CHILTON®
ASIAN
MECHANICAL SERVICE
Acura - Mazda
2005 Edition

THOMSON

DELMAR LEARNING

Australia • Canada • Mexico • Singapore • Spain • United Kingdom • United States

THOMSON

DELMAR LEARNING

Chilton
Asian Mechanical Service
2005 Edition

Acura-Mazda

Vice President, Technology and Trades SBU:
Alar Elken

Executive Director, Professional Business Unit:
Gregory L. Clayton

Publisher, Professional Business Unit:
David Koontz

Marketing Director:
Beth A. Lutz

Marketing Specialist:
Brian McGrath

Marketing Coordinator:
Marissa Mariella

Production Director:
Mary Ellen Black

Production Editor:
Elizabeth Hough

Editorial Assistant:
Christine Wade

Publishing Coordinator
Paula Baillie

Editors:
Dennis Bailey
Terry Blomquist
Timothy A. Crain
Matthew Frederick
George Heinrich
Thomas A. Mellon
Richard J. Rivele
Christine L. Sheeky
Jon Wallace

Cover Design
Melinda Possinger

ISBN: 1-4018-6716-2

ISSN: 1548-0887

NOTICE TO THE READER

Table of Contents

Model Index

EDITORIAL POLICY

Manufacturer and Model Coverage

This manual does not cover every Asian make and model that is currently available on the market. Rather, the Chilton editorial staff makes judicious decisions as to which models warrant coverage, based on which vehicles are serviced by most technicians.

Model Year Information

This manual is published toward the end of the year prior to the edition year. Every effort is made to gather current data from the Original Vehicle Manufacturers (OEMs) when they publish it. Different OEMs choose to release their new model information at different times of the year. Indeed, the same OEM can publish information early one season and late the next season. As a result, not all models are equally current when each edition of this manual is published.

Although information in this manual is based on industry sources and is as complete as possible at the time of publication, some vehicle manufacturers may make changes which cannot be included here. Information on very late models may not be available in some circumstances. While striving for total accuracy, the publisher cannot assume responsibility for any errors, changes, or omissions that may occur in the compilation of this data.

Safety Notice

Proper service and repair procedures are vital to the safe, reliable operation of all motor vehicles, as well as the personal safety of those performing the repairs. This manual outlines procedures for servicing and repairing vehicles using safe, effective methods. The procedures contain many NOTES, WARNINGS and CAUTIONS which should be followed along with standard safety procedures to reduce the possibility of personal injury or improper service which could damage the vehicle or compromise its safety.

Repair procedures, tools, parts, and technician skill and experience vary widely. It is not possible to anticipate all conceivable ways or conditions under which vehicles may be serviced, or to provide cautions for all possible hazards that may result. Standard and accepted safety precautions and equipment should be used when handling toxic or flammable fluids, and safety goggles or other protection should be used during cutting, grinding, chiseling, prying, or any other process that can cause material removal or projectiles.

Some procedures require the use of tools specially designed for a specific purpose. Before substituting another tool or procedure, you must be completely satisfied that neither your personal safety, nor the performance of the vehicle will be endangered.

LOCATING AND USING THE INFORMATION

Organization

To find where a particular model section or procedure is located, look in the Table of Contents. Main topics are listed with the page number on which they may be found. Following the main topics is an alphabetical listing of all of the procedures within the section and their page numbers.

Part Numbers

Part numbers listed in this book are not recommendations by the publisher for any product by brand name. They are references that can be used with interchanges manuals and aftermarket supplier catalogs to locate each brand supplier's discrete part number.

Special Tools

Special tools are recommended by the vehicle manufacturer to perform specific jobs. When necessary, special tools are referred to in the text by the part number of the tool manufacturer. These tools may be purchased, under the appropriate part number, from your local dealer or regional distributor, or an equivalent tool can be purchased locally from a tool supplier or parts outlet. Before substituting any tool for the one recommended, read the previous Safety Notice.

ACKNOWLEDGEMENT

The publisher would like to express appreciation to the following vehicle manufacturers for their assistance in producing this publication. No further reproduction or distribution of the material in this manual is allowed without the expressed written permission of the vehicle manufacturers and the publisher.

American Honda Motor Co., including Acura and Honda Division
Fuji Heavy Industries Ltd., including Subaru
Isuzu Motors Ltd.
Hyundai Group, including Hyundai and Kia Motor
Mazda Motor Corp.
Mitsubishi Motors North America, Inc.
Nissan North America, including Infiniti and Nissan Division
Suzuki Motor Corp.
Toyota Motor Sales USA, including Lexus and Toyota Division

ACURA

3.2CL • 3.2TL • 3.5RL • Integra • NSX • RSX • TSX

1

SPECIFICATION AND MAINTENANCE CHARTS

ENGINE AND VEHICLE IDENTIFICATION

Engine							Model Year	
Code	Liters (cc)	Cu. In.	Cyl.	Fuel Sys.	Engine Type	Eng. Mfg.	Code ①	Year
B18B1	1.8 (1834)	112	4	PGM-FI	DOHC	Honda	1	2001
B18C1	1.8 (1797)	110	4	PGM-FI	DOHC	Honda	2	2002
B18C5	1.8 (1797)	110	4	PGM-FI	DOHC	Honda	3	2003
K20A2	2.0 (1999)	122	4	PGM-FI	DOHC	Honda	4	2004
K20A3	2.0 (1999)	122	4	PGM-FI	DOHC	Honda	5	2005
K24A4	2.4 (2354)	144	4	PGM-FI	DOHC	Honda		
C30A1	3.0 (2977)	183	6	PGM-FI	DOHC	Honda		
J30A1	3.0 (2997)	183	6	PGM-FI	SOHC	Honda		
C32A6	3.2 (3206)	196	6	PGM-FI	SOHC	Honda		
C32B1	3.2 (3206)	196	6	PGM-FI	DOHC	Honda		
J32A1	3.2 (3210)	196	6	PGM-FI	SOHC	Honda		
J32A3	3.2 (3210)	196	6	PGM-FI	SOHC	Honda		
C35A1	3.5 (3474)	211	6	PGM-FI	SOHC	Honda		

PGM-FI: Programmed Fuel Injection
DOHC: Double Overhead Camshaft
SOHC: Single Overhead Camshaft

① 10th digit of the Vehicle Identification Number (VIN)

67162-ATSX-C01

GENERAL ENGINE SPECIFICATIONS

Year	Model	Engine Displacement Liters	Engine ID	Net Horsepower @ rpm	Net Torque @ rpm (ft. lbs.)	Bore x Stroke (in.)	Com-pression Ratio	Oil Pressure @ rpm
2001	Integra	1.8	B18B1/①	140@6300	127@5200	3.19x3.50	9.2:1	50@3000
	Integra GSR	1.8	B18C1/②	170@7600	128@6200	3.19x3.43	10.0:1	50@3000
	NSX	3.0	C30A1/NA1	252@6600	210@5300	3.54x3.07	10.2:1	50@3000
	NSX	3.2	C32B1/NA2	290@7100	224@5500	3.66x3.07	10.2:1	50@3000
	3.2TL	3.2	J32A1/UA5	225@5500	216@5000	3.50x3.39	9.8:1	71@3000
	3.2CL	3.2	J32A2/YA4	260@6100	232@3500	3.50x3.39	10.5:1	71@3000
	3.5RL	3.5	C35A1/KA9	210@5200	224@2800	3.54x3.58	9.6:1	50@3000
2002	RSX	2.0	K20A3	160@6500	141@4000	3.39x3.39	9.8:1	44@3000
	RSX ③	2.0	K20A2	200@7400	142@6000	3.39x3.39	11.0:1	44@3000
	NSX	3.0	C30A1/NA1	252@6600	210@5300	3.54x3.07	10.2:1	50@3000
	NSX	3.2	C32B1/NA2	290@7100	224@5500	3.66x3.07	10.2:1	50@3000
	3.2TL	3.2	J32A1/UA5	225@5500	216@5000	3.50x3.39	9.8:1	71@3000
	3.2CL	3.2	J32A2/YA4	260@6100	232@3500	3.50x3.39	10.5:1	71@3000
	3.5RL	3.5	C35A1/KA9	210@5200	224@2800	3.54x3.58	9.6:1	50@3000
2003	RSX	2.0	K20A3	160@6500	141@4000	3.39x3.39	9.8:1	44@3000
	RSX ③	2.0	K20A2	200@7400	142@6000	3.39x3.39	11.0:1	44@3000
	NSX	3.0	C30A1/NA1	252@6600	210@5300	3.54x3.07	10.2:1	50@3000
	NSX	3.2	C32B1/NA2	290@7100	224@5500	3.66x3.07	10.2:1	50@3000
	3.2TL	3.2	J32A1/UA5	225@5500	216@5000	3.50x3.39	9.8:1	71@3000
	3.2CL	3.2	J32A2/YA4	260@6100	232@3500	3.50x3.39	10.5:1	71@3000
	3.5RL	3.5	C35A1/KA9	210@5200	224@2800	3.54x3.58	9.6:1	50@3000
2004	RSX	2.0	K20A3	160@6500	141@4000	3.39x3.39	9.8:1	44@3000
	RSX ③	2.0	K20A2	200@7400	142@6000	3.39x3.39	11.0:1	44@3000
	TSX	2.4	K24A2	200@6800	166@4500	3.42X3.89	10.5:1	50@3000
	NSX	3.0	C30A1/NA1	252@6600	210@5300	3.54x3.07	10.2:1	50@3000
	NSX	3.2	C32B1/NA2	290@7100	224@5500	3.66x3.07	10.2:1	50@3000
	3.2TL	3.2	J32A1/UA5	225@5500	216@5000	3.50x3.39	9.8:1	71@3000
	3.5RL	3.5	C35A1/KA9	210@5200	224@2800	3.54x3.58	9.6:1	50@3000

PGM-FI: Programmed Fuel Injection

① DB7: 4 door
  DC4: 3 door

② DB8: 4 door (Except Type R)
  DC2: 3 door

③ Type-S

67162-ATSX-C02

ENGINE TUNE-UP SPECIFICATIONS

Year	Engine Displacement Liters	Engine ID/VIN	Spark Plug Gap (in.)	Ignition Timing (deg.) MT	Ignition Timing (deg.) AT	Fuel Pump (psi)	Idle Speed (rpm) MT	Idle Speed (rpm) AT	Valve Clearance In.	Valve Clearance Ex.
2001	1.8	B18B1/①	0.039-0.043	16B	16B	40-47 ②	700-800	700-800	0.003-0.005	0.006-0.008
	1.8	B18C1/③	0.051	16B	16B	48-55 ②	700-800	—	0.006-0.007	0.007-0.008
	3.0	C30A1/NA1	0.043-0.047	15B	15B	47-53 ②	750-850	730-830	0.006-0.007	0.007-0.008
	3.2	C32B1/NA2	0.043-0.047	—	15B	47-53 ②	—	750-850	0.006-0.007	0.007-0.008
	3.2	J32A1/UA5	0.039-0.043	—	10B	41-48 ②	—	630-730	0.008-0.009	0.011-0.013
	3.2	J32A2/YA4	0.039-0.043	—	10B	41-48 ②	—	700-800	0.008-0.009	0.011-0.013
	3.5	C35A1/KA9	0.039-0.043	—	15B	43-50 ②	—	600-700	HYD	HYD
2002	2.0	K20A3	0.039-0.043	8B	8B	47-52 ②	600-700	600-700	0.003-0.005	0.006-0.008
	2.0	K20A2	0.039-0.043	8B	8B	47-52 ②	650-750	650-750	0.006-0.007	0.007-0.008
	3.0	C30A1/NA1	0.043-0.047	15B	15B	47-53 ②	750-850	730-830	0.006-0.007	0.007-0.008
	3.2	C32B1/NA2	0.043-0.047	—	15B	47-53 ②	—	750-850	0.006-0.007	0.007-0.008
	3.2	J32A1/UA5	0.039-0.043	—	10B	41-48 ②	—	630-730	0.008-0.009	0.011-0.013
	3.2	J32A2/YA4	0.039-0.043	—	10B	41-48 ②	—	700-800	0.008-0.009	0.011-0.013
	3.5	C35A1/KA9	0.039-0.043	—	15B	43-50 ②	—	600-700	HYD	HYD
2003	2.0	K20A3	0.039-0.043	8B	8B	47-52 ②	600-700	600-700	0.003-0.005	0.006-0.008
	2.0	K20A2	0.039-0.043	8B	8B	47-52 ②	650-750	650-750	0.006-0.007	0.007-0.008
	3.0	C30A1/NA1	0.043-0.047	15B	15B	47-53 ②	750-850	730-830	0.006-0.007	0.007-0.008
	3.2	C32B1/NA2	0.043-0.047	—	15B	47-53 ②	—	750-850	0.006-0.007	0.007-0.008
	3.2	J32A1/UA5	0.039-0.043	—	10B	41-48 ②	—	630-730	0.008-0.009	0.011-0.013
	3.2	J32A2/YA4	0.039-0.043	—	10B	41-48 ②	—	700-800	0.008-0.009	0.011-0.013
	3.5	C35A1/KA9	0.039-0.043	—	15B	43-50 ②	—	600-700	HYD	HYD

67162-ATSX-C03

ENGINE TUNE-UP SPECIFICATIONS

Year	Engine Displacement Liters	Engine ID/VIN	Spark Plug Gap (in.)	Ignition Timing (deg.)		Fuel Pump (psi)	Idle Speed (rpm)		Valve Clearance	
				MT	AT		MT	AT	In.	Ex.
2004	2.0	K20A3	0.039-0.043	8B	8B	47-52 ②	600-700	600-700	0.003-0.005	0.006-0.008
	2.0	K20A2	0.039-0.043	8B	8B	47-52 ②	650-750	650-750	0.006-0.007	0.007-0.008
	2.4	K24A2	0.039-0.043	8B	8B	48-55 ②	650-750	750-850	0.008-0.010	0.011-0.013
	3.0	C30A1/NA1	0.043-0.047	15B	15B	47-53 ②	750-850	730-830	0.006-0.007	0.007-0.008
	3.2	C32B1/NA2	0.043-0.047	—	15B	47-53 ②	—	750-850	0.006-0.007	0.007-0.008
	3.2	J32A1/UA5	0.039-0.043	—	10B	41-48 ②	—	630-730	0.008-0.009	0.011-0.013
	3.2	J32A2/YA4	0.039-0.043	—	10B	41-48 ②	—	700-800	0.008-0.009	0.011-0.013
	3.5	C35A1/KA9	0.039-0.043	—	15B	43-50 ②	—	600-700	HYD	HYD

NOTE: The Vehicle Emission Control Information label reflects specification changes during production and must be used if they differ from this chart.

B: Before Top Dead Center

HYD: Hydraulic

① DB7: 4 door
   DC4: 3 door

② At idle, pressure regulator vacuum hose disconnected

③ DB8: 4 door (Except Type R)
   DC2: 3 door

67162-ATSX-C04

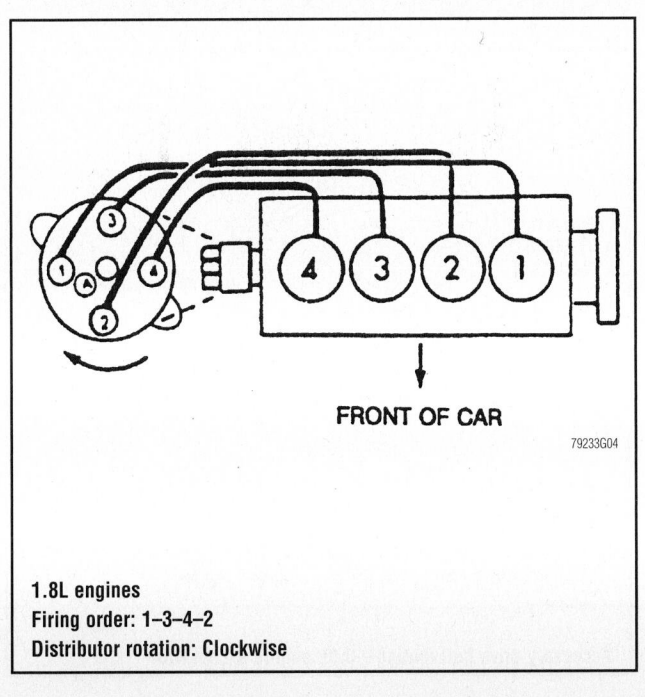

1.8L engines
Firing order: 1-3-4-2
Distributor rotation: Clockwise

79233G04

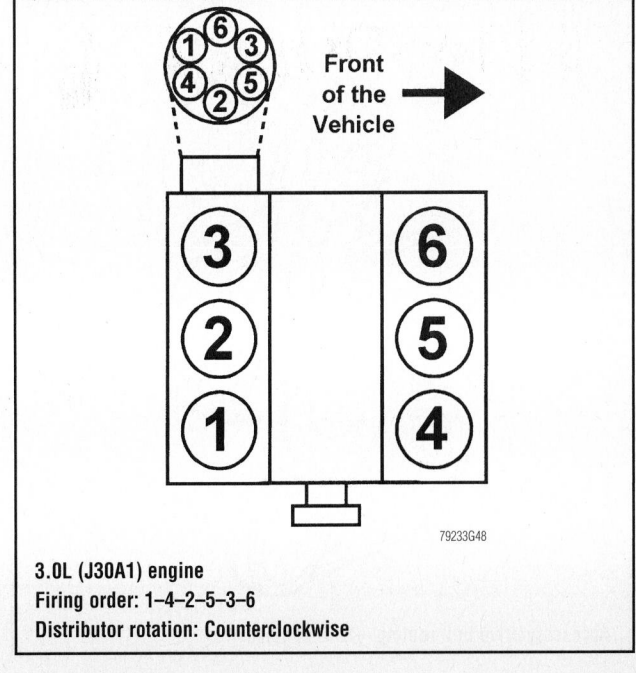

3.0L (J30A1) engine
Firing order: 1-4-2-5-3-6
Distributor rotation: Counterclockwise

79233G48

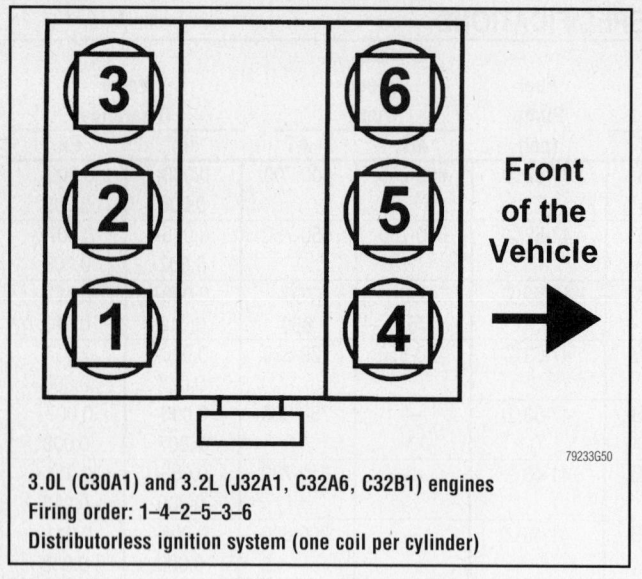

Front of the Vehicle →

79233G50

3.0L (C30A1) and 3.2L (J32A1, C32A6, C32B1) engines
Firing order: 1–4–2–5–3–6
Distributorless ignition system (one coil per cylinder)

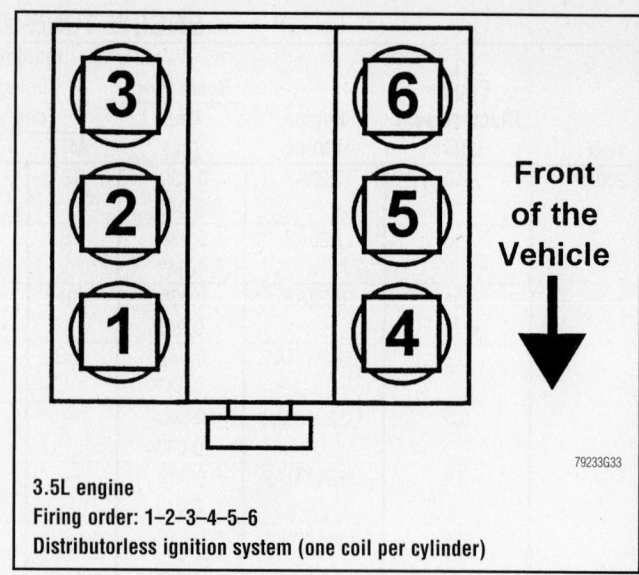

Front of the Vehicle ↓

79233G33

3.5L engine
Firing order: 1–2–3–4–5–6
Distributorless ignition system (one coil per cylinder)

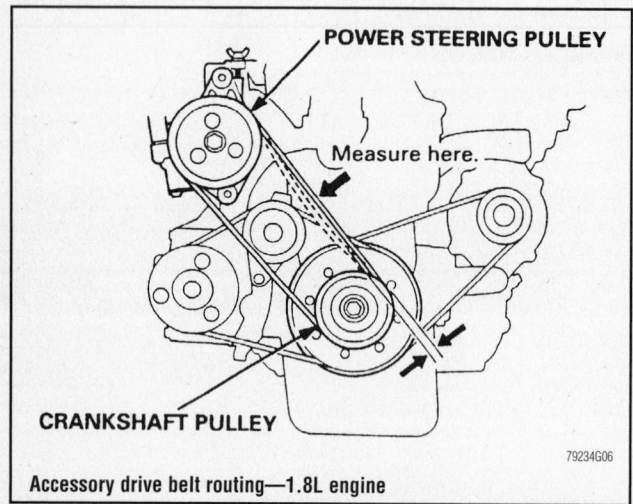

POWER STEERING PULLEY

Measure here.

CRANKSHAFT PULLEY

79234G06

Accessory drive belt routing—1.8L engine

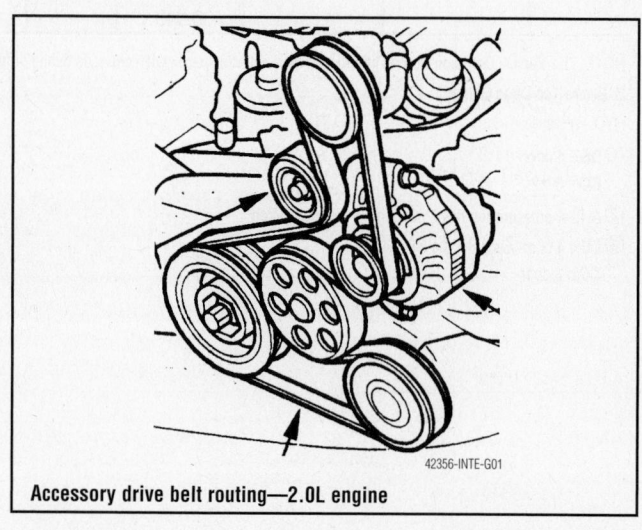

42356-INTE-G01

Accessory drive belt routing—2.0L engine

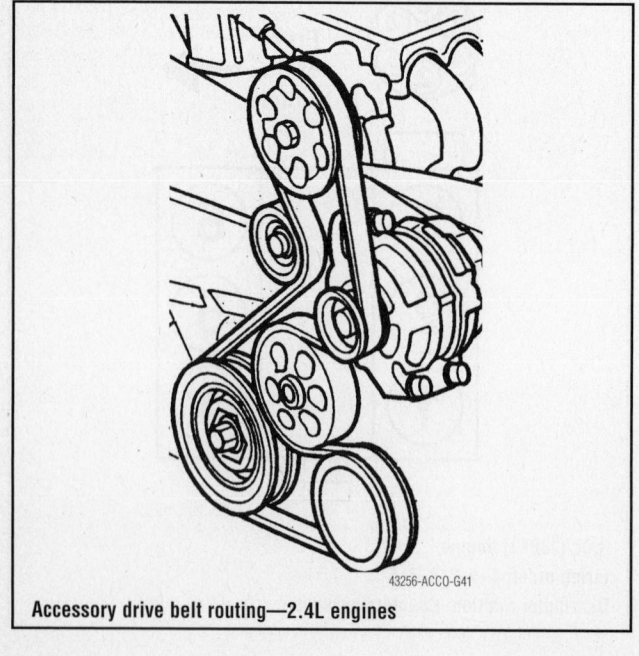

43256-ACCO-G41

Accessory drive belt routing—2.4L engines

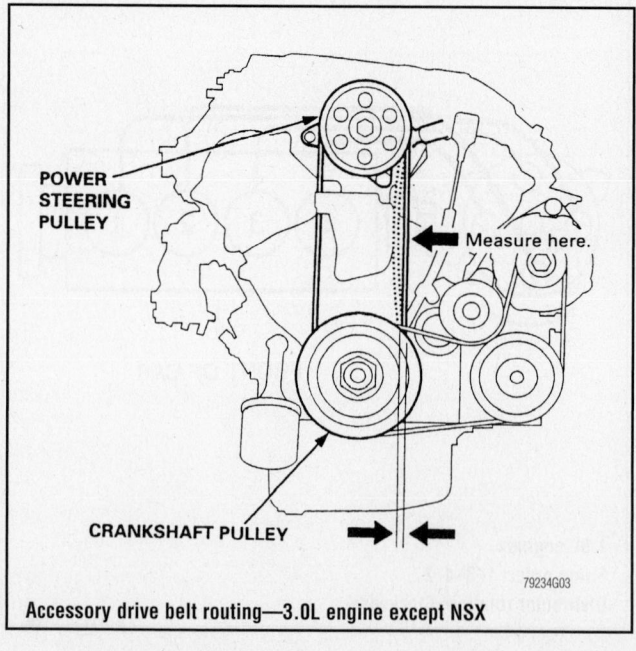

POWER STEERING PULLEY

Measure here.

CRANKSHAFT PULLEY

79234G03

Accessory drive belt routing—3.0L engine except NSX

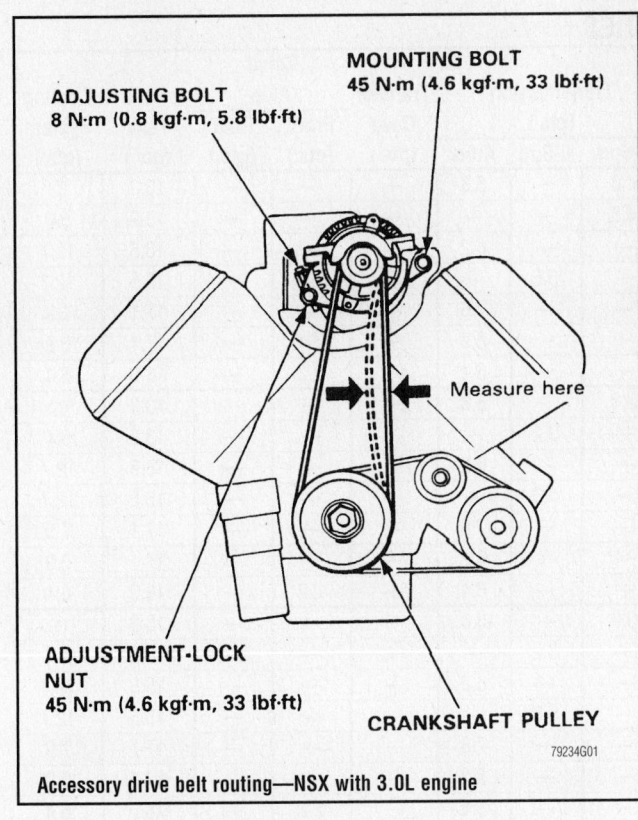

ADJUSTING BOLT
8 N·m (0.8 kgf·m, 5.8 lbf·ft)

MOUNTING BOLT
45 N·m (4.6 kgf·m, 33 lbf·ft)

Measure here

ADJUSTMENT-LOCK
NUT
45 N·m (4.6 kgf·m, 33 lbf·ft)

CRANKSHAFT PULLEY

79234G01

Accessory drive belt routing—NSX with 3.0L engine

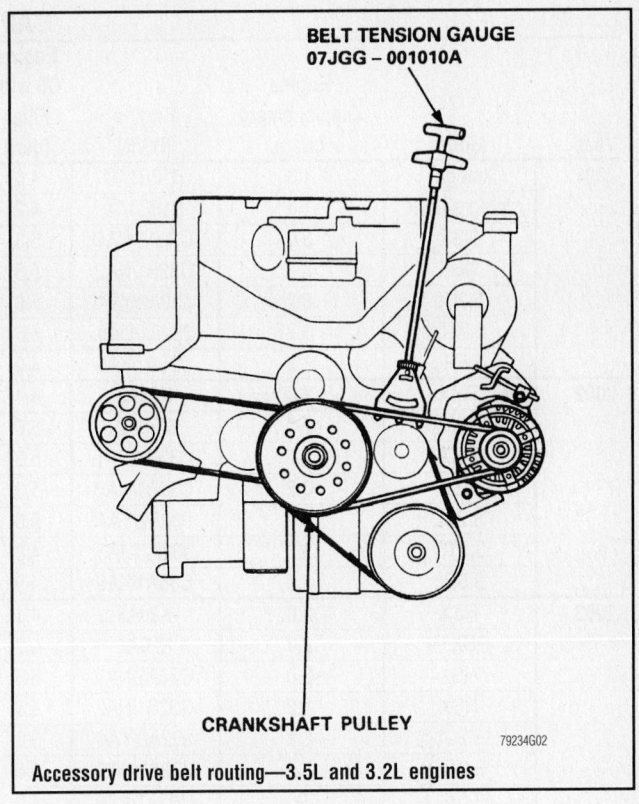

BELT TENSION GAUGE
07JGG – 001010A

CRANKSHAFT PULLEY

79234G02

Accessory drive belt routing—3.5L and 3.2L engines

CAPACITIES

Year	Model	Engine Displacement Liters	Engine ID/VIN	Engine Oil with Filter (qts.)	Transmission (pts.) 5-Spd	6-Spd	Auto.	Transfer Case (pts.)	Drive Axle Front (pts.)	Rear (pts.)	Fuel Tank (gal.)	Cooling System (qts.)
2001	Integra	1.8	B18B1/①	4.0	4.6	—	5.8	—	—	—	13.2	②
	Integra GSR	1.8	B18C1/③	4.2	4.6	—	—	—	—	—	13.2	5.0
	NSX	3.0	C30A1/NA1	5.3	—	—	6.2	—	—	—	18.5	12.7
	NSX	3.2	C32B1/NA2	5.3	—	5.6	—	—	—	—	18.5	12.7
	3.2CL	3.2	J32A2/YA4	5.0	—	—	7.6	—	—	—	17.1	7.9
	3.2TL	3.2	J32A1/UA5	4.6	—	—	6.2	—	—	—	17.1	5.9
	3.5RL	3.5	C35A1/KA9	4.9	—	—	6.4	—	2.2	—	18.0	6.4
2002	RSX	2.0	K20A3	4.4	3.2	—	5.8	—	—	—	13.2	⑤
	RSX ④	2.0	K20A2	5.0	—	3.2	—	—	—	—	13.2	5.4
	NSX	3.0	C30A1/NA1	5.3	—	—	6.2	—	—	—	18.5	12.7
	NSX	3.2	C32B1/NA2	5.3	—	5.6	—	—	—	—	18.5	12.7
	3.2CL	3.2	J32A2/YA4	5.0	—	—	7.6	—	—	—	17.1	7.9
	3.2TL	3.2	J32A1/UA5	4.6	—	—	6.2	—	—	—	17.1	5.9
	3.5RL	3.5	C35A1/KA9	4.9	—	—	6.4	—	2.2	—	18.0	6.4
2003	RSX	2.0	K20A3	4.4	3.2	—	5.8	—	—	—	13.2	⑤
	RSX ④	2.0	K20A2	5.0	—	3.2	—	—	—	—	13.2	5.4
	NSX	3.0	C30A1/NA1	5.3	—	—	6.2	—	—	—	18.5	12.7
	NSX	3.2	C32B1/NA2	5.3	—	5.6	—	—	—	—	18.5	12.7
	3.2CL	3.2	J32A2/YA4	5.0	—	—	7.6	—	—	—	17.1	7.9
	3.2TL	3.2	J32A1/UA5	4.6	—	—	6.2	—	—	—	17.1	5.9
	3.5RL	3.5	C35A1/KA9	4.9	—	—	6.4	—	2.2	—	18.0	6.4
2004	RSX	2.0	K20A3	4.4	3.2	—	5.8	—	—	—	13.2	⑤
	RSX ④	2.0	K20A2	5.0	—	3.2	—	—	—	—	13.2	5.4
	TSX	2.4	K24A2	4.4	—	4.2	6.0	—	—	—	17.1	⑥
	NSX	3.0	C30A1/NA1	5.3	—	—	6.2	—	—	—	18.5	12.7
	NSX	3.2	C32B1/NA2	5.3	—	5.6	—	—	—	—	18.5	12.7
	3.2TL	3.2	J32A1/UA5	4.6	—	—	6.2	—	—	—	17.1	5.9
	3.5RL	3.5	C35A1/KA9	4.9	—	—	6.4	—	2.2	—	18.0	6.4

NOTE: All capacities are approximate. Add fluid gradually and ensure a proper fluid level is obtained.

NOTE: Capacities given are service, not overhaul capacities

① DB7: 4 door
　DC4: 3 door

② Automatic transmission: 5.0
　Manual transmission: 4.6

③ DB8: 4 door (Except Type R)
　DC2: 3 door

④ Type-S

⑤ Automatic transmission: 5.3
　Manual Transmission: 5.4

⑥ Automatic transmission: 5.6
　Manual Transmission: 5.7

67162-ATSX-C05

VALVE SPECIFICATIONS

Year	Engine Displacement Liters	Engine ID/VIN	Seat Angle (deg.)	Face Angle (deg.)	Spring Test Pressure (lbs. @ in.)	Spring Installed Height (in.)	Stem-to-Guide Clearance (in.)		Stem Diameter (in.)	
							Intake	Exhaust	Intake	Exhaust
2001	1.8	B18B1/①	45	45	NA	NA	0.0010-0.0020	0.0020-0.0030	0.2591-0.2594	0.2579-0.2583
	1.8	B18C1/②	45	45	NA	NA	0.0010-0.0022	0.0020-0.0031	0.2156-0.2159	0.2146-0.2150
	3.0	C30A1/NA1	45	45	NA	NA	0.0010-0.0020	0.0020-0.0030	0.2156-0.2159	0.2146-0.2150
	3.2	C32B1/NA2	45	45	NA	NA	0.0010-0.0029	0.0020-0.0030	0.2156-0.2159	0.2146-0.2150
	3.2	J32A2/YA4	45	45	NA	NA	0.0008-0.0018	0.0022-0.0031	0.2159-0.2163	0.2146-0.2150
	3.2	J32A1/UA5	45	45	NA	NA	0.0008-0.0018	0.0022-0.0031	0.2159-0.2163	0.2146-0.2150
	3.5	C35A1/KA9	45	45	NA	NA	0.0010-0.0020	0.0020-0.0030	0.2157-0.2161	0.2146-0.2150
2002	2.0	K20A3	45	45	NA	NA	0.0012-0.0022	0.0022-0.0031	0.2156-0.2159	0.2146-0.2150
	2.0	K20A4	45	45	NA	NA	0.0012-0.0022	0.0022-0.0031	0.2156-0.2159	0.2146-0.2150
	3.0	C30A1/NA1	45	45	NA	NA	0.0010-0.0020	0.0020-0.0030	0.2156-0.2159	0.2146-0.2150
	3.2	C32B1/NA2	45	45	NA	NA	0.0010-0.0029	0.0020-0.0030	0.2156-0.2159	0.2146-0.2150
	3.2	J32A2/YA4	45	45	NA	NA	0.0008-0.0018	0.0022-0.0031	0.2159-0.2163	0.2146-0.2150
	3.2	J32A1/UA5	45	45	NA	NA	0.0008-0.0018	0.0022-0.0031	0.2159-0.2163	0.2146-0.2150
	3.5	C35A1/KA9	45	45	NA	NA	0.0010-0.0020	0.0020-0.0030	0.2157-0.2161	0.2146-0.2150
2003	2.0	K20A3	45	45	NA	NA	0.0012-0.0022	0.0022-0.0031	0.2156-0.2159	0.2146-0.2150
	2.0	K20A4	45	45	NA	NA	0.0012-0.0022	0.0022-0.0031	0.2156-0.2159	0.2146-0.2150
	3.0	C30A1/NA1	45	45	NA	NA	0.0010-0.0020	0.0020-0.0030	0.2156-0.2159	0.2146-0.2150
	3.2	C32B1/NA2	45	45	NA	NA	0.0010-0.0029	0.0020-0.0030	0.2156-0.2159	0.2146-0.2150
	3.2	J32A2/YA4	45	45	NA	NA	0.0008-0.0018	0.0022-0.0031	0.2159-0.2163	0.2146-0.2150
	3.2	J32A1/UA5	45	45	NA	NA	0.0008-0.0018	0.0022-0.0031	0.2159-0.2163	0.2146-0.2150
	3.5	C35A1/KA9	45	45	NA	NA	0.0010-0.0020	0.0020-0.0030	0.2157-0.2161	0.2146-0.2150

67162-ATSX-C06

VALVE SPECIFICATIONS

Year	Engine Displacement Liters	Engine ID/VIN	Seat Angle (deg.)	Face Angle (deg.)	Spring Test Pressure (lbs. @ in.)	Spring Installed Height (in.)	Stem-to-Guide Clearance (in.)		Stem Diameter (in.)	
							Intake	Exhaust	Intake	Exhaust
2004	2.0	K20A3	45	45	NA	NA	0.0012-0.0022	0.0022-0.0031	0.2156-0.2159	0.2146-0.2150
	2.0	K20A4	45	45	NA	NA	0.0012-0.0022	0.0022-0.0031	0.2156-0.2159	0.2146-0.2150
	2.4	K24A2	45	45	NA	NA	0.0012-0.0022	0.0022-0.0031	0.2156-0.2159	0.2146-0.2150
	3.0	C30A1/NA1	45	45	NA	NA	0.0010-0.0020	0.0020-0.0030	0.2156-0.2159	0.2146-0.2150
	3.2	J32A2/YA4	45	45	NA	NA	0.0008-0.0018	0.0022-0.0031	0.2159-0.2163	0.2146-0.2150
	3.2	J32A1/UA5	45	45	NA	NA	0.0008-0.0018	0.0022-0.0031	0.2159-0.2163	0.2146-0.2150
	3.2	C32B1/NA2	45	45	NA	NA	0.0010-0.0029	0.0020-0.0030	0.2156-0.2159	0.2146-0.2150
	3.5	C35A1/KA9	45	45	NA	NA	0.0010-0.0020	0.0020-0.0030	0.2157-0.2161	0.2146-0.2150

NA: Not Available

① DB7: 4 door
DC4: 3 door

② DB8: 4 door (Except Type R)
DC2: 3 door

67162-ATSX-C07

CRANKSHAFT AND CONNECTING ROD SPECIFICATIONS
All measurements are given in inches.

Year	Engine Displacement Liters	Engine ID/VIN	Crankshaft				Connecting Rod		
			Main Brg. Journal Dia.	Main Brg. Oil Clearance	Shaft End-play	Thrust on No.	Journal Diameter	Oil Clearance	Side Clearance
2001	1.8	B18B1/①	②	③	0.0040-0.0180	4	1.7707-1.7717	0.0008-0.0015	0.0060-0.0160
	1.8	B18C1/④	⑤	⑥	0.0040-0.0180	4	1.7707-1.7717	0.0013-0.0020	0.0060-0.0160
	3.0	C30A1/NA1	2.5187-2.5197	0.0009-0.0019	0.0040-0.0110	3	2.0463-2.0742	0.0016-0.0025	0.0060-0.0120
	3.2	C32B1/NA2	2.5187-2.5197	0.0009-0.0019	0.0040-0.0110	3	2.0463-2.0742	0.0016-0.0025	0.0060-0.0120
	3.2	J32A1/UA5	2.8337-2.8346	0.0008-0.0017	0.0040-0.0140	3	2.1644-2.1654	0.0008-0.0017	0.0060-0.0140
	3.2	J32A2/YA4	2.8337-2.8346	0.0008-0.0017	0.0040-0.0140	3	2.1644-2.1654	0.0008-0.0017	0.0060-0.0140
	3.5	C35A1/KA9	2.6762-2.6772	0.0008-0.0017	0.0040-0.0110	3	2.1248-2.1257	0.0009-0.0018	0.0060-0.0120
2002	2.0	K20A3	2.1648-2.1648	⑦	0.0040-0.0140	4	1.7707-1.7717	0.0008-0.0019	0.0060-0.0160
	2.0	K20A2	2.1646-2.1655	⑦	0.0040-0.0140	4	1.8888-1.8898	0.0013-0.0024	0.0060-0.0160
	3.0	C30A1/NA1	2.5187-2.5197	0.0009-0.0019	0.0040-0.0110	3	2.0463-2.0472	0.0016-0.0025	0.0060-0.0120
	3.2	C32B1/NA2	2.5187-2.5197	0.0009-0.0019	0.0040-0.0110	3	2.0463-2.0472	0.0016-0.0025	0.0060-0.0120
	3.2	J32A1/UA5	2.8337-2.8346	0.0008-0.0017	0.0040-0.0140	3	2.1644-2.1654	0.0008-0.0017	0.0060-0.0140
	3.2	J32A2/YA4	2.8337-2.8346	0.0008-0.0017	0.0040-0.0140	3	2.1644-2.1654	0.0008-0.0017	0.0060-0.0140
	3.5	C35A1/KA9	2.6762-2.6772	0.0008-0.0017	0.0040-0.0110	3	2.1248-2.1257	0.0009-0.0018	0.0060-0.0120
2003	2.0	K20A3	2.1648-2.1648	⑦	0.0040-0.0140	4	1.7707-1.7717	0.0008-0.0019	0.0060-0.0160
	2.0	K20A2	2.1646-2.1655	⑦	0.0040-0.0140	4	1.8888-1.8898	0.0013-0.0024	0.0060-0.0160
	3.0	C30A1/NA1	2.5187-2.5197	0.0009-0.0019	0.0040-0.0110	3	2.0463-2.0472	0.0016-0.0025	0.0060-0.0120
	3.2	C32B1/NA2	2.5187-2.5197	0.0009-0.0019	0.0040-0.0110	3	2.0463-2.0472	0.0016-0.0025	0.0060-0.0120
	3.2	J32A1/UA5	2.8337-2.8346	0.0008-0.0017	0.0040-0.0140	3	2.1644-2.1654	0.0008-0.0017	0.0060-0.0140
	3.2	J32A2/YA4	2.8337-2.8346	0.0008-0.0017	0.0040-0.0140	3	2.1644-2.1654	0.0008-0.0017	0.0060-0.0140
	3.5	C35A1/KA9	2.6762-2.6772	0.0008-0.0017	0.0040-0.0110	3	2.1248-2.1257	0.0009-0.0018	0.0060-0.0120

67162-ATSX-C08

CRANKSHAFT AND CONNECTING ROD SPECIFICATIONS
All measurements are given in inches.

Year	Engine Displacement Liters	Engine ID/VIN	Crankshaft			Thrust on No.	Connecting Rod		
			Main Brg. Journal Dia.	Main Brg. Oil Clearance	Shaft End-play		Journal Diameter	Oil Clearance	Side Clearance
2004	2.0	K20A3	2.1648-2.1648	⑦	0.0040-0.0140	4	1.7707-1.7717	0.0008-0.0019	0.0060-0.0160
	2.0	K20A2	2.1646-2.1655	⑦	0.0040-0.0140	4	1.8888-1.8898	0.0013-0.0024	0.0060-0.0160
	2.4	K24A4	⑧	⑦	0.0040-0.0140	4	1.8888-1.8898	0.0013-0.0026	0.0060-0.0160
	3.0	C30A1/NA1	2.5187-2.5197	0.0009-0.0019	0.0040-0.0110	3	2.0463-2.0472	0.0016-0.0025	0.0060-0.0120
	3.2	C32B1/NA2	2.5187-2.5197	0.0009-0.0019	0.0040-0.0110	3	2.0463-2.0472	0.0016-0.0025	0.0060-0.0120
	3.2	J32A1/UA5	2.8337-2.8346	0.0008-0.0017	0.0040-0.0140	3	2.1644-2.1654	0.0008-0.0017	0.0060-0.0140
	3.2	J32A2/YA4	2.8337-2.8346	0.0008-0.0017	0.0040-0.0140	3	2.1644-2.1654	0.0008-0.0017	0.0060-0.0140
	3.5	C35A1/KA9	2.6762-2.6772	0.0008-0.0017	0.0040-0.0110	3	2.1248-2.1257	0.0009-0.0018	0.0060-0.0120

① DB7: 4 door
DC4: 3 door

② Nos. 1, 2, 4 and 5: 2.1644-2.1654
No. 3: 2.1642-2.1651

③ Nos. 1,2,4, and 5: 0.0009-0.0017
No. 3: 0.0012-0.0019

④ DB8: 4 door (Except Type R)
DC2: 3 door

⑤ Nos. 1, 2, 4 and 5: 2.1644-2.1654
No. 3: 2.1643-2.1653

⑥ Nos. 1, 2, 4 and 5: 0.0009-0.0017
No. 3: 0.0012-0.0019

⑦ Nos. 1, 2, 4 and 5: 0.0007-0.0016
No. 3: 0.0010-0.0019

⑧ Nos. 1, 2, 4 and 5: 2.1648-2.1657
No. 3: 2.1644-2.1654

67162-ATSX-C09

PISTON AND RING SPECIFICATIONS
All measurements are given in inches

Year	Engine Displacement Liters	Engine ID/VIN	Piston Clearance	Ring Gap			Ring Side Clearance		
				Top Compression	Bottom Compression	Oil Control	Top Compression	Bottom Compression	Oil Control
2001	1.8	B18B1/①	0.0004-0.0016	0.0080-0.0140	0.0160-0.0220	0.0080-0.0200	0.0018-0.0028	0.0018-0.0026	NA
	1.8	B18C1/②	0.0004-0.0016	0.0080-0.0140	0.0160-0.0220	0.0080-0.0200	0.0018-0.0028	0.0018-0.0026	NA
	3.0	C30A1/NA1	0.0002-0.0014	0.0100-0.0160	0.0140-0.0200	0.0080-0.0280	0.0012-0.0022	0.0012-0.0022	NA
	3.0	J30A1/YA2	0.0006-0.00160	0.0080-0.0140	0.0160-0.0220	0.0080-0.0280	0.0014-0.0025	0.0012-0.0022	NA
	3.2	C32B1/NA2	0.0002-0.0012	0.0080-0.0120	0.0140-0.0200	0.0080-0.0280	0.0014-0.0026	0.0012-0.0024	NA
	3.2	J32A1/UA5	0.0006-0.00160	0.0080-0.0140	0.0160-0.0220	0.0080-0.0280	0.0014-0.0024	0.0012-0.0022	NA
	3.2	J32A2/YA4	0.0006-0.00160	0.0080-0.0140	0.0160-0.0220	0.0080-0.0280	0.0014-0.0024	0.0012-0.0022	NA
	3.5	C35A1/KA	0.0010-0.0020	0.0100-0.0160	0.0160-0.0220	③	0.0022-0.0031	0.0012-0.0022	NA
2002	2.0	K20A3	0.0008-0.0016	0.0080-0.0014	0.0160-0.0220	0.0100-0.0260	0.0014-0.0024	0.0012-0.0022	NA
	2.0	K20A2	0.0008-0.0016	0.0080-0.0014	0.0200-0.0260	0.0080-0.0280	0.0016-0.0026	0.0018-0.0028	NA
	3.0	C30A1/NA1	0.0002-0.0014	0.0100-0.0160	0.0140-0.0200	0.0080-0.0280	0.0012-0.0022	0.0012-0.0022	NA
	3.0	J30A1/YA2	0.0006-0.00160	0.0080-0.0140	0.0160-0.0220	0.0080-0.0280	0.0014-0.0025	0.0012-0.0022	NA
	3.2	C32B1/NA2	0.0002-0.0012	0.0080-0.0120	0.0140-0.0200	0.0080-0.0280	0.0014-0.0026	0.0012-0.0024	NA
	3.2	J32A1/UA5	0.0006-0.00160	0.0080-0.0140	0.0160-0.0220	0.0080-0.0280	0.0014-0.0024	0.0012-0.0022	NA
	3.2	J32A2/YA4	0.0006-0.00160	0.0080-0.0140	0.0160-0.0220	0.0080-0.0280	0.0014-0.0024	0.0012-0.0022	NA
	3.5	C35A1/KA	0.0010-0.0020	0.0100-0.0160	0.0160-0.0220	③	0.0022-0.0031	0.0012-0.0022	NA
2003	2.0	K20A3	0.0008-0.0016	0.0080-0.0014	0.0160-0.0220	0.0100-0.0260	0.0014-0.0024	0.0012-0.0022	NA
	2.0	K20A2	0.0008-0.0016	0.0080-0.0014	0.0200-0.0260	0.0080-0.0280	0.0016-0.0026	0.0018-0.0028	NA
	3.0	C30A1/NA1	0.0002-0.0014	0.0100-0.0160	0.0140-0.0200	0.0080-0.0280	0.0012-0.0022	0.0012-0.0022	NA
	3.0	J30A1/YA2	0.0006-0.00160	0.0080-0.0140	0.0160-0.0220	0.0080-0.0280	0.0014-0.0025	0.0012-0.0022	NA
	3.2	C32B1/NA2	0.0002-0.0012	0.0080-0.0120	0.0140-0.0200	0.0080-0.0280	0.0014-0.0026	0.0012-0.0024	NA
	3.2	J32A1/UA5	0.0006-0.00160	0.0080-0.0140	0.0160-0.0220	0.0080-0.0280	0.0014-0.0024	0.0012-0.0022	NA
	3.2	J32A2/YA4	0.0006-0.00160	0.0080-0.0140	0.0160-0.0220	0.0080-0.0280	0.0014-0.0024	0.0012-0.0022	NA
	3.5	C35A1/KA	0.0010-0.0020	0.0100-0.0160	0.0160-0.0220	③	0.0022-0.0031	0.0012-0.0022	NA

67162-ATSX-C10

PISTON AND RING SPECIFICATIONS
All measurements are given in inches

Year	Engine Displacement Liters	Engine ID/VIN	Piston Clearance	Ring Gap			Ring Side Clearance		
				Top Compression	Bottom Compression	Oil Control	Top Compression	Bottom Compression	Oil Control
2004	2.0	K20A3	0.0008-0.0016	0.0080-0.0014	0.0160-0.0220	0.0100-0.0260	0.0014-0.0024	0.0012-0.0022	NA
	2.0	K20A2	0.0008-0.0016	0.0080-0.0014	0.0200-0.0260	0.0080-0.0280	0.0016-0.0026	0.0018-0.0028	NA
	2.4	K24A2	0.0008-0.0016	0.0080-0.0014	0.0200-0.0260	0.0080-0.0280	0.0018-0.0028	0.0016-0.0026	NA
	3.0	C30A1/NA1	0.0002-0.0014	0.0100-0.0160	0.0140-0.0200	0.0080-0.0280	0.0012-0.0022	0.0012-0.0022	NA
	3.0	J30A1/YA2	0.0006-0.00160	0.0080-0.0140	0.0160-0.0220	0.0080-0.0280	0.0014-0.0025	0.0012-0.0022	NA
	3.2	C32B1/NA2	0.0002-0.0012	0.0080-0.0120	0.0140-0.0200	0.0080-0.0280	0.0014-0.0026	0.0012-0.0024	NA
	3.2	J32A1/UA5	0.0006-0.00160	0.0080-0.0140	0.0160-0.0220	0.0080-0.0280	0.0014-0.0024	0.0012-0.0022	NA
	3.2	J32A2/YA4	0.0006-0.00160	0.0080-0.0140	0.0160-0.0220	0.0080-0.0280	0.0014-0.0024	0.0012-0.0022	NA
	3.5	C35A1/KA	0.0010-0.0020	0.0100-0.0160	0.0160-0.0220	③	0.0022-0.0031	0.0012-0.0022	NA

NA; Not Applicable

① DB7: 4 door
DC4: 3 door

② DB8: 4 door
DC2: 3 door (Except Type R)

③ RIKEN: 0.0080-0.0280 inches
TEIKOKU: 0.0080-0.0200 inches

67162-ATSX-C11

TORQUE SPECIFICATIONS
All readings in ft. lbs.

Year	Engine Displacement Liters	Engine ID/VIN	Cylinder Head Bolts	Main Bearing Bolts	Rod Bearing Bolts	Crankshaft Damper Bolts	Flywheel Bolts	Manifold Intake	Manifold Exhaust	Spark Plugs	Oil Pan Drain Plug
2001	1.8	B18B1/①	②	56	③	130	④	17	23	13	33
	1.8	B18C1/⑤	②	⑥	⑦	130	76	17	23	13	33
	3.0	C30A1/NA1	56	⑧	⑨	181	④	16	25	13	33
	3.2	C32B1/NA2	56	⑩	29	181	76	16	22	13	33
	3.2	C35A1/KA9	56	⑪	33	181	54	16	22	13	29
	3.2	J32A2/YA4	⑫	⑬	⑭	181	54	16	23	13	29
	3.5	J32A1/UA5	⑫	⑬	⑭	181	54	16	23	13	33
2002	2.0	K20A3	⑮	⑯	14	181	⑰	16	33	13	33
	2.0	K20A2	⑮	⑯	22	181	90	16	33	⑱	29
	3.0	C30A1/NA1	56	⑦	⑧	181	⑨	16	25	13	33
	3.2	C32B1/NA2	56	⑩	29	181	76	16	22	13	33
	3.2	C35A1/KA9	56	⑪	33	181	54	16	22	13	29
	3.2	J32A2/YA4	⑫	⑬	⑭	181	54	16	23	13	29
	3.5	J32A1/UA5	⑫	⑬	⑭	181	54	16	23	13	33
2003	2.0	K20A3	⑮	⑯	14	181	⑰	16	33	13	33
	2.0	K20A2	⑮	⑯	22	181	90	16	33	⑱	29
	3.0	C30A1/NA1	56	⑦	⑧	181	⑨	16	25	13	33
	3.2	C32B1/NA2	56	⑩	29	181	76	16	22	13	33
	3.2	C35A1/KA9	56	⑪	33	181	54	16	22	13	29
	3.2	J32A2/YA4	⑫	⑬	⑭	181	54	16	23	13	29
	3.5	J32A1/UA5	⑫	⑬	⑭	181	54	16	23	13	33
2004	2.0	K20A3	⑮	⑯	14	181	⑰	16	33	13	33
	2.0	K20A2	⑮	⑯	22	181	90	16	33	⑱	29
	2.4	K24A2	⑮	⑯	16	⑲	76	16	33	13	33
	3.0	C30A1/NA1	56	⑦	⑧	181	⑨	16	25	13	33
	3.2	C32B1/NA2	56	⑩	29	181	76	16	22	13	29
	3.2	C35A1/KA9	56	⑪	33	181	54	16	22	13	29
	3.2	J32A2/YA4	⑫	⑬	⑭	181	54	16	23	13	29
	3.5	J32A1/UA5	⑫	⑬	⑭	181	54	16	23	13	33

① DB7: 4 door
  DC4: 3 door

② Step 1: 22 ft. lbs.
  Step 2: 63 ft. lbs.

③ Step 1: 14 ft. lbs.
  Step 2: 23 ft. lbs.

④ Manual transmission: 76 ft. lbs.
  Automatic transmission: 54 ft. lbs.

⑤ DB8: 4 door (Except Type R)
  DC2: 3 door

⑥ Step 1: 22 ft. lbs.
  Step 2: Cap Nos. 1, 5: 56 ft. lbs.
       Cap Nos. 2, 3 and 4: 49 ft. lbs.

⑦ Step 1: 14 ft. lbs.
  Step 2: 33 ft. lbs.

⑧ Cap bolts: 29 ft. lbs.
  Cap bridge bolts: 48 ft. lbs.

⑨ 14 ft. lbs. plus 116 degrees

⑩ Step 1: Cap bolts 48 ft. lbs.
  Step 2: Side bolts 36 ft. lbs.

⑪ Step 1: Outer (9mm) 29 ft. lbs.
  Step 2: Inner (11mm) 56 ft. lbs.
  Step 3: Side (10mm) 36 ft. lbs.

⑫ Step 1: 29 ft. lbs.
  Step 2: 51 ft. lbs.
  Step 3: 72.3 ft. lbs.

⑬ Step 1: Cap bolts 56 ft. lbs.
  Step 2: Side bolts 36 ft. lbs.

⑭ Step 1: 14 ft. lbs.
  Step 2: Rotate 90 degrees

⑮ Step 1: 29 ft. lbs.
  Step 2: Rotate 90 degrees
  Step 3: Rotate an additional 90 degrees
  Step 4 (new bolts only): additional 90 degrees

⑯ Step 1: 22 ft. lbs.
  Step 2: Rotate 56 degrees

⑰ Manual trans.: 54 ft. lbs.
  Auto trans.: 76 ft. lbs.

⑱ NGK type IFRG-11KS & Denso type SK22PRM11S: 18 ft. lbs.
  All others: 13 ft. lbs.

⑲ Step 1: 36 ft. lbs.
  Step 2: Rotate 90 degrees

WHEEL ALIGNMENT

Year	Model		Caster Range (+/-Deg.)	Caster Preferred Setting (Deg.)	Camber Range (+/-Deg.)	Camber Preferred Setting (Deg.)	Toe-in (in.)
2001	3.2 TL	F	1.00	+2.80	1.00	0	0 +/- 0.06
		R	—	—	0.50	-0.50	0.06 +/- 0.06
	3.2 CL	F	1.00	+2.80	1.00	0	0 +/- 0.06
		R	—	—	1.00	-0.50	0.06 +/- 0.06
	3.5 RL	F	1.00	+2.81	1.00	0	0 +/- 0.06
		R	—	—	1.00	-0.50	0.06 +/- 0.06
	Integra ①	F	1.00	+1.16	1.00	+0.16	0 +/- 0.16
		R	—	—	1.00	+0.75	0.16 +/- 0.16
	Integra ②	F	1.00	+1.16	1.00	+0.16	0 +/- 0.06
		R	—	—	1.00	+0.50	0.06 +/- 0.06
	NSX ③	F	0.25	+8.00	0.17	-0.33	0.14 +/- 0.04
		R	—	—	0.50	-1.50	0.18 +/- 0.06
	NSX ④	F	0.25	+8.23	0.17	+0.50	0.14 +/- 0.04
		R	—	—	0.50	-2.00	0.18 +/- 0.06
2002	3.2 TL	F	1.00	+2.80	1.00	0	0 +/- 0.06
		R	—	—	0.50	-0.50	0.06 +/- 0.06
	3.2 CL	F	1.00	+2.80	1.00	0	0 +/- 0.06
		R	—	—	1.00	-0.50	0.06 +/- 0.06
	3.5 RL	F	1.00	+2.81	1.00	0	0 +/- 0.06
		R	—	—	1.00	-0.50	0.06 +/- 0.06
	RSX ③	F	1.00	+1.16	1.00	+0.16	0 +/- 0.16
		R	—	—	1.00	+0.75	0.16 +/- 0.16
	RSX ④	F	1.00	+1.16	1.00	+0.16	0 +/- 0.06
		R	—	—	1.00	+0.50	0.06 +/- 0.06
	NSX ③	F	0.25	+8.00	0.17	-0.33	0.14 +/- 0.04
		R	—	—	0.50	-1.50	0.18 +/- 0.06
	NSX ④	F	0.25	+8.23	0.17	+0.50	0.14 +/- 0.04
		R	—	—	0.50	-2.00	0.18 +/- 0.06
2003	3.2 TL	F	1.00	+2.80	1.00	0	0 +/- 0.06
		R	—	—	0.50	-0.50	0.06 +/- 0.06
	3.2 CL	F	1.00	+2.80	1.00	0	0 +/- 0.06
		R	—	—	1.00	-0.50	0.06 +/- 0.06
	3.5 RL	F	1.00	+2.81	1.00	0	0 +/- 0.06
		R	—	—	1.00	-0.50	0.06 +/- 0.06
	RSX ③	F	1.00	+1.16	1.00	+0.16	0 +/- 0.16
		R	—	—	1.00	+0.75	0.16 +/- 0.16
	RSX ④	F	1.00	+1.16	1.00	+0.16	0 +/- 0.06
		R	—	—	1.00	+0.50	0.06 +/- 0.06
	NSX ③	F	0.25	+8.00	0.17	-0.33	0.14 +/- 0.04
		R	—	—	0.50	-1.50	0.18 +/- 0.06
	NSX ④	F	0.25	+8.23	0.17	+0.50	0.14 +/- 0.04
		R	—	—	0.50	-2.00	0.18 +/- 0.06

67162-ATSX-C13

WHEEL ALIGNMENT

Year	Model			Caster		Camber		Toe-in (in.)
				Range (+/-Deg.)	Preferred Setting (Deg.)	Range (+/-Deg.)	Preferred Setting (Deg.)	
2004	3.2 TL	F		1.00	+2.80	1.00	0	0 +/- 0.06
		R		—	—	0.50	-0.50	0.06 +/- 0.06
	3.5 RL	F		1.00	+2.81	1.00	0	0 +/- 0.06
		R		—	—	1.00	-0.50	0.06 +/- 0.06
	RSX ③	F		1.00	+1.16	1.00	+0.16	0 +/- 0.16
		R		—	—	1.00	+0.75	0.16 +/- 0.16
	RSX ④	F		1.00	+1.16	1.00	+0.16	0 +/- 0.06
		R		—	—	1.00	+0.50	0.06 +/- 0.06
	NSX ③	F		0.25	+8.00	0.17	-0.33	0.14 +/- 0.04
		R		—	—	0.50	-1.50	0.18 +/- 0.06
	NSX ④	F		0.25	+8.23	0.17	+0.50	0.14 +/- 0.04
		R		—	—	0.50	-2.00	0.18 +/- 0.06
	TSX	F		0.45	+3.13	0.45	0	0 +/- 0.08
		R		—	—	0.30	-1.00	0.08 +/- 0.08

① Except Type R
② Type R
③ Except Type S
④ Type S

67162-ATSX-C14

TIRE, WHEEL AND BALL JOINT SPECIFICATIONS

| Year | Model | OEM Tires | | Tire Pressures (psi) | | Wheel Size | Ball Joint Inspection | Lug Nut (ft. lbs.) |
		Standard	Optional	Front	Rear			
2001	Integra	P195/55VR15	None	32	30	6-JJ	NS	80
	3.2TL	P205/60VR16	None	29	29	6-JJ	NS	80
	3.2CL	P205/60VR16	None	29	29	6-JJ	NS	80
	3.5RL	P225/55/R16	None	32	29	6-JJ	NS	80
	NSX	①	None	32	29	6-JJ	NS	80
2002	RSX	P205/55R16	None	32	30	6-JJ	NS	80
	3.2TL	P205/60VR16	None	29	29	6-JJ	NS	80
	3.2CL	P205/60VR16	None	29	29	6-JJ	NS	80
	3.5RL	P225/55/R16	None	32	29	6-JJ	NS	80
	NSX	①	None	32	29	6-JJ	NS	80
2003	RSX	P205/55R16	None	32	30	6-JJ	NS	80
	3.2TL	P205/60VR16	None	29	29	6-JJ	NS	80
	3.2CL	P205/60VR16	None	29	29	6-JJ	NS	80
	3.5RL	P225/55/R16	None	32	29	6-JJ	NS	80
	NSX	①	None	32	29	6-JJ	NS	80
2004	RSX	P205/55R16	None	32	30	6-JJ	NS	80
	3.2TL	P205/60VR16	None	29	29	6-JJ	NS	80
	3.5RL	P225/55/R16	None	32	29	6-JJ	NS	80
	TSX	P215/50/R17	None	32	30	6-JJ	NS	80
	NSX	①	None	32	29	6-JJ	NS	80

OEM: Original Equipment Manufacturer

PSI: Pounds Per Square Inch

STD: Standard

OPT: Optional

NS: Not Specified by manufacturer

① Front tires: P215/40R17

Rear tires: P255/40R17

67162-ATSX-C15

BRAKE SPECIFICATIONS
All measurements in inches unless noted

Year	Model		Brake Disc Original Thickness	Brake Disc Minimum Thickness	Brake Disc Maximum Runout	Brake Drum Diameter Original Inside Diameter	Brake Drum Diameter Max. Wear Limit	Brake Drum Diameter Maximum Machine Diameter	Minimum Lining Thickness Front	Minimum Lining Thickness Rear	Brake Caliper Bracket Bolts (ft. lbs.)	Brake Caliper Mounting Bolts (ft. lbs.)
2001	3.5RL	F	0.910	0.830	0.004	—	—	—	0.06	—	80	36
		R	0.350	0.300	0.004	—	—	—	—	0.06	28	17
	3.2TL	F	1.100	1.020	0.004	—	—	—	0.06	—	80	36
		R	0.350	0.310	0.004	①	6.693 ②	②	—	③	41	17
	3.2CL	F	1.100	1.020	0.004	—	—	—	0.06	—	80	36
		R	0.350	0.310	0.004	①	6.693 ②	②	—	③	41	17
	Integra	F	0.830	0.750	0.004	—	—	—	0.06	—	—	24
		R	0.350	0.310	0.004	—	—	—	—	0.06	—	24
	NSX	F	1.100	1.020	0.004	—	—	—	0.06	—	80	36
		R	0.830	0.750	0.004	—	—	—	—	0.06	80	36
2002	3.5RL	F	0.910	0.830	0.004	—	—	—	0.06	—	80	36
		R	0.350	0.300	0.004	—	—	—	—	0.06	28	17
	3.2TL	F	1.100	1.020	0.004	—	—	—	0.06	—	80	36
		R	0.350	0.310	0.004	①	6.693 ②	②	—	③	41	17
	3.2CL	F	1.100	1.020	0.004	—	—	—	0.06	—	80	36
		R	0.350	0.310	0.004	①	6.693 ②	②	—	③	41	17
	RSX	F	0.830	0.750	0.004	—	—	—	0.06	—	—	24
		R	0.350	0.310	0.004	—	—	—	—	0.06	—	24
	RSX Type-S	F	0.980	0.910	0.004	—	—	—	0.06	—	—	24
		R	0.350	0.310	0.004	—	—	—	—	0.06	—	24
	NSX	F	1.100	1.020	0.004	—	—	—	0.06	—	80	36
		R	0.830	0.750	0.004	—	—	—	—	0.06	80	36
2003	3.5RL	F	0.910	0.830	0.004	—	—	—	0.06	—	80	36
		R	0.350	0.300	0.004	—	—	—	—	0.06	28	17
	3.2TL	F	1.100	1.020	0.004	—	—	—	0.06	—	80	36
		R	0.350	0.310	0.004	①	6.693 ②	②	—	③	41	17
	3.2CL	F	1.100	1.020	0.004	—	—	—	0.06	—	80	36
		R	0.350	0.310	0.004	①	6.693 ②	②	—	③	41	17
	RSX	F	0.830	0.750	0.004	—	—	—	0.06	—	—	24
		R	0.350	0.310	0.004	—	—	—	—	0.06	—	24
	RSX Type-S	F	0.980	0.910	0.004	—	—	—	0.06	—	—	24
		R	0.350	0.310	0.004	—	—	—	—	0.06	—	24
	NSX	F	1.100	1.020	0.004	—	—	—	0.06	—	80	36
		R	0.830	0.750	0.004	—	—	—	—	0.06	80	36
2004	3.5RL	F	0.910	0.830	0.004	—	—	—	0.06	—	80	36
		R	0.350	0.300	0.004	—	—	—	—	0.06	28	17
	3.2TL	F	1.100	1.020	0.004	—	—	—	0.06	—	80	36
		R	0.350	0.310	0.004	①	6.693 ②	②	—	③	41	17
	RSX	F	0.830	0.750	0.004	—	—	—	0.06	—	—	24
		R	0.350	0.310	0.004	—	—	—	—	0.06	—	24
	RSX Type-S	F	0.980	0.910	0.004	—	—	—	0.06	—	—	24
		R	0.350	0.310	0.004	—	—	—	—	0.06	—	24
	TSX	F	0.980	0.910	0.004	—	—	—	0.06	—	—	24
		R	0.390	0.310	0.004	—	—	—	—	0.06	—	24
	NSX	F	1.100	1.020	0.004	—	—	—	0.06	—	80	36
		R	0.830	0.750	0.004	—	—	—	—	0.06	80	36

NA: Not Available
F: Front
R: Rear
① Rear parking brake drum: 6.693 inches

② Rear parking brake drum maximum diameter: 6.732 inches
③ Rear pad: 0.06 inches
   Rear parking brake shoes: 0.04 inches

67162-ATSX-C16

SCHEDULED MAINTENANCE INTERVALS
ACURA—3.2CL, 3.2TL, 3.5RL, INTEGRA, RSX, NSX & TSX

TO BE SERVICED	TYPE OF SERVICE	VEHICLE MILEAGE INTERVAL (x1000)												
		7.5	15	22.5	30	37.5	45	52.5	60	67.5	75	82.5	90	97.5
Engine oil	R	✓	✓	✓	✓	✓	✓	✓	✓	✓	✓	✓	✓	✓
Rear brake discs, calipers & pads	S/I		✓		✓		✓		✓		✓		✓	
Rotate tires	S/I	✓	✓	✓	✓	✓	✓	✓	✓	✓	✓	✓	✓	✓
A/C filter	R		✓		✓		✓		✓		✓		✓	
A/C filter (3.5RL)	R				✓				✓				✓	
Brake hoses & lines (including ABS)	S/I		✓		✓		✓		✓		✓		✓	
Cooling system hoses & connections	S/I		✓		✓		✓		✓		✓		✓	
Driveshaft boots	S/I		✓		✓		✓		✓		✓		✓	
Exhaust system	S/I		✓		✓		✓		✓		✓		✓	
Front brake discs & calipers	S/I		✓		✓		✓		✓		✓		✓	
Fuel pipes, hoses & connections	S/I		✓		✓		✓		✓		✓		✓	
Suspension components	S/I		✓		✓		✓		✓		✓		✓	
Suspension mounting bolts	S/I		✓		✓		✓		✓		✓		✓	
Tie rods, steering gear box & boots	S/I		✓		✓		✓		✓		✓		✓	
Steering operation, tie rod ends, steering gearbox & boots	S/I		✓		✓				✓				✓	
Valve clearance (NSX)	S/I				✓				✓				✓	
Parking brake	S/I		✓		✓		✓		✓		✓		✓	
Air cleaner element	R				✓				✓				✓	
Automatic transmission fluid	R				✓				✓				✓	
Brake fluid (including ABS) (Integra & RSX)	R				✓				✓				✓	
Brake fluid (including ABS) (3.5RL)	R								✓				✓	
Brake fluid (including ABS) (3.2TL & NSX)	R						✓				✓			
Front differential fluid (3.2TL & 3.5RL)	R								✓				✓	
Manual transmission fluid	R				✓				✓				✓	
ABS operation	S/I				✓				✓				✓	
Drive belt(s)	S/I				✓				✓				✓	
Spark plugs (Integra exc. GSR & RSX exc. Type S)	R				✓				✓				✓	
Spark plugs (3.2TL, Integra GSR & NSX)	R								✓				✓	
Spark plugs (3.5.RL)	R								✓					
Engine coolant	R						✓				✓			

67162-ATSX-C17

SCHEDULED MAINTENANCE INTERVALS
ACURA—3.2CL, 3.2TL, 3.5RL, INTEGRA, RSX, NSX & TSX

TO BE SERVICED	TYPE OF SERVICE	VEHICLE MILEAGE INTERVAL (x1000)												
		7.5	15	22.5	30	37.5	45	52.5	60	67.5	75	82.5	90	97.5
ABS high pressure hose (NSX)	R								✓					
Fuel filter	R								✓					
PCV valve	S/I								✓					
Timing belt (except as noted below)	R												✓	
Timing belt & timing balancer belt (3.5RL) ①	R													
Transmission fluid	R												✓	
Distributor, ignition cap & rotor (Integra)	S/I								✓					
Idle speed (3.2TL, Integra & NSX)	S/I								✓					
Idle speed (3.5RL) ②	S/I		✓		✓		✓		✓		✓		✓	
Ignition wires	S/I		✓		✓				✓				✓	
TWC converter heat shield	S/I		✓		✓		✓		✓		✓		✓	
Water pump	S/I				✓				✓				✓	
Water pump (3.5RL) ②	S/I	✓			✓		✓		✓		✓		✓	

R: Replace   S/I: Service or Inspect

① Replace at 105,000 miles.

② Service or inspect at 105,000 miles.

FREQUENT OPERATION MAINTENANCE (SEVERE SERVICE)

If a vehicle is operated under any of the following conditions it is considered severe service:

- Extremely dusty areas.

-50% or more of the vehicle operation is in 32°C (90°F) or higher temperatures, or constant operation in temperatures below 0°C (32°F).

-Prolonged idling (vehicle operation in stop and go traffic).

-Frequent short running periods (engine does not warm to normal operating temperatures).

-Police, taxi, delivery usage or trailer towing usage.

Oil & oil filter: change every 3750 miles.

Brake hoses & lines (including ABS) (3.5RL): service or inspect every 7500 miles.

Cooling system hoses & connections (3.5RL): service or inspect every 7500 miles.

Driveshaft boots (3.2TL & 3.5RL): check every 7500 miles.

Exhaust system (3.5RL): check every 7500 miles.

Brake discs, calipers & pads: service or inspect every 7500 miles.

Fuel pipes, hoses & connections (3.5RL): check every 7500 miles.

Power steering system: service or inspect every 7500 miles.

Suspension components: service or inspect every 7500 miles.

Tie rod ends, steering gear box & boots (3.2TL & 3.5RL): service or inspect every 7500 miles.

Air cleaner element (NSX): service or inspect every 7500 miles.

Air cleaner element (except NSX): service or inspect every 15,000 miles.

Front differential fluid (3.2TL, 3.5RL): replace every 15,000 miles.

Transmission fluid (NSX): replace every 15,000 miles.

Transmission fluid (3.2TL & 3.5RL): replace every 30,000 miles

Timing belt (3.2TL): replace every 60,000 miles.

Water pump (3.2TL): service or inspect every 60,000 miles.

67162-ATSX-C18

PRECAUTIONS

Before servicing any vehicle, please be sure to read all of the following precautions, which deal with personal safety, prevention of component damage and important points to take into consideration when servicing a motor vehicle:

• Never open, service or drain the radiator or cooling system when the engine is hot; serious burns can occur from the steam and hot coolant.

• Observe all applicable safety precautions when working around fuel. Whenever servicing the fuel system, always work in a well-ventilated area. Do not allow fuel spray or vapors to come in contact with a spark, open flame, or excessive heat (a hot drop light, for example). Keep a dry chemical fire extinguisher near the work area. Always keep fuel in a container specifically designed for fuel storage; also, always properly seal fuel containers to avoid the possibility of fire or explosion. Refer to the additional fuel system precautions later in this section.

• Fuel injection systems often remain pressurized, even after the engine has been turned **OFF**. The fuel system pressure must be relieved before disconnecting any fuel lines. Failure to do so may result in fire and/or personal injury.

• Brake fluid often contains polyglycol ethers and polyglycols. Avoid contact with the eyes and wash your hands thoroughly after handling brake fluid. If you do get brake fluid in your eyes, flush your eyes with clean, running water for 15 minutes. If eye irritation persists, or if you have taken brake fluid internally, seek medical assistance IMMEDIATELY.

• The EPA warns that prolonged contact with used engine oil may cause a number of skin disorders, including cancer! You should make every effort to minimize your exposure to used engine oil. Protective gloves should be worn when changing oil. Wash your hands and any other exposed skin areas as soon as possible after exposure to used engine oil. Soap and water, or waterless hand cleaner should be used.

• All new vehicles are now equipped with an air bag system. The system must be disabled before performing service on or around system components, steering column, instrument panel components, wiring and sensors. Failure to follow safety and disabling procedures could result in accidental air bag deployment, possible personal injury and unnecessary system repairs.

• Always wear safety goggles when working with, or around, the air bag system. When carrying a non-deployed air bag, be sure the bag and trim cover are pointed away from your body. When placing a non-deployed air bag on a work surface, always face the bag and trim cover upward, away from the surface. This will reduce the motion of the module if it is accidentally deployed. Refer to the additional air bag system precautions later in this section.

• Clean, high quality brake fluid from a sealed container is essential to the safe and proper operation of the brake system. You should always buy the correct type of brake fluid for your vehicle. If the brake fluid becomes contaminated, completely flush the system with new fluid. Never reuse any brake fluid. Any brake fluid that is removed from the system should be discarded. Also, do not allow any brake fluid to come in contact with a painted surface; it will damage the paint.

• Never operate the engine without the proper amount and type of engine oil; doing so WILL result in severe engine damage.

• Timing belt maintenance is extremely important! Many models utilize an interference type, non-freewheeling engine. If the timing belt breaks, the valves in the cylinder head may strike the pistons, causing potentially serious (also time consuming and expensive) engine damage. Refer to the maintenance interval charts in the front of this section for the recommended replacement interval for the timing belt and to the timing belt procedure for belt replacement and inspection.

• Disconnecting the negative battery cable on some vehicles may interfere with the functions of the on-board computer system(s) and may require the computer to undergo a relearning process once the negative battery cable is reconnected.

• When servicing drum brakes, only disassemble and assemble one side at a time, leaving the remaining side intact for reference.

• Only an MVAC-trained, EPA-certified automotive technician should service the air conditioning system or its components.

• The radio may contain a coded theft protection circuit. Always obtain the code number before disconnecting the battery.

ENGINE REPAIR

➡Disconnecting the negative battery cable on some vehicles may interfere with the functions of the on board computer system. The computer may undergo a relearning process once the negative battery cable is reconnected.

Distributor

REMOVAL & INSTALLATION

1.8L Engines

1. Before servicing the vehicle, refer to the precautions in the beginning of this section.
2. Remove or disconnect the following:
   • Negative battery cable
   • Engine wiring harness and connectors from the distributor

   • Spark plug wires from the distributor cap
3. If removing the ignition coil, remove the distributor cap, rotor, and cap seal, then remove the leak cover.
4. Remove or disconnect the following:
   • 2 screws to disconnect the wires from the coil
   • Ignition coil screws and slide the ignition coil out of the distributor housing
   • Distributor hold-down bolts
   • Distributor from the cylinder head

To install:

5. Use a new O-ring coated with engine oil, on the distributor housing.
6. Slip the distributor into position.

➡Lugs on the end of the distributor and the matching grooves in the camshaft end are offset to eliminate any possibility of installing the distributor 180 degrees out of time.

7. Install the hold-down bolts and hand tighten them.
8. Slide the ignition coil into the distributor housing and install the 2 mounting screws.
9. Install or connect the following:
   • 2 wires to the coil and install the 2 screws.
   • Distributor leak cover, rotor, cap seal, and cap
   • Engine wiring harness and connector to distributor
   • Spark plug wires
   • Negative battery cable
10. Set the timing, then tighten the hold down bolts to 16 ft. lbs. (22 Nm).

3.0L Engines

1. Before servicing the vehicle, refer to the precautions in the beginning of this section.
2. Remove or disconnect the following:
   - The negative battery cable
   - The spark plug and coil wires from the distributor cap
   - Harness connector(s) from the distributor
   - Distributor mounting bolts
   - Distributor from the cylinder head

To install:

3. Install or connect the following:
   - New O–ring coated with engine oil, on the distributor housing
   - Distributor
   - Mounting bolts and tighten them to 13 ft. lbs. (18 Nm)
   - Spark plug and coil wires
   - Negative battery cable
4. Check the ignition timing with a timing light. The timing marks are located on the crankshaft pulley and lower timing cover. If the timing is not within specification replace the PCM module.

Alternator

REMOVAL & INSTALLATION

1.8L and 3.5L Engines

1. Before servicing the vehicle, refer to the precautions in the beginning of this section.
2. Remove or disconnect the following:
   - Both battery cables
   - 4 prong connector and the black wire from the rear of the alternator
   - Alternator adjusting bolt(s)
   - Mounting bolt(s)
   - Alternator belt
   - Alternator assembly

To install:

3. Install or connect the following:
   - Alternator assembly
   - Mounting bolts and tighten them to 33 ft. lbs. (44 Nm)
   - Alternator adjusting bolt, hand tight
   - Alternator belt
4. Adjust alternator belt to tension.
5. Tighten the 8 x 1.25mm locknut/lock bolt(s) to 16 ft. lbs. (22 Nm).
6. Install or connect the following:
   - 4 prong connector and the black wire to the rear of the alternator
   - Both battery cables

❄❄ WARNING

Be sure to adjust the alternator belt to the proper tension or alternator bearing failure may occur.

➡ **The Powertrain Control Module (PCM) idle memory must be reset after reconnecting the battery. Start the engine and hold it at 3000 rpm until the cooling fan comes on. Then allow the engine to idle for about 5 minutes with all accessories OFF and with the transmission in Park or Neutral.**

2.0L Engines

1. Before servicing the vehicle, refer to the precautions in the beginning of this section.
2. Remove or disconnect the following:
   - Both battery cables
   - Drive belt
   - Auto tensioner
   - Connector and the black wire from the rear of the alternator
   - Mounting bolt(s)
   - Alternator assembly

To install:

3. Install or connect the following:
   - Alternator assembly
   - Mounting bolts and tighten them to 16 ft. lbs. (22 Nm)
   - Connector and the black wire to the rear of the alternator
   - Auto tensioner
   - Drive belt
   - Both battery cables

❄❄ WARNING

Be sure to adjust the alternator belt to the proper tension or alternator bearing failure may occur.

➡ **The Powertrain Control Module (PCM) idle memory must be reset after reconnecting the battery. Start the engine and hold it at 3000 rpm until the cooling fan comes on. Then allow the engine to idle for about 5 minutes with all accessories OFF and with the transmission in Park or Neutral.**

2.4L Engine

1. Before servicing the vehicle, refer to the precautions in the beginning of this section.
2. Note the radio security code and the radio presets.
3. Remove or disconnect the following:
   - Negative battery cable, then the positive

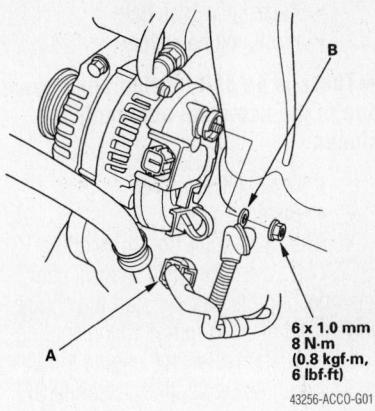

6 x 1.0 mm
8 N·m
(0.8 kgf·m,
6 lbf·ft)

43256-ACCO-G01

Alternator mounting—2.4L engine

- Drive belt
- Auto-tensioner
- Connectors from the alternator
- 3 alternator mounting bolts and the alternator

To install:

- Alternator and 3 mounting bolts. Torque the bolts to 16 ft. lbs. (22 Nm).
- Electrical connectors
- Auto-tensioner
- Drive belt
- Positive, then negative battery cables
4. Enter the security code and radio presets

3.0L Engine

1. Before servicing the vehicle, refer to the precautions in the beginning of this section.
2. Remove or disconnect the following:
   - Battery cover
   - Both battery cables
   - Engine cover
   - Accessory drive belt
   - Fan motor connector and A/C compressor clutch wiring connector from the fan shroud
   - A/C condenser fan shroud assembly
   - Ground cable
   - 4 prong connector
   - Alternator and bracket assembly

To install:

3. Install or connect the following:
   - Alternator and bracket assembly and tighten the 10 x 1.25 mounting bolts to 33 ft. lbs. (44 Nm) and the 8 x 1.25mm locknut/lock bolt(s) to 16 ft. lbs. (22 Nm)
   - 4 prong connector
   - Ground cable
   - A/C condenser fan shroud assembly

- Accessory drive belt
- Front engine cover

➡ **There is no belt tension adjustment due to the use of an automatic tensioner.**

- Both battery cables
- Battery cover

4. The Powertrain Control Module (PCM) idle memory must be reset after reconnecting the battery. Start the engine and hold it at 3000 rpm until the cooling fan comes on. Then allow the engine to idle for about 5 minutes with all accessories OFF and with the transmission in Park or Neutral.

3.2L Engine

1. Before servicing the vehicle, refer to the precautions in the beginning of this section.

2. Remove or disconnect the following:
- Both battery cables
- Adjusting bolts
- Accessory drive belt
- Mounting bolts

3. Turn the alternator 90° in a counter–clockwise direction.
- Alternator
- 4 prong connector
- Harness clip and bracket assembly
- Black wire from the terminal
- Alternator from the vehicle

To install:

4. Install or connect the following:
- Alternator
- Black wire
- Harness clip and bracket assembly and t tighten the bolt to 104 inch lbs. (12 Nm)
- 4 prong connector
- Mounting bolts and tighten the 10 x 1.25mm bolts to 33 ft. lbs. (44 Nm), the 8 x 1.25mm locknut/lock bolt(s) to 16 ft. lbs. (22 Nm) and the 6 x 1.0mm 104 inch lbs. (12 Nm)
- Accessory drive belt

➡ **There is no belt tension adjustment due to the use of an automatic tensioner.**

5. The Powertrain Control Module (PCM) idle memory must be reset after reconnecting the battery. Start the engine and hold it at 3000 rpm until the cooling fan comes on. Then allow the engine to idle for about 5 minutes with all accessories OFF and with the transmission in Park or Neutral.

6. Connect the positive, then the negative battery cable.

Ignition Timing

ADJUSTMENT

1.8L Engines

1. Before servicing the vehicle, refer to the precautions in the beginning of this section.

2. Start the engine and hold the engine speed at 3000 rpm, until the radiator fan comes on. The engine should be at idle speed and at normal operating temperature. Be sure all electrical devices (radio, air conditioning, lights, etc.) are turned OFF.

3. Locate the Service Check (SCS) connector:
- 1.8L engines: behind the right kick panel

4. Connect the SCS service connector (part number 07PAZ–0010100) to the service check connector.

5. Connect a timing light to No. 1 ignition wire and point the light toward the pointer on the timing belt cover.

6. Check the idle speed and adjust if necessary.

7. The red mark on the crankshaft pulley should be aligned with the pointer on the timing belt cover.

➡ **The white mark on the crank pulley is Top Dead Center (TDC).**

8. Adjust the ignition timing by loosening the distributor mounting bolts and rotating the distributor housing to adjust the timing. Set as follows:
- 1.8L (Except Type R): 16 degrees Before Top Dead Center (BTDC) at 700–800 rpm
- 1.8L (Type R): 16 degrees BTDC at 750–850 rpm

9. Tighten the distributor bolts to 17 ft. lbs. (24 Nm) and recheck the timing.

10. Remove the SCS service connector from the service check connector.

2.0L, 2.4L and 3.0L Engines

➡ **The ignition timing is controlled by the Powertrain Control Module (PCM) and can be checked for diagnostic purposes. If the timing is out of specification, all mechanical and electrical systems should be checked for proper operation before replacing the PCM.**

1. Before servicing the vehicle, refer to the precautions in the beginning of this section.

2. To check the ignition timing, start the engine and allow it to fast idle at 3000 rpm with all electrical accessories off and the

transmission in **N** or **P**. Allow the engine to warm up and reach normal operating temperature. The engine cooling fan should cycle at least 1 time.

3. Locate the Service Check (SCS) connector under the glove box. Connect the service connector tool part number 07PAZ–0010100 to the SCS terminals.

4. Check the idle speed and adjust if necessary.

5. Connect a timing light to the No. 1 plug wire. While engine idles, point the light toward the pointer on the timing belt cover.

6. Inspect the ignition timing at idle. The specification is 6–10 degrees Before Top Dead Center (BTDC) at idle for the 2.0L engines, or 8–12 degrees BTDC at 700–800 for 3.0L engines

➡ **All mechanical and electrical systems should checked for proper operation before replacing the PCM.**

7. If the ignition timing is incorrect, replace the PCM.

8. Remove the service connector.

3.2L and 3.5L Engines

➡ **The ignition timing is controlled by the Powertrain Control Module (PCM) and can be checked for diagnostic purposes. If the timing is out of specification, all mechanical and electrical systems should checked for proper operation before replacing the PCM.**

1. Before servicing the vehicle, refer to the precautions in the beginning of this section.

2. To check the ignition timing, start the engine and allow it to fast idle at 3000 rpm with all electrical accessories off and the transmission in **N** or **P**. Allow the engine to warm up and reach normal operating temperature. The engine cooling fan should cycle at least 1 time.

3. Locate the Service Check (SCS) connector under the glove box and connect the service connector tool part number 07PAZ–0010100 to it.

4. Check the idle speed and adjust if necessary.

5. Connect a timing light to the No. 1 plug wire. With the engine idling at normal operating temperature point the timing light toward the pointer on the timing belt cover.

6. Inspect the ignition timing. The specifications are as follows:
- 3.2L: 13–17 degrees Before Top Dead Center (BTDC) at 590–690 rpm
- 3.5L: 13–17 degrees BTDC at 700–800 rpm

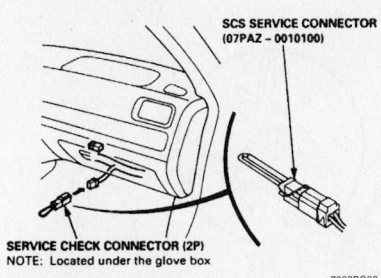

Service check connector—3.2TL and 3.5RL

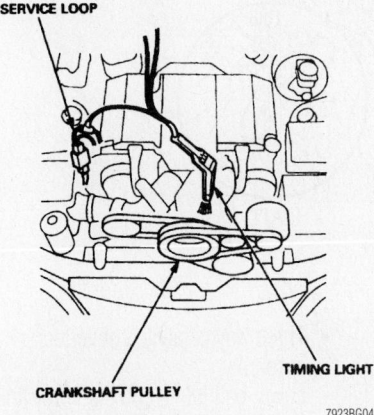

Timing light attachment—3.2TL

➡All mechanical and electrical systems should checked for proper operation before replacing the PCM.

7. If the ignition timing is incorrect, replace the PCM.

Only replace the PCM as a last resort.

8. Remove the timing light.

9. Disconnect the special tool (SCS service connector) from the service check connector.

Engine Assembly

REMOVAL & INSTALLATION

1.8L Engines

1. Before servicing the vehicle, refer to the precautions in the beginning of this section.

2. Relieve the fuel system pressure.

3. Remove or disconnect the following:
- Both battery cables
- Hood
- Strut brace (if equipped)
- Battery cables from the under–hood fuse/relay box and under–hood Antilock Brake (ABS) System fuse/relay box

- Air cleaner assembly and mounting bracket
- Evaporative emission (EVAP) control canister hose and vacuum hose from the intake manifold
- Brake booster vacuum hose
- Fuel return hose
- Engine wiring harness connectors on the right side of the engine compartment
- Fuel feed hose
- Throttle cable

➡Be careful not to bend the cable when removing it. Replace the cable if it gets kinked.

- Engine wiring harness connectors on the left side of the engine compartment
- Cruise control actuator
- Engine ground cable
- Power steering belt
- Power steering pump without disconnecting the hoses
- Air conditioning compressor belt
- Clutch slave cylinder and pipe/hose assembly (if equipped). Do not disconnect the pipe/hose assembly.
- Transmission ground cable and hose clamp
- Front wheels
- Lower splash shield

4. Drain the engine coolant, engine oil, and transmission fluid into sealable containers. Reinstall the drain plugs using new washers. Be careful not to overtighten the drain plugs.

5. Remove or disconnect the following:
- Upper and lower radiator hoses and the heater hoses from the engine
- Transmission oil cooler hoses (if equipped)
- Radiator assembly
- Air conditioning compressor without disconnecting the hoses
- Heated Oxygen (HO2S) sensor connector
- Front exhaust pipe from the exhaust manifold
- Shift rod and extension rod from the transaxle (if equipped)
- Shift cable cover, then disconnect the shift cable from the transaxle (if equipped)
- Damper fork
- Lower ball joints
- Halfshafts from the transaxle

6. Attach a hoist to the engine.
- Left and right front engine mounts and brackets
- Rear engine mount and bracket
- Side engine mount

- Transmission mount

7. Check that the engine is completely clear of all vacuum, fuel and coolant hoses, and electrical wiring.

8. Raise the engine and transaxle assembly all the way and remove it from the vehicle.

9. Separate the engine and transaxle.

To install:

10. Installation is the reverse of the removal procedure, while using the following torque values:
- Transaxle bolts: 47 ft. lbs. (64 Nm) (if equipped with a manual transaxle), or 43 ft. lbs. (59 Nm) (if equipped with an automatic transaxle)
- Rear mounting bracket: 87 ft. lbs. (118 Nm)
- Upper mounting bolts: 54 ft. lbs. (74 Nm)
- Torque converter bolts: 104 inch lbs. (12 Nm) (if equipped)
- Transmission mount bolts: 47 ft. lbs. (64 Nm)
- Engine side mount bolt: 47 ft. lbs. (64 Nm)
- Rear mount bracket bolts: 40 ft. lbs. (54 Nm)
- Right front mount/bracket bolts: 47 ft. lbs. (64 Nm)
- Damper fork bolts: 47 ft. lbs. (64 Nm)
- Lower ball joints nuts: 40 ft. lbs. (54 Nm)
- Shift cable bolt: 10 ft. lbs. (14 Nm)
- Shift rod and extension rod bolt: 16 ft. lbs. (22 Nm)
- Front exhaust pipe nuts: 40 ft. lbs. (54 Nm)
- A/C compressor bolts: 17 ft. lbs. (24 Nm)
- Power steering pump bolts: 17 ft. lbs. (24 Nm)
- Cruise control actuator bolts: 17 ft. lbs. (24 Nm)

11. Fill the engine with oil and the transaxle with fluid.

12. Fill and bleed (if necessary) the air from the cooling system.

13. Connect the positive, then the negative battery cable and enter the radio security code.

14. Switch the ignition **ON** but do not engage the starter. The fuel pump should run for approximately 2 seconds, building pressure within the lines. Switch the ignition **OFF**, then **ON** 2 or 3 more times to build full system pressure. Check for fuel leaks.

15. Start the engine, allowing it to idle. Check the hoses and lines carefully for any sign of leakage.

16. Check the timing and idle speed.

17. After the engine has warmed up fully and the fan(s) have come on at least once, recheck the engine for fluid leaks. Switch the engine OFF.

18. Adjust the belts and throttle cable as necessary.

2.0L Engines

1. Before servicing the vehicle, refer to the precautions in the beginning of this section.

2. Obtain the anti–theft code for the radio.

3. Drain the engine oil, coolant, and transmission oil (or fluid) into sealable containers and carefully reinstall the drain plugs using new sealing washers.

4. Properly relieve the fuel system pressure.

5. Remove or disconnect the following:
- Negative and positive battery cables
- Battery
- Intake manifold cover
- Intake Air Temperature (IAT) sensor connector
- Breather hose
- Air cleaner housing
- Intake air duct
- Battery cable from the underhood fuse/relay box, then the harness clamps and ground cable
- Throttle cover, if equipped

6. Fully open the throttle link and cruise control link by hand, then remove the cables from the links. Loosen the locknuts and remove the cables from the bracket.
- Engine Control Module (ECM)/Powertrain Control Module (PCM) connectors and main wire harness connector
- Harness clamps and grommet, then pull the engine wire harness through the bulkhead
- Fuel feed hose
- Evaporative Emission (EVAP) canister hose and brake booster vacuum hose
- Clutch slave cylinder and clutch line bracket mounting bolt, if equipped with a manual transaxle
- Shift cable and select cable, on manual transaxles
- Drive belt
- Power steering pump and position aside without disconnecting the hoses
- Bolt holding the power steering hose bracket

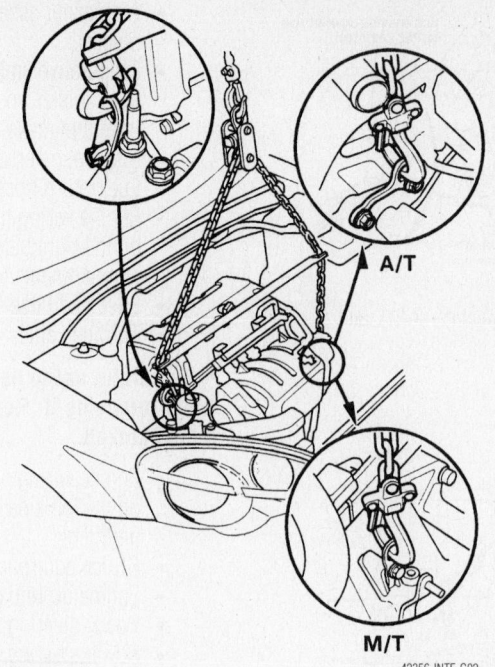

A/T

M/T

42356-INTE-G02

Attach a suitable engine tilt hanger to the engine

- Radiator cap
- Front wheels and tires
- Splash shield
- Air Fuel (A/F) ratio sensor connector
- Secondary Heated Oxygen (HO2) sensor connector

- Three way catalytic converter assembly
- Lower ball joints and stabilizer links
- Halfshafts. Coat all machined surfaces with clean engine oil, then tie plastic bags over the halfshaft ends to protect them.

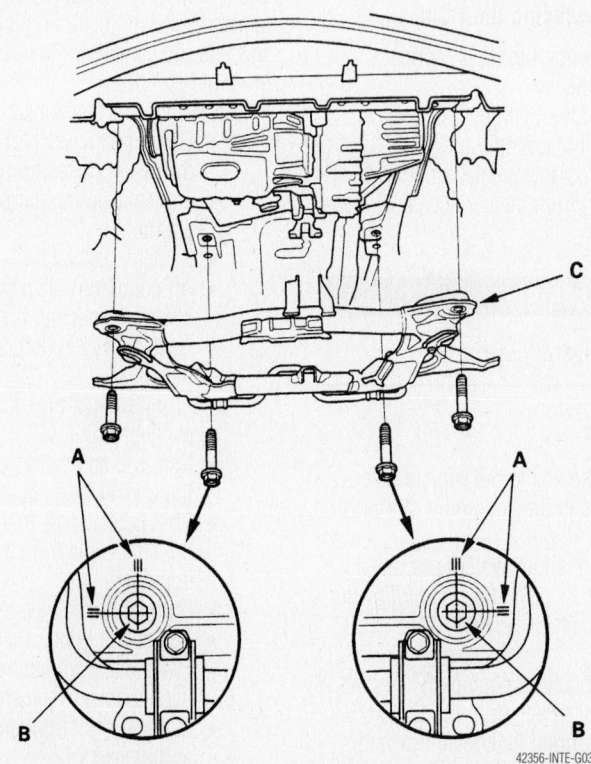

42356-INTE-G03

Make alignment marks on the reference lines (A) that align with the centers of the rear subframe mounting bolts (B), then remove the front subframe (C)

- Shift cable holder and the shift cable cover. To avoid damaging the control lever joint, be sure to remove the bolts holding the shift cable cover, on automatic transaxles.
- Spring clip and control pin, then the shift cable from the control lever, on automatic transaxles
- Lower hose
- Automatic Transaxle Fluid (ATF) filter mounting bolt, if equipped
- ATF cooler hoses, then plug the hoses and lines
- Upper hose and heater hoses
- Radiator

7. Attach a suitable engine tilt hanger to the engine.
- Transaxle mount bracket support bolts/nuts
- Upper bracket mounting bolt and nut

8. Make sure the hoist brackets are positioned properly, then raise the hoist to its full height.
- Rear mount mounting bolts
- Front mount bracket mounting bolt

9. Make alignment marks on the reference lines that align with the centers of the rear subframe mounting bolts, then remove the subframe.
- Compressor clutch connector and position the compressor aside without disconnect the refrigerant lines
- Any remaining electrical connectors, vacuum, fuel or coolant hoses

10. Slowly lower the engine about 6 in. (15 cm). Check that all hoses and wires are disconnected from the engine/transaxle.

11. Lower the engine all the way, then remove the chain hoist from the engine.

12. Remove the engine from under the vehicle.

To install:

13. Installation is the reverse of the removal procedure, while using the following torque values:
- Engine mount bracket: 33 ft. lbs. (44 Nm)
- A/C compressor bracket: 33 ft. lbs. (44 Nm)
- A/C compressor bolts: 16 ft. lbs. (22 Nm)
- New subframe bolts: 76 ft. lbs. (103 Nm)
- Rear mount mounting bolts: 43 ft. lbs. (59 Nm)
- Upper bracket mounting bolt: 40 ft. lbs. (54 Nm)
- Support bolts/nuts: 40 ft. lbs. (54 Nm)

- Front engine mount bracket mounting bolts: 47 ft. lbs. (64 Nm)
- A/T shift control cable: 7.2 ft. lbs. (9.8 Nm)
- New catalytic converter self-locking nuts: 25 ft. lbs. (33 Nm)
- Catalytic converter bolt: 16 ft. lbs. (22 Nm)
- ATF filter mounting bolt: 104 inch lbs. (12 Nm)
- Power steering pump bolts: 16 ft. lbs. (22 Nm)
- Power steering hose bracket bolt: 7.2 ft. lbs. (9.8 Nm)
- M/T cable bolts: 7.2 ft. lbs. (9.8 Nm)
- Clutch slave cylinder mounting bolts: 17 ft. lbs. (24 Nm)
- Clutch line bracket mounting bolt: 7.2 ft. lbs. (9.8 Nm)
- Battery cable harness clamp bolt: 104 inch lbs. (12 Nm)
- Air cleaner housing bolts: 104 inch lbs. (12 Nm)
- Intake manifold cover bolts: 104 inch lbs. (12 Nm)

14. Fill the engine with oil and the transaxle with fluid.

15. Fill and bleed (if necessary) the air from the cooling system.

16. Connect the positive, then the negative battery cable and enter the radio security code.

17. Switch the ignition **ON** but do not engage the starter. The fuel pump should run for approximately 2 seconds, building pressure within the lines. Switch the ignition **OFF**, then **ON** 2 or 3 more times to build full system pressure. Check for fuel leaks.

18. Start the engine, allowing it to idle. Check the hoses and lines carefully for any sign of leakage.

19. Check the timing and idle speed.

20. After the engine has warmed up fully and the fan(s) have come on at least once, recheck the engine for fluid leaks. Switch the engine OFF.

21. Adjust the belts and throttle cable as necessary.

2.4L Engines

1. Before servicing the vehicle, refer to the precautions in the beginning of this section.

2. Obtain the anti-theft code for the radio.

3. Remove the splash shield in order to drain the vehicle's fluids.

4. Drain the engine oil, coolant, and transmission oil (or fluid) into sealable containers and carefully reinstall the drain plugs using new sealing washers.

5. Properly relieve the fuel system pressure.

6. Remove or disconnect the following:
- Negative and positive battery cables
- Battery

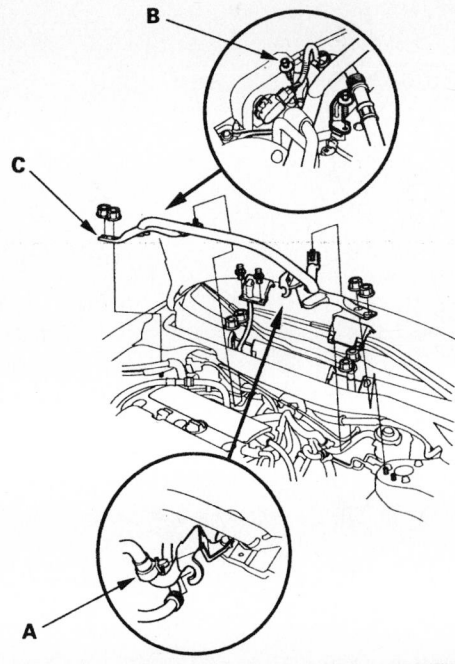

67162-ATSX-G01

To remove the strut brace (C), you must remove the harness clamp (A) and bolt (B)—2.4L engine

- Air cleaner housing
- Harness clamps and harness bracket
- Battery base retainers and base
- Battery cables from the underhood fuel/relay box and detach the electrical connector
- 2 bolts holding the underhood fuse/relay box
- Harness clamp and bolt and strut brace
- Quick-connect fitting cover and fuel feed hose
- Evaporative emission (EVAP) canister hose and brake booster vacuum hose
- Engine Control Module/Powertrain Control Module (ECM/PCM) connectors and main wire harness
- Accelerator Pedal Position (APP) sensor connector
- Harness clamps and grommet, then pull the engine wire harness through the bulkhead
- Clutch slave cylinder, shift cable and select cable, if M/T
- Drive belt
- Power steering pump and the hose clamp, but do NOT disconnect the fluid lines
- A/C compressor, but do not disconnect the A/C hoses
- Front wheels and tires
- Stabilizer links
- Damper fork
- Lower control arm ball joints

- Driveshafts. Coat all machined surfaces with clean engine oil, then tie plastic bags over the driveshaft ends.
- Shift control cable, if A/T
- Exhaust pipe "A"
- Nuts holding the transmission lower front mount and transmission lower rear mount
- Automatic Transmission Fluid cooler hoses
- Upper radiator hose and heater hoses

7. To disconnect the lower radiator hose, perform the following:

   a. Clean dirt from the quick-connector, radiator and lower radiator hose.

   b. Pull out the lock by hand, and wiggle the quick-connector to remove it from the radiator. Do not use tools to remove the quick-connector, do it by hand.

8. Remove or disconnect the following:
- Ground cable and upper bracket

9. Proper attach a chain hoist to the engine.

10. For vehicles with a M/T, remove the ground cable, transmission upper mount/bracket assembly and clutch line clamp bracket

11. For vehicles with an A/T, remove the ground cable, then remove the bracket plate, transmission upper bracket and transmission upper mount.

12. Remove or disconnect the following:
- Vacuum hose from the vacuum line
- Front mount stop and vacuum hose

clamp bracket and the front mount bolt
- Rear mount stop and the rear mount bolt

13. Make sure that the engine/transmission assembly is totally free of all vacuum lines, wiring and fuel and coolant hoses.

14. Slowly raise the engine about 6 in. (150mm), then recheck that all wires/hoses are disconnected from the engine/transmission assembly.

15. Raise the engine/transmission assembly all the way and remove it from the vehicle.

To install:

16. Install the accessory brackets to the engine and tighten to the following torque specifications:
- Side engine mount bracket: 33 ft. lbs. (45 Nm)
- Rear mount bracket (use NEW bolts): 65 ft. lbs. (88 Nm)
- Front mount bracket (use NEW bolts): 47 ft. lbs. (64 Nm)
- A/C compressor bracket: 33 ft. lbs. (45 Nm)

➡ **You must tighten the mounting bolts/support nuts in the order listed below, following the tightening sequences when given. Failure to follow the proper sequence may cause excessive noise and vibration and may also reduce the life of the bushings.**

17. Installation is the reverse of the removal procedure, while using the following torque values:
- Front mount bolt (use a NEW bolt): 40 ft. lbs. (54 Nm)
- Front mount stop nuts: 58 ft. lbs. (78 Nm).
- Rear mount bolt (use a NEW bolt): 40 ft. lbs. (54 Nm)
- Rear mount stop nuts: 51 ft. lbs. (69 Nm)
- 2 top upper bracket bolts: 40 ft. lbs. (54 Nm)

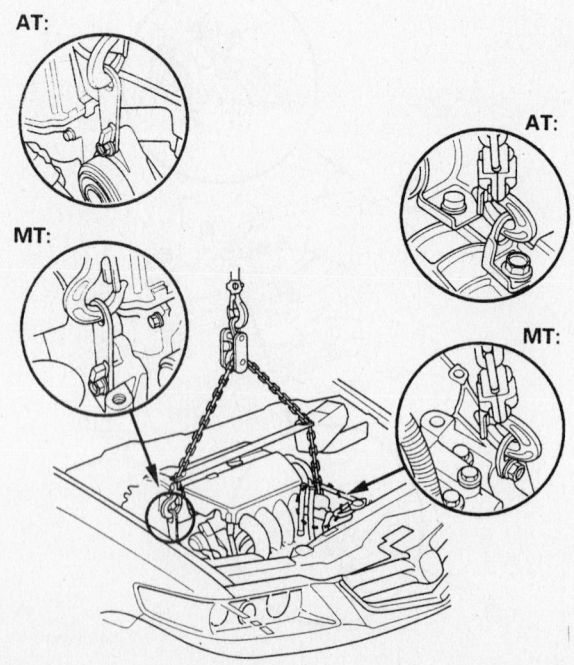

Make sure the chain hoist is properly attached to the engine—2.4L engine

67162-ATSX-G02

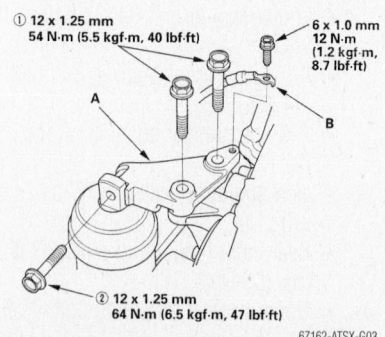

Proper tightening sequence for the upper bracket bolts—2.4L engine

67162-ATSX-G03

- Side upper bracket bolt: 47 ft. lbs. (64 Nm)
- Ground cable bolt: 104 inch lbs. (12 Nm)

18. If equipped with a M/T, loosen the transmission upper mount bolt, then install the transmission upper mount/bracket assembly, clutch line clamp bracket and ground cable.

19. If equipped with an A/T, install the transmission upper mount, transmission upper bracket and ground cable, then loosely install the transmission upper bracket plate.

20. On M/T equipped vehicles, tighten a NEW transmission upper mount bolt to 40 ft. lbs. (54 Nm). On A/T equipped vehicles, tighten NEW transmission upper mount bracket plate mounting bolts to 40 ft. lbs. (54 Nm) and NEW transmission upper mount bolt to 40 ft. lbs. (54 Nm).

21. The remainder of installation is the reverse of the removal procedure. Note the following tightening specifications:

- Transmission lower front and rear mounts: 33 ft. lbs. (44 Nm)
- Exhaust pipe "A" self-locking nuts (use NEW nuts): 25 ft. lbs.
- Exhaust pipe "A" self-locking bolt (use a NEW bolt): 16 ft. lbs. (22 Nm)
- A/C compressor mounting bolts: 16 ft. lbs. (22 Nm)
- Power steering pump mounting bolts: 16 ft. lbs. (22 Nm)
- Power steering hose clamp bolt: 104 inch lbs. (12 Nm)
- Strut brace nuts: 16 ft. lbs. (22 Nm)
- Harness clamp bolt: 104 inch lbs. (12 Nm)
- Underhood fuse/relay box bolts: 104 inch lbs. (12 Nm)
- Battery base bolts: 16 ft. lbs. (22 Nm)
- Harness bracket bolt: 104 inch lbs. (12 Nm)

22. Fill the engine with oil and the transmission with fluid.

23. Fill and bleed (if necessary) the air from the cooling system.

24. Connect the positive, then the negative battery cable and enter the radio security code.

25. Switch the ignition **ON** but do not engage the starter. The fuel pump should run for approximately 2 seconds, building pressure within the lines. Switch the ignition **OFF**, then **ON** 2 or 3 more times to build full system pressure. Check for fuel leaks.

26. Start the engine, allowing it to idle. Check the hoses and lines carefully for any sign of leakage.

27. Check the timing and idle speed.

28. After the engine has warmed up fully and the fan(s) have come on at least once, recheck the engine for fluid leaks. Switch the engine OFF.

29. Adjust the belts and throttle cable as necessary.

3.0L Engines

1. Before servicing the vehicle, refer to the precautions in the beginning of this section.

2. Obtain the anti-theft code for the radio.

3. Drain the engine oil, coolant, and transmission oil (or fluid) into sealable containers and carefully reinstall the drain plugs using new sealing washers.

4. Properly relieve the fuel system pressure.

5. Remove or disconnect the following:
- Air cleaner assembly
- Hood support struts and support the hood in a vertical position
- Negative battery cable and then the positive cable
- Strut brace
- Battery, battery tray and the engine ground cable
- Accelerator and cruise control cables from the throttle body and bracket
- Battery cables from the underhood Antilock Brake System (ABS) fuse/relay box and underhood fuse/relay box assemblies
- Engine wiring harness on the right side of the engine
- Evaporative emission (EVAP) control canister hose
- Brake booster vacuum hose
- Fuel feed and return hoses
- Engine wiring harness on the left side of the engine
- Accessory drive belt(s)
- Power steering pump leaving the hoses attached
- Shift cable (if equipped with a manual transmission)
- Clutch slave cylinder leaving (if equipped) the hydraulic line attached
- Reverse light switch connector (if equipped with a manual transmission)
- Cruise control vacuum tank (if equipped)
- Front wheels
- Lower splash shield
- Center support beam
- Heated Oxygen (HO$_2$S) sensor connector

- Front exhaust pipe from the manifold
- Shift selector cable and cover (if equipped with an automatic transmission)
- Damper fork
- Lower ball joints
- Driveshafts from the transaxle
- Crankshaft pulley (if equipped with 3.0L engine)
- VTEC/oil filter housing (if equipped with 3.0L engine)
- Upper and lower radiator hoses
- Transmission oil cooler hoses (if equipped)
- Radiator
- Air conditioning compressor leaving the hoses attached
- Heater hoses

6. Attach a suitable engine lifting hoist to the engine lifting hooks and secure the engine.
- Front engine mount
- Rear engine mount
- Side engine mount
- Transmission mount

7. Lift the engine slightly and check that all hoses, cables and wires have been properly disconnected.

8. Carefully raise the engine/transmission from the vehicle.

To install:

9. Installation is the reverse of the removal procedure, while using the following torque values:
- Transmission mount bolts: 47 ft. lbs. (64 Nm)
- Front engine mount bolts: 47 ft. lbs. (64 Nm)
- Side engine mount bolts: 47 ft. lbs. (64 Nm)
- Rear engine mount bolts: 47 ft. lbs. (64 Nm)
- A/C compressor bolts: 16 ft. lbs. (22 Nm)
- VTEC/oil filter housing bolts (if equipped with 3.0L engine): 16 ft. lbs. (22 Nm)
- Crankshaft pulley bolt (if equipped with 3.0L engine): 181 ft. lbs. (245 Nm)
- Lower ball joint nuts: 40 ft. lbs. (54 Nm)
- Damper fork bolts: 47 ft. lbs. (64 Nm)
- Front exhaust pipe nuts: 40 ft. lbs. (54 Nm)
- Center support beam bolts: 37 ft. lbs. (50 Nm)
- Splash shield bolts: 84 inch lbs. (9.5 Nm)
- Power steering pump bolts: 16 ft. lbs. (22 Nm)

• Strut brace bolts (if equipped): 16 ft. lbs. (22 Nm)

10. Fill the engine with oil and the transmission with fluid.

11. Fill and bleed (if necessary) the air from the cooling system.

12. Connect the positive, then the negative battery cable and enter the radio security code.

13. Switch the ignition **ON** but do not engage the starter. The fuel pump should run for approximately 2 seconds, building pressure within the lines. Switch the ignition **OFF**, then **ON** 2 or 3 more times to build full system pressure. Check for fuel leaks.

14. Start the engine, allowing it to idle. Check the hoses and lines carefully for any sign of leakage.

15. Check the timing and idle speed.

16. After the engine has warmed up fully and the fan(s) have come on at least once, recheck the engine for fluid leaks. Switch the engine OFF.

17. Adjust the belts and throttle cable as necessary.

3.2L Engine

1. Before servicing the vehicle, refer to the precautions in the beginning of this section.

2. Do not remove the hood. Disconnect the hood support strut and reconnect it to hold the hood in a vertical position.

3. Properly relieve the fuel pressure.

4. Drain the engine oil, coolant, transmission fluid and the differential fluid.

5. Remove or disconnect the following:

• Negative battery cable, then the positive battery cable
• battery and battery tray
• Air cleaner assembly
• Water bypass hoses (on 3.2CL)
• Traction Control System (TCS) control valve actuator connector (on 3.2CL)
• TCS control valve angel sensor connector (on 3.2CL)
• Throttle cable and cruise control cable from the throttle and bracket
• Left side engine wiring harness connectors
• Fuel feed and return hoses
• Brake booster vacuum hose
• Battery cables from the under–hood fuse/relay box
• Under–hood fuse/relay box
• Powertrain Control Module (PCM) electrical connectors
• Accessory drive belts
• Power steering pump without disconnecting the hoses

• Vehicle Speed sensor (VSS) connector, then remove the VSS/power steering sensor leaving the fluid hoses attached
• Front wheels
• Lower splash shield
• Heated Oxygen (HO2S) sensor connector
• Front exhaust pipe from the manifold
• Front damper forks
• Lower ball joints from the steering knuckles
• Halfshafts from the differential and the intermediate shaft
• Shift cable cover
• Shift cable with the control lever from the transaxle
• Power steering hose clamps and the engine mount control vacuum hose
• Upper and lower radiator hoses
• Heater hoses
• Transmission fluid cooler hoses
• Ground cable
• Power steering hose clamp from the rear beam assembly

6. Attach a suitable chain hoist to the engine lifting hooks and support the engine.

➡**The engine and transmission assembly is removed by lowering it from the vehicle. Be sure the vehicle is in a position that will allow the engine and transmission assembly enough clearance to be moved from the vehicle once it is lowered away from the vehicle.**

7. Remove or disconnect the following:

• Side, rear and front engine mount support fasteners
• Front suspension radius rod bolts

8. Make alignment marks on the front beam and remove the front beam.

9. Remove or disconnect the following:

• Air conditioning compressor leaving the hoses attached
• Rear mounts from the engine and transmission

10. Check that all hoses, cables and wires have been properly disconnected and slowly lower the engine about 6 inches (150 mm). Recheck that all hoses, cables and wires have been properly disconnected.

11. Carefully lower the engine/transmission assembly from the vehicle.

To install:

12. Installation is the reverse of the removal procedure, while using the following torque values:

• Transmission rear mount bolts: 28 ft. lbs. (38 Nm)

• Air conditioning compressor bolts: 16 ft. lbs. (22 Nm)
• Front beam nuts: 28 ft. lbs. (38 Nm)
• Front beam bolts: 76 ft. lbs. (103 Nm)
• Rear bolts: 28 ft. lbs. (38 Nm)
• Front suspension radius rod bolts: 119 ft. lbs. (162 Nm)
• Front engine mount nut: 40 ft. lbs. (54 Nm)
• Rear engine mount nut: 40 ft. lbs. (54 Nm)
• Rear engine mount bolt: 28 ft. lbs. (38 Nm)
• Side engine mount bracket bolts: 33 ft. lbs. (44 Nm)
• Side engine mount through bolt: 40 ft. lbs. (54 Nm)
• Lower ball joint nuts: 40 ft. lbs. (54 Nm)
• Damper fork bolts: 47 ft. lbs. (64 Nm)
• Front exhaust pipe nuts: 40 ft. lbs. (54 Nm)
• Shift cable and control lever bolts: 10 ft. lbs. (14 Nm)
• Power steering pump bolts: 16 ft. lbs. (22 Nm)

13. Fill the engine with oil and the transmission with fluid.

14. Fill and bleed (if necessary) the air from the cooling system.

15. Connect the positive, then the negative battery cable and enter the radio security code.

16. Switch the ignition **ON** but do not engage the starter. The fuel pump should run for approximately 2 seconds, building pressure within the lines. Switch the ignition **OFF**, then **ON** 2 or 3 more times to build full system pressure. Check for fuel leaks.

17. Start the engine, allowing it to idle. Check the hoses and lines carefully for any sign of leakage.

18. Check the timing and idle speed.

19. After the engine has warmed up fully and the fan(s) have come on at least once, recheck the engine for fluid leaks. Switch the engine OFF.

20. Adjust the belts and throttle cable as necessary.

3.5L Engine

1. Before servicing the vehicle, refer to the precautions in the beginning of this section.

2. Move the front passenger's seat forward.

3. Relieve the fuel system pressure.

Bracket Bolts Torque Specifications:

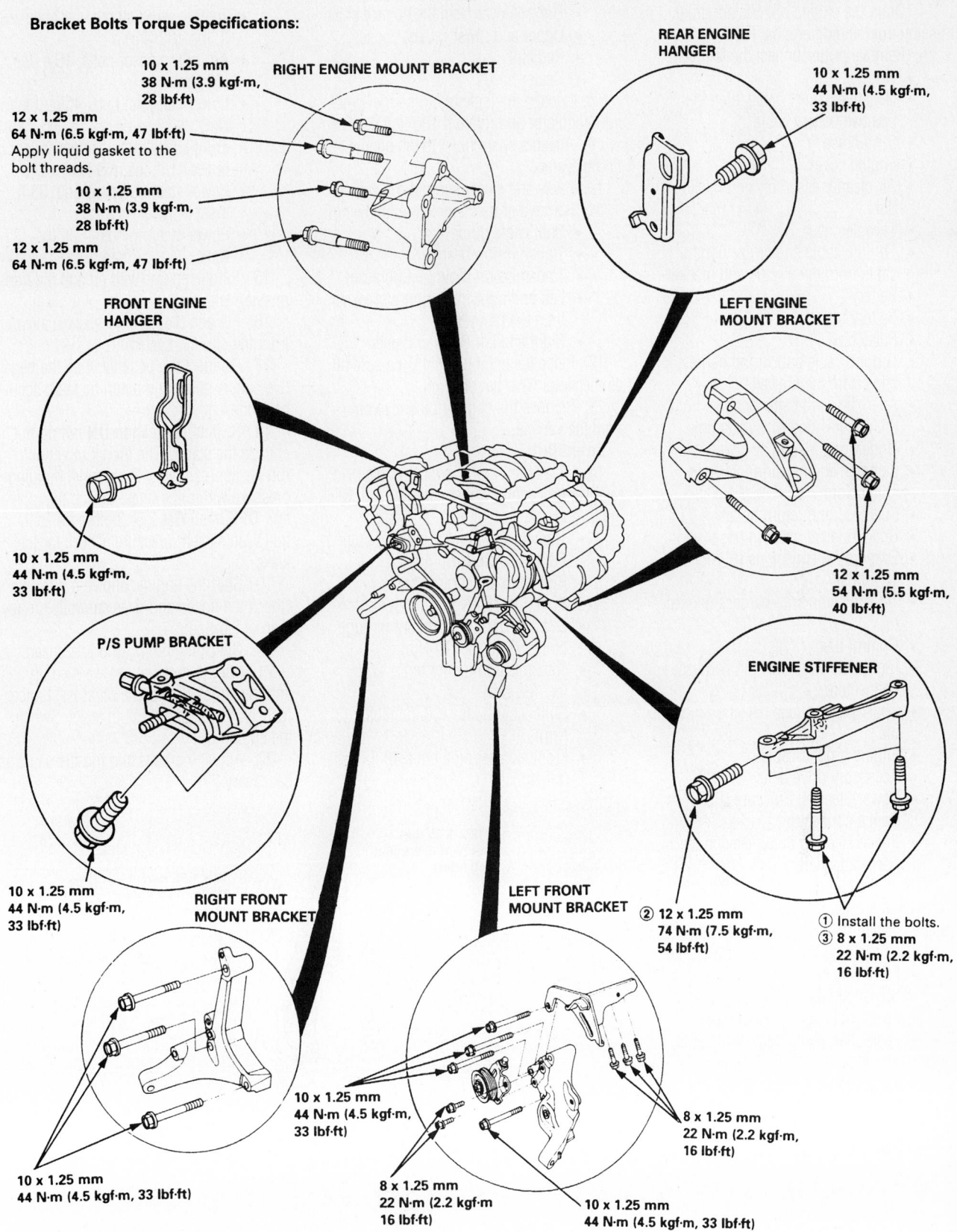

10 x 1.25 mm
38 N·m (3.9 kgf·m,
28 lbf·ft)

RIGHT ENGINE MOUNT BRACKET

12 x 1.25 mm
64 N·m (6.5 kgf·m, 47 lbf·ft)
Apply liquid gasket to the
bolt threads.

10 x 1.25 mm
38 N·m (3.9 kgf·m,
28 lbf·ft)

12 x 1.25 mm
64 N·m (6.5 kgf·m, 47 lbf·ft)

REAR ENGINE HANGER

10 x 1.25 mm
44 N·m (4.5 kgf·m,
33 lbf·ft)

FRONT ENGINE HANGER

LEFT ENGINE MOUNT BRACKET

10 x 1.25 mm
44 N·m (4.5 kgf·m,
33 lbf·ft)

12 x 1.25 mm
54 N·m (5.5 kgf·m,
40 lbf·ft)

P/S PUMP BRACKET

ENGINE STIFFENER

10 x 1.25 mm
44 N·m (4.5 kgf·m,
33 lbf·ft)

RIGHT FRONT MOUNT BRACKET

LEFT FRONT MOUNT BRACKET

② **12 x 1.25 mm**
74 N·m (7.5 kgf·m,
54 lbf·ft)

① Install the bolts.
③ **8 x 1.25 mm**
22 N·m (2.2 kgf·m,
16 lbf·ft)

10 x 1.25 mm
44 N·m (4.5 kgf·m,
33 lbf·ft)

8 x 1.25 mm
22 N·m (2.2 kgf·m,
16 lbf·ft)

10 x 1.25 mm
44 N·m (4.5 kgf·m, 33 lbf·ft)

8 x 1.25 mm
22 N·m (2.2 kgf·m
16 lbf·ft)

10 x 1.25 mm
44 N·m (4.5 kgf·m, 33 lbf·ft)

7923BG77

View of the engine mounting bracket showing torque specifications—3.5RL

4. Drain the engine oil, coolant, transmission fluid and differential fluid.

5. Remove or disconnect the following:
- Hood
- Negative battery cable, then the positive battery cable
- Strut brace
- Engine cover
- Air cleaner assembly and intake duct
- Throttle cover
- Throttle cable and cruise control cable from the throttle and bracket
- Battery
- Battery tray
- Relay box
- Ground cable and wiring harness clips from the firewall
- Alternator and battery cables from the under–hood fuse/relay box
- Underhood fuse/relay box
- Left side engine wiring harness connectors
- Fuel feed and return hoses
- Brake booster vacuum hose
- Evaporative emissions (EVAP) canister hose
- Transmission sub–harness connector
- Control box
- Right side engine wiring harness connectors
- Spark plug voltage detection module
- Engine ground cables
- Accessory drive belts
- Power Steering Pressure (PSP) switch connector
- Power steering pump leaving the hoses attached

6. Pull the carpet back under the front passengers seat and detach the secondary Heated Oxygen (HO$_2$S) sensor connector.

7. Remove or disconnect the following:
- Front wheels
- Splash shield
- Front suspension damper forks
- Lower ball joints from the steering knuckles
- Halfshafts from the transmission
- Air conditioning compressor without disconnecting the hoses
- Vehicle Speed Sensor (VSS) leaving the hoses attached
- Transmission stop collars
- Front exhaust pipe from the vehicle
- Wire harness cover and grommet
- Three way catalytic converter
- Converter heat shield
- Transmission fluid cooler lines
- Shift cable cover
- Shift cable from the transmission

- Control lever from the control shaft
- Upper and lower radiator hoses
- Radiator
- Heater hoses

8. Loosen the locknut on the fuel pressure regulator and rotate it 180 degrees.

9. Attach a chain hoist to the engine lifting eyelets.

10. Raise and safely support the vehicle.

11. Remove or disconnect the following:
- Shift cable guide
- Transmission beam
- Transmission mount and bracket
- Left and right front mount brackets from the mounts
- Right and left engine mounts

12. Raise the engine slightly, be sure all connections have be removed.

13. Remove the engine/transmission from the vehicle.

To install:

14. Installation is the reverse of the removal procedure, while using the following torque values:
- Transmission mount bracket bolts: 28 ft. lbs. (38 Nm)
- Right and left engine mount nuts: 47 ft. lbs. (64 Nm)
- Left and right front mount through bolts: 52 ft. lbs. (74 Nm)
- Transmission mount bolts: 28 ft. lbs. (38 Nm)
- Shift cable bolts: 108 inch lbs. (12 Nm)
- Front exhaust pipe nuts: 40 ft. lbs. (54 Nm)

- Transmission stop collar bolts: 28 ft. lbs. (38 Nm)
- A/C compressor bolts: 16 ft. lbs. (22 Nm)
- Lower ball joint nuts: 40 ft. lbs. (54 Nm)
- Front suspension damper fork bolts: 41 ft. lbs. (69 Nm)
- Power steering pump bolt: 33 ft. lbs. (44 Nm)
- Power steering nut: 16 ft. lbs. (22 Nm)

15. Fill the engine with oil and the transmission with fluid.

16. Fill and bleed (if necessary) the air from the cooling system.

17. Connect the positive, then the negative battery cable and enter the radio security code.

18. Switch the ignition **ON** but do not engage the starter. The fuel pump should run for approximately 2 seconds, building pressure within the lines. Switch the ignition **OFF**, then **ON** 2 or 3 more times to build full system pressure. Check for fuel leaks.

19. Start the engine, allowing it to idle. Check the hoses and lines carefully for any sign of leakage.

20. Check the timing and idle speed.

21. After the engine has warmed up fully and the fan(s) have come on at least once, recheck the engine for fluid leaks. Switch the engine OFF.

22. Adjust the belts and throttle cable as necessary.

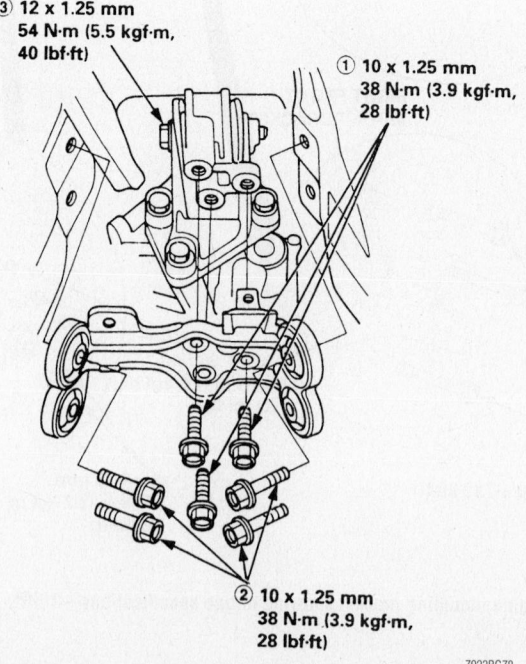

③ 12 x 1.25 mm 54 N·m (5.5 kgf·m, 40 lbf·ft)

① 10 x 1.25 mm 38 N·m (3.9 kgf·m, 28 lbf·ft)

② 10 x 1.25 mm 38 N·m (3.9 kgf·m, 28 lbf·ft)

7923BG78

Transmission beam bolt tightening sequence and torque specifications—3.5RL

Water Pump

REMOVAL & INSTALLATION

1.8L Engines

1. Before servicing the vehicle, refer to the precautions in the beginning of this section.
2. Disconnect the negative battery cable.
3. Drain the engine coolant.
4. Remove or disconnect the following:
   • Timing belt
   • Camshaft pulleys
   • Rear timing belt cover
   • 5 water pump mounting bolts and remove the water pump

To install:

5. Install or connect the following:
   • Water pump and tighten the bolts to 108 inch lbs. (12 Nm)
   • Rear timing belt cover
   • Camshaft pulleys
   • Timing belt
6. Fill the engine with coolant and bleed the air from the cooling system.
7. Connect the negative battery cable and enter the radio security code.
8. Run the engine and check for cooling system leaks.

2.0L Engines

1. Before servicing the vehicle, refer to the precautions in the beginning of this section.
2. Disconnect the negative battery cable.
3. Drain the engine coolant.
4. Remove or disconnect the following:
   • Drive belt
   • Crankshaft pulley
   • For the K20A3 (base) engine, 6 bolts securing the water pump
   • For the K20A2 (Type–S) engine, oil cooler joint pipe and the 7 bolts securing the water pump
   • Water pump
5. Clean and inspect the O–ring groove and mating surfaces.

To install:

6. Install or connect the following:
   • Water pump with a new O–ring. Tighten the mounting bolts to 104 inch lbs. (12 Nm).
   • Crankshaft pulley
   • Drive belt
7. Fill the engine with coolant and bleed the air from the cooling system.
8. Connect the negative battery cable and enter the radio security code.
9. Run the engine and check for cooling system leaks.

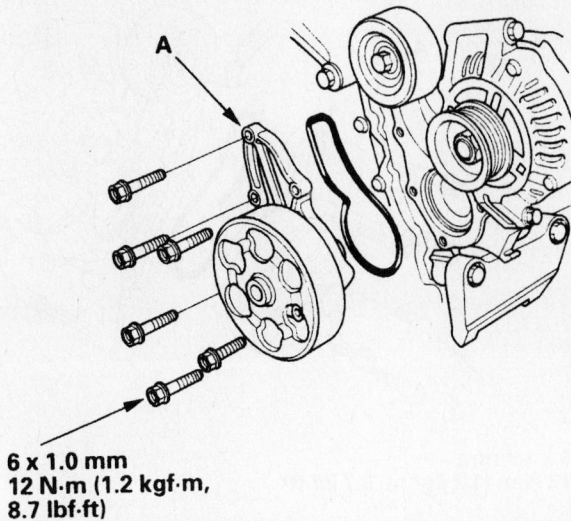

6 x 1.0 mm
12 N·m (1.2 kgf·m, 8.7 lbf·ft)

42356-INTE-G04

Water pump mounting—2.0L (K20A3) engine

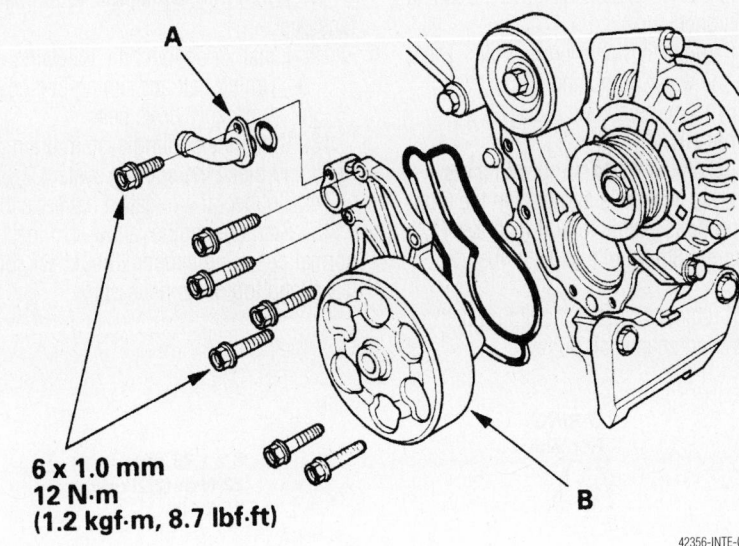

6 x 1.0 mm
12 N·m
(1.2 kgf·m, 8.7 lbf·ft)

42356-INTE-G05

Water pump mounting—2.0L (K20A2) engine

2.4L Engine

1. Before servicing the vehicle, refer to the precautions in the beginning of this section.
2. Drain the cooling system.
3. Remove or disconnect the following:
   • Crankshaft pulley
   • 6 water pump mounting bolts, then the pump and O-ring seal
4. Clean the seal groove and mating surfaces.
5. Install or connect the following:
   • Water pump, with a new seal. Tighten the bolts to 104 inch lbs. (12 Nm).
   • Crankshaft pulley
6. Fill the cooling system.
7. Start the engine and check for leaks.

3.0L, 3.2L and 3.5L Engines

➡ **Perform this service operation with the engine cold.**

1. Before servicing the vehicle, refer to the precautions in the beginning of this section.
2. Disconnect the negative battery cable.
3. Remove the front splash panel.
4. Drain the cooling system.
5. Remove the timing belt. Inspect the timing belt for any signs of damage or oil and coolant contamination. Replace the timing belt if there is any doubt about its condition.
6. On 3.5 RL models, remove the left camshaft pulley and back cover.
7. On 2000–03 3.2 TL models and

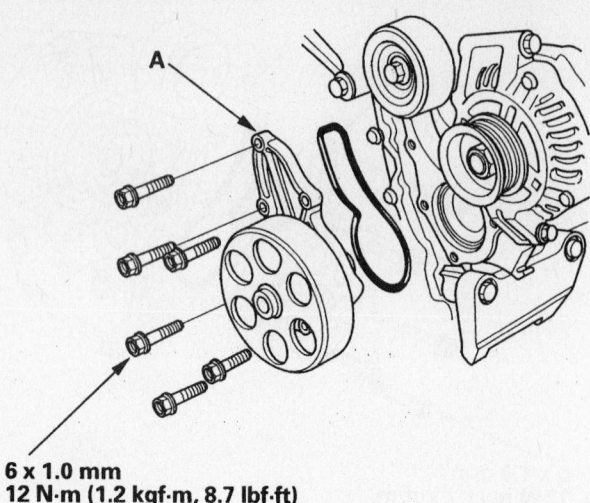

6 x 1.0 mm
12 N·m (1.2 kgf·m, 8.7 lbf·ft)

43256-ACCO-G02

Water pump mounting—2.4L engine

2001–03 3.2 CL models, remove the timing belt tensioner.

8. Remove the water pump bolts. Then, remove the water pump and sprocket assembly from the engine block.

To install:

9. Install the water pump with a new O–ring. Use new bolts and tighten the 6mm mounting bolts evenly to 104 inch lbs. (12 Nm) and the 8mm bolts to 16 ft. lbs. (22 Nm).

10. If removed, install the timing belt rear cover and camshaft pulley.

11. If removed, install the timing belt tensioner.

12. Install or connect the following:
   • Timing belt and timing belt covers
   • Accessory drive belts

13. Close the cooling system drain plug. Refill and bleed the cooling system.

14. Connect the negative battery cable.

15. Start the engine, allow it to reach normal operating temperature, check for leaks, and top off as necessary.

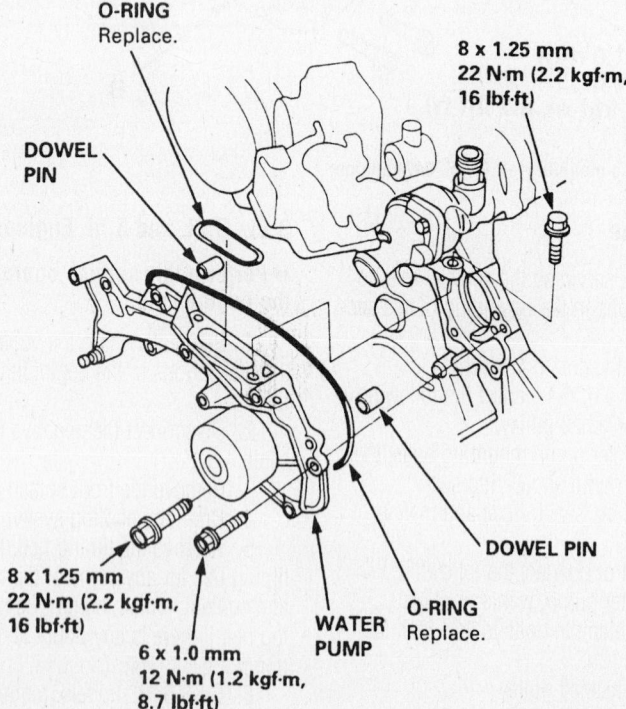

O-RING Replace.

DOWEL PIN

8 x 1.25 mm
22 N·m (2.2 kgf·m, 16 lbf·ft)

DOWEL PIN

8 x 1.25 mm
22 N·m (2.2 kgf·m, 16 lbf·ft)

6 x 1.0 mm
12 N·m (1.2 kgf·m, 8.7 lbf·ft)

WATER PUMP

O-RING Replace.

7923BG79

Water pump mounting and bolt torque specifications—3.5L engine

Heater Core

REMOVAL AND INSTALLATION

Integra and RSX

➡ Be sure to acquire the anti–theft code for the radio; then, write down the frequencies for the preset buttons.

1. Before servicing the vehicle, refer to the precautions in the beginning of this section.

2. Disconnect the negative battery cable.

✻✻ CAUTION

After disconnecting the negative battery cable, wait for at least 3 minutes for the SRS module to deplete its energy.

3. Drain the cooling system into a clean container for reuse.

4. Remove or disconnect the following:
   • Heater hoses from the heater core
   • Heater housing-to-cowl nut, located in the engine compartment

5. Remove the instrument panel by removing or disconnecting the following:
   • Front seats
   • Front and rear consoles
   • Lower dashboard-to-instrument panel screws and the lower dashboard, located on the driver's side
   • Knee bolster
   • Glove box
   • 4 glove box frame-to-instrument panel bolts and the frame
   • Clock
   • Moon roof switch
   • Stereo radio/cassette
   • Lower steering column cover clamps and the cover
   • Wrap a shop towel around the steering column to prevent damage to the column.
   • Steering column-to-instrument panel nuts/bolts and lower the steering column
   • SRS module-to-instrument panel nuts (located on the driver's side); then, carefully, remove the SRS module and disconnect the electrical connector
   • Air mix control cable and electrical connectors, located at the center of the instrument panel
   • Antenna lead
   • Electrical connectors, located at the under–dash fuse/relay box

- Pry out the access panels, from both sides of the instrument panel
- Instrument panel-to-chassis bolts and remove the instrument panel
- Wiring harness clips; then, remove the 4 heater housing-to-blower motor duct screws and the duct, (models not equipped with air conditioning only)

6. If equipped with air conditioning, remove the evaporator housing removing or disconnecting the following:

- Discharge and recover the air conditioning system refrigerant
- Refrigerant lines from the evaporator core, discard the O–rings and plug the openings to prevent contamination
- Cut the insulation pad (at the firewall) at the lower evaporator housing-to-chassis location
- Thermostat electrical connector, located at the evaporator housing
- Wiring harness clips from the evaporator housing
- Evaporator housing-to-chassis screws, the nut and bolt
- Drain hose
- Evaporator housing, carefully
- 2 SRS support beam bolts, nut and the SRS beam, located at the passenger's side
- Mode control motor connector and the wiring harness clip from the heater housing
- Heater housing-to-chassis nuts and the heater housing

7. Remove the damper arm cover-to-heater housing screw and the cover.

8. Remove or disconnect the following:

- Damper arm link; then, remove the damper arm-to-heater housing screw and the arm
- 2 heater core cover-to-heater housing screws and the cover
- Pipe clamp screw and the clamp
- Heater core from the heater housing

To install:

9. Install or connect the following:

- Heater core to the heater housing
- Pipe clamp screw and the clamp
- Heater core cover and the 2 cover-to-heater housing screws
- Damper arm-to-heater housing screw and the arm; then, connect the damper arm link
- Damper arm cover and the cover-to-heater housing screw
- Heater housing and the heater housing-to-chassis nuts
- Mode control motor connector and

▲ : Bolt, nut locations

A▲ : Bolt, 2

8 x 1.25 mm
22 N·m (2.2 kgf·m, 16 lbf·ft)

B▲ : Nut, 2

8 x 1.25 mm
13 N·m (1.3 kgf·m, 9 lbf·ft)
Replace.

STEERING COLUMN

CLAMP

STEERING JOINT COVER

93112GP5

View of the steering column and related components—Integra

▲ : Nut locations, 4

6 x 1.0 mm
9.8 N·m
(1.0 kgf·m,
7.2 lbf·ft)

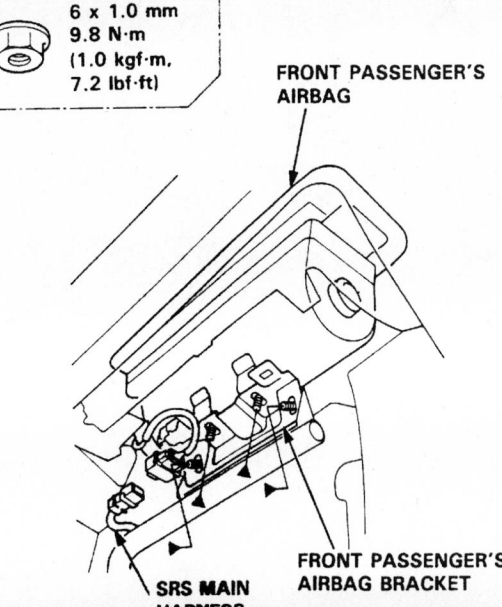

FRONT PASSENGER'S AIRBAG

FRONT PASSENGER'S AIRBAG BRACKET

SRS MAIN HARNESS

93112GP6

View of the passenger's side SRS module and related components—Integra

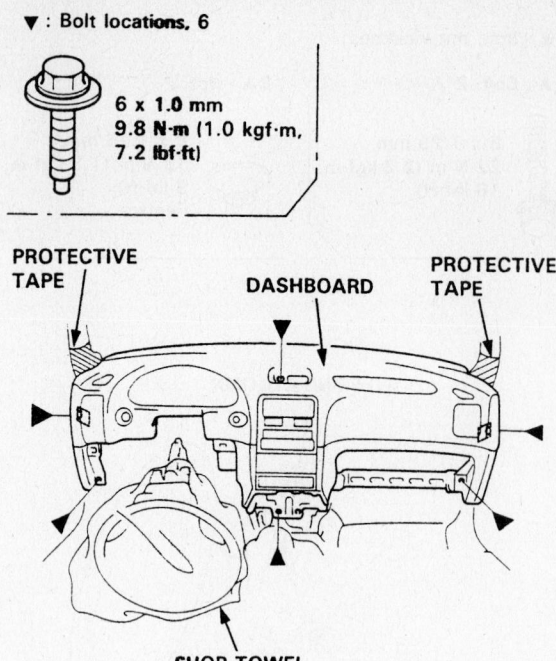

▼ : Bolt locations, 6

6 x 1.0 mm
9.8 N·m (1.0 kgf·m, 7.2 lbf·ft)

PROTECTIVE TAPE

DASHBOARD

PROTECTIVE TAPE

SHOP TOWEL

93112GP7

View of the instrument panel fastener locations—Integra

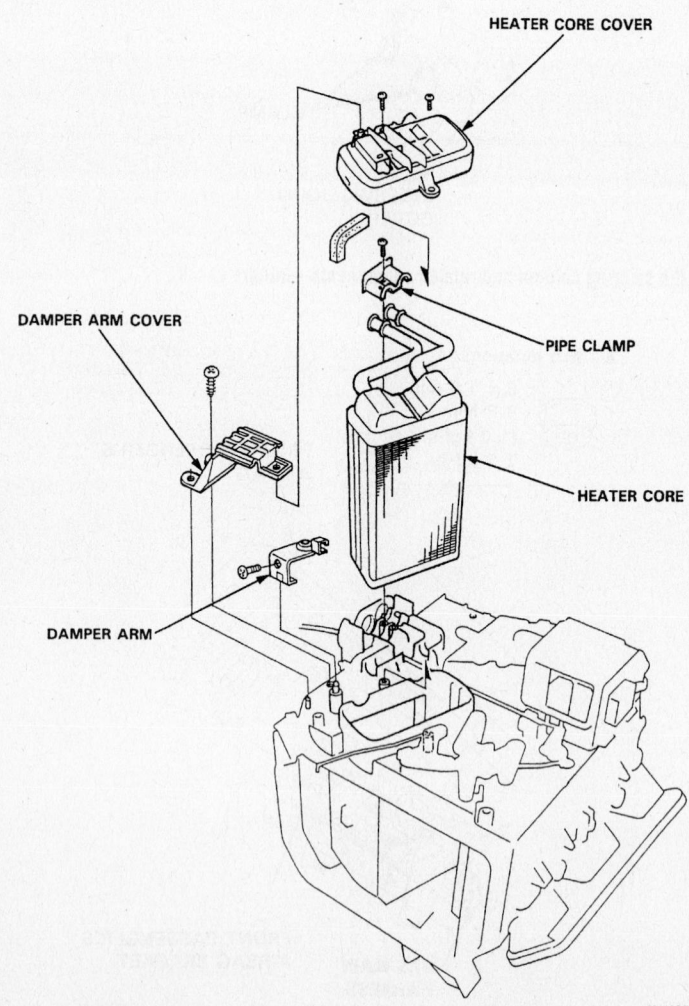

HEATER CORE COVER

DAMPER ARM COVER

PIPE CLAMP

HEATER CORE

DAMPER ARM

93112GP8

Exploded view of the heater core, heater housing and related components—Integra

the wiring harness clip to the heater housing
- SRS support beam, nut and the 2 SRS beam bolts on the passengers side

10. If equipped with air conditioning, install the evaporator housing by installing or connecting the following:
- Evaporator housing
- Drain hose
- 4 evaporator housing-to-chassis screws, the nut and bolt
- Wiring harness clips to the evaporator housing
- Thermostat electrical connector at the evaporator housing
- Refrigerant lines to the evaporator core making sure to use new O–rings
- Heater housing-to-blower motor duct and the 4 duct screws; then, connect the wiring harness clips (models not equipped with air conditioning only)

11. Install the instrument panel by installing or connecting the following:
- instrument panel and the instrument panel-to-chassis bolts
- access panels on both sides of the instrument panel
- Electrical connectors at the under-dash fuse/relay box
- Antenna lead
- Air mix control cable and electrical connectors at the center of the instrument panel
- SRS module electrical connector (located on the passenger's side) and torque the SRS module-to-instrument panel nuts to 84 inch lbs. (9.8 Nm)
- Steering column and torque the steering column-to-instrument panel nuts to 108 ft. lbs. (13 Nm) and the bolts to 16 ft. lbs. (22 Nm)
- Lower steering column cover and the cover clamps
- Stereo radio/cassette
- Moon roof switch
- Clock
- Glove box frame and the 4 frame-to-instrument panel bolts
- Glove box
- Knee bolster
- Lower dashboard and the lower dashboard-to-instrument panel screws on the drivers side
- Front and rear consoles
- Front seats
- Heater housing-to-cowl nut (located in the engine compart-

ment) and torque to 16 ft. lbs. (22 Nm)
- Heater hoses to the heater core

12. Refill the cooling system.

13. Connect the negative battery cable.

14. Evacuate, charge and leak test the air conditioning system refrigerant.

15. Operate the engine to normal operating temperatures; then, check the climate control operation and check for leaks.

3.2CL

➡ Be sure to acquire the anti-theft code for the radio; then, write down the frequencies for the preset buttons.

1. Before servicing the vehicle, refer to the precautions in the beginning of this section.

2. Disconnect the negative battery cable.

※※ CAUTION

After disconnecting the negative battery cable, wait for at least 3 minutes for the SRS module to deplete its energy.

3. Drain the cooling system into a clean container for reuse.

4. Remove or disconnect the following:

- Heater hoses from the heater core
- 2 heater housing-to-cowl nuts, located in the engine compartment

5. Place the front wheel in the straight-ahead position.

6. Remove the SRS module and the steering wheel by removing or disconnecting the following:

- Access panel-to-steering wheel screws and the panel
- SRS module electrical connector
- SRS module-to-steering wheel covers from both sides of the steering wheel

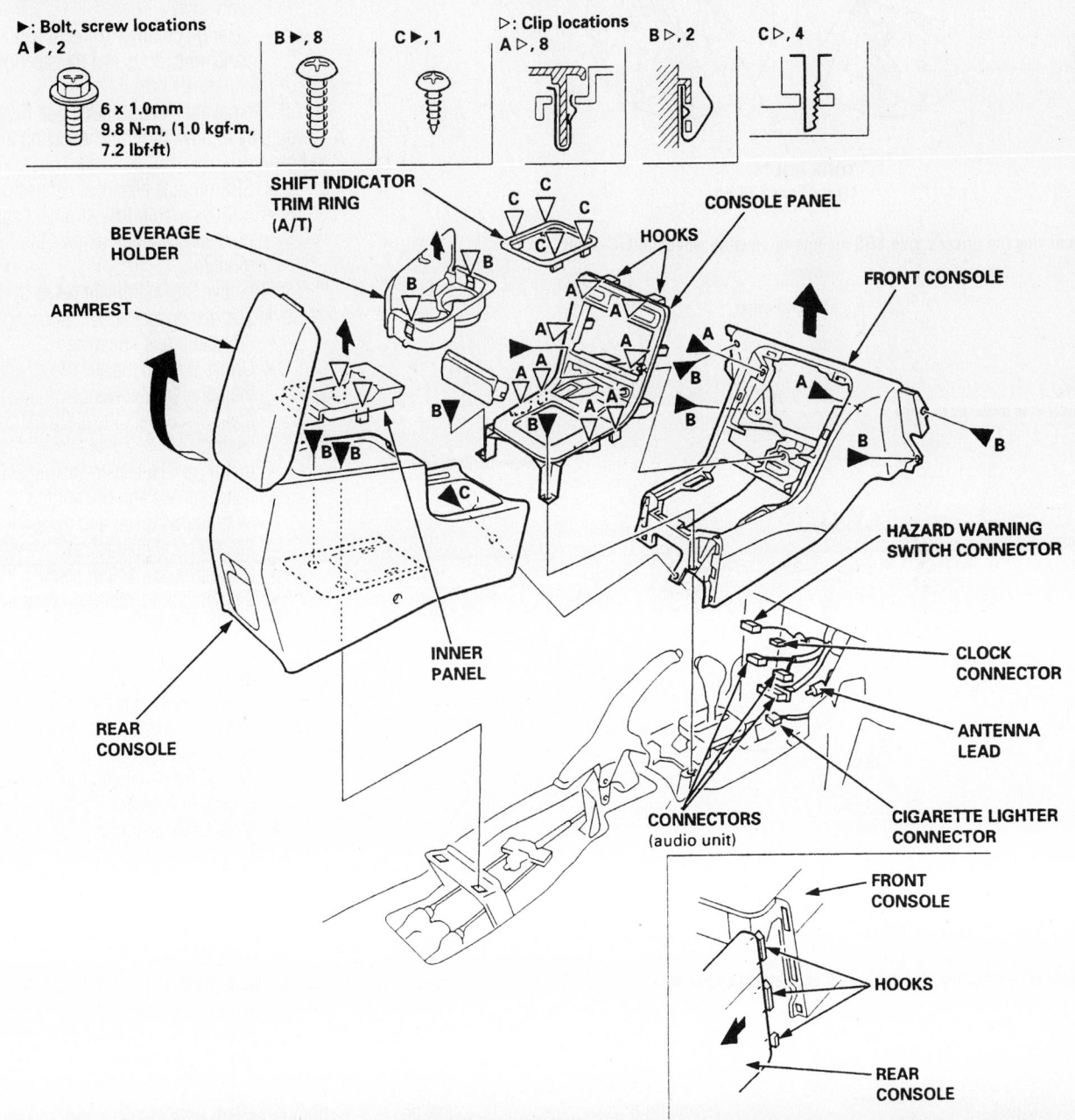

View of the front and rear consoles and related components—3.2CL

93112GP9

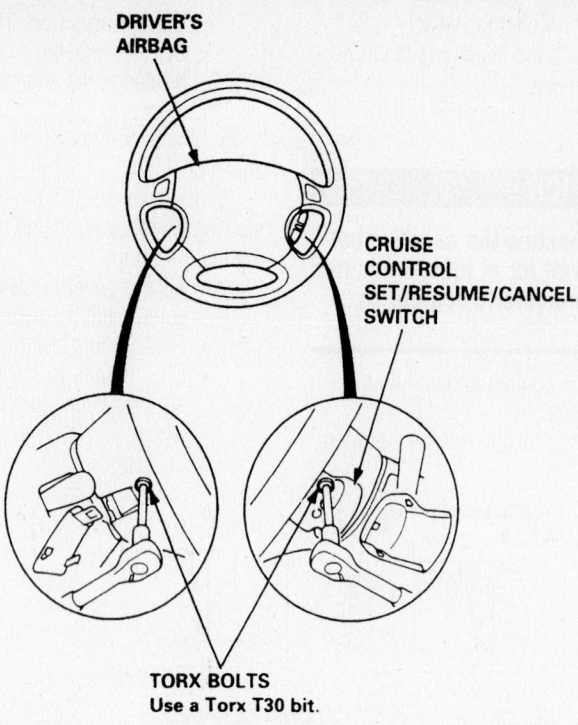

DRIVER'S AIRBAG

CRUISE CONTROL SET/RESUME/CANCEL SWITCH

TORX BOLTS
Use a Torx T30 bit.

93112GP0

Removing the driver's side SRS module-to-steering wheel bolts—3.2CL

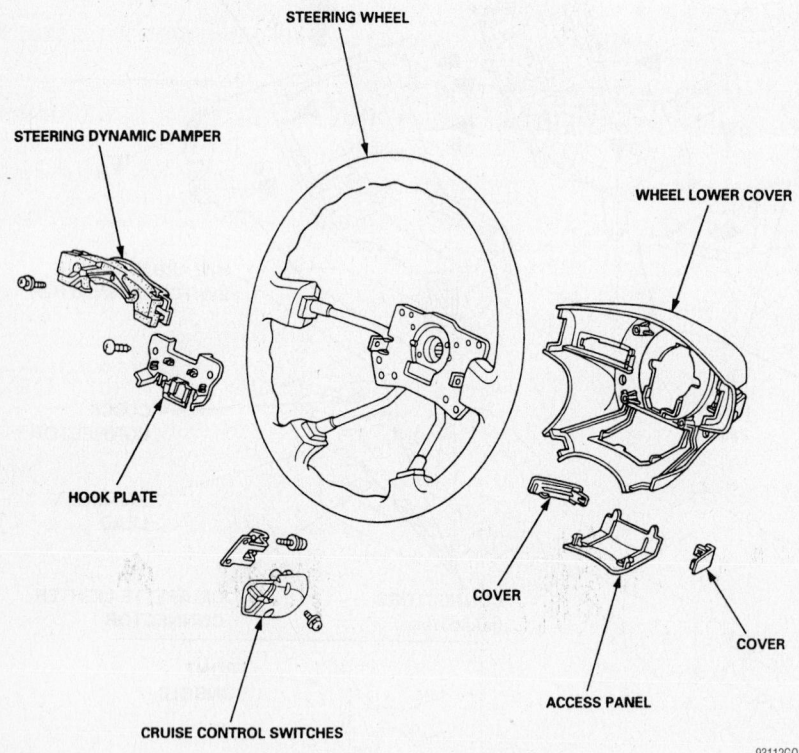

STEERING WHEEL

STEERING DYNAMIC DAMPER

WHEEL LOWER COVER

HOOK PLATE

COVER

COVER

ACCESS PANEL

CRUISE CONTROL SWITCHES

93112GQ1

Exploded view of the steering wheel and related components—3.2CL

- Both SRS module-to-steering wheel bolts, using a T30 Torx®bit
- SRS module from the steering wheel
- Electrical connectors from the steering wheel

- Steering wheel from the steering column

7. Remove the steering column by removing or disconnecting the following:
- Coin pocket, the lower instrument panel-to-dash screw and the lower

instrument panel from the driver's side
- Electrical connector and the air hose at the knee bolster; then, remove the knee bolster-to-dash bolts and the knee bolster
- Steering column cover screws and the covers
- Combination switch-to-steering column screws, disconnect the electrical connectors and remove the combination switch
- Ignition switch connectors
- Clamps, clips and the steering joint cover
- Steering joint from the steering column shaft
- Steering column-to-instrument panel nuts/bolts and the steering column

8. Remove the passenger's side SRS module by removing or disconnecting the following:
- SRS module electrical connector
- 5 SRS module-to-instrument panel nuts and carefully remove the SRS module

9. Remove the instrument panel by removing or disconnecting the following:
- Front and rear consoles
- Cruise control master switch and the panel brightness controller; then, disconnect the electrical connectors
- Instrument cluster-to-instrument panel screws and pry out the instrument cluster and disconnect the electrical connectors
- Glove box-to-damper screw
- 2 glove box-to-dash screws and the glove box
- Audio unit-to-instrument panel fasteners and the audio unit
- 3 passenger's dashboard panel screws (located on the passengers side); then, carefully pry out the dashboard panel and the climate control unit as an assembly
- Side defogger trim from both sides of the instrument panel
- Instrument panel electrical connectors, the recirculation control motor connector and the resistor connector
- Instrument panel-to-chassis bolts and the instrument panel
- Steering hanger beam-to-chassis nuts, bolts and the beam

10. Discharge and recover the air conditioning system refrigerant.

11. Remove the evaporator housing by removing or disconnecting the following:

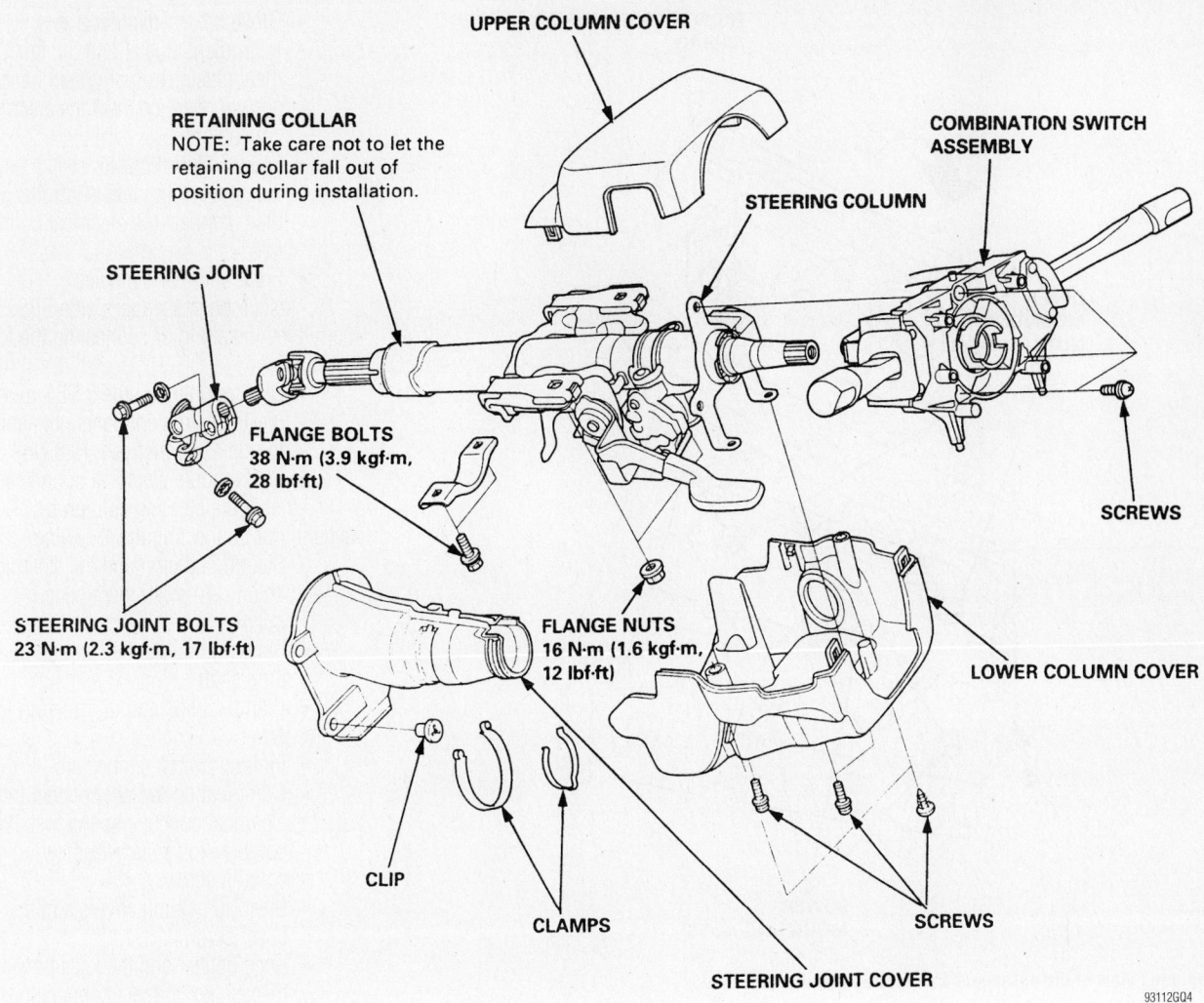

UPPER COLUMN COVER

RETAINING COLLAR
NOTE: Take care not to let the retaining collar fall out of position during installation.

STEERING JOINT

STEERING COLUMN

COMBINATION SWITCH ASSEMBLY

FLANGE BOLTS
38 N·m (3.9 kgf·m, 28 lbf·ft)

SCREWS

STEERING JOINT BOLTS
23 N·m (2.3 kgf·m, 17 lbf·ft)

FLANGE NUTS
16 N·m (1.6 kgf·m, 12 lbf·ft)

LOWER COLUMN COVER

CLIP

CLAMPS

SCREWS

STEERING JOINT COVER

93112GQ4

Exploded view of the steering wheel column and related components—3.2CL

- Refrigerant lines from the evaporator core. Discard the O–rings and plug the openings to prevent contamination.
- Thermostat electrical connector at the evaporator housing
- Wiring harness clips from the evaporator housing
- Evaporator housing-to-chassis screws, the nut and bolt
- Drain hose
- Evaporator housing
12. Remove or disconnect the following:
- Wiring harness clip; then, disconnect the mode control motor and the air mix control motor connectors
- Heater housing-to-chassis bolt and the heater housing
- Vent/defroster duct-to-heater housing screws and the duct
- Heater core pipe clamp-to-heater housing screw and the clamp
- Heater core clamp-to-heater housing screw and the heater core

To install:
13. Install or connect the following:
- Heater core clamp and the heater core-to-heater housing screw
- Heater core pipe clamp and the clamp-to-heater housing screw
- Vent/defroster duct and the duct-to-heater housing screws
- Heater housing and the heater housing-to-chassis bolt
- Wiring harness clip; then, connect the mode control motor and the air mix control motor connectors
14. Install the evaporator housing by installing or connecting the following:
- Evaporator housing
- Drain hose
- Evaporator housing-to-chassis screws, the nut and bolt
- Wiring harness clips to the evaporator housing
- Thermostat electrical connector, at the evaporator housing

- Refrigerant lines to the evaporator core, make sure to use new O–rings
- Steering hanger beam and the beam-to-chassis nuts and bolts
15. Install the instrument panel by installing or connecting the following:
- Instrument panel and the instrument panel-to-chassis bolts
- Instrument panel electrical connectors, the recirculation control motor connector and the resistor connector
- Side defogger trim at both sides of the instrument panel
- Dashboard panel and the climate control unit as an assembly; then, install the 3 passenger's dashboard panel screws at the passenger's side
- Audio unit and the audio unit-to-instrument panel fasteners
- Glove box and the 2 glove box-to-dash screws

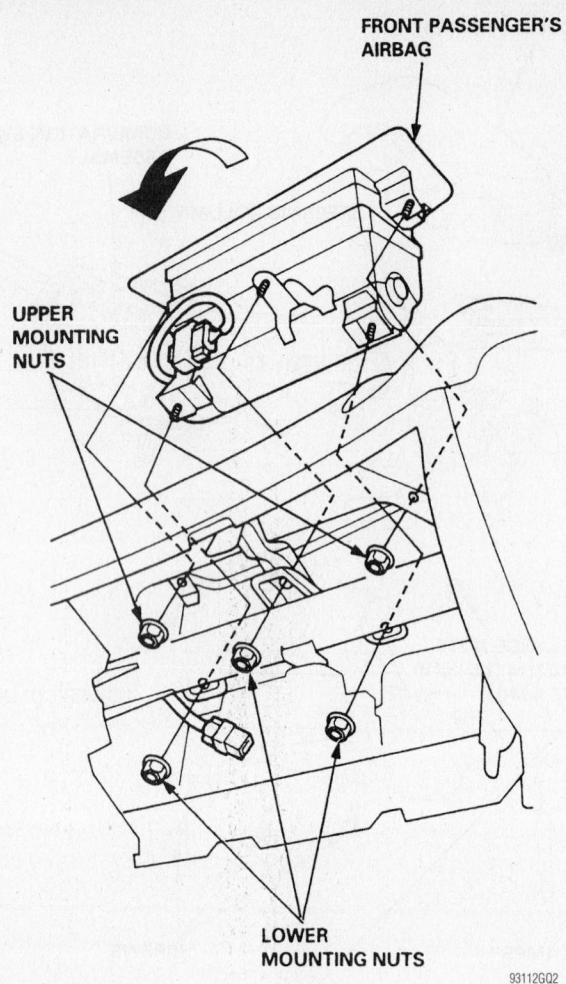

FRONT PASSENGER'S AIRBAG

UPPER MOUNTING NUTS

LOWER MOUNTING NUTS

93112GQ2

Exploded view of the passenger's side SRS module—3.2CL

- Glove box-to-damper screw
- Instrument cluster and the instrument cluster-to-instrument panel screws; then, connect the electrical connectors
- Cruise control master switch and the panel brightness controller; then, connect the electrical connectors
- Front and rear consoles

16. Install the passenger's side SRS module by installing or connecting the following:

- SRS module and the 5 SRS module-to-instrument panel nuts and torque to 84 inch lbs. (9.8 Nm)
- SRS module electrical connector

17. Install the steering column by installing or connecting the following:

- Steering column and the steering column-to-instrument panel nuts/bolts
- Steering joint to the steering column shaft
- Clamps, clips and the steering joint cover
- Ignition switch connectors
- Combination switch, connect the electrical connectors and install the combination switch-to-steering column screws
- Steering column covers and the cover screws
- Knee bolster and the knee bolster-to-dash bolts; then, connect the

▶: Bolt locations, 6

6 x 1.0 mm
9.8 N·m (1.0 kgf·m, 7.2 lbf·ft)

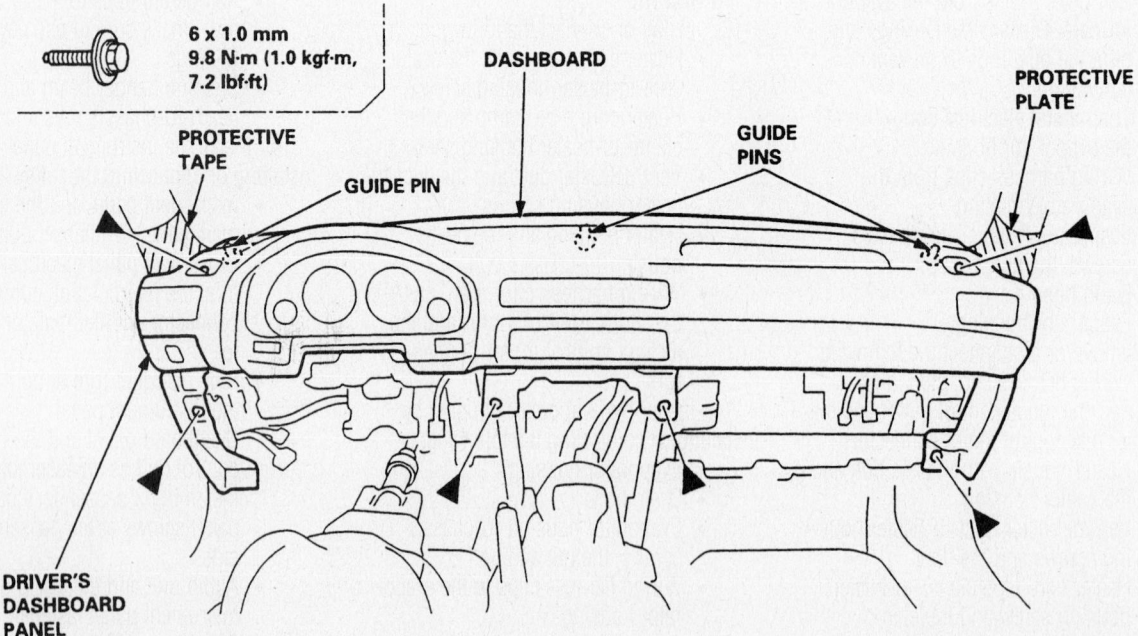

PROTECTIVE TAPE

GUIDE PIN

DASHBOARD

GUIDE PINS

PROTECTIVE PLATE

DRIVER'S DASHBOARD PANEL

93112GQ3

View of the dashboard bolt locations—3.2CL

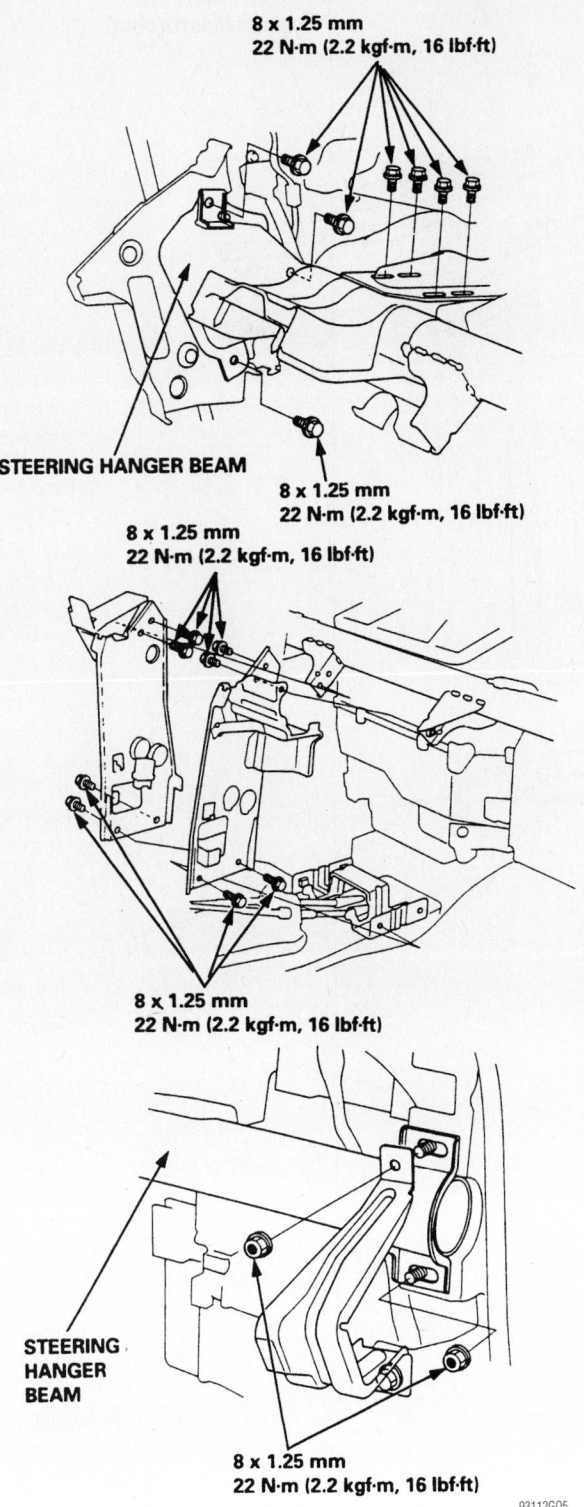

View of the steering hanger beam and related components—3.2CL

8 x 1.25 mm
22 N·m (2.2 kgf·m, 16 lbf·ft)

STEERING HANGER BEAM

8 x 1.25 mm
22 N·m (2.2 kgf·m, 16 lbf·ft)

8 x 1.25 mm
22 N·m (2.2 kgf·m, 16 lbf·ft)

8 x 1.25 mm
22 N·m (2.2 kgf·m, 16 lbf·ft)

STEERING HANGER BEAM

8 x 1.25 mm
22 N·m (2.2 kgf·m, 16 lbf·ft)

93112GQ5

electrical connector and the air hose.
- Lower instrument panel, the lower instrument panel-to-dash screw and the coin pocket

18. Install the SRS module and the steering wheel by installing or connecting the following:

- Steering wheel to the steering column and torque the nut to 28 ft. lbs. (38 Nm)
- Electrical connectors to the steering wheel
- SRS module to the steering wheel
- Both SRS module-to-steering wheel bolts and torque to 84 inch

lbs. (9.8 Nm) using a T30 Torx bit
- SRS module-to-steering wheel covers
- SRS module electrical connector
- Access panel-to-steering wheel screws and the panel.
- 2 heater housing-to-cowl nuts located in the engine compartment
- Heater hoses to the heater core

19. Refill the cooling system.
20. Connect the negative battery cable.
21. Evacuate, charge and leak test the air conditioning system.
22. Operate the engine to normal operating temperatures; then, check the climate control operation and check for leaks.

3.2TL

➡ Be sure to acquire the anti–theft code for the radio; then, write down the frequencies for the preset buttons.

1. Before servicing the vehicle, refer to the precautions in the beginning of this section.
2. Disconnect the negative battery cable.

✳✳ CAUTION

After disconnecting the negative battery cable, wait for at least 3 minutes for the SRS module to deplete its energy.

3. Drain the cooling system into a clean container for reuse.
4. Disconnect the heater hoses from the heater core.
5. In the engine compartment, remove the heater housing-to-cowl nut.
6. Remove the instrument panel by removing or disconnecting the following:
- Clips and remove the dashboard side cover, located at the driver's side
- Lower dashboard cover screw, detach the clips and remove the lower dashboard cover
- Console panel screws and the console panel
- Glove box-to-instrument panel screw; then, remove the damper from the glove box
- Glove box stop located at each side while holding the glove box
- Both glove box-to-instrument panel screws and the glove box
- Clips and remove the dashboard side cover located at the passenger's side
- Glove box cover-to-instrument panel screws, disconnect the elec-

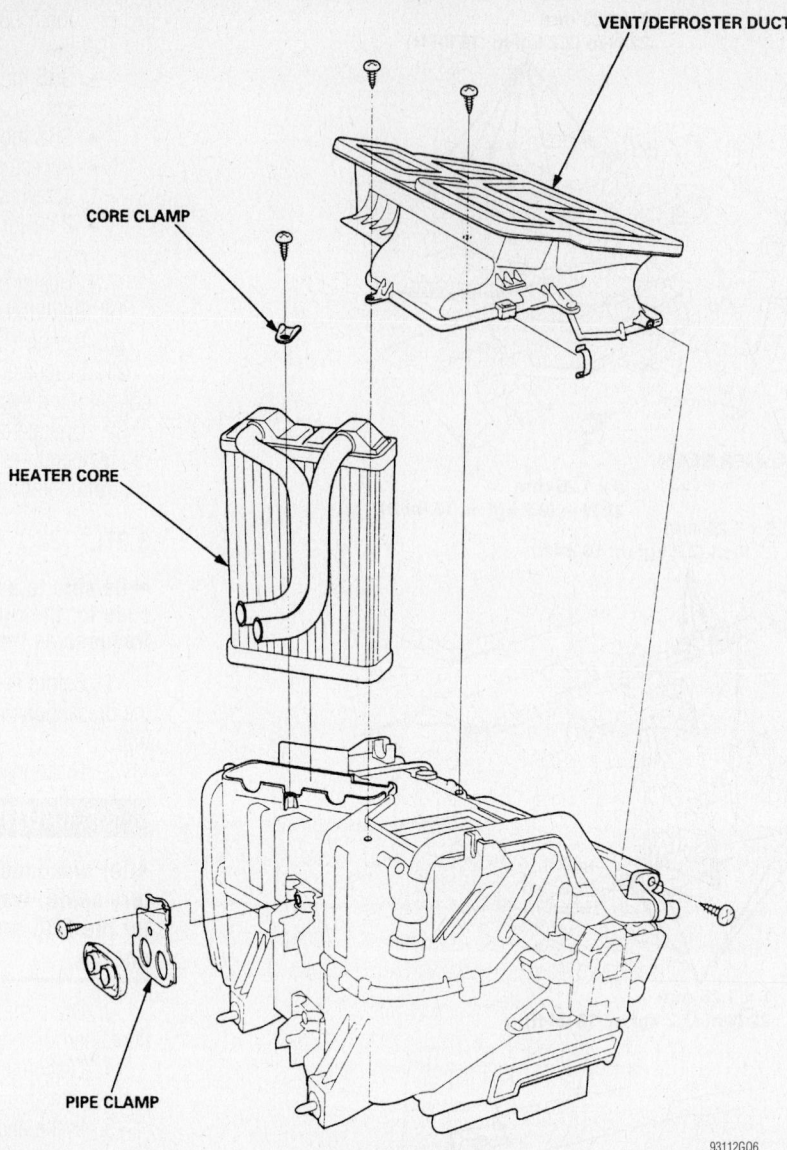

VENT/DEFROSTER DUCT

CORE CLAMP

HEATER CORE

PIPE CLAMP

93112GQ6

Exploded view of the heater core, heater housing and related components—3.2CL

trical connectors and remove the glove box cover
- Rear console and the front console cover
- 4 audio unit-to-instrument panel screws, disconnect the electrical connectors and remove the audio unit
- Pry out the front pillar trim at both sides
- Bolts at the rear vent duct, disconnect the clips and the OBD–II connector; then, detach the tabs and remove the rear vent duct
- Steering column cover screws and the covers
- Steering column electrical connectors
- Steering joint cover clamp, screws and the cover

- Steering column-to-instrument panel nuts/bolts and lower the steering column
- Instrument panel's electrical connectors and clips
- Instrument panel-to-chassis bolts and the instrument panel

7. Discharge and recover the air conditioning system refrigerant.

8. Remove the evaporator housing by removing or disconnecting the following:
- Refrigerant lines from the evaporator core. Discard the O–rings and plug the openings to prevent contamination.
- Electrical connectors from the evaporator housing
- Evaporator housing-to-chassis screws, nut and bolts; then, remove the evaporator housing

9. Remove or disconnect the following:
- Heater housing-to-chassis bolts and the heater housing
- Mode control motor-to-heater housing screws, the linkage and the motor
- Bracket-to-heater housing screws and the brackets
- Upper-to-lower heater housing screws and separate the housings
- Heater core

To install:
10. Install or connect the following:
- Heater core
- Upper-to-lower heater housing and install the heater housing screws
- Brackets and the bracket-to-heater housing screws
- Mode control motor, the linkage and the motor-to-heater housing screws

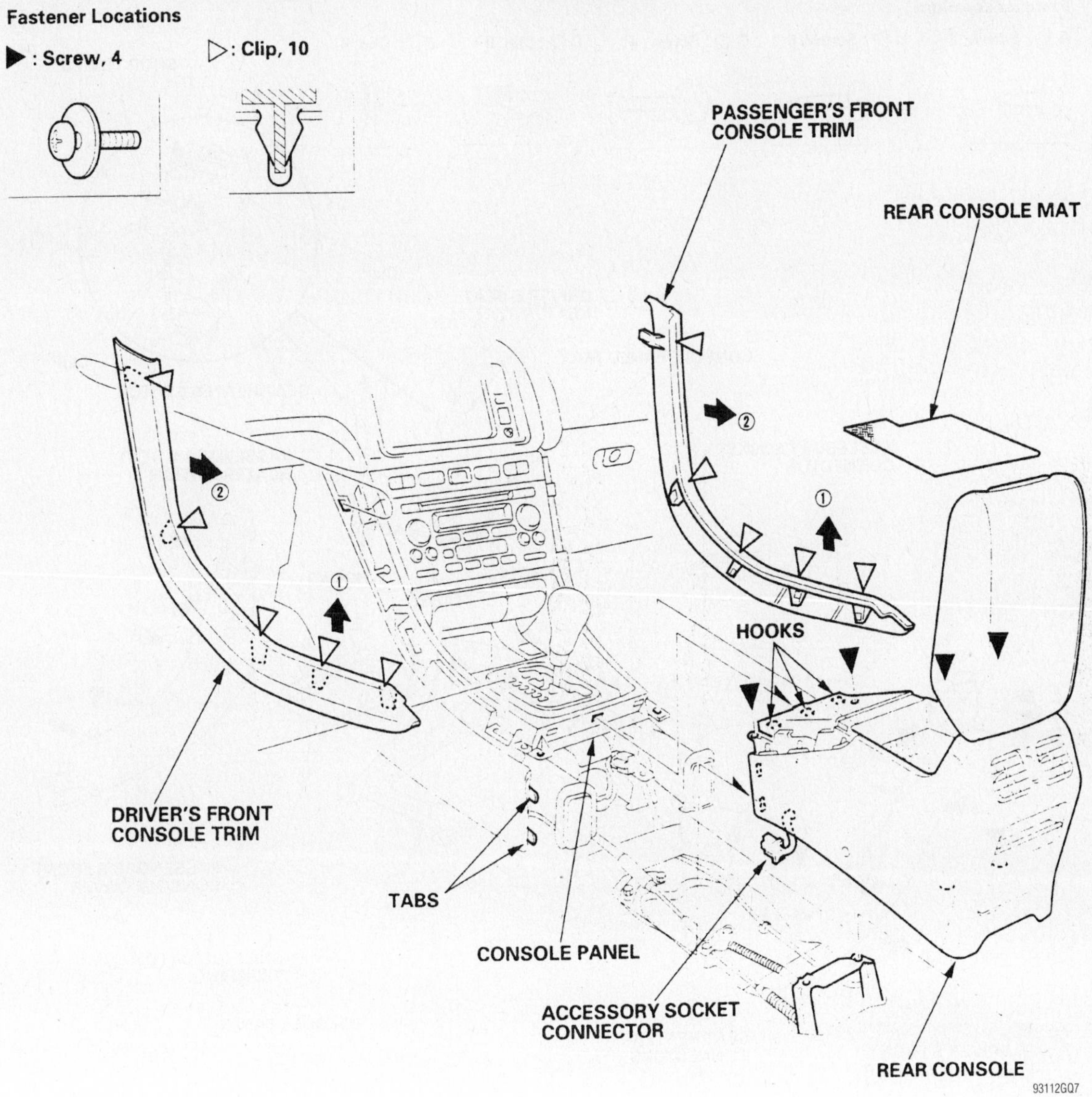

Fastener Locations

▶ : Screw, 4   ▷ : Clip, 10

PASSENGER'S FRONT
CONSOLE TRIM

REAR CONSOLE MAT

DRIVER'S FRONT
CONSOLE TRIM

HOOKS

TABS

CONSOLE PANEL

ACCESSORY SOCKET
CONNECTOR

REAR CONSOLE

93112GQ7

Exploded view of the rear console and related components—3.2TL

- Heater housing and the heater housing-to-chassis bolts
11. Install the evaporator housing by installing or connecting the following:
  - Evaporator housing and the evaporator housing-to-chassis screws, nut and bolts
  - Electrical connectors to the evaporator housing
  - Refrigerant lines to the evaporator core, make sure to use new O–rings
12. Install the instrument panel by installing or connecting the following:
  - Instrument panel and the instrument panel-to-chassis bolts
  - Instrument panel's electrical connectors and clips
  - Steering column and torque the steering column-to-instrument

panel nuts to 12 ft. lbs. (16 Nm) and bolts to 17 ft. lbs. (23 Nm)
  - Steering joint cover and the cover clamp and screws
  - Steering column electrical connectors
  - Steering column covers and the cover screws
  - Rear vent duct, connect the clips and the OBD–II connector, attach the tabs and install the bolts
  - Front pillar trim. On both sides.
  - Audio unit, connect the electrical connectors and install the 4 audio unit-to-instrument panel screws
  - Rear console and the front console cover
  - Glove box cover, connect the electrical connectors and install the glove box cover-to-instrument panel screws

- Dashboard side cover and attach the clips at the passenger's side
- Both glove box and the glove box-to-instrument panel screws
- Glove box stop (located at each side), while holding the glove box
- Damper to the glove box and the glove box-to-instrument panel screw
- Console panel and the console panel screws
- Lower dashboard cover, attach the clips and install the lower dashboard cover screw
- Dashboard side cover and attach the clips at the driver's side
- Heater housing-to-cowl nut located in the engine compartment
- Heater hoses to the heater core
13. Refill the cooling system.

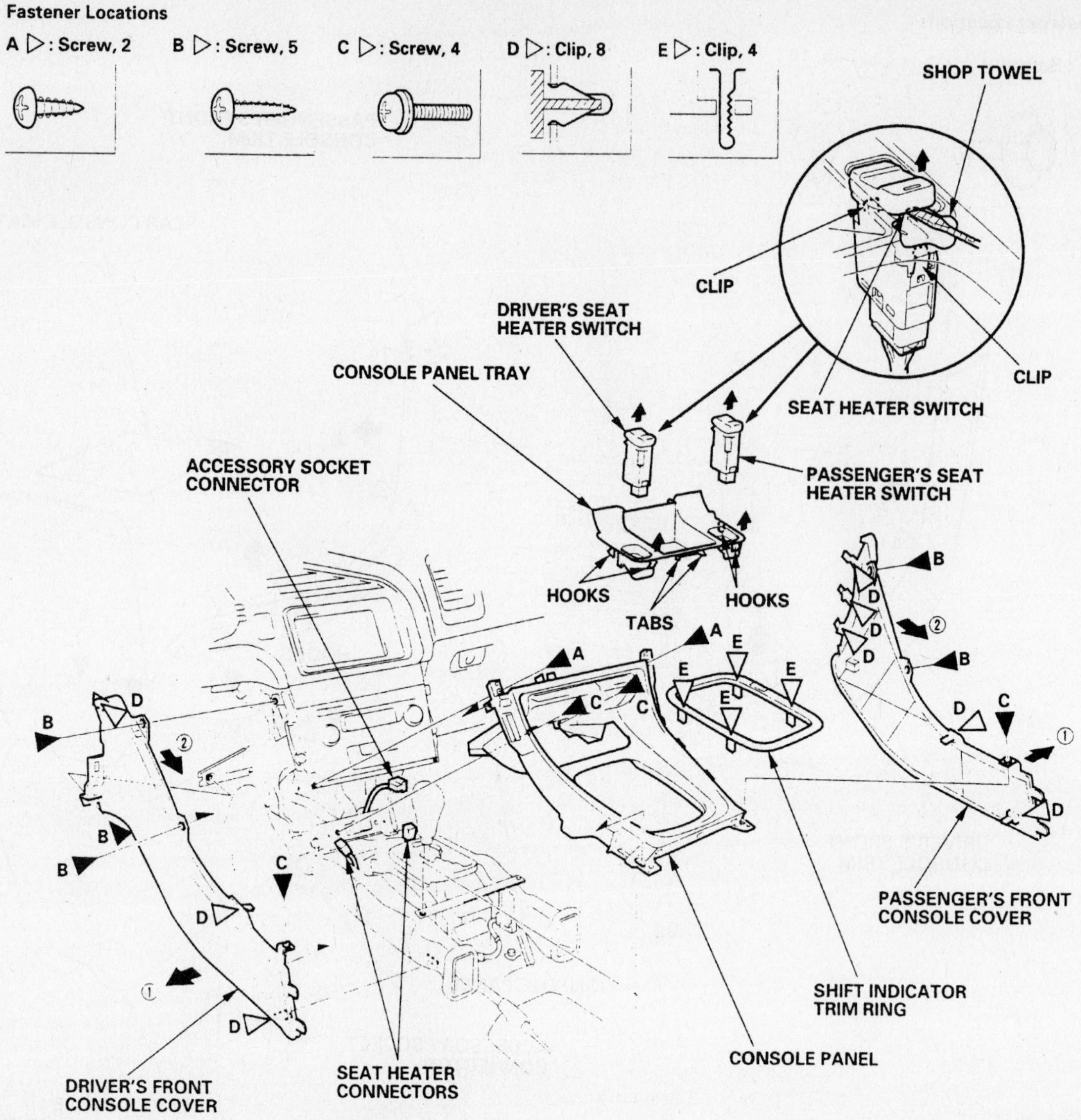

Fastener Locations

A ▷: Screw, 2   B ▷: Screw, 5   C ▷: Screw, 4   D ▷: Clip, 8   E ▷: Clip, 4

SHOP TOWEL

CLIP

CLIP

DRIVER'S SEAT
HEATER SWITCH

SEAT HEATER SWITCH

CONSOLE PANEL TRAY

PASSENGER'S SEAT
HEATER SWITCH

ACCESSORY SOCKET
CONNECTOR

HOOKS

TABS

HOOKS

B

D
D
D

②

B

D   C

①

D

PASSENGER'S FRONT
CONSOLE COVER

A

A

C   C

E   E   E

E

B

D

②

B

B

C

D

①

D

SHIFT INDICATOR
TRIM RING

CONSOLE PANEL

DRIVER'S FRONT
CONSOLE COVER

SEAT HEATER
CONNECTORS

93112GQ8

Exploded view of the console panel and related components—3.2TL

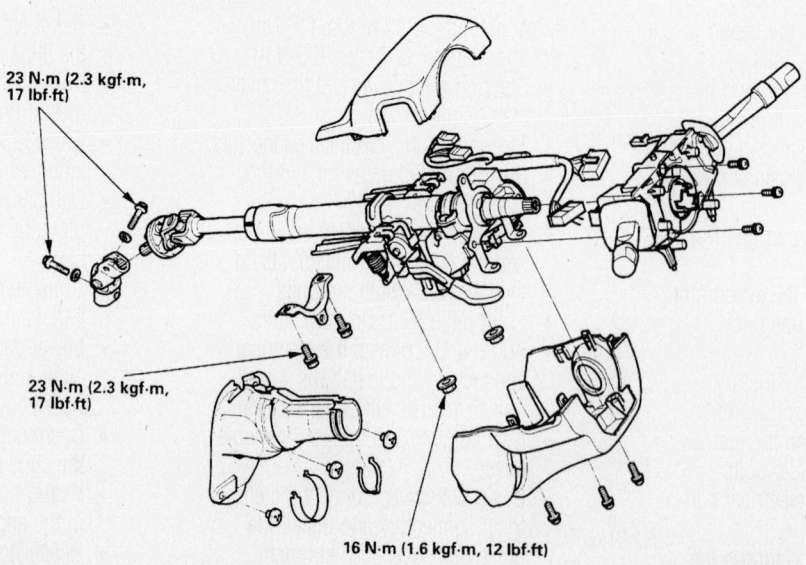

23 N·m (2.3 kgf·m,
17 lbf·ft)

23 N·m (2.3 kgf·m,
17 lbf·ft)

16 N·m (1.6 kgf·m, 12 lbf·ft)

93112GQ9

Exploded view of the steering column and related components—3.2TL

Fastener Locations

B ▶ : Bolt, 6   C ▶ : Bolt, 3   D ▶ : Bolt, 1   E ▶ : Bolt, 3   F ▶ : Bolt, 1

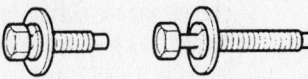  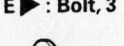

8 x 1.25 mm
22 N·m (2.2 kgf·m,
16 lbf·ft)

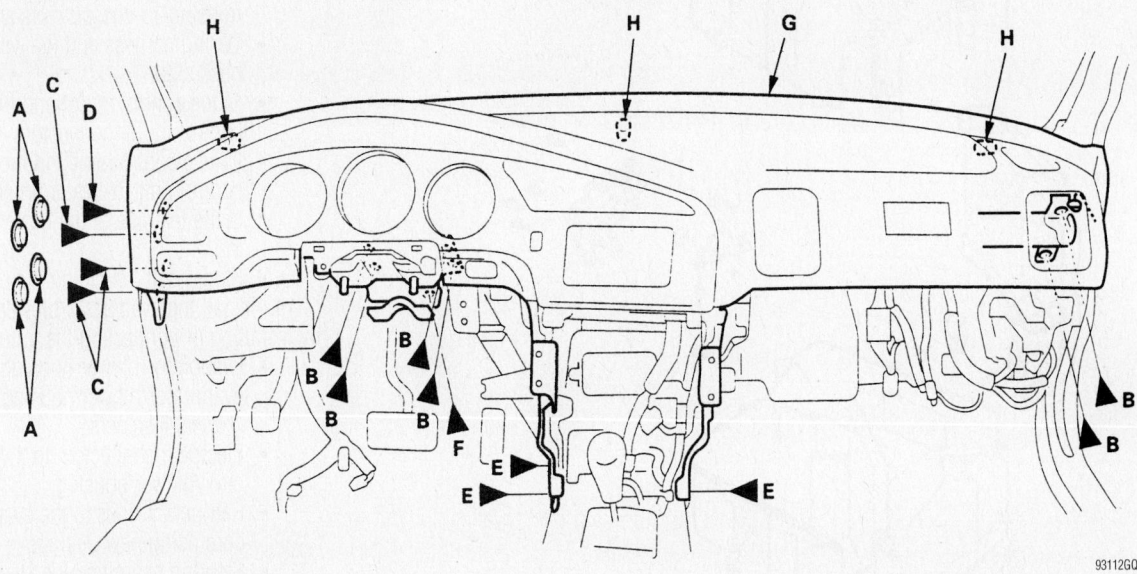

93112GQ0

View of the instrument panel and bolt locations—3.2TL

14. Connect the negative battery cable.

15. Evacuate, charge and leak test the air conditioning system.

16. Operate the engine to normal operating temperatures; then, check the climate control operation and check for leaks.

3.5RL

➡**Be sure to acquire the anti–theft code for the radio; then, write down the frequencies for the preset buttons.**

1. Before servicing the vehicle, refer to the precautions in the beginning of this section.

2. Disconnect the negative battery cable.

> ❊❊ **CAUTION**

After disconnecting the negative battery cable, wait for at least 3 minutes for the SRS module to deplete its energy.

3. Drain the cooling system into a clean container for reuse.

4. Disconnect the heater hoses from the heater core.

5. Place the front wheel in the straight–ahead position.

6. Remove the SRS module and the steering wheel by removing or disconnecting the following:

- Access panel-to-steering wheel screws and the panel
- SRS module electrical connector
- SRS module-to-steering wheel covers, located at both sides of the steering wheel
- Both SRS module-to-steering wheel bolts using a T30 Torx bit
- SRS module from the steering wheel
- Electrical connectors from the steering wheel
- Steering wheel from the steering column

7. Remove the passenger's side SRS module by removing or disconnecting the following:

- SRS module electrical connector
- 2 SRS module-to-instrument panel nuts and carefully remove the SRS module

> ❊❊ **CAUTION**

Store the SRS module in a safe place with the front facing upward.

8. Remove the instrument panel by removing or disconnecting the following:

- Console panel and the rear console
- Center air vent using a suitable prytool
- 4 climate control unit/audio assem-

bly-to-instrument panel bolts; then, disconnect the electrical connectors and remove the climate control unit/audio assembly
- Lower instrument panel cover at the passenger's side
- The stop located at each side of the glove box
- Damper clip and the electrical connector while holding the glove box
- Glove box-to-instrument panel bolts and the glove box
- 3 glove box back cover screws, disconnect the clips and remove the cover
- Lower carpet at both sides of the console
- Lower instrument panel cover and the kick panel at the driver's side
- Steering column cover screws and the covers
- Combination switch-to-steering column screws, the combination switch and disconnect the electrical connectors
- Steering joint cover clamp, screws and the cover
- Steering column-to-instrument panel nuts and bolts; then, lower the steering column
- Instrument panel wiring harness

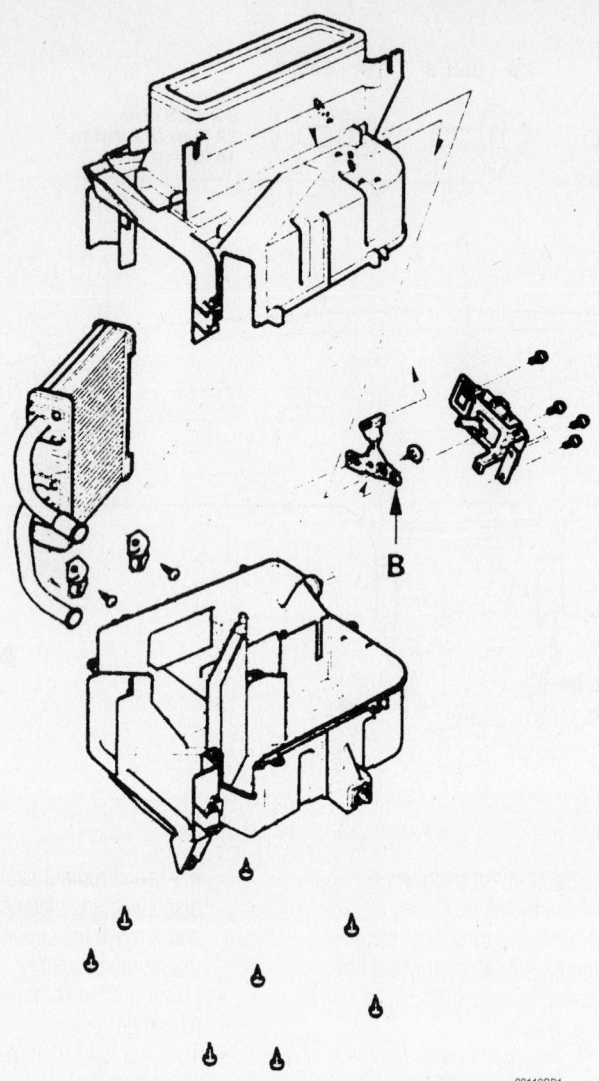

93112GR1

Exploded view of the heater core, heater housing and related components—3.2TL

connectors, the clip and the air hose
- Instrument panel-to-chassis bolts, the screws and the instrument panel
- Steering beam-to-chassis bolts at the left side
- Steering beam-to-chassis nuts, the bolt and the steering hanger beam at the right side

9. Discharge and recover the air conditioning system refrigerant.

10. Remove the evaporator/blower housing by removing or disconnecting the following:
- Refrigerant lines from the evaporator core
- Electrical connectors from the evaporator/blower housing
- Evaporator/blower housing-to-chassis fasteners and the evaporator/blower housing

- Clips and the floor heater duct

11. At the left side of the heater housing, remove the cruise control unit -to-heater housing bolt and the cruise control unit.

12. Remove or disconnect the following:
- Mode control motor and the air mix control motor connectors
- Wiring harness clips and the wiring harness
- Heater housing-to-chassis nuts, the bolt and the heater housing
- Vent/defroster duct-to-heater housing screws and the duct
- Heater core-to-heater housing pipe clamp screws and the clamps
- Heater core-to-heater housing clamp screws and the clamp
- Heater core from the heater housing

To install:

13. Install or connect the following:
- Heater core to the heater housing

- Heater core-to-heater housing clamp and the clamp screws
- Heater core-to-heater housing pipe clamps and the clamp screws
- Vent/defroster duct and the duct-to-heater housing screws
- Heater housing and the heater housing-to-chassis nuts and bolt
- Wiring harness and the wiring harness clips
- Mode control motor and the air mix control motor connectors
- Cruise control unit and the cruise control unit-to-heater housing bolt at the left side of the heater housing
- Floor heater duct and the clips

14. Install the evaporator/blower housing by installing or connecting the following:
- Evaporator/blower housing and the evaporator/blower housing-to-chassis fasteners
- Electrical connectors to the evaporator/blower housing
- Refrigerant lines to the evaporator core using new O–rings
- Steering beam and the steering hanger beam-to-chassis nuts
- Steering beam-to-chassis bolts

15. Install the instrument panel by installing or connecting the following:
- Instrument panel and the instrument panel-to-chassis bolts and the screws
- Instrument panel wiring harness connectors, the clip and the air hose
- Lower the steering column and torque the steering column-to-instrument panel nuts to 12 ft. lbs. (16 Nm) and bolts to 16 ft. lbs. (22 Nm)
- Steering joint cover and the cover clamp and screws
- Combination switch, the combination switch-to-steering column screws and connect the electrical connectors
- Steering column covers and the cover screws
- Lower instrument panel cover and the kick panel at the driver's side
- Lower carpet at both sides of the console
- Glove box back cover, connect the clips and install the 3 cover screws
- Glove box and the glove box-to-instrument panel bolts
- Damper clip and the electrical connector while holding the glove box
- The stop at each side of the glove box

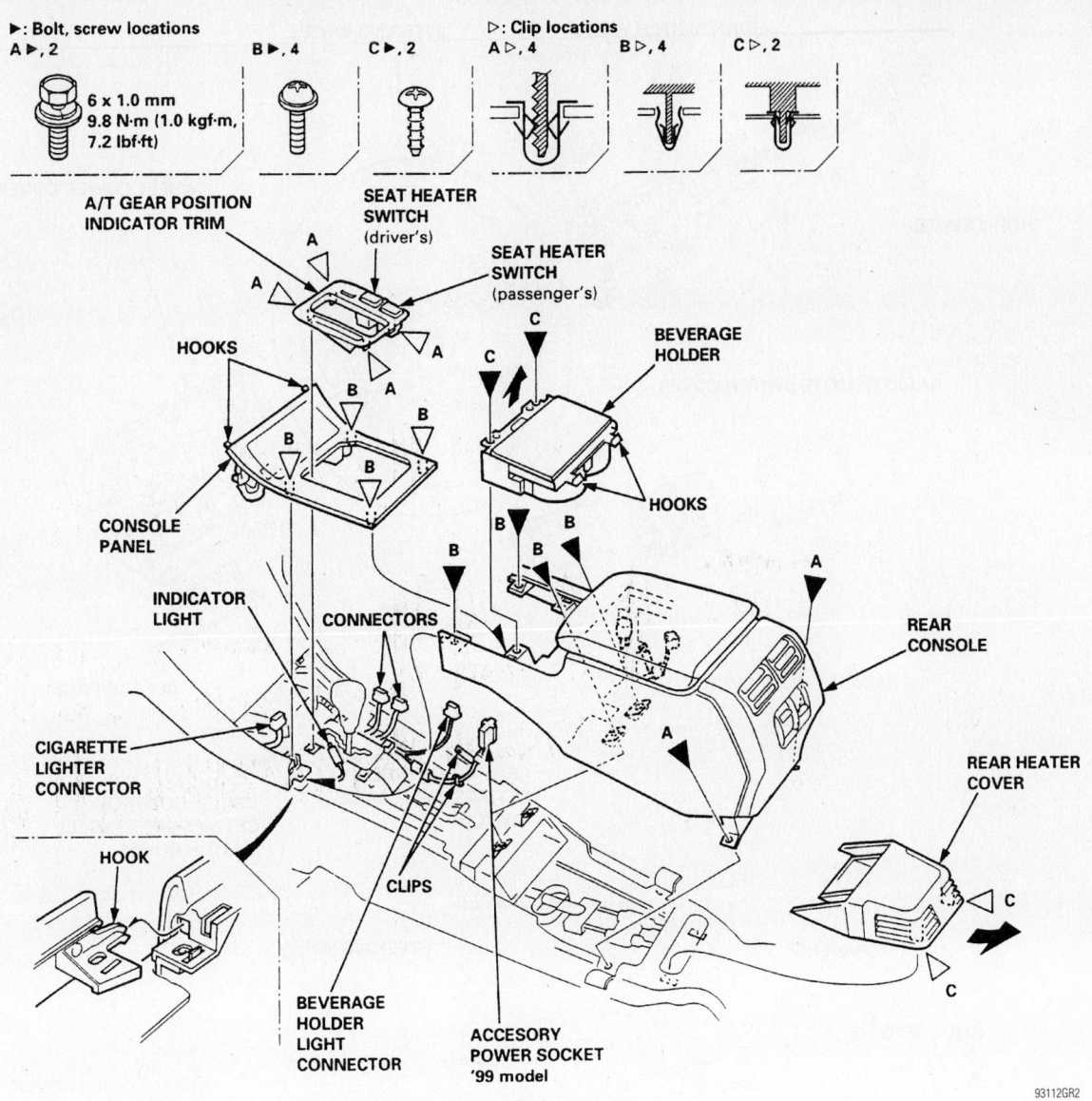

► : Bolt, screw locations
A ► , 2   B ► , 4   C ► , 2

▷ : Clip locations
A ▷ , 4   B ▷ , 4   C ▷ , 2

6 x 1.0 mm
9.8 N·m (1.0 kgf·m,
7.2 lbf·ft)

A/T GEAR POSITION
INDICATOR TRIM

SEAT HEATER
SWITCH
(driver's)

SEAT HEATER
SWITCH
(passenger's)

BEVERAGE
HOLDER

HOOKS

HOOKS

CONSOLE
PANEL

INDICATOR
LIGHT

CONNECTORS

REAR
CONSOLE

CIGARETTE
LIGHTER
CONNECTOR

REAR HEATER
COVER

HOOK

CLIPS

BEVERAGE
HOLDER
LIGHT
CONNECTOR

ACCESORY
POWER SOCKET
'99 model

93112GR2

Exploded view of the console panel, rear console and related components—3.5RL

- Lower instrument panel cover
- Climate control unit/audio assembly; then, connect the electrical connectors and install the 4 climate control unit/audio assembly-to-instrument panel bolts
- Center air vent.
- Console panel and the rear console

16. Install the passenger's side SRS module by installing or connecting the following:

- SRS module and the 2 SRS module-to-instrument panel nuts; then, torque the nuts to 84 inch lbs. (9.8 Nm)
- SRS module electrical connector

17. Install the SRS module and the steering wheel by installing or connecting the following:

- Steering wheel to the steering col-

umn and torque the nut to 36 ft. lbs. (49 Nm)
- Electrical connectors to the steering wheel
- SRS module to the steering wheel
- Both SRS module-to-steering wheel bolts and torque to 84 inch lbs. (9.8 Nm) using a T30 Torx bit
- SRS module-to-steering wheel covers
- SRS module electrical connector
- Access panel-to-steering wheel screws and the panel

18. Connect the heater hoses to the heater core.

19. Refill the cooling system.

20. Connect the negative battery cable.

21. Evacuate, charge and leak test the air conditioning system.

22. Operate the engine to normal operat-

ing temperatures; then, check the climate control operation and check for leaks.

NSX

✵✵ CAUTION

Before starting service procedures on components, especially under the instrument panel and/or near the steering column, disable the SRS system. In addition, the vehicle may be equipped with a radio anti–theft code (5 digits). Get this code from the customer before disconnecting the battery.

1. Before servicing the vehicle, refer to the precautions in the beginning of this section.

2. If equipped with SRS or an anti–theft radio, perform the following procedures:

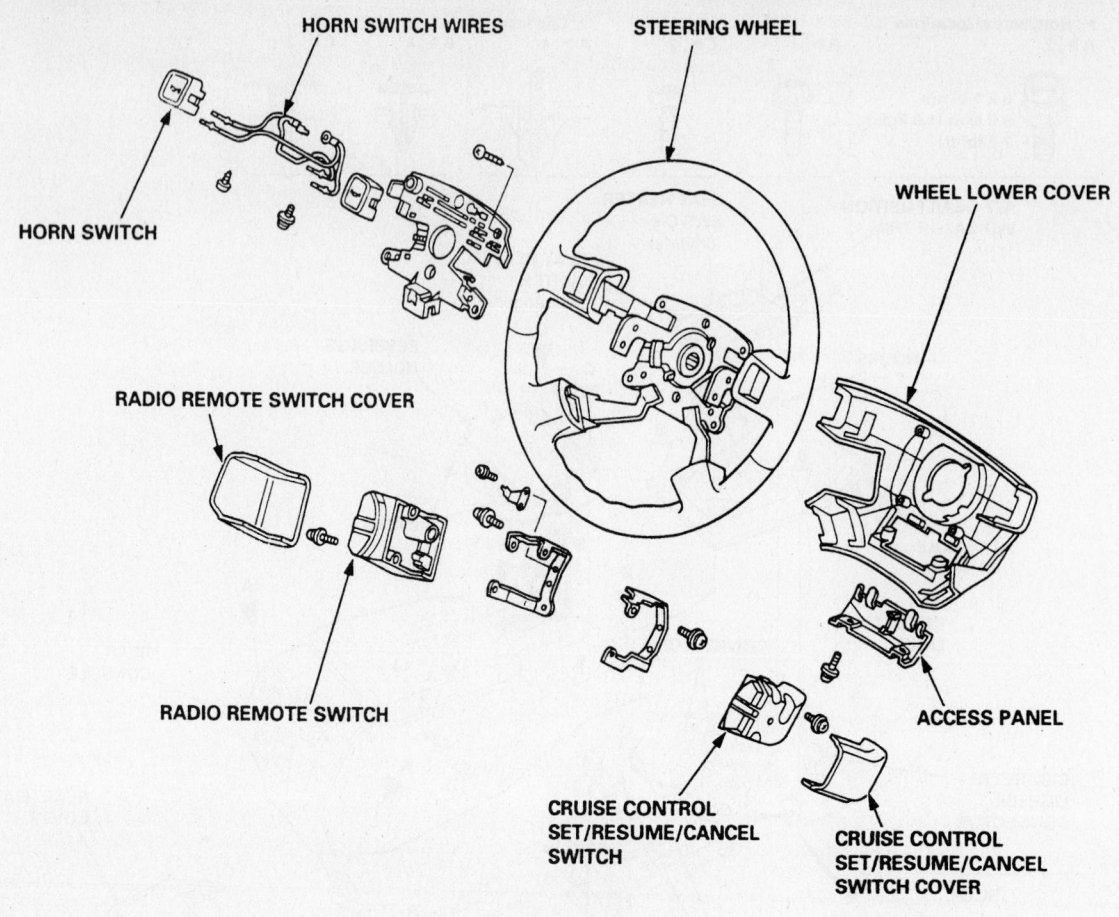

HORN SWITCH WIRES

STEERING WHEEL

HORN SWITCH

WHEEL LOWER COVER

RADIO REMOTE SWITCH COVER

RADIO REMOTE SWITCH

ACCESS PANEL

CRUISE CONTROL SET/RESUME/CANCEL SWITCH

CRUISE CONTROL SET/RESUME/CANCEL SWITCH COVER

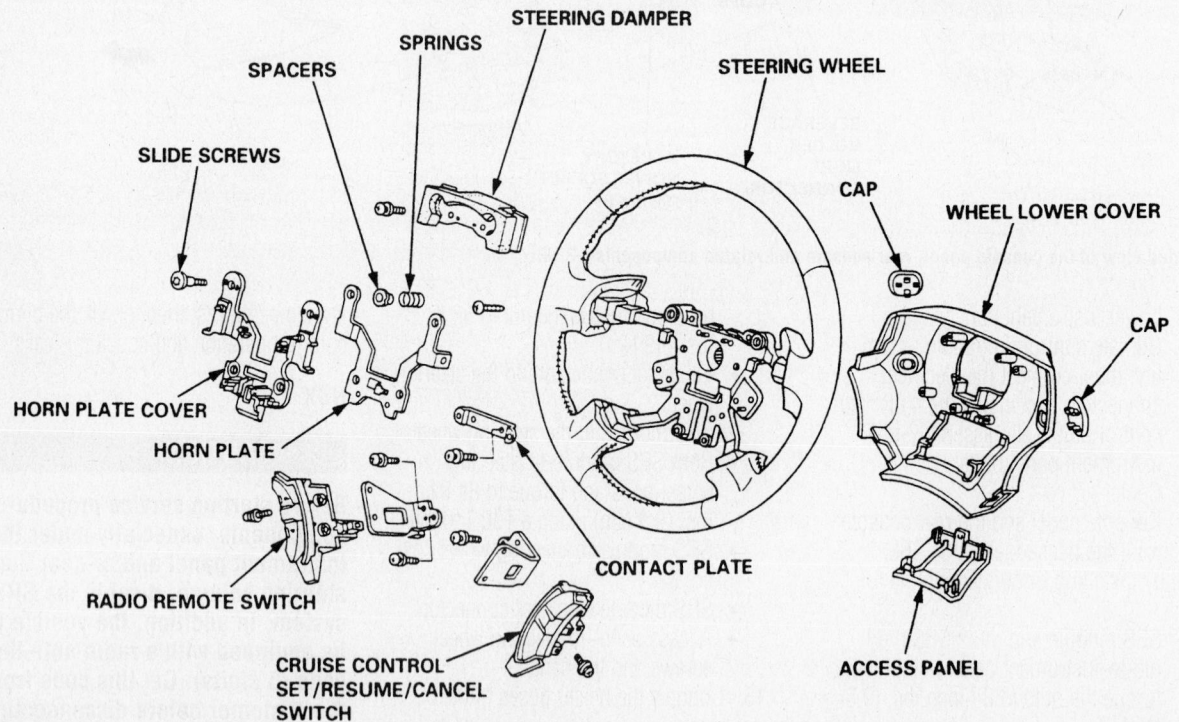

SLIDE SCREWS

SPACERS

SPRINGS

STEERING DAMPER

STEERING WHEEL

CAP

WHEEL LOWER COVER

CAP

HORN PLATE COVER

HORN PLATE

RADIO REMOTE SWITCH

CONTACT PLATE

CRUISE CONTROL SET/RESUME/CANCEL SWITCH

ACCESS PANEL

93112GR3

Exploded view of the SRS module, the steering wheel and related components—3.5RL

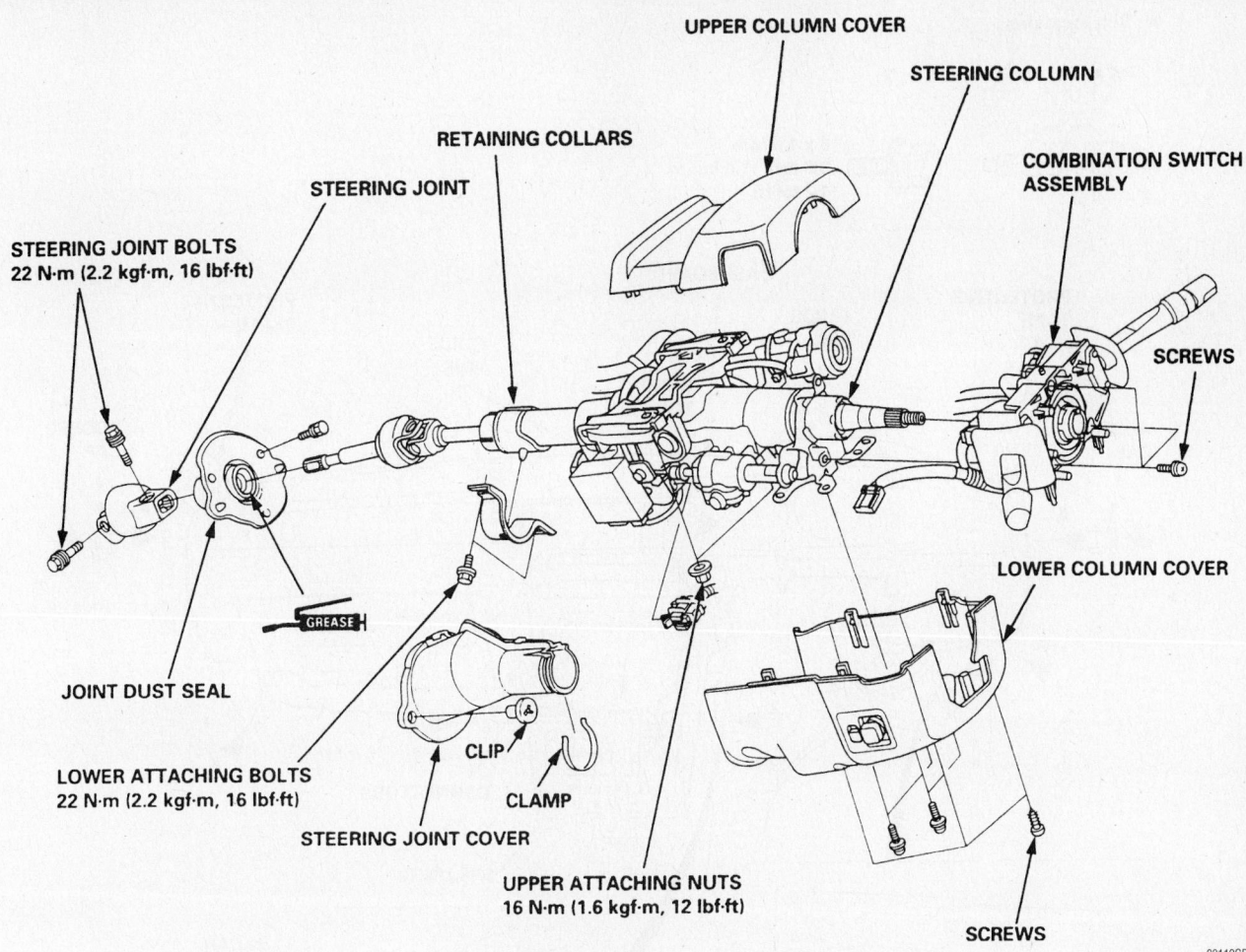

UPPER COLUMN COVER

STEERING COLUMN

RETAINING COLLARS

COMBINATION SWITCH ASSEMBLY

STEERING JOINT

STEERING JOINT BOLTS
22 N·m (2.2 kgf·m, 16 lbf·ft)

SCREWS

GREASE

JOINT DUST SEAL

LOWER COLUMN COVER

LOWER ATTACHING BOLTS
22 N·m (2.2 kgf·m, 16 lbf·ft)

CLIP

CLAMP

STEERING JOINT COVER

UPPER ATTACHING NUTS
16 N·m (1.6 kgf·m, 12 lbf·ft)

SCREWS

93112GR4

Exploded view of the steering column and related components—3.5RL

- Disconnect the negative battery cable, then disconnect the positive battery cable.
- Remove the access panel beneath the air bag on the under-side of the steering wheel center housing.
- Disconnect the connector between the air bag and the cable reel and install the short connector (red) that is provided. Install this connector on the air bag side of the connector.
- The system is now disabled. When repairs are complete, unplug the red short connector, reconnect the air bag and cable reel connector and replace the access panel.
- Reconnect the battery. When the word "CODE" appears, re-enter the 5-digit code in the radio, if equipped.
- When the ignition is turned to the **II** position, the SRS light on the instrument cluster should come ON for about 6 seconds, then go out. If so, the system is okay.

✳✳ WARNING

Do not use electrically powered test equipment on this system or related circuits. All SRS wiring is covered with a special yellow cable housing for identification.

✳✳ CAUTION

All SRS wiring harnesses are covered with yellow outer insulation. Before disconnecting any part of the SRS wire harness, install short connection to prevent accidental discharge of the air bag system. If the SRS harness assembly should have an open circuit or damaged wiring, replace the entire affected harness.

3. Disconnect the negative battery cable.

4. Properly drain the cooling system. Properly discharge the air conditioning system using an approved refrigerant recover/recycling system.

5. Remove or disconnect the following:
- Blower assembly
- Refrigerant lines from their connection at the firewall. Cap all opening immediately to keep dirt and moisture from contaminating the system.

6. Remove the dashboard by removing or disconnecting the following:
- Front seats
- Knee bolster, pad, dashboard stay and center armrest
- Clock, center air vent and center console panel
- Heater/air conditioning control panel and the radio assembly
- Right dashboard lower panel. Remove the glove box
- Lower the steering column
- Instrument cluster bezel and the instrument cluster
- Air vents from both ends of the dashboard
- U–plate at the foot of the center console, rearward of the parking brake

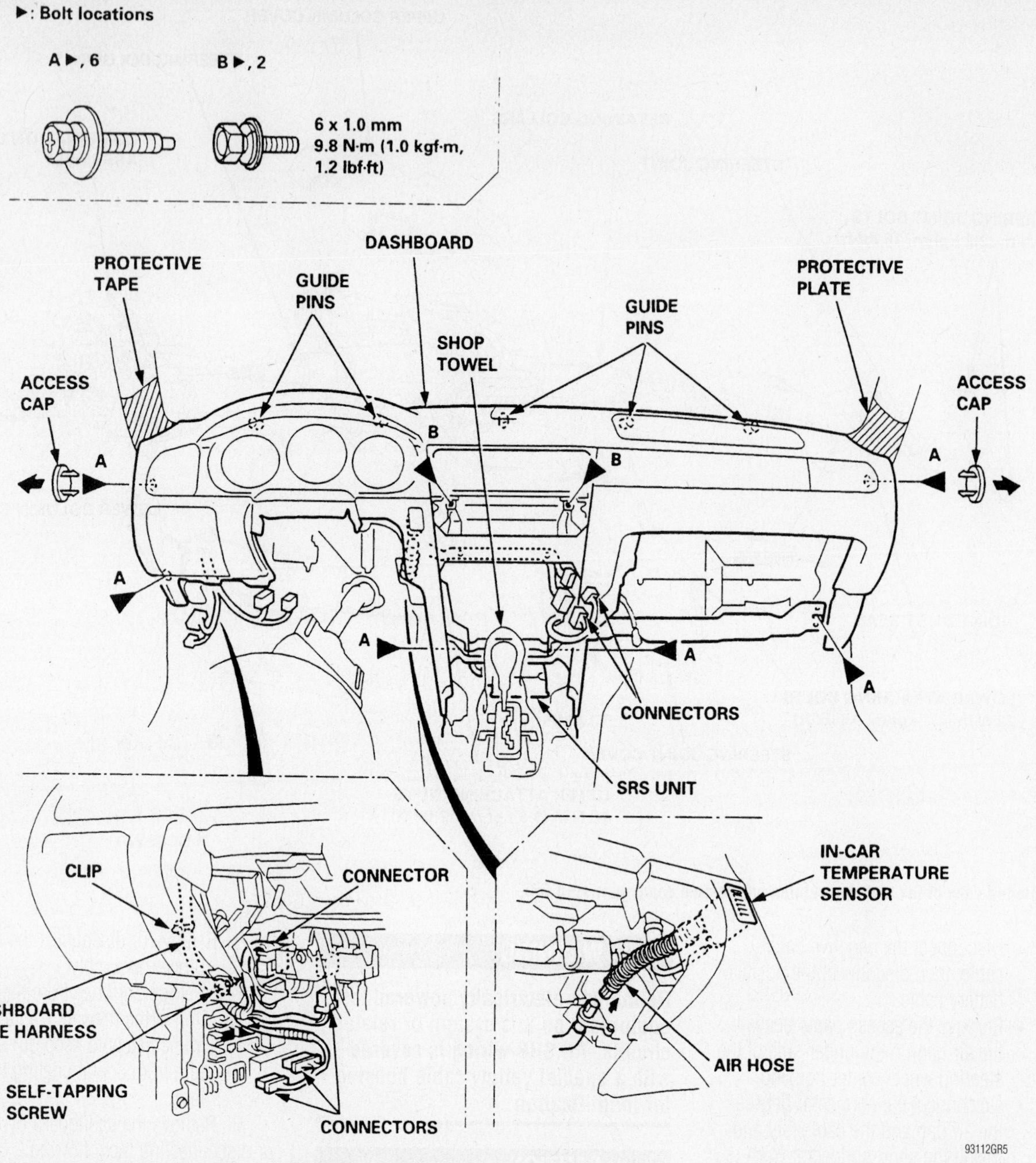

▶: Bolt locations

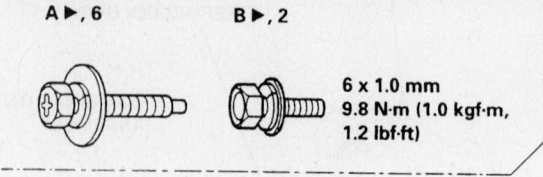

A ▶, 6   B ▶, 2

6 x 1.0 mm
9.8 N·m (1.0 kgf·m,
1.2 lbf·ft)

PROTECTIVE
TAPE

DASHBOARD

GUIDE
PINS

SHOP
TOWEL

GUIDE
PINS

PROTECTIVE
PLATE

ACCESS
CAP

ACCESS
CAP

A

B

B

A

A

A

A

CONNECTORS

SRS UNIT

CLIP

CONNECTOR

IN-CAR
TEMPERATURE
SENSOR

DASHBOARD
WIRE HARNESS

SELF-TAPPING
SCREW

CONNECTORS

AIR HOSE

93112GR5

View of the instrument panel bolt locations—3.5RL

- Defrost outlet grille, remove the mounting bolts and lift out the dashboard. Note the position of the center guide pin for reinstallation.
7. Remove or disconnect the following:
- Heater duct
- Connectors from the control u nit and from the evaporator temperature sensor, then remove the control unit and bracket
- Sound system speaker
- Connectors from the control unit and from the evaporator temperature sensor, then remove the control unit and bracket

- Sound system speaker
- Actuator connectors and sensor connectors from the heater-evaporator unit
- 2 under-dash mounting bolts and remove the heater-evaporator unit through the passenger's door
8. The unit can now be disassembled and the heater core removed.

To install:
9. Installation is the reverse of the removal procedures.
10. Add 0.3 oz. of extra refrigerant oil if the evaporator core was replaced.

11. Adjust the heater valve cable, if needed.
12. Refill the cooling system and bleed the system by performing the following procedure:

✳✳ WARNING

Failure to properly bleed the air from the cooling system could cause engine damage.

13. Turn the ignition switch **ON** and slowly set the climate control temperature knob to **90**. This will allow the coolant in the heater to drain out with the rest of the system.

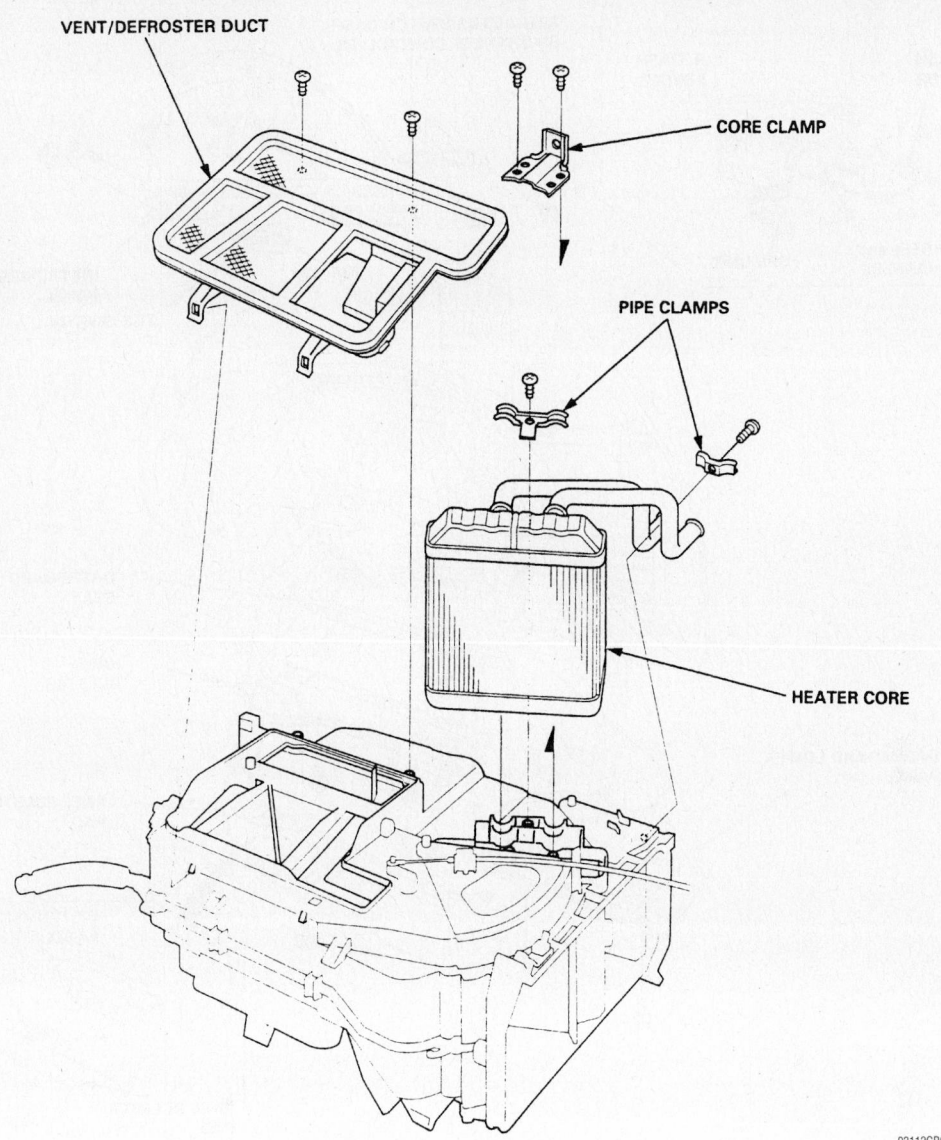

VENT/DEFROSTER DUCT

CORE CLAMP

PIPE CLAMPS

HEATER CORE

93112GR6

Exploded view of the heater core, heater housing and related components—3.5RL

14. Open the hood, rear hatch and the engine cover.

15. Remove the cover protecting the water pipes and shift cables on the underside of the car.

16. Carefully loosen the coolant reserve tank cap. Loosen the drain plug at the bottom of the radiator. Remove the 2 drain bolts from the water pipes. Install rubber hoses to the drain bolts at the front and rear of the engine under the cylinder bank and loosen the drain bolts to drain the coolant.

17. Coolant will drain more quickly if all the air bleed bolts, plug and cap are opened. Be sure the coolant reserve tank has drained completely before opening the air bleed bolts.

18. Using new washers on the water pipe drain bolts, install the drain bolts and radiator drain plug.

19. Open all 4 air bleeder bolts (radiator, heater pipe, water pipe and engine thermostat cover).

20. Using approved coolant in a 50/50 mixture, fill the coolant reserve tank. Tighten the bleeders in sequence; thermostat cover bleed bolt, radiator bleed plug, heater pipe bleed cap, and water pipe bleed bolt as coolant runs out in a steady stream with no bubbles.

21. After tightening all the bleed bolts, fill the coolant tank to the MAX line. Loosen the thermostat bleed bolt to remove any remaining air.

22. When bleeding is complete, tighten the thermostat bolt and fill the coolant reserve tank to the MAX line again. Install the tank cap to the 1st detent.

23. Start the engine and run it to normal operating temperatures (thermostat opens and radiator cooling fan runs).

24. Turn the engine **OFF**, check and adjust the coolant to the MAX line if needed.

25. Install the coolant tank cap securely and install the car's undercover.

26. Evacuate, charge and leak test the air conditioning system.

27. Connect the negative battery cable.

TSX

➡**Be sure to acquire the anti–theft code for the radio; then, write down the frequencies for the preset buttons.**

1. Before servicing the vehicle, refer to the precautions in the beginning of this section.

2. Disconnect the negative battery cable.

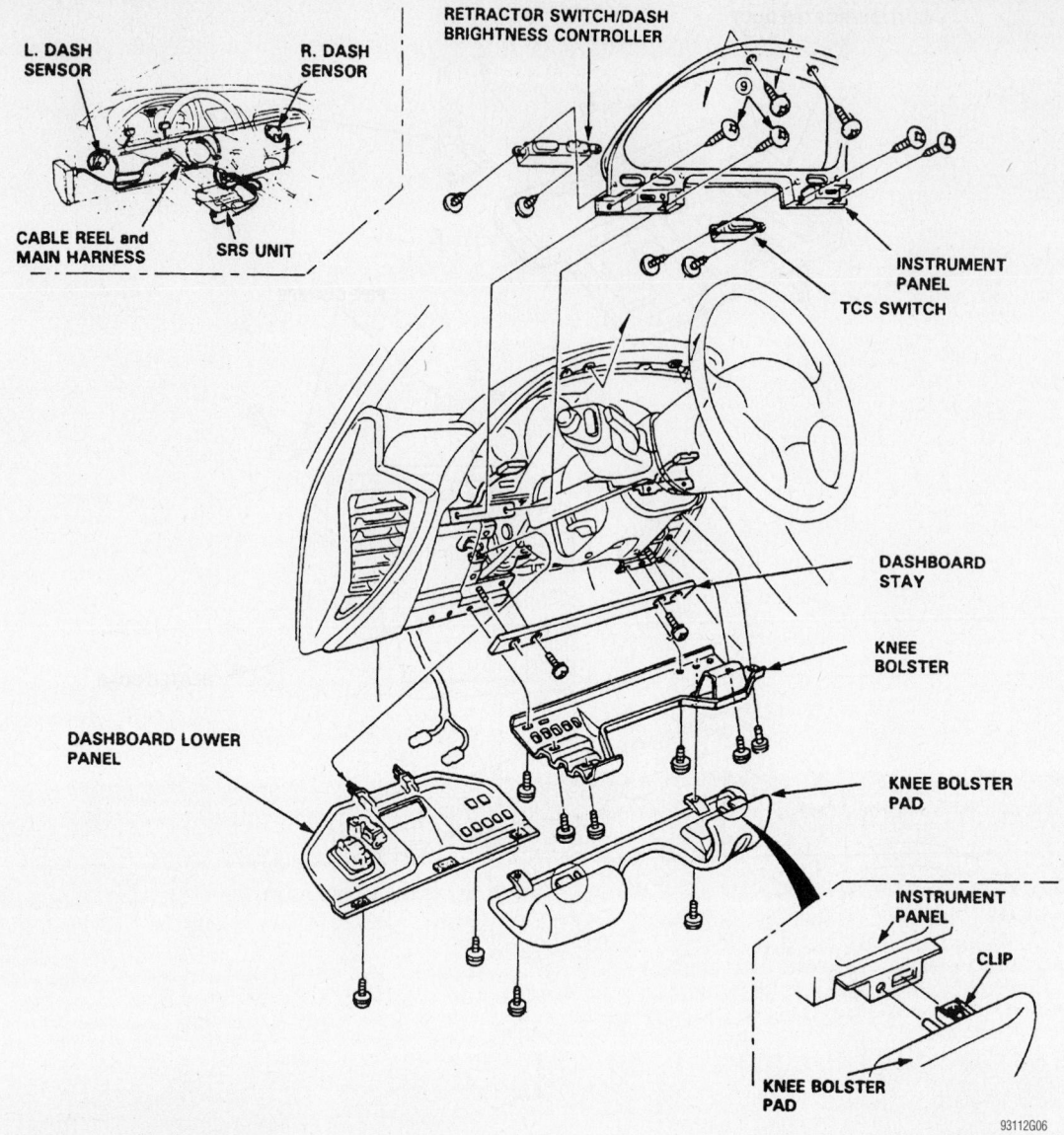

Exploded view of the dashboard and related components—NSX

✳✳ CAUTION

After disconnecting the negative battery cable, wait for at least 3 minutes for the SRS module to deplete its energy.

3. Properly drain the cooling system. Properly discharge the air conditioning system using an approved refrigerant recover/recycling system.

4. Remove or disconnect the following:
- Suction and receiver lines from the evaporator core

5. From under the hood, open the cable clamp and disconnect the heater valve cable from the heater valve arm. Turn the heater valve arm to the fully opened position.

6. Position a drain pan under the hoses. Slide the hose clamps back Remove the bolt and water valve bracket, then disconnect the inlet and outlet heater hoses from the heater unit.

✳✳ WARNING

Do not let any coolant that drains from the hoses to contact any painted surfaces or electrical parts. Immediately wipe up and coolant that spills.

7. Remove or disconnect the following:
- Mounting nut from the heater unit. Do not damage or bend the fuel lines or brake lines
- Dashboard
- Footrest and footrest bracket

8. Remove or disconnect the connectors, wire harness clips, connector clips and wire harness, after disconnecting the following:
- Driver's side mix control motor
- Evaporator temperature sensor
- Power transistor
- Mode control
- Passenger's air mix control motor

9. Remove or disconnect the following:
- Heater ducts
- Mounting nuts and blower/heater unit
- Self-tapping screws and the joint ducts "A" and "B"
- Self-tapping screws and the passenger's heater outlet and heater core cover
- Self-tapping screws, heater pipe brackets, grommets and heater pipe brackets
- Heater core, carefully, being sure not to bend the inlet and outlet pipes.

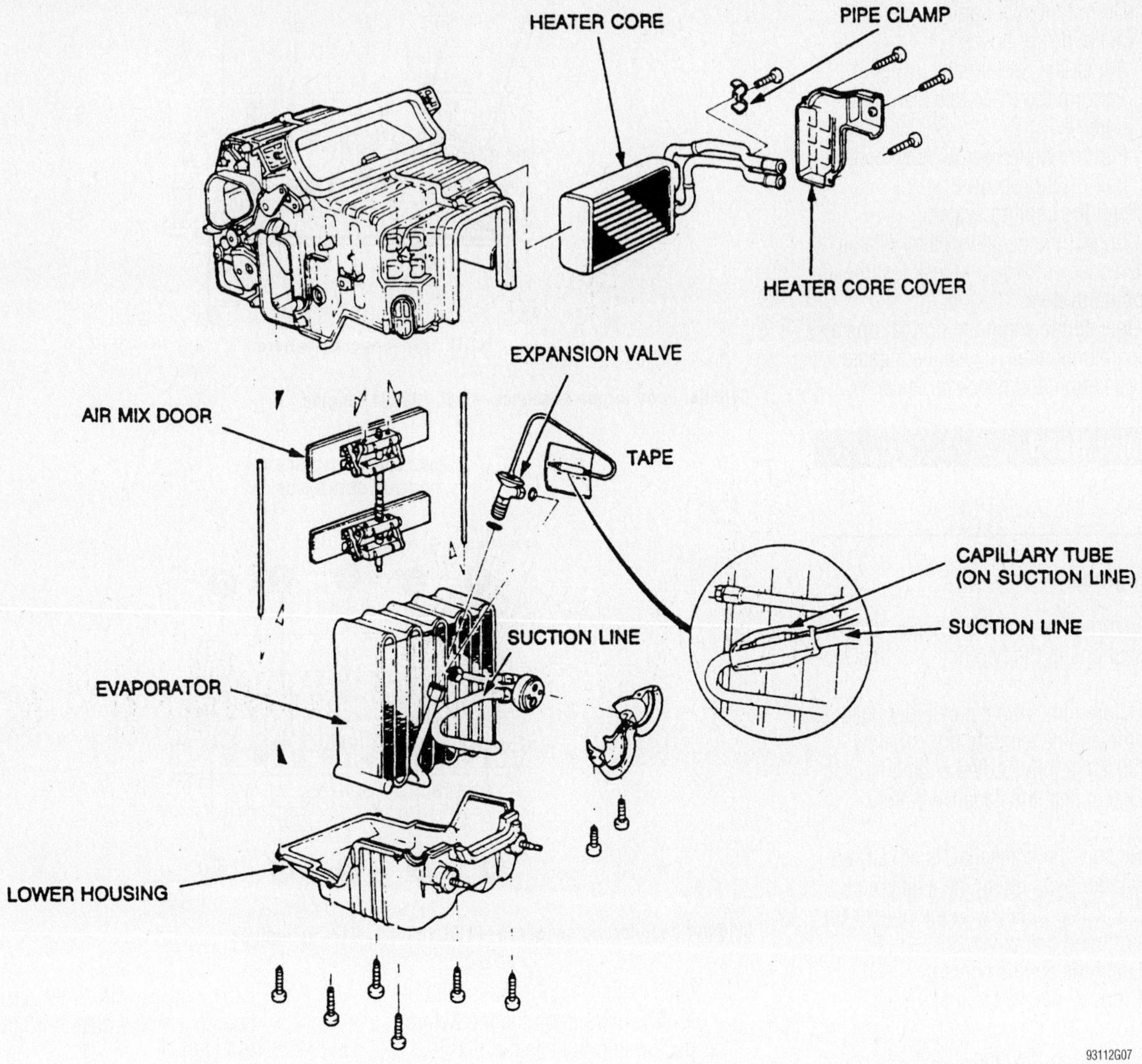

HEATER CORE

PIPE CLAMP

HEATER CORE COVER

EXPANSION VALVE

AIR MIX DOOR

TAPE

CAPILLARY TUBE
(ON SUCTION LINE)

SUCTION LINE

SUCTION LINE

EVAPORATOR

LOWER HOUSING

93112G07

Exploded view of the heater core/evaporator assembly—NSX

To install:

10. Installation is the reverse of the removal procedure, noting the following points:

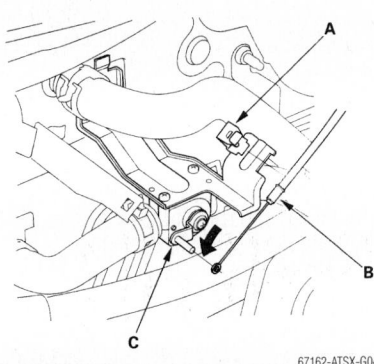

67162-ATSX-G04

Open the heater valve cable clamp (A), then disconnect the cable (B) from the heater valve arm. Turn the heater valve arm (C) to the fully opened position—TSX

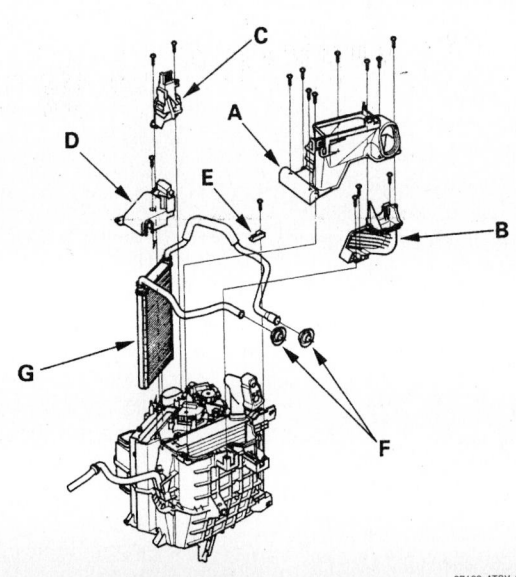

67162-ATSX-G05

Exploded view of the passenger heater outlet (C), heater core cover (D), heater pipe brackets (E), grommets (F) and heater core (G)—TSX

- Do not interchange the inlet and outlet heater hoses
- Adjust the heater valve cable
- Perform the PCM idle learn procedure
- Perform the power window control unit reset procedure

11. Refill the cooling system.

12. Connect the negative battery cable.

13. Evacuate, charge and leak test the air conditioning system.

14. Operate the engine to normal operating temperatures; then, check the climate control operation and check for leaks.

Cylinder Head

REMOVAL & INSTALLATION

1.8L Engines

1. Before servicing the vehicle, refer to the precautions in the beginning of this section.

2. Before removing the cylinder head, be sure the engine temperature is below 100°F (38°C); a fully cooled engine is best.

3. Disconnect the negative battery cable.

4. Be sure the crankshaft is at TDC on No. 1 cylinder by aligning the white mark on the crankshaft pulley with the pointer on the lower timing belt cover.

5. Drain the engine coolant.

6. Properly relieve the fuel system pressure.

7. Remove or disconnect the following:
- Cylinder head cover
- Crankshaft pulley
- Middle and lower timing belt covers
- Timing belt
- Camshaft pulleys
- Back cover
- Exhaust manifold cover
- Exhaust manifold and bracket

8. Loosen the locknuts and adjusting screws, then remove the camshaft holder bolts. Remove the camshaft holders, camshafts and rocker arms.

9. Remove the cylinder head bolts, then remove the cylinder head. To prevent warpage, unscrew the bolts in the reverse of the torque sequence, ⅓ turn at a time. Repeat the sequence until all bolts are loosened.

To install:

10. Install the cylinder head with a new gasket. Be sure to pay attention to the following points:
- Be sure the No. 1 cylinder is at top

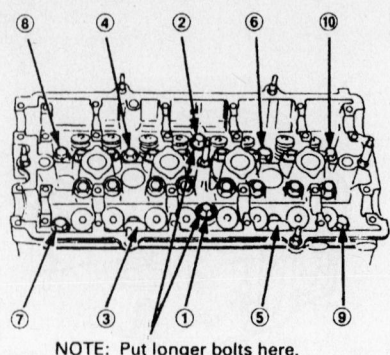

NOTE: Put longer bolts here.

7923BG07

Cylinder head torque sequence—1.8L (B18B1) engine

CYLINDER HEAD BOLT TORQUE SEQUENCE

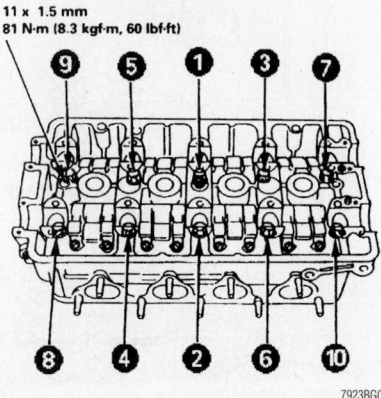

11 x 1.5 mm
81 N·m (8.3 kgf·m, 60 lbf·ft)

7923BG08

Cylinder head torque sequence—1.8L (B18C1, B18C5) engines

dead center and the camshaft pulley **UP** mark is on the top before positioning the head in place.
- The cylinder head dowel pins and oil control orifice must be cleaned and aligned.
- Replace the washer if damaged or deteriorated.
- Apply engine oil to the cylinder head bolts and the washers.
- Use the longer cylinder head bolts at the No. 1 and No. 2 positions.

11. Tighten the cylinder head bolts in 2 steps. In the first step tighten all bolts in sequence to 22 ft. lbs. (29 Nm), then in the second step tighten all bolts in the same sequence to 63 ft. lbs. (85 Nm).

12. Install or connect the following:
- Intake manifold with a new gasket and torque the bolts in a criss–cross pattern to 17 ft. lbs. (24 Nm)
- Intake manifold bracket and torque the bolts to 17 ft. lbs. (24 Nm)
- Exhaust manifold and tighten the new self–locking nuts in a criss-cross pattern in to 23 ft. lbs. (31 Nm)

- Exhaust pipe with a new gasket and tighten the new nuts to 40 ft. lbs. (54 Nm)

13. Be sure that the keyways on the camshafts are facing up and that the rocker arms are in their original position. The valve locknuts should be loosened and the adjusting screw backed off before installation.

14. Place the rocker arms on the pivot bolts and the valve stems.

15. Install the camshafts, then install the camshaft seals with the open side facing in. Install the rubber cap with liquid gasket applied. If the rubber cap has 2 horizontal marks, align the marks with the cylinder head upper surface.

16. Apply liquid gasket to the cylinder head mating surfaces of the No. 1 and No. 6 intake and exhaust camshaft holders and install them, along with No. 2, 3, 4 and 5. Be sure to pay attention to the following points:

17. Do not apply oil to the holder mating surface of camshaft seals.

➡️I or E marks are stamped on the camshaft holders.

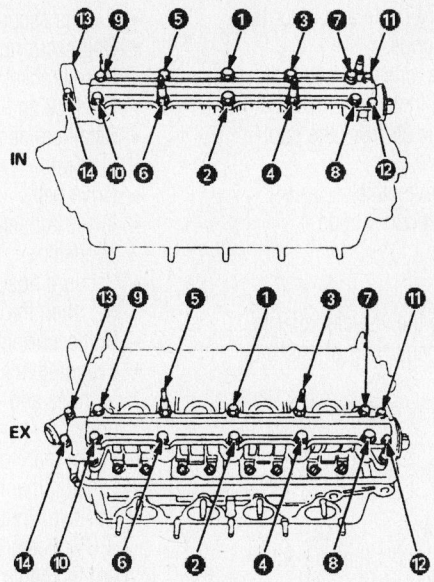

1 – **10**: 8 x 1.25 mm 27 N·m (2.8 kgf·m, 20 lbf·ft)
11 – **14**: 6 x 1.0 mm 9.8 N·m (1.0 kgf·m, 7.2 lbf·ft)

7923BG09

Camshaft plate torque sequence—1.8L engines

18. The arrows marked on the camshaft holders should point to the timing belt.

19. Tighten the camshaft holders temporarily. Be sure that the rocker arms are properly positioned on the valve stems.

20. Tighten each bolt in 2 steps to ensure that the rockers do not bind on the valves. Tighten the 6mm bolts to 86 inch lbs. (9.8 Nm) and the 8mm bolts to 20 ft. lbs. (27 Nm) working from the middle outward.

21. Install the keys into the camshaft grooves. To set the No. 1 piston at TDC, align the holes on the camshaft with the holes in the No. 1 camshaft holders and insert 5.0mm pin punches into the holes.

22. Install or connect the following:
- Rear timing belt cover and torque the bolts to 84 inch lbs. (9.5 Nm)
- Camshaft pulleys and tighten the retaining bolts to 27 ft. lbs. (37 Nm) (B18B1 engine) or 41 ft. lbs. (56 Nm) (B18C1 and B18C5 engines)
- Timing belt and adjust the tension
- Timing belt cover(s) and torque the bolts to 84 inch lbs. (9.5 Nm)

23. Adjust the valve clearance.
- Cylinder head cover and torque the nuts to 86 inch lbs. (9.8 Nm)
- Engine side mount and torque the mounting bolts to 38 ft. lbs. (52 Nm) and the through bolt to 54 ft. lbs. (74 Nm)
- Crankshaft pulley and torque the bolt to 130 ft. lbs. (177 Nm)

24. Connect the negative battery cable and enter the radio security code.

25. After installation, check to see that all hoses and wires are installed correctly.

26. Fill and bleed the air from the cooling system.

27. Attach the negative battery cable.

28. Enter the radio security code.

2.0L Engines

1. Before servicing the vehicle, refer to the precautions in the beginning of this section.

2. Disconnect the negative battery cable.

3. Drain the coolant.

4. Relieve the fuel system pressure.

5. Remove or disconnect the following:
- Fuel feed hose, K20A3 engine only
- Drive belt
- Intake manifold
- Water bypass hose
- Exhaust manifold
- Timing chain

6. Disconnect the following engine wire harness connectors and wire harness clamps from the cylinder head:
- Fuel injector connectors, K20A3 engine
- Engine Coolant Temperature (ECT) sensor connector
- Exhaust and intake Camshaft Position (CMP) sensor connectors

7. Remove or disconnect the following:
- Upper radiator hose and heater hose
- Harness holder from the bracket
- Connecting pipe mounting bolt and water bypass line mounting bolts
- Water bypass hose
- Rocker arm assembly
- Cylinder head bolts, loosening in sequence, ⅓ turn at a time until removed
- Cylinder head

To install:

8. Clean and inspect the cylinder head and mating surfaces.

9. Install a new cylinder head gasket and dowel pins on the cylinder block

10. Set the crankshaft to Top Dead Center (TDC). Align the TDC mark on the crankshaft sprocket with the pointer on the cylinder block.

11. Measure the diameter of each cylinder head bolt at point A and point B, as shown in the illustration. If either specification is less than 0.42 in. (10.6mm), you must replace the cylinder head bolt.

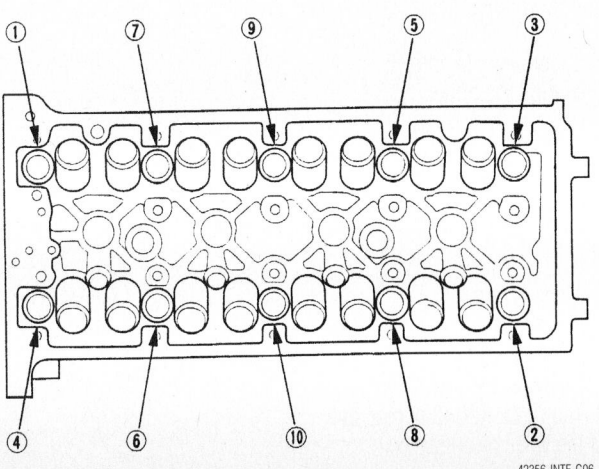

42356-INTE-G06

Cylinder head bolt loosening sequence—2.0L engines

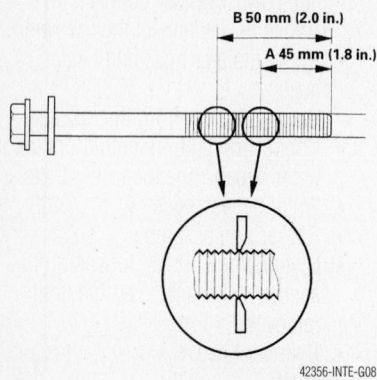

42356-INTE-G07

Align the TDC mark on the crankshaft sprocket (A) with the pointer (B) on the cylinder block—2.0L engines

B 50 mm (2.0 in.)
A 45 mm (1.8 in.)

42356-INTE-G08

Measure the diameter of the head bolts to determine if they are reusable—2.0L engine

12. Carefully position the cylinder heads on the engine.

13. Lubricate the cylinder head bolts with clean engine oil.

14. Tighten the cylinder head bolts in 3 steps. Be sure to follow the tightening torque sequence:

   a. Step 1: 29 ft. lbs. (39 Nm).
   b. Step 2: 90 degrees.
   c. Step 3: 90 degrees.
   d. Step 4 (new bolts only): 90 degrees.

15. Install or connect the following:
- Rocker arm assembly
- Water bypass hose
- Connecting pipe mounting bolt and water bypass line mounting bolts
- Harness holder on the bracket
- Upper radiator hose and heater hose
- Water bypass hose
- Intake manifold
- Exhaust manifold
- Timing chain
- Fuel feed hose, K20A3 engine only

16. Adjust the valve clearance.
- Drive belt

- Negative battery cable and enter the radio security code.

17. After installation, check to see that all hoses and wires are installed correctly.

18. Fill and bleed the air from the cooling system.

19. Attach the negative battery cable.

20. Enter the radio security code.

2.4L Engine

1. Before servicing the vehicle, refer to the precautions in the beginning of this section.

2. Obtain the security code for the radio.

3. Disconnect the negative battery cable.

4. Drain the coolant.

5. Remove or disconnect the following:
- Intake manifold cover
- 4 ignition coils

- 2 bolts securing the vacuum line
- Bolt securing the power steering hose bracket
- Dipstick and breather hose
- Retainers and cylinder head cover
- Fuel line
- Drive belt
- Intake Air Temperature (IAT) sensor connector
- Vacuum hose (B) and breather pipe (C), then the intake air duct (D)
- Bolt securing the connecting pipe
- Evaporative emission (EVAP) canister hose and brake booster vacuum hose
- Intake manifold
- Exhaust manifold
- Positive Crankcase Ventilation (PCV) hose, vacuum hose and ground cable
- Upper radiator hose, heater hoses and water bypass hose

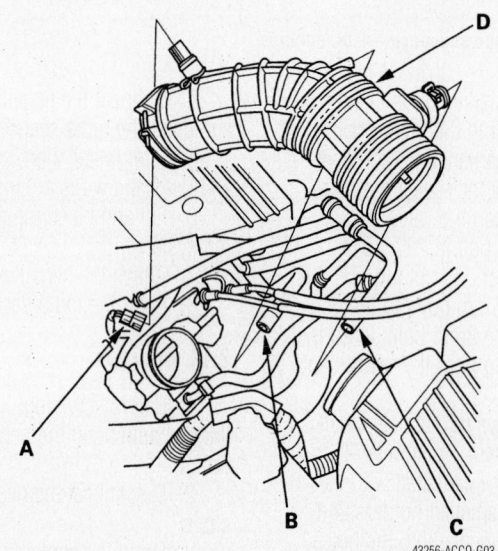

43256-ACCO-G03

Remove the vacuum hose (B), breather pipe (C), then remove the intake air duct (D)—2.4L engine

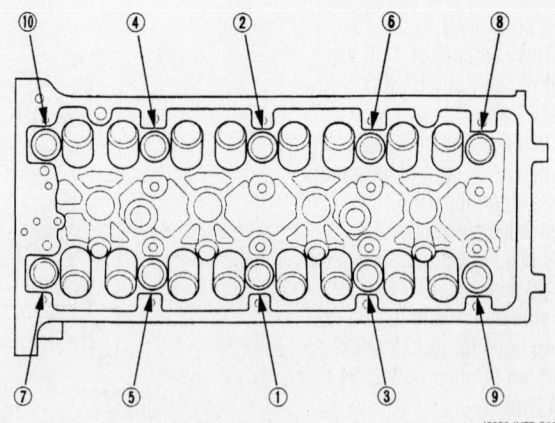

42356-INTE-G09

Cylinder head bolt tightening sequence—2.0L engines

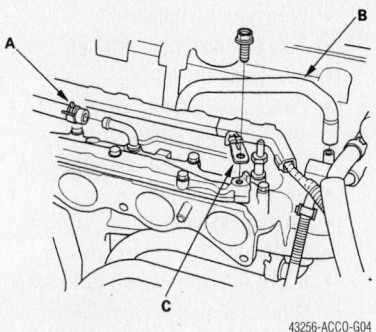

43256-ACCO-G04

Disconnect the PCV hose (A), vacuum hose (B) and ground cable (C)—2.4L engine

- Engine wire harness connectors and wire harness clamps from the cylinder head
- 4 injector connectors
- Engine Coolant Temperature (ECT) sensor connector
- Camshaft Position (CMP) sensor connectors
- Exhaust Gas Recirculation (EGR) valve connector
- VTEC solenoid valve connector
- Engine Oil Pressure (EOP) sensor connector
- 2 bolts securing the EVAP canister purge valve bracket and the bolt securing the harness bracket
- Timing chain
- Rocker arm assembly
- Cylinder head bolts, in sequence, ⅓ turn at a time until completely loosened
- Cylinder head. Discard the gasket.

To install:

6. Be sure all cylinder head and block gasket surfaces are clean. Check the cylinder head for warpage. If warpage is less than 0.002 in. (0.05mm), cylinder head resurfacing is not required. Maximum resurface limit is 0.008 in. (0.2mm) based

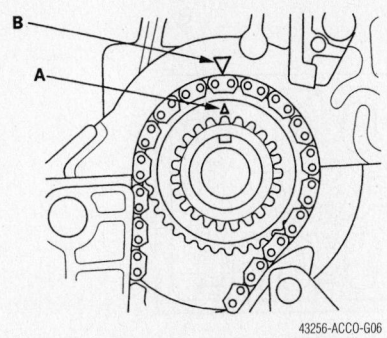

43256-ACCO-G06

Set the crankshaft to TDC by aligning the mark (A) on the crankshaft sprocket with the pointer (B) on the cylinder block—2.4L engine

on a cylinder head height of 3.94 in. (100mm).

7. Install or connect the following:
- New gasket and dowel pins on the cylinder block

8. Set the crankshaft to Top Dead Center (TDC). Align the TDC mark (A) on the crankshaft sprocket with the pointer (B) on the cylinder block.

9. Measure the diameter of each cylinder head bolt at points A & B, as shown in the illustration. If either diameter is less than 0.42 in. (10.6mm), replace the head bolt

10. Apply engine oil to the threads and under the bolt heads of all of the bolts.

11. Install the cylinder head. Tighten the bolts in sequence as follows:
   a. Step 1: 29 ft. lbs. (39 Nm).
   b. Step 2: Plus 90 degrees.
   c. Step 3: Plus 90 degrees.
   d. Step 4: If using new cylinder head bolts, add an additional 90 degrees.

12. Install or connect the following:
- Rocker arm assembly
- Timing chain
- 2 bolts securing the EVAP canister

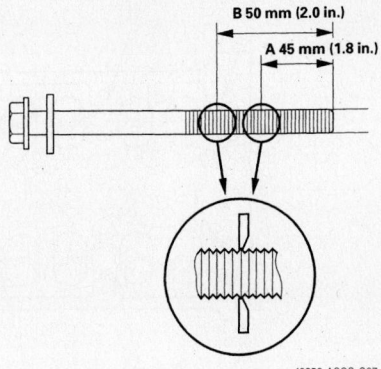

43256-ACCO-G07

You must measure the cylinder head bolts to see if they can be reused or need to be replaced—2.4L engine

purge valve bracket and tighten to 16 ft. lbs. (22 Nm)
- Bolt securing the harness bracket and tighten to 104 ft. lbs. (12 Nm)
- Upper radiator hose, heater hoses and water bypass hose
- PCV hose, vacuum hose and ground cable
- Exhaust manifold
- Intake manifold
- EVAP canister hose and brake booster vacuum hose
- Fuel line
- Bolt securing the connecting pipe and tighten to 16 ft. lbs. (22 Nm)
- Intake air duct, IAT sensor connector, vacuum hose and breather pipe
- Cylinder head cover gasket in the groove of the cylinder head cover
- Apply liquid gasket P/N 08718-0009 or equivalent on the chain cover and No. 5 rocker shaft holder mating areas. The parts must be installed within 5 minutes of applying liquid gasket.
- Spark plug seals on the spark plug tubes
- Cylinder head cover on the cylinder head, then slide the cover back and forth gently to seat it
- Cover washers
- Cylinder head cover bolts. Torque, in sequence, in 2 or 3 steps to 104 inch lbs. (12 Nm).
- Dipstick and breather hose
- Bolt securing the power steering hose bracket
- 2 bolts securing the vacuum line
- 4 ignition coils
- Intake manifold cover, and tighten the retainers to 104 inch lbs. (12 Nm)
- All of the remaining hoses, tubes,

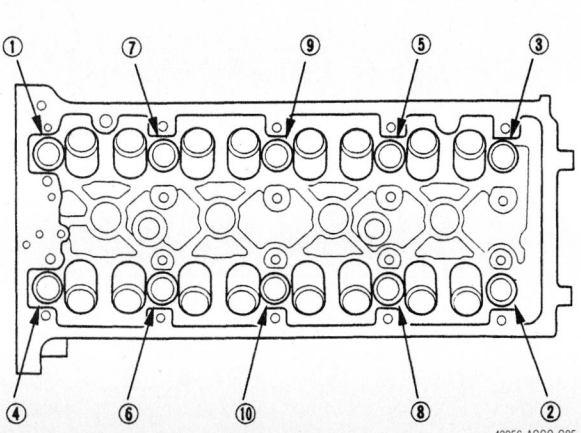

43256-ACCO-G05

Cylinder head bolt loosening sequence—2.4L engine

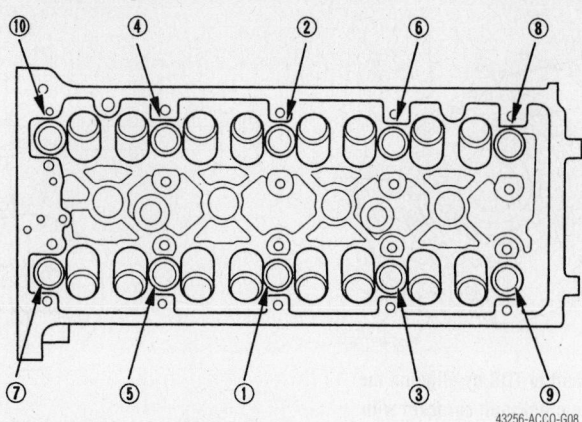

Cylinder head bolt tightening sequence—2.4L engine

43256-ACCO-G08

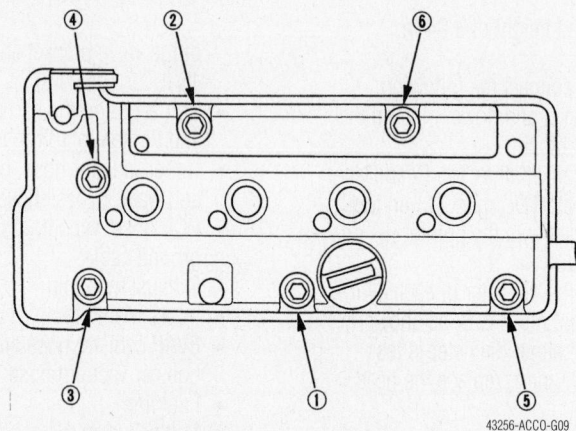

Cylinder head cover bolt tightening sequence—2.4L engine

43256-ACCO-G09

and connectors are installed correctly.

13. Fill the cooling system.

14. Connect the negative battery cable and enter the radio security code.

15. Start the engine and check carefully for any leaks.

3.0L, 3.2L and 3.5L Engines

1. Before servicing the vehicle, refer to the precautions in the beginning of this section.

2. Disconnect the negative battery cable.

3. Drain the coolant.

4. Relieve the fuel system pressure.

5. Remove or disconnect the following:

- Engine covers
- Strut brace
- Water bypass hose
- Traction Control System (TCS) control valve from the throttle body (if equipped)
- Evaporative Emissions (EVAP) canister hose from the throttle body
- Intake air duct
- Upper engine covers

- Accelerator and cruise control cables from the throttle body
- Fuel feed and return hoses
- Brake booster vacuum hose
- PCV hose
- Intake Manifold Runner Control (IMRC) actuator

- Wire harness holder
- Side engine mount bracket
- Accessory drive belts
- Power steering pump without disconnecting the lines
- Ground cable from the engine
- Alternator
- Spark plug wires
- Distributor
- Intake Air Temperature (IAT) sensor connector
- Idle Air Control (IAC) valve connector
- Throttle Position (TPS) sensor connector
- Manifold Absolute Pressure (MAP) sensor connector
- Engine Coolant Temperature (ECT) sensor connector
- Radiator fan switch connectors
- Crankshaft Position (CKP) sensor connector
- Top Dead Center (TDC) sensor connector
- Exhaust Gas Recirculation (EGR) valve connector
- Engine oil pressure switch connector
- Ignition coils
- Intake manifold
- Fuel injector connectors
- Fuel rails
- Intake Air Bypass (IAB) control valve vacuum hoses
- Heater hoses
- Upper and lower radiator hoses
- Exhaust manifolds
- Water passage assembly
- Crankshaft pulley
- Front timing belt cover

6. Set the engine to TDC by aligning

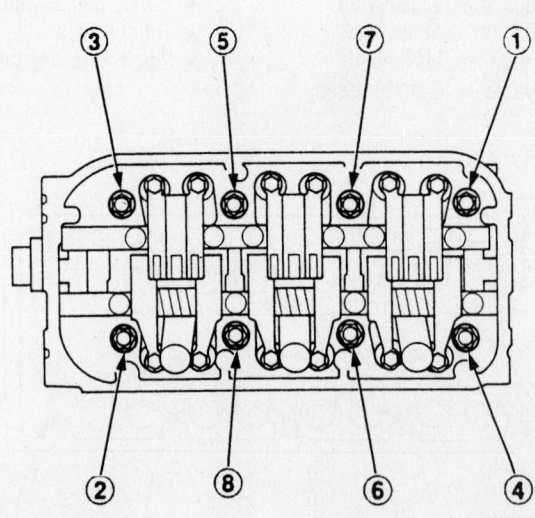

7923BG61

Loosen the cylinder head bolts in the sequence shown to prevent damage to the head—3.0L engine

the marks on the crankshaft and camshaft pulleys.

- Timing belt
- Camshaft pulleys
- Rear timing belt covers

7. Loosen each cylinder head bolt ⅓ turn at a time in the reverse order of the tightening sequence.

8. Remove the cylinder heads and the oil control orifices.

To install:

9. Install the oil control orifices (A) using new O–rings (B).

10. If removed, install the dowel pins (C).

11. Position new cylinder head gaskets (D) on the cylinder block.

12. If moved, set the crankshaft and camshaft pulleys to TDC by aligning the marks on the pulley and oil pump.

13. Carefully position the cylinder heads on the engine.

14. Lubricate the cylinder head bolts with clean engine oil.

15. Tighten the cylinder head bolts in 3 steps. Be sure to follow the tightening torque sequence:

   a. Step 1: 29 ft. lbs. (39 Nm).
   b. Step 2: 51 ft. lbs. (69 Nm).
   c. Step 3: 72 ft. lbs. (98 Nm).

16. Install or connect the following:

- Exhaust manifolds with new gaskets and torque the nuts to 23 ft. lbs. (31 Nm)
- Rear timing belt covers and torque the bolts to 16 ft. lbs. (22 Nm)

17. Install the camshaft pulleys and torque to:

   a. 3.0L and 3.2L engines: 67 ft. lbs. (90 Nm).
   b. 3.5L engines: 23 ft. lbs. (31 Nm.

18. Install or connect the following:

- Timing belt
- Front timing belt cover and torque the bolts to 108 inch lbs. (12 Nm)

19. Check and adjust the valve clearance.

- Crankshaft pulley and torque the bolt to 181 ft. lbs. (245 Nm)
- Water passage assembly and torque the bolts to 16 ft. lbs. (22 Nm)
- Intake manifold with new gaskets and O–rings and tighten the bolts to 16 ft. lbs. (22 Nm)
- Cylinder head cover and tighten the bolts to 108 inch lbs. (12 Nm)
- Ignition coils and torque the bolts to 108 inch lbs. (12 Nm)
- Exhaust manifolds and torque the bolts to 23 ft. lbs. (31 Nm)
- Upper and lower radiator hoses

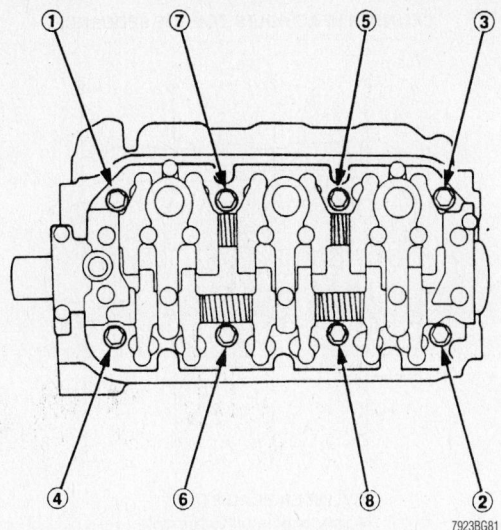

Loosen the cylinder head bolts in the sequence shown—3.5L engine

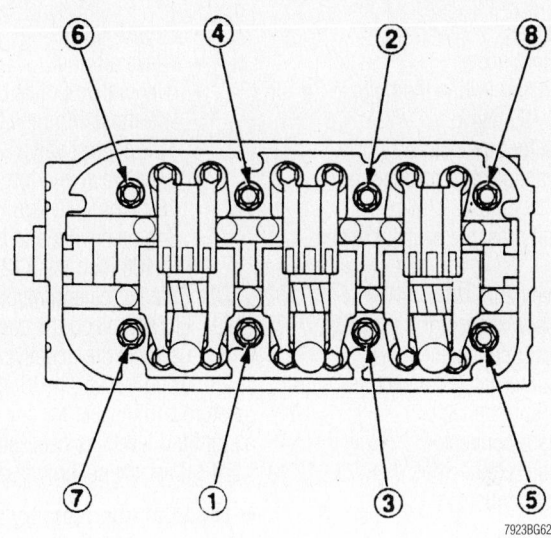

Cylinder head torque sequence—3.0L engine

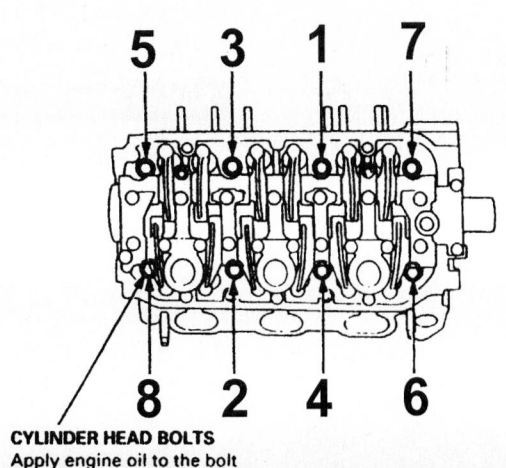

CYLINDER HEAD BOLTS
Apply engine oil to the bolt threads.

Cylinder head torque sequence—3.2L engine

CYLINDER HEAD BOLTS TORQUE SEQUENCE

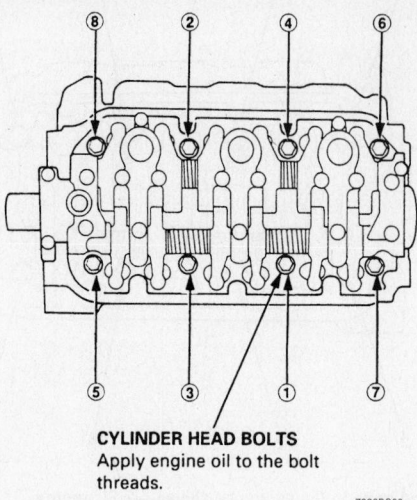

CYLINDER HEAD BOLTS
Apply engine oil to the bolt threads.

7923BG82

Cylinder head torque sequence—3.5L engine

- Heater hoses
- IAB vacuum hoses
- Fuel rails and torque the bolts to 84 inch lbs. (9.5 Nm)
- Fuel injector connectors
- Intake manifold and torque the bolts to 16 ft. lbs. (22 Nm)
- Engine oil pressure switch connector
- EGR valve connector
- TDC sensor connector
- CKP sensor connector
- Radiator fan switch connectors
- MAP sensor connector
- ECT sensor connector
- TPS sensor connector
- IAC sensor connector
- IAT sensor connector
- Spark plug wires
- Distributor
- Alternator and torque the upper bolt to 16 ft. lbs. (22 Nm) and the lower bolt to 33 ft. lbs. (44 Nm)
- Ground cable
- Power steering pump and torque the bolts to 17 ft. lbs. (24 Nm)
- Accessory drive belts
- Side engine mount and torque the mounting bolts to 33 ft. lbs. (44 Nm) and the through bolt to 40 ft. lbs. (54 Nm)
- Wire harness holder
- IMRC actuator
- PCV hose
- Brake booster vacuum hose
- Fuel feed and return lines and torque the fitting to 16 ft. lbs. (22 Nm)
- Accelerator and cruise control cables

- Engine covers
- Intake air duct
- Evaporative Emissions (EVAP) canister hose from the throttle body
- TCS control valve
- Strut brace and torque the bolts to 16 ft. lbs. (22 Nm)
- Engine covers and torque the bolts to 108 inch lbs. (12 Nm)
20. Change the engine oil and filter.
21. Fill and bleed the cooling system.
22. Connect the negative battery cable.
23. Bring the engine to operating temperature and inspect for any fluid leaks. Top off all fluid levels as necessary.
24. Enter the security code for the radio.

➡**The PCM idle memory must be reset after reconnecting the battery. Start the engine and hold it at 3000 rpm until the cooling fan comes on. Then allow the engine to idle for about 5 minutes with all accessories OFF and with the transmission in Park or Neutral.**

Rocker Arms/Shafts

REMOVAL & INSTALLATION

1.8L (B18B1) Engine

1. Before servicing the vehicle, refer to the precautions in the beginning of this section.
2. Remove or disconnect the following:
- Negative battery cable
- Spark plug wires
- Cylinder head cover
- Timing belt cover
- Timing belt

- Distributor
3. Install 5.0mm pin punches to the No.1 camshaft holders, then remove the camshaft sprockets.
4. Loosen the valve adjusters to remove as much spring tension as possible.
5. Remove the pin punches from the camshaft holders.
6. To remove the camshaft bearing caps, loosen each bolt 2 turns at a time in a crisscross pattern to avoid damage to the valves or rockers. Mark the caps so they can be replaced in their original position.
7. Lift the camshafts from the cylinder head, wipe them clean and inspect the lift ramps. Replace the camshafts and rockers if the lobes are pitted, scored or excessively worn.
8. Label the rocker arms before removing so they can be installed in their original locations.
9. Remove the rocker arms.
To install:
10. Check the following before installing the camshafts:
   a. Be certain the keyways on the camshafts are facing UP (No. 1 cylinder at TDC).
   b. The valve adjuster locknuts should be loosened and the adjusting screws backed off before installation.
11. Lubricate the rocker arms and camshafts with clean engine oil.
12. Place the rocker arms on the pivot bolts and the valve stems, making sure that the rocker arms are in their original positions.
13. Install the camshaft seals with the open side (spring) facing in. Lubricate the lip of the seal.
14. Be sure the keyways on the camshafts are facing up and install the camshafts to the cylinder head.
15. Apply liquid gasket to the head mating surfaces of the No. 1 and No. 6 camshaft holders, then install them along with Nos. 2, 3, 4 and 5 camshaft holders. The arrows stamped on the holders should point toward the timing belt. Do not apply oil to the holder mating surface where the camshaft seals are housed.
16. Tighten the camshaft holders temporarily and be sure that the rocker arms are properly positioned.
17. Press the oil seals into the No.1 camshaft holders with a seal driver.
18. Tighten the bolts in a crisscross pattern to 84 inch lbs. (10 Nm). Check that the rockers do not bind on the valves.
19. Install the cylinder head plug to the end of the cylinder head. If the plug has alignment marks, align the marks.

20. Install the rear timing belt cover and tighten the bolts to 108 inch lbs. (12 Nm).

21. Install 5.0mm pin punches to the No.1 camshaft holders, then install the camshaft pulley keys onto the grooves in the camshafts.

22. Push the camshaft pulleys onto the camshafts, then tighten the retaining bolts to 27 ft. lbs. (38 Nm).

23. Install the timing belt and timing belt covers. Remove the pin punches from the camshaft holders.

24. Adjust the valves and pour oil over the camshafts and rocker arms.

25. Apply liquid gasket to the rubber seal at the 8 corners of the recesses.

26. Install the cylinder head cover and engine ground cable. Be sure the contact surfaces are clean and do not touch surfaces where liquid gasket has been applied.

27. Tighten the cylinder head cover nuts in 2–3 steps. In the final step, tighten all nuts in sequence, to 84 inch lbs. (10 Nm).

28. Install the distributor to the cylinder

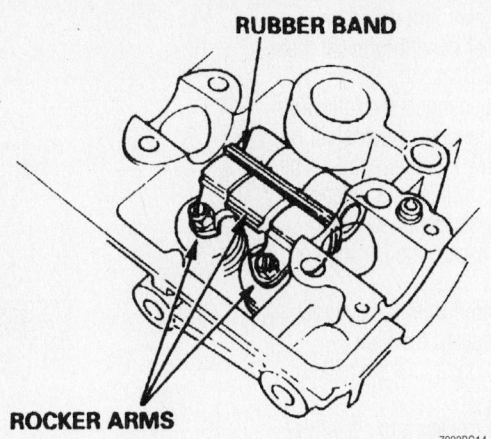

RUBBER BAND

ROCKER ARMS

7923BG14

Rocker arms with rubber band installed—1.8L (B18B1, B18C1 and B18C5) engines

head and reconnect the spark plug wires to the spark plugs.

29. Connect the negative battery cable and enter the radio security code.

30. Change the engine oil. Wait at least 20 minutes for the sealant to cure before filling the engine with oil.

1.8L (B18C1, B18C5) Engine and NSX

1. Before servicing the vehicle, refer to the precautions in the beginning of this section.

2. Remove or disconnect the following:
- Negative battery cable
- Cylinder head from the vehicle

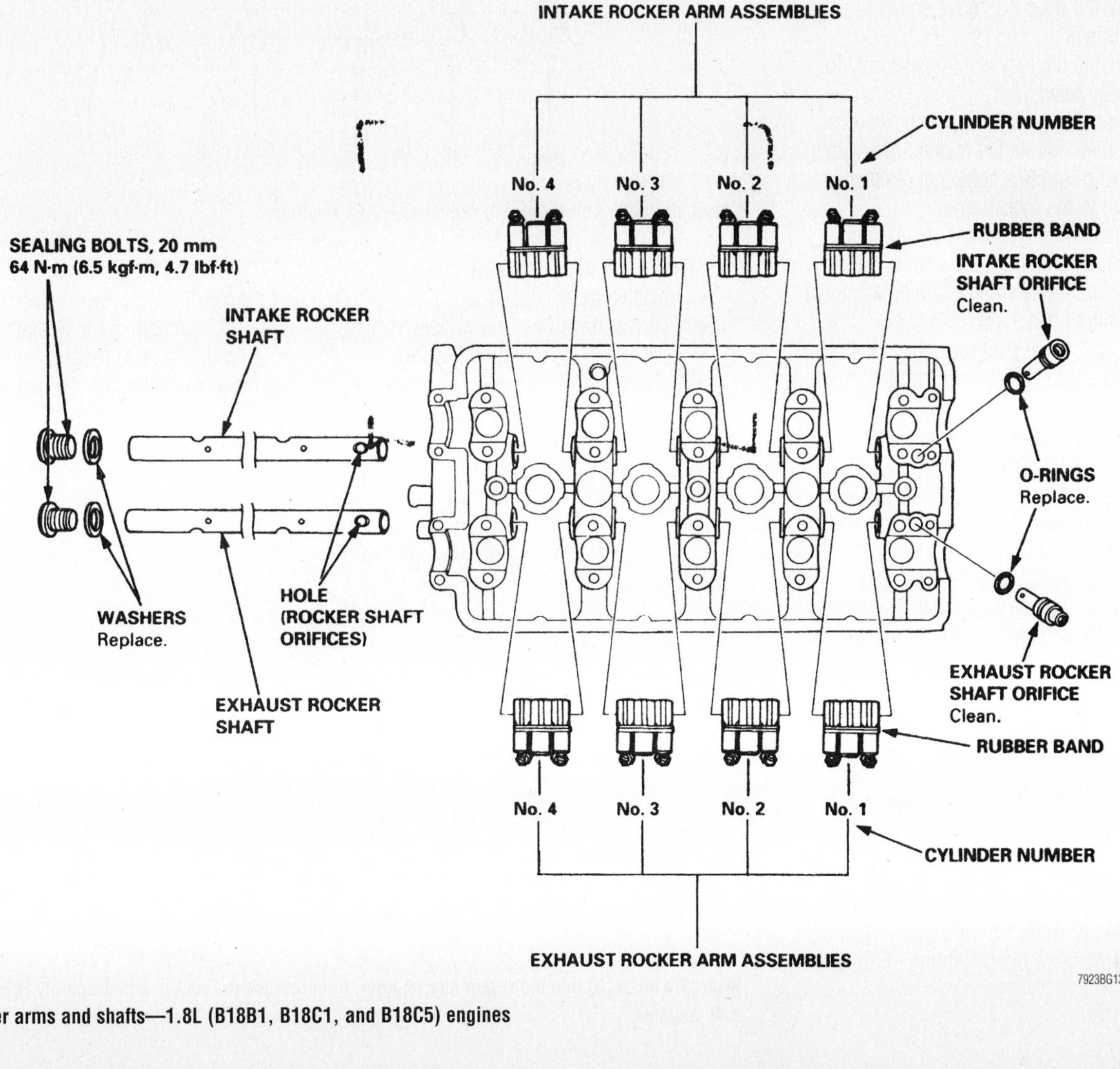

INTAKE ROCKER ARM ASSEMBLIES

CYLINDER NUMBER

No. 4   No. 3   No. 2   No. 1

RUBBER BAND

INTAKE ROCKER SHAFT ORIFICE Clean.

SEALING BOLTS, 20 mm 64 N·m (6.5 kgf·m, 4.7 lbf·ft)

INTAKE ROCKER SHAFT

O-RINGS Replace.

WASHERS Replace.

HOLE (ROCKER SHAFT ORIFICES)

EXHAUST ROCKER SHAFT

EXHAUST ROCKER SHAFT ORIFICE Clean.

RUBBER BAND

No. 4   No. 3   No. 2   No. 1

CYLINDER NUMBER

EXHAUST ROCKER ARM ASSEMBLIES

7923BG13

Rocker arms and shafts—1.8L (B18B1, B18C1, and B18C5) engines

3. Hold each rocker arm assembly together with a rubber band to prevent them from separating.

4. Remove or disconnect the following:
- Intake and exhaust rocker shaft orifices from the cylinder head—The rocker shaft orifices are different and should be identified when removed. Discard the O–rings on the orifices.
- VTEC solenoid from the cylinder head and discard the filter

5. Insert 12mm bolts into the rocker arm shafts.

6. Remove each rocker arm set while slowly pulling out the rocker arm shaft.

➡Tag each rocker arm set to assure installation in their original locations.

7. Inspect the rocker arm pistons. If they do not move smoothly, replace the rocker arm assembly.

8. Remove the lost motion assembly from the cylinder head. Inspect the lost motion assembly by pushing the plunger with your finger. Replace the lost motion assembly if it does not move smoothly.

To install:

9. Install the lost motion assembly to the cylinder head.

10. Apply engine oil to the rocker arm pistons, then bundle the rocker arms with a rubber band. Apply a light coat of clean engine oil to the rocker arms.

11. Position the rocker arms in their original locations, if they are being reused. If new assembles are being used place them in the cylinder head.

12. Lightly coat the rocker arm shafts with clean engine oil, then install the rocker arm shafts into the cylinder head. A 12mm bolt can be installed into the end of the rocker arm shafts to aid in their installation. Be sure to install the shafts in the proper positions. Remove the 12mm bolts from the rocker arm shafts.

13. Clean and install the rocker arm shaft orifices with new O–rings. If the holes in the rocker arm shafts are not aligned screw a 12mm bolt into the end of the shaft to position the it.

14. Install the sealing bolts with new washers and tighten them to 47 ft. lbs. (64 Nm).

15. Install the cylinder head into the vehicle.

2.0L Engines

1. Before servicing the vehicle, refer to the precautions in the beginning of this section.

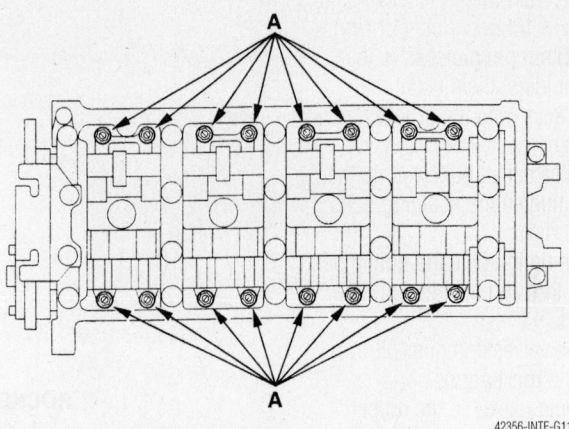

Loosen the rocker arm adjusting screws (A)—2.0L engines

42356-INTE-G11

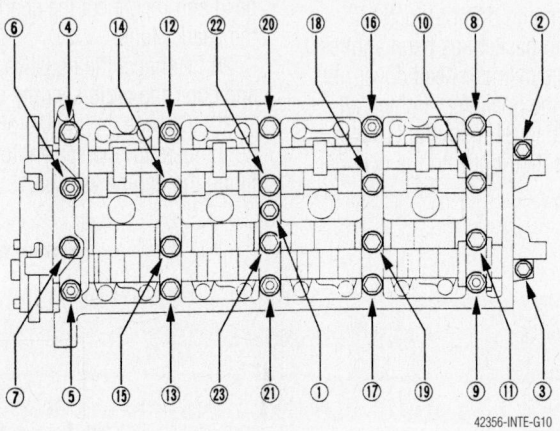

Camshaft holder bolt loosening sequence—2.0L engines

42356-INTE-G10

2. Remove or disconnect the following:
- Negative battery cable
- Cylinder head from the vehicle
- Loosen the rocker arm adjusting screws
- Camshaft holder bolts, 2 turns at a time in a criss–cross pattern
- Timing chain guide B, camshaft holders and camshafts

3. Insert the bolts into the rocker shaft

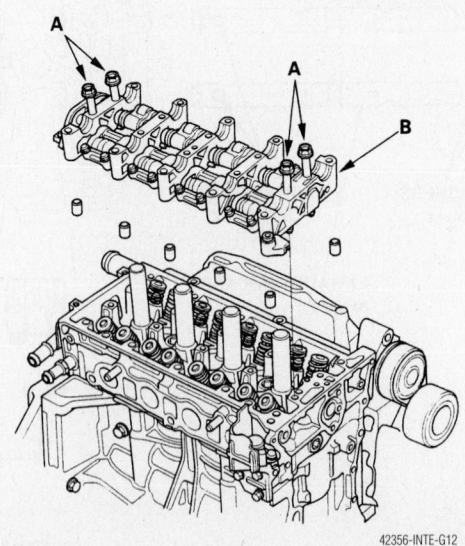

Insert the bolts (A) into the rocker shaft holder, then remove the rocker arm assembly (B)—2.0L engines

42356-INTE-G12

holder, then remove the rocker arm assembly.

4. Disassemble the rocker shafts as necessary.

To install:

5. Clean and dry the No. 5 rocker shaft holder mating surface.

6. Apply liquid gasket part No. 08718–0009 evenly to the cylinder head mating surface of the No. 5 rocker shaft holder and install within 5 minutes.

7. Reassemble the rocker arm assembly.

8. Insert the bolts into the rocker shaft holder, then install the rocker arm assembly on the cylinder head. Remove the bolts from the rocker shaft holder.

9. Make sure the punch marks on the VTC actuator and exhaust camshaft sprocket are facing up, then set the camshafts in the holder.

10. Set the camshaft holders and timing chain guide B in place.

11. Tighten the bolts, in sequence, to the following specifications:

- 8mm bolts: 16 ft. lbs. (22 Nm)
- 6mm bolts (nos. 21, 22 & 23): 104 inch lbs. (12 Nm)

12. Install or connect the following:

- Timing chain, then adjust the valve clearance
- Cylinder head
- Negative battery cable

13. Check for proper engine and valve train operation.

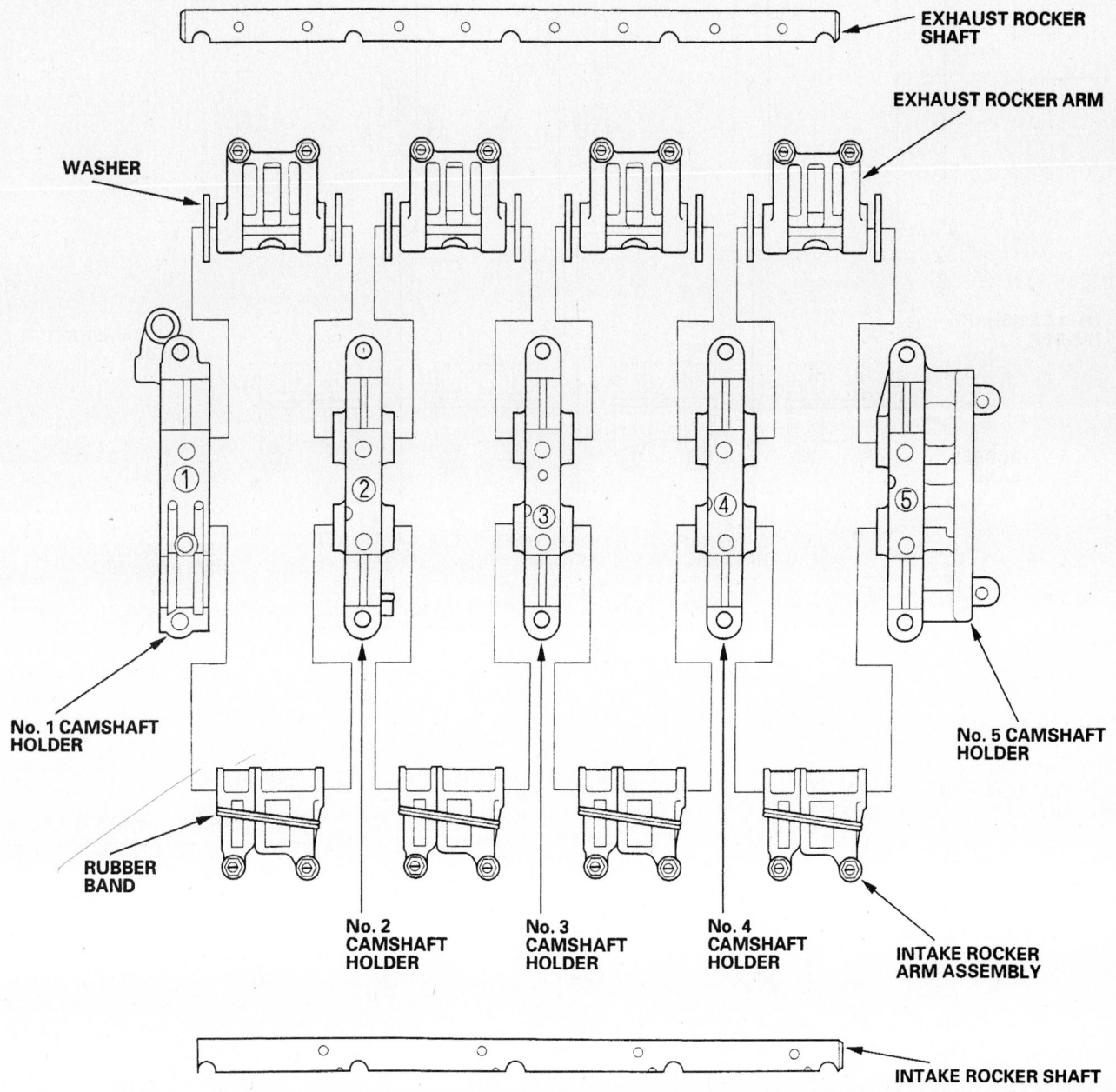

Rocker arm disassembly—2.0L (K20A3) engine

42356-INTE-G13

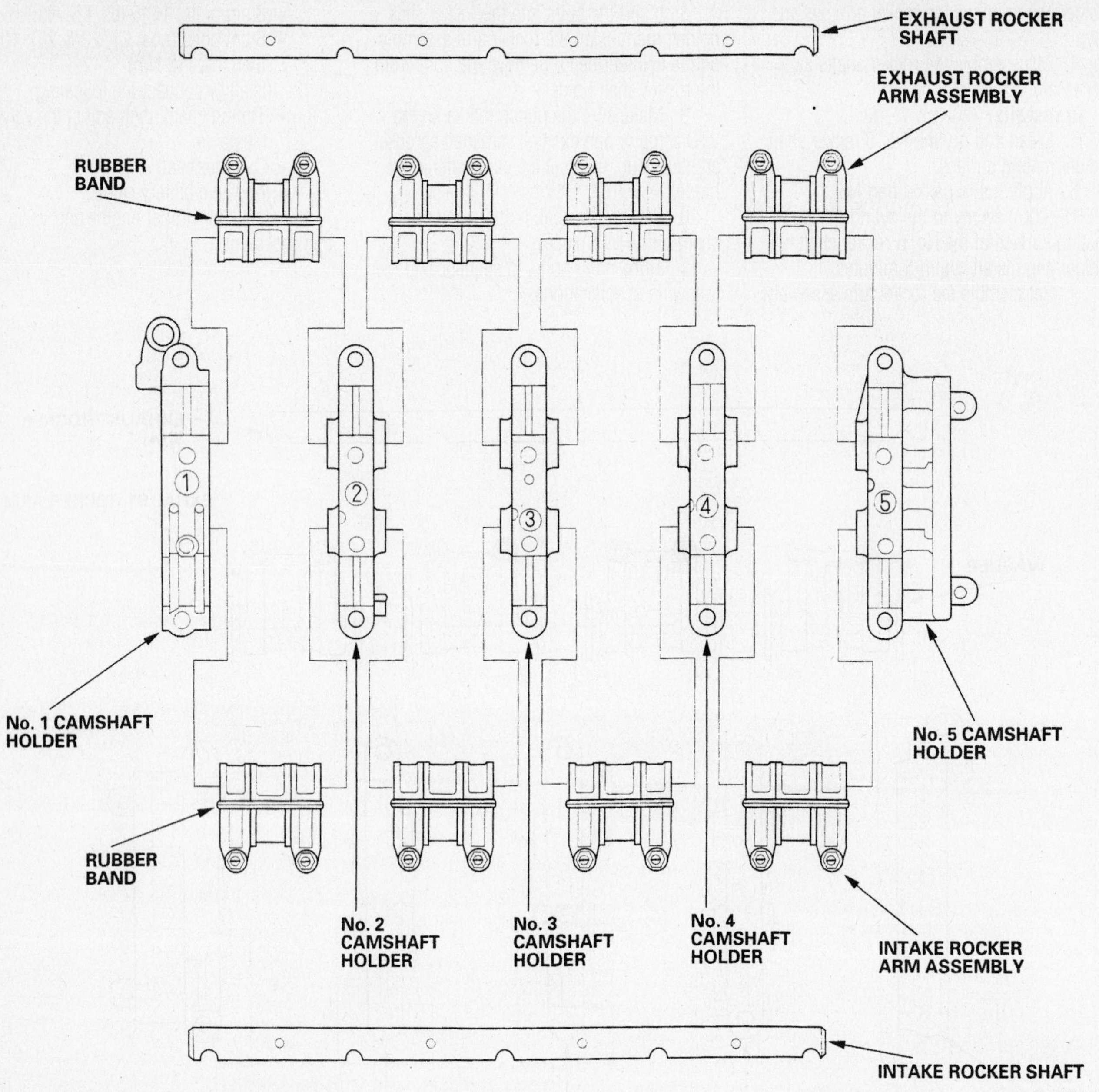

EXHAUST ROCKER SHAFT

EXHAUST ROCKER ARM ASSEMBLY

RUBBER BAND

No. 1 CAMSHAFT HOLDER

No. 5 CAMSHAFT HOLDER

RUBBER BAND

No. 2 CAMSHAFT HOLDER

No. 3 CAMSHAFT HOLDER

No. 4 CAMSHAFT HOLDER

INTAKE ROCKER ARM ASSEMBLY

INTAKE ROCKER SHAFT

42356-INTE-G14

Rocker arm disassembly—2.0L (K20A2) engine

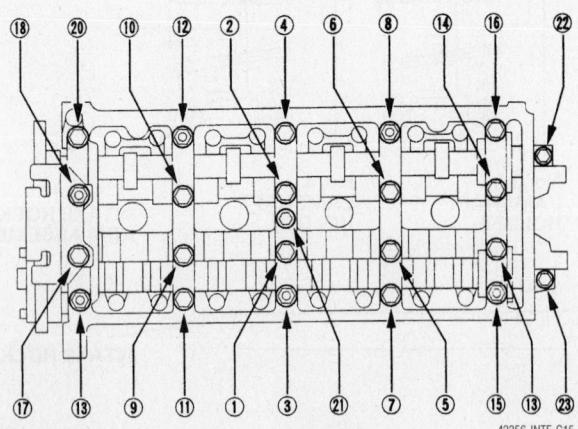

42356-INTE-G15

Rocker arm assembly tightening sequence—2.0L engines

2.4L Engine

1. Before servicing the vehicle, refer to the precautions in the beginning of this section.

2. Remove or disconnect the following:
   - Timing chain
   - Loosen the rocker arm adjusting screws
   - Camshaft holder bolts, two turns at a time in sequence
   - Timing chain guide (B), camshaft and camshafts

3. Insert the bolts (A) into the rocker shaft holder, then remove the rocker arm assembly (B)

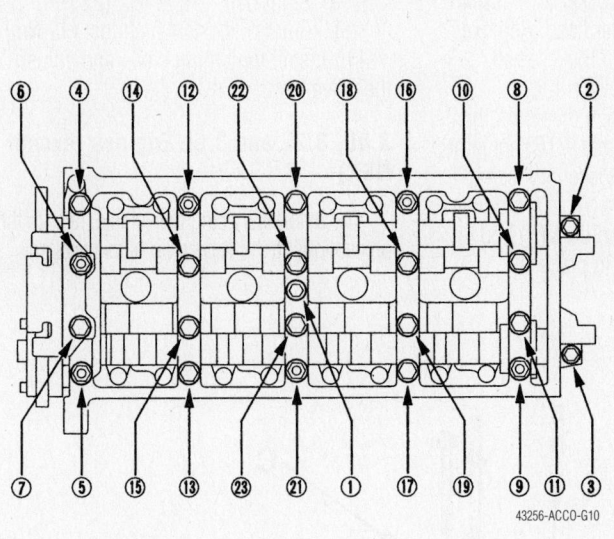

Camshaft holder bolt loosening sequence—2.4L engine

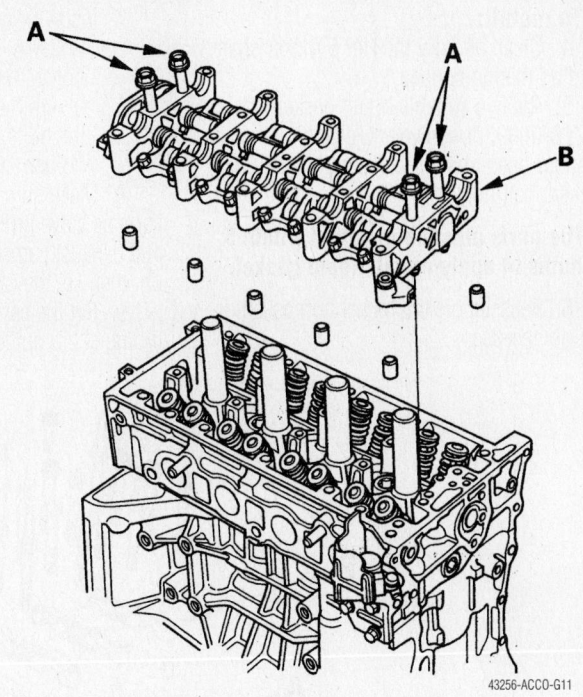

Insert the bolts (A) into the rocker shaft holder, then remove the rocker arm assembly (B)—2.4L engine

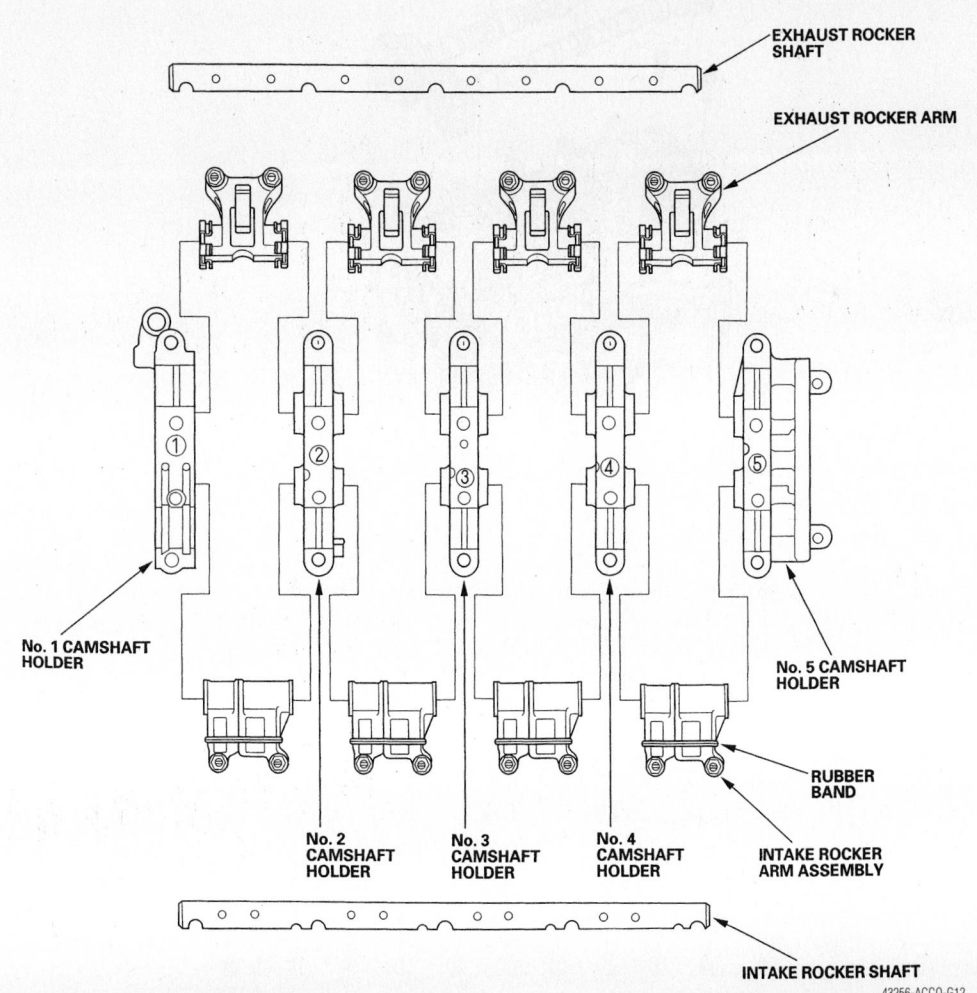

Exploded view of the rocker arms and related components—2.4L engine

To install:

4. Clean and dry the No. 5 rocker shaft holding mating surface.

5. Apply a suitable liquid gasket P/N 08718-0009, or equivalent, evenly to the cylinder head mating surface of the No. 5 rocker shaft holder.

➡ **The parts must be installed within 5 minutes of applying the liquid gasket.**

6. Reassemble the rocker arm assembly, as necessary.

7. Install or connect the following
- Bolts (A) into the rocker shaft holder, then the rocker arm assembly on the cylinder head. Remove the bolts from the rocker shaft holder.

8. Make sure the punch marks on the variable valve timing control (VTC) actuator and exhaust camshaft sprocket are facing up, then set the camshafts (A) in the holder

9. Set the camshaft holders (B) and timing chain guide B (C) in place.

10. Tighten the bolts, in sequence, to the following specification:
- a. 8mm bolts: 16 ft. lbs. (22 Nm)
- b. 6mm bolts: 104 inch lbs. (12 Nm)

11. Install the timing chain and adjust the valve lash.

3.0L, 3.2L and 3.5L Engines (except NSX)

1. Before servicing the vehicle, refer to the precautions in the beginning of this section.

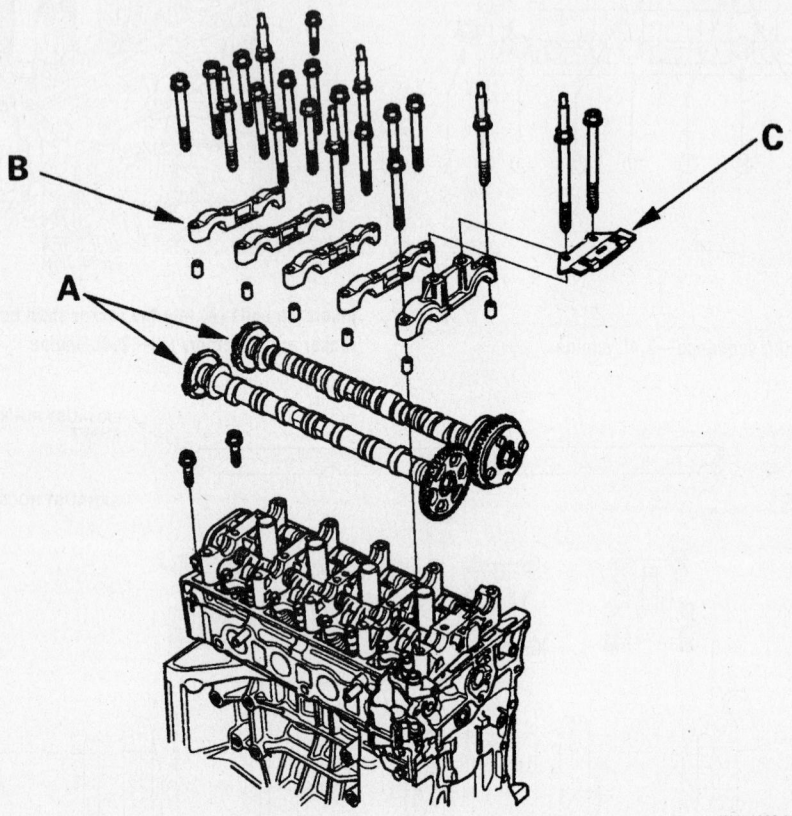

43256-ACCO-G13

When installing the camshafts (A) make sure the punch marks on the VTC actuator and exhaust cam sprockets are facing up—2.4L engine

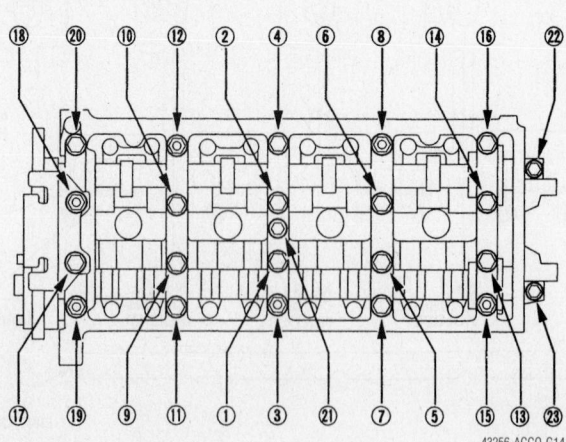

43256-ACCO-G14

Rocker arm assembly bolt tightening sequence—2.4L engine

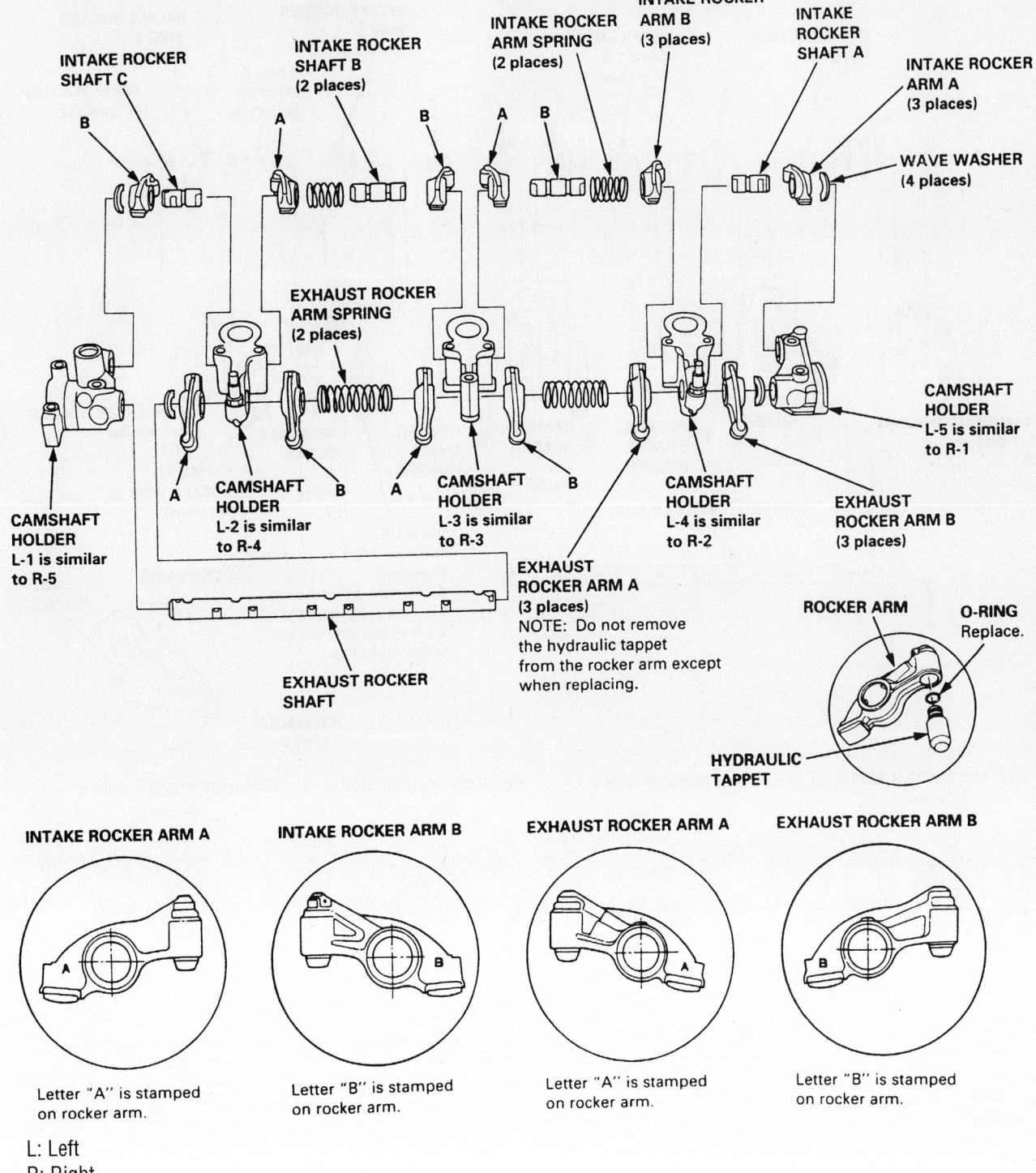

Exploded view of the rocker arms and related components—3.5L engine

L: Left
R: Right

2. Disconnect the negative battery cable.

3. Remove the cylinder head from the vehicle.

4. Loosen the rocker shaft holder bolts 1 turn at a time in the opposite of the installation sequence. Following this procedure will prevent the camshafts and rocker assemblies from warping.

5. After all bolts are loose, remove the rocker arm shafts as an assembly with the bolts still in the holders.

6. If the rocker shafts are to be disassembled, note that each rocker arm has a letter **A** or **B** stamped into the side. Before disassembling the rocker arms, make a note of the position of each letter so the arms can be reassembled the same way.

7. For 3.2L and 3.5L engines, do not remove the hydraulic tappets from the rocker arms unless they are to be replaced. Handle the rocker arms carefully so the oil does not drain out of the tappets.

8. Lift the camshafts from the cylinder

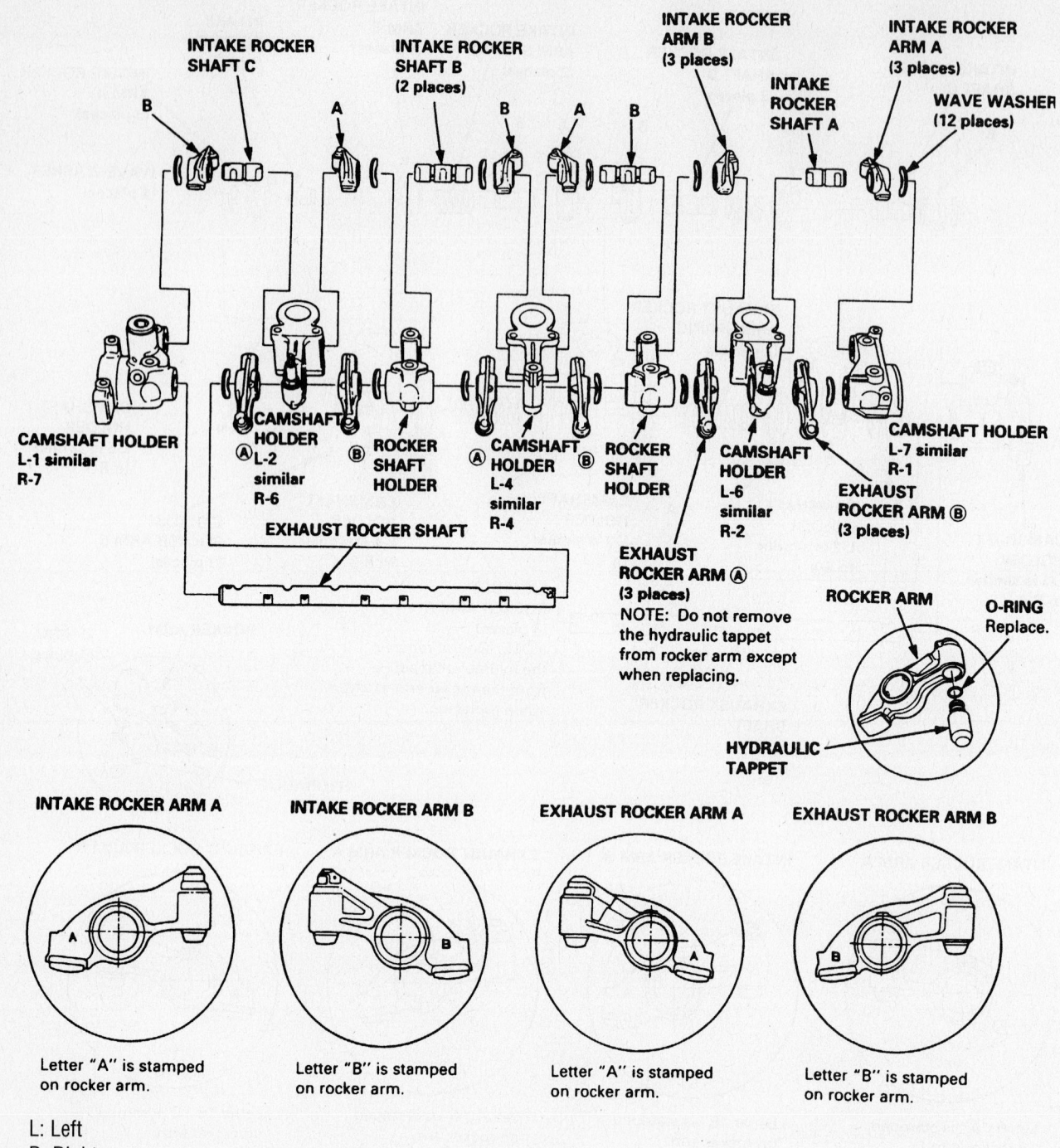

INTAKE ROCKER SHAFT C

INTAKE ROCKER SHAFT B (2 places)

INTAKE ROCKER ARM B (3 places)

INTAKE ROCKER ARM A (3 places)

INTAKE ROCKER SHAFT A

WAVE WASHER (12 places)

CAMSHAFT HOLDER L-1 similar R-7

CAMSHAFT HOLDER Ⓐ L-2 similar R-6

ROCKER SHAFT HOLDER Ⓑ

CAMSHAFT HOLDER Ⓐ L-4 similar R-4

ROCKER SHAFT HOLDER Ⓑ

CAMSHAFT HOLDER L-6 similar R-2

CAMSHAFT HOLDER L-7 similar R-1

EXHAUST ROCKER ARM Ⓑ (3 places)

EXHAUST ROCKER SHAFT

EXHAUST ROCKER ARM Ⓐ (3 places)

NOTE: Do not remove the hydraulic tappet from rocker arm except when replacing.

ROCKER ARM

O-RING Replace.

HYDRAULIC TAPPET

INTAKE ROCKER ARM A

Letter "A" is stamped on rocker arm.

INTAKE ROCKER ARM B

Letter "B" is stamped on rocker arm.

EXHAUST ROCKER ARM A

Letter "A" is stamped on rocker arm.

EXHAUST ROCKER ARM B

Letter "B" is stamped on rocker arm.

L: Left
R: Right

7923BG84

Exploded view of the rocker arms and related components—3.2L engine

head, wipe them clean and inspect the lift ramps. Replace the camshafts and rockers if the lobes are pitted, scored, or excessively worn.

To install:

9. Lubricate the camshaft and its journals with fresh engine oil.

10. Place a new camshaft seal on the end of the camshaft. The spring side of the seal must face in. Lubricate the journals and set the camshaft in place on the head.

11. Install the camshaft onto the cylinder head with the keyway pointed up.

12. Apply liquid gasket to the mounting surfaces of the camshaft end holders.

13. Set the rocker arm assemblies in place and start all the cam holder bolts. Be sure the rocker arms are properly positioned

and turn each bolt in sequence 2 turns at a time until the holders are seated on the head. Follow this procedure to avoid damaging the camshaft and rocker assemblies.

14. When all the camshaft and rocker holders are seated, tighten the bolts in the same sequence. Tighten the 8mm bolts to 16 ft. lbs. (22 Nm) and the 6mm bolts to 104 inch lbs. (12 Nm).

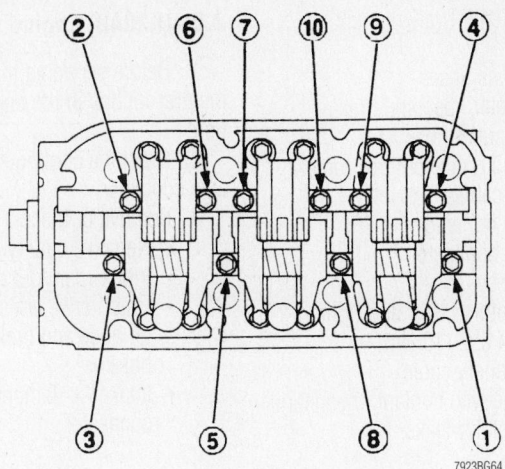

Be sure to loosen the rocker arm shaft bolts in the correct order as shown—3.0L engine

7923BG64

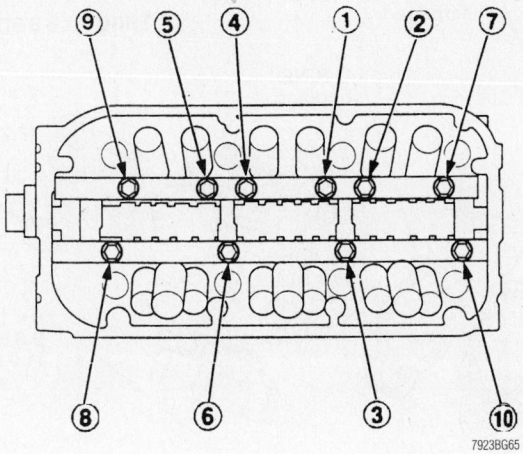

Tighten the bolts 2 turns at a time in the sequence shown—3.0L engine

7923BG65

15. Install or connect the following:
  • Cylinder head
  • Negative battery cable
16. Check for proper engine and valve train operation.

Intake Manifold

REMOVAL & INSTALLATION

1.8L (B18B1, B18C1 and B18C5) Engines

1. Before servicing the vehicle, refer to the precautions in the beginning of this section.
2. Disconnect the negative battery cable.
3. Drain the cooling system into a sealable container.
4. Remove the strut brace (if equipped).
5. Properly relieve the fuel pressure.
6. Remove or disconnect the following:
  • Intake air duct
  • Fuel feed and return hoses
  • PCV hose
  • Brake booster vacuum hose
  • Throttle cable from the throttle body, take great care not to kink or damage the cable

7. Label and disconnect all the emission vacuum hoses from the intake manifold.
8. Label and disconnect the wiring connected to the intake manifold. Disconnect sensors as needed; release wiring retainers and clips.
9. Remove or disconnect the following:
  • Water bypass hoses from the manifold
  • Attaching the intake manifold to the support bracket
  • Nuts attaching the intake manifold to the cylinder head in a crisscross pattern, beginning from the center and moving out to both ends
  • Intake manifold

To install:
10. Installation is the reverse of the removal procedure, while using the following torque values:

  • Intake manifold nuts:17 ft. lbs. (23 Nm)
  • Manifold support bracket bolts: 17 ft. lbs. (23 Nm)
  • Strut brace nuts: 17 ft. lbs. (23 Nm)

2.0L (K20A3) Engines

1. Before servicing the vehicle, refer to the precautions in the beginning of this section.
2. Drain the cooling system into a sealable container.
3. Remove or disconnect the following:
  • Negative battery cable
  • Intake manifold cover
  • Intake Air Temperature (IAT) sensor connector
  • Breather hose
  • Air cleaner housing
  • Throttle cover. Fully open the throttle link and cruise control link by hand, then remove the throttle cable and cruise control cable from the links
  • Locknut, loosen only, then the cables from the bracket
  • Evaporative emission (EVAP) canister hose and brake booster vacuum hose
  • Water bypass hoses, then plug them
  • Intake Manifold Runner Control (IMRC) valve actuator control solenoid valve connector
  • Positive Crankcase Ventilation (PCV) hose
  • IMRC valve control solenoid valve mounting bolt
  • Front bumper
  • Hood switch connector
  • A/C line bracket mounting bolt
  • Intake airduct mounting bolt and harness clamps
  • Upper bracket and cushion mounting bolts, then the bulkhead
  • Idle Air Control (IAC) valve connector
  • Throttle Position (TP) sensor connector
  • Manifold Absolute Pressure (MAP) sensor connector
  • Evaporative Emission (EVAP) canister purge valve connector
  • Intake Manifold Runner Control (IMRC) valve position sensor connector
  • Intake manifold and O-rings

To install:
4. Install or connect the following:
  • Intake manifold, with new O-rings. Tighten the bolts/nuts in a

criss–cross pattern in 2 or 3 steps to 16 ft. lbs. (22 Nm).
- Bulkhead, and upper bracket and cushion mounting bolts
- Hood switch connector
- A/C line bracket mounting bolt and tighten to 104 inch lbs. (10 Nm)
- Intake airduct mounting bolt and tighten to 104 inch lbs. (10 Nm)
- Harness clamps
- Front bumper
- IMRC valve actuator control solenoid valve connector
- PCV hose and IMRC valve actuator

control solenoid valve mounting bolt
- Water bypass hoses
- EVAP canister hose and brake booster vacuum hose
- Throttle and cruise control cables. Adjust the cables.
- Air cleaner housing
- IAT sensor connector
- Breather hose
- Intake manifold cover and tighten to 104 inch lbs. (10 Nm)
- Negative battery cable

5. Fill the engine with coolant, then start the engine and check for leaks.

2.0L (K20A2) Engine

1. Before servicing the vehicle, refer to the precautions in the beginning of this section.

2. Drain the cooling system into a sealable container.

3. Remove or disconnect the following:
- Negative battery cable
- Intake manifold cover
- Evaporative emission (EVAP) canister hose and brake booster vacuum hose
- Intake Air Temperature (IAT) sensor connector

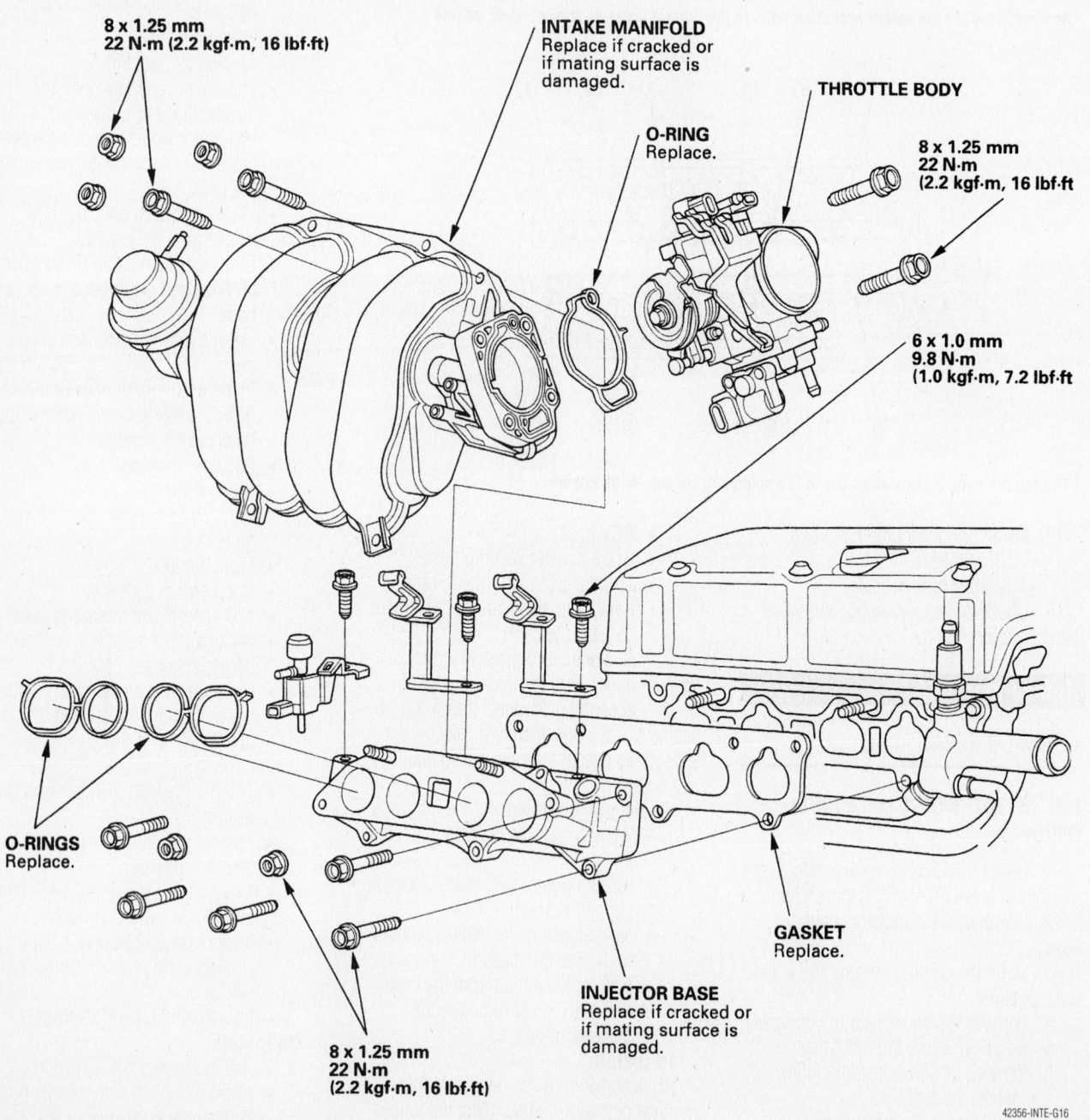

Exploded view of the intake manifold—2.0L (K20A3) engine

42356-INTE-G16

- Breather hose
- Air cleaner housing
- Throttle cover. Fully open the throttle link and cruise control link by hand, then remove the throttle cable and cruise control cable from the links.
- Locknut, loosen only, then the cables from the bracket
- Water bypass hoses, then plug them

4. Relieve the fuel system pressure.
   - Fuel feed hose
   - Positive Crankcase Ventilation (PCV) hose, harness holder mounting bolt and harness clamp mounting bolt

- Idle Air Control (IAC) valve connector
- Throttle Position (TP) sensor connector
- Manifold Absolute Pressure (MAP) sensor connector
- Evaporative Emission (EVAP) canister purge valve connector
- Intake Manifold Runner Control (IMRC) valve position sensor connector
- 2 bolts securing the intake manifold and brackets
- Intake manifold mounting bolts and nuts
- 2 stud bolts and the intake manifold. Discard the gasket(s).

To install:
5. Install or connect the following:
   - Intake manifold with a new gasket, then tighten the 2 stud bolts to 16 ft. lbs. (22 Nm). Tighten the intake manifold bolts/nuts in a criss-cross pattern, beginning with the inner bolt to 16 ft. lbs. (22 Nm).
   - 2 bolts securing the intake manifold brackets and tighten to 16 ft. lbs. (22 Nm).
   - PCV hose, harness holder mounting bolt and harness clamp mounting bolt. Tighten the bolts to 104 inch lbs. (10 Nm).
   - Fuel feed hose
   - Water bypass hoses

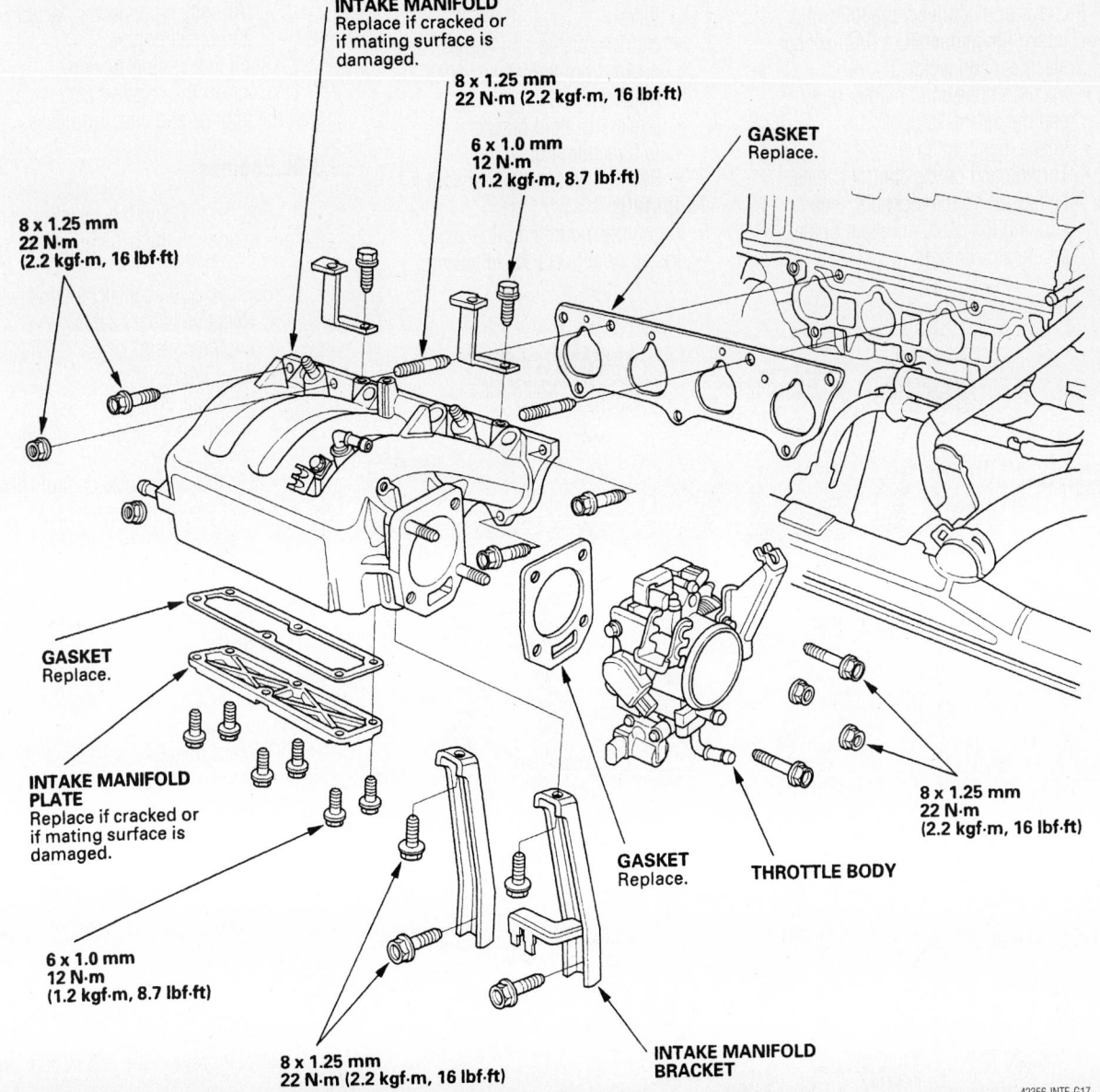

Exploded view of the intake manifold—2.0L (K20A2) engine

42356-INTE-G17

- Throttle and cruise control cables. Adjust the cables.
- Air cleaner housing
- IAT sensor connector
- Breather hose
- EVAP canister hose, brake booster vacuum hose and vacuum hoses
- Intake manifold cover and tighten to 104 inch lbs. (10 Nm)
- Negative battery cable

6. Fill the engine with coolant, then start the engine and check for leaks.

2.4L Engine

1. Before servicing the vehicle, refer to the precautions in the beginning of this section.
2. Disconnect the negative battery cable.
3. Drain the engine coolant into a sealable container.
4. Remove or disconnect the following:
- Intake Air Temperature (IAT) sensor electrical connector
- Vacuum hose and breather pipe and the air intake duct
- Intake manifold cover
- Throttle and cruise control cables by loosening the locknuts, then slipping the cable ends out of the accelerator linkage.

➡Do not bend the cables during removal. Always replace any throttle or cruise control cables that get kinked during removal.

- Evaporative emission (EVAP) canister hose and brake booster vacuum hose
- Idle Air Control (IAC) valve connectors
- Throttle Position (TP) sensor connector
- Manifold Absolute Pressure (MAP) sensor connector
- Necessary engine wire harness connectors and wire harness clamps from the intake manifold
- Bolt securing the harness holder and remove the harness clamps
- Water bypass hoses, then plug them
- Harness clamp and harness connector from the intake manifold bracket
- Intake manifold bracket
- A/T vacuum hose
- Retainer and intake manifold

To install:

5. Clean the mounting surfaces.
6. Install or connect the following:

- New gasket
- Intake manifold. Tighten the bolts, in a criss-cross pattern beginning with the inner bolt, to 16 ft. lbs. (22 Nm).
- A/T vacuum hose
- Intake manifold bracket
- Harness clamp and connector to the intake manifold bracket
- Water bypass hoses
- Bolt securing the harness holder and tighten to 104 inch lbs. (12 Nm)
- Harness clamps
- EVAP canister hose and brake booster vacuum hose
- Throttle and cruise control cables
- Intake manifold cover
- Intake air duct
- IAT sensor connector, vacuum hose and breather pipe

7. Refill the cooling system.
8. Connect the negative battery cable, start the engine, and check for leaks.

3.0L Engines

1. Before servicing the vehicle, refer to the precautions in the beginning of this section.
2. Remove or disconnect the following:
- Negative battery cable
- Engine covers
- Strut brace
- Intake air duct
- Throttle body and intake manifold covers
- Throttle and cruise control cables
- Fuel feed and return hoses
- Brake booster vacuum hose
- Electrical connectors from the manifold
- Intake manifold

To install:

3. Installation is the reverse of the removal procedure, while using the following torque values:
- Intake manifold bolts: 16 ft. lbs. (22 Nm)
- Strut brace bolts: 16 ft. lbs. (22 Nm)

3.2L and 3.5L Engines

1. Before servicing the vehicle, refer to the precautions in the beginning of this section.
2. Disconnect the negative battery cable.
3. Drain the cooling system.
4. Properly relieve the fuel system pressure.
5. Remove or disconnect the following:
- Intake air duct

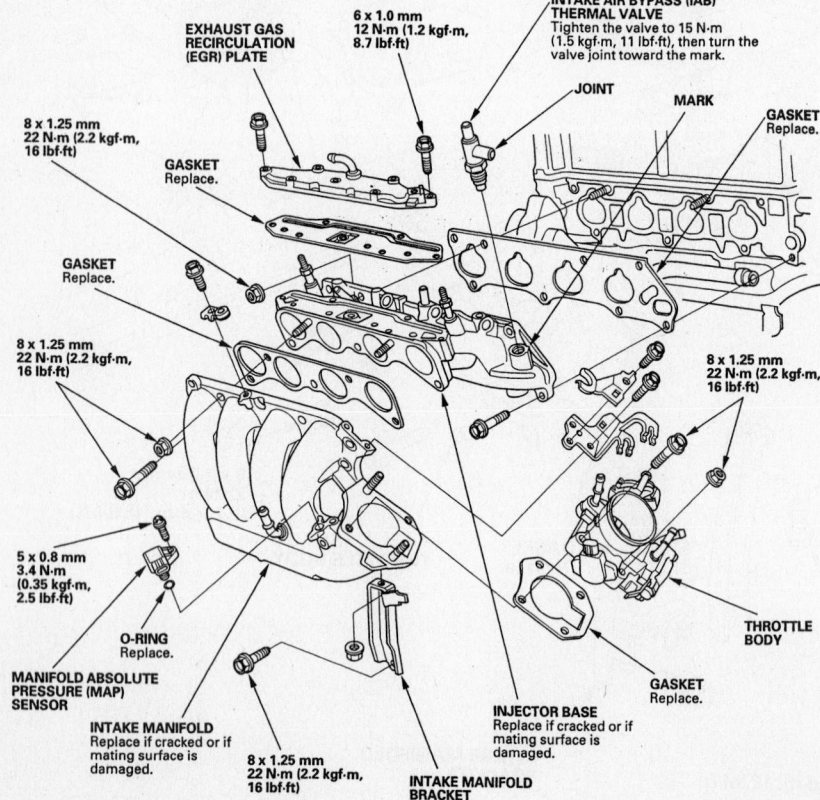

EXHAUST GAS RECIRCULATION (EGR) PLATE

6 x 1.0 mm
12 N·m (1.2 kgf·m, 8.7 lbf·ft)

INTAKE AIR BYPASS (IAB) THERMAL VALVE
Tighten the valve to 15 N·m (1.5 kgf·m, 11 lbf·ft), then turn the valve joint toward the mark.

JOINT **MARK**

GASKET Replace.

8 x 1.25 mm
22 N·m (2.2 kgf·m, 16 lbf·ft)

GASKET Replace.

GASKET Replace.

8 x 1.25 mm
22 N·m (2.2 kgf·m, 16 lbf·ft)

8 x 1.25 mm
22 N·m (2.2 kgf·m, 16 lbf·ft)

5 x 0.8 mm
3.4 N·m (0.35 kgf·m, 2.5 lbf·ft)

O-RING Replace.

MANIFOLD ABSOLUTE PRESSURE (MAP) SENSOR

INTAKE MANIFOLD
Replace if cracked or if mating surface is damaged.

INJECTOR BASE
Replace if cracked or if mating surface is damaged.

8 x 1.25 mm
22 N·m (2.2 kgf·m, 16 lbf·ft)

INTAKE MANIFOLD BRACKET

THROTTLE BODY

GASKET Replace.

43256-ACC0-G15

Exploded view of the intake manifold and related components—2.4L engine

- Strut brace (if equipped)
- Intake manifold cover
- Water bypass hose
- Traction Control System (TCS) actuator (if equipped)
- Evaporative Emission (EVAP) canister hose
- Throttle and cruise control cables
- Brake booster vacuum hose
- Upper intake manifold cover
- Intake manifold nuts and bolts in a crisscross pattern, beginning from the center and moving out

6. Verify that all vacuum lines are disconnected and remove the intake manifold and throttle body as a unit.

7. Inspect the manifold for cracks, flatness, or other damage; replace any damaged parts. If the intake manifold is to be replaced, transfer all the necessary components to the new manifold.

To install:

- Intake fasteners:16 ft. lbs. (22 Nm)
- Upper intake manifold cover fasteners: 108 inch lbs. (12 Nm)
- TCS actuator bolts: 16 ft. lbs. (22 Nm)
- Strut brace bolts: 16 ft. lbs. (22 Nm)

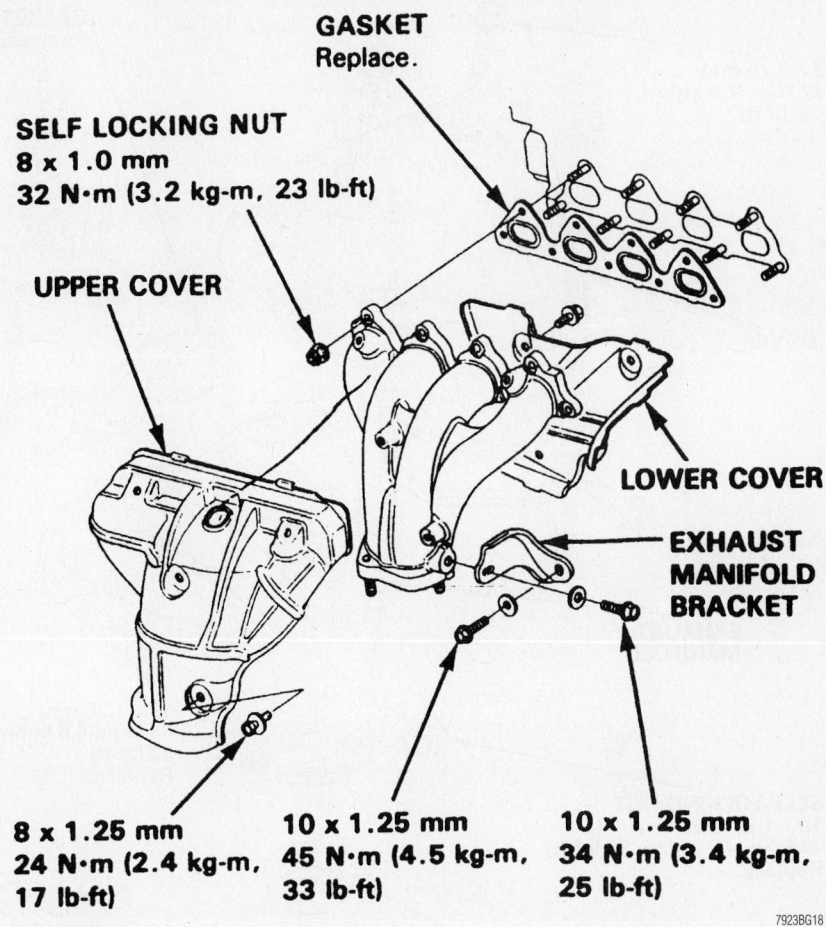

GASKET
Replace.

SELF LOCKING NUT
8 x 1.0 mm
32 N·m (3.2 kg-m, 23 lb-ft)

UPPER COVER

LOWER COVER

EXHAUST MANIFOLD BRACKET

8 x 1.25 mm
24 N·m (2.4 kg-m, 17 lb-ft)

10 x 1.25 mm
45 N·m (4.5 kg-m, 33 lb-ft)

10 x 1.25 mm
34 N·m (3.4 kg-m, 25 lb-ft)

7923BG18

Exhaust manifold—1.8L engines

Exhaust Manifold

REMOVAL & INSTALLATION

1.8L Engines

1. Before servicing the vehicle, refer to the precautions in the beginning of this section.

2. Remove or disconnect the following:
- Negative battery cable
- Exhaust manifold cover
- Front exhaust pipe from the manifold
- Exhaust manifold support bracket
- Exhaust manifold from the engine
- Rear cover from the exhaust manifold (if necessary)

To install:

➡**Use new exhaust manifold nuts for installation.**

3. Install or connect the following:
- Rear heat shield bolts: 17 ft. lbs. (24 Nm)
- Exhaust manifold nuts: 23 ft. lbs. (31 Nm)
- Bracket bolts: 33 ft. lbs. (44 Nm)
- Front exhaust pipe nuts: 40 ft. lbs. (54 Nm)
- Exhaust manifold heat shield bolts: 17 ft. lbs. (24 Nm)

2.0L Engine

1. Before servicing the vehicle, refer to the precautions in the beginning of this section.

2. Remove or disconnect the following:
- VTEC solenoid valve
- Intermediate shaft cover
- Cover and exhaust manifold bracket
- Mounting nuts and bolts and exhaust manifold. Discard the gasket.

To install:

3. Install or connect the following:
- Exhaust manifold with a new gasket. Tighten the retainers to 33 ft. lbs. (45 Nm), in a criss–cross pattern starting at the inner bolt.
- Cover and exhaust manifold bracket
- Intermediate shaft cover
- VTEC solenoid valve

2.4L Engine

1. Before servicing the vehicle, refer to the precautions in the beginning of this section.

2. Raise and safely support the vehicle.

3. Remove or disconnect the following:
- VTEC solenoid valve
- Driveshaft heat shield
- Cover and exhaust manifold bracket
- Exhaust manifold

To install:

4. Clean the mounting surfaces.

5. Install or connect the following:
- New gasket on the cylinder head
- Exhaust manifold. Tighten the nuts, in a criss-cross pattern starting with the inner nut, to 33 ft. lbs. (45 Nm).
- Exhaust manifold bracket and cover
- Driveshaft heat shield
- VTEC solenoid valve

3.0L Engine

1. Before servicing the vehicle, refer to the precautions in the beginning of this section.

2. Remove or disconnect the following:
- Manifold cover
- Exhaust pipe from the manifold to be removed
- Mounting nuts and the exhaust manifold

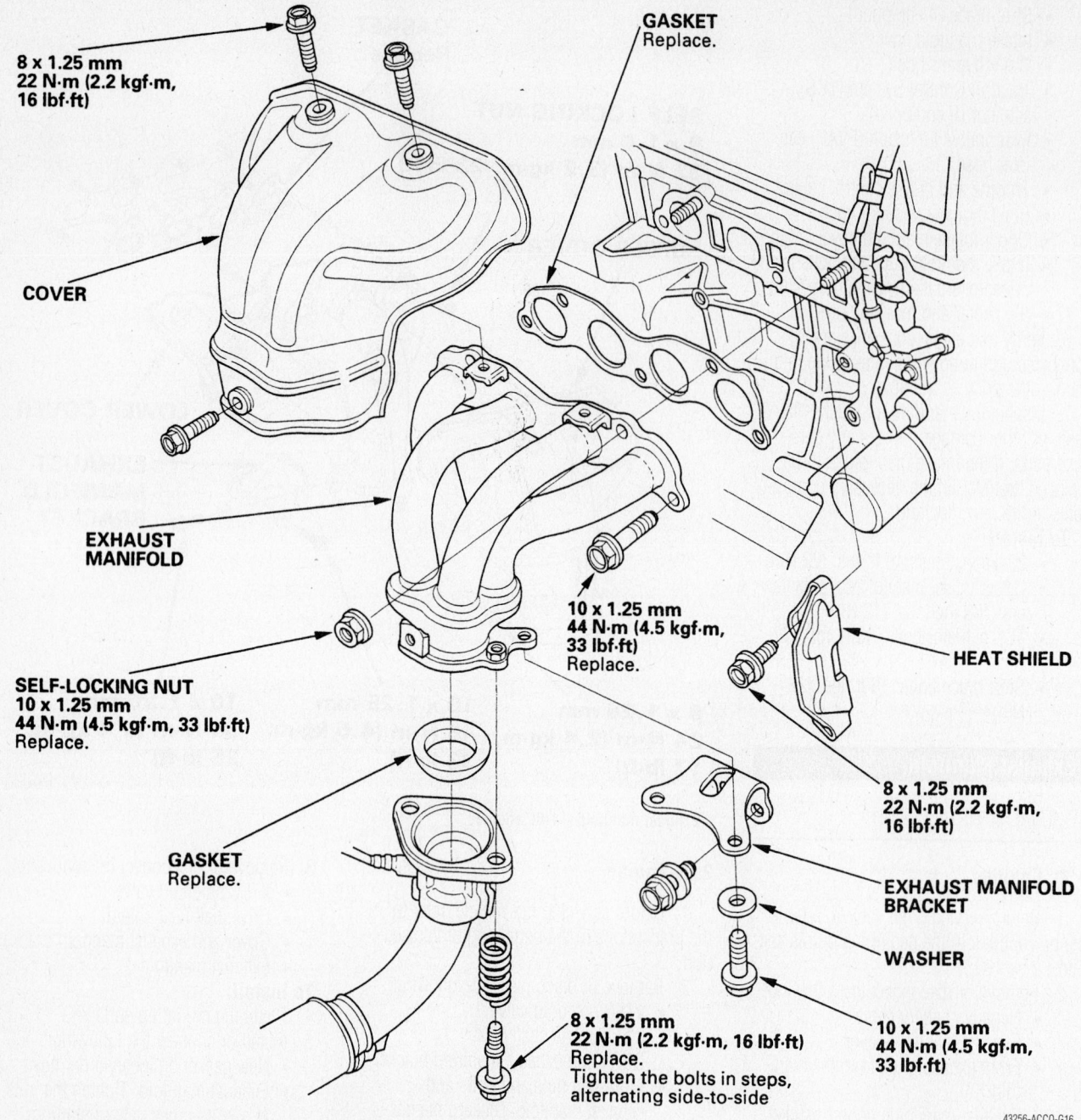

8 x 1.25 mm
22 N·m (2.2 kgf·m,
16 lbf·ft)

GASKET
Replace.

COVER

EXHAUST
MANIFOLD

SELF-LOCKING NUT
10 x 1.25 mm
44 N·m (4.5 kgf·m, 33 lbf·ft)
Replace.

10 x 1.25 mm
44 N·m (4.5 kgf·m,
33 lbf·ft)
Replace.

HEAT SHIELD

8 x 1.25 mm
22 N·m (2.2 kgf·m,
16 lbf·ft)

GASKET
Replace.

EXHAUST MANIFOLD
BRACKET

WASHER

8 x 1.25 mm
22 N·m (2.2 kgf·m, 16 lbf·ft)
Replace.
Tighten the bolts in steps,
alternating side-to-side

10 x 1.25 mm
44 N·m (4.5 kgf·m,
33 lbf·ft)

43256-ACCO-G16

Exploded view of the exhaust manifold and related components—2.4L engine

To install:

3. Install or connect the following:
- Exhaust manifold with a new gasket and tighten the nuts to 23 ft. lbs. (31 Nm)
- Exhaust pipe to the manifold using a new gasket and tighten the nuts to 40 ft. lbs. (54 Nm)
- Manifold cover and tighten the bolts to 16 ft. lbs. (22 Nm)

3.2L and 3.5L Engines

1. Before servicing the vehicle, refer to the precautions in the beginning of this section.

2. Remove or disconnect the following:
- Negative battery cable
- Exhaust manifold covers
- Small heat shields from the cylinder heads (if equipped)
- Exhaust pipe from the manifold
- Heated Oxygen (HO2S) sensors
- Exhaust manifold nuts in a crisscross pattern starting from the center of the manifold
- Exhaust manifold

To install:

3. Install or connect the following:
- Exhaust manifold with a new gas-

ket and new nuts and tighten the nuts in a crisscross pattern starting from the center to 22 ft. lbs. (30 Nm)
- Small heat shields and tighten the attaching bolts to 16 ft. lbs. (22 Nm) (if equipped)
- Exhaust pipe to the manifold with a new gasket and tighten the nuts to 40 ft. lbs. (55 Nm)
- HO2S sensor and tighten it to 33 ft. lbs. (45 Nm)
- Manifold covers and tighten the bolts to 16 ft. lbs. (22 Nm)

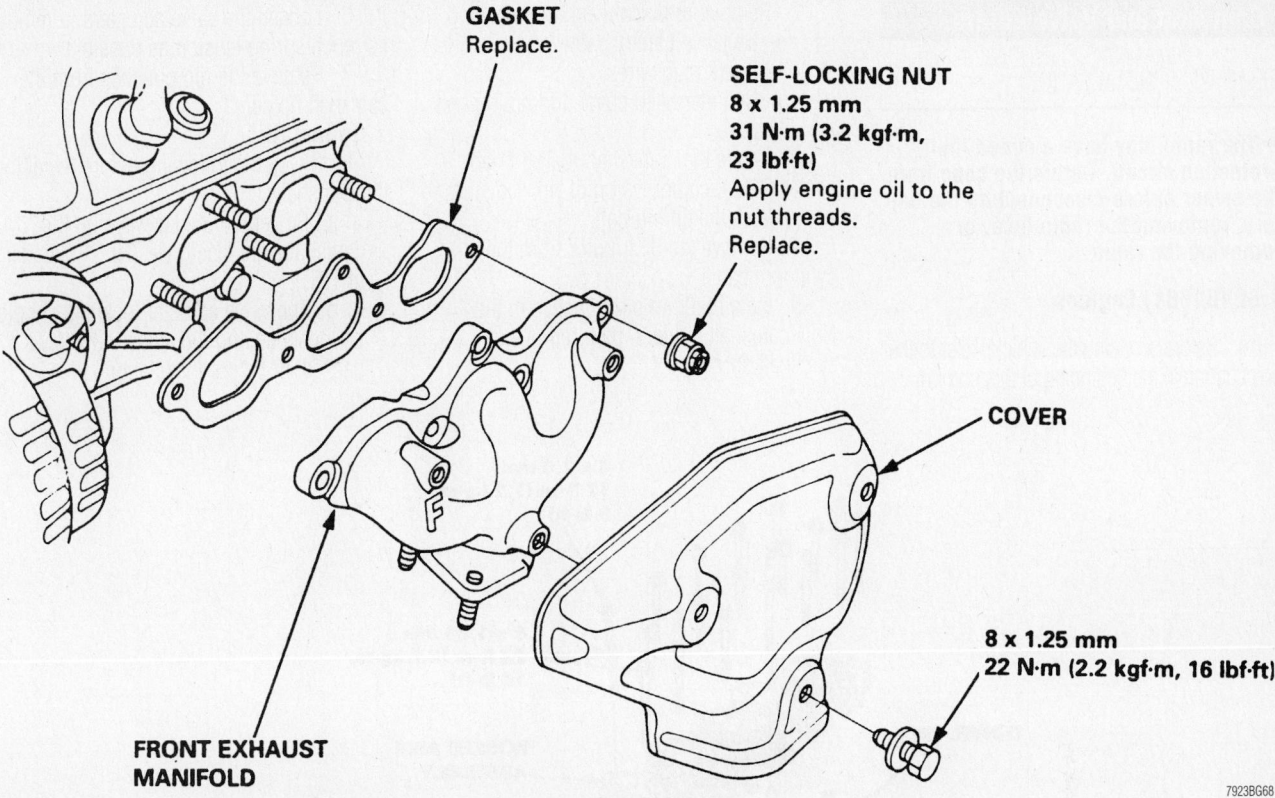

GASKET
Replace.

SELF-LOCKING NUT
8 x 1.25 mm
31 N·m (3.2 kgf·m,
23 lbf·ft)
Apply engine oil to the
nut threads.
Replace.

COVER

8 x 1.25 mm
22 N·m (2.2 kgf·m, 16 lbf·ft)

**FRONT EXHAUST
MANIFOLD**

7923BG68

Exploded view of the front exhaust manifold mounting—3.0L engine

4. Verify that all vacuum lines and wiring are properly connected.
5. Reconnect the negative battery cable.
6. Start the engine and check for leaks.

Front Crankshaft Seal

REMOVAL & INSTALLATION

1. Before servicing the vehicle, refer to the precautions in the beginning of this section.
2. Disconnect negative cable at the battery.
3. Raise and safely support the vehicle. Drain the engine oil and properly dispose of it.
4. Be sure the crankshaft is at TDC on No. 1 cylinder by aligning the white mark on the crankshaft pulley with the pointer on the lower timing belt cover.
5. Remove or disconnect the following:
   • Crankshaft pulley
   • Cylinder head cover
   • Timing belt cover
   • Timing belt
   • Crankshaft Speed Fluctuation (CKF) sensor (if equipped)
   • Timing belt drive gear from the crankshaft
6. Using a suitable prytool, carefully remove the seal.

To install:
7. Apply a light coat of oil to the seal lip.
8. Position the seal, then using a seal driver, install the seal into the housing.
9. Install or connect the following:
   • Timing belt drive gear
   • Timing belt
   • Timing belt cover
   • Cylinder head cover
   • CKF sensor and tighten the attach-

ing bolts to 96 inch lbs. (11 Nm) (if equipped)
   • Crankshaft pulley
10. Lower the vehicle and check and fill the engine with oil as necessary.
11. Connect the negative battery cable and enter the radio security code.
12. Run the engine and check for leaks.
13. Turn off engine and check the oil level. Top off the oil level if necessary.

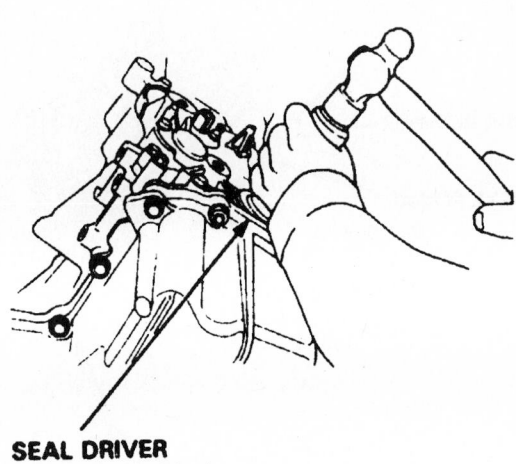

SEAL DRIVER
Install seal with the part number side facing out.

7923BG19

Installing the seal

Camshaft and Valve Lifters

REMOVAL & INSTALLATION

➡ **The radio may have a coded theft protection circuit. Obtain the code from the owner before disconnecting the battery, removing the radio fuse, or removing the radio.**

1.8L (B18B1) Engines

1. Before servicing the vehicle, refer to the precautions in the beginning of this section.

2. Remove or disconnect the following:
- Negative battery cable
- Spark plug wires
- Cylinder head cover and timing belt cover

3. Rotate the crankshaft to Top Dead center (TDC), compression of No. 1 piston and remove the timing belt.

4. Remove the distributor from the cylinder head.

5. Install 5.0mm pin punches to the No.1 camshaft holders, then remove the camshaft sprockets.

6. Loosen the valve adjusters to remove as much spring tension as possible.

7. Remove the pin punches from the camshaft holders.

To install:

8. Check the following before installing the camshafts:

a. Be certain the keyways on the camshafts are facing UP (No. 1 cylinder at TDC).

b. The valve adjuster locknuts should be loosened and the adjusting screws backed off before installation.

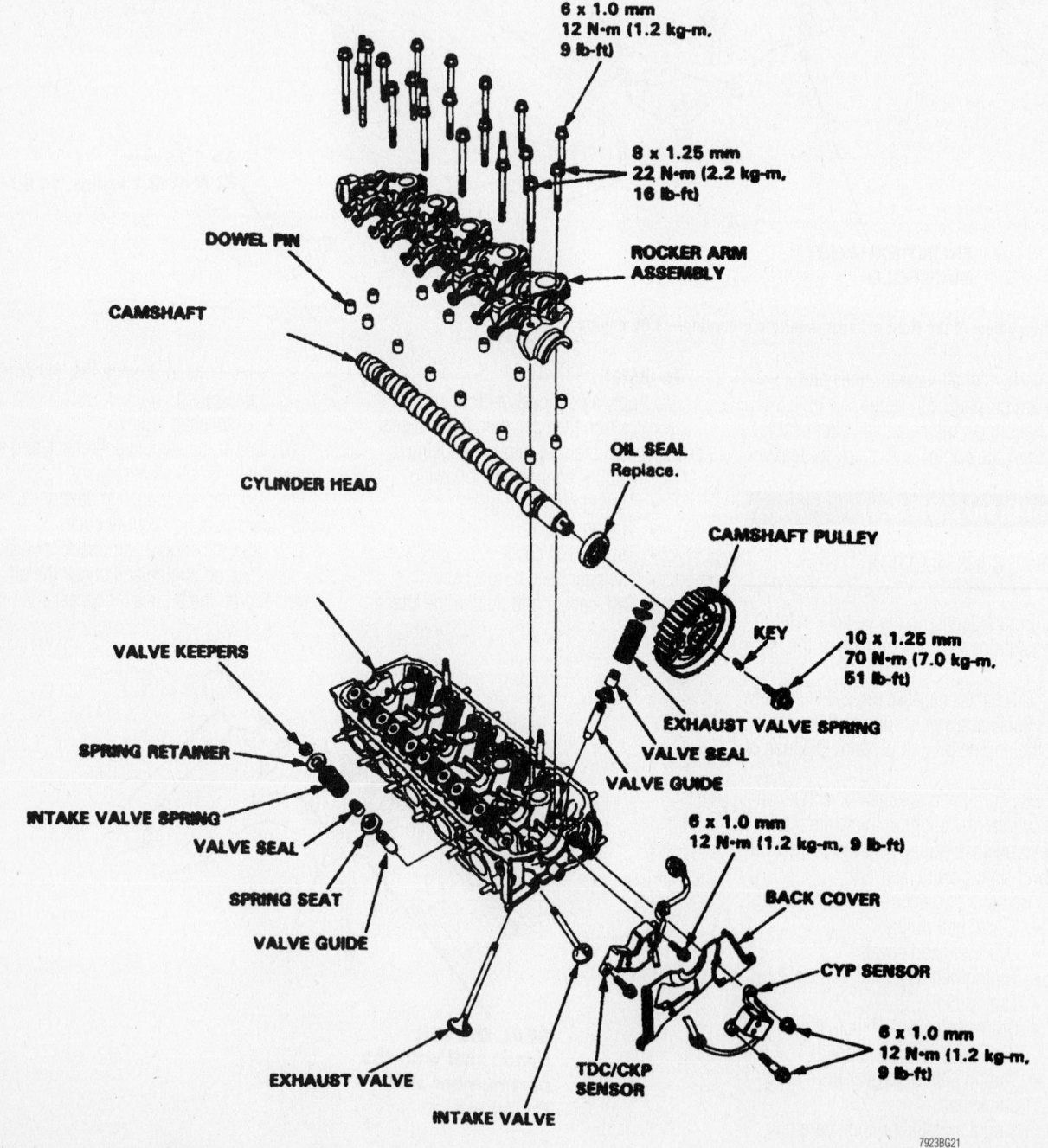

**6 x 1.0 mm
12 N·m (1.2 kg-m, 9 lb-ft)**

**8 x 1.25 mm
22 N·m (2.2 kg-m, 16 lb-ft)**

ROCKER ARM ASSEMBLY

DOWEL PIN

CAMSHAFT

CYLINDER HEAD

OIL SEAL
Replace.

CAMSHAFT PULLEY

KEY

**10 x 1.25 mm
70 N·m (7.0 kg-m, 51 lb-ft)**

EXHAUST VALVE SPRING

VALVE SEAL

VALVE GUIDE

VALVE KEEPERS

SPRING RETAINER

INTAKE VALVE SPRING

VALVE SEAL

SPRING SEAT

VALVE GUIDE

**6 x 1.0 mm
12 N·m (1.2 kg-m, 9 lb-ft)**

BACK COVER

CYP SENSOR

**6 x 1.0 mm
12 N·m (1.2 kg-m, 9 lb-ft)**

EXHAUST VALVE

INTAKE VALVE

TDC/CKP SENSOR

7923BG21

Exploded view of the cylinder head—1.8L (B18B1) engine

9. Lubricate the rocker arms and camshafts with clean oil.

10. Place the rocker arms on the pivot bolts and the valve stems, making sure that the rocker arms are in their original positions.

11. Install the camshaft seals with the open side (spring) facing in. Lubricate the lip of the seal.

12. Be sure the keyways on the camshafts are facing up and install the camshafts to the cylinder head.

13. Apply liquid gasket to the head mating surfaces of the No. 1 and No. 6 camshaft holders, then install them along with No. 2, 3, 4 and 5 camshaft holders. The arrows stamped on the holders should point toward the timing belt. Do not apply oil to the holder mating surface where the camshaft seals are housed.

14. Tighten the camshaft holders temporarily and be sure that the rocker arms are properly positioned.

15. Press the oil seals into the No.1 camshaft holders with a seal driver.

16. Tighten the bolts in a crisscross pattern to 84 inch lbs. (10 Nm). Check that the rockers do not bind on the valves.

17. Install the cylinder head plug to the end of the cylinder head. If the plug has alignment marks, align the marks with the cylinder head upper surface.

18. If equipped with a timing belt back cover, install the cover and tighten the bolts to 84 inch lbs. (10 Nm).

19. Install 5.0mm pin punches to the No.1 camshaft holders, then install the camshaft pulley keys onto the grooves in the camshafts.

20. Push the camshaft pulleys onto the camshafts, then tighten the retaining bolts to 27 ft. lbs. (38 Nm).

21. Install the timing belt and timing belt covers. Remove the pin punches from the camshaft holders.

22. Adjust the valves and pour oil over the camshafts and rocker arms.

23. Install or connect the following:
- Cylinder head cover and engine ground cable
- Distributor to the cylinder head and reconnect the spark plug wires to the spark plugs
- Negative battery cable and enter the radio security code

24. Change the engine oil. Wait at least 20 minutes for the sealant to cure before filling the engine with oil.

1.8L (B18C1, B18C5) Engines

1. Before servicing the vehicle, refer to the precautions in the beginning of this section.

2. Disconnect the negative battery cable.

3. Be sure the crankshaft is at TDC/compression on No. 1 cylinder by aligning the white mark on the crankshaft pulley with the pointer on the lower timing belt cover.

4. Remove or disconnect the following:
- Strut brace
- Cylinder head cover, timing belt cover, and timing belt
- Camshaft pulleys and back cover

5. Loosen the rocker arm locknuts and adjusting screws.

6. Remove the camshaft holder bolts, then, remove the camshaft holder plates, the camshaft holders, and camshafts.

To install:

7. Be sure that the keyways on the camshafts are facing up and that the rocker arms are in their original position. The valve locknuts should be loosened and the adjusting screw backed off before installation

8. Install or connect the following:
- Camshafts
- Camshaft seals with the open side facing in
- Rubber cap with liquid gasket applied
- O–ring and the dowel pin to the oil passage of the No. 3 camshaft holder

9. Apply liquid gasket to the head of the mating surfaces of the No. 1 and No. 5 camshaft holders, then install them, along with No. 2, 3, and 4. Be sure to pay attention to the following points:
- Do not apply oil to the holder mating surface of camshaft seals
- The arrows marked on the camshaft holders should point to the timing belt

10. Tighten the camshaft holders temporarily. Be sure that the rocker arms are properly positioned on the valve stems.

11. Tighten the camshaft holder bolts in 2 steps, following the proper sequence, to

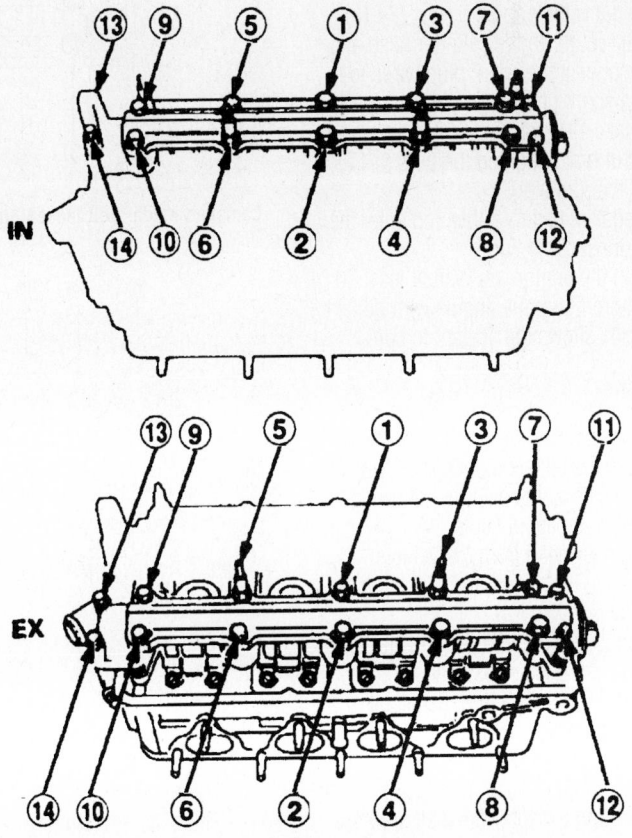

1-10 : 8 x 1.25 mm 27 N·m (2.8 kgf-m, 20 lbf·ft)
11-14 : 6 x 1.0 mm 9.8 N·m (1.0 kgf-m, 7.2 lbf·ft)

7923BG20

Camshaft holder plates torque sequence—1.8L (B18C1, B18C5) engines

ensure that the rockers do not bind on the valves. Tighten the 8 x 1.25mm bolts to 20 ft. lbs. (27 Nm). Tighten the 6 x 1.0mm bolts to 84 inch lbs. (10 Nm).

12. Install or connect the following:
- Keys into the camshaft grooves—To set the No. 1 piston at TDC, align the holes on the camshaft with the holes in the No. 1 camshaft holders and insert 5.0mm pin punches into the holes.
- Back cover and push the camshaft pulleys onto the camshafts, then tighten the retaining bolts to 27 ft. lbs. (37 Nm)
- Timing belt and adjust the tension, then install the timing belt covers

13. Adjust the valve clearance.
14. Install or connect the following:
- Cylinder head cover, be sure that the seal and groove are thoroughly clean first
- Engine side mount, tighten the 2 new nuts and new bolt to the engine to 38 ft. lbs. (52 Nm) and tighten the bolt attaching the mount to the vehicle to 54 ft. lbs. (74 Nm)
- Distributor to the cylinder head and reconnect the spark plug wires to the spark plugs
- Intake air duct
- Strut brace, tighten the nuts to 17 ft. lbs. (24 Nm)
- Negative battery cable and enter the radio security code

15. Drain the engine oil. Wait at least 20 minutes before filling the engine with oil; the time delay allows the sealant to cure.

2.0L Engine

1. Before servicing the vehicle, refer to the precautions in the beginning of this section.
2. Remove or disconnect the following:
- Negative battery cable
- Cylinder head from the vehicle
- Loosen the rocker arm adjusting screws
- Camshaft holder bolts, 2 turns at a time in a criss–cross pattern
- Timing chain guide B, camshaft holders and camshafts

To install:
3. Make sure the punch marks on the VTC actuator and exhaust camshaft sprocket are facing up, then set the camshafts in the holder.
4. Set the camshaft holders and timing chain guide B in place.
5. Tighten the bolts, in sequence, to the following specifications:
- 8mm bolts: 16 ft. lbs. (22 Nm)

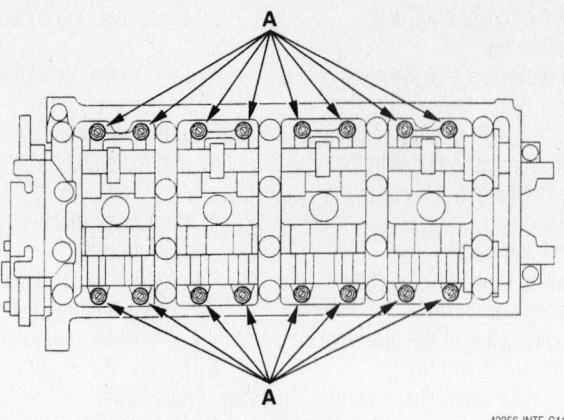

Loosen the rocker arm adjusting screws (A)—2.0L engines

42356-INTE-G11

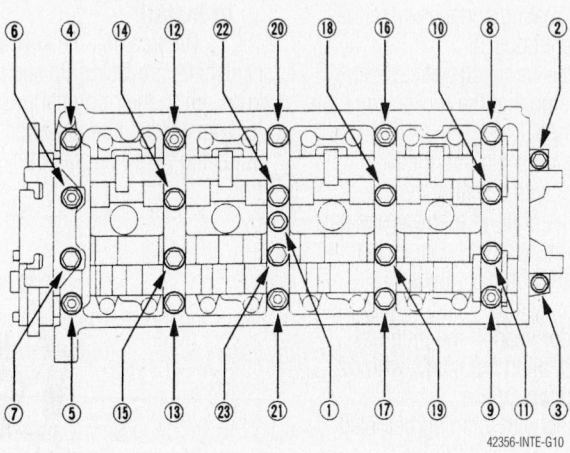

Camshaft holder bolt loosening sequence—2.0L engines

42356-INTE-G10

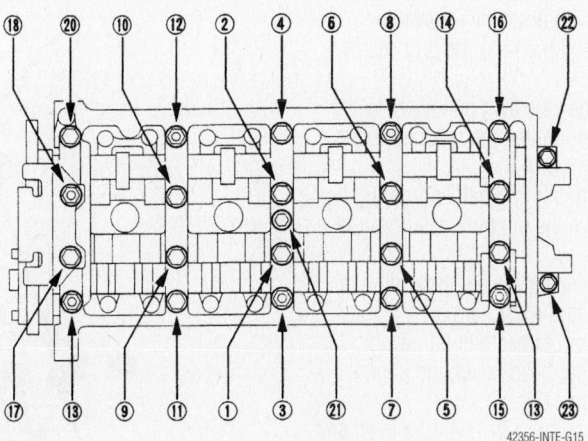

Rocker arm assembly tightening sequence—2.0L engines

42356-INTE-G15

- 6mm bolts (nos. 21, 22 & 23): 104 inch lbs. (12 Nm)
6. Install or connect the following:
- Timing chain, then adjust the valve clearance
- Cylinder head
- Negative battery cable
7. Check for proper engine and valve train operation.

2.4L Engine

1. Before servicing the vehicle, refer to the precautions in the beginning of this section.
2. Disconnect the negative battery cable.
3. Remove or disconnect the following:
- Timing chain

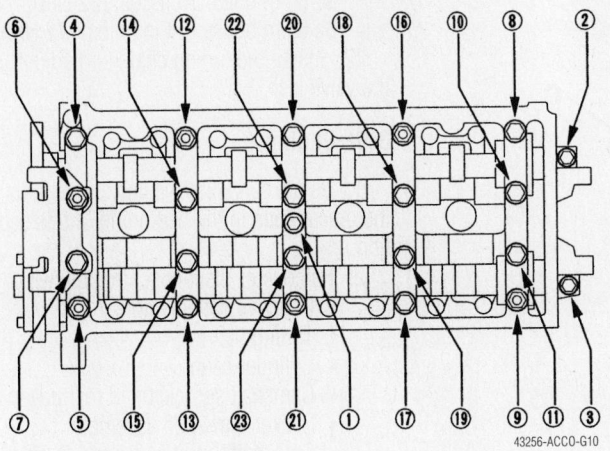

Camshaft holder bolt loosening sequence—2.4L engine

43256-ACCO-G10

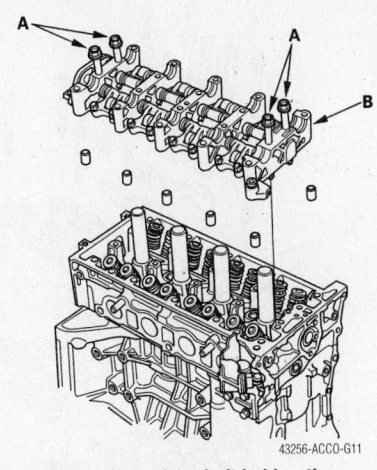

Insert the bolts (A) into the rocker shaft holder, then remove the rocker arm assembly (B)—2.4L engine

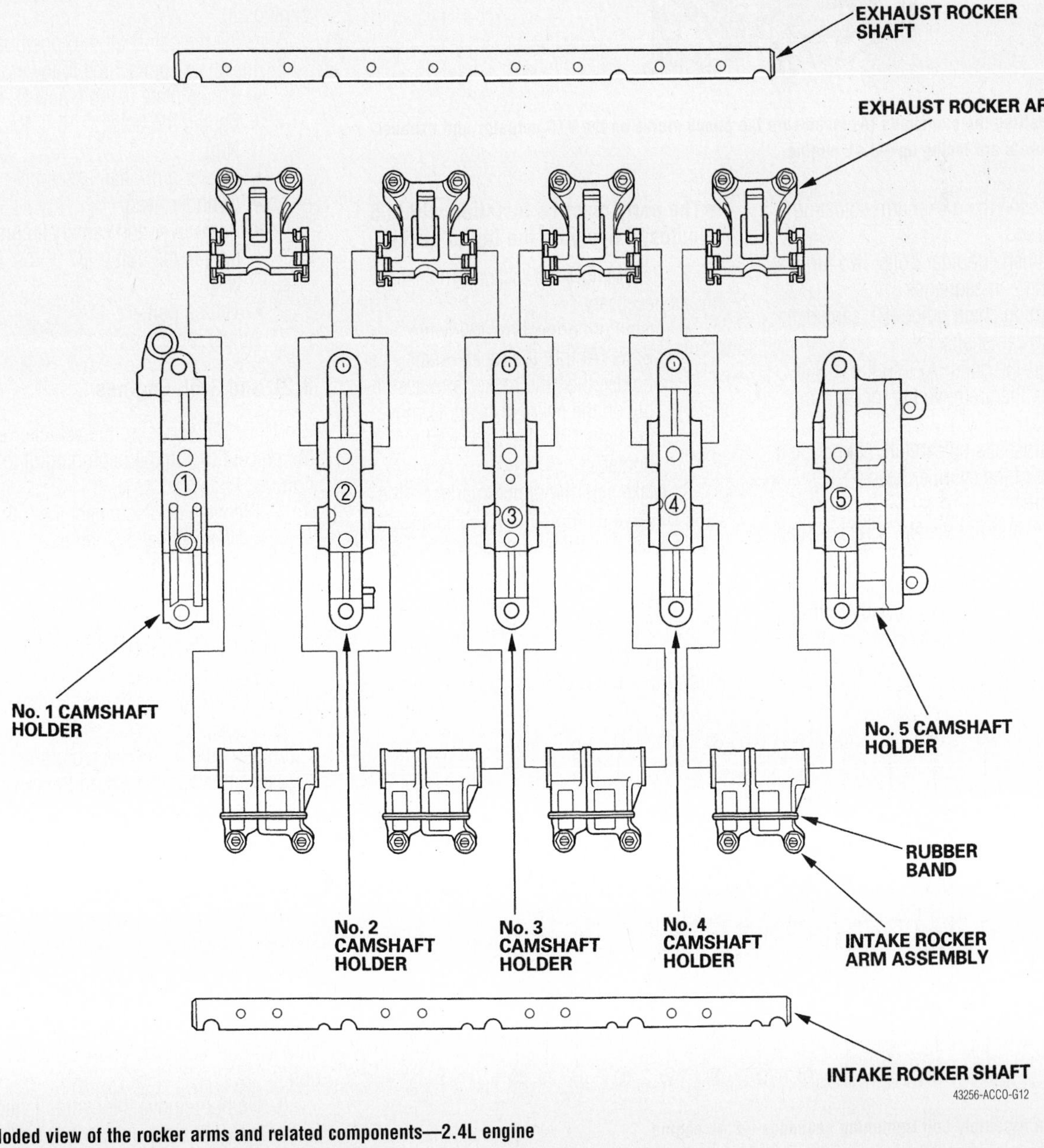

43256-ACCO-G11

EXHAUST ROCKER SHAFT

EXHAUST ROCKER ARM

No. 1 CAMSHAFT HOLDER

No. 5 CAMSHAFT HOLDER

No. 2 CAMSHAFT HOLDER

No. 3 CAMSHAFT HOLDER

No. 4 CAMSHAFT HOLDER

RUBBER BAND

INTAKE ROCKER ARM ASSEMBLY

INTAKE ROCKER SHAFT

43256-ACCO-G12

Exploded view of the rocker arms and related components—2.4L engine

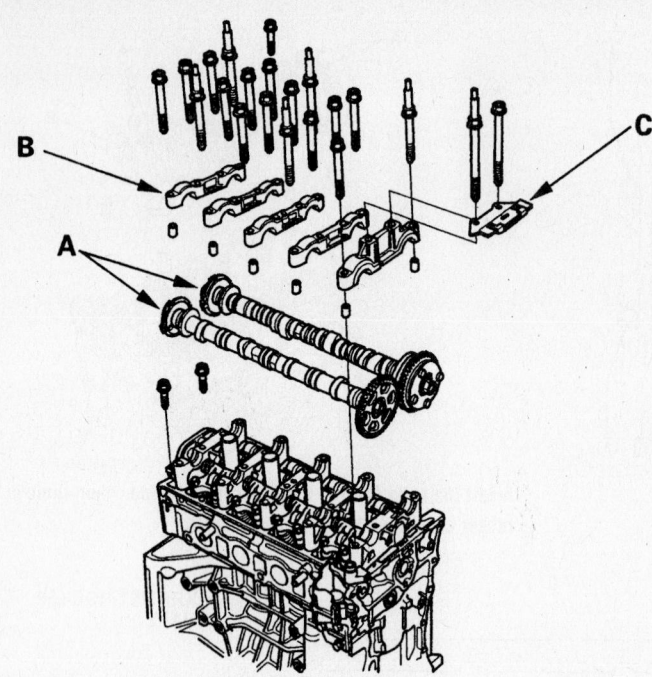

43256-ACCO-G13

When installing the camshafts (A) make sure the punch marks on the VTC actuator and exhaust cam sprockets are facing up—2.4L engine

- Loosen the rocker arm adjusting screws
- Camshaft holder bolts, two turns at a time in sequence
- Timing chain guide (B), camshaft and camshafts

4. Insert the bolts (A) into the rocker shaft holder, then remove the rocker arm assembly (B)

- Camshafts by carefully lifting them out of the cylinder head

To install:

5. Clean and dry the No. 5 rocker shaft holding mating surface.

6. Apply a suitable liquid gasket P/N 08718-0009, or equivalent, evenly to the cylinder head mating surface of the No. 5 rocker shaft holder.

➡**The parts must be installed within 5 minutes of applying the liquid gasket.**

7. Reassemble the rocker arm assembly, as necessary.

8. Install or connect the following

- Bolts (A) into the rocker shaft holder, then the rocker arm assembly on the cylinder head. Remove the bolts from the rocker shaft holder.

9. Make sure the punch marks on the variable valve timing control (VTC) actuator and exhaust camshaft sprocket are facing up, then set the camshafts (A) in the holder

10. Set the camshaft holders (B) and timing chain guide B (C) in place.

11. Tighten the bolts, in sequence, to the following specification:

a. 8mm bolts: 16 ft. lbs. (22 Nm)
b. 6mm bolts: 104 inch lbs. (12 Nm)

12. Install the timing chain and adjust the valve lash.

3.0L Engine

1. Before servicing the vehicle, refer to the precautions in the beginning of this section.

2. Remove or disconnect the following:
- Negative battery cable
- Timing belt
- Cylinder head
- Camshaft sprocket and rear cover
- Rocker arm/shaft assembly
- Camshaft thrust cover and O–ring

3. Pull out the camshaft.

To install:

4. Lubricate the camshaft with clean engine oil.

5. Slide the camshaft into position.

6. Install or connect the following:
- Thrust plate using a new O–ring, tighten the bolts to 16 ft. lbs. (22 Nm)
- Rocker arm/shaft assembly
- Cylinder head
- Rear cover and camshaft sprocket, tighten the bolt to 67 ft. lbs. (90 Nm)
- Timing belt

7. Adjust the valves.

3.2L and 3.5L Engines

1. Before servicing the vehicle, refer to the precautions in the beginning of this section.

2. Remove or disconnect the following:
- Negative battery cable
- Timing belt covers and cylinder head covers

3. Rotate the crankshaft to Top Dead Center (TDC) for the No. 1 piston and remove the timing belt.

4. Remove the camshaft sprockets.

5. Loosen the rocker shaft holder bolts 1 turn at a time in the reverse of the torque sequence to avoid damaging the valves, camshafts, or rocker assemblies.

6. After all bolts are loose, remove the rocker arm shafts as an assembly with the bolts still in the holders.

7. If the rocker shafts are to be disassembled, note that each rocker arm has a letter **A** or **B** stamped into the side. Before disassembling the rocker arms, make a note of the position of each letter so that the arms can be reassembled in the same position.

8. Do not remove the hydraulic tappets from the rocker arms unless they are to be

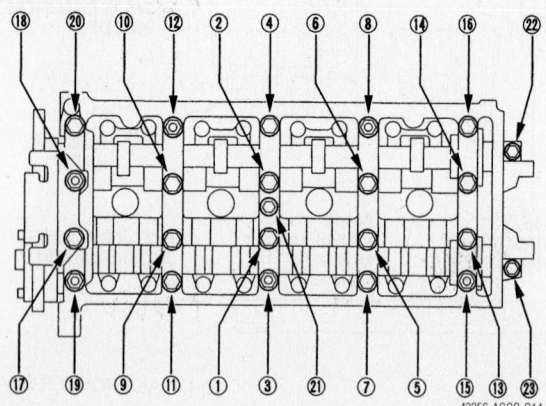

43256-ACCO-G14

Rocker arm assembly bolt tightening sequence—2.4L engine

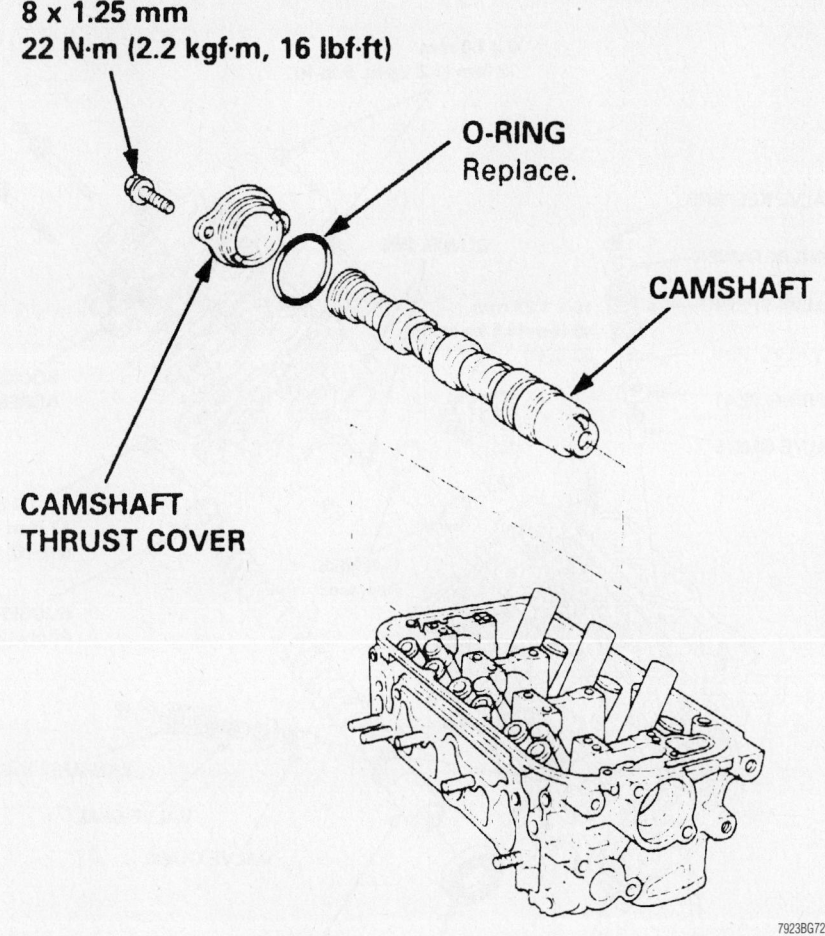

8 x 1.25 mm
22 N·m (2.2 kgf·m, 16 lbf·ft)

O-RING
Replace.

CAMSHAFT

CAMSHAFT
THRUST COVER

7923BG72

Camshaft installation—3.0L engine

replaced. Handle the rocker arms carefully so the oil does not drain out of the tappets.

9. Lift the camshafts from the cylinder head, wipe them clean and inspect the lift ramps. Replace the camshafts and rockers if the lobes are pitted, scored, or excessively worn.

To install:

10. Place a new seal on the end of the camshaft, lubricate the journals and set the camshaft in place on the head.

➡**The pin hole in the front of the camshaft designates the top position.**

11. Apply liquid gasket to the mounting surfaces of the camshaft end holders.

12. Set the rocker arm assemblies in place and start all of the camshaft holder bolts. Be sure the rocker arms are properly positioned and turn each bolt in sequence 2 turns at a time until the holders are seated on the head to avoid damaging the valves or rocker assemblies.

13. When all the camshaft and rocker holders are seated, tighten the bolts in the same sequence. Tighten the 8mm bolts to 16 ft. lbs. (22 Nm) and the 6mm bolts to 104 inch lbs. (12 Nm).

14. Install or connect the following:
- Camshaft pulleys and tighten the bolts to 23 ft. lbs. (32 Nm)
- Timing belt and pour oil over the camshafts
- Cylinder head cover and reassemble accessory components

15. Verify that all electrical connections and vacuum lines are connected.

16. Reconnect the negative battery cable.

17. Run the engine and check for leaks and proper operation..

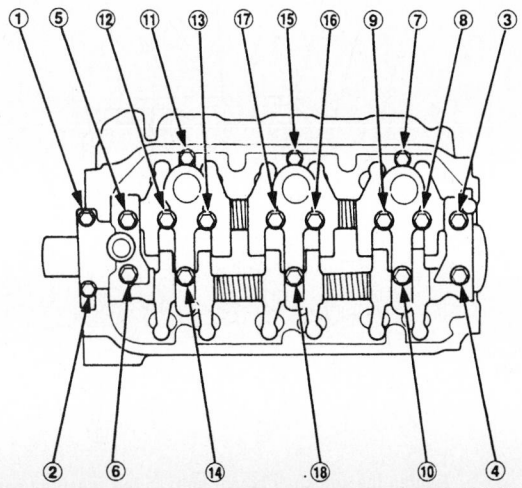

7923BG85

Loosen the camshaft holder bolts in the specified sequence—3.5L engine

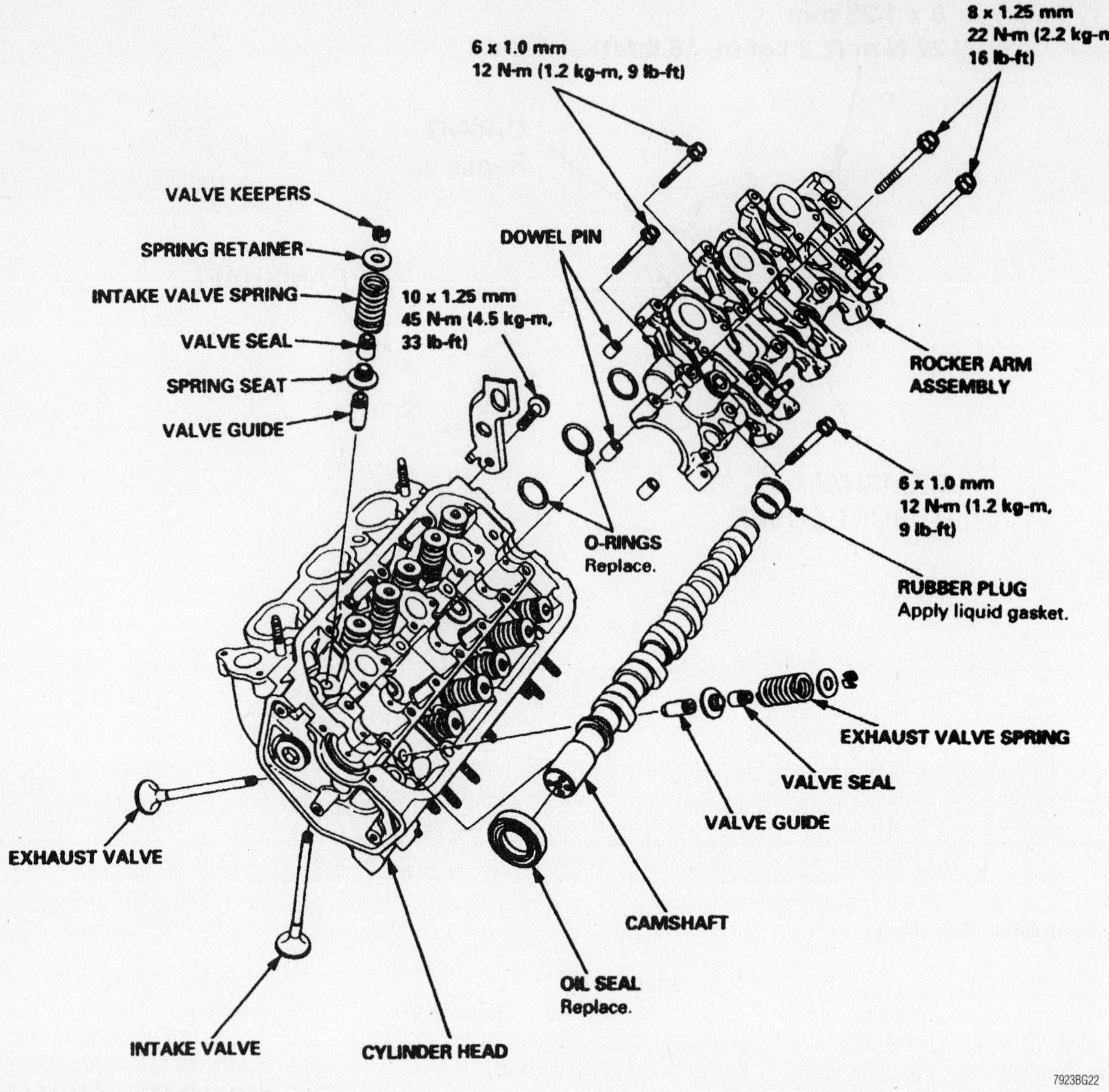

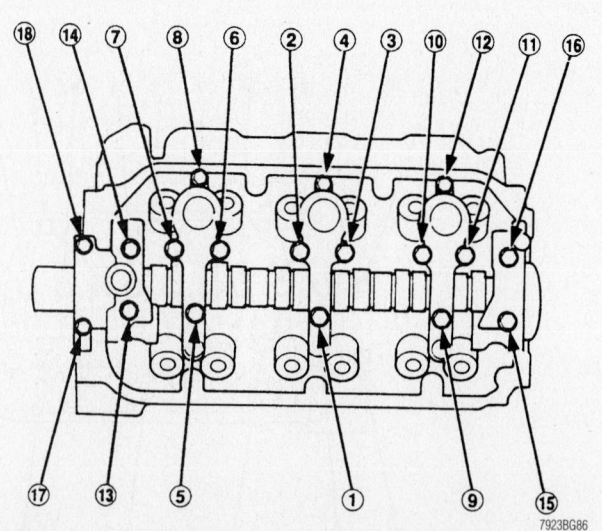

6 x 1.0 mm
12 N·m (1.2 kg-m, 9 lb-ft)
8 x 1.25 mm
22 N·m (2.2 kg-m, 16 lb-ft)
VALVE KEEPERS
SPRING RETAINER
INTAKE VALVE SPRING
VALVE SEAL
SPRING SEAT
VALVE GUIDE
DOWEL PIN
10 x 1.25 mm
45 N·m (4.5 kg-m, 33 lb-ft)
ROCKER ARM ASSEMBLY
6 x 1.0 mm
12 N·m (1.2 kg-m, 9 lb-ft)
O-RINGS
Replace.
RUBBER PLUG
Apply liquid gasket.
EXHAUST VALVE SPRING
VALVE SEAL
VALVE GUIDE
EXHAUST VALVE
CAMSHAFT
OIL SEAL
Replace.
INTAKE VALVE
CYLINDER HEAD
7923BG22

Camshaft and rocker arm assembly—3.2L engines

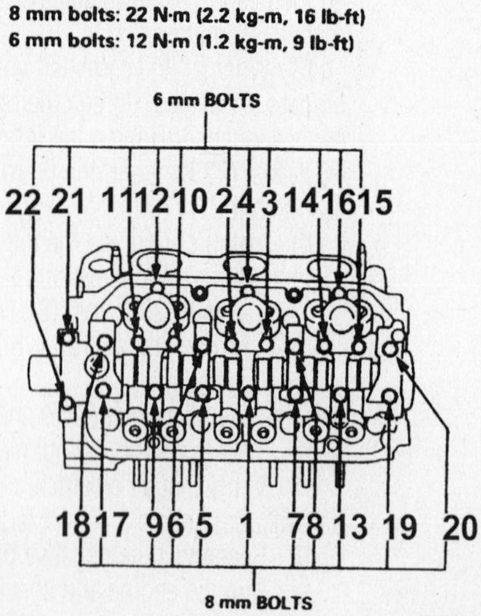

Specified torque:
8 mm bolts: 22 N·m (2.2 kg-m, 16 lb-ft)
6 mm bolts: 12 N·m (1.2 kg-m, 9 lb-ft)
6 mm BOLTS
22 21 11 12 10 2 4 3 14 16 15
18 17 9 6 5 1 7 8 13 19 20
8 mm BOLTS
7923BG23

Camshaft holder bolt tightening sequence—3.2L engine

18 14 7 8 6 2 4 3 10 12 11 16
17 13 5 1 9 15
7923BG86

Tighten the camshaft holder bolts in the specified sequence—3.5L engine

Valve Lash

ADJUSTMENT

1.8L and 2.0L Engines

➡While all valve adjustments must be as accurate as possible, it is better to have the valve adjustment slightly loose rather than too tight. Burned valves may result from overly tight adjustments. Perform the valve adjustment for each cylinder in the same sequence as the firing order: 1–3–4–2.

1. Before servicing the vehicle, refer to the precautions in the beginning of this section.

2. Be sure the engine is cold; cylinder head temperature must be below 100° F (38° C). Overnight cold is best.

3. Remove the cylinder head cover.

4. Remove the upper timing belt cover, 1.8L engines only

5. Set the No. 1 cylinder to Top Dead Center (TDC).

a. On 1.8L engines, the word **UP** should appear at the top and the TDC grooves on the pulley should align with the cylinder head surface or the mark on the rear belt cover.

b. On 2.0L engines, the punch mark (arrow) on the Variable Valve Timing Control (VTC) actuator and the punch mark on the exhaust camshaft sprockets should be at the top. Align the TDC marks on the VTC actuator and exhaust camshaft sprocket.

6. For 1.8L engines, valve clearances are:

- B18B1 engine: Intake— 0.003–0.005 in. (0.08–0.12mm) Exhaust—0.006–0.008 in. (0.16–0.20mm)
- B18C1 and B18C5 VTEC engine: Intake—0.006–0.007 in. (0.15–0.19mm) Exhaust— 0.007–0.008 in. (0.17–0.20mm)

7. For 2.0L engines, valve clearances are:

- K20A3 engine: Intake— 0.008–0.010 in. (0.21–0.25mm) Exhaust—0.011–0.013 in. (0.28–0.32mm)
- K20A2 engine: Intake— 0.008–0.010 in. (0.21–0.25mm) Exhaust—0.010–0.011 in. (0.25–0.29mm)

8. With the No. 1 cylinder at TDC, adjust the valves of the No. 1 cylinder by performing the following procedures:

a. Hold the rocker arm against the

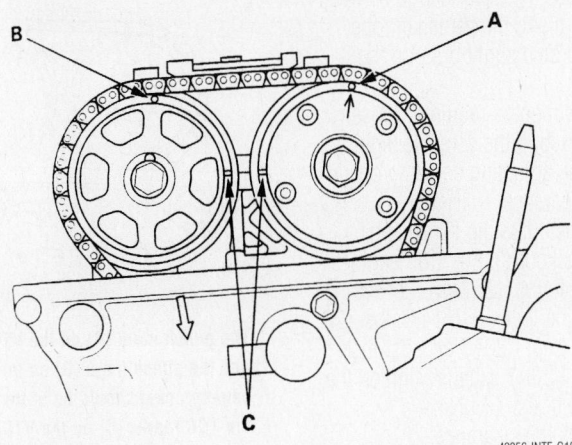

42356-INTE-G18

The punch mark (A) on the VTC actuator and the punch mark (B) on the exhaust camshaft sprockets should be at the top. Align the TDC marks (C) on the VTC actuator and exhaust camshaft sprocket.

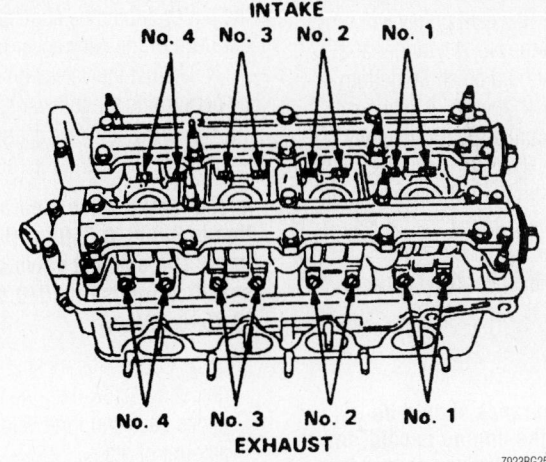

7923BG25

Valve arrangement—1.8L (B18B1, B18C1, and B18C5) engines

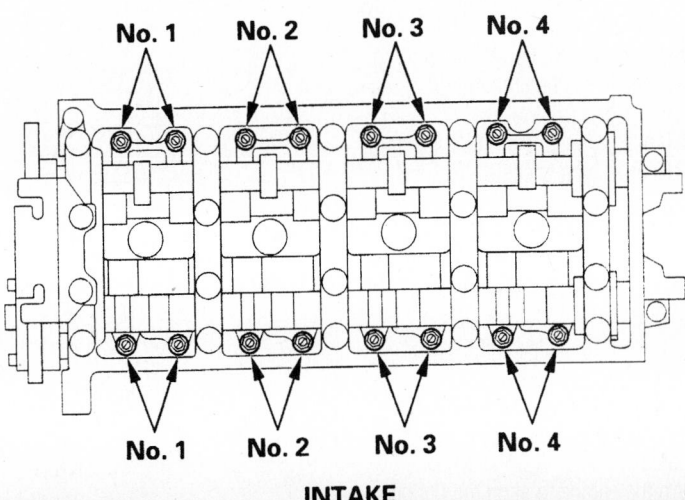

42356-INTE-G19

Valve arrangement—2.0L engines

valve and place the feeler gauge between the rocker arm and the camshaft lobe. There should be a slight drag on the feeler gauge.

b. If adjustment is required, loosen the valve adjusting the screw locknut.

c. Turn the adjusting screw to obtain the proper clearance.

d. Hold the adjusting screw and tighten the locknut(s). For 1.8L engines tighten to 18 ft. lbs. (25 Nm). For the 2.0L tighten the locknut to 14 ft. lbs. (20 Nm) for all except K20A3 exhaust locknuts. For the K20A3 exhaust, tighten the locknut to 10 ft. lbs. (14 Nm).

e. Recheck the clearance.

9. Turn the crankshaft 180 degrees counterclockwise; the cam pulley will turn 90 degrees. With the No. 3 cylinder at TDC, the **UP** marks should be at the exhaust side. Adjust the valves on the No. 3 cylinder.

10. Turn the crankshaft 180 degrees counterclockwise; the cam pulley will turn 90 degrees. With the No. 4 cylinder at TDC, both **UP** marks should be at the bottom. Adjust the valves on the No. 4 cylinder.

11. Turn the crankshaft 180 degrees counterclockwise. The No. 2 cylinder will now be on TDC and the **UP** marks should be at the intake side. Adjust the valves on the No. 2 cylinder.

12. Install the cylinder head cover and upper timing belt cover.

2.4L Engine

➡**The valve clearance should be adjusted when the engine is cold, the cylinder head temperature should be less than 100°F (38°C).**

➡**The radio may contain a coded theft protection circuit. Always obtain the code number before disconnecting the battery.**

1. Before servicing the vehicle, refer to the precautions in the beginning of this section.

2. Remove or disconnect the following:
• Negative battery cable

➡**Label the wires before disconnecting them.**

• Spark plug wires from the spark plugs
• Positive Crankcase Ventilation (PCV) hose
• Cylinder head cover. Replace the rubber seals if damaged or deteriorated.

3. Turn the engine to align the timing marks and set cylinder No.1 to TDC. The white mark on the crankshaft pulley should

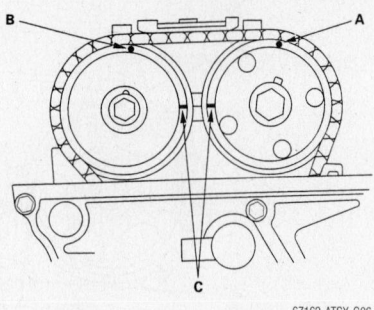

67162-ATSX-G06

The punch mark (A) on the VTC actuator and the punch mark (B) on the exhaust cam sprocket should be at the top. Align the TDC marks (C) on the VTC actuator and exhaust cam sprockets—2.4L engine

align with the pointer on the timing belt cover. The words **UP** embossed on the camshaft pulley should be aligned in the upward position. The marks on the edge of the pulley should be aligned with the cylinder head or the back cover upper edge.

4. Adjust the valves on cylinder No. 1 by performing the following:

a. Insert a feeler gauge in between the camshaft lobe and the rocker arm.

➡**The intake valve clearance specification is 0.008–0.010 in. (0.21–0.25 mm). The exhaust valve clearance specification is 0.010–0.011 in. (0.25–0.29mm).**

b. Loosen the locknut and turn the adjusting screw until the feeler gauge slides back and forth with a slight amount of drag.

c. Tighten the locknut to 14 ft. lbs. (20 Nm) and recheck the valve clearance. Repeat the valve adjustment if necessary.

5. Rotate the crankshaft 180° counterclockwise (the camshaft pulleys will turn

90°) The **UP** arrow marks should be pointing to the exhaust side of the cylinder head.

6. Adjust the valves on cylinder No. 3 by performing the following:

a. Insert a feeler gauge in between the camshaft lobe and the rocker arm.

b. Loosen the locknut and turn the adjusting screw until the feeler gauge slides back and forth with a slight amount of drag.

c. Tighten the locknut to 14 ft. lbs. (20 Nm) and recheck the valve clearance. Repeat the valve adjustment if necessary.

7. Rotate the crankshaft 180° counterclockwise (the camshaft pulleys will turn 90°) to bring No. 4 piston to TDC. The **UP** arrow marks should be pointing down, toward the crankshaft.

8. Adjust the valves on cylinder No. 4 by performing the following:

a. Insert a feeler gauge in between the camshaft lobe and the rocker arm.

b. Loosen the locknut and turn the adjusting screw until the feeler gauge slides back and forth with a slight amount of drag.

c. Tighten the locknut to 14 ft. lbs. (20 Nm) and recheck the valve clearance. Repeat the valve adjustment if necessary.

9. Rotate the crankshaft 180° counterclockwise (the camshaft pulleys will turn 90°) to bring piston No. 2 to TDC. The **UP** arrow marks should be pointing to the intake side of the cylinder head.

10. Adjust the valves on cylinder No. 2 by performing the following:

a. Insert a feeler gauge in between the camshaft lobe and the rocker arm.

b. Loosen the locknut and turn the adjusting screw until the feeler gauge slides back and forth with a slight amount of drag.

c. Tighten the locknut to 14 ft. lbs.

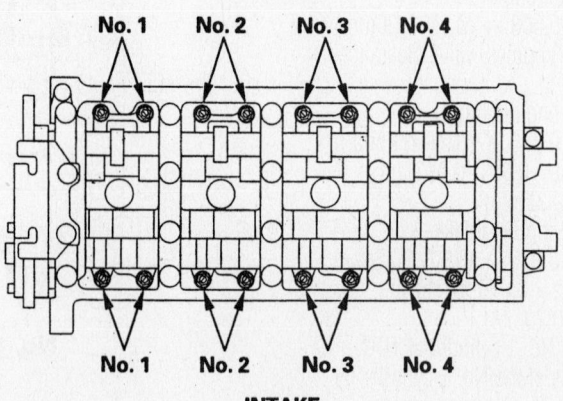

43256-ACCO-G17

Valve clearance adjusting screw locations

(20 Nm) and recheck the valve clearance. Repeat the valve adjustment if necessary.

11. Install the cylinder head cover gasket cover to the groove of the cylinder head cover. Before installing the gasket, thoroughly clean the seal and the groove. Seat the recesses for the camshaft first, then work it into the groove around the outside edges. Be sure the gasket is seated securely in the corners of the recesses.

12. Apply liquid gasket to the 4 corners of the recesses of the cylinder head cover gasket. Do not install the parts if 5 minutes or more have elapsed since applying liquid gasket. After assembly, wait at least 20 minutes before filling the engine with oil.

13. Install or connect the following:
- Cylinder head (valve) cover. Tighten the bolts attaching to 84 inch lbs. (10 Nm).
- Spark plug wires to the correct spark plugs.
- Positive, then the negative battery cable and enter the radio security code.

3.0L Engine

1. Before servicing the vehicle, refer to the precautions in the beginning of this section.

2. Remove or disconnect the following:
- Cylinder head cover
- Upper front timing belt cover

3. Rotate the crankshaft so the No. 1 piston is at Top Dead Center (TDC) on compression to adjust the valves for the No. 1 cylinder.

4. Loosen the locknuts and adjust the screws until a slight drag can be felt with the feeler gage when the gage is placed between the valve and rocker arm tip as shown. The specifications are as follows:
- Intake: 0.006–0.007 in. (0.15–0.18mm)
- Exhaust: 0.007–0.008 in. (0.18–0.20mm)

5. Rotate the crankshaft clockwise until the No. 4 on the camshaft sprocket is near the pointer on the rear cover. This is the No. 4 cylinder firing position.

6. Adjust the valves for the No. 4 cylinder while the sprocket is in this position. Tighten the locknuts to 14 ft. lbs. (20 Nm).

7. Continue to rotate the crankshaft and adjust the valves for cylinders 3 and 2.

8. Install the timing belt and cylinder head covers.

3.2L and 3.5L Engines

These engines are equipped with hydraulic valve lash adjusters on the rocker

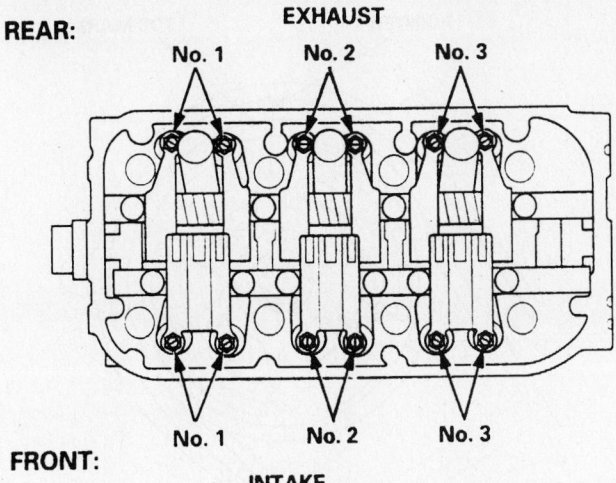

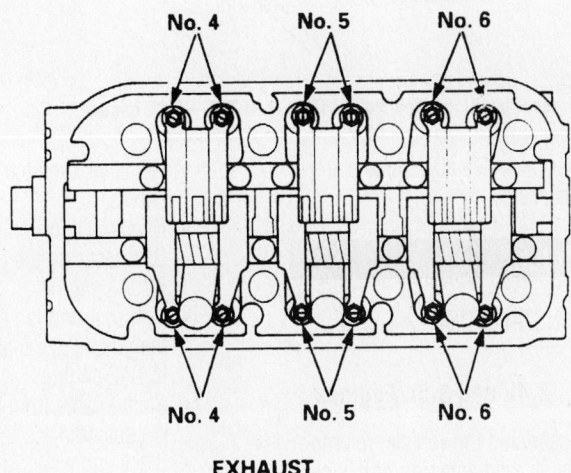

Adjusting screw locations for valve lash adjustment—3.0L engine

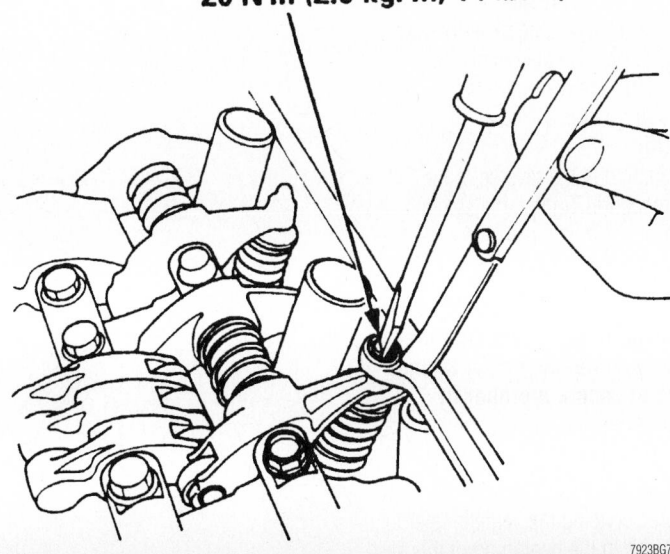

LOCKNUT
7 x 0.75 mm
20 N·m (2.0 kgf·m, 14 lbf·ft)

Slide the feeler gauge between the valve and rocker arm while turning the adjusting screw—3.0L engine

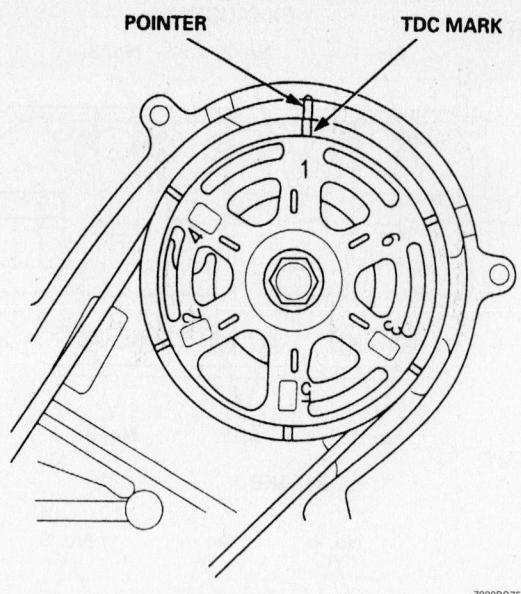

POINTER   TDC MARK

Camshaft sprocket position when No. 1 piston is at TDC—3.0L engine

7923BG75

arms. No valve clearance adjustments are possible or necessary.

Starter Motor

REMOVAL & INSTALLATION

Except 2.0L, 2.4L and 3.5L Engines

1. Before servicing the vehicle, refer to the precautions in the beginning of this section.
2. Disconnect the negative battery cable.
3. On the 1.8L engine, remove the intake air duct.
4. On 3.0L engines, remove the automatic transmission cooler hose from the bracket on the starter motor.
5. Remove or disconnect the following:
   - Starter electrical connectors
   - 2 starter mounting bolts
   - Starter motor

To install:

6. Installation is the reverse of the removal procedure. Tighten the starter motor bolts to 33 ft. lbs. (44 Nm).

➡ When installing the starter cable, be sure to place the closed loop connector over the stud on the starter with the crimped side of the connector facing up. This is to ensure a proper fit against the stud.

2.0L Engine

1. Before servicing the vehicle, refer to the precautions in the beginning of this section.
2. Remove or disconnect the following:
   - Negative, then the positive battery cables
   - Knock Sensor (KS) connector
   - Bolt securing the harness bracket, K20A3 engine
   - Bolt securing the harness bracket, then the intake manifold brackets, K20A2 engine
   - Starter cable from the B terminal on the solenoid
   - Black and white wire from the S terminal
   - 2 starter mounting bolts and starter

To install:

3. Installation is the reverse of the removal procedure, noting the following:
   - Make sure the crimped side of the

B terminal connector ring terminal is facing out
   - Tighten the top starter mounting bolt to 33 ft. lbs. (44 Nm) and the bottom bolt to 47 ft. lbs. (64 Nm)

2.4L Engine

➡ The factory sound system has a coded theft protection system. It is recommended that you know your reset code before you begin.

1. Before servicing the vehicle, refer to the precautions in the beginning of this section.
2. Remove or disconnect the following:
   - Negative then the positive battery cables
   - Intake manifold
   - Starter cable from the B terminal
   - Black/white wire from the S (solenoid) terminal
   - Harness clamp and holder
   - Two bolts that mount the starter to the transaxle assembly
   - Starter

To install:

3. Install in the reverse order of removal.

➡ When installing the heavy gauge starter cable, make sure the crimped side of the terminal end is facing out.

4. Enter the anti-theft code and radio presets.

3.5L Engine

➡ This procedure requires the use of an engine hoist to lift the engine slightly.

1. Before servicing the vehicle, refer to the precautions in the beginning of this section.

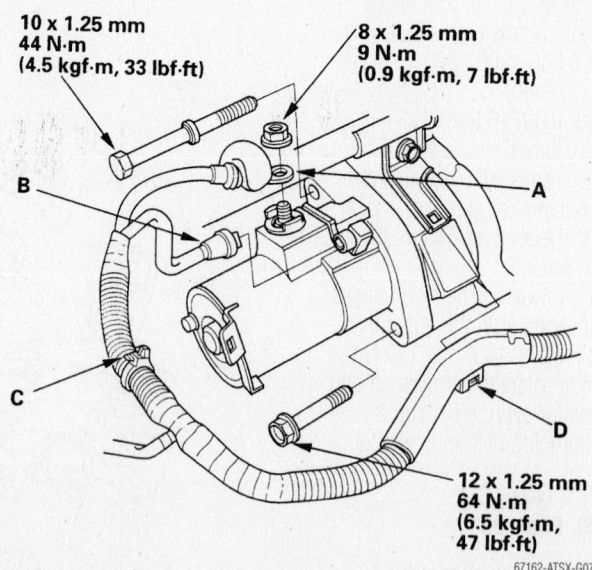

10 x 1.25 mm
44 N·m
(4.5 kgf·m, 33 lbf·ft)

8 x 1.25 mm
9 N·m
(0.9 kgf·m, 7 lbf·ft)

B

A

C

D

12 x 1.25 mm
64 N·m
(6.5 kgf·m, 47 lbf·ft)

67162-ATSX-G07

Starter mounting—2.4L engine

2. Remove or disconnect the following:
- Both battery cables
- Battery and tray
- Alternator and belt
- Left exhaust manifold cover
- Left damper fork
- Left lower ball joint from the suspension
- Left drive shaft
- Transmission stop collar
- Exhaust system Y–pipe
- Front mounting bolts

3. Attach a suitable engine hoist and slightly lift the engine.
- Left engine mount bracket
- Starter electrical connectors
- Starter motor

To install:

4. Install or connect the following:
- Starter motor and tighten the bolts to 33 ft. lbs. (44 Nm)
- Starter electrical connectors

➡ **Upon installation of the starter cable and the black/white wire, make sure that the crimped side of the connector is facing up.**

5. Lower the engine onto the motor mount and tighten the nut to 47 ft. lbs. (64 Nm) and the bolts to 40 ft. lbs. (54 Nm)

6. Install or connect the following:
- Exhaust system Y–pipe with new gaskets and tighten the 10mm bolts and nuts to 40 ft. lbs. (54 Nm) and the 8mm nuts to 16 ft. lbs. (22 Nm)
- Transmission stop collar and tighten the bolts to 28 ft. lbs. (38 Nm)
- Left drive shaft
- Damper fork
- Left lower ball joint
- Left exhaust manifold cover
- Alternator and belt
- Battery tray
- Battery and battery cables

Oil Pan

REMOVAL & INSTALLATION

1.8L Engines

1. Before servicing the vehicle, refer to the precautions in the beginning of this section.
2. Raise and safely support the vehicle.
3. Drain the engine oil.
4. Remove or disconnect the following:
- Negative battery cable
- Splash shield
- Heated Oxygen (HO2S) sensor connector

- Front exhaust pipe from the vehicle
- Center beam from the subframe

5. Loosen the oil pan bolts in a criss-cross pattern. To remove the oil pan, lightly tap the corners of the oil pan with a rubber or plastic faced mallet.

To install:

6. Apply liquid gasket to the oil pan mating surface where the oil pump and the right side cover meet the engine block.
7. Install or connect the following:
- Oil pan with a new gasket
- Oil pan nuts and bolts and torque them in sequence to 10 ft. lbs. (14 Nm)

✳✳ CAUTION

Excessive tightening can cause distortion of the oil pan gasket and oil leakage.

- Oil drain plug with a new gasket and torque it to 33 ft. lbs. (44 Nm)
- Front exhaust pipe using new gaskets and locknuts and tighten the manifold nuts to 40 ft. lbs. (54 Nm) and the others to 16 ft. lbs. (22 Nm)
- HO2S sensor connector

- Lower splash shield
8. Fill the engine with oil.
9. Connect the negative battery cable and enter the radio security code.
10. Run the engine and check for leaks.
11. Turn off engine and check the oil level. Top off the oil level if necessary.

2.0L Engine

K20A3 ENGINE

1. Before servicing the vehicle, refer to the precautions in the beginning of this section.
2. Raise and safely support the vehicle.
3. Drain the engine oil.
4. Attach a chain hoist to the engine.
5. Remove or disconnect the following:
- Lower ball joints
- Rear mount mounting bolts
- Front mount mounting bolts
- Automatic Transaxle Fluid (ATF) filter mounting bolt, if equipped with an automatic transaxle
6. Make alignment marks on the reference lines that align with the centers of the rear subframe mounting bolts.
- Front subframe
- Bolts/nuts securing the oil pan

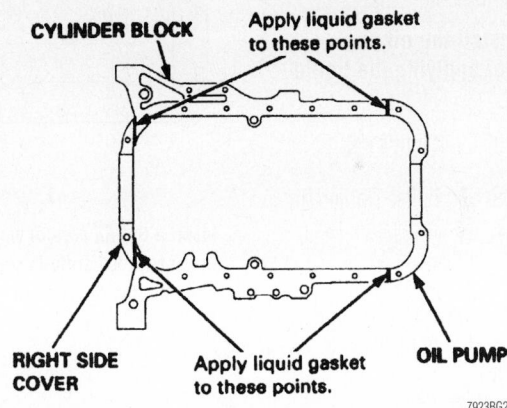

Apply liquid gasket to the oil pan as shown—1.8L (B18B1, B18C1 and B18C5) engines

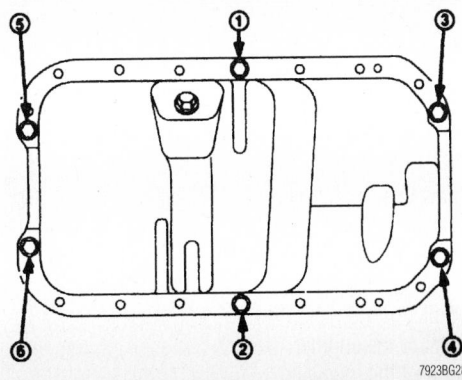

Oil pan bolt tightening sequence—1.8L (B18B1, B18C1 and B18C5) engines

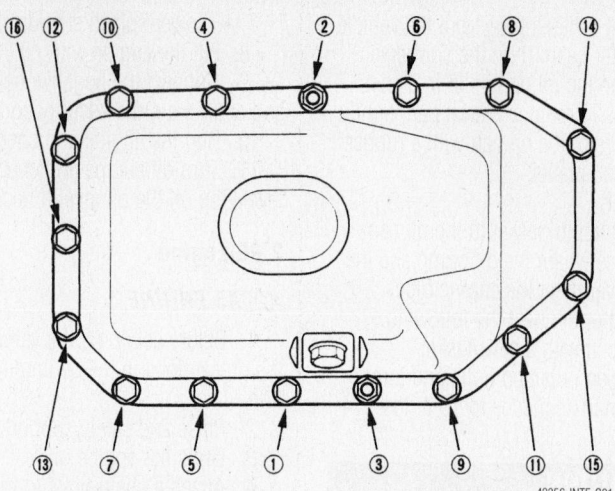

Oil pan tightening sequence—2.0L (K20A3) engine

- Oil pan by driving an oil pan seal cutter between the oil pan and cylinder block, then cutting the oil pan seal by striking the side of the cutter to slide the cutter along the oil pan

To install:

7. Thoroughly clean the mating surfaces, bolts and bolt holes. Apply liquid gasket part no. 08718–0009 evenly to the cylinder block mating surface of the oil pan and inner threads of the bolt holes.

➡**Make sure to install the oil pan within 5 minutes of applying the liquid gasket.**

8. Install or connect the following:
   - Oil pan
   - Oil pan mounting bolts. Tighten, in sequence, in 2 or 3 steps to 104 inch lbs. (12 Nm).
   - Subframe. Align the reference lines on the subframe with the bolt head center, tighten the bolts to 76 ft. lbs. (103 Nm).
   - ATF filter mounting bolt
   - Front mounting bolt
   - Rear mounting bolts
   - Lower ball joints
   - Negative battery cable

9. Wait 30 minutes, then fill the engine with oil. Start the engine and check for leaks.

K20A2 ENGINE

1. Before servicing the vehicle, refer to the precautions in the beginning of this section.
2. Raise and safely support the vehicle.
3. Drain the engine oil.
4. Attach a chain hoist to the engine.
5. Remove or disconnect the following:
   - Lower ball joints

- Rear mount mounting bolts
- Front mount mounting bolts

6. Make alignment marks on the reference lines that align with the centers of the rear subframe mounting bolts.

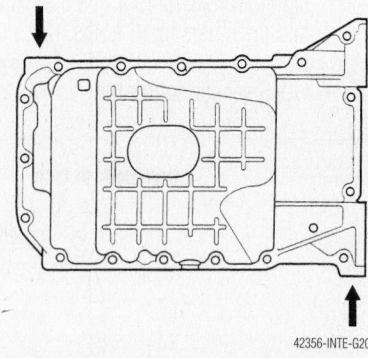

Insert a flat tip prytool where indicated by the arrows to carefully separate the oil pan from the block—2.0 (K20A2) engine

- Front subframe
- Clutch cover and 2 bolts securing the transaxle
- Bolts/nuts securing the oil pan
- Oil pan by inserting a flat tip prytool where shown in the illustration and separate the oil pan from the block

To install:

7. Thoroughly clean the mating surfaces, bolts and bolt holes. Apply liquid gasket part no. 08718–0009 evenly to the cylinder block mating surface of the oil pan and inner threads of the bolt holes.

➡**Make sure to install the oil pan within 5 minutes of applying the liquid gasket.**

8. Install or connect the following:
   - Oil pan
   - Oil pan mounting bolts. Tighten, in sequence, in 2 or 3 steps to 104 inch lbs. (12 Nm).
   - Clutch cover. Tighten the bolts to 104 inch lbs. (12 Nm).
   - 2 bolts securing the transaxle and tighten to 47 ft. lbs. (64 Nm)
   - Subframe. Align the reference lines on the subframe with the bolt head center, tighten the bolts to 76 ft. lbs. (103 Nm).
   - Front mounting bolt
   - Rear mounting bolts
   - Lower ball joints
   - Negative battery cable

9. Wait 30 minutes, then fill the engine with oil. Start the engine and check for leaks.

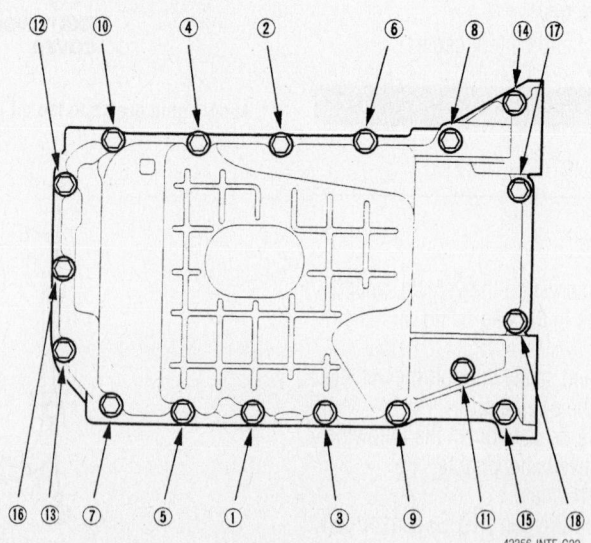

Oil pan tightening sequence—2.0L (K20A2) engine

2.4L Engine

1. Before servicing the vehicle, refer to the precautions in the beginning of this section.

2. Remove or disconnect the following:
- Negative battery cable
- Engine oil
- Front tire and wheel assemblies
- Splash shield
- Stabilizer links
- Right side damper fork
- Right side lower ball joint
- Right side halfshaft. Coat all machined surfaces with clean engine oil and secure a plastic bag over the end of the halfshaft.

3. From the engine compartment, remove the front mount stop and front mount bolt
- Rear mount stop and rear mount bolt
- Ground cable and upper bracket
- Bolt holding the side engine mount bracket and attach Engine Hanger Plate EQS00BRSX0, or equivalent.

4. Use a jack to lift the engine 1.2–2.4 in. (30–60mm).
- Stiffener
- Oil pan bolts and nuts

5. Hammer a seal cutter between the engine block and oil pan to break the seal.

6. Remove the oil pan.

To install:

7. Clean the oil pan flange and engine block mounting surface.

8. Install or connect the following:
- Sealant to the oil pan flange. Be sure to apply sealant toward the inside of the bolt holes.
- Oil pan on the engine. Tighten the bolts in sequence, in 2 or 3 steps, to 104 inch lbs. (12 Nm).
- Stiffener. Torque the retaining bolts to 33 ft. lbs. (45 Nm)

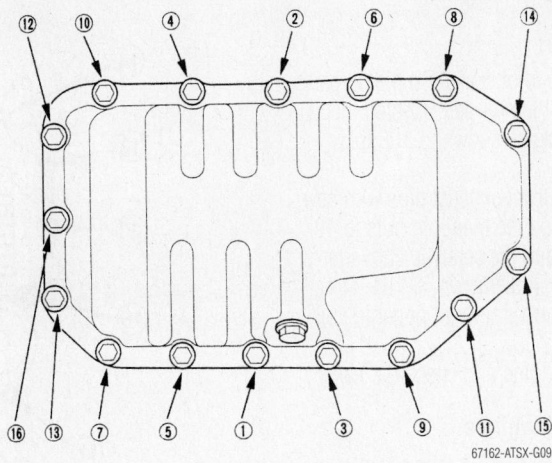

Oil pan bolt tightening sequence—2.4L engine

9. Remove the engine hanger tool and tighten the side engine mount bracket bolt to 33 ft. lbs. (45 Nm).
- Upper bracket and ground cable
- New set ring on the end of the driveshaft, then install the driveshaft. Make sure each ring "clicks" into plate in the differential.
- Right side lower ball joint
- Right side damper fork
- Stabilizer links
- Splash shield
- Tire and wheel assemblies
- Front mount bolt and front mount stop
- Rear mount bolt and rear mount stop

⁂ WARNING

Operating the engine without the proper amount and type of engine oil will result in severe engine damage.

10. After 30 minutes, fill the engine with the correct amount of oil.

11. Connect the negative battery cable.

12. Start the engine and check for leaks.

3.0L Engine

1. Before servicing the vehicle, refer to the precautions in the beginning of this section.

2. Raise and safely support the vehicle.

3. Drain the engine oil.

4. Remove or disconnect the following:
- Front exhaust pipe

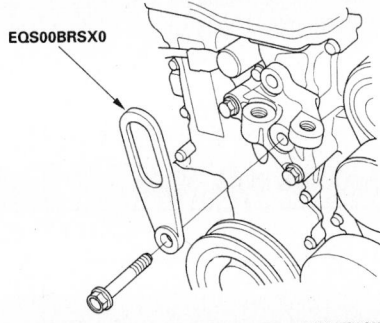

EQS00BRSX0

67162-ATSX-G08

Proper installation of Engine Hanger Plate EQS00BRSX0—2.4L engine

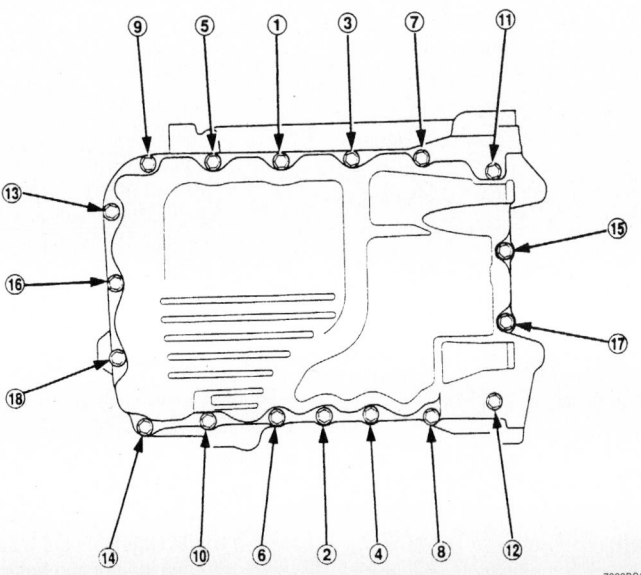

Oil pan bolt tightening sequence—3.0L engine

- Oil pan mounting bolts
- Oil pan

To install:

5. Apply a bead of sealant to the oil pan flange and install the oil pan. Tighten the bolts in the sequence shown to 108 inch lbs. (12 Nm).

6. Install the front exhaust pipe with new gaskets and torque the manifold nuts to 40 ft. lbs. (54 Nm) and the catalytic converter nuts to 25 ft. lbs. (33 Nm).

7. Add the correct amount of engine oil to the crankcase.

8. Start the engine and check for leaks.

3.2L and 3.5L Engines

1. Before servicing the vehicle, refer to the precautions in the beginning of this section.

2. Drain the engine oil.

3. Drain the differential oil.

4. Remove or disconnect the following:
- Negative battery cable
- Accessory drive belts
- Power steering pump without disconnecting the lines
- Exhaust manifold covers
- Front wheels
- Splash shield
- Strut forks
- Lower ball joints
- Halfshafts from the differential
- Intermediate shaft
- Vehicle speed/power steering speed sensor without disconnecting the hoses
- Lower plate from the rear beam
- A/C compressor without disconnecting the lines

5. Attach a chain hoist to the engine.
- Left engine mount bracket
- Right engine mount bracket
- Right engine mount
- 36mm sealing bolt on the transaxle. Ensure that the transaxle is in 1st gear (manual) or **P** (automatic).
- Extension shaft from the differential with an extension shaft puller
- Differential mounting bolts and the 26mm shim, then remove the differential
- Rear engine stiffener
- Flywheel cover or the torque converter covers
- Oil pan

➡**Do not lose the dowel pins from the oil pan**

To install:

6. Apply liquid gasket to the cylinder block. Be sure that the mating surfaces are

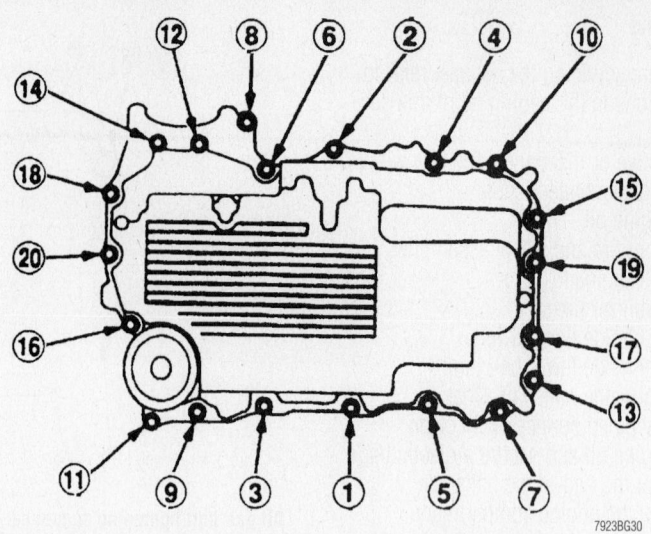

Be sure to tighten the oil pan bolts in the sequence shown—3.2L engines

clean and dry before installing the liquid gasket. Do not apply liquid gasket to the O–ring grooves.

7. Install or connect the following:
- Oil pan with new O–rings and torque the bolts to 16 ft. lbs. (22 Nm)

➡**Be sure the dowel pins are still in place.**

- Flywheel or torque converter cover and torque the bolts to 108 inch lbs. (12 Nm)
- Rear engine stiffener and tighten the bolt attaching the engine stiffener to the transaxle first, to 47 ft. lbs. (64 Nm), then tighten the bolts to the engine block to 16 ft. lbs. (22 Nm)
- Differential to the engine and torque the bolts to 47 ft. lbs. (64 Nm)

➡**Be sure to install the shim in the original location**

- Air conditioning compressor to the engine block and tighten the mounting bolts to 16 ft. lbs. (22 Nm)
- New set ring to the extension shaft—Using an extension shaft installer tool, install the shaft to the differential.
- Extension shaft with a new set ring

8. Fill the secondary gear with super high temperature grease. Applying sealer to the threads of the 36mm sealing bolt, then install the bolt and tighten it to 58 ft. lbs. (78 Nm).

- Right engine mount and torque the bolts to 28 ft. lbs. (38 Nm)
- Right engine mount bracket and torque the nut and bolts to 47 ft. lbs. (64 Nm)

- Left engine mount bracket and torque the bolts to 40 ft. lbs. (54 Nm)
- A/C compressor and torque the bolts to 16 ft. lbs. (22 Nm)
- Lower plate and torque the bolts to 28 ft. lbs. (38 Nm)
- Vehicle speed/power steering speed sensor and torque the bolts to 108 inch lbs. (12 Nm)
- Intermediate shaft and the halfshafts
- Lower ball joints and tighten the nuts to 40 ft. lbs. (54 Nm)
- Strut forks and torque the bolts to 51 ft. lbs. (69 Nm)
- Engine splash shield and tighten the bolts to 84 inch lbs. (9.5 Nm)
- Front wheels
- Exhaust manifold covers and torque the bolts to 16 ft. lbs. (22 Nm)
- Power steering pump and torque the bolt to 33 ft. lbs. (44 Nm) and the nut to 16 ft. lbs. (22 Nm)
- Accessory drive belts
- Negative battery cable

9. Fill the engine with oil.

10. Fill the differential with oil.

11. Run the engine and check for leaks.

12. Check the front wheel alignment.

Oil Pump

REMOVAL & INSTALLATION

Except RSX, NSX & TSX

1. Before servicing the vehicle, refer to the precautions in the beginning of this section.

2. Drain the engine oil.

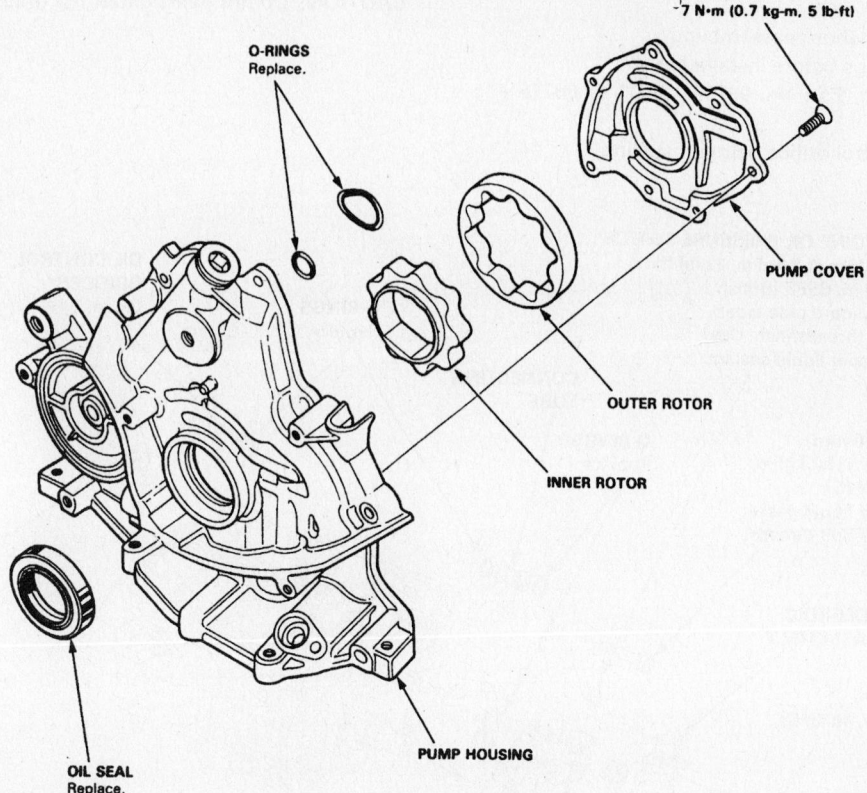

O-RINGS
Replace.

6 x 1.0 mm
7 N·m (0.7 kg-m, 5 lb-ft)

PUMP COVER

OUTER ROTOR

INNER ROTOR

PUMP HOUSING

OIL SEAL
Replace.

7923BG31

Exploded view of the oil pump—1.8L (B18B1) engines

3. Be sure the crankshaft is at Top Dead Center (TDC) on the No. 1 cylinder.

4. Remove or disconnect the following:
- Negative battery cable
- Cylinder head cover
- Timing belt cover
- Timing belt
- Crankshaft Position (CKP) sensor (if necessary)
- Crankshaft timing belt gear
- Oil pan
- Pickup screen
- Oil pump from the front of the engine

➥Any time the oil pump is removed, the front oil seal should be replaced.

To install:

5. Install a new oil seal in the oil pump.

6. Apply liquid gasket to the mounting surface of the oil pump.

7. Install the oil pump, using new O–rings. For all engines, except the 1.8L (B18B1, B18C1, B18C5) engines, tighten the 6mm bolts to 108 inch lbs. (12 Nm) and the 8mm bolts to 16 ft. lbs. (22 Nm). For 1.8L (B18B1, B18C1, B18C5) engines, tighten the 8 x 1.25mm bolts to 17 ft. lbs. (24 Nm), tighten the 6 x 1.0mm bolts to 96 inch lbs. (11 Nm).

✳✳ WARNING

The B18B1, B18C1 and B18C5 engines use different oil pumps, be sure that you have the correct oil pump. Match the crankshaft timing mark on the new oil pump with the timing mark on the old oil pump, because the timing marks are in different locations. If an oil pump is used with the timing mark in the wrong position the pistons may contact the valves.

8. Install or connect the following:
- Oil pump pickup screen
- Oil pan and tighten the bolts to 108 inch lbs. (12 Nm)
- Crankshaft timing belt gear
- Timing belt
- Crankshaft Position (CKP) sensor and torque the bolt to 108 inch lbs. (12 Nm)
- Timing belt cover
- Cylinder head cover
- Negative battery cable

9. Wait at least 30 minutes after completion of procedure before refilling the engine with oil. The waiting period is to allow the silicone sealant (liquid gasket) to cure.

10. Refill the engine with oil.

11. Start the engine and check the engine for leaks.

12. Turn off engine and check the oil level. Top off the oil level if necessary.

RSX

1. Before servicing the vehicle, refer to the precautions in the beginning of this section.

2. Drain the engine oil.

3. Remove or disconnect the following:
- Oil pan
- Oil pump chain tensioner
- Oil pump

To install:

4. Install or connect the following:
- Oil pump
- Oil pump chain tensioner
- Oil pan

5. Wait 30 minutes, then fill the engine with oil. Start the engine and check for leaks.

NSX

1. Before servicing the vehicle, refer to the precautions in the beginning of this section.

2. Drain the engine oil.

NOTE:
- Use new O-rings when reassembling.
- Apply oil to O-rings before installation.
- Use liquid gasket, Part No. 08718 – 0001 or 08718 – 0003.
- Clean the oil control orifice before installing.

CAUTION: Do not overtighten the drain bolt.

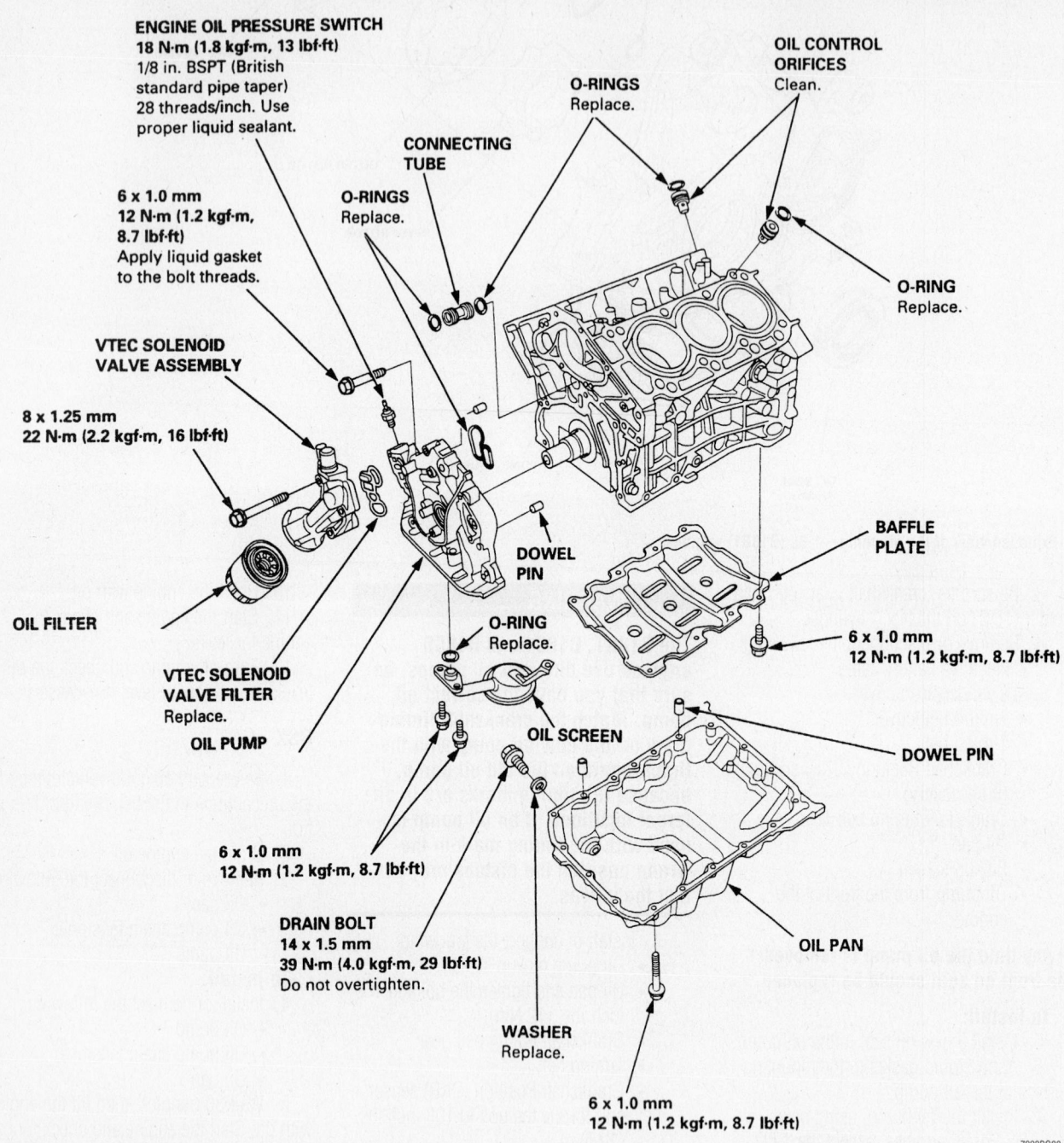

ENGINE OIL PRESSURE SWITCH
18 N·m (1.8 kgf·m, 13 lbf·ft)
1/8 in. BSPT (British standard pipe taper) 28 threads/inch. Use proper liquid sealant.

6 x 1.0 mm
12 N·m (1.2 kgf·m, 8.7 lbf·ft)
Apply liquid gasket to the bolt threads.

CONNECTING TUBE

O-RINGS
Replace.

O-RINGS
Replace.

OIL CONTROL ORIFICES
Clean.

O-RING
Replace.

VTEC SOLENOID VALVE ASSEMBLY

8 x 1.25 mm
22 N·m (2.2 kgf·m, 16 lbf·ft)

OIL FILTER

DOWEL PIN

O-RING
Replace.

VTEC SOLENOID VALVE FILTER
Replace.

OIL PUMP

OIL SCREEN

BAFFLE PLATE

6 x 1.0 mm
12 N·m (1.2 kgf·m, 8.7 lbf·ft)

DOWEL PIN

6 x 1.0 mm
12 N·m (1.2 kgf·m, 8.7 lbf·ft)

DRAIN BOLT
14 x 1.5 mm
39 N·m (4.0 kgf·m, 29 lbf·ft)
Do not overtighten.

OIL PAN

WASHER
Replace.

6 x 1.0 mm
12 N·m (1.2 kgf·m, 8.7 lbf·ft)

7923BG88

Lubrication system—3.0L engine

3. Remove or disconnect the following:
- Timing belt
- Oil level indicator tube
- Oil filter assembly
- Front exhaust manifold (if equipped with a manual transmission)
- Oil pan
- Oil screen
- Baffle plate
- Oil pass pipe and joint
- Oil pump
- Oil seal from the oil pump

To install:
4. Install a new oil seal in the oil pump
5. Apply liquid gasket to the mounting surface of the oil pump.
6. Install or connect the following:

NOTE:
- Use new O-rings when reassembling.
- Apply oil to O-rings before installation.
- Use liquid gasket, Part No. 08718 – 0001 or 08718 – 0003.
- Clean the oil control orifice before installing.
- Remove the balancer shaft

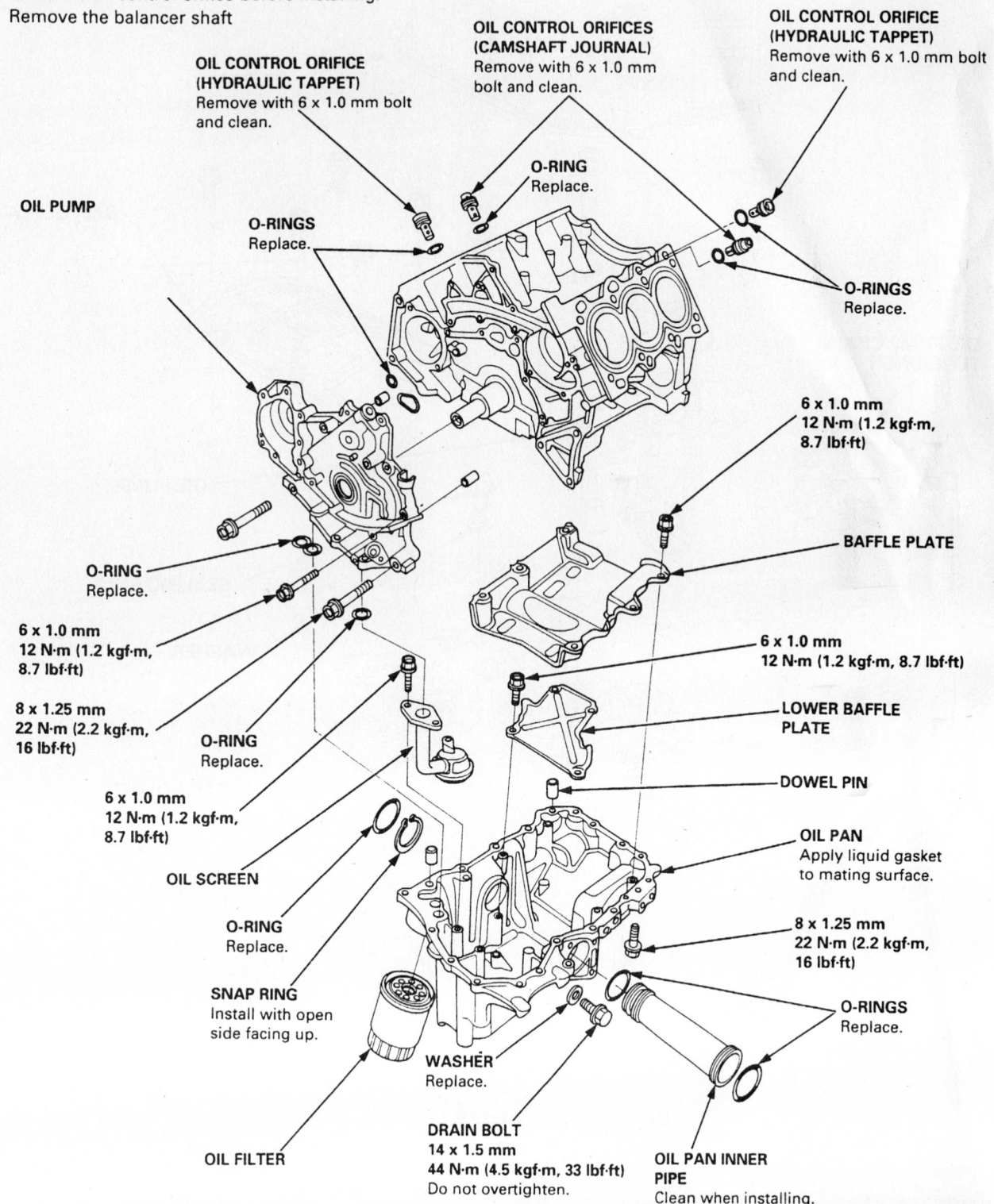

OIL CONTROL ORIFICE
(HYDRAULIC TAPPET)
Remove with 6 x 1.0 mm bolt
and clean.

OIL CONTROL ORIFICES
(CAMSHAFT JOURNAL)
Remove with 6 x 1.0 mm
bolt and clean.

OIL CONTROL ORIFICE
(HYDRAULIC TAPPET)
Remove with 6 x 1.0 mm bolt
and clean.

O-RING
Replace.

O-RINGS
Replace.

OIL PUMP

O-RINGS
Replace.

O-RINGS
Replace.

6 x 1.0 mm
12 N·m (1.2 kgf·m,
8.7 lbf·ft)

BAFFLE PLATE

O-RING
Replace.

6 x 1.0 mm
12 N·m (1.2 kgf·m,
8.7 lbf·ft)

6 x 1.0 mm
12 N·m (1.2 kgf·m, 8.7 lbf·ft)

8 x 1.25 mm
22 N·m (2.2 kgf·m,
16 lbf·ft)

LOWER BAFFLE
PLATE

O-RING
Replace.

DOWEL PIN

6 x 1.0 mm
12 N·m (1.2 kgf·m,
8.7 lbf·ft)

OIL SCREEN

OIL PAN
Apply liquid gasket
to mating surface.

O-RING
Replace.

8 x 1.25 mm
22 N·m (2.2 kgf·m,
16 lbf·ft)

SNAP RING
Install with open
side facing up.

O-RINGS
Replace.

WASHER
Replace.

OIL FILTER

DRAIN BOLT
14 x 1.5 mm
44 N·m (4.5 kgf·m, 33 lbf·ft)
Do not overtighten.

OIL PAN INNER
PIPE
Clean when installing.

7923BG89

Exploded view of the lubrication system—3.5L engine

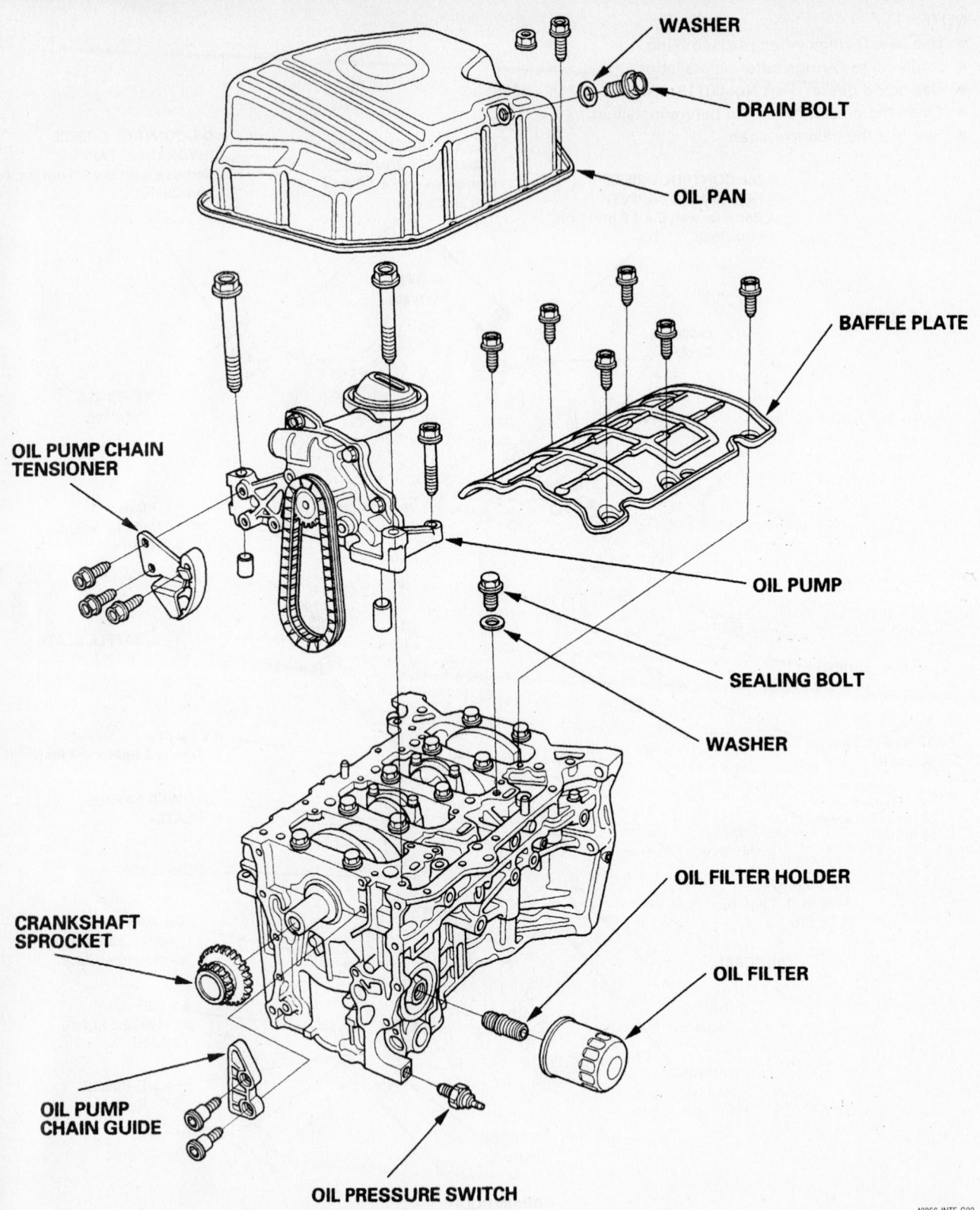

WASHER

DRAIN BOLT

OIL PAN

BAFFLE PLATE

OIL PUMP CHAIN
TENSIONER

OIL PUMP

SEALING BOLT

WASHER

OIL FILTER HOLDER

CRANKSHAFT
SPROCKET

OIL FILTER

OIL PUMP
CHAIN GUIDE

OIL PRESSURE SWITCH

42356-INTE-G23

Exploded view of the lubrication system—2.0L (K20A3) engine shown

- Oil pump and torque the bolts to 16 ft. lbs. (22 Nm)
- Oil pass pipe and joint with new O–rings and torque the bolts to 108 inch lbs. (12 Nm)
- Baffle plate and torque the bolts to 108 inch lbs. (12 Nm)
- Oil screen and torque the bolts to 108 inch lbs. (12 Nm)
- Oil pan and torque the bolts to 16 ft. lbs. (22 Nm)
- Front exhaust manifold (if removed) and torque the nuts to 23 ft. lbs. (31 Nm)
- Oil filter assembly
- Oil level indicator tube
- Timing belt

7. Refill the engine oil.
8. Start the engine and check for leaks.

TSX

1. Before servicing the vehicle, refer to the precautions in the beginning of this section.
2. Raise and safely support the vehicle.
3. Drain the engine oil.
4. Turn the crankshaft to position the No. 1 piston at Top Dead Center (TDC) on the compression stroke.
5. Remove or disconnect the following:
   - Oil pan
   - Oil pump chain tensioner, and discard

6. Secure the rear balancer shaft by inserting a 6mm pin into the maintenance hole in the lower balancer shaft holder and through the rear balancer shaft
   - Oil pump sprocket mounting bolt
   - Oil pump sprocket and oil pump

To install:

7. Make sure the No. 1 piston is still at TDC.

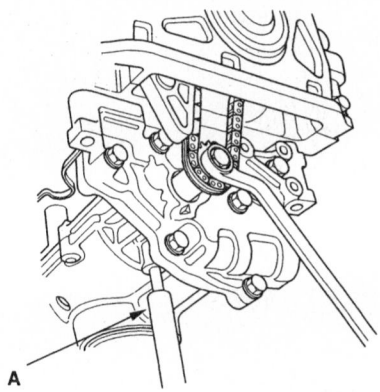

Insert a 6mm pin into the maintenance hole in the lower balancer shaft holder and through the rear balancer shaft

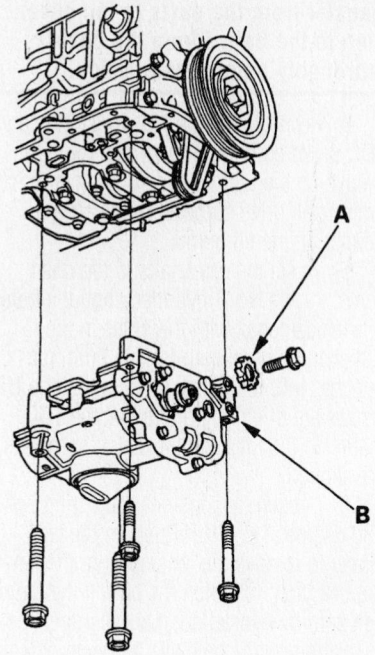

43256-ACCO-G21

Exploded view of the oil pump sprocket (A) and oil pump (B)—2.4L engine

8. Align the dowel pin on the rear balancer shaft with the mark on the oil pump
9. Secure the rear balancer shaft by inserting a 6mm pin into the maintenance hole in the lower balancer shaft holder and through the rear balancer shaft

10. Install or connect the following:
    - Engine oil to the threads of the oil pump sprocket mounting bolt
    - Oil pump and pump sprocket, loosely. Remove the 6mm pin.
    - Oil pump mounting bolts and tighten as shown in the illustration

11. Squeeze a new oil pump chain tensioner then install the set clip on it. The set clip is provided with the oil pump chain tensioner
    - New oil pump chain tensioner and tighten the bolts to 8.7 ft. lbs. (12 Nm). Remove the set clip from the tensioner.
    - Oil pan

12. Fill the crankcase with the proper amount of new engine oil.

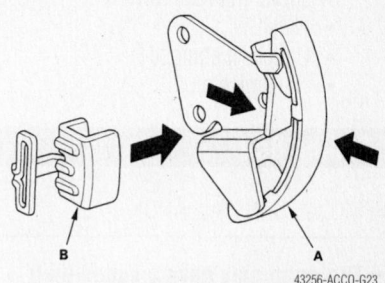

43256-ACCO-G23

Squeeze a new oil pump chain tensioner (A) then install the set clip (B) on it—2.4L engine

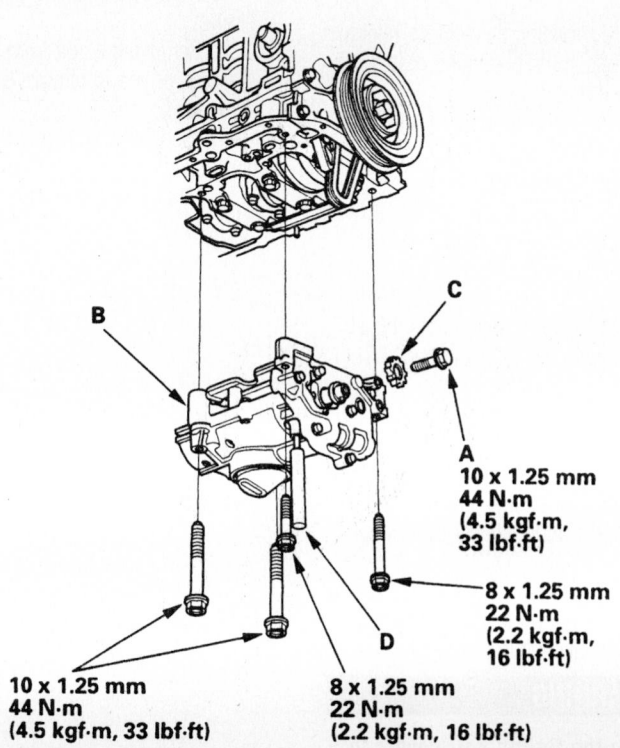

A
10 x 1.25 mm
44 N·m
(4.5 kgf·m, 33 lbf·ft)

B
8 x 1.25 mm
22 N·m
(2.2 kgf·m, 16 lbf·ft)

10 x 1.25 mm
44 N·m
(4.5 kgf·m, 33 lbf·ft)

8 x 1.25 mm
22 N·m
(2.2 kgf·m, 16 lbf·ft)

43256-ACCO-G22

Oil pump tightening specifications—2.4L engine

Rear Main Seal

REMOVAL & INSTALLATION

1. Before servicing the vehicle, refer to the precautions in the beginning of this section.
2. Remove or disconnect the following:
   - Transaxle
   - Clutch (if equipped)
   - Flexplate
   - Crankshaft seal by prying it out of the retainer

To install:

3. Install or connect the following:
   - Clean engine oil to the lip of the new seal
   - New seal into the retainer using an appropriate seal driver
   - Flywheel
   - Clutch (if equipped)
   - Transmission

Timing Belt

REMOVAL & INSTALLATION

➡**The radio may have a coded theft protection circuit. Obtain the code before disconnecting the battery, removing the radio fuse, or removing the radio.**

1.8L (B18B1 and B18C1) Engines

1. Before servicing the vehicle, refer to the precautions in the beginning of this section.
2. Turn the crankshaft pulley until cylinder No. 1 is set to Top Dead Center (TDC) on the compression stroke. The white crankshaft pulley mark should be aligned with the pointer on the lower timing belt cover.
3. Remove or disconnect the following:
   - All necessary components to gain access to the cylinder head and timing belt covers
   - Cylinder head and timing belt covers
   - Crankshaft pulley bolt and the crankshaft pulley. Bolt a pulley holder (holder attachment tool part No. 07MAB–PY3010A and holder handle tool part No. 07JAB–001020A) will be needed to keep the crankshaft from turning.

✳✳ WARNING

Do not use the timing belt covers to store small parts. Grease or oil can

transfer from the parts to the cover, then to the belt. Clean the covers thoroughly before installation.

4. Recheck that the No. 1 piston is at TDC on its compression stroke. Align the groove on the toothed side of the crankshaft timing belt drive sprocket to the arrow pointer on the oil pump.
5. To set the camshafts to top dead center for the No. 1 cylinder, align the hole in each camshaft with the holes in the No. 1 camshaft holders, then push 5.0mm pin punches into the holes. Be sure that the **UP** arrows are pointing up and that the TDC marks on the intake and exhaust sprockets are aligned.
6. Loosen the tensioner adjusting bolt 180 degrees (½ turn). Push on the tensioner to remove the tension from the timing belt, then retighten the bolt. If the timing belt is to be reinstalled, mark the direction of rotation on the belt with a crayon or white paint.
7. Remove the timing belt from the sprockets.

➡**Be sure the water pump pulley turns counterclockwise freely. Check for signs of seal leakage; a small amount of weeping from the bleed hole is normal.**

8. If necessary, remove the timing belt tensioner by performing the following:
   a. Remove the timing belt tensioner spring.
   b. Remove the bolt from the timing belt tensioner and remove the tensioner.

To install:

9. If the timing belt tensioner was removed, perform the following:
   a. Position the timing belt tensioner

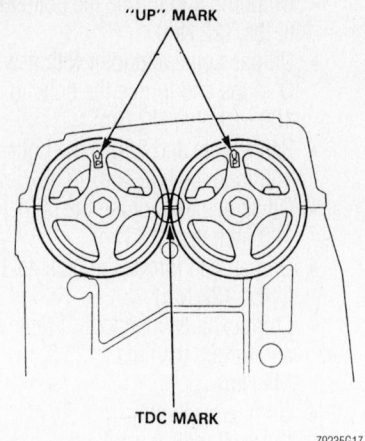

Camshaft timing belt sprocket alignment marks for TDC—1.8L (B18B1 and B18C1) engines

on the engine and install the attaching bolt loosely.
   b. Install the timing belt tensioner spring.
   c. Push the tensioner down, then snug the tensioner bolt to hold this position.

➡**Before reinstallation, check every component for cleanliness. All covers, pulleys, shields, etc. must be completely free of grease and oil.**

➡**Install the timing belt in the correct sequence. Also, if installing the old belt, be sure it is turning the same direction.**

10. Install the timing belt first to the crankshaft pulley, then to the adjuster, then to the water pump pulley, the exhaust camshaft and finally to the intake camshaft pulley.

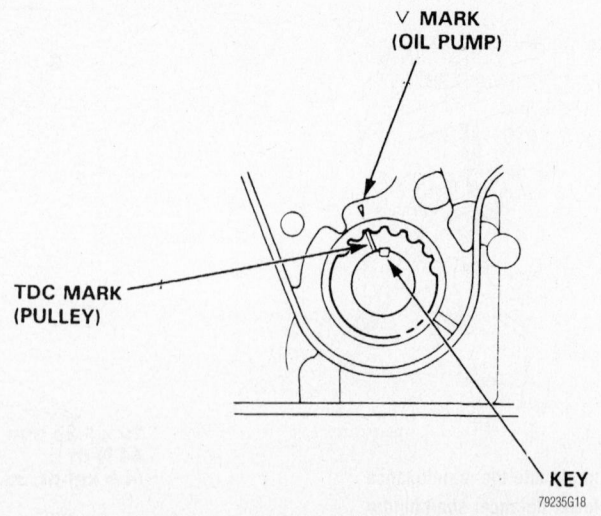

Crankshaft timing belt sprocket alignment marks for TDC—1.8L (B18B1 and B18C1) engines

11. Install the lower belt cover. Install the crankshaft pulley, tightening the bolt to 130 ft. lbs. (177 Nm). Lubricate the threads and the flange of the bolt with engine oil before installation.

12. Loosen the adjusting bolt, allowing the adjuster to tension the belt. Retighten the bolt to 40 ft. lbs. (54 Nm).

13. Remove the pin punches from the camshafts.

14. Rotate the crankshaft 4–6 turns counterclockwise. This allows the belt to equalize tension across all of the pulleys.

15. Once again, set the engine to TDC compression for cylinder No. 1. Check that all timing marks for the cam and crankshaft are properly aligned. If any mark is out of alignment, remove the timing belt and reinstall it.

16. Loosen the adjusting bolt 180 degrees (½ turn). Rotate the crankshaft counterclockwise until the camshaft pulleys have moved 3 teeth. Retighten the adjusting bolt to 40 ft. lbs. (54 Nm).

17. Check the torque of the crankshaft pulley bolt.

18. Install or connect the following:
- Other timing belt covers
- Rubber seal in the groove of the cylinder head cover. Be sure that the seal and groove are thoroughly clean first.
- Liquid gasket to the rubber seal at the eight corners of the recesses. Do not install the parts if 20 minutes or more have elapsed since applying the liquid gasket. Instead, reapply liquid gasket after removing the old residue.
- Cylinder head cover and all other applicable components. Tighten the cylinder head cover nuts in 2 steps to 88 inch lbs. (10 Nm).

3.2L (C32A6) Engines

➡**Under normal driving conditions, the timing belt and timing balancer belt are to be replaced at 105,000 miles (168,000 km).**

1. Before servicing the vehicle, refer to the precautions in the beginning of this section.

2. Be sure to acquire the anti–theft code for the radio and the frequencies for the radio's preset buttons.

3. Remove or disconnect the following:
- Negative battery cable
- Left wheel well splash shield
- Power steering pump belt by loosening the power steering pump's adjusting bolt, locknut and mounting

- Alternator belt by loosening the alternator adjusting bolt, locknut and mounting nut
- Alternator terminal and connector

4. To remove the engine-to-chassis center bracket, located at the front of the engine, perform the following procedure:

a. Using a floor jack, position a cushion between the jack and the oil pan.

b. Raise the engine slightly to take the weight off of the center bracket.

c. Remove the center bracket-to-engine bolts, the center bracket-to-chassis bolts and the center bracket.

5. Remove or disconnect the following:
- TCS upper and lower brackets
- TCS throttle sensor and actuator connectors and remove the TCS control valve assembly
- Oil pressure switch connector, the engine ground cable and the

engine wire harness cover, from the front of the engine
- Dipstick and the dipstick tube-to-engine bolt; then, pull the tube from its O–ring mount. Discard the O–ring.

6. Turn the engine to align the timing marks and set cylinder No. 1 to Top Dead Center (TDC) on the compression stroke. The white mark on the crankshaft pulley should align with the pointer on the timing belt cover. Remove the inspection caps on the upper timing belt covers to check the alignment of the timing marks. The pointers for the camshafts should align with the marks on the camshaft sprockets.

7. Using the Holder Handle and the 50mm Offset Holder Attachment tool 07MAB–PY3010A and a 19mm socket, secure the crankshaft pulley and remove the crankshaft pulley bolt.

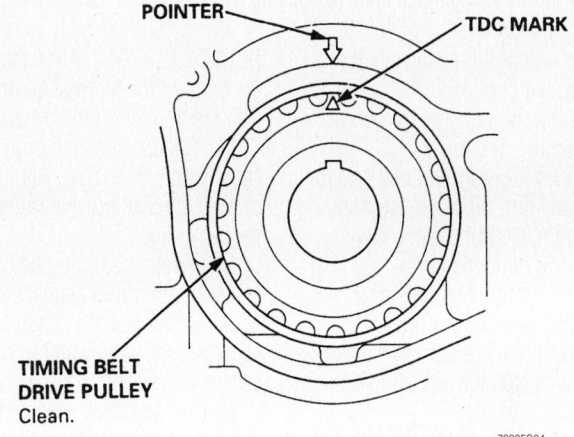

Crankshaft timing belt sprocket alignment mark locations—3.2TL 3.2L (C32A6) engines

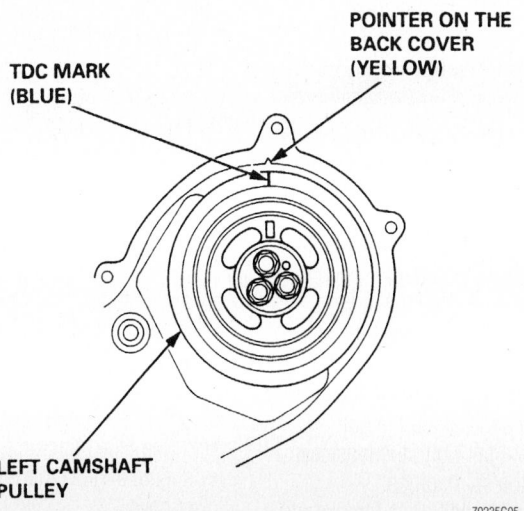

Left camshaft sprocket alignment mark positioning for TDC—3.2TL 3.2L (C32A6) engines

RIGHT:

Right camshaft sprocket alignment mark positioning for TDC—3.2TL 3.2L (C32A6) engines

- Washer and pull the crankshaft pulley from the engine
- Necessary components for access to the timing belt covers
- Upper and lower timing belt covers. Clean any dirt, oil or grease from the covers. Do not use the covers for storing removed items.

8. Loosen the timing belt tensioner adjusting bolt 180 degrees (½ turn). Push on the tensioner to remove tension from the timing belt, then tighten the adjusting bolt.

- Timing belt

9. If necessary, remove the timing belt tensioner by performing the following:

   a. Remove the spring from the tensioner.

   b. Remove the bolt mounting the tensioner, then remove the tensioner.

To install:

✷✷ CAUTION

Do not rotate the crankshaft pulley or camshaft pulleys with the timing belt removed. The pistons may hit the valves and cause damage.

10. If necessary, install the timing belt tensioner by performing the following:

   a. Install the tensioner and the attaching bolt.

   b. Move the tensioner its full deflection to the left and tighten the bolt.

   c. Install the spring to the tensioner.

11. Remove the spark plugs.

12. Set the timing belt drive (crankshaft) sprocket so that the No. 1 piston is at Top Dead Center (TDC). Align the TDC mark on the tooth of the timing belt drive sprocket with the pointer on the oil pump.

13. Set the camshaft pulleys so that the No. 1 piston is at TDC. Align the TDC mark on the camshaft pulleys to the pointers on the back covers.

14. Install the timing belt on the sprockets in the following sequence: drive sprocket (crankshaft), tensioner pulley, left camshaft sprocket, water pump pulley, right camshaft sprocket.

15. Loosen, then retighten the timing belt adjuster bolt to tension the timing belt.

16. Install the lower timing belt cover.

17. Install the crankshaft pulley and finger–tighten the bolt and washer. Using the Holder Handle and the 50mm Offset Holder Attachment tool 07MAB–PY3010A and a 19mm socket with a torque wrench, tighten the crankshaft pulley bolt to 181 ft. lbs. (245 Nm).

18. Rotate the crankshaft 5–6 turns clockwise so that the timing belt positions itself properly on the sprockets.

19. Set cylinder No. 1 to TDC by aligning the timing marks. If the timing marks do not align, remove the timing belt, then adjust the components and reinstall the timing belt.

20. Rotate the crankshaft clockwise enough to move the camshaft pulley nine teeth (the blue mark on the crankshaft pulley should align with the pointer on the lower cover).

21. Loosen the timing belt adjusting bolt 180 degrees (½ turn), then tighten the bolt to 31 ft. lbs. (42 Nm).

22. Install the upper timing belt covers, then install all applicable components. When installing the center bracket, tighten the bolts attaching the brackets to 40 ft. lbs. (54 Nm), then the mount through–bolt to 40 ft. lbs. (54 Nm).

23. To complete the installation, reverse the removal procedures. Adjust the tension of the drive belts.

3.0L (J30A1) and 3.2L (J32A1) Engines

1. Before servicing the vehicle, refer to the precautions in the beginning of this section.

2. Remove or disconnect the following:

- Negative battery cable
- Ignition coil cover
- Front tire/wheel assemblies
- Splash shield from under the vehicle
- Drive belts

3. To remove the engine-to-chassis side mount, located at the front of the engine, perform the following procedure:

   a. Using a floor jack, position a cushion between the jack and the oil pan.

   b. Raise the engine slightly to take the weight off of the side mount.

   c. Remove the side mount-to-engine bracket bolt, the side mount-to-chassis bolts and the side mount.

4. Remove the dipstick and the dipstick tube-to-engine bolt; then, pull the tube from its O–ring mount. Discard the O–ring.

5. Turn the engine to align the timing marks and set cylinder No. 1 to Top Dead Center (TDC) on the compression stroke. The white mark on the crankshaft pulley should align with the pointer on the timing belt cover. Remove the inspection caps on the upper timing belt covers to check the alignment of the timing marks. The pointers for the camshaft pulleys should align with the marks on rear upper cover mark.

6. Using the Holder Handle and the 50mm Offset Holder Attachment tool 07MAB–PY3010A and a 19mm socket, secure the crankshaft pulley and remove the crankshaft pulley bolt.

7. Remove or disconnect the following:

- Washer and pull the crankshaft pulley from the engine
- All necessary components for access to the timing belt covers
- Upper and lower timing belt covers. Clean any dirt, oil or grease from the covers. Do not use the covers for storing removed items.

- One of the battery clamp bolts and grind a 45° bevel on the threaded end. Screw the battery clamp bolt into hole provided at the base of the right cylinder head to hold the timing belt adjuster in it's current position; tighten the bolt by hand, Do NOT use a wrench.
- Engine mount bracket

8. At the base of the left cylinder head, loosen the idler pulley bolt about 5–6 turns; then, remove the timing belt.

To install:

✳✳ CAUTION

Do not rotate the crankshaft pulley or camshaft pulleys with the timing belt removed. The pistons may hit the valves and cause damage.

9. Remove the spark plugs.

10. Set the timing belt drive (crankshaft) sprocket so that the No. 1 piston is at Top Dead Center (TDC). Align the TDC mark on the tooth of the timing belt drive sprocket with the pointer on the oil pump.

11. Set the camshaft pulleys so that the No. 1 piston is at TDC. Align the TDC mark on the camshaft pulleys to the pointers on the back covers.

12. Remove the battery clamp bolt from the back cover. Remove the auto-tensioner.

13. Service the auto-tensioner by performing the following procedure:

a. Position the auto-tensioner in a soft jawed vise with the maintenance bolt facing upward. DO NOT grip the body of the auto-tensioner.

b. Remove the maintenance bolt.

c. Be careful not to spill the oil from inside the tensioner. If oil is spilled, replenish it; the total capacity is 0.22 fl. oz. (6.5 ml).

d. Using Stopper tool 14540–P8A–A01, position it on the auto-tensioner while turning the internal screw.

e. Insert a flat–blade screwdriver into the maintenance hole and turn it clockwise to compress the bottom.

✳✳ WARNING

Be careful not to damage the threads or the gasket contact surface with the screwdriver.

f. Using a new gasket, reinstall the maintenance bolt and torque it to 6 ft. lbs. (8 Nm).

g. Make sure that no oil is leaking around the maintenance bolt and install

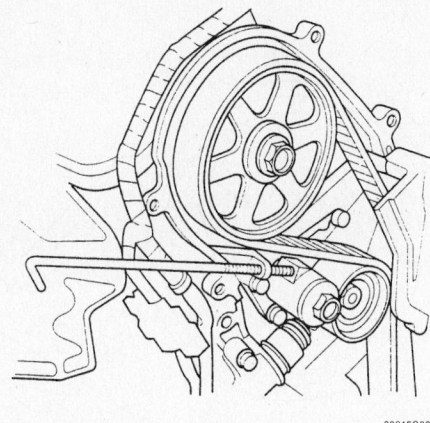

Using battery clamp bolt to hold timing belt adjuster in position—3.0L (J30A1) and 3.2L (J32A1) engines

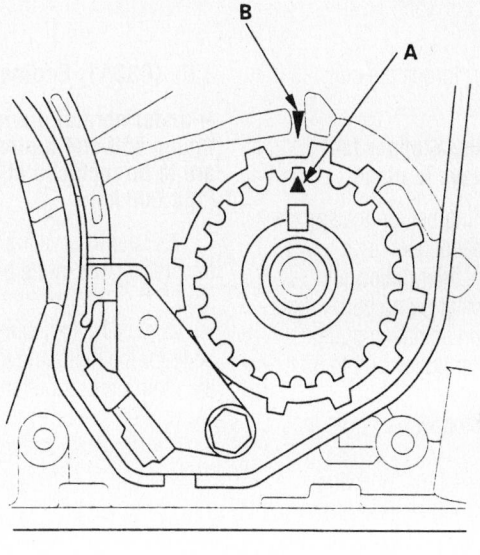

Crankshaft sprocket alignment mark positioning for TDC—3.0L (J30A1) and 3.2L (J32A1) engines

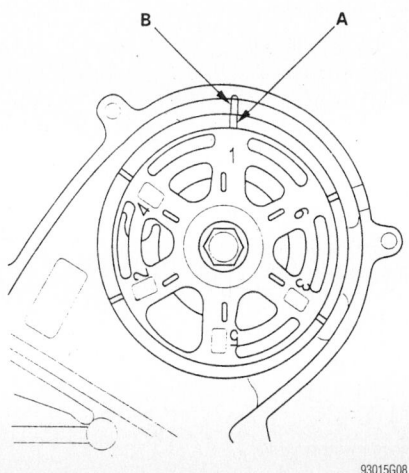

Left camshaft sprocket alignment mark positioning for TDC—3.0L (J30A1) and 3.2L (J32A1) engines

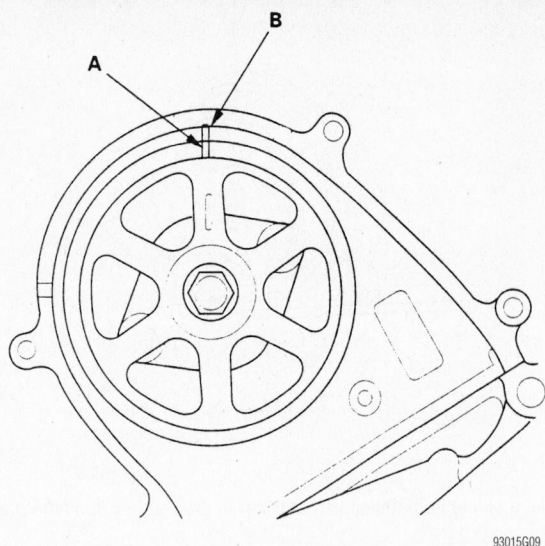

Right camshaft sprocket alignment mark positioning for TDC—3.0L (J30A1) and 3.2L (J32A1) engines

the auto-tensioner; torque the bolts 33 ft. lbs. (44 Nm).

➡**Make sure that the Stopper tool 14540–P8A–A01 stays in place.**

14. Install the timing belt on the sprockets in the following sequence: drive sprocket (crankshaft), idler pulley, left camshaft sprocket, water pump pulley, right camshaft sprocket and adjusting pulley.

15. Torque the idler pulley bolt to 33 ft. lbs. (44 Nm).

16. Remove the Stopper tool from the auto-tensioner.

17. Install or connect the following:
- Engine mount-to-engine and torque the No. 10 bolts to 33 ft. lbs. (44 Nm) and the No. 6 bolt to 104 inch lbs. (12 Nm)
- Lower and upper timing belt covers
- Crankshaft pulley and finger–tighten the bolt and washer. Using the Holder Handle and the 50mm Offset Holder Attachment tool 07MAB–PY3010A and a 19mm socket with a torque wrench, tighten the crankshaft pulley bolt to181 ft. lbs. (245 Nm).

18. Rotate the crankshaft 5–6 turns clockwise so that the timing belt positions itself properly on the sprockets.

19. Set cylinder No. 1 to TDC by aligning the timing marks. If the timing marks do not align, remove the timing belt, then adjust the components and reinstall the timing belt.

20. Install all applicable components.

21. To complete the installation, reverse the removal procedures. Adjust the tension of the drive belts.

3.5L (C35A1) Engines

➡**Under normal driving conditions, the timing belt and timing balancer belt are to be replaced at 105,000 miles (168,000 km).**

1. Before servicing the vehicle, refer to the precautions in the beginning of this section.

2. Be sure to acquire the anti–theft code for the radio and the frequencies for the radio's preset buttons.

3. Remove or disconnect the following:
- Negative battery cable

- Strut brace located at the top rear of the engine, if necessary

4. Rotate the crankshaft pulley so that the No. 1 piston is at Top Dead Center (TDC) of its compression stroke.
- Top engine cover-to-engine bolts and the cover
- Air intake duct and air cleaner housing
- Alternator and A/C compressor drive belts
- TCS control valve upper and lower brackets
- Power steering belt
- TCS throttle sensor connector, the TCS throttle actuator connector and the Throttle Position (TP) sensor
- TCS control valve assembly.
- Vehicle Speed Sensor (VSS) harness connector and remove the wire harness holder
- Breather and vacuum hoses
- Ignition Control Module (ICM) bracket from the right timing belt cover
- Idler pulley bracket from the left side of the crankshaft pulley
- Dipstick and the dipstick tube-to-engine bolt; then, pull the tube from its O–ring mount. Discard the O–ring.

5. Using the Holder Handle and the 50mm Offset Holder Attachment tool 07MAB–PY3010A and a 19mm socket, secure the crankshaft pulley and remove the crankshaft pulley bolt.

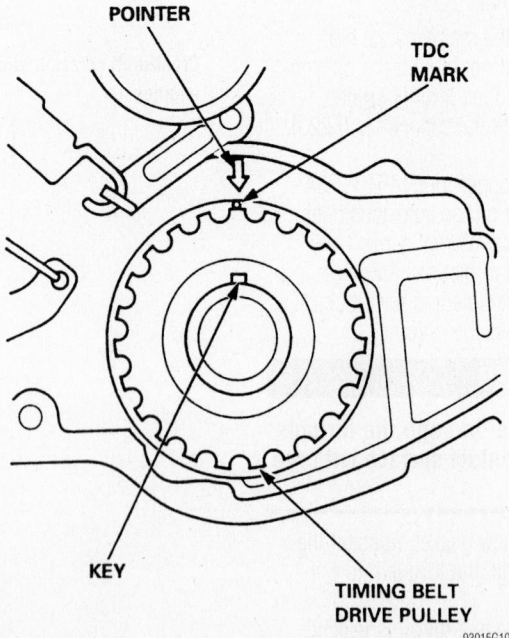

Crankshaft timing belt pulley alignment mark locations—3.5L (C35A1) engine

- Washer and pull the crankshaft pulley from the engine
- Upper and lower timing belt cover

6. Loosen the balancer belt adjusting nut 180° (½) turn. Push the tensioner to relieve the tension from the balancer belt; then, retighten the adjusting bolt. Remove the balancer belt.

7. Loosen the timing belt adjusting nut 180° (½) turn. Push the tensioner to relieve the tension from the timing belt; then, retighten the adjusting bolt. Remove the timing belt.

To install:

8. Remove the spark plugs.

9. Remove the balancer belt drive pulley and the timing belt guide plate from the crankshaft.

10. Clean the upper and lower timing belt covers.

11. Position the timing belt pulley so the No. 1 piston is at TDC of its compression stroke. Align the mark on the pulley (near keyway) with the pointer mark on the oil pump.

12. Adjust the camshaft pulley so that the No. 1 piston is the TDC of the compression stroke. Align the TDC marks on the pulley with the upper surface of the back cover; the arrow mark on the left camshaft pulley and the "1" on the right camshaft pulley should be facing the back cover pointer.

13. Install the timing belt in the following sequence:

   a. Crankshaft timing belt pulley sprocket.

   b. Adjusting pulley.

   c. Left camshaft pulley.

   d. Water pump pulley.

   e. Right camshaft pulley.

❊❊ WARNING

Make sure that the camshaft and crankshaft pulleys are at TDC.

➡ **For easier installation, turn the right camshaft pulley about ½ tooth from TDC.**

14. Loosen and retighten the timing belt adjusting bolt to tension the timing belt.

15. Install the lower cover and the crankshaft pulley.

16. Rotate the crankshaft pulley about 5–6 turns clockwise to position the timing belt on the pulleys.

17. To adjust the timing belt tension, perform the following procedure:

   a. Make sure that the No. 1 piston is at TDC of its compression stroke.

   b. Rotate the crankshaft clockwise ten

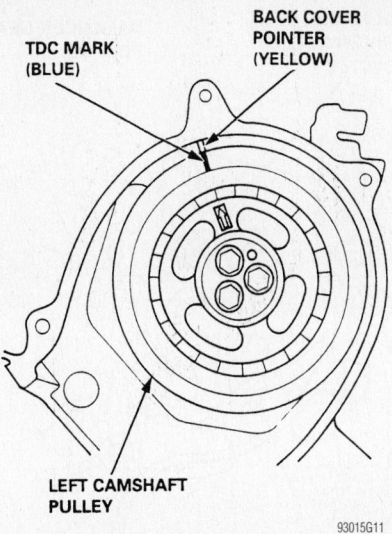

Left camshaft timing belt pulley alignment mark locations—3.5L (C35A1) engine

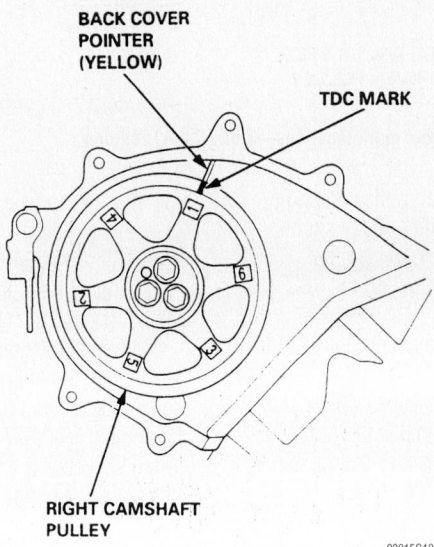

Right camshaft timing belt pulley alignment mark locations—3.5L (C35A1) engine

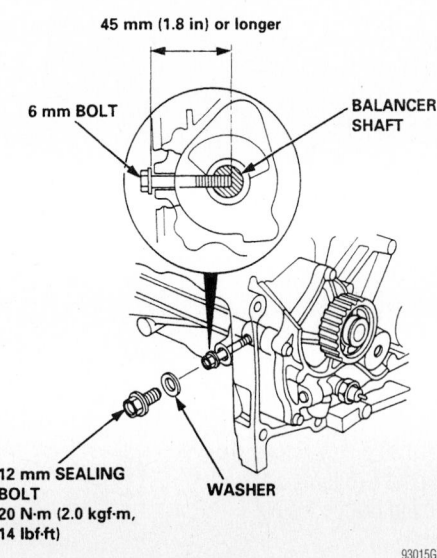

Securing the balancer shaft—3.5L (C35A1) engine

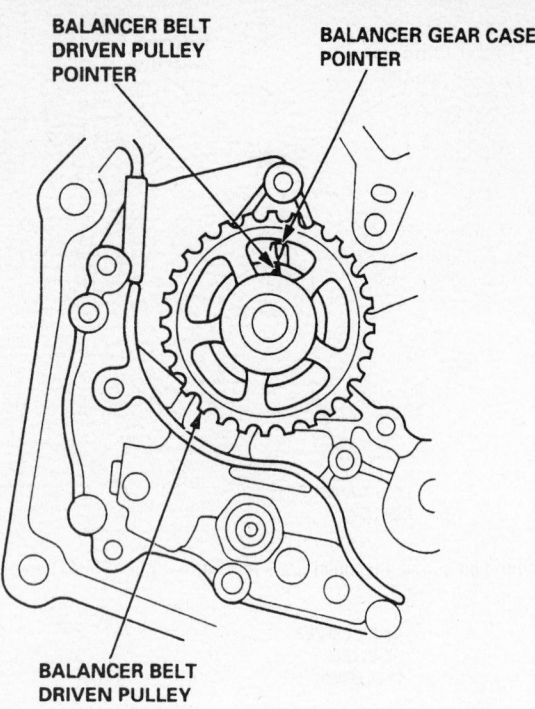

BALANCER BELT
DRIVEN PULLEY
POINTER

BALANCER GEAR CASE
POINTER

BALANCER BELT
DRIVEN PULLEY

93015G14

Balancer shaft alignment mark locations—3.5L (C35A1) engine

teeth on the camshaft pulley; the blue mark on the crankshaft pulley should align with the lower cover pointer.

c. Loosen the adjusting nut 180° (½ turn).

d. Tighten the adjusting nut to 31 ft. lbs. (42 Nm).

18. Remove the crankshaft pulley and the lower cover; then, install the timing belt guide plate and the balancer belt drive pulley.

19. Align the balancer shaft pulley by performing the following procedure:

a. Using a 6 x 45mm bolt, insert it into the maintenance hole and the balancer shaft.

b. Align the pointer on the balancer belt pulley with the pointer on the balancer gear case.

20. Adjust the timing belt drive pulley so that the No. 1 piston is at TDC of the compression stroke.

21. Install the balancer belt drive pulley and the balancer belt.

22. Loosen and retighten the balancer adjuster bolt to place tension on the balancer belt.

23. Remove the 6mm bolt and install the sealing bolt in the maintenance hole using a new washer.

24. Install the crankshaft pulley. Rotate the crankshaft pulley about 5–6 turns clockwise to position the timing belt on the pulleys.

25. Loosen the balancer belt adjuster

bolt 180° (½ turn) and retighten the bolt to 33 ft. lbs. (44 Nm).

26. Remove the crankshaft pulley.

27. Install the upper and lower timing belt covers and the crankshaft pulley.

28. Install the crankshaft pulley and finger–tighten the bolt and washer. Using the Holder Handle and the 50mm Offset Holder Attachment tool 07MAB–PY3010A and a 19mm socket with a torque wrench, tighten the crankshaft pulley bolt to 181 ft. lbs. (245 Nm).

29. Make sure the crankshaft and camshaft pulleys are at TDC.

➡If the camshaft or crankshaft pulley is not at TDC, remove the timing belt and re–perform the adjustment procedure.

30. To complete the installation, reverse the removal procedures. Adjust the tension of the drive belts.

Timing Chain

REMOVAL & INSTALLATION

2.0L Engines

1. Before servicing the vehicle, refer to the precautions in the beginning of this section.

2. Rotate the crankshaft pulley so that the No. 1 piston is at Top Dead Center (TDC) of its compression stroke.

3. Remove or disconnect the following:
- Front wheels and tires
- Splash shield
- Drive belt
- Cylinder head cover

4. Hold the crankshaft pulley with holder handle 07JAB–001020A and holder attachment 07NAB–001040A, then remove the pulley bolt with a 19mm socket and breaker bar.
- Oil cooler hose joint pipe from the water pump, K20A2 engines
- Crankshaft Position (CKP) sensor connector and Variable Valve Timing Control (VTC) oil control solenoid valve connector
- VTC oil control solenoid valve

5. Support the engine under the oil pan with a jack and block of wood.
- Ground cable and upper bracket

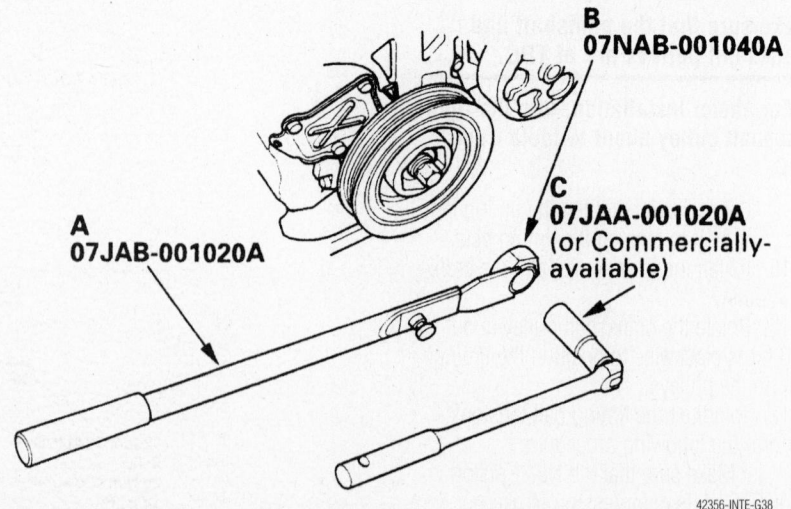

B
07NAB-001040A

A
07JAB-001020A

C
07JAA-001020A
(or Commercially-
available)

42356-INTE-G38

Hold the crankshaft pulley with holder handle (A) and holder attachment (B)—2.0L engines

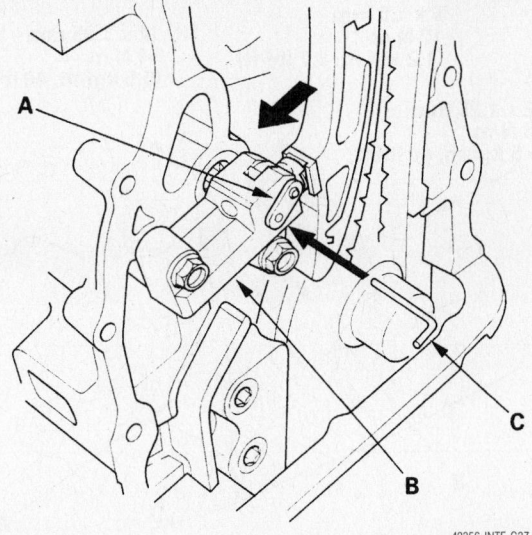

Align the holes on the lock (A) and auto-tensioner (B), then insert a 0.06 in. (1.5mm) pin (C) into the hole–2.0L engine

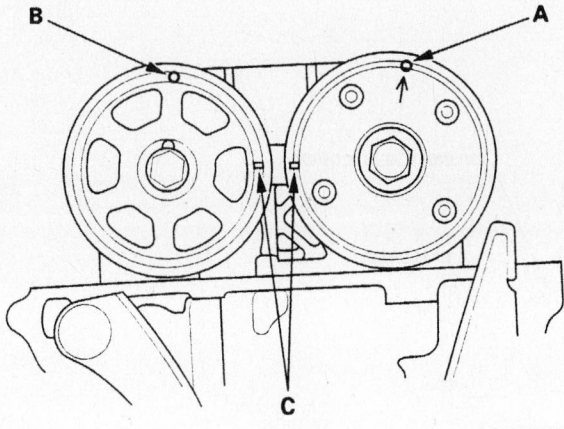

The punch mark (arrow) (A) on the VTC actuator and punch mark (B) on the exhaust camshaft sprocket should be at the top. Align the TDC marks (C)on the VTC actuator and exhaust camshaft sprocket

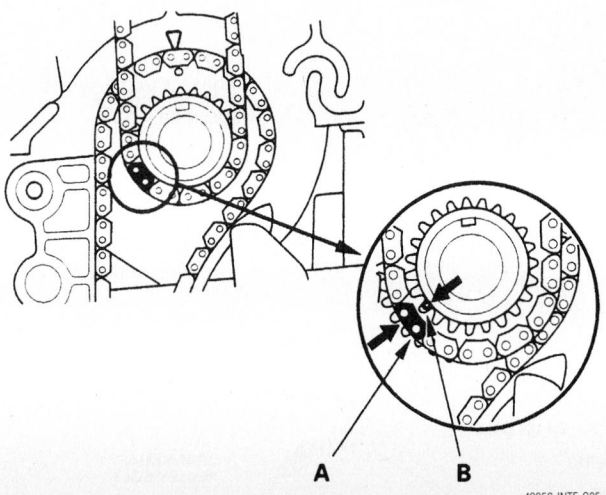

Install the timing chain on the crankshaft sprocket with the colored piece aligned with the punch mark on the crankshaft sprocket—2.0L engines

- Side engine mounting bracket
- Retaining bolts and timing chain case

6. Loosely install the crankshaft pulley.

7. Turn the crankshaft counterclockwise to compress the auto-tensioner.

8. Align the holes on the lock and auto-tensioner, then insert a 0.06 in. (1.5mm) pin into the hole. Turn the crankshaft clockwise to hold the pin.

9. Auto-tensioner
- Timing chain guide B
- Timing chain guide A and tensioner arm
- Timing chain

➡**Do not let the timing chain near any magnetic fields.**

To install:

10. Set the crankshaft to TDC. Align the TDC mark on the crankshaft sprocket with the pointer on the cylinder block.

11. Set the camshafts to TDC. The punch mark (arrow) on the VTC actuator and punch mark on the exhaust camshaft sprocket should be at the top. Align the TDC marks on the VTC actuator and exhaust camshaft sprocket.

12. Install or connect the following:
- Timing chain on the crankshaft sprocket with the colored piece aligned with the punch mark on the crankshaft sprocket
- Timing chain on the VTC actuator and exhaust camshaft sprocket with the punch marks aligned with the 2 colored pieces
- Timing chain guide A and tighten the bolts to 104 inch lbs. (12 Nm)
- Tensioner arm and tighten the bolts to 16 ft. lbs. (22 Nm)
- Auto-tensioner. Tighten the bolts to 104 inch lbs. (12 Nm).
- Timing chain guide B and tighten the bolts to 16 ft. lbs. (22 Nm)

13. Remove the pin from the auto-tensioner.

14. Check the timing chain case oil seal for damage and replace if necessary.

15. Remove the oil liquid gasket material from the chain case mating surfaces, bolts and bolt holes.

16. Apply liquid gasket part no. 08718–0009, evenly to the cylinder block mating surface chain case and the inner threads of the holes. Apply liquid gasket to the cylinder block upper surface contact areas on the chain case and to the oil pan mating surface of the chain case and inner threads of the holes.

17. Install or connect the following:
- New O–ring on the timing chain

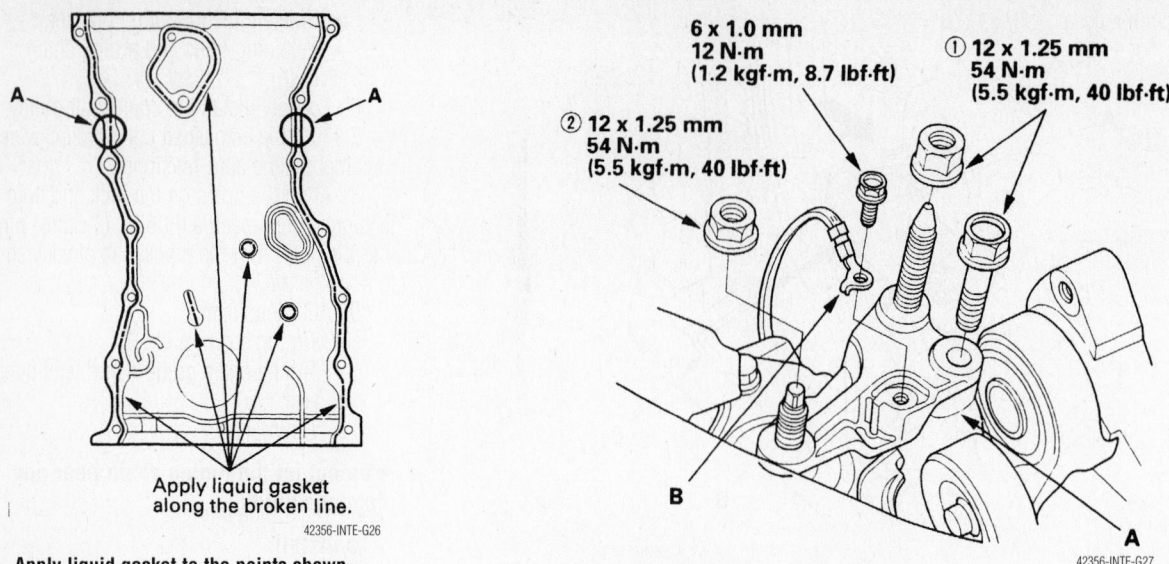

A ← → A

Apply liquid gasket along the broken line.

42356-INTE-G26

Apply liquid gasket to the points shown—2.0L engines

6 x 1.0 mm
12 N·m
(1.2 kgf·m, 8.7 lbf·ft)

① 12 x 1.25 mm
54 N·m
(5.5 kgf·m, 40 lbf·ft)

② 12 x 1.25 mm
54 N·m
(5.5 kgf·m, 40 lbf·ft)

B

A

42356-INTE-G27

Upper bracket tightening sequence and specifications—2.0L engines

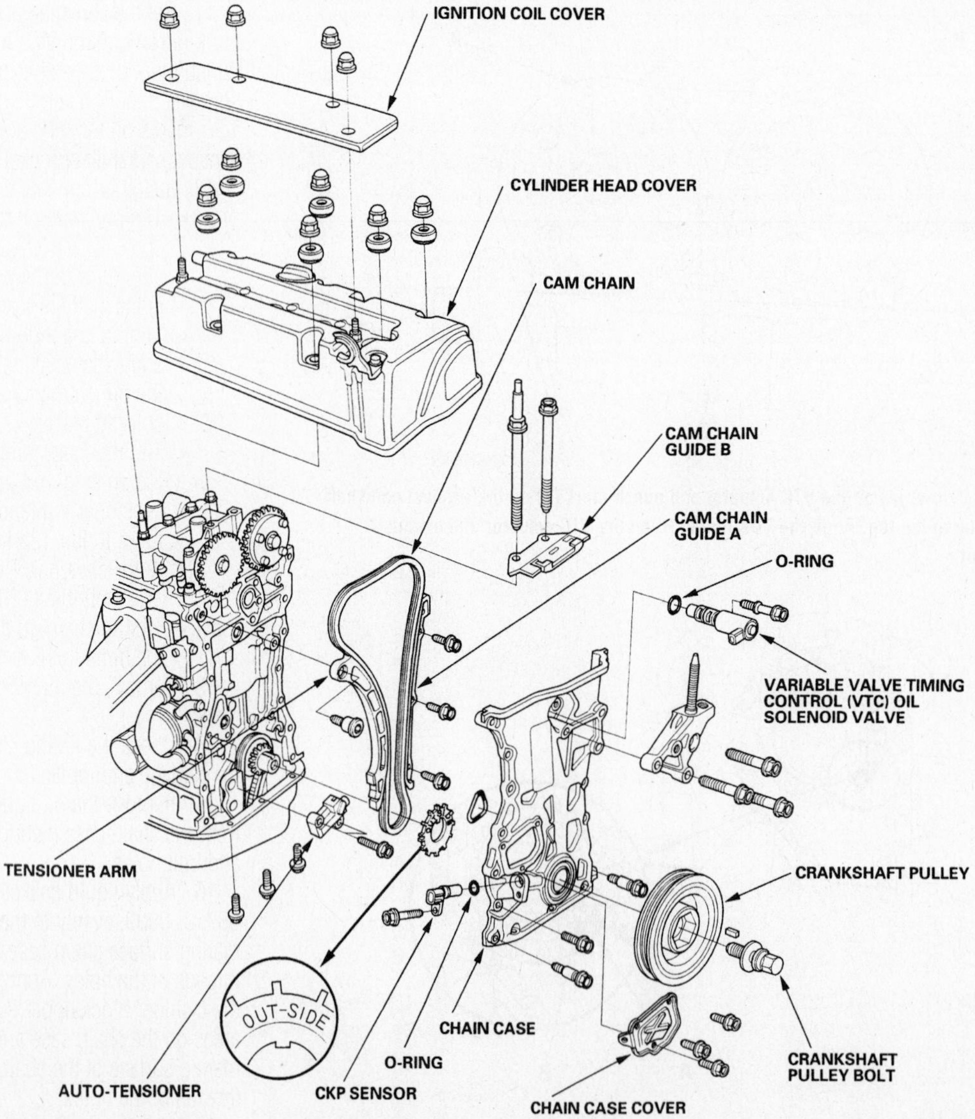

IGNITION COIL COVER

CYLINDER HEAD COVER

CAM CHAIN

CAM CHAIN GUIDE B

CAM CHAIN GUIDE A

O-RING

VARIABLE VALVE TIMING CONTROL (VTC) OIL SOLENOID VALVE

TENSIONER ARM

CRANKSHAFT PULLEY

OUT-SIDE

AUTO-TENSIONER

CKP SENSOR

O-RING

CHAIN CASE

CHAIN CASE COVER

CRANKSHAFT PULLEY BOLT

42356-INTE-G28

Exploded view of the timing (cam) chain and related components—2.0L engines

case. Set the edge of the chain case to the edge of the oil pan, then install the case on the cylinder block. Tighten the case bolts to 104 inch lbs. (12 Nm).

✳✳ WARNING

When installing the chain case, do not slide the bottom surface on the oil pan mounting surface.

- Side engine mount bracket and tighten to 33 ft. lbs. (44 m)
- Upper bracket, then tighten the bolts/nuts in sequence as shown in the illustration
- Ground cable
- VTC coil control solenoid valve
- CKP sensor connector and VTC oil control solenoid valve connector
- Oil cooler hose using a new O–ring

18. Clean the crankshaft pulley and pulley bolt, then apply lubrication to the pulley bolt and washer.

- Crankshaft pulley and hold the pulley with the holder handle and holder attachment. Using a 19mm socket and torque wrench, tighten the pulley bolt to 181 ft. lbs. (245 Nm).
- Cylinder head cover
- Drive belt
- Splash shield
- Front wheels and tires

2.4L Engine

1. Before servicing the vehicle, refer to the precautions in the beginning of this section.
2. Set the engine to Top Dead Center (TDC).
3. Drain the cooling system.
4. Relieve the fuel system pressure.
5. Turn the crankshaft pulley so its Top Dead Center (TDC) mark lines up with the pointer.
6. Remove or disconnect the following:

- Negative battery cable
- Front tires and wheels
- Splash shield
- Drive belt
- Cylinder head cover

➡**Make sure the No. 1 piston TDC marks on the VTC actuator and exhaust camshaft are aligned.**

- Crankshaft pulley
- Crankshaft Position (CKP) sensor connector
- Variable Valve Timing Control (VTC) oil control solenoid valve connector
- VTC oil control solenoid valve

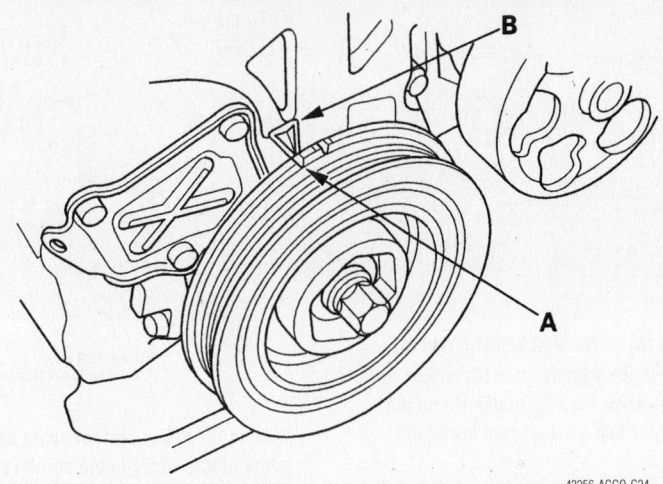

Turn the crankshaft pulley so the TDC mark (A) is aligned with the pointer (B)—2.4L engine

7. Support the engine with a suitable jack with a wooden block under the oil pan.

- Ground cable and upper bracket
- Side engine mount bracket
- Chain cover/case

8. Loosely install the crankshaft pulley. Turn the crankshaft counterclockwise to compress the auto-tensioner.

9. Align the holes on the lock (A) and the auto-tensioner (B), then place a 1.5mm pin into the holes. Turn the crankshaft clockwise to secure the pin.

10. Remove or disconnect the following:

- Auto-tensioner
- Timing chain guide B (top guide)
- Timing chain guide A and tensioner arm
- Timing chain

✳✳ WARNING

Do not place the timing chain near any magnetic fields.

To install:

11. Set the crankshaft to TDC. Align the TDC mark (A) on the crankshaft sprocket with the pointer (B) on the cylinder block.

12. Set the camshafts to TDC. The punch

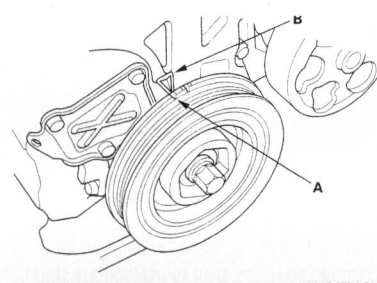

Turn the crankshaft pulley so its Top Dead Center (TDC) mark (A) line up with the pointer (B)—2.4L engine

mark (A) on the VTC actuator and the punch mark (B) on the exhaust camshaft (C) should be at the top. Align the TDC marks (C) on the VTC actuator and exhaust camshaft sprockets.

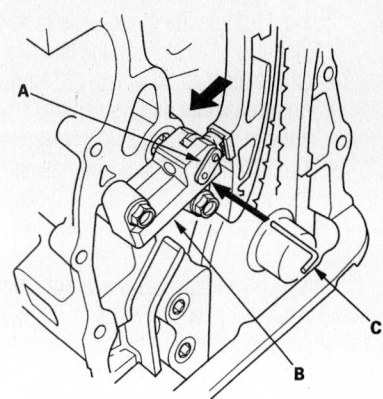

Align the holes on the lock (A) and the auto-tensioner (B), then place a 1.5mm pin into the holes. Turn the crankshaft clockwise to secure the pin

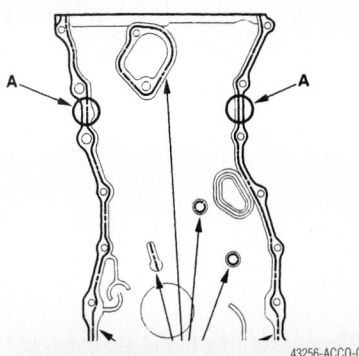

Set the crankshaft to TDC. Align the TDC mark (A) on the crankshaft sprocket with the pointer (B) on the cylinder block

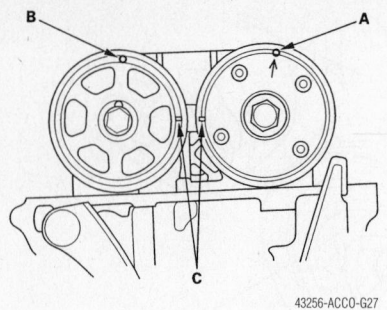

43256-ACCO-G27

The mark (A) on the VTC actuator and the mark (B) on the exhaust cam (C) should be at the top. Align the TDC marks (C) on the VTC actuator and exhaust cam sprockets

13. Install or connect the following:
- Timing chain the crankshaft sprocket with the colored link of the chain aligned with the mark on the crank sprocket
- Timing chain on the VTC actuator and exhaust camshaft sprocket with the punch marks aligned with the center of the 2 colored links
- Timing chain guide A and tensioner arm. Tighten the guide bolts to 104 inch lbs. (12 Nm) and the tensioner arm retainer to 16 ft. lbs. (22 Nm).
- Auto-tensioner and tighten the bolts to 104 inch lbs. (12 Nm)
- Timing chain guide B and tighten the retainers to 16 ft. lbs. (22 Nm)

14. Remove the pin from the auto-tensioner.
15. Inspect the chain cover seal for damage and replace if necessary. Clean and dry the chain cover mating surfaces.
16. Install or connect the following:
- Liquid gasket, P/N 08718-0009 evenly to the cylinder block mating surface of the timing chain cover and the inner threads of the holes
- Liquid gasket to the cylinder block upper surface contact areas on the chain cover and the oil pan mating surface of the chain cover in the inner threads of the holes

➡Make sure to install the components within 5 minutes of applying the sealer.

- New O-ring the timing chain cover. Set the edge of the cover to the edge of the oil pan, then install the cover on the engine block. Tighten the retainers to 104 inch lbs. (12 Nm).

➡When installing the chain case, do not slide the bottom surface on the oil pan mounting surface.

- Side engine mounting bracket and

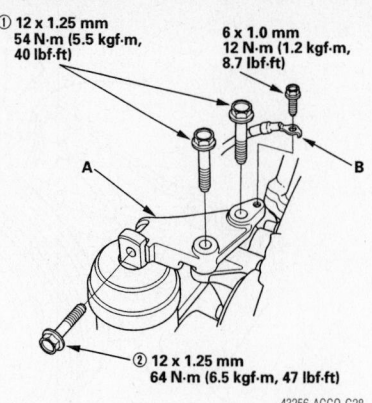

① 12 x 1.25 mm
54 N·m (5.5 kgf·m, 40 lbf·ft)

6 x 1.0 mm
12 N·m (1.2 kgf·m, 8.7 lbf·ft)

② 12 x 1.25 mm
64 N·m (6.5 kgf·m, 47 lbf·ft)

43256-ACCO-G28

Tighten the upper bracket upper bolt/nuts in the proper order to the correct specification—2.4L engine

tighten the retainers to 33 ft. lbs. (44 Nm)
- Upper bracket, then tighten the bolts/nuts as shown in the illustration
- Ground cable
- VTC oil control solenoid valve
- CKP sensor and VTC oil control solenoid valve connectors
- Crankshaft pulley
- Cylinder head cover
- Drive belt
- Splash shield
- Wheels and tires

17. Fill the engine cooling system and connect the negative battery cable.

Piston and Ring

POSITIONING

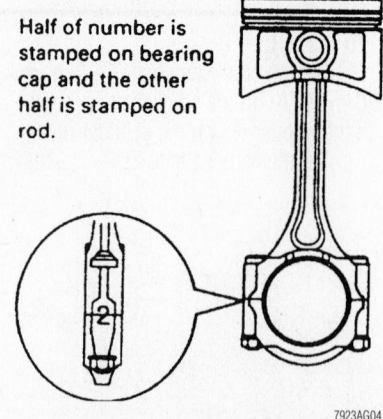

Half of number is stamped on bearing cap and the other half is stamped on rod.

7923AG04

Before removing the caps from the connecting rods, be sure to matchmark them as shown

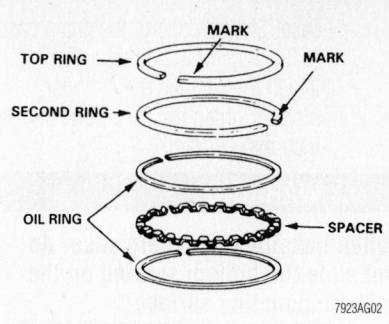

7923AG02

Piston ring positioning

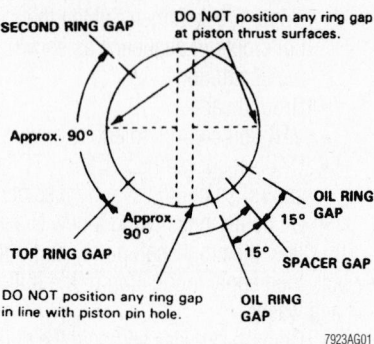

7923AG01

Piston ring end-gap spacing—except 2.0L engines

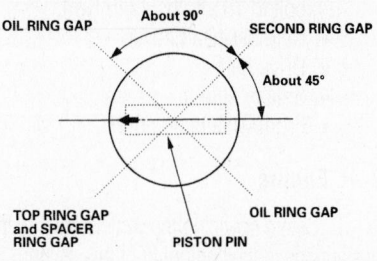

42356-INTE-G30

Piston ring end-gap spacing—2.0L engines

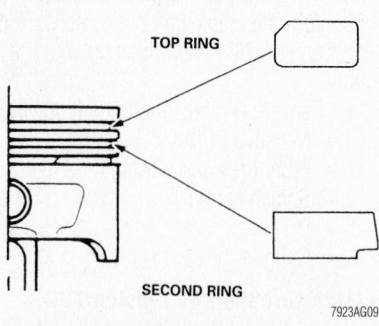

7923AG09

Compression ring locations—1.8L engines

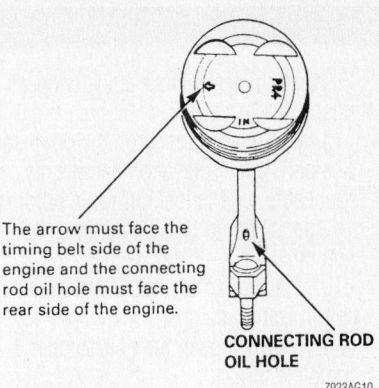

The arrow must face the timing belt side of the engine and the connecting rod oil hole must face the rear side of the engine.

CONNECTING ROD OIL HOLE

7923AG10

Piston/connecting rod assembly-to-engine orientation—1.8L (B18B1) engine

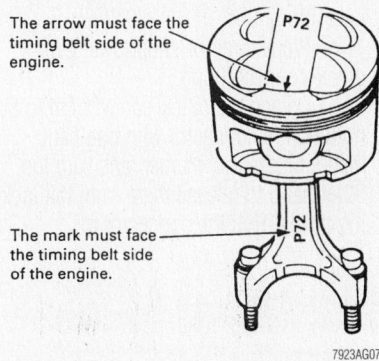

The arrow must face the timing belt side of the engine.

The mark must face the timing belt side of the engine.

7923AG07

Piston/connecting rod assembly-to-engine orientation—1.8L (B18C1) engine

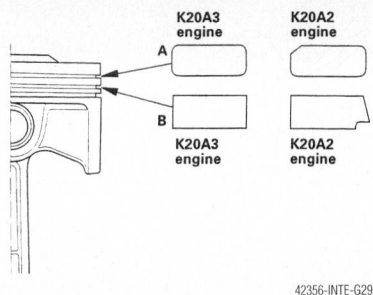

K20A3 engine   K20A2 engine

A
B

K20A3 engine   K20A2 engine

42356-INTE-G29

Compression ring locations—2.0L engines

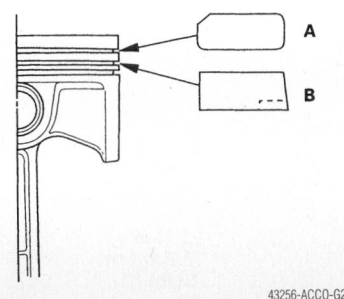

A

B

43256-ACCO-G29

Compression ring locations—2.4L (K24A2) engines

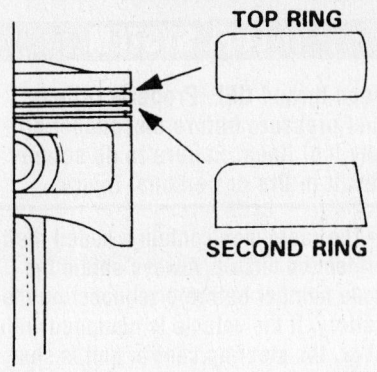

TOP RING

SECOND RING

7923AG08

Compression ring locations—3.0L (J30A1) and 3.2L (C32A1) engines

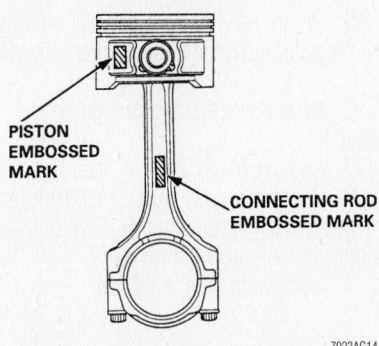

PISTON EMBOSSED MARK

CONNECTING ROD EMBOSSED MARK

7923AG14

Piston-to-connecting rod assembly—3.0L (J30A1) engine

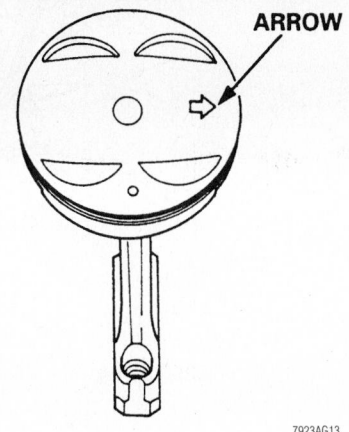

ARROW

7923AG13

Piston directional arrow location—3.0L (J30A1) engine

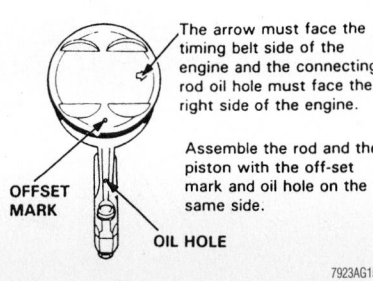

The arrow must face the timing belt side of the engine and the connecting rod oil hole must face the right side of the engine.

Assemble the rod and the piston with the off-set mark and oil hole on the same side.

OFFSET MARK

OIL HOLE

7923AG15

Piston/connecting rod assembly-to-engine orientation—3.5L engine

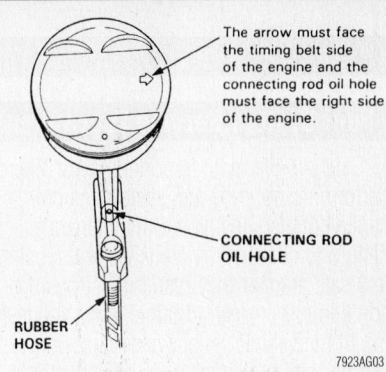

The arrow must face the timing belt side of the engine and the connecting rod oil hole must face the right side of the engine.

CONNECTING ROD OIL HOLE

RUBBER HOSE

7923AG03

Piston/connecting rod assembly-to-engine orientation—3.2L (C32A6) and (J32A1) engines

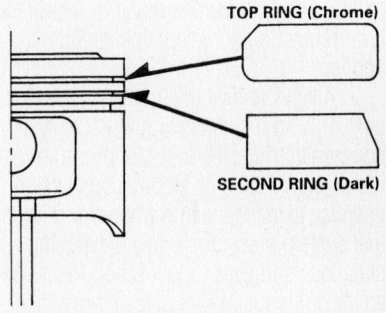

TOP RING (Chrome)

SECOND RING (Dark)

7923AG16

Compression ring locations—3.5L engine

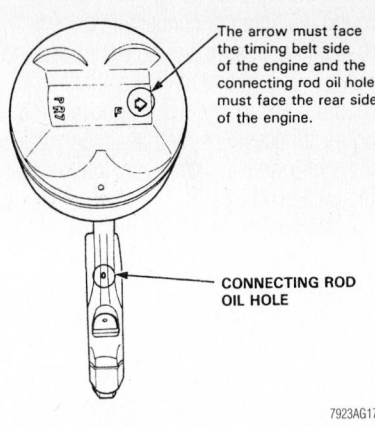

The arrow must face the timing belt side of the engine and the connecting rod oil hole must face the rear side of the engine.

CONNECTING ROD OIL HOLE

7923AG17

Piston/connecting rod assembly-to-engine orientation—3.0L (C30A1) and 3.2L (C32B1) engines

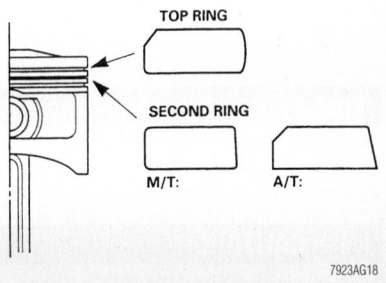

TOP RING

SECOND RING

M/T: A/T:

7923AG18

Compression ring locations—3.0L (C30A1), (C32A6), (J32A1) and 3.2L (C32B1) engines

FUEL SYSTEM

Fuel System Service Precautions

Safety is the most important factor when performing not only fuel system maintenance but also any type of maintenance. Failure to conduct maintenance and repairs in a safe manner may result in serious personal injury or death. Maintenance and testing of the vehicle's fuel system components can be accomplished safely and effectively by adhering to the following rules and guidelines.

• To avoid the possibility of fire and personal injury, always disconnect the negative battery cable unless the repair or test procedure requires that battery voltage be applied.

• Always relieve the fuel system pressure prior to disconnecting any fuel system component (injector, fuel rail, pressure regulator, etc.), fitting or fuel line connection. Exercise extreme caution whenever relieving fuel system pressure, to avoid exposing skin, face and eyes to fuel spray. Please be advised that fuel under pressure may penetrate the skin or any part of the body that it contacts.

• Always place a shop towel or cloth around the fitting or connection prior to loosening to absorb any excess fuel due to spillage. Ensure that all fuel spillage (should it occur) is quickly removed from engine surfaces. Ensure that all fuel soaked cloths or towels are deposited into a suitable waste container.

• Always keep a dry chemical (Class B) fire extinguisher near the work area.

• Do not allow fuel spray or fuel vapors to come into contact with a spark or open flame.

• Always use a back–up wrench when loosening and tightening fuel line connection fittings. This will prevent unnecessary stress and torsion to fuel line piping. Always follow the proper torque specifications.

• Always replace worn fuel fitting O–rings with new. Do not substitute fuel hose or equivalent, where fuel pipe is installed.

Fuel System Pressure

RELIEVING

❋❋ CAUTION

The fuel injection system remains under pressure after the engine has been turned OFF. Properly relieve fuel pressure before disconnecting any fuel lines. Failure to do so may result in fire or personal injury.

➡The radio may contain a coded theft protection circuit. Always obtain the code number before disconnecting the battery. If the vehicle is equipped with 4WS, the steering control unit is shut down when the battery is disconnected. After connecting the battery, turn the steering wheel lock-to-lock to reset the steering control unit.

1. Before servicing the vehicle, refer to the precautions in the beginning of this section.

2. Disconnect the negative battery cable.

3. Remove the kick panel, then remove the PGM-FI main relay (FUEL PUMP) from the under-dash fuse/relay box. Start the engine and let it run until it stalls.

4. Remove the fuel filler cap.

5. On vehicles without quick connect fittings:

a. Use a box wrench on the 6mm service bolt or fuel pulsation damper, as applicable, on the fuel rail while holding the special banjo bolt with another wrench.

b. Place a rag or shop towel over the bolt/pulsation damper.

c. Slowly loosen the bolt/damper 1 complete turn.

d. On vehicles with quick-connect fittings:

e. Remove the quick-connect fitting cover.

f. Clean any dirt from the quick-connect fitting.

g. Place a rag or shop towel over quick-connect fitting.

h. Detach the quick-connect fitting by holding the connector with one hand, then squeeze the retainer tabs with the other hand to release them from the locking pawls. Pull the connector off.

K20A2 engine

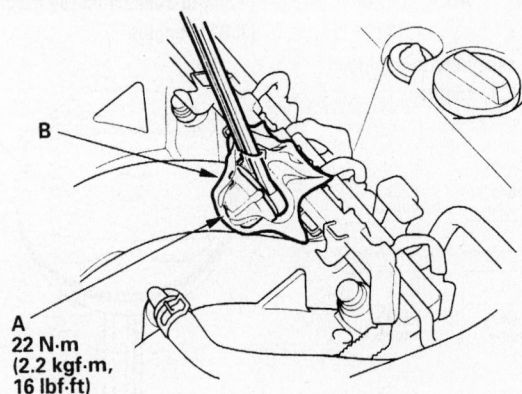

A
22 N·m
(2.2 kgf·m,
16 lbf·ft)

K20A3 engine

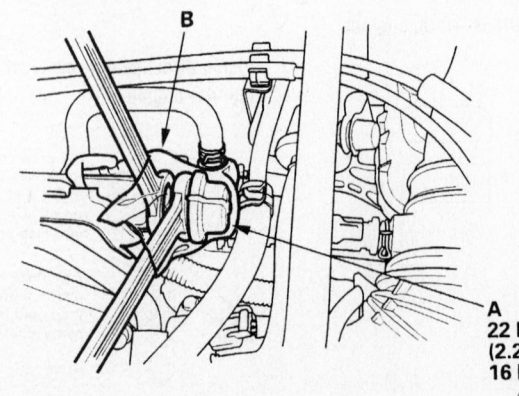

A
22 N·m
(2.2 kgf·m,
16 lbf·ft)

42356-INTE-G31

View of the fuel pulsation damper (A)—2.0L engines

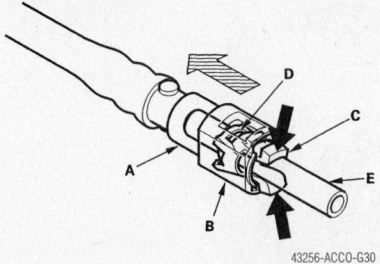

Hold the quick-connect (A) connector (B) with one hand, then squeeze the retainer tabs (C) with the other hand to release them from the locking pawls (D)

✳✳ CAUTION

Do not allow fuel spray or fuel vapors to come in contact with a spark or open flame. Keep a dry chemical fire extinguisher nearby. Never store fuel in an open container due to risk of fire or explosion.

➡ **A fuel pressure gauge may be attached at the 6mm service bolt/pulsation damper or quick-connect location. Always replace the washer between the service bolt and the banjo bolt whenever the service bolt is loosened.**

6. If equipped, remove the service bolt/damper and install a new washer. Tighten the 6mm service bolt to 104 inch lbs. (12 Nm). If equipped with a fuel pulsation damper, tighten to 16 ft. lbs. (22 Nm). Don't overtighten the service bolts, their threads may strip and cause leaks.

7. If equipped with a quick-connect fitting, connect the fitting, making sure the locking pawls are properly engaged,

8. Clean up any fuel spilled on the engine and intake manifold.

9. Install the fuel pump relay to the underdash fuel/relay box and install the kick-panel cover.

10. Install the fuel filler cap.

11. Properly dispose of the fuel soaked rag or shop towel.

12. Reconnect the negative battery cable.

13. Turn the ignition **ON**, but don't start the engine. Repeat this 2 or 3 times to pressurize the fuel system. Check for fuel leaks.

14. Enter the radio security code.

Fuel Filter

REMOVAL & INSTALLATION

Except Fuel Pump-Mounted Filters

1. Before servicing the vehicle, refer to the precautions in the beginning of this section.

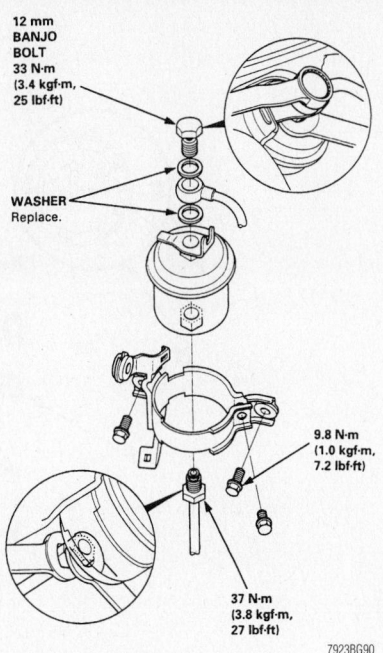

Fuel filter assembly

2. Disconnect the negative battery cable.

3. Properly relieve the fuel pressure.

4. Wrap a shop towel around the filter fittings. Use a properly sized wrench to slowly loosen the fuel line fittings.

5. Remove or disconnect the following:
   • Fuel pipes from the fuel filter
   • Fuel filter clamp
   • Filter from the vehicle

✳✳ WARNING

It is very important that ALL of the fuel line banjo bolt washers be replaced every time the banjo bolts are loosened. If the washers are not replaced, the fuel lines will leak pressurized fuel, causing the risk of fire or explosion.

To install:

6. Install or connect the following:
   • New filter in position and tighten the clamp mounting bolt to 89 inch lbs. (10 Nm)
   • Banjo bolt with new washers and tighten it to 25 ft. lbs. (33 Nm)
   • Fuel feed line and tighten the fitting to 27 ft. lbs. (37 Nm)
   • Negative battery cable and enter the radio security code

7. Start the vehicle and check for leaks.

Fuel Pump-Mounted Filter

1. Before servicing the vehicle, refer to the precautions in the beginning of this section.

2. Properly relieve the fuel pressure.

3. Remove or disconnect the following:
   • Negative battery cable
   • Fuel pump
   • Fuel filter set

To install:

4. Install or connect the following:
   • Fuel filter set, using a new base gasket and new O-rings

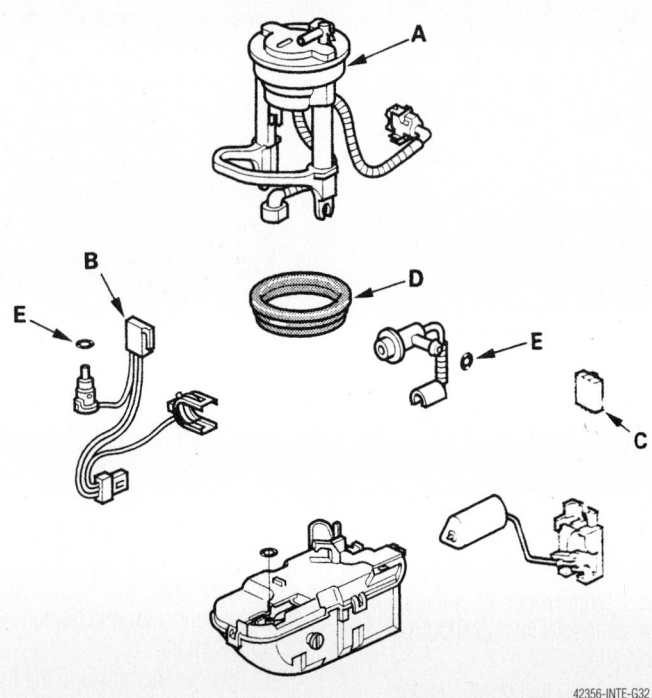

View of the fuel pump–mounted fuel filter (A)—2.0L engine shown, others similar

- Fuel pump. When installing the fuel gauge sending unit, make sure the is secure and the connector is locked into place.
- Negative battery cable

Fuel Pump

REMOVAL & INSTALLATION

1.8L and 2.0L Engines

1. Before servicing the vehicle, refer to the precautions in the beginning of this section.
2. Disconnect the negative battery cable.
3. Properly relieve the fuel system pressure.
4. Remove the rear seat to gain access to the fuel pump access panel.
5. Remove the maintenance access cover.
6. Disconnect the electrical connector from the fuel pump.
7. If equipped with quick–connect fittings, hold the fuel line connector with one hand and press down the retainer tabs with the other hand, and then pull the connector off. Check the contact area of the pipe for dirt or damage, clean or replace the pipe or pump as required. Remove the old retainer from the pipe and discard. Cover the connector and pipe with plastic bags to prevent damage and keep foreign material out.
8. Remove the fuel pump mounting nuts, then remove the fuel pump from the fuel tank.

To install:

9. Installation is the reverse of the removal procedure. Tighten the fuel pump mounting nuts to 53 inch lbs. (6 Nm).

2.4L Engine

1. Before servicing the vehicle, refer to the precautions in the beginning of this section.
2. Properly relieve the fuel pressure.
3. Remove or disconnect the following:
   - Negative battery cable, if not already done
   - Fuel fill cap
   - Trunk floor
   - Access panel from the floor
   - Fuel pump connector
   - Quick-connect fitting from the fuel tank unit
   - Fuel pump/sender assembly locknut using special tool 07XAA-001010A,
   - Locknut and fuel pump/sender assembly

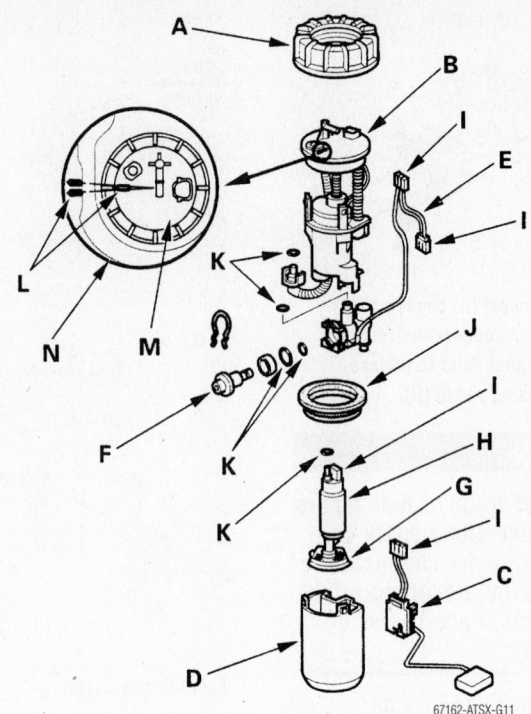

Exploded view of the fuel filter (B), fuel gauge sending unit (C), case (D), wire harness (E), fuel pressure regulator (F) and related components—2.4L engine

67162-ATSX-G11

- Fuel filter, fuel gauge sending unit, case, wire harness and fuel pressure regulator

To install:

4. Installation is the reverse of the removal procedure, noting the following points:

   a. When connecting the wire harness, make sure the connection is secure and the connectors are firmly locked into place. Do not bend or twist the fuel gauge sending unit more than necessary.

   b. Install the pump assembly components in the reverse order of removal, with a new base gasket and new O-rings. Tighten the fuel pump/sender locknut to 69 ft. lbs. (93 Nm) with special tool 07XAA-001010A.

   c. When installing the fuel tank unit, align the marks on the unit and the fuel tank.

3.0L Engines

1. Before servicing the vehicle, refer to the precautions in the beginning of this section.
2. Properly relieve the fuel system pressure.
3. Lower the fuel tank and detach the electrical connector and fuel lines from the pump assembly.
4. Remove the nuts and the fuel pump from the tank.

To install:

5. Installation is the reverse of the removal procedure. Tighten the fuel pump mounting nuts to 53 inch lbs. (6 Nm).

3.2L Engines

1. Before servicing the vehicle, refer to the precautions in the beginning of this section.
2. Remove or disconnect the following:
   - Negative battery cable
   - Left rear wheel
   - Tank drain bolt and drain the fuel into an approved container
   - Pump and float wiring connectors located under the trunk floor
   - Fuel hose and pipe covers from the inside of the quarter panel

3. Support the tank with a transmission jack, remove the straps and lower the tank out of the vehicle. If it sticks on the undercoating, carefully pry it free using a blunt or wooden instrument as a lever.
4. Disconnect the fuel line by removing the banjo bolt or uncoupling the quick–connect fittings.
5. Remove the fuel pump mounting nuts. Remove the fuel pump from the fuel tank.

To install:

6. Installation is the reverse of the removal procedure, while using the following torque values:

- Fuel pump mounting nuts: 48 inch lbs. (6 Nm)
- Fuel tank strap bolts: 28 ft. lbs. (38 Nm)
- Fuel tank drain bolt: 36 ft. lbs. (49–50 Nm)

3.5L Engine

1. Before servicing the vehicle, refer to the precautions in the beginning of this section.
2. Properly relieve the fuel system pressure.
3. Remove or disconnect the following:
   - Rear seat cushion
   - Access panel from the floor
   - Fuel line and wiring from the fuel pump assembly

- Mounting nuts and the fuel pump from the fuel tank

To install:

4. Installation is the reverse of the removal procedure. Tighten the fuel pump mounting nuts to 53 inch lbs. (6 Nm).

Fuel Injector

REMOVAL & INSTALLATION

✳✳ CAUTION

Fuel injection systems remain under pressure, even after the engine has been turned OFF. The fuel system pressure must be relieved before disconnecting any fuel lines. Failure to

do so may result in fire and/or personal injury. Observe all applicable safety precautions when working around fuel. Whenever servicing the fuel system, always work in a well-ventilated area. Do not allow fuel spray or vapors to come in contact with a spark or open flame. Keep a dry chemical fire extinguisher near the work area. Always keep fuel in a container specifically designed for fuel storage; also, always properly seal fuel containers to avoid the possibility of fire or explosion.

1. Disconnect the negative battery cable.
2. Relieve the fuel system pressure.
3. Remove or disconnect the following:
   - Intake manifold cover, if equipped

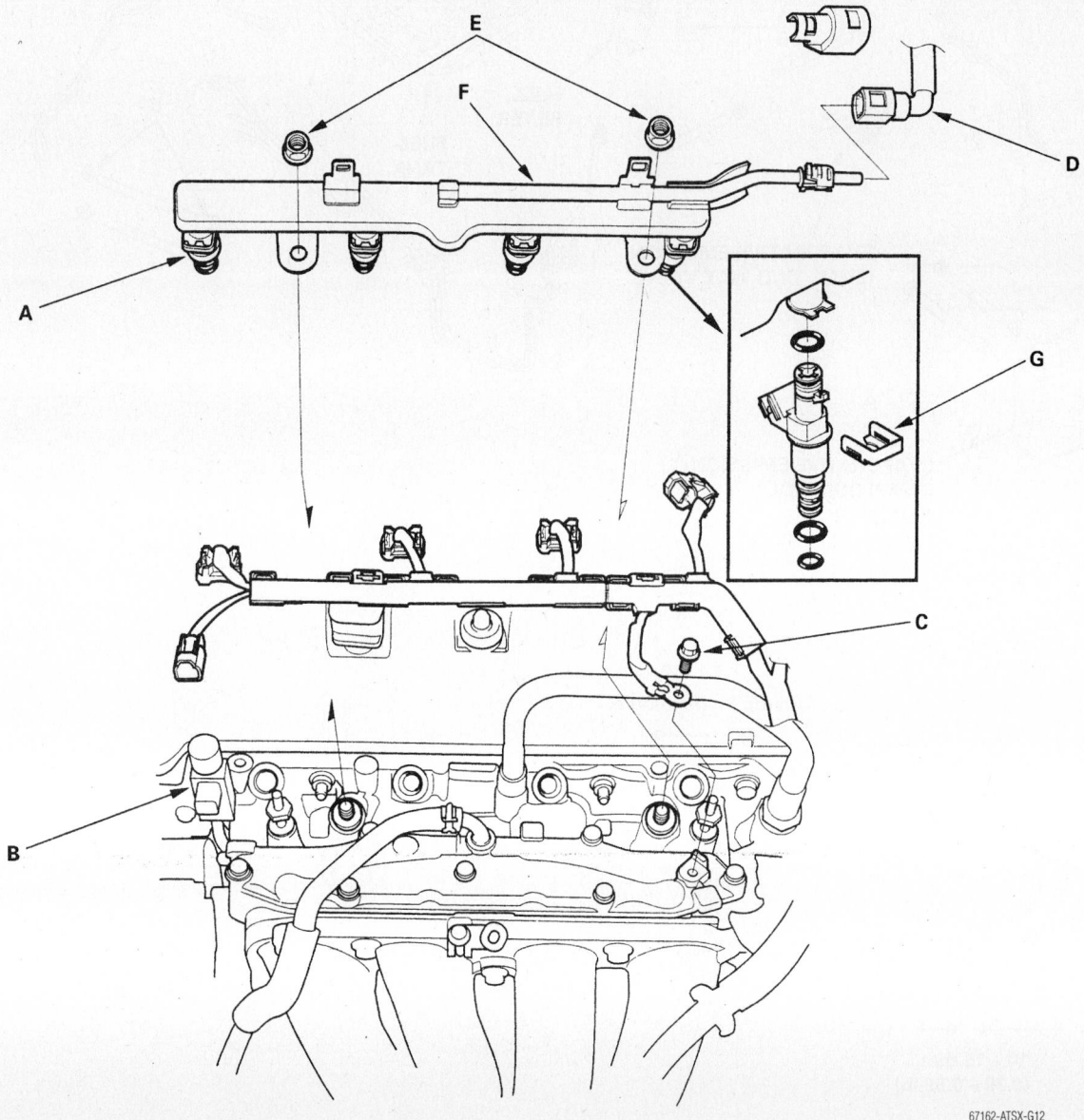

Exploded view of the fuel rail and injectors—2.4L engine shown

NOTE: Check all hose clamps and retighten if necessary.

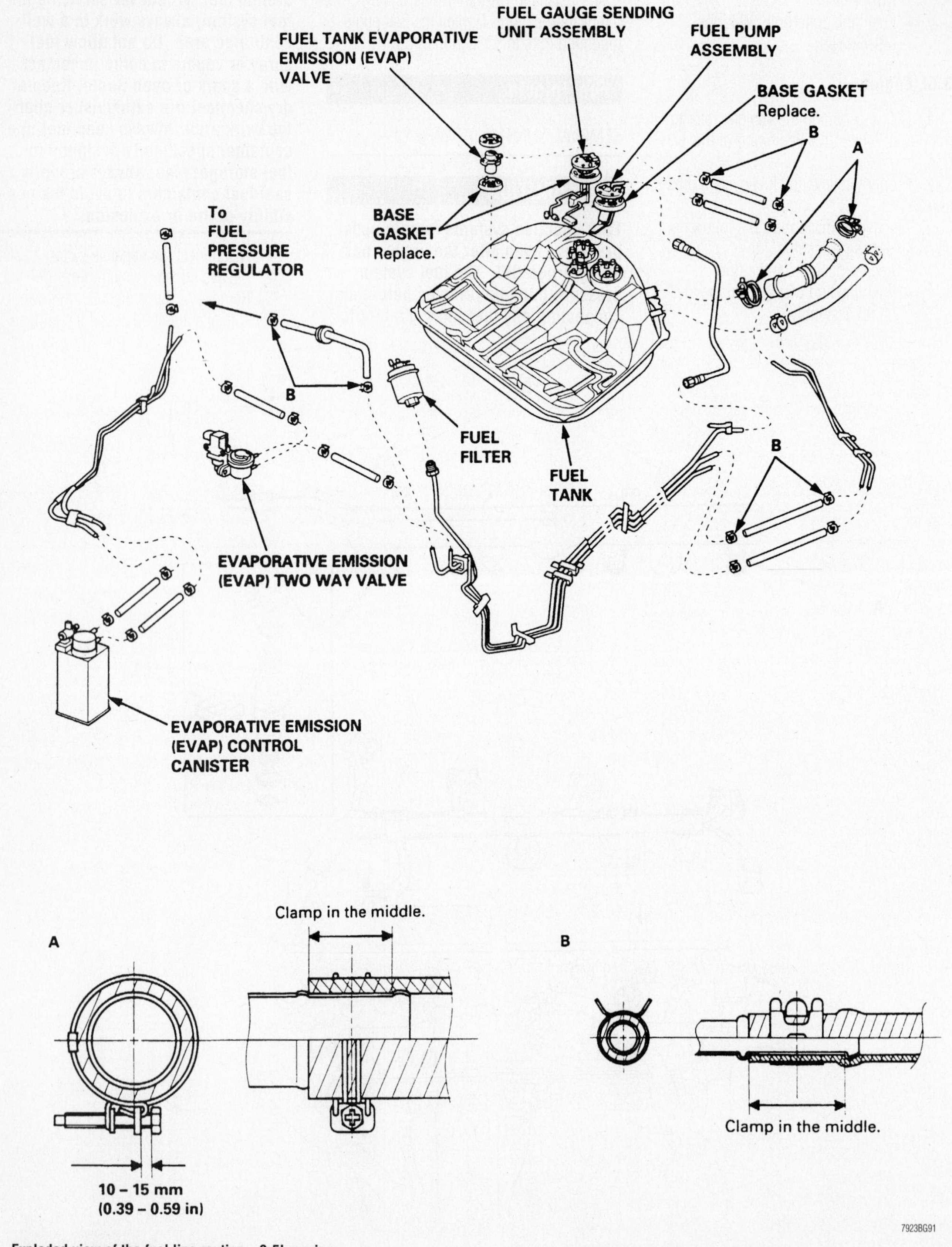

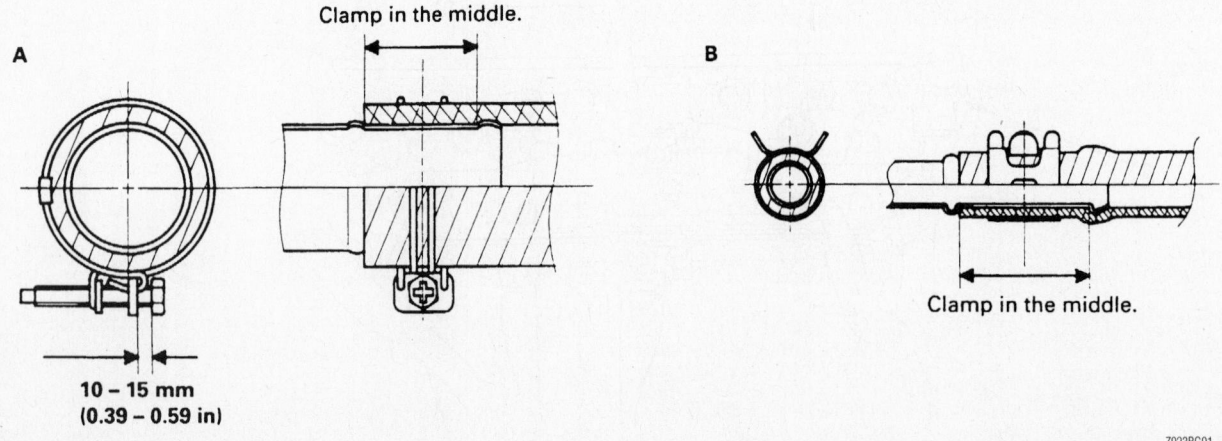

Exploded view of the fuel line routing—3.5L engine

- Fuel injector electrical connectors
- Ground cable bolt, if necessary
- Fuel line quick-connect fittings
- Fuel feed line from the fuel rail
- Vacuum hose and fuel return line from the fuel pressure regulator
- Fuel rail mounting nuts and fuel rail
- Fuel injector retaining clip

4. Grasp the fuel injectors body and pull up while gently rocking the fuel injector from side to side.

5. Once removed, inspect the fuel injector cap and body for signs of deterioration. Replace as required.

6. Remove the O-rings and discard.

To install:

7. Install or connect the following:
- Apply a small amount of clean engine oil to the new O-rings and install them onto each injector
- Injectors into the fuel rail
- Injector retaining clips
- Fuel rail and injectors into injector base

- Fuel rail mounting nuts and tighten to 16 ft. lbs. (22 Nm)
- Ground cable bolt, if removed
- Fuel feed line and quick-connect fittings
- Vacuum hose and fuel return line to the fuel pressure regulator
- Fuel injector electrical connectors
- Intake manifold cover, if equipped
- Negative battery cable

8. Run the engine at idle for 2 minutes, then turn the engine **OFF** and check for fuel leaks and proper operation.

DRIVE TRAIN

Transaxle Assembly

REMOVAL & INSTALLATION

Manual

INTEGRA

1. Before servicing the vehicle, refer to the precautions in the beginning of this section.

2. Disconnect the negative battery cable, then the positive battery cable.

3. Drain the transaxle oil. Install the drain plug with a new washer.

4. Remove or disconnect the following:
- Air cleaner housing and the intake air tube
- Backup light switch connector
- Transaxle ground wire
- Lower radiator hose clamp from the transaxle hanger
- Wiring harness clips
- Starter motor electrical connectors
- Vehicle Speed (VSS) sensor electrical connector
- Clutch pipe bracket and slave cylinder

✳✳ CAUTION

Do not operate the clutch pedal once the slave cylinder has been removed.

- 3 upper transaxle mounting bolts and the lower starter mounting bolt
- Engine splash shield
- Heated Oxygen (HO$_2$S) sensor connector
- Front exhaust pipe from the vehicle
- Ball joints from the lower control arms
- Right strut fork
- Both halfshafts
- Intermediate shaft
- Extension rod

- Shift rod
- Front and rear engine stiffeners
- Clutch cover
- Right front mount/bracket

5. Place a transmission jack under the transaxle and a jackstand under the engine.
- Transaxle mount
- Rear mounting bracket
- Transaxle mounting bolts
- Transaxle from the vehicle

To install:

6. Installation is the reverse of the removal procedure, while using the following torque values:
- Transaxle mounting bolts: 47 ft. lbs. (64 Nm)
- Rear mounting bracket bolts: 87 ft. lbs. (118 Nm)
- Transaxle mount fasteners: 47 ft. lbs. (64 Nm)
- Transaxle mount through bolt: 54 ft. lbs. (74 Nm)
- Starter motor bolts: 33 ft. lbs. (44 Nm)
- Right front mount/bracket self-locking bolt: 61 ft. lbs. (83 Nm)
- Right front bracket and mount bolts, except self-locking bolt: 33 ft. lbs. (44 Nm)
- Clutch cover 6mm bolts: 108 inch lbs. (12 Nm)
- Clutch cover 8mm bolt: 17 ft. lbs. (24 Nm)
- Clutch cover 12mm bolt: 42 ft. lbs. (57 Nm)
- Front and rear engine stiffener transaxle bolts: 42 ft. lbs. (57 Nm)
- Front and rear engine stiffener engine bolts: 17 ft. lbs. (24 Nm)
- Extension rod bolt: 16 ft. lbs. (22 Nm)
- Intermediate shaft mounting bolts: 29 ft. lbs. (39 Nm)

- Halfshafts
- Ball joint nuts: 36–43 ft. lbs. (49–59 Nm)
- Right strut fork pinch bolt: 32 ft. lbs. (43 Nm)
- Right strut fork lower nut and bolt: 47 ft. lbs. (64 Nm)
- Exhaust pipe nuts: 40 ft. lbs. (54 Nm)
- Catalytic converter nuts: 25 ft. lbs. (33 Nm)
- Slave cylinder bolts: 16 ft. lbs. (22 Nm)

RSX

1. Before servicing the vehicle, refer to the precautions in the beginning of this section.

2. Disconnect the negative battery cable, then the positive battery cable.

3. Drain the transaxle oil. Install the drain plug with a new washer.

4. Remove or disconnect the following:
- Intake manifold cover
- Air cleaner housing
- Intake air duct
- Battery tray
- Transmission ground cable
- Back-up light switch connector
- Vehicle Speed Sensor (VSS) connector
- Reverse lockout solenoid connector
- Cable bracket and cable from the top of the transaxle. Remove the cables and bracket together to avoid bending the cables.
- Harness clips
- Slave cylinder, being careful not to bend the clutch line. Do not depress the clutch pedal after the slave cylinder has been removed.
- Engine wire harness cover by lifting up on the locktab, then sliding the harness forward off the air cleaner housing mounting bracket

- Water pipe mounting bolt and lower the pipe slightly
- Loosen the air cleaner housing bracket mounting bolt, then the mounting bolt
- Brake booster and Evaporative emission (EVAP) line brake mounting bolts and attach special tool EQS00BRSX0 to the threaded hole in the cylinder head

5. Install the engine support hanger to the vehicle and attach the hook to the special tool.

- 2 upper transmission mounting bolts
- Transmission mount bracket and transmission mounting bolt
- Air cleaner bracket
- Splash shield
- Three–way catalytic converter
- Halfshafts
- Intermediate shaft
- Front engine mount bracket mounting bolt
- 3 bolts securing the transaxle rear mount

6. Support the subframe with the subframe adapter and a jack.

7. Make reference marks of the installed position on the front suspension subframe and mounting bolts, then remove the subframe.

8. Remove or disconnect the following:
- Clutch cover. The cover will have 2 bolts on 5–speed transaxles and 3 bolts on 6–speed transaxles.
- Front engine mount
- Transaxle mounting bolts, after placing a jack under the transaxle. 5–speed models will have 4 lower bolts and 6–speed models will have 2 rear and 2 lower bolts.
- Harness bracket and intake manifold bracket, 6–speed transaxles only
- 2 front transaxle mounting bolts, 5–speed transaxles only

9. Pull the transaxle away from the engine until the mainshaft clears the clutch pressure plate, then lower the transmission on the jack.

- Transaxle rear mount and rear mount bracket
- Boot, release fork and release bearing from the transaxle

10. Installation is the reverse of the removal procedure, while using the following torque values:
- Transaxle rear mount bracket bolts: 47 ft. lbs. (64 Nm)
- 2 front transaxle mounting bolts (5–speed only): 47 ft. lbs. (64 Nm)

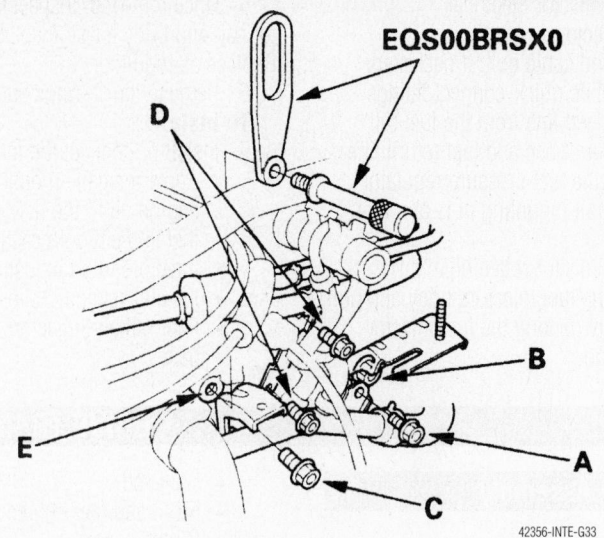

Attach the special tool to the threaded hole (E) in the cylinder head —RSX

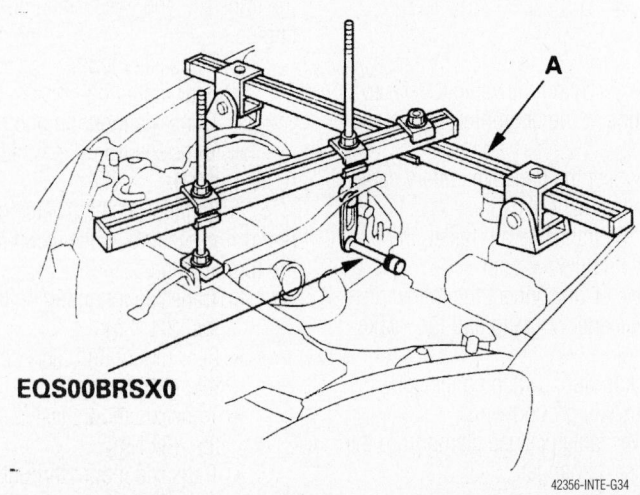

Install the engine support hanger (A) to the vehicle and attach the hook to the special tool—RSX

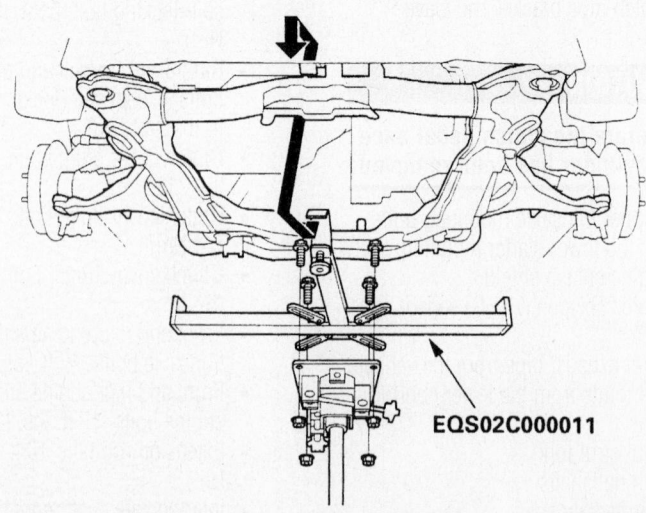

Support the subframe with the subframe adapter and a jack—RSX

- 4 lower transaxle mounting bolts (5–speed only): 47 ft. lbs. (64 Nm)
- 2 rear transaxle mounting bolts (6–speed only): 47 ft. lbs. (64 Nm)
- 2 lower transaxle mounting bolts (6–speed only): 33 ft. lbs. (44 Nm)
- Front engine mount bolts: 47 ft. lbs. (64 Nm)
- Clutch cover bolts: 9 ft. lbs. (12 Nm)
- Subframe mounting bolts: 76 ft. lbs. (98 Nm)
- 3 transaxle rear mount mounting bolts: 43 ft. lbs. (59 Nm)
- Transaxle mount bracket and transmission mounting bolt: 40 ft. lbs. (54 Nm)
- Air cleaner bracket bolts: 9 ft. lbs. (12 Nm)
- 2 upper transaxle mounting bolts: 47 ft. lbs. (64 Nm)
- 2 Slave cylinder mounting bolts: 16 ft. lbs. (22 Nm)
- Transmission ground cable: 16 ft. lbs. (22 Nm)

NSX

1. Before servicing the vehicle, refer to the precautions in the beginning of this section.
2. Drain the transaxle fluid.
3. Remove or disconnect the following:
   - Negative battery cable, then the positive battery cable
   - Strut bar
   - Air cleaner assembly
   - Control box
   - Transmission ground cable
   - Back up light switch electrical connector
   - Neutral position switch electrical connector
   - Differential speed sensor electrical connector
   - Reverse lockout solenoid electrical connector
   - Vehicle Speed (VSS) sensor electrical connector
   - Starter motor
   - Transmission mount
   - Parking brake cable holders from the rear beam rod
   - Rear beam rod
   - Parking brake cable holder from the rear sub frame
   - Wheel sensor wire clamps from the lower control arms
   - Toe control arms from the side beams
   - Strut forks
   - Lower control arm from the side beam

- Half shafts from the differential
- Intermediate shaft from the differential
- Lower cover
- Change wire bracket
- Upper cover
- Shift and select cables
- Slave cylinder
- Release fork from the clutch release hanger

4. Attach a chain hoist to the transmission hangers
   - Front engine mounting bolts on the transmission side
   - Transmission mounting bolts and stiffener
   - Transmission housing mounting bolts
   - Transmission from the vehicle

To install:

5. Installation is the reverse of the removal procedure, while using the following torque values:
   - Transmission and engine stiffener bolts: 47 ft. lbs. (64 Nm)
   - Front engine mounting bolts: 43 ft. lbs. (60 Nm)
   - Slave cylinder bolts: 16 ft. lbs. (22 Nm)
   - Upper cover bolts: 108 inch lbs. (12 Nm)
   - Change wire bracket bolts: 19 ft. lbs. (25 Nm)
   - Lower cover bolts: 108 inch lbs. (12 Nm)
   - Lower control arm bolts: 90 ft. lbs. (123 Nm)
   - Strut fork bolts: 69 ft. lbs. (93 Nm)
   - Toe control arms and torque the bolts to 69 ft. lbs. (93 Nm)
   - Parking brake cable holder bolts: 16 ft. lbs. (22 Nm)
   - Rear beam rod bolts: 43 ft. lbs. (59 Nm)
   - Transmission mount bolts: 43 ft. lbs. (59 Nm)
   - Starter motor bolts: 54 ft. lbs. (74 Nm)
   - Control box bolt: 16 ft. lbs. (22 Nm)
   - Strut bar bolts: 16 ft. lbs. (22 Nm)

TSX

➡This procedure requires the use of the following special tools, or their equivalents: Engine hanger/adapter VSB02C000015, Engine support hanger A & Reds AAR-T-12566 and Subframe adapter VSB02C000016.

1. Before servicing the vehicle, refer to the precautions in the beginning of this section.

2. Note the radio security code and the radio presets.
3. Remove or disconnect the following:
4. Drain the transaxle fluid.
5. Remove or disconnect the following:
   - Negative battery cable, then the positive battery cable
   - Air cleaner housing assembly
   - Battery base
   - Clutch slave cylinder, but do NOT bend the clutch line. Do NOT depress the clutch pedal once the slave cylinder is removed.
   - Cable bracket and cables from the top of the transaxle housing
   - Countershaft speed sensor connector
   - Back-up light switch connector
   - Reverse solenoid lockout connector
   - Secondary Heater Oxygen Sensor (HO2S) connector and the bracket
   - Engine front mount stop and the engine mount front mounting bolt
   - Engine rear mount stop
   - Right and left side steering gearbox mounting bolts
   - Transaxle upper mounting bolts

6. Attach Engine hanger/adapter VSB02C000015, or equivalent to the threaded holes in the cylinder head.

7. Install the engine support hanger to the vehicle, then attach the hook to the special tool.

8. Remove or disconnect the following:
   - Transaxle mount bracket and ground cable. Do NOT drop the mount collar.
   - Splash shield
   - Exhaust pipe "A"
   - Front stabilizer link

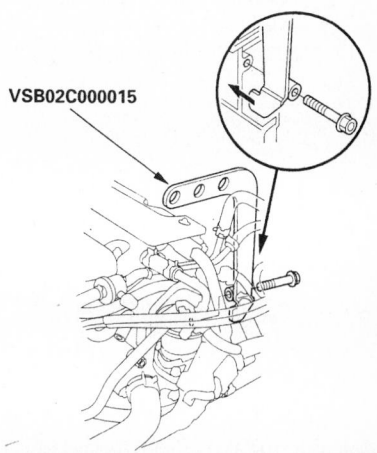

VSB02C000015

67162-ATSX-G13

Attach Engine hanger/adapter VSB02C000015, or equivalent to the threaded holes in the cylinder head—TSX

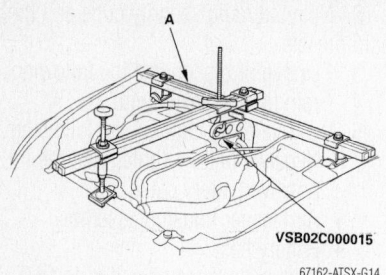

VSB02C000015

67162-ATSX-G14

Install the engine support hanger to the vehicle, then attach the hook to the special tool—TSX

- Damper fork
- Knuckle from the lower control arm
- Inboard joint of the driveshaft
- Intermediate shaft
- Heat shield
- Power steering line bracket mounting bolt
- Power steering line from the subframe. There are 2 clamps and 1 mounting bracket bolt.
- Rear engine mount lower mounting bolts
- Transaxle lower mount mounting nuts
- Middle subframe mounting bolts

9. Matchmark the ends of the subframe to the edge of the stiffeners.

10. Support the subframe with the subframe adapter tool and a suitable jack.

11. Suspend the steering gearbox with a suitable support, then remove the front suspension subframe stays and subframe.

12. Remove or disconnect the following:
- Front engine mount upper bracket
- Shift cable bracket, rear engine mount upper mounting bolt and rear engine mount
- Rear engine mount upper bracket
- Clutch cover

13. Place a transmission jack under the

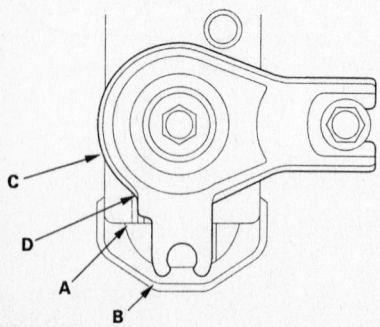

67162-ATSX-G15

Matchmark (A) the ends of the subframe (B) to the edge (C) of the stiffeners (D)—TSX

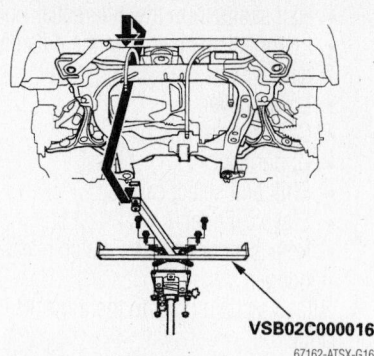

VSB02C000016

67162-ATSX-G16

Support the subframe with the subframe adapter tool and a suitable jack —TSX

transaxle, then remove the lower transaxle mounting bolts.

14. Pull the transaxle away from the engine until the mainshaft clears the clutch pressure plate, then carefully lower the transaxle using the jack.

15. Remove or disconnect the following:
- Transmission lower front mount and lower rear mount
- Release fork boot from the clutch housing
- Release fork from the clutch housing by squeezing the release fork set spring with pliers
- Release bearing

To install:

16. Make sure the 2 dowel pins are installed in the clutch housing.

17. Apply super high temp urea grease (P/N 087989002) to the release fork and release bearing, then install the release fork, release bearing and boot.

18. Make sure the mount bracket collars are installed in the transaxle, then install the transmission lower front and rear mounts. Tighten the front mount bolts to 43 ft. lbs. (59 Nm) and the lower rear mount bolts to 33 ft. lbs. (45 Nm).

19. Place the transaxle on the transaxle jack and raise it to engine level.

20. Install or connect the following:
- Transmission mounting lower bolts and tighten to 47 ft. lbs. (64 Nm)
- Clutch cover and tighten the retainers to 33 ft. lbs. (45 Nm)
- Rear engine mount upper bracket and tighten NEW bolts to 65 ft. lbs. (90 Nm)
- Rear engine mount
- New rear engine mount upper mounting bolt and tighten to 40 ft. lbs. (54 Nm)
- Shift cable bracket and tighten the bolt to 16 ft. lbs. (22 Nm)
- Front engine mount upper bracket and tighten NEW bolts to 47 ft. lbs. (64 Nm)

21. Support the subframe with the subframe adapter tool and a jack.

22. Install the front suspension subframe and subframe stays.

23. Align the matchmarks with the edge of both rear stiffeners and tighten the rear subframe mounting bolts, then the front bolts, as shown in the accompanying figure.

24. Install or connect the following:
- New middle subframe mounting bolts. Torque the side (short) bolts to 36 ft. lbs. (49 Nm) and the bot-

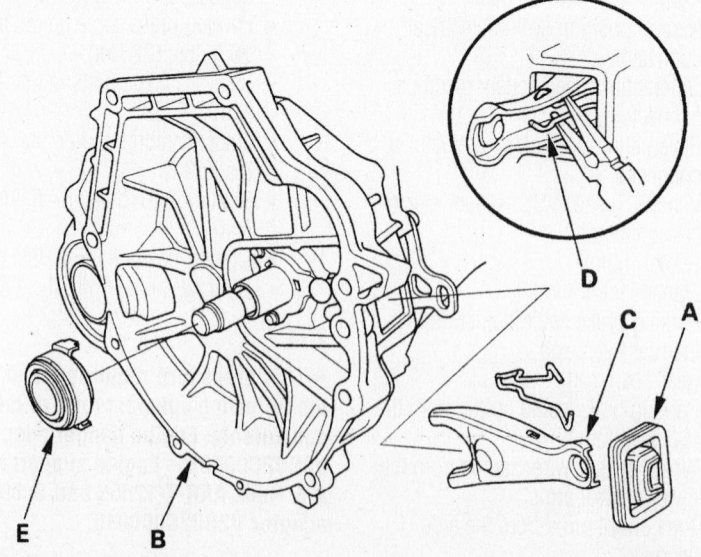

67162-ATSX-G17

View of the release fork boot (A), clutch housing (B), release fork (C), set spring (D) and release bearing (E)—TSX

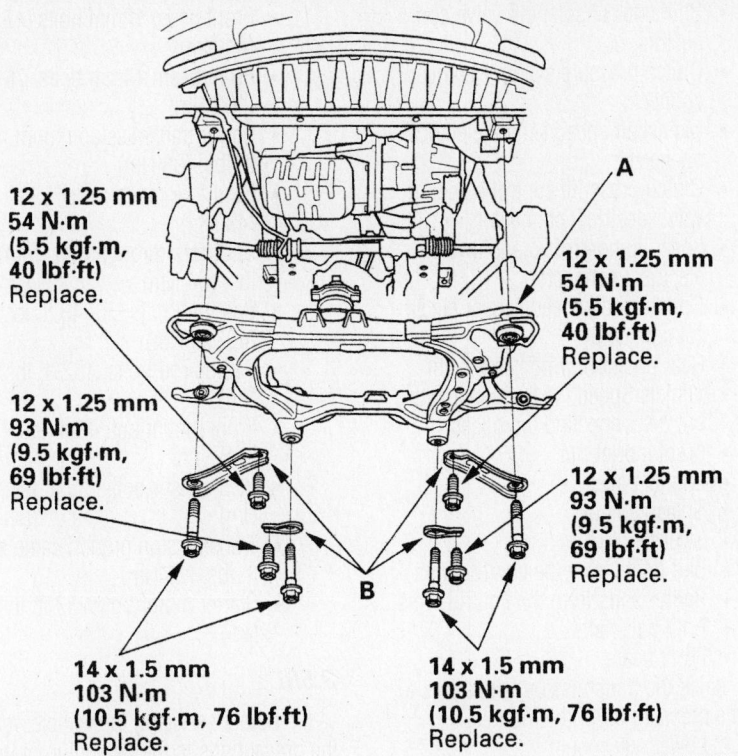

12 x 1.25 mm
54 N·m
(5.5 kgf·m,
40 lbf·ft)
Replace.

12 x 1.25 mm
54 N·m
(5.5 kgf·m,
40 lbf·ft)
Replace.

A

12 x 1.25 mm
93 N·m
(9.5 kgf·m,
69 lbf·ft)
Replace.

12 x 1.25 mm
93 N·m
(9.5 kgf·m,
69 lbf·ft)
Replace.

B

14 x 1.5 mm
103 N·m
(10.5 kgf·m, 76 lbf·ft)
Replace.

14 x 1.5 mm
103 N·m
(10.5 kgf·m, 76 lbf·ft)
Replace.

67162-ATSX-G18

Subframe bolt tightening specifications—TSX

tom (long) bolt to 33 ft. lbs. (44 Nm).
- Transaxle lower mount mounting nuts and tighten to 33 ft. lbs. (45 Nm)
- New rear engine mount lower mounting bolts and tighten to 33 ft. lbs. (45 Nm)
- Power steering line to the subframe with the 2 clamps and bolt. Torque the bolt to 7.2 ft. lbs. (9.8 Nm).
- Power steering line bracket mounting bolt and tighten to 7.2 ft. lbs. (9.8 Nm)
- Heat shield. Tighten the retainers to 7.2 ft. lbs. (9.8 Nm).
- Intermediate shaft
- Driveshaft inboard joint
- Knuckle onto the lower arm
- Damper fork
- Front stabilizer
- Exhaust pipe "A" with new gaskets. Torque the bolt to 16 ft. lbs. (22 Nm) and the nut to 25 ft. lbs. (33 Nm).
- Splash shield

25. Make sure the mount bracket collars are installed in the transaxle, then install the transaxle mount bracket and ground cable. Torque the mount bracket bolts to 40 ft. lbs. (54 Nm) and the ground cable bolt to 7.2 ft. lbs. (9.8 Nm).
26. Install or connect the following:

- Transmission upper mounting bolts and tighten to 47 ft. lbs. (64 Nm)
- New steering gearbox mounting bolts. Tighten the right side bolts to 28 ft. lbs. (38 Nm) and the left side bolts to 43 ft. lbs. (59 Nm).
- Engine rear mount stop and tighten the nuts to 51 ft. lbs. (69 Nm)
- New engine mount front mounting bolt and tighten to 40 ft. lbs. (54 Nm)
- Engine front mount stop and tighten the nuts to 58 ft. lbs. (78 Nm)
- Bracket and secondary HO2S sensor connector
- Reverse lockout solenoid, back-up light switch and countershaft speed sensor connectors
- Cable bracket and cables. Tighten the bolts to 20 ft. lbs. (27 Nm). Apply a light coat of super high temp grease to the cable ends, then install new cotter pins.

27. Apply super high temp urea grease to the end of the cylinder rod, then install the slave cylinder. Tighten the slave cylinder mounting bolts to 16 ft. lbs. (22 Nm) and the fluid line bolt to 7.2 ft. lbs. (9.8 Nm). Do not bend the clutch line.

- Battery base
- Air cleaner housing
- Battery

28. Fill the transaxle with fluid.
29. Connect the positive and negative battery cables.
30. Program the station presets, if necessary.
31. Test drive and check clutch operation and front wheel alignment.

Automatic

INTEGRA & RSX

1. Before servicing the vehicle, refer to the precautions in the beginning of this section.
2. Drain the transaxle fluid.
3. Remove or disconnect the following:

- Negative battery cable, then the positive battery cable
- Air cleaner housing assembly with intake air duct
- Starter motor cables and holder
- Transaxle ground cable from the transaxle hanger
- Lockup control solenoid valve connector
- Shift control solenoid valve connector
- Harness clamp on the lockup control solenoid harness from the harness stay
- Vehicle Speed (VSS) sensor electrical connector
- Main shaft speed sensor electrical connector
- Counter shaft speed sensor electrical connector
- Upper transaxle mounting bolts
- Splash shield
- Front wheels
- Ball joints from the lower control arms
- Right strut fork
- Both halfshafts from the vehicle
- Heated Oxygen (HO2S) sensor connector
- Front exhaust pipe from the vehicle
- Intermediate shaft
- Shift cable cover
- Shift cable by removing the control lever

✳✳ WARNING

Do not bend the shift control cable when removing it.

- Right front mount/bracket
- End of the throttle control cable from the throttle control drum
- Transmission oil cooler hoses from the joint pipes
- Engine stiffener
- Torque converter cover

- 8 drive plate bolts
4. Support the transaxle.
  - Transaxle mounting bolts and rear engine mounting bolts
  - Transaxle from the engine
  - Starter motor from the transaxle (if necessary)

To install:

5. Installation is the reverse of the removal procedure, while using the following torque values:
  - Starter motor bolts: 33 ft. lbs. (45 Nm)
  - Transaxle mounting bolts: 47 ft. lbs. (64 Nm)
  - Rear engine mount bolts: 87 ft. lbs. (118 Nm)
  - Transmission mount bolt: 54 ft. lbs. (74 Nm)
  - Transmission mount nuts: 47 ft. lbs. (64 Nm)
  - Drive plate bolts: 108 inch lbs. (12 Nm)
  - Torque converter cover 6mm bolts: 108 inch lbs. (12 Nm)
  - Torque converter cover 10mm bolt: 33 ft. lbs. (44 Nm)
  - Engine stiffener transaxle bolt: 32 ft. lbs. (43 Nm)
  - Engine stiffener engine bolts: 17 ft. lbs. (24 Nm)
  - Right front mount/bracket 12mm bolts: 47 ft. lbs. (64 Nm)
  - Right front mount/bracket 10mm bolts: 33 ft. lbs. (44 Nm)
  - Control lever bolt: 10 ft. lbs. (14 Nm)
  - Intermediate shaft mounting bolts: 29 ft. lbs. (39 Nm)
  - Front exhaust pipe manifold nuts: 40 ft. lbs. (54 Nm)
  - Catalytic converter nuts: 16 ft. lbs. (22 Nm)
  - Strut fork pinch bolt: 32 ft. lbs. (43 Nm)
  - Strut fork lower bolt: 47 ft. lbs. (64 Nm)
  - Splash shield bolts: 108 inch lbs. (12 Nm)

3.2CL AND 3.2TL

1. Before servicing the vehicle, refer to the precautions in the beginning of this section.
2. Shift the transmission into **P**.
3. Drain the transmission fluid.
4. Remove or disconnect the following:
  - Both battery cables
  - Battery and the tray
  - Intake air duct
  - Transmission oil cooler hoses
  - Starter motor
  - Transmission ground cable

- Shift control solenoid valve connectors
- Clutch pressure switch electrical connector
- Mainshaft speed sensor electrical connector
- Clutch pressure control solenoid valve electrical connector
- Lock–up control solenoid valve electrical connector
- Countershaft speed sensor electrical connector
- Gear position switch connector
- Vehicle Speed (VSS) sensor without disconnecting the hoses
- Front mount nut
- Engine cover
- Splash shield
- Strut forks
- Ball joints from the control arms
- Radius rods from the control arms
- Both halfshafts
- Front beam

5. Raise the transmission with a jack to take the pressure off of the mounts.
  - Lower rear mount
  - Intermediate shaft
  - Shift cable holder
  - Shift cable cover
  - Shift cable with the control lever
  - Torque converter cover
  - Drive plate bolts
  - Engine stiffener
  - Front mount bracket
  - Transmission-to-engine bolts
  - Transmission from the vehicle

To install:

6. Installation is the reverse of the removal procedure, while using the following torque values:
  - Transmission mounting bolts: 47 ft. lbs. (64 Nm)
  - Front mount bracket bolts: 28 ft. lbs. (38 Nm)
  - Engine stiffener bolts: 28 ft. lbs. (38 Nm)
  - Drive plate bolts: 20 ft. lbs. (26 Nm)
  - Torque converter cover bolts: 108 inch lbs. (12 Nm)
  - Control lever bolt: 10 ft. lbs. (14 Nm)
  - Cable cover bolts: 16 ft. lbs. (22 Nm)
  - Shift cable holder bolts: 84 inch lbs. (9.5 Nm)
  - Intermediate shaft bolts: 30 ft. lbs. (39 Nm)
  - Lower transmission mount bolts: 28 ft. lbs. (38 Nm)
  - Front beam 10mm bolts: 28 ft. lbs. (38 Nm)

- Front beam 12mm bolts: 47 ft. lbs. (64 Nm)
- Front beam 14mm bolts: 76 ft. lbs. (103 Nm)
- Lower transmission mount nuts: 28 ft. lbs. (38 Nm)
- Strut fork pinch bolts: 32 ft. lbs. (43 Nm)
- Strut fork through bolts to 47 ft. lbs. (64 Nm)
- Ball joint nuts: 36–43 ft. lbs. (49–59 Nm)
- Radius rod bolts: 132 ft. lbs. (179 Nm)
- Front mount nut: 40 ft. lbs. (54 Nm)
- VSS sensor bolts: 20 ft. lbs. (26 Nm)
- Transmission ground cable bolt: 20 ft. lbs. (26 Nm)
- Starter motor bolts: 33 ft. lbs. (44 Nm)

3.5RL

1. Before servicing the vehicle, refer to the precautions in the beginning of this section.
2. Disconnect the negative, then the positive battery cables.
3. Shift the transaxle into **P**.
4. Drain the transmission.
5. Remove or disconnect the following:
  - Strut brace
  - Control box
  - Transaxle sub–harness connectors, and remove the sub–harness clamp
  - 3 bolts securing the transaxle dipstick pipe bracket
  - Upper transaxle mounting bolts

6. Pull the carpet back under the passenger seat to expose the secondary Heated Oxygen (HO2S) sensor connector. Detach the connector and push it out from the inside of the vehicle.

7. Remove or disconnect the following:
  - Transmission stop collars
  - Front exhaust pipe from the vehicle
  - HO2S sensor wiring harness cover and grommet
  - Catalytic converter
  - Exhaust heat shield from the floor of the vehicle
  - Transaxle oil cooler hoses
  - Shift solenoid valve electrical connector
  - Shift cable cover from the transaxle
  - Shift cable holder from the holder base
  - Control lever from the control shaft
  - Transaxle dipstick pipe from the torque converter housing
  - Range switch connector

- Lower plate from under the steering gear, then re-install the 2 steering gear mounting bolts
- Shift cable guide bracket from the transaxle beam

8. Raise the transmission slightly to take the weight off of the mounts.

- Transaxle beam
- Rear transaxle mount bracket and the mount
- Exhaust pipe bracket
- Sealing bolt from the differential
- Extension shaft from the differential
- Transmission-to-engine bolts
- Engine stiffener
- Torque converter covers
- Drive plate bolts
- Transmission from the vehicle

To install:

9. Installation is the reverse of the removal procedure, while using the following torque values:

- Drive plate bolts: Step 1: 108 inch lbs. (12 Nm). Step 2: 20 ft. lbs. (26 Nm)
- Torque converter cover bolts: 108 inch lbs. (12 Nm)
- Transaxle housing 8mm bolts: 16 ft. lbs. (22 Nm)
- Transaxle housing 12mm bolts: 47 ft. lbs. (64 Nm)
- Engine stiffener bolts: 16 ft. lbs. (22 Nm)
- Transaxle beam bolts: 28 ft. lbs. (38 Nm)
- Rear transaxle mount bracket bolts: 28 ft. lbs. (38 Nm)
- Rear transaxle mount bolts: 40 ft. lbs. (54 Nm)
- Shift cable guide bolt: 84 inch lbs. (9.5 Nm)
- Differential sealing bolt: 58 ft. lbs. (78 Nm)
- Lower plate bolts: 28 ft. lbs. (38 Nm)
- Steering gear bolts: 43 ft. lbs. (59 Nm)
- Shift control lever nut: 12 ft. lbs. (16 Nm)
- Shift cable holder bolts: 108 inch lbs. (12 Nm)
- Shift cable cover bolts: 108 inch lbs. (12 Nm)
- Exhaust heat shield bolts: 84 inch lbs. (9.5 Nm)
- Catalytic converter nuts: 25 ft. lbs. (33 Nm)
- Front exhaust pipe manifold nuts: 40 ft. lbs. (54 Nm)
- Transmission stop collar bolts: 28 ft. lbs. (38 Nm)
- Control box mounting bolts: 108 inch lbs. (12 Nm)

- Strut brace bolts: 16 ft. lbs. (22 Nm)

NSX

1. Before servicing the vehicle, refer to the precautions in the beginning of this section.

2. Drain the transmission.

3. Remove or disconnect the following:

- Both battery cables
- Strut brace
- Air cleaner assembly
- Control box
- Vehicle Speed (VSS) sensor electrical connector
- Transmission ground cable
- Starter cables
- Lock-up solenoid valve electrical connector
- Shift control solenoid valve electrical connector
- Transaxle oil cooler
- Starter motor
- Upper transmission housing and mounting bolts
- Parking brake cable holders from the rear beam
- Rear beam
- Front exhaust pipe
- Toe control arms from the side beams
- Damper forks
- Lower control arms from the side beams
- Both halfshafts and the intermediate shaft
- Shift cable cover and holder
- Shift cable from the control lever
- Torque converter cover
- Drive plate bolts

4. Place a jack under the transmission and raise it just enough to take the weight off of the mounts.

- Front engine mount bolts on the transaxle side
- Rear transaxle mount bolts
- Transaxle-to-engine bolts
- Transmission from the vehicle

To install:

5. Installation is the reverse of the removal procedure, while using the following torque values:

- Transaxle mounting bolts: 54 ft. lbs. (74 Nm)
- Rear transaxle mount bolts: 43 ft. lbs. (59 Nm)
- Front engine mount bolts: 43 ft. lbs. (59 Nm)
- Drive plate bolts: 108 inch lbs. (12 Nm)
- Torque converter cover bolts: 108 inch lbs. (12 Nm)

- Shift cable cover bolts: 6 ft. lbs. (8 Nm)
- Intermediate shaft bolts: 16 ft. lbs. (22 Nm)
- Intermediate shaft heat shield bolts: 84 inch lbs. (9.5 Nm)
- Lower control arm bolts: 90 ft. lbs. (123 Nm)
- Damper fork bolts: 69 ft. lbs. (93 Nm)
- Toe control arm bolts: 69 ft. lbs. (93 Nm)
- Front exhaust pipe manifold nuts: 40 ft. lbs. (54 Nm)
- Catalytic converter nuts: 25 ft. lbs. (33 Nm)
- Rear beam 10mm bolts: 43 ft. lbs. (59 Nm)
- Rear beam nuts and 12mm bolts: 69 ft. lbs. (93 Nm)
- Parking brake cable holder bolts: 16 ft. lbs. (22 Nm)
- Starter motor bolts: 33 ft. lbs. (44 Nm)
- Transaxle oil cooler bolts: 13 ft. lbs. (18 Nm)
- Strut bar bolts: 28 ft. lbs. (38 Nm)

TSX

➡**This procedure requires the use of the following special tools, or their equivalents: Engine hanger/adapter VSB02C000015, Engine support hanger A & Reds AAR-T-12566 and Front subframe adapter EQS02C00011.**

1. Before servicing the vehicle, refer to the precautions in the beginning of this section.

2. Note the radio security code and the radio presets.

3. Remove or disconnect the following:

4. Remove the splash shield.

5. Drain the transaxle fluid. Reinstall the drain plug with a new washer and tighten the plug to 36 ft. lbs. (49 Nm).

6. Remove or disconnect the following:

- Negative battery cable, then the positive battery cable
- Battery tray
- Air cleaner housing assembly
- Battery base
- A/T clutch pressure control solenoid valve A connector, 2nd clutch transmission fluid pressure switch connector
- Harness clams from the clamp brackets
- Transmission range switch connector from its bracket and disconnect it
- AF sensor connector from its bracket and disconnect it

- Mainshaft speed sensor connector
- Countershaft speed sensor connector and harness clamps from the clamp brackets
- 3rd clutch transmission fluid pressure switch connector
- Harness clamp from the clamp bracket
- Shift solenoid harness connector A/T clutch pressure control solenoid valve B and C connectors
- Harness clamp from the clamp bracket
- ATF cooler hoses from the lines. Plug the hoses and lines to prevent fluid from leaking out. Check for signs of leakage at the hose joints.
- ATF cooler hose from the hose clamp
- ATF cooler hose from the line and plug the hose
- Ground cable, transaxle upper mount bracket plate and upper mount bracket
- Harness clamp and hose
- Bolts holding the hose clamps
- Mounting nuts and strut brace

7. Remove the clamp bracket to attach Engine hanger/adapter VSB02C000015. Attach the tool to the threaded holes in the cylinder head.

8. Install the engine support hanger to the vehicle, then attach the hook to the special tool.

9. Install the Engine support hanger AAR-T-12566, or equivalent to the vehicle. Attach the hook to the special tool adapter as shown in the accompanying figure.

Tighten the wing nut by hand and lift and support the engine.

10. Remove or disconnect the following:
- Vacuum hose from the clamp and hose from the vacuum line
- Front mount stop and clamp bracket
- Front mount bolt

11. Insert a 6mm hex wrench in the top of the ball joint pin, remove the nuts, then separate the stabilizer link from the lower control arms.
- Cotter pins, castle nuts, damper pinch bolt, self-locking nut, bolt

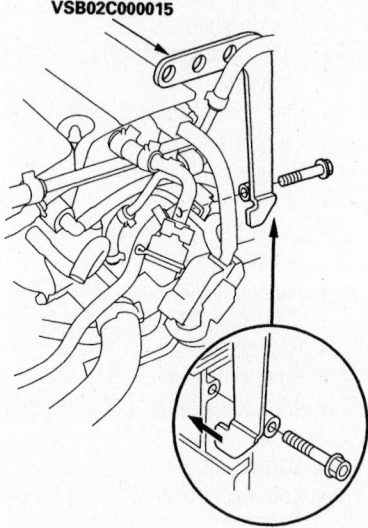

VSB02C000015

67162-ATSX-G22

Attach Engine hanger/adapter VSB02C000015, or equivalent to the threaded holes in the cylinder head—TSX

and damper forks, then separate the ball joints from the lower control arms
- Exhaust pipe "A" and its mount
- Steering gearbox heat shield and bolt securing the power steering fluid line bracket
- Power steering line from the clips from the front subframe
- Engine stiffener
- Driveplate bolts, while rotating the crankshaft pulley
- 3 bolts from the shift cable holder, then remove the cover
- Spring clip and control pin and separate the cable from the control lever. Do not bend the cable more than necessary.
- Rear mount stop
- Power steering fluid line bracket from the front subframe
- Steering gearbox mounting bolts and stiffener
- Steering gearbox mounting bracket bolts
- Damper and rear mount base bracket bolts
- Transmission lower mount nuts
- Both mid mounts

12. Matchmark the ends of the subframe to the edge of the stiffeners.

13. Attach the special tool to the subframe with hanging the hook of the special tool over the front of the subframe, then tighten the special tool screw.

14. Raise the jack and line up the slots in the arms with the bolt holes on the corner of the jack base, then attach them with the bolts securely.

15. Remove the 4 bolts holding the stiffeners and 4 bolts holding the front subframe. Lower the front subframe while sliding the steering gearbox out to clear the gearbox mounting bracket on the subframe.

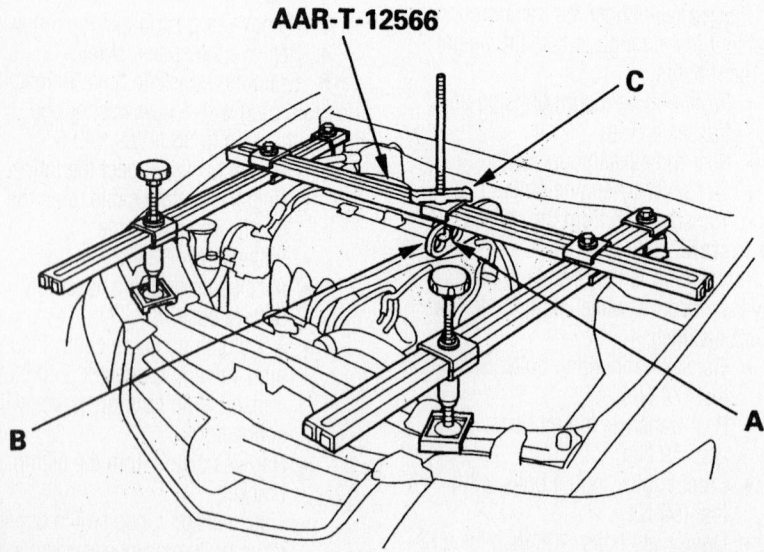

AAR-T-12566

67162-ATSX-G23

Attach the hook (A) to the special tool adapter (B), then tighten the wing nut (C) by hand and lift and support the engine

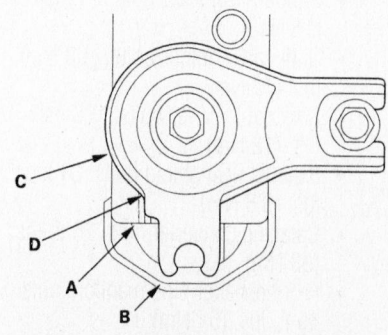

67162-ATSX-G15

Matchmark (A) the ends of the subframe (B) to the edge (C) of the stiffeners (D)—TSX

RSX

✳✳ CAUTION

The Supplemental Restraint System (SRS, air bag system) must be disarmed before any of its components are disconnected or removed. Failing to disable the SRS before servicing its components may cause accidental deployment of the air bag, resulting in unnecessary repairs and possible personal injury.

1. Before servicing the vehicle, refer to the precautions in the beginning of this section.
2. Turn the ignition switch **OFF**.
3. Disconnect the negative battery cable, then wait 3 minutes.
4. For the driver air bag:
  a. Remove the access panel from the steering wheel and disconnect the driver's airbag 4P connector from the cable reel.
5. For the passenger air bag:
  a. Remove the glove box.
  b. Disconnect the front passenger's 4P connector from the dashboard wire harness.
6. For the side airbag:
  a. Disconnect both side airbag 2P connectors from the floor wire harness.
7. For the seat belt tensioners:
  a. Remove the B–pillar trim panels.
  b. Disconnect both seat belt tensioner 2P connectors from the floor wire harness.
8. For the seat belt buckle tensioners:
  a. Disconnect both seat belt buckle tensioner 4P connectors.

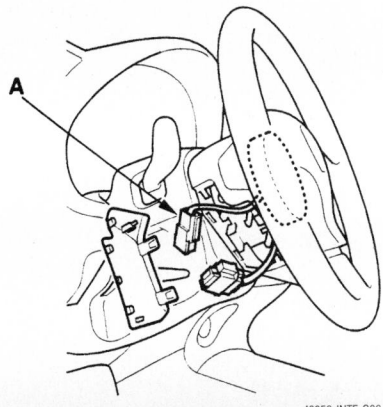

Remove the access panel from the steering wheel and disconnect the driver's airbag 4P connector from the cable reel—RSX

42356-INTE-G36

To enable:
9. For the seat belt buckle tensioners:
  a. Connect both seat belt buckle tensioner 4P connectors.
10. For the seat belt tensioners:
  a. Connect both seat belt tensioner 2P connectors to the floor wire harness.
  b. Install the B–pillar trim panels.
11. For the side airbag:
  a. Connect both side airbag 2P connectors to the floor wire harness.
12. For the passenger air bag:
  a. Connect the front passenger's 4P connector to the dashboard wire harness.
  b. Install the glove box.
13. For the driver air bag:
  a. Connect the driver's airbag 4P connector to the cable reel. Install the access panel to the steering wheel.
14. Reconnect the negative battery cable.
15. Turn the ignition switch to the **ON** position, but don't start the engine. The SRS indicator light should turn on for 6 seconds, then turn off. If the SRS indicator light doesn't come on, or stays on longer than 6 seconds, the system fault must be diagnosed.
16. Enter the radio security code.

TSX

✳✳ CAUTION

The Supplemental Restraint System (SRS, air bag system) must be disarmed before any of its components are disconnected or removed. Failing to disable the SRS before servicing its components may cause accidental deployment of the air bag, resulting in unnecessary repairs and possible personal injury.

1. Before servicing the vehicle, refer to the precautions in the beginning of this section.
2. Turn the ignition switch **OFF**.
3. Disconnect the negative battery cable, then wait 3 minutes.
4. For the driver air bag:
  a. Remove the access panel from the steering wheel and disconnect the driver's airbag 4P connector from the cable reel.
5. For the passenger air bag:
  a. Remove the glove box.
  b. Disconnect the front passenger's 4P connector from the dashboard wire harness.
6. For the side airbag:
  a. Disconnect both side airbag 2P connectors from the floor wire harness.

7. For the side curtain air bag:
  a. Disconnect the side curtain airbag 2P connector from the side curtain airbag subharness.
8. For the seat belt tensioners:
  a. Disconnect both seat belt tensioner 2P connectors from the floor wire harness.

To enable:
9. For the seat belt tensioners:
  a. Connect both seat belt tensioner 2P connectors to the floor wire harness.
  b. Install the B–pillar trim panels.
10. For the side curtain air bag:
  a. Disconnect the side curtain airbag 2P connector from the side curtain airbag subharness.
11. For the side airbag:
  a. Connect both side airbag 2P connectors to the floor wire harness.
12. For the passenger air bag:
  a. Connect the front passenger's 4P connector to the dashboard wire harness.
  b. Install the glove box.
13. For the driver air bag:
  a. Connect the driver's airbag 4P connector to the cable reel. Install the access panel to the steering wheel.
14. Reconnect the negative battery cable.
15. Turn the ignition switch to the **ON** position, but don't start the engine. The SRS indicator light should turn on for 6 seconds, then turn off. If the SRS indicator light doesn't come on, or stays on longer than 6 seconds, the system fault must be diagnosed.
16. Enter the radio security code.

3.2CL and 3.2TL

✳✳ CAUTION

The SRS must be disarmed before any of its components are disconnected or removed. Failing to disable the SRS before servicing its components may cause accidental deployment of the air bag, resulting in unnecessary repairs and possible personal injury.

1. Before servicing the vehicle, refer to the precautions in the beginning of this section.
2. Turn the ignition switch to the **LOCK** position. Remove the key.
3. Disconnect the negative and positive battery cables.
4. Always wait at least 3 minutes after disconnecting the battery before working around the air bags.
5. Remove or disconnect the following:

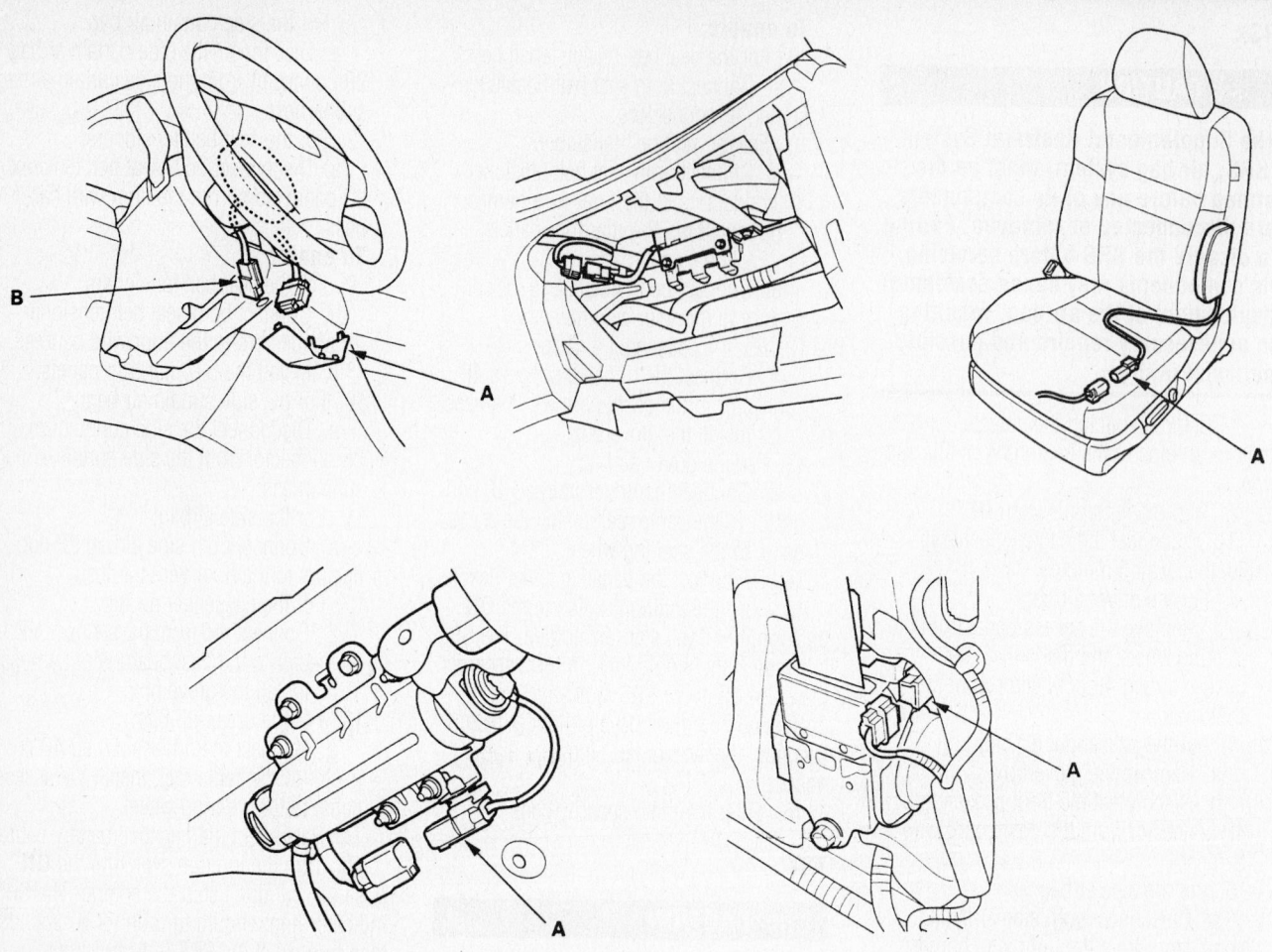

Locations of the SRS electrical connections—TSX

42356-ATSX-G21

- Steering wheel lower access cover
- Clip securing the air bag module/cable reel connection to the steering column
- Air bag and cable reel connection—Immediately install the red shorting connector onto the air bag module connector.

➡ **The driver's side air bag connection contains a spring–contact self–disabling device. A shorting connector doesn't need to be installed on the driver's air bag connector.**

6. After servicing has been completed, couple the air bag and cable reel connectors.

7. Install or connect the following:
- Clip securing the air bag/cable reel connection to the steering column
- Access cover

To enable:

8. Reconnect the positive and negative battery cables.

9. Turn the ignition switch to the **ON** position, but don't start the engine. The SRS indicator light should turn on for 6 seconds, then turn off. If the SRS indicator light doesn't come on, or stays on longer than 6 seconds, the system fault must be diagnosed.

10. Enter the radio security code.

3.5RL

1. Before servicing the vehicle, refer to the precautions in the beginning of this section.

2. Disconnect the negative battery cable, then the positive cable.

3. Wait 3 minutes for the air bag reserve power to discharge before preceding with work.

To enable:

4. Reconnect the positive and negative battery cables.

5. Turn the ignition switch to the **ON** position, but don't start the engine. The SRS indicator light should turn on for 6 seconds, then turn off. If the SRS indicator light doesn't come on, or stays on longer than 6 seconds, the system fault must be diagnosed.

6. Enter the radio security code.

Rack and Pinion Steering Gear

REMOVAL & INSTALLATION

Power

3.2CL, INTEGRA AND RSX

1. Before servicing the vehicle, refer to the precautions in the beginning of this section.

2. Note the radio security code and the radio presets.

3. Drain the power steering fluid.

4. Remove or disconnect the following:
- Both battery cables
- Wheels
- Steering wheel lower access panel
- Supplemental Inflatable Restraint (SIR) electrical connector
- Steering wheel side panels
- Air bag
- Horn electrical connector
- Cruise control electrical connector
- Audio remote switch and navigation guide switch, if equipped
- Steering wheel

- Steering joint cover
- Steering joint lower bolt and pull the joint toward the column
- Tie rods from the steering knuckles
- Shift linkage (if equipped with a manual transmission)
- Heated Oxygen (HO$_2$S) sensor electrical connector

- Catalytic converter from the vehicle
- Return line from the steering gear
- Rear beam brace rod
- Left tie rod end, then slide the steering gear all of the way to the right
- 2 lines from the valve body unit on the steering gear

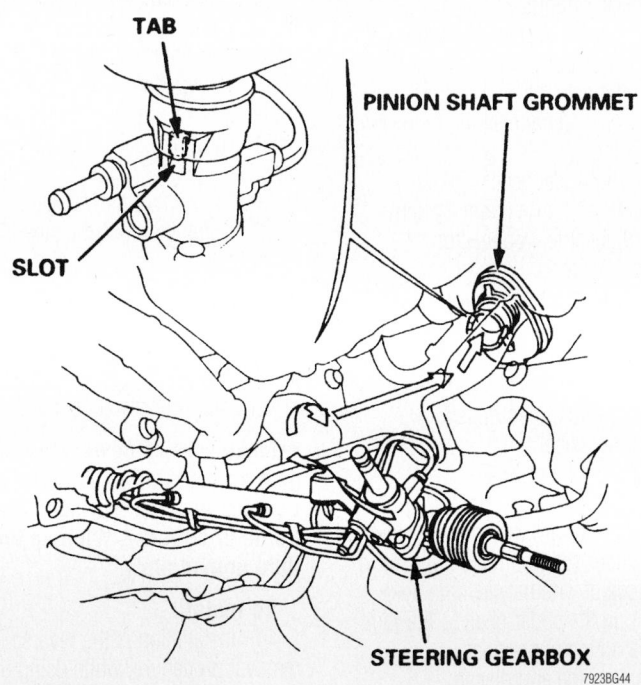

SHIFT CABLE HOLDER

SHIFT CABLE

8 x 1.25 mm
22 N·m (2.2 kgf·m, 16 lbf·ft)

SHIFT CABLE COVER

CONTROL SHAFT

CONTROL LEVER

6 x 1.0 mm
14 N·m (1.4 kgf·m, 10 lbf·ft)

LOCK WASHER
Replace.

6 x 1.0 mm
12 N·m (1.2 kgf·m, 8.7 lbf·ft)

7923BG43

Automatic transaxle shift cable attachment—Integra

TAB

PINION SHAFT GROMMET

SLOT

STEERING GEARBOX

7923BG44

Installing the steering gear—Integra

After disconnecting the hose and pipe, plug or cap the hose and pipe to prevent foreign materials from entering the valve body unit.

➡**Do not loosen the cylinder pipes between the valve body unit and the cylinder.**

5. Remove the steering gear mounting bolts.

6. Pull the steering gear all the way down to clear the pinion shaft from the bulkhead, and remove the pinion shaft grommet.

7. Slide the rack all of the way to the right, then place the left rack end below the rear beam.

8. Move the steering gear assembly to the left, and tilt the left side down to remove it from the vehicle.

To install:

9. Installation is the reverse of the removal procedure, while using the following torque values:
- Left steering gear mounting bolts: 28 ft. lbs. (38 Nm)
- Right steering gear mounting bolts: 43 ft. lbs. (58 Nm)

➡**After installing the steering gear, check the air hose connections for interference with adjacent parts.**

- Intermediate shaft pinch bolt: 16 ft. lbs. (22 Nm)
- Rear beam brace rod bolts: 28 ft. lbs. (38 Nm)
- Catalytic converter nuts: 25 ft. lbs. (33 Nm)
- Tie rod end nuts: 33 ft. lbs. (44 Nm)
- Steering wheel nut: 36 ft. lbs. (49 Nm)
- Air bag bolts: 84 inch lbs. (9.5 Nm)

3.2TL AND 3.5RL

1. Before servicing the vehicle, refer to the precautions in the beginning of this section.

2. Note the radio security code and the radio presets.

3. Drain the power steering fluid.

4. Remove or disconnect the following:
- Negative then the positive battery cables
- Steering wheel lower access panel
- Supplemental Inflatable Restraint (SIR) electrical connector
- Steering wheel side panels
- Air bag

- Horn electrical connector
- Cruise control electrical connector
- Audio remote switch and navigation guide switch, if equipped
- Steering wheel
- Steering joint bolts then move the joint toward the column
- Front wheels
- Tie rods from the steering knuckles
- Splash guard
- Feed line from the valve body
- Line mounting clamps
- Feed line from the line mounting cushions
- Sensor inlet line and both return lines from the hoses
- Steering gear mounting brackets
- Steering gear from the vehicle

To install:

5. Installation is the reverse of the removal procedure, while using the following torque values:
- Line mounting clamps bolts: 84 inch lbs. (9.5 Nm)
- Feed line bolts: 96 inch lbs. (11 Nm)
- Mounting bracket bolts: 28 ft. lbs. (38 Nm)
- Splash shield short bolts: 28 ft. lbs. (38 Nm)
- Splash shield long bolts: 43 ft. lbs. (59 Nm)
- Steering joint pinch bolt: 16 ft. lbs. (22 Nm)
- Steering wheel nut: 36 ft. lbs. (49 Nm)
- Air bag bolts: 84 inch lbs. (9.5 Nm)

NSX

1. Before servicing the vehicle, refer to the precautions in the beginning of this section.
2. Note the radio security code and the radio presets.
3. Drain the power steering fluid.
4. Remove or disconnect the following:
- Negative then the positive battery cables
- Battery
- Steering joint cover
- Steering joint bolts then move the joint toward the column to separate it
- Wheels
- Tie rods from the steering knuckles
- Spare tire
- Spare tire holder plate
- Spare tire holder
- Floor under cover
- Terminal guard
- Ground cable

- Wires from the steering gear terminals
- Radiator pipe bracket at the front compartment bulkhead
- Radiator pipe bracket at the floor

5. Support the steering gear and the front crossbeam
- Steering gear and front crossbeam bolts and nuts
- Steering gear and crossbeam from the vehicle

To install:

6. Installation is the reverse of the removal procedure, while using the following torque values:
- Steering gear and crossbeam fasteners: 43 ft. lbs. (59 Nm)
- Radiator pipe bracket bolts: 16 ft. lbs. (22 Nm)
- Ground cable bolt: 84 inch lbs. (9.5 Nm)
- Terminal guard bolts: 84 inch lbs. (9.5 Nm)
- Floor under cover bolts: 84 inch lbs. (9.5 Nm)
- Spare tire holder bolt: 18 ft. lbs. (25 Nm)
- Steering joint bolts: 16 ft. lbs. (22 Nm)

TSX

1. Before servicing the vehicle, refer to the precautions in the beginning of this section.
2. Note the radio security code and the radio presets.
3. Disconnect the negative battery cable and wait at least 3 minutes.
4. Make sure the front wheels are in the straight ahead position.
5. Drain the power steering fluid.
6. Remove or disconnect the following:
- Wheels
- Steering wheel access panel
- Supplemental Restraint System (SRS) electrical connector
- 2 Torx® screws
- Horn electrical connector
- Air bag
- Cruise control electrical connector
- Audio remote switch and navigation guide switch, if equipped
- Steering wheel
- Steering joint cover A
- Steering joint lower bolt and pull the joint toward the column. Hold the slider shaft on the column with a piece of wire between the yoke joint on the slider shaft to the joint yoke on the upper shaft.
- Center guide and discard
- Steering joint cover B. Do not dam-

age the mating surfaces of the cover and pinion shaft grommet. Replace the cover seal if necessary.
- Tie rods from the steering knuckles
- Power steering heat baffle
- Feed line holder mounting from the front suspension subframe
- Feed line holder mounting bolt and return hose from the gearbox mounting bracket
- Shift linkage (if equipped with a manual transmission)

7. Place some rags under the fluid line connections and cover the gearbox mounting part to protect it from power steering fluid. Loosen the flare nut and disconnect the fluid and return lines.

✳✳ CAUTION

After disconnecting the fluid and return lines, plug or cap the hose and pipe to prevent foreign materials from entering the valve body unit.

➡**Do not loosen the cylinder pipes between the valve body unit and the cylinder.**

8. Remove or disconnect the following:
- Front suspension subframe left mid mount
- Steering gear mounting bolts from the left gearbox mount, then remove the steering stiffener plate
- 2 flange bolts from the right side of the gearbox, then remove the gearbox mounting bracket, and cushion

9. Move the steering gear toward the front, and remove the pinion shaft grommet from the top of the valve body unit.
10. Put vinyl tape on the brake lines to protect them from the pinion shaft.
11. Move the steering gear to the driver's side and rotate it so the pinion shaft points toward the front of the vehicle.
12. Carefully move the steering gearbox toward the driver's side of the vehicle until the pinion shaft clears the wheelwell opening. Do not damage the brake lines with the pinion shaft.
13. Remove the steering gear from the driver's side wheelwell opening.

➡**Make sure no power steering fluid gets on the gearbox mount cushions, gearbox housing, surfaces of the subframe or stiffener. Wipe up any spilled fluid immediately.**

To install:

14. Installation is the reverse of the removal procedure, while using the following torque values:

- Left steering gear mounting bracket flange bolts: 28 ft. lbs. (38 Nm)
- Right steering gear mounting bolts: 43 ft. lbs. (58 Nm)

➡ **After installing the steering gear, check the air hose connections for interference with adjacent parts.**

- Return line flare nut: 21 ft. lbs. (28 Nm)
- Feed line flare nut: 31 ft. lbs. (42 Nm)

- Front suspension subframe mid mount bottom (long) bolt: 33 ft. lbs. (44 Nm)
- Front suspension subframe mid mount side (short) bolts: 36 ft. lbs. (49 Nm)
- Steering joint to pinion shaft bolt: 21 ft. lbs. (29 Nm)
- Tie rod end nuts: 32 ft. lbs. (43 Nm)
- Steering wheel nut: 29 ft. lbs. (39 Nm)
- Air bag bolts: 86 inch lbs. (9.8 Nm)

Strut

REMOVAL & INSTALLATION

Front

EXCEPT NSX & TSX

1. Before servicing the vehicle, refer to the precautions in the beginning of this section.
2. Support the lower suspension arm with a jack.

CAUTION:
- Replace the self-locking nuts after removal.
- The vehicle should be on the ground before any bolts or nuts connected to rubber mounts or bushings are tightened.
- Torque the castle nut to the lower torque specification, then tighten it only far enough to align the slot with the pin hole. Do not align the nut by loosening.

NOTE: Wipe off the grease before tightening the nut at the ball joint.

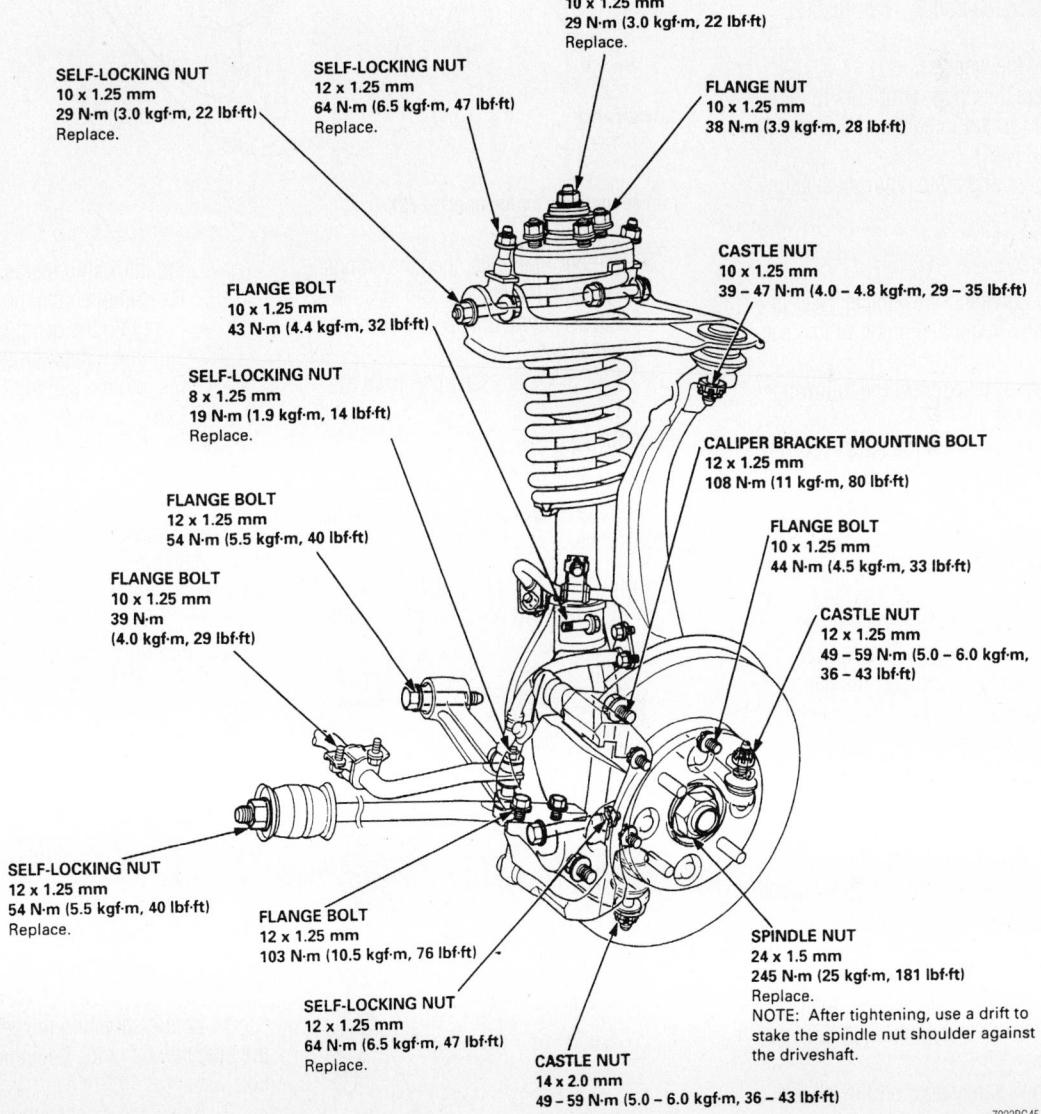

Front suspension showing the torque specifications

7923BG45

3. Remove or disconnect the following:
- Wheel(s)
- Brake hose from the strut
- Strut fork
- Upper strut mounting nuts
- Strut from the vehicle

To install:

4. Install the strut and loosely install the mounting nuts

5. Install the strut fork, do not tighten the bolts at this time

➡**All suspension nuts and bolts should be tightened with the vehicle on the ground, or with a floor jack supporting the vehicle's weight.**

6. Tighten the pinch bolt to 32 ft. lbs. (43 Nm) for all models except Integra. For Integra, tighten to 16 ft. lbs. (22 Nm).

7. Tighten the lower strut fork bolt to 47 ft. lbs. (64 Nm).

8. Install the brake hose to the strut fork and tighten the bolt to 16 ft. lbs. (22 Nm).

9. Install the wheels.

10. Tighten the upper strut nuts to 28 ft. lbs. (38 Nm) (except Integra and 3.2 models 37 ft. lbs. [50 Nm])

11. Measure and adjust the wheel alignment.

NSX

1. Before servicing the vehicle, refer to the precautions in the beginning of this section.

2. Remove or disconnect the following:
- Wheel(s)
- Brake hose from the strut
- Lower strut mounting bolt
- Upper strut mounting nuts
- Strut from the vehicle

To install:

3. Install or connect the following:
- Strut to the vehicle and loosely install the lower mounting bolt
- New upper mounting nuts hand tight
- Brake hose and torque the bolt to 16 ft. lbs. (22 Nm)
- Wheel(s)

4. Lower the vehicle.

5. Torque the upper mounting nuts to 32 ft. lbs. (43 Nm) and the lower mounting bolt to 69 ft. lbs. (93 Nm).

TSX

1. Before servicing the vehicle, refer to the precautions in the beginning of this section.

2. Raise and safely support the vehicle.

3. Remove or disconnect the following:
- Front wheels

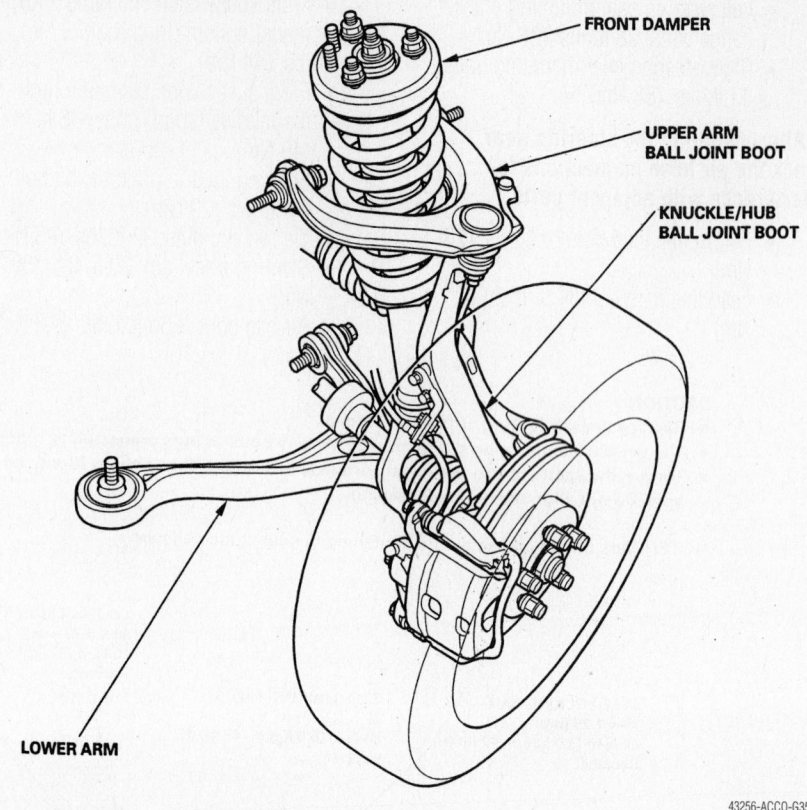

Front suspension components—TSX

43256-ACCO-G35

- Damper fork bolts, then the damper fork from the damper and lower arm
- 2 8mm flange nuts and 3 10mm flange nuts
- Strut (damper assembly) from the vehicle

To install:

➡**Use new self-locking bolts when installing the struts and assembling the damper forks.**

4. Install or connect the following:
- Strut into the vehicle with the align-

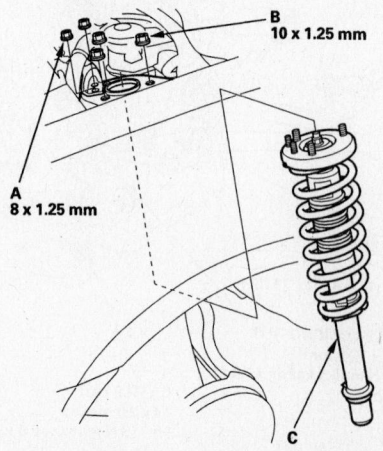

Front strut (damper) mounting—TSX

43256-ACCO-G36

ing tab facing inside, if equipped. Hand-tighten the mounting nuts.
- Strut into the damper fork. The alignment mark on the strut tube fits into the slot on the damper fork.
- Pinch bolt and damper fork bolt/nut. Only hand-tighten these bolts.
- Front wheels and lower the vehicle.

5. With all 4 of the vehicle's wheels on the ground, tighten the damper fork nut to 47 ft. lbs. (65 Nm) while holding the damper fork bolt. Tighten the damper fork pinch bolt to 32 ft. lbs. (44 Nm). Tighten the damper assembly 8mm bolts to 16 ft. lbs. (22 Nm) and the 10mm bolts to 37 ft. lbs. (50 Nm).

6. Tighten the wheel nuts to 80 ft. lbs. (110 Nm).

7. Check and adjust the vehicle's front end alignment.

Rear

3.2TL

1. Before servicing the vehicle, refer to the precautions in the beginning of this section.

2. Raise and safely support the vehicle and remove the rear wheels.

3. Remove the rear seat:

a. Remove the lower cushion bolts located under the armrest and near the floor.

b. Pull the rear of the lower cushion up and lift it forward to release it from the clips.

c. Pull down the trunk bulkhead trim and release the armrest lid clips.

d. Remove the bolts from the top and bottom of the back cushion, then lift it up and forward to disengage the securing hooks.

4. Place a floor jack under the lower arm and slightly compress the spring.

5. Remove the upper mounting nuts and the lower flange bolt.

6. Lower the jack to remove the strut. Be sure and mark the right and left struts so they can be reinstalled on the proper sides.

To install:

7. Install the strut into the vehicle. Loosely install the mounting nuts and mounting bolt, but do not tighten them until the weight of the vehicle is on the suspension.

8. Raise the rear suspension with a floor jack until the weight of the vehicle is on the strut. Tighten the upper mounting nuts to 28 ft. lbs. (39 Nm), then tighten the

CAUTION:
- Replace the self-locking nuts after removal.
- The vehicle should be on the ground before any bolts or nuts connected to rubber mounts or bushings are tightened.
- Torque the castle nut to the lower torque specification, then tighten it only far enough to align the slot with the pin hole. Do not align the nut by loosening.

NOTE: Wipe off the grease before tightening the nut at the ball joint.

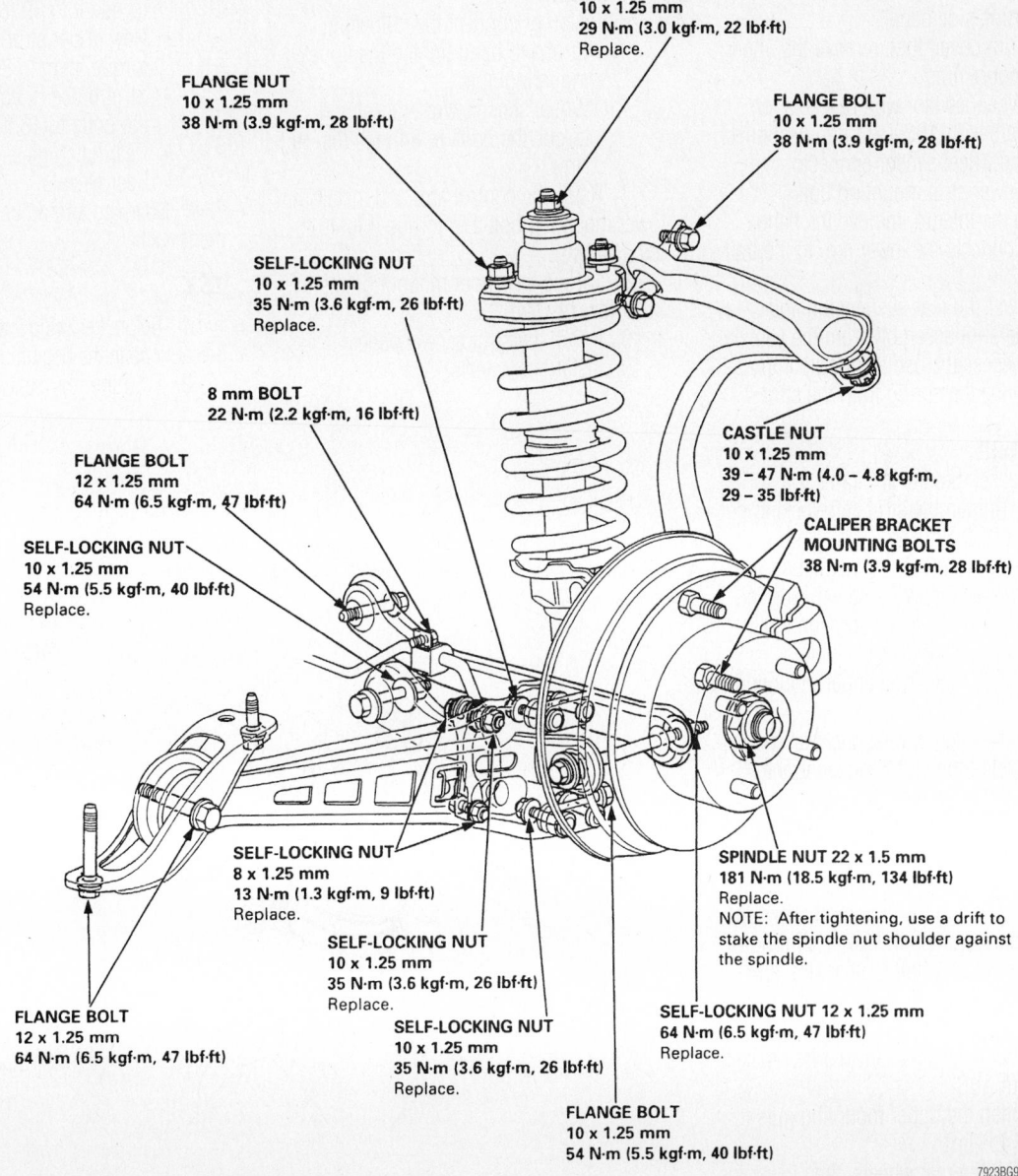

SELF-LOCKING NUT
10 x 1.25 mm
29 N·m (3.0 kgf·m, 22 lbf·ft)
Replace.

FLANGE NUT
10 x 1.25 mm
38 N·m (3.9 kgf·m, 28 lbf·ft)

FLANGE BOLT
10 x 1.25 mm
38 N·m (3.9 kgf·m, 28 lbf·ft)

SELF-LOCKING NUT
10 x 1.25 mm
35 N·m (3.6 kgf·m, 26 lbf·ft)
Replace.

8 mm BOLT
22 N·m (2.2 kgf·m, 16 lbf·ft)

FLANGE BOLT
12 x 1.25 mm
64 N·m (6.5 kgf·m, 47 lbf·ft)

SELF-LOCKING NUT
10 x 1.25 mm
54 N·m (5.5 kgf·m, 40 lbf·ft)
Replace.

CASTLE NUT
10 x 1.25 mm
39 – 47 N·m (4.0 – 4.8 kgf·m, 29 – 35 lbf·ft)

CALIPER BRACKET MOUNTING BOLTS
38 N·m (3.9 kgf·m, 28 lbf·ft)

SELF-LOCKING NUT
8 x 1.25 mm
13 N·m (1.3 kgf·m, 9 lbf·ft)
Replace.

SELF-LOCKING NUT
10 x 1.25 mm
35 N·m (3.6 kgf·m, 26 lbf·ft)
Replace.

SELF-LOCKING NUT
10 x 1.25 mm
35 N·m (3.6 kgf·m, 26 lbf·ft)
Replace.

FLANGE BOLT
12 x 1.25 mm
64 N·m (6.5 kgf·m, 47 lbf·ft)

FLANGE BOLT
10 x 1.25 mm
54 N·m (5.5 kgf·m, 40 lbf·ft)

SPINDLE NUT 22 x 1.5 mm
181 N·m (18.5 kgf·m, 134 lbf·ft)
Replace.
NOTE: After tightening, use a drift to stake the spindle nut shoulder against the spindle.

SELF-LOCKING NUT 12 x 1.25 mm
64 N·m (6.5 kgf·m, 47 lbf·ft)
Replace.

7923BG92

Rear suspension showing the torque specifications

lower mounting bolts to 40 ft. lbs. (55 Nm). Be careful not to pinch the ABS speed sensor wire between the strut and bracket.

9. Install the rear wheels and lower the vehicle.

10. Install the rear seat back and torque the bolts to 84 inch lbs. (9.5 Nm).

11. Install the armrest lid.

12. Install the lower seat cushion and torque the bolts to 84 inch lbs. (9.5 Nm).

3.5RL

1. Before servicing the vehicle, refer to the precautions in the beginning of this section.

2. Raise and safely support the vehicle and remove the rear wheels.

3. Remove or disconnect the following:
- Upper strut mount cover from the rear panel, just below the speaker—On sedans, remove the trunk side panel.
- Trim cover, then remove the upper mount nuts
- Wheel sensor wire brackets, on cars with ABS—do not disconnect the wheel sensor connector
- Lower strut mounting bolt

4. On the Integra, remove the flange bolt that connects the lower arm to the trailing arm.

5. Lower the rear suspension and remove the strut assembly from the vehicle.

6. If necessary, use a spring compressor to remove the spring from the strut assembly.

To install:

7. Reassemble the strut and coil spring assembly. Tighten the strut self–locking nut to 22 ft. lbs. (30 Nm).

8. Lower the rear suspension and position the strut assembly in the vehicle. The nut welded to the lower strut mounting should face the front of the vehicle.

9. Loosely install the upper mounting nuts.

10. On the Integra, raise the rear suspension and install the bolt connecting the lower arm to the trailing arm.

11. Install the strut lower mounting bolt.

12. Raise the vehicle until the vehicle just lifts off the safety stand and tighten the lower strut bolt and lower control arm bolt. On the Integra, tighten the lower strut bolt and the control arm bolt to 40 ft. lbs. (54 Nm). In the 3.5RL, tighten the lower strut mounting bolt to 76 ft. lbs. (103 Nm).

13. Install the wheel sensor wire bracket on cars with ABS.

14. Tighten the upper mounting nuts to 36 ft. lbs. (49 Nm).

15. Install the rear wheels, then lower the vehicle.

16. Install the trim panel, or the trunk side panel.

17. Check the vehicle's alignment.

INTEGRA AND RSX

1. Before servicing the vehicle, refer to the precautions in the beginning of this section.

2. Raise and safely support the vehicle.

3. Support the control arm with a jack.

4. Remove or disconnect the following:
- Rear wheels
- Strut access panel
- Upper mounting nuts
- Wheel sensor wire brackets (if equipped)
- Lower mounting bolt

5. Lower the jack and remove the strut from the vehicle.

To install:

6. Install or connect the following:
- Strut and hand tighten the upper mounting nuts
- Wheel sensor wire bracket and torque the bolts to 84 inch lbs. (9.5 Nm)

7. Raise the control arm and install the lower mounting bolt and torque it to 40 ft. lbs. (54 Nm).

8. Torque the upper mounting nuts to 36 ft. lbs. (49 Nm).

9. Install the strut access panel.

10. Install the wheels.

NSX

1. Before servicing the vehicle, refer to the precautions in the beginning of this section.

2. Remove or disconnect the following:
- Rear wheels
- Lower rear hatch glass trim
- Upper strut mounting nuts
- Rear strut brace (NSX–T model only)
- Lower strut mounting nut
- Stabilizer link
- Strut from the vehicle

To install:

3. Install or connect the following:
- Strut
- Stabilizer link and torque the nut (at the stabilizer bar) to 61 ft. lbs. (83 Nm)
- Lower strut mounting bolt and torque it to 69 ft. lbs. (93 Nm)
- New upper strut mounting nuts and torque them to 39 ft. lbs. (53 Nm)
- Strut brace (if equipped) and torque the bolts to 16 ft. lbs. (22 Nm)
- Rear hatch glass trim
- Rear wheels

4. Measure and adjust the wheel alignment.

TSX

1. Before servicing the vehicle, refer to the precautions in the beginning of this section.

2. Fold the rear seat forward.

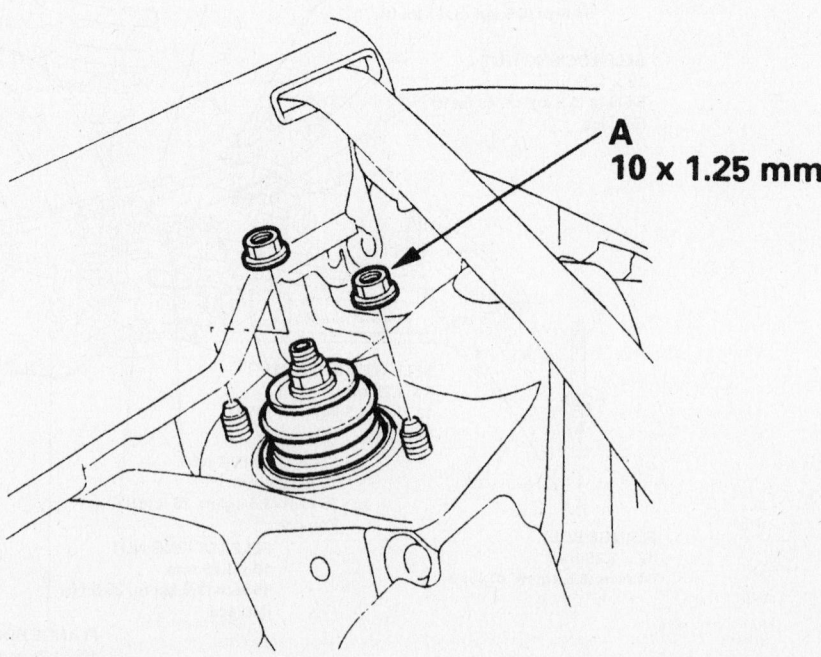

Rear strut upper mounting nut locations—TSX

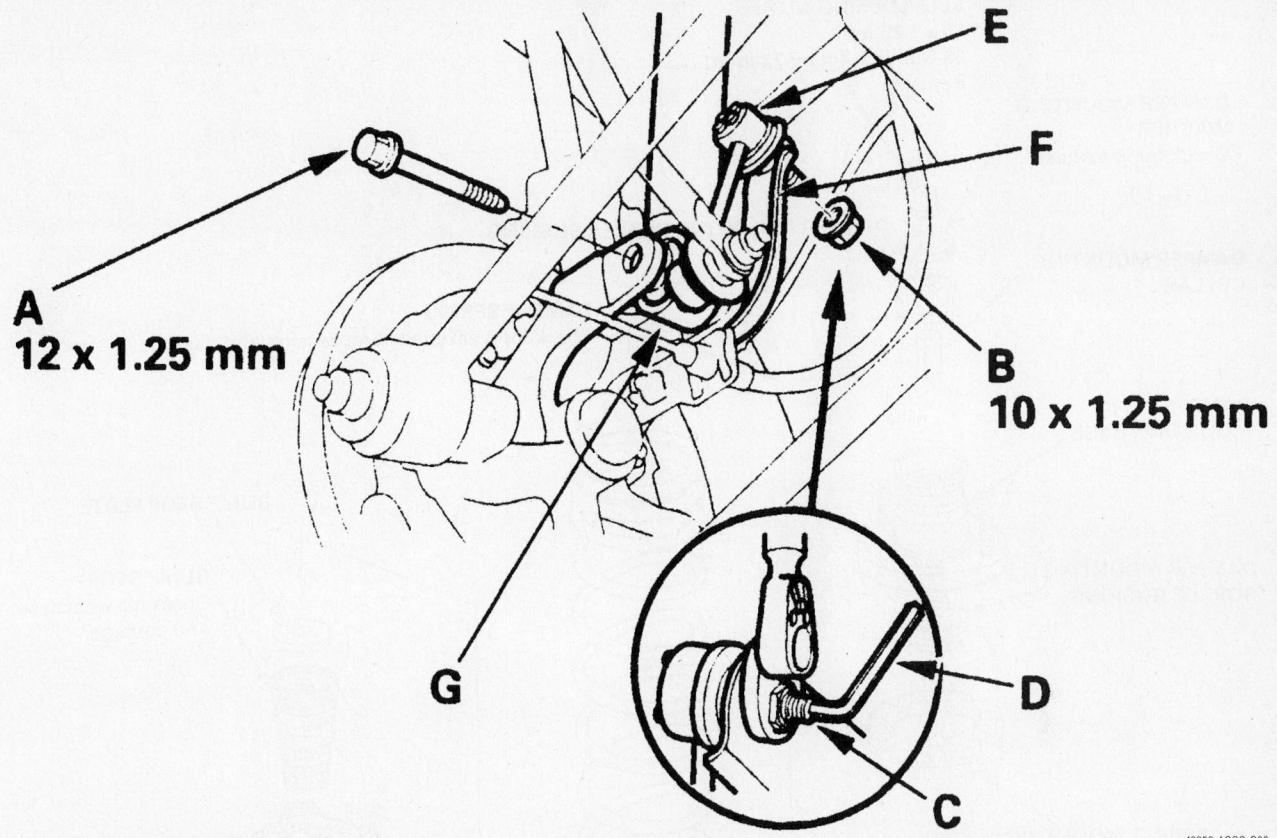

43256-ACCO-G38

Remove the lower flange (A) bolt from the knuckle, then remove the flange nut (B) while holding the joint pin (C) with a hex wrench (D) and disconnect the stabilizer link (E) from the bracket (F)

3. Remove or disconnect the following:
- Rear shelf cover
- Seat side bolster cushions. The side bolster cushions are secured by 1 screw at the bottom, and 2 clips at the top.
- Strut mount cap, if equipped, and upper strut mounting nuts

4. Raise and safely support the vehicle.
- Rear wheels, then support the knuckle with a floor jack
- Strut lower flange bolt from the knuckle
- Strut flange nut while holding the joint pin with a hex wrench
- Stabilizer link from the bracket
- Strut, while lowering rear suspension

➡ **The left and right struts are different, so be sure to mark them L & R if you are removing both struts before continuing.**

To install:

➡ **Use new self-locking nuts when installing the strut.**

5. Lower the rear suspension.
6. Install or connect the following:
- Strut into the upper mount. Only hand-tighten the upper mounting nuts.
- Strut into position on the knuckle, then loosely install the flange bolt on the bottom of the strut
- Stabilizer link on the bracket and loosely install the flange nut

7. Place a jack under the lower strut mount. Raise the jack until the weight of the vehicle is on the jack.

8. With the suspension under load, tighten the lower mount bolt to 43 ft. lbs. (59 Nm).

9. While holding the joint pin with a cotter pin, tighten the flange nut to 29 ft. lbs. (39 Nm).

10. Tighten the upper nuts to 37 ft. lbs. (50 Nm).

11. Install or connect the following:
- Rear wheel(s). Lower the vehicle to the ground. Tighten the wheel nuts to 80 ft. lbs. (110 Nm).
- Rear seat side bolsters and fold the seat back into place.
- Rear bulkhead cover

12. Check and adjust the vehicle's rear wheel alignment.

Coil Spring

REMOVAL & INSTALLATION

1. Before servicing the vehicle, refer to the precautions in the beginning of this section.

2. Raise and support the vehicle and remove the front wheels.

3. Remove the strut (damper).

4. Place the strut assembly in a coil spring compressor.

5. Compress the coil spring and remove the locking nut from the top of the strut.

6. Release the pressure from the spring compressor.

7. Remove the coil spring and related pieces from the strut.

To install:

➡ **Use new self-locking nuts and bolts when assembling the strut.**

8. Install the strut, coil spring and related components on the spring compressor.

9. Compress the spring.

10. Install the mounting washer, and loosely install a new self-locking nut.

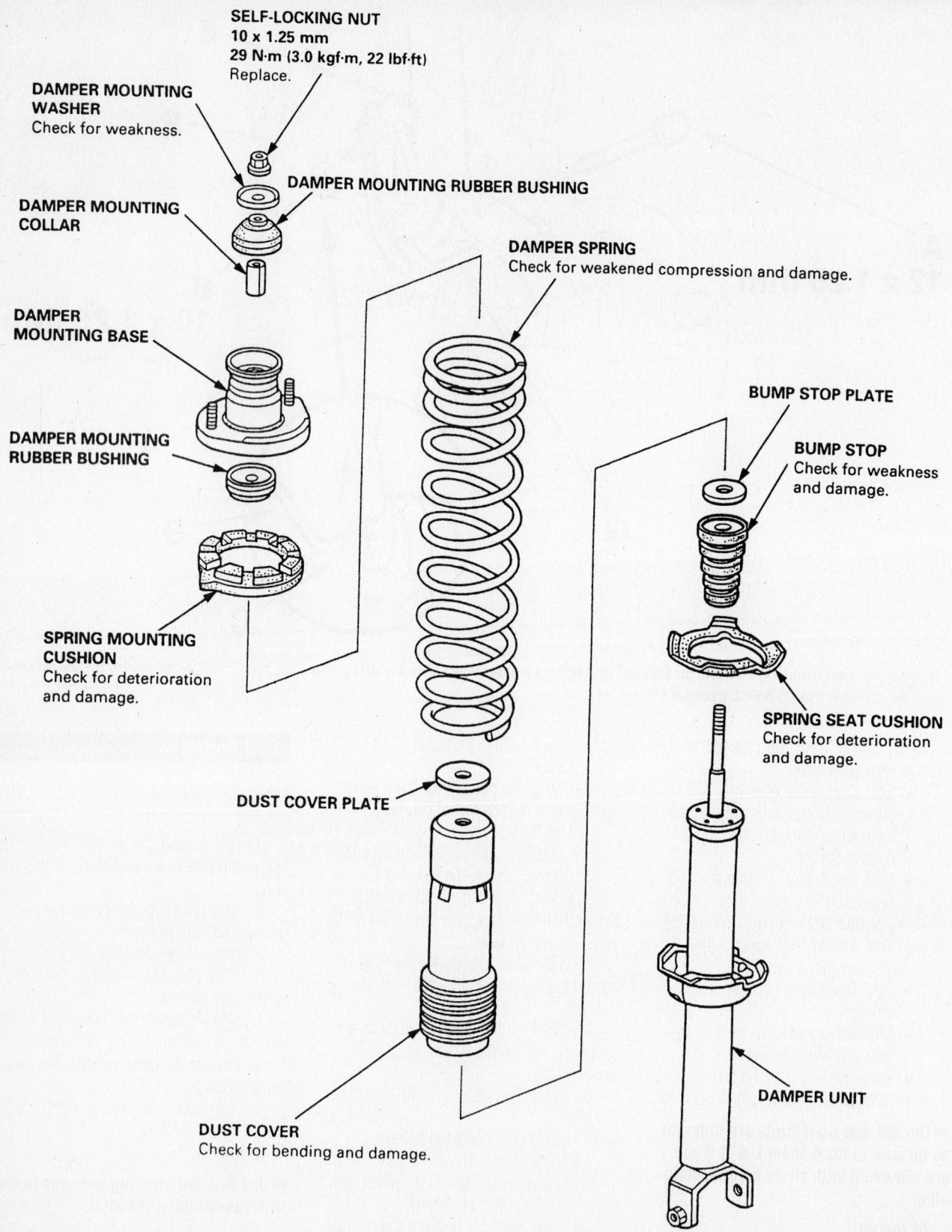

SELF-LOCKING NUT
10 x 1.25 mm
29 N·m (3.0 kgf·m, 22 lbf·ft)
Replace.

DAMPER MOUNTING WASHER
Check for weakness.

DAMPER MOUNTING RUBBER BUSHING

DAMPER MOUNTING COLLAR

DAMPER SPRING
Check for weakened compression and damage.

DAMPER MOUNTING BASE

DAMPER MOUNTING RUBBER BUSHING

BUMP STOP PLATE

BUMP STOP
Check for weakness and damage.

SPRING MOUNTING CUSHION
Check for deterioration and damage.

SPRING SEAT CUSHION
Check for deterioration and damage.

DUST COVER PLATE

DUST COVER
Check for bending and damage.

DAMPER UNIT

7923BG93

Exploded view of the rear strut (damper)

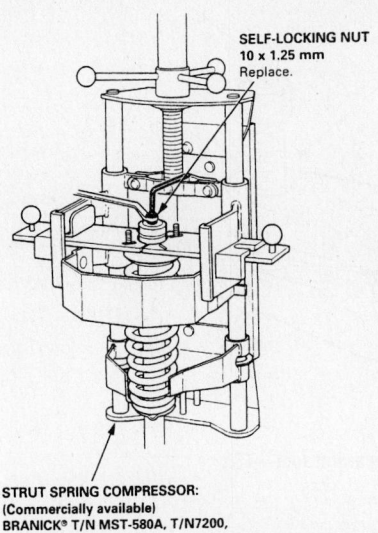

SELF-LOCKING NUT
10 x 1.25 mm
Replace.

STRUT SPRING COMPRESSOR:
(Commercially available)
BRANICK® T/N MST-580A, T/N7200,
or equivalent

7923BG94

Compress the coil spring until the spring moves away from the seat and use a hex wrench to hold the piston rod while removing the nut

11. Hold the strut piston rod with a hex wrench and tighten the self–locking nut to 22 ft. lbs. (30 Nm).

12. Install the strut in the vehicle.

13. Check and adjust the vehicle's front wheel alignment.

Upper Ball Joint

REMOVAL & INSTALLATION

➡The upper ball joint cannot be removed from the control arm. If the ball joint is damaged, the upper arm assembly must be replaced.

Lower Ball Joint

REMOVAL & INSTALLATION

Integra & RSX

1. Before servicing the vehicle, refer to the precautions in the beginning of this section.

2. Raise and support the vehicle safely.

3. Remove or disconnect the following:
- Front wheels
- Axle nut
- Brake hose from the knuckle
- Brake caliper
- Brake rotor
- Wheel sensor wire bracket
- Wheel sensor from the knuckle
- Tie rod from the knuckle
- Lower ball joint from the knuckle

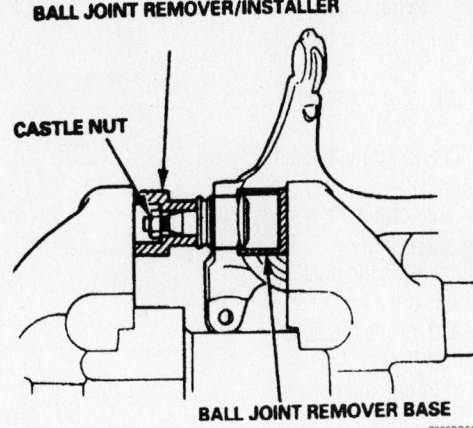

BALL JOINT REMOVER/INSTALLER

CASTLE NUT

BALL JOINT REMOVER BASE

7923BG50

Use a vise or press to remove the ball joint from the steering knuckle

- Upper ball joint from the knuckle
- Steering knuckle
- Boot by prying off the snapring

4. Press the ball joint from the knuckle.

To install:

5. Place the ball joint in position by hand. Install the ball joint into the tool and press in the new ball joint.

✳✳ WARNING

After installing the boot, check the ball joint pin tapered section for grease contamination and wipe it if necessary.

6. Install or connect the following:
- Ball joint boot and snapring

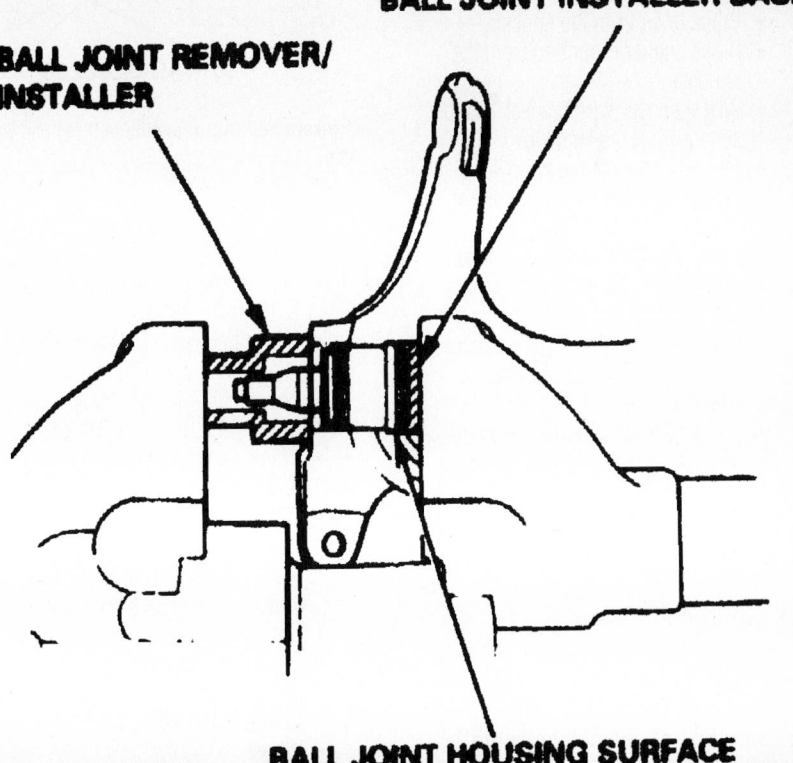

BALL JOINT INSTALLER BASE

BALL JOINT REMOVER/
INSTALLER

BALL JOINT HOUSING SURFACE

7923BG51

Press the new ball joint into the steering knuckle

- Steering knuckle to the vehicle
- Lower ball joint and torque the nut to 40 ft. lbs. (54 Nm)
- Tie rod and torque the nut to 32 ft. lbs. (43 Nm)
- Upper ball joint and torque the nut to 32 ft. lbs. (43 Nm)
- Wheel sensor and torque the bolts to 84 inch lbs. (9.5 Nm)
- Brake rotor and torque the bolts to 84 inch lbs. (9.5 Nm)
- Brake caliper and torque the bolts to 80 ft. lbs. (108 Nm)
- Brake hose and torque the bolts to 84 inch lbs. (9.5 Nm)
- New axle nut and torque it to 134 ft. lbs. (181 Nm)
- Front wheels

7. Check the front wheel alignment and adjust if necessary.

3.2TL, 3.2CL, 3.5RL, NSX and TSX

➡**The lower ball joint cannot be removed from the steering knuckle.**

1. Before servicing the vehicle, refer to the precautions in the beginning of this section.
2. Raise and safely support the vehicle
3. Remove or disconnect the following:
- Wheels
- Axle nut
- Brake hose from the knuckle
- Brake caliper mounting from the knuckle
- Wheel sensor wire bracket and the sensor from the knuckle
- Tie rod end from the knuckle
- Lower ball joint from the control arm
- Upper ball joint from the knuckle
- Knuckle and hub by sliding the assembly off the halfshaft. Tap the end of the halfshaft with a plastic mallet to release it from the knuckle.
- Hub and rotor assembly from the knuckle
- Splash guard from the knuckle

To install:
4. Install or connect the following:
- Splash guard and torque the bolts to 84 inch lbs. (9.5 Nm)
- Hub assembly and tighten the self–locking bolts to 33 ft. lbs. (45 Nm)

➡**Be sure that all the hub bolts are properly tightened to avoid warpage of the brake disc.**

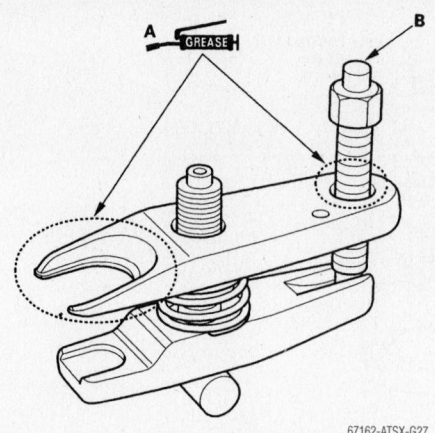

Apply grease to the A & B areas on the ball joint separator tool—TSX

67162-ATSX-G27

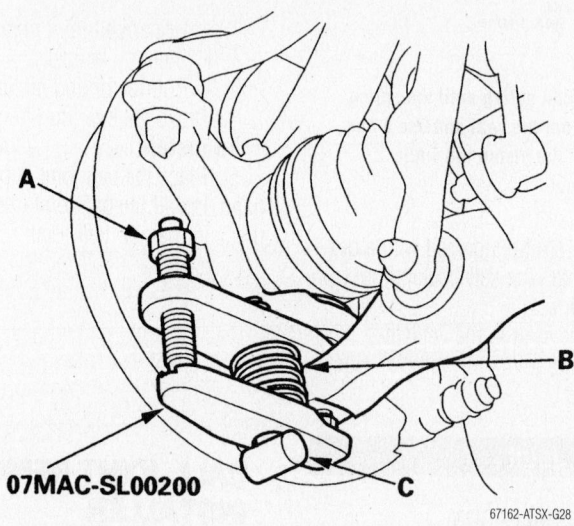

07MAC-SL00200

67162-ATSX-G28

Pressure bolt (A), adjusting bolt (B) and proper position of the head of the adjusting bolt (C)—TSX

- Knuckle and hub assembly onto the halfshaft
- Tie rod and torque the nut to 36–43 ft. lbs. (49–59 Nm)
- Upper ball joint and torque the nut to 29–35 ft. lbs. (39–47 Nm)
- Lower ball joint and torque the nut to 36–43 ft. lbs. (49–59 Nm)
- Wheel sensor and torque the bolts to 16 ft. lbs. (22 Nm)
- Wheel sensor wire and torque the bolts to 84 inch lbs. (9.5 Nm)
- Brake caliper and torque the bolts to 80 ft. lbs. (108 Nm)
- Brake hoses and torque the bolts to 84 inch lbs. (9.5 Nm)
- New axle nut and torque it to 181 ft. lbs. (245 Nm)
- Wheel
5. Measure and adjust the wheel alignment.

Upper Control Arm

REMOVAL & INSTALLATION

1. Before servicing the vehicle, refer to the precautions in the beginning of this section.
2. Raise and safely support the vehicle.
3. Remove or disconnect the following:
- Front wheel
- Front damper/strut
- Wheel speed sensor bracket from the upper control arm
- Cotter pin from the upper ball joint and loosen the nut
- Upper ball joint from the knuckle, using a suitable separator tool
- Upper control arm bolts or nuts, as applicable
- Upper control arm from the vehicle

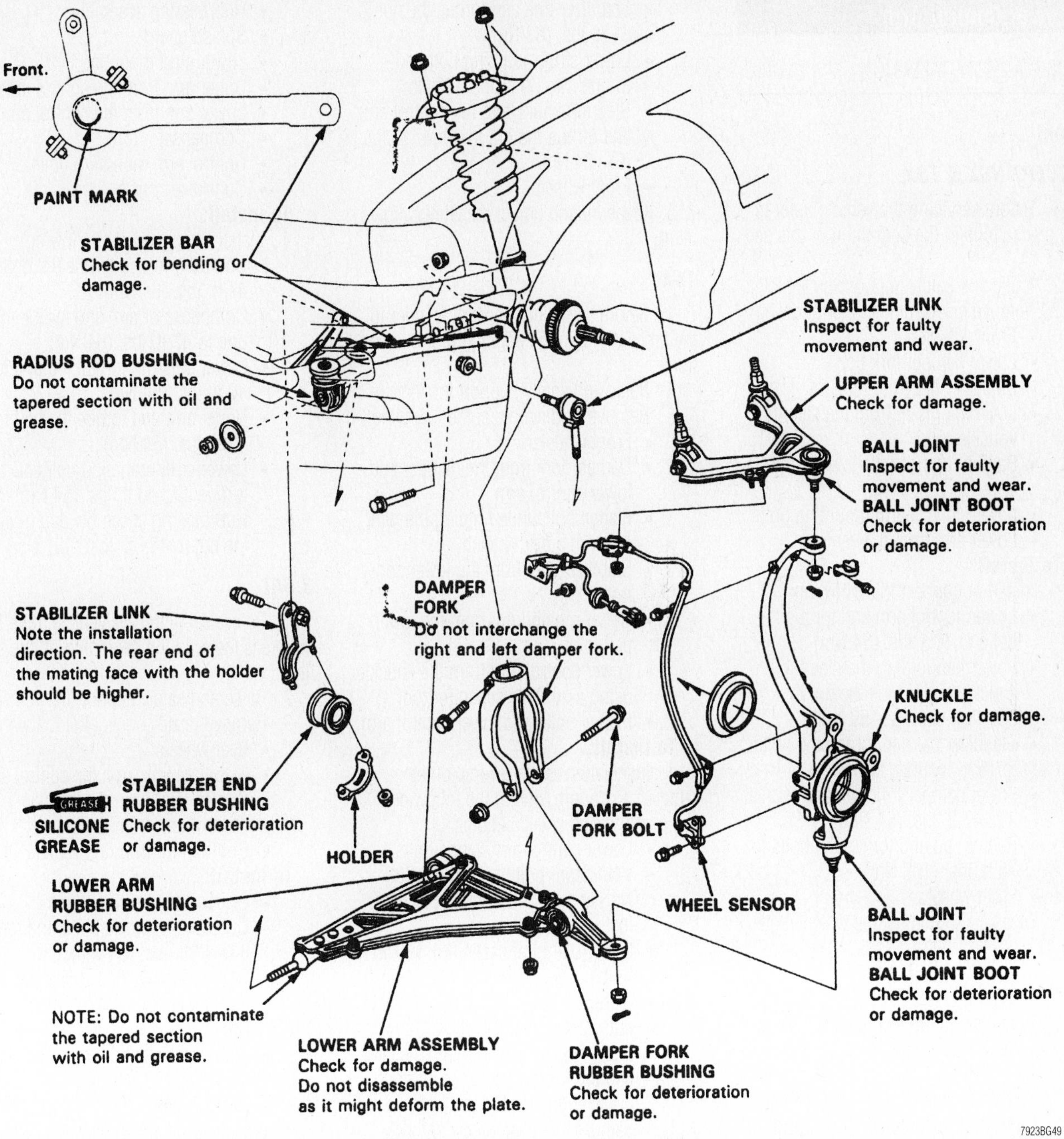

Front.

PAINT MARK

STABILIZER BAR
Check for bending or damage.

RADIUS ROD BUSHING
Do not contaminate the tapered section with oil and grease.

STABILIZER LINK
Note the installation direction. The rear end of the mating face with the holder should be higher.

SILICONE GREASE

STABILIZER END RUBBER BUSHING
Check for deterioration or damage.

LOWER ARM RUBBER BUSHING
Check for deterioration or damage.

HOLDER

NOTE: Do not contaminate the tapered section with oil and grease.

LOWER ARM ASSEMBLY
Check for damage.
Do not disassemble as it might deform the plate.

DAMPER FORK
Do not interchange the right and left damper fork.

DAMPER FORK BOLT

DAMPER FORK RUBBER BUSHING
Check for deterioration or damage.

WHEEL SENSOR

STABILIZER LINK
Inspect for faulty movement and wear.

UPPER ARM ASSEMBLY
Check for damage.

BALL JOINT
Inspect for faulty movement and wear.
BALL JOINT BOOT
Check for deterioration or damage.

KNUCKLE
Check for damage.

BALL JOINT
Inspect for faulty movement and wear.
BALL JOINT BOOT
Check for deterioration or damage.

7923BG49

A common upper control arm and ball joint assembly—early model vehicle

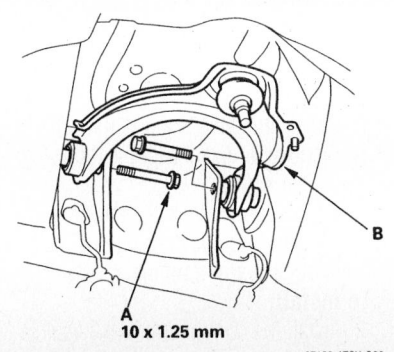

A
10 x 1.25 mm

67162-ATSX-G26

Upper control arm (B) and bolt (A) mounting—TSX shown

To install:
4. Install or connect the following:
   - Upper control arm
   - Upper control arm bolts: 23 ft. lbs. (31 Nm)
   - Upper control arm-to-chassis nuts and torque them to 47 ft. lbs. (64 Nm)
   - Ball joint to the steering knuckle and torque the nut to 29–35 ft. lbs. (39–47 Nm)
   - Front wheel
5. Check the front wheel alignment and adjust if necessary.

CONTROL ARM BUSHING REPLACEMENT

➡The bushings are an integral part of the control arm and are not replaceable. If they are damaged the control arm should be replaced.

Lower Control Arm

REMOVAL & INSTALLATION

Front

EXCEPT NSX & TSX

1. Before servicing the vehicle, refer to the precautions in the beginning of this section.
2. Raise and safely support the vehicle.
3. Remove or disconnect the following:
   - Front wheels
   - Lower damper fork bolt
   - Stabilizer bar from the arm
   - Lower ball joint from the steering knuckle
   - Radius rod from the lower control arm
   - Lower control arm mounting bolts
   - Lower arm from the vehicle

To install:

4. Install or connect the following:
   - Lower control arm and torque the bolts to 40 ft lbs. (54 Nm)
   - Lower ball joint to the steering knuckle and torque the nut to 36–43 ft. lbs. (49–59 Nm)
   - Stabilizer bar and torque the bolts to 16 ft. lbs. (22 Nm)
   - Lower damper fork bolt and torque it to 47 ft. lbs. (64 Nm)
   - Radius rod and torque the bolts to 76 ft. lbs. (103 Nm)
   - Front wheels
5. Measure and adjust the wheel alignment.

NSX

1. Before servicing the vehicle, refer to the precautions in the beginning of this section.
2. Raise and safely support the vehicle.
3. Remove or disconnect the following:
   - Front wheels
   - Steering knuckle from the control arm
   - Lower strut mounting bolt
   - Stabilizer link from the control arm
   - Ball joint from the compliance pivot assembly
   - Control arm adjusting bolt
   - Control arm from the vehicle

To install:

4. Install or connect the following:
   - Lower control arm to the vehicle and torque the adjusting bolt to 90 ft. lbs. (123 Nm)
   - Ball joint to the compliance pivot assembly and torque the nut to 40–47 ft. lbs. (54–64 Nm)

- Stabilizer link and torque the nut to 61 ft. lbs. (83 Nm)
- Lower strut mounting bolt and torque it to 69 ft. lbs. (93 Nm)
- Steering knuckle to the control arm and torque the nut to 40–47 ft. lbs. (54–64 Nm)
- Front wheels

5. Measure and adjust the wheel alignment.

TSX

1. Before servicing the vehicle, refer to the precautions in the beginning of this section.
2. Raise and safely support the vehicle.
3. Remove or disconnect the following:
   - Front wheels
   - Damper fork from the damper and lower control arm
   - Flange nut, while holding the joint pin with a hex wrench
   - Stabilizer link from the lower control arm
   - Cotter pin and nut from the lower ball joint
   - Lower control arm from the knuckle using a suitable separator tool
   - Flange bolts and lower control arm

To install:

4. Installation is the reverse of the removal procedure, noting the following steps and torque specifications:
   - Lower control arm flange bolts: 14x1.5mm bolt to 61 ft. lbs. (83 Nm) and 12x1.25mm bolt to 47 ft. lbs. (64 Nm)
   - Lower ball joint castle nut: 58–65 ft. lbs. (78–88 Nm)
   - Stabilizer link-to-control arm flange nut: 22 ft. lbs. (29 Nm)
   - Insert the damper fork into the damper lower end so the aligning tab is aligned with the slot in the damper fork. Use a new damper fork mounting nut
   - Lower damper fork mounting bolt: 47 ft. lbs. (64 Nm)
   - Upper damper fork mounting bolt: 32 ft. lbs. (44 Nm)
   - Check and adjust the front wheel alignment

Rear

INTEGRA AND RSX

1. Before servicing the vehicle, refer to the precautions in the beginning of this section.
2. Raise and safely support the vehicle.
3. Remove or disconnect the following:
   - Rear wheels

- Hub/bearing assembly
- Splash guard
- Lower strut mounting bolt
- Upper arm from the control arm
- Lower arm from the control arm
- Compensator arm
- Control arm mounting bolts
- Control arm from the vehicle

To install:

4. Install or connect the following:
   - Control arm and torque the bolts to 47 ft. lbs. (64 Nm)
   - Compensator arm and torque the bolt to 47 ft. lbs. (64 Nm)
   - Lower arm and torque the bolt to 40 ft. lbs. (54 Nm)
   - Upper arm and torque the bolt to 40 ft. lbs. (54 Nm)
   - Lower strut mounting bolt and torque it to 40 ft. lbs. (54 Nm)
   - Hub/bearing assembly and torque the nut to 134 ft. lbs. (181 Nm)

3.5RL

1. Before servicing the vehicle, refer to the precautions in the beginning of this section.
2. Raise and safely support the vehicle.
3. Remove or disconnect the following:
   - Rear wheels
   - Stabilizer link from the control arm
   - Control arm from the knuckle
   - Control arm from the bracket
   - Control arm from the vehicle

To install:

- Control arm to the vehicle
- Control arm bracket bolt and torque it to 47 ft. lbs. (64 Nm)
- Control arm to the knuckle and torque the nuts to 47 ft. lbs. (64 Nm)
- Stabilizer link and torque the nut to 22 ft. lbs. (29 Nm)
- Rear wheels

4. Measure and adjust the wheel alignment.

3.2CL AND 3.2TL

1. Before servicing the vehicle, refer to the precautions in the beginning of this section.
2. Raise and safely support the vehicle.
3. Remove or disconnect the following:
   - Rear wheels
   - Control arm from the knuckle
   - Control arm from the subframe
   - Control arm from the vehicle

To install:

4. Install the control arm to the vehicle and torque the subframe nut to 40 ft. lbs. (54 Nm) and the knuckle nut to 43 ft. lbs. (59 Nm).

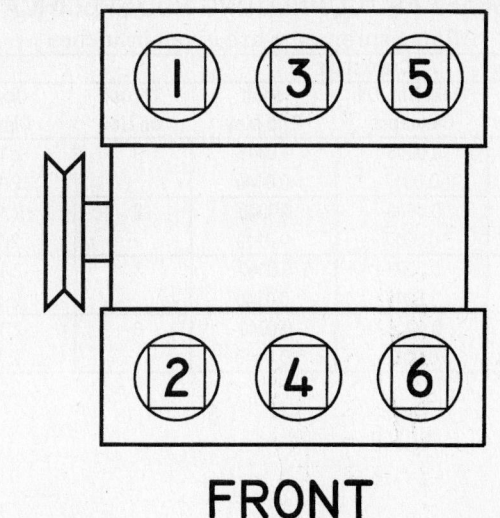

FRONT

9308MG32

3.5L Engine
Firing Order: 1–2–3–4–5–6
Distributorless ignition system (One coil per cylinder)

CAPACITIES

Year	Model	Engine Displacement Liters (VIN)	Engine Oil with Filter (qts.)	Transmission (pts.) 5-Spd	Transmission (pts.) Auto.	Transfer Case (pts.)	Drive Axle Front (pts.)	Drive Axle Rear (pts.)	Fuel Tank (gal.)	Cooling System (qts.)
2001	MDX	3.5 (J35A3)	5.0	—	9.0	—	—	—	19.2	8.0
2002	MDX	3.5 (J35A3)	5.0	—	9.0	—	—	—	19.2	8.0
2003	MDX	3.5 (J35A3)	5.0	—	9.0	—	—	—	19.2	8.0
2004-05	MDX	3.5 (J35A3)	5.0	—	9.0	—	—	—	19.2	8.0

NOTE: All capacities are approximate. Add fluid gradually and check to be sure a proper fluid level is obtained.

67162-AMDX-C04

VALVE SPECIFICATIONS

Year	Engine Displacement Liters (cc)	Seat Angle (deg.)	Face Angle (deg.)	Spring Test Pressure (lbs. @ in.)	Spring Installed Height (in.)	Stem-to-Guide Clearance (in.) Intake	Stem-to-Guide Clearance (in.) Exhaust	Stem Diameter (in.) Intake	Stem Diameter (in.) Exhaust
2001	3.5 (J35A3)	45	45	NA	①	0.0008-0.0018	0.0022-0.0031	0.2159-0.2163	0.2146-0.2150
2002	3.5 (J35A3)	45	45	NA	①	0.0008-0.0018	0.0022-0.0031	0.2159-0.2163	0.2146-0.2150
2003	3.5 (J35A3)	45	45	NA	①	0.0008-0.0018	0.0022-0.0031	0.2159-0.2163	0.2146-0.2150
2004-05	3.5 (J35A3)	45	45	NA	①	0.0008-0.0018	0.0022-0.0031	0.2159-0.2163	0.2146-0.2150

NA: Not Available

① Valve spring free length:
   Intake: 1.9713 in.
   Exhaust: 2.1060 in.

67162-AMDX-C05

CRANKSHAFT AND CONNECTING ROD SPECIFICATIONS
All measurements are given in inches

Year	Engine Displacement Liters (cc)	Crankshaft Main Brg. Journal Dia.	Main Brg. Oil Clearance	Shaft End-play	Thrust on No.	Connecting Rod Journal Diameter	Oil Clearance	Side Clearance
2001	3.5 (J35A3)	2.8337-2.8346	0.0008-0.0017	0.0040-0.0140	3	2.1644-2.1654	0.0008-0.0017	0.0060-0.0140
2002	3.5 (J35A3)	2.8337-2.8346	0.0008-0.0017	0.0040-0.0140	3	2.1644-2.1654	0.0008-0.0017	0.0060-0.0140
2003	3.5 (J35A3)	2.8337-2.8346	0.0008-0.0017	0.0040-0.0140	3	2.1644-2.1654	0.0008-0.0017	0.0060-0.0140
2004-05	3.5 (J35A3)	2.8337-2.8346	0.0008-0.0017	0.0040-0.0140	3	2.1644-2.1654	0.0008-0.0017	0.0060-0.0140

67162-AMDX-C06

PISTON AND RING SPECIFICATIONS
All measurements are given in inches

Year	Engine Displacement Liters (cc)	Piston Clearance	Ring Gap Top Compression	Bottom Compression	Oil Control	Ring Side Clearance Top Compression	Bottom Compression	Oil Control
2001	3.5 (J35A3)	0.0006-0.0016	0.0080-0.0140	0.0160-0.0220	0.0080-0.0280	0.0014-0.0024	0.0012-0.0022	NA
2002	3.5 (J35A3)	0.0006-0.0016	0.0080-0.0140	0.0160-0.0220	0.0080-0.0280	0.0014-0.0024	0.0012-0.0022	NA
2003	3.5 (J35A3)	0.0006-0.0016	0.0080-0.0140	0.0160-0.0220	0.0080-0.0280	0.0014-0.0024	0.0012-0.0022	NA
2004-05	3.5 (J35A3)	0.0006-0.0016	0.0080-0.0140	0.0160-0.0220	0.0080-0.0280	0.0014-0.0024	0.0012-0.0022	NA

NA: Not Available

67162-AMDX-C07

TORQUE SPECIFICATIONS
All readings in ft. lbs.

Year	Engine Displacement Liters (VIN)	Cylinder Head Bolts	Main Bearing Bolts	Rod Bearing Bolts	Crankshaft Damper Bolts	Flywheel Bolts	Manifold Intake	Exhaust	Spark Plugs	Oil Pan Drain Plug
2001	3.5 (J35A1)	①	②	③	181	54	16	23	13	29
2002	3.5 (J35A1)	①	②	③	181	54	16	23	13	29
2003	3.5 (J35A1)	①	②	③	181	54	16	23	13	29
2004-05	3.5 (J35A1)	①	②	③	181	54	16	23	13	29

NOTE: Dip main bearing bolts and crankshaft damper bolt in clean engine oil prior to tightening.

① Step 1: 29 ft. lbs.
Step 2: 51 ft. lbs.
Step 3: 72 ft. lbs.

② 11mm bolt 56 ft. lbs.
10mm bolt 36 ft. lbs.

③ Step 1: 14 ft. lbs.
Step 2: 90 degrees

67162-AMDX-C08

WHEEL ALIGNMENT

Year	Model		Caster Range (+/-Deg.)	Caster Preferred Setting (Deg.)	Camber Range (+/-Deg.)	Camber Preferred Setting (Deg.)	Toe-in (in.)
2001	MDX	F	1.00	1.00	1.00	30	0+/-1/16
		R	—	—	0	30	0+/-1/16
2002	MDX	F	1.00	1.00	1.00	30	0+/-1/16
		R	—	—	0	30	0+/-1/16
2003	MDX	F	1.00	1.00	1.00	30	0+/-1/16
		R	—	—	0	30	0+/-1/16
2004-05	MDX	F	1.00	1.00	1.00	30	0+/-1/16
		R	—	—	0	30	0+/-1/16

67162-AMDX-C09

TIRE, WHEEL AND BALL JOINT SPECIFICATIONS

Year	Model	OEM Tires Standard	OEM Tires Optional	Tire Pressures (psi) Front	Tire Pressures (psi) Rear	Wheel Size	Ball Joint Inspection	Lug Nut
2001	MDX	P235/65R17	None	32	32	R17	NS	80
2002	MDX	P235/65R17	None	32	32	R17	NS	80
2003	MDX	P235/65R17	None	32	32	R17	NS	80
2004-05	MDX	P235/65R17	None	32	32	R17	NS	80

OEM: Original Equipment Manufacturer

PSI: Pounds Per Square Inch

NS: Not specified by manufacturer

67162-AMDX-C10

BRAKE SPECIFICATIONS
Acura MDX
All measurements in inches unless noted

Year	Model		Brake Disc Original Thickness	Brake Disc Minimum Thickness	Maximum Runout	Brake Drum Diameter Original Inside Diameter	Max. Wear Limit	Maximum Machine Diameter	Minimum Lining Thickness Front	Minimum Lining Thickness Rear	Brake Caliper Bracket Bolts (ft. lbs.)	Brake Caliper Mounting Bolts (ft. lbs.)
2001	MDX	F	1.100	1.020	0.004	—	—	—	0.060	—	80	27
		R	0.430	0.350	0.004	—	—	—	—	0.41	41	27
2002	MDX	F	1.100	1.020	0.004	—	—	—	0.060	—	80	27
		R	0.430	0.350	0.004	—	—	—	—	0.41	41	27
2003	MDX	F	1.100	1.020	0.004	—	—	—	0.060	—	80	27
		R	0.430	0.350	0.004	—	—	—	—	0.41	41	27
2004-05	MDX	F	1.100	1.020	0.004	—	—	—	0.060	—	80	27
		R	0.430	0.350	0.004	—	—	—	—	0.41	41	27

F: Front

R: Rear

67162-AMDX-C11

SCHEDULED MAINTENANCE INTERVALS
ACURA—MDX

TO BE SERVICED	TYPE OF SERVICE	VEHICLE MILEAGE INTERVAL (x1000)															
		7.5	15	22.5	30	37.5	45	52.5	60	67.5	75	82.5	90	97.5	105	112.5	120
Accessory drive belts	I & A				✓				✓				✓				✓
Air cleaner element	R				✓				✓				✓				✓
Brake fluid	R	Every 3 years															
Brake hoses & lines (including ABS)	I		✓		✓		✓		✓		✓		✓		✓		✓
Cooling system hoses & connections	I		✓		✓		✓		✓		✓		✓		✓		✓
Engine coolant ①	R						✓						✓				
Engine oil	R	✓	✓	✓	✓	✓	✓	✓	✓	✓	✓	✓	✓	✓	✓	✓	✓
Engine oil and coolant levels	I	Inspect at each fuel stop															
Engine oil filter	R		✓		✓		✓		✓		✓		✓		✓		✓
Exhaust system	I		✓		✓		✓		✓		✓		✓		✓		✓
Fluid levels and condition	I		✓		✓		✓		✓		✓		✓		✓		✓
Front and rear brakes	I		✓		✓		✓		✓		✓		✓		✓		✓
Fuel lines & connection	I		✓		✓		✓		✓		✓		✓		✓		✓
Halfshaft boots	I		✓		✓		✓		✓		✓		✓		✓		✓
Idle speed	I & A														✓		
Parking brake system	I & A		✓		✓		✓		✓		✓		✓		✓		✓
Rear differential fluid	R	✓			✓		✓		✓				✓				✓
Rotate and inspect tires	I	✓	✓	✓	✓	✓	✓	✓	✓	✓	✓	✓	✓	✓	✓	✓	✓
Spark plugs	R														✓		
Supplemental Restrain system (SRS)	I	Inspect the SRS 10 years after production															
Suspension components	I		✓		✓		✓		✓		✓		✓		✓		✓
Tie rod ends, steering gear box & boots	I		✓		✓		✓		✓		✓		✓		✓		✓
Timing belt	R														✓		
Transmission fluid	R						✓				✓				✓		
Valve clearance	I	Adjust if valves are noisy															
Water pump	S/I														✓		

R: Replace   I: Inspect   A: Adjust

① Every 12,000 miles or 10 years, then every 60,000 miles or 5 years

FREQUENT OPERATION MAINTENANCE (SEVERE SERVICE)

If a vehicle is operated under any of the following conditions it is considered severe service:

- Towing a trailer or using a camper or car-top carrier.
- Repeated short trips of less than 5 miles in temperatures below freezing, or trips of less than 10 miles in any temperature.
- Extensive idling or low-speed driving for long distances as in heavy commercial use, such as delivery, taxi or police cars.
- Operating on rough, muddy or salt-covered roads.
- Operating on unpaved or dusty roads.
- Driving in extremely hot (over 90°) conditions.

Air cleaner element: replace every 15,000 miles

Engine oil and filter: replace every 3750 miles or 6 months, whichever occurs first.

Timing belt: replace every 60,000 miles if the vehicle is regularly driven in temperatures above 110°F or below -20°F, or if frequently towing a trailer.

Transmission fluid: replace every 30,000 miles.

Rear differential fluid: replace every 60,000 miles.

Front and rear brakes: inspect every 7500 miles or 6 months, whichever occurs first.

Locks and hinges: lubricate every 15,000 miles.

Tie rods, steering gear box, boots: inspect every 7500 miles or 6 months, whichever occurs first.

Suspension components: inspect every 7500 miles or 6 months, whichever occurs first.

Halfshaft boots: inspect every 7500 miles or 6 months, whichever occurs first.

PRECAUTIONS

Before servicing any vehicle, please be sure to read all of the following precautions, which deal with personal safety, prevention of component damage, and important points to take into consideration when servicing a motor vehicle:

• Never open, service or drain the radiator or cooling system when the engine is hot; serious burns can occur from the steam and hot coolant.

• Observe all applicable safety precautions when working around fuel. Whenever servicing the fuel system, always work in a well-ventilated area. Do not allow fuel spray or vapors to come in contact with a spark, open flame or excessive heat (a hot drop light, for example). Keep a dry chemical fire extinguisher near the work area. Always keep fuel in a container specifically designed for fuel storage; also, always properly seal fuel containers to avoid the possibility of fire or explosion. Refer to the additional fuel system precautions later in this section.

• Fuel injection systems often remain pressurized, even after the engine has been turned **OFF**. The fuel system pressure must be relieved before disconnecting any fuel lines. Failure to do so may result in fire and/or personal injury.

• Brake fluid often contains polyglycol ethers and polyglycols. Avoid contact with the eyes and wash your hands thoroughly after handling brake fluid. If you do get brake fluid in your eyes, flush your eyes with clean, running water for 15 minutes. If eye irritation persists, or if you have taken brake fluid internally, IMMEDIATELY seek medical assistance.

• The EPA warns that prolonged contact with used engine oil may cause a number of skin disorders, including cancer! You should make every effort to minimize your exposure to used engine oil. Protective gloves should be worn when changing oil. Wash your hands and any other exposed skin areas as soon as possible after exposure to used engine oil. Soap and water, or waterless hand cleaner should be used.

• All new vehicles are now equipped with an air bag system. The system must be disabled before performing service on or around system components, steering column, instrument panel components, wiring and sensors. Failure to follow safety and disabling procedures could result in accidental air bag deployment, possible personal injury and unnecessary system repairs.

• Always wear safety goggles when working with, or around, the air bag system. When carrying a non-deployed air bag, be sure the bag and trim cover are pointed away from your body. When placing a non-deployed air bag on a work surface, always face the bag and trim cover upward, away from the surface. This will reduce the motion of the module if it is accidentally deployed. Refer to the additional air bag system precautions later in this section.

• Clean, high quality brake fluid from a sealed container is essential to the safe and proper operation of the brake system. You should always buy the correct type of brake fluid for your vehicle. If the brake fluid becomes contaminated, completely flush the system with new fluid. Never reuse any brake fluid. Any brake fluid that is removed from the system should be discarded. Also, do not allow any brake fluid to come in contact with a painted surface; it will damage the paint.

• Never operate the engine without the proper amount and type of engine oil; doing so WILL result in severe engine damage.

• Timing belt maintenance is extremely important! Many models utilize an interference-type, non-freewheeling engine. If the timing belt breaks, the valves in the cylinder head may strike the pistons, causing potentially serious (also time-consuming and expensive) engine damage.

• Disconnecting the negative battery cable on some vehicles may interfere with the functions of the on-board computer system(s) and may require the computer to undergo a relearning process once the negative battery cable is reconnected.

• When servicing drum brakes, only disassemble and assemble one side at a time, leaving the remaining side intact for reference.

• Only an MVAC-trained, EPA-certified automotive technician should service the air conditioning system or its components.

ENGINE REPAIR

➡**Disconnecting the negative battery cable on some vehicles may interfere with the functions of the on board computer system. The computer may undergo a relearning process once the negative battery cable is reconnected.**

Distributor

The MDX is equipped with a Distributorless Ignition System (DIS).

Alternator

REMOVAL

1. Before servicing the vehicle, refer to the precautions in the beginning of this section.
2. Remove or disconnect the following:
   • Negative battery cable

• Intake manifold and ignition coil covers
• Accessory drive belt
• Alternator wiring harness connectors
• Alternator mounting bolts
• Wiring harness clamp
• Alternator

INSTALLATION

1. Install or connect the following:
   • Alternator
   • Wiring harness clamp. Tighten the bolt to 105 inch lbs. (12 Nm).
   • Alternator mounting bolts. Tighten the 10mm bolt to 33 ft. lbs. (44 Nm) and the 8mm bolt to 16 ft. lbs. (22 Nm).
   • Alternator wiring harness connectors. Tighten the battery terminal nut to 105 inch lbs. (12 Nm).

• Accessory drive belt
• Intake manifold and ignition coil covers
• Negative battery cable

Ignition Timing

ADJUSTMENT

The MDX is equipped with a Distributorless Ignition System (DIS). The ignition timing is controlled by the Powertrain Control module (PCM). No adjustment is necessary.

Engine Assembly

REMOVAL & INSTALLATION

➡**The engine and transaxle are removed from the vehicle as a unit.**

1. Before servicing the vehicle, refer to the precautions in the beginning of this section.
2. Drain the cooling system.
3. Drain the power steering system.
4. Drain the transaxle fluid.
5. Drain the engine oil.
6. Relieve fuel system pressure.
7. Remove or disconnect the following:
- Negative battery cable
- Battery and tray
- Intake and ignition coil covers
- Air intake duct
- Left engine wire harness connectors
- Relay bracket
- Starter cable and harness clamp
- Accelerator cable
- Cruise control cable
- Fuel lines
- EVAP canister hose

8. Remove the drivers side center console lower panel and pull back the cover to access steering joint cover.
- Steering joint bolt
- Powertrain Control Module (PCM) connectors
- Heated Oxygen (HO$_2$S) sensor connector and grommet. Pull the PCM harness through the firewall.
- Brake booster vacuum line
- Power steering hose's clamps and clips
- Fuse/Relay box battery cable
- Accessory drive belts
- Front wheels
- Splash shield
- Front sub-frame stiffener
- Exhaust front pipe
- Propeller shaft
- Shift control cable
- Transfer assembly
- Ball joints
- Stabilizer bar links
- Power steering hose and pressure switch connector
- Transaxle lower front mount
- Transaxle lower rear mount
- A/C compressor
- Heater hoses
- Radiator hoses
- Ground cable
- Transaxle oil cooler lines
- Radiator

9. Attach a hoist to the engine lifting eyes and support the powertrain weight.
10. Remove or disconnect the following:
- Side engine mount bracket
- Front mount bracket support nut
11. Matchmark the front subframe to the mounting points.
12. Remove or disconnect the following:

- Front subframe
- All remaining hoses and electrical connections
13. Lower the powertrain away from the vehicle.

To install:
14. Raise the powertrain into position.
15. Installation is the reverse of removal but please note the following steps:
- A/C compressor bolts to 16 ft. lbs. (22 Nm)
- Front subframe. Use new bolts and tighten the 14mm bolts to 76 ft. lbs. (103 Nm). Tighten the front brace bolts to 54 ft. lbs. (74 Nm) and the rear brace bolts to 86 ft. lbs. (117 Nm).
- Transaxle lower front mount nuts to 28 ft. lbs. (38 Nm)
- Transaxle lower rear mount bolts to 28 ft. lbs. (38 Nm)
- Front mount bracket support nut to 40 ft. lbs. (54 Nm)
- Side engine mount bracket bolts to

33 ft. lbs. (44 Nm) and the through bolt to 40 ft. lbs. (54 Nm)
16. Fill the engine crankcase to the correct level.
17. Fill the transaxle to the correct level.
18. Fill the cooling system.
19. Fill the power steering system.
20. Start the engine and check for leaks.
21. Check the wheel alignment and adjust as necessary.

Water Pump

REMOVAL & INSTALLATION

1. Before servicing the vehicle, refer to the precautions in the beginning of this section.
2. Drain the cooling system.
3. Remove or disconnect the following:
- Negative battery cable
- Accessory drive belts
- Front cover
- Timing belt

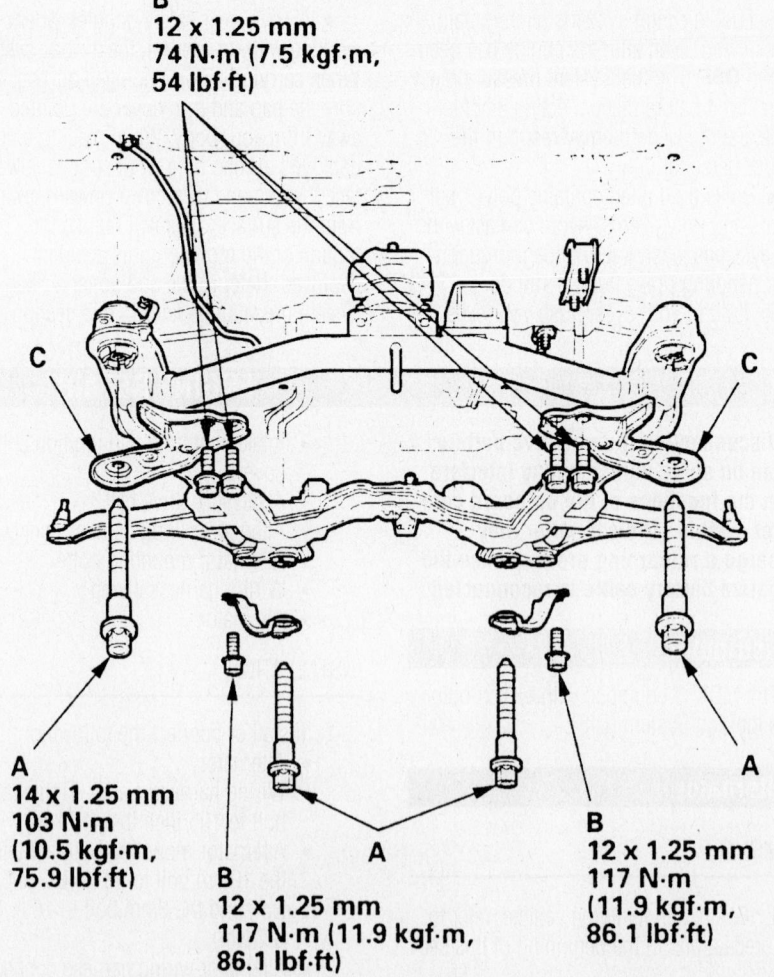

B
12 x 1.25 mm
74 N·m (7.5 kgf·m, 54 lbf·ft)

A
14 x 1.25 mm
103 N·m (10.5 kgf·m, 75.9 lbf·ft)

B
12 x 1.25 mm
117 N·m (11.9 kgf·m, 86.1 lbf·ft)

B
12 x 1.25 mm
117 N·m (11.9 kgf·m, 86.1 lbf·ft)

9302MG69

Sub-frame fastener locations and tightening torque—MDX

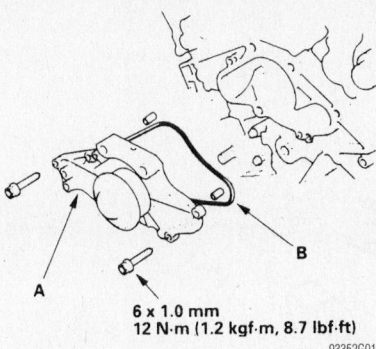

6 x 1.0 mm
12 N·m (1.2 kgf·m, 8.7 lbf·ft)

93352G01

Exploded view of the water pump mounting

- Timing belt tensioner
- Water pump

To install:

4. Install or connect the following:
   - Water pump. Use a new O-ring seal and tighten the bolts to 105 inch lbs. (12 Nm).
   - Timing belt tensioner
   - Timing belt
   - Front cover
   - Accessory drive belts
   - Negative battery cable
5. Fill the cooling system.
6. Start the engine and check for leaks.

Heater Core

REMOVAL & INSTALLATION

1. Before servicing the vehicle, refer to the precautions in the beginning of this section.
2. Drain the cooling system.
3. Remove or disconnect the following:
   - Negative battery cable
4. Recover the refrigerant using approved equipment.
   - Heater valve cable from the valve arm. Turn the valve arm to the fully opened position.
   - Heater hoses from the heater unit
   - Mounting nut from the heater unit. Be careful not to bend or damage fuel or brake lines.
5. Remove the dashboard as follows:
   a. Remove the center console by unlatching the clips.
   b. Remove the dashboard lower cover screw, gently pull down on the cover to disengage the clips and disconnect the electrical connections.
   c. Remove the dashboard side cover by gently pulling and turning to unfasten the clips.
   d. While holding the glove box, remove the box stop from each side, then disconnect the lock from the damper.

e. Remove the glove ox bolts and the glove box.
   f. Remove the shift lever assembly.
   g. Remove the front door trim, lick panel and A-pillar trim from both sides.
   h. Remove the cap from the front pillar corner trim. Unfasten the screw, slide the trim upward along the pillar and remove it. Remove the remaining clips from the body.
   i. On the drivers side, remove the fuel/relay box nut and pull out the box.
   j. Remove the steering column
   k. On the passenger side remove the fuse/relay bolt and pull out the box.
   l. Disconnect all electrical connections from the dashboard.
   m. If equipped with a navigation system, pull back the carpet, remove the harness cushions and then pull out the GPS harness.
   n. Remove all harness and connector clips.
   o. Remove all the bolts and lift up on the dashboard to release the dashboard and steering hanger beam from the guide pins.
   p. Remove the dashboard through the door.
6. Remove the evaporator as follows:
   a. Disconnect the receiver and suction lines from the evaporator.
   b. Remove the mounting nuts and plug the lines to avoid system contamination.
   c. Remove the plastic brace and glove box frame.
   d. Disconnect the wire harness and evaporator temperature sensor connector.
   e. Remove the self-tapping screws, the nuts and the evaporator.
7. Remove or disconnect the following:
   - Mounting bolts and the heater unit
   - Self-tapping screws and the clamp, then pull the heater core from the case being careful not to bend the pipes

To install:

8. Install or connect the following:
   - Heater core in the case
   - Clamp and the screws
   - Heater unit and tighten the bolts to 7 ft. lbs. (10 Nm)
   - Evaporator in the reverse order of removal. Tighten all the retainers to 7 ft. lbs. (10 Nm) .
9. Install the dashboard in the reverse order of removal keeping in mind the following points:
   a. Make sure the dashboard is seated properly and that the wiring harness and steering hanger beam wire harness are not pinched.

b. Referring to the accompanying illustration, tighten bolts **(A)** to 7 ft. lbs. (10 Nm). Tighten all the other bolts to 16 ft. lbs. (22 Nm). Apply thread lock to the **B** bolts before installation.
   c. Ensure that all electrical connectors are properly connected.
10. Install or connect the following:
    - Mounting nut to the heater unit and tighten to 9 ft. lbs. (13 Nm)
    - Heater hoses
11. Connect the heater valve cable and adjust as follows:
    a. In the engine compartment, open the cable clamp (A), then disconnect the heater valve cable (B) from the valve arm (C).
    b. Under the dashboard, disconnect the valve cable housing from the clamp (A) and the cable (B) from the mix control linkage (C).
    c. Set the temperature control button to the MAX COOL position with the ignition switch in the on position.

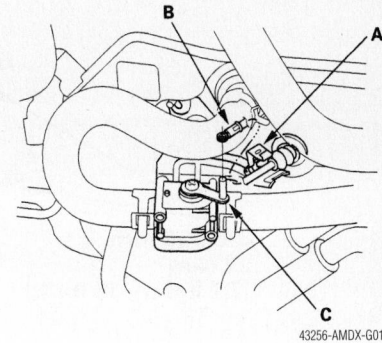

43256-AMDX-G01

In the engine compartment, open the cable clamp (A), then disconnect the heater valve cable (B) from the valve arm (C)

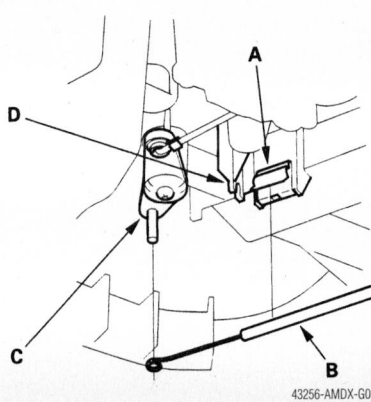

43256-AMDX-G02

Under the dashboard, disconnect the valve cable housing from the clamp (A) and the cable (B) from the mix control linkage (C)

d. Attach the valve cable (B) to the mix control linkage (C) as shown in the illustration, hold the end of the cable housing against the stop, then snap the cable housing into the clamp. Sin the engine compartment, turn the valve arm (C) to the fully closed position as shown in the accompanying illustration and hold it there. Attach the cable (B) to the vale arm and pull gently on the cable housing to take up the slack, then install the cable housing into the clamp (A).

12. Fill the cooling system
13. Connect the battery cable.

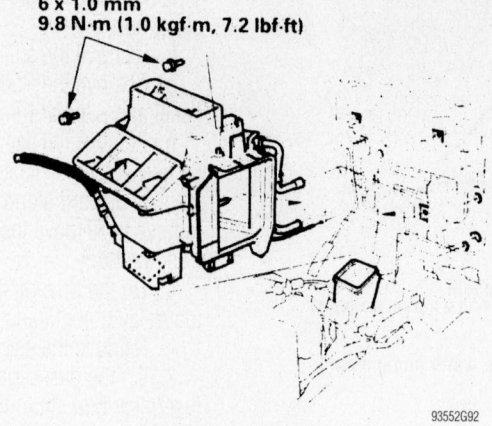

6 x 1.0 mm
9.8 N·m (1.0 kgf·m, 7.2 lbf·ft)

93552G92

Exploded view of the evaporator mounting—Acura MDX

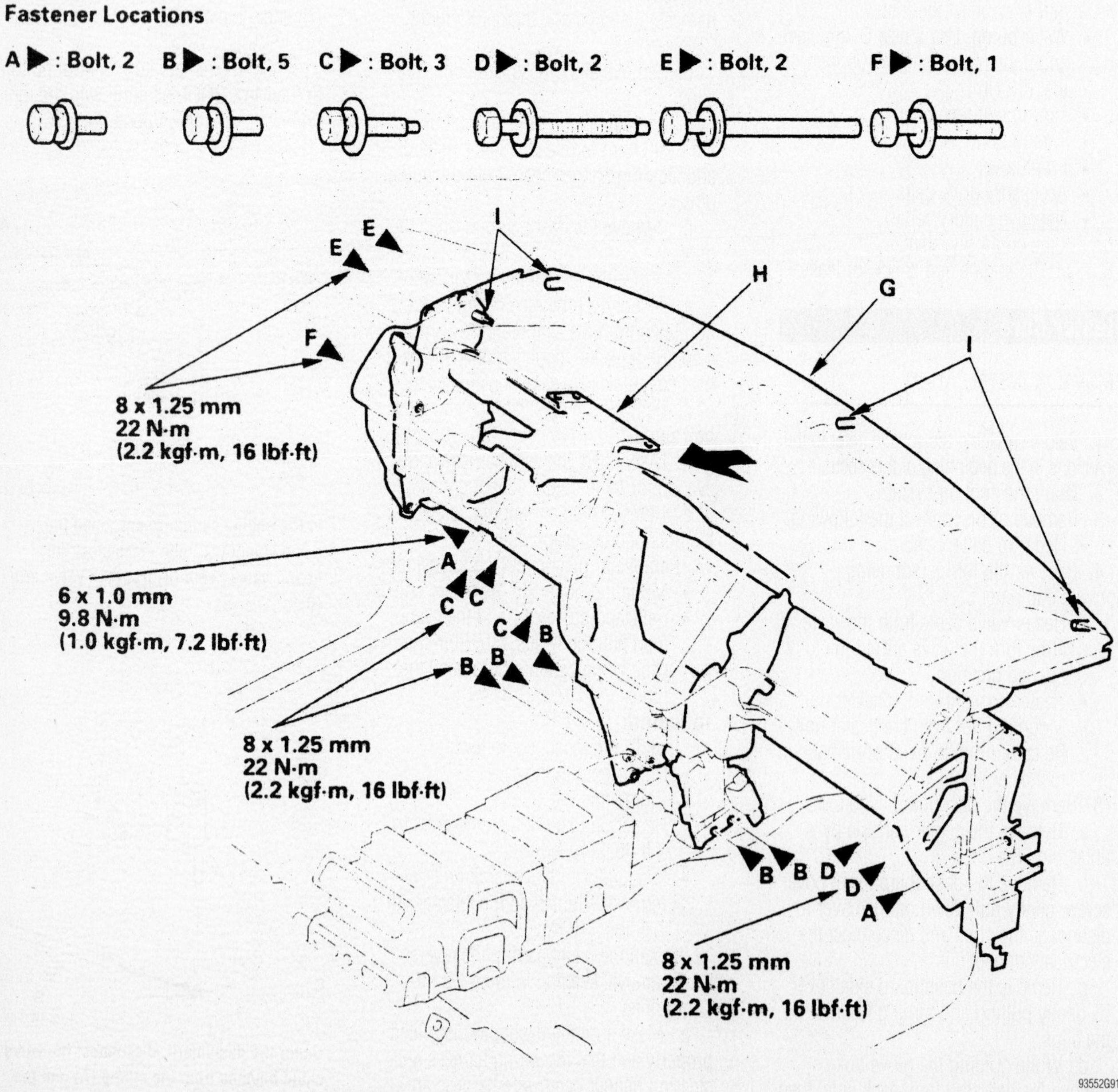

Fastener Locations

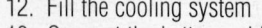

A ▶ : Bolt, 2   B ▶ : Bolt, 5   C ▶ : Bolt, 3   D ▶ : Bolt, 2   E ▶ : Bolt, 2   F ▶ : Bolt, 1

8 x 1.25 mm
22 N·m
(2.2 kgf·m, 16 lbf·ft)

6 x 1.0 mm
9.8 N·m
(1.0 kgf·m, 7.2 lbf·ft)

8 x 1.25 mm
22 N·m
(2.2 kgf·m, 16 lbf·ft)

8 x 1.25 mm
22 N·m
(2.2 kgf·m, 16 lbf·ft)

93552G91

Exploded view of the dashboard mounting—Acura MDX

Cylinder Head

REMOVAL & INSTALLATION

1. Before servicing the vehicle, refer to the precautions in the beginning of this section.
2. Drain the cooling system.
3. Relieve the fuel system pressure.
4. Remove or disconnect the following:
   - Negative battery cable
   - Ignition coil covers
   - Intake manifold cover
   - Air intake tube
   - Accelerator cable
   - Cruise control cable
   - EVAP canister, breather, water bypass hoses and the bypass pipe bolt
   - Fuel lines
   - Brake booster vacuum line
   - Intake manifold vacuum line
   - Positive Crankcase Ventilation (PCV) valve and hose
   - Power steering hose clamp
   - Intake Manifold Runner Control (IMRC) actuator
   - Wiring harness holder and joint connector
5. Support the engine with a jack and a block of wood.
   - Side engine mount bracket
   - Accessory drive belts
   - Power steering pump
   - Alternator
   - Intake Air Temperature (IAT) sensor connector
   - Idle Air Control (IAC) valve connector
   - Throttle Position (TP) sensor connector
   - Manifold Absolute Pressure (MAP) sensor connector
   - Evaporative Emission (EVAP) canister purge valve connector
   - Engine Coolant Temperature (ECT) sensor connector
   - Radiator fan switch connectors
   - ECT gauge sending unit connector
   - Crankshaft Position (CKP) sensor connector
   - Top Dead Center (TDC) sensor connector
   - Exhaust Gas Recirculation (EGR) connector
   - Variable Valve Timing and Valve Lift Electronic Control (VTEC) solenoid valve connector
   - VTEC oil pressure switch connector
   - Oil pressure switch connector
   - Ignition coils
   - Intake manifold
   - Fuel injector connectors
   - Fuel supply manifold

   - Fuel injection air control valve vacuum lines
   - Front cover
   - Timing belt
   - Radiator hoses
   - Heater hoses
   - Front and rear exhaust manifolds
   - Coolant cross-over pipe
   - Valve covers
6. Loosen the cylinder head bolts in

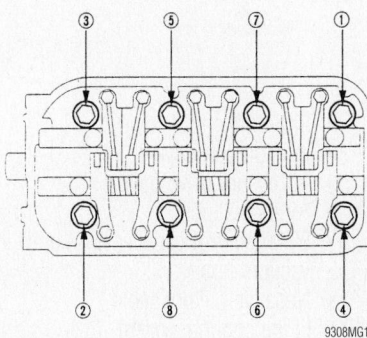

Cylinder head bolt loosening sequence—MDX

sequence and ⅓ turns until all bolts are loose.
7. Remove the cylinder head.
To install:
8. Align the crankshaft and camshaft sprocket TDC marks as shown.

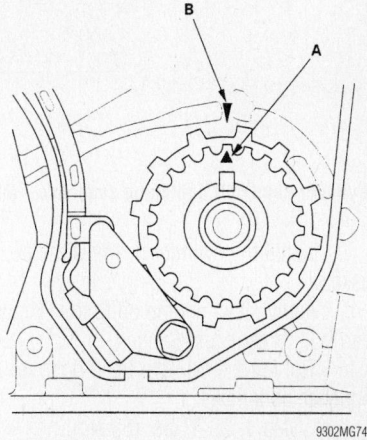

9302MG74

Crankshaft timing belt sprocket TDC marks. Align sprocket mark (A) with pointer (B)—MDX

FRONT:

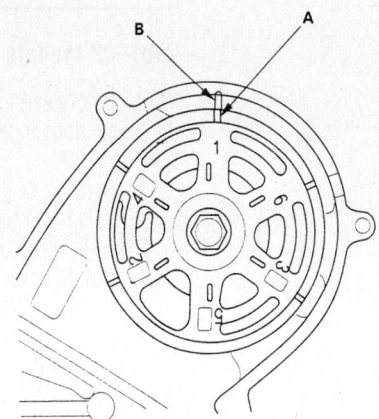

REAR:

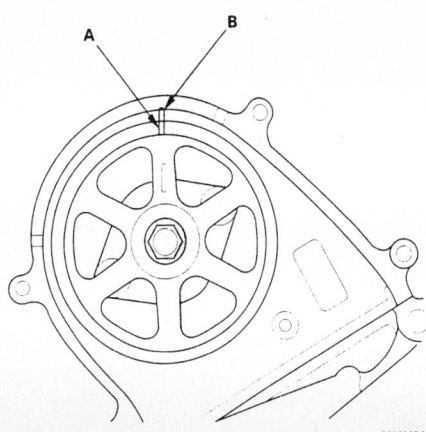

9302MG85

Camshaft TDC marks. Align sprocket mark (A) with the back cover pointer (B)—MDX

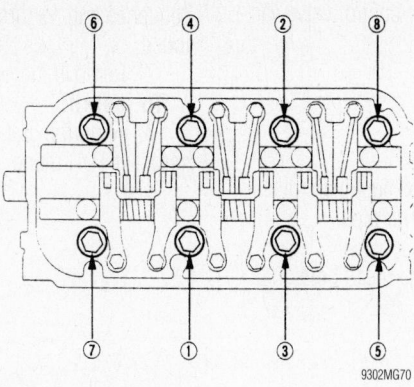

Cylinder head bolt tightening sequence—MDX

9. Install the cylinder heads with new gaskets.

10. Apply clean engine oil to the cylinder head bolt threads and flanges.

11. Tighten the cylinder head bolts in sequence as follows:
   a. Step 1: 29 ft. lbs. (39 Nm).
   b. Step 2: 51 ft. lbs. (69 Nm).
   c. Step 3: 72 ft. lbs. (98 Nm).

12. Install or connect the following:
   • Valve covers
   • Coolant cross-over pipe
   • Front and rear exhaust manifolds
   • Heater hoses
   • Radiator hoses
   • Timing belt
   • Front cover
   • Fuel injection air control valve vacuum lines
   • Fuel supply manifold
   • Fuel injector connectors
   • Intake manifold
   • Ignition coils
   • Oil pressure switch connector
   • VTEC oil pressure switch connector
   • VTEC solenoid valve connector
   • EGR connector
   • TDC sensor connector
   • CKP sensor connector
   • ECT gauge sending unit connector
   • Radiator fan switch connectors
   • ECT sensor connector
   • MAP sensor connector
   • TP sensor connector
   • IAC valve connector
   • IAT sensor connector
   • Alternator
   • Power steering hose clamp
   • Power steering pump
   • Side engine mount bracket
   • PCV valve and hose
   • Intake manifold vacuum line
   • Brake booster vacuum line
   • Fuel lines
   • Cruise control cable
   • Accelerator cable

   • Intake manifold cover
   • Ignition coil covers
   • Accessory drive belts
   • Air intake tube
   • EVAP control canister hose and vacuum hose
   • Negative battery cable

13. Fill the cooling system.

14. Start the engine and check for leaks.

Rocker Arms/Shafts

REMOVAL & INSTALLATION

2001–02 Models

1. Before servicing the vehicle, refer to the precautions in the beginning of this section.

2. Remove or disconnect the following:
   • Negative battery cable
   • Air intake tube
   • Ignition coil covers
   • Intake manifold cover
   • Intake manifold
   • Valve cover

3. Loosen the valve adjuster locknuts and screws so that all valves are closed.

4. Loosen the rocker arm shaft bolts evenly in sequence.

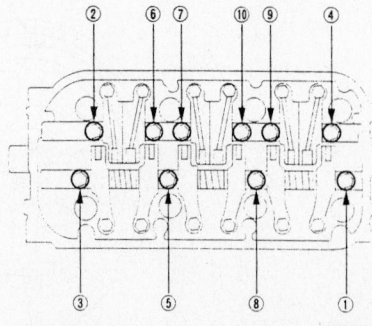

Rocker shaft loosening sequence— 2001–02 models

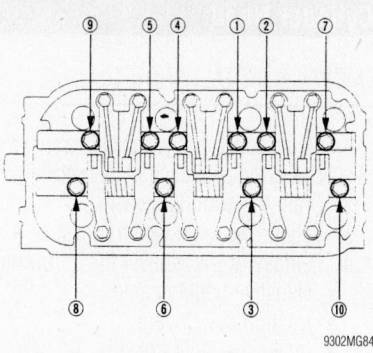

Rocker shaft tightening sequence— 2001–02 models

5. Remove the rocker arms and shafts from the vehicle as an assembly.

➡ **Keep all valvetrain components in order for assembly.**

6. Remove the rocker arms and springs from the rocker arm shafts.

To install:

7. Assemble the rocker arms and springs to the rocker arm shafts in their original positions.

8. Install the rocker arm assemblies. Tighten the bolts in sequence and in multiple passes to 17 ft. lbs. (24 Nm).

9. Adjust the valve clearance.

10. Install or connect the following:
   • Valve covers
   • Intake manifold
   • Intake manifold cover
   • Ignition coil covers
   • Air intake tube
   • Negative battery cable

11. Start the engine and check for leaks.

2003–05 Models

1. Before servicing the vehicle, refer to the precautions in the beginning of this section.

2. Remove or disconnect the following:
   • Negative battery cable
   • Intake manifold cover
   • Ignition coil covers
   • Ignition coils
   • Dipstick
   • 2 bolts attaching the harness holder
   • Positive Crankcase ventilation (PCV) hose
   • Injector electrical connections
   • Power steering hose bracket bolt and the harness holder bolts
   • Harness clamps and the breather hose
   • Valve cover
   • Rocker arm adjusting screws. refer to the illustration for location.

3. remove the rocker arm assembly as follows:

   a. Unscrew the rocker shaft bolts 2 turns at a time in a criss-cross pattern to avoid damaging the vales or rocker assembly.

   b. Do not remove the rocker shaft bolts. These bolts keep the springs and rocker arms on the shafts.

4. Loosen the valve adjuster locknuts and screws so that all valves are closed.

5. Remove the rocker arms and shafts from the vehicle as an assembly.

➡**Keep all valvetrain components in order for assembly.**

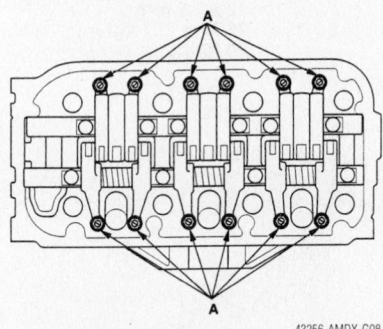

Rocker arm shaft adjusting screw locations—2003–05 models

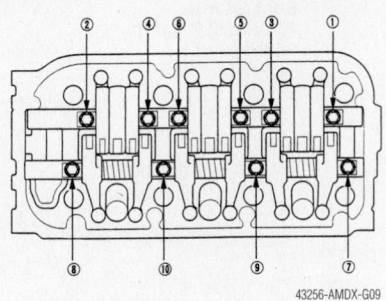

43256-AMDX-G09

Rocker arm shaft loosening sequence—2003–05 models

6. Remove the rocker arms and springs from the rocker arm shafts.

To install:

7. Assemble the rocker arms and springs to the rocker arm shafts in their original positions.

8. Install the rocker arm assemblies. Tighten the bolts in sequence and in multiple passes to 17 ft. lbs. (24 Nm).

9. Adjust the valve clearance.

10. Install or connect the following:
- Valve covers
- Harness clamps and the breather hose
- Power steering hose bracket bolt and the harness holder bolts

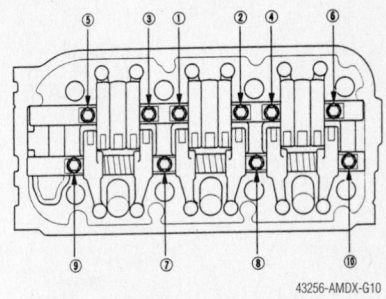

43256-AMDX-G10

Rocker shaft tightening sequence—2003–05 models

- Injector electrical connections
- PCV hose
- 2 bolts attaching the harness holder
- Dipstick
- Ignition coils and torque the retainers to 9 ft. lbs. (12 Nm)
- Ignition coil covers
- Intake manifold cover
- Negative battery cable

Intake Manifold

REMOVAL & INSTALLATION

2001–2002 Models

1. Before servicing the vehicle, refer to the precautions in the beginning of this section.

2. Remove or disconnect the following:
- Negative battery cable
- Evaporative Emissions (EVAP) control canister hose and vacuum hose
- Air intake tube
- Intake manifold cover
- Accelerator cable
- Cruise control cable
- Brake booster vacuum line
- Intake manifold vacuum line
- Positive Crankcase Ventilation (PCV) valve and hose
- Intake Air Temperature (IAT) sensor connector
- Idle Air Control (IAC) valve connector
- Throttle Position (TP) sensor connector
- Manifold Absolute Pressure (MAP) sensor connector
- Intake manifold

To install:

3. Install or connect the following:
- New intake manifold gasket
- Intake manifold. Tighten the fasteners in sequence and in several passes to 16 ft. lbs. (22 Nm).
- MAP sensor connector

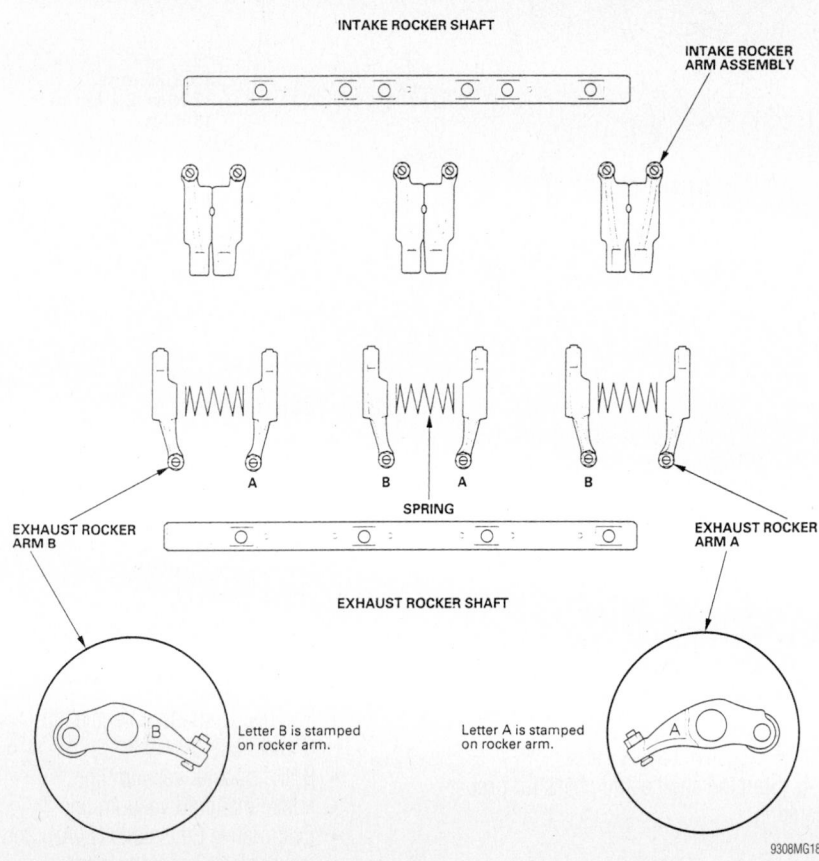

Exploded view of the rocker arms and shafts—MDX

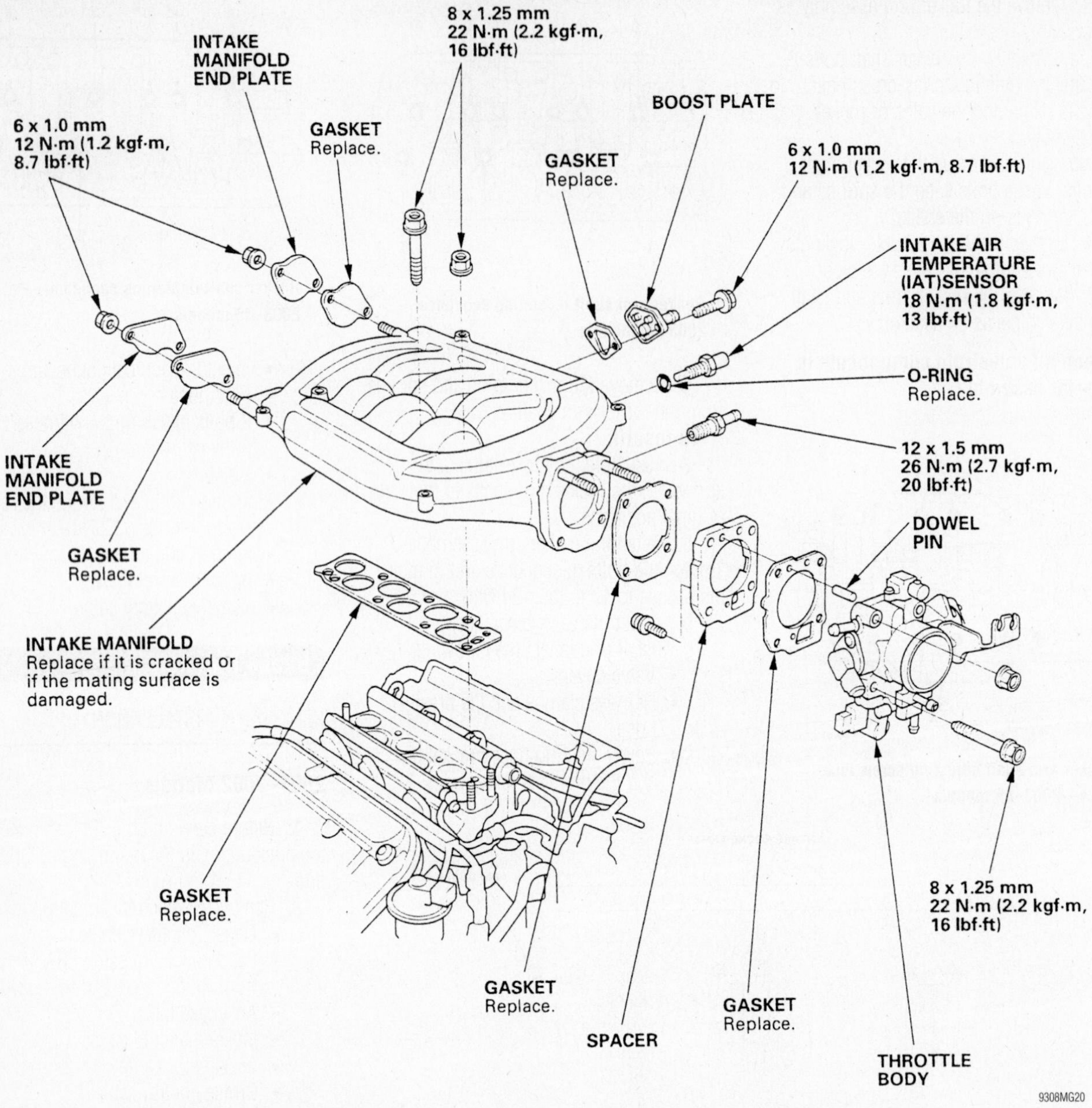

6 x 1.0 mm
12 N·m (1.2 kgf·m,
8.7 lbf·ft)

INTAKE MANIFOLD END PLATE

GASKET
Replace.

8 x 1.25 mm
22 N·m (2.2 kgf·m,
16 lbf·ft)

BOOST PLATE

GASKET
Replace.

6 x 1.0 mm
12 N·m (1.2 kgf·m, 8.7 lbf·ft)

INTAKE AIR TEMPERATURE (IAT) SENSOR
18 N·m (1.8 kgf·m, 13 lbf·ft)

O-RING
Replace.

INTAKE MANIFOLD END PLATE

12 x 1.5 mm
26 N·m (2.7 kgf·m, 20 lbf·ft)

GASKET
Replace.

DOWEL PIN

INTAKE MANIFOLD
Replace if it is cracked or if the mating surface is damaged.

GASKET
Replace.

8 x 1.25 mm
22 N·m (2.2 kgf·m, 16 lbf·ft)

GASKET
Replace.

GASKET
Replace.

SPACER

THROTTLE BODY

9308MG20

Exploded view of the intake manifold—2001–02 models

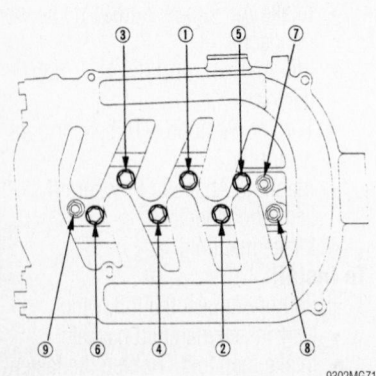

9302MG71

Intake manifold torque sequence—MDX

- TP sensor connector
- IAC valve connector
- IAT sensor connector
- PCV valve and hose
- Intake manifold vacuum line
- Brake booster vacuum line
- Cruise control cable
- Accelerator cable
- Intake manifold cover
- Air intake tube
- EVAP control canister hose and vacuum hose
- Negative battery cable

4. Start the engine and check for proper operation.

2003–2005 Models

1. Before servicing the vehicle, refer to the precautions in the beginning of this section.

2. Remove or disconnect the following:
   - Negative battery cable
   - Intake manifold cover
   - Intake Air Temperature (IAT) sensor 2 connector
   - Air intake tube
   - Positive Crankcase Ventilation (PCV) hose
   - Brake booster vacuum line
   - Intake manifold vacuum line
   - Evaporative Emissions (EVAP) control canister hose and clamp bracket

- Water bypass hoses from the throttle body and plug the hoses
3. Remove the following electrical connections and clamps from the manifold:
  - IAT sensor 1 connector
  - Throttle Position (TP) sensor connector
  - Manifold Absolute Pressure (MAP) sensor connector

- EVAP canister purge valve connector
- Intake Manifold Runner Control (IMRC) actuator connector
4. Remove or disconnect the following:
  - Upper cover bolts and nuts in the sequence illustrated using two or three passes
  - Intake manifold bolts in the sequence illustrated

- Intake manifold and spacer
To install:
5. Install or connect the following:
  - New intake manifold gasket and spacer
  - Intake manifold. Tighten the fasteners in sequence and in several passes to 16 ft. lbs. (22 Nm).

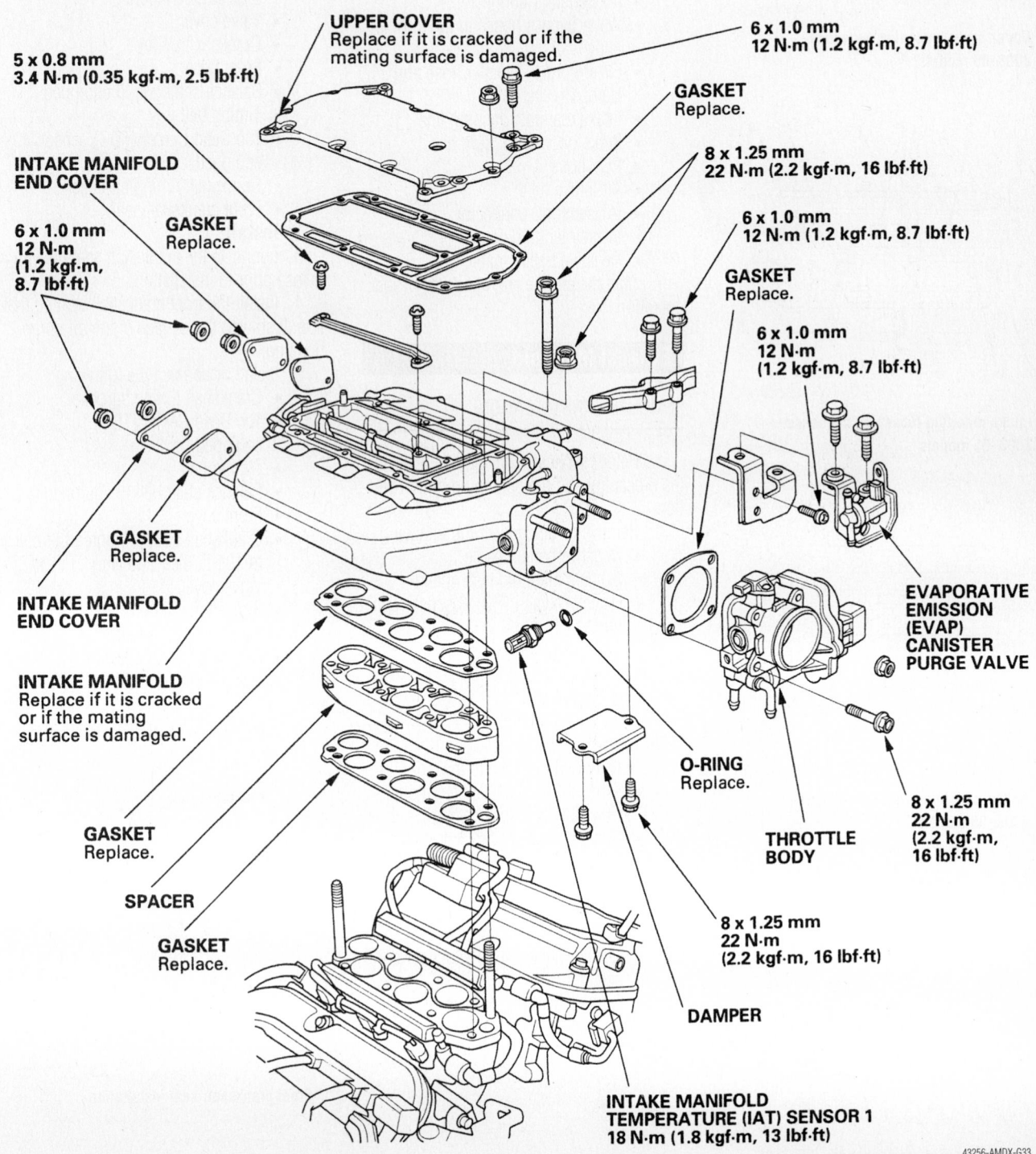

5 x 0.8 mm
3.4 N·m (0.35 kgf·m, 2.5 lbf·ft)

INTAKE MANIFOLD END COVER

6 x 1.0 mm 12 N·m (1.2 kgf·m, 8.7 lbf·ft)

GASKET Replace.

GASKET Replace.

INTAKE MANIFOLD END COVER

INTAKE MANIFOLD Replace if it is cracked or if the mating surface is damaged.

GASKET Replace.

SPACER

GASKET Replace.

UPPER COVER Replace if it is cracked or if the mating surface is damaged.

6 x 1.0 mm 12 N·m (1.2 kgf·m, 8.7 lbf·ft)

GASKET Replace.

8 x 1.25 mm 22 N·m (2.2 kgf·m, 16 lbf·ft)

6 x 1.0 mm 12 N·m (1.2 kgf·m, 8.7 lbf·ft)

GASKET Replace.

6 x 1.0 mm 12 N·m (1.2 kgf·m, 8.7 lbf·ft)

EVAPORATIVE EMISSION (EVAP) CANISTER PURGE VALVE

O-RING Replace.

THROTTLE BODY

8 x 1.25 mm 22 N·m (2.2 kgf·m, 16 lbf·ft)

8 x 1.25 mm 22 N·m (2.2 kgf·m, 16 lbf·ft)

DAMPER

INTAKE MANIFOLD TEMPERATURE (IAT) SENSOR 1 18 N·m (1.8 kgf·m, 13 lbf·ft)

43256-AMDX-G33

Exploded view of the intake manifold—2003–05 models

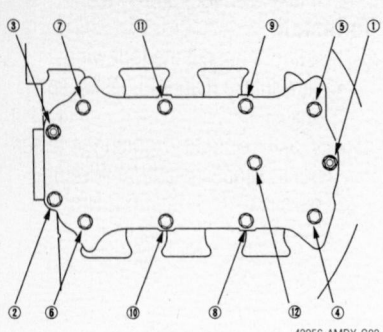

Upper cover loosening sequence—2003–05 models

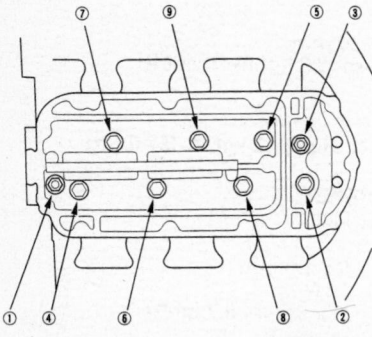

Intake manifold loosening sequence—2003–05 models

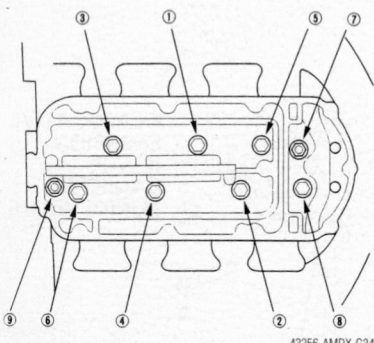

Intake manifold torque sequence—2003–05 models

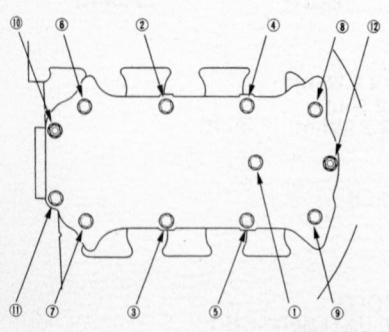

Upper cover torque sequence—2003–05 models

- Upper cover bolts and nuts in the sequence illustrated using two or three passes to 9 ft. lbs. (12 Nm)
6. Connect the following electrical connections and clamps to the manifold:
  - Intake Manifold Runner Control (IMRC) actuator connector
  - EVAP canister purge valve connector
  - MAP sensor connector
  - TP sensor connector
  - IAT sensor 1 connector
  - Water bypass hoses to the throttle body
  - EVAP control canister hose and clamp bracket
  - Intake manifold vacuum line
  - Brake booster vacuum line
  - PCV hose
  - Air intake tube
  - IAT sensor 2 connector
  - Intake manifold cover
  - Negative battery cable
7. Start the engine and check for proper operation.

Exhaust Manifold

REMOVAL & INSTALLATION

1. Before servicing the vehicle, refer to the precautions in the beginning of this section.
2. Remove or disconnect the following:
   - Negative battery cable
   - Exhaust manifold heat shield
   - Heated Oxygen (HO$_2$S) sensor connector
   - Exhaust front pipe
   - Exhaust manifold bracket, if equipped
   - Exhaust manifold

To install:
3. Install or connect the following:
   - Exhaust manifold. Tighten the fasteners to 23 ft. lbs. (31 Nm).
   - Exhaust manifold bracket, if equipped. Tighten the bolts to 33 ft. lbs. (44 Nm).
   - Exhaust front pipe. Tighten the nuts to 40 ft. lbs. (55 Nm).
   - Heated Oxygen (HO$_2$S) sensor connector
   - Exhaust manifold heat shield
   - Negative battery cable

Front Crankshaft Seal

REMOVAL & INSTALLATION

1. Before servicing the vehicle, refer to the precautions in the beginning of this section.
2. Remove or disconnect the following:
   - Negative battery cable
   - Accessory drive belts
   - Side engine mount
   - Valve cover
   - Crankshaft pulley
   - Front cover
   - Balance shaft belt, if equipped
   - Timing belt
   - Top Dead Center (TDC) sensor, if equipped
   - Crankshaft timing sprocket
   - Front crankshaft seal

To install:
3. Lubricate the crankshaft seal lip with grease prior to installation.
4. Install the front crankshaft seal so that it is flush with the surface of the oil pump housing.
5. Install or connect the following:
   - Crankshaft timing sprocket
   - Top Dead Center (TDC) sensor, if equipped
   - Timing belt
   - Balance shaft belt, if equipped
   - Front cover
   - Crankshaft pulley. Tighten the bolt to 181 ft. lbs. (245 Nm).
   - Valve cover
   - Side engine mount
   - Accessory drive belts
   - Negative battery cable
6. Check the engine oil level and add if necessary.
7. Start the engine and check for leaks.

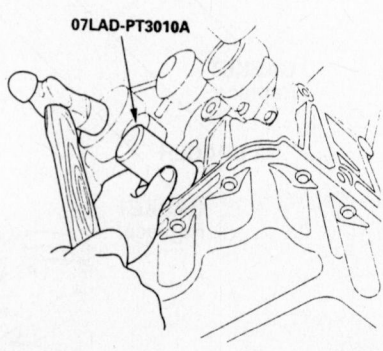

Front crankshaft seal installation

Camshaft

REMOVAL & INSTALLATION

2001–02 Models

1. Before servicing the vehicle, refer to the precautions in the beginning of this section.
2. Remove or disconnect the following:
   - Negative battery cable
   - Air intake tube
   - Accessory drive belts
   - Front cover
   - Timing belt
   - Camshaft sprockets
   - Timing belt rear covers
   - Ignition coil covers
   - Intake manifold cover
   - Intake manifold
   - Valve cover
   - Rocker arms and shaft assembly
   - Camshaft thrust cover
   - Camshaft

To install:

➥Use new O-rings, seals and gaskets when installing the camshaft.

3. Install or connect the following:
   - Camshaft

- Camshaft thrust cover. Tighten the bolts to 16 ft. lbs. (22 Nm).
- Rocker arms and shaft assembly
- Valve cover
- Intake manifold
- Intake manifold cover
- Ignition coil covers
- Timing belt rear covers
- Camshaft sprockets. Tighten the bolts to 67 ft. lbs. (90 Nm).
- Timing belt
- Front cover
- Accessory drive belts
- Air intake tube
- Negative battery cable
4. Start the engine and check for leaks.

2003–05 Models

FRONT

1. Before servicing the vehicle, refer to the precautions in the beginning of this section.
2. Remove or disconnect the following:
   - Negative and positive battery cables
   - Battery
3. Drain the coolant.
   - Exhaust Gas Recirculation (EGR) vale
   - Timing belt
   - Rocker arm assembly

- Front camshaft pulley
- Thrust plate and camshaft

To install:

4. Install or connect the following:
   - Camshaft using a new O-ring. Tighten the thrust plate to 16 ft. lbs. (22 Nm).
   - Front camshaft pulley
   - Rocker arm assembly
   - Timing belt
   - Exhaust Gas Recirculation (EGR) vale
   - Battery
   - Positive, then negative battery cables
5. Fill the cooling system.
   - Camshaft
6. Start the engine and check for leaks.

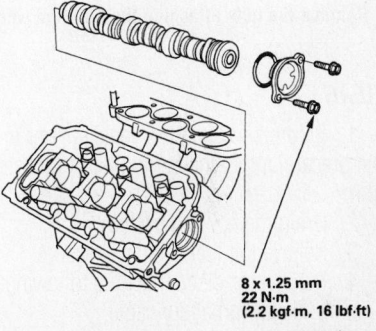

Front camshaft assembly—2003–05 models

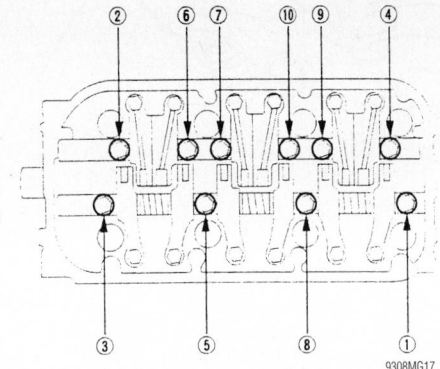

Rocker arm shaft loosening sequence—2001–02 models

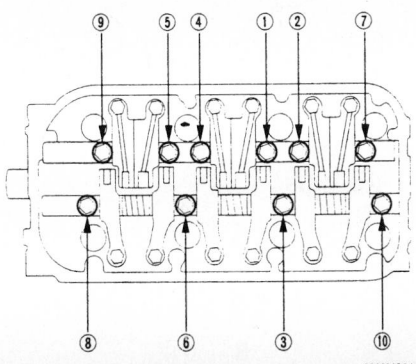

Rocker shaft tightening sequence—2001–02 models

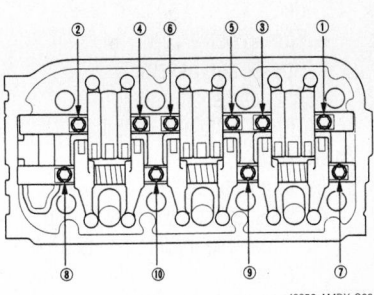

Rocker arm shaft loosening sequence—2003–05 models

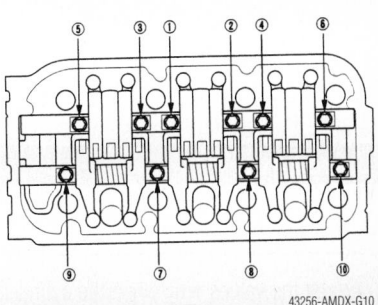

Rocker shaft tightening sequence—2003–05 models

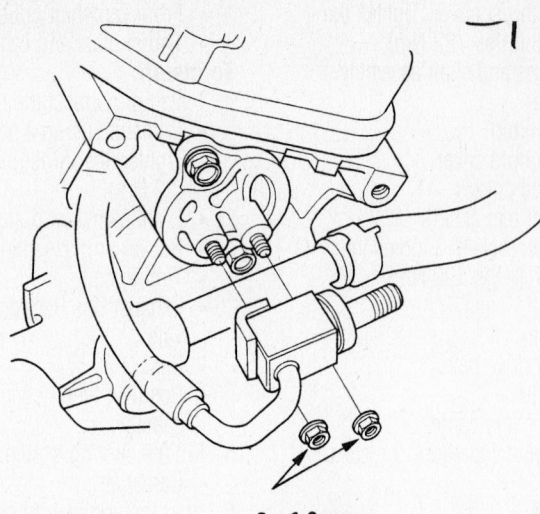

6 x 1.0 mm
12 N·m (1.2 kgf·m, 8.7 lbf·ft)

43256-AMDX-G07

Remove the nuts attaching the fuel line when removing the rear camshaft—2003–05 models

REAR

1. Before servicing the vehicle, refer to the precautions in the beginning of this section.
2. Drain the cooling system.
3. Relieve the fuel system pressure.
4. Remove or disconnect the following:
   - Negative battery cable
   - Under-hood fuse box
   - Fuel feed hose
   - Nuts securing the fuel line
   - Brake lines from the master cylinder
   - Timing belt
   - Rocker arm assembly
   - Rear camshaft pulley
   - Thrust plate and camshaft

To install:
5. Install or connect the following:
   - Camshaft using a new O-ring. Tighten the thrust plate to 16 ft. lbs. (22 Nm).
   - Rear camshaft pulley
   - Rocker arm assembly
   - Timing belt
   - Brake lines to the master cylinder
   - Nuts securing the fuel line
   - Fuel feed hose
   - Under-hood fuse box
   - Negative battery cable

Valve Lash

ADJUSTMENT

Adjust the valves only when the cylinder head temperature is less than 100°F (38°C).
1. Before servicing the vehicle, refer to the precautions in the beginning of this section.
2. Remove or disconnect the following:
   - Negative battery cable
   - Air intake tube
   - Intake manifold
   - Valve cover
3. Rotate the crankshaft so that the valves to be adjusted are closed and the rocker arm is contacting the camshaft lobe base circle.
4. Measure the valve clearance. If adjustment is necessary, loosen the locknut and turn the adjusting screw as necessary to achieve the correct valve clearance.
5. The correct valve clearance is:
   - Intake valves: 0.008–0.009 inches (0.20–0.24mm)
   - Exhaust valves: 0.011–0.013 inches (0.28–0.32mm)
6. After adjustment, tighten the locknuts to 14 ft. lbs. (20 Nm).

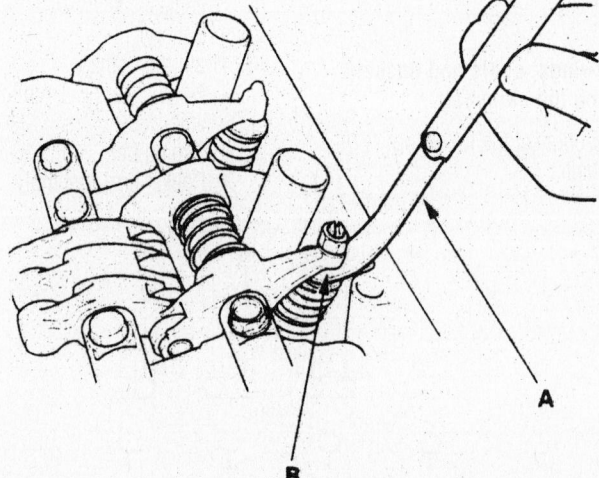

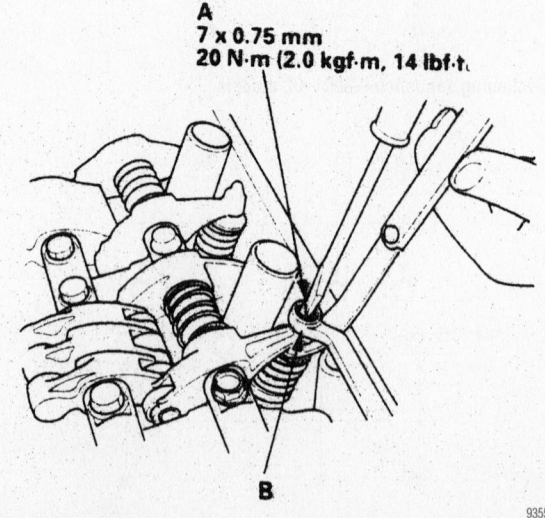

A
7 x 0.75 mm
20 N·m (2.0 kgf·m, 14 lbf·t

93552G04

Inspect the valve clearance, adjust to specification and tighten the retainer to specification

Adjusting screw locations:

EXHAUST

REAR:

No. 1   No. 2   No. 3

No. 1   No. 2   No. 3

INTAKE

FRONT:

No. 4   No. 5   No. 6

No. 4   No. 5   No. 6

EXHAUST

93552G03

Valve adjusting retainer locations

7. Install or connect the following:
   - Valve cover
   - Intake manifold
   - Air intake tube
   - Negative battery cable
8. Start the engine and check for proper operation.

Starter Motor

REMOVAL & INSTALLATION

2001–02 Models

1. Before servicing the vehicle, refer to the precautions in the beginning of this section.
2. Remove or disconnect the following:
   - Negative battery cable and wait at least 3 minutes
   - Transmission fluid cooler line clamp
   - Starter wiring harness connectors
   - Starter motor

To install:
3. Install or connect the following:

- Starter motor. Tighten the 10mm bolt to 33 ft. lbs. (44 Nm) and the 12mm bolt to 47 ft. lbs. (64 Nm).
- Starter wiring harness connectors. Tighten the battery cable nut to 79 inch lbs. (9 Nm).
- Transmission fluid cooler line clamp
- Negative battery cable

2003–05 models

1. Before servicing the vehicle, refer to the precautions in the beginning of this section.
2. Remove or disconnect the following:
   - Negative battery cable and wait at least 3 minutes
   - Battery and battery tray
   - Harness clamps
   - Starter wiring harness connectors
   - Starter motor

To install:
3. Install or connect the following:
   - Starter motor. Tighten the bolts to 54 ft. lbs. (74 Nm).
   - Starter wiring harness connectors. Tighten the battery cable nut to 79 inch lbs. (9 Nm).
   - Harness clamps
   - Battery tray
   - Negative battery cable

Oil Pan

REMOVAL & INSTALLATION

1. Before servicing the vehicle, refer to the precautions in the beginning of this section.
2. Drain the engine oil.
3. Remove or disconnect the following:
   - Negative battery cable
   - Front splash shield
   - Heated Oxygen (HO_2S) sensor connector
   - Exhaust front pipe
   - Torque converter cover, if equipped with an automatic transaxle
   - Oil pan

To install:
4. Install the oil pan. Apply liquid gasket as shown.

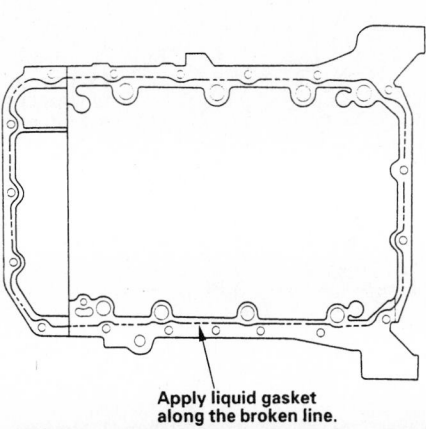

Apply liquid gasket along the broken line.

9302MG75

Apply liquid gasket to the inner threads of the bolt holes and the engine block along the area indicated by the broken line—MDX

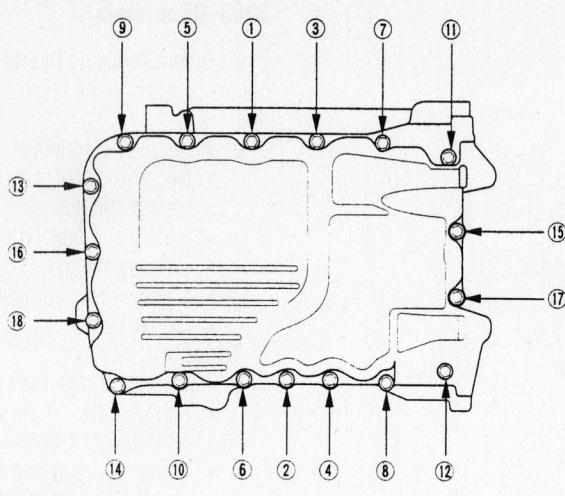

Oil pan tightening sequence—MDX

5. Tighten the bolts in sequence to 105 inch lbs. (12 Nm) using several passes.
6. Install or connect the following:
   - Torque converter cover, if removed
   - Exhaust front pipe
   - Subframe center beam, if removed
   - HO2S sensor connector
   - Front splash shield
   - Negative battery cable

Oil Pump

REMOVAL & INSTALLATION

1. Before servicing the vehicle, refer to the precautions in the beginning of this section.
2. Drain the engine oil.
3. Remove or disconnect the following:
   - Negative battery cable
   - Accessory drive belts
   - Front cover

   - Timing belt
   - Timing belt idler pulley
   - Crankshaft Position (CKP) sensor
   - Crankshaft timing sprocket
   - Variable Valve Timing and Valve Lift Electronic Control (VTEC) solenoid valve connector
   - Oil filter adapter
   - Oil pan
   - Oil pump pickup tube
   - Oil pump

To install:

➡**Use new gaskets and O-ring seals for assembly.**

4. Apply liquid gasket to the oil pump and to the bolt hole threads.
5. Install or connect the following:
   - Oil pump. Tighten the bolts to 105 inch lbs. (12 Nm).
   - Oil pump pickup tube. Tighten the bolts to 105 inch lbs. (12 Nm).

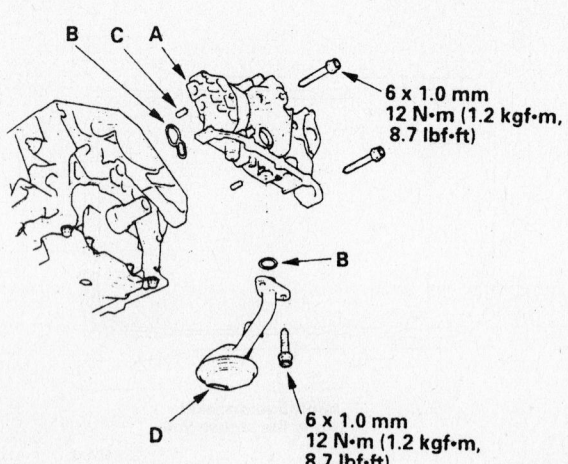

Exploded view of the oil pump assembly

- Oil pan
- Oil filter adapter
- VTEC solenoid valve connector
- Crankshaft timing sprocket
- CKP sensor
- Timing belt idler pulley
- Timing belt
- Front cover
- Accessory drive belts
- Negative battery cable

6. Fill the crankcase to the correct level.
7. Start the engine and check for leaks.

Rear Main Seal

REMOVAL & INSTALLATION

1. Before servicing the vehicle, refer to the precautions in the beginning of this section.
2. Remove or disconnect the following:
   - Transaxle
   - Clutch pressure plate and disc, if equipped
   - Flywheel
   - Oil seal

To install:

3. Install or connect the following:
   - Oil seal. Drive the seal square into the seal case.
   - Flywheel. Tighten the bolts in a crossing pattern to 54 ft. lbs. (73 Nm).
   - Clutch pressure plate and disc, if equipped
   - Transaxle

4. Check the fluid levels.
5. Start the engine and check for leaks.

Timing Belt

REMOVAL & INSTALLATION

1. Before servicing the vehicle, refer to the precautions in the beginning of this section.
2. Turn the crankshaft so the white mark aligns with the pointer.
3. Make sure the number 1 piston is at Top Dead center (TDC).
4. Remove or disconnect the following:
   - Negative battery cable
   - Wheels and splash shield
   - Drive belt

5. Support the engine with a block of wood and a jack under the oil pan.
   - Upper side engine mount

6. Remove the crankshaft pulley using holder tool shown in the accompanying illustration and a breaker and socket, loosen the 19mm bolt and remove the pulley.

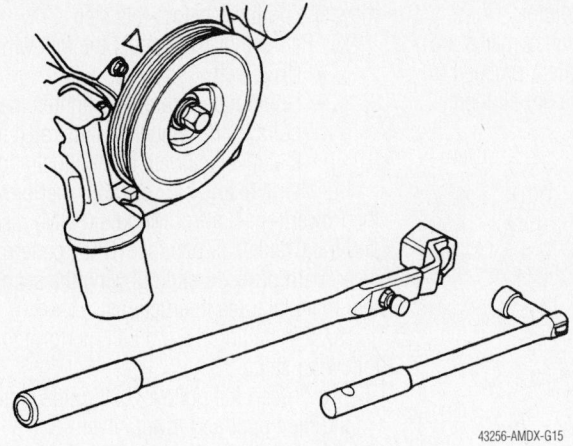

Remove the crankshaft pulley using holder tool shown

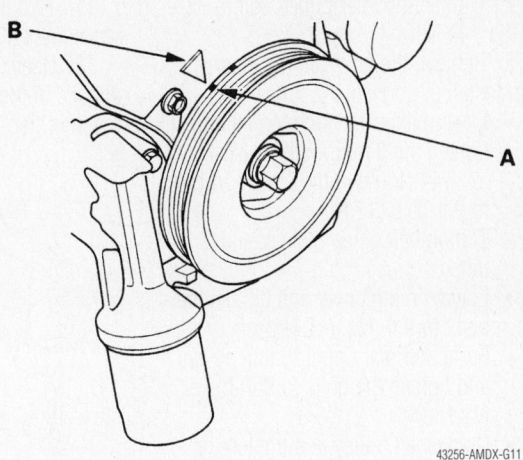

Turn the crankshaft so the white mark (A) aligns with the pointer (B)

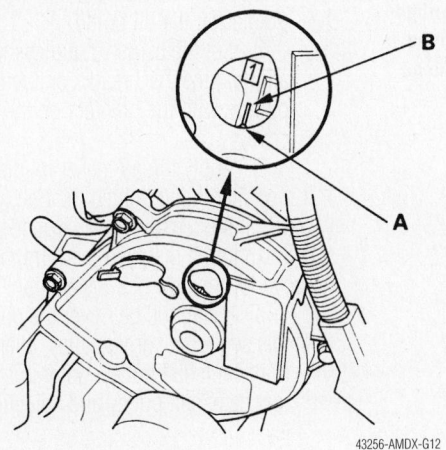

Make sure the number 1 piston is at top dead center (A) on the front camshaft pulley and pointer (B)

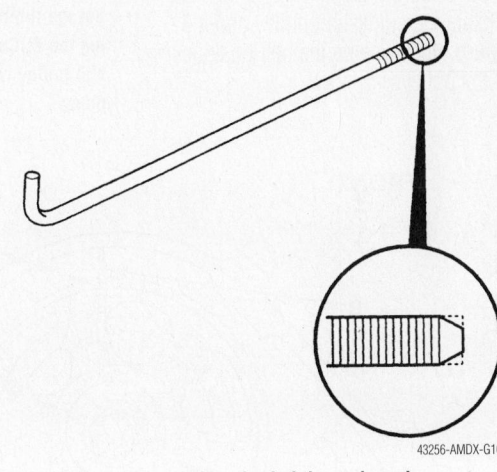

Remove a battery clamp bolt and grind the end as shown

- Front upper cover, rear upper cover and the lower cover
- One of the battery clamp bolts and grind the end as illustrated

7. Screw the battery clamp bolt as illustrated to hold the belt adjuster in position. Do not use a wrench, hand tighten only.

- Lower side engine mount
- Idler pulley bolt and the pulley
- Timing belt

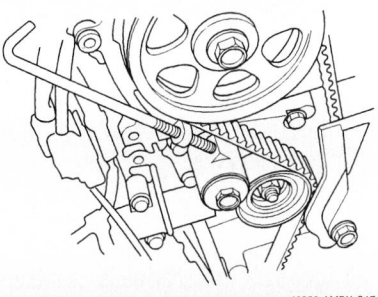

Install the battery clamp bolt as shown to hold the belt adjuster in position

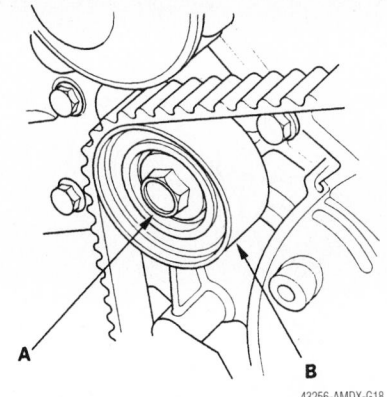

Remove the idler pulley bolt (A), pulley (B) and the timing belt

To install:

8. If installing a new belt, perform the following steps:

   a. Clean the pulleys, belt guide plate and the upper and lower covers.

   b. Set the timing belt drive pulley to TDC by aligning the TDC mark on the tooth of the belt drive pulley with the pointer on the oil pump.

   c. Set the camshaft pulleys to TDC by aligning the TDC marks on the camshaft pulleys with the pointers on the back covers.

   d. Remove the battery clamp bolt.

   e. Remove the belt tensioner.

   f. Align the holes on the rod and housing of the tensioner.

   g. Using a press or other suitable device, slowly compress the tensioner and insert a 0.08 inch (2mm) pin through the housing and rod.

   h. Install the tensioner making sure the pin is still installed.

   i. Apply thread locker to idler pulley bolt then hand tighten the bolt.

   j. Install the belt over the pulleys in this sequence; drive pulley, idler pulley, front camshaft pulley, water pump pulley, rear camshaft pulley and adjusting pulley.

k. Tighten the idler pulley bolt to 33 ft. lbs. (44 Nm).

l. Remove the pin from the tensioner.

9. Install or connect the following:
- Lower half of the side mount and tighten the 3 long bolts to 33 ft. lbs. (44 Nm) and the one short bolt to 9 ft. lbs. (12 Nm)
- Timing belt guide plate as illustrated
- Lower timing cover and tighten the bolts to 9 ft. lbs. (12 Nm)
- Front and rear upper timing covers and tighten the bolts to 9 ft. lbs. (12 Nm)
- Crankshaft pulley and tighten the bolts to 181 ft. lbs. (245 Nm), using the holding tool to prevent the unit from turning

10. Rotate the crankshaft pulley about 5 or 6 degrees clockwise so the belt positions itself on the pulleys.

11. Turn the crankshaft pulley so the white mark aligns with the pointer.

12. Check the camshaft pulley marks are aligned. If the marks are aligned, proceed to the next step. If the marks are not aligned,

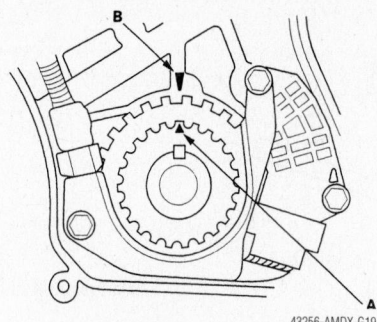

43256-AMDX-G19

Set the timing belt pulley to TDC by aligning the TDC mark (A) on the tooth of the belt pulley with the pointer (B) on the oil pump

remove the timing belt and reinstall using the steps outlined before this step.

13. Remove or disconnect the following:
- Drive belt
- Upper side mount and tighten the bolts in the sequence illustrated to the specifications in the illustration

14. Using a suitable scan tool, perform the Powertrain Control Module (PCM) reset and the Crankshaft position (CKP) pattern clear/learn procedures, following the scan tool manufactures instructions.

15. If installing the old belt, perform the following steps:

a. Clean the pulleys, belt guide plate and the upper and lower covers.

b. Set the timing belt drive pulley to TDC by aligning the TDC mark on the tooth of the belt drive pulley with the pointer on the oil pump.

c. Set the camshaft pulleys to TDC by aligning the TDC marks on the camshaft pulleys with the pointers on the back covers.

d. Apply thread locker to idler pulley bolt then hand tighten the bolt.

16. If the tensioner was extended and the belt cannot be installed, perform the steps above for the new belt installation.

a. Install the belt over the pulleys in this sequence; drive pulley, idler pulley, front camshaft pulley, water pump pulley, rear camshaft pulley and adjusting pulley.

b. Tighten the idler pulley bolt to 33 ft. lbs. (44 Nm).

c. Remove the battery clamp bolt.

17. Install or connect the following:
- Lower half of the side mount and tighten the 3 long bolts to 33 ft. lbs. (44 Nm) and the one short bolt to 9 ft. lbs. (12 Nm)
- Timing belt guide plate as illustrated
- Lower timing cover and tighten the bolts to 9 ft. lbs. (12 Nm)
- Front and rear upper timing covers and tighten the bolts to 9 ft. lbs. (12 Nm)
- Crankshaft pulley and tighten the bolts to 181 ft. lbs. (245 Nm), using the holding tool to prevent the

18. Rotate the crankshaft pulley about 5 or 6 degrees clockwise so the belt positions itself on the pulleys.

19. Turn the crankshaft pulley so the white mark aligns with the pointer.

20. Check the camshaft pulley marks are aligned. If the marks are aligned, proceed to the next step. If the marks are not aligned,

FRONT:

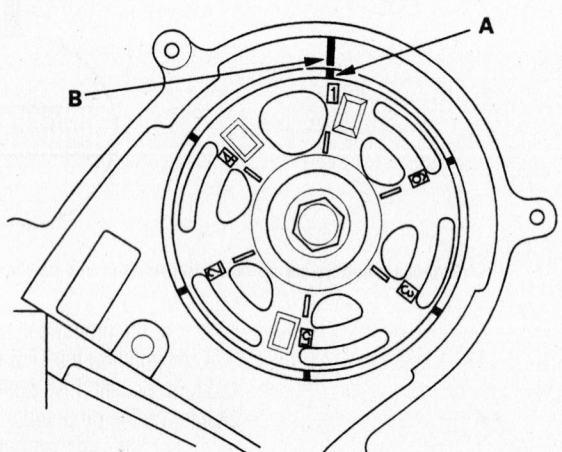

REAR:

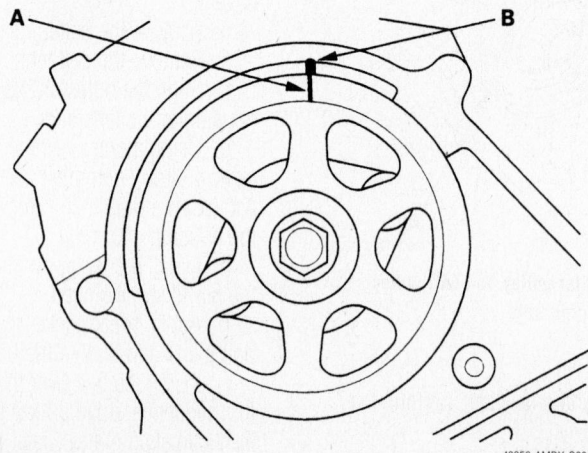

43256-AMDX-G20

Set the camshaft pulleys to TDC by aligning the TDC marks (A) on the camshaft pulleys with the pointers (B) on the back covers

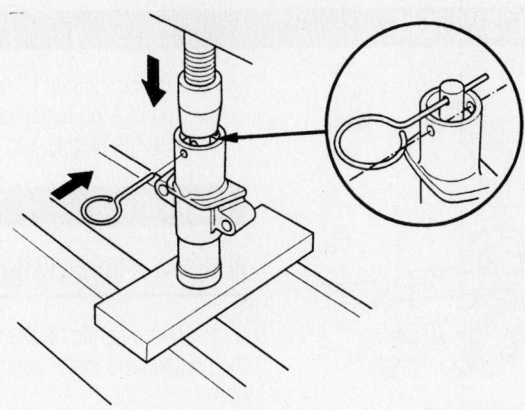

Insert a 0.08 inch (2mm) pin through the tensioner housing and rod

-1 Drive pulley (A).
-2 Idler pulley (B).
-3 Front camshaft pulley (C).
-4 Water pump pulley (D).
-5 Rear camshaft pulley (E).
-6 Adjusting pulley (F).

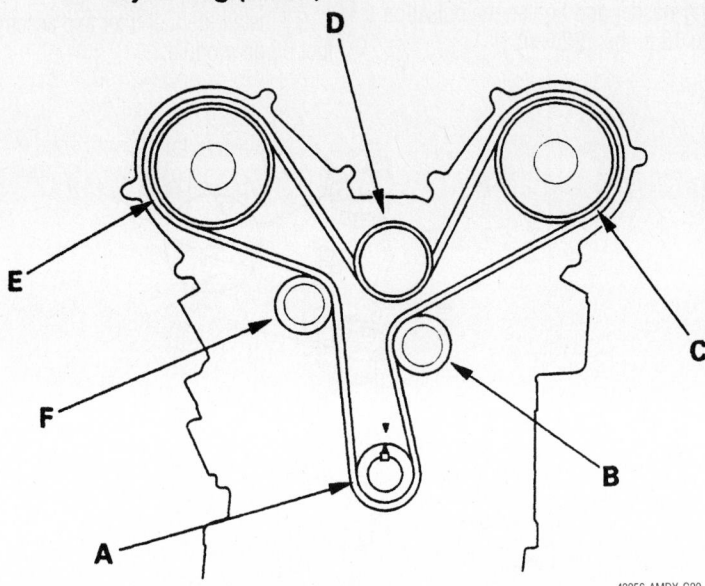

Route the belt as shown in the sequence listed

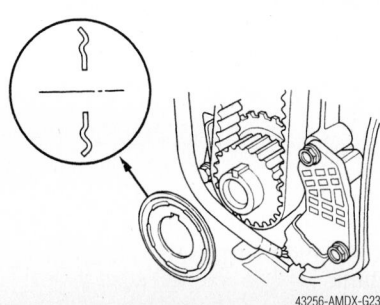

Install the timing belt guide plate as shown

FRONT CAMSHAFT PULLEY:

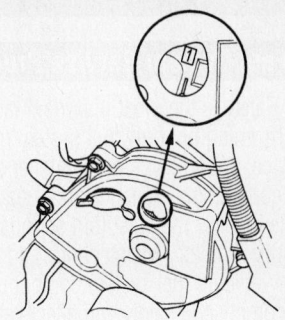

REAR CAMSHAFT PULLEY:

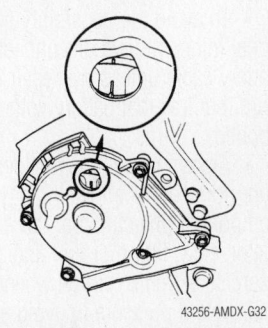

Check that the camshaft pulley marks are aligned as shown

Piston and Ring

POSITIONING

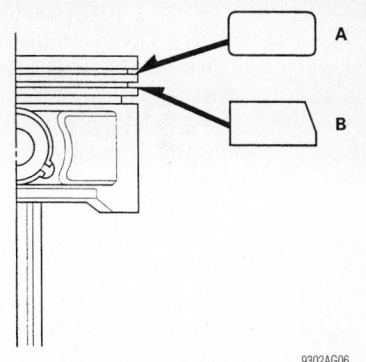

Compression ring identification—3.5L engine

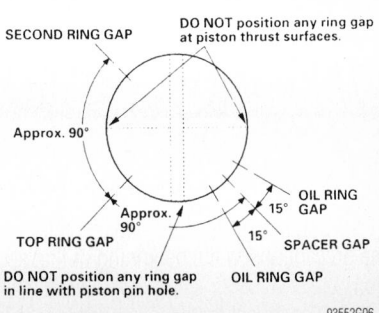

Ring end gap positioning

remove the timing belt and reinstall using the steps outlined before this step.

21. Remove or disconnect the following:
   • Drive belt
   • Upper side mount and tighten the bolts in the sequence illustrated to the specifications in the illustration

22. Using a suitable scan tool, perform the Powertrain Control Module (PCM) reset and the Crankshaft position (CKP) pattern clear/learn procedures, following the scan tool manufactures instructions.

FUEL SYSTEM

Fuel System Service Precautions

Safety is the most important factor when performing not only fuel system maintenance, but any type of maintenance. Failure to conduct maintenance and repairs in a safe manner may result in serious personal injury or death. Maintenance and testing of the vehicle's fuel system components can be accomplished safely and effectively by adhering to the following rules and guidelines:

• To avoid the possibility of fire and personal injury, always disconnect the negative battery cable unless the repair or test procedure requires that battery voltage be applied.

• Always relieve the fuel system pressure prior to disconnecting any fuel system component (injector, fuel rail, pressure regulator, etc.), fitting or fuel line connection. Exercise extreme caution whenever relieving fuel system pressure to avoid exposing skin, face and eyes to fuel spray. Please be advised that fuel under pressure may penetrate the skin or any part of the body that it contacts.

• Always place a shop towel or cloth around the fitting or connection prior to loosening to absorb any excess fuel due to spillage. Ensure that all fuel spillage (should it occur) is quickly removed from engine surfaces. Ensure that all fuel soaked cloths or towels are deposited into a suitable waste container.

• Always keep a dry chemical (Class B) fire extinguisher near the work area.

• Do not allow fuel spray or fuel vapors to come into contact with a spark or open flame.

• Always use a backup wrench when loosening and tightening fuel line connection fittings. This will prevent unnecessary stress and torsion to fuel line piping. Always follow the proper torque specifications.

• Always replace worn fuel fitting O-rings with new. Do not substitute fuel hose or equivalent, where fuel pipe is installed.

Fuel System Pressure

RELIEVING

1. Before servicing the vehicle, refer to the precautions in the beginning of this section.
2. Disconnect the negative battery cable.
3. Remove the fuel filler cap.

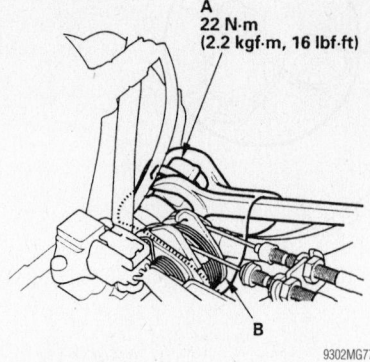

A
22 N·m
(2.2 kgf·m, 16 lbf·ft)

B

9302MG77

Use a wrench on the fuel pulsation damper (A). Place a rag over the damper (B) when relieving residual fuel pressure—MDX

4. Place a shop towel over the fuel pulsation damper.
5. Loosen the fuel pulsation damper 1 turn.
6. When service is completed, replace the sealing washer and tighten the pulsation damper to 16 ft. lbs. (22 Nm).

7. Replace the fuel filler cap.
8. Connect the negative battery cable.
9. Start the engine and check for leaks.

Fuel Filter

REMOVAL & INSTALLATION

1. Before servicing the vehicle, refer to the precautions in the beginning of this section.
2. Relieve the fuel system pressure.
3. Remove or disconnect the following:
   • Negative battery cable
   • Driver's side second row seat and cut the carpet along the dotted line. Be careful not to cut the wiring harness under the carpet.
   • Access panel
   • Fuel pump module
4. Disassemble the fuel pump module and remove the fuel filter.
   To install:
5. Install the fuel filter and assemble the fuel pump module.

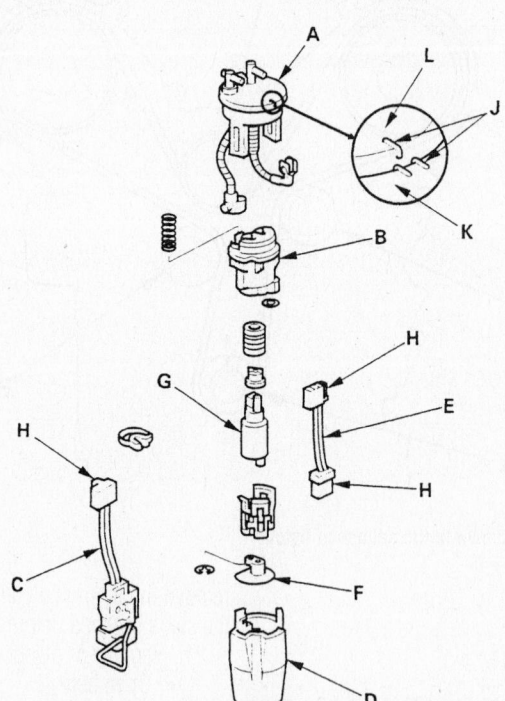

A. Bracket
B. Fuel filter
C. Fuel gauge sender
D. Case
E. Wire harness
F. Suction filter
G. Fuel pump
H. Connectors
J. Alignment marks
K. Fuel tank
L. Fuel pump module

9308MG26

Exploded view of the fuel pump module—MDX

6. Install or connect the following:
- Fuel pump module
- Access panel
- Carpet and seat
- Negative battery cable
7. Start the engine and check for leaks.

Fuel Pump

REMOVAL & INSTALLATION

1. Before servicing the vehicle, refer to the precautions in the beginning of this section.
2. Relieve the fuel system pressure.
3. Remove or disconnect the following:
- Negative battery cable
- Driver's side second row seat and cut the carpet along the dotted line. Be careful not to cut the wiring harness under the carpet.
- Access panel
- Fuel pump module wiring connector
- Fuel supply and return lines

- Fuel pump locknut
- Fuel pump module

To install:

4. Install or connect the following:
- Fuel pump module. Use a new seal and align the matchmarks.
- Fuel pump locknut
- Fuel supply and return lines
- Fuel pump module wiring connector
- Access panel
- Carpet and seat
- Negative battery cable
5. Start the engine and check for leaks.

Fuel Injector

REMOVAL & INSTALLATION

1. Before servicing the vehicle, refer to the precautions in the beginning of this section.
2. Relieve the fuel system pressure.
3. Remove or disconnect the following:
- Negative battery cable

- Intake manifold
- Fuel lines
- Fuel injector connectors
- Fuel pressure regulator vacuum line
- Fuel supply manifold
4. Separate the fuel injectors from the fuel supply manifold.

To install:

5. Install the fuel injectors to the fuel supply manifold with new cushion rings and O-rings.
6. Install new seal rings to the intake manifold.
7. Install or connect the following:
- Fuel supply manifold and injector assembly. Tighten the bolts to 86 inch lbs. (10 Nm).
- Fuel pressure regulator vacuum line
- Fuel injector connectors
- Fuel lines
- Intake manifold
- Negative battery cable
8. Start the engine and check for leaks.

DRIVE TRAIN

Transaxle Assembly

REMOVAL & INSTALLATION

1. Before servicing the vehicle, refer to the precautions in the beginning of this section.
2. Drain the transaxle.
3. Drain the power steering system.
4. Remove the engine appearance covers.
5. Remove the drivers side center console lower panel and pull back the cover to access steering joint cover.
6. Remove or disconnect the following:
- Steering joint bolt
- Steering joint from the steering gearbox pinion shaft
- Air intake assembly

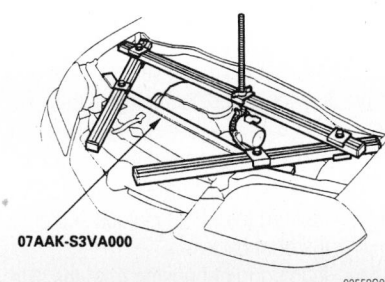

07AAK-S3VA000

93552G08

Support the engine while removing the transaxle—MDX

- Battery
- Battery tray
- Power steering pump hose and the clamp bolt
- Transmission breather tube
- Cooler hose from the clamp on the starter
- Transaxle oil cooler lines
- Starter motor
- Shift control solenoid valve connectors
- Transaxle ground cable
- 8P connector from the bracket and the connector
- Clutch pressure switch connectors
- Joint connector and transmission range switch connector from the brackets
- Countershaft speed sensor connector
- Heated Oxygen (HO2S) sensor connectors
- Transmission housing mounting bolts
- Nut from the front mount and the ground cable from the engine
- Bulkhead cover, windshield wiper arms, cowl cover sealing and cover
- Install a support fixture to the engine lifting eyes.
- Splash shield
- Front sub-frame stiffener
- Exhaust front pipe

- Lower control arms from the knuckle
- Stabilizer bar links
- Tie rod ends from the knuckle
- Left driveshaft from the differential
- Right driveshaft from the intermediate shaft
- Propeller shaft from the companion flange
- Shift cable cover and holder
- Shift control cable and lever
7. Install a 6 x 1 x 14mm bolt and nut on the cable cover, then reinstall the cable cover to the torque converter housing. If this is not done, the bolt head of the cable cover may prevent torque converter removal.
- Transfer assembly
- Engine-to-torque converter bolts
- Power steering pressure switch connection
- Power steering hose clamp, then the hose from the pipe at the subframe
- Transmission lower mount nuts
8. Matchmark the front subframe to the vehicle body.
- Rear mount bracket bolts
9. Support the sub-frame with a 4 x 4 x 50 inch piece of wood and a jack.
- Sub-frame
- Transaxle lower mounts
- Driveshafts from the differential and intermediate shaft

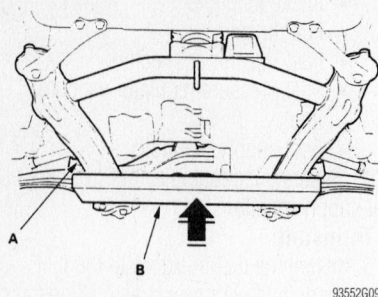

Support the sub-frame with a 4 x 4 x 50 inch piece of wood and a jack

93552G09

- Intermediate shaft
- Transmission front mount bracket
- Transmission flange bolts
- Transmission

To install:

➡**Use new circlips, split pins and self-locking nuts for assembly.**

10. Installation is the reverse of removal. Please note the following specifications:
- Transmission housing bolts and harness clamp bolts to 47 ft. lbs. (64 Nm)
- Transmission housing bolts to 40 ft. lbs. (54 Nm)
- Front mount bracket bolts to 28 ft. lbs. (38 Nm)
- Intermediate shaft bolts to 29 ft. lbs. (39 Nm)

11. Raise the subframe into position and align the match marks. Tighten the subframe bolts to 76 ft. lbs. (103 Nm). Tighten the front subframe bracket bolts to 54 ft. lbs. (74 Nm) and the rear bracket bolts to 86 ft. lbs. (117 Nm).
- Rear engine mount bolts to 28 ft. lbs. (38 Nm)
- Engine-to-torque converter bolts. Tighten the 6 x 1 mm bolts to 105 inch lbs. (12 Nm), 10 x 1.25mm bolt to 28 ft. lbs. (38 Nm).
- Front motor mount nut to 40 ft. lbs. (54 Nm)

12. Fill the transaxle to the correct level.
13. Start the engine and check for leaks.
14. Check the wheel alignment and adjust as necessary.

Transfer Assembly

REMOVAL & INSTALLATION

1. Before servicing the vehicle, refer to the precautions in the beginning of this section.
2. Drain the transmission fluid.

3. Remove or disconnect the following:
- Negative battery cable
- Heated Oxygen (HO₂S) sensor connectors
- Front sub-frame stiffener
- Exhaust front pipe
- Breather tube bracket bolt, then the tube from the breather pipe
- Propeller shaft from the transfer assembly
- Transfer assembly bolts and the assembly

To install:

4. Install or connect the following:
- New O-ring on the transfer cover
- Dowel pin on the assembly
- Transfer assembly and tighten the bolts to 33 ft. lbs. (44 Nm) on 2001–02 models and 38 ft. lbs. (51 Nm) on 2003–05 models
- Propeller shaft
- Breather tube bracket, attach the tube with the dot facing outwards and tighten the bolt to 9 ft. lbs. (12 Nm)
- Exhaust front pipe
- Front sub-frame stiffener and tighten the bolts to 40 ft. lbs. (54 Nm)
- Heated Oxygen (HO₂S) sensor connectors
- Negative battery cable

Halfshaft

REMOVAL & INSTALLATION

1. Before servicing the vehicle, refer to the precautions in the beginning of this section.
2. Drain the transaxle if removing the left halfshaft. If is not necessary to drain the fluid if removing the right halfshaft.
3. Remove or disconnect the following:
- Negative battery cable
- Front wheels
- Spindle nut
- Stabilizer bar link
- Lower ball joint

4. Pry the inboard joint from the transaxle or intermediate shaft.
5. Remove the outer CV-joint stub shaft from the hub by tapping the stub shaft with a plastic hammer.

To install:

➡**Use new circlips, split pins and self-locking nuts for assembly.**

6. Install the outer CV-joint stub shaft into the hub.
7. Install the inner CV-joint to the

transaxle or intermediate shaft until the circlip locks in the retaining groove.
8. Install or connect the following:
- Lower ball joint. Tighten the nut to 43–51 ft. lbs. (59–69 Nm).
- Stabilizer bar link. Tighten the nut to 58 ft. lbs. (78 Nm).
- Spindle nut. Tighten the nut to 181 ft. lbs. (245 Nm) on 2001–02 models, or 210 ft. lbs. (285 Nm) on 2003–05 models.
- Front wheels
- Negative battery cable

9. Fill the transaxle to the correct level and check for leaks.

CV-Joint

OVERHAUL

Front

OUTBOARD JOINT

1. Before servicing the vehicle, refer to the precautions in the beginning of this section.
2. Remove or disconnect the following:
- Axle halfshaft from the vehicle and place it in a vise
- Outboard joint boot clamps and push the boot back
- Outboard joint by driving it off the axle shaft with a brass drift and hammer
- Outboard joint boot

To install:

➡**Use new circlips and boot clamps for assembly.**

3. Install the outboard joint boot and clamps to the axle shaft.
4. Fill the outboard joint with grease. Install the outboard joint to the axle shaft. Tap the stub shaft with a brass hammer to seat the circlip.
5. Fill the outboard joint boot with grease and install the boot clamps.
6. Install the axle halfshaft to the vehicle.

INBOARD JOINT

1. Before servicing the vehicle, refer to the precautions in the beginning of this section.
2. Remove or disconnect the following:
- Axle halfshaft from the vehicle.
- Inboard joint boot clamps and push the boot back
- Inboard joint housing from the axle
- Rollers from the spider

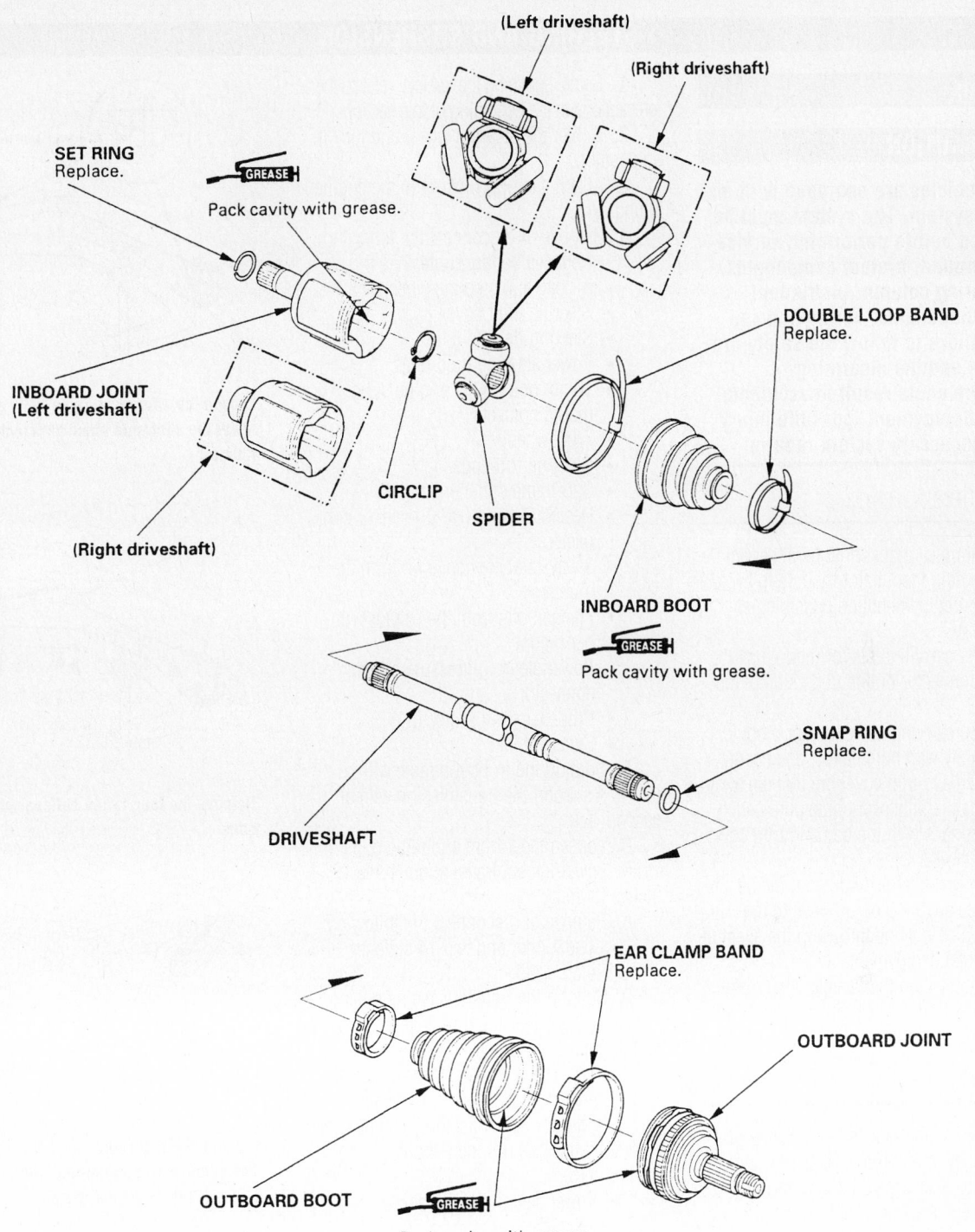

Front axle exploded view—MDX

- Snapring and the spider from the axle shaft
- Inboard joint boot

To install:

➡ **Use new circlips and boot clamps for assembly.**

3. Install or connect the following:
   - Inboard joint boot and clamps to the axle shaft
   - Spider with a new snapring
   - Rollers to the spider

4. Fill the joint housing with grease and install it.

5. Fill the inboard joint boot with grease and install the boot clamps.

6. Install the axle halfshaft to the vehicle.

STEERING AND SUSPENSION

Air Bag

※※ CAUTION

Some vehicles are equipped with an air bag system. The system must be disarmed before performing service on, or around, system components, the steering column, instrument panel components, wiring and sensors. Failure to follow the safety precautions and the disarming procedure could result in accidental air bag deployment, possible injury and unnecessary system repairs.

PRECAUTIONS

Several precautions must be observed when handling the inflator module to avoid accidental deployment and possible personal injury.

• Never carry the inflator module by the wires or connector on the underside of the module.

• When carrying a live inflator module, hold securely with both hands, and ensure that the bag and trim cover are pointed away.

• Place the inflator module on a bench or other surface with the bag and trim cover facing up.

• With the inflator module on the bench, never place anything on or close to the module which may be thrown in the event of an accidental deployment.

Before servicing the vehicle, also make sure to refer to the precautions in the beginning of this section as well.

DISARMING

Disconnect and isolate the negative battery cable. Wait 3 minutes for the system capacitor to discharge before performing any service.

Power Rack and Pinion Steering Gear

REMOVAL & INSTALLATION

※※ WARNING

Do not permit the steering wheel to turn whenever the steering gear is disconnected from the steering column. Damage to the air bag wiring can result.

1. Before servicing the vehicle, refer to the precautions in the beginning of this section.
2. Center the steering wheel and lock it in position.
3. Attach a support fixture to the engine lifting eyes.
4. Remove or disconnect the following:
   • Negative battery cable
   • Air bag and steering wheel
   • Steering joint cover
   • Steering flexible joint
   • Power steering fluid lines
   • 10mm bolt on the engine side mount bracket
   • Front wheels
   • Outer tie rod ends
   • Sub-frame stiffener
   • Heated Oxygen (HO$_2$S) sensor connectors
   • 3 way catalytic converter from the mufflers
   • Flange bolts from the exhaust rubber mount
   • Power steering pressure switch connector
   • Propeller shaft protector
   • Splash shield
5. Support the front subframe with a jack and support the transmission with a second jack.
6. Loosen the 14mm subframe bolts.
7. Lower the subframe about 1 3/16 inches (30mm).
8. Remove or disconnect the following:
   • Two 12mm and two 14 stiffener plate bolts
9. Support the transfer case by raising the transmission jack and remove the two 12mm bolts.
   • Two 14mm bolts and the rear stiffener plats from the sub-frame
10. Lower the transmission jack until the front subframe has dropped about 1 15/16 inch (50mm).
    • Power steering line brackets
    • Feed line
    • Return hose
    • Two 10mm bolts from the right side gearbox
    • Mounting bracket and cushion
    • Two 10mm bolts from the left side gearbox
11. Lower the transmission jack until the front subframe has dropped about 3 15/16 inch (100mm).
    • Gearbox stiffener bracket
12. Slide the gearbox between the body and front sub-frame towards the left and from the vehicle.

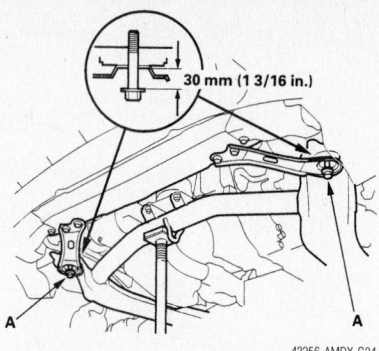

Loosen the 14mm subframe bolts and lower the subframe about 1³⁄₁₆ inches (30mm)

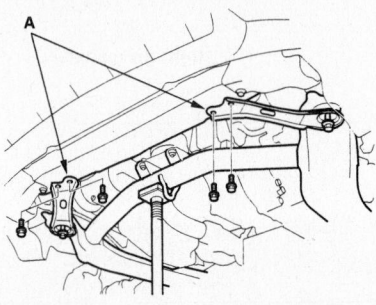

Remove the four 12mm stiffener plate bolts

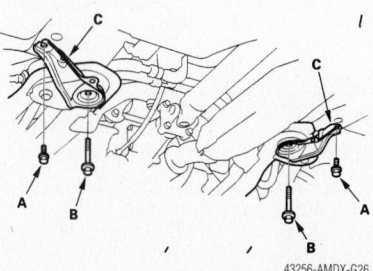

Remove the two 12mm bolts (A), then the 14mm bolts (B) and the rear stiffener plates (C) from the sub-frame

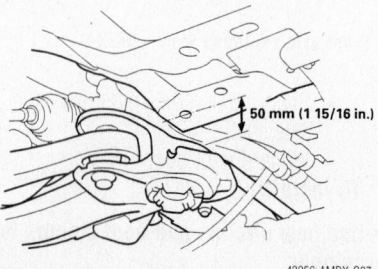

Lower the transmission jack until the front subframe has dropped about 1¹⁵⁄₁₆ inch (50mm)

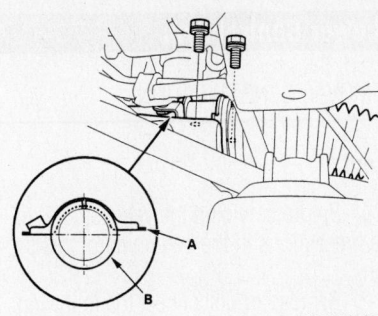

43256-AMDX-G28

Remove the two 10mm bolts from the right side gearbox and the mounting bracket and cushion

To install:

13. Position the steering gear in the vehicle.

14. Install or connect the following:
- Left steering gear mounting bolts. Tighten the bolts to 43 ft. lbs. (58 Nm).
- Right steering gear mounting bracket. Tighten the bolts to 29 ft. lbs. (39 Nm).
- Return hose
- Feed line
- Power steering line mounting brackets and tighten the bolts to 7 ft. lbs. (10 Nm)

15. Raise the subframe into position. Tighten the 14mm bolts to 76 ft. lbs. (103 Nm) and the 12mm bolts to 86 ft. lbs. (117 Nm).

16. Install or connect the following:
- Front stiffener plates. Tighten the 14mm bolts to 76 ft. lbs. (103 Nm) and the 12mm bolts to 54 ft. lbs. (74 Nm).
- Splash shield
- Propeller shaft protector
- Power steering pressure switch
- 3 way catalytic converter and mufflers. Tighten the nuts to 25 ft. lbs. (33 Nm)
- Rubber exhaust mount and tighten the bolts to 28 ft. lbs. (38 Nm)
- HO_2S sensor connectors
- Sub-frame stiffener plate
- 10mm flange bolts on the engine side mount bracket to 33 ft. lbs. (44 Nm)
- Power steering hoses
- Outer tie rod ends
- Front wheels. Position the wheels straight-ahead.
- Steering flexible joint. Tighten the pinch bolts to 16 ft. lbs. (22 Nm).
- Steering joint cover
- Negative battery cable

17. Fill the power steering system.

18. Check the wheel alignment and adjust as necessary.

Strut

REMOVAL & INSTALLATION

Front

1. Before servicing the vehicle, refer to the precautions in the beginning of this section.

2. Remove or disconnect the following:
- Front wheel
- Wheel speed sensor wiring bracket
- Brake hose bracket
- Stabilizer bar link
- Strut pinch bolts
- Upper mount nuts
- Strut

To install:
3. Install or connect the following:
- Strut. Tighten the upper mount nuts to 43 ft. lbs. (59 Nm).

- Strut pinch bolts. Tighten the nuts to 116 ft. lbs. (157 Nm).
- Stabilizer bar link. Tighten the nut to 58 ft. lbs. (78 Nm).
- Brake hose bracket
- Wheel speed sensor wiring bracket
- Front wheel

4. Check the wheel alignment and adjust as necessary.

Shock Absorber

REMOVAL & INSTALLATION

Rear

1. Before servicing the vehicle, refer to the precautions in the beginning of this section.

2. Support the vehicle under the lower control arm.

3. Remove or disconnect the following:
- Rear wheel
- Upper shock absorber flange bolt

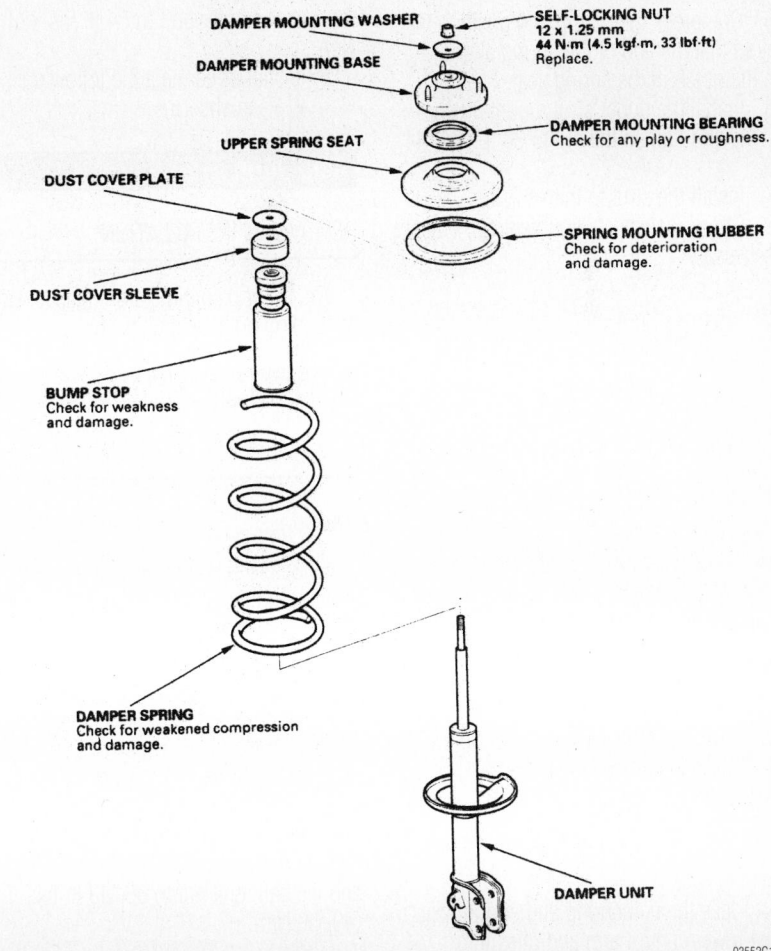

93552G11

Exploded view of the front strut assembly

- Lower shock absorber nut
- Shock absorber

To install:

4. Install or connect the following:
- Shock absorber. Tighten the fasteners to 47 ft. lbs. (64 Nm).
- Rear wheel

Coil Spring

REMOVAL & INSTALLATION

Front

1. Before servicing the vehicle, refer to the precautions in the beginning of this section.

2. Remove the strut from the vehicle and install in a strut spring compressor. Compress the spring until the end of the spring comes away from the spring seat.

3. Remove the upper strut mount, spring seat and related components.

4. Remove the coil spring from the strut spring compressor.

To install:

➡ **Use a new self-locking nut.**

5. Compress the spring and position the strut so that the end of the spring aligns with the notch in the spring seat.

6. Install the upper strut mounting components and tighten the nut to 33 ft. lbs. (44 Nm).

7. Install the strut to the vehicle.

8. Check the wheel alignment and adjust as necessary.

Rear

1. Before servicing the vehicle, refer to the precautions in the beginning of this section.

2. Support the vehicle under the lower control arm.

3. Remove or disconnect the following:
- Rear wheel
- Stabilizer link from the lower arm
- Wheel speed sensor wiring harness from the lower arm. Do not disconnect the connector.
- Upper shock absorber flange bolt
- Lower control arm bolts

4. Lower the floor jack and remove the coil spring and spring seats.

To install:

➡ **Use new self-locking nuts for assembly.**

5. Place the coil spring and spring seats on the lower control arm and raise into position. Tighten the inboard bolt to 61 ft.

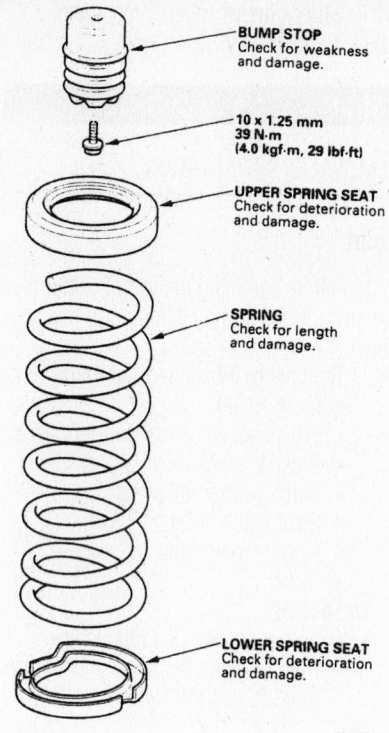

BUMP STOP
Check for weakness and damage.

10 x 1.25 mm
39 N·m
(4.0 kgf·m, 29 lbf·ft)

UPPER SPRING SEAT
Check for deterioration and damage.

SPRING
Check for length and damage.

LOWER SPRING SEAT
Check for deterioration and damage.

93552G12

Exploded view of the rear spring assembly

lbs. and the outer bolt to 54 ft. lbs. (74 Nm).

6. Install or connect the following:
- Rear wheel

Ball Joint

REMOVAL & INSTALLATION

The lower ball joints are replaced with the control arms as an assembly.

Upper Control Arm

REMOVAL & INSTALLATION

Rear

1. Before servicing the vehicle, refer to the precautions in the beginning of this section.

2. Support the control arm at the knuckle.

3. Remove or disconnect the following:
- Wheel
- Upper ball joint from the knuckle
- Upper arm bolt and the arm

4. Installation is the reverse of removal. Tighten the arm bolt to 47 ft. lbs. (64 Nm) and the ball joint nut to 36–43 ft. lbs. (49–59 Nm).

Lower Control Arm

REMOVAL & INSTALLATION

Front

1. Before servicing the vehicle, refer to the precautions in the beginning of this section.

2. Remove or disconnect the following:
- Front wheel
- Lower ball joint
- Front inner flange bolt
- Rear inner flange bolt
- Lower control arm

To install:

➡ **Use a new split pin for assembly.**

3. Install or connect the following:
- Lower control arm. Tighten the inner flange bolts to 69 ft. lbs. (93 Nm).
- Lower ball joint. Tighten the nut to 43–51 ft. lbs. (59–69 Nm).
- Front wheel

4. Check the wheel alignment and adjust as necessary.

Rear

LOWER ARM (A)

1. Before servicing the vehicle, refer to the precautions in the beginning of this section.

2. Remove or disconnect the following:
- Lower arm mounting bolt and nut
- Lower arm

3. Installation is the reverse of removal. Tighten the bolt to 105 ft. lbs. (142 Nm) and the nut to 47 ft. lbs. (64 Nm).

LOWER ARM (B)

1. Before servicing the vehicle, refer to the precautions in the beginning of this section.

2. Support the control arm with a jack.

3. Remove or disconnect the following:
- Wheel

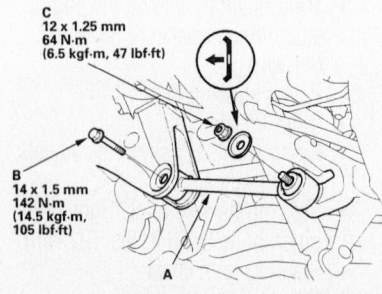

C
12 x 1.25 mm
64 N·m
(6.5 kgf·m, 47 lbf·ft)

B
14 x 1.5 mm
142 N·m
(14.5 kgf·m, 105 lbf·ft)

A

43256-AMDX-G29

Rear lower arm (A) mounting

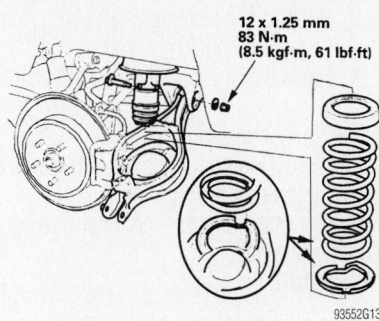

12 x 1.25 mm
83 N·m
(8.5 kgf·m, 61 lbf·ft)

93552G13

Rear lower arm (B) mounting

- Stabilizer link from the lower arm
- Wheel speed sensor wiring harness from the lower arm. Do not disconnect the connector.
- Flange bolts that attaches the lower arm to the knuckle
4. Spring assembly
   - Inner nuts and bolts and the arm
5. Install or connect the following:
   - Arm, inner bolt and loosely install the nut
   - Spring assembly
6. Raise the arm into position and install the flange bolt.
7. Raise the rear suspension with a floor jack to load the vehicle weight.

- Tighten the flange bolt to 54 ft. lbs. (74 Nm) and the inner nut and bolt to 61 ft. lbs. (83 Nm).
- Wheel speed sensor harness
- Wheel
8. Check the vehicle alignment.

CONTROL ARM BUSHING REPLACEMENT

The control arm bushings are serviced with the control arms as an assembly.

Wheel Bearings

ADJUSTMENT

The wheel bearings are sealed units and are not adjustable.

REMOVAL & INSTALLATION

Front

1. Before servicing the vehicle, refer to the precautions in the beginning of this section.
2. Remove or disconnect the following:
   - Front wheel
   - Brake hose mounting bolt

- Brake caliper
- Wheel speed sensor
- Spindle nut
- Brake rotor
- Outer tie rod end
- Lower ball joint
- Steering knuckle
3. Press the hub out of the wheel bearing.
   - Splash guard
   - Snapring and press the wheel bearing out of the steering knuckle
4. If necessary, press the inner bearing race off of the hub.

To install:

➡**Use a new ball joint nut, split pin, snapring and spindle nut for assembly.**

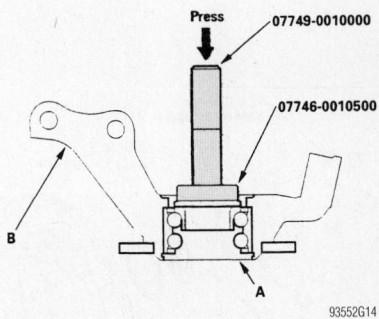

Press 07749-0010000
07746-0010500
B
A
93552G14

Press the wheel bearing out of the knuckle

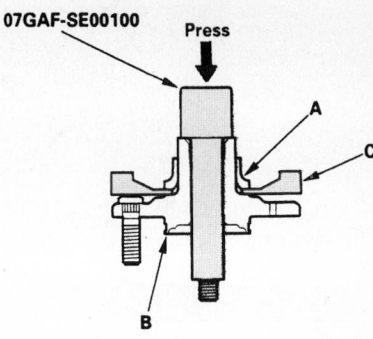

07GAF-SE00100   Press
A
C
B
93552G15

Press the wheel bearing inner race from the hub

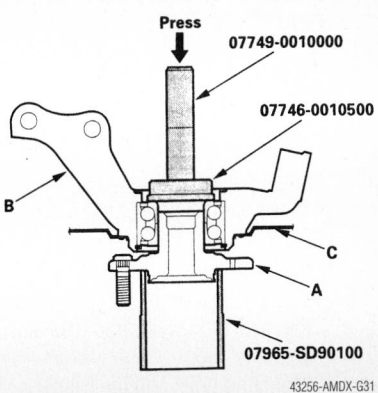

Press 07749-0010000
07746-0010500
B
C
A
07965-SD90100
43256-AMDX-G31

Press the wheel bearing into the knuckle

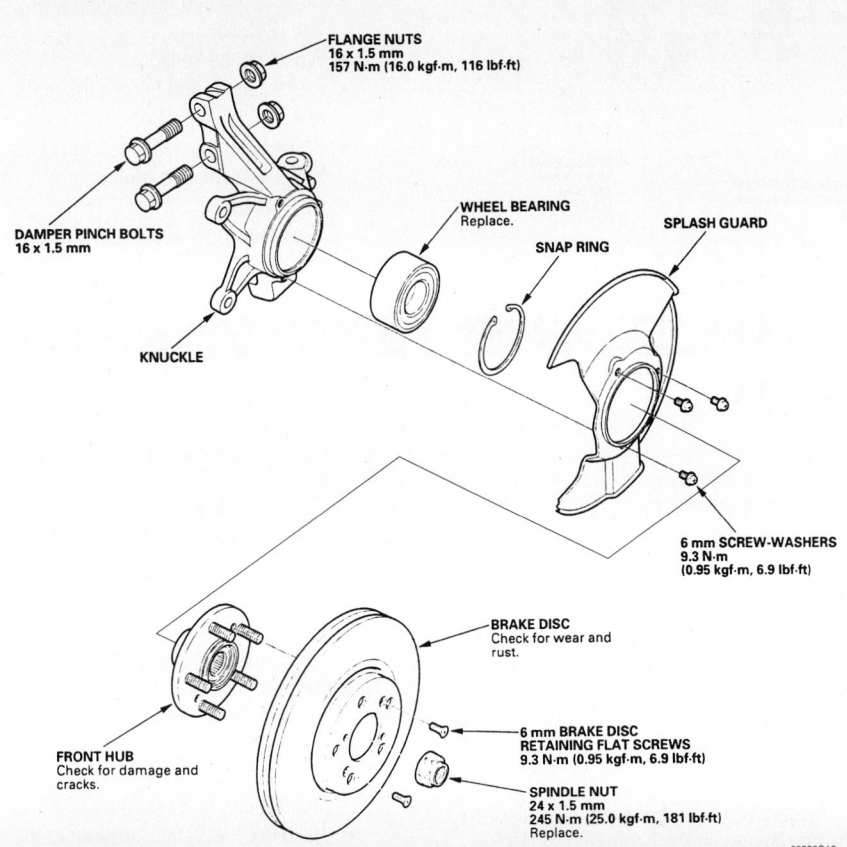

FLANGE NUTS
16 x 1.5 mm
157 N·m (16.0 kgf·m, 116 lbf·ft)

DAMPER PINCH BOLTS
16 x 1.5 mm

KNUCKLE

WHEEL BEARING
Replace.

SNAP RING

SPLASH GUARD

6 mm SCREW-WASHERS
9.3 N·m
(0.95 kgf·m, 6.9 lbf·ft)

BRAKE DISC
Check for wear and rust.

FRONT HUB
Check for damage and cracks.

6 mm BRAKE DISC
RETAINING FLAT SCREWS
9.3 N·m (0.95 kgf·m, 6.9 lbf·ft)

SPINDLE NUT
24 x 1.5 mm
245 N·m (25.0 kgf·m, 181 lbf·ft)
Replace.

93552G16

Front wheel bearing assembly, spindle nut torque shown is for 2001–02, on 2003–04 torque is 210 ft. lbs. (285 Nm)

5. Press the bearing into the steering knuckle and install the snapring.

6. Install the splash guard.

7. Press the hub into the bearing.

8. Install or connect the following:
- Steering knuckle. Tighten the ball joint nut to 43–51 ft. lbs. (59–69 Nm) and the damper flange bolts to 116 ft. lbs. (157 Nm).
- Outer tie rod end. Tighten the nut to 40 ft. lbs. (54 Nm).
- Wheel speed sensor, if equipped
- Brake caliper and rotor
- Brake hose
- Spindle nut. Tighten the nut to 181 ft. lbs. (245 Nm) on 2001–02 models. On 2003–05 models, torque to 210 ft. lbs. (285 Nm).
- Front wheel

9. Check the wheel alignment and adjust as necessary.

Rear

1. Before servicing the vehicle, refer to the precautions in the beginning of this section.

2. Remove or disconnect the following:
- Rear wheel
- Brake hose bracket mounting bolts from the trailing arm and the knuckle
- Brake caliper
- Wheel speed sensor
- Spindle nut
- Brake rotor
- Upper ball joint
- Lower arm (A)
- Lower arm (B) from the trailing arm

3. Support the lower arm (B)
- Steering knuckle

4. Press the hub out of the wheel bearing.
- Splash guard
- Snapring and press the wheel bearing out of the steering knuckle

5. If necessary, press the inner bearing race off of the hub.

To install:

➡**Use a new ball joint nut, split pin, snapring and spindle nut for assembly.**

6. Press the bearing into the steering knuckle and install the snapring.

7. Install the splash guard.

8. Press the hub into the bearing.

9. Install or connect the following:
- Steering knuckle. Tighten the flange bolt to 54 ft. lbs. (74 Nm) and the lower shock nut to 47 ft. lbs. (64 Nm)
- Lower arm (B) to the trailing arm and tighten the bolts to 47 ft. lbs. (64 Nm)
- Lower arm (A)
- Upper ball joint and tighten the nut to 40 ft. lbs. (54 Nm)
- Brake rotor and tighten the screws to 7 ft. lbs. (9 Nm)
- Spindle nut and tighten to 181 ft, lbs. (245 Nm)
- Wheel speed sensor
- Brake caliper and tighten the bolts to 41 ft. lbs. (55 Nm)
- Brake hose bracket mounting bolts to the knuckle and trailing arm
- Rear wheel

10. Check the wheel alignment and adjust as necessary.

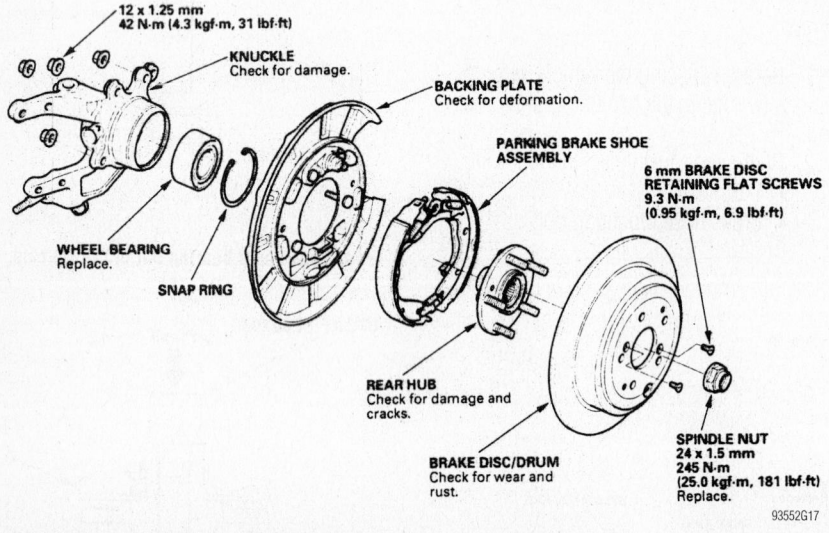

12 x 1.25 mm
42 N·m (4.3 kgf·m, 31 lbf·ft)

KNUCKLE
Check for damage.

BACKING PLATE
Check for deformation.

PARKING BRAKE SHOE ASSEMBLY

6 mm BRAKE DISC RETAINING FLAT SCREWS
9.3 N·m
(0.95 kgf·m, 6.9 lbf·ft)

WHEEL BEARING
Replace.

SNAP RING

REAR HUB
Check for damage and cracks.

BRAKE DISC/DRUM
Check for wear and rust.

SPINDLE NUT
24 x 1.5 mm
245 N·m
(25.0 kgf·m, 181 lbf·ft)
Replace.

93552G17

Exploded view of the rear wheel bearing assembly

BRAKES

Brake Caliper

REMOVAL & INSTALLATION

Front

1. Before servicing the vehicle, refer to the precautions in the beginning of this section.

2. Remove some fluid from the reservoir with a suction pump.

3. Remove or disconnect the following:
- Front wheels
- Banjo bolt and disconnect the brake hose from the caliper. Plug the hose to prevent fluid loss and contamination.
- Mounting bolts and the caliper from its mounting bracket

To Install:

4. Install or connect the following:
- Caliper over the pads and onto its mounting bracket. Torque both caliper bolts to 27 ft. lbs. (36 Nm).
- Brake hose to the caliper using new sealing washers. Carefully torque the banjo bolt to 25 ft. lbs. (34 Nm).

5. Fill the reservoir with fluid and bleed the brakes.
- Front wheels

Rear

1. Before servicing the vehicle, refer to the precautions in the beginning of this section.

2. Remove some fluid from the reservoir with a suction pump.

3. Remove or disconnect the following:
- Rear wheels
- Banjo bolt and disconnect the brake hose from the caliper. Plug the hose to prevent fluid loss and contamination.
- 2 caliper mounting bolts and the caliper from its mounting bracket

To Install:

4. Install or connect the following:
- Caliper over the pads and onto its mounting bracket. Tighten the caliper bolts to 27 ft. lbs. (37 Nm).
- Brake hose with new sealing washers. Tighten the banjo bolt to 25 ft. lbs. (34 Nm).

Front Brake Caliper Overhaul

⚠CAUTION

Frequent inhalation of brake pad dust, regardless of material composition, could be hazardous to your health.
- Avoid breathing dust particles.
- Never use an air hose or brush to clean brake assemblies. Use an OSHA-approved vacuum cleaner.

Remove, disassemble, inspect, reassemble, and install the caliper, and note these items:

- Do not spill brake fluid on the vehicle; it may damage the paint; if brake fluid gets on the paint, wash it off immediately with water.
- To prevent dripping brake fluid, cover disconnected hose joints with rags or shop towels.
- Clean all parts in brake fluid and air dry; blow out all passages with compressed air.
- Before reassembling, check that all parts are free of dirt and other foreign particles.
- Replace parts with new ones as specified in the illustration.
- Make sure no dirt or other foreign matter gets in the brake fluid.
- Make sure no grease or oil gets on the brake discs or pads.
- When reusing pads, always reinstall them in their original positions to prevent loss of braking efficiency.
- Do not reuse drained brake fluid. Use only clean Genuine Honda DOT 3 Brake Fluid. Non-Honda brake fluid can cause corrosion and shorten the life of the system.
- Coat the piston, piston seal groove, and caliper bore with clean brake fluid.
- Replace all rubber parts with new ones.
- After installing the caliper, check the brake hose and line for leaks, interference, and twisting.

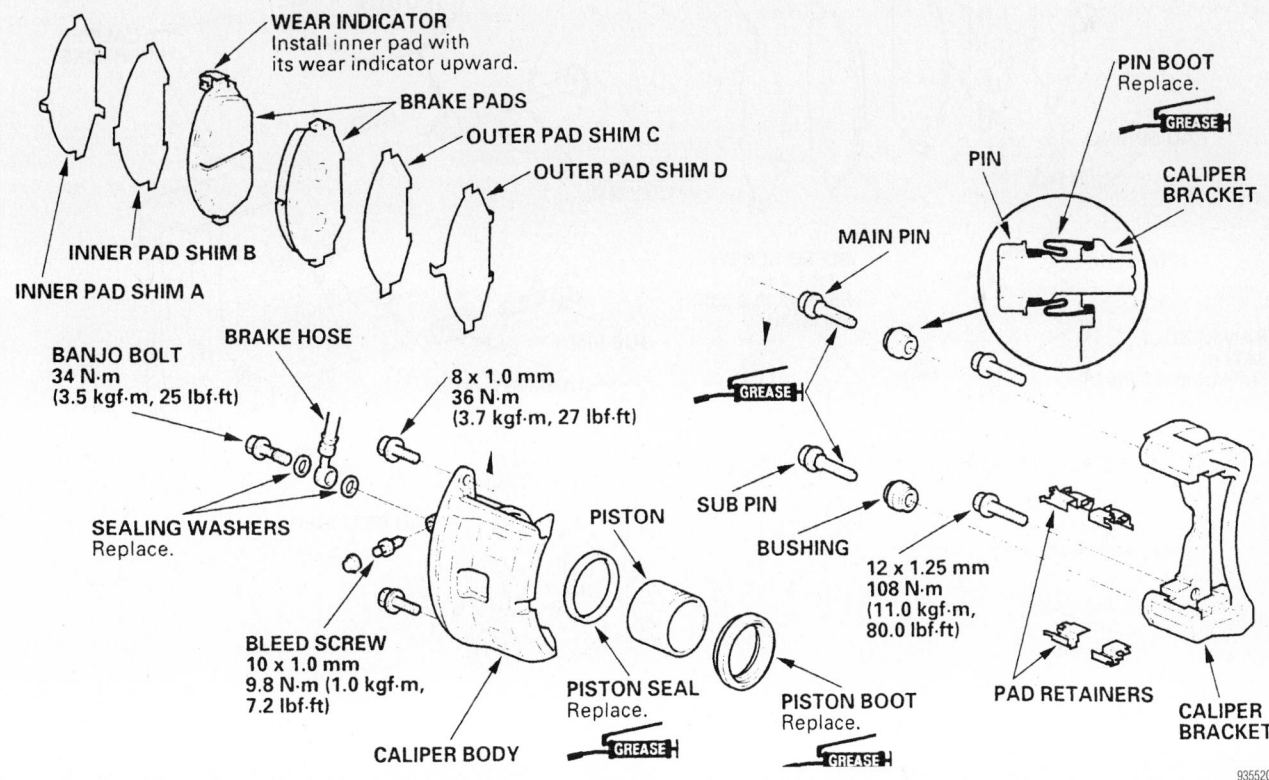

Exploded view of the front caliper components—2001–02 models

93552GZZ

⚠️CAUTION

Frequent inhalation of brake pad dust, regardless of material composition, could be hazardous to your health.
• Avoid breathing dust particles.
• Never use an air hose or brush to clean brake assemblies. Use an OSHA-approved vacuum cleaner.

Remove, disassemble, inspect, reassemble, and install the caliper, and note these items:

• Do not spill brake fluid on the vehicle; it may damage the paint; if brake fluid gets on the paint, wash it off immediately with water.
• To prevent dripping brake fluid, cover disconnected hose joints with rags or shop towels.
• Clean all parts in brake fluid and air dry; blow out all passages with compressed air.
• Before reassembling, check that all parts are free of dirt and other foreign particles.
• Replace parts with new ones as specified in the illustration.
• Make sure no dirt or other foreign matter gets in the brake fluid.
• Make sure no grease or oil gets on the brake discs or pads.
• When reusing pads, always reinstall them in their original positions to prevent loss of braking efficiency.
• Do not reuse drained brake fluid. Use only clean Honda DOT 3 Brake Fluid. Non-Honda brake fluid can cause corrosion and shorten the life of the system.
• Coat the piston, piston seal groove, and caliper bore with clean brake fluid.
• Replace all rubber parts with new ones.
• After installing the caliper, check the brake hose and line for leaks, interference, and twisting.

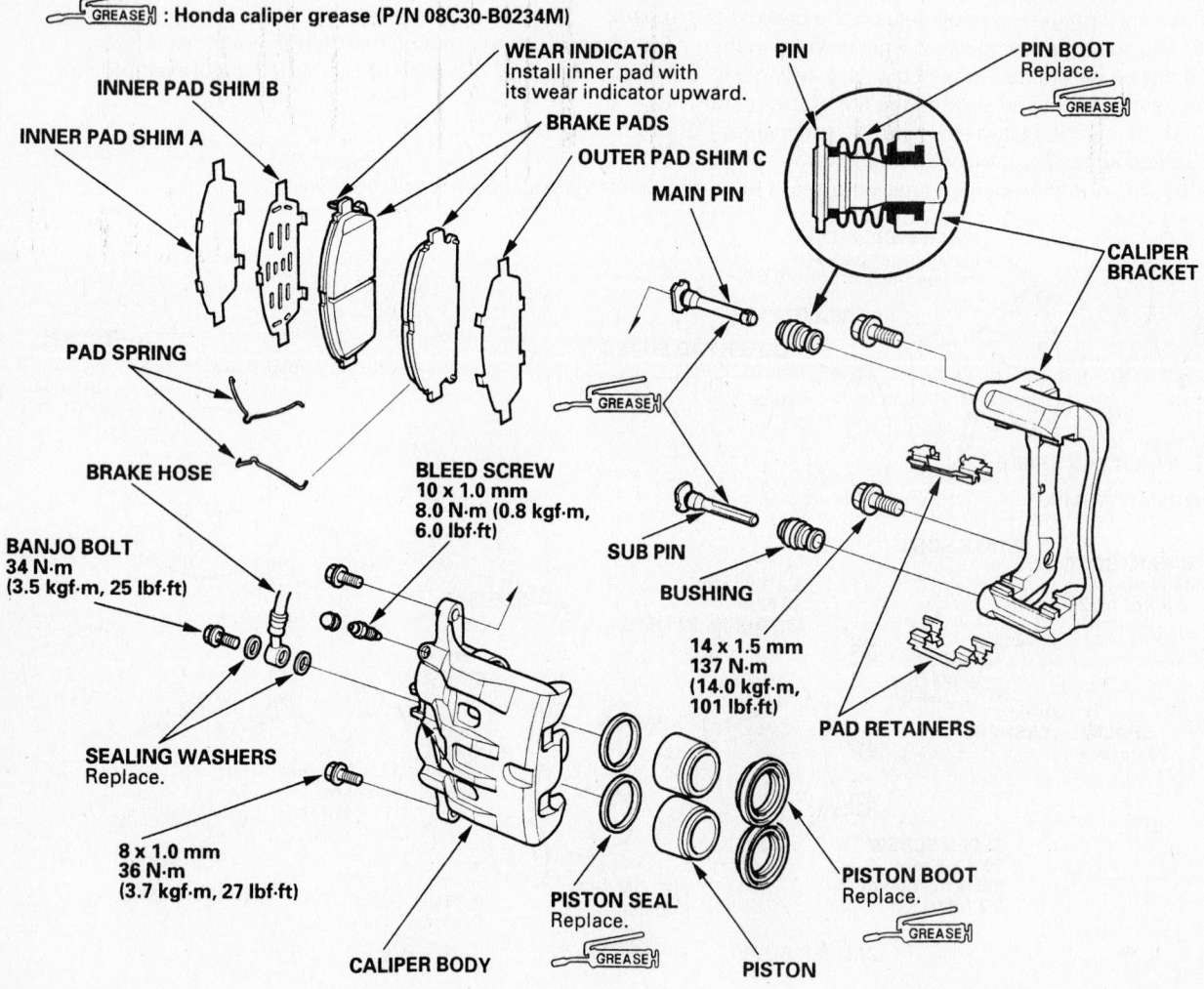

Exploded view of the front caliper components—2003–05 models

43256-AMDX-G30

Rear Brake Caliper Overhaul

> ⚠️**CAUTION**
>
> Frequent inhalation of brake pad dust, regardless of material composition, could be hazardous to your health.
> • Avoid breathing dust particles.
> • Never use an air hose or brush to clean brake assemblies. Use an OSHA-approved vacuum cleaner.

Remove, disassemble, inspect, reassemble, and install the caliper, and note these items:

• Do not spill brake fluid on the vehicle; It may damage the paint; If brake fluid gets on the paint, wash it off immediately with water.
• To prevent dripping brake fluid, cover disconnected hose joints with rags or shop towels.
• Clean all parts in brake fluid and air dry; blow out all passages with compressed air.
• Before reassembling, check that all parts are free of dirt and other foreign particles.
• Replace parts with new ones as specified in the illustration.
• Make sure no dirt or other foreign matter gets into the brake fluid.
• Make sure no grease or oil gets on the brake discs or pads.
• When reusing pads, always reinstall them in their original positions to prevent loss of braking efficiency.
• Do not reuse drained brake fluid. Use only clean Genuine Honda DOT 3 Brake Fluid. Non-Honda brake fluid can cause corrosion and shorten the life of the system.
• Coat the piston, piston seal groove, and caliper bore with clean brake fluid.
• Replace all rubber parts with new ones.
• After installing the caliper, check the brake hose and line for leaks, interference, and twisting.

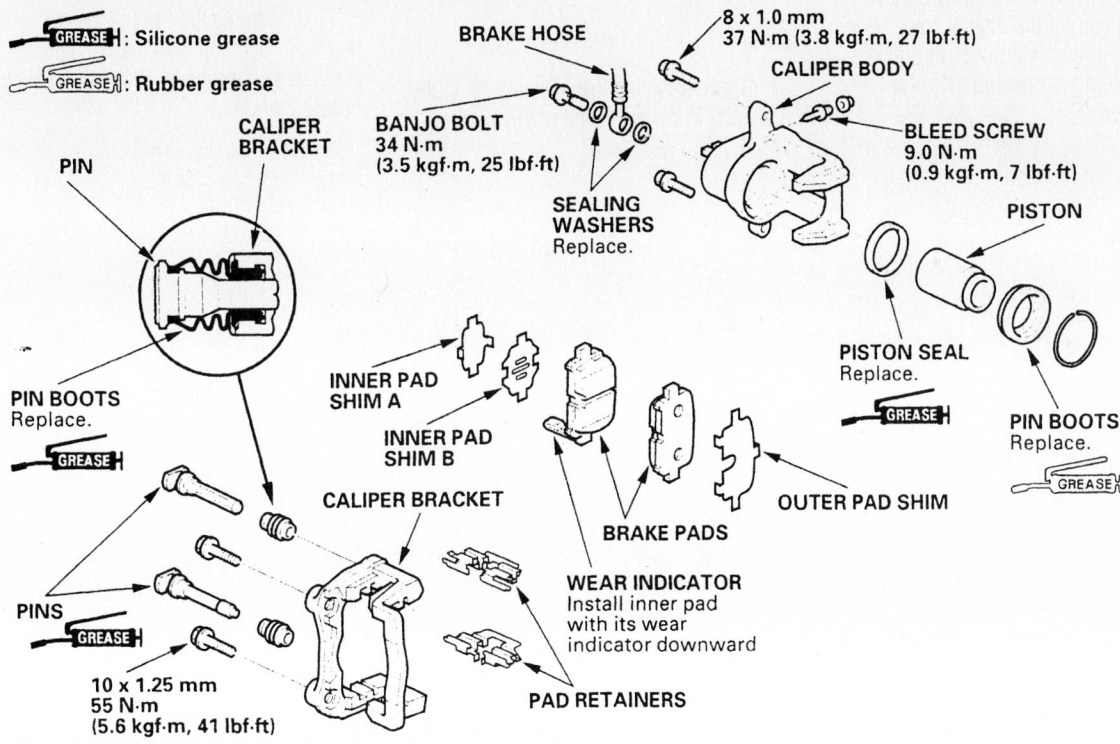

Exploded view of the rear caliper components—MDX

93352GZB

5. Fill the reservoir with fluid and bleed the brake system. Adjust the parking brake if necessary.
- Rear wheels

Disc Brake Pads

REMOVAL & INSTALLATION

Front

1. Before servicing the vehicle, refer to the precautions in the beginning of this section.
2. Remove or disconnect the following:
- Front wheels
3. Remove a small amount of brake fluid from the reservoir using a suction pump.
- Brake hose clamp from the knuckle by unfastening the retaining bolts
- Lower caliper retaining bolt and pivot the caliper upward, off of the pads
- Pad shim and pad retainers
- Disc brake pads from the caliper

To install:

4. Clean the caliper thoroughly; remove any rust from the lip of the disc or rotor. Check the brake rotor for grooves or cracks. If any heavy scoring is present, the rotor must be replaced.

5. Install or connect the following:
- Pad retainers. Apply molybdenum brake grease to both surfaces of the shims and the back of the disc brake pads.
- Pads and shims. The pad with the wear indicator goes in the inboard position.
6. Push in the caliper piston so the caliper will fit over the pads. This is most easily accomplished with a pad spreader or large C-clamp.
- Caliper down into position and tighten the mounting bolt to 27 ft. lbs. (37 Nm)
- Brake hose to the knuckle, if removed
- Wheels
7. Add brake fluid to the master cylinder reservoir and install the cap.
8. Depress the brake pedal several times and make sure that the movement feels normal. The first brake pedal application may result in a very long pedal action due to the pistons being retracted. Always make several brake applications before starting the vehicle. Bleed the system if necessary.

Rear

1. Before servicing the vehicle, refer to the precautions in the beginning of this section.

2. Remove a small amount of brake fluid from the reservoir using a suction pump.
3. Remove or disconnect the following:
- Rear wheels
- 2 caliper mounting bolts and the caliper from the bracket
- Pads, shims, and pad retainers

To install:

4. Clean the caliper thoroughly; remove any dirt or dust. Check the brake rotor for grooves or cracks and machine or replace, as necessary.
5. Install or connect the following:
- Pad retainers. Apply molybdenum brake grease to both surfaces of the shims and the back of the disc brake pads.
- Pads and shims. The wear retainer on the inboard pad faces down.
6. Use a suitable tool to push caliper piston into its bore and enable the caliper to fit over the pads. Lubricate the piston boot with silicon grease. Avoid twisting the boot.
- Brake caliper and tighten the mounting bolts to 27 ft. lbs. (37 Nm)
- Rear wheels
7. Add brake fluid to the master cylinder reservoir. Depress the brake pedal several times to seat the pads. Bleed the brakes if necessary.

HONDA

Accord • Civic • Prelude • S2000

SPECIFICATION AND MAINTENANCE CHARTS

ENGINE AND VEHICLE IDENTIFICATION

Engine							Model Year	
Code	Liters	Cu. In. (cc)	Cyl.	Fuel Sys.	Eng. Mfg.		Code	Year
D17A1	1.7	101.7 (1668)	4	PGM-FI	Honda		1	2001
D17A2	1.7	101.7 (1668)	4	PGM-FI	Honda		2	2002
D17A6	1.7	101.7 (1668)	4	PGM-FI	Honda		3	2003
F20C1	2.0	121.9 (1997)	4	PGM-FI	Honda		4	2004
F23A1	2.3	137 (2254)	4	PGM-FI	Honda		5	2005
F23A4	2.3	137 (2254)	4	PGM-FI	Honda			
F23A5	2.3	137 (2254)	4	PGM-FI	Honda			
H22A4	2.2	132 (2157)	4	PGM-FI	Honda			
K24A4	2.4	144 (2354)	4	PGM-FI	Honda			
J30A1	3.0	183 (2997)	6	PGM-FI	Honda			
J30A4	3.0	183 (2997)	6	PGM-FI	Honda			

PGM-FI: Programmed Fuel Injection

67162-ACC0-C01

GENERAL ENGINE SPECIFICATIONS

Year	Model	Engine Displacement Liters	Engine ID/VIN	Net Horsepower @ rpm	Net Torque @ rpm (ft. lbs.)	Bore X Stroke (in.)	Com-pression Ratio	Oil Pressure @ rpm
2001	Civic DX	1.7	D17A1	115@6100	110X4500	2.95X3.72	9.5:1	50@3000
	Civic LX	1.7	D17A1	115@6100	110X4500	2.95X3.72	9.5:1	50@3000
	Civic EX	1.7	D17A2	127@6100	117X4500	2.95X3.72	9.9:1	50@3000
	Civic HX	1.7	D17A6	127@6100	117X4500	2.95X3.72	9.9:1	50@3000
	Prelude	2.2	H22A4	①	158@5500	3.43X3.57	10.0:1	50@3000
	Prelude SH	2.2	H22A4	①	158@5500	3.43X3.57	10.0:1	50@3000
	Accord Coupe (EX, LX)	2.3	F23A1	150@5700	152@4900	3.39X3.82	9.3:1	50@3000
	Accord Coupe (EX, LX)	2.3	F23A4	150@5700	152@4900	3.39X3.82	9.3:1	50@3000
	Accord Sedan (DX)	2.3	F23A5	150@5700	152@4900	3.39X3.82	9.3:1	50@3000
	Accord Sedan (EX, LX)	2.3	F23A1	150@5700	152@4900	3.39X3.82	9.3:1	50@3000
	Accord Sedan (EX, LX)	2.3	F23A4	150@5700	152@4900	3.39X3.82	9.3:1	50@3000
	Accord Coupe (EX, LX)	3.0	J30A1	200@5500	195@4700	3.39X3.39	9.4:1	50@3000
	Accord Sedan (EX, LX)	3.0	J30A1	200@5500	195@4700	3.39X3.39	9.4:1	50@3000
	S2000	3.0	F20C1	240@8300	153@7500	3.43X3.31	11:01	85@3000
2002	Civic DX	1.7	D17A1	115@6100	110X4500	2.95X3.72	9.5:1	50@3000
	Civic LX	1.7	D17A1	115@6100	110X4500	2.95X3.72	9.5:1	50@3000
	Civic EX	1.7	D17A2	127@6100	117X4500	2.95X3.72	9.9:1	50@3000
	Civic HX	1.7	D17A6	127@6100	117X4500	2.95X3.72	9.9:1	50@3000
	Accord Coupe (EX, LX)	2.3	F23A1	150@5700	152@4900	3.39X3.82	9.3:1	50@3000
	Accord Coupe (EX, LX)	2.3	F23A4	150@5700	152@4900	3.39X3.82	9.3:1	50@3000
	Accord Sedan (DX)	2.3	F23A5	150@5700	152@4900	3.39X3.82	9.3:1	50@3000
	Accord Sedan (EX, LX)	2.3	F23A1	150@5700	152@4900	3.39X3.82	9.3:1	50@3000
	Accord Sedan (EX, LX)	2.3	F23A4	150@5700	152@4900	3.39X3.82	9.3:1	50@3000
	Accord Coupe (EX, LX)	3.0	J30A1	200@5500	195@4700	3.39X3.39	9.4:1	50@3000
	Accord Sedan (EX, LX)	3.0	J30A1	200@5500	195@4700	3.39X3.39	9.4:1	50@3000
	S2000	2.0	F20C1	240@8300	153@7500	3.43X3.31	11:01	85@3000
2003	Civic DX	1.7	D17A1	115@6100	110X4500	2.95X3.72	9.5:1	50@3000
	Civic LX	1.7	D17A1	115@6100	110X4500	2.95X3.72	9.5:1	50@3000
	Civic EX	1.7	D17A2	127@6100	117X4500	2.95X3.72	9.9:1	50@3000
	Civic HX	1.7	D17A6	127@6100	117X4500	2.95X3.72	9.9:1	50@3000
	Accord Coupe (EX, LX)	2.4	K24A4	160@5500	161@4500	3.42X3.89	9.7:1	50@3000
	Accord Sedan (DX)	2.4	K24A4	160@5500	161@4500	3.42X3.89	9.7:1	50@3000
	Accord Sedan (EX, LX)	2.4	K24A4	160@5500	161@4500	3.42X3.89	9.7:1	50@3000
	Accord Coupe (EX, LX)	3.0	J30A4	240@6250	212@5000	3.38X3.38	10.0:1	50@3000
	Accord Sedan (EX, LX)	3.0	J30A4	240@6250	212@5000	3.39X3.39	10.0:1	50@3000
	S2000	2.0	F20C1	240@8300	153@7500	3.43X3.31	11:01	85@3000
2004	Civic DX	1.7	D17A1	115@6100	110X4500	2.95X3.72	9.5:1	50@3000
	Civic LX	1.7	D17A1	115@6100	110X4500	2.95X3.72	9.5:1	50@3000
	Civic EX	1.7	D17A2	127@6100	117X4500	2.95X3.72	9.9:1	50@3000
	Civic HX	1.7	D17A6	127@6100	117X4500	2.95X3.72	9.9:1	50@3000
	Accord Coupe (EX, LX)	2.4	K24A4	160@5500	161@4500	3.42X3.89	9.7:1	50@3000
	Accord Sedan (DX)	2.4	K24A4	160@5500	161@4500	3.42X3.89	9.7:1	50@3000
	Accord Sedan (EX, LX)	2.4	K24A4	160@5500	161@4500	3.42X3.89	9.7:1	50@3000
	Accord Coupe (EX, LX)	3.0	J30A4	240@6250	212@5000	3.38X3.38	10.0:1	50@3000
	Accord Sedan (EX, LX)	3.0	J30A4	240@6250	212@5000	3.39X3.39	10.0:1	50@3000
	S2000	2.0	F20C1	240@8300	153@7500	3.43X3.31	11:01	85@3000

PGM-FI: Programmed Fuel Injection

① Manual transaxle: 195@7000

Automatic transaxle: 190@6600

ENGINE TUNE-UP SPECIFICATIONS

Year	Engine Displacement Liters	Engine ID/VIN	Spark Plugs Gap (in.)	Ignition Timing (deg.) MT	AT	Fuel Pump (psi)	Idle Speed (rpm) MT	AT	Valve Clearance In.	Ex.
2001	1.7	D17A1	0.039-0.043	8B	8B	40-47	650-750	650-750	0.007-0.009	0.009-0.011
	1.7	D17A2	0.039-0.043	8B	8B	40-47	650-750	650-750	0.007-0.009	0.009-0.011
	1.7	D17A6	0.039-0.043	8B	8B	40-47	650-750	650-750	0.007-0.009	0.009-0.011
	2.0	F20C1	0.039-0.043	5B	5B	47-54	750-850	—	0.008-0.010	0.010-0.011
	2.3	F23A1	0.039-0.043	12B	12B	40-47	650-750	650-750	0.009-0.011	0.011-0.013
	2.3	F23A4	0.039-0.043	12B	12B	40-47	650-750	650-750	0.009-0.011	0.011-0.013
	2.3	F23A5	0.039-0.043	12B	12B	40-47	650-750	650-750	0.009-0.011	0.011-0.013
	2.2	H22A4	0.039-0.043	15B	15B	47-54	650-750	650-750	0.006-0.007	0.007-0.008
	3.0	J30A1	0.039-0.043	—	10B	41-48	—	630-730	0.008-0.009	0.011-0.013
2002	1.7	D17A1	0.039-0.043	8B	8B	40-47	650-750	650-750	0.007-0.009	0.009-0.011
	1.7	D17A2	0.039-0.043	8B	8B	40-47	650-750	650-750	0.007-0.009	0.009-0.011
	1.7	D17A6	0.039-0.043	8B	8B	40-47	650-750	650-750	0.007-0.009	0.009-0.011
	2.0	F20C1	0.039-0.043	5B	5B	47-54	750-850	—	0.008-0.010	0.010-0.011
	2.3	F23A1	0.039-0.043	12B	12B	40-47	650-750	650-750	0.009-0.011	0.011-0.013
	2.3	F23A4	0.039-0.043	12B	12B	40-47	650-750	650-750	0.009-0.011	0.011-0.013
	2.3	F23A5	0.039-0.043	12B	12B	40-47	650-750	650-750	0.009-0.011	0.011-0.013
	3.0	J30A1	0.039-0.043	—	10B	41-48	—	630-730	0.008-0.009	0.011-0.013
2003	1.7	D17A1	0.039-0.043	8B	8B	40-47	650-750	650-750	0.007-0.009	0.009-0.011
	1.7	D17A2	0.039-0.043	8B	8B	40-47	650-750	650-750	0.007-0.009	0.009-0.011
	1.7	D17A6	0.039-0.043	8B	8B	40-47	650-750	650-750	0.007-0.009	0.009-0.011
	2.0	F20C1	0.039-0.043	5B	5B	47-54	750-850	—	0.008-0.010	0.010-0.011
	2.4	K24A4	0.039-0.043	8B	8B	①	650-750	750-850	0.008-0.010	0.011-0.013
	3.0	J30A4	0.039-0.043	—	10B	41-48	—	630-730	0.008-0.009	0.011-0.013

ENGINE TUNE-UP SPECIFICATIONS

Year	Engine Displacement Liters	Engine ID/VIN	Spark Plugs Gap (in.)	Ignition Timing (deg.)		Fuel Pump (psi)	Idle Speed (rpm)		Valve Clearance	
				MT	AT		MT	AT	In.	Ex.
2004	1.7	D17A1	0.039-0.043	8B	8B	40-47	650-750	650-750	0.007-0.009	0.009-0.011
	1.7	D17A2	0.039-0.043	8B	8B	40-47	650-750	650-750	0.007-0.009	0.009-0.011
	1.7	D17A6	0.039-0.043	8B	8B	40-47	650-750	650-750	0.007-0.009	0.009-0.011
	2.0	F20C1	0.039-0.043	5B	5B	47-54	750-850	—	0.008-0.010	0.010-0.011
	2.4	K24A4	0.039-0.043	8B	8B	①	650-750	750-850	0.008-0.010	0.011-0.013
	3.0	J30A4	0.039-0.043	—	10B	41-48	—	630-730	0.008-0.009	0.011-0.013

NOTE: The Vehicle Emission Control Information label often reflects specification changes made during production. The label figures must be used if they differ from those in this chart.

B: Before Top Dead Center

① Except SULEV: 48-55 psi
   SULEV: 47-54

67162-ACCO-C04

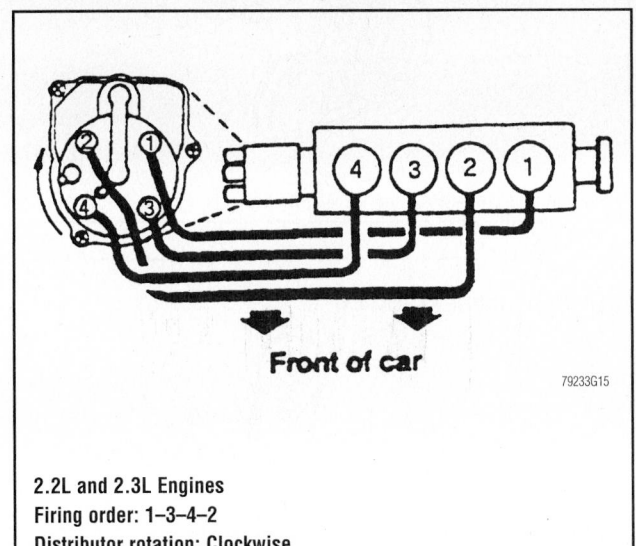

2.2L and 2.3L Engines
Firing order: 1–3–4–2
Distributor rotation: Clockwise

79233G15

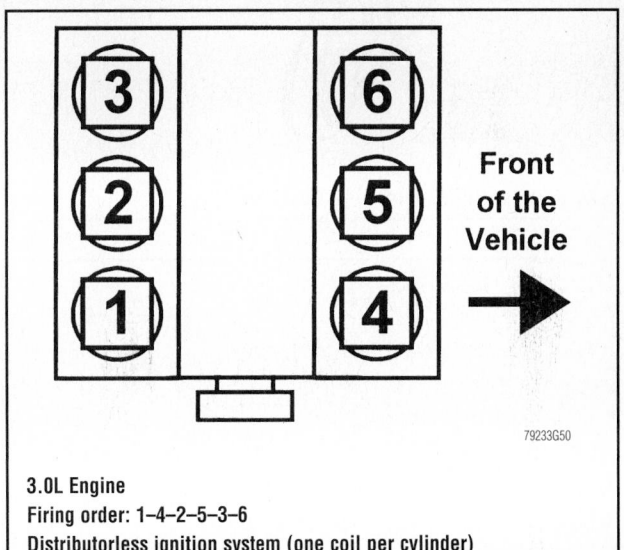

Front of the Vehicle

3.0L Engine
Firing order: 1–4–2–5–3–6
Distributorless ignition system (one coil per cylinder)

79233G50

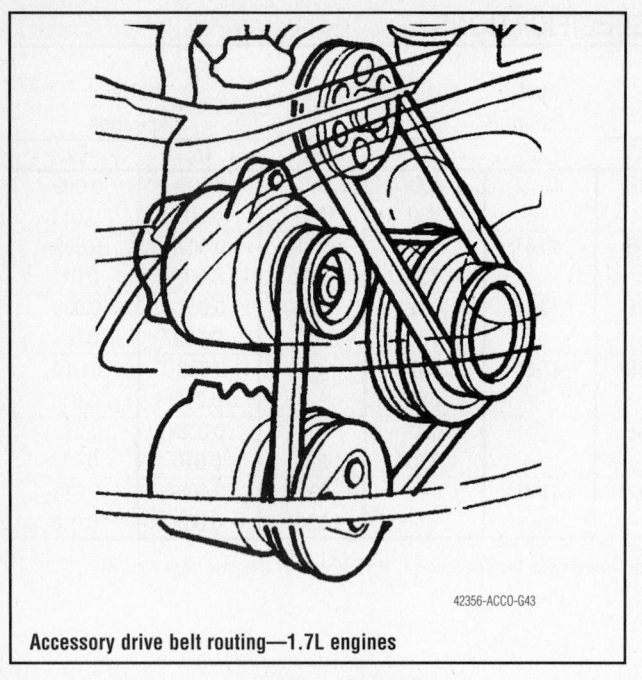

Accessory drive belt routing—1.7L engines

42356-ACCO-G43

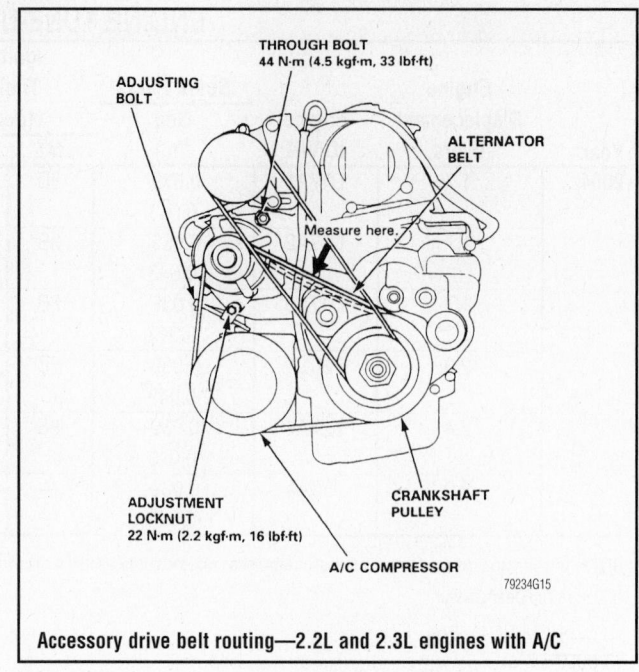

Accessory drive belt routing—2.2L and 2.3L engines with A/C

79234G15

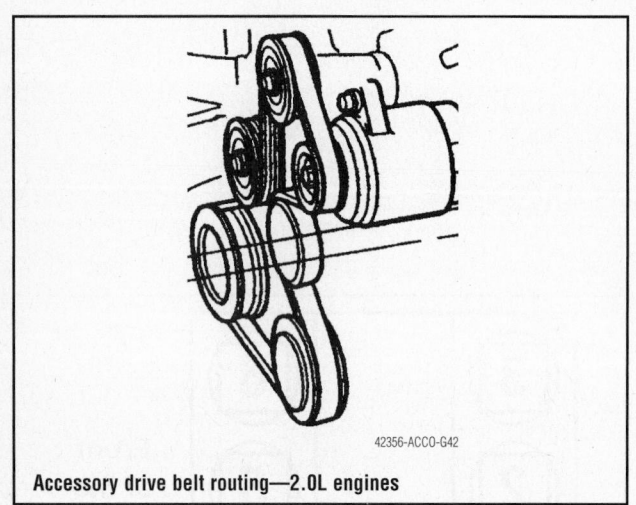

Accessory drive belt routing—2.0L engines

42356-ACCO-G42

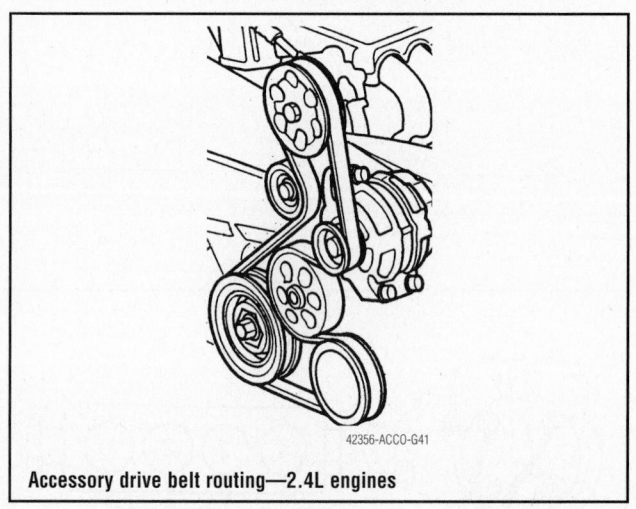

Accessory drive belt routing—2.4L engines

42356-ACCO-G41

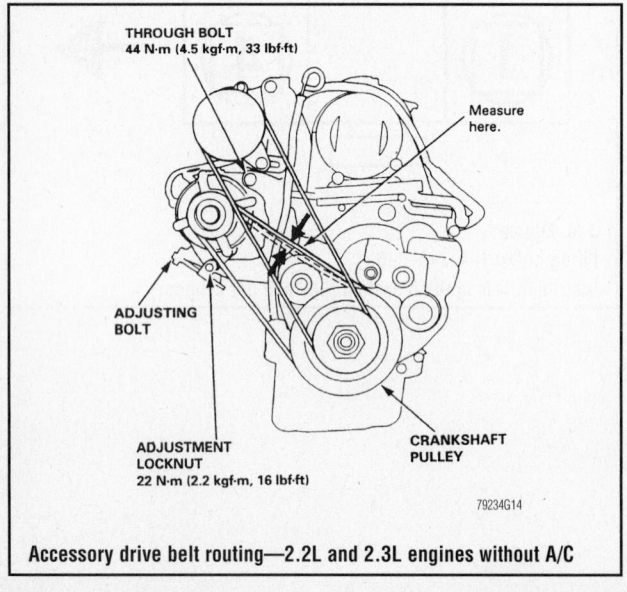

Accessory drive belt routing—2.2L and 2.3L engines without A/C

79234G14

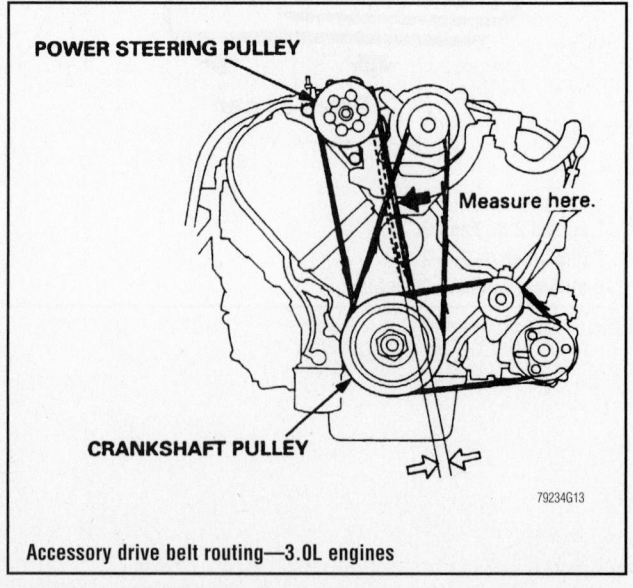

Accessory drive belt routing—3.0L engines

79234G13

CAPACITIES

Year	Model	Engine Displacement Liters	Engine ID	Engine Oil with Filter	Transmission (pts.) 5-Spd	Transmission (pts.) Auto.	Drive Axle Front (pts.)	Drive Axle Rear (pts.)	Fuel Tank (gal.)	Cooling System (qts.)
2001	Accord Coupe (EX, LX)	2.3	F23A1	4.0	4.0	5.0	—	—	17.0	①
	Accord Coupe (EX, LX)	2.3	F23A4	4.5	4.0	5.0	—	—	17.0	①
	Accord Coupe (EX, LX)	3.0	J30A1	4.6	—	6.2	—	—	17.1	5.9
	Accord Sedan (DX)	2.3	F23A5	4.5	4.0	5.2	—	—	17.0	①
	Accord Sedan (EX, LX)	2.3	F23A1	4.0	4.0	5.0	—	—	17.0	①
	Accord Sedan (EX, LX)	2.3	F23A4	4.5	4.0	5.2	—	—	17.0	①
	Accord Sedan (EX, LX)	3.0	J30A1	4.6	—	6.2	—	—	17.1	5.9
	Civic	1.7	D17A1	3.7	3.2	5.8	—	—	11.9	5.4
	Civic	1.7	D17A2	3.7	3.2	5.8	—	—	11.9	5.4
	Civic	1.7	D17A6	3.7	3.2	5.8	—	—	11.9	5.4
	S2000	2.0	F20C1	5.1	3.12	—	—	—	13.2	6.9
	Prelude	2.2	H22A4	5.1	4.0	—	—	—	15.9	4.6
	Prelude SH	2.2	H22A4	5.1	4.0	—	—	—	15.9	4.6
2002	Accord Coupe (EX, LX)	2.3	F23A1	4.0	4.0	5.0	—	—	17.0	①
	Accord Coupe (EX, LX)	2.3	F23A4	4.5	4.0	5.0	—	—	17.0	①
	Accord Coupe (EX, LX)	3.0	J30A1	4.6	—	6.2	—	—	17.1	5.9
	Accord Sedan (DX)	2.3	F23A5	4.5	4.0	5.2	—	—	17.0	①
	Accord Sedan (EX, LX)	2.3	F23A1	4.0	4.0	5.0	—	—	17.0	①
	Accord Sedan (EX, LX)	2.3	F23A4	4.5	4.0	5.2	—	—	17.0	①
	Accord Sedan (EX, LX)	3.0	J30A1	4.6	—	6.2	—	—	17.1	5.9
	Civic	1.7	D17A1	3.7	3.2	5.8	—	—	11.9	5.4
	Civic	1.7	D17A2	3.7	3.2	5.8	—	—	11.9	5.4
	Civic	1.7	D17A6	3.7	3.2	5.8	—	—	11.9	5.4
	S2000	2.0	F20C1	5.1	3.12	—	—	—	13.2	6.9
2003	Accord Coupe (EX, LX)	2.4	K24A4	4.4	4.0	6.0	—	—	17.1	②
	Accord Coupe (EX, LX)	3.0	J30A4	4.6	—	6.2	—	—	17.1	7.1
	Accord Sedan (DX)	2.4	K24A4	4.4	4.0	6.0	—	—	17.1	②
	Accord Sedan (EX, LX)	2.4	K24A4	4.4	4.0	6.0	—	—	17.1	②
	Accord Sedan (EX, LX)	3.0	J30A4	4.6	—	6.2	—	—	17.1	7.1
	Civic	1.7	D17A1	3.7	3.2	5.8	—	—	11.9	5.4
	Civic	1.7	D17A2	3.7	3.2	5.8	—	—	11.9	5.4
	Civic	1.7	D17A6	3.7	3.2	5.8	—	—	11.9	5.4
	S2000	2.0	F20C1	5.1	3.12	—	—	—	13.2	6.9
2004	Accord Coupe (EX, LX)	2.4	K24A4	4.4	4.0	6.0	—	—	17.1	②
	Accord Coupe (EX, LX)	3.0	J30A4	4.6	—	6.2	—	—	17.1	7.1
	Accord Sedan (DX)	2.4	K24A4	4.4	4.0	6.0	—	—	17.1	②
	Accord Sedan (EX, LX)	2.4	K24A4	4.4	4.0	6.0	—	—	17.1	②
	Accord Sedan (EX, LX)	3.0	J30A4	4.6	—	6.2	—	—	17.1	7.1
	Civic	1.7	D17A1	3.7	3.2	5.8	—	—	11.9	5.4
	Civic	1.7	D17A2	3.7	3.2	5.8	—	—	11.9	5.4
	Civic	1.7	D17A6	3.7	3.2	5.8	—	—	11.9	5.4
	S2000	2.0	F20C1	5.1	3.12	—	—	—	13.2	6.9

NOTE: All capacities are approximate. Add fluid gradually and ensure a proper fluid level is obtained.

NOTE: Capacities given are service, not overhaul capacities

① Automatic Transaxle: 5.7
  Manual Transaxle: 5.8

② Automatic Transaxle: 5.3
  Manual Transaxle: 5.4

VALVE SPECIFICATIONS

Year	Engine Displacement Liters	Engine ID/VIN	Seat Angle (deg.)	Face Angle (deg.)	Spring Test Pressure (lbs. @ in.)	Spring Installed Height (in.)	Stem-to-Guide Clearance (in.)		Stem Diameter (in.)	
							Intake	Exhaust	Intake	Exhaust
2001	1.7	D17A1	45	45	NA	NA	0.0008-0.0020	0.0020-0.0031	0.2157-0.2161	0.2146-0.2150
	1.7	D17A2	45	45	NA	NA	0.0008-0.0020	0.0020-0.0031	0.2157-0.2161	0.2146-0.2150
	1.7	D17A6	45	45	NA	NA	0.0008-0.0020	0.0020-0.0031	0.2157-0.2161	0.2146-0.2150
	2.0	F20C1	45	45	NA	NA	0.0010-0.0020	0.0020-0.0030	0.2157-0.2162	0.2146-0.2150
	2.3	F23A1	45	45	NA	NA	0.0008-0.0018	0.0022-0.0031	0.2159-0.2163	0.2146-0.2150
	2.3	F23A4	45	45	NA	NA	0.0008-0.0018	0.0022-0.0031	0.2159-0.2163	0.2146-0.2150
	2.3	F23A5	45	45	NA	NA	0.0008-0.0018	0.0022-0.0031	0.2159-0.2163	0.2146-0.2150
	2.2	H22A4	45	45	NA	NA	0.0010-0.0022	0.0020-0.0031	0.2156-0.2159	0.2156-0.2159
	3.0	J30A1	45	45	NA	NA	0.0008-0.0018	0.0022-0.0031	0.2159-0.2163	0.2146-0.2150
2002	1.7	D17A1	45	45	NA	NA	0.0008-0.0020	0.0020-0.0031	0.2157-0.2161	0.2146-0.2150
	1.7	D17A2	45	45	NA	NA	0.0008-0.0020	0.0020-0.0031	0.2157-0.2161	0.2146-0.2150
	1.7	D17A6	45	45	NA	NA	0.0008-0.0020	0.0020-0.0031	0.2157-0.2161	0.2146-0.2150
	2.0	F20C1	45	45	NA	NA	0.0010-0.0020	0.0020-0.0030	0.2157-0.2162	0.2146-0.2150
	2.3	F23A1	45	45	NA	NA	0.0008-0.0018	0.0022-0.0031	0.2159-0.2163	0.2146-0.2150
	2.3	F23A4	45	45	NA	NA	0.0008-0.0018	0.0022-0.0031	0.2159-0.2163	0.2146-0.2150
	2.3	F23A5	45	45	NA	NA	0.0008-0.0018	0.0022-0.0031	0.2159-0.2163	0.2146-0.2150
	3.0	J30A1	45	45	NA	NA	0.0008-0.0018	0.0022-0.0031	0.2159-0.2163	0.2146-0.2150
2003	1.7	D17A1	45	45	NA	NA	0.0008-0.0020	0.0020-0.0031	0.2157-0.2161	0.2146-0.2150
	1.7	D17A2	45	45	NA	NA	0.0008-0.0020	0.0020-0.0031	0.2157-0.2161	0.2146-0.2150
	1.7	D17A6	45	45	NA	NA	0.0008-0.0020	0.0020-0.0031	0.2157-0.2161	0.2146-0.2150
	2.0	F20C1	45	45	NA	NA	0.0010-0.0020	0.0020-0.0030	0.2157-0.2162	0.2146-0.2150
	2.4	K24A4	45	45	NA	NA	0.0012-0.0022	0.0022-0.0031	0.2156-0.2159	0.2146-0.2150
	3.0	J30A4	45	45	NA	NA	0.0008-0.0018	0.0022-0.0031	0.2159-0.2163	0.2146-0.2150

67162-ACCO-C09

VALVE SPECIFICATIONS

Year	Engine Displacement Liters	Engine ID/VIN	Seat Angle (deg.)	Face Angle (deg.)	Spring Test Pressure (lbs. @ in.)	Spring Installed Height (in.)	Stem-to-Guide Clearance (in.)		Stem Diameter (in.)	
							Intake	Exhaust	Intake	Exhaust
2004	1.7	D17A1	45	45	NA	NA	0.0008-0.0020	0.0020-0.0031	0.2157-0.2161	0.2146-0.2150
	1.7	D17A2	45	45	NA	NA	0.0008-0.0020	0.0020-0.0031	0.2157-0.2161	0.2146-0.2150
	1.7	D17A6	45	45	NA	NA	0.0008-0.0020	0.0020-0.0031	0.2157-0.2161	0.2146-0.2150
	2.0	F20C1	45	45	NA	NA	0.0010-0.0020	0.0020-0.0030	0.2157-0.2162	0.2146-0.2150
	2.4	K24A4	45	45	NA	NA	0.0012-0.0022	0.0022-0.0031	0.2156-0.2159	0.2146-0.2150
	3.0	J30A4	45	45	NA	NA	0.0008-0.0018	0.0022-0.0031	0.2159-0.2163	0.2146-0.2150

NA: Not Available

67162-ACCO-C10

CRANKSHAFT AND CONNECTING ROD SPECIFICATIONS

All measurements are given in inches.

Year	Engine Displacement Liters (cc)	Engine ID/VIN	Crankshaft				Connecting Rod		
			Main Brg. Journal Dia.	Main Brg. Oil Clearance	Shaft End-play	Thrust on No.	Journal Diameter	Oil Clearance	Side Clearance
2001	1.7	D17A1	2.1644-2.1654	0.0007- ① 0.0014	0.004-0.014	4	1.7707-1.7717	0.0009-0.0017	0.0006-0.0016
	1.7	D17A2	2.1644-2.1654	0.0007- ① 0.0014	0.004-0.014	4	1.7707-1.7717	0.0009-0.0017	0.0006-0.0016
	1.7	D17A6	2.1644-2.1654	0.0007- ① 0.0014	0.004-0.014	4	1.7707-1.7717	0.0009-0.0017	0.0006-0.0016
	2.0	F20C1	2.1644-2.1654	0.0007-0.0016	0.004-0.014	4	1.8888-1.8898	0.0012-0.0020	0.0006-0.0016
	2.3	F23A1	②	②	0.004-0.018	4	1.7707-1.7717	0.0008-0.0024	0.0006-0.0016
	2.3	F23A4	②	③	0.004-0.018	4	1.7707-1.7717	0.0008-0.0024	0.0006-0.0018
	2.3	F23A5	②	③	0.004-0.018	4	1.7707-1.7717	0.0008-0.0024	0.0006-0.0016
	2.2	H22A4	④	⑤	0.004-0.018	4	1.8888-1.8898	0.0011-0.0020	0.0006-0.0016
	3.0	J30A1	2.8337-2.8346	0.0008-0.0020	0.004-0.018	3	2.0857-2.0866	0.0008-0.0020	0.0006-0.0018
2002	1.7	D17A1	2.1644-2.1654	0.0007- ① 0.0014	0.004-0.014	4	1.7707-1.7717	0.0009-0.0017	0.0006-0.0016
	1.7	D17A2	2.1644-2.1654	0.0007- ③ 0.0014	0.004-0.014	4	1.7707-1.7717	0.0009-0.0017	0.0006-0.0016
	1.7	D17A6	2.1644-2.1654	0.0007- ① 0.0014	0.004-0.014	4	1.7707-1.7717	0.0009-0.0017	0.0006-0.0016
	2.0	F20C1	2.1644-2.1654	0.0007-0.0016	0.004-0.014	4	1.8888-1.8898	0.0012-0.0020	0.0006-0.0016
	2.3	F23A1	②	③	0.004-0.018	4	1.7707-1.7717	0.0008-0.0024	0.0006-0.0016
	2.3	F23A4	②	③	0.004-0.018	4	1.7707-1.7717	0.0008-0.0024	0.0006-0.0018
	2.3	F23A5	②	③	0.004-0.018	4	1.7707-1.7717	0.0008-0.0024	0.0006-0.0016
	3.0	J30A1	2.8337-2.8346	0.0008-0.0020	0.004-0.018	3	2.0857-2.0866	0.0008-0.0020	0.0006-0.0018
2003	1.7	D17A1	2.1644-2.1654	0.0007- ① 0.0014	0.004-0.014	4	1.7707-1.7717	0.0009-0.0017	0.0006-0.0016
	1.7	D17A2	2.1644-2.1654	0.0007- ① 0.0014	0.004-0.014	4	1.7707-1.7717	0.0009-0.0017	0.0006-0.0016
	1.7	D17A6	2.1644-2.1654	0.0007- ① 0.0014	0.004-0.014	4	1.7707-1.7717	0.0009-0.0017	0.0006-0.0016
	2.0	F20C1	2.1644-2.1654	0.0007-0.0016	0.004-0.014	4	1.8888-1.8898	0.0012-0.0020	0.0006-0.0016
	2.4	K24A4	④	⑤	0.004-0.014	4	1.8888-1.8898	0.0002-0.0006	0.0006-0.0016
	3.0	J30A4	2.8337-2.8346	0.0008-0.0020	0.004-0.018	3	2.0857-2.0866	0.0008-0.0020	0.0006-0.0018

CRANKSHAFT AND CONNECTING ROD SPECIFICATIONS
All measurements are given in inches.

Year	Engine Displacement Liters (cc)	Engine ID/VIN	Crankshaft				Connecting Rod		
			Main Brg. Journal Dia.	Main Brg. Oil Clearance	Shaft End-play	Thrust on No.	Journal Diameter	Oil Clearance	Side Clearance
2004	1.7	D17A1	2.1644-2.1654	0.0007- ① 0.0014	0.004-0.014	4	1.7707-1.7717	0.0009-0.0017	0.0006-0.0016
	1.7	D17A2	2.1644-2.1654	0.0007- ① 0.0014	0.004-0.014	4	1.7707-1.7717	0.0009-0.0017	0.0006-0.0016
	1.7	D17A6	2.1644-2.1654	0.0007- ① 0.0014	0.004-0.014	4	1.7707-1.7717	0.0009-0.0017	0.0006-0.0016
	2.0	F20C1	2.1644-2.1654	0.0007-0.0016	0.004-0.014	4	1.8888-1.8898	0.0012-0.0020	0.0006-0.0016
	2.4	K24A4	④	⑤	0.004-0.014	4	1.8888-1.8898	0.0002-0.0006	0.0006-0.0016
	3.0	J30A4	2.8337-2.8346	0.0008-0.0020	0.004-0.018	3	2.0857-2.0866	0.0008-0.0020	0.0006-0.0018

① Journals 1 and 5
  Journals 2, 3 and 4: 0.0009 - 0.0019
② Journals 1, 2 and 4: 2.1646 - 2.1655
  Journal 3: 2.1644 - 2.1654
  Journal 5: 2.1650 - 2.1660
③ Journals 1, 2 and 4: 0.0008 - 0.0020
  Journal 3: 0.0010 - 0.0022
  Journal 5: 0.0004 - 0.0016
④ Journals 1 and 2: 1.9676 - 1.9685
  Journal 3: 1.9674 - 1.9683
  Journal 4: 1.9679 - 1.9688
  Journal 5: 1.9680 - 1.9690

⑤ Journals 1 and 2: 0.0008 - 0.0020
  Journal 3: 0.0010 - 0.0022
  Journal 4: 0.0005 - 0.0020
  Journal 5: 0.0004 - 0.0016
⑥ Journals 1, 2 4 and 5: 2.1648 - 2.1657
  Journal 3: 2.1644 - 2.1654
⑦ Journals 1, 2 4 and 5: 0.0007-0.0016
  Journal 3: 0.0010-0.0019

67162-ACCO-C06

PISTON AND RING SPECIFICATIONS
All measurements are given in inches.

Year	Engine Displacement Liters	Engine ID/VIN	Piston Clearance	Ring Gap			Ring Side Clearance		
				Top Compression	Bottom Compression	Oil Control	Top Compression	Bottom Compression	Oil Control
2001	1.7	D17A1	0.0004-0.0016	0.006-0.024	0.012-0.024	0.008-0.031	0.0014-0.0024	0.0012-0.0022	N/A
	1.7	D17A2	0.0004-0.0016	0.006-0.024	0.012-0.024	0.008-0.031	0.0014-0.0024	0.0012-0.0022	N/A
	1.7	D17A6	0.0004-0.0016	0.006-0.024	0.012-0.024	0.008-0.031	0.0014-0.0024	0.0012-0.0022	N/A
	2.0	F20C1	0.0002-0.0011	0.010-0.024	0.024-0.035	0.008-0.031	0.0018-0.0035	0.0016-0.0028	N/A
	2.3	F23A1	0.0008-0.0020	0.008-0.024	0.016-0.028	0.008-0.031	0.0014-0.0050	0.0012-0.0050	N/A
	2.3	F23A4	0.0008-0.0020	0.008-0.024	0.016-0.028	0.008-0.031	0.0014-0.0050	0.0012-0.0050	N/A
	2.3	F23A5	0.0008-0.0020	0.008-0.024	0.016-0.028	0.008-0.031	0.0014-0.0050	0.0012-0.0050	N/A
	2.2	H22A4	0.0002-0.0020	0.010-0.024	0.024-0.035	0.008-0.031	0.0022-0.0050	0.0016-0.0050	N/A
	3.0	J30A1	0.0006-0.0030	0.008-0.024	0.016-0.028	0.008-0.031	0.0014-0.0050	0.0012-0.0050	N/A
2002	1.7	D17A1	0.0004-0.0016	0.006-0.024	0.012-0.024	0.008-0.031	0.0014-0.0024	0.0012-0.0022	N/A
	1.7	D17A2	0.0004-0.0016	0.006-0.024	0.012-0.024	0.008-0.031	0.0014-0.0024	0.0012-0.0022	N/A
	1.7	D17A6	0.0004-0.0016	0.006-0.024	0.012-0.024	0.008-0.031	0.0014-0.0024	0.0012-0.0022	N/A
	2.0	F20C1	0.0002-0.0011	0.010-0.024	0.024-0.035	0.008-0.031	0.0018-0.0035	0.0016-0.0028	N/A
	2.3	F23A1	0.0008-0.0020	0.008-0.024	0.016-0.028	0.008-0.031	0.0014-0.0050	0.0012-0.0050	N/A
	2.3	F23A4	0.0008-0.0020	0.008-0.024	0.016-0.028	0.008-0.031	0.0014-0.0050	0.0012-0.0050	N/A
	2.3	F23A5	0.0008-0.0020	0.008-0.024	0.016-0.028	0.008-0.031	0.0014-0.0050	0.0012-0.0050	N/A
	3.0	J30A1	0.0006-0.0030	0.008-0.024	0.016-0.028	0.008-0.031	0.0014-0.0050	0.0012-0.0050	N/A
2003	1.7	D17A1	0.0004-0.0016	0.006-0.024	0.012-0.024	0.008-0.031	0.0014-0.0024	0.0012-0.0022	N/A
	1.7	D17A2	0.0004-0.0016	0.006-0.024	0.012-0.024	0.008-0.031	0.0014-0.0024	0.0012-0.0022	N/A
	1.7	D17A6	0.0004-0.0016	0.006-0.024	0.012-0.024	0.008-0.031	0.0014-0.0024	0.0012-0.0022	N/A
	2.0	F20C1	0.0002-0.0011	0.010-0.024	0.024-0.035	0.008-0.031	0.0018-0.0035	0.0016-0.0028	N/A
	2.4	K24A4	0.0008-0.0016	0.008-0.0014	0.016-0.022	0.008-0.028	0.0018-0.0028	0.0020-0.0030	N/A
	3.0	J30A4	0.0006-0.0030	0.008-0.024	0.016-0.028	0.008-0.031	0.0014-0.0050	0.0012-0.0050	N/A

PISTON AND RING SPECIFICATIONS
All measurements are given in inches.

Year	Engine Displacement Liters	Engine ID/VIN	Piston Clearance	Ring Gap			Ring Side Clearance		
				Top Compression	Bottom Compression	Oil Control	Top Compression	Bottom Compression	Oil Control
2004	1.7	D17A1	0.0004-0.0016	0.006-0.024	0.012-0.024	0.008-0.031	0.0014-0.0024	0.0012-0.0022	N/A
	1.7	D17A2	0.0004-0.0016	0.006-0.024	0.012-0.024	0.008-0.031	0.0014-0.0024	0.0012-0.0022	N/A
	1.7	D17A6	0.0004-0.0016	0.006-0.024	0.012-0.024	0.008-0.031	0.0014-0.0024	0.0012-0.0022	N/A
	2.0	F20C1	0.0002-0.0011	0.010-0.024	0.024-0.035	0.008-0.031	0.0018-0.0035	0.0016-0.0028	N/A
	2.4	K24A4	0.0008-0.0016	0.008-0.0014	0.016-0.022	0.008-0.028	0.0018-0.0028	0.0020-0.0030	N/A
	3.0	J30A4	0.0006-0.0030	0.008-0.024	0.016-0.028	0.008-0.031	0.0014-0.0050	0.0012-0.0050	N/A

NA: Not Available

67162-ACCO-C08

TORQUE SPECIFICATIONS
All readings in ft. lbs.

Year	Engine Displacement Liters	Engine ID/VIN	Cylinder Head Bolts	Main Bearing Bolts	Rod Bearing Bolts	Crankshaft Damper Bolts	Flywheel Bolts	Manifold Intake	Manifold Exhaust	Spark Plugs	Oil Pan Drain Plug
2001	1.7	D17A1	①	②	24	③	87	16	23	13	33
	1.7	D17A2	①	②	24	③	87	16	23	13	29
	1.7	D17A6	①	②	24	③	87	16	23	13	29
	2.0	F20C1	④	⑤	⑥	181	94	16	23	13	29
	2.2	H22A4	⑦	54	34	181	③	16	23	13	33
	2.3	F23A1	④	⑧	⑨	181	③	16	23	13	33
	2.3	F23A4	④	⑧	⑨	181	③	16	23	13	33
	2.3	F23A5	④	⑧	⑨	181	③	16	23	13	33
	3.0	J30A1	⑦	⑩	⑨	181	③	16	23	13	29
2002	1.7	D17A1	①	②	24	⑨	87	16	23	13	33
	1.7	D17A2	①	②	24	⑨	87	16	23	13	29
	1.7	D17A6	①	②	24	⑨	87	16	23	13	29
	2.0	F20C1	④	⑤	⑥	181	94	16	23	13	29
	2.3	F23A1	④	⑧	⑨	181	③	16	23	13	33
	2.3	F23A4	④	⑧	⑨	181	③	16	23	13	33
	2.3	F23A5	④	⑧	⑨	181	③	16	23	13	33
	3.0	J30A1	⑦	⑩	⑨	181	③	16	23	13	29
2003	1.7	D17A1	①	②	24	③	87	16	23	13	33
	1.7	D17A2	①	②	24	⑨	87	16	23	13	29
	1.7	D17A6	①	②	24	⑨	87	16	23	13	29
	2.0	F20C1	④	⑤	⑥	181	94	16	23	13	29
	2.4	K24A4	⑪	⑫	16	⑬	③	16	33	13	33
	3.0	J30A4	⑦	⑩	⑨	181	③	16	23	13	29
2004	1.7	D17A1	①	②	24	③	87	16	23	13	33
	1.7	D17A2	①	②	24	⑨	87	16	23	13	29
	1.7	D17A6	①	②	24	⑨	87	16	23	13	29
	2.0	F20C1	④	⑤	⑥	181	94	16	23	13	33
	2.4	K24A4	⑪	⑫	16	⑬	③	16	33	13	33
	3.0	J30A4	⑦	⑩	⑨	181	③	16	23	13	29

① Step 1: 14 ft. lbs.
Step 2: 36 ft. lbs.
Step 3: 49 ft. lbs.
Step 4: Bolts 1-2, retorque
to 49 ft. lbs.

② Step 1: 18 ft. lbs.
Step 2: 38 ft. lbs.

③ Automatic transaxle: 54 ft. lbs.
Manual transaxle: 76 ft. lbs.

④ Step 1: 22 ft. lbs.
Step 2: Rotate 90 degrees
Step 3: Rotate 90 degrees
Step 4: If new bolt rotate
additional 90 degrees

⑤ Step 1: Bearing cap bolts to 22 ft. lbs.
Step 2: Bearing cap bolts plus 60 degrees
Step 3: 8mm bolts to 16 ft. lbs.

⑥ Step 1: 18 ft. lbs.
Step 2: Plus 90 degrees

⑦ Step 1: 29 ft. lbs.
Step 2: 51 ft. lbs.
Step 3: 72.3 ft. lbs.

⑧ Step 1: 11mm bolts, 29 ft. lbs.
Step 2: 11mm bolts, 58 ft. lbs.
Step 3: 6mm bolts, 8.7 ft. lbs.

⑨ Step 1: 14 ft. lbs.
Step 2: Rotate 90 degrees

⑩ Step 1: Cap bolts, 56 ft. lbs.
Step 2: Side bolts, 36 ft. lbs.

⑪ Step 1: 29 ft. lbs.
Step 2: Rotate 90 degrees
Step 3: Rotate 90 degrees
Step 4: If new bolt rotate
additional 90 degrees

⑫ Step 1: 22 ft. lbs.
Step 2: Rotate 56 degrees

⑬ Step 1: 36 ft. lbs.
Step 2: Rotate 90 degrees

67162-ACCO-C12

WHEEL ALIGNMENT

Year	Model		Caster Range (+/-Deg.)	Caster Preferred Setting (Deg.)	Camber Range (+/-Deg.)	Camber Preferred Setting (Deg.)	Toe-in (in.)
2001	Accord	F	1.00	+2.80	1.00	+0.06	0 +/- 0.03
		R	—	—	0.50	-0.50	0.03 +/- 0.03
	Civic	F	—	+1.66	—	0	0.03
		R	—	—	—	-1.00	0.03
	Prelude	F	1.00	+2.66	1.00	0	0 +/- 0.03
		R	—	—	1.00	-0.45	0.03 +/- 0.03
	S2000	F	0.75	+6.00	0.50	-0.50	0 +/- 0.03
		R	—	—	0.50	-1.50	0.12 +/- 0.03
2002	Accord	F	1.00	+2.80	1.00	+0.06	0 +/- 0.03
		R	—	—	0.50	-0.50	0.03 +/- 0.03
	Civic	F	—	+1.66	—	0	0.03
		R	—	—	—	-1.00	0.03
	S2000	F	0.75	+6.00	0.50	-0.50	0 +/- 0.03
		R	—	—	0.50	-1.50	0.12 +/- 0.03
2003	Accord	F	1.00	+2.80	1.00	+0.06	0 +/- 0.03
		R	—	—	0.50	-0.50	0.03 +/- 0.03
	Civic	F	—	+1.66	—	0	0.03
		R	—	—	—	-1.00	0.03
	S2000	F	0.75	+6.00	0.50	-0.50	0 +/- 0.03
		R	—	—	0.50	-1.50	0.12 +/- 0.03
2004	Accord	F	1.00	+2.80	1.00	+0.06	0 +/- 0.03
		R	—	—	0.50	-0.50	0.03 +/- 0.03
	Civic	F	—	+1.66	—	0	0.03
		R	—	—	—	-1.00	0.03
	S2000	F	0.75	+6.00	0.50	-0.50	0 +/- 0.03
		R	—	—	0.50	-1.50	0.12 +/- 0.03

67162-ACCO-C13

TIRE, WHEEL AND BALL JOINT SPECIFICATIONS

| Year | Model | OEM Tires | | Tire Pressures (psi) | | Wheel Size | Ball Joint Inspection | Lug Nut (ft. lbs.) |
		Standard	Optional	Front	Rear			
2001	Civic	P185/65R14	P195/55VR15	30	29	5-J	NS	80
	Accord DX	P195/70SR14	None	32	32	5-J	NS	80
	Accord EX, LX 4-cyl.	P195/65HR15	None	32	32	6.5	NS	80
	Accord LX V6	P205/65VR15	None	30	30	5-J	NS	80
	Prelude	P205/50VR16	None	32	32	6.5-JJ	NS	80
	S2000	P225/50VR16	None	32	32	5-J	NS	80
2002	Civic	P185/65R14	P195/55VR15	30	29	5-J	NS	80
	Accord DX	P195/70SR14	None	32	32	5-J	NS	80
	Accord EX, LX 4-cyl.	P195/65HR15	None	32	32	6.5	NS	80
	Accord LX V6	P205/65VR15	None	30	30	5-J	NS	80
	S2000	P225/50VR16	None	32	32	5-J	NS	80
2003	Civic	P185/65R14	P195/55VR15	30	29	5-J	NS	80
	Accord DX	P195/65R15	None	32	32	5-J	NS	80
	Accord LX 4-cyl.	P205/65R15	None	32	32	6.5	NS	80
	Accord EX 4-cyl.	P205/60R16	None	32	32	6.5	NS	80
	Accord LX V6	P205/65VR15	None	30	30	5-J	NS	80
	Accord EX V6	P205/60R16	None	32	32	5-J	NS	80
	S2000	P225/50VR16	None	32	32	5-J	NS	80
2004	Civic	P185/65R14	P195/55VR15	30	29	5-J	NS	80
	Accord DX	P195/65R15	None	32	32	5-J	NS	80
	Accord LX 4-cyl.	P205/65R15	None	32	32	6.5	NS	80
	Accord EX 4-cyl.	P205/60R16	None	32	32	6.5	NS	80
	Accord LX V6	P205/65VR15	None	30	30	5-J	NS	80
	Accord EX V6	P205/60R16	None	32	32	5-J	NS	80
	S2000	P225/50VR16	None	32	32	5-J	NS	80

OEM: Original Equipment Manufacturer

PSI: Pounds Per Square Inch

STD: Standard

OPT: Optional

NS: Not specified by manufacturer

67162-ACCO-C14

BRAKE SPECIFICATIONS
All measurements in inches unless noted

Year	Model		Brake Disc Original Thickness	Brake Disc Minimum Thickness	Brake Disc Maximum Runout	Brake Drum Diameter Original Inside Diameter	Brake Drum Diameter Max. Wear Limit	Brake Drum Diameter Maximum Machine Diameter	Minimum Lining Thickness Front	Minimum Lining Thickness Rear	Brake Caliper Bracket Bolts (ft. lbs.)	Brake Caliper Mounting Bolts (ft. lbs.)
2001	Accord	F	0.910	0.830	0.004	—	—	—	0.060	—	—	①
		R	0.358	0.310	0.004	8.66	8.70	8.70	—	0.080 ②	—	17
	Civic	F	0.840	0.750	0.004	—	—	—	0.060	—	—	①
		R	—	—	—	7.87	7.91	7.91	—	0.080	—	—
	Prelude	F	0.910	0.830	0.004	—	—	—	0.060	—	83	①
		R	0.390	0.320	0.004	—	—	—	—	0.060	—	17
	S2000	F	0.990	0.910	0.004	—	—	—	0.060	—	80	24
		R	0.476	0.390	0.004	—	—	—	—	0.060	41	17
2002	Accord	F	0.910	0.830	0.004	—	—	—	0.060	—	—	①
		R	0.358	0.310	0.004	8.66	8.70	8.70	—	0.080 ②	—	17
	Civic	F	0.840	0.750	0.004	—	—	—	0.060	—	—	①
		R	—	—	—	7.87	7.91	7.91	—	0.080	—	—
	S2000	F	0.990	0.910	0.004	—	—	—	0.060	—	80	24
		R	0.476	0.390	0.004	—	—	—	—	0.060	41	17
2003	Accord	F	0.910	0.830	0.004	—	—	—	0.060	—	—	①
		R	0.400	0.310	0.004	8.66	8.70	8.70	—	0.080 ②	—	17
	Civic	F	0.840	0.750	0.004	—	—	—	0.060	—	—	①
		R	—	—	—	7.87	7.91	7.91	—	0.080	—	—
	S2000	F	0.990	0.910	0.004	—	—	—	0.060	—	80	24
		R	0.476	0.390	0.004	—	—	—	—	0.060	41	17
2004	Accord	F	0.910	0.830	0.004	—	—	—	0.060	—	—	①
		R	0.400	0.310	0.004	8.66	8.70	8.70	—	0.080 ②	—	17
	Civic	F	0.840	0.750	0.004	—	—	—	0.060	—	—	①
		R	—	—	—	7.87	7.91	7.91	—	0.080	—	—
	S2000	F	0.990	0.910	0.004	—	—	—	0.060	—	80	24
		R	0.476	0.390	0.004	—	—	—	—	0.060	41	17

NA: Not Available

F: Front

R: Rear

① Calipers with long pins beyond bolt threads, 54 ft. lbs.
   Calipers with no pin beyond threads, 20 ft. lbs.

② With rear disc: 0.060

67162-ACCO-C15

SCHEDULED MAINTENANCE INTERVALS
Honda—Civic, Accord, Prelude & S2000

TO BE SERVICED	TYPE OF SERVICE	VEHICLE MILEAGE INTERVAL (x1000)												
		7.5	15	22.5	30	37.5	45	52.5	60	67.5	75	82.5	90	97.5
Engine oil & filter	R	✓	✓	✓	✓	✓	✓	✓	✓	✓	✓	✓	✓	✓
Front brake pads	S/I	✓	✓	✓	✓	✓	✓	✓	✓	✓	✓	✓	✓	✓
Rotate tires	S/I	✓	✓	✓	✓	✓	✓	✓	✓	✓	✓	✓	✓	✓
Cooling system, hoses & connections	S/I		✓		✓		✓		✓		✓		✓	
Driveshaft boots	S/I		✓		✓		✓		✓		✓		✓	
Exhaust system	S/I		✓		✓		✓		✓		✓		✓	
Front brake discs & calipers	S/I		✓		✓		✓		✓		✓		✓	
Front wheel alignment	S/I		✓		✓		✓		✓		✓		✓	
Front & rear wheel alignment (Prelude w/4WS)	S/I		✓		✓		✓		✓		✓		✓	
Fuel pipes, hoses & connections	S/I		✓		✓		✓		✓		✓		✓	
Parking brake adjustment	S/I		✓		✓		✓		✓		✓		✓	
Power steering system	S/I		✓		✓		✓		✓		✓		✓	
Rear brake discs, calipers & pads	S/I		✓		✓		✓		✓		✓		✓	
Suspension components	S/I		✓		✓		✓		✓		✓		✓	
Suspension mounting bolts	S/I		✓		✓		✓		✓		✓		✓	
Tie rods, steering gear box & boots	S/I		✓		✓		✓		✓		✓		✓	
Valve clearance (Prelude VTEC) ①	S/I		✓		✓		✓		✓		✓		✓	
Valve clearance (Accord L4, Civic & Prelude non-VTEC)	S/I				✓				✓				✓	
Parking brake	S/I		✓		✓				✓				✓	
Air cleaner element	R				✓				✓				✓	
Transmission fluid (Civic CVT)	R				✓		✓		✓		✓		✓	
Transmission fluid (A/T or M/T) (except as noted below)	R				✓				✓				✓	
Transmission fluid (Prelude L4)	R												✓	
Brake fluid (including ABS) (Accord V6)	R				✓				✓				✓	
Brake fluid (including ABS) (Accord L4, Civic, & Prelude)	R						✓						✓	
Spark plugs (non-VTEC)	R				✓				✓				✓	
Spark plugs (VTEC) ①	R								✓				✓	
ABS operation	S/I				✓				✓				✓	
Alternator drive belt	S/I				✓				✓					
Power steering pump belt	S/I				✓				✓				✓	
Rear brake drums, wheel cylinders & linings (except Prelude)	S/I				✓				✓				✓	
Engine coolant	R						✓				✓			

67162-ACCO-C16

SCHEDULED MAINTENANCE INTERVALS
Honda—Civic, Accord, Prelude & S2000

TO BE SERVICED	TYPE OF SERVICE	VEHICLE MILEAGE INTERVAL (x1000)												
		7.5	15	22.5	30	37.5	45	52.5	60	67.5	75	82.5	90	97.5
ABS high pressure hose	R								✓					
Fuel filter	R								✓					
Timing belt	R												✓	
Timing balancer belt	R												✓	
Distributor, ignition cap & rotor	S/I								✓					
Idle speed	S/I								✓					
Ignition wires	S/I								✓					
PCV valve	S/I								✓					
TWC converter heat shield	S/I								✓					
Water pump	S/I												✓	

R: Replace   S/I: Service or Inspect

① S2000: 105,000 miles

FREQUENT OPERATION MAINTENANCE (SEVERE SERVICE)

If a vehicle is operated under any of the following conditions it is considered severe service:

- Extremely dusty areas.

- 50% or more of the vehicle operation is in 32°C (90°F) or higher temperatures, or constant operation in temperatures below 0°C (32°F).

- Prolonged idling (vehicle operation in stop and go traffic).

- Frequent short running periods (engine does not warm to normal operating temperatures).

- Police, taxi, delivery usage or trailer towing usage.

Oil & oil filter: change every 3750 miles.

Driveshaft boots: service or inspect every 7500 miles.

Front brake discs & calipers, & rear brake discs, calipers & pads: service or inspect every 7500 miles.

Power steering system: service or inspect every 7500 miles.

Suspension components: service or inspect every 7500 miles.

Tie rods, steering gear box & boots: service or inspect every 7500 miles.

Air cleaner element: service or inspect every 15,000 miles.

Transmission fluid (Accord V6 & Civic CVT): replace every 15,000 miles.

Transmission fluid (Accord L4, Civic, & Prelude): replace every 30,000 miles.

Timing balancer belt: replace every 60,000 miles.

Timing belt: replace every 60,000 miles.

Water pump: service or inspect every 60,000 miles.

67162-ACCO-C17

PRECAUTIONS

Before servicing any vehicle, please be sure to read all of the following precautions, which deal with personal safety, prevention of component damage, and important points to take into consideration when servicing a motor vehicle:

• Never open, service or drain the radiator or cooling system when the engine is hot; serious burns can occur from the steam and hot coolant.

• Observe all applicable safety precautions when working around fuel. Whenever servicing the fuel system, always work in a well-ventilated area. Do not allow fuel spray or vapors to come in contact with a spark, open flame, or excessive heat (a hot drop light, for example). Keep a dry chemical fire extinguisher near the work area. Always keep fuel in a container specifically designed for fuel storage; also, always properly seal fuel containers to avoid the possibility of fire or explosion. Refer to the additional fuel system precautions later in this section.

• Fuel injection systems often remain pressurized, even after the engine has been turned **OFF**. The fuel system pressure must be relieved before disconnecting any fuel lines. Failure to do so may result in fire and/or personal injury.

• Brake fluid often contains polyglycol ethers and polyglycols. Avoid contact with the eyes and wash your hands thoroughly after handling brake fluid. If you do get brake fluid in your eyes, flush your eyes with clean, running water for 15 minutes. If eye irritation persists, or if you have taken brake fluid internally, IMMEDIATELY seek medical assistance.

• The EPA warns that prolonged contact with used engine oil may cause a number of skin disorders, including cancer! You should make every effort to minimize your exposure to used engine oil. Protective gloves should be worn when changing oil. Wash your hands and any other exposed skin areas as soon as possible after exposure to used engine oil. Soap and water, or waterless hand cleaner should be used.

• All new vehicles are now equipped with an air bag system. The system must be disabled before performing service on or around system components, steering column, instrument panel components, wiring and sensors. Failure to follow safety and disabling procedures could result in accidental air bag deployment, possible personal injury, and unnecessary system repairs.

• Always wear safety goggles when working with, or around, the air bag system. When carrying a non-deployed air bag, be sure the bag and trim cover are pointed away from your body. When placing a non-deployed air bag on a work surface, always face the bag and trim cover upward, away from the surface. This will reduce the motion of the module if it is accidentally deployed. Refer to the additional air bag system precautions later in this section.

• Clean, high quality brake fluid from a sealed container is essential to the safe and proper operation of the brake system. You should always buy the correct type of brake fluid for your vehicle. If the brake fluid becomes contaminated, completely flush the system with new fluid. Never reuse any brake fluid. Any brake fluid that is removed from the system should be discarded. Also, do not allow any brake fluid to come in contact with a painted surface; it will damage the paint.

• Never operate the engine without the proper amount and type of engine oil; doing so WILL result in severe engine damage.

• Timing belt maintenance is extremely important! Many models utilize an interference-type, non-freewheeling engine. If the timing belt breaks, the valves in the cylinder head may strike the pistons, causing potentially serious (also time-consuming and expensive) engine damage. Refer to the maintenance interval charts in the front of this section for the recommended replacement interval for the timing belt, and to the timing belt procedure in this section for belt replacement and inspection.

• Disconnecting the negative battery cable on some vehicles may interfere with the functions of the on-board computer system(s) and may require the computer to undergo a relearning process once the negative battery cable is reconnected.

• When servicing drum brakes, only disassemble and assemble one side at a time, leaving the remaining side intact for reference.

ENGINE REPAIR

➡Disconnecting the negative battery cable on some vehicles may interfere with the functions of the on board computer systems and may require the computer to undergo a relearning process, once the negative battery cable is reconnected.

Alternator

REMOVAL & INSTALLATION

1.7L Engines

1. Before servicing the vehicle, refer to the precautions in the beginning of this section.
2. Remove or disconnect the following:
   • Negative battery cable
   • Accessory drive belts

   • Power steering pump
   • 4P connector and battery terminal wire
   • Alternator bolts
   • Alternator

To install:
   • Alternator and tighten the bolts to 33 ft. lbs. (44 Nm)
   • 4P connector and battery terminal wire. Tighten the battery terminal wire nut to 108 inch lbs. (12 Nm).
   • Power steering pump
   • Accessory drive belts
   • Negative battery cable

2.0L Engine

1. Before servicing the vehicle, refer to the precautions in the beginning of this section.
2. Remove or disconnect the following:

   • Negative battery cable
   • Accessory drive belt
   • 4P connector and battery terminal wire
   • Alternator bolts
   • Alternator

To install:
   • Alternator and tighten the bolts to 33 ft. lbs. (44 Nm)
   • 4P connector and battery terminal wire. Tighten the battery terminal wire nut to 108 inch lbs. (12 Nm).
   • Accessory drive belt
   • Negative battery cable

2.2L Engine

1. Before servicing the vehicle, refer to the precautions in the beginning of this section.
2. Remove or disconnect the following:

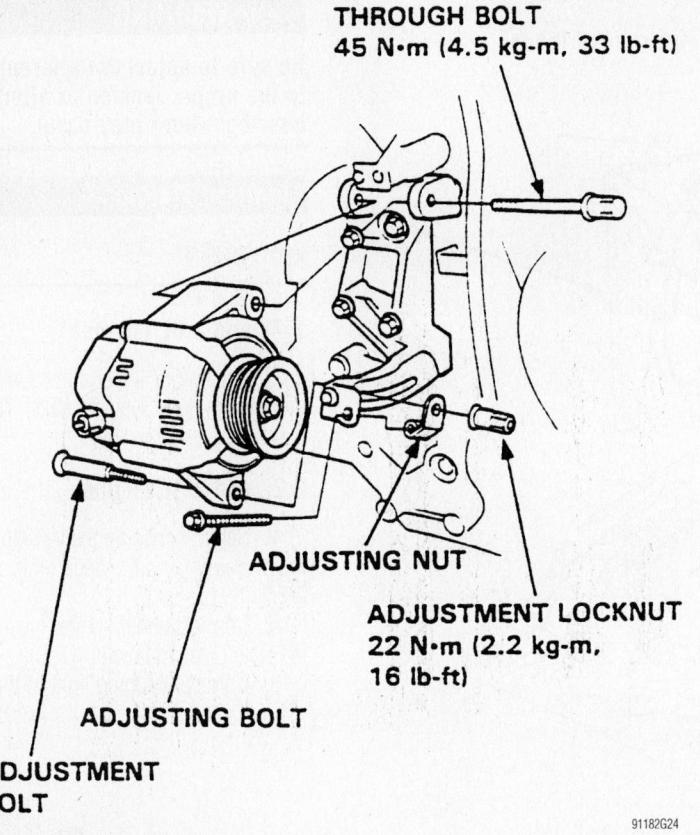

THROUGH BOLT
45 N•m (4.5 kg-m, 33 lb-ft)

ADJUSTING NUT

ADJUSTMENT LOCKNUT
22 N•m (2.2 kg-m,
16 lb-ft)

ADJUSTING BOLT

ADJUSTMENT
BOLT

91182G24

Alternator mounting bolt locations

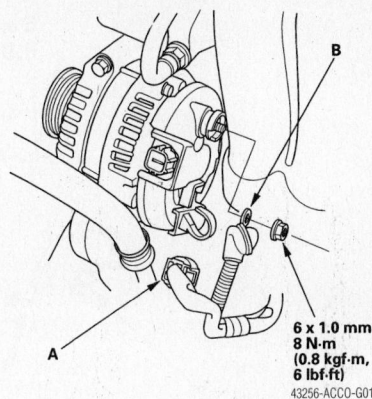

6 x 1.0 mm
8 N•m
(0.8 kgf·m,
6 lbf·ft)
43256-ACCO-G01

Alternator mounting—2.4L engine

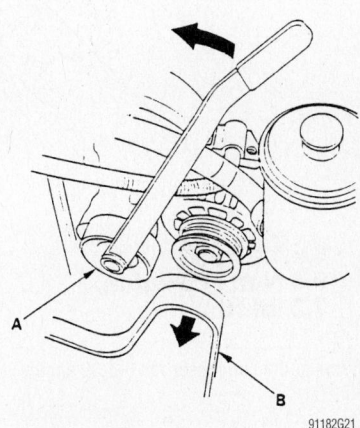

91182G21

Relieve the belt tension by pulling back on the tensioner—3.0L engine

- Negative battery cable, then the positive
- Power steering pump
- Cruise control actuator but do not remove the cable

3. Loosen the through bolt and then loosen adjusting bolt.
- Alternator belt
- Adjusting bolt
- Through bolt and then the alternator

To install:
4. Installation is the reverse of removal.
5. Adjust the alternator belt tension.

2.3L Engines

1. Before servicing the vehicle, refer to the precautions in the beginning of this section.
2. Note the radio security code and the radio presets.
3. Remove or disconnect the following:
- Negative battery cable, then the positive
- 4P connector and battery terminal wire from the alternator
- Adjusting bolt, locknut and the mounting bolt
- Alternator belt
- Alternator from the bracket

To install:
4. Installation is the reverse of removal.
5. Adjust the alternator belt tension.
6. Enter the anti-theft code for the radio.

2.4L Engine

1. Before servicing the vehicle, refer to the precautions in the beginning of this section.
2. Note the radio security code and the radio presets.
3. Remove or disconnect the following:
- Negative battery cable, then the positive
- Drive belt
- Auto-tensioner
- Connectors from the alternator
- 3 alternator mounting bolts and the alternator

To install:
- Alternator and 3 mounting bolts. Torque the bolts to 16 ft. lbs. (22 Nm).
- Electrical connectors
- Auto-tensioner
- Drive belt
- Positive, then negative battery cables

4. Enter the security code and radio presets

3.0L Engine

1. Before servicing the vehicle, refer to the precautions in the beginning of this section.
2. Note the radio security code and the radio presets.
3. Remove or disconnect the following:
- Negative battery cable, then the positive
- Alternator belt tension by pulling back on the adjuster and then remove the belt
- Condenser fan motor connector from the shroud
- Condenser fan assembly
- Four prong connector from the rear of the alternator
- Alternator mounting bolts
- Wiring harness clamp
- Alternator assembly

To install:
4. Alternator installation is the reverse of the removal procedure.
5. Connect the positive battery cable, then the negative battery cable. Enter the radio security code and station presets.

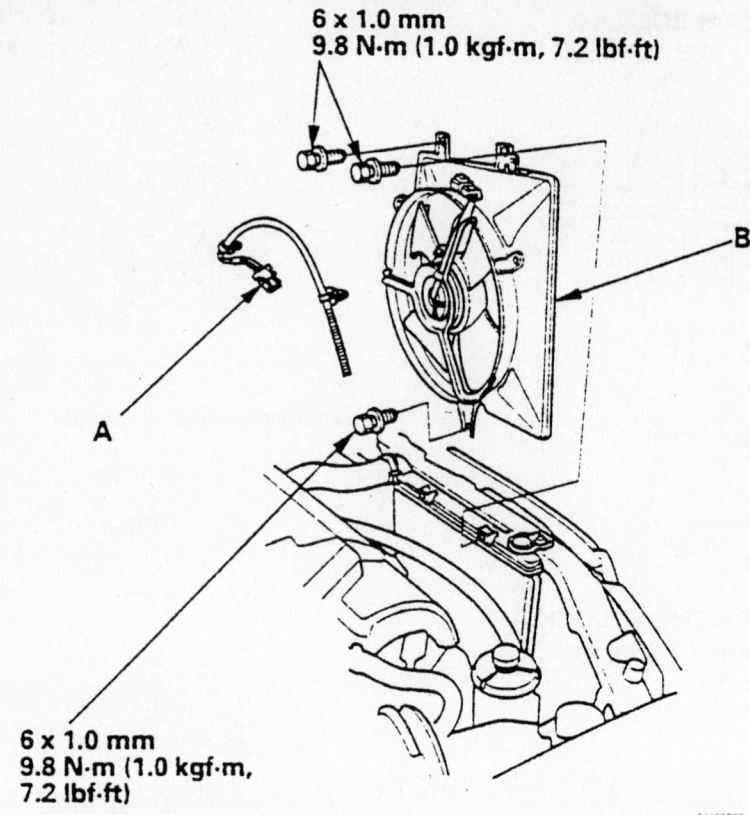

6 x 1.0 mm
9.8 N·m (1.0 kgf·m, 7.2 lbf·ft)

B

A

6 x 1.0 mm
9.8 N·m (1.0 kgf·m,
7.2 lbf·ft)

91182G22

Remove the condenser fan—3.0L engine

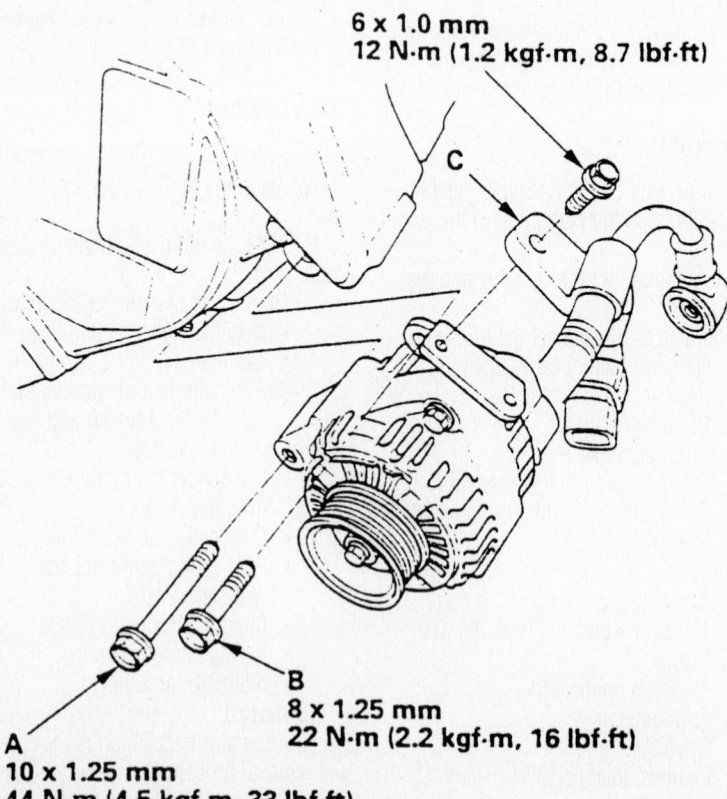

6 x 1.0 mm
12 N·m (1.2 kgf·m, 8.7 lbf·ft)

C

B
8 x 1.25 mm
22 N·m (2.2 kgf·m, 16 lbf·ft)

A
10 x 1.25 mm
44 N·m (4.5 kgf·m, 33 lbf·ft)

91182G23

Torque the alternator bolts to the specs shown—3.0L engine

❄❄ **WARNING**

Be sure to adjust the alternator belt to the proper tension or alternator bearing failure may occur.

Ignition Timing

ADJUSTMENT

1.7L and 2.0L Engines

These engines are equipped with Distributorless Ignition Systems (DIS). No adjustment is necessary.

2.2L and 2.3L Engines

1. Before servicing the vehicle, refer to the precautions in the beginning of this section.
2. Connect a PGM tester (scan tool) to the Data Link Connector (DLC).
3. Connect a timing light to the No. 1 ignition cable.

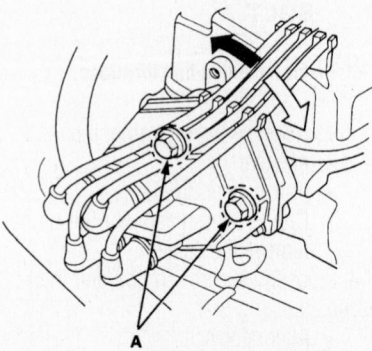

A
22 N·m (2.2 kgf·m, 16 lbf·ft)

7923FG04

Distributor hold-down bolt locations—2.3L (F23A1 and F23A4) engine

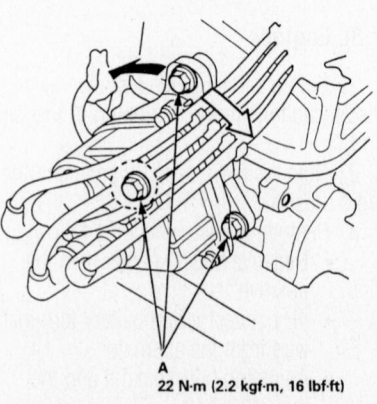

A
22 N·m (2.2 kgf·m, 16 lbf·ft)

7923FG05

Distributor hold-down bolt locations—2.3L (F23A5) engine

4. Start the engine and allow it to warm up until the electric fan comes on.

5. Be sure to turn off all accessories.

6. Verify the idle speed is 650–750 rpm.

7. Point the light at the timing belt cover near the crankshaft pulley and read the timing. Correct timing is 10–14° Before Top Dead Center (BTDC) for both automatic and manual transmissions. If necessary, loosen the distributor hold-down bolt and rotate the distributor slightly to adjust the timing. Turn it counterclockwise to advance and clockwise to retard the timing.

8. Tighten the hold-down bolt to 16 ft. lbs. (22 Nm). Recheck the timing after the bolt is tight to confirm the correct timing.

9. Disconnect the PGM tester.

2.4L and 3.0L Engines

The ignition timing is only adjustable by the Powertrain Control Module (PCM), but the ignition base timing can be checked by performing the following:

1. Before servicing the vehicle, refer to the precautions in the beginning of this section.

2. Connect a PGM tester (scan tool) to the data link connector.

3. Connect a timing light to the No. 1 ignition cable.

4. Start the engine and allow it to warm up until the electric fan comes on.

5. Be sure to turn off all accessories.

6. Verify that the idle speed is 630–730 rpm.

7. Point the light at the timing belt cover near the crankshaft pulley and read the timing. Correct timing is 8–12° Before Top Dead Center (BTDC). If the ignition timing is different from the specification, replace the PCM.

Engine Assembly

REMOVAL & INSTALLATION

➡ **The original radio contains a coded anti-theft circuit. Obtain the security code number before disconnecting the battery cables.**

Civic

1. Before servicing the vehicle, refer to the precautions in the beginning of this section.

2. Disconnect the negative and positive battery cables. Wait at least 3 minutes before working around the air bags.

➡ **The engine and transaxle are removed from the vehicle as 1 unit.**

3. Support the hood as far open as possible. If the hood is to be removed, first matchmark the hinge plates with a felt-tipped marker.

4. Remove the battery from the vehicle. Unbolt and remove the battery tray.

5. Disconnect the battery and alternator cables from the underhood fuse and relay box on the right shock tower.

6. Remove the lower right kick panel to expose the Powertrain Control Module (PCM).

7. Label and disconnect the 5 wiring harness connections from the PCM.

8. Unbolt the main wiring harness retainer from the rear of the fuse and relay box on the right side of the bulkhead. Carefully pull the grommet out of its bulkhead opening. Next, pull the PCM harness and connectors through the opening. Be careful not to damage the wiring, insulation, or connectors.

9. Relieve the fuel pressure:

   a. Loosen the fuel filler cap.

   b. Use a box-end wrench and a flare nut wrench to hold the fuel filter banjo fitting.

   c. Place a shop towel over the fuel filter to catch the fuel spray.

   d. Slowly loosen the fuel filter service bolt 1 full turn.

   e. Clean up any spilled fuel.

10. Remove or disconnect the following:
- Intake air duct and air cleaner
- Intake Air Temperature (IAT) sensor connector from the air cleaner case, if equipped
- Fuel feed hose from the fuel filter
- Fuel return hose from the fuel rail
- Intake manifold/throttle body vacuum hoses
- Brake booster vacuum hose
- Evaporative emissions (EVAP) canister vacuum hose
- Power Steering Pressure (PSP) switch and detach its clamp from the bracket below the brake booster
- Transaxle ground cable
- Radiator hose bracket

11. Loosen the throttle cable locknut, then disconnect the cable from the throttle body linkage. Move the cable aside without kinking it.

12. Loosen the power steering pump mounting bolts. Slip power steering belt off its pulleys. Unbolt the steering pump and move it out of the work area. Don't disconnect the hydraulic hoses.

➡ **Label the connectors before detaching them.**

13. Remove or disconnect the following:
- Engine wiring harness connectors

at the left side of the engine compartment

14. Drain the coolant from the radiator and engine block.

15. Remove or disconnect the following:
- Upper and lower radiator hoses
- Heater hoses from the cylinder head

16. If equipped with a CVT transaxle, loosen the shift cable locknut. Remove the spring clip and washers and disconnect the shift cable from its linkage. Be careful not to kink the cable or damage its boot.

17. Remove or disconnect the following:
- Hydraulic line brackets from the top of the transaxle case, if equipped with a manual transaxle

18. Attach a chain hoist to the engine lifting brackets. Don't raise the hoist to lift the engine yet.

19. Raise the vehicle and support it safely. Remove the front wheels.

20. Remove or disconnect the following:
- Engine splash shield

21. Drain the engine oil.

22. Drain the fluid from the transaxle.

23. Remove or disconnect the following:
- Left front engine mount bracket from the shock tower, if equipped with air conditioning

24. Loosen the compressor idler pulley and adjusting bolt. Slip the belt around the engine mount stud to remove it.

25. Remove or disconnect the following:
- Compressor mounting bolts to separate the compressor from its mounting plate. Move the compressor out of the work area. Do not disconnect the air conditioning refrigerant lines.

26. Remove or disconnect the following:
- ATF cooler lines, if equipped. Plug the cooler lines to prevent fluid leakage and contamination
- Slave cylinder from the transaxle case without disconnecting its hydraulic line, if equipped with manual transaxle
- Front exhaust pipe from the exhaust manifold and catalytic converter. Unbolt its hanger bracket and remove the exhaust pipe.
- Shift cable from the transaxle control shaft, if equipped with automatic transaxle
- Shift rod and extension rod from the transaxle, if equipped with manual transaxle
- Strut damper fork
- Steering knuckle ball joint from the lower control arm using a ball joint separator

27. Pry the inboard CV-joints from the transaxle. Then, move the halfshafts away from the transaxle and wire them to the undercarriage of the vehicle. Tie plastic bags over the inboard CV-joints to prevent damage to the boots and splined shafts.

28. Raise the hoist slightly to take up the weight of the engine and transaxle assembly.

29. Disconnect the engine mounts in the following order:

a. Unbolt and remove the left front engine mount.

b. Unbolt and remove the right front engine mount and bracket assembly.

c. Remove the rear engine mount through-bolt. Then, unbolt the rear mount bracket from the engine block.

30. If necessary, lower the vehicle slightly to gain access to the side engine and transaxle mounts. Do not release the tension on the chain hoist. The engine must be securely supported.

31. Unbolt the side engine mount bracket from the engine block bracket and mount damper.

32. Unbolt the transaxle mount bracket from the transaxle case. Then, unbolt the mount from the shock tower.

33. Raise the chain hoist to lift the engine a few inches off of its mounts.

34. Verify that all electrical, vacuum, and fuel lines have been disconnected.

35. Raise the engine and transaxle assembly and remove it from the vehicle.

To install:

➡**Use new self-locking nuts and gaskets when installing the front exhaust pipe and when assembling the front suspension. Use new set rings on the inboard CV-joint splined shafts.**

36. Lower the engine and transaxle assembly into the vehicle.

37. Install and connect the engine and transaxle mounts and brackets. Use new self-locking nuts and color-coded bolts. At this point, only tighten the mounting nuts and bolts by hand.

38. Before installing the left front engine mount, fit the air conditioning compressor back into place and install the compressor belt. Tighten the compressor bolts to 17 ft. lbs. (24 Nm).

➡**Failure to tighten the bolts in the proper sequence can cause excessive noise and vibration and reduce bushing life. Be sure to check that the bushings are not twisted or offset.**

39. The engine and transaxle mount and bracket fasteners must be tightened in the proper sequence with the weight of the engine resting upon them. This step is important for engine mount pre-loading. Tighten the engine mount bolts in the following sequence:

a. Transaxle mount bolts: 47 ft. lbs. (64 Nm); or 28 ft. lbs. (38 Nm) for CVT-equipped vehicles.

b. Side engine mount bracket nuts: 54 ft. lbs. (74 Nm).

c. Rear mount bracket bolts: 61 ft. lbs. (83 Nm); or 43 ft. lbs. (59 Nm) for CVT-equipped vehicles.

d. Rear mount through-bolt: 43 ft. lbs. (59 Nm).

e. Transaxle mount bracket nuts or bolts: 47 ft. lbs. (64 Nm).

f. Transaxle mount through-bolt: 54 ft. lbs. (74 Nm).

g. Right front mount bracket bolts: 33 ft. lbs. (44 Nm).

h. Right front mount carrier bolts: 33 ft. lbs. (44 Nm).

i. Left front mount: stud: 61 ft. lbs. (85 Nm); carrier bolts: 33 ft. lbs. (44 Nm); nut: 43 ft. lbs. (59 Nm).

40. Remove the chain hoist from the engine lifting hooks.

41. Install or connect the following:

- New set rings on the inboard splined shafts of each halfshaft. Check that the set ring on each inboard CV-joint clicks into place when the halfshafts are installed into the transaxle.

- Damper fork and reconnect the lower ball joint. When the weight of the vehicle is resting on its suspension, tighten the pinch bolt to 32 ft. lbs. (44 Nm) and the fork bolt to 47 ft. lbs. (65 Nm). Tighten the ball joint castle nut to 36–43 ft. lbs. (50–60 Nm). Next, tighten the castle nut only enough to install a new cotter pin.

- Slave cylinder, if equipped. Tighten the slave cylinder mounting bolts to 16 ft. lbs. (22 Nm). If the clutch hydraulic line was disconnected, air must be bled from the system.

- Transaxle shift and extension rods to the linkage at the transaxle case, if equipped. Install a new 8mm spring pin into the shift rod linkage. Then, install the retainer clip and boot. Tighten the extension rod bolt to 16 ft. lbs. (22 Nm).

- Shift cable to the control shaft, if equipped with an automatic transaxle. Use a new lockwasher and tighten the lockbolt to 10 ft. lbs. (14 Nm). Tighten the shift

cable cover bolts to 16 ft. lbs. (22 Nm). Install the shift cable cover and tighten its bolts to 16 ft. lbs. (22 Nm).

42. Install the front exhaust pipe using new self-locking nuts:

a. If equipped with the D16Y8 engine, tighten the converter flange nuts to 16 ft. lbs. (22 Nm).

b. Tighten the exhaust manifold nuts to 40 ft. lbs. (55 Nm).

c. If equipped with the D16Y5 or D16Y7 engine, tighten the converter flange nuts to 25 ft. lbs. (33 Nm).

43. Install or connect the following:

- Tighten the exhaust flange bolts to 16 ft. lbs. (22 Nm).

- ATF cooler lines. If the rubber cooler lines are cracked or stressed, they must be replaced.

- Engine splash shield

44. Refill the engine with fresh oil.

45. Refill the transaxle with the proper fluid.

46. Lower the vehicle.

47. If equipped, fit the clutch hydraulic line brackets back into place. Tighten the 8mm bolts to 17 ft. lbs. (24 Nm). Tighten the 6mm bolts to 96 inch lbs. (11 Nm).

48. If equipped with a CVT transaxle, reconnect the shift cable to the linkage. Use new plastic washers and a new spring clip. Tighten the locknut to 22 ft. lbs. (29 Nm).

49. Adjust the alternator and air conditioning compressor belt tensions.

50. Install or connect the following:

- Upper and lower radiator hoses and the heater hoses

- Power steering pump into its mounts. Adjust the pump belt tension, then tighten the mounting bolts to 17 ft. lbs. (24 Nm).

- PSP switch connector and attach its harness clamp

- Intake manifold/throttle body vacuum hoses

- Brake booster vacuum hose

- EVAP canister vacuum hose

- Fuel line fittings to the fuel filter and fuel rail. Use new sealing washers. Tighten the banjo fittings to 25 ft. lbs. (33 Nm), and the service bolts to 11 ft. lbs. (15 Nm). Don't overtighten the fittings

- Throttle cable and adjust its deflection to 10–12mm (0.39–0.47 in.).

51. Feed the PCM harness through the hole in the bulkhead. Apply sealant to the grommet, then install the retainer.

52. Install or connect the following:

- Engine wiring harness and ground cables that were disconnected dur-

ing removal. Be sure the grounds are free of corrosion to ensure good contact.
- Fuse and relay box back into position
- Battery and alternator cables
- Air cleaner case and air intake duct
- IAT connector
- 5 PCM connectors and kick panel
- Battery tray and the battery

53. Verify that all wiring harnesses and grounds, vacuum lines, fuel lines have been reconnected.

54. Refill the radiator with fresh coolant.

55. If it was removed, install the hood. Reconnect the windshield washer tubing. After installation, check to be sure that the hood, fender, and grille panel gaps are equal.

56. Reconnect the positive and negative battery cables.

57. Turn the ignition switch to the **ON** position, but don't start the engine. Then, turn the ignition **OFF**. Repeat this procedure 2 or 3 times and check for fuel leaks.

58. Start the engine and allow it to warm up to its normal operating temperature.

59. Bleed the air from the cooling system with the heater valve open.

60. Check the throttle cable deflection and operation.

61. Check and adjust the ignition timing.

62. Shut the engine off and check the drive belt adjustments.

63. Check all fluid levels and top up as necessary.

64. Check and adjust the front wheel alignment.

65. Road test the vehicle.

Prelude

1. Before servicing the vehicle, refer to the precautions in the beginning of this section.

2. Secure the hood as far open as possible.

3. Remove or disconnect the following:
- Negative battery cable, then the positive battery cable
- Radiator cap

4. Raise and safely support the vehicle. Remove the front wheels and the engine splash shield.

5. Drain the engine coolant into a sealable container.

6. Drain the transaxle fluid into a sealable container. Install the drain plug with a new gasket.

7. Lower the vehicle to a working level.

8. Remove or disconnect the following:
- Air intake duct and the air cleaner case

- PAIR vacuum tank and bracket
- Battery and the battery base
- Battery cable and starter cable harnesses from the body

9. Relieve the pressure from the fuel system.

❈❈ CAUTION

The fuel injection system remains under pressure after the engine has been turned OFF. Properly relieve fuel pressure before disconnecting any fuel lines. Failure to do so may result in fire or personal injury.

10. Remove or disconnect the following:
- Fuel feed hose from the fuel rail and the fuel return line from the fuel pressure regulator
- Injector resistor connector on the left side of the engine compartment
- Throttle cable by loosening the locknut, then slip the cable end out of the throttle linkage. Take care not to bend the cable when removing it. Always replace any kinked cable with a new one.
- Engine wiring harness connectors, terminal, and clamps on the right side of the engine
- Power cable from the under-hood fuse/relay box
- Brake booster vacuum hose and emissions control vacuum tubes from the intake manifold
- Cruise control actuator electrical connector and vacuum tube, then the actuator
- Engine ground cable from the body side
- Power steering pump drive belt, then the pump
- Air conditioning condenser fan, then install a protector plate on the radiator
- Alternator mounting bolt, nut, and adjusting nut, then the alternator drive belt
- Air conditioning compressor electrical connector
- Compressor without disconnecting the air conditioning hoses. Support the compressor with a strong wire out of the way.
- Upper and lower radiator hoses, and the heater hoses from the engine
- Transaxle ground cable
- Cooler hoses, if equipped with an automatic transaxle

11. If equipped with a manual transaxle perform the following:

a. Disconnect the shift cable and the select cable from the transaxle. Do not bend the cables when removing them. Replace any kinked cable with a new one.

b. Remove the clutch slave cylinder and the pipe/hose assembly. Do not depress the clutch pedal once the slave cylinder has been removed.

c. Remove the clutch damper assembly.

12. Remove or disconnect the following:
- Vehicle Speed/Power Steering Speed (VSS/PSS) sensor assembly. Do not disconnect the hoses
- Nuts attaching the exhaust pipe to the exhaust manifold and the catalytic converter
- Bolts from the exhaust pipe hanger, then the exhaust pipe and discard the gaskets

13. If equipped with an automatic transaxle, remove the shift cable cover, then disconnect the shift cable. Do not bend the cable and replace the cable if it becomes kinked.

14. Remove or disconnect the following:
- Left and the right side damper forks
- Lower ball joints from the lower control arms

15. Pry the halfshafts from the transaxle. Cover the inner CV-joints with plastic bags to protect them.

16. Swing the halfshafts under the fender out of the way.

17. Attach an engine hoist to the engine lifting points and raise the hoist to remove all slack from the chain.

18. Remove or disconnect the following:
- Rear engine mount bracket
- Front engine mount bracket
- Left side engine mount
- Transaxle mount and the mount bracket

19. Check that the engine is completely free of vacuum hoses, fuel and coolant hoses, and electrical wiring.

20. Slowly raise the engine approximately 6 in. (150mm). Check once again that all hoses and wires have been disconnected from the engine.

21. Raise the engine all the way and remove it from the vehicle.

22. Remove the transaxle.

23. If equipped with a manual transaxle, remove the clutch cover (pressure plate) and clutch disc.

24. Mount the engine on an engine stand, making sure the mounting bolts are tight. If an engine stand is not available, support the engine in an upright position with blocks. Never leave an engine hanging from a lift or hoist.

To install:

25. Install or connect the following:
- Clutch disc and pressure plate to the flywheel for manual transaxle vehicles
- Transaxle
- Engine into position and lower it into the car, aligning the mounts and bushings.

➡ **When installing the engine mounts and vibration dampers in the following steps, they must be tightened to the correct tension in the correct order if they are to damp vibration properly.**

- Side engine mount and the through-bolt. Do not tighten the through-bolt at this time
- Nut and bolt attaching the side mount to the engine: 40 ft. lbs. (55 Nm)
- Transaxle mount and through-bolt. Do not tighten the through-bolt at this time
- Rear engine mount and new bolts attaching the mount to the engine: 40 ft. lbs. (55 Nm)
- New rear engine mount through-bolt: 47 ft. lbs. (65 Nm)
- Front mount and the 3 bolts attaching the mount to the engine assembly, only snug the bolts in place
- New through-bolt to the front mount: 47 ft. lbs. (65 Nm)
- Nuts to the transaxle mount: 28 ft. lbs. (39 Nm)

26. Tighten the side engine mount through-bolt to 47 ft. lbs. (65 Nm).

27. Tighten the transaxle mount through-bolt to 47 ft. lbs. (65 Nm).

28. Tighten the 3 bolts attaching the front mount to the engine to 28 ft. lbs. (39 Nm).

29. Remove the hoist equipment from the engine.

30. Install or connect the following:
- New spring clips to the inner CV-joints
- Halfshafts into the transaxle. Be sure that the inner joint spring clips click into place
- Lower ball joints to the lower control arms. Tighten the nuts to 36–43 ft. lbs. (50–60 Nm). Install a new cotter pin to the ball joint stud.

31. Using new self-locking bolts, attach the damper forks to the struts and tighten to 32 ft. lbs. (44 Nm). Tighten the new nut and bolt attaching the damper fork to the lower control arm to 47 ft. lbs. (65 Nm).

32. If equipped with an automatic transaxle, connect the shift cable to the transaxle. Install a new lockwasher and tighten the attaching bolt to 84 inch lbs. (10 Nm). Install the shift cable cover. Tighten the shift cable cover attaching bolts to 13 ft. lbs. (18 Nm).

33. Install or connect the following:
- Exhaust pipe with new gaskets
- New nuts attaching the exhaust pipe to the exhaust manifold: 40 ft. lbs. (55 Nm)
- New nuts attaching the exhaust pipe to the catalytic converter: 25 ft. lbs. (34 Nm)
- New attaching bolts to the exhaust pipe hanger: 13 ft. lbs. (18 Nm)
- VSS/PSS, connect the electrical connector, and tighten the mounting bolt to 13 ft. lbs. (18 Nm)

34. If equipped with a manual transaxle perform the following:
   a. Install the clutch damper assembly and tighten the attaching bolts to 16 ft. lbs. (22 Nm).
   b. Install the clutch slave cylinder and the pipe/hose assembly and tighten the slave cylinder mounting bolts to 16 ft. lbs. (22 Nm).
   c. Connect the shift cable and the select cable to the transaxle. Adjust the shift cable and select cable.

35. Install or connect the following:
- Cooler hoses, if equipped with an automatic transaxle
- Transaxle ground cable
- Upper and lower radiator hoses and heater hoses to the engine
- Air conditioning compressor, connect the electrical connector. Tighten the mounting bolts to 16 ft. lbs. (22 Nm).
- Alternator drive belt, then adjust

36. Remove the protector plate from the radiator and install the air conditioning condenser fan.

37. Install or connect the following:
- Power steering pump and drive belt. Adjust the drive belt tension, then tighten the attaching nuts and bolts to 16 ft. lbs. (22 Nm).
- Engine ground cable to the body
- Cruise control actuator, electrical connector and vacuum tube. Tighten the mounting bolts to 84 inch lbs. (10 Nm)
- Brake booster vacuum hose and the emissions control vacuum tubes to the intake manifold
- Engine wiring harness connectors, terminal, and clamps
- Throttle cable, then adjust
- Injector resistor connector on the left side of the engine compartment

- Fuel return hose to the regulator
- Fuel feed hose to the fuel rail with new washers. Tighten the cap nut to 16 ft. lbs. (22 Nm)
- Battery and starter cables to the body
- Battery base and the battery. Tighten the battery base attaching bolts to 16 ft. lbs. (22 Nm)
- PAIR vacuum tank and bracket. Tighten the mounting bolts to 96 inch lbs. (10 Nm)
- Air cleaner duct and housing
- Engine splash shield and the front wheels

38. Lower the vehicle.

39. Fill the engine with oil and the transaxle with fluid.

40. Fill and bleed the air from the cooling system.

41. Connect the positive, then the negative battery cable and enter the radio security code.

42. Switch the ignition **ON** but do not engage the starter. The fuel pump should run for approximately 2 seconds, building pressure within the lines. Switch the ignition **OFF**, then **ON** 2 or 3 more times to build full system pressure. Check for fuel leaks.

43. Start the engine, allowing it to idle. Check the hoses and lines carefully for any sign of leakage.

44. Check the timing and idle speed.

45. After the engine has warmed up fully and the fan(s) have come on at least once, recheck the engine for fluid leaks. Switch the engine OFF.

46. Adjust the belts and throttle cable as necessary.

47. If equipped with 4WS, start the engine and turn the steering wheel lock-to-lock to reset the 4WS control unit.

48. Road test the vehicle, then loosen and retighten the 3 bolts attaching the front engine mount to the engine. Tighten the bolts to 28 ft. lbs. (39 Nm).

Accord

1. Before servicing the vehicle, refer to the precautions in the beginning of this section.

2. Obtain the anti-theft code for the radio, then disconnect the battery cables. Be sure to disconnect the negative cable first.

3. Remove the air intake duct.

4. Secure the hood in the open position with a long prop rod such as P/N 74145-S84-A00.

5. Remove or disconnect the following:
- Both battery cables and the con-

nector from the underhood relay box
- Battery and tray, on the 3.0L engine
- Bolt securing the relay box to the body
- Accelerator and cruise control cables from the throttle body and bracket

6. Properly relieve the fuel system pressure.
- Fuel hoses from the fuel rail
- Brake booster vacuum and evaporative emissions (EVAP) canister hoses
- Vacuum hose from the canister
- Hose securing the power steering hose on the engine
- Power steering pump belt, then remove the pump and position it out of the way. Use wire if necessary.
- Powertrain Control Module (PCM) connectors from the control module. Remove the grommet and pull the connectors through.
- Wiring harness connectors at the right side of the engine compartment for 2.3L and 2.4L engines and on the left side for the 3.0L engine.

7. On the 2.3L and 2.4L engine, remove the starter cable (A) and clamp (B). Remove the ground cable (C) and back-up light switch connectors (D). On the 3.0L engine, remove the starter wiring from the engine compartment attaching points.

8. On vehicles with a manual transaxle, disconnect the shift and select cables from the transaxle. Remove the slave cylinder mounting bolts and position the cylinder out of the way. Be sure not to bend the line.

9. Remove or disconnect the following:
- Rear engine mount through-bolt and stiffener
- Front engine mount bracket mounting bolts and loosen the through-bolt
- Radiator cap

10. Raise and safely support the vehicle.

11. Remove or disconnect the following:
- Front tires
- Engine under cover

12. Loosen the radiator drain plug and drain the coolant.

13. Drain the transaxle oil or fluid, then reinstall the plug using a new washer.

14. Drain the engine oil, then reinstall the plug using a new washer.

15. Lower the vehicle and remove the upper and lower radiator hoses and heater hoses from the engine.

16. On vehicles with an automatic transaxle, disconnect the ATF fluid cooler lines.

17. Remove the air conditioning compressor from the engine and position it to the side without disconnecting the hoses.

18. Raise the vehicle and remove the front exhaust pipe.

19. On vehicles with automatic transaxle, remove the 2 bolts (A) for the shift cable holder (B), then remove the shift cable cover (C). To prevent damage to the linkage, be sure to remove the shift cable holder before removing the bolts for the cover.

20. Remove or disconnect the following:
- Lockbolt (D) from the control lever (E), then the shift cable (F) with the control lever
- Through-bolt securing the bottom of the shock absorber to the control arm
- Halfshafts
- Rear engine mounting bracket
- 2 flange bolts from each of the radius rods

21. Mark the location of the front beams

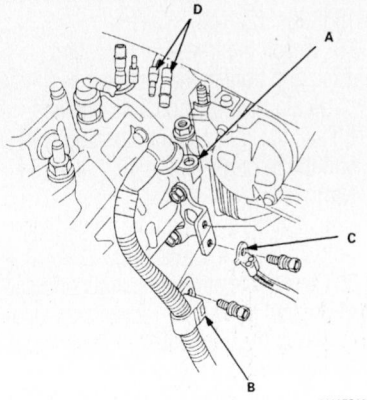

Starter cable, clamp, ground cable and back-up light switch connector locations—Accord with 2.3L engine

7923FG06

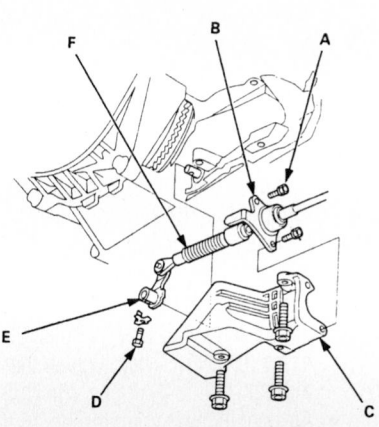

Automatic transaxle linkage components—2.3L Accord

7923FG07

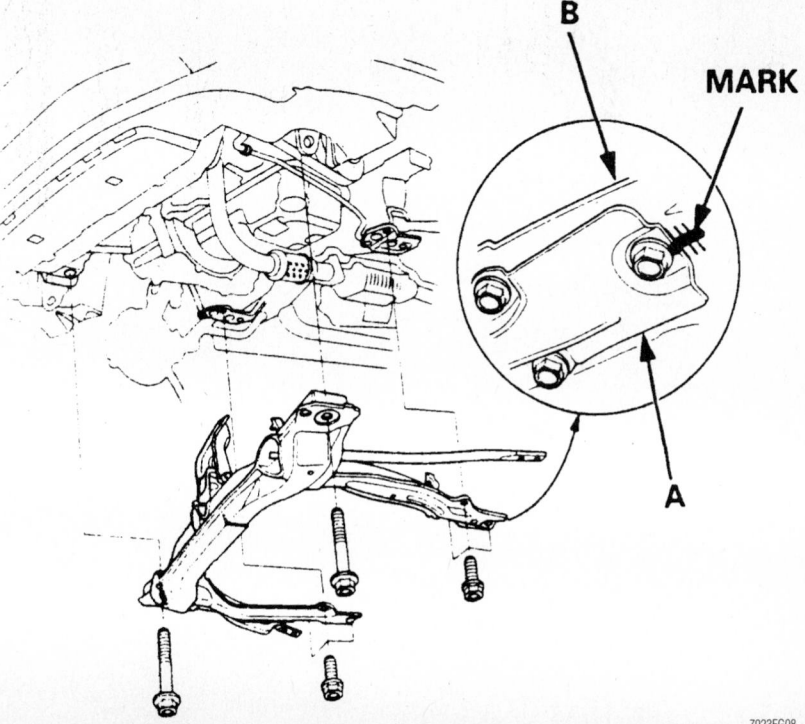

Be sure to mark the location of the front beams (A) on the rear beams (B) before removing the subframe—2.3L Accord

7923FG08

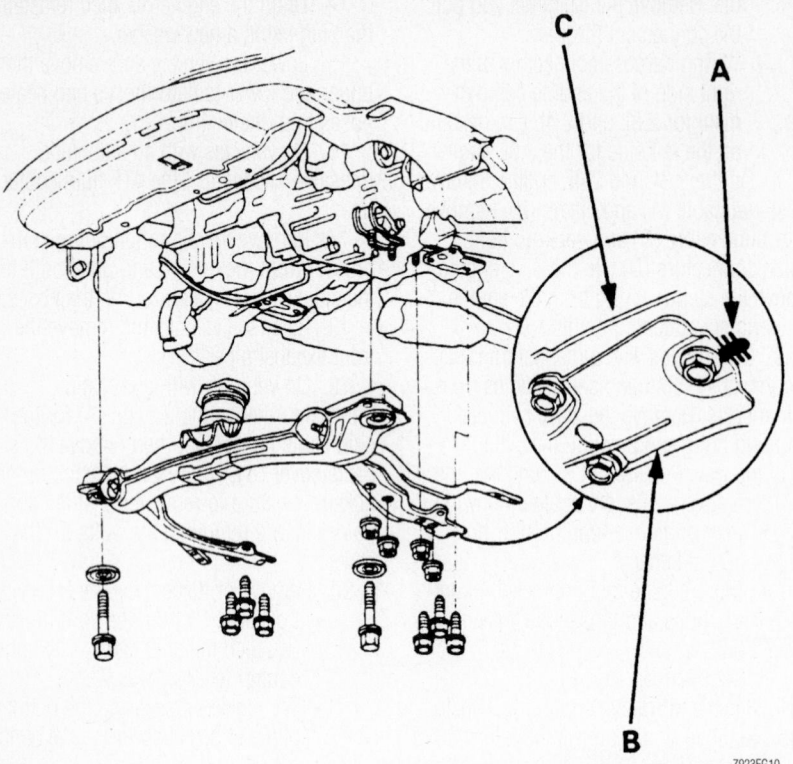

Mark the location of the front beams (A) on the rear beams (B) before removing the subframe—3.0L Accord

7923FG10

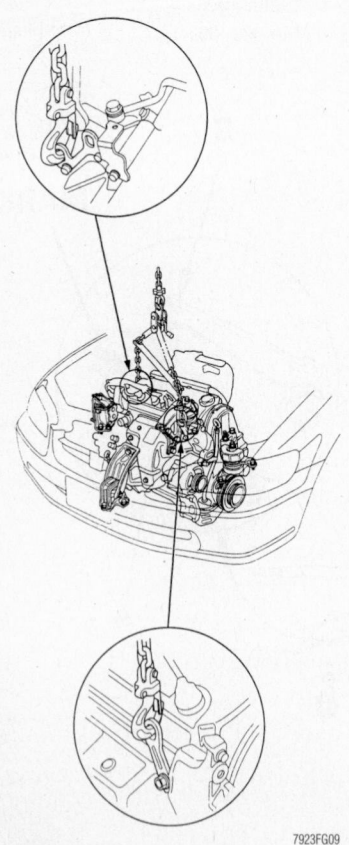

7923FG09

Engine lifting points—2.3L Accord shown, 2.4L similar

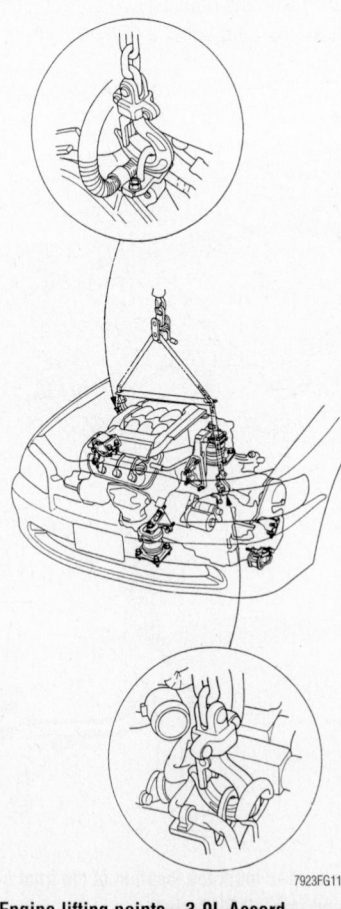

7923FG11

Engine lifting points—3.0L Accord

(A) on the rear beams (B). Remove the 4 bolts and the subframe.

22. Lower the vehicle about half way and attach a chain hoist to the engine lifting points as shown. Apply slight upward pressure to the engine/transaxle assembly.

23. Remove the remaining engine and transaxle mounting brackets.

24. Lower the engine about 6 in. (150mm) and check that the engine/transaxle is free of any hoses, cables or wiring.

25. Lower the assembly completely and remove it from under the vehicle.

To install:

26. Lift the engine into position and install the engine mounting brackets. Tighten the retainers as follows:

   a. On the 2.3L and 2.4L engines, tighten the engine mounting bolts and nuts to 40 ft. lbs. (54 Nm).

   b. On the 3.0L, tighten the bolts to 28 ft. lbs. (38 Nm).

27. On the 3.0L engine, install the air conditioning compressor. Tighten the bolts to 16 ft. lbs. (22 Nm).

28. Install the transaxle mounting bracket, and tighten the retainers as follows:

   a. On the 2.3L and 2.4L engines, tighten the nuts to 28 ft. lbs. (38 Nm) and the through-bolt to 40 ft. lbs. (54 Nm).

   b. On the 3.0L engine, tighten the bolts to 28 ft. lbs. (38 Nm).

29. Install the sub-frame in its original position, and tighten the retainers as follows:

   a. On the 2.3L and 2.4L engines, tighten the rear bolts to 47 ft. lbs. (64 Nm) and the front bolts to 76 ft. lbs. (103 Nm).

   b. On the 3.0L engine, tighten the rear bolts to 40 ft. lbs. (54 Nm), front bolts to 76 ft. lbs. (103 Nm) and the nuts to 28 ft. lbs. (38 Nm).

30. On the 2.3L and 2.4L engines, install or connect the following:

- Radius rod bolts: 119 ft. lbs. (162 Nm)
- Rear mount bracket: 40 ft. lbs. (54 Nm)
- Stiffener. Tighten the through-bolt to 47 ft. lbs. (64 Nm) for manual transaxles or the nut and bolt to 28 ft. lbs. (38 Nm) for automatic transaxles.
- 3 front mounting bracket bolts: 28 ft. lbs. (38 Nm). Then, tighten the through-bolt to 47 ft. lbs. (64 Nm).
- Air conditioning compressor: 16 ft. lbs. (22 Nm)

31. On the 3.0L engine, install or connect the following:

- Radius rod bolts: 119 ft. lbs. (162 Nm)
- Front mounting bracket support nut: 40 ft. lbs. (54 Nm)
- Rear mounting bracket nut and bolt. Tighten the nut to 40 ft. lbs. (54 Nm) and the bolt to 28 ft. lbs. (38 Nm).
- Side mounting bracket. Tighten the bolts to 40 ft. lbs. (54 Nm) and the through-bolt to 40 ft. lbs. (54 Nm).
- Exhaust system
- Shift linkage, if equipped with an automatic transaxle

32. The remainder of the installation is the reverse of the removal.

33. Refill and bleed the cooling system.

✷✷ WARNING

Operating the engine without the proper amount and type of engine oil will result in severe engine damage.

34. Fill the engine with the correct amount of oil.

35. Install the battery if removed. Start the engine and check for leaks.

S2000

1. Before servicing the vehicle, refer to the precautions in the beginning of this section.

2. Drain the cooling system.

3. Relieve the fuel system pressure.

4. Remove or disconnect the following:
- Battery
- Front wheels
- Transmission
- Engine Control Module (ECM) connectors and the main wire harness connector. Pass the connectors through the cowl panel.
- Vacuum tank
- Throttle cable
- Electrical Power Steering (EPS) control unit
- Battery cable at the main underhood fuse/relay box
- Battery cable at the auxiliary fuse box
- Ground cable and harness clamps
- Fuel lines
- Brake booster vacuum line
- Evaporative Emissions (EVAP) canister hose
- Front motor mount and bracket
- Heater hoses
- Radiator hoses
- Left motor mount
- Right motor mount bracket

5. Carefully raise the engine out of the vehicle.

To install:

6. Installation is the reverse of the removal procedure, while using the following torque values:
- Right motor mount bracket bolts: 28 ft. lbs. (38 Nm)
- Motor mount nuts: 40 ft. lbs. (54 Nm)
- Front motor mount bolts: 16 ft. lbs. (22 Nm)

Water Pump

REMOVAL & INSTALLATION

2.2L and 2.3L Engines

➡**The original radio contains a coded anti-theft circuit. Obtain the security code number before disconnecting the battery cables.**

1. Before servicing the vehicle, refer to the precautions in the beginning of this section.

2. Remove or disconnect the following:
- Negative battery cable

3. Drain the cooling system.

4. Remove or disconnect the following:
- Accessory drive belts, the valve cover, and the upper timing belt cover

5. Set the timing at Top Dead Center (TDC)/compression for No. 1 piston.

6. Remove or disconnect the following:
- Crankshaft pulley and lower timing belt cover
- Timing belt. Replace the timing belt if it is contaminated with oil or coolant or shows any signs of wear and damage.
- Crankshaft Speed Fluctuation (CKF) sensor bracket and move the sensor out of the way, if equipped. Cover the sensor with a shop towel to keep coolant off of it.
- Water pump from the engine block.

To install:

7. Clean the water pump and O-ring mating surfaces before installation.

8. Install or connect the following:
- Water pump with a new O-ring. Coat only the bolt threads with liquid gasket and tighten them to 108 inch lbs. (12 Nm).
- Timing belt. Be sure it is fitted and adjusted properly.
- CKF sensor, if equipped, and tighten the bracket bolts to 108 inch lbs. (12 Nm).
- Lower belt cover and crankshaft pulley
- Upper timing belt cover, the valve cover, and the accessory drive belts

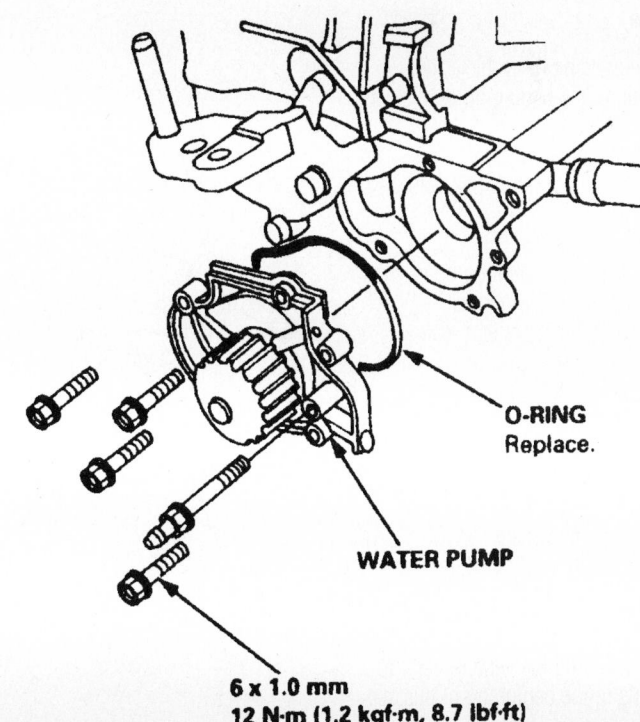

O-RING
Replace.

WATER PUMP

6 x 1.0 mm
12 N·m (1.2 kgf·m, 8.7 lbf·ft)

7923FG12

Water pump—2.2L and 2.3L engines

9. Be sure the cooling system drain plug is closed. Refill and bleed the cooling system.

10. Connect the negative battery cable and enter the radio security code.

11. Start the engine, allow it to reach normal operating temperature, and check for coolant leaks.

12. If equipped with 4WS, and turn the steering wheel lock-to-lock to reset the 4WS control unit.

2.0L Engine

1. Before servicing the vehicle, refer to the precautions in the beginning of this section.

2. Drain the cooling system.

3. Remove or disconnect the following:
- Negative battery cable
- Accessory drive belt
- Water pump pulley
- Water pump

To install:

4. Install or connect the following:
- Water pump with a new O ring seal. Tighten the 8mm bolts to 16 ft. lbs. (22 Nm) and the 10mm bolt to 33 ft. lbs. (44 Nm).
- Water pump pulley and tighten the bolts to 10 ft. lbs. (14 Nm)
- Accessory drive belt
- Negative battery cable

5. Fill the cooling system.

6. Start the engine and check for leaks.

2.4L Engine

1. Before servicing the vehicle, refer to the precautions in the beginning of this section.

2. Drain the cooling system.

3. Remove or disconnect the following:
- Crankshaft pulley
- 6 water pump mounting bolts, then the pump and O-ring seal

4. Clean the seal groove and mating surfaces.

5. Install or connect the following:
- Water pump, with a new seal. Tighten the bolts to 104 inch lbs. (12 Nm).
- Crankshaft pulley

6. Fill the cooling system.

7. Start the engine and check for leaks.

3.0L Engine

1. Before servicing the vehicle, refer to the precautions in the beginning of this section.

2. Remove or disconnect the following:

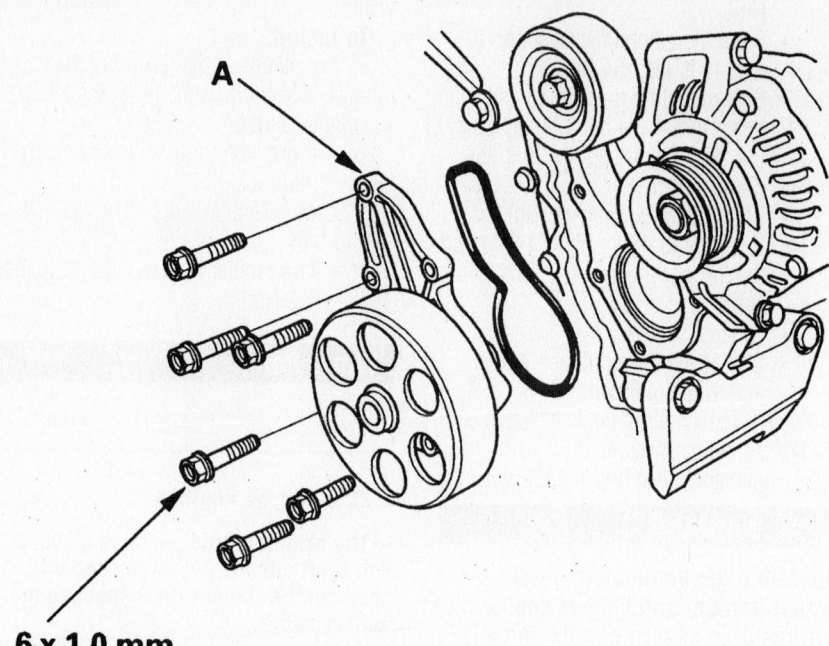

6 x 1.0 mm
12 N·m (1.2 kgf·m, 8.7 lbf·ft)

43256-ACCO-G02

Water pump mounting—2.4L engine

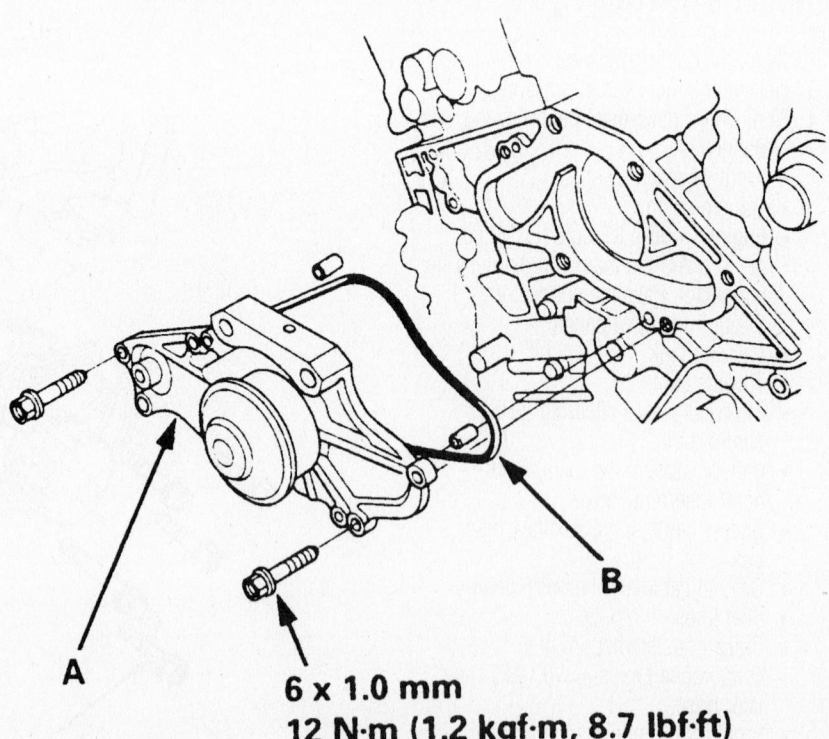

6 x 1.0 mm
12 N·m (1.2 kgf·m, 8.7 lbf·ft)

7923FG14

Exploded view of the water pump mounting—3.0L engine

- Timing belt.
- Timing belt tensioner
- 5 water pump mounting bolts, then remove the pump and seal

To install:

3. Clean the seal groove and mating surfaces.
4. Install or connect the following:
   - Water pump, with a new seal. Tighten the bolts to 104 inch lbs. (12 Nm).
   - Timing belt tensioner
   - Timing belt.
5. Refill the cooling system.
6. Start the engine and check for leaks.
7. Top off the cooling system if necessary after the engine has cooled.

Heater Core

REMOVAL AND INSTALLATION

Accord

➡Make sure to acquire the anti-theft code form the radio and write down the frequencies for the radio's preset buttons.

1. Disconnect the negative battery cable.

CAUTION

After disconnecting the negative battery cable, wait for at least 3 minutes for the air bag module to deplete its energy before working the on the instrument panel or steering wheel.

2. Drain the cooling system into a clean container for reuse.
3. In the engine compartment, open the heater valve cable clamp and disconnect the cable from the heater valve arm. Then, turn the heater valve to the fully opened position.

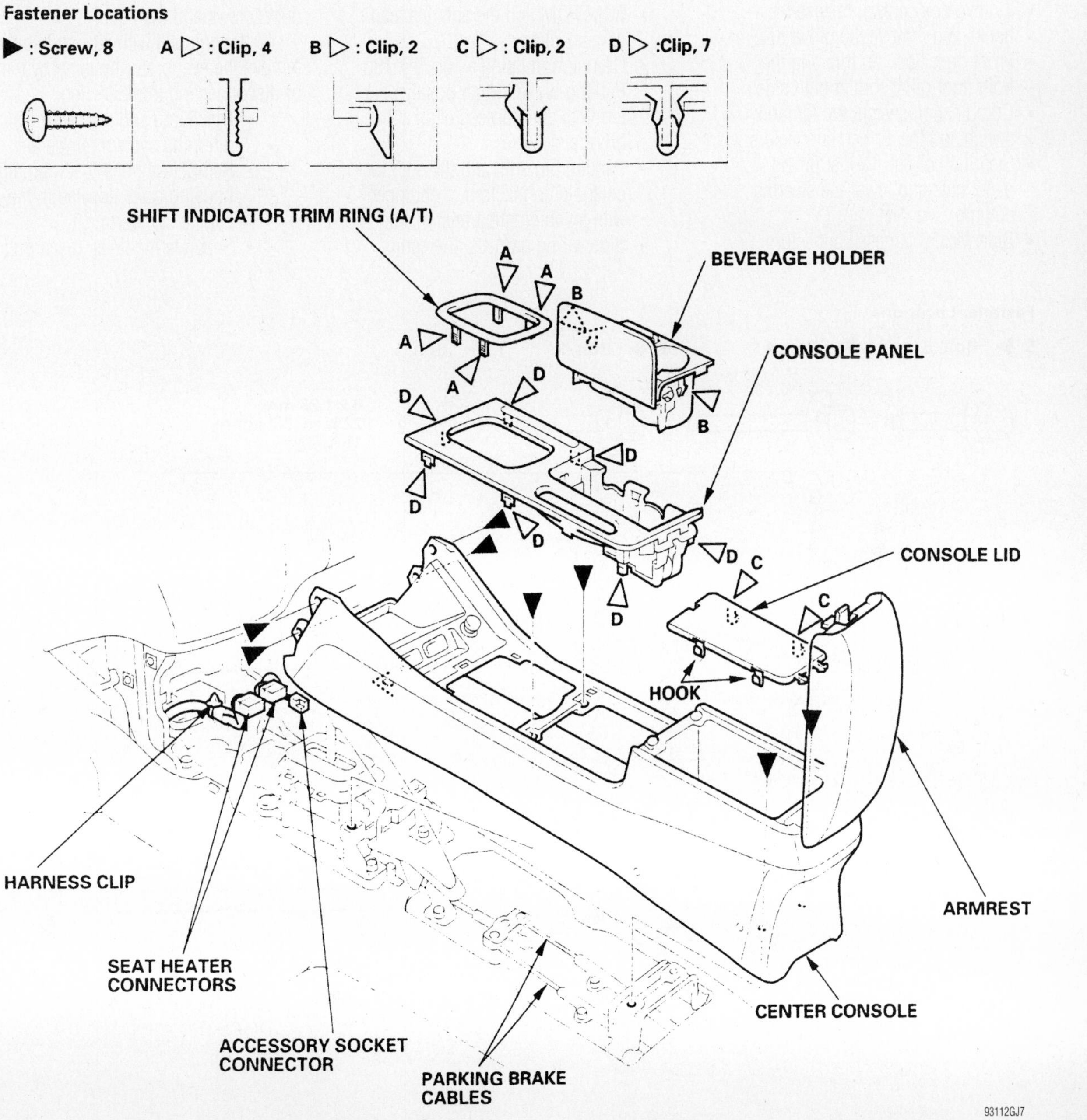

Fastener Locations
▶ : Screw, 8
A ▷ : Clip, 4
B ▷ : Clip, 2
C ▷ : Clip, 2
D ▷ :Clip, 7
SHIFT INDICATOR TRIM RING (A/T)
BEVERAGE HOLDER
CONSOLE PANEL
CONSOLE LID
HOOK
HARNESS CLIP
ARMREST
SEAT HEATER CONNECTORS
CENTER CONSOLE
ACCESSORY SOCKET CONNECTOR
PARKING BRAKE CABLES
93112GJ7

Exploded view of the center console and related components—Accord

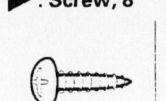

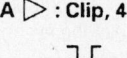

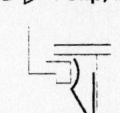

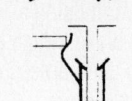

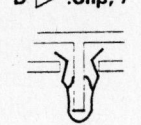

4. Remove or disconnect the following:
- Heater hoses from the heater core
- Heater housing-to-chassis nut

➡ **When removing the heater housing nut, be careful not to damage or bend the fuel lines, the brake lines etc.**

- Center console

5. Remove the instrument panel by removing or disconnecting the following:
- Center lower cover
- Passenger's side lower cover
- Lower glove box screws
- Hooks at the top inner side of the glove box by placing a flat tipped screwdriver in the cap notch and pry out the cap
- 4 glove box screws, release the hooks and remove the glove box
- Front door opening trim and the front door pillar trim at both sides
- Combination switch, the ignition switch and the air bag connectors
- Steering column-to-instrument panel nuts and lower the steering column
- Side wiring harness connectors, cabin wire harness and door wire harness from the fuse box at the driver's side
- Instrument panel wiring harness connector
- Wiring harness, the steering hanger beam wiring harness and the brake switch connectors under the dash
- Clutch switch connectors, if equipped with a manual transmission
- Pull back the carpet at the passenger's side
- SRS wiring harness, the ECM/PCM, the air mix control motor, the evaporator temperature sensor and the antenna lead
- ECM/PCM and the antenna lead harness clips
- Ground bolt using a Torx®bit T30
- Parking brake switch positive (+) terminal and harness clips at the driver's side
- Parking pin shift and the shift lock solenoid connectors, if equipped with an automatic transmission
- Side wiring harness, the cabin wiring harness, the roof wiring harness and the door wiring harness from the passenger's fuse box at the passenger's side
- Instrument panel wiring harness
- Instrument pane harness, the blower motor and the recirculation control motor connectors under the passenger's side
- Harness, the harness holder and the connector clips
- Instrument panel-to-chassis cap and bolts
- Instrument panel from the guide pins and remove it

6. If not equipped with air conditioning, remove the blower motor-to-heater housing duct screws and the duct.

7. If equipped with air conditioning, remove the evaporator housing by removing or disconnecting the following:
- Discharge and recover the air conditioning system refrigerant
- Refrigerant lines-to-evaporator housing bolts, located in the engine compartment
- Separate the lines, discard the O-

Fastener Locations

B ▶ : Bolt, 6   C ▶ : Bolt, 4   D ▶ : Bolt, 3   E ▶ : Bolt, 1

8 x 1.25 mm
22 N·m (2.2 kgf·m, 16 lbf·ft)

View of the instrument panel bolt locations—Accord

93112GJ8

rings and plug the openings to prevent contamination
- Evaporator housing's temperature sensor connector
- Evaporator housing-to-chassis nut/bolts and remove the evaporator housing

8. Remove or disconnect the following:
- Heater housing-to-chassis bolts and the heater housing
- Heater core-to-heater housing bracket screws and the brackets
- Heater core from the heater housing

To install:
9. Install or connect the following:
- Heater core to the heater housing
- Heater core-to-heater housing bracket and the bracket screws
- Heater housing and the heater housing-to-chassis bolts

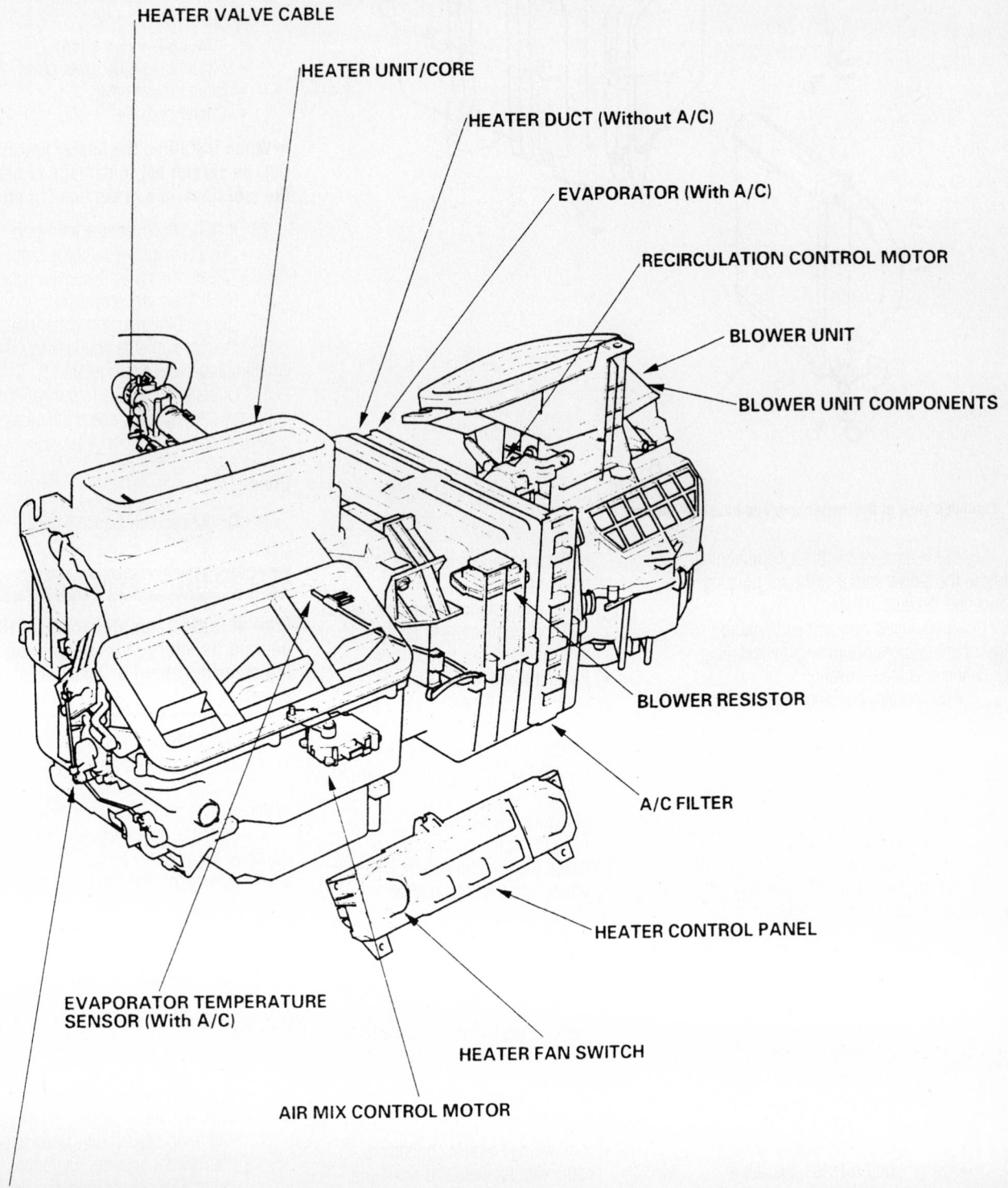

HEATER VALVE CABLE

HEATER UNIT/CORE

HEATER DUCT (Without A/C)

EVAPORATOR (With A/C)

RECIRCULATION CONTROL MOTOR

BLOWER UNIT

BLOWER UNIT COMPONENTS

BLOWER RESISTOR

A/C FILTER

HEATER CONTROL PANEL

HEATER FAN SWITCH

AIR MIX CONTROL MOTOR

EVAPORATOR TEMPERATURE SENSOR (With A/C)

MODE CONTROL MOTOR

93112GJ0

View of the heater housing, evaporator housing and related components—Accord

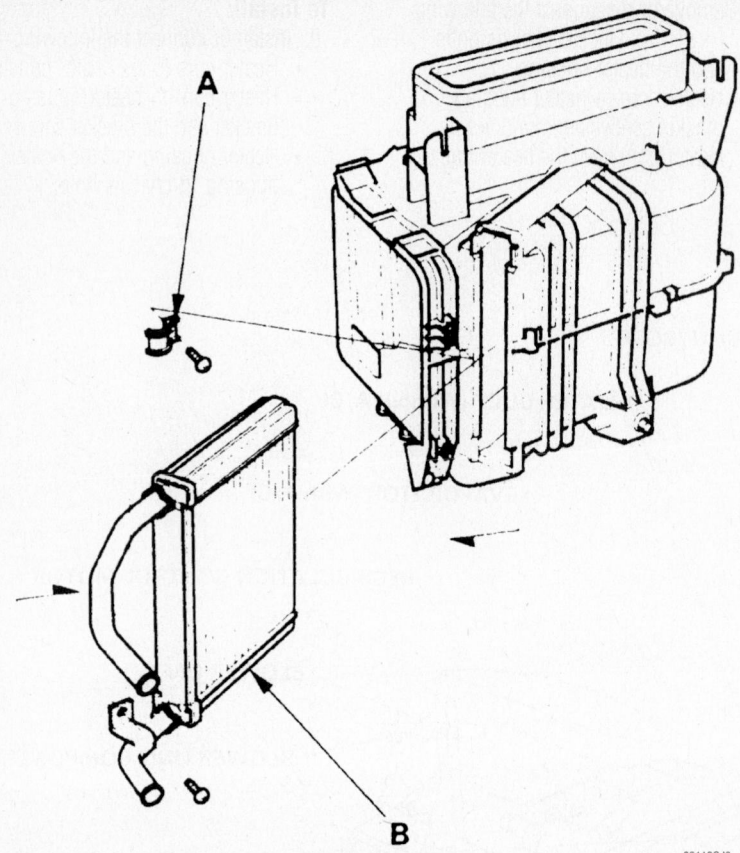

A

B

93112GJ9

Exploded view of the heater core and housing—Accord

10. If not equipped with air conditioning, install the blower motor-to-heater housing duct and the duct screws.

11. If equipped with air conditioning, install the evaporator housing by installing or connecting the following:
- Evaporator housing and the evaporator housing-to-chassis nut/bolts
- Evaporator housing's temperature sensor connector
- Refrigerant lines using new O-rings
- Refrigerant lines-to-evaporator housing bolts, located in the engine compartment

12. Install the instrument panel by installing or connecting the following:
- Instrument panel to the guide pins
- Instrument panel-to-chassis cap and bolts
- Harness holder and the connector clips
- Instrument pane harness, the blower motor and the recirculation control motor connectors under the passenger's side
- Instrument panel wiring harness
- Side wiring harness, the cabin wiring harness, the roof wiring harness and the door wiring harness

to the passenger's fuse box at the passenger's side
- Parking pin shift and the shift lock solenoid connectors, if equipped with an automatic transmission
- Parking brake switch positive (+) terminal and harness clips a the driver's side
- Ground bolt using a Torx bit T30
- ECM/PCM and the antenna lead harness clips
- SRS wiring harness, the ECM/PCM, the air mix control motor, the evaporator temperature sensor and the antenna lead
- Move back the carpet.
- Clutch switch connectors, if equipped with a manual transmission
- Wiring harness, the steering hanger beam wiring harness and the brake switch connectors, located under the dash
- Instrument panel wiring harness connector
- Side wiring harness connectors, cabin wire harness and door wire harness from the fuse box, located at the driver's side

- Steering column and the steering column-to-instrument panel nuts
- Combination switch, the ignition switch and the air bag connectors
- Front door pillar trim and the front door opening trim on both sides
- Glove box and the 4 glove box screws
- Cap at the top inner side of the glove box
- Lower glove box screws
- Passenger's side lower cover
- Center lower cover
- Center console

➡When installing the heater housing nut, be careful not to damage or bend the fuel lines, the brake lines or etc.

13. Install or connect the following:
- Heater housing-to-chassis nut
- Heater hoses to the heater core
14. Refill the cooling system.
15. Connect the negative battery cable.
16. Evacuate, charge and leak test the air conditioning system refrigerant.
17. Operate the engine to normal operating temperatures; then, check the climate control operation and check for leaks.

Civic

1. Disconnect the negative battery cable.

✳✳ CAUTION

Wait at least 3 minutes for the SRS to deplete its energy before working on the steering wheel or instrument panel.

2. In the engine compartment, remove the heater valve cable clamp; then, disconnect the heater valve cable and rotate the heater valve to the fully open position.

3. Drain the engine coolant into a clean container for reuse.

4. Remove or disconnect the following:
- Heater hoses from the heater unit
- Heater housing-to-chassis nut
- Instrument panel.

5. If not equipped with air conditioning, remove the wiring harness from the heater duct; then, remove the 2 screws and the heater duct.

6. If equipped with air conditioning, remove or disconnect the following:
- Discharge and recover the air conditioning system refrigerant
- Refrigerant lines-to-evaporator bolts. Disconnect the lines. Discard the O-rings. Plug the openings to prevent contamination

- Thermostat electrical connector and the wiring harness from the evaporator
- 4 evaporator housing screws, the bolt and nut
- Drain hose and remove the evaporator housing

7. On 2001 models, disconnect the electrical connectors from the blower motor, the power transistor, the blower motor high relay and the recirculation control motor.

8. Remove or disconnect the following:
- Electrical connectors from the mode control motor and the air mix

control motor (for 2001); then, remove the electrical harness clips and wiring harness from the heater housing
- Heater duct clip
- 2 heater housing-to-chassis nuts and the heater housing

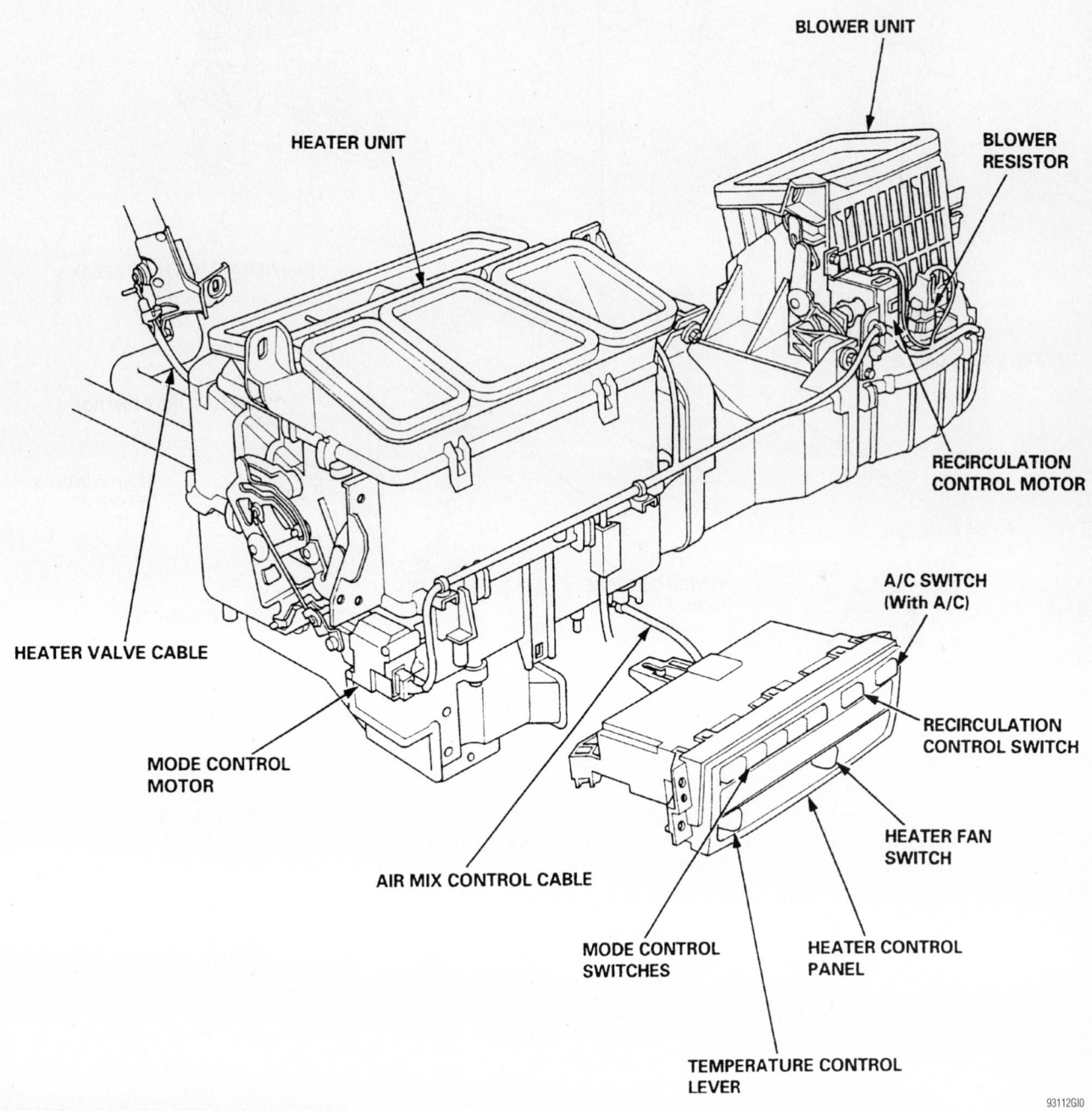

BLOWER UNIT

HEATER UNIT

BLOWER RESISTOR

RECIRCULATION CONTROL MOTOR

A/C SWITCH (With A/C)

RECIRCULATION CONTROL SWITCH

HEATER VALVE CABLE

MODE CONTROL MOTOR

AIR MIX CONTROL CABLE

HEATER FAN SWITCH

MODE CONTROL SWITCHES

HEATER CONTROL PANEL

TEMPERATURE CONTROL LEVER

93112GI0

View of the heater housing, evaporator housing and related components—Civic

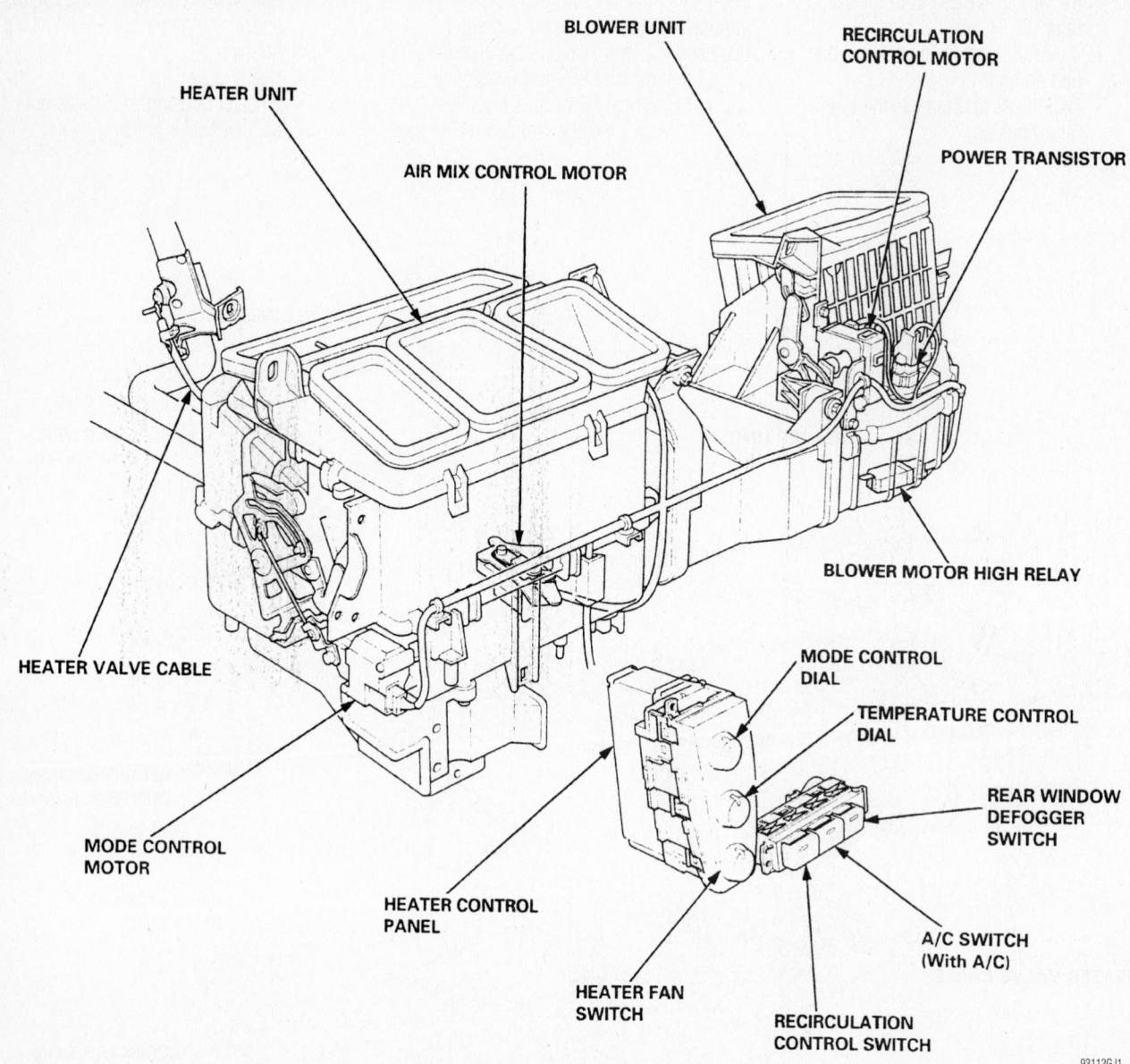

View of the heater housing, evaporator housing and related components—Civic

- Heater core cover-to-heater housing screws, the cover, the clamp and the heater core

To install:

9. Install or connect the following:
- Heater core, the clamp, the cover and the heater core cover-to-heater housing screws
- Heater housing and the 2 heater housing-to-chassis nuts
- Heater duct clip
- Electrical connectors to the mode control motor and the air mix control motor (for 2001); then, install the electrical harness clips and wiring harness to the heater housing

10. On 2001 models, connect the electrical connectors to the blower motor, the power transistor, the blower motor high relay and the recirculation control motor.

11. If not equipped with air conditioning, install the heater duct, the 2 screws and the wiring harness to the heater duct.

12. If equipped with air conditioning, install or connect the following:
- Drain hose and install the evaporator housing
- 4 evaporator housing screws, the bolt and nut
- Thermostat electrical connector and the wiring harness to the evaporator
- Refrigerant lines-to-evaporator

bolts, connect the lines using new O-rings
- Evacuate, charge and leak test the air conditioning system refrigerant.

13. Install or connect the following:
- Instrument panel
- Heater housing-to-chassis nut
- Heater hoses to the heater unit

14. Refill the cooling system.

15. In the engine compartment, install the heater valve cable clamp and connect the heater valve cable.

16. Connect the negative battery cable.

17. Operate the engine to normal operating temperatures; then, check the climate control operation and check for leaks.

Prelude

➡Make sure to acquire the anti-theft code form the radio and write down the frequencies for the radio's preset buttons.

1. Disconnect the negative battery cable.

✳✳ CAUTION

Wait at least 3 minutes for the SRS to deplete its energy before working on the steering wheel or instrument panel.

2. In the engine compartment, remove the heater valve cable clamp; then, disconnect the heater valve cable and rotate the heater valve to the fully open position.

3. Drain the engine coolant into a clean container for reuse.

4. Remove or disconnect the following:
   • Heater hoses from the heater unit
   • Heater housing-to-chassis nuts
   • Instrument panel
   • Steering hanger beam mounting bolts and the steering hanger beam

5. Discharge and recover the air conditioning system refrigerant

6. Remove the evaporator housing by removing or disconnecting the following:
   • Refrigerant lines. Discard the O-rings. Plug the openings to prevent contamination.
   • Thermostat electrical connector
   • Drain hose and remove the evaporator housing.

7. Remove or disconnect the following:
   • Electrical connector from the mode control motor
   • Wiring harness clips from the heater housing
   • Heater housing-to-chassis nuts and the heater housing
   • Heater housing screws and separate the housings
   • Heater core from the heater housing

To install:

8. Install or connect the following:
   • Heater core to the heater housing
   • Assemble the housings and install the heater housing screws
   • Heater housing and the heater housing-to-chassis nuts
   • Wiring harness clips to the heater housing
   • Electrical connector to the mode control motor

9. Install the evaporator housing:
   • Evaporator housing and connect the drain hose

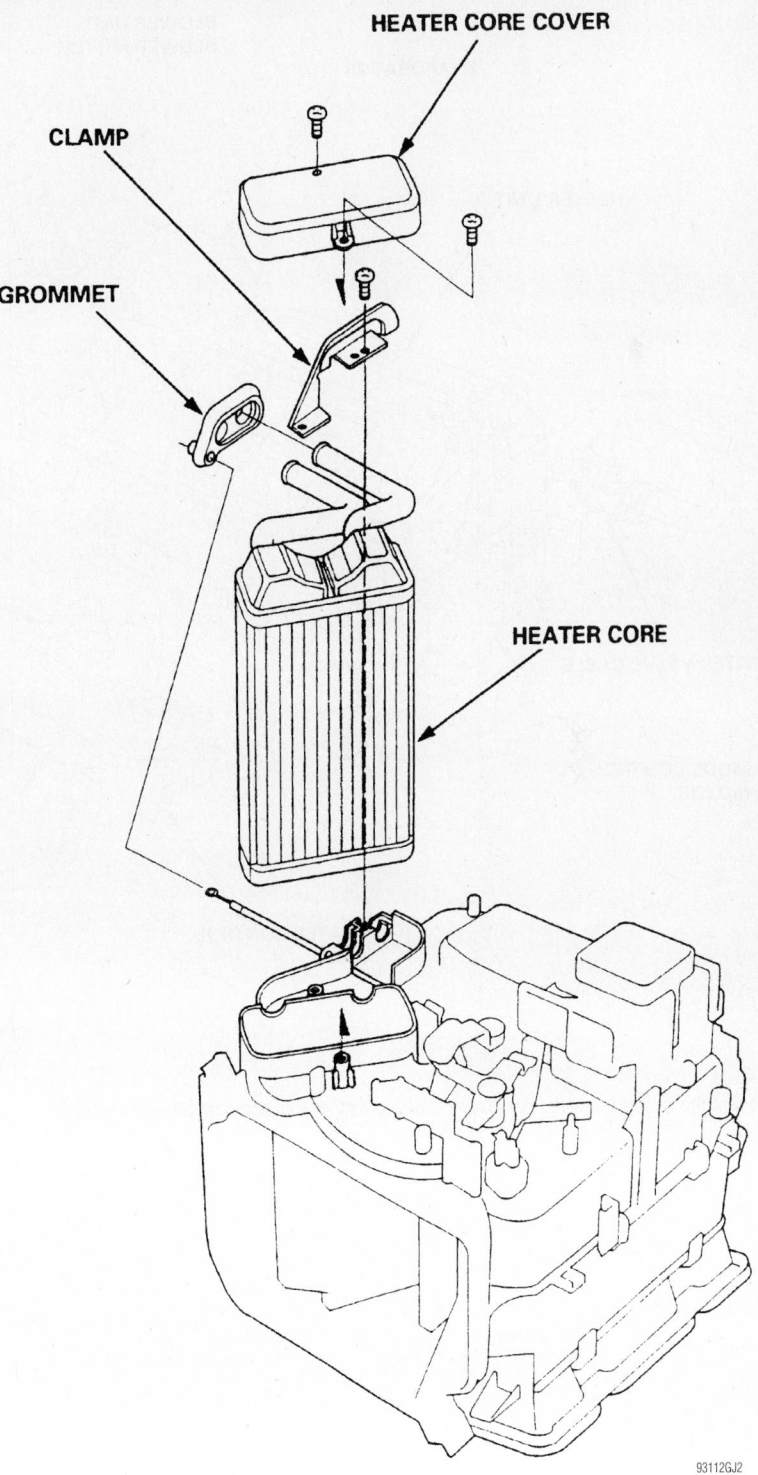

Exploded view of the heater core—Civic

   • Evaporator housing-to-chassis screws, the bolt and nuts
   • Thermostat electrical connector and the wiring harness to the evaporator
   • Refrigerant lines using new O-rings
   • Evacuate, charge and leak test the air conditioning system refrigerant

10. Install or connect the following:
   • Steering hanger beam and the steering hanger beam mounting bolts
   • Instrument panel
   • Heater housing-to-chassis nuts
   • Heater hoses to the heater unit

11. Refill the cooling system.

12. In the engine compartment, Install the heater valve cable clamp; then, connect the heater valve cable.

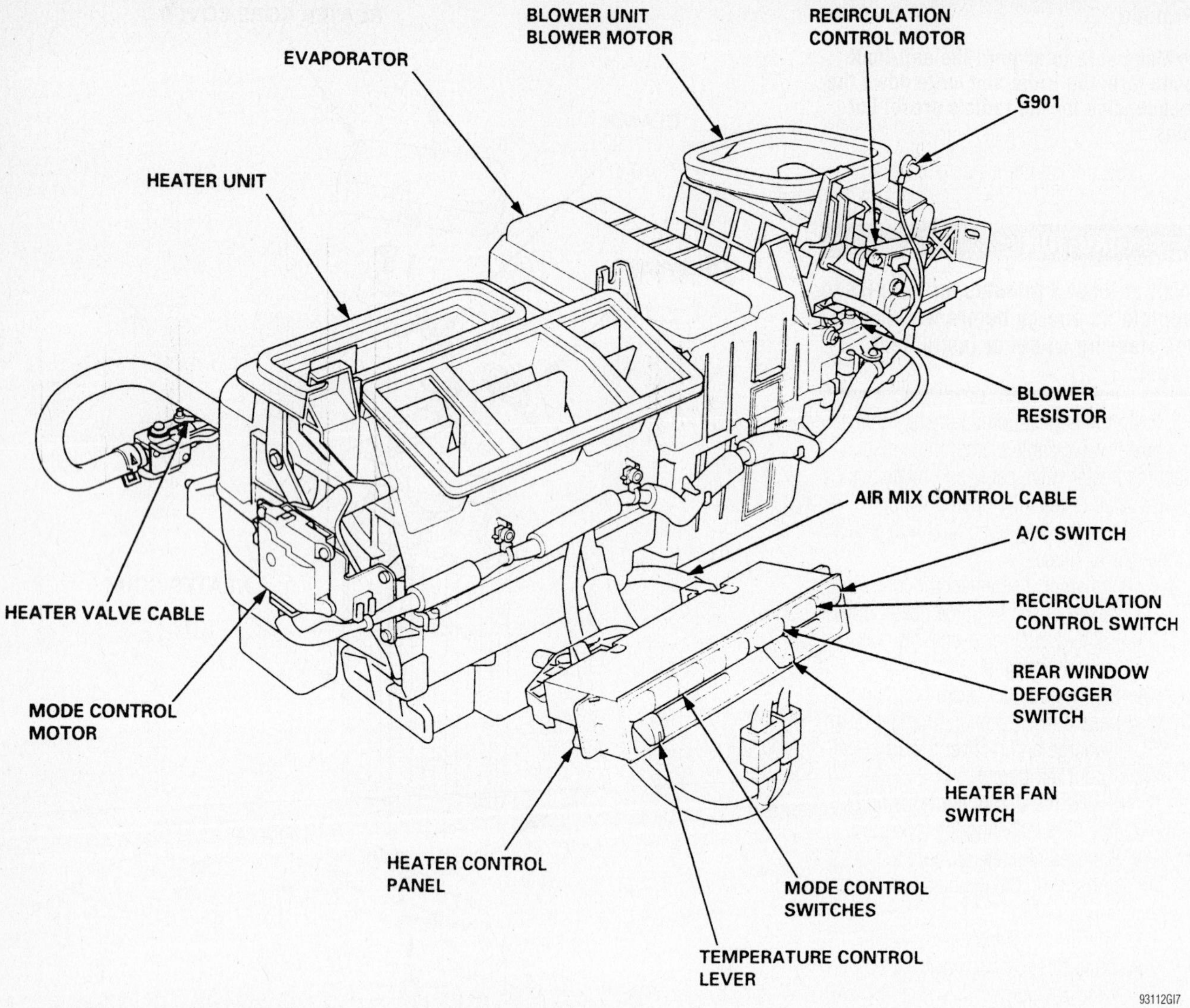

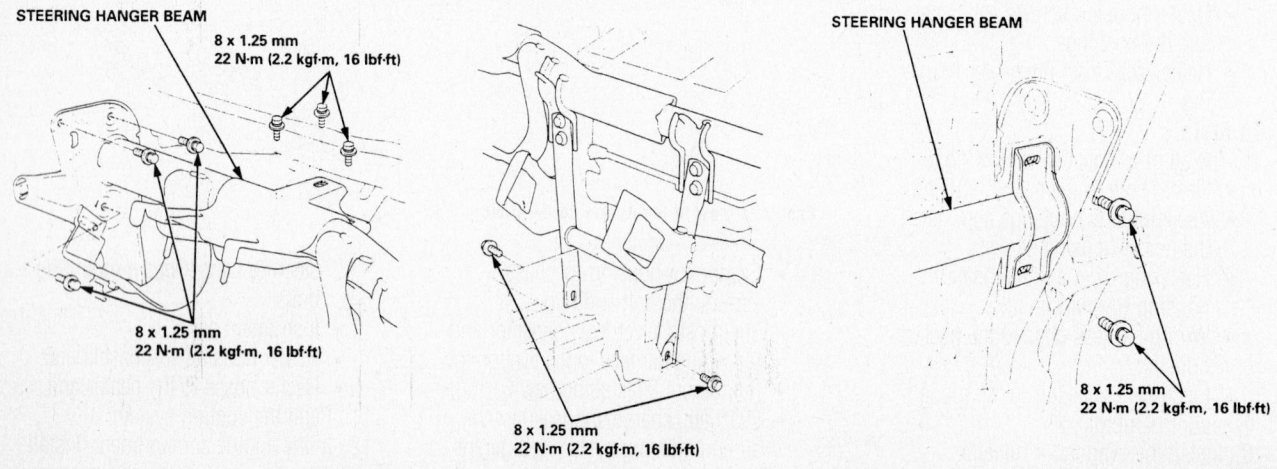

BLOWER UNIT
BLOWER MOTOR

RECIRCULATION
CONTROL MOTOR

G901

EVAPORATOR

HEATER UNIT

BLOWER
RESISTOR

AIR MIX CONTROL CABLE

A/C SWITCH

RECIRCULATION
CONTROL SWITCH

HEATER VALVE CABLE

REAR WINDOW
DEFOGGER
SWITCH

HEATER FAN
SWITCH

MODE CONTROL
MOTOR

HEATER CONTROL
PANEL

MODE CONTROL
SWITCHES

TEMPERATURE CONTROL
LEVER

93112GI7

View of the heater housing, evaporator housing and related components—Prelude

STEERING HANGER BEAM

STEERING HANGER BEAM

8 x 1.25 mm
22 N·m (2.2 kgf·m, 16 lbf·ft)

8 x 1.25 mm
22 N·m (2.2 kgf·m, 16 lbf·ft)

8 x 1.25 mm
22 N·m (2.2 kgf·m, 16 lbf·ft)

8 x 1.25 mm
22 N·m (2.2 kgf·m, 16 lbf·ft)

93112GI8

View of the steering hanger beam and related components—Prelude

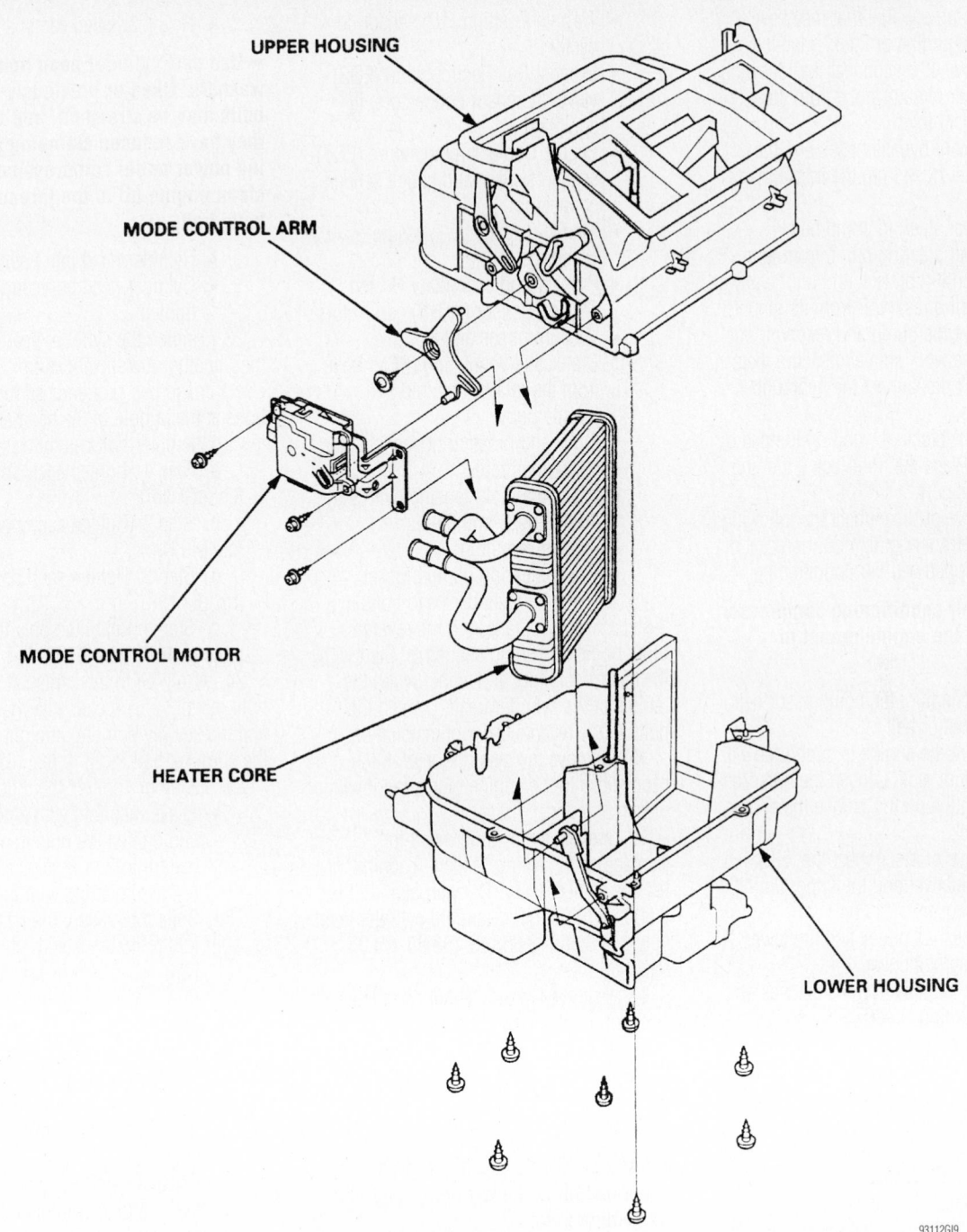

UPPER HOUSING

MODE CONTROL ARM

MODE CONTROL MOTOR

HEATER CORE

LOWER HOUSING

93112GI9

Exploded view of the heater core, the heater housing and related components—Prelude

13. Connect the negative battery cable.

14. Operate the engine to normal operating temperatures; then, check the climate control operation and check for leaks.

Cylinder Head

REMOVAL & INSTALLATION

➡**The radio may contain a coded theft protection circuit. Always obtain the code number before disconnecting the** **battery. If the vehicle is equipped with 4WS, the steering control unit is shut down when the battery is disconnected. After connecting the battery, turn the steering wheel lock-to-lock to reset the steering control unit.**

1.7L Engines

1. Before servicing the vehicle, refer to the precautions in the beginning of this section.

2. Be sure the cylinder head is cool to the touch before beginning the removal pro-

cedure. The coolant temperature must be below 100°F (38°C).

3. Remove or disconnect the following:
   • Negative battery cable

4. Drain the cooling system.

5. Remove or disconnect the following:

➡**Label the wires before disconnecting them.**

   • Ignition wires
   • Air intake duct and the air cleaner assembly

6. Relieve the fuel pressure.

7. Clean up any fuel that may have spilled on the engine or intake manifold.

8. Remove or disconnect the following:
- Upper radiator hose from the coolant inlet
- Coolant bypass hoses and the heater hose from the intake manifold
- Power steering pump belt
- Power steering pump from its mounting bracket and lift the power steering reservoir from its mount. Move the pump and reservoir out of the work area and secure them. Don't disconnect the hydraulic lines.

9. Place a block of wood on the pad of a floor jack. Place the floor jack under the engine for support.

10. Remove or disconnect the following:
- Left-front engine mount bracket, if equipped with air conditioning

➡**Slip the air conditioning compressor belt around the engine mount to remove it.**

- Air conditioning compressor belt
- Alternator belt

11. Be sure the engine is supported with the padded floor jack. Loosen the nuts from left side engine mount. Remove the engine mount bracket.

12. Remove or disconnect the following:
- Valve cover and the upper timing belt cover
- Crankshaft pulley and the lower timing belt cover
- Dipstick tube from its catches on the timing cover
- Timing belt.

13. With the timing belt removed, inspect the water pump and replace it if necessary.

14. Remove or disconnect the following:
- Distributor, if equipped
- Camshaft sprocket
- Fuel lines from the intake manifold fuel rail. Immediately plug the lines to prevent fuel leakage and contamination
- Throttle cable from the linkage by first loosening its locknut, then slipping it out of its holder.

➡**Label the electrical connectors, before disconnecting them.**

- Fuel injector wiring harness connectors
- VTEC solenoid valve and pressure switch connectors, if equipped
- Idle Air Control (IAC) valve connector

- Throttle Position (TP) sensor connector
- Exhaust Gas Recirculation (EGR) valve lift sensor connectors, if equipped
- Engine Coolant Temperature (ECT) sensor, switch, and gauge sender connectors
- Manifold Absolute Pressure (MAP) sensor connector
- Primary and secondary Heated Oxygen Sensor (HO2S) connectors
- Vacuum hoses and Positive Crankcase Ventilation (PCV) hose from the intake manifold and throttle body
- EVAP and breather hoses from the intake manifold
- Intake manifold together with the throttle body and plenum
- Exhaust manifold
- Power steering pump bracket

15. Loosen the cylinder head bolts in a 3-step crisscross pattern in the reverse order of the tightening sequence. Start with the outermost bolts and work toward the middle of the cylinder head. Loosen the bolts in the reverse order of installation.

16. Remove the cylinder head. If the head sticks to the engine block, tap it with a plastic or wooden mallet.

17. Inspect the cylinder head for warpage and cracking. Repair, machine, or replace as necessary. The warpage limit is 0.002 in. (0.05mm). Standard cylinder head height is 3.659–3.663 in. (92.95–93.05 mm).

18. Remove the old cylinder head gasket and thoroughly clean the mating surfaces.

19. Cover the engine block with a sheet of plastic to keep out dust and foreign objects.

To install:

➡**Use new O-ring, seals, and gaskets when installing the cylinder head and its components.**

20. Be sure the cylinder head and the engine block surfaces are clean, level, and straight.

21. Be sure the cylinder head dowel pins and control orifice are aligned. Clean the oil control orifice and reinstall it with a new O-ring.

22. Install or connect the following:
- New head gasket onto the engine block
- Camshaft, if removed, with the keyway facing up so that the engine will remain at Top Dead Center (TDC)/compression for the No. 1 cylinder

- New lubricated camshaft seal

➡**Use new cylinder head bolts and washers. Used or previously-tightened bolts may be stretched, and therefore they have reduced clamping and sealing power under compression. Apply clean engine oil to the threads of each head bolt.**

- Cylinder head into position
- Cylinder head bolts and hand-tighten

23. Tighten the cylinder head bolts to their final torque specification in 4 steps. Use a crisscross sequence starting with the bolts at the middle of the head and working toward the outer bolts as follows:
   a. Step 1: Tighten each bolt to 14 ft. lbs. (20 Nm).
   b. Step 2: Tighten each bolt to 36 ft. lbs. (49 Nm).
   c. Step 3: Tighten each bolt to 49 ft. lbs. (67 Nm).
   d. Step 4: Retighten only the 2 center bolts to 49 ft. lbs. (67 Nm).

24. Apply oil to the camshaft sprocket bolt. Install the sprocket with the UP mark and the keyway pointing straight up. Tighten the sprocket bolt to 27 ft. lbs. (37 Nm).

25. Install or connect the following:
- Intake manifold with a new gasket, and tighten the nuts in a crisscross pattern in 2 or 3 steps to 17 ft. lbs. (24 Nm) starting with the inner nuts
- Bolts that secure the intake manifold to its bracket: 17 ft. lbs. (24 Nm)
- Power steering pump bracket: 33 ft. lbs. (44 Nm)
- Exhaust manifold with a new gasket. Apply anti-seize paste to the studs, and tighten the nuts to 23 ft. lbs. (31 Nm) in a crisscross sequence.
- Exhaust manifold to the front exhaust pipe. Tighten the self-locking nuts to 25 ft. lbs. (33 Nm). On vehicles with the D16Y8 engine, tighten the nuts to 40 ft. lbs. (55 Nm).

26. Verify that the engine is at Top Dead Center (TDC)/compression for the No. 1 cylinder.

27. Install or connect the following:
- Timing belt. After the timing belt has been properly tensioned, tighten the adjusting bolt to 33 ft. lbs. (44 Nm).
- Lower timing belt cover
- Crankshaft pulley; tighten its bolt to 134 ft. lbs. (181 Nm)
- Dipstick tube back into its catches

28. Adjust the valves. If equipped with a VTEC engine, also check the rocker arms for free and smooth motion..

29. If equipped with a VTEC engine, remove the VTEC solenoid valve and its filter. Install a new filter, then reinstall the VTEC solenoid valve and tighten its bolt to 108 inch lbs. (12 Nm).

30. Install or connect the following:
    • Distributor, if equipped

31. Be sure all the spark plug tube sealing gaskets are fully seated.

32. Install or connect the following:
    • New gasket onto to the valve cover. Apply liquid gasket to the corner recesses of the gasket. Don't let the sealant cure before installing the valve cover onto the cylinder head.
    • Valve cover. Gently wiggle the valve cover to be sure it is fully seated. Tighten the valve cover bolts in a crisscross pattern to 84 inch lbs. (10 Nm).
    • New spark plugs
    • Ignition wires
    • Upper radiator hose, heater hoses, and intake manifold coolant bypass hoses
    • Intake manifold vacuum lines, PCV, EVAP canister, and breather hoses
    • Fuel lines to the fuel rail. Use new sealing washers on the banjo fitting. Carefully tighten the banjo fitting to 21 ft. lbs. (28 Nm) for the D16Y5 engine, or to 16 ft. lbs. (22 Nm) for all other engines. Tighten the service bolt to 10–11 ft. lbs. (13–15 Nm).
    • Throttle cable. Adjust its tension so the cable has a deflection of 10–12mm (0.39–0.47 in.).

33. Installation of the remaining components is the reverse of removal.

2.0L Engine

1. Before servicing the vehicle, refer to the precautions in the beginning of this section.

2. Drain the cooling system.

3. Relieve the fuel system pressure.

4. Remove or disconnect the following:
    • Negative battery cable
    • Air cleaner housing
    • Accessory drive belt
    • Throttle cable
    • Fuel lines
    • Brake booster vacuum hose
    • Evaporative Emissions (EVAP) canister hose
    • Intake manifold bracket and air hose

• Water outlet housing
• Water bypass hose
• Fuel injector connectors
• Intake Air Temperature (IAT) sensor connector
• Idle Air Control (IAC) valve connector
• Throttle Position (TP) sensor connector
• Manifold Absolute Pressure (MAP) sensor connector
• Engine Coolant Temperature (ECT) sensor connector
• Heated Oxygen (HO$_2$S) sensor connector
• VTEC solenoid valve connector
• VTEC pressure switch connector
• Crankshaft Position (CKP) sensor connector
• Exhaust manifold cover
• Exhaust manifold heat shield
• Exhaust manifold
• Oil level dipstick
• Positive Crankcase Ventilation (PCV) valve and hose
• Ignition coil cover
• Ignition coils
• Intake manifold

• Valve cover
• Timing chain auto-tensioner
• Camshafts
• Timing chain idler gear
• Cylinder head. Loosen the bolts in sequence and in ⅓ turns.

To install:

5. Install the cylinder head with a new gasket. Tighten the bolts in sequence as follows:
    a. Step 1: 22 ft. lbs. (29 Nm).
    b. Step 2: Plus 90 degrees.
    c. Step 3: Plus 90 degrees.
    d. Step 4: If using new cylinder head bolts, add an additional 90 degrees.

6. The remainder of the installation is the reverse of the removal procedure.

2.2L Engine

1. Before servicing the vehicle, refer to the precautions in the beginning of this section.

2. Disconnect the negative battery cable.

3. Bring the No. 1 cylinder to TDC.

4. Drain the engine coolant into a sealable container.

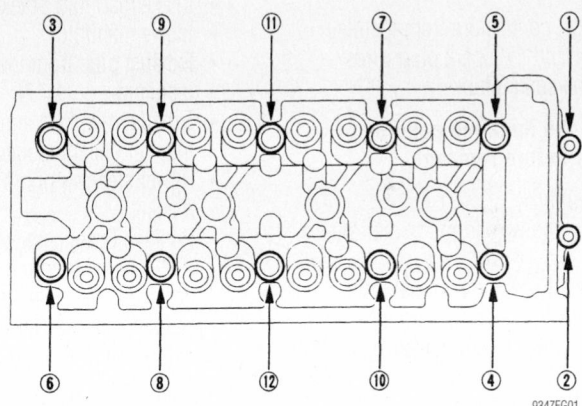

Cylinder head loosening sequence—2.0L engine

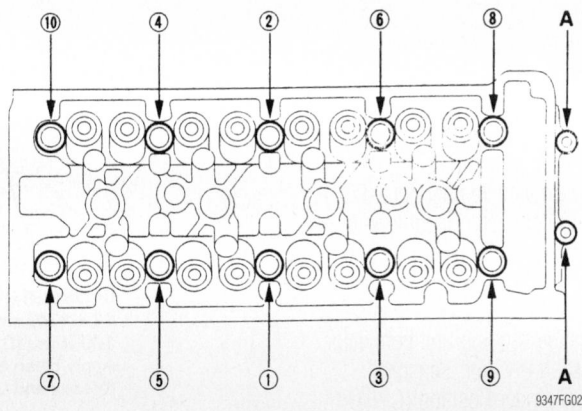

Cylinder head torque sequence—2.0L engine

5. Relieve the fuel system pressure.
6. Remove or disconnect the following:
- Vacuum hose, breather hose and air intake duct
- Water bypass hose from the cylinder head
- Fuel feed and return hose from the fuel rail
- Evaporative emissions (EVAP) control canister hose from the intake manifold
- Brake booster vacuum hose from the intake manifold
- Vacuum hose mount, on automatic transaxle equipped vehicles
- Throttle cable from the throttle body
- Throttle control cable from the throttle body, on automatic transaxle equipped vehicles

➡ **Be careful not to bend the cable when removing. Do not use pliers to remove the cable from the linkage. Always replace a kinked cable with a new one.**

- Ignition coil

➡ **Label the connectors before disconnecting them.**

- Electrical connectors from the distributor and the spark plug wires from the spark plugs

➡ **Matchmark the installed position of the distributor before removal.**

- Distributor
- Ignition coil wire from the distributor
- Connector and the terminal from the alternator, then the engine wiring harness from the valve cover
- Fuel injector connectors
- Intake Air Temperature (IAT) sensor connector, if equipped
- Idle Air Control (IAC) valve connector
- Throttle Position (TP) sensor connector
- Exhaust Gas Recirculation (EGR) valve lift sensor
- Ground cable terminals
- Engine Coolant Temperature (ECT) switch B connector, if equipped
- Heated Oxygen Sensor (HO$_2$S) connector
- ECT sensor
- ECT gauge sending unit connector
- Crankshaft Position Sensor (CKP)/Cylinder Position (CYP) sensor connector, if equipped

- Vehicle Speed Sensor (VSS) connector
- ECT switch **A** connector
- Upper radiator hose and the heater inlet hose from the cylinder head
- Lower radiator hose and heater outlet hose from the intake manifold
- Bypass hose from the thermostat housing and intake manifold
- Thermostat. Discard the O-rings.

➡ **Tag all vacuum hoses before disconnecting them.**

- Emissions vacuum hoses from the intake manifold assembly
- Cruise control actuator electrical connector and the vacuum tube, then the actuator
- Engine ground cable from the body
- Mounting bolts and drive belt from the power steering pump. Pull the pump away from the mounting bracket without disconnecting the hoses. Support the pump out of the way.
7. Raise and safely support the vehicle.
8. Remove or disconnect the following:
- Front wheel and tire assemblies
- Splash shield
- Intake manifold bracket bolts
- Intake manifold
- Exhaust pipe from the exhaust manifold
- Exhaust manifold and the exhaust manifold heat insulator
- Power steering pump mounting bracket
- Positive Crankcase Ventilation (PCV) hose, then remove the cylinder head cover. Replace the rubber seals if damages or deteriorated.
- Timing belt.

- Cylinder head bolts in the reverse order of installation

➡ **To prevent warpage, unscrew the bolts in sequence 1/3 turn at a time. Repeat the sequence until all bolts are loosened.**

9. Separate the cylinder head from the engine block with a suitable flat bladed pry-tool.

To install:

10. Be sure all cylinder head and block gasket surfaces are clean. Check the cylinder head for warpage. If warpage is less than 0.002 in. (0.05mm), cylinder head resurfacing is not required. Maximum resurface limit is 0.008 in. (0.2mm) based on a cylinder head height of 3.94 in. (100mm).
11. Always use a new head gasket.
12. The **UP** mark on the camshaft pulley should be at the top.
13. Be sure the No. 1 cylinder is at Top Dead Center (TDC).
14. Clean the oil control orifice and install a new O-ring. Install and align the cylinder head dowel pins and oil control jet.
15. Install the bolts that secure the intake manifold to its bracket but do not tighten them.
16. Position the camshaft correctly.
17. Install the cylinder head, then tighten the cylinder head bolts sequentially in 3 steps:
   a. Step 1: 29 ft. lbs. (40 Nm).
   b. Step 2: 51 ft. lbs. (70 Nm).
   c. Step 3: 72 ft. lbs. (100 Nm).
18. Install or connect the following:
- Intake manifold and tighten the nuts in a crisscross pattern, in 2 or 3 steps, beginning with the inner nuts. Final torque should be

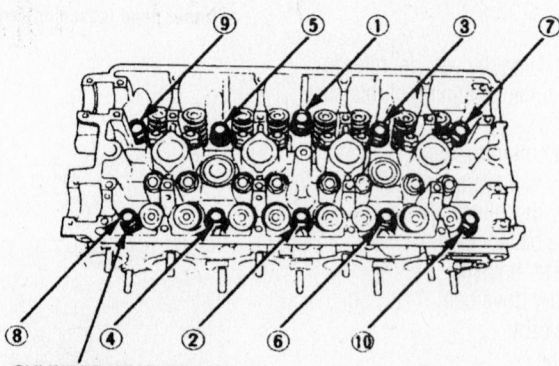

CYLINDER HEAD BOLTS
12 x 1.25 mm
100 N·m (10.0 kg-m, 72 lb-ft)
Apply clean engine oil bolt threads and under bolt heads.

7923FG19

Cylinder head torque sequence—2.2L engine

16 ft. lbs. (22 Nm). Always use a new intake manifold gasket.

- Intake manifold bracket to the intake manifold: 16 ft. lbs. (22 Nm)
- Heat insulator to the cylinder head and the block
- Power steering pump mounting bracket to the cylinder head. Tighten the 2 10mm bolts to 36 ft. lbs. (50 Nm). Torque the 8mm bolt to 16 ft. lbs. (22 Nm).
- Exhaust manifold and tighten the nuts in a crisscross pattern in 2 or 3 steps, beginning with the inner nut. Final torque should be 23 ft. lbs. (32 Nm). Always use a new exhaust manifold gasket.
- Exhaust manifold bracket, then the exhaust pipe, bracket and upper shroud

➡ **Be sure the camshaft sprocket and the crankshaft pulleys are aligned to TDC.**

- Timing belt.
- Splash shield and the front wheels

19. Lower the vehicle.

20. Check and adjust the valves, as necessary.

21. Tighten the crankshaft pulley bolt to 181 ft. lbs. (250 Nm).

22. Installation of the remaining components is the reverse of removal.

23. Fill the cooling system.

24. Connect the negative battery cable and enter the radio security code.

25. Start the engine and check carefully for any leaks.

26. Check the ignition timing and tighten the distributor bolts to 13 ft. lbs. (18 Nm).

27. If equipped with engine and turn the steering wheel lock-to-lock to reset the 4WS control unit.

2.3L Engines

1. Before servicing the vehicle, refer to the precautions in the beginning of this section.

2. Drain the cooling system.

3. Relieve the fuel system pressure.

4. Remove or disconnect the following:
- Negative battery cable
- Air intake duct
- Throttle and cruise control cables
- Positive Crankcase Ventilation (PCV) valve and hose
- Fuel lines
- Brake booster vacuum hose
- Evaporative Emissions (EVAP) canister hoses

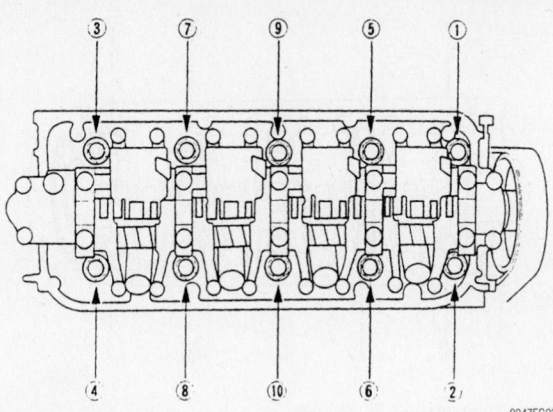

9347FG03

Cylinder head loosening sequence—2.3L engines

- Water bypass hoses
- Accessory drive belts
- Power steering pump
- Alternator wiring harness
- Radiator hoses
- Heater hoses
- Fuel injector connectors
- Intake Air Temperature (IAT) sensor connector
- Idle Air Control (IAC) valve connector
- Throttle Position (TP) sensor connector
- Manifold Absolute Pressure (MAP) sensor connector
- Heated Oxygen (HO$_2$S) sensor connector (F23A1, F23A5 engines)
- Air/Fuel ratio sensor connector (F23A4 engine)
- Engine Coolant Temperature (ECT) sensor connector
- Radiator fan switch connector
- Coolant temperature gauge sender connector
- Exhaust Gas Recirculation (EGR) valve connector
- Crankshaft Position (CKP) sensor connector
- VTEC solenoid valve connector (F23A1, F23A4 engines)
- VTEC oil pressure switch connector (F23A1, F23A4 engines)
- Distributor
- Front motor mount bracket
- Valve cover
- Timing belt
- Camshaft pulley and back cover
- Intake manifold
- Exhaust manifold
- Cylinder head. Loosen the bolts in sequence and in ⅓ turns.

To install:

5. Set the crankshaft pulley to Top Dead Center (TDC).

6. Set the camshaft pulley to TDC.

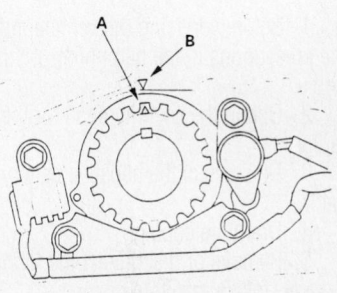

9347FG04

Set the crankshaft to TDC by aligning pointers A and B—2.3L engines

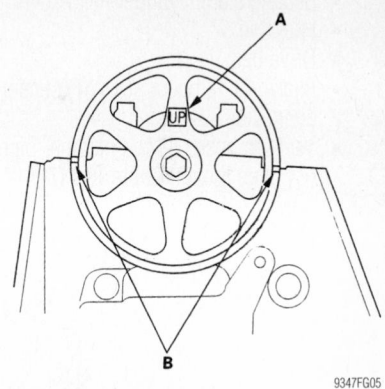

9347FG05

Align the camshaft pulley as shown prior to cylinder head installation—2.3L engines

7. Install the cylinder head with a new gasket. Tighten the bolts in sequence as follows:
   a. Step 1: 22 ft. lbs. (29 Nm).
   b. Step 2: Plus 90 degrees.
   c. Step 3: Plus 90 degrees.
   d. Step 4: If using new cylinder head bolts, add an additional 90 degrees.

8. The remainder of the installation is the reverse of the removal procedure.

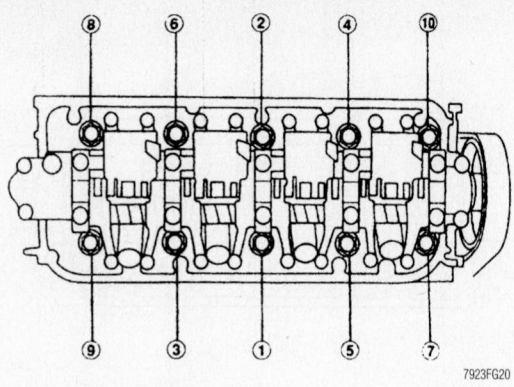

7923FG20

Cylinder head torque sequence—2.3L engines

2.4L Engine

1. Before servicing the vehicle, refer to the precautions in the beginning of this section.

2. Obtain the security code for the radio.

3. Disconnect the negative battery cable.

4. Drain the coolant.

5. Remove or disconnect the following:
- Intake manifold cover
- 4 ignition coils
- 2 bolts securing the vacuum line
- Bolt securing the power steering hose bracket
- Dipstick and breather hose
- Retainers and cylinder head cover
- Fuel line
- Drive belt
- Intake Air Temperature (IAT) sensor connector
- Vacuum hose (B) and breather pipe (C), then the intake air duct (D)

- Bolt securing the connecting pipe
- Evaporative emission (EVAP) canister hose and brake booster vacuum hose
- Intake manifold
- Exhaust manifold
- Positive Crankcase Ventilation

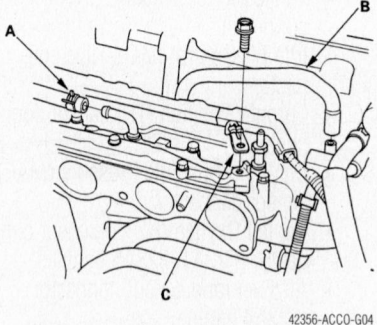

42356-ACCO-G04

Disconnect the PCV hose (A), vacuum hose (B) and ground cable (C)—2.4L engine

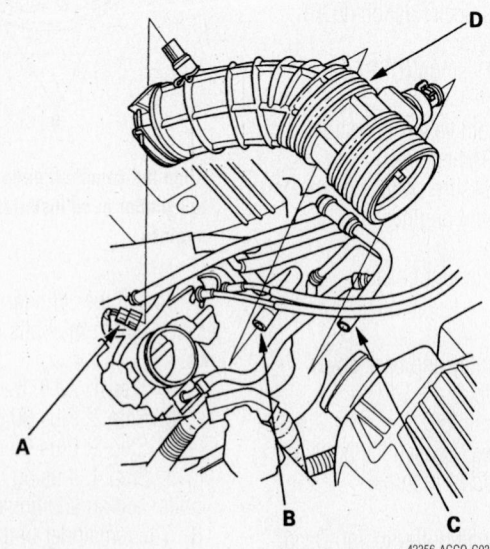

42356-ACCO-G03

Remove the vacuum hose (B), breather pipe (C), then remove the intake air duct (D)—2.4L engine

(PCV) hose, vacuum hose and ground cable
- Upper radiator hose, heater hoses and water bypass hose
- Engine wire harness connectors and wire harness clamps from the cylinder head
- 4 injector connectors
- Engine Coolant Temperature (ECT) sensor connector
- Camshaft Position (CMP) sensor connectors
- Exhaust Gas Recirculation (EGR) valve connector
- VTEC solenoid valve connector
- Engine Oil Pressure (EOP) sensor connector
- 2 bolts securing the EVAP canister purge valve bracket and the bolt securing the harness bracket
- Timing chain
- Rocker arm assembly
- Cylinder head bolts, in sequence, ⅓ turn at a time until completely loosened
- Cylinder head. Discard the gasket.

To install:

6. Be sure all cylinder head and block gasket surfaces are clean. Check the cylinder head for warpage. If warpage is less than 0.002 in. (0.05mm), cylinder head resurfacing is not required. Maximum resurface limit is 0.008 in. (0.2mm) based on a cylinder head height of 3.94 in. (100mm).

7. Install or connect the following:
- New gasket and dowel pins on the cylinder block

8. Set the crankshaft to Top Dead Center (TDC). Align the TDC mark (A) on the crankshaft sprocket with the pointer (B) on the cylinder block.

9. Measure the diameter of each cylinder head bolt at points A & B, as shown in the illustration. If either diameter is less than 0.42 in. (10.6mm), replace the head bolt

10. Apply engine oil to the threads and under the bolt heads of all of the bolts.

11. Install the cylinder head. Tighten the bolts in sequence as follows:
   a. Step 1: 29 ft. lbs. (39 Nm).
   b. Step 2: Plus 90 degrees.
   c. Step 3: Plus 90 degrees.
   d. Step 4: If using new cylinder head bolts, add an additional 90 degrees.

12. Install or connect the following:
- Rocker arm assembly
- Timing chain
- 2 bolts securing the EVAP canister purge valve bracket and tighten to 16 ft. lbs. (22 Nm)

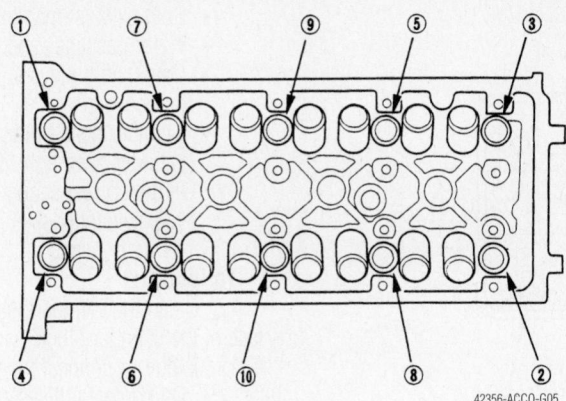

Cylinder head bolt loosening sequence—2.4L engine

42356-ACCO-G05

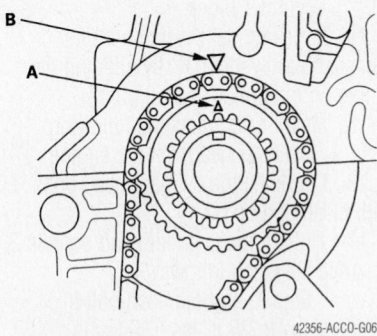

42356-ACCO-G06

Set the crankshaft to TDC by aligning the mark (A) on the crankshaft sprocket with the pointer (B) on the cylinder block—2.4L engine

- Bolt securing the harness bracket and tighten to 104 ft. lbs. (12 Nm)
- Upper radiator hose, heater hoses and water bypass hose
- PCV hose, vacuum hose and ground cable
- Exhaust manifold
- Intake manifold
- EVAP canister hose and brake booster vacuum hose

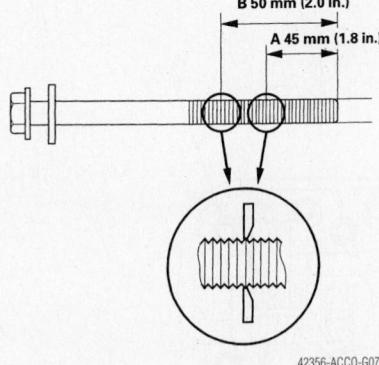

42356-ACCO-G07

You must measure the cylinder head bolts to see if they can be reused or need to be replaced—2.4L engine

- Fuel line
- Bolt securing the connecting pipe and tighten to 16 ft. lbs. (22 Nm)
- Intake air duct, IAT sensor connector, vacuum hose and breather pipe
- Cylinder head cover gasket in the groove of the cylinder head cover
- Apply liquid gasket P/N 08718-0009 or equivalent on the chain cover and No. 5 rocker shaft holder

mating areas. The parts must be installed within 5 minutes of applying liquid gasket.

- Spark plug seals on the spark plug tubes
- Cylinder head cover on the cylinder head, then slide the cover back and forth gently to seat it
- Cover washers
- Cylinder head cover bolts. Torque, in sequence, in 2 or 3 steps to 104 inch lbs. (12 Nm).
- Dipstick and breather hose
- Bolt securing the power steering hose bracket
- 2 bolts securing the vacuum line
- 4 ignition coils
- Intake manifold cover, and tighten the retainers to 104 inch lbs. (12 Nm)
- All of the remaining hoses, tubes, and connectors are installed correctly.

13. Fill the cooling system.

14. Connect the negative battery cable and enter the radio security code.

15. Start the engine and check carefully for any leaks.

3.0L Engine

1. Before servicing the vehicle, refer to the precautions in the beginning of this section.

2. Obtain the security code for the radio.

3. Disconnect the negative battery cable.

4. Drain the coolant.

5. Remove or disconnect the following:
- Evaporative emissions (EVAP) canister hose from the throttle body
- Air intake duct
- Upper engine covers
- Accelerator and cruise control cables from the throttle body
- Spark plug wire holder, cover, and intake manifold covers

6. Properly relieve the fuel system pressure.

7. Remove or disconnect the following:
- Fuel hoses from the supply rail
- Brake booster vacuum hose
- Positive Crankcase Ventilation (PCV) hose
- Breather hose
- Water bypass hose
- Vacuum hose from the throttle body
- Ground cable from the engine
- Alternator belt

8. Support the engine with a jack and a block of wood and remove the side engine mounting bracket.

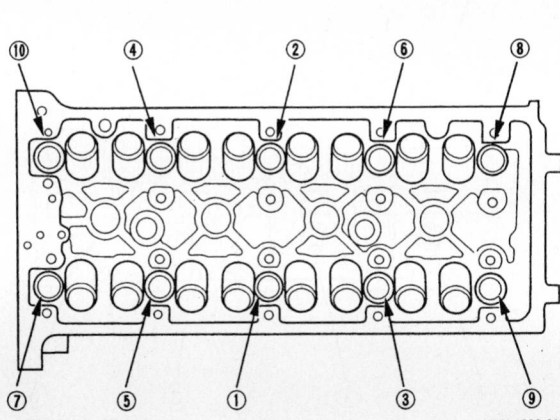

Cylinder head bolt tightening sequence—2.4L engine

42356-ACCO-G08

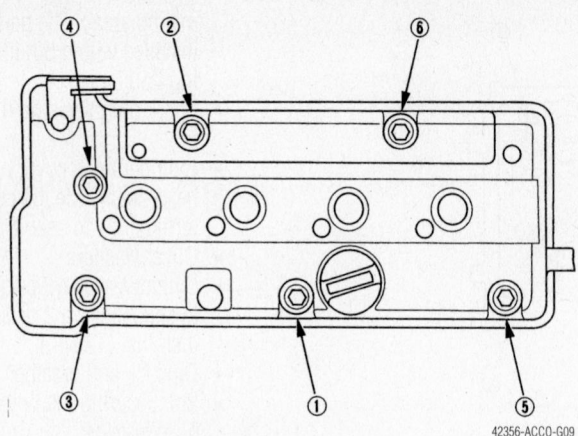

Cylinder head cover bolt tightening sequence—2.4L engine

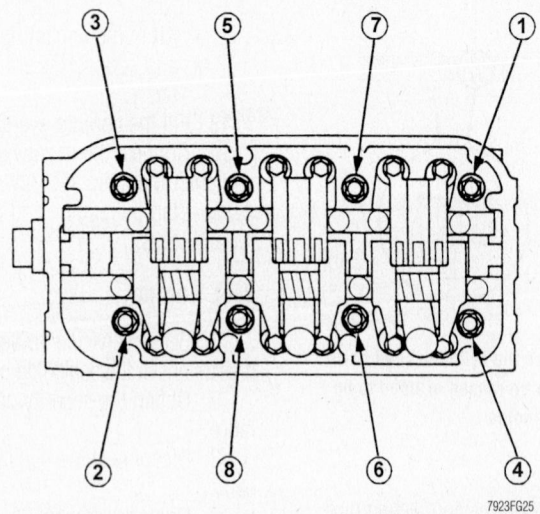

Loosen the cylinder head bolts in the sequence shown to prevent damage to the head—3.0L engine

9. Remove or disconnect the following:
- Power steering pump without disconnecting the hoses
- Alternator
- Wiring harness connectors from the components on the engine that may interfere with removing the cylinder head
- Distributor and spark plug wires
- Intake manifold
- Connectors from the fuel injectors
- Fuel supply rails
- Vacuum hoses from the fuel control valve

10. Set the engine to Top Dead Center (TDC) by aligning the marks on the crankshaft and camshaft pulleys.

11. Remove or disconnect the following:
- Timing belt
- Upper and lower radiator hoses
- Heater hoses

- Both exhaust manifolds
- Water passage assembly
- Camshaft pulleys and rear timing belt covers
- Cylinder head bolts; loosen each cylinder head bolt ⅓ turn at a time in the correct sequence. This will take several passes.
- Cylinder heads

To install:

12. Clean the cylinder head and the surface of the cylinder block.

13. Install or connect the following:
- Oil control orifices and install them using new o-rings
- Dowel pins, if removed
- New cylinder head gaskets on the cylinder block

14. If moved, set the crankshaft and camshaft pulleys to TDC by aligning the marks on the pulley and oil pump.

15. Install or connect the following:
- Cylinder heads on the engine

16. Lubricate the cylinder head bolts with clean engine oil.

17. Tighten the cylinder head bolts in 3 separate steps, as follows:
   a. Step 1: Tighten each bolt in sequence to 29 ft. lbs. (39 Nm).
   b. Step 2: Tighten each bolt in sequence to 51 ft. lbs. (69 Nm).
   c. Step 3: Tighten each bolt a third time in sequence to a final torque of 72 ft. lbs. (98 Nm).

➡If any cylinder head bolt makes noise while being tightened, loosen the bolts and begin the tightening sequence again.

18. Install or connect the following:
- Exhaust manifolds

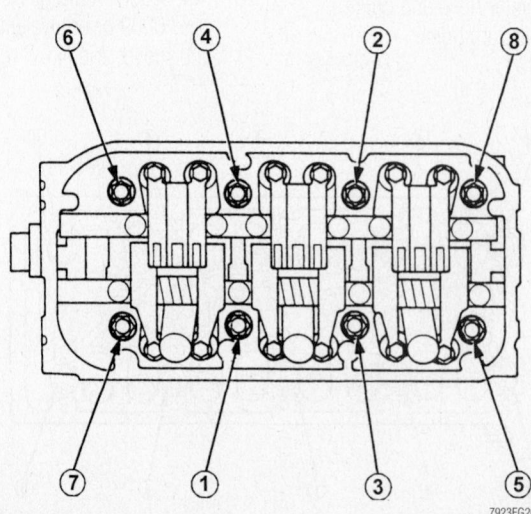

Cylinder head torque sequence—3.0L engine

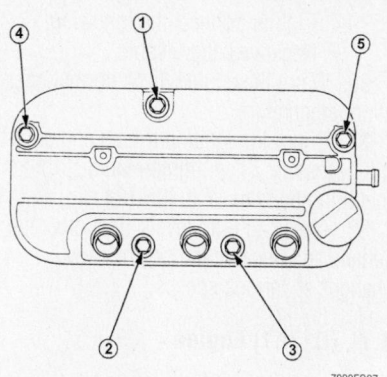

Tighten the cylinder head cover bolts in the sequence shown—3.0L engine

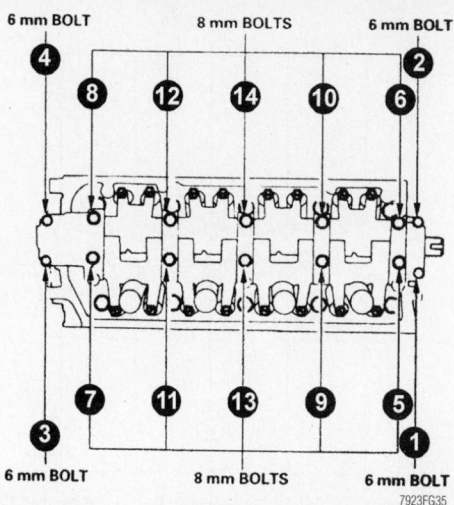

Rocker arm/shaft bolt loosening sequence—1.7L (D17A2 & D17A6) engine

• Timing belt

19. Check and adjust the valve clearance if necessary.

20. Install or connect the following:
   • Cylinder head cover. Tighten the bolts in sequence to 108 inch lbs. (12 Nm).
   • Water passage. Be sure to use new gaskets and o-rings. Tighten the bolts to 16 ft. lbs. (22 Nm).
   • Intake manifold
   • All of the remaining hoses, tubes, and connectors are installed correctly.
   • Negative battery cable

21. Enter the security code for the radio.

Rocker Arms/Shafts

REMOVAL & INSTALLATION

1.7L (D17A2 & D17A6) Engines

1. Before servicing the vehicle, refer to the precautions in the beginning of this section.

2. Remove or disconnect the following:
   • Negative battery cable

➡**Label the wires before disconnecting them.**

   • Ignition wires
   • Spark plugs and note their cylinder assignments.
   • Valve cover

3. Rotate the crankshaft to set the No. 1 cylinder to Top Dead Center (TDC) for the compression stroke. The white TDC mark on the crankshaft pulley aligns with the pointers on the lower timing cover.

4. Remove or disconnect the following:
   • Distributor, if equipped

5. Loosen the valve adjusting screws

6. Remove or disconnect the following:
   • Variable Timing Electronic Control (VTEC) solenoid valve connector

7. Loosen the camshaft holder bolts 2 turns at a time in a crisscross pattern to prevent damaging the valves or rocker assembly.

8. Remove or disconnect the following:
   • Rocker arm and shaft assemblies together with the camshaft holders. Do not remove the rocker shaft bolts yet. The bolts keep the bearing caps, springs, and rocker arms in place on the shafts.

➡**The rocker arms and shafts are an assembly. They must be removed from the engine as a unit. Always follow the torque sequence carefully when installing the rocker shaft assembly.**

   • Camshaft holder bolts from the rocker arm and shaft assembly

9. Bundle the intake rocker arm assemblies with rubber bands so they don't separate when the intake rocker shaft is removed.

10. Disassemble the rocker arm and shaft assemblies. Label the parts as they are removed from the shafts to ensure reinstallation in the original location.

11. Disassemble the rocker arm assemblies taking care not to mix up any of the parts. Inspect the rocker arm synchronizing and timing pistons by pushing them with your fingers. If the pistons don't move smoothly in the rocker arm bores, replace the rocker arm assembly.

12. Apply oil to the synchronizing pistons, timing piston, and timing spring and reassemble the rocker arms. Bundle the rocker arm assemblies with rubber bands to prevent the parts from separating.

13. Inspect the timing plates and return springs which are located on the camshaft holders. Set each timing plate and return spring so that the C-shaped upper arm of the plate is position parallel to the top of the camshaft holder.

To install:

14. Verify that the engine is set at Top Dead Center (TDC)/compression for the No. 1 cylinder.

15. Lubricate the camshaft journals and lobes. Coat the rocker shafts and camshaft holders with oil.

16. Remove the oil control orifice. Thoroughly clean it and reinstall it with a new O-ring.

17. Install or connect the following:
   • New camshaft seal if necessary

18. Assemble the rocker arm and shaft assemblies. Be sure the intake shaft collars and exhaust shaft springs are in the proper locations.

19. After the rocker arms and shafts are assembled, cut the rubber bands and remove them from the intake rockers. Be sure that no rubber band fragments are left in the engine.

20. Install or connect the following:
   • Fresh oil to the threads of camshaft holder bolts
   • Camshaft holder bolts
   • Liquid gasket to the cylinder head mating surfaces of the No. 1 and No. 5 camshaft holders. Do not allow the sealant to cure before installation.
   • Rocker arm and shaft assembly in place
   • Rocker and bolts, hand-tight

21. Tighten each bolt 2 turns at a time in the crisscross sequence so that the rockers are evenly tightened and don't bind on the valves. Tighten the 8mm rocker arm bolts to 14 ft. lbs. (20 Nm). Tighten the 6mm bolts to 108 inch lbs. (12 Nm).

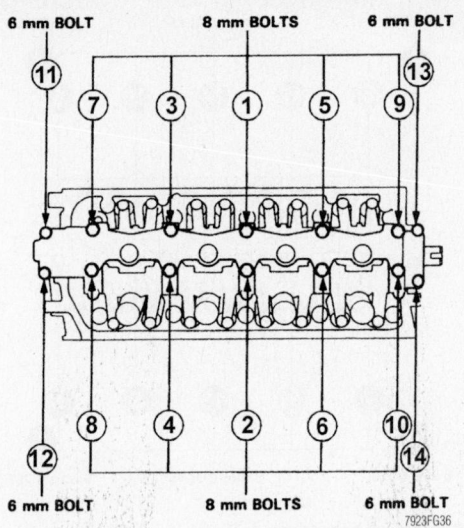

6 mm BOLT   8 mm BOLTS   6 mm BOLT

6 mm BOLT   8 mm BOLTS   6 mm BOLT

7923FG36

Rocker arm/shaft bolt torque sequence—1.7L (D17A2 & D17A6) engines

22. Starting with the No. 1 cylinder at Top Dead Center (TDC)/compression, adjust the valve clearances. After the clearance has been reached, tighten the locknuts to 14 ft. lbs. (20 Nm). Set the No. 3, No. 4, and No. 2 cylinders at Top Dead Center (TDC)/compression, and adjust their valve clearances.

- Intake: 0.007–0.009 in. (0.18–0.22mm)
- Exhaust: 0.009–0.011 in. (0.23–0.27mm)

23. Remove the VTEC solenoid valve, then remove the filter. Install a new VTEC solenoid valve filter. Tighten the solenoid valve bolts to 108 inch lbs. (12 Nm) and reconnect the solenoid valve connector.

24. Rotate the crankshaft to set the No. 1 cylinder at Top Dead Center (TDC)/compression. Then, manually inspect the operation of each of the VTEC intake rocker arms:

  a. Move the No. 1 cylinder's secondary intake rocker arm up and down.

  b. Verify that the secondary intake rocker arm moves independently of the primary intake rocker arm.

  c. Repeat the rocker arm inspection for the other 3 cylinders with each cylinder set at Top Dead Center (TDC)/compression.

25. Rotate the crankshaft back to TDC/compression for the No. 1 cylinder. Install the distributor, if equipped, but do not tighten the mounting bolts yet.

26. Tighten the crankshaft pulley bolt to 134 ft. lbs. (181 Nm).

27. Install or connect the following:
- Valve cover. Be sure the gasket is in good condition, and apply sealant to the corners where the gasket meets the camshaft holders.
- Spark plugs and the ignition wires

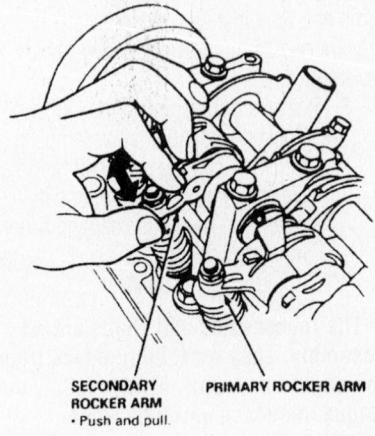

SECONDARY ROCKER ARM
• Push and pull.

PRIMARY ROCKER ARM

7923FG37

VTEC rocker arm inspection—1.7L (D17A2 & D17A6) engines

28. Drain the engine oil and remove the oil filter. Install a new oil filter and refill the engine with fresh oil.

29. Install or connect the following:
- Negative battery cable

30. Warm the engine up to normal operating temperature.

31. Check the ignition timing and adjust it if necessary. Then, tighten the distributor mounting bolts to 17 ft. lbs. (24 Nm).

32. Check all fluid levels. Test drive the vehicle and observe the engine RPM changes at various speeds.

1.7L (D17A1) Engine

1. Before servicing the vehicle, refer to the precautions in the beginning of this section.

2. Remove or disconnect the following:
- Negative battery cable

➡ **Label the wires before disconnecting them.**

- Ignition wires and spark plugs
- Valve cover and the upper timing belt cover

3. Set the No. 1 cylinder to Top Dead Center (TDC) for the compression stroke. Verify that the TDC marks are correctly aligned. Once the engine is set in this position, it must not be disturbed.

4. Remove or disconnect the following:
- Distributor, if equipped

5. Loosen the valve adjusting screws.

6. Cover the timing belt with a clean shop towel to protect it from engine oil. If the belt is contaminated with oil, it must be replaced.

7. Remove or disconnect the following:
- Camshaft holder bolts. Unscrew the bolts 2 turns at a time in a criss-cross pattern to prevent damaging the valves, camshaft, or rocker arm assembly.

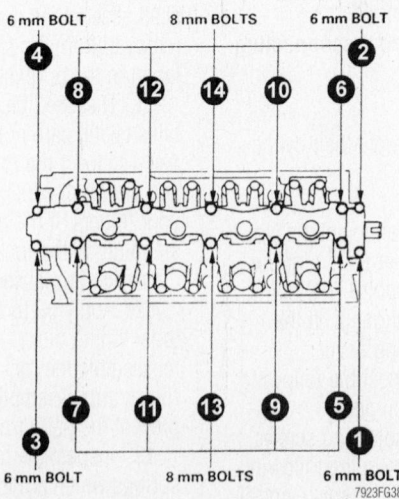

6 mm BOLT   8 mm BOLTS   6 mm BOLT

6 mm BOLT   8 mm BOLTS   6 mm BOLT

7923FG38

Rocker arm/shaft bolt loosening sequence—1.7L (D17A1) engines

➡The rocker arms and shafts are an assembly; they must be removed from the engine as a unit. To prevent warpage, always follow the torque sequence carefully when removing or installing the rocker shaft assembly.

• Rocker arm and shaft assemblies. Do not remove the camshaft holder bolts. The bolts keep the camshaft bearing caps, springs, and rocker arms in place on the shafts.

8. If the rocker arms or shafts are to be replaced, identify the parts as they are removed from the shafts to ensure reinstallation in the original location.

To install:

9. Verify that the engine is set to TDC/compression for the No. 1 cylinder. The camshaft keyway faces up when the engine is at Top Dead Center (TDC)/compression.

10. Lubricate the camshaft journals and lobes with clean engine oil. Install a new camshaft seal if necessary.

11. Remove the oil control orifice. Thoroughly clean it and install it with a new O-ring.

12. Assemble the rocker arms, shafts, and camshaft bearing caps.

13. Apply sealant to the mating surfaces of the No. 1 and No. 5 camshaft bearing caps. Do not allow the sealant to cure before the rocker arm assembly is installed.

14. Set the rocker arm assembly in place. Apply engine oil to the holder bolt threads, then loosely install the bolts. Tighten each bolt in a 2 step crisscross pattern to ensure that the rockers do not bind on the valves. Tighten the 8mm bolts to 14 ft. lbs. (20 Nm). Tighten the 6mm bolts to 104 inch lbs. (12 Nm).

15. Verify that the engine is at Top Dead Center (TDC)/compression for the No. 1 piston and install the distributor, if equipped.

16. Adjust the valves and tighten the locknuts to 14 ft. lbs. (20 Nm).

17. Install the valve cover and upper timing belt cover.

18. Reconnect the negative battery cable.

19. Check the ignition timing and adjust if necessary. Tighten the distributor mounting bolts to 17 ft. lbs. (24 Nm).

2.0L Engine

1. Before servicing the vehicle, refer to the precautions in the beginning of this section.

2. Remove or disconnect the following:

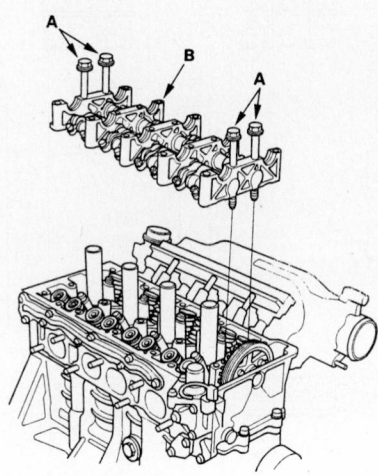

9347FG07

Use bolts (A) to hold the rocker arm assembly (B) together while removing or replacing the assembly—2.0L engine

• Negative battery cable
• Valve cover
• Camshafts

3. Replace the front and rear camshaft holder bolts as shown, and remove the rocker arm assembly.

To install:

➡Keep all valvetrain components in order. If reused, they must be installed in their original locations.

4. Installation is the reverse of the removal procedure.

2.2L Engine

1. Before servicing the vehicle, refer to the precautions in the beginning of this section.

2. Remove or disconnect the following:
• Negative battery cable
• Air intake duct
• Positive Crankcase Ventilation (PCV) hose
• Cylinder head cover. Replace the rubber seals if damaged or deteriorated.
• Timing belt upper cover

3. Bring the No. 1 cylinder to TDC. The white mark on the crankshaft pulley should align with the pointer on the timing belt cover. The words **UP** embossed on the camshaft pulley should be aligned in the upward position. The marks on the edge of the pulley should be aligned with the cylinder head or the back cover upper edge. Once in this position, the engine must NOT be turned or disturbed.

4. Remove or disconnect the following:
• Electrical connectors from the distributor and the spark plug wires from the spark plugs.

➡Matchmark the installed position distributor before removing it.

• Distributor from the cylinder head
• Power steering pump drive belt

5. Mark the rotation of the timing belt if it is to be used again. Loosen the timing belt adjusting bolt ¾ to 1 turn, then release the tension on the timing belt. Push the tensioner to release tension from the belt, then tighten the adjusting bolt.

6. Remove or disconnect the following:
• Timing belt from the camshaft sprocket

❋❋ WARNING

Do not crimp or bend the timing belt more than 90°, or less than 1 inch (25mm) in diameter

6 mm BOLT   8 mm BOLTS   6 mm BOLT

⑪   ⑬

⑦   ③   ①   ⑤   ⑨

⑧   ④   ②   ⑥   ⑩

⑫   ⑭

6 mm BOLT   8 mm BOLTS   6 mm BOLT

7923FG39

Rocker arm/shaft bolt tightening sequence—1.7L (D17A1) engines

NOTE:
- Identify parts as they are removed to ensure reinstallation in their original locations.
- Inspect rocker shafts and rocker arms
- Rocker arms must be installed in the same position if reused.
- Prior to reassembling, clean all the parts in solvent, dry them, and apply lubricant to any contact points.
- Bundle the rocker arms with rubber bands to keep them together as a set.

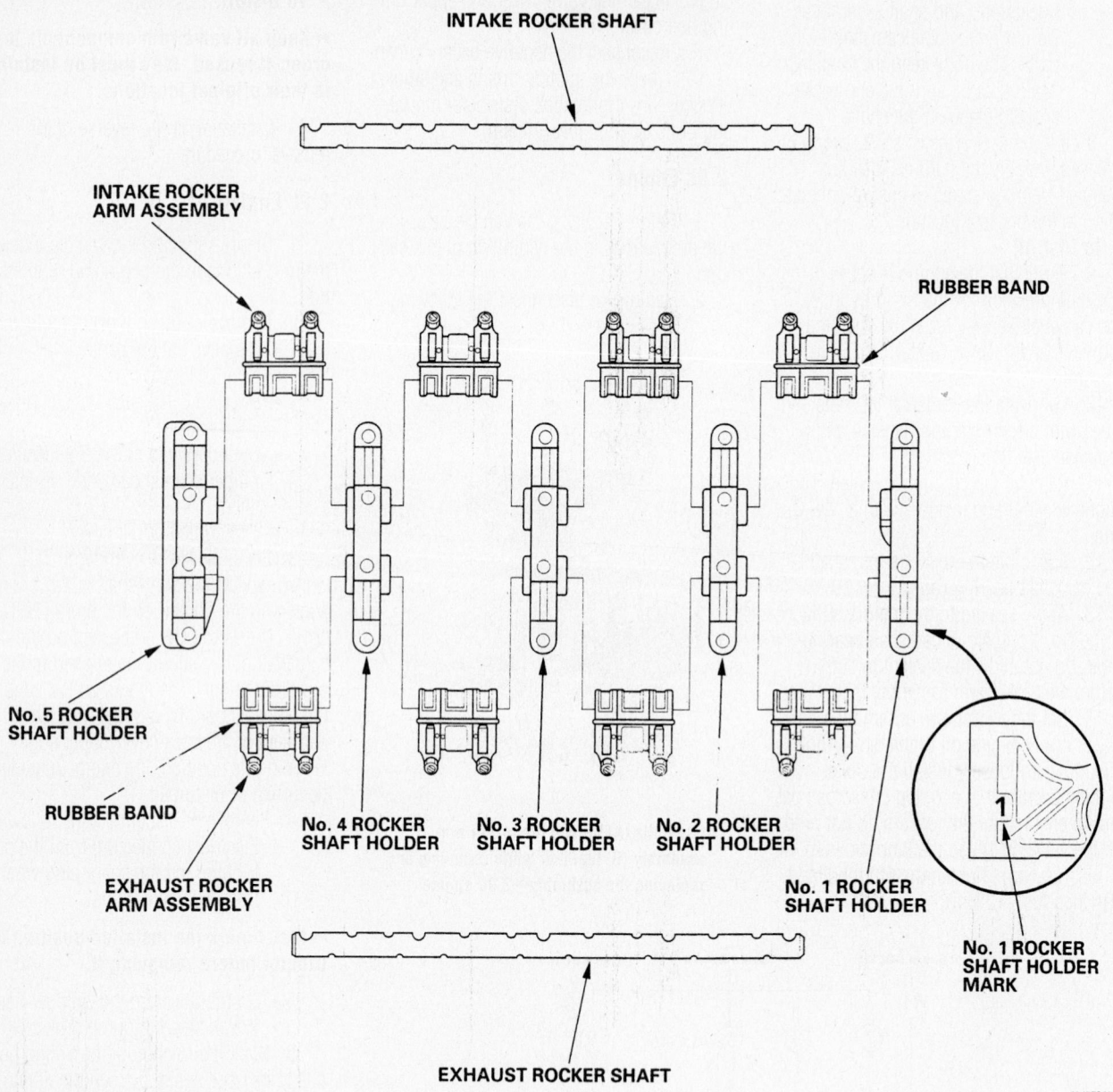

INTAKE ROCKER SHAFT

INTAKE ROCKER ARM ASSEMBLY

RUBBER BAND

No. 5 ROCKER SHAFT HOLDER

RUBBER BAND

EXHAUST ROCKER ARM ASSEMBLY

No. 4 ROCKER SHAFT HOLDER

No. 3 ROCKER SHAFT HOLDER

No. 2 ROCKER SHAFT HOLDER

No. 1 ROCKER SHAFT HOLDER

No. 1 ROCKER SHAFT HOLDER MARK

EXHAUST ROCKER SHAFT

9347FG08

Exploded view of the rocker arm assembly—2.0L engine

➡Ensure the words UP embossed on the camshaft pulley is aligned in the upward position.

- Camshaft sprocket bolt, sprocket and sprocket key
- Timing belt back cover
- Valve adjusting screws
- Camshaft holder attaching bolts. Loosen the bolts 2 turns at a time in the proper sequence to prevent

damaging the valves or rocker arm assemblies.

➡When removing the rocker arm assembly, do not remove the camshaft holder bolts. The bolts will keep the camshaft holders, springs, and the rocker arms on the shafts.

- Camshaft holders and rocker arm assembly. If the rocker arm and shaft assembly needs to be disas-

sembled for service, note the location of the components as they are removed. Install a rubber band around the VTEC rocker arm assemblies to keep them from coming apart during disassembly of the rocker arm assembly. The rocker arms must be installed in the same position if reused.
- Camshaft from the cylinder head and discard the seal.

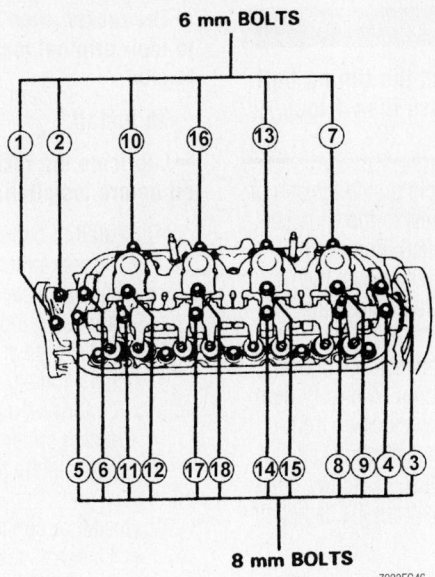

6 mm BOLTS

8 mm BOLTS

7923FG46

Rocker arm assembly bolt removal sequence—2.2L (H22A4) engines

- Oil control orifice

To install:

7. Wipe the camshaft and the camshaft journals clean, then lubricate both surfaces, and install the camshaft.

8. Turn the camshaft so that its keyway is facing up (No. 1 cylinder will be at Top Dead Center (TDC)).

9. Clean the oil control orifice and install a new O-ring, then install the oil control orifice.

10. Reassemble the rocker arm and shaft assembly, if it was disassembled. Lubricate the rocker arm and shaft assembly with clean oil, then apply liquid gasket to the head mating surfaces of the No. 1 and No. 6 camshaft holders.

11. Set the camshaft holders and rocker arm assembly in place, then loosely install the attaching bolts.

12. Apply clean oil to the camshaft oil seal lip and the seal guide (part # 07NAG-PT0010A), then install the seal to the seal guide. Install the seal guide to the camshaft, then the installer cup (part # 07NAF-PT0010A), and the installer shaft (part # 07NAF-PT0020A). Tighten the nut on the installer shaft to press the seal into the cylinder head.

13. Tighten the camshaft holder bolts 2 turns at a time in the proper sequence. The final torque for the 8mm bolts is 16 ft. lbs. (22 Nm) and the final torque for the 6mm bolts is 108 inch lbs. (12 Nm).

14. Install or connect the following:

- Timing belt back cover and a new gasket, if necessary. Tighten the bolt toward the exhaust manifold to 84 inch lbs. (10 Nm) and tighten the bolt toward the intake manifold to 108 inch lbs. (12 Nm).
- Camshaft sprocket key to the

camshaft, then the camshaft sprocket and bolt. Tighten the bolt to 27 ft. lbs. (37 Nm).

15. Ensure the words **UP** embossed on the camshaft pulley is aligned in the upward position, then install the timing belt onto the camshaft sprocket. Loosen, then tighten the timing belt adjusting nut.

16. Rotate the crankshaft pulley 5 or 6 turns to position the timing belt on the pulleys.

17. Set the No. 1 cylinder to TDC and loosen the timing belt adjusting nut 1 turn. Turn the crankshaft counterclockwise until the cam pulley has moved 3 teeth; this creates tension on the timing belt. Loosen, then tighten the adjusting nut, and tighten it to 33 ft. lbs. (45 Nm).

18. Adjust the valves.

19. Tighten the crankshaft pulley bolt to 181 ft. lbs. (245 Nm).

20. Install or connect the following:

- Upper timing belt cover. Tighten the bolt toward the exhaust manifold to 84 inch lbs. (10 Nm) and tighten the bolt toward the intake manifold to 108 inch lbs. (12 Nm).
- Cylinder head cover gasket cover to the groove of the cylinder head cover. Before installing the gasket, thoroughly clean the seal and the groove. Seat the recesses for the camshaft first, then work it into the groove around the outside edges. Be sure the gasket is seated securely in the corners of the recesses.

21. Apply liquid gasket to the 4 corners of the recesses of the cylinder head cover gasket. Do not install the parts if 5 minutes or more have elapsed since applying liquid gasket. After assembly, wait at least 20 minutes before filling the engine with oil.

22. Install or connect the following:

- Cylinder head (valve) cover. Tighten the bolts attaching the cylinder head cover in the proper sequence to 84 inch lbs. (10 Nm).
- PCV hose to the cylinder head cover
- Power steering belt, then adjust the belt.
- Distributor to the cylinder head. Snug the attaching bolts until the timing has been checked and adjusted.
- Spark plug wires and the distributor electrical connectors
- Air intake duct

23. Drain the oil from the engine into a sealable container. Install the drain plug and refill the engine with clean oil.

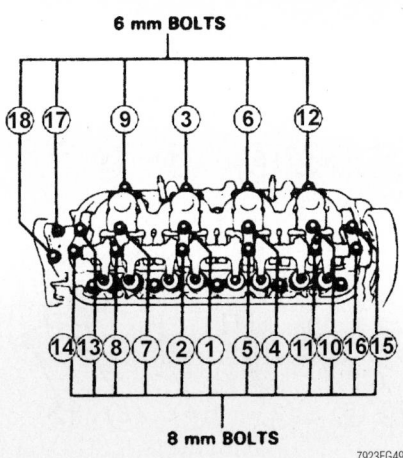

6 mm BOLTS

8 mm BOLTS

7923FG49

Rocker arm assembly torque sequence—2.2L (H22A4) engines

24. Connect the negative battery cable and enter the radio security code.

25. Start the engine and check carefully for any leaks.

26. Check and adjust the ignition timing as necessary, then tighten the distributor bolts to 13 ft. lbs. (18 Nm).

2.3L Engines

1. Before servicing the vehicle, refer to the precautions in the beginning of this section.

2. Disconnect the negative battery cable.

3. Turn the crankshaft so the No. 1 piston is at Top Dead Center (TDC).

➡ **The No. 1 piston is at top dead center when the pointer on the block aligns with the white painted mark on the flywheel (manual transaxle) or driveplate (automatic transaxle).**

4. Remove or disconnect the following:
   - Air intake duct
   - Engine ground cable from the cylinder head cover
   - Connector and the terminal from the alternator
   - Engine wiring harness from the valve cover
   - Ignition coil

➡ **Label all electrical connectors before detaching them.**

   - Electrical connectors from the distributor and the spark plug wires from the spark plugs.

➡ **Matchmark the installed position if the distributor before removal.**

   - Distributor from the cylinder head
   - Ignition coil wire from the distributor
   - Positive Crankcase Ventilation (PCV) hose
   - Cylinder head cover. Replace the rubber seals if damaged or deteriorated.
   - Timing belt middle cover

5. Ensure the words **UP** embossed on the camshaft pulleys are aligned in the upward position.

6. Mark the rotation of the timing belt if it is to be used again. Loosen the timing belt adjusting nut ½ turn, then release the tension on the timing belt. Push the tensioner to release tension from the belt, then tighten the adjusting nut.

7. Remove the timing belt from the camshaft sprockets.

❊❊ WARNING

Do not crimp or bend the timing belt more than 90°, or less than 1 inch (25mm) in diameter

---

8. Insert a 5.0mm pin punch in each of the camshaft caps, nearest to the sprockets, through the holes provided. Remove the camshaft sprocket attaching bolts, then remove the sprockets. Do not lose the sprocket keys.

9. Remove or disconnect the following:
   - Side engine mount bracket B, then the timing belt back cover from behind the camshaft sprockets.
   - Rocker arm adjusting screws, then the pin punches from the camshaft caps

➡ **Note the camshaft holders locations for ease of installation. Loosen the bolts in the reverse order of the holder bolts torque sequence.**

   - Camshaft holders
   - Camshafts from the cylinder head, then discard the camshaft seals.
   - Rubber cap from the head, located at the end of the intake camshaft.
   - Rocker arms from the cylinder head. Note the locations of the rocker arms.

➡ **The rocker arms have to be installed to their original locations if being reused.**

To install:

➡ **Lubricate the rocker arms with clean oil before installation.**

10. Install or connect the following:
   - Rocker arms on the pivot bolts and the valve stems. If the rocker arms are being reused, install them to their original locations. The locknuts and adjustment screws should be loosened before installing the rocker arms.

11. Lubricate the camshafts with clean oil.

12. Install or connect the following:
   - Camshaft seals to the end of the camshafts that the timing belt sprockets attach to. The open side (spring) should be facing into the cylinder head when installed.

➡ **Be sure the keyways on the camshafts are facing up and install the camshafts to the cylinder head.**

   - Rubber plug to the cylinder head at the end of the intake camshaft

13. Apply liquid gasket to the head mating surfaces of the No. 1 and No. 6

Specified torque:
Except Intake ⑤, ⑦. Exhaust ⑥, ⑧:
      10 N·m (1.0 kg-m, 7 lb-ft)
Intake ⑤, ⑦. Exhaust ⑥, ⑧:
      12 N·m (1.2 kg-m, 9 lb-ft)

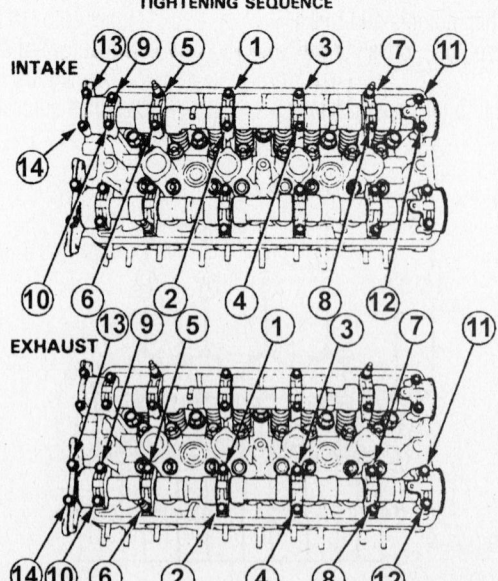

Camshaft holders torque sequence—2.3L engines

7923FG45

camshaft holders, then install them along with No. 2, 3, 4 and 5. **I** or **E** marks are stamped on the camshaft holders to identify them as Intake or Exhaust side holders. The arrows stamped on the holders should point toward the timing belt.

14. Snug the camshaft holders in place.

15. Press the camshaft seals securely into place.

16. Tighten the camshaft holder bolts in 2 steps, following the proper sequence, to ensure that the rockers do not bind on the valves. Tighten all the bolts, except the 4 studs, to 84 inch lbs. (10 Nm). Tighten the studs (number 5 and 7 bolts in the correct sequence) to 108 inch lbs. (12 Nm).

17. Install or connect the following:
- Timing belt back cover.
- Side engine mount bracket B. Tighten the bolt attaching the bracket to the cylinder head to 33 ft. lbs. (45 Nm). Tighten the bolts attaching the bracket to the side engine mount to 16 ft. lbs. (22 Nm).

18. Insert a 5.0mm pin punch in each of the camshaft caps, nearest to the pulleys, through the holes provided. Install the keys into the camshaft grooves.

19. Push the camshaft sprockets onto the camshafts, then tighten the retaining bolts to 27 ft. lbs. (38 Nm).

20. Ensure the words **UP** embossed on the camshaft pulleys are aligned in the upward position. Install the timing belt to the camshaft sprockets, then remove the 2, 5.0mm pin punches from the camshaft bearing caps.

21. Loosen, then tighten the timing belt adjuster nut.

22. Turn the crankshaft counterclockwise until the cam pulley has moved 3 teeth; this creates tension on the timing belt. Loosen, then tighten the adjusting nut and tighten it to 33 ft. lbs. (45 Nm).

23. Adjust the valves.

24. Tighten the crankshaft pulley bolt to 181 ft. lbs. (250 Nm).

25. Install or connect the following:
- Middle timing belt cover and tighten the attaching bolts to 108 inch lbs. (12 Nm).
- Cylinder head cover and tighten the cap nuts to 84 inch lbs. (10 Nm).
- PCV hose to the cylinder head cover
- Distributor. Snug the attaching bolts until the timing has been checked and adjusted.
- Spark plug wires, then the distributor electrical connectors
- Ignition coil wire to the distributor.

- Ignition coil
- Alternator wiring harness to the cylinder head cover
- Terminal and connector to the alternator
- Engine ground cable to the cylinder head cover
- Air intake duct

26. Drain the oil from the engine into a sealable container. Install the drain plug and refill the engine with clean oil.

27. Connect the negative battery cable and enter the radio security code.

28. Start the engine and check carefully for any leaks.

29. Check and adjust the ignition timing. Tighten the distributor bolts to 13 ft. lbs. (18 Nm).

30. If equipped with 4WS, turn the steering wheel lock-to-lock to reset the 4WS control unit.

2.4L Engine

1. Before servicing the vehicle, refer to the precautions in the beginning of this section.

2. Remove or disconnect the following:
- Timing chain
- Loosen the rocker arm adjusting screws
- Camshaft holder bolts, two turns at a time in sequence
- Timing chain guide (B), camshaft and camshafts

3. Insert the bolts (A) into the rocker shaft holder, then remove the rocker arm assembly (B)

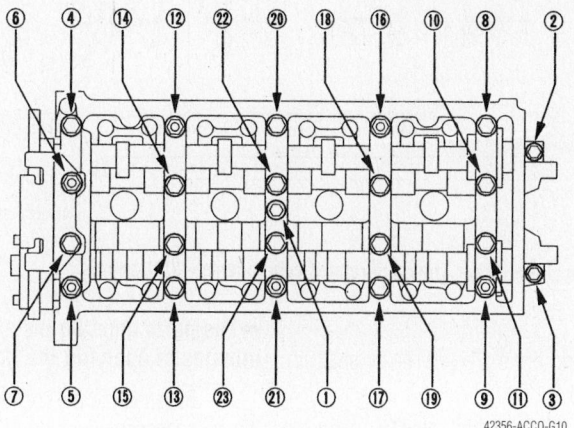

42356-ACCO-G10

Camshaft holder bolt loosening sequence—2.4L engine

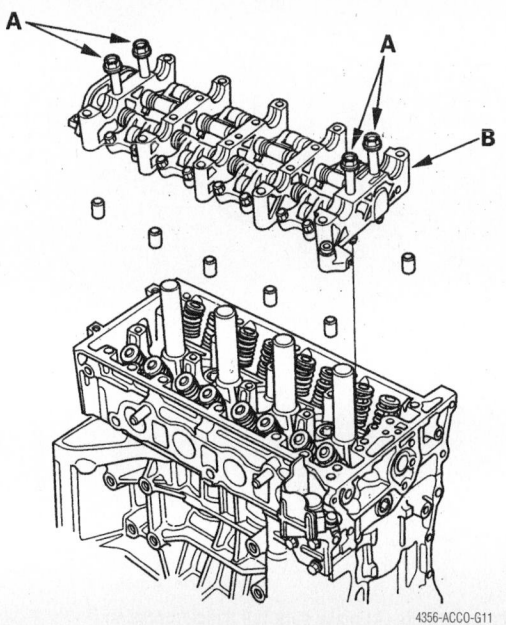

4356-ACCO-G11

Insert the bolts (A) into the rocker shaft holder, then remove the rocker arm assembly (B)—2.4L engine

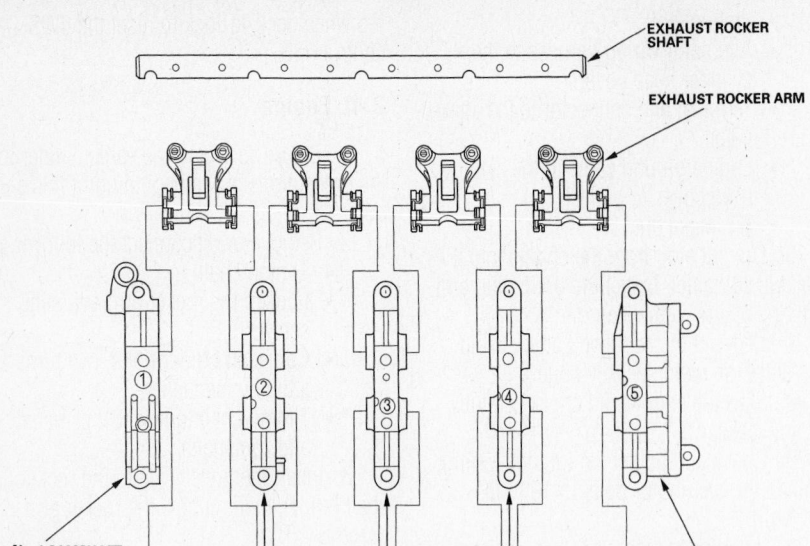

Exploded view of the rocker arms and related components—2.4L engine

To install:

4. Clean and dry the No. 5 rocker shaft holding mating surface.

5. Apply a suitable liquid gasket P/N 08718-0009, or equivalent, evenly to the cylinder head mating surface of the No. 5 rocker shaft holder.

➡ **The parts must be installed within 5 minutes of applying the liquid gasket.**

6. Reassemble the rocker arm assembly, as necessary.

7. Install or connect the following
• Bolts (A) into the rocker shaft

holder, then the rocker arm assembly on the cylinder head. Remove the bolts from the rocker shaft holder.

8. Make sure the punch marks on the variable valve timing control (VTC) actuator and exhaust camshaft sprocket are facing up, then set the camshafts (A) in the holder

9. Set the camshaft holders (B) and timing chain guide B (C) in place.

10. Tighten the bolts, in sequence, to the following specification:
    a. 8mm bolts: 16 ft. lbs. (22 Nm)
    b. 6mm bolts: 8.7 ft. lbs. (12 Nm)

11. Install the timing chain and adjust the valve lash.

3.0L Engine

1. Before servicing the vehicle, refer to the precautions in the beginning of this section.

2. Remove or disconnect the following:

• Cylinder head cover
• Jam nuts and screws
• Rocker arm shaft bolts, 2 turns at a time in the sequence shown.

3. Lift the rocker arm assembly from the cylinder head. Leave the bolts in the shafts to retain the rocker arms and springs.

To install:

4. Clean all parts in solvent, dry with compressed air and lubricate with clean engine oil.

5. Place the rocker arm assemblies on

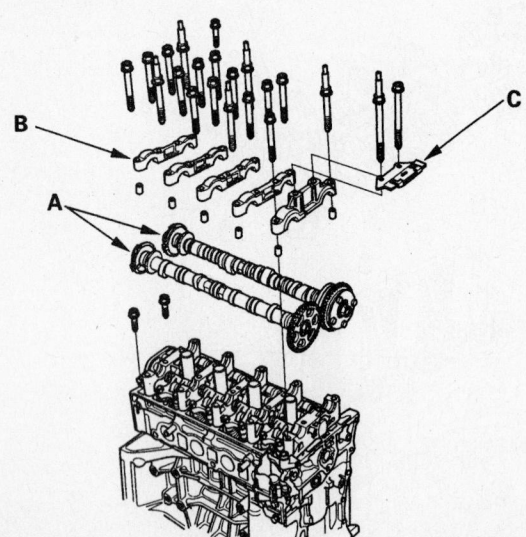

When installing the camshafts (A) make sure the punch marks on the VTC actuator and exhaust cam sprockets are facing up—2.4L engine

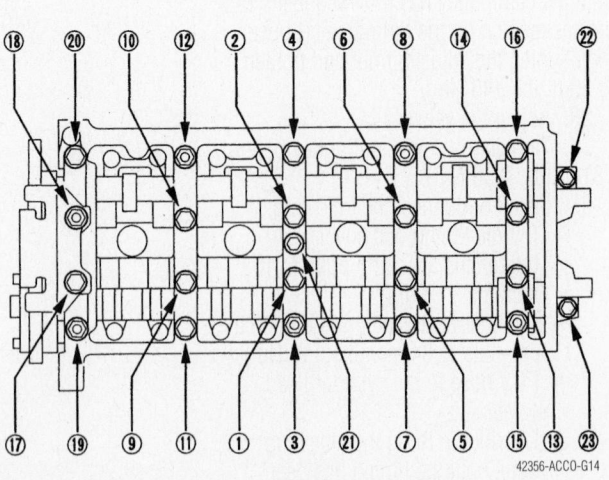

Rocker arm assembly bolt tightening sequence—2.4L engine

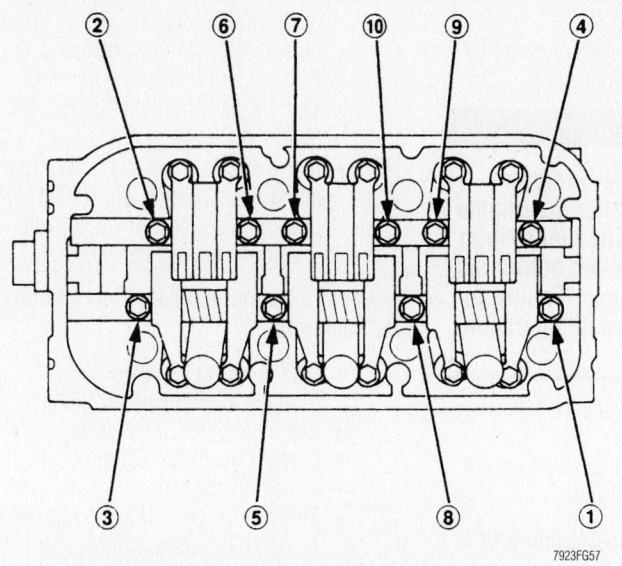

Be sure to loosen the rocker arm shaft bolts in the correct order as shown—3.0L engine

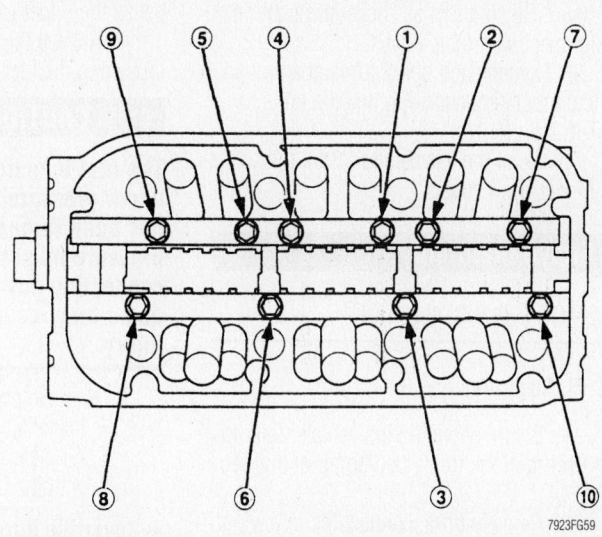

Tighten the bolts 2 turns at a time in the sequence shown—3.0L engine

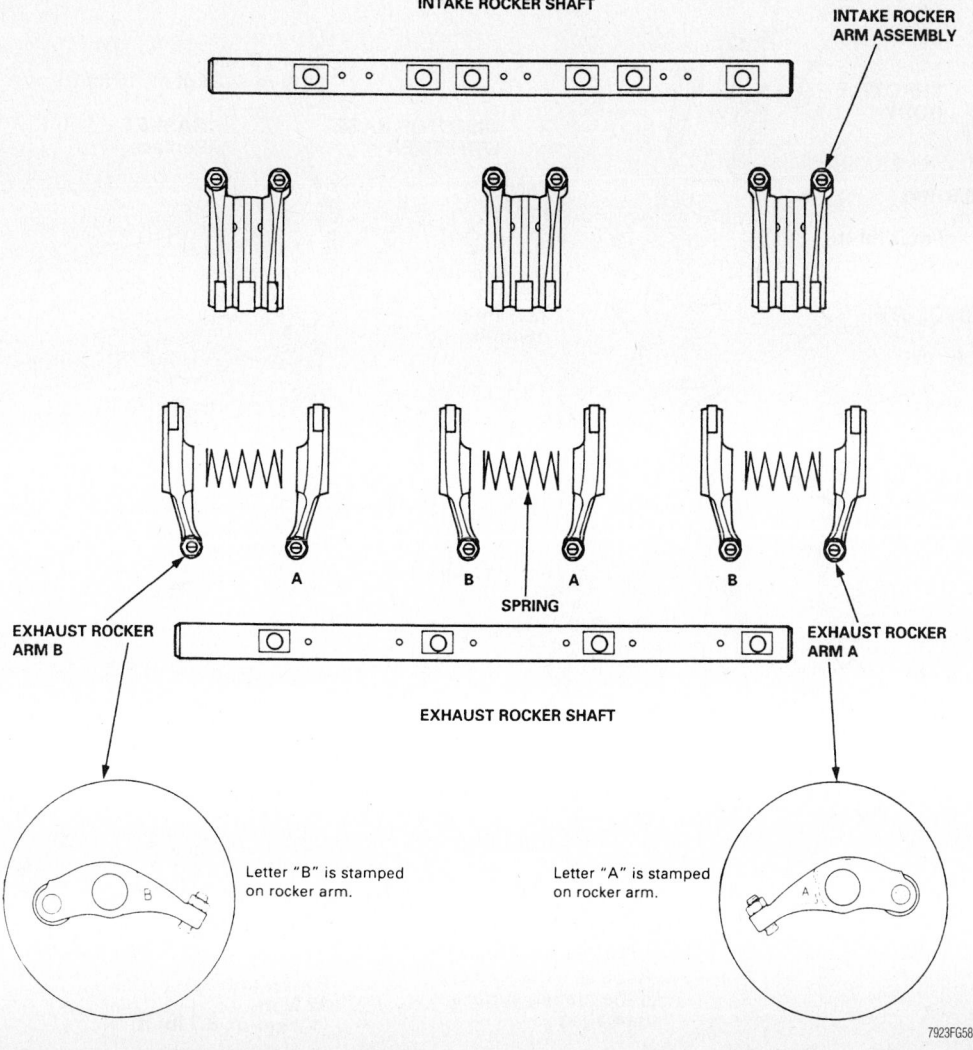

Exploded view of the rocker arms and related components—3.0L engine

the cylinder head and install the bolts loosely. Be sure that all rocker arms are in alignment with their valves.

6. Tighten each bolt 2 turns at a time in the correct sequence. Tighten the bolts to 17 ft. lbs. (24 Nm).

7. Adjust the valves and install the cylinder head covers.

Intake Manifold

REMOVAL & INSTALLATION

1.7L Engines

1. Before servicing the vehicle, refer to the precautions in the beginning of this section.

2. Remove or disconnect the following:
   • Negative battery cable

3. Drain the cooling system to a level below the upper radiator hose.

4. Relieve the fuel system pressure by loosening the fuel filter service bolt.

✳✳ CAUTION

The fuel injection system remains under pressure even after the engine has been turned off. The fuel system pressure must be relieved before disconnecting any fuel lines. Failure to do so may result in fire and personal injury.

5. Remove or disconnect the following:
   • Intake air duct
   • Air cleaner assembly, if equipped with the D16Y7 engine

➡ Cover the throttle body opening to keep dirt out.

• Fuel line from the fuel rail. Clean up any spilled fuel.
• Fuel injector wiring harnesses
• Fuel rail and injectors
• Throttle cable from the linkage at the throttle body
• Intake manifold cooling hoses. Use a drain pan to catch any spilled coolant. Also, be sure no coolant spills on electrical connections.

➡ Label all electrical connectors before detaching them.

• Engine wiring harness connectors from the intake manifold sensors
• Idle Air Control (IAC) valve
• Exhaust Gas Recirculation (EGR valve), if equipped
• Throttle Position (TP) and Manifold Absolute Pressure (MAP) sensor
• Manifold from its support bracket

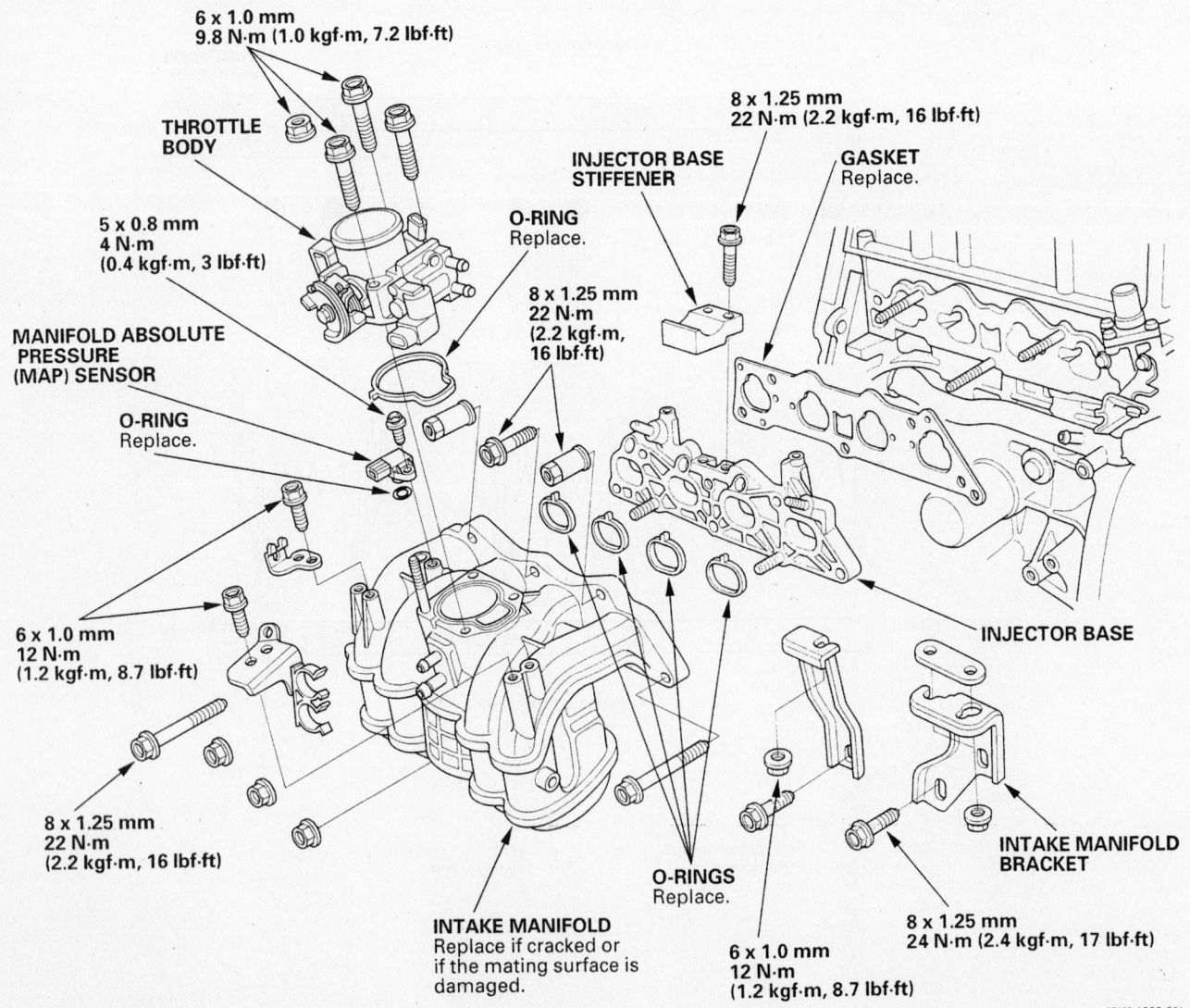

Exploded view of the intake manifold and related components—1.7L engines

- Intake manifold nuts in a crisscross pattern.
- Intake manifold assembly from the vehicle.

To install:

➡**Use new gaskets when installing the intake manifold. Use new O-rings when installing manifold sensors and components. Use new sealing washers when reconnecting the fuel lines.**

6. Clean all gasket mating surfaces.
7. Install or connect the following:
- New intake manifold gaskets
- Intake manifold

8. Tighten the intake manifold nuts in 2 or 3 steps in a crisscross pattern starting with the inside nuts. Tighten the nuts to 16 ft. lbs. (22 Nm).
9. Install or connect the following:
- Support bracket bolts: 17 ft. lbs. (24 Nm)
- Fuel rail and injectors
- Fuel line using new washers
- EGR valve and tighten its nuts to 15 ft. lbs. (21 Nm).
- IAC valve. Tighten its mounting bolts to 16 ft. lbs. (22 Nm).
- Fuel injector wiring harnesses
- Intake manifold wiring harnesses
- Intake manifold cooling hoses
- Throttle cable
- Intake air duct and air cleaner assembly

10. Refill and bleed the cooling system.
11. Connect the negative battery cable.
12. Verify that all sensors, valves, and vacuum lines are installed and connected properly. Be sure there are no loose electrical connections.
13. Turn the ignition on and off several times without starting the engine to pressurize the fuel system. Run the engine and check for proper operation. Check for vacuum leaks.
14. After the engine has warmed up, check the operation of the throttle cable and adjust it if necessary.

2.0L Engine

1. Before servicing the vehicle, refer to the precautions in the beginning of this section.
2. Drain the cooling system.
3. Relieve the fuel system pressure.
4. Remove or disconnect the following:
- Negative battery cable
- Cooling hoses from the intake manifold
- Vacuum hoses and electrical connectors from the manifold and throttle body

- Throttle cable from the throttle body
- Fuel rail and fuel injectors
- Intake manifold support brackets
- Intake manifold

To install:

5. Installation is the reverse of the removal procedure, while using the following torque values:
- Intake manifold fasteners: 16 ft. lbs. (22 Nm)
- Throttle body fasteners: 16 ft. lbs. (22 Nm)
- Intake manifold bracket bolts: 33 ft. lbs. (44 Nm)

2.2L and 2.3L Engines

1. Before servicing the vehicle, refer to the precautions in the beginning of this section.
2. Disconnect the negative battery cable.
3. Drain the engine coolant into a sealable container.
4. Remove or disconnect the following:
- Cooling hoses from the intake manifold

➡**Label all vacuum hoses and electrical connectors before detaching them.**

- Vacuum hoses and electrical connectors from the manifold and throttle body
- Connector from the Exhaust Gas Recirculation (EGR) valve. Position the wiring harnesses out of the way.
- Throttle cable from the throttle body
5. Relieve the fuel pressure.

6. Remove or disconnect the following:
- Fuel rail and fuel injectors
- Thermostat housing from the intake manifold and the connecting pipe by pulling and twisting the housing. Discard the O-rings.

➡**It may be necessary to remove the upper intake manifold plenum and throttle body assembly in order to access the nuts securing the manifold to the head.**

- Intake manifold support bracket bolts and the bracket. If it is necessary to access it from under the vehicle; raise and support the vehicle safely.
7. While supporting the intake manifold, remove the nuts attaching the intake manifold to the cylinder head, then remove the manifold. Remove the old gasket from the cylinder head.
8. Clean any old gasket material from the cylinder head and the intake manifold. Check and clean the FIA chamber on the cylinder head.

To install:

9. Install or connect the following:
- New gasket
- Intake manifold, and support the manifold.
- Support bracket to the manifold. Tighten the retaining bolt to 16 ft. lbs. (22 Nm).
10. Starting with the inner or center nuts, tighten the nuts, in a crisscross pattern, to the correct torque. The tension must be even across the entire face of the manifold if leaks are to be prevented. Correct torque is 16 ft. lbs. (22 Nm).

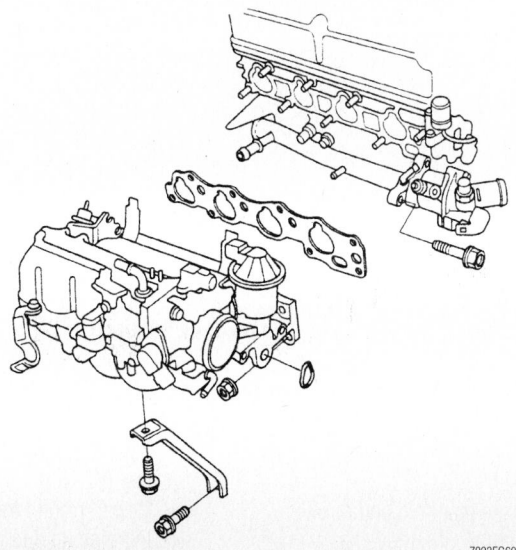

7923FG60

Intake manifold and related components—2.3L engine

NOTE: Use new O-rings and gaskets when reassembling.

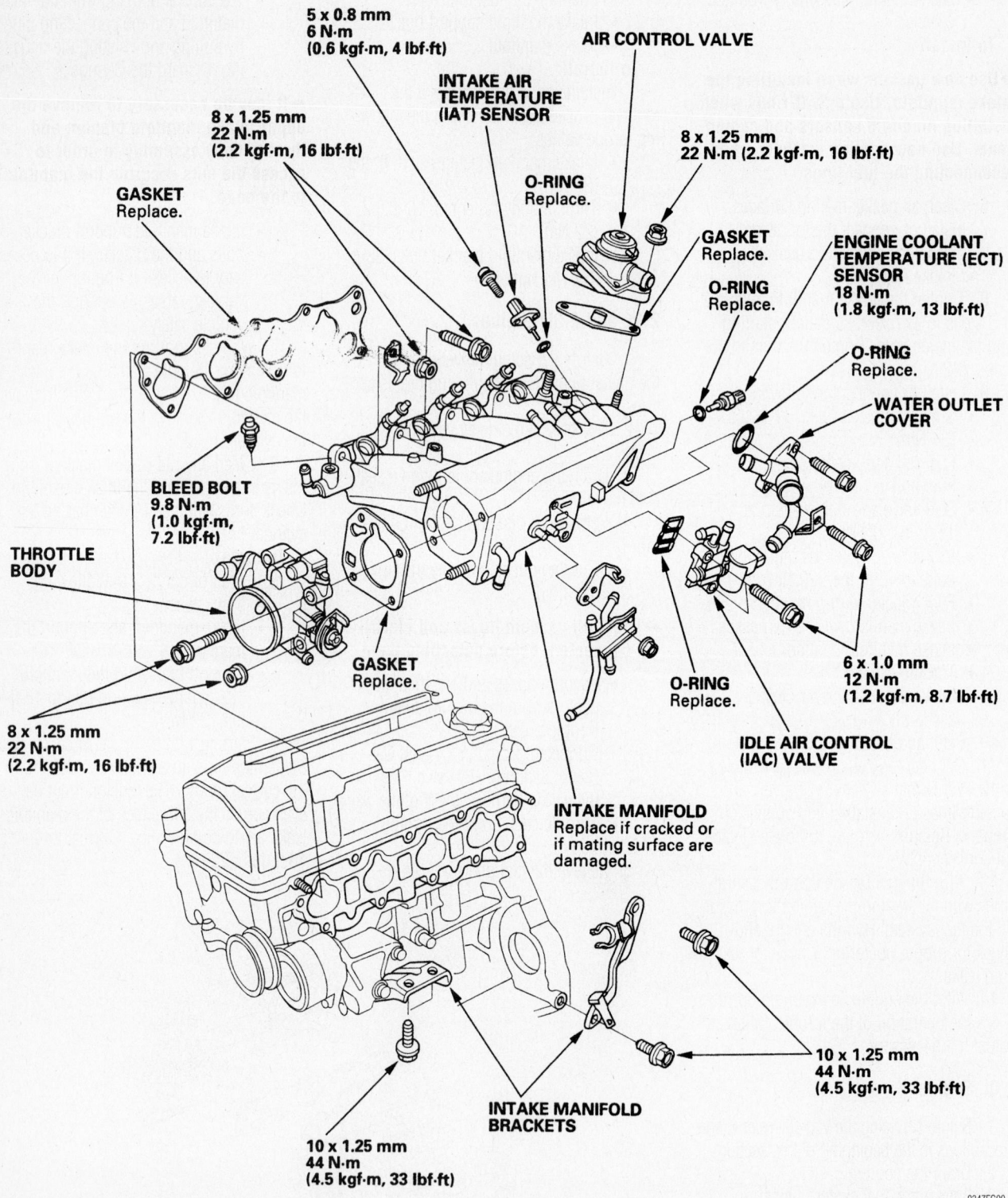

5 x 0.8 mm
6 N·m
(0.6 kgf·m, 4 lbf·ft)

AIR CONTROL VALVE

INTAKE AIR
TEMPERATURE
(IAT) SENSOR

8 x 1.25 mm
22 N·m
(2.2 kgf·m, 16 lbf·ft)

8 x 1.25 mm
22 N·m (2.2 kgf·m, 16 lbf·ft)

O-RING
Replace.

GASKET
Replace.

GASKET
Replace.

O-RING
Replace.

ENGINE COOLANT
TEMPERATURE (ECT)
SENSOR
18 N·m
(1.8 kgf·m, 13 lbf·ft)

O-RING
Replace.

WATER OUTLET
COVER

BLEED BOLT
9.8 N·m
(1.0 kgf·m,
7.2 lbf·ft)

THROTTLE
BODY

GASKET
Replace.

O-RING
Replace.

6 x 1.0 mm
12 N·m
(1.2 kgf·m, 8.7 lbf·ft)

IDLE AIR CONTROL
(IAC) VALVE

8 x 1.25 mm
22 N·m
(2.2 kgf·m, 16 lbf·ft)

INTAKE MANIFOLD
Replace if cracked or
if mating surface are
damaged.

INTAKE MANIFOLD
BRACKETS

10 x 1.25 mm
44 N·m
(4.5 kgf·m, 33 lbf·ft)

10 x 1.25 mm
44 N·m
(4.5 kgf·m, 33 lbf·ft)

9347FG09

Exploded view of the intake manifold—2.0L engine

11. Install or connect the following:
   • New gasket
   • Upper intake manifold and throttle
     body assembly, if removed as a
     separate unit. Tighten the nuts and

bolts holding the chamber to 16 ft.
lbs. (22 Nm).
   • New O-ring to the coolant connect-
     ing pipe, and to the thermostat
     housing.

   • Housing to the coolant pipe and the
     intake manifold. Tighten the mount-
     ing bolts to 16 ft. lbs. (22 Nm).
   • Throttle cable and adjust.
   • Fuel rail/injector assembly

- Fuel lines
- Wiring harnesses and the electrical connectors
- Vacuum hoses

12. Fill and bleed the air from the cooling system.

13. Connect the negative battery cable and enter the radio security code.

14. Start the engine and check carefully for any leaks of fuel, coolant or vacuum. Check the manifold gasket areas carefully for any leakage of vacuum.

15. If equipped with 4WS, turn the steering wheel lock-to-lock to reset the 4WS control unit.

2.4L Engine

1. Before servicing the vehicle, refer to the precautions in the beginning of this section.

2. Disconnect the negative battery cable.

3. Drain the engine coolant into a sealable container.

4. Remove or disconnect the following:
- Intake Air Temperature (IAT) sensor electrical connector
- Vacuum hose and breather pipe and the air intake duct
- Intake manifold cover
- Throttle and cruise control cables

by loosening the locknuts, then slipping the cable ends out of the accelerator linkage.

→Do not bend the cables during removal. Always replace any throttle or cruise control cables that get kinked during removal.

- Evaporative emission (EVAP) canister hose and brake booster vacuum hose
- Idle Air Control (IAC) valve connectors
- Throttle Position (TP) sensor connector

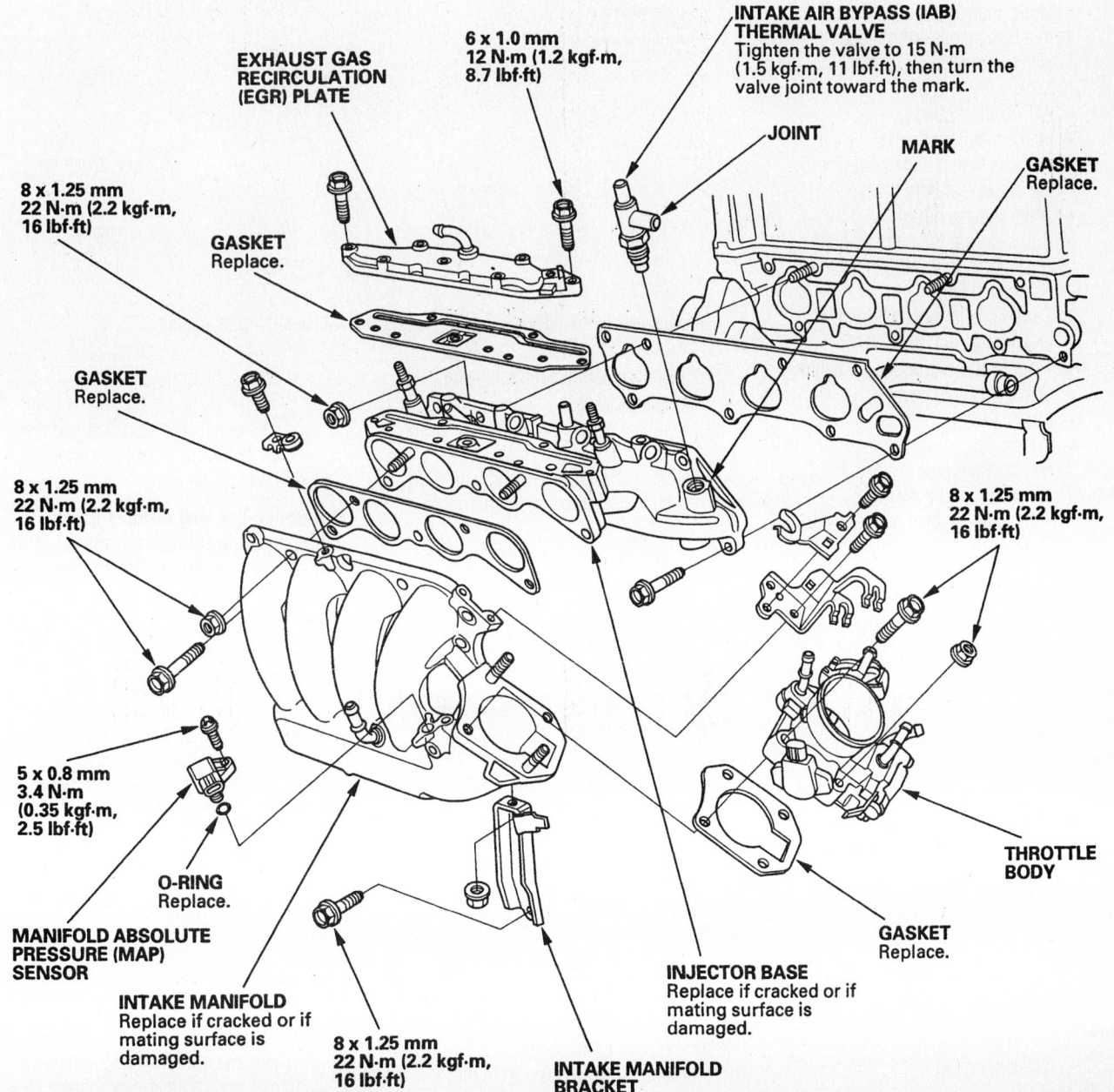

Exploded view of the intake manifold and related components—2.4L engine

42356-ACCO-G15

- Manifold Absolute Pressure (MAP) sensor connector
- Necessary engine wire harness connectors and wire harness clamps from the intake manifold
- Bolt securing the harness holder and remove the harness clamps
- Water bypass hoses, then plug them
- Harness clamp and harness connector from the intake manifold bracket
- Intake manifold bracket
- A/T vacuum hose
- Retainer and intake manifold

To install:
5. Clean the mounting surfaces.
6. Install or connect the following:
- New gasket
- Intake manifold. Tighten the bolts, in a criss-cross pattern beginning with the inner bolt, to 16 ft. lbs. (22 Nm).
- A/T vacuum hose
- Intake manifold bracket
- Harness clamp and connector to the intake manifold bracket
- Water bypass hoses
- Bolt securing the harness holder and tighten to 8.7 ft. lbs. (12 Nm)
- Harness clamps
- EVAP canister hose and brake booster vacuum hose
- Throttle and cruise control cables
- Intake manifold cover
- Intake air duct
- IAT sensor connector, vacuum hose and breather pipe
7. Refill the cooling system.
8. Connect the negative battery cable, start the engine, and check for leaks.

3.0L Engine

1. Before servicing the vehicle, refer to the precautions in the beginning of this section.
2. Obtain the security code for the radio.
3. Disconnect the negative battery cable.
4. Drain the coolant.
5. Remove or disconnect the following:
- Evaporative emissions (EVAP) canister hose from the throttle body.
- Air intake duct
- Upper engine covers
- Accelerator and cruise control cables from the throttle body.

➡**Ensure that all components have been removed from the intake manifold.**

- Intake manifold

To install:
6. Clean the mounting surfaces.

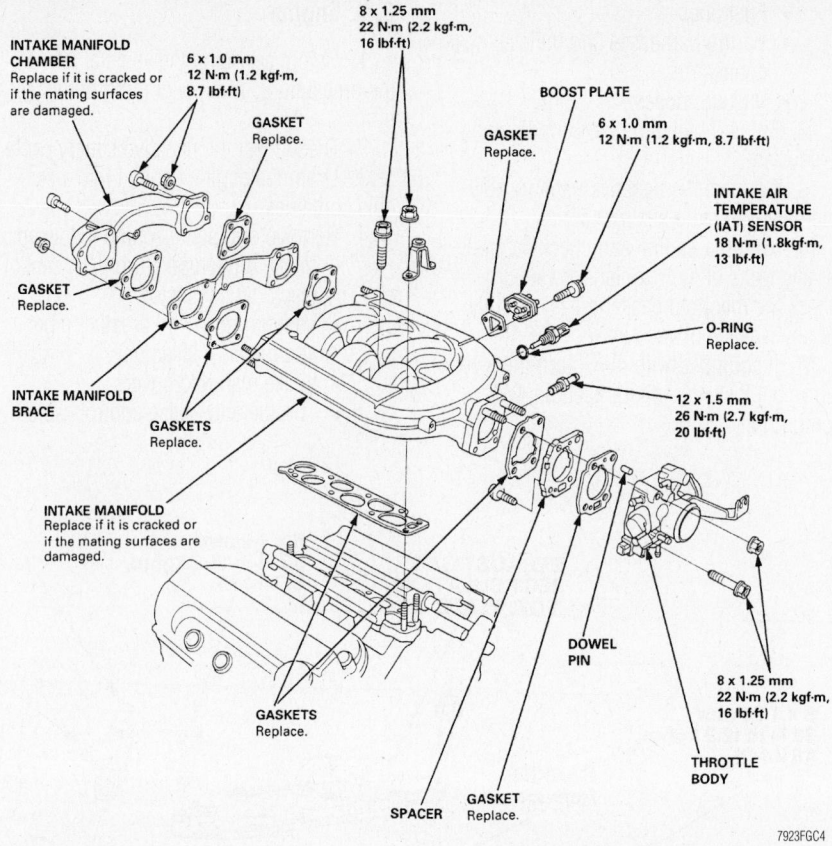

8 x 1.25 mm
22 N·m (2.2 kgf·m, 16 lbf·ft)

INTAKE MANIFOLD CHAMBER Replace if it is cracked or if the mating surfaces are damaged.

6 x 1.0 mm 12 N·m (1.2 kgf·m, 8.7 lbf·ft)

GASKET Replace.

BOOST PLATE

GASKET Replace.

6 x 1.0 mm 12 N·m (1.2 kgf·m, 8.7 lbf·ft)

INTAKE AIR TEMPERATURE (IAT) SENSOR 18 N·m (1.8kgf·m, 13 lbf·ft)

GASKET Replace.

O-RING Replace.

INTAKE MANIFOLD BRACE

GASKETS Replace.

12 x 1.5 mm 26 N·m (2.7 kgf·m, 20 lbf·ft)

INTAKE MANIFOLD Replace if it is cracked or if the mating surfaces are damaged.

DOWEL PIN

8 x 1.25 mm 22 N·m (2.2 kgf·m, 16 lbf·ft)

THROTTLE BODY

GASKETS Replace.

SPACER

GASKET Replace.

7923FGC4

Exploded view of the intake manifold and related components—3.0L engine

7. Install or connect the following:
- New gasket
- Intake manifold. Tighten the bolts to 16 ft. lbs. (22 Nm).
- All removed hoses and wiring on the intake manifold and throttle body.
- Engine covers
- Intake air duct
8. Refill the cooling system.
9. Connect the negative battery cable, start the engine, and check for leaks.

Exhaust Manifold

REMOVAL & INSTALLATION

1.7L Engines

1. Before servicing the vehicle, refer to the precautions in the beginning of this section.
2. Disconnect the negative battery cable.
3. Raise and support the front of the vehicle and block the rear wheels.
4. Remove or disconnect the following:
- Front exhaust pipe from the exhaust manifold/catalytic converter

- Exhaust manifold support brackets if their bolts are accessible from this angle. The splash shield may be removed for better access.
5. Lower the vehicle.

➡**Remove any rust or dirt from the exhaust manifold before removal. This will prevent dirt from entering the exhaust pipes.**

6. Remove or disconnect the following:
- Manifold heat shield
- Heated Oxygen Sensor (HO2S) harness
- HO2S, using an oxygen sensor socket or box end wrench to unscrew the sensor from the manifold. Handle the sensor carefully.
- Exhaust manifold brackets
- Exhaust manifold and separate it from the cylinder head
- Exhaust manifold and gasket

To install:

➡**Use new gaskets and self-locking nuts when installing the exhaust manifold.**

7. Clean the gasket mating surfaces of the manifold and cylinder head ports.
8. Install or connect the following:

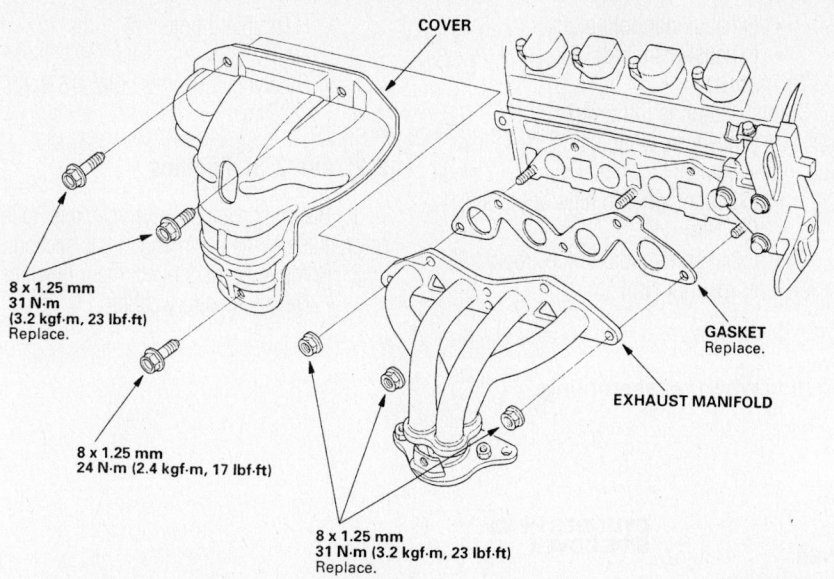

8 x 1.25 mm
31 N·m
(3.2 kgf·m, 23 lbf·ft)
Replace.

COVER

8 x 1.25 mm
24 N·m (2.4 kgf·m, 17 lbf·ft)

8 x 1.25 mm
31 N·m (3.2 kgf·m, 23 lbf·ft)
Replace.

GASKET
Replace.

EXHAUST MANIFOLD

67162-ACCO-G02

Exhaust manifold components—1.7L (D17A2) engine

bolts to 33 ft. lbs. (45 Nm) for all other engines

9. Carefully coat only the threads of the oxygen sensor body with anti-seize paste. Don't get any anti-seize on the sensor probe.

10. Install or connect the following:
  - HO2S and carefully tighten it to 33 ft. lbs. (45 Nm)
  - Heat shield and tighten the bolts to 16 ft. lbs. (22 Nm)
  - HO2S connector

11. Raise and support the front of the vehicle and block the rear wheels.

12. Install or connect the following:
  - Front exhaust pipe and the exhaust manifold/catalytic converter. Tighten the self-locking nuts to 40 ft. lbs. (55 Nm), if the converter is not attached to the manifold. If the converter is attached, tighten to 25 ft. lbs., (34 Nm). Install any manifold brackets and tighten them to 33 ft. lbs. (45 Nm).
  - Splash shield if it was removed

13. Lower the vehicle and connect the negative battery cable.

14. Run the engine and check for exhaust leaks.

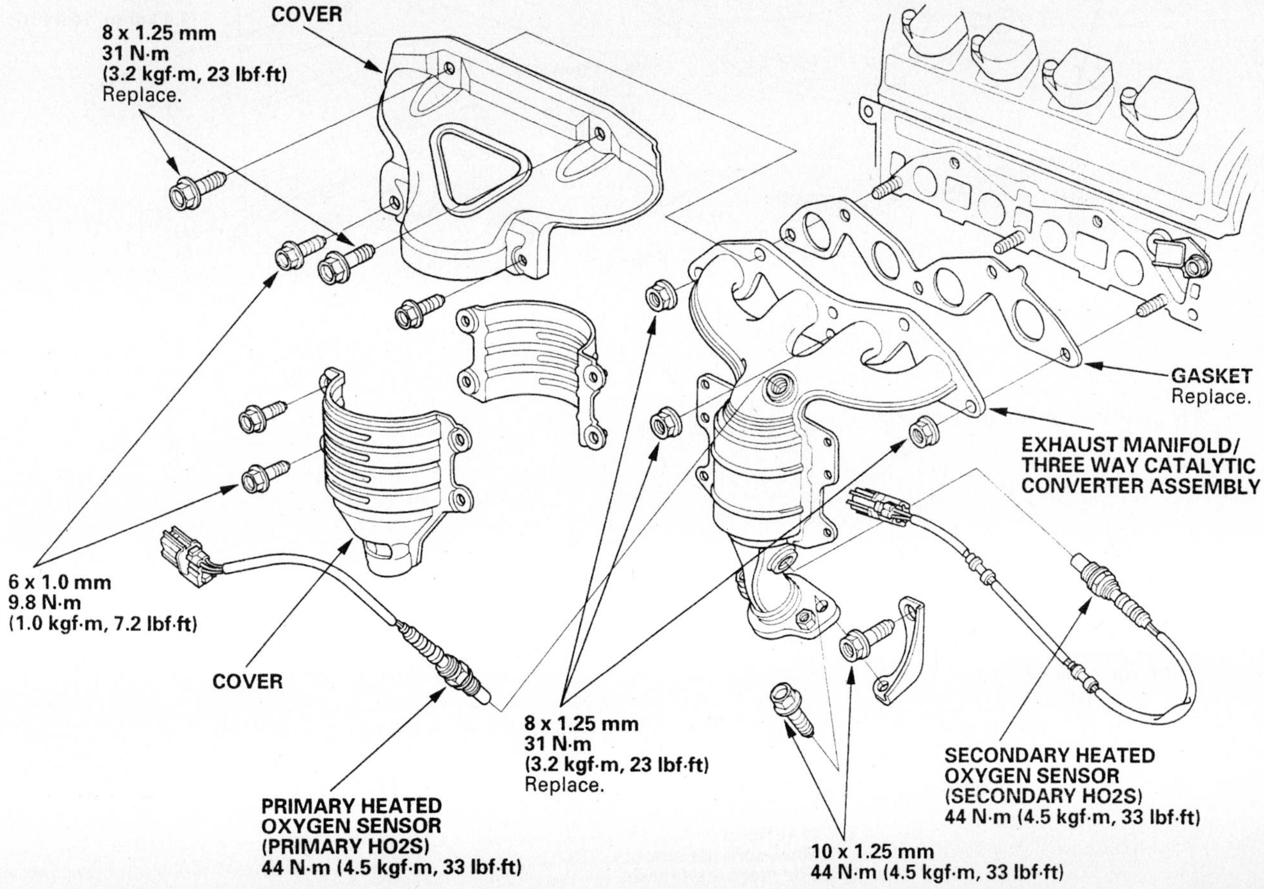

- New gasket onto the cylinder head
- New gaskets onto the exhaust pipe flange
- Exhaust manifold. Apply anti-seize paste to the studs. Tighten the self-locking nuts to 23 ft. lbs. (32 Nm) in a crisscross pattern starting in the center of the manifold and working outward.
- Manifold brackets and tighten their

8 x 1.25 mm
31 N·m
(3.2 kgf·m, 23 lbf·ft)
Replace.

COVER

6 x 1.0 mm
9.8 N·m
(1.0 kgf·m, 7.2 lbf·ft)

COVER

PRIMARY HEATED
OXYGEN SENSOR
(PRIMARY HO2S)
44 N·m (4.5 kgf·m, 33 lbf·ft)

8 x 1.25 mm
31 N·m
(3.2 kgf·m, 23 lbf·ft)
Replace.

10 x 1.25 mm
44 N·m (4.5 kgf·m, 33 lbf·ft)

GASKET
Replace.

EXHAUST MANIFOLD/
THREE WAY CATALYTIC
CONVERTER ASSEMBLY

SECONDARY HEATED
OXYGEN SENSOR
(SECONDARY HO2S)
44 N·m (4.5 kgf·m, 33 lbf·ft)

67162-ACCO-G03

Exhaust manifold components—1.7L (D17A1 & D17A6) engines

2.0L Engine

1. Before servicing the vehicle, refer to the precautions in the beginning of this section.

2. Remove or disconnect the following:
- Negative battery cable
- Catalytic converter
- Exhaust manifold heat shields
- Heated Oxygen (HO2S) sensor connector
- Exhaust manifold bracket
- Exhaust manifold

To install:

3. Installation is the reverse of the removal procedure, while using the following torque values:
- Exhaust manifold nuts: 23 ft. lbs. (31 Nm)
- Exhaust manifold bracket bolts: 33 ft. lbs. (44 Nm)
- Heat shield bolts: 16 ft. lbs. (22 Nm)
- Catalytic converter bolts: 16 ft. lbs. (22 Nm)

2.2L and 2.3L Engines

1. Before servicing the vehicle, refer to the precautions in the beginning of this section.

2. Remove or disconnect the following:
- Negative battery cable

NOTE: Use new gaskets and self-locking nuts when reassembling.

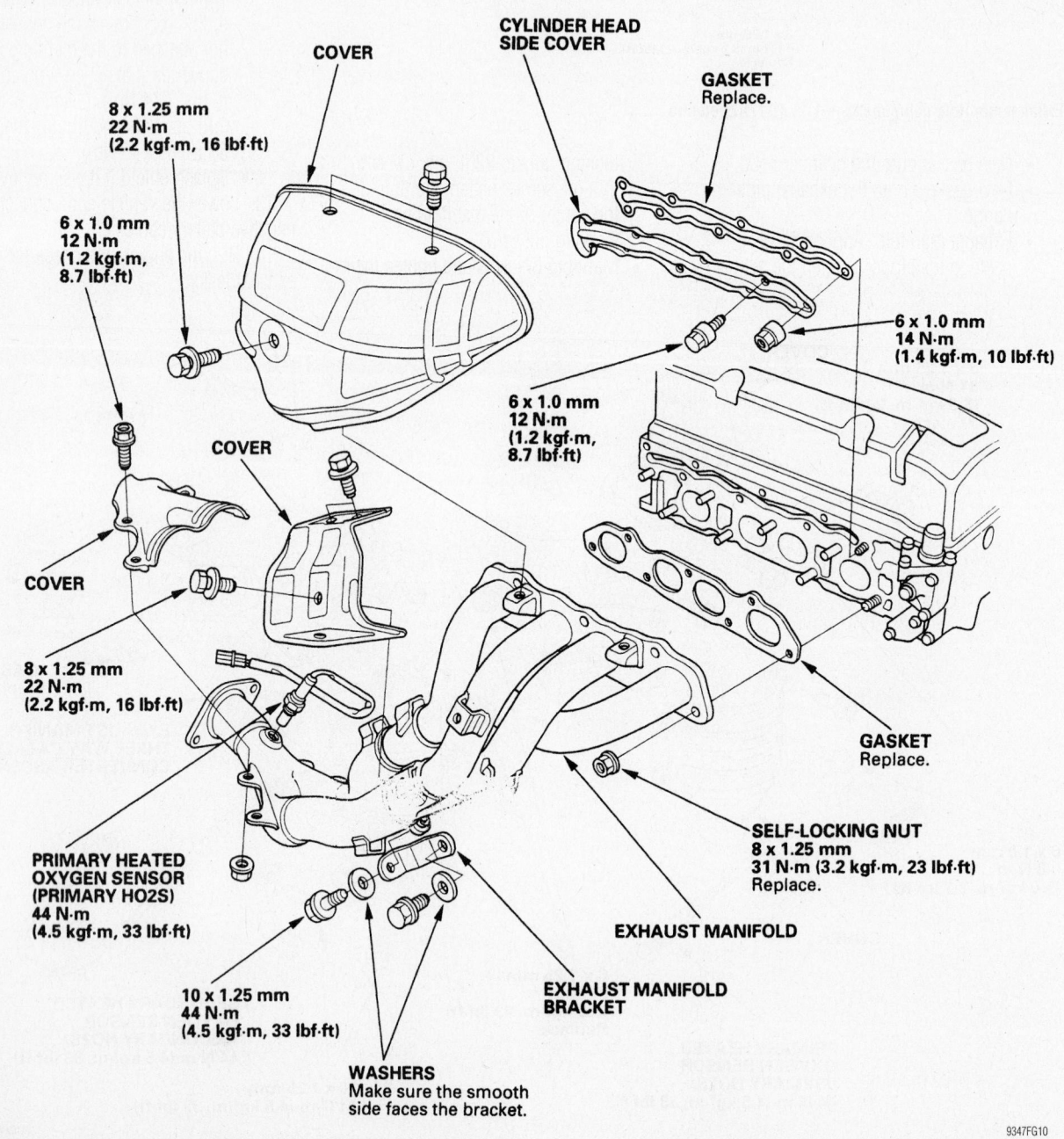

COVER

CYLINDER HEAD SIDE COVER

GASKET
Replace.

8 x 1.25 mm
22 N·m
(2.2 kgf·m, 16 lbf·ft)

6 x 1.0 mm
12 N·m
(1.2 kgf·m, 8.7 lbf·ft)

6 x 1.0 mm
14 N·m
(1.4 kgf·m, 10 lbf·ft)

6 x 1.0 mm
12 N·m
(1.2 kgf·m, 8.7 lbf·ft)

COVER

COVER

GASKET
Replace.

8 x 1.25 mm
22 N·m
(2.2 kgf·m, 16 lbf·ft)

SELF-LOCKING NUT
8 x 1.25 mm
31 N·m (3.2 kgf·m, 23 lbf·ft)
Replace.

EXHAUST MANIFOLD

PRIMARY HEATED OXYGEN SENSOR (PRIMARY HO2S)
44 N·m
(4.5 kgf·m, 33 lbf·ft)

10 x 1.25 mm
44 N·m
(4.5 kgf·m, 33 lbf·ft)

EXHAUST MANIFOLD BRACKET

WASHERS
Make sure the smooth side faces the bracket.

9347FG10

Exploded view of the exhaust manifold—2.0L engine

3. Safely raise and support the vehicle.
4. Remove or disconnect the following:
   - Oxygen Sensor (O_2S) connector, if it is located in the manifold.
   - Exhaust manifold upper cover
   - Heat insulator from the manifold, if equipped with air conditioning.
   - Nuts attaching the exhaust manifold to the front exhaust pipe.
   - Pipe from the manifold and discard the gasket. Support the pipe with wire; do not allow it to hang by itself.
   - Exhaust manifold bracket(s) bolts and bracket(s).
   - Exhaust manifold attaching nuts, using a crisscross pattern (starting from the center).
   - Manifold and discard the gasket. Clean the manifold and cylinder head mating surfaces.
   - Lower manifold cover from the manifold, if equipped.

To install:
5. Install or connect the following:
   - Lower manifold cover, if equipped, and tighten the attaching bolts to 16 ft. lbs. (22 Nm).
   - New gasket
   - Exhaust manifold into position and support it
   - New nuts snug on the studs
   - Support bracket(s) below the manifold. Tighten the bracket(s) mounting bolts to 33 ft. lbs. (44 Nm).

6. Starting with the manifold inner or center nuts, tighten the nuts in a crisscross pattern to the correct torque. The tension must be even across the entire face of the manifold if leaks are to be prevented. Tighten the nuts to 23 ft. lbs. (31 Nm).

7. Install or connect the following:
   - Heat insulator to the manifold, if equipped with air conditioning. Tighten the attaching bolts to 84 inch lbs. (10 Nm) on Prelude mod-

els and 108 inch lbs. (12 Nm) on Accord models.
   - Upper manifold cover and tighten the bolts to 16 ft. lbs. (22 Nm).
   - Oxygen sensor connector, if detached
   - Front exhaust pipe using new gaskets and nuts. Tighten the exhaust pipe attaching nuts to 40 ft. lbs. (55 Nm).
   - Negative battery cable and enter the radio security code.

8. Start the engine and check for exhaust leaks.

9. If equipped with 4WS, turn the steering wheel lock-to-lock to reset the 4WS control unit.

2.4L Engine

1. Before servicing the vehicle, refer to the precautions in the beginning of this section.

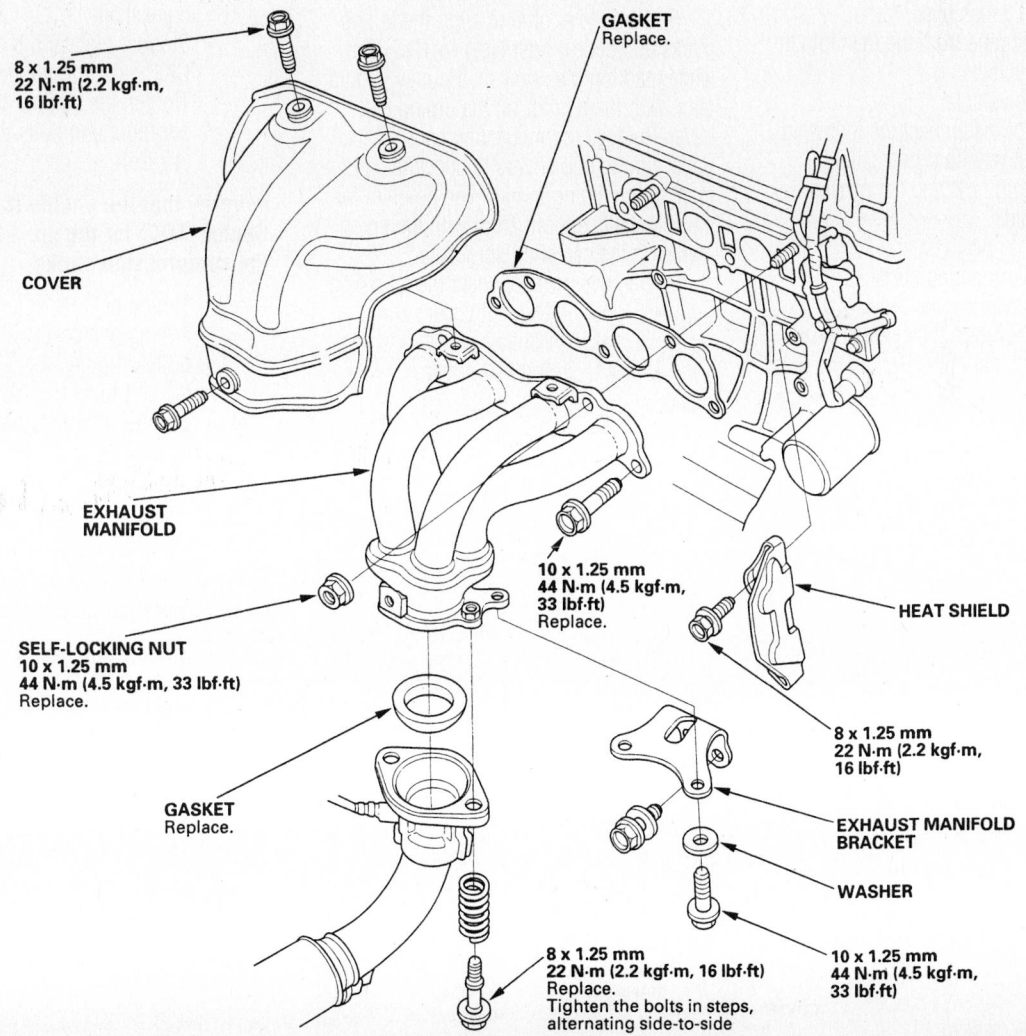

8 x 1.25 mm
22 N·m (2.2 kgf·m, 16 lbf·ft)

GASKET
Replace.

COVER

EXHAUST MANIFOLD

10 x 1.25 mm
44 N·m (4.5 kgf·m, 33 lbf·ft)
Replace.

HEAT SHIELD

SELF-LOCKING NUT
10 x 1.25 mm
44 N·m (4.5 kgf·m, 33 lbf·ft)
Replace.

8 x 1.25 mm
22 N·m (2.2 kgf·m, 16 lbf·ft)

EXHAUST MANIFOLD BRACKET

GASKET
Replace.

WASHER

8 x 1.25 mm
22 N·m (2.2 kgf·m, 16 lbf·ft)
Replace.
Tighten the bolts in steps, alternating side-to-side.

10 x 1.25 mm
44 N·m (4.5 kgf·m, 33 lbf·ft)

42356-ACCO-G16

Exploded view of the exhaust manifold and related components—2.4L engine

2. Raise and safely support the vehicle.
3. Remove or disconnect the following:
   - VTEC solenoid valve
   - Driveshaft heat shield
   - Cover and exhaust manifold bracket
   - Exhaust manifold

To install:

4. Clean the mounting surfaces.
5. Install or connect the following:
   - New gasket on the cylinder head
   - Exhaust manifold. Tighten the nuts, in a criss-cross pattern starting with the inner nut, to 33 ft. lbs. (45 Nm).
   - Exhaust manifold bracket and cover
   - Driveshaft heat shield
   - VTEC solenoid valve

3.0L Engine

1. Before servicing the vehicle, refer to the precautions in the beginning of this section.
2. Raise and safely support the vehicle.
3. Remove or disconnect the following:
   - Engine undercover
   - Exhaust pipe from the manifold to be removed
4. Lower the vehicle.
5. Remove or disconnect the following:
   - Exhaust manifold heat shield
   - Mounting nuts and the exhaust manifold.

To install:

6. Clean the mounting surfaces.
7. Install or connect the following:
   - New gasket on the cylinder head
   - Exhaust manifold. Tighten the nuts to 23 ft. lbs. (31 Nm).

- Heat shield. Tighten the bolts to 16 ft. lbs. (22 Nm).

8. Raise the vehicle and connect the exhaust pipe to the manifold using a new gasket. Tighten the nuts to 40 ft. lbs. (54 Nm).

Front Crankshaft Seal

REMOVAL & INSTALLATION

➡**The original radio may contain a coded anti-theft circuit. Obtain the security code number before disconnecting the battery cables.**

1. Before servicing the vehicle, refer to the precautions in the beginning of this section.
2. Disconnect the negative battery cable.
3. Safely raise and support the vehicle.
4. Remove or disconnect the following:
   - Splash shield
   - Engine accessory drive belts
5. Turn the engine to align the timing marks and set cylinder No.1 to TDC. The white mark on the crankshaft pulley should align with the pointer on the timing belt cover. Remove the inspection caps on the upper timing belt covers to check the alignment of the timing marks. The pointers for the camshafts should align with the green marks on the camshaft sprockets.
6. Remove or disconnect the following:
   - Upper timing belt covers and crankshaft pulley
   - Lower timing belt cover

➡**Mark the direction of the timing belt rotation if it is to be reinstalled.**

- Timing belt
- Crankshaft Position (CKP) sensor from the oil pump, if equipped
- Stopper plate
- Timing belt sprocket from the crankshaft. Do not lose the sprocket key.
- Seal from the front of the engine, using a suitable seal removal tool

To install:

7. Clean the seal mounting surfaces on the engine block.
8. Apply a thin coat of grease on the crankshaft and seal lips.
9. Install or connect the following:
   - Seal with the part number facing out. Use a seal driver to seat the seal against the oil pump. Clean any excess grease off the crankshaft and be sure the seal lip is not distorted.
   - Timing belt sprocket and key to the crankshaft
   - Stopper plate and if equipped, the CKP sensor to the oil pump. Tighten the stopper plate and sensor mounting bolts to 108 inch lbs. (12 Nm).

➡**Verify that the engine is at Top Dead Center (TDC) for the no. 1 cylinder on the compression stroke.**

- Timing belt
- Timing belt covers and crankshaft pulley. Tighten the crankshaft pulley bolt to 181 ft. lbs. (245 Nm), with the aid of a crank pulley holder.
- Accessory drive belts, then adjust.

➡**Verify that all engine components that may have been removed have been reinstalled correctly.**

- Splash shield and lower the vehicle.
- Negative battery cable

10. Top up the engine oil if necessary.
11. Run the engine and check for leaks.

Camshaft

REMOVAL & INSTALLATION

1.7L Engines

1. Before servicing the vehicle, refer to the precautions in the beginning of this section.

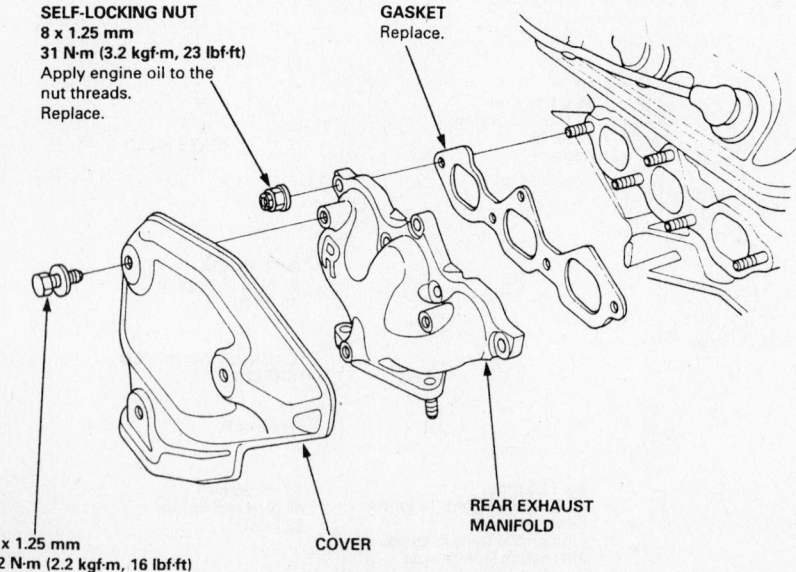

SELF-LOCKING NUT
8 x 1.25 mm
31 N·m (3.2 kgf·m, 23 lbf·ft)
Apply engine oil to the nut threads.
Replace.

GASKET
Replace.

8 x 1.25 mm
22 N·m (2.2 kgf·m, 16 lbf·ft)

COVER

REAR EXHAUST MANIFOLD

7923FG93

Exploded view of the rear exhaust manifold mounting—3.0L engine

2. Remove or disconnect the following components:
- Negative battery cable
- Ignition wires
- Valve cover and the upper timing belt cover.

3. Rotate the crankshaft to set the No. 1 cylinder at Top Dead Center (TDC) for the compression stroke. Once the engine is in this position, it shouldn't be disturbed.

4. Remove or disconnect the following components:
- Timing belt. If the timing belt is contaminated with oil or coolant, it must be replaced. If the timing belt is to be reused, mark its direction of rotation.
- Distributor, if equipped
- Camshaft sprocket and its key. Remove the upper rear timing cover.
- Rocker arm locknuts and the valve adjusting screws.
- Camshaft holder bolts in a 2-step crisscross sequence, starting at the edges and working toward the center of the cylinder head.
- Rocker arm and shaft assembly. Leave the camshaft holder bolts in the camshaft holders to hold the rocker arm and shaft assembly together.

5. Wrap rubber bands around the VTEC rocker arm assemblies so that they do not separate.

6. Store the rocker arm and shaft assembly away from your work area. Cover the assembly with shop towels or a sheet of plastic to protect it from dust.

7. Lift the camshaft from the cylinder head. Remove the camshaft seal.

8. Inspect the camshaft journals and lobes for signs of scoring or other damage.

To install:

9. Remove the oil control orifice. Thoroughly clean it and reinstall it with a new O-ring.

10. Clean and inspect the camshaft bearing caps in the cylinder head.

11. Lubricate the lobes and journals of the camshaft prior to installation.

12. Install or connect the following:
- Camshaft with the keyway facing up so that the camshaft will be at Top Dead Center (TDC)/compression for the No. 1 cylinder.
- New, lightly lubricated, camshaft seal

13. Install the rocker arm and shaft assembly as follows:
   a. Remove the rubber bands from the VTEC rocker arms.

   b. Lubricate the rocker arm contact surfaces.

   c. Apply liquid gasket to the head mating surfaces of the No. 1 and No. 5 camshaft holders. Don't allow the sealant to cure before installing the rocker arm assembly.

   d. Set the rocker arm and shaft assembly in place. If equipped, install the lost motion assembly holder.

   e. Coat the threads of the camshaft holder bolts with clean oil and loosely install them.

   f. Tighten each bolt 2 turns at a time in the crisscross sequence to ensure that the rockers and camshaft holder do not bind on the camshaft journals.

   g. Tighten the 8mm camshaft holder bolts to 14 ft. lbs. (20 Nm), and the 6mm camshaft holder bolts to 108 inch lbs. (12 Nm).

14. Install or connect the following:
- Camshaft sprocket and key. Tighten the retaining bolt to 27 ft. lbs. (38 Nm).

➥**Verify that the engine remains at Top Dead Center (TDC)/compression for the No. 1 cylinder.**

- Distributor, if equipped
- Timing belt. Tighten the tensioner bolt to 33 ft. lbs. (44 Nm) once the belt has been properly tensioned.
- Lower timing cover. Tighten the crankshaft pulley bolt to 134 ft. lbs. (181 Nm).

15. Adjust the valves.

16. Manually inspect the VTEC rocker arms for smooth motion.

17. Be sure all the spark plug tube sealing gaskets are fully seated.

18. Apply liquid gasket to the corner recesses of a new valve cover gasket.

19. Install or connect the following:
- Gasket to the valve cover. Don't allow the sealant to cure before installation.
- Valve cover. Gently wiggle the valve cover to be sure it is fully seated. Tighten the valve cover bolts in a crisscross pattern to 84 inch lbs. (10 Nm).
- Ignition wires

20. Refill the engine with fresh oil and install a new filter.

21. Reconnect the battery cable.

22. Warm the engine up to normal operating temperature. Check for oil leaks.

23. Check the ignition timing and adjust it if necessary. Then, tighten the distributor mounting bolts to 17 ft. lbs. (24 Nm).

2.0L Engine

1. Before servicing the vehicle, refer to the precautions in the beginning of this section.

2. Loosen the valve adjustment screws so that all valves are closed and all rocker arms are loose.

3. Remove or disconnect the following:
- Negative battery cable
- Valve cover
- Camshaft bearing caps
- Camshafts

To install:

4. Set the engine to Top Dead Center (TDC) so that the timing chain sprocket timing marks are aligned with the cylinder head surface as shown.

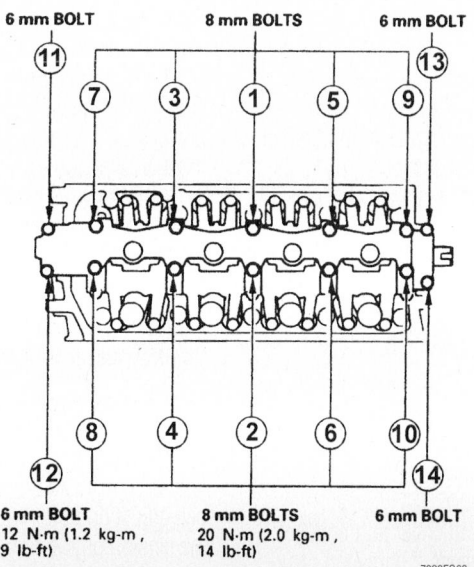

Camshaft holder bolt tightening sequence—1.7L engines

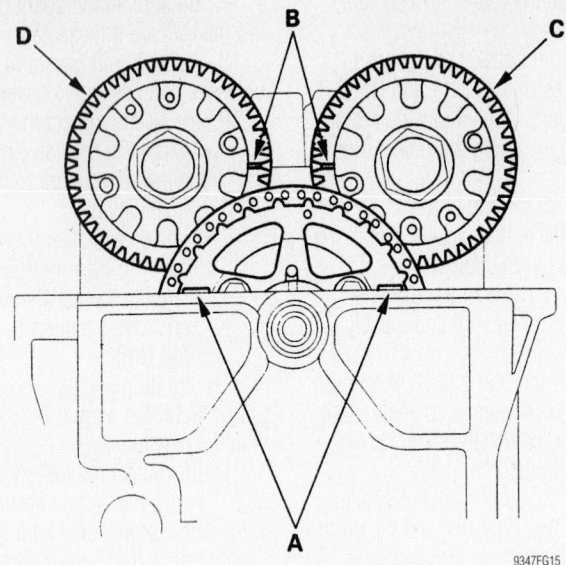

Timing chain sprocket alignment marks (A) and camshaft sprocket alignment marks (B)—2.0L engine

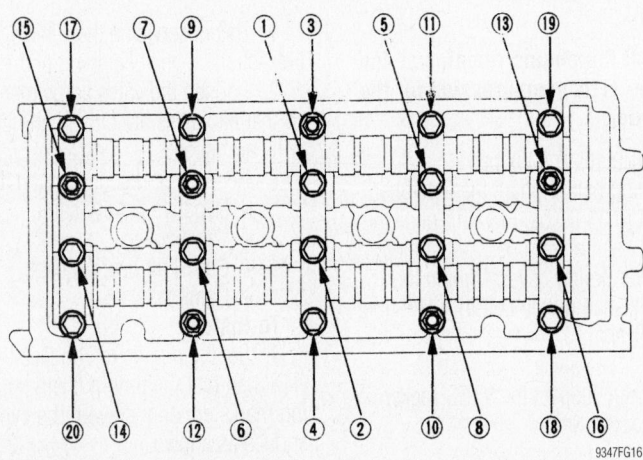

Camshaft bearing cap torque sequence—2.0L engine

5. Install or connect the following:
- Camshafts with the sprocket timing marks aligned as shown
- Camshaft bearing caps and tighten the bolts in sequence to 16 ft. lbs. (22 Nm). Adjust the valve clearance.
- Valve cover
- Negative battery cable

2.2L Engine

1. Before servicing the vehicle, refer to the precautions in the beginning of this section.
2. Remove or disconnect the following:
- Negative battery cable
- Air intake duct
- Cylinder head cover and replace the rubber seals if damaged or deteriorated.

- Timing belt upper cover
3. Turn the engine to align the timing marks and set cylinder No.1 to TDC/compression. Once in this position, the engine must NOT be turned or disturbed.
4. Remove or disconnect the following:
- Electrical connectors from the distributor and the spark plug wires from the spark plugs.

➡ Matchmark the installed position of the distributor before removing it.

- Distributor from the cylinder head
- Power steering pump drive belt
5. Mark the rotation of the timing belt if it is to be used again. Loosen the timing belt adjusting bolt ¾ to 1 turn, then release the tension on the timing belt. Push the tensioner to release tension from the belt, then tighten the adjusting bolt.

6. Remove or disconnect the following:
- Timing belt from the camshaft sprocket

✳✳ WARNING

Do not crimp or bend the timing belt more than 90°, or less than 1 inch (25mm) in diameter

➡ Ensure the words UP embossed on the camshaft pulley is aligned in the upward position before removing the sprocket bolt.

- Camshaft sprocket bolt, sprocket and sprocket key
- Timing belt back cover
- Valve adjusting screws
- Camshaft holder attaching bolts 2 turns at a time, in the proper sequence to prevent damaging the valves or rocker arm assemblies

➡ When removing the rocker arm assembly, do not remove the camshaft holder bolts. The bolts will keep the camshaft holders, springs, and the rocker arms on the shafts.

- Camshaft holders and rocker arm assembly. If the rocker arm and shaft assembly needs to be disassembled for service, note the location of the components as they are removed. The rocker arms must be installed in the same position if reused
- Camshaft from the cylinder head and discard the seal

To install:
7. Wipe the camshaft and the camshaft journals clean, then lubricate both surfaces,
8. Install or connect the following:
- Camshaft
9. Turn the camshaft so that its keyway is facing up (No. 1 cylinder will be at Top Dead Center (TDC)).
10. Reassemble the rocker arm and shaft assembly if it was disassembled. Lubricate the rocker arm and shaft assembly with clean oil, then apply liquid gasket to the head mating surfaces of the No. 1 and No. 6 camshaft holders.
11. Install or connect the following:
- Camshaft holders and rocker arm assembly in place, then loosely install the attaching bolts
- Clean oil to the camshaft oil seal lip and the seal guide (part # 07NAG-PT0010A)
- Seal to the seal guide
- Seal guide to the camshaft, then the installer cup (part # 07NAF-

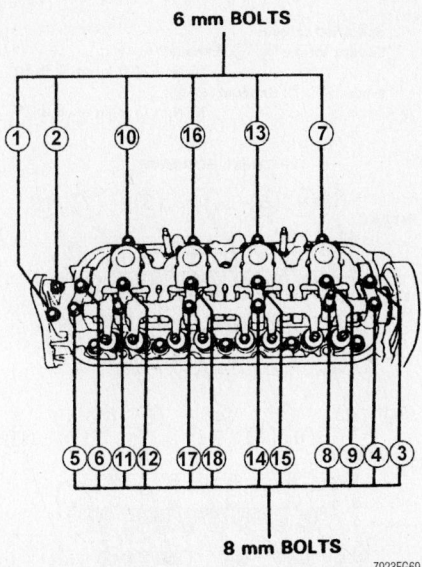

6 mm BOLTS

8 mm BOLTS

7923FG69

Rocker arm assembly bolt removal sequence—2.2L (H22A4) engines

PT0010A) and the installer shaft (part # 07NAF-PT0020A). Tighten the nut on the installer shaft to press the seal into the cylinder head.

12. Tighten the camshaft holder bolts 2 turns at a time in the proper sequence. The final torque for the 8mm bolts is 16 ft. lbs. (22 Nm) and the final torque for the 6mm bolts is 108 inch lbs. (12 Nm).

13. Install or connect the following:
   • Timing belt back cover and tighten the attaching bolt to 108 inch lbs. (12 Nm)
   • Camshaft sprocket key to the camshaft, then the camshaft sprocket. Install the bolt and tighten it to 27 ft. lbs. (37 Nm)

➡**Ensure the words UP embossed on the camshaft pulley is aligned in the upward position, before installing the timing belt.**

   • Timing belt onto the camshaft sprocket. Loosen, then tighten the timing belt adjusting nut.

14. Rotate the crankshaft pulley 5 or 6 turns to position the timing belt on the pulleys.

15. Set the No. 1 cylinder to TDC and loosen the timing belt adjusting nut 1 turn. Turn the crankshaft counterclockwise until the cam pulley has moved 3 teeth; this creates tension on the timing belt. Loosen the timing belt adjusting nut, then tighten it to 33 ft. lbs. (45 Nm).

16. Adjust the valves.

17. Tighten the crankshaft pulley bolt to 181 ft. lbs. (245 Nm)

18. Install or connect the following:
   • Upper timing belt cover and tighten the bolt to 108 inch lbs. (12 Nm)
   • Cylinder head cover gasket in the groove on the cylinder head cover. Before installing the gasket thoroughly clean the seal and the groove. Seat the recesses for the camshaft first, then work it into the groove around the outside edges. Be sure the gasket is seated securely in the corners of the recesses.
   • Cylinder head (valve) cover and tighten the cap nuts to 84 inch lbs. (10 Nm).
   • Power steering belt, then adjust
   • Distributor to the cylinder head. Snug the attaching bolts until the timing has been checked and adjusted.
   • Spark plug wires and the distributor electrical connectors
   • Ignition coil wire to the distributor
   • Air intake duct

19. Drain the oil from the engine into a sealable container. Install the drain plug and refill the engine with clean oil.

20. Connect the negative battery cable and enter the radio security code.

21. Start the engine and check carefully for any leaks.

22. Check and adjust the ignition timing as necessary, then tighten the distributor bolts to 16 ft. lbs. (22 Nm).

2.3L Engines

1. Before servicing the vehicle, refer to the precautions in the beginning of this section.

2. Disconnect the negative battery cable.

3. Turn the crankshaft so the No. 1 piston is at Top Dead Center (TDC).

➡**The No. 1 piston is at Top Dead Center (TDC) when the pointer on the block aligns with the white painted mark on the flywheel (manual transaxle) or driveplate (automatic transaxle).**

4. Remove or disconnect the following:
   • Air intake duct
   • Engine ground cable from the cylinder head cover
   • Connector and the terminal from the alternator
   • Engine wiring harness from the valve cover
   • Ignition coil

➡**Tag all electrical connectors before disconnecting them.**

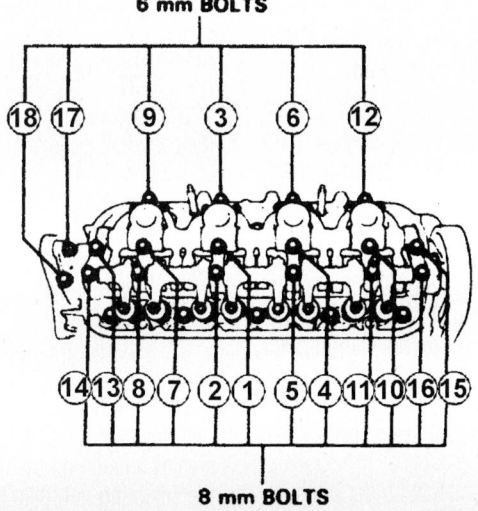

6 mm BOLTS

8 mm BOLTS

7923FG71

Rocker arm assembly torque sequence—2.2L (H22A4) engines

- Electrical connectors from the distributor
- Spark plug wires from the spark plugs
- Distributor from the cylinder head
- Ignition coil wire from the distributor
- Positive Crankcase Ventilation (PCV) hose
- Cylinder head cover. Replace the rubber seals if damaged or deteriorated.
- Timing belt middle cover

5. Ensure the words **UP** embossed on the camshaft pulleys are aligned in the upward position.

6. Mark the rotation of the timing belt if it is to be used again. Loosen the timing belt adjusting nut ½ turn, then release the tension on the timing belt. Push the tensioner to release tension from the belt, then tighten the adjusting nut.

7. Remove or disconnect the following:
- Timing belt from the camshaft sprockets

✳✳ WARNING

Do not crimp or bend the timing belt more than 90°, or less than 1 inch (25mm) in diameter

8. Insert a 5.0mm pin punch in each of the camshaft caps nearest to the sprockets through the holes provided.

9. Remove or disconnect the following:
- Camshaft sprocket attaching bolts, then the sprockets. Do not lose the sprocket keys.
- Side engine mount bracket B, then the timing belt back cover from behind the camshaft sprockets
- Rocker arm adjusting screws, then the pin punches from the camshaft caps
- Camshaft holders, note the holders locations for ease of installation. Loosen the bolts in the reverse order of the installation.
- Camshafts from the cylinder head, then discard the camshaft seals
- Rubber cap from the head, located at the end of the intake camshaft

10. Remove the rocker arms from the cylinder head. Note the locations of the rocker arms.

➡ **The rocker arms have to be installed to their original locations if being reused.**

To install:

11. Lubricate the rocker arms with clean oil.

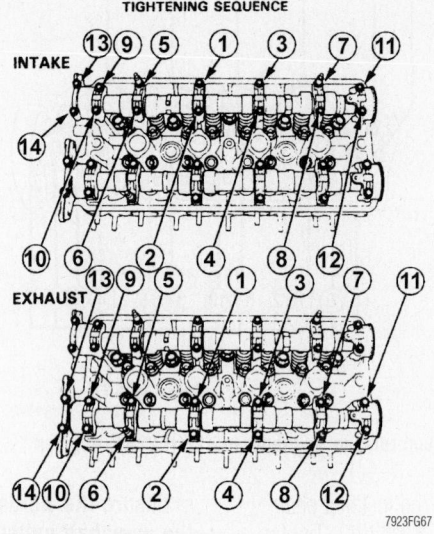

Specified torque:
Except Intake ⑤, ⑦. Exhaust ⑥, ⑧:
   10 N·m (1.0 kg-m, 7 lb-ft)
Intake ⑤, ⑦. Exhaust ⑥, ⑧:
   12 N·m (1.2 kg-m, 9 lb-ft)

Camshaft holders torque sequence—2.3L engines

12. Install or connect the following:
- Rocker arms on the pivot bolts and the valve stems. If the rocker arms are being reused, install them to their original locations. The locknuts and adjustment screws should be loosened before installing the rocker arms.

13. Lubricate the camshafts with clean oil.

14. Install or connect the following:
- Camshaft seals to the end of the camshafts that the timing belt sprockets attach to. The open side (spring) should be facing into the cylinder head when installed.

➡ **Be sure the keyways on the camshafts are facing up and install the camshafts to the cylinder head.**

- Rubber plug to the cylinder head at the end of the intake camshaft

15. Apply liquid gasket to the head mating surfaces of the No. 1 and No. 6 camshaft holders, then install them along with No. 2, 3, 4 and 5. **I** or **E** marks are stamped on the camshaft holders to identify them as Intake or Exhaust side holders. The arrows stamped on the holders should point toward the timing belt.

16. Snug the camshaft holders in place.

17. Press the camshaft seals securely into place.

18. Tighten the camshaft holder bolts in 2 steps, following the proper sequence, to ensure that the rockers do not bind on the valves. Tighten all the bolts, except the 4 studs, to 84 inch lbs. (10 Nm). Tighten the studs (number 5 and 7 bolts in the correct sequence) to 108 inch lbs. (12 Nm).

19. Install or connect the following:
- Timing belt back cover
- Side engine mount bracket B. Tighten the bolt attaching the bracket to the cylinder head to 33 ft. lbs. (45 Nm). Tighten the bolts attaching the bracket to the side engine mount to 16 ft. lbs. (22 Nm).
- 5.0mm pin punch in each of the camshaft caps, nearest to the pulleys, through the holes provided.
- Keys into the camshaft grooves

20. Push the camshaft sprockets onto the camshafts, then tighten the retaining bolts to 27 ft. lbs. (38 Nm).

21. Ensure the words **UP** embossed on the camshaft pulleys are aligned in the upward position.

22. Install or connect the following:
- Timing belt to the camshaft sprockets, then remove the 2, 5.0mm pin punches from the camshaft bearing caps

23. Loosen, then tighten the timing belt adjuster nut.

24. Turn the crankshaft counterclockwise until the cam pulley has moved 3 teeth; this creates tension on the timing belt. Loosen the adjusting nut, then tighten it to 33 ft. lbs. (45 Nm).

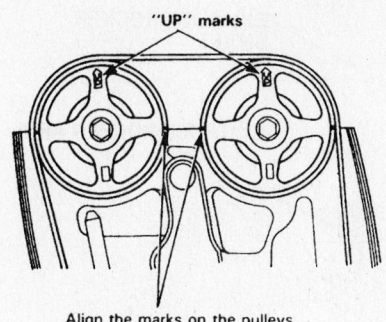

Camshaft sprockets alignment—2.3L engines

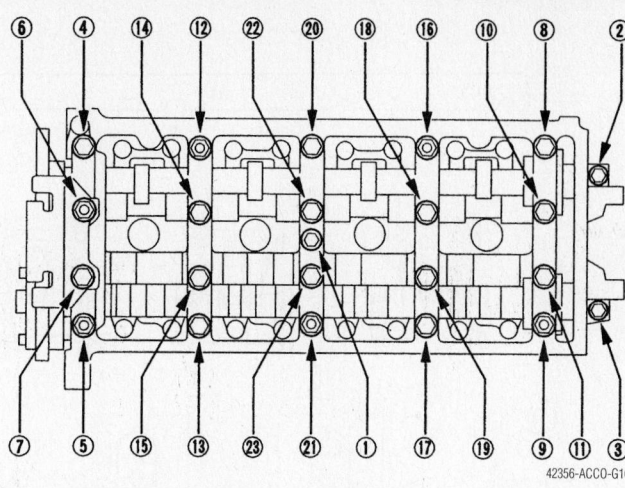

Camshaft holder bolt loosening sequence—2.4L engine

25. Adjust the valves.

26. Tighten the crankshaft pulley bolt to 181 ft. lbs. (250 Nm).

27. Install or connect the following:
- Middle timing belt cover and tighten the attaching bolts to 108 inch lbs. (12 Nm)
- Cylinder head cover and tighten the cap nuts to 84 inch lbs. (10 Nm)
- PCV hose to the cylinder head cover
- Distributor to the cylinder head, snug the attaching bolts until the timing has been checked and adjusted
- Spark plug wires and distributor electrical connectors
- Ignition coil wire to the distributor
- Ignition coil
- Alternator wiring harness to the cylinder head cover
- Terminal and connector to the alternator
- Engine ground cable to the cylinder head cover
- Air intake duct

28. Drain the oil from the engine into a sealable container. Install the drain plug and refill the engine with clean oil.

29. Connect the negative battery cable and enter the radio security code.

30. Start the engine and check carefully for any leaks.

31. Check and adjust the ignition timing. Tighten the distributor bolts to 13 ft. lbs. (18 Nm).

32. If equipped with 4WS, turn the steering wheel lock-to-lock to reset the 4WS control unit.

2.4L Engine

1. Before servicing the vehicle, refer to the precautions in the beginning of this section.

2. Disconnect the negative battery cable.

3. Remove or disconnect the following:
- Timing chain
- Loosen the rocker arm adjusting screws
- Camshaft holder bolts, two turns at a time in sequence
- Timing chain guide (B), camshaft and camshafts

4. Insert the bolts (A) into the rocker shaft holder, then remove the rocker arm assembly (B)
- Camshafts by carefully lifting them out of the cylinder head

To install:

5. Clean and dry the No. 5 rocker shaft holding mating surface.

6. Apply a suitable liquid gasket P/N 08718-0009, or equivalent, evenly to the cylinder head mating surface of the No. 5 rocker shaft holder.

➡**The parts must be installed within 5 minutes of applying the liquid gasket.**

7. Reassemble the rocker arm assembly, as necessary.

8. Install or connect the following
- Bolts (A) into the rocker shaft holder, then the rocker arm assembly on the cylinder head. Remove the bolts from the rocker shaft holder.

9. Make sure the punch marks on the variable valve timing control (VTC) actuator and exhaust camshaft sprocket are facing up, then set the camshafts (A) in the holder

10. Set the camshaft holders (B) and timing chain guide B (C) in place.

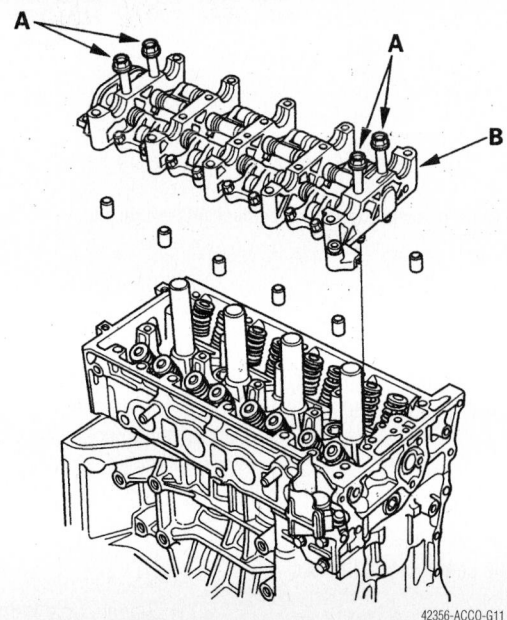

Insert the bolts (A) into the rocker shaft holder, then remove the rocker arm assembly (B)—2.4L engine

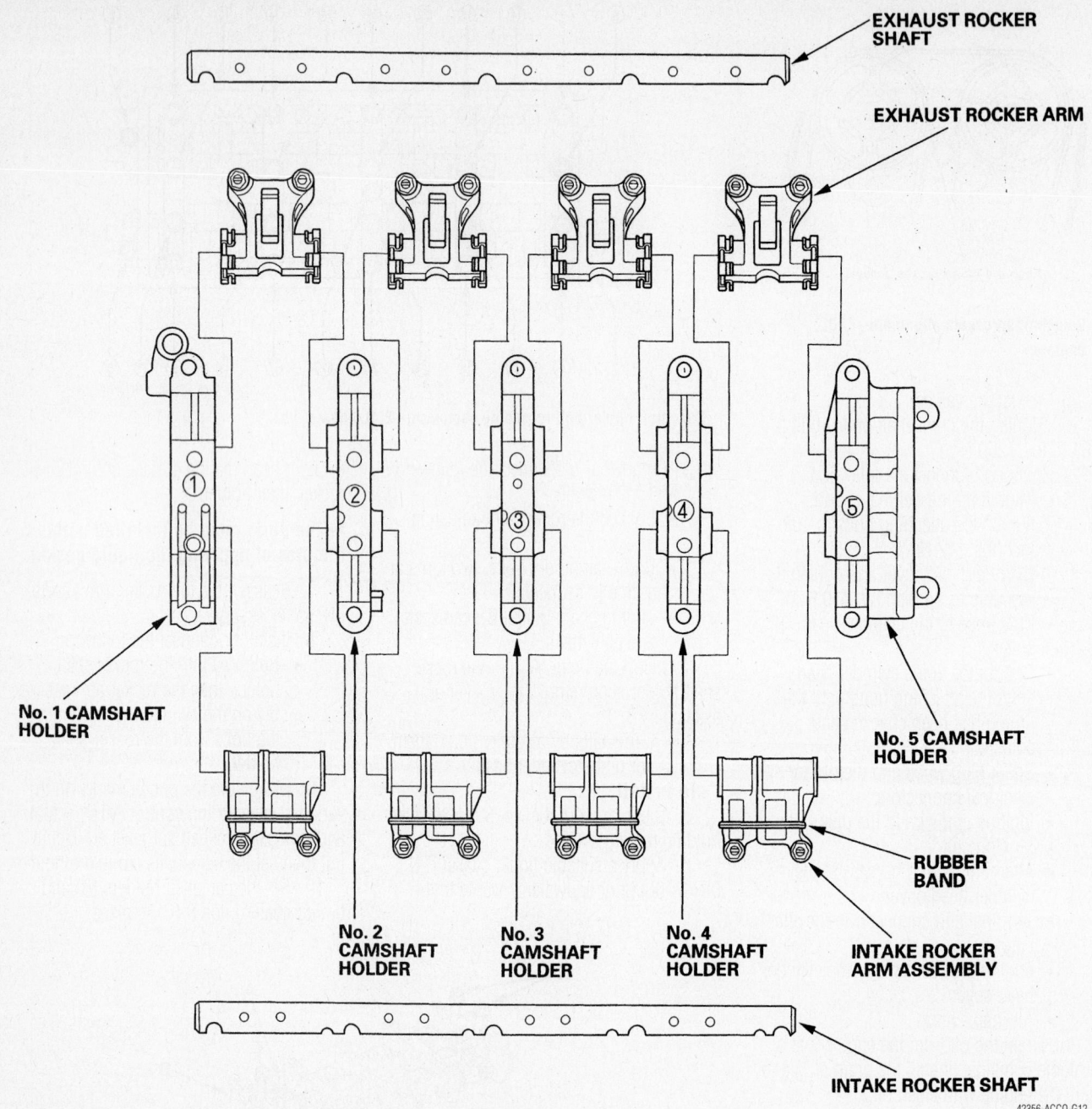

EXHAUST ROCKER SHAFT

EXHAUST ROCKER ARM

No. 1 CAMSHAFT HOLDER

No. 5 CAMSHAFT HOLDER

No. 2 CAMSHAFT HOLDER

No. 3 CAMSHAFT HOLDER

No. 4 CAMSHAFT HOLDER

RUBBER BAND

INTAKE ROCKER ARM ASSEMBLY

INTAKE ROCKER SHAFT

42356-ACCO-G12

Exploded view of the rocker arms and related components—2.4L engine

11. Tighten the bolts, in sequence, to the following specification:
   a. 8mm bolts: 16 ft. lbs. (22 Nm)
   b. 6mm bolts: 8.7 ft. lbs. (12 Nm)
12. Install the timing chain and adjust the valve lash.

3.0L Engine

1. Before servicing the vehicle, refer to the precautions in the beginning of this section.
2. Disconnect the negative battery cable.

3. Turn the engine to align the timing marks and set cylinder No.1 to TDC. The white mark on the crankshaft pulley should align with the pointer on the timing belt cover. Remove the inspection caps on the upper timing belt covers to check the alignment of the timing marks. The pointers for the camshafts should align with the green marks on the camshaft pulleys.
4. Remove or disconnect the following:
   • Intake air duct
   • Starter cable from the strut brace
   • Strut brace
   • Intake manifold covers

   • Breather hose from the cylinder head cover
   • Positive Crankcase Ventilation (PCV) hose from the cylinder head cover.
   • Air conditioning compressor belt
   • Alternator drive belt
   • Power steering drive belt

➡Label all electrical connectors before detaching them.

   • Electrical connectors from the distributor and spark plug wires from the spark plugs

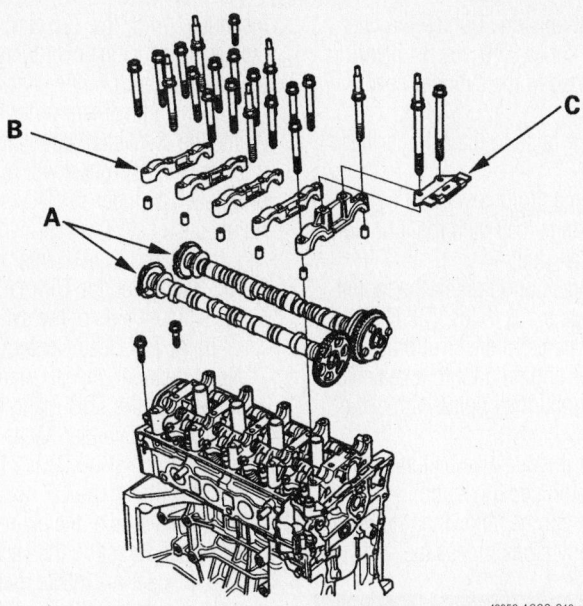

42356-ACCO-G13

When installing the camshafts (A) make sure the punch marks on the VTC actuator and exhaust cam sprockets are facing up—2.4L engine

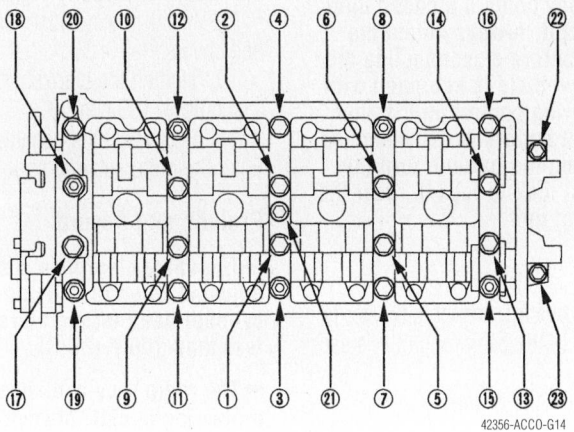

42356-ACCO-G14

Rocker arm assembly bolt tightening sequence—2.4L engine

Specified torque:
8 x 1.25 mm
24 N·m (2.4 kgf·m, 17 lbf·ft)
Apply engine oil to the bolt threads and flange

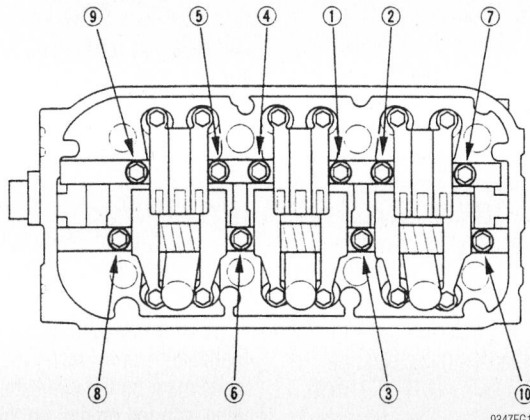

9347FG11

Camshaft holders torque sequence—3.0L engine

- Distributor from the cylinder head
- Cylinder head covers and the side covers
- Timing belt covers and the timing belt.
- Camshaft sprockets and the timing belt back covers.
- Bolts attaching the camshaft holder plates and camshaft holders in the opposite order of the installation sequence.
- Camshaft holder plates, camshaft holders, and the dowel pins. Discard the O-rings.
- Camshafts from the cylinder heads and the rubber cap from the rear cylinder head. Discard the camshaft seals.

To install:

5. Apply clean engine oil to the rocker arms and the camshafts.

6. Loosen the exhaust rocker arm adjusting screws and locknuts.

7. Be sure the rocker arms are properly positioned on the valve stems. Advance the crankshaft 30° from TDC to prevent interference between the pistons and valves, then install the camshafts. Position the rear camshaft on the cylinder head so the cam is not pushing on any valves.

8. Apply liquid gasket around the rubber cap, then install it to the cylinder head.

9. Install or connect the following:
- Camshaft seals to the camshafts with the open side (spring) facing in.
- Liquid gasket the cylinder head and camshaft holder mating surfaces
- Camshaft holders and the camshaft plates with the dowel pins
- New O-rings to the camshaft holder plates
- Clean oil to the camshaft holder bolts
- Bolts and tighten them in the proper sequence to 17 ft. lbs. (24 Nm)
- Timing belt back covers and tighten the attaching bolts to 108 inch lbs. (12 Nm).
- Camshaft sprockets and tighten the attaching bolts to 23 ft. lbs. (31 Nm).

10. Set the camshaft sprockets so that the No. 1 piston is at Top Dead Center (TDC). Align the TDC marks (green mark) on the camshaft pulleys to the pointers on the back covers.

11. Turn the crankshaft counterclockwise to set it at Top Dead Center (TDC). Align the TDC mark on the tooth of the timing belt drive pulley with the pointer on the oil pump.

12. Install or connect the following:
- Timing belt and timing belt covers.

13. Set No. 1 cylinder to TDC.

14. Tighten the valve adjusting screws for No. 1, No. 2 and No. 4 cylinders. Tighten the screw until it contacts the valve, then tighten the screw 1⅛ turns. Hold the screw in place and tighten the locknut to 14 ft. lbs. (20 Nm).

15. Rotate the crankshaft pulley 1 turn clockwise, then tighten the adjusting screws for No. 3, No. 5, and No. 6 cylinders. Tighten the screw until it contacts the valve, then tighten the screw 1⅛ turns. Hold the screw in place and tighten the locknut to 14 ft. lbs. (20 Nm).

16. Install or connect the following:
- Cylinder head cover gasket into the groove of the cylinder head cover. Seat the recesses for the camshaft first, then work it into the groove around the outside edges.

➡**Before installing the cylinder head cover gasket, thoroughly clean the seal groove.**

- Liquid gasket to the cylinder head cover gasket at the 4 corners of the recesses. Use a shop towel and wipe the cylinder heads where the cylinder head covers will come in contact.
- Cylinder head covers, hold the gasket in the groove by placing your fingers on the camshaft contacting surfaces. With the cylinder head cover on the cylinder heads, slide the covers slightly back and forth to seat the cylinder head cover gaskets. Replace the washers if damaged or deteriorated

17. Tighten the cylinder head cover bolts in 2 or 3 steps. In the final step, tighten all the bolts, in sequence, to 11 ft. lbs. (15 Nm).

18. Install or connect the following:
- Cylinder head side covers with new O-rings and tighten the bolts to 108 inch lbs. (12 Nm).
- Distributor to the cylinder head and tighten the mounting bolt to 16 ft. lbs. (22 Nm).
- Spark plug wires to the correct spark plugs and the distributor electrical connectors.
- Power steering belt, and adjust the tension.
- Alternator belt and adjust the tension. Tighten the alternator mounting nut and bolt to 16 ft. lbs. (22 Nm).
- Air conditioning belt, and adjust

the belt tension. Tighten the idler center nut to 33 ft. lbs. (44 Nm).
- PCV hose to the cylinder head cover
- Breather hose to the cylinder head cover
- Intake manifold cover and tighten the bolts to 108 inch lbs. (12 Nm).
- Intake air duct
- Strut brace and tighten the mounting bolts to 16 ft. lbs. (22 Nm).
- Starter cable to the strut brace

19. Drain the engine oil into a sealable container, then refill the engine with clean oil.

20. Connect the negative battery cable and enter the radio security code.

21. Start the engine, allowing it to idle and check for any signs of leakage.

Valve Clearance

ADJUSTMENT

➡**The radio may contain a coded theft protection circuit. Always obtain the code number before disconnecting the battery. If the vehicle is equipped with 4WS, the steering control unit is shut down when the battery is disconnected. After connecting the battery, turn the steering wheel lock-to-lock to reset the steering control unit.**

Civic

1. Before servicing the vehicle, refer to the precautions in the beginning of this section.

2. Disconnect the negative battery cable.

3. Remove the cylinder head cover and the upper timing belt cover.

4. Rotate the crankshaft to align the white TDC mark on the crankshaft pulley with the pointer on the cover for the No. 1 cylinder compression stroke. Be sure the **UP** mark on the camshaft sprocket is up and the TDC marks align with the edge of the cylinder head.

5. Hold a No. 1 cylinder rocker arm against the camshaft and use a feeler gauge to check the clearance at the valve stem. Except on B16A2 engines, intake valve clearance should be 0.007–0.009 in. (0.18–0.26mm), exhaust valve clearance should be 0.009–0.011 in. (0.23–0.27mm). On B16A2 engines, the intake valve clearance should be 0.006–0.007 in. (0.15–0.19mm), exhaust valve clearance should be 0.007–0.008 in. (0.17–0.21mm). Loosen the locknut and turn the adjusting screw to adjust the clearance. Tighten the

locknut to 14 ft. lbs. (20 Nm) and recheck the clearance. Don't overtighten the locknut, the aluminum rockers will strip easily.

6. The adjustment order is 1–3–4–2. Rotate the crankshaft counterclockwise 180° (the camshaft sprocket will rotate 90°) to bring each cylinder to TDC/compression. Adjust each set of valves.
   a. At Top Dead Center (TDC) for the No. 3 cylinder, the UP mark is pointed to the exhaust side of the cylinder head.
   b. At Top Dead Center (TDC) for the No. 4 cylinder, the UP mark is pointed down, and the TDC marks align with the edge of the cylinder head.
   c. At Top Dead Center (TDC) for the No. 2 cylinder, the UP mark is pointed to the intake side of the cylinder head.

7. After adjusting the valves of a VTEC engine, inspect the intake rocker arms for smooth and independent motion.

8. Apply sealant to the edges of the valve cover gasket where it meets the camshaft holders. Be sure the spark plug tube seals are properly seated.

9. Install the cylinder head and timing belt covers.

10. Tighten the crankshaft pulley bolt to 134 ft. lbs. (185 Nm).

11. Reconnect the negative battery cable. Enter the radio security code.

Prelude and Accord

➡**The valve clearance should be adjusted when the engine is cold, the cylinder head temperature should be less than 100°F (38°C).**

➡**The radio may contain a coded theft protection circuit. Always obtain the code number before disconnecting the battery.**

1. Before servicing the vehicle, refer to the precautions in the beginning of this section.

2. Remove or disconnect the following:
- Negative battery cable

➡**Label the wires before disconnecting them.**

- Spark plug wires from the spark plugs
- Positive Crankcase Ventilation (PCV) hose
- Cylinder head cover. Replace the rubber seals if damaged or deteriorated.

3. Turn the engine to align the timing marks and set cylinder No.1 to TDC. The white mark on the crankshaft pulley should align with the pointer on the timing belt cover. The words **UP** embossed on the

camshaft pulley should be aligned in the upward position. The marks on the edge of the pulley should be aligned with the cylinder head or the back cover upper edge.

4. Adjust the valves on cylinder No. 1 by performing the following:

    a. Insert a feeler gauge in between the camshaft lobe and the rocker arm.

➡**The intake valve clearance specification is 0.009–0.011 in (0.24–0.28mm) for 2001–02 vehicles or 0.008–0.010 in. (0.21–0.25mm) for 2003–04 vehicles. The exhaust valve clearance specification is 0.011–0.013 in. (0.27–0.32mm).**

    b. Loosen the locknut and turn the adjusting screw until the feeler gauge slides back and forth with a slight amount of drag.

    c. Tighten the locknut to 14 ft. lbs. (20 Nm) for intake or 10 ft. lbs. (14 Nm) for exhaust and recheck the valve clearance. Repeat the valve adjustment if necessary.

5. Rotate the crankshaft 180° counterclockwise (the camshaft pulleys will turn 90°) The **UP** arrow marks should be pointing to the exhaust side of the cylinder head.

6. Adjust the valves on cylinder No. 3 by performing the following:

    a. Insert a feeler gauge in between the camshaft lobe and the rocker arm.

    b. Loosen the locknut and turn the adjusting screw until the feeler gauge slides back and forth with a slight amount of drag.

    c. Tighten the locknut to 14 ft. lbs. (20 Nm) for intake or 10 ft. lbs. (14 Nm) for exhaust and recheck the valve clearance. Repeat the valve adjustment if necessary.

7. Rotate the crankshaft 180° counterclockwise (the camshaft pulleys will turn 90°) to bring No. 4 piston to TDC. The **UP** arrow marks should be pointing down, toward the crankshaft.

8. Adjust the valves on cylinder No. 4 by performing the following:

    a. Insert a feeler gauge in between the camshaft lobe and the rocker arm.

    b. Loosen the locknut and turn the adjusting screw until the feeler gauge slides back and forth with a slight amount of drag.

    c. Tighten the locknut to 14 ft. lbs. (20 Nm) for intake or 10 ft. lbs. (14 Nm) for exhaust and recheck the valve clearance. Repeat the valve adjustment if necessary.

9. Rotate the crankshaft 180° counterclockwise (the camshaft pulleys will turn 90°) to bring piston No. 2 to TDC. The **UP** arrow marks should be pointing to the intake side of the cylinder head.

10. Adjust the valves on cylinder No. 2 by performing the following:

    a. Insert a feeler gauge in between the camshaft lobe and the rocker arm.

    b. Loosen the locknut and turn the adjusting screw until the feeler gauge slides back and forth with a slight amount of drag.

    c. Tighten the locknut to 14 ft. lbs. (20 Nm) and recheck the valve clearance. Repeat the valve adjustment if necessary.

11. Install the cylinder head cover gasket cover to the groove of the cylinder head cover. Before installing the gasket, thoroughly clean the seal and the groove. Seat the recesses for the camshaft first, then work it into the groove around the outside edges. Be sure the gasket is seated securely in the corners of the recesses.

12. Apply liquid gasket to the 4 corners of the recesses of the cylinder head cover gasket. Do not install the parts if 5 minutes or more have elapsed since applying liquid gasket. After assembly, wait at least 20 minutes before filling the engine with oil.

13. Install or connect the following:

- Cylinder head (valve) cover. Tighten the bolts attaching to 84 inch lbs. (10 Nm).
- Spark plug wires to the correct spark plugs.
- Positive, then the negative battery cable and enter the radio security code.

14. If equipped with 4WS, start the engine and turn the steering wheel lock-to-lock to reset the 4WS control unit.

S2000

➡**The valve clearance is checked with the engine COLD.**

1. Before servicing the vehicle, refer to the precautions at the beginning of this section.

2. Disconnect the negative battery cable.

3. Remove the valve covers

4. Rotate the crankshaft so that the camshaft lobe is directed away from the valve tappet to be measured.

5. Insert a feeler gauge under the camshaft lobe at a 90 degree angle to the camshaft. Clearance for the intake valves should be 0.008–0.010 inch (0.21–0.25mm). Clearance for the exhaust valves should be 0.010–0.011 inch (0.25–0.29mm).

6. If adjustment is necessary, loosen the locknut and turn the adjusting screw until the clearance is correct.

7. Tighten the locknut to 14 ft. lbs. (20 Nm) and recheck the valve clearance.

8. Repeat for each valve requiring adjustment.

Starter Motor

REMOVAL & INSTALLATION

1.7L Engines

1. Before servicing the vehicle, refer to the precautions in the beginning of this section.

2. Remove or disconnect the following:

- Negative battery cable
- Air intake resonator
- Starter motor wiring harness
- Starter motor

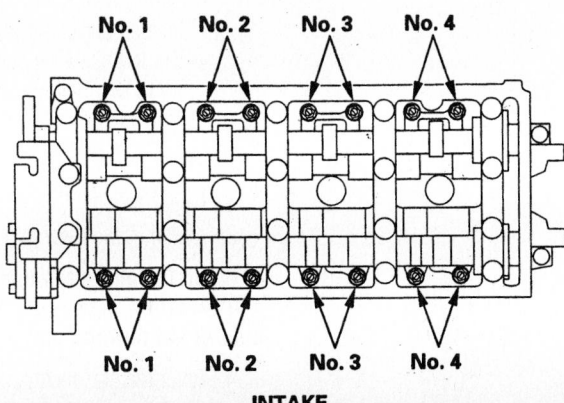

EXHAUST

No. 1   No. 2   No. 3   No. 4

No. 1   No. 2   No. 3   No. 4

INTAKE

42356-ACCO-G17

Valve clearance adjusting screw locations

To install:

3. Install or connect the following:
- Starter motor and tighten the bolts to 33 ft. lbs. (44 Nm)
- Starter motor wiring harness and tighten the battery cable terminal bolt to 84 inch lbs. (9 Nm)
- Negative battery cable

2.0L Engine

1. Before servicing the vehicle, refer to the precautions in the beginning of this section.
2. Remove or disconnect the following:
- Negative battery cable
- Accessory drive belt and tensioner
- Alternator
- Starter motor wiring harness
- Starter motor

To install:

3. Install or connect the following:
- Starter motor and tighten the bolts to 33 ft. lbs. (44 Nm)
- Starter motor wiring harness and tighten the battery cable terminal bolt to 84 inch lbs. (9 Nm)
- Alternator
- Accessory drive belt tensioner and tighten the bolts to 16 ft. lbs. (22 Nm)

- Accessory drive belt
- Negative battery cable

2.2L and 2.3L Engines

➡ **The factory sound system has a coded theft protection system. It is recommended that you know your reset code before you begin.**

1. Before servicing the vehicle, refer to the precautions in the beginning of this section.
2. Remove or disconnect the following:
- Negative battery cable
- Wiring from harness
- Lower radiator hose from the bracket on the starter motor
- Starter cable from terminal B located on the back of the solenoid
- Black/white wire from the S (solenoid) terminal
- Two bolts that mount the starter to the transaxle assembly
- Starter

To install:

3. Install in the reverse order of removal.

➡ **When installing the heavy gauge starter cable, make sure the crimped side of the terminal end is facing out.**

4. Enter the anti-theft code and radio presets.

2.4L Engine

➡ **The factory sound system has a coded theft protection system. It is recommended that you know your reset code before you begin.**

1. Before servicing the vehicle, refer to the precautions in the beginning of this section.
2. Remove or disconnect the following:
- Negative then the positive battery cables
- Intake manifold
- Starter cable from the B terminal
- Black/white wire from the S (solenoid) terminal
- Harness clamp and holder
- Two bolts that mount the starter to the transaxle assembly
- Starter

To install:

3. Install in the reverse order of removal.

➡ **When installing the heavy gauge starter cable, make sure the crimped side of the terminal end is facing out.**

4. Enter the anti-theft code and radio presets.

3.0L Engine

➡ **The factory sound system has a coded theft protection system. It is recommended that you know your reset code before you begin.**

1. Before servicing the vehicle, refer to the precautions in the beginning of this section.
2. Remove or disconnect the following:
- Negative then the positive battery cables
- Automatic Transmission Fluid (ATF) cooler
- Starter cable from terminal B located on the back of the solenoid
- Black/white wire from the S (solenoid) terminal
- Two bolts that mount the starter to the transaxle assembly
- Starter

To install:

3. Install in the reverse order of removal.

➡ **When installing the heavy gauge starter cable, make sure the crimped side of the terminal end is facing out.**

4. Enter the anti-theft code and radio presets.

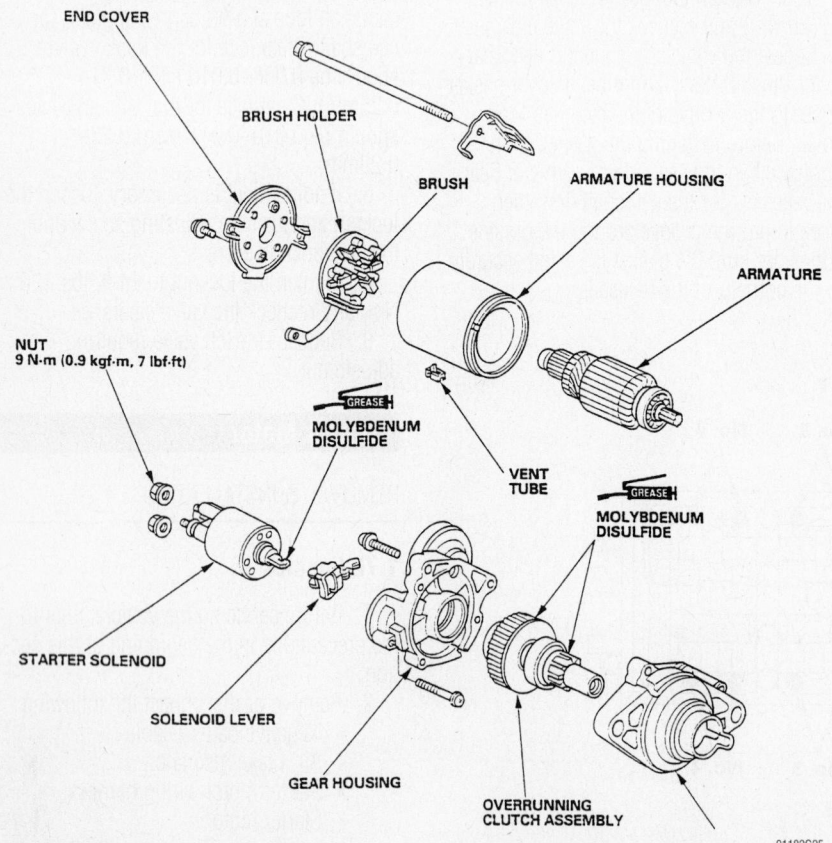

END COVER

BRUSH HOLDER

BRUSH

ARMATURE HOUSING

ARMATURE

NUT
9 N·m (0.9 kgf·m, 7 lbf·ft)

GREASE
MOLYBDENUM DISULFIDE

VENT TUBE

GREASE
MOLYBDENUM DISULFIDE

STARTER SOLENOID

SOLENOID LEVER

GEAR HOUSING

OVERRUNNING CLUTCH ASSEMBLY

91182G25

Exploded view of a typical Honda Starter

M/T:

**10 x 1.25 mm
44 N·m
(4.5 kgf·m, 33 lbf·ft)**

**8 x 1.25 mm
9 N·m
(0.9 kgf·m, 7 lbf·ft)**

**12 x 1.25 mm
64 N·m
(6.5 kgf·m,
47 lbf·ft)**

A/T:

**10 x 1.25 mm
44 N·m
(4.5 kgf·m, 33 lbf·ft)**

**8 x 1.25 mm
9 N·m
(0.9 kgf·m, 7 lbf·ft)**

**12 x 1.25 mm
64 N·m
(6.5 kgf·m,
47 lbf·ft)**

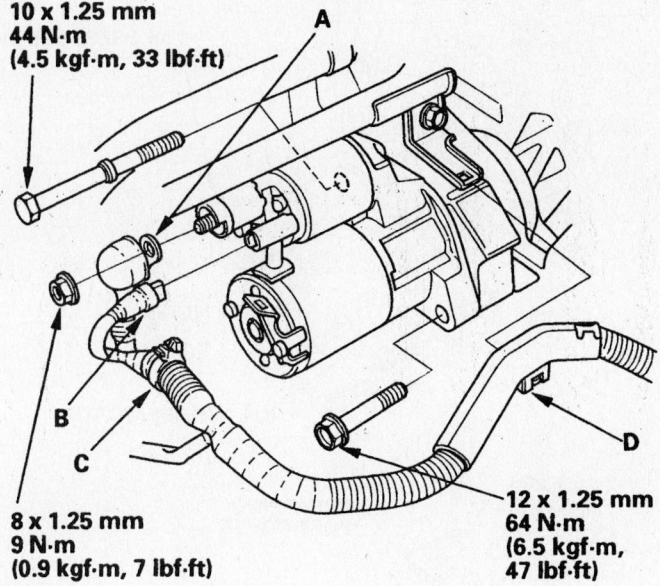

42356-ACCO-G18

Starter mounting—2.4L engine

Oil Pan

REMOVAL & INSTALLATION

➡The radio may contain a coded theft protection circuit. Always obtain the code number before disconnecting the battery. If the vehicle is equipped with 4WS, the steering control unit is shut down when the battery is disconnected. After connecting the battery, turn the steering wheel lock-to-lock to reset the steering control unit.

1.7L Engines

1. Before servicing the vehicle, refer to the precautions in the beginning of this section.
2. Disconnect the negative battery cable.
3. Raise and safely support the vehicle, then drain the oil.
4. Remove or disconnect the following:
   - Lower splash panel
   - Nuts and bolts connecting the exhaust pipe to the catalytic converter. Discard the gasket and the locknuts.
   - Nuts attaching the exhaust pipe to the exhaust hanger
   - Locknuts attaching the exhaust pipe to the exhaust manifold, then discard the nuts
   - Remove the exhaust pipe from the vehicle. Discard the exhaust gaskets.
   - Oil pan bolts in a crisscross pattern
   - Oil pan. Lightly tap the corners of the oil pan with a rubber or plastic faced mallet. Clean off all the old gasket material.
5. Inspect the oil screen and pick-up tube for damaged and clogging. If the screen and tube are clogged with oil residue, they should be thoroughly cleaned or replaced.

To install:

6. Install or connect the following:
   - Oil screen and tube with a new

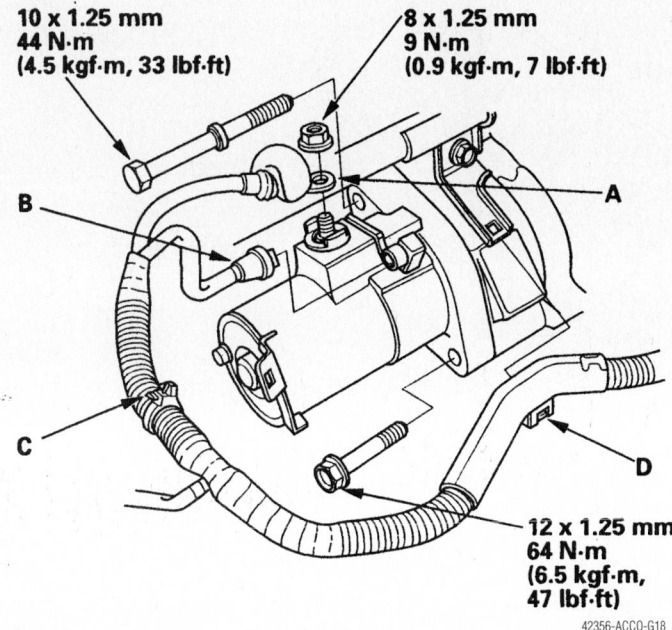

91182G27

Location of starter wiring—3.0L engine

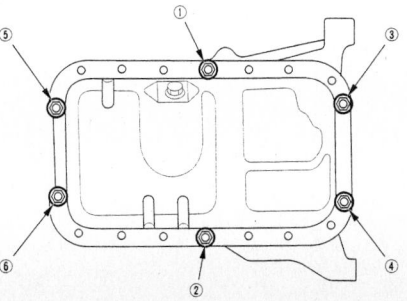

67162-ACCO-G06

Oil pan bolt tightening sequence—Civic with steel oil pan

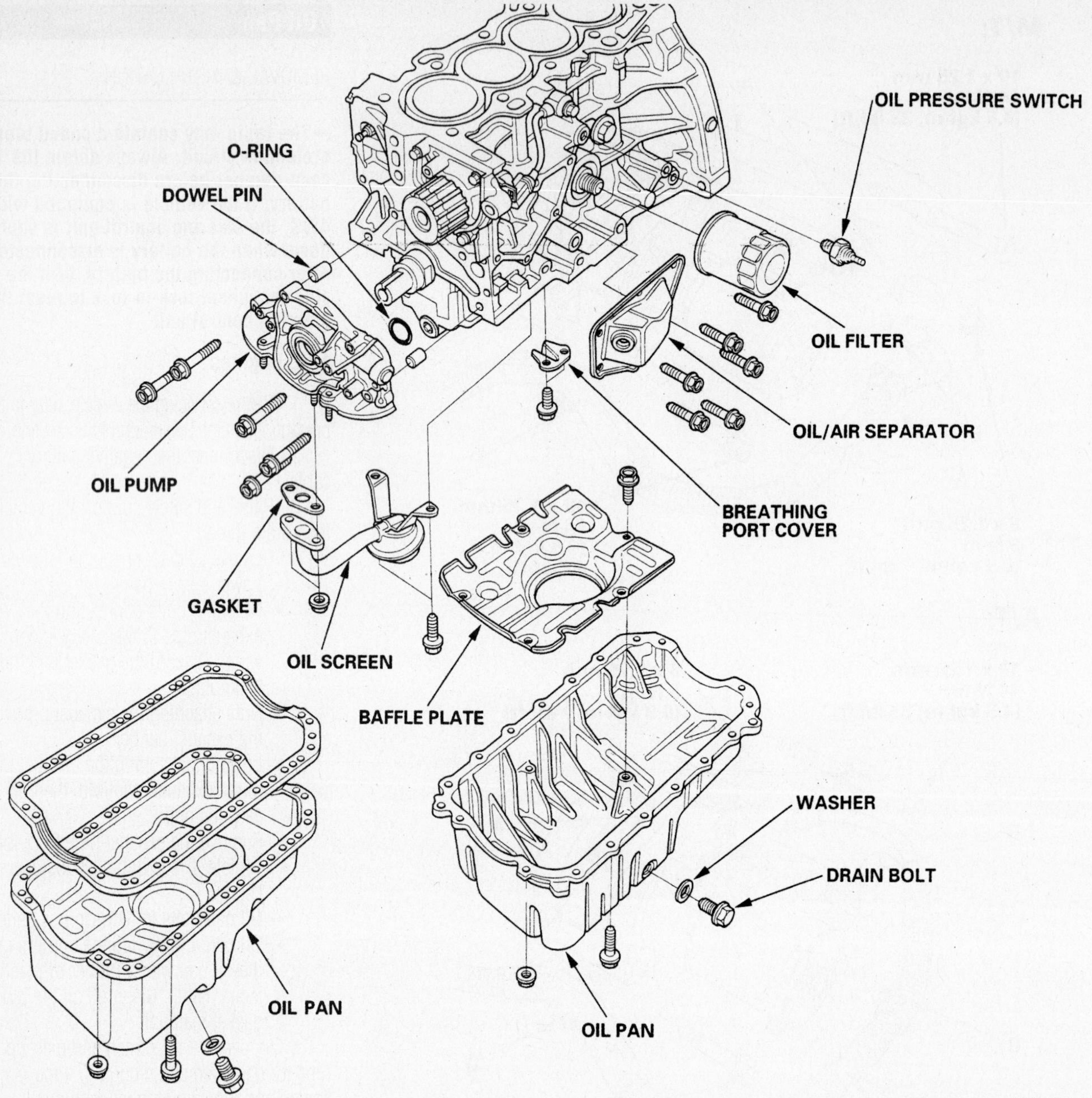

OIL PRESSURE SWITCH
O-RING
DOWEL PIN
OIL FILTER
OIL/AIR SEPARATOR
OIL PUMP
BREATHING PORT COVER
GASKET
OIL SCREEN
BAFFLE PLATE
WASHER
DRAIN BOLT
OIL PAN
OIL PAN
67162-ACCO-G04

Oil pan and oil screen—1.7L engine

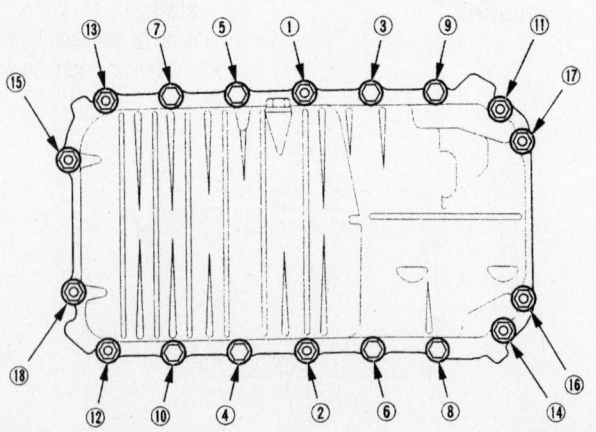

13
7
5
1
3
9
11
15
17
18
16
12
10
4
2
6
8
14
67162-ACCO-G05

Oil pan bolt tightening sequence—Civic with aluminum oil pan

gasket, if removed. Tighten the mounting nuts and bolts to 96 inch lbs. (11 Nm).

- Liquid gasket to the oil pan mating surface where the oil pump and the right side cover meet the engine block
- Oil pan gasket to the oil pan
- Oil pan
- Center and end mounting nuts and bolts. Evenly hand-tighten the oil pan nuts and bolts.

7. Tighten the oil pan mounting nuts and bolts in a 3-step clockwise pattern starting with the center bolt next to the oil drain plug. The final torque value for the nuts and bolts is 8.7 ft. lbs. (12 Nm).

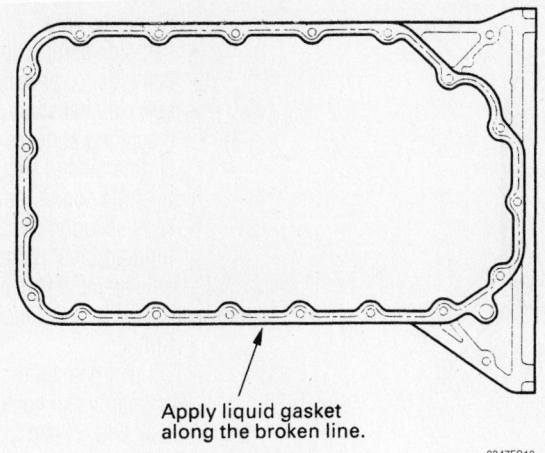

Apply liquid gasket
along the broken line.

9347FG12

Apply liquid gasket along the broken line—2.0L engine

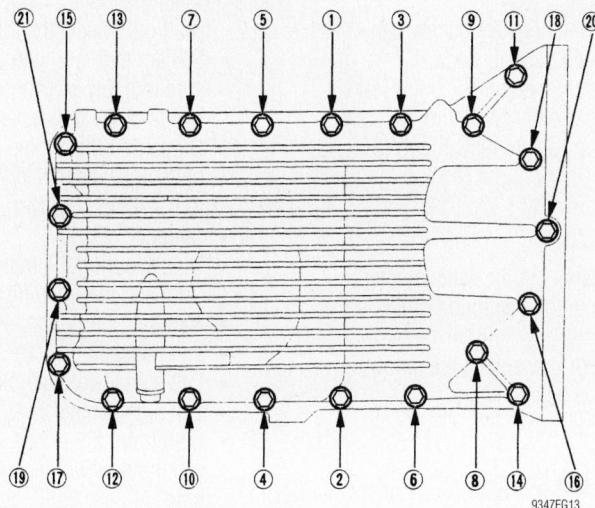

9347FG13

Oil pan torque sequence—2.0L engine

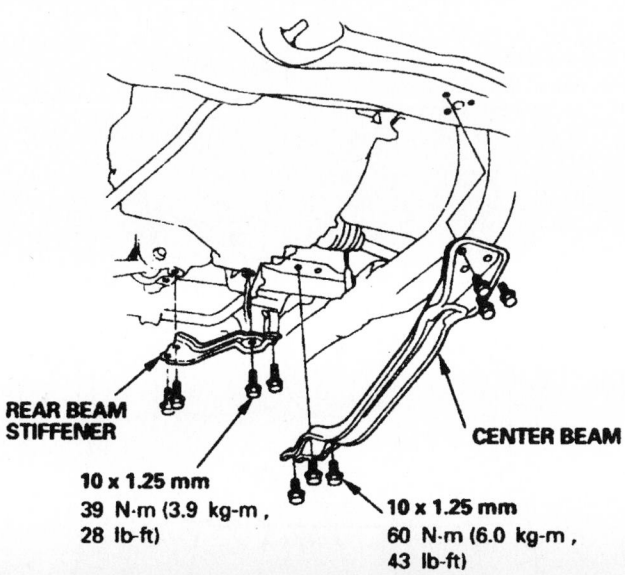

REAR BEAM
STIFFENER

CENTER BEAM

10 x 1.25 mm
39 N·m (3.9 kg-m,
28 lb-ft)

10 x 1.25 mm
60 N·m (6.0 kg-m,
43 lb-ft)

7923FG84

To gain access to the oil pan, remove the center beam—2.2L and 2.3L engines

➡**Excessive tightening can cause distortion of the oil pan gasket and oil leakage.**

8. Install or connect the following:
- Oil drain plug with a new crush washer. Tighten the plug to 33 ft. lbs. (44 Nm).
- Exhaust pipe using new gaskets and locknuts. Tighten the nuts attaching the exhaust pipe to the exhaust manifold to 40 ft. lbs. (54 Nm). Tighten the nuts attaching the exhaust pipe to the catalytic converter and the exhaust pipe hanger to 16 ft. lbs. (22 Nm).
- Lower splash panel

9. Lower the vehicle.

10. Refill the engine with clean oil.

11. Connect the negative battery cable and enter the radio security code.

12. Run the engine and check for leaks.

13. Turn off the engine and check the oil level. Top off the oil level if necessary.

2.0L Engine

1. Before servicing the vehicle, refer to the precautions in the beginning of this section.

2. Drain the engine oil.

3. Remove the oil pan bolts and the oil pan.

To install:

4. Remove old liquid gasket material from the oil pan mating surfaces and the bolt holes.

5. Clean and dry the oil pan mating surfaces.

6. Apply liquid gasket (PN 08718-0009) as shown.

7. Install the oil pan. Tighten the bolts in sequence and in two or three steps to 104 inch lbs. (12 Nm).

8. Fill the crankcase to the correct level.

9. Start the engine and check for leaks.

2.2L and 2.3L Engines

1. Before servicing the vehicle, refer to the precautions in the beginning of this section.

2. Disconnect the negative battery cable.

3. Raise and safely support the vehicle.

4. Drain the engine oil into a sealable container. Install the drain bolt with a new gasket and tighten to 33 ft. lbs. (44 Nm).

5. Remove or disconnect the following:
- Front wheels and the splash shield
- Center beam
- Oxygen (O_2S) sensor electrical connector
- Bolts from the support bracket on the exhaust pipe

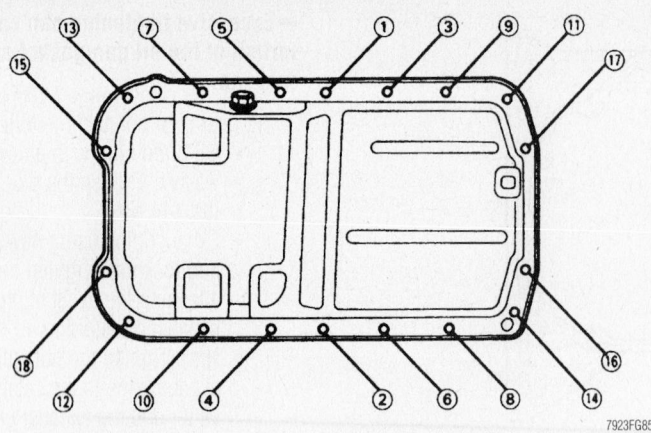

Oil pan mounting bolt tightening sequence—2.2L (H22A4) engines

- Nuts attaching the exhaust pipe to the exhaust manifold and the catalytic converter
- Exhaust pipe and discard the gaskets
- Converter cover, if equipped with an automatic transaxle
- Clutch cover, if equipped with a manual transaxle
- Oil pan nuts and bolts (in a criss-cross pattern) and the oil pan; if necessary, use a mallet to tap the corners of the oil pan. DO NOT pry on the pan to get it loose

6. Clean the oil pan mounting surface of old gasket material and engine oil.

To install:

7. Install or connect the following:
- New oil pan gasket to the oil pan. Apply liquid gasket to the corners of the curved section of the gasket.
- Oil pan to the engine
- Oil pan nuts and bolts and tighten the nuts and bolts in sequence. Tighten the nuts and bolts in 2 steps to 10 ft. lbs. (14 Nm).
- Torque converter cover or clutch cover, as applicable. Tighten the bolts to 108 inch lbs. (12 Nm).
- Exhaust pipe with new gaskets and new locknuts. Tighten the nuts attaching the exhaust pipe to the manifold to 40 ft. lbs. (54 Nm) and tighten the nuts attaching the exhaust pipe to the catalytic converter to 25 ft. lbs. (33 Nm).
- Bolts to the exhaust pipe support bracket and tighten to 13 ft. lbs. (18 Nm)
- O$_2$S electrical connector
- Center beam and tighten the mounting bolts to 43 ft. lbs. (60 Nm)
- Splash shield and tighten the mounting bolts to 84 inch lbs. (10 Nm)

- Front wheels

8. Lower the vehicle and fill the engine with oil.

9. Connect the negative battery cable and enter the radio security code.

10. Start the engine and check for leaks.

11. If equipped with 4WS, turn the steering wheel lock-to-lock to reset the 4WS control unit.

2.4L Engine

1. Before servicing the vehicle, refer to the precautions in the beginning of this section.

2. Remove or disconnect the following:
- Negative battery cable
- Engine oil
- Battery
- Air cleaner housing
- Harness clamp
- Battery base
- Clutch slave cylinder and clutch line bracket mounting bolt, if manual transaxle
- Ground cable and the transaxle upper mount bracket assembly
- Front and rear mount stops and mount bolts

- Front tire and wheel assemblies
- Stabilizer links
- Left side damper fork
- Left side lower ball joint
- Left side halfshaft. Coat all machined surfaces with clean engine oil and secure a plastic bag over the end of the halfshaft.
- Nuts securing the transaxle lower front and rear mounts

3. Use a suitable jack to lift the transaxle about 1.6–2.2 inches.
- Stiffener
- Oil pan bolts and nuts

4. Hammer a seal cutter between the engine block and oil pan to break the seal.

5. Remove the oil pan.

To install:

6. Clean the oil pan flange and engine block mounting surface.

7. Install or connect the following:
- Sealant to the oil pan flange. Be sure to apply sealant toward the inside of the bolt holes.
- Oil pan on the engine. Tighten the bolts in sequence, in 2 or 3 steps, to 8.7 ft. lbs. (12 Nm).
- Stiffener
- Nuts securing the transaxle lower front and rear mountings
- New set ring on the left halfshaft
- Left halfshaft
- Left side lower ball joint
- Left side damper fork
- Stabilizer
- Front mount bolt and front mount stop
- Rear mount bolt and rear mount stop

8. Loosen the mount bolt then install the transaxle upper mount/bracket assembly, clutch line clamp bracket (if M/T) and ground cable.
- Transaxle upper mount bolt if M/T
- Transaxle upper mount bracket

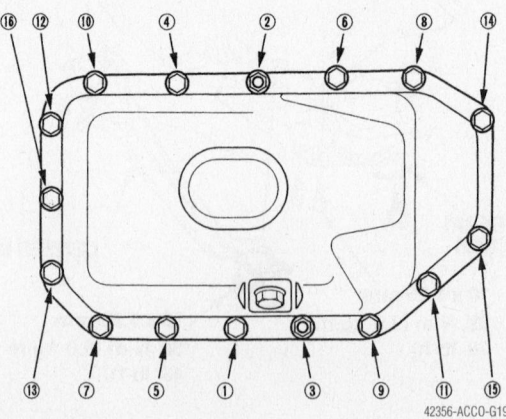

Oil pan bolt tightening sequence—2.4L engine

plate mounting bolts, then the upper mount bolt, if A/T
- Clutch slave cylinder and clutch line brake mounting bolt, if M/T
- Battery base
- Harness clamp
- Air cleaner housing
- Battery

☀☀ WARNING

Operating the engine without the proper amount and type of engine oil will result in severe engine damage.

9. Refill the engine with the correct amount of oil.
10. Connect the negative battery cable.
11. Start the engine and check for leaks.

3.0L Engine

1. Before servicing the vehicle, refer to the precautions in the beginning of this section.
2. Remove or disconnect the following:
   - Negative battery cable
3. Raise and safely support the vehicle.
4. Remove or disconnect the following:
   - Undercover
5. Drain the engine oil and replace the drain plug.
6. Remove or disconnect the following:
   - Front exhaust pipe
   - Oil pan mounting bolts
7. Hammer a seal cutter between the engine block and oil pan to break the seal.
8. Remove the oil pan.

To install:

9. Clean the oil pan flange and engine block mounting surface.
10. Install or connect the following:
    - Sealant to the oil pan flange. Be sure to apply sealant toward the inside of the bolt holes.

- Oil pan on the engine. Tighten the bolts in sequence to 10 ft. lbs. (14 Nm).
- Exhaust pipe
- Undercover
11. Lower the vehicle.

☀☀ WARNING

Operating the engine without the proper amount and type of engine oil will result in severe engine damage.

12. Refill the engine with the correct amount of oil.
13. Connect the negative battery cable.
14. Start the engine and check for leaks.

Oil Pump

REMOVAL & INSTALLATION

➡**The original radio may contain a coded anti-theft circuit. Always obtain the security code number before disconnecting the battery cables.**

1.7L (D17A1, D17A2) Engines

1. Before servicing the vehicle, refer to the precautions in the beginning of this section.
2. Disconnect the negative battery cable.
3. Raise and safely support the vehicle.
4. Drain the engine oil.
5. Rotate the crankshaft to set the No. 1 cylinder to TDC for the compression stroke. The white TDC mark on the crankshaft pulley should align with the TDC pointers on the lower timing cover.

➡**Mark the direction of the timing belt rotation if it is to be reinstalled.**

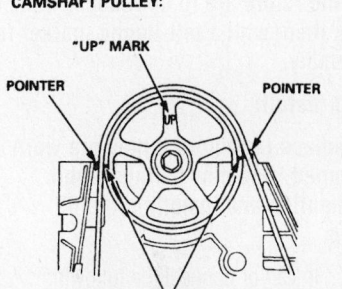

CRANKSHAFT PULLEY:

TDC MARK (WHITE)

CAMSHAFT PULLEY:

"UP" MARK
POINTER   POINTER
TDC MARK

7923FG88

Crankshaft and camshaft TDC marks—1.7L engines

6. Remove or disconnect the following:
   - Accessory drive belts and the crankshaft pulley
   - Valve cover and the upper and lower timing belt covers

➡**Cover the rocker arm and shaft assemblies with a towel or sheet of plastic to keep out dirt and foreign objects.**

- Dipstick and its tube from the oil pump housing

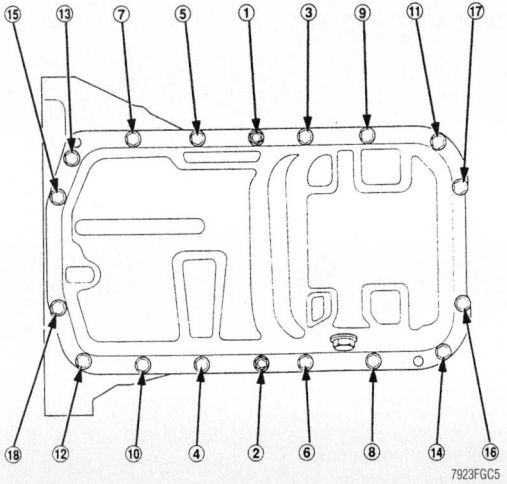

Oil pan mounting bolt tightening sequence—3.0L engine

7923FGC5

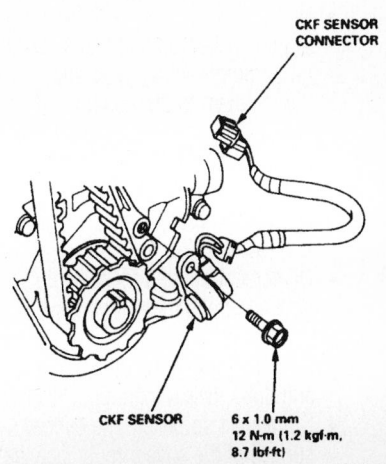

CKF SENSOR CONNECTOR

CKF SENSOR

6 x 1.0 mm
12 N·m (1.2 kgf·m, 8.7 lbf·ft)

7923FG89

CKF sensor location—1.7L engines

- Timing belt
- Crankshaft Speed Fluctuation (CKF) sensor from the oil pump cover, and position it out of the way so that it will not come in contact with oil or become damaged.
- Crankshaft sprocket
- Oil pan
- Oil screen and pick-up tube from the oil pump housing and crankshaft buttress. If the screen and pick-up tube are blocked with oil residue, clean or replace them as necessary.
- Oil pump assembly

➡**If the rotors are to be reused, match-mark them with a felt-tipped marker for assembly.**

To install:

➡**Replace the rotors if they are worn or damaged. Use new O-rings when assembling and installing the oil pump.**

7. Install or connect the following:
   - Rotors back into their original positions. Be sure they move without binding. Pack the rotor cavity with petroleum jelly to prevent oil starvation damage when the engine is initially started.

8. Assemble the oil pump and tighten the rotor cover bolts to 60 inch lbs. (7 Nm).

9. Be sure all gasket mating surfaces are clean prior to installation.

10. Install or connect the following:
   - New crankshaft oil seal into the oil pump housing.
   - Liquid gasket to the cylinder block mating surface of the block.
   - Light coat of oil to the crankshaft seal lip.
   - New O-ring on the cylinder block
   - Oil pump
   - Liquid gasket to the threads of the oil pump mounting bolts and tighten them to 96 inch lbs. (11 Nm).
   - Lightly lubricated relief valve piston and spring. Tighten the sealing bolt (with a new crush washer) to 29 ft. lbs. (39 Nm).
   - Oil screen. Tighten the fastening nuts and bolts to 96 inch lbs. (11 Nm).
   - Oil pan. Tighten the oil pan nuts and bolts to 108 inch lbs. (12 Nm).
   - Crankshaft sprocket. The concave surface of the spacer must face the engine block.

➡**Verify that the engine is at Top Dead Center (TDC)/compression for the No. 1 cylinder.**

- Timing belt Tighten the tensioner adjusting bolt to 33 ft. lbs. (44 Nm).
- CKF sensor. Tighten the sensor mounting bolt to 108 inch lbs. (12 Nm).
- Upper and lower timing belt covers and valve cover. Be sure all rubber seals and gaskets are properly seated.
- Dipstick tube with a new O-ring.
- Crankshaft pulley bolt: 134 ft. lbs. (181 Nm).
- New oil filter. Refill the engine with fresh oil.

11. Slowly rotate the engine several times by hand to prime the oil pump and verify that the timing belt has been installed and tensioned correctly.

12. Install and adjust the accessory drive belts.

13. Connect the negative battery cable.

14. Run the engine and check for proper oil pressure.

15. Check for leaks. Top up the engine oil if necessary.

2.0L Engine

1. Before servicing the vehicle, refer to the precautions in the beginning of this section.

2. Drain the engine oil.

3. Remove or disconnect the following:
   - Negative battery cable
   - Oil pan
   - Timing chain
   - Oil pump chain guide and tensioner
   - Baffle plate
   - Oil pump, chain, and crankshaft sprocket

To install:

➡**Use a new oil pump chain tensioner for assembly.**

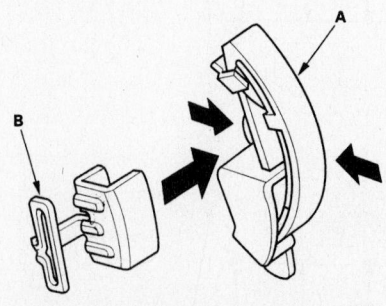

9347FG14

Compress the oil pump chain tensioner (A) and install the retaining clip (B)—2.0L engine

4. Compress the oil pump chain tensioner and install the supplied retaining clip.

5. Install or connect the following:
   - Oil pump, chain, and crankshaft sprocket. Tighten the 8mm bolts to 16 ft. lbs. (22 Nm) and the 6mm bolt to 104 inch lbs. (12 Nm).
   - Baffle plate and tighten the bolts to 104 inch lbs. (12 Nm)
   - Oil pump chain guide and tensioner. Tighten the 8mm bolt to 16 ft. lbs. (22 Nm) and the 6mm bolts to 104 inch lbs. (12 Nm).
   - Timing chain
   - Oil pan
   - Negative battery cable

6. Fill the crankcase to the correct level.

7. Start the engine and check for leaks.

2.2L and 2.3L Engines

1. Before servicing the vehicle, refer to the precautions in the beginning of this section.

2. Disconnect the negative battery cable.

3. Drain the engine oil into a sealable container.

4. Turn the engine to align the timing marks and set cylinder No.1 to TDC. The white mark on the crankshaft pulley should align with the pointer on the timing belt cover.

5. Remove or disconnect the following:
   - Valve cover and upper timing belt cover
   - Power steering pump belt and the alternator belt, also the air conditioning belt if so equipped
   - Crankshaft pulley and the lower timing belt cover
   - Balancer belt and the timing belt. Be sure to mark the rotation of the timing belt if it is going to be reused.
   - Timing belt and balancer belt tensioners
   - Crankshaft Position (CKP) sensor, if equipped
   - Timing belt drive pulley and key from the crankshaft
   - Balancer driven pulley, by inserting a suitable tool into the maintenance hole in the front balancer shaft.

6. Align the rear timing balancer pulley using a 6 x 100mm bolt or rod. Mark the bolt or rod at a point 2.9 in. (74mm) from the end. Remove the bolt from the maintenance hole on the side of the block; insert the bolt/rod into the hole. Align the 74mm mark with the face of the hole. This pin will hold the shaft in place.

7. Remove or disconnect the following:
- Balancer gear case and the dowel pins. Discard the O-ring.
- Balancer driven gear attaching bolt and the balancer driven gear
- Oil pan and the oil screen. Discard the screen gasket.
- Oil pump mounting bolts and oil pump assembly
- Dowel pins from the engine and clean the oil pump mating surfaces of old gasket material and oil. Discard the O-rings.

To install:

8. Install the 2 dowel pins and new O-rings to the cylinder block.

9. Be sure that the mating surfaces are clean and dry. Apply a liquid gasket evenly in a narrow bead, centered on the mating surface. Once the sealant is applied, do not wait longer than 20 minutes to install the parts; the sealant will become ineffective. After final assembly, wait at least 30 minutes before adding oil to the engine, giving the sealant time to set. To prevent leakage of oil, apply a suitable thread sealer to the inner threads of the bolt holes.

10. Install or connect the following:
- Oil pump to the engine block. Tighten the mounting bolts to 108 inch lbs. (12 Nm).
- Oil screen. Tighten the screen mounting bolts and nuts to 108 inch lbs. (12 Nm).
- Oil pan
- Balancer driven pulley to the front balancer belt, hold the balancer shaft in place with a suitable tool. Tighten the attaching bolt to 22 ft. lbs. (29 Nm).
- Balancer driven gear to the rear balancer shaft. Tighten the bolt to 18 ft. lbs. (25 Nm).

➡**Before installing the balancer driven gear and the gear case, apply molybdenum disulfide (lithium grease) to the thrust surfaces of the balancer gears.**

11. Align the groove on the pulley edge to the pointer on the balancer gear case.

12. Install or connect the following:
- Balancer gear case to the engine and the mounting bolts and nut. The rear balancer shaft is being held in place with a 6 x 100mm bolt. Tighten the mounting bolts and nut to 18 ft. lbs. (25 Nm).

13. Check the alignment of the pointer on the balancer pulley to the pointer on the oil pump.

14. Install or connect the following:
- Drive pulley to the crankshaft

- CKP sensor. Tighten the mounting bolts to 108 inch lbs. (12 Nm).
- Timing belt tensioners
- Timing belt and the balancer belt
- Crankshaft pulley and the lower timing belt cover
- Drive belts for the alternator, power steering, and air conditioning compressor. Tension the belts properly.
- Valve cover and upper timing belt cover

15. Refill the engine with clean, fresh oil.

16. Connect the negative battery cable and enter the radio security code.

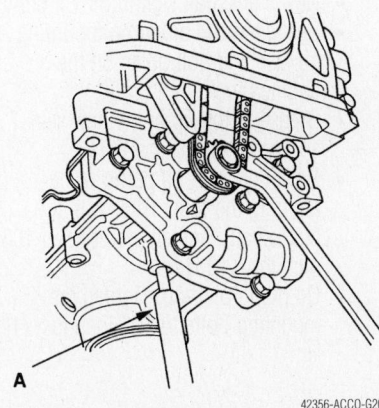

42356-ACCO-G20

Insert a 6mm pin into the maintenance hole in the lower balancer shaft holder and through the rear balancer shaft

2.4L Engine

1. Before servicing the vehicle, refer to the precautions in the beginning of this section.
2. Raise and safely support the vehicle.
3. Drain the engine oil.

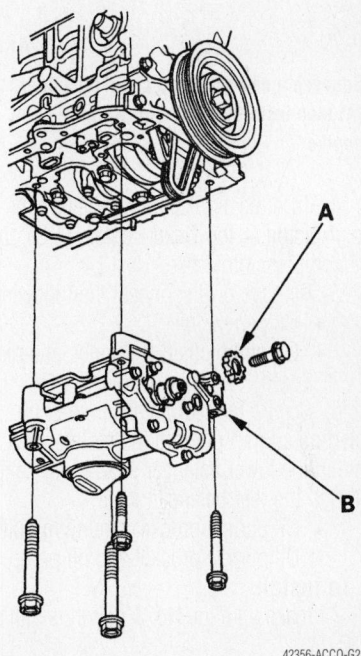

42356-ACCO-G21

Exploded view of the oil pump sprocket (A) and oil pump (B)—2.4L engine

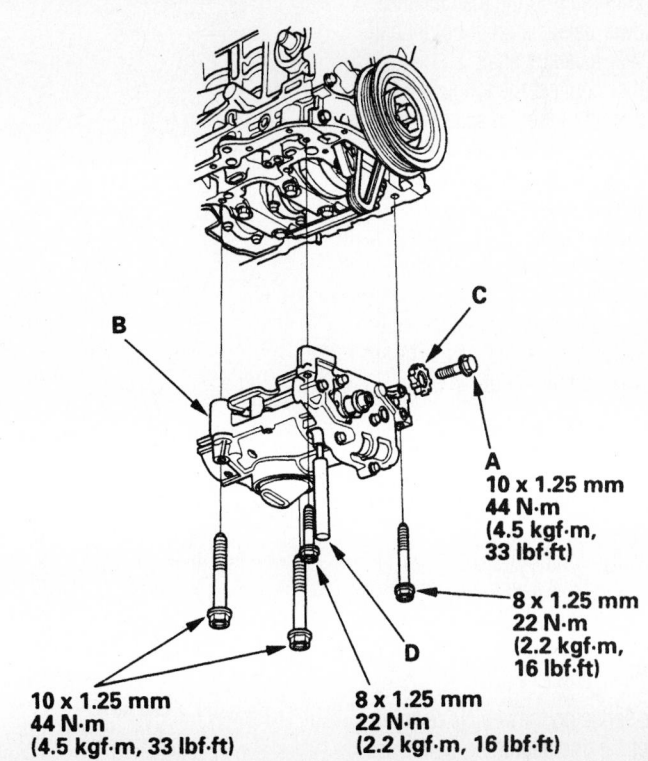

A 10 x 1.25 mm 44 N·m (4.5 kgf·m, 33 lbf·ft)
8 x 1.25 mm 22 N·m (2.2 kgf·m, 16 lbf·ft)
10 x 1.25 mm 44 N·m (4.5 kgf·m, 33 lbf·ft)
8 x 1.25 mm 22 N·m (2.2 kgf·m, 16 lbf·ft)

42356-ACCO-G22

Oil pump tightening specifications—2.4L engine

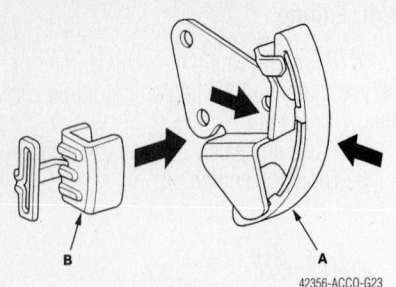

42356-ACCO-G23

Squeeze a new oil pump chain tensioner (A) then install the set clip (B) on it—2.4L engine

4. Turn the crankshaft to position the No. 1 piston at Top Dead Center (TDC) on the compression stroke.

5. Remove or disconnect the following:
- Oil pan
- Oil pump chain tensioner, and discard

6. Secure the rear balancer shaft by inserting a 6mm pin into the maintenance hole in the lower balancer shaft holder and through the rear balancer shaft
- Oil pump sprocket mounting bolt
- Oil pump sprocket and oil pump

To install:

7. make sure the No. 1 piston is still at TDC.

8. Align the dowel pin on the rear balancer shaft wit the mark on the oil pump

9. Secure the rear balancer shaft by inserting a 6mm pin into the maintenance hole in the lower balancer shaft holder and through the rear balancer shaft

10. Install or connect the following:
- Engine oil to the threads of the oil pump sprocket mounting bolt
- Oil pump and pump sprocket, loosely. Remove the 6mm pin.
- Oil pump mounting bolts and tighten as shown in the illustration

11. Squeeze a new oil pump chain tensioner then install the set clip on it. The set clip is provided with the oil pump chain tensioner
- New oil pump chain tensioner and tighten the bolts to 8.7 ft. lbs. (12 Nm). Remove the set clip from the tensioner.
- Oil pan

12. Fill the crankcase with the proper amount of new engine oil.

3.0L Engine

1. Before servicing the vehicle, refer to the precautions in the beginning of this section.

2. Raise and safely support the vehicle.

3. Drain the engine oil.

4. Turn the crankshaft to position the No. 1 piston at Top Dead Center (TDC) on the compression stroke.

5. Remove or disconnect the following:
- Timing belt
- Idler pulley
- Crankshaft Position (CKP) sensor
- Variable Timing Electronic Control (VTEC) solenoid valve
- Oil filter
- Oil pan and pick-up
- Oil pump assembly

To install:

6. Install or connect the following:
- New crankshaft seal in the oil pump
- Sealant to the oil pump mounting surface and bolt holes on the engine block
- Grease to the lip of the new seal and engine oil to the O-ring
- Dowel pin and oil pump while aligning the inner rotor with the crankshaft. Tighten the bolts to 108 inch lbs. (12 Nm).
- Oil pump pick-up. Tighten the mounting bolts to 108 inch lbs. (12 Nm).

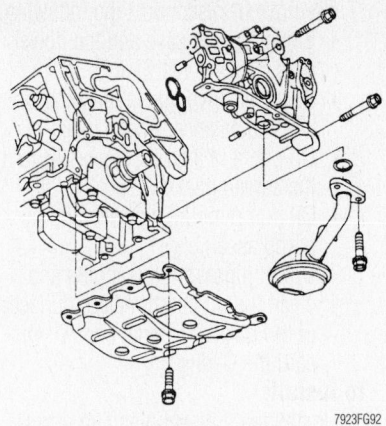

7923FG92

Oil pump mounting—3.0L engine

- Oil pan, VTEC solenoid, oil filter, CKP, and idler pulley
- Timing belt

✻✻ WARNING

Operating the engine without the proper amount and type of engine oil will result in severe engine damage.

7. Fill the crankcase with the proper amount of new engine oil.

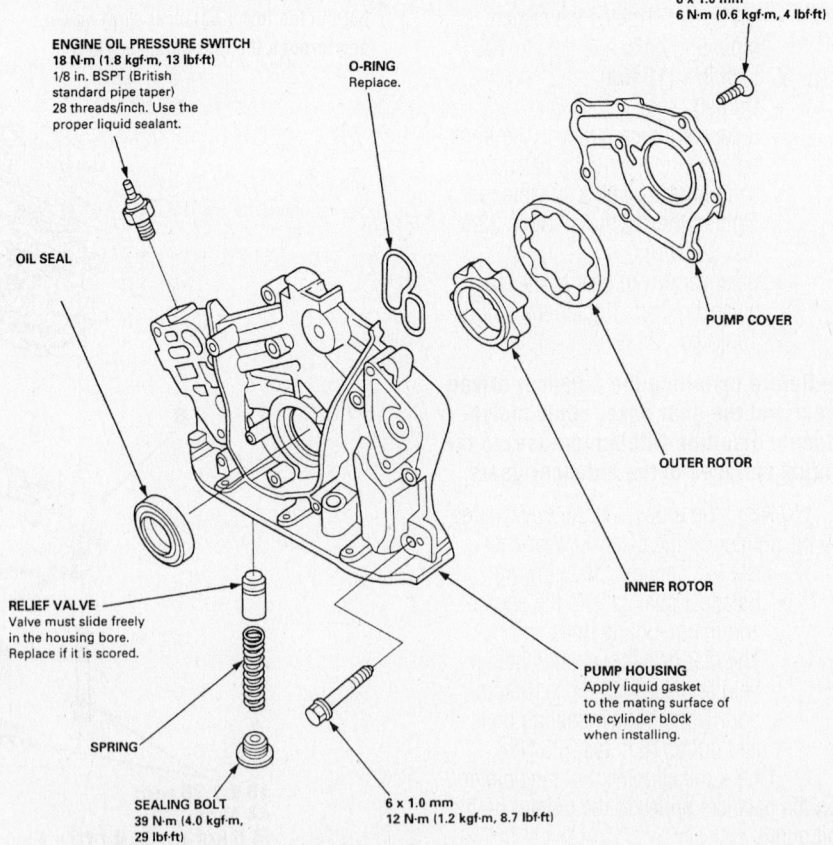

ENGINE OIL PRESSURE SWITCH
18 N·m (1.8 kgf·m, 13 lbf·ft)
1/8 in. BSPT (British standard pipe taper)
28 threads/inch. Use the proper liquid sealant.

O-RING
Replace.

OIL SEAL

6 x 1.0 mm
6 N·m (0.6 kgf·m, 4 lbf·ft)

PUMP COVER

OUTER ROTOR

INNER ROTOR

PUMP HOUSING
Apply liquid gasket to the mating surface of the cylinder block when installing.

RELIEF VALVE
Valve must slide freely in the housing bore. Replace if it is scored.

SPRING

SEALING BOLT
39 N·m (4.0 kgf·m, 29 lbf·ft)

6 x 1.0 mm
12 N·m (1.2 kgf·m, 8.7 lbf·ft)

7923FG91

Exploded view of the oil pump—3.0L engine

Rear Main Seal

REMOVAL & INSTALLATION

1. Before servicing the vehicle, refer to the precautions in the beginning of this section.

2. Remove the transmission.

3. Remove the driveplate from the crankshaft.

4. Carefully pry the crankshaft seal out of the retainer.

To install:

5. Apply clean engine oil to the lip of the new seal.

6. Install the seal onto the crankshaft and into the retainer using the appropriate seal driver.

7. Install the driveplate and the transmission.

Timing Belt, Sprockets, Front Cover and Seal

REMOVAL & INSTALLATION

1.7L Engines

1. Before servicing the vehicle, refer to the precautions in the beginning of this section.

2. Rotate the crankshaft to set the engine at Top Dead Center (TDC) on the compression stroke for the No. 1 piston. The white mark on the crankshaft pulley should align with the pointers on the timing cover. Once the engine is in this position, it must not be disturbed.

3. Remove or disconnect the following:

- Necessary components for access to the cylinder head and timing belt covers. Cover the rocker arm and shaft assemblies with a towel or sheet of plastic to keep out dust and foreign objects.
- Timing belt covers

4. Loosen the timing belt adjusting bolt 180 degrees (½ turn). Push the tensioner pulley down to release the belt tension. After releasing the tension, retighten the tensioner pulley bolt until snug.

➡**Do not remove the tensioner pulley unless it is to be replaced.**

- Timing belt. Mark the direction of the belt's rotation if it is to be reinstalled.

To install:

➡**Inspect the water pump when replacing the timing belt; the manufacturer**

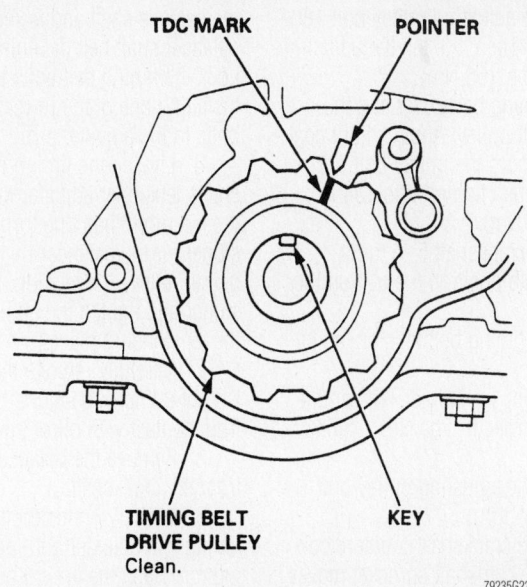

TDC alignment mark locations for the crankshaft sprocket—1.7L SOHC engines

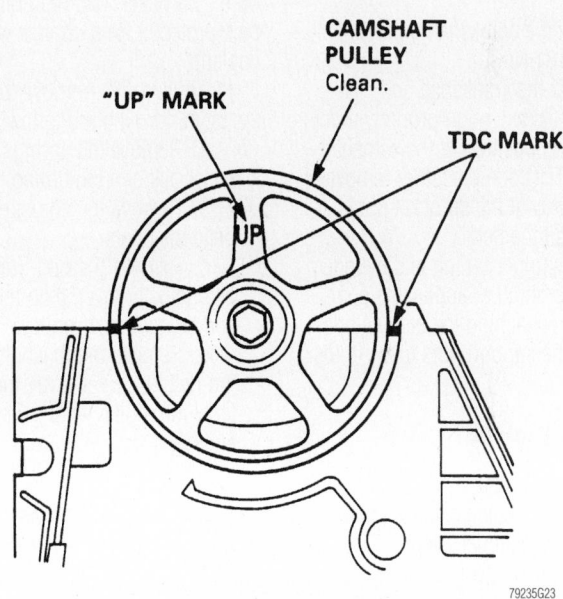

Single camshaft timing belt sprocket TDC mark positioning for timing belt installation—1.7L SOHC engines

recommends replacing the water pump at the timing belt's service interval. Replace the timing belt if it shows any signs of wear, or if it is contaminated with oil or coolant.

5. Verify that the timing is set at TDC on the compression stroke for the No. 1 cylinder as follows:

a. The groove in the crankshaft sprocket must align with the pointer on the oil pump.

b. The TDC marks on the camshaft sprockets must align with the pointer located between the sprockets. The TDC marks will also be in line with the upper surface of the head.

c. On other engines, the TDC mark on the camshaft sprocket must align with the pointer on the back cover.

d. The **UP** mark on the camshaft sprocket must point up.

6. Install or connect the following:

- Timing belt onto the crankshaft sprocket, then around the adjusting pulley and water pump sprocket, and finally over the camshaft sprocket.

7. Loosen the adjusting pulley bolt 180 degrees (½ turn). Then, tighten the adjusting bolt to 40 ft. lbs. (55 Nm).

• Lower timing belt cover and the crankshaft pulley. Apply a light coat of fresh oil to the pulley bolt threads, then tighten it to 134 ft. lbs. (181 Nm).

8. Rotate the crankshaft 5–6 turns counterclockwise to position the belt on the sprockets.

9. Adjust the timing belt tension, as follows:

a. Set the No. 1 piston at TDC on the compression stroke for the No. 1 cylinder.

b. Loosen the adjusting pulley bolt 180 degrees (½ turn).

c. Rotate the crankshaft counterclockwise so that the camshaft sprocket moves 3 teeth from the TDC/compression position.

d. Tighten the adjusting bolt to 33 ft. lbs. (45 Nm).

e. Tighten the crankshaft pulley to 134 ft. lbs. (181 Nm).

10. Verify that the crankshaft and camshaft sprockets will align properly at the TDC/compression position. If the camshaft pulley is not at TDC/compression, remove the timing belt, adjust the sprocket positions and reinstall the belt.

11. Install the upper timing and cylinder head covers, and all other applicable components. When reattaching the side engine mount, tighten the support nuts to 54 ft. lbs. (75 Nm).

2.2L and 2.3L Engines

1. Before servicing the vehicle, refer to the precautions in the beginning of this section.

2. Remove the cylinder head (valve) and upper timing belt covers.

3. Turn the engine to align the timing marks and set cylinder No. 1 to Top Dead Center (TDC). The white mark on the crankshaft sprocket should align with the pointer on the timing belt cover. The words **UP** embossed on the camshaft sprocket should be aligned in the upward position. The marks on the edge of the sprocket should be aligned with the cylinder head or the back cover upper edge. Once in this position, the engine must NOT be turned or disturbed.

4. Remove all necessary components for access to the lower timing belt cover, then remove the cover.

5. There are two belts in this system; the one running to the camshaft sprocket is the timing belt. The other, shorter one drives

the balance shaft and is referred to as the balancer shaft belt or timing balancer belt. Lock the timing belt adjuster in position, by installing one of the lower timing belt cover bolts to the adjuster arm.

6. Loosen the timing belt and balancer shafts tensioner adjuster nut, do not loosen the nut more than one turn. Push the tensioner for the balancer belt away from the belt to relieve the tension. Hold the tensioner and tighten the adjusting nut to hold the tensioner in place.

7. Carefully remove the balancer belt. Do not crimp or bend the belt; protect it from contact with oil or coolant.

8. Remove the balancer belt sprocket from the crankshaft.

9. Loosen the lockbolt installed to the timing belt adjuster and loosen the adjusting nut. Push the timing belt adjuster to remove the tension on the timing belt, then tighten the adjuster nut.

10. Remove the timing belt by sliding it off the sprockets. Do not crimp or bend the belt; protect it from contact with oil or coolant.

11. If defective, remove the belt tensioners by performing the following:

a. Remove the springs from the balancer belt and the timing belt tensioners.

b. Remove the adjusting nut from the belt tensioners.

c. Remove the bolt from the balancer belt adjuster lever, then remove the lever and the tensioner pulley.

d. Remove the lockbolt from the timing belt tensioner lever, then remove the tensioner pulley and lever from the engine.

12. This is an excellent time to check or replace the water pump. Even if the timing belt is only being replaced as part of a good maintenance schedule, consider replacing the pump at the same time.

To install:

13. If the water pump is to be replaced, install a new O-ring and make certain it is properly seated. Install the water pump and tighten the mounting bolts to 106 inch lbs. (12 Nm).

14. If the tensioners were removed, perform the following procedures:

a. Install the timing belt tensioner lever and the tensioner pulley.

b. Install the balancer belt pulley and adjuster lever.

c. Install the adjusting nut and the bolt to the balancer belt adjuster lever.

d. Install the springs to the tensioners.

e. Install the lockbolt to the timing belt tensioner, then move it its full deflection and tighten the lockbolt.

f. Move the balancer it's full deflection and tighten the adjusting nut to hold its position.

15. The pointer on the crankshaft sprocket should be aligned with the pointer on the oil pump; the camshaft sprocket must be aligned so that the word **UP** is at the top of the sprocket and the marks on the edge of the sprocket are aligned with the surfaces of the head or the back cover upper edge.

16. Install the timing belt on the sprockets in the following sequence: crankshaft sprocket, tensioner sprocket, water pump sprocket and camshaft sprocket.

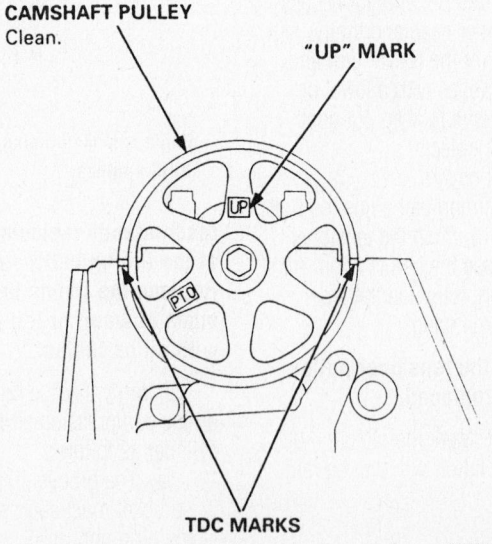

CAMSHAFT PULLEY
Clean.

"UP" MARK

TDC MARKS

79235G28

Position the camshaft sprocket as indicated for timing belt installation—Honda 2.2L and 2.3L engines

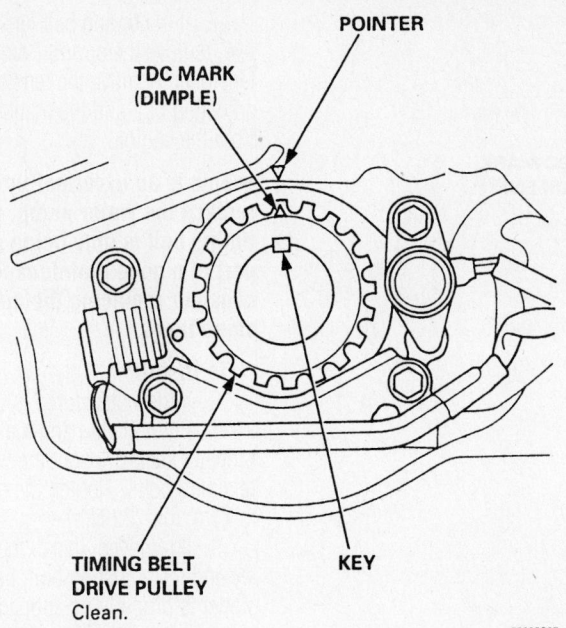

Before installing the timing belt, ensure the crankshaft sprocket marks are properly aligned—Honda 2.2L and 2.3L engines

17. Check the timing marks to be sure that they did not move.

18. Loosen, then retighten the timing belt adjusting nut; this will apply the proper amount of tension to the timing belt.

19. Install the timing balancer belt drive sprocket and the lower timing belt cover.

20. Install the crankshaft pulley and bolt, tighten the bolt to 181 ft. lbs. (245 Nm). Rotate the crankshaft sprocket 5–6 turns to position the timing belt on the sprockets.

21. Set the No. 1 cylinder to TDC and loosen the timing belt adjusting nut one turn. Turn the crankshaft counterclockwise until the cam sprocket has moved 3 teeth; this creates tension on the timing belt.

22. Tighten the timing belt adjusting nut.

23. Set the crankshaft sprocket and the camshaft sprocket to TDC. If the sprockets do not align, remove the belt to realign the marks, then install the belt.

24. Remove the crankshaft pulley and the lower cover.

25. With the timing marks aligned, lock the timing belt adjuster in place with one of the lower cover mounting bolts.

26. Loosen the adjusting nut and ensure the timing balancer belt adjuster moves freely.

27. Align the rear timing balancer sprocket using a 6 x 100mm bolt or rod. Mark the bolt or rod at a point 2.9 in. (74mm) from the end. Remove the bolt from the maintenance hole on the side of the block; insert the bolt/rod into the hole and align the 2.9 in. (74mm) mark with the face

of the hole. This will hold the shaft in place during installation.

28. Align the groove on the front balancer shaft sprocket with the pointer on the oil pump.

29. Install the balancer belt. Once the belts are in place, be sure that all the engine alignment marks are still correct. If not, remove the belts, realign the engine and reinstall the belts. Once the belts are properly installed, slowly loosen the adjusting nut, allowing the tensioner to move against

the belt. Remove the bolt from the maintenance hole and reinstall the bolt and washer.

30. Install the crankshaft pulley, then turn the crankshaft sprocket 1 turn counterclockwise and tighten the timing belt adjusting nut to 33 ft. lbs. (45 Nm).

31. Remove the crankshaft pulley and the bolt locking the timing belt adjuster in place.

32. Install the lower and upper timing belt covers, and all applicable components. When installing the crankshaft pulley, coat the threads and seating face of the pulley bolt with engine oil, then install and tighten the bolt to 181 ft. lbs. (250 Nm).

33. Install the cylinder head cover gasket cover to the groove of the cylinder head cover. Before installing the gasket thoroughly clean the seal and the groove. Seat the recesses for the camshaft first, then work it into the groove around the outside edges. Be sure the gasket is seated securely in the corners of the recesses.

34. Apply liquid gasket to the four corners of the recesses of the cylinder head cover gasket. Do not install the parts if 5 minutes or more have elapsed since applying liquid gasket. After assembly, wait at least 20 minutes before filling the engine with oil.

35. Install the cylinder head (valve) cover and all other applicable components.

3.0L Engines

1. Before servicing the vehicle, refer to the precautions in the beginning of this section.

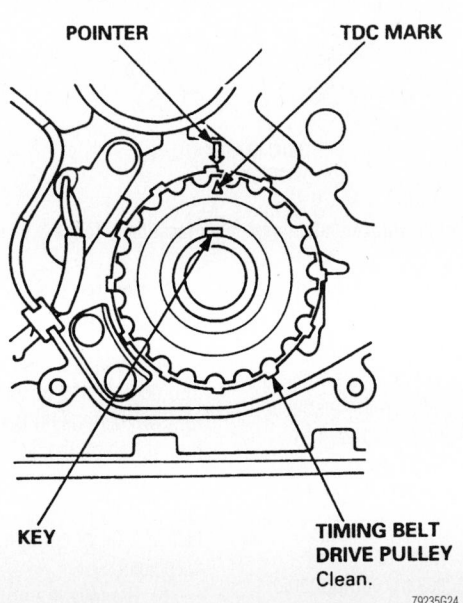

Crankshaft timing belt sprocket alignment mark locations—Honda 3.0L engine

FRONT:

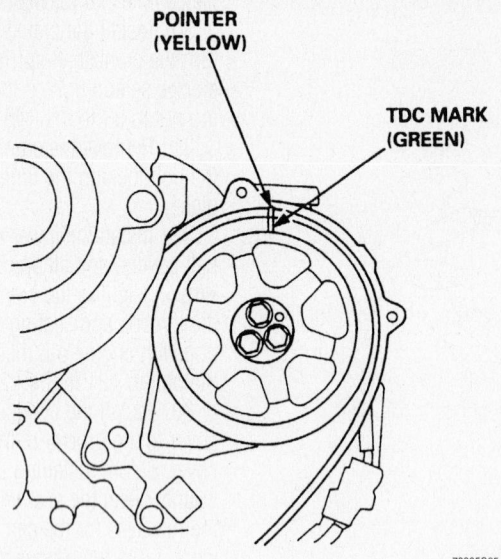

Left camshaft timing belt sprocket alignment mark location—Honda 3.0L engine

REAR:

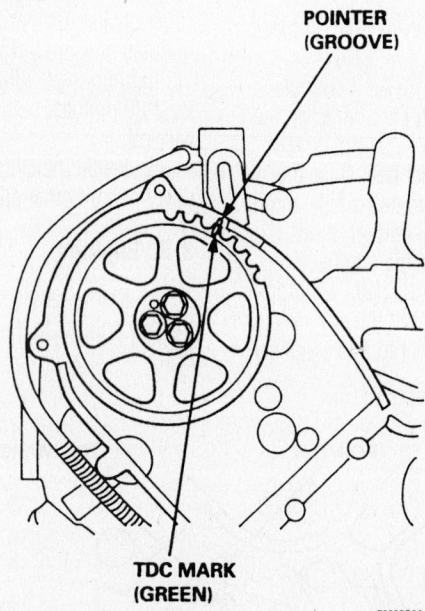

Rear camshaft timing belt sprocket alignment mark location—Honda 3.0L engine

2. Turn the engine to align the timing marks and set cylinder No. 1 to Top Dead Center (TDC). The white mark on the crankshaft pulley should align with the pointer on the timing belt cover. Remove the inspection caps on the upper timing belt covers to check the alignment of the timing marks. The pointers for the camshafts should align with the green marks on the camshaft sprockets.

3. Remove all necessary components for access to the timing belt covers, then remove the covers.

➡**Do not use the covers to store removed items.**

4. Loosen the timing belt adjuster bolt 180 degrees (½ turn). Push the tensioner to remove the tension from the timing belt, then retighten the adjusting bolt.

5. Remove the timing belt. Do not crimp or bend the belt; protect it from contact with oil or coolant. Slide the belt off the sprockets.

6. Remove the bolts attaching the camshaft sprockets to the camshafts, then remove the sprockets.

7. If the timing belt tensioner is defective, remove the spring from the timing belt tensioner. Remove the tensioner pulley adjusting bolt and the adjuster assembly from the engine.

➡**This is an excellent time to check or replace the water pump. Even if the timing belt is only being replaced as part of a good maintenance schedule, consider replacing the pump at the same time.**

To install:

8. If the water pump is to be replaced, install a new O-ring and make certain it is properly seated. Install the water pump and retaining bolts. Tighten the mounting bolts to 16 ft. lbs. (22 Nm).

9. If removed, install the tensioner pulley and the adjusting bolt, be sure the tensioner is properly positioned on its pivot pin. Install the spring to the tensioner, then push the tensioner to its full deflection and tighten the adjusting bolt.

10. Set the timing belt drive sprocket so that the No. 1 piston is at TDC. Align the TDC mark on the tooth of the timing belt drive sprocket with the pointer on the oil pump.

11. Set the camshaft sprockets so that the No. 1 piston is at TDC. Align the TDC marks (green mark) on the camshaft sprockets to the pointers on the back covers.

12. Install the timing belt onto the sprockets in the following sequence: crankshaft sprocket, tensioner pulley, front camshaft sprocket, water pump pulley and rear camshaft sprocket.

13. Loosen, then retighten the timing belt adjuster bolt to tension the timing belt.

14. Install the lower timing belt cover.

15. Install the crankshaft sprocket and the crankshaft pulley bolt. Tighten the bolt to 181 ft. lbs. (245 Nm) with the aid of the crank pulley holder.

16. Rotate the crankshaft five or six turns clockwise so that the timing belt positions on the sprockets.

17. Set cylinder No. 1 to TDC by aligning the timing marks. If the timing marks do not align, remove the timing belt, then adjust the components and reinstall the timing belt.

18. Loosen the timing belt adjusting bolt 180 degrees (½ turn) and retighten the adjusting bolt. Tighten the adjusting bolt to 31 ft. lbs. (42 Nm).

19. Install the upper timing belt cover and all other applicable components. When installing the side engine mount to the engine, use 3 new attaching bolts. Tighten the new bolts to 40 ft. lbs. (54 Nm).

Timing Chain, Sprockets, Front Cover and Seal

REMOVAL & INSTALLATION

2.0L Engine

1. Before servicing the vehicle, refer to the precautions in the beginning of this section.

2. Set the engine to Top Dead Center (TDC).

3. Drain the cooling system.

4. Relieve the fuel system pressure.

5. Remove or disconnect the following:
   - Negative battery cable
   - Air cleaner housing
   - Vacuum tank
   - Accessory drive belt
   - Water bypass hose and tube
   - Water pump pulley
   - Accessory drive belt tensioner
   - Alternator
   - Idler pulley
   - Throttle cable
   - Intake manifold
   - Exhaust manifold

6. Remove the timing chain auto tensioner as follows:

   a. Remove the end cover and nozzle from the auto tensioner.

   b. Use a 5 x 0.8mm bolt and locknut as shown to compress the auto-tensioner.

   c. Remove the timing chain auto-tensioner.
   - Valve cover
   - Camshafts
   - Timing chain idler gear
   - Crankshaft sprocket
   - Front crankshaft seal
   - Front cover
   - Oil pump chain guide

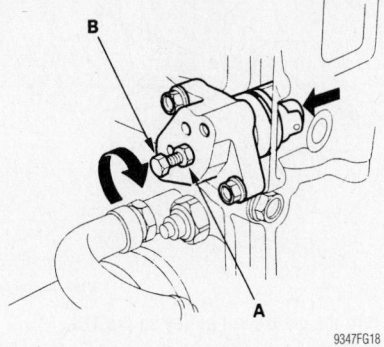

Use a 5 x 0.8mm bolt (B) and locknut (A) to compress the auto-tensioner—2.0L engine

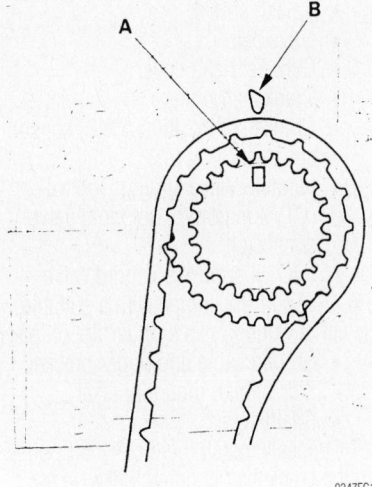

Align the sprocket key (A) with the cylinder block pointer (B) to set the engine to TDC—2.0L engine

   - Timing chain

To install:

7. Ensure that the crankshaft sprocket is set to TDC.

8. Install the timing chain with the col-ored link aligned with the crankshaft sprocket punch mark as shown.

9. Install the timing chain idler sprocket so that the two colored links are aligned with the sprocket punch mark, and the TDC marks are aligned with the cylinder head surface as shown. Tighten the idler sprocket bolt to 36 ft. lbs. (49 Nm).

10. Prepare the timing chain auto-tensioner for installation as follows:

    a. Clamp the auto-tensioner in a soft-jawed vise.

    b. Tighten the 5 x 0.8mm bolt to compress the tensioner until a set pin can be inserted.

    c. Remove the 5 x 0.8mm bolt and install the nozzle and cover. Tighten the nozzle to 48 inch lbs. (5 Nm) and the cover bolts to 104 inch lbs. (12 Nm).

    d. Install the auto-tensioner with new O ring seals and tighten the bolts to 104 inch lbs. (12 Nm).

    e. Remove the service bolt from the cylinder head and remove the set pin.

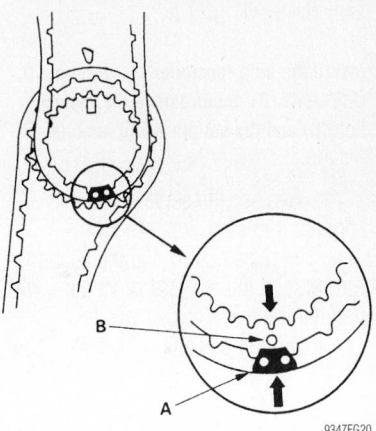

Install the timing chain with the colored link (A) aligned with the crankshaft sprocket punch mark (B)—2.0L engine

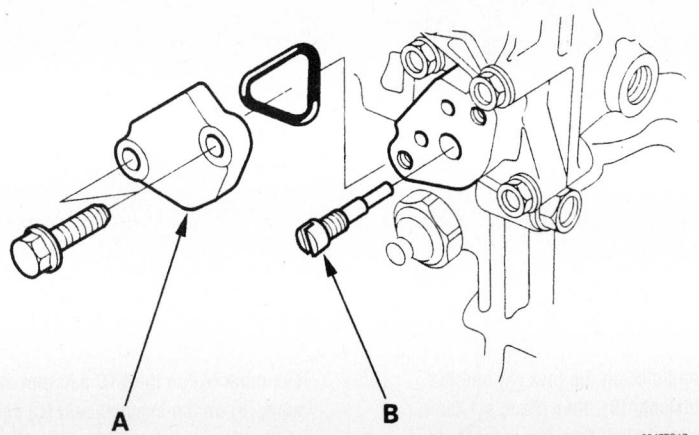

Timing chain auto-tensioner cover (A) and nozzle (B)—2.0L engine

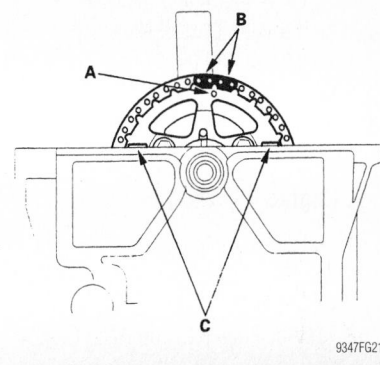

Timing chain idler sprocket punch mark (A), colored links (B) and TDC marks (C) in proper alignment—2.0L engine

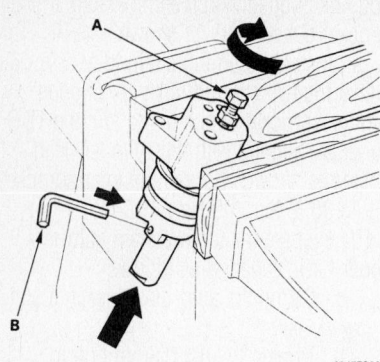

9347FG22

Tighten the 5 x).8mm bolt (A) and insert the set pin (B)—2.0L engine

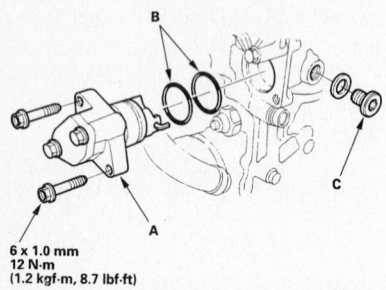

6 x 1.0 mm
12 N·m
(1.2 kgf·m, 8.7 lbf·ft)

9347FG23

Install the auto-tensioner (A) with new O ring seals (B), then remove the service bolt (C) and the set pin—2.0L engine

f. Replace the service bolt and tighten it to 22 ft. lbs. (29 Nm).

11. The remainder of the installation is the reverse of the removal procedure, while using the following torque values:

- Oil pump chain guide bolts: 104 inch lbs. (12 Nm)
- Front cover: 10mm bolts to 33 ft. lbs. (44 Nm) and the 6mm bolts to 104 inch lbs. (12 Nm)
- Crankshaft sprocket bolt: 181 ft. lbs. (245 Nm)
- Idler pulley: 10mm bolts to 33. Ft. lbs. (44 Nm) and the 6mm bolt to 104 inch lbs. (12 Nm)
- Water pump pulley bolts: 10 ft. lbs. (14 Nm)
- Bypass tube bolts: 104 inch lbs. (12 Nm)

2.4L Engine

1. Before servicing the vehicle, refer to the precautions in the beginning of this section.

2. Set the engine to Top Dead Center (TDC).

3. Drain the cooling system.

4. Relieve the fuel system pressure.

5. Remove or disconnect the following:

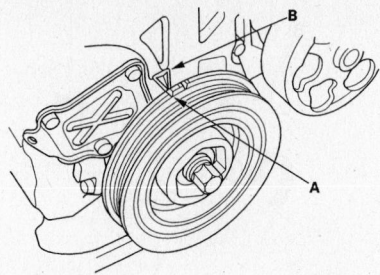

42356-ACCO-G24

Turn the crankshaft pulley so the TDC mark (A) is aligned with the pointer (B)— 2.4L engine

- Negative battery cable
- Front tires and wheels
- Splash shield
- Drive belt
- Cylinder head cover
- Crankshaft pulley
- Crankshaft Position (CKP) sensor connector
- Variable Valve Timing Control (CTV) oil control solenoid valve connector
- VTC oil control solenoid valve

6. Support the engine with a suitable jack with a wooden block under the oil pan.
- Ground cable and upper bracket
- Side engine mount bracket
- Chain cover

7. Loosely install the crankshaft pulley. Turn the crankshaft counterclockwise to compress the auto-tensioner.

8. Align the holes on the lock (A) and the auto-tensioner (B), then place a 1.5mm pin into the holes. Turn the crankshaft clockwise to secure the pin.

9. Remove or disconnect the following:

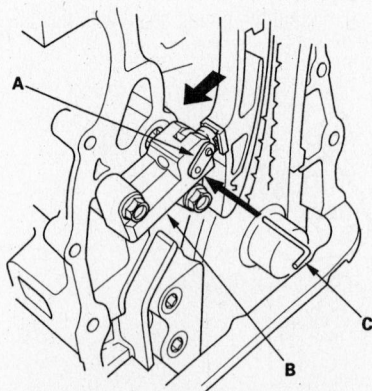

42356-ACCO-G25

Align the holes on the lock (A) and the auto-tensioner (B), then place a 1.5mm pin into the holes. Turn the crankshaft clockwise to secure the pin

- Auto-tensioner
- Timing chain guide B (top guide)
- Timing chain guide A and tensioner arm
- Timing chain

✳✳ WARNING

Do not let the timing chain near any magnetic fields.

To install:

10. Set the crankshaft to TDC. Align the TDC mark (A) on the crankshaft sprocket with the pointer (B) on the cylinder block.

11. Set the camshafts to TDC. The punch mark (A) on the VTC actuator and the punch mark (B) on the exhaust camshaft (C) should be at the top. Align the TDC marks (C) on the VTC actuator and exhaust camshaft sprockets.

12. Install or connect the following:
- Timing chain the crankshaft sprocket with the colored link of the chain aligned with the mark on the crank sprocket
- Timing chain on the VTC actuator and exhaust camshaft sprocket with

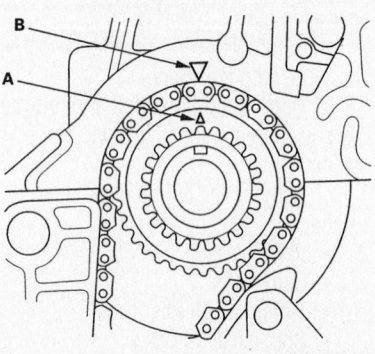

42356-ACCO-G26

Set the crankshaft to TDC. Align the TDC mark (A) on the crankshaft sprocket with the pointer (B) on the cylinder block

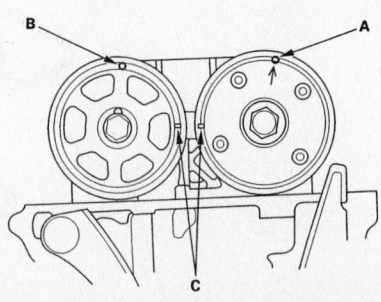

42356-ACCO-G27

The mark (A) on the VTC actuator and the mark (B) on the exhaust cam (C) should be at the top. Align the TDC marks (C) on the VTC actuator and exhaust cam sprockets

the punch marks aligned with the center of the 2 colored links
- Timing chain guide A and tensioner arm. Tighten the guide bolts to 8.7 ft. lbs. (12 Nm) and the tensioner arm retainer to 16 ft. lbs. (22 Nm).
- Auto-tensioner and tighten the bolts to 8.7 ft. lbs. (12 Nm)
- Timing chain guide B and tighten the retainers to 16 ft. lbs. (22 Nm)

13. Remove the pin from the auto-tensioner.

14. Inspect the chain cover seal for damage and replace if necessary. Clean and dry the chain cover mating surfaces.

15. Install or connect the following:
- Liquid gasket, P/N 08718-0009 evenly to the cylinder block mating surface of the timing chain cover and the inner threads of the holes
- Liquid gasket to the cylinder block upper surface contact areas on the chain cover and the oil pan mating surface of the chain cover in the inner threads of the holes

➥**Make sure to install the components within 5 minutes of applying the sealer.**

- New O-ring the timing chain cover. Set the edge of the cover to the edge of the oil pan, then install the cover on the engine block. Tighten the retainers to 8.7 ft. lbs. (12 Nm).

➥**When installing the chain case, do not slide the bottom surface on the oil pan mounting surface.**

- Side engine mounting bracket and tighten the retainers to 33 ft. lbs. (44 Nm)
- Upper bracket, then tighten the bolts/nuts as shown in the illustration
- Ground cable

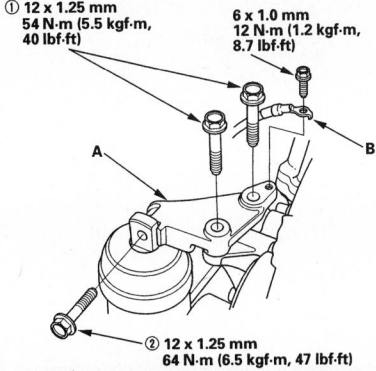

① 12 x 1.25 mm
54 N·m (5.5 kgf·m, 40 lbf·ft)

⑥ x 1.0 mm
12 N·m (1.2 kgf·m, 8.7 lbf·ft)

A   B

② 12 x 1.25 mm
64 N·m (6.5 kgf·m, 47 lbf·ft)

42356-ACCO-G28

Tighten the upper bracket upper bolt/nuts in the proper order to the correct specification—2.4L engine

- VTC oil control solenoid valve
- CKP sensor and VTC oil control solenoid valve connectors
- Crankshaft pulley
- Cylinder head cover
- Drive belt
- Splash shield

16. Fill the engine cooling system and connect the negative battery cable.

Piston and Ring

POSITIONING

When assembling the pistons, piston rings and connecting rods, and when installing these assemblies into the engine block, it is vitally important to ensure that these three components are properly positioned with respect to each other. Often times the engine block is designed so that if a connecting rod or piston is installed backwards, or in the wrong bank of cylinders, internal engine damage may occur once the engine is started. The piston ring end-gap spacing that is recommended by the engine manufacturer is often with the purpose of increased compression pressures during the engine break-in period. Failure to properly space the piston ring end-gaps may lead to increased oil consumption and extended break-in time. Therefore, always be sure to position the pistons, rings and connecting rods as shown in the accompanying illustrations.

✱✱ WARNING

Always be sure to matchmark the connecting rods and caps prior to disassembly so that they may be reassembled with their original coun-

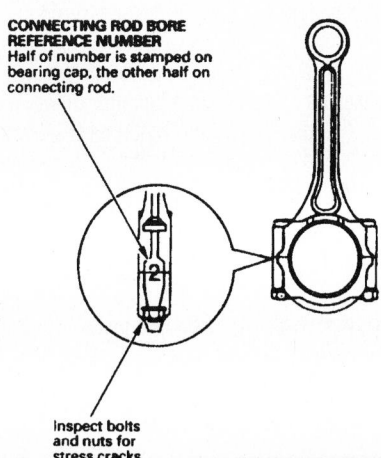

CONNECTING ROD BORE REFERENCE NUMBER
Half of number is stamped on bearing cap, the other half on connecting rod.

Inspect bolts and nuts for stress cracks.

7923AG27

Honda engines—before removing the caps from the connecting rods, be sure to matchmark them as shown

terparts. If the caps are not installed on their original connecting rods, the assemblies will most likely need machining to avoid bearing, connecting rod and/or crankshaft damage.

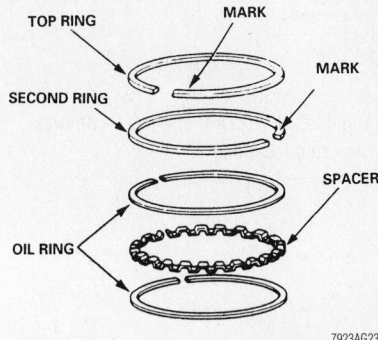

TOP RING   MARK
MARK
SECOND RING
SPACER
OIL RING

7923AG23

Honda engines—piston ring positioning

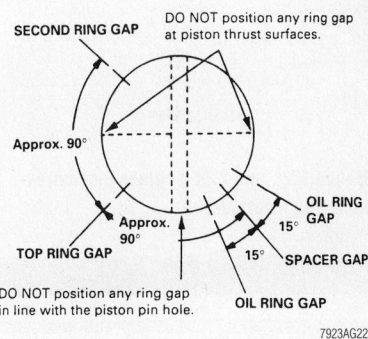

SECOND RING GAP   DO NOT position any ring gap at piston thrust surfaces.

Approx. 90°

Approx. 90°

TOP RING GAP

OIL RING GAP
15°
SPACER GAP
15°
OIL RING GAP

DO NOT position any ring gap in line with the piston pin hole.

7923AG22

Honda engines—piston ring end-gap spacing

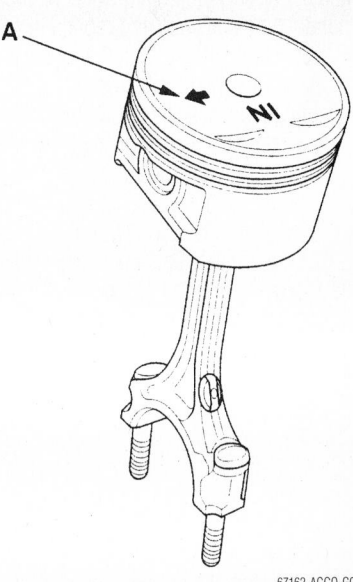

A

67162-ACCO-G07

Honda 1.7L engines—piston/connecting rod assembly-to-engine orientation. The arrow (A) must face the timing belt side of the engine

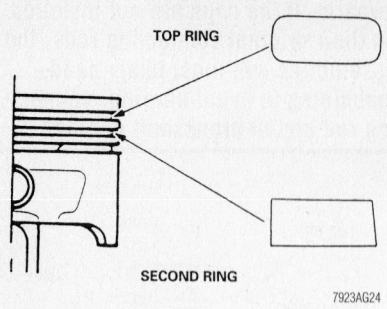

Honda 2.2L (H22A4) engines—compression ring locations

7923AG24

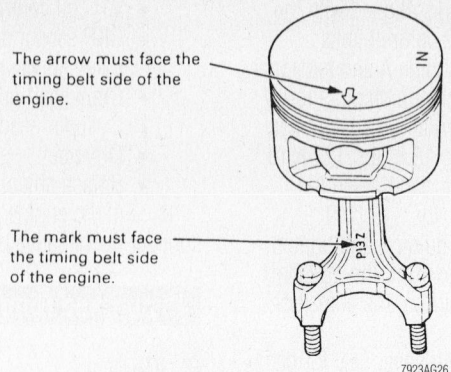

The arrow must face the timing belt side of the engine.

The mark must face the timing belt side of the engine.

7923AG26

Honda 2.2L (H22A4) engines—piston/connecting rod assembly-to-engine orientation

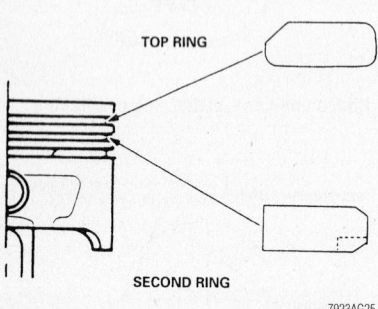

Honda 2.2L and 2.3L engines—compression ring locations

7923AG25

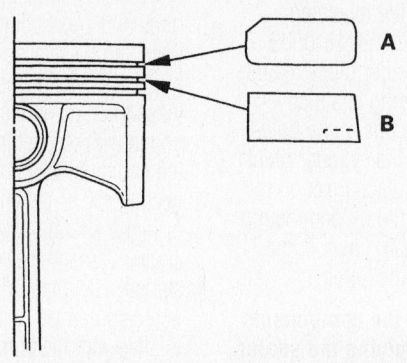

42356-ACCO-G29

Honda 2.4 engine—compression ring locations

FUEL SYSTEM

Fuel System Service Precautions

Safety is the most important factor when performing not only fuel system maintenance but any type of maintenance. Failure to conduct maintenance and repairs in a safe manner may result in serious personal injury or death. Maintenance and testing of the vehicle's fuel system components can be accomplished safely and effectively by adhering to the following rules and guidelines.

• To avoid the possibility of fire and personal injury, always disconnect the negative battery cable unless the repair or test procedure requires that battery voltage be applied.

• Always relieve the fuel system pressure prior to disconnecting any fuel system component (injector, fuel rail, pressure regulator, etc.), fitting or fuel line connection. Exercise extreme caution whenever relieving fuel system pressure, to avoid exposing skin, face and eyes to fuel spray. Please be advised that fuel under pressure may penetrate the skin or any part of the body that it contacts.

• Always place a shop towel or cloth around the fitting or connection prior to loosening to absorb any excess fuel due to spillage. Ensure that all fuel spillage (should it occur) is quickly removed from engine surfaces. Ensure that all fuel soaked cloths or towels are deposited into a suitable waste container.

• Always keep a dry chemical (Class B) fire extinguisher near the work area.

• Do not allow fuel spray or fuel vapors to come into contact with a spark or open flame.

• Always use a back-up wrench when loosening and tightening fuel line connection fittings. This will prevent unnecessary stress and torsion to fuel line piping. Always follow the proper torque specifications.

• Always replace worn fuel fitting O-rings with new. Do not substitute fuel hose or equivalent, where fuel pipe is installed.

Fuel System Pressure

RELIEVING

✳✳ CAUTION

The fuel injection system remains under pressure after the engine has been turned OFF. Properly relieve fuel pressure before disconnecting any fuel lines. Failure to do so may result in fire or personal injury.

➡**The radio may contain a coded theft protection circuit. Always obtain the code number before disconnecting the battery. If the vehicle is equipped with 4WS, the steering control unit is shut down when the battery is disconnected. After connecting the battery, turn the steering wheel lock-to-lock to reset the steering control unit.**

1. Before servicing the vehicle, refer to the precautions in the beginning of this section.

2. Disconnect the negative battery cable.

3. Remove the kick panel, then remove the PGM-FI main relay (FUEL PUMP) from the under-dash fuse/relay box. Start the engine and let it run until it stalls.

4. Remove the fuel filler cap.

5. On vehicles without quick connect fittings:

   a. Use a box wrench to loosen the 6mm service bolt while holding the special banjo bolt with another wrench. It is either located on the fuel filter or the fuel rail.

b. Place a rag or shop towel over the 6mm service bolt.

c. Slowly loosen the 6mm service bolt 1 complete turn.

d. On vehicles with quick-connect fittings:

e. Remove the quick-connect fitting cover.

f. Clean any dirt from the quick-connect fitting.

g. Place a rag or shop towel over quick-connect fitting.

h. Detach the quick-connect fitting by holding the connector with one hand, then squeeze the retainer tabs with the other hand to release them from the locking pawls. Pull the connector off.

※※ CAUTION

Do not allow fuel spray or fuel vapors to come in contact with a spark or open flame. Keep a dry chemical fire extinguisher nearby. Never store fuel in an open container due to risk of fire or explosion.

➡ **A fuel pressure gauge may be attached at the 6mm service bolt or quick-connect location. Always replace the washer between the service bolt and the banjo bolt whenever the service bolt is loosened.**

6. If equipped, remove the service bolt and install a new washer. Tighten the 6mm service bolt to 108 inch lbs. (12 Nm). Don't overtighten the service bolts, their threads may strip and cause leaks.

7. If equipped with a quick-connect fitting, connect the fitting, making sure the locking pawls are properly engaged.

8. Clean up any fuel spilled on the engine and intake manifold.

9. Install the fuel pump relay to the underdash fuel/relay box and install the kick-panel cover.

10. Install the fuel filler cap.

11. Reconnect the negative battery cable.

12. Turn the ignition **ON**, but don't start the engine. Repeat this 2 or 3 times to pressurize the fuel system. Check for fuel leaks.

13. Enter the radio security code.

14. If equipped with 4WS, start the engine and turn the steering wheel lock-to-lock to reset the 4WS control unit.

Fuel Filter

REMOVAL & INSTALLATION

Prelude

➡ **The radio may contain a coded theft protection circuit. Always obtain the code number before disconnecting the battery. If the vehicle is equipped with 4WS, the steering control unit is shut down when the battery is disconnected. After connecting the battery, turn the steering wheel lock-to-lock to reset the steering control unit.**

1. Before servicing the vehicle, refer to the precautions in the beginning of this section.

2. Disconnect the negative battery cable.

3. Place a shop towel under and around the fuel rail, then relieve the fuel pressure.

※※ CAUTION

Do not allow fuel spray or fuel vapors to come in contact with a spark or open flame. Keep a dry chemical fire extinguisher nearby. Never store fuel in an open container due to risk of fire or explosion.

4. Remove or disconnect the following:
- 12mm banjo bolt and the fuel feed pipe from the fuel filter. Discard the washers.
- Fuel filter clamp and the fuel filter.

To install:

5. Install or connect the following:
- Fuel filter on the bracket and the filter clamp. Tighten the clamp bolts to 84 inch lbs. (10 Nm).

➡ **Clean the fuel fittings thoroughly before reconnecting them.**

- Fuel feed pipe to the filter and tighten the fitting to 28 ft. lbs. (38 Nm).
- Fuel outlet pipe to the filter using new gaskets around the fitting. Tighten the banjo bolt to 20 ft. lbs. (28 Nm).
- Negative battery cable and enter the radio security code.

6. Turn the ignition **ON** and check for fuel leaks.

7. If equipped with 4WS, start the engine and turn the steering wheel lock-to-lock to reset the 4WS control unit.

Accord & Civic

1. Before servicing the vehicle, refer to the precautions in the beginning of this section.

2. Relieve the fuel system pressure.

3. Remove or disconnect the following:
- Fuel pump from the tank
- Fuel filter from the pump module

To install:

4. Install or connect the following:
- New filter on the pump module
- Fuel pump in the tank

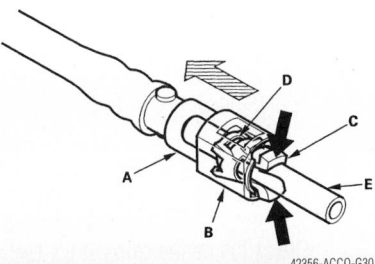

42356-ACCO-G30

Hold the quick-connect (A) connector (B) with one hand, then squeeze the retainer tabs (C) with the other hand to release them from the locking pawls (D)

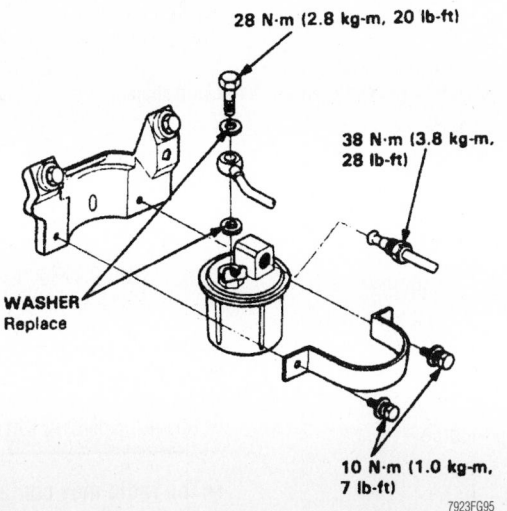

28 N·m (2.8 kg-m, 20 lb-ft)

38 N·m (3.8 kg-m, 28 lb-ft)

WASHER Replace

10 N·m (1.0 kg-m, 7 lb-ft)

7923FG95

Fuel filter mounting—Prelude

Except SULEV model:

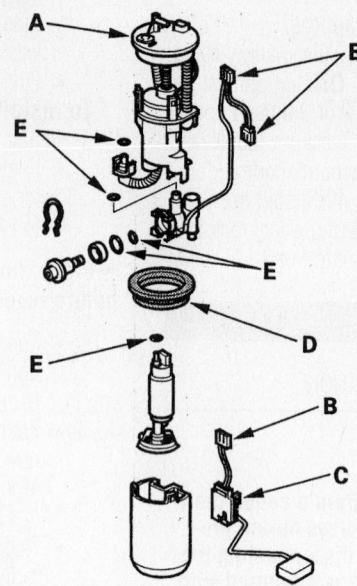

SULEV model:

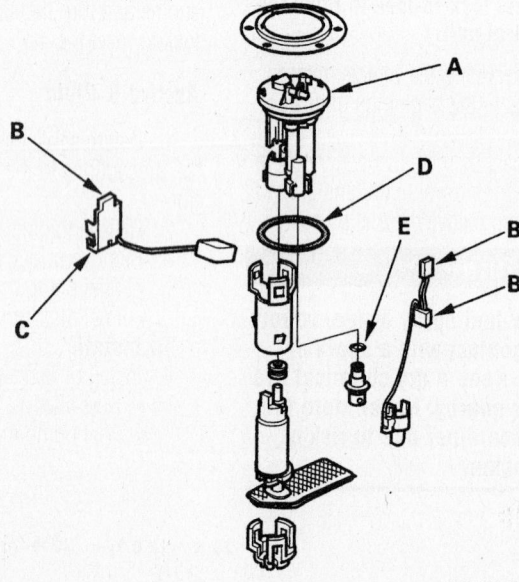

42356-ACCO-G31

Fuel filter (A) and related components—2003-04 Accord shown

S2000

1. Before servicing the vehicle, refer to the precautions in the beginning of this section.
2. Relieve the fuel system pressure.
3. Remove or disconnect the following:
   - Negative battery cable
   - Rear package tray
   - Access panel
   - Fuel pump module
   - Fuel filter

To install:

4. Install or connect the following:

- Fuel filter
- Fuel pump module and tighten the bolts to 36 inch lbs. (4 Nm)
- Access panel
- Rear package tray
- Negative battery cable

Fuel Pump

REMOVAL & INSTALLATION

➡The radio may contain a coded theft protection circuit. Always obtain the code number before disconnecting the battery. If the vehicle is equipped with 4WS, the steering control unit is shut down when the battery is disconnected. After connecting the battery, turn the steering wheel lock-to-lock to reset the steering control unit.

Civic

✳✳ CAUTION

The fuel injection system remains under pressure, even after the engine has been turned OFF. The fuel system pressure must be relieved before disconnecting any fuel lines. Failure to follow this procedure may result in fire, explosion, or personal injury.

1. Before servicing the vehicle, refer to the precautions in the beginning of this section.
2. Disconnect the negative battery cable.
3. Relieve the fuel pressure.
4. Remove or disconnect the following:
   - Rear seat cushions
   - Fuel pump access panel
   - 2-wire fuel pump harness

➡Clean the fuel line fittings before disconnecting them.

- Fuel line and the hose from the fuel pump.
- Fuel pump bolts and fuel pump from the fuel tank. Allow the fuel in the pump drain into the tank before removing the pump from the vehicle.
- Fuel pump motor from its bracket.

To install:

➡Use new sealing washers when reconnecting the fuel line banjo bolt.

5. Install or connect the following:
   - Fuel pump into the fuel tank with a new O-ring. Then, tighten the mounting nuts to 48 inch lbs. (6 Nm).
   - Hose and the fuel line. Carefully tighten the banjo bolt to 20 ft. lbs. (28 Nm).
   - Fuel pump harness
   - Fuel filler cap
   - Fuel filter service bolt: 11 ft. lbs. (15 Nm)
   - Battery cable and turn the ignition switch **ON** and **OFF** several times to pressurize the fuel system. Check the connections at the fuel pump for any leaks. Check the fuel filter service bolt for leaks.
   - Fuel pump access cover
   - Rear seat cushions or rear com-

partment trim. Be sure the clips are properly seated.

Prelude

❋❋ CAUTION

The fuel injection system remains under pressure after the engine has been turned OFF. Properly relieve fuel pressure before disconnecting any fuel lines. Failure to do so may result in fire or personal injury.

1. Before servicing the vehicle, refer to the precautions in the beginning of this section.
2. Disconnect the negative battery terminal.
3. Relieve the fuel pressure.
4. Remove or disconnect the following:
   • Carpet in the luggage area.
   • Fuel pump maintenance access cover in the floor.
   • Electrical connector at the pump unit.

➡ **Label the fuel lines before disconnecting them.**

   • Fuel lines. Discard the washers from the fuel feed connection.
   • Retaining nuts holding the pump
   • Pump up and out of the tank.

➡ **The pump sits on an angle and may require some manipulation to remove. If the pump still won't come out, loosen the fuel tank mounting nuts under the car, slide the tank downward a bit to give more clearance at the top.**

To install:

5. Install or connect the following:
   • New sealing ring
   • Fuel pump, making certain it is correctly seated and not wedged or jammed.
   • Retaining nuts and tighten them evenly and alternately to 48 inch lbs. (6 Nm).
   • Fuel lines. Make certain the clamp is secure; use new ones if necessary.
   • New washers to the fuel feed connection before installing the attaching bolt. Tighten the fuel feed attaching bolt to 20 ft. lbs. (28 Nm).
   • Fuel pump connector
   • Negative battery cable and enter the radio security code.
6. Switch the ignition **ON** but do not engage the starter. The fuel pump should

run for approximately 2 seconds, building pressure within the lines. Switch the ignition **OFF**, then **ON** 2 or 3 more times to build full system pressure. Check for fuel leaks.

7. If equipped with 4WS, start the engine and turn the steering wheel lock-to-lock to reset the 4WS control unit.
8. Install the maintenance access cover and seal or gasket, if used.
9. Reposition the carpeting in the luggage compartment.

Accord

1. Before servicing the vehicle, refer to the precautions in the beginning of this section.
2. Relieve the fuel pressure.
3. Remove or disconnect the following:
   • Negative battery cable
   • Spare tire cover, if equipped
   • Trunk floor
   • Access panel from the floor
   • 5-pin connector from the pump assembly.
   • Quick-connect connections from the pump assembly.
   • Mounting bolts and the fuel tank unit
   • Strainer case, fuel gauge sending unit and wire harness, if necessary

To install:

4. Assembly the strainer case, fuel gauge sending unit and wire harness, if necessary
5. Install or connect the following:
   • Fuel pump, using a new gasket

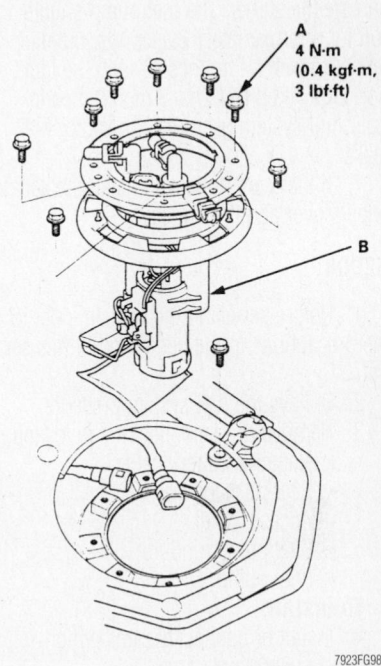

Fuel pump assembly mounting—Accord

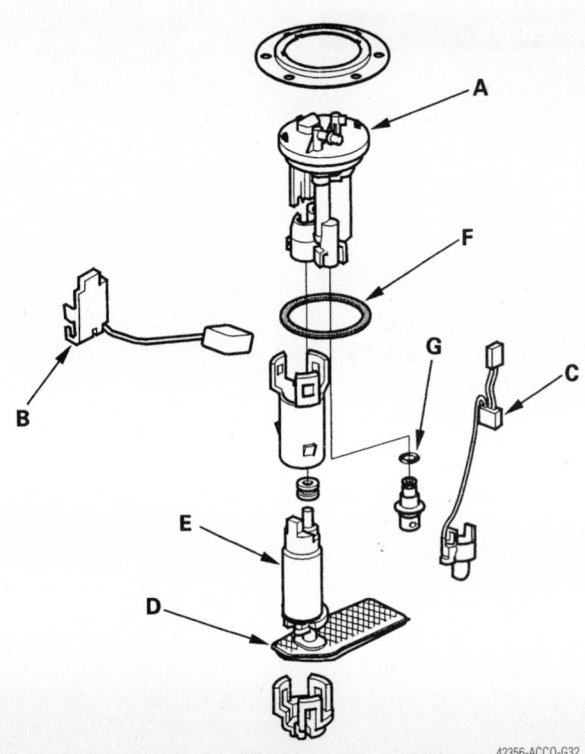

Exploded view of the strainer case (A), fuel gauge sending unit (B), wire harness (C), suction filter (D) and fuel pump (E)—Accord

- Quick-connect fuel lines to the pump assembly
- Fuel cap
- 5-pin connector to the pump
- Negative battery cable and enter the radio security code

6. Switch the ignition **ON** but do not engage the starter. The fuel pump should run for approximately 2 seconds, building pressure within the lines. Switch the ignition **OFF**, then **ON** 2 or 3 more times to build full system pressure. Check for fuel leaks.

7. If there are no leaks, install the access panel cover and the tire cover.

S2000

1. Before servicing the vehicle, refer to the precautions in the beginning of this section.

2. Relieve the fuel system pressure.

3. Remove or disconnect the following:
- Negative battery cable
- Rear package tray
- Access panel
- Fuel pump module
- Fuel pump

To install:

4. Install or connect the following:
- Fuel pump

- Fuel pump module and tighten the bolts to 36 inch lbs. (4 Nm)
- Access panel
- Rear package tray
- Negative battery cable

Fuel Injector

REMOVAL & INSTALLATION

1. Disconnect the negative battery cable.

2. Relieve the fuel system pressure.

3. Remove the fuel rail assembly

4. Carefully pull the injectors from the intake manifold.

5. Discard the seal rings, cushion rings and O-rings.

To install:

6. Slide new cushion rings onto the injectors.

7. Coat new O-rings with clean engine oil and put them on the injectors.

8. Insert the injectors into the fuel rail. Be sure to align the center line on the injector with the mark on the fuel rail.

9. Coat new seal rings with clean engine oil and insert them into the intake manifold.

10. Install the fuel rail assembly.

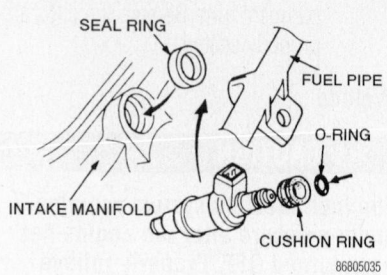

Always use new cushion rings, seal rings and O-rings

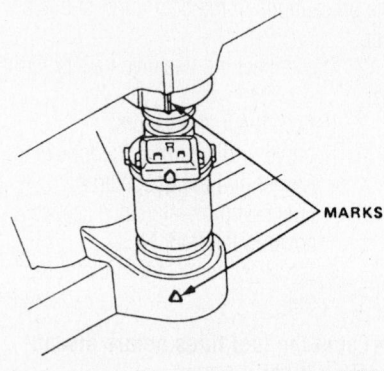

Be sure to align the center line on the injector with the mark on the fuel rail

---

DRIVE TRAIN

Transaxle Assembly

REMOVAL & INSTALLATION

Manual

CIVIC

✳✳ WARNING

Use only genuine Honda manual transaxle fluid (MTF)-it is specially formulated for use in Honda transaxles. If Honda MTF is not available, API SG/SJ 10W-30 or 10W-40 motor oil may be used as a temporary lubricant. However, motor oil will cause increased transaxle wear and shifting effort. Refill the transaxle with Honda MTF as soon as possible.

1. Before servicing the vehicle, refer to the precautions in the beginning of this section.

2. Remove or disconnect the following:
- Negative and positive battery cables

- Resonator, the air cleaner box, and the air intake duct.
- Starter cables and the transaxle ground cable.
- Back-up light switch connection
- Upper radiator hose out of its bracket.
- Vehicle Speed Sensor (VSS) connector.
- Clutch fluid line bracket
- Slave cylinder. It isn't necessary to disconnect the clutch fluid line.

3. Raise and safely support the vehicle.

4. Drain the transaxle fluid.

5. Remove or disconnect the following:
- Front wheels
- Strut pinch bolt and fork bolt
- Lower ball joint from the steering knuckle using a ball joint remover.
- Halfshaft inboard joints out of the transaxle case. Swing the steering knuckles out to free the halfshafts from the transaxle.

6. Tie the halfshafts up and out of the way with wire so that the joints will not be stressed. Tie plastic bags over the inboard joints to prevent damage to the CV-boots and splined shafts.

7. Remove or disconnect the following:
- Shift rod and extension rod from the transaxle case. Drive the shift rod retaining pin out with a pin punch.
- Front exhaust pipe
- Engine-to-transaxle stiffener brackets and the clutch cover plate

8. Attach a lifting chain to the engine and lift slightly to ease the tension on the mounts.

9. Remove or disconnect the following:
- Splash shield from underneath the vehicle.
- Right-front mount/bracket assembly

10. Place a jack under the transaxle to support its weight.

11. Remove or disconnect the following:
- Transaxle side mount and bracket
- Starter's lower mounting bolt and the upper 3 transaxle case bolts
- 3 rear transaxle mount bracket bolts, then the lower 3 transaxle case bolts.
- Pull the transaxle away from the engine until it clears the mainshaft. Lower the transaxle out of the

vehicle Be careful not to bend the clutch hydraulic line.

To install:

➡**Use new self-locking nuts and color-coded bolts when installing the transaxle and suspension components.**

12. Apply high temperature grease to the mainshaft splines, release fork contact points, and throw-out bearing. The manufacturer recommends part No. 08798-9002, Honda Super High temp Urea Grease.

13. Place the transaxle on a transaxle jack and raise it to the level of the engine.

14. Align the transaxle and engine. Be sure the transaxle case dowel pins are securely seated, and fit the transaxle onto the engine. Install the upper and lower transaxle case bolts and the 14mm rear mount bolts and washers. Only hand-tighten them at this time.

15. Raise the transaxle and install the side mount. Tighten the upper and lower transaxle case bolts to 47 ft. lbs. (64 Nm). Tighten the 14mm rear mount bracket bolts to 61 ft. lbs. (84 Nm).

16. First, tighten the transaxle side mount bracket nuts and bolt to 47 ft. lbs. (64 Nm) each. Next, tighten the mount bushing bolts to 47 ft. lbs. (64 Nm). Finally, tighten the through-bolt to 54 ft. lbs. (74 Nm).

17. Install or connect the following:
- Right-front mount/bracket assembly. Use 3 new 12mm bolt and washers, and tighten them to 47 ft. lbs. (64 Nm). Tighten the 2 10mm bolts to 33 ft. lbs. (45 Nm).
- Clutch cover
- Engine-to-transaxle stiffener brackets and tighten the 8mm bolts to 17 ft. lbs. (24 Nm). Tighten the 10mm bolts to 33 ft. lbs. (44 Nm).

18. Once the transaxle is bolted to the engine, and the transaxle mounts are securely tightened, the engine lifting chain may be removed.

19. Install or connect the following:
- Shift rod with a new spring pin and clip. Then, fit the shift rod boot back into place. Connect the torque rod and tighten the bolt to 16 ft. lbs. (22 Nm).
- Front exhaust pipe. Use new self-locking nuts and gaskets. Tighten the rear flange nuts to 16 ft. lbs. (22 Nm). Tighten the front flange nuts to 40 ft. lbs. (54 Nm) for D16Y8 and D16Y7 engines, or 25 ft. lbs. (33Nm) for all others.
- New set rings on the inboard CV-joint splines

- Halfshafts into the transaxle case and intermediate shaft. The inboard joints must snap into place.
- Lower ball joint and damper fork
- Front wheels
- Slave cylinder and clutch pipe stay. Coat the slave cylinder's tip with high temperature grease. Be sure it snaps into the release fork. Tighten the slave cylinder mounting bolts to 16 ft. lbs. (22 Nm).
- VSS connector and back-up light switch connectors
- Wiring harness clamps and starter cables.
- Resonator, air cleaner box, and air intake duct. Fit the upper radiator hose back into its bracket.

20. Lower the vehicle and tighten the strut pinch bolts to 32 ft. lbs. (44 Nm). Tighten the fork bolts to 47 ft. lbs. (65 Nm). Tighten the ball joint castle nuts to 40 ft. lbs. (55 Nm), then tighten them only enough to install new cotter pins.

21. Turn the breather cap so that the arrow with the **F** mark points toward the front of the vehicle.

22. Refill the transaxle with the Honda MTF fluid.

23. Reconnect the positive and negative battery cables.

24. Bleed the clutch hydraulic system.

25. Check the clutch and transaxle for smooth operation.

26. Check and adjust the front wheel alignment.

PRELUDE

1. Before servicing the vehicle, refer to the precautions in the beginning of this section.

2. Shift the transaxle to **R**.

3. Remove or disconnect the following:
- Negative and positive battery cables
- Battery
- Intake duct, air cleaner case and battery base
- Vacuum tank and bracket. Do not disconnect the hoses.
- Starter wires and the starter.

4. Loosen, but do not remove the 2 upper transaxle mounting bolts.

5. Remove or disconnect the following:
- Transaxle ground cable and the back-up light switch wire.
- Engine harness clamp
- Shift cables from the transaxle case, leaving them attached to their bracket, and wire them safely out of the work area.

- Vehicle Speed Sensor (VSS) connector. Leave the sensor hoses connected and remove the sensor from the transaxle case.
- Slave cylinder mounting bolts.
- Slave cylinder from the release fork and move it out of the work area, leaving the hydraulic line connected to the slave cylinder

➡**Do not depress the clutch pedal once the slave cylinder has been removed. Be careful not to kink the metal hydraulic line.**

6. Raise and safely support the vehicle. Drain the transaxle fluid.

7. Remove or disconnect the following:
- Clutch damper mounting bolts and raise the clutch damper.
- Remove the rear engine mount bracket stay, if equipped.
- Front wheels
- Cotter pins and lower arm ball joint nuts
- Ball joints and lower arms using a press-type ball joint tool.
- Damper fork bolt and the radius rod on the right side of the vehicle only.

8. Use a suitable tool to pry the right and left halfshafts out of the differential and the intermediate shaft. Pull on the inboard joint and remove the right and left halfshafts.

9. Remove or disconnect the following:
- Intermediate shaft from the differential. Tie plastic bags over the halfshaft inboard joints to prevent damage to the boots and splines. Wire the halfshafts to the underbody of the vehicle so that their weight doesn't hang on their outboard joints.
- Center beam and remove the clutch cover
- Rear beam stiffener and the intake manifold stay.
- 3 rear engine mount bracket bolts.

10. Place a transaxle jack under the transaxle and raise the transaxle just enough to take its weight off the mounts.

11. Remove or disconnect the following:
- Transaxle mount and mount bracket.
- 2 upper transaxle housing mounting bolts and the 3 lower transaxle housing bolts.

12. Pull the transaxle away from the engine to clear the mainshaft.

13. Lower the transaxle from the vehicle.

To install:

➡ **Use new self-locking nuts and set rings when assembling the front suspension components and halfshafts. Use new self-locking bolts when installing the center beam and rear engine mount bracket. These fasteners can be purchased from a Honda dealer.**

14. Be sure the dowel pins are installed into the transaxle case.

15. Apply heavy duty high temperature grease (use Honda part number 08798–9002) to the mainshaft splines, release fork bolt and paws, and the throw-out bearing. Install the bearing and release fork. Be sure the release fork snaps into place.

16. Raise the transaxle into position with a transaxle jack.

17. Install or connect the following:
- 3 lower and 2 upper transaxle mounting bolts and evenly tighten them to 47 ft. lbs. (65 Nm).
- Transaxle mount and mount bracket
- Through-bolt and tighten temporarily. Be sure the engine is level. First tighten the 3 bracket-to-mount nuts and 2 bolts to 28 ft. lbs. (39 Nm). Then, tighten the through-bolt to 47 ft. lbs. (65 Nm).
- 3 new rear engine mount bracket bolts on the engine side and tighten them to 40 ft. lbs. (55 Nm).
- Rear beam stiffener and tighten the bolts to 28 ft. lbs. (39 Nm).
- Intake manifold stay and tighten the bolts to 16 ft. lbs. (22 Nm).
- Clutch cover and tighten the bolts to 108 inch lbs. (12 Nm).
- Center beam and tighten the bolts to 43 ft. lbs. (60 Nm).
- Intermediate shaft. Tighten its mounting bolts to 28 ft. lbs. (39 Nm).
- New set rings onto the halfshaft inboard joint splines.
- Halfshafts, making sure that they lock into place.
- Radius rod and damper fork. Only hand-tighten their fasteners at this time.
- Ball joint to the lower arm. Tighten the castle nut to 36–43 ft. lbs. (50–60 Nm). Then, only tighten the nut enough to install a new cotter pin.
- Rear engine mount bracket stay, if equipped. Tighten the nut to 15 ft. lbs. (21 Nm) and the bolt to 28 ft. lbs. (39 Nm).
- Clutch damper and tighten the bolts to 16 ft. lbs. (22 Nm).

- Front wheels

18. Lower the vehicle.

19. Use a floor jack placed under the right front control arm to raise the vehicle enough so that its weight is supported by the jack. Tighten the radius rod mounting bolts to 76 ft. lbs. (105 Nm) and the radius rod nut to 32 ft. lbs. (44 Nm). Tighten the damper pinch bolt to 32 ft. lbs. (44 Nm). Tighten the damper fork bolt to 47 ft. lbs. (65 Nm). After pre-loading the suspension, lower the vehicle and remove the floor jack.

20. Coat the tip of the slave cylinder with heavy duty high temperature grease.

21. Install or connect the following:
- Clutch hose pipe and clutch slave cylinder to the transaxle housing. Be sure the slave cylinder snaps into the release fork. Tighten the slave cylinder mounting bolts to 16 ft. lbs. (22 Nm).
- Speed sensor. Tighten the mounting bolt to 14 ft. lbs. (19 Nm).
- Shift cable and select cable to the shift arm lever.
- Shift cable assembly onto the transaxle case. Tighten the cable bracket mounting bolts to 16 ft. lbs. (22 Nm).
- New cotter pins
- Back-up light switch coupler and the transaxle ground cable.
- Harness clamp
- Starter. Tighten the 10mm bolt to 32 ft. lbs. (45 Nm) and the 12mm bolt to 54 ft. lbs. (75 Nm).
- Starter wires

22. Loosen the 3 front engine mount bracket bolts. Tighten them to 28 ft. lbs. (39 Nm).

23. Install or connect the following:
- Vacuum tank and its bracket
- Air cleaner case and intake duct

24. Fill the transaxle with the proper type and quantity of oil.

25. Install or connect the following:
- Battery base stay and the battery base. Tighten the battery base bolts to 16 ft. lbs. (22 Nm).

26. Install or connect the following:
- Battery and connect the battery cables.

27. Check the clutch pedal free-play.

28. Start the vehicle and check the transaxle and clutch for smooth operation.

29. On Preludes equipped with 4WS, and turn the steering wheel lock-to-lock to reset the steering control unit.

30. Check and adjust the front wheel alignment.

31. Enter the radio security code.

ACCORD

1. Before servicing the vehicle, refer to the precautions in the beginning of this section.

2. Shift the transaxle into **R**.

3. Remove or disconnect the following:
- Negative and positive battery cables and the battery.
- Idle Air Control (IAC) solenoid connector
- Intake duct, resonator, air cleaner case, and battery base.
- Starter wires and starter
- Transaxle ground cable and the back-up light switch wire.
- Cable stay, then the cables from the top housing of the transaxle
- Both cables and the stay together
- Vehicle Speed Sensor (VSS). Leave the speed sensor hoses connected.
- Shift cable bracket
- Shift and select cables from the top of the transaxle case. Leave the cables and bracket together, and wire them out of the work area.
- Mounting bolts and clutch slave cylinder with the clutch pipe and pushrod.
- Mounting bolt and clutch hose joint with the clutch pipe and clutch hose.

➡ **Do not depress the clutch pedal once the slave cylinder has been removed. Be careful not to kink the metal hydraulic lines.**

- 2 upper transaxle case bolts

4. Raise and safely support the vehicle.

5. Remove or disconnect the following:
- Front wheels
- Engine splash shield

6. Drain the transaxle fluid.

7. Remove or disconnect the following:
- Clutch damper bracket and raise it out of the way.
- Subframe center beam
- Cotter pins and lower arm ball joint nuts
- Ball joints and lower arms using a press type ball joint tool.
- Right damper fork bolt
- Right damper pinch bolt, then separate the damper fork and damper.
- Radius rod bolts and nut, then the right radius rod.
- Right and left halfshafts from the differential and the intermediate shaft, using a suitable prytool.
- Left halfshaft
- Intermediate shaft from the differential by removing its 3 bearing shaft mounting bolts.

8. Swing the right halfshaft out and wire it up inside the right fender well. Tie plastic bags over the inboard CV-joints to protect the boots and splines from damage.

9. Remove or disconnect the following:
- Engine stiffener and the clutch cover
- Intake manifold bracket
- Rear engine mount bracket
- 3 rear engine mount bracket mounting bolts, then discard.

10. Place a transaxle jack under the transaxle. Raise the transaxle just enough to take the weight off the its mounts.

➡**A chain hoist may be attached the transaxle lifting hooks to steady it and aid in lowering it from the vehicle.**

11. Remove or disconnect the following:
- Transaxle housing mounting bolt on the engine side
- Transaxle mount bolt and loosen the mount bracket nuts.
- 3 transaxle housing mounting bolts, then the transaxle from the vehicle.

To install:

➡**Use new self-locking nuts when assembling the front suspension. Install new set rings onto the inboard CV-joints. Use new self-locking bolts when installing transaxle rear mount bracket (the bolts are color coded by type). New fasteners are available from a Honda dealer.**

12. Be sure the 2 dowel pins are installed into the transaxle case.

13. Apply heavy duty high temperature grease (use Honda part No. 08798–9002) to the release bearing, mainshaft splines, and the release fork pawls. Install the release fork and release bearing.

14. Raise the transaxle into position.

15. Install or connect the following:
- 3 lower transaxle case bolts and tighten to 47 ft. lbs. (65 Nm).
- Transaxle mount and mount bracket
- Through-bolt and tighten temporarily. Be sure the engine is level and tighten the 3 mount bracket nuts to 40 ft. lbs. (55 Nm). Tighten the through-bolt to 47 ft. lbs. (65 Nm).
- Upper transaxle case bolts on the engine side and tighten to 47 ft. lbs. (65 Nm).
- 3 new rear engine bracket mounting bolts and tighten to 40 ft. lbs. (55 Nm).
- Intake manifold bracket and tighten the bolts to 16 ft. lbs. (22 Nm).
- Clutch cover and tighten the bolts to 108 inch lbs. (12 Nm).

- Subframe center beam with new self-locking bolts. Evenly tighten the bolts to 37 ft. lbs. (50 Nm).
- Engine stiffener plate, if equipped. Loosely install the mounting bolts. Tighten the stiffener-to-transaxle case mounting bolt to 28 ft. lbs. (39 Nm), then tighten the 2 stiffener-to-engine block mounting bolts to 28 ft. lbs. (39 Nm) beginning with the bolt closest to the transaxle.
- Radius rod and the damper fork. Hand-tighten all the fasteners.
- Intermediate shaft and tighten its mounting bolts to 28 ft. lbs. (39 Nm).
- New set ring on the end of each halfshaft.
- Right and left halfshafts. Turn the right and left steering knuckle fully outward and slide the axle into the differential, until the set ring is felt engaging the differential side gear.
- Lower control arm ball joints. Tighten the castle nuts to 40 ft. lbs. (50 Nm). Then, tighten them only enough to install a new cotter pin.
- Clutch damper and tighten its mounting bolts to 16 ft. lbs. (22 Nm).
- Front wheels. Lower the vehicle.

16. Place a floor jack under the right front knuckle, and raise the jack until it is supporting the vehicle's weight.

17. Tighten the radius rod mounting bolts to 76 ft. lbs. (105 Nm) and the radius rod nut to 32 ft. lbs. (44 Nm). Tighten the damper fork nut while holding the damper fork bolt to 40 ft. lbs. (55 Nm). Tighten the damper pinch bolt to 32 ft. lbs. (44 Nm).

18. Coat the tip of the slave cylinder with high temperature grease. Install the clutch hose joint and clutch slave cylinder to the transaxle housing. Be sure the slave cylinders tip snaps into the release fork. Tighten the slave cylinder mounting bolts to 16 ft. lbs. (22 Nm).

19. Install or connect the following:
- Speed sensor. Tighten the mounting bolt to 13 ft. lbs. (18 Nm).
- Shift cable and select cable to the shift arm lever. Tighten the cable bracket mounting bolts to 20 ft. lbs. (27 Nm).
- New cotter pins
- Back-up light switch connector
- Starter. Tighten the 10mm bolt to 32 ft. lbs. (45 Nm) and the 12mm bolt to 54 ft. lbs. (75 Nm).
- Starter wires
- Transaxle ground cable

20. Fill the transaxle with the proper type and quantity of oil.

21. Install or connect the following:
- Air cleaner case and the resonator, then the intake duct.
- Battery tray bracket and battery tray and tighten the bolts to 16 ft. lbs. (22 Nm).
- Battery and the battery cables.

22. Check the clutch pedal free-play.

23. Check and adjust the front wheel alignment.

24. Road test the vehicle and check the transaxle for smooth operation.

25. Loosen the 3 front engine mount bracket mounting bolts, then retighten them to 28 ft. lbs. (38 Nm).

26. Enter the radio security code.

S2000

1. Before servicing the vehicle, refer to the precautions in the beginning of this section.

2. Remove or disconnect the following:
- Battery cables and the battery
- Shift lever knob
- Center console
- Shift lever boot
- Shift lever
- Air cleaner housing
- Steering shaft from the steering gear box
- Alternator
- A/C compressor
- Exhaust manifold heat shields
- Upper starter mounting bolt
- Upper intake manifold bracket mounting bolt
- Suction valve hose
- Camshaft Position (CMP) sensor connectors
- Splash shield
- Steering gear box electrical connector
- Torque sensor connector
- Intake manifold bracket
- Heated Oxygen (HO$_2$S) sensor connectors
- Catalytic converter
- Exhaust manifold
- Driveshaft
- Shifter boot holder bolts
- Clutch slave cylinder
- Clutch release fork
- Lower transmission flange bolts

3. Support the front subframe with a floor jack and remove the two center mounting bolts.

4. Loosen the four outer mounting bolts 3 inches (75 mm).

5. Lower the front subframe until it is supported by the loosened bolts.

6. Support the transmission with the floor jack.

7. Remove or disconnect the following:
- Rear transmission mount
- Speed sensor connector and wiring harness
- Upper transmission flange bolts
- Transmission

To install:

➡**Use new subframe bolts for assembly.**

8. Installation is the reverse of the removal procedure, while using the following torque values:
- Transmission flange bolts: 47 ft. lbs. (64 Nm)
- Rear transmission mount bolts: 28 ft. lbs. (38 Nm)
- Subframe mounting bolts: 14mm bolts to 85 ft. lbs. (116 Nm) and 12mm bolts to 43 ft. lbs. (59 Nm)
- Clutch slave cylinder bolts: 16 ft. lbs. (22 Nm)
- A/C compressor bolts: 33 ft. lbs. (44 Nm)
- Steering shaft pinch bolt: 16 ft. lbs. (22 Nm)
- Shift lever bolts: 86 inch lbs. (10 Nm)

Automatic

CIVIC

1. Before servicing the vehicle, refer to the precautions in the beginning of this section.

2. Remove or disconnect the following:
- Negative and positive battery cables
- Resonator, the air cleaner box, and the air intake duct.
- Starter cables and the transaxle ground cable
- Engine wiring harness clip

➡**Label all electrical connectors before removing them.**

- Lock-up control solenoid connector
- Vehicle Speed Sensor and countershaft speed sensor connectors
- Upper transaxle case bolts and the rear engine mounting bolt

3. Raise and safely support the vehicle.

4. Remove or disconnect the following:
- Front wheels

5. Drain the automatic transaxle fluid. Then, install the drain plug with a new crush washer. Note the color, consistency, and odor of the drained fluid.

6. Remove or disconnect the following:
- Front splash shield

- Shift control and linear solenoid connectors
- Mainshaft speed sensor connector
- Strut pinch bolt and fork bolt
- Lower ball joint using a ball joint remover.

7. Pry the halfshaft inboard joints out of the transaxle case and intermediate shaft. Swing the steering knuckles out to free the halfshafts from the transaxle.

8. Tie the halfshafts up and out of the way with wire. Tie plastic bags over the inboard joints to prevent damage to the CV-boots and splined shafts.

9. Remove or disconnect the following:
- Front exhaust pipe
- Shift cable cover
- Shift cable from the transaxle control shaft. Move the shift cable out of the way, and tie it up with wire.
- Automatic Transaxle Fluid (ATF) cooler hoses from the cooler lines. Cap the lines to prevent fluid lose and contamination.
- Right-front mount and bracket assembly
- Engine stiffener and the torque converter cover plate.
- 8 torque converter-to-driveplate bolts 1 at a time by rotating the crankshaft pulley.

➡**There are no gear teeth on the driveplate; the starter motor engages a ring gear on the inner edge of the torque converter.**

10. After unbolting the torque converter from the driveplate, rotate the crankshaft to set the engine at Top Dead Center (TDC)/compression for the No. 1 cylinder.

11. Remove or disconnect the following:
- Ignition wires
- Distributor, if equipped

12. Attach a lifting chain to the engine and lift slightly to ease the tension on the mounts.

13. Place a transaxle jack under the transaxle and remove the transaxle side mount and bracket.

14. With the transaxle supported, remove the transaxle rear mount bracket bolts and transaxle case bolts.

15. Pull the transaxle away from the engine until it clears the locating dowel pins. Carefully lower the transaxle from the vehicle with the torque converter angled upward so it doesn't drop out of the transaxle.

16. Remove the torque converter from the transaxle. Inspect the ring gear teeth for breakage and inspect the converter's hub for burrs and scoring. Check the condition of

the converter fluid. Replace the torque converter if necessary.

17. Inspect the transaxles front oil pump bearing and seal for signs of leakage and scoring. Inspect the mainshaft for burrs, scoring, and roughness.

18. With the transaxle removed, carefully inspect the driveplate for stress cracks, enlarged bolt holes, and other defects. Replace it if necessary.

To install:

➡**Use new self-locking nuts and color-coded bolts when installing the transaxle and suspension components.**

19. Flush the transaxle cooler lines to remove any contaminated fluid and residual clutch material:

a. Use a pressurized flusher (Honda J38405-A or equivalent). Use only Honda flushing fluid (Honda J35944–20); other fluids may damage the system.

b. Fill the flusher with 21 ounces of fluid. Pressurize the flusher to 80–120 psi, following the procedure on the fluid container and flusher.

c. Clamp the discharge hose of the flusher to the cooler return line. Clamp the drain hose to the cooler inlet line and route it into a bucket or drain tank.

d. Connect the flusher to air and water lines. The air line use a water trap to keep excess moisture out.

e. Open the flusher water valve and flush the cooler for 10 seconds.

f. Depress the flusher trigger to mix flushing fluid with the water. Flush for 2 minutes, turning the air valve on and off for 5 seconds every 15–20 seconds to create a surging action.

g. After finishing 1 flushing cycle, reverse the hose and flush in the opposite direction.

h. Dry the cooler lines with compressed air for 2 minutes or longer to remove all excess moisture from the system.

20. Install or connect the following:
- Starter motor onto the transaxle case and tighten its mounting bolts to 33 ft. lbs. (45 Nm).
- Torque converter with a new hub O-ring.

21. Place the transaxle on a transaxle jack and raise it to the level of the engine.

22. Align the transaxle and engine. Install the transaxle case bolts. Install new 14mm rear mount bolts and washers.

23. Raise the transaxle and install the side mount. Tighten the case bolts to 47 ft. lbs. (64 Nm). Tighten all of the 14mm rear mount bolts to 61 ft. lbs. (85 Nm).

24. Install or connect the following:
- Transaxle side mount and bracket. Tighten the bracket nuts to 47 ft. lbs. (64 Nm). Tighten the mount through-bolt to 54 ft. lbs. (74 Nm).

25. Remove the transaxle jack.

26. Rotate the crankshaft and install the torque converter-to-driveplate bolts. Tighten the bolts to 108 inch lbs. (12 Nm) in a crisscross pattern. Tighten the bolts to the specification in 2 steps.

27. Rotate the crankshaft to reset the engine at Top Dead Center (TDC)/compression for the No. 1 cylinder. After the engine is set at Top Dead Center (TDC), it must not be disturbed until the distributor, if equipped, has been reinstalled.

28. Install or connect the following:
- Torque converter cover and tighten the bolts to 108 inch lbs. (12 Nm).
- Engine stiffener and tighten the 8mm bolts to 17 ft. lbs. (24 Nm). Tighten the 10mm bolts to 33 ft. lbs. (45 Nm).
- Right-front mount and bracket assembly. Tighten the 10mm bolt to 33 ft. lbs. (44 Nm), and the 12mm bolts to 47 ft. lbs. (64 Nm).

29. Remove the lifting chain and chain hooks.

30. Verify that the engine is at Top Dead Center (TDC)/compression for the No. 1 cylinder. Align the tabs on the distributor drive with the grooves on the end of the camshaft. Install the distributor and hand-tighten the mounting bolts. Reconnect the ignition wires.

31. Tighten the crankshaft pulley to 134 ft. lbs. (181 Nm).

32. Install or connect the following:
- Transaxle cooler lines
- New set rings on the inboard CV-joint splines
- Halfshafts into the transaxle case and intermediate shaft. The inboard joints must snap into place.
- Lower ball joint and damper fork
- Front wheels
- Shift cable linkage to the transaxle control shaft.
- New lockwasher and tighten the linkage bolt to 10 ft. lbs. (14 Nm).
- Shift cable cover and tighten its bolt to 16 ft. lbs. (22 Nm).
- Front exhaust pipe. Use new self-locking nuts and gaskets. Tighten the rear flange nuts to 16 ft. lbs. (22 Nm), and the front flange nuts to 47 ft. lbs. (64 Nm).
- VSS and countershaft speed sensor connectors
- Lock-up control solenoid connector.

- Shift control and linear solenoid connectors
- Mainshaft speed sensor connector
- Wiring harness clamps and starter cables
- Resonator, air cleaner box, and air intake duct
- Front splash shield

33. Lower the vehicle and tighten the strut pinch bolts to 32 ft. lbs. (44 Nm). Tighten the fork bolts to 47 ft. lbs. (65 Nm). Tighten the ball joint castle nuts to 40 ft. lbs. (55 Nm), then tighten them only enough to install new cotter pins.

34. Refill the transaxle with fresh ATF. Use only Honda Premium ATF or DEXRON® II or III ATF. Reconnect the positive and negative battery cables.
   a. Leave the flusher drain hose attached to the cooler return line.
   b. With the transaxle in park, run the engine for 30 seconds, or until approximately 1 quart of fluid is discharged. Immediately shut the engine off. This completes the cooler flushing process.
   c. Remove the drain hose and reconnect the cooler return line.
   d. Refill the transaxle to the proper level.

35. Check shift cable and throttle cable adjustments.

36. Check the ignition timing. Rotate the distributor counterclockwise to advance the timing, or clockwise to retard the timing. When the timing has been set, tighten the distributor mounting bolts to 13 ft. lbs. (18 Nm).

37. Start the engine and shift through all the gears 3 times.

38. Let the engine warm up to operating temperature and check the fluid level with the transaxle in the **P** or **N** position.

39. Check and adjust the front wheel alignment.

40. Road test the vehicle. Recheck the transaxle fluid level.

PRELUDE

1. Before servicing the vehicle, refer to the precautions in the beginning of this section.

2. Disconnect both cables from the battery.

3. Shift the transaxle into **N**.

4. Remove or disconnect the following:
- Battery hold-down and the battery

5. Drain the transaxle fluid and reinstall the drain plug with a new crush washer.

6. Remove or disconnect the following:
- Air intake duct, air cleaner case, and resonator
- Connector from the vacuum tank

and the vacuum tank and tank bracket. Do not remove the vacuum tube from the vacuum tank.
- Transaxle-to-body ground cable
- Battery base with the ground cable and the battery base stay.
- Lock-up control solenoid valve and shift control solenoid valve connectors.
- Throttle control cable from the throttle control lever.
- Countershaft speed sensor connector
- Vehicle Speed Sensor (VSS) connector
- Rear stiffener, then remove the VSS and Power Steering Sensor (PSS).

➡**Do not disconnect the power steering pressure hoses from the VSS and PSS.**

- Automatic Transaxle Fluid (ATF) cooler hoses at the joint pipes. Turn the ends of the cooler hoses upward to prevent fluid loss. Plug the joint pipes.
- Starter motor
- Upper transaxle housing mounting bolts
- Front engine mount bracket bolts
- Transaxle mount

7. Raise and support the vehicle safely.

8. Remove or disconnect the following:
- Front wheels
- Splash shield, subframe center beam and rear beam stiffener.
- Cotter pins and castle nuts from the lower ball joints. Use a press-type ball joint tool to separate the ball joints from the lower arm.
- Damper fork bolts, then separate the damper fork and the damper.

9. Use a suitable prytool to separate the right and left halfshafts from the differential.

10. Pull on the inboard joint and remove the right and left halfshafts. Tie plastic bags over the halfshaft ends to protect the boots and splined shafts from damage.

11. Remove or disconnect the following:
- Right damper pinch bolt, then separate the right damper fork from the strut.
- Right radius rod bolts and nut, then the radius rod
- Torque converter cover and the shift cable cover.
- Control lever lockbolt and the shift cable with the lever. Do not bend the shift control cable during removal. Wire the cable to the underbody of the vehicle our of the work area.

- Driveplate bolts while rotating the crankshaft.

12. Place a transaxle jack below the transaxle and raise it enough to take the weight off the mounts.

13. Remove or disconnect the following:
- Intake manifold bracket
- Lower transaxle housing mounting bolts and lower rear engine mounting bolts.

14. Pull the transaxle away from the engine until it clears the dowel pins. Lower the transaxle out of the vehicle.

To install:

→**Use new self-locking nuts when assembling the front suspension components. Use new set rings on the half-shaft inboard joints. Use new self-locking bolts for the subframe beams. These fasteners are available from a Honda dealer.**

15. Flush the transaxle cooling lines before installing the transaxle. Use a pressurized flushing canister, such as Honda tool No. J38405-A, or its equivalent. Use only biodegradable flushing fluid, Honda part No. J35944–20.

a. Fill the flusher with 21 ounces of fluid. Pressurize the flusher to 80–120 psi, following the procedure on the fluid container and flusher.

b. Clamp the discharge hose of the flusher to the cooler return line. Clamp the drain hose to the cooler inlet line and route it into a bucket or drain tank.

c. Connect the flusher to air and water lines. Open the flusher water valve and flush the cooler for 10 seconds.

d. Depress the flusher trigger to mix flushing fluid with the water. Flush for 2 minutes, turning the air valve on and off for 5 seconds every 15–20 seconds.

e. After finishing 1 flushing cycle, reverse the hose and flush in the opposite direction.

f. Dry the cooler lines with compressed air so that no moisture remains in the cooler lines.

16. Install the starter motor onto the transaxle case. Install the torque converter with a new hub O-ring. Tighten the starter bolts to 33 ft. lbs. (45 Nm).

17. Place the transaxle on a transaxle jack and raise it to the level of the engine.

18. Align the transaxle to the engine and install the transaxle housing mounting bolts and lower rear engine mounting bolts. Tighten the rear engine mounting bolts to 40 ft. lbs. (55 Nm) and the transaxle mounting bolts to 47 ft. lbs. (65 Nm). Install the

intake manifold bracket and tighten the bolts to 16 ft. lbs. (22 Nm).

19. Tighten the front engine mount bracket bolts to 28 ft. lbs. (39 Nm).

20. Install or connect the following:
- Transaxle mount. Tighten the bolt to 47 ft. lbs. (65 Nm) and the nuts to 28 ft. lbs. (39 Nm).

21. Remove the transaxle jack.

22. Install or connect the following:
- Torque converter to the driveplate and the mounting bolts. Turn the crankshaft to rotate the driveplate. Tighten the bolts in 2 steps, first to 50 inch lbs. (6 Nm) in a crisscross pattern and finally to 108 inch lbs. (12 Nm). Check for free rotation after tightening the last bolt.
- Shift cable onto the control shaft and tighten the lockbolt to 10 ft. lbs. (14 Nm).
- Torque converter cover and the shift cable cover.
- New set ring onto the inboard joint of each halfshaft.
- Damper fork bolts and ball joint nuts to the lower arms. Tighten the ball joint nut to 47 ft. lbs. (65 Nm)
- New cotter pin
- Radius rod and the damper fork. Only hand-tighten the radius rod and damper fork fasteners at this point.

23. Turn the right steering knuckle fully outward and slide the axle into the differential until the spring clip is felt engaging the differential side gear. Repeat the procedure on the left side.

24. Install or connect the following:
- Subframe rear beam stiffener and the center beam. Tighten the stiffener bolts to 28 ft. lbs. (39 Nm). Tighten the subframe center beam bolts to 43 ft. lbs. (60 Nm).
- Front wheels and lower the vehicle.

25. Use a floor jack to place the weight of the vehicle onto the right front knuckle. Tighten the radius rod bolts to 76 ft. lbs. (105 Nm) and the nut to 40 ft. lbs. (55 Nm). Tighten the damper pinch bolt to 32 ft. lbs. (44 Nm). Tighten the nut to 47 ft. lbs. (65 Nm) while holding the damper fork bolt. Remove the floor jack.

26. Install or connect the following:
- Speedometer sensor. Tighten the sensor bolt to 108 inch lbs. (12 Nm).
- ATF cooler hoses to the joint pipes
- Lock-up control solenoid and shift control solenoid valve connectors.
- VSS and PSS sensor connectors
- Starter motor cables and battery base and base stay.

- Ground cables on the body and on the transaxle.
- Vacuum tank, tank bracket, and tank connector.
- Resonator, air cleaner case, and air intake duct.

27. Refill the transaxle with ATF. Use only Honda Premium ATF or DEXRON®II ATF. Connect the negative and positive battery cables.

a. Leave the flusher drain hose attached to the cooler return line.

b. With the transaxle in park, run the engine for 30 seconds, or until approximately 1 quart of fluid is discharged. This completes the cooler flushing process.

c. Remove the drain hose and reconnect the cooler return line.

d. Refill the transaxle to the proper level with ATF.

28. Start the engine, set the parking brake, and shift the transaxle through all gears 3 times. Check for proper control cable adjustment.

29. On Preludes equipped with 4WS, and turn the steering wheel lock-to-lock to reset the steering control unit.

30. Check and adjust the front wheel alignment.

31. Let the engine reach operating temperature with the transaxle in **N** or **P**, then turn the engine OFF and check the fluid level

32. After road testing the vehicle, loosen the front engine mount bolts, and tighten them to 28 ft. lbs. (39 Nm).

33. Enter the radio security code.

ACCORD—2.3L AND 2.4L ENGINES

1. Before servicing the vehicle, refer to the precautions in the beginning of this section.

2. Remove or disconnect the following:
- Negative, then the positive battery cables
- Battery

3. Shift the transaxle into **N**.

4. Remove or disconnect the following:
- Air intake hose, air cleaner housing, and the resonator assembly.
- Battery base and the base stay
- Throttle cable from the throttle control lever.
- Transaxle ground cable and the speed sensor connectors.
- Solenoid valve connectors
- Lock-up control solenoid valve and shift control solenoid valve connectors.
- Transaxle cooler hoses from the joint pipes and plug the hoses.

- Starter cables and starter
- Countershaft Speed Sensor (CSS) and Vehicle Speed Sensor (VSS) connectors

5. Install a hoist to the engine.

6. Remove or disconnect the following:
- 4 upper bolts attaching the transaxle to the engine block.
- 3 bolts attaching the front engine mount bracket to the engine.
- Transaxle mount

7. Raise and safely support the vehicle. Remove the front wheels.

8. Drain the transaxle fluid and reinstall the drain plug with a new washer.

9. Remove or disconnect the following:
- Splash shield
- Subframe center beam
- Cotter pins and lower arm ball joint nuts, then separate the ball joints from the lower arms using a suitable tool.
- Right damper pinch bolt, then separate the damper fork and damper.
- Bolts and nut, then the right radius rod.

10. Using a small prying device, carefully pry the right and left halfshafts out of the differential. Remove the right and left halfshafts. Tie plastic bags over the halfshaft ends to prevent damage to the CV boots and splines.

11. Remove or disconnect the following:
- Bolts mounting the intermediate shaft, then the intermediate shaft from the differential.
- Torque converter cover and shift cable cover.
- Shift control cable by removing the lockbolt.
- Shift cable lever from the control shaft. Don't disconnect the control lever from the shift cable. Wire the shift cable out of the work area and be careful not to kink it.
- 8 drive plate bolts, one at a time while rotating the crankshaft pulley.

12. Place a suitable jack under the transaxle and raise the jack just enough to take weight off of the mounts.

13. Remove or disconnect the following:
- Intake manifold bracket.
- Transaxle housing mounting bolts
- Mounting bolts from the rear engine mount bracket.
- 4 transaxle housing mounting bolts and 3 mount bracket nuts.

14. Pull the transaxle away from the engine until it clears the 14mm dowel pins, then lower it using the jack.

To install:

➡**Use new self-locking nuts when assembling the front suspension components. Install new set rings onto the halfshaft inboard joint splines. Replace any color-coded self-locking bolts.**

15. Flush the transaxle cooler lines before installing the transaxle. Use a pressurized flushing unit such as Honda J38405-A or equivalent. Use only Honda biodegradable flushing fluid, Honda J35944-20. Other fluids will damage the automatic transmission cooling system.

a. Fill the flusher with 21 ounces of fluid. Pressurize the flusher to 80–120 psi, following the procedure on the fluid container and flusher.

b. Clamp the discharge hose of the flusher to the cooler return line. Clamp the drain hose to the cooler inlet line and route it into a bucket or drain tank.

c. Connect the flusher to air and water lines. Open the flusher water valve and flush the cooler for 10 seconds. The air line should be equipped with a water trap to keep the system dry.

d. Depress the flusher trigger to mix flushing fluid with the water. Flush for 2 minutes, turning the air valve on and off for 5 seconds every 15–20 seconds to create a surging action.

e. After finishing 1 flushing cycle, reverse the hose and flush in the opposite direction following the same steps.

f. Dry the cooler lines with compressed air so that no moisture is left in the cooler system.

16. Be sure the 2, 14mm dowel pins are installed into the torque converter housing.

17. Install or connect the following:
- Torque converter onto the transaxle mainshaft with a new hub O-ring.
- Starter motor onto the transaxle case and tighten the mounting bolts to 33 ft. lbs. (44 Nm).
- Transaxle and transaxle housing mounting bolts: 47 ft. lbs. (65 Nm)
- Rear engine mounting bolts: 40 ft. lbs. (54 Nm)
- Intake manifold bracket and tighten the bolts to 16 ft. lbs. (22 Nm).
- Upper bolts attaching the transaxle to the engine: 47 ft. lbs. (64 Nm)
- Front engine mount bracket bolts: 28 ft. lbs. (38 Nm)
- Transaxle mount and the nuts and bolt that attach the mount. Tighten the nuts first to 28 ft. lbs. (38 Nm), then tighten the bolt to 47 ft. lbs. (64 Nm).

18. Remove the jack from the transaxle.

19. Install or connect the following:
- Torque converter to the drive plate with the 8 bolts. Tighten the bolts in 2 steps in a crisscross pattern: first to 54 inch lbs. (6 Nm), and finally to 108 inch lbs. (12 Nm). Check for free rotation after tightening the last bolt.
- Shift control cable and control cable holder. Tighten the shift cable lockbolt to 10 ft. lbs. (14 Nm). Tighten the shift cable cover bolts to 13 ft. lbs. (18 Nm).
- Torque converter cover and tighten the bolts to 108 inch lbs. (12 Nm).

20. Remove the engine hoist.

21. Install or connect the following:
- Radius rod and damper fork
- Intermediate shaft into the differential and tighten the mounting bolts to 28 ft. lbs. (38 Nm).
- New set ring on the end of each halfshaft.

22. Turn the right steering knuckle fully outward and slide the axle into the differential until the set ring snaps into the differential side gear. Repeat the procedure on the left side.

23. Install or connect the following:
- Damper fork bolts and ball joint nuts to the lower arms: 40 ft. lbs. (55 Nm) with a new cotter pin.
- Subframe center beam and tighten the center beam bolts to 28 ft. lbs. (39 Nm).
- Splash shield
- Front wheels and lower the vehicle.
- Speed sensor connector

24. Support the right front knuckle with a floor jack, until the weight of the vehicle is held by the jack. Tighten the damper fork pinch bolt to 32 ft. lbs. (44 Nm). Tighten the radius rod bolts to 76 ft. lbs. (105 Nm), and the radius rod nut to 32 ft. lbs. (44 Nm). Hold the damper fork bolt with a wrench, and tighten the nut to 40 ft. lbs. (55 Nm). Remove the floor jack.

25. Install or connect the following:
- Cables to the starter
- Throttle control cable
- Lock-up control solenoid valve and shift control solenoid valve connectors
- Speed sensor connectors and the transaxle ground cable
- Transaxle cooler inlet hose to the joint pipe. Attach a drain hose to the return line.
- Battery base stay and the battery base

26. Install the resonator assembly, the air cleaner assembly, and the air intake hose.

- Battery, positive then the negative battery cables

27. Refill the transaxle with ATF. Use only Honda Premium ATF or DEXRON®II ATF.

a. With the flusher drain hose attached to the cooler return line.

b. Place the transaxle in **P**, run the engine for 30 seconds, or until approximately 1 quart of fluid is discharged. Immediately shut off the engine. This completes the cooler flushing process.

c. Remove the drain hose and reconnect the cooler return line.

d. Refill the transaxle to the proper level with ATF.

28. Start the engine, set the parking brake, and shift the transaxle through all gears 3 times. Check for proper shift cable adjustment.

29. Let the engine reach operating temperature with the transaxle in **P** or **N**. Then, shut off the engine and check the fluid level.

30. Road test the vehicle.

31. After road testing the vehicle, loosen the front engine mount bracket bolts, then retighten them to 28 ft. lbs. (39 Nm).

32. Check and adjust the vehicle's front end alignment.

33. Enter the radio security code.

ACCORD—3.0L ENGINES

1. Before servicing the vehicle, refer to the precautions in the beginning of this section.

2. Remove or disconnect the following:
- Negative, then the positive battery cables
- Battery and tray
- Clamps securing the battery cables to the base.
- Intake air duct and the air cleaner assembly

3. Raise the vehicle and drain the transaxle fluid. Replace the drain plug with a new washer.

4. Remove or disconnect the following:
- Starter wiring and harness clamps, then the breather and radiator hoses from the retainer.
- Wiring connectors from the transaxle assembly
- Cooler lines; point them up to prevent fluid drainage.
- Bolt and nut, then the rear stiffener.
- Bolts securing the transaxle to the engine.
- Front mounting bracket bolts
- Engine under cover
- Lower shock absorber mounting and the lower ball joints from the control arms.

- Bolts securing the radius rods to the lower arms.
- Halfshafts. Keep the splined ends of the shafts clean.

➡**Matchmark the installed position of the sub-frame on the main-frame before removing it.**

- Sub-frame from the main-frame
- Engine brace from the rear of the engine
- Shift cable cover, bracket and cable
- 8 bolts securing the drive plate to the torque converter

5. Attach a chain hoist to the engine and raise it slightly.

6. Place a jack under the transaxle.

7. Remove or disconnect the following:
- Transaxle mount bracket
- Intake manifold support bracket
- Rear mount bracket

8. Pull the transaxle back slightly until it comes off the dowels and lower it from the vehicle. Do not let the torque converter fall out of the transaxle.

To install:

9. Install or connect the following:
- Torque converter using a new O-ring, if removed.
- Dowel pins in the torque converter housing.
- Transaxle to the engine and the rear mount bracket. Tighten the 8mm bolt to 16 ft. lbs. (22 Nm) and the 12mm bolts to 40 ft. lbs. (54 Nm).
- Transaxle-to-engine bolts: 47 ft. lbs. (64 Nm)
- Breather tube with the dot facing up
- Transaxle mount bracket. Tighten the nuts to 28 ft. lbs. (38 Nm) and the through-bolt to 40 ft. lbs. (54 Nm).
- Driveplate-to-torque converter bolts: 108 inch lbs. (12 Nm) in a crisscross pattern.
- Shift cable, bracket and cover
- Engine brace on the rear of the engine.
- Halfshafts
- Sub-frame after aligning the matchmarks. Tighten the rear bolts to 47 ft. lbs. (64 Nm) and the front bolts to 76 ft. lbs. (103 Nm).
- Front mount. Tighten the bolts to 28 ft. lbs. (38 Nm).
- Shock absorbers and the radius rods to the lower control arms.
- Engine under cover
- All the wiring connectors
- Starter wiring and harness clamps
- Battery
- Air cleaner assembly and intake duct

10. Refill the transaxle with Genuine Honda® premium automatic transmission fluid.

Clutch

REMOVAL & INSTALLATION

1. Before servicing the vehicle, refer to the precautions in the beginning of this section.

2. Remove or disconnect the following:
- Negative battery cable

3. Raise and safely support the vehicle.

4. Remove or disconnect the following:
- Transmission from the vehicle. Matchmark the flywheel and clutch for reassembly.

5. Use a flywheel ring-gear holder to lock the flywheel in position.

6. Remove or disconnect the following:
- Pressure plate bolts, 2 turns at a time working in a crisscross pattern to prevent warping the pressure plate.
- Pressure plate and clutch disc

7. Inspect the flywheel, disc, and pressure plate for wear, cracks, and warpage. Light scoring of the flywheel may be polished out; gouges, warpage, burn marks, cracks, or chipped teeth require replacement of the flywheel.

➡**If the flywheel is to be removed, but is going to be reused, matchmark it to the engine block prior to removal. Aligning the matchmarks upon reassembly will preserve driveline balance.**

8. Inspect the flywheel's ball bearing: turn the inner race of the bearing with your finger, and be sure it turns smoothly and quietly. If the bearing is loose or noisy, or exhibits rough motion, replace it.

9. Remove or disconnect the following:
- Release fork boot. Squeeze the release fork retaining spring to disengage the fork from its pivot.
- Release fork from the clutch housing.
- Release bearing. Spin the bearing by hand to check its degree of play. Replace the release bearing if it has excessive play or is leaking grease.

10. Inspect the rear main bearing oil seal for signs of leakage. If necessary, replace the seal to prevent oil leakage onto the clutch's friction surfaces.

To install:

11. If necessary, drive out the flywheel bearing, then use a suitably-sized bearing

driver to install a new one. Use a crisscross pattern to tighten the flywheel mounting bolts in several steps to 87 ft. lbs. (118 Nm) for vehicles with SOHC engines. If equipped with the B16A2 engine, tighten the flywheel bolts to 76 ft. lbs. (105 Nm). For 2.0L engines, tighten the flywheel bolts to 94 ft. lbs. (127 Nm).

12. Install or connect the following:
- Clutch disc and pressure plate by aligning the dowels on the flywheel with the dowel holes in the pressure plate. If a new pressure plate is not being installed, align the matchmarks that were made during removal.
- Pressure plate bolts, hand-tight

13. Insert a suitable clutch disc alignment tool into the splined hole in the clutch disc. Align the clutch and pressure plate.

14. Tighten the pressure plate bolts in a crisscross pattern 2 turns at a time to prevent warping the pressure plate. The final torque is 19 ft. lbs. (26 Nm).

15. Remove the alignment tool and ring gear holder.

16. Coat the mainshaft with heavy-duty high-temperature grease. The manufacturer recommends part No. 08798–9002, Honda super high-temp urea grease.

17. Coat the release fork pawls and the inner race of the release bearing with high temperature grease and install them into the clutch housing. Be sure the release fork retainer spring snaps into place on the pivot. The bearing and fork must fit together properly and slide back and forth smoothly.

18. Coat the tip of the slave cylinder with grease. Install the release fork boot.

19. Install or connect the following:
- Transmission, making sure the mainshaft is properly aligned with the clutch disc splines, and the transmission case dowels are properly aligned with the engine block.
- Transmission case bolts: 47 ft. lbs. (65 Nm), sequentially

20. Bleed the clutch hydraulic system.

21. Adjust the clutch pedal free-play.

22. Verify that all engine and transaxle components are installed and connected properly.

23. Reconnect the negative battery cable.

24. Road test the vehicle.

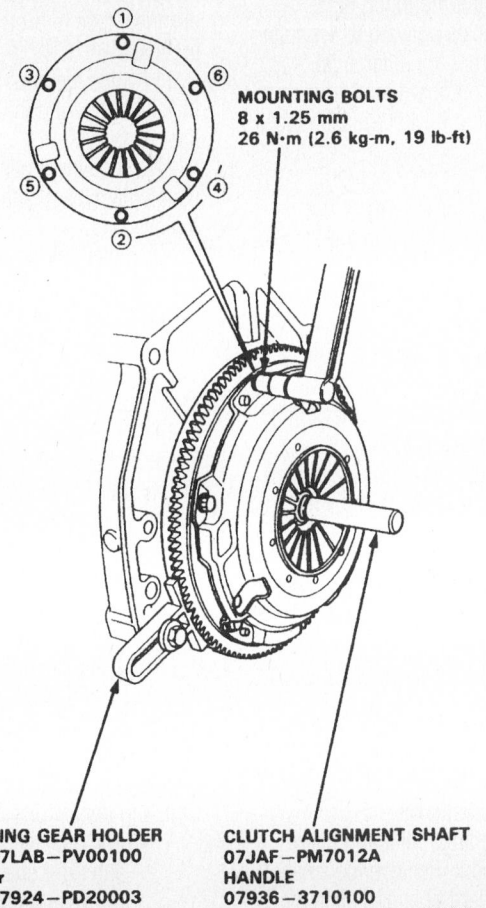

RING GEAR HOLDER
07LAB–PV00100
or
07924–PD20003

CLUTCH ALIGNMENT SHAFT
07JAF–PM7012A
HANDLE
07936–3710100

MOUNTING BOLTS
8 x 1.25 mm
26 N·m (2.6 kg-m, 19 lb-ft)

7923FG99

Clutch alignment tools and pressure plate torque sequence

Hydraulic Clutch System

BLEEDING

1. Before servicing the vehicle, refer to the precautions in the beginning of this section.

2. Fill the clutch master cylinder reservoir with clean DOT 3 or 4 brake fluid.

3. Attach a rubber tube to the clutch slave cylinder bleed screw. Route the tube into a container of clean brake fluid.

4. Loosen the bleed screw.

5. Slowly pump the clutch pedal until the fluid draining from the slave cylinder is free of air bubbles.

6. Tighten the bleed screw to 72–84 inch lbs. (8–10 Nm).

7. Refill the clutch master cylinder reservoir with brake fluid.

Halfshaft

REMOVAL & INSTALLATION

Except S2000

1. Before servicing the vehicle, refer to the precautions in the beginning of this section.

2. Loosen the front spindle nut.

3. Raise and safely support the vehicle.

4. Remove or disconnect the following:
- Front wheels and the spindle nut

5. Drain the transaxle fluid and install the drain plug with a new washer. If the halfshaft to be removed is installed into the intermediate shaft, the transaxle fluid does not need to be drained.
- Flange nut, while holding the stabilizer ball joint pin with a hex wrench
- Front stabilizer link from the lower arm
- Damper fork nut and damper pinch bolt(s)
- Damper fork
- Cotter pin and castle nut from the lower arm ball joint.

6. Install a hex nut flush onto the ball joint stud to prevent the ball joint tool from damaging the stud threads.

7. Using a ball joint tool, separate the lower arm from the knuckle.

8. Pull the knuckle outward.

9. Remove or disconnect the following:
- Heat shield, if equipped
- Halfshaft outboard joint from the hub by tapping it with a plastic hammer.

- Inner CV-joint away from the transaxle case to force the halfshaft set ring out of the groove.
- Halfshaft from the differential case or intermediate shaft by pulling on the inboard CV-joint

➡**Do not pull on the halfshaft as the CV-joint may come apart. Use care when prying out the assembly and pull it straight to avoid damaging the differential oil seal or intermediate shaft oil or dust seals.**

To install:

10. Replace the differential oil seal or intermediate shaft seal if either were damaged during removal.

11. Install or connect the following:
- New set rings on the ends of the halfshafts

12. For the 2003–04 Accord, apply 0.02–0.04 oz. of grease to the whole splined surface of the right side halfshaft. Then, remove the grease from the splined grooves at intervals of 2–3 splines and from the set ring groove to let air bleed from the intermediate shaft.
- Halfshafts and be sure the set ring locks in the differential gear groove and the halfshaft bottoms in the differential or intermediate shaft.
- Outboard joint into the hub. Be sure the splines mesh together and the joint is fully seated into the hub.
- Heat shield, if equipped
- Ball joint stud into the lower control arm.

13. Torque the new castle nut to the lower of the following specifications, then tighten only enough to install a new cotter pin. Never loosen the nut to install the cotter pin.

    a. 2001–02 Civic: 36–43 ft. lbs. (49–59 Nm)

    b. 2003–04 Civic: 43–51 ft. lbs. (59–69 Nm)

    c. Prelude: 36–43 ft. lbs. (49–59 Nm)

    d. 2001–02 Accord: 36–43 ft. lbs. (49–59 Nm)

    e. 2003–04 Accord: 65–72 ft. lbs. (88–98 Nm)

14. Install or connect the following:
- Damper fork into position. Make sure the aligning tab is lined up with the slot in the damper fork. Tighten the retainers loosely
- Front stabilizer link to the lower arm. Hold the stabilizer link ball joint pin with a hex wrench and tighten the new flange nut to 22 ft. lbs. (30 Nm).

- Tighten the upper damper pinch bolt to 32 ft. lbs. (44 Nm) and the fork nut to 47 ft. lbs. (65 Nm).
- Front wheels
- New spindle nut; don't tighten it yet.

15. Lower the vehicle.

16. Tighten the spindle nut to 134 ft. lbs. (181 Nm) for the Civic, and 4-cylinder Accord with A/T. For all other models, tighten the spindle nut to 181 ft. lbs. (245 Nm) and stake its tab. Tighten the wheel nuts to 80 ft. lbs. (110 Nm).

17. Fill the transaxle with the proper type and quantity of fluid.

18. Warm the engine up, check the transaxle fluid level, and road test the vehicle.

S2000

1. Before servicing the vehicle, refer to the precautions in the beginning of this section.

2. Remove or disconnect the following:
- Rear wheel
- Spindle nut
- Lower ball joint
- Wheel speed sensor harness
- Inboard joint mounting bolts

3. Pull the knuckle outward to separate the inboard joint from the differential.

4. Remove the outboard joint from the wheel hub by tapping it with a plastic-faced hammer.

To install:

5. Installation is the reverse of the removal procedure, while using the following torque values:
- Inboard joint mounting bolts: 61 ft. lbs. (83 Nm)
- Lower ball joint nut: 43–51 ft. lbs. (69–78 Nm)
- Wheel speed sensor harness bolts: 88 inch lbs. (10 Nm)
- Spindle nut: 181 ft. lbs. (245 Nm)
- Wheel lug nuts: 80 ft. lbs. (108 Nm)

CV-Joints

OVERHAUL

1. Before servicing the vehicle, refer to the precautions in the beginning of this section.

2. Remove the halfshaft.

3. Remove the large retaining band from the inboard boot. Remove the smaller band from the inboard boot and slide the boot off the joint.

4. Carefully remove the stub end of the

inboard joint. Check the splines for cracks, wear or damage. Check the inside bore for any sign of wear.

5. Remove and discard the snapring from the end of the halfshaft. This will allow removal of the spider assembly.

6. Mark the rollers, spider and the stub end of the axle so that all parts may be reassembled in the same position. Remove the rollers from the spider.

7. Remove the second snapring from the shaft. Remove the joint boot. If equipped, remove the dynamic damper from the shaft.

8. If the outer joint's boot is to be replaced, remove the boot clamps and slide the boot off the joint, then off the shaft. Hold the outer joint and swivel the end. If the joint is noisy, it must be replaced. The replacement joint will come with a new shaft; the inner joint must be assembled onto the shaft.

9. Clean and inspect all disassembled parts. Any sign of wear requires replacement.

To install:

10. Thoroughly pack the inboard and outboard joints with moly grease. Use only moly grease; other lubricants will not last. Wrap the splines of the shaft in vinyl or electrical tape to protect the boots as they are installed.

11. Slide the boot for the outer joint over the shaft and onto the joint. Do not install the bands yet.

12. Slide the inner boot onto the shaft. Install the dynamic damper if it was removed.

13. Install the inboard snapring on the shaft. Install the rollers and bearing races on the spider shafts. Hold the shaft upright, then slide the spider assembly into the inboard shaft joint. Install the outer snapring.

14. Slide the boots over both joints. Position the small end of the boot so that the band will be centered between the locating humps on the shaft. Install the band; bend both sets of locking tabs. Once the band is in place, expand and compress the boots once or twice; allow the boots to return to their normal size and length.

15. Adjust the length of the halfshaft to the specifications listed below, then adjust the boots to halfway between full compression and full extension. Make sure the ends of the boots seat in the groove of the driveshaft and joint. Correct shaft lengths are:
- 2001–02 Accord with M/T—Left and right shafts: 19.1–19.3 in. (486–491mm).
- 2001–02 Accord with A/T—Left shaft: 33.3–33.5 in. (845–850mm).

Right shaft: 19.1–19.9 in. (486–491mm).
- 2003–04 Accord (4 cyl.) with M/T—Left shaft: 21.42–21.61 in. (544–549mm).Right shaft: 18.62–18.82 in. (473–478mm).
- 2003–04 Accord (4 cyl.) with A/T—Left shaft: 21.61–21.81 in. (549–554mm). Right shaft: 33.43–33.62 in. (849–854mm).
- 2003–04 Accord (V6)—Left shaft: 21.81–22.01 in. (554–559mm). Right shaft: 20.16–20.35 in. (512–517mm)
- Civic—see accompanying illustration.
- Prelude—Right: 20.0–20.1 in. (507.9–512.9mm). Left with M/T: 20.5–20.7 in. (520.9–525.9mm). Left with A/T: 33.9–34.1 in. (862.9–867.9mm).
- S2000—Left halfshaft: 22.8–23 inches (579–584 mm). Right half-shaft—24.6—24.8 inches (624–629 mm).

16. Install new boot bands on the large ends of the boots. Be sure to bend both sets of locking tabs. Lightly tap the doubled-over portion of the band(s) to reduce the height. Do NOT hit the boot.

17. Position the dynamic damper so that it is 0.1–1.2 in. (3–7mm) from the CV-boot. Install a new retaining band in the same fashion as the boot bands.

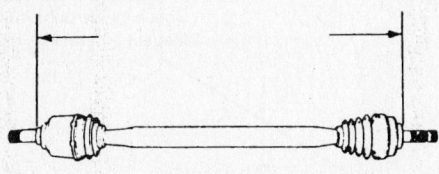

LEFT DRIVESHAFT

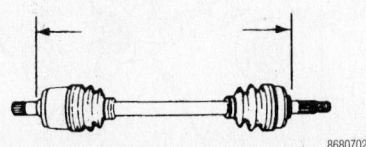

RIGHT DRIVESHAFT

Halfshafts must be set to the correct length before installing boot bands

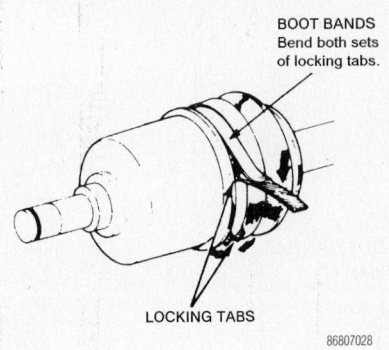

BOOT BANDS
Bend both sets of locking tabs.

LOCKING TABS

Always use new boot bands

Left driveshaft:
  Japan-produced models:
    M/T: 788—793 mm (31.0—31.2 in.)
    A/T: 792—797 mm (31.2—31.4 in.)
  U.S.-and Canada-produced models:
    M/T: 788.2—793.2 mm (31.0—31.2 in.)
    A/T: 791.7—796.7 mm (31.2—31.4 in.)

Right driveshaft:
  Japan-produced models:
    '01-02 models: 502—507 mm (19.8—20.0 in.)
    '03 model M/T: 497—502 mm (19.6—19.8 in.)
    '03 model A/T: 502—507 mm (19.8—20.0 in.)
  U.S.-and Canada-produced models:
    501.2—506.2 mm (19.7—19.9 in.)

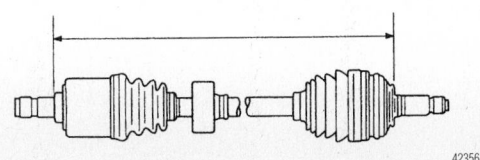

Halfshaft length specifications—Civic

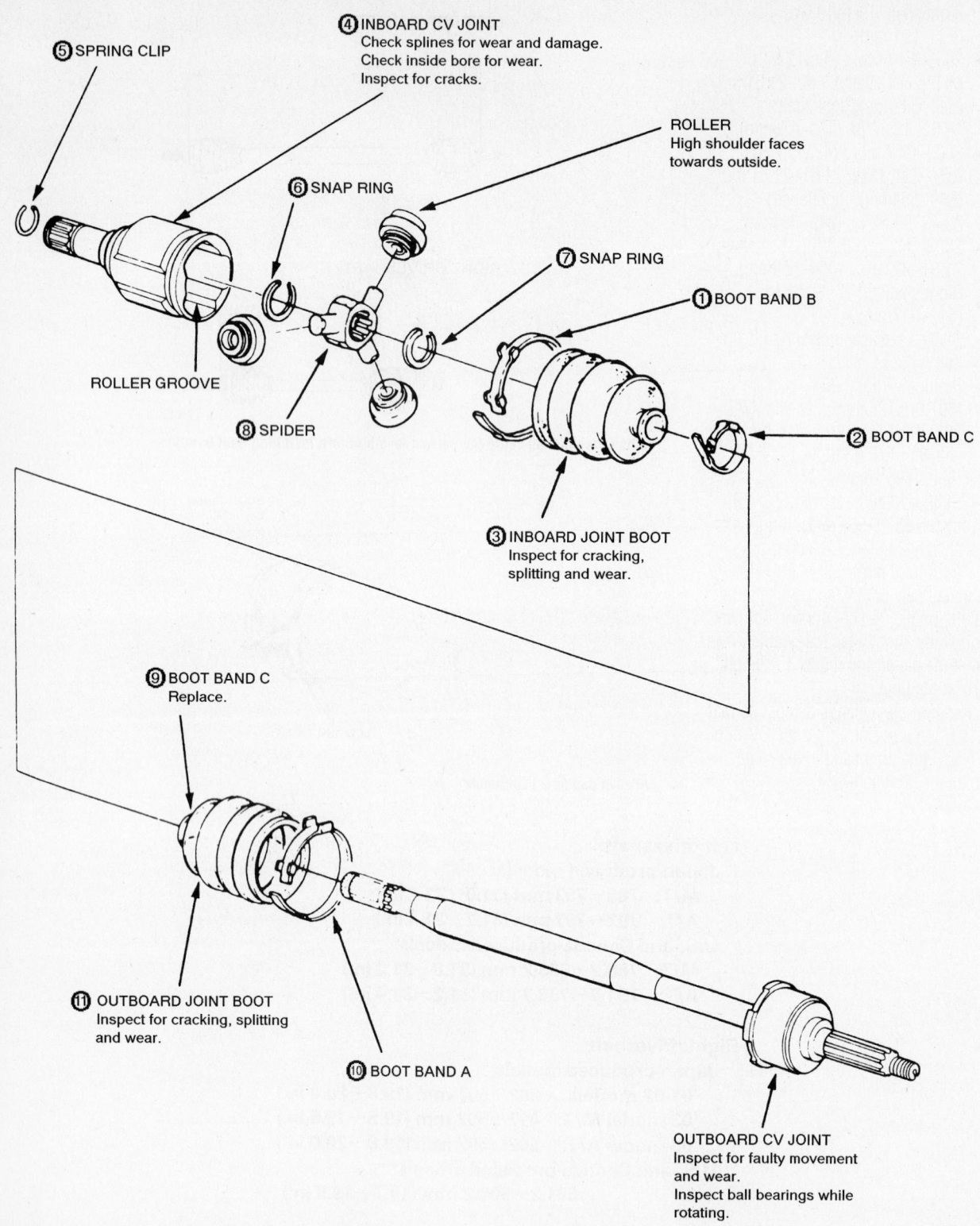

⑤ SPRING CLIP

④ INBOARD CV JOINT
Check splines for wear and damage.
Check inside bore for wear.
Inspect for cracks.

ROLLER
High shoulder faces
towards outside.

⑥ SNAP RING

⑦ SNAP RING

① BOOT BAND B

ROLLER GROOVE

② BOOT BAND C

⑧ SPIDER

③ INBOARD JOINT BOOT
Inspect for cracking,
splitting and wear.

⑨ BOOT BAND C
Replace.

⑪ OUTBOARD JOINT BOOT
Inspect for cracking, splitting
and wear.

⑩ BOOT BAND A

OUTBOARD CV JOINT
Inspect for faulty movement
and wear.
Inspect ball bearings while
rotating.

86807026

Exploded view of the halfshaft

18. Install a new snapring on the inboard end of the joint, then install the half-shaft.

Pinion Seal

REMOVAL & INSTALLATION

S2000

1. Before servicing the vehicle, refer to the precautions in the beginning of this section.
2. Drain the axle housing fluid.
3. Remove or disconnect the following:
   - Negative battery cable
   - Rear wheels
   - Driveshaft
   - Brake calipers and pads

➡**The brake calipers and pads must be removed so that there is no additional drag when measuring pinion bearing preload.**

4. Use an inch lb. torque wrench and measure and record the amount of torque required to maintain pinion rotation through several revolutions.
5. Remove or disconnect the following:
   - Pinion flange
   - Pinion seal
   - Pinion bearing
   - Collapsible spacer

To install:

➡**Use a new collapsible spacer and flange nut for assembly.**

6. Install or connect the following:
   - Collapsible spacer
   - Pinion bearing
   - Pinion seal
   - Pinion flange
7. Rotate the pinion flange occasionally while tightening the flange nut to make sure the pinion bearings seat correctly.
8. Tighten the flange nut to 94 ft. lbs. (127 Nm) and then measure bearing preload torque.

9. Continue tightening the flange nut to achieve the bearing preload torque originally measured. Do not exceed 210 ft. lbs. (284 Nm) flange nut torque.
10. If using new pinion bearings, add 8–12 inch lbs. (0.88–1.37 Nm) to the originally measured bearing preload.

✳✳ CAUTION

Never loosen the pinion nut to reduce bearing preload. If it is necessary to reduce bearing preload, install a new collapsible spacer and pinion nut.

11. Install or connect the following:
    - Driveshaft
    - Brake calipers and pads
    - Wheels
    - Negative battery cable
12. Fill the differential with gear lubricant and check for leaks.

STEERING AND SUSPENSION

Air Bag

The air bag modules must be disabled if they, or any other part of the SRS, must be serviced or disconnected. Failing to disable the SRS before servicing its components may cause accidental air bag deployment and possible personal injury.

PRECAUTIONS

Several precautions must be observed when handling the inflator module to avoid accidental deployment and possible personal injury.

- Never carry the inflator module by the wires or connector on the underside of the module.
- When carrying a live inflator module, hold securely with both hands, and ensure that the bag and trim cover are pointed away.
- Place the inflator module on a bench or other surface with the bag and trim cover facing up.
- With the inflator module on the bench, never place anything on or close to the module which may be thrown in the event of an accidental deployment.

DISARMING

➡**The radio may contain a coded theft protection circuit. Always obtain the code number before disconnecting the battery.**

Driver's Side

1. Before servicing the vehicle, refer to the precautions in the beginning of this section.
2. Remove or disconnect the following:
   - Negative and positive battery cables

✳✳ CAUTION

Always wait at least 3 minutes after disconnecting the battery before working around the air bag.

- Steering wheel lower access cover
- Clip securing the air bag module/cable reel connection to the steering column.

➡**Spring-loaded air bag connectors contain a self-disabling contact. A shorting connector doesn't need to be installed on the driver's air bag connector.**

3. Uncouple the spring-loaded connectors:
   a. Hold the connector body, not the wiring.
   b. Pull the spring-loaded locking sleeve toward its stop while holding the opposite half of the connector.
   c. After releasing the locking sleeve, uncouple the connectors.

Passenger's Side

1. Before servicing the vehicle, refer to the precautions in the beginning of this section.
2. Remove or disconnect the following:
   - Negative and positive battery cables

✳✳ CAUTION

Always wait at least 3 minutes after disconnecting the battery before working around the air bag.

- Glove box door and frame
- Lower mounting brackets that may cover the air bag connection, if equipped.
- Passenger's air bag connector. Pull the spring-loaded sleeve toward the stop while holding the opposite half of the connector and pull the connector apart.

REARMING

Driver's Side

1. After servicing has been completed, couple the air bag and cable reel connectors. Press the sleeve side of the connector into the pawl side until the sleeve locks the connectors together.
2. Install or connect the following:
   - Clip securing the air bag/cable reel connection to the steering column.

- Access cover
- Positive and negative battery cables

3. Turn the ignition switch to the **ON** position, but don't start the engine. The air bag indicator light should turn on for 6 seconds, then turn off. If the air bag indicator light doesn't come on, or stays on longer than 6 seconds, the system fault must be diagnosed.

4. Enter the radio security code.

Passenger's Side

1. After servicing has been completed, immediately couple the air bag and cable reel connectors.

2. Install or connect the following:
- Any lower mounting brackets that may have been removed.
- Glove box frame and glove box door
- Positive and negative battery cables

3. Turn the ignition switch to the **ON** position, but don't start the engine. The air bag indicator light should turn on for 6 seconds, then turn off. If the air bag indicator light doesn't come on, or stays on longer than 6 seconds, the system fault must be diagnosed.

4. Enter the radio security code.

Rack and Pinion Steering Gear

REMOVAL & INSTALLATION

Manual

CIVIC

✳✳ CAUTION

The air bag must be disabled before removing the steering wheel to center the cable reel. Failure to disarm the air bag system may cause accidental air bag deployment, resulting in unnecessary air bag system repairs and the risk of personal injury.

1. Before servicing the vehicle, refer to the precautions in the beginning of this section.

2. Position the front wheels straight ahead. Lock the steering column and remove the ignition key.

3. Remove or disconnect the following:
- Negative, then the positive battery cables.

4. Disable the air bag.

5. Remove or disconnect the following:
- Steering joint cover

- Upper and lower steering joint bolts

6. Raise and support the vehicle safely.

7. Remove or disconnect the following:
- Front wheels
- Tie-rod end cotter pins and castle nuts
- Tie-rod ends from the steering knuckles, using a ball joint tool.
- Left tie-rod end and slide the rack all the way to the right.
- Self-locking nuts, then separate the catalytic converter or front exhaust pipe from the rear exhaust pipes.
- Catalytic converter or front exhaust pipe

8. If equipped with a manual transaxle, remove or disconnect the following:
- Shift lever extension rod from the clutch housing
- Pin retainer out of the way, then drive out the spring pin
- Shift rod

9. If equipped with an automatic transaxle, remove or disconnect the following:
- Shift cable bracket and holder
- Shift cable from the control shaft. Suspend the cable from the underbody with a piece of wire.

10. Remove or disconnect the following:
- Steering rack stiffener plate
- Steering rack mounting bracket
- Steering rack from the pinion shaft, by pulling the rack down.

11. Drop the steering rack far enough to permit the end of the pinion shaft and the grommet to come out of the hole in the bulkhead.

12. Slide the gearbox to the right until the left tie rod clears the subframe, then drop it down and out of the vehicle to the left.

To install:

➡ **Use new self-locking nuts and gaskets when installing the catalytic converter.**

13. Install or connect the following:
- Steering rack into position
- Pinion shaft grommet, insert the pinion through the hole in the bulkhead.
- Steering rack mounting cushion, bracket, and bolts. The arrow on the bracket faces the front of the vehicle. Tighten the bracket bolts to 28 ft. lbs. (39 Nm).
- Steering rack stiffener plate. Tighten the steering rack mounting bolts to 43 ft. lbs. (59 Nm). Tighten the stiffener plate bolts to 28 ft. lbs. (39 Nm).

14. Center the rack ends within their steering strokes.

15. Install or connect the following:
- Tie rod ends onto the rack ends
- Tie rod ends to the steering knuckles, then the castle nuts.
- Front wheels
- Catalytic converter using new gaskets and self-locking nuts. Tighten the front nuts to 16 ft. lbs. (22 Nm), and the rear nuts to 25 ft. lbs. (34 Nm).

16. If equipped with a manual transaxle, install or connect the following:
- Shift linkage with a new spring pin and clip.
- Extension rod and tighten its bolt to 16 ft. lbs. (22 Nm).

17. If equipped with an automatic transaxle, install or connect the following:
- Shift cable and brackets. Tighten the bracket bolts to 108 inch lbs. (12 Nm). Tighten the cable lockbolt to 10 ft. lbs. (14 Nm). Tighten the cable holder bolts to 16 ft. lbs. (22 Nm).

18. Verify that the rack is centered within its strokes. Lower the vehicle.

19. Center the air bag cable reel as follows:
  a. Remove the steering wheel.
  b. Turn the cable reel clockwise until it stops.
  c. Turn the steering wheel counterclockwise, approximately 2 turns, until the arrow on the label points straight up.
  d. Install the steering wheel.

20. During steering wheel installation, verify that the slot on the steering wheel shaft engages with the tabs on the turn signal canceling sleeve. The pins on the cable reel fit into the holes on the steering wheel body. Install a new steering wheel nut and tighten it to 36 ft. lbs. (50 Nm).

21. Line up the bolt hole in the steering joint with the groove in the pinion shaft. Slip the joint onto the pinion shaft. Pull the joint up and down to be sure the splines are fully seated. Tighten the joint bolts to 16 ft. lbs. (22 Nm).

➡ **Connect the steering joint and pinion shaft with the cable reel and steering rack centered. Verify that the lower joint bolt is securely seated in the pinion shaft groove. If the steering wheel and rack are not centered, reposition the serrations at the lower end of the steering joint.**

22. Install or connect the following:
- Steering joint cover

23. Tighten the ball joint castle nuts to

29–35 ft. lbs. (40–48 Nm). Then, tighten them only enough to install new cotter pins.

24. Enable the air bag.

25. Install or connect the following:
- Steering wheel's lower access cover
- Negative and positive battery cables

26. Turn the ignition switch to the **ON** position. The air bag indicator light should come on for 6 seconds, then turn off. This light sequence indicates that the air bag system is enabled and functioning normally. If the air bag light stays on longer, or doesn't turn on, the system must be diagnosed.

27. Check the front wheel alignment and steering wheel spoke angle. Make adjustments by turning the left and right tie-rod ends equally.

28. Road test the vehicle.

Power

➡The radio may contain a coded theft protection circuit. Always obtain the code number before disconnecting the battery. If the vehicle is equipped with 4WS, the steering control unit is shut down when the battery is disconnected. After connecting the battery, turn the steering wheel lock-to-lock to reset the steering control unit.

CIVIC

✳✳ CAUTION

The air bag must be disabled before removing the steering wheel to center the cable reel. Failure to disarm the air bag system may cause accidental air bag deployment, resulting in unnecessary air bag system repairs and the risk of personal injury.

1. Before servicing the vehicle, refer to the precautions in the beginning of this section.

2. Remove or disconnect the following:
- Power steering reservoir off of its mount
- Inlet hose

3. Insert a length of tubing into the inlet hose and route the tubing into a drain container.

4. With the engine running at idle, turn the steering wheel lock-to-lock several times until fluid stops running out of the hose.

5. Position the front wheels straight ahead. Shut off the engine and lock the steering column and remove the ignition key. Reconnect the reservoir inlet hose.

6. Remove or disconnect the following:
- Negative and positive battery cables. Wait 3 minutes before working around the air bags.
- Steering wheel lower access cover

7. Uncouple the air bag connector from the cable reel connector as follows:
   a. Hold the cable reel connector. With your other hand, slide the spring-loaded sleeve toward the stop tab on the air bag connector.
   b. Separate the 2 connectors. There is no need to install a shorting connector, as the connectors are automatically grounded when they are uncoupled.

8. Remove or disconnect the following:
- Steering joint cover, then the upper and lower steering joint bolts.

9. Raise and support the vehicle safely.

10. Remove or disconnect the following:
- Front wheels
- Tie rod end cotter pins and castle nuts
- Tie rod ends from the steering knuckles, using a ball joint tool

11. If equipped with a manual transaxle, remove or disconnect the following:
- Shift lever extension rod from the clutch housing
- Pin retainer out of the way, then drive out the spring pin
- Shift rod

12. If equipped with an automatic transaxle, remove or disconnect the following:
- Shift cable bracket and holder
- Shift cable from the control shaft. Suspend the cable from the underbody with a piece of wire.

13. Remove or disconnect the following:
- Self-locking nuts and separate the catalytic converter from the exhaust pipes.
- Catalytic converter
- Hydraulic line and hose from the rack valve body using a flare nut wrench.
- Left tie rod end and slide the rack all the way to the right.
- Steering rack mounting bolts
- Steering rack from the pinion shaft by pulling the rack downward

14. Drop the gearbox far enough to permit the end of the pinion shaft to come out of the hole in the frame channel.

15. Slide the gearbox to the right until the left tie rod clears the subframe, then drop it down and out of the vehicle to the left.

To install:

➡Use new self-locking nuts when installing the catalytic converter.

✳✳ WARNING

Use only genuine Honda power steering fluid. Any other type or brand of fluid will damage the power steering pump.

16. Install or connect the following:
- Steering rack into position
- Pinion shaft grommet, insert the pinion through the hole in the bulkhead.
- Rack mounting bolts. Tighten the bracket bolts to 28 ft. lbs. (39 Nm). Tighten the mounting bolt under the valve body to 43 ft. lbs. (59 Nm).
- 2 hydraulic lines to the rack valve body. Carefully tighten the hydraulic line fitting to 28 ft. lbs. (38 Nm). Securely tighten the return hose clamp.

17. Center the rack ends within their steering strokes.

18. Install or connect the following:
- Tie rod ends onto the rack ends
- Tie rod ends to the steering knuckles, then the castle nuts.
- Front wheels
- Catalytic converter using new gaskets and self-locking nuts. Tighten the front nuts to 16 ft. lbs. (22 Nm), and the rear nuts to 25 ft. lbs. (34 Nm).

19. If equipped with a manual transaxles, install or connect the following:
- Shift linkage with a new spring pin and clip.
- Extension rod and tighten its bolt to 16 ft. lbs. (22 Nm).

20. If equipped with an automatic transaxles, install or connect the following:
- Shift cable and brackets. Tighten the bracket bolts to 108 inch lbs. (12 Nm). Tighten the cable lockbolt to 10 ft. lbs. (14 Nm). Tighten the cable holder bolts to 16 ft. lbs. (22 Nm).

21. Verify that the rack is centered within its strokes. Lower the vehicle.

22. Center the air bag cable reel as follows:
   a. Remove the steering wheel.
   b. Turn the cable reel clockwise until it stops.
   c. Turn the steering wheel counterclockwise (approximately 2 turns) until the arrow on the label points straight up.
   d. Install the steering wheel.

23. During steering wheel installation, verify that the slot on the steering wheel shaft engages with the tabs on the turn sig-

nal canceling sleeve. The pins on the cable reel fit into the holes on the steering wheel body. Install a new steering wheel nut and tighten it to 36 ft. lbs. (50 Nm).

24. Line up the bolt hole in the steering joint with the groove in the pinion shaft. Slip the joint onto the pinion shaft. Pull the joint up and down to be sure the splines are fully seated. Tighten the joint bolts to 16 ft. lbs. (22 Nm).

➡Connect the steering joint and pinion shaft with the cable reel and steering rack centered. Verify that the lower joint bolt is securely seated in the pinion shaft groove. If the steering wheel and rack are not centered, reposition the serrations at the lower end of the steering joint.

25. Install or connect the following:
   • Steering joint cover
26. Tighten the ball joint castle nuts to 29–35 ft. lbs. (40–48 Nm). Then, tighten them only enough to install new cotter pins.
27. Install or connect the following:
   • Air bag and cable reel connectors: Be sure the connectors fit squarely together. Then, press the connectors to couple them. The spring-loaded sleeve will lock into place as the 2 connectors are coupled.
   • Steering wheel lower access cover
   • Negative and positive battery cables
28. Turn the ignition switch to the **ON** position. The air bag indicator light should come on for 6 seconds, then turn off. This light sequence indicates that the air bag system is enabled and functioning normally. If the air bag light stays on longer, or doesn't turn on, the system must be diagnosed.
29. Be sure the reservoir inlet line has been reconnected. Fill the reservoir to the upper line with Honda power steering fluid. Run the engine at idle and turn the steering wheel lock-to-lock several times to bleed air from the system and fill the rack valve body. Recheck the fluid level and add more if necessary.
30. Check the power steering system for leaks.
31. Check the front wheel alignment and steering wheel spoke angle. Make adjustments by turning the left and right tie rod ends equally.
32. Road test the vehicle.

PRELUDE

➡The electronic neutral check must be performed on 4WS equipped Preludes any time the steering rack, steering wheel, or steering column is removed, and before the wheels are aligned.

1. Before servicing the vehicle, refer to the precautions in the beginning of this section.
2. Remove or disconnect the following:
   • Power steering reservoir off of its mount
   • Inlet hose
3. Insert a length of tubing into the inlet hose and route the tubing into a drain container.
4. With the engine running at idle, turn the steering wheel lock-to-lock several times until fluid stops running out of the hose. Shut off the engine.
5. Position the front wheels straight ahead. Lock the steering column with the ignition key. Reconnect the reservoir inlet hose.
6. Remove or disconnect the following:
   • Negative battery cable
   • Steering joint cover, then the upper and lower steering joint bolts.
7. Raise and support the vehicle safely.
8. Remove or disconnect the following:
   • Front wheels
   • Tie rod end cotter pins and castle nuts. Install a 12mm nut onto the end of the ball joint stud to protect the threads from damage.
   • Tie rod ends from the steering knuckles, using a ball joint tool
   • Heated Oxygen Sensor (HO$_2$S) sensor connector
   • Self-locking nuts, then separate the catalytic converter from the exhaust pipe.
   • Exhaust pipe from the intake manifold
   • Exhaust pipe from the vehicle
9. If equipped with an automatic transaxle, remove or disconnect the following:
   • Remove the shift cable cover
   • Shift cable, and wire it up and out of the way.

➡Clean any oil or dirt off of the valve body with solvent.

   • Center beam from the subframe
   • Valve body shield
   • 4 hydraulic lines from the rack valve body, using a flare nut wrench. Plug the lines to keep dirt and moisture out.
10. On models with 4-Wheel Steering (4WS) remove or disconnect the following:
   • Carefully cut the wire tie securing the cover to the front sub-steering angle sensor
   • Cover
   • Sensor wiring harness from the 2 securing clamps
   • Sensor connector from the 4WS steering main wiring harness.
11. Remove or disconnect the following:
   • Steering joint bolt, then slide the pinion shaft out of the joint.
   • Left mounting bracket, then the right mounting brackets.
   • Left tie rod end and slide the rack all the way to the right.
12. Pull the steering rack down to release it from the pinion shaft.
13. Slide the steering rack to the right until the left tie rod clears the subframe, then drop it down and out of the vehicle to the left.

   To install:

➡Use new gaskets and self-locking nuts when installing the exhaust pipe.

✱✱ WARNING

Use only genuine Honda power steering fluid. Any other type or brand of fluid will damage the power steering pump.

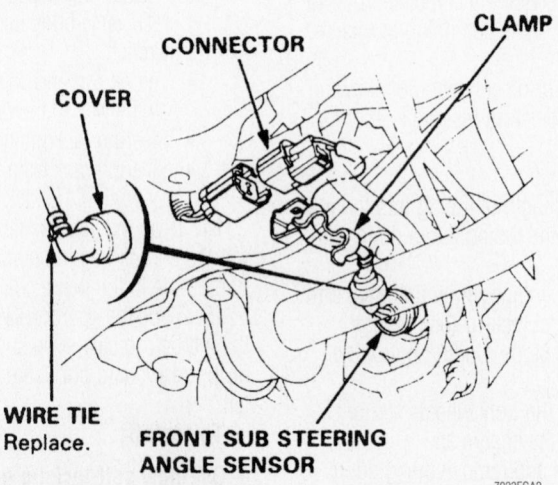

4WS front sub-steering angle sensor—Prelude

7923FGA2

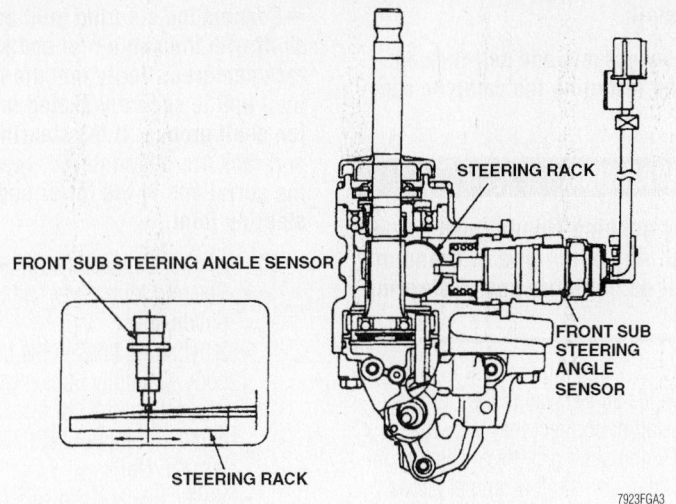

Front sub-steering angle sensor, harness, and steering rack—Prelude

14. Install or connect the following:
- Steering rack into position
- Pinion shaft grommet and insert the pinion through the hole in the firewall.
- Right and left mounting brackets. Tighten the short bolts to 28 ft. lbs. (39 Nm), and the long bolts to 32 ft. lbs. (44 Nm).

15. Center the rack ends within their steering strokes.

16. Center the air bag cable reel as follows:

  a. Turn the steering wheel clockwise until it stops.

  b. Turn the steering wheel counterclockwise until the yellow gear tooth lines up with the alignment mark on the lower column cover.

17. Line up the bolt hole in the steering joint with the groove in the pinion shaft. Slip the joint onto the pinion shaft. Pull the joint up and down to be sure the splines are fully seated. Tighten the joint bolts to 16 ft. lbs. (22 Nm).

➡**Connect the steering joint and pinion shaft with the cable reel and steering rack centered. Verify that the lower joint bolt is securely seated in the pinion shaft groove. If the steering wheel and rack are not centered, reposition the serrations at the lower end of the steering joint.**

18. Install or connect the following:
- 4 hydraulic lines to the rack valve body. Carefully tighten the 12mm fittings to 108 inch lbs. (13 Nm), the 14mm inlet fitting to 28 ft. lbs. (37 Nm), and the 17mm oil cooler fitting to 21 ft. lbs. (29 Nm).
- Front sub-steering angle sensor to the 4WS harness.

- Wire back into its clamps, making sure that it doesn't interfere with the stabilizer bar.
- Sensor cover with a new wire tie
- Valve body shield
- Center beam. Use new self-locking bolts and tighten them to 43 ft. lbs. (60 Nm).

19. If equipped with an automatic transaxle, install or connect the following:
- Shift cable and tighten the locknut to 10 ft. lbs. (14 Nm).
- Cable holder and tighten its bolts to 13 ft. lbs. (18 Nm).

20. Install or connect the following:
- Catalytic converter using new gaskets and self-locking nuts. Tighten the exhaust manifold nuts to 40 ft. lbs. (55 Nm), and the rear nuts to 25 ft. lbs. (34 Nm).
- HO2S connector

- Tie rod ends onto the rack ends
- Tie rod ends to the steering knuckles, then the castle nuts.
- Front wheels

21. Verify that the rack is centered within its strokes. Lower the vehicle.

22. Install the steering joint cover.

23. Tighten the ball joint castle nuts to 36–43 ft. lbs. (50–60 Nm). Then, tighten them only enough to install new cotter pins.

24. Reconnect the negative battery cable.

25. Be sure the reservoir inlet line has been reconnected. Fill the reservoir to the upper line with Honda power steering fluid. Run the engine at idle and turn the steering wheel lock-to-lock several times to bleed air from the system and fill the rack valve body. Recheck the fluid level and add more if necessary.

26. Check the power steering system for leaks.

27. On Preludes without 4WS, check and adjust the front wheel alignment. On Preludes with 4WS, the electronic neutral check must be performed on the 4WS system.

ACCORD

1. Before servicing the vehicle, refer to the precautions in the beginning of this section.

2. Lift the power steering reservoir off of its mount and disconnect the inlet hose.

3. Insert a length of tubing into the inlet hose and route the tubing into a drain container.

4. With the engine running at idle, turn the steering wheel lock-to-lock several times until fluid stops running out of the hose. Immediately shut off the engine.

5. Position the front wheels straight ahead. Lock the steering column with the

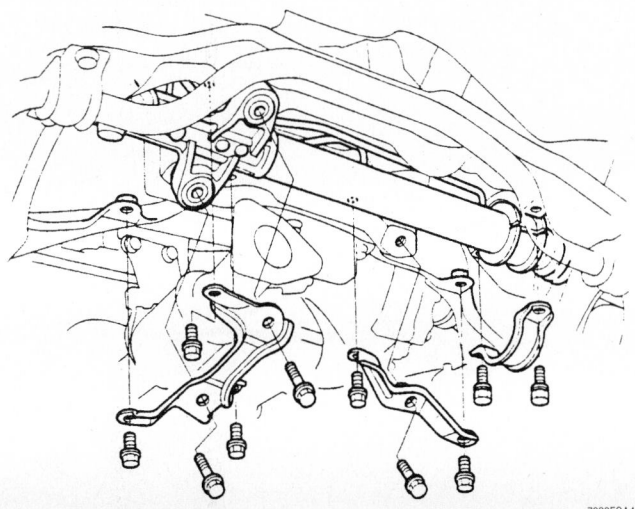

Power rack and pinion steering gear mounting—Accord

ignition key. Reconnect the reservoir inlet hose.

6. Remove or disconnect the following:
   - Negative battery cable
   - Steering joint cover and the upper and lower steering joint bolts.
7. Raise and support the vehicle safely.
8. Remove or disconnect the following:
   - Front wheels
   - Tie rod end cotter pins and castle nuts
   - Tie rod ends from the steering knuckles, using a ball joint tool.
   - Left tie rod end and slide the rack all the way to the right.
   - Heated Oxygen Sensor (HO2S) connector
   - Self-locking nuts, then separate the catalytic converter from the exhaust pipe.
   - Catalytic converter
   - Shift linkage from the transaxle case, if equipped with a manual transaxle.
   - Shift cable cover and cable (wire it up and out of the way), if equipped with an automatic transaxle.
   - 2 hydraulic lines from the rack valve body, using a flare nut wrench. Plug the lines to keep dirt and moisture out. Carefully move the disconnected lines to the rear of the rack assembly so that they are not damaged when the rack is removed.
   - Rack stiffener plate, then the steering rack mounting bolts.
9. Pull the steering rack down to release it from the pinion shaft.
10. Drop the steering rack far enough to permit the end of the pinion shaft to come out of the hole in the frame channel.
11. Slide the steering rack to the right until the left tie rod clears the subframe, then drop it down and out of the vehicle to the left.

To install:

➡️ **Use new gaskets and self-locking nuts when installing the catalytic converter.**

✳✳ WARNING

Use only genuine Honda power steering fluid. Any other type or brand of fluid will damage the power steering pump.

12. Before installing the rack & pinion, slide the ends all the way to the right.
13. Install or connect the following:
    - Pinion shaft grommet. The lug on the pinion shaft grommet aligns with the slot on the valve body.
    - Steering rack into position
    - Pinion shaft grommet and insert the pinion through the hole in the bulkhead.
    - Rack mounting bolts. Tighten the bracket bolts to 28 ft. lbs. (39 Nm). Tighten the stiffener plate mounting bolts to 32 ft. lbs. (43 Nm).
14. Center the rack ends within their steering strokes.
15. Center the air bag cable reel, as follows:
    a. Turn the steering wheel left approximately 150°, to check the cable reel position with the indicator.
    b. If the cable reel is centered, the yellow gear tooth lines up with the alignment mark on the cover.
    c. Return the steering wheel right approximately 150° to position the steering wheel in the straight-ahead position.
16. Line up the bolt hole in the steering joint with the groove in the pinion shaft. Slip the joint onto the pinion shaft. Pull the joint up and down to be sure the splines are fully seated. Tighten the joint bolts to 16 ft. lbs. (22 Nm).

➡️ **Connect the steering joint and pinion shaft with the cable reel and steering rack centered. Verify that the lower joint bolt is securely seated in the pinion shaft groove. If the steering wheel and rack are not centered, reposition the serrations at the lower end of the steering joint.**

17. Install or connect the following:
    - Steering joint cover and the rack & pinion cover
    - 2 hydraulic lines to the rack valve body. Carefully tighten the 14mm inlet fitting to 27 ft. lbs. (37 Nm) and the 16mm outlet fitting to 21 ft. lbs. (28 Nm).
    - Shift cable and the select cable to the transaxle with new cotter pins, if equipped with a manual transaxle.
    - Shift cable to the transaxle using a new lockwasher, if equipped with an automatic transaxle. Tighten the lockbolt to 10 ft. lbs. (14 Nm).
    - Catalytic converter using new gaskets and self-locking nuts. Tighten the front nuts to 16 ft. lbs. (22 Nm), and the rear nuts to 25 ft. lbs. (34 Nm).
    - HO2S sensor connector
    - Tie rod ends onto the rack ends
    - Tie rod ends to the steering knuckles, then the castle nuts.
18. Tighten the ball joint castle nuts to 29–35 ft. lbs. (40–48 Nm). Then, tighten them only enough to install new cotter pins.
19. Install the front wheels.
20. Lower the vehicle.
21. Reconnect the negative battery cable.
22. Be sure the reservoir inlet line has been reconnected. Fill the reservoir to the upper line with Honda power steering fluid. Run the engine at idle and turn the steering wheel lock-to-lock several times to bleed air from the system and fill the rack valve body. Recheck the fluid level and add more if necessary.
23. Check the power steering system for leaks.
24. Check the front wheel alignment and steering wheel spoke angle. Make adjustments by turning the left and right tie rod ends equally.
25. Road test the vehicle.

S2000

1. Before servicing the vehicle, refer to the precautions in the beginning of this section.
2. Remove or disconnect the following:
   - Negative battery cable

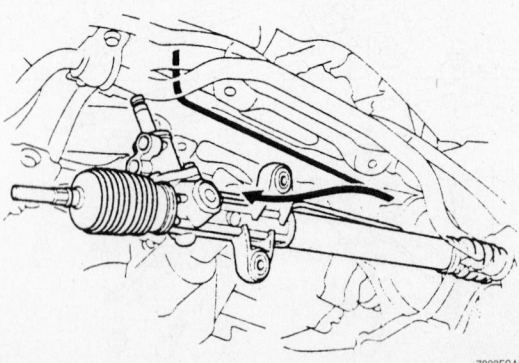

7923FGA5

Move the steering rack to the right, then down and out of the vehicle—Accord

- Front wheels
- Driver's air bag
- Steering wheel
- Steering coupler
- Outer tie rod ends
- Splash shield
- Stabilizer bar brackets
- Steering gear wiring connectors
- Steering gear mounting bolts

3. Move the steering gear forward and to the right to remove the steering gear.

To install:

4. Installation is the reverse of the removal procedure, while using the following torque values:

- Steering gear mounting bolts: 33 ft. lbs. (44 Nm)
- Steering gear ground cable bolt: 88 inch lbs. (10 Nm)
- Stabilizer bar bracket bolts: 61 ft. lbs. (83 Nm)
- Splash shield bolts: 88 inch lbs. (10 Nm)
- Outer tie rod end nuts: 40 ft. lbs. (54 Nm)
- Steering coupler pinch bolts: 16 ft. lbs. (22 Nm)

Strut

REMOVAL & INSTALLATION

Front

CIVIC

1. Before servicing the vehicle, refer to the precautions in the beginning of this section.
2. Raise and safely support the vehicle.
3. Remove or disconnect the following:
   - Front wheels
   - Brake hose brackets from the bottom of the strut tube. Do not disconnect the brake hoses.

➡**Some Civic models may not have brake hose brackets on their struts. In these cases, there is no need to unbolt the brackets.**

- Damper pinch bolt
- Damper fork nut and bolt
- Damper fork
- 2 strut mounting bolts from the shock tower
- Strut from the vehicle

To install:

➡**Use new self-locking nuts when installing the strut.**

4. Install or connect the following:
   - Strut into the vehicle. Hand-tighten

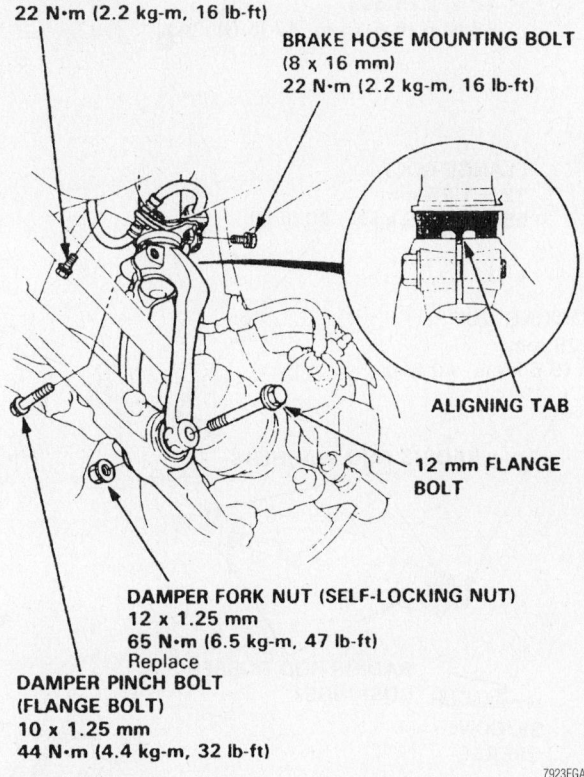

BRAKE HOSE MOUNTING BOLT
(8 x 20 mm)
22 N·m (2.2 kg-m, 16 lb-ft)

BRAKE HOSE MOUNTING BOLT
(8 x 16 mm)
22 N·m (2.2 kg-m, 16 lb-ft)

ALIGNING TAB

12 mm FLANGE
BOLT

DAMPER FORK NUT (SELF-LOCKING NUT)
12 x 1.25 mm
65 N·m (6.5 kg-m, 47 lb-ft)
Replace

DAMPER PINCH BOLT
(FLANGE BOLT)
10 x 1.25 mm
44 N·m (4.4 kg-m, 32 lb-ft)

7923FGA6

Damper fork components—Civic

the strut mounting bolts. The alignment mark on the strut tube faces away from the wheel.
   - Damper fork onto the strut and lower control arm
   - Pinch and fork bolts
   - Brake hose brackets to the strut tube and tighten them to 16 ft. lbs. (22 Nm).
   - Front wheels and lower the vehicle.

5. Tighten the strut mount bolts to 36 ft. lbs. (50 Nm).
6. Tighten the pinch bolt to 32 ft. lbs. (44 Nm). Tighten the damper fork nut to 47 ft. lbs. (65 Nm).
7. Tighten the wheel nuts to 80 ft. lbs. (110 Nm).
8. Check the vehicle's front end alignment and adjust it if necessary.

PRELUDE AND ACCORD

1. Before servicing the vehicle, refer to the precautions in the beginning of this section.
2. Raise and safely support the vehicle.
3. Remove or disconnect the following:
   - Front wheels
   - Brake hose clamp bolts from the strut
   - Damper fork bolts, then the damper fork from the damper and lower arm

   - 3 strut mounting nuts or 2 8mm flange nuts and 3 10mm flange nuts, as applicable
   - Strut (damper assembly) from the vehicle

To install:

➡**Use new self-locking bolts when installing the struts and assembling the damper forks.**

4. Install or connect the following:
   - Strut into the vehicle with the aligning tab facing inside, if equipped. Hand-tighten the mounting nuts.
   - Strut into the damper fork. The alignment mark on the strut tube fits into the slot on the damper fork.
   - Pinch bolt and damper fork bolt/nut. Only hand-tighten these bolts.
   - Front wheels and lower the vehicle.

5. With all 4 of the vehicle's wheels on the ground, tighten the damper fork nut to 47 ft. lbs. (65 Nm) while holding the damper fork bolt. Tighten the damper fork pinch bolt to 32 ft. lbs. (44 Nm). Tighten the strut mounting nuts to 28 ft. lbs. (39 Nm) for 2001–02 models. For 2003–04 models,

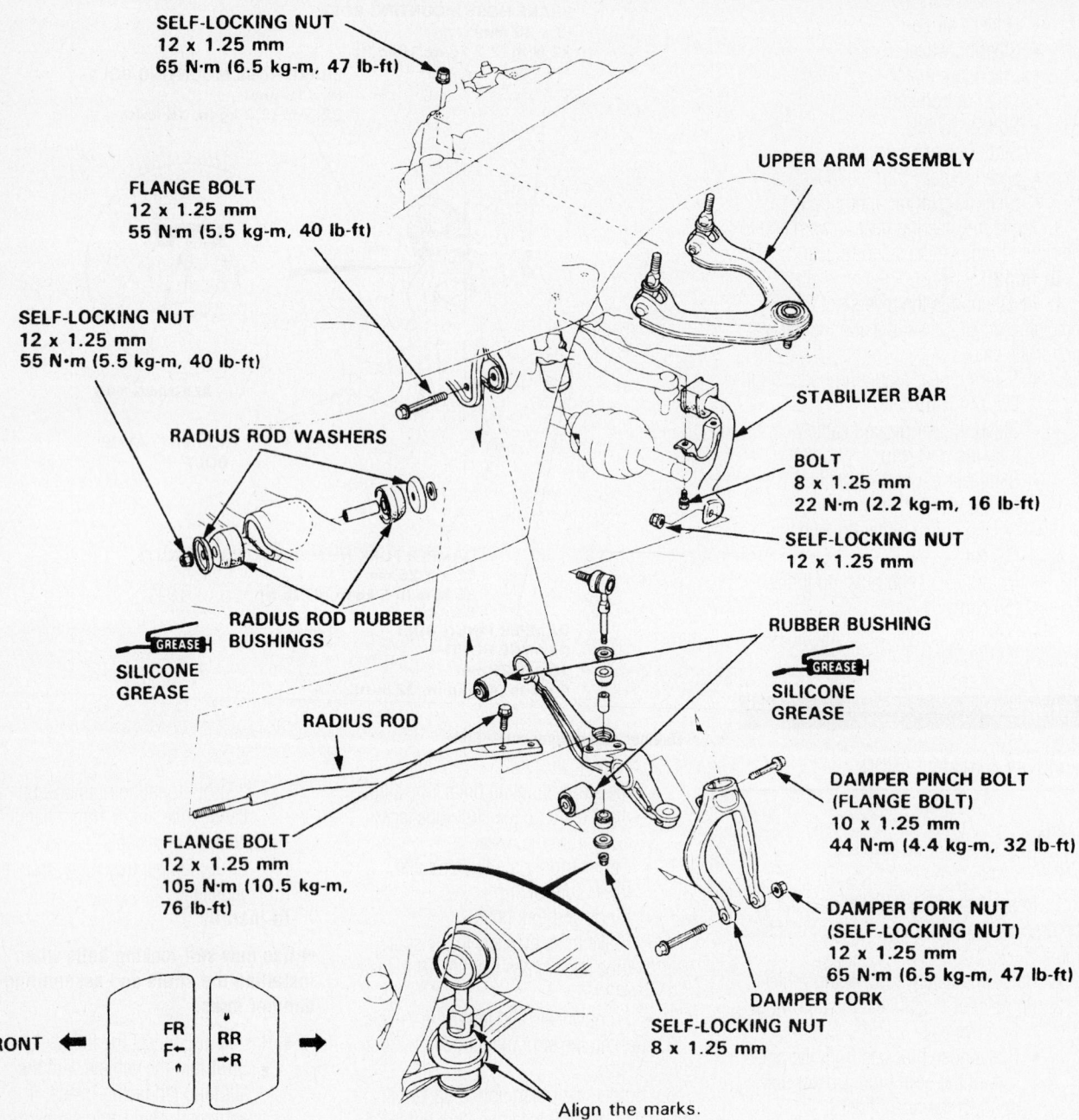

SELF-LOCKING NUT
12 x 1.25 mm
65 N·m (6.5 kg-m, 47 lb-ft)

FLANGE BOLT
12 x 1.25 mm
55 N·m (5.5 kg-m, 40 lb-ft)

UPPER ARM ASSEMBLY

SELF-LOCKING NUT
12 x 1.25 mm
55 N·m (5.5 kg-m, 40 lb-ft)

RADIUS ROD WASHERS

STABILIZER BAR

BOLT
8 x 1.25 mm
22 N·m (2.2 kg-m, 16 lb-ft)

SELF-LOCKING NUT
12 x 1.25 mm

RADIUS ROD RUBBER BUSHINGS

GREASE

SILICONE GREASE

RUBBER BUSHING

GREASE

SILICONE GREASE

RADIUS ROD

DAMPER PINCH BOLT (FLANGE BOLT)
10 x 1.25 mm
44 N·m (4.4 kg-m, 32 lb-ft)

FLANGE BOLT
12 x 1.25 mm
105 N·m (10.5 kg-m, 76 lb-ft)

DAMPER FORK NUT (SELF-LOCKING NUT)
12 x 1.25 mm
65 N·m (6.5 kg-m, 47 lb-ft)

DAMPER FORK

SELF-LOCKING NUT
8 x 1.25 mm

FRONT ◄── | FR F← | RR ►R | ──►

Align the marks.

7923FGA7

Front suspension components—Prelude and 2001–02 Accord

tighten the 8mm bolts to 16 ft. lbs. (22 Nm) and the 10mm bolts to 37 ft. lbs. (50 Nm).

6. Tighten the wheel nuts to 80 ft. lbs. (110 Nm).

7. Check and adjust the vehicle's front end alignment. On Preludes equipped with 4WS, the electronic neutral check must be performed before aligning all 4 wheels.

S2000

1. Before servicing the vehicle, refer to the precautions in the beginning of this section.

2. Remove or disconnect the following:
   • Front wheel
   • Lower ball joint
   • Brake caliper bracket bolt
   • Upper strut mount nuts
   • Lower flange bolt
   • Strut

To install:

3. Installation is the reverse of the removal procedure, while using the following torque values:
   • Lower flange bolt: 47 ft. lbs. (64 Nm)

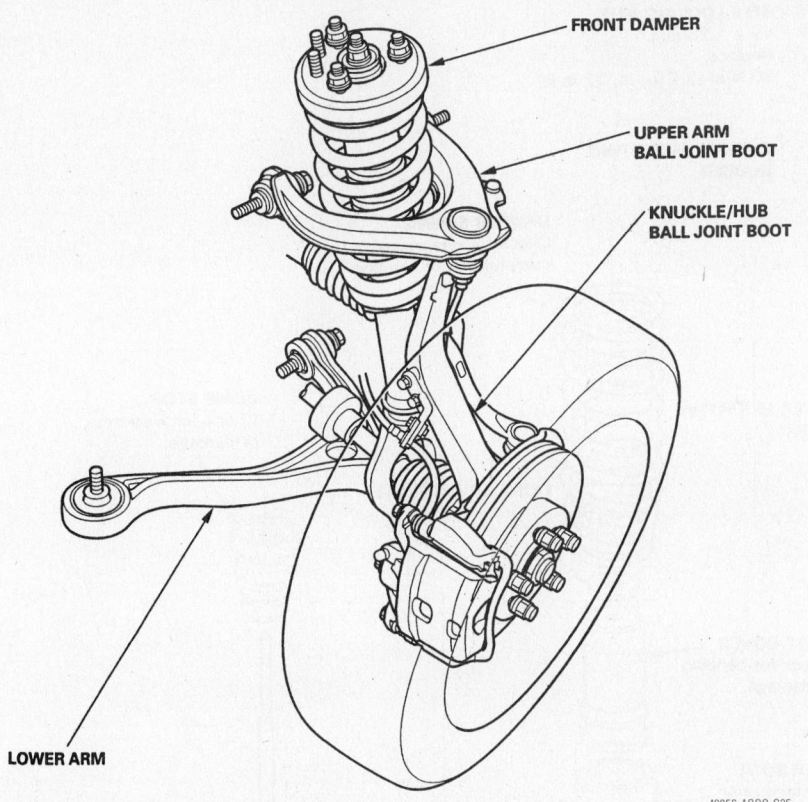

Front suspension components—2003–04 Accord

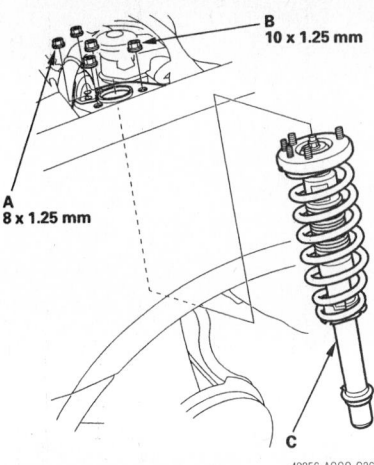

**Front strut (damper) mounting—2003
Accord shown, earlier models similar**

- Upper mount nuts: 36 ft. lbs. (49 Nm)
- Lower ball joint nut: 43–51 ft. lbs. (59–69 Nm)
- Brake caliper bracket bolt: 16 ft. lbs. (22 Nm)

Rear

CIVIC

> ### ✳✳ CAUTION
>
> Removing rear suspension components may make the vehicle front-heavy and cause it to tip forward when raised on a hoist. Use under-lift support stands, or place additional weight in the trunk of the vehicle before hoisting it.

1. Before servicing the vehicle, refer to the precautions in the beginning of this section.

2. Remove the interior or trunk trim pieces that cover the strut mount, as follows:

   a. **Sedan and Coupe models:** Fold down the upper rear seat cushion. Carefully pry out the clips that secure the trunk and shock tower trim to the body. Remove the trunk trim to expose the strut mounts.

   b. **Hatchback models:** Fold down the rear seat. Unbolt and remove the rear side shelf/speaker grille assemblies. Disconnect and remove the speaker. Carefully pry out the clips and remove the screws to remove shock tower trim panel.

3. Raise and support the vehicle.
4. Remove or disconnect the following:
   - Rear wheels
   - 2 upper mounting nuts
   - Wheel sensor bracket from the lower control arm.
   - Lower strut bolt and the knuckle flange bolt.
   - Strut from the vehicle

To install:

➡ **All suspension nuts and bolts should be tightened with the vehicle on the ground. Alternatively, raise the lower control arm with a floor jack until the jack is supporting the weight of the vehicle. This method pre-loads the suspension and allows room to work.**

5. Install or connect the following:
   - Strut into the vehicle with the locknut facing the front of the vehicle. Hand-tighten the upper mounting nuts.
   - Wheel sensor bracket onto the lower control arm. Tighten the bolts to 84 inch lbs. (10 Nm).
   - Knuckle flange bolt and the lower strut bolt. Hand-tighten the bolts.
   - Wheels and lower the vehicle.
   - Upper mounting nuts to 36 ft. lbs. (50 Nm). Tighten the knuckle flange bolt and strut bolts to 40 ft. lbs. (55 Nm). Tighten the wheel nuts to 80 ft. lbs. (110 Nm).
   - Trunk side trim panels

6. Check and adjust the vehicle's rear wheel alignment.

PRELUDE

1. Before servicing the vehicle, refer to the precautions in the beginning of this section.

2. Raise and safely support the vehicle.
3. Remove or disconnect the following:
   - Trunk side trim and the 2 top strut nuts.
   - Upper ball joint cover
   - Cotter pin and upper ball joint nut

4. Fit a 10mm nut on the ball joint and separate the ball joint and the knuckle by using a ball joint removal tool.

5. Remove or disconnect the following:
   - Lower strut mounting bolt and lower the suspension.
   - Strut from the vehicle

To install:

➡ **Use new self-locking nuts when installing the rear struts.**

6. Install or connect the following:
   - Strut; loosely install the lower mounting bolt. Do not tighten.

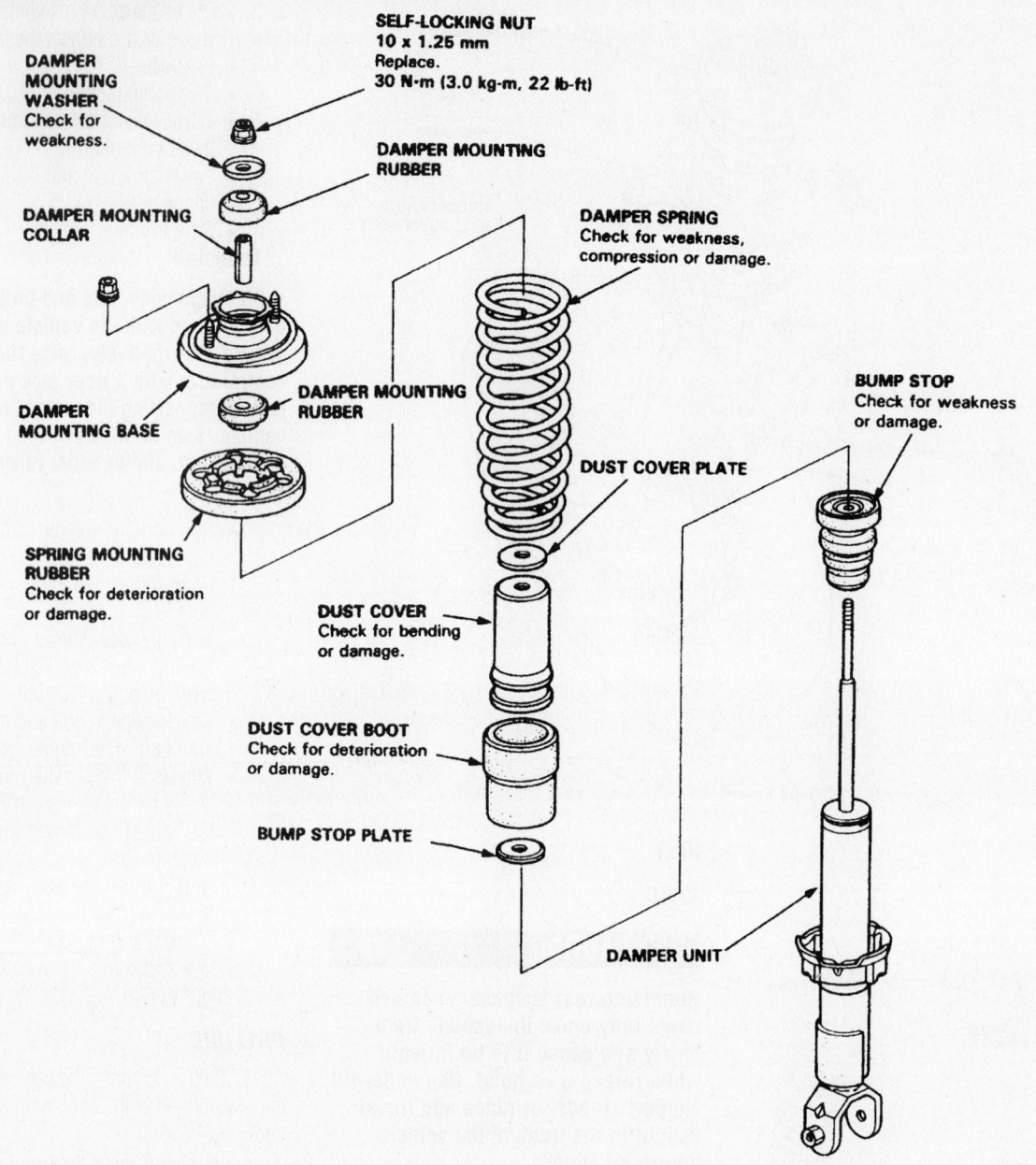

SELF-LOCKING NUT
10 x 1.25 mm
Replace.
30 N·m (3.0 kg-m, 22 lb-ft)

DAMPER MOUNTING WASHER
Check for weakness.

DAMPER MOUNTING RUBBER

DAMPER MOUNTING COLLAR

DAMPER SPRING
Check for weakness, compression or damage.

DAMPER MOUNTING RUBBER

BUMP STOP
Check for weakness or damage.

DAMPER MOUNTING BASE

DUST COVER PLATE

SPRING MOUNTING RUBBER
Check for deterioration or damage.

DUST COVER
Check for bending or damage.

DUST COVER BOOT
Check for deterioration or damage.

BUMP STOP PLATE

DAMPER UNIT

7923FGA9

Exploded view of the rear suspension strut—Civic

- Upper strut mounting bolts: 28 ft. lbs. (39 Nm).
- Upper arm and knuckle and tighten the castle nut to 29–35 ft. lbs. (40–48 Nm).
- Upper ball joint cover

7. Raise the rear suspension with a floor jack until the weight is on the strut.

8. Tighten the lower strut mounting bolt to 47 ft. lbs. (65 Nm).

9. Install the rear wheels and lower the vehicle.

10. Tighten the rear wheel nuts to 80 ft. lbs. (110 Nm).

11. Check and adjust the vehicle's rear wheel alignment.

ACCORD

1. Before servicing the vehicle, refer to the precautions in the beginning of this section.

2. Fold the rear seat forward.

3. Remove or disconnect the following:
- Rear bulkhead cover
- Side bolster cushions. The side bolster cushions are secured by 1 screw at the bottom, and 2 clips at the top.
- Strut mount cap, if equipped, and upper strut mounting nuts

4. Raise and safely support the vehicle.
- Rear wheels, then support the knuckle with a floor jack

- Strut lower flange bolt from the knuckle
- Strut flange nut while holding the joint pin with a hex wrench
- Stabilizer link from the bracket
- Strut, while lowering rear suspension

➡ **The left and right struts are different, so be sure to mark them L & R if you are removing both struts before continuing.**

To install:

➡ **Use new self-locking nuts when installing the strut.**

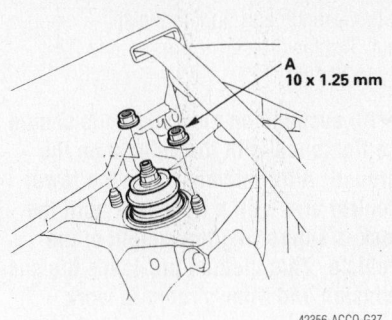

A
10 x 1.25 mm

42356-ACCO-G37

Rear strut upper mounting nut locations—Accord

5. Lower the rear suspension.
6. Install or connect the following:
   - Strut into the upper mount. Only hand-tighten the upper mounting nuts.
   - Strut into position on the knuckle, then loosely install the flange bolt on the bottom of the strut
   - Stabilizer link on the bracket and loosely install the flange nut
7. Place a jack under the lower strut mount. Raise the jack until the weight of the vehicle is on the jack.
8. With the suspension under load, tighten the lower mount bolt to 40 ft. lbs. (55 Nm) for 2001–02 vehicles or to 43 ft. lbs. (59 Nm) for 2003–04 vehicles.
9. While holding the joint pin with a cotter pin, tighten the flange nut to 29 ft. lbs. (39 Nm).
10. Tighten the upper nuts to 28 ft. lbs. (39 Nm) for 2001–02 vehicles or to 37 ft. lbs. (50 Nm) for 2003–04 vehicles.
11. Install or connect the following:
    - Rear wheel(s). Lower the vehicle to the ground. Tighten the wheel nuts to 80 ft. lbs. (110 Nm).

- Rear seat side bolsters and fold the seat back into place.
- Rear bulkhead cover
12. Check and adjust the vehicle's rear wheel alignment.

S2000

1. Before servicing the vehicle, refer to the precautions in the beginning of this section.
2. Remove or disconnect the following:
   - Rear wheel
   - Spare tire
   - Upper mount flange nuts
   - Lower flange bolt
   - Strut

To install:

3. Installation is the reverse of the removal procedure, while using the following torque values:
   - Lower flange bolt: 47 ft. lbs. (64 Nm)
   - Upper mount flange nuts: 36 ft. lbs. (49 Nm)

Coil Spring

REMOVAL & INSTALLATION

Civic

FRONT

1. Before servicing the vehicle, refer to the precautions in the beginning of this section.
2. Raise and safely support the vehicle.
3. Remove or disconnect the following:
   - Front wheels
   - Brake hose brackets from the bottom of the strut tube. Do not disconnect the brake hoses.

➡**Some Civic models may not have brake hose brackets on their struts.**

- Damper fork pinch bolt and flange bolt, then the damper fork.
- Strut's upper mounting nuts
- Strut assembly from the vehicle.
4. Install a spring compressor onto the strut assembly and tighten the compressor according to the manufacturer's instructions.
5. Remove the locking nut from the top of the shock absorber piston. Disassemble the strut and remove the coil spring.

To install:

➡**Use new self-locking nuts when assembling the strut.**

6. Install or connect the following:
   - Spring compressor onto the coil spring
7. Assemble the lower strut mounts, dust covers, coil spring, and upper strut mount onto the shock absorber. Position the strut bearing mounting studs so that they will line up with the mounting holes in the shock tower.
8. Install or connect the following:
   - Mounting washer, and a new self-locking nut (loosely).
9. Hold the shock absorber piston with a hex wrench and tighten the self-locking nut. Tighten the self-locking nut to 22 ft. lbs. (30 Nm).

➡**All suspension nuts and bolts should be tightened with the vehicle on the ground.**

10. Install or connect the following:
    - Strut assembly into the vehicle. Tighten the upper mounting nuts to 36 ft. lbs. (50 Nm).
    - Damper fork. Tighten the pinch bolt to 32 ft. lbs. (44 Nm), and the fork bolt to 47 ft. lbs. (64 Nm).
    - Brake hose clamps. Tighten them to 16 ft. lbs. (22 Nm).
    - Wheel, and tighten the wheel nuts to 80 ft. lbs. (110 Nm).
11. Check and adjust the vehicle's front wheel alignment.

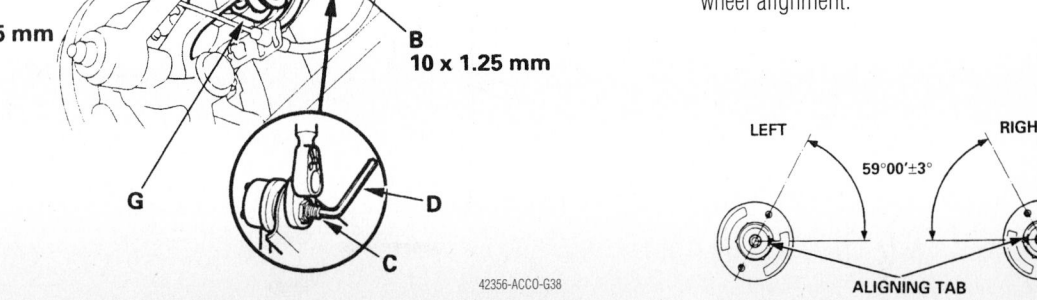

A
12 x 1.25 mm

B
10 x 1.25 mm

42356-ACCO-G38

Remove the lower flange (A) bolt from the knuckle, then remove the flange nut (B) while holding the joint pin (C) with a hex wrench (D) and disconnect the stabilizer link (E) from the bracket (F)

LEFT   **RIGHT**

59°00'±3°

ALIGNING TAB

7923FGB2

Strut bearing installation direction—Civic

REAR

⁑ CAUTION

Removing rear suspension components may make the vehicle front-heavy and cause it to tip forward when raised on a hoist. Use under-lift support stands, or place additional weight in the trunk of the vehicle before hoisting it.

1. Before servicing the vehicle, refer to the precautions in the beginning of this section.

2. Remove the interior or trunk trim pieces that cover the strut mount:

   a. **Sedan and Coupe models:** Fold down the upper rear seat cushion. Carefully pry out the clips that secure the trunk and shock tower trim to the body. Remove the trunk trim to expose the strut mounts.

   b. **Hatchback models:** Fold down the rear seat. Unbolt and remove the rear side shelf/speaker grille assemblies. Disconnect and remove the speaker. Carefully pry out the clips and remove the screws to remove shock tower trim panel.

3. Raise and safely support the vehicle.

4. Remove or disconnect the following:

- 2 strut mounting bolts
- Wheel sensor brackets from the lower control arm. Do not disconnect the sensor.

5. Support the lower control arm with a floor jack.

6. Remove or disconnect the following:

- Strut mounting flange bolt and the knuckle flange bolt, then lower the floor jack
- Strut from the vehicle.

7. Install a spring compressor onto the strut assembly and tighten the compressor according to the manufacturer's instructions.

8. Remove the locking nut from the top of the shock absorber. Disassemble the strut and remove the coil spring.

To install:

➡ **Use new self-locking nuts when assembling the strut.**

9. Install or connect the following:

- Spring compressor onto the coil spring.
- Upper and lower strut mounts, dust covers, and coil spring onto the shock absorber.
- Mounting washer, and a new self-locking nut (loosely)

10. Hold the shock absorber piston with

a hex wrench and tighten the self-locking nut. Tighten the self-locking nut to 22 ft. lbs. (30 Nm).

➡ **All suspension nuts and bolts should be tightened with the vehicle on the ground. Alternatively, raise the lower control arm with a floor jack until the jack is supporting the weight of the vehicle. This method pre-loads the suspension and allows room to work.**

11. Install or connect the following:

- Strut assembly into the vehicle. Tighten the upper mounting nuts to 36 ft. lbs. (50 Nm).
- Shock mounting bolt at the knuckle and tighten to 40 ft. lbs. (55 Nm).
- Knuckle flange bolt and tighten it to 40 ft. lbs. (55 Nm).
- Wheel sensor brackets
- Wheel, and tighten the wheel nuts to 80 ft. lbs. (110 Nm).
- Trunk side trim

12. Check and adjust the rear wheel alignment.

Accord and Prelude

FRONT

1. Before servicing the vehicle, refer to the precautions in the beginning of this section.

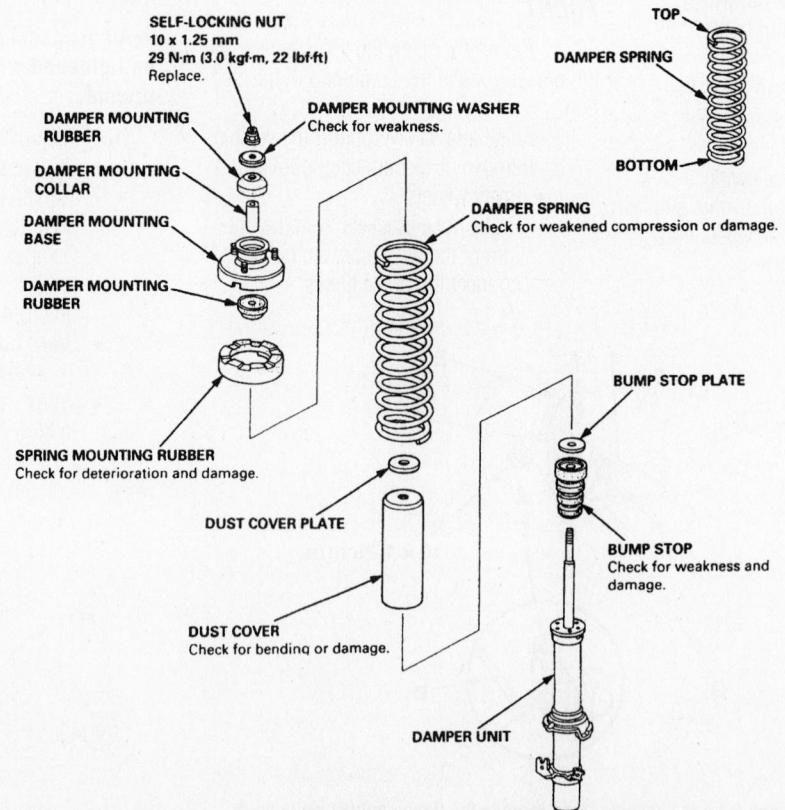

SELF-LOCKING NUT
10 x 1.25 mm
29 N·m (3.0 kgf·m, 22 lbf·ft)
Replace.

DAMPER MOUNTING WASHER
Check for weakness.

DAMPER MOUNTING RUBBER

DAMPER MOUNTING COLLAR

DAMPER MOUNTING BASE

DAMPER MOUNTING RUBBER

SPRING MOUNTING RUBBER
Check for deterioration and damage.

DUST COVER PLATE

DUST COVER
Check for bending or damage.

TOP

DAMPER SPRING

BOTTOM

DAMPER SPRING
Check for weakened compression or damage.

BUMP STOP PLATE

BUMP STOP
Check for weakness and damage.

DAMPER UNIT

7923FGB3

Coil spring, strut cartridge, and strut mount components—2001–02 Accord and Prelude shown, 2003–04 similar

2. Raise and safely support the vehicle.

3. Remove or disconnect the following:
- Strut from the vehicle

4. Place the strut in vice and install a spring compressor onto the coil spring. Follow the spring compressor manufacturer's instructions.

5. Compress the spring and remove the self-locking nut from the top of the strut. Disassemble the strut mounts and remove the coil spring.

6. Inspect the strut mounts for wear and damage. Replace any damaged or worn parts.

To install:

➡ **Use new self-locking nuts when assembling and installing the struts.**

7. Install or connect the following:
- Spring compressor onto the coil spring. Set the spring onto the strut cartridge. The flat part of the coil spring is its top.

8. Assemble the strut mount and washer onto the strut. Tighten the self-locking nut to 22 ft. lbs. (29 Nm). Remove the spring compressor.

9. Install or connect the following:
- Strut

10. Check and adjust the vehicle's front wheel alignment. On Preludes equipped with 4WS, the electronic neutral check must be performed before all 4 wheels are aligned.

REAR—ACCORD

1. Before servicing the vehicle, refer to the precautions in the beginning of this section.

2. Remove or disconnect the following:
- Strut

3. Place the strut in a vice and install a spring compressor onto the coil spring. Follow the spring compressor manufacturer's instructions.

4. Compress the spring and remove the self-locking nut from the strut. Disassemble the strut mounts and remove the coil spring.

5. Inspect the strut mounts for wear and damage. Replace any damaged or worn parts.

To install:

➡ **Use new self-locking nuts when assembling and installing the struts.**

6. Install or connect, the following:
- Spring compressor onto the coil spring. Set the spring onto the strut cartridge. The flat part of the coil spring is its top.
- Strut mount and washer onto the strut. Tighten the self-locking nut to

22 ft. lbs. (29 Nm). Remove the spring compressor.
- Strut into the vehicle. Hand-tighten the mounting nuts.
- Strut into position on the knuckle.
- Mounting bolt

7. Place a jack under the lower strut mount. Raise the jack until the weight of the vehicle is on the jack.

8. With the suspension under load, tighten the lower mount bolt to 40 ft. lbs.

(55 Nm). Tighten the upper nuts to 28 ft. lbs. (39 Nm).

9. Install or connect, the following:
- Rear wheel. Lower the vehicle to the ground.
- Wheel nuts to 80 ft. lbs. (110 Nm).
- Rear seat side bolsters and fold the seat back into place.

10. Check and adjust the vehicle's rear wheel alignment.

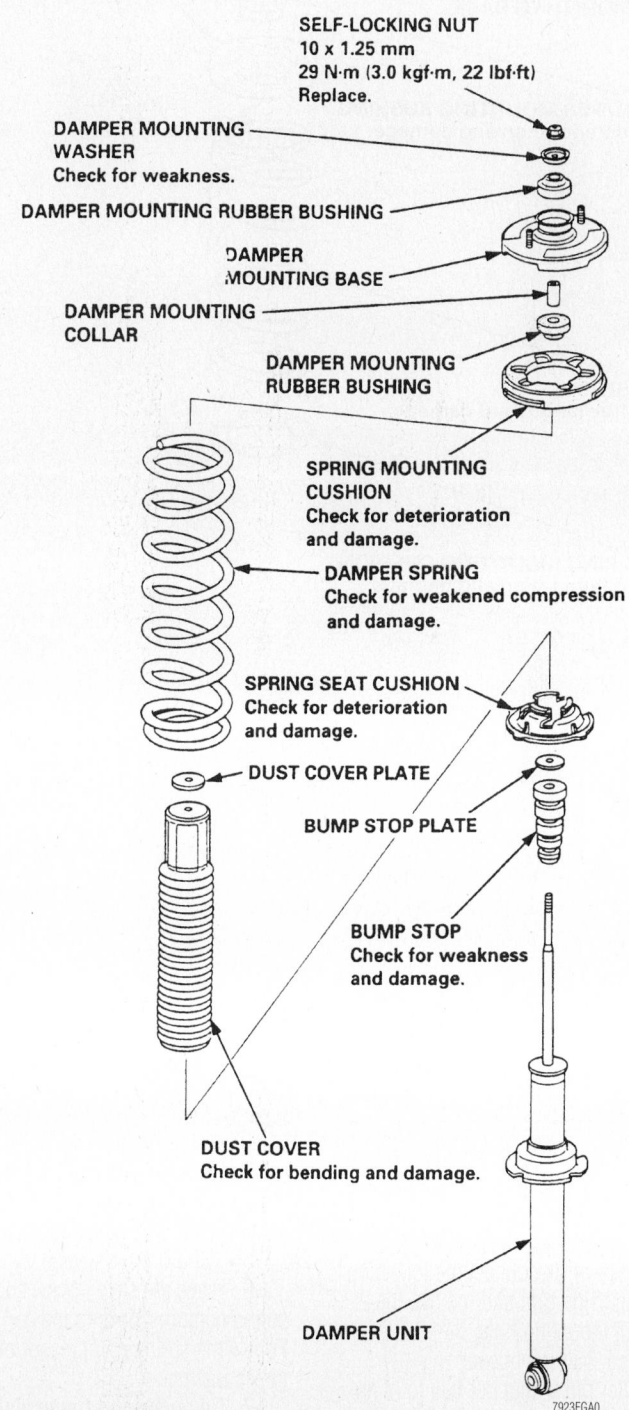

SELF-LOCKING NUT
10 x 1.25 mm
29 N·m (3.0 kgf·m, 22 lbf·ft)
Replace.

DAMPER MOUNTING WASHER
Check for weakness.

DAMPER MOUNTING RUBBER BUSHING

DAMPER MOUNTING BASE

DAMPER MOUNTING COLLAR

DAMPER MOUNTING RUBBER BUSHING

SPRING MOUNTING CUSHION
Check for deterioration and damage.

DAMPER SPRING
Check for weakened compression and damage.

SPRING SEAT CUSHION
Check for deterioration and damage.

DUST COVER PLATE

BUMP STOP PLATE

BUMP STOP
Check for weakness and damage.

DUST COVER
Check for bending and damage.

DAMPER UNIT

7923FGA0

Exploded view of the rear suspension strut assembly—Accord

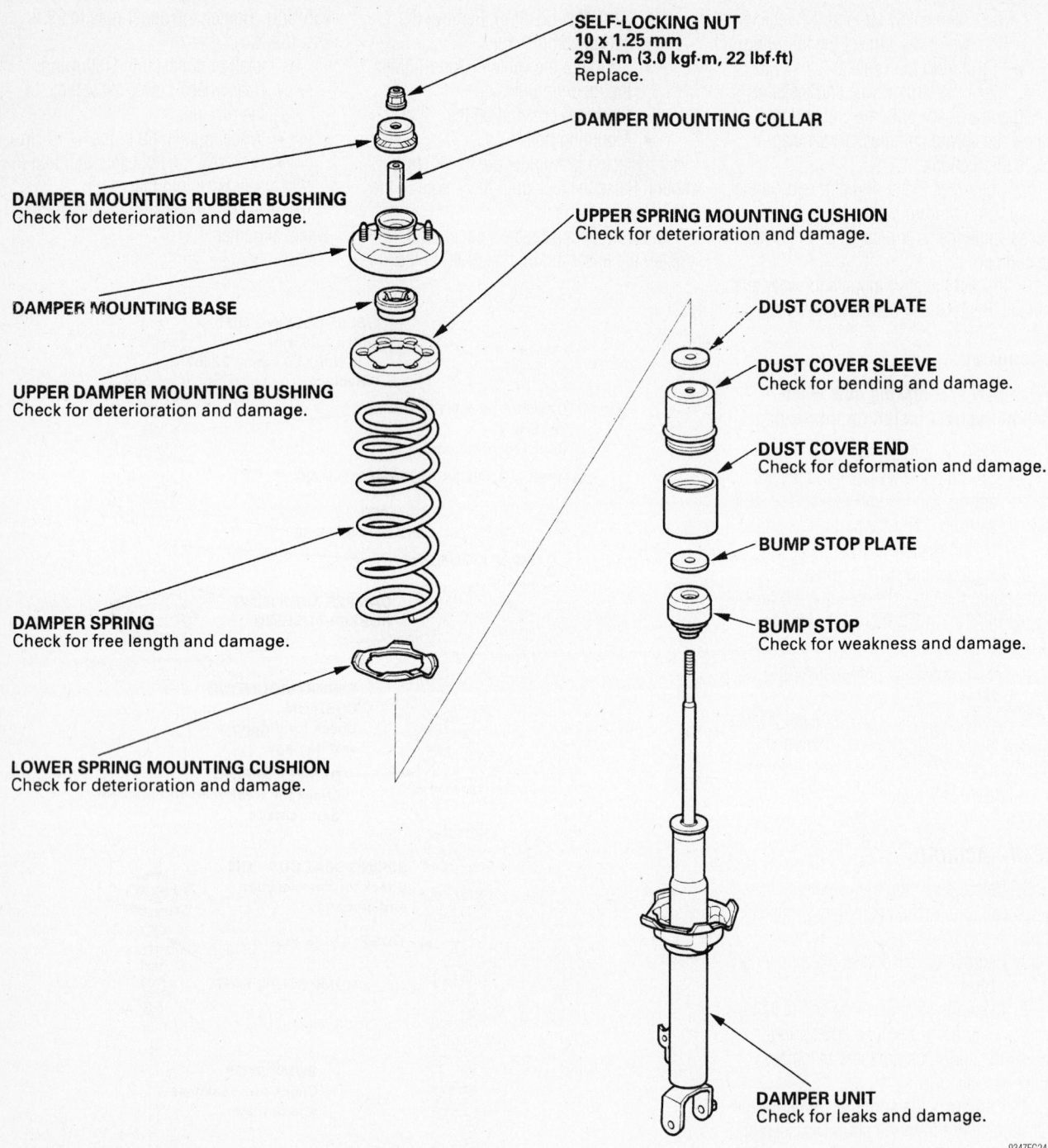

SELF-LOCKING NUT
10 x 1.25 mm
29 N·m (3.0 kgf·m, 22 lbf·ft)
Replace.

DAMPER MOUNTING COLLAR

DAMPER MOUNTING RUBBER BUSHING
Check for deterioration and damage.

UPPER SPRING MOUNTING CUSHION
Check for deterioration and damage.

DAMPER MOUNTING BASE

DUST COVER PLATE

DUST COVER SLEEVE
Check for bending and damage.

UPPER DAMPER MOUNTING BUSHING
Check for deterioration and damage.

DUST COVER END
Check for deformation and damage.

BUMP STOP PLATE

DAMPER SPRING
Check for free length and damage.

BUMP STOP
Check for weakness and damage.

LOWER SPRING MOUNTING CUSHION
Check for deterioration and damage.

DAMPER UNIT
Check for leaks and damage.

9347FG24

Exploded view of the strut and spring assembly—S2000—front shown

REAR—PRELUDE

1. Before servicing the vehicle, refer to the precautions in the beginning of this section.

2. Raise and safely support the vehicle.

3. Remove or disconnect the following:
   - Trunk side trim and remove the 2 strut mounting nuts.
   - Upper ball joint cover
   - Cotter pin and upper ball joint nut

4. Fit a 10mm nut on the ball joint and separate the ball joint and the knuckle by using a ball joint removal tool.

5. Remove or disconnect the following:
   - Lower strut mounting bolt and lower the suspension.
   - Strut from the vehicle.

6. Place the strut in vice and install a spring compressor onto the coil spring. Follow the spring compressor manufacturer's instructions.

7. Compress the spring and remove the self-locking nut from the strut. Disassemble the strut mounts and remove the coil spring.

8. Inspect the strut mounts for wear and damage. Replace any damaged or worn parts.

To install:

➡ **Use new self-locking nuts when installing the rear struts.**

9. Install or connect the following:
   - Spring compressor onto the coil spring. Set the spring onto the strut cartridge. The flat part of the coil spring is its top.

- Strut mount and washer onto the strut. Tighten the self-locking nut to 22 ft. lbs. (29 Nm). Remove the spring compressor.
- Strut to the vehicle and the lower mounting bolt (loosely). Do not tighten at this time.
- Upper strut mounting bolts. Tighten the bolts to 28 ft. lbs. (39 Nm).
- Upper arm and knuckle and tighten the castle nut to 29–35 ft. lbs. (40–48 Nm).
- Upper ball joint cover

10. Raise the rear suspension with a floor jack until the weight is on the strut.

11. Tighten the lower strut mounting bolt to 47 ft. lbs. (65 Nm).

12. Install or connect the following:
- Rear wheels and lower the vehicle.
- Rear wheel nuts to 80 ft. lbs. (110 Nm)
- Trunk trim

13. Check and adjust the vehicle's rear wheel alignment. On Preludes equipped with 4WS, the electronic neutral check must be performed before all 4 wheels are aligned.

S2000

FRONT AND REAR

1. Before servicing the vehicle, refer to the precautions at the beginning of this section.

2. Remove the strut from the vehicle.

3. Compress the coil spring using a suitable spring compressor until the spring comes away from the seat.

4. Remove the center nut and slowly release the spring compressor.

To install:

5. Compress the spring and install it on the strut.

6. Install the lower washer and mounting bracket.

7. Install the upper washer and a new nut. Tighten the nut to 22 ft. lbs. (29 Nm).

8. Install the strut assembly in the vehicle.

Upper Ball Joint

REMOVAL & INSTALLATION

Front and Rear

ALL MODELS

The upper ball joint cannot be removed from the upper control arm. If the ball joint is faulty or worn, the entire control arm

must be replaced. If the upper ball joint boot is damaged and the ball joint itself is still usable, the boot can be replaced.

Upper Control Arm

REMOVAL & INSTALLATION

Front

CIVIC

1. Before servicing the vehicle, refer to the precautions in the beginning of this section.

2. Raise and support the vehicle safely.

3. Remove or disconnect the following:
- Front wheels
- Damper fork from the lower control arm
- Strut mounting nuts, then the strut from the vehicle
- Upper ball joint from the steering knuckle using a suitable ball joint remover.
- Self-locking nuts, then the upper arm from the vehicle.
- Upper arm bolts to separate the

control arm from its anchor bolt assembly. Inspect the bushings for signs of deterioration and replace them if they are damaged.

4. Place the upper control arm anchor bolt assembly into a vice and drive out the upper arm bushings.

To install:

➡Use new self-locking nuts when assembling the anchor bolts and when installing the control arm into the vehicle.

5. Drive the new upper arm bushings into the upper arm anchor bolts. Center the bushing in the anchor bolt so that equal amounts of the bushing sleeve protrude on either side.

6. Install or connect the following:
- Anchor bolt assembly onto the control arm. Align the marks on the arm and anchor assembly. Tighten the nuts to 22 ft. lbs. (30 Nm).
- Upper control arm assembly into the shock tower.
- Strut into the vehicle
- Damper fork bolt and nut
- Steering arm and upper ball joint

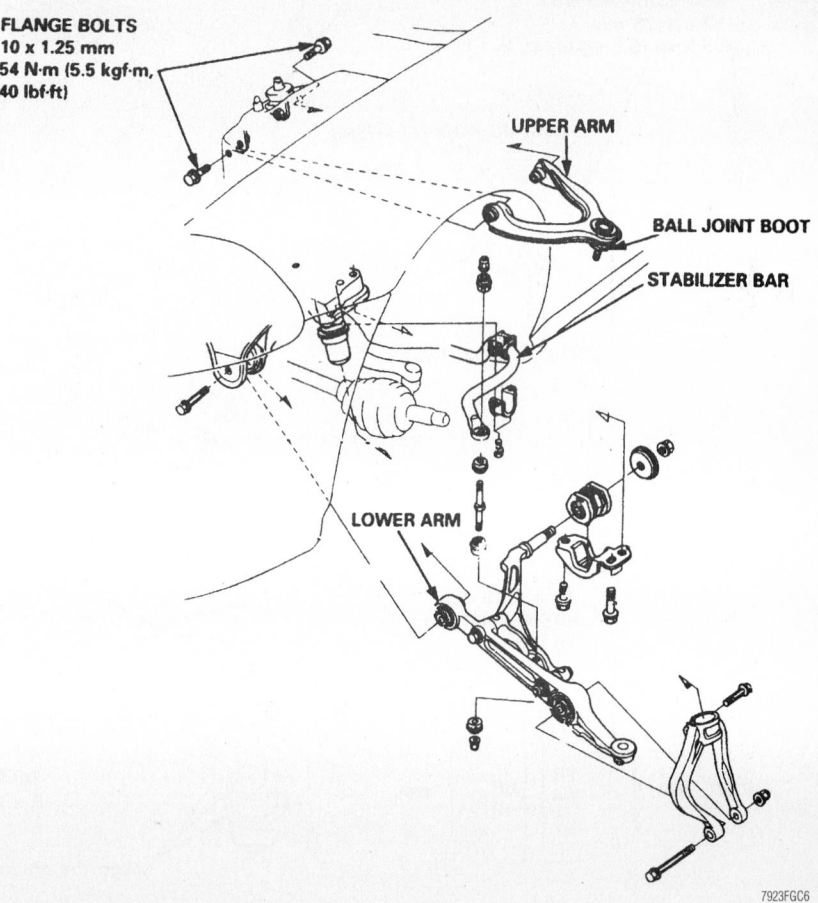

FLANGE BOLTS
10 x 1.25 mm
54 N·m (5.5 kgf·m, 40 lbf·ft)

UPPER ARM

BALL JOINT BOOT

STABILIZER BAR

LOWER ARM

Front suspension components—Civic

7923FGC6

- Front wheels. Lower the vehicle to the ground.
  7. Torque the strut mounting nuts to 36 ft. lbs. (50 Nm).
  8. Torque the upper control arm mounting nuts to 47 ft. lbs. (65 Nm).
  9. Torque the damper fork nut to 47 ft. lbs. (65 Nm).
  10. Torque the upper ball joint castle nut to 29–35 ft. lbs. (40–48 Nm). Then, tighten the nut only enough to install a new cotter pin.
  11. Tighten the wheel nuts to 80 ft. lbs. (108 Nm).
  12. Check the vehicle's front end alignment and adjust it if necessary. Road test the vehicle.

ACCORD AND PRELUDE—2001–02 VEHICLES

➡Do not disassemble the upper arm. If the ball joint or bushings are faulty, or the upper arm is damaged, the entire upper arm must be replaced.

1. Before servicing the vehicle, refer to the precautions in the beginning of this section.
2. Raise and support the vehicle safely.
3. Remove or disconnect the following:
   - Front wheels. Support the lower control arm assembly with a floor jack.
   - Upper ball joint from the steering knuckle using a ball joint separator tool.
   - Self-locking nuts from the upper arm anchor bolts.
   - Upper arm from the vehicle

➡Do not disassemble the upper arm. If the ball joint or bushings are faulty, or the upper arm is damaged, the entire upper arm must be replaced.

To install:

➡Use new self-locking nuts when installing the upper arm and strut.

4. Install or connect the following:
   - Upper control arm assembly into the strut tower.
   - Upper ball joint
   - Front wheels and lower the vehicle.

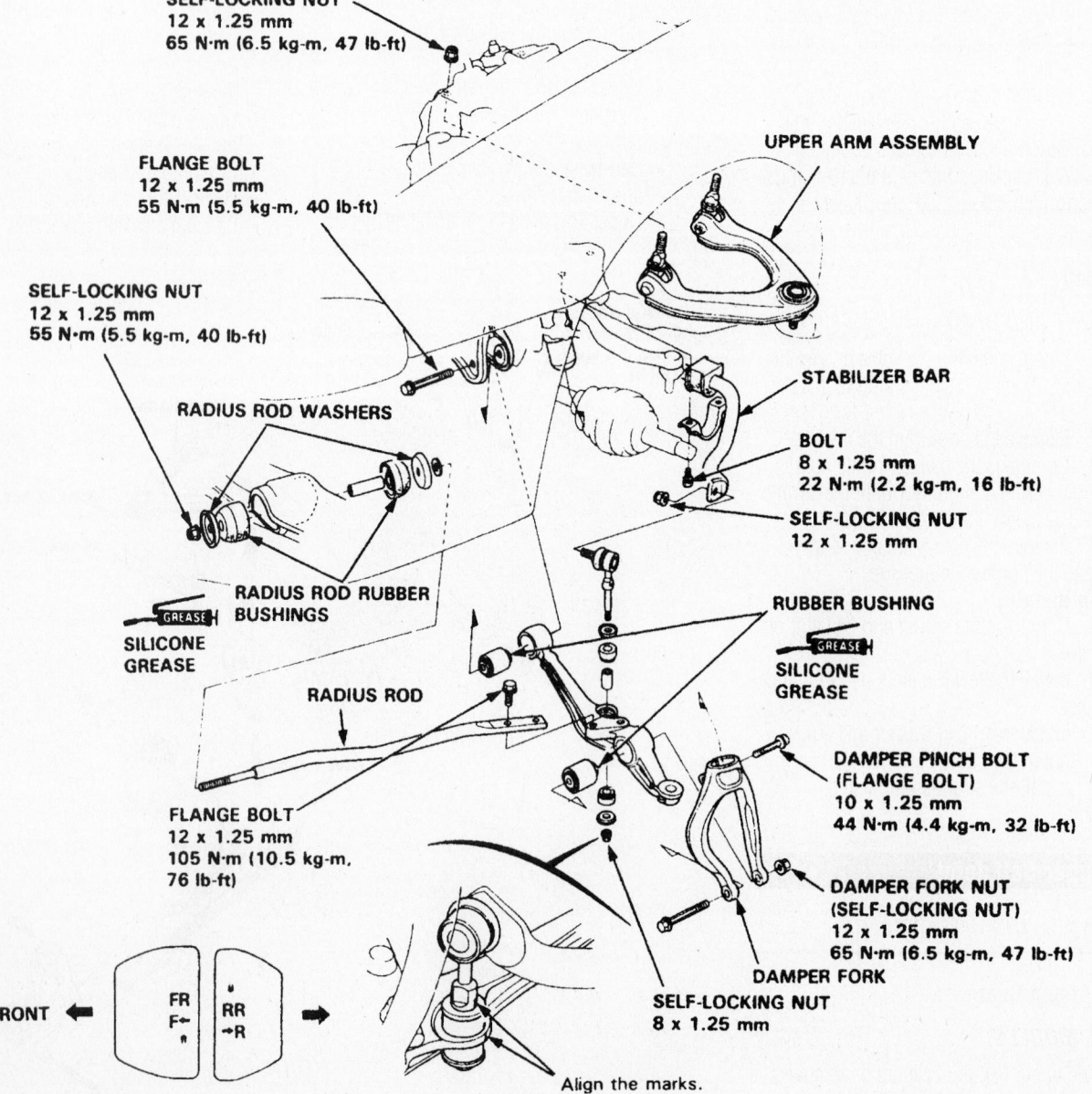

SELF-LOCKING NUT
12 x 1.25 mm
65 N·m (6.5 kg-m, 47 lb-ft)

UPPER ARM ASSEMBLY

FLANGE BOLT
12 x 1.25 mm
55 N·m (5.5 kg-m, 40 lb-ft)

SELF-LOCKING NUT
12 x 1.25 mm
55 N·m (5.5 kg-m, 40 lb-ft)

RADIUS ROD WASHERS

STABILIZER BAR

BOLT
8 x 1.25 mm
22 N·m (2.2 kg-m, 16 lb-ft)

SELF-LOCKING NUT
12 x 1.25 mm

RADIUS ROD RUBBER BUSHINGS

RUBBER BUSHING

GREASE
SILICONE GREASE

GREASE
SILICONE GREASE

RADIUS ROD

DAMPER PINCH BOLT
(FLANGE BOLT)
10 x 1.25 mm
44 N·m (4.4 kg-m, 32 lb-ft)

FLANGE BOLT
12 x 1.25 mm
105 N·m (10.5 kg-m, 76 lb-ft)

DAMPER FORK NUT
(SELF-LOCKING NUT)
12 x 1.25 mm
65 N·m (6.5 kg-m, 47 lb-ft)

DAMPER FORK

SELF-LOCKING NUT
8 x 1.25 mm

FRONT ◀

FR
F←
RR
→R

Align the marks.

7923FGC7

Front suspension components—Prelude and Accord

5. With all 4 of the vehicle's wheels on the ground, torque the upper control arm nuts to 47 ft. lbs. (65 Nm). Torque the castle nut to 32 ft. lbs. (44 Nm); then, only tighten it only enough to install a new cotter pin.

6. Tighten the wheel nuts to 80 ft. lbs. (110 Nm).

7. Check and adjust the vehicle's front end alignment. On Preludes equipped with 4WS, the electronic neutral check must be performed before all 4 wheels are aligned.

ACCORD—2003–04 VEHICLES

1. Before servicing the vehicle, refer to the precautions in the beginning of this section.

2. Remove or disconnect the following:
- Front wheel
- Front strut (damper) assembly
- Wheel Speed Sensor (WSS) from the upper arm
- Cotter pin from the upper arm ball joint, then loosen the nut
- Upper ball joint from the knuckle
- Upper control arm mounting bolts
- Upper control arm

To install:

3. Install or connect the following:
- Upper control arm and loosely install the mounting bolts
- Upper ball joint to the knuckle. Tighten the nut to 29 ft. lbs. (39 Nm), then tighten it up to 35 ft. lbs. (47 Nm) to align the holes to install a new cotter pin. Never loosen the nut to install the cotter pin.

4. To tighten the upper control arm bolts, insert a rod about 6mm in diameter by 300mm long into the positioning holes and place the upper arm on the rock to position it. Tighten the bolts to 23 ft .lbs. (31 Nm).
- WSS to the upper arm
- Front strut (damper) assembly

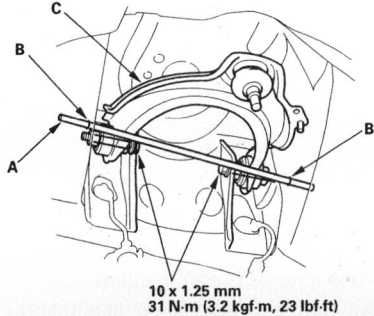

10 x 1.25 mm
31 N·m (3.2 kgf·m, 23 lbf·ft)

42356-ACCO-G39

Insert a rod (A) into the positioning holes (B) and place the upper control arm (C) on the rod to position it before tightening the bolts—2003–04 Accord

- Front wheel

5. Tighten the wheel nuts to 80 ft. lbs. (110 Nm).

6. Check and adjust the vehicle's front end alignment.

S2000

1. Before servicing the vehicle, refer to the precautions in the beginning of this section.

2. Remove or disconnect the following:
- Front wheel
- Wheel speed sensor harness
- Upper ball joint
- Inner flange bolts and the upper control arm

To install:

3. Installation is the reverse of the removal procedure, while using the following torque values:
- Inner flange bolts: 76 ft. lbs. (103 Nm)
- Upper ball joint nut: 36–43 ft. lbs. (49–59 Nm)
- Wheel speed sensor harness bolts: 88 inch lbs. (10 Nm)

Rear

CIVIC

✳✳ CAUTION

Removing rear suspension components may make the vehicle front-heavy and cause it to tip forward when raised on a hoist. Use under-lift support stands, or place additional weight in the trunk of the vehicle before hoisting it.

1. Before servicing the vehicle, refer to the precautions in the beginning of this section.

2. Raise and safely support the vehicle.

3. Remove or disconnect the following:
- Rear wheels

4. Support the lower control arm with a floor jack.

5. Remove or disconnect the following:
- Upper control arm from the trailing arm
- Upper control arm flange bar from its vehicle body mount
- Upper control arm

6. Inspect the upper control arm and bushings for signs of wear and distortion. The bushings are replaced as follows:

a. Press the bushings out of the upper control arm using suitably sized press fixtures.

b. Matchmark the bolt flange bar to the body of the upper control arm.

c. Lubricate the new bushings with silicon grease before installation.

d. Press the new bushings into the control arm. Make sure the bolt flange bar matchmarks align. The leading edges of the control arm bushings must be flush with the edges of the control arm body.

To install:

➡**Use new self-locking nuts and color-coded bolts when assembling suspension components.**

7. Install or connect the following:
- Control arm to its body mount. Hand-tighten the flange bolts.
- Control arm to the trailing arm. Hand-tighten the flange bolt.
- Rear wheel and lower the vehicle.

8. Torque the bolts with the vehicle on the ground. Tighten the control arm bolts-to-body to 29 ft. lbs. (40 Nm). Tighten the control arm-to-trailing arm bolt to 40 ft. lbs. (55 Nm).

9. Check and adjust the vehicle's rear wheel alignment.

10. Tighten the wheel nuts to 80 ft. lbs. (110 Nm).

PRELUDE

1. Before servicing the vehicle, refer to the precautions in the beginning of this section.

2. Raise and support the vehicle safely.

3. Remove or disconnect the following:
- Rear wheels. Support the knuckle and lower control arm assembly with a jack.
- Upper ball joint from the knuckle using a ball joint separator tool.
- Trunk side trim
- 2 strut mounting nuts
- Self-locking nuts from the upper arm anchor bolts
- Upper arm from the vehicle

➡**Do not disassemble the upper arm. If the ball joint or bushings are faulty, or the upper arm is damaged, the entire upper arm must be replaced.**

To install:

➡**Use new self-locking nuts when installing the upper arm and strut.**

4. Install or connect the following:
- Upper control arm assembly into the strut tower.
- Upper ball joint
- Rear wheels and lower the vehicle.

5. With all 4 of the vehicle's wheels on the ground, torque the upper control arm nuts to 47 ft. lbs. (65 Nm). Torque the castle

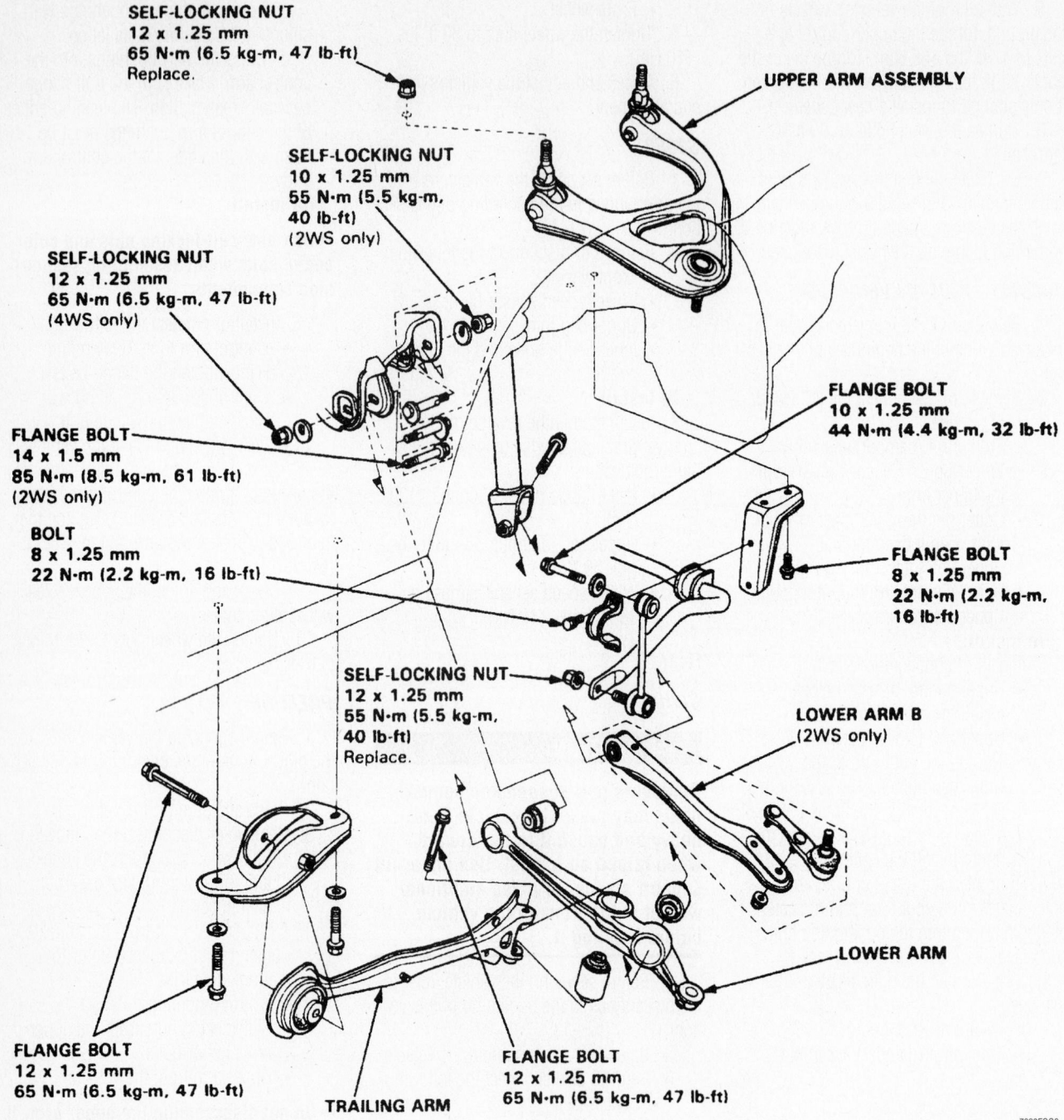

SELF-LOCKING NUT
12 x 1.25 mm
65 N·m (6.5 kg-m, 47 lb-ft)
Replace.

SELF-LOCKING NUT
10 x 1.25 mm
55 N·m (5.5 kg-m,
40 lb-ft)
(2WS only)

SELF-LOCKING NUT
12 x 1.25 mm
65 N·m (6.5 kg-m, 47 lb-ft)
(4WS only)

UPPER ARM ASSEMBLY

FLANGE BOLT
14 x 1.5 mm
85 N·m (8.5 kg-m, 61 lb-ft)
(2WS only)

BOLT
8 x 1.25 mm
22 N·m (2.2 kg-m, 16 lb-ft)

FLANGE BOLT
10 x 1.25 mm
44 N·m (4.4 kg-m, 32 lb-ft)

FLANGE BOLT
8 x 1.25 mm
22 N·m (2.2 kg-m,
16 lb-ft)

SELF-LOCKING NUT
12 x 1.25 mm
55 N·m (5.5 kg-m,
40 lb-ft)
Replace.

LOWER ARM B
(2WS only)

LOWER ARM

FLANGE BOLT
12 x 1.25 mm
65 N·m (6.5 kg-m, 47 lb-ft)

TRAILING ARM

FLANGE BOLT
12 x 1.25 mm
65 N·m (6.5 kg-m, 47 lb-ft)

7923FGC8

Rear suspension components—Prelude

nut to 32 ft. lbs. (44 Nm); then, only tighten it only enough to install a new cotter pin.

6. Tighten the wheel nuts to 80 ft. lbs. (110 Nm).

7. Put the trunk side trim back into position.

8. Check and adjust the vehicle's rear end wheel alignment. On Preludes equipped with 4WS, the electronic neutral check must be performed before all 4 wheels are aligned.

ACCORD

1. Raise and safely support the vehicle.
2. Remove or disconnect the following:
   - Rear wheels
3. Support the knuckle and lower control arm with a floor jack to compress the strut.
4. Remove or disconnect the following:
   - Castle nut cap, cotter pin, and castle nut from the upper ball joint. Use a ball joint separator tool to separate the ball joint from the knuckle.
   - Upper control arm
5. Check upper control arm and bushing for signs of wear and damage. Replace the upper control arm if the ball joint is faulty.

To install:

➡Use new self-locking nuts when assembling suspension components.

6. Install or connect the following:
   - Upper arm into the vehicle
   - Mounting bolts and only hand-tighten them.

- Upper arm to the knuckle.
- Castle nut at the ball joint to 32 ft. lbs. (44 Nm). Tighten the castle nut only enough to install a new cotter pin.
- Castle nut cap
- Rear wheels and lower the vehicle.

7. Tighten the upper mounting bolts to 28 ft. lbs. (39 Nm).

8. Tighten the wheel nuts to 80 ft. lbs. (110 Nm).

9. Check and adjust the vehicle's rear wheel alignment.

S2000

1. Before servicing the vehicle, refer to the precautions in the beginning of this section.

2. Remove or disconnect the following:
- Rear wheel
- Wheel speed sensor harness
- Upper ball joint
- Inner flange bolts and the upper control arm

To install:

3. Installation is the reverse of the removal procedure, while using the following torque values:
- Inner flange bolts: 98 ft. lbs. (132 Nm)
- Upper ball joint nut: 36–43 ft. lbs. (49–59 Nm)
- Wheel speed sensor harness bolts: 88 inch lbs. (10 Nm)

Lower Ball Joint

REMOVAL & INSTALLATION

Civic

➡ **The steering knuckle must be removed from the vehicle for the ball joint to be replaced. The following special tools or their equivalents are needed to press the ball joint in and out of the knuckle: ball joint installer base tool 07965-SB00200, ball joint installer/remover tool 07965-SB00100, and ball joint remover base tool 07965-SH20200. A large vise will be required to hold the knuckle and the press tools. A ball joint clip guide tool 07974-SA50700 or 07GAG-SD40700 is used to install the retaining clip on the joint boot.**

1. Before servicing the vehicle, refer to the precautions in the beginning of this section.

2. Remove or disconnect the following:
- Steering knuckle assembly from the vehicle.

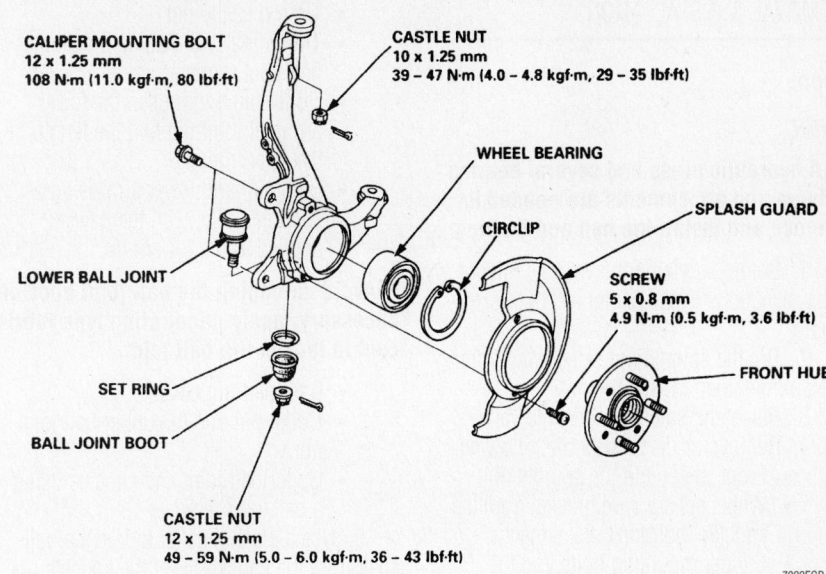

CALIPER MOUNTING BOLT
12 x 1.25 mm
108 N·m (11.0 kgf·m, 80 lbf·ft)

CASTLE NUT
10 x 1.25 mm
39 – 47 N·m (4.0 – 4.8 kgf·m, 29 – 35 lbf·ft)

WHEEL BEARING

CIRCLIP

SPLASH GUARD

SCREW
5 x 0.8 mm
4.9 N·m (0.5 kgf·m, 3.6 lbf·ft)

FRONT HUB

LOWER BALL JOINT

SET RING

BALL JOINT BOOT

CASTLE NUT
12 x 1.25 mm
49 – 59 N·m (5.0 – 6.0 kgf·m, 36 – 43 lbf·ft)

7923FGB4

Knuckle components—Civic

- Ball joint boot snapring and the boot.
- Snapring out of the groove in the ball joint body.

3. Install the ball joint removal tool onto the ball joint with the large end facing out. Install the ball joint nut to attach the tool to the joint.

4. Position the removal base tool on the ball joint and set the assembly in a large vise. Press the ball joint out of the steering knuckle.

To install:

5. Install or connect the following:
- New ball joint into the hole of the steering knuckle.
- Ball joint installer tool over the ball joint with the small end facing out.
- Installation base tool on the ball joint and set the assembly in a large vise. Press the ball joint into the steering knuckle.

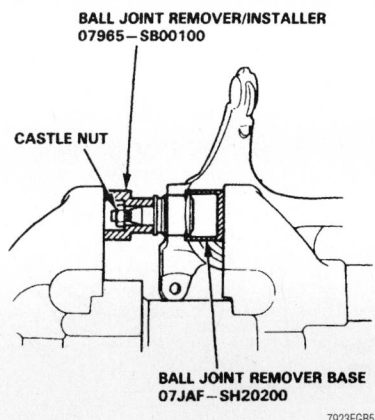

BALL JOINT REMOVER/INSTALLER
07965 — SB00100

CASTLE NUT

BALL JOINT REMOVER BASE
07JAF — SH20200

7923FGB5

Ball joint removal tools—Civic

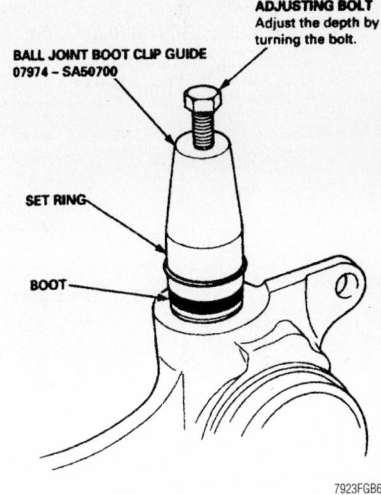

ADJUSTING BOLT
Adjust the depth by turning the bolt.

BALL JOINT BOOT CLIP GUIDE
07974 – SA50700

SET RING

BOOT

7923FGB6

Ball joint boot clip guide—Civic

- Snapring in the groove of the ball joint.

6. Adjust the boot clip tool with the adjusting bolt until the end of the tool aligns with the groove on the boot. Slide the clip over the tool and into position.

7. Install the ball joint stud in the steering knuckle. Tighten the nut to 44 ft. lbs. (60 Nm).

Wheel Bearings

ADJUSTMENT

The wheel bearings are not adjustable or repairable and should be replaced if found defective.

REMOVAL & INSTALLATION

Front

CIVIC

➡ **A hydraulic press and several bearing drivers and attachments are needed to remove and install the hub and bearing.**

1. Before servicing the vehicle, refer to the precautions in the beginning of this section.

2. Pry the spindle nut stake away from the spindle, then loosen the nut.

3. Raise and safely support the vehicle.

4. Remove or disconnect the following:
- Front wheel and the spindle nut
- Wheel sensor wire bracket from the knuckle, but don't disconnect it.
- Caliper mounting bolts and the caliper. Support the caliper out of the way with a length of wire. Do not let the caliper hang from the brake hose.
- 6mm brake disc retaining screws. Screw 2, 12mm bolts into the disc to push it away from the hub.

- Tie rod castle nut
- Tie rod ball joint using a suitable ball joint remover.
- Cotter pin and loosen the lower arm ball joint nut half the length of the joint threads.
- Ball joint and lower arm using a suitable puller with the pawls applied to the lower arm.

➡ **Avoid damaging the ball joint boot. If necessary, apply penetrating type lubricant to loosen the ball joint.**

- Ball joint nut cover
- Cotter pin and the upper ball joint nut.
- Upper ball joint and knuckle using a ball joint remover.

5. Use a plastic mallet to free the half-shaft from the knuckle. Pull the knuckle out to remove it.

➡ **A new wheel bearing must be used when the hub is removed.**

6. Place the knuckle in a press and use a base and pilot to press the hub assembly out of the wheel bearing.

7. Remove the knuckle ring seal and circlip. Remove the splash guard from the knuckle.

8. Press the wheel bearing out of the knuckle using a driving attachment.

To install:

9. Clean the knuckle and hub assembly and inspect them for damage.

10. Install or connect the following:
- New wheel bearing into the hub using a driving tool.
- Circlip in the outer groove of the knuckle.
- Splash guard
- Hub assembly into the steering knuckle using a base and a driving and guide tool.
- Knuckle ring seal
- Knuckle onto the spindle
- Knuckle onto the upper and lower ball joints and tighten the castle nuts.
- Tie rod ball joint onto the steering knuckle.

11. Tighten the upper ball joint nut and tie rod nut to 29–35 ft. lbs. (40–48 Nm) and

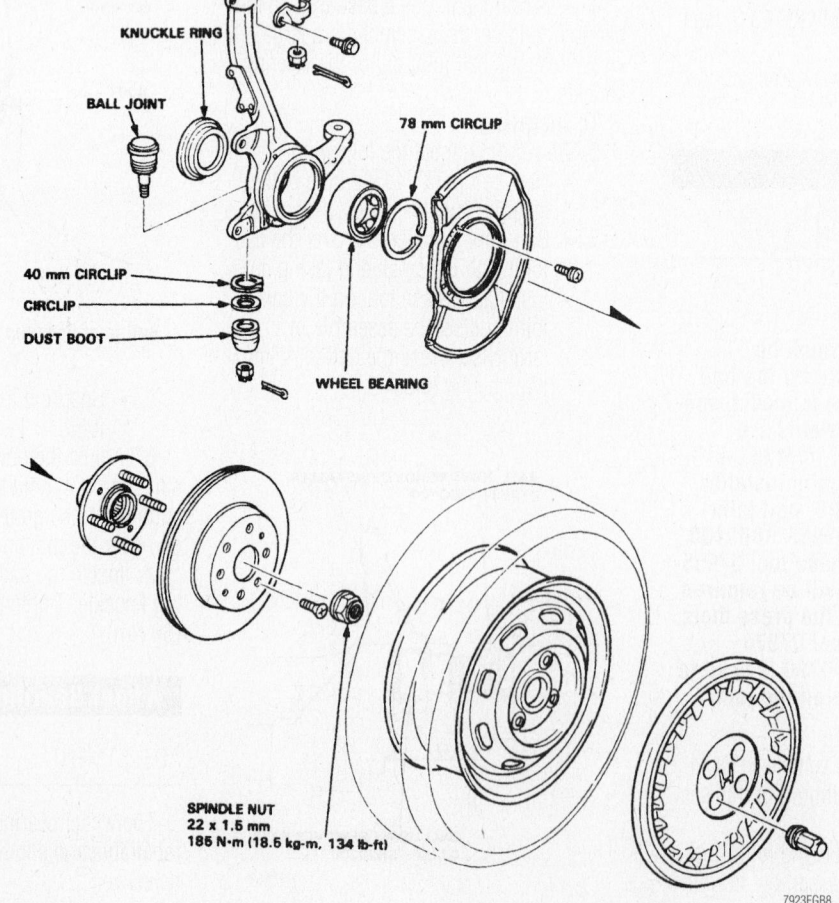

Wheel hub, bearing and steering knuckle components—Prelude and 2001–02 Accord

the lower ball joint castle nut to 36–43 ft. lbs. (50–60 Nm).

12. Install or connect the following:

- Anti-lock Brake System (ABS) wheel sensor wire brackets onto the knuckle. Tighten the mounting bolts to 84 inch lbs. (10 Nm).
- Brake disc; use 2 lug nuts to evenly draw the disc onto the hub.
- Retainer screws: 84 inch lbs. (10 Nm).
- Spindle washer and nut. Don't tighten the nut until the vehicle is on the ground.
- Brake caliper and tighten the bolts to 80 ft. lbs. (110 Nm).

- Front wheels and lower the vehicle.

13. Tighten the spindle nut to 134 ft. lbs. (185 Nm), stake the nut, and install the grease cap.

14. Check and adjust the vehicle's front wheel alignment.

➡ Avoid damaging the ball joint boot. If necessary, apply penetrating-type lubricant to loosen the ball joint.

PRELUDE AND ACCORD

➡ Once the hub has been removed, the wheel bearings must be replaced. A hydraulic press and bearing drivers must be used to remove and install the bearing.

1. Before servicing the vehicle, refer to the precautions in the beginning of this section.

2. Pry the spindle nut stake away from the spindle and loosen the nut. Do not tighten or loosen a spindle nut unless the vehicle is sitting on all 4 wheels. The torque required is high enough to cause the vehicle to fall off the stands even when properly supported.

3. Raise and safely support the vehicle.

4. Remove or disconnect the following:

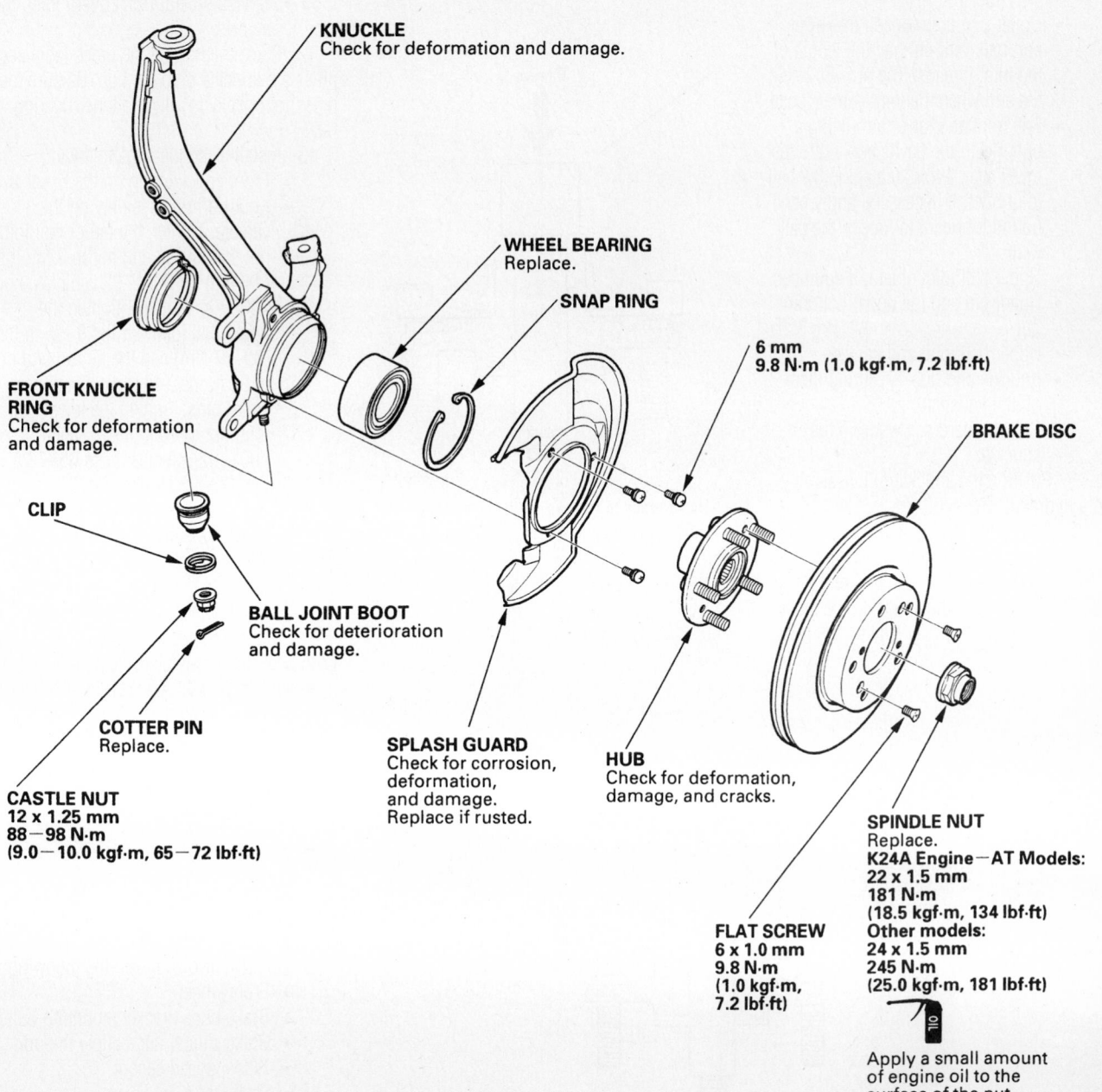

KNUCKLE
Check for deformation and damage.

WHEEL BEARING
Replace.

SNAP RING

6 mm
9.8 N·m (1.0 kgf·m, 7.2 lbf·ft)

BRAKE DISC

FRONT KNUCKLE RING
Check for deformation and damage.

CLIP

BALL JOINT BOOT
Check for deterioration and damage.

COTTER PIN
Replace.

CASTLE NUT
12 x 1.25 mm
88–98 N·m
(9.0–10.0 kgf·m, 65–72 lbf·ft)

SPLASH GUARD
Check for corrosion, deformation, and damage. Replace if rusted.

HUB
Check for deformation, damage, and cracks.

FLAT SCREW
6 x 1.0 mm
9.8 N·m
(1.0 kgf·m, 7.2 lbf·ft)

SPINDLE NUT
Replace.
K24A Engine—AT Models:
22 x 1.5 mm
181 N·m
(18.5 kgf·m, 134 lbf·ft)
Other models:
24 x 1.5 mm
245 N·m
(25.0 kgf·m, 181 lbf·ft)

Apply a small amount of engine oil to the surface of the nut.

42356-ACCO-G40

Wheel hub, bearing and steering knuckle components—2003–04 Accord

- Wheel and the spindle nut
- Caliper mounting bolts and the caliper. Support the caliper out of the way with a length of wire. Do not let the caliper hang from the brake hose.
- 6mm brake disc retaining screws. Screw 2, 8 x 1.25mm bolts into the disc to push it away from the hub.

➡**Turn each bolt 2 turns at a time to prevent cocking the brake disc.**

- Cotter pin from the tie rod castle nut, then the nut.
- Tie rod ball joint using a ball joint remover, then lift the tie rod out of the knuckle.
- Cotter pin, then loosen the lower arm ball joint nut half the length of the joint threads. The nut will retain the arm when the joint comes loose.
- Ball joint and lower arm using a puller with the pawls applied to the lower arm. Avoid damaging the ball joint boot. If necessary, apply penetrating lubricant to loosen the ball joint.
- Upper ball joint shield, if equipped.
- Cotter pin and the upper ball joint nut.
- Upper ball joint and knuckle
- Knuckle and hub by sliding them off the halfshaft.
- Splash guard screws from the knuckle.

5. Position the knuckle/hub assembly in a hydraulic press.

6. Remove or disconnect the following:
- Hub from the knuckle using a driver of the proper diameter while supporting the knuckle. The inner bearing race may stay on the hub.

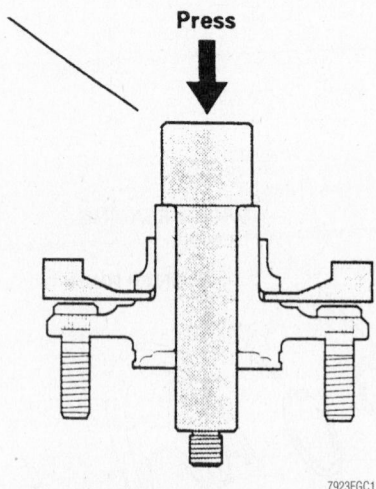

Press the hub out of the knuckle—Prelude and Accord

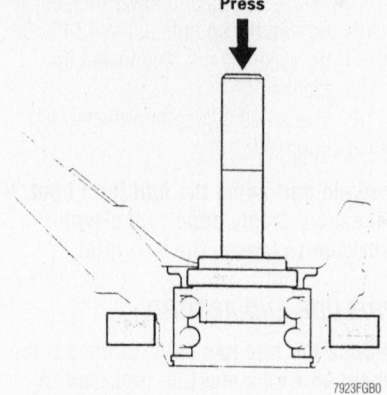

Press the bearing out of the knuckle—Prelude and Accord

Use a press to remove the inner bearing race from the hub—Prelude and Accord

- Splash guard and snapring from the knuckle.

7. Press the wheel bearing out of the knuckle while supporting the knuckle.

8. If necessary, remove the outboard bearing inner race from the hub using a bearing puller.

To install:

9. Clean the knuckle and hub thoroughly.

10. Press a new wheel bearing into the knuckle. Be sure the press tool contacts only the outer bearing race and properly support the knuckle so it is stable.

11. Install or connect the following:
- Snapring
- Splash shield. Don't overtighten the screws.

12. Place the hub on the press table and press the knuckle onto the hub. Be sure the press tool contacts only the inner bearing race.

13. Install or connect the following:
- Front knuckle ring on the knuckle
- Knuckle/hub assembly on the vehicle. Tighten the upper ball joint nut and tie rod end nut to 32 ft. lbs. (44 Nm) for 2001–02 vehicles. For 2003–04 vehicles, tighten the upper ball joint nut to 29–35 ft. lbs. (39–47 Nm) and the tie rod end nut to 32 ft. lbs. (44 Nm). Install new cotter pins. Tighten the lower ball joint nut to 40 ft. lbs. (55 Nm) for 2001–02 vehicles, or to 65–72 ft. lbs. (88–98 Nm) for 2003–04 vehicles and install a new cotter pin.
- Brake disc and caliper. Tighten the caliper bracket bolts to 80 ft. lbs. (110 Nm).
- Front wheels and lower the vehicle.

14. Tighten the spindle nut to 181 ft. lbs. (245 Nm) for all models except 2003–04 4-cylinder Accord with A/T. For the 2003–04 4-cylinder A/T Accord, tighten the spindle nut to 134 ft. lbs. (181 Nm). Tighten the wheel nuts to 80 ft. lbs. (110 Nm).

15. Check and adjust the vehicle's front wheel alignment.

S2000

1. Before servicing the vehicle, refer to the precautions in the beginning of this section.

2. Remove or disconnect the following:
- Front wheel
- Brake hose bracket mounting bolts
- Brake caliper and caliper support
- Wheel speed sensor
- Brake rotor
- Outer tie rod end
- Upper and lower ball joints

- Steering knuckle from the vehicle
- Dust cover
- Spindle nut
- Wheel speed pulse ring

3. Mount the steering knuckle in a press and press the hub out of the wheel bearing.

4. Remove the splash guard and the wheel bearing snapring.

5. Press the wheel bearing out of the steering knuckle.

To install:

6. Installation is the reverse of the removal procedure, while using the following torque values:

- Splash guard screws: 48 inch lbs. (5 Nm)
- Spindle nut: 242 ft. lbs. (329 Nm)
- Upper ball joint nut: 36–43 ft. lbs. (49–59 Nm)
- Lower ball joint nut: 43–51 ft. lbs. (56–69 Nm)
- Outer tie rod end nut: 40 ft. lbs. (54 Nm)
- Brake caliper support bolts: 83 ft. lbs. (113 Nm)

Rear

CIVIC

1. Before servicing the vehicle, refer to the precautions in the beginning of this section.

2. Remove or disconnect the following:

- Hub dust cap and loosen the spindle nut.

3. Raise and safely support the vehicle.

4. Remove or disconnect the following:

- Rear wheels.
- 2 brake rotor or drum retaining screws
- Brake drum, if equipped with drum brakes.

5. If equipped with disc brakes, remove or disconnect the following:

- Caliper shield and brake hose bracket
- Caliper bracket and hang the caliper out of the way with a piece of wire.
- Brake rotor

6. Remove or disconnect the following:

- Hub assembly from the spindle

7. Clean the hub assembly in solvent.

8. Inspect the hub assembly for any signs of wear or damage. If the wheel bearings are damaged, the hub assembly must be replaced.

To install:

9. Clean the spindle and the brake rotor/drum mounting surfaces.

10. Install or connect the following:

- Hub assembly onto the spindle
- Spindle washer
- Brake rotor or brake drum. Apply anti-seize paste to the retaining screws and tighten them to 84 inch

lbs. (10 Nm). Don't overtighten the retaining screws.

11. If equipped with disc brakes, install or connect the following:

- Brake caliper and tighten the mounting bolts to 28 ft. lbs. (39 Nm).
- Brake hose bracket onto its mount.
- Caliper dust shield and tighten the bolts to 84 inch lbs. (10 Nm).
- New spindle nut and wheel assembly.

12. Lower the vehicle.

13. Tighten the spindle nut to 134 ft. lbs. (185 Nm). Tighten the wheel nuts to 80 ft. lbs. (110 Nm). Stake the spindle nut with a punch. If the dust cap was bent during removal, install a new one.

PRELUDE AND ACCORD

➡The rear wheel bearing and hub unit are replaced as a unit.

1. Before servicing the vehicle, refer to the precautions in the beginning of this section.

2. Loosen the spindle nut.

3. Raise the vehicle and support it safely.

4. Remove or disconnect the following:

- Rear wheels
- Brake disc retaining screws
- Brake hose brackets from the knuckle

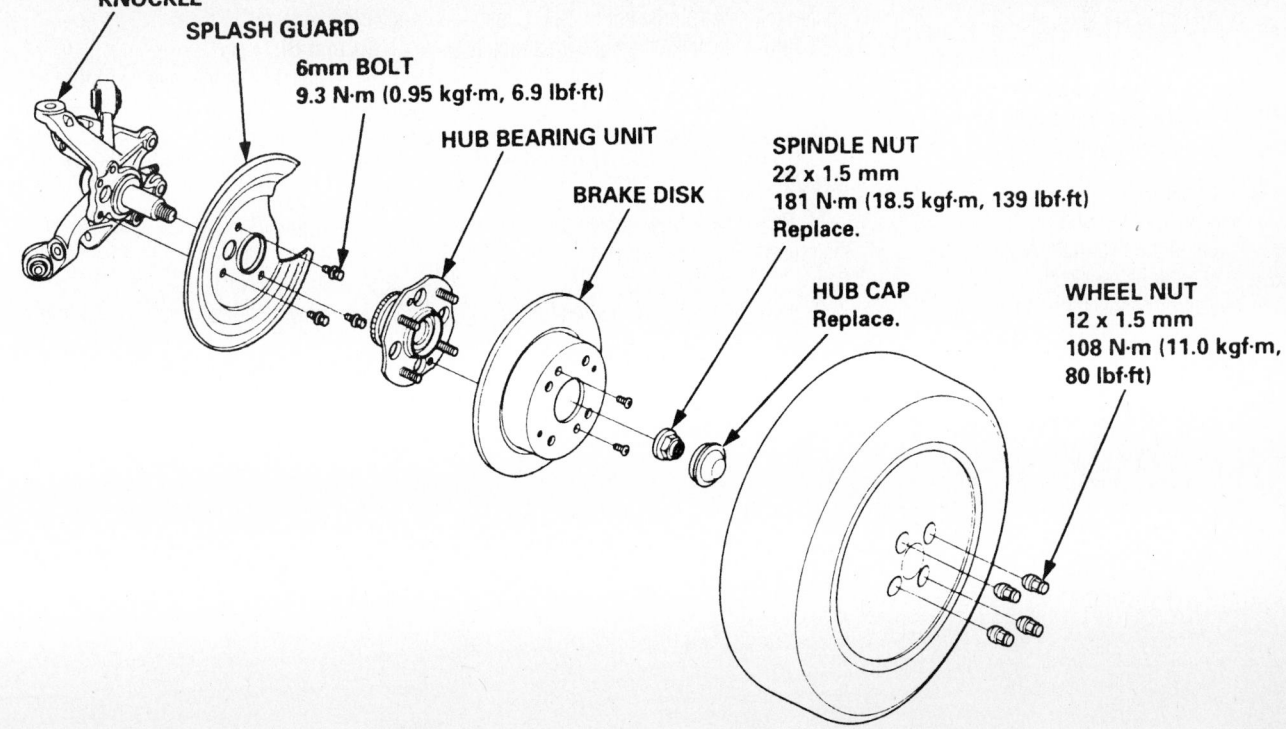

KNUCKLE

SPLASH GUARD

6mm BOLT
9.3 N·m (0.95 kgf·m, 6.9 lbf·ft)

HUB BEARING UNIT

BRAKE DISK

SPINDLE NUT
22 x 1.5 mm
181 N·m (18.5 kgf·m, 139 lbf·ft)
Replace.

HUB CAP
Replace.

WHEEL NUT
12 x 1.5 mm
108 N·m (11.0 kgf·m, 80 lbf·ft)

Exploded view of the hub unit, drum brakes—Accord

7923FGC2

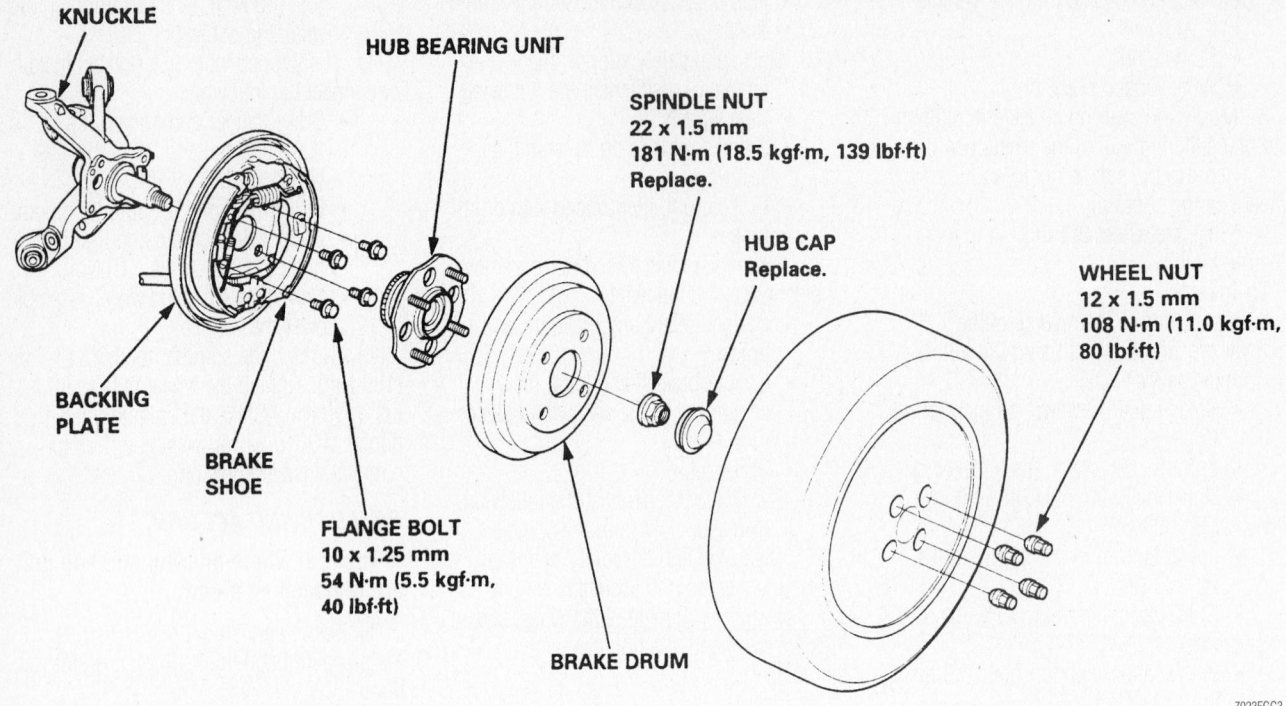

KNUCKLE

HUB BEARING UNIT

SPINDLE NUT
22 x 1.5 mm
181 N·m (18.5 kgf·m, 139 lbf·ft)
Replace.

HUB CAP
Replace.

WHEEL NUT
12 x 1.5 mm
108 N·m (11.0 kgf·m, 80 lbf·ft)

BACKING
PLATE

BRAKE
SHOE

FLANGE BOLT
10 x 1.25 mm
54 N·m (5.5 kgf·m, 40 lbf·ft)

BRAKE DRUM

7923FGC3

Hub unit, disc brakes—Accord and Prelude

- Caliper bracket mounting bolts and hang the caliper out of the way with a piece of wire.
- Brake disc. If the disc is frozen on the hub, screw 2, 8 x 1.25mm bolts evenly into the disc to push it away from the hub.
- Spindle nut and pull the hub unit off of the spindle.

➥**Clean the backing plate and the mating surfaces of the brake disc and hub with brake cleaner. Clean the spindle, washer, and hub with solvent.**

To install:

5. Inspect the hub unit for signs of damage or wear. If the bearings are worn, the entire unit must be replaced.

6. Install or connect the following:
- Hub unit and spindle washer onto the spindle.
- Spindle nut but do not tighten it.
- Brake disc and tighten the retaining screws to 84 inch lbs. (10 Nm).

- Brake caliper and tighten the mounting bolts to 28 ft. lbs. (39 Nm).
- Brake hose brackets onto the knuckle and tighten the bolts to 16 ft. lbs. (22 Nm).
- Rear wheels and lower the vehicle.

7. With the vehicle on the ground, tighten the new spindle nut to 134 ft. lbs. (185 Nm), then stake the nut with a punch.

8. Tighten the wheel nuts to 80 ft. lbs. (110 Nm).

9. Test the operation of the brakes.

S2000

1. Before servicing the vehicle, refer to the precautions in the beginning of this section.

2. Remove or disconnect the following:
- Rear wheel
- Brake caliper support bracket
- Wheel speed sensor
- Spindle nut
- Brake rotor

- Control arm
- Upper and lower ball joints
- Spindle from the vehicle

3. Mount the steering knuckle in a press and press the hub out of the wheel bearing.

4. Remove the splash guard and the wheel bearing snapring.

5. Press the wheel bearing out of the steering knuckle.

To install:

6. Installation is the reverse of the removal procedure, while using the following torque values:
- Splash guard screws: 48 inch lbs. (5 Nm)
- Spindle nut: 181 ft. lbs. (245 Nm)
- Upper ball joint nut: 36–43 ft. lbs. (49–59 Nm)
- Lower ball joint nut: 51–58 ft. lbs. (68–78 Nm)
- Control arm ball joint nut: 36–43 ft. lbs. (49–59 Nm)
- Brake caliper support bolts: 41 ft. lbs. (55 Nm)

BRAKES

Brake Caliper

REMOVAL & INSTALLATION

Civic

FRONT

➡ **Two distinct types of front calipers are used on these vehicles. The caliper body will be marked either 5410 or 2056 depending on type. Servicing procedures are similar, but the different calipers use different pads.**

1. Before servicing the vehicle, refer to the precautions in the beginning of this section.

2. Remove or disconnect the following:
   • Front wheels
   • Banjo bolt and brake hose from the caliper
   • Mounting bolts, and then the caliper from its mounting bracket
   • Caliper mounting bracket from the steering knuckle, if necessary for servicing

To install:
3. Install or connect the following:
   • Caliper mounting bracket was removed, if removed
   • Caliper pins with new pin boots, after applying brake seal grease to the pins and
   • Apply anti-seize paste to the caliper

mounting bolts and tighten them to 80 ft. lbs. (108 Nm)
   • Brake pads
   • Caliper over the pads and onto its mounting bracket

4. On vehicles equipped with type 2056 calipers, torque the top caliper bolt to 25 ft. lbs. (35 Nm). Torque the lower bolt to 20–24 ft. lbs. (27–32 Nm).

5. On vehicles equipped with type 5410 calipers, torque both caliper bolts to 24 ft. lbs. (33 Nm).

6. Install or connect the following:
   • Brake hose to the caliper using new sealing washers. Carefully torque the banjo bolt to 25 ft. lbs. (35 Nm).

• Coat the piston, piston seal, and caliper bore with clean brake fluid.
• Replace all rubber parts with new ones whenever disassembled.

⎓ GREASE : Use recommended rubber grease in the caliper seal set.

⎓ GREASE : Use recommended seal grease in the caliper seal set.

BRAKE PADS

WEAR INDICATOR
Install inner pad with its wear indicator upward.

OUTER PAD SHIM

PIN BOOTS
Replace.

CALIPER PIN B

CALIPER PIN A

CALIPER BRACKET MOUNTING BOLT
108 N·m (11.0 kgf·m, 80 lbf·ft)

CALIPER BRACKET

PAD RETAINERS

CALIPER BOLTS
32 N·m (3.3 kgf·m, 24 lbf·ft)

BLEED SCREW
9 N·m (0.9 kgf·m, 6.5 lbf·ft)

CALIPER BODY

PISTON

PISTON SEAL
Replace.

PISTON BOOT
Replace.

BOOT CLIP
Replace.

93016G05

Exploded view of the front brakes—Honda 5410 Type shown

- Reservoir with fresh brake fluid and bleed the brake system
- Front wheels

REAR

1. Before servicing the vehicle, refer to the precautions in the beginning of this section.

2. Remove or disconnect the following:
- Rear wheels
- Caliper shield
- Lock pin and clevis pin from the parking brake cable
- Cable securing clip
- Parking brake cable from the caliper
- Banjo bolt and the brake hose from the caliper
- 2 caliper mounting bolts
- Caliper from its mounting bracket
- Caliper mounting bracket from the

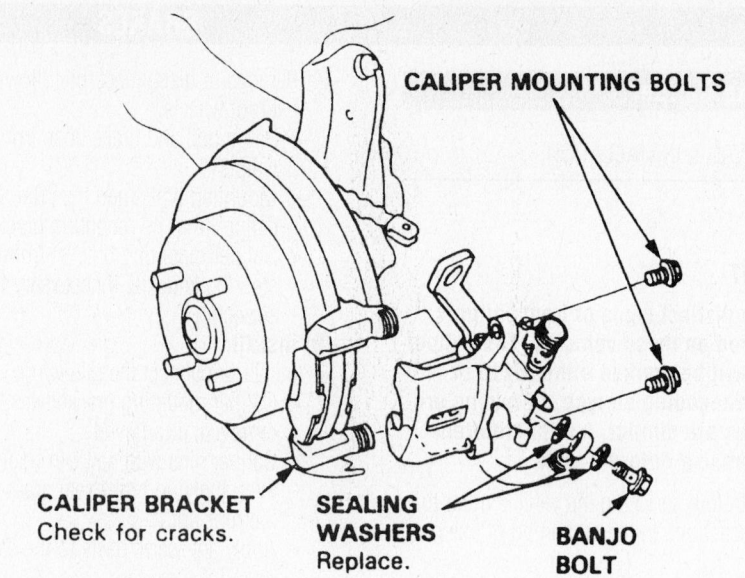

CALIPER MOUNTING BOLTS

CALIPER BRACKET
Check for cracks.

SEALING WASHERS
Replace.

BANJO BOLT

93016G06

Rear brake caliper mounting—Civic

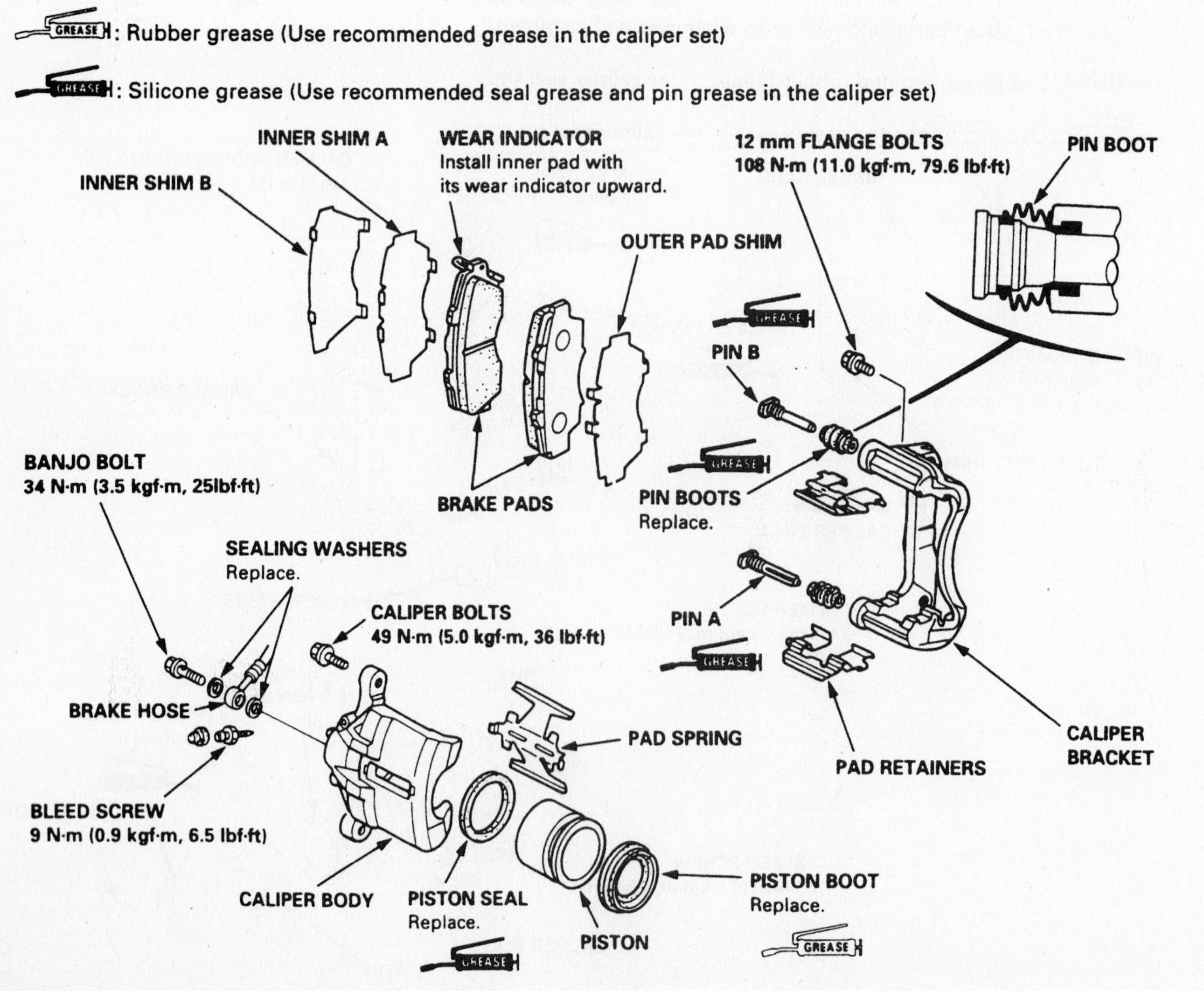

GREASE : Rubber grease (Use recommended grease in the caliper set)

GREASE : Silicone grease (Use recommended seal grease and pin grease in the caliper set)

INNER SHIM A

INNER SHIM B

WEAR INDICATOR
Install inner pad with its wear indicator upward.

OUTER PAD SHIM

12 mm FLANGE BOLTS
108 N·m (11.0 kgf·m, 79.6 lbf·ft)

PIN BOOT

PIN B

BRAKE PADS

PIN BOOTS
Replace.

BANJO BOLT
34 N·m (3.5 kgf·m, 25 lbf·ft)

SEALING WASHERS
Replace.

CALIPER BOLTS
49 N·m (5.0 kgf·m, 36 lbf·ft)

PIN A

BRAKE HOSE

BLEED SCREW
9 N·m (0.9 kgf·m, 6.5 lbf·ft)

PAD SPRING

PAD RETAINERS

CALIPER BRACKET

CALIPER BODY

PISTON SEAL
Replace.

PISTON

PISTON BOOT
Replace.

93016G07

Exploded view of the front brakes—Accord V6 shown

Sedan:

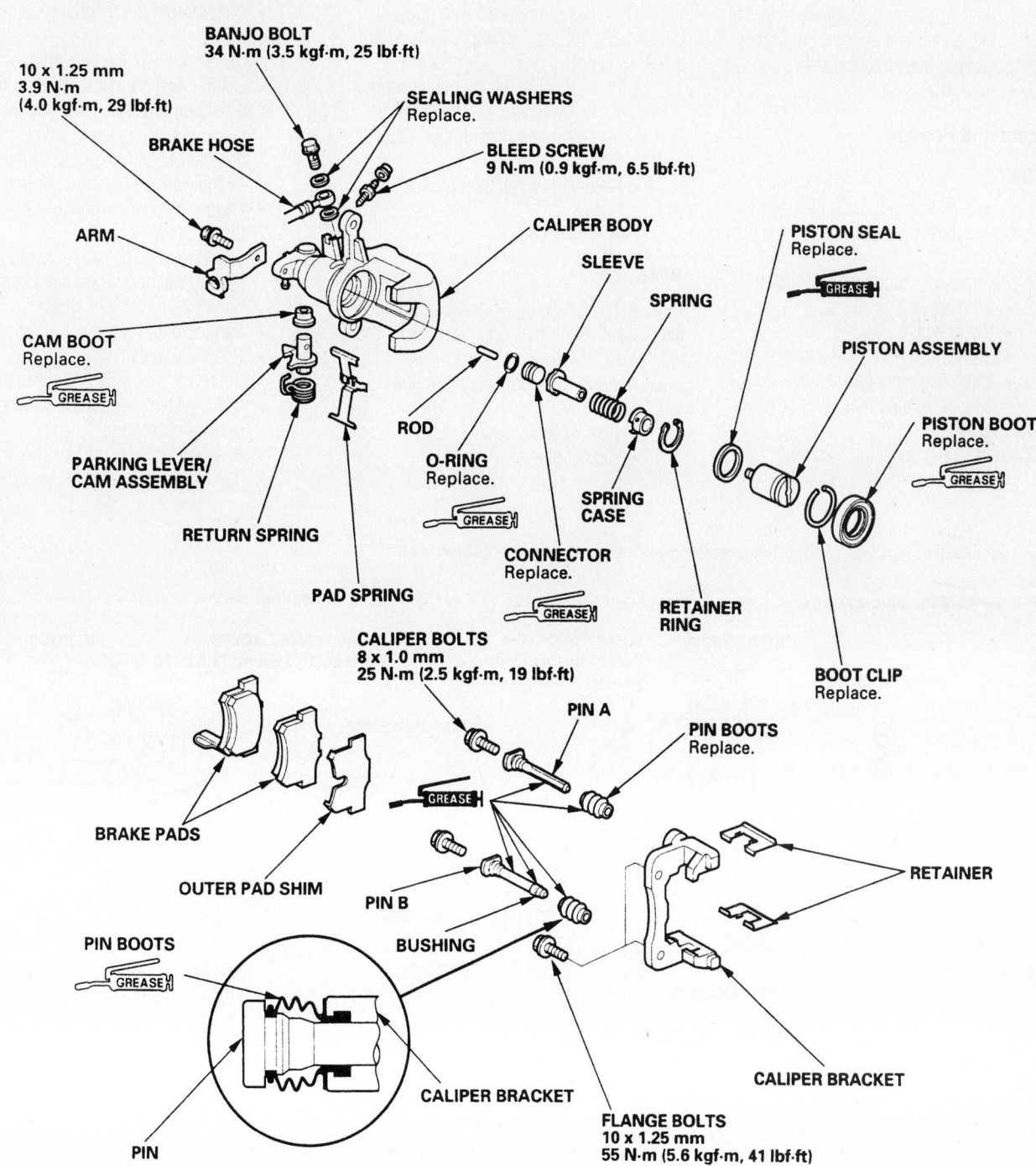

: Silicone grease

: Rubber grease

BANJO BOLT
34 N·m (3.5 kgf·m, 25 lbf·ft)

10 x 1.25 mm
3.9 N·m
(4.0 kgf·m, 29 lbf·ft)

SEALING WASHERS
Replace.

BRAKE HOSE

BLEED SCREW
9 N·m (0.9 kgf·m, 6.5 lbf·ft)

ARM

CALIPER BODY

SLEEVE

SPRING

PISTON SEAL
Replace.

PISTON ASSEMBLY

CAM BOOT
Replace.

ROD

O-RING
Replace.

SPRING CASE

PISTON BOOT
Replace.

**PARKING LEVER/
CAM ASSEMBLY**

CONNECTOR
Replace.

RETURN SPRING

RETAINER RING

BOOT CLIP
Replace.

PAD SPRING

CALIPER BOLTS
8 x 1.0 mm
25 N·m (2.5 kgf·m, 19 lbf·ft)

PIN A

PIN BOOTS
Replace.

BRAKE PADS

OUTER PAD SHIM

PIN B

RETAINER

PIN BOOTS

BUSHING

PIN

CALIPER BRACKET

CALIPER BRACKET

FLANGE BOLTS
10 x 1.25 mm
55 N·m (5.6 kgf·m, 41 lbf·ft)

93016G08

Rear disc brakes—Accord Sedan shown

trailing arm, if necessary for servicing

To install:
3. Install or connect the following:
   • Caliper mounting bracket, if removed.

• Caliper pins with new pin boots, after applying brake seal grease to the pins
• Anti-seize paste to the caliper bracket mounting bolts and tighten them to 28 ft. lbs. (39 Nm)

• Brake pads
4. Rotate the caliper piston clockwise into place in the cylinder and then align the groove in the piston with the tab on inner pad.
• Caliper over the pads and onto its

mounting bracket. Tighten the caliper bolts to 17 ft. lbs. (23 Nm).
- Parking brake cable, after greasing the parking brake linkage
- Caliper shield
- Reservoir with fresh brake fluid and bleed the brake system. Adjust the parking brake if necessary.
- Rear wheels

Accord and Prelude

FRONT

1. Before servicing the vehicle, refer to the precautions in the beginning of this section.
2. Remove or disconnect the following:
   - Front wheels
   - Banjo bolt
   - Brake hose from the caliper
   - Mounting bolts and remove the caliper from its mounting bracket

To install:
3. Install or connect the following:

- Caliper over the pads and onto its mounting bracket
4. On vehicles equipped with long caliper pins, torque the top caliper bolts to 54 ft. lbs. (74 Nm).
5. On vehicles equipped with short caliper bolts, torque the caliper bolts to 36 ft. lbs. (50 Nm).
   - Brake hose to the caliper using new sealing washers. Carefully torque the banjo bolt to 25 ft. lbs. (35 Nm).
   - Reservoir with fluid and bleed the brakes
   - Front wheels

REAR

1. Before servicing the vehicle, refer to the precautions in the beginning of this section.
2. Remove or disconnect the following:
   - Rear wheels
   - Caliper shield

- Parking brake cable from the caliper
- Banjo bolt
- Brake hose from the caliper
- 2 caliper mounting bolts
- Caliper from the mounting bracket

To install:
3. Install or connect the following:
   - Caliper over the pads and onto the mounting bracket. Rotate the piston clockwise into place in the cylinder and then align the groove in the piston with the tab on inner pad.
   - Caliper bolts and tighten to 17 ft. lbs. (23 Nm)
   - Brake hose to the caliper using new sealing washers. Torque the banjo bolt to 25 ft. lbs. (35 Nm).
   - Parking brake cable
   - Caliper shield
   - Reservoir with fluid and bleed the brake system. Adjust the parking brake if necessary.
   - Rear wheels

GREASE: Rubber grease (Use recommended grease in the caliper set)

GREASE: Silicone grease (Use recommended seal grease and pin grease in the caliper set)

INNER SHIM A

INNER SHIM B

WEAR INDICATOR
Install inner pad with its wear indicator upward.

12 mm FLANGE BOLTS
108 N·m (11.0 kgf·m, 79.6 lbf·ft)

PIN BOOT

OUTER PAD SHIM

PIN B

BRAKE PADS

PIN BOOTS
Replace.

BANJO BOLT
34 N·m (3.5 kgf·m, 25 lbf·ft)

SEALING WASHERS
Replace.

CALIPER BOLTS
49 N·m (5.0 kgf·m, 36 lbf·ft)

PIN A

CALIPER BRACKET

BRAKE HOSE

PAD SPRING

PAD RETAINERS

BLEED SCREW
9 N·m (0.9 kgf·m, 6.5 lbf·ft)

CALIPER BODY

PISTON SEAL
Replace.

PISTON

PISTON BOOT
Replace.

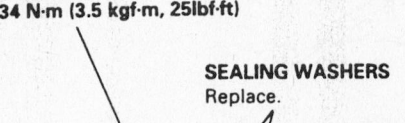

93016G07

Exploded view of the front brakes—Accord V6 shown

A
34 N·m
(3.5 kgf·m,
25 lbf·ft)

E
23 N·m (2.3 kgf·m, 17 lbf·ft)

D

B

C

F

42356-ACCO-G44

Rear caliper mounting—S2000

S2000

FRONT

1. Before servicing the vehicle, refer to the precautions in the beginning of this section.
2. Remove or disconnect the following:
   - Front wheels
   - Banjo bolt
   - Brake hose from the caliper
   - Mounting bolts and remove the caliper from its mounting bracket

To install:

3. Install or connect the following:
   - Caliper over the pads and onto its mounting bracket
4. On vehicles equipped with long caliper pins, torque the top caliper bolts to 54 ft. lbs. (74 Nm).
5. On vehicles equipped with short caliper bolts, torque the caliper bolts to 36 ft. lbs. (50 Nm).
   - Brake hose to the caliper using new sealing washers. Carefully torque the banjo bolt to 25 ft. lbs. (35 Nm).
   - Reservoir with fluid and bleed the brakes
   - Front wheels

REAR

1. Before servicing the vehicle, refer to the precautions in the beginning of this section.
2. Remove or disconnect the following:
   - Rear wheels
   - Caliper shield
   - Brake hose mounting bracket from caliper
   - Cable clip from the parking brake cable

   - Parking brake cable from the parking brake arm
   - Banjo bolt and brake hose
   - Brake hose sealing washers and discard
   - 2 caliper bolts, while holding the caliper pin with a wrench
   - Caliper from the bracket

To install:

3. Install or connect the following:
   - Caliper onto the bracket
   - Caliper bolts and tighten to 17 ft. lbs. (23 Nm)
   - Brake hose, with new sealing washers
   - Banjo bolt and tighten to 25 ft. lbs. (34 Nm)
   - Parking brake cable
   - Cable clip
   - Brake hose mounting bracket
   - Caliper shield
   - Rear wheels

Disc Brake Pads

REMOVAL & INSTALLATION

Civic

FRONT

➡**Two distinct types of front caliper are used on these vehicles. The caliper body will be marked either 5410 or 2056, according to type. Servicing procedures are similar, but the different calipers use different pads.**

1. Before servicing the vehicle, refer to the precautions in the beginning of this section.
2. Use a suction pump to remove some

brake fluid from the master cylinder reservoir.
3. Remove or disconnect the following:
   - Front wheels
   - Brake hose clamp from the steering knuckle
   - Lower caliper retaining bolt and pivot the caliper up
   - Disc brake pads, shims, and pad retainers from the caliper

To install:

4. Install or connect the following:
   - Pad retainers. Apply brake grease to the inner side of the shims and the back of the disc brake pads.
   - Pads, shims, and pad retainers. Make sure the wear indicator on the inner pad is facing up.
5. Compress the caliper piston with a suitable tool so that the caliper will fit over the pads.
6. Pivot the caliper down into position. Install the caliper bolts and tighten them as follows:
   - 5410 calipers: 24 ft. lbs. (33 Nm)
   - 2056 calipers: 25 ft. lbs. (35 Nm) (top) and 20 ft. lbs. (27 Nm) (bottom)
7. Fill the reservoir with clean brake fluid.
8. Install the front wheels.

REAR

1. Before servicing the vehicle, refer to the precautions in the beginning of this section.
2. Use a suction pump to remove some brake fluid from the master cylinder reservoir.
3. Remove or disconnect the following:
   - Rear wheels

- Caliper dust shield
- 2 caliper mounting bolts
- Caliper from the bracket and hang it out of the way with a piece of wire
- Pads, shims, and pad retainers from the caliper

To install:

4. Check the brake rotor for grooves or cracks and machine or replace if necessary.

5. Install or connect the following:
- Pad retainers. Apply brake grease to the inner side of the shims and to the back of the disc brake pads.
- Pads, shims, and pad retainers

6. Rotate the caliper piston clockwise into the caliper bore enough to allow the caliper to fit over the brake pads. Lubricate the piston boot with silicon grease. Avoid twisting the piston boot.
- Brake caliper. Align the groove in the piston with the tab on the inner pad. Tighten the mounting bolts to 17 ft. lbs. (23 Nm).
- Master cylinder reservoir with clean brake fluid
- Rear wheels

Accord and Prelude

FRONT

1. Before servicing the vehicle, refer to the precautions in the beginning of this section.

2. Remove or disconnect the following:
- Small amount of brake fluid from the reservoir using a suction pump
- Wheels
- Brake hose clamp from the strut or knuckle by removing the retaining bolts
- Lower caliper retaining bolt and pivot the caliper upward
- Pad shim and pad retainers
- Disc brake pads from the caliper

To install:

3. Check the brake rotor for grooves or cracks. If any heavy scoring is present, the rotor must be replaced.

4. Install or connect the following:
- Pad retainers. Apply a disc brake pad lubricant to both surfaces of the shims and the back of the disc brake pads.
- Pads and shims. The pad with the wear indicator goes in the inboard position.

5. Compress the caliper piston so the caliper will fit over the pads.
- Caliper by pivoting it down into position and tighten the mounting bolt to 36 ft. lbs. (49 Nm)

- Brake hose to the strut or knuckle, if removed
- Wheels
- Brake fluid to the master cylinder reservoir and install the cap

REAR

1. Before servicing the vehicle, refer to the precautions in the beginning of this section.

2. Remove or disconnect the following:
- Small amount of brake fluid from the reservoir using a suction pump
- Rear wheels
- Dust shield
- 2 caliper mounting bolts
- Caliper from the bracket
- Pads, shims and pad retainers

To install:

3. Clean the caliper thoroughly; remove any dirt or dust. Check the brake rotor for grooves or cracks and machine or replace, as necessary.

4. Install or connect the following:
- Pad retainers. Apply a disc brake pad lubricant to both surfaces of the shims and the back of the disc brake pads.
- Pads and shims. The wear retainer on the inboard pad faces down.

5. Use a suitable tool to rotate the caliper piston clockwise into the caliper bore, enough to enable the caliper to fit over the pads. Lubricate the piston boot with silicon grease, and avoid twisting the boot.
- Brake caliper, aligning the cutout in the piston with the tab on the inner pad. Tighten the mounting bolts to 17 ft. lbs. (23 Nm).
- Wheels
- Brake fluid to the master cylinder reservoir

S2000

FRONT

1. Before servicing the vehicle, refer to the precautions in the beginning of this section.

2. Remove or disconnect the following:
- Small amount of brake fluid from the reservoir using a suction pump
- Wheels

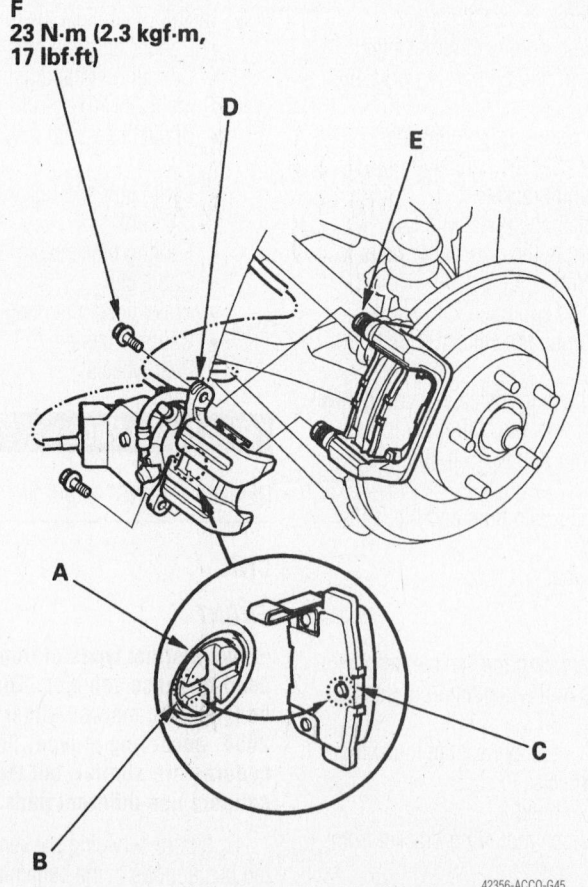

F
23 N·m (2.3 kgf·m, 17 lbf·ft)

Rotate the caliper piston (A) clockwise into the caliper bore, then align the cutout (B) in the piston with the tab (C) on the inner pad by turning the piston back—S2000

42356-ACCO-G45

- Lower caliper bolt, while holding the pin with a wrench. Pivot the caliper up for access to the pads.
- Pad shim and pad retainers
- Disc brake pads from the caliper

To install:

3. Check the brake rotor for grooves or cracks. If any heavy scoring is present, the rotor must be replaced.

4. Install or connect the following:
- Pad retainers. Apply a disc brake pad lubricant to both surfaces of the shims and the back of the disc brake pads.
- Pads and shims. The pad with the wear indicator goes in the inboard position.

5. Compress the caliper piston so the caliper will fit over the pads.
- Caliper by pivoting it down into position and tighten the mounting bolt to 24 ft. lbs. (32 Nm)
- Wheels
- Brake fluid to the master cylinder reservoir and install the cap

REAR

1. Before servicing the vehicle, refer to the precautions in the beginning of this section.

2. Remove or disconnect the following:
- Small amount of brake fluid from the reservoir using a suction pump
- Rear wheels
- 2 caliper mounting bolts, while holding the caliper pin with a wrench
- Caliper from the bracket
- Pads, shims and pad retainers

To install:

3. Clean the caliper thoroughly; remove any dirt or dust. Check the brake rotor for grooves or cracks and machine or replace, as necessary.

4. Install or connect the following:
- Pad retainers. Apply a disc brake pad lubricant to both surfaces of the shims and the back of the disc brake pads.
- Pads and shims. The wear retainer on the inboard pad faces upward.

5. Use a suitable tool to rotate the caliper piston clockwise into the caliper bore, then align the cutout in the piston with the tab on the inner pad by turning the piston back. Lubricate the piston boot with silicon grease, and avoid twisting the boot.
- Brake caliper
- Caliper bolts, and tighten to 17 ft. lbs. (23 Nm), while holding the caliper pin with a wrench
- Wheels
- Brake fluid to the master cylinder reservoir

Brake Drums

REMOVAL & INSTALLATION

Accord and Civic

1. Before servicing the vehicle, refer to the precautions in the beginning of this section.

2. Remove or disconnect the following:
- Rear wheels
- Rear brake drum

To install:

3. Make certain the brake shoes are adjusted to allow the drum clearance during installation. Fit the drum into position.

4. Install the rear wheels.

Brake Shoes

REMOVAL & INSTALLATION

Accord and Civic

1. Before servicing the vehicle, refer to the precautions in the beginning of this section.

2. Remove or disconnect the following:
- Rear wheels and brake drums
- Upper return spring from the brake shoes

3. Push the retainer springs and turn the tension pins to release the shoes from the backing plate.
- Lower the brake shoe assembly and remove the lower return spring
- Brake shoe assembly from the backing plate
- Parking brake cable from the brake shoe lever
- Upper return spring, self-adjuster lever, and self-adjuster spring. Separate the brake shoes.
- Wave washer, parking brake lever, and pivot pin from the brake shoe by removing the U-clip.

To install:

4. Install or connect the following:
- Brake cylinder grease to the sliding surface of the pivot pin and insert the pin into the brake shoe.

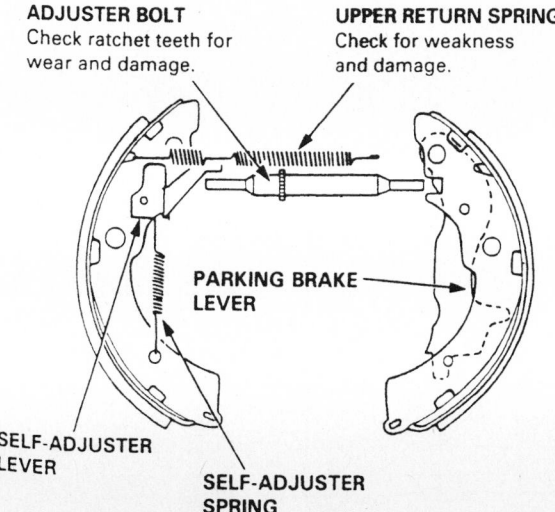

ADJUSTER BOLT
Check ratchet teeth for wear and damage.

UPPER RETURN SPRING
Check for weakness and damage.

PARKING BRAKE LEVER

SELF-ADJUSTER LEVER

SELF-ADJUSTER SPRING

Rear drum brakes—Civic shown

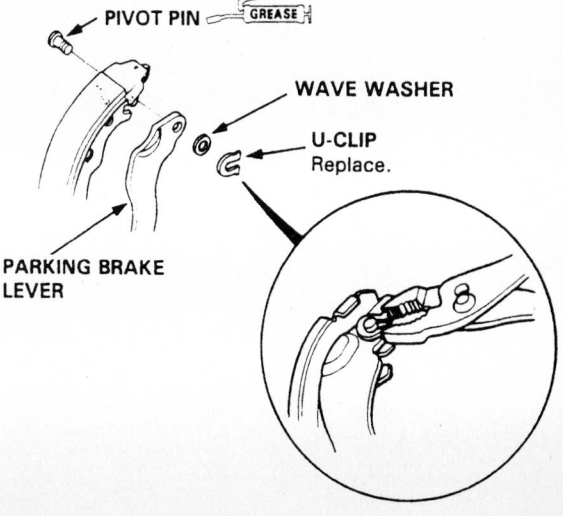

PIVOT PIN **GREASE**

WAVE WASHER

U-CLIP
Replace.

PARKING BRAKE LEVER

93016G09

- Parking brake lever and wave washer on the pivot pin and pinch the U-clip with a pair of pliers to secure the pivot pin to the shoe
- Parking brake cable to the parking brake lever
- Adjuster spring to the adjuster lever first, then to the brake shoe

- Adjuster bolt/clevis assembly and the upper return spring
- Brake shoes to the backing plate
- Lower return spring, the tension pins and retaining springs
- Connect the upper return spring

5. Turn the adjuster bolt to force the brake shoes out until the brake drum will not easily go on. Back off the adjuster bolt just enough that the brake drum will go on and turn easily.

6. Install the wheels.

7. Depress the brake pedal several times to set the self-adjusting brake. Adjust the parking brake.

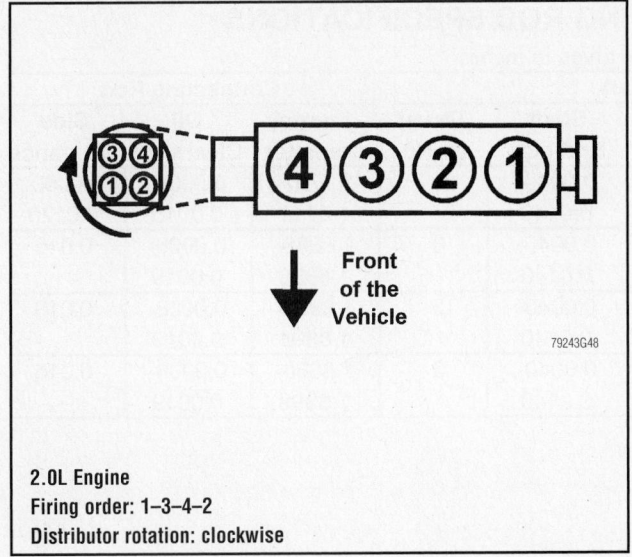

2.0L Engine
Firing order: 1–3–4–2
Distributor rotation: clockwise

79243G48

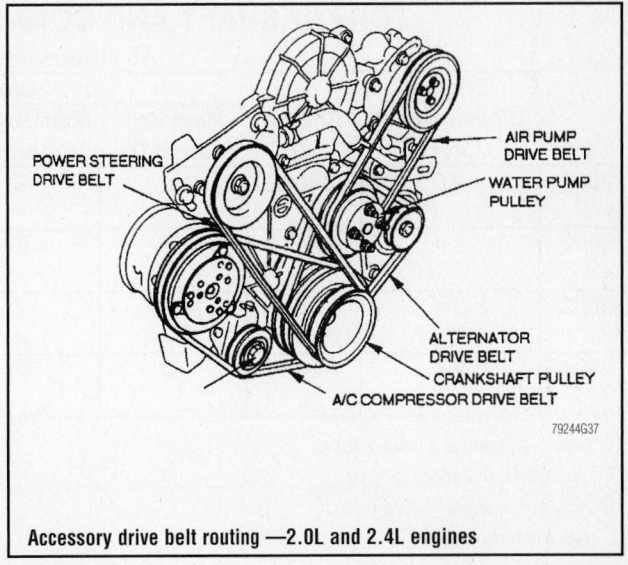

Accessory drive belt routing —2.0L and 2.4L engines

79244G37

CAPACITIES

Year	Model	Engine Displacement Liters	Engine ID	Engine Oil with Filter (qts.)	Transmission (pts.) 5-Spd	Auto.	Transfer Case (pts.)	Drive Axle Front (pts.)	Rear (pts.)	Fuel Tank (gal.)	Cooling System (qts.)
2001	CR-V	2.0	B20Z2	4.0	3.6	①	②	②	2.2	15.3	4.1
2002	CR-V	2.4	K24A1	4.4	4.0	③	②	②	2.2	15.3	5.8
2003	CR-V	2.4	K24A1	4.4	4.0	③	②	②	2.2	15.3	5.8
2004	CR-V	2.4	K24A1	4.4	4.0	③	②	②	2.2	15.3	5.8

NOTE: All capacities are approximate. Add fluid gradually and check to be sure a proper fluid level is obtained.

① 4WD: 6.2
   2WD: 5.8

② Included in transaxle refill figure

③ 4WD: 6.6 pts.
   2WD: 6.2 pts.

67162-HCRV-C04

VALVE SPECIFICATIONS

Year	Engine Displacement Liters	Engine ID	Seat Angle (deg.)	Face Angle (deg.)	Spring Test Pressure (lbs. @ in.)	Spring Installed Height (in.)	Stem-to-Guide Clearance (in.) Intake	Exhaust	Stem Diameter (in.) Intake	Exhaust
2001	2.0	B20Z2	45	45	NA	①	0.0010-0.0020	0.0020-0.0030	0.2591-0.2594	0.2579-0.2583
2002	2.4	K24A1	NA	NA	NA	①	0.0012-0.0022	0.0022-0.0031	0.2156-0.2159	0.2146-0.2150
2003	2.4	K24A1	NA	NA	NA	①	0.0012-0.0022	0.0022-0.0031	0.2156-0.2159	0.2146-0.2150
2004	2.4	K24A1	NA	NA	NA	①	0.0012-0.0022	0.0022-0.0031	0.2156-0.2159	0.2146-0.2150

NA: Not Available

① Valve spring free length:
   Intake: 1.668 in.
   Exhaust: 1.745 in.

67162-HCRV-C05

CRANKSHAFT AND CONNECTING ROD SPECIFICATIONS

All measurements are given in inches

Year	Engine Displacement Liters	Engine ID	Crankshaft				Connecting Rod		
			Main Brg. Journal Dia.	Main Brg. Oil Clearance	Shaft End-play	Thrust on No.	Journal Diameter	Oil Clearance	Side Clearance
2001	2.0	B20Z2	①	②	0.0040-0.0140	4	1.7707-1.7717	0.0008-0.0015	0.0060-0.0120
2002	2.4	K24A1	③	④	0.0040-0.0140	3	1.8888-1.8898	0.0008-0.0019	0.016
2003	2.4	K24A1	③	④	0.0040-0.0140	3	1.8888-1.8898	0.0008-0.0019	0.016
2004	2.4	K24A1	③	④	0.0040-0.0140	3	1.8888-1.8898	0.0008-0.0019	0.016

① Nos. 1, 2, 4 and 5: 2.1644-2.1654
   No. 3: 2.1642-2.1651

② Nos. 1, 2, 4 and 5: 0.0009-0.0017
   No. 3: 0.0012-0.0019

③ Except No. 3: 2.1648-2.1657
   No. 3: 2.1644-2.1654

④ Except No. 3: 0.0007-0.0016
   No. 3: 0.0010-0.0019

67162-HCRV-C06

PISTON AND RING SPECIFICATIONS

All measurements are given in inches

Year	Engine Displacement Liters	Engine ID	Piston Clearance	Ring Gap			Ring Side Clearance		
				Top Compression	Bottom Compression	Oil Control	Top Compression	Bottom Compression	Oil Control
2001	2.0	B20Z2	0.0004-0.0016	0.0080-0.0120	0.0160-0.0220	0.0080-0.0200	0.0022-0.0031	0.0014-0.0024	NA
2002	2.4	K24A1	0.0008-0.0016	0.0080-0.0140	0.0160-0.0220	0.0080-0.0280	0.0018-0.0028	0.0020-0.0030	NA
2003	2.4	K24A1	0.0008-0.0016	0.0080-0.0140	0.0160-0.0220	0.0080-0.0280	0.0018-0.0028	0.0020-0.0030	NA
2004	2.4	K24A1	0.0008-0.0016	0.0080-0.0140	0.0160-0.0220	0.0080-0.0280	0.0018-0.0028	0.0020-0.0030	NA

NA: Not Applicable

67162-HCRV-C07

TORQUE SPECIFICATIONS
All readings in ft. lbs.

Year	Engine Displacement Liters	Engine ID	Cylinder Head Bolts	Main Bearing Bolts	Rod Bearing Bolts	Crankshaft Damper Bolts	Flywheel Bolts	Manifold Intake	Manifold Exhaust	Spark Plugs	Oil Pan Drain Plug
2001	2.0	B20Z2	①	②	23	130	54	17	23	13	33
2002	2.4	K24A1	③	④	⑤	181	90	16	33	13	33
2003	2.4	K24A1	③	④	⑤	181	90	16	33	13	33
2004	2.4	K24A1	③	④	⑤	⑥	90	16	33	13	33

NOTE: Dip main bearing bolts and crankshaft damper bolt in clean engine oil prior to tightening.

① Step 1: 22 ft. lbs.
  Step 2: 63 ft. lbs.

② Step 1: 18 ft. lbs.
  Step 2: 56 ft. lbs.

③ Step 1: 29 ft. lbs.
  Step 2: +90 degrees
  Step 3: +90 degrees
  Step 4: NEW BOLT ONLY +90 degrees

④ 22 ft. lbs. +56 degrees

⑤ 14 ft. lbs. +90 degrees

⑥ 36 ft. lbs. +90 degrees

67162-HCRV-C08

WHEEL ALIGNMENT

Year	Model		Caster Range (+/-Deg.)	Caster Preferred Setting (Deg.)	Camber Range (+/-Deg.)	Camber Preferred Setting (Deg.)	Toe-in (in.)
2001	CR-V	F	1.00	+2.10	1.00	0	0+/-0.12
		R	—	—	1.00	-1.00	0.08+/-0.08
2002	CR-V	F	1.00	+1.75	0.75	0	0+/-0.08
		R	—	—	0.75	-1.00	0.08+/-0.08
2003	CR-V	F	1.00	+1.75	0.75	0	0+/-0.08
		R	—	—	0.75	-1.00	0.08+/-0.08
2004	CR-V	F	1.00	+1.75	0.75	0	0+/-0.08
		R	—	—	0.75	-1.00	0.08+/-0.08

67162-HCRV-C09

TIRE, WHEEL AND BALL JOINT SPECIFICATIONS

| Year | Model | OEM Tires | | Tire Pressures (psi) | | Wheel Size | Ball Joint Inspection | Lugnut Torque (ft. lbs.) |
		Standard	Optional	Front	Rear			
2001	CR-V	P205/70R15	None	26	26	6JJ	NS	80
2002	CR-V	P205/70R15	None	26	26	6JJ	NS	80
2003	CR-V	P205/70R15	None	26	26	6JJ	NS	80
2004	CR-V	P205/70R15	None	26	26	6JJ	NS	80

OEM: Original Equipment Manufacturer

PSI: Pounds Per Square Inch

NS: Not specified by manufacturer

67162-HCRV-C10

BRAKE SPECIFICATIONS

All measurements in inches unless noted

| Year | Model | | Brake Disc | | | Brake Drum Diameter | | | Minimum Lining Thickness | | Brake Caliper | |
			Original Thickness	Minimum Thickness	Maximum Runout	Original Inside Diameter	Max. Wear Limit	Maximum Machine Diameter	Front	Rear	Bracket Bolts (ft. lbs.)	Mounting Bolts (ft. lbs.)
2001	CR-V	F	0.929	0.830	0.004	—	—	—	0.060	—	80	36
		R	—	—	—	8.66	8.70	8.70	—	0.080	—	—
2002	CR-V	F	0.910	0.830	0.004	—	—	—	0.060	—	80	25
		R	0.350	0.280	0.004	—	—	—	0.040	—	41	16
2003	CR-V	F	0.910	0.830	0.004	—	—	—	0.060	—	80	25
		R	0.350	0.280	0.004	—	—	—	0.040	—	41	16
2004	CR-V	F	0.910	0.830	0.004	—	—	—	0.060	—	80	25
		R	0.350	0.280	0.004	—	—	—	0.040	—	41	16

F: Front

R: Rear

67162-HCRV-C11

SCHEDULED MAINTENANCE INTERVALS

2001 HONDA—CRV

TO BE SERVICED	SERVICE	VEHICLE MILEAGE INTERVAL (x1000)															
		8	15	22.5	30	37.5	45	52.5	60	67.5	75	82.5	90	97.5	105	112.5	120
Accessory drive belts	I & A				✓				✓				✓				✓
Air cleaner element	R				✓				✓				✓				✓
Air conditioning filter	R				✓				✓				✓				✓
Brake fluid	R					✓							✓				
Brake hoses & lines	I		✓		✓		✓		✓		✓		✓		✓		✓
Cooling system	I		✓		✓		✓		✓		✓		✓		✓		✓
Engine coolant	R					✓							✓				
Engine oil	R	✓	✓	✓	✓	✓	✓	✓	✓	✓	✓	✓	✓	✓	✓	✓	✓
Engine oil and coolant levels	I	Inspect at each fuel stop															
Engine oil filter	R		✓		✓		✓		✓		✓		✓		✓		✓
Exhaust system	I		✓		✓		✓		✓		✓		✓		✓		✓
Fluid levels and condition	I		✓		✓		✓		✓		✓		✓		✓		✓
Front and rear brakes	I		✓		✓		✓		✓		✓		✓		✓		✓
Fuel lines & connection	I		✓		✓		✓		✓		✓		✓		✓		✓
Halfshaft boots	I		✓		✓		✓		✓		✓		✓		✓		✓
Idle speed	I & A												✓				
Parking brake system	I & A		✓		✓		✓		✓		✓		✓				✓
Rear differential fluid	R												✓				
Rotate and inspect tires	I	✓	✓	✓	✓	✓	✓	✓	✓	✓	✓	✓	✓	✓	✓	✓	✓
Spark plugs	R				✓				✓				✓				✓
Supplemental Restrain system	I	Inspect the SRS 10 years after production															
Suspension components	I		✓		✓		✓		✓		✓		✓		✓		✓
Tie rod ends, steering gear box & boots	I		✓		✓		✓		✓		✓		✓		✓		✓
Timing balancer belt ①	R														✓		
Timing belt	R														✓		
Transmission fluid	R					✓					✓				✓		
Valve clearance	I			✓					✓				✓				✓
Water pump	S/I														✓		

R: Replace   I: Inspect   A: Adjust

FREQUENT OPERATION MAINTENANCE (SEVERE SERVICE)

If a vehicle is operated under any of the following conditions it is considered severe service:

- Towing a trailer or using a camper or car-top carrier.
- Repeated short trips of less than 5 miles in temperatures below freezing, or trips of less than 10 miles in any temperature.
- Extensive idling or low-speed driving for long distances as in heavy commercial use, such as delivery, taxi or police cars.
- Operating on rough, muddy or salt-covered roads.
- Operating on unpaved or dusty roads.
- Driving in extremely hot (over 90°) conditions.

Air cleaner element: replace every 15,000 miles

Engine oil and filter: replace every 3750 miles or 6 months, whichever occurs first.

Timing belt: replace every 60,000 miles if the vehicle is regularly driven in temperatures above 110°F or below -20°F.

Transmission fluid: replace every 30,000 miles.

Rear differential fluid: replace every 60,000 miles.

Front and rear brakes: inspect every 7500 miles or 6 months, whichever occurs first.

Locks and hinges: lubricate every 15,000 miles.

Tie rods, steering gear box, boots: inspect every 7500 miles or 6 months, whichever occurs first.

Suspension components: inspect every 7500 miles or 6 months, whichever occurs first.

Halfshaft boots: inspect every 7500 miles or 6 months, whichever occurs first.

SCHEDULED MAINTENANCE INTERVALS
2002-04 HONDA—CRV

TO BE SERVICED	TYPE OF SERVICE	10	20	30	40	50	60	70	80	90	100	110	120
		VEHICLE MILEAGE INTERVAL (x1000)											
Accessory drive belts	I & A			✓			✓			✓			✓
Air cleaner element	R			✓			✓			✓			✓
Air conditioning filter	R			✓			✓			✓			✓
Brake fluid	R											✓	
Brake hoses & lines (including ABS)	I		✓		✓		✓		✓		✓		
Cooling system hoses & connections	I		✓		✓		✓		✓		✓		
Engine coolant	R												✓
Engine oil	R	✓	✓	✓	✓	✓	✓	✓	✓	✓	✓	✓	✓
Engine oil and coolant levels	I	Inspect at each fuel stop											
Engine oil filter	R		✓		✓		✓		✓		✓		
Exhaust system	I		✓		✓		✓		✓		✓		
Fluid levels and condition	I		✓		✓		✓		✓		✓		
Front and rear brakes	I		✓		✓		✓		✓		✓		
Fuel lines & connection	I		✓		✓		✓		✓		✓		
Halfshaft boots	I		✓		✓		✓		✓		✓		
Idle speed	I & A											✓	
Parking brake system	I & A		✓		✓		✓		✓		✓		
Rear differential fluid	R										✓		
Rotate and inspect tires	I	✓	✓	✓	✓	✓	✓	✓	✓	✓	✓	✓	✓
Spark plugs	R											✓	
Suspension components	I		✓		✓		✓		✓		✓		
Tie rod ends, steering gear box & boots	I		✓		✓		✓		✓		✓		
Transmission fluid	R												✓
Valve clearance	I											✓	

R: Replace   I: Inspect   A: Adjust

FREQUENT OPERATION MAINTENANCE (SEVERE SERVICE)

If a vehicle is operated under any of the following conditions it is considered severe service:

- Towing a trailer or using a camper or car-top carrier.
- Repeated short trips of less than 5 miles in temperatures below freezing, or trips of less than 10 miles in any temperature.
- Extensive idling or low-speed driving for long distances as in heavy commercial use, such as delivery, taxi or police cars.
- Operating on rough, muddy or salt-covered roads.
- Operating on unpaved or dusty roads.
- Driving in extremely hot (over 90°) conditions.

Air cleaner element: replace every 15,000 miles

Engine oil and filter: replace every 3750 miles or 6 months, whichever occurs first.

Timing belt: replace every 60,000 miles if the vehicle is regularly driven in temperatures above 110°F or below -20°F.

Transmission fluid: replace every 30,000 miles.

Rear differential fluid: replace every 60,000 miles.

Front and rear brakes: inspect every 7500 miles or 6 months, whichever occurs first.

Locks and hinges: lubricate every 15,000 miles.

Tie rods, steering gear box, boots: inspect every 7500 miles or 6 months, whichever occurs first.

Suspension components: inspect every 7500 miles or 6 months, whichever occurs first.

Halfshaft boots: inspect every 7500 miles or 6 months, whichever occurs first.

67162-HCRV-C13

PRECAUTIONS

Before servicing any vehicle, please be sure to read all of the following precautions, which deal with personal safety, prevention of component damage, and important points to take into consideration when servicing a motor vehicle:

- Never open, service or drain the radiator or cooling system when the engine is hot; serious burns can occur from the steam and hot coolant.

- Observe all applicable safety precautions when working around fuel. Whenever servicing the fuel system, always work in a well-ventilated area. Do not allow fuel spray or vapors to come in contact with a spark, open flame, or excessive heat (a hot drop light, for example). Keep a dry chemical fire extinguisher near the work area. Always keep fuel in a container specifically designed for fuel storage; also, always properly seal fuel containers to avoid the possibility of fire or explosion. Refer to the additional fuel system precautions later in this section.

- Fuel injection systems often remain pressurized, even after the engine has been turned **OFF**. The fuel system pressure must be relieved before disconnecting any fuel lines. Failure to do so may result in fire and/or personal injury.

- Brake fluid often contains polyglycol ethers and polyglycols. Avoid contact with the eyes and wash your hands thoroughly after handling brake fluid. If you do get brake fluid in your eyes, flush your eyes with clean, running water for 15 minutes. If eye irritation persists, or if you have taken brake fluid internally, seek medical assistance IMMEDIATELY.

- The EPA warns that prolonged contact with used engine oil may cause a number of skin disorders, including cancer. You should make every effort to minimize your exposure to used engine oil. Protective gloves should be worn when changing oil. Wash your hands and any other exposed skin areas as soon as possible after exposure to used engine oil. Soap and water, or waterless hand cleaner should be used.

- All new vehicles are now equipped with an air bag system. The system must be disabled before performing service on or around system components, steering column, instrument panel components, wiring and sensors. Failure to follow safety and disabling procedures could result in accidental air bag deployment, possible personal injury and unnecessary system repairs.

- Always wear safety goggles when working with, or around, the air bag system. When carrying a non-deployed air bag, be sure the bag and trim cover are pointed away from your body. When placing a non-deployed air bag on a work surface, always face the bag and trim cover upward, away from the surface. This will reduce the motion of the module if it is accidentally deployed. Refer to the additional air bag system precautions later in this section.

- Clean, high quality brake fluid from a sealed container is essential to the safe and proper operation of the brake system. You should always buy the correct type of brake fluid for your vehicle. If the brake fluid becomes contaminated, completely flush the system with new fluid. Never reuse any brake fluid. Any brake fluid that is removed from the system should be discarded. Also, do not allow any brake fluid to come in contact with a painted surface; it will damage the paint.

- Never operate the engine without the proper amount and type of engine oil; doing so WILL result in severe engine damage.

- Timing belt maintenance is extremely important. Many models utilize an interference-type, non-freewheeling engine. If the timing belt breaks, the valves in the cylinder head may strike the pistons, causing potentially serious (also time-consuming and expensive) engine damage. Refer to the maintenance interval charts for the recommended replacement interval for the timing belt, and to the timing belt section for belt replacement and inspection.

- Disconnecting the negative battery cable on some vehicles may interfere with the functions of the on-board computer system(s) and may require the computer to undergo a relearning process once the negative battery cable is reconnected.

- When servicing drum brakes, only disassemble and assemble one side at a time, leaving the remaining side intact for reference.

- Only an MVAC-trained, EPA-certified automotive technician should service the air conditioning system or its components.

ENGINE REPAIR

➡**Disconnecting the negative battery cable on some vehicles may interfere with the functions of the on board computer system. The computer may undergo a relearning process once the negative battery cable is reconnected.**

Distributor

The 2.4L engine does not have a distributor.

REMOVAL & INSTALLATION

1. Before servicing the vehicle, refer to the precautions in the beginning of this section.
2. Remove or disconnect the following:
   - Negative battery cable
   - Cruise control cable

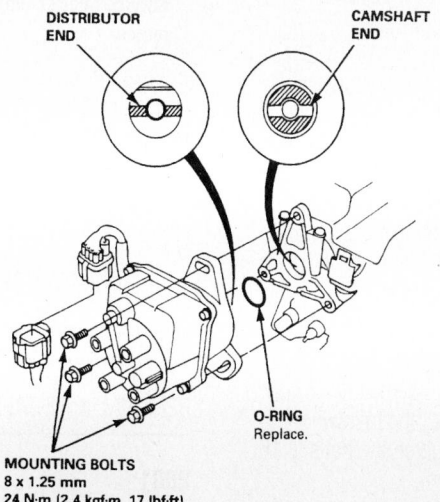

Exploded view of the distributor mounting—CR-V

7924MG02

- Air intake duct
- Distributor harness connector
- Spark plug wires
- Distributor

To install:

3. Install or connect the following:
- Distributor. Use a new O-ring seal.
- Spark plug wires
- Distributor harness connector
- Air intake duct
- Cruise control cable
- Negative battery cable

4. Set the ignition timing and tighten the mounting bolts to 13 ft. lbs. (18 Nm).

Alternator

REMOVAL

2001

1. Before servicing the vehicle, refer to the precautions in the beginning of this section.
2. Remove or disconnect the following:
- Negative battery cable
- Accessory drive belts
- Alternator wiring harness connectors
- Alternator

2002–04

1. Before servicing the vehicle, refer to the precautions in the beginning of this section.
2. Remove or disconnect the following:
- Negative battery cable
- Front cover
- Accessory drive belt
- The 3 bolts holding the alternator
- Alternator wiring harness connectors
- Alternator

INSTALLATION

2001

1. Install or connect the following:
- Alternator. Tighten the mounting nut to 33 ft. lbs. (44 Nm) and the adjustment locknut to 17 ft. lbs. (24 Nm).
- Alternator wiring harness connectors. Tighten the battery terminal nut to 70 inch lbs. (8 Nm).
- Accessory drive belts
- Negative battery cable

2002–04

1. Install or connect the following:
- Alternator. Tighten the bolts to 16 ft. lbs. (22 Nm).
- Alternator wiring harness connectors. Tighten the battery terminal nut to 70 inch lbs. (8 Nm).

- Accessory drive belts
- Front cover
- Negative battery cable

Ignition Timing

ADJUSTMENT

Adjustment is not possible on the 2.4L engine.

➡**Timing adjustments are made with the engine at operating temperature.**

1. Before servicing the vehicle, refer to the precautions in the beginning of this section.
2. Short the **2P** Service Check connector.
3. Connect a timing light to the No. 1 ignition wire.

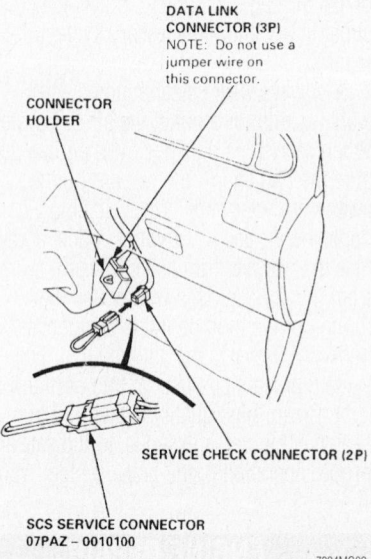

Service Check connector and shorting jumper

4. The timing should be 13–17 degrees Before Top Dead Center (BTDC) (red timing mark on crankshaft pulley) at 650–750 rpm.
5. Adjust the timing as necessary and tighten the distributor bolts to 13 ft. lbs. (18 Nm).
6. Remove the **2P** connector jumper.

Engine Assembly

REMOVAL & INSTALLATION

2001

➡**The engine and transaxle are removed from the vehicle as a unit.**

1. Before servicing the vehicle, refer to the precautions in the beginning of this section.
2. Drain the cooling system.
3. Drain the transaxle fluid.
4. Drain the engine oil.
5. Relieve fuel system pressure.
6. Remove or disconnect the following:
- Negative battery cable
- Fuse/Relay box battery cables
- Battery and tray
- Air intake assembly
- Powertrain Control Module (PCM) connectors and grommet. Pull the PCM harness through the firewall.
- Left engine wire harness connectors
- Cruise control actuator
- Power steering pump belt and pump
- A/C compressor drive belt
- Fuel lines
- Brake booster vacuum line
- Accelerator cable
- Power Steering Pressure (PSP) switch
- Splash shield
- Radiator hoses
- Heater hoses
- Heated Oxygen Sensor (HO_2S) connector
- Exhaust front pipe
- Right damper fork
- Lower ball joints

7. Separate the inner CV-joints from the transaxle and support the axle halfshafts out of the work area with safety wire.

8. If equipped with a manual transaxle, remove or disconnect the following:
- Clutch slave cylinder

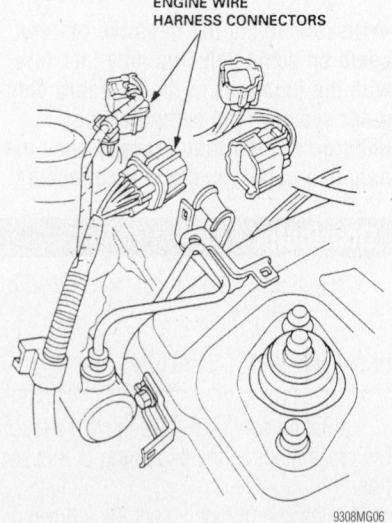

Left engine wire harness connectors—CR-V

- Clutch hose bracket
- Transaxle ground cable
- Shift cables

9. If equipped with an automatic transaxle, remove or disconnect the following:
- Shift cable cover
- Shift cable
- Transaxle ground cable and hose clamp
- Transaxle fluid cooler lines

10. For all vehicles, remove or disconnect the following:
- A/C hose clamp
- Radiator
- A/C compressor
- Rear driveshaft, if equipped

11. Attach a hoist to the engine lifting eyes and support the powertrain weight.

12. Remove or disconnect the following:
- Left front mount and bracket
- Right front mount and bracket
- Rear mount bracket through bolt
- Upper bracket
- Transaxle mount and bracket

13. Lift the powertrain away from the vehicle.

To install:

➡**Use new self-locking nuts and color-coded self-locking bolts when installing the engine mounts and suspension components.**

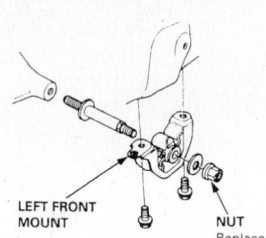
M/T:

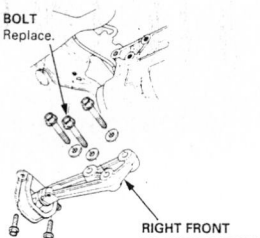
A/T:

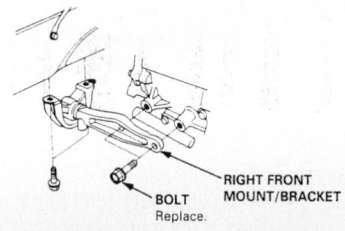
Front mounts—CR-V

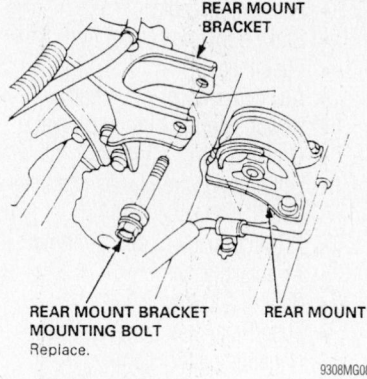
Rear mount—CR-V

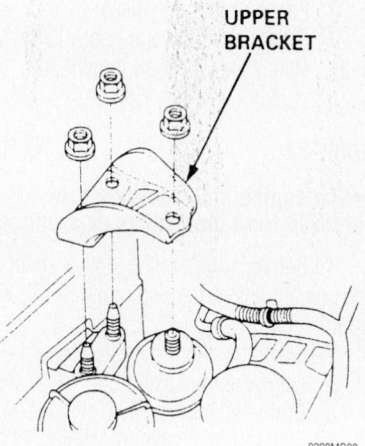
Upper bracket—CR-V

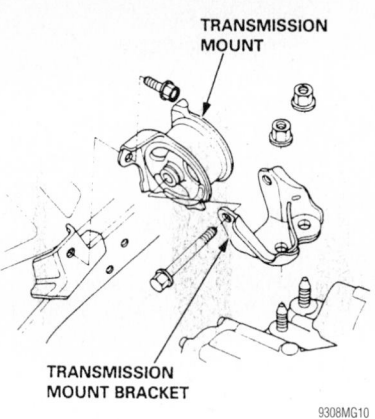
Transaxle mount—CR-V

➡**Do not tighten the engine or transaxle mount fasteners until instructed to do so.**

14. Lower the powertrain into position.
15. Install or connect the following:
- Transaxle mount and bracket. Tighten the frame mounting bolts to 47 ft. lbs. (64 Nm).
- Upper bracket. Tighten the nuts in sequence to 54 ft. lbs. (74 Nm).

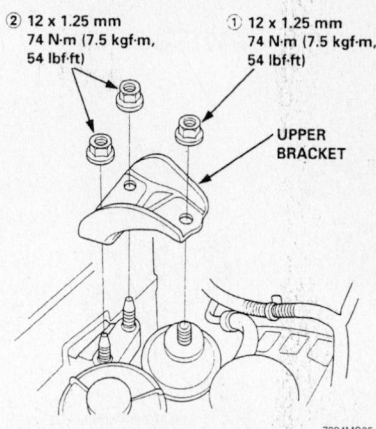
Upper bracket tightening sequence—CR-V

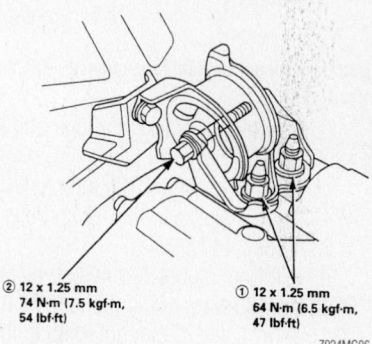
Transaxle mount fastener tightening sequence—CR-V

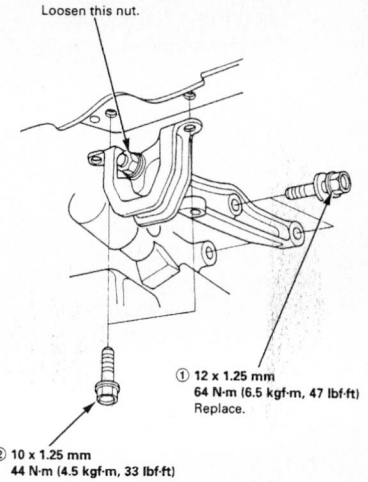
Right front mount tightening sequence—CR-V

- Rear mount bracket through bolt
- Right front mount and bracket
- Left front mount and bracket

16. Tighten the remaining mount fasteners as follows:

a. Transaxle mount fasteners to 47 ft. lbs. (64 Nm) and the through bolt to 54 ft. lbs. (74 Nm).

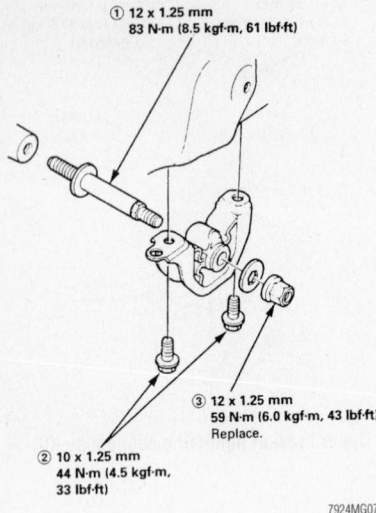

① 12 x 1.25 mm
83 N·m (8.5 kgf·m, 61 lbf·ft)

③ 12 x 1.25 mm
59 N·m (6.0 kgf·m, 43 lbf·ft)
Replace.

② 10 x 1.25 mm
44 N·m (4.5 kgf·m,
33 lbf·ft)

7924MG07

Left front mount tightening sequence—CR-V

b. Rear mount bracket through bolt to 43 ft. lbs. (59 Nm).

c. Right front mount 12mm bolts to 47 ft. lbs. (64 Nm) and the 10mm bolts to 33 ft. lbs. (44 Nm).

d. Left front mount 12mm stud bolt to 61 ft. lbs. (83 Nm), 10mm bolts to 33 ft. lbs. (44 Nm), and 12mm nut to 43 ft. lbs. (59 Nm).

e. Right front mount 12mm nut to 43 ft. lbs. (59 Nm).

17. Install or connect the following:
- Rear driveshaft, if equipped
- A/C compressor
- Radiator
- A/C hose clamp

18. If equipped with a manual transaxle, install or connect the following:
- Shift cables
- Transaxle ground cable
- Clutch hose bracket
- Clutch slave cylinder

19. If equipped with an automatic transaxle, install or connect the following:
- Transaxle fluid cooler lines
- Transaxle ground cable and hose clamp
- Shift cable
- Shift cable cover

20. For all vehicles, install or connect the following:
- Axle halfshafts
- Lower ball joints
- Right damper fork
- Exhaust front pipe
- HO2S connector
- Heater hoses
- Radiator hoses
- Splash shield
- PSP switch

- Accelerator cable
- Brake booster vacuum line
- Fuel lines
- A/C compressor drive belt
- Power steering pump and belt
- Cruise control actuator
- Left engine wire harness connectors
- PCM connectors and grommet
- Air intake assembly
- Battery and tray
- Fuse/Relay box battery cables
- Negative battery cable

21. Fill the engine crankcase to the correct level.

22. Fill the transaxle to the correct level.

23. Fill the cooling system.

24. Start the engine and check for leaks.

25. Check the wheel alignment and adjust as necessary.

2002–04

➡**The engine and transaxle are removed from the vehicle as a unit.**

1. Before servicing the vehicle, refer to the precautions in the beginning of this section.

2. Drain the cooling system.

3. Drain the transaxle fluid.

4. Drain the engine oil.

5. Relieve fuel system pressure.

6. Remove or disconnect the following:
- Negative battery cable
- Fuse/Relay box battery cables
- Battery and tray
- Intake manifold cover
- IAT sensor connector
- Breather hose
- Intake duct
- Cables from the power distribution center
- Throttle and cruise cables
- Powertrain Control Module (PCM) connectors and grommet. Pull the PCM harness through the firewall.
- Fuel lines
- EVAP canister
- Brake booster vacuum line
- Clutch slave cylinder
- Clutch hose bracket
- Shift cables
- Drive belt
- Power steering pump, leaving the hoses connected

7. Attach a hoist to the engine lifting eyes and support the powertrain weight.

8. Remove or disconnect the following:
- Splash shield
- Wheels
- Catalytic converter
- Rear driveshaft

- Stabilizer links
- Right damper fork
- Halfshafts
- Shift cable
- Radiator
- Upper bracket
- Transaxle mount and bracket
- Front mount bolt
- Rear mount bracket bolts. Match-mark the sub-frame mounting bolt centers.

There is a special tool necessary for sub-frame removal. The Honda tool number is EQS02C000011. Attach the tool as explained in the tool instructions, attach a floor jack with adapter, remove the 4 sub-frame bolts and lower the sub-frame.

9. Remove or disconnect the following:
- A/C compressor without disconnecting the hoses

10. Check that all hoses and wires are disconnected.

11. Lower the engine about 6 inches and recheck all clearances.

12. Lower the engine all the way.

13. Remove the chain hoist.

To install:

14. Installation is the reverse of removal. Observe the following torques:
- Front engine mount bracket bolts: 33 ft. lbs. (44Nm)
- A/C compressor bracket: 33 ft. lbs. (44Nm)
- Stiffener 10mm bolts: 33 ft. lbs. (44Nm); 6mm bolts 9 ft. lbs. (12 Nm)
- A/C compressor bolts: 33 ft. lbs. (44 Nm)
- Subframe front bolt: 47 ft. lbs. (64 Nm)
- Subframe rear bolts: 43 ft. lbs. (59 Nm)
- Upper bracket bolt and nut: 40 ft. lbs. (54 Nm)
- Transmission mount bracket support bolts/nuts: 40 ft. lbs. (54 Nm)
- PS pump bolts: 16 ft. lbs. (22 Nm)
- Intake manifold cover: 9 ft. lbs. (12 Nm)

➡**Use new self-locking nuts and color-coded self-locking bolts when installing the engine mounts and suspension components.**

➡**Do not tighten the engine or transaxle mount fasteners until instructed to do so.**

15. Lower the powertrain into position.

16. Install or connect the following:
- Transaxle mount and bracket. Tighten the frame mounting bolts to 47 ft. lbs. (64 Nm).

- Upper bracket. Tighten the nuts in sequence to 54 ft. lbs. (74 Nm).
- Rear mount bracket through bolt
- Right front mount and bracket
- Left front mount and bracket

17. Tighten the remaining mount fasteners as follows:

a. Transaxle mount fasteners to 47 ft. lbs. (64 Nm) and the through bolt to 54 ft. lbs. (74 Nm).

b. Rear mount bracket through bolt to 43 ft. lbs. (59 Nm).

c. Right front mount 12mm bolts to 47 ft. lbs. (64 Nm) and the 10mm bolts to 33 ft. lbs. (44 Nm).

d. Left front mount 12mm stud bolt to 61 ft. lbs. (83 Nm), 10mm bolts to 33 ft. lbs. (44 Nm), and 12mm nut to 43 ft. lbs. (59 Nm).

e. Right front mount 12mm nut to 43 ft. lbs. (59 Nm).

18. Install or connect the following:
- Rear driveshaft, if equipped
- A/C compressor
- Radiator
- A/C hose clamp

19. If equipped with a manual transaxle, install or connect the following:
- Shift cables
- Transaxle ground cable
- Clutch hose bracket
- Clutch slave cylinder

20. If equipped with an automatic transaxle, install or connect the following:
- Transaxle fluid cooler lines
- Transaxle ground cable and hose clamp
- Shift cable
- Shift cable cover

21. For all vehicles, install or connect the following:
- Axle halfshafts
- Lower ball joints
- Right damper fork
- Exhaust front pipe
- HO2S connector
- Heater hoses
- Radiator hoses
- Splash shield
- PSP switch
- Accelerator cable
- Brake booster vacuum line
- Fuel lines
- A/C compressor drive belt
- Power steering pump and belt
- Cruise control actuator
- Left engine wire harness connectors
- PCM connectors and grommet
- Air intake assembly
- Battery and tray
- Fuse/Relay box battery cables

- Negative battery cable

22. Fill the engine crankcase to the correct level.

23. Fill the transaxle to the correct level.

24. Fill the cooling system.

25. Start the engine and check for leaks.

26. Check the wheel alignment and adjust as necessary.

Water Pump

REMOVAL & INSTALLATION

2001

1. Before servicing the vehicle, refer to the precautions in the beginning of this section.

2. Drain the cooling system.

3. Remove or disconnect the following:
- Negative battery cable
- Accessory drive belts
- Front cover
- Timing belt
- Water pump

To install:

4. Install or connect the following:
- Water pump. Use a new O-ring seal and tighten the bolts to 105 inch lbs. (12 Nm).
- Timing belt
- Front cover
- Accessory drive belts
- Negative battery cable

5. Fill the cooling system.

6. Start the engine and check for leaks.

2002-04

1. Before servicing the vehicle, refer to the precautions in the beginning of this section.

2. Drain the cooling system.

3. Remove or disconnect the following:
- Negative battery cable
- Accessory drive belt

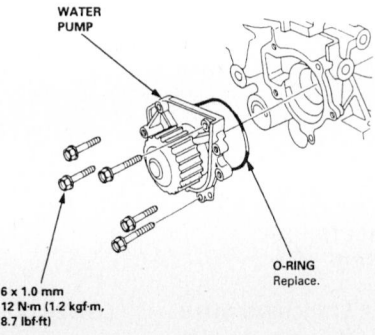

WATER PUMP

6 x 1.0 mm
12 N·m (1.2 kgf·m,
8.7 lbf·ft)

O-RING
Replace.

7924MG10

Exploded view of the water pump mounting

- Crankshaft pulley
- Water pump (6 bolts)

Installation is the reverse of removal. Use new O-rings. Torque the bolts to 104 inch lbs. (12 Nm).

Heater Core

REMOVAL & INSTALLATION

1. Disconnect the negative battery cable.

2. Drain the cooling system into a clean container for reuse.

3. In the engine compartment, open the heater valve cable clamp and disconnect the cable from the heater valve arm. Then, turn the heater valve to the fully opened position.

4. Disconnect the heater hoses from the heater core.

5. Remove the heater housing-to-chassis nut.

➡ **When removing the heater housing nut, be careful not to damage or bend the fuel lines, the brake lines, etc.**

6. Remove the instrument panel by performing the following procedure:

a. Remove the driver's side lower instrument panel cover screws, disengage the clips and remove the lower cover.

b. Remove the knee bolster bolts and the knee bolster.

c. Remove the glove box stops from each side of the glove box.

d. Remove the glove box-to-instrument panel bolts and the glove box.

e. Remove the lower console cover by disengaging the 4 clips and removing the cover.

f. Remove the 6 center pocket-to-instrument panel screws; then, insert a flat tipped screwdriver at the upper right side corner of the center pocket, push down on the top of the hook and remove the center pocket/beverage holder assembly.

g. Remove the center instrument panel lower cover screws and disengage the clips on the upper left side; then, disconnect the electrical connectors and remove the cover.

h. Gently, push the power window switch from the instrument panel's lower cover opening by hand. Disconnect the electrical connectors and remove the power window switch.

i. Close the driver's side air vent; then, gently, push out the clips and pull out the vent. Disconnect the electrical connectors and remove the vent.

j. Gently, push out the driver's side defogger trim; then, disconnect the electrical connector and remove the side defogger trim.

k. At the base of the steering wheel, remove the access panel and disconnect the air bag electrical connector.

l. Remove the steering column covers screws and the covers.

m. Remove the steering column-to-instrument panel nuts/bolts and lower the steering column.

n. Remove the instrument panel side covers.

o. Disconnect the wiring harness connector and remove the nuts.

p. Move the under-dash fuse/relay box.

q. Disconnect the antenna connector and the harness clips.

r. Remove the connector holder from the instrument panel frame.

s. Remove the control unit/relay bracket from behind the center of the instrument panel.

t. Remove the passenger's side lower instrument panel cover.

u. Disconnect the connectors and the harness clips.

v. Remove the instrument panel-to-chassis bolts.

w. Using an assistant, remove the instrument panel.

7. Remove the evaporator housing by performing the following procedure:

a. Discharge and recover the air conditioning system refrigerant.

b. In the engine compartment, remove the refrigerant lines-to-evaporator housing bolts.

c. Separate the lines, discard the grommets and plug the openings to prevent contamination.

d. Disconnect the evaporator housing's temperature sensor connector.

e. Remove the evaporator housing-to-chassis screws/nut and remove the evaporator housing.

8. Disconnect the mode control motor and the air mix control motor electrical connectors and remove the wiring harness clips and the wiring harness from the heater housing.

9. Remove the heater duct clip, the heater housing-to-chassis nuts and the heater housing.

10. Remove the heater core cover screws and the cover.

11. Remove the heater core pipe clamp screws and the clamp.

12. Remove the heater core from the heater housing.

To install:

13. Install the heater core in the heater housing.

14. Install the heater core pipe clamp and the clamp screws.

15. Install the heater core cover and the cover screws.

16. Install the heater housing, the heater housing-to-chassis nuts and the heater duct clip.

17. Install the wiring harness clips and the wiring harness to the heater housing and connect the mode control motor and the air mix control motor electrical connectors.

18. Install the evaporator housing by performing the following procedure:

a. Install the evaporator housing and the evaporator housing-to-chassis screws/nut.

b. Connect the evaporator housing's temperature sensor connector.

c. Using new grommets, connect the refrigerant lines.

d. In the engine compartment, install

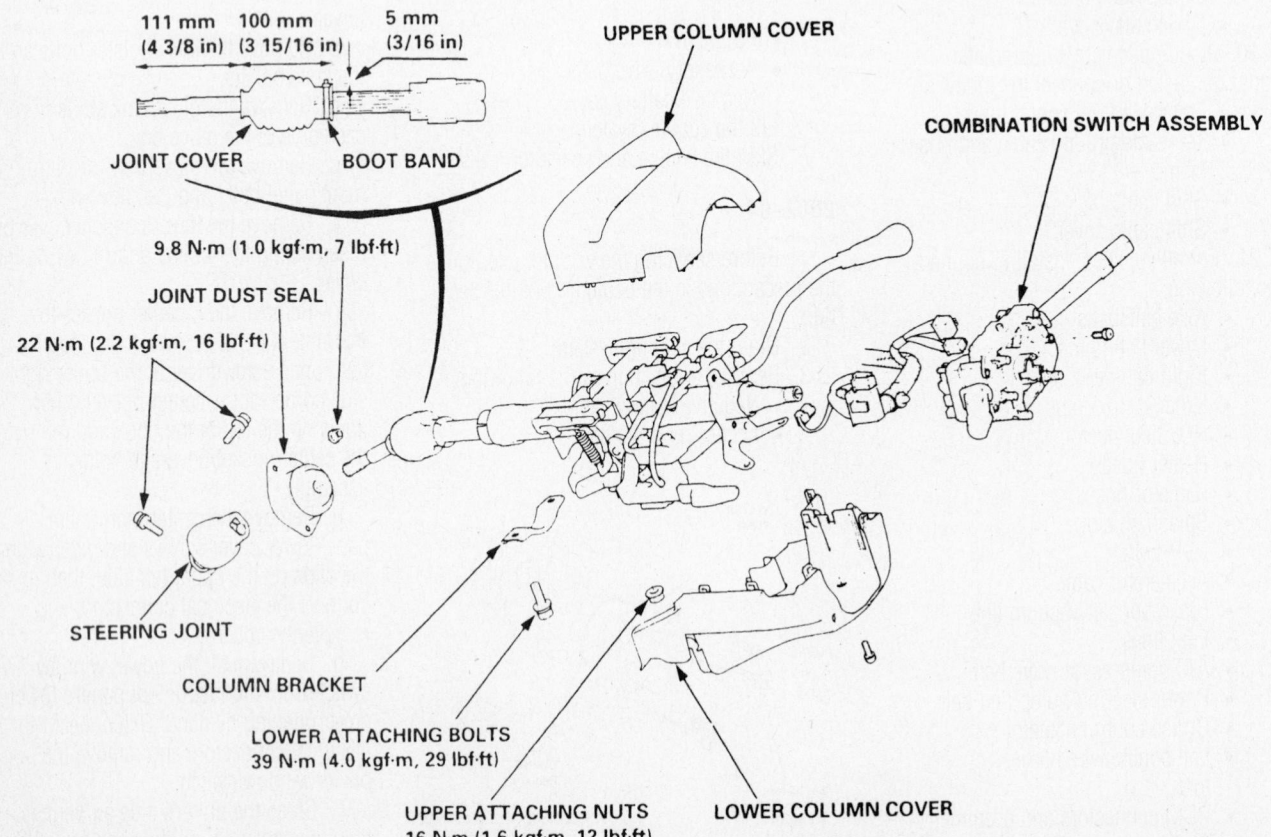

111 mm (4 3/8 in) 100 mm (3 15/16 in) 5 mm (3/16 in)

JOINT COVER   BOOT BAND

UPPER COLUMN COVER

COMBINATION SWITCH ASSEMBLY

9.8 N·m (1.0 kgf·m, 7 lbf·ft)

JOINT DUST SEAL

22 N·m (2.2 kgf·m, 16 lbf·ft)

STEERING JOINT

COLUMN BRACKET

LOWER ATTACHING BOLTS
39 N·m (4.0 kgf·m, 29 lbf·ft)

UPPER ATTACHING NUTS
16 N·m (1.6 kgf·m, 12 lbf·ft)

LOWER COLUMN COVER

93113GI2

Exploded view of the steering column and related components

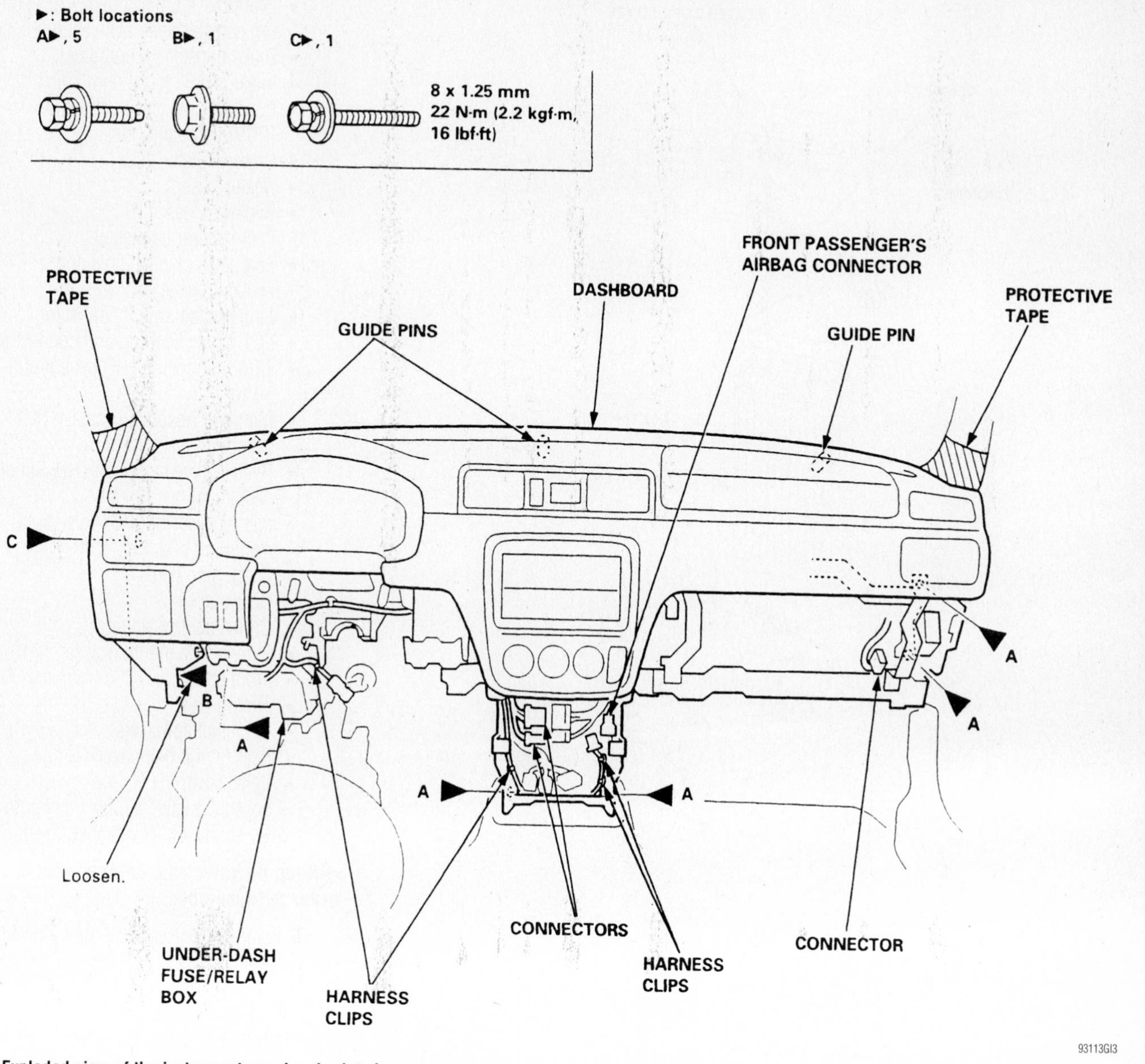

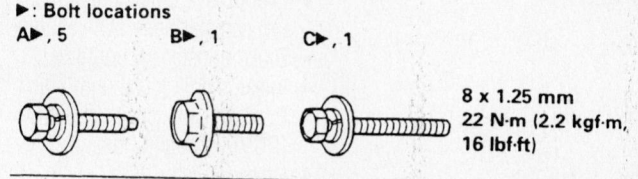

▶: Bolt locations
A▶, 5   B▶, 1   C▶, 1

8 x 1.25 mm
22 N·m (2.2 kgf·m,
16 lbf·ft)

PROTECTIVE
TAPE

GUIDE PINS

DASHBOARD

FRONT PASSENGER'S
AIRBAG CONNECTOR

GUIDE PIN

PROTECTIVE
TAPE

C ▶

B

A

A ▶

A

A ▶   ▶ A

A

A

Loosen.

UNDER-DASH
FUSE/RELAY
BOX

HARNESS
CLIPS

CONNECTORS

HARNESS
CLIPS

CONNECTOR

93113GI3

Exploded view of the instrument panel and related components

the refrigerant lines-to-evaporator housing bolts.

19. Install the instrument panel by performing the following procedure:

a. Using an assistant, install the
instrument panel.

b. Install the instrument panel-to-
chassis bolts.

c. Connect the connectors and the
harness clips.

d. Install the passenger's side lower
instrument panel cover.

e. Install the control unit/relay bracket
to the center of the instrument panel.

f. Install the connector holder to the
instrument panel frame.

g. Connect the antenna connector and
the harness clips.

h. Install the under-dash fuse/relay box.

i. Connect the wiring harness connector and install the nuts.

j. Install the instrument panel side
covers.

k. Install the steering column and the
column-to-instrument panel nuts/bolts.
Torque the nuts to 12 ft. lbs. (16 Nm)
and the bolts to 29 ft. lbs. (39 Nm).

l. Install the steering column covers
and the cover screws.

m. At the base of the steering wheel,
connect the air bag electrical connector
and install the access panel.

n. Connect the electrical connector
and install the driver's side defogger trim.

o. Connect the electrical connectors
and install driver's side air vent.

p. Connect the electrical connectors
and install the power window switch to
the instrument panel's lower cover opening.

q. Install the center instrument panel
lower cover and engage the clips on the
upper left side; then, connect the electrical connectors and install the cover
screws.

r. Install the center pocket/beverage
holder assembly and the 6 center pocket-
to-instrument panel screws.

s. Install the lower console cover by
engaging the 4 clips.

t. Install the glove box and the glove
box-to-instrument panel bolts.

u. Install the glove box stops to each
side of the glove box.

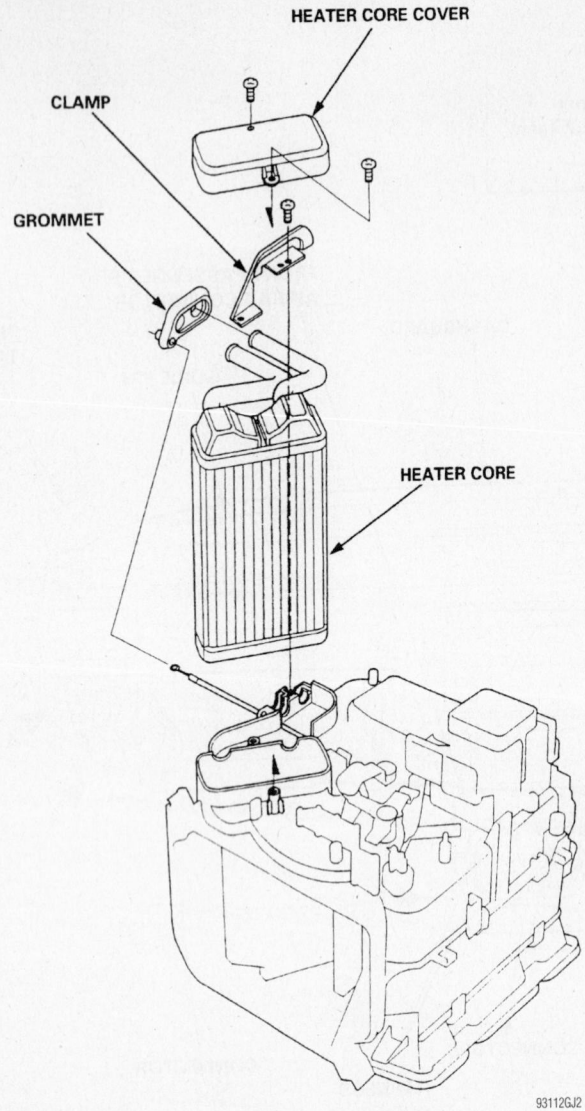

HEATER CORE COVER

CLAMP

GROMMET

HEATER CORE

93112GJ2

Exploded view of the heater core and housing

v. Install the knee bolster and the knee bolster bolts.

w. Install the driver's side lower instrument panel cover, engage the clips and install the lower cover screws.

➡When installing the heater housing nut, be careful not to damage or bend the fuel lines, the brake lines or etc.

20. Install the heater housing-to-chassis nut.

21. Connect the heater hoses to the heater core.

22. In the engine compartment, connect the cable to the heater valve arm and close the heater valve cable clamp.

23. Refill the cooling system.

24. Connect the negative battery cable.

25. Evacuate and charge and leak test the air conditioning system refrigerant.

26. Run the engine to normal operating temperatures; then, check the climate control operation and check for leaks.

Cylinder Head

REMOVAL & INSTALLATION

2001

1. Before servicing the vehicle, refer to the precautions in the beginning of this section.

2. Drain the cooling system.

3. Relieve the fuel system pressure.

4. Remove or disconnect the following:
- Negative battery cable
- Air intake assembly
- Accessory drive belts
- Power steering pump and bracket
- Accelerator cable
- Fuel lines
- Evaporative Emissions (EVAP) control canister hose and vacuum hose
- Brake booster vacuum line
- Intake manifold vacuum line
- Positive Crankcase Ventilation (PCV) valve and hose
- Upper radiator hose
- Heater hose
- Bypass hoses
- Fuel injector connectors
- Engine Coolant Temperature (ECT) sensor connector
- Radiator fan switch connector
- ECT gauge sending unit connector
- Throttle Position (TP) sensor connector
- Manifold Absolute Pressure (MAP) sensor connector
- Heated Oxygen Sensor (HO$_2$S) connector
- Idle Air Control (IAC) valve connector
- Spark plug wires
- Distributor
- Cruise control actuator
- Engine side mount bracket
- Front cover
- Timing belt.
- Camshaft sprockets
- Rear timing belt cover
- Valve cover

5. Loosen the valve adjuster locknuts and screws so that all valves are closed.

➡Keep all valvetrain components in order for assembly.

6. Remove or disconnect the following:
- Camshafts
- Rocker arms
- Exhaust front pipe
- Exhaust manifold bracket
- Intake manifold

7. Loosen the cylinder head bolts in sequence and ⅓ turns until all bolts are loose.

8. Remove the cylinder head.

To install:

9. Install the cylinder head with a new gasket.

10. Apply clean engine oil to the bolt threads and under the bolt heads.

11. Tighten the cylinder head bolts in sequence as follows:
   a. Step 1: 22 ft. lbs. (29 Nm)
   b. Step 2: 63 ft. lbs. (85 Nm)

12. Install or connect the following:
- Intake manifold. Tighten the bolts to 17 ft. lbs. (24 Nm).
- Exhaust manifold bracket
- Exhaust front pipe
- Rocker arms in their original positions

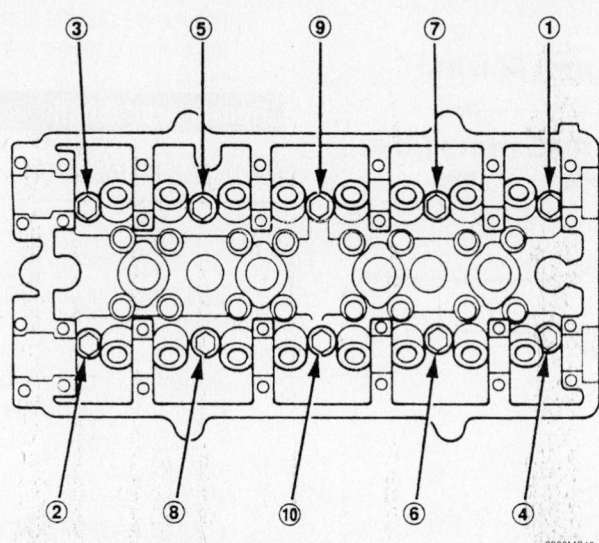

Cylinder head bolt loosening sequence—2.0L

9308MG13

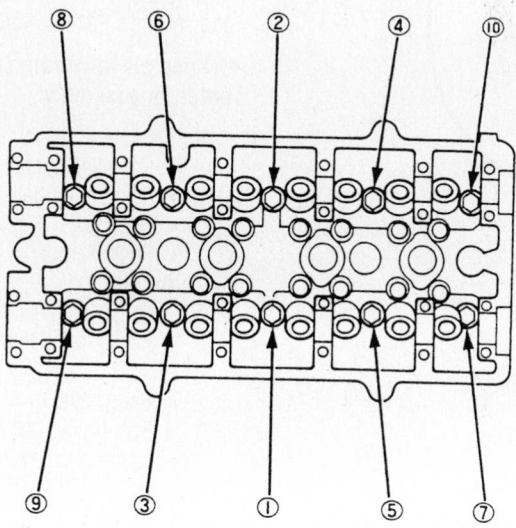

Cylinder head bolt tightening sequence—2.0L

9308MG14

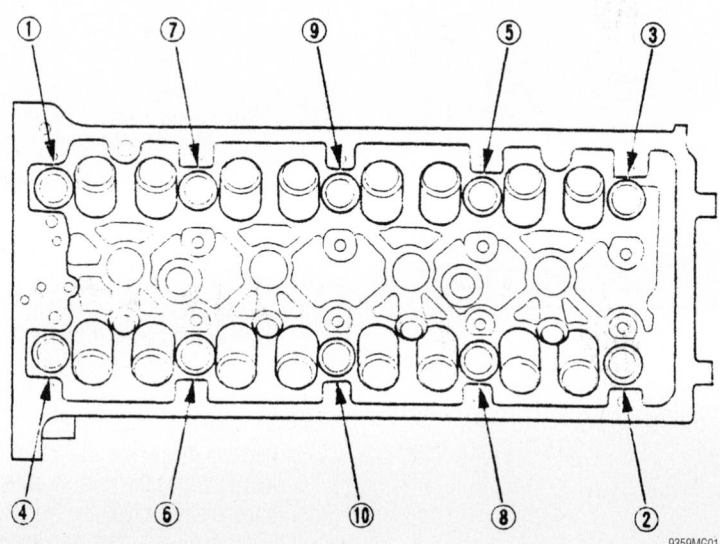

Cylinder head bolt loosening sequence—2.4L

9359MG01

- Camshafts
- Rear timing belt cover
- Camshaft sprockets. Tighten the bolts to 27 ft. lbs. (37 Nm).
- Timing belt
13. Adjust the valve clearance.
14. Install or connect the following:
    - Valve cover
    - Front cover
    - Engine side mount bracket
    - Cruise control actuator
    - Distributor
    - Spark plug wires
    - IAC valve connector
    - HO2S connector
    - MAP sensor connector
    - TP sensor connector
    - ECT gauge sending unit connector
    - Radiator fan switch connector
    - ECT sensor connector
    - Fuel injector connectors
    - Bypass hoses
    - Heater hose
    - Upper radiator hose
    - PCV valve and hose
    - Intake manifold vacuum line
    - Brake booster vacuum line
    - EVAP control canister hose and vacuum hose
    - Fuel lines
    - Accelerator cable
    - Power steering pump and bracket
    - Accessory drive belts
    - Air intake assembly
    - Negative battery cable
15. Fill the cooling system.
16. Start the engine and check for leaks.

2002–04

1. Before servicing the vehicle, refer to the precautions in the beginning of this section.
2. Drain the cooling system.
3. Relieve the fuel system pressure.
4. Remove or disconnect the following:
   - Negative battery cable
   - Accessory drive belt
   - Fuel lines
   - Intake manifold
   - Bypass hoses
   - Exhaust manifold
   - Cam chain
   - Engine wiring harness connectors
   - Upper radiator hose
   - Heater hose
   - Brake booster vacuum line
   - Intake manifold vacuum line
   - Rocker arms
5. Loosen the cylinder head bolts in sequence and ⅓ turns until all bolts are loose.

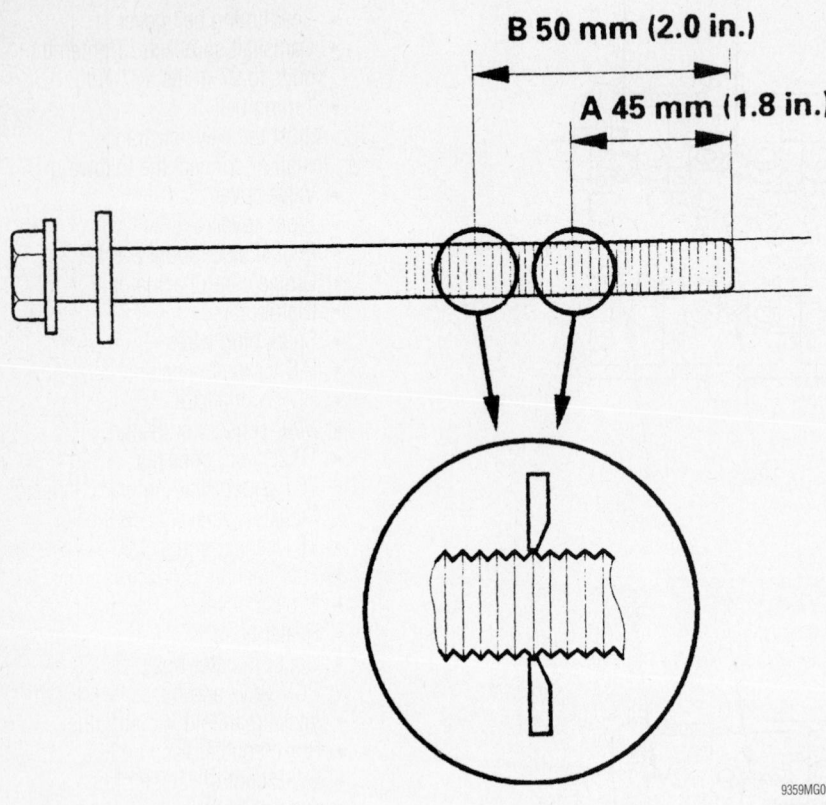

B 50 mm (2.0 in.)

A 45 mm (1.8 in.)

9359MG02

Cylinder head bolt inspection—2.4L

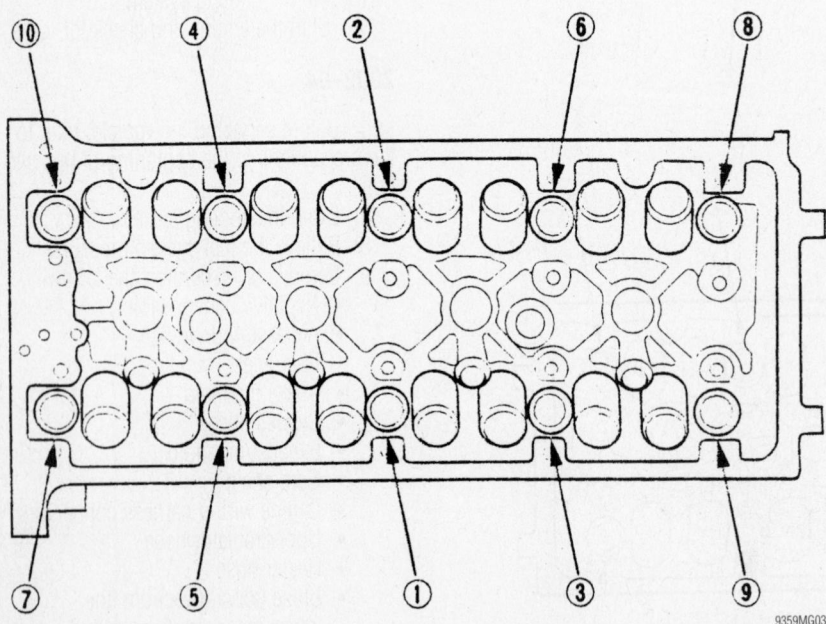

9359MG03

Cylinder head bolt torque sequence—2.4L

6. Remove the cylinder head.
7. Installation is the reverse of removal.

Rocker Arms/Shafts

REMOVAL & INSTALLATION

2.0L

1. Before servicing the vehicle, refer to the precautions in the beginning of this section.
2. Remove or disconnect the following:
   - Negative battery cable
   - Spark plug wires
   - Distributor
   - Valve cover
   - Accessory drive belts
   - Front cover
   - Timing belt.
3. Loosen the valve adjuster locknuts and screws so that all valves are closed.

➡ **Keep all valvetrain components in order for assembly.**

4. Remove or disconnect the following:
   - Camshaft sprockets
   - Rear timing belt cover
   - Camshafts
   - Rocker arms

To install:

5. Install or connect the following:
   - Rocker arms in their original positions
   - Camshafts
   - Rear timing belt cover
   - Camshaft sprockets. Tighten the bolts to 27 ft. lbs. (37 Nm).
   - Timing belt
6. Adjust the valve clearance.
7. Install or connect the following:
   - Front cover
   - Accessory drive belts
   - Valve cover
   - Distributor
   - Spark plug wires
   - Negative battery cable
8. Start the engine and check for leaks.

2.4L

1. Remove the camshaft chain
2. Loosen the rocker adjusting screws
3. Remove the camshaft holder bolts 2 turns at a time in the sequence shown.
4. Remove the camshaft chain guide, camshaft holders and camshafts.
5. Remove the rocker arm assembly.
6. Installation is the reverse of removal. Prior to installation, clean the No.5 rocker shaft holder mating surface and apply RTV gasket sealer to the mounting point on the

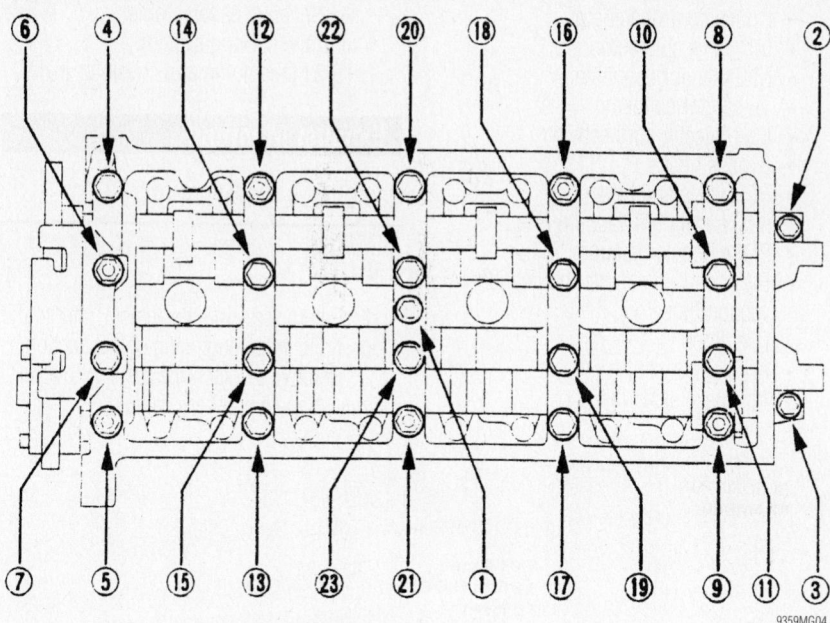

Camshaft holder bolt loosening sequence—2.4L

9359MG04

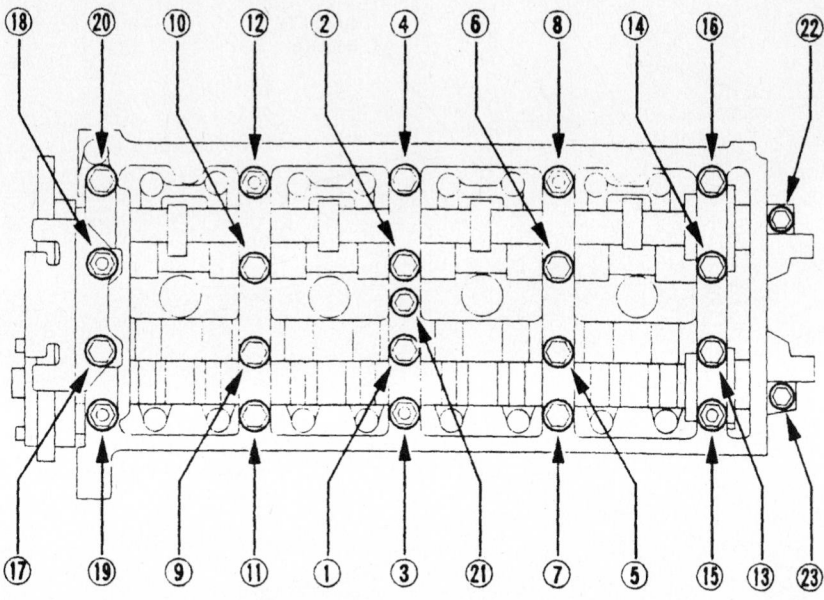

Camshaft holder bolt torque—2.4L

9359MG05

head. See the illustration for the torque sequence. Torque the 8mm bolts to 16 ft. lbs. (22 Nm) and the 6mm bolts to 9 ft. lbs. (12 Nm).

Intake Manifold

REMOVAL & INSTALLATION

2.0L

1. Before servicing the vehicle, refer to the precautions in the beginning of this section.

2. Drain the cooling system.
3. Relieve the fuel system pressure.
4. Remove or disconnect the following:
   - Negative battery cable
   - Air intake assembly
   - Intake manifold resonator chamber and bracket
   - Accelerator cable
   - Fuel lines
   - Evaporative Emissions (EVAP) control canister hose and vacuum hose
   - Brake booster vacuum line
   - Intake manifold vacuum line

   - Positive Crankcase Ventilation (PCV) valve and hose
   - Bypass hoses
   - Fuel injector connectors
   - Throttle Position (TP) sensor connector
   - Manifold Absolute Pressure (MAP) sensor connector
   - Idle Air Control (IAC) valve connector
   - Cruise control actuator
   - Intake manifold brackets
   - Intake manifold

To install:

5. Install or connect the following:
   - New intake manifold gasket
   - Intake manifold. Tighten the fasteners to 17 ft. lbs. (23 Nm).
   - Intake manifold brackets
   - Cruise control actuator
   - IAC valve connector
   - MAP sensor connector
   - TP sensor connector
   - Fuel injector connectors
   - Bypass hoses
   - PCV valve and hose
   - Intake manifold vacuum line
   - Brake booster vacuum line
   - EVAP control canister hose and vacuum hose
   - Fuel lines
   - Accelerator cable
   - Intake manifold resonator chamber and bracket
   - Air intake assembly
   - Negative battery cable
6. Fill the cooling system.
7. Start the engine and check for leaks.

2.4L

1. Before servicing the vehicle, refer to the precautions in the beginning of this section.

2. Drain the cooling system.
3. Relieve the fuel system pressure.
4. Remove or disconnect the following:
   - Negative battery cable
   - Air intake assembly
   - Accelerator cable
   - Cruise control cable
   - Evaporative Emissions (EVAP) control canister hose and vacuum hose
   - Brake booster vacuum line
   - Bypass hoses
   - Intake manifold resonator chamber and bracket, for 2001 models
   - Fuel lines
   - Intake manifold vacuum line
   - Positive Crankcase Ventilation (PCV) valve and hose
   - Fuel injector connectors

- Throttle Position (TP) sensor connector
- Manifold Absolute Pressure (MAP) sensor connector
- Idle Air Control (IAC) valve connector
- Intake manifold brackets
- Intake manifold

To install:

5. Install or connect the following:
- New intake manifold gasket
- Intake manifold. Tighten the fasteners to 16 ft. lbs. (22Nm).
- Intake manifold brackets

- Cruise control actuator
- IAC valve connector
- MAP sensor connector
- TP sensor connector
- Fuel injector connectors
- Bypass hoses
- PCV valve and hose
- Intake manifold vacuum line
- Brake booster vacuum line
- EVAP control canister hose and vacuum hose
- Fuel lines
- Accelerator cable
- Air intake assembly

- Negative battery cable
6. Fill the cooling system.
7. Start the engine and check for leaks.

Exhaust Manifold

REMOVAL & INSTALLATION

2.0L

1. Before servicing the vehicle, refer to the precautions in the beginning of this section.
2. Remove or disconnect the following:
- Negative battery cable

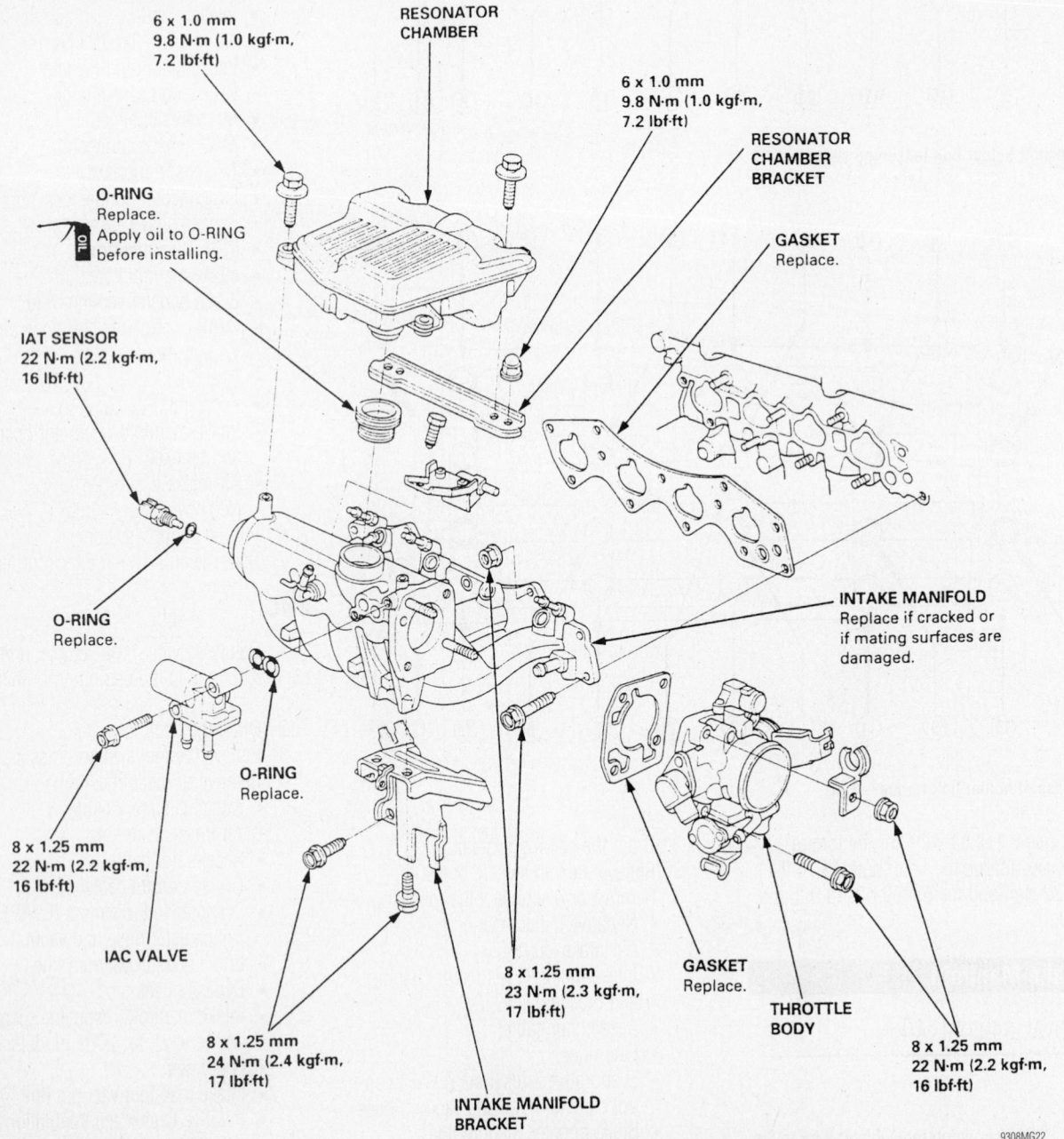

6 x 1.0 mm
9.8 N·m (1.0 kgf·m, 7.2 lbf·ft)

RESONATOR CHAMBER

6 x 1.0 mm
9.8 N·m (1.0 kgf·m, 7.2 lbf·ft)

RESONATOR CHAMBER BRACKET

O-RING
Replace.
Apply oil to O-RING before installing.

GASKET
Replace.

IAT SENSOR
22 N·m (2.2 kgf·m, 16 lbf·ft)

O-RING
Replace.

INTAKE MANIFOLD
Replace if cracked or if mating surfaces are damaged.

8 x 1.25 mm
22 N·m (2.2 kgf·m, 16 lbf·ft)

O-RING
Replace.

IAC VALVE

8 x 1.25 mm
24 N·m (2.4 kgf·m, 17 lbf·ft)

INTAKE MANIFOLD BRACKET

8 x 1.25 mm
23 N·m (2.3 kgf·m, 17 lbf·ft)

GASKET
Replace.

THROTTLE BODY

8 x 1.25 mm
22 N·m (2.2 kgf·m, 16 lbf·ft)

9308MG22

Intake manifold exploded view—2001 CR-V

- Exhaust manifold heat shield
- Heated Oxygen Sensor (HO2S) connector
- Exhaust front pipe
- Exhaust manifold bracket, if equipped
- Exhaust manifold

To install:

3. Install or connect the following:
- Exhaust manifold. Tighten the fasteners to 23 ft. lbs. (31 Nm).
- Exhaust manifold bracket, if equipped. Tighten the bolts to 33 ft. lbs. (44 Nm).
- Exhaust front pipe. Tighten the nuts to 40 ft. lbs. (55 Nm).
- Heated Oxygen Sensor (HO2S) connector
- Exhaust manifold heat shield
- Negative battery cable

2.4L

1. Before servicing the vehicle, refer to the precautions in the beginning of this section.
2. Remove or disconnect the following:
- Negative battery cable
- VTEC solenoid valve
- Driveshaft heat shield
- Exhaust manifold heat shield
- Exhaust front pipe
- Exhaust manifold bracket, if equipped
- Exhaust manifold

To install:

3. Install or connect the following:
- Exhaust manifold. Tighten the fasteners to 3 ft. lbs. (44m).
- Exhaust manifold bracket, if equipped. Tighten the bolts to 33 ft. lbs. (44 Nm).
- Exhaust front pipe. Tighten the nuts to 16t. lbs. (22m).
- Exhaust manifold heat shield
- Driveshaft heat shield
- VTEC solenoid valve
- Negative battery cable

Front Crankshaft Seal

REMOVAL & INSTALLATION

For the 2.4L engine, see the Timing Chain Removal & Installation procedure.

2.0L

1. Before servicing the vehicle, refer to the precautions in the beginning of this section.
2. Remove or disconnect the following:
- Negative battery cable
- Accessory drive belts

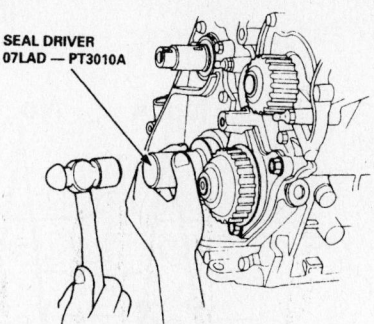

SEAL DRIVER
07LAD — PT3010A

7924MG48

Front crankshaft seal installation–2.0L

- Side engine mount
- Valve cover
- Crankshaft pulley
- Front cover
- Balance shaft belt, if equipped
- Timing belt
- Top Dead Center (TDC) sensor, if equipped
- Crankshaft timing sprocket
- Front crankshaft seal

To install:

3. Lubricate the crankshaft seal lip with grease prior to installation.
4. Install the front crankshaft seal so that it is flush with the surface of the oil pump housing.
5. Install or connect the following:
- Crankshaft timing sprocket
- Top Dead Center (TDC) sensor, if equipped

- Timing belt.
- Balance shaft belt, if equipped
- Front cover
- Crankshaft pulley. Tighten the bolt to 130 ft. lbs. (177 Nm).
- Valve cover
- Side engine mount
- Accessory drive belts
- Negative battery cable

6. Check the engine oil level and add if necessary.
7. Start the engine and check for leaks.

Camshaft

REMOVAL & INSTALLATION

2.0L

1. Before servicing the vehicle, refer to the precautions in the beginning of this section.
2. Remove or disconnect the following:
- Negative battery cable
- Spark plug wires
- Distributor
- Valve cover
- Accessory drive belts
- Front cover
- Timing belt

3. Loosen the valve adjuster locknuts and screws so that all valves are closed.

➡**Keep all valvetrain components in order for assembly.**

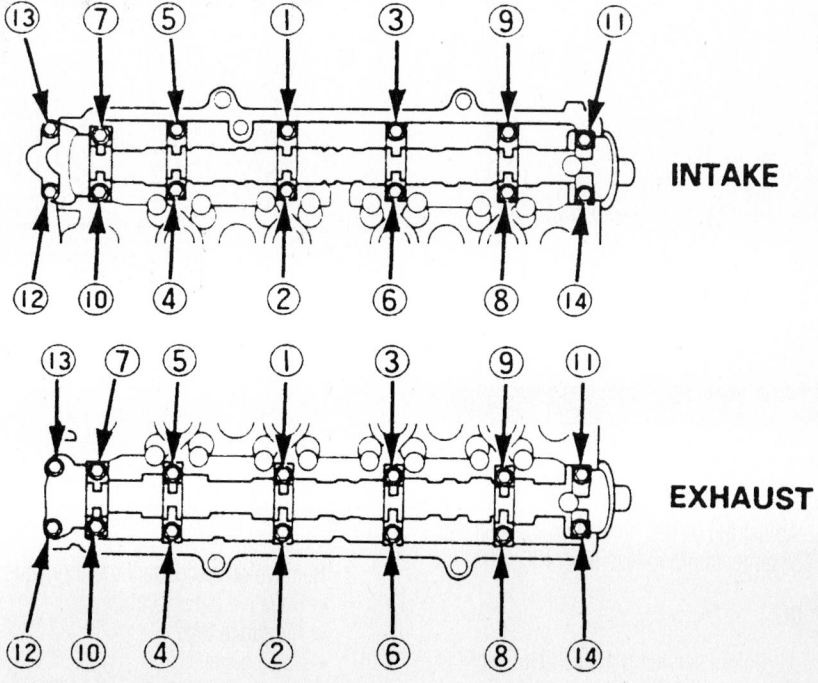

7924MG16

Camshaft bearing tightening sequence—2.0L

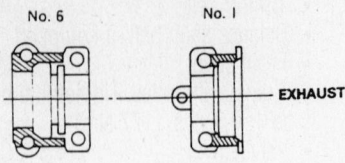

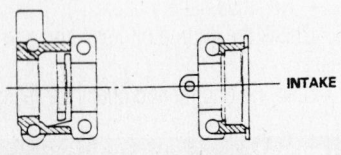

Apply liquid gasket to the shaded areas.

9302MG72

Apply liquid gasket to the shaded areas of the camshaft journals—2.0L

4. Remove or disconnect the following:
   - Camshaft sprockets
   - Rear timing belt cover
   - Camshafts

To install:

➡ **Use new O-rings, seals and gaskets when installing the camshaft.**

5. Install or connect the following:
   - Camshafts. Tighten the bearing cap bolts in sequence to 86 inch lbs. (10 Nm).
   - Rear timing belt cover
   - Camshaft sprockets. Tighten the bolts to 27 ft. lbs. (37 Nm).
   - Timing belt
6. Adjust the valve clearance.
7. Install or connect the following:
   - Front cover
   - Accessory drive belts
   - Valve cover
   - Distributor
   - Spark plug wires
   - Negative battery cable
8. Start the engine and check for leaks.

2.4L

See the Rocker Arm Shaft Removal & Installation procedure.

Valve Lash

ADJUSTMENT

Adjust the valves only when the cylinder head temperature is less than 100°F (38°C).

2.0L

1. Before servicing the vehicle, refer to the precautions in the beginning of this section.

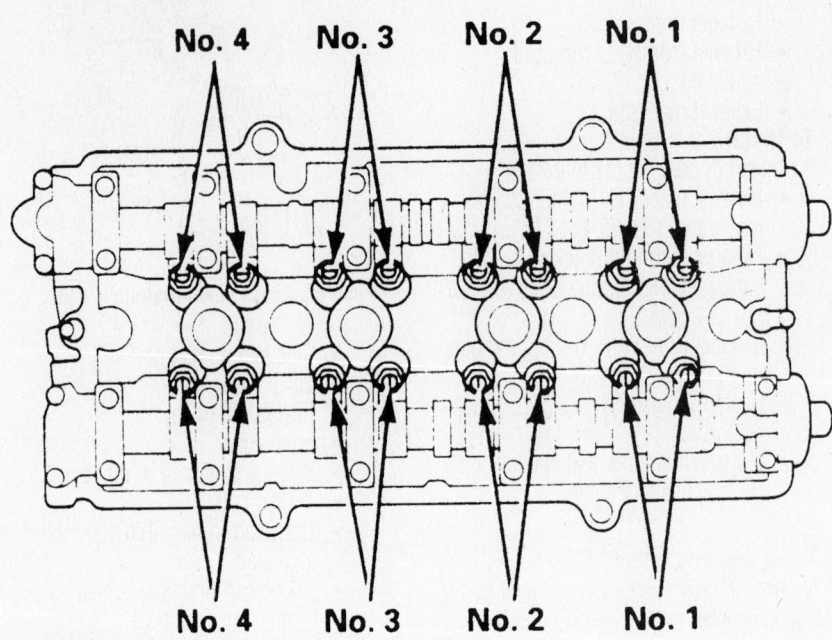

Intake and exhaust valve identification—2.0L

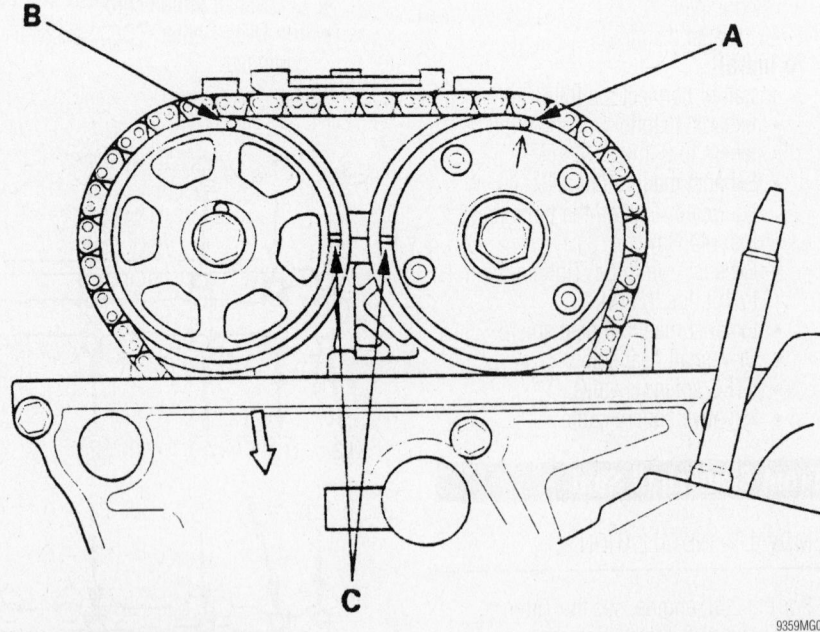

Align the timing marks—2.4L

2. Remove or disconnect the following:
   - Negative battery cable
   - Air intake tube
   - Valve cover
3. Rotate the crankshaft so that the valves to be adjusted are closed and the rocker arm is contacting the camshaft lobe base circle.
4. Measure the valve clearance. If adjustment is necessary, loosen the locknut and turn the adjusting screw as necessary to achieve the correct valve clearance.

5. The correct valve clearance is:
- Intake valves: 0.003–0.005 inches (0.08–0.12mm)
- Exhaust valves: 0.006–0.008 inches (0.16–0.20mm)

6. After adjustment, tighten the locknuts to 18 ft. lbs. (24 Nm).

7. Install or connect the following:
- Valve cover
- Air intake tube
- Negative battery cable

8. Start the engine and check for proper operation.

2.4L

1. Before servicing the vehicle, refer to the precautions in the beginning of this section.

2. Remove or disconnect the following:
- Negative battery cable
- cylinder head cover

3. Set the timing marks as shown in the illustration with N0.1 at TDC. Check all clearances. Intake should be 0.008–0.010 in.; exhaust should be 0.011–0.013 in. Intake locknut torque is 14 ft. lbs.; exhaust is 10 ft. lbs.

4. Rotate the crankshaft 180 degrees clockwise and recheck No.3.

5. Rotate the crankshaft 180 degrees clockwise and recheck No.4.

6. Rotate the crankshaft 180 degrees clockwise and recheck No.2.

Starter Motor

REMOVAL & INSTALLATION

2.0L

1. Before servicing the vehicle, refer to the precautions in the beginning of this section.

2. Remove or disconnect the following:
- Negative battery cable
- Starter wiring harness connectors
- Starter motor

To install:
3. Install or connect the following:
- Starter motor. Tighten the bolts to 33 ft. lbs. (44 Nm).
- Starter wiring harness connectors. Tighten the battery cable nut to 79 inch lbs. (9 Nm).
- Negative battery cable

2.4L

1. Before servicing the vehicle, refer to the precautions in the beginning of this section.

2. Remove or disconnect the following:
- Negative battery cable
- Knock sensor connector
- Starter wiring harness connectors
- Starter motor

To install:
3. Install or connect the following:
- Starter motor. Tighten the upper bolt to 33 ft. lbs. (44 Nm); the lower bolt to 47 ft. lbs. (64 Nm).
- Starter wiring harness connectors. Tighten the battery cable nut to 84 inch lbs. (9 Nm).
- Knock sensor connector
- Negative battery cable

Timing Belt

REMOVAL & INSTALLATION

2.0L

1. Disconnect the negative battery cable.

2. Position crankshaft so that No. 1 piston is at Top Dead Center (TDC).

3. Remove the splash guard.

4. Remove the accessory drive belts.

5. If equipped, remove the cruise control actuator.

6. Place a piece of wood between the oil pan and the jack, support the engine with a jack.

7. Remove upper engine bracket.

8. Remove the valve cover.

9. Remove the timing belt covers.

10. Loosen the adjusting bolt 180 degrees. Release the tension from the belt by pushing on the tensioner, then retighten the adjusting bolt.

11. Remove the timing belt.

To install:
12. Be sure the timing marks are properly aligned.

13. Install the timing belt on the pulleys following this sequence:
   a. Crankshaft pulley.
   b. Adjusting pulley.
   c. Water pump pulley.
   d. Exhaust camshaft pulley.
   e. Intake camshaft pulley.

14. Loosen and retighten the adjusting bolt to allow tension to be applied to the belt.

15. Install the lower and middle timing covers.

16. Install the crankshaft pulley and tighten the bolt to 130 ft. lbs. (177 Nm).

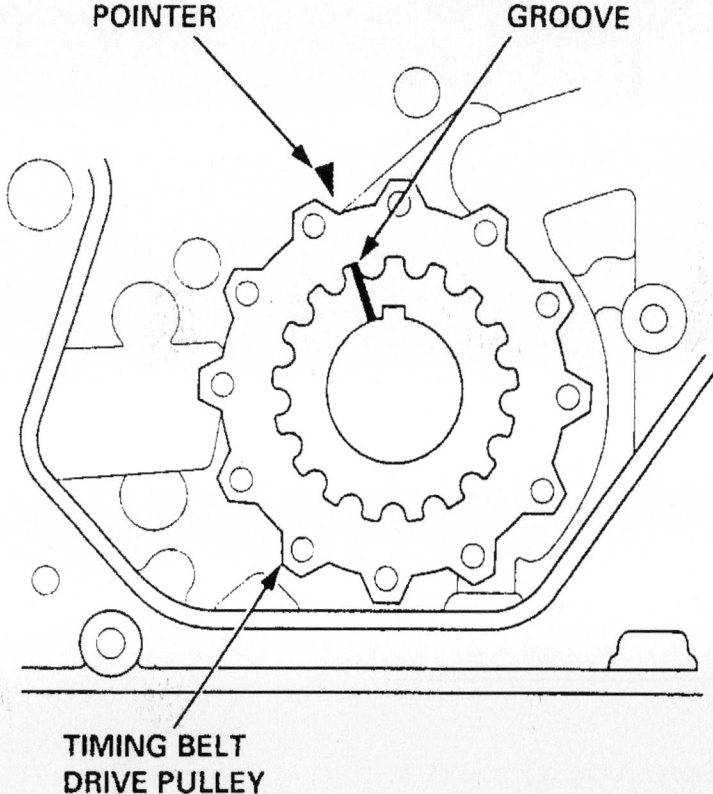

POINTER   **GROOVE**

TIMING BELT DRIVE PULLEY

79245G44

Crankshaft timing mark will be easier to verify when clean —2.0L (B20B4) engine

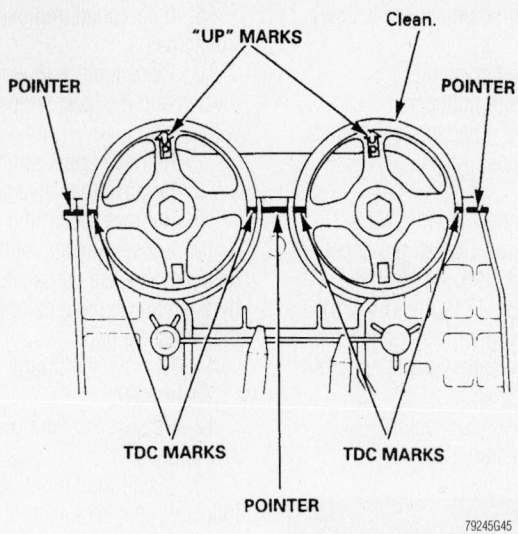

Intake and exhaust camshaft timing marks properly aligned at TDC —2.0L (B20B4) engine

79245G45

❄❄ WARNING

If any binding is felt when adjusting the timing belt tension by turning the crankshaft, STOP turning the engine, because the pistons may be hitting the valves.

17. Rotate the crankshaft about 5–6 times counterclockwise to seat the timing belt.

18. Position the No. 1 piston to TDC.

19. Loosen the adjusting bolt ½ turn.

20. Rotate the crankshaft counterclockwise 3 teeth on the camshaft pulley.

21. Tighten the adjusting bolt to 40 ft. lbs. (54Nm).

22. Retighten the crankshaft pulley bolt to 130 ft. lbs. (177 Nm).

23. Install the valve cover.

24. Install the engine mounting bracket, then remove the jack.

25. If removed, install the cruise control actuator.

26. Install the accessory drive belts.

27. Install the splash guard.

28. Connect the negative battery cable.

29. Check the engine operation and road test.

Timing Chain and Front Seal

REMOVAL & INSTALLATION

2.4L

1. Before servicing the vehicle, refer to the precautions in the beginning of this section.

2. Drain the engine oil.

3. Align the timing marks at TDC No.1.

4. Remove or disconnect the following:
- Negative battery cable
- Front splash shield
- Drive belt
- Cylinder head cover
- Crankshaft pulley
- CKP sensor
- VTC oil control connector
- VTC oil control solenoid valve

5. Support the engine with a block of wood and jack.

6. Remove or disconnect the following:
- Ground cable
- Upper support bracket
- Side engine mount
- Chain case

7. Loosely install the crank pulley. Turn the crankshaft counterclockwise to com-

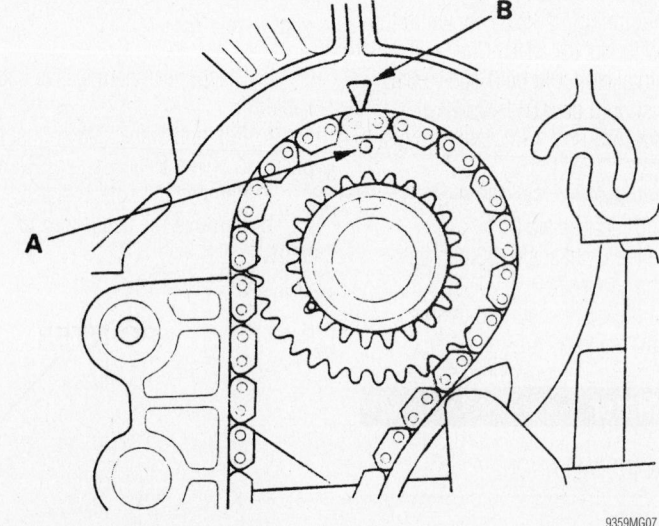

9359MG07

Align the crankshaft timing marks—2.4L

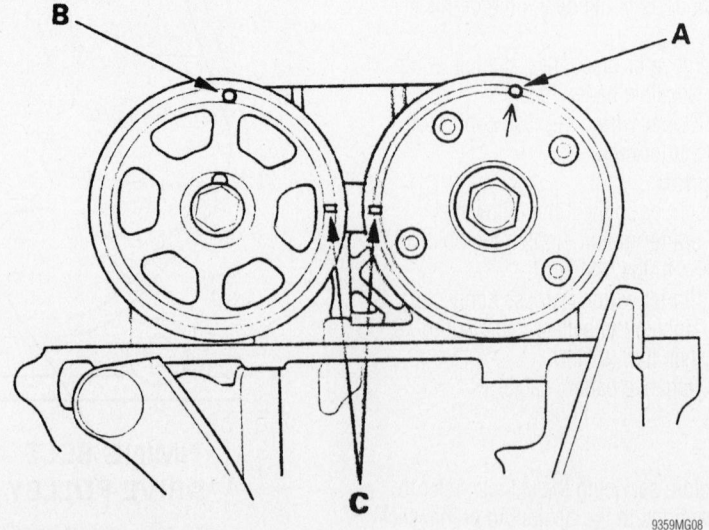

9359MG08

Align the camshaft timing marks—2.4L

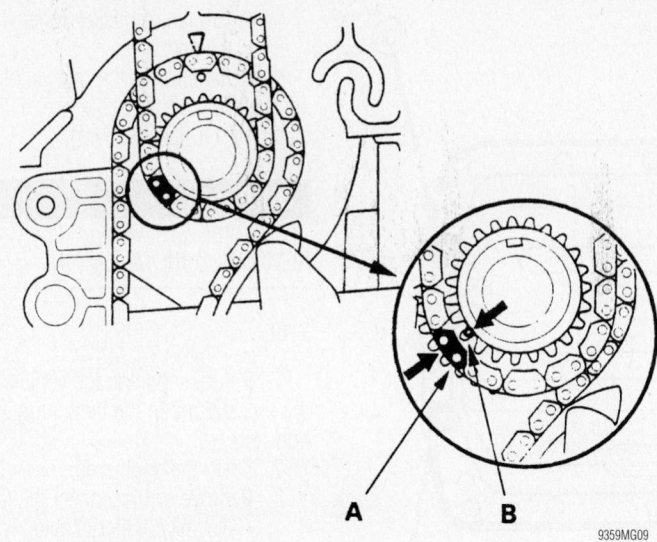

Chain installed on crankshaft—2.4L

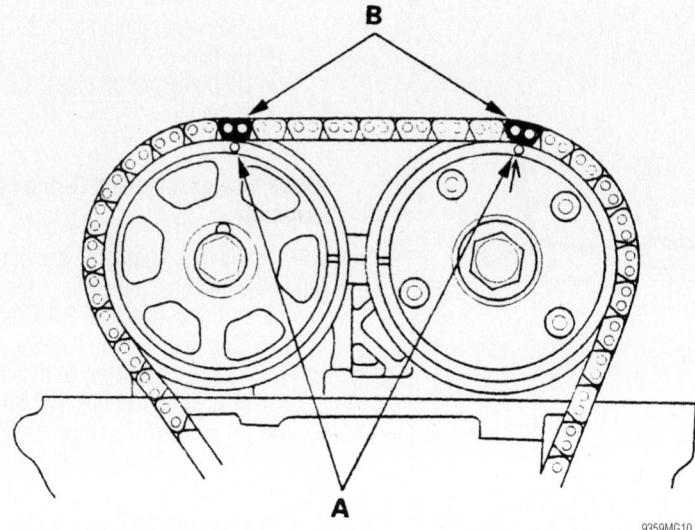

Chain installed on camshafts—2.4L

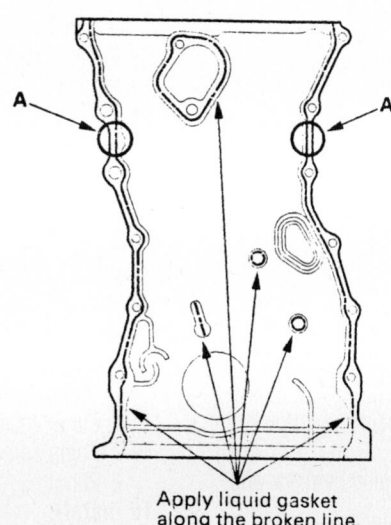

Apply liquid gasket
along the broken line.

Chain case sealer application—2.4L

press the auto-tensioner. Align the holes on the lock and auto-tensioner and insert a 1.5mm pin into the holes. Turn the crank clockwise to hold the pin.

8. Remove the auto-tensioner.
9. Remove the chain guides.
10. Remove the tensioner arm.
11. Remove the chain.
12. With the case removed, drive out the old seal and install a new one.

To install:

13. Set the crankshaft to TDC.
14. Set the camshafts to TDC.
15. Install the chain on the sprocket with the colored piece A aligned with the punch mark B.

16. The remainder of installation is the reverse of removal. See the accompanying illustration for sealer application. Observe the following torques:

- Chain guide: 9 ft. lbs. (12 Nm)
- Tensioner arm: 16 ft. lbs. (22 Nm)
- Auto-tensioner: 9 ft. lbs. (12 Nm)
- Upper chain guide: 16 ft. lbs. (22 Nm)
- Case: 9 ft. lbs. (12 Nm)
- Side mount: 33 ft. lbs. (44 Nm)
- Upper bracket: 40 ft. lbs. (54 Nm)

Oil Pan

REMOVAL & INSTALLATION

2.0L

1. Before servicing the vehicle, refer to the precautions in the beginning of this section.
2. Drain the engine oil.
3. Remove or disconnect the following:
   - Negative battery cable
   - Front splash shield
   - Heated Oxygen Sensor (HO2S) connector
   - Exhaust front pipe
   - Torque converter cover, if equipped with an automatic transaxle
   - Oil pan

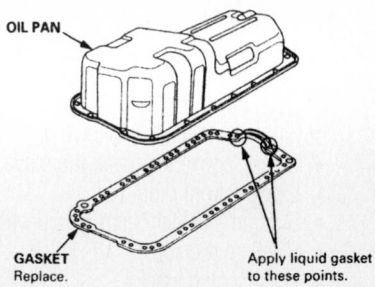

Oil pan gasket installation–2.0L

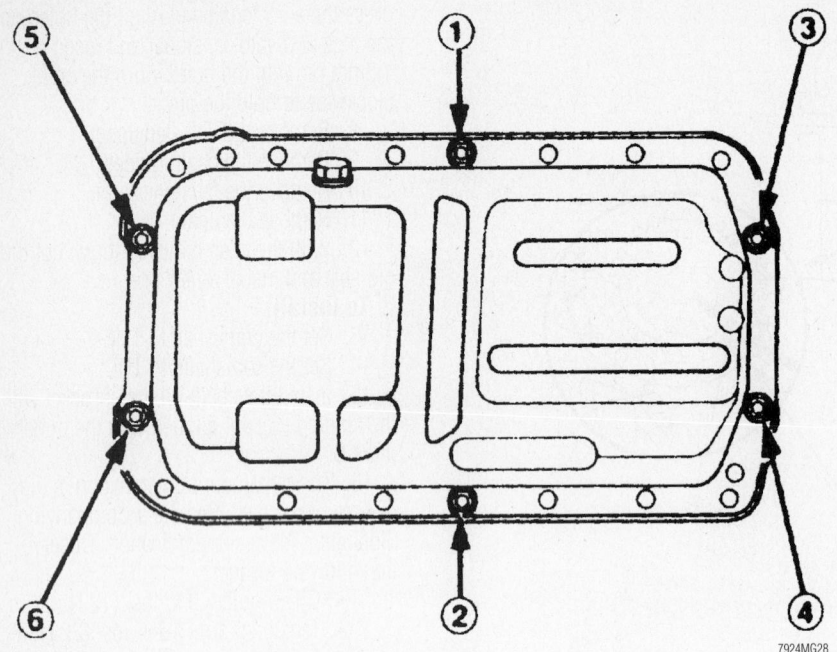

Oil pan fastener tightening sequence—2.0L

7924MG28

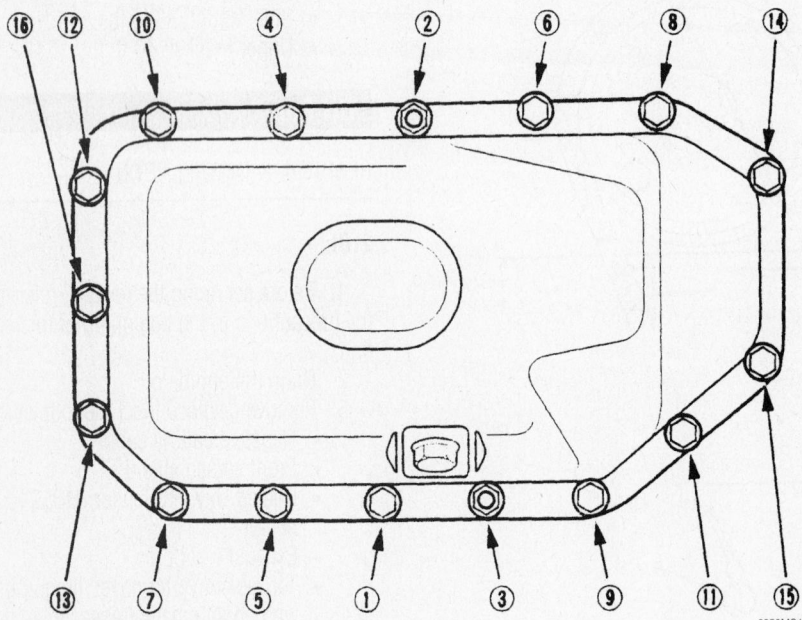

Oil pan fastener tightening sequence—2.4L

9359MG12

To install:

4. Install the oil pan.
5. Tighten the bolts in sequence to 105 inch lbs. (12 Nm).
6. Install or connect the following:

*   Torque converter cover, if removed
*   Exhaust front pipe
*   Subframe center beam, if removed
*   HO$_2$S connector
*   Front splash shield
*   Negative battery cable

2.4L

1. Before servicing the vehicle, refer to the precautions in the beginning of this section.
2. Drain the engine oil.
3. Remove or disconnect the following:

*   Subframe. See engine Removal and Installation.
*   With MT, the stiffener
*   Oil pan bolts

*   Oil pan. A gasket cutter will be needed.

4. Installation is the reverse of removal. Torque the bolts, in sequence, in 2 or 3 steps, to 9 ft. lbs. (12 Nm).

Oil Pump

REMOVAL & INSTALLATION

2.0L

1. Before servicing the vehicle, refer to the precautions in the beginning of this section.
2. Drain the engine oil.
3. Remove or disconnect the following:

*   Negative battery cable
*   Accessory drive belts
*   Front cover
*   Timing belt.
*   Crankshaft timing sprocket
*   Oil pan
*   Oil pump pickup tube
*   Oil pump

To install:

➡**Use new gaskets and O-ring seals for assembly.**

4. Apply liquid gasket to the oil pump and to the bolt hole threads.
5. Install or connect the following:

*   Oil pump. Tighten the 8mm bolts to 17 ft. lbs. (24 Nm) and the 6mm bolts to 86 inch lbs. (10 Nm).
*   Oil pump pickup tube. Tighten the fasteners to 86 inch lbs. (10 Nm).
*   Oil pan
*   Crankshaft timing sprocket
*   Timing belt.
*   Front cover
*   Accessory drive belts
*   Negative battery cable

6. Fill the crankcase to the correct level.
7. Start the engine and check for leaks.

2.4L

1. Before servicing the vehicle, refer to the precautions in the beginning of this section.
2. Drain the engine oil.
3. Remove or disconnect the following:

*   Negative battery cable
*   Oil pan
*   Pump chain
*   Pump sprocket
*   Pump

To install:

4. Make sure that No.1 piston is at TDC.

5. Align the dowel pin on the rear balance shaft with the mark on the pump.

6. Install the pump and sprocket loosely.

7. Remove the balance shaft holding pin.

8. Torque the 10mm mounting bolts to 33 ft. lbs. (44 Nm); the 8mm bolts to 16 ft. lbs. (22 Nm).

9. Torque the pulley bolt to 33 ft. lbs. (44 Nm).

10. Torque the tensioner bolts to 9 ft. lbs. (12 nm).

Rear Main Seal

REMOVAL & INSTALLATION

1. Before servicing the vehicle, refer to the precautions in the beginning of this section.

2. Remove or disconnect the following:
- Transaxle
- Clutch pressure plate and disc, if equipped
- Flywheel
- Oil seal

To install:

3. Install or connect the following:
- Oil seal. Drive the seal square into the seal case.

- Flywheel. Tighten the bolts in a crossing pattern to 54 ft. lbs. (73 Nm).
- Clutch pressure plate and disc, if equipped
- Transaxle

4. Check the fluid levels.

5. Start the engine and check for leaks.

Piston and Ring

POSITIONING

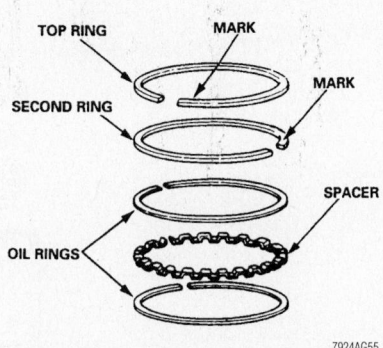

Piston ring positioning and top mark location

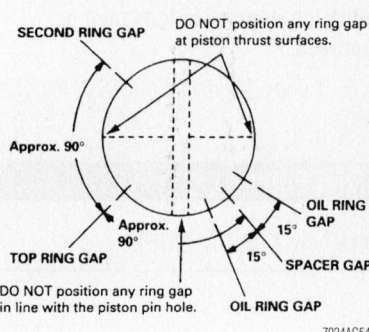

Piston ring end-gap spacing

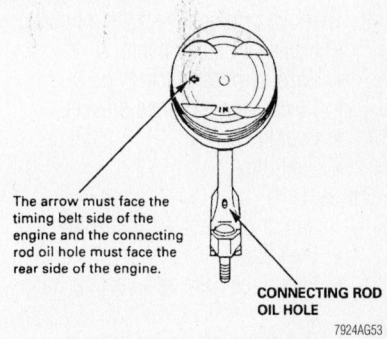

Piston and connecting rod assembly

FUEL SYSTEM

Fuel System Service Precautions

Safety is the most important factor when performing not only fuel system maintenance, but any type of maintenance. Failure to conduct maintenance and repairs in a safe manner may result in serious personal injury or death. Maintenance and testing of the vehicle's fuel system components can be accomplished safely and effectively by adhering to the following rules and guidelines:

- To avoid the possibility of fire and personal injury, always disconnect the negative battery cable unless the repair or test procedure requires that battery voltage be applied.
- Always relieve the fuel system pressure prior to disconnecting any fuel system component (injector, fuel rail, pressure regulator, etc.), fitting or fuel line connection. Exercise extreme caution whenever relieving fuel system pressure to avoid exposing skin, face and eyes to fuel spray. Please be advised that fuel under pressure may penetrate the skin or any part of the body that it contacts.
- Always place a shop towel or cloth around the fitting or connection prior to loosening to absorb any excess fuel due to spillage. Ensure that all fuel spillage

(should it occur) is quickly removed from engine surfaces. Ensure that all fuel soaked cloths or towels are deposited into a suitable waste container.

- Always keep a dry chemical (Class B) fire extinguisher near the work area.
- Do not allow fuel spray or fuel vapors to come into contact with a spark or open flame.
- Always use a backup wrench when loosening and tightening fuel line connection fittings. This will prevent unnecessary stress and torsion to fuel line piping. Always follow the proper torque specifications.
- Always replace worn fuel fitting O-rings with new. Do not substitute fuel hose or equivalent, where fuel pipe is installed.

Fuel System Pressure

RELIEVING

2.0L

1. Before servicing the vehicle, refer to the precautions in the beginning of this section.

2. Disconnect the negative battery cable.

3. Remove the fuel filler cap.

4. Hold the fuel rail inlet banjo bolt with a flare nut wrench. Hold the service bolt with a box end wrench.

5. Place a shop towel over the fitting to absorb leakage.

6. Loosen the service bolt 1 turn.

7. When repairs are complete, replace the sealing washers and tighten the service bolt to 25 ft. lbs. (33 Nm).

8. Install the fuel filler cap.

9. Connect the negative battery cable.

10. Start the engine and check for leaks.

2.4L

1. Before servicing the vehicle, refer to the precautions in the beginning of this section.

2. Disconnect the negative battery cable.

3. Remove the fuel filler cap.

4. Remove the engine cover.

5. Using a back up wrench and shop towel, turn the fuel pulsation damper one complete turn, slowly.

➡️Replace all washers whenever the pulsation damper is loosened or removed.

6. Tighten the damper to 16 ft. lbs. (22 Nm).

Fuel Filter

REMOVAL & INSTALLATION

2.0L

1. Before servicing the vehicle, refer to the precautions in the beginning of this section.
2. Relieve the fuel system pressure.
3. Remove or disconnect the following:
   - Negative battery cable
   - Wire harness bracket
   - Power steering hose bracket
   - Fuel lines
   - Fuel filter

To install:

4. Install or connect the following:
   - Fuel filter
   - Fuel lines. Use new sealing washers.
   - Power steering hose bracket
   - Wire harness bracket

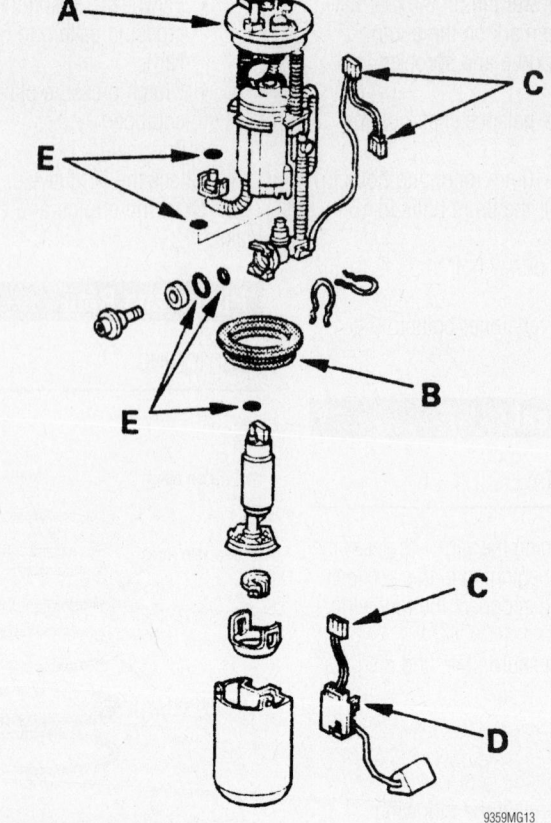

Exploded view of the fuel filter mounting–2.4L

SPECIAL BOLT
22 N·m
(2.2 kgf·m, 16 lbf·ft)

WASHER
Replace.

9.8 N·m
(1.0 kgf·m, 7.2 lbf·ft)

37 N·m
(3.8 kgf·m, 27 lbf·ft)

7924MG31

Exploded view of the fuel filter mounting–2.0L

- Negative battery cable
5. Start the engine and check for leaks.

2.4L

➡️The fuel filter should be replaced whenever the fuel pressure drops below 48 psi, after making sure that the fuel pump and fuel pressure regulator are okay.

1. Before servicing the vehicle, refer to the precautions in the beginning of this section.
2. Relieve the fuel system pressure.
3. Remove or disconnect the following:
   - Negative battery cable
   - Fuel pump
   - Fuel filter carrier (A)
   - Fuel filter

To install:

4. Install or connect the following:
   - Fuel filter
   - Fuel lines
   - New gasket (B)
   - New o-rings (E)
   - Connectors (C)
   - Sending unit (D)
5. Start the engine and check for leaks.

Fuel Pump

REMOVAL & INSTALLATION

2.4L

1. Before servicing the vehicle, refer to the precautions in the beginning of this section.
2. Relieve the fuel system pressure.
3. Remove or disconnect the following:
   - Negative battery cable
   - Left rear seat cushion
   - Base frame cover
   - Access panel
   - Fuel pump module wiring connector
   - Fuel supply and return lines
   - Fuel pump module

To install:

4. Install or connect the following:
   - Fuel pump module. Use a new seal and tighten the nuts to 52 inch lbs. (6 Nm).
   - Fuel supply and return lines
   - Fuel pump module wiring connector
   - Access panel
   - Base frame cover
   - Left rear seat cushion
   - Negative battery cable
5. Start the engine and check for leaks.

2.4L

1. Before servicing the vehicle, refer to the precautions in the beginning of this section.
2. Relieve the fuel system pressure.

3. Remove or disconnect the following:
   - Negative battery cable
   - Fuel filler cap
   - Access panel, under the rear seats
   - Fuel pump connector
   - Fuel supply and return lines
   - Fuel pump locknut
   - Fuel pump module
4. Installation is the reverse of removal.

Fuel Injector

REMOVAL & INSTALLATION

2.0L

1. Before servicing the vehicle, refer to the precautions in the beginning of this section.
2. Relieve the fuel system pressure.
3. Remove or disconnect the following:
   - Negative battery cable
   - Air intake resonator
   - Injector connectors
   - Positive Crankcase Ventilation (PCV) valve and hose
   - Fuel pressure regulator vacuum line
   - Fuel lines
   - Fuel supply manifold
   - Fuel injectors

To install:

4. Install the fuel injectors to the fuel supply manifold with new cushion rings and O-rings.
5. Install new seal rings to the intake manifold.
6. Install or connect the following:

- Fuel supply manifold and injector assembly. Tighten the nuts to 105 inch lbs. (12 Nm).
- Fuel lines
- Fuel pressure regulator vacuum line
- PCV valve and hose
- Injector connectors
- Air intake resonator
- Negative battery cable
7. Start the engine and check for leaks.

2.4L

1. Before servicing the vehicle, refer to the precautions in the beginning of this section.
2. Relieve the fuel system pressure.
3. Remove or disconnect the following:
   - Negative battery cable
   - Injector connectors, ground cable and harness holder
   - Fuel line
   - Fuel supply manifold
   - Fuel injectors

To install:

4. Install the fuel injectors to the fuel supply manifold with new O-rings coated with clean engine oil.
5. Install new seal rings, coated with clean engine oil, in the intake manifold.
6. Install or connect the following:
   - Fuel supply manifold and injector assembly. Tighten the nuts to 16 ft. lbs. (22 Nm).
   - Fuel lines
   - Injector connectors
   - Negative battery cable
7. Start the engine and check for leaks.

DRIVE TRAIN

Transaxle Assembly

REMOVAL & INSTALLATION

Automatic Transaxle

2001

1. Before servicing the vehicle, refer to the precautions in the beginning of this section.
2. Drain the transaxle.
3. Remove or disconnect the following:
   - Battery
   - Battery tray
   - Air intake assembly
   - Starter motor
   - Transaxle ground cable
   - Clutch pressure control solenoid valve connector
   - Mainshaft speed sensor connector

- Clutch pressure switch connectors
- Shift control solenoid valve connectors
- Lockup control solenoid connector
- Countershaft speed sensor connector
- Transaxle oil cooler lines
- Gear position switch connector
- Engine splash shield
- Front wheels
- Subframe center beam
- Rear driveshaft, if equipped
- Front motor mount bracket
- Lower ball joints
- Lower damper fork bolts
4. Separate the inner CV-joints from the transaxle and intermediate shaft and support the axle halfshafts out of the work area with safety wire.
5. Remove or disconnect the following:

- Right damper fork
- Right radius rod
- Intermediate shaft
- Shift cable holder and shift cable
- Engine stiffener
- Torque converter
- Transaxle mount
- Intake manifold bracket
- Rear mount bracket
- Transaxle flange bolts
- Transaxle

To install:

➡**Use new circlips, split pins and self-locking nuts for assembly.**

6. Install or connect the following:
   - Transaxle. Tighten the flange bolts to 47 ft. lbs. (64 Nm).
   - Rear mount bracket. Tighten the bolts to 40 ft. lbs. (54 Nm).

- Intake manifold bracket. Tighten the bolts to 16 ft. lbs. (22 Nm).
- Transaxle mount. Tighten the stud bolt and nuts to 28 ft. lbs. (38 Nm) and the through bolt to 40 ft. lbs. (54 Nm).
- Front motor mount bracket. Tighten the bolts to 28 ft. lbs. (38 Nm).
- Torque converter. Tighten the drive-plate bolts to 105 inch lbs. (12 Nm).
- Engine stiffener. Tighten the bolts to 33 ft. lbs. (44 Nm).
- Shift cable holder and shift cable
- Intermediate shaft. Tighten the bolts to 29 ft. lbs. (39 Nm).
- Axle halfshafts
- Right damper fork. Tighten the pinch bolt to 32 ft. lbs. (43 Nm).
- Right radius rod. Tighten the bolts to 76 ft. lbs. (103 Nm) and the nut to 32 ft. lbs. (43 Nm).
- Lower damper fork bolts. Tighten the nut to 47 ft. lbs. (64 Nm).
- Lower ball joints. Tighten the nut to 36–43 ft. lbs. (49–59 Nm).
- Rear driveshaft, if equipped
- Subframe center beam. Tighten the bolts to 37 ft. lbs. (50 Nm).
- Front wheels
- Engine splash shield
- Gear position switch connector
- Transaxle oil cooler lines
- Countershaft speed sensor connector
- Lockup control solenoid connector
- Shift control solenoid valve connectors
- Clutch pressure switch connectors
- Mainshaft speed sensor connector
- Clutch pressure control solenoid valve connector
- Transaxle ground cable
- Starter motor. Tighten the bolts to 33 ft. lbs. (44 Nm).
- Air intake assembly
- Battery tray
- Battery

7. Fill the transaxle to the correct level.
8. Start the engine and check for leaks.
9. Check the wheel alignment and adjust as necessary.

2002–04

1. Before servicing the vehicle, refer to the precautions in the beginning of this section.
2. Drain the transaxle.
3. Remove or disconnect the following:
- Battery
- Battery tray
- Air intake assembly
- Splash shield
- Transaxle ground cable
- Starter motor

- Clutch pressure control solenoid valve connector
- Mainshaft speed sensor connector
- Clutch pressure switch connectors
- Shift control solenoid valve connectors
- Lockup control solenoid connector
- Countershaft speed sensor connector
- Transaxle oil cooler lines
- Engine wiring harness from the air cleaner bracket
- Water pipe mounting bolt
- Brake booster and EVAP line mounting bolts. Attach an engine support hanger to the head to support the weight of the engine.
- Stabilizer link from the lower arm
- Lower arms from the knuckles
- Torque converter nuts
- Shift cable
- Front mount bolt and nut
- Rear mount bolts
- Subframe (see the Engine Removal and Installation procedure)
- Rear driveshaft, if equipped

4. Separate the inner CV-joints from the transaxle and intermediate shaft and support the axle halfshafts out of the work area with safety wire.
- Intermediate shaft
- Engine stiffener
- Transaxle mount bolts and nuts
- Transaxle

5. Installation is the reverse of removal. Observe the following torques:
- Air cleaner housing bracket bolt: 16 ft. lbs. (22 Nm)
- Front mount bolts: 47 ft. lbs. (64 Nm)
- Rear mount bracket bolts: 40 ft. lb. (54 Nm)
- Transmission-to-engine bolts: 47 ft. lbs. (64 Nm)
- Upper transmission mount bolt: 40 ft. lbs. (54 Nm)
- Upper transmission mount nuts: 40 ft. lbs. (54 Nm)
- Rear driveshaft bolts: 24 ft. lbs. (32 Nm)
- Subframe bolts: 76 ft. lbs. (103 Nm)

Manual Transaxle

2001

1. Before servicing the vehicle, refer to the precautions in the beginning of this section.
2. Remove or disconnect the following:
- Negative battery cable
- Air intake assembly
- Clutch slave cylinder and hose bracket

- Starter motor
- Transaxle ground cable
- Reverse lamp switch connector
- Wire harness bracket
- Shift cables and bracket
- Vehicle Speed Sensor (VSS) connector
- Splash shield
- Heated Oxygen Sensor (HO2S) connector
- Exhaust front pipe
- Rear driveshaft
- Lower ball joints
- Right damper fork

3. Separate the inner CV-joints from the transaxle and intermediate shaft and support the axle halfshafts out of the work area with safety wire.
4. Remove or disconnect the following:
- Intermediate shaft
- Rear engine stiffener
- Clutch housing cover
- Right front mount and bracket
- Transaxle mount and bracket
- Rear engine mounting bolts
- Transaxle flange bolts
- Transaxle

To install:

➡ **Use new circlips, split pins and self-locking nuts for assembly.**

5. Install or connect the following:
- Transaxle. Tighten the flange bolts to 47 ft. lbs. (64 Nm) and the rear engine mounting bolts to 61 ft. lbs. (83 Nm).
- Transaxle mount and bracket. Tighten the bracket bolts to 47 ft. lbs. (64 Nm) and the through bolt to 54 ft. lbs. (74 Nm).
- Right front mount and bracket. Tighten the mount bolts to 33 ft. lbs. (44 Nm) and the bracket bolts to 47 ft. lbs. (64 Nm).
- Clutch housing cover. Tighten the 12mm bolts to 22 ft. lbs. (29 Nm) and the 6mm bolts to 105 inch lbs. (12 Nm).
- Rear engine stiffener. Tighten the 8mm bolts to 18 ft. lbs. (24 Nm) and the 10mm bolt to 33 ft. lbs. (44 Nm).
- Intermediate shaft. Tighten the bolts to 29 ft. lbs. (39 Nm).
- Axle halfshafts
- Right damper fork. Tighten the pinch bolt to 32 ft. lbs. (43 Nm) and the nut to 47 ft. lbs. (64 Nm).
- Lower ball joints. Tighten the nut to 36–43 ft. lbs. (49–59 Nm).
- Rear driveshaft. Tighten the bolts to 24 ft. lbs. (32 Nm).
- Exhaust front pipe

- HO$_2$S connector
- Splash shield
- VSS connector
- Shift cables and bracket. Tighten the bracket bolts to 20 ft. lbs. (27 Nm).
- Wire harness bracket
- Reverse lamp switch connector
- Transaxle ground cable
- Starter motor. Tighten the bolts to 32 ft. lbs. (44 Nm).
- Clutch slave cylinder and hose bracket
- Air intake assembly
- Negative battery cable

6. Fill the transaxle to the correct level.

7. Start the engine and check for leaks.

8. Check the wheel alignment and adjust as necessary.

2002–04

1. Before servicing the vehicle, refer to the precautions in the beginning of this section.

2. Remove or disconnect the following:
- Negative battery cable
- Air intake assembly
- Transaxle ground cable
- Vehicle Speed Sensor (VSS) connector
- Splash shield
- Shift cables and bracket
- Clutch slave cylinder and hose bracket
- Wire harness bracket
- Water pipe mounting bolt
- Brake booster and EVAP line mounting bolts. Attach an engine support hanger to the head to support the weight of the engine.
- Upper transmission mounting bolts
- Transaxle mount and bracket

3. Separate the inner CV-joints from the transaxle and intermediate shaft and support the axle halfshafts out of the work area with safety wire.

4. Remove or disconnect the following:
- Intermediate shaft
- Right front mount and bracket
- Rear engine mounting bolts
- Rear driveshaft
- Subframe (see the Engine Removal and Installation procedure)
- Clutch housing cover
- Transaxle

5. Installation is the reverse of removal. Observe the following torques:
- Transaxle rear mount and bracket. Tighten the bracket bolts to 40 ft. lbs. (54 Nm) and the through bolt to 47 ft. lbs. (64 Nm)
- Transaxle. Tighten the flange bolts to 47 ft. lbs. (64 Nm)

- Front mount and bracket. Tighten the bolts to 47 ft. lbs. (64 Nm).
- Clutch housing cover. Tighten the bolts to 29 ft. lbs. (39 Nm)
- Subframe: 76 ft. lbs. (98 Nm)

Clutch

ADJUSTMENTS

The CR-V is equipped with a hydraulic clutch system. No adjustment is necessary.

REMOVAL & INSTALLATION

1. Before servicing the vehicle, refer to the precautions in the beginning of this section.

2. Remove or disconnect the following:
- Negative battery cable
- Transaxle
- Pressure plate. Loosen the bolts evenly in a crossing pattern.
- Clutch disc

To install:

3. Install the clutch disc and pressure plate. Tighten the pressure plate bolts in a crossing pattern and in several steps to 19 ft. lbs. (25 Nm).

4. Install or connect the following:
- Transaxle
- Negative battery cable

Hydraulic Clutch System

BLEEDING

1. Before servicing the vehicle, refer to the precautions in the beginning of this section.

2. Attach a hose to the bleeder screw and suspend the other end in a container of clean brake fluid.

3. Open the bleeder screw.

4. Slowly pump the clutch pedal until no more air bubbles appear at the bleeder hose.

5. Tighten the bleeder screw to 70 inch lbs. (8 Nm).

6. Refill the clutch master cylinder as necessary.

7. Check for leaks and proper clutch operation.

Transfer Assembly

REMOVAL & INSTALLATION

1. Before servicing the vehicle, refer to the precautions in the beginning of this section.

2. Drain the transaxle fluid.

3. Remove or disconnect the following:
- Negative battery cable
- Heated Oxygen Sensor (HO$_2$S) connector on 2001 models
- Exhaust front pipe on 2001 models
- Rear driveshaft
- Transfer assembly and bracket

To install:

4. Install the transfer assembly and bracket with a new O-ring seal. Tighten the 10mm bolts to 33 ft. lbs. (44 Nm) and the 8mm bolts to 17 ft. lbs. (24 Nm).

5. Install or connect the following:
- Rear driveshaft. Tighten the bolts to 24 ft. lbs. (32 Nm).
- Exhaust front pipe
- HO$_2$S connector
- Negative battery cable

6. Fill the transaxle to the correct level and check for leaks.

Halfshaft

REMOVAL & INSTALLATION

Front

2001

1. Before servicing the vehicle, refer to the precautions in the beginning of this section.

2. Drain the transaxle.

3. Remove or disconnect the following:
- Negative battery cable
- Front wheels
- Damper fork
- Lower ball joint
- Spindle nut

4. Pry the inboard joint from the transaxle or intermediate shaft.

5. Remove the outer CV-joint stub shaft from the hub by tapping the stub shaft with a plastic hammer.

To install:

➡**Use new circlips, split pins and self-locking nuts for assembly.**

6. Install the outer CV-joint stub shaft into the hub.

7. Install the inner CV-joint to the transaxle or intermediate shaft until the circlip locks in the retaining groove.

8. Install or connect the following:
- Lower ball joint. Tighten the nut to 36–43 ft. lbs. (49–59 Nm).
- Damper fork. Tighten the pinch bolt to 32 ft. lbs. (43 Nm) and the nut to 47 ft. lbs. (64 Nm).
- Spindle nut. Tighten the nut to 181 ft. lbs. (245 Nm).
- Front wheels

- Negative battery cable

9. Fill the transaxle to the correct level and check for leaks.

2002–04

1. Before servicing the vehicle, refer to the precautions in the beginning of this section.

2. Drain the transaxle.

3. Remove or disconnect the following:
- Negative battery cable
- Front wheels
- Stabilizer bar
- Lower ball joint
- Spindle nut

4. On the left side, pry the inboard joint from the case with a prybar.

5. On the right side, drive the inboard shaft off the intermediate shaft with a drift and hammer.

6. Installation is the reverse of removal. Observe the following torques:
- Ball stud nuts: 40 ft. lbs. (54 Nm)
- Stabilizer link nuts: 29 ft. lbs. (39 Nm)
- Spindle nut: 181 ft. lbs. (245 Nm)

Rear

1. Before servicing the vehicle, refer to the precautions in the beginning of this section.

2. Drain the differential.

3. Remove or disconnect the following:
- Negative battery cable
- Rear wheels
- Spindle nut

4. Pry the inboard joint from the differential.

5. Remove the outer CV-joint stub shaft from the hub by tapping the stub shaft with a plastic hammer.

To install:

➡**Use new circlips and self-locking nuts for assembly.**

6. Install the outer CV-joint stub shaft into the hub.

7. Install the inner CV-joint to the differential until the circlip locks in the retaining groove.

8. Install or connect the following:

- Spindle nut. Tighten the nut to 134 ft. lbs. (181 Nm).
- Rear wheels
- Negative battery cable

9. Fill the differential to the correct level and check for leaks.

CV-Joint

OVERHAUL

Front

OUTBOARD JOINT

1. Before servicing the vehicle, refer to the precautions in the beginning of this section.

2. Remove or disconnect the following:
- Axle halfshaft from the vehicle and place it in a vise
- Outboard joint boot clamps and push the boot back
- Outboard joint by driving it off the axle shaft with a brass drift and hammer
- Outboard joint boot

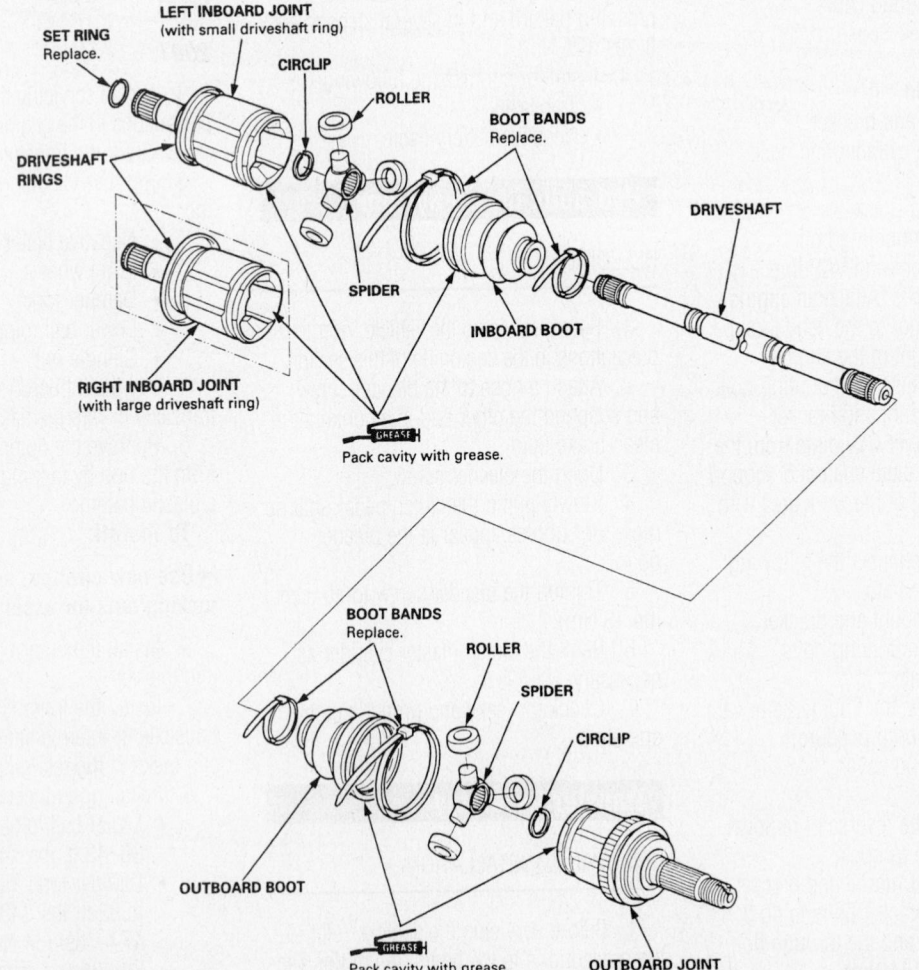

Exploded view of the rear axle—CR-V

To install:

➡**Use new circlips and boot clamps for assembly.**

3. Install the outboard joint boot and clamps to the axle shaft.

4. Fill the outboard joint with grease. Install the outboard joint to the axle shaft. Tap the stub shaft with a brass hammer to seat the circlip.

5. Fill the outboard joint boot with grease and install the boot clamps.

6. Install the axle halfshaft to the vehicle.

INBOARD JOINT

1. Before servicing the vehicle, refer to the precautions in the beginning of this section.

2. Remove or disconnect the following:
- Axle halfshaft from the vehicle.
- Inboard joint boot clamps and push the boot back
- Inboard joint housing from the axle
- Rollers from the spider
- Snapring and the spider from the axle shaft
- Inboard joint boot

To install:

➡**Use new circlips and boot clamps for assembly.**

3. Install or connect the following:
- Inboard joint boot and clamps to the axle shaft
- Spider with a new snapring
- Rollers to the spider

4. Fill the joint housing with grease and install it.

5. Fill the inboard joint boot with grease and install the boot clamps.

6. Install the axle halfshaft to the vehicle.

Rear

1. Before servicing the vehicle, refer to the precautions in the beginning of this section.

2. Remove or disconnect the following:
- Axle halfshaft from the vehicle
- Joint boot clamps and push the boot back
- Joint housing from the axle
- Rollers from the spider
- Snapring and the spider from the axle shaft
- Joint boot

To install:

➡**Use new circlips and boot clamps for assembly.**

3. Install or connect the following:
- Joint boot and clamps to the axle shaft
- Spider with a new snapring
- Rollers to the spider

4. Fill the joint housing with grease and install it.

5. Fill the joint boot with grease and install the boot clamps.

6. Install the axle halfshaft to the vehicle.

Pinion Seal

REMOVAL & INSTALLATION

1. Before servicing the vehicle, refer to the precautions in the beginning of this section.

2. Remove or disconnect the following:
- Driveshaft
- Companion flange
- Pinion seal

To install:

➡**Use a new locknut and O-ring for assembly.**

3. Install or connect the following:
- Pinion seal. Drive the seal square into the bore.
- Companion flange. Tighten the locknut to 87 ft. lbs. (118 Nm).
- Driveshaft. Tighten the flange bolts to 24 ft. lbs. (32 Nm).

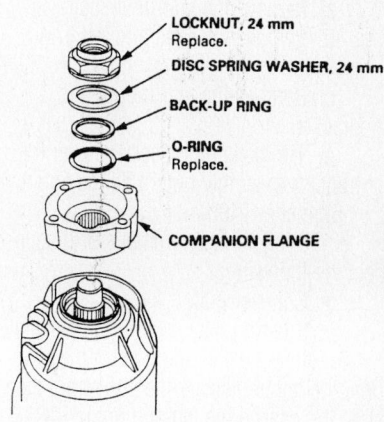

9308MG31

Exploded view of the rear differential pinion components—CR-V

STEERING AND SUSPENSION

Air Bag

✳✳ CAUTION

Some vehicles are equipped with an air bag system. The system must be disarmed before performing service on, or around, system components, the steering column, instrument panel components, wiring and sensors. Failure to follow the safety precautions and the disarming procedure could result in accidental air bag deployment, possible injury and unnecessary system repairs.

PRECAUTIONS

Several precautions must be observed when handling the inflator module to avoid accidental deployment and possible personal injury.

- Never carry the inflator module by the wires or connector on the underside of the module.

- When carrying a live inflator module, hold securely with both hands, and ensure that the bag and trim cover are pointed away.

- Place the inflator module on a bench or other surface with the bag and trim cover facing up.

- With the inflator module on the bench, never place anything on or close to the module which may be thrown in the event of an accidental deployment.

Before servicing the vehicle, also make sure to refer to the precautions in the beginning of this section as well.

DISARMING

Disconnect and isolate the negative battery cable. Wait 3 minutes for the system capacitor to discharge before performing any service.

Power Rack and Pinion Steering Gear

REMOVAL & INSTALLATION

2001

✳✳ WARNING

Do not permit the steering wheel to turn whenever the steering gear is disconnected from the steering column. Damage to the air bag wiring can result.

1. Before servicing the vehicle, refer to the precautions in the beginning of this section.

2. Center the steering wheel and lock it in position.

3. Remove or disconnect the following:

- Negative battery cable
- Air bag and steering wheel
- Steering flexible joint
- Front wheels
- Outer tie rod ends
- Heated Oxygen Sensor (HO2S) connector
- Exhaust front pipe
- Transmission shift cable
- Power steering fluid lines

4. For 4 wheel drive vehicles, perform the following:

a. Remove the rear driveshaft.

b. Remove the right and left front mounts.

c. Remove the rear mount and bracket.

d. Tilt the engine back to lower the transfer assembly output flange about 1.57 inches (40mm).

5. For all vehicles, remove or disconnect the following:

- Stiffener plate
- Steering gear mounting brackets

6. Slide the steering gear to the right until the left end clears the subframe, then slide the gear to the left and out of the vehicle.

To install:

7. Position the steering gear in the vehicle, right end first.

8. Install or connect the following:

- Steering gear mounting brackets. Tighten the bolts to 29 ft. lbs. (39 Nm).
- Stiffener plate. Tighten the plate bolts to 28 ft. lbs. (38 Nm) and the steering bear bolt to 32 ft. lbs. (43 Nm).

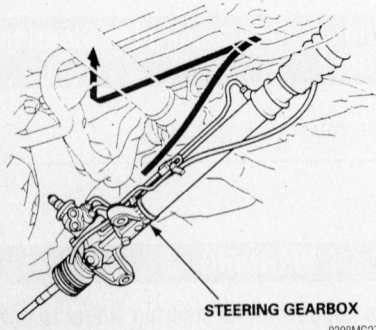

STEERING GEARBOX

9308MG27

Rack and pinion steering gear installation—2001

➡**Use new self-locking nuts and bolts for engine mount installation.**

9. For 4 wheel drive vehicles, install or connect the following:

- Rear mount and bracket. Tighten the upper bracket bolt to 43 ft. lbs. (59 Nm), the lower bracket bolt to 61 ft. lbs. (83 Nm), the mount attaching bolts to 47 ft. lbs. (64 Nm), and the mount through bolt to 43 ft. lbs. (59 Nm).
- Left and right front mounts. Tighten the bolts to 33 ft. lbs. (44 Nm) and the nuts to 43 ft. lbs. (59 Nm).
- Rear driveshaft. Tighten the bolts to 24 ft. lbs. (32 Nm).

10. For all vehicles, install or connect the following:

- Power steering fluid lines
- Transmission shift cable
- Exhaust front pipe
- HO2S connector
- Outer tie rod ends
- Front wheels. Position the wheels straight ahead.
- Steering flexible joint. Tighten the pinch bolts to 16 ft. lbs. (22 Nm).
- Steering wheel and air bag
- Negative battery cable

11. Fill the power steering system.

12. Check the wheel alignment and adjust as necessary.

2002–04

⁂ WARNING

Do not permit the steering wheel to turn whenever the steering gear is disconnected from the steering column. Damage to the air bag wiring can result.

1. Before servicing the vehicle, refer to the precautions in the beginning of this section.

2. Center the steering wheel and lock it in position.

3. Remove or disconnect the following:

- Negative battery cable
- Air bag and steering wheel
- Front wheels
- Driver's side dashboard lower cover and undercover
- Air cleaner housing
- Steering joint bolts
- Tie rod ends
- Steering hoses
- Left side flange bolts
- Mounting brackets

4. Lower the unit so the pinion shaft points outward. Remove the pinion shaft

grommet. The steering gear is removed through the driver's side.

5. Installation is the reverse of removal. Observe the following torques:

- Mounting bracket and side flange bolts: 46 ft. lbs. (62 Nm)
- Supply line flare nut: 27 ft. lbs. (37 Nm)
- Tie rod ball stud nuts: 32 ft. lbs. (43 Nm)
- Steering joint bolts: 21 ft. lbs. (28 Nm)

Strut

REMOVAL & INSTALLATION

Front

2001

1. Before servicing the vehicle, refer to the precautions in the beginning of this section.

2. Remove or disconnect the following:

- Front wheel
- Brake hose retainer
- Damper fork
- Strut

To install:

➡**Use new self-locking fasteners for assembly.**

3. Install or connect the following:

- Strut. Tighten the mounting nuts to 43 ft. lbs. (59 Nm).
- Damper fork. Tighten the pinch bolt to 32 ft. lbs. (43 Nm) and the lower bolt to 47 ft. lbs. (64 Nm).
- Brake hose retainer
- Front wheel

2002–04

1. Before servicing the vehicle, refer to the precautions in the beginning of this section.

2. Remove or disconnect the following:

- Front wheel
- Tie rod end
- Brake hose retainer
- ABS sensor
- Strut

To install:

➡**Use new self-locking fasteners for assembly.**

3. Install or connect the following:

- Strut. Tighten the upper mounting nuts to 33 ft. lbs. (44 Nm).
- Tighten the pinch bolts to 116 ft. lbs. (157 Nm)
- ABS sensor

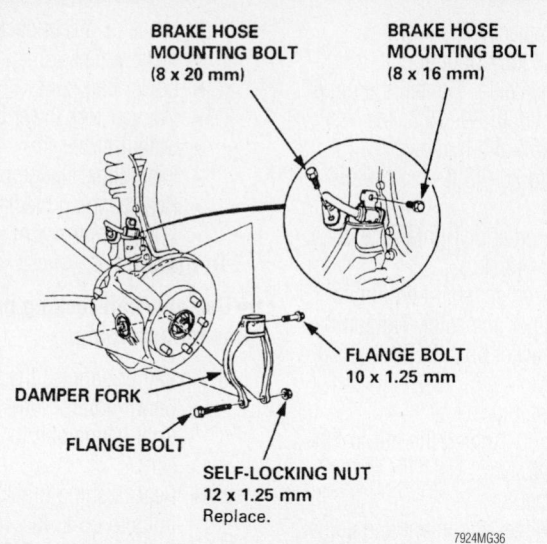

BRAKE HOSE
MOUNTING BOLT
(8 x 20 mm)

BRAKE HOSE
MOUNTING BOLT
(8 x 16 mm)

FLANGE BOLT
10 x 1.25 mm

DAMPER FORK

FLANGE BOLT

SELF-LOCKING NUT
12 x 1.25 mm
Replace.

7924MG36

Identification of some of the front suspension components

- Tie rod end
- Brake hose retainer
- Front wheel

Rear

1. Before servicing the vehicle, refer to the precautions in the beginning of this section.

2. Support the vehicle under the lower control arm.

3. Remove or disconnect the following:

- Rear wheel
- Interior access panel
- Damper cap
- Upper strut mount nuts
- Lower strut flange bolt
- Strut

To install:

4. Install or connect the following:
- Strut. Tighten the nuts to 36 ft. lbs. (49 Nm) for 2001; 54 ft. lbs. (74 Nm) for 2002–04; the bolt to 40 ft. lbs. (54 Nm) for 2001; 69 ft. lbs. (93 Nm) for 2002–04.
- Damper cap
- Interior access panel
- Rear wheel

Coil Spring

REMOVAL & INSTALLATION

Front

1. Before servicing the vehicle, refer to the precautions in the beginning of this section.

2. Remove the strut from the vehicle and install in a strut spring compressor. Com-

press the spring until the end of the spring comes away from the spring seat.

3. Remove the upper strut mount, spring seat and related components.

4. Remove the coil spring from the strut spring compressor.

To install:

➡ **Use a new self-locking nut.**

5. Compress the spring and position the strut so that the end of the spring aligns with the notch in the spring seat.

6. Install the upper strut mounting components and tighten the nut to 22 ft. lbs. (29 Nm) for 2001; 33 ft. lbs. (44 Nm) for 2002–04.

7. Install the strut to the vehicle.

8. Check the wheel alignment and adjust as necessary.

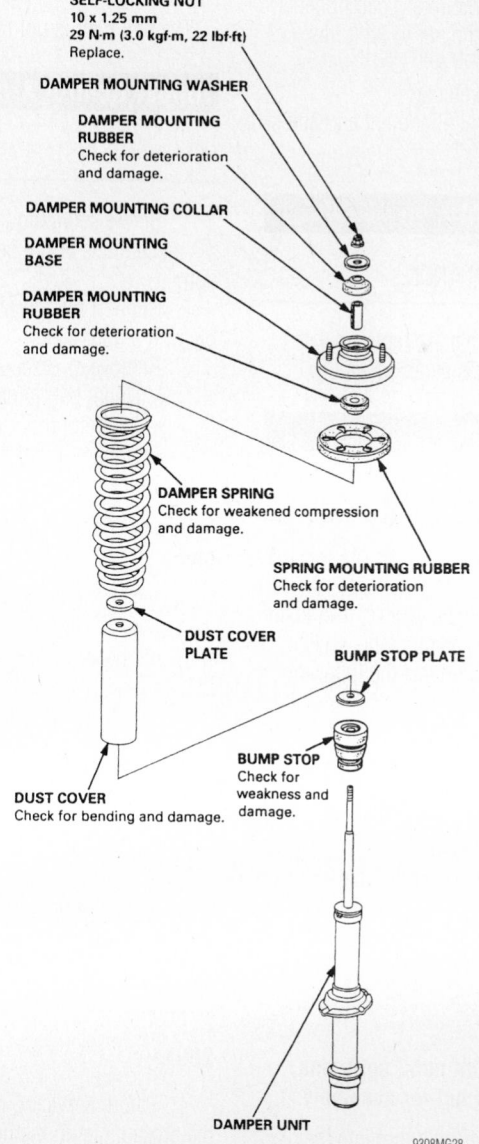

SELF-LOCKING NUT
10 x 1.25 mm
29 N·m (3.0 kgf·m, 22 lbf·ft)
Replace.

DAMPER MOUNTING WASHER

DAMPER MOUNTING
RUBBER
Check for deterioration
and damage.

DAMPER MOUNTING COLLAR

DAMPER MOUNTING
BASE

DAMPER MOUNTING
RUBBER
Check for deterioration
and damage.

DAMPER SPRING
Check for weakened compression
and damage.

SPRING MOUNTING RUBBER
Check for deterioration
and damage.

DUST COVER
PLATE

BUMP STOP PLATE

BUMP STOP
Check for
weakness and
damage.

DUST COVER
Check for bending and damage.

DAMPER UNIT

9308MG28

Front strut and spring exploded view—2001 shown

Rear

1. Before servicing the vehicle, refer to the precautions in the beginning of this section.

2. Remove the strut from the vehicle and install in a strut spring compressor. Compress the spring until the end of the spring comes away from the spring seat.

3. Remove or disconnect the following:
- Upper strut mount, spring seat and related components
- Coil spring from the strut spring compressor

To install:

➡**Use a new self-locking nut.**

4. Compress the spring and position the strut so that the end of the spring aligns with the notch in the spring seat.

5. Install or connect the following:
- Upper strut mounting components and tighten the nut to 22 ft. lbs. (29 Nm).
- Strut to the vehicle

6. Check the wheel alignment and adjust as necessary.

Upper Ball Joint

REMOVAL & INSTALLATION

The upper ball joints are replaced with the upper control arms as an assembly.

Lower Ball Joint

REMOVAL & INSTALLATION

2001

1. Before servicing the vehicle, refer to the precautions in the beginning of this section.

2. Remove or disconnect the following:
- Front wheel
- Spindle nut
- Brake hose bracket
- Brake caliper and rotor
- Wheel speed sensor, if equipped
- Outer tie rod end
- Upper and lower ball joints
- Steering knuckle
- Lower ball joint boot and set ring

3. Press the lower ball joint out of the steering knuckle.

To install:

➡**Use new ball joint nuts, split pins, and a new spindle nut for assembly.**

4. Press the lower ball joint into the steering knuckle.

5. Install or connect the following:
- Lower ball joint boot and set ring
- Steering knuckle. Tighten the upper ball joint nut to 29–35 ft. lbs. (39–47 Nm) and the lower ball joint nut to 36–43 ft. lbs. (49–59 Nm).
- Outer tie rod end. Tighten the nut to 32 ft. lbs. (43 Nm).
- Wheel speed sensor, if equipped
- Brake caliper and rotor. Tighten the caliper bracket bolts to 80 ft. lbs. (108 Nm).
- Brake hose bracket
- Spindle nut. Tighten the nut to 181 ft. lbs. (245 Nm).
- Front wheel

6. Check the wheel alignment and adjust as necessary.

2002–04

The ball joint is not replaceable.

Upper Control Arm

REMOVAL & INSTALLATION

1. Before servicing the vehicle, refer to the precautions in the beginning of this section.

2. Support the lower control arm assembly with a floor jack.

3. Remove or disconnect the following:
- Upper ball joint
- Inner control arm flange bolts
- Upper control arm

To install:

➡**Use new self-locking nuts for assembly.**

4. Install the upper control arm. Tighten the ball joint nut to 29–35 ft. lbs. (39–47 Nm) and the inner flange bolts to 40 ft. lbs. (54 Nm).

CONTROL ARM BUSHING REPLACEMENT

The upper control arm bushings are serviced with the upper control arm as an assembly.

Lower Control Arm

REMOVAL & INSTALLATION

2001

1. Before servicing the vehicle, refer to the precautions in the beginning of this section.

2. Remove or disconnect the following:
- Front wheel
- Lower ball joint
- Damper fork lower bolt
- Stabilizer bar link
- Front inner flange bolt
- Rear bushing bracket bolts
- Lower control arm

To install:

➡**Use new self-locking nuts and split pins for assembly.**

3. Install or connect the following:
- Lower control arm. Tighten the front flange bolt to 76 ft. lbs. (103 Nm).
- Rear bushing bracket. Tighten the bolts to 66 ft. lbs. (89 Nm).
- Stabilizer bar link. Tighten the nut to 22 ft. lbs. (29 Nm).
- Damper fork lower bolt. Tighten the nut to 47 ft. lbs. (64 Nm).
- Lower ball joint. Tighten the nut to 36–43 ft. lbs. (49–59 Nm).
- Front wheel

4. Check the wheel alignment and adjust as necessary.

2002–04

1. Before servicing the vehicle, refer to the precautions in the beginning of this section.

2. Remove or disconnect the following:
- Front wheel
- Stabilizer link
- Lower arm from the knuckle
- Lower arm

3. Installation is the reverse of removal. Observe the following torques:
- Lower arm bolts: 61 ft. lbs. (83 nm)
- Ball stud nut: 51 ft. lbs. (69 Nm)
- Stabilizer link: 29 ft. lbs. (39 Nm)

CONTROL ARM BUSHING REPLACEMENT

The lower control arm front inner bushing and the damper fork bushing are serviced with the control arm as an assembly.

REAR INNER BUSHING

1. Before servicing the vehicle, refer to the precautions in the beginning of this section.

2. Remove or disconnect the following:
- Front wheel
- Rear bushing bracket
- Rear bushing

To install:

➡**Use a new self-locking nut for assembly.**

3. Install or connect the following:
- Rear bushing. Tighten the nut to 61 ft. lbs. (83 Nm).
- Rear bushing bracket. Tighten the bolts to 66 ft. lbs. (89 Nm).
- Front wheel

4. Check the wheel alignment and adjust as necessary.

Wheel Bearings

ADJUSTMENT

The wheel bearings are sealed units and are not adjustable.

REMOVAL & INSTALLATION

Front

2001

1. Before servicing the vehicle, refer to the precautions in the beginning of this section.

2. Remove or disconnect the following:
- Front wheel
- Spindle nut
- Brake hose bracket
- Brake caliper and rotor
- Wheel speed sensor, if equipped
- Outer tie rod end
- Upper and lower ball joints
- Steering knuckle

3. Press the hub out of the wheel bearing.

4. Remove the splash guard.

5. Remove the snapring and press the wheel bearing out of the steering knuckle.

6. If necessary, press the inner bearing race off of the hub.

To install:

➡**Use new ball joint nuts, split pins, snapring and a new spindle nut for assembly.**

7. Press the bearing into the steering knuckle and install the snapring.

8. Install the splash guard.

9. Press the hub into the bearing.

10. Install or connect the following:
- Steering knuckle. Tighten the upper ball joint nut to 29–35 ft. lbs. (39–47 Nm) and the lower ball joint nut to 36–43 ft. lbs. (49–59 Nm).
- Outer tie rod end. Tighten the nut to 32 ft. lbs. (43 Nm).
- Wheel speed sensor, if equipped
- Brake caliper and rotor. Tighten the caliper bracket bolts to 80 ft. lbs. (108 Nm).

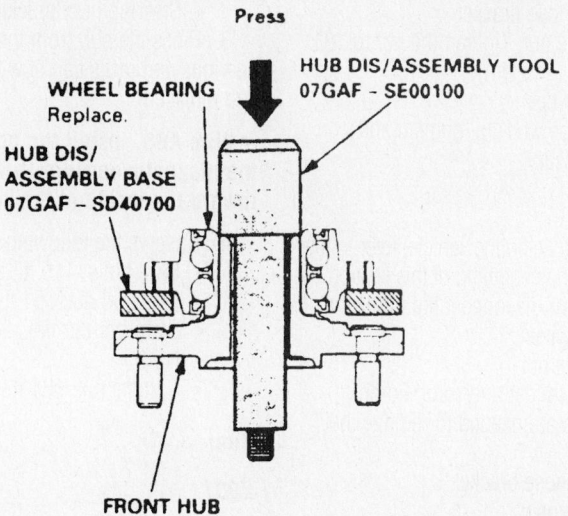

Removing the hub from the wheel bearing using the disassembly tools

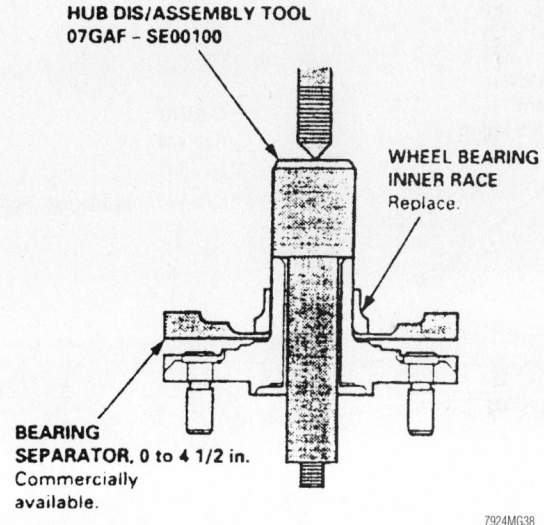

Pressing out the wheel bearing inner race

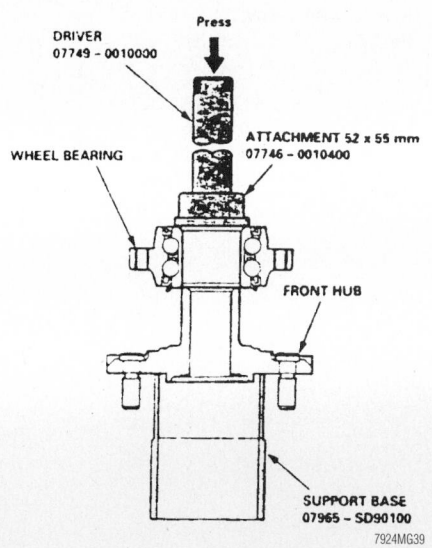

Utilizing the hub support base and driving attachment tools to install the new wheel bearing

- Brake hose bracket
- Spindle nut. Tighten the nut to 181 ft. lbs. (245 Nm).
- Front wheel

11. Check the wheel alignment and adjust as necessary.

2002-04

1. Before servicing the vehicle, refer to the precautions in the beginning of this section.
2. Remove or disconnect the following:
- Front wheel
- Spindle nut
- Brake caliper and rotor. Forcing screws are needed to remove the rotor.
- Brake hose bracket
- ABS sensor
- Stabilizer link
- Lower arm from the knuckle
- Strut-to-knuckle bolts

- Steering hub/knuckle assembly

3. Press the hub from the knuckle. The bearings and races can now be pressed out and replaced.

➡**With ABS, install the bearing with the magnetic encoder (brown color) toward the inside of the knuckle.**

4. Observe the following torques:
- Strut bolts: 116 ft. lbs. (157 Nm)
- Ball stud nuts: 51 ft. lbs. (69 Nm)
- Stabilizer bar link: 29 ft. lbs. (39 Nm)
- Spindle nut: 181 ft. lbs. (245 Nm)

Rear

2001

1. Before servicing the vehicle, refer to the precautions in the beginning of this section.

2. Remove or disconnect the following:
- Rear wheel
- Brake drum
- Spindle nut
- Brake shoes and parking brake cable
- Brake fluid line
- Wheel sensor, if equipped
- Wheel bearing flange bolts
- Hub, backing plate and bearing assembly

3. Press the hub out of the wheel bearing.
4. If necessary, press the inner bearing race off of the hub.
5. Remove the backing plate from the bearing assembly.

To install:

➡**Use a new spindle nut for assembly.**

6. Use a new O-ring and install the backing plate to the bearing assem-

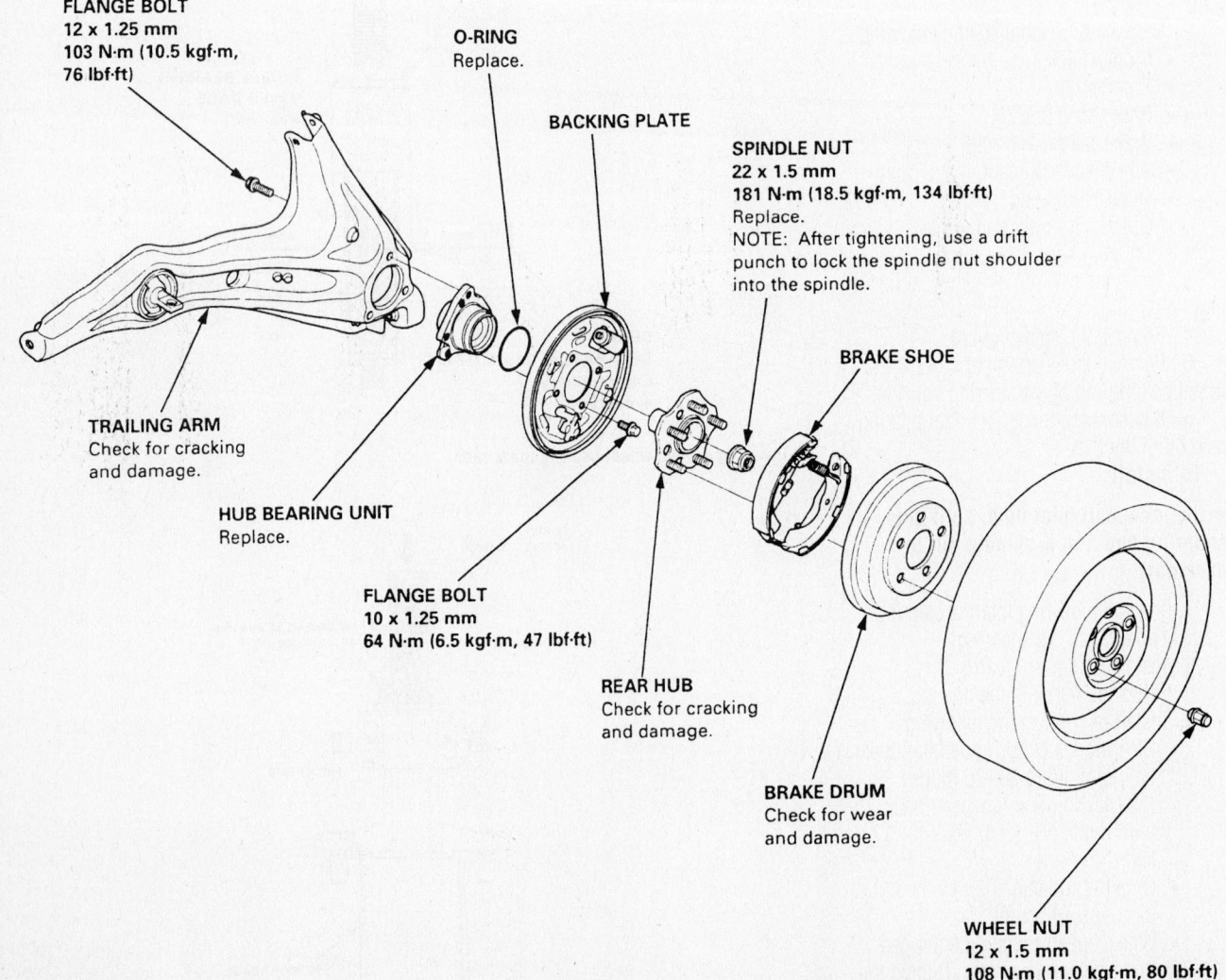

FLANGE BOLT
12 x 1.25 mm
103 N·m (10.5 kgf·m, 76 lbf·ft)

O-RING
Replace.

BACKING PLATE

SPINDLE NUT
22 x 1.5 mm
181 N·m (18.5 kgf·m, 134 lbf·ft)
Replace.
NOTE: After tightening, use a drift punch to lock the spindle nut shoulder into the spindle.

BRAKE SHOE

TRAILING ARM
Check for cracking and damage.

HUB BEARING UNIT
Replace.

FLANGE BOLT
10 x 1.25 mm
64 N·m (6.5 kgf·m, 47 lbf·ft)

REAR HUB
Check for cracking and damage.

BRAKE DRUM
Check for wear and damage.

WHEEL NUT
12 x 1.5 mm
108 N·m (11.0 kgf·m, 80 lbf·ft)

7924MG43

Exploded view of the rear hub and wheel bearing components—CR-V

bly. Tighten the bolts to 47 ft. lbs. (64 Nm).

7. Press the hub into the bearing assembly.

8. Install or connect the following:
- Hub, backing plate and bearing assembly. Tighten the flange bolts to 76 ft. lbs. (103 Nm).
- Brake fluid line
- Parking brake cable and brake shoes
- Wheel sensor, if equipped
- Spindle nut. Tighten the nut to 134 ft. lbs. (181 Nm).
- Brake drum
- Rear wheel

9. Bleed the brake system.

2002–04

1. Before servicing the vehicle, refer to the precautions in the beginning of this section.

2. Remove or disconnect the following:
- Rear wheel
- Brake caliper
- Rotor
- Spindle nut
- Axle shaft (2wd)
- Parking brake shoes
- Parking brake cable
- Wheel sensor, if equipped

3. Support the trailing arm.

4. Remove or disconnect the following:
- Upper arm from the knuckle

5. Matchmark the trailing arm cam

adjusting bolt and cam. Remove the bolt. Discard the nut.

6. Remove the flange bolt.

7. Remove the knuckle assembly.

8. Press the hub from the knuckle. The bearings and races can now be pressed out and replaced.

➡**With ABS, install the bearing with the magnetic encoder (brown color) toward the inside of the knuckle.**

9. Observe the following torques:
- Flange bolt: 69 ft. lbs. (93 Nm)
- Cam bolts: 43 ft. lbs. (59 Nm)
- Spindle nut: 134 ft. lbs. (181 Nm)
- Caliper mounting bolts: 41 ft. lbs. (55 Nm)

BRAKES

Brake Caliper

REMOVAL & INSTALLATION

2001

FRONT

1. Raise and safely support the vehicle. Remove the front wheels.

2. Remove some brake fluid from the master cylinder reservoir.

3. Disconnect and plug the brake fluid line from the caliper.

4. Remove the brake caliper mounting bolt and guide bolt and remove the caliper from the mount. The brackets can be remove for additional work space.

5. Remove the brake pads and clips

from the caliper. Inspect the brake pads for wear; replace them, if necessary.

To install:

6. Install the brake pads and clips onto the caliper.

7. If the caliper bracket was removed, tighten the bolts to 103–126 ft. lbs. (139–171 Nm).

8. Install the caliper on the mounting bracket. Torque the caliper-to-mounting bracket bolts to 20–27 ft. lbs. (27–37 Nm).

9. Connect the fluid line to the caliper using new washers. Torque the brake line banjo fitting to 26 ft. lbs. (35 Nm).

✴✴ WARNING

Be sure the hook end of the flexible brake line is positioned in the anti-rotation cavity.

10. Refill the master cylinder reservoir and bleed the brake system.

11. Install the front wheels and lower the vehicle.

REAR

1. Raise and safely support the vehicle. Remove the rear wheels.

2. Remove some fluid from the master cylinder reservoir.

3. Disconnect and plug the brake fluid line from the caliper.

➡**Discard the parking brake cable mounting pin after removal.**

4. If equipped with caliper actuated parking brakes; remove the mounting pin from the parking brake cable and disconnect the parking cable from the disc caliper.

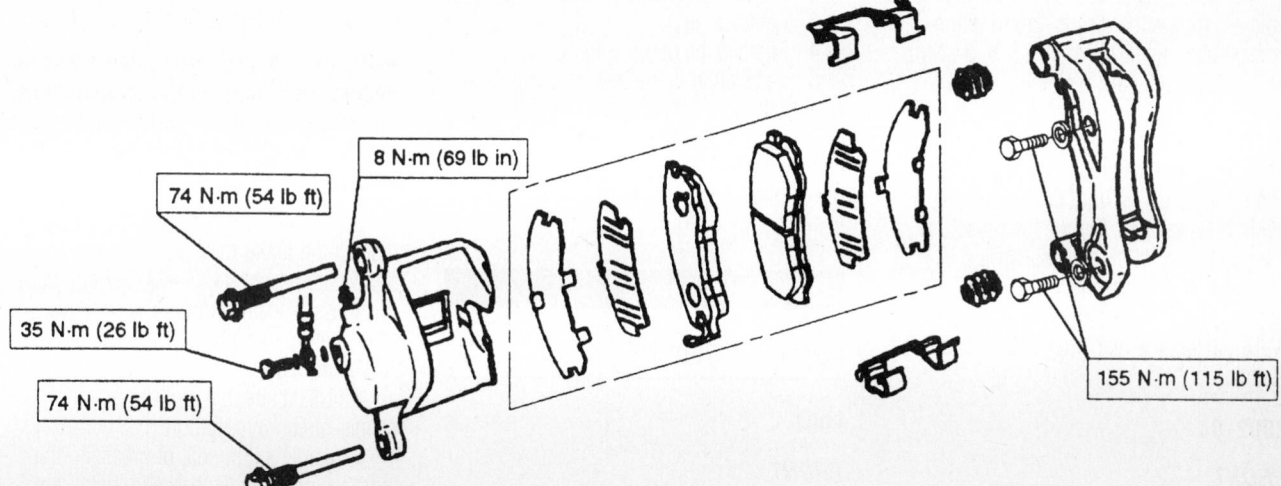

93026G54

Exploded view of the front caliper components

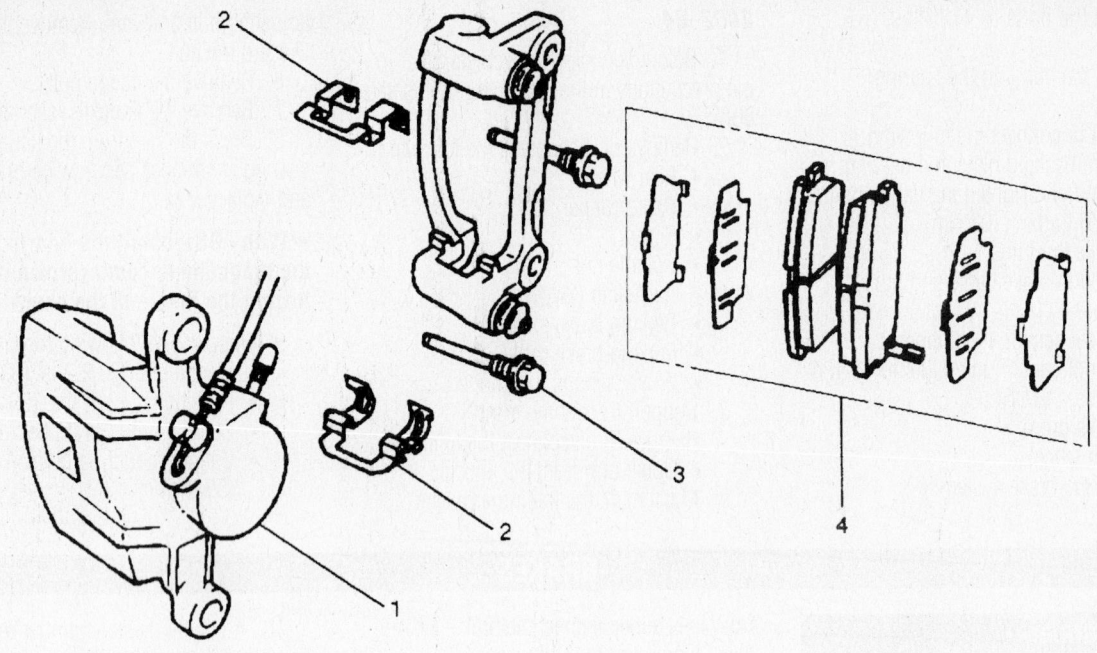

(1) Caliper Assembly
(2) Clip
(3) Lock Bolt
(4) Pad Assembly

93026G55

Exploded view of the rear caliper components

5. Remove the brake caliper mounting bolt and guide bolt and remove the caliper from the mount.

6. Remove the brake pads and clips from the caliper. Inspect the brake pads for wear; replace them, if necessary.

To install:

7. Install the brake pads and clips onto the caliper.

8. If the mounting bracket was removed, tighten the bolts to 69–84 ft. lbs. (93–114 Nm).

9. Install the caliper on the mounting bracket. Torque the caliper-to-mounting bracket bolts to 12–17 ft. lbs. (16–24 Nm), or 32 ft. lbs. (43 Nm) on vehicles with shoe-type parking brakes.

10. Connect the parking brake cable to the caliper and install a new mounting pin.

11. Connect the fluid line to the caliper using new washers. Torque the brake line banjo fitting to 26 ft. lbs. (35 Nm).

12. Refill the master cylinder reservoir and bleed the brake system.

13. Install the rear wheels and lower the vehicle.

2002–04

FRONT

1. Remove the upper and lower bolts.
2. Lift off the caliper.

3. Remove the pad springs.
4. Remove the pads and shims.
5. Remove the pad retainers.
6. Installation is the reverse of removal. Coat both sides of the shims and the backs of the pads with brake grease. Torque the bolts to 25 ft. lbs. (34 Nm). If the hose was disconnected, torque the banjo bolt to 25 ft. lbs. (34 Nm).

REAR

1. Remove the caliper pin bolts.
2. Lift off the caliper and suspend it safely.
3. Remove the pads and shims.
4. Remove the pad retainers.
5. Installation is the reverse of removal. Coat both sides of the shims and the backs of the pads with brake grease. Torque the bolts to 16 ft. lbs. (22 Nm). If the hose was disconnected, torque the banjo bolt to 16 ft. lbs. (22 Nm).

Disc Brake Pads

REMOVAL & INSTALLATION

2001

FRONT

Most disc brake pads are equipped with wear indicators. If a squealing noise occurs

from the brakes while driving, check the pad wear indicator plate. If there is evidence of the indicator plate contacting the brake disc, the brake pad should be replaced.

1. Remove ½ of the volume of brake fluid from the master cylinder to prevent overflow when the caliper piston is compressed.

2. Raise and safely support the vehicle.
3. Remove the wheel and tire assemblies.
4. Remove the brake caliper without disconnecting the brake line. Support the caliper with a length of wire. Do not let the caliper hang from the brake hose.

➡On some disc brake systems it is not necessary to remove the caliper when installing new brake pads. Remove the lower slide bolt and rotate the caliper upward to remove the pads.

5. Remove the brake pads and shims. Inspect the brake rotor and machine or replace as necessary. Check the minimum thickness (specification is cast into the rotor) before machining.

To install:

6. Use a suitable tool to push the caliper piston into its bore.

7. Apply a thin coat of grease to the rear face of the brake pad and install the shim. Install the brake pads.

8. Install the calipers. Lubricate the

caliper bolts and boots. Tighten the caliper mounting bolts to 24 ft. lbs. (33 Nm).

9. Install the wheel and tire assemblies and lower the vehicle.

10. Apply the brakes several times to seat the pads before moving the vehicle. Check the fluid in the master cylinder and add as necessary.

REAR

1. Raise and safely support the vehicle. Remove the rear wheels.

2. Remove the brake caliper mounting bolts and remove the caliper without disconnecting the brake fluid line. Support the caliper so it does not hang on the brake line.

3. Remove the brake pads and retaining clips from the caliper.

4. If equipped with caliper-activated parking brakes; use tool J-37617 or equivalent, to rotate the piston clockwise until it retracts into the bore. Align the notches of the piston face so the centerline of the notches is perpendicular to the centerline of the mounting bosses.

To install:

5. Install the new brake pads and clips in the caliper and install the caliper in the mounting bracket.

6. Tighten the caliper mounting bolts to

12–17 ft. lbs. (16–24 Nm), or 32 ft. lbs. (43 Nm) on vehicles with shoe-type parking brakes.

7. Install the rear wheels. Check the brake fluid level.

8. Pump the brake pedal until pressure is felt before moving the vehicle.

2002–04

FRONT

1. Remove the lower bolt.
2. Pivot the caliper up and hold the pads.
3. Remove the pad springs.
4. Remove the pads and shims.
5. Remove the pad retainers.
6. Installation is the reverse of removal. Coat both sides of the shims and the backs of the pads with brake grease. Torque the lower bolt to 25 ft. lbs. (34 Nm).

REAR

1. Remove the caliper pin bolts.
2. Lift off the caliper and suspend it safely.
3. Remove the pads and shims.
4. Remove the pad retainers.
5. Installation is the reverse of removal. Coat both sides of the shims and the backs of the pads with brake grease. Torque the bolts to 16 ft. lbs. (22 Nm).

Brake Drums

REMOVAL & INSTALLATION

1. Raise and safely support the vehicle. Release the parking brake.

2. Remove the rear wheels.

3. Use chalk to mark the brake drum to one of the wheel studs as an index mark for reinstallation.

4. Remove the retaining screw that holds the brake drum to the axle flange.

5. Pull the brake drum from the axle flange.

To install:

6. Align the index mark and install the brake drum to the axle flange.

7. Install the retaining screw to secure the brake drum to the axle flange.

8. Install the rear wheels.

Rear Brake Shoes

REMOVAL & INSTALLATION

1. Raise and safely support the vehicle.
2. Remove the rear wheels.
3. Remove the brake drums.
4. Remove the brake return springs.

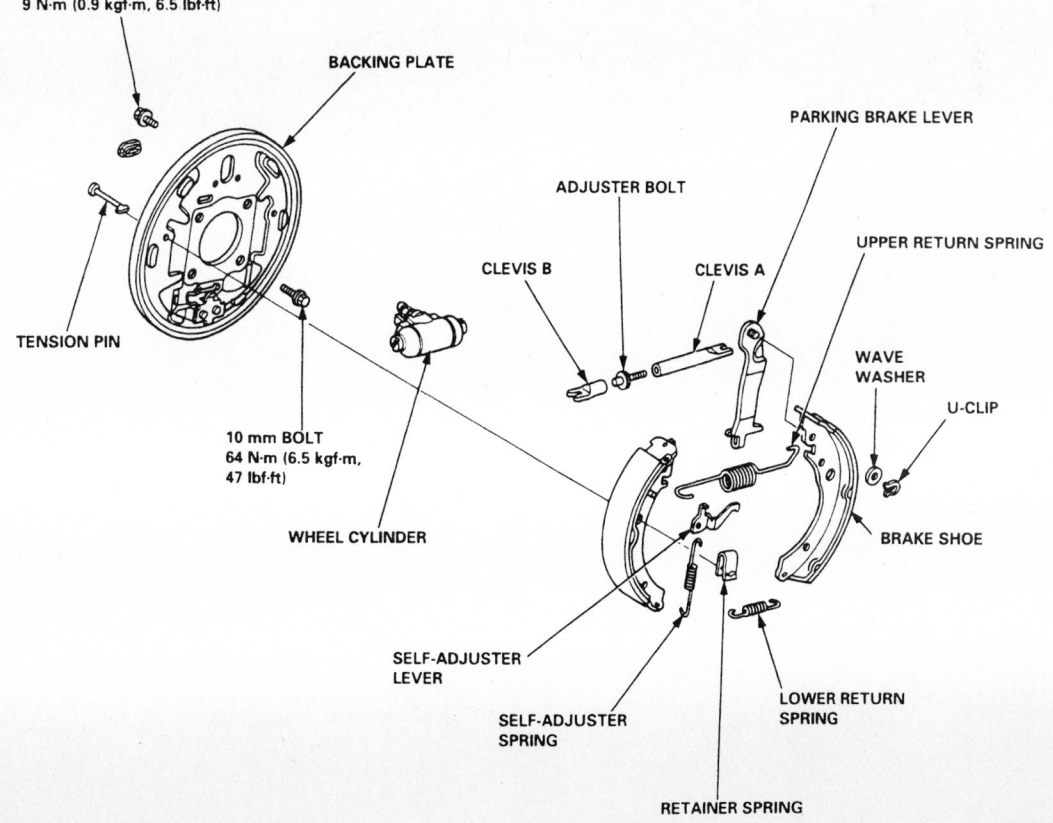

Exploded view of the rear drum brakes

93026G59

5. Remove the leading shoe holding pin and spring, and then the leading shoe.

6. Remove the self-adjuster and the adjuster lever.

7. Remove the trailing shoe holding pin and spring.

8. Disconnect the parking brake cable from the trailing shoe and remove the trailing shoe. Remove the parking brake lever from the trailing shoe.

To install:

9. Attach the parking brake lever to the trailing shoe.

10. Connect the parking brake cable to the parking brake lever.

11. Apply a thin coat of high temperature grease to the shoe contact points on the brake backing plate contact surface (B), and self-adjuster (D).

12. Position the trailing shoe on the backing plate and install the hold-down pin, spring, and retainer. Be careful not to stretch the return spring when fitting the shoes onto the backing plate.

13. Connect the upper return spring and the leading shoe to the trailing shoe and position the leading brake shoe on the backing plate.

14. Install the adjuster assembly and the hold-down pin, spring, and retainer.

15. Use a brake spring tool to install the lower return spring.

16. Install the self-adjuster lever and adjuster spring.

17. Adjust the shoe-to-drum clearance to 0.0098–0.0157 in. (0.25–0.40mm) and install the brake drum.

18. Check the brake drum for scoring or other wear. Machine or replace as necessary. Check the maximum brake drum diameter specification when machining.

19. Install the rear wheels. Lower the vehicle.

20. Road-test the vehicle.

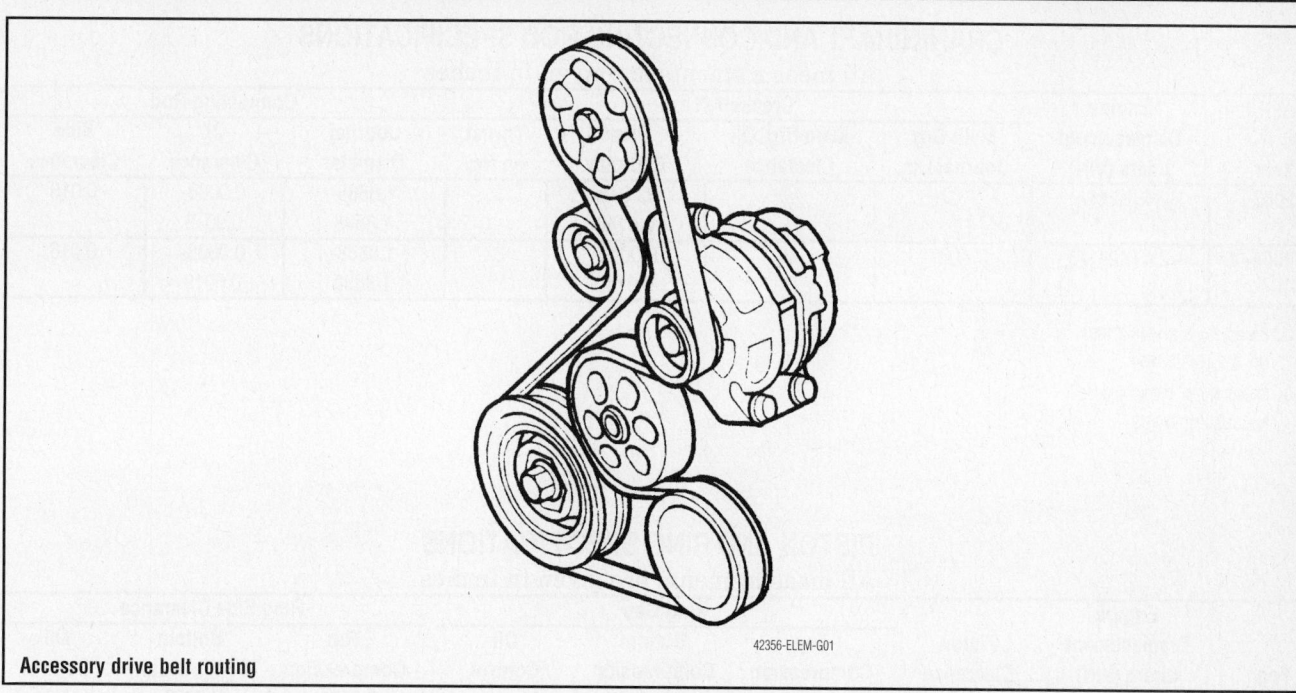

Accessory drive belt routing

42356-ELEM-G01

CAPACITIES

Year	Model	Engine Displacement Liters (VIN)	Engine Oil with Filter (qts.)	Transmission (pts.) 5-Spd	Transmission (pts.) Auto.	Transfer Case (pts.)	Drive Axle Front (pts.)	Drive Axle Rear (pts.)	Fuel Tank (gal.)	Cooling System (qts.)
2003	Element	2.4 (K241A)	4.4	4.0	①	②	②	2.2	15.9	③
2004-05	Element	2.4 (K241A)	4.4	4.0	①	②	②	2.2	15.9	③

NOTE: All capacities are approximate. Add fluid gradually and check to be sure a proper fluid level is obtained.

① 2WD: 6.6 pts. for fluid change, 15.2 for overhaul
    4WD: 6.3 pts. for fluid change, 13.8 for overhaul

② Included in transaxle refill figure

②③ Manual trans: 7.6 qts
    Auto trans: 7.5 qts.

67162-ELEM-C04

VALVE SPECIFICATIONS

Year	Engine Displacement Liters (VIN)	Seat Angle (deg.)	Face Angle (deg.)	Spring Test Pressure (lbs. @ in.)	Spring Installed Height (in.)	Stem-to-Guide Clearance (in.) Intake	Stem-to-Guide Clearance (in.) Exhaust	Stem Diameter (in.) Intake	Stem Diameter (in.) Exhaust
2003	2.4 (K241A)	NA	NA	NA	①	0.0012-0.0022	0.0022-0.0031	0.2156-0.2159	0.2146-0.2150
2004-05	2.4 (K241A)	NA	NA	NA	①	0.0012-0.0022	0.0022-0.0031	0.2156-0.2159	0.2146-0.2150

NA: Not Available

① Valve spring free length:
    Intake: 1.668 in.
    Exhaust: 1.745 in.

67162-ELEM-C05

CRANKSHAFT AND CONNECTING ROD SPECIFICATIONS
All measurements are given in inches

Year	Engine Displacement Liters (VIN)	Crankshaft				Connecting Rod		
		Main Brg. Journal Dia.	Main Brg. Oil Clearance	Shaft End-play	Thrust on No.	Journal Diameter	Oil Clearance	Side Clearance
2003	2.4 (K241A)	①	②	0.0040-0.0140	3	1.8888-1.8898	0.0008-0.0019	0.016
2004-05	2.4 (K241A)	①	②	0.0040-0.0140	3	1.8888-1.8898	0.0008-0.0019	0.016

① Except No. 3: 2.1648-2.1657
   No. 3: 2.1644-2.1654

② Except No. 3: 0.0007-0.0016
   No. 3: 0.0010-0.0019

67162-ELEM-C06

PISTON AND RING SPECIFICATIONS
All measurements are given in inches

Year	Engine Displacement Liters (VIN)	Piston Clearance	Ring Gap			Ring Side Clearance		
			Top Compression	Bottom Compression	Oil Control	Top Compression	Bottom Compression	Oil Control
2003	2.4 (K241A)	0.0008-0.0016	0.0080-0.0140	0.0160-0.0220	0.0080-0.0280	0.0018-0.0028	0.0020-0.0030	NA
2004-05	2.4 (K241A)	0.0008-0.0016	0.0080-0.0140	0.0160-0.0220	0.0080-0.0280	0.0018-0.0028	0.0020-0.0030	NA

NA: Not Applicable

67162-ELEM-C07

TORQUE SPECIFICATIONS
All readings in ft. lbs.

Year	Engine Displacement Liters (VIN)	Cylinder Head Bolts	Main Bearing Bolts	Rod Bearing Bolts	Crankshaft Damper Bolts	Flywheel Bolts	Manifold		Spark Plugs	Oil Pan Drain Plug
							Intake	Exhaust		
2003	2.4 (K241A)	①	②	③	181	76	16	33	13	33
2004-05	2.4 (K241A)	①	②	③	181	76	16	33	13	33

NOTE: Dip main bearing bolts and crankshaft damper bolt in clean engine oil prior to tightening.

① Step 1: 29 ft. lbs.
   Step 2: +90 degrees
   Step 3: +90 degrees
   Step 4: NEW BOLT ONLY +90 degrees

② 22 ft. lbs. +56 degrees

③ 14 ft. lbs. +90 degrees

67162-ELEM-C08

WHEEL ALIGNMENT

Year	Model		Caster Range (+/-Deg.)	Caster Preferred Setting (Deg.)	Camber Range (+/-Deg.)	Camber Preferred Setting (Deg.)	Toe-in (in.)
2003	Element	F	1.00	+1.75	0.75	0	0+/-0.08
		R	—	—	0.75	-1.00	0.08+/-0.08
2004-05	Element	F	1.00	+1.75	0.75	0	0+/-0.08
		R	—	—	0.75	-1.00	0.08+/-0.08

67162-ELEM-C09

TIRE, WHEEL AND BALL JOINT SPECIFICATIONS

Year	Model	OEM Tires Standard	OEM Tires Optional	Tire Pressures Front	Tire Pressures Rear	Wheel Size	Ball Joint Inspection	Lug Nut
2003	Element	P215/70R16	None	26	26	6JJ	NS	80
2004-05	Element	P215/70R16	None	26	26	6JJ	NS	80

OEM: Original Equipment Manufacturer
PSI: Pounds Per Square Inch
NS: Not specified by manufacturer

67162-ELEM-C10

BRAKE SPECIFICATIONS
HONDA ELEMENT
All measurements in inches unless noted

Year	Model		Brake Disc Original Thickness	Brake Disc Minimum Thickness	Brake Disc Maximum Runout	Original Inside Diameter	Max. Wear Limit	Maximum Machine Diameter	Lining Front	Lining Rear	Bracket Bolts (ft. lbs.)	Mounting Bolts (ft. lbs.)
2003	Element	F	0.910	0.830	0.004	—	—	—	0.060	—	80	25
		R	0.350	0.280	0.004	—	—	—	0.040	—	41	16
2004-05	Element	F	0.910	0.830	0.004	—	—	—	0.060	—	80	25
		R	0.350	0.280	0.004	—	—	—	0.040	—	41	16

F: Front
R: Rear

67162-ELEM-C11

SCHEDULED MAINTENANCE INTERVALS
HONDA—ELEMENT

TO BE SERVICED	TYPE OF SERVICE	VEHICLE MILEAGE INTERVAL (x1000)											
		10	20	30	40	50	60	70	80	90	100	110	120
Accessory drive belts	I & A			✓			✓			✓			✓
Air cleaner element	R			✓			✓			✓			✓
Air conditioning filter	R			✓			✓			✓			✓
Brake fluid	R											✓	
Brake hoses & lines (including ABS)	I		✓		✓		✓		✓		✓		
Cooling system hoses & connections	I		✓		✓		✓		✓		✓		
Engine coolant	R												✓
Engine oil	R	✓	✓	✓	✓	✓	✓	✓	✓	✓	✓	✓	✓
Engine oil and coolant levels	I	Inspect at each fuel stop											
Engine oil filter	R		✓		✓		✓		✓		✓		
Exhaust system	I		✓		✓		✓		✓		✓		
Fluid levels and condition	I		✓		✓		✓		✓		✓		
Front and rear brakes	I		✓		✓		✓		✓		✓		
Fuel lines & connection	I		✓		✓		✓		✓		✓		
Halfshaft boots	I		✓		✓		✓		✓		✓		
Idle speed	I & A											✓	
Parking brake system	I & A		✓		✓		✓		✓		✓		
Rear differential fluid	R										✓		
Rotate and inspect tires	I	✓	✓	✓	✓	✓	✓	✓	✓	✓	✓	✓	✓
Spark plugs	R											✓	
Suspension components	I		✓		✓		✓		✓		✓		
Tie rod ends, steering gear box & boots	I		✓		✓		✓		✓		✓		
Transmission fluid	R												✓
Valve clearance	I											✓	

R: Replace   I: Inspect   A: Adjust

FREQUENT OPERATION MAINTENANCE (SEVERE SERVICE)

If a vehicle is operated under any of the following conditions it is considered severe service:

- Towing a trailer or using a camper or car-top carrier.
- Repeated short trips of less than 5 miles in temperatures below freezing, or trips of less than 10 miles in any temperature.
- Extensive idling or low-speed driving for long distances as in heavy commercial use, such as delivery, taxi or police cars.
- Operating on rough, muddy or salt-covered roads.
- Operating on unpaved or dusty roads.
- Driving in extremely hot (over 90°) conditions.

Air cleaner element: replace every 15,000 miles

Engine oil and filter: replace every 3750 miles or 6 months, whichever occurs first.

Timing belt: replace every 60,000 miles if the vehicle is regularly driven in temperatures above 110°F or below -20°F.

Transmission fluid: replace every 30,000 miles.

Rear differential fluid: replace every 60,000 miles.

Front and rear brakes: inspect every 7500 miles or 6 months, whichever occurs first.

Locks and hinges: lubricate every 15,000 miles.

Tie rods, steering gear box, boots: inspect every 7500 miles or 6 months, whichever occurs first.

Suspension components: inspect every 7500 miles or 6 months, whichever occurs first.

Halfshaft boots: inspect every 7500 miles or 6 months, whichever occurs first.

PRECAUTIONS

Before servicing any vehicle, please be sure to read all of the following precautions, which deal with personal safety, prevention of component damage, and important points to take into consideration when servicing a motor vehicle:

• Never open, service or drain the radiator or cooling system when the engine is hot; serious burns can occur from the steam and hot coolant.

• Observe all applicable safety precautions when working around fuel. Whenever servicing the fuel system, always work in a well-ventilated area. Do not allow fuel spray or vapors to come in contact with a spark, open flame, or excessive heat (a hot drop light, for example). Keep a dry chemical fire extinguisher near the work area. Always keep fuel in a container specifically designed for fuel storage; also, always properly seal fuel containers to avoid the possibility of fire or explosion. Refer to the additional fuel system precautions later in this section.

• Fuel injection systems often remain pressurized, even after the engine has been turned **OFF**. The fuel system pressure must be relieved before disconnecting any fuel lines. Failure to do so may result in fire and/or personal injury.

• Brake fluid often contains polyglycol ethers and polyglycols. Avoid contact with the eyes and wash your hands thoroughly after handling brake fluid. If you do get brake fluid in your eyes, flush your eyes with clean, running water for 15 minutes. If eye irritation persists, or if you have taken brake fluid internally, seek medical assistance IMMEDIATELY.

• The EPA warns that prolonged contact with used engine oil may cause a number of skin disorders, including cancer. You should make every effort to minimize your exposure to used engine oil. Protective gloves should be worn when changing oil. Wash your hands and any other exposed skin areas as soon as possible after exposure to used engine oil. Soap and water, or waterless hand cleaner should be used.

• All new vehicles are now equipped with an air bag system. The system must be disabled before performing service on or around system components, steering column, instrument panel components, wiring and sensors. Failure to follow safety and disabling procedures could result in accidental air bag deployment, possible personal injury and unnecessary system repairs.

• Always wear safety goggles when working with, or around, the air bag system. When carrying a non-deployed air bag, be sure the bag and trim cover are pointed away from your body. When placing a non-deployed air bag on a work surface, always face the bag and trim cover upward, away from the surface. This will reduce the motion of the module if it is accidentally deployed. Refer to the additional air bag system precautions later in this section.

• Clean, high quality brake fluid from a sealed container is essential to the safe and proper operation of the brake system. You should always buy the correct type of brake fluid for your vehicle. If the brake fluid becomes contaminated, completely flush the system with new fluid. Never reuse any brake fluid. Any brake fluid that is removed from the system should be discarded. Also, do not allow any brake fluid to come in contact with a painted surface; it will damage the paint.

• Never operate the engine without the proper amount and type of engine oil; doing so WILL result in severe engine damage.

• Timing belt maintenance is extremely important. Many models utilize an interference-type, non-freewheeling engine. If the timing belt breaks, the valves in the cylinder head may strike the pistons, causing potentially serious (also time-consuming and expensive) engine damage.

• Disconnecting the negative battery cable on some vehicles may interfere with the functions of the on-board computer system(s) and may require the computer to undergo a relearning process once the negative battery cable is reconnected.

• When servicing drum brakes, only disassemble and assemble one side at a time, leaving the remaining side intact for reference.

• Only an MVAC-trained, EPA-certified automotive technician should service the air conditioning system or its components.

ENGINE REPAIR

➡Disconnecting the negative battery cable on some vehicles may interfere with the functions of the on board computer system. The computer may undergo a relearning process once the negative battery cable is reconnected.

Distributor

The 2.4L engine does not have a distributor.

Alternator

REMOVAL

1. Before servicing the vehicle, refer to the precautions in the beginning of this section.
2. Remove or disconnect the following:

• Negative, then the positive battery cables
• Accessory drive belt
• Auto-tensioner
• Alternator wiring harness connectors and harness clamp
• Positive Crankcase Ventilation (PCV) valve
• 3 bolts holding the alternator
• Alternator

INSTALLATION

1. Install or connect the following:
• Alternator. Tighten the bolts to 16 ft. lbs. (22 Nm).
• PCV valve
• Alternator wiring harness connectors and harness clamp
• Auto tensioner
• Accessory drive belt
• Negative battery cable

Ignition Timing

ADJUSTMENT

Adjustment is not possible on the 2.4L engine.

Engine Assembly

REMOVAL & INSTALLATION

➡The engine and transaxle are removed from the vehicle as a unit.

1. Before servicing the vehicle, refer to the precautions in the beginning of this section.
2. Drain the cooling system.
3. Drain the transaxle fluid.
4. Drain the engine oil.
5. Relieve fuel system pressure.

6. Remove or disconnect the following:
- Negative battery cable
- Fuse/Relay box battery cables
- Battery and tray
- Intake manifold cover
- IAT sensor connector
- Breather hose
- Intake duct
- Cables from the power distribution center
- Throttle and cruise cables
- Powertrain Control Module (PCM) connectors and grommet. Pull the PCM harness through the firewall.
- Fuel lines
- EVAP canister
- Brake booster vacuum line
- Clutch slave cylinder
- Clutch hose bracket
- Shift cables
- Drive belt
- Power steering pump, leaving the hoses connected

7. Attach a hoist to the engine lifting eyes and support the powertrain weight.
8. Remove or disconnect the following:
- Splash shield
- Wheels
- Catalytic converter
- Rear driveshaft
- Stabilizer links
- Right damper fork
- Halfshafts
- Shift cable
- Radiator
- Upper bracket
- Transaxle mount and bracket
- Front mount bolt
- Rear mount bracket bolts. Match-mark the sub-frame mounting bolt centers.

There is a special tool necessary for sub-frame removal. The Honda tool number is EQS02C000011. Attach the tool as explained in the tool instructions, attach a floor jack with adapter, remove the 4 sub-frame bolts and lower the sub-frame.

9. Remove or disconnect the following:
- A/C compressor without disconnecting the hoses
10. Check that all hoses and wires are disconnected.
11. Lower the engine about 6 inches and recheck all clearances.
12. Lower the engine all the way.
13. Remove the chain hoist.

To install:
14. Installation is the reverse of removal. Observe the following torques:
- Front engine mount bracket bolts: 33 ft. lbs. (44Nm)

- A/C compressor bracket: 33 ft. lbs. (44Nm)
- Stiffener 10mm bolts: 33 ft. lbs. (44Nm); 6mm bolts 9 ft. lbs. (12 Nm)
- A/C compressor bolts: 33 ft. lbs. (44 Nm)
- Subframe front bolt: 47 ft. lbs. (64 Nm)
- Subframe rear bolts: 43 ft. lbs. (59 Nm)
- Upper bracket bolt and nut: 40 ft. lbs. (54 Nm)
- Transmission mount bracket support bolts/nuts: 40 ft. lbs. (54 Nm)
- PS pump bolts: 16 ft. lbs. (22 Nm)
- Intake manifold cover: 9 ft. lbs. (12 Nm)

➡**Use new self-locking nuts and color-coded self-locking bolts when installing the engine mounts and suspension components.**

➡**Do not tighten the engine or transaxle mount fasteners until instructed to do so.**

15. Lower the powertrain into position.
16. Install or connect the following:
- Transaxle mount and bracket. Tighten the frame mounting bolts to 47 ft. lbs. (64 Nm).
- Upper bracket. Tighten the nuts in sequence to 54 ft. lbs. (74 Nm).
- Rear mount bracket through bolt
- Right front mount and bracket
- Left front mount and bracket

17. Tighten the remaining mount fasteners as follows:
a. Transaxle mount fasteners to 47 ft. lbs. (64 Nm) and the through bolt to 54 ft. lbs. (74 Nm).
b. Rear mount bracket through bolt to 43 ft. lbs. (59 Nm).
c. Right front mount 12mm bolts to 47 ft. lbs. (64 Nm) and the 10mm bolts to 33 ft. lbs. (44 Nm).
d. Left front mount 12mm stud bolt to 61 ft. lbs. (83 Nm), 10mm bolts to 33 ft. lbs. (44 Nm), and 12mm nut to 43 ft. lbs. (59 Nm).
e. Right front mount 12mm nut to 43 ft. lbs. (59 Nm).

18. Install or connect the following:
- Rear driveshaft, if equipped
- A/C compressor
- Radiator
- A/C hose clamp

19. If equipped with a manual transaxle, install or connect the following:
- Shift cables
- Transaxle ground cable
- Clutch hose bracket

- Clutch slave cylinder
20. If equipped with an automatic transaxle, install or connect the following:
- Transaxle fluid cooler lines
- Transaxle ground cable and hose clamp
- Shift cable
- Shift cable cover

21. For all vehicles, install or connect the following:
- Axle halfshafts
- Lower ball joints
- Right damper fork
- Exhaust front pipe
- HO2S connector
- Heater hoses
- Radiator hoses
- Splash shield
- PSP switch
- Accelerator cable
- Brake booster vacuum line
- Fuel lines
- A/C compressor drive belt
- Power steering pump and belt
- Cruise control actuator
- Left engine wire harness connectors
- PCM connectors and grommet
- Air intake assembly
- Battery and tray
- Fuse/Relay box battery cables
- Negative battery cable

22. Fill the engine crankcase to the correct level.
23. Fill the transaxle to the correct level.
24. Fill the cooling system.
25. Start the engine and check for leaks.
26. Check the wheel alignment and adjust as necessary.

Water Pump

REMOVAL & INSTALLATION

1. Before servicing the vehicle, refer to the precautions in the beginning of this section.
2. Drain the cooling system.
3. Remove or disconnect the following:
- Negative battery cable
- Accessory drive belt
- Crankshaft pulley
- Water pump (6 bolts)

To install:
4. Clean the water pump mating surfaces.
5. Install or connect the following:
- Water pump with a new O-ring. Torque the bolts to 7.8 ft. lbs. (12 Nm).
- Crankshaft pulley

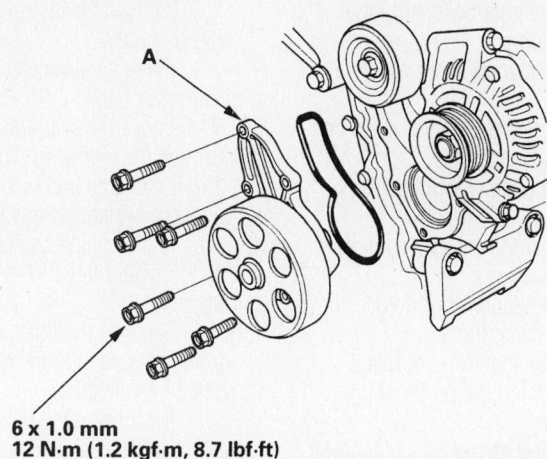

A

6 x 1.0 mm
12 N·m (1.2 kgf·m, 8.7 lbf·ft)

42356-ELEM-G02

Exploded view of the water pump mounting

- Accessory drive belt
- Negative battery cable
6. Refill the engine cooling system.

Heater Core

REMOVAL & INSTALLATION

1. Before servicing the vehicle, refer to the precautions in the beginning of this section.

2. Disconnect the negative battery cable.

3. Drain the cooling system into a clean container for reuse.

4. In the engine compartment, open the heater valve cable clamp and disconnect the cable from the heater valve arm. Then, turn the heater valve to the fully opened position.

5. Remove or disconnect the following:
- Heater hoses from the heater core
- Heater housing-to-chassis nut

➡**When removing the heater housing nut, be careful not to damage or bend the fuel lines, the brake lines, etc.**

6. Remove the instrument panel by performing the following procedure:
 a. Remove the driver's side lower instrument panel cover screws, disengage the clips and remove the lower cover.
 b. Remove the knee bolster bolts and the knee bolster.
 c. Remove the glove box stops from each side of the glove box.
 d. Remove the glove box-to-instrument panel bolts and the glove box.
 e. Remove the lower console cover by disengaging the 4 clips and removing the cover.
 f. Remove the 6 center pocket-to-instrument panel screws; then, insert a flat tipped screwdriver at the upper right side corner of the center pocket, push down on the top of the hook and remove the center pocket/beverage holder assembly.
 g. Remove the center instrument panel lower cover screws and disengage the clips on the upper left side; then, disconnect the electrical connectors and remove the cover.

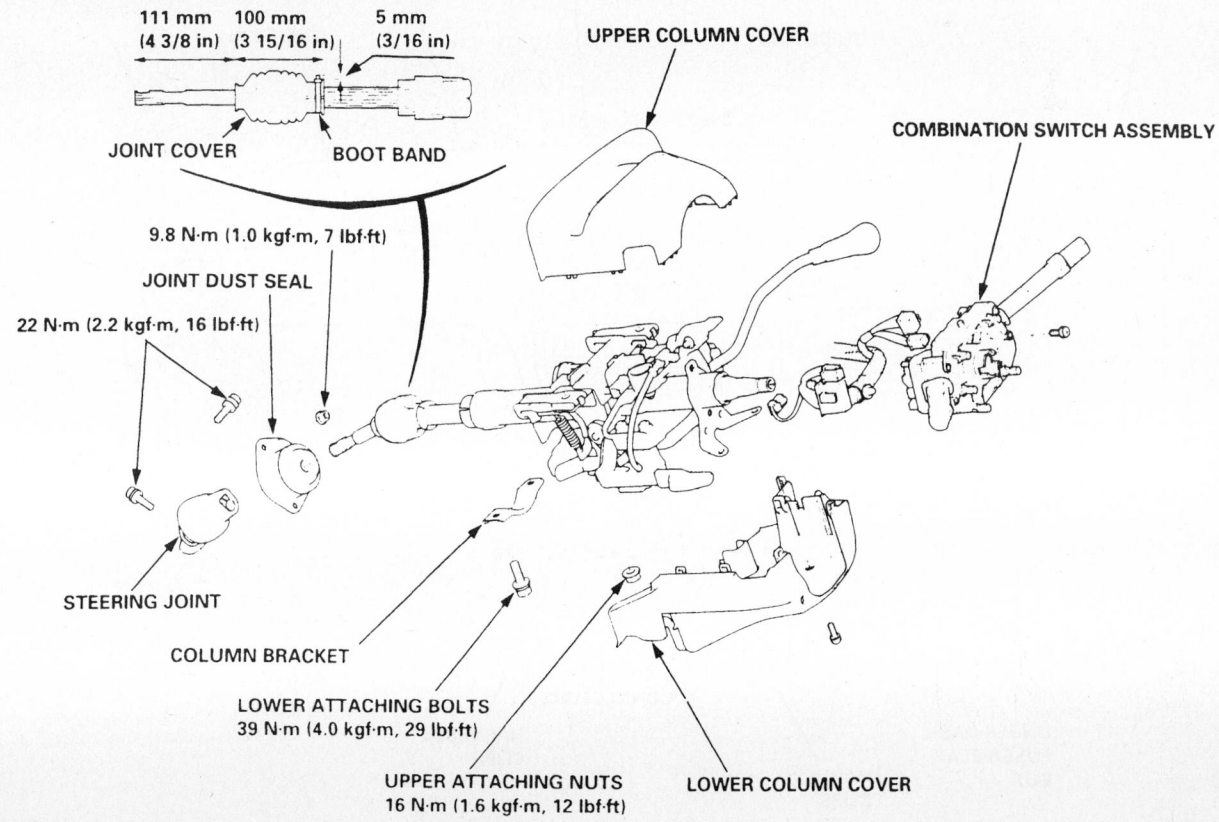

111 mm   100 mm   5 mm
(4 3/8 in) (3 15/16 in) (3/16 in)

JOINT COVER   BOOT BAND

UPPER COLUMN COVER

COMBINATION SWITCH ASSEMBLY

9.8 N·m (1.0 kgf·m, 7 lbf·ft)

JOINT DUST SEAL

22 N·m (2.2 kgf·m, 16 lbf·ft)

STEERING JOINT

COLUMN BRACKET

LOWER ATTACHING BOLTS
39 N·m (4.0 kgf·m, 29 lbf·ft)

UPPER ATTACHING NUTS
16 N·m (1.6 kgf·m, 12 lbf·ft)

LOWER COLUMN COVER

93113GI2

Exploded view of the steering column and related components

h. Gently, push the power window switch from the instrument panel's lower cover opening by hand. Disconnect the electrical connectors and remove the power window switch.

i. Close the driver's side air vent; then, gently, push out the clips and pull out the vent. Disconnect the electrical connectors and remove the vent.

j. Gently, push out the driver's side defogger trim; then, disconnect the electrical connector and remove the side defogger trim.

k. At the base of the steering wheel, remove the access panel and disconnect the air bag electrical connector.

l. Remove the steering column covers screws and the covers.

m. Remove the steering column-to-instrument panel nuts/bolts and lower the steering column.

n. Remove the instrument panel side covers.

o. Disconnect the wiring harness connector and remove the nuts.

p. Move the under-dash fuse/relay box.

q. Disconnect the antenna connector and the harness clips.

r. Remove the connector holder from the instrument panel frame.

s. Remove the control unit/relay bracket from behind the center of the instrument panel.

t. Remove the passenger's side lower instrument panel cover.

u. Disconnect the connectors and the harness clips.

v. Remove the instrument panel-to-chassis bolts.

w. Using an assistant, remove the instrument panel.

7. Remove the evaporator housing by performing the following procedure:

a. Discharge and recover the air conditioning system refrigerant.

b. In the engine compartment, remove the refrigerant lines-to-evaporator housing bolts.

c. Separate the lines, discard the grommets and plug the openings to prevent contamination.

d. Disconnect the evaporator housing's temperature sensor connector.

e. Remove the evaporator housing-to-chassis screws/nut and remove the evaporator housing.

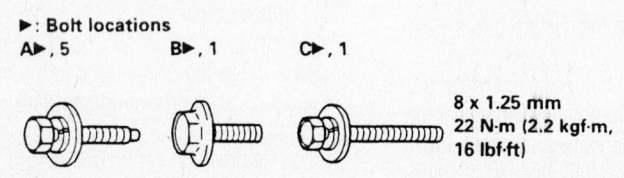

▶: Bolt locations

A▶, 5   B▶, 1   C▶, 1

8 x 1.25 mm
22 N·m (2.2 kgf·m,
16 lbf·ft)

PROTECTIVE TAPE

GUIDE PINS

DASHBOARD

FRONT PASSENGER'S AIRBAG CONNECTOR

GUIDE PIN

PROTECTIVE TAPE

C ▶

▶ A

▶ A

B

▶ A

A ▶

▶ A

Loosen.

UNDER-DASH FUSE/RELAY BOX

HARNESS CLIPS

CONNECTORS

HARNESS CLIPS

CONNECTOR

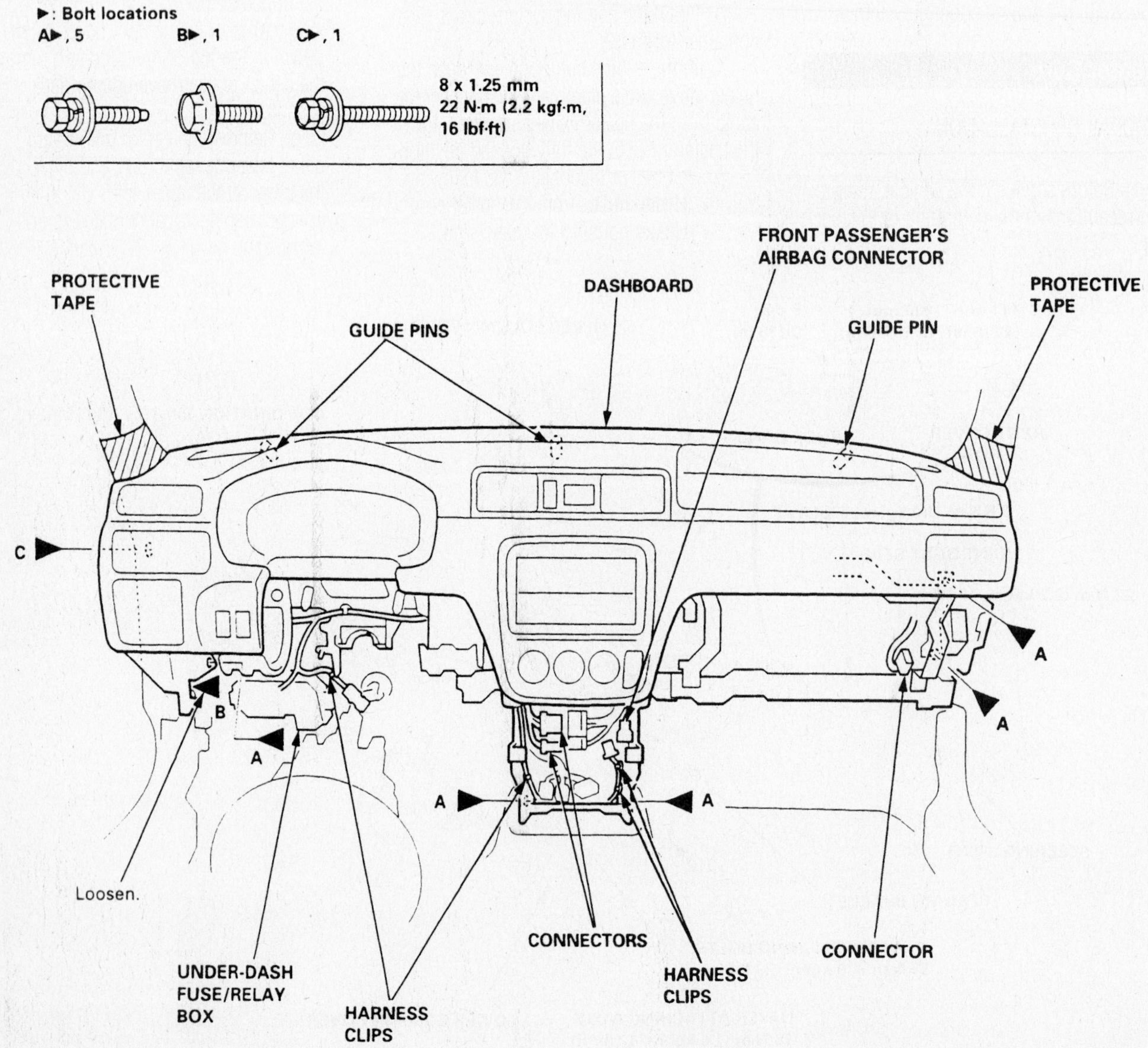

Exploded view of the instrument panel and related components

93113GI3

8. Disconnect the mode control motor and the air mix control motor electrical connectors and remove the wiring harness clips and the wiring harness from the heater housing.

9. Remove the heater duct clip, the heater housing-to-chassis nuts and the heater housing.

10. Remove the heater core cover screws and the cover.

11. Remove the heater core pipe clamp screws and the clamp.

12. Remove the heater core from the heater housing.

To install:

13. Install the heater core in the heater housing.

14. Install the heater core pipe clamp and the clamp screws.

15. Install the heater core cover and the cover screws.

16. Install the heater housing, the heater housing-to-chassis nuts and the heater duct clip.

17. Install the wiring harness clips and the wiring harness to the heater housing and connect the mode control motor and the air mix control motor electrical connectors.

18. Install the evaporator housing by performing the following procedure:

a. Install the evaporator housing and the evaporator housing-to-chassis screws/nut.

b. Connect the evaporator housing's temperature sensor connector.

c. Using new grommets, connect the refrigerant lines.

d. In the engine compartment, install the refrigerant lines-to-evaporator housing bolts.

19. Install the instrument panel by performing the following procedure:

a. Using an assistant, install the instrument panel.

b. Install the instrument panel-to-chassis bolts.

c. Connect the connectors and the harness clips.

d. Install the passenger's side lower instrument panel cover.

e. Install the control unit/relay bracket to the center of the instrument panel.

f. Install the connector holder to the instrument panel frame.

g. Connect the antenna connector and the harness clips.

h. Install the under-dash fuse/relay box.

i. Connect the wiring harness connector and install the nuts.

j. Install the instrument panel side covers.

k. Install the steering column and the column-to-instrument panel nuts/bolts. Torque the nuts to 12 ft. lbs. (16 Nm) and the bolts to 29 ft. lbs. (39 Nm).

l. Install the steering column covers and the cover screws.

m. At the base of the steering wheel, connect the air bag electrical connector and install the access panel.

n. Connect the electrical connector and install the driver's side defogger trim.

o. Connect the electrical connectors and install driver's side air vent.

p. Connect the electrical connectors and install the power window switch to the instrument panel's lower cover opening.

q. Install the center instrument panel lower cover and engage the clips on the upper left side; then, connect the electrical connectors and Install the cover screws.

r. Install the center pocket/beverage holder assembly and the 6 center pocket-to-instrument panel screws.

s. Install the lower console cover by engaging the 4 clips.

t. Install the glove box and the glove box-to-instrument panel bolts.

u. Install the glove box stops to each side of the glove box.

v. Install the knee bolster and the knee bolster bolts.

w. Install the driver's side lower instrument panel cover, engage the clips and install the lower cover screws.

➡**When installing the heater housing nut, be careful not to damage or bend the fuel lines, the brake lines or etc.**

20. Install the heater housing-to-chassis nut.

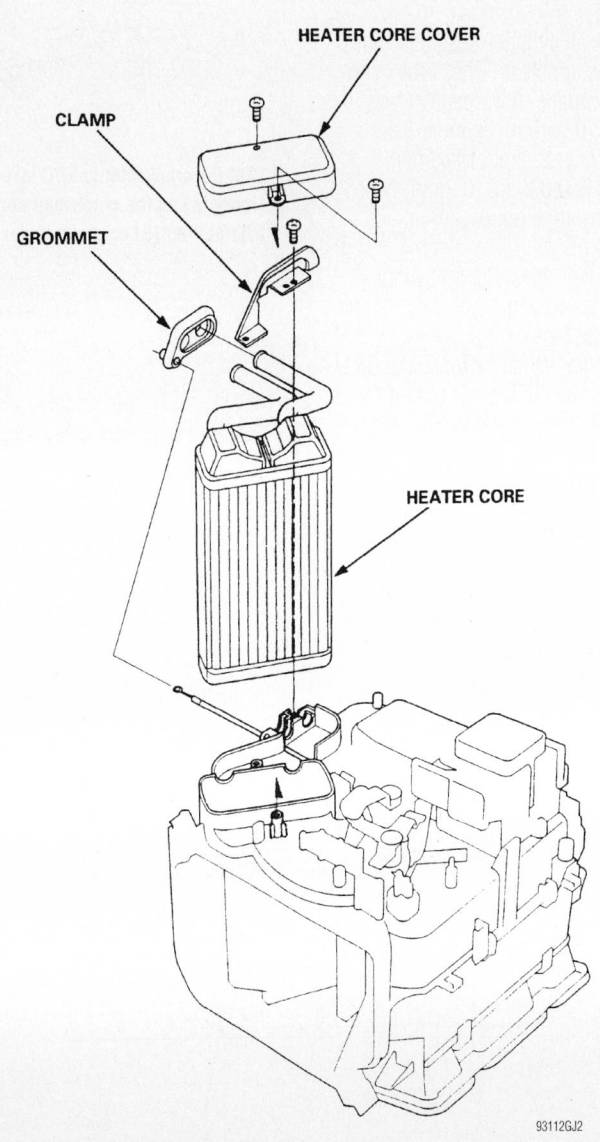

HEATER CORE COVER

CLAMP

GROMMET

HEATER CORE

93112GJ2

Exploded view of the heater core and housing

21. Connect the heater hoses to the heater core.

22. In the engine compartment, connect the cable to the heater valve arm and close the heater valve cable clamp.

23. Refill the cooling system.

24. Connect the negative battery cable.

25. Evacuate and charge and leak test the air conditioning system refrigerant.

26. Run the engine to normal operating temperatures; then, check the climate control operation and check for leaks.

Cylinder Head

REMOVAL & INSTALLATION

1. Before servicing the vehicle, refer to the precautions in the beginning of this section.

2. Drain the cooling system.

3. Relieve the fuel system pressure.

4. Remove or disconnect the following:
- Negative battery cable
- Accessory drive belt
- Intake Air Temperature (IAT) sensor connector
- Vacuum hoses and breather pipe and air intake duct
- Fuel feed hose
- Bolt securing the connecting pipe support bracket to the engine block
- Evaporative emission (EVAP) canister hose and brake booster vacuum hose
- Intake manifold
- Exhaust manifold
- Cam chain
- Positive Crankcase Ventilation (PCV) hose and ground cable
- Upper radiator hose, heater hoses and water bypass hose

5. Remove the following engine wire harness connectors and wire harness clamps from the cylinder head:
- Four injector connector
- Engine Coolant Temperature (ECT) sensor connector
- Camshaft Position (CMP) sensor A & B (intake & exhaust) connectors
- VTEC solenoid valve connector
- Engine Oil Pressure (EOP) sensor connector

6. Remove or disconnect the following:
- 3 bolts holding the EVAP canister purge valve bracket and remove the two bolts (B) securing the harness bracket

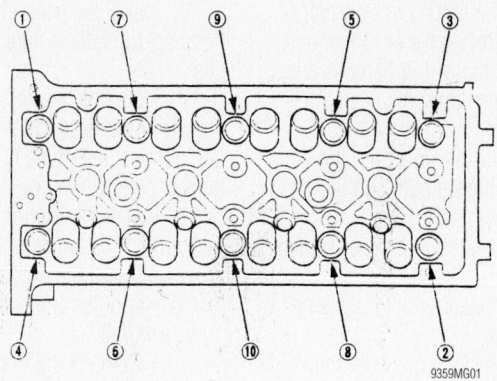

Cylinder head bolt loosening sequence

- Timing (cam) chain
- Rocker arm assembly

7. Loosen the cylinder head bolts in sequence and ⅓ turns until all bolts are loose.

8. Remove the cylinder head.

To install:

9. Be sure all cylinder head and block gasket surfaces are clean. Check the cylinder head for warpage. If warpage is less than 0.002 in. (0.05mm), cylinder head resurfacing is not required. Maximum resurface limit is 0.008 in. (0.2mm) based on a cylinder head height of 3.94 in. (100mm).

10. Install or connect the following:
- New gasket and dowel pins on the cylinder block

11. Set the crankshaft to Top Dead Center (TDC). Align the TDC mark (A) on the crankshaft sprocket with the pointer (B) on the cylinder block.

12. Measure the diameter of each cylinder head bolt at points A & B, as shown in the illustration. If either diameter is less than 0.42 in. (10.6mm), replace the head bolt.

13. Apply engine oil to the threads and under the bolt heads of all of the bolts.

14. Install the cylinder head. Tighten the bolts in sequence as follows:

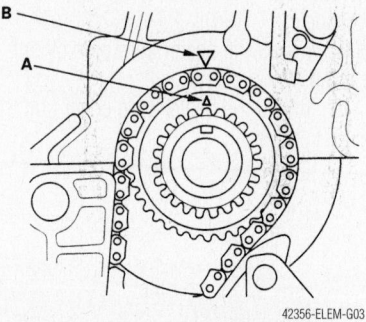

Set the crankshaft to TDC by aligning the mark (A) on the crankshaft sprocket with the pointer (B) on the cylinder block

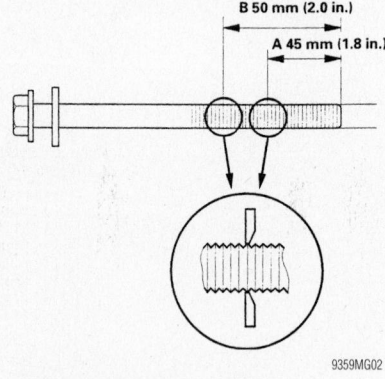

Cylinder head bolt inspection

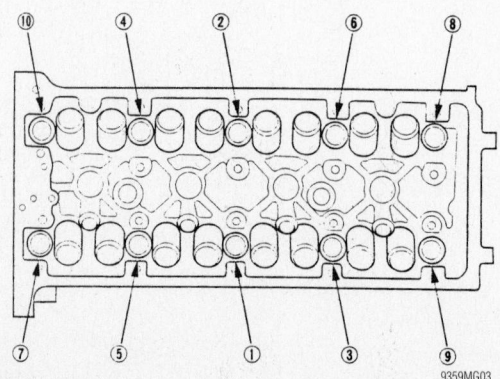

Cylinder head bolt torque sequence

a. Step 1: 29 ft. lbs. (39 Nm).

b. Step 2: Plus 90 degrees.

c. Step 3: Plus 90 degrees.

d. Step 4: If using new cylinder head bolts, add an additional 90 degrees.

15. The remainder of installation is the reverse of removal.

16. Fill the cooling system.

17. Connect the negative battery cable and enter the radio security code.

18. Start the engine and check carefully for any leaks.

Rocker Arms/Shafts

REMOVAL & INSTALLATION

1. Before servicing the vehicle, refer to the precautions in the beginning of this section.

2. Remove or disconnect the following:
- Timing (cam) chain
- Loosen the rocker arm adjusting screws
- Camshaft holder bolts, two turns at a time in sequence
- Timing chain guide (B), camshaft holders and camshafts

3. Insert the bolts (A) into the rocker shaft holder, then remove the rocker arm assembly (B)

To install:

4. Clean and dry the No. 5 rocker shaft holding mating surface.

5. Apply a suitable liquid gasket P/N 08718-0009, or equivalent, evenly to the cylinder head mating surface of the No. 5 rocker shaft holder.

➡**The parts must be installed within 5 minutes of applying the liquid gasket.**

6. Reassemble the rocker arm assembly, as necessary.

7. Install or connect the following
- Bolts (A) into the rocker shaft holder, then the rocker arm assembly on the cylinder head. Remove the bolts from the rocker shaft holder.

8. Make sure the punch marks on the variable valve timing control (VTC) actuator and exhaust camshaft sprocket are facing up, then set the camshafts (A) in the holder

9. Set the camshaft holders (B) and timing chain guide B (C) in place.

10. Tighten the bolts, in sequence, to the following specification:

a. 8mm bolts: 16 ft. lbs. (22 Nm)

b. 6mm bolts: 8.7 ft. lbs. (12 Nm)

11. Install the timing chain and adjust the valve lash.

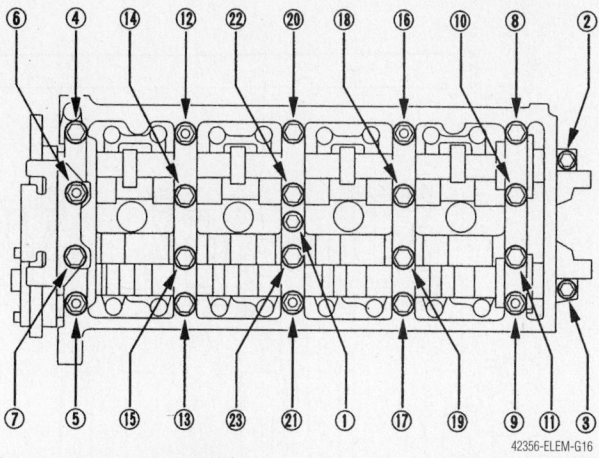

Camshaft holder bolt loosening sequence

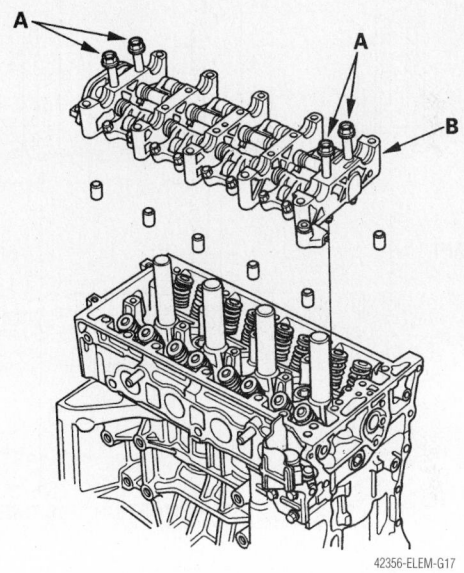

Insert the bolts (A) into the rocker shaft holder, then remove the rocker arm assembly (B)

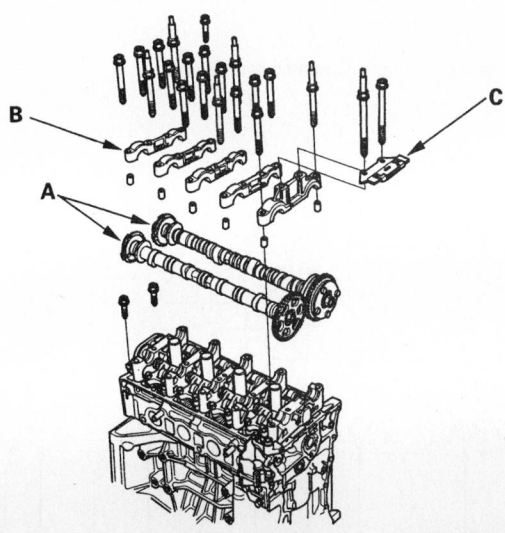

When installing the camshafts (A) make sure the punch marks on the VTC actuator and exhaust cam sprockets are facing up

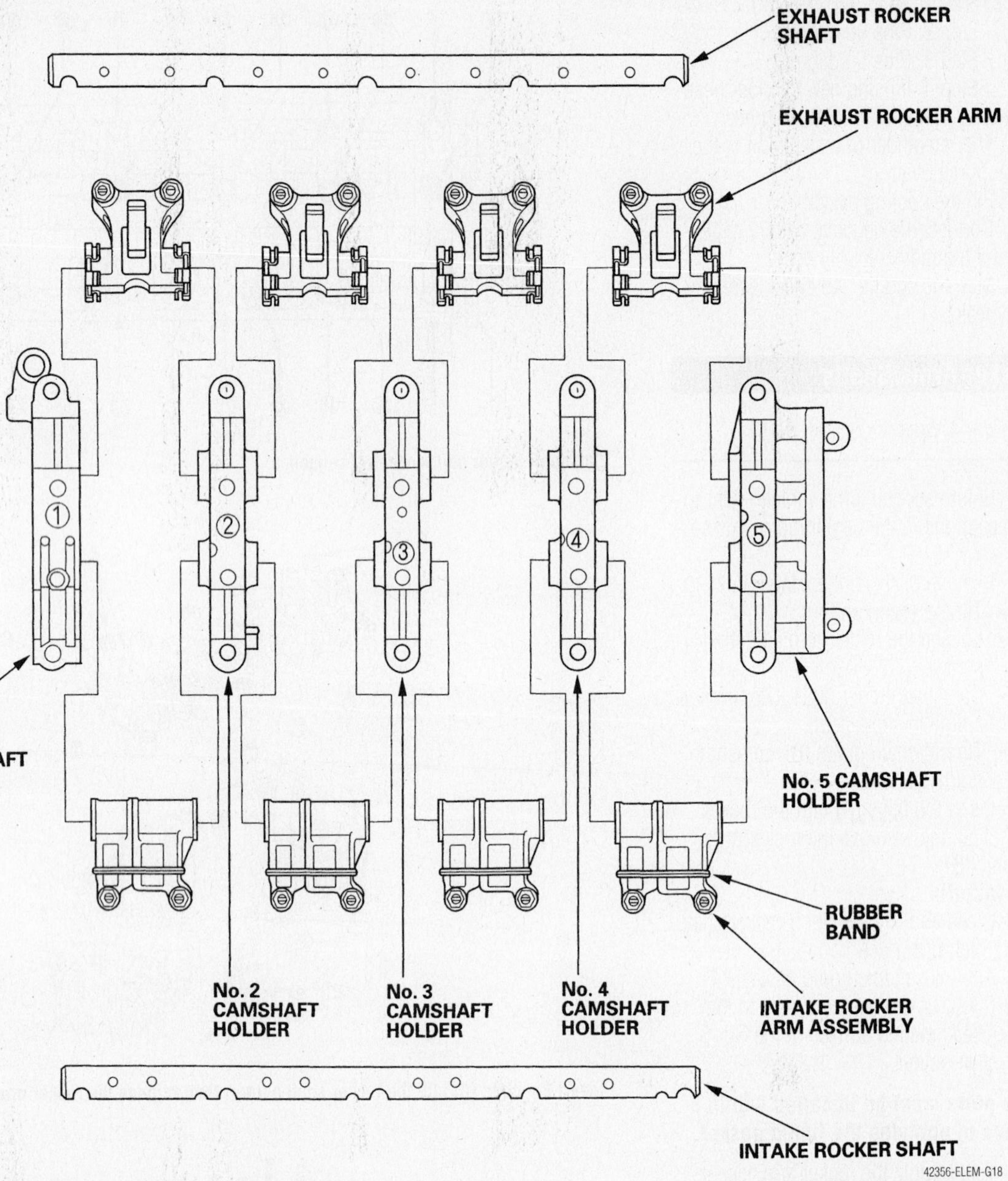

EXHAUST ROCKER
SHAFT

EXHAUST ROCKER ARM

No. 1 CAMSHAFT
HOLDER

No. 5 CAMSHAFT
HOLDER

No. 2
CAMSHAFT
HOLDER

No. 3
CAMSHAFT
HOLDER

No. 4
CAMSHAFT
HOLDER

RUBBER
BAND

INTAKE ROCKER
ARM ASSEMBLY

INTAKE ROCKER SHAFT

42356-ELEM-G18

Exploded view of the rocker arms and related components

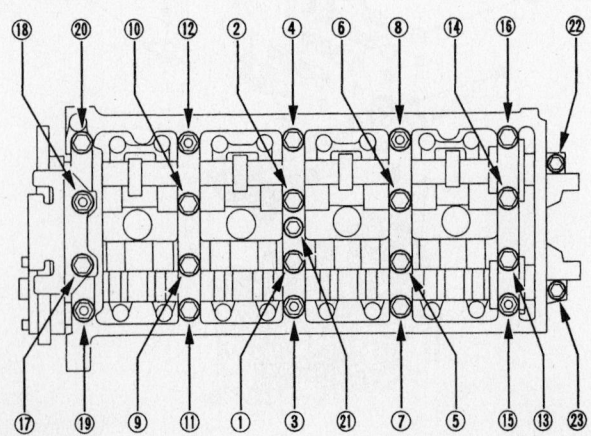

42356-ELEM-G20

Rocker arm assembly bolt tightening sequence

Intake Manifold

REMOVAL & INSTALLATION

1. Before servicing the vehicle, refer to the precautions in the beginning of this section.

2. Disconnect the negative battery cable.

3. Drain the engine coolant into a sealable container.

4. Remove or disconnect the following:
- Intake Air Temperature (IAT) sensor electrical connector
- Vacuum hose and breather pipe and the air intake duct

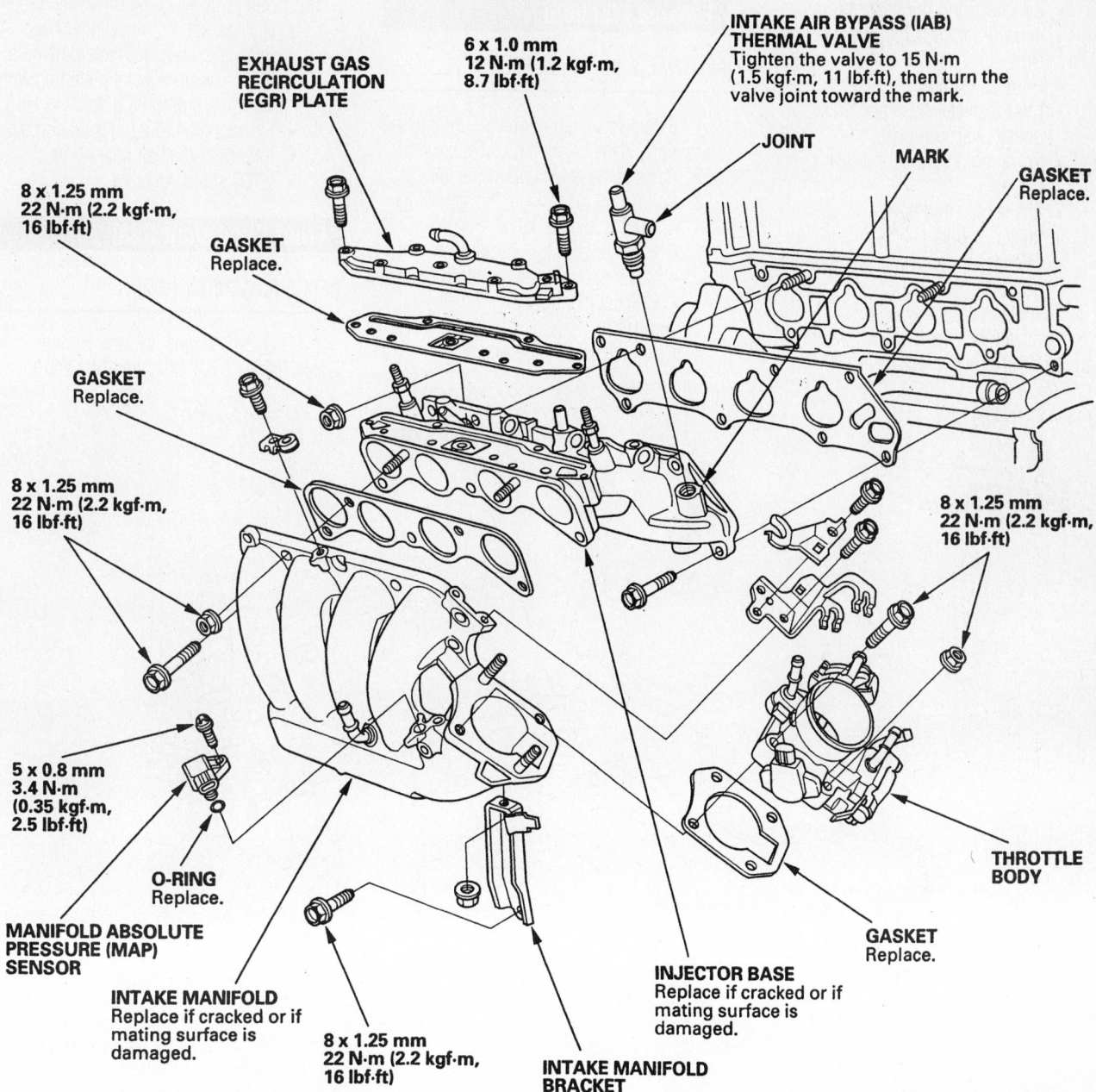

EXHAUST GAS RECIRCULATION (EGR) PLATE

6 x 1.0 mm
12 N·m (1.2 kgf·m, 8.7 lbf·ft)

INTAKE AIR BYPASS (IAB) THERMAL VALVE
Tighten the valve to 15 N·m (1.5 kgf·m, 11 lbf·ft), then turn the valve joint toward the mark.

JOINT

MARK

GASKET
Replace.

8 x 1.25 mm
22 N·m (2.2 kgf·m, 16 lbf·ft)

GASKET
Replace.

GASKET
Replace.

8 x 1.25 mm
22 N·m (2.2 kgf·m, 16 lbf·ft)

8 x 1.25 mm
22 N·m (2.2 kgf·m, 16 lbf·ft)

5 x 0.8 mm
3.4 N·m (0.35 kgf·m, 2.5 lbf·ft)

O-RING
Replace.

MANIFOLD ABSOLUTE PRESSURE (MAP) SENSOR

INTAKE MANIFOLD
Replace if cracked or if mating surface is damaged.

8 x 1.25 mm
22 N·m (2.2 kgf·m, 16 lbf·ft)

INTAKE MANIFOLD BRACKET

INJECTOR BASE
Replace if cracked or if mating surface is damaged.

THROTTLE BODY

GASKET
Replace.

42356-ELEM-G21

Exploded view of the intake manifold and related components

- Intake manifold cover
- Throttle and cruise control cables by loosening the locknuts, then slipping the cable ends out of the accelerator linkage.

➡ **Do not bend the cables during removal. Always replace any throttle or cruise control cables that get kinked during removal.**

- Evaporative emission (EVAP) canister hose and brake booster vacuum hose
- Idle Air Control (IAC) valve connectors

- Throttle Position (TP) sensor connector
- Manifold Absolute Pressure (MAP) sensor connector
- Necessary engine wire harness connectors and wire harness clamps from the intake manifold
- Bolt securing the harness holder and remove the harness clamps
- Water bypass hoses, then plug them
- Harness clamp and harness connector from the intake manifold bracket
- Intake manifold bracket

- A/T vacuum hose
- Retainer and intake manifold

To install:
5. Clean the mounting surfaces.
6. Install or connect the following:
   - New gasket
   - Intake manifold. Tighten the bolts, in a criss-cross pattern beginning with the inner bolt, to 16 ft. lbs. (22 Nm).
   - A/T vacuum hose
   - Intake manifold bracket
   - Harness clamp and connector to the intake manifold bracket
   - Water bypass hoses

- Bolt securing the harness holder and tighten to 8.7 ft. lbs. (12 Nm)
- Harness clamps
- EVAP canister hose and brake booster vacuum hose
- Throttle and cruise control cables
- Intake manifold cover
- Intake air duct
- IAT sensor connector, vacuum hose and breather pipe

7. Refill the cooling system.
8. Connect the negative battery cable, start the engine, and check for leaks.

Exhaust Manifold

REMOVAL & INSTALLATION

1. Before servicing the vehicle, refer to the precautions in the beginning of this section.
2. Raise and safely support the vehicle.
3. Remove or disconnect the following:
   - VTEC solenoid valve
   - Intermediate shaft heat cover
   - Cover and exhaust manifold bracket
   - Exhaust manifold

To install:
4. Clean the mounting surfaces.

5. Install or connect the following:
   - New gasket on the cylinder head
   - Exhaust manifold. Tighten the nuts, in a criss-cross pattern starting with the inner nut, to 33 ft. lbs. (45 Nm).
   - Exhaust manifold bracket and cover
   - Intermediate shaft heat cover
   - VTEC solenoid valve

Front Crankshaft Seal

REMOVAL & INSTALLATION

For the 2.4L engine, see the Timing Chain Removal & Installation procedure.

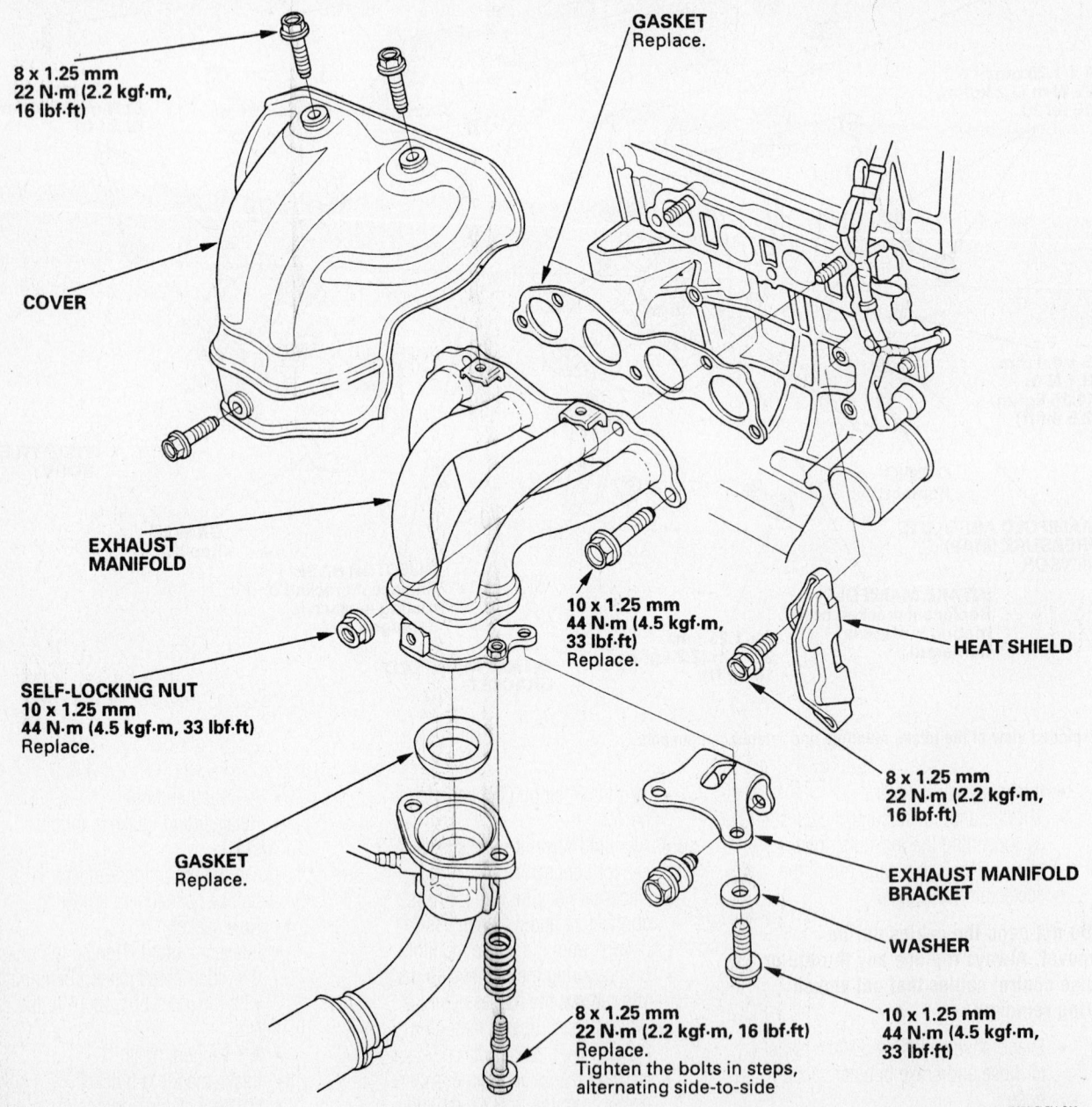

GASKET
Replace.

8 x 1.25 mm
22 N·m (2.2 kgf·m, 16 lbf·ft)

COVER

EXHAUST MANIFOLD

SELF-LOCKING NUT
10 x 1.25 mm
44 N·m (4.5 kgf·m, 33 lbf·ft)
Replace.

10 x 1.25 mm
44 N·m (4.5 kgf·m, 33 lbf·ft)
Replace.

HEAT SHIELD

GASKET
Replace.

8 x 1.25 mm
22 N·m (2.2 kgf·m, 16 lbf·ft)

EXHAUST MANIFOLD BRACKET

WASHER

8 x 1.25 mm
22 N·m (2.2 kgf·m, 16 lbf·ft)
Replace.
Tighten the bolts in steps, alternating side-to-side

10 x 1.25 mm
44 N·m (4.5 kgf·m, 33 lbf·ft)

42356-ELEM-G22

Exploded view of the exhaust manifold and related components

Camshaft

REMOVAL & INSTALLATION

See the Rocker Arm Shaft Removal & Installation procedure.

Valve Lash

ADJUSTMENT

Adjust the valves only when the cylinder head temperature is less than 100°F (38°C).

1. Before servicing the vehicle, refer to the precautions in the beginning of this section.

2. Remove or disconnect the following:
   • Negative battery cable
   • Cylinder head cover

3. Set the timing marks as shown in the illustration with NO.1 at TDC. Check all clearances. Intake should be 0.008–0.010 in.; exhaust should be 0.011–0.013 in. Intake locknut torque is 14 ft. lbs. (19 Nm); exhaust is 10 ft. lbs. (14 Nm).

4. Rotate the crankshaft 180 degrees clockwise and recheck No.3.

5. Rotate the crankshaft 180 degrees clockwise and recheck No.4.

6. Rotate the crankshaft 180 degrees clockwise and recheck No.2.

Starter Motor

REMOVAL & INSTALLATION

➡ The factory sound system has a coded theft protection system. It is recommended that you know your reset code before you begin.

1. Before servicing the vehicle, refer to the precautions in the beginning of this section.

2. Remove or disconnect the following:
   • Negative then the positive battery cables
   • Intake manifold
   • Starter cable from the B terminal
   • Black/white wire from the S (solenoid) terminal
   • Harness clamp and holder
   • Two bolts that mount the starter to the transaxle assembly
   • Starter

To install:

3. Install in the reverse order of removal.

➡ **When installing the heavy gauge starter cable, make sure the crimped side of the terminal end is facing out.**

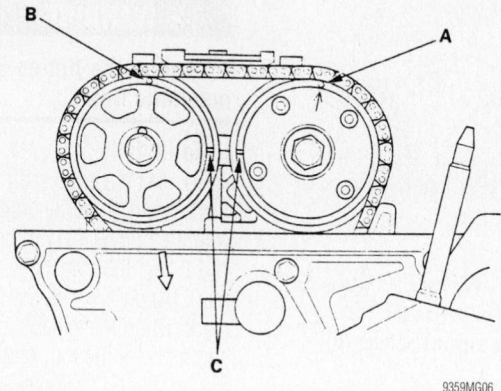

Align the timing marks

9359MG06

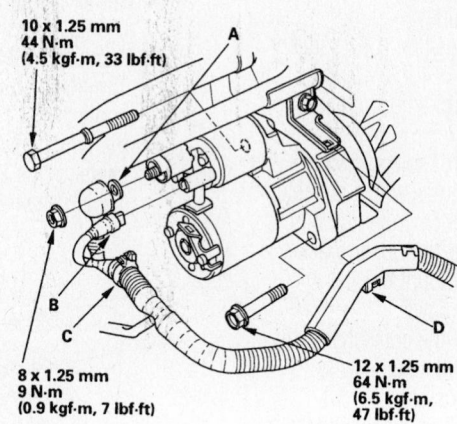

M/T:

10 x 1.25 mm
44 N·m
(4.5 kgf·m, 33 lbf·ft)

8 x 1.25 mm
9 N·m
(0.9 kgf·m, 7 lbf·ft)

12 x 1.25 mm
64 N·m
(6.5 kgf·m, 47 lbf·ft)

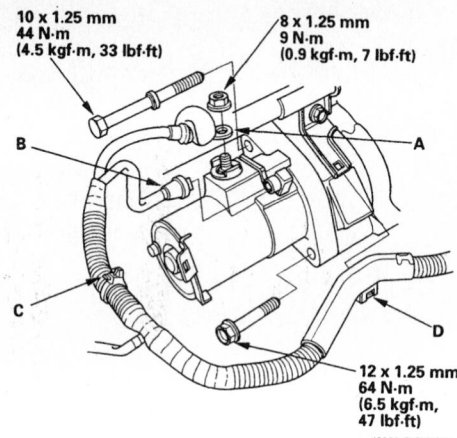

A/T:

10 x 1.25 mm
44 N·m
(4.5 kgf·m, 33 lbf·ft)

8 x 1.25 mm
9 N·m
(0.9 kgf·m, 7 lbf·ft)

12 x 1.25 mm
64 N·m
(6.5 kgf·m, 47 lbf·ft)

42356-ELEM-G23

Starter mounting

4. Enter the anti-theft code and radio presets.

Timing Chain and Front Seal

REMOVAL & INSTALLATION

1. Before servicing the vehicle, refer to the precautions in the beginning of this section.

2. Set the engine to Top Dead Center (TDC).

3. Drain the cooling system.

4. Relieve the fuel system pressure.

5. Remove or disconnect the following:
   • Negative battery cable
   • Front tires and wheels
   • Splash shield
   • Drive belt
   • Cylinder head cover. Check that the

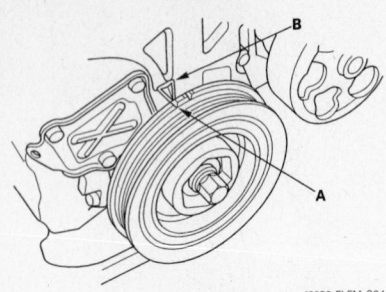

42356-ELEM-G24

Turn the crankshaft pulley so the TDC mark (A) is aligned with the pointer (B)

No. 1 piston TDC marks on the Variable Valve Timing Control (VTC) actuator and exhaust camshaft sprocket are aligned
- Crankshaft pulley
- Crankshaft Position (CKP) sensor connector
- VTC oil control solenoid valve connector
- VTC oil control solenoid valve

6. Support the engine with a suitable jack with a wooden block under the oil pan.
- Ground cable and upper bracket
- Side engine mount bracket
- Chain (case) cover

7. Loosely install the crankshaft pulley. Turn the crankshaft counterclockwise to compress the auto-tensioner.

8. Align the holes on the lock (A) and the auto-tensioner (B), then place a 1.5mm pin into the holes. Turn the crankshaft clockwise to secure the pin.

9. Remove or disconnect the following:
- Auto-tensioner
- Timing chain guide B (top guide)
- Timing chain guide A and tensioner arm
- Timing chain

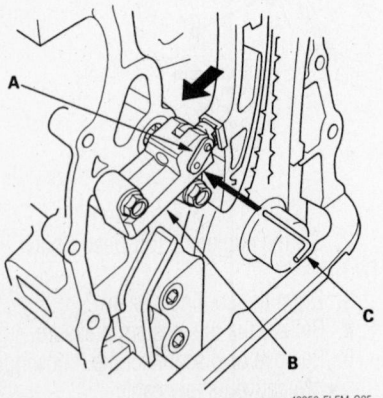

42356-ELEM-G25

Align the holes on the lock (A) and the auto-tensioner (B), then place a 1.5mm pin into the holes. Turn the crankshaft clockwise to secure the pin

Do not let the timing chain near any magnetic fields.

To install:

10. Set the crankshaft to TDC. Align the TDC mark (A) on the crankshaft sprocket with the pointer (B) on the cylinder block.

11. Set the camshafts to TDC. The punch mark (A) on the VTC actuator and the punch mark (B) on the exhaust camshaft (C) should be at the top. Align the TDC marks (C) on the VTC actuator and exhaust camshaft sprockets.

12. Install or connect the following:
- Timing chain the crankshaft sprocket with the colored link of the chain aligned with the mark on the crank sprocket
- Timing chain on the VTC actuator and exhaust camshaft sprocket with the punch marks aligned with the center of the 2 colored links
- Timing chain guide A and tensioner

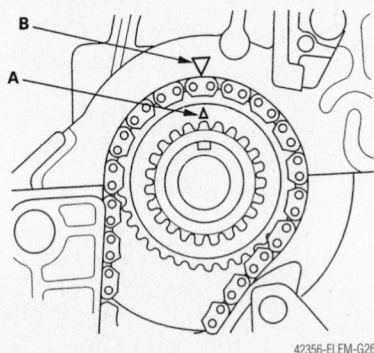

42356-ELEM-G26

Set the crankshaft to TDC. Align the TDC mark (A) on the crankshaft sprocket with the pointer (B) on the cylinder block

arm. Tighten the guide bolts to 8.7 ft. lbs. (12 Nm) and the tensioner arm retainer to 16 ft. lbs. (22 Nm).
- Auto-tensioner and tighten the bolts to 8.7 ft. lbs. (12 Nm)
- Timing chain guide B and tighten the retainers to 16 ft. lbs. (22 Nm)

13. Remove the pin from the auto-tensioner.

14. Inspect the chain cover seal for damage and replace if necessary. Clean and dry the chain cover mating surfaces.

15. Install or connect the following:
- Liquid gasket, P/N 08718-0009 evenly to the cylinder block mating surface of the timing chain cover and the inner threads of the holes
- Liquid gasket to the cylinder block upper surface contact areas on the chain cover and the oil pan mating surface of the chain cover in the inner threads of the holes

➡ **Make sure to install the components within 5 minutes of applying the sealer.**

- New O-ring the timing chain cover. Set the edge of the cover to the edge of the oil pan, then install the cover on the engine block. Tighten the retainers to 8.7 ft. lbs. (12 Nm).

➡ **When installing the chain case, do not slide the bottom surface on the oil pan mounting surface.**

- Side engine mounting bracket and tighten the retainers to 33 ft. lbs. (44 Nm)
- Upper bracket, then tighten the bolts/nuts as shown in the illustration
- Ground cable
- VTC oil control solenoid valve

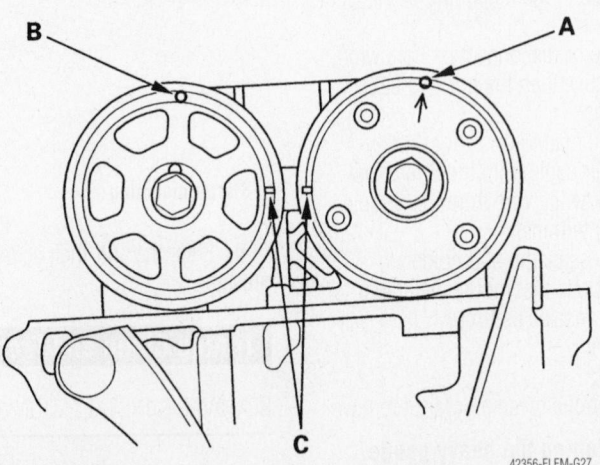

42356-ELEM-G27

The mark (A) on the VTC actuator and the mark (B) on the exhaust cam (C) should be at the top. Align the TDC marks (C) on the VTC actuator and exhaust cam sprockets

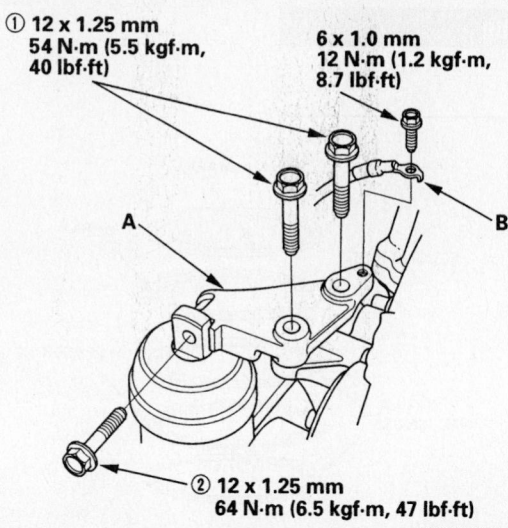

① 12 x 1.25 mm
54 N·m (5.5 kgf·m, 40 lbf·ft)

6 x 1.0 mm
12 N·m (1.2 kgf·m, 8.7 lbf·ft)

A

B

② 12 x 1.25 mm
64 N·m (6.5 kgf·m, 47 lbf·ft)

42356-ELEM-G28

Tighten the upper bracket upper bolt/nuts in the proper order to the correct specification

- CKP sensor and VTC oil control solenoid valve connectors
- Crankshaft pulley
- Cylinder head cover
- Drive belt
- Splash shield

16. Fill the engine cooling system and connect the negative battery cable.

Oil Pan

REMOVAL & INSTALLATION

1. Before servicing the vehicle, refer to the precautions in the beginning of this section.
2. Drain the engine oil.
3. Remove or disconnect the following:
   - Subframe. See engine Removal and Installation.
   - With MT, the stiffener
   - Oil pan bolts

- Oil pan. A gasket cutter will be needed.

4. Installation is the reverse of removal. Torque the bolts, in sequence, in 2 or 3 steps, to 9 ft. lbs. (12 Nm).

Oil Pump

REMOVAL & INSTALLATION

1. Before servicing the vehicle, refer to the precautions in the beginning of this section.
2. Drain the engine oil.
3. Set the No. 1 piston to Top Dead Center (TDC).
4. Remove or disconnect the following:
   - Negative battery cable
   - Oil pan
   - Oil pump chain tensioner and discard
5. Insert a 6mm pin driver into the maintenance hole in the lower balance shaft holder and through the rear balancer shaft to hold the rear balancer shaft.
6. Loosen the oil pump sprocket mounting bolt.
   - Oil pump sprocket
   - Oil pump

To install:

7. Make sure that No.1 piston is at TDC.

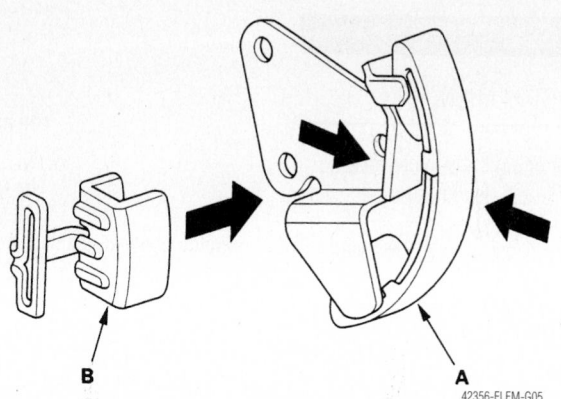

B

A

42356-ELEM-G05

Squeeze the new oil pump chain tensioner (A) then install the set clip (A) on it as shown. The clip is supplied with the new tensioner

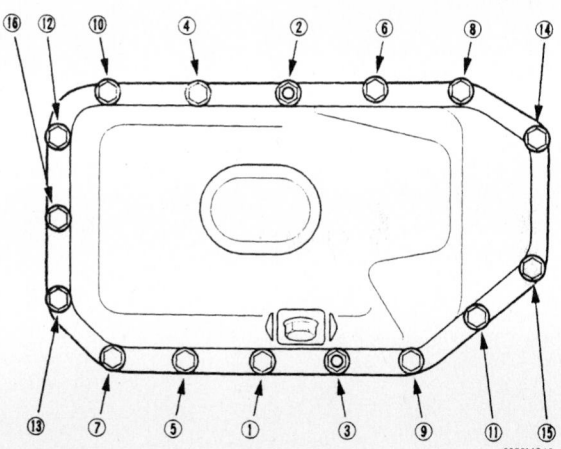

⑯ ⑫ ⑩ ④ ② ⑥ ⑧ ⑭

⑬ ⑦ ⑤ ① ③ ⑨ ⑪ ⑮

9359MG12

Oil pan fastener tightening sequence

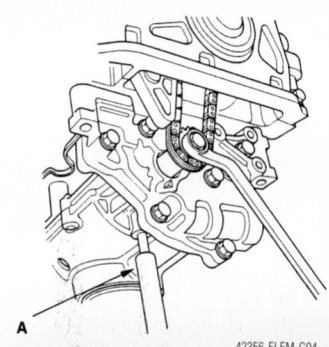

A

42356-ELEM-G04

Insert a 6mm pin into the maintenance hole in the lower balance shaft holder, through the rear balancer shaft to hold the shaft, then loosen the sprocket mounting bolt

8. Align the dowel pin on the rear balance shaft with the mark on the pump.

9. Insert a 6mm pin into the maintenance hole in the lower balance shaft holder, through the rear balancer shaft to hold the shaft.

10. Install or connect the following:
   - Engine oil to the threads of the oil pump sprocket mounting bolt
   - Oil pump and sprocket loosely

11. Remove the balance shaft holding pin.

12. Torque the 10mm mounting bolts to 33 ft. lbs. (44 Nm); the 8mm bolts to 16 ft. lbs. (22 Nm).

13. Torque the pulley bolt to 33 ft. lbs. (44 Nm).

14. Squeeze the new oil pump chain tensioner then install the set clip on it as shown in the illustration.

15. Install or connect the following:
   - New oil pump chain tensioner and torque the bolts to 9 ft. lbs. (12 Nm). Remove the set clip from the tensioner.
   - Oil pan

16. Fill the engine with oil.

Rear Main Seal

REMOVAL & INSTALLATION

1. Before servicing the vehicle, refer to the precautions in the beginning of this section.

2. Remove or disconnect the following:
   - Transaxle
   - Clutch pressure plate and disc, if equipped
   - Flywheel
   - Oil seal

To install:

3. Install or connect the following:
   - Oil seal. Drive the seal square into the seal case.
   - Flywheel. Tighten the bolts in a crossing pattern to 76 ft. lbs. (103 Nm).
   - Clutch pressure plate and disc, if equipped
   - Transaxle

4. Check the fluid levels.

5. Start the engine and check for leaks.

Piston and Ring

POSITIONING

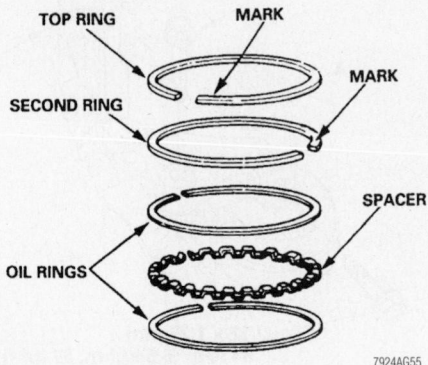

Piston ring positioning and top mark location

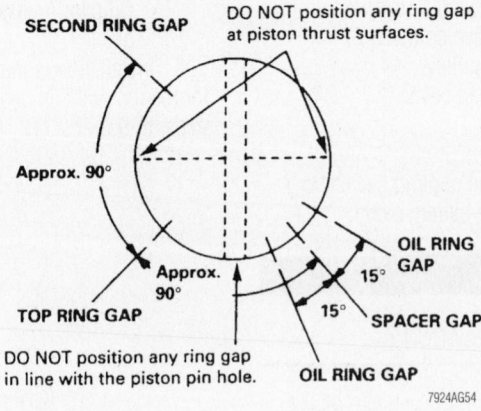

Piston ring end-gap spacing

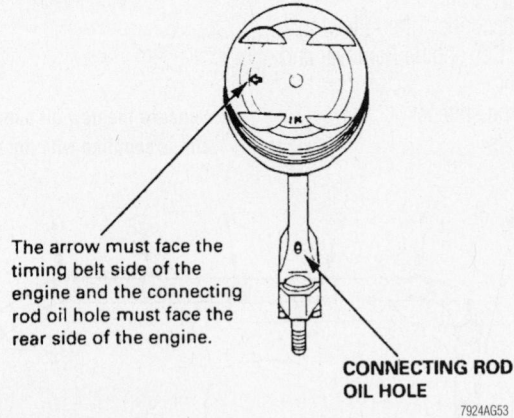

Piston and connecting rod assembly

FUEL SYSTEM

Fuel System Service Precautions

Safety is the most important factor when performing not only fuel system maintenance, but any type of maintenance. Failure to conduct maintenance and repairs in a safe manner may result in serious personal injury or death. Maintenance and testing of the vehicle's fuel system components can be accomplished safely and effectively by adhering to the following rules and guidelines:

• To avoid the possibility of fire and personal injury, always disconnect the negative battery cable unless the repair or test procedure requires that battery voltage be applied.

• Always relieve the fuel system pressure prior to disconnecting any fuel system component (injector, fuel rail, pressure regulator, etc.), fitting or fuel line connection. Exercise extreme caution whenever relieving fuel system pressure to avoid exposing skin, face and eyes to fuel spray. Please be advised that fuel under pressure may penetrate the skin or any part of the body that it contacts.

• Always place a shop towel or cloth around the fitting or connection prior to loosening to absorb any excess fuel due to spillage. Ensure that all fuel spillage (should it occur) is quickly removed from engine surfaces. Ensure that all fuel soaked cloths or towels are deposited into a suitable waste container.

• Always keep a dry chemical (Class B) fire extinguisher near the work area.

• Do not allow fuel spray or fuel vapors to come into contact with a spark or open flame.

• Always use a backup wrench when loosening and tightening fuel line connection fittings. This will prevent unnecessary stress and torsion to fuel line piping. Always follow the proper torque specifications.

• Always replace worn fuel fitting O-rings with new. Do not substitute fuel hose or equivalent, where fuel pipe is installed.

Fuel System Pressure

RELIEVING

✳✳ CAUTION

The fuel injection system remains under pressure after the engine has been turned OFF. Properly relieve

fuel pressure before disconnecting any fuel lines. Failure to do so may result in fire or personal injury.

➡**The radio may contain a coded theft protection circuit. Always obtain the code number before disconnecting the battery.**

1. Before servicing the vehicle, refer to the precautions in the beginning of this section.

2. Disconnect the negative battery cable.

3. Remove the glove box, then remove the PGM-FI main relay (FUEL PUMP) from the fuse/relay box. Start the engine and let it run until it stalls.

4. Turn the engine OFF.

5. Remove the fuel filler cap.

6. Remove the quick-connect fitting cover.

7. Clean any dirt from the quick-connect fitting.

8. Place a rag or shop towel over quick-connect fitting.

9. Detach the quick-connect fitting by holding the connector with one hand, then squeeze the retainer tabs with the other hand to release them from the locking pawls. Pull the connector off.

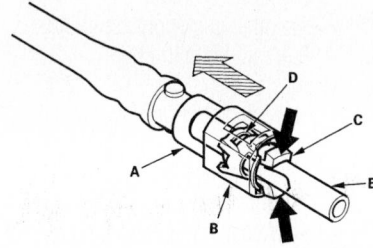

42356-ELEM-G29

Hold the quick-connect (A) connector (B) with one hand, then squeeze the retainer tabs (C) with the other hand to release them from the locking pawls (D)

✳✳ CAUTION

Do not allow fuel spray or fuel vapors to come in contact with a spark or open flame. Keep a dry chemical fire extinguisher nearby. Never store fuel in an open container due to risk of fire or explosion.

➡**A fuel pressure gauge may be attached at the quick-connect location.**

10. Connect the quick-connect fitting, making sure the locking pawls are properly engaged,

11. Clean up any fuel spilled on the engine and intake manifold.

12. Install the fuel pump relay to the underdash fuel/relay box and install the glove box.

13. Install the fuel filler cap.

14. Reconnect the negative battery cable.

15. Turn the ignition **ON**, but don't start the engine. Repeat this 2 or 3 times to pressurize the fuel system. Check for fuel leaks.

16. Enter the radio security code.

Fuel Filter

REMOVAL & INSTALLATION

➡**The fuel filter should be replaced whenever the fuel pressure drops below 48 psi, after making sure that the fuel pump and fuel pressure regulator are okay.**

1. Before servicing the vehicle, refer to the precautions in the beginning of this section.

2. Relieve the fuel system pressure.

3. Remove or disconnect the following:
   • Negative battery cable
   • Fuel pump
   • Fuel filter carrier (A)
   • Fuel filter

To install:

4. Install or connect the following:
   • Fuel filter
   • Fuel lines
   • New gasket (B)

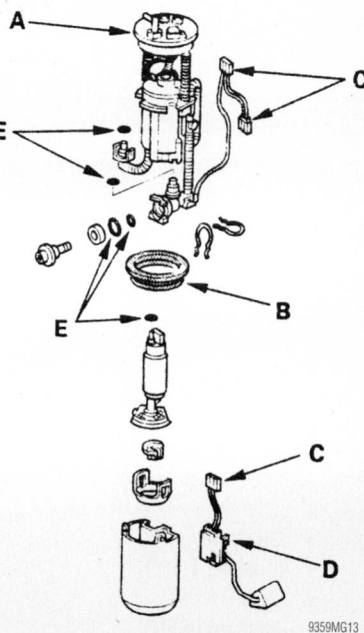

9359MG13

Exploded view of the fuel filter mounting

- New o-rings (E)
- Connectors (C)
- Sending unit (D)
5. Start the engine and check for leaks.

Fuel Pump

REMOVAL & INSTALLATION

1. Before servicing the vehicle, refer to the precautions in the beginning of this section.
2. Relieve the fuel system pressure.
3. Remove or disconnect the following:
   - Negative battery cable
   - Fuel filler cap
   - Center console, then both track floor covers and sill trims.
4. Fold back the floor covering until you can get to the access panel
   - Access panel from the floor
   - Fuel pump connector
   - Fuel supply and return line quick-connect fittings
   - Fuel pump locknut, using special tool No. 07XAA-001010A
   - Fuel pump sending assembly
5. Installation is the reverse of removal.

Fuel Injector

REMOVAL & INSTALLATION

1. Before servicing the vehicle, refer to the precautions in the beginning of this section.
2. Relieve the fuel system pressure.
3. Remove or disconnect the following:
   - Negative battery cable
   - Engine cover

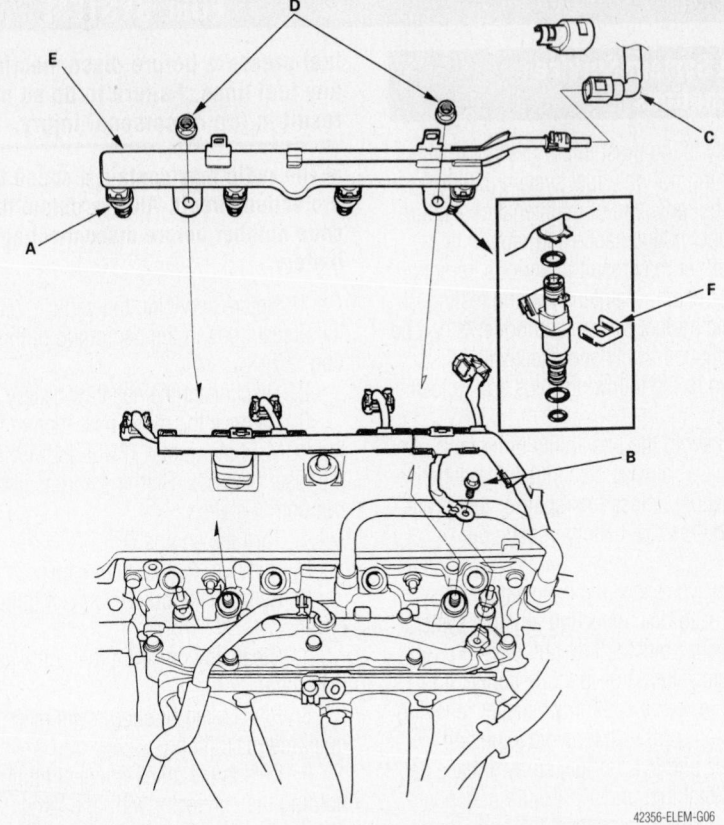

42356-ELEM-G06

Exploded view of the fuel rail (E), injectors (A) and related components

- Injector connectors, ground cable and harness holder
- Fuel line quick-connect fittings
- Fuel rail mounting nuts
- Injector clip(s) from the injector(s)
- Fuel injectors from the fuel rail

To install:
4. Install or connect the following:
   - Injectors to the fuel rail with new O-rings coated with clean engine oil.

- Injector clips
- Injectors in the injector base
- Fuel rail and injector assembly. Tighten the nuts to 16 ft. lbs. (22 Nm).
- Ground cable bolt
- Injector connectors
- Fuel lines
- Negative battery cable
5. Start the engine and check for leaks.

DRIVE TRAIN

Transaxle Assembly

REMOVAL & INSTALLATION

Automatic Transaxle

1. Before servicing the vehicle, refer to the precautions in the beginning of this section.
2. Drain the transaxle.
3. Remove or disconnect the following:
   - Battery
   - Battery tray
   - Air intake assembly
   - Splash shield
   - Transaxle ground cable
   - Starter motor

- Clutch pressure control solenoid valve connector
- Mainshaft speed sensor connector
- Clutch pressure switch connectors
- Shift control solenoid valve connectors
- Lockup control solenoid connector
- Countershaft speed sensor connector
- Transaxle oil cooler lines
- Engine wiring harness from the air cleaner bracket
- Water pipe mounting bolt
- Brake booster and EVAP line mounting bolts. Attach an engine support hanger to the head to support the weight of the engine.
- Stabilizer link from the lower arm

- Lower arms from the knuckles
- Torque converter nuts
- Shift cable
- Front mount bolt and nut
- Rear mount bolts
- Subframe (see the Engine Removal and Installation procedure)
- Rear driveshaft, if equipped
4. Separate the inner CV-joints from the transaxle and intermediate shaft and support the axle halfshafts out of the work area with safety wire.
   - Intermediate shaft
   - Engine stiffener
   - Transaxle mount bolts and nuts
   - Transaxle
5. Installation is the reverse of removal. Observe the following torques:

- Air cleaner housing bracket bolt: 16 ft. lbs. (22 Nm)
- Front mount bolts: 47 ft. lbs. (64 Nm)
- Rear mount bracket bolts: 40 ft. lb. (54 Nm)
- Transmission-to-engine bolts: 47 ft. lbs. (64 Nm)
- Upper transmission mount bolt: 40 ft. lbs. (54 Nm)
- Upper transmission mount nuts: 40 ft. lbs. (54 Nm)
- Rear driveshaft bolts: 24 ft. lbs. (32 Nm)
- Subframe bolts: 76 ft. lbs. (103 Nm)

Manual Transaxle

1. Before servicing the vehicle, refer to the precautions in the beginning of this section.
2. Remove or disconnect the following:
   - Negative battery cable
   - Air intake assembly
   - Transaxle ground cable
   - Vehicle Speed Sensor (VSS) connector
   - Splash shield
   - Shift cables and bracket
   - Clutch slave cylinder and hose bracket
   - Wire harness bracket
   - Water pipe mounting bolt
   - Brake booster and EVAP line mounting bolts. Attach an engine support hanger to the head to support the weight of the engine.
   - Upper transmission mounting bolts
   - Transaxle mount and bracket
3. Separate the inner CV-joints from the transaxle and intermediate shaft and support the axle halfshafts out of the work area with safety wire.
4. Remove or disconnect the following:
   - Intermediate shaft
   - Right front mount and bracket
   - Rear engine mounting bolts
   - Rear driveshaft
   - Subframe (see the Engine Removal and Installation procedure)
   - Clutch housing cover
   - Transaxle
5. Installation is the reverse of removal. Observe the following torques:
   - Transaxle rear mount and bracket. Tighten the bracket bolts to 40 ft. lbs. (54 Nm) and the through bolt to 47 ft. lbs. (64 Nm)
   - Transaxle. Tighten the flange bolts to 47 ft. lbs. (64 Nm)
   - Front mount and bracket. Tighten the bolts to 47 ft. lbs. (64 Nm).

- Clutch housing cover. Tighten the bolts to 29 ft. lbs. (39 Nm)
- Subframe: 76 ft. lbs. (98 Nm)

Clutch

ADJUSTMENTS

The Element is equipped with a hydraulic clutch system. No adjustment is necessary.

REMOVAL & INSTALLATION

1. Before servicing the vehicle, refer to the precautions in the beginning of this section.
2. Remove or disconnect the following:
   - Negative battery cable
   - Transaxle
   - Pressure plate. Loosen the bolts evenly in a crossing pattern.
   - Clutch disc

To install:

3. Install the clutch disc and pressure plate. Tighten the pressure plate bolts in a crisscross pattern, in several steps to 19 ft. lbs. (26 Nm).
4. Install or connect the following:
   - Transaxle
   - Negative battery cable

Hydraulic Clutch System

BLEEDING

1. Before servicing the vehicle, refer to the precautions in the beginning of this section.
2. Attach a hose to the bleeder screw and suspend the other end in a container of clean brake fluid.
3. Open the bleeder screw.
4. Slowly pump the clutch pedal until no more air bubbles appear at the bleeder hose.
5. Tighten the bleeder screw to 70 inch lbs. (8 Nm).
6. Refill the clutch master cylinder as necessary.
7. Check for leaks and proper clutch operation.

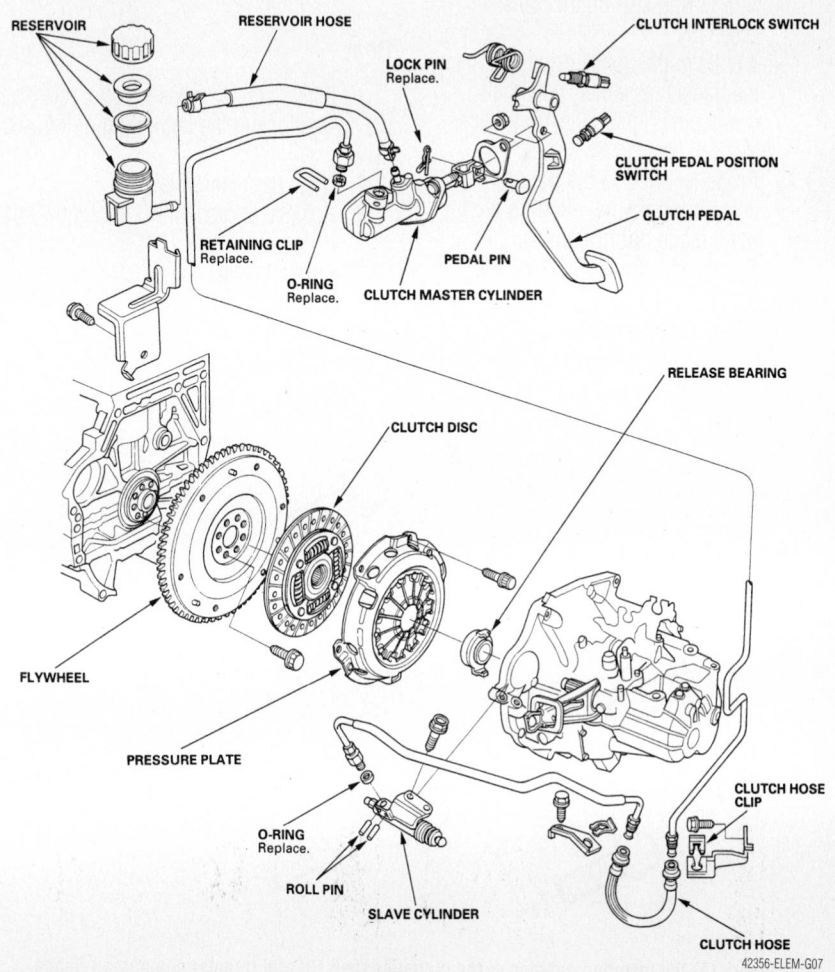

Exploded view of the clutch system components

42356-ELEM-G07

Transfer Assembly

REMOVAL & INSTALLATION

1. Before servicing the vehicle, refer to the precautions in the beginning of this section.
2. Drain the transaxle fluid. Install the drain plug with a new gasket and tighten to 36 ft. lbs. (49 Nm).
3. Disconnect the negative battery cable.
4. Matchmark the installed position of the propeller shaft and transfer companion flange.
5. Remove or disconnect the following:

- Propeller shaft from the transfer assembly
- Mounting bolts and transfer assembly

To install:

6. Clean the transfer assembly mating surfaces, then apply clean transmission fluid to the mating surfaces.
7. Install or connect the following:

- New O-ring seal on the transfer assembly
- 4 bolts in the transfer housing, then the transfer assembly with the dowel pin. Tighten the 10mm bolts to 33 ft. lbs. (44 Nm).
- Propeller shaft to the transfer companion flange, aligning the mark made during removal. Tighten the 8mm bolts to 24 ft. lbs. (33 Nm).
- Negative battery cable

8. Fill the transaxle to the correct level and check for leaks.

Halfshaft

REMOVAL & INSTALLATION

Front

1. Before servicing the vehicle, refer to the precautions in the beginning of this section.
2. Drain the transaxle.
3. Remove or disconnect the following:

- Negative battery cable
- Front wheels
- Stabilizer bar
- Lower ball joint
- Spindle nut

4. On the left side, pry the inboard joint from the case with a prybar.
5. On the right side, drive the inboard shaft off the intermediate shaft with a drift and hammer.
6. Installation is the reverse of removal. Observe the following torques:

- Ball stud nuts: 40 ft. lbs. (54 Nm)
- Stabilizer link nuts: 29 ft. lbs. (39 Nm)
- Spindle nut: 181 ft. lbs. (245 Nm)

Rear

1. Before servicing the vehicle, refer to the precautions in the beginning of this section.
2. Drain the differential.
3. Remove or disconnect the following:

- Negative battery cable
- Rear wheels
- Spindle nut

4. Pry the inboard joint from the differential.

5. Remove the outer CV-joint stub shaft from the hub by tapping the stub shaft with a plastic hammer.

To install:

➡ **Use new circlips and self-locking nuts for assembly.**

6. Install the outer CV-joint stub shaft into the hub.
7. Install the inner CV-joint to the differential until the circlip locks in the retaining groove.
8. Install or connect the following:

- Spindle nut. Tighten the nut to 134 ft. lbs. (181 Nm).
- Rear wheels
- Negative battery cable

9. Fill the differential to the correct level and check for leaks.

CV-Joint

OVERHAUL

Front

OUTBOARD JOINT

1. Before servicing the vehicle, refer to the precautions in the beginning of this section.
2. Remove or disconnect the following:

- Axle halfshaft from the vehicle and place it in a vise
- Outboard joint boot clamps and push the boot back
- Outboard joint by driving it off the axle shaft with a brass drift and hammer
- Outboard joint boot

To install:

➡ **Use new circlips and boot clamps for assembly.**

3. Install the outboard joint boot and clamps to the axle shaft.
4. Fill the outboard joint with grease. Install the outboard joint to the axle shaft. Tap the stub shaft with a brass hammer to seat the circlip.
5. Fill the outboard joint boot with grease and install the boot clamps.
6. Install the axle halfshaft to the vehicle.

INBOARD JOINT

1. Before servicing the vehicle, refer to the precautions in the beginning of this section.
2. Remove or disconnect the following:

- Axle halfshaft from the vehicle.
- Inboard joint boot clamps and push the boot back

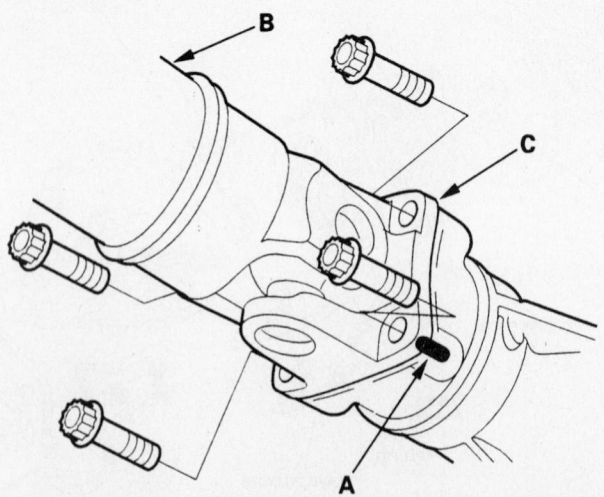

Matchmark (A) the installed position of the propeller shaft (B) and transfer companion flange (C)

42356-ELEM-G08

- Inboard joint housing from the axle
- Rollers from the spider
- Snapring and the spider from the axle shaft
- Inboard joint boot

To install:

➥**Use new circlips and boot clamps for assembly.**

3. Install or connect the following:
   - Inboard joint boot and clamps to the axle shaft
   - Spider with a new snapring
   - Rollers to the spider

4. Fill the joint housing with grease and install it.

5. Fill the inboard joint boot with grease and install the boot clamps.

6. Install the axle halfshaft to the vehicle.

Rear

1. Before servicing the vehicle, refer to the precautions in the beginning of this section.

2. Remove or disconnect the following:
   - Axle halfshaft from the vehicle
   - Joint boot clamps and push the boot back
   - Joint housing from the axle

- Rollers from the spider
- Snapring and the spider from the axle shaft
- Joint boot

To install:

➥**Use new circlips and boot clamps for assembly.**

3. Install or connect the following:
   - Joint boot and clamps to the axle shaft
   - Spider with a new snapring
   - Rollers to the spider

4. Fill the joint housing with grease and install it.

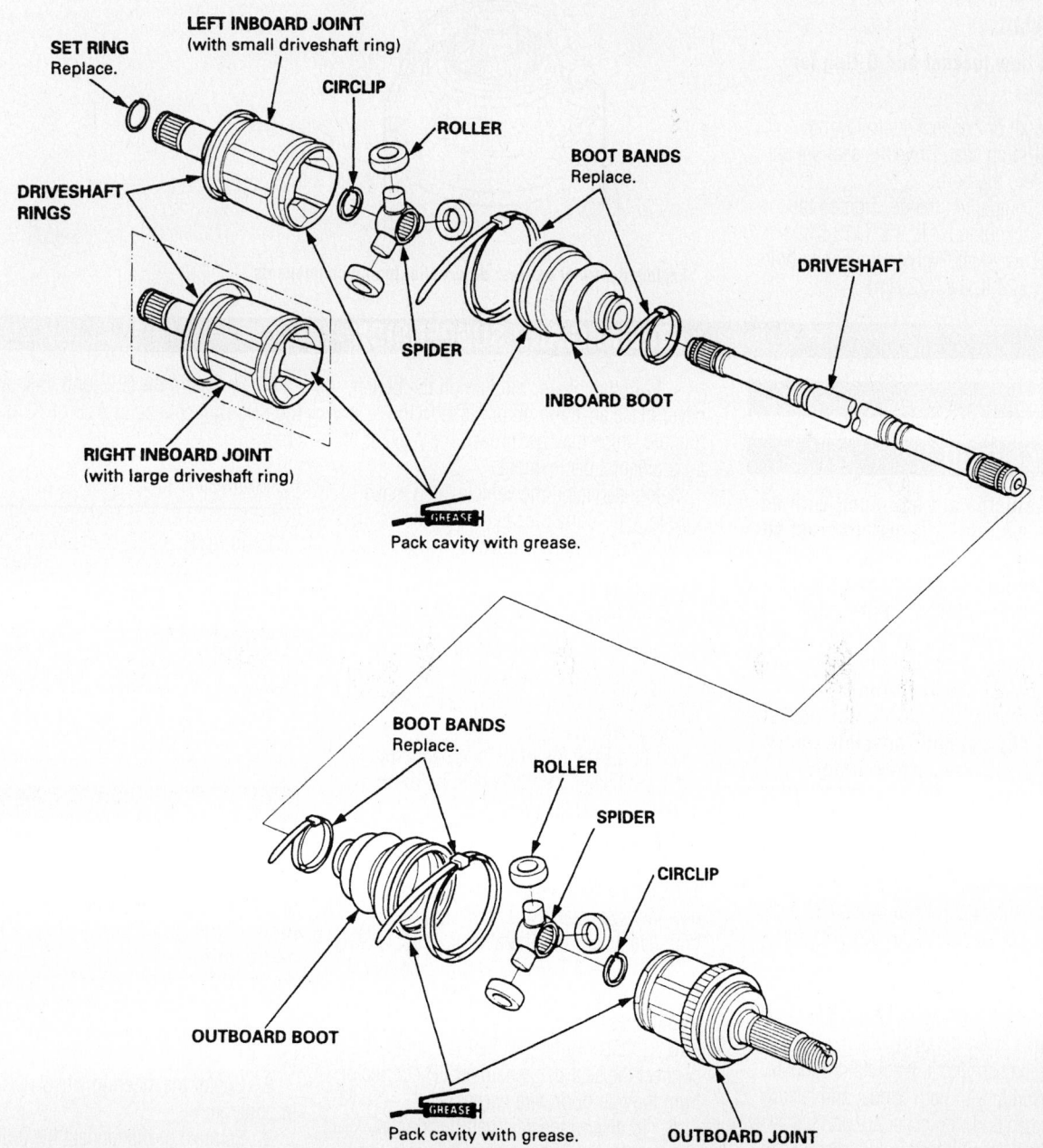

Exploded view of the rear axle

9308MG30

5. Fill the joint boot with grease and install the boot clamps.

6. Install the axle halfshaft to the vehicle.

Pinion Seal

REMOVAL & INSTALLATION

1. Before servicing the vehicle, refer to the precautions in the beginning of this section.

2. Remove or disconnect the following:
- Driveshaft
- Companion flange
- Pinion seal

To install:

➡**Use a new locknut and O-ring for assembly.**

3. Install or connect the following:
- Pinion seal. Drive the seal square into the bore.
- Companion flange. Tighten the locknut to 87 ft. lbs. (118 Nm).
- Driveshaft. Tighten the flange bolts to 24 ft. lbs. (32 Nm).

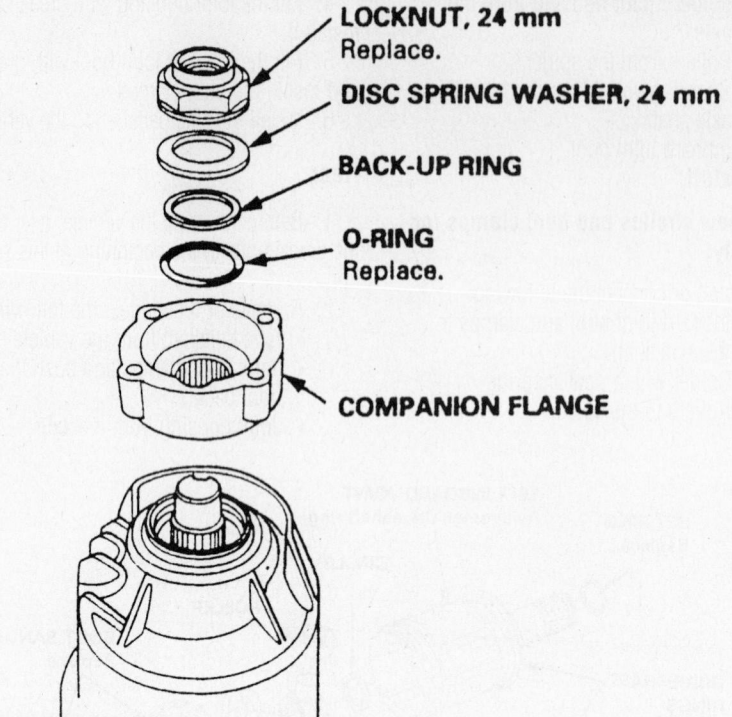

LOCKNUT, 24 mm
Replace.

DISC SPRING WASHER, 24 mm

BACK-UP RING

O-RING
Replace.

COMPANION FLANGE

9308MG31

Exploded view of the rear differential pinion components

STEERING AND SUSPENSION

Air Bag

※ CAUTION

Some vehicles are equipped with an air bag system. The system must be disarmed before performing service on, or around, system components, the steering column, instrument panel components, wiring and sensors. Failure to follow the safety precautions and the disarming procedure could result in accidental air bag deployment, possible injury and unnecessary system repairs.

PRECAUTIONS

Several precautions must be observed when handling the inflator module to avoid accidental deployment and possible personal injury.
- Never carry the inflator module by the wires or connector on the underside of the module.
- When carrying a live inflator module, hold securely with both hands, and ensure that the bag and trim cover are pointed away.
- Place the inflator module on a bench or other surface with the bag and trim cover facing up.

- With the inflator module on the bench, never place anything on or close to the module which may be thrown in the event of an accidental deployment.

Before servicing the vehicle, also make sure to refer to the precautions in the beginning of this section as well.

DISARMING

1. Disconnect and isolate the negative battery cable. Wait 3 minutes for the system capacitor to discharge before performing any service.

2. To disarm the driver's airbag, remove the access panel from the steering wheel, then disconnect the driver's airbag 4P connector from the cable reel.

3. To disarm the front passenger's airbag, remove the glove box, then disconnect the passenger's airbag 4P connector from dashboard wire harness B.

4. To disarm the side airbag, disconnect the side airbag 2P connector from the floor wire harness.

5. To disarm the seat belt tensioner, disconnect the seat belt tensioner 2P connector from the rear door wire harness.

6. To disarm the seat belt buckle tensioner, disconnect the seat belt buckle tensioner 4P connector.

7. To disarm the SRS unit, disconnect the SRS unit connector A, B or C, as applicable.

REARMING

1. To rearm, connect the electrical connector(s) as necessary, then connect the negative battery cable.

Power Rack and Pinion Steering Gear

REMOVAL & INSTALLATION

※ WARNING

Do not permit the steering wheel to turn whenever the steering gear is disconnected from the steering column. Damage to the air bag wiring can result.

1. Before servicing the vehicle, refer to the precautions in the beginning of this section.

2. Center the steering wheel and lock it in position.

3. Remove or disconnect the following:
- Negative battery cable
- Air bag and steering wheel

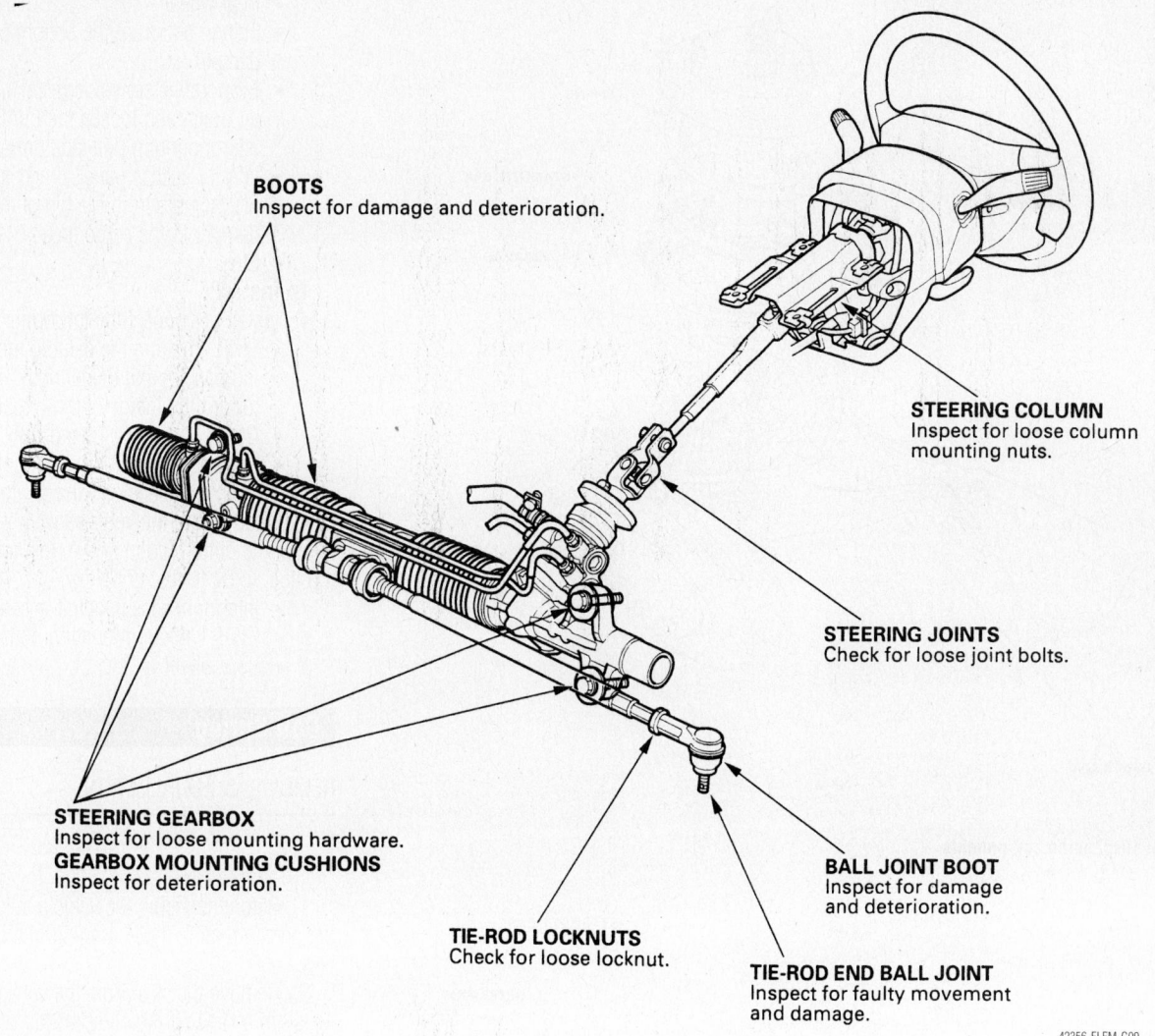

BOOTS
Inspect for damage and deterioration.

STEERING COLUMN
Inspect for loose column
mounting nuts.

STEERING JOINTS
Check for loose joint bolts.

STEERING GEARBOX
Inspect for loose mounting hardware.
GEARBOX MOUNTING CUSHIONS
Inspect for deterioration.

BALL JOINT BOOT
Inspect for damage
and deterioration.

TIE-ROD LOCKNUTS
Check for loose locknut.

TIE-ROD END BALL JOINT
Inspect for faulty movement
and damage.

42356-ELEM-G09

Power steering gear and related components

- Front wheels
- Driver's side dashboard lower cover
  and undercover
- Air cleaner housing
- Steering joint bolts
- Tie rod ends
- Steering hoses
- Left side flange bolts
- Mounting brackets

4. Lower the unit so the pinion shaft
points outward. Remove the pinion shaft
grommet. The steering gear is removed
through the driver's side.

5. Installation is the reverse of removal.
Observe the following torques:

- Mounting bracket and side flange
  bolts: 46 ft. lbs. (62 Nm)
- Supply line flare nut: 27 ft. lbs. (37
  Nm)
- Tie rod ball stud nuts: 32 ft. lbs.
  (43 Nm)
- Steering joint bolts: 21 ft. lbs. (28
  Nm)

Strut (Damper)

REMOVAL & INSTALLATION

Front

1. Before servicing the vehicle, refer to
the precautions in the beginning of this section.

2. Remove or disconnect the following:
- Front wheel
- Tie rod end
- Brake hose retainer
- ABS sensor harness bracket and
  brake hose bracket. Do not discon-
  nect the wheel sensor connector.
- Pinch bolts from the damper, while
  holding the nuts
- Flange nuts from the top of the
  damper
- Strut (damper), after lowering the
  lower control arm

To install:

→**Use new self-locking fasteners for
assembly.**

3. Install or connect the following:
- Strut (damper). Tighten the upper
  mounting nuts to 33 ft. lbs. (44
  Nm).
- Tighten the pinch bolts to 116 ft.
  lbs. (157 Nm)
- ABS sensor
- Tie rod end
- Brake hose retainer
- Front wheel

Rear

1. Before servicing the vehicle, refer to
the precautions in the beginning of this section.

2. Support the vehicle under the lower
control arm.

3. Remove or disconnect the following:

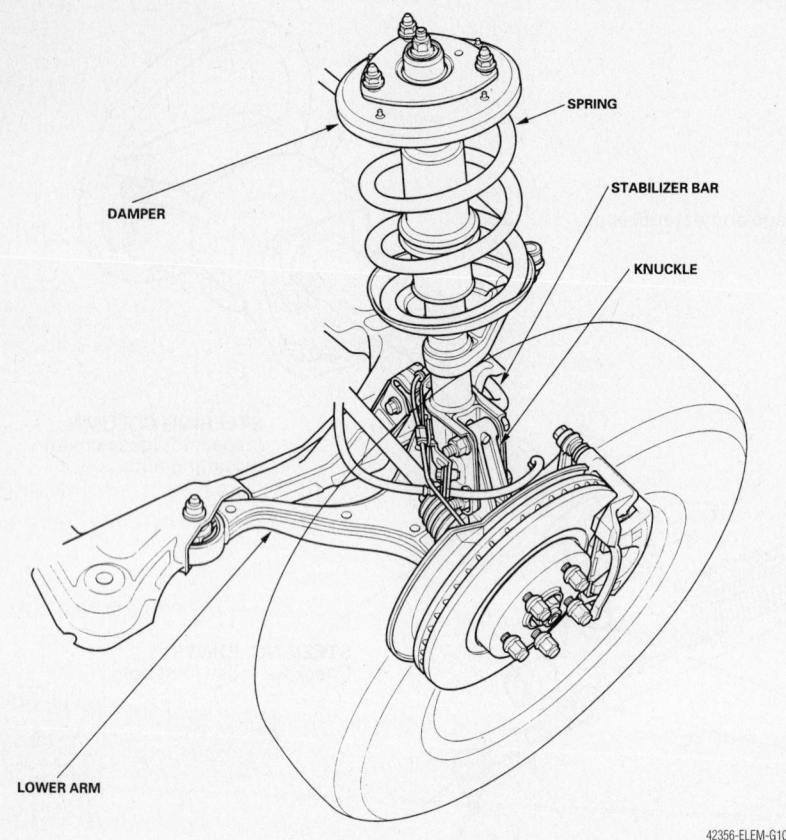

Front suspension components

42356-ELEM-G10

Labels: SPRING, STABILIZER BAR, KNUCKLE, DAMPER, LOWER ARM

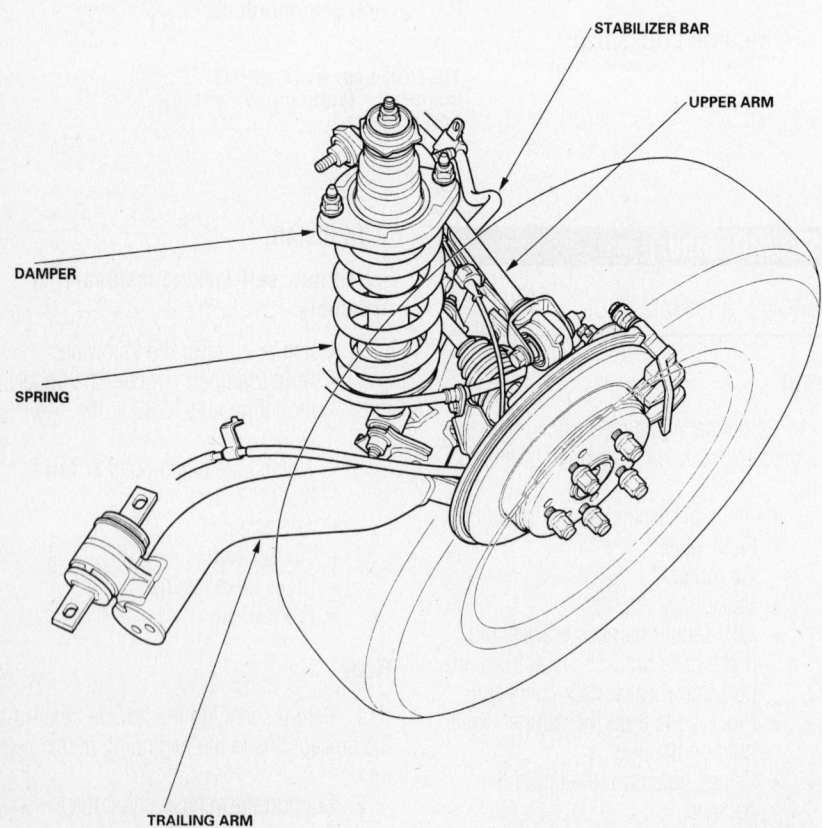

Rear suspension components

42356-ELEM-G12

Labels: STABILIZER BAR, UPPER ARM, DAMPER, SPRING, TRAILING ARM

- Rear wheel
- Flange bolt from the bottom of the damper (strut)
- Evaporative emission (EVAP) canister bolts, and loosen the EVAP canister mounting (left side only)
- Interior access panel, if necessary
- Flange nuts from the top of the damper in the cargo area
- Strut

To install:

4. Install or connect the following:
   - Strut. Position the damper mounting base so the indent mark is toward the inside of the vehicle,
   - Upper flange nuts, hand-tight only
   - Bottom flange bolt, hand-tight only

5. With the suspension raised with a jack to load it with the vehicles weight, tighten the bottom bolt to 69 ft. lbs. and the top nuts to 54 ft. lbs. (74 Nm).
   - Interior access panel, if necessary
   - EVAP canister mounting bolts
   - Rear wheel

Coil Spring

REMOVAL & INSTALLATION

Front

1. Before servicing the vehicle, refer to the precautions in the beginning of this section.

2. Remove the strut from the vehicle and install in a strut spring compressor. Compress the spring until the end of the spring comes away from the spring seat.

3. Remove the upper strut mount, spring seat and related components.

4. Remove the coil spring from the strut spring compressor.

To install:

➡**Use a new self-locking nut.**

5. Compress the spring and position the strut so that the end of the spring aligns with the notch in the spring seat.

6. Install the upper strut mounting components and tighten the nut to 33 ft. lbs. (44 Nm).

7. Install the strut to the vehicle.

8. Check the wheel alignment and adjust as necessary.

Rear

1. Before servicing the vehicle, refer to the precautions in the beginning of this section.

2. Remove the strut from the vehicle and install in a strut spring compressor. Com-

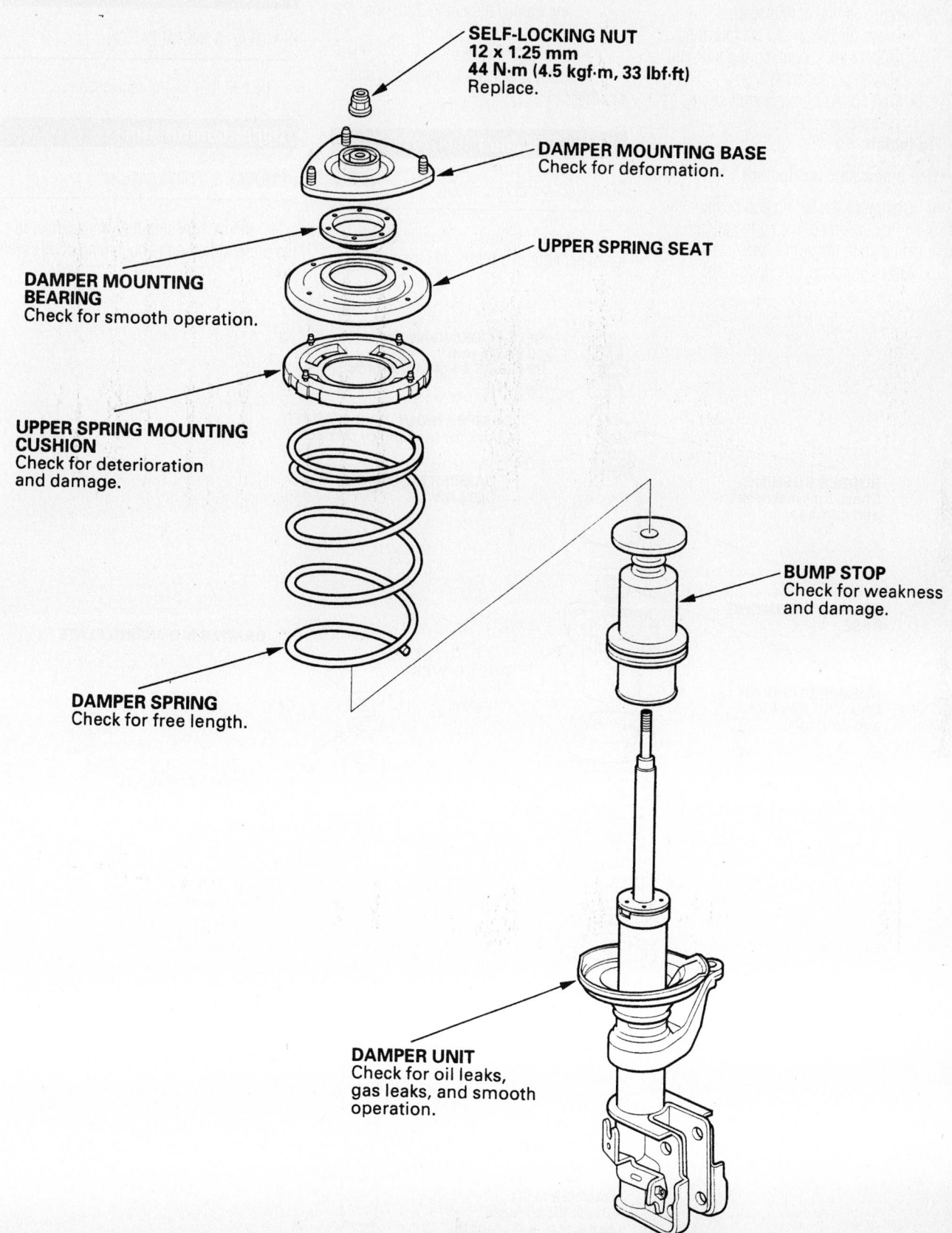

SELF-LOCKING NUT
12 x 1.25 mm
44 N·m (4.5 kgf·m, 33 lbf·ft)
Replace.

DAMPER MOUNTING BASE
Check for deformation.

UPPER SPRING SEAT

**DAMPER MOUNTING
BEARING**
Check for smooth operation.

**UPPER SPRING MOUNTING
CUSHION**
Check for deterioration
and damage.

BUMP STOP
Check for weakness
and damage.

DAMPER SPRING
Check for free length.

DAMPER UNIT
Check for oil leaks,
gas leaks, and smooth
operation.

42356-ELEM-G11

Exploded view of the front strut (damper and spring) assembly

press the spring until the end of the spring comes away from the spring seat.

3. Remove or disconnect the following:
   • Upper strut mount, spring seat and related components
   • Coil spring from the strut spring compressor

To install:

➡**Use a new self-locking nut.**

4. Compress the spring and position the strut so that the end of the spring aligns with the notch in the spring seat.

5. Install or connect the following:

   • Upper strut mounting components and tighten the nut to 22 ft. lbs. (29 Nm).
   • Strut to the vehicle

6. Check the wheel alignment and adjust as necessary.

Upper Ball Joint

REMOVAL & INSTALLATION

The upper ball joints are replaced with the upper control arms as an assembly.

Lower Ball Joint

REMOVAL & INSTALLATION

The ball joint is not replaceable.

Upper Control Arm

REMOVAL & INSTALLATION

1. Before servicing the vehicle, refer to the precautions in the beginning of this section.

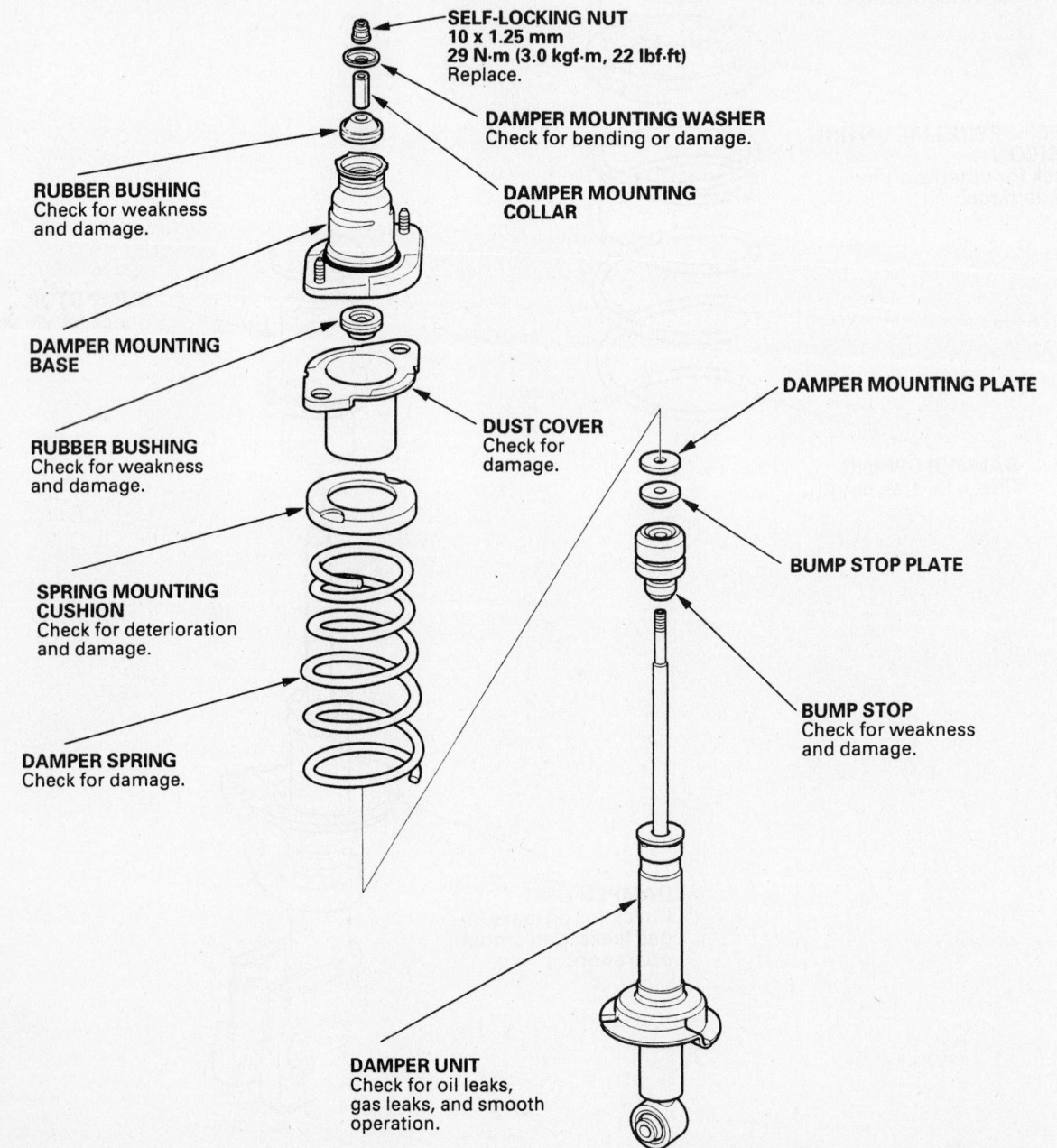

SELF-LOCKING NUT
10 x 1.25 mm
29 N·m (3.0 kgf·m, 22 lbf·ft)
Replace.

DAMPER MOUNTING WASHER
Check for bending or damage.

DAMPER MOUNTING COLLAR

RUBBER BUSHING
Check for weakness and damage.

DAMPER MOUNTING BASE

RUBBER BUSHING
Check for weakness and damage.

DUST COVER
Check for damage.

SPRING MOUNTING CUSHION
Check for deterioration and damage.

DAMPER SPRING
Check for damage.

DAMPER MOUNTING PLATE

BUMP STOP PLATE

BUMP STOP
Check for weakness and damage.

DAMPER UNIT
Check for oil leaks, gas leaks, and smooth operation.

42356-ELEM-G13

Exploded view of the strut (damper and spring) assembly

2. Support the lower control arm assembly with a floor jack.

3. Remove or disconnect the following:
- Upper ball joint
- Inner control arm flange bolts.
- Upper control arm

To install:

➡**Use new self-locking nuts for assembly.**

4. Install the upper control arm. Tighten the ball joint nut to 29–35 ft. lbs. (39–47 Nm) and the inner flange bolts to 40 ft. lbs. (54 Nm).

CONTROL ARM BUSHING REPLACEMENT

The upper control arm bushings are serviced with the upper control arm as an assembly.

Lower Control Arm

REMOVAL & INSTALLATION

1. Before servicing the vehicle, refer to the precautions in the beginning of this section.

2. Remove or disconnect the following:

- Front wheel
- Stabilizer link
- Lower arm from the knuckle
- Lower arm

3. Installation is the reverse of removal. Observe the following torques:
- Lower arm bolts: 61 ft. lbs. (83 Nm)
- Ball stud nut: 51 ft. lbs. (69 Nm)
- Stabilizer link: 29 ft. lbs. (39 Nm)

CONTROL ARM BUSHING REPLACEMENT

The lower control arm front inner bushing and the damper fork bushing are serviced with the control arm as an assembly.

REAR INNER BUSHING

1. Before servicing the vehicle, refer to the precautions in the beginning of this section.

2. Remove or disconnect the following:
- Front wheel
- Rear bushing bracket
- Rear bushing

To install:

➡**Use a new self-locking nut for assembly.**

3. Install or connect the following:
- Rear bushing. Tighten the nut to 61 ft. lbs. (83 Nm).
- Rear bushing bracket. Tighten the bolts to 66 ft. lbs. (89 Nm).
- Front wheel

4. Check the wheel alignment and adjust as necessary.

Wheel Bearings

ADJUSTMENT

The wheel bearings are sealed units and are not adjustable.

REMOVAL & INSTALLATION

Front

1. Before servicing the vehicle, refer to the precautions in the beginning of this section.

2. Remove or disconnect the following:
- Front wheel
- Spindle nut
- Brake caliper and rotor. Forcing screws are needed to remove the rotor.

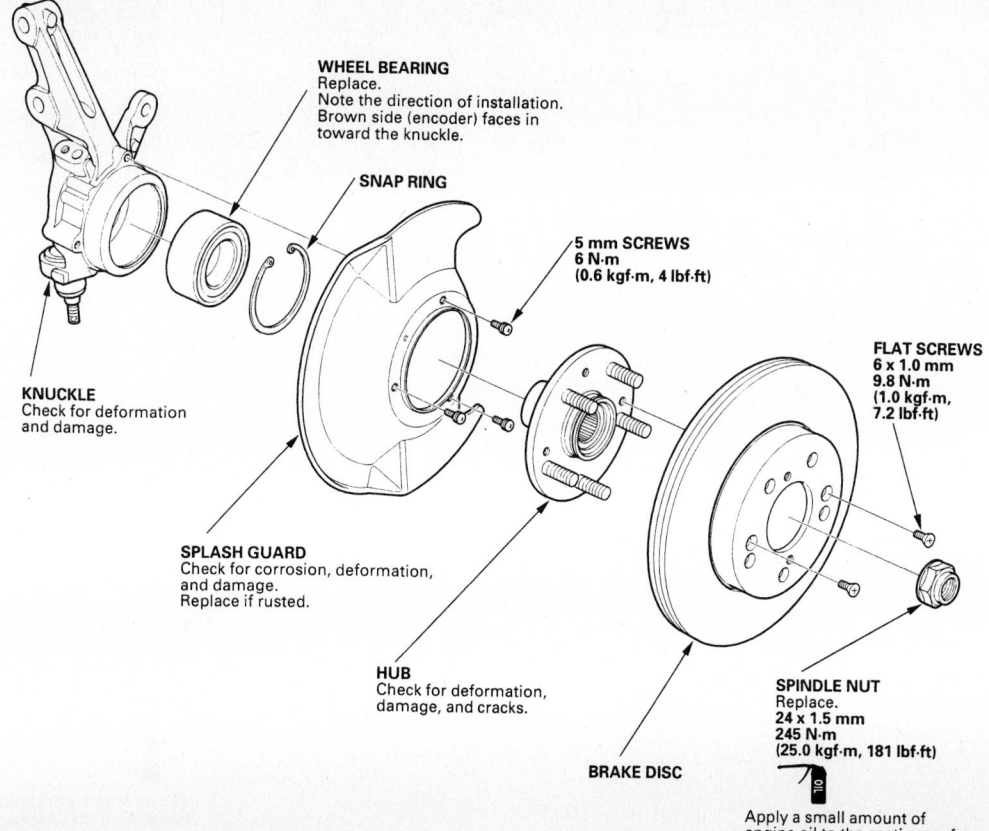

WHEEL BEARING
Replace.
Note the direction of installation.
Brown side (encoder) faces in toward the knuckle.

SNAP RING

5 mm SCREWS
6 N·m
(0.6 kgf·m, 4 lbf·ft)

FLAT SCREWS
6 x 1.0 mm
9.8 N·m
(1.0 kgf·m, 7.2 lbf·ft)

KNUCKLE
Check for deformation and damage.

SPLASH GUARD
Check for corrosion, deformation, and damage.
Replace if rusted.

HUB
Check for deformation, damage, and cracks.

BRAKE DISC

SPINDLE NUT
Replace.
24 x 1.5 mm
245 N·m
(25.0 kgf·m, 181 lbf·ft)

Apply a small amount of engine oil to the seating surface.

42356-ELEM-G14

Exploded view of the front hub, wheel bearing and related components

- Brake hose bracket
- ABS sensor
- Stabilizer link
- Lower arm from the knuckle
- Strut-to-knuckle bolts
- Steering hub/knuckle assembly

3. Press the hub from the knuckle. The bearings and races can now be pressed out and replaced.

➡**With ABS, install the bearing with the magnetic encoder (brown color) toward the inside of the knuckle.**

4. Observe the following torques:
- Strut bolts: 116 ft. lbs. (157 Nm)
- Ball stud nuts: 51 ft. lbs. (69 Nm)
- Stabilizer bar link: 29 ft. lbs. (39 Nm)

Rear

1. Before servicing the vehicle, refer to the precautions in the beginning of this section.
2. Remove or disconnect the following:
- Rear wheel
- Brake caliper
- Rotor
- Spindle nut
- Axle shaft (2wd)
- Parking brake shoes
- Parking brake cable
- Wheel sensor, if equipped
3. Support the trailing arm.
4. Remove or disconnect the following:
- Upper arm from the knuckle
5. Matchmark the trailing arm cam

adjusting bolt and cam. Remove the bolt. Discard the nut.
6. Remove the flange bolt.
7. Remove the knuckle assembly.
8. Press the hub from the knuckle. The bearings and races can now be pressed out and replaced.

➡**With ABS, install the bearing with the magnetic encoder (brown color) toward the inside of the knuckle.**

9. Observe the following torques:
- Flange bolt: 69 ft. lbs. (93 Nm)
- Cam bolts: 43 ft. lbs. (59 Nm)
- Spindle nut: 134 ft. lbs. (181 Nm)
- Caliper mounting bolts: 41 ft. lbs. (55 Nm)

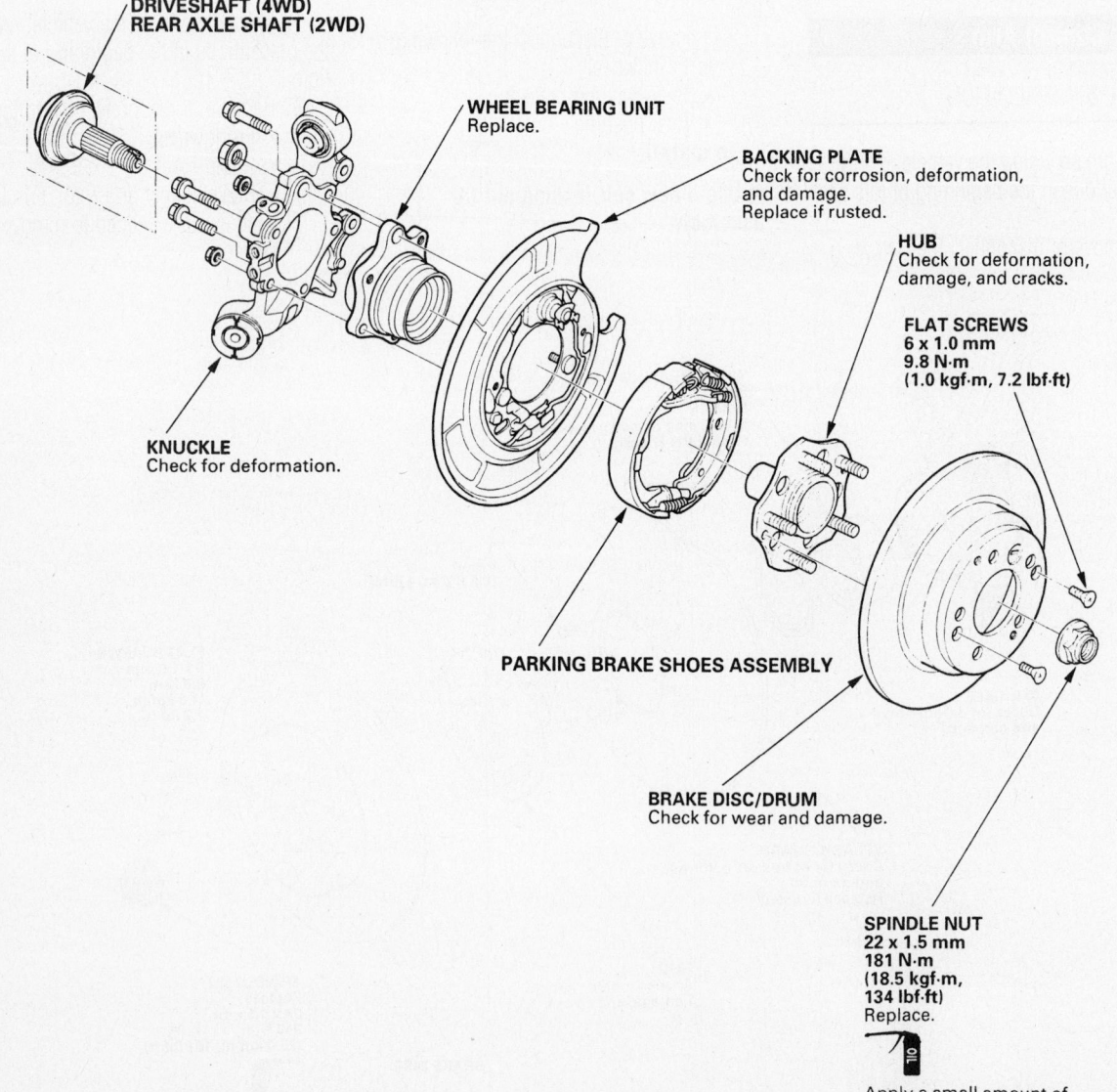

DRIVESHAFT (4WD)
REAR AXLE SHAFT (2WD)

WHEEL BEARING UNIT
Replace.

BACKING PLATE
Check for corrosion, deformation, and damage.
Replace if rusted.

HUB
Check for deformation, damage, and cracks.

FLAT SCREWS
6 x 1.0 mm
9.8 N·m
(1.0 kgf·m, 7.2 lbf·ft)

KNUCKLE
Check for deformation.

PARKING BRAKE SHOES ASSEMBLY

BRAKE DISC/DRUM
Check for wear and damage.

SPINDLE NUT
22 x 1.5 mm
181 N·m
(18.5 kgf·m, 134 lbf·ft)
Replace.

Apply a small amount of engine oil to the seating surface.

42356-ELEM-G15

Exploded view of the rear hub, wheel bearing and related components

BRAKES

Brake Caliper

REMOVAL & INSTALLATION

Front

1. Remove the upper and lower bolts.
2. Lift off the caliper.
3. Remove the pad springs.
4. Remove the pads and shims.
5. Remove the pad retainers.
6. Installation is the reverse of removal. Coat both sides of the shims and the backs of the pads with brake grease. Torque the bolts to 25 ft. lbs. (34 Nm). If the hose was disconnected, torque the banjo bolt to 25 ft. lbs. (34 Nm).

Rear

1. Remove the caliper pin bolts.
2. Lift off the caliper and suspend it safely.
3. Remove the pads and shims.
4. Remove the pad retainers.
5. Installation is the reverse of removal. Coat both sides of the shims and the backs of the pads with brake grease. Torque the bolts to 16 ft. lbs. (22 Nm). If the hose was disconnected, torque the banjo bolt to 16 ft. lbs. (22 Nm).

Disc Brake Pads

REMOVAL & INSTALLATION

Front

1. Remove the lower bolt.
2. Pivot the caliper up and hold the pads.
3. Remove the pad springs.
4. Remove the pads and shims.
5. Remove the pad retainers.
6. Installation is the reverse of removal. Coat both sides of the shims and the backs of the pads with brake grease. Torque the lower bolt to 25 ft. lbs. (34 Nm).

Rear

1. Remove the caliper pin bolts.
2. Lift off the caliper and suspend it safely.
3. Remove the pads and shims.
4. Remove the pad retainers.
5. Installation is the reverse of removal. Coat both sides of the shims and the backs of the pads with brake grease. Torque the bolts to 16 ft. lbs. (22 Nm).

Brake Drums

REMOVAL & INSTALLATION

1. Raise and safely support the vehicle. Release the parking brake.
2. Remove the rear wheels.
3. Use chalk to mark the brake drum to one of the wheel studs as an index mark for reinstallation.
4. Remove the retaining screw that holds the brake drum to the axle flange.
5. Pull the brake drum from the axle flange.

To install:

6. Align the index mark and install the brake drum to the axle flange.
7. Install the retaining screw to secure the brake drum to the axle flange.
8. Install the rear wheels.

Rear Brake Shoes

REMOVAL AND INSTALLATION

1. Raise and safely support the vehicle.
2. Remove the rear wheels.
3. Remove the brake drums.
4. Remove the brake return springs.

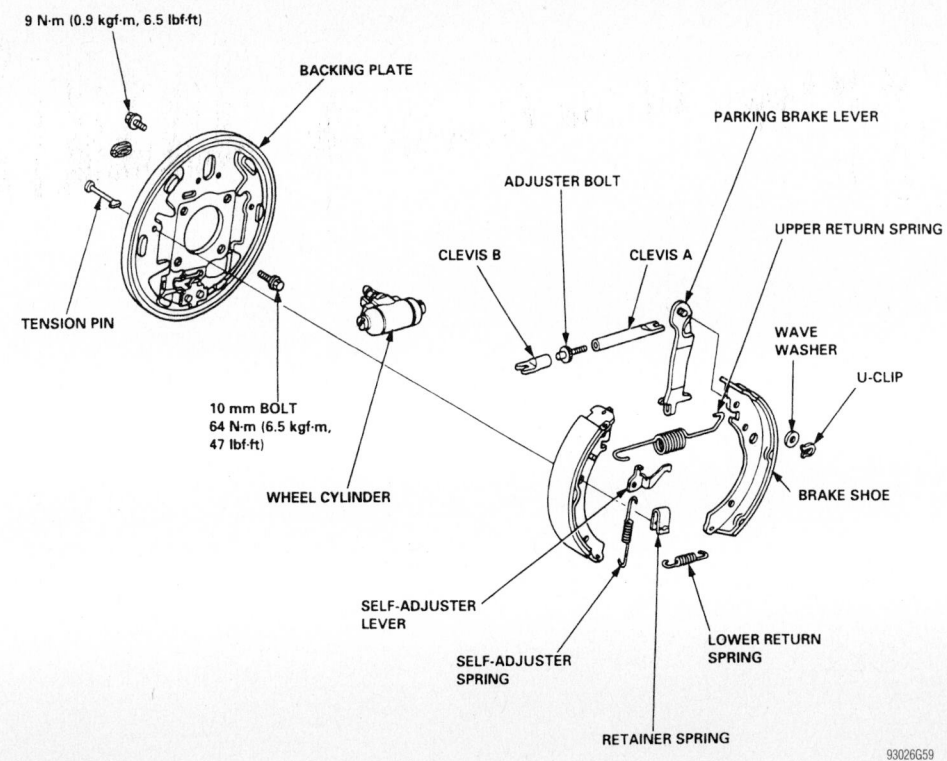

9 N·m (0.9 kgf·m, 6.5 lbf·ft)

BACKING PLATE

PARKING BRAKE LEVER

ADJUSTER BOLT

UPPER RETURN SPRING

CLEVIS B   CLEVIS A

TENSION PIN

WAVE WASHER

U-CLIP

10 mm BOLT
64 N·m (6.5 kgf·m, 47 lbf·ft)

WHEEL CYLINDER

BRAKE SHOE

SELF-ADJUSTER LEVER

SELF-ADJUSTER SPRING

LOWER RETURN SPRING

RETAINER SPRING

93026G59

Exploded view of the rear drum brakes

5. Remove the leading shoe holding pin and spring, and then the leading shoe.

6. Remove the self-adjuster and the adjuster lever.

7. Remove the trailing shoe holding pin and spring.

8. Disconnect the parking brake cable from the trailing shoe and remove the trailing shoe. Remove the parking brake lever from the trailing shoe.

To install:

9. Attach the parking brake lever to the trailing shoe.

10. Connect the parking brake cable to the parking brake lever.

11. Apply a thin coat of high temperature grease to the shoe contact points on the brake backing plate contact surface (B), and self-adjuster (D).

12. Position the trailing shoe on the backing plate and install the hold-down pin, spring, and retainer. Be careful not to stretch the return spring when fitting the shoes onto the backing plate.

13. Connect the upper return spring and the leading shoe to the trailing shoe and position the leading brake shoe on the backing plate.

14. Install the adjuster assembly and the hold-down pin, spring, and retainer.

15. Use a brake spring tool to install the lower return spring.

16. Install the self-adjuster lever and adjuster spring.

17. Adjust the shoe-to-drum clearance to 0.0098–0.0157 in. (0.25–0.40mm) and install the brake drum.

18. Check the brake drum for scoring or other wear. Machine or replace as necessary. Check the maximum brake drum diameter specification when machining.

19. Install the rear wheels. Lower the vehicle.

20. Road-test the vehicle.

SPECIFICATION AND MAINTENANCE CHARTS

ENGINE AND VEHICLE IDENTIFICATION CHART

		Engine Code					Model Year	
Code	Liters (cc)	Cu. In.	Cyl.	Fuel Sys.	Engine Type	Eng. Mfg.	Code ①	Year
ECA1	1.0 (999)	61	3	SMFI	SOHC	Honda	1	2001
							2	2002
							3	2003
							4	2004
							5	2005

SOHC: Single Overhead Cam

SMFI: Sequential Multi-port Fuel Injection

① 10th position of VIN

67162-INSI-C01

GENERAL ENGINE SPECIFICATIONS

Year	Model	Engine Displacement Liters	Engine ID/VIN	Net Horsepower @ rpm	Net Torque @ rpm (ft. lbs.)	Bore x Stroke (in.)	Compression Ratio	Oil Pressure @ rpm
2001	Insight	1.0	ECA1	71@5700	89@2000	2.83x3.21	①	50@3000
2002	Insight	1.0	ECA1	71@5700	89@2000	2.83x3.21	①	50@3000
2003	Insight	1.0	ECA1	71@5700	89@2000	2.83x3.21	①	50@3000
2004	Insight	1.0	ECA1	71@5700	89@2000	2.83x3.21	①	50@3000

① Man. Trans. 10.8:1

Auto. Trans. 10.3:1

67162-INSI-C02

ENGINE TUNE-UP SPECIFICATIONS

Year	Engine Displacement Liters	Engine ID/VIN	Spark Plug Gap (in.)	Ignition Timing (deg.) MT	Ignition Timing (deg.) AT	Fuel Pump (psi)	Idle Speed (rpm) MT	Idle Speed (rpm) AT	Valve Clearance (in.) In.	Valve Clearance (in.) Ex.
2001	1.0	ECA1	0.039-0.043	10-14B	10-14B	40-47	850-950	850-950	0.007-0.009	0.008-0.01
2002	1.0	ECA1	0.039-0.043	10-14B	10-14B	40-47	850-950	850-950	0.007-0.009	0.008-0.01
2003	1.0	ECA1	0.039-0.043	10-14B	10-14B	40-47	850-950	850-950	0.007-0.009	0.008-0.01
2004	1.0	ECA1	0.039-0.043	10-14B	10-14B	40-47	850-950	850-950	0.007-0.009	0.008-0.01

NOTE: The Vehicle Emission Control Information label often reflects changes made during production and must be used if they differ from this chart.

NOTE: The fuel pressure readings are given with the vacuum hose connected to the regulator and the engine running

B: Before top dead center

67162-INSI-C03

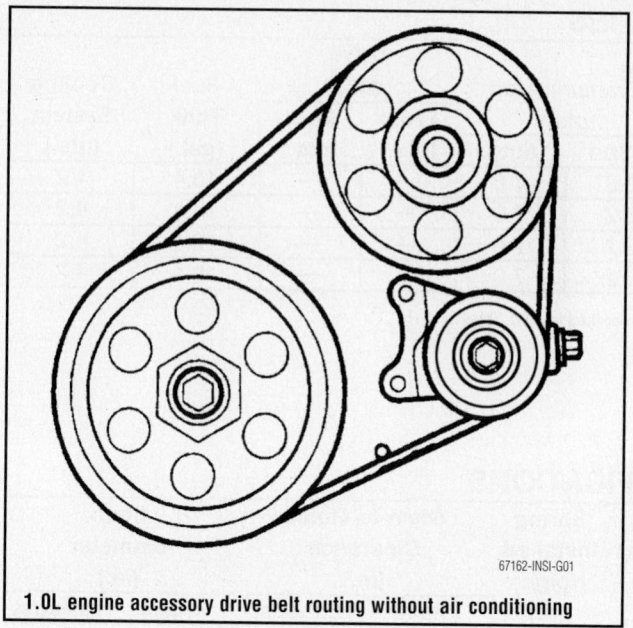

67162-INSI-G01

1.0L engine accessory drive belt routing without air conditioning

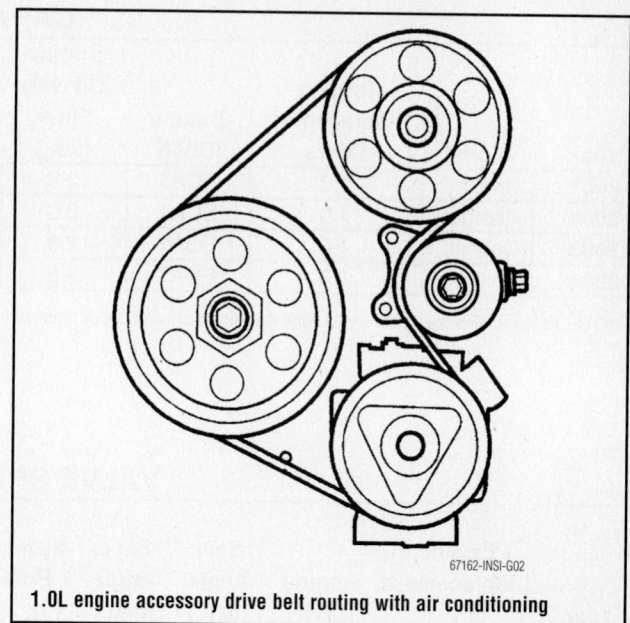

67162-INSI-G02

1.0L engine accessory drive belt routing with air conditioning

CAPACITIES

Year	Model	Engine Displacement Liters	Engine ID/VIN	Engine Oil with Filter (qts.)	Transmission (pts.) 5-Spd	Auto.	Drive Axle Front (pts.)	Rear (pts.)	Fuel Tank (gal.)	Cooling System (qts.)
2001	Insight	1.0	ECA1	2.6	3.2	7.0	—	—	10.4	4.2
2002	Insight	1.0	ECA1	2.6	3.2	7.0	—	—	10.4	4.2
2003	Insight	1.0	ECA1	2.6	3.2	7.0	—	—	10.4	4.2
2004	Insight	1.0	ECA1	2.6	3.2	7.0	—	—	10.4	4.2

NOTE: All capacities are approximate. Add fluid gradually and check to be sure a proper fluid level is obtained.

67162-INSI-C04

VALVE SPECIFICATIONS

Year	Engine Displacement Liters	Engine ID/VIN	Seat Angle (deg.)	Face Angle (deg.)	Spring Test Pressure (lbs. @ in.)	Spring Installed Height (in.)	Stem-to-Guide Clearance (in.) Intake	Exhaust	Stem Diameter (in.) Intake	Exhaust
2001	1.0	ECA1	45	45	NA	①	0.0010-0.0020	0.0020-0.0031	0.2157-0.2161	0.2146-0.2150
2002	1.0	ECA1	45	45	NA	①	0.0010-0.0020	0.0020-0.0031	0.2157-0.2161	0.2146-0.2150
2003	1.0	ECA1	NA	NA	NA	①	0.0010-0.0020	0.0020-0.0031	0.2157-0.2161	0.2146-0.2150
2004	1.0	ECA1	NA	NA	NA	①	0.0010-0.0020	0.0020-0.0031	0.2157-0.2161	0.2146-0.2150

NA: Not Available

① Valve spring free length:
  Intake: 2.689 in.
  Exhaust: 2.800 in.

67162-INSI-C05

CRANKSHAFT AND CONNECTING ROD SPECIFICATIONS

All measurements are given in inches

Year	Engine Displacement Liters	Engine ID/VIN	Crankshaft Main Brg. Journal Dia.	Main Brg. Oil Clearance	Shaft End-play	Thrust on No.	Connecting Rod Journal Diameter	Oil Clearance	Side Clearance
2001	1.0	ECA1	①	②	0.0040-0.0140	NA	1.4164-1.4173	0.0008-0.0015	NA
2002	1.0	ECA1	①	②	0.0040-0.0140	NA	1.4164-1.4173	0.0008-0.0015	NA
2003	1.0	ECA1	①	②	0.0040-0.0140	NA	1.4164-1.4173	0.0008-0.0015	NA
2004	1.0	ECA1	①	②	0.0040-0.0140	NA	1.4164-1.4173	0.0008-0.0015	NA

NA: Not available

① Nos. 1 and 4: 1.5741-1.5750
  Nos.2 and 3:1.5739-1.5748

② Nos. 1 and 4: 0.0006-0.0013
  Nos. 2 and 3: 0.0008-0.0015

67162-INSI-C06

PISTON AND RING SPECIFICATIONS

All measurements are given in inches

Year	Engine Displacement Liters	Engine ID/VIN	Piston Clearance	Ring Gap			Ring Side Clearance		
				Top Compression	Bottom Compression	Oil Control	Top Compression	Bottom Compression	Oil Control
2001	1.0	ECA1	0.0002-0.0017	0.0060-0.0120	0.0140-0.0200	0.0080-0.0200	0.0022-0.0031	0.0012-0.0022	NA
2002	1.0	ECA1	0.0002-0.0017	0.0060-0.0120	0.0140-0.0200	0.0080-0.0200	0.0022-0.0031	0.0012-0.0022	NA
2003	1.0	ECA1	0.0002-0.0017	0.0060-0.0120	0.0140-0.0200	0.0080-0.0280	0.0022-0.0031	0.0012-0.0022	NA
2004	1.0	ECA1	0.0002-0.0017	0.0060-0.0120	0.0140-0.0200	0.0080-0.0280	0.0022-0.0031	0.0012-0.0022	NA

NA: Not Available

67162-INSI-C07

TORQUE SPECIFICATIONS

All readings in ft. lbs.

Year	Engine Displacement Liters	Engine ID/VIN	Cylinder Head Bolts	Main Bearing Bolts	Rod Bearing Bolts	Crankshaft Damper Bolts	Flywheel Bolts	Manifold		Spark Plugs	Oil Pan Drain Plug
								Intake	Exhaust		
2001	1.0	ECA1	①	②	③	④	33	16	17	17	29
2002	1.0	ECA1	①	②	③	④	33	16	17	17	29
2003	1.0	ECA1	①	②	③	④	33	16	17	17	29
2004	1.0	ECA1	①	②	③	④	33	16	17	17	29

NOTE: Dip connecting rod and main bearing bolts and crankshaft damper bolt in clean engine oil prior to tightening.

① Step 1: 29 ft. lbs.
Step 2: +90 degrees
Step 3: 6 mm bolts to 106 inch lbs.

② Step 1: 18 ft. lbs.
Step 2: +60 degrees

③ Step 1: 87 inch lbs.
Step 2: +90 degrees

④ Step 1: 14 ft. lbs.
Step 2: +90 degrees

67162-INSI-C08

WHEEL ALIGNMENT

Year	Model		Caster		Camber		Toe-in (in.)
			Range (+/-Deg.)	Preferred Setting (Deg.)	Range (+/-Deg.)	Preferred Setting (Deg.)	
2001	Insight	F	1.00	+2.00	1.00	0	0+/- 0.08
		R	—	—	1.00	-1.00	0.12+/- 0.12
2002	Insight	F	1.00	+2.00	1.00	0	0+/- 0.08
		R	—	—	1.00	-1.00	0.12+/- 0.12
2003	Insight	F	1.00	+2.00	1.00	0	0+/- 0.08
		R	—	—	1.00	-1.00	0.12+/- 0.12
2004	Insight	F	1.00	+2.00	1.00	0	0+/- 0.08
		R	—	—	1.00	-1.00	0.12+/- 0.12

67162-INSI-C09

TIRE, WHEEL AND BALL JOINT SPECIFICATIONS

Year	Model	OEM Tires Standard	OEM Tires Optional	Tire Pressures (psi) Front	Tire Pressures (psi) Rear	Wheel Size	Ball Joint Inspection	Lug Nuts
2001	Insight	P165/65R14 78S	None	38	35	NS	NS	80
2002	Insight	P165/65R14 78S	None	38	35	NS	NS	80
2003	Insight	P165/65R14 78S	None	38	35	NS	NS	80
2004	Insight	P165/65R14 78S	None	38	35	NS	NS	80

OEM: Original Equipment Manufacturer

PSI: Pounds Per Square Inch

NS: Not specified by manufacturer

67162-INSI-C10

BRAKE SPECIFICATIONS
All measurements in inches unless noted

Year	Model		Brake Disc Original Thickness	Brake Disc Minimum Thickness	Brake Disc Maximum Runout	Brake Drum Diameter Original Inside Diameter	Brake Drum Diameter Max. Wear Limit	Brake Drum Diameter Maximum Machine Diameter	Minimum Lining Thickness Front	Minimum Lining Thickness Rear	Brake Caliper Bracket Bolts (ft. lbs.)	Brake Caliper Mounting Bolts (ft. lbs.)
2001	Insight	F	0.670	0.590	0.002	—	—	—	0.080	—	80	36
		R	—	—	—	7.08	7.12	7.12	—	0.040	—	—
2002	Insight	F	0.670	0.590	0.002	—	—	—	0.080	—	80	36
		R	—	—	—	7.08	7.12	7.12	—	0.040	—	—
2003	Insight	F	0.670	0.590	0.002	—	—	—	0.080	—	80	36
		R	—	—	—	7.08	7.12	7.12	—	0.040	—	—
2004	Insight	F	0.670	0.590	0.002	—	—	—	0.080	—	80	36
		R	—	—	—	7.08	7.12	7.12	—	0.040	—	—

F: Front

R: Rear

67162-INSI-C11

SCHEDULED MAINTENANCE INTERVALS

2001-03 HONDA—Insight

TO BE SERVICED	TYPE OF SERVICE	VEHICLE MILEAGE INTERVAL (x1000)															
		7.5	15	22.5	30	37.5	45	52.5	60	67.5	75	82.5	90	97.5	105	112.5	120
Accessory drive belts	I & A				✓				✓				✓				✓
Air cleaner element	R				✓				✓				✓				✓
Air conditioning filter	R				✓				✓				✓				✓
Brake fluid	R	Replace every 3 years															
Brake hoses & lines	I		✓		✓		✓		✓		✓		✓		✓		✓
Cooling system	I		✓		✓		✓		✓		✓		✓		✓		✓
Engine coolant	R														✓		
Engine oil	R	✓	✓	✓	✓	✓	✓	✓	✓	✓	✓	✓	✓	✓	✓	✓	✓
Engine oil and coolant levels	I	Inspect at each fuel stop															
Engine oil filter	R		✓		✓		✓		✓		✓		✓		✓		✓
Exhaust system	I		✓		✓		✓		✓		✓		✓		✓		✓
Fluid levels and condition	I		✓		✓		✓		✓		✓		✓		✓		✓
Front and rear brakes	I		✓		✓		✓		✓		✓		✓		✓		✓
Fuel lines & connection	I		✓		✓		✓		✓		✓		✓		✓		✓
Halfshaft boots	I		✓		✓		✓		✓		✓		✓		✓		✓
Idle speed	I & A														✓		
Parking brake system	I & A		✓		✓		✓		✓		✓		✓		✓		✓
Rotate and inspect tires	I	✓	✓	✓	✓	✓	✓	✓	✓	✓	✓	✓	✓	✓	✓	✓	✓
Spark plugs	R		✓						✓				✓				✓
Supplemental Restrain system	I	Inspect the SRS 10 years after production															
Suspension components	I		✓		✓		✓		✓		✓		✓		✓		✓
Tie rod ends, steering gear box & boots	I		✓		✓		✓		✓		✓		✓		✓		✓
Automatic transmission fluid	R				✓				✓				✓				✓
Manual transmission fluid	R																✓
Valve clearance	I				✓				✓				✓				✓

R: Replace   I: Inspect   A: Adjust

FREQUENT OPERATION MAINTENANCE (SEVERE SERVICE)

If a vehicle is operated under any of the following conditions it is considered severe service:

- Towing a trailer or using a camper or car-top carrier.
- Repeated short trips of less than 5 miles in temperatures below freezing, or trips of less than 10 miles in any temperature.
- Extensive idling or low-speed driving for long distances as in heavy commercial use, such as delivery, taxi or police cars.
- Operating on rough, muddy or salt-covered roads.
- Operating on unpaved or dusty roads.
- Driving in extremely hot (over 90°) conditions.

Air cleaner element: replace every 15,000 miles

Engine oil and filter: replace every 3750 miles or 6 months, whichever occurs first.

Automatic transmission fluid: replace every 15,000 miles.

Front and rear brakes: inspect every 7500 miles or 6 months, whichever occurs first.

Locks and hinges: lubricate every 15,000 miles.

Tie rods, steering gear box, boots: inspect every 7500 miles or 6 months, whichever occurs first.

Suspension components: inspect every 7500 miles or 6 months, whichever occurs first.

Halfshaft boots: inspect every 7500 miles or 6 months, whichever occurs first.

SCHEDULED MAINTENANCE INTERVALS

2004 HONDA—Insight

TO BE SERVICED	TYPE OF SERVICE	VEHICLE MILEAGE INTERVAL (x1000)															
		7.5	15	22.5	30	37.5	45	52.5	60	67.5	75	82.5	90	97.5	105	112.5	120
Accessory drive belts	I & A				✓				✓				✓				✓
Air cleaner element	R				✓				✓				✓				✓
Air conditioning filter	R				✓				✓				✓				✓
Brake fluid	R	Replace every 3 years															
Brake hoses & lines	I		✓		✓		✓		✓		✓		✓		✓		✓
Cooling system	I		✓		✓		✓		✓		✓		✓		✓		✓
Engine coolant	R								✓								
Engine oil	R	✓	✓	✓	✓	✓	✓	✓	✓	✓	✓	✓	✓	✓	✓	✓	✓
Engine oil and coolant levels	I	Inspect at each fuel stop															
Engine oil filter	R		✓		✓		✓		✓		✓		✓		✓		✓
Exhaust system	I		✓		✓		✓		✓		✓		✓		✓		✓
Fluid levels and condition	I		✓		✓		✓		✓		✓		✓		✓		✓
Front and rear brakes	I		✓		✓		✓		✓		✓		✓		✓		✓
Fuel lines & connection	I		✓		✓		✓		✓		✓		✓		✓		✓
Halfshaft boots	I		✓		✓		✓		✓		✓		✓		✓		✓
Idle speed	I & A														✓		
Parking brake system	I & A		✓		✓		✓		✓		✓		✓		✓		✓
Rotate and inspect tires	I	✓	✓	✓	✓	✓	✓	✓	✓	✓	✓	✓	✓	✓	✓	✓	✓
Spark plugs	R																✓
Supplemental Restrain system	I	Inspect the SRS 10 years after production															
Suspension components	I		✓		✓		✓		✓		✓		✓		✓		✓
Tie rod ends, steering gear box & boots	I		✓		✓		✓		✓		✓		✓		✓		✓
Automatic transmission fluid	R				✓				✓				✓				✓
Manual transmission fluid	R																✓
Valve clearance	I																✓

R: Replace   I: Inspect   A: Adjust

FREQUENT OPERATION MAINTENANCE (SEVERE SERVICE)

If a vehicle is operated under any of the following conditions it is considered severe service:

- Towing a trailer or using a camper or car-top carrier.
- Repeated short trips of less than 5 miles in temperatures below freezing, or trips of less than 10 miles in any temperature.
- Extensive idling or low-speed driving for long distances as in heavy commercial use, such as delivery, taxi or police cars.
- Operating on rough, muddy or salt-covered roads.
- Operating on unpaved or dusty roads.
- Driving in extremely hot (over 90°) conditions.

Air cleaner element: replace every 15,000 miles

Engine oil and filter: replace every 3750 miles or 6 months, whichever occurs first.

Automatic transmission fluid: replace every 15,000 miles.

Front and rear brakes: inspect every 7500 miles or 6 months, whichever occurs first.

Locks and hinges: lubricate every 15,000 miles.

Tie rods, steering gear box, boots: inspect every 7500 miles or 6 months, whichever occurs first.

Suspension components: inspect every 7500 miles or 6 months, whichever occurs first.

Halfshaft boots: inspect every 7500 miles or 6 months, whichever occurs first.

67162-INSI-C13

PRECAUTIONS

Before servicing any vehicle, please be sure to read all of the following precautions, which deal with personal safety, prevention of component damage, and important points to take into consideration when servicing a motor vehicle:

• Never open, service or drain the radiator or cooling system when the engine is hot; serious burns can occur from the steam and hot coolant.

• Observe all applicable safety precautions when working around fuel. Whenever servicing the fuel system, always work in a well-ventilated area. Do not allow fuel spray or vapors to come in contact with a spark, open flame, or excessive heat (a hot drop light, for example). Keep a dry chemical fire extinguisher near the work area. Always keep fuel in a container specifically designed for fuel storage; also, always properly seal fuel containers to avoid the possibility of fire or explosion. Refer to the additional fuel system precautions later in this section.

• Fuel injection systems often remain pressurized, even after the engine has been turned **OFF**. The fuel system pressure must be relieved before disconnecting any fuel lines. Failure to do so may result in fire and/or personal injury.

• Brake fluid often contains polyglycol ethers and polyglycols. Avoid contact with the eyes and wash your hands thoroughly after handling brake fluid. If you do get brake fluid in your eyes, flush your eyes with clean, running water for 15 minutes. If eye irritation persists, or if you have taken brake fluid internally, seek medical assistance IMMEDIATELY.

• The EPA warns that prolonged contact with used engine oil may cause a number of skin disorders, including cancer. You should make every effort to minimize your exposure to used engine oil. Protective gloves should be worn when changing oil. Wash your hands and any other exposed skin areas as soon as possible after exposure to used engine oil. Soap and water, or waterless hand cleaner should be used.

• All new vehicles are now equipped with an air bag system. The system must be disabled before performing service on or around system components, steering column, instrument panel components, wiring and sensors. Failure to follow safety and disabling procedures could result in accidental air bag deployment, possible personal injury and unnecessary system repairs.

• Always wear safety goggles when working with, or around, the air bag system. When carrying a non-deployed air bag, be sure the bag and trim cover are pointed away from your body. When placing a non-deployed air bag on a work surface, always face the bag and trim cover upward, away from the surface. This will reduce the motion of the module if it is accidentally deployed. Refer to the additional air bag system precautions later in this section.

• Clean, high quality brake fluid from a sealed container is essential to the safe and proper operation of the brake system. You should always buy the correct type of brake fluid for your vehicle. If the brake fluid becomes contaminated, completely flush the system with new fluid. Never reuse any brake fluid. Any brake fluid that is removed from the system should be discarded. Also, do not allow any brake fluid to come in contact with a painted surface; it will damage the paint.

• Never operate the engine without the proper amount and type of engine oil; doing so WILL result in severe engine damage.

• Timing belt maintenance is extremely important. Many models utilize an interference-type, non-freewheeling engine. If the timing belt breaks, the valves in the cylinder head may strike the pistons, causing potentially serious (also time-consuming and expensive) engine damage. Refer to the maintenance interval charts for the recommended replacement interval for the timing belt, and to the timing belt section for belt replacement and inspection.

• Disconnecting the negative battery cable on some vehicles may interfere with the functions of the on-board computer system(s) and may require the computer to undergo a relearning process once the negative battery cable is reconnected.

• When servicing drum brakes, only disassemble and assemble one side at a time, leaving the remaining side intact for reference.

• Only an MVAC-trained, EPA-certified automotive technician should service the air conditioning system or its components.

• The Insight uses an Integrated Motor Assist (IMA) system which has an Auto-stop system that shuts down the engine under certain conditions to improve fuel economy when the vehicle is at a stop. BEFORE servicing the vehicle, turn the ignition switch **OFF** and remove the key so the engine cannot be started. Before performing any service, turn the battery switch module switch **OFF** and wait 5 minutes before disconnect the negative battery cable or working on the system. See the IMA system in this article for procedures on how to disable the IMA system.

• Many sections of the Insight are built with aluminum alloys. When replacing components on aluminum parts, always use Dacro type nuts and bolts recommended by Honda. These bolts can be identified with a gray coating. Some of these bolts have a green coating on the thread section of the bolt for easier installation. These are called "torquer" bolts. Always tighten these bolts with a torque wrench to the specified torque.

ENGINE REPAIR

➡Disconnecting the negative battery cable on some vehicles may interfere with the functions of the on board computer system. The computer may undergo a relearning process once the negative battery cable is reconnected.

Distributor

The 1.0L engine does not use a distributor.

Alternator

REMOVAL

The 1.0L engine has a DC electric converter and therefore does not use a standard alternator.

Ignition Timing

ADJUSTMENT

Adjustment is not possible on the 1.0L engine. Use the following procedure to inspect the timing.

INSPECTION

1. Connect the Honda Diagnostic System (HDS) to the Data Link Connector (DLC) and check for Diagnostic Trouble Codes (DTC's). If any DTC's are present, diagnose and repair the cause.
2. Start the engine and hold it at 3000 rpm until the cooling fan comes on, then return to idle.
3. Place the HDS tester in the "SCS" mode.
4. Connect a timing light to the no. 1 ignition coil wire.
5. With all accessories off, check the ignition timing on the oil pump pointer at the crankshaft pulley. Timing should be 10–14 degrees BTDC.
6. If the timing is off, check the cam timing. If the cam timing is okay, check the Electronic Control Module (ECM) for the latest software and update, or substitute a known good ECM and check the timing again.
7. If the timing is okay with a substitute ECM, replace the original ECM.

Engine Assembly

REMOVAL & INSTALLATION

➡The engine, electric motor and transaxle are removed from the vehicle as a unit.

1. Before servicing the vehicle, refer to the precautions in the beginning of this section. Also see the Integrated Motor Assembly (IMA) system disabling before proceeding with any procedure.
2. Turn the battery module switch off and measure the voltage.
3. Disable the Integrated Motor Assembly (IMA) system.
4. Drain the cooling system, engine oil and transmission fluids.
5. Relieve the fuel system pressure.
6. Disconnect the battery cables and remove the battery.
7. Remove the battery box.
8. Remove the engine appearance cover.
9. Remove the breather pipe and brake booster vacuum line bracket, then remove the air cleaner and intake air duct.
10. Disconnect the throttle cable linkage.
11. Remove or disconnect the following:
   • Brake booster vacuum hose
   • EVAP canister hose
   • Starter ground and positive cables
12. Mark the location and disconnect the "U", "V" and "W" phase power cables from the electric motor.
13. Disconnect the Electronic Control Module (ECM) and the main wiring harness connectors.

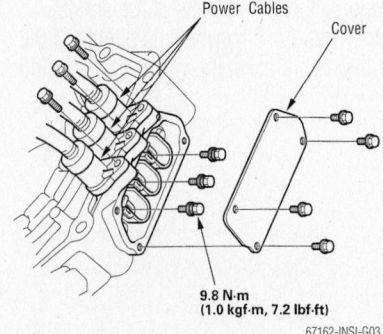

Electric motor power cable locations—Insight

14. Remove or disconnect the following:
   • Fuel feed and return lines
   • Accessory drive belts
   • Clutch slave cylinder
   • Transmission shift cables
   • Electric motor power cable clamps
   • Raise and support the vehicle
   • Engine splash shields and brackets
   • Lower control arm ball joints
   • Axle halfshafts
   • A/C compressor without disconnect refrigerant lines
   • Oxygen sensor and catalytic converter
   • Motor power cable and cable holder
   • Radiator and heater hoses
   • Shift cables
   • Transmission oil cooler lines on automatic transmission models
15. Attach a suitable engine chain hoist device to the engine.
16. Remove the rear and side engine mounting nuts and bolts.
17. Remove the transaxle mount nuts and bolts.
18. Ensure that all vacuum hoses, wiring harnesses and fuel lines are free from interfering with the engine removal.
19. Place a suitable rolling jack under the engine assembly, and lower the chain hoist until the engine assembly is on the jack.
20. Remove the engine/electric motor/transmission from the vehicle.
To install:

➡Use new self-locking nuts and color-coded self-locking bolts when installing the engine mounts and suspension components.

➡Do not tighten the engine or transaxle mount fasteners until instructed to do so.

21. Raise the powertrain into position.
22. Install or connect the following:
   • Transaxle mount and bracket. Tighten the bolts and nut to 41 ft. lbs. (64 Nm) on manual transmission, or 76 ft. lbs. (103 Nm) on automatic transmission. Loosen the mount through bolt.
   • Side engine mount. Tighten the bolt and nuts to 38 ft. lbs. (52 Nm). Loosen the mount through bolt

- Upper bracket. Tighten the nuts in sequence to 54 ft. lbs. (74 Nm).
- Raise the vehicle
- Rear mount. Tighten the bolts to 54 ft. lbs. (74 Nm). Loosen the through bolt.
- Lower the vehicle and remove the chain hoist.
- Tighten the transaxle mount through bolt to 40 ft. lbs. (54 Nm).
- Tighten the side mount through bolt to 40 ft. lbs. (54 Nm).
- Tighten the rear mount through bolt to 66 ft. lbs. (89 Nm).
- Motor power cable and cable holder
- Left engine undercover
- Oxygen sensor and catalytic converter
- A/C compressor
- Axle halfshafts
- Lower control arm ball joints
- Engine splash shields and brackets
- Lower the vehicle
- Electric motor power cable clamps
- Transmission oil cooler lines
- Shift cables
- Accessory drive belts
- Radiator and heater hoses
- ECM and main wiring harness connectors
- Motor power cables
- Fuel feed and return lines
- Starter ground and positive cables
- EVAP canister hose
- Brake booster vacuum hose
- Throttle cable linkage
- Clutch slave cylinder
- Air cleaner and intake duct
- Breather pipe and brake booster vacuum line bracket
- Battery box
- Battery cables
- Fill the cooling system, engine oil and transmission fluids
- Engine appearance cover

23. Start the engine and hold at 3500 rpm with no load for 10 minutes.

24. Check that the IMA battery level indicator reads full

25. Check the engine idle.

Water Pump

REMOVAL & INSTALLATION

1. Before servicing the vehicle, refer to the precautions in the beginning of this section. Also see the Integrated Motor Assembly (IMA) system disabling before proceeding with any procedure.

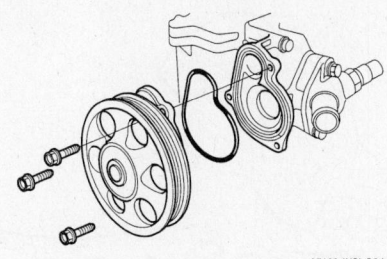

67162-INSI-G04

Exploded view of the water pump mounting—Insight

2. Drain the cooling system.
3. Remove or disconnect the following:
   - Negative battery cable
   - Accessory drive belt
   - Water pump

To install:

4. Install or connect the following:
   - Water pump. Use a new seal and tighten the bolts to 105 inch lbs. (12 Nm).
   - Accessory drive belt
   - Negative battery cable
5. Fill the cooling system.
6. Start the engine and check for leaks.

Heater Core

REMOVAL & INSTALLATION

1. Before servicing the vehicle, refer to the precautions in the beginning of this section. Also see the Integrated Motor Assembly (IMA) system disabling before proceeding with any procedure.

2. Disconnect the negative battery cable and wait for 3 minutes.

3. Place the front wheels in the straight ahead position.

4. Drain the cooling system into a clean container for reuse.

5. In the engine compartment, open the heater valve cable clamp and disconnect the cable from the heater valve arm. Then, turn the heater valve to the fully opened position.

6. Disconnect the heater hoses from the heater core.

7. Remove the heater housing-to-chassis nut.

➡ **When removing the heater housing nut, be careful not to damage or bend the fuel lines, the brake lines, etc.**

8. Remove the front center floor console.
9. Remove the glove box
10. Remove the center lower cover.
11. Remove the radio.
12. Remove the steering wheel access cover and disconnect the driver air bag connector.

13. Remove the steering wheel cover and remove the driver air bag.

14. Loosen the steering wheel nut, then use a puller and remove the steering wheel.

15. Remove the steering column covers.

16. Disconnect the cable wheel connectors and remove the cable reel.

17. Disconnect the combination switch and ignition switch connectors.

18. Remove the steering joint pinch bolts.

19. Remove the steering column attaching nuts and remove the steering column.

20. Disconnect the passenger air bag connector.

21. Remove the 3 air bag mounting nuts, then pry carefully with a screwdriver and lift the air bag out of the dashboard.

22. Remove the A pillar trim on both sides.

23. From under and behind the dashboard, disconnect and label all electrical connectors.

24. Remove the driver coin tray.

25. Remove the dashboard attaching screws and bolts as shown.

26. Carefully lift up on the dashboard and disengage it from the holders.

27. Remove the dashboard through the door opening.

28. Discharge and recover the air conditioning system refrigerant.

29. In the engine compartment, remove the refrigerant lines-to-evaporator housing bolts.

30. Separate the lines, discard the grommets and plug the openings to prevent contamination.

31. Disconnect the evaporator housing's temperature sensor connector.

32. Remove the evaporator housing-to-chassis screws/nut and remove the evaporator housing.

33. Disconnect the mode control motor and the air mix control motor electrical connectors and remove the wiring harness clips and the wiring harness from the heater housing.

34. Remove the heater duct clip, the heater housing-to-chassis nuts and the heater housing.

35. Remove the heater core cover screws and the cover.

36. Remove the heater core pipe clamp screws and the clamp.

37. Remove the heater core from the heater housing.

To install:

38. Install the heater core in the heater housing.

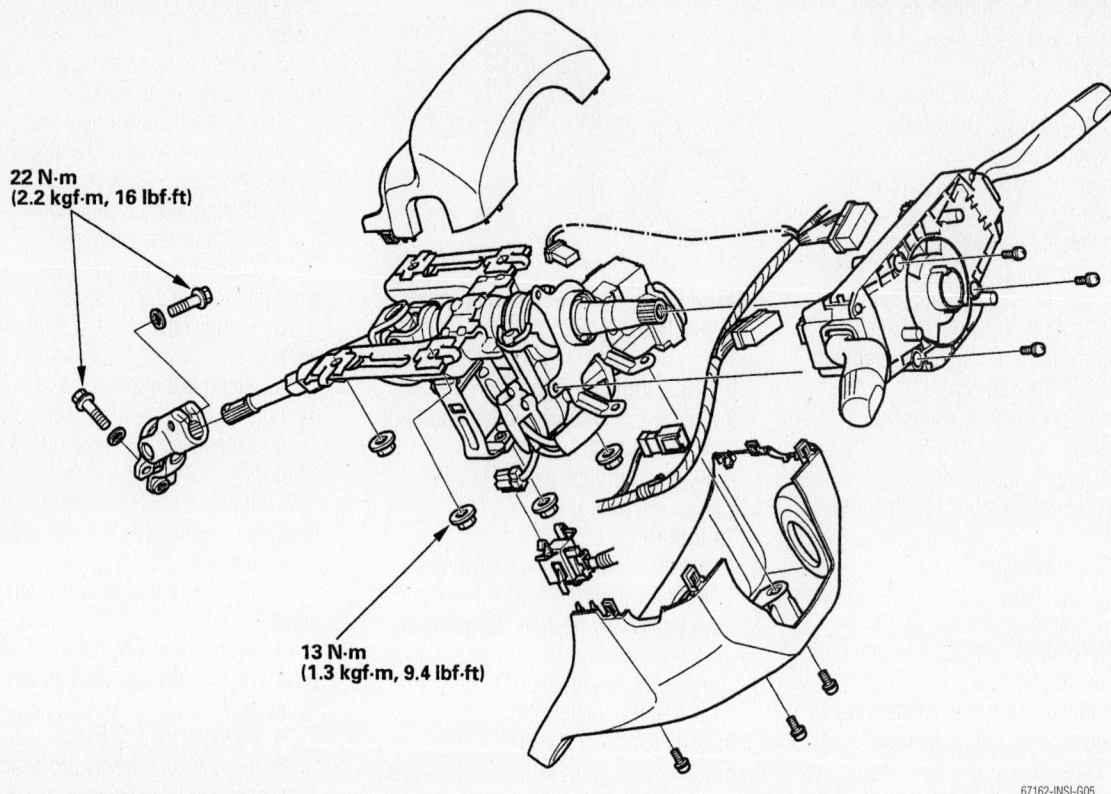

**22 N·m
(2.2 kgf·m, 16 lbf·ft)**

**13 N·m
(1.3 kgf·m, 9.4 lbf·ft)**

67162-INSI-G05

Exploded view of the steering column and related components—Insight

Fastener Locations

A ▶ : Bolt, 7   B ▶ : Screw, 2

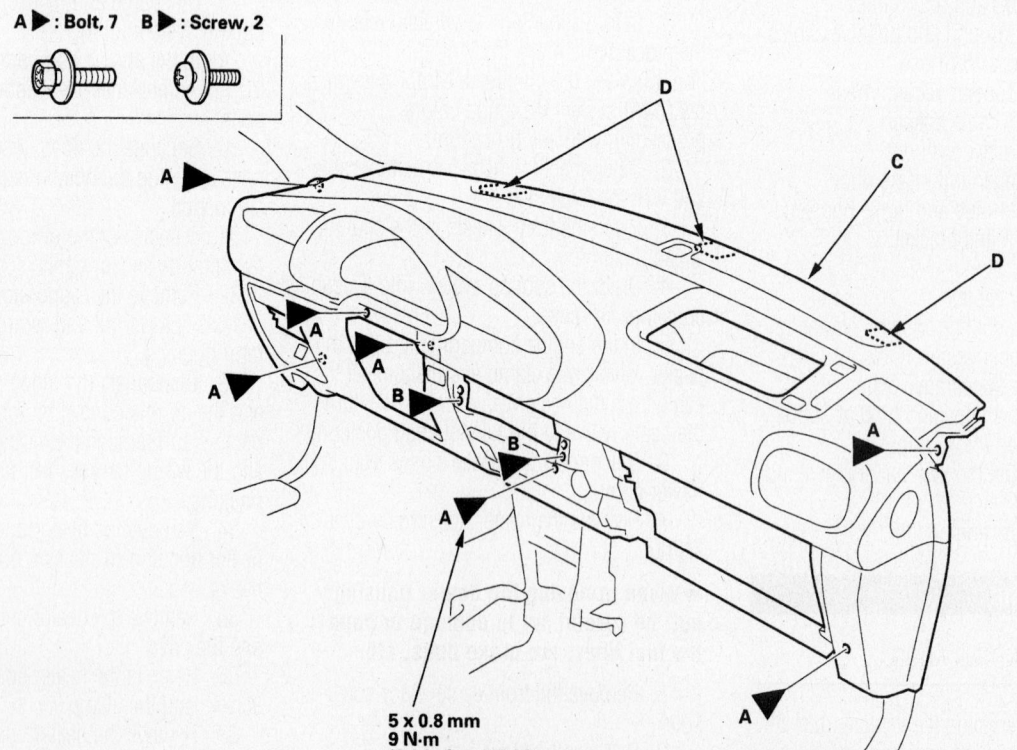

**5 x 0.8 mm
9 N·m
(0.9 kgf·m, 7 lbf·ft)**

67162-INSI-G06

Exploded view of the instrument panel—Insight

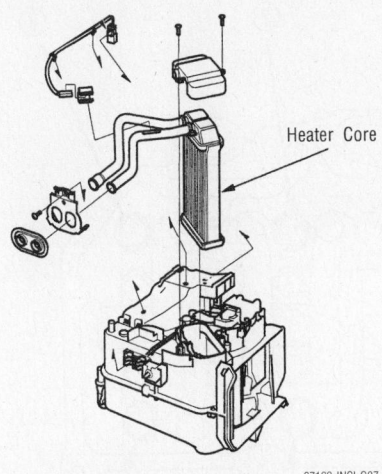

Exploded view of the heater core and housing—Insight

39. Install the heater core pipe clamp and the clamp screws.

40. Install the heater core cover and the cover screws.

41. Install the heater housing, the heater housing-to-chassis nuts and the heater duct clip.

42. Install the wiring harness clips and the wiring harness to the heater housing and connect the mode control motor and the air mix control motor electrical connectors.

43. Install the evaporator housing and the evaporator housing-to-chassis screws/nut.

44. Connect the evaporator housing's temperature sensor connector.

45. Using new grommets, connect the refrigerant lines.

46. In the engine compartment, install the refrigerant lines-to-evaporator housing bolts.

47. Using an assistant, install the instrument panel making sure it is engaged in the holders.

48. Install the driver coin tray.

49. From under and behind the dashboard, reconnect all electrical connectors.

50. Install the A pillar trim on both sides.

51. Place the passenger air bag in the dashboard and tighten 3 mounting nuts.

52. Connect the passenger air bag connector.

53. Install the steering column.

54. Insert the upper end of the steering joint into the steering shaft and line up the bolt hole with the flat on the shaft.

55. Insert the lower end of the joint into the pinion shaft and line up the bolt hole with the groove around the shaft.

56. Install the lower bolt and pull up on the joint to ensure it is securely seated.

57. Install the upper bolt and tighten both bolts to 16 ft. lbs. (22 Nm).

58. Install the steering column covers.

59. Connect the combination switch and ignition switch connectors.

60. Install the cable and center it by rotating it clockwise until it stops. Rotate it counterclockwise about 2 and one half turns until the arrow mark on the reel point straight up.

61. Install the steering wheel on the shaft making sure the wheel hub engages the pins on the cable reel and the tabs of the turn signal canceling sleeve.

62. Install the steering wheel bolt nut and tighten it to 29 ft. lbs. (39 Nm).

63. Install the driver air bag and tighten the bolts to 87 inch lbs. (9.8 Nm).

64. Connect the driver air bag connector and install the steering wheel access cover.

65. Remove the radio.

66. Install the center lower cover.

67. Install the glove box

68. Install the front center floor console.

69. Install the heater housing-to-chassis nut.

70. Connect the heater hoses to the heater core.

71. In the engine compartment, connect the cable to the heater valve arm and close the heater valve cable clamp.

72. Refill the cooling system.

73. Connect the negative battery cable.

74. Evacuate and charge and leak test the air conditioning system refrigerant.

75. Run the engine to normal operating temperatures; then, check the climate control operation and check for leaks.

Cylinder Head

REMOVAL & INSTALLATION

1. Before servicing the vehicle, refer to the precautions in the beginning of this section. Also see the IMA Integrated Motor Assembly (IMA) system disabling before proceeding with any procedure.

2. Drain the cooling system and engine oil.

3. Relieve the fuel system pressure.

4. Remove or disconnect the following:
- Negative and then positive battery cables
- Accessory drive belt
- Engine appearance cover
- Breather pipe and brake booster vacuum line
- Air cleaner and intake air duct
- Throttle cable linkage
- Brake booster vacuum hose
- EVAP canister hose
- Heater hoses
- Cylinder head cover
- Intake manifold
- Radiator hoses
- Catalytic converter
- Dipstick
- Cam chain tensioner by loosening the bolt in sequence one turn at a time.
- Cylinder head plug
- Timing chain

5. Hold the camshaft with an open end wrench, then remove the camshaft sprocket bolt.

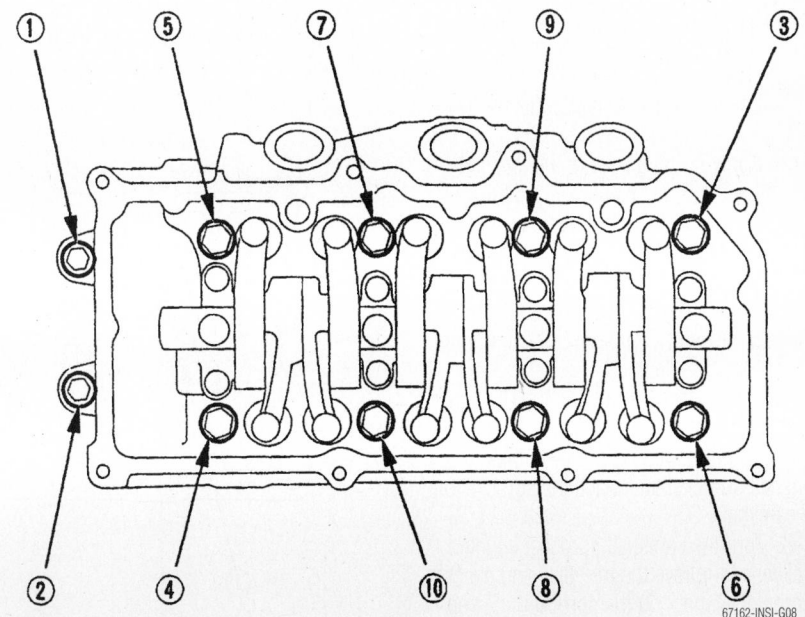

Cylinder head bolt loosening sequence—Insight

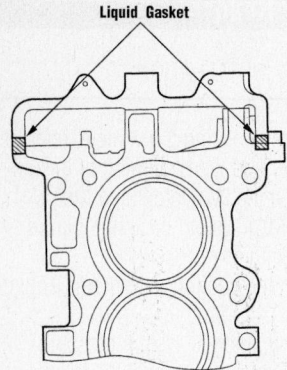

Liquid Gasket

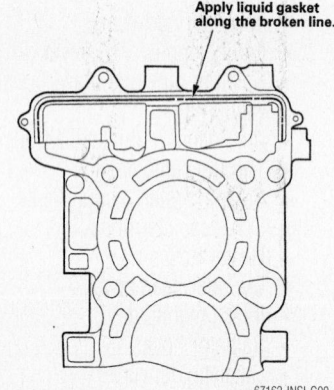

Apply liquid gasket along the broken line.

67162-INSI-G09

Engine block sealant application areas—Insight

6. Remove the camshaft sprocket.

7. Remove the engine mount bracket bolt on the cylinder head side.

8. Loosen the cylinder head bolts in sequence and ⅓ turns until all bolts are loose.

9. Remove the cylinder head.

To install:

10. Clean the cylinder head and block mating surfaces.

11. Ensure the crankshaft pulley is at TDC.

12. Apply liquid gasket to the areas of the engine block as shown.

13. Install the cylinder head.

14. Coat the cylinder head bolt threads with clean engine oil.

15. Tighten the cylinder head bolts in the sequence shown to 29 ft. lbs. (39 Nm), plus an additional 90 degrees.

16. Tighten the 6 mm bolts to 105 inch lbs. (12 Nm).

17. Install the timing chain on the camshaft sprocket and place the sprocket on the camshaft.

18. Turn the camshaft sprocket counterclockwise to relieve the free play and check that the TDC mark on the sprocket lines up with the cylinder head surface.

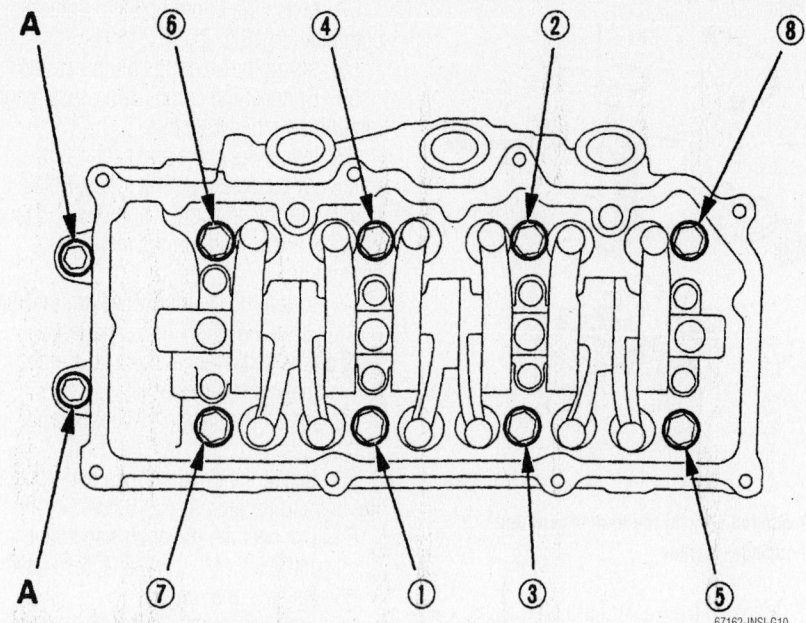

Cylinder head bolt tightening sequence—Insight

67162-INSI-G10

19. If the mark is not aligned correctly, remove the sprocket from the camshaft and reposition the timing chain until the mark is aligned.

20. Hold the camshaft with an open end wrench and tighten the sprocket bolt to 41 ft. lbs. (56 Nm).

21. Install a new cylinder head plug.

22. Press the rod to pump the oil out of the timing chain tensioner.

23. Install a new o-ring to the tensioner spacer and a new gasket on the tensioner and then tighten the bolts and nuts equally to 105 inch lbs. (12 Nm), while pressing the tensioner against the head.

24. Adjust the valve clearance.

25. The remainder of the installation is the reverse of the removal procedure.

26. Start the engine and hold at 3500 rpm with no load for 10 minutes.

27. Check that the IMA battery level indicator reads full

28. Check the engine idle.

Rocker Arms/Shafts

REMOVAL & INSTALLATION

1. Before servicing the vehicle, refer to the precautions in the beginning of this section. Also see the Integrated Motor Assembly (IMA) system disabling before proceeding with any procedure.

2. Remove the valve cover.

3. Loosen the rocker adjusting screws

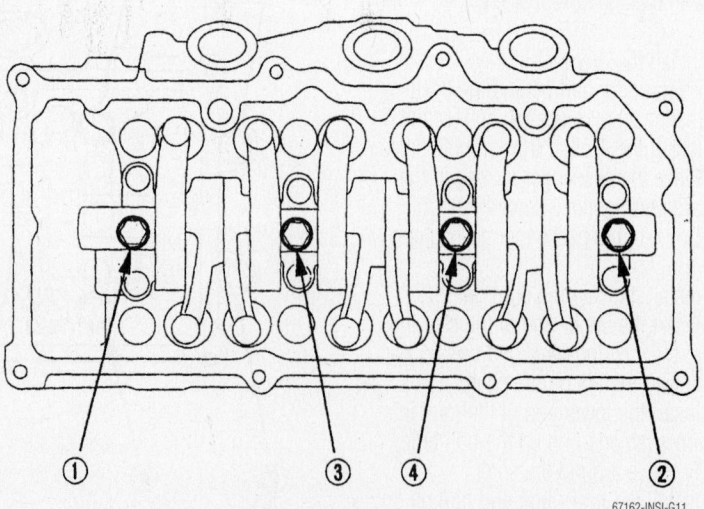

67162-INSI-G11

Rocker arm shaft bolt loosening sequence—Insight

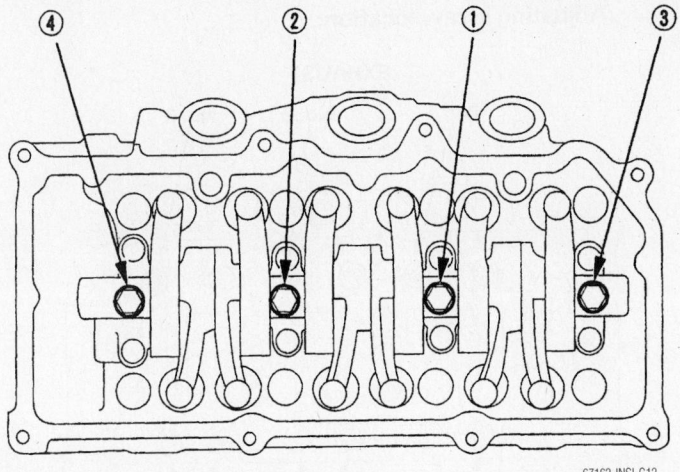

67162-INSI-G12

Rocker arm shaft bolt tightening sequence—Insight

4. Remove the rocker shaft bolts 2 turns at a time in the sequence shown.

5. Remove the rocker arm assembly.

6. Installation is the reverse of removal. Tighten the bolts in sequence as shown to 16 ft. lbs. (22 Nm).

Intake Manifold

REMOVAL & INSTALLATION

1. Before servicing the vehicle, refer to the precautions in the beginning of this section. Also see the Integrated Motor Assembly (IMA) system disabling before proceeding with any procedure.

2. Drain the cooling system and engine oil.

3. Relieve the fuel system pressure.

4. Remove or disconnect the following:
- Negative and then positive battery cables
- Accessory drive belt
- Engine appearance cover

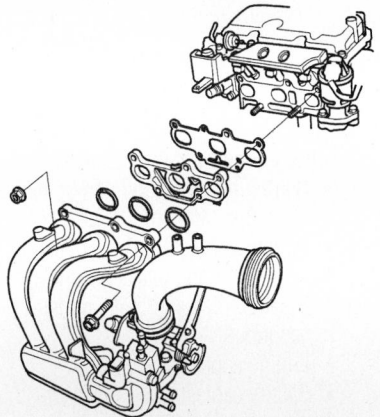

67162-INSI-G13

Intake manifold exploded view—Insight

- Breather pipe and brake booster vacuum line
- Air cleaner and intake air duct
- Throttle cable linkage
- Brake booster vacuum hose
- EVAP canister hose
- Heater hoses

5. Disconnect and label all the electrical connectors that would interfere with the valve cover removal, then remove the valve cover.

6. Remove the intake manifold.

To install:

7. Install or connect the following:
- New intake manifold gasket and o-rings
- Intake manifold. Tighten the fasteners to 16 ft. lbs. (22 Nm).
- Valve cover
- Electrical connectors
- Heater hoses
- EVAP canister hose
- Brake booster vacuum hose
- Throttle cable linkage
- Air cleaner and intake air duct
- Breather pipe and brake booster vacuum line
- Engine appearance cover
- Accessory drive belt
- Negative and then positive battery cables

8. Fill the cooling system.

9. Start the engine and check for leaks.

Exhaust Manifold

REMOVAL & INSTALLATION

1. Before servicing the vehicle, refer to the precautions in the beginning of this section. Also see the Integrated Motor Assembly (IMA) system disabling before proceeding with any procedure.

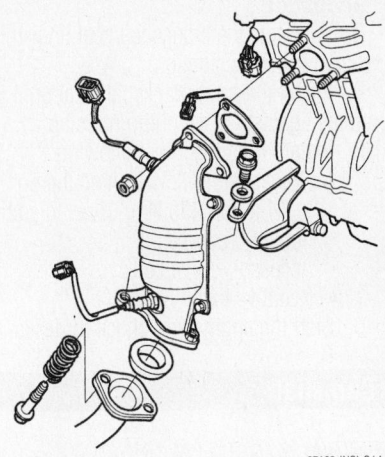

67162-INSI-G14

Exploded view of the exhaust system—Insight

2. Remove or disconnect the following:
- Negative battery cable
- Exhaust manifold heat shield
- Heated Oxygen Sensor (HO_2S) connector
- Air fuel ratio sensor connector
- Exhaust front pipe
- Exhaust manifold bracket, if equipped
- Exhaust manifold

To install:

3. Install or connect the following:
- Exhaust manifold. Tighten the new fasteners to 17 ft. lbs. (24 Nm).
- Exhaust manifold bracket, if equipped. Tighten the bolts to 33 ft. lbs. (44 Nm).
- Exhaust front pipe. Tighten the bolts alternately to 16 ft. lbs. (22 Nm).
- Heated Oxygen Sensor (HO_2S) connector
- Air fuel ratio sensor connector
- Exhaust manifold heat shield
- Negative battery cable

Front Crankshaft Seal

REMOVAL & INSTALLATION

1. Before servicing the vehicle, refer to the precautions in the beginning of this section. Also see the Integrated Motor Assembly (IMA) system disabling before proceeding with any procedure.

2. Remove or disconnect the following:
- Negative battery cable
- Accessory drive belt
- Crankshaft pulley
- Front crankshaft seal

To install:

3. Lubricate the crankshaft seal lip with grease prior to installation.

4. Install the front crankshaft seal until it bottoms against the oil pump housing.

5. Install or connect the following:
- Crankshaft pulley. Tighten the bolt to 14 ft. lbs. (20 Nm), plus an additional 90 degrees.
- Accessory drive belt
- Negative battery cable

6. Start the engine and check for leaks.

Camshaft

REMOVAL & INSTALLATION

1. Before servicing the vehicle, refer to the precautions in the beginning of this section. Also see the Integrated Motor Assembly (IMA) system disabling before proceeding with any procedure.

2. Remove or disconnect the following:
- Negative battery cable
- Accessory drive belt
- Valve cover
- Camshaft sprocket
- Rocker arms and shaft

➡ **Keep all valvetrain components in order for assembly.**

3. Remove the camshaft holder bolts in reverse of the tightening sequence.

4. Remove the camshaft.

To install:
- Camshafts. Tighten the holder bolts in sequence to 105 inch lbs. (12 Nm).
- Rocker arms and shaft
- Camshaft sprocket
- Valve cover
- Accessory drive belt
- Negative battery cable

5. Adjust the valve clearance.

6. Start the engine and check for leaks.

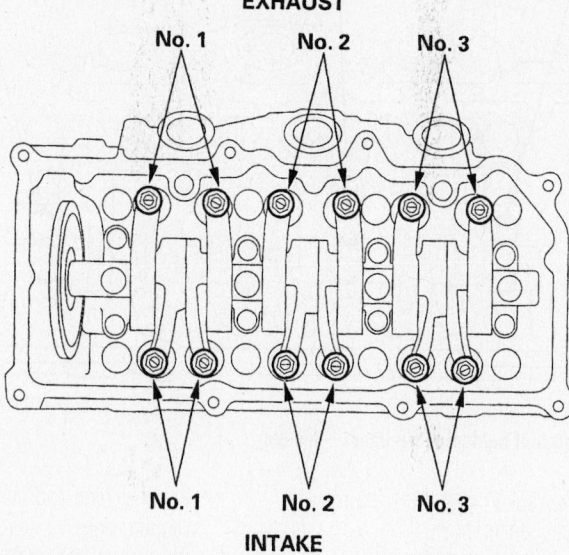

Adjusting screw location:

Valve adjustment locations—Insight

67162-INSI-G16

Valve Lash

ADJUSTMENT

Adjust the valves only when the cylinder head temperature is less than 100°F (38°C).

1. Before servicing the vehicle, refer to the precautions in the beginning of this section. Also see the Integrated Motor Assembly (IMA) system disabling before proceeding with any procedure.

2. Remove or disconnect the following:
- Negative battery cable
- Valve cover

3. Set the No. 1 piston at TDC. The TDC mark on the camshaft sprocket should line up with the cylinder head surface.

4. Check the clearance on No. 1 cylinder. Intake should be 0.007–0.009 in.; exhaust should be 0.008–0.010 in. Tighten the locknut.

5. Rotate the crankshaft 240 degrees clockwise and check No.3.

6. Rotate the crankshaft 240 degrees clockwise and check No. 2.

Starter Motor

REMOVAL & INSTALLATION

1. Before servicing the vehicle, refer to the precautions in the beginning of this section. Also see the Integrated Motor Assembly (IMA) system disabling before proceeding with any procedure.

2. Remove or disconnect the following:
- Negative and then positive battery cable
- Engine appearance cover
- Breather pipe and brake booster vacuum line
- Air cleaner and intake air duct
- Starter wiring harness connectors
- Starter motor

To install:

3. Install or connect the following:
- Starter motor. Tighten the bolts to 33 ft. lbs. (44 Nm).
- Starter wiring harness connectors
- Air cleaner and intake air duct
- Breather pipe and brake booster vacuum line
- Engine appearance cover
- Battery cables

4. Remove the No. 15 (40A) fuse from the under hood relay box.

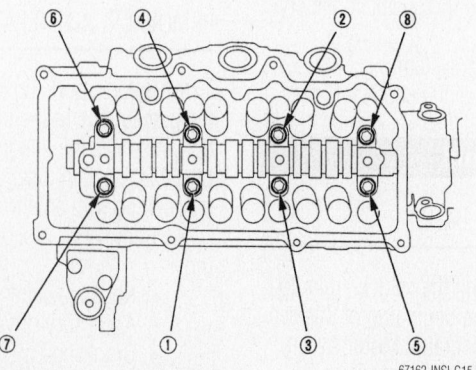

Camshaft holder tightening sequence—Insight

67162-INSI-G15

5. Start the engine and run at 3500 rpm for 10 minutes.

6. Check that the battery module indicator reads full.

7. Install the No. 15 fuse.

Timing Chain, Oil Pan and Oil Pump

REMOVAL & INSTALLATION

1. Before servicing the vehicle, refer to the precautions in the beginning of this section. Also see the Integrated Motor Assembly (IMA) system disabling before proceeding with any procedure.

2. Disconnect the negative battery cable.

3. Remove the engine.

4. Remove the cylinder head.

5. Remove the idler pulley bracket mounting bolt.

6. Remove the water pump.

7. Remove the crankshaft pulley.

8. On 2001–03 models, remove the engine oil cooler bypass hoses.

9. Remove the oil pan.

10. Remove the oil screen and oil pump.

11. Remove the timing chain.

12. Remove the crankshaft sprocket.

13. Remove the timing chain tensioning arm and chain guide.

14. Remove the Crankshaft Position Sensor (CKP) pulser plate.

To install:

15. Install the CKP sensor plate.

16. Install the timing chain guide and tighten the bolt to 105 inch lbs. (12 Nm).

17. Install the timing chain tensioning arm and tighten the bolts to 16 ft. lbs. (22 Nm).

18. Set the crankshaft sprocket so no. 1 piston is at TDC. Align the TDC mark (A) on the crankshaft pulser with the pointer on the block.

19. Align the timing chain colored link (A) with the punch mark (B) on the crankshaft sprocket.

20. Clean and dry the oil pump mating surfaces.

21. Apply liquid gasket to the oil pump

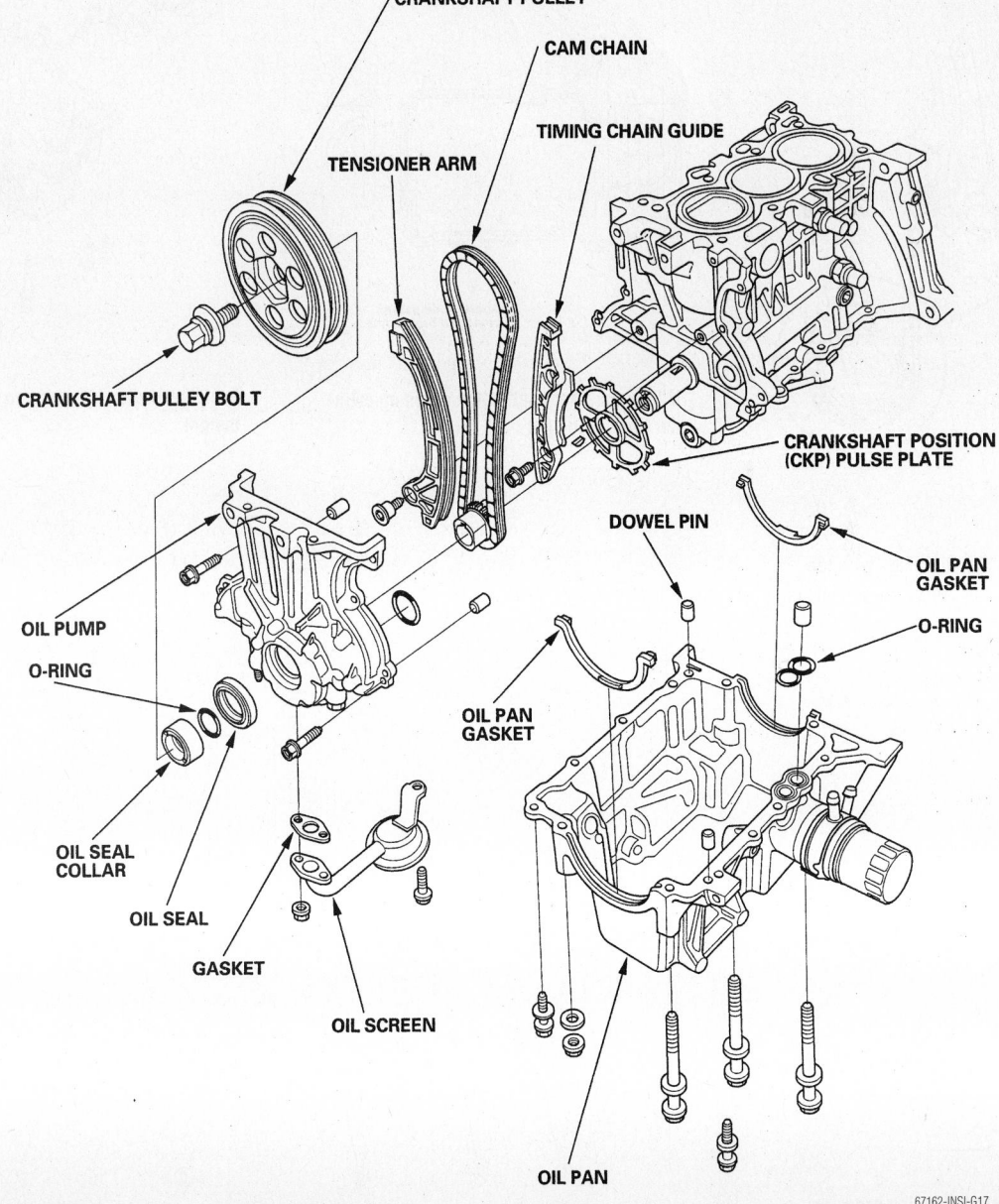

Exploded view of the timing chain components—Insight

67162-INSI-G17

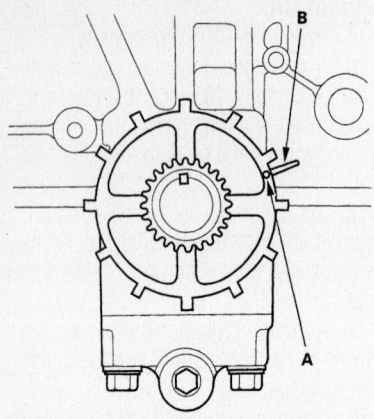

Aligning the crankshaft sprocket timing marks—Insight

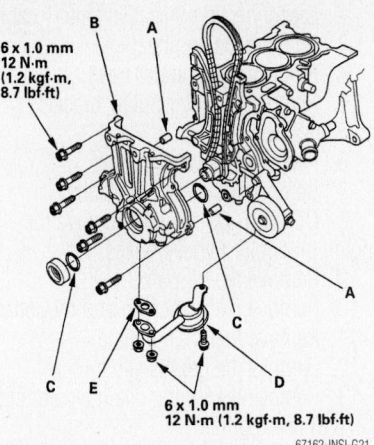

6 x 1.0 mm
12 N·m
(1.2 kgf·m,
8.7 lbf·ft)

6 x 1.0 mm
12 N·m (1.2 kgf·m, 8.7 lbf·ft)

67162-INSI-G21

Exploded view of the oil pump installation—Insight

mating surface on the cylinder block and the bolt holes as shown.

22. Install the dowel pins (A) and align the inner rotor with the crankshaft and install the oil pump (B) using new o-rings (C).

23. Install the oil screen (D) using a new gasket (E).

24. Clean and dry the oil pan mating surfaces.

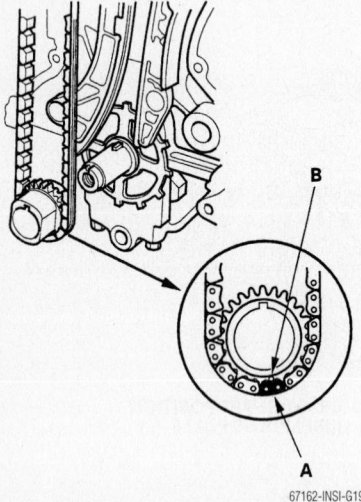

Aligning the timing chain link with the crankshaft sprocket timing mark—Insight

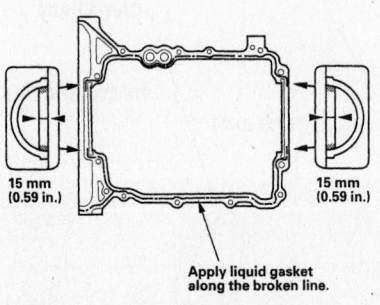

15 mm (0.59 in.)   15 mm (0.59 in.)

Apply liquid gasket along the broken line.

67162-INSI-G22

Oil pan sealant application areas on cylinder block—Insight

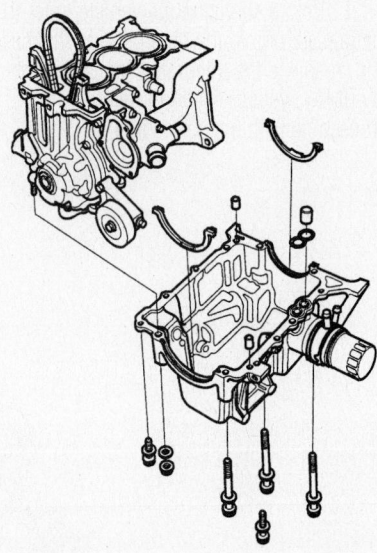

67162-INSI-G23

Exploded view of the oil pan installation—Insight

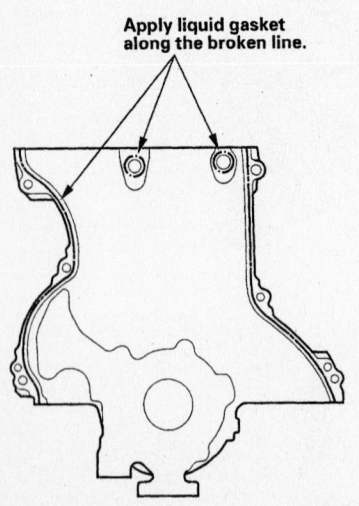

Apply liquid gasket along the broken line.

67162-INSI-G20

Oil pump sealant application areas on cylinder block—Insight

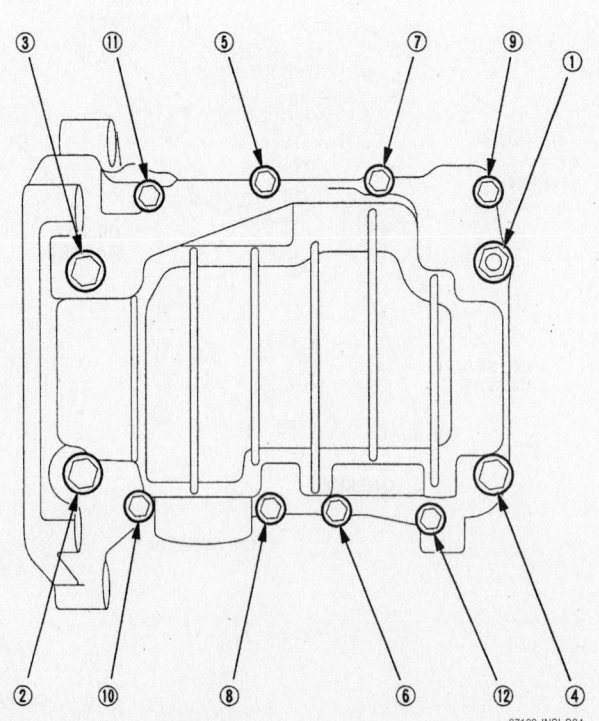

67162-INSI-G24

Oil pan tightening sequence—Insight

25. Apply liquid gasket to the oil pan mating surface on the cylinder block and the bolt holes as shown.

26. Using new gaskets and o-rings, install the oil pan. Tighten the bolts in sequence as shown. Tighten M8 bolts to 16 ft. lbs. (22 Nm). Tighten the M6 bolts to 105 inch lbs. (12 Nm).

27. Lubricate the crankshaft pulley bolt with clean engine oil and install the pulley. Tighten the bolt to 14 ft. lbs. (20 Nm), plus an additional 90°.

28. Install or connect the following:
- On 2000–03 models, the engine oil cooler bypass hoses
- Crankshaft pulley
- Water pump
- Idler pulley bracket bolt
- Cylinder head
- Engine
- Connect the negative battery cable.
- Check the engine operation and road test.

Rear Main Seal

REMOVAL & INSTALLATION

1. Remove or disconnect the following:
- Transaxle
- Clutch pressure plate and disc, if equipped
- Flywheel
- Oil seal

To install:

2. Install or connect the following:
- Oil seal. Drive the seal square into the seal case.
- Flywheel. Tighten the bolts in a crossing pattern to 33 ft. lbs. (44 Nm).
- Clutch pressure plate and disc, if equipped
- Transaxle

3. Check the fluid levels.

4. Start the engine and check for leaks.

Piston and Ring

POSITIONING

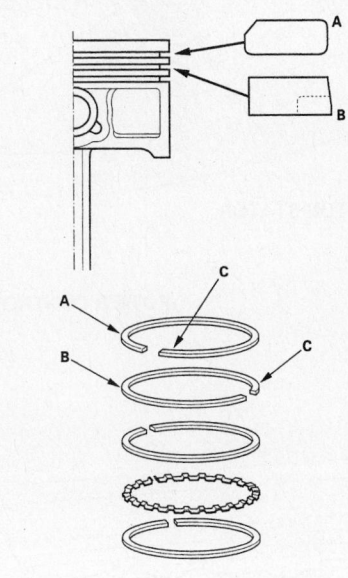

Piston Ring Dimensions:

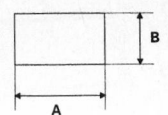

Top Ring (Standard):
A: 2.3 mm (0.09 in.)
B: 1.0 mm (0.04 in.)

Second Ring (Standard):
A: 3.0 mm (0.12 in.)
B: 1.2 mm (0.05 in.)

67162-INSI-G25

Piston ring positioning and top mark location

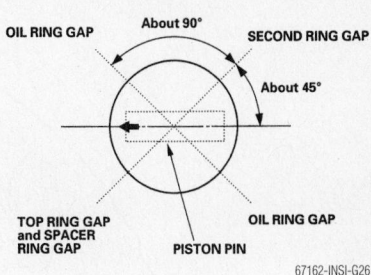

67162-INSI-G26

Piston ring end-gap spacing

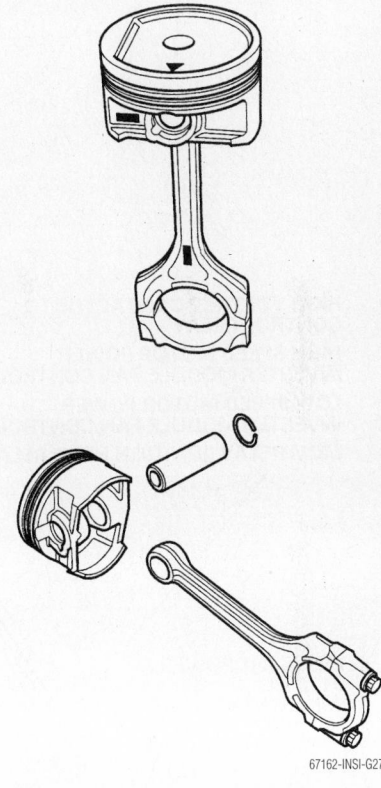

67162-INSI-G27

Piston and connecting rod assembly

INTEGRATED MOTOR ASSEMBLY (IMA) SYSTEM

IMA Service Precautions

Before servicing any vehicle, please be sure to read all of the following precautions, which deal with personal safety, prevention of component damage, and important points to take into consideration when servicing the integrated motor assembly:

- The Integrated Motor Assembly (IMA) system uses 144 volt circuits. Always shut off electrical circuits and isolate the IMA system and related parts before servicing the system.

- The high voltage cables and their covers are identified by an Orange coloring. Caution labels are attached to high voltage and related parts. Be careful not to touch these cables and parts without adequate protective gear.

- If the 12 volt battery has been discharged or disconnected or the Motor Control Module (MCM) has been reset, the IMA battery level indicator will not show the state of charge when the engine is started. Start the engine and hold the idle at 3500 rpm with no load until the battery level indicator reads full charge.

- Wear insulated gloves with no tears, pin holes or other damage when inspecting or servicing the IMA.

- Turn the battery module switch **OFF** and secure the switch in the **OFF** position with the locking cover before servicing the IMA system.

- Wait for 5 minutes or more after turning the battery module switch off, then disconnect the negative battery cable, It takes about 5 minutes for the PDU capacitor to discharge.

- Before disconnecting the high voltage cables, make sure the voltage between the terminals is less than 30 volts when checked with a voltmeter.

- When servicing parts without the Orange insulating color, always use insulated tools to prevent shorts.

- The rotor assembly contains very strong magnets and should be handled with care. People with pacemakers or other magnetically sensitive medical devices should not handle the rotor assembly.

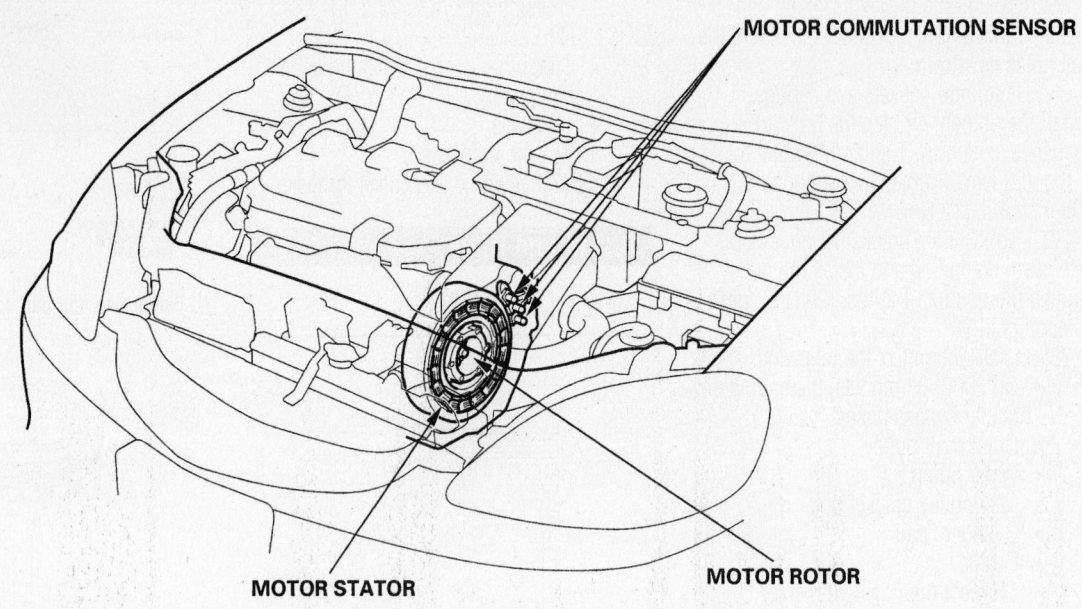

MOTOR COMMUTATION SENSOR

MOTOR STATOR

MOTOR ROTOR

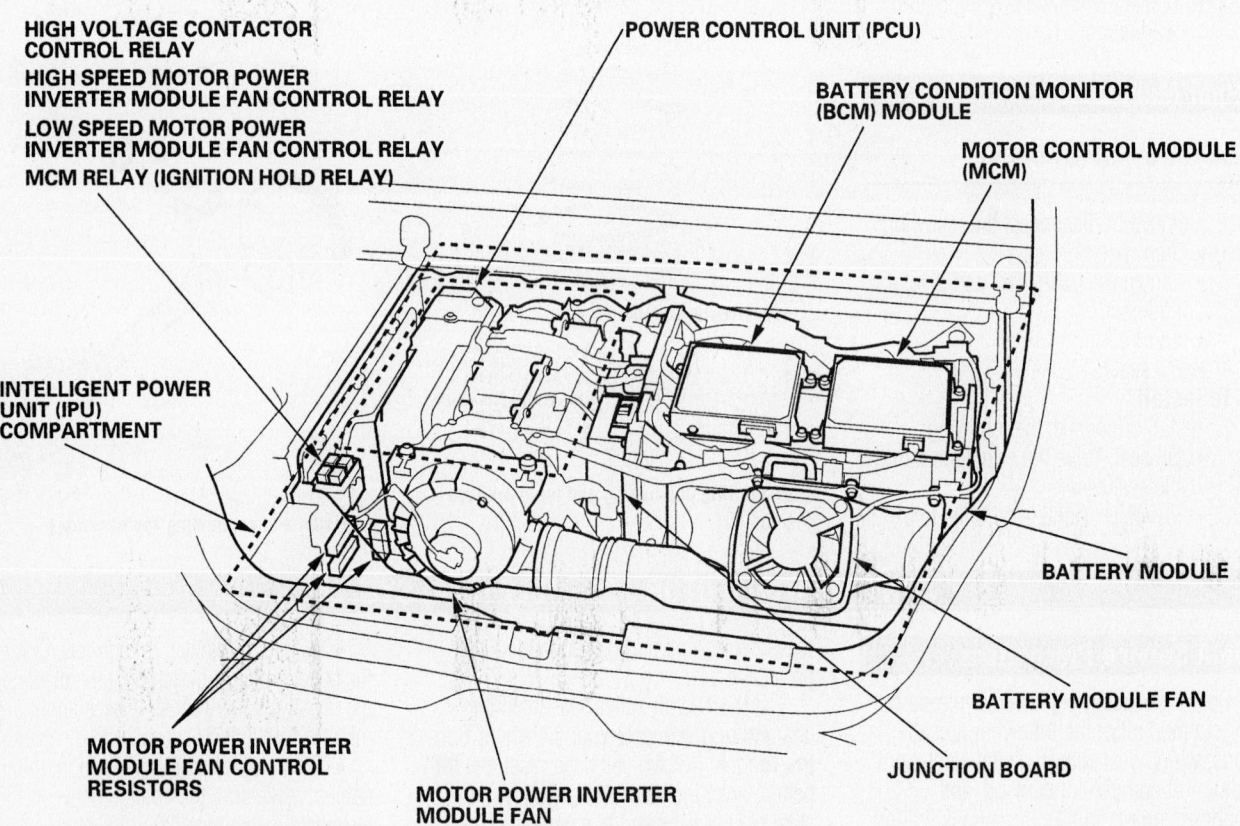

HIGH VOLTAGE CONTACTOR
CONTROL RELAY

HIGH SPEED MOTOR POWER
INVERTER MODULE FAN CONTROL RELAY

LOW SPEED MOTOR POWER
INVERTER MODULE FAN CONTROL RELAY

MCM RELAY (IGNITION HOLD RELAY)

POWER CONTROL UNIT (PCU)

BATTERY CONDITION MONITOR
(BCM) MODULE

MOTOR CONTROL MODULE
(MCM)

INTELLIGENT POWER
UNIT (IPU)
COMPARTMENT

BATTERY MODULE

MOTOR POWER INVERTER
MODULE FAN CONTROL
RESISTORS

MOTOR POWER INVERTER
MODULE FAN

BATTERY MODULE FAN

JUNCTION BOARD

67162-INSI-G28

IMA system component locations—Insight

- Keep the rotor away from magnetically sensitive devices.
- After disconnecting the high voltage terminals or other parts, isolate them with insulated parts.
- Attach a sign saying "WORKING ON HIGH VOLTAGE PARTS. DO NOT TOUCH," to the steering wheel as a safety measure.

Disabling The IMA System Power

1. Turn the ignition switch off.
2. Remove the luggage compartment floor mat.
3. Remove the battery module cover and the locking cover.

4. Turn the battery module switch **OFF** and install the locking cover.
5. Wait at least 5 minutes to allow the Intelligent Power Unit (IPU) capacitor to discharge.
6. Remove the right side shelf support.
7. Remove the luggage compartment cover clips and the cover.

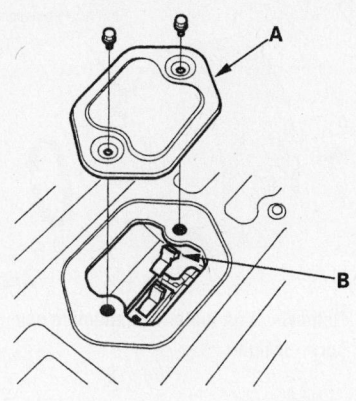

IMA battery module switch and locking
cover location—Insight

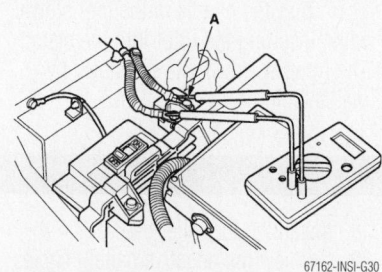

Measuring IMA battery voltage—Insight

8. Measure the voltage at the junction
board terminals (A). There should be less
than 30 volts. If more than 30 volts are
present, there is a problem with the system.
DO NOT proceed with any procedure without doing a Diagnostic Trouble Code (DTC)
troubleshooting to determine the cause.

Battery Module and Power Control Unit

REMOVAL & INSTALLATION

1. Before servicing the vehicle, refer to
the precautions in the beginning of this section.
2. Disable the IMA system power.
3. Remove the foam inserts.
4. Remove the mid frame (A), the front
and rear braces (B) and disconnect the high
voltage cables (C).
5. Wrap the cables with insulating tape.
6. Remove the air duct mounting bolt
and push the air duct forward.
7. Remove the battery module mounting bolts.
8. Disconnect the 3 connectors at the
left side and secure the connectors.
9. Install battery lifting tool 07YAK-
001010A on the battery module.
10. With the aid of an assistant, lift the

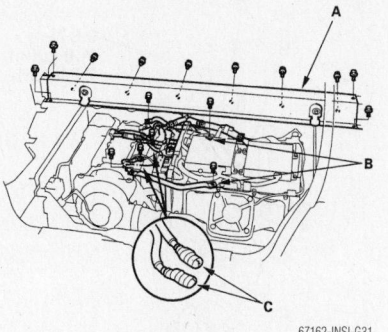

Removing the battery module braces and
high voltage cables—Insight

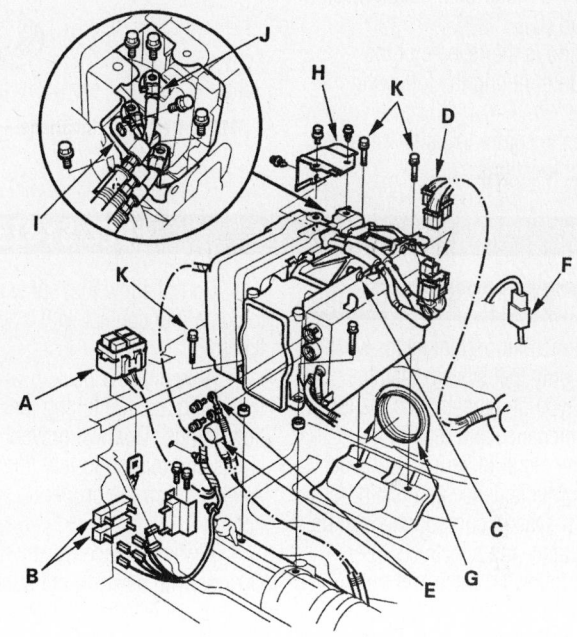

Removing the power control unit—Insight

battery module from the vehicle and place
on a flat surface.
11. To remove the Power Control Unit
(PCU), disconnect the connector (A) from
the cooling fan assembly (B) and remove
the clip from the fan shroud (C).
12. Remove the 4 fan bolts (D) and
remove the fan bracket (E).
13. Remove the fan duct (F) from the fan
and remove the fan.
14. Lift the relay pack (A) from its holder
and disconnect the harness from the resistors (B).
15. Disconnect the Y condenser ground
(C).
16. Disconnect the connector (D) from
the PCU and then disconnect the battery
cables (E).
17. Disconnect the high voltage DC-DC
converter 2P connector (F).

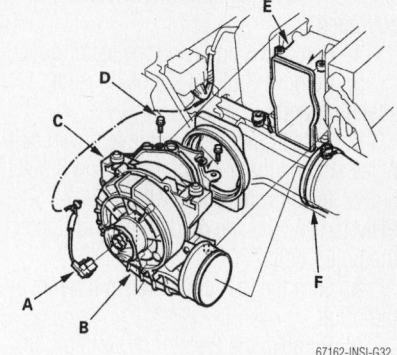

Removing the cooling fan assembly—
Insight

18. Pull away the intake duct (G).
19. Remove the PCU terminal cover (H)
and remove the clip (J) and bolts that hold
the three phase cables (J) in place.
20. Remove the 4 PCU mounting bolts
(K).
21. Remove the PCU and place it on a
flat surface.
To install:
22. Installation is the reverse of the
removal procedure.

IMA Motor

REMOVAL & INSTALLATION

1. Before servicing the vehicle, refer to
the precautions in the beginning of this section.
2. Disable the IMA system power.

3. Remove the transaxle and clutch, if equipped.

4. Remove 8 bolts and the stator cover.

5. Remove 3 of the opposing rotor attaching bolts.

6. Install guide pins in the removed bolt holes and then remove the remaining 3 bolts.

7. Install rotor puller tool 07YAC-PHM1010A over the guide pins and tighten the puller bolts.

8. Turn the tool clockwise and remove the rotor.

9. Remove the 2 stator mounting bolts.

10. Remove the stator.

11. Remove the cover (A) and the motor power cables (B).

12. Remove the motor commutation sensors (A), (B) and (C).

13. Installation is the reverse of the removal procedure, noting the following items:

a. Install the communication sensors in the correct locations.

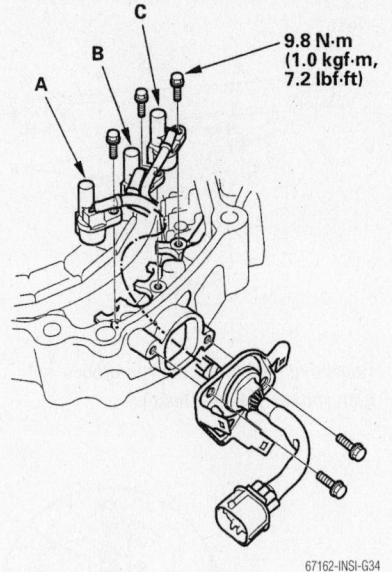

9.8 N·m (1.0 kgf·m, 7.2 lbf·ft)

67162-INSI-G34

IMA power cable locations—Insight

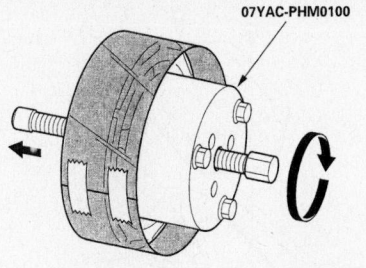

07YAC-PHM0100

67162-INSI-G35

Removing IMA motor commutation sensors—Insight

b. Install the motor power cables in the correct locations.

c. Set the rotor on the special tool and install the rotor with the end of the tool extended.

d. Turn the handle of the tool slowly when inserting the rotor into the stator. The rotor is drawn into the stator by magnetic force.

FUEL SYSTEM

Fuel System Service Precautions

Safety is the most important factor when performing not only fuel system maintenance, but any type of maintenance. Failure to conduct maintenance and repairs in a safe manner may result in serious personal injury or death. Maintenance and testing of the vehicle's fuel system components can be accomplished safely and effectively by adhering to the following rules and guidelines:

• To avoid the possibility of fire and personal injury, always disconnect the negative battery cable unless the repair or test procedure requires that battery voltage be applied.

• Always relieve the fuel system pressure prior to disconnecting any fuel system component (injector, fuel rail, pressure regulator, etc.), fitting or fuel line connection. Exercise extreme caution whenever relieving fuel system pressure to avoid exposing skin, face and eyes to fuel spray. Please be advised that fuel under pressure may penetrate the skin or any part of the body that it contacts.

• Always place a shop towel or cloth around the fitting or connection prior to loosening to absorb any excess fuel due to spillage. Ensure that all fuel spillage (should it occur) is quickly removed from engine surfaces. Ensure that all fuel soaked cloths or towels are deposited into a suitable waste container.

• Always keep a dry chemical (Class B) fire extinguisher near the work area.

• Do not allow fuel spray or fuel vapors to come into contact with a spark or open flame.

• Always use a backup wrench when loosening and tightening fuel line connection fittings. This will prevent unnecessary stress and torsion to fuel line piping. Always follow the proper torque specifications.

• Always replace worn fuel fitting O-rings with new. Do not substitute fuel hose or equivalent, where fuel pipe is installed.

Fuel System Pressure

RELIEVING

1. Before servicing the vehicle, refer to the precautions in the beginning of this section. Also see the IMA system disabling before proceeding with any procedure.

2. Disconnect the negative battery cable.

3. Remove the fuel filler cap.

4. Hold the fuel rail inlet banjo bolt with a flare nut wrench.

5. Place a shop towel over the fitting to absorb leakage.

6. Loosen the service bolt 1 turn.

7. When repairs are complete, replace the sealing washers and tighten the service bolt to 16 ft. lbs. (22 Nm) on 2001–03 models, or 105 inch lbs. (12 Nm) on 2004 models.

8. Install the fuel filler cap.

9. Connect the negative battery cable.

10. Start the engine and check for leaks.

Fuel Filter and Pump

REMOVAL & INSTALLATION

1. Before servicing the vehicle, refer to the precautions in the beginning of this section. Also see the IMA system disabling before proceeding with any procedure.

2. Relieve the fuel system pressure.

3. Drain the fuel tank into a suitable approved container.

4. Remove or disconnect the following:

• Negative battery cable
• Muffler
• Place a support under the fuel tank
• Fuel tank shield and straps
• Fuel supply and return lines
• Fuel pump connector
• Fuel tank
• Fuel pump unit locknut
• Fuel pump
• Fuel pump filter

To install:

5. Installation is the reverse of the removal procedure noting the following items:

a. Install a new fuel pump base gasket.

b. Do not push on the suction filter.

c. Ensure the wiring harness connections are secure.

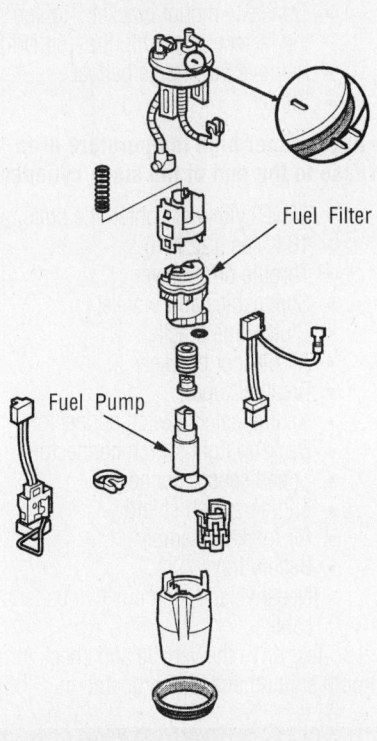

Exploded view of the fuel pump module assembly—Insight; A/T shown M/T similar

d. When installing the fuel pump align the marks on the fuel tank with the marks on the fuel pump.

e. Tighten the fuel tank straps bolts to 28 ft. lbs. (38 Nm).

Fuel Injector

REMOVAL & INSTALLATION

1. Before servicing the vehicle, refer to the precautions in the beginning of this sec-tion. Also see the IMA system disabling before proceeding with any procedure.

2. Relieve the fuel system pressure.

3. Remove or disconnect the following:
- Negative battery cable
- Fuel rail cover
- Injector connectors
- Vacuum hose and fuel return line from the pressure regulator
- Fuel rail nuts
- PCV valve
- Fuel rail
- Fuel injectors

To install:

4. Install new cushion rings (A) on the fuel injectors (B).

5. Coat new O-rings (C) with clean engine oil.

6. Install the fuel injectors to the fuel rail (D).

7. Coat new seal rings (E) with clean engine oil and press them in.

8. Install or connect the following:
- Fuel rail. Tighten the nuts to 105 inch lbs. (12 Nm).
- Fuel line to rail with new washers
- Vacuum line and fuel return line
- PCV valve
- Negative battery cable

9. Turn the ignition on but do not start the vehicle. Repeat this 2 or 3 times.

10. Start the engine and check for leaks.

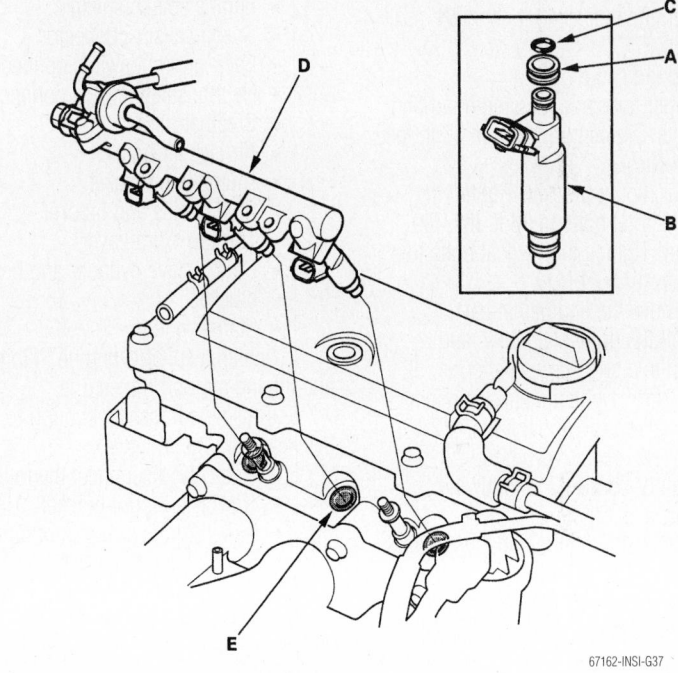

Exploded view of the fuel injector assembly—Insight

DRIVE TRAIN

Transaxle Assembly

REMOVAL & INSTALLATION

Automatic Transaxle

1. Before servicing the vehicle, refer to the precautions in the beginning of this section. Also see the IMA system disabling before proceeding with any procedure.

2. Drain the transaxle.

3. Remove or disconnect the following:
- Negative and then positive battery cables
- Battery tray
- Air intake assembly
- Engine splash shield
- Starter motor
- Transaxle ground cable
- Solenoid valve connector
- Input shaft speed sensor connector
- Shift cable and bracket
- Range switch connector
- Output shaft speed sensor connectors
- Upper transaxle housing mounting bolts
- Front wheels
- Axle shaft nuts
- Lower arms from knuckle
- Axle shafts from differential
- Transaxle oil cooler hoses and cap the openings
- Rear mount and bracket
- Drive plate bolts through the starter opening

4. Install a suitable engine lifting device and lift and support the engine.

5. Raise the transaxle enough to take the weight off the mounts.

6. Install a transmission jack.

7. Remove the transaxle ground cable from the transaxle mount bracket, then remove the mount bracket.

8. Remove the heat shield and exhaust manifold bracket.

9. Remove the lower transmission housing mounting bolts.

10. Pull the transaxle back until it clears the dowel pins and remove the transaxle.

To install:

➡ **Use new circlips, split pins and self-locking nuts for assembly.**

11. Install or connect the following:
- Transaxle
- Lower transaxle housing mounting bolts. Tighten the bolts to 33 ft. lbs. (44 Nm).
- Exhaust manifold bracket and heat shield. Tighten the bolts to 33 ft. lbs. (44 Nm).
- Transaxle mount bracket. Tighten the nuts to 76 ft. lbs. (103 Nm), then the bolt to 40 ft. lbs. (54 Nm).
- Ground cable
- Upper transaxle housing mounting bolts. Tighten the bolts to 47 ft. lbs. (64 Nm).
- Rear mount bracket. Tighten the horizontal bolts to 36 ft. lbs. (49 Nm). Tighten the vertical bolts to 54 ft. lbs. (74 Nm).
- Remove the engine hangar

12. Install the drive plate bolts and tighten them in a criss-cross pattern to 52 inch lbs. (6 Nm). Tighten them again in a criss cross pattern to 105 inch lbs. (12 Nm).

13. Install or connect the following:
- Transaxle oil cooler hoses
- Axle shafts to differential using new circlips
- Lower arms to knuckle. Tighten the nut to 40–47 ft. lbs. (54–64 Nm).
- Axle shaft nut. Tighten the nut to 134 ft. lbs. (181 Nm).
- Front wheels
- Splash shield
- Output shaft speed sensor connectors
- Range switch connector
- Shift cable and bracket
- Input shaft speed sensor connector
- Transaxle ground cable
- Starter motor
- Engine splash shield
- Air intake assembly
- Battery tray
- Negative and then positive battery cables

14. Check the engine timing.

15. Fill the transaxle to the correct level.

16. Start the engine and check for leaks.

17. Check the wheel alignment and adjust as necessary.

18. Connect the Honda Diagnostic System (HDS) tool to the data link connector.

19. Follow the prompts on the tester to perform the Start Clutch Control system calibration procedure.

Manual Transaxle

1. Before servicing the vehicle, refer to the precautions in the beginning of this section. Also see the IMA system disabling before proceeding with any procedure.

2. Drain the transaxle.

3. Remove or disconnect the following:
- Negative and then positive battery cables
- Battery tray
- Air intake assembly
- Engine splash shield
- Speed sensor connector
- Back-up light switch connector
- Neutral safety switch connector
- Breather tube
- Air cleaner bracket
- Clutch line bracket
- Shift cables and bracket
- Throttle drum cover
- Clutch slave cylinder and hose bracket
- Starter motor

4. Install a suitable engine lifting device and lift and support the engine.

5. Raise the transaxle enough to take the weight off the mounts.

6. Remove or disconnect the following:
- Transaxle mount bracket
- Upper 2 transaxle mounting bolts
- Splash shield bracket
- Front wheels
- Axle shafts

7. Install a transmission jack.

8. Remove the rear engine mount bracket.

9. Remove the 4 lower transmission mounting bolts

10. Pull the transaxle back until it clears the dowel pins and remove the transaxle.

To install:

➡ **Use new circlips, split pins and self-locking nuts for assembly.**

11. Install or connect the following:
- Transaxle
- Lower transaxle housing mounting bolts. Tighten the bolts to 47 ft. lbs. (65 Nm).
- Rear engine mount bracket. Tighten the bolts to 36 ft. lbs. (49 Nm). Tighten the through bolt to 66 ft. lbs. (89 Nm).
- Upper transaxle housing mounting bolts. Tighten the bolts to 47 ft. lbs. (65 Nm).

- Transaxle mount bracket. Tighten the fasteners to 40 ft. lbs. (54 Nm).
- Remove the engine hangar
- Starter

➡ **Apply super high temperature urea grease to the end of the slave cylinder.**

- Slave cylinder. Tighten the bolts to 16 ft. lbs. (22 Nm).
- Throttle drum cover
- Shift cables and bracket
- Clutch line bracket
- Air cleaner bracket
- Breather tube
- Neutral safety switch connector
- Back-up light switch connector
- Speed sensor connector
- Engine splash shield
- Air intake assembly
- Battery tray
- Negative and then positive battery cables

12. Test drive the vehicle and check for smooth shifting and clutch operation.

Clutch

ADJUSTMENTS

The Insight is equipped with a hydraulic clutch system. No adjustment is necessary.

REMOVAL & INSTALLATION

1. Before servicing the vehicle, refer to the precautions in the beginning of this section. Also see the IMA system disabling before proceeding with any procedure.

2. Remove or disconnect the following:
- Negative battery cable
- Transaxle
- Pressure plate. Loosen the bolts evenly in a crossing pattern.
- Clutch disc

To install:

3. Install the clutch disc and pressure plate. Tighten the pressure plate bolts in a crossing pattern and in several steps to 19 ft. lbs. (25 Nm).

4. Install or connect the following:
- Transaxle
- Negative battery cable

Hydraulic Clutch System

BLEEDING

1. Before servicing the vehicle, refer to the precautions in the beginning of this section.

2. Attach a hose to the bleeder screw and suspend the other end in a container of clean brake fluid.

3. Open the bleeder screw.

4. Slowly pump the clutch pedal until no more air bubbles appear at the bleeder hose.

5. Tighten the bleeder screw to 70 inch lbs. (8 Nm).

6. Refill the clutch master cylinder as necessary.

7. Check for leaks and proper clutch operation.

Halfshaft

REMOVAL & INSTALLATION

1. Before servicing the vehicle, refer to the precautions in the beginning of this section. Also see the IMA system disabling before proceeding with any procedure.

2. Drain the transaxle.

3. Remove or disconnect the following:
- Negative battery cable
- Front wheels
- Spindle nut
- Stabilizer link
- Brake hose clamp
- Wheel speed sensor clamp

4. Turn the front knuckle outward.

5. Tap the halfshaft inward with a plastic hammer.

6. Remove the lower ball joint.

7. Remove the inboard joint heat cover.

8. Pry the inboard joint from the transaxle or bearing support.

9. Pull the axle shaft straight out to avoid damaging the oil seal.

To install:

➡**Use new circlips, split pins and self-locking nuts for assembly.**

10. Install the new circlip into the circlip groove of the driveshaft

11. Install the inner CV-joint to the transaxle or bearing support until the circlip locks in the retaining groove.

12. Install or connect the following:
- Inboard joint heat cover
- Outboard joint into the front hub
- Lower ball joint. Tighten the nut to 40–47 ft. lbs. (54–64 Nm).
- Wheel speed sensor clamp
- Brake hose clamp
- Stabilizer link
- Axle shaft nut. Tighten the nut to 134 ft. lbs. (181 Nm).
- Front wheels
- Battery cables

13. Fill the transaxle to the correct level and check for leaks.

CV-Joint

OVERHAUL

Outboard Joint

1. Before servicing the vehicle, refer to the precautions in the beginning of this section. Also see the IMA system disabling before proceeding with any procedure.

2. Remove or disconnect the following:
- Axle halfshaft from the vehicle and place it in a vise
- Outboard joint boot clamps and push the boot back
- Outboard joint by using a slide hammer to pull it off the driveshaft
- Snap ring
- Outboard joint boot

To install:

➡**Use new circlips and boot clamps for assembly.**

3. Install the outboard joint boot and clamps to the axle shaft.

4. Install a new snap ring on the driveshaft groove.

5. Insert the driveshaft into the joint until the snap ring closes.

6. Seat the joint by dropping it straight down onto a hard surface from about 4 inches. DO NOT use a hammer to seat the joint.

7. Fill the outboard joint with grease.

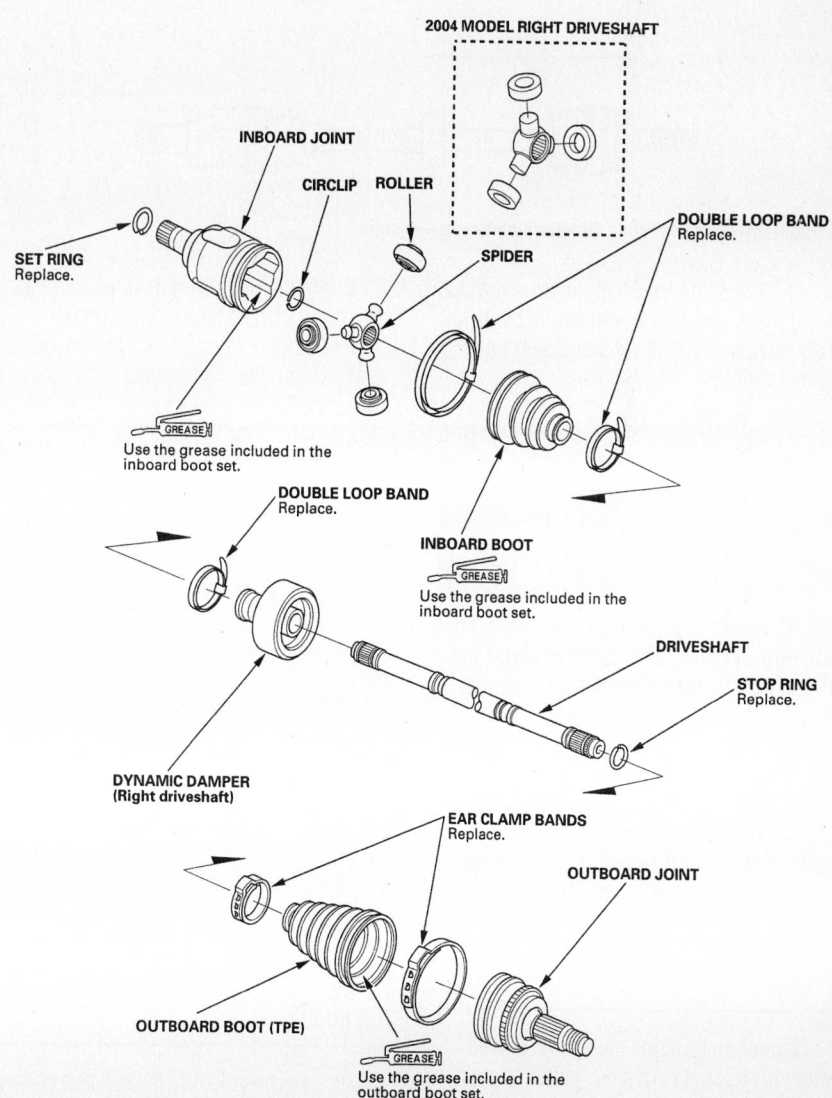

Exploded view of the driveshaft/CV-joint assembly—Insight

67162-INSI-G38

Left driveshaft: 481−486 mm (18.9−19.1 in.)
Right driveshaft: 761−766 mm (30.0−30.2 in.)

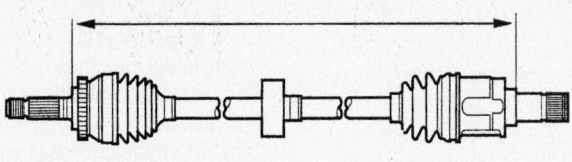

67162-INSI-G39

Drive shaft assembled length dimensions—Insight

Standard: 415.5−419.5 mm (16.4−16.5 in.)

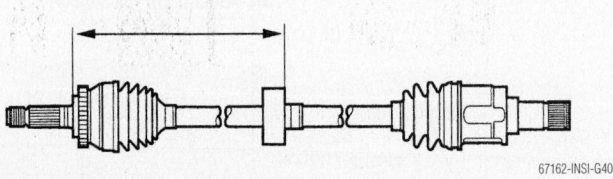

67162-INSI-G40

Right driveshaft dynamic damper dimensions—Insight

8. Adjust the length of the driveshaft to the dimensions shown. Adjust the boots to half-way between full extension and full compression.

9. On the right driveshaft, position the dynamic damper to the length shown.
10. Fit the boot ends onto the driveshaft.
11. Crimp the clamps onto the boots.

12. Install the axle halfshaft to the vehicle.

Inboard Joint

1. Before servicing the vehicle, refer to the precautions in the beginning of this section. Also see the IMA system disabling before proceeding with any procedure.
2. Remove or disconnect the following:
   - Axle halfshaft from the vehicle.
   - Circlip from the joint
   - Inboard joint boot clamps and push the boot back
   - Inboard joint housing from the axle
   - Rollers from the spider
   - Snapring and the spider from the axle shaft
   - Inboard joint boot

To install:

➡**Use new circlips and boot clamps for assembly.**

3. Install or connect the following:
   - Inboard joint boot and clamps to the axle shaft
   - Spider with a new snapring
   - Rollers to the spider
4. Fill the joint housing with grease and install it.
5. Fill the inboard joint boot with grease and install the boot clamps.
6. Install the axle halfshaft to the vehicle.

STEERING AND SUSPENSION

Air Bag

✳✳ CAUTION

Some vehicles are equipped with an air bag system. The system must be disarmed before performing service on, or around, system components, the steering column, instrument panel components, wiring and sensors. Failure to follow the safety precautions and the disarming procedure could result in accidental air bag deployment, possible injury and unnecessary system repairs.

PRECAUTIONS

Several precautions must be observed when handling the inflator module to avoid accidental deployment and possible personal injury.

- Never carry the inflator module by the wires or connector on the underside of the module.
- When carrying a live inflator module, hold securely with both hands, and ensure that the bag and trim cover are pointed away.
- Place the inflator module on a bench or other surface with the bag and trim cover facing up.
- With the inflator module on the bench, never place anything on or close to the module which may be thrown in the event of an accidental deployment.

Before servicing the vehicle, also make sure to refer to the precautions in the beginning of this section as well.

DISARMING

Disconnect and isolate the negative battery cable. Wait 3 minutes for the system capacitor to discharge before performing any service.

Electronic Power Steering Gear

REMOVAL & INSTALLATION

➡**Some steering assembly bolts are Dacro coated. Always replace them with the same size and specified bolts.**

✳✳ WARNING

Do not permit the steering wheel to turn whenever the steering gear is disconnected from the steering column. Damage to the air bag wiring can result.

1. Before servicing the vehicle, refer to the precautions in the beginning of this section. Also see the IMA system disabling before proceeding with any procedure.
2. Center the steering wheel and lock it in position.

3. Raise and support the vehicle.

4. Remove or disconnect the following:
- Negative battery cable
- Air bag and steering wheel
- Steering flexible joint
- Front wheels
- Battery box
- Front damper base beam in the engine compartment
- Unlock the locking tabs on the tie rod stop plate
- Bolts, stop plate, tie rod plate and washers
- Tie rod ends from center of gear
- Steering gear ground cable and electronic connectors
- Steering gear mounting bolts

5. Slide the steering gear to the passenger side until the left end clears the subframe, then slide the gear to the driver side and out of the vehicle.

To install:

6. Installation is the reverse of the removal procedure, noting the following:

a. Use Dacro bolts to mount the steering gear and tighten them to 43 ft. lbs. (58 Nm).

b. Tighten the stop plate bolts to 80 ft. lbs. (108 Nm). Ensure the bolt head flats line up with the lock tabs, then bend the lock tabs over the bolt heads.

Strut Damper and Spring

REMOVAL & INSTALLATION

Front

1. Before servicing the vehicle, refer to the precautions in the beginning of this section. Also see the IMA system disabling before proceeding with any procedure.

2. Remove or disconnect the following:
- Front wheel
- Tie rod end nuts
- Upper ball joint
- Stabilizer link
- Brake hose retainer

- Speed sensor bracket
- Lower damper pinch bolts
- Strut top nut
- Strut and damper

To install:

➡ **Use new self-locking fasteners for assembly.**

3. Install or connect the following:
- Strut. Loosely tighten the mounting nuts and pinch bolts
- Stabilizer link and hand tighten the nut.

4. Place a jack under the lower arm and raise the suspension to load it.

5. Tighten the strut top nuts to 40 ft. lbs. (54 Nm).

6. Tighten the lower pinch bolts to 72 ft. lbs. (98 Nm).

7. Tighten the stabilizer link nut to 22 ft. lbs. (29 Nm).

8. Tighten the tie rod nut to 47 ft. lbs. (64 Nm).

9. Install the speed sensor bracket and brake line retainer.

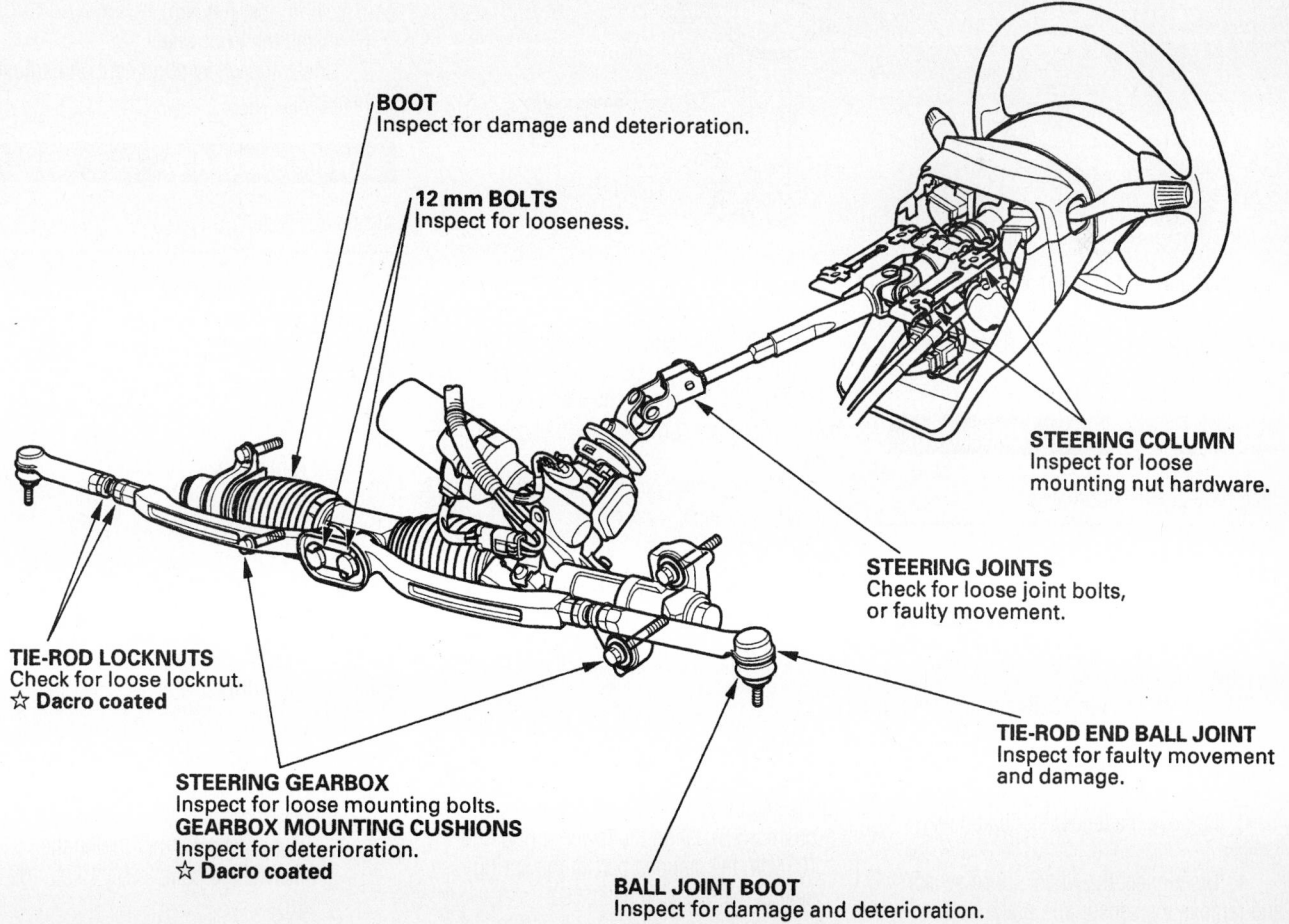

BOOT
Inspect for damage and deterioration.

12 mm BOLTS
Inspect for looseness.

STEERING COLUMN
Inspect for loose mounting nut hardware.

STEERING JOINTS
Check for loose joint bolts, or faulty movement.

TIE-ROD LOCKNUTS
Check for loose locknut.
☆ **Dacro coated**

STEERING GEARBOX
Inspect for loose mounting bolts.
GEARBOX MOUNTING CUSHIONS
Inspect for deterioration.
☆ **Dacro coated**

TIE-ROD END BALL JOINT
Inspect for faulty movement and damage.

BALL JOINT BOOT
Inspect for damage and deterioration.

67162-INSI-G41

Exploded view of steering gear —Insight

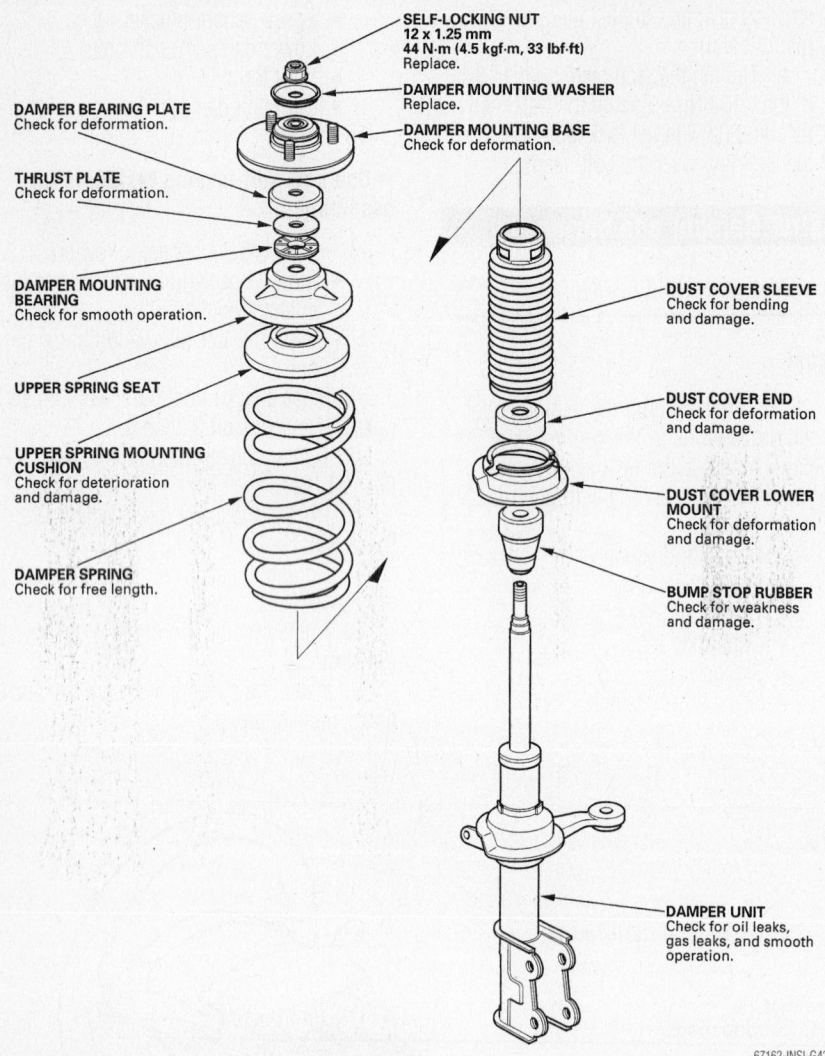

SELF-LOCKING NUT
12 x 1.25 mm
44 N·m (4.5 kgf·m, 33 lbf·ft)
Replace.

DAMPER MOUNTING WASHER
Replace.

DAMPER MOUNTING BASE
Check for deformation.

DAMPER BEARING PLATE
Check for deformation.

THRUST PLATE
Check for deformation.

DAMPER MOUNTING BEARING
Check for smooth operation.

UPPER SPRING SEAT

UPPER SPRING MOUNTING CUSHION
Check for deterioration and damage.

DAMPER SPRING
Check for free length.

DUST COVER SLEEVE
Check for bending and damage.

DUST COVER END
Check for deformation and damage.

DUST COVER LOWER MOUNT
Check for deformation and damage.

BUMP STOP RUBBER
Check for weakness and damage.

DAMPER UNIT
Check for oil leaks, gas leaks, and smooth operation.

67162-INSI-G42

Exploded view of front strut and damper—Insight

10. Install the front wheels and check the wheel alignment.

Coil Spring and Shock Absorber

REMOVAL & INSTALLATION

Rear

1. Before servicing the vehicle, refer to the precautions in the beginning of this section. Also see the IMA system disabling before proceeding with any procedure.

2. Raise and support the rear of the vehicle.

3. Remove the fender skirts and rear wheels.

4. Disconnect the wheel speed sensors and remove the sensor and brake line brackets.

5. Place a jack under each end of the rear axle beam.

6. Remove the lower shock absorber mounting bolts.

7. Slowly lower the jacks and remove the coil springs, spring cushion and bump stop.

8. Remove the shock upper bolts and remove the shock.

To install:

➡Use a new self-locking nut.

9. Installation is the reverse of the removal procedure, noting the following items:

   a. Install the shock and upper mounting bolts.

   b. Align the upper spring end (A) with the stepped part of the upper spring cushion (B). Align the lower spring end (C) with the stepped part of the spring seat of the rear axle beam (D).

   c. Install the lower shock bolt loosely, raise and load the suspension with the jacks then tighten the lower bolts to 43 ft.

lbs. (59 Nm). Tighten the upper bolts to 40 ft. lbs. (54 Nm).

10. Check the wheel alignment and adjust as necessary.

Upper or Lower Ball Joint

REMOVAL & INSTALLATION

1. Before servicing the vehicle, refer to the precautions in the beginning of this section. Also see the IMA system disabling before proceeding with any procedure.

2. Remove or disconnect the following:

- Front wheel
- Ball joints nuts and cotter pin

3. Press the ball joint out of the steering arm or suspension arm.

To install:

➡Use new ball joint nuts and cotter pins.

4. Press the ball joint into the steering arm or suspension arm.

5. Install the ball joint nuts. Tighten the nuts to 47 ft. lbs. (64 Nm).

6. Install the front wheel.

7. Check the wheel alignment and adjust as necessary.

Lower Control Arm

REMOVAL & INSTALLATION

1. Before servicing the vehicle, refer to the precautions in the beginning of this section. Also see the IMA system disabling before proceeding with any procedure.

2. Remove or disconnect the following:

- Front wheel
- Lower ball joint
- Lower arm cover
- Front inner flange bolt and nut
- Lower control arm

To install:

➡Use new self-locking nuts and split pins for assembly.

3. Loosely install the mounting bolts and raise and load the suspension before final tightening of the bolts.

4. Install or connect the following:

- Lower control arm. Tighten the inner flange bolts to 51 ft. lbs. (69 Nm).
- Upper flange bolt. Tighten the bolt to 54 ft. lbs. (74 Nm).
- Lower arm cover

- Lower ball joint. Tighten the nuts to 47 ft. lbs. (64 Nm).
- Front wheel

5. Check the wheel alignment and adjust as necessary.

Steering Knuckle, Hub and Bearing

ADJUSTMENT

The wheel bearings are not adjustable.

REMOVAL & INSTALLATION

Front

1. Before servicing the vehicle, refer to the precautions in the beginning of this section. Also see the IMA system disabling before proceeding with any procedure.

2. Remove or disconnect the following:
- Front wheel
- Spindle nut
- Brake hose bracket
- Brake caliper and rotor
- Wheel speed sensor, if equipped
- Lower ball joint
- Steering knuckle

3. Press the hub out of the knuckle.

4. Press the inner bearing race out of the hub.

5. Remove the splash guard.

6. Remove the bearing unit from the knuckle.

To install:

➡**Use new ball joint nuts, split pins, snapring and a new spindle nut for assembly.**

7. Install the hub unit and tighten the bolts to 43 ft. lbs. (59 Nm).

8. Install the splash guard.

9. Press the bearing unit into the hub.

10. Install the splash guard.

11. Press the hub into the bearing.

12. Install or connect the following:
- Steering knuckle. Tighten the knuckle-to-damper unit bolts to 72 ft. lbs. (98 Nm)
- Lower ball joint. Tighten the nut 40–47 ft. lbs. (54–64 Nm).
- Wheel speed sensor, if equipped
- Brake caliper and rotor. Tighten the caliper bracket bolts to 80 ft. lbs. (108 Nm).
- Brake hose bracket
- Spindle nut. Tighten the nut to 134 ft. lbs. (181 Nm).
- Front wheel

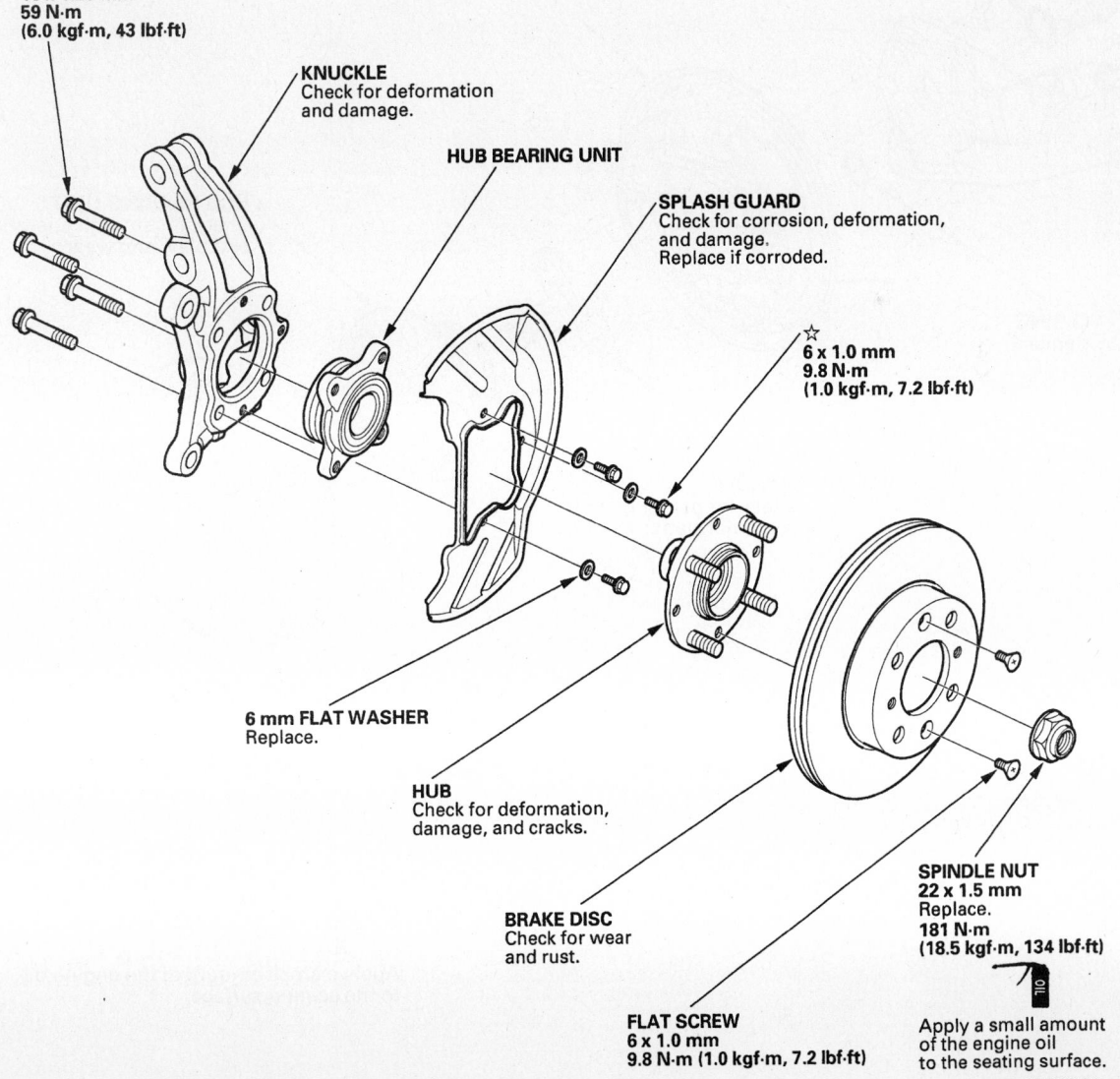

10 x 1.25 mm
59 N·m
(6.0 kgf·m, 43 lbf·ft)

KNUCKLE
Check for deformation and damage.

HUB BEARING UNIT

SPLASH GUARD
Check for corrosion, deformation, and damage.
Replace if corroded.

☆
6 x 1.0 mm
9.8 N·m
(1.0 kgf·m, 7.2 lbf·ft)

6 mm FLAT WASHER
Replace.

HUB
Check for deformation, damage, and cracks.

BRAKE DISC
Check for wear and rust.

FLAT SCREW
6 x 1.0 mm
9.8 N·m (1.0 kgf·m, 7.2 lbf·ft)

SPINDLE NUT
22 x 1.5 mm
Replace.
181 N·m
(18.5 kgf·m, 134 lbf·ft)

Apply a small amount of the engine oil to the seating surface.

67162-INSI-G43

Exploded view of front steering knuckle, hub and wheel bearing—Insight

13. Check the wheel alignment and adjust as necessary.

Rear

1. Before servicing the vehicle, refer to the precautions in the beginning of this section. Also see the IMA system disabling before proceeding with any procedure.

2. Remove or disconnect the following:
- Rear wheel
- Fender skirts
- Spindle nut
- Brake drum
- Hub and bearing assembly

To install:

➡ Use a new spindle nut for assembly.

3. Install or connect the following:
- Hub and bearing assembly. Tighten the flange bolts to 76 ft. lbs. (103 Nm).
- Spindle nut. Tighten the nut to 119 ft. lbs. (162 Nm).
- Brake drum
- Rear wheel

Rear Axle Beam

REMOVAL & INSTALLATION

1. Before servicing the vehicle, refer to the precautions in the beginning of this section. Also see the IMA system disabling before proceeding with any procedure.

2. Place a jack at each end of the rear axle beam.

3. Remove or disconnect the following:

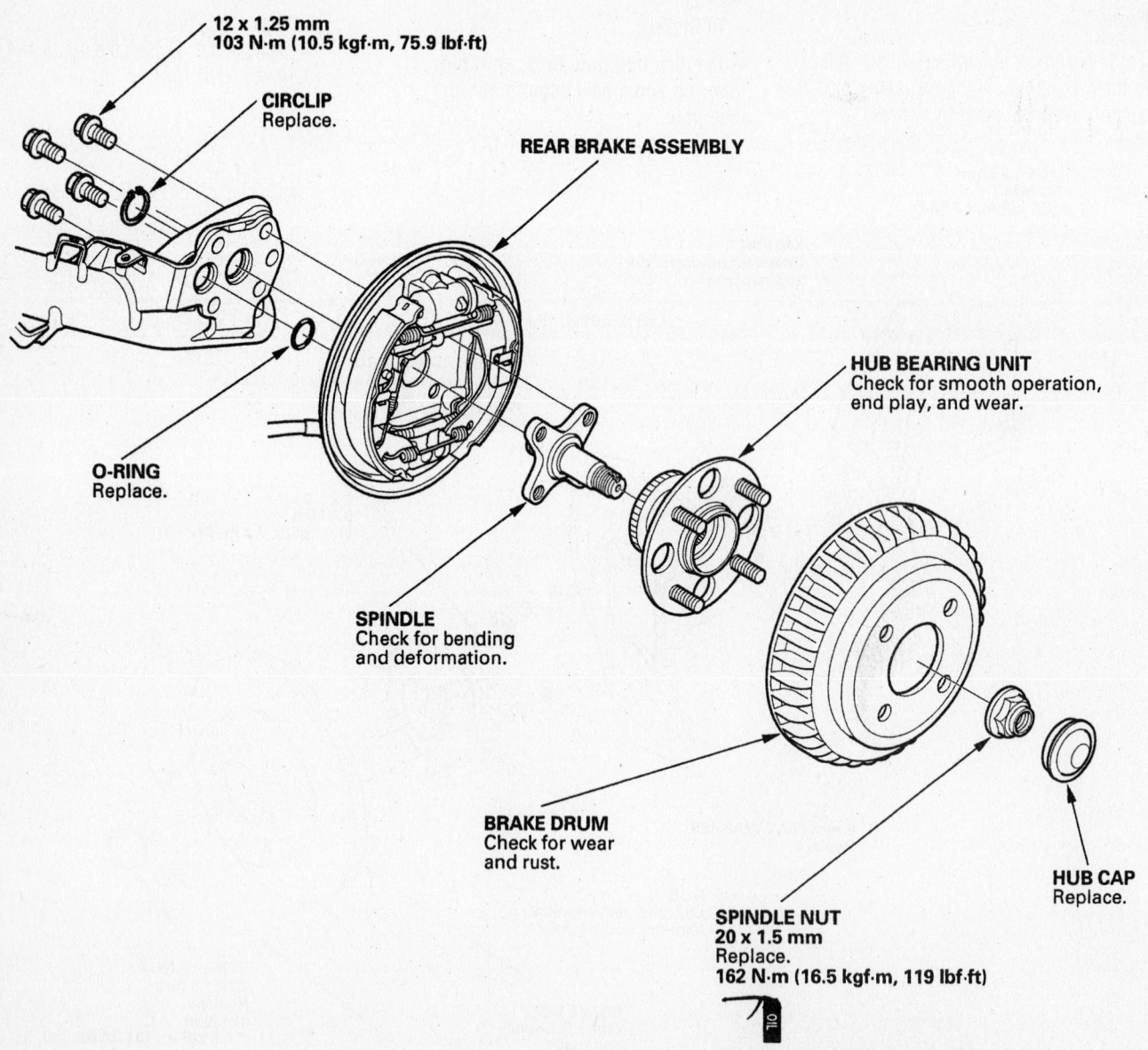

12 x 1.25 mm
103 N·m (10.5 kgf·m, 75.9 lbf·ft)

CIRCLIP
Replace.

REAR BRAKE ASSEMBLY

HUB BEARING UNIT
Check for smooth operation, end play, and wear.

O-RING
Replace.

SPINDLE
Check for bending and deformation.

BRAKE DRUM
Check for wear and rust.

SPINDLE NUT
20 x 1.5 mm
Replace.
162 N·m (16.5 kgf·m, 119 lbf·ft)

Apply a small amount of the engine oil to the seating surface.

HUB CAP
Replace.

Exploded view of rear hub and bearing—Insight

- Rear wheel
- Fender skirts
- Wheel speed sensors and remove the sensor and brake line brackets
- Parking brake cable brackets
- Speed sensor harnesses
- Lower shock bolts
- Coil springs
- Spring bump stop

- Spindle nut
- Brake drum
- Bearing hub
- Brake assembly
- Axle beam

To install:

4. Installation is the reverse of the removal procedure, noting the following items:

a. Tighten the axle beam bolts to 43 ft. lbs. (59 Nm).

b. Install the lower shock bolt loosely, raise and load the suspension with the jacks then tighten the lower bolts to 43 ft. lbs. (59 Nm). Tighten the upper bolts to 40 ft. lbs. (54 Nm).

5. Check the wheel alignment and adjust as necessary.

BRAKES

Brake Caliper

REMOVAL & INSTALLATION

1. Before servicing the vehicle, refer to the precautions in the beginning of this section. Also see the IMA system disabling before proceeding with any procedure.

2. Remove or disconnect the following:
- Wheel
- Brake hose bracket
- Caliper mounting bolts

- Caliper

3. Installation is the reverse of removal. Torque the bolts to 16 ft. lbs. (22 Nm).

Disc Brake Pads

REMOVAL & INSTALLATION

1. Remove the lower bolt.
2. Pivot the caliper up and hold the pads.
3. Remove the pad springs.
4. Remove the pads and shims.
5. Remove the pad retainers.
6. Installation is the reverse of removal.

Coat both sides of the shims and the backs of the pads with brake grease. Torque the lower bolt to 16 ft. lbs. (22 Nm).

Brake Drums

REMOVAL & INSTALLATION

1. Raise and safely support the vehicle. Release the parking brake.
2. Remove the rear wheels.
3. Use chalk to mark the brake drum to one of the wheel studs as an index mark for reinstallation.

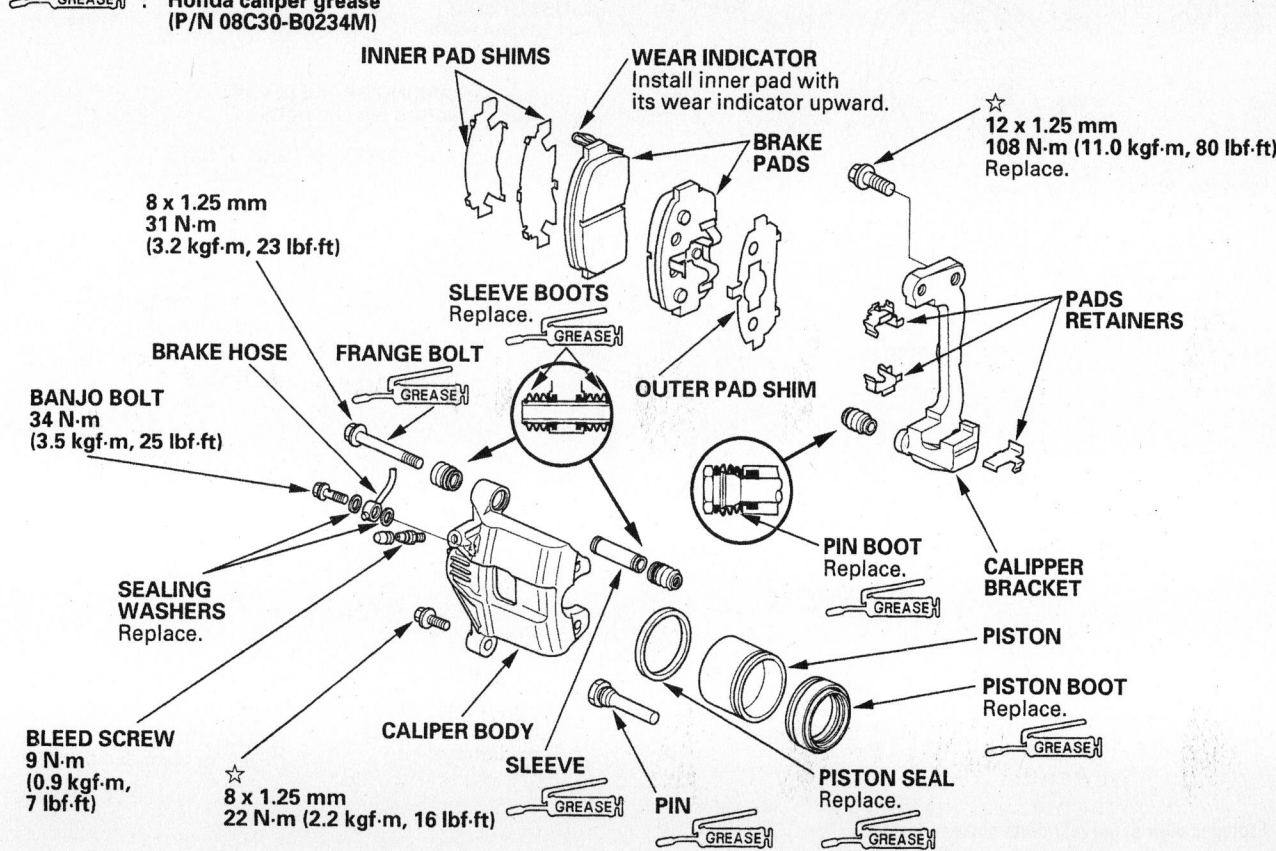

GREASE : **Honda caliper grease (P/N 08C30-B0234M)**

INNER PAD SHIMS

WEAR INDICATOR
Install inner pad with its wear indicator upward.

BRAKE PADS

☆
12 x 1.25 mm
108 N·m (11.0 kgf·m, 80 lbf·ft)
Replace.

8 x 1.25 mm
31 N·m
(3.2 kgf·m, 23 lbf·ft)

SLEEVE BOOTS
Replace.
GREASE

OUTER PAD SHIM

PADS RETAINERS

BRAKE HOSE   FRANGE BOLT
GREASE

BANJO BOLT
34 N·m
(3.5 kgf·m, 25 lbf·ft)

PIN BOOT
Replace.
GREASE

CALIPPER BRACKET

SEALING WASHERS
Replace.

PISTON

PISTON BOOT
Replace.
GREASE

BLEED SCREW
9 N·m
(0.9 kgf·m, 7 lbf·ft)

☆
8 x 1.25 mm
22 N·m (2.2 kgf·m, 16 lbf·ft)

CALIPER BODY

SLEEVE
GREASE

PIN
GREASE

PISTON SEAL
Replace.
GREASE

Exploded view of the front disc brakes—Insight

67162-INSI-G45

4. Remove the retaining screw that holds the brake drum to the axle flange.

5. Pull the brake drum from the axle flange.

To install:

6. Align the index mark and install the brake drum to the axle flange.

7. Install the retaining screw to secure the brake drum to the axle flange.

8. Install the rear wheels.

Rear Brake Shoes

REMOVAL & INSTALLATION

1. Raise and safely support the vehicle.
2. Remove the rear wheels.
3. Remove the brake drums.
4. Remove the brake return springs.
5. Remove the leading shoe holding pin and spring, and then the leading shoe.

6. Remove the self-adjuster and the adjuster lever.

7. Remove the trailing shoe holding pin and spring.

8. Disconnect the parking brake cable from the trailing shoe and remove the trailing shoe. Remove the parking brake lever from the trailing shoe.

To install:

9. Attach the parking brake lever to the trailing shoe.

10. Connect the parking brake cable to the parking brake lever.

11. Apply a thin coat of high temperature grease to the shoe contact points on the brake backing plate contact surface (B), and self-adjuster (D).

12. Position the trailing shoe on the backing plate and install the hold-down pin, spring, and retainer. Be careful not to stretch the return spring when fitting the shoes onto the backing plate.

13. Connect the upper return spring and the leading shoe to the trailing shoe and position the leading brake shoe on the backing plate.

14. Install the adjuster assembly and the hold-down pin, spring, and retainer.

15. Use a brake spring tool to install the lower return spring.

16. Install the self-adjuster lever and adjuster spring.

17. Check the brake drum for scoring or other wear. Machine or replace as necessary. Check the maximum brake drum diameter specification when machining.

18. Install the rear wheels. Lower the vehicle.

19. Road-test the vehicle.

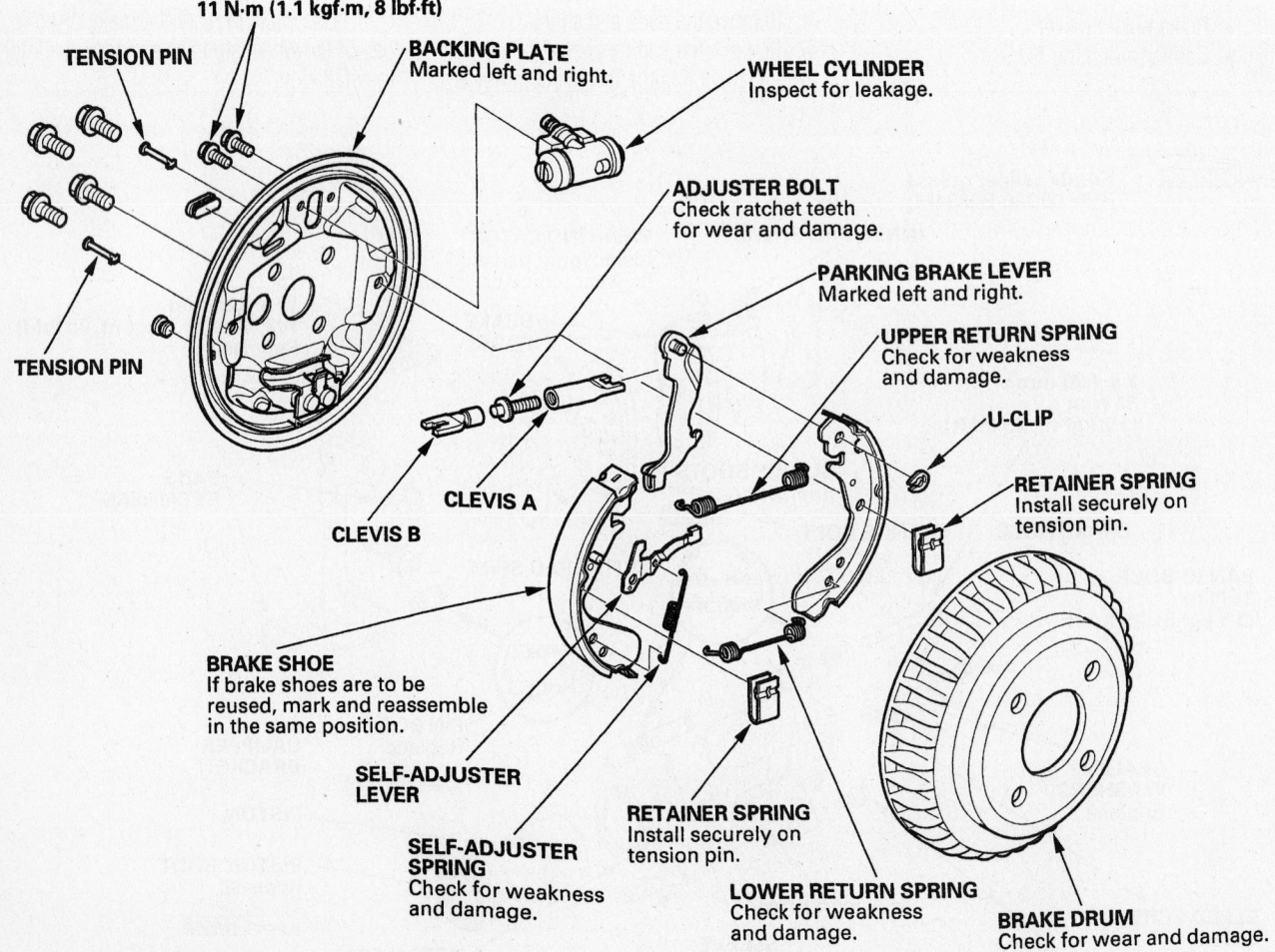

11 N·m (1.1 kgf·m, 8 lbf·ft)

TENSION PIN

BACKING PLATE
Marked left and right.

WHEEL CYLINDER
Inspect for leakage.

TENSION PIN

ADJUSTER BOLT
Check ratchet teeth for wear and damage.

PARKING BRAKE LEVER
Marked left and right.

UPPER RETURN SPRING
Check for weakness and damage.

U-CLIP

CLEVIS A

CLEVIS B

RETAINER SPRING
Install securely on tension pin.

BRAKE SHOE
If brake shoes are to be reused, mark and reassemble in the same position.

SELF-ADJUSTER LEVER

SELF-ADJUSTER SPRING
Check for weakness and damage.

RETAINER SPRING
Install securely on tension pin.

LOWER RETURN SPRING
Check for weakness and damage.

BRAKE DRUM
Check for wear and damage.

67162-INSI-G46

Exploded view of the rear drum brakes—Insight

SPECIFICATIONS AND MAINTENANCE CHARTS

ENGINE AND VEHICLE IDENTIFICATION CHART

Code	Liters (cc)	Cu. In.	Cyl.	Fuel Sys.	Engine Type	Eng. Mfg.
J35A1	3.5 (3471)	212	6	SMFI	SOHC	Honda
J35A4	3.5 (3471)	212	6	SMFI	SOHC	Honda

Code ①	Year
1	2001
2	2002
3	2003
4	2004
5	2005

SOHC: Single Overhead Cam

SMFI: Sequential Multi-port Fuel Injection

① 10th position of VIN

67162-ODYS-C01

GENERAL ENGINE SPECIFICATIONS

Year	Model	Engine Displacement Liters	Engine ID	Net Horsepower @ rpm	Net Torque @ rpm (ft. lbs.)	Bore x Stroke (in.)	Compression Ratio	Oil Pressure @ rpm
2001	Odyssey	3.5	J35A1	240@5200	242@3500	3.50x3.66	9.4:1	71@3000
2002	Odyssey	3.5	J35A4	240@5200	242@3500	3.50x3.66	10.0:1	71@3000
2003	Odyssey	3.5	J35A4	250@5200	242@4500	3.50x3.66	10.0:1	71@3000
2004	Odyssey	3.5	J35A4	250@5200	242@4500	3.50x3.66	10.0:1	71@3000

SMFI: Sequential Multi-port Fuel Injection

67162-ODYS-C02

ENGINE TUNE-UP SPECIFICATIONS

Year	Engine Displacement Liters	Engine ID	Spark Plug Gap (in.)	Ignition Timing (deg.) MT	Ignition Timing (deg.) AT	Fuel Pump (psi)	Idle Speed (rpm) MT	Idle Speed (rpm) AT	Valve Clearance (in.) In.	Valve Clearance (in.) Ex.
2001	3.5	J35A1	0.039-0.043	—	10B	43-50	—	680-780	0.008-0.009	0.011-0.013
2002	3.5	J35A4	0.039-0.043	—	10B	48-54	—	680-780	0.008-0.009	0.011-0.013
2003	3.5	J35A4	0.039-0.040	—	10B	41-48	—	680-780	0.008-0.009	0.011-0.013
2004	3.5	J35A4	0.039-0.040	—	10B	41-48	—	680-780	0.008-0.009	0.011-0.013

NOTE: The Vehicle Emission Control Information label often reflects changes made during production and must be used if they differ from this chart.

NOTE: The fuel pressure readings are given with the vacuum hose connected to the regulator and the engine running

B: Before top dead center

67162-ODYS-C03

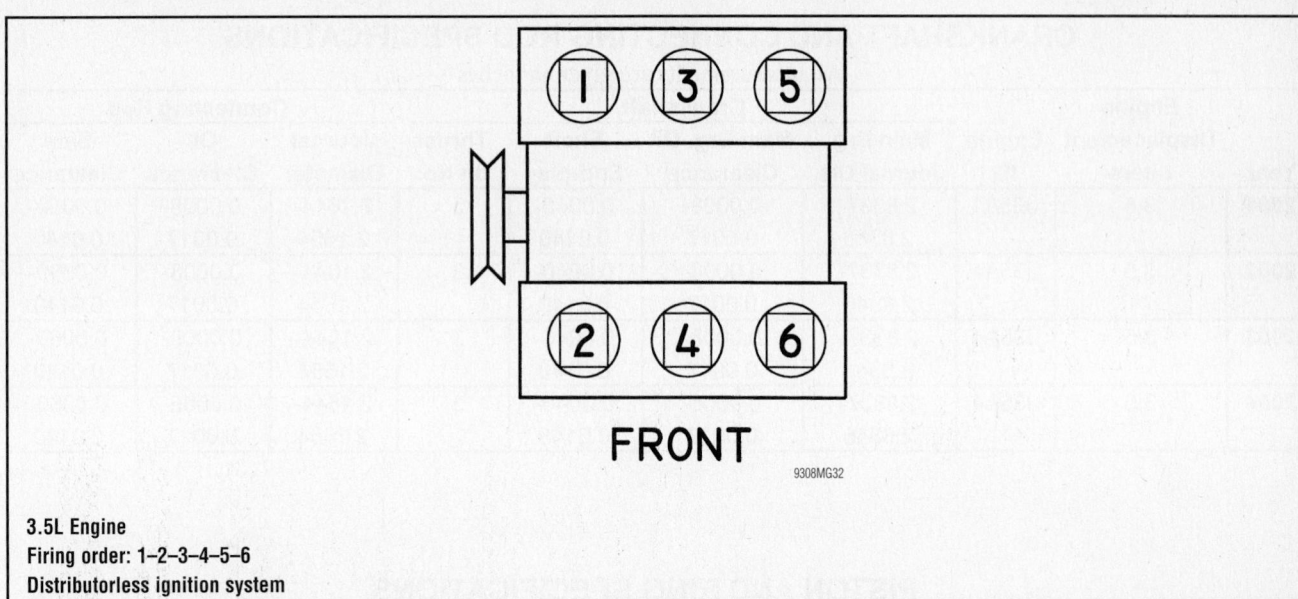

3.5L Engine
Firing order: 1–2–3–4–5–6
Distributorless ignition system

9308MG32

CAPACITIES

Year	Model	Engine Displacement Liters	Engine ID	Engine Oil with Filter (qts.)	Transmission (qts.) 5-Spd	Auto.	Fuel Tank (gal.)	Cooling System (qts.)
2001	Odyssey	3.5	J35A1	4.6	—	6.2	20.0	7.0
2002	Odyssey	3.5	J35A4	4.6	—	8.3	20.0	7.0
2003	Odyssey	3.5	J35A4	4.6	—	8.3	20.0	8.0
2004	Odyssey	3.5	J35A4	4.6	—	8.3	20.0	8.0

NOTE: All capacities are approximate. Add fluid gradually and check to be sure a proper fluid level is obtained.

67162-ODYS-C04

VALVE SPECIFICATIONS

Year	Engine Displacement Liters	Engine ID	Seat Angle (deg.)	Face Angle (deg.)	Spring Test Pressure (lbs. @ in.)	Spring Installed Height (in.)	Stem-to-Guide Clearance (in.) Intake	Exhaust	Stem Diameter (in.) Intake	Exhaust
2001	3.5	J35A1	45	45	NA	①	0.0008-0.0018	0.0022-0.0031	0.2159-0.2163	0.2146-0.2150
2002	3.5	J35A4	45	45	NA	②	0.0008-0.0018	0.0022-0.0031	0.2159-0.2163	0.2146-0.2150
2003	3.5	J35A4	45	45	NA	②	0.0008-0.0018	0.0022-0.0031	0.2159-0.2163	0.2146-0.2150
2004	3.5	J35A4	45	45	NA	②	0.0008-0.0018	0.0022-0.0031	0.2159-0.2163	0.2146-0.2150

NA: Not Available

① Valve spring free length:
   Intake: 1.9713 in.
   Exhaust: 2.1060 in.

② Valve spring free length:
   Intake: 2.0290 in.
   Exhaust: 2.1000 in.

67162-ODYS-C05

CRANKSHAFT AND CONNECTING ROD SPECIFICATIONS

All measurements are given in inches

Year	Engine Displacement Liters	Engine ID	Crankshaft				Connecting Rod		
			Main Brg. Journal Dia.	Main Brg. Oil Clearance	Shaft End-play	Thrust on No.	Journal Diameter	Oil Clearance	Side Clearance
2001	3.5	J35A1	2.8337-2.8346	0.0008-0.0017	0.0040-0.0140	3	2.1644-2.1654	0.0008-0.0017	0.0060-0.0140
2002	3.5	J35A4	2.8337-2.8346	0.0008-0.0017	0.0040-0.0140	3	2.1644-2.1654	0.0008-0.0017	0.0060-0.0140
2003	3.5	J35A4	2.8337-2.8346	0.0008-0.0017	0.0040-0.0140	3	2.1644-2.1654	0.0008-0.0017	0.0060-0.0140
2004	3.5	J35A4	2.8337-2.8346	0.0008-0.0017	0.0040-0.0140	3	2.1644-2.1654	0.0008-0.0017	0.0060-0.0140

67162-ODYS-C06

PISTON AND RING SPECIFICATIONS

All measurements are given in inches

Year	Engine Displacement Liters	Engine ID	Piston Clearance	Ring Gap			Ring Side Clearance		
				Top Compression	Bottom Compression	Oil Control	Top Compression	Bottom Compression	Oil Control
2001	3.5	J35A1	0.0006-0.0016	0.0080-0.0140	0.0160-0.0220	0.0080-0.0280	0.0014-0.0024	0.0012-0.0022	NA
2002	3.5	J35A4	0.0006-0.0016	0.0080-0.0140	0.0160-0.0220	0.0080-0.0280	0.0022-0.0031	0.0012-0.0022	NA
2003	3.5	J35A4	0.0006-0.0016	0.0080-0.0140	0.0160-0.0220	0.0080-0.0280	0.0022-0.0031	0.0012-0.0022	NA
2004	3.5	J35A4	0.0006-0.0016	0.0080-0.0140	0.0160-0.0220	0.0080-0.0280	0.0022-0.0031	0.0012-0.0022	NA

NA: Not Applicable

67162-ODYS-C07

TORQUE SPECIFICATIONS

All readings in ft. lbs.

Year	Engine Displacement Liters	Engine ID	Cylinder Head Bolts	Main Bearing Bolts	Rod Bearing Bolts	Crankshaft Damper Bolts	Flywheel Bolts	Manifold		Spark Plugs	Oil Pan Drain Plug
								Intake	Exhaust		
2001	3.5	J35A1	①	②	③	181	54	16	23	13	29
2002	3.5	J35A4	①	②	③	181	54	16	23	13	29
2003	3.5	J35A4	①	②	③	181	54	16	23	13	29
2004	3.5	J35A4	①	②	③	181	54	16	23	13	29

NOTE: Dip main bearing bolts and crankshaft damper bolt in clean engine oil prior to tightening.

① Step 1: 29 ft. lbs.
Step 2: 51 ft. lbs.
Step 3: 72 ft. lbs.

② 11mm bolt 56 ft. lbs.
10mm bolt 36 ft. lbs.

③ Step 1: 14 ft. lbs.
Step 2: 90 degrees

67162-ODYS-C08

WHEEL ALIGNMENT

Year	Model		Caster Range (+/-Deg.)	Caster Preferred Setting (Deg.)	Camber Range (+/-Deg.)	Camber Preferred Setting (Deg.)	Toe-in (in.)
2001	Odyssey	F	1.00	+2.07	1.00	0	0+/-0.08
		R	—	—	0.75	-0.50	0+/-0.08
2002	Odyssey	F	1.00	+2.07	1.00	0	0+/-0.08
		R	—	—	0.75	-0.50	0+/-0.08
2003	Odyssey	F	1.00	+2.07	1.00	0	0+/-0.08
		R	—	—	0.75	-0.50	0+/-0.08
2004	Odyssey	F	1.00	+2.07	1.00	0	0+/-0.08
		R	—	—	0.75	-0.50	0+/-0.08

67162-ODYS-C09

TIRE, WHEEL AND BALL JOINT SPECIFICATIONS

Year	Model	OEM Tires Standard	OEM Tires Optional	Tire Pressures (psi) Front	Tire Pressures (psi) Rear	Wheel Size	Ball Joint Inspection	Lugnut Torque (ft. lbs.)
2001	Odyssey	P215/65R16	None	32	32	6JJ	NS	80
2002	Odyssey	P215/65R16	None	32	32	6JJ	NS	80
2003	Odyssey	P225/60R16	None	32	32	6.5JJ	NS	80
2004	Odyssey	P225/60R16	None	32	32	6.5JJ	NS	80

OEM: Original Equipment Manufacturer

PSI: Pounds Per Square Inch

STD: Standard

OPT: Optional

NS: Not specified by manufacturer

67162-ODYS-C10

BRAKE SPECIFICATIONS

All measurements in inches unless noted

Year	Model		Brake Disc Original Thickness	Brake Disc Minimum Thickness	Brake Disc Maximum Runout	Brake Drum Diameter Original Inside Diameter	Brake Drum Diameter Max. Wear Limit	Brake Drum Diameter Maximum Machine Diameter	Minimum Lining Thickness Front	Minimum Lining Thickness Rear	Brake Caliper Bracket Bolts (ft. lbs.)	Brake Caliper Mounting Bolts (ft. lbs.)
2001	Odyssey	F	1.100	1.020	0.004	—	—	—	0.060	—	80	27
		R	—	—	—	9.996	10.04	—	—	0.080	—	—
2002	Odyssey	F	1.100	1.020	0.004	—	—	—	0.060	—	80	27
		R	0.440	0.350	0.004	8.268 ①	8.272	—	0.060	0.040	41	27
2003	Odyssey	F	1.100	1.020	0.004	—	—	—	0.060	—	80	20
		R	0.440	0.350	0.004	8.268 ①	8.272	—	0.060	0.040	41	27
2004	Odyssey	F	1.100	1.020	0.004	—	—	—	0.060	—	80	20
		R	0.440	0.350	0.004	8.268 ①	8.272	—	0.060	0.040	41	27

F: Front

R: Rear

① Parking brake drum

67162-ODYS-C11

SCHEDULED MAINTENANCE INTERVALS
Honda Odyssey

TO BE SERVICED	TYPE OF SERVICE	VEHICLE MILEAGE INTERVAL (x1000)															
		7.5	15	22.5	30	37.5	45	52.5	60	67.5	75	82.5	90	97.5	105	112	120
Accessory drive belts	I & A				✓				✓				✓				✓
Air cleaner element	R				✓				✓				✓				✓
Air conditioning filter	R				✓				✓				✓				✓
Brake fluid	R					✓							✓				
Brake hoses & lines	I		✓		✓		✓		✓		✓		✓		✓		✓
Cooling system hoses	I		✓		✓		✓		✓		✓		✓		✓		✓
Engine coolant	R					✓							✓				
Engine oil	R	✓	✓	✓	✓	✓	✓	✓	✓	✓	✓	✓	✓	✓	✓	✓	✓
Engine oil and coolant levels	I	Inspect at each fuel stop															
Engine oil filter	R		✓		✓		✓		✓		✓		✓		✓		✓
Exhaust system	I		✓		✓		✓		✓		✓		✓		✓		✓
Fluid levels and condition	I		✓		✓		✓		✓		✓		✓		✓		✓
Front and rear brakes	I		✓		✓		✓		✓		✓		✓		✓		✓
Fuel lines & connection	I		✓		✓		✓		✓		✓		✓		✓		✓
Halfshaft boots	I		✓		✓		✓		✓		✓		✓		✓		✓
Idle speed	I & A													✓			
Parking brake system	I & A		✓		✓		✓		✓		✓		✓		✓		✓
Rear differential fluid	R												✓				
Rotate and inspect tires	I	✓	✓	✓	✓	✓	✓	✓	✓	✓	✓	✓	✓	✓	✓	✓	✓
Spark plugs	R				✓				✓				✓				✓
Supplemental Restraint system	I	Inspect the SRS 10 years after production															
Suspension components	I		✓		✓		✓		✓		✓		✓		✓		✓
Steering components	I		✓		✓		✓		✓		✓		✓		✓		✓
Timing balancer belt	R													✓			
Timing belt	R													✓			
Transmission fluid	R						✓				✓				✓		
Valve clearance	I				✓				✓				✓				✓
Water pump	S/I													✓			

R: Replace   I: Inspect   A: Adjust

FREQUENT OPERATION MAINTENANCE (SEVERE SERVICE)

If a vehicle is operated under any of the following conditions it is considered severe service:

- Towing a trailer or using a camper or car-top carrier.
- Repeated short trips of less than 5 miles in temperatures below freezing, or trips of less than 10 miles in any temperature.
- Extensive idling or low-speed driving for long distances as in heavy commercial use, such as delivery, taxi or police cars.
- Operating on rough, muddy or salt-covered roads.
- Operating on unpaved or dusty roads.
- Driving in extremely hot (over 90°) conditions.

Air cleaner element: replace every 15,000 miles

Engine oil and filter: replace every 3750 miles or 6 months, whichever occurs first.

Timing belt: replace every 60,000 miles if the vehicle is regularly driven in temperatures above 110°F or below -20°F.

Transmission fluid: replace every 30,000 miles.

Rear differential fluid: replace every 60,000 miles.

Front and rear brakes: inspect every 7500 miles or 6 months, whichever occurs first.

Locks and hinges: lubricate every 15,000 miles.

PRECAUTIONS

Before servicing any vehicle, please be sure to read all of the following precautions, which deal with personal safety, prevention of component damage, and important points to take into consideration when servicing a motor vehicle:

• Never open, service or drain the radiator or cooling system when the engine is hot; serious burns can occur from the steam and hot coolant.

• Observe all applicable safety precautions when working around fuel. Whenever servicing the fuel system, always work in a well-ventilated area. Do not allow fuel spray or vapors to come in contact with a spark, open flame, or excessive heat (a hot drop light, for example). Keep a dry chemical fire extinguisher near the work area. Always keep fuel in a container specifically designed for fuel storage; also, always properly seal fuel containers to avoid the possibility of fire or explosion. Refer to the additional fuel system precautions later in this section.

• Fuel injection systems often remain pressurized, even after the engine has been turned **OFF**. The fuel system pressure must be relieved before disconnecting any fuel lines. Failure to do so may result in fire and/or personal injury.

• Brake fluid often contains polyglycol ethers and polyglycols. Avoid contact with the eyes and wash your hands thoroughly after handling brake fluid. If you do get brake fluid in your eyes, flush your eyes with clean, running water for 15 minutes. If eye irritation persists, or if you have taken brake fluid internally, seek medical assistance IMMEDIATELY.

• The EPA warns that prolonged contact with used engine oil may cause a number of skin disorders, including cancer. You should make every effort to minimize your exposure to used engine oil. Protective gloves should be worn when changing oil. Wash your hands and any other exposed skin areas as soon as possible after exposure to used engine oil. Soap and water, or waterless hand cleaner should be used.

• All new vehicles are now equipped with an air bag system. The system must be disabled before performing service on or around system components, steering column, instrument panel components, wiring and sensors. Failure to follow safety and disabling procedures could result in accidental air bag deployment, possible personal injury and unnecessary system repairs.

• Always wear safety goggles when working with, or around, the air bag system. When carrying a non-deployed air bag, be sure the bag and trim cover are pointed away from your body. When placing a non-deployed air bag on a work surface, always face the bag and trim cover upward, away from the surface. This will reduce the motion of the module if it is accidentally deployed. Refer to the additional air bag system precautions later in this section.

• Clean, high quality brake fluid from a sealed container is essential to the safe and proper operation of the brake system. You should always buy the correct type of brake fluid for your vehicle. If the brake fluid becomes contaminated, completely flush the system with new fluid. Never reuse any brake fluid. Any brake fluid that is removed from the system should be discarded. Also, do not allow any brake fluid to come in contact with a painted surface; it will damage the paint.

• Never operate the engine without the proper amount and type of engine oil; doing so WILL result in severe engine damage.

• Timing belt maintenance is extremely important. Many models utilize an interference-type, non-freewheeling engine. If the timing belt breaks, the valves in the cylinder head may strike the pistons, causing potentially serious (also time-consuming and expensive) engine damage. Refer to the maintenance interval charts in the front of this manual for the recommended replacement interval for the timing belt, and to the timing belt section for belt replacement and inspection.

• Disconnecting the negative battery cable on some vehicles may interfere with the functions of the on-board computer system(s) and may require the computer to undergo a relearning process once the negative battery cable is reconnected.

• When servicing drum brakes, only disassemble and assemble one side at a time, leaving the remaining side intact for reference.

• Only an MVAC-trained, EPA-certified automotive technician should service the air conditioning system or its components.

ENGINE REPAIR

➡**Disconnecting the negative battery cable on some vehicles may interfere with the functions of the on board computer system. The computer may undergo a relearning process once the negative battery cable is reconnected.**

Distributor

The 3.5L engine is equipped with a Distributorless Ignition System (DIS).

REMOVAL & INSTALLATION

Alternator

REMOVAL

1. Before servicing the vehicle, refer to the precautions in the beginning of this section.

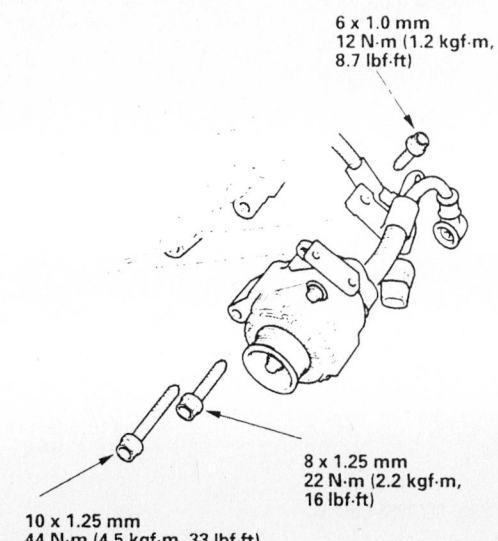

6 x 1.0 mm
12 N·m (1.2 kgf·m, 8.7 lbf·ft)

8 x 1.25 mm
22 N·m (2.2 kgf·m, 16 lbf·ft)

10 x 1.25 mm
44 N·m (4.5 kgf·m, 33 lbf·ft)

9358MG01

Exploded view of the alternator mounting

2. Remove or disconnect the following:
   - Negative battery cable
   - Accessory drive belt
   - Alternator wiring harness connectors
   - Alternator mounting bolts
   - Wiring harness clamp
   - Alternator

INSTALLATION

1. Install or connect the following:
   - Alternator
   - Wiring harness clamp. Tighten the bolt to 105 inch lbs. (12 Nm).
   - Alternator mounting bolts. Tighten the 10mm bolt to 33 ft. lbs. (44 Nm) and the 8mm bolt to 16 ft. lbs. (22 Nm).
   - Alternator wiring harness connectors. Tighten the battery terminal nut to 105 inch lbs. (12 Nm).
   - Accessory drive belt
   - Negative battery cable

Ignition Timing

ADJUSTMENT

The 3.5L engine is equipped with a Distributorless Ignition System (DIS). The ignition timing is controlled by the Powertrain Control module (PCM). No adjustment is necessary.

Engine Assembly

REMOVAL & INSTALLATION

➡**The engine and transaxle are removed from the vehicle as a unit.**

1. Before servicing the vehicle, refer to the precautions in the beginning of this section.
2. Drain the cooling system.
3. Drain the transaxle fluid.
4. Drain the engine oil.
5. Relieve fuel system pressure.
6. Remove or disconnect the following:
   - Negative battery cable
   - Intake manifold cover and ignition coil cover on 2002–04 models
   - Evaporative Emissions (EVAP) control canister or vacuum hose, whichever applies
   - Air intake duct
   - Battery
   - Left engine wire harness connectors
   - Relay bracket

- Battery tray
- Accessory drive belts
- Accelerator cable
- Cruise control cable
- Fuel lines
- Brake booster vacuum line
- Vacuum supply hose
- Powertrain Control Module (PCM) connectors and grommet. Pull the PCM harness through the firewall.
- Fuse/Relay box battery cable
- Ground cable
- Power steering pump
- Starter cable and harness clamp
- Radiator hoses
- Heater hoses
- Bypass hose
- Transaxle oil cooler lines
- Front wheels
- Splash shield
- Heated Oxygen Sensor (HO2S) connector
- Exhaust front pipe
- Stabilizer bar links
- Lower ball joints

7. Separate the inner CV-joints from the transaxle and support the axle halfshafts out of the work area with safety wire.
8. Remove or disconnect the following:
   - Shift cable bracket

- Shift cable cover
- Shift control lever with cable attached
- Power steering hose clamp and clips
- Transaxle lower front mount
- Transaxle lower rear mount
- Steering rack and pinion gear. Support the steering gear with safety wire.

9. Attach a hoist to the engine lifting eyes and support the powertrain weight.
10. Remove or disconnect the following:
    - Side engine mount bracket
    - Front mount bracket support nut
11. Matchmark the front subframe to the mounting points.
12. Remove or disconnect the following:
    - Front subframe
    - A/C compressor
13. Lower the powertrain away from the vehicle.

To install:

14. Raise the powertrain into position.
15. Install or connect the following:
    - A/C compressor. Tighten the bolts to 16 ft. lbs. (22 Nm).
    - Front subframe. Use new bolts and tighten the 14mm bolts to 76 ft. lbs. (103 Nm). Tighten the front

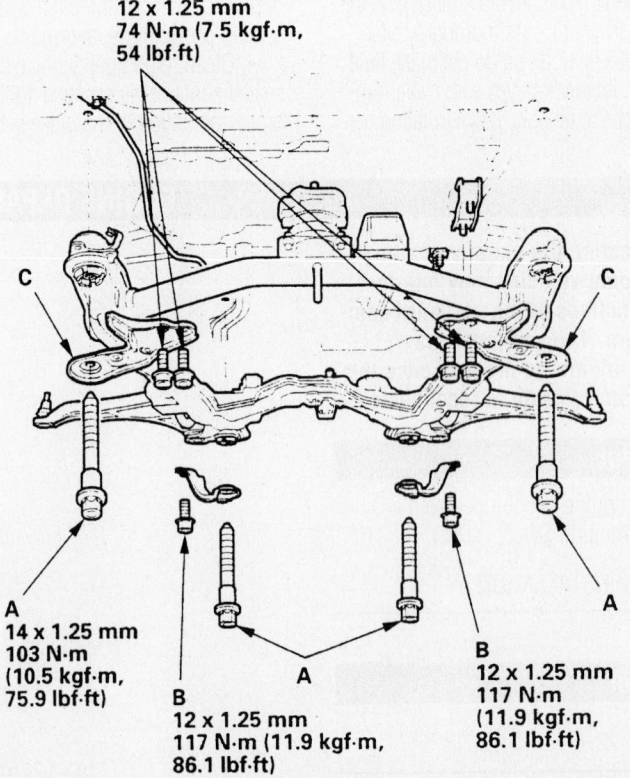

B
12 x 1.25 mm
74 N·m (7.5 kgf·m, 54 lbf·ft)

C   **C**

A   **A**
14 x 1.25 mm   **12 x 1.25 mm**
103 N·m   **117 N·m**
(10.5 kgf·m,   **(11.9 kgf·m,**
75.9 lbf·ft)   **86.1 lbf·ft)**

B
12 x 1.25 mm
117 N·m (11.9 kgf·m, 86.1 lbf·ft)

9302MG69

Sub-frame fastener locations and tightening torque—2001

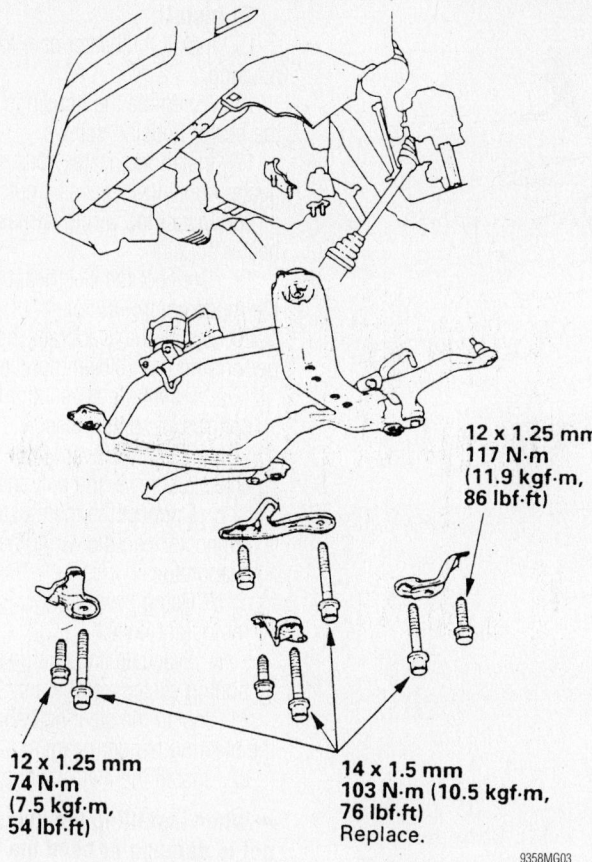

12 x 1.25 mm
117 N·m
(11.9 kgf·m,
86 lbf·ft)

12 x 1.25 mm
74 N·m
(7.5 kgf·m,
54 lbf·ft)

14 x 1.5 mm
103 N·m (10.5 kgf·m,
76 lbf·ft)
Replace.

9358MG03

Sub-frame fastener locations and tightening torque—2002–04

brace bolts to 54 ft. lbs. (74 Nm) and the rear brace bolts to 86 ft. lbs. (117 Nm).
- Transaxle lower front mount. Tighten the nuts to 28 ft. lbs. (38 Nm).
- Transaxle lower rear mount. Tighten the bolts to 28 ft. lbs. (38 Nm).
- Front mount bracket support nut. Tighten the nut to 40 ft. lbs. (54 Nm).
- Side engine mount bracket. Tighten the bracket bolts to 33 ft. lbs. (44 Nm) and the through bolt to 40 ft. lbs. (54 Nm).
- Steering rack and pinion gear. Tighten the bolts to 29 ft. lbs. (39 Nm).
- Power steering hose clamp and clips
- Shift control lever with cable attached
- Shift cable cover
- Shift cable bracket
- Axle halfshafts. Use new circlips.
- Lower ball joints. Tighten the nuts to 43–51 ft. lbs. (59–69 Nm).
- Stabilizer bar links. Tighten the nuts to 58 ft. lbs. (78 Nm).
- Exhaust front pipe

- HO2S connector
- Splash shield
- Front wheels
- Transaxle oil cooler lines
- Radiator hoses
- Heater hoses
- Bypass hose
- Starter cable and harness clamp
- Power steering pump
- Ground cable
- Fuse/Relay box battery cable
- PCM connectors and grommet
- Vacuum supply hose
- Brake booster vacuum line
- Cruise control cable
- Accelerator cable
- Accessory drive belts
- Fuel lines
- Battery tray
- Relay bracket
- Left engine wire harness connectors
- Battery
- Air intake duct
- EVAP control canister hose
- Negative battery cable
16. Fill the engine crankcase to the correct level.
17. Fill the transaxle to the correct level.
18. Fill the cooling system.

19. Start the engine and check for leaks.
20. Check the wheel alignment and adjust as necessary.

Heater Core

REMOVAL & INSTALLATION

➡**Make sure to acquire the anti-theft code for the radio and write down the frequencies for the radio's preset buttons.**

1. Disconnect the negative battery cable.

✳✳ **CAUTION**

Wait at least 3 minutes for the air bag to deplete its energy before working on the steering wheel or instrument panel.

2. In the engine compartment, remove the heater valve cable clamp; then, disconnect the heater valve cable and rotate the heater valve to the fully open position.
3. Drain the engine coolant into a clean container for reuse.
4. Disconnect the heater hoses from the heater unit.
5. Remove the heater housing-to-chassis nuts.
6. Remove the center console.
7. Remove the instrument panel.
8. Remove the steering hanger beam mounting bolts and the steering hanger beam.
9. Remove the evaporator housing by performing the following procedure:
   a. Discharge and recover the air conditioning system refrigerant.
   b. Remove the refrigerant lines. Discard the O-rings. Plug the openings to prevent contamination.
   c. Disconnect the thermostat electrical connector and the wiring harness from the evaporator.
   d. Remove the evaporator housing-to-chassis screws, the bolt and nuts.
   e. Disconnect the drain hose and remove the evaporator housing.
10. Disconnect the electrical connector from the mode control motor.
11. Remove the wiring harness clips from the heater housing.
12. Remove the heater housing-to-chassis nuts and the heater housing.
13. Remove the heater housing screws and separate the housings.
14. Remove the heater core from the heater housing.

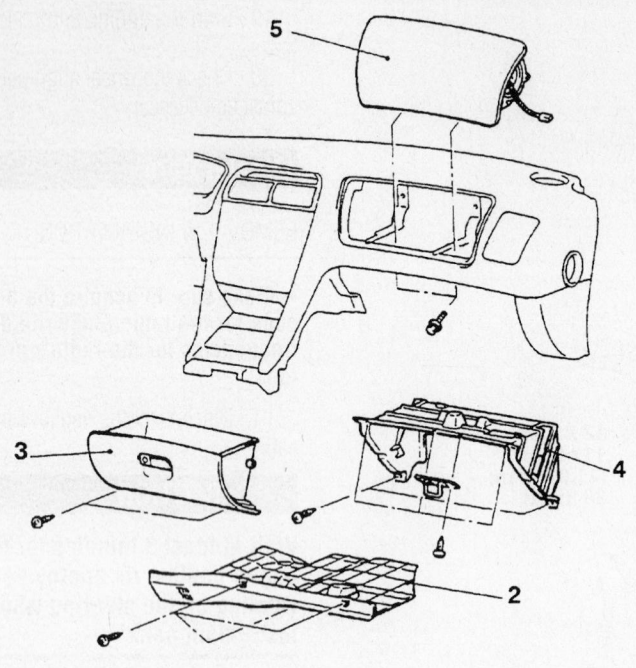

2. UNDERCOVER
3. GLOVE BOX ASSEMBLY
4. GLOVE BOX CASE
5. AIR BAG MODULE

93112GG1

View of the heater housing, evaporator housing and related components

To install:

15. Install the heater core to the heater housing.

16. Assemble the housings and install the heater housing screws.

17. Install the heater housing and the heater housing-to-chassis nuts.

18. Install the wiring harness clips to the heater housing.

19. Connect the electrical connector to the mode control motor.

20. Install the evaporator housing by performing the following procedure:

   a. Install the evaporator housing and connect the drain hose.

   b. Install the evaporator housing-to-chassis screws, the bolt and nuts.

   c. Connect the thermostat electrical connector and the wiring harness to the evaporator.

   d. Using new O-rings, install the refrigerant lines.

   e. Evacuate and charge the air conditioning system refrigerant.

21. Install the steering hanger beam and the steering hanger beam mounting bolts.

22. Install the instrument panel.

➡**When installing the nuts, be careful not to damage or bend the fuel lines, the brake lines or etc.**

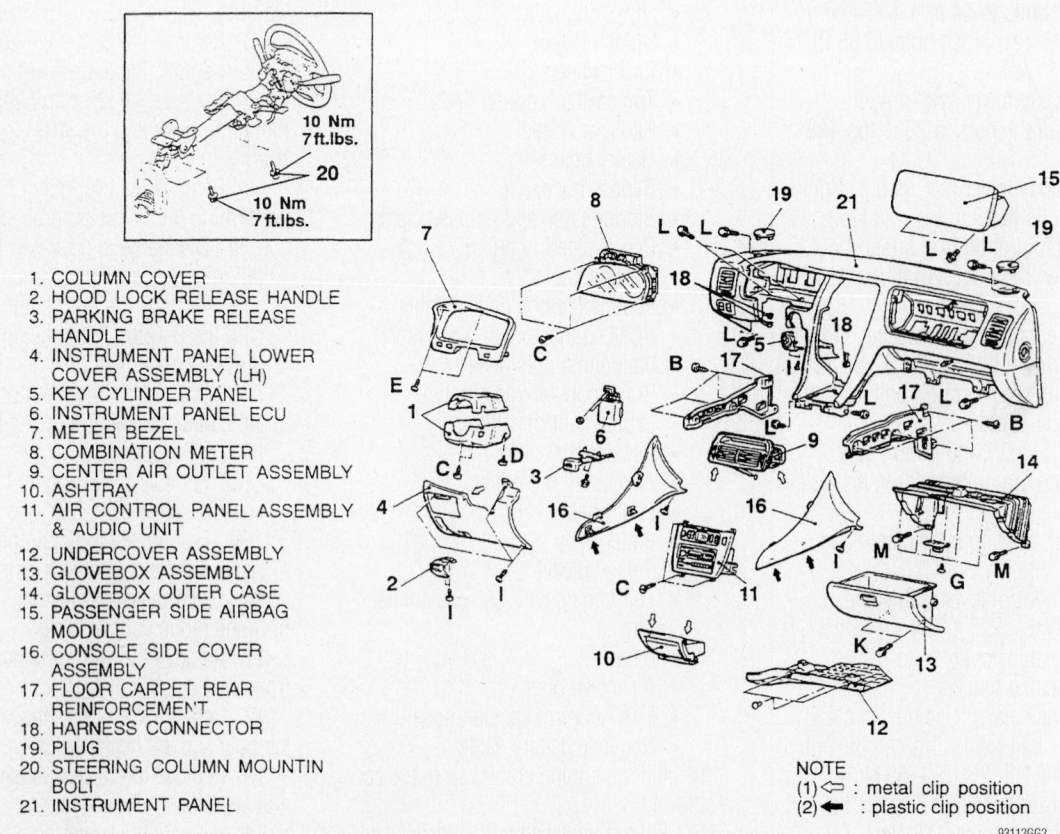

1. COLUMN COVER
2. HOOD LOCK RELEASE HANDLE
3. PARKING BRAKE RELEASE HANDLE
4. INSTRUMENT PANEL LOWER COVER ASSEMBLY (LH)
5. KEY CYLINDER PANEL
6. INSTRUMENT PANEL ECU
7. METER BEZEL
8. COMBINATION METER
9. CENTER AIR OUTLET ASSEMBLY
10. ASHTRAY
11. AIR CONTROL PANEL ASSEMBLY & AUDIO UNIT
12. UNDERCOVER ASSEMBLY
13. GLOVEBOX ASSEMBLY
14. GLOVEBOX OUTER CASE
15. PASSENGER SIDE AIRBAG MODULE
16. CONSOLE SIDE COVER ASSEMBLY
17. FLOOR CARPET REAR REINFORCEMENT
18. HARNESS CONNECTOR
19. PLUG
20. STEERING COLUMN MOUNTIN BOLT
21. INSTRUMENT PANEL

NOTE
(1) ⇐ : metal clip position
(2) ⬅ : plastic clip position

93112GG2

View of the steering hanger beam and related components

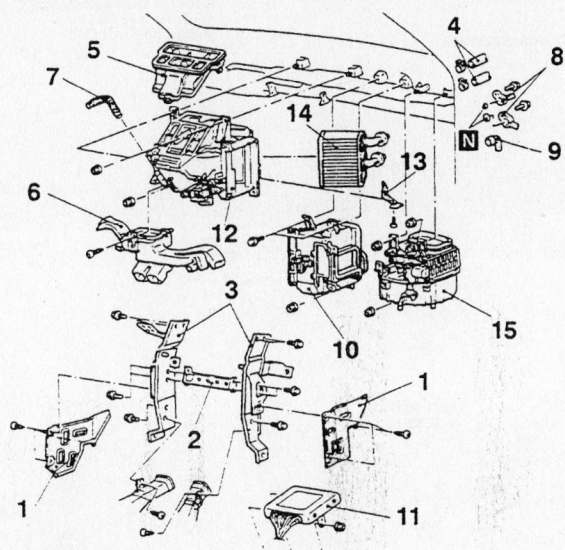

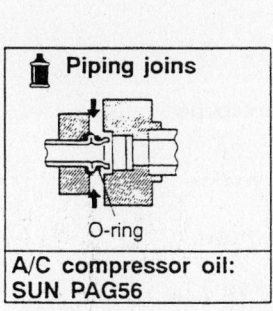

Piping joins

O-ring

**A/C compressor oil:
SUN PAG56**

1. FLOOR CARPET FRONT
   REINFORCEMENT
3. ECU BRACKET
4. CENTER STAY ASSEMBLY
5. HEATER HOSE CONNECTION
6. CENTER DUCT ASSEMBLY
7. FOOT DISTRIBUTION DUCT
8. BREATHER HOSE
9. SUCTION PIPE, LIQUID PIPE B
   AND COOLING UNIT
   CONNECTION
10. DRAIN HOSE
11. EVAPORATOR
12. ENGINE CONTROL MODULE
13. HEATER UNIT
14. HEATER CORE SUPPORT
15. HEATER CORE

93112GG3

Exploded view of the heater core, the heater housing and related components

23. Install the heater housing-to-chassis nuts.

24. Connect the heater hoses to the heater unit.

25. Refill the cooling system.

26. In the engine compartment, Install the heater valve cable clamp; then, connect the heater valve cable.

27. Connect the negative battery cable.

28. Run the engine to normal operating temperatures; then, check the climate control operation and check for leaks.

Water Pump

REMOVAL & INSTALLATION

1. Before servicing the vehicle, refer to the precautions in the beginning of this section.

2. Drain the cooling system.

3. Remove or disconnect the following:
   - Negative battery cable
   - Accessory drive belts
   - Front cover
   - Timing belt. Refer to the Timing Belt unit repair section.
   - Timing belt tensioner
   - Water pump

To install:

4. Install or connect the following:
   - Water pump. Use a new O-ring seal and tighten the bolts to 105 inch lbs. (12 Nm).
   - Timing belt tensioner
   - Timing belt
   - Front cover
   - Accessory drive belts
   - Negative battery cable

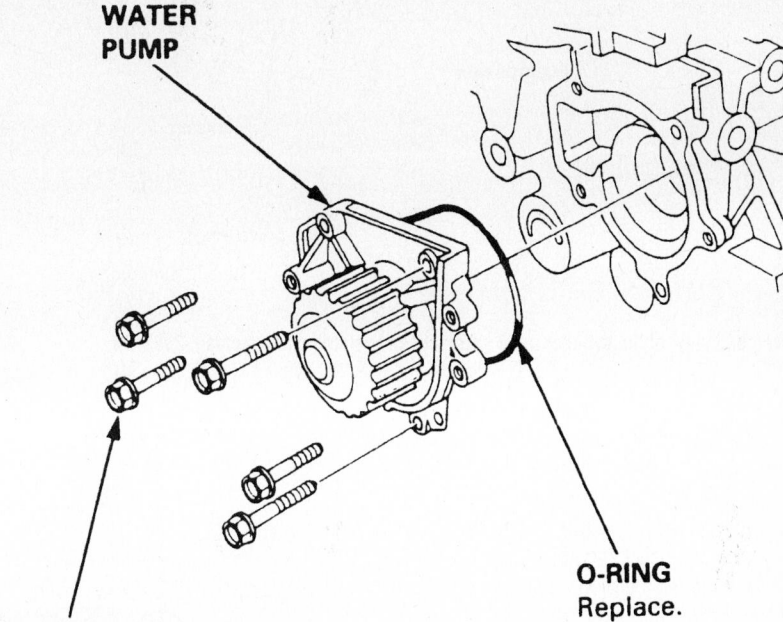

WATER PUMP

**6 x 1.0 mm
12 N·m (1.2 kgf·m,
8.7 lbf·ft)**

O-RING
Replace.

7924MG10

Exploded view of the water pump mounting

5. Fill the cooling system.
6. Start the engine and check for leaks.

Cylinder Head

REMOVAL & INSTALLATION

1. Before servicing the vehicle, refer to the precautions in the beginning of this section.

2. Drain the cooling system.
3. Relieve the fuel system pressure.
4. Remove or disconnect the following:
   - Negative battery cable
   - Accessory drive belts
   - Evaporative Emissions (EVAP) control canister hose and vacuum hose
   - Air intake tube
   - Ignition coil covers
   - Intake manifold cover
   - Accelerator cable

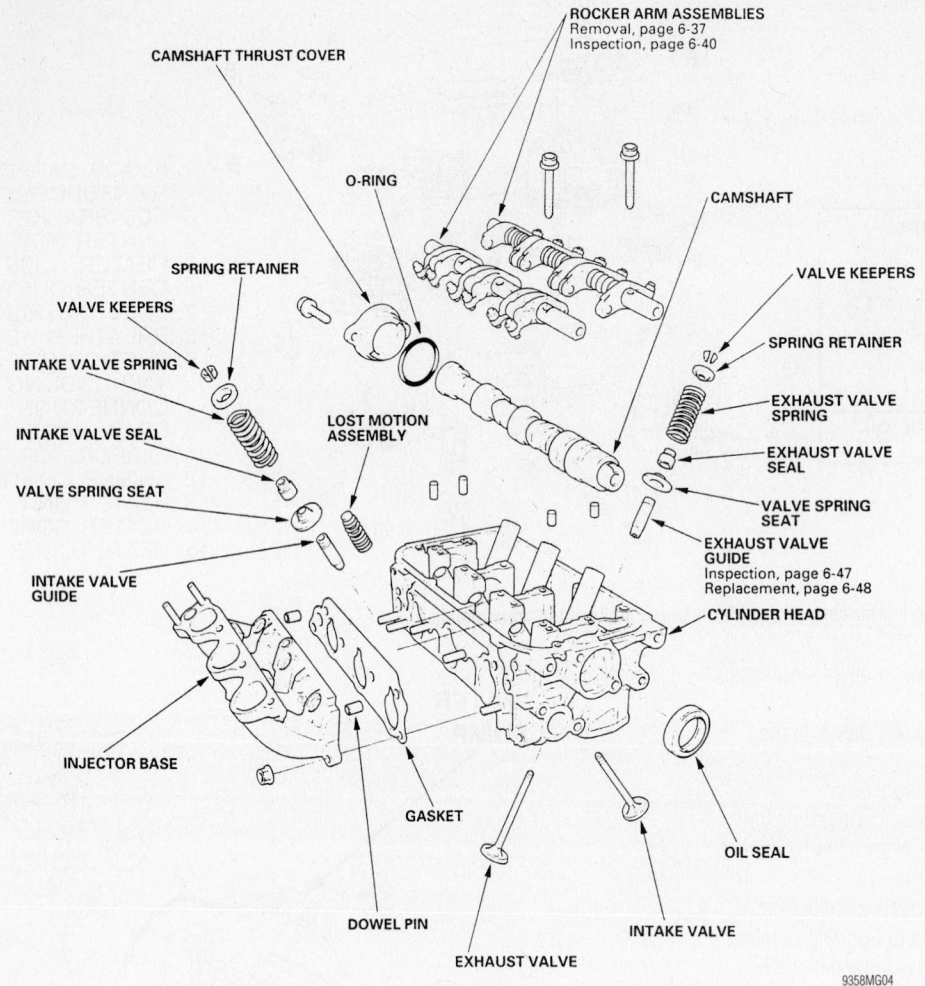

ROCKER ARM ASSEMBLIES
Removal, page 6-37
Inspection, page 6-40

CAMSHAFT THRUST COVER

O-RING

CAMSHAFT

SPRING RETAINER

VALVE KEEPERS

VALVE KEEPERS

INTAKE VALVE SPRING

SPRING RETAINER

INTAKE VALVE SEAL

LOST MOTION ASSEMBLY

EXHAUST VALVE SPRING

VALVE SPRING SEAT

EXHAUST VALVE SEAL

INTAKE VALVE GUIDE

VALVE SPRING SEAT

EXHAUST VALVE GUIDE
Inspection, page 6-47
Replacement, page 6-48

INJECTOR BASE

CYLINDER HEAD

GASKET

OIL SEAL

DOWEL PIN

INTAKE VALVE

EXHAUST VALVE

9358MG04

Exploded view of the cylinder head assembly and related components—2002–04

- Cruise control cable
- Fuel lines
- Brake booster vacuum line
- Intake manifold vacuum line
- Positive Crankcase Ventilation (PCV) valve and hose
- Side engine mount bracket
- Power steering pump
- Power steering hose clamp
- Alternator
- Intake Air Temperature (IAT) sensor connector
- Idle Air Control (IAC) valve connector
- Throttle Position (TP) sensor connector
- Manifold Absolute Pressure (MAP) sensor connector
- Engine Coolant Temperature (ECT) sensor connector
- Radiator fan switch connectors
- ECT gauge sending unit connector
- Crankshaft Position (CKP) sensor connector
- Top Dead Center (TDC) sensor connector

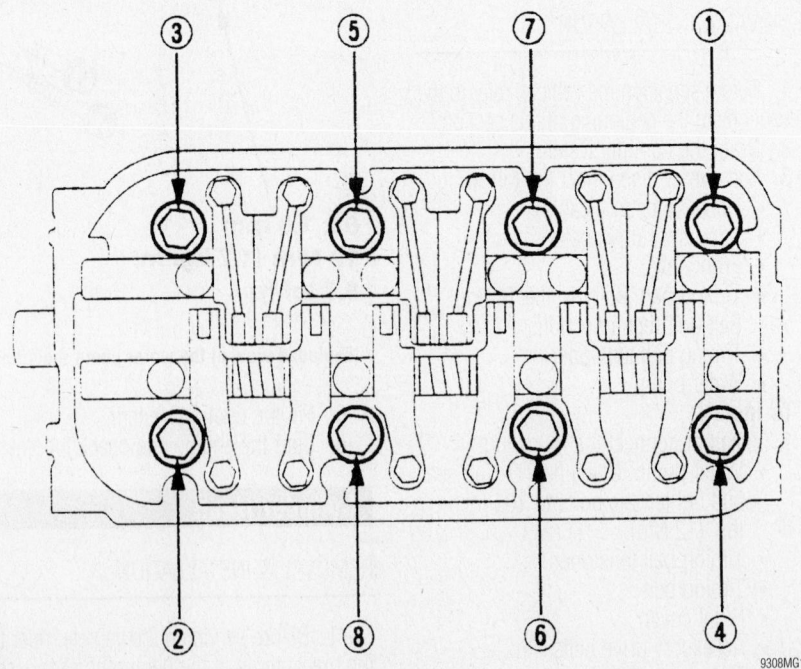

9308MG12

Cylinder head bolt loosening sequence

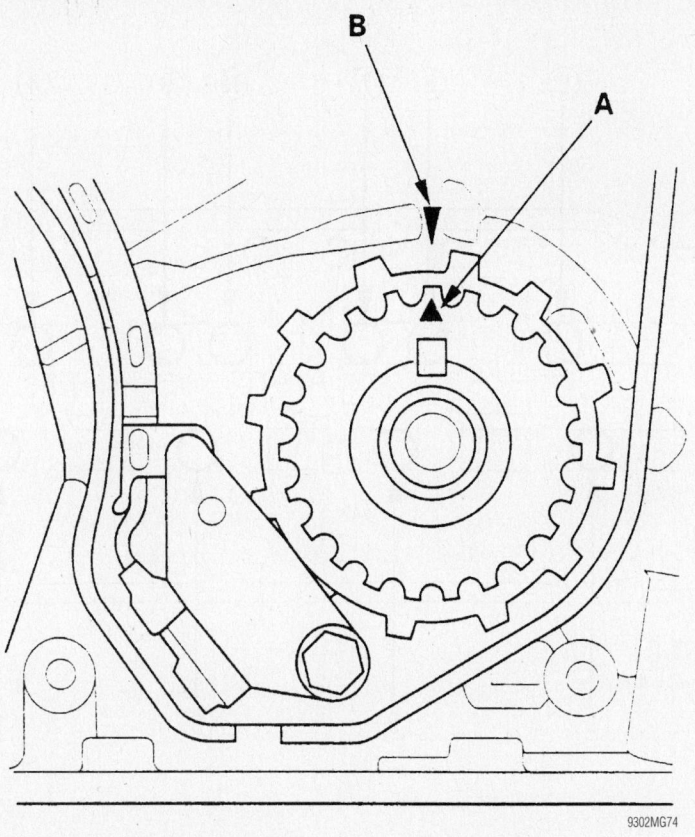

Crankshaft timing belt sprocket TDC marks. Align sprocket mark (A) with pointer (B)

- Exhaust Gas Recirculation (EGR) connector
- Variable Valve Timing and Valve Lift Electronic Control (VTEC) solenoid valve connector
- VTEC oil pressure switch connector
- Oil pressure switch connector
- Ignition coils
- Intake manifold
- Fuel injector connectors
- Fuel supply manifold
- Fuel injection air control valve vacuum lines
- Front cover
- Timing belt. Refer to the Timing Belt unit repair section.
- Radiator hoses
- Heater hoses
- Front and rear exhaust manifolds
- Coolant cross-over pipe
- Valve covers

5. Loosen the cylinder head bolts in sequence and ⅓ turns until all bolts are loose.

6. Remove the cylinder head.

To install:

7. Align the crankshaft and camshaft sprocket TDC marks as shown.

8. Install the cylinder heads with new gaskets.

FRONT:

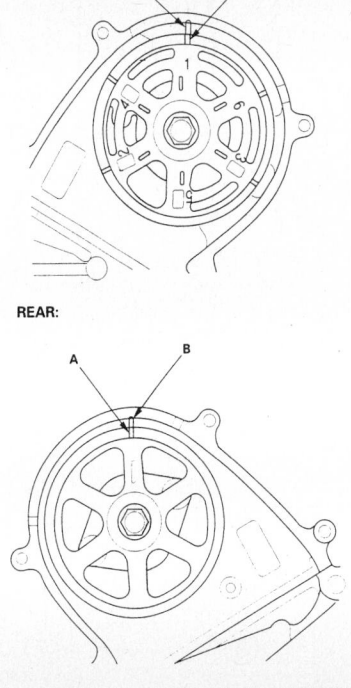

REAR:

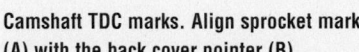

Camshaft TDC marks. Align sprocket mark (A) with the back cover pointer (B)

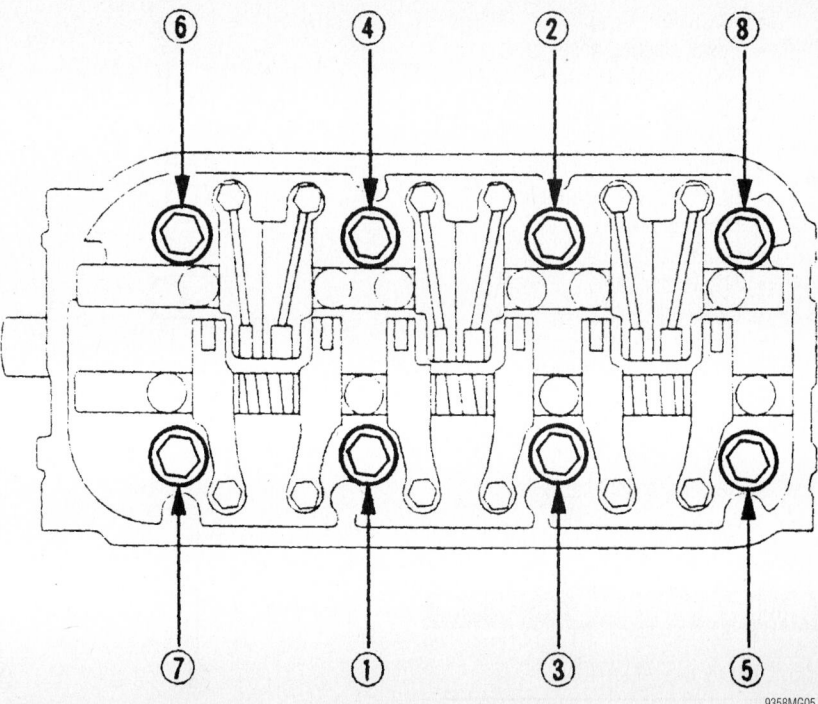

Cylinder head bolt torque sequence

9. Apply clean engine oil to the cylinder head bolt threads and flanges.

10. Tighten the cylinder head bolts in sequence as follows:
   a. Step 1: 29 ft. lbs. (39 Nm)
   b. Step 2: 51 ft. lbs. (69 Nm)
   c. Step 3: 72 ft. lbs. (98 Nm)

11. Install or connect the following:
   • Valve covers
   • Coolant cross-over pipe
   • Front and rear exhaust manifolds
   • Heater hoses
   • Radiator hoses
   • Timing belt
   • Front cover
   • Fuel injection air control valve vacuum lines
   • Fuel supply manifold
   • Fuel injector connectors
   • Intake manifold
   • Ignition coils
   • Oil pressure switch connector
   • VTEC oil pressure switch connector
   • VTEC solenoid valve connector
   • EGR connector
   • TDC sensor connector
   • CKP sensor connector
   • ECT gauge sending unit connector
   • Radiator fan switch connectors
   • ECT sensor connector
   • MAP sensor connector
   • TP sensor connector
   • IAC valve connector
   • IAT sensor connector
   • Alternator
   • Power steering hose clamp
   • Power steering pump
   • Side engine mount bracket
   • PCV valve and hose
   • Intake manifold vacuum line
   • Brake booster vacuum line
   • Fuel lines
   • Cruise control cable
   • Accelerator cable
   • Intake manifold cover
   • Ignition coil covers
   • Accessory drive belts
   • Air intake tube
   • EVAP control canister hose and vacuum hose
   • Negative battery cable

12. Fill the cooling system.

13. Start the engine and check for leaks.

Rocker Arms/Shafts

REMOVAL & INSTALLATION

1. Before servicing the vehicle, refer to the precautions in the beginning of this section.

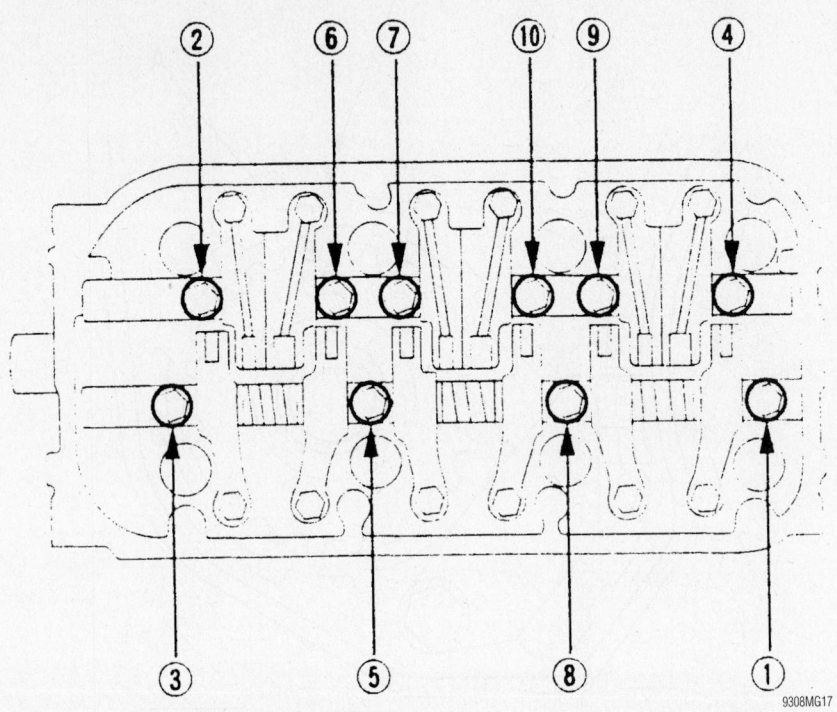

Rocker arm shaft loosening sequence—2001

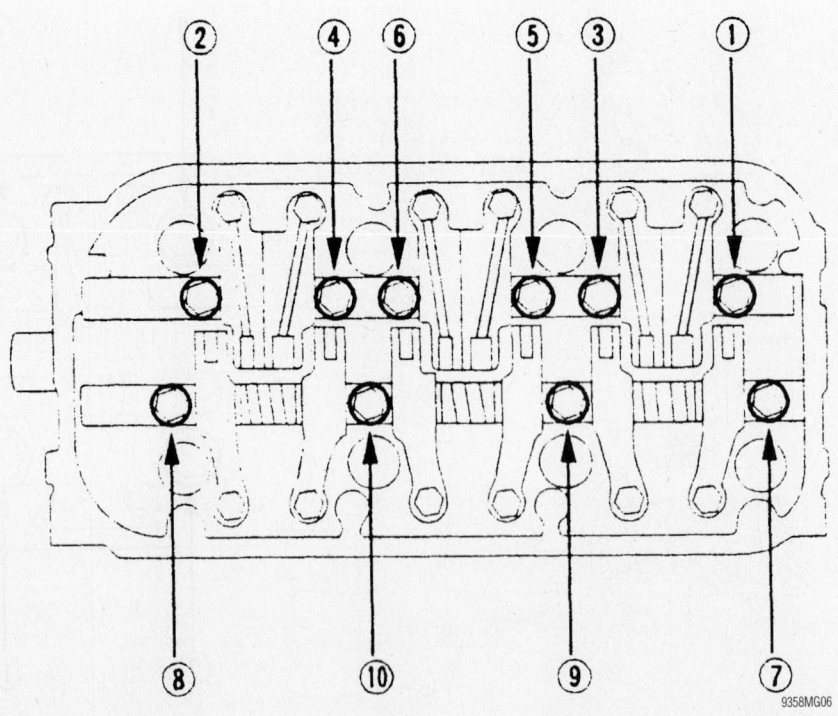

Rocker arm shaft loosening sequence—2002–04

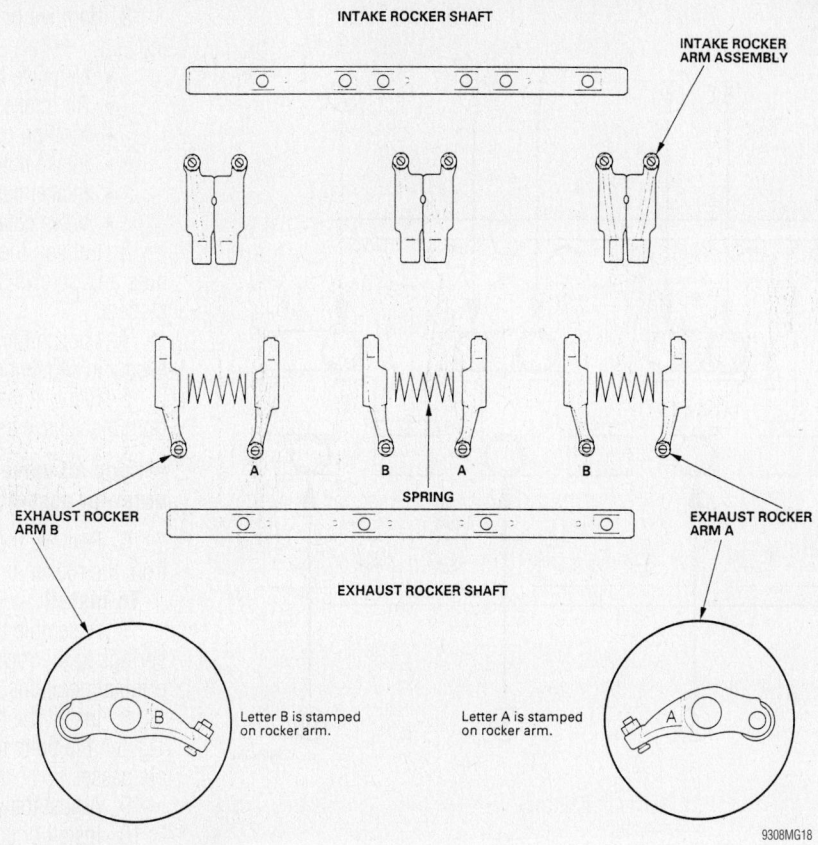

Exploded view of the rocker arms and shafts—2001

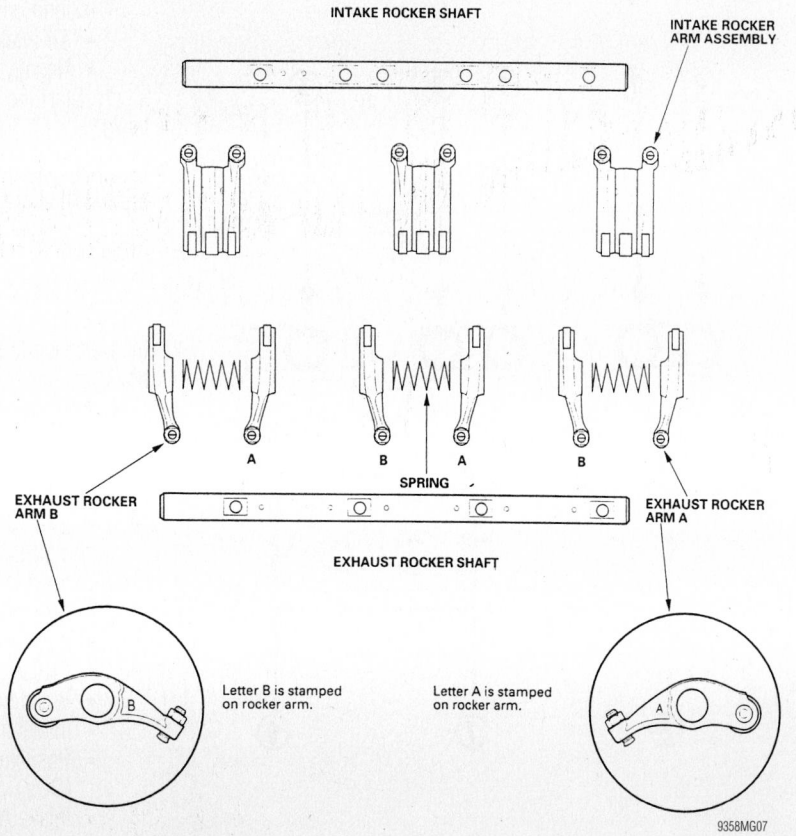

Exploded view of the rocker arms and shafts—2002–04

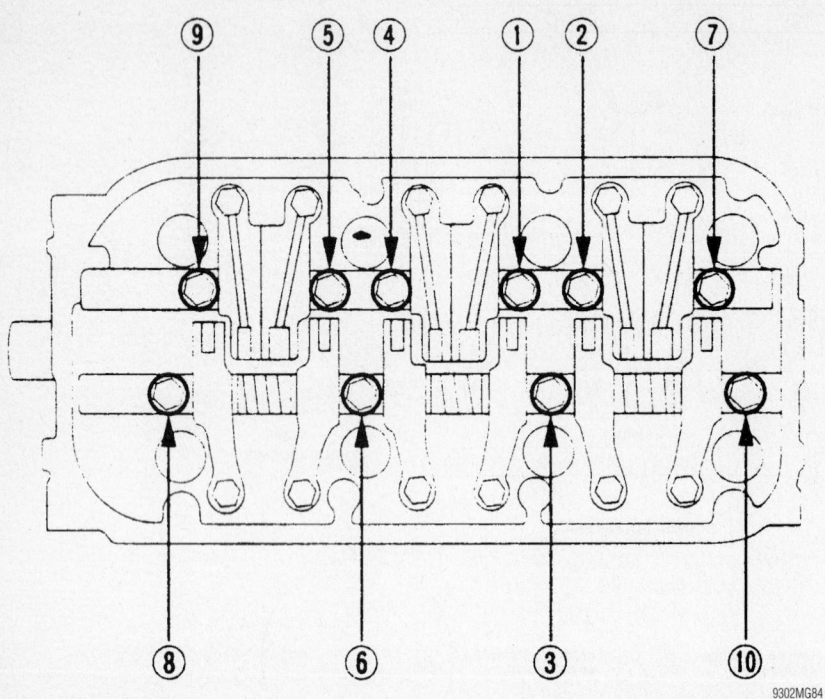

Rocker shaft tightening sequence—2001

9302MG84

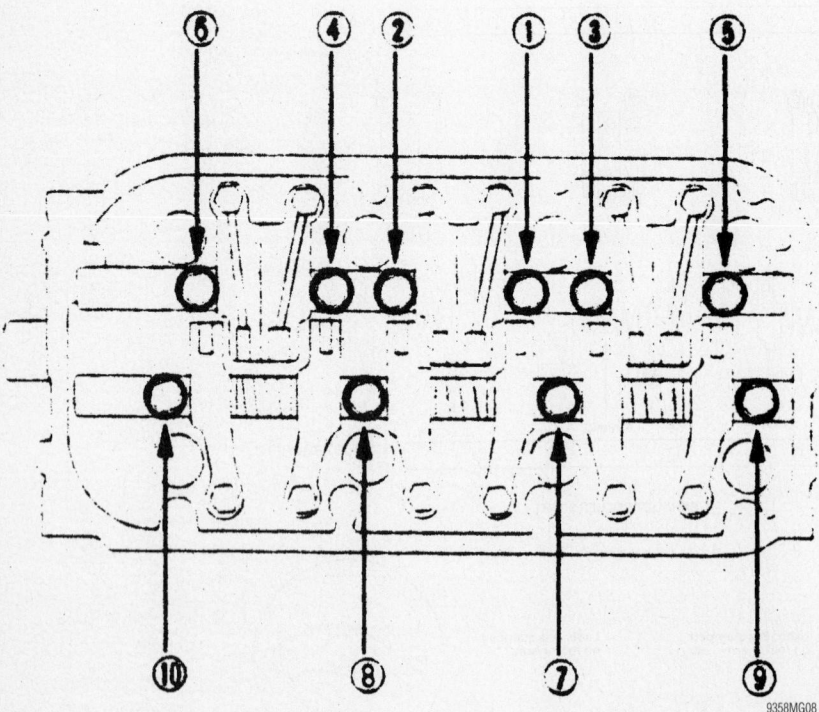

Rocker shaft tightening sequence—2002–04

9358MG08

2. Remove or disconnect the following:

- Negative battery cable
- Air intake tube
- Ignition coil covers
- Intake manifold cover
- Intake manifold
- Valve cover

3. Loosen the valve adjuster locknuts and screws so that all valves are closed.

4. Loosen the rocker arm shaft bolts evenly in sequence.

5. Remove the rocker arms and shafts from the vehicle as an assembly.

➡ **Keep all valvetrain components in order for assembly.**

6. Remove the rocker arms and springs from the rocker arm shafts.

To install:

7. Assemble the rocker arms and springs to the rocker arm shafts in their original positions.

8. Install the rocker arm assemblies. Tighten the bolts in sequence and in multiple passes to 17 ft. lbs. (24 Nm).

9. Adjust the valve clearance.

10. Install or connect the following:

- Valve covers
- Intake manifold
- Intake manifold cover
- Ignition coil covers
- Air intake tube
- Negative battery cable

11. Start the engine and check for leaks.

Intake Manifold

REMOVAL & INSTALLATION

1. Before servicing the vehicle, refer to the precautions in the beginning of this section.

2. Remove or disconnect the following:

- Negative battery cable
- Intake manifold cover
- Evaporative Emissions (EVAP) control canister hose or vacuum hose
- Air intake tube
- Accelerator cable
- Cruise control cable
- Brake booster vacuum line
- Intake manifold vacuum line
- Positive Crankcase Ventilation (PCV) valve and hose
- Intake Air Temperature (IAT) sensor connector

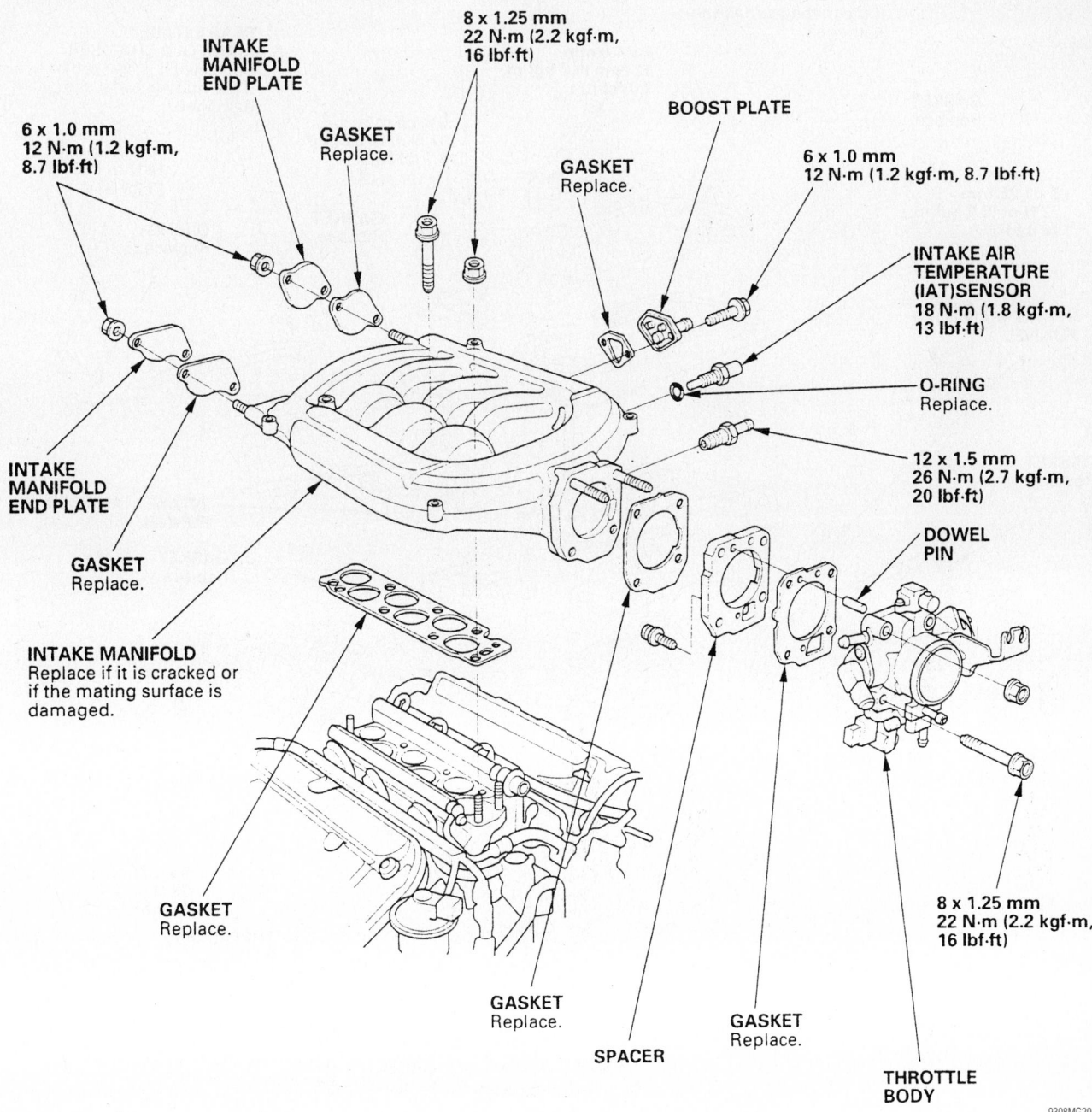

6 x 1.0 mm
12 N·m (1.2 kgf·m,
8.7 lbf·ft)

INTAKE
MANIFOLD
END PLATE

GASKET
Replace.

8 x 1.25 mm
22 N·m (2.2 kgf·m,
16 lbf·ft)

GASKET
Replace.

BOOST PLATE

6 x 1.0 mm
12 N·m (1.2 kgf·m, 8.7 lbf·ft)

INTAKE AIR
TEMPERATURE
(IAT)SENSOR
18 N·m (1.8 kgf·m,
13 lbf·ft)

O-RING
Replace.

12 x 1.5 mm
26 N·m (2.7 kgf·m,
20 lbf·ft)

INTAKE
MANIFOLD
END PLATE

GASKET
Replace.

DOWEL
PIN

INTAKE MANIFOLD
Replace if it is cracked or
if the mating surface is
damaged.

GASKET
Replace.

GASKET
Replace.

SPACER

GASKET
Replace.

8 x 1.25 mm
22 N·m (2.2 kgf·m,
16 lbf·ft)

THROTTLE
BODY

9308MG20

Exploded view of the intake manifold—2001

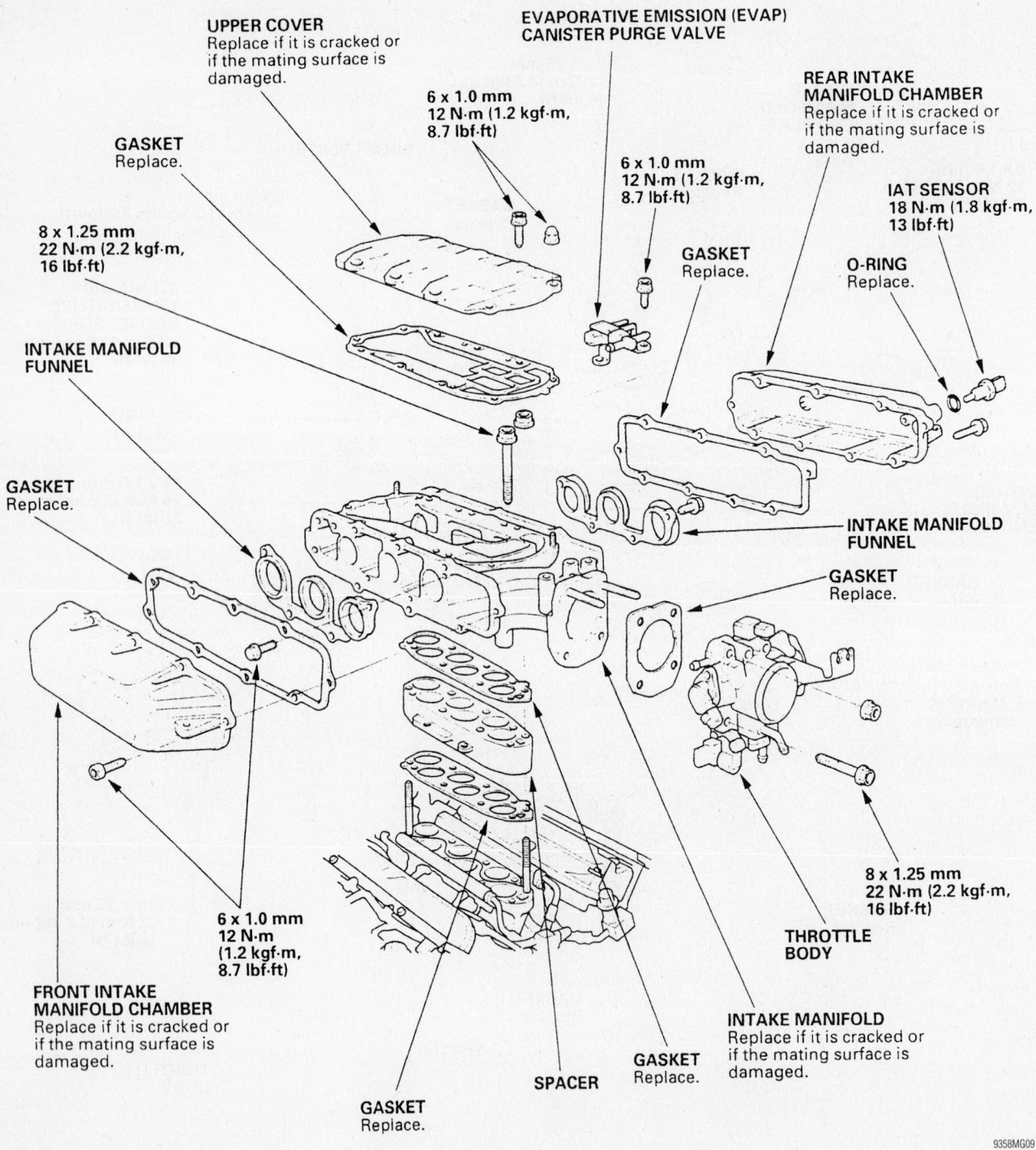

UPPER COVER
Replace if it is cracked or
if the mating surface is
damaged.

EVAPORATIVE EMISSION (EVAP)
CANISTER PURGE VALVE

REAR INTAKE
MANIFOLD CHAMBER
Replace if it is cracked or
if the mating surface is
damaged.

6 x 1.0 mm
12 N·m (1.2 kgf·m,
8.7 lbf·ft)

6 x 1.0 mm
12 N·m (1.2 kgf·m,
8.7 lbf·ft)

GASKET
Replace.

IAT SENSOR
18 N·m (1.8 kgf·m,
13 lbf·ft)

8 x 1.25 mm
22 N·m (2.2 kgf·m,
16 lbf·ft)

GASKET
Replace.

O-RING
Replace.

INTAKE MANIFOLD
FUNNEL

INTAKE MANIFOLD
FUNNEL

GASKET
Replace.

GASKET
Replace.

THROTTLE
BODY

8 x 1.25 mm
22 N·m (2.2 kgf·m,
16 lbf·ft)

6 x 1.0 mm
12 N·m
(1.2 kgf·m,
8.7 lbf·ft)

FRONT INTAKE
MANIFOLD CHAMBER
Replace if it is cracked or
if the mating surface is
damaged.

GASKET
Replace.

SPACER

GASKET
Replace.

INTAKE MANIFOLD
Replace if it is cracked or
if the mating surface is
damaged.

9358MG09

Exploded view of the intake manifold—2002–04

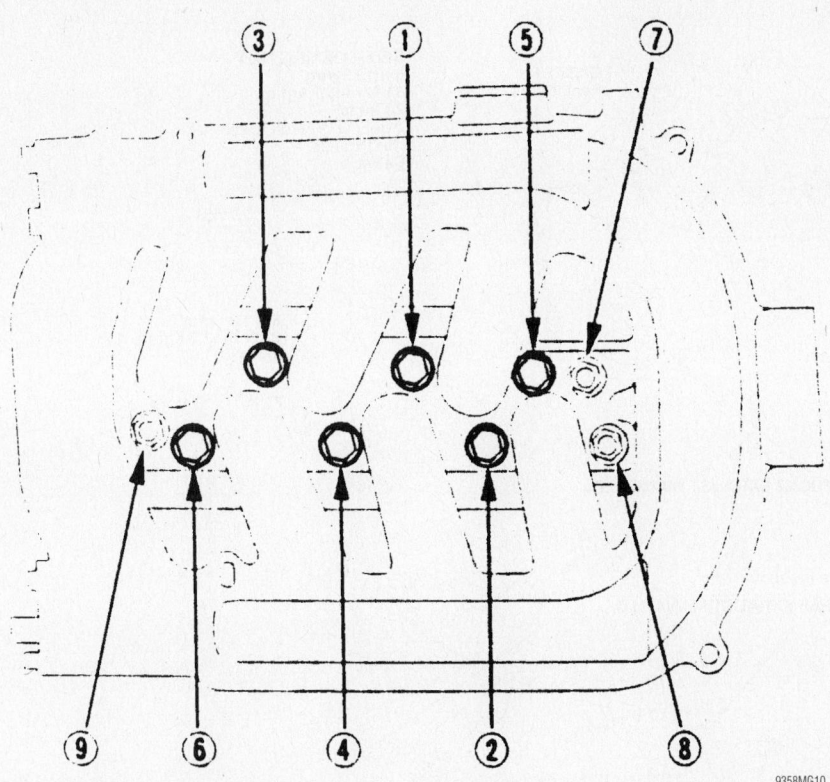

Intake manifold torque sequence—2001

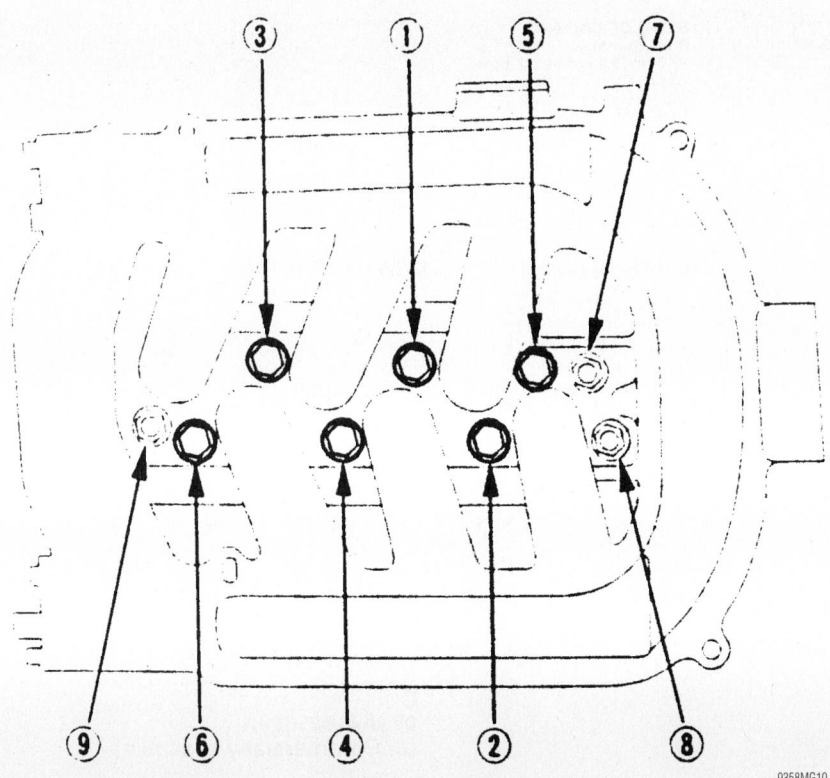

Intake manifold torque sequence—2002–04

- Idle Air Control (IAC) valve connector
- Throttle Position (TP) sensor connector
- Manifold Absolute Pressure (MAP) sensor connector
- Intake manifold

To install:

3. Install or connect the following:
   - New intake manifold gasket
   - Intake manifold. Tighten the fasteners in sequence and in several passes to 16 ft. lbs. (22 Nm).
   - MAP sensor connector
   - TP sensor connector
   - IAC valve connector
   - IAT sensor connector
   - PCV valve and hose
   - Intake manifold vacuum line
   - Brake booster vacuum line
   - Cruise control cable
   - Accelerator cable
   - Intake manifold cover
   - Air intake tube
   - EVAP control canister hose and vacuum hose
   - Negative battery cable

4. Start the engine and check for proper operation.

Exhaust Manifold

REMOVAL & INSTALLATION

1. Before servicing the vehicle, refer to the precautions in the beginning of this section.

2. Remove or disconnect the following:
   - Negative battery cable
   - Exhaust manifold heat shield
   - Heated Oxygen Sensor (HO$_2$S) connector
   - Exhaust front pipe
   - Exhaust manifold bracket, if equipped
   - Exhaust manifold

To install:

3. Install or connect the following:
   - Exhaust manifold. Tighten the fasteners to 23 ft. lbs. (31 Nm).
   - Exhaust manifold bracket, if equipped. Tighten the bolts to 33 ft. lbs. (44 Nm).
   - Exhaust front pipe. Tighten the nuts to 40 ft. lbs. (55 Nm).
   - HO$_2$S connector
   - Exhaust manifold heat shield
   - Negative battery cable

FRONT:

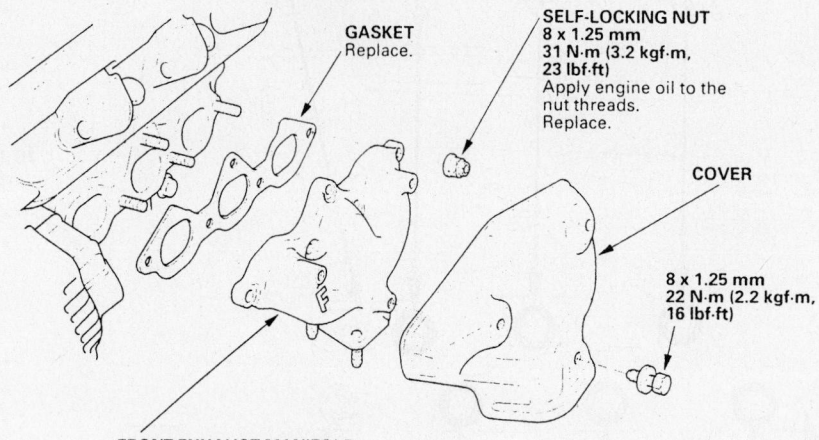

GASKET
Replace.

SELF-LOCKING NUT
8 x 1.25 mm
31 N·m (3.2 kgf·m,
23 lbf·ft)
Apply engine oil to the
nut threads.
Replace.

COVER

8 x 1.25 mm
22 N·m (2.2 kgf·m,
16 lbf·ft)

FRONT EXHAUST MANIFOLD

REAR:

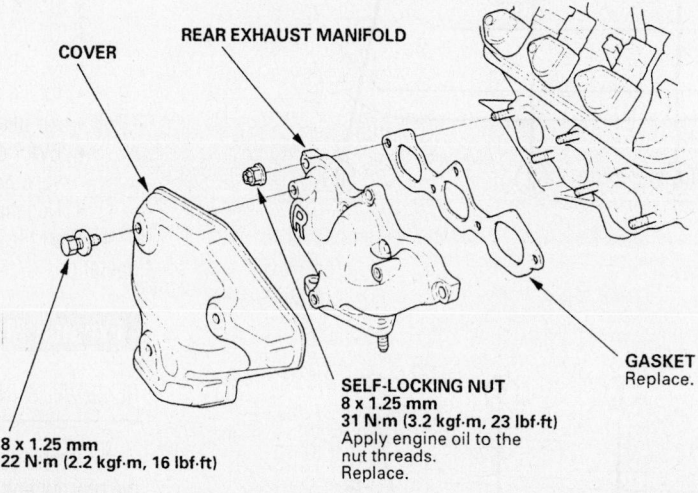

COVER

REAR EXHAUST MANIFOLD

GASKET
Replace.

8 x 1.25 mm
22 N·m (2.2 kgf·m, 16 lbf·ft)

SELF-LOCKING NUT
8 x 1.25 mm
31 N·m (3.2 kgf·m, 23 lbf·ft)
Apply engine oil to the
nut threads.
Replace.

9358MG12

Exploded view of the exhaust manifolds

Front Crankshaft Seal

REMOVAL & INSTALLATION

1. Before servicing the vehicle, refer to the precautions in the beginning of this section.

2. Remove or disconnect the following:

- Negative battery cable
- Accessory drive belts
- Side engine mount
- Valve cover
- Crankshaft pulley using the tools in the accompanying illustration
- Front cover
- Balance shaft belt, if equipped
- Timing belt. Refer to the Timing Belt unit repair section.
- Top Dead Center (TDC) sensor, if equipped

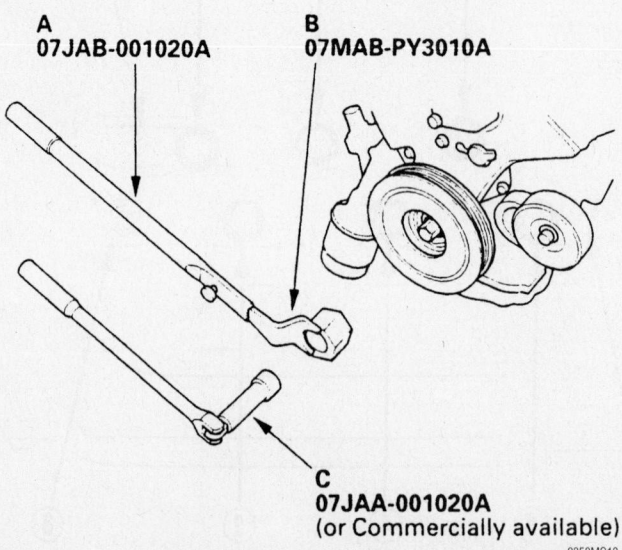

A
07JAB-001020A

B
07MAB-PY3010A

C
07JAA-001020A
(or Commercially available)

9358MG13

Remove the crankshaft pulley using tools (A) Holder handle 07JAB-001020A, (B) attachment 07MAB-PY3010A and (C) a breaker bar and 19mm socket

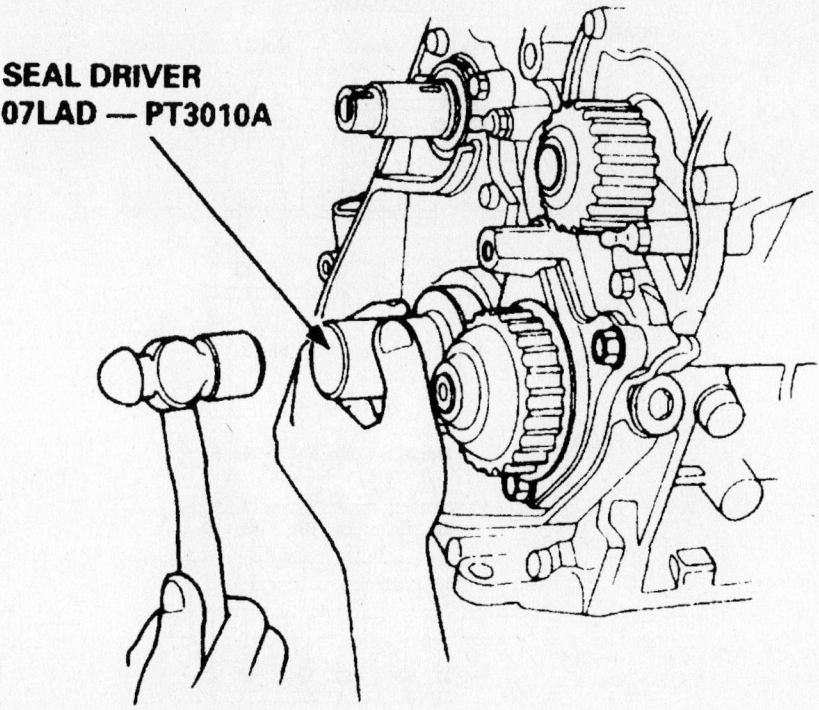

**SEAL DRIVER
07LAD — PT3010A**

7924MG48

Typical front crankshaft seal installation

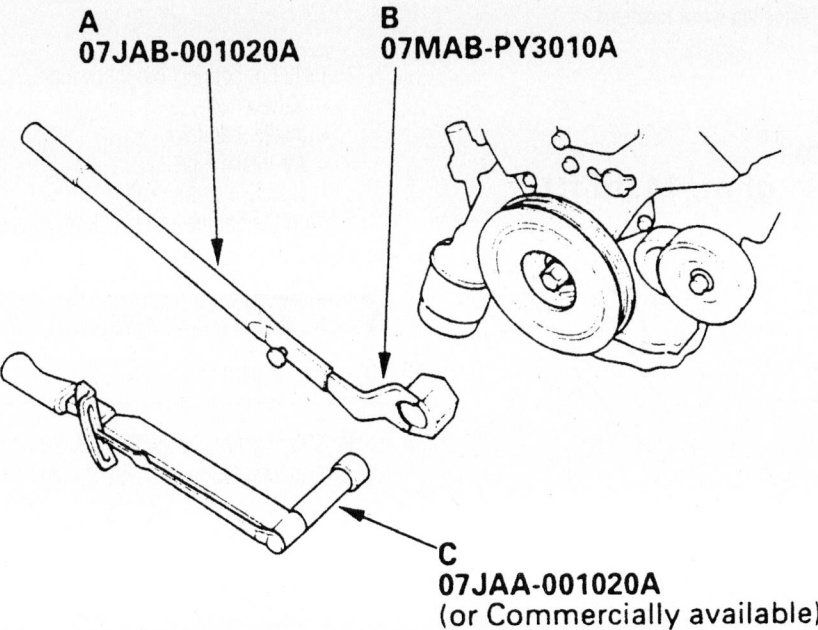

**A
07JAB-001020A**

**B
07MAB-PY3010A**

**C
07JAA-001020A
(or Commercially available)**

9358MG14

Install the crankshaft pulley using tools (A) Holder handle 07JAB-001020A, (B) attachment 07MAB-PY3010A and (C) a torque wrench and 19mm socket

- Crankshaft timing sprocket
- Front crankshaft seal

To install:

3. Lubricate the crankshaft seal lip with grease prior to installation.

4. Install the front crankshaft seal so that

it is flush with the surface of the oil pump housing.

5. Install or connect the following:
- Crankshaft timing sprocket
- Top Dead Center (TDC) sensor, if equipped

- Timing belt. Refer to the Timing Belt unit repair section.
- Balance shaft belt, if equipped
- Front cover
- Crankshaft pulley. Tighten the bolt to 181 ft. lbs. (245 Nm) using the tools in the accompanying illustration.
- Valve cover
- Side engine mount
- Accessory drive belts
- Negative battery cable

6. Check the engine oil level and add if necessary.

7. Start the engine and check for leaks.

Camshaft

REMOVAL & INSTALLATION

1. Before servicing the vehicle, refer to the precautions in the beginning of this section.

2. Remove or disconnect the following:
- Negative battery cable
- Air intake tube
- Accessory drive belts
- Front cover
- Timing belt. Refer to the Timing Belt unit repair section.
- Camshaft sprockets
- Timing belt rear covers
- Ignition coil covers
- Intake manifold cover
- Intake manifold
- Valve cover
- Rocker arms and shaft assembly
- Camshaft thrust cover
- Camshaft

To install:

➡**Use new O-rings, seals and gaskets when installing the camshaft.**

3. Install or connect the following:
- Camshaft
- Camshaft thrust cover. Tighten the bolts to 16 ft. lbs. (22 Nm).
- Rocker arms and shaft assembly
- Valve cover
- Intake manifold
- Intake manifold cover
- Ignition coil covers
- Timing belt rear covers
- Camshaft sprockets. Tighten the bolts to 67 ft. lbs. (90 Nm).
- Timing belt
- Front cover
- Accessory drive belts
- Air intake tube
- Negative battery cable

4. Start the engine and check for leaks.

Valve Lash

ADJUSTMENT

Adjust the valves only when the cylinder head temperature is less than 100°F (38°C).

1. Before servicing the vehicle, refer to the precautions in the beginning of this section.

2. Remove or disconnect the following:
   - Negative battery cable
   - Air intake tube
   - Intake manifold
   - Valve cover

3. Rotate the crankshaft so that the valves to be adjusted are closed and the rocker arm is contacting the camshaft lobe base circle.

4. Measure the valve clearance. If adjustment is necessary, loosen the locknut and turn the adjusting screw as necessary to achieve the correct valve clearance.

5. The correct valve clearance is:
   - Intake valves: 0.008–0.009 inches (0.20–0.24mm)
   - Exhaust valves: 0.011–0.013 inches (0.28–0.32mm)

6. After adjustment, tighten the locknuts to 14 ft. lbs. (20 Nm).

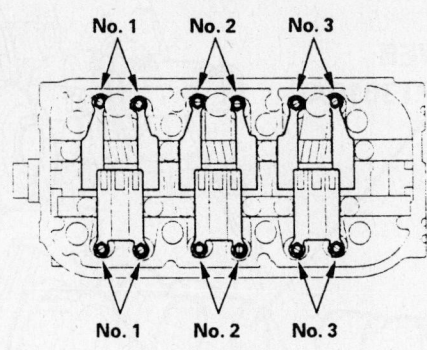

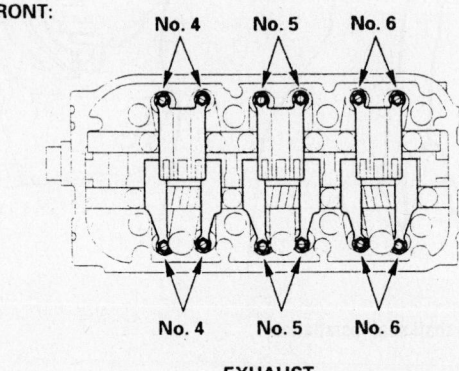

Adjusting screw locations

9358MG16

7. Install or connect the following:
   - Valve cover
   - Intake manifold
   - Air intake tube
   - Negative battery cable

8. Start the engine and check for proper operation.

Starter Motor

REMOVAL & INSTALLATION

1. Before servicing the vehicle, refer to the precautions in the beginning of this section.

2. Remove or disconnect the following:
   - Negative battery cable
   - Transmission fluid cooler line clamp
   - Starter wiring harness connectors
   - Starter motor

To install:

3. Install or connect the following:
   - Starter motor. Tighten the bolts to 33 ft. lbs. (44 Nm).
   - Starter motor. Tighten the upper bolt to 33 ft. lbs. (44 Nm) and the lower bolt to 47 ft. lbs. (64 Nm).
   - Starter wiring harness connectors.

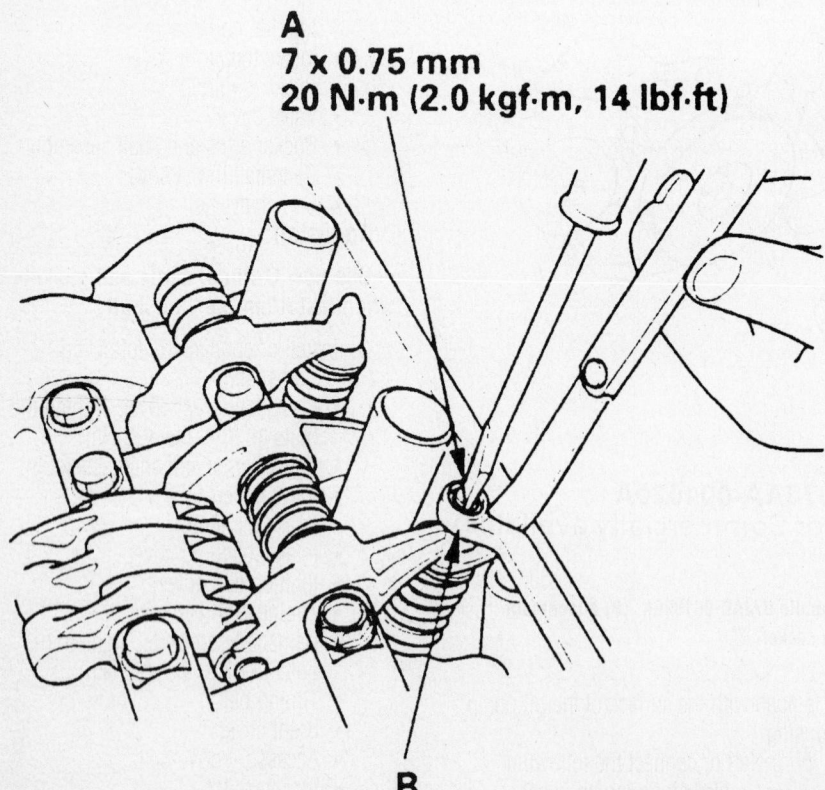

A
7 x 0.75 mm
20 N·m (2.0 kgf·m, 14 lbf·ft)

B

9358MG15

After adjustment tighten the locknut to specification

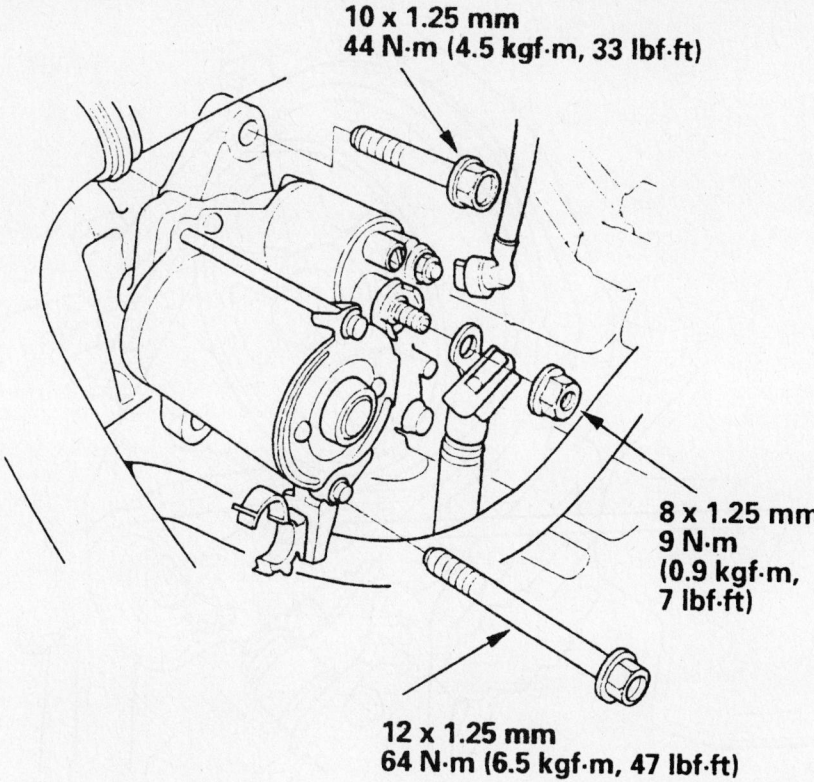

10 x 1.25 mm
44 N·m (4.5 kgf·m, 33 lbf·ft)

8 x 1.25 mm
9 N·m
(0.9 kgf·m,
7 lbf·ft)

12 x 1.25 mm
64 N·m (6.5 kgf·m, 47 lbf·ft)

9358MG02

Exploded view of the starter mounting

Tighten the battery cable nut to 79 inch lbs. (9 Nm).
- Transmission fluid cooler line clamp
- Negative battery cable

Timing Belt and Front Cover

REMOVAL & INSTALLATION

➡The radio may contain a coded theft protection circuit. Always make note the code number before disconnecting the battery.

1. Disconnect the negative battery terminal.
2. Turn the crankshaft so the white mark on the crankshaft pulley aligns with the pointer on the oil pump housing cover.
3. Open the inspection plugs on the upper timing belt covers and check that the camshaft sprocket marks align with the upper cover marks.

✳✳ WARNING

Align the camshaft and crankshaft sprockets with their alignment marks before removing the timing belt. Failure to align the timing marks correctly may result in valve damage.

4. Raise and safely support the vehicle and remove both front tires/wheels.
5. Remove the front lower splash shield.
6. Move the alternator tensioner with a Belt Tensioner Release Arm tool YA9317, or equivalent, to release tension from the belt and remove the alternator drive belt.
7. Remove the alternator belt tensioner release arm.
8. Loosen the power steering pump adjustment nut, adjustment locknut and mounting bolt, then remove the power steering pump with the hoses attached.
9. Support the weight of the engine by placing a wood block on a floor jack and carefully lift on the oil pan.
10. Remove the bolts from the side engine mount bracket and remove the bracket.
11. Remove the dipstick, the dipstick tube and discard the O-ring.
12. Hold the crankshaft pulley with the Handle tool 07JAB-001020A and Crankshaft Holding tool 07MAB-PY3010A, or equivalent. While holding the crankshaft pulley, remove the crankshaft pulley bolt using a heavy duty ¾ in. (19mm) socket and breaker bar.

FRONT CAMSHAFT PULLEY:

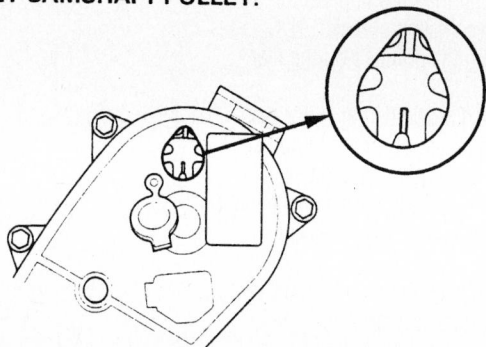

REAR CAMSHAFT PULLEY:

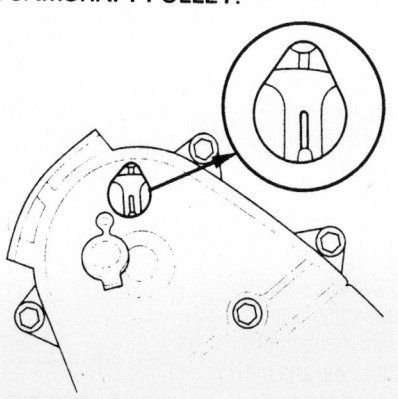

93025G04

Crankshaft and camshaft timing marks at Top Dead Center (TDC)

13. Remove the crankshaft pulley, the upper timing belt covers and the lower timing belt cover.

14. Remove one of the battery clamp fasteners from the battery tray and grind a 45 degree bevel on the threaded end of the battery clamp bolt.

15. Screw in the battery hold-down bolt into the threaded bracket just above the auto-tensioner (automatic timing belt adjuster) and tighten the bolt hand-tight to hold the auto-tensioner adjuster in its current position.

16. Remove the engine mount bracket bolts and the bracket.

17. Loosen the timing belt idler pulley bolt (located on the right side across from the auto-tensioner pulley) about 5–6 revolutions and remove the timing belt.

To install:

18. Clean the timing belt sprockets and the timing belt covers.

✱✱ WARNING

Align the camshaft and crankshaft sprockets with their alignment marks before installing the timing belt. Failure to align the timing marks correctly may result in valve damage.

19. Align the timing mark on the crankshaft sprocket with the oil pump pointer.

20. Align the camshaft sprocket TDC timing marks with the pointers on the rear cover.

21. If installing a new belt or if the auto-tensioner has extended or if the timing belt cannot be reinstalled easily, the auto-tensioner must be collapsed before installation of the timing belt, perform the following procedures:

    a. Remove the battery hold-down bolt from the auto-tensioner bracket.

    b. Remove the timing belt auto-tensioner bolts and the auto-tensioner.

    c. Secure the auto-tensioner in a soft jawed vise, clamping onto the flat surface of one of the mounting bolt holes with the maintenance bolt facing upward.

    d. Remove the maintenance bolt and use caution not to spill oil from the tensioner assembly.

    e. Should oil spill from the tensioner, be sure the tensioner is filled with 0.22 ounces (6.5 ml) of fresh engine oil.

    f. Using care not to damage the threads or the gasket sealing surface, insert a flat-blade screwdriver through the tensioner maintenance hole and turn the screwdriver clockwise to compress the auto-tensioner bottom while the Ten-

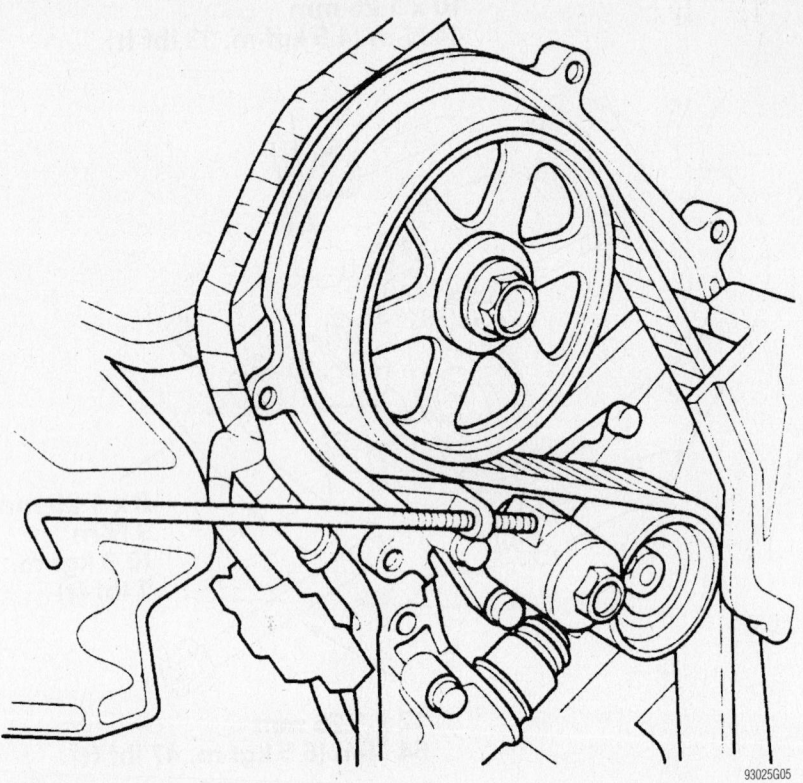

Battery hold-down bolt installed to hold auto-tensioner

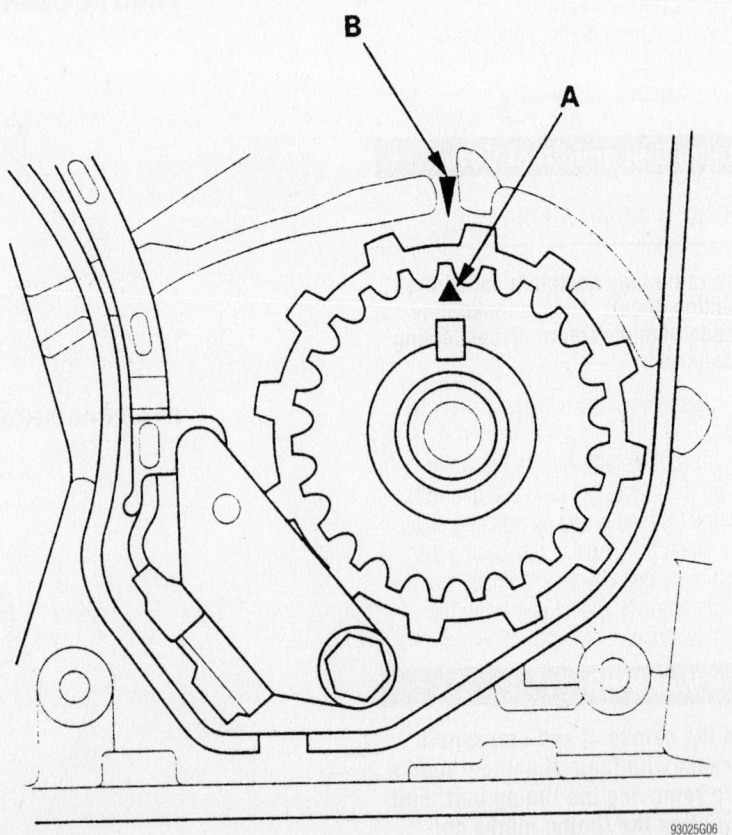

Crankshaft sprocket Top Dead Center (TDC) mark

sioner Holder tool 14540-P8A-A01, or
equivalent, is installed on the auto-tensioner assembly.

g. Install the auto-tensioner maintenance bolt with a new gasket and tighten
to a torque 72 inch lbs. (8 Nm).

h. Install the auto-tensioner on the
engine with the tensioner holder tool
installed and torque the mounting bolts
to 104 inch lbs. (12 Nm).

22. Install the timing belt in a counterclockwise pattern starting with the crankshaft drive sprocket. Install the timing belt
counterclockwise in the following sequence:

- Crankshaft drive sprocket.
- Idler pulley.
- Left side camshaft sprocket.
- Water pump.
- Right side camshaft sprocket.
- Auto-tensioner adjustment pulley.

23. Torque the timing belt idler pulley
bolt to 33 ft. lbs. (44 Nm).

24. Remove the auto-tensioner holding
tool to allow the tensioner to extend.

25. Install the engine mount bracket to
the engine and torque the bolts to 33 ft. lbs.
(44 Nm).

26. Install the lower timing belt cover
and both upper timing belt covers.

27. Hold the crankshaft pulley with special tools 07JAB-001020A handle and
07MAB-PY3010A crankshaft holding tool,
or equivalent tools. While holding the
crankshaft pulley, install the crankshaft pulley bolt using a heavy duty ¾ in. (19mm)
socket and a commercially available torque
wrench and torque the bolt to 181 ft. lbs.
(245 Nm).

✳✳ WARNING

**If any binding is felt while moving
the crankshaft pulley, STOP turning
the crankshaft pulley immediately
because the pistons may be hitting
the valves.**

28. Rotate the crankshaft pulley clockwise 5–6 revolutions to allow the timing
belt to be seated in the pulleys.

29. Move the crankshaft pulley to the
white TDC mark and inspect the camshaft
TDC marks to ensure proper timing of the
camshafts.

✳✳ WARNING

**If the timing marks do not align, the
timing belt removal and installation
procedure must be performed again.**

30. Install the engine dipstick tube using
a new O-ring.

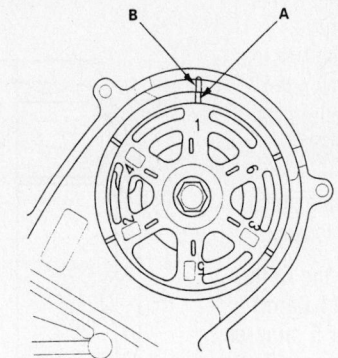
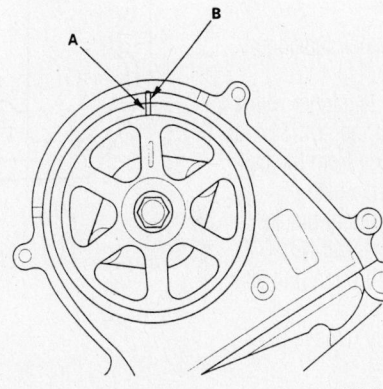

Camshaft sprocket Top Dead Center (TDC) mark

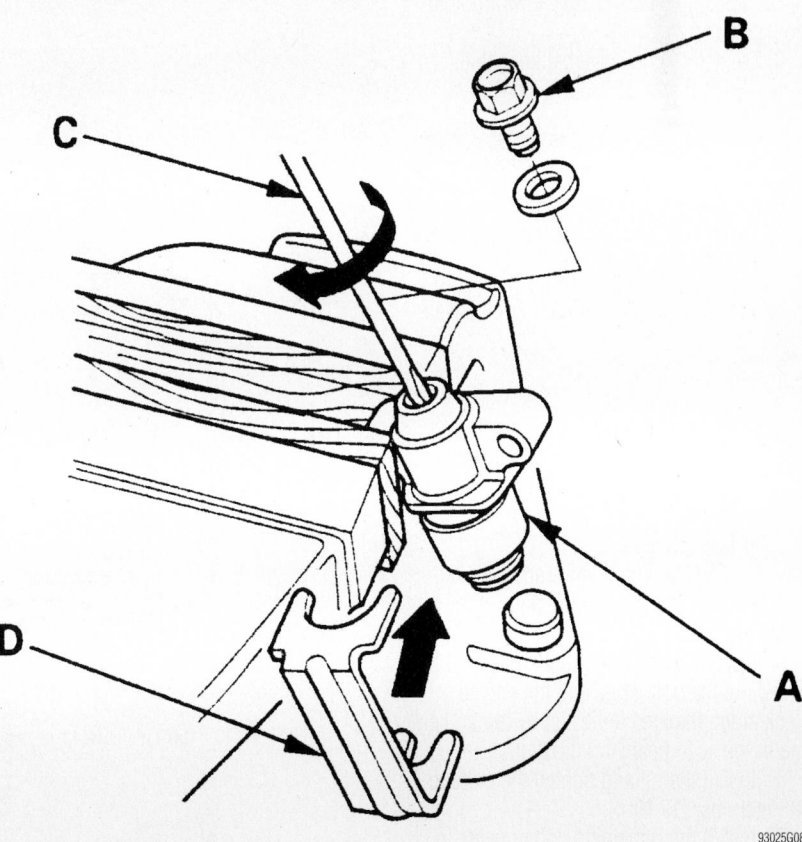

Adjusting the auto-tensioner

31. Install the power steering pump, and loosely install the mounting bolt, adjustment locknut and adjustment nut.

32. Adjust the power steering belt to a tension such that a 22 lb. (98 N) pull halfway between the 2 drive pulleys will allow the belt to move 0.51–0.65 in. (13.0–16.5mm).

33. Tighten the power steering pump mounting bolt and adjustment locknut.

➡**If a new belt is used, set the deflection to 0.33–0.43 in. (8.5–11.0mm) and after engine has run for 5 minutes, readjust the new belt to the used belt specification.**

34. Install the alternator belt tensioner arm.

35. Move the alternator tensioner with a Belt Tensioner Release Arm tool YA9317, or equivalent, to release tension from the belt and install the alternator drive belt.

36. Install both engine mount bracket bolts and torque to 33 ft. lbs. (44 Nm).

37. Install the bushing through bolt and tighten to 40 ft. lbs. (54 Nm).

38. Release and carefully remove the floor jack.

39. Install the front lower splash shield.

40. Install both front tires/wheels.

41. Carefully lower the vehicle.

42. Install the battery hold-down bolt in the battery tray.

43. Install the negative battery cable.

44. Enter the radio security code.

Oil Pan

REMOVAL & INSTALLATION

1. Before servicing the vehicle, refer to the precautions in the beginning of this section.

2. Drain the engine oil.

3. Set the engine at Top Dead Center (TDC).

4. Remove or disconnect the following:
- Negative battery cable
- Timing belt
- Idler pulley
- VTEC solenoid valve and oil filter
- Oil pan bolts and the pan

To install:

5. Apply liquid gasket to the inner threads of the bolt holes and the engine block along the area indicated by the broken line in the accompanying illustration.

6. Install the oil and tighten the bolts to 105 inch lbs. (12 Nm).

7. Install the remaining components in the reverse order of removal.

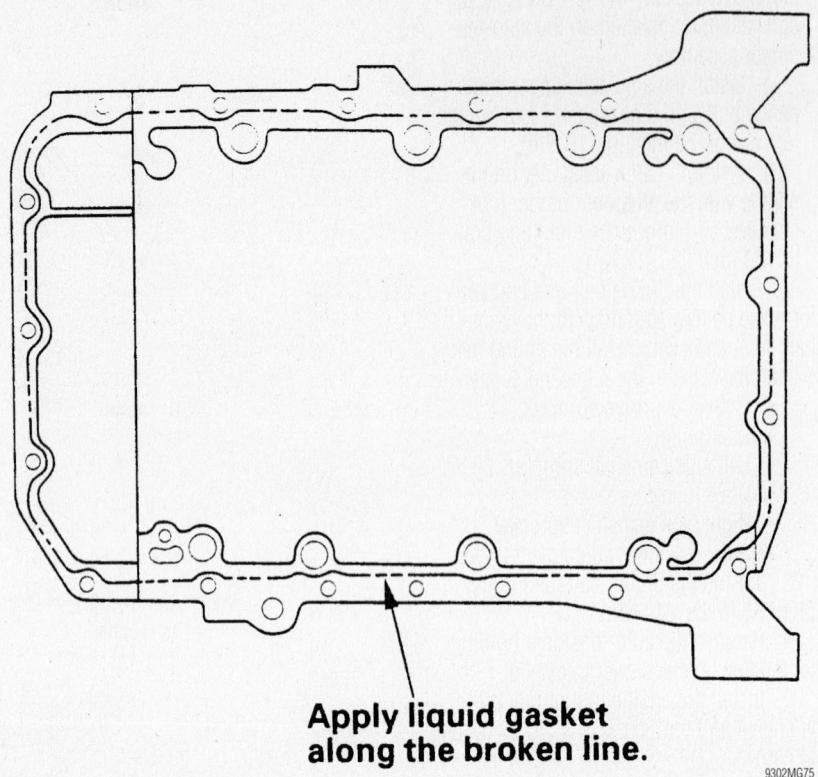

Apply liquid gasket along the broken line.

9302MG75

Apply liquid gasket to the inner threads of the bolt holes and the engine block along the area indicated by the broken line

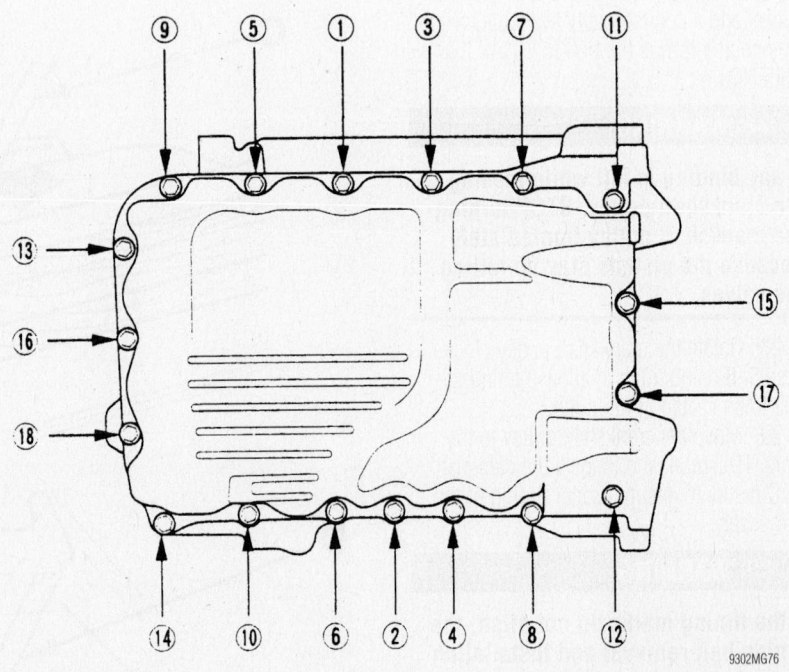

9302MG76

Oil pan tightening sequence

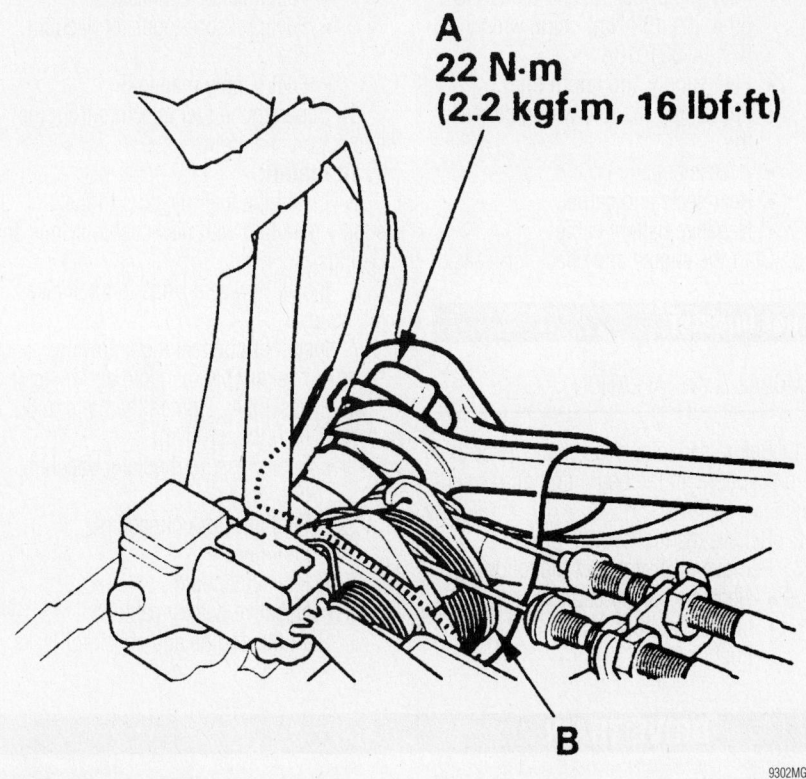

A
22 N·m
(2.2 kgf·m, 16 lbf·ft)

B

9302MG77

Use a wrench on the fuel pulsation damper (A). Place a rag over the damper (B) when relieving residual fuel pressure

2. Disconnect the negative battery cable.
3. Remove the fuel filler cap.
4. Place a shop towel over the fuel pulsation damper.
5. Loosen the fuel pulsation damper 1 turn.
6. When service is completed, replace the sealing washer and tighten the pulsation damper to 16 ft. lbs. (22 Nm).
7. Replace the fuel filler cap.
8. Connect the negative battery cable.
9. Start the engine and check for leaks.

Fuel Filter

REMOVAL & INSTALLATION

1. Before servicing the vehicle, refer to the precautions in the beginning of this section.
2. Relieve the fuel system pressure.
3. Remove or disconnect the following:
   - Negative battery cable
   - Rear seats and carpet
   - Access panel
   - Fuel lines
   - Fuel pump locknut using wrench 07XAA-001010A as shown in the accompanying illustration.
   - Fuel pump module

4. Disassemble the fuel pump module and remove the fuel filter.

To install:
5. Install the fuel filter and assemble the fuel pump module.
6. Install or connect the following:
   - Fuel pump module
   - Fuel pump locknut and tighten to 69 ft. lbs. (93 Nm) using wrench 07XAA-001010A
   - Fuel lines
   - Access panel
   - Rear seats and carpet
   - Negative battery cable
7. Start the engine and check for leaks.

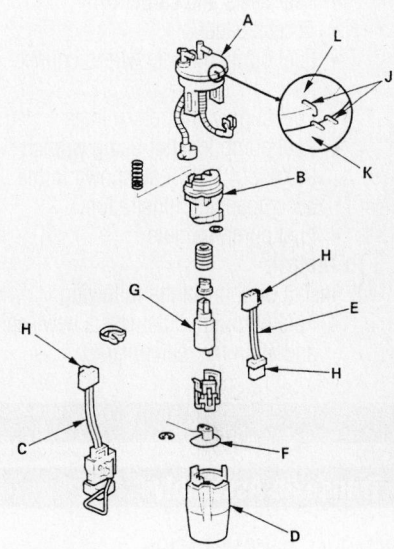

A. Bracket	F. Suction filter
B. Fuel filter	G. Fuel pump
C. Fuel gauge sender	H. Connectors
D. Case	J. Alignment marks
E. Wire harness	K. Fuel tank
	L. Fuel pump module

9308MG26

Exploded view of the fuel pump module

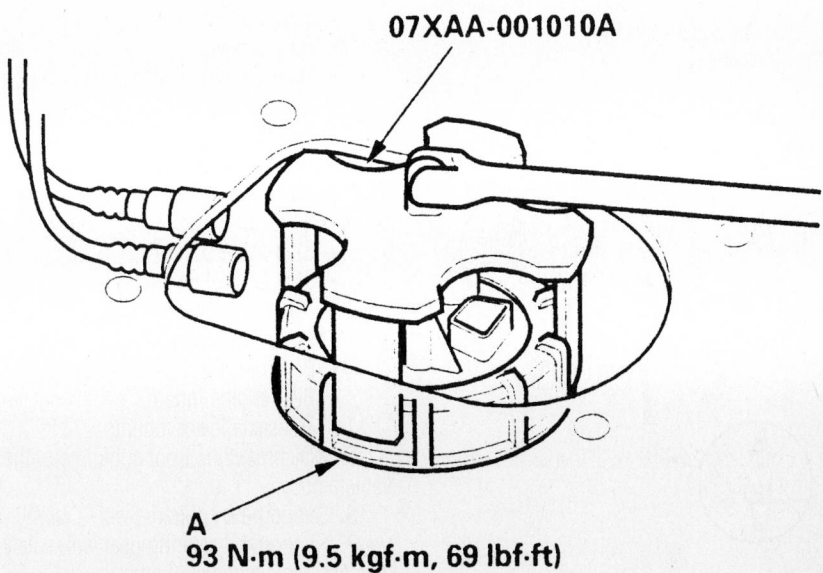

07XAA-001010A

A
93 N·m (9.5 kgf·m, 69 lbf·ft)

9358MG19

Use wrench 07XAA-001010A to remove and install the fuel pump locknut

Fuel Pump

REMOVAL & INSTALLATION

1. Before servicing the vehicle, refer to the precautions in the beginning of this section.
2. Relieve the fuel system pressure.
3. Remove or disconnect the following:
   - Negative battery cable
   - Rear seats and carpet
   - Access panel
   - Fuel pump module wiring connector
   - Fuel supply and return lines
   - Fuel pump locknut using wrench 07XAA-001010A as shown in the accompanying illustration.
   - Fuel pump module

To install:

4. Install or connect the following:
   - Fuel pump module. Use a new seal and align the matchmarks.
   - Fuel pump locknut and tighten to 69 ft. lbs. (93 Nm) using wrench 07XAA-001010A
   - Fuel supply and return lines
   - Fuel pump module wiring connector
   - Access panel
   - Rear seats and carpet
   - Negative battery cable
5. Start the engine and check for leaks.

Fuel Injector

REMOVAL & INSTALLATION

1. Before servicing the vehicle, refer to the precautions in the beginning of this section.
2. Relieve the fuel system pressure.
3. Remove or disconnect the following:
   - Negative battery cable
   - Intake manifold
   - Fuel lines
   - Fuel injector connectors
   - Fuel pressure regulator vacuum line
   - Fuel supply manifold
4. Separate the fuel injectors from the fuel supply manifold.

To install:

5. Install the fuel injectors to the fuel supply manifold with new cushion rings and O-rings.
6. Install new seal rings to the intake manifold.
7. Install or connect the following:
   - Fuel supply manifold and injector assembly. Tighten the bolts to 86 inch lbs. (10 Nm).
   - Fuel pressure regulator vacuum line
   - Fuel injector connectors
   - Fuel lines
   - Intake manifold
   - Negative battery cable
8. Start the engine and check for leaks.

DRIVE TRAIN

Transaxle Assembly

REMOVAL & INSTALLATION

Automatic Transaxle

1. Before servicing the vehicle, refer to the precautions in the beginning of this section.
2. Drain the transaxle.
3. Remove the engine appearance covers and install a support fixture to the engine lifting eyes.
4. Remove or disconnect the following:
   - Air intake assembly
   - Battery
   - Battery tray
   - Transaxle oil cooler lines

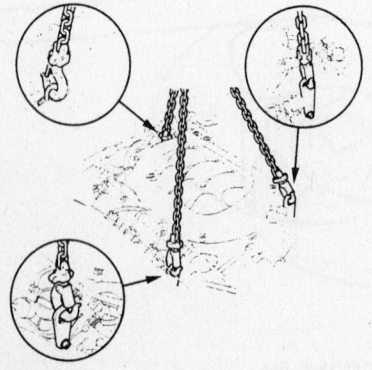

Support the engine while removing the transaxle

- Starter motor
- Transaxle ground cable
- Shift control solenoid valve connectors
- Clutch pressure switch connectors
- Mainshaft speed sensor connector
- Pressure control solenoid valve connectors
- Connector bracket
- Wiring harness cover
- Countershaft speed sensor connector
- Gear position switch connector
- Front motor mount
- Vacuum tube
- Splash shield
- Heated Oxygen Sensor (HO2S) connectors
- Exhaust front pipe
- Stabilizer bar links
- Lower ball joints
- Shift cable bracket
- Shift cable cover
- Shift control lever
- Torque converter
- Power steering hose bracket
- Power steering gear and brace
- Rear engine mount
- Transaxle lower mounts

5. Matchmark the front subframe to the vehicle body.
6. Support the subframe with a jack.
7. Support the steering gear with safety wire and remove the subframe.
8. Separate the inner CV-joints from the transaxle and intermediate shaft and support

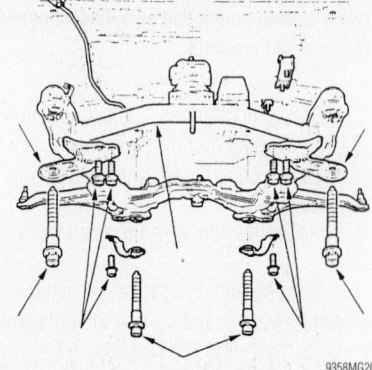

Remove the subframe bolts

the axle halfshafts out of the work area with safety wire.

9. Remove or disconnect the following:
   - Intermediate shaft
   - Transaxle flange bolts
   - Transaxle

To install:

➡**Use new circlips, split pins and self-locking nuts for assembly.**

10. Install or connect the following:
   - Transaxle. Tighten the flange bolts to 47 ft. lbs. (64 Nm).
   - Intermediate shaft. Tighten the bolts to 29 ft. lbs. (39 Nm).
   - Axle halfshafts
11. Raise the subframe into position and align the matchmarks. Tighten the subframe bolts to 76 ft. lbs. (103 Nm). Tighten the

front subframe bracket bolts to 54 ft. lbs. (74 Nm) and the rear bracket bolts to 86 ft. lbs. (117 Nm).

12. Install or connect the following:
- Transaxle lower mounts. Tighten the nuts to 28 ft. lbs. (38 Nm).
- Rear engine mount. Tighten the bolts to 28 ft. lbs. (38 Nm).
- Power steering gear. Tighten the bolts to 29 ft. lbs. (39 Nm).
- Power steering gear brace. Tighten the bolts to 43 ft. lbs. (58 Nm).
- Power steering hose bracket
- Torque converter. Tighten the bolts to 105 inch lbs. (12 Nm).
- Shift control lever
- Shift cable cover
- Shift cable bracket
- Lower ball joints. Tighten the nuts to 43–51 ft. lbs. (59–69 Nm).
- Stabilizer bar links. Tighten the nuts to 58 ft. lbs. (78 Nm).
- Exhaust front pipe
- HO₂S connectors
- Splash shield
- Vacuum tube
- Front motor mount. Tighten the nut to 40 ft. lbs. (54 Nm).
- Gear position switch connector
- Countershaft speed sensor connector
- Wiring harness cover
- Connector bracket
- Pressure control solenoid valve connectors
- Mainshaft speed sensor connector
- Clutch pressure switch connectors
- Shift control solenoid valve connectors
- Transaxle ground cable
- Starter motor
- Transaxle oil cooler lines
- Battery tray
- Battery
- Air intake assembly

13. Fill the transaxle to the correct level.
14. Start the engine and check for leaks.
15. Check the wheel alignment and adjust as necessary.

Halfshaft

REMOVAL & INSTALLATION

1. Before servicing the vehicle, refer to the precautions in the beginning of this section.
2. Drain the transaxle.
3. Remove or disconnect the following:
- Negative battery cable
- Front wheels

➡**If removing the left halfshaft, drain the transmission fluid, it is not necessary to drain the fluid if removing the right halfshaft.**

- Stabilizer bar link
- Lower ball joint
- Spindle nut

4. Pry the inboard joint from the transaxle or intermediate shaft.
5. Remove the outer CV-joint stub shaft from the hub by tapping the stub shaft with a plastic hammer.

To install:

➡**Use new circlips, split pins and self-locking nuts for assembly.**

6. Install the outer CV-joint stub shaft into the hub.
7. Install the inner CV-joint to the transaxle or intermediate shaft until the circlip locks in the retaining groove.
8. Install or connect the following:
- Lower ball joint. Tighten the nut to 43–51 ft. lbs. (59–69 Nm).
- Stabilizer bar link. Tighten the nut to 58 ft. lbs. (78 Nm).
- Spindle nut. Tighten the nut to 181 ft. lbs. (245 Nm).

- Front wheels
- Negative battery cable

9. Fill the transaxle to the correct level and check for leaks.

CV-Joint

OVERHAUL

OUTBOARD JOINT

1. Before servicing the vehicle, refer to the precautions in the beginning of this section.
2. Remove or disconnect the following:

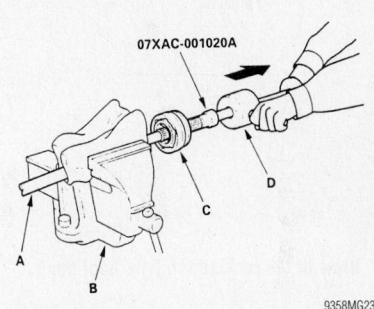

Removing the outboard joint using tool 07XAC-001020A

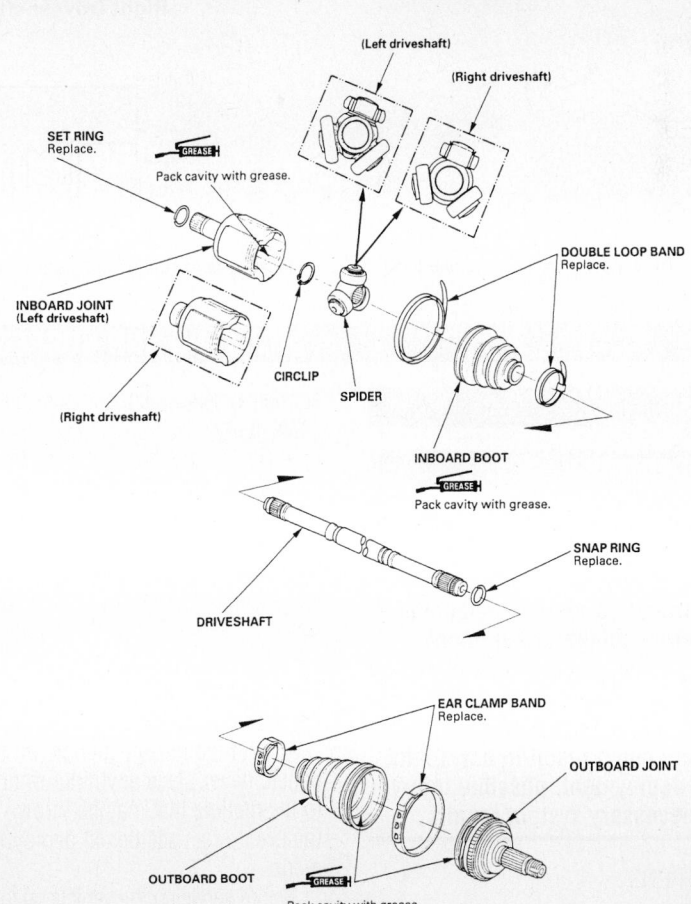

Front axle exploded view

- Axle halfshaft from the vehicle and place it in a vise
- Outboard joint boot clamps and push the boot back
- Outboard joint by driving it off the axle shaft with a brass drift and hammer or tool 07XAC-001020A, if available
- Outboard joint boot

To install:

➡ Use new circlips and boot clamps for assembly.

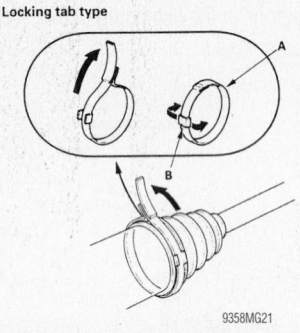

Locking tab type

9358MG21

View of the locking tab type boot band

Double loop type

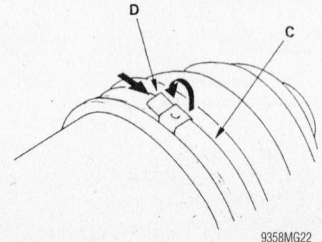

9358MG22

View of the double loop tab type boot band

3. Install the outboard joint boot and clamps to the axle shaft.
4. Fill the outboard joint with grease. Install the outboard joint to the axle shaft. Tap the stub shaft with a brass hammer to seat the circlip.
5. Fill the outboard joint boot with grease and install the boot clamps.
6. Install the axle halfshaft to the vehicle.

INBOARD JOINT

1. Before servicing the vehicle, refer to the precautions in the beginning of this section.
2. Remove or disconnect the following:
   - Axle halfshaft from the vehicle.
   - Inboard joint boot clamps and push the boot back
   - Inboard joint housing from the axle
   - Rollers from the spider

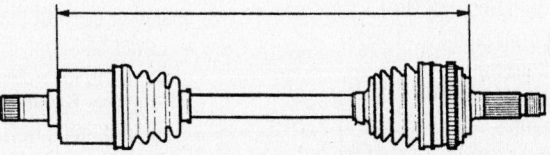

Right Driveshaft: 555.8—560.8 mm (21.9—22.1 in.)

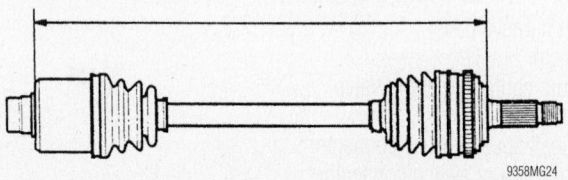

9358MG24

Make sure to adjust the length of the driveshafts as shown

- Snapring and the spider from the axle shaft
- Inboard joint boot

To install:

➡ Use new circlips and boot clamps for assembly.

3. Install or connect the following:
   - Inboard joint boot and clamps to the axle shaft
   - Spider with a new snapring
   - Rollers to the spider
4. Fill the joint housing with grease and install it.
5. Fill the inboard joint boot with grease and install the boot clamps. Make sure to adjust the length of the driveshafts as shown in the accompanying illustration.
6. Install the axle halfshaft to the vehicle.

STEERING AND SUSPENSION

Air Bag

❊❊ CAUTION

Some vehicles are equipped with an air bag system. The system must be disarmed before performing service on, or around, system components, the steering column, instrument panel components, wiring and sensors. Failure to follow the safety precautions and the disarming procedure could result in accidental air bag deployment, possible injury and unnecessary system repairs.

PRECAUTIONS

Several precautions must be observed when handling the inflator module to avoid accidental deployment and possible personal injury.

- Never carry the inflator module by the wires or connector on the underside of the module.
- When carrying a live inflator module, hold securely with both hands, and ensure that the bag and trim cover are pointed away.
- Place the inflator module on a bench or other surface with the bag and trim cover facing up.
- With the inflator module on the bench, never place anything on or close to the module that may be thrown in the event of an accidental deploy0 ment.

Before servicing the vehicle, also make sure to refer to the precautions in the beginning of this section as well.

DISARMING

Disconnect and isolate the negative battery cable. Wait 3 minutes for the system capacitor to discharge before performing any service.

Power Rack and Pinion Steering Gear

REMOVAL & INSTALLATION

❊❊ WARNING

Do not permit the steering wheel to turn whenever the steering gear is disconnected from the steering column. Damage to the air bag wiring can result.

1. Before servicing the vehicle, refer to the precautions in the beginning of this section.

2. Center the steering wheel and lock it in position.

3. Attach a support fixture to the engine lifting eyes.

4. Remove or disconnect the following:
- Negative battery cable
- Drivers side air bag assembly, if equipped
- Steering joint cover
- Steering flexible joint
- Front wheels
- Outer tie rod ends
- Splash shield
- Heated Oxygen Sensor (HO$_2$S) connectors
- Exhaust front pipe
- Fuel feed line
- Power steering fluid lines
- Rear engine mount

5. Support the front subframe with a jack.

6. Loosen the 14mm subframe bolts and remove the 12mm stiffener plate bolts.

7. Lower the subframe about 1 ³⁄₁₆ inches (30mm).

8. Remove or disconnect the following:
- Right steering gear mounting bracket
- Left steering gear mounting bolts
- Steering gear

To install:

9. Position the steering gear in the vehicle.

10. Install or connect the following:
- Left steering gear mounting bolts. Tighten the bolts to 43 ft. lbs. (58 Nm).
- Right steering gear mounting bracket. Tighten the bolts to 29 ft. lbs. (39 Nm).

11. Raise the subframe into position. Tighten the 14mm bolts to 76 ft. lbs. (103 Nm) and the 12mm bolts to 54 ft. lbs. (74 Nm).

12. Install or connect the following:
- Rear engine mount
- Power steering fluid lines
- Exhaust front pipe
- HO$_2$S connectors
- Splash shield
- Outer tie rod ends
- Front wheels. Position the wheels straight-ahead.
- Steering flexible joint. Tighten the pinch bolts to 16 ft. lbs. (22 Nm).
- Steering joint cover
- Negative battery cable

13. Fill the power steering system.

14. Check the wheel alignment and adjust as necessary.

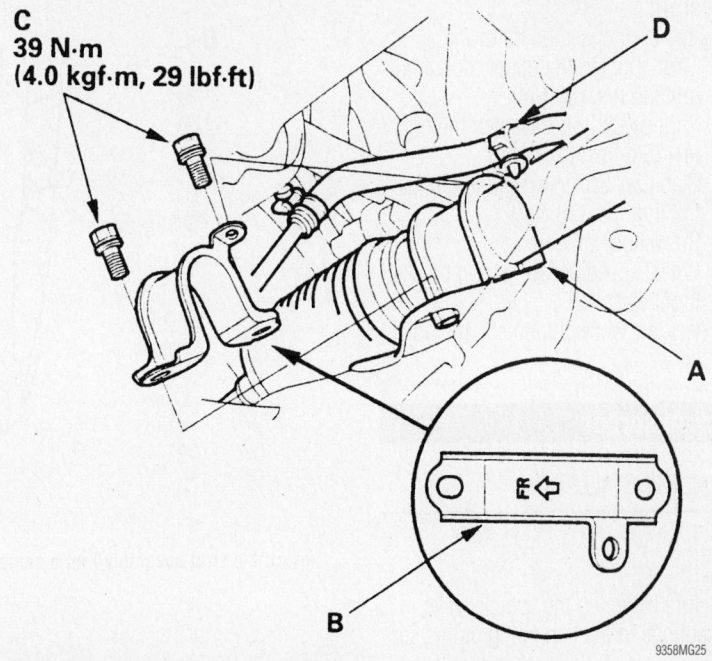

C
39 N·m
(4.0 kgf·m, 29 lbf·ft)

9358MG25

Make sure the right steering gear mounting bracket is oriented properly

Strut

REMOVAL & INSTALLATION

Front

1. Before servicing the vehicle, refer to the precautions in the beginning of this section.

2. Remove or disconnect the following:
- Front wheel
- Wheel speed sensor wiring bracket
- Brake hose bracket
- Stabilizer bar link
- Strut pinch bolts
- Upper mount nuts
- Strut

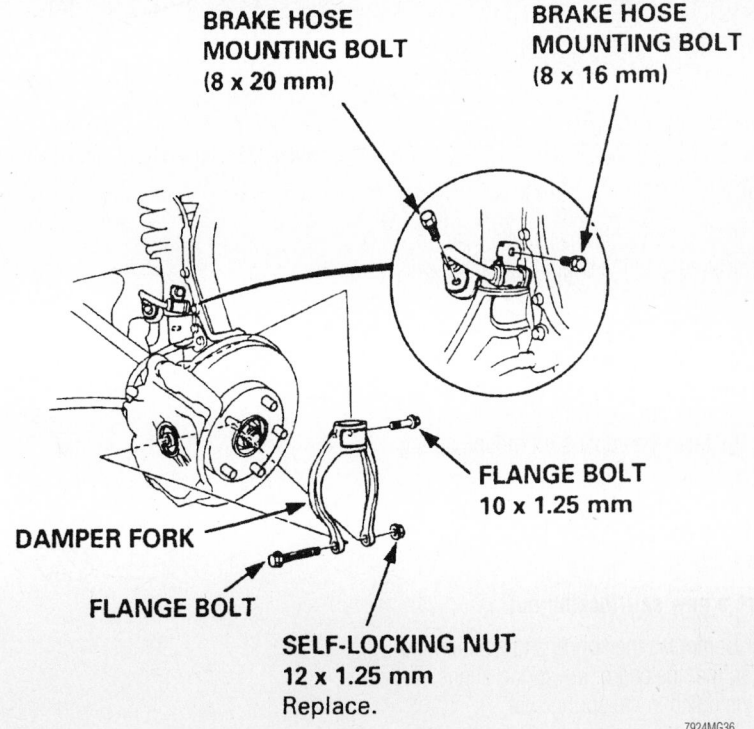

BRAKE HOSE MOUNTING BOLT (8 x 20 mm)

BRAKE HOSE MOUNTING BOLT (8 x 16 mm)

FLANGE BOLT 10 x 1.25 mm

DAMPER FORK

FLANGE BOLT

SELF-LOCKING NUT 12 x 1.25 mm Replace.

7924MG36

Identification of some of the front suspension components

To install:

3. Install or connect the following:
   - Strut. Tighten the upper mount nuts to 43 ft. lbs. (59 Nm).
   - Strut pinch bolts. Tighten the nuts to 116 ft. lbs. (157 Nm).
   - Stabilizer bar link. Tighten the nut to 58 ft. lbs. (78 Nm).
   - Brake hose bracket
   - Wheel speed sensor wiring bracket
   - Front wheel

4. Check the wheel alignment and adjust as necessary.

Shock Absorber

REMOVAL & INSTALLATION

Rear

1. Before servicing the vehicle, refer to the precautions in the beginning of this section.

2. Support the vehicle under the lower control arm.

3. Remove or disconnect the following:
   - Rear wheel
   - Upper shock absorber flange bolt
   - Lower shock absorber nut
   - Shock absorber

To install:

4. Install or connect the following:
   - Shock absorber. Tighten the fasteners to 47 ft. lbs. (64 Nm).
   - Rear wheel

Coil Spring

REMOVAL & INSTALLATION

Front

1. Before servicing the vehicle, refer to the precautions in the beginning of this section.

2. Remove the strut from the vehicle and install in a strut spring compressor. Compress the spring until the end of the spring comes away from the spring seat.

3. Remove the upper strut mount, spring seat and related components.

4. Remove the coil spring from the strut spring compressor.

To install:

➤**Use a new self-locking nut.**

5. Compress the spring and position the strut so that the end of the spring aligns with the notch in the spring seat.

6. Install the upper strut mounting components and tighten the nut to 22 ft. lbs. (29

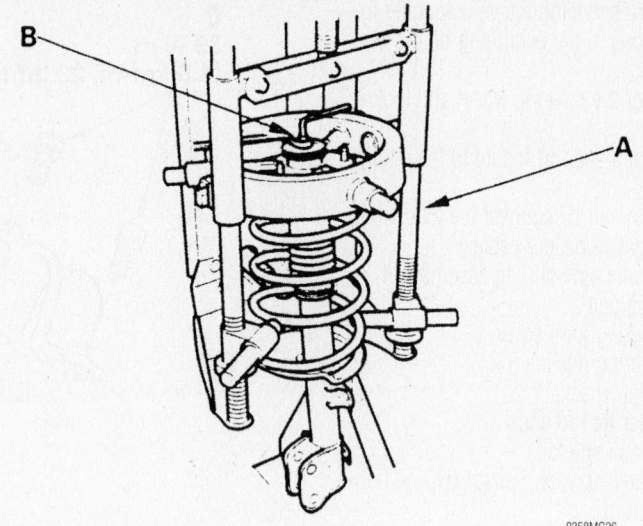

Install the strut assembly into a compressor

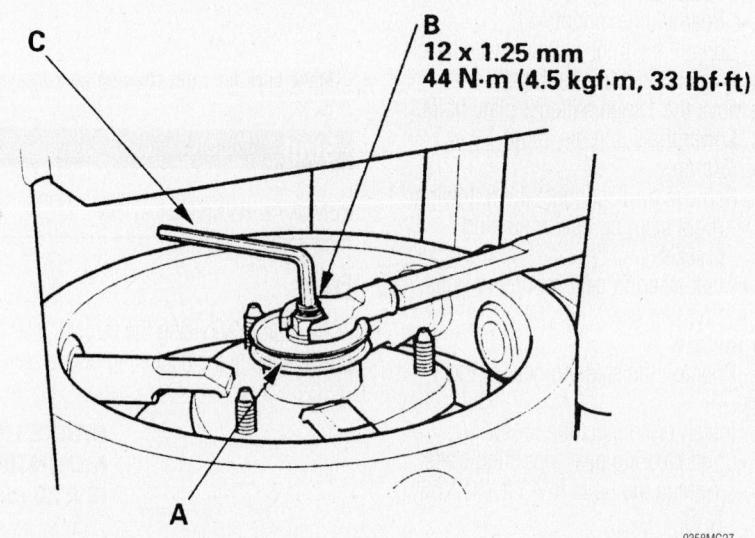

B
12 x 1.25 mm
44 N·m (4.5 kgf·m, 33 lbf·ft)

Tighten the strut nut to specification

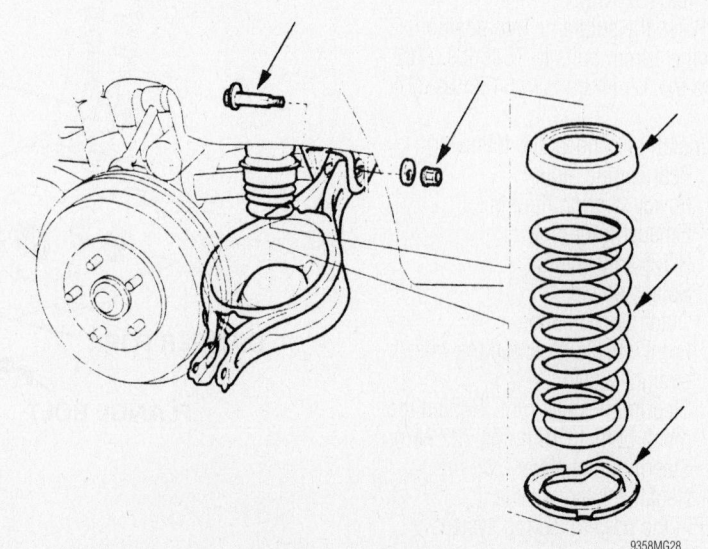

Exploded view of the rear spring assembly

Nm) for vehicles with 4 cylinder engines or to 33 ft. lbs. (44 Nm) for vehicles with 6 cylinder engines.

7. Install the strut to the vehicle.

8. Check the wheel alignment and adjust as necessary.

Rear

1. Before servicing the vehicle, refer to the precautions in the beginning of this section.

2. Support the vehicle under the lower control arm.

3. Remove or disconnect the following:
- Rear wheel
- Upper shock absorber flange bolt
- Lower shock absorber nut
- Shock absorber
- Wheel speed sensor wiring harness
- Lower control arm bolts

4. Lower the floor jack and remove the coil spring and spring seats.

To install:

→**Use new self-locking nuts for assembly.**

5. Place the coil spring and spring seats on the lower control arm and raise into position. Tighten the inboard bolt to 61 ft. lbs. and the outer bolt to 47 ft. lbs. (64 Nm).

6. Install or connect the following:
- Shock absorber. Tighten the fasteners to 47 ft. lbs. (64 Nm).
- Rear wheel

Upper Ball Joint

REMOVAL & INSTALLATION

The upper ball joints are replaced with the upper control arms as an assembly.

Lower Ball Joint

REMOVAL & INSTALLATION

The lower ball joints are replaced with the lower control arms as an assembly.

Upper Control Arm

REMOVAL & INSTALLATION

1. Before servicing the vehicle, refer to the precautions in the beginning of this section.

2. Support the lower control arm assembly with a floor jack.

3. Remove or disconnect the following:

- Upper ball joint
- Inner control arm flange bolts.
- Upper control arm

To install:

→**Use new self-locking nuts for assembly.**

4. Install the upper control arm. Tighten the ball joint nut to 29–35 ft. lbs. (39–47 Nm) and the inner flange nuts to 47 ft. lbs. (64 Nm).

CONTROL ARM BUSHING REPLACEMENT

The upper control arm bushings are serviced with the upper control arm as an assembly.

Lower Control Arm

REMOVAL & INSTALLATION

1. Before servicing the vehicle, refer to the precautions in the beginning of this section.

2. Remove or disconnect the following:

- Front wheel
- Lower ball joint
- Front inner flange bolt

- Rear inner flange bolt
- Lower control arm

To install:

→**Use a new split pin for assembly.**

3. Install or connect the following:
- Lower control arm. Tighten the inner flange bolt (A) to 69 ft. lbs. (93 Nm) and bolt (B) to 90 ft. lbs. (122 Nm) (69 ft. lbs. for 2004 models). Refer to the accompanying illustration for the locations of bolts A and B.
- Lower ball joint. Tighten the nut to 43–51 ft. lbs. (59–69 Nm).
- Front wheel

4. Check the wheel alignment and adjust as necessary.

CONTROL ARM BUSHING REPLACEMENT

The lower control arm bushings are serviced with the lower control arm as an assembly.

REAR INNER BUSHING

1. Before servicing the vehicle, refer to the precautions in the beginning of this section.

2. Remove or disconnect the following:

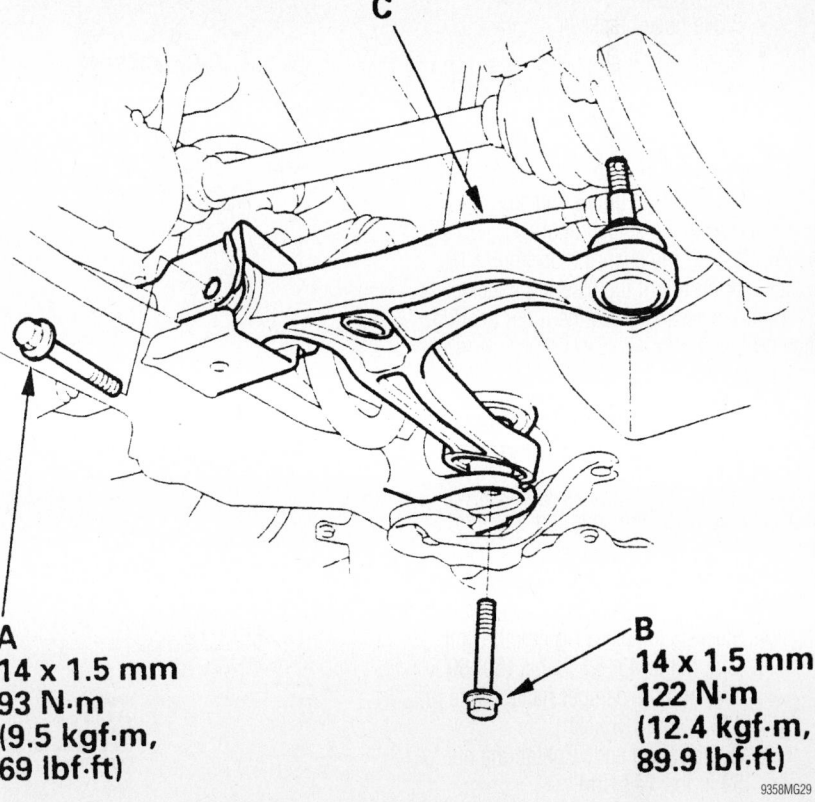

C

A
14 x 1.5 mm
93 N·m
(9.5 kgf·m,
69 lbf·ft)

B
14 x 1.5 mm
122 N·m
(12.4 kgf·m,
89.9 lbf·ft)

9358MG29

Tighten the lower control arm bolts to the specifications shown

- Front wheel
- Rear bushing bracket
- Rear bushing

To install:

➡ **Use a new self-locking nut for assembly.**

3. Install or connect the following:
   - Rear bushing. Tighten the nut to 61 ft. lbs. (83 Nm).
   - Rear bushing bracket. Tighten the bolts to 66 ft. lbs. (89 Nm).
   - Front wheel
4. Check the wheel alignment and adjust as necessary.

Wheel Bearings

ADJUSTMENT

The wheel bearings are sealed units and are not adjustable.

REMOVAL & INSTALLATION

Front

1. Before servicing the vehicle, refer to the precautions in the beginning of this section.
2. Remove or disconnect the following:
   - Front wheel
   - Spindle nut
   - Brake hose bracket
   - Brake caliper and rotor
   - Wheel speed sensor, if equipped
   - Outer tie rod end
   - Lower ball joint
   - Steering knuckle
3. Press the hub out of knuckle.
4. Remove the splash guard.
5. Remove the snapring and press the wheel bearing out of the steering knuckle.
6. If necessary, press the inner bearing race off of the hub.

To install:

➡ **Use a new ball joint nut, split pin, snapring and spindle nut for assembly.**

7. Press the bearing into the steering knuckle and install the snapring.
8. Install the splash guard.
9. Press the hub into the bearing.
10. Install or connect the following:
   - Steering knuckle. Tighten the ball joint nut to 43–51 ft. lbs. (59–69 Nm) and the damper flange bolts to 116 ft. lbs. (157 Nm).
   - Outer tie rod end. Tighten the nut to 32 ft. lbs. (43 Nm).
   - Wheel speed sensor, if equipped
   - Brake caliper and rotor. Tighten the

caliper bracket bolts to 80 ft. lbs. (108 Nm).
   - Brake hose bracket
   - Spindle nut. Tighten the nut to 181 ft. lbs. (245 Nm).
   - Front wheel
11. Check the wheel alignment and adjust as necessary.

Rear

WITH DRUM BRAKES

1. Before servicing the vehicle, refer to the precautions in the beginning of this section.
2. Remove or disconnect the following:
   - Rear wheel
   - Brake drum,
   - Spindle cap and nut
   - Hub and bearing assembly

To install:

➡ **Use a new spindle nut for assembly.**

3. Install or connect the following:
   - Hub and bearing assembly. Tighten the spindle nut to 181 ft. lbs. (245 Nm).

- Spindle cap
- Brake drum
- Rear wheel

WITH DISC BRAKES

1. Before servicing the vehicle, refer to the precautions in the beginning of this section.
2. Remove or disconnect the following:
   - Rear wheel
   - Brake caliper and rotor
   - Spindle cap, nut and washer
   - Hub and bearing assembly

To install:

➡ **Use a new spindle nut for assembly.**

3. Install or connect the following:
   - Hub and bearing assembly. Tighten the spindle nut to 181 ft. lbs. (245 Nm).
   - Spindle cap
   - Brake caliper and rotor. Tighten the caliper bracket bolts to 28 ft. lbs. (38 Nm).
   - Rear wheel

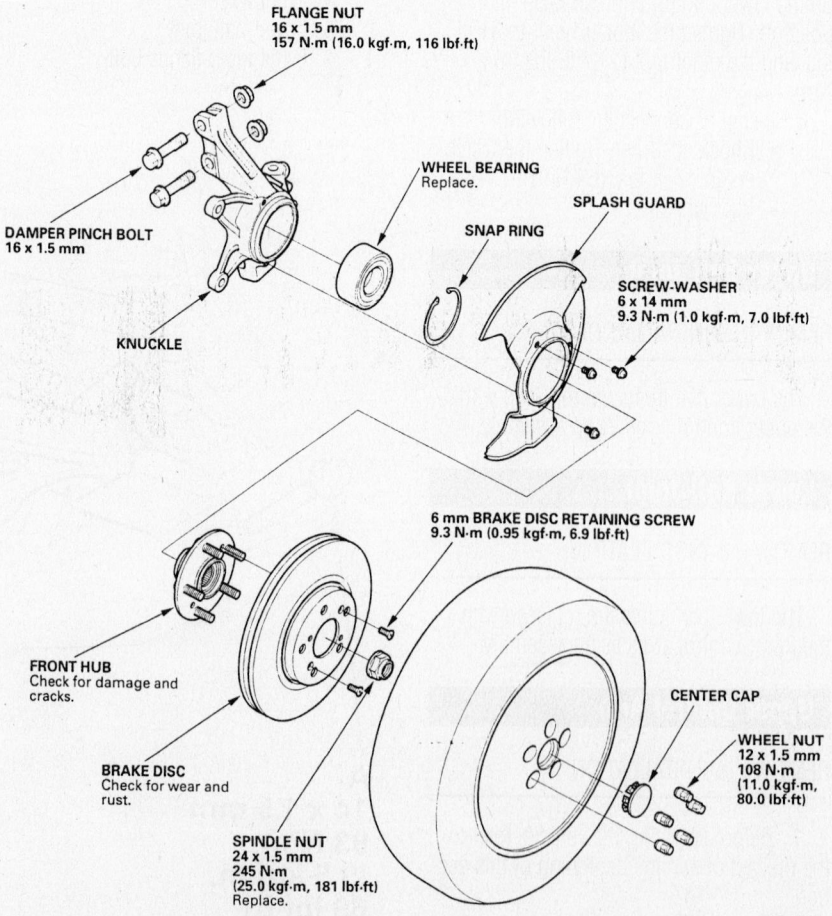

FLANGE NUT
16 x 1.5 mm
157 N·m (16.0 kgf·m, 116 lbf·ft)

DAMPER PINCH BOLT
16 x 1.5 mm

KNUCKLE

WHEEL BEARING
Replace.

SNAP RING

SPLASH GUARD

SCREW-WASHER
6 x 14 mm
9.3 N·m (1.0 kgf·m, 7.0 lbf·ft)

6 mm BRAKE DISC RETAINING SCREW
9.3 N·m (0.95 kgf·m, 6.9 lbf·ft)

FRONT HUB
Check for damage and cracks.

BRAKE DISC
Check for wear and rust.

SPINDLE NUT
24 x 1.5 mm
245 N·m
(25.0 kgf·m, 181 lbf·ft)
Replace.

CENTER CAP

WHEEL NUT
12 x 1.5 mm
108 N·m
(11.0 kgf·m,
80.0 lbf·ft)

9358MG32

Exploded view of the front hub/knuckle assembly

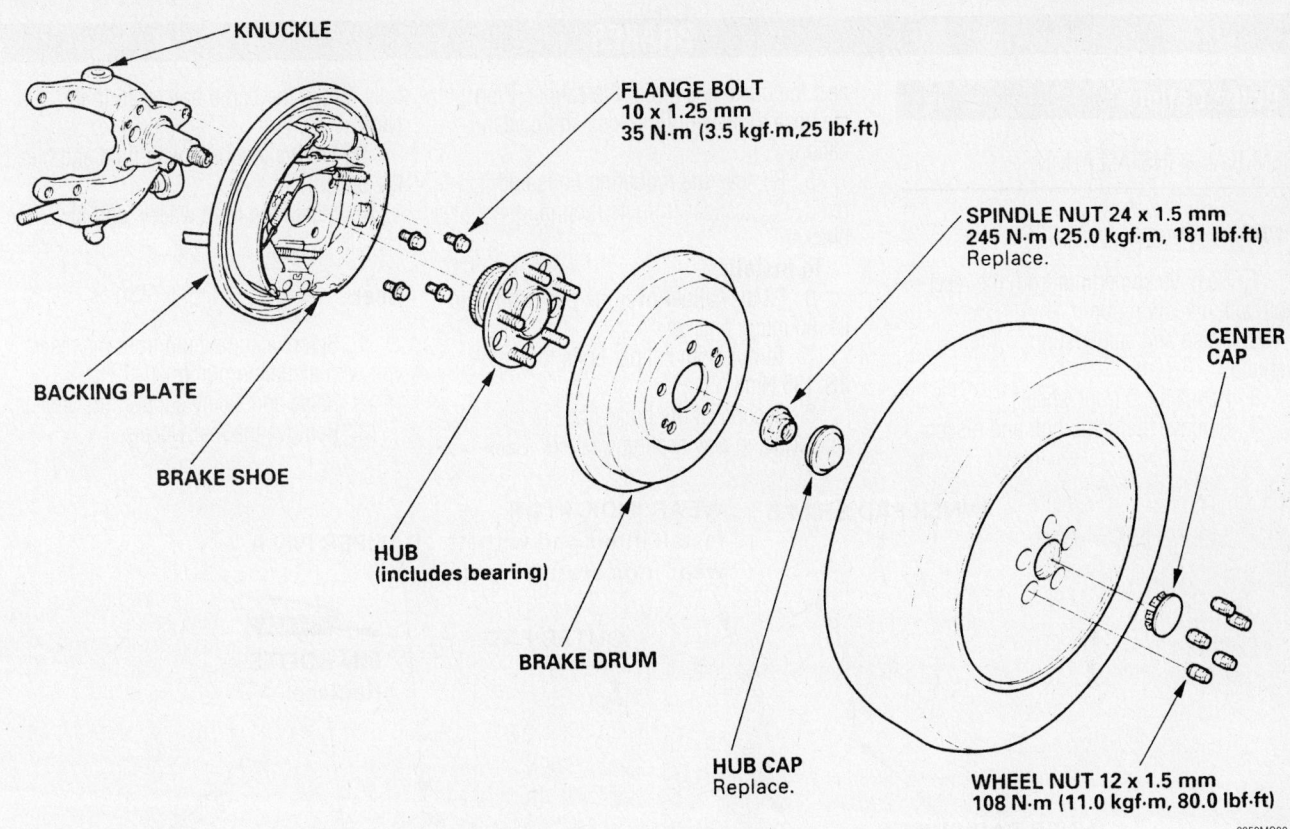

KNUCKLE

FLANGE BOLT
10 x 1.25 mm
35 N·m (3.5 kgf·m, 25 lbf·ft)

SPINDLE NUT 24 x 1.5 mm
245 N·m (25.0 kgf·m, 181 lbf·ft)
Replace.

CENTER
CAP

BACKING PLATE

BRAKE SHOE

HUB
(includes bearing)

BRAKE DRUM

HUB CAP
Replace.

WHEEL NUT 12 x 1.5 mm
108 N·m (11.0 kgf·m, 80.0 lbf·ft)

9358MG30

Exploded view of the rear hub/bearing assembly on models equipped with drum brakes

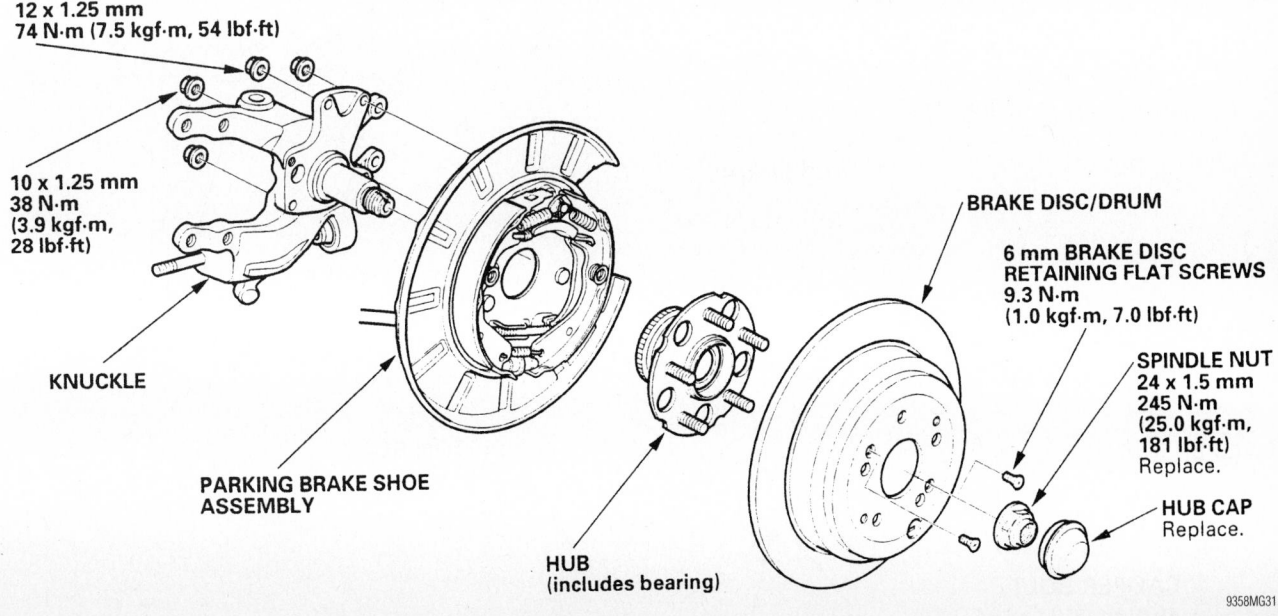

12 x 1.25 mm
74 N·m (7.5 kgf·m, 54 lbf·ft)

10 x 1.25 mm
38 N·m
(3.9 kgf·m,
28 lbf·ft)

KNUCKLE

BRAKE DISC/DRUM

6 mm BRAKE DISC
RETAINING FLAT SCREWS
9.3 N·m
(1.0 kgf·m, 7.0 lbf·ft)

SPINDLE NUT
24 x 1.5 mm
245 N·m
(25.0 kgf·m,
181 lbf·ft)
Replace.

HUB CAP
Replace.

PARKING BRAKE SHOE
ASSEMBLY

HUB
(includes bearing)

9358MG31

Exploded view of the rear hub/bearing assembly on models equipped with disc brakes

BRAKES

Brake Caliper

REMOVAL & INSTALLATION

Front

1. Remove some fluid from the reservoir with a suction pump.
2. Raise and safely support the vehicle.
3. Remove the front wheels.
4. Remove the banjo bolt and disconnect the brake hose from the caliper. Plug the hose to prevent fluid loss and contamination.
5. Remove the mounting bolts and remove the caliper from its mounting bracket.

To Install:

6. Fit the caliper over the pads and onto its mounting bracket.
7. Torque both caliper bolts to 27 ft. lbs. (36 Nm).
8. Reconnect the brake hose to the caliper using new sealing washers. Carefully torque the banjo bolt to 25 ft. lbs. (35 Nm).
9. Fill the reservoir with fluid and bleed the brakes.
10. Install the front wheels and lower the vehicle.

Rear

1. Remove some fluid from the reservoir with a suction pump.
2. Raise and safely support the vehicle.
3. Remove the rear wheels.

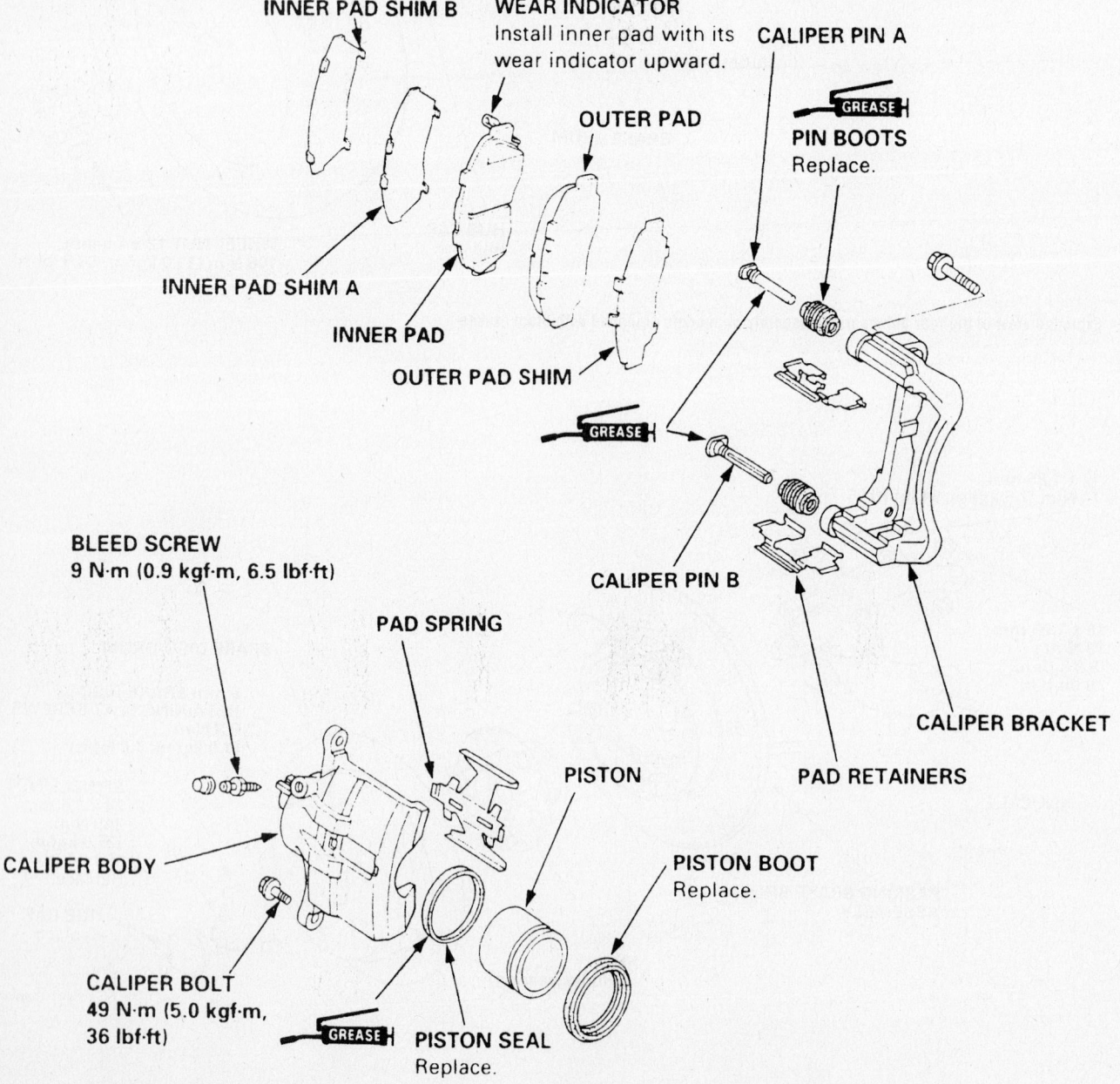

INNER PAD SHIM B

WEAR INDICATOR
Install inner pad with its wear indicator upward.

CALIPER PIN A

PIN BOOTS
Replace.

OUTER PAD

INNER PAD SHIM A

INNER PAD

OUTER PAD SHIM

CALIPER PIN B

BLEED SCREW
9 N·m (0.9 kgf·m, 6.5 lbf·ft)

PAD SPRING

CALIPER BRACKET

PISTON

PAD RETAINERS

CALIPER BODY

PISTON BOOT
Replace.

CALIPER BOLT
49 N·m (5.0 kgf·m, 36 lbf·ft)

GREASE

PISTON SEAL
Replace.

93026G57

Exploded view of the front caliper components—Odyssey

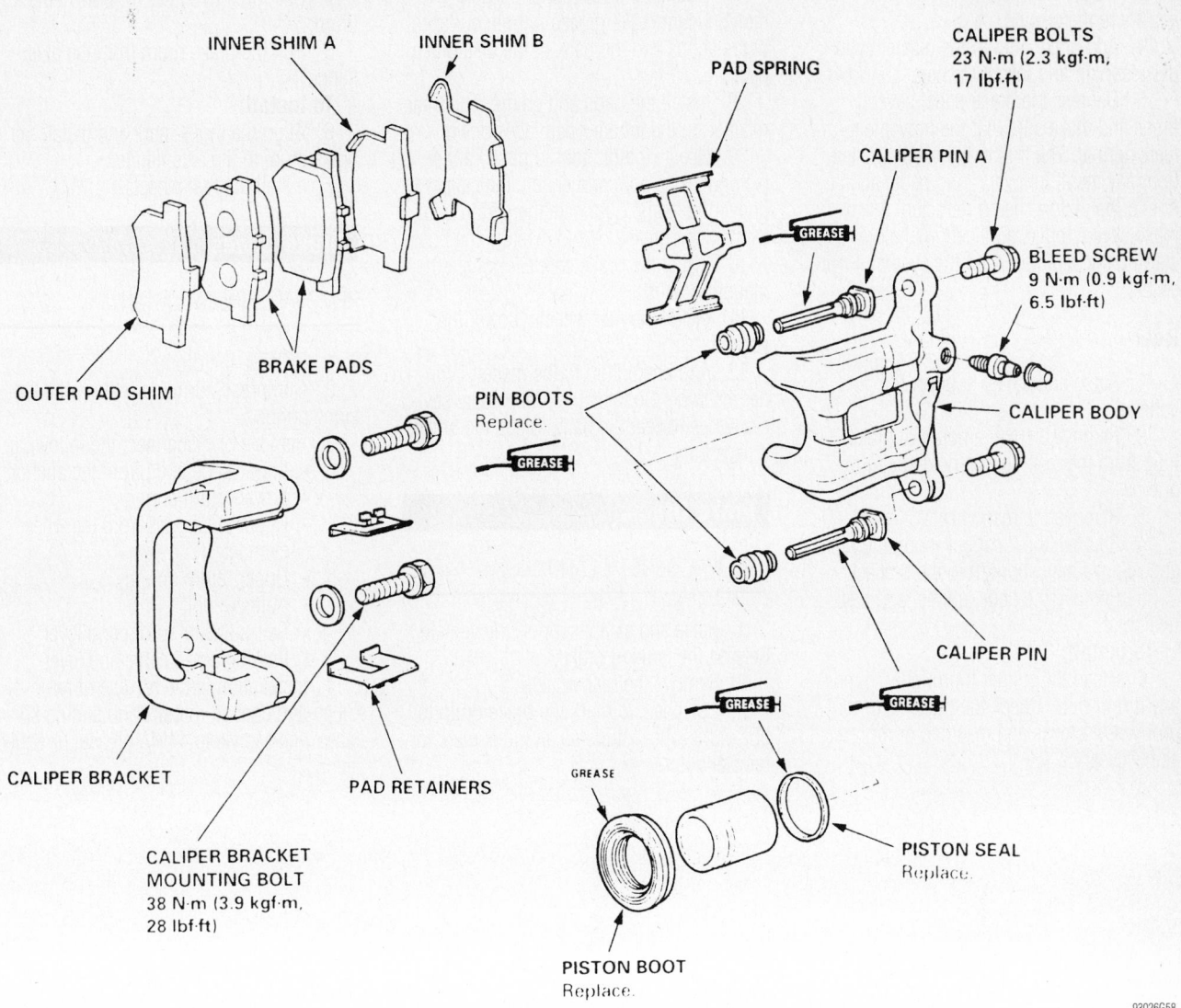

INNER SHIM A INNER SHIM B PAD SPRING CALIPER BOLTS 23 N·m (2.3 kgf·m, 17 lbf·ft)

CALIPER PIN A

BLEED SCREW 9 N·m (0.9 kgf·m, 6.5 lbf·ft)

OUTER PAD SHIM BRAKE PADS PIN BOOTS Replace. CALIPER BODY

GREASE

CALIPER PIN

CALIPER BRACKET PAD RETAINERS GREASE

CALIPER BRACKET MOUNTING BOLT 38 N·m (3.9 kgf·m, 28 lbf·ft) GREASE

PISTON SEAL Replace.

PISTON BOOT Replace.

93026G58

Exploded view of the rear caliper components—Odyssey

4. Remove the banjo bolt and disconnect the brake hose from the caliper. Plug the hose to prevent fluid loss and contamination.

5. Remove the 2 caliper mounting bolts. Remove the caliper from its mounting bracket.

To Install:

6. Fit the caliper over the pads and onto its mounting bracket.

7. Tighten the caliper bolts to 27 ft. lbs. (36 Nm).

8. Reconnect the brake hose with new sealing washers. Tighten the banjo bolt to 17 ft. lbs. (34 Nm).

9. Fill the reservoir with fluid and bleed the brake system. Adjust the parking brake if necessary.

10. Install the rear wheels and lower the vehicle.

Disc Brake Pads

REMOVAL & INSTALLATION

Front

1. Raise and support the vehicle safely.
2. Remove the front wheels.
3. Remove a small amount of brake fluid from the reservoir using a suction pump.
4. Unbolt the brake hose clamp from the knuckle by removing the retaining bolts.
5. Remove the lower caliper retaining bolt and pivot the caliper upward, off of the pads.
6. Remove the pad shim and pad retainers. Remove the disc brake pads from the caliper.

To install:

7. Clean the caliper thoroughly; remove any rust from the lip of the disc or rotor. Check the brake rotor for grooves or cracks. If any heavy scoring is present, the rotor must be replaced.

8. Install the pad retainers. Apply molybdenum brake grease to both surfaces of the shims and the back of the disc brake pads.

9. Install the pads and shims. The pad with the wear indicator goes in the inboard position.

10. Push in the caliper piston so the caliper will fit over the pads. This is most easily accomplished with a pad spreader or large C-clamp.

11. Pivot the caliper down into position and tighten the mounting bolt.

12. Connect the brake hose to the knuckle, if removed.

13. Install the wheel and lower the vehicle to the ground.

14. Add brake fluid to the master cylinder reservoir and install the cap.

15. Depress the brake pedal several times and make sure that the movement feels normal. The first brake pedal application may result in a very long pedal action due to the pistons being retracted. Always make several brake applications before starting the vehicle. Bleed the system if necessary.

Rear

1. Raise and safely support the vehicle.

2. Remove a small amount of brake fluid from the reservoir using a suction pump.

3. Remove the rear wheels.

4. Remove the 2 caliper mounting bolts and remove the caliper from the bracket.

5. Remove the pads, shims, and pad retainers.

To install:

6. Clean the caliper thoroughly; remove any dirt or dust. Check the brake rotor for grooves or cracks and machine or replace, as necessary.

7. Install the pad retainers. Apply molybdenum brake grease to both surfaces of the shims and the back of the disc brake pads.

8. Install the pads and shims. The wear retainer on the inboard pad faces down.

9. Use a suitable tool to push caliper piston into its bore and enable the caliper to fit over the pads. Lubricate the piston boot with silicon grease. Avoid twisting the boot.

10. Install the brake caliper. Tighten the mounting bolts.

11. Install the rear wheels. Lower the vehicle.

12. Add brake fluid to the master cylinder reservoir. Depress the brake pedal several times to seat the pads. Bleed the brakes if necessary.

Brake Drums

REMOVAL & INSTALLATION

1. Raise and safely support the vehicle. Release the parking brake.

2. Remove the rear wheels.

3. Use chalk to mark the brake drum to one of the wheel studs as an index mark for reinstallation.

4. Use forcing screws to remove the drum.

5. Pull the brake drum from the axle flange.

To install:

6. Align the index mark and install the brake drum to the axle flange.

7. Install the rear wheels.

Brake Shoes

REMOVAL & INSTALLATION

1. Remove the drum.

2. Compress, turn and remove the tension springs.

3. Remove or disconnect the following:
   - Lower shoe ends from the anchor
   - Lower return spring
   - Upper shoe ends from the wheel cylinder
   - Upper return spring
   - Adjuster bolt
   - Self-adjuster spring and lever
   - Parking brake cable and lever

4. Installation is the reverse of removal. Clean and coat all threads and sliding surfaces with Molykote 44MA grease, or equivalent.

HONDA

Pilot

SPECIFICATION AND MAINTENANCE CHARTS

ENGINE AND VEHICLE IDENTIFICATION CHART

Engine Code							Model Year	
Code	Liters (cc)	Cu. In.	Cyl.	Fuel Sys.	Engine Type	Eng. Mfg.	Code ①	Year
J35A4	3.5 (3471)	212	6	SMFI	SOHC	Honda	3	2003
							4	2004
							5	2005

SOHC: Single Overhead Cam

SMFI: Sequential Multi-port Fuel Injection

① 10th position of VIN

67162-HPIL-C01

GENERAL ENGINE SPECIFICATIONS

Year	Model	Engine Displacement Liters (VIN)	Net Horsepower @ rpm	Net Torque @ rpm (ft. lbs.)	Bore x Stroke (in.)	Com-pression Ratio	Oil Pressure @ rpm
2003	Pilot	3.5 (J35A4)	210@5200	229@4300	3.50x3.66	9.4:1	71@3000
2004-05	Pilot	3.5 (J35A4)	210@5200	229@4300	3.50x3.66	9.4:1	71@3000

SMFI: Sequential Multi-port Fuel Injection

67162-HPIL-C02

ENGINE TUNE-UP SPECIFICATIONS

Year	Engine Displacement Liters (VIN)	Spark Plug Gap (in.)	Ignition Timing (deg.)		Fuel Pump (psi)	Idle Speed (rpm)		Valve Clearance (in.)	
			MT	AT		MT	AT	In.	Ex.
2003	3.5 (J35A4)	0.039-0.043	—	8-12B	32-40	—	680-780	0.008-0.009	0.011-0.013
2004-05	3.5 (J35A4)	0.039-0.043	—	8-12B	32-40	—	680-780	0.008-0.009	0.011-0.013

NOTE: The Vehicle Emission Control Information label often reflects changes made during production and must be used if they differ from this chart.

NOTE: The fuel pressure readings are given with the vacuum hose connected to the regulator and the engine running

B: Before top dead center

67162-HPIL-C03

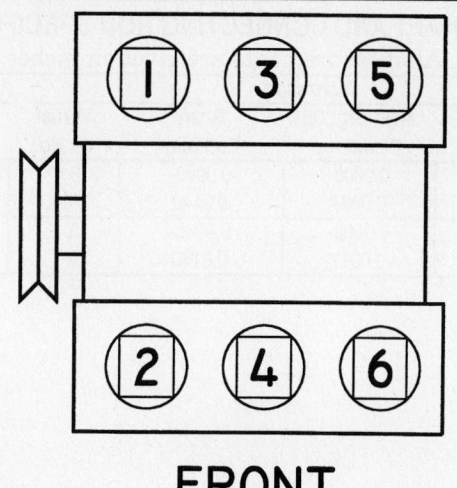

FRONT

9308MG32

3.5L Engine
Firing Order: 1–2–3–4–5–6
Distributorless ignition system (One coil per cylinder)

CAPACITIES

Year	Model	Engine Displacement Liters (VIN)	Engine Oil with Filter (qts.)	Transmission (pts.) 5-Spd	Transmission (pts.) Auto.	Transfer Case (pts.)	Drive Axle Front (pts.)	Drive Axle Rear (pts.)	Fuel Tank (gal.)	Cooling System (qts.)
2003	Pilot	3.5 (J35A4)	5.0	—	①	—	—	5.5	19.2	8.0
2004-05	Pilot	3.5 (J35A4)	5.0	—	①	—	—	5.5	19.2	8.0

NOTE: All capacities are approximate. Add fluid gradually and check to be sure a proper fluid level is obtained.

① Without filter change: 6.4 pts;

   With filter Change: 16.4 pts.

67162-HPIL-C04

VALVE SPECIFICATIONS

Year	Engine Displacement Liters (VIN)	Seat Angle (deg.)	Face Angle (deg.)	Spring Test Pressure (lbs. @ in.)	Spring Installed Height (in.)	Stem-to-Guide Clearance (in.) Intake	Stem-to-Guide Clearance (in.) Exhaust	Stem Diameter (in.) Intake	Stem Diameter (in.) Exhaust
2003	3.5 (J35A4)	45	45	NA	①	0.0008-0.0018	0.0022-0.0031	0.2159-0.2163	0.2146-0.2150
2004-05	3.5 (J35A4)	45	45	NA	①	0.0008-0.0018	0.0022-0.0031	0.2159-0.2163	0.2146-0.2150

NA: Not Available

① Valve spring free length:

   Intake: 1.9713 in.

   Exhaust: 2.1060 in.

67162-HPIL-C05

CRANKSHAFT AND CONNECTING ROD SPECIFICATIONS
All measurements are given in inches

Year	Engine Displacement Liters (VIN)	Crankshaft				Connecting Rod		
		Main Brg. Journal Dia.	Main Brg. Oil Clearance	Shaft End-play	Thrust on No.	Journal Diameter	Oil Clearance	Side Clearance
2003	3.5 (J35A4)	2.8337-2.8346	0.0008-0.0017	0.0040-0.0140	3	2.1644-2.1654	0.0008-0.0017	0.0060-0.0140
2004-05	3.5 (J35A4)	2.8337-2.8346	0.0008-0.0017	0.0040-0.0140	3	2.1644-2.1654	0.0008-0.0017	0.0060-0.0140

67162-HPIL-C06

PISTON AND RING SPECIFICATIONS
All measurements are given in inches

Year	Engine Displacement Liters (VIN)	Piston Clearance	Ring Gap			Ring Side Clearance		
			Top Compression	Bottom Compression	Oil Control	Top Compression	Bottom Compression	Oil Control
2003	3.5 (J35A4)	0.0006-0.0016	0.0080-0.0140	0.0160-0.0220	0.0080-0.0280	0.0014-0.0024	0.0012-0.0022	NA
2004-05	3.5 (J35A4)	0.0006-0.0016	0.0080-0.0140	0.0160-0.0220	0.0080-0.0280	0.0014-0.0024	0.0012-0.0022	NA

NA: Not Available

67162-HPIL-C07

TORQUE SPECIFICATIONS
All readings in ft. lbs.

Year	Engine Displacement Liters (VIN)	Cylinder Head Bolts	Main Bearing Bolts	Rod Bearing Bolts	Crankshaft Damper Bolts	Flywheel Bolts	Manifold		Spark Plugs	Oil Pan Drain Plug
							Intake	Exhaust		
2003	3.5 (J35A4)	①	②	③	181	54	16	23	13	29
2004-05	3.5 (J35A4)	①	②	③	181	54	16	23	13	29

NOTE: Dip main bearing bolts and crankshaft damper bolt in clean engine oil prior to tightening.

① Step 1: 29 ft. lbs.
Step 2: 51 ft. lbs.
Step 3: 72 ft. lbs.

② 11mm bolt 56 ft. lbs.
10mm bolt 36 ft. lbs.

③ Step 1: 14 ft. lbs.
Step 2: 90 degrees

67162-HPIL-C08

WHEEL ALIGNMENT

Year	Model		Caster Range (+/-Deg.)	Caster Preferred Setting (Deg.)	Camber Range (+/-Deg.)	Camber Preferred Setting (Deg.)	Toe-in (in.)
2003	Pilot	F	1.00	1.00	1.00	30	0+/-1/16
		R	—	—	0	30	0+/-1/16
2004-05	Pilot	F	1.00	1.00	1.00	30	0+/-1/16
		R	—	—	0	30	0+/-1/16

67162-HPIL-C09

TIRE, WHEEL AND BALL JOINT SPECIFICATIONS

Year	Model	OEM Tires Standard	OEM Tires Optional	Tire Pressures (psi) Front	Tire Pressures (psi) Rear	Wheel Size	Ball Joint Inspection	Lug Nut
2003	Pilot	P235/65R17	None	32	32	R17	NS	80
2004-05	Pilot	P235/65R17	None	32	32	R17	NS	80

OEM: Original Equipment Manufacturer

PSI: Pounds Per Square Inch

NS: Not specified by manufacturer

67162-HPIL-C10

BRAKE SPECIFICATIONS
Honda Pilot
All measurements in inches unless noted

Year	Model		Brake Disc Original Thickness	Brake Disc Minimum Thickness	Brake Disc Maximum Runout	Brake Drum Diameter Original Inside Diameter	Max. Wear Limit	Maximum Machine Diameter	Minimum Lining Thickness Front	Minimum Lining Thickness Rear	Brake Caliper Bracket Bolts (ft. lbs.)	Brake Caliper Mounting Bolts (ft. lbs.)
2003	Pilot	F	1.100	1.020	0.004	—	—	—	0.060	—	80	27
		R	0.430	0.350	0.004	—	—	—	—	0.41	41	27
2004-05	Pilot	F	1.100	1.020	0.004	—	—	—	0.060	—	80	27
		R	0.430	0.350	0.004	—	—	—	—	0.41	41	27

F: Front

R: Rear

67162-HPIL-C11

SCHEDULED MAINTENANCE INTERVALS
HONDA—PILOT

TO BE SERVICED	TYPE OF SERVICE	VEHICLE MILEAGE INTERVAL (x1000)															
		7.5	15	22.5	30	37.5	45	52.5	60	67.5	75	82.5	90	97.5	105	112.5	120
Accessory drive belts	I & A				✓				✓				✓				✓
Air cleaner element	R				✓				✓				✓				✓
Brake fluid	R	Every 3 years															
Brake hoses & lines (including ABS)	I		✓		✓		✓		✓		✓		✓		✓		✓
Cooling system hoses & connections	I		✓		✓		✓		✓		✓		✓		✓		✓
Engine coolant ①	R						✓						✓				
Engine oil	R	✓	✓	✓	✓	✓	✓	✓	✓	✓	✓	✓	✓	✓	✓	✓	✓
Engine oil and coolant levels	I	Inspect at each fuel stop															
Engine oil filter	R		✓		✓		✓		✓		✓		✓		✓		✓
Exhaust system	I		✓		✓		✓		✓		✓		✓		✓		✓
Fluid levels and condition	I		✓		✓		✓		✓		✓		✓		✓		✓
Front and rear brakes	I		✓		✓		✓		✓		✓		✓		✓		✓
Fuel lines & connection	I		✓		✓		✓		✓		✓		✓		✓		✓
Halfshaft boots	I		✓		✓		✓		✓		✓		✓		✓		✓
Idle speed	I & A														✓		
Parking brake system	I & A		✓		✓		✓		✓		✓		✓		✓		✓
Rear differential fluid	R	✓			✓				✓				✓				✓
Rotate and inspect tires	I	✓	✓	✓	✓	✓	✓	✓	✓	✓	✓	✓	✓	✓	✓	✓	✓
Spark plugs	R														✓		
Supplemental Restrain system (SRS)	I	Inspect the SRS 10 years after production															
Suspension components	I		✓		✓		✓		✓		✓		✓		✓		✓
Tie rod ends, steering gear box & boots	I		✓		✓		✓		✓		✓		✓		✓		✓
Timing belt	R														✓		
Transmission fluid	R							✓			✓				✓		
Valve clearance	I	Adjust if valves are noisy															
Water pump	S/I														✓		

R: Replace   I: Inspect   A: Adjust

① Every 12,000 miles or 10 years, then every 60,000 miles or 5 years

FREQUENT OPERATION MAINTENANCE (SEVERE SERVICE)

If a vehicle is operated under any of the following conditions it is considered severe service:

- Towing a trailer or using a camper or car-top carrier.
- Repeated short trips of less than 5 miles in temperatures below freezing, or trips of less than 10 miles in any temperature.
- Extensive idling or low-speed driving for long distances as in heavy commercial use, such as delivery, taxi or police cars.
- Operating on rough, muddy or salt-covered roads.
- Operating on unpaved or dusty roads.
- Driving in extremely hot (over 90°) conditions.

Air cleaner element: replace every 15,000 miles

Engine oil and filter: replace every 3750 miles or 6 months, whichever occurs first.

Timing belt: replace every 60,000 miles if the vehicle is regularly driven in temperatures above 110°F or below -20°F, or if frequently towing a trailer.

Transmission fluid: replace every 30,000 miles.

Rear differential fluid: replace every 60,000 miles.

Front and rear brakes: inspect every 7500 miles or 6 months, whichever occurs first.

Locks and hinges: lubricate every 15,000 miles.

Tie rods, steering gear box, boots: inspect every 7500 miles or 6 months, whichever occurs first.

Suspension components: inspect every 7500 miles or 6 months, whichever occurs first.

Halfshaft boots: inspect every 7500 miles or 6 months, whichever occurs first.

PRECAUTIONS

Before servicing any vehicle, please be sure to read all of the following precautions, which deal with personal safety, prevention of component damage, and important points to take into consideration when servicing a motor vehicle:

• Never open, service or drain the radiator or cooling system when the engine is hot; serious burns can occur from the steam and hot coolant.

• Observe all applicable safety precautions when working around fuel. Whenever servicing the fuel system, always work in a well-ventilated area. Do not allow fuel spray or vapors to come in contact with a spark, open flame or excessive heat (a hot drop light, for example). Keep a dry chemical fire extinguisher near the work area. Always keep fuel in a container specifically designed for fuel storage; also, always properly seal fuel containers to avoid the possibility of fire or explosion. Refer to the additional fuel system precautions later in this section.

• Fuel injection systems often remain pressurized, even after the engine has been turned **OFF**. The fuel system pressure must be relieved before disconnecting any fuel lines. Failure to do so may result in fire and/or personal injury.

• Brake fluid often contains polyglycol ethers and polyglycols. Avoid contact with the eyes and wash your hands thoroughly after handling brake fluid. If you do get brake fluid in your eyes, flush your eyes with clean, running water for 15 minutes. If eye irritation persists, or if you have taken brake fluid internally, IMMEDIATELY seek medical assistance.

• The EPA warns that prolonged contact with used engine oil may cause a number of skin disorders, including cancer! You should make every effort to minimize your exposure to used engine oil. Protective gloves should be worn when changing oil. Wash your hands and any other exposed skin areas as soon as possible after exposure to used engine oil. Soap and water, or waterless hand cleaner should be used.

• All new vehicles are now equipped with an air bag system. The system must be disabled before performing service on or around system components, steering column, instrument panel components, wiring and sensors. Failure to follow safety and disabling procedures could result in accidental air bag deployment, possible personal injury and unnecessary system repairs.

• Always wear safety goggles when working with, or around, the air bag system. When carrying a non-deployed air bag, be sure the bag and trim cover are pointed away from your body. When placing a non-deployed air bag on a work surface, always face the bag and trim cover upward, away from the surface. This will reduce the motion of the module if it is accidentally deployed. Refer to the additional air bag system precautions later in this section.

• Clean, high quality brake fluid from a sealed container is essential to the safe and proper operation of the brake system. You should always buy the correct type of brake fluid for your vehicle. If the brake fluid becomes contaminated, completely flush the system with new fluid. Never reuse any brake fluid. Any brake fluid that is removed from the system should be discarded. Also, do not allow any brake fluid to come in contact with a painted surface; it will damage the paint.

• Never operate the engine without the proper amount and type of engine oil; doing so WILL result in severe engine damage.

• Timing belt maintenance is extremely important! Many models utilize an interference-type, non-freewheeling engine. If the timing belt breaks, the valves in the cylinder head may strike the pistons, causing potentially serious (also time-consuming and expensive) engine damage.

• Disconnecting the negative battery cable on some vehicles may interfere with the functions of the on-board computer system(s) and may require the computer to undergo a relearning process once the negative battery cable is reconnected.

• When servicing drum brakes, only disassemble and assemble one side at a time, leaving the remaining side intact for reference.

• Only an MVAC-trained, EPA-certified automotive technician should service the air conditioning system or its components.

ENGINE REPAIR

➡**Disconnecting the negative battery cable on some vehicles may interfere with the functions of the on board computer system. The computer may undergo a relearning process once the negative battery cable is reconnected.**

Distributor

The Pilot is equipped with a Distributorless Ignition System (DIS).

Alternator

REMOVAL

1. Before servicing the vehicle, refer to the precautions in the beginning of this section.

2. Remove or disconnect the following:
   • Negative battery cable
   • Accessory drive belt
   • Intake manifold and ignition coil covers
   • Alternator wiring harness connectors
   • Alternator mounting bolts
   • Wiring harness clamp
   • Alternator

INSTALLATION

1. Install or connect the following:
   • Alternator
   • Wiring harness clamp. Tighten the bolt to 105 inch lbs. (12 Nm).
   • Alternator mounting bolts. Tighten the 10mm bolt to 33 ft. lbs. (44 Nm) and the 8mm bolt to 16 ft. lbs. (22 Nm).
   • Alternator wiring harness connectors. Tighten the battery terminal nut to 105 inch lbs. (12 Nm).
   • Accessory drive belt
   • Intake manifold and ignition coil covers
   • Negative battery cable

Ignition Timing

ADJUSTMENT

The Pilot is equipped with a Distributorless Ignition System (DIS). The ignition timing is controlled by the Powertrain Control module (PCM). No adjustment is necessary.

Engine Assembly

REMOVAL & INSTALLATION

➡**The engine and transaxle are removed from the vehicle as a unit.**

1. Before servicing the vehicle, refer to the precautions in the beginning of this section.
2. Drain the cooling system.
3. Drain the power steering system.
4. Drain the transaxle fluid.
5. Drain the engine oil.
6. Relieve fuel system pressure.
7. Remove or disconnect the following:
   - Negative battery cable
   - Battery
   - Intake and ignition coil covers
   - Air intake duct
   - Left engine wire harness connectors
   - Relay bracket
   - Battery and tray
   - Starter cable and harness clamp
   - Accelerator cable
   - Cruise control cable
   - Fuel lines
   - EVAP canister hose
8. Remove the drivers side center console lower panel and pull back the cover to access steering joint cover.
   - Steering joint bolt
   - Powertrain Control Module (PCM) connectors
   - Heated Oxygen (HO2S) sensor connector and grommet. Pull the PCM harness through the firewall.
   - Brake booster vacuum line
   - Clamps and clips from power steering hoses
   - Fuse/relay box battery cable
   - Accessory drive belts
   - Front wheels
   - Splash shield
   - Front sub-frame stiffener
   - Exhaust front pipe
   - Propeller shaft
   - Shift control cable
   - Transfer assembly
   - Ball joints
   - Stabilizer bar links
   - Halfshafts
   - Power steering hose and pressure switch connector
   - Transaxle lower front mount
   - Transaxle lower rear mount
   - A/C compressor
   - Heater hoses
   - Radiator hoses
   - Ground cable
   - Transaxle oil cooler lines

   - Radiator
9. Attach a hoist to the engine lifting eyes and support the powertrain weight.
10. Remove or disconnect the following:
    - Side engine mount bracket
    - Front mount bracket support nut
11. Matchmark the front subframe to the mounting points.
12. Remove or disconnect the following:
    - Front subframe
    - All remaining hoses and electrical connections
13. Lower the powertrain away from the vehicle.

To install:

14. Raise the powertrain into position.
15. Installation is the reverse of removal but please note the following steps:
    - A/C compressor bolts to 16 ft. lbs. (22 Nm)
    - Front subframe. Use new bolts and tighten the 14mm bolts to 76 ft. lbs. (103 Nm). Tighten the front brace bolts to 54 ft. lbs. (74 Nm) and the rear brace bolts to 86 ft. lbs. (117 Nm).

    - Transaxle lower front mount nuts to 28 ft. lbs. (38 Nm)
    - Transaxle lower rear mount bolts to 28 ft. lbs. (38 Nm)
    - Front mount bracket support nut to 40 ft. lbs. (54 Nm)
    - Side engine mount bracket bolts to 33 ft. lbs. (44 Nm) and the through bolt to 40 ft. lbs. (54 Nm)
16. Fill the engine crankcase to the correct level.
17. Fill the transaxle to the correct level.
18. Fill the cooling system.
19. Fill the power steering system.
20. Start the engine and check for leaks.
21. Check the wheel alignment and adjust as necessary.

Water Pump

REMOVAL & INSTALLATION

1. Before servicing the vehicle, refer to the precautions in the beginning of this section.

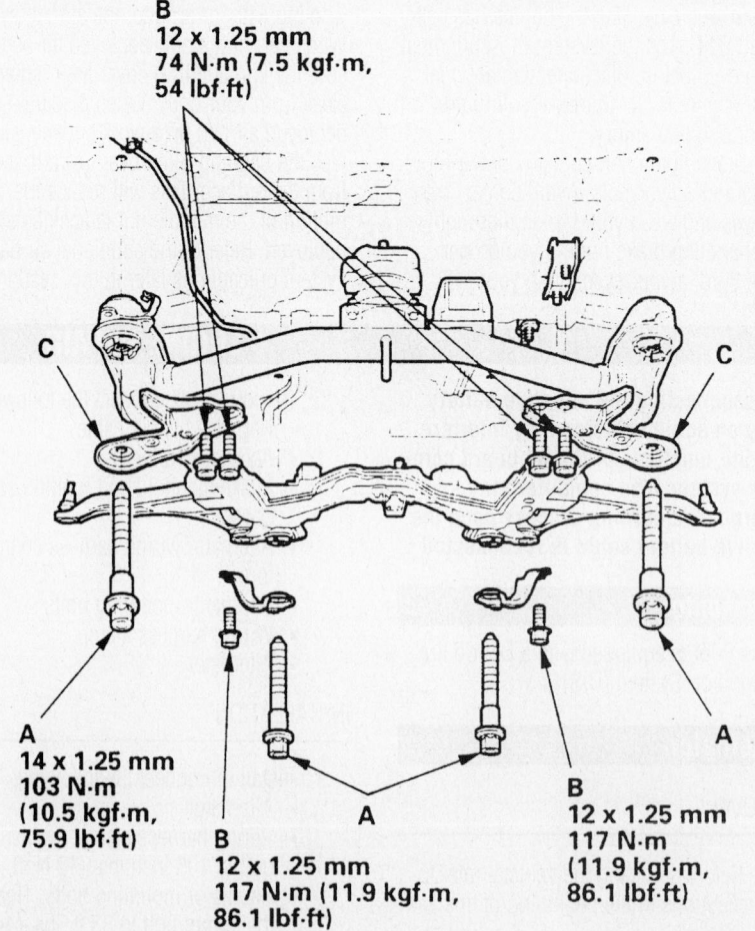

B
12 x 1.25 mm
74 N·m (7.5 kgf·m, 54 lbf·ft)

C

A
14 x 1.25 mm
103 N·m (10.5 kgf·m, 75.9 lbf·ft)

B
12 x 1.25 mm
117 N·m (11.9 kgf·m, 86.1 lbf·ft)

B
12 x 1.25 mm
117 N·m (11.9 kgf·m, 86.1 lbf·ft)

9302MG69

Sub-frame fastener locations and tightening torque—Pilot

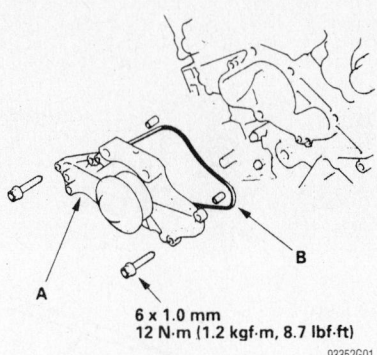

6 x 1.0 mm
12 N·m (1.2 kgf·m, 8.7 lbf·ft)

93352G01

Exploded view of the water pump mounting

2. Drain the cooling system.
3. Remove or disconnect the following:
   • Negative battery cable
   • Accessory drive belts
   • Front cover
   • Timing belt
   • Timing belt tensioner
   • Water pump

To install:

4. Install or connect the following:
   • Water pump. Use a new O-ring seal and tighten the bolts to 105 inch lbs. (12 Nm).
   • Timing belt tensioner
   • Timing belt
   • Front cover
   • Accessory drive belts
   • Negative battery cable
5. Fill the cooling system.
6. Start the engine and check for leaks.

Heater Core

REMOVAL & INSTALLATION

1. Before servicing the vehicle, refer to the precautions in the beginning of this section.
2. Drain the cooling system.
3. Remove or disconnect the following:
   • Negative battery cable
4. Recover the refrigerant using approved equipment.
   • Heater valve cable from the valve arm. Turn the valve arm to the fully opened position.
   • Heater hoses from the heater unit
   • Mounting nut from the heater unit. Be careful not to bend or damage fuel or brake lines.
5. Remove the dashboard as follows:
   a. Remove the center console by unlatching the clips.
   b. Remove the dashboard lower cover screw, gently pull down on the cover to

disengage the clips and disconnect the electrical connections.
   c. Remove the dashboard side cover by gently pulling and turning to unfasten the clips.
   d. Remove the right kick panel.
   e. While holding the glove box, remove the box stop from each side, then disconnect the lock from the damper.
   f. Remove the glove box bolts and the glove box.
   g. Remove the front door trim, kick panels and A-pillar trim from both sides.
   h. Remove the cap from the front pil-lar corner trim. Unfasten the screw, slide the trim upward along the pillar and remove it. Remove the remaining clips from the body.
   i. On the drivers side, remove the fuel/relay box nut and pull out the box.
   j. Remove the steering column
   k. On the passenger side remove the fuse/relay bolt and pull out the box.
   l. Disconnect all electrical connections from the dashboard.
   m. If equipped with a navigation system, remove the passenger seat, pull back the carpet, remove the harness

Fastener Locations

A ▶ : Bolt, 2   B ▶ : Bolt, 1   C ▶ : Bolt, 3

D ▶ : Bolt, 5   E ▶ : Bolt, 2   F ▶ : Bolt, 2

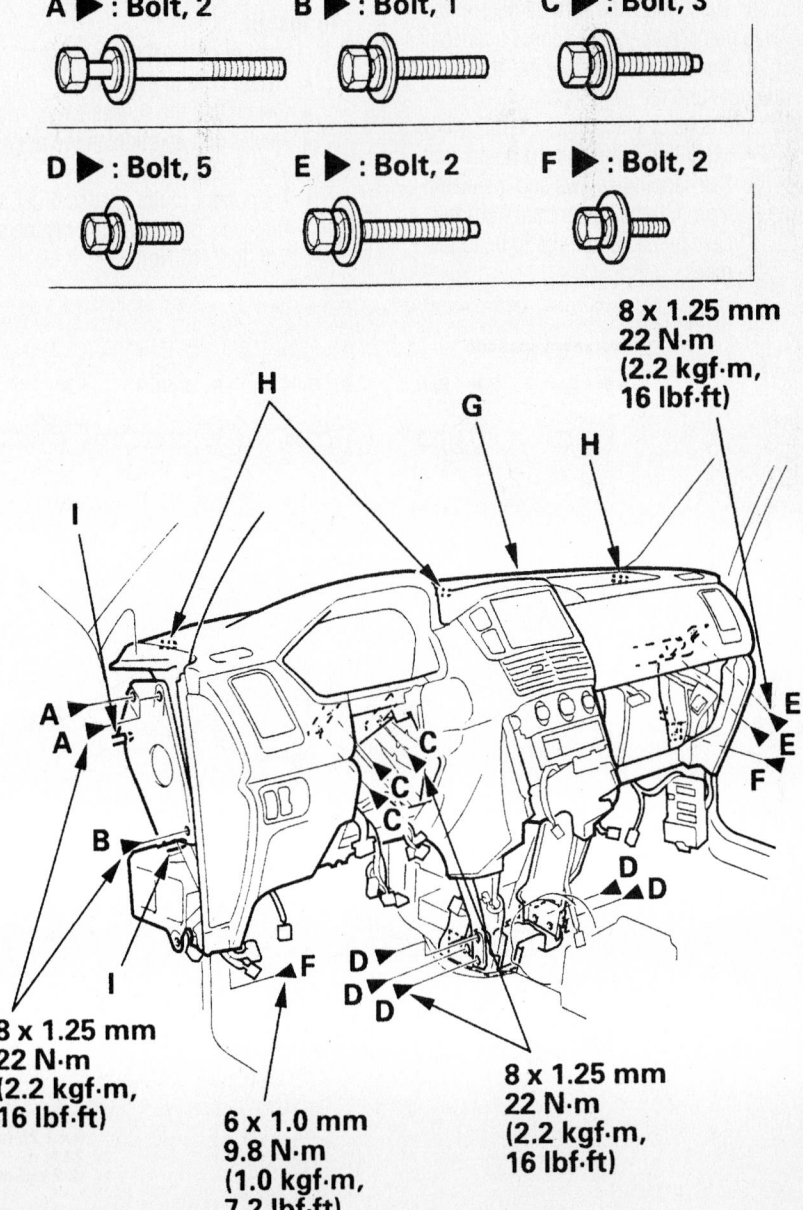

8 x 1.25 mm
22 N·m
(2.2 kgf·m,
16 lbf·ft)

8 x 1.25 mm
22 N·m
(2.2 kgf·m,
16 lbf·ft)

6 x 1.0 mm
9.8 N·m
(1.0 kgf·m,
7.2 lbf·ft)

8 x 1.25 mm
22 N·m
(2.2 kgf·m,
16 lbf·ft)

Tighten the dashboard bolts as illustrated

42356-HPIL-G03

cushions and then pull out the GPS harness.

n. Remove all harness and connector clips.

o. Remove all the bolts and lift up on the dashboard to release the dashboard and steering hanger beam from the guide pins.

p. Remove the dashboard through the door.

6. Remove the evaporator as follows:

a. Disconnect the receiver and suction lines from the evaporator.

b. Remove the mounting nuts and plug the lines to avoid system contamination.

c. Remove the plastic brace and glove box frame.

d. Disconnect the wire harness and evaporator temperature sensor connector.

e. Remove the self-tapping screws, the nuts and the evaporator.

7. Remove or disconnect the following:

- Mounting bolts and the heater unit
- Self-tapping screws and the clamp, then pull the heater core from the case being careful not to bend the pipes

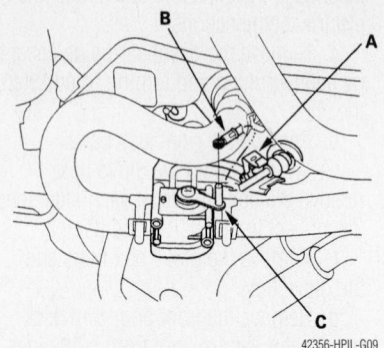

42356-HPIL-G09

In the engine compartment, open the cable clamp (A), then disconnect the heater valve cable (B) from the valve arm (C)

To install:

8. Install or connect the following:

- Heater core in the case
- Clamp and the screws
- Heater unit and tighten the bolts to 7 ft. lbs. (10 Nm)
- Evaporator in the reverse order of removal. Tighten all the retainers to 7 ft. lbs. (10 Nm) .

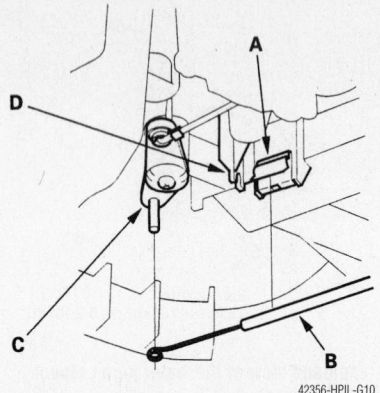

42356-HPIL-G10

Under the dashboard, disconnect the valve cable housing from the clamp (A) and the cable (B) from the mix control linkage (C)

9. Install the dashboard in the reverse order of removal keeping in mind the following points:

a. Make sure the dashboard is seated properly and that the wiring harness and steering hanger beam wire harness are not pinched.

b. Referring to the accompanying illustration, tighten bolts (A, B, C, D

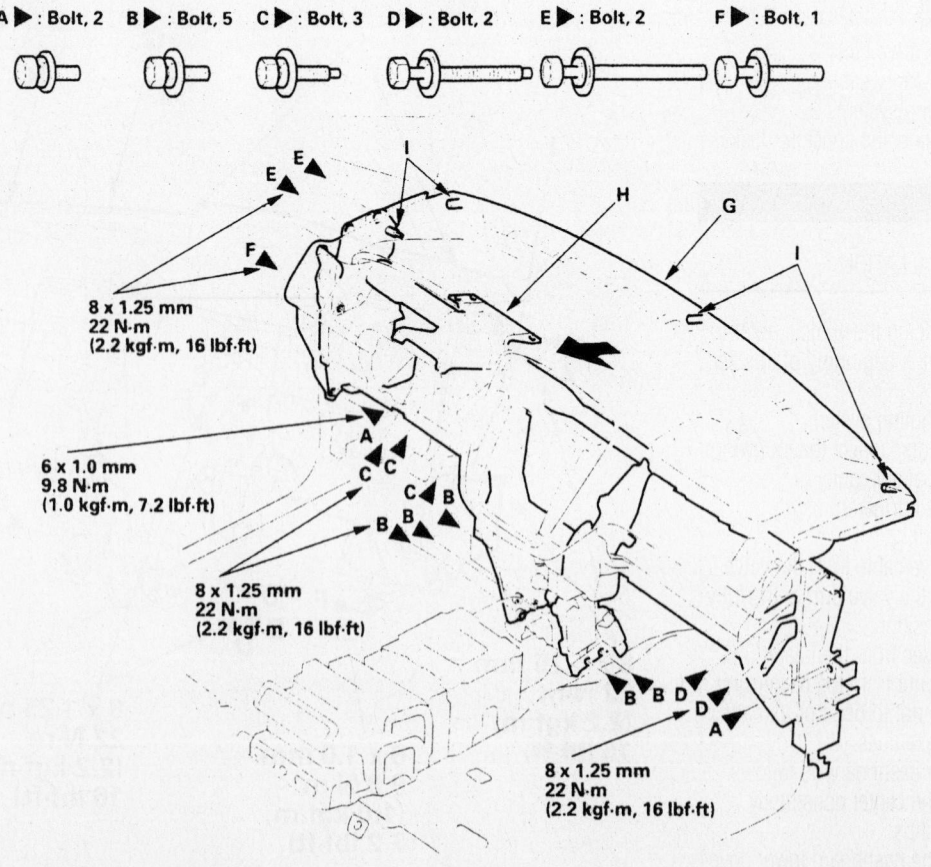

Fastener Locations

A ▶ : Bolt, 2   B ▶ : Bolt, 5   C ▶ : Bolt, 3   D ▶ : Bolt, 2   E ▶ : Bolt, 2   F ▶ : Bolt, 1

8 x 1.25 mm
22 N·m
(2.2 kgf·m, 16 lbf·ft)

6 x 1.0 mm
9.8 N·m
(1.0 kgf·m, 7.2 lbf·ft)

8 x 1.25 mm
22 N·m
(2.2 kgf·m, 16 lbf·ft)

8 x 1.25 mm
22 N·m
(2.2 kgf·m, 16 lbf·ft)

93552G91

Exploded view of the dashboard mounting—Pilot

6 x 1.0 mm
9.8 N·m (1.0 kgf·m, 7.2 lbf·ft)

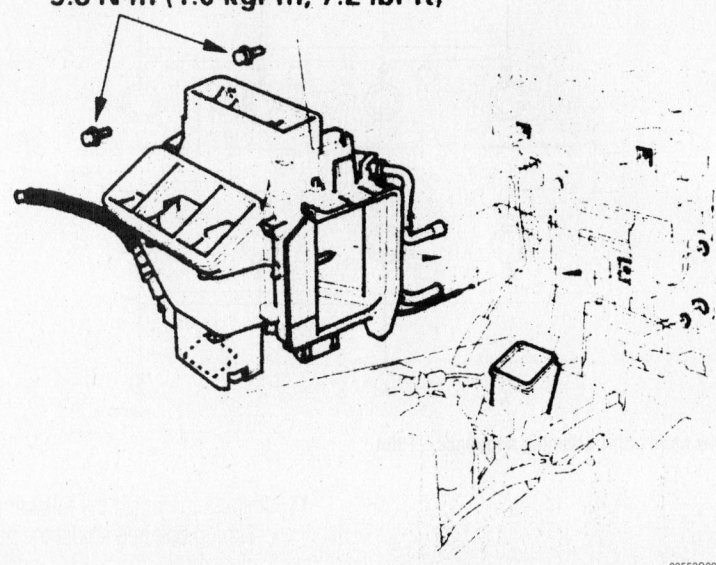

93552G92

Exploded view of the evaporator mounting—Pilot

and E) to 16 ft. lbs. (22 Nm). Tighten bolts **F** to 7 ft. lbs. (10 Nm). Apply thread lock to the **B** bolts before installation.

   c. Ensure that all electrical connectors are properly connected.

10. Install or connect the following:
- Mounting nut to the heater unit and tighten to 7 ft. lbs. (10 Nm)
- Heater hoses

11. Connect the heater valve cable and adjust as follows:

   a. In the engine compartment, open the cable clamp (A), then disconnect the heater valve cable (B) from the valve arm (C).

   b. Under the dashboard, disconnect the valve cable housing from the clamp (A) and the cable (B) from the mix control linkage (C).

   c. Set the temperature control button to the MAX COOL position with the ignition switch in the on position.

   d. Attach the valve cable (B) to the mix control linkage (C) as shown in the illustration, hold the end of the cable housing against the stop, then snap the cable housing into the clamp.

   e. In the engine compartment, turn the valve arm (C) to the fully closed position as shown in the accompanying illustration and hold it there. Attach the cable (B) to the valve arm and pull gently on the cable housing to take up the slack, then install the cable housing into the clamp (A).

12. Fill the cooling system

13. Connect the battery cable.

Cylinder Head

REMOVAL & INSTALLATION

1. Before servicing the vehicle, refer to the precautions in the beginning of this section.
2. Drain the cooling system.
3. Relieve the fuel system pressure.
4. Remove or disconnect the following:
- Negative battery cable
- Alternator belt
- Intake manifold cover
- Ignition coil covers
- Power steering belt and pump
- Power steering hose clamp
- Alternator
- Fuel feed and return lines
- EVAP canister hose
- Intake manifold
- Ignition coils
- Timing belt
- Fuel injector connectors
- Engine Coolant Temperature (ECT) sensor connector
- Radiator fan switch connectors
- Crankshaft Position (CKP) sensor connector
- Camshaft Position (CMP) sensor connector
- Exhaust Gas Recirculation (EGR) connector
- Valve Lift Electronic Control (VTEC) solenoid valve connector and oil pressure switch connections
- Oil pressure switch connector
- Vacuum hoses from the intake air bypass control valve
- Fuel rails
- Heater hose
- Radiator hoses
- Ground cable
- Exhaust manifolds
- Water passage
- Camshaft pulleys and back covers
- Valve covers

5. Loosen the cylinder head bolts in sequence and ⅓ turns until all bolts are loose.

6. Remove the cylinder head.

To install:

7. Align the crankshaft and camshaft sprocket TDC marks as shown.

8. Install the cylinder heads with new gaskets.

9. Apply clean engine oil to the cylinder head bolt threads and flanges.

10. Tighten the cylinder head bolts in sequence as follows:

   a. Step 1: 29 ft. lbs. (39 Nm).
   b. Step 2: 51 ft. lbs. (69 Nm).
   c. Step 3: 72 ft. lbs. (98 Nm).

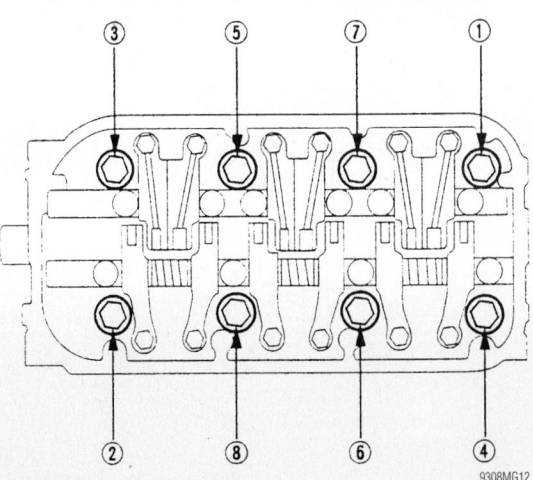

9308MG12

Cylinder head bolt loosening sequence—Pilot

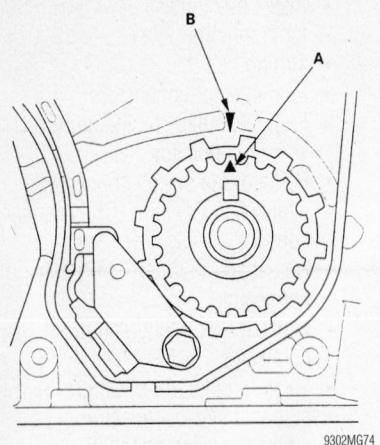

Crankshaft timing belt sprocket TDC marks. Align sprocket mark (A) with pointer (B)—Pilot

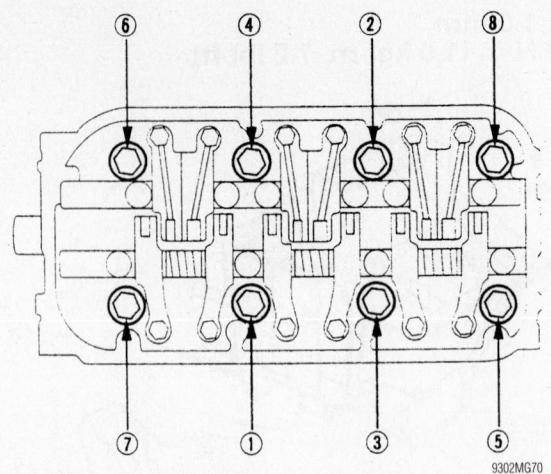

Cylinder head bolt tightening sequence—Pilot

FRONT:

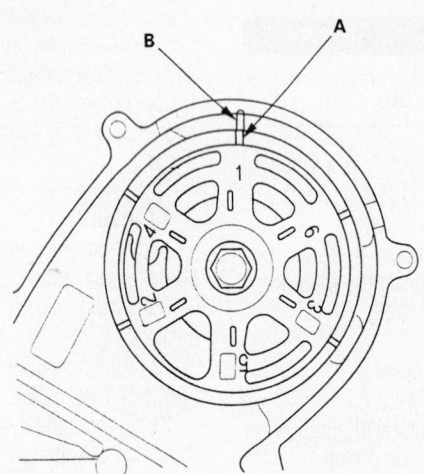

REAR:

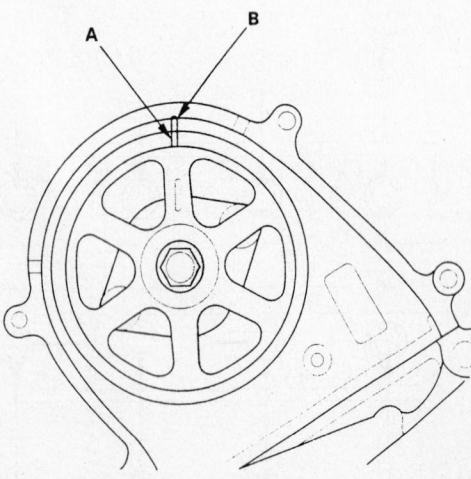

Camshaft TDC marks. Align sprocket mark (A) with the back cover pointer (B)—Pilot

11. Install or connect the following:
- Timing belt and adjust the valve clearance
- Valve covers
- Exhaust manifolds
- Water passage
- Fuel rails
- Vacuum hoses to the intake air bypass control valve
- Radiator hoses
- Heater hose
- Oil pressure switch connector
- VTEC solenoid valve connector and oil pressure switch connections
- EGR connector
- CMP sensor connector
- CKP sensor connector
- Radiator fan switch connectors
- ECT sensor connector
- Fuel injector connectors
- Ignition coils
- Intake manifold
- EVAP canister hose
- Fuel feed and return lines
- Alternator
- Power steering hose clamp
- Power steering pump and belt
- Ground cable
- Ignition coil covers
- Intake manifold cover
- Alternator belt
- Negative battery cable
12. Fill the cooling system.
13. Start the engine and check for leaks.

Rocker Arms/Shafts

REMOVAL & INSTALLATION

1. Before servicing the vehicle, refer to the precautions in the beginning of this section.

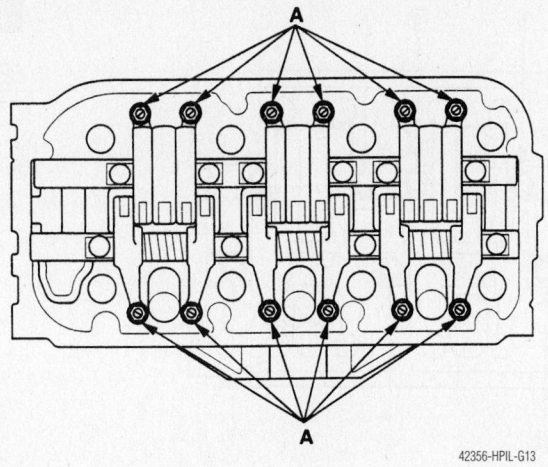

42356-HPIL-G13

Rocker arm shaft adjusting screw locations

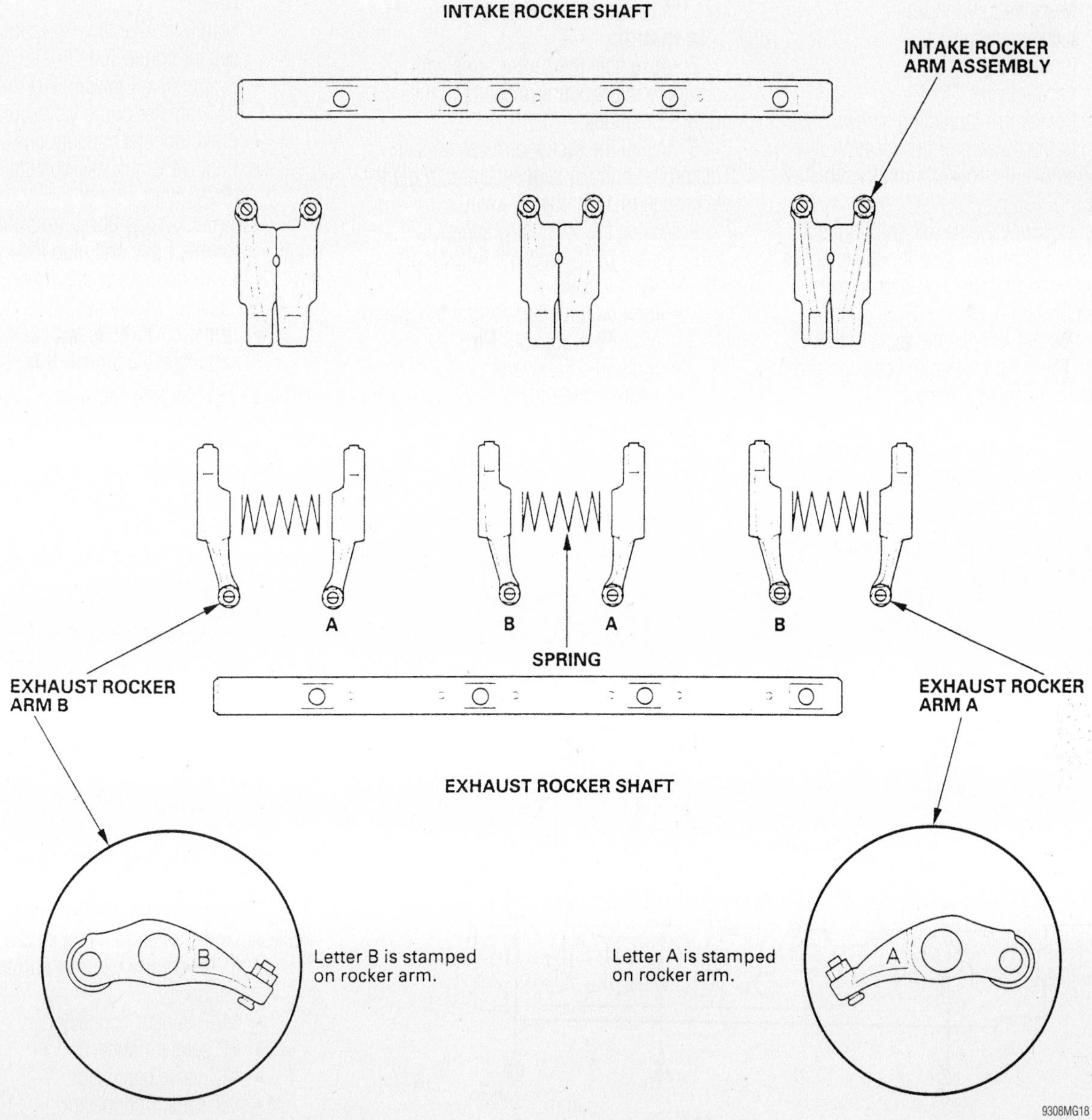

INTAKE ROCKER SHAFT

INTAKE ROCKER
ARM ASSEMBLY

A   B   A   B

EXHAUST ROCKER
ARM B

SPRING

EXHAUST ROCKER
ARM A

EXHAUST ROCKER SHAFT

Letter B is stamped
on rocker arm.

Letter A is stamped
on rocker arm.

9308MG18

Exploded view of the rocker arms and shafts—Pilot

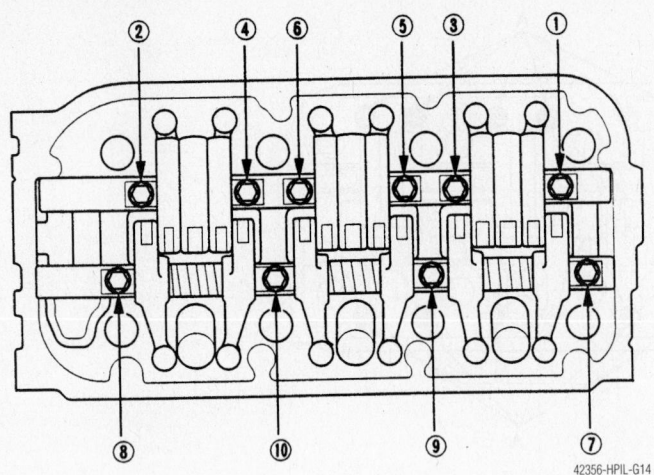

42356-HPIL-G14

Rocker arm shaft loosening sequence

2. Remove or disconnect the following:
- Negative battery cable
- Intake manifold
- Ignition coils
- Valve cover
- Rocker arm adjusting screws. refer to the illustration for location.

3. Remove the rocker arm assembly as follows:

a. Unscrew the rocker shaft bolts 2 turns at a time in a criss-cross pattern to avoid damaging the valves or rocker assembly.

b. Do not remove the rocker shaft bolts. These bolts keep the springs and rocker arms on the shafts.

4. Loosen the valve adjuster locknuts and screws so that all valves are closed.

5. Remove the rocker arms and shafts from the vehicle as an assembly.

➡ **Keep all valvetrain components in order for assembly.**

6. Remove the rocker arms and springs from the rocker arm shafts.

To install:

7. Assemble the rocker arms and springs to the rocker arm shafts in their original positions.

8. Install the rocker arm assemblies. Tighten the bolts in sequence and in multiple passes to 17 ft. lbs. (24 Nm).

9. Adjust the valve clearance.

10. Install or connect the following:
- Valve covers
- Ignition coils and torque the retainers to 9 ft. lbs. (12 Nm)
- Ignition coil covers
- Intake manifold
- Negative battery cable

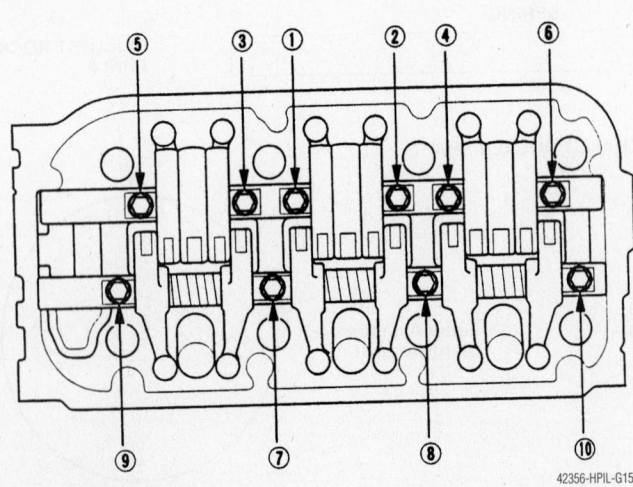

42356-HPIL-G15

Rocker shaft tightening sequence

Intake Manifold

REMOVAL & INSTALLATION

1. Before servicing the vehicle, refer to the precautions in the beginning of this section.

2. Remove or disconnect the following:
- Negative battery cable
- Intake manifold cover
- Air intake tube
- Throttle and cruise control cables

3. Remove the following electrical connections and clamps from the manifold:
- Intake Air Temperature (IAT) sensor connector
- Idle Air Control (IAC) valve connector
- Throttle Position (TP) sensor connector
- Manifold Absolute Pressure (MAP) sensor connector
- Evaporative Emissions (EVAP) control canister purge valve connector
- Brake booster vacuum line
- Positive Crankcase Ventilation (PCV) hose
- Water bypass hoses from the throttle body and plug the hoses
- EVAP control canister hose

4. Remove or disconnect the following:
- Upper cover bolts and nuts in the sequence illustrated using two or three passes
- Intake manifold bolts using two or three passes
- Intake manifold and spacer

To install:

5. Install or connect the following:
- New intake manifold gasket and spacer
- Intake manifold. Tighten the fasteners in sequence and in several passes to 16 ft. lbs. (22 Nm).
- Upper cover bolts and nuts in the sequence illustrated using two or three passes to 9 ft. lbs. (12 Nm)
- EVAP control canister hose
- Water bypass hoses toe throttle body
- Brake booster vacuum line
- PCV hose

6. Connect the following electrical connections and clamps to the manifold:
- EVAP control canister purge valve connector
- MAP sensor connector
- TP sensor connector
- IAC valve connector
- IAT sensor connector
- Throttle and cruise control cables

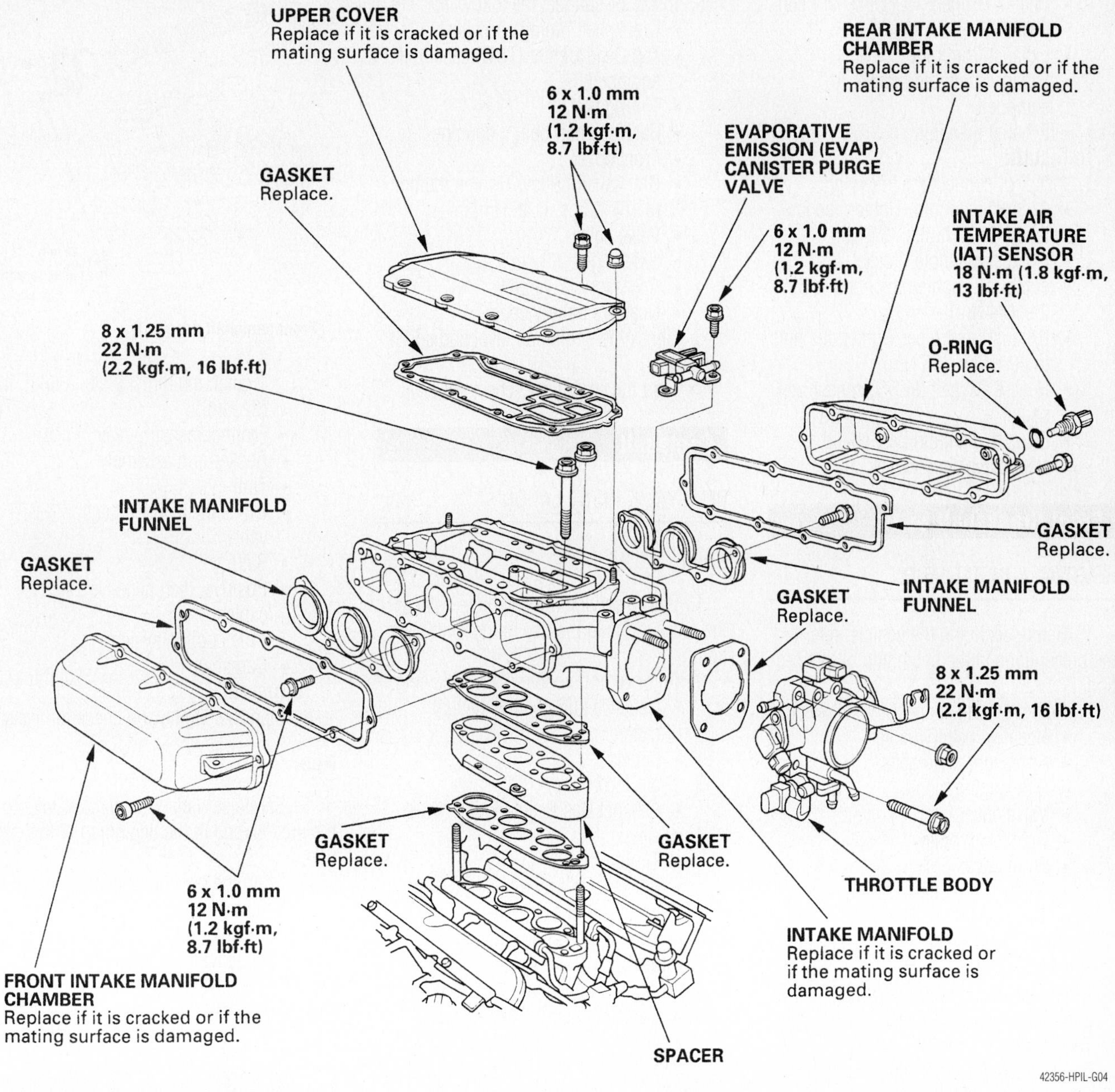

UPPER COVER
Replace if it is cracked or if the mating surface is damaged.

6 x 1.0 mm
12 N·m
(1.2 kgf·m, 8.7 lbf·ft)

REAR INTAKE MANIFOLD CHAMBER
Replace if it is cracked or if the mating surface is damaged.

GASKET
Replace.

EVAPORATIVE EMISSION (EVAP) CANISTER PURGE VALVE

6 x 1.0 mm
12 N·m
(1.2 kgf·m, 8.7 lbf·ft)

INTAKE AIR TEMPERATURE (IAT) SENSOR
18 N·m (1.8 kgf·m, 13 lbf·ft)

8 x 1.25 mm
22 N·m
(2.2 kgf·m, 16 lbf·ft)

O-RING
Replace.

INTAKE MANIFOLD FUNNEL

GASKET
Replace.

GASKET
Replace.

INTAKE MANIFOLD FUNNEL

GASKET
Replace.

GASKET
Replace.

8 x 1.25 mm
22 N·m
(2.2 kgf·m, 16 lbf·ft)

GASKET
Replace.

GASKET
Replace.

6 x 1.0 mm
12 N·m
(1.2 kgf·m, 8.7 lbf·ft)

THROTTLE BODY

INTAKE MANIFOLD
Replace if it is cracked or if the mating surface is damaged.

FRONT INTAKE MANIFOLD CHAMBER
Replace if it is cracked or if the mating surface is damaged.

SPACER

42356-HPIL-G04

Exploded view of the intake manifold

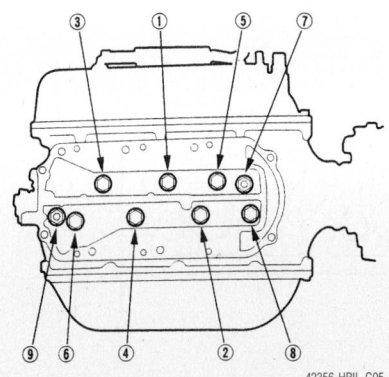

42356-HPIL-G05

Intake manifold torque sequence

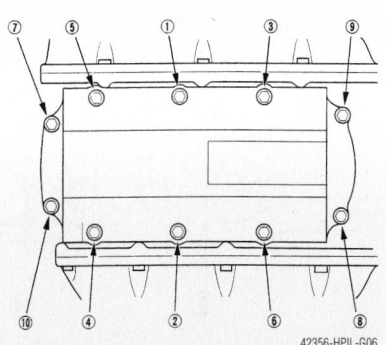

42356-HPIL-G06

Upper cover torque sequence

- Air intake tube
- Intake manifold cover
- Negative battery cable

7. Start the engine and check for proper operation.

Exhaust Manifold

REMOVAL & INSTALLATION

1. Before servicing the vehicle, refer to the precautions in the beginning of this section.

2. Remove or disconnect the following:
- Negative battery cable
- Exhaust manifold heat shield

- Heated Oxygen (HO2S) sensor connector
- Exhaust front pipe
- Exhaust manifold bracket, if equipped
- Exhaust manifold

To install:

3. Install or connect the following:
- Exhaust manifold. Tighten the fasteners to 23 ft. lbs. (31 Nm).
- Exhaust manifold bracket, if equipped. Tighten the bolts to 33 ft. lbs. (44 Nm).
- Exhaust front pipe. Tighten the nuts to 40 ft. lbs. (55 Nm).
- Heated Oxygen (HO2S) sensor connector
- Exhaust manifold heat shield
- Negative battery cable

Front Crankshaft Seal

REMOVAL & INSTALLATION

1. Before servicing the vehicle, refer to the precautions in the beginning of this section.

2. Remove or disconnect the following:
- Negative battery cable
- Accessory drive belts
- Side engine mount
- Valve cover
- Crankshaft pulley
- Front cover
- Balance shaft belt, if equipped
- Timing belt
- Top Dead Center (TDC) sensor, if equipped
- Crankshaft timing sprocket
- Front crankshaft seal

To install:

3. Lubricate the crankshaft seal lip with grease prior to installation.

4. Install the front crankshaft seal so that it is flush with the surface of the oil pump housing.

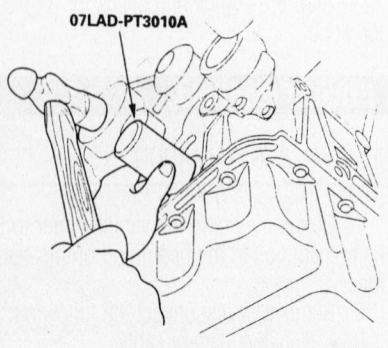

07LAD-PT3010A

93352G02

Front crankshaft seal installation

5. Install or connect the following:
- Crankshaft timing sprocket
- Top Dead Center (TDC) sensor, if equipped
- Timing belt
- Balance shaft belt, if equipped
- Front cover
- Crankshaft pulley. Tighten the bolt to 181 ft. lbs. (245 Nm).
- Valve cover
- Side engine mount
- Accessory drive belts
- Negative battery cable

6. Check the engine oil level and add if necessary.

7. Start the engine and check for leaks.

Camshaft

REMOVAL & INSTALLATION

Front

1. Before servicing the vehicle, refer to the precautions in the beginning of this section.

2. Remove or disconnect the following:
- Negative and positive battery cables
- Battery

3. Drain the coolant.
- Exhaust Gas Recirculation (EGR) valve
- Timing belt
- Rocker arm assembly
- Front camshaft pulley
- Thrust plate and camshaft

To install:

4. Install or connect the following:
- Camshaft using a new O-ring.

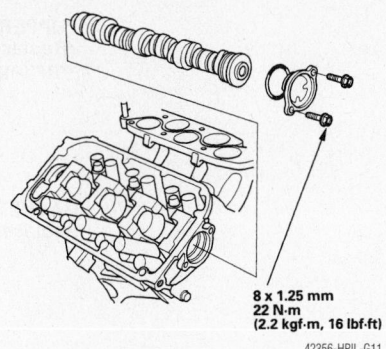

**8 x 1.25 mm
22 N·m
(2.2 kgf·m, 16 lbf·ft)**

42356-HPIL-G11

Front camshaft assembly

Tighten the thrust plate to 16 ft. lbs. (22 Nm).
- Front camshaft pulley
- Rocker arm assembly
- Timing belt
- Exhaust Gas Recirculation (EGR) valve
- Battery
- Positive, then negative battery cables

5. Fill the cooling system.
- Camshaft

To install:

6. Start the engine and check for leaks.

Rear

1. Before servicing the vehicle, refer to the precautions in the beginning of this section.

2. Drain the cooling system.

3. Relieve the fuel system pressure.

4. Remove or disconnect the following:
- Negative battery cable
- Under-hood fuse box
- Fuel feed hose

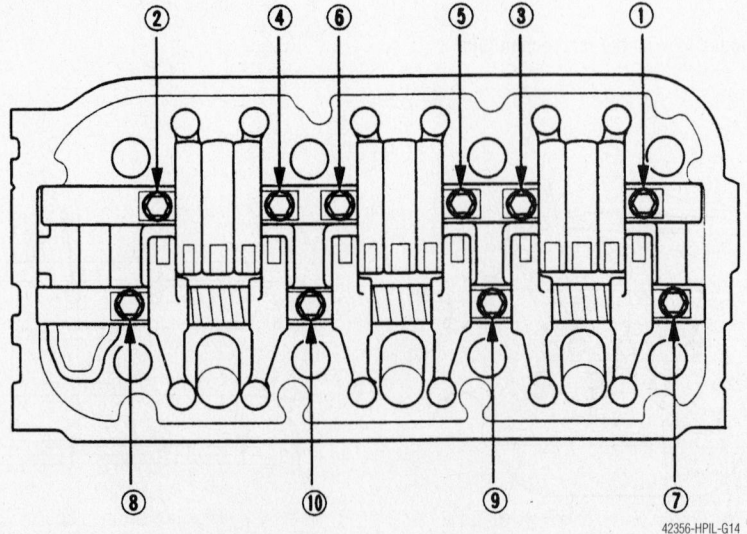

42356-HPIL-G14

Rocker arm shaft loosening sequence

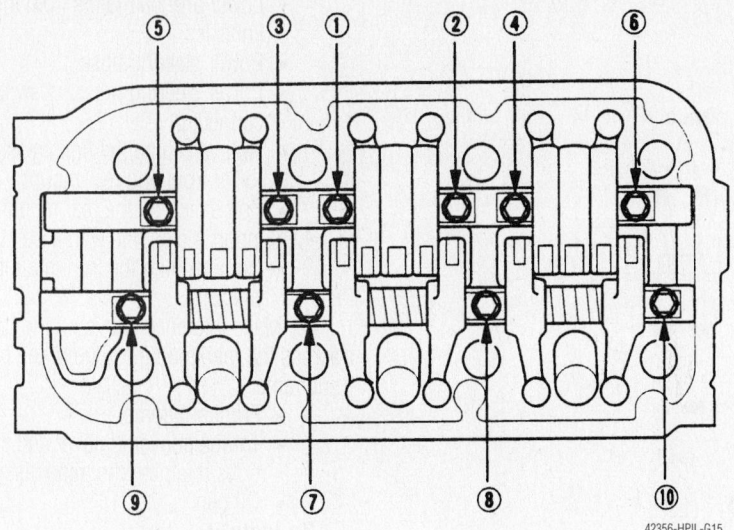

Rocker shaft tightening sequence

42356-HPIL-G15

ADJUSTMENT

Adjust the valves only when the cylinder head temperature is less than 100°F (38°C).

1. Before servicing the vehicle, refer to the precautions in the beginning of this section.

2. Remove or disconnect the following:
   - Negative battery cable
   - Air intake tube
   - Intake manifold
   - Valve cover

3. Rotate the crankshaft so that the valves to be adjusted are closed and the rocker arm is contacting the camshaft lobe base circle.

4. Measure the valve clearance. If adjustment is necessary, loosen the locknut and turn the adjusting screw as necessary to achieve the correct valve clearance.

5. The correct valve clearance is:
   - Intake valves: 0.008–0.009 inches (0.20–0.24mm)

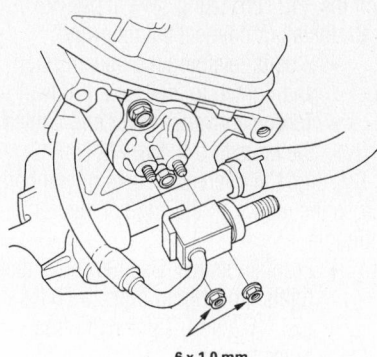

6 x 1.0 mm
12 N·m (1.2 kgf·m, 8.7 lbf·ft)

42356-HPIL-G12

Remove the nuts attaching the fuel line when removing the rear camshaft

- Nuts securing the fuel line
- Brake lines from the master cylinder
- Timing belt
- Rocker arm assembly
- Rear camshaft pulley
- Thrust plate and camshaft

To install:

5. Install or connect the following:
   - Camshaft using a new O-ring. Tighten the thrust plate to 16 ft. lbs. (22 Nm).
   - Rear camshaft pulley
   - Rocker arm assembly
   - Timing belt
   - Brake lines to the master cylinder
   - Nuts securing the fuel line
   - Fuel feed hose
   - Under-hood fuse box
   - Negative battery cable

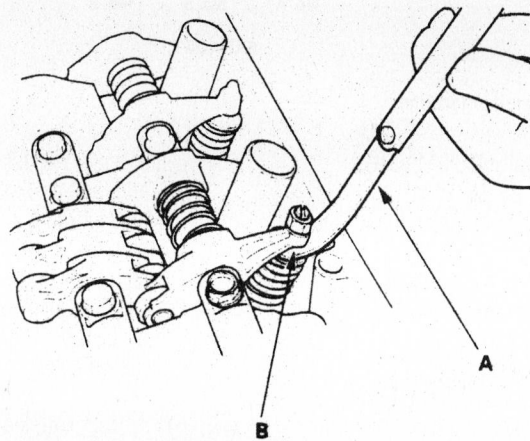

A

B

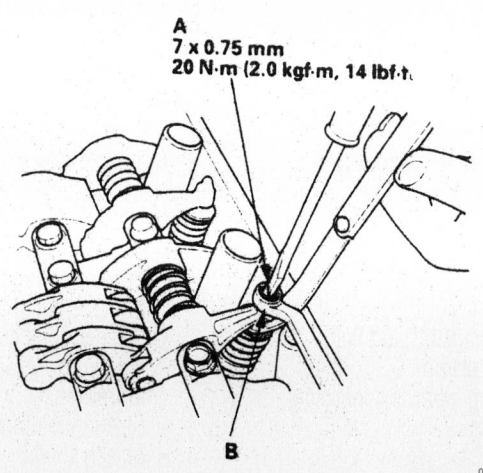

A
7 x 0.75 mm
20 N·m (2.0 kgf·m, 14 lbf·t)

B

93552G04

Inspect the valve clearance, adjust to specification and tighten the retainer to specification

Adjusting screw locations:

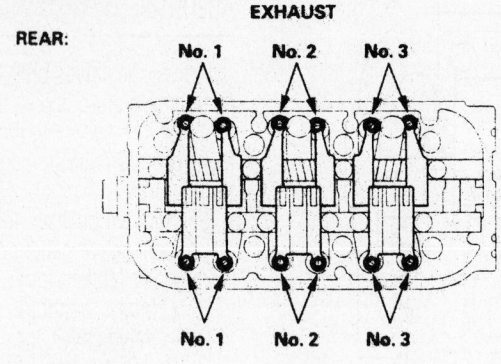

REAR:

EXHAUST

No. 1   No. 2   No. 3

No. 1   No. 2   No. 3

INTAKE

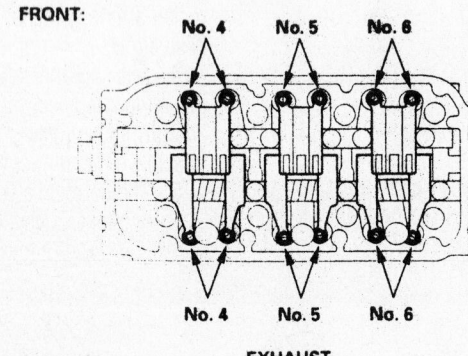

FRONT:

No. 4   No. 5   No. 6

No. 4   No. 5   No. 6

EXHAUST

93552G03

Valve adjusting retainer locations

- Exhaust valves: 0.011–0.013 inches (0.28–0.32mm)
6. After adjustment, tighten the locknuts to 14 ft. lbs. (20 Nm).
7. Install or connect the following:
  - Valve cover
  - Intake manifold
  - Air intake tube
  - Negative battery cable
8. Start the engine and check for proper operation.

Starter Motor

REMOVAL & INSTALLATION

1. Before servicing the vehicle, refer to the precautions in the beginning of this section.
2. Remove or disconnect the following:
  - Negative battery cable and wait at least 3 minutes
  - Unlock the transmission fluid cooler hose clamp
  - Starter wiring harness connectors
  - Starter motor
To install:
3. Install or connect the following:
  - Starter motor. Tighten the 10mm

bolt to 33 ft. lbs. (44 Nm) and the 12mm bolt to 47 ft. lbs. (64 Nm).
- Starter wiring harness connectors. Tighten the battery cable nut to 79 inch lbs. (9 Nm).
- Lock the transmission fluid cooler hose clamp
- Negative battery cable

Oil Pan

REMOVAL & INSTALLATION

1. Before servicing the vehicle, refer to the precautions in the beginning of this section.
2. Drain the engine oil and power steering system.
3. Remove or disconnect the following:
  - Negative battery cable
  - Power steering pump outlet hose from the pump and the hose clamp
  - Steering joint cover, mark the steering joint-to-gearbox pinion shaft for reference
  - Steering joint from the pinion shaft
  - Splash shield
  - Transfer assembly
  - Tie rod ends from the knuckles

- Lower arm ball joints from the knuckles
- Power steering hose
- Power steering pressure switch connector
- Nuts attaching the transmission lower front and rear mount
- Bolt attaching the rear mount
4. Support the engine with a hoist.
  - Nut attaching the front mount bracket
5. Make reference marks on the body across the marks on the edge of the front subframe.
  - Front subframe
  - Torque converter cover and the 2 bolts retaining the transmission
  - Oil pan
To install:
6. Install the oil pan. Apply liquid gasket as shown.
7. Tighten the bolts in sequence to 105 inch lbs. (12 Nm) using several passes.
8. Install or connect the following:
  - 2 bolts retaining the transmission and tighten to 28 ft. lbs. (38 Nm)
  - Torque converter cover and tighten the bolts to 9 ft. lbs. (12 Nm)
9. Align the reference marks on the body across the marks on the edge of the front subframe.
  - Front subframe. Use new bolts and tighten the 14mm bolts to 76 ft. lbs. (103 Nm). Tighten the front brace bolts to 54 ft. lbs. (74 Nm) and the rear brace bolts to 86 ft. lbs. (117 Nm).
  - Nut attaching the front mount bracket to 40 ft. lbs. (54 Nm)
  - Bolt attaching the rear mount to 28 ft. lbs. (38 Nm)
  - Nuts attaching the transmission lower front and rear mount to 28 ft. lbs. (38 Nm)

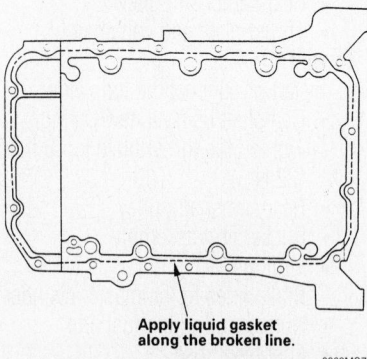

Apply liquid gasket along the broken line.

9302MG75

Apply liquid gasket to the inner threads of the bolt holes and the engine block along the area indicated by the broken line— Pilot

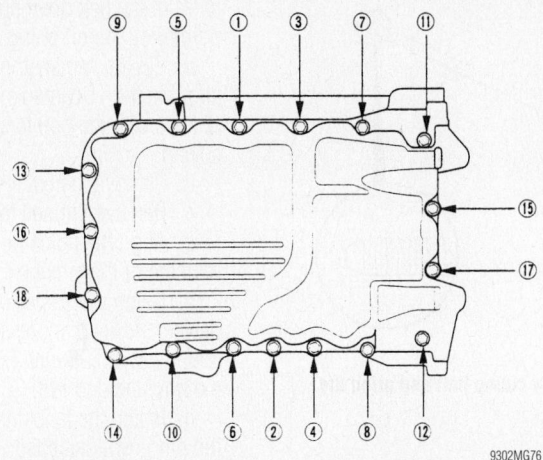

Oil pan tightening sequence—Pilot

- Power steering pressure switch connector
- Power steering hose
- Lower arm ball joints to the knuckles
- Tie rod ends to the knuckles
- Transfer assembly
- Splash shield
- Steering joint to the pinion shaft
- Steering joint cover
- Power steering pump outlet hose clamp and hose
- Negative battery cable

Oil Pump

REMOVAL & INSTALLATION

1. Before servicing the vehicle, refer to the precautions in the beginning of this section.
2. Drain the engine oil.
3. Turn the crankshaft and place the engine at Top Dead Center (TDC).
4. Remove or disconnect the following:
   - Negative battery cable
   - Accessory drive belts
   - Front cover
   - Timing belt
   - Timing belt idler pulley
   - Crankshaft Position (CKP) sensor
   - Crankshaft timing sprocket
   - Variable Valve Timing and Valve Lift Electronic Control (VTEC) solenoid valve connector
   - Oil filter adapter
   - Oil pan
   - Oil pump pickup tube
   - Oil pump

To install:

➡**Use new gaskets and O-ring seals for assembly.**

5. Apply liquid gasket to the oil pump and to the bolt hole threads.
6. Install or connect the following:
   - Oil pump. Tighten the bolts to 9 ft. lbs. (12 Nm).
   - Oil pump pickup tube. Tighten the bolts to 9 ft. lbs. (12 Nm).
   - Oil pan
   - Oil filter adapter
   - VTEC solenoid valve connector
   - Crankshaft timing sprocket
   - CKP sensor
   - Timing belt idler pulley
   - Timing belt
   - Front cover
   - Accessory drive belts
   - Negative battery cable
7. Fill the crankcase to the correct level.
8. Start the engine and check for leaks.

Rear Main Seal

REMOVAL & INSTALLATION

1. Before servicing the vehicle, refer to the precautions in the beginning of this section.
2. Remove or disconnect the following:
   - Transaxle
   - Flywheel
   - Oil seal

To install:
3. Install or connect the following:
   - Oil seal. Drive the seal square into the seal case.
   - Flywheel. Tighten the bolts in a crossing pattern to 54 ft. lbs. (73 Nm).
   - Transaxle
4. Check the fluid levels.
5. Start the engine and check for leaks.

Timing Belt

REMOVAL & INSTALLATION

1. Before servicing the vehicle, refer to the precautions in the beginning of this section.
2. Turn the crankshaft so the white mark aligns with the pointer.
3. Make sure the number 1 piston is at Top Dead center (TDC).
4. Remove or disconnect the following:
   - Negative battery cable
   - Wheels and splash shield

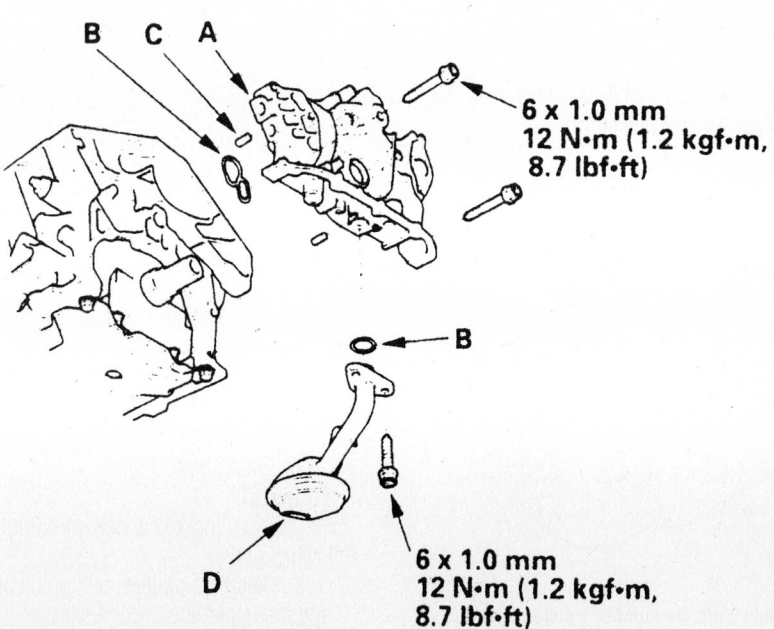

**6 x 1.0 mm
12 N·m (1.2 kgf·m, 8.7 lbf·ft)**

**6 x 1.0 mm
12 N·m (1.2 kgf·m, 8.7 lbf·ft)**

Exploded view of the oil pump assembly

- Drive belts

5. Support the engine with a block of wood and a jack under the oil pan.

- Upper side engine mount
- Dipstick tube

6. Remove the crankshaft pulley using holder tool shown in the accompanying illustration and a breaker and socket, loosen the 19mm bolt and remove the pulley.

- Front upper cover, rear upper cover and the lower cover
- One of the battery clamp bolts and grind the end as illustrated

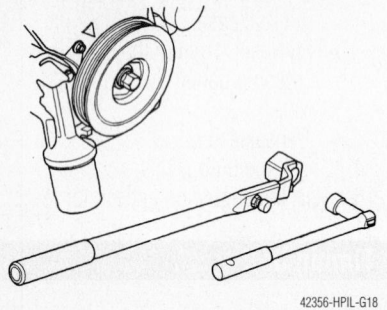

42356-HPIL-G18

Remove the crankshaft pulley using holder tool shown

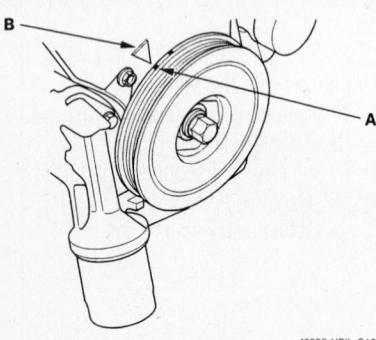

42356-HPIL-G16

Turn the crankshaft so the white mark (A) aligns with the pointer (B)

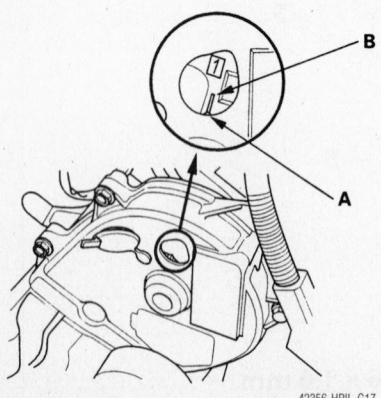

42356-HPIL-G17

Make sure the number 1 piston is at top dead center (A) on the front camshaft pulley and pointer (B)

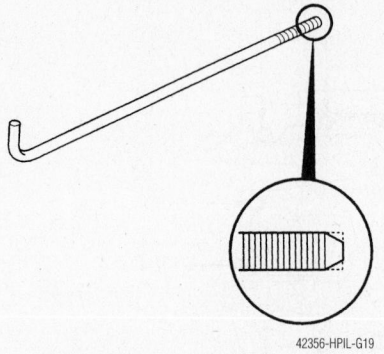

42356-HPIL-G19

Remove a battery clamp bolt and grind the end as shown

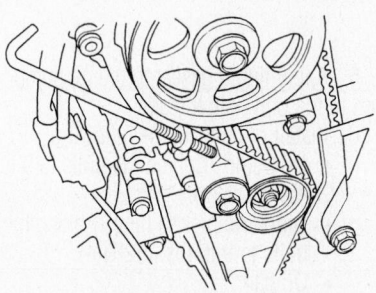

42356-HPIL-G20

Install the battery clamp bolt as shown to hold the belt adjuster in position

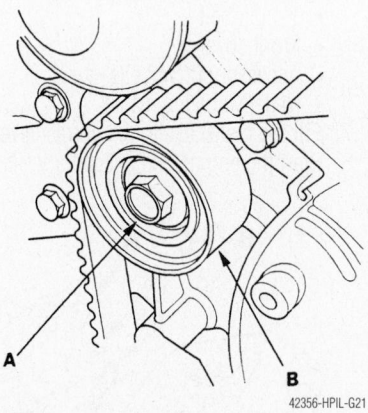

42356-HPIL-G21

Remove the idler pulley bolt (A), pulley (B) and the timing belt

7. Screw the battery clamp bolt as illustrated to hold the belt adjuster in position. Do not use a wrench, hand tighten only.

- Lower side engine mount
- Idler pulley bolt and the pulley
- Timing belt

To install:

8. If installing a new belt, perform the following steps:

   a. Clean the pulleys, belt guide plate and the upper and lower covers.

   b. Set the timing belt drive pulley to TDC by aligning the TDC mark on the tooth of the belt drive pulley with the pointer on the oil pump.

   c. Set the camshaft pulleys to TDC by aligning the TDC marks on the camshaft pulleys with the pointers on the back covers.

   d. Remove the battery clamp bolt.

   e. Remove the belt tensioner.

   f. Align the holes on the rod and housing of the tensioner.

   g. Using a press or other suitable device, slowly compress the tensioner and insert a 0.08 inch (2mm) pin through the housing and rod.

   h. Install the tensioner making sure the pin is still installed.

   i. Apply thread locker to idler pulley bolt then hand tighten the bolt.

   j. Install the belt over the pulleys in this sequence; drive pulley, idler pulley, front camshaft pulley, water pump pulley, rear camshaft pulley and adjusting pulley.

   k. Tighten the idler pulley bolt to 33 ft. lbs. (44 Nm).

   l. Remove the pin from the tensioner.

9. Install or connect the following:

- Lower half of the side mount and tighten the 3 long bolts to 33 ft. lbs. (44 Nm) and the one short bolt to 9 ft. lbs. (12 Nm)
- Timing belt guide plate as illustrated
- Lower timing cover and tighten the bolts to 9 ft. lbs. (12 Nm)
- Front and rear upper timing covers and tighten the bolts to 9 ft. lbs. (12 Nm)
- Crankshaft pulley and tighten the bolts to 181 ft. lbs. (245 Nm), using the holding tool to prevent the unit from turning

10. Rotate the crankshaft pulley about 5 or 6 degrees clockwise so the belt positions itself on the pulleys.

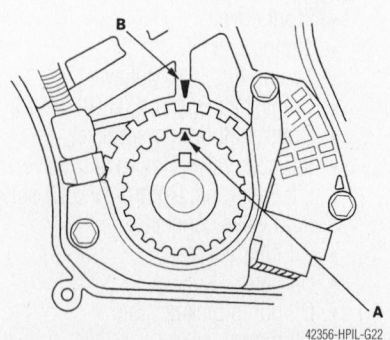

42356-HPIL-G22

Set the timing belt pulley to TDC by aligning the TDC mark (A) on the tooth of the belt pulley with the pointer (B) on the oil pump

FRONT:

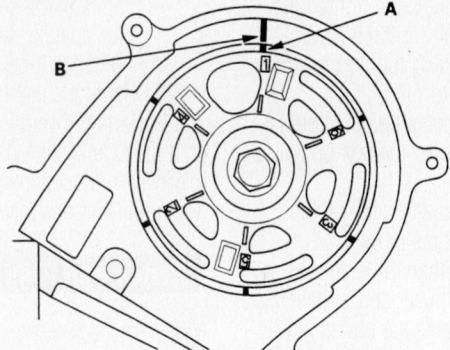

REAR:

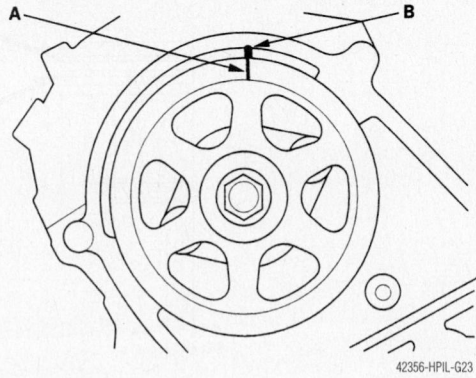

42356-HPIL-G23

Set the camshaft pulleys to TDC by aligning the TDC marks (A) on the camshaft pulleys with the pointers (B) on the back covers

11. Turn the crankshaft pulley so the white mark aligns with the pointer.

12. Check the camshaft pulley marks are aligned. If the marks are aligned, proceed to the next step. If the marks are not aligned, remove the timing belt and reinstall using the steps outlined before this step.

13. Remove or disconnect the following:
- Drive belt
- Upper side mount and tighten the bolts in the sequence illustrated to the specifications in the illustration

14. Using a suitable scan tool, perform the Powertrain Control Module (PCM) reset

and the Crankshaft position (CKP) pattern clear/learn procedures, following the scan tool manufactures instructions.

15. If installing the old belt, perform the following steps:

a. Clean the pulleys, belt guide plate and the upper and lower covers.

b. Set the timing belt drive pulley to TDC by aligning the TDC mark on the tooth of the belt drive pulley with the pointer on the oil pump.

c. Set the camshaft pulleys to TDC by aligning the TDC marks on the camshaft pulleys with the pointers on the back covers.

d. Apply thread locker to idler pulley bolt then hand tighten the bolt.

16. If the tensioner was extended and the belt cannot be installed, perform the steps above for the new belt installation.

a. Install the belt over the pulleys in this sequence; drive pulley, idler pulley, front camshaft pulley, water pump pulley, rear camshaft pulley and adjusting pulley.

b. Tighten the idler pulley bolt to 33 ft. lbs. (44 Nm).

c. Remove the battery clamp bolt.

17. Install or connect the following:
- Lower half of the side mount and tighten the 3 long bolts to 33 ft. lbs. (44 Nm) and the one short bolt to 9 ft. lbs. (12 Nm)

1 Drive pulley (A).
2 Idler pulley (B).
3 Front camshaft pulley (C).
4 Water pump pulley (D).
5 Rear camshaft pulley (E).
6 Adjusting pulley (F).

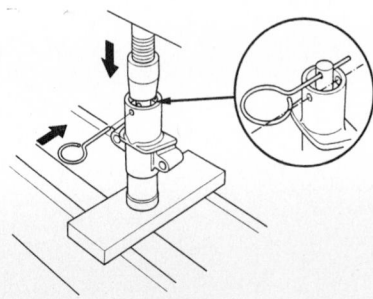

42356-HPIL-G24

Insert a 0.08 inch (2mm) pin through the tensioner housing and rod

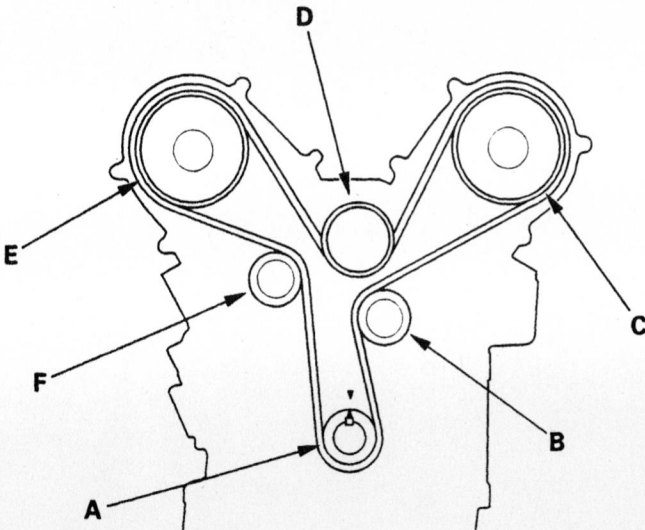

42356-HPIL-G25

Route the belt as shown in the sequence listed

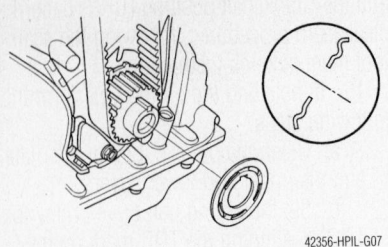

42356-HPIL-G07

Install the timing belt guide plate as shown

- Timing belt guide plate as illustrated
- Lower timing cover and tighten the bolts to 9 ft. lbs. (12 Nm)
- Front and rear upper timing covers

FRONT CAMSHAFT PULLEY:

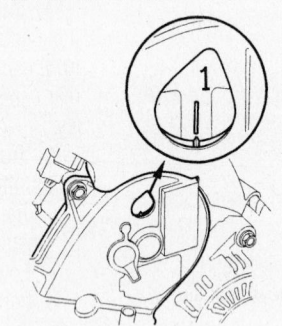

REAR CAMSHAFT PULLEY:

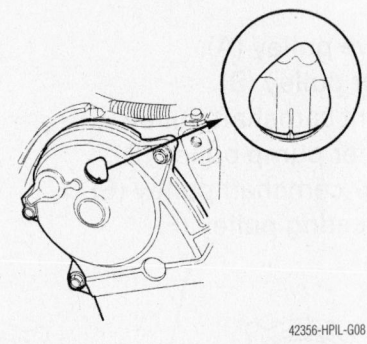

42356-HPIL-G08

Check that the camshaft pulley marks are aligned as shown

and tighten the bolts to 9 ft. lbs. (12 Nm)
- Crankshaft pulley and tighten the bolts to 181 ft. lbs. (245 Nm), using the holding tool to prevent the unit from turning.

18. Rotate the crankshaft pulley about 5 or 6 degrees clockwise so the belt positions itself on the pulleys.

19. Turn the crankshaft pulley so the white mark aligns with the pointer.

20. Check the camshaft pulley marks are aligned. If the marks are aligned, proceed to the next step. If the marks are not aligned, remove the timing belt and reinstall using the steps outlined before this step.

21. Install or connect the following:

- Drive belt
- Upper side mount and tighten the bolts in the sequence illustrated to the specifications in the illustration
- Dipstick tube

22. Using a suitable scan tool, perform the Powertain Control Module (PCM) reset and the Crankshaft position (CKP) pattern clear/learn procedures, following the scan tool manufactures instructions.

Piston and Ring

POSITIONING

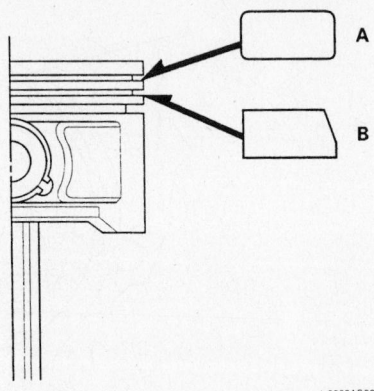

9302AG06

Compression ring identification—3.5L engine

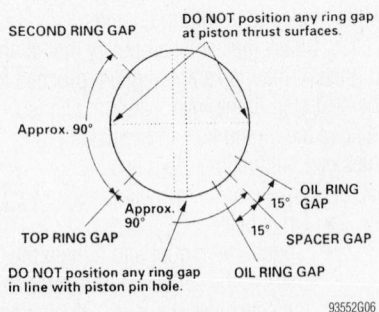

93552G06

Ring end gap positioning

FUEL SYSTEM

Fuel System Service Precautions

Safety is the most important factor when performing not only fuel system maintenance, but any type of maintenance. Failure to conduct maintenance and repairs in a safe manner may result in serious personal injury or death. Maintenance and testing of the vehicle's fuel system components can be accomplished safely and effectively by adhering to the following rules and guidelines:

• To avoid the possibility of fire and personal injury, always disconnect the negative battery cable unless the repair or test procedure requires that battery voltage be applied.

• Always relieve the fuel system pressure prior to disconnecting any fuel system component (injector, fuel rail, pressure regulator, etc.), fitting or fuel line connection. Exercise extreme caution whenever relieving fuel system pressure to avoid exposing skin, face and eyes to fuel spray. Please be advised that fuel under pressure may penetrate the skin or any part of the body that it contacts.

• Always place a shop towel or cloth around the fitting or connection prior to loosening to absorb any excess fuel due to spillage. Ensure that all fuel spillage (should it occur) is quickly removed from engine surfaces. Ensure that all fuel soaked cloths or towels are deposited into a suitable waste container.

• Always keep a dry chemical (Class B) fire extinguisher near the work area.

• Do not allow fuel spray or fuel vapors to come into contact with a spark or open flame.

• Always use a backup wrench when loosening and tightening fuel line connection fittings. This will prevent unnecessary stress and torsion to fuel line piping. Always follow the proper torque specifications.

• Always replace worn fuel fitting O-rings with new. Do not substitute fuel hose or equivalent, where fuel pipe is installed.

Fuel System Pressure

RELIEVING

1. Before servicing the vehicle, refer to the precautions in the beginning of this section.

2. Disconnect the negative battery cable.

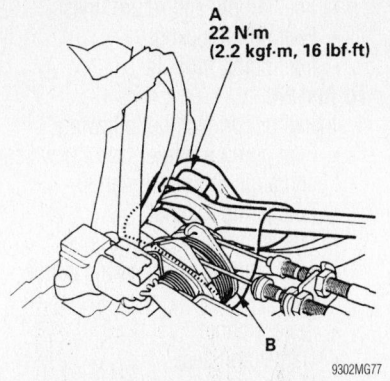

A
22 N·m
(2.2 kgf·m, 16 lbf·ft)

B

9302MG77

Use a wrench on the fuel pulsation damper (A). Place a rag over the damper (B) when relieving residual fuel pressure—Pilot

3. Remove the fuel filler cap.
4. Remove the intake manifold cover.
5. Place a shop towel over the fuel pulsation damper.

6. Loosen the fuel pulsation damper 1 turn.
7. When service is completed, replace the sealing washer and tighten the pulsation damper to 16 ft. lbs. (22 Nm).
8. Replace the fuel filler cap.
9. Connect the negative battery cable.
10. Start the engine and check for leaks.

Fuel Filter

REMOVAL & INSTALLATION

1. Before servicing the vehicle, refer to the precautions in the beginning of this section.
2. Relieve the fuel system pressure.
3. Remove or disconnect the following:
   • Negative battery cable
   • Driver's side second row seat and cut the carpet along the dotted line. Be careful not to cut the wiring harness under the carpet.

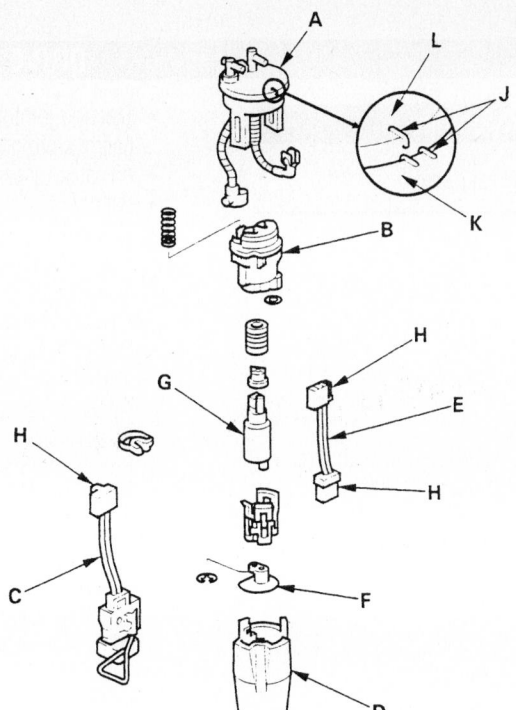

A. Bracket	F. Suction filter
B. Fuel filter	G. Fuel pump
C. Fuel gauge sender	H. Connectors
D. Case	J. Alignment marks
E. Wire harness	K. Fuel tank
	L. Fuel pump module

9308MG26

Exploded view of the fuel pump module—Pilot

- Access panel
- Fuel pump module

4. Disassemble the fuel pump module and remove the fuel filter.

To install:

5. Install the fuel filter and assemble the fuel pump module.

6. Install or connect the following:
- Fuel pump module
- Access panel
- Carpet and seat
- Negative battery cable

7. Start the engine and check for leaks.

Fuel Pump

REMOVAL & INSTALLATION

1. Before servicing the vehicle, refer to the precautions in the beginning of this section.

2. Relieve the fuel system pressure.

3. Remove or disconnect the following:
- Negative battery cable
- Driver's side second row seat and cut the carpet along the dotted line. Be careful not to cut the wiring harness under the carpet.

- Access panel
- Fuel pump module wiring connector
- Fuel supply and return lines
- Fuel pump locknut
- Fuel pump module

To install:

4. Install or connect the following:
- Fuel pump module. Use a new seal and align the matchmarks.
- Fuel pump locknut
- Fuel supply and return lines
- Fuel pump module wiring connector
- Access panel
- Carpet and seat
- Negative battery cable

5. Start the engine and check for leaks.

Fuel Injector

REMOVAL & INSTALLATION

1. Before servicing the vehicle, refer to the precautions in the beginning of this section.

2. Relieve the fuel system pressure.

3. Remove or disconnect the following:
- Negative battery cable
- Intake manifold
- Fuel lines
- Fuel injector connectors
- Fuel pressure regulator vacuum line
- Fuel supply manifold

4. Separate the fuel injectors from the fuel supply manifold.

To install:

5. Install the fuel injectors to the fuel supply manifold with new cushion rings and O-rings.

6. Install new seal rings to the intake manifold.

7. Install or connect the following:
- Fuel supply manifold and injector assembly. Tighten the bolts to 86 inch lbs. (10 Nm).
- Fuel pressure regulator vacuum line
- Fuel injector connectors
- Fuel lines
- Intake manifold
- Negative battery cable

8. Start the engine and check for leaks.

DRIVE TRAIN

Transaxle Assembly

REMOVAL & INSTALLATION

1. Before servicing the vehicle, refer to the precautions in the beginning of this section.

2. Drain the transaxle.

3. Drain the power steering system.

4. Remove the engine appearance covers.

5. Remove the drivers side center console lower panel and pull back the cover to access steering joint cover.

6. Remove or disconnect the following:
- Steering joint bolt

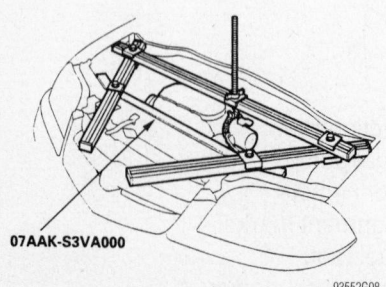

07AAK-S3VA000

93552G08

Support the engine while removing the transaxle—Pilot

- Steering joint from the steering gearbox pinion shaft
- Air intake assembly
- Battery
- Battery tray
- Power steering pump hose and the clamp bolt
- Transmission breather tube
- Cooler hose from the clamp on the starter
- Transaxle oil cooler lines
- Starter motor
- Shift control solenoid valve connectors
- Transaxle ground cable
- 8P connector from the bracket and the connector
- Clutch pressure switch connectors
- Joint connector and transmission range switch connector from the brackets
- Countershaft speed sensor connector
- Heated Oxygen (HO2S) sensor connectors
- Transmission housing mounting bolts
- Nut from the front mount and the ground cable from the engine
- Bulkhead cover, windshield wiper arms, cowl cover sealing and cover

- Install a support fixture to the engine lifting eyes.
- Front sub-frame stiffener
- Primary HO2S sensor clamp bracket from the transmission and harness from the clamp
- Exhaust front pipe
- Lower control arms from the knuckle
- Stabilizer bar links
- Tie rod ends from the knuckle
- Left driveshaft from the differential
- Right driveshaft from the intermediate shaft
- Propeller shaft from the companion flange
- Shift cable cover and holder
- Shift control cable and lever

7. Install a 6 x 1 x 14mm bolt and nut on the cable cover, then reinstall the cable cover to the torque converter housing. If this is not done, the bolt head of the cable cover may prevent torque converter removal.

- Transfer assembly
- Engine-to-torque converter bolts
- Power steering pressure switch connection
- Power steering hose clamp, then the hose from the pipe at the sub-frame
- Transmission lower mount nuts

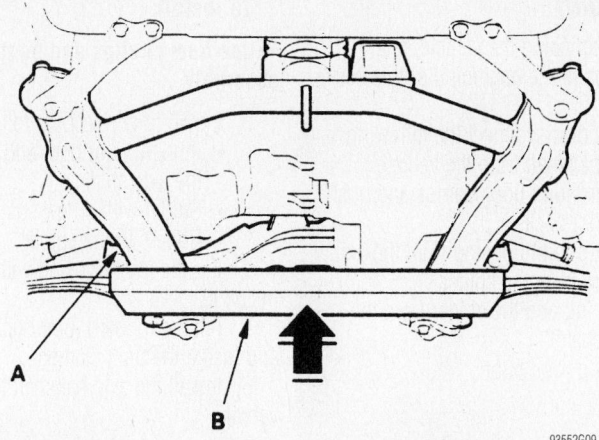

Support the sub-frame with a 4 x 4 x 50 inch piece of wood and a jack

8. Matchmark the front subframe to the vehicle body.
- Rear mount bracket bolts

9. Support the sub-frame with a 4 x 4 x 50 inch piece of wood and a jack.
- Sub-frame
- Transaxle lower mounts
- Driveshafts from the differential and intermediate shaft
- Intermediate shaft
- Transmission front mount bracket
- Transmission flange bolts
- Transmission

To install:

➡**Use new circlips, split pins and self-locking nuts for assembly.**

10. Installation is the reverse of removal. Please note the following specifications:
- Transmission housing bolts and harness clamp bolts to 47 ft. lbs. (64 Nm)
- Transmission housing bolts to 40 ft. lbs. (54 Nm)
- Front mount bracket bolts to 28 ft. lbs. (38 Nm)
- Intermediate shaft bolts to 29 ft. lbs. (39 Nm)
- Transfer assembly bolts to 33 ft. lbs. (44 Nm)

11. Raise the subframe into position and align the matchmarks. Tighten the subframe bolts to 76 ft. lbs. (103 Nm). Tighten the front subframe bracket bolts to 54 ft. lbs. (74 Nm) and the rear bracket bolts to 86 ft. lbs. (117 Nm).
- Rear engine mount bolts to 28 ft. lbs. (38 Nm)
- Engine-to-torque converter bolts. Tighten the 6 x 1 mm bolts to 105 inch lbs. (12 Nm), 10 x 1.25mm bolt to 28 ft. lbs. (38 Nm).
- Front motor mount nut to 40 ft. lbs. (54 Nm)

12. Fill the transaxle to the correct level.
13. Start the engine and check for leaks.
14. Check the wheel alignment and adjust as necessary.

Transfer Assembly

REMOVAL & INSTALLATION

1. Before servicing the vehicle, refer to the precautions in the beginning of this section.
2. Drain the transmission fluid.
3. Remove or disconnect the following:
- Negative battery cable
- Heated Oxygen (HO$_2$S) sensor connectors
- Front sub-frame stiffener
- Exhaust front pipe
- Breather tube bracket bolt, then the tube from the breather pipe
- Propeller shaft from the transfer assembly
- Transfer assembly bolts and the assembly

To install:

4. Install or connect the following:
- New O-ring on the transfer cover
- Dowel pin on the assembly
- Transfer assembly and tighten the bolts to 33 ft. lbs. (44 Nm) in a star pattern
- Propeller shaft
- Breather tube bracket, attach the tube with the dot facing outwards and tighten the bolt to 9 ft. lbs. (12 Nm)
- Exhaust front pipe
- Front sub-frame stiffener and tighten the bolts to 40 ft. lbs. (54 Nm)
- Heated Oxygen (HO$_2$S) sensor connectors
- Negative battery cable

Halfshaft

REMOVAL & INSTALLATION

1. Before servicing the vehicle, refer to the precautions in the beginning of this section.
2. Drain the transaxle if removing the left halfshaft. If is not necessary to drain the fluid if removing the right halfshaft.
3. Remove or disconnect the following:
- Negative battery cable
- Front wheels
- Spindle nut
- Stabilizer bar link
- Lower ball joint

4. Pry the inboard joint from the transaxle or intermediate shaft.
5. Remove the outer CV-joint stub shaft from the hub by tapping the stub shaft with a plastic hammer.

To install:

➡**Use new circlips, split pins and self-locking nuts for assembly.**

6. Install the outer CV-joint stub shaft into the hub.
7. Install the inner CV-joint to the transaxle or intermediate shaft until the circlip locks in the retaining groove.
8. Install or connect the following:
- Lower ball joint. Tighten the nut to 47 ft. lbs. (64 Nm).
- Stabilizer bar link. Tighten the nut to 58 ft. lbs. (78 Nm).
- Spindle nut. Tighten the nut to 181 ft. lbs. (245 Nm).
- Front wheels
- Negative battery cable

9. Fill the transaxle to the correct level and check for leaks.

CV-Joint

OVERHAUL

Front

OUTBOARD JOINT

1. Before servicing the vehicle, refer to the precautions in the beginning of this section.
2. Remove or disconnect the following:
- Axle halfshaft from the vehicle and place it in a vise
- Outboard joint boot clamps and push the boot back
- Outboard joint by driving it off the axle shaft with a brass drift and hammer
- Outboard joint boot

To install:

→**Use new circlips and boot clamps for assembly.**

3. Install the outboard joint boot and clamps to the axle shaft.

4. Fill the outboard joint with grease. Install the outboard joint to the axle shaft. Tap the stub shaft with a brass hammer to seat the circlip.

5. Fill the outboard joint boot with grease and install the boot clamps.

6. Install the axle halfshaft to the vehicle.

INBOARD JOINT

1. Before servicing the vehicle, refer to the precautions in the beginning of this section.

2. Remove or disconnect the following:
- Axle halfshaft from the vehicle.
- Inboard joint boot clamps and push the boot back
- Inboard joint housing from the axle
- Rollers from the spider
- Snapring and the spider from the axle shaft
- Inboard joint boot

To install:

→**Use new circlips and boot clamps for assembly.**

3. Install or connect the following:
- Inboard joint boot and clamps to the axle shaft
- Spider with a new snapring
- Rollers to the spider

4. Fill the joint housing with grease and install it.

5. Fill the inboard joint boot with grease and install the boot clamps.

6. Install the axle halfshaft to the vehicle.

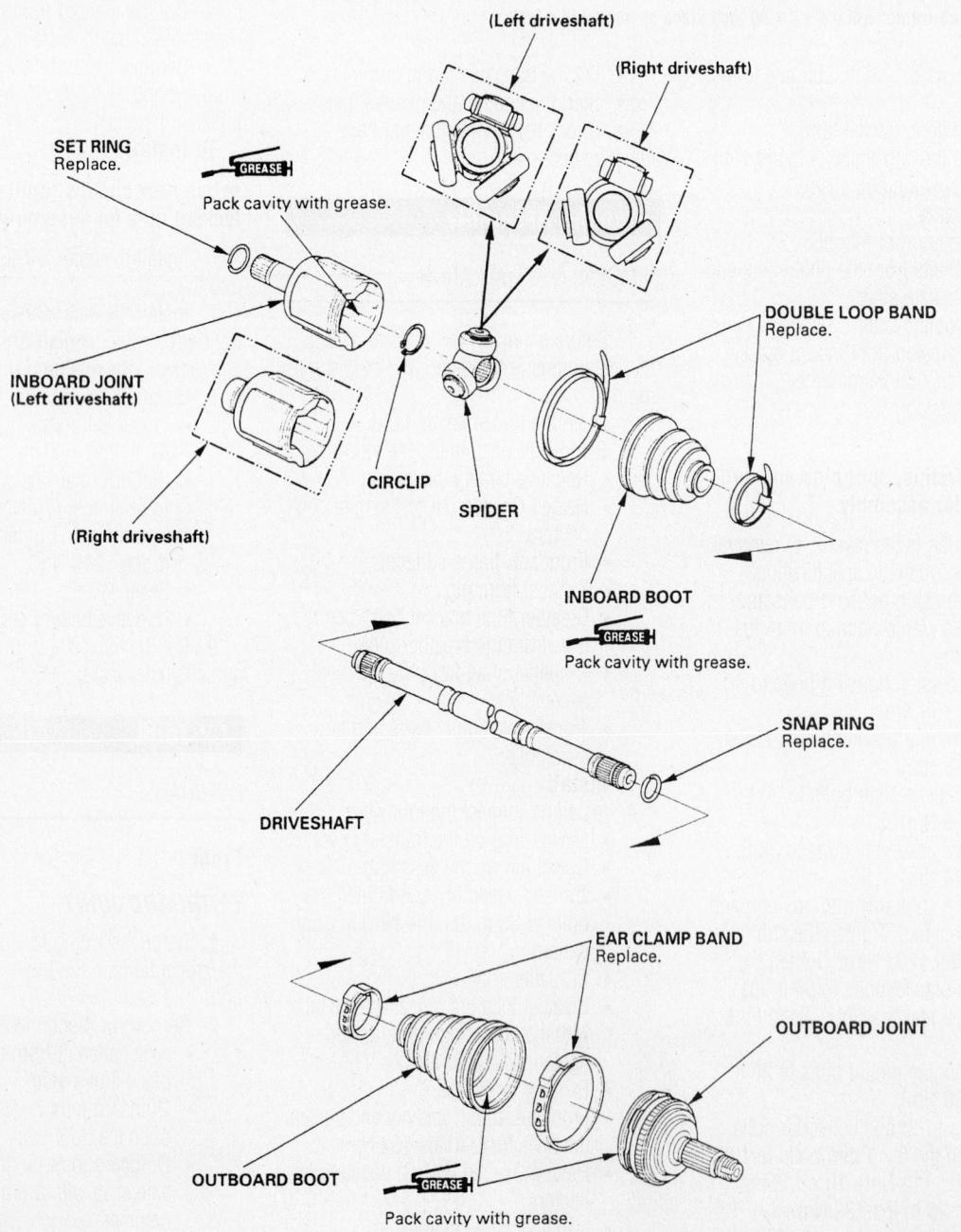

Front axle exploded view—Pilot

9308MG29

STEERING AND SUSPENSION

Air Bag

✳✳ CAUTION

Some vehicles are equipped with an air bag system. The system must be disarmed before performing service on, or around, system components, the steering column, instrument panel components, wiring and sensors. Failure to follow the safety precautions and the disarming procedure could result in accidental air bag deployment, possible injury and unnecessary system repairs.

PRECAUTIONS

Several precautions must be observed when handling the inflator module to avoid accidental deployment and possible personal injury.

• Never carry the inflator module by the wires or connector on the underside of the module.

• When carrying a live inflator module, hold securely with both hands, and ensure that the bag and trim cover are pointed away.

• Place the inflator module on a bench or other surface with the bag and trim cover facing up.

• With the inflator module on the bench, never place anything on or close to the module which may be thrown in the event of an accidental deployment.

Before servicing the vehicle, also make sure to refer to the precautions in the beginning of this section as well.

DISARMING

Disconnect and isolate the negative battery cable. Wait 3 minutes for the system capacitor to discharge before performing any service.

Power Rack and Pinion Steering Gear

REMOVAL & INSTALLATION

✳✳ WARNING

Do not permit the steering wheel to turn whenever the steering gear is disconnected from the steering column. Damage to the air bag wiring can result.

1. Before servicing the vehicle, refer to the precautions in the beginning of this section.
2. Center the steering wheel and lock it in position.
3. Attach a support fixture to the engine lifting eyes.
4. Remove or disconnect the following:
   • Negative battery cable
   • Air bag and steering wheel
   • Steering joint cover
   • Steering flexible joint
   • Power steering fluid lines
   • 10mm bolt on the engine side mount bracket
   • Front wheels
   • Outer tie rod ends
   • Sub-frame stiffener
   • Heated Oxygen (HO$_2$S) sensor connectors
   • 3 way catalytic converter from the mufflers
   • Flange bolts from the exhaust rubber mount
   • Power steering pressure switch connector
   • Propeller shaft protector
   • Splash shield
5. Support the front subframe with a jack and support the transmission with a second jack.
6. Loosen the 14mm subframe bolts.
7. Lower the subframe about 1 ³/₁₆ inches (30mm).
8. Remove or disconnect the following:
   • Two 12mm and two 14 stiffener plate bolts
9. Support the transfer case by raising the transmission jack and remove the two 12mm bolts.
   • Two 14mm bolts and the rear stiffener plats from the sub-frame

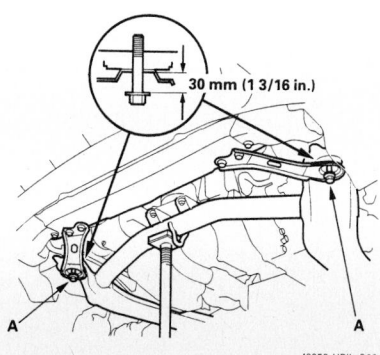

Loosen the 14mm subframe bolts and lower the subframe about 1 ³/₁₆ inches (30mm)

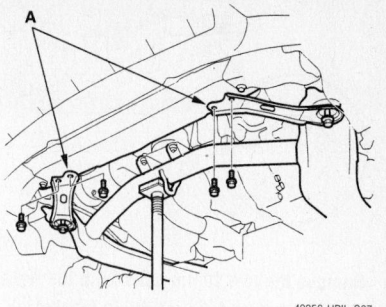

Remove the Four 12mm stiffener plate bolts

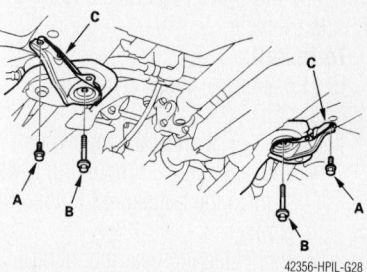

Remove the Two 12mm bolts (A), then the 14mm bolts (B) and the rear stiffener plates (C) from the sub-frame

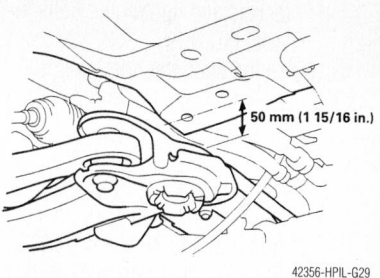

Lower the transmission jack until the front subframe has dropped about 1 ¹⁵/₁₆ inch (50mm)

10. Lower the transmission jack until the front subframe has dropped about 1 ¹⁵/₁₆ inch (50mm).
   • Power steering line brackets
   • Feed line
   • Return hose
   • Two 10mm bolts from the right side gearbox
   • Mounting bracket and cushion
   • Two 10mm bolts from the left side gearbox
11. Lower the transmission jack until the front subframe has dropped about 3 ¹⁵/₁₆ inch (100mm).
   • Gearbox stiffener bracket

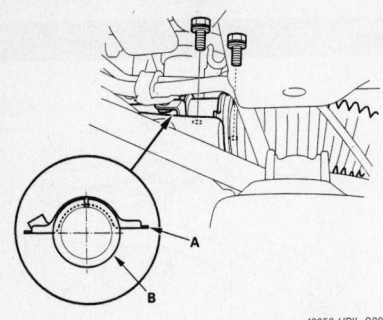

42356-HPIL-G30

Remove the two 10mm bolts from the right side gearbox and the mounting bracket and cushion

12. Slide the gearbox between the body and front sub-frame towards the left and from the vehicle.

To install:

13. Position the steering gear in the vehicle.

14. Install or connect the following:
- Left steering gear mounting bolts. Tighten the bolts to 43 ft. lbs. (58 Nm).
- Right steering gear mounting bracket. Tighten the bolts to 29 ft. lbs. (39 Nm).
- Return hose
- Feed line
- Power steering line mounting brackets and tighten the bolts to 7 ft. lbs. (10 Nm)

15. Raise the subframe into position. Tighten the 14mm bolts to 76 ft. lbs. (103 Nm) and the 12mm bolts to 86 ft. lbs. (117 Nm).

16. Install or connect the following:
- Front stiffener plates. Tighten the 14mm bolts to 76 ft. lbs. (103 Nm) and the 12mm bolts to 54 ft. lbs. (74 Nm).
- Splash shield
- Propeller shaft protector
- Power steering pressure switch
- 3 way catalytic converter and mufflers. Tighten the nuts to 25 ft. lbs. (33 Nm)
- Rubber exhaust mount and tighten the bolts to 28 ft. lbs. (38 Nm)
- HO2S sensor connectors
- Sub-frame stiffener plate
- 10mm flange bolts on the engine side mount bracket to 33 ft. lbs. (44 Nm)
- Power steering hoses
- Outer tie rod ends
- Front wheels. Position the wheels straight-ahead.
- Steering flexible joint. Tighten the pinch bolts to 16 ft. lbs. (22 Nm).

- Steering joint cover
- Negative battery cable

17. Fill the power steering system.

18. Check the wheel alignment and adjust as necessary.

Strut

REMOVAL & INSTALLATION

Front

1. Before servicing the vehicle, refer to the precautions in the beginning of this section.

2. Remove or disconnect the following:
- Front wheel
- Wheel speed sensor wiring bracket
- Brake hose bracket
- Stabilizer bar link
- Strut pinch bolts
- Upper mount nuts
- Strut

To install:

3. Install or connect the following:
- Strut. Tighten the upper mount nuts to 43 ft. lbs. (59 Nm).
- Strut pinch bolts. Tighten the nuts to 116 ft. lbs. (157 Nm).
- Stabilizer bar link. Tighten the nut to 58 ft. lbs. (78 Nm).
- Brake hose bracket
- Wheel speed sensor wiring bracket
- Front wheel

4. Check the wheel alignment and adjust as necessary.

Shock Absorber

REMOVAL & INSTALLATION

Rear

1. Before servicing the vehicle, refer to the precautions in the beginning of this section.

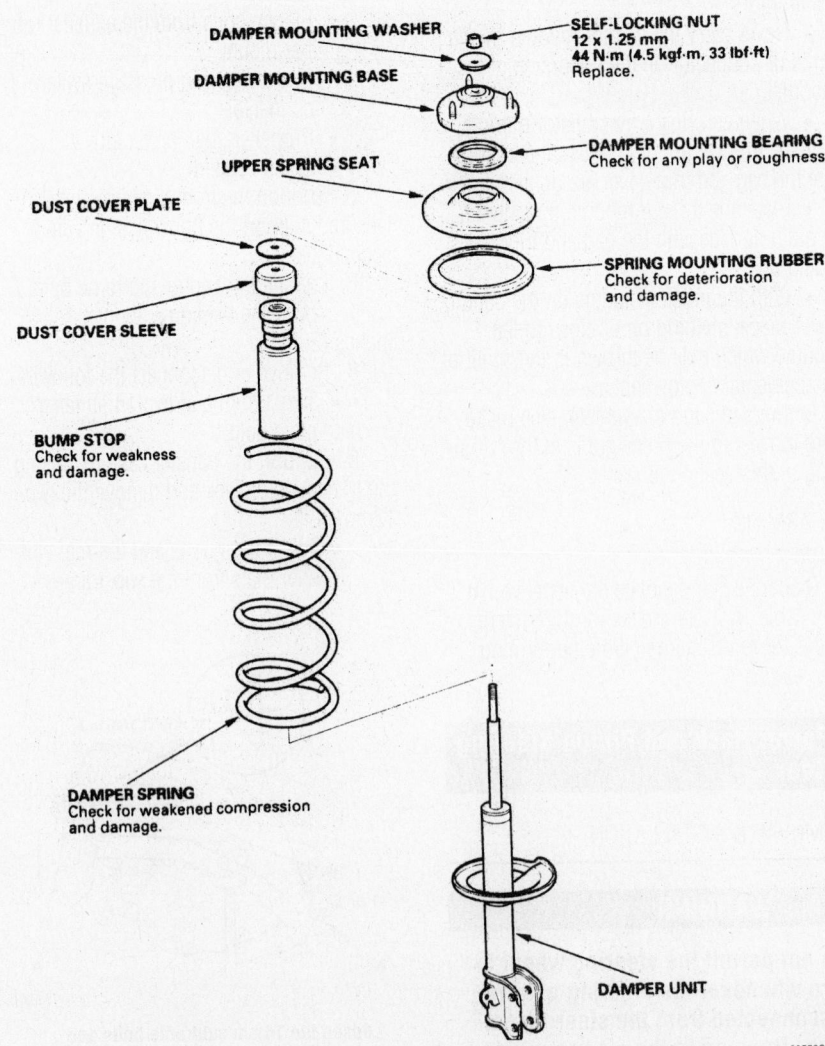

Exploded view of the front strut assembly

93552G11

2. Support the vehicle under the lower control arm.

3. Remove or disconnect the following:
- Rear wheel
- Upper shock absorber flange bolt
- Lower shock absorber nut
- Shock absorber

To install:

4. Install or connect the following:
- Shock absorber. Tighten the fasteners to 47 ft. lbs. (64 Nm).
- Rear wheel

Coil Spring

REMOVAL & INSTALLATION

Front

1. Before servicing the vehicle, refer to the precautions in the beginning of this section.

2. Remove the strut from the vehicle and install in a strut spring compressor. Compress the spring until the end of the spring comes away from the spring seat.

3. Remove the upper strut mount, spring seat and related components.

4. Remove the coil spring from the strut spring compressor.

To install:

➡ **Use a new self-locking nut.**

5. Compress the spring and position the strut so that the end of the spring aligns with the notch in the spring seat.

6. Install the upper strut mounting components and tighten the nut to 33 ft. lbs. (44 Nm).

7. Install the strut to the vehicle.

8. Check the wheel alignment and adjust as necessary.

Rear

1. Before servicing the vehicle, refer to the precautions in the beginning of this section.

2. Support the vehicle under the lower control arm.

3. Remove or disconnect the following:
- Rear wheel
- Stabilizer link from the lower arm
- Wheel speed sensor wiring harness from the lower arm. Do not disconnect the connector.
- Upper shock absorber flange bolt
- Lower control arm bolts

4. Lower the floor jack and remove the coil spring and spring seats.

To install:

➡ **Use new self-locking nuts for assembly.**

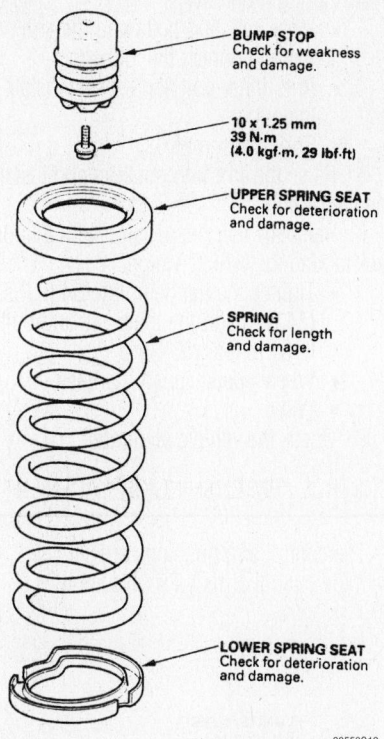

BUMP STOP
Check for weakness and damage.

**10 x 1.25 mm
39 N·m
(4.0 kgf·m, 29 lbf·ft)**

UPPER SPRING SEAT
Check for deterioration and damage.

SPRING
Check for length and damage.

LOWER SPRING SEAT
Check for deterioration and damage.

93552G12

Exploded view of the rear spring assembly

5. Place the coil spring and spring seats on the lower control arm and raise into position. Tighten the inboard bolt to 61 ft. lbs. and the outer bolt to 54 ft. lbs. (74 Nm).

6. Install or connect the following:
- Rear wheel

Ball Joint

REMOVAL & INSTALLATION

The lower ball joints are replaced with the control arms as an assembly.

Upper Control Arm

REMOVAL & INSTALLATION

Rear

1. Before servicing the vehicle, refer to the precautions in the beginning of this section.

2. Support the control arm at the knuckle.

3. Remove or disconnect the following:
- Wheel
- Upper ball joint from the knuckle
- Upper arm bolt and the arm

4. Installation is the reverse of removal. Tighten the arm bolt to 47 ft. lbs. (64 Nm)

and the ball joint nut to 36–43 ft. lbs. (49–59 Nm).

Lower Control Arm

REMOVAL & INSTALLATION

Front

1. Before servicing the vehicle, refer to the precautions in the beginning of this section.

2. Remove or disconnect the following:
- Front wheel
- Lower ball joint
- Front inner flange bolt
- Rear inner flange bolt
- Lower control arm

To install:

➡ **Use a new split pin for assembly.**

3. Install or connect the following:
- Lower control arm. Tighten the inner flange bolts to 69 ft. lbs. (93 Nm).
- Lower ball joint. Tighten the nut to 43–51 ft. lbs. (59–69 Nm).
- Front wheel

4. Check the wheel alignment and adjust as necessary.

Rear

LOWER ARM (A)

1. Before servicing the vehicle, refer to the precautions in the beginning of this section.

2. Remove or disconnect the following:
- Lower arm mounting bolt and nut
- Lower arm

3. Installation is the reverse of removal. Tighten the bolt to 105 ft. lbs. (142 Nm) and the nut to 47 ft. lbs. (64 Nm).

LOWER ARM (B)

1. Before servicing the vehicle, refer to the precautions in the beginning of this section.

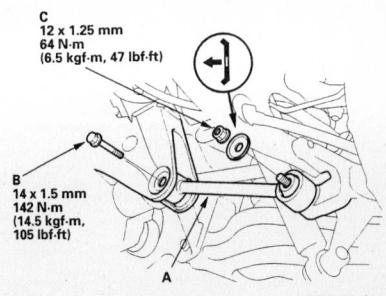

**C
12 x 1.25 mm
64 N·m
(6.5 kgf·m, 47 lbf·ft)**

**B
14 x 1.5 mm
142 N·m
(14.5 kgf·m,
105 lbf·ft)**

A

42356-HPIL-G31

Rear lower arm (A) mounting

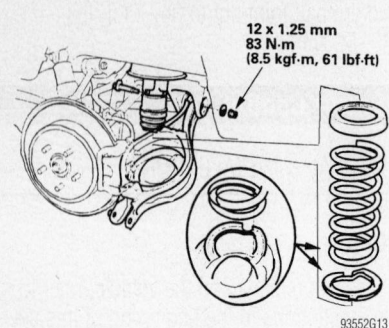

12 x 1.25 mm
83 N·m
(8.5 kgf·m, 61 lbf·ft)

93552G13

Rear lower arm (B) mounting

2. Support the control arm with a jack.
3. Remove or disconnect the following:
   - Wheel
   - Stabilizer link from the lower arm
   - Wheel speed sensor wiring harness from the lower arm. Do not disconnect the connector.
   - Flange bolts that attaches the lower arm to the knuckle

4. Spring assembly
   - Inner nuts and bolts and the arm
5. Install or connect the following:
   - Arm, inner bolt and loosely install the nut
   - Spring assembly
6. Raise the arm into position and install the flange bolt.
7. Raise the rear suspension with a floor jack to load the vehicle weight.
   - Tighten the flange bolt to 54 ft. lbs. (74 Nm) and the inner nut and bolt to 61 ft. lbs. (83 Nm).
   - Wheel speed sensor harness
   - Wheel
8. Check the vehicle alignment.

CONTROL ARM BUSHING REPLACEMENT

The control arm bushings are serviced with the control arms as an assembly.

Wheel Bearings

ADJUSTMENT

The wheel bearings are sealed units and are not adjustable.

REMOVAL & INSTALLATION

Front

1. Before servicing the vehicle, refer to the precautions in the beginning of this section.
2. Remove or disconnect the following:
   - Front wheel
   - Brake hose mounting bolt
   - Brake caliper
   - Wheel speed sensor
   - Spindle nut
   - Brake rotor
   - Outer tie rod end

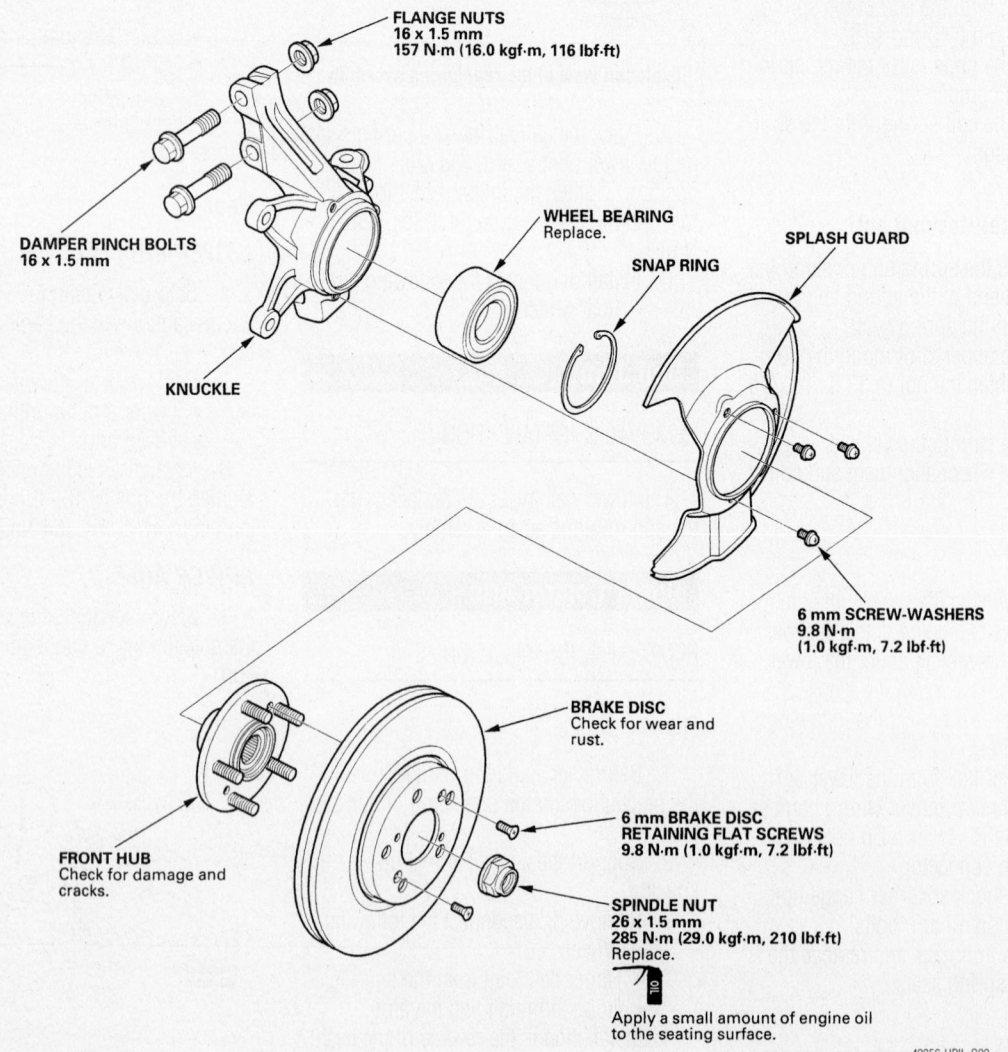

FLANGE NUTS
16 x 1.5 mm
157 N·m (16.0 kgf·m, 116 lbf·ft)

DAMPER PINCH BOLTS
16 x 1.5 mm

KNUCKLE

WHEEL BEARING
Replace.

SNAP RING

SPLASH GUARD

6 mm SCREW-WASHERS
9.8 N·m
(1.0 kgf·m, 7.2 lbf·ft)

BRAKE DISC
Check for wear and rust.

FRONT HUB
Check for damage and cracks.

6 mm BRAKE DISC
RETAINING FLAT SCREWS
9.8 N·m (1.0 kgf·m, 7.2 lbf·ft)

SPINDLE NUT
26 x 1.5 mm
285 N·m (29.0 kgf·m, 210 lbf·ft)
Replace.

Apply a small amount of engine oil to the seating surface.

42356-HPIL-G02

Front wheel bearing assembly

- Lower ball joint
- Steering knuckle
3. Press the hub out of the wheel bearing.
- Splash guard
- Snapring and press the wheel bearing out of the steering knuckle

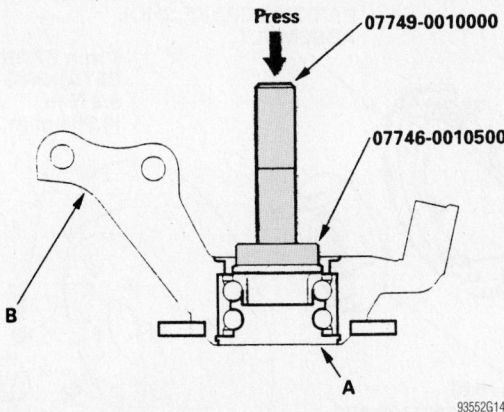

Press the wheel bearing out of the knuckle

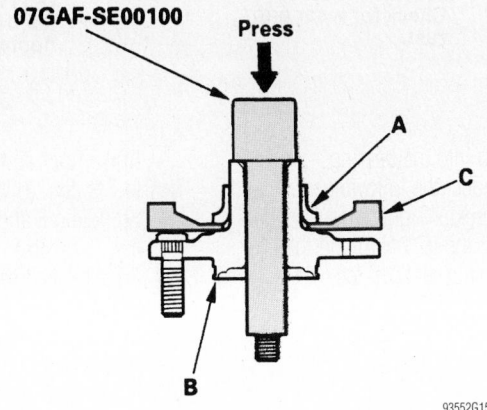

Press the wheel bearing inner race from the hub

4. If necessary, press the inner bearing race off of the hub.

To install:

➡**Use a new ball joint nut, split pin, snapring and spindle nut for assembly.**

5. Press the bearing into the steering knuckle and install the snapring.
6. Install the splash guard.
7. Press the hub into the bearing.
8. Install or connect the following:
- Steering knuckle. Tighten the ball joint nut to 43–51 ft. lbs. (59–69 Nm) and the damper flange bolts to 116 ft. lbs. (157 Nm).
- Outer tie rod end. Tighten the nut to 40 ft. lbs. (54 Nm).
- Wheel speed sensor, if equipped
- Brake caliper and rotor
- Brake hose
- Spindle nut. Tighten the nut to 210 ft. lbs. (285 Nm).
- Front wheel
9. Check the wheel alignment and adjust as necessary.

Rear

1. Before servicing the vehicle, refer to the precautions in the beginning of this section.
2. Remove or disconnect the following:
- Rear wheel
- Brake hose bracket mounting bolts from the trailing arm and the knuckle
- Brake caliper
- Wheel speed sensor
- Spindle nut
- Brake rotor
- Upper ball joint
- Lower arm (A)
- Lower arm (B) from the trailing arm
3. Support the lower arm (B)
- Steering knuckle
4. Press the hub out of the wheel bearing.

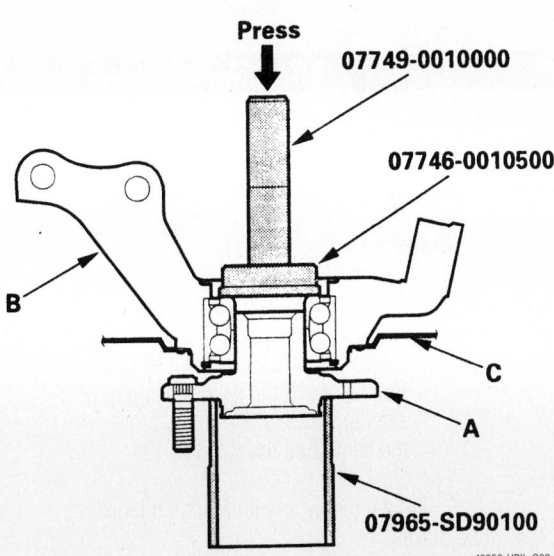

Press the wheel bearing into the knuckle

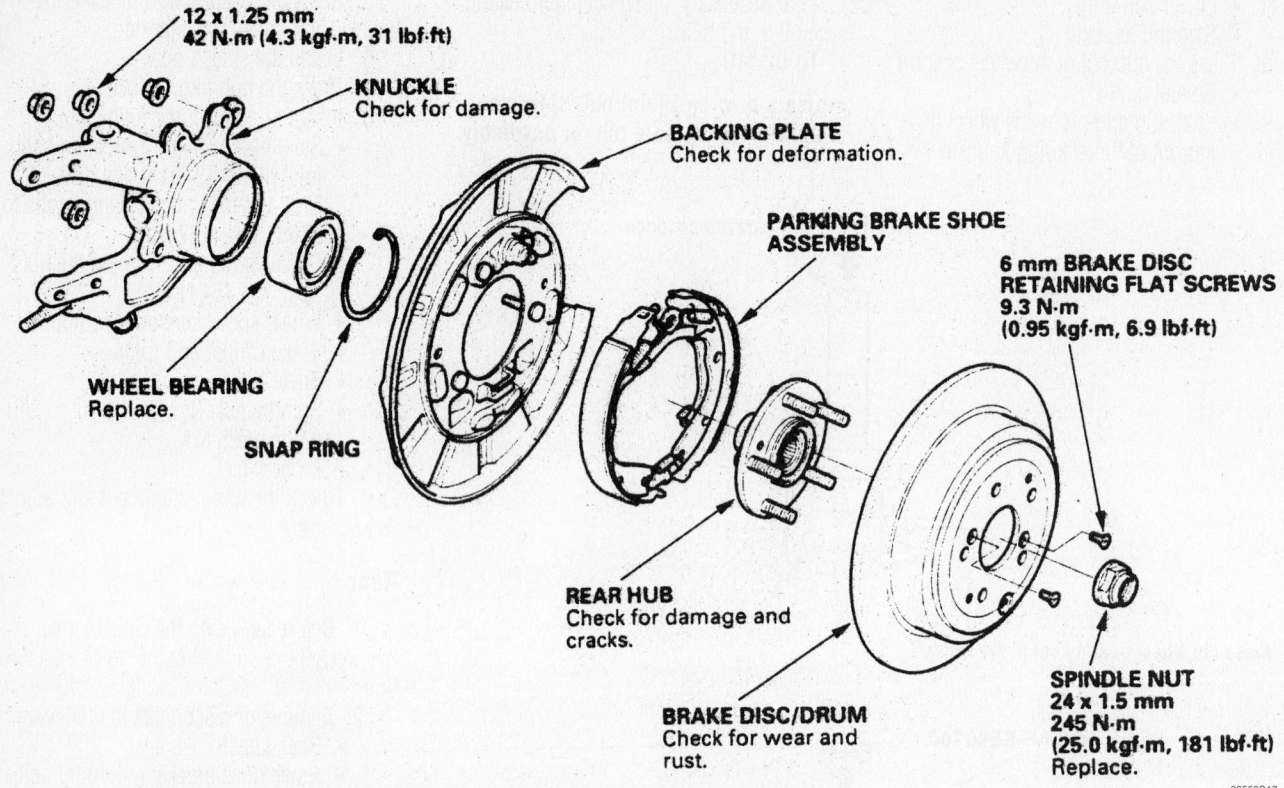

12 x 1.25 mm
42 N·m (4.3 kgf·m, 31 lbf·ft)

KNUCKLE
Check for damage.

BACKING PLATE
Check for deformation.

PARKING BRAKE SHOE ASSEMBLY

6 mm BRAKE DISC RETAINING FLAT SCREWS
9.3 N·m
(0.95 kgf·m, 6.9 lbf·ft)

WHEEL BEARING
Replace.

SNAP RING

REAR HUB
Check for damage and cracks.

BRAKE DISC/DRUM
Check for wear and rust.

SPINDLE NUT
24 x 1.5 mm
245 N·m
(25.0 kgf·m, 181 lbf·ft)
Replace.

93552G17

Exploded view of the rear wheel bearing assembly

- Splash guard
- Snapring and press the wheel bearing out of the steering knuckle

5. If necessary, press the inner bearing race off of the hub.

To install:

➡**Use a new ball joint nut, split pin, snapring and spindle nut for assembly.**

6. Press the bearing into the steering knuckle and install the snapring.
7. Install the splash guard.

8. Press the hub into the bearing.
9. Install or connect the following:
   - Steering knuckle. Tighten the flange bolt to 54 ft. lbs. (74 Nm) and the lower shock nut to 47 ft. lbs. (64 Nm)
   - Lower arm (B) to the trailing arm and tighten the bolts to 47 ft. lbs. (64 Nm)
   - Lower arm (A)
   - Upper ball joint and tighten the nut to 40 ft. lbs. (54 Nm)

   - Brake rotor and tighten the screws to 7 ft. lbs. (9 Nm)
   - Spindle nut and tighten to 181 ft. lbs. (245 Nm)
   - Wheel speed sensor
   - Brake caliper and tighten the bolts to 41 ft. lbs. (55 Nm)
   - Brake hose bracket mounting bolts to the knuckle and trailing arm
   - Rear wheel
10. Check the wheel alignment and adjust as necessary.

BRAKES

Brake Caliper

REMOVAL & INSTALLATION

Front

1. Before servicing the vehicle, refer to the precautions in the beginning of this section.
2. Remove some fluid from the reservoir with a suction pump.
3. Remove or disconnect the following:
   - Front wheels
   - Banjo bolt and disconnect the brake hose from the caliper. Plug

the hose to prevent fluid loss and contamination.
   - Mounting bolts and the caliper from its mounting bracket

To Install:
4. Install or connect the following:
   - Caliper over the pads and onto its mounting bracket. Torque both caliper bolts to 27 ft. lbs. (36 Nm).
   - Brake hose to the caliper using new sealing washers. Carefully torque the banjo bolt to 25 ft. lbs. (34 Nm).
5. Fill the reservoir with fluid and bleed the brakes.
   - Front wheels

Rear

1. Before servicing the vehicle, refer to the precautions in the beginning of this section.
2. Remove some fluid from the reservoir with a suction pump.
3. Remove or disconnect the following:
   - Rear wheels
   - Banjo bolt and disconnect the brake hose from the caliper. Plug the hose to prevent fluid loss and contamination.
   - 2 caliper mounting bolts and the caliper from its mounting bracket

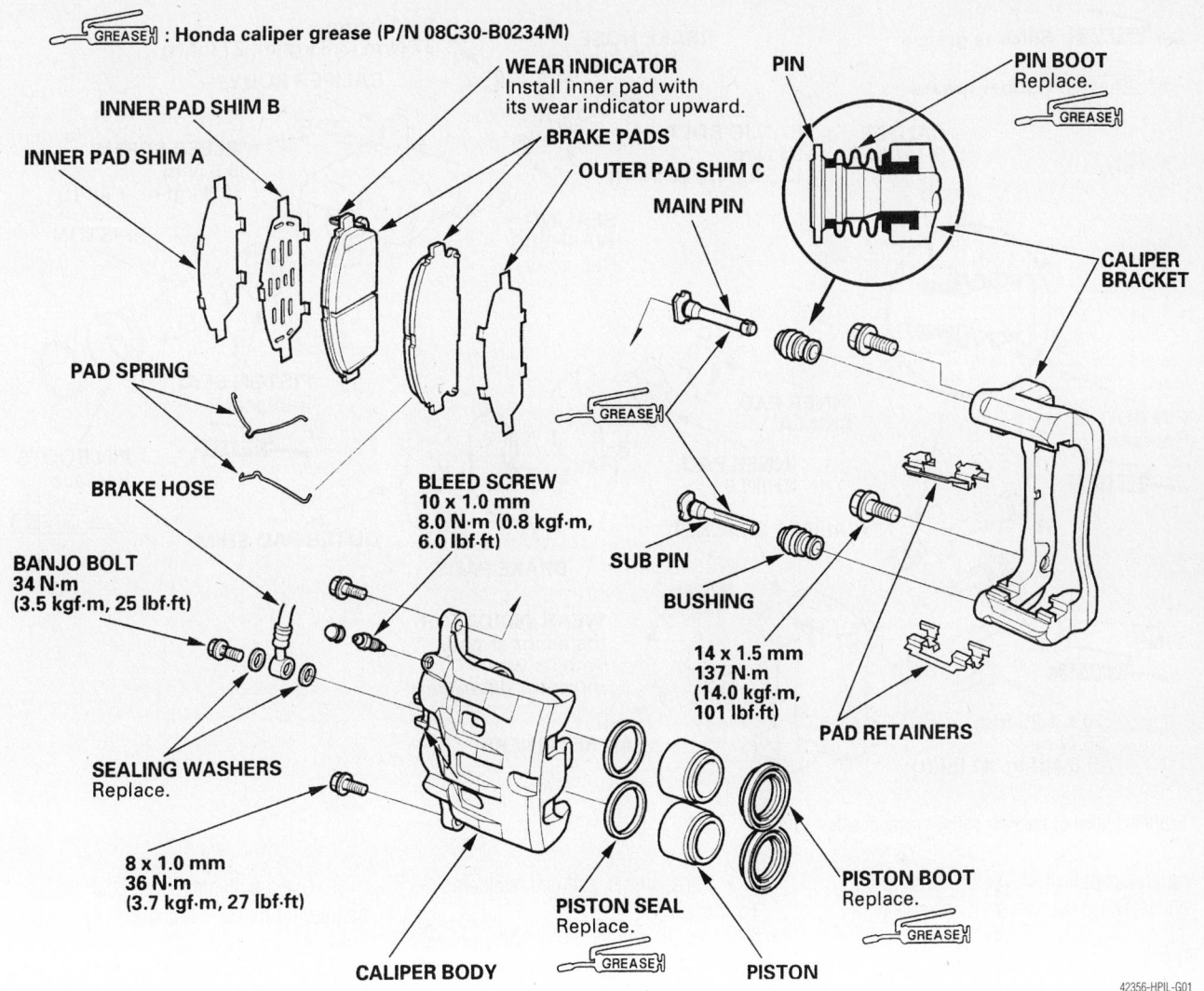

⊏GREASE⊐ : Honda caliper grease (P/N 08C30-B0234M)

INNER PAD SHIM B

INNER PAD SHIM A

WEAR INDICATOR
Install inner pad with
its wear indicator upward.

BRAKE PADS

OUTER PAD SHIM C

MAIN PIN

PIN

PIN BOOT
Replace.
⊏GREASE⊐

CALIPER
BRACKET

PAD SPRING

BRAKE HOSE

BLEED SCREW
10 x 1.0 mm
8.0 N·m (0.8 kgf·m,
6.0 lbf·ft)

⊏GREASE⊐

SUB PIN

BUSHING

14 x 1.5 mm
137 N·m
(14.0 kgf·m,
101 lbf·ft)

PAD RETAINERS

BANJO BOLT
34 N·m
(3.5 kgf·m, 25 lbf·ft)

SEALING WASHERS
Replace.

8 x 1.0 mm
36 N·m
(3.7 kgf·m, 27 lbf·ft)

CALIPER BODY

PISTON SEAL
Replace.
⊏GREASE⊐

PISTON

PISTON BOOT
Replace.
⊏GREASE⊐

42356-HPIL-G01

Exploded view of the front caliper components

To Install:

4. Install or connect the following:
   - Caliper over the pads and onto its mounting bracket. Tighten the caliper bolts to 27 ft. lbs. (37 Nm).
   - Brake hose with new sealing washers. Tighten the banjo bolt to 25 ft. lbs. (34 Nm).

5. Fill the reservoir with fluid and bleed the brake system. Adjust the parking brake if necessary.
   - Rear wheels

Disc Brake Pads

REMOVAL & INSTALLATION

Front

1. Before servicing the vehicle, refer to the precautions in the beginning of this section.

2. Remove or disconnect the following:
   - Front wheels

3. Remove a small amount of brake fluid from the reservoir using a suction pump.
   - Brake hose clamp from the knuckle by unfastening the retaining bolts
   - Lower caliper retaining bolt and pivot the caliper upward, off of the pads
   - Pad springs while holding the pads
   - Pad shim and pad retainers
   - Disc brake pads from the caliper

To install:

4. Clean the caliper thoroughly; remove any rust from the lip of the disc or rotor. Check the brake rotor for grooves or cracks. If any heavy scoring is present, the rotor must be replaced.

5. Install or connect the following:
   - Pad retainers. Apply molybdenum brake grease to both surfaces of the shims and the back of the disc brake pads.

   - Pads and shims. The pad with the wear indicator goes in the inboard position.
   - Pad springs while holding the pads

6. Push in the caliper piston so the caliper will fit over the pads. This is most easily accomplished with a pad spreader or large C-clamp.
   - Caliper down into position and tighten the mounting bolt to 27 ft. lbs. (37 Nm)
   - Brake hose to the knuckle, if removed
   - Wheels

7. Add brake fluid to the master cylinder reservoir and install the cap.

8. Depress the brake pedal several times and make sure that the movement feels normal. The first brake pedal application may result in a very long pedal action due to the pistons being retracted. Always make sev-

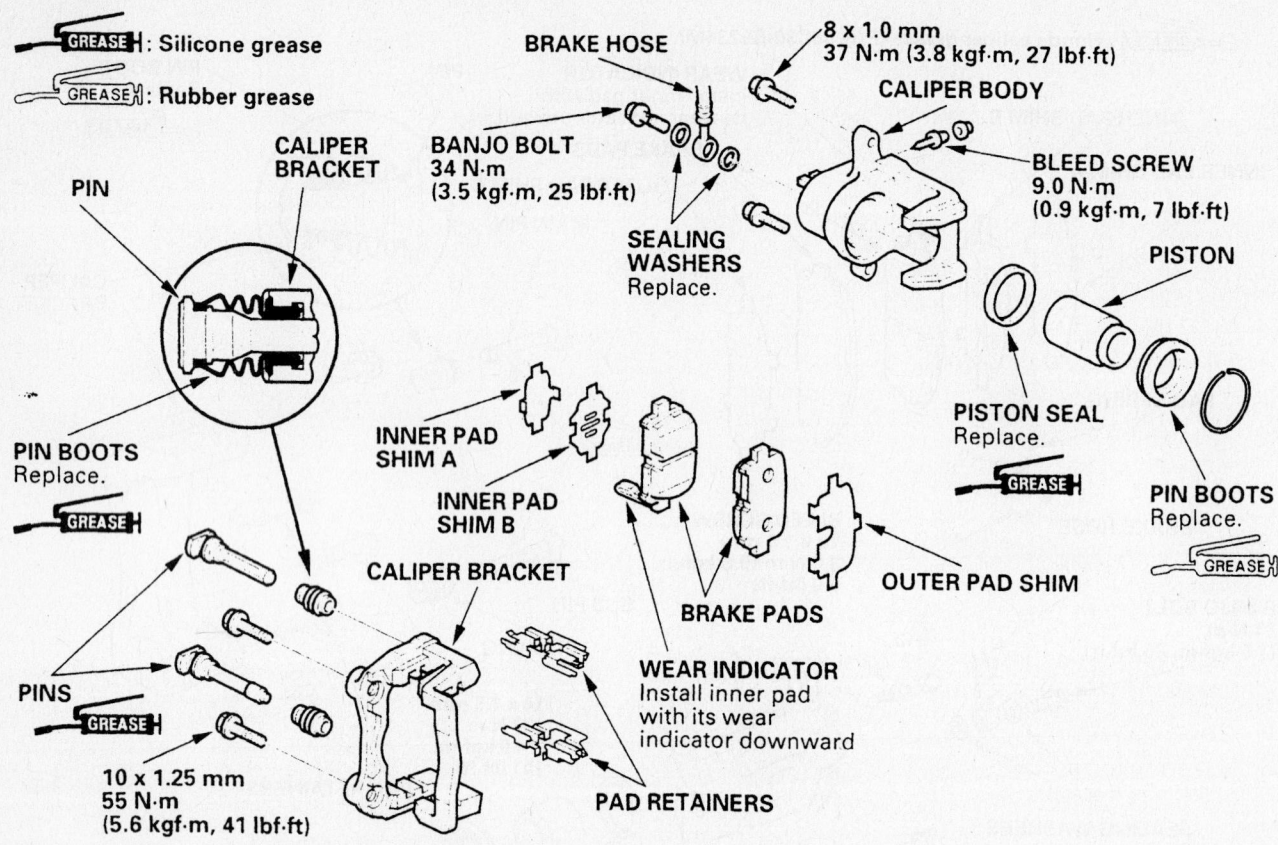

Exploded view of the rear caliper components—Pilot

eral brake applications before starting the vehicle. Bleed the system if necessary.

Rear

1. Before servicing the vehicle, refer to the precautions in the beginning of this section.

2. Remove a small amount of brake fluid from the reservoir using a suction pump.

3. Remove or disconnect the following:
   - Rear wheels
   - 2 caliper mounting bolts and the caliper from the bracket

   - Pads, shims, and pad retainers

To install:

4. Clean the caliper thoroughly; remove any dirt or dust. Check the brake rotor for grooves or cracks and machine or replace, as necessary.

5. Install or connect the following:
   - Pad retainers. Apply molybdenum brake grease to both surfaces of the shims and the back of the disc brake pads.
   - Pads and shims. The wear retainer on the inboard pad faces down.

6. Use a suitable tool to push caliper piston into its bore and enable the caliper to fit over the pads. Lubricate the piston boot with silicon grease. Avoid twisting the boot.
   - Brake caliper and tighten the mounting bolts to 27 ft. lbs. (37 Nm)
   - Rear wheels

7. Add brake fluid to the master cylinder reservoir. Depress the brake pedal several times to seat the pads. Bleed the brakes if necessary.

HYUNDAI

Accent • Elantra • Sonata • Tiburon • XG 300

9

SPECIFICATION AND MAINTENANCE CHARTS

ENGINE AND VEHICLE IDENTIFICATION

Code	Liters (cc)	Cu. In.	Cyl.	Fuel Sys.	Engine Type	Eng. Mfg.
G	1.5 (1495)	91.17	4	MPFI	DOHC	Hyundai
F ①	2.0 (1975)	120.52	4	MPFI	DOHC	Hyundai
D ②	2.0 (1975)	120.52	4	MPFI	DOHC	Hyundai
S	2.4 (2351)	143.46	4	MPFI	DOHC	Hyundai
V	2.5 (2493)	152.13	6	MPFI	DOHC	Hyundai
H ③	2.7 (2656)	164.30	6	MPFI	DOHC	Hyundai
F ④	2.7 (2656)	164.30	6	MPFI	DOHC	Hyundai
D	3.0 (2972)	181.40	6	MPFI	DOHC	Hyundai
E	3.5 (3496)	211.60	6	MPFI	DOHC	Hyundai

Model Year	
Code	Year
1	2001
2	2002
3	2003
4	2004
5	2005

MPFI: Multi-Point Fuel Injection

SOHC: Single Overhead Camshaft

DOHC: Double Overhead Camshafts

① 2001 Tiburon

② Elantra and 2003–04 Tiburon

③ Sonata

④ Tiburon

67162-ELAN-C01

GENERAL ENGINE SPECIFICATIONS

Year	Engine Displacement Liters	Engine ID/VIN	Net Horsepower @ rpm	Net Torque @ rpm (ft. lbs.)	Bore x Stroke (in.)	Compression Ratio	Oil Pressure @ rpm
2001	1.5	G	92@5500	97@4000	2.97 x 3.29	10.0:1	21@Idle
	2.0	F	140@6000	133@4800	3.23 x 3.68	10.3:1	24@Idle
	2.0	D	140@6000	133@4800	3.23 x 3.68	10.3:1	24@Idle
	2.4	S	137@6000	129@4000	3.41 x 3.94	10.0:1	12@Idle
	2.5	V	142@5000	168@2500	3.59 x 2.99	10.0:1	12@Idle
	3.0	D	192@6000	178@4800	3.59 x 2.99	10.0:1	12@Idle
2002	1.5	G	92@5500	97@4000	2.97 x 3.29	10.0:1	21@Idle
	2.0	D	140@6000	133@4800	3.23 x 3.68	10.3:1	24@Idle
	2.4	S	137@6000	129@4000	3.41 x 3.94	10.0:1	12@Idle
	2.7	H	172@6000	181@4000	3.41 x 2.95	10.0:1	12@Idle
	3.5	E	194@5500	216@3500	3.66 x 3.38	10.0:1	12@Idle
2003	1.5	G	92@5500	97@4000	2.97 x 3.29	10.0:1	21@Idle
	2.0	D	140@6000	133@4800	3.23 x 3.68	10.3:1	24@Idle
	2.4	S	137@6000	129@4000	3.41 x 3.94	10.0:1	12@Idle
	2.7	H	172@6000	181@4000	3.41 x 2.95	10.0:1	12@Idle
	2.7	F	172@6000	181@4000	3.41 x 2.95	10.0:1	12@Idle
	3.5	E	194@5500	216@3500	3.66 x 3.38	10.0:1	12@Idle
2004	1.5	G	92@5500	97@4000	2.97 x 3.29	10.0:1	21@Idle
	2.0	D	140@6000	133@4800	3.23 x 3.68	10.3:1	24@Idle
	2.4	S	137@6000	129@4000	3.41 x 3.94	10.0:1	12@Idle
	2.7	H	172@6000	181@4000	3.41 x 2.95	10.0:1	12@Idle
	2.7	F	172@6000	181@4000	3.41 x 2.95	10.0:1	12@Idle
	3.5	E	194@5500	216@3500	3.66 x 3.38	10.0:1	12@Idle

GASOLINE ENGINE TUNE-UP SPECIFICATIONS

Year	Engine Displacement Liters	Engine ID/VIN	Spark Plugs Gap (in.)	Ignition Timing (deg.) MT	AT	Fuel Pump (psi)	Idle Speed (rpm) MT	AT	Valve Clearance In.	Ex.
2001	1.5	G	0.039-0.043	6-16B	6-16B	43	700-900	700-900	HYD	HYD
	2.0	F	0.039-0.043	5-15B	5-15B	43	700-900	700-900	HYD	HYD
	2.0	D	0.039-0.043	5-15B	5-15B	43	700-900	700-900	HYD	HYD
	2.4	S	0.039-0.043	3-7B	3-7B	48	650-850	650-850	HYD	HYD
	2.5	V	0.039-0.043	7-17B	7-17B	48	600-800	600-800	HYD	HYD
	3.0	D	0.039-0.043	—	3-7B	48	600-800	600-800	HYD	HYD
2002	1.5	G	0.039-0.043	6-16B	6-16B	43	700-900	700-900	HYD	HYD
	2.0	D	0.039-0.043	5-15B	5-15B	43	700-900	700-900	HYD	HYD
	2.4	S	0.039-0.043	3-7B	3-7B	48	650-850	650-850	HYD	HYD
	2.7	H	0.039-0.043	7-17B	7-17B	48	600-800	600-800	HYD	HYD
	3.5	E	0.039-0.043	—	3-7B	48	600-800	600-800	HYD	HYD
2003	1.5	G	0.039-0.043	6-16B	6-16B	43	700-900	700-900	HYD	HYD
	2.0	D	0.039-0.043	5-15B	5-15B	43	700-900	700-900	HYD	HYD
	2.4	S	0.039-0.043	3-7B	3-7B	48	650-850	650-850	HYD	HYD
	2.7	H	0.039-0.043	7-17B	7-17B	48	600-800	600-800	HYD	HYD
	2.7	F	0.039-0.043	7-17B	7-17B	48	600-800	600-800	HYD	HYD
	3.5	E	0.039-0.043	—	3-7B	48	600-800	600-800	HYD	HYD
2004	1.5	G	0.039-0.043	6-16B	6-16B	43	700-900	700-900	HYD	HYD
	2.0	D	0.039-0.043	5-15B	5-15B	43	700-900	700-900	HYD	HYD
	2.4	S	0.039-0.043	3-7B	3-7B	48	650-850	650-850	HYD	HYD
	2.7	H	0.039-0.043	7-17B	7-17B	48	600-800	600-800	HYD	HYD
	2.7	F	0.039-0.043	7-17B	7-17B	48	600-800	600-800	HYD	HYD
	3.5	E	0.039-0.043	—	3-7B	48	600-800	600-800	HYD	HYD

HYD: Hydraulic Valve Lifters

B: Before Top Dead Center

67162-ELAN-C03

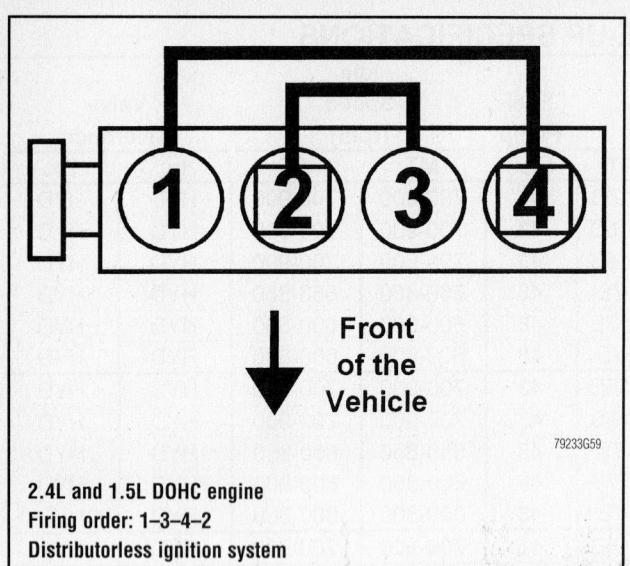

2.4L and 1.5L DOHC engine
Firing order: 1–3–4–2
Distributorless ignition system

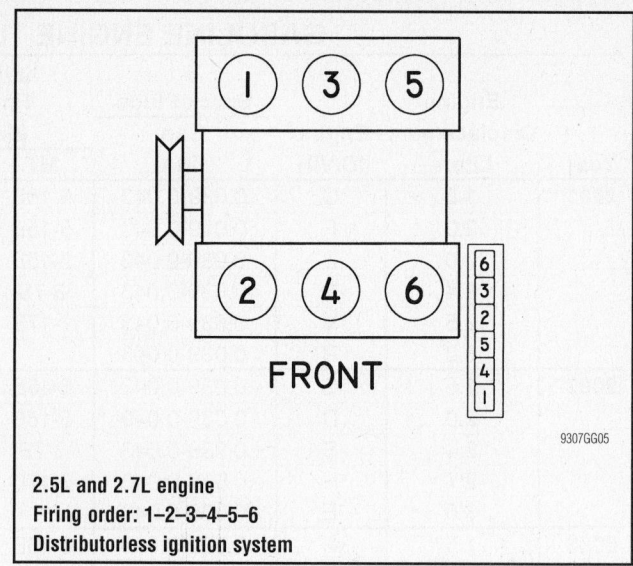

2.5L and 2.7L engine
Firing order: 1–2–3–4–5–6
Distributorless ignition system

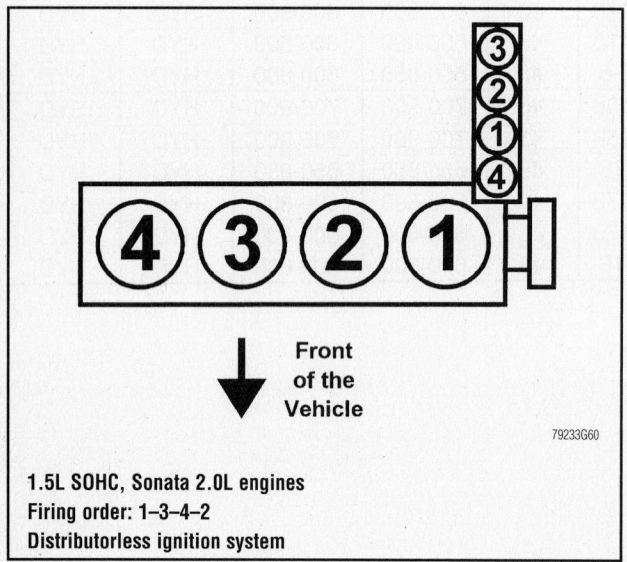

1.5L SOHC, Sonata 2.0L engines
Firing order: 1–3–4–2
Distributorless ignition system

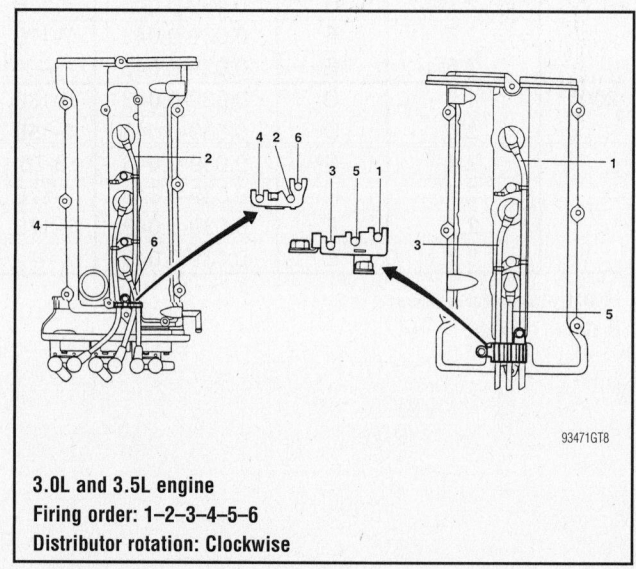

3.0L and 3.5L engine
Firing order: 1–2–3–4–5–6
Distributor rotation: Clockwise

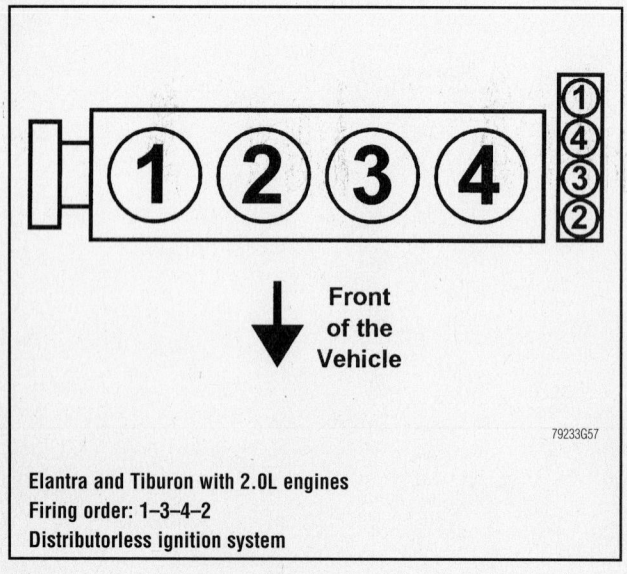

Elantra and Tiburon with 2.0L engines
Firing order: 1–3–4–2
Distributorless ignition system

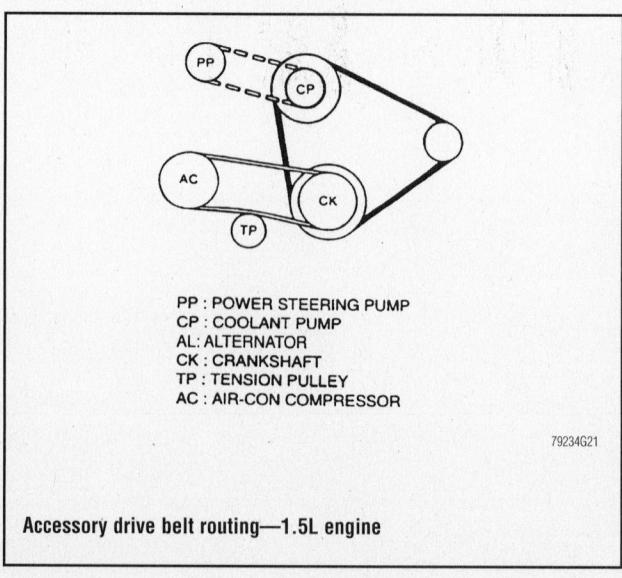

PP : POWER STEERING PUMP
CP : COOLANT PUMP
AL: ALTERNATOR
CK : CRANKSHAFT
TP : TENSION PULLEY
AC : AIR-CON COMPRESSOR

Accessory drive belt routing—1.5L engine

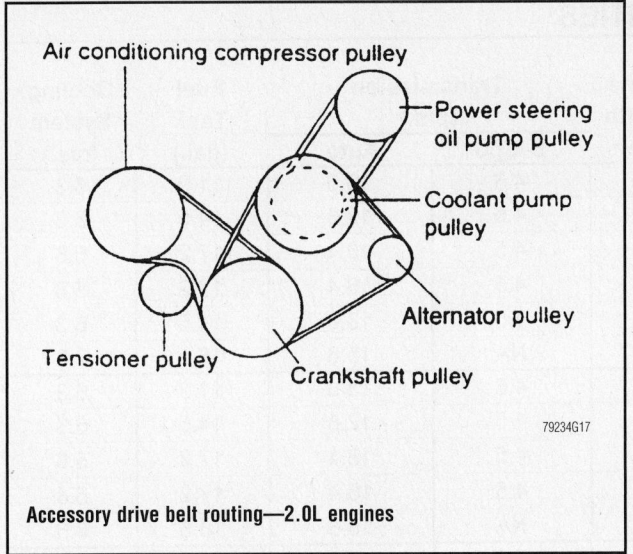

Air conditioning compressor pulley

Power steering oil pump pulley

Coolant pump pulley

Alternator pulley

Tensioner pulley

Crankshaft pulley

79234G17

Accessory drive belt routing—2.0L engines

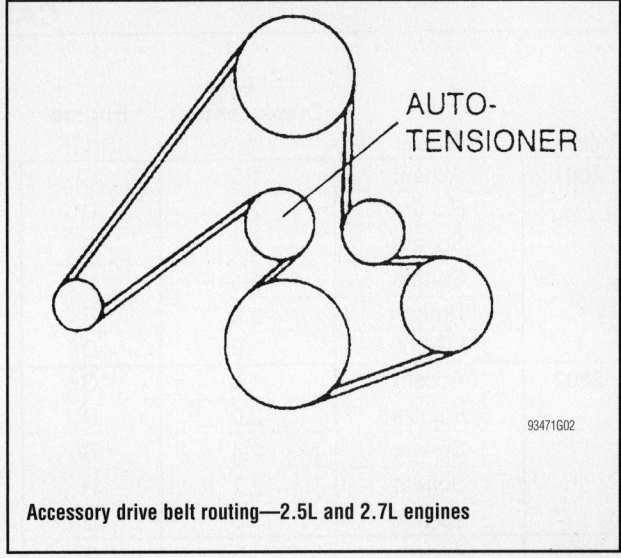

AUTO-TENSIONER

93471G02

Accessory drive belt routing—2.5L and 2.7L engines

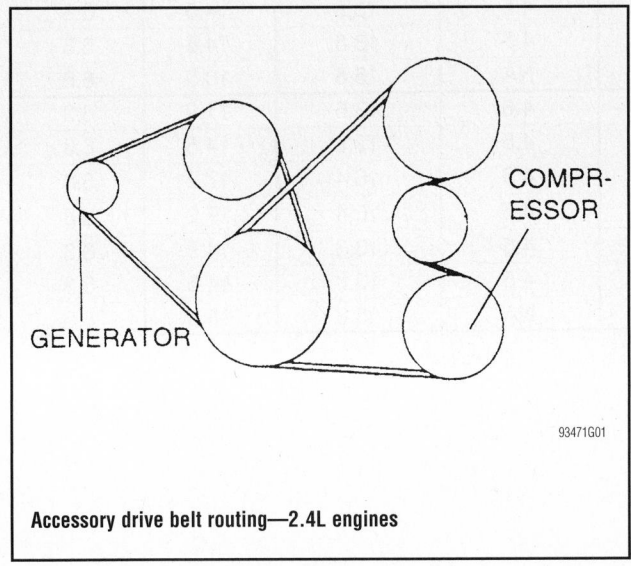

COMPR-ESSOR

GENERATOR

93471G01

Accessory drive belt routing—2.4L engines

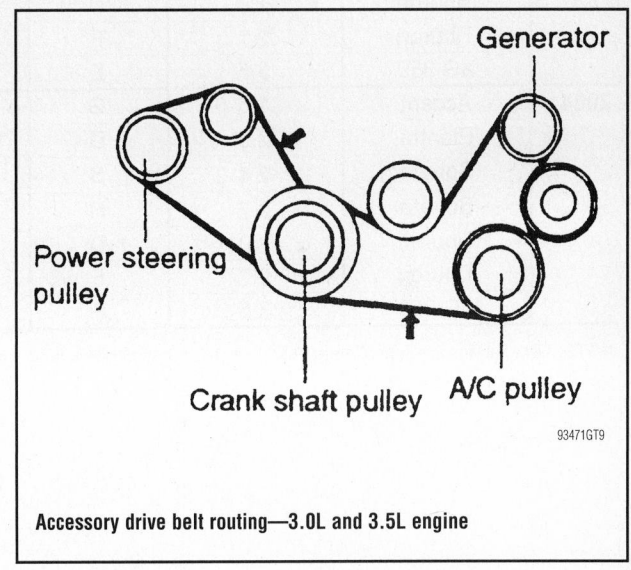

Generator

Power steering pulley

Crank shaft pulley

A/C pulley

93471GT9

Accessory drive belt routing—3.0L and 3.5L engine

CAPACITIES

Year	Model	Engine Displacement Liters	Engine ID/VIN	Engine Oil with Filter	Transmission (pts.)		Fuel Tank (gal.)	Cooling System (qts.)
					5–Spd	Auto.		
2001	Accent	1.5	G	3.5	4.6	13.6	11.9	6.3
	Elantra	2.0	D	4.2	4.5	12.8	14.5	6.3
	Sonata	2.4	S	4.2	4.5	16.4	17.2	5.8
	Sonata	2.5	V	4.5	4.5	16.4	17.2	5.8
	Tiburon	2.0	F	4.2	4.5	13.8	14.5	6.3
	XG 300	3.0	D	4.3	NA	15.8	18.5	7.3
2002	Accent	1.5	G	3.5	4.6	13.6	11.9	6.3
	Elantra	2.0	D	4.2	4.5	12.8	14.5	6.3
	Sonata	2.4	S	4.2	4.5	16.4	17.2	5.8
	Sonata	2.7	H	4.5	4.5	16.4	17.2	5.8
	XG 350	3.5	E	4.3	NA	15.8	18.5	8.6
2003	Accent	1.5	G	3.5	4.6	13.6	11.9	6.3
	Elantra	2.0	D	4.2	4.5	12.8	14.5	6.3
	Sonata	2.4	S	4.2	4.5	16.4	17.2	5.8
	Sonata	2.7	H	4.5	4.5	16.4	17.2	5.8
	Tiburon	2.0	D	4.2	4.5	13.8	14.5	6.3
	Tiburon	2.7	F	4.2	4.5	13.8	14.5	6.3
	XG 350	3.5	E	4.3	NA	15.8	18.5	8.6
2004	Accent	1.5	G	3.5	4.6	13.6	11.9	6.3
	Elantra	2.0	D	4.2	4.5	12.8	14.5	6.3
	Sonata	2.4	S	4.2	4.5	16.4	17.2	5.8
	Sonata	2.7	H	4.5	4.5	16.4	17.2	5.8
	Tiburon	2.0	D	4.2	4.5	13.8	14.5	6.3
	Tiburon	2.7	F	4.2	4.5	13.8	14.5	6.3
	XG 350	3.5	E	4.3	NA	15.8	18.5	8.6

NOTE: All capacities are approximate. Add fluid gradually and check to be sure a proper fluid level is obtained.

NA: Not Available

67162-ELAN-C04

CRANKSHAFT AND CONNECTING ROD SPECIFICATIONS

All measurements are given in inches.

Year	Engine Displacement Liters	Engine ID/VIN	Crankshaft				Connecting Rod		
			Main Brg. Journal Dia.	Main Brg. Oil Clearance	Shaft End-play	Thrust on No.	Journal Diameter	Oil Clearance	Side Clearance
2001	1.5	G	2.2440	0.0011-0.0018	0.0019-0.0068	3	1.7700	0.0009-0.0016	0.0039-0.0098
	2.0	F	2.2400	0.0011-0.0018	0.0023-0.0100	3	1.7700	0.0009-0.0016	0.0039-0.0098
	2.0	D	2.2400	0.0011-0.0018	0.0023-0.0100	3	1.7700	0.0009-0.0016	0.0039-0.0098
	2.4	S	2.2434-2.2442	0.0007-0.0014 ①	0.0020-0.0098	3	1.7709-1.7717	0.0008-0.0020	0.0040-0.0098
	2.5	V	2.4402-2.4409	0.0002-0.0009	0.0028-0.0098	3	1.8891-1.8898	0.0007-0.0014	0.0039-0.0098
	3.0	D	2.3617-2.3620	0.0007-0.0014	0.002-0.0098	3	1.9677-1.9685	0.0009-0.0020	0.0039-0.0098
2002	1.5	G	2.2440	0.0011-0.0018	0.0019-0.0068	3	1.7700	0.0009-0.0016	0.0039-0.0098
	2.0	D	2.2400	0.0011-0.0018	0.0023-0.0100	3	1.7700	0.0009-0.0016	0.0039-0.0098
	2.4	S	2.2434-2.2442	0.0007-0.0014 ①	0.0020-0.0098	3	1.7709-1.7717	0.0008-0.0020	0.0040-0.0098
	2.7	H	2.4402-2.4409	0.0002-0.0009	0.0028-0.0098	3	1.8891-1.8898	0.0007-0.0014	0.0039-0.0098
	3.5	E	2.5190-2.5197	0.00086-0.00157	0.0020-0.0098	3	2.1650-2.1653	0.0010-0.0017	0.0039-0.0098
2003	1.5	G	2.2440	0.0011-0.0018	0.0019-0.0068	3	1.7700	0.0009-0.0016	0.0039-0.0098
	2.0	D	2.2400	0.0011-0.0018	0.0023-0.0100	3	1.7700	0.0009-0.0016	0.0039-0.0098
	2.4	S	2.2434-2.2442	0.0007-0.0014 ①	0.0020-0.0098	3	1.7709-1.7717	0.0008-0.0020	0.0040-0.0098
	2.7	H	2.4402-2.4409	0.0002-0.0009	0.0028-0.0098	3	1.8891-1.8898	0.0007-0.0014	0.0039-0.0098
	2.7	F	2.4402-2.4409	0.0002-0.0009	0.0028-0.0098	3	1.8891-1.8898	0.0007-0.0014	0.0039-0.0098
	3.5	E	2.5190-2.5197	0.00086-0.00157	0.0020-0.0098	3	2.1650-2.1653	0.0010-0.0017	0.0039-0.0098
2004	1.5	G	2.2440	0.0011-0.0018	0.0019-0.0068	3	1.7700	0.0009-0.0016	0.0039-0.0098
	2.0	D	2.2400	0.0011-0.0018	0.0023-0.0100	3	1.7700	0.0009-0.0016	0.0039-0.0098
	2.4	S	2.2434-2.2442	0.0007-0.0014 ①	0.0020-0.0098	3	1.7709-1.7717	0.0008-0.0020	0.0040-0.0098
	2.7	H	2.4402-2.4409	0.0002-0.0009	0.0028-0.0098	3	1.8891-1.8898	0.0007-0.0014	0.0039-0.0098
	2.7	F	2.4402-2.4409	0.0002-0.0009	0.0028-0.0098	3	1.8891-1.8898	0.0007-0.0014	0.0039-0.0098
	3.5	E	2.5190-2.5197	0.00086-0.00157	0.0020-0.0098	3	2.1650-2.1653	0.0010-0.0017	0.0039-0.0098

① No. 3: 0.0009 - 0.0016

VALVE SPECIFICATIONS

Year	Engine Displacement Liters	Engine ID/VIN	Seat Angle (deg.)	Face Angle (deg.)	Spring Test Pressure (lbs. @ in.)	Spring Installed Height (in.)	Stem-to-Guide Clearance (in.)		Stem Diameter (in.)	
							Intake	Exhaust	Intake	Exhaust
2001	1.5	G	45	45	54@1.358	1.358	0.0012-0.0024	0.0014-0.0026	0.3920	0.3960
	2.0	F	45	45	56@1.457	1.358	0.0008-0.0019	0.0019-0.0033	0.2348-0.2354	0.2334-0.2342
	2.0	D	45	45	56@1.457	1.358	0.0008-0.0019	0.0019-0.0033	0.2348-0.2354	0.2334-0.2342
	2.4	S	45	45-45.5	56@1.457	1.358	0.0008-0.0019	0.0020-0.0033	0.2585-0.2891	0.2571-0.2579
	2.5	V	45	45	49@1.378	1.378	0.0009-0.0020	0.0014-0.0026	0.2348-0.2354	0.2343-0.2348
	3.0	D	45	45-45.5	74@1.591	1.826	0.0009-0.0020	0.0020-0.0033	0.258-0.2594	0.257-0.2580
2002	1.5	G	45	45	54@1.358	1.358	0.0012-0.0024	0.0014-0.0026	0.3920	0.3960
	2.0	F	45	45	56@1.457	1.358	0.0008-0.0019	0.0019-0.0033	0.2348-0.2354	0.2334-0.2342
	2.4	S	45	45-45.5	56@1.457	1.358	0.0008-0.0019	0.0020-0.0033	0.2585-0.2891	0.2571-0.2579
	2.7	H	45	45-45.5	49@1.378	1.378	0.0009-0.0020	0.0014-0.0026	0.2348-0.2354	0.2343-0.2348
	3.5	E	45	45-45.5	53@1.492	1.826	0.0009-0.0020	0.0020-0.0033	0.258-0.2590	0.257-0.258
2003	1.5	G	45	45	54@1.358	1.358	0.0012-0.0024	0.0014-0.0026	0.3920	0.3960
	2.0	D	45	45	56@1.457	1.358	0.0008-0.0019	0.0019-0.0033	0.2348-0.2354	0.2334-0.2342
	2.4	S	45	45-45.5	56@1.457	1.358	0.0008-0.0019	0.0020-0.0033	0.2585-0.2891	0.2571-0.2579
	2.7	H	45	45	49@1.378	1.378	0.0009-0.0020	0.0014-0.0026	0.2348-0.2354	0.2343-0.2348
	3.5	E	45	45-45.5	53@1.492	1.826	0.0009-0.0020	0.0020-0.0033	0.258-0.2590	0.257-0.258
2004	1.5	G	45	45	54@1.358	1.358	0.0012-0.0024	0.0014-0.0026	0.3920	0.3960
	2.0	D	45	45	56@1.457	1.358	0.0008-0.0019	0.0019-0.0033	0.2348-0.2354	0.2334-0.2342
	2.4	S	45	45-45.5	56@1.457	1.358	0.0008-0.0019	0.0020-0.0033	0.2585-0.2891	0.2571-0.2579
	2.7	H	45	45	49@1.378	1.378	0.0009-0.0020	0.0014-0.0026	0.2348-0.2354	0.2343-0.2348
	3.5	E	45	45-45.5	53@1.492	1.826	0.0009-0.0020	0.0020-0.0033	0.258-0.2590	0.257-0.258

67162-ELAN-C06

PISTON AND RING SPECIFICATIONS

All measurements are given in inches.

| Year | Engine Displacement Liters | Engine ID/VIN | Piston Clearance | Ring Gap | | | Ring Side Clearance | | |
				Top Compression	Bottom Compression	Oil Control	Top Compression	Bottom Compression	Oil Control
2001	1.5	G	0.0008-0.0016	0.008-0.020	0.008-0.020	0.008-0.039	0.0016-0.0033	0.0016-0.0033	snug
	2.0	F	0.0008-0.0016	0.009-0.015	0.013-0.019	0.008-0.024	0.0015-0.0031	0.0012-0.0027	snug
	2.0	D	0.0008-0.0016	0.009-0.015	0.013-0.019	0.008-0.024	0.0015-0.0031	0.0012-0.0027	snug
	2.4	S	0.0008-0.0012	0.010-0.014	0.006-0.022	0.004-0.016	0.0008-0.0024	0.0008-0.0024	snug
	2.5	V	0.0008-0.0012	0.008-0.014	0.015-0.020	0.008-0.028	0.0016-0.0031	0.0012-0.0028	snug
	3.0	D	0.0008-0.0016	0.012-0.018	0.010-0.016	0.008-0.028	0.0012-0.0035	0.0008-0.0024	snug
2002	1.5	G	0.0008-0.0016	0.008-0.020	0.008-0.020	0.008-0.039	0.0016-0.0033	0.0016-0.0033	snug
	2.0	D	0.0008-0.0016	0.009-0.015	0.013-0.019	0.008-0.024	0.0015-0.0031	0.0012-0.0027	snug
	2.4	S	0.0008-0.0012	0.010-0.014	0.006-0.022	0.004-0.016	0.0008-0.0024	0.0008-0.0024	snug
	2.7	H	0.0004-0.0012	0.008-0.014	0.015-0.020	0.008-0.028	0.0016-0.0031	0.0012-0.0028	snug
	3.5	E	0.0012-0.0020	0.0078-0.012	0.0177-0.0236	0.0079-0.0276	0.0016-0.0031	0.0008-0.0024	snug
2003	1.5	G	0.0008-0.0016	0.008-0.020	0.008-0.020	0.008-0.039	0.0016-0.0033	0.0016-0.0033	snug
	2.0	D	0.0008-0.0016	0.009-0.015	0.013-0.019	0.008-0.024	0.0015-0.0031	0.0012-0.0027	snug
	2.4	S	0.0008-0.0012	0.010-0.014	0.006-0.022	0.004-0.016	0.0008-0.0024	0.0008-0.0024	snug
	2.7	H	0.0004-0.0012	0.008-0.014	0.015-0.020	0.008-0.028	0.0016-0.0031	0.0012-0.0028	snug
	2.7	F	0.0004-0.0012	0.008-0.014	0.015-0.020	0.008-0.028	0.0016-0.0031	0.0012-0.0028	snug
	3.5	E	0.0012-0.0020	0.0078-0.012	0.0177-0.0236	0.0079-0.0276	0.0016-0.0031	0.0008-0.0024	snug
2004	1.5	G	0.0008-0.0016	0.008-0.020	0.008-0.020	0.008-0.039	0.0016-0.0033	0.0016-0.0033	snug
	2.0	D	0.0008-0.0016	0.009-0.015	0.013-0.019	0.008-0.024	0.0015-0.0031	0.0012-0.0027	snug
	2.4	S	0.0008-0.0012	0.010-0.014	0.006-0.022	0.004-0.016	0.0008-0.0024	0.0008-0.0024	snug
	2.7	H	0.0004-0.0012	0.008-0.014	0.015-0.020	0.008-0.028	0.0016-0.0031	0.0012-0.0028	snug
	2.7	F	0.0004-0.0012	0.008-0.014	0.015-0.020	0.008-0.028	0.0016-0.0031	0.0012-0.0028	snug
	3.5	E	0.0012-0.0020	0.0078-0.012	0.0177-0.0236	0.0079-0.0276	0.0016-0.0031	0.0008-0.0024	snug

67162-ELAN-C07

TORQUE SPECIFICATIONS
All readings in ft. lbs.

Year	Engine Displacement Liters	Engine ID/VIN	Cylinder Head Bolts	Main Bearing Bolts	Rod Bearing Bolts	Crankshaft Damper Bolts	Flywheel Bolts	Manifold Intake	Manifold Exhaust	Spark Plugs	Oil Pan Drain Plug
2001	1.5	G	①	40-44	25-28	110-118	94-101	11-14	11-14	18	30-33
	2.0	F	②	③	34-39	125-133	88-95	11-14	17-22	18	30-33
	2.0	D	②	③	34-39	125-133	88-95	11-14	17-22	18	30-33
	2.4	S	④	⑤	⑤	80-94	94-101	⑥	⑦	15-22	25-33
	2.5	V	①	⑧	⑨	130-138	53-55	14-15	18-22	15-22	25-33
	3.0	D	75-82	65-72	⑩	NA	NA	9-10	20-40	15-22	26-32
2002	1.5	G	①	40-44	25-28	110-118	94-101	11-14	11-14	18	30-33
	2.0	D	②	③	34-39	125-133	88-95	11-14	17-22	18	30-33
	2.4	S	④	⑤	⑤	80-94	94-101	⑥	⑦	15-22	25-33
	2.7	H	①	⑧	⑨	130-138	53-55	14-15	18-22	15-22	25-33
	3.5	E	75-82	65-72	⑩	NA	NA	9-10	20-40	15-22	26-32
2003	1.5	G	①	40-44	25-28	110-118	94-101	11-14	11-14	18	30-33
	2.0	D	②	③	34-39	125-133	88-95	11-14	17-22	18	30-33
	2.4	S	④	⑤	⑤	80-94	94-101	⑥	⑦	15-22	25-33
	2.7	H	①	⑧	⑨	130-138	53-55	14-15	18-22	15-22	25-33
	2.7	F	①	⑧	⑨	130-138	53-55	14-15	18-22	15-22	25-33
	3.5	E	75-82	65-72	⑩	NA	NA	9-10	20-40	15-22	26-32
2004	1.5	G	①	40-44	25-28	110-118	94-101	11-14	11-14	18	30-33
	2.0	D	②	③	34-39	125-133	88-95	11-14	17-22	18	30-33
	2.4	S	④	⑤	⑤	80-94	94-101	⑥	⑦	15-22	25-33
	2.7	H	①	⑧	⑨	130-138	53-55	14-15	18-22	15-22	25-33
	2.7	F	①	⑧	⑨	130-138	53-55	14-15	18-22	15-22	25-33
	3.5	E	75-82	65-72	⑩	NA	NA	9-10	20-40	15-22	26-32

① Step 1: 18 ft. lbs.
Step 2: Plus 60-64 degrees
Step 3: Plus 45-49 degrees

② Step 1: M10 bolts to 22 ft. lbs. and M12 bolts to 26 ft. lbs.
Step 2: Plus 60-65 degrees
Step 3: Plus 60-65 degrees

③ Step 1: 20-24 ft. lbs.
Step 2: Plus 60-65 degrees

④ Step 1: 14 ft. lbs.
Step 2: Plus 90 degrees
Step 3: Loosen to 0 ft. lbs.
Step 4: 14 ft. lbs.
Step 5: Plus 90 degrees
Step 6: Plus 90 degrees

⑤ 13-16 ft. lbs. plus 90-94 degrees

⑥ M8: 11-14 ft. lbs.
M10: 13-18 ft. lbs.
Nuts: 22-30 ft. lbs.

⑦ M8: 18-22 ft. lbs.
M10: 25-40 ft. lbs.

⑧ M10: 20-24 ft. lbs. plus 90-94 degrees
M7: 10-13 ft. lbs. plus 90-94 degrees

⑨ 12-14 ft. lbs. plus 90-94 degrees

⑩ 26 ft. lbs. + 90 degrees

67162-ELAN-C08

WHEEL ALIGNMENT

Year	Model		Caster Range (+/-Deg.)	Caster Preferred Setting (Deg.)	Camber Range (+/-Deg.)	Camber Preferred Setting (Deg.)	Toe-in (in.)
2001	Accent	F	0.50	+1.80	0.50	0	0 +/- 0.12
		R	—	—	0.50	-0.68	0.12 +/- 0.08
	Elantra	F	0.50	+2.35	2.00	+4.00	0 +/- 0.12
		R	—	—	0.50	-0.70	0.14 +/- 0.02
	Sonata	F	1.00	+3.25	0.50	0	0 +/- 0.12
		R	—	—	0.50	-0.50	0.08 +/- 0.08
	Tiburon	F	0.50	+2.45	2.00	+4.00	0 +/- 0.12
		R	—	—	0.50	-0.90	0.14 +/- 0.02
	XG 300	F	1.00	+3.15	0.50	0	0 +/- 0.12
		R	—	—	0.50	-0.50	0.08 +/- 0.08
2002	Accent	F	0.50	+1.80	0.50	0	0 +/- 0.12
		R	—	—	0.50	-0.68	0.12 +/- 0.08
	Elantra	F	0.50	+2.35	2.00	+4.00	0 +/- 0.12
		R	—	—	0.50	-0.70	0.14 +/- 0.02
	Sonata	F	1.00	+3.25	0.50	0	0 +/- 0.12
		R	—	—	0.50	-0.50	0.08 +/- 0.08
	XG 350	F	1.00	+3.15	0.50	0	0 +/- 0.12
		R	—	—	0.50	-0.50	0.08 +/- 0.08
2003	Accent	F	0.50	+1.80	0.50	0	0 +/- 0.12
		R	—	—	0.50	-0.68	0.12 +/- 0.08
	Elantra	F	0.50	+2.35	2.00	+4.00	0 +/- 0.12
		R	—	—	0.50	-0.70	0.14 +/- 0.02
	Sonata	F	1.00	+3.25	0.50	0	0 +/- 0.12
		R	—	—	0.50	-0.50	0.08 +/- 0.08
	Tiburon	F	0.50	+2.45	2.00	+4.00	0 +/- 0.12
		R	—	—	0.50	-0.90	0.14 +/- 0.02
	XG 350	F	1.00	+3.15	0.50	0	0 +/- 0.12
		R	—	—	0.50	-0.50	0.08 +/- 0.08
2004	Accent	F	0.50	+1.80	0.50	0	0 +/- 0.12
		R	—	—	0.50	-0.68	0.12 +/- 0.08
	Elantra	F	0.50	+2.35	2.00	+4.00	0 +/- 0.12
		R	—	—	0.50	-0.70	0.14 +/- 0.02
	Sonata	F	1.00	+3.25	0.50	0	0 +/- 0.12
		R	—	—	0.50	-0.50	0.08 +/- 0.08
	Tiburon	F	0.50	+2.45	2.00	+4.00	0 +/- 0.12
		R	—	—	0.50	-0.90	0.14 +/- 0.02
	XG 350	F	1.00	+3.15	0.50	0	0 +/- 0.12
		R	—	—	0.50	-0.50	0.08 +/- 0.08

67162-ELAN-C09

TIRE, WHEEL AND BALL JOINT SPECIFICATIONS

| Year | Model | OEM Tires | | Tire Pressures (psi) | | Wheel Size | Ball Joint Inspection | Lug Nut Torque ① |
		Standard	Optional	Front	Rear			
2001	Accent	P155/80R13	P175/70R13 P175/65R14	30	30	5-J	②	65-80
	Elantra	P195/60R14	None	30	30	5.5-JJ	②	65-80
	Sonata	P195/70R14	P205/60HR15	30	30	Std: 5.5-JJ Opt: 6-JJ	②	65-80
	Tiburon	P195/60R14	None	30	30	5.5JJ	②	67-81
	XG 300	P205/65R15	None	30		6.0J	②	67-81
2002	Accent	P155/80R13	P175/70R13 P175/65R14	30	30	5-J	②	67-81
	Elantra	P195/60R14	None	30	30	5.5-JJ	②	65-80
	Sonata	P195/70R14	P205/60HR15	30	30	Std: 5.5-JJ Opt: 6-JJ	②	67-81
	XG 350	P205/60R16	None	30		6.0J	②	67-81
2003	Accent	P155/80R13	P175/70R13 P175/65R14	30	30	5-J	②	65-80
	Elantra	P195/60R14	None	30	30	5.5-JJ	②	67-81
	Sonata	P195/70R14	P205/60HR15	30	30	Std: 5.5-JJ Opt: 6-JJ	②	67-81
	Tiburon	P195/60R14	None	30	30	5.5JJ	②	67-81
	XG 350	P205/60R16	None	30		6.0J	②	65-80
2004	Accent	P155/80R13	P175/70R13 P175/65R14	30	30	5-J	②	65-80
	Elantra	P195/60R14	None	30	30	5.5-JJ	②	67-81
	Sonata	P195/70R14	P205/60HR15	30	30	Std: 5.5-JJ Opt: 6-JJ	②	67-81
	Tiburon	P195/60R14	None	30	30	5.5JJ	②	67-81
	XG 350	P205/60R16	None	30		6.0J	②	67-81

OEM: Original Equipment Manufacturer

PSI: Pounds Per Square Inch

STD: Standard

OPT: Optional

① Ft. Lbs.

② Replace if any measurable movement is found.

67162-ELAN-C10

BRAKE SPECIFICATIONS
All measurements in inches unless noted

| Year | Model | | Brake Disc | | | Brake Drum Diameter | | | Minimum Lining Thickness | | Brake Caliper | |
			Original Thickness	Minimum Thickness	Maximum Run-out	Original Inside Diameter	Max. Wear Limit	Maximum Machine Diameter	Front	Rear	Bracket Bolts (ft. lbs.)	Mounting Bolts (ft. lbs.)
2001	Accent	F	0.750	0.670	0.002	—	—	—	0.039	—	50	①
		R	—	—	—	7.100	—	7.165	—	0.039	—	—
	Elantra	F	0.750	0.670	0.002	—	—	—	0.039	—	50	①
		R	—	—	—	8.000	—	8.079	—	0.039	—	—
	Elantra w/rear disc	F	0.750	0.670	0.002	—	—	—	0.039	—	50	①
		R	0.354	NA	NA	—	—	—	—	0.031	—	23
	Sonata	F	0.945	0.787	0.002	—	—	—	0.079	—	51-63	16-24
		R	—	—	—	9.000	—	8.936	—	0.059	—	—
	Sonata w/rear disc	F	0.945	0.880	0.003	—	—	—	0.079	—	51-63	16-24
		R	0.390	0.350	0.005	—	—	—	—	0.079	—	23
	XG 300	F	0.413	0.096	0.002	—	—	—	0.079	—	51-63	16-24
		R	0.390	0.080	NA	—	—	—	—	—	51-63	16-24
2002	Accent	F	0.750	0.670	0.002	—	—	—	0.039	—	50	①
		R	—	—	—	7.100	—	7.165	—	0.039	—	—
	Elantra	F	0.750	0.670	0.002	—	—	—	0.039	—	50	①
		R	—	—	—	8.000	—	8.079	—	0.039	—	—
	Elantra w/rear disc	F	0.750	0.670	0.002	—	—	—	0.039	—	50	①
		R	0.354	NA	NA	—	—	—	—	0.031	—	23
	Sonata	F	0.945	0.787	0.002	—	—	—	0.079	—	51-63	16-24
		R	—	—	—	9.000	—	8.936	—	0.059	—	—
	Sonata w/rear disc	F	0.945	0.880	0.003	—	—	—	0.079	—	51-63	16-24
		R	0.390	0.350	0.005	—	—	—	—	0.079	—	23
	Tiburon	F	0.866	0.787	0.002	—	—	—	0.079	—	55	①
		R	—	—	—	8.000	—	8.079	—	0.059	—	—
	Tiburon w/rear disc	F	0.866	0.787	0.002	—	—	—	0.079	—	55	①
		R	0.354	NA	NA	—	—	—	—	0.031	—	23
	XG 350	F	0.413	0.096	0.002	—	—	—	0.079	—	51-63	16-24
		R	0.390	0.080	NA	—	—	—	—	—	51-63	16-24
2003	Accent	F	0.750	0.670	0.002	—	—	—	0.039	—	50	①
		R	—	—	—	7.100	—	7.165	—	0.039	—	—
	Elantra	F	0.750	0.670	0.002	—	—	—	0.039	—	50	①
		R	—	—	—	8.000	—	8.079	—	0.039	—	—
	Elantra w/rear disc	F	0.750	0.670	0.002	—	—	—	0.039	—	50	①
		R	0.354	NA	NA	—	—	—	—	0.031	—	23
	Sonata	F	0.945	0.787	0.002	—	—	—	0.079	—	51-63	16-24
		R	—	—	—	9.000	—	8.936	—	0.059	—	—
	Sonata w/rear disc	F	0.945	0.880	0.003	—	—	—	0.079	—	51-63	16-24
		R	0.390	0.350	0.005	—	—	—	—	0.079	—	23
	Tiburon	F	0.866	0.787	0.002	—	—	—	0.079	—	55	①
		R	—	—	—	8.000	—	8.079	—	0.059	—	—
	Tiburon w/rear disc	F	0.866	0.787	0.002	—	—	—	0.079	—	55	①
		R	0.354	NA	NA	—	—	—	—	0.031	—	23
	XG 350	F	0.413	0.096	0.002	—	—	—	0.079	—	51-63	16-24
		R	0.390	0.080	NA	—	—	—	—	—	51-63	16-24

BRAKE SPECIFICATIONS

All measurements in inches unless noted

Year	Model		Brake Disc Original Thickness	Brake Disc Minimum Thickness	Brake Disc Maximum Run-out	Brake Drum Diameter Original Inside Diameter	Brake Drum Diameter Max. Wear Limit	Brake Drum Diameter Maximum Machine Diameter	Minimum Lining Thickness Front	Minimum Lining Thickness Rear	Brake Caliper Bracket Bolts (ft. lbs.)	Brake Caliper Mounting Bolts (ft. lbs.)
2004	Accent	F	0.750	0.670	0.002	—	—	—	0.039	—	50	①
		R	—	—	—	7.100	—	7.165	—	0.039	—	—
	Elantra	F	0.750	0.670	0.002	—	—	—	0.039	—	50	①
		R	—	—	—	8.000	—	8.079	—	0.039	—	—
	Elantra w/rear disc	F	0.750	0.670	0.002	—	—	—	0.039	—	50	①
		R	0.354	NA	NA	—	—	—	—	0.031	—	23
	Sonata	F	0.945	0.787	0.002	—	—	—	0.079	—	51-63	16-24
		R	—	—	—	9.000	—	8.936	—	0.059	—	—
	Sonata w/rear disc	F	0.945	0.880	0.003	—	—	—	0.079	—	51-63	16-24
		R	0.390	0.350	0.005	—	—	—	—	0.079	—	23
	Tiburon	F	0.866	0.787	0.002	—	—	—	0.079	—	55	①
		R	—	—	—	8.000	—	8.079	—	0.059	—	—
	Tiburon w/rear disc	F	0.866	0.787	0.002	—	—	—	0.079	—	55	①
		R	0.354	NA	NA	—	—	—	—	0.031	—	23
	XG 350	F	0.413	0.096	0.002	—	—	—	0.079	—	51-63	16-24
		R	0.390	0.080	NA	—	—	—	—	—	51-63	16-24

NA: Not Available

F: Front

R: Rear

① Upper: 28 ft. lbs.
   Lower: 19 ft. lbs.

67162-ELAN-C12

SCHEDULED MAINTENANCE INTERVALS
HYUNDAI—ACCENT, ELANTRA, SONATA, TIBURON, XG300 & XG350

TO BE SERVICED	TYPE OF SERVICE	VEHICLE MILEAGE INTERVAL (x1000)												
		7.5	15	22.5	30	37.5	45	52.5	60	67.5	75	82.5	90	97.5
Engine oil & filter	R	✓	✓	✓	✓	✓	✓	✓	✓	✓	✓	✓	✓	✓
Automatic transaxle fluid	S/I		✓		✓		✓		✓		✓		✓	
Brake pads, calipers & rotors	S/I		✓		✓		✓		✓		✓		✓	
Driveshafts & boots	S/I		✓		✓		✓		✓		✓		✓	
Wheel bearing grease	S/I				✓				✓				✓	
Air cleaner filter	R				✓				✓				✓	
Automatic transaxle fluid & filter	R				✓				✓				✓	
Brake fluid	R				✓				✓				✓	
Engine coolant	R				✓				✓				✓	
Fuel hose, vapor hose & fuel filter cap	S/I							✓						
Spark plugs	R				✓				✓				✓	
Spark plugs (Sonata 3.0L V6)	R								✓					
Bolts & nuts on chassis & body (Accent)	S/I				✓				✓				✓	
Drive belts	S/I				✓				✓				✓	
Exhaust pipe connections, muffler & suspension bolts	S/I				✓				✓				✓	
Manual transaxle oil	S/I				✓				✓				✓	
Rear brake drums, linings & parking brake	S/I				✓				✓				✓	
Steering gear rack, linkage & boots	S/I				✓				✓				✓	
Suspension ball joints & dust covers (Accent)	S/I				✓				✓				✓	
Timing belt (Accent & Elantra)	S/I				✓				✓				✓	
Timing belt (except Accent & Elantra)	R								✓					
Fuel filter	R							✓						
Fuel lines & connections	S/I								✓					
Vacuum & crankcase ventilation hoses	S/I								✓					

R: Replace   S/I: Service or Inspect

FREQUENT OPERATION MAINTENANCE (SEVERE SERVICE)

If a vehicle is operated under any of the following conditions it is considered severe service

- Extremely dusty areas.

- 50% or more of the vehicle operation is in 32°C (90°F) or higher temperatures, or constant operation in temperatures below 0°C (32°F).

- Prolonged idling (vehicle operation in stop and go traffic).

- Frequent short running periods (engine does not warm to normal operating temperatures).

- Police, taxi, delivery usage or trailer towing usage.

Oil & oil filter: change every 3000 miles.

Brake pads, calipers & rotors: service or inspect every 7500 miles.

Driveshaft boots: service or inspect every 7500 miles

Steering gear rack, linkage & boots: service or inspect every 7500 miles.

Air cleaner filter: service or inspect every 15,000 miles.

Automatic transaxle fluid & filter: replace every 15,000 miles.

Rear brake drums & linings: service or inspect every 15,000 miles.

Spark plugs: replace every 24,000 miles.

67162-ELAN-C13

PRECAUTIONS

Before servicing any vehicle, please be sure to read all of the following precautions, which deal with personal safety, prevention of component damage, and important points to take into consideration when servicing a motor vehicle:

• Never open, service or drain the radiator or cooling system when the engine is hot; serious burns can occur from the steam and hot coolant.

• Observe all applicable safety precautions when working around fuel. Whenever servicing the fuel system, always work in a well-ventilated area. Do not allow fuel spray or vapors to come in contact with a spark, open flame, or excessive heat (a hot drop light, for example). Keep a dry chemical fire extinguisher near the work area. Always keep fuel in a container specifically designed for fuel storage; also, always properly seal fuel containers to avoid the possibility of fire or explosion. Refer to the additional fuel system precautions later in this section.

• Fuel injection systems often remain pressurized, even after the engine has been turned **OFF**. The fuel system pressure must be relieved before disconnecting any fuel lines. Failure to do so may result in fire and/or personal injury.

• Brake fluid often contains polyglycol ethers and polyglycols. Avoid contact with the eyes and wash your hands thoroughly after handling brake fluid. If you do get brake fluid in your eyes, flush your eyes with clean, running water for 15 minutes. If

eye irritation persists, or if you have taken brake fluid internally, IMMEDIATELY seek medical assistance.

• The EPA warns that prolonged contact with used engine oil may cause a number of skin disorders, including cancer. You should make every effort to minimize your exposure to used engine oil. Protective gloves should be worn when changing oil. Wash your hands and any other exposed skin areas as soon as possible after exposure to used engine oil. Soap and water, or waterless hand cleaner should be used.

• All new vehicles are now equipped with an air bag system. The system must be disabled before performing service on or around system components, steering column, instrument panel components, wiring and sensors. Failure to follow safety and disabling procedures could result in accidental air bag deployment, possible personal injury and unnecessary system repairs.

• Always wear safety goggles when working with, or around, the air bag system. When carrying a non-deployed air bag, be sure the bag and trim cover are pointed away from your body. When placing a non-deployed air bag on a work surface, always face the bag and trim cover upward, away from the surface. This will reduce the motion of the module if it is accidentally deployed. Refer to the additional air bag system precautions later in this section.

• Clean, high quality brake fluid from a sealed container is essential to the safe and

proper operation of the brake system. You should always buy the correct type of brake fluid for your vehicle. If the brake fluid becomes contaminated, completely flush the system with new fluid. Never reuse any brake fluid. Any brake fluid that is removed from the system should be discarded. Also, do not allow any brake fluid to come in contact with a painted surface; it will damage the paint.

• Never operate the engine without the proper amount and type of engine oil; doing so WILL result in severe engine damage.

• Timing belt maintenance is extremely important. Many models utilize an interference-type, non-freewheeling engine. If the timing belt breaks, the valves in the cylinder head may strike the pistons, causing potentially serious (also time-consuming and expensive) engine damage. Refer to the maintenance interval charts in the front of this section for the recommended replacement interval for the timing belt, and to the timing belt procedure for belt replacement and inspection.

• Disconnecting the negative battery cable on some vehicles may interfere with the functions of the on-board computer system(s) and may require the computer to undergo a relearning process once the negative battery cable is reconnected.

• When servicing drum brakes, only disassemble and assemble one side at a time, leaving the remaining side intact for reference.

ENGINE REPAIR

➡**Disconnecting the negative battery cable on some vehicles may interfere with the functions of the on board computer system. The computer may undergo a relearning process once the negative battery cable is reconnected.**

Distributor

All engines except the 3.0L and 3.5L (VIN T) V6 are equipped with a Distributorless Ignition System (DIS).

REMOVAL

1. Before servicing the vehicle, refer to the precautions in the beginning of this section.
2. Remove or disconnect the following:
   • Distributor cap

• Distributor electrical connectors
3. Matchmark the rotor to the distributor housing.
4. Matchmark the distributor housing to the engine.
5. Unbolt and remove the distributor.

INSTALLATION

Timing Not Disturbed

1. Before servicing the vehicle, refer to the precautions in the beginning of this section.
2. Install or connect the following:
   • Distributor, with the rotor-to-housing and the housing-to-engine matchmarks aligned
   • Distributor electrical connectors
   • Distributor cap

3. Check the ignition timing and adjust as necessary.

Timing Disturbed

1. Before servicing the vehicle, refer to the precautions in the beginning of this section.
2. Set the crankshaft to Top Dead Center (TDC) of the compression stroke for the No. 1 cylinder.
3. Align the timing marks on the distributor gear and the distributor housing.
4. Install or connect the following:
   • Distributor, by aligning the groove of the installation flange with the center of the installation stud
   • Distributor electrical connectors
   • Distributor cap
5. Check the ignition timing and adjust as necessary.

Alternator

REMOVAL

1. Before servicing the vehicle, refer to the precautions in the beginning of this section.

2. Remove or disconnect the following:
   - Negative battery cable
   - Alternator drive belt
   - Alternator wiring harness connectors
   - Alternator. It may be necessary to raise the radiator

INSTALLATION

1. Before servicing the vehicle, refer to the precautions in the beginning of this section.

2. Install or connect the following:
   - Alternator and reposition the radiator, if raised
   - Alternator wiring harness connectors
   - Alternator drive belt

3. Adjust the alternator belt and torque the bolts to the following specifications:
   a. Accent: Pivot bolt to 14–18 ft. lbs. (19–25 Nm) and adjustment bolt to 14–20 ft. lbs. (19–28 Nm).
   b. 2.0L, 2.5L and 2.7L Sonata, Elantra and Tiburon: Pivot bolt to 14–18 ft. lbs. (19–25 Nm) and adjustment bolt to 105–132 inch lbs. (12–15 Nm).
   c. 2.4L Sonata: Pivot bolt to 26–41 ft. lbs. (34–54 Nm) and the adjustment bolt to 14–18 ft. lbs. (19–25 Nm).
   d. XG 300 and XG 350: Pivot bolt to 14–18 ft. lbs. (19–25 Nm) and

adjustment bolt to 11–16 ft. lbs. (15–22 Nm).

4. Install the radiator mounting bolts, if removed.

5. Connect the negative battery cable.

Ignition Timing

ADJUSTMENT

These engines are equipped with a Distributorless Ignition System (DIS). No adjustment is necessary.

Engine Assembly

REMOVAL & INSTALLATION

➡ **Hyundai recommends that the engine and transaxle be removed as a single unit on all models.**

1. Before servicing the vehicle, refer to the precautions in the beginning of this section.

2. Drain the cooling system.

3. Drain the transaxle.

4. Drain the engine oil.

5. Relieve fuel system pressure.

6. Remove or disconnect the following:
   - Battery
   - Hood
   - Air intake assembly
   - Accessory drive belts
   - Engine wiring harness connectors
   - Reverse lamp switch connector, if equipped
   - Speedometer cable
   - Alternator harness connectors
   - Oil pressure gauge sender connector
   - Radiator hoses
   - Cooling fan
   - Fuel lines
   - Control cable, if equipped
   - Brake booster vacuum line
   - Intake manifold vacuum lines
   - Heater hoses
   - Accelerator cable
   - Cruise control cable, if equipped
   - Engine ground cable

7. If equipped with a manual transaxle, disconnect or remove the following:
   - Clutch cable
   - Select control valve connector
   - Shift linkage rods

8. If equipped with an automatic transaxle, disconnect or remove the following:
   - Transaxle oil cooler lines

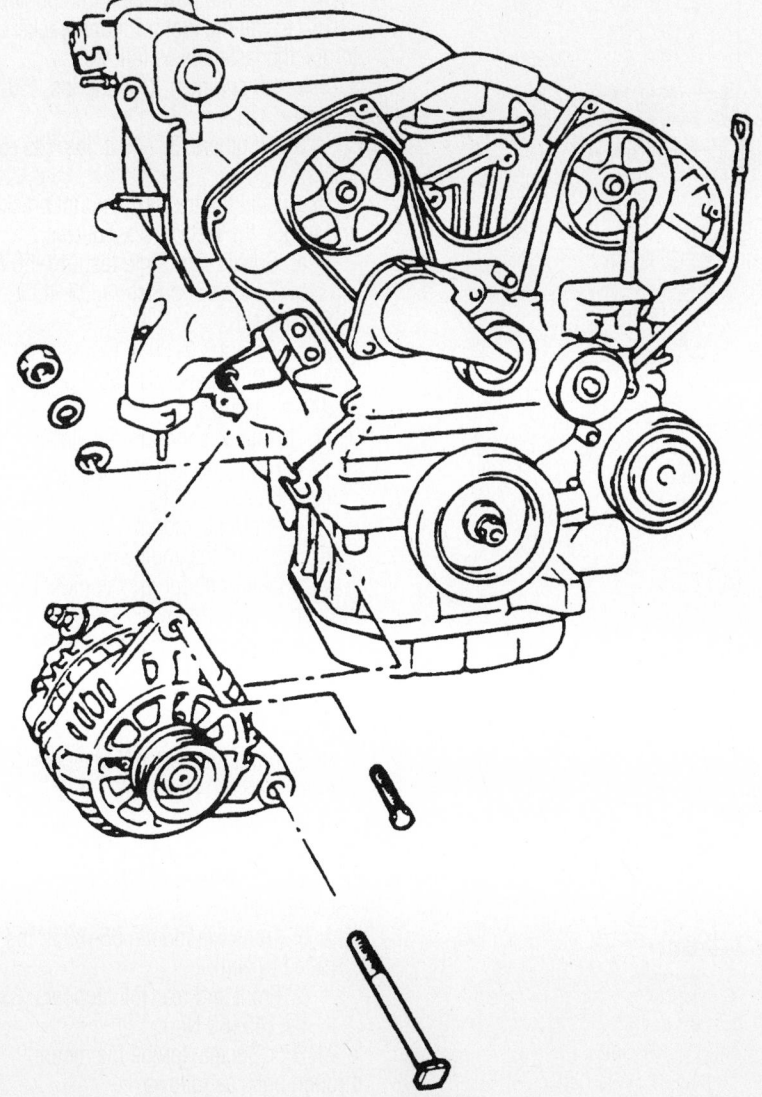

9347KG01

Exploded view of the alternator—XG 300 and XG 350

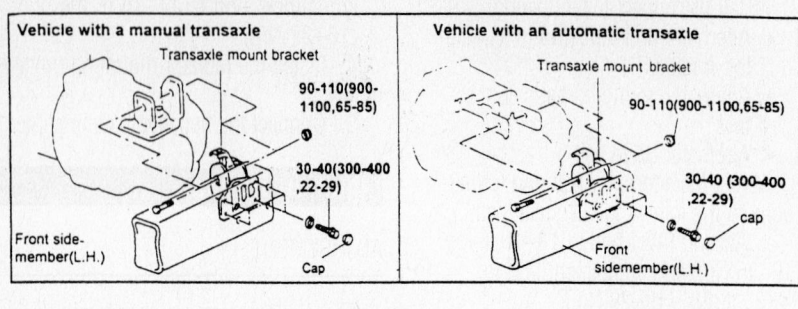

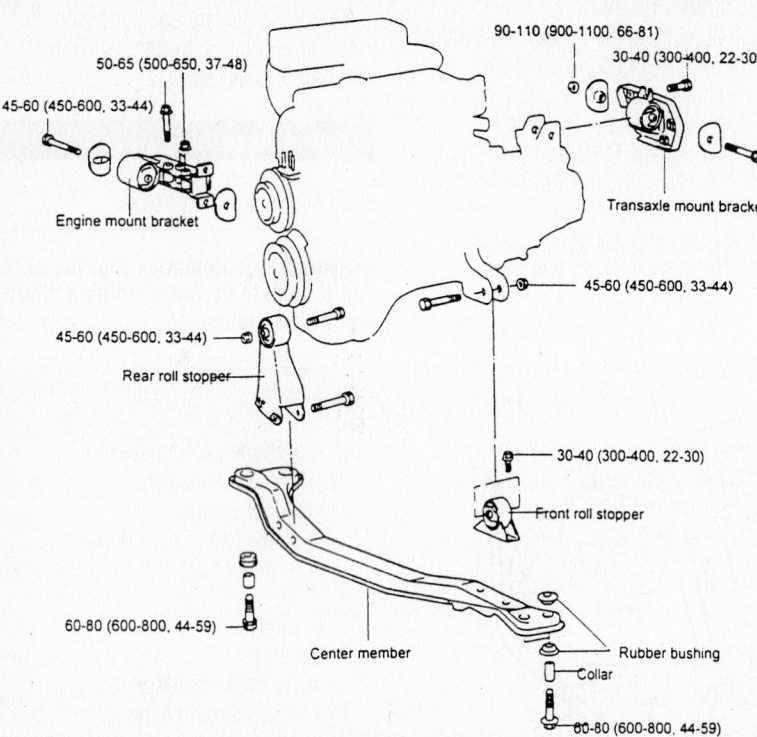

Exploded view of the engine mounts and torque specifications—Accent

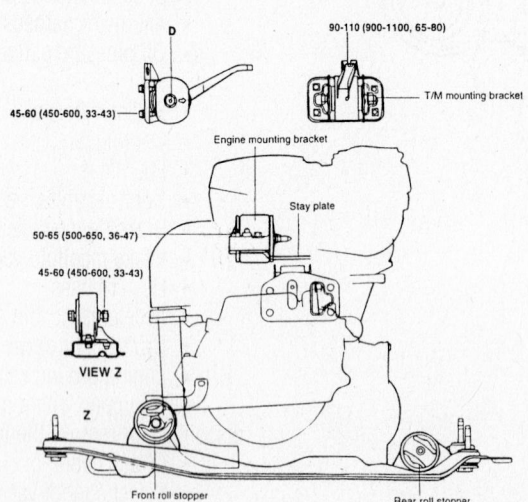

TORQUE: Nm (kg.cm, lb.ft)

Exploded view of the engine mounts and torque specifications—Tiburon and Elantra

- Shift cable
- Transaxle wiring connectors

9. For all vehicles, remove or disconnect the following:
- Radiator
- Power steering pump
- A/C compressor, if equipped
- Exhaust front pipe
- Lower ball joints
- Stabilizer bar links

10. Separate the inner CV-joints from the transaxle and suspend the halfshafts out of the work area with safety wire.

11. Attach a hoist to the engine lifting eyes.

12. Remove or disconnect the following:
- Front and rear roll stoppers
- Engine mount and bracket
- Transaxle mount and bracket

13. Lift the powertrain out of the vehicle.

To install:

14. Lower the powertrain into position.

15. Install the motor mount bracket and torque the fasteners as follows:
    a. V6 engines: 43–58 ft. lbs. (60–80 Nm).
    b. All others: 37–48 ft. lbs. (45–60 Nm).

16. Install the transaxle mount bracket and torque the fasteners as follows:
    a. Sonata: 29–36 ft. lbs. (40–50 Nm).
    b. Tiburon and Elantra: 33–43 ft. lbs. (45–60 Nm).
    c. Accent: 22–30 ft. lbs. (30–40 Nm).
    d. XG 300 and XG 350: 65–79 ft. lbs. (90–110 Nm).

17. Install or connect the following:
- Front and rear roll stoppers
- Engine mount
- Transaxle mount

18. Remove the engine hoist.

19. For Accent, torque the mount through bolts as follows:
    a. Engine mount: 33–43 ft. lbs. (45–60 Nm).
    b. Transaxle mount: 65–80 ft. lbs. (90–110 Nm).
    c. Front and rear roll stoppers: 33–43 ft. lbs. (45–60 Nm).

20. For Elantra and Tiburon, torque the mount through bolts as follows:
    a. Engine mount: 36–47 ft. lbs. (50–65 Nm).
    b. Transaxle mount: 65–80 ft. lbs. (90–110 Nm).
    c. Front and rear roll stoppers: 33–43 ft. lbs. (45–60 Nm).

21. For Sonata, torque the mount through bolts as follows:
    a. 4 cylinder engine mount: 43–58 ft. lbs. (60–80 Nm).

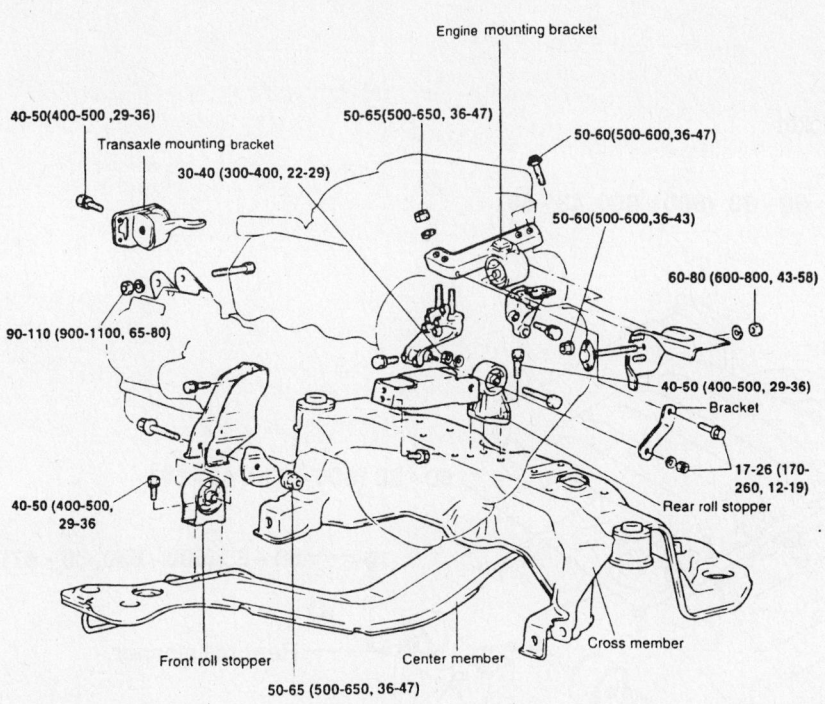

Exploded view of the engine mounts and torque specifications—4 cylinder Sonata

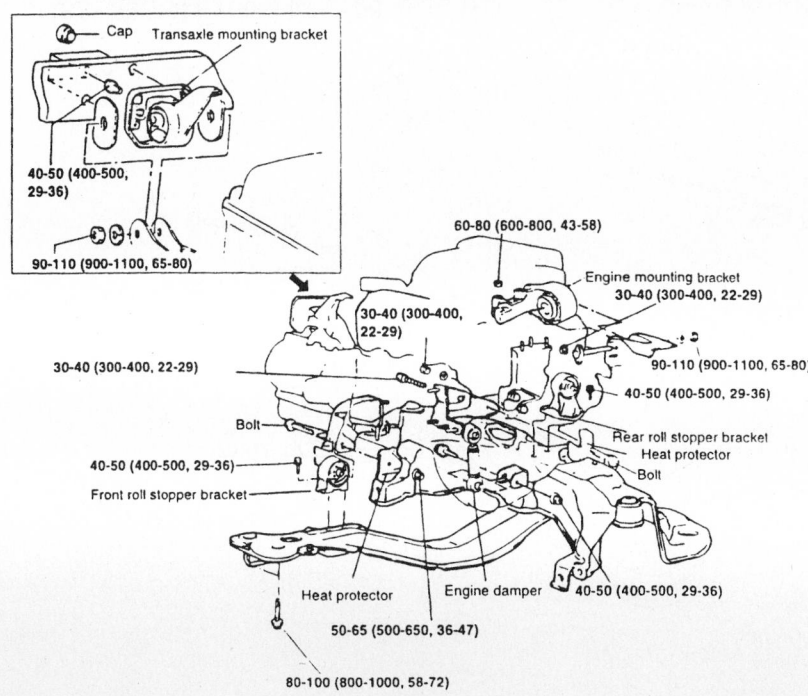

Exploded view of the engine mounts and torque specifications—V6 Sonata

b. V6 engine mount: 65–80 ft. lbs. (90–110 Nm).

c. Transaxle mount: 65–80 ft. lbs. (90–110 Nm).

d. Front roll stopper: 36–47 ft. lbs. (50–65 Nm).

e. Rear roll stopper: 22–29 ft. lbs. (30–40 Nm).

22. For XG 300 and XG 350, torque the mount through bolts as follows:

a. Front roll stopper: 36–47 ft. lbs. (50–65 Nm).

b. Rear roll stopper: 36–47 ft. lbs. (50–65 Nm).

23. Install or connect the following:
- Axle halfshafts using new circlips
- Stabilizer bar links
- Lower ball joints
- Exhaust front pipe
- A/C compressor, if equipped
- Power steering pump
- Radiator

24. If equipped with a manual transaxle, install or connect the following:
- Clutch cable
- Select control valve connector
- Shift linkage rods

25. If equipped with an automatic transaxle, install or connect the following:
- Transaxle oil cooler lines
- Shift cable
- Transaxle wiring connectors

26. For all vehicles, install or connect the following:
- Engine ground cable
- Cruise control cable, if equipped
- Accelerator cable
- Heater hoses
- Intake manifold vacuum lines
- Brake booster vacuum line
- Fuel lines
- Cooling fan
- Radiator hoses
- Oil pressure gauge sender connector
- Alternator harness connectors
- Speedometer cable
- Reverse lamp switch connector
- Engine wiring harness connectors
- Accessory drive belts
- Air intake assembly
- Hood
- Battery

27. Fill the engine with clean oil.

28. Fill the transaxle to the correct level.

29. Fill the cooling system to the proper level.

30. Start the engine and check for leaks.

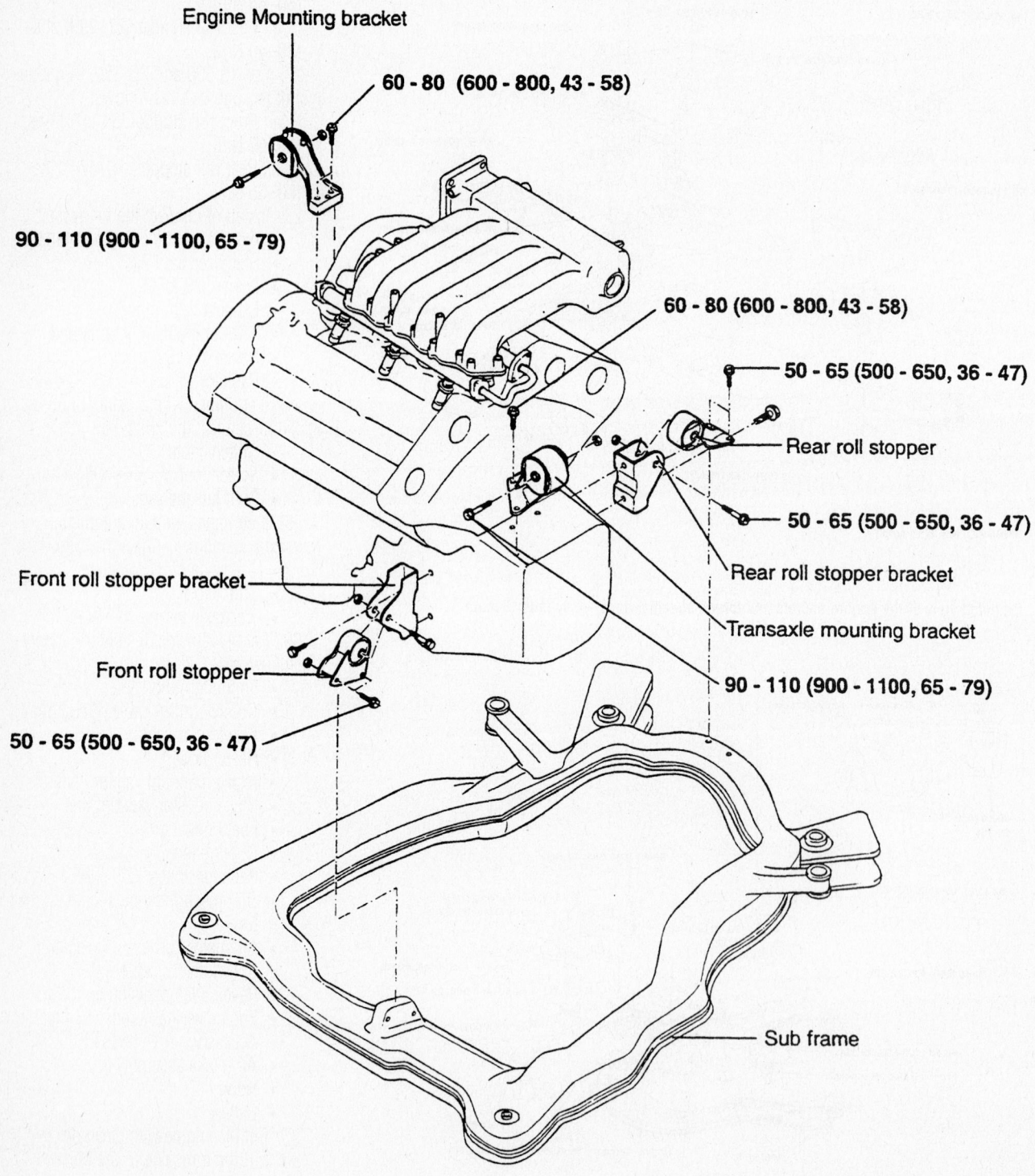

Engine Mounting bracket

60 - 80 (600 - 800, 43 - 58)

90 - 110 (900 - 1100, 65 - 79)

60 - 80 (600 - 800, 43 - 58)

50 - 65 (500 - 650, 36 - 47)

Rear roll stopper

50 - 65 (500 - 650, 36 - 47)

Rear roll stopper bracket

Transaxle mounting bracket

90 - 110 (900 - 1100, 65 - 79)

Front roll stopper bracket

Front roll stopper

50 - 65 (500 - 650, 36 - 47)

Sub frame

9347KG02

Exploded view of the engine mounts and torque specifications—XG 300 and XG 350

Water Pump

REMOVAL & INSTALLATION

Except XG 300 and XG 350

1. Before servicing the vehicle, refer to the precautions in the beginning of this section.

2. Drain the cooling system.

3. Remove or disconnect the following:
- Negative battery cable
- Accessory drive belts
- Radiator hose
- Bypass hose, if equipped
- Water pump pulley
- Front cover
- Timing belt
- Alternator bracket
- Water pump

➡️**The water pump bolts are different lengths. Note the bolt location for assembly.**

To install:

4. Install the water pump with new gaskets and O-rings. Tighten the bolts to the following specifications:

a. 1.5L SOHC engine: 105–132 inch lbs. (12–15 Nm).

b. 1.5L DOHC engine: 14–20 ft. lbs. (20–27 Nm).

c. 2.0L, and 2.4L engines: 14–20 ft. lbs. (20–27 Nm).

d. 2.5L and 2.7L engine: 11–16 ft. lbs. (15–22 Nm).

5. Install or connect the following:
- Alternator bracket
- Timing belt and front cover
- Water pump pulley
- Bypass hose, if equipped
- Radiator hose
- Accessory drive belts
- Negative battery cable

6. Fill the cooling system to the proper level.

7. Start the engine and check for leaks.

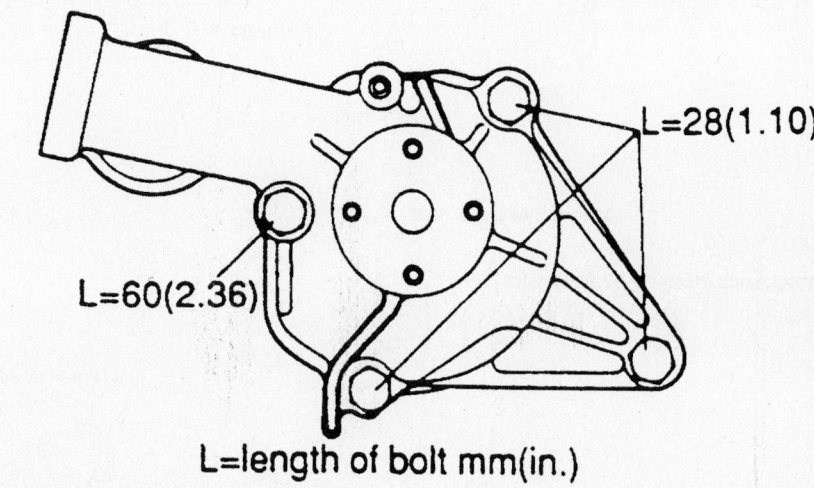

L=28(1.10)

L=60(2.36)

L=length of bolt mm(in.)

7923GG19

Water pump bolt lengths—1.5L engines

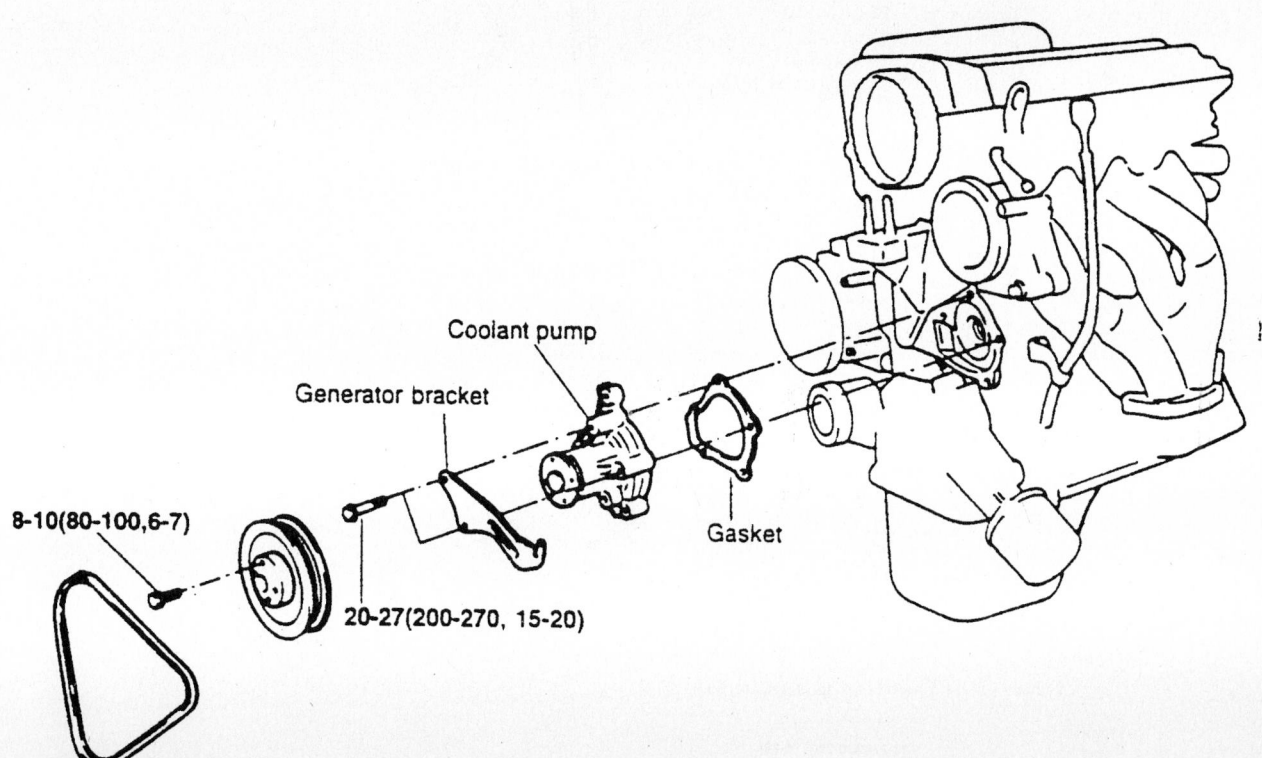

Coolant pump

Generator bracket

8-10(80-100,6-7)

20-27(200-270, 15-20)

Gasket

7923GG17

Exploded view of the water pump assembly—1.5L engines

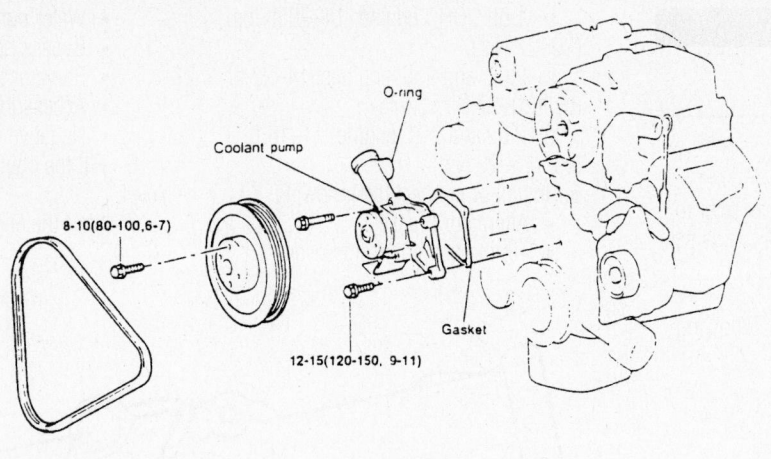

TORQUE : Nm (kg.cm, lb.ft)

7923GG20

Water pump assembly—2.0L engines

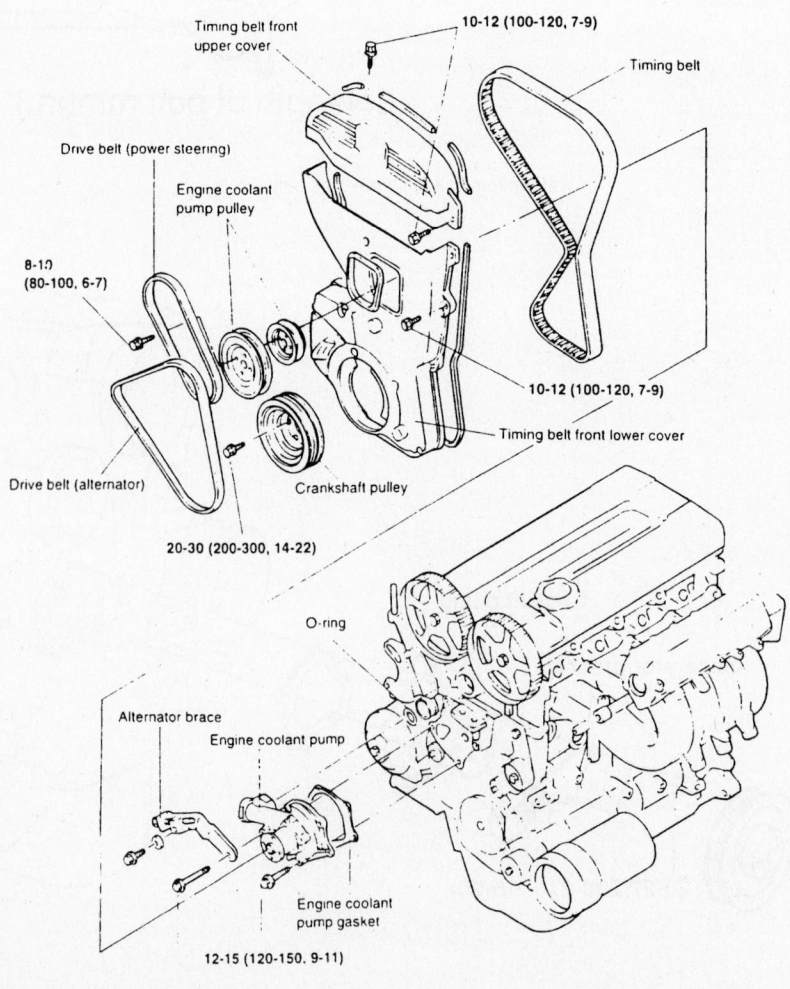

TORQUE : Nm (kg.cm, lb.ft)

7923GG21

Exploded view of the water pump assembly and related components—2.0L engines

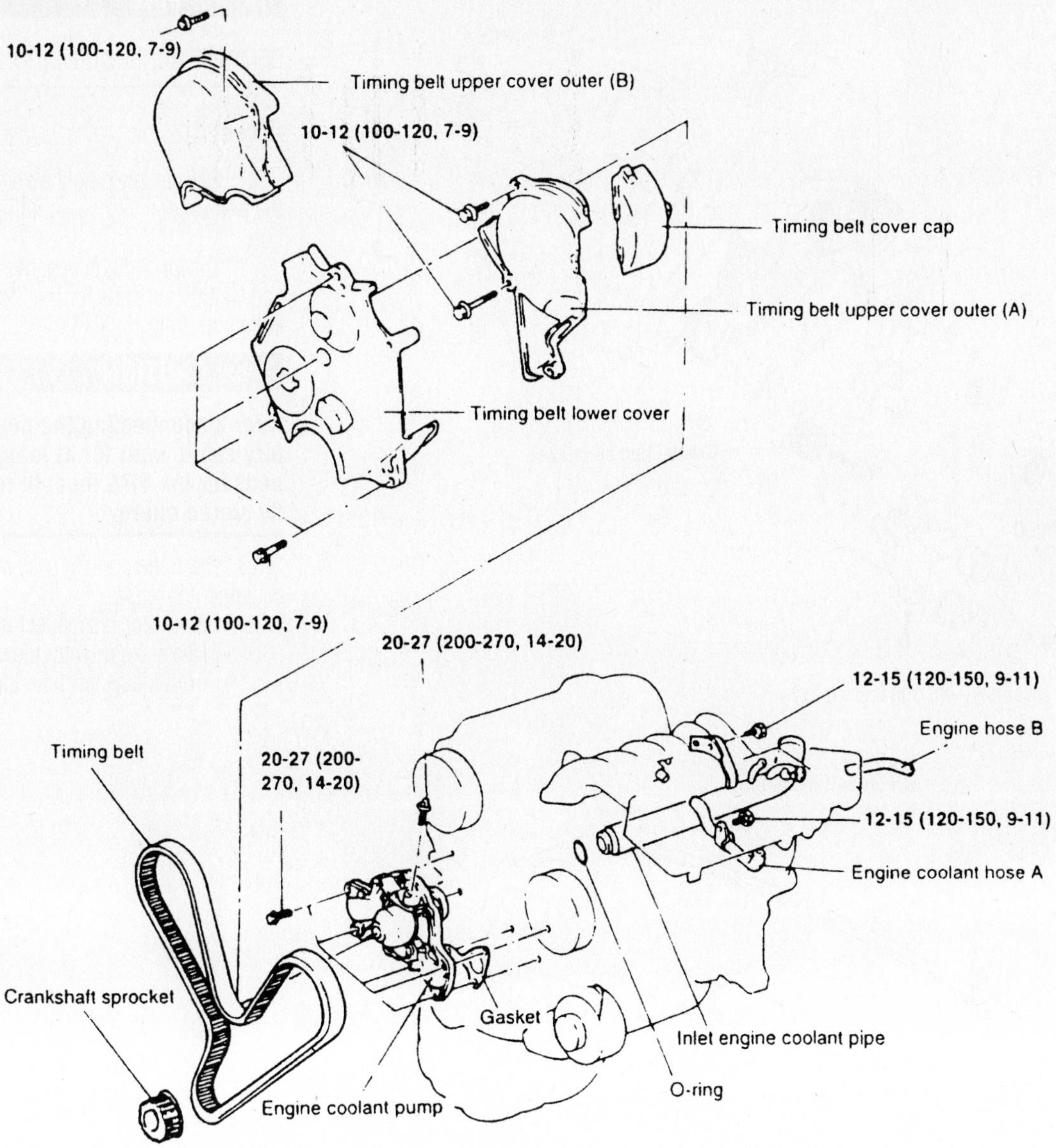

10-12 (100-120, 7-9)

Timing belt upper cover outer (B)

10-12 (100-120, 7-9)

Timing belt cover cap

Timing belt upper cover outer (A)

Timing belt lower cover

10-12 (100-120, 7-9)

20-27 (200-270, 14-20)

12-15 (120-150, 9-11)

Engine hose B

Timing belt

20-27 (200-270, 14-20)

12-15 (120-150, 9-11)

Engine coolant hose A

Crankshaft sprocket

Gasket

Inlet engine coolant pipe

O-ring

Engine coolant pump

TORQUE : Nm (kg.cm, lb.ft)

7923GG22

Water pump assembly—3.0L and 3.5L engines

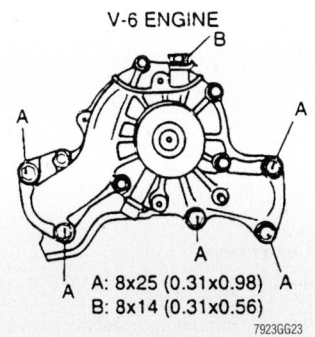

V-6 ENGINE
B

A   A

A   A

A: 8x25 (0.31x0.98) A
B: 8x14 (0.31x0.56)

7923GG23

Water pump bolt lengths—3.0L and 3.5L engines

XG 300 and XG 350

1. Before servicing the vehicle, refer to the precautions in the beginning of this section.
2. Drain the cooling system.
3. Remove or disconnect the following:

- Negative battery cable
- Accessory drive belts
- Water pump pulley
- Timing belt, tensioner and idler pulley
- Water pump

To install:

4. Clean all mating surfaces of any residual gasket material.
5. Install or connect the following:

- Water pump with a new gasket and torque the bolts to 16 ft. lbs. (22 Nm)
- Tensioner and timing belt and idler pulley
- Water pump pulley and drive belts
- Negative battery cable

6. Fill the cooling system to the proper level.
7. Start the vehicle, check for leaks and repair if necessary.

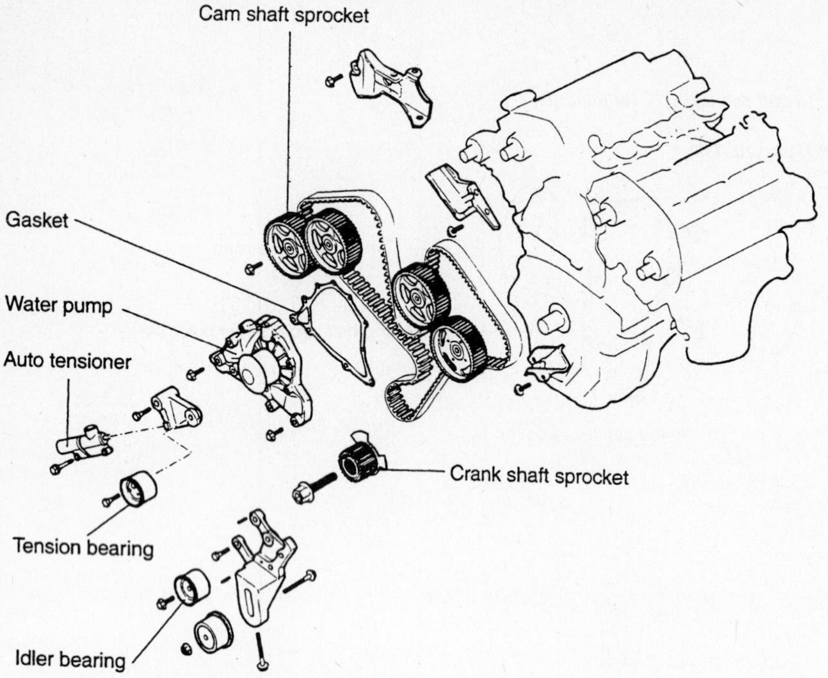
Cam shaft sprocket
Gasket
Water pump
Auto tensioner
Crank shaft sprocket
Tension bearing
Idler bearing
9347KG03

Water pump assembly—XG 300 and XG 350

Heater Core

REMOVAL AND INSTALLATION

Accent

1. Before servicing the vehicle, refer to the precautions in the beginning of this section.
2. Disconnect the negative battery cable and wait 90 seconds for the SRS memory battery to drain.

⁂ CAUTION

After disconnecting the negative battery cable, wait for at least 30 seconds for the SRS module to deplete its stored energy.

3. Drain the cooling system into a clean container for reuse.
4. Remove or disconnect the following:
   - Heater hoses with the vacuum hose from the heater housing

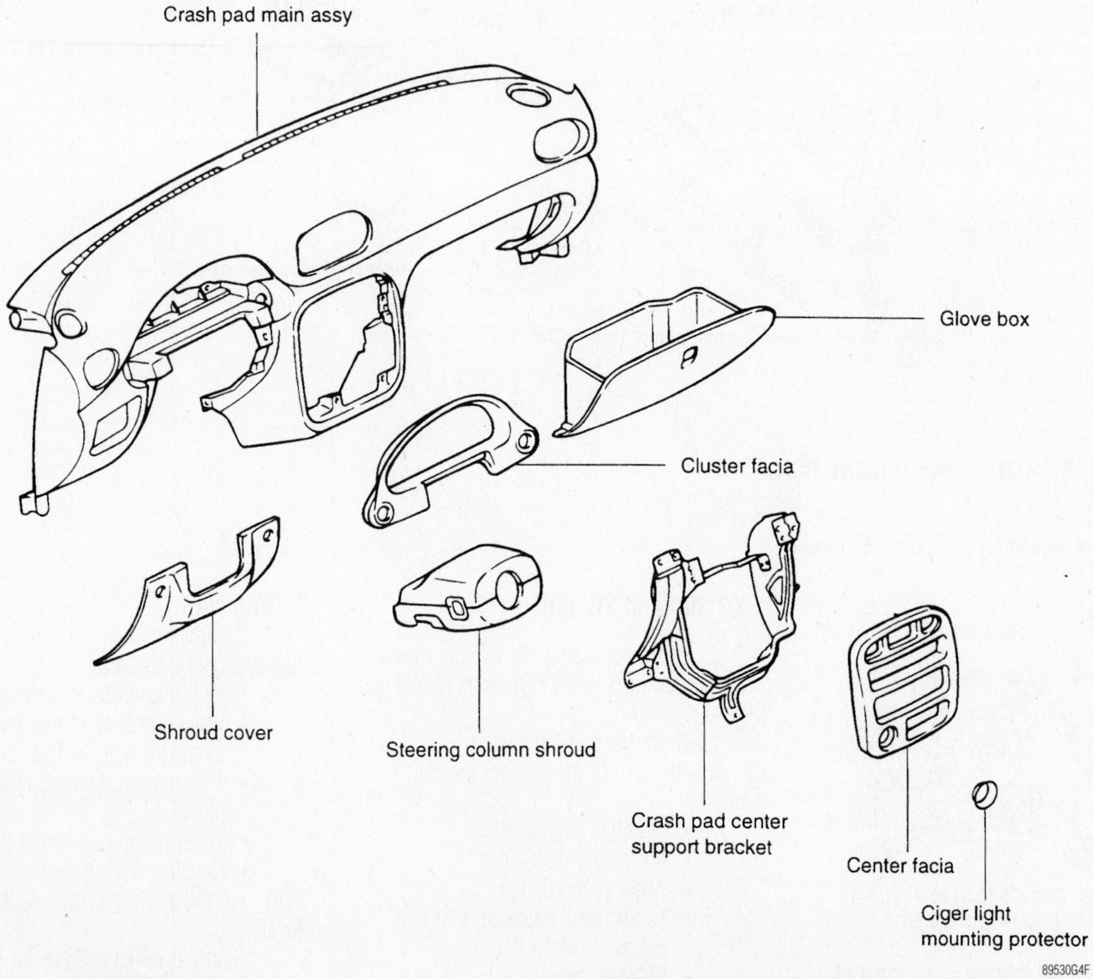
Crash pad main assy
Glove box
Cluster facia
Shroud cover
Steering column shroud
Crash pad center
support bracket
Center facia
Ciger light
mounting protector
89530G4F

Instrument panel assembly—Accent

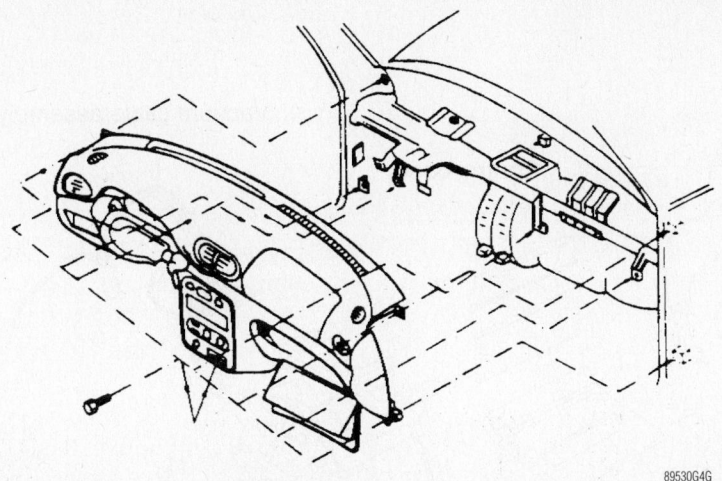

Instrument panel screw locations—Accent

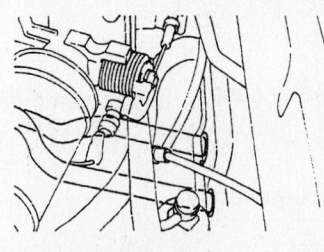

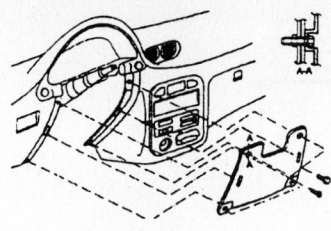

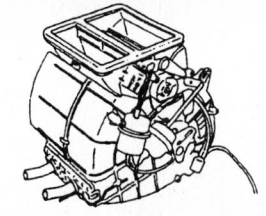

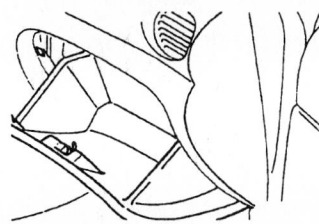

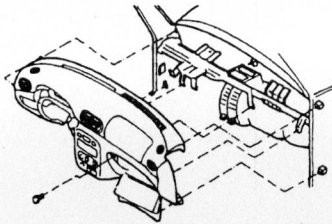

View of the heater housing assembly and related components—Accent

- Discharge and recover the air conditioning system refrigerant
- Suction and discharge hoses from the evaporator assembly

5. Remove the SRS module and the steering wheel:
  - Steering wheel-to-SRS module nuts
  - SRS module from the steering wheel and disconnect the electrical connector
  - Steering wheel-to-steering column nut
  - Steering wheel from the steering column

6. Remove or disconnect the following:
  - Multi-function switch assembly
  - Front and rear console assemblies
  - Lower left side crash pad
  - Center fascia panel and disconnect the connectors and vacuum connector from the heater control assembly
  - Heater control assembly and the audio system
  - Glove box
  - 4 mounting bolts from the passenger air bag mounting bracket, if equipped
  - Main crash pad assembly
  - Cables from the heater housing and the thermostatic switch connector from the evaporator housing
  - Any remaining connectors
  - Main crash pad assembly
  - 3 evaporator mounting bolts (or nuts)
  - Evaporator housing
  - 3 mounting bolts from the heater housing
  - Heater housing

7. Disassemble the heater housing by removing or disconnecting the following:

- Vacuum motor-to-heater housing bolts (2 for each vacuum motor)
- Vacuum motor rod end connection and remove the vacuum motors
- Heater housing cover clips
- Cover and the heater core

To install:

8. Assemble the heater housing by installing or connecting the following:
  - Heater core and the cover
  - Heater housing cover clips
  - Vacuum motor rod end connection and install the vacuum motors
  - Vacuum motor-to-heater housing bolts (2 for each vacuum motor)

9. Install or connect the following:
  - Heater housing
  - 3 mounting bolts to the heater housing
  - Evaporator housing
  - 3 evaporator mounting bolts (or nuts)
  - Main crash pad assembly
  - Any remaining connectors
  - Cables to the heater housing and the thermostatic switch connector to the evaporator housing
  - Main crash pad assembly
  - 4 mounting bolts to the passenger air bag mounting bracket, if equipped
  - Glove box
  - Heater control assembly and the audio system
  - Connectors and vacuum connector to the heater control assembly. Install the center fascia panel.
  - Lower left side crash pad
  - Front and rear console assemblies

10. Install the SRS module and the steering wheel by installing or connecting the following:
  - Steering wheel to the steering column
  - Steering wheel-to-steering column nut and torque to 30–37 ft. lbs. (40–50 Nm)
  - SRS module to the steering wheel and connect the electrical connector
  - Steering wheel-to-SRS module nuts

11. Install or connect the following:
  - Multi-function switch assembly
  - Suction and discharge hoses to the evaporator assembly
  - Heater hoses with the vacuum hose to the heater housing

12. Refill the cooling system.
13. Connect the negative battery cable.
14. Evacuate, charge and leak test the air conditioning system refrigerant.

* Heater Assembly

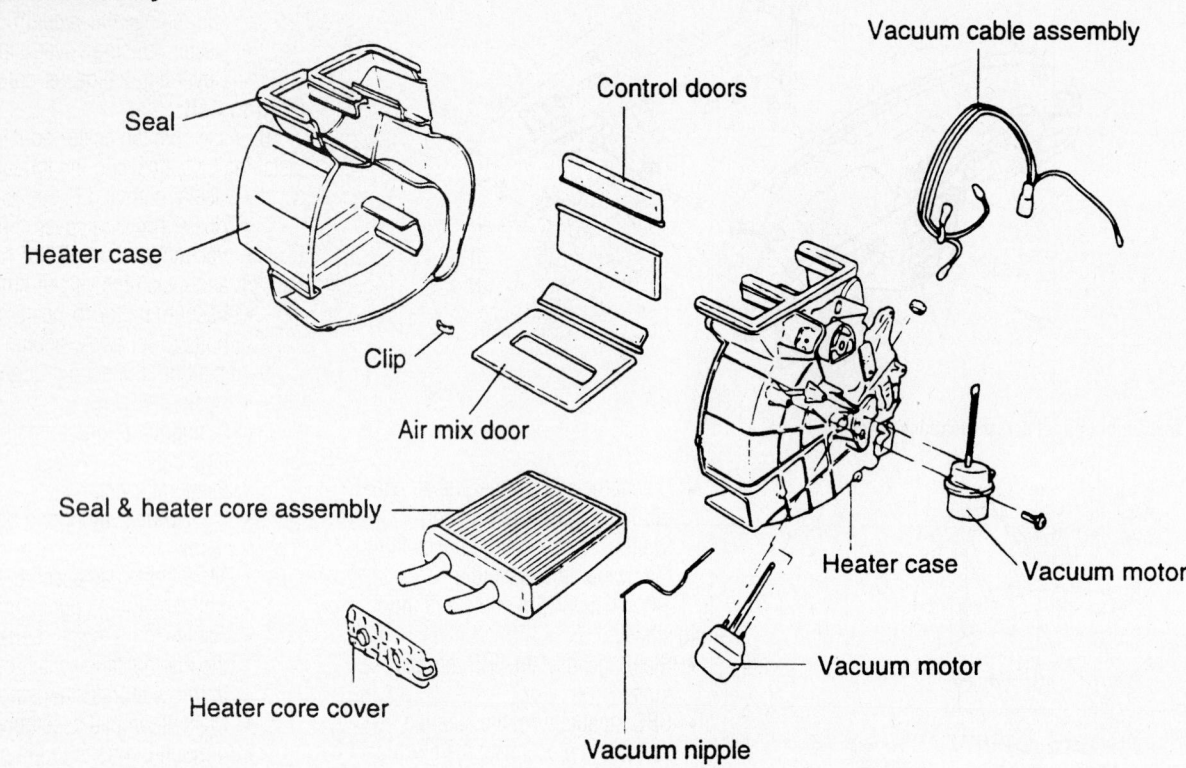

Seal

Control doors

Vacuum cable assembly

Heater case

Clip

Air mix door

Seal & heater core assembly

Heater core cover

Heater case

Vacuum motor

Vacuum motor

Vacuum nipple

* Vacuum Source Lines

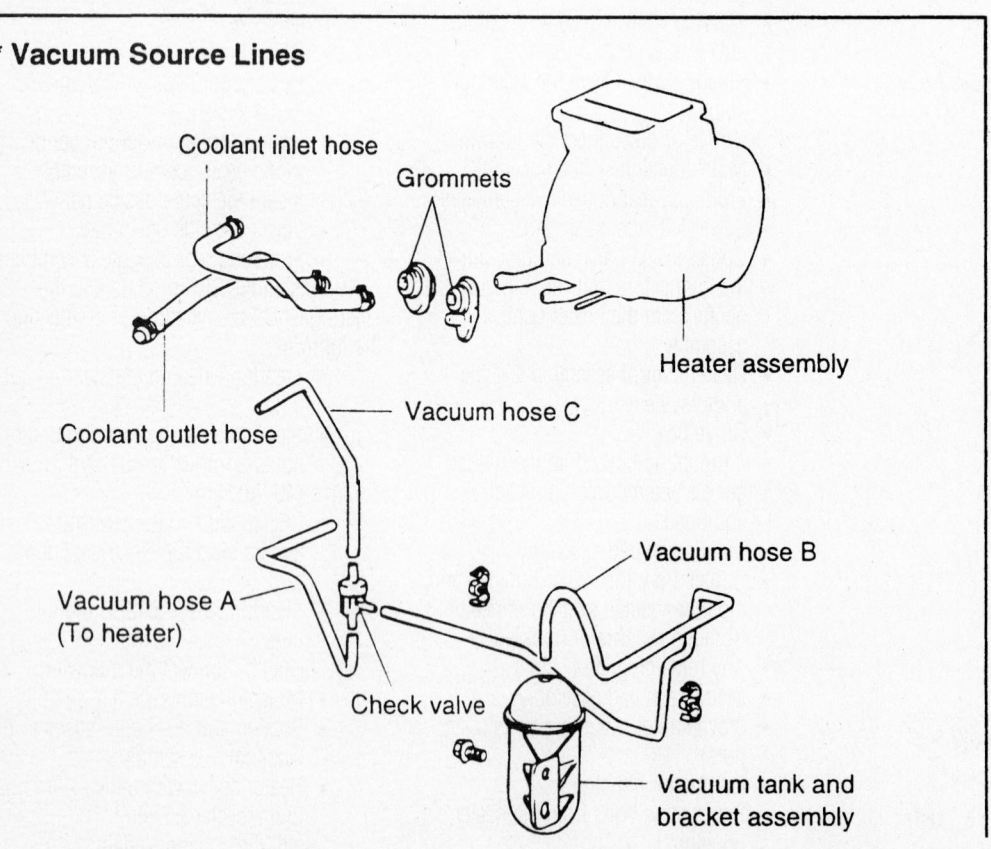

Coolant inlet hose

Grommets

Heater assembly

Coolant outlet hose

Vacuum hose C

Vacuum hose B

Vacuum hose A
(To heater)

Check valve

Vacuum tank and
bracket assembly

93112GP4

Exploded view of the heater core, heater housing and related components—Accent

15. Operate the engine to normal operating temperatures; then, check the climate control operation and check for leaks.

Elantra

1. Before servicing the vehicle, refer to the precautions in the beginning of this section.

2. Disconnect the negative battery cable.

> **※ CAUTION**
>
> **After disconnecting the negative battery cable, wait for at least 30 seconds for the SRS module to deplete its stored energy.**

3. Discharge and recover the air conditioning system refrigerant.

4. Drain the engine coolant into a clean container for reuse.

5. Disconnect the heater hoses from the heater core. Plug the openings.

6. Disconnect the vacuum line from the heater housing vacuum nipple.

7. Remove the SRS module and the steering wheel by removing or disconnecting the following:

- Steering wheel-to-SRS module nuts
- SRS module from the steering wheel and disconnect the electrical connector
- Steering wheel-to-steering column nut
- Steering wheel from the steering column

8. Remove the instrument panel by removing or disconnecting the following:

- Steering column cover screws and the covers
- Instrument panel lower cover at the driver's side
- Multi-function switch and disconnect the electrical connector at the steering column
- Instrument panel cluster fascia panel
- Instrument cluster-to-instrument panel screws, disconnect the electrical connectors and remove the instrument cluster
- Side fascia panel and disconnect the mirror control connector
- Hood release mounting screws
- Rheostat and the upper console cover

- Heater control cable
- Electrical connectors at the center of the instrument panel and remove the center fascia panel assembly
- Console
- Radio-to-chassis screws and the radio
- Glove box screws and the glove box
- Glove box striker screws and the upper glove box cover
- Defroster nozzle
- Loosen the speedometer drive gear sleeve and disconnect the speedometer cable from the instrument panel
- Passenger's side SRS module connector
- Instrument panel-to-chassis bolts
- Any remaining electrical connectors
- Ventilation ducts from the instrument panel
- Instrument panel

9. Remove or disconnect the following:

- Front right side heating duct from the heater housing
- Pull back the carpet and remove the right side console mounting bracket

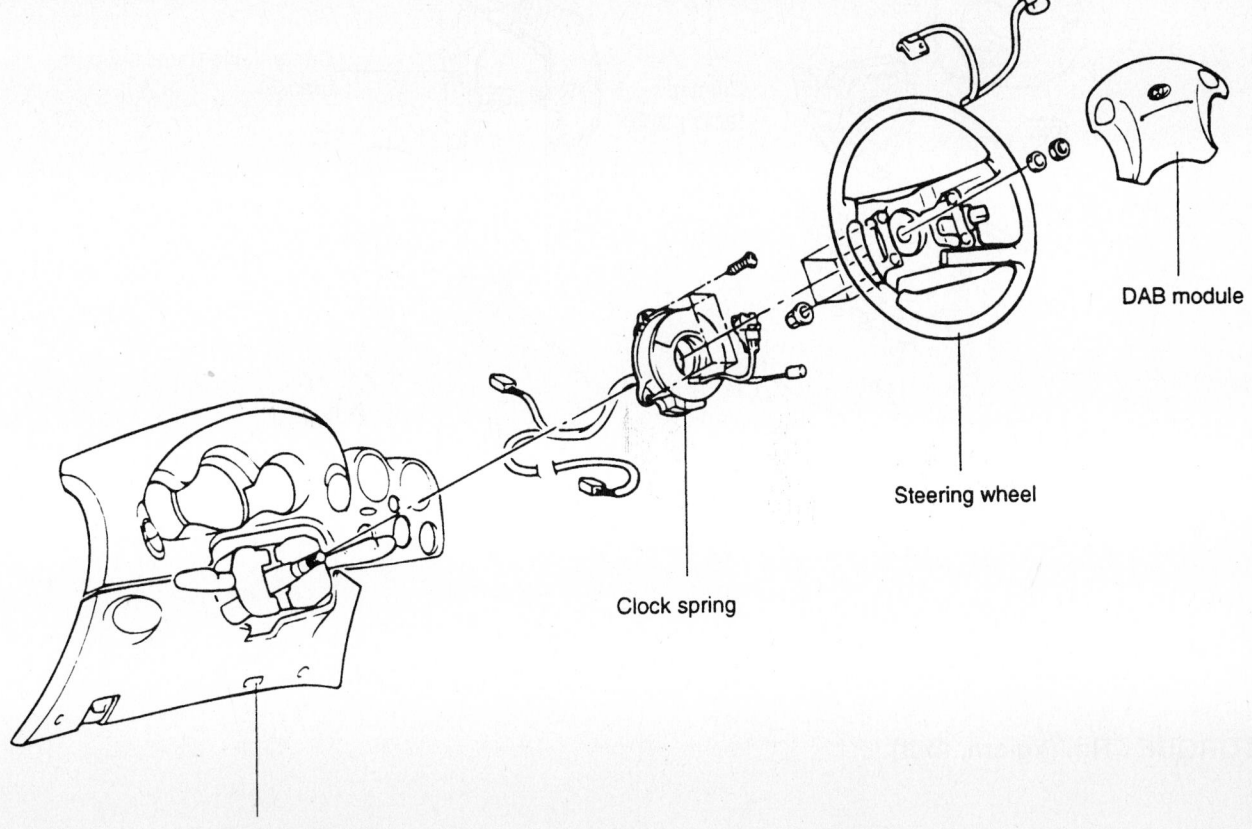

DAB module

Steering wheel

Clock spring

Data link connector

93112G07

Exploded view of the SRS module, steering wheel and related components—Elantra

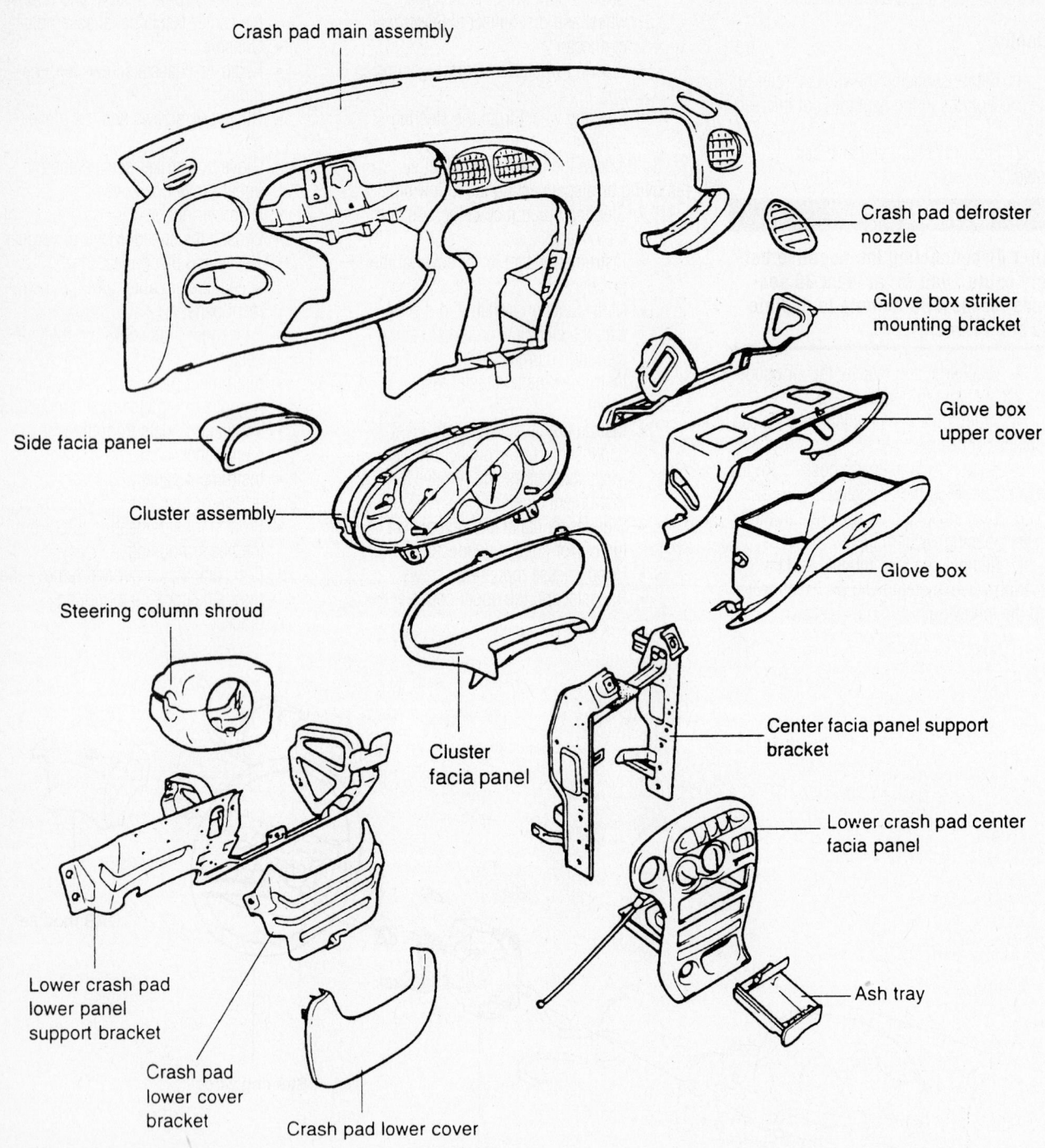

Crash pad main assembly

Crash pad defroster nozzle

Glove box striker mounting bracket

Glove box upper cover

Side facia panel

Cluster assembly

Glove box

Steering column shroud

Cluster facia panel

Center facia panel support bracket

Lower crash pad center facia panel

Lower crash pad lower panel support bracket

Crash pad lower cover bracket

Crash pad lower cover

Ash tray

TORQUE : Nm (kg·cm, lb·ft)

93112G08

Exploded view of the instrument panel and related components—Elantra

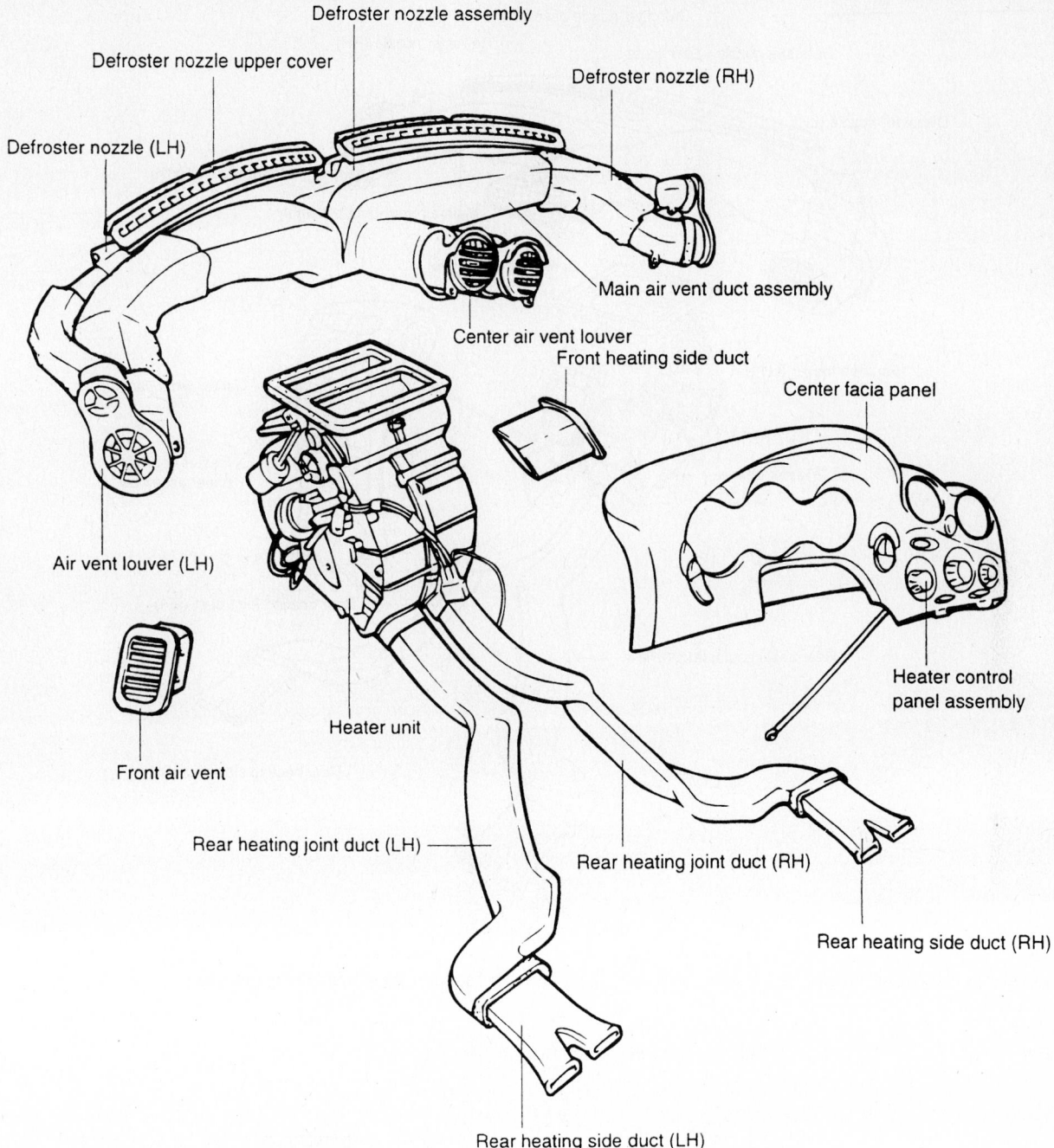

Defroster nozzle assembly

Defroster nozzle upper cover

Defroster nozzle (RH)

Defroster nozzle (LH)

Main air vent duct assembly

Center air vent louver

Front heating side duct

Center facia panel

Air vent louver (LH)

Front air vent

Heater unit

Heater control panel assembly

Rear heating joint duct (LH)

Rear heating joint duct (RH)

Rear heating side duct (RH)

Rear heating side duct (LH)

93112G00

Exploded view of the heater housing, center fascia, distribution ducts and related components—Elantra Coupe

- Front left side duct from the heater housing
- Pull back the carpet and remove the left side console mounting bracket
- Rear heating duct from the heater housing
- Control modules electrical connectors at the center fascia panel support bracket
- Center fascia panel support bracket screws, bolts and/or nuts; then, remove the center fascia panel support bracket

- Center support bars
- Glove box support bracket-to-instrument panel bolts and the bracket

10. Remove the evaporator housing by removing or disconnecting the following:
- Refrigerant lines from the evaporator housing and discard the O-rings
- Thermostatic switch connector
- Evaporator housing upper and lower bolts
- Evaporator housing

11. Remove or disconnect the following:
- Heater housing-to-chassis bolts and the housing
- Vacuum motor-to-heater housing bolts (2 for each vacuum motor)
- Vacuum motor rod end connection and remove the vacuum motors
- Heater housing cover clips
- Cover and the heater core

To install:

12. Assemble the heater housing by installing or connecting the following:
- Heater core and the cover

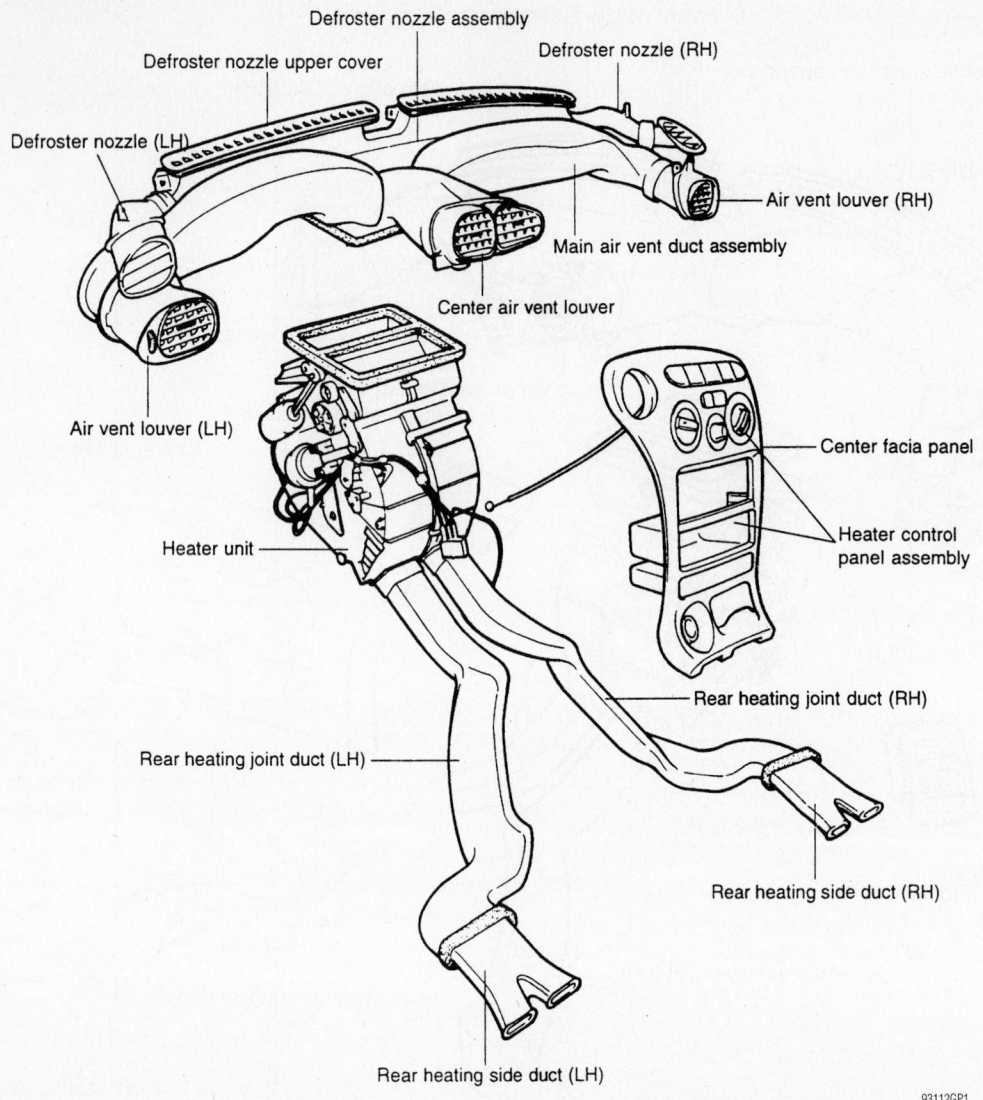

Defroster nozzle assembly

Defroster nozzle upper cover

Defroster nozzle (RH)

Defroster nozzle (LH)

Air vent louver (RH)

Main air vent duct assembly

Center air vent louver

Air vent louver (LH)

Center facia panel

Heater control panel assembly

Heater unit

Rear heating joint duct (RH)

Rear heating joint duct (LH)

Rear heating side duct (RH)

Rear heating side duct (LH)

93112GP1

Exploded view of the heater housing, center fascia, distribution ducts and related components—Elantra Sedan and Wagon

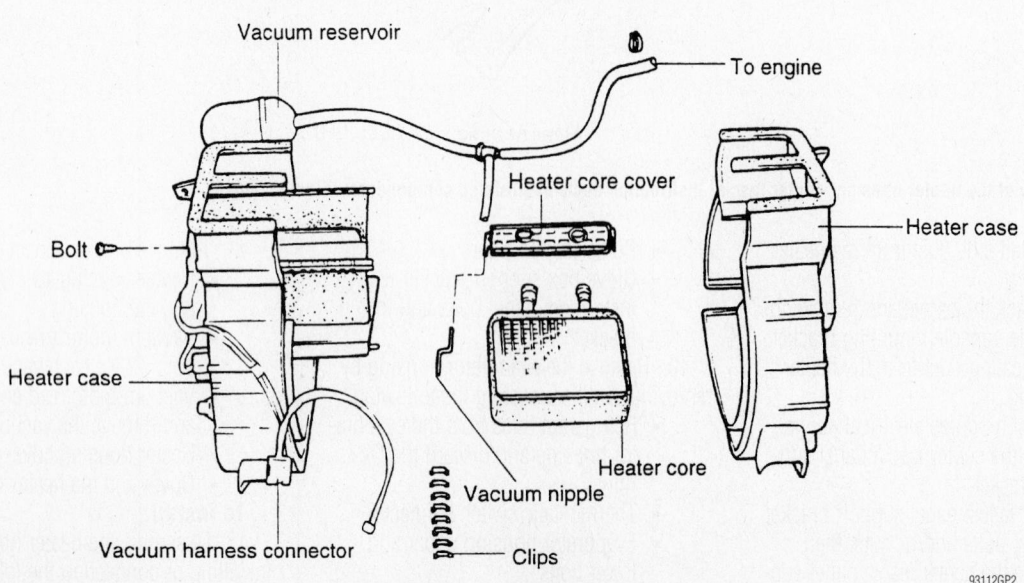

Vacuum reservoir

To engine

Heater core cover

Heater case

Bolt

Heater case

Heater core

Heater case

Vacuum nipple

Vacuum harness connector

Clips

93112GP2

Exploded view of the heater core and heater housing and related components—Elantra

- Heater housing cover clips
- Vacuum motor rod end connection and install the vacuum motors
- Vacuum motor-to-heater housing bolts (2 for each vacuum motor)
13. Install or connect the following:
- Heater housing and the housing-to-chassis bolts
- Evaporator housing
- Evaporator housing upper and lower bolts
- Thermostatic switch connector
- Connect the refrigerant lines to the evaporator housing
14. Install or connect the following:
- Glove box support bracket and the bracket-to-instrument panel bolts
- Center support bars
- Center fascia panel support bracket;

then, install the center fascia panel support bracket screws, bolts and/or nuts
- Center fascia panel support bracket, connect the control modules electrical connectors
- Rear heating duct to the heater housing
- Left side console mounting bracket and install the carpet
- Front left side duct to the heater housing
- Right side console mounting bracket and install the carpet
- Front right side heating duct to the heater housing
15. Install the instrument panel by installing or connecting the following:
- Instrument panel

- Ventilation ducts to the instrument panel
- Instrument panel-to-chassis bolts
- Passenger's side SRS module connector
- Speedometer cable to the instrument panel and tighten the speedometer drive gear sleeve
- Defroster nozzle
- Upper glove box cover and the glove box striker screws
- Glove box and the glove box screws
- Radio and the radio-to-chassis screws
- Console
- Electrical connectors and install the center fascia panel assembly
- Heater control cable

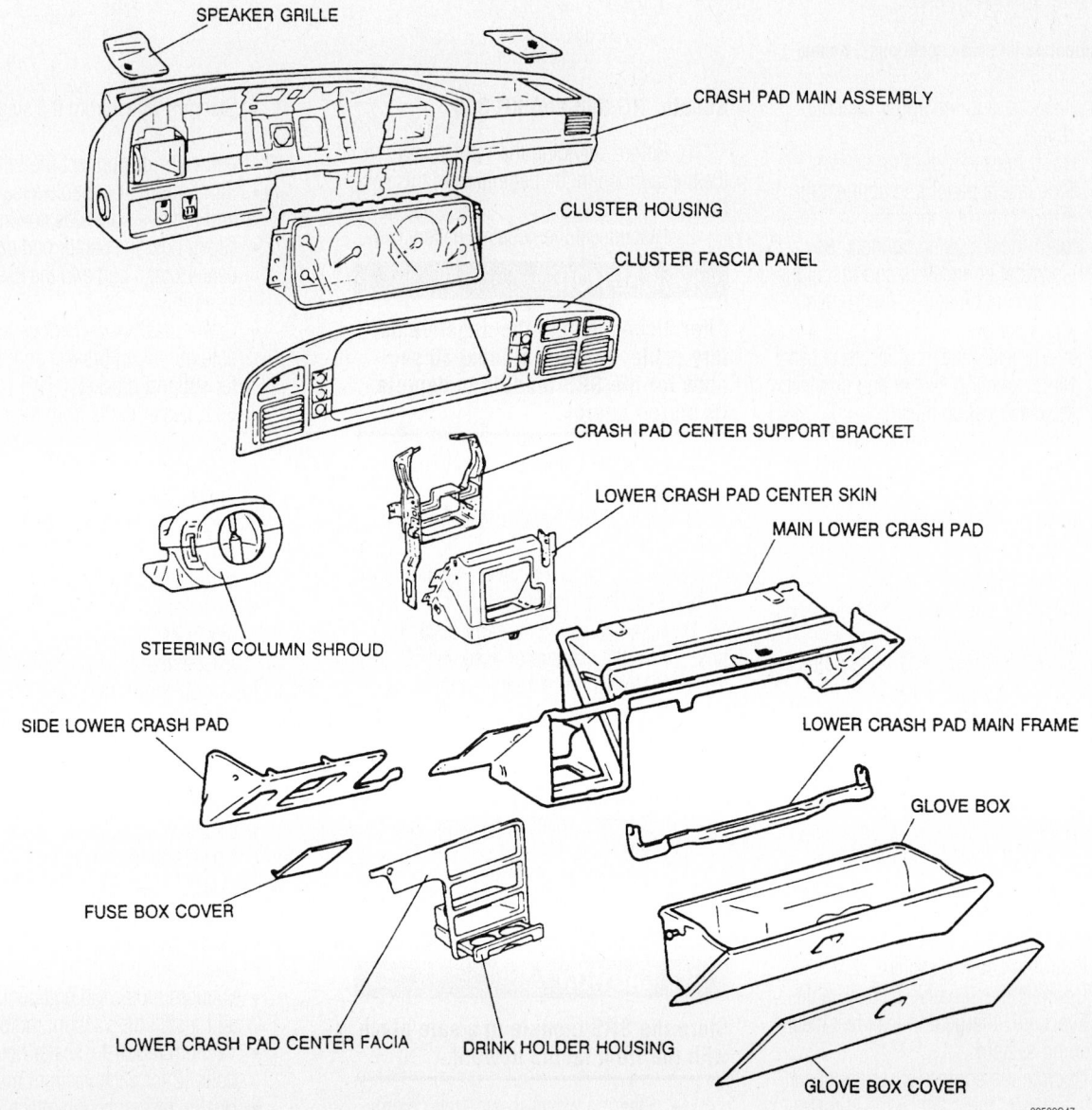

SPEAKER GRILLE

CRASH PAD MAIN ASSEMBLY

CLUSTER HOUSING

CLUSTER FASCIA PANEL

CRASH PAD CENTER SUPPORT BRACKET

LOWER CRASH PAD CENTER SKIN

MAIN LOWER CRASH PAD

STEERING COLUMN SHROUD

SIDE LOWER CRASH PAD

LOWER CRASH PAD MAIN FRAME

GLOVE BOX

FUSE BOX COVER

LOWER CRASH PAD CENTER FACIA

DRINK HOLDER HOUSING

GLOVE BOX COVER

89530G47

Instrument panel assembly—Sonata

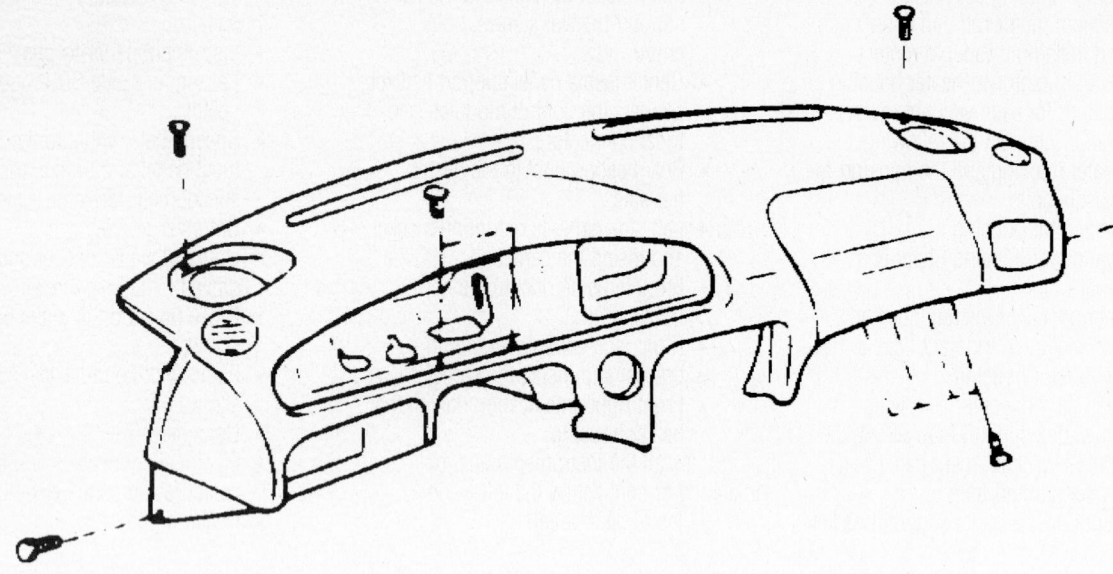

89530G48

Instrument panel screw locations—Sonata

- Rheostat and the upper console cover
- Hood release mounting screws
- Side fascia panel and connect the mirror control connector
- Instrument cluster, connect the electrical connectors and install the instrument cluster-to-instrument panel screws
- Instrument panel cluster fascia panel
- Multi-function switch and connect the electrical connector
- Instrument panel lower cover
- Steering column cover and the cover screws

16. Install the SRS module and the steering wheel by installing or connecting the following:
- Steering wheel to the steering column
- Steering wheel-to-steering column nut and torque to 30–37 ft. lbs. (40–50 Nm)
- SRS module to the steering wheel and connect the electrical connector
- Steering wheel-to-SRS module nuts

17. Connect the vacuum line to the heater housing vacuum nipple.
18. Connect the heater hoses to the heater core.
19. Refill the cooling system.
20. Connect the negative battery cable.
21. Evacuate, charge and leak test the air conditioning system.
22. Operate the engine to normal operating temperatures; then, check the climate control operation and check for leaks.

Sonata, XG 300 and XG 350

1. Before servicing the vehicle, refer to the precautions in the beginning of this section.
2. Disconnect the negative battery cable.

✳✳ CAUTION

After disconnecting the negative battery cable, wait for at least 30 seconds for the SRS module to deplete its stored energy.

3. Drain the cooling system into a clean container for reuse.
4. Remove the heater hoses from the heater housing.
5. Discharge and recover the air conditioning system refrigerant.
6. Remove the suction and discharge hoses from the evaporator assembly. Cap the hoses to minimize contamination.
7. Remove the evaporator drain hose.
8. Remove the SRS module and the steering wheel by removing or disconnecting the following:
- Steering wheel-to-SRS module nuts
- SRS module from the steering wheel and disconnect the electrical connector

✳✳ CAUTION

Store the SRS module in a safe place with the front facing upward.

- Steering wheel-to-steering column nut

- Steering wheel from the steering column

9. Remove or disconnect the following:
- Front and rear console assembly and remove both side covers
- Glove box, the center pad cover, the center crash pad and the cassette assembly
- Lower crash pad. Remove the console mounting bracket and the center support bracket
- Rear heater ducts from the heater housing
- Control assembly
- Blower speed control actuator connector and the blend door actuator connector, if equipped with semi-automatic temperature control
- 4 retaining bolts and remove the heater assembly

10. Disassemble the heater housing by removing or disconnecting the following:
- Vacuum motor-to-heater housing bolts (2 for each vacuum motor)
- Vacuum motor rod end connection and remove the vacuum motors
- Heater housing cover clips
- Cover and the heater core

To install:

11. Install or connect the following:
- Heater core and the cover
- Heater housing cover clips
- Vacuum motor rod end connection and install the vacuum motors
- Vacuum motor-to-heater housing bolts (2 for each vacuum motor)
- Heater assembly and attach it to the dash panel with the mounting bolts

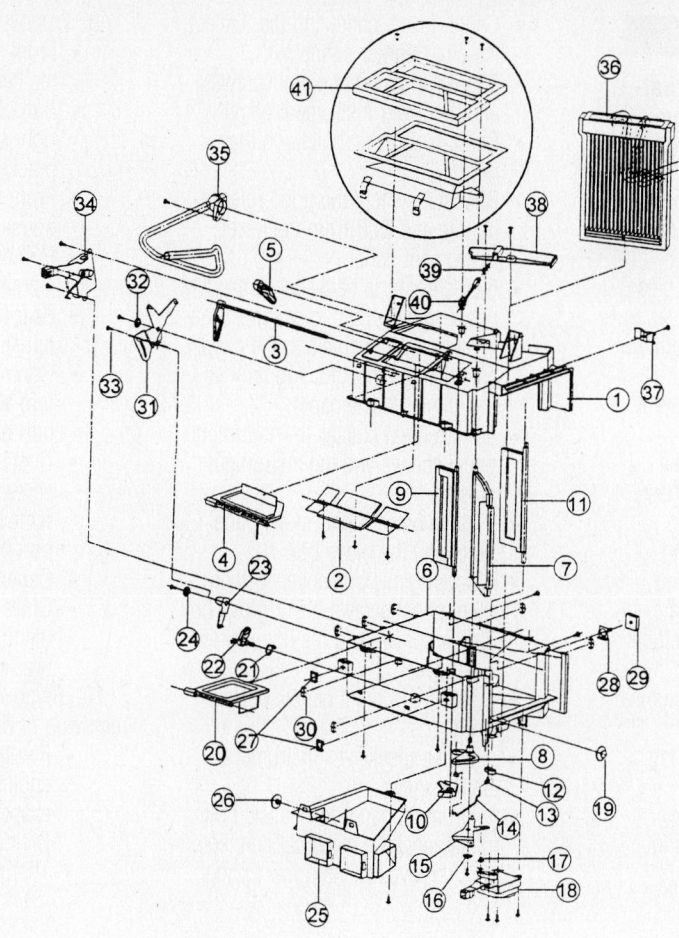

1 Case-heater upper	11 Door ass'y-by pass	21 Spring	31 Cam-mode
2 Door ass'y-vent	12 Arm-By pass door	22 Arm-floor door	32 Spring-washer
3 Shaft ass'y-vent door	13 Holder-rod link	23 Lever-floor door	33 Holder-rod link
4 Door ass'y-defrost	14 Link	24 Spring washer	34 Mode actuator
5 Arm defrost door	15 Lever-temp. door	25 Duct-floor	35 Aspirator & hose ass'y
6 Case-heater lower	16 Spring washer	26 Guide bush	36 Heater core
7 Door ass'y-temp.	17 Guide bush	27 U-nut	37 Clip
8 Arm-temp. door	18 Blend door actuator	28 Clip & Bolt ass'y	38 Cover-heater core
9 Door ass'y (A)-temp. door	(For AUTO A/C only)	29 Seal (A)-heater to D/panel	39 Stopper
10 Arm (A)-temp. door (A)	19 Guide bush	30 Clip	40 Sensor
	20 Door ass'y-floor		41 Plenum duct ass'y

93112GP3

Exploded view of the heater housing assembly—Sonata

- Heater control assembly. Connect the ducts to the heater housing
- Console mounting bracket and the center support bracket
- Lower crash pad and both side covers
- Front and rear console assembly

12. Install the SRS module and the steering wheel by installing or connecting the following:
- Steering wheel to the steering column

- Steering wheel-to-steering column nut and torque to 30–37 ft. lbs. (40–50 Nm)
- SRS module to the steering wheel and connect the electrical connector
- Steering wheel-to-SRS module nuts
- Evaporator tubes, the heater hoses and the drain hose

13. Refill the cooling system.
14. Connect the negative battery cable.
15. Evacuate, charge and leak test the air conditioning system.

16. Operate the engine to normal operating temperatures; then, check the climate control operation and check for leaks.

Tiburon

1. Before servicing the vehicle, refer to the precautions in the beginning of this section.

2. Disconnect the negative battery cable.

❄❄ CAUTION

After disconnecting the negative battery cable, wait for at least 30 seconds for the SRS module to deplete its stored energy.

3. Discharge and recover the air conditioning system refrigerant.

4. Drain the engine coolant into a clean container for reuse.

5. Disconnect the heater hoses from the heater core. Plug the openings.

6. Disconnect the vacuum line from the heater housing vacuum nipple.

7. Remove the SRS module and the steering wheel by removing or disconnecting the following:
- Steering wheel-to-SRS module nuts
- SRS module from the steering wheel and disconnect the electrical connector
- Steering wheel-to-steering column nut
- Steering wheel from the steering column

8. Remove the instrument panel by removing or disconnecting the following:
- Upper console cover

- Center fascia panel and disconnect the cigar lighter connector
- 3 lower instrument panel screws and the lower instrument panel
- Radio-to-bracket bolts and the radio
- Rheostat switch, the hood release handle and DLC from the lower instrument panel
- 5 cluster fascia panel-to-instrument panel screws; then, disconnect the heater control cable and the cluster electrical connectors and remove the cluster fascia panel
- 4 instrument cluster-to-instrument panel screws and the instrument cluster
- 2 glove box-to-instrument panel bolts and the glove box
- 4 upper glove box cover-to-instrument panel screws, the 2 glove box striker screws and the upper glove box cover
- Upper instrument panel speaker grille
- 2 upper speaker-to-instrument panel screws
- Instrument panel-to-chassis bolts, disconnect the electrical connectors and remove the instrument panel

9. Remove or disconnect the following:
- Front right side heating duct from the heater housing
- Pull back the carpet and remove the right side console mounting bracket
- Front left side duct from the heater housing
- Pull back the carpet and remove the left side console mounting bracket
- Rear heating duct from the heater housing
- Control modules electrical connectors at the center fascia panel support bracket
- Center fascia panel support bracket screws, bolts and/or nuts; then, remove the center fascia panel support bracket
- Center support bars
- Glove box support bracket-to-instrument panel bolts and the bracket

10. Remove the evaporator housing by removing or disconnecting the following:
- Refrigerant lines (located in the engine compartment), from the evaporator housing and discard the O-rings
- Thermostatic switch connector

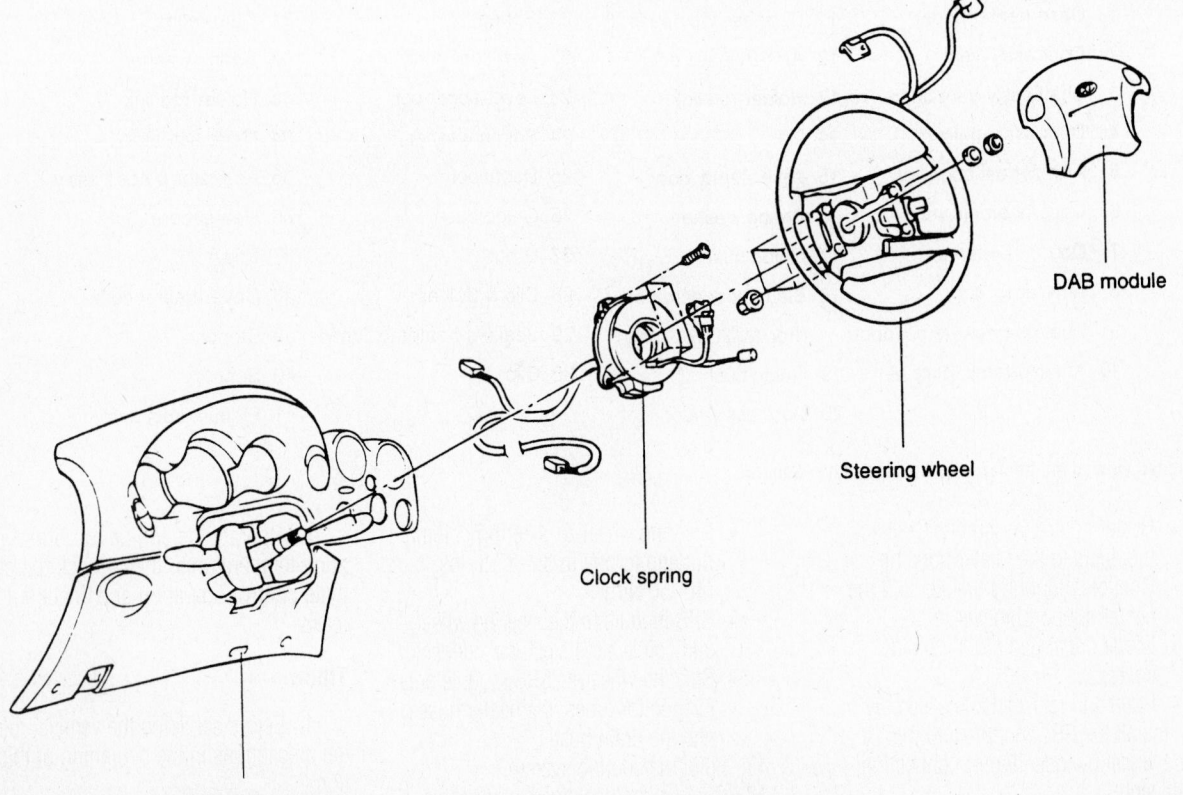

DAB module

Steering wheel

Clock spring

Data link connector

93112G07

Exploded view of the SRS module, steering wheel and related components—Tiburon

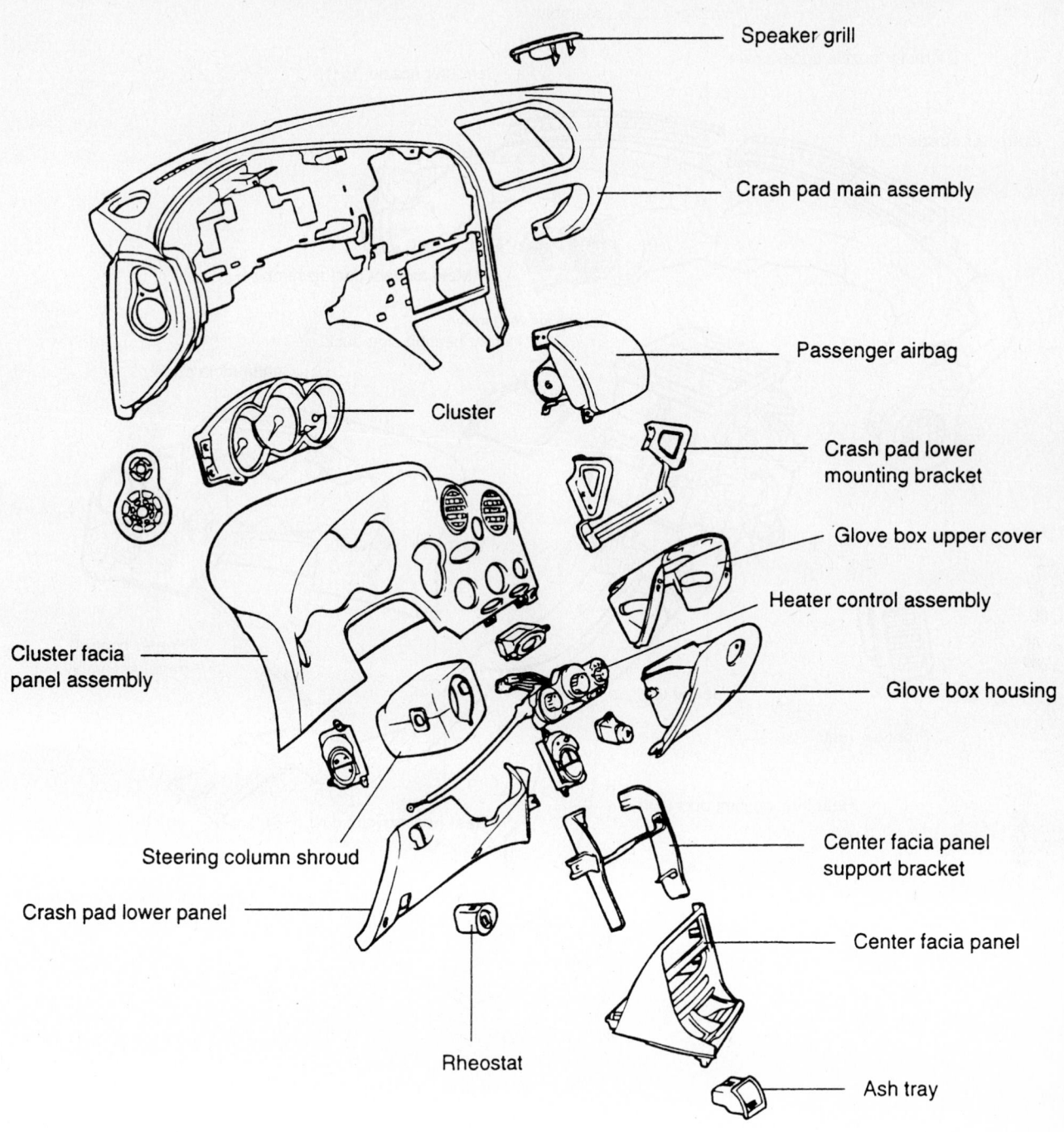

Speaker grill

Crash pad main assembly

Passenger airbag

Cluster

Crash pad lower mounting bracket

Glove box upper cover

Heater control assembly

Cluster facia panel assembly

Glove box housing

Steering column shroud

Center facia panel support bracket

Crash pad lower panel

Center facia panel

Rheostat

Ash tray

93112G09

Exploded view of the instrument panel and related components—Tiburon

- Evaporator housing upper and lower bolts
- Evaporator housing
- Heater housing-to-chassis bolts and the housing

11. Disassemble the heater housing by removing or disconnecting the following:
- Vacuum motor-to-heater housing bolts (2 for each vacuum motor)
- Vacuum motor rod end connection and remove the vacuum motors

- Heater housing cover clips
- Cover and the heater core

To install:

12. Assemble the heater housing by installing or connecting the following:
- Heater core and the cover
- Heater housing cover clips
- Vacuum motor rod end connection and install the vacuum motors
- Vacuum motor-to-heater housing bolts (2 for each vacuum motor)

- Heater housing and the housing-to-chassis bolts

13. Install the evaporator housing by installing or connecting the following:
- Evaporator housing
- Evaporator housing upper and lower bolts
- Thermostatic switch connector
- Connect the refrigerant lines to the evaporator housing using new O-rings

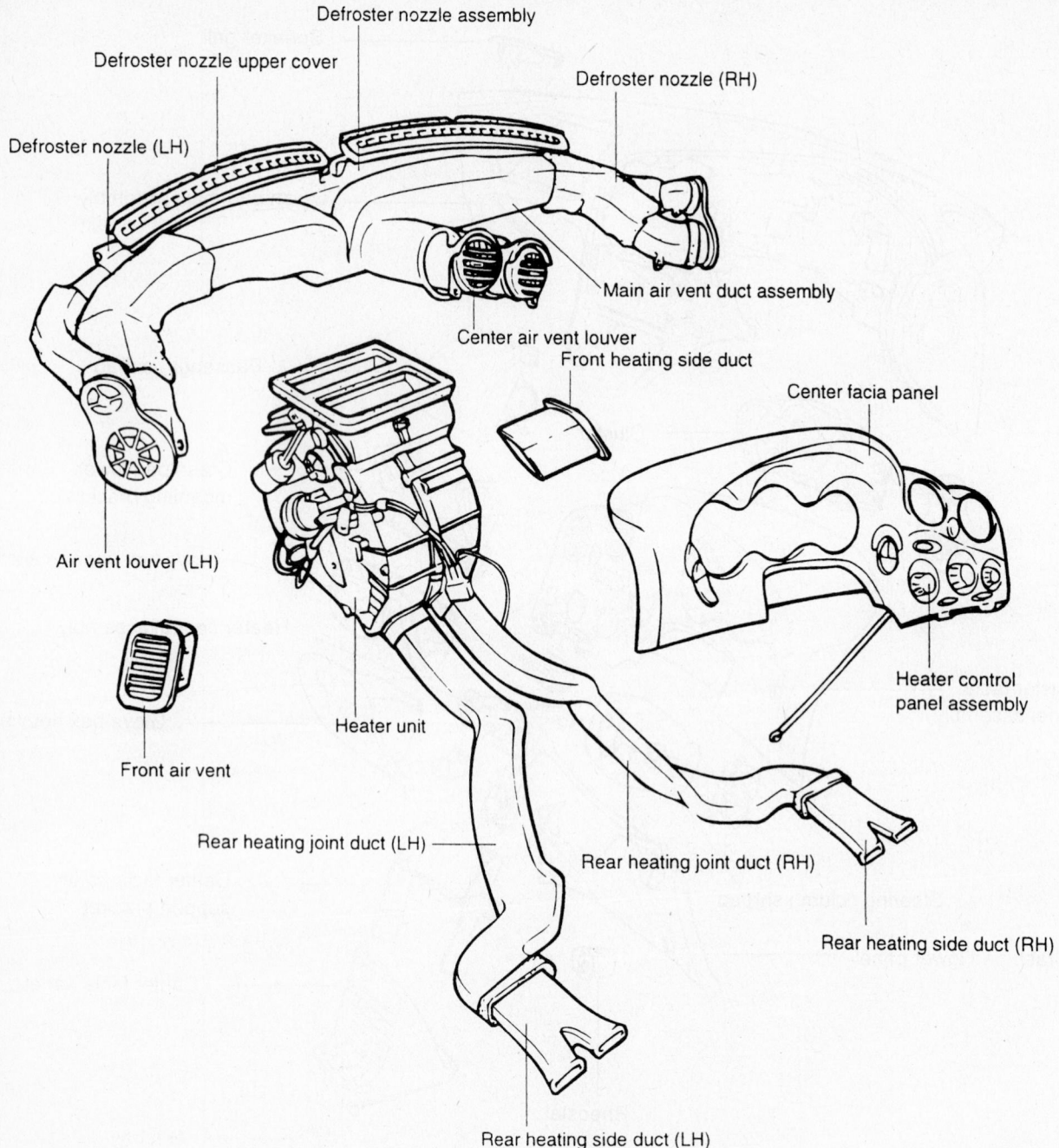

Defroster nozzle assembly

Defroster nozzle upper cover

Defroster nozzle (RH)

Defroster nozzle (LH)

Main air vent duct assembly

Center air vent louver

Front heating side duct

Center facia panel

Air vent louver (LH)

Heater control panel assembly

Front air vent

Heater unit

Rear heating joint duct (LH)

Rear heating joint duct (RH)

Rear heating side duct (RH)

Rear heating side duct (LH)

93112G00

Exploded view of the heater housing, center fascia, distribution ducts and related components—Tiburon Coupe

14. Install or connect the following:
  - Glove box support bracket and the bracket-to-instrument panel bolts
  - Center support bars
  - Center fascia panel support bracket; then, install the center fascia panel support bracket screws, bolts and/or nuts
  - Control modules electrical connectors at the center fascia panel support bracket
  - Rear heating duct to the heater housing

  - Left side console mounting bracket and install the carpet
  - Front left side duct to the heater housing
  - Right side console mounting bracket and install the carpet
  - Front right side heating duct to the heater housing

15. Install the instrument panel by installing or connecting the following:
  - Instrument panel, connect the electrical connectors and install the instrument panel-to-chassis bolts

  - 2 upper speaker-to-instrument panel screws
  - Upper instrument panel speaker grille
  - Upper glove box cover, the 2 glove box striker screws and the 4 upper glove box cover-to-instrument panel screws
  - Glove box and the 2 glove box-to-instrument panel bolts
  - Instrument cluster and the 4 instrument cluster-to-instrument panel screws

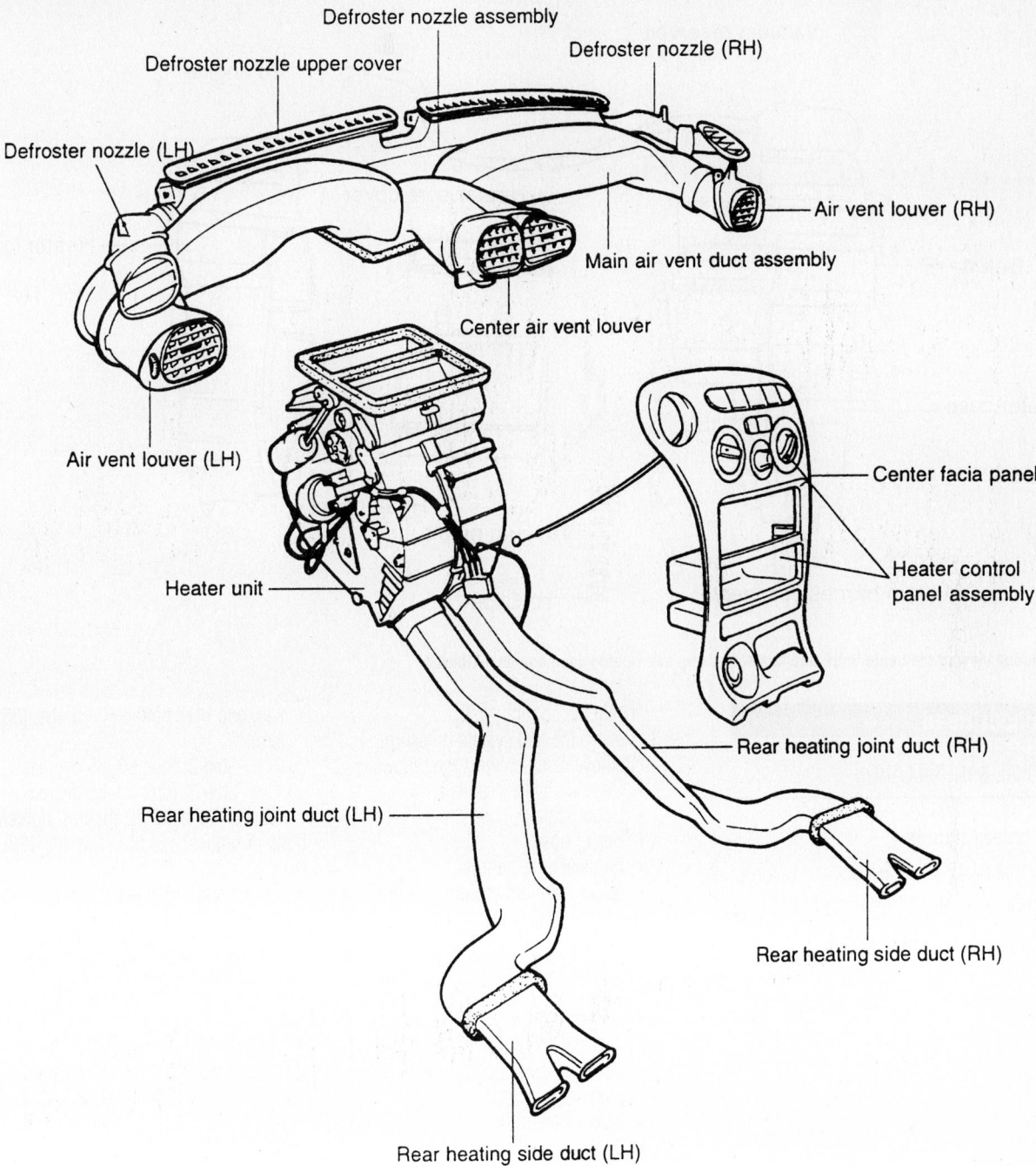

Defroster nozzle assembly

Defroster nozzle upper cover

Defroster nozzle (RH)

Defroster nozzle (LH)

Air vent louver (RH)

Main air vent duct assembly

Center air vent louver

Air vent louver (LH)

Center facia panel

Heater control panel assembly

Heater unit

Rear heating joint duct (RH)

Rear heating joint duct (LH)

Rear heating side duct (RH)

Rear heating side duct (LH)

93112GP1

Exploded view of the heater housing, center fascia, distribution ducts and related components—Tiburon Sedan and Wagon

- Cluster fascia panel; then, connect the heater control cable and the cluster electrical connectors and install the 5 cluster fascia panel-to-instrument panel screws
- Rheostat switch, the hood release handle and DLC to the lower instrument panel
- Radio and the radio-to-bracket bolts
- Lower instrument panel and the 3 lower instrument panel screws
- Cigar lighter connector and install the center fascia panel

- Upper console cover
16. Install the SRS module and the steering wheel by installing or connecting the following:
- Steering wheel to the steering column
- Steering wheel-to-steering column nut and torque to 30–37 ft. lbs. (40–50 Nm)
- SRS module to the steering wheel and connect the electrical connector
- Steering wheel-to-SRS module nuts

17. Install or connect the following:
- Vacuum line to the heater housing vacuum nipple
- Heater hoses to the heater core
18. Refill the cooling system.
19. Connect the negative battery cable.
20. Evacuate, charge and leak test the air conditioning system.
21. Operate the engine to normal operating temperatures; then, check the climate control operation and check for leaks.

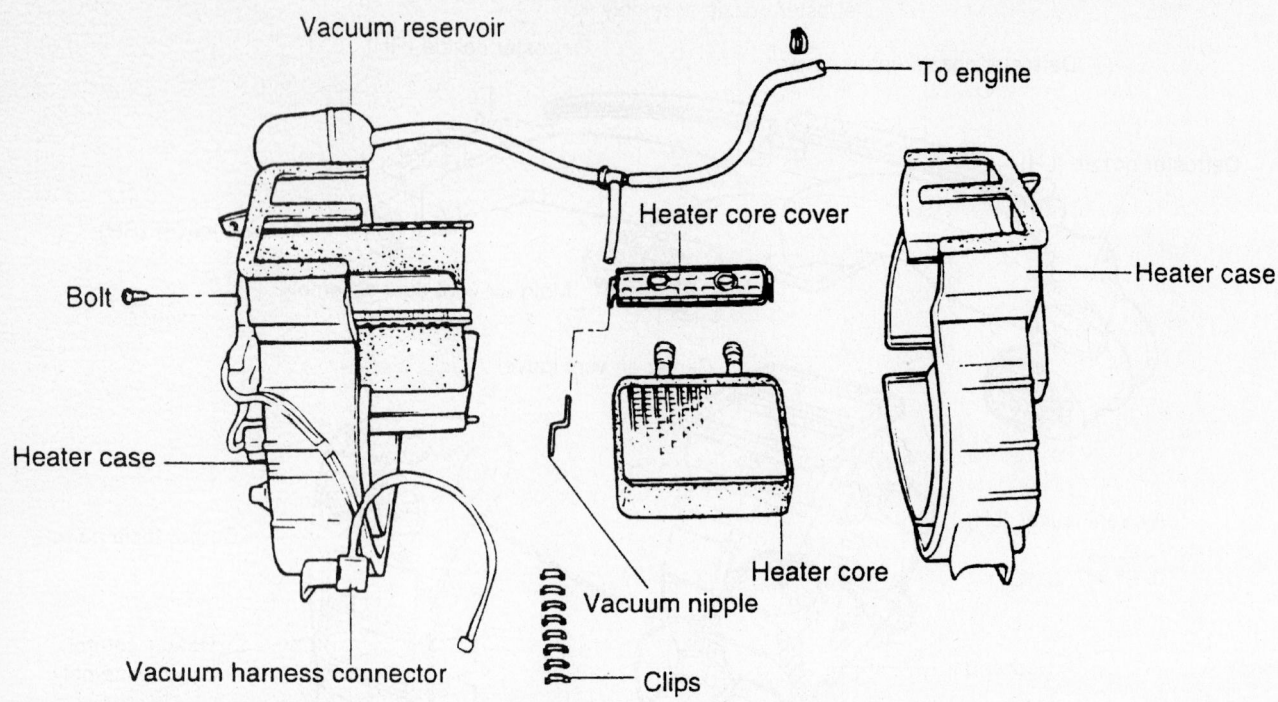

Exploded view of the heater core and heater housing and related components—Tiburon

Cylinder Head

REMOVAL & INSTALLATION

4 Cylinder Engines

1. Before servicing the vehicle, refer to the precautions in the beginning of this section.

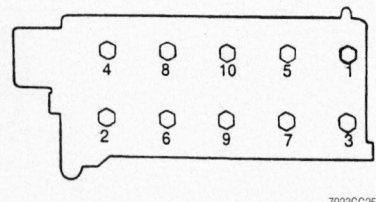

Cylinder head bolt loosening sequence—1.5L, Elantra and Tiburon 2.0L engines

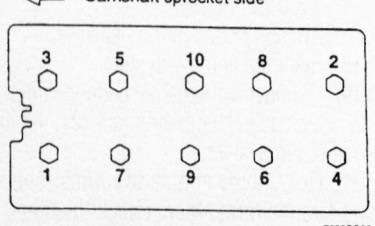

Cylinder head bolt loosening sequence—Sonata 2.0L and 2.4L engines

2. Drain the cooling system.
3. Relieve the fuel system pressure.
4. Remove or disconnect the following:
   - Negative battery cable
   - Upper radiator hose
   - Heater hose
   - Air cleaner assembly
   - Intake manifold vacuum lines
   - Engine control wiring harness connectors
   - Spark plug wires
   - Distributor, if equipped
   - Ignition coil
   - Accessory drive belts
   - Power steering pump and bracket
   - Fuel lines
   - Intake manifold
   - Exhaust manifold
   - Front cover
   - Timing belt
   - Valve cover bolts
   - Cylinder head by loosening the bolts in sequence
   - Cylinder head and discard the gasket

To install:

5. Install the cylinder head with a new gasket.
6. For 1.5L engines, tighten the bolts in sequence to 51–54 ft. lbs. (71–75 Nm).
7. For 2.0L Elantra and Tiburon engines, tighten the bolts in sequence as follows:
   a. Step 1: M10 bolts to 22 ft. lbs. (30

Nm) and M12 bolts to 26 ft. lbs. (35 Nm).
   b. Step 2: Plus 60–65 degrees.
   c. Step 3: Plus 60–65 degrees.
8. For Sonata 2.0L engines, tighten the bolts in sequence to 65–72 ft. lbs. (90–100 Nm).
9. For 2.4L engines, tighten the bolts in sequence as follows:
   a. Step 1: 14 ft. lbs. (20 Nm).
   b. Step 2: Plus 90 degrees.

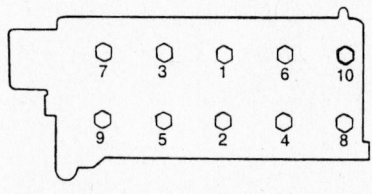

Cylinder head torque sequence—1.5L, and Elantra and Tiburon 2.0L Engines

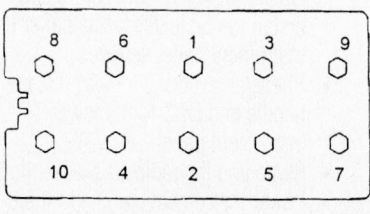

Cylinder head torque sequence—Sonata 2.0L and 2.4L Engines

c. Step 3: Loosen all bolts in reverse of tightening order.
d. Step 4: 14 ft. lbs. (20 Nm).
e. Step 5: Plus 90 degrees.
f. Step 6: Plus 90 degrees.

10. Install or connect the following:
- Valve cover
- Timing belt and front cover
- Exhaust manifold
- Intake manifold
- Fuel lines
- Power steering pump and bracket
- Accessory drive belts
- Ignition coil
- Distributor, if equipped
- Spark plug wires
- Engine control wiring harness connectors
- Intake manifold vacuum lines
- Air cleaner assembly
- Heater hose
- Upper radiator hose
- Negative battery cable

11. Fill the cooling system.
12. Start the engine and check for leaks.

V6 Engines

2.5L AND 2.7L ENGINE

1. Before servicing the vehicle, refer to the precautions in the beginning of this section.
2. Drain the cooling system.
3. Relieve the fuel system pressure.
4. Remove or disconnect the following:
- Negative battery cable
- Accessory drive belts
- Air intake assembly
- A/C compressor
- Alternator
- Power steering pump
- Front covers
- Timing belt
- Engine control wiring harness connectors
- Intake manifold vacuum lines
- Spark plug wires
- Distributor

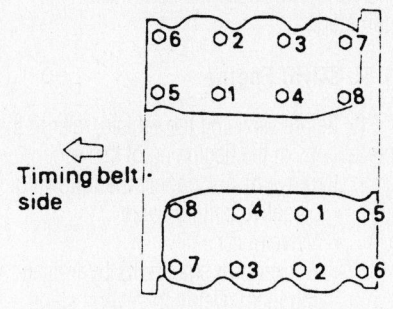

Timing belt side

9307GG02

Cylinder head torque sequence—2.5L and 2.7L Sonata

- Fuel lines
- Intake manifold
- Exhaust manifolds
- Valve cover bolts
- Cylinder heads by loosening the bolts in sequence
- Cylinder head and discard the gasket

To install:
5. Install or connect the following:
- New head gaskets
- Cylinder heads
6. For 2.5L and 2.7L engines, tighten the bolts in sequence as follows:
a. Step 1: 18 ft. lbs. (25 Nm).
b. Step 2: Plus 60–64 degree turn.
c. Step 3: Plus 45–49 degree turn.
7. Install or connect the following:
- Valve covers
- Exhaust manifolds
- Intake manifold
- Fuel lines
- Distributor
- Spark plug wires
- Intake manifold vacuum lines
- Engine control wiring harness connectors
- Timing belt and front covers
- Power steering pump
- Alternator
- A/C compressor
- Air intake assembly

- Accessory drive belts
- Negative battery cable
8. Fill the cooling system.
9. Start the engine and check for leaks.

3.0L AND 3.5L AND 3.5L ENGINE

1. Before servicing the vehicle, refer to the precautions in the beginning of this section.
2. Drain the cooling system.
3. Relieve the fuel system pressure.
4. Remove or disconnect the following:
- Negative battery cable
- Upper radiator hose
- Breather hose
- Air intake hose
- Vacuum hose
- Fuel hoses
- Intake manifold
- Spark plug wires
- Ignition coil
- Upper and lower timing belt covers
- Timing belt
- Camshaft sprockets
- Heat protector and exhaust manifold
- Water pump
- Rocker arm cover
- Camshafts
- Cylinder head and discard the gasket
5. Clean all mating surfaces of any residual gasket material.

To install:
6. Install or connect the following:
- New gasket with the identification mark facing the cylinder head
- Cylinder head and torque the bolts, in sequence, to 82 ft. lbs. (115 Nm)
- Camshafts
- Rocker arm cover and torque the bolts to 89 inch lbs. (10 Nm)
- Water pump and torque the bolts to 16 ft. lbs. (22 Nm)
- Heat protector and exhaust manifold and torque the bolts to 14 ft. lbs. (19 Nm)

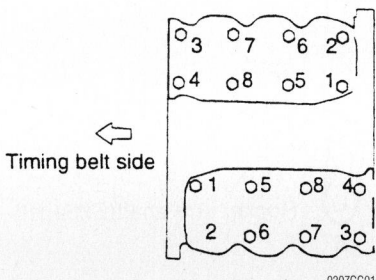

Timing belt side

9307GG01

Cylinder head bolt loosening sequence—2.5L and 2.7L engine

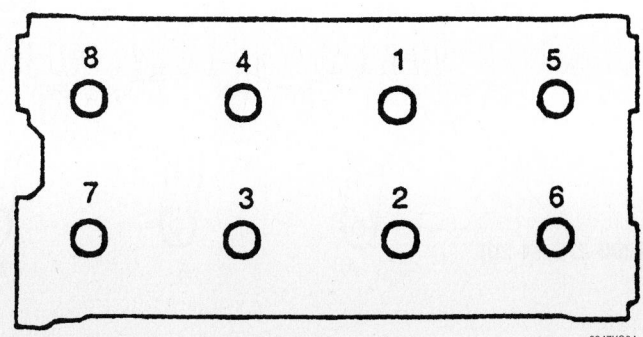

9347KG04

Cylinder head torque sequence—3.0L and 3.5L engines

- Camshaft sprockets
- Timing belt
- Upper and lower timing belt covers
- Ignition coil and spark plug wires
- Intake manifold and torque the bolts to 10 ft. lbs. (15 Nm)
- Fuel hoses
- Vacuum hose
- Air intake hose
- Breather hose
- Upper radiator hose
- Negative battery cable

7. Fill the cooling system to the proper level.

8. Start the vehicle, check for leaks and repair if necessary.

Rocker Arms/Shafts

REMOVAL & INSTALLATION

Except 1.5L SOHC and Sonata 2.0L Engines

These engines are not equipped with rocker arms. The camshaft acts directly on the valves through hydraulic lash adjusters.

1.5L SOHC Engine

1. Before servicing the vehicle, refer to the precautions in the beginning of this section.
2. Remove or disconnect the following:
   - Negative battery cable
   - Valve cover
   - Rocker arm shaft bolts by loosening them evenly in several steps
   - Rocker arm and shaft assemblies

➡Keep all valvetrain components in order for assembly.

   - Rocker arms and springs from the shafts

To install:
3. Install or connect the following:
   - Rocker arms and springs in their original positions
   - Rocker arm and shaft assemblies and torque the bolts evenly to 14–20 ft. lbs. (20–26 Nm)
   - Valve cover
   - Negative battery cable

Sonata 2.0L Engine

1. Before servicing the vehicle, refer to the precautions in the beginning of this section.
2. Remove or disconnect the following:
   - Negative battery cable
   - Valve cover
   - Accessory drive belts
   - Front cover
   - Timing belt
   - Camshaft
   - Rocker arms

➡Keep all valvetrain components in order for assembly.

To install:
3. Install or connect the following:
   - Rocker arms in their original positions
   - Camshaft
   - Timing belt
   - Front cover
   - Accessory drive belts
   - Valve cover
   - Negative battery cable

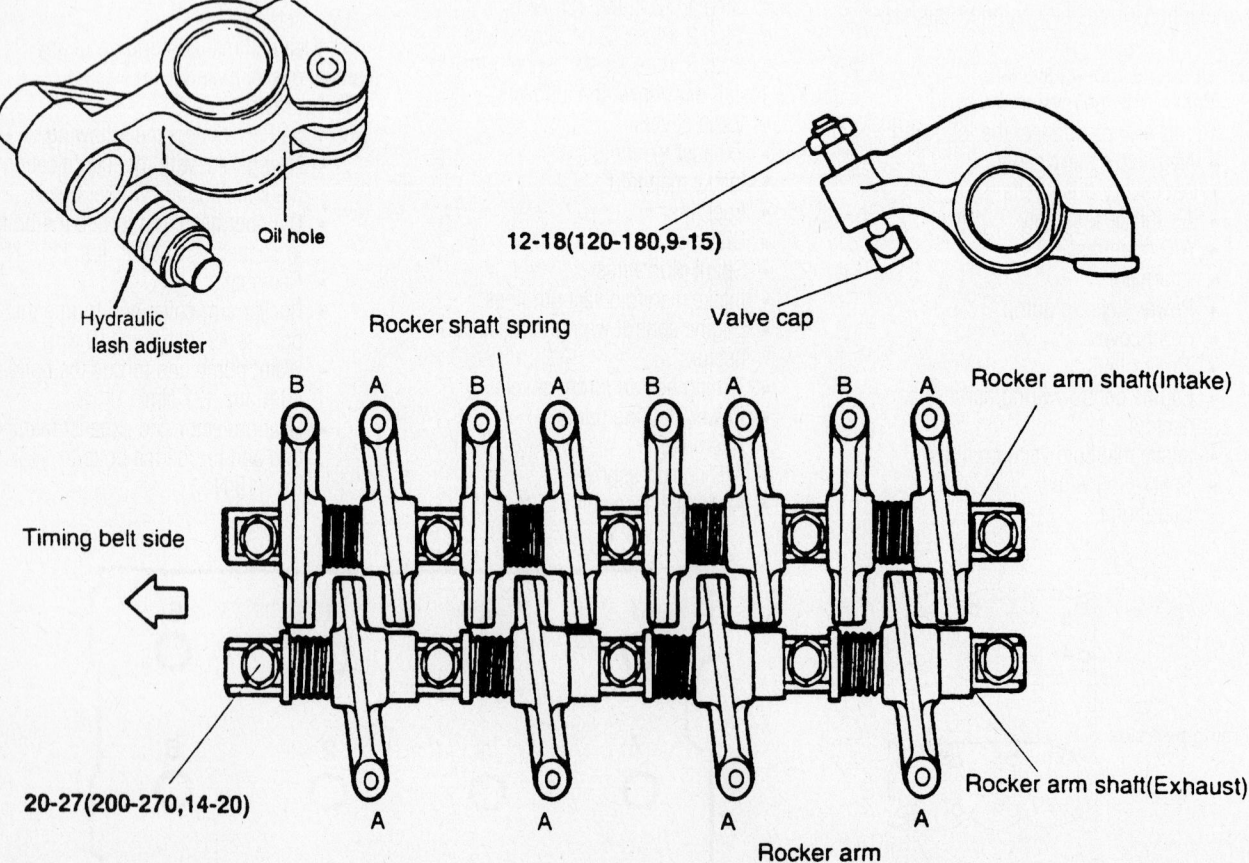

Oil hole

Hydraulic lash adjuster

12-18(120-180,9-15)

Valve cap

Rocker shaft spring

B A B A B A B A

Rocker arm shaft(Intake)

Timing belt side

20-27(200-270,14-20)

A A A A

Rocker arm shaft(Exhaust)

Rocker arm

Rocker assembly components and arrangement. Rockers marked "A" and "B" must be returned to their original positions—1.5L SOHC engines

7923GG28

Intake Manifold

REMOVAL & INSTALLATION

4 Cylinder Engines

1. Before servicing the vehicle, refer to the precautions in the beginning of this section.
2. Relieve the fuel system pressure.
3. Drain the cooling system.
4. Remove or disconnect the following:

- Negative battery cable
- Air intake hose
- Accelerator cable
- Upper radiator hose
- Engine control wiring harness connectors
- Throttle body
- Positive Crankcase Ventilation (PCV) valve and hose
- Brake booster vacuum line
- Intake manifold vacuum hoses
- Fuel lines
- Surge tank
- Fuel injector connectors
- Fuel supply manifold
- Heater hose
- Engine Coolant Temperature (ECT) sensor connector
- Spark plug wires
- Thermostat housing
- Distributor, if equipped
- Ignition coil
- Intake manifold bracket
- Intake manifold and discard the gasket

To install:

5. Install or connect the following:

- Intake manifold using a new gasket and torque the nuts to 11–14 ft. lbs. (15–20 Nm), starting from the center and working outwards
- Intake manifold bracket and torque the bolts to 13–18 ft. lbs. (18–25 Nm)
- Ignition coil
- Distributor, if equipped
- Thermostat housing and torque the bolts to 12–14 ft. lbs. (17–20 Nm)
- Spark plug wires
- ECT sensor connector

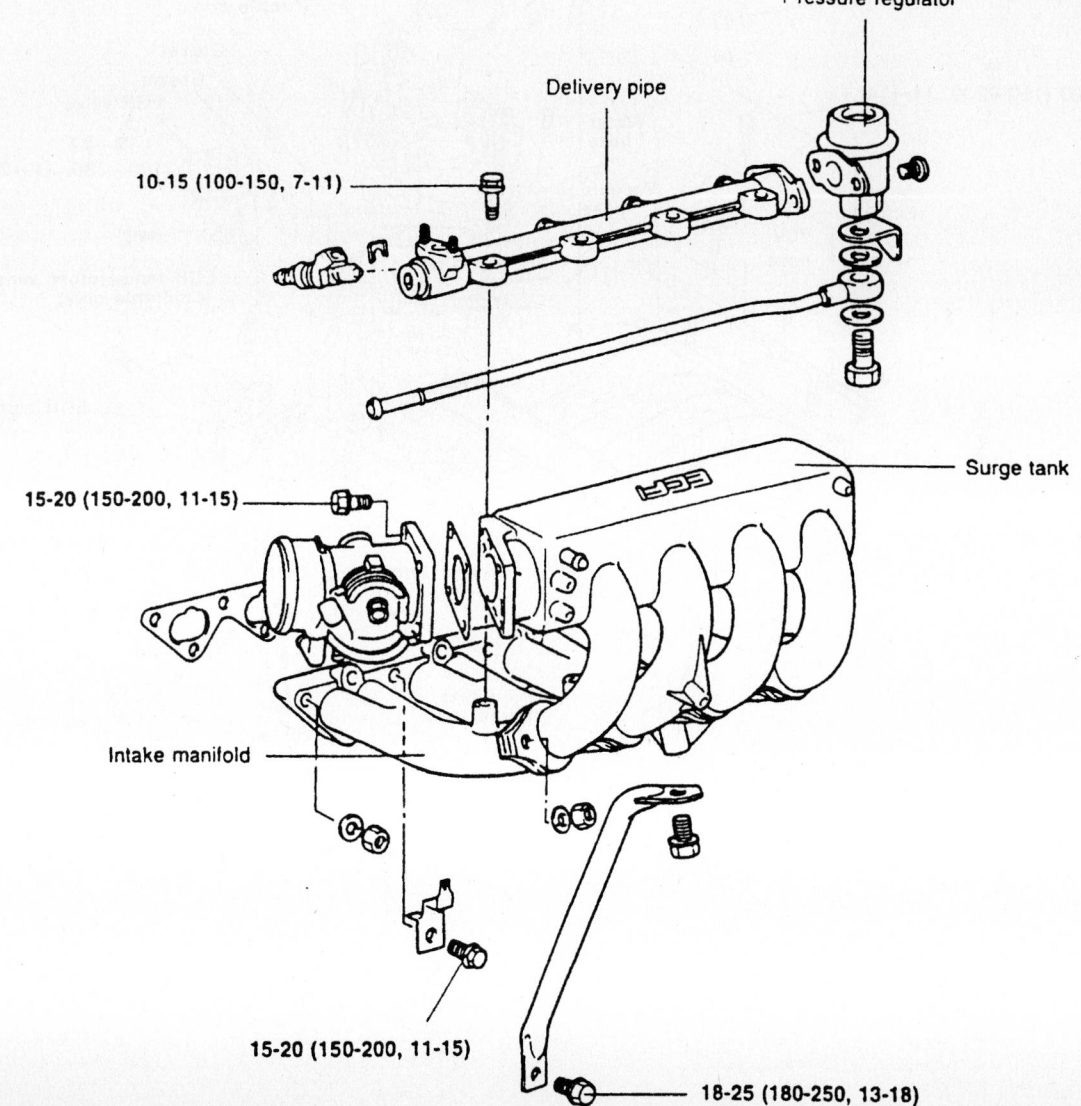

Pressure regulator

Delivery pipe

10-15 (100-150, 7-11)

Surge tank

15-20 (150-200, 11-15)

Intake manifold

15-20 (150-200, 11-15)

18-25 (180-250, 13-18)

TORQUE : Nm (kg.cm, lb.ft)

7923GG34

Surge tank and intake manifold components—1.5L engine

- Heater hose
- Fuel supply manifold
- Fuel injector connectors
- Surge tank using a new gasket and torque the bolts to 11–16 ft. lbs. (15–22 Nm)
- Fuel lines
- Intake manifold vacuum hoses
- Brake booster vacuum line
- PCV valve and hose
- Throttle body
- Engine control wiring harness connectors

- Upper radiator hose
- Accelerator cable
- Air intake hose
- Negative battery cable
6. Fill the cooling system.
7. Start the engine and check for leaks.

V6 Engines

1. Before servicing the vehicle, refer to the precautions in the beginning of this section.
2. Relieve the fuel system pressure.
3. Drain the cooling system.
4. Remove or disconnect the following:

- Negative battery cable
- Air intake hose
- Accelerator cable
- Upper radiator hose
- Engine control wiring harness
- Throttle body
- Positive Crankcase Ventilation (PCV) valve and hose
- Intake manifold vacuum hoses
- Exhaust Gas Recirculation (EGR) pipe
- Surge tank
- Fuel lines

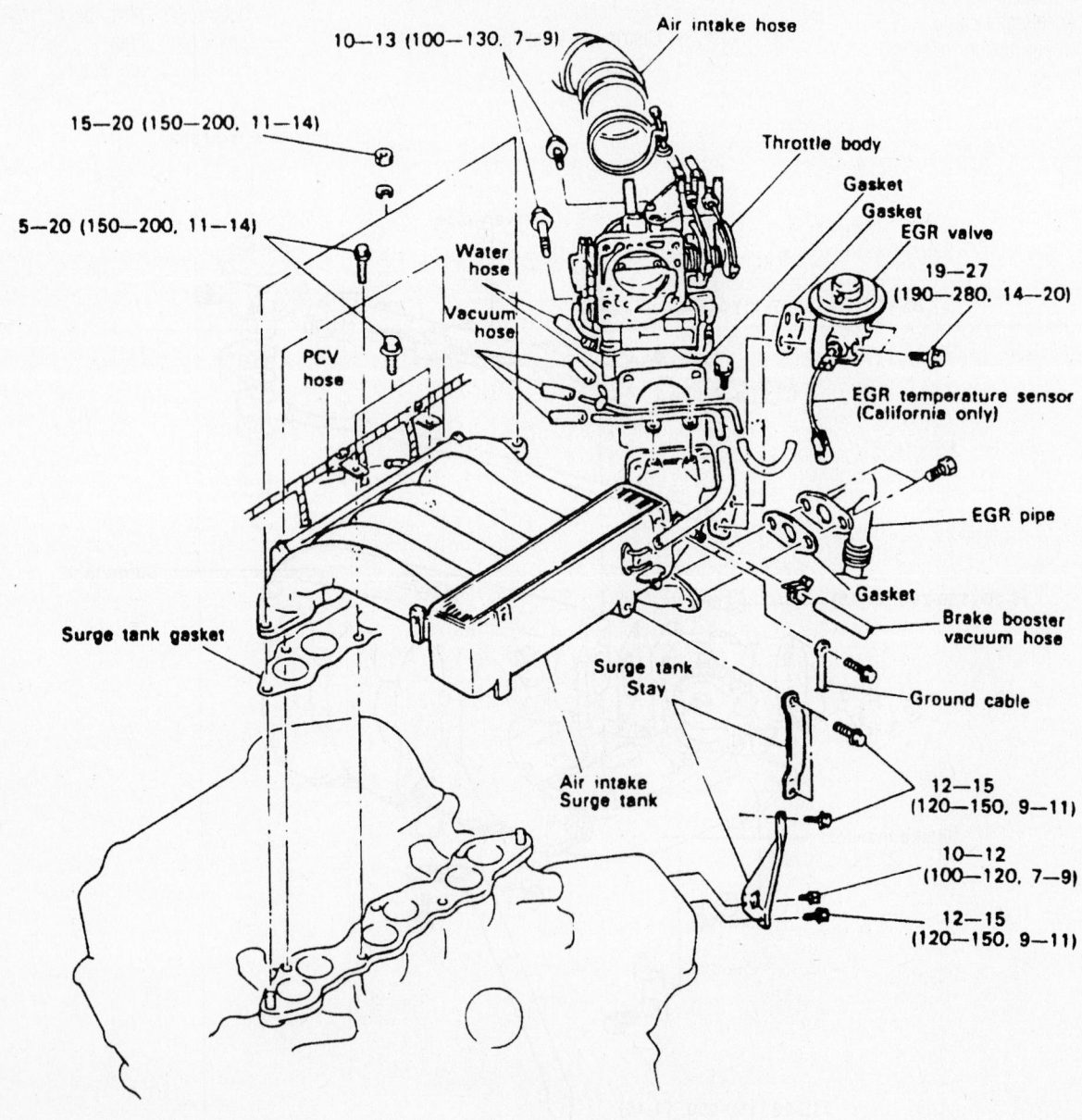

TORQUE : Nm (kg.cm, lb.ft)

Surge tank and intake manifold components—Sonata 2.0L and 2.4L engines

7923GG38

- Fuel injector connectors
- Fuel supply manifold
- Thermostat housing
- Intake manifold and discard the gasket

To install:

5. Install or connect the following:
   - Intake manifold using a new gasket and torque the nuts to 11–14 ft. lbs. (15–20 Nm) for 3.0L and 3.5L engines or to 14–15 ft. lbs. (15–20 Nm) for 2.5L and 2.7L engines starting from the center and working outwards
   - Thermostat housing and torque the bolts to 12–14 ft. lbs. (17–20 Nm)
   - Fuel supply manifold and torque the bolts to 84–108 inch lbs. (10–13 Nm)
   - Fuel injector connectors
   - Fuel lines
   - Surge tank and torque the bolts to 11–14 ft. lbs. (15–20 Nm)
   - EGR pipe
   - Intake manifold vacuum hoses
   - PCV valve and hose

➡ **One throttle body bolt is shorter than the rest. This bolt is installed in the upper left hole when viewed from the front of the throttle body.**

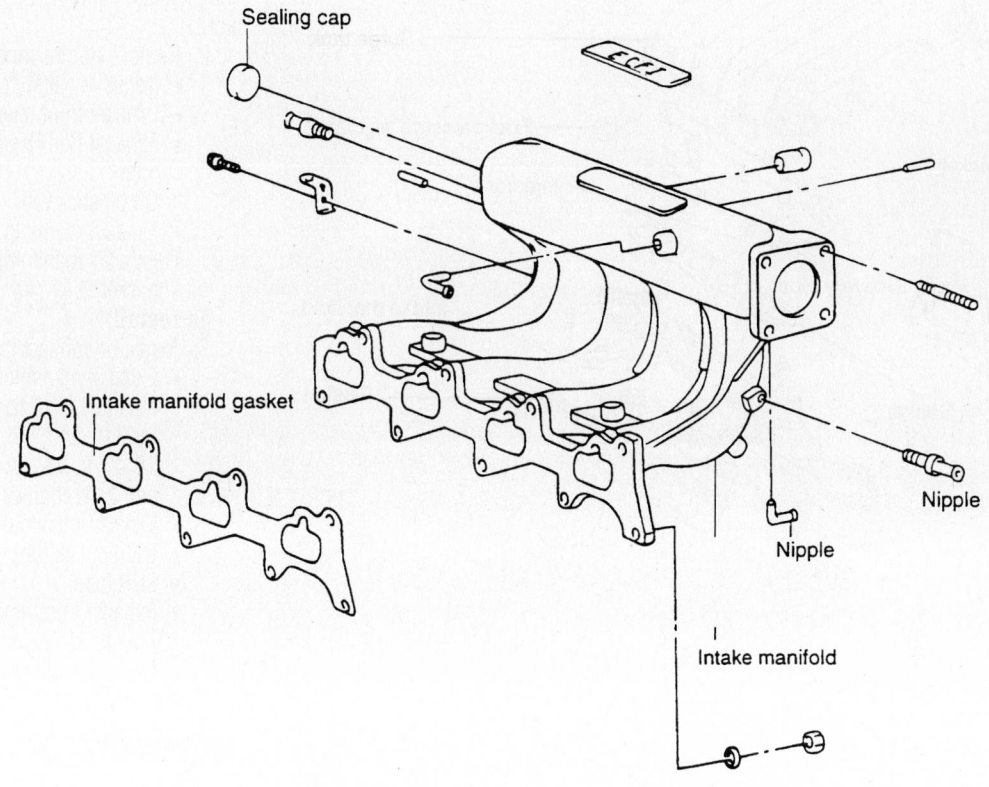

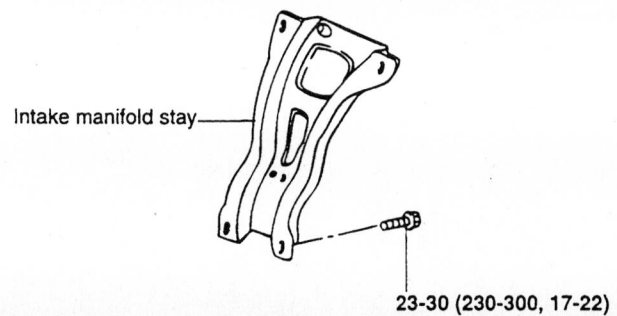

23-30 (230-300, 17-22)

TORQUE : Nm (kg.cm, lb.ft)

7923GG39

Surge tank and intake manifold components—Elantra and Tiburon 2.0L engine

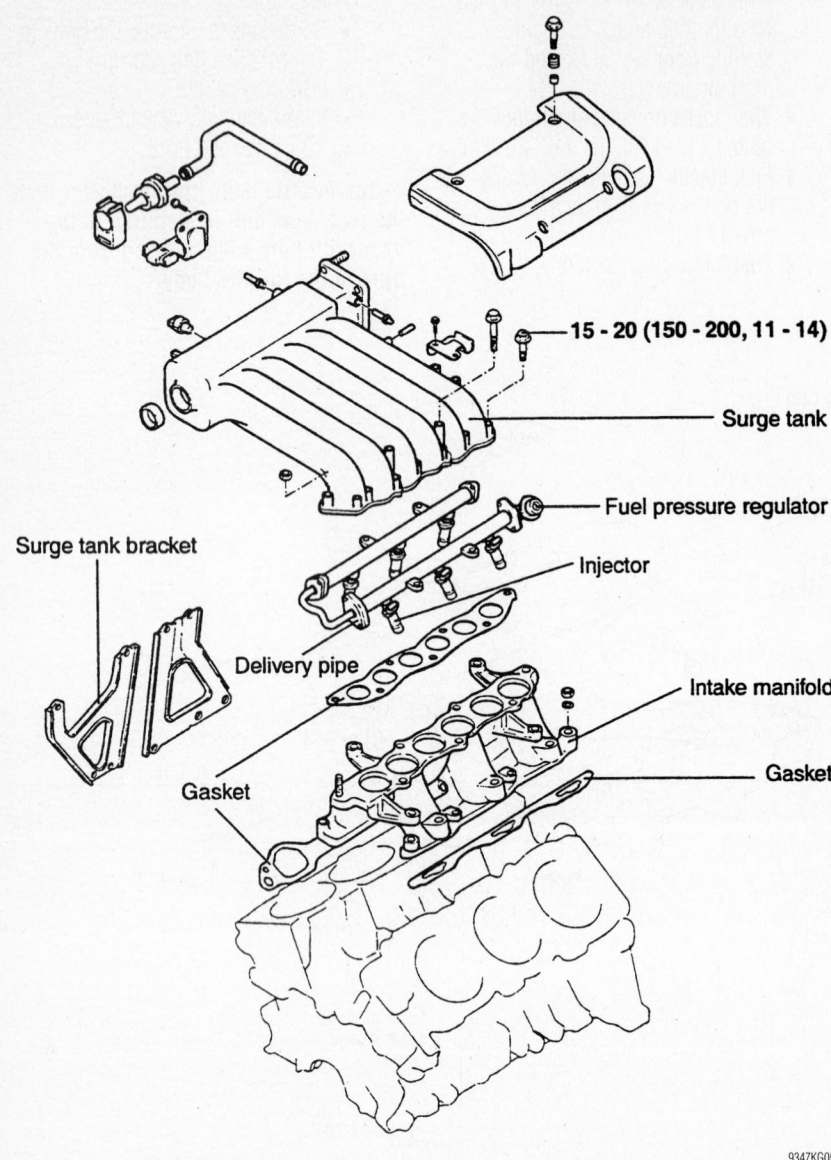

15 - 20 (150 - 200, 11 - 14)

Surge tank

Fuel pressure regulator

Injector

Surge tank bracket

Delivery pipe

Intake manifold

Gasket

Gasket

9347KG05

Surge tank and intake manifold components—XG 300 and XG 350

- Throttle body and torque the bolts to 11–16 ft. lbs. (15–22 Nm)
- Engine control wiring harness
- Upper radiator hose
- Accelerator cable
- Air intake hose
- Negative battery cable
6. Fill the cooling system.
7. Start the engine and check for leaks.

Exhaust Manifold

REMOVAL & INSTALLATION

4 Cylinder Engines

1. Before servicing the vehicle, refer to the precautions in the beginning of this section.

2. Remove or disconnect the following:

- Negative battery cable
- Heated Oxygen (HO2S) sensor
- Exhaust manifold heat shield
- Exhaust front pipe
- Exhaust manifold and discard the gasket

To install:

3. Install the exhaust manifold with a new gasket and torque the nuts to the following specifications:

a. 1.5L engine: 11–15 ft. lbs. (15–20 Nm).

b. Sonata 2.0L engines: 18–22 ft. lbs. (25–30 Nm).

c. Elantra and Tiburon 2.0L engine: 32–41 ft. lbs. (43–50 Nm).

d. 2.4L engine: M8 fasteners to

18–22 ft. lbs. (25–30 Nm) and M10 fasteners to 25–40 ft. lbs. (34–55 Nm).

4. Install or connect the following:

- Exhaust front pipe
- Exhaust manifold heat shield
- HO2S sensor
- Negative battery cable

5. Start the engine and check for leaks.

6 Cylinder Engines

SONATA

1. Before servicing the vehicle, refer to the precautions in the beginning of this section.

2. Remove or disconnect the following:

- Negative battery cable
- Exhaust front pipe
- Exhaust Gas Recirculation (EGR) tube
- Oil dipstick tube
- Exhaust manifold heat shields
- Exhaust manifolds and discard the gasket

To install:

3. Install or connect the following:

- Exhaust manifolds and torque the nuts to 11–16 ft. lbs. (15–22 Nm) for 3.0L and 3.5L engines or to 18–22 ft. lbs. (25–30 Nm) for 2.5L and 2.7L engines
- Exhaust manifold heat shields
- Oil dipstick tube
- EGR tube
- Exhaust front pipe and torque the nuts to 22–29 ft. lbs. (30–40 Nm)
- Negative battery cable

4. Start the engine and check for leaks.

XG 300 AND XG 350

1. Before servicing the vehicle, refer to the precautions in the beginning of this section.

2. Remove or disconnect the following:

- Negative battery cable
- Heat protector
- Heated Oxygen (HO2S) sensor
- Exhaust manifold and discard the gaskets

3. Clean and residual gasket material from all mating surfaces.

To install:

4. Install or connect the following:

- Exhaust manifold with a new gasket and torque the bolts to 22 ft. lbs. (30 Nm)
- Heat protector and torque the bolts to 11 ft. lbs. (15 Nm)
- HO2S sensor
- Negative battery cable

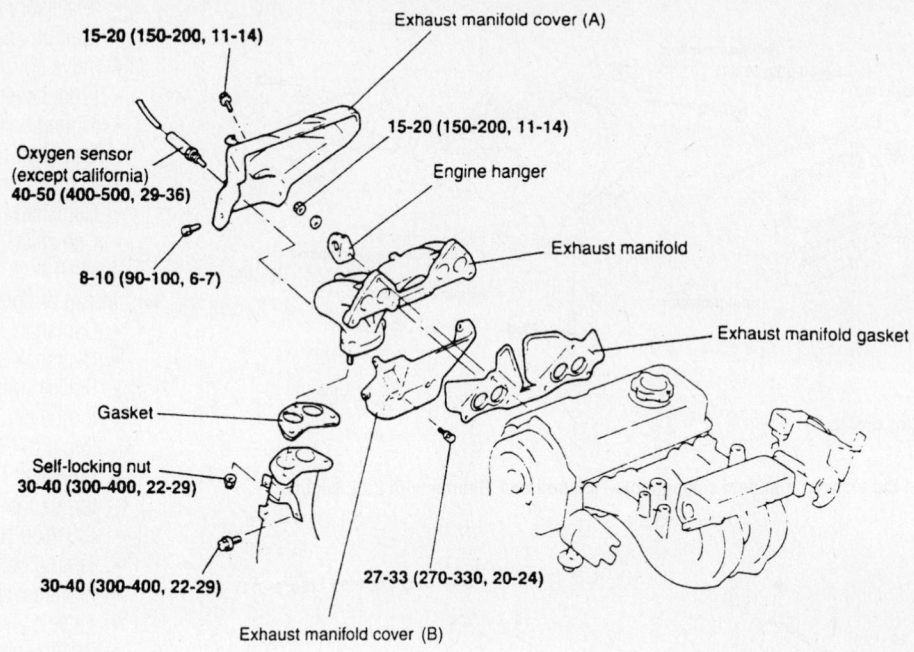

TORQUE : Nm (kg.cm, lb.ft)

7923GG41

Exploded view of the exhaust manifold and related components—1.5L engines

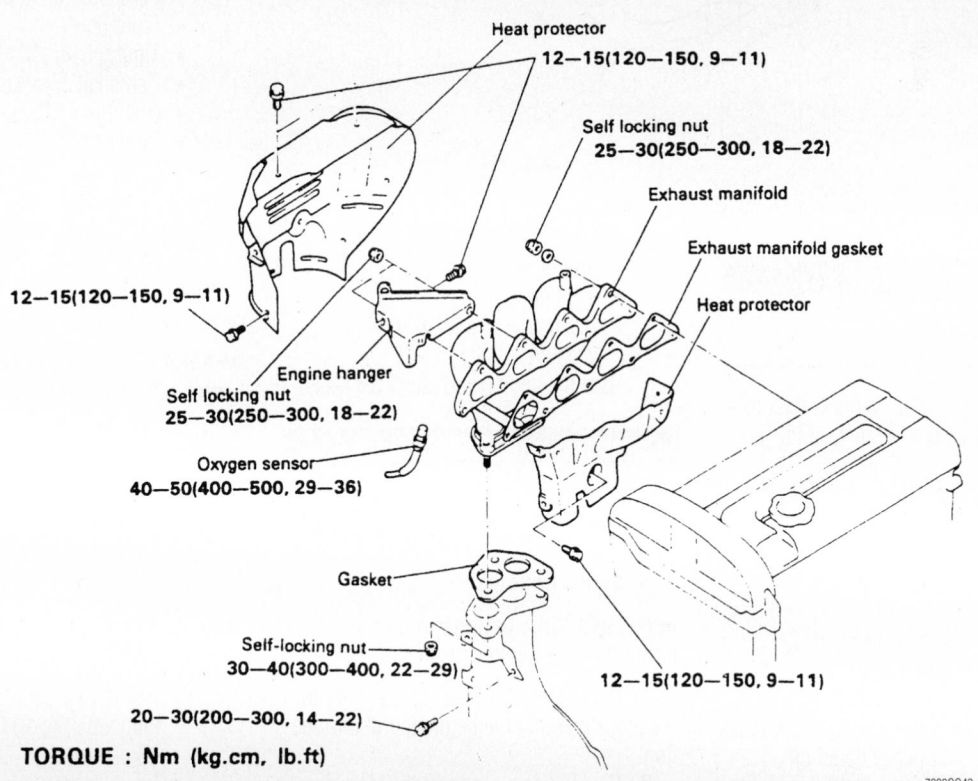

TORQUE : Nm (kg.cm, lb.ft)

7923GG42

Exploded view of the exhaust manifold components—Sonata 2.0L engines

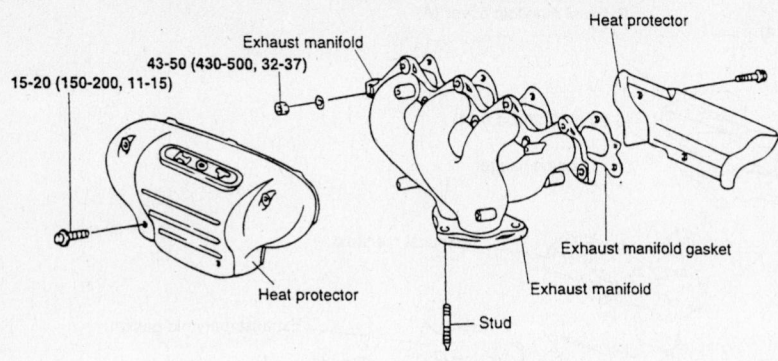

TORQUE: Nm (kg.cm, lb.ft)

7923GG43

Exploded view of the exhaust manifold components—Elantra and Tiburon with 2.0L engine

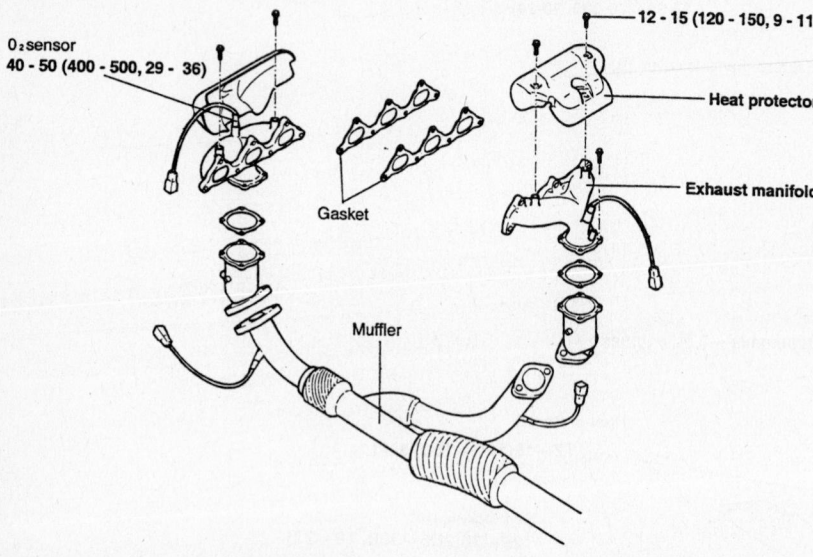

9347KG07

Exploded view of the exhaust manifold—XG 300 and XG 350

Front Crankshaft Seal

REMOVAL & INSTALLATION

1. Before servicing the vehicle, refer to the precautions in the beginning of this section.
2. Remove or disconnect the following:
   - Negative battery cable
   - Accessory drive belts
   - Front cover
   - Timing belt
   - Crankshaft timing sprocket
   - Front crankshaft seal

To install:

3. Install the front crankshaft seal so that it is flush with the oil pump housing.
4. Install or connect the following:
   - Crankshaft timing sprocket
   - Timing belt
   - Front cover
   - Accessory drive belts
   - Negative battery cable
5. Start the engine and check for leaks.

Camshaft and Valve Lifters

REMOVAL & INSTALLATION

1.5L SOHC Engines

➡**The hydraulic lash adjusters are housed in the rocker arms.**

1. Before servicing the vehicle, refer to the precautions in the beginning of this section.
2. Remove or disconnect the following:
   - Negative battery cable

- Accessory drive belts
- Ignition coil
- Valve cover
- Front cover
- Timing belt
- Camshaft timing belt sprocket
- Rocker arm and shaft assembly
- Camshaft bearing caps
- Camshaft

To install:

3. Install or connect the following:
   - Camshaft
   - Camshaft bearing caps
   - Rocker arm and shaft assembly and torque the bolts evenly to 14–20 ft. lbs. (20–26 Nm)
   - Camshaft timing belt sprocket and torque the bolt to 58–72 ft. lbs. (80–100 Nm)
   - Timing belt
   - Front cover
   - Valve cover
   - Ignition coil
   - Accessory drive belts
   - Negative battery cable

1.5L DOHC, 2.0L and 2.4L Engines

1. Before servicing the vehicle, refer to the precautions in the beginning of this section.
2. Remove or disconnect the following:
   - Negative battery cable
   - Accessory drive belts
   - Front cover
   - Timing belt
   - Camshaft sprocket
   - Distributor, if equipped
   - Camshaft position sensor, if equipped
   - Valve cover
   - Camshaft bearing caps and timing chain
   - Intake and exhaust camshafts
   - Hydraulic lash adjusters

➡**Keep all valvetrain components in order for assembly.**

To install:

3. Install or connect the following:
   - Hydraulic lash adjusters in their original positions
   - Intake and exhaust camshafts with the secondary chain aligned as shown
   - Camshaft bearing caps and timing chain and torque the bolts to 15 ft. lbs. (21 Nm) for 2.0L (VIN P) engines or to 10 ft. lbs. (14 Nm) for all others
   - Valve cover
   - Distributor, if equipped

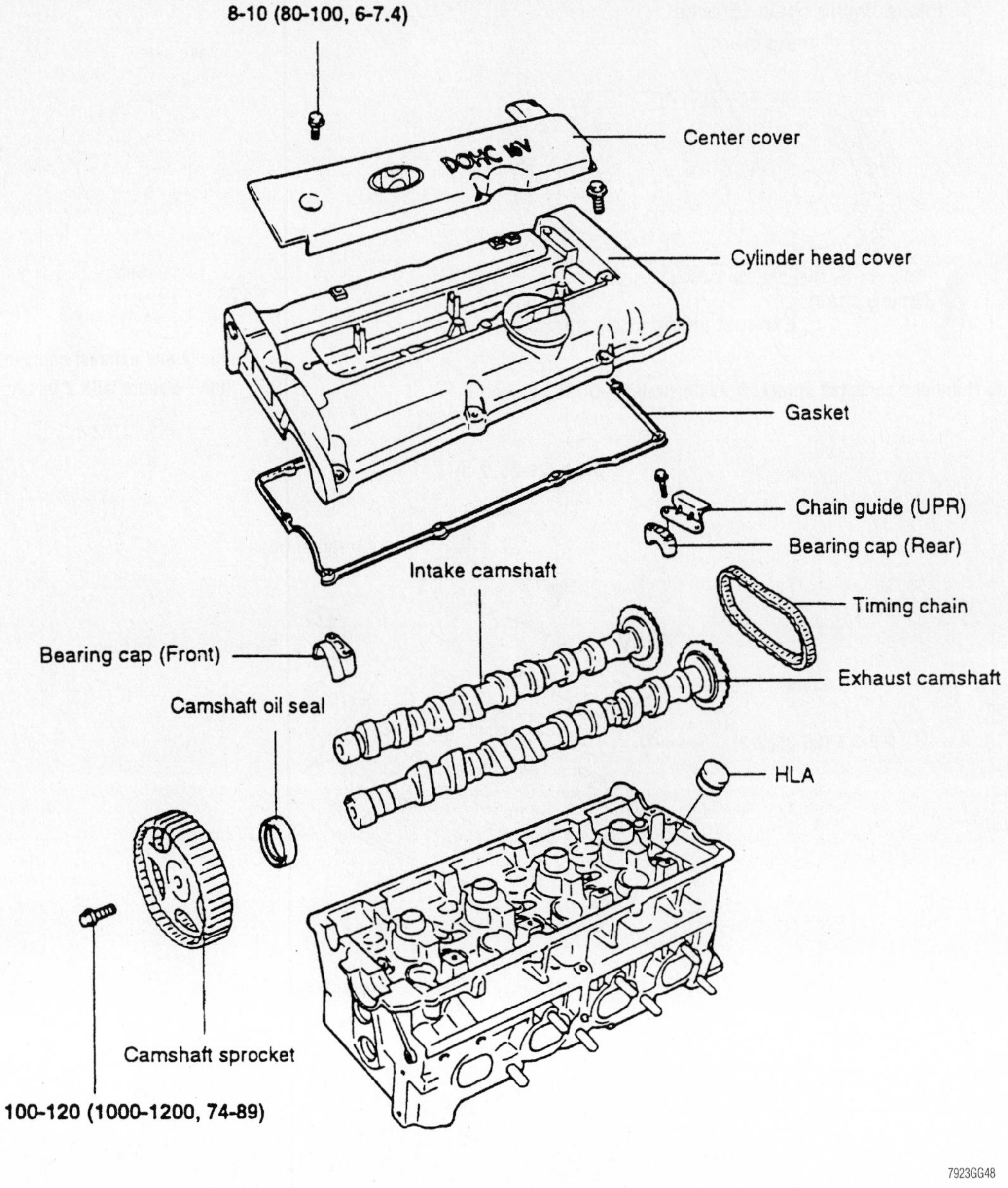

8-10 (80-100, 6-7.4)

Center cover

Cylinder head cover

Gasket

Chain guide (UPR)

Bearing cap (Rear)

Timing chain

Exhaust camshaft

Intake camshaft

Bearing cap (Front)

Camshaft oil seal

HLA

Camshaft sprocket

100-120 (1000-1200, 74-89)

7923GG48

Camshaft assembly components—1.5L DOHC, and Elantra and Tiburon 2.0L engines

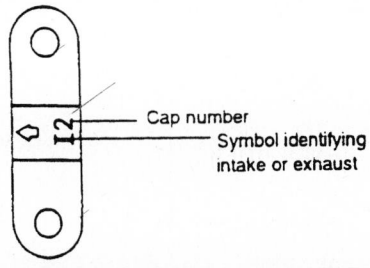

Cap number

Symbol identifying
intake or exhaust

7923GG49

The camshaft bearing caps are identified
with a letter and number stamp. The letter
indicates either intake or exhaust and the
number is sequential from the cylinder
head end opposite the timing chain—1.5L
DOHC, and Elantra and Tiburon 2.0L
engines

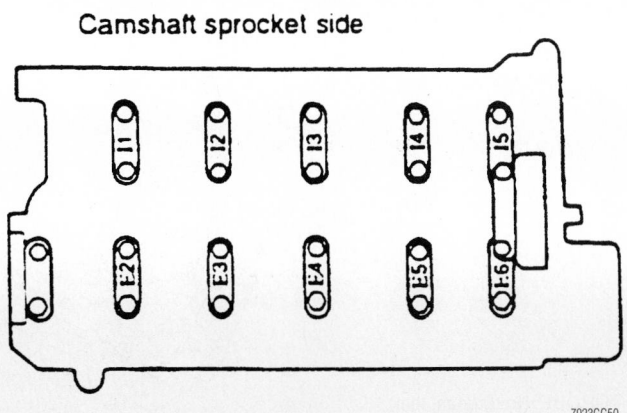

Camshaft sprocket side

I1 I2 I3 I4 I5

E2 E3 E4 E5 E6

7923GG50

The camshaft bearing caps are arranged on the cylinder head as illustrated—1.5L DOHC, and
Elantra and Tiburon 2.0L engines

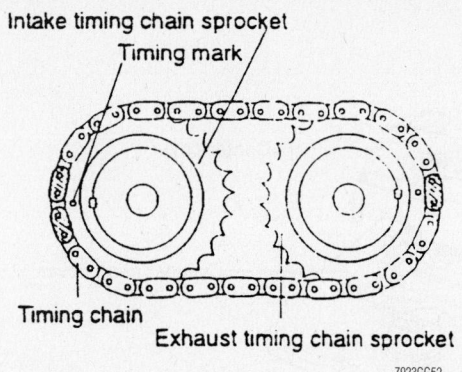

Align the timing chain and camshaft sprockets as illustrated—DOHC engines

Intake timing chain sprocket
Timing mark
Timing chain
Exhaust timing chain sprocket

7923GG52

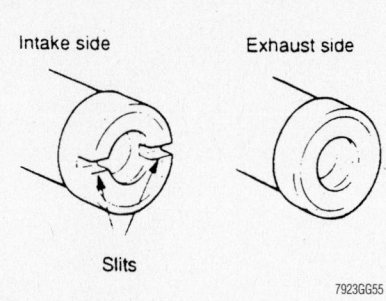

Intake and exhaust camshaft identification—Sonata with 2.0L engine

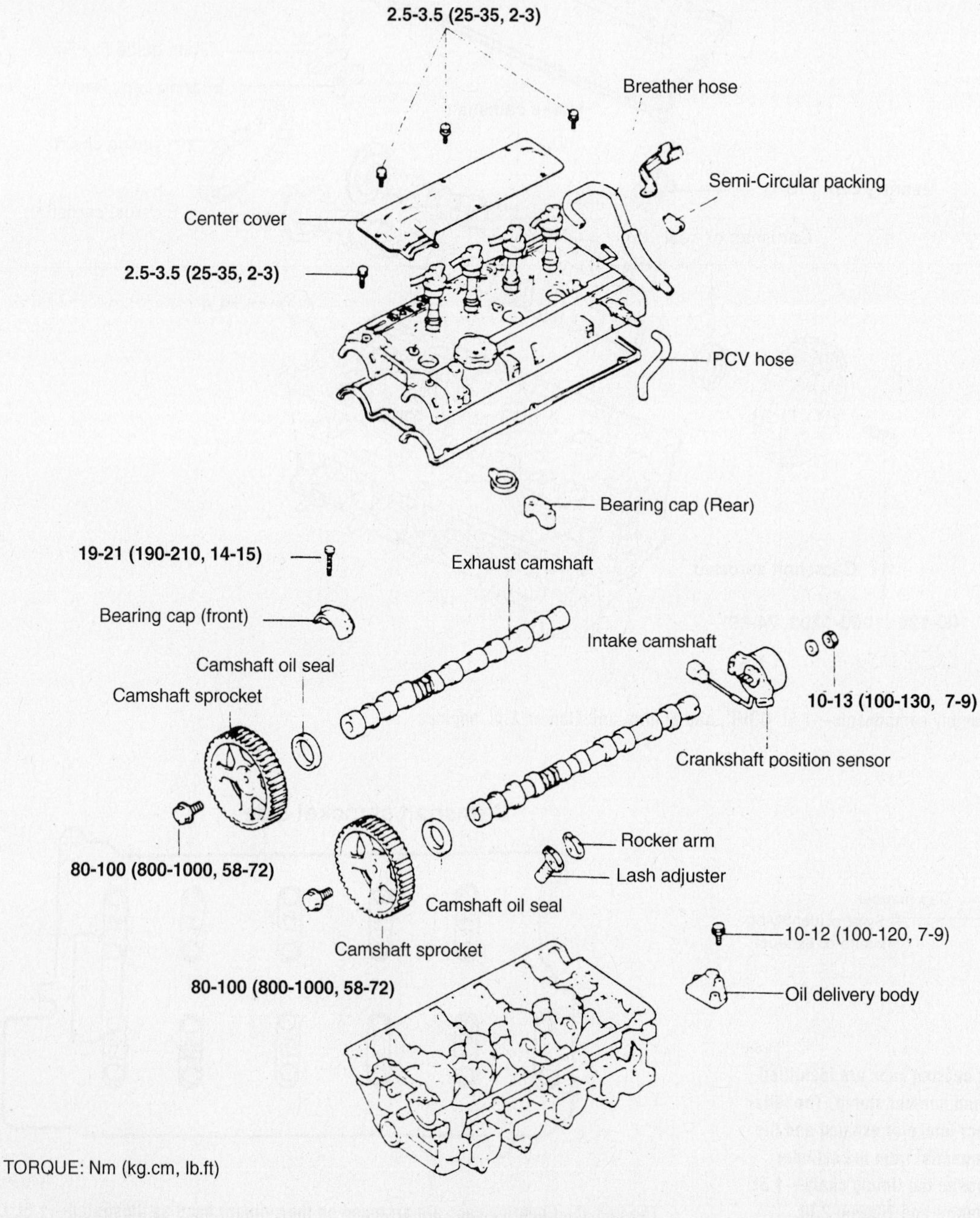

Intake side
Exhaust side
Slits

7923GG55

2.5-3.5 (25-35, 2-3)

Breather hose

Semi-Circular packing

Center cover

2.5-3.5 (25-35, 2-3)

PCV hose

Bearing cap (Rear)

19-21 (190-210, 14-15)

Exhaust camshaft

Bearing cap (front)

Intake camshaft

Camshaft oil seal

Camshaft sprocket

10-13 (100-130, 7-9)

Crankshaft position sensor

80-100 (800-1000, 58-72)

Rocker arm
Lash adjuster

Camshaft oil seal

Camshaft sprocket

10-12 (100-120, 7-9)

Oil delivery body

80-100 (800-1000, 58-72)

TORQUE: Nm (kg.cm, lb.ft)

7923GG54

Exploded view of the camshaft and rocker arm assembly components—Sonata with 2.0L engine

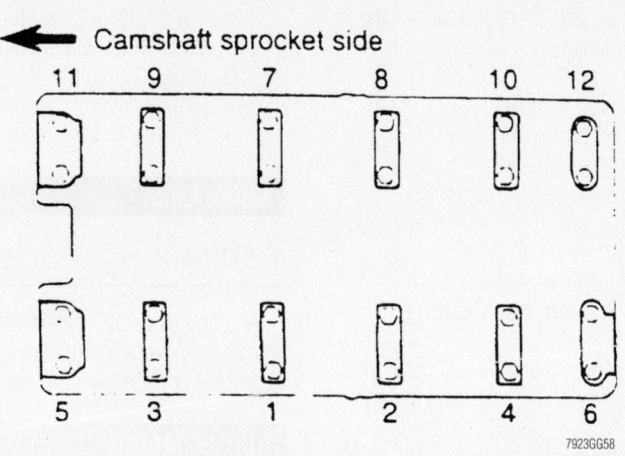

← Camshaft sprocket side

11 9 7 8 10 12

5 3 1 2 4 6

7923GG58

Bearing cap torque sequence—Sonata with 2.0L engine

- Camshaft position sensor, if equipped
- Camshaft sprocket and torque the bolt to 60–74 ft. lbs. (80–100 Nm)
- Timing belt
- Front cover
- Accessory drive belts
- Negative battery cable

2.5L and 2.7L Engine

1. Before servicing the vehicle, refer to the precautions in the beginning of this section.
2. Remove or disconnect the following:
   - Negative battery cable
   - Accessory drive belts

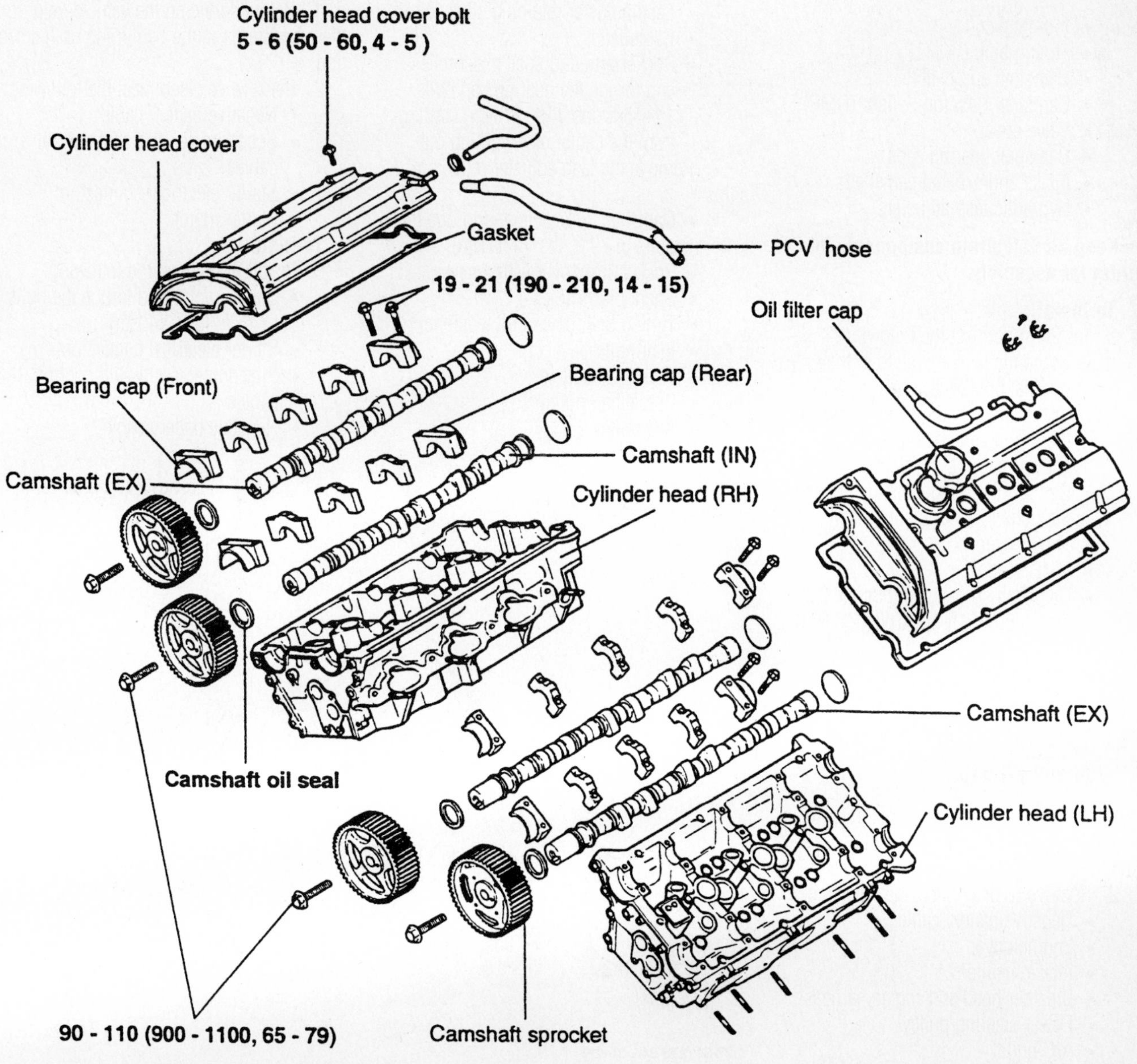

Cylinder head cover bolt
5 - 6 (50 - 60, 4 - 5)

Cylinder head cover

Gasket

PCV hose

Oil filter cap

19 - 21 (190 - 210, 14 - 15)

Bearing cap (Front)

Bearing cap (Rear)

Camshaft (EX)

Camshaft (IN)

Cylinder head (RH)

Camshaft oil seal

Camshaft (EX)

Cylinder head (LH)

90 - 110 (900 - 1100, 65 - 79)

Camshaft sprocket

9347KG08

Exploded view of the camshaft and related components—XG 300 and XG 350

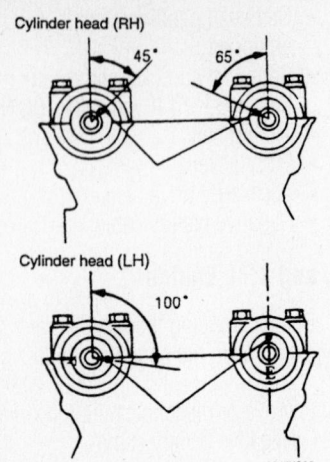

Cylinder head (RH)

45° 65°

Cylinder head (LH)

100°

9347KG09

Install the camshaft dowel pin as shown—XG 300 and XG 350

- Front cover
- Timing belt
- Camshaft sprocket
- Camshaft Position Sensor (CMP)
- Valve cover
- Camshaft bearing caps
- Intake and exhaust camshafts
- Hydraulic lash adjusters

➡**Keep all valvetrain components in order for assembly.**

To install:

3. Install or connect the following:
- Hydraulic lash adjusters in their original positions
- Intake and exhaust camshafts with the secondary chain aligned as shown
- Camshaft bearing caps and torque the bolts to 10 ft. lbs. (14 Nm)
- Valve cover
- CMP sensor
- Camshaft sprocket and torque the bolt to 60–74 ft. lbs. (80–100 Nm)
- Timing belt
- Front cover
- Accessory drive belts
- Negative battery cable

XG 300 and XG 350

1. Before servicing the vehicle, refer to the precautions in the beginning of this section.
2. Remove or disconnect the following:
- Negative battery cable
- Engine cover
- Intake manifold
- Breather hose and engine harness
- Power steering pulley
- A/C pulley
- Crankshaft pulley
- Idler pulley and tensioner pulley

- Timing belt cover and loosen the auto tensioner
- Timing belt
- Spark plug cables
- Rocker arm cover
- Camshaft sprockets
- Camshaft bearing caps
- Camshafts

To install:

3. Rotate the crankshaft so that the No. 1 cylinder is at the Top Dead Center (TDC) position.
4. Make certain that the rocker arm is installed properly on the lash adjuster and valve.
5. Install the camshaft dowel pin.
6. Install or connect the following:
- Hydraulic lash adjusters in their original locations
- Lash adjuster retaining clips
- Camshafts
- Rocker arm and shaft assemblies and torque the bearing cap bolts 14–15 ft. lbs. (19–21 Nm) starting from the center and working out
7. Remove the lash adjuster retaining clips.
- Camshaft sprockets and torque the bolts to 79 ft. lbs. (110 Nm)
- Rocker arm cover with a new gasket
- Spark plug cables
- Timing belt, cover and tensioner
- Idler pulley
- Tensioner pulley
- Crankshaft pulley
- A/C pulley

- Power steering pump pulley
- Breather hose and engine harness
- Intake manifold
- Engine cover
- Negative battery cable

Valve Lash

ADJUSTMENT

All engines use hydraulic valve lash adjusters. Valve lash adjustments are not necessary or possible on these engines.

Starter Motor

REMOVAL & INSTALLATION

1. Before servicing the vehicle, refer to the precautions in the beginning of this section.
2. Remove or disconnect the following:
- Negative battery cable
- Speedometer and shift control cables
- Starter electrical connectors
- Starter motor

To install:

3. Install or connect the following:
- Starter motor and torque the bolts to 25 ft. lbs. (34 Nm)
- Starter electrical connectors
- Speedometer and shift control cables
- Negative battery cable

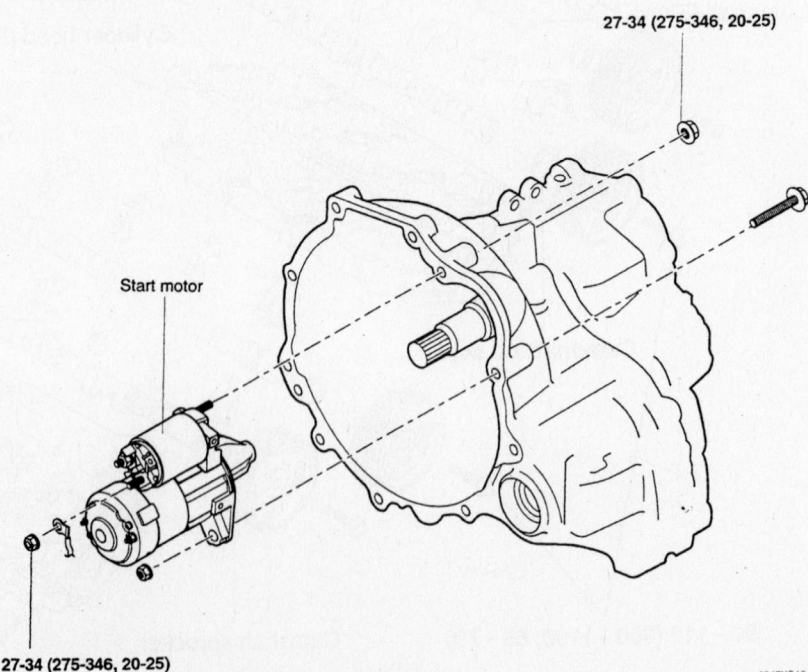

27-34 (275-346, 20-25)

Start motor

27-34 (275-346, 20-25)

Exploded view of the starter—XG 300 and XG 350

9347KG10

bolt slightly with the tensioner moved as far as possible away from the water pump.

11. Install the free end of the spring into the locating tang on the front case.

12. Position the timing belt over the crankshaft sprocket, then over the camshaft sprocket. Slip the back of the belt over the tensioner wheel.

13. Turn the camshaft sprocket in the opposite of its normal direction of rotation until the straight side of the belt is tight and be sure the timing marks align.

➡️**If the timing marks are not properly aligned, shift the belt 1 tooth at a time in the appropriate direction until they are aligned.**

14. Loosen the tensioner mounting bolts so the tensioner works, without the interference of any friction, under spring pressure. Be sure the belt follows the curve of the camshaft pulley so the teeth are engaged all the way around. Correct the path of the belt, if necessary.

15. Tighten the tensioner adjusting bolt, then the tensioner pivot bolt to 15–18 ft. lbs. (20–26 Nm).

➡️**Bolts must be tightened in the stated order or tension won't be correct.**

16. Turn the crankshaft 1 turn clockwise until timing marks again align to seat the belt.

17. Loosen both tensioner attaching bolts and let the tensioner position itself under spring tension. Retighten the bolts.

18. Check belt tension by putting a finger on the water pump side of the tensioner wheel and pull the belt toward the water pump. The belt should move toward the pump until the teeth are approximately ½ of the way across the head of the tensioner adjusting bolt. Retension the belt, if necessary.

19. Install or connect the following:
- Timing belt cover
- Crankshaft pulley
- A/C compressor belt
- Water pump pulley
- V belt
- Negative battery cable
- Engine with coolant

2.0L Engines

1. Before servicing the vehicle, refer to the precautions in the beginning of this section.

2. Remove or disconnect the following:
- Negative battery cable
- Engine coolant
- Water pump pulley bolts
- Alternator bolt, loosen only
- Water pump pulley and drive belts

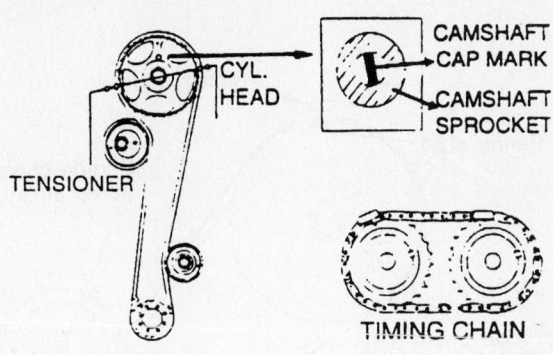

Proper alignment of the timing belt alignment marks for belt removal and installation— Hyundai 2.0L engines

- Crankshaft pulley
- Timing belt cover(s)

3. Rotate the crankshaft clockwise and align the timing marks so No. 1 piston will be at Top Dead Center (TDC) of the compression stroke.

4. Remove the timing belt tensioner and idler pulley.

5. Mark the timing belt with an arrow showing direction of rotation.

6. Remove the timing belt.

To install:

7. Align the timing marks of the camshaft sprocket and check that the crankshaft timing marks are still in alignment.

8. Install the timing belt tensioner.

9. Install the idler pulley, if equipped. Tighten bolt to 32–41 ft. lbs. (43–55 Nm).

10. Position the timing belt over the camshaft sprocket, then over the crankshaft sprocket.

11. Tension the timing belt and tighten the tensioner pulley bolt to 32–41 ft. lbs. (43–55 Nm). When properly tensioned, the timing belt should deflect 0.16–0.24 in. (4–6mm) when a force of 5 lbs. (2.2kg) is placed on the longest span of the belt.

12. Turn the crankshaft sprocket one turn clockwise and realign the crankshaft sprocket timing mark.

13. Recheck the belt tension and adjust as necessary.

14. Install or connect the following:
- Timing belt cover(s)
- Crankshaft pulley
- A/C compressor belt
- Water pump pulley
- V belt
- Negative battery cable
- Engine with coolant

3.0L and 3.5L Engine

1. Before servicing the vehicle, refer to the precautions in the beginning of this section.

2. Before servicing the vehicle, refer to the precautions in the beginning of this section.

3. Remove or disconnect the following:
- Negative battery cable
- Engine coolant
- Water pump pulley bolts
- Alternator bolt, loosen only
- Water pump pulley and drive belts
- Crankshaft pulley
- Timing belt cover(s)

4. Turn the crankshaft until the timing marks on the camshaft sprocket and cylinder head are aligned.

5. Loosen the timing belt tensioner bolt and turn the tensioner counterclockwise as far as it will go. Tighten the adjusting bolt.

6. Mark the timing belt with an arrow showing direction of rotation.

7. Remove the timing belt.

8. If defective, remove the timing belt tensioner.

To install:

9. If necessary, install the timing belt tensioner.

10. Attach the top of the tensioner spring on the engine coolant pump pin. Ensure the hook on the pin is facing down and the hook on the tensioner is facing away from the engine

11. Rotate the timing belt tensioner to the extreme counterclockwise position. Temporarily lock the tensioner in place.

12. Align the timing marks of the camshaft and crankshaft sprockets.

13. Install the timing belt on the crankshaft sprocket, then onto the rear camshaft sprocket.

14. Route the belt to the coolant pump pulley, the front camshaft sprocket and the timing belt tensioner.

15. Apply force counterclockwise to the rear camshaft sprocket with tension on the tight side of the belt and check that timing marks are aligned.

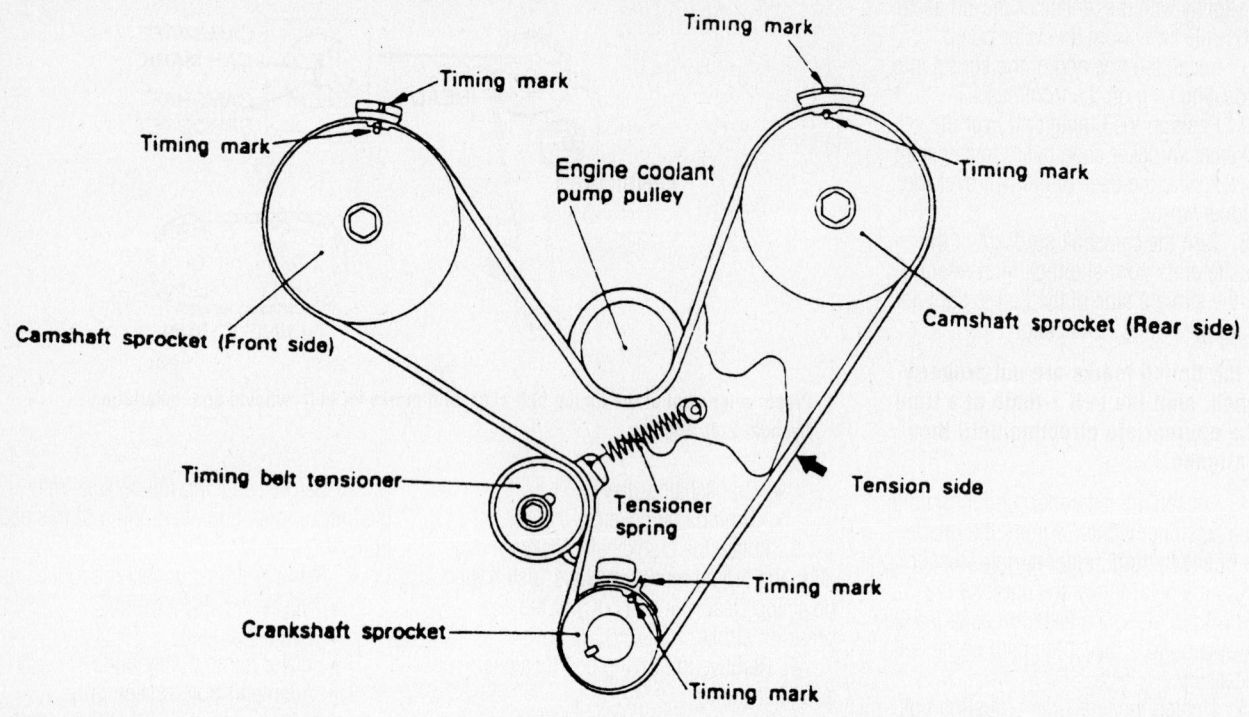

Timing belt sprocket alignment mark positioning for belt removal and installation—3.0L and 3.5L engine

16. Loosen the tensioner bolt 1–2 turns and tighten the timing belt to a tension of 57–84 lbs. (260–380 N).

17. Turn the crankshaft two turns clockwise.

18. Readjust the sprocket timing marks and tighten the tensioner bolts.

19. Install the timing belt covers.

20. Install the crankshaft pulley and tighten to 108–116 ft. lbs. (150–160 Nm).

21. Install or connect the following:
- Timing belt cover(s)
- Crankshaft pulley
- A/C compressor belt
- Water pump pulley
- V belt
- Negative battery cable
- Engine with coolant

Piston and Ring

POSITIONING

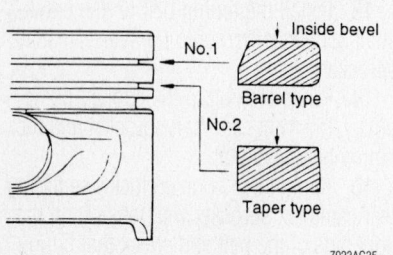

Compression ring identification

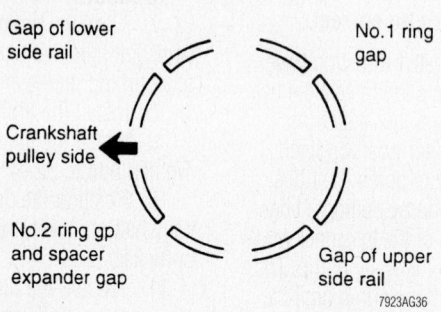

Piston ring end-gap spacing

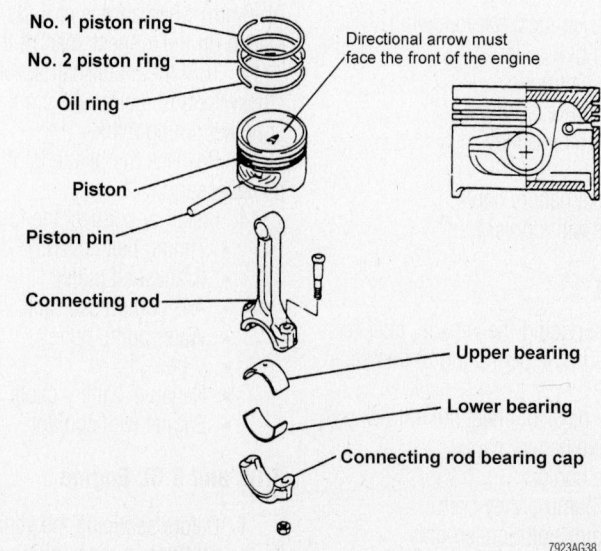

Piston and connecting rod assembly

FUEL SYSTEM

Fuel System Service Precautions

Safety is the most important factor when performing not only fuel system maintenance, but any type of maintenance. Failure to conduct maintenance and repairs in a safe manner may result in serious personal injury or death. Maintenance and testing of the vehicle's fuel system components can be accomplished safely and effectively by adhering to the following rules and guidelines.

• To avoid the possibility of fire and personal injury, always disconnect the negative battery cable unless the repair or test procedure requires that battery voltage be applied.

• Always relieve the fuel system pressure prior to disconnecting any fuel system component (injector, fuel rail, pressure regulator, etc.), fitting or fuel line connection. Exercise extreme caution whenever relieving the fuel system pressure, to avoid exposing skin, face and eyes to fuel spray. Please be advised that fuel under pressure may penetrate the skin or any part of the body that it contacts.

• Always place a shop towel or cloth around the fitting or connection prior to loosening to absorb any excess fuel due to spillage. Ensure that all fuel spillage (should it occur) is quickly removed from engine surfaces. Ensure that all fuel soaked cloths or towels are deposited into a suitable waste container.

• Always keep a dry chemical (Class B) fire extinguisher near the work area.

• Do not allow fuel spray or fuel vapors to come into contact with a spark or open flame.

• Always use a back-up wrench when loosening and tightening fuel line connection fittings. This will prevent unnecessary stress and torsion to fuel line piping. Always follow the proper torque specifications.

• Always replace worn fuel fitting O-rings with new. Do not substitute fuel hose where fuel pipe is installed.

Fuel System Pressure

RELIEVING

1. Before servicing the vehicle, refer to the precautions in the beginning of this section.
2. Remove or disconnect the following:
   • Rear seat cushion
   • Access panel
   • Fuel pump module connector
3. Start the engine and allow it to run until it stalls.
4. Turn the ignition switch to the **OFF** position.
5. Disconnect the negative battery cable.
6. Attach the fuel pump harness connector.

Fuel Filter

REMOVAL & INSTALLATION

➡**The fuel filter is located underneath the car, near the fuel tank.**

1. Before servicing the vehicle, refer to the precautions in the beginning of this section.
2. Relieve the fuel system pressure.
3. Remove or disconnect the following:
   • Negative battery cable
   • Fuel supply and pressure lines
   • Fuel filter bracket, if equipped
   • Fuel filter

To install:
4. Install or connect the following:
   • Fuel filter and torque the mounting bolts to 18–25 ft. lbs. (25–35 Nm)

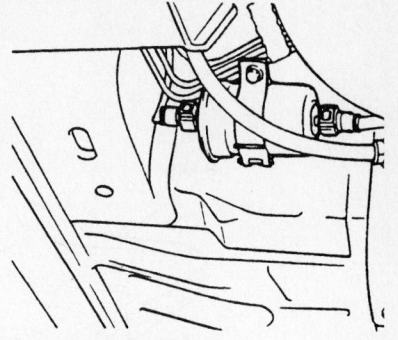

9347KG12

Remove the fuel filter

   • Fuel filter bracket, if equipped
   • Fuel supply and pressure lines
   • Negative battery cable
5. Start the engine, check leaks and repair if necessary.

Fuel Pump

REMOVAL & INSTALLATION

1. Before servicing the vehicle, refer to the precautions in the beginning of this section.
2. Relieve the fuel system pressure.

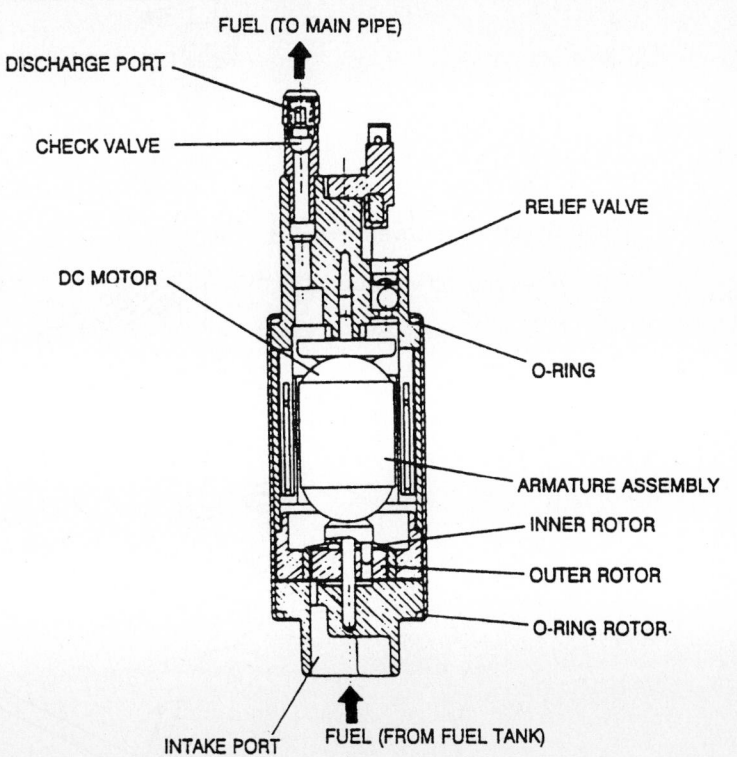

Cut away view of the electric fuel pump

7923GG81

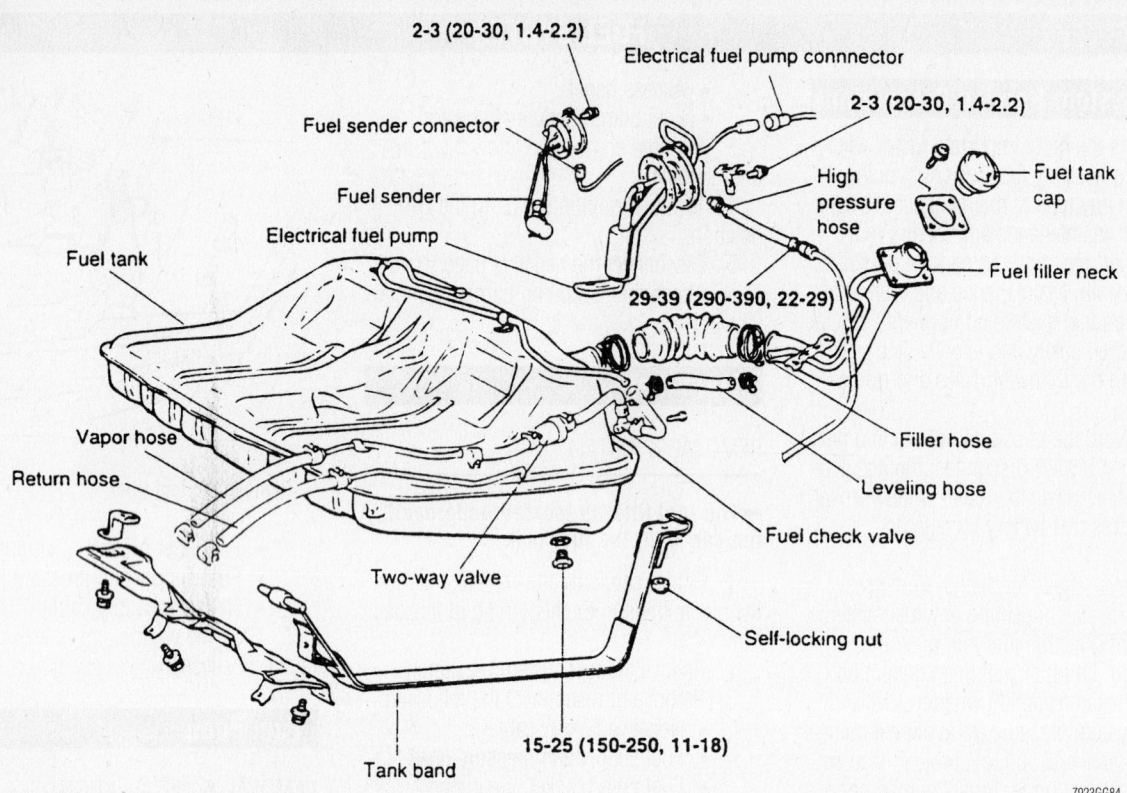

2-3 (20-30, 1.4-2.2)
Electrical fuel pump connnector
Fuel sender connector
2-3 (20-30, 1.4-2.2)
Fuel sender
Fuel tank cap
High pressure hose
Electrical fuel pump
Fuel tank
Fuel filler neck
29-39 (290-390, 22-29)
Vapor hose
Filler hose
Return hose
Leveling hose
Fuel check valve
Two-way valve
Self-locking nut
15-25 (150-250, 11-18)
Tank band

7923GG84

Exploded view of the fuel tank and fuel pump assembly—Sonata

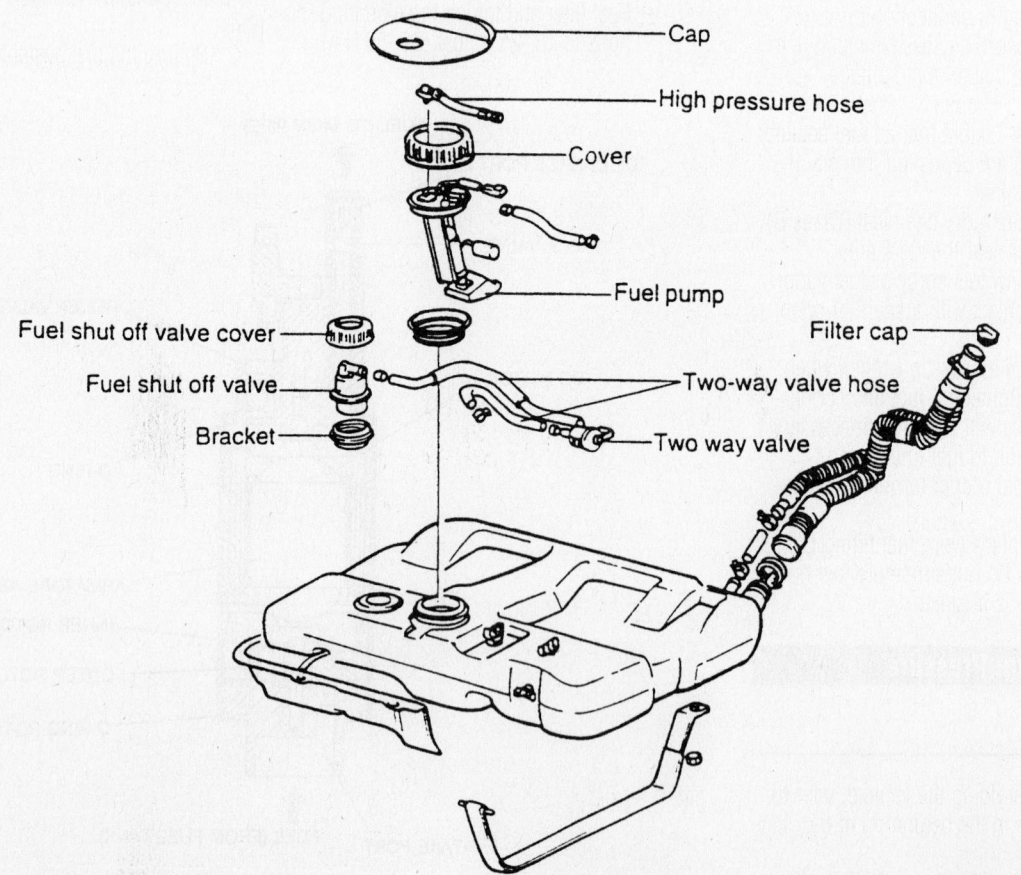

Cap
High pressure hose
Cover
Fuel pump
Fuel shut off valve cover
Filter cap
Fuel shut off valve
Two-way valve hose
Bracket
Two way valve

7923GG85

Exploded view of the fuel tank and fuel pump assembly—Accent

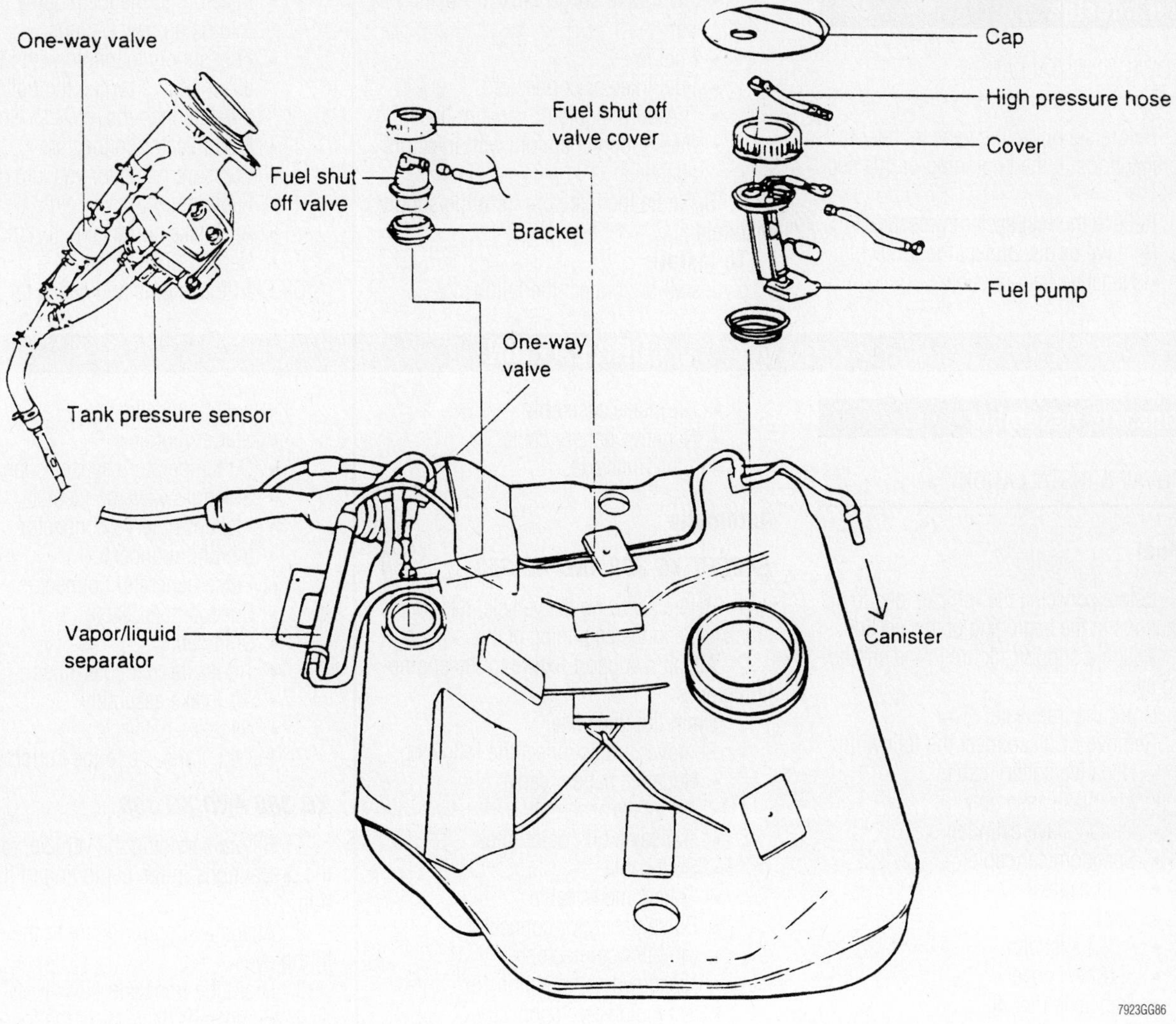

One-way valve

Fuel shut off valve

Fuel shut off valve cover

Bracket

Tank pressure sensor

One-way valve

Cap

High pressure hose

Cover

Fuel pump

Vapor/liquid separator

Canister

7923GG86

Fuel tank and fuel pump assembly—Tiburon and Elantra

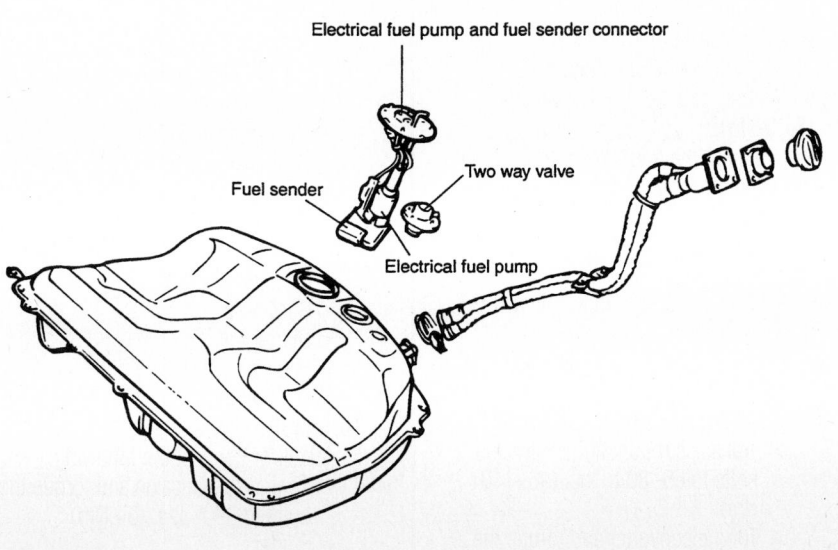

Electrical fuel pump and fuel sender connector

Fuel sender

Two way valve

Electrical fuel pump

9347KG13

Fuel tank and fuel pump assembly—XG 300 and XG 350

3. Drain the fuel tank.
4. Remove or disconnect the following:
   • Negative battery cable
   • Fuel supply, return and vapor lines
   • Fuel fill and vent hoses
   • Fuel pump module connector
   • Fuel level sender connector
   • Fuel tank straps
   • Fuel tank
   • Fuel pump module

To install:

5. Install or connect the following:
   • Fuel pump module and torque the mounting bolts to 12–24 inch lbs. (1–3 Nm)
   • Fuel tank
   • Fuel tank straps
   • Fuel level sender connector
   • Fuel pump module connector
   • Fuel fill and vent hoses
   • Fuel supply, return and vapor lines
   • Negative battery cable

6. Fill the tank with fuel and check for proper fuel pump operation.

Fuel Injector

REMOVAL & INSTALLATION

1. Before servicing the vehicle, refer to the precautions in the beginning of this section.
2. Relieve the fuel system pressure.
3. Remove or disconnect the following:
   - Negative battery cable
   - Air intake surge tank, if necessary
   - Fuel lines
   - Fuel injector connectors
   - Pressure regulator vacuum line
   - Fuel supply manifold with injectors attached
4. Separate the injectors from the supply manifold.

To install:

5. Install or connect the following:
   - Injectors to the fuel supply manifold using new O-rings
   - Fuel supply manifold with injectors attached and torque the bolts to 84–132 inch lbs. (10–15 Nm)
   - Fuel injector connectors
   - Pressure regulator vacuum line
   - Fuel lines
   - Air intake surge tank, if removed
   - Negative battery cable
6. Start the engine and check for leaks.

DRIVE TRAIN

Transaxle Assembly

REMOVAL & INSTALLATION

Manual

1. Before servicing the vehicle, refer to the precautions in the beginning of this section.
2. Attach a support fixture to the engine lifting eyes.
3. Drain the transaxle.
4. Remove or disconnect the following:
   - Negative battery cable
   - Air intake assembly
   - Clutch slave cylinder
   - Speedometer cable
   - Shift cables
   - Starter motor
   - Axle halfshafts
   - Flywheel cover
   - Transaxle mount
   - Transaxle flange bolts
   - Transaxle

To install:

➡**Use new circlips, split pins and self-locking fasteners for assembly.**

5. Position the transaxle to the engine and tighten the flange bolts to the following specifications:
   a. M8 bolts: 72–84 inch lbs. (8–10 Nm).
   b. M10 bolts: 22–25 ft. lbs. (30–35 Nm).
   c. M12 bolts: 32–39 ft. lbs. (43–55 Nm).
6. Install or connect the following:
   - Transaxle mount and torque the bolts to 65–80 ft. lbs. (90–110 Nm)
   - Flywheel cover and torque the bolts to 72–84 inch lbs. (8–10 Nm)
   - Axle halfshafts
   - Starter motor
   - Shift cables
   - Speedometer cable
   - Clutch slave cylinder
   - Air intake assembly
   - Negative battery cable
7. Fill the transaxle.

Automatic

EXCEPT XG 300 AND XG 350

1. Before servicing the vehicle, refer to the precautions in the beginning of this section.
2. Attach a support fixture to the engine lifting eyes.
3. Drain the transaxle.
4. Remove or disconnect the following:
   - Negative battery cable
   - Air intake assembly
   - Transaxle oil cooler lines
   - Shift cable
   - Speedometer cable
   - Pulse generator connector
   - Inhibitor connector
   - Kickdown servo connector
   - Solenoid valve connector
   - Oil temperature sensor connector
   - Starter motor
   - Axle halfshafts
   - Flywheel cover
   - Torque converter
   - Transaxle mount
   - Transaxle flange bolts
   - Transaxle

To install:

5. Position the transaxle to the engine and tighten the flange bolts to the following specifications:
   a. M8 bolts: 72–84 inch lbs. (8–10 Nm).
   b. M10 bolts: 22–25 ft. lbs. (30–35 Nm).
   c. M12 bolts: 32–39 ft. lbs. (43–55 Nm).
6. Install or connect the following:
   - Transaxle mount and torque the bolts to 65–80 ft. lbs. (90–110 Nm)
   - Torque converter and torque the bolts to 34–39 ft. lbs. (46–53 Nm)
   - Flywheel cover and torque the bolts to 72–84 inch lbs. (8–10 Nm)
   - Axle halfshafts
   - Starter motor
   - Oil temperature sensor connector
   - Solenoid valve connector
   - Kickdown servo connector
   - Inhibitor connector
   - Pulse generator connector
   - Speedometer cable
   - Shift cable
   - Transaxle oil cooler lines
   - Air intake assembly
   - Negative battery cable
7. Fill the transaxle to the correct level.

XG 300 AND XG 350

1. Before servicing the vehicle, refer to the precautions in the beginning of this section.
2. Attach a support fixture to the engine lifting eyes.
3. Drain the transaxle.
4. Remove or disconnect the following:
   - Negative battery cable
   - Air cleaner assembly
   - Control cable
   - Speedometer sensor connector
   - Transaxle range switch connector
   - Solenoid connector
   - Oil temperature sensor connector
   - Oil cooler lines
   - Steering gear
   - Sway bar
   - Tie rod end
   - Ball joints and drive shafts
   - Steering U-joint and return tube mounting bolts
   - Subframe
   - Starter
   - Engine-to-transaxle bolts
   - Transaxle

To install:

5. Install or connect the following:
   - Engine-to-transaxle and torque the bolts to 39 ft. lbs. (54 Nm)
   - Starter
   - Subframe and torque the subframe to transaxle bolts to 58 ft. lbs. (80

Nm) and the roll stopper bolts to 38 ft. lbs. (55 Nm)
- Steering U-joint and return tube mounting bolts
- Ball joints and driveshafts
- Tie rod end
- Sway bar
- Gear box
- Oil cooler lines
- Oil temperature sensor connector
- Solenoid connector
- Transaxle range switch connector
- Speedometer sensor connector and torque to19 ft. lbs. (26 Nm)
- Control cable and torque the bracket bolt to 18 ft. lbs. (25 Nm)
- Air cleaner assembly
- Negative battery cable

6. Fill the transaxle fluid to the proper level.

7. Start the vehicle, check for leaks and repair if necessary.

Clutch

ADJUSTMENTS

These vehicles are equipped with a hydraulic clutch system. No adjustment is necessary.

REMOVAL & INSTALLATION

1. Before servicing the vehicle, refer to the precautions in the beginning of this section.

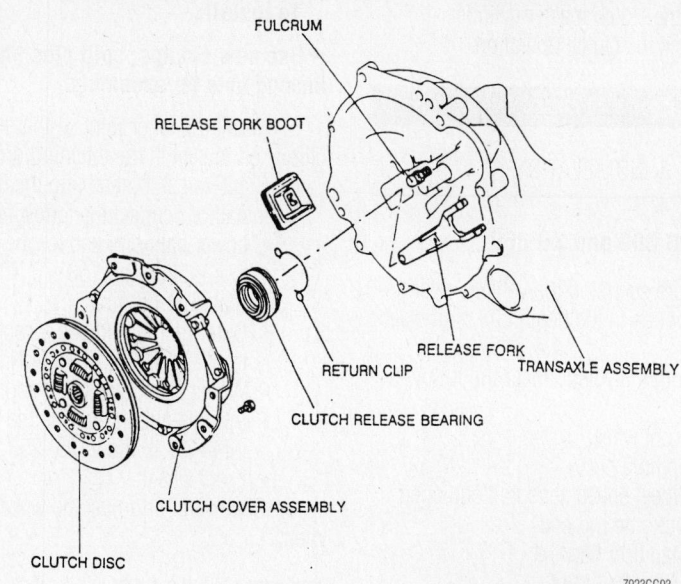

Exploded view of the clutch disc and pressure plate components

2. Remove or disconnect the following:
- Transaxle
- Pressure plate bolts by loosening them evenly in a crossing pattern
- Pressure plate and clutch disc

To install:
3. Install or connect the following:
- Clutch disc on the flywheel
- Pressure plate and torque the bolts evenly in a crossing pattern to 11–15 ft. lbs. (15–21 Nm)
- Transaxle

Hydraulic Clutch System

BLEEDING

1. Connect a hose to the bleeder screw and place the other end of hose into a container of clean brake fluid. Open the bleeder screw.

2. Have an assistant pump the clutch pedal slowly until no air bubbles are present at the bleeder screw.

3. Close the bleeder screw.

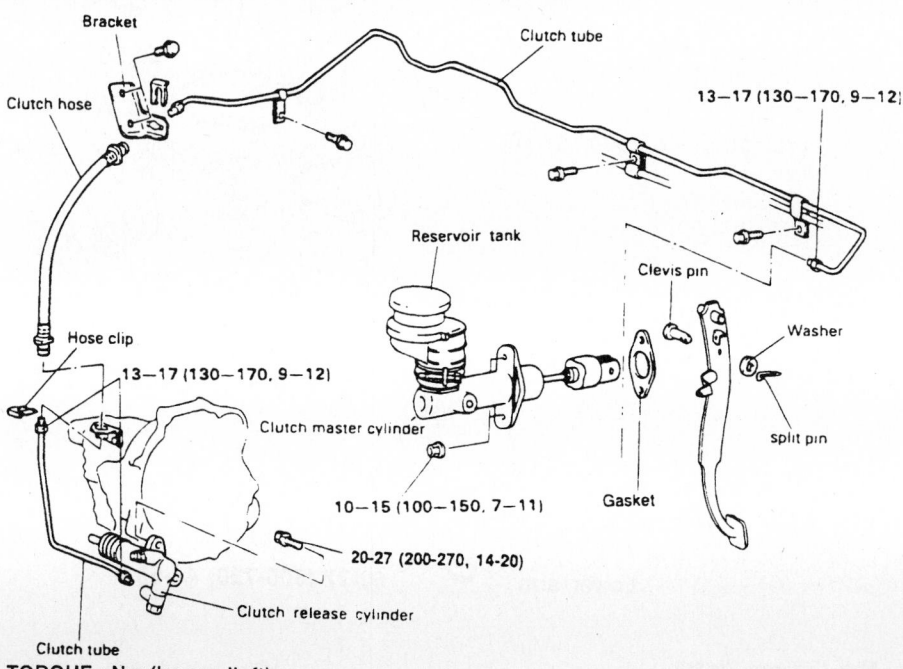

TORQUE : Nm (kg.cm, lb.ft)

Exploded view of the clutch hydraulic system

4. Fill the clutch master cylinder.

5. Check the clutch operation.

Halfshaft

REMOVAL & INSTALLATION

Except XG 300 and XG 350

1. Before servicing the vehicle, refer to the precautions in the beginning of this section.

2. Remove or disconnect the following:

- Front wheel
- Spindle nut
- Wheel speed sensor, if equipped
- Outer tie rod end
- Stabilizer bar link
- Lower ball joint

3. Press the stub shaft out of the hub.

4. Pry the inner joint out of the transaxle or intermediate shaft.

To install:

➡Use new circlips, split pins and self-locking nuts for assembly.

5. Install the inner joint so that the circlip is felt to seat in the retaining groove.

6. Guide the stub shaft into the hub.

7. Install or connect the following:

- Lower ball joint and torque the nut to 43–52 ft. lbs. (58–70 Nm)
- Stabilizer bar link
- Outer tie rod end and torque the nut to 17–25 ft. lbs. (23–34 Nm)
- Wheel speed sensor, if equipped
- Spindle nut and torque the nut to 144–187 ft. lbs. (195–253 Nm)
- Front wheel

8. Check and/or adjust the wheel alignment.

XG 300 and XG 350

1. Before servicing the vehicle, refer to the precautions in the beginning of this section.

2. Drain the transaxle fluid.

3. Remove or disconnect the following:

- Front wheel
- Split pin and halfshaft nut
- Ball joint from the steering knuckle
- Halfshaft from the axle hub

4. Install a prybar between the center bearing bracket and the halfshaft and pry the halfshaft from the transaxle.

5. Remove the center bearing bracket bolts and install a prybar between bracket and the engine and disconnect the bracket from the engine.

6. Remove the inner shaft from the transaxle.

To install:

7. Install or connect the following:

- Inner halfshaft to the transaxle
- Center bearing bracket and torque the bolt to 36 ft. lbs. (50 Nm)
- Halfshaft to the axle hub
- Ball joint to the steering knuckle and torque 88 ft. lbs. (110 Nm)

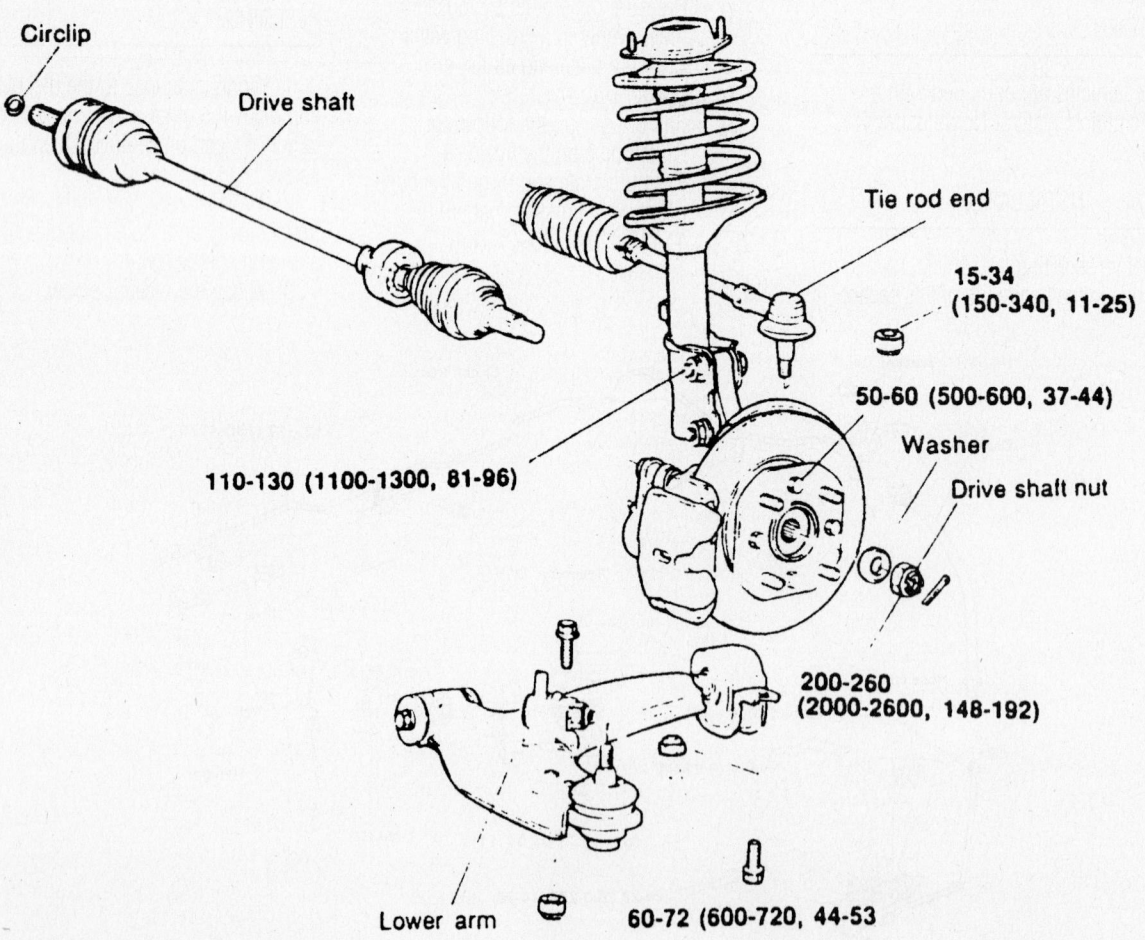

Circlip

Drive shaft

Tie rod end

15-34
(150-340, 11-25)

50-60 (500-600, 37-44)

Washer

Drive shaft nut

110-130 (1100-1300, 81-96)

200-260
(2000-2600, 148-192)

Lower arm

60-72 (600-720, 44-53)

TORQUE : Nm (kg.cm, lb.ft)

Halfshaft components—except V6 Sonata

7923GG94

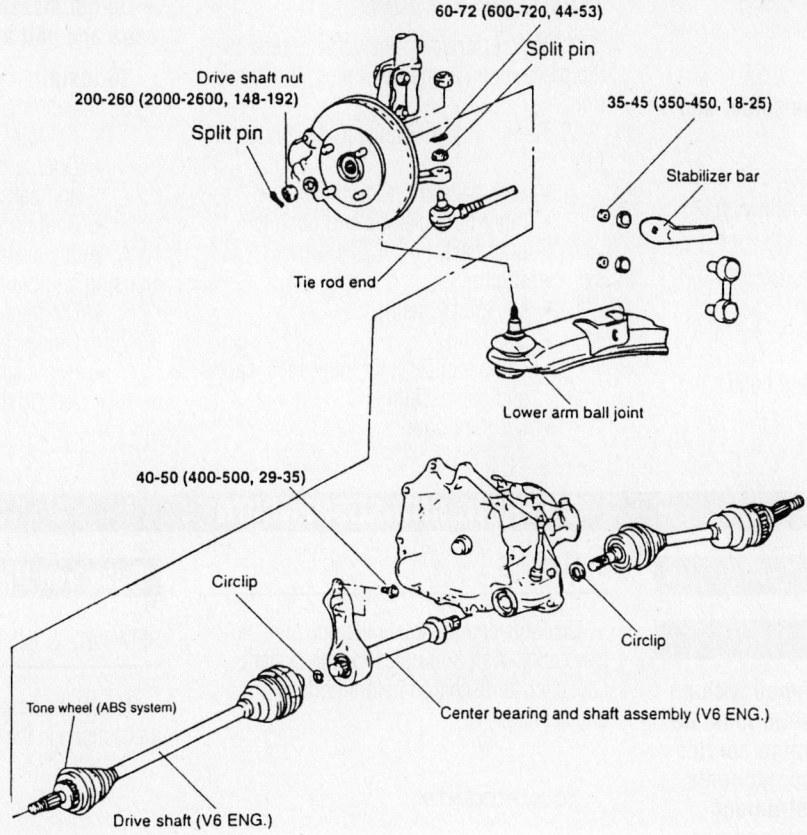

60-72 (600-720, 44-53)

Split pin

Drive shaft nut
200-260 (2000-2600, 148-192)

Split pin

35-45 (350-450, 18-25)

Stabilizer bar

Tie rod end

Lower arm ball joint

40-50 (400-500, 29-35)

Circlip

Circlip

Tone wheel (ABS system)

Center bearing and shaft assembly (V6 ENG.)

Drive shaft (V6 ENG.)

7923GG95

Halfshaft components—V6 Sonata

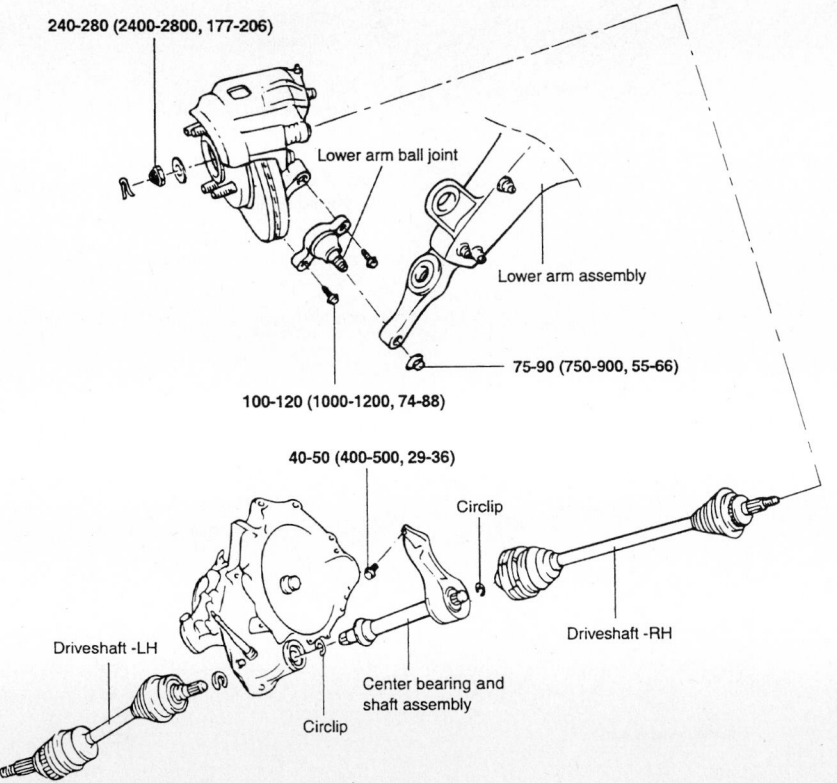

240-280 (2400-2800, 177-206)

Lower arm ball joint

Lower arm assembly

75-90 (750-900, 55-66)

100-120 (1000-1200, 74-88)

40-50 (400-500, 29-36)

Circlip

Driveshaft -LH

Driveshaft -RH

Center bearing and
shaft assembly

Circlip

9347KG14

Exploded view of the halfshaft assembly—XG 300 and XG 350

- Split pin and halfshaft nut and torque the nut to 206 ft. lbs. (280 Nm)
- Front wheel

8. Fill the transaxle fluid to the proper level.

9. Check and/or adjust the wheel alignment.

CV-Joints

OVERHAUL

Outer CV-Joint

The outer CV-joint is serviced with the axle halfshaft as an assembly. The outer CV-joint boot may be replaced by first removing the inner joint.

Inner CV-Joint

TRIPOD JOINT

1. Before servicing the vehicle, refer to the precautions in the beginning of this section.

2. Remove or disconnect the following:
- Axle halfshaft from the vehicle
- Inner boot clamps and push the boot back
- CV-joint housing

- Snapring
- Spider and rollers
- CV-joint boot

➡**Do not disassemble the spider and rollers.**

To install:
3. Install or connect the following:
  - CV-joint boot
  - Spider and rollers
  - Snapring
4. Apply clean grease to the CV-joint housing and the boot.
  - CV-joint housing and tighten the boot clamps
  - Axle to the vehicle

DOUBLE OFFSET JOINT

1. Before servicing the vehicle, refer to the precautions in the beginning of this section.
2. Remove or disconnect the following:

  - Axle halfshaft from the vehicle
  - Inner boot clamps and push the boot back
  - Circlip
  - CV-joint housing
  - Snapring
  - Double offset joint inner race, cage and ball assembly
  - CV-joint boot

➡**Do not disassemble the inner race, cage and ball assembly.**

To install:
3. Install or connect the following:
  - CV-joint boot
  - Double offset joint inner race, cage and ball assembly
  - Snapring
4. Apply clean grease to the CV-joint housing and the boot.
  - CV-joint housing
  - Circlip
  - Boot clamps by tightening them
  - Axle to the vehicle

STEERING AND SUSPENSION

Air Bag

✳✳ CAUTION

Some vehicles are equipped with an air bag system. The system must be disarmed before performing service on, or around, system components, the steering column, instrument panel components, wiring and sensors. Failure to follow the safety precautions and the disarming procedure could result in accidental air bag deployment, possible injury and unnecessary system repairs.

PRECAUTIONS

Several precautions must be observed when handling the inflator module to avoid accidental deployment and possible personal injury.

- Never carry the inflator module by the wires or connector on the underside of the module.
- When carrying a live inflator module, hold securely with both hands, and ensure that the bag and trim cover are pointed away.
- Place the inflator module on a bench or other surface with the bag and trim cover facing up.
- With the inflator module on the bench, never place anything on or close to the module which may be thrown in the event of an accidental deployment.

Before servicing the vehicle, also make sure to refer to the precautions in the beginning of this section as well.

DISARMING

Disconnect and isolate the negative battery cable. Wait 3 minutes for the system capacitor to discharge before performing any service.

COMPONENTS

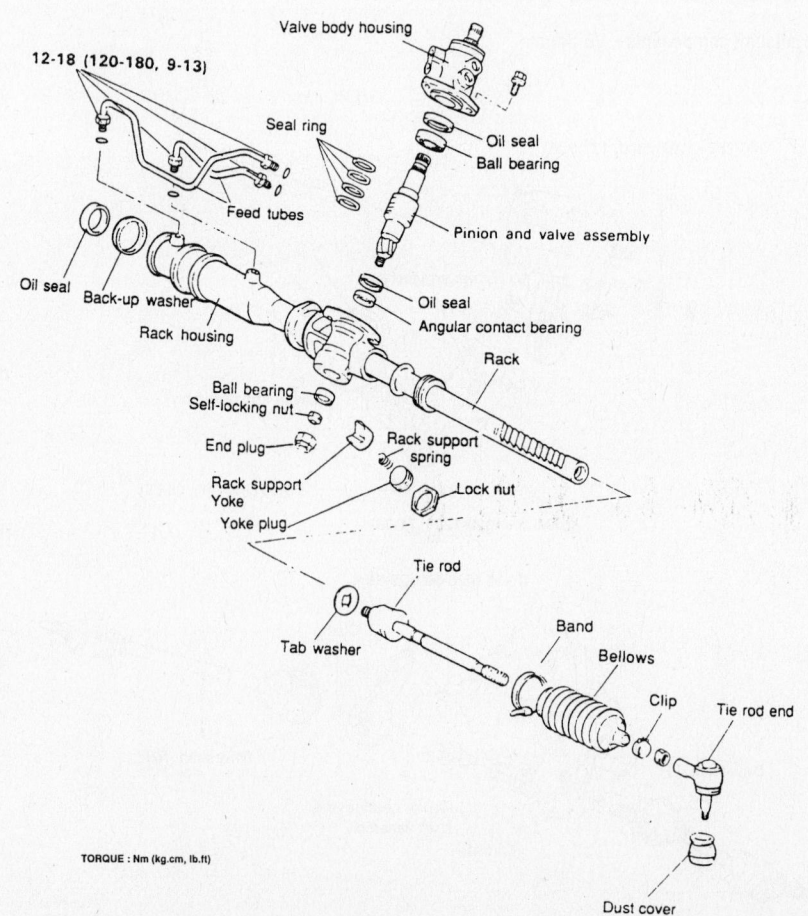

12-18 (120-180, 9-13)

Valve body housing
Seal ring
Oil seal
Ball bearing
Feed tubes
Pinion and valve assembly
Oil seal
Oil seal
Angular contact bearing
Back-up washer
Rack housing
Rack
Ball bearing
Self-locking nut
End plug
Rack support spring
Rack support Yoke
Lock nut
Yoke plug
Tie rod
Band
Tab washer
Bellows
Clip
Tie rod end
Dust cover

TORQUE : Nm (kg.cm, lb.ft)

Exploded view of the rack and pinion assembly

7923GG98

Rack and Pinion Steering Gear

REMOVAL & INSTALLATION

1. Before servicing the vehicle, refer to the precautions in the beginning of this section.

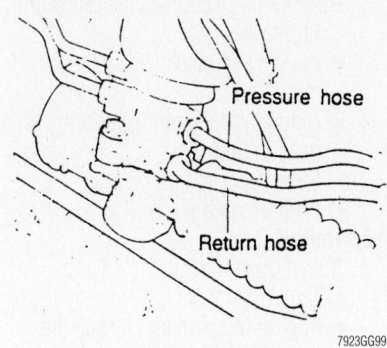

Pressure and return hose location on the rack

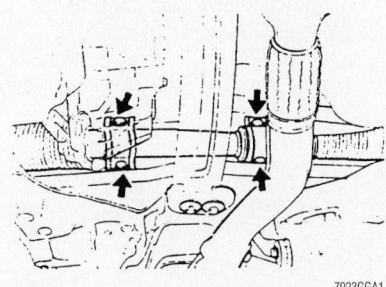

Power rack and pinion mounting bolt locations

2. Remove or disconnect the following:
- Negative battery cable
- Front wheels
- Outer tie rod ends
- Steering column flexible coupler
- Power steering fluid hoses, if equipped with power steering
- Subframe center beam
- Exhaust front pipe
- Left lower control arm
- Stabilizer bar
- Steering gear

To install:

3. Install or connect the following:
- Steering gear and torque the bolts to 44–59 ft. lbs. (60–80 Nm)
- Stabilizer bar
- Left lower control arm
- Exhaust front pipe
- Subframe center beam
- Power steering fluid hoses, if equipped with power steering
- Steering column flexible coupler and torque the bolt to 11–14 ft. lbs. (15–19 Nm)
- Outer tie rod ends and torque the nuts to 17–25 ft. lbs. (23–34 Nm)
- Front wheels
- Negative battery cable

4. Fill the power steering system.
5. Start the engine and check for leaks.

Struts

REMOVAL & INSTALLATION

Front

2000 VEHICLES

1. Before servicing the vehicle, refer to the precautions in the beginning of this section.

2. Remove or disconnect the following:
- Front wheel
- Brake hose bracket
- Steering knuckle pinch bolts
- Upper strut mount
- Strut

To install:

3. Position the strut to the vehicle and tighten the fasteners to the following specifications:

a. Steering knuckle pinch bolts, for Accent and Sonata: 65–76 ft. lbs. (95–105 Nm).

b. Steering knuckle pinch bolts, for Elantra and Tiburon: 80–94 ft. lbs. (110–130 Nm).

c. Upper strut mount nuts, for Accent: 14–22 ft. lbs. (20–30 Nm).

d. Upper strut mount nuts, for Elantra and Tiburon: 25–33 ft. lbs. (35–45 Nm).

e. Upper strut mount nuts, for Sonata: 18–25 ft. lbs. (25–34 Nm).

4. Install or connect the following:
- Brake hose bracket
- Front wheel

5. Check and/or adjust the wheel alignment.

2001–03 ACCENT, ELANTRA AND TIBURON

1. Before servicing the vehicle, refer to the precautions in the beginning of this section.

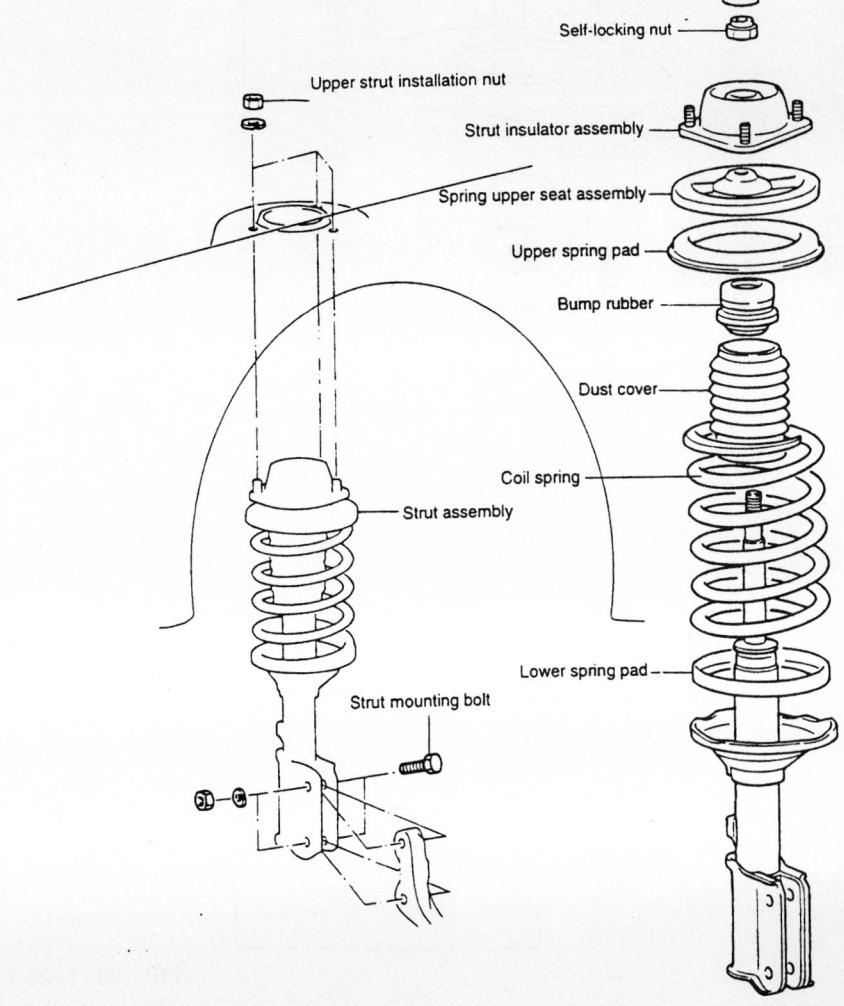

Exploded view of the strut assembly components

2. Remove or disconnect the following:
- Front wheel
- Brake hose bracket
- Strut upper mounting bolts

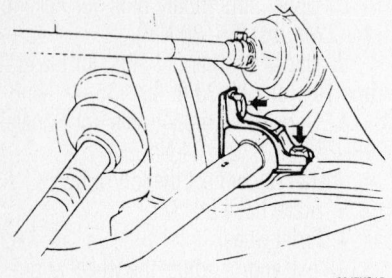

Exploded view of the strut assembly

- Lower mounting bolts
- Strut assembly

To install:

3. Install or connect the following:
- Strut assembly and torque the lower bolts to 66 ft. lbs. (90 Nm)
- Strut upper mounting bolts and torque to 22 ft. lbs. (30 Nm)
- Brake hose bracket
- Front wheel

4. Check and/or adjust the wheel alignment.

SONATA, XG 300 AND XG 350

1. Before servicing the vehicle, refer to the precautions in the beginning of this section.

2. Remove or disconnect the following:
- Front wheel
- Brake hose bracket from the mounting fork
- Mounting fork and lower arm connecting bolt
- Strut upper mounting nuts
- Strut assembly

To install:

3. Install or connect the following:
- Strut assembly
- Fork to the strut and torque the bolts to 59 ft. lbs. (80 Nm)
- Fork to the lower arm and torque the bolt to 88 ft. lbs. (120 Nm)
- Upper strut mounting nuts and torque to 36 ft. lbs. (50 Nm)

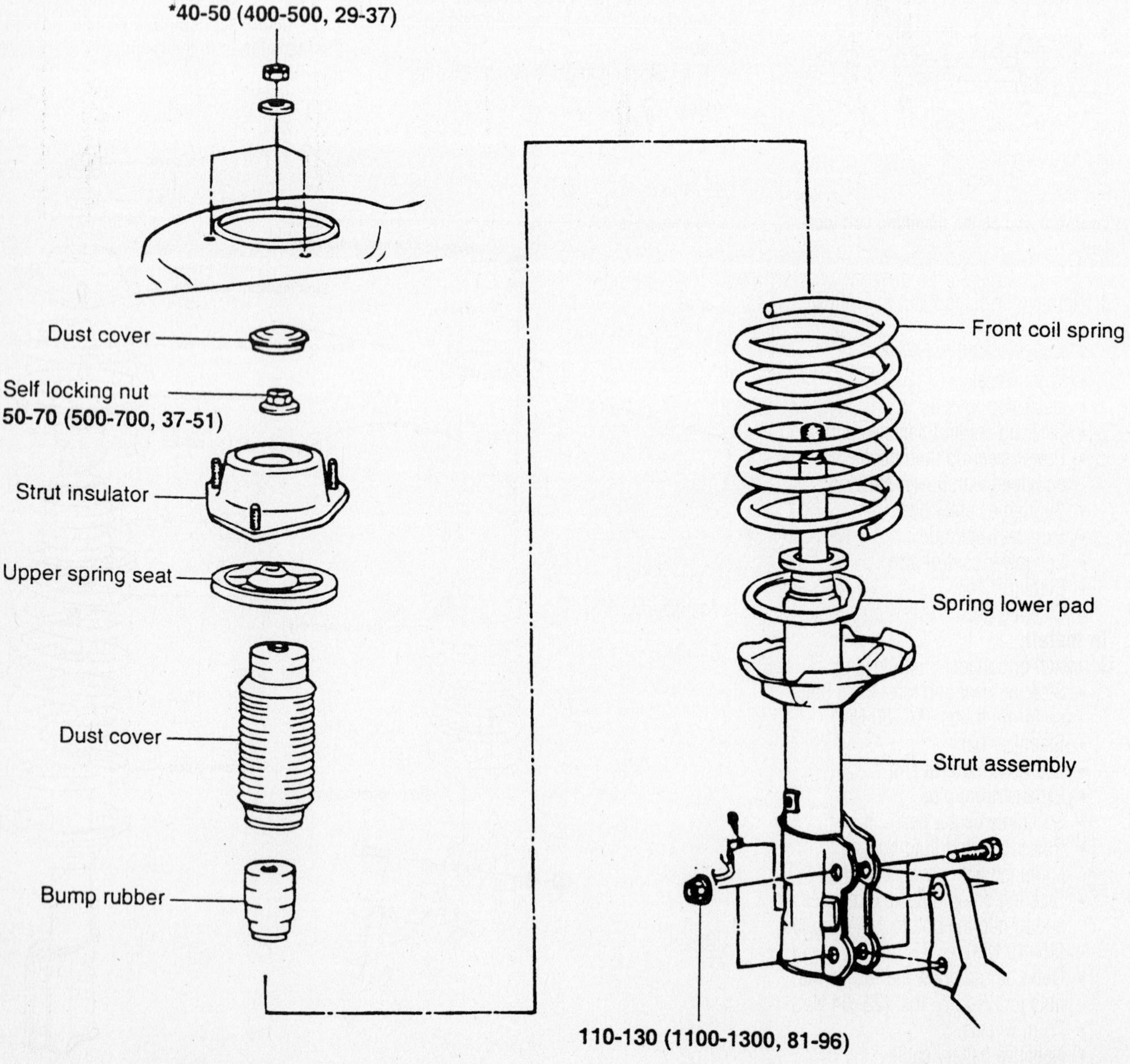

*40-50 (400-500, 29-37)

Dust cover

Self locking nut
50-70 (500-700, 37-51)

Strut insulator

Upper spring seat

Dust cover

Bump rubber

Front coil spring

Spring lower pad

Strut assembly

110-130 (1100-1300, 81-96)

Exploded view of the front strut assembly

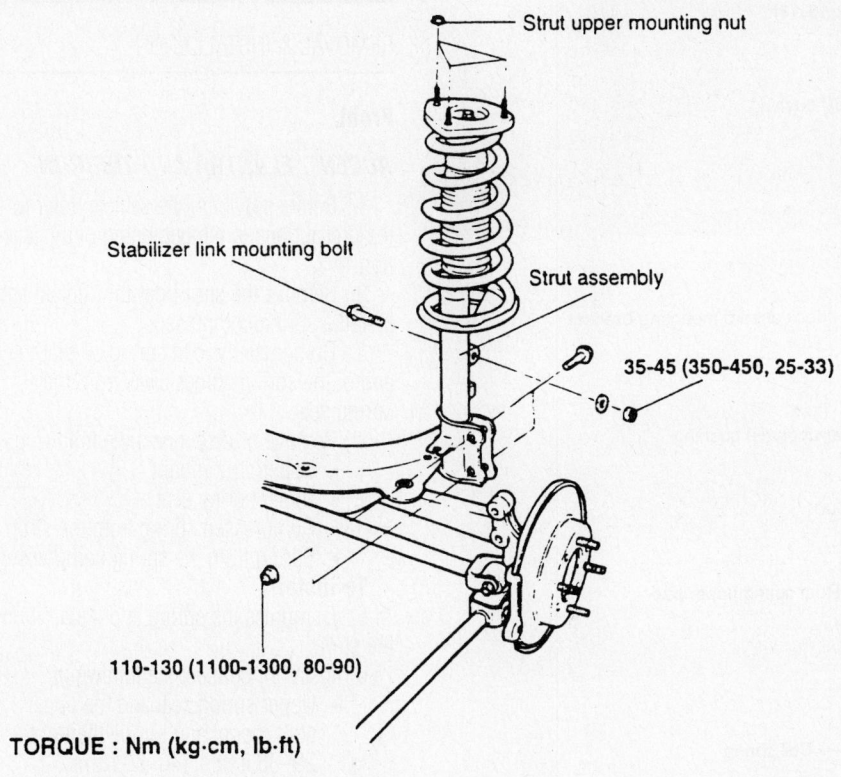

Strut upper mounting nut

Stabilizer link mounting bolt

Strut assembly

35-45 (350-450, 25-33)

110-130 (1100-1300, 80-90)

TORQUE : Nm (kg·cm, lb·ft)

7923GGA3

Exploded view of the rear strut assembly

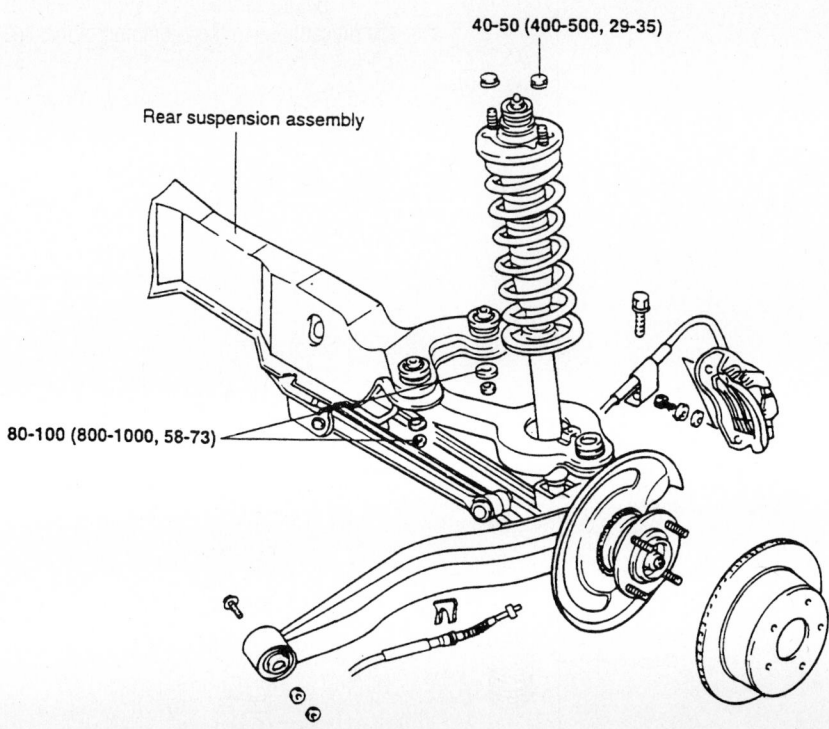

Rear suspension assembly

40-50 (400-500, 29-35)

80-100 (800-1000, 58-73)

7923GGA9

Rear suspension components—Sonata

- Brake hose bracket
- Front wheel

4. Check and/or adjust the wheel alignment.

Rear

2000 MODELS

1. Before servicing the vehicle, refer to the precautions in the beginning of this section.

2. Remove or disconnect the following:

- Rear wheel
- Upper strut mount access panel
- Upper strut mount nuts
- Wheel speed sensor connector
- Stabilizer bar link
- Hub knuckle pinch bolts
- Strut

To install:

3. Install or connect the following:

- Strut and torque the pinch bolts to 80–90 ft. lbs. (110–130 Nm) and the upper mount nuts to 14–22 ft. lbs. (20–30 Nm)
- Stabilizer bar link and torque the bolt to 25–33 ft. lbs. (35–45 Nm)
- Wheel speed sensor connector
- Upper strut mount access panel
- Rear wheel

4. Check and/or adjust the wheel alignment.

2001–03 ACCENT, ELANTRA AND TIBURON

1. Before servicing the vehicle, refer to the precautions in the beginning of this section.

2. Remove or disconnect the following:

- Rear seatback assembly and wheel house cover
- Rear wheel
- Upper mounting nuts
- Brake hose and wheel speed sensor connectors
- Carrier mounting nuts
- Strut assembly

To install:

3. Install or connect the following:

- Strut assembly and torque the carrier mounting nuts to 66 ft. lbs. (90 Nm)
- Brake hose and wheel speed sensor connectors
- Upper mounting nuts and torque the nuts to 22 ft. lbs. (30 Nm) on the Accent and to 37 ft. lbs. (50 Nm) for the Elantra

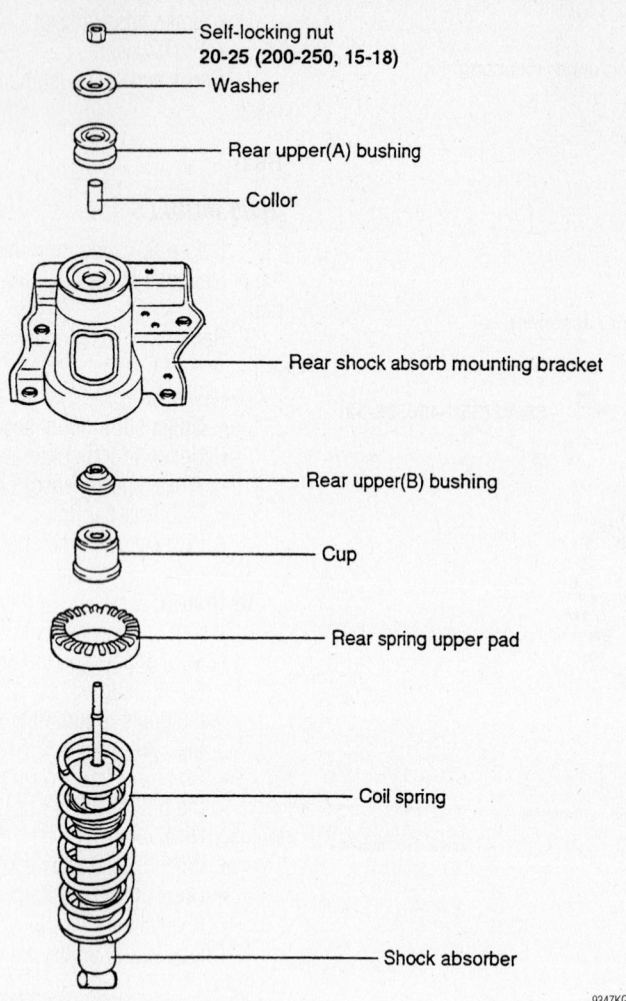

- Self-locking nut
  20-25 (200-250, 15-18)
- Washer
- Rear upper(A) bushing
- Collor
- Rear shock absorb mounting bracket
- Rear upper(B) bushing
- Cup
- Rear spring upper pad
- Coil spring
- Shock absorber

9347KG17

Exploded view of the rear strut assembly

- Wheel house cover and wheel
- Rear seatback

SONATA, XG 300 AND XG 350

1. Before servicing the vehicle, refer to the precautions in the beginning of this section.
2. Remove or disconnect the following:

- Rear wheel
- Lower mounting bolt
- Upper arm and rear carrier bolt
- Strut mounting bracket
- Strut assembly

To install:

3. Install or connect the following:

- Strut assembly and mounting bracket and torque the bolt to 36 ft. lbs. (50 Nm)
- Upper arm and rear carrier bolt and torque the bolt to 88 ft. lbs. (120 Nm)
- Lower mounting bolt
- Rear wheel

Coil Spring

REMOVAL & INSTALLATION

Front

ACCENT, ELANTRA AND TIBURON

1. Before servicing the vehicle, refer to the precautions in the beginning of this section.
2. Remove the strut from the vehicle and install a spring compressor.
3. Compress the coil spring so that the end of the spring comes away from the spring seat.
4. Remove or disconnect the following:

- Upper strut mount
- Upper spring seat
- Compressed spring from the strut
- Spring from the spring compressor

To install:

5. Compress the spring and install it on the strut.
6. Install or connect the following:

- Upper spring seat and the upper strut mount and torque the nut to 29–36 ft. lbs. (40–50 Nm)
- Strut to the vehicle

7. Check and/or adjust the wheel alignment.

SONATA, XG 300 AND XG 350

1. Before servicing the vehicle, refer to the precautions in the beginning of this section.

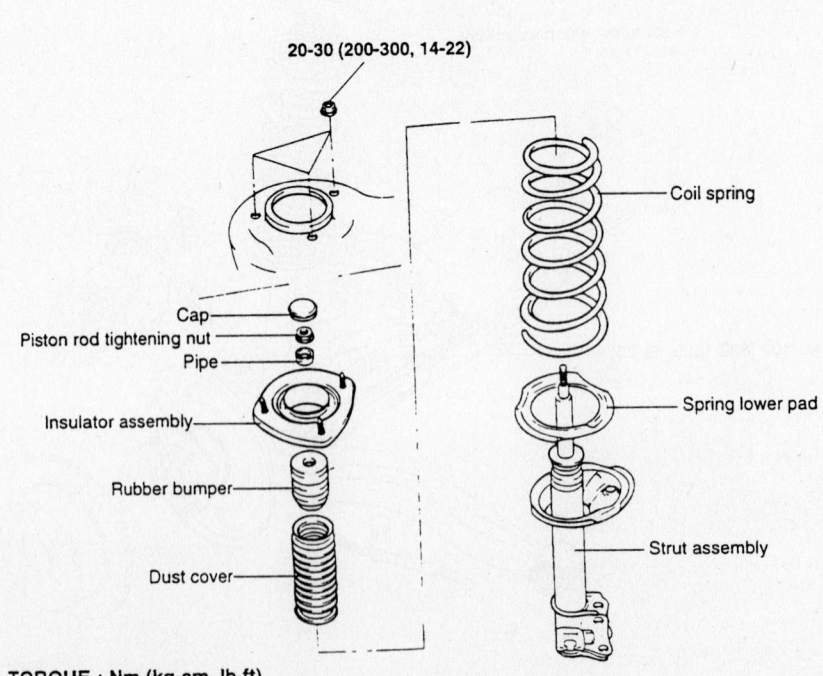

20-30 (200-300, 14-22)

- Cap
- Piston rod tightening nut
- Pipe
- Insulator assembly
- Rubber bumper
- Dust cover
- Coil spring
- Spring lower pad
- Strut assembly

TORQUE : Nm (kg·cm, lb·ft)

Rear strut components

7923GGA4

2. Remove the strut from the vehicle and install a Spring Compressor Tool, such as J38402.

3. Compress the coil spring so that the end of the spring comes away from the spring seat.

4. Remove or disconnect the following:
- Self locking nut

5. Install the Compressor Tool, J38402.

6. Remove the bracket, spring pad and coil spring.

To install:

7. Compress the coil spring with Compressor Tool J38402.

8. Install or connect the following:
- Coil spring to the strut
- Dust cover, upper spring pad, bushing and hand tighten the lock nut

9. Remove the compressor tool when the coil spring is properly aligned and torque the lock nut to 18 ft. lbs. (25 Nm).

10. Install the strut assembly.

Upper Ball Joint

The upper ball joints are replaced with the upper control arms as an assembly.

Lower Ball Joint

REMOVAL & INSTALLATION

Bolt-On Type

1. Before servicing the vehicle, refer to the precautions in the beginning of this section.

2. Remove or disconnect the following:
- Front wheel
- Ball joint stud from the knuckle
- Ball joint from the lower control arm

To install:

➡**Use a new split pin for assembly.**

3. Install or connect the following:
- Ball joint and torque the mounting bolts to 69–87 ft. lbs. (95–120 Nm)
- Stud nut and torque to 43–52 ft. lbs. (60–72 Nm)
- Front wheel

4. Check and/or adjust the wheel alignment.

Press-In Type

1. Before servicing the vehicle, refer to the precautions in the beginning of this section.

2. Remove or disconnect the following:

- Front wheel
- Lower control arm
- Ball joint dust cover

3. Press the ball joint out of the lower control arm.

To install:

4. Press the ball joint into the control arm.

5. Install or connect the following:
- Ball joint dust cover
- Lower control arm and torque the stud nut to 43–52 ft. lbs. (60–72 Nm)
- Front wheel

6. Check and/or adjust the wheel alignment.

Upper Control Arm

REMOVAL & INSTALLATION

Sonata, XG 300 and XG 350

1. Before servicing the vehicle, refer to the precautions in the beginning of this section.

2. Support the lower control arm assembly with a floor jack.

3. Remove or disconnect the following:
- Front wheel
- Ball joint nut, loosen only
- Upper arm ball joint from the steering knuckle with Special Tool 09568-34000

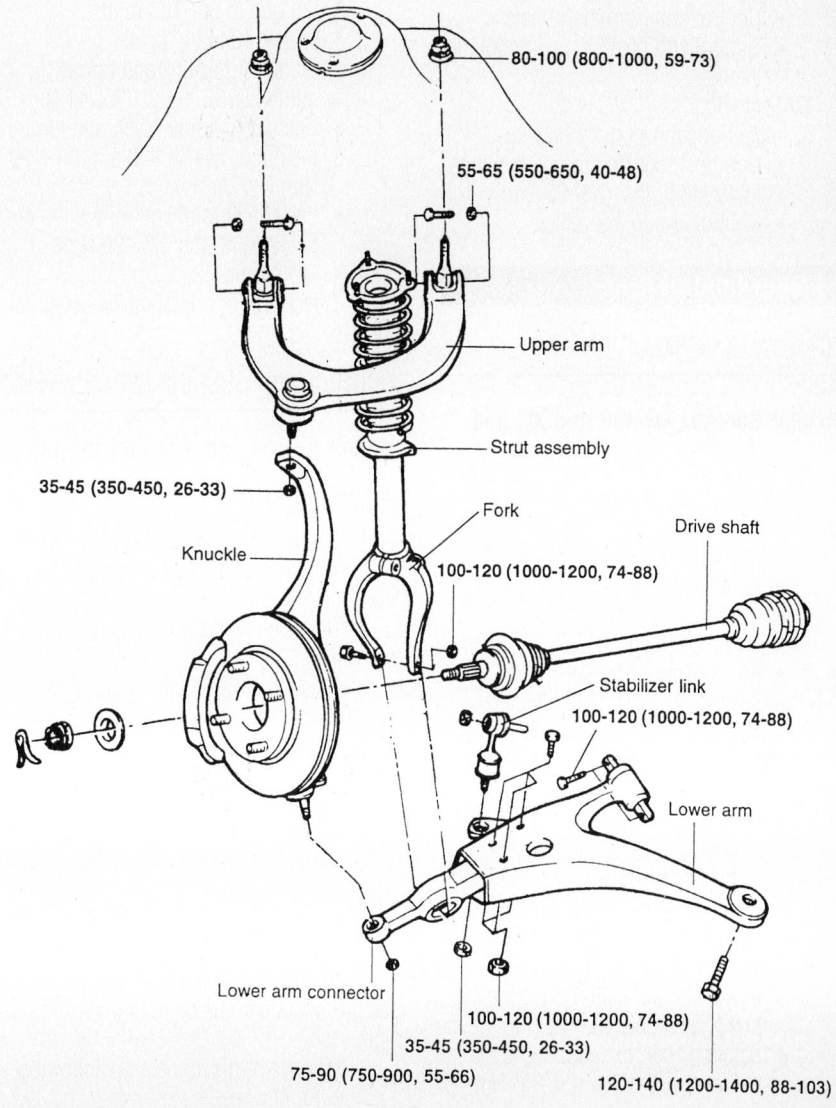

Exploded view of the upper control arm assembly—Sonata

9347KG18

- Wheel house panel nuts
- Upper arm assembly
- Upper arm shaft

To install:

4. Install or connect the following:
- Upper control arm shaft
- Upper control arm assembly and torque the bolts to 73 ft. lbs. (100 Nm)
- Wheel house panel nuts and torque the nuts to 48 ft. lbs. (65 Nm)
- Upper arm ball joint to the steering knuckle and torque the bolts to 33 ft. lbs. (45 Nm)
- Front wheel

CONTROL ARM BUSHING REPLACEMENT

1. Before servicing the vehicle, refer to the precautions in the beginning of this section.
2. Remove or disconnect the following:
- Control arm from the vehicle
- Control arm bushings by unbolting them

To install:

3. Install or connect the following:
- New bushings and torque the bolts to 40–48 ft. lbs. (55–65 Nm)
- Control arm to the vehicle

Lower Control Arm

REMOVAL & INSTALLATION

Except Sonata, XG 300 and XG 350

1. Before servicing the vehicle, refer to the precautions in the beginning of this section.
2. Remove or disconnect the following:
- Front wheel
- Stabilizer bar link
- Lower ball joint
- Rear bushing bracket
- Front bolt
- Lower control arm

To install:

3. Install or connect the following:
- Lower control arm and torque the front bolt to 72–87 ft. lbs. (100–120 Nm)
- Rear bushing bracket. Tighten the bolts to 58–72 ft. lbs. (80–100 Nm)
- Lower ball joint and torque the nut to 43–52 ft. lbs. (60–72 Nm)
- Stabilizer bar link and torque the nut to 25–33 ft. lbs. (35–45 Nm)
- Front wheel
4. Check and/or adjust the wheel alignment.

Sonata, XG 300 and XG 350

1. Before servicing the vehicle, refer to the precautions in the beginning of this section.
2. Remove or disconnect the following:
- Front wheel
- Lower ball joint nut, loosen only
- Lower arm ball joint from the lower arm connector with Special Tool 09445-21000
- Ball joint
- Fork from the lower arm connector
- Stabilizer bar link
- Control arm inner bushing bolts
- Lower control arm

To install:

3. Install or connect the following:
- Lower control arm and torque the front bushing bolts to 74–88 ft. lbs. (100–120 Nm) and the rear bushing bolt to 88–103 ft. lbs. (120–140 Nm)
- Stabilizer bar link and torque the nut to 26–33 ft. lbs. (35–45 Nm)
- Damper fork lower bolt and torque the nut to 74–88 ft. lbs. (100–120 Nm)
- Lower ball joint and torque the nut to 55–66 ft. lbs. (75–90 Nm)
- Front wheel
4. Check and/or adjust the wheel alignment.

CONTROL ARM BUSHING REPLACEMENT

Except Sonata, XG 300 and XG 350

FRONT BUSHING

1. Before servicing the vehicle, refer to the precautions in the beginning of this section.
2. Remove the lower control arm from the vehicle.
3. Press the front bushing out of control arm.

To install:

4. Lubricate the front bushing with soap and press into the control arm.
5. Install the control arm to the vehicle.
6. Check and/or adjust the wheel alignment.

REAR BUSHING

1. Before servicing the vehicle, refer to the precautions in the beginning of this section.
2. Remove or disconnect the following:
- Front wheel
- Rear bushing bracket
- Rear bushing nut
- Rear bushing

To install:

3. Install or connect the following:
- Rear bushing and torque the nut to 25–33 ft. lbs. (35–45 Nm)
- Rear bushing bracket and torque the bolts to 58–72 ft. lbs. (80–100 Nm)
- Front wheel
4. Check and/or adjust the wheel alignment.

Sonata, XG 300 and XG 350

FRONT BUSHING

The front control arm bushing is serviced with the control arm as an assembly.

REAR BUSHING AND DAMPER FORK BUSHING

1. Before servicing the vehicle, refer to the precautions in the beginning of this section.
2. Remove the control arm from the vehicle.
3. Press the rear bushing and the damper fork bushing out of the control arm.

To install:

4. Press the rear bushing and the damper fork bushing into the control arm.
5. Install the control arm to the vehicle.
6. Check and/or adjust the wheel alignment.

Wheel Bearings

ADJUSTMENT

Front

The front wheel bearing is a sealed unit and is not adjustable.

Rear

WITH REAR DRUM BRAKES

1. Before servicing the vehicle, refer to the precautions in the beginning of this section.
2. Remove the rear wheels.
3. Loosen the spindle nut.
4. Torque the nut to 108–145 ft. lbs. (150–200 Nm). Check for correct bearing end-play by placing a dial indicator on the hub surface and moving the hub outward. Note the movement of the gauge and compare it to the desired reading of 0.008 in. (0.2mm) or less. If end-play exceeds the desired reading, retighten the rear hub bearing nut and recheck the end-play. If the reading is still excessive, replace the hub unit.
5. If end-play is correct, check the start-

ing torque by attaching a spring balance to the hub lug bolts and pulling at a 90 degree angle while noting the required force to turn the hub. If the force required is above the desired reading of 5 lbs. (2.3 kg) or less, loosen the nut and again tighten to the desired torque. Recheck the starting torque. If the torque is still above the desired reading, replace the rear bearings.

6. Install the rear wheels.

WITH REAR DISC BRAKES

The rear wheel bearing is an integral part of the rear hub. No adjustment is possible.

REMOVAL & INSTALLATION

Front

1. Before servicing the vehicle, refer to the precautions in the beginning of this section.
2. Remove or disconnect the following:
   - Front wheel
   - Brake caliper
   - Lower ball joint
   - Spindle nut
   - Knuckle pinch bolts
   - Steering knuckle
3. Press the hub out of the wheel bearing.
4. Press the wheel bearings out of the steering knuckle.
5. If necessary, press the inner race off the hub.

To install:

6. Press the wheel bearings into the steering knuckle.
7. Install the outer grease seal and press the hub into the wheel bearings.
8. Install or connect the following:
   - Inner grease seal
   - Steering knuckle and torque the knuckle pinch bolts to 65–76 ft. lbs. (95–105 Nm)

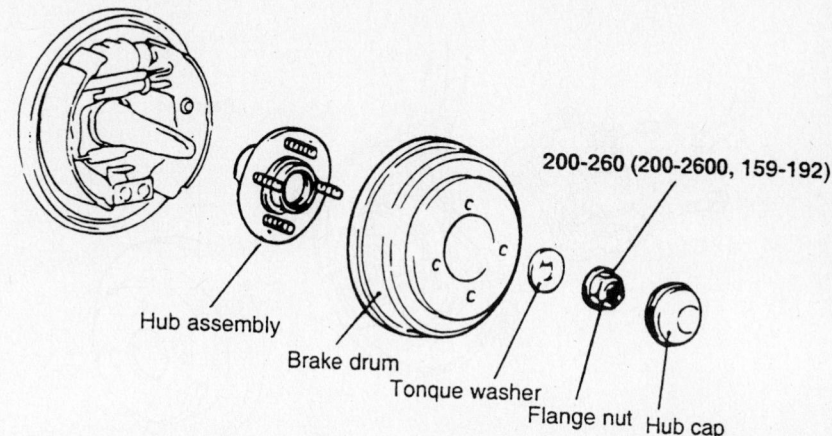

200-260 (200-2600, 159-192)

Hub assembly

Brake drum

Tonque washer

Flange nut Hub cap

9347KG20

Exploded view of the rear hub assembly—with drum brakes

- Lower ball joint and torque the stud nut to 43–52 ft. lbs. (60–72 Nm)
- Spindle nut and torque the nut to 144–187 ft. lbs. (195–253 Nm)
- Brake caliper and torque the bracket bolts to 50 ft. lbs. (68 Nm)
- Front wheel

9. Check and/or adjust the wheel alignment.

Rear

DRUM BRAKES

1. Before servicing the vehicle, refer to the precautions in the beginning of this section.
2. Remove or disconnect the following:
   - Rear wheel
   - Speed sensor, if equipped
   - Grease cap
   - Flange nut
   - Outer bearing
   - Brake drum
   - Inner grease seal
   - Inner bearing

3. Drive the bearing races out of the drum hub.

To install:

4. Install the inner and outer bearing races.
5. Apply grease to the bearings and to the cavity in the hub.
6. Install or connect the following:
   - Inner bearing
   - Inner grease seal
   - Brake drum
   - Outer bearing
   - Flange nut and torque the nut to 159–192 ft. lbs. (200–260 Nm)
   - Grease cap
   - Wheel speed sensor, if equipped
   - Rear wheel

DISC BRAKES

1. Before servicing the vehicle, refer to the precautions in the beginning of this section.
2. Release the parking brake.
3. Remove or disconnect the following:
   - Rear wheel
   - Wheel speed sensor, if equipped
   - Brake caliper and rotor
   - Rear axle hub bolts
   - Tone wheel with Tool 09445-21000
   - Carrier assembly
   - Nut after unstaking it
4. Press out the rear axle hub.
5. Remove the bearing inner race with Tool 09445-21000.
6. Remove the bushings from the carrier with Tools 09453-33000B and 09545-21100.

To install:

7. Press in the bushings to the carrier with Tools 09453-33000B and 09545-21100.

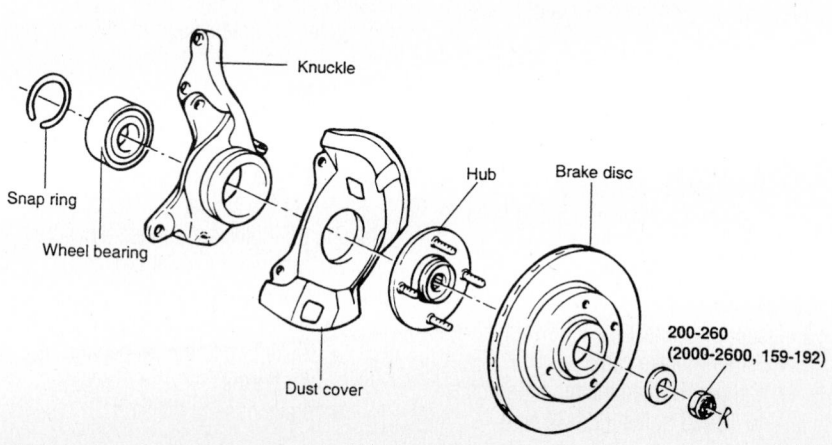

Knuckle

Snap ring

Wheel bearing

Dust cover

Hub   Brake disc

200-260
(2000-2600, 159-192)

9347KG19

Exploded view of the front hub assembly

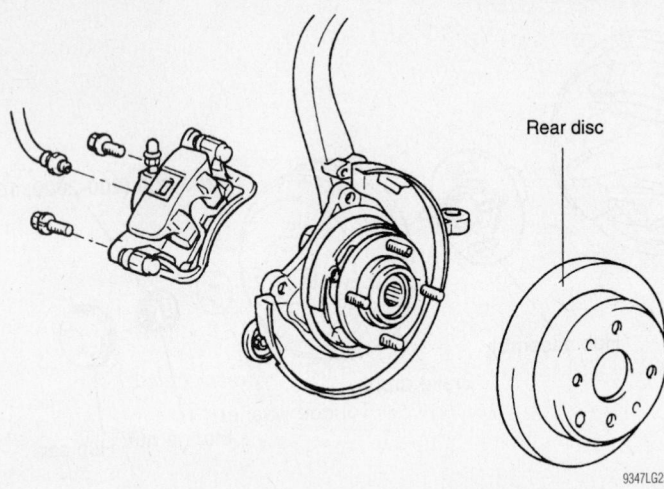

Rear disc

9347LG25

Exploded view of the rear wheel bearing assembly—with disc brakes

8. Press in the bearing to the hub with Tool 09221-21000.

9. Tighten the flange nut to meet the concave portion of the spindle.

10. Press in the tone wheel with Tool 09221-21000. Torque the nut to 191 ft. lbs. (260 Nm).

11. Install or connect the following:
- Hub and bearing assembly to the backing plate and torque the bolts to 88 ft. lbs. (120 Nm)
- Brake caliper and rotor
- Wheel speed sensor, if equipped
- Rear wheel

BRAKES

Brake Caliper

REMOVAL & INSTALLATION

Accent

1. Before servicing the vehicle, refer to the precautions in the beginning of this section.

2. Remove or disconnect the following:
- Front wheels
- Brake line at the caliper
- Brake pads
- Pin and sleeve boots
- Lower caliper bolt and raise the caliper up and out to remove it

To install:

3. Install or connect the following:
- Caliper onto its mounting and install the lower mounting bolt. Torque the bolt to 16–24 ft. lbs. (22–32 Nm).
- Pin boots, sleeve boots and brake pads
- Brake line to the caliper with 2 new metal gaskets. Torque the brake line union bolt to 18–22 ft. lbs. (25–30 Nm).

4. Bleed the system.
- Front wheels

Elantra

FRONT

1. Before servicing the vehicle, refer to the precautions in the beginning of this section.

2. Remove or disconnect the following:
- Front wheels
- Brake line at the caliper
- Brake pads
- Pin and sleeve boots
- Lower caliper bolt and raise the caliper up and out to remove it

To install:

3. Install or connect the following:
- Caliper onto its mounting
- Lower mounting bolt. Torque the bolt to 16–24 ft. lbs. (22–32 Nm).
- Pin boots, sleeve boots and brake pads
- Brake line to the caliper with 2 new metal gaskets. Torque the brake line union bolt to 18–22 ft. lbs. (24–30 Nm).

4. Bleed the system.
- Front wheels

REAR

1. Before servicing the vehicle, refer to the precautions in the beginning of this section.

2. Remove or disconnect the following:
- Center console and loosen the parking brake adjustment
- Wheels
- Brake hose
- Caliper assembly mounting bolts
- Caliper
- Parking brake cable

To install:

3. Install or connect the following:
- Parking brake cable
- Caliper. Tighten the mounting bolts to 16–23 ft. lbs. (22–32 Nm).

- Brake hose
- Master cylinder with clean fluid and bleed the hydraulic system
- Wheel. Adjust the parking brake.
- Center console

Sonata, XG 300 and XG 350

FRONT

1. Before servicing the vehicle, refer to the precautions in the beginning of this section.

2. Remove or disconnect the following:
- Front wheels
- Brake tube from the brake hose
- Brake hose clip
- Brake hose from the strut
- Brake line from the caliper
- Small retaining pin from the lower part of the caliper

3. Swing the caliper up until it clears the rotor and pads.

4. Slide the caliper inboard until the locating pin disengages from its groove in the caliper. Pull the caliper from the locating pin.

To install:

5. Lubricate the locating pin bore with white silicone compound and mount the caliper onto the locating pin.

6. Lower the caliper until the small retaining pin holes are aligned. Install a new retaining pin into the lower part of the caliper. Tighten the pin.

7. Install or connect the following:
- Brake line to the caliper and bleed the brakes
- Brake hose to the strut

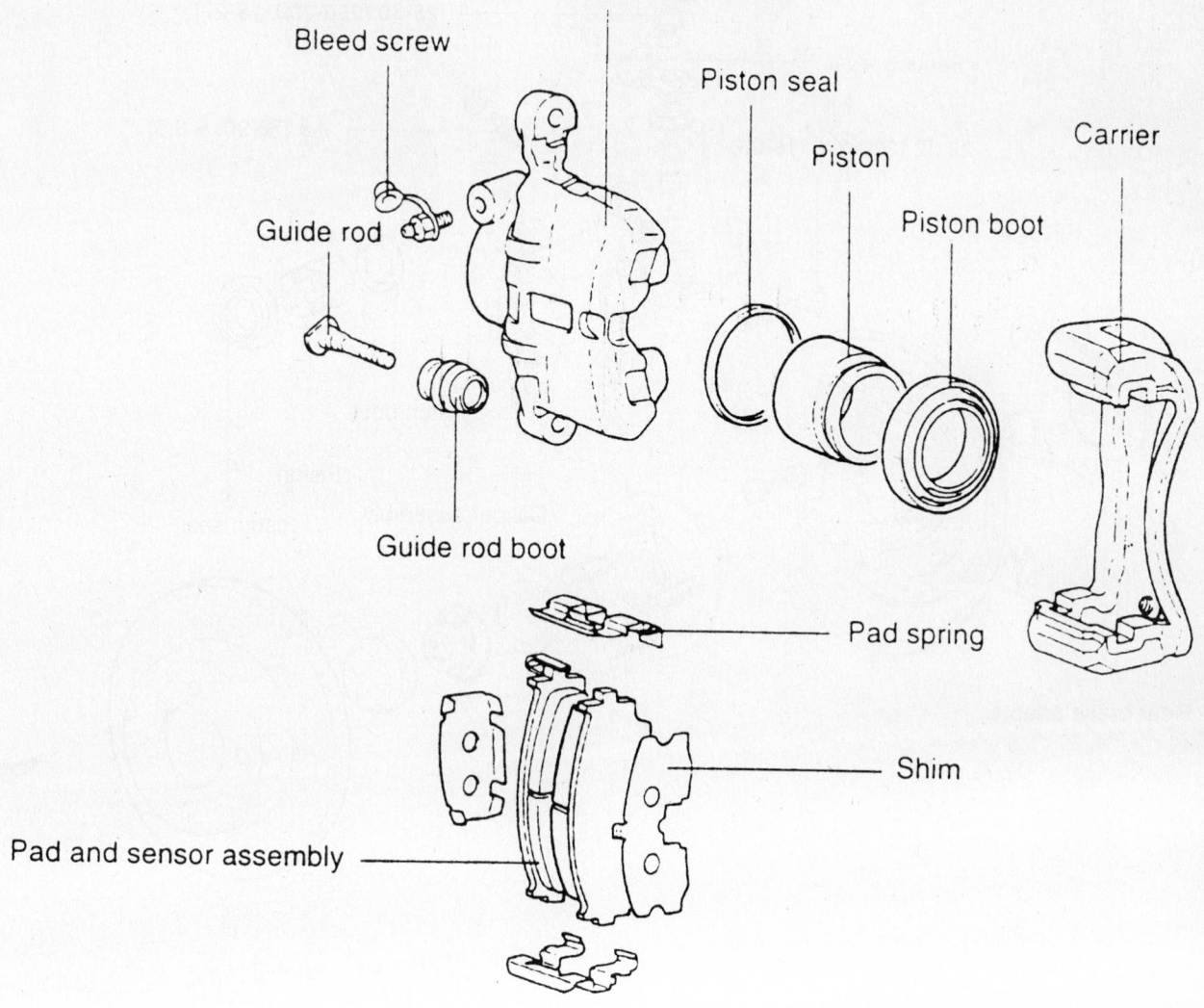

Front caliper—Sonata shown

- Brake hose clip
- Brake tube to the brake hose
- Front wheels

REAR

1. Before servicing the vehicle, refer to the precautions in the beginning of this section.
2. Remove or disconnect the following:
   - Center console and loosen the parking brake adjustment
   - Wheels
   - Brake hose
   - Caliper assembly mounting bolts
   - Caliper
   - Parking brake cable

To install:

3. Install or connect the following:

- Parking brake cable
- Caliper. Tighten the mounting bolts to 16–23 ft. lbs. (22–32 Nm).
- Brake hose
- Master cylinder with clean fluid and bleed the hydraulic system
- Wheel. Adjust the parking brake.
- Center console

Tiburon

FRONT

1. Before servicing the vehicle, refer to the precautions in the beginning of this section.
2. Remove or disconnect the following:
   - Wheels
   - Brake hose

- Caliper
- Caliper support

To install:

3. Install or connect the following:
   - Caliper support and tighten the bolts to 44–63 ft. lbs. (69–85 Nm)
   - Caliper. Tighten the guide rod bolts to 16–24 ft. lbs. (22–32 Nm).
   - Brake hose and tighten to 18–22 ft. lbs. (25–30 Nm)
   - Wheels
4. Fill the master cylinder with clean brake fluid and bleed the hydraulic system.

REAR

1. Before servicing the vehicle, refer to the precautions in the beginning of this section.

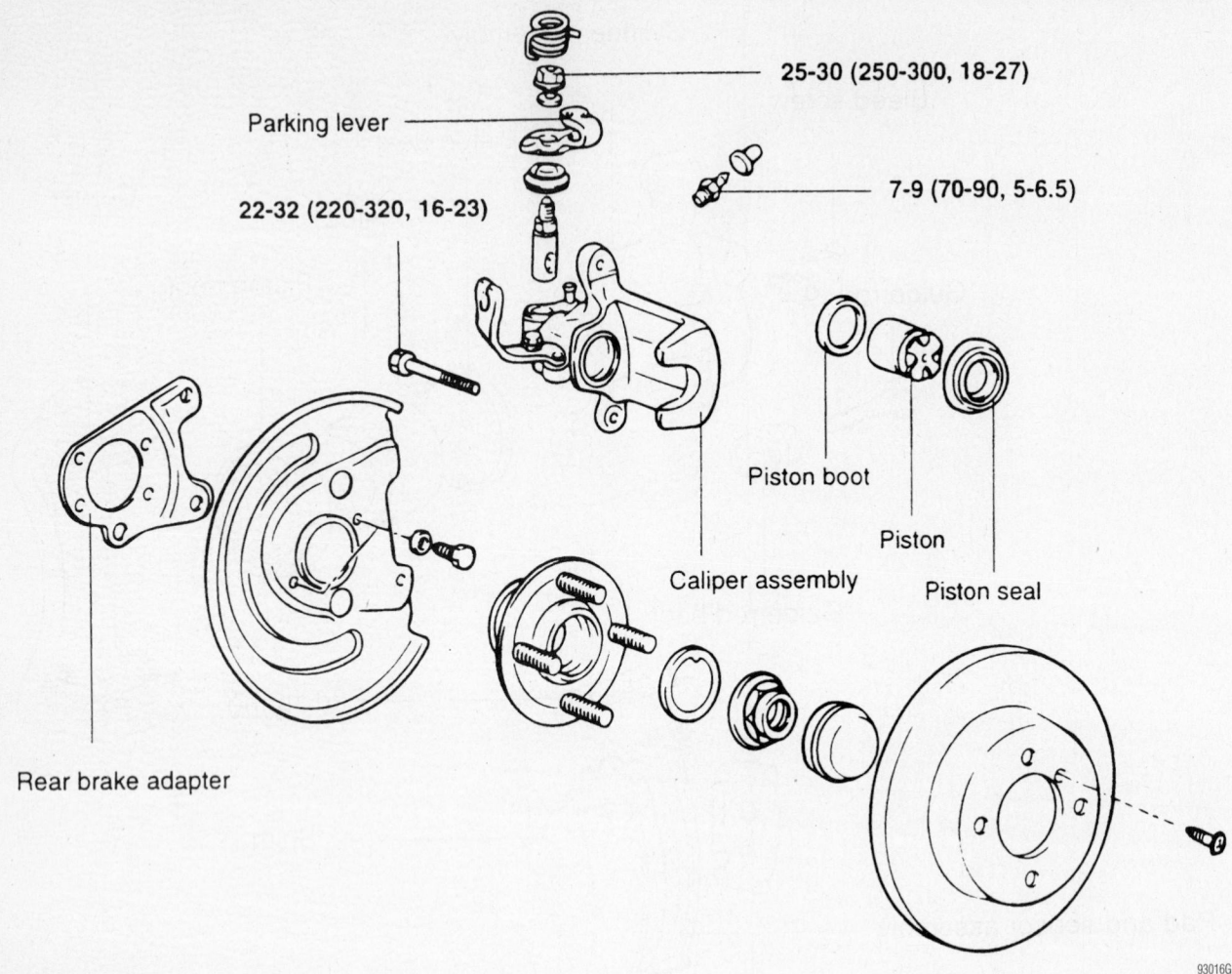

25-30 (250-300, 18-27)

Parking lever

7-9 (70-90, 5-6.5)

22-32 (220-320, 16-23)

Piston boot

Piston

Caliper assembly

Piston seal

Rear brake adapter

93016G11

Rear caliper—Elantra shown

2. Remove or disconnect the following:
   • Wheels
   • Brake hose
   • Caliper
   • Parking brake cable

To install:

3. Install or connect the following:
   • Parking brake cable
   • Caliper. Tighten the mounting bolts to 16–24 ft. lbs. (22–32 Nm).
   • Brake hose

4. Fill the master cylinder with clean brake fluid and bleed the hydraulic system.
   • Wheels

Disc Brake Pads

REMOVAL & INSTALLATION

Accent

1. Before servicing the vehicle, refer to the precautions in the beginning of this section.

2. Remove or disconnect the following:
   • Front wheels
   • Lower caliper mounting bolt and rotate the caliper upward
   • Pads from the caliper support
   • Pad clips, if necessary

To install:

3. Install or connect the following:
   • Pad clips
   • Pads onto the pad clips

4. Compress the caliper piston using a C-clamp. Rotate the caliper downward and install the mounting bolt.
   • Wheels

Elantra

FRONT

1. Before servicing the vehicle, refer to the precautions in the beginning of this section.

2. Remove or disconnect the following:
   • Front wheels

   • Lower caliper mounting bolt and rotate the caliper upward
   • Pads from the caliper support
   • Pad clips

To install:

3. Install or connect the following:
   • Pad clips
   • Pads onto the pad clips

4. Compress the caliper piston using a C-clamp. Rotate the caliper downward and install the mounting bolt.
   • Wheels

REAR

1. Before servicing the vehicle, refer to the precautions in the beginning of this section.

2. Remove or disconnect the following:
   • ½ of the fluid from the brake master cylinder
   • Wheels
   • Caliper mounting bolts
   • Caliper
   • Brake pads and retaining clips

To install:

3. Install or connect the following:
- New pads and retainers

4. Compress the caliper piston using special tool 09580-3400.
- Caliper. Tighten the mounting bolts to 16–24 ft. lbs. (22–32 Nm).
- Master cylinder with clean brake fluid and bleed the hydraulic system
- Wheels

Sonata, Tiburon and XG 300 and XG 350

FRONT

1. Before servicing the vehicle, refer to the precautions in the beginning of this section.

2. Remove or disconnect the following:
- ½ of the fluid from the brake master cylinder
- Front wheels
- Small retaining pin from the lower part of the caliper

3. Swing the caliper up until it clears the rotor and pads.
- Pads and anti-squeal spring from the caliper support

To install:

4. Install or connect the following:
- Pads and anti-rattle spring

5. Compress the caliper piston using special tool 09580-3400.
- Caliper. Tighten the mounting bolts to 16–24 ft. lbs. (22–32 Nm).
- Master cylinder with clean brake fluid and bleed the hydraulic system
- Wheels

REAR

1. Before servicing the vehicle, refer to the precautions in the beginning of this section.

2. Remove or disconnect the following:
- ½ of the fluid from the brake master cylinder
- Rear wheels
- Caliper mounting bolts and remove the caliper
- Brake pads and retaining clips

To install:

3. Install or connect the following:
- New pads and retainers

4. Compress the caliper piston using special tool 09580-3400.
- Caliper. Tighten the mounting bolts to 16–24 ft. lbs. (22–32 Nm).

- Master cylinder with clean brake fluid and bleed the hydraulic system
- Wheels

Brake Drums

REMOVAL & INSTALLATION

1. Before servicing the vehicle, refer to the precautions in the beginning of this section.

2. Remove or disconnect the following:
- Wheels
- Dust cap, cotter pin, nut lock, wheel bearing nut and washer from the spindle
- Outer wheel bearing
- Drum with the inner wheel bearing from the spindle

To install:

3. Install or connect the following:
- Lubricated inner wheel bearing
- New grease seal
- Drum to the spindle
- Lubricated outer wheel bearing, washer and nut. Adjust the bearing preload as required.
- Nut lock and a new cotter pin

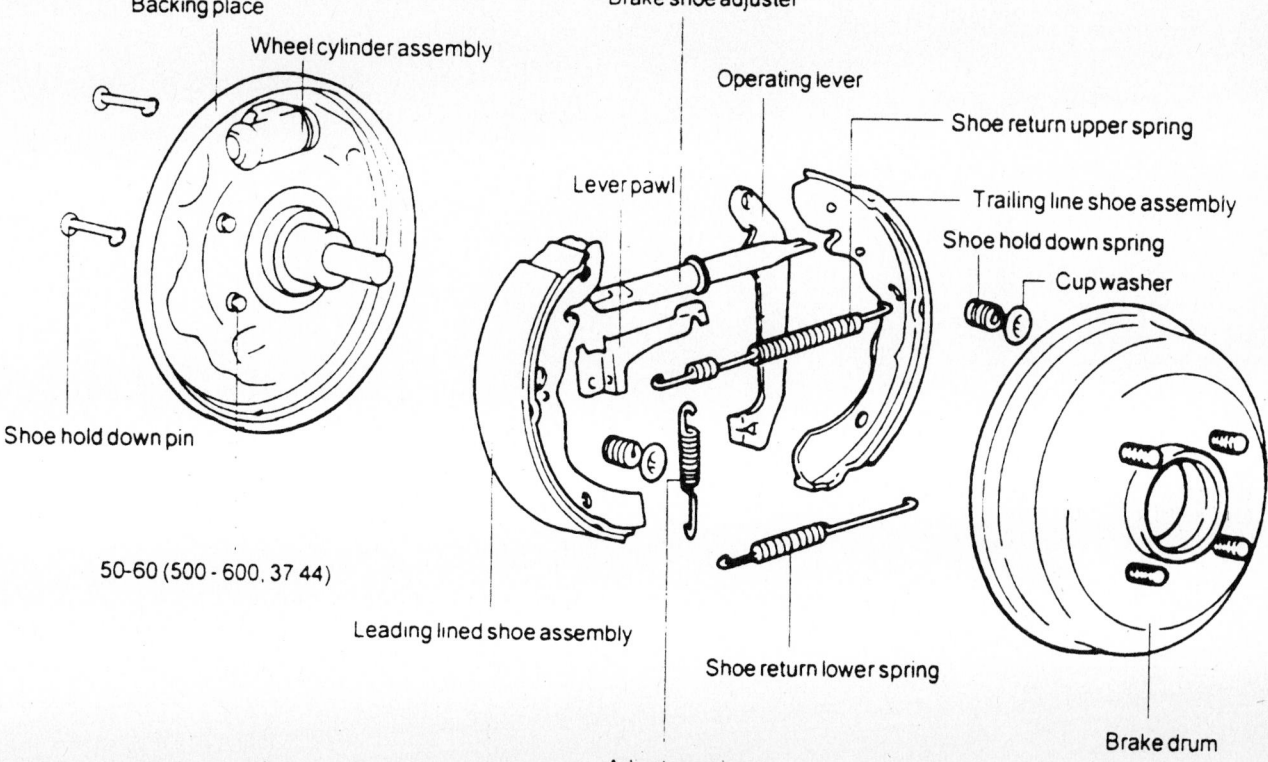

50-60 (500 - 600, 37 44)

Backing place
Wheel cylinder assembly
Shoe hold down pin
Brake shoe adjuster
Operating lever
Lever pawl
Shoe return upper spring
Trailing line shoe assembly
Shoe hold down spring
Cup washer
Leading lined shoe assembly
Shoe return lower spring
Adjuster spring
Brake drum

93016G12

Rear drum brakes—Accent shown

- Grease cap
- Wheels. Adjust the rear brakes as required.

Brake Shoes

REMOVAL & INSTALLATION

1. Before servicing the vehicle, refer to the precautions in the beginning of this section.
2. Remove or disconnect the following:
   - Wheels
   - Brake drum

- Self-adjuster spring and the adjuster lever
- Spread the shoes and remove the adjuster strut
- Shoe-to-shoe spring and the hold-down springs
- Primary brake shoe
- Horseshoe clip and the parking brake lever from the secondary brake shoe

To install:

3. Clean the backing plate with brake cleaning solvent.
4. Install or connect the following:
   - Light coating of lithium grease to

the friction points on the backing plate
- Primary shoe on the backing plate
- Hold-down spring and pin
- Parking brake lever to the secondary shoe
- Secondary shoe to the backing plate
- Adjuster strut assembly and the adjuster lever and spring
- Lower shoe-to-shoe spring
- Brake drum and the wheel

5. Adjust the brake shoes.

SPECIFICATION AND MAINTENANCE CHARTS

ENGINE AND VEHICLE IDENTIFICATION CHART

Engine Code								Model Year	
Code	Liters (cc)	Cu. In.	Cyl.	Fuel Sys.	Engine Type	Eng. Mfg.		Code ①	Year
B	2.4 (2351)	120	4	MFI	DOHC	Hyundai		1	2001
D	2.7 (2656)	120	6	MFI	DOHC	Hyundai		2	2002
E	3.5 (3496)	120	6	MFI	DOHC	Hyundai		3	2003
								4	2004
								5	2005

DOHC: Double Overhead Cam

MFI: Multi-port Fuel Injection

① 8th position of VIN

67162-HSFE-C01

GENERAL ENGINE SPECIFICATIONS

Year	Model	Engine Displacement Liters (VIN)	Net Horsepower @ rpm	Net Torque @ rpm (ft. lbs.)	Bore x Stroke (in.)	Com-pression Ratio	Oil Pressure @ rpm
2001	Santa Fe	2.4 (B)	149@5500	156@3000	3.41x3.94	10:01	①
		2.7 (D)	181@6000	177@4000	3.41x2.95	10:01	②
2002	Santa Fe	2.4 (B)	149@5500	156@3000	3.41x3.94	10:01	①
		2.7 (D)	181@6000	177@4000	3.41x2.95	10:01	②
2003	Santa Fe	2.4 (B)	149@5500	156@3000	3.41x3.94	10:01	①
		2.7 (D)	181@6000	177@4000	3.41x2.95	10:01	②
2004-05	Santa Fe	2.4 (B)	149@5500	156@3000	3.41x3.94	10:01	①
		2.7 (D)	181@6000	177@4000	3.41x2.95	10:01	②
		3.5 (E)	195@5500	219@3500	3.66x3.77	10:01	③

MFI: Multi-port Fuel Injection

① 11.6 Psi (80 kPa) @ idle.

② 7.3 Psi (50 kPa) or more @ idle.

③ 11.4 Psi (80 kPa) or more @ 700 RPM with engine oil temperature at 204 degrees F (95 degrees C)

67162-HSFE-C02

ENGINE TUNE-UP SPECIFICATIONS

Year	Engine Displacement Liters (VIN)	Spark Plug Gap (in.)	Ignition Timing (deg.)		Fuel Pump (psi)	Idle Speed (rpm)		Valve Clearance (in.)	
			MT	AT		MT	AT	In.	Ex.
2001	2.4 (B)	0.039-0.043	2-12B	2-12B	37	625-825	625-825	NA	NA
	2.7 (D)	0.040-0.043	7-19B	7-19B	37	625-825	625-825	NA	NA
2002	2.4 (B)	0.039-0.043	2-12B	2-12B	37	625-825	625-825	NA	NA
	2.7 (D)	0.040-0.043	7-19B	7-19B	37	625-825	625-825	NA	NA
2003	2.4 (B)	0.039-0.043	2-12B	2-12B	37	625-825	625-825	NA	NA
	2.7 (D)	0.040-0.043	7-19B	7-19B	37	625-825	625-825	NA	NA
2004-05	2.7 (D)	0.040-0.043	7-19B	7-19B	37	625-825	625-825	NA	NA
	3.5 (E)	0.039-0.043	①	①	37	②	②	NA	NA

NOTE: The Vehicle Emission Control Information label often reflects changes made during production and must be used if they differ from this chart.

NOTE: The fuel pressure readings are given with the vacuum hose connected to the regulator and the engine running

B: Before top dead center

HYD: Hydraulic

NA: Not Availible

① The timing is controlled by the Powertrain Control Module (PCM) and is not adjustable.

② 600-700 with the A/C OFF or 750-850 with the A/C ON

67162-HSFE-C03

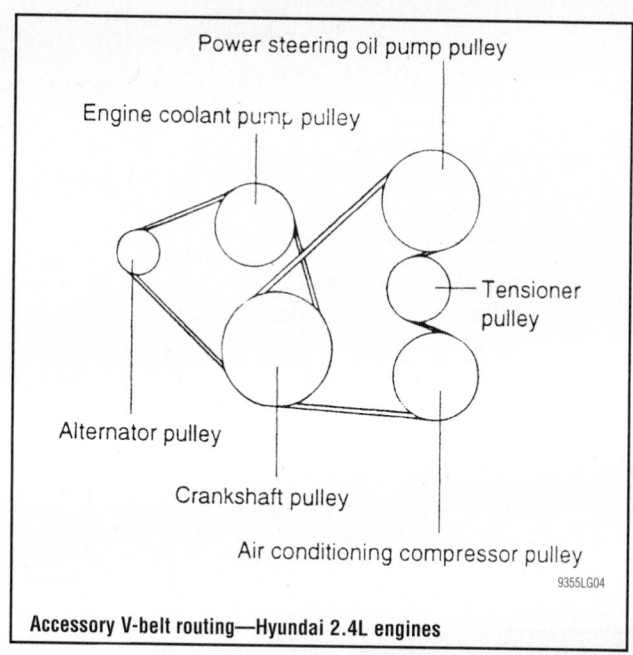

Accessory V-belt routing—Hyundai 2.4L engines

9355LG04

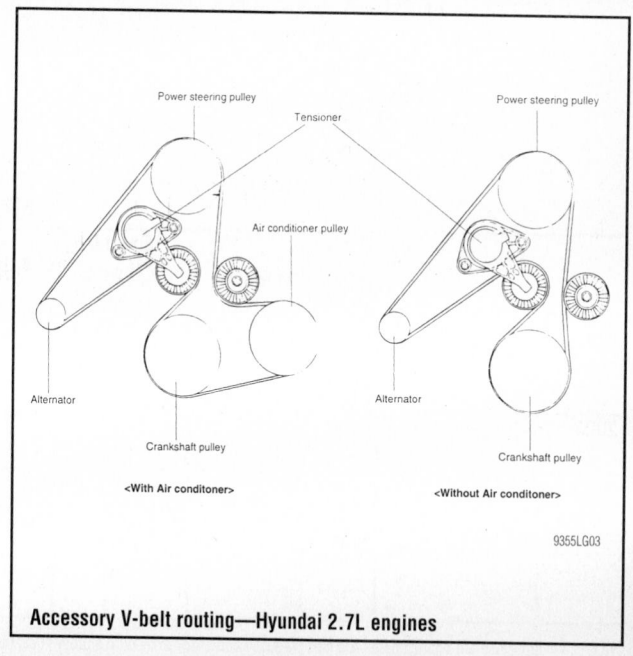

Accessory V-belt routing—Hyundai 2.7L engines

9355LG03

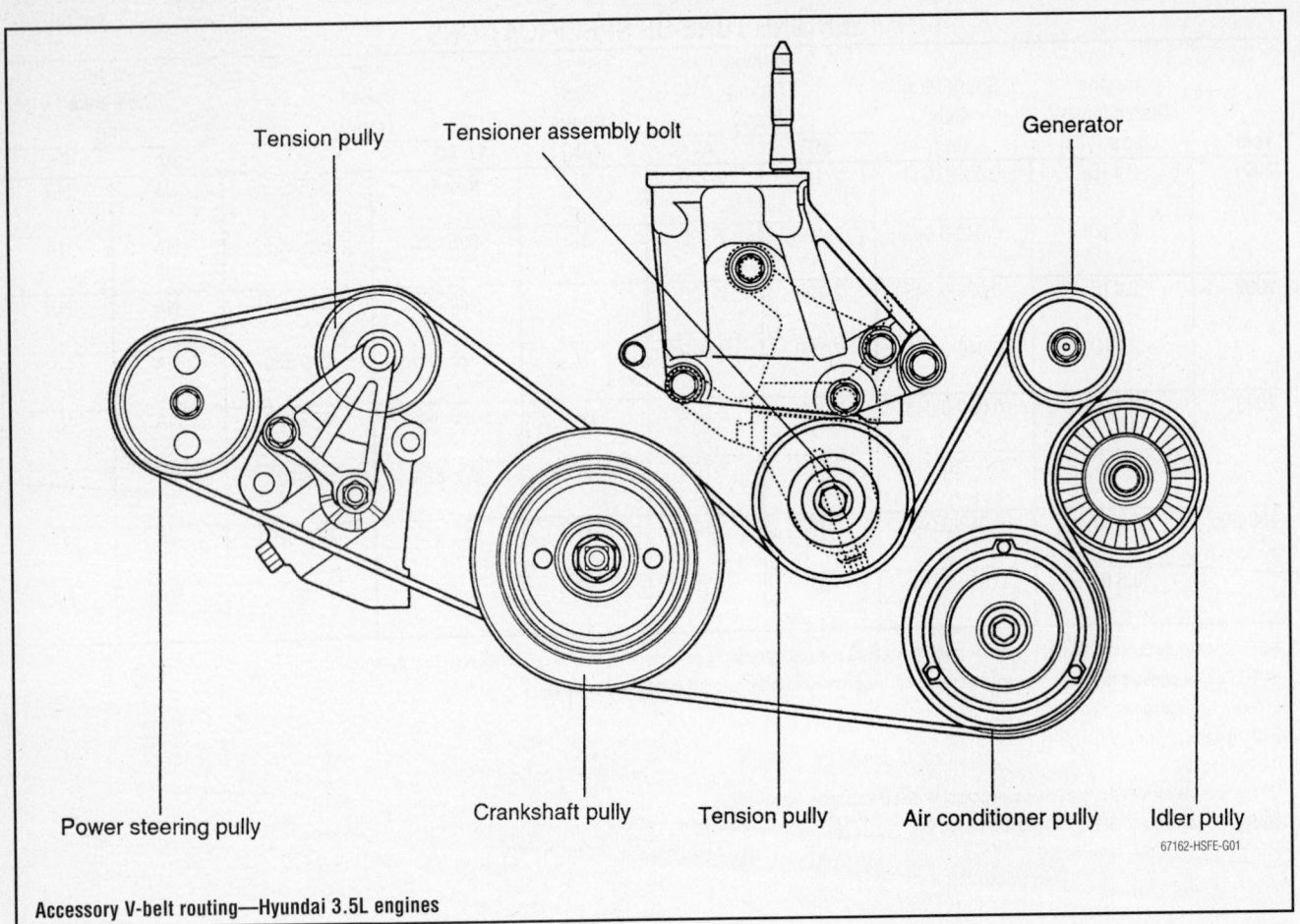

Accessory V-belt routing—Hyundai 3.5L engines

67162-HSFE-G01

CAPACITIES

Year	Model	Engine Displacement Liters (VIN)	Engine Oil with Filter (qts.)	Transmission (qts.)		Transfer Case (pts.)	Drive Axle		Fuel Tank (gal.)	Cooling System (qts.)
				Man.	Auto.		Front (pts.)	Rear (pts.)		
2001	Santa Fe	2.4 (B)	4.53	2.3	8.2	—	—	2.2	17.2	7.35
		2.7 (D)	4.76	2.3	8.2	—	—	2.2	17.2	8.94
2002	Santa Fe	2.4 (B)	4.53	2.3	8.2	—	—	2.2	17.2	7.35
		2.7 (D)	4.76	2.3	8.2	—	—	2.2	17.2	8.94
2003	Santa Fe	2.4 (B)	4.53	2.3	8.2	—	—	2.2	17.2	7.35
		2.7 (D)	4.76	2.3	8.2	—	—	2.2	17.2	8.94
2004-05	Santa Fe	2.4 (B)	4.53	2.3	8.2	—	—	2.2	19.0	7.71
		2.7 (D)	4.76	2.3	8.2	—	—	2.2	19.0	9.09
		3.5 (E)	4.55	2.3	8.98	—	—	2.2	19.0	8.66

NOTE: All capacities are approximate. Add fluid gradually and check to be sure a proper fluid level is obtained.

67162-HSFE-C04

VALVE SPECIFICATIONS

Year	Engine Displacement Liters (VIN)	Seat Angle (deg.)	Face Angle (deg.)	Spring Test Pressure (lbs. @ in.)	Spring Free Length (in.)	Stem-to-Guide Clearance (in.)		Stem Diameter (in.)	
						Intake	Exhaust	Intake	Exhaust
2001	2.4 (B)	45	45	NA	1.804	0.0008-0.0019	0.0020-0.0030	0.2580-0.2590	0.2571-0.2579
	2.7 (D)	45	45	NA	1.670	0.008-0.0020	0.0014-0.0026	0.2350-0.2354	0.2340-0.2350
2002	2.4 (B)	45	45	NA	1.804	0.0008-0.0019	0.0020-0.0030	0.2580-0.2590	0.2571-0.2579
	2.7 (D)	45	45	NA	1.670	0.008-0.0020	0.0014-0.0026	0.2350-0.2354	0.2340-0.2350
2003	2.4 (B)	45	45	NA	1.804	0.0008-0.0019	0.0020-0.0030	0.2580-0.2590	0.2571-0.2579
	2.7 (D)	45	45	NA	1.670	0.008-0.0020	0.0014-0.0026	0.2350-0.2354	0.2340-0.2350
2004-05	2.4 (B)	45	45	NA	1.804	0.0008-0.0019	0.0020-0.0030	0.2580-0.2590	0.2571-0.2579
	2.7 (D)	45	45	NA	1.670	0.008-0.0020	0.0014-0.0026	0.2350-0.2354	0.2340-0.2350
	3.5 (E)	43.5-44	43.5-44	NA	1.826	0.009-0.0020	0.0020-0.0033	0.2580-0.2590	0.2570-0.2580

NA: Not Available

67162-HSFE-C05

CRANKSHAFT AND CONNECTING ROD SPECIFICATIONS
All measurements are given in inches

Year	Engine Displacement Liters (VIN)	Crankshaft				Connecting Rod		
		Main Brg. Journal Dia.	Main Brg. Oil Clearance	Shaft End-play	Thrust on No.	Journal Diameter	Oil Clearance	Side Clearance
2001	2.4 (B)	2.2434-2.2441	①	0.0020-0.0098	3	2.2434-2.2411	0.0007-0.0014	0.004-0.0098
	2.7 (D)	2.4402-2.4409	0.0002-0.0009	0.0024-0.0094	3	2.2434-2.2411	0.0007-0.0014	0.0039-0.0098
2002	2.4 (B)	2.2434-2.2441	①	0.0020-0.0098	3	2.2434-2.2411	0.0007-0.0014	0.004-0.0098
	2.7 (D)	2.4402-2.4409	0.0002-0.0009	0.0024-0.0094	3	2.2434-2.2411	0.0007-0.0014	0.0039-0.0098
2003	2.4 (B)	2.2434-2.2441	①	0.0020-0.0098	3	2.2434-2.2411	0.0007-0.0014	0.004-0.0098
	2.7 (D)	2.4402-2.4409	0.0002-0.0009	0.0024-0.0094	3	2.2434-2.2411	0.0007-0.0014	0.0039-0.0098
2004-05	2.4 (B)	2.2434-2.2441	①	0.0020-0.0098	3	2.2434-2.2411	0.0007-0.0014	0.004-0.0098
	2.7 (D)	2.4402-2.4409	0.0002-0.0009	0.0024-0.0094	3	2.2434-2.2411	0.0007-0.0014	0.0039-0.0098
	3.5 (E)	2.5190-2.5197	0.00086-0.00157	0.0020-0.0098	3	NA	0.0001-0.0017	0.0039-0.0098

NA: Not Available

① Nos. 1, 2, 4 and 5: 0.0007-0.0014
    No. 3: 0.0009-0.0016

67162-HSFE-C06

PISTON AND RING SPECIFICATIONS
All measurements are given in inches

Year	Engine Displacement Liters (VIN)	Piston Clearance	Ring Gap			Ring Side Clearance		
			Top Compression	Bottom Compression	Oil Control	Top Compression	Bottom Compression	Oil Control
2001	2.4 (B)	0.0008-0.0016	0.0098-0.0138	0.0157-0.0216	0.0039-0.0157	0.0012-0.0028	0.0008-0.0024	0.0024-0.0059
	2.7 (D)	0.0004-0.0012	0.0079-0.0138	0.0146-0.0205	0.0079-0.0276	0.0016-0.0031	0.0012-0.0028	NA
2002	2.4 (B)	0.0008-0.0016	0.0098-0.0138	0.0157-0.0216	0.0039-0.0157	0.0012-0.0028	0.0008-0.0024	0.0024-0.0059
	2.7 (D)	0.0004-0.0012	0.0079-0.0138	0.0146-0.0205	0.0079-0.0276	0.0016-0.0031	0.0012-0.0028	NA
2003	2.4 (B)	0.0008-0.0016	0.0098-0.0138	0.0157-0.0216	0.0039-0.0157	0.0012-0.0028	0.0008-0.0024	0.0024-0.0059
	2.7 (D)	0.0004-0.0012	0.0079-0.0138	0.0146-0.0205	0.0079-0.0276	0.0016-0.0031	0.0012-0.0028	NA
2004-05	2.4 (B)	0.0008-0.0016	0.0098-0.0138	0.0157-0.0216	0.0039-0.0157	0.0012-0.0028	0.0008-0.0024	0.0024-0.0059
	2.7 (D)	0.0004-0.0012	0.0079-0.0138	0.0146-0.0205	0.0079-0.0276	0.0016-0.0031	0.0012-0.0028	NA
	3.5 (E)	0.0012-0.0020	0.0078-0.0118	0.0157-0.0216	0.0079-0.0276	0.0016-0.0031	0.0008-0.0024	NA

NA: Not Applicable

67162-HSFE-C07

TORQUE SPECIFICATIONS
All readings in ft. lbs.

Year	Engine Displacement Liters (VIN)	Cylinder Head Bolts	Main Bearing Bolts	Rod Bearing Bolts	Crankshaft Damper Bolts	Flywheel Bolts	Manifold		Spark Plugs	Oil Pan Drain Plug
							Intake	Exhaust		
2001	2.4 (B)	①	②	③	58-72	94-101	④	⑤	15-22	25-33
	2.7 (D)	⑥	⑦	⑧	65-80	53-56	14-15	18-22	15-22	25-33
2002	2.4 (B)	①	②	③	58-72	94-101	④	⑤	15-22	25-33
	2.7 (D)	⑥	⑦	⑧	65-80	53-56	14-15	18-22	15-22	25-33
2003	2.4 (B)	①	②	③	58-72	94-101	④	⑤	15-22	25-33
	2.7 (D)	⑥	⑦	⑧	65-80	53-56	14-15	18-22	15-22	25-33
2004-05	2.4 (B)	①	②	③	58-72	94-101	④	⑤	15-22	25-33
	2.7 (D)	⑥	⑦	⑧	65-80	53-56	14-15	18-22	15-22	25-33
	3.5 (E)	75-82	52-59	⑨	130-138	NA	14-16	29-33	15-22	25-33

NOTE: Dip main bearing bolts and crankshaft damper bolt in clean engine oil prior to tightening.

NA: Not Available

① If using used parts:
Step 1: 14 ft. lbs. (20 Nm).
Step 2: plus an additional 90 degrees.
Step 3: plus an additional 90 degrees.
If using new parts:
Step 1: 46 ft. lbs. (64 Nm)
Step 2: Release the bolts.
Step 3: 14 ft. lbs. (20 Nm)
Step 4: plus an additional 90 degrees.
Step 5: plus an additional 90 degrees.

② 15 ft. lbs. Plus 90 degrees

③ 14 ft. lbs. Plus 90 degrees

④ Bolt (M8): 11-14 ft. lbs.
Nut: 22-30 ft. lbs.

⑤ Bolt (M8): 18-2 ft. lbs.
Bolt (M10): 25-40

⑥ Step 1: 14 ft. lbs.
Step 2: plus an additional 90 degrees.
Step 3: plus an additional 90 degrees.

⑥ Step 1: 18 ft. lbs.
Step 2: plus an additional 58<en dash>62 degrees.
Step 3: plus an additional 43<en dash>47 degrees.

⑦ Bolt (M10): 10-12 ft. lbs.
Bolt (M7): 7-9

⑧ 12-15 ft. lbs. Plus 90-94 degrees

⑨ 26 ft. lbs. Plus 90 degrees

67162-HSFE-C08

[2.4 I4]

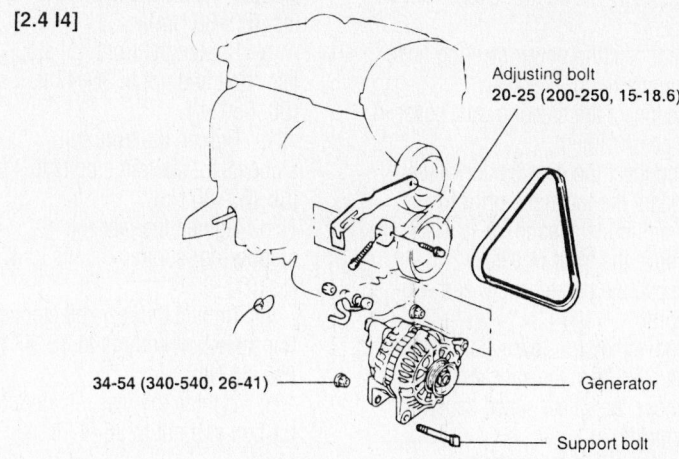

Adjusting bolt
20-25 (200-250, 15-18.6)

34-54 (340-540, 26-41) — Generator

Support bolt

[2.7 V6]

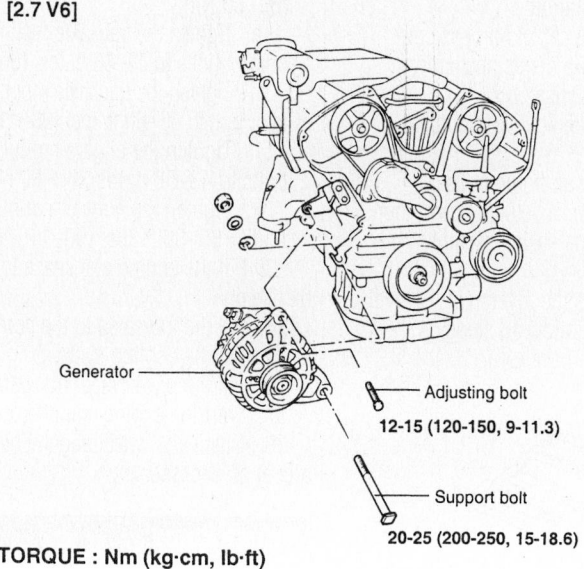

Generator

Adjusting bolt
12-15 (120-150, 9-11.3)

Support bolt
20-25 (200-250, 15-18.6)

TORQUE : Nm (kg·cm, lb·ft)

Exploded view of the alternator mounting for both engines used in the Santa Fe

Ignition Timing

ADJUSTMENT

The Santa Fe is equipped with a Distributorless Ignition System (DIS). The ignition timing is controlled by the Powertrain Control module (PCM). No adjustment is necessary.

Engine Assembly

REMOVAL & INSTALLATION

Except 3.5L Engine

1. Before servicing the vehicle, refer to the precautions in the beginning of this section.

2. Remove the battery and air cleaner assembly.
3. Drain the cooling system.
4. Drain the engine oil.
5. Drain the transaxle fluid.
6. Relieve the fuel system pressure.
7. Disconnect the following electrical connections:
   • Starter
   • Alternator
   • Throttle Position Sensor (TPS)
   • Power steering switch connector
   • Oil pressure gauge connector
   • Back-up lamp switch connector
   • A/T solenoid inhibitor switch connector
   • Coolant Temperature Sensor (CTS)
   • Ignition coil
   • Idle Speed Control (ISC) valve connector

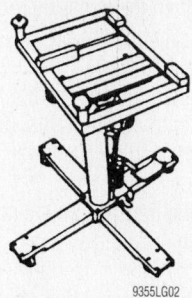

Attach the special tool to the transmission jack and support the transmission

   • Manifold Absolute Pressure (MAP) sensor
   • Oxygen (O_2S) sensor connector
8. If equipped with an automatic transmission, disconnect the oil cooler lines.
9. Remove or disconnect the following:
   • Radiator hoses from the engine
   • Radiator
   • Engine ground
   • Brake vacuum hose
   • Heater hoses at the engine
   • Throttle cable at the engine
   • Cruise control cable at the engine, if equipped
   • Main fuel line at the supply/return pipe
   • Speedometer cable at the transaxle
   • Clutch or control cable from the transaxle
   • Power steering hoses from the pump
   • Steering dust cover in the engine compartment
   • Gear box universal joint bolt
   • Front wheel
   • Brake caliper and support with wire
   • Strut lower bolt
   • Front muffler bolts
   • Transaxle control rod and extension rod, if equipped with a manual transmission
10. Support the transmission with a jack using the special attachment shown in the accompanying illustration.
11. Make sure all cable, harness connectors and hoses are disconnected from the engine and transmission.
   • Engine and transaxle mounting brackets
   • Sub frame bolts
   • Drive shaft
12. Lower the engine and transaxle assembly enough so the front and rear roll stoppers can be removed.
13. Remove the engine assembly
14. Installation is the reverse of removal but please note the following steps:
   • Tighten the roll stopper bolts to 36–47 ft. lbs. (50–65 Nm)

- Tighten the transaxle mounting bracket bolts to 65–80 ft. lbs. (90–110 Nm)
- Tighten the engine mount bracket bolts to 43–58 ft. lbs. (60–80 Nm)

15. Fill the engine crankcase to the correct level.

16. Fill the transaxle to the correct level.

17. Fill the cooling system.

18. Fill the power steering system.

19. Start the engine and check for leaks.

20. Check the wheel alignment and adjust as necessary.

3.5L Engine

1. Before servicing the vehicle, refer to the precautions in the beginning of this section.

2. Remove the battery and engine cover assembly.

3. Remove the battery stay.

4. Remove the air cleaner.

5. Drain the cooling system.

6. Drain the engine oil.

7. Drain the transaxle fluid.

8. Relieve the fuel system pressure.

9. Disconnect the following electrical connections:

- Alternator
- Starter
- Power steering switch connector
- Oil pressure gauge connector
- A/C switch
- Fuel injector connectors
- Back-up lamp switch connector
- A/T solenoid inhibitor switch connector
- Ignition coils
- Power TR selector connector
- Idle Speed Control (ISC) valve connector
- AFS and ATS connectors
- Oxygen (O_2S) sensor connector

10. Remove any remaining electrical connections that would interfere with engine removal.

11. If equipped with an automatic transmission, disconnect the oil cooler lines.

12. Disconnect the radiator hoses from the engine.

13. Disconnect the engine and transaxle grounds.

14. Disconnect the brake booster vacuum hose.

15. Disconnect the heater hoses from the engine.

16. Disconnect the fuel delivery and return lines.

17. Disconnect the speedometer cable from the transaxle.

18. Disconnect the control cable from the transaxle.

19. Disconnect the power steering hose from the engine mount bracket.

20. Disconnect the steering dust cover in the engine compartment.

21. Disconnect the gear box universal joint bolt. Mark the locations prior to removal to aid in installation.

22. Remove the front wheel.

23. Remove the brake caliper and support with wire.

24. Remove the strut lower bolt and disconnect the strut from the knuckle.

25. Remove the wheel speed sensor from the knuckle.

26. Remove the front muffler bolts.

27. Support the transmission with a jack using the special attachment shown in the accompanying illustration.

28. Make sure all cable, harness connectors and hoses are disconnected from the engine and transmission.

29. Remove the engine and transaxle mounting brackets.

30. Remove the sub frame bolts.

31. Lower the transaxle side down, then lift the engine and transaxle assembly from the vehicle.

32. Installation is the reverse of removal but please note the following steps:

a. Tighten the front roll

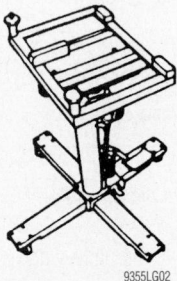

9355LG02

Attach the special tool to the transmission jack and support the transmission

stopper–to–transaxle bolts to 43–58 ft. lbs. (60–80 Nm).

b. Tighten the front roll stopper insulator bolt and nut to 36–47 ft. lbs. (50–65 Nm).

c. Tighten the front roll stopper–to–subframe bolts to 43–58 ft. lbs. (60–80 Nm).

d. Tighten the rear roll stopper–to–subframe bolts to 43–58 ft. lbs. (60–80 Nm).

e. Tighten the rear roll stopper–to–transaxle bolt and nut to 36–47 ft. lbs. (50–65 Nm).

f. Tighten the rear roll stopper insulator bolt and nut to 36–47 ft. lbs. (50–65 Nm).

g. Tighten the transaxle mounting sub–bracket bolts to 43–58 ft. lbs. (60–80 Nm).

h. Tighten the transaxle mounting bracket bolts to 43–58 ft. lbs. (60–80 Nm).

i. Tighten the transaxle mounting insulator bolt to 65–80 ft. lbs. (90–110 Nm).

j. Tighten the engine mount bracket bolts to 43–58 ft. lbs. (60–80 Nm).

k. Tighten the engine mount insulator bolt to 65–80 ft. lbs. (90–110 Nm).

33. Fill the engine crankcase to the correct level.

34. Fill the transaxle to the correct level.

35. Fill the cooling system.

36. Fill the power steering system.

37. Start the engine and check for leaks.

38. Check the wheel alignment and adjust as necessary.

Water Pump

REMOVAL & INSTALLATION

2.4L Engine

1. Before servicing the vehicle, refer to the precautions in the beginning of this section.

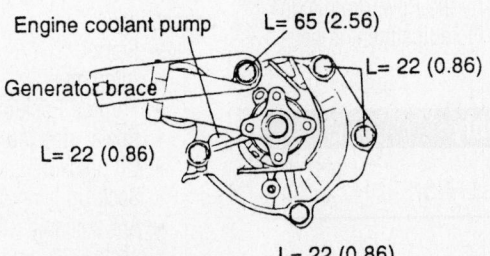

[DOHC ENGINE]

Engine coolant pump   L= 65 (2.56)

L= 22 (0.86)

Generator brace

L= 22 (0.86)

L= 22 (0.86)

L=Length of bolt mm (in.)

9355LG05

Make sure the water pump bolts are positioned in their original positions as the bolts are different lengths—2.4L engine

2. Drain the cooling system.
3. Remove or disconnect the following:
   - Negative battery cable
   - Water pump inlet pipe
   - Drive belt and water pump pulley
   - Timing belt covers
   - Timing belt tensioner
   - Water pump bolts
   - Alternator brace
   - Water pump and gasket
4. Clean the gasket mating surfaces.

To install:
5. Install or connect the following:
   - New O-ring onto the groove on the front end of the coolant pipe and wet the O-ring with water
   - Water pump and gasket
   - Bolts and alternator bracket. Tighten the bolts to 14–20 ft. lbs. (20–27 Nm).
   - Timing belt tensioner
   - Timing belt covers
   - Water pump pulley and drive belt
   - Water pump inlet pipe
   - Negative battery cable
6. Fill the cooling system.
7. Start the engine and check for leaks.

2.7L Engine

1. Before servicing the vehicle, refer to the precautions in the beginning of this section.
2. Drain the cooling system.

3. Remove or disconnect the following:
   - Negative battery cable
   - Drive belt and water pump pulley
   - Timing belt covers
   - Timing belt tensioner
   - Idler pulley
   - Water pump bolts
   - Water pump and gasket

4. Clean the gasket mating surfaces.
To install:
5. Install or connect the following:
   - Water pump and gasket
   - Bolts and alternator bracket. Tighten the bolts to 11–16 ft. lbs. (15–22 Nm).
   - Idler pulley

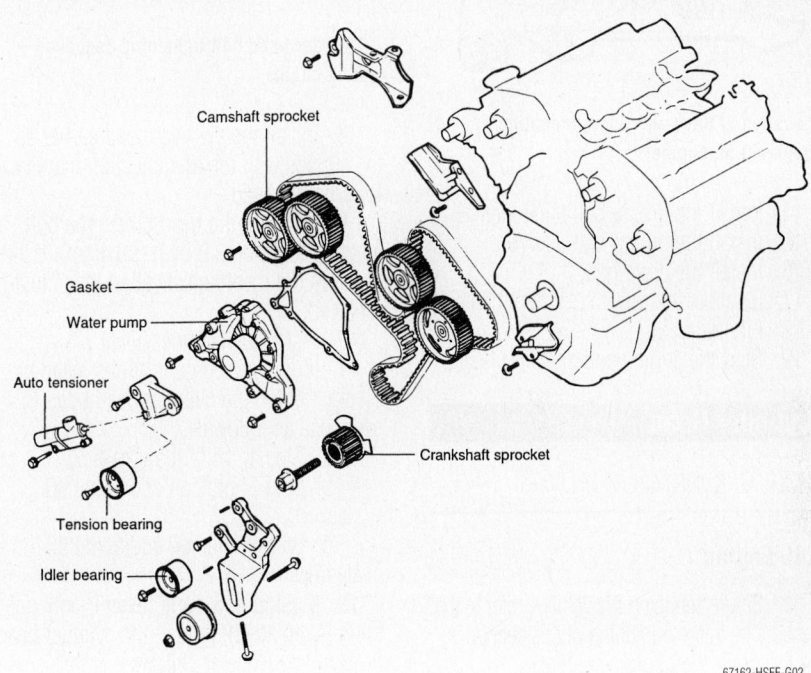

67162-HSFE-G02

Exploded view of the water pump mounting and related components—3.5L engine

- Timing belt tensioner
- Timing belt covers
- Water pump pulley and drive belt
- Negative battery cable
6. Fill the cooling system.
7. Start the engine and check for leaks.

3.5L Engine

1. Before servicing the vehicle, refer to the precautions in the beginning of this section.
2. Drain the cooling system.
3. Disconnect the negative battery cable.
4. Remove the drive belt.
5. Remove the timing belt covers.
6. Remove the timing belt tensioner.
7. Remove the idler pulley.
8. Remove the water pump bolts.
9. Remove the water pump and gasket.
10. Clean the gasket mating surfaces.
To install:
11. Install the water pump and gasket.
12. Install the bolts. Tighten the M8 bolts to 11–16 ft. lbs. (15–22 Nm) and the M10 bolt to 24–36 ft. lbs. (33–50 Nm).
13. Install the idler pulley.

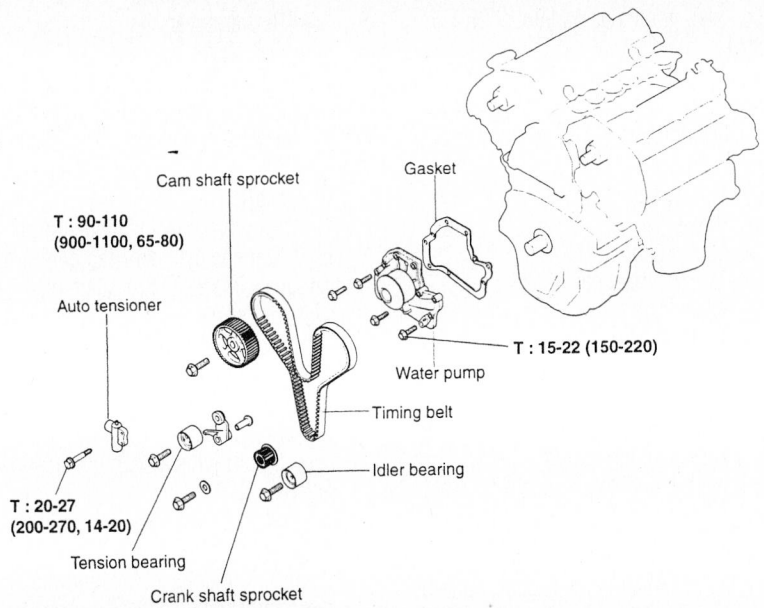

TORQUE : Nm (kg·cm, lb·ft)

9355LG06

Exploded view of the water pump mounting and related components—2.7L engine

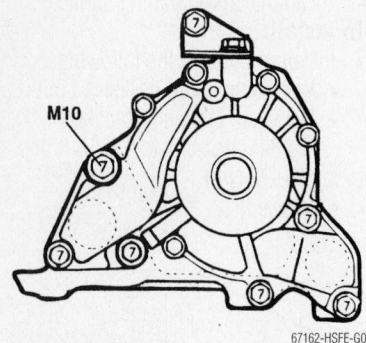

Location of the water pump mounting bolts—3.5L engine

14. Install the timing belt tensioner.
15. Install the timing belt covers.
16. Install the drive belt.
17. Connect the negative battery cable.
18. Fill the cooling system.
19. Start the engine and check for leaks.

Cylinder Head

REMOVAL & INSTALLATION

2.4L Engine

1. Before servicing the vehicle, refer to the precautions in the beginning of this section.
2. Drain the cooling system.
3. Relieve the fuel system pressure.
4. Remove or disconnect the following:
   - Negative battery cable
   - All necessary electrical connections, hoses and cables
   - Air cleaner
   - Intake manifold
   - Ignition coil
   - Timing belt
   - Exhaust manifold
   - Rocker cover
   - Camshafts
5. Loosen the cylinder head bolts in the sequence illustrated
6. Remove the cylinder head and gasket.

To install:
7. Clean the gasket mating surfaces

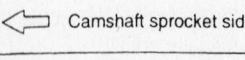

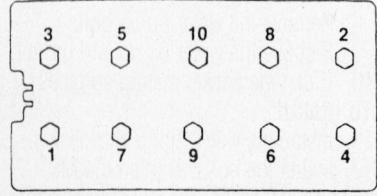

Cylinder head bolt loosening sequence—2.4L engine

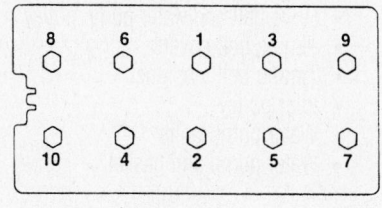

Cylinder head bolt tightening sequence—2.4L engine

8. Install the cylinder head gasket so the surface with the identification mark faces towards the head.
9. Measure the head bolts. The bolt length should be 3.9 inch (99.4mm). If the bolts do not meet specification they must be replaced.
10. Install the cylinder head.
11. If using used parts (bolts, head or block), tighten the cylinder head bolts in sequence as follows:
   a. Step 1: 14 ft. lbs. (20 Nm).
   b. Step 2: plus an additional 90 degrees.
   c. Step 3: plus an additional 90 degrees.
12. If using new parts (even if only one thing is replaced), tighten the cylinder head bolts in sequence as follows:
   a. Step 1: 46 ft. lbs. (64 Nm).
   b. Step 2: release the bolts.
   c. Step 3: 14 ft. lbs. (20 Nm).
   d. Step 4: plus an additional 90 degrees.
   e. Step 5: plus an additional 90 degrees.
13. Install or connect the following:
   - Camshafts
   - Timing belt
   - Rocker cover
   - Ignition coil
   - Exhaust manifold
   - Intake manifold
   - Air cleaner
   - All necessary electrical connections, hoses and cables
   - Negative battery cable
14. Fill the cooling system.
15. Start the engine and check for leaks.

2.7L Engine

1. Before servicing the vehicle, refer to the precautions in the beginning of this section.
2. Drain the cooling system.
3. Relieve the fuel system pressure.
4. Remove or disconnect the following:
   - Negative battery cable

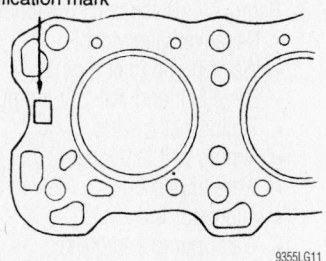

Location of the cylinder head gasket identification mark—2.7L engine

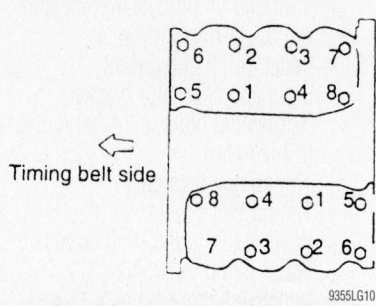

Cylinder head bolt tightening sequence—2.7L engine

   - All necessary electrical connections, hoses and cables
   - Air cleaner
   - Intake manifold
   - Ignition coil
   - Timing belt
   - Exhaust manifold
   - Rocker cover
   - Camshafts
5. Loosen the cylinder head bolts in the reverse order of the tightening sequence.
6. Remove the cylinder head and gasket.

To install:
7. Clean the gasket mating surfaces
8. Install the cylinder head gasket so the surface with the identification mark faces towards the head.
9. Install the cylinder head.
10. Tighten the cylinder head bolts in sequence as follows:
   a. Step 1: 18 ft. lbs. (25 Nm).
   b. Step 2: plus an additional 58–62 degrees.
   c. Step 3: plus an additional 43–47 degrees.
11. Install or connect the following:
   - Camshafts
   - Timing belt
   - Rocker cover
   - Ignition coil
   - Exhaust manifold
   - Intake manifold

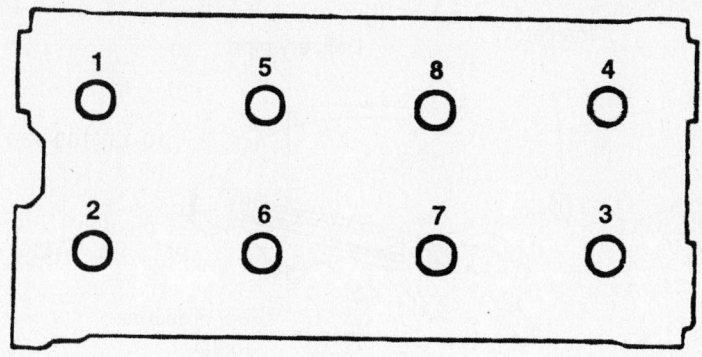

Cylinder head bolt loosening sequence—3.5L engine

67162-HSFE-G04

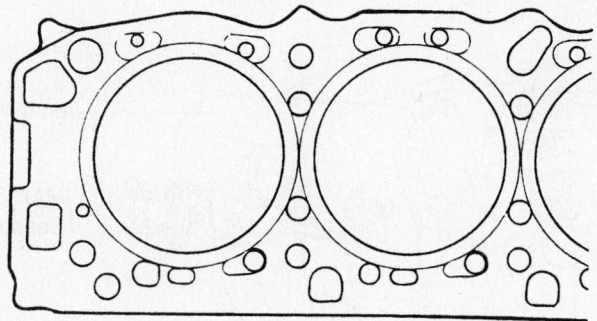

Location of the cylinder head gasket identification mark—3.5L engine

67162-HSFE-G05

- Air cleaner
- All necessary electrical connections, hoses and cables
- Negative battery cable
12. Fill the cooling system.
13. Start the engine and check for leaks.

3.5L Engine

1. Before servicing the vehicle, refer to the precautions in the beginning of this section.
2. Drain the cooling system.
3. Relieve the fuel system pressure.
4. Remove the engine from the vehicle.
5. Disconnect the spark plug wires.
6. Remove the ignition coil.
7. Remove the timing belt covers.
8. Remove the timing belt and camshaft sprockets.
9. Remove the heat shield and exhaust manifold.
10. Remove the water pump pulley and valve cover.
11. Remove the intake and exhaust camshafts.
12. Loosen the cylinder head bolts in the sequence shown using a 12mm socket in 2–3 steps.
13. Remove the cylinder head and gasket.

To install:
14. Clean the gasket mating surfaces
15. Install the cylinder head gasket so the surface with the identification mark faces towards the head.
16. Install the cylinder head.
17. Tighten the cylinder head bolts to 75–82 ft. lbs. (105–115 Nm).
18. Install the intake and exhaust camshafts.
19. Install the water pump pulley and valve cover.
20. Install the heat shield and exhaust manifold.
21. Install the timing belt and camshaft sprockets.
22. Install the timing belt covers.
23. Install the ignition coil.
24. Connect the spark plug wires.
25. Install the engine in the vehicle.
26. Fill the cooling system.
27. Change the oil and filter.
28. Start the engine and check for leaks.

Rocker Arms/Shafts

REMOVAL & INSTALLATION

Refer to the camshaft removal and installation procedure.

Intake Manifold

REMOVAL & INSTALLATION

2.4L Engine

1. Before servicing the vehicle, refer to the precautions in the beginning of this section.
2. Remove or disconnect the following:
   - Negative battery cable
   - Air breather hose from the throttle body
   - Throttle cable
   - Engine coolant hose and throttle body
   - Positive Crankcase Ventilation (PCV) valve and brake booster vacuum hose
   - Vacuum hose connector
   - Injector cover
   - High pressure fuel hose
   - Fuel injector harness connector
   - Delivery pipe with the injectors and the pressure regulator as an assembly
   - Intake manifold stay
   - Intake manifold

To install:
3. Install or connect the following:

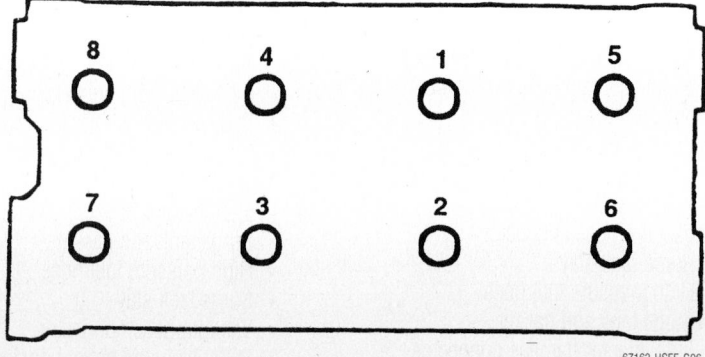

Cylinder head bolt tightening sequence—3.5L engine

67162-HSFE-G06

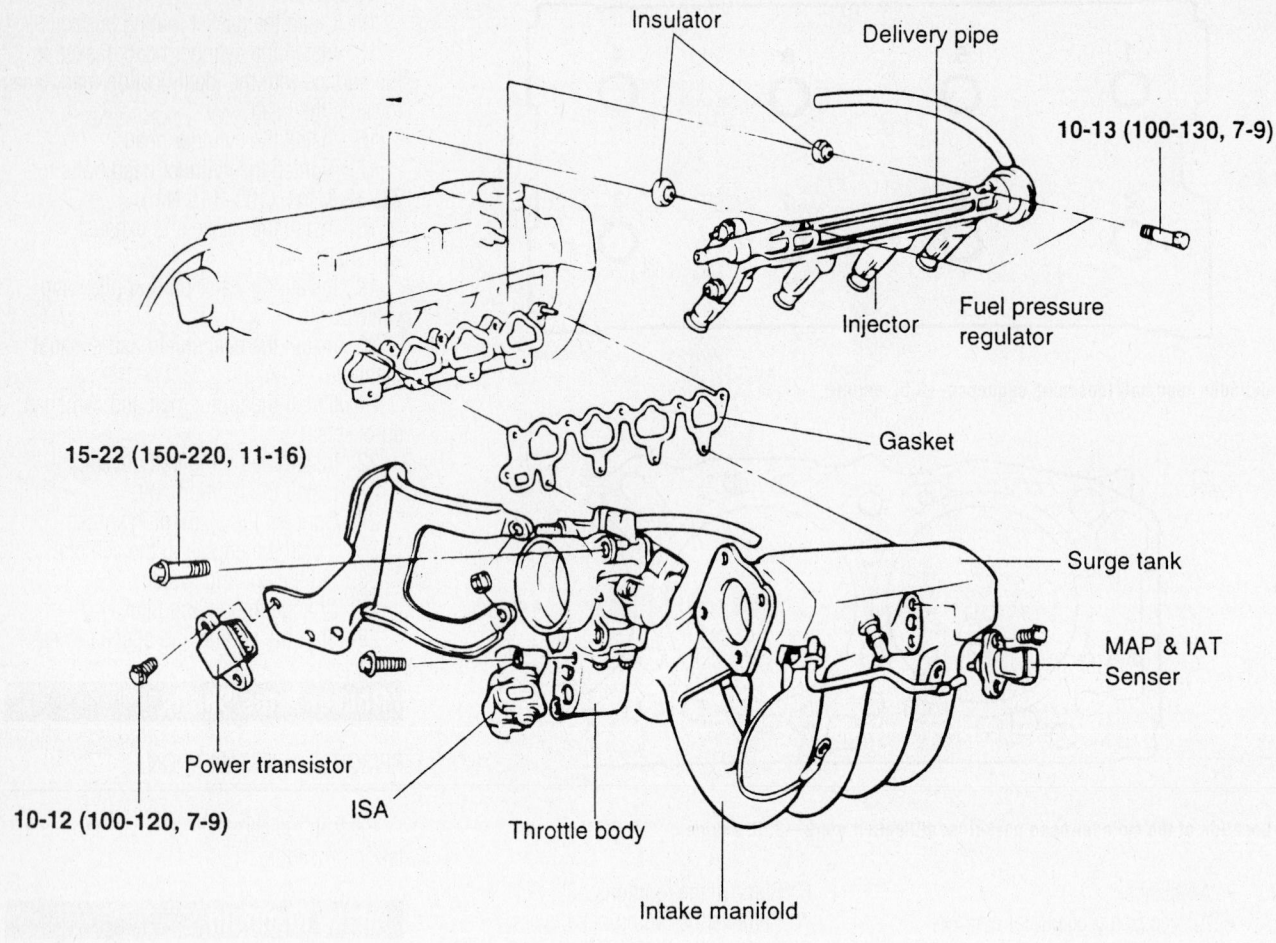

Insulator

Delivery pipe

10-13 (100-130, 7-9)

Injector

Fuel pressure regulator

Gasket

15-22 (150-220, 11-16)

Surge tank

MAP & IAT Senser

Power transistor

ISA

10-12 (100-120, 7-9)

Throttle body

Intake manifold

TORQUE : Nm (kg·cm, lb·ft)

9355LG12A

Exploded view of the intake manifold—2.4L engine

- New intake manifold gasket
- Intake manifold and bolts and nuts. Tighten the bolts to 11–14 ft. lbs. (15–20 Nm) and the nuts to 22–30 ft. lbs. (30–42 Nm).
- Delivery pipe and injector assembly
- Intake manifold stay and tighten the bolts to 13–18 ft. lbs. (18–25 Nm)
- Fuel injector harness connector
- High pressure fuel hose
- Injector cover
- Vacuum hose connector
- PCV valve and brake booster vacuum hose
- Throttle body and engine coolant hose
- Throttle cable
- Air breather hose from the throttle body
- Negative battery cable

4. Start the engine and check for proper operation.

2.7L Engine

1. Before servicing the vehicle, refer to the precautions in the beginning of this section.

2. Remove or disconnect the following:

- Negative battery cable
- Air breather hose from the throttle body
- Throttle and cruise control cables
- Engine coolant hose and throttle body
- Positive Crankcase Ventilation (PCV) valve and brake booster vacuum hose
- Vacuum hose connector
- Surge tank stay
- High pressure fuel hose
- Surge tank and gasket
- Fuel injector harness connector
- Delivery pipe with the injectors and

the pressure regulator as an assembly
- Coolant Temperature Sensor (CTS) electrical connector
- Intake manifold

To install:

3. Install or connect the following:

- New intake manifold gasket
- Intake manifold and tighten the bolts to 14–15 ft. lbs. (19–21 Nm)
- CTS electrical connector
- Delivery pipe with the injectors and the pressure regulator as an assembly
- Fuel injector harness connector
- Surge tank and gasket
- High pressure fuel hose
- Surge tank stay
- Vacuum hose connector
- PCV valve and brake booster vacuum hose

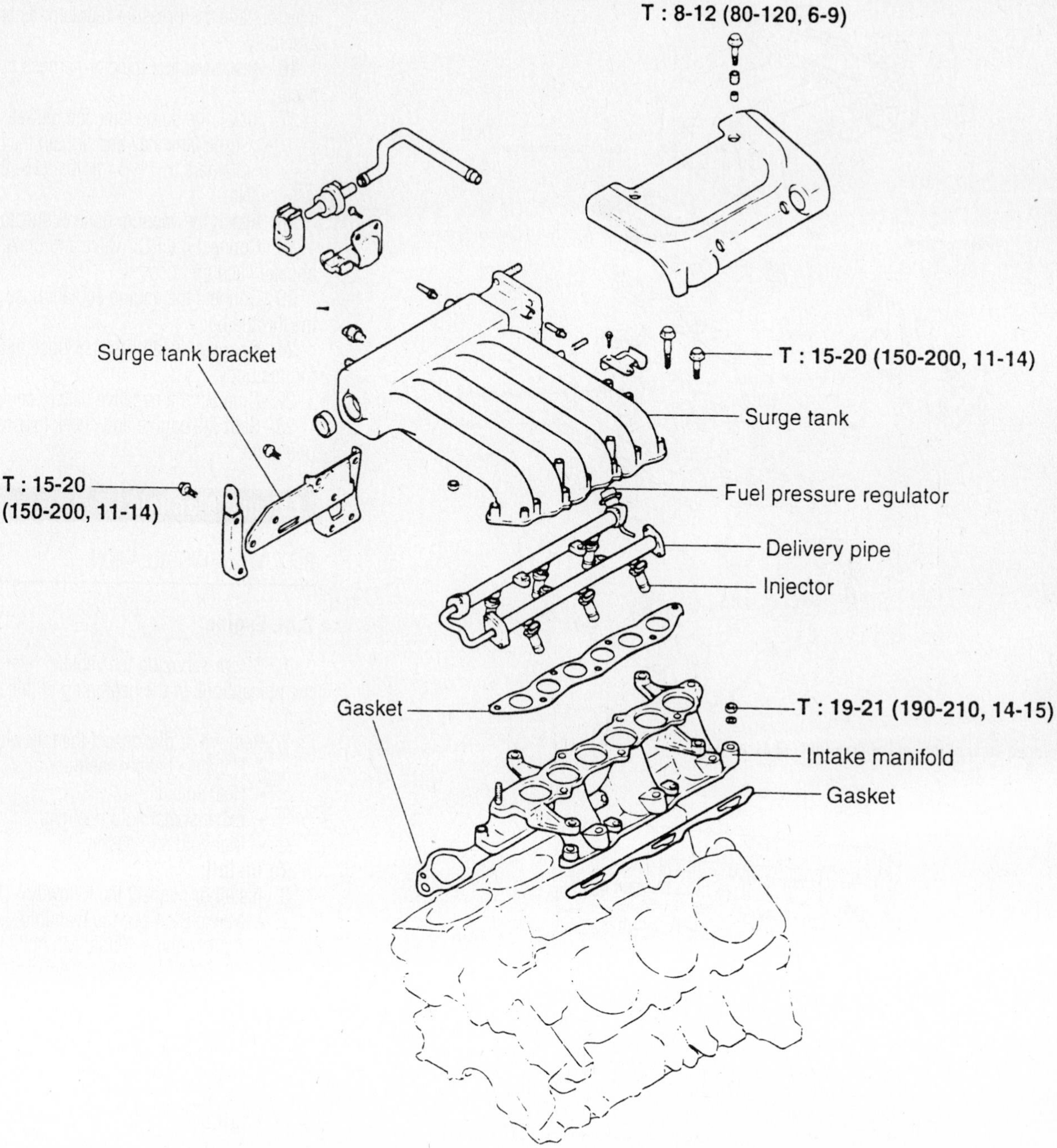

T : 8-12 (80-120, 6-9)

Surge tank bracket

T : 15-20 (150-200, 11-14)

Surge tank

T : 15-20
(150-200, 11-14)

Fuel pressure regulator

Delivery pipe

Injector

Gasket

T : 19-21 (190-210, 14-15)

Intake manifold

Gasket

9355LG12

Exploded view of the intake manifold—2.7L engine

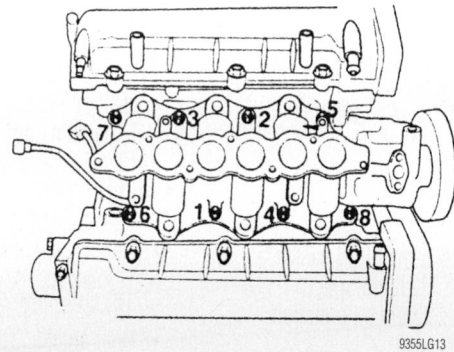

9355LG13

Intake manifold torque sequence—2.7L engine

- Engine coolant hose and throttle body
- Throttle and cruise control cables
- Air breather hose from the throttle body
- Negative battery cable

4. Start the engine and check for proper operation.

3.5L Engine

1. Before servicing the vehicle, refer to the precautions in the beginning of this section.

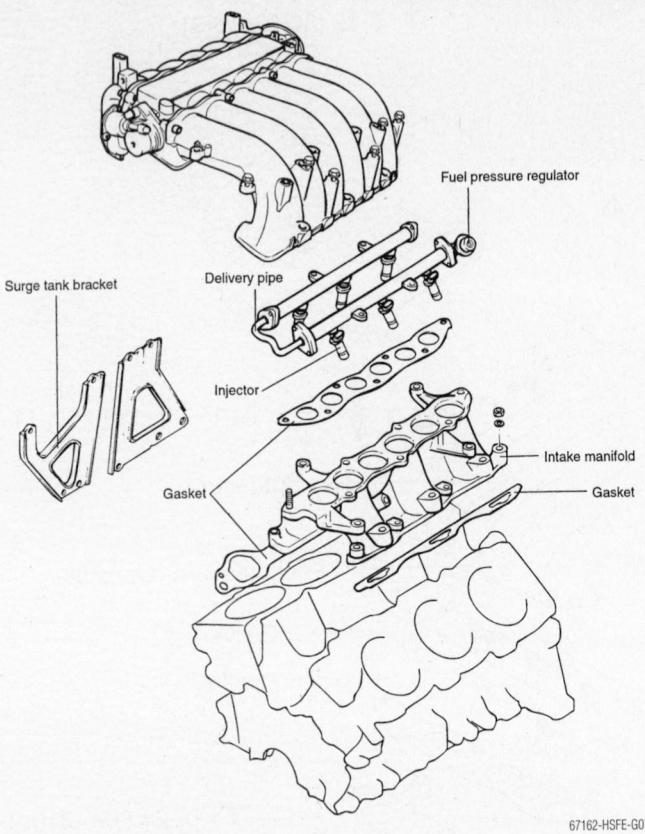

Exploded view of the intake manifold—3.5L engine

67162-HSFE-G07

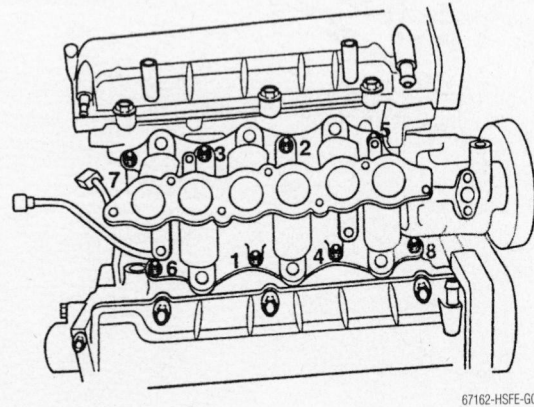

Intake manifold torque sequence—3.5L engine

67162-HSFE-G08

2. Disconnect the negative battery cable.

3. Remove the air breather hose from the throttle body.

4. Remove the Positive Crankcase Ventilation (PCV) valve and brake booster vacuum hoses.

5. Disconnect the vacuum hose connections.

6. Remove the surge tank stay.

7. Remove the surge tank and gasket.

8. Disconnect the fuel injector harness connector.

9. Remove the delivery pipe with the injectors and the pressure regulator as an assembly.

10. Disconnect the Coolant Temperature Sensor (CTS) electrical connector.

11. Remove the intake manifold bolts, manifold and gasket.

To install:

12. Install a new intake manifold gasket.

13. Install the intake manifold and tighten the bolts in the sequence illustrated to 14–16 ft. lbs. (20–23 Nm)

14. Attach the CTS electrical connector.

15. Install the delivery pipe with the injectors and the pressure regulator as an assembly.

16. Attach the fuel injector harness connector.

17. Install the surge tank and gasket.
- Surge tank stay and tighten the retainers to 11–14 ft. lbs. (15–20 Nm).

18. Attach the vacuum hose connectors.

19. Connect the PCV valve and brake booster vacuum hoses.

20. Connect the engine coolant hose to the throttle body.

21. Connect the air breather hose to the throttle body.

22. Connect the negative battery cable.

23. Start the engine and check for proper operation.

Exhaust Manifold

REMOVAL & INSTALLATION

2.4L Engine

1. Before servicing the vehicle, refer to the precautions in the beginning of this section.

2. Remove or disconnect the following:
- Negative battery cable
- Heat shield
- Exhaust manifold retainers
- Manifold and gasket

To install:

3. Install or connect the following:
- New gasket and the manifold. Tighten the manifold M8 bolts to 18–22 ft. lbs. (25–30 Nm) and the M10 bolts to 25–40 ft. lbs. (35–55 Nm).
- Heat shield
- Negative battery cable

2.7L Engine

1. Before servicing the vehicle, refer to the precautions in the beginning of this section.

2. Remove or disconnect the following:
- Negative battery cable
- Heat shield
- Exhaust manifold retainers
- Manifold and gasket

To install:

3. Install or connect the following:
- New gasket and the manifold. Tighten the manifold bolts to 18–22 ft. lbs. (25–30 Nm).
- Heat shield
- Negative battery cable

3.5L Engine

1. Before servicing the vehicle, refer to the precautions in the beginning of this section.

2. In necessary, disconnect the Oxygen ($O_2$2) sensor connector and remove the sensor.

3. Remove the front muffler.

4. Remove the heat shield.

5. Remove the exhaust manifold retainers.

6. Remove the manifold and gasket.

To install:

7. Install a new gasket and the manifold. Tighten the manifold bolts to 29–33 ft. lbs. (40–45 Nm).

8. Install the heat shield.

9. If removed, install the O_2 sensor, tighten to 29–36 ft. lbs. (40–50 Nm) and attach the electrical connector.

10. Install the front muffler.

Camshaft

REMOVAL & INSTALLATION

2.4L Engine

1. Before servicing the vehicle, refer to the precautions in the beginning of this section.

2. Drain the cooling system.

3. Remove or disconnect the following:
- Negative battery cable
- Breather hose from the air cleaner and rocker cover
- Air cleaner
- Timing belt cover
- Rocker cover and the Crankshaft Position (CKP) sensor
- Camshaft sprocket bolts and the sprockets
- Timing belt
- Camshaft bearing cap bolts using several passes
- Camshaft bearing caps, camshafts, rocker arms and lash adjusters

To install:

4. Inspect all parts for wear and damage.

5. Install or connect the following:
- Camshafts and apply engine oil to the journals. Do not install the rocker arms yet.

➡ **The exhaust camshaft has a slit on the rear end for the CKP sensor.**

- Bearing caps. The caps are marked **I** for intake and **E** for exhaust, they also contain the cap number. For example **I2** would be intake cap number 2.

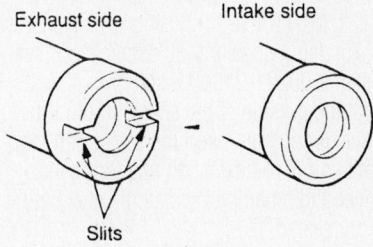

Exhaust side   Intake side

Slits

The exhaust camshaft has a slit on the rear end for the CKP sensor—2.4L engine

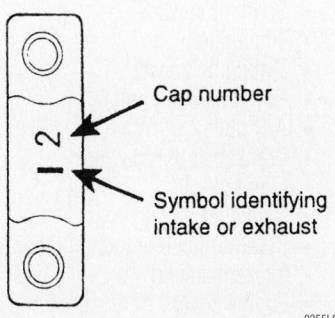

Cap number

Symbol identifying intake or exhaust

The caps are marked I for intake and E for exhaust, they also contain the cap number. For example I2 would be intake cap number 2—2.4L engine

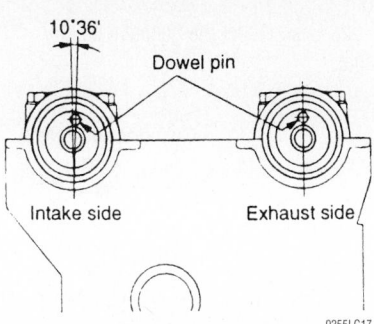

10°36'   Dowel pin

Intake side   Exhaust side

Make sure the dowel pins on the ends of camshaft sprockets are facing up—2.4L engine

6. Make sure the camshafts can turn freely, then remove the caps and the camshafts.

7. Make sure the dowel pins on the ends of camshaft sprockets are facing up.
- Rocker arms
- Camshafts and bearing caps. Tighten the bearing cap bolts uniformly to 14–15 ft. lbs. (19–21 Nm).

8. Using special tools, camshaft oil seal installer and guide 09221-21000, 09221-21100 install the oil seal. Coat the outside of the seal with oil prior to installation, then slide the seal along the front end of the

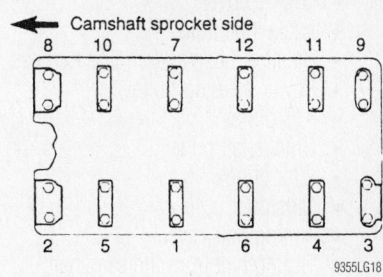

Camshaft sprocket side
8 10 7 12 11 9
2 5 1 6 4 3

Camshaft bearing cap locations—2.4L engines

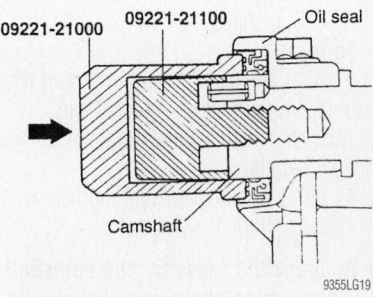

09221-21000   09221-21100   Oil seal

Camshaft

Installing the camshaft oil seal—2.4L engine

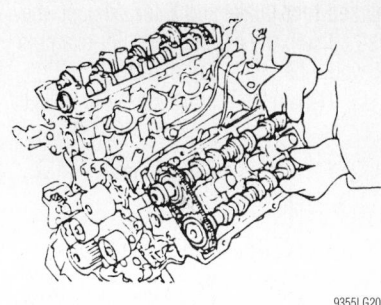

Remove the camshafts from the head—2.7L engine

camshaft and using the driver and a hammer install the seal until it is full seated.
- Camshaft sprockets and bolts. Tighten the bolts to 58–72 ft. lbs. (80–100 Nm).
- CKP sensor and rocker cover
- Breather hose to the air cleaner and rocker cover
- Air cleaner
- Negative battery cable

9. Start the engine and check for leaks.

2.7L Engine

1. Before servicing the vehicle, refer to the precautions in the beginning of this section.

2. Remove or disconnect the following:
- Negative battery cable

- Engine cover
- Intake manifold
- Breather hose and engine harness
- Power steering pulley
- A/C pulley
- Crankshaft pulley
- Idler pulley
- Tensioner pulley
- Timing belt cover
- Timing belt from the camshaft sprocket(s)
- Spark plug wires
- Rocker arm cover
- Camshaft sprockets
- Camshaft bearing caps
- Camshafts

To install:

3. Align the camshaft timing chain with the intake timing chain sprocket and exhaust sprocket as shown in the accompanying illustration.

4. Lubricate the camshaft journals with oil and install them.

➡**To check the press fit, the camshaft (IN) and timing chain sprocket should be separable by a force greater than 1000kg minimum at room temperature.**

5. Install the bearing caps. The caps are marked **I** for intake and **E** for exhaust, they also contain the cap number. For example **I2** would be intake cap number 2.

6. Tighten the bearing caps M10 bolts to

10–12 ft. lbs. (14–16 Nm) and the M7 bolts to 7–9 ft. lbs. (10–12 Nm) using several passes.

7. Using special tool, camshaft oil seal installer 09221-21000 install the oil seal. Coat the outside of the seal with oil prior to installation, then insert the seal along the front end of the camshaft and using the driver and a hammer install the seal until it is full seated.

8. Install or connect the following:
- Camshaft sprocket and tighten the bolt to 65–80 ft. lbs. (90–110 Nm)
- Rocker cover
- Spark plug wires
- Timing belt
- Timing belt cover
- Power steering pulley
- A/C pulley
- Crankshaft pulley
- Idler pulley
- Tensioner pulley
- Breather hose and engine harness
- Intake manifold
- Engine cover
- Negative battery cable

3.5L Engine

1. Before servicing the vehicle, refer to the precautions in the beginning of this section.

2. Disconnect the negative battery cable.

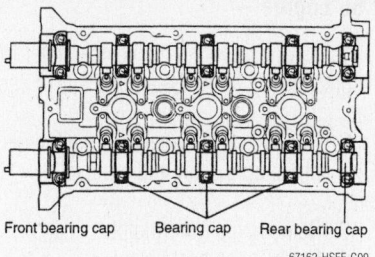

Remove the camshaft bearing caps—3.5L engine

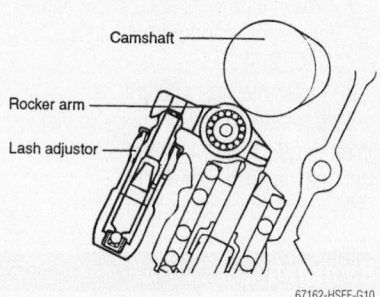

Camshaft and related component positioning—3.5L engine

3. Remove the engine cover.
4. Remove the intake manifold.
5. Disconnect the breather hose and engine harness.
6. Remove the timing belt.

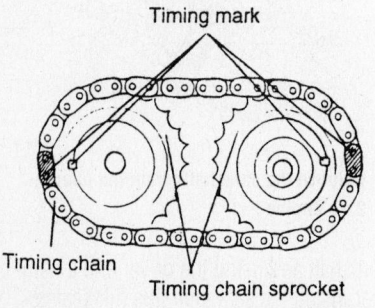

Align the camshaft timing chain with the intake timing chain sprocket and exhaust sprocket—2.7L engine

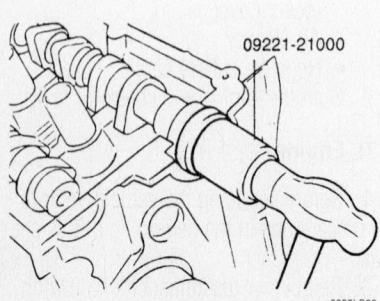

Install the camshaft oil seal—2.7L engine

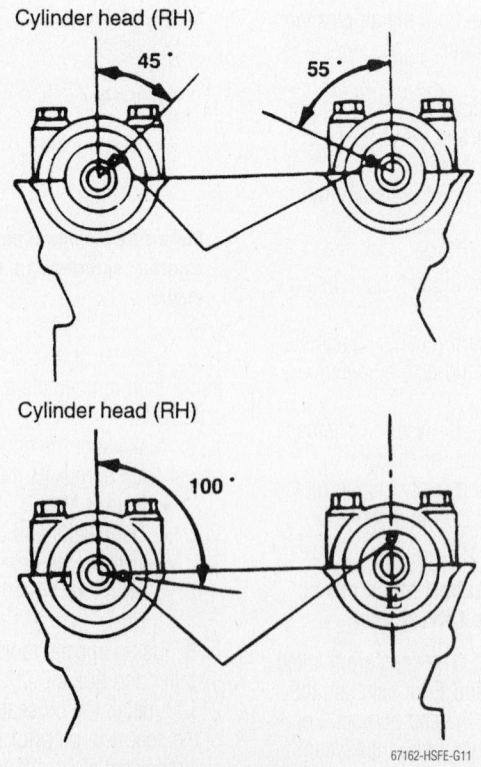

Install the camshaft dowel pin as shown—3.5L engine

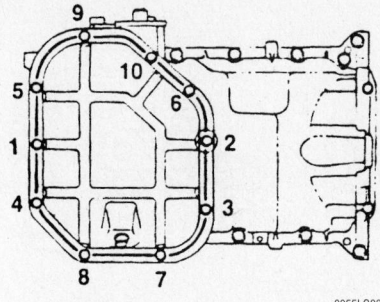

Lower oil pan bolt torque sequence—2.7L engine

9355LG33

- Lower oil pan bolts and the pan
- Upper oil pan bolts and the pan

4. Clean the gasket mating surfaces

To install:

5. Apply 0.16 inch (4mm) of sealant to the lower oil pan groove. Install the pan within 15 minutes of sealant installation.

6. Install the upper oil pan and the tighten the bolts in sequence as follows:

    a. 0.937 x 1.4961 inch (10 x 38mm) bolt to 22–30 ft. lbs. (30–42 Nm).

    b. 0.3150 x 0.866a inch (8 x 22mm) bolt 14–20 inch (19–28 Nm).

    c. 6.7519 inch (171.5mm) bolt to 4–5 ft. lbs. (5–7 Nm).

    d. 6.7520 inch (152.5mm) bolt to 4–5 ft. lbs. (5–7 Nm).

7. Install the lower pan and tighten the bolts to 7–9 ft. lbs. (10–12 Nm).

8. Connect the negative battery cable.

9. Refill the crankcase with oil.

3.5L Engine

1. Before servicing the vehicle, refer to the precautions in the beginning of this section.

2. Drain the engine oil.

3. Remove the oil pressure switch.

4. Disconnect the negative battery cable.

5. Remove the lower oil pan bolts and the pan.

6. Remove the upper oil pan bolts and the pan.

7. Clean the gasket mating surfaces

To install:

8. Apply 0.16 inch (4mm) of sealant to the lower oil pan groove. Install the pan within 15 minutes of sealant installation.

9. Install the upper oil pan and the tighten the bolts in sequence as follows:

    a. 0.7087 inch (6 x 18mm), 6.004 inch (6 x 152.5mm) bolts identified with either * or ** in the illustration to 4–5 ft. lbs. (5–7 Nm).

    b. 1.4961 inch (10 x 38mm) bolts

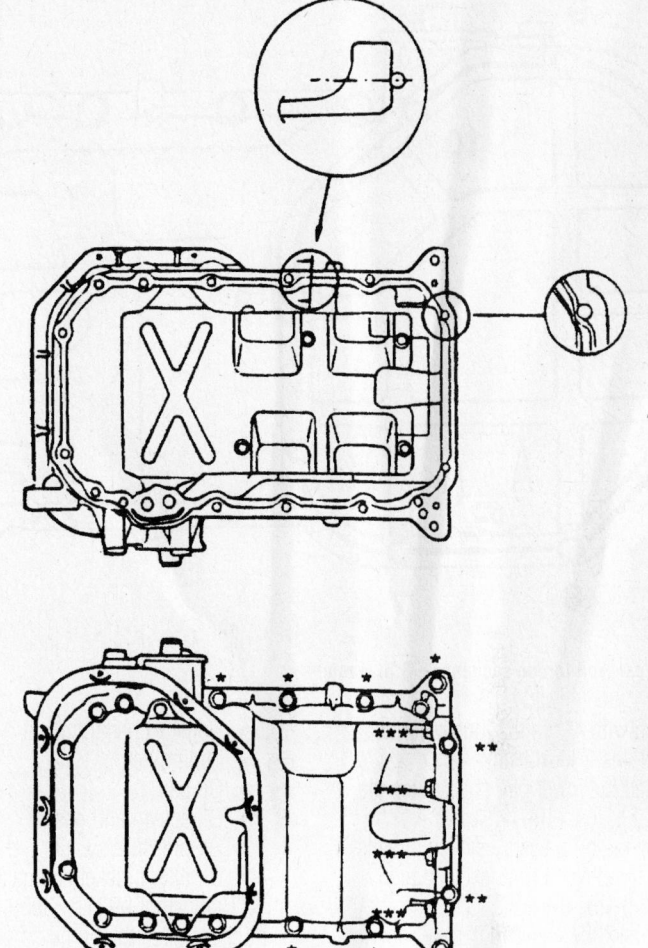

Apply 0.16 inch (4mm) of sealant to the lower oil pan groove—3.5L engine

67162-HSFE-G16

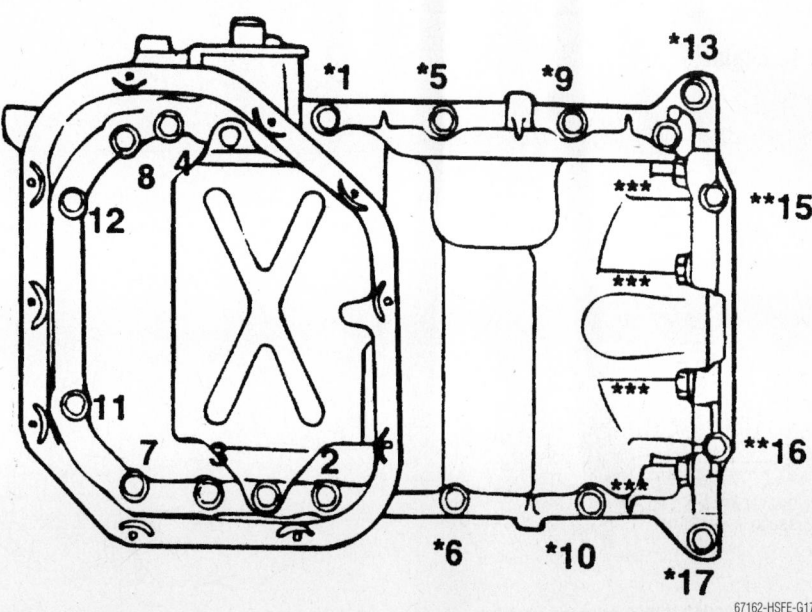

67162-HSFE-G17

Upper oil pan torque sequence (tighten bolts indicated with * or ** in the illustration to 4–5 ft. lbs. (5–7 Nm), and bolts indicated with a * to 22–30 ft. lbs. (30–42 Nm)—3.5L engine**

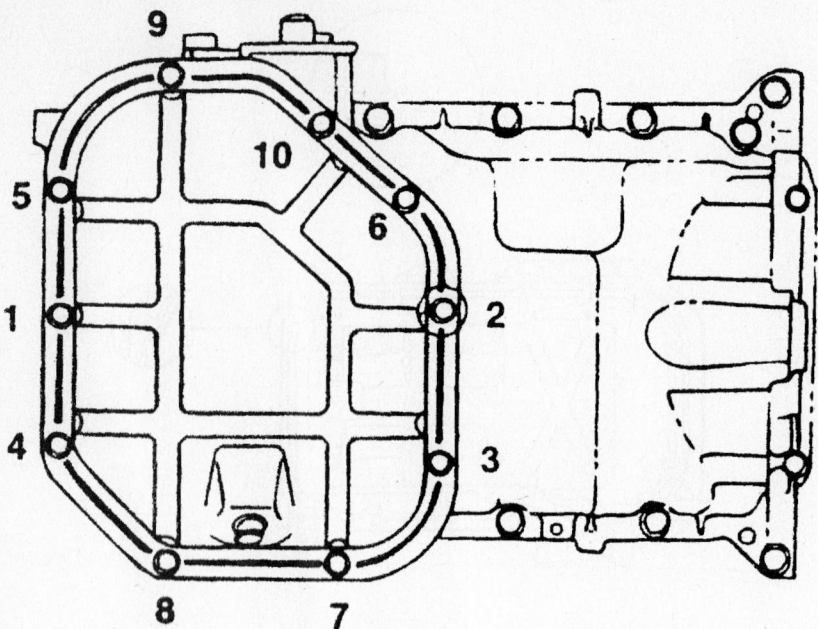

Lower oil pan bolt torque sequence—3.5L engine

67162-HSFE-G18

identified with *** in the illustration to 22–30 ft. lbs. (30–42 Nm).

10. Install the lower pan and tighten the bolts to 7–9 ft. lbs. (10–12 Nm).

11. Coat the oil pressure switch threads with Three bond No. 1104E sealant and tighten to 6 ft. lbs. (8 Nm).

12. Connect the negative battery cable.

13. Refill the crankcase with oil.

Oil Pump

REMOVAL & INSTALLATION

2.4L Engine

1. Before servicing the vehicle, refer to the precautions in the beginning of this section.

2. Drain the engine oil.

3. Remove or disconnect the following:

- Negative battery cable
- Timing belt
- Oil pan
- Oil screen and gasket
- Oil pressure switch
- Oil filter bracket and gasket

4. Using Tool 09213-33000, remove the plug cap from the oil pump portion of the case.

- Plug from the left side of the block

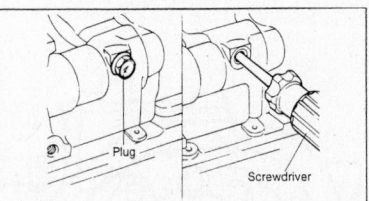

Insert an 0.32 inch (8mm) screwdriver at least 2.4 inch (60mm) into the plug hole—2.4L engine

9355LG25

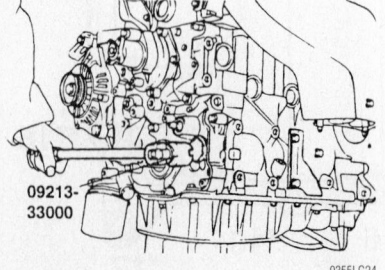

Use Tool 09213-33000, remove the plug cap from the oil pump portion of the case—2.4L engine

9355LG24

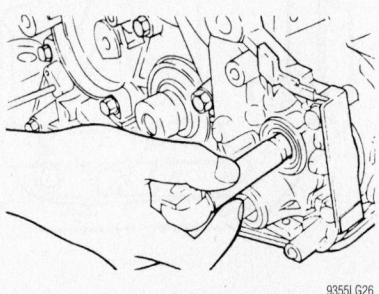

Remove the pump driven gear the left counter balance shaft bolt—2.4L engine

9355LG26

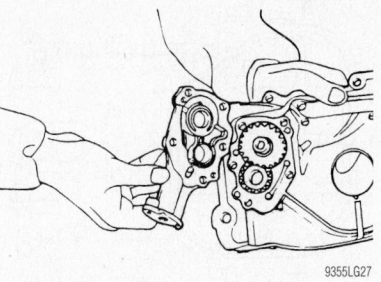

Remove the oil pump cover from the case—2.4L engine

9355LG27

and insert a 0.32 inch (8mm) screwdriver into the plug hole. The screwdriver must be inserted at least 2.4 inch (60mm).

- Pump driven gear the left counter balance shaft bolt
- Front case bolts (noting the bolt length and location), the case and gasket.
- Two counter balance shafts from the block
- Oil pump cover from the case
- Oil pump gears from the case
- Screwdriver from the plug hole

To install:

5. Install the oil pump gears.

6. Inspect the tip clearance of the gears using a feeler gauge. The specifications are as follows:

a. Standard value drive gear: 0.0063–0.0083 inch (0.16–0.21mm).

b. Standard value driven gear: 0.0071–0.0083 inch (0.18–0.21mm).

c. Limit drive gear: 0.0098 inch (0.25mm).

d. Limit driven gear: 0.0098 inch (0.25mm).

7. Inspect the side clearance of the gears using a feeler gauge. The specifications are as follows:

a. Standard value drive gear: 0.0031–0.0055 inch (0.08–0.14mm).

b. Standard value driven gear: 0.0024–0.0047 inch (0.06–0.12mm).

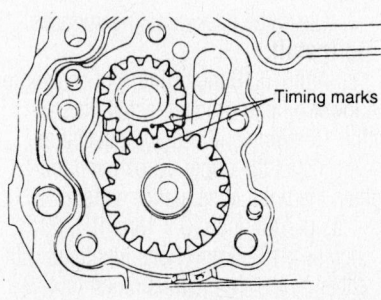

Apply engine oil to both gears and align the gear timing marks—2.4L engine

9355LG28

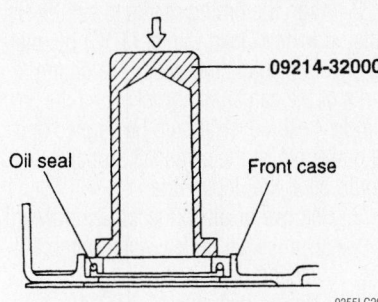

Using crankshaft front oil seal install Tool 09214-32000, install the oil seal into the case—2.4L engine

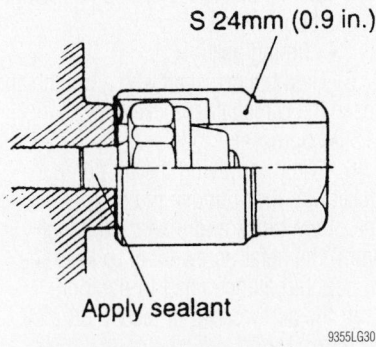

Front case bolt length and location—2.4L engine

c. Limit drive gear: 0.0098 inch (0.25mm).
d. Limit driven gear: 0.0098 inch (0.25mm).
8. Apply engine oil to both gears and align the gear timing marks.
9. Install the oil pump case.
10. Using crankshaft front oil seal install Tool 09214-32000, install the oil seal into the case.
11. Place special tool 09214-32100 on the front of the crankshaft and apply a coat of oil to the outside of the tool to aid in case installation.
12. Install or connect the following:
- New front case gasket and temporarily tighten the flange bolts
- Front case and tighten the bolts to 14–20 ft. lbs. (20–27 Nm), making sure the correct length bolt is installed in the correct location.
13. Insert an 0.32 inch (8mm) screwdriver into the plug hole. The screwdriver must be inserted at least 2.4 inch (60mm). Verify the shaft is in place and install the bolt.
- New O-ring on the groove on the front case
- Plug case and tighten to 14–20 ft. lbs. (20–27 Nm)
- Oil screen and gasket

- Oil pan
- Oil pressure switch using a 24mm deep socket. Apply Three bond 1104 sealant to the threads before installation and tighten to 6–9 ft. lbs. (8–12 Nm).
- Timing belt
- Negative battery cable
14. Fill the crankcase to the correct level.
15. Start the engine and check for leaks.

2.7L Engine

1. Before servicing the vehicle, refer to the precautions in the beginning of this section.
2. Drain the engine oil.
3. Remove or disconnect the following:
- Negative battery cable
- Oil pressure switch
- Oil filter and pans
- Oil screen and gasket
- Oil filter bracket and gasket
- Oil relief valve plug from the pump case
- Oil pump case

To install:
4. Install the oil pump gears.
5. Inspect the side clearance of the gears using a feeler gauge. The specifications are as follows:
   a. Standard body clearance: 0.0039–0.0071 inch (0.100–0.181mm).
   b. Standard side clearance: 0.0016–0.0037 inch (0.040–0.095mm).
6. Install the oil pump case with a new gasket. Tighten the bolt to 9–11 ft. lbs. (12–15 Nm) and the screw to 6–9 ft. lbs. (8–12 Nm).
7. Install a new oil seal into the pump as tightly as possible.
8. Using crankshaft front oil seal install Tool 09214-33000, install the oil seal into the case.
9. Install the relief plunger and spring and tighten the valve plug to 29–36 ft. lbs. (40–50 Nm).

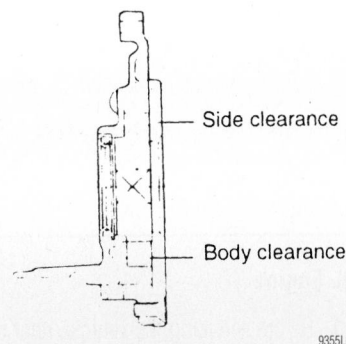

Check the oil pump gears side and body clearance–2.7L

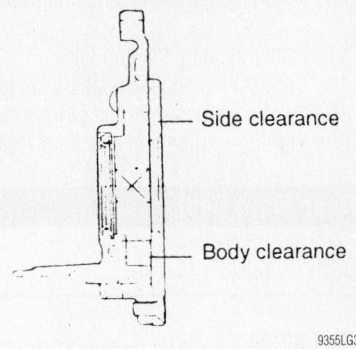

Check the oil pump gears side and body clearance–3.5L

10. Install the oil screen and a new gasket.
11. Install the oil pans and filter.
12. Connect the negative battery cable.
13. Fill the crankcase to the correct level.
14. Start the engine and check for leaks.

3.5L Engine

1. Before servicing the vehicle, refer to the precautions in the beginning of this section.
2. Drain the engine oil.
3. Disconnect the negative battery cable.
4. Remove the oil pressure switch.
5. Remove the oil filter and pans.
6. Remove the oil screen and gasket.
7. Remove the oil filter bracket and gasket.
8. Remove the oil relief valve plug from the pump case.
9. Remove the oil pump case.

To install:
10. Install the oil pump gears.
11. Inspect the side clearance of the gears using a feeler gauge. The specifications are as follows:
   a. Standard body clearance: 0.0039–0.0071 inch (0.100–0.181mm).
   b. Standard side clearance: 0.0016–0.0037 inch (0.040–0.095mm).
   c. Oil tip clearance: 0.0024–0.0071 inch (0.06–0.18mm).
12. Install the oil pump case with a new gasket. Tighten the bolt to 11–14 ft. lbs. (15–20 Nm) and the screw to 6–9 ft. lbs. (8–12 Nm).
13. Install a new oil seal into the pump as tightly as possible.
14. Using crankshaft front oil seal install Tool 09214-33000, install the oil seal into the case.
15. Install the relief plunger and spring and tighten the valve plug to 29–36 ft. lbs. (40–50 Nm).

16. Install the oil screen and a new gasket.

17. Install the oil pans and filter.

18. Connect the negative battery cable.

19. Fill the crankcase to the correct level.

20. Start the engine and check for leaks.

Rear Main Seal

REMOVAL & INSTALLATION

2.4L Engine

1. Before servicing the vehicle, refer to the precautions in the beginning of this section.

2. Remove or disconnect the following:
   - Transaxle
   - Clutch pressure plate and disc, if equipped
   - Flywheel
   - Oil seal case bolts and the case
   - Oil seal

To install:

3. Install or connect the following:
   - Oil seal. Drive the seal square into the seal case.
   - Oil seal case so that the oil hole in the separator may be directed downwards. Tighten the bolts to 7–9 ft. lbs. (10–12 Nm).
   - Flywheel

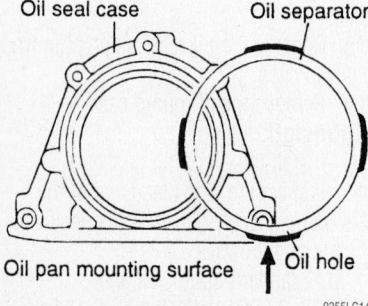

Position the oil seal case so that the oil hole in the separator may be directed downwards—2.4L engine

9355LG14

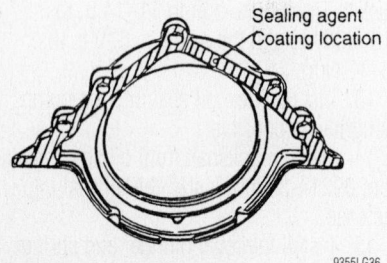

Apply sealant to the areas shown—2.7L and 3.5L engines

9355LG36

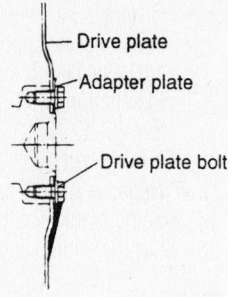

(A/T)
Drive plate
Adapter plate
Drive plate bolt

9355LG37

Install the drive plate and adapter plate— 2.7L and 3.5L engines

- Clutch pressure plate and disc, if equipped
- Transaxle

4. Check the fluid levels.

5. Start the engine and check for leaks.

2.7L And 3.5L Engines

1. Before servicing the vehicle, refer to the precautions in the beginning of this section.

2. Remove or disconnect the following:
   - Transaxle
   - Clutch pressure plate and disc, if equipped
   - Flywheel
   - Drive plate and adapter plate
   - Oil seal case bolts and the case
   - Oil seal

To install:

3. Install or connect the following:
   - Oil seal. Drive the seal square into the seal case.
   - Oil seal case so that the oil hole in the separator may be directed downwards. Tighten the bolts to 7–9 ft. lbs. (10–12 Nm).
   - Drive plate and adapter plate. Tighten the bolt to 53–56 ft. lbs. (73–77 Nm).
   - Flywheel
   - Clutch pressure plate and disc, if equipped
   - Transaxle

4. Check the fluid levels.

5. Start the engine and check for leaks.

Timing Belt

REMOVAL & INSTALLATION

2.4L Engine

1. Before servicing the vehicle, refer to the precautions in the beginning of this section.

2. Align the timing marks to set the No. 1 piston to Top Dead Center (TDC) by rotating the crankshaft clockwise. The timing marks of the camshaft sprocket and the cylinder head cover should be aligned and the dowel pin of the camshaft sprocket should be at the upper side.

3. Remove or disconnect the following:
   - Crankshaft pulley, water pump pulley and drive belt
   - Timing belt cover
   - Auto tensioner

4. Mark the timing belt is being reused, mark an arrow on the belt noting the direction of rotation or the front of the engine to make sure the belt is reinstalled in its original position.
   - Timing belt

5. Hold the camshaft with a wrench and loosen the camshaft sprocket bolts.
   - Sprockets

6. When removing the oil pump socket nut, first remove the plug at the side of the block and insert a 0.3 inch (8mm) diameter screwdriver to keep the left counterbalance shaft in position. Insert the screwdriver at least 2.36 inch (60 mm).
   - Oil pump sprocket nut and the sprocket
   - Loosen the right counterbalance shaft sprocket bolt until you can loosen it by hand
   - Tensioner **B** and timing belt **B**. Refer to the accompanying illustration for tensioner and belt identification.

✳✳ CAUTION

Do not attempt to loosen bolts while holding the sprocket with pliers or any tool after removing timing belt B.

   - Crankshaft sprocket **B** from the crankshaft

To install:

7. Install or connect the following:
   - Crankshaft sprocket **B** to the crankshaft

✳✳ CAUTION

Pay attention to the direction of the flange. If it is installed in the wrong direction, the belt will break.

8. Apply engine oil to the outer surface of the spacer lightly and install the spacer to the right counterbalance shaft. Be sure to install spacer correctly.
   - Counterbalance shaft sprocket onto the right counterbalance shaft and

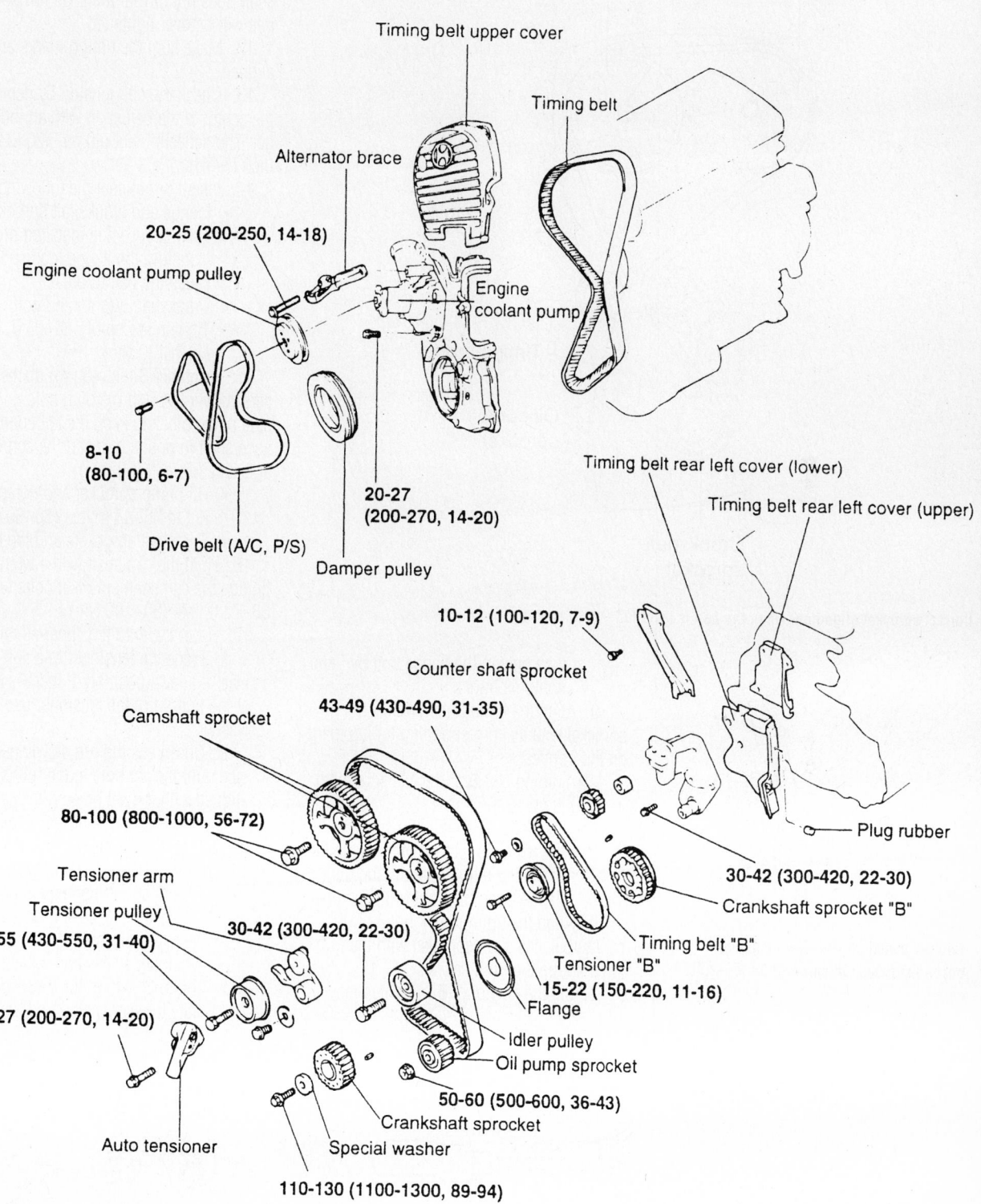

Timing belt upper cover

Timing belt

Alternator brace

20-25 (200-250, 14-18)

Engine coolant pump pulley

Engine coolant pump

8-10 (80-100, 6-7)

Drive belt (A/C, P/S)

20-27 (200-270, 14-20)

Damper pulley

Timing belt rear left cover (lower)

Timing belt rear left cover (upper)

10-12 (100-120, 7-9)

Counter shaft sprocket

43-49 (430-490, 31-35)

Plug rubber

30-42 (300-420, 22-30)

Crankshaft sprocket "B"

Camshaft sprocket

80-100 (800-1000, 56-72)

Tensioner arm

Tensioner pulley

43-55 (430-550, 31-40)

30-42 (300-420, 22-30)

Timing belt "B"

Tensioner "B"

15-22 (150-220, 11-16)

Flange

20-27 (200-270, 14-20)

Idler pulley

Oil pump sprocket

Auto tensioner

50-60 (500-600, 36-43)

Crankshaft sprocket

Special washer

110-130 (1100-1300, 89-94)

TORQUE : Nm (kg·cm, lb·ft)

9355LG73

Exploded view of the timing belt assembly and related components—2.4L engine

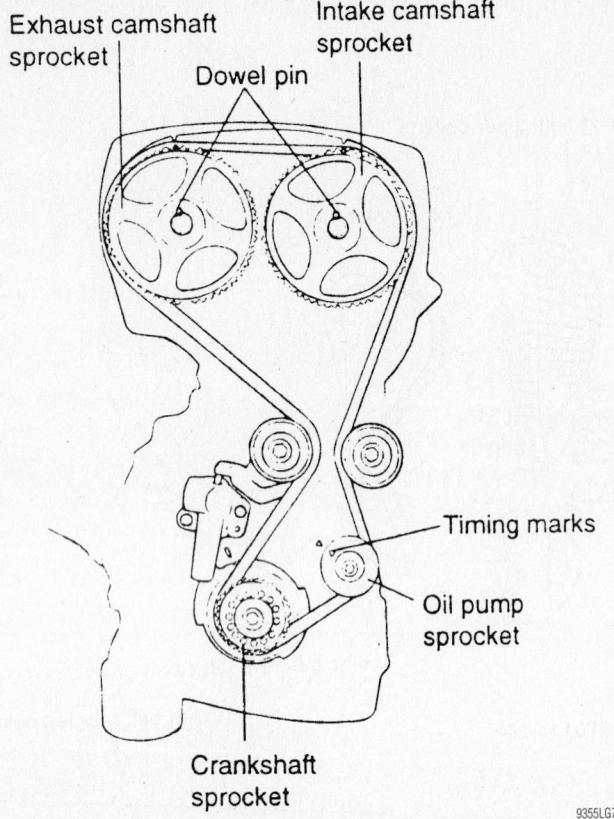

Correct sprocket alignment when the belt is installed—2.4L engine

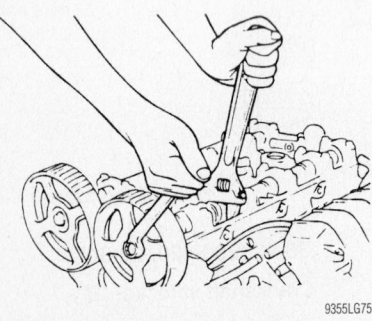

Hold the camshaft with a wrench and loosen the camshaft sprocket bolts—2.4L engine

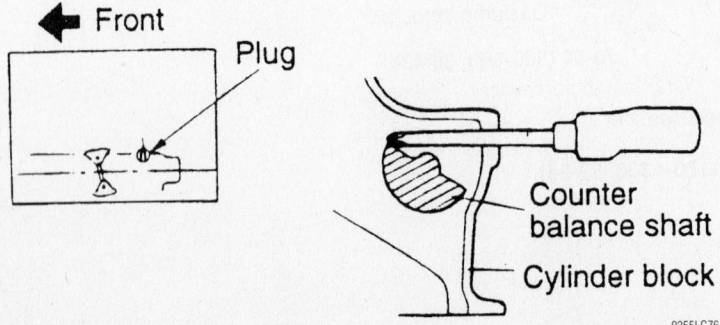

Insert a 0.3 inch (8mm) diameter screwdriver to keep the left counterbalance shaft in position—2.4L engine

then tighten the flange bolt by hand until it is tight

9. Align the timing mark on each sprocket with its corresponding timing mark on the front case.

- Timing belt **B** and make sure there is no slack
- Tensioner **B** so that the center of the pulley is located on the left side of the mounting bolt and the pulley flange faces the front of the engine

10. Align the timing mark on the right counterbalance shaft sprocket with the timing mark on the front case.

11. Lift the tensioner **B** to tighten tensioner **B** so that its tension side is pulled tight. Tighten the bolt on tensioner **B**. As the bolt is being tightened, make sure the shaft does not turn. If the shaft turns, the belt will be over tightened.

12. Make sure the timing marks are aligned.

13. Check the belt tension by depressing the center of the belt span with an index finger. The deflection should be 0.20–0.28 inch (5–7mm).

14. Install or connect the following:
- Flange and crankshaft sprocket making sure it is installed properly. Installing the flange incorrectly will cause the belt to break.
- Crankshaft washer and bolt. Tighten the bolt to 80–94 ft. lbs. (110–130 Nm).

15. Insert a 0.3 inch (8mm) diameter screwdriver through the plug hole on the left side of the block to keep the left counterbalance shaft in position. Insert the screwdriver at least 2.36 inch (60 mm).
- Oil pump sprocket and tighten the nut to 36–43 ft. lbs. (50–60 Nm)
- Camshaft sprockets and the bolts

16. Hold the camshaft with a wrench and tighten the camshaft sprocket bolts to 56–72 ft. lbs. (80–100 Nm).

17. Reset the auto tensioner as follows:
a. Place the tensioner in a soft jawed vice in a level position. If there is a plug at the bottom of the tensioner use a plain washer.

b. Compress the rod slowly using the vice until the set hole in the rod is aligned with the set hole on the cylinder.

c. Insert a set pin through the body and rod and leave the pin installed.

18. Install or connect the following:
- Tensioner and tighten the bolts to 14–20 ft. lbs. (20–27 Nm). Leave the set pin in place.
- Tensioner pulley and tighten the bolt to 31–40 ft. lbs. (43–55 Nm)

19. Rotate the camshaft sprockets so that

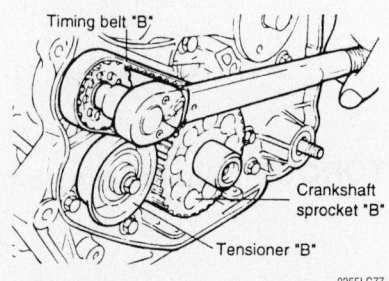

Remove tensioner B and timing belt B—2.4L engine

the dowel pin is at the upper side. Set the timing mark of the sprocket correctly.

➡**Before installing the belt, the timing mark of the camshaft sprocket doers not coincide with that of the rocker cover, do not rotate the cam sprocket more than 2 teeth of the sprocket in any direction. Rotating the sprocket more that 2 teeth may make the valve and piston contact each other. If it is necessary to rotate the sprocket more that 2 teeth, rotate the crankshaft sprocket counterclockwise first based on the timing mark. After the camshaft sprocket is properly timed, return the crankshaft to TDC.**

20. Align the crankshaft sprocket timing marks.

21. Align the pump sprocket timing marks.

22. Install the timing belt counterclockwise around the tensioner pulley and crankshaft sprocket. Hold the belt onto the tensioner pulley using your hand.

23. Pull the belt around the oil pump sprocket using your other hand.

24. Install the belt around the right-hand idler pulley, then the intake camshaft sprocket.

25. Turn the exhaust camshaft sprocket one tooth clockwise to align its timing mark with the cylinder top surface, then pull the belt around the exhaust camshaft sprocket.

26. Raise the tensioner pulley gently so that the belt does not sag and temporarily tighten the pulley center bolt.

27. Recheck that all timing marks are correct.

28. Remove the set pin from the auto tensioner.

29. Rotate the crankshaft two turns clockwise and let it sit for around 15 minutes. After 15 minutes, measure the auto tensioner protrusion **A** (the distance between the tensioner arm and tensioner) as shown in the accompanying illustration. The specification should be 0.24–0.35 inch (6–9mm).

30. Install the timing covers. Tighten the bolts as shown in the accompanying illustration to specification as shown in the accompanying illustration.

2.7L Engine

1. Before servicing the vehicle, refer to the precautions in the beginning of this section.

2. Remove or disconnect the following:
   • Engine cover

3. Using a 16mm wrench, rotate the tensioner arm clockwise about 14 degrees and remove the drive belt.
   • Power steering pump pulley, idler pulley, tensioner pulley and crankshaft pulley

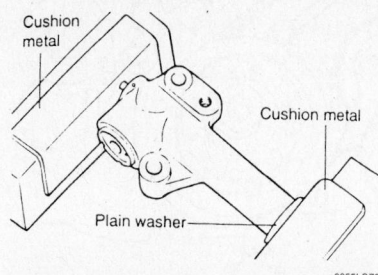

Place the tensioner in a soft jawed vice in a level position. If there is a plug at the bottom of the tensioner use a plain washer—2.4L engine

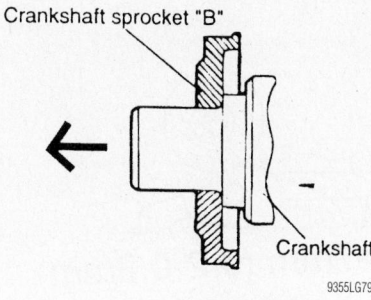

Pay attention to the direction of the flange, if it is installed in the wrong direction, the belt will break—2.4L engine

• Upper and lower timing covers
• Timing belt tensioner

4. Rotate the crankshaft clockwise and align the timing mark to set the No. 1 cylinder to Top Dead Center (TDC). Make sure the timing marks of the camshaft sprocket

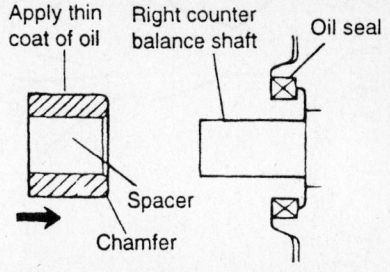

Apply engine oil to the outer surface of the spacer lightly and install the spacer to the right counterbalance shaft. Be sure to install spacer correctly—2.4L engine

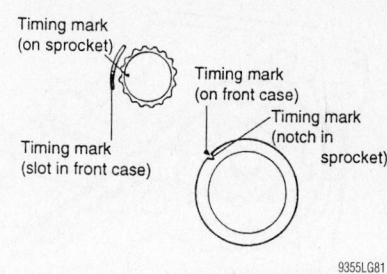

Align the timing mark on each sprocket with its corresponding timing mark on the front case—2.4L engine

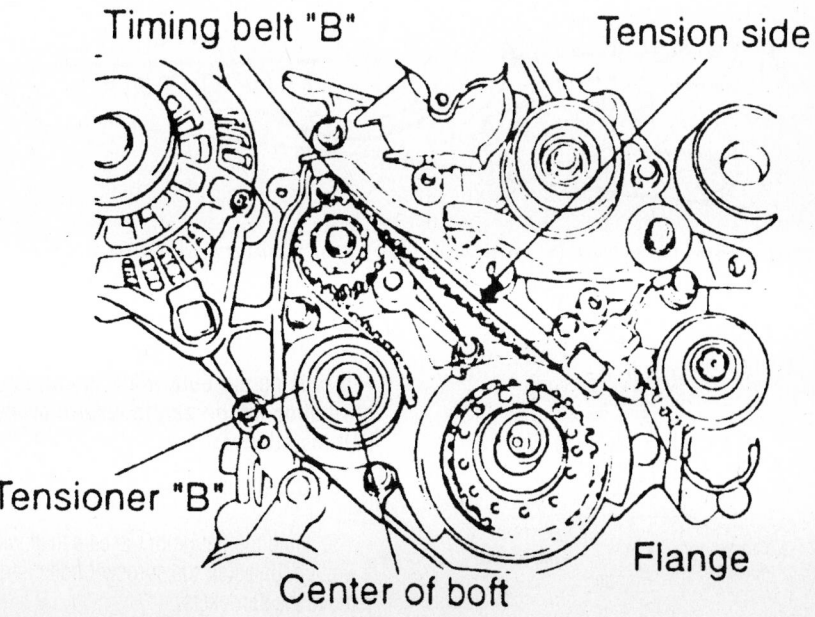

Align the timing mark on the right counterbalance shaft sprocket with the timing mark on the front case—2.4L engine

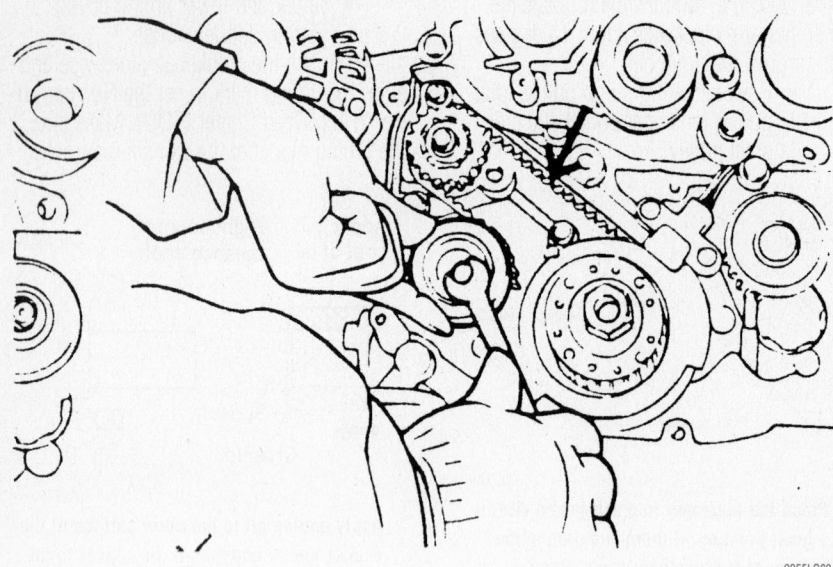

Lift the tensioner B to tighten tensioner B so that its tension side is pulled tight—2.4L engine

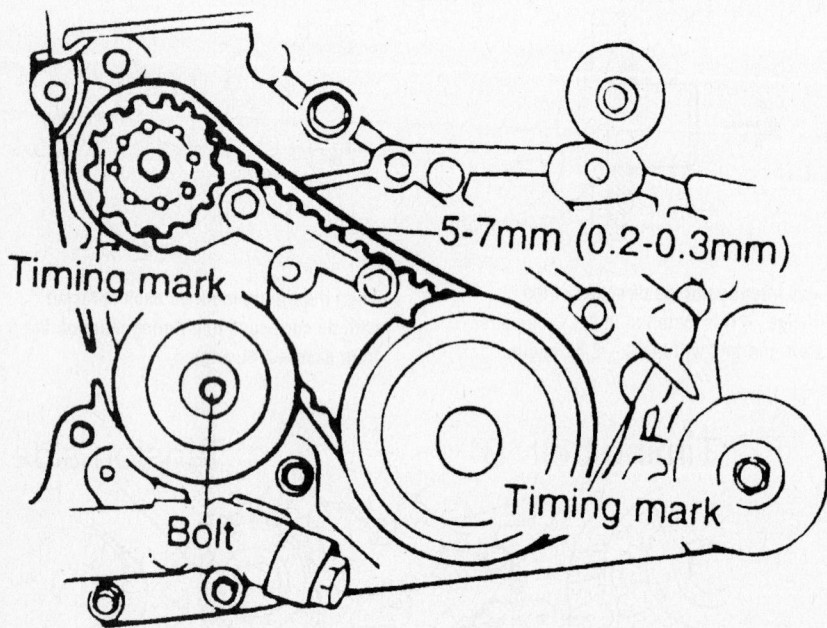

Make sure the timing belt B marks are aligned and the belt tensioner is correct—2.4L engine

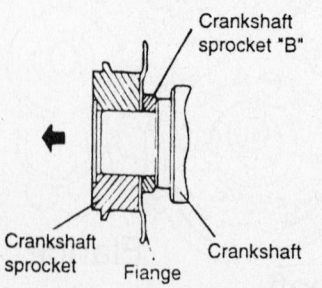

Install the flange and crankshaft sprocket making sure it is installed properly—2.4L engine

and cylinder head cover should align with each other.

➡If reusing the belt, mark the direction of rotation on the belt to ensure proper belt installation.

5. Unbolt the tensioner and remove the belt.

6. Hold the flange of the camshaft with a wrench, unfasten the sprocket bolts and remove the sprockets.

To install:

7. Install or connect the following:

- Idler pulley on the water pump boss
- Idler pulley to the roll pin that is pressed in the water pump boss
- Tensioner arm and plain washer to the block
- Tensioner pulley to the tensioner arm
- Camshaft sprockets and align the timing marks. Tighten the bolts to 65–80 ft. lbs. (90–110 Nm).

8. Compress the tensioner and install a set pin to keep the plunger in position.

- Tensioner and tighten the bolt to 14–20 ft. lbs. (20–27 Nm)

9. Align the sprocket timing marks and install the belt in the following order:

- Crankshaft sprocket
- Idler pulley
- Camshaft sprocket on the left hand side
- Water pump pulley
- Camshaft sprocket on the right hand side
- Tensioner pulley

10. Remove the tensioner set pin.

11. Adjust the timing belt tension as follows:

  a. Rotate the crankshaft 2 turns clockwise and measure the projected length of the auto tensioner at TDC after being installed for 5 minutes.

  b. The projected length should be 6–8mm.

Make sure the timing marks are in their specified positions or remove and reinstall the belt.

12. Install or connect the following:

- Upper and lower timing belt covers
- Power steering pump pulley, idler pulley, tensioner pulley and crankshaft pulley
- Drive belt
- Engine cover

3.5L Engine

1. Before servicing the vehicle, refer to the precautions in the beginning of this section.

2. Remove the passenger side wheel.

3. Remove the side cover.

4. Remove the engine cover.

5. Disconnect the power steering hose from the engine mount bracket.

6. Support the engine with a jack and a wooden block and remove the engine mount bracket.

7. Disconnect the connectors from the upper timing belt cover.

8. Remove the upper timing belt cover.

9. Remove the drive belt.

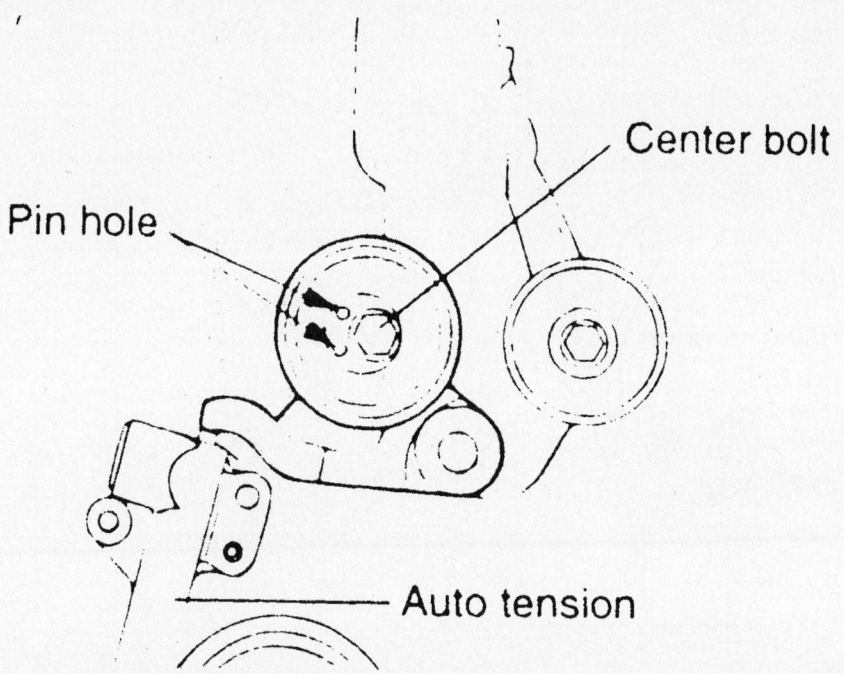

Install the tensioner pulley—2.4L engine

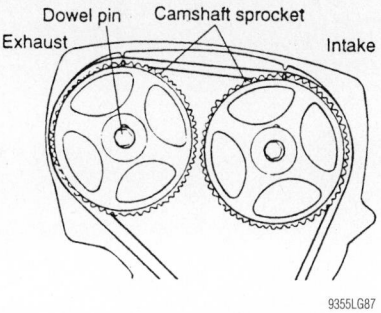

Rotate the camshaft sprockets so that the dowel pin is at the upper side, set the timing mark of the sprocket correctly—2.4L engine

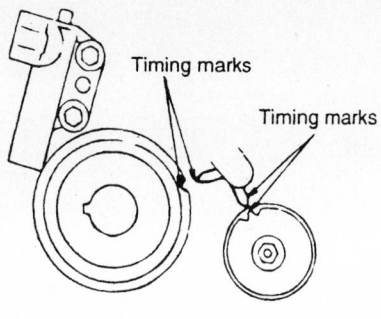

Align the oil pump sprocket timing marks—2.4L engine

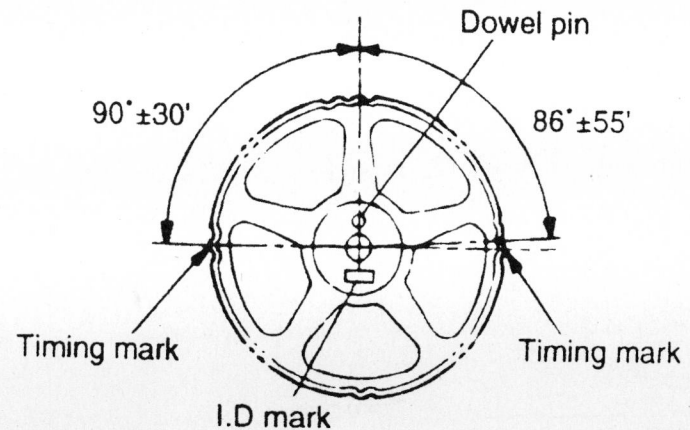

Camshaft sprocket alignment—2.4L engine

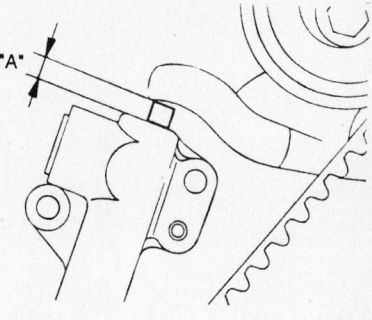

Measure the auto tensioner protrusion A —2.4L engine

10. Remove the 2 alternator mounting nuts and 2 bolts attaching the engine support bracket.

11. Loosen the 5 engine support bracket bolts.

12. Remove the alternator.

13. Remove the engine support bracket by moving the engine up and down slightly and remove the bracket upwards.

14. Remove the auto tensioner.

15. Rotate the crankshaft clockwise and align the timing mark to set the No. 1 cylinder to Top Dead Center (TDC). Make sure the timing marks of the camshaft sprocket and cylinder head cover should align with each other.

➡**If reusing the belt, mark the direction of rotation on the belt to ensure proper belt installation.**

16. Unbolt the tensioner and remove the belt.

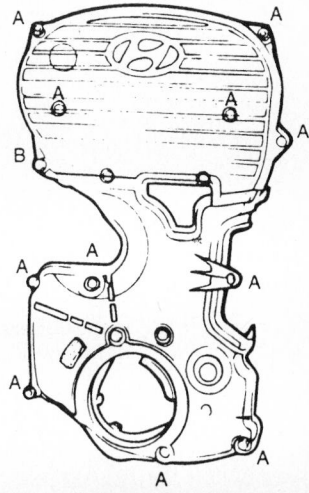

A:8-10 N.m (80-100 kg.cm, 6-7 lb.ft)
B:10-12 N.m (100-120 kg.cm, 7-9 lb.ft)

Timing cover bolt location and torque specifications A —2.4L engine

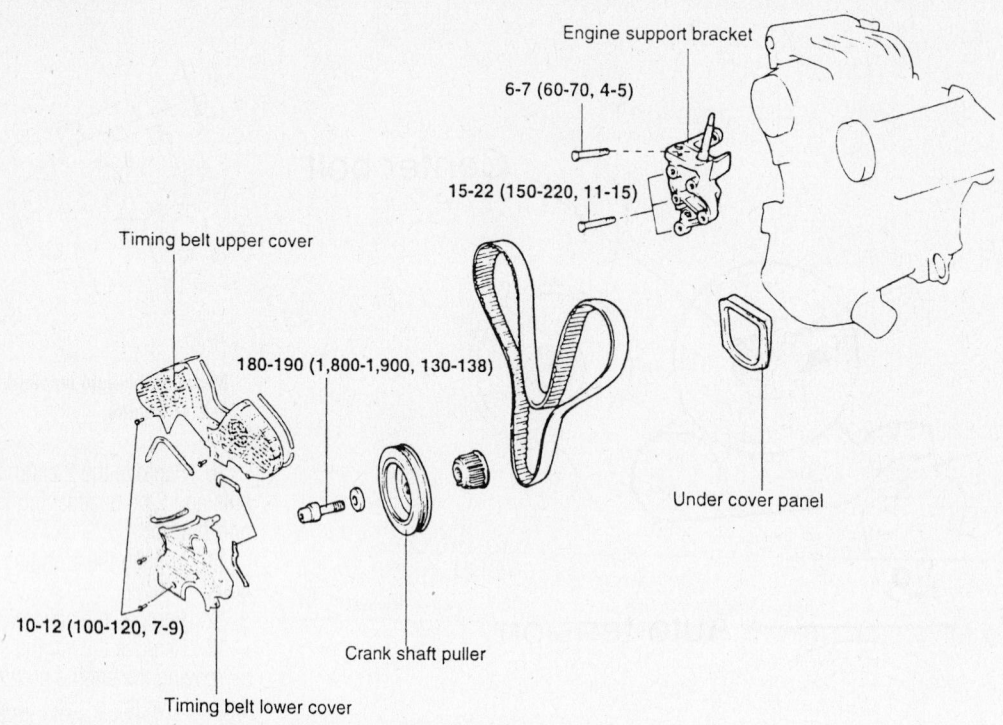

TORQUE : Nm (kg.cm, lb.ft)

Exploded view of the timing belt assembly—2.7L engine

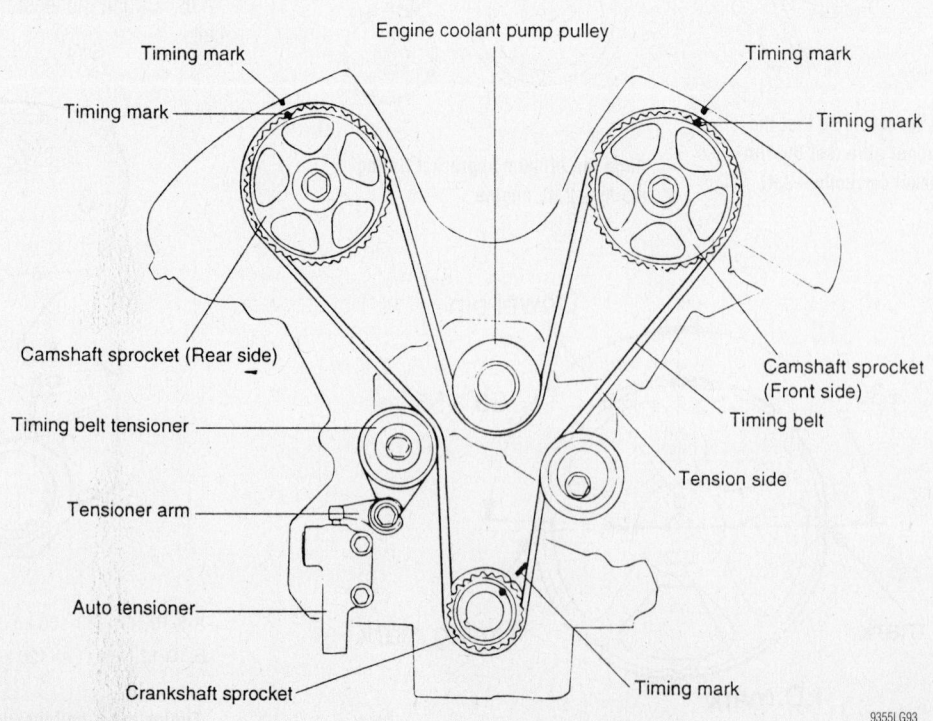

Correct timing marks alignment with the timing belt installed—2.7L engine

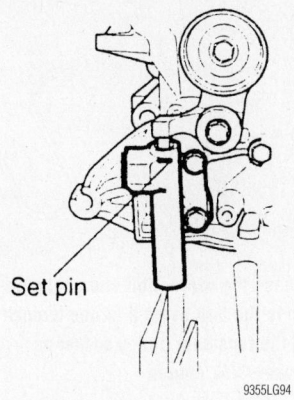

Remove the set pin from the auto tensioner—2.7L engine

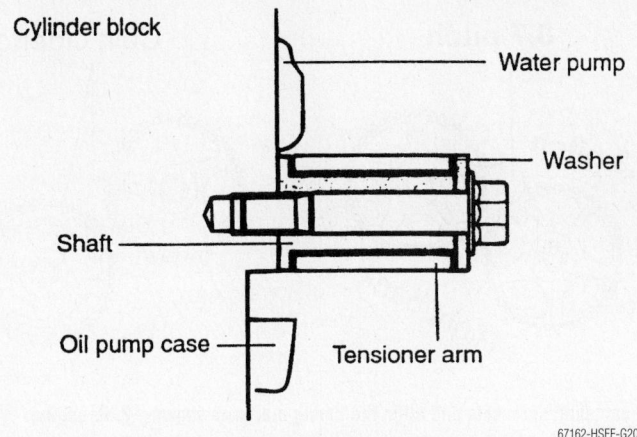

Install the tensioner arm, shaft and washer to the block—3.5L engine

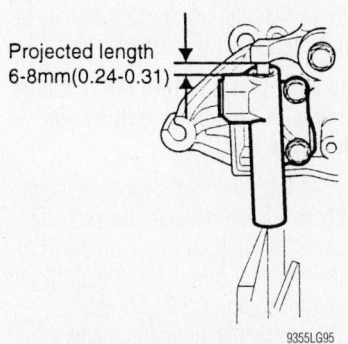

The projected length of the auto tensioner at Top Dead Center (TDC) should be 6–8mm—2.7L engine

17. Hold the flange of the camshaft with a wrench, unfasten the sprocket bolts and remove the sprockets.

To install:

18. Install the idler pulley to the engine support bracket.

19. Install the tensioner arm, shaft and washer to the block. Tighten to 25–40 ft. lbs. (35–55 Nm).

20. Install the crankshaft sprocket and align the timing mark as illustrated. Tighten the bolts to 60–74 ft. lbs. (80–100 Nm). Be careful not to bend the crankshaft sensing blade.

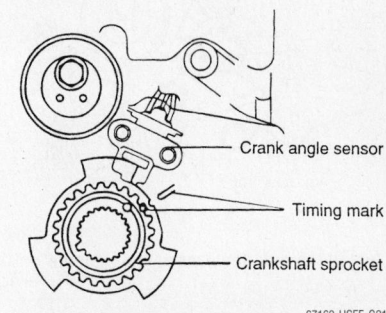

After installing the crankshaft sprocket, align the timing mark as shown—3.5L engine

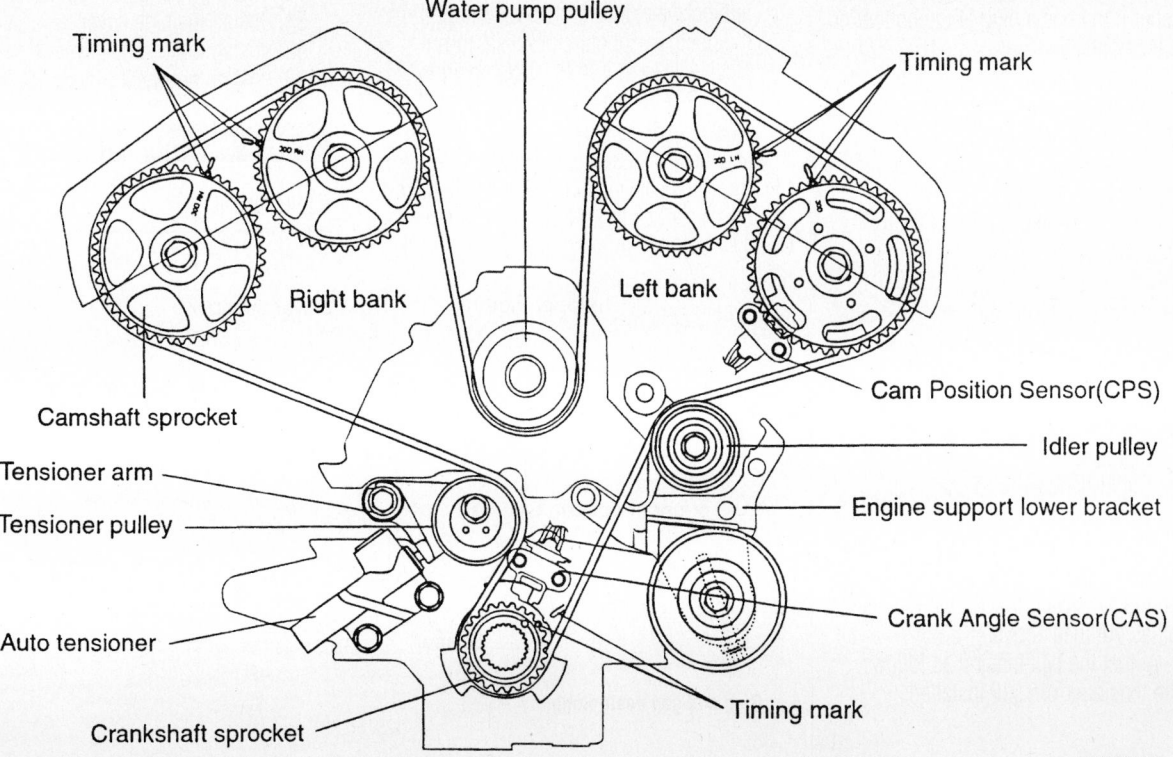

Exploded view of the timing belt assembly—3.5L engine

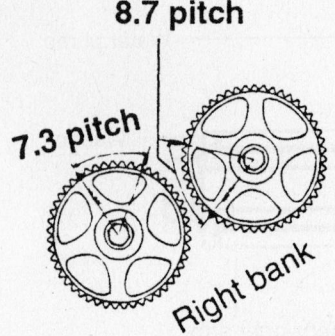

Install the camshaft sprockets and align the timing marks as shown—3.5L engine

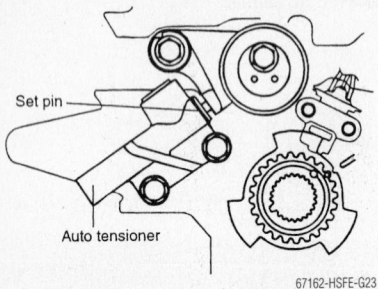

Install the auto tensioner—3.5L engine

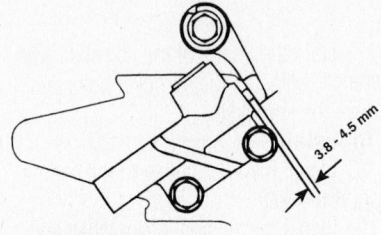

Insert a 3.8—4.5mm pin through the tensioner body and rod when resetting the tensioner—3.5L engine

21. Install the auto tensioner to the oil pump case. If the auto tensioner is in its fully extended position, perform the following:

   a. Place the tensioner in a soft jawed vise in a level position. Use a plain washer if there is a plug at the bottom of the tensioner.

   b. Slowly compress the rod until the set hole in the rod is aligned with the set hole in the tensioner.

   c. Insert a 3.8—4.5mm pin through the tensioner body and rod.

22. Align the timing marks of each sprocket and install the timing belt in this order:

- Crankshaft sprocket
- Idler pulley
- Left hand exhaust camshaft sprocket
- Right hand intake camshaft sprocket
- Right hand exhaust camshaft sprocket
- Tensioner pulley

23. Make sure the engine is at TDC or remove the belt reset to TDC and reinstall the belt as outlined above.

24. Adjust the belt tension as follows with the tensioner pin still installed:

➡**Be careful not to turn the tensioner pulley with the center bolt as you tighten the bolt.**

   a. Turn the crankshaft ¼ turn counterclockwise, turn it clockwise to fit it in the position counterclockwise, turn it clockwise to fit it in the TDC position.

   b. Release the center bolt and apply tension to the belt using a torque wrench and the tensioner pulley socket as shown in the accompanying illustration and tighten the center bolt to 3.71 ft. lbs. (5 Nm).

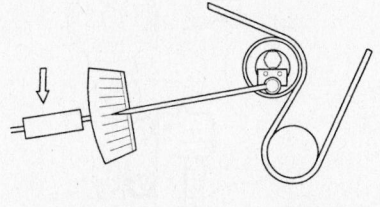

Release the center bolt and apply tension to the belt using a torque wrench and the tensioner pulley socket as shown—3.5L engine

   c. Tighten the auto tensioner bolt to 31–39 ft. lbs. (43–55 Nm) and pull out

25. Adjust the belt tension as follows with the tensioner pin removed:

➡**Be careful not to turn the tensioner pulley with the center bolt as you tighten the bolt.**

   a. Turn the crankshaft 2 rotations clockwise and measure the projected load of the auto tensioner in the TDC position after 5 minutes. The projected length should be 3.8–4.5mm.

26. Check that all timing marks are aligned.

27. Install the engine support bracket.

28. Install the timing belt cover.

29. Install the alternator and drive belt.

30. Install the engine mount bracket.

31. Install the side cover.

32. Install the engine cover.

33. Install the wheel.

Piston and Ring

POSITIONING

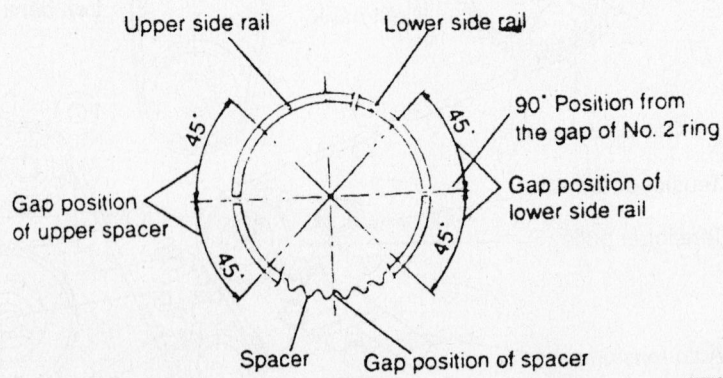

Ring end gap positioning

FUEL SYSTEM

Fuel System Service Precautions

Safety is the most important factor when performing not only fuel system maintenance, but any type of maintenance. Failure to conduct maintenance and repairs in a safe manner may result in serious personal injury or death. Maintenance and testing of the vehicle's fuel system components can be accomplished safely and effectively by adhering to the following rules and guidelines:

• To avoid the possibility of fire and personal injury, always disconnect the negative battery cable unless the repair or test procedure requires that battery voltage be applied.

• Always relieve the fuel system pressure prior to disconnecting any fuel system component (injector, fuel rail, pressure regulator, etc.), fitting or fuel line connection. Exercise extreme caution whenever relieving fuel system pressure to avoid exposing skin, face and eyes to fuel spray. Please be advised that fuel under pressure may penetrate the skin or any part of the body that it contacts.

• Always place a shop towel or cloth around the fitting or connection prior to loosening to absorb any excess fuel due to spillage. Ensure that all fuel spillage (should it occur) is quickly removed from engine surfaces. Ensure that all fuel soaked cloths or towels are deposited into a suitable waste container.

• Always keep a dry chemical (Class B) fire extinguisher near the work area.

• Do not allow fuel spray or fuel vapors to come into contact with a spark or open flame.

• Always use a backup wrench when loosening and tightening fuel line connection fittings. This will prevent unnecessary stress and torsion to fuel line piping. Always follow the proper torque specifications.

• Always replace worn fuel fitting O-rings with new. Do not substitute fuel hose or equivalent, where fuel pipe is installed.

Fuel System Pressure

RELIEVING

1. Before servicing the vehicle, refer to the precautions in the beginning of this section.
2. Remove the fuel filler cap.
3. Remove the fuel pump fuse and crank the engine until it stalls.

4. Disconnect the negative battery cable.
5. Replace the fuse.

Fuel Filter

The fuel filter is part of the fuel pump assembly located in the tank.

REMOVAL & INSTALLATION

1. Before servicing the vehicle, refer to the precautions in the beginning of this section.
2. Relieve the fuel system pressure.
3. Disconnect the negative battery cable.
4. Remove the two fitting nuts while holding the filter.
5. Remove the filter mounting bolts and the filter from the clamp.
6. Installation is the reverse of removal. Tighten the fitting nuts to 22–29 ft. lbs. (30–40 Nm).
7. Connect the negative battery cable.
8. Start the engine and check for leaks.

Fuel Pump

REMOVAL & INSTALLATION

1. Before servicing the vehicle, refer to the precautions in the beginning of this section.
2. Relieve the fuel system pressure.
3. Remove or disconnect the following:
   • Negative battery cable
   • Fuel pump connector
   • Fuel feed and return lines from the pump assembly
   • Pump cover screws
   • Pump assembly from the tank
To install:
4. Install or connect the following:
   • Pump into the tank
   • Pump cover screws
   • Fuel feed and return lines to the pump assembly
   • Fuel pump connector
   • Negative battery cable
5. Start the engine and check for leaks.

Fuel Injector

REMOVAL & INSTALLATION

1. Before servicing the vehicle, refer to the precautions in the beginning of this section.
2. Relieve the fuel system pressure.

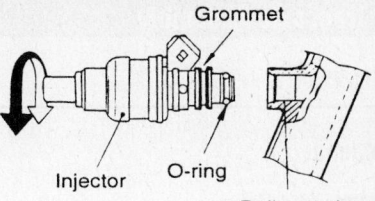

Install the injector using a twisting motion

3. Remove or disconnect the following:
   • Negative battery cable
   • Air breather hose from the throttle body
   • Throttle cable
   • Engine coolant hose and throttle body
   • Positive Crankcase Ventilation (PCV) valve and brake booster vacuum hose
   • Vacuum hose connector
   • Injector cover
   • High pressure fuel hose
   • Fuel injector harness connector
   • Delivery pipe with the injectors and the pressure regulator as an assembly
   • Injector from the delivery pipe
   • Injector O-ring, grommet and discard
To install:
4. Install a new grommet and O-ring
5. Apply a coating of spindle oil or gasoline to the injector O-ring
6. Install the injector into the delivery pipe while turning the injector left and right making sure the injector turns smoothly. If the injector does not turn smoothly check for a jammed O-ring, remove the injector and reinsert it again.
7. Install or connect the following:
   • Delivery pipe and injector assembly
   • Intake manifold stay and tighten the bolts to 13–18 ft. lbs. (18–25 Nm)
   • Fuel injector harness connector
   • High pressure fuel hose
   • Injector cover
   • Vacuum hose connector
   • PCV valve and brake booster vacuum hose
   • Throttle body and engine coolant hose
   • Throttle cable
   • Air breather hose from the throttle body
   • Negative battery cable
8. Start the engine and check for proper operation.

DRIVE TRAIN

Transaxle Assembly

REMOVAL & INSTALLATION

Manual

2001–03 Models

1. Before servicing the vehicle, refer to the precautions in the beginning of this section.
2. Drain the transaxle.
3. Remove or disconnect the following:
   - Negative battery cable
   - Air cleaner duct
   - Air cleaner and air flow hose
   - Back-up light connector
   - Clutch line and clip
   - Clutch release cylinder
   - Speedometer cable
   - Gear select or shift cables
   - Starter motor mounting bolts
   - Transaxle upper bolts
4. Attach an engine support fixture to the engine.
   - Transaxle mounting bracket and insulator
   - Front wheels
   - Tie rod end, lower ball joint and drive shaft
   - Gear box u-joint bolt and the return tube mounting bolts
   - Front muffler
   - Sub-frame mounting bolts and the frame
   - Transaxle front and rear mounting brackets
   - Transaxle side mounting bolts
5. Using a suitable jack, remove the transaxle from the vehicle.

To install:
6. Installation is the reverse of removal keeping in mind the following torques:
   - Transaxle case bolts to 15–20 ft. lbs. (20–27 Nm)
   - Transaxle mounting sub-bracket nut 43–58 ft. lbs. (60–80 Nm)
   - Transaxle mounting bracket bolts 29–40 ft. lbs. (40–55 Nm)
   - Transaxle mounting insulator bolt 65–80 ft. lbs. (90–110 Nm)
   - Clutch release cylinder retainers to 11–16 ft. lbs. (15–22 Nm)
7. Fill the transaxle to the correct level.
8. Start the engine and check for leaks.
9. Check the wheel alignment and adjust as necessary.

2004 Models

1. Before servicing the vehicle, refer to the precautions in the beginning of this section.
2. Drain the transaxle.
3. Remove the battery and tray.
4. Remove the air cleaner duct.
5. Remove the air cleaner and air flow hose.
6. Disconnect the back-up light connector, Crankshaft Position (CKP) sensor and oil pressure switch connectors.
7. Disconnect the speedometer cable.
8. Remove the clutch release cylinder bolt.
9. Remove the cotter pin from the shift cable at the transaxle.
10. Remove the clip from the shift cable on the transaxle side.
11. Remove the clip from the select cable at the transaxle side.
12. Separate the steering column shaft joint.
13. Separate the power steering pump hose.
14. Separate the hose after removing the clip from the power steering return hose.
15. Remove the upper transaxle bolt.
16. Remove the starter motor.
17. Install an engine support fixture.
18. Remove the front wheels and calipers.
19. Disconnect the tie rod end, wheel speed sensor, and knuckle mounting bolt.
20. Remove the transaxle mounting bracket and insulator.
21. Remove the front roll stopper insulator bolt, upper and lower stopper bolts and the front roll stopper.
22. Remove the rear roll stopper insulator bolt, and stopper bolt and the roll stopper.
23. Remove the drive shaft.
24. Remove the subframe bolts and subframe.
25. Install a transaxle jack.
26. Remove the transaxle lower bolts.
27. Remove the transaxle–to–engine bolts and the transaxle.

To install:
28. Installation is the reverse of removal keeping in mind the following steps and torques:
   a. Transaxle case bolts to 15–20 ft. lbs. (20–27 Nm)
   b. Tighten the transaxle mounting bracket bolts to 43–58 ft. lbs. (60–80 Nm).
   c. Transaxle mounting bracket bolts 29–40 ft. lbs. (40–55 Nm)
   d. Transaxle mounting insulator bolt 65–80 ft. lbs. (90–110 Nm)
   e. Tighten the front roll stopper–to–transaxle bolts to 43–58 ft. lbs. (60–80 Nm).
   f. Tighten the front roll stopper insulator bolt and nut to 36–47 ft. lbs. (50–65 Nm).
   g. Tighten the front roll stopper–to–subframe bolts to 43–58 ft. lbs. (60–80 Nm).
   h. Tighten the rear roll stopper–to–subframe bolts to 43–58 ft. lbs. (60–80 Nm).
   i. Tighten the rear roll stopper–to–transaxle bolt and nut to 36–47 ft. lbs. (50–65 Nm).
   j. Tighten the rear roll stopper insulator bolt and nut to 36–47 ft. lbs. (50–65 Nm).
   k. Clutch release cylinder retainers to 11–16 ft. lbs. (15–22 Nm)
29. Fill the transaxle to the correct level.
30. Start the engine and check for leaks.
31. Check the wheel alignment and adjust as necessary.

Automatic

2001–03 MODELS

1. Before servicing the vehicle, refer to the precautions in the beginning of this section.
2. Drain the transaxle.
3. Remove or disconnect the following:
   - Negative battery cable
   - Air cleaner assembly
   - Control cable
   - Speedometer sensor connector
   - Transaxle range switch, solenoid, and oil temperature sensor connectors
   - Oil cooler hose
4. Attach an engine support fixture to the engine.
   - Gear box, stabilizer bar, tie rod end, lower ball joint and drive shaft
   - Gear box u-joint bolt and the return tube mounting bolts
   - Sub-frame mounting bolts and the frame
   - Starter motor
   - Transaxle mounting bolts
5. Using a suitable jack, remove the transaxle from the vehicle.

To install:
6. Installation is the reverse of removal keeping in mind the following torques:
   - Transaxle case bolts to 15–20 ft. lbs. (20–27 Nm)

- Transaxle mounting sub-bracket nut 43–58 ft. lbs. (60–80 Nm)
- Transaxle mounting bracket bolts 29–40 ft. lbs. (40–55 Nm)
- Transaxle mounting insulator bolt 65–80 ft. lbs. (90–110 Nm)

7. Adjust the control cable as follows:

a. Move the shift lever and the transaxle range switch to the **N** position and install the cable.

b. When attaching the control cable to the bracket, make sure the clip is installs so it contacts the cable.

c. Adjust the nut to remove any free-play in the cable and make sure the lever moves freely.

8. Fill the transaxle to the correct level.

9. Start the engine and check for leaks.

10. Check the wheel alignment and adjust as necessary.

2004 Models

1. Before servicing the vehicle, refer to the precautions in the beginning of this section.

2. Drain the transaxle.

3. Remove the battery and tray.

4. Remove the air cleaner duct.

5. Remove the air cleaner and air flow hose.

6. Disconnect the back-up light connector, Crankshaft Position (CKP) sensor and oil pressure switch connectors.

7. Disconnect the speedometer cable.

8. Remove the cotter pin from the shift cable at the transaxle.

9. Remove the clip from the shift cable on the transaxle side.

10. Remove the clip from the select cable at the transaxle side.

11. Separate the steering column shaft joint.

12. Separate the power steering pump hose.

13. Separate the hose after removing the clip from the power steering return hose.

14. Remove the upper transaxle bolt.

15. Remove the starter motor.

16. Install an engine support fixture.

17. Remove the front wheels and calipers.

18. Disconnect the tie rod end, wheel speed sensor, and knuckle mounting bolt.

19. Remove the transaxle mounting bracket and insulator.

20. Remove the front roll stopper insulator bolt, upper and lower stopper bolts and the front roll stopper.

21. Remove the rear roll stopper insulator bolt, and stopper bolt and the roll stopper.

22. Remove the drive shaft.

23. Remove the subframe bolts and subframe.

24. Install a transaxle jack.

25. Remove the transaxle lower bolts.

26. Remove the transaxle–to–engine bolts and the transaxle.

To install:

27. Installation is the reverse of removal keeping in mind the following steps and torques:

a. Transaxle case bolts to 15–20 ft. lbs. (20–27 Nm)

b. Tighten the transaxle mounting bracket bolts to 43–58 ft. lbs. (60–80 Nm).

c. Transaxle mounting bracket bolts 29–40 ft. lbs. (40–55 Nm)

d. Transaxle mounting insulator bolt 65–80 ft. lbs. (90–110 Nm)

e. Tighten the front roll stopper–to–transaxle bolts to 43–58 ft. lbs. (60–80 Nm).

f. Tighten the front roll stopper insulator bolt and nut to 36–47 ft. lbs. (50–65 Nm).

g. Tighten the front roll stopper–to–subframe bolts to 43–58 ft. lbs. (60–80 Nm).

h. Tighten the rear roll stopper–to–subframe bolts to 43–58 ft. lbs. (60–80 Nm).

i. Tighten the rear roll stopper–to–transaxle bolt and nut to 36–47 ft. lbs. (50–65 Nm).

j. Tighten the rear roll stopper insulator bolt and nut to 36–47 ft. lbs. (50–65 Nm).

k. Move the shift lever and the transaxle range switch to the **N** position and install the cable.

l. When attaching the control cable to the bracket, make sure the clip is installs so it contacts the cable.

m. Adjust the nut to remove any free-play in the cable and make sure the lever moves freely.

28. Fill the transaxle to the correct level.

29. Start the engine and check for leaks.

30. Check the wheel alignment and adjust as necessary.

Transfer Case

REMOVAL & INSTALLATION

1. Before servicing the vehicle, refer to the precautions in the beginning of this section.

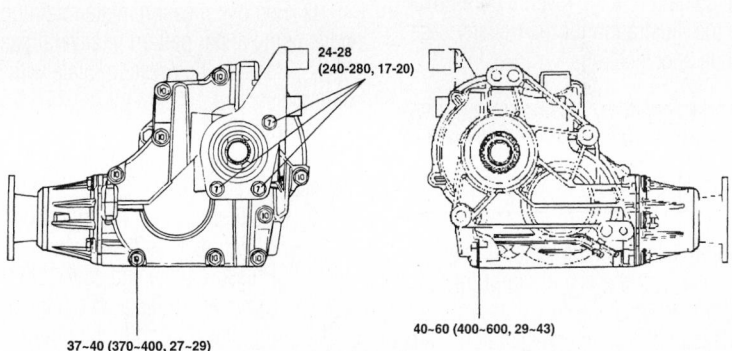

<Right view> <Left view>

24-28
(240-280, 17-20)

37~40 (370~400, 27~29)

40-60 (400-600, 29-43)

<Rear view>

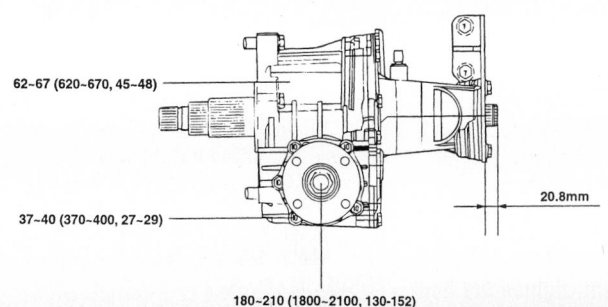

62~67 (620~670, 45-48)

20.8mm

37~40 (370~400, 27~29)

180~210 (1800~2100, 130-152)

TORQUE : N·m (kg·cm, lb·ft)

Transfer case bolt torque spoliations

2. Attach an engine support fixture to the engine.

3. Drain the transfer case fluid.

4. Disconnect the negative battery cable.

5. Disconnect the Oxygen (O2S) sensor connector and remove the sensor connector from the mounting bracket.

6. Remove the O2S sensor wire bracket.

7. Remove the alternator wire terminal mounting nut.

8. Remove the right hand wheel assembly.

9. Remove the right hand side engine cover.

10. Remove the lower ball joint mounting bolt and disconnect the ball joint from the knuckle.

11. Remove the drive shaft from the transfer case and support with a piece of wire.

12. Remove the exhaust heat shield.

13. Remove the exhaust manifold.

14. Remove the propeller shaft.

15. Remove the transfer case mounting bracket.

16. Remove the transfer case mounting bolts.

17. Using a flat bladed prytool, remove the transfer case from its mounting by prying it from side–to–side.

18. Remove the transfer case from the vehicle.

19. Installation is the reverse of removal. Refer to the illustration for the transfer case bolt torque specifications.

Clutch

ADJUSTMENTS

1. Before servicing the vehicle, refer to the precautions in the beginning of this section.

2. Measure the clutch pedal height from the face of the pedal pad to the floorboard. The proper measurement (8.597 inch (218.9mm).

3. Measure the clutch pedal clevis pin play. The measurement should be 0.04–0.11 inch (1–3mm).

4. Adjust the height and clevis pin if necessary as follows:

a. Turn and adjust the bolt, then secure it by tightening the lock nut.

➡**After adjustment, tighten the bolt until it reaches the pedal stopper and tighten the lock nut.**

b. Turn the push rod until the proper specification is reached and tighten the lock nut.

➡**When adjusting the clevis pin play or the pedal height, make sure not to push the push rod towards the master cylinder.**

c. After the adjustments are made, check that the clutch pedal free play is 0.2–0.5 inch (6–13mm). The free play is measured at the face of the pedal pad

5. If the adjustments do not bring the pedal into specifications, check the hydraulic system for air or a faulty component.

REMOVAL & INSTALLATION

1. Before servicing the vehicle, refer to the precautions in the beginning of this section.

2. Disconnect the negative battery cable.

3. Remove the transmission assembly.

4. If the pressure plate is attached to the flywheel, remove the release bearing using snap-ring pliers as follows:

a. Insert the snap-ring pliers under the wave washer and place it in the center of the snap-ring

b. Spread the snap-ring by pushing down on the bearing assembly.

c. Once the snap-ring is fully expanded, remove the release bearing

5. Install a clutch disc alignment tool.

6. Remove pressure plate retaining bolts using a star pattern in several passes.

7. Remove the pressure plate with clutch disc.

To install:

8. Apply multi-purpose grease to clutch splines.

9. Install or connect the following:
- Clutch plate to the flywheel
- Pressure plate and cover. Do not torque the bolts at this time.

10. Align the bearing to the release fork and install it until it is fully engaged.

11. Torque the pressure plate bolts using a star pattern torque sequence to 11–16 ft. lbs. (15–22 Nm) on models with a single mass flywheel or 14–19 ft. lbs. (20–27 Nm) on models with a dual mass flywheel.

12. Install transmission.

13. Connect the negative battery cable.

Hydraulic Clutch System

BLEEDING

Bleeding air from the hydraulic clutch system is necessary whenever any part of the system has been disconnected or the fluid level (in the reservoir) has been allowed to fall so low that air has been drawn into the master cylinder.

✲✲ WARNING

NEVER use fluid that has been bled from a clutch system to fill the master cylinder reservoir, as it may be aerated, contain excessive moisture and/or be contaminated in some other way.

1. Before servicing the vehicle, refer to the precautions in the beginning of this section.

2. Fill the clutch master cylinder reservoir with new hydraulic clutch fluid.

3. Attach a hose to the bleeder on the clutch actuator and submerge the other end of the hose in a container of hydraulic clutch fluid.

4. Have an assistant slowly depress and hold the clutch pedal.

5. Loosen the bleeder to purge air.

6. Tighten the bleeder.

7. Repeat the above 3 steps until all air is completely purged from the system.

8. Refill the clutch master cylinder reservoir.

Halfshaft

REMOVAL & INSTALLATION

Front

1. Before servicing the vehicle, refer to the precautions in the beginning of this section.

2. Drain the transaxle.

3. Remove or disconnect the following:
- Negative battery cable
- Aluminum wheel cover
- Split pin and halfshaft nut
- Wheel Speed Sensor (WSS) from the bracket, if equipped
- Brake hose from the bracket
- Knuckle from the strut by removing the flange bolts
- Halfshaft from the hub by tapping the end with a plastic mallet
- Halfshaft from the transaxle using a pry bar.

4. Insert a plug in the transaxle opening.

To install:

➡**Use new circlips, split pins and self-locking nuts for assembly.**

5. Remove the plug from the transaxle opening.

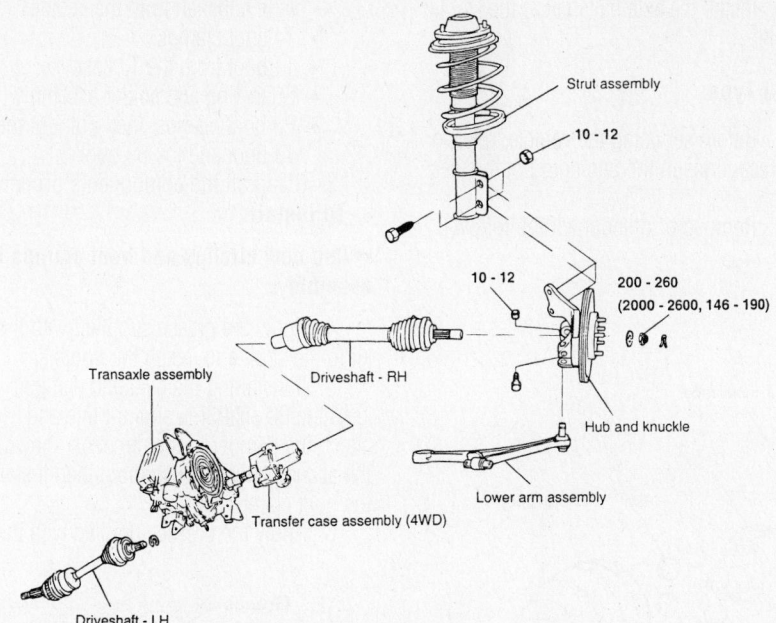

Strut assembly

10 - 12

10 - 12

200 - 260
(2000 - 2600, 146 - 190)

Transaxle assembly

Driveshaft - RH

Hub and knuckle

Transfer case assembly (4WD)

Lower arm assembly

Driveshaft - LH

TORQUE : Nm (kg·cm, lb·ft)

9355LG40

Exploded view of the halfshaft mounting and related components

6. Coat the halfshaft splines and sliding surfaces with gear oil.

7. Make sure the gap of the circlip is facing downwards.

8. Install the inner CV-joint to the transaxle until the circlip locks in the retaining groove. Pull on the shaft by hand to make sure it is properly engaged.

9. Install or connect the following:
- Halfshaft into the hub
- Knuckle to the strut
- Flange bolts and tighten to 74–88 ft. lbs. (100–120 Nm)
- Halfshaft nut and tighten to 146–190 ft. lbs. (200–260 Nm)
- New split pin
- Brake hose to the bracket
- WSS to the bracket, if equipped
- Aluminum wheel cover
- Negative battery cable

10. Fill the transaxle to the correct level and check for leaks.

Rear

1. Before servicing the vehicle, refer to the precautions in the beginning of this section.

2. Remove or disconnect the following:
- Rear wheel
- Split pin and nut

- Spare tire and support the hanger of the main muffler to avoid interfering with the carrier during right hand shaft removal

3. Matchmark the propeller rubber coupling and differential flange, then remove the bolts and nuts.

4. Support the differential carrier with a jack and remove the differential carrier mounting nuts and bolts.
- Shaft from the carrier by inserting a prybar between the carrier and the shaft
- Differential carrier to the rear after lowering the jack
- Shaft from the axle hub using a plastic mallet
- Shaft from the vehicle

To install:

5. Install or connect the following:
- Shaft into the axle hub
- Differential carrier to the rear using a jack
- Shaft to the carrier
- Differential carrier mounting nuts and bolts. Tighten the carrier nuts and bolts to 51–58 ft. lbs. (70–80 Nm) and the carrier rear bracket bolt to 58–73 ft. lbs. (80–100 Nm).
- Propeller shaft

- Spare tire and remove the support from the main muffler hanger
- New shaft nut and tighten to 146–253 ft. lbs. (200–260 Nm)
- New split pin
- Rear wheel

CV-Joint

OVERHAUL

DOJ-BJ Type

The DOJ-BJ type joint is used on both the front and rear of the vehicle and the following overhaul procedure is used.

1. Before servicing the vehicle, refer to the precautions in the beginning of this section.

➡**Do not disassemble the BJ assembly.**

2. Remove or disconnect the following:
- Axle halfshaft from the vehicle and place it in a vise
- DOJ boot clamps and push the boot back from the outer race
- Circlip with a flat bladed prytool
- Driveshaft from the DOJ outer race
- Snap-ring and take out the inner race, cage and balls as an assembly

3. Clean the outer race, cage and balls without disassembling them.
- BJ boot clamps
- DOJ and BJ boots

To install:

➡**Use new circlips and boot clamps for assembly.**

4. Apply some of the grease supplied with the kit to the halfshaft.

5. Install the boots.

6. Apply the grease supplied with the kit to the inner race and cage.

7. Install the cage so that it is offset on the race.

8. Apply the grease supplied with the kit to the cage and install the balls

9. Install the chamfered side of the cage as shown in the accompanying illustration, then insert the inner race onto the shaft and install the snap-ring.

10. Apply the grease supplied with the kit to the BJ outer race (67–73 grams of grease in the joint and 62–68 grams of grease in the boot); and install the outer race onto the shaft.

11. Apply the grease supplied with the kit to the DOJ outer race (62–68 grams in the joint and 37–43 grams in the boot); and install the circlip.

12. Tighten the DOJ boot clamps.
13. Apply the grease supplied with the kit to the BJ.
14. Install the boots.
15. Tighten the BJ boot clamps

➡ **To control the volume of air in the DOJ boot, make sure the distance between the boot bands is as shown in the accompanying illustration.**

16. Install the axle halfshaft to the vehicle.

TJ-BJ Type

1. Before servicing the vehicle, refer to the precautions in the beginning of this section.
2. Remove or disconnect the following:

- Axle halfshaft from the vehicle.
- TJ boot clamps
- TJ boot from the TJ case
- Snap-ring and spider assembly
- BJ boot clamps, then pull out the TJ boot and the BJ boot
3. Clean all the components properly

To install:

➡ **Use new circlips and boot clamps for assembly.**

4. Apply the grease supplied with the kit to the shaft and install the boots.
5. If installing the dynamic damper, keep the BJ shaft in a straight line and attach the damper in the direction shown in the accompanying illustration, then install the boot clamp.
6. Apply the grease supplied with the

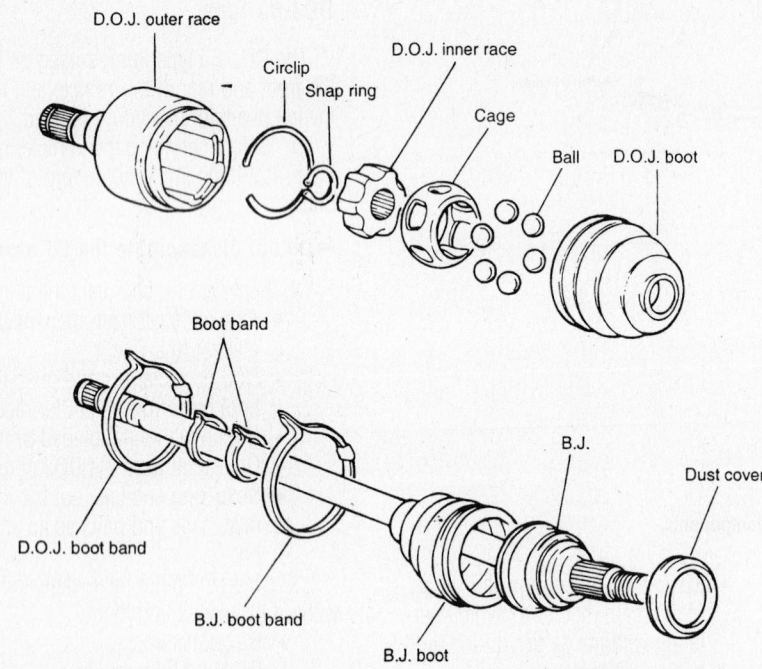

TORQUE : Nm (kg·cm, lb·ft)

9355LG41

Exploded view of the DOJ-BJ type CV-joint assembly

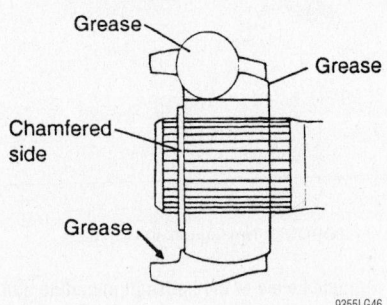

Install the chamfered side of the cage as shown

9355LG46

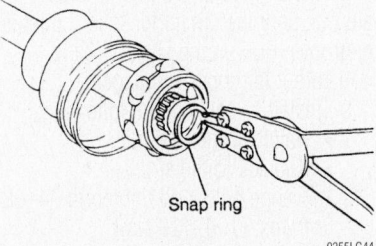

Snap ring

9355LG44

Remove snap-ring and take out the inner race, cage and balls as an assembly

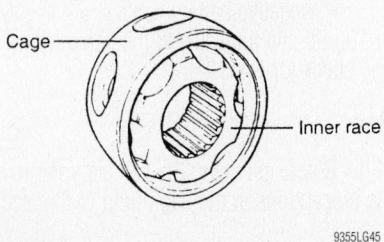

Cage

Inner race

9355LG45

Install the cage so that it is offset on the race

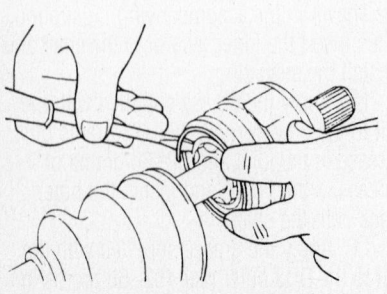

9355LG43

Remove the circlip with a suitable tool

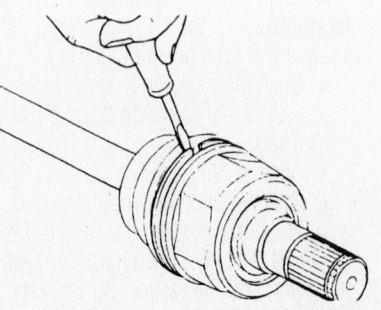

9355LG42

Remove the boot clamps from the DOJ

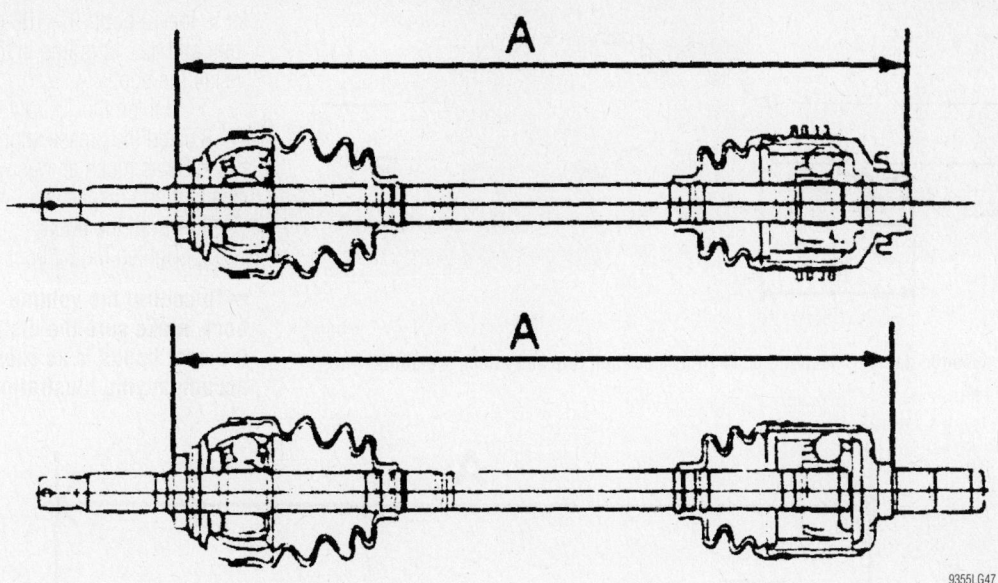

9355LG47

To control the volume of air in the DOJ boot, make sure the distance between the boot bands is as shown

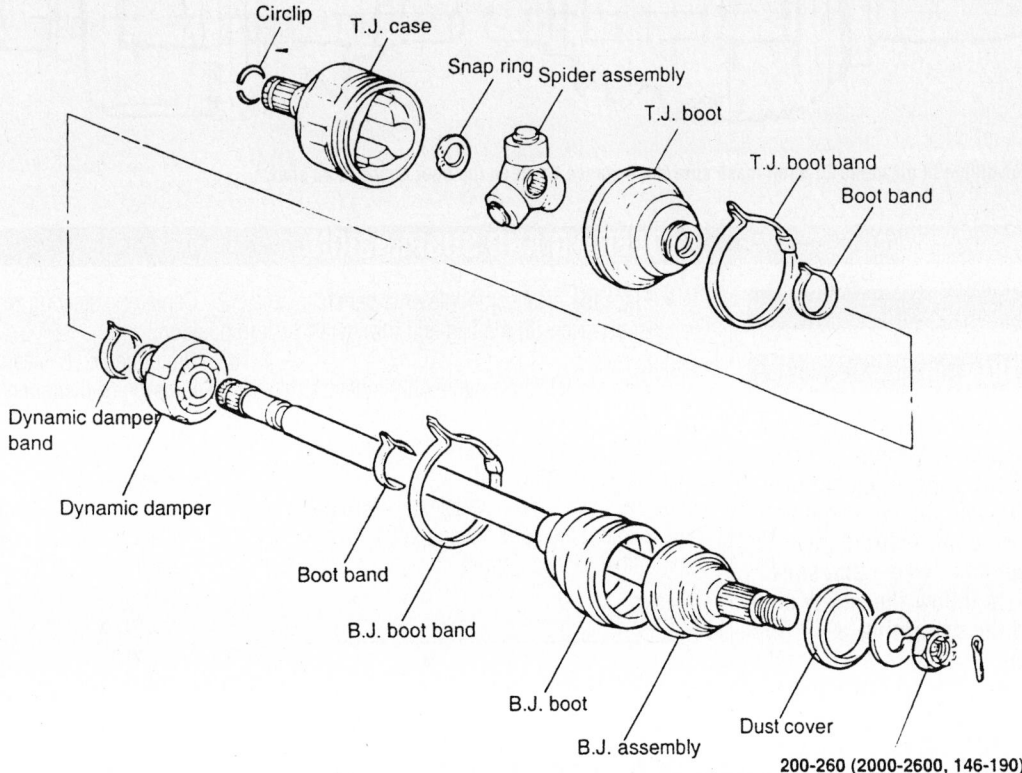

Circlip
T.J. case
Snap ring
Spider assembly
T.J. boot
T.J. boot band
Boot band
Dynamic damper band
Dynamic damper
Boot band
B.J. boot band
B.J. boot
B.J. assembly
Dust cover
200-260 (2000-2600, 146-190)

TORQUE : Nm (kg·cm, lb·ft)

9355LG48

Exploded view of the TJ-BJ type CV assembly

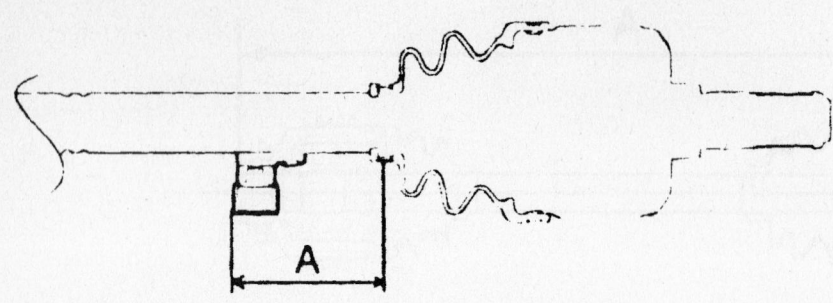

If installing the dynamic damper, keep the BJ shaft in a straight line and attach the damper in the direction shown

kit to the TJ boot (97–103 grams in the joint and 42–48 grams in the boot), then install the boot.

7. Tighten the TJ boot clamps.

8. Add the grease supplied with the kit to the BJ as much as was wiped out during cleaning and inspection.

9. Install the boots.

10. Tighten the BJ boot clamps.

➡ **To control the volume of air in the TJ boot, make sure the distance between the boot bands is as shown in the accompanying illustration**

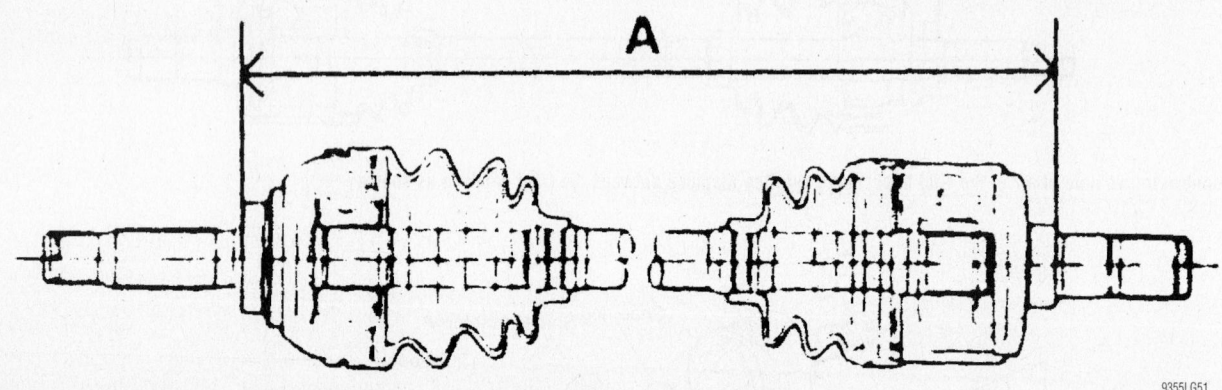

To control the volume of air in the TJ boot, make sure the distance between the boot bands is as shown

STEERING AND SUSPENSION

Air Bag

✳✳ CAUTION

Some vehicles are equipped with an air bag system. The system must be disarmed before performing service on, or around, system components, the steering column, instrument panel components, wiring and sensors. Failure to follow the safety precautions and the disarming procedure could result in accidental air bag deployment, possible injury and unnecessary system repairs.

PRECAUTIONS

Several precautions must be observed when handling the inflator module to avoid accidental deployment and possible personal injury.

• Never carry the inflator module by the wires or connector on the underside of the module.

• When carrying a live inflator module, hold securely with both hands, and ensure that the bag and trim cover are pointed away.

• Place the inflator module on a bench or other surface with the bag and trim cover facing up.

• With the inflator module on the bench, never place anything on or close to the module which may be thrown in the event of an accidental deployment.

Before servicing the vehicle, also make sure to refer to the precautions in the beginning of this section as well.

DISARMING

1. Turn the wheel to the straight ahead position.

2. Disconnect the negative battery cable and wait at least 30 seconds for the air bag energy to deplete.

3. After work has been completed, connect the battery cable.

Power Rack and Pinion Steering Gear

REMOVAL & INSTALLATION

1. Before servicing the vehicle, refer to the precautions in the beginning of this section.

2. Center the steering wheel and lock it in position.

3. Drain the power steering system.

4. Remove or disconnect the following:

• Negative battery cable
• Pressure and return hoses
• Joint assembly connecting bolt
• Tie rod end from the knuckle
• Feed tube
• Gear box mounting bolts
• Gear box assembly with the rubber mounts

To install:

5. Install or connect the following:

• Gear box assembly with the rubber mounts
• Gear box mounting bolts and tighten to 66–81 ft. lbs. (90–110 Nm)
• Feed tube and tighten to 7–11 ft. lbs. (10–16 Nm)
• Tie rod end from the knuckle and tighten to 18–25 ft. lbs. (24–34 Nm)
• Joint assembly connecting bolt and tighten to 11–14 ft. lbs. (15–20 Nm)
• Pressure and return hoses. Tighten

the fittings to 9–13 ft. lbs. (12–18 Nm).
  • Negative battery cable
6. Fill the power steering system.
7. Check the wheel alignment and adjust as necessary.

Strut

REMOVAL & INSTALLATION

Front

1. Before servicing the vehicle, refer to the precautions in the beginning of this section.

2. Remove or disconnect the following:
  • Front wheel
  • Brake hose clip from the strut mounting bracket
  • Wheel Speed Sensor (WSS) from the knuckle
  • Stabilizer bar link
  • Strut-to-knuckle bolts
  • 3 upper mounting nuts
  • Strut

To install:

3. Install or connect the following:
  • Strut. Tighten the upper mount nuts to 30–37 ft. lbs. (40–50 Nm)

on 2001–03 models or 33–44 ft. Lbs. (45–60 Nm) on 2004 models..
  • Strut-to-knuckle bolts and tighten to 74–88 ft. lbs. (100–120 Nm)
  • Stabilizer bar link and tighten to 30–37 ft. lbs. (30–50 Nm)
  • WSS from the knuckle
  • Brake hose clip to the strut mounting bracket
  • Front wheel

4. Check the wheel alignment and adjust as necessary.

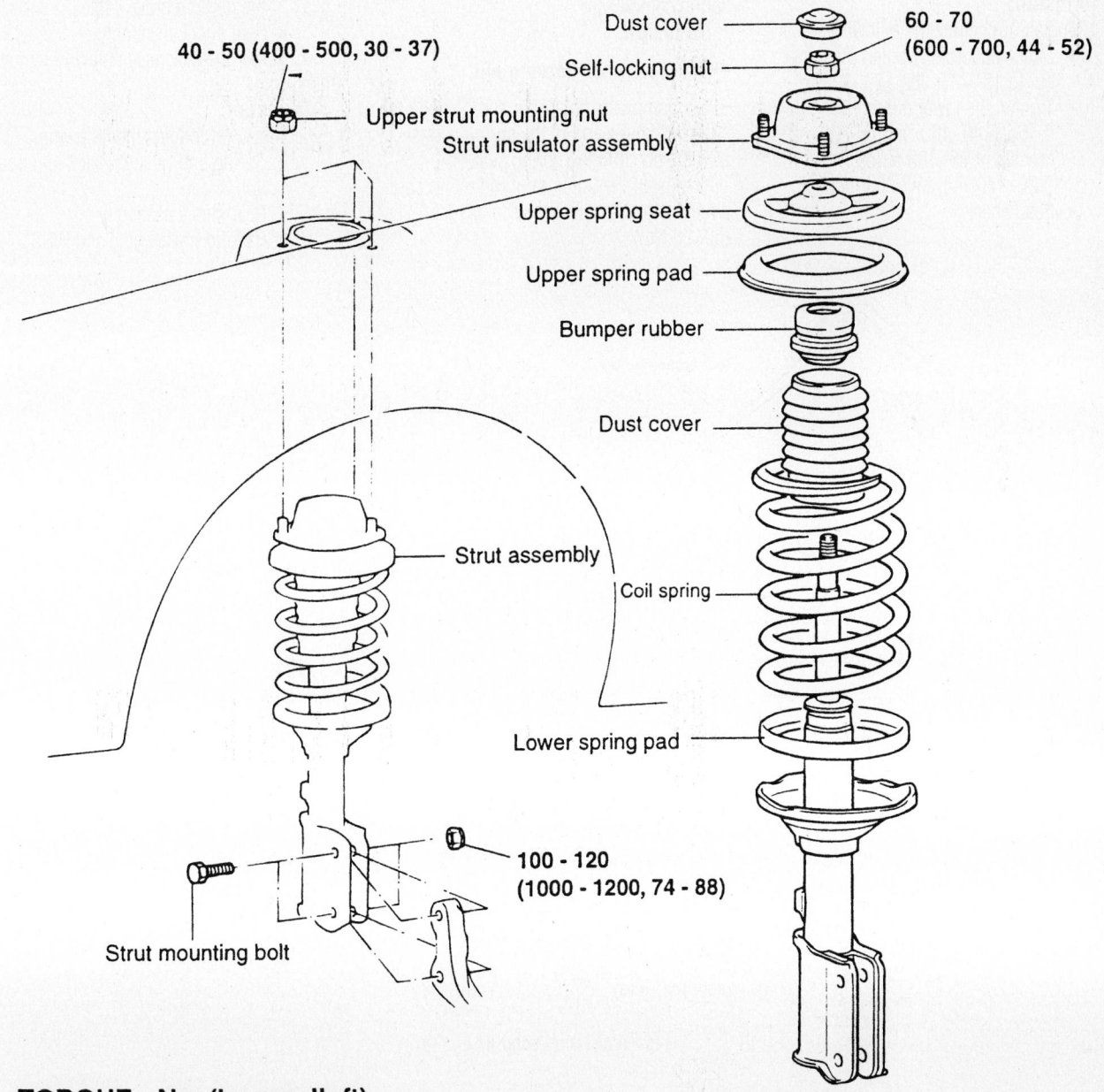

40 - 50 (400 - 500, 30 - 37)

Upper strut mounting nut

Dust cover

60 - 70
(600 - 700, 44 - 52)

Self-locking nut

Strut insulator assembly

Upper spring seat

Upper spring pad

Bumper rubber

Dust cover

Strut assembly

Coil spring

Lower spring pad

100 - 120
(1000 - 1200, 74 - 88)

Strut mounting bolt

TORQUE : Nm (kg·cm, lb·ft)

9355LG52

Exploded view of the front strut assembly and mounting

Shock Absorber

REMOVAL & INSTALLATION

Rear

1. Before servicing the vehicle, refer to the precautions in the beginning of this section.

2. Support the vehicle under the lower control arm.

3. Remove or disconnect the following:
   - Rear wheel
   - Upper shock absorber upper nut
   - Lower shock absorber nut
   - Shock absorber

To install:

4. Install or connect the following:
   - Shock absorber. Tighten the upper nut to 15–22 ft. lbs. (20–30 Nm)
   - Tighten the lower nut to 104–118 ft. lbs. (140–160 Nm) on 2001–03 models or 88–103 ft. lbs. (120–140 Nm) on 2004 models.
   - Rear wheel

Coil Spring

REMOVAL & INSTALLATION

Front

1. Before servicing the vehicle, refer to the precautions in the beginning of this section.

2. Remove the strut from the vehicle and install in a strut spring compressor. Compress the spring until the end of the spring comes away from the spring seat.

3. Remove the upper strut mount, spring seat and related components.

4. Remove the coil spring from the strut spring compressor.

To install:

➡ **Use a new self-locking nut.**

5. Compress the spring and position the strut so that the end of the spring aligns with the notch in the spring seat.

6. Install the upper strut mounting components and tighten the nut to 44–52 ft. lbs. (60–70 Nm).

7. Install the strut to the vehicle.

8. Check the wheel alignment and adjust as necessary.

Rear

1. Before servicing the vehicle, refer to the precautions in the beginning of this section.

2. Support the vehicle under the lower control arm.

3. Remove or disconnect the following:
   - Rear wheel
   - Flange nut and brake caliper assembly
   - Parking brake assembly
   - Wheel Speed Sensor (WSS) and the parking brake cable
   - Rear shock assembly

4. Lower the jack assembly and remove the spring.

To install:

5. Install or connect the following:
   - Spring and raise the jack into position
   - Rear shock assembly
   - Parking brake cable and WSS

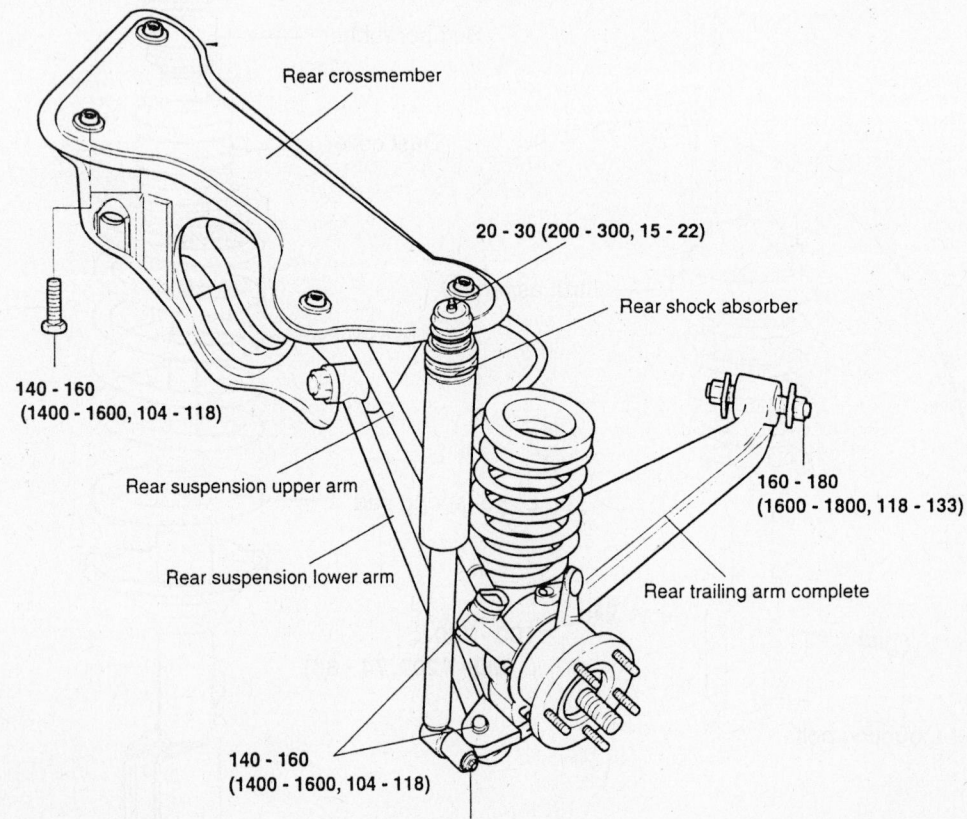

Rear crossmember

20 - 30 (200 - 300, 15 - 22)

Rear shock absorber

140 - 160
(1400 - 1600, 104 - 118)

Rear suspension upper arm

160 - 180
(1600 - 1800, 118 - 133)

Rear suspension lower arm

Rear trailing arm complete

140 - 160
(1400 - 1600, 104 - 118)

140 - 160 (1400 - 1600, 104 - 118)

TORQUE : Nm (kg·cm, lb·ft)

9355LG53

Exploded view of the rear suspension assembly

- Parking brake assembly
- Flange nut and brake caliper assembly
- Rear wheel

Ball Joint

REMOVAL & INSTALLATION

1. Before servicing the vehicle, refer to the precautions in the beginning of this section.
2. Remove the control arm.
3. Using tools 09551-3100 and 09216-21100, remove the bushing.
4. Using a suitable prytool and remove the dust cover from the ball joint.
5. Remove the snap-ring.
6. Remove the ball joint from the arm using a plastic hammer.

To install:

7. Install the ball joint.
8. Install the bushings into the arm using the appropriate tools. Make sure that the ball joint flange is supported while pressing down on the bushing until the flange touches the arm surface.
9. Install the ball joint snap-ring. Be careful to keep the snap-ring expansion as small as possible during installation.
10. Apply multi-purpose grease to the dust cover lip and inside of the cover.
11. Using tool 09545-11000, install the dust cover until it is completely seated on the snap ring.
12. Install the control arm.

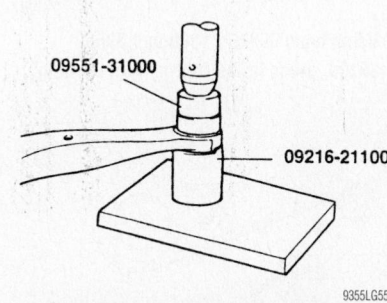

09551-31000
09216-21100
9355LG55

Removing the bushing from the control arm using the appropriate removal and installation tools

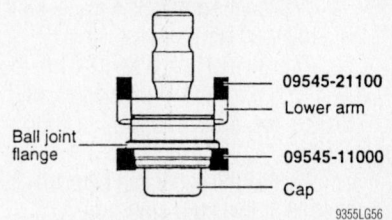

09545-21100
Lower arm
Ball joint flange
09545-11000
Cap
9355LG56

Installing the ball joint dust cover

Lower Control Arm

REMOVAL & INSTALLATION

Front

1. Before servicing the vehicle, refer to the precautions in the beginning of this section.
2. Remove or disconnect the following:
   - Front wheel
   - Ball joint-to-knuckle bolt
   - Sub frame bolts and the frame
   - Lower arm bolts and the arm
3. Installation is the reverse of removal. Tighten the fasteners as follows:

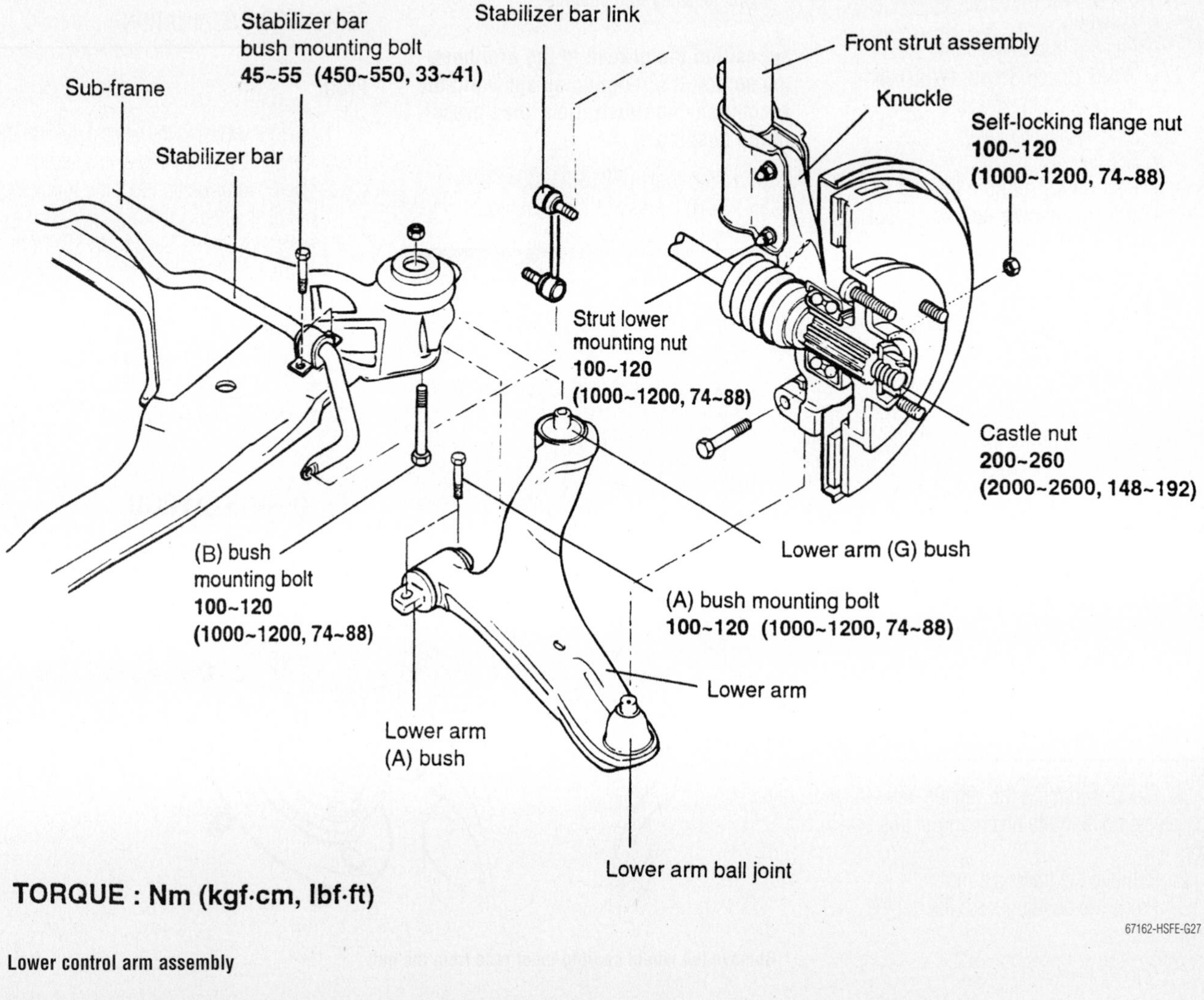

Stabilizer bar bush mounting bolt
45~55 (450~550, 33~41)
Stabilizer bar link
Front strut assembly
Sub-frame
Knuckle
Self-locking flange nut
100~120
(1000~1200, 74~88)
Stabilizer bar
Strut lower mounting nut
100~120
(1000~1200, 74~88)
Castle nut
200~260
(2000~2600, 148~192)
Lower arm (G) bush
(B) bush mounting bolt
100~120
(1000~1200, 74~88)
(A) bush mounting bolt
100~120 (1000~1200, 74~88)
Lower arm
Lower arm
(A) bush
Lower arm ball joint
TORQUE : Nm (kgf·cm, lbf·ft)
67162-HSFE-G27

Lower control arm assembly

a. Tighten the arm bolt **A** to 74–88 ft. lbs. (100–120 Nm).

b. Tighten and bolt **B** to 66–81 ft. lbs. (90–110 Nm) on 2001–03 models, or 74–88 ft. lbs. (100–120 Nm) on 2004 models.

c. Tighten the sub frame bolts to 118–148 ft. lbs. (160–200 Nm).

d. Tighten the ball joint-to-knuckle bolt to 74–88 ft. lbs. (100–120 Nm).

4. Check the wheel alignment and adjust as necessary.

Trailing Arm

REMOVAL & INSTALLATION

Rear

1. Before servicing the vehicle, refer to the precautions in the beginning of this section.

2. Support the vehicle under the lower control arm.

3. Remove or disconnect the following:
- Rear wheel
- Flange nut and brake caliper assembly
- Parking brake assembly
- Wheel Speed Sensor (WSS) and the parking brake cable
- Rear shock assembly
- Spring
- Rear driveshaft from the rear axle
- Upper and lower arm using tool 09517-43001.
- Trailing arm bolt and the arm

To install:

4. Install or connect the following:
- Trailing arm and tighten the trailing arm bolt to 118–133 (160–180 Nm)
- Upper and lower arm. Tighten the ball joint nut to 104–11 ft. lbs. (140–118 Nm).
- Spring and raise the jack into position
- Rear shock assembly
- Parking brake cable and WSS
- Parking brake assembly
- Flange nut and brake caliper assembly
- Rear wheel

TRAILING ARM BUSHING REPLACEMENT

1. Before servicing the vehicle, refer to the precautions in the beginning of this section.

2. Remove the trailing arm.

3. Press the bushing from the trailing arm.

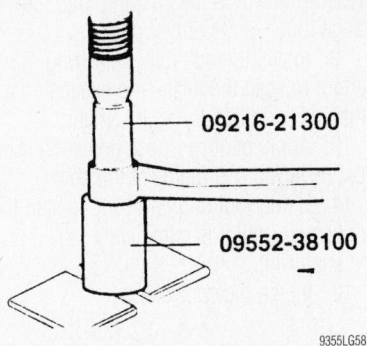

Using tools 09216-21300 and 09552-38100, press fit the trailing arm bushing

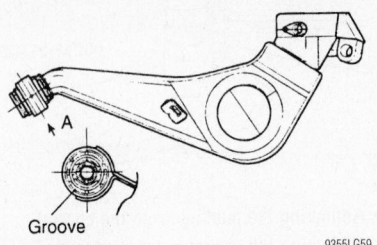

Position the groove in the arm bushing so that it is aligned as shown, before pressing the bushing into position

➡Position the groove in the arm bushing so that it is aligned as shown in the accompanying illustration, then press fit the bushing.

4. Using tools 09216-21300 and 09552-38100, press fit the bushing.

Wheel Bearings

ADJUSTMENT

The wheel bearings are sealed units and are not adjustable.

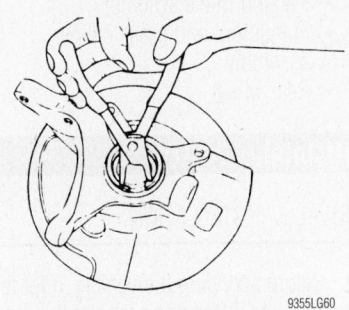

Remove the snap-ring from the hub

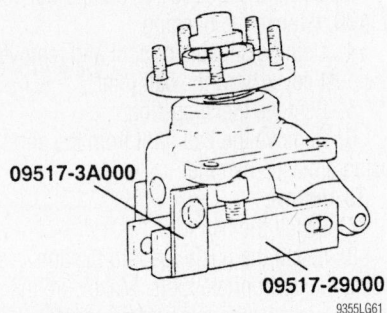

Remove the hub from the knuckle

REMOVAL & INSTALLATION

Front

1. Before servicing the vehicle, refer to the precautions in the beginning of this section.

2. Remove or disconnect the following:
- Front wheel
- Wheel Speed Sensor (WSS) from the knuckle
- Brake caliper and suspend it aside using wire
- Split pin and nut from the axle
- Strut from the knuckle
- Tie rod end from the knuckle

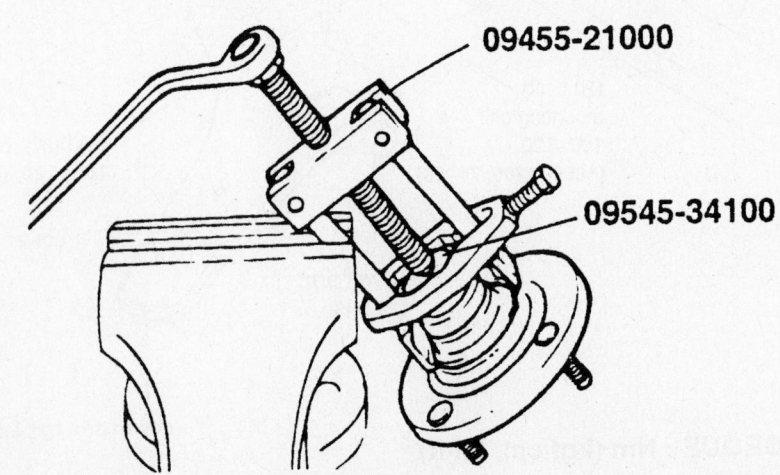

Remove the wheel bearing inner race from the hub

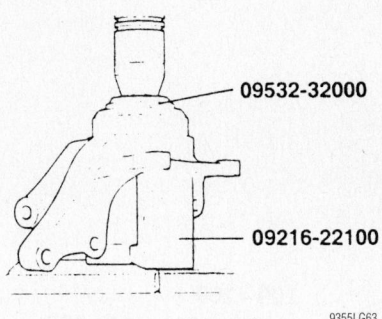

Remove the wheel bearing outer race from the knuckle

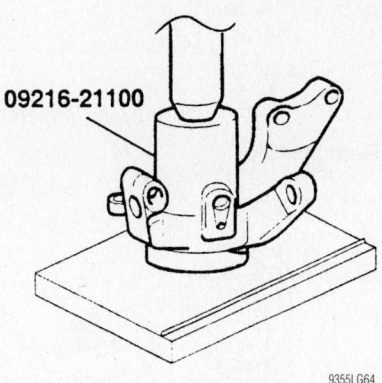

Install the bearing onto the knuckle

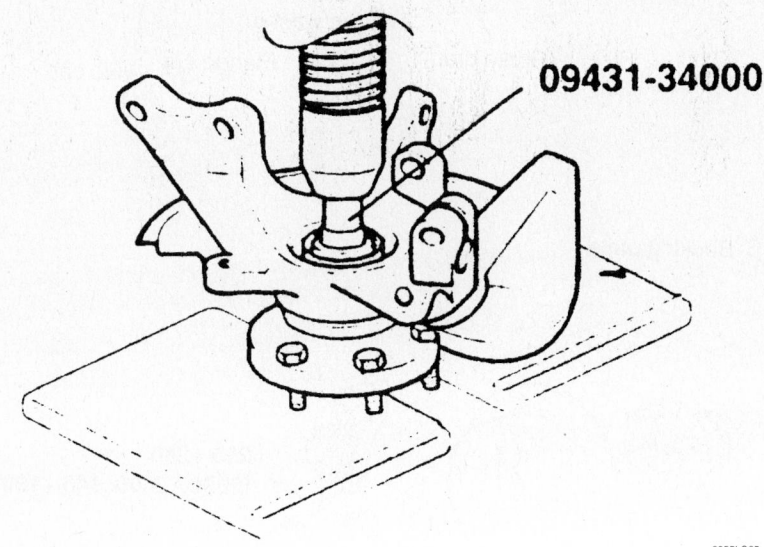

Press the hub onto the knuckle

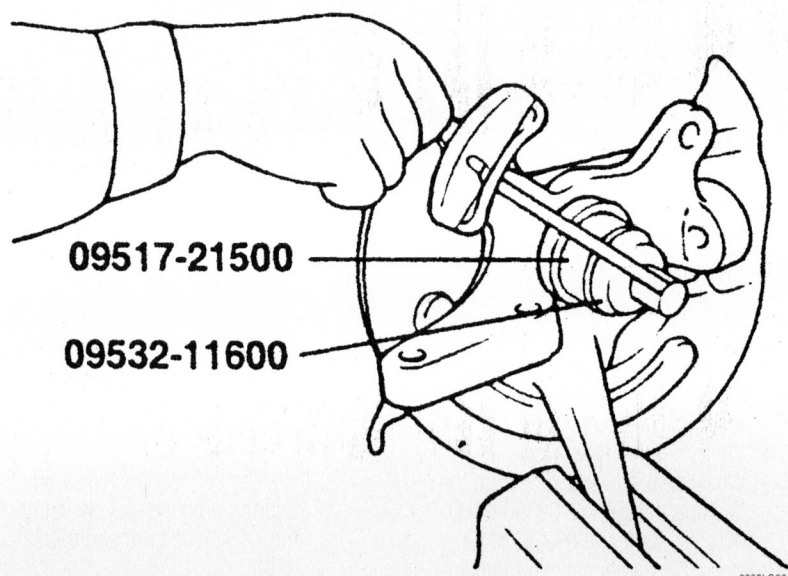

Check the wheel bearing starting torque

- Lower ball joint bolt
- Axle shaft from the knuckle using a plastic hammer
- Brake disc
- Knuckle assembly
- Snap-ring from the hub
- Hub from the knuckle by installing tools 09517-3A00 and 09517-2900, then tighten the nut of the tool to separate the tool from the knuckle
- Wheel bearing inner race from the hub using tools 09455-2100 and 09545-34100
- Wheel bearing outer race from the knuckle using tools 09532-3200 and 09216-22100

3. Check all components for wear or damage and replace as necessary.

To install:

4. Apply a thin coat of multi-purpose grease to the surface on the knuckle and bearing.

➡**Do not press against the outer race of the bearing as this can cause bearing damage and always use a new bearing kit.**

5. Install or connect the following:
- Bearing onto the knuckle using tool 09216-21100.
- Snap-ring into the groove of the knuckle
- Backing plate onto the knuckle

➡**Do not press against the outer race of the bearing as this can cause bearing damage and always use a new bearing kit.**

- Hub onto the knuckle by pressing it into position using tool 09431-3400

6. Rotate the bearing several times to seat the bearing.

7. Measure the wheel bearing torque using an inch lb. torque wrench. The measurement is 16.64 inch lbs. (1.88 Nm).

8. Measure the end-play of the hub using a dial gauge. The specification is 0.003-0.008mm on 2001–03 models or

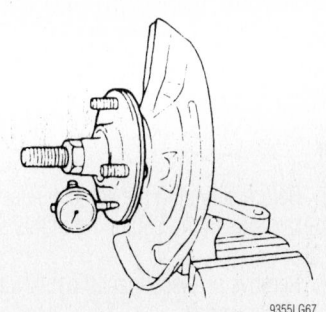

Check the hub end end-play

<DRUM BRAKE>

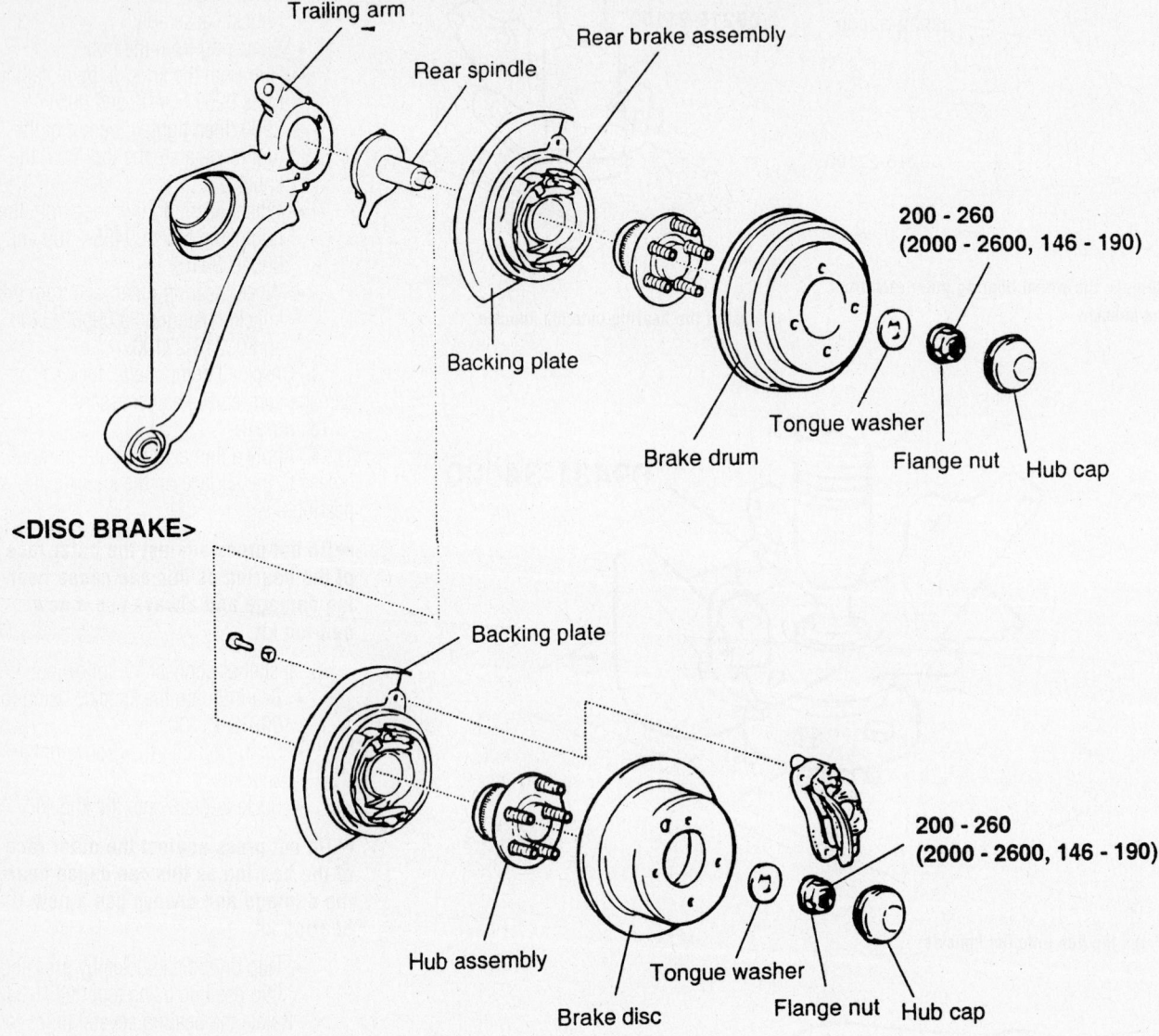

200 - 260
(2000 - 2600, 146 - 190)

<DISC BRAKE>

200 - 260
(2000 - 2600, 146 - 190)

TORQUE : Nm (kg·cm, lb·ft)

9355LG68

Exploded view of the rear hub assembly

0.0025–0.0035 inch (0.064–0.088mm) on 2004 models.

9. Install the remaining components in the reverse order of removal.

10. Check the wheel alignment and adjust as necessary.

Rear

1. Before servicing the vehicle, refer to the precautions in the beginning of this section.

2. Remove or disconnect the following:
• Rear wheel

• Flange nut and washer
• Drum or rotor
• Brake line
• Parking brake assembly
• Parking brake cable
• Spindle bolts and the spindle
• Rear hub from the housing using tool 09517-43001
• Wheel bearing snap-ring
• Wheel bearing inner race from the housing using tools 09500-2100, 09527-33000 and 09216-22100

3. Inspect the components for damage and replace as necessary.

To install:

4. Apply a thin coat of multi-purpose grease to the surface on the housing and bearing.

➡**Do not press against the outer race of the bearing as this can cause bearing damage and always use a new bearing kit.**

5. Install or connect the following:
• Bearing onto the spindle using tools 09216-21100 and 09532-3200
• Snap-ring

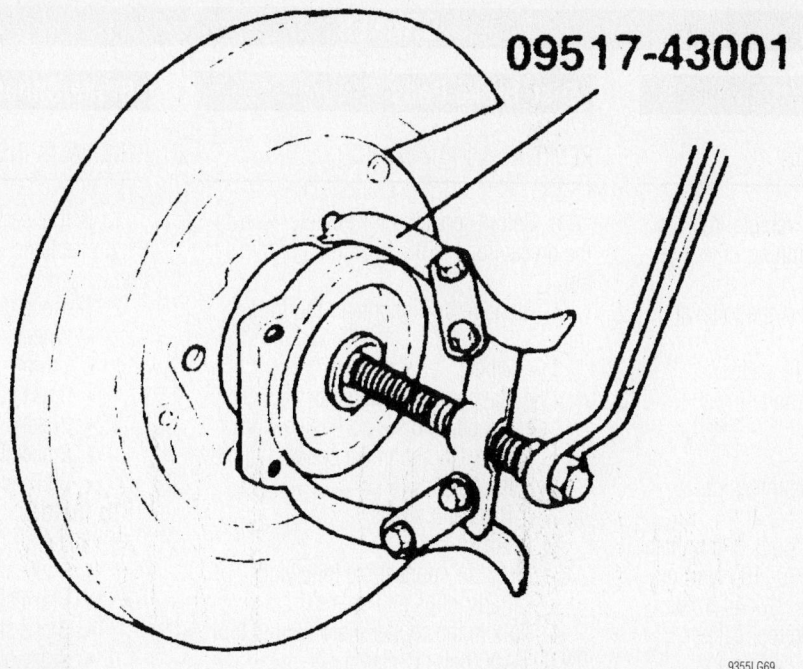

09517-43001

9355LG69

Removing the rear hub from housing

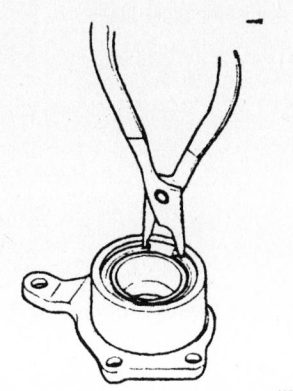

9355LG70

Removing the wheel bearing snap-ring

- Backing plate, then press the hub onto the housing using tool 09517-21500
6. Rotate the bearing several times to seat the bearing.

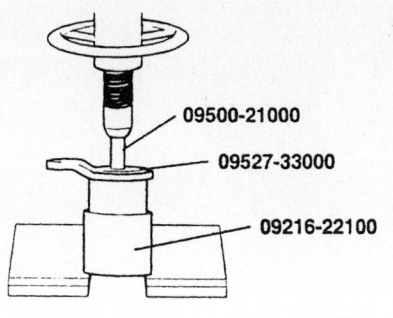

09500-21000

09527-33000

09216-22100

9355LG71

Removing the wheel bearing inner race from housing

7. Measure the wheel bearing torque using an inch lb. torque wrench. The measurement is 16.64 inch lbs. (1.88 Nm) on 2001–03 models, or 0.73 inch (0.99 Nm) on 2004 models.

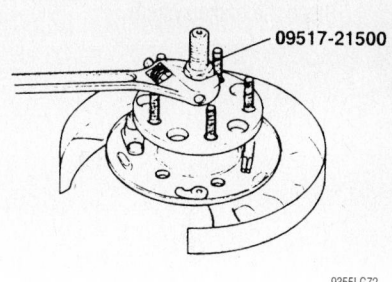

09517-21500

9355LG72

Press the hub onto the housing

8. Measure the end-play of the hub using a dial gauge. The specification is 0.003-0.008mm.
9. Install the remaining components in the reverse order of removal.
10. Check the wheel alignment and adjust as necessary.

BRAKES

Brake Caliper

REMOVAL & INSTALLATION

1. Before servicing the vehicle, refer to the precautions in the beginning of this section.
2. Remove or disconnect the following:
   - Wheel
   - Brake hose from the caliper
   - Caliper mounting bolts
   - Caliper

To install:

3. Install or connect the following:
   - Caliper
   - Caliper mounting bolts and tighten to 58–73 ft. lbs. (80–100 Nm) on the front caliper, or 37–44 ft. lbs. 50–60 Nm) on the rear caliper.
   - Brake hose to the caliper and tighten the fitting too 18–22 ft. lbs. (25–30 Nm)
   - Wheel

4. Bleed the brake system.

Disc Brake Pads

REMOVAL & INSTALLATION

1. Before servicing the vehicle, refer to the precautions in the beginning of this section.
2. Remove or disconnect the following:
   - Wheel
   - Caliper mounting bolts
   - Caliper and support aside with wire. Do not let the caliper hang by the hose.
   - Pads and shims

To install:

3. Install or connect the following:
   - Pads, clips and shims.

4. Bottom the caliper piston using tool 09581-11000 or a C-clamp
   - Caliper mounting bolts and tighten to 16–24 ft. lbs. (22–32Nm)
   - Wheel

Brake Shoes

REMOVAL & INSTALLATION

1. Before servicing the vehicle, refer to the precautions in the beginning of this section.
2. Remove or disconnect the following:
   - Wheel
   - Drum
   - Brake shoe hold-down spring
   - Brake strut
   - Brake shoe return spring
   - Brake shoes

To install:

3. Install or connect the following:
   - Brake shoes
   - Brake shoe return spring
   - Brake strut
   - Brake shoe hold-down spring
   - Brake drum
   - Wheel

4. Adjust the brake shoes.

SPECIFICATION AND MAINTENANCE CHARTS

ENGINE AND VEHICLE IDENTIFICATION

		Engine					Model Year	
Code ①	Liters (cc)	Cu. In.	Cyl.	Fuel Sys.	Engine Type	Eng. Mfg.	Code ②	Year
VQ35DE	3.5 (3498)	183	6	SFI	DOHC	Nissan	3	2003
VK45DE	4.5 (4494)	274	8	SFI	DOHC	Nissan	4	2004
							5	2005

NA: Information not available

SFI: Sequential Fuel Injection

DOHC: Double Overhead Camshaft

① Stamped on the upper rear of the engine block, just behind a cylinder head.

② 10th digit of the Vehicle Identification Number (VIN)

67162-FX35-C01

GENERAL ENGINE SPECIFICATIONS

Year	Model	Engine Displacement Liters	Engine Series ID	Net Horsepower @ rpm	Net Torque @ rpm (ft. lbs.)	Bore x Stroke (in.)	Compression Ratio	Oil Pressure @ rpm
2003	FX35	3.5	VQ35DE	280@6200	270@4800	3.76 x 3.21	10.3:1	43@2000
	FX45	4.5	VK45DE	315@6400	329@4000	3.66 x 3.26	10.5:1	43@2000
2004	FX35	3.5	VQ35DE	280@6200	270@4800	3.76 x 3.21	10.3:1	43@2000
	FX45	4.5	VK45DE	315@6400	329@4000	3.66 x 3.26	10.5:1	43@2000
2005	FX35	3.5	VQ35DE	280@6200	270@4800	3.76 x 3.21	10.3:1	43@2000
	FX45	4.5	VK45DE	315@6400	329@4000	3.66 x 3.26	10.5:1	43@2000

67162-FX35-C02

ENGINE TUNE-UP SPECIFICATIONS

Year	Engine Displacement Liters	Engine ID	Spark Plug Gap (in.)	Ignition Timing (deg.) ①	Fuel Pump (psi)	Idle Speed (rpm)	Valve Clearance ② Intake	Valve Clearance ② Exhaust
2003	3.5	VQ35DE	0.043	10-20B	51	600-700	0.010-0.013	0.011-0.015
	4.5	VK45DE	0.043	7-17B	51	600-700	0.010-0.013	0.011-0.015
2004	3.5	VQ35DE	0.043	10-20B	51	600-700	0.010-0.013	0.011-0.015
	4.5	VK45DE	0.043	7-17B	51	600-700	0.010-0.013	0.011-0.015
2005	3.5	VQ35DE	0.043	10-20B	51	600-700	0.010-0.013	0.011-0.015
	4.5	VK45DE	0.043	7-17B	51	600-700	0.010-0.013	0.011-0.015

NOTE: The Vehicle Emission Control Information label often reflects specification changes made during production.

The label figures must be used if they differ from those in this chart.

B: Before top dead center

① With terminals TC and CG of DLC3 connected

② Engine cold - approximately 68°F (20°C)

67162-FX35-C03

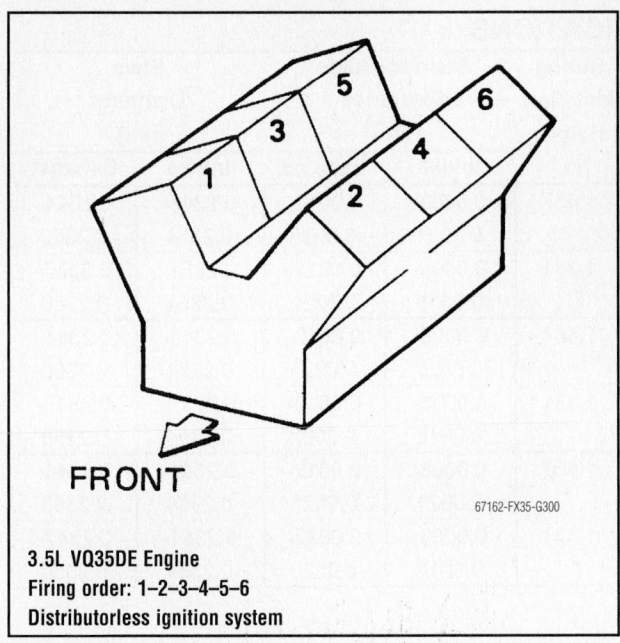

3.5L VQ35DE Engine
Firing order: 1–2–3–4–5–6
Distributorless ignition system

67162-FX35-G300

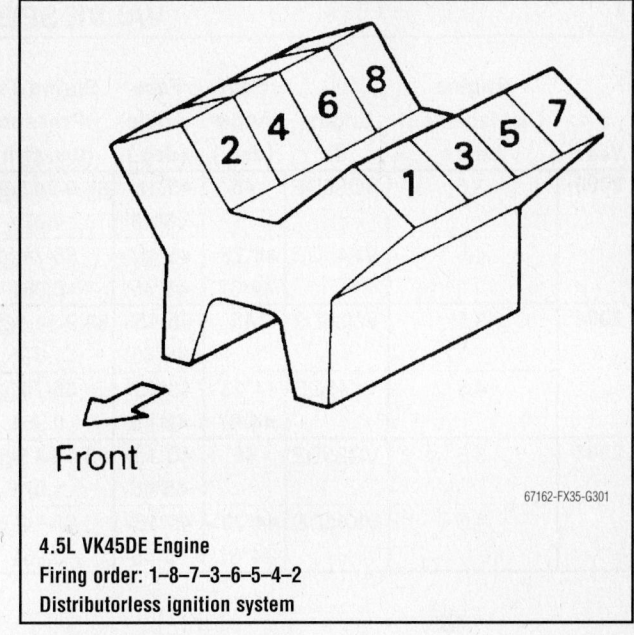

4.5L VK45DE Engine
Firing order: 1–8–7–3–6–5–4–2
Distributorless ignition system

67162-FX35-G301

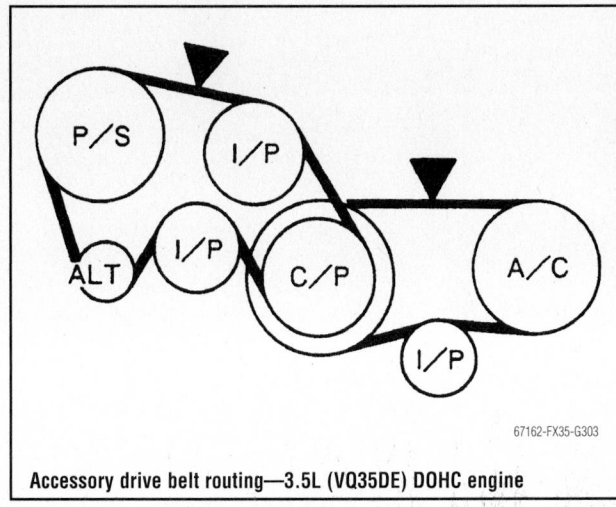

Accessory drive belt routing—3.5L (VQ35DE) DOHC engine

67162-FX35-G303

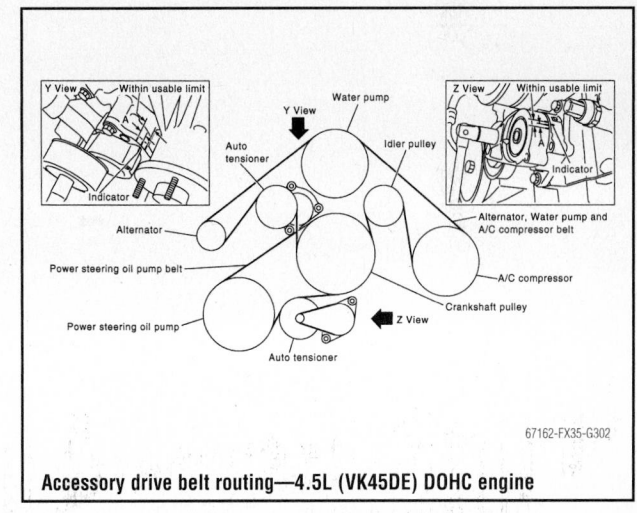

Accessory drive belt routing—4.5L (VK45DE) DOHC engine

67162-FX35-G302

CAPACITIES

Year	Model	Engine Displacement Liters	Engine ID	Engine Oil with Filter (qts.)	Transaxle (pts.) ①	Transfer Case (pts.)	Rear Drive Axle (pts.)	Fuel Tank (gal.)	Cooling System (qts.)
2003	FX35	3.5	VQ35DE	5.0	10.9	2.6	3.0	23.8	8.6
	FX45	4.5	VK45DE	7.5	10.9	2.6	3.0	23.8	10.0
2004	FX35	3.5	VQ35DE	5.0	10.9	2.6	3.0	23.8	8.6
	FX45	4.5	VK45DE	7.5	10.9	2.6	3.0	23.8	10.0
2005	FX35	3.5	VQ35DE	5.0	10.9	2.6	3.0	23.8	8.6
	FX45	4.5	VK45DE	7.5	10.9	2.6	3.0	23.8	10.0

① After draining, add the following amounts, then, fill to the cold full line.

67162-FX35-C04

VALVE SPECIFICATIONS

Year	Engine Displacement Liters	Engine ID	Seat Angle (deg.)	Face Angle (deg.)	Spring Test Pressure (lbs. @ in.)	Spring Installed Height (in.)	Stem-to-Guide Clearance (in.)		Stem Diameter (in.)	
							Intake	Exhaust	Intake	Exhaust
2003	3.5	VQ35DE	45	45°15'-45°45'	83.9-94.5@1.071	1.457	0.0008-0.0021	0.0012-0.0025	0.2348-0.2354	0.2344-0.2350
	4.5	VK45DE	44°23'-44°67'	45°15'-45°45'	65-74@0.961	1.331	0.0008-0.0018	0.0012-0.0022	0.2351-0.2354	0.2347-0.2350
2004	3.5	VQ35DE	45	45°15'-45°45'	83.9-94.5@1.071	1.457	0.0008-0.0021	0.0012-0.0025	0.2348-0.2354	0.2344-0.2350
	4.5	VK45DE	44°23'-44°67'	45°15'-45°45'	65-74@0.961	1.331	0.0008-0.0018	0.0012-0.0022	0.2351-0.2354	0.2347-0.2350
2005	3.5	VQ35DE	45	45°15'-45°45'	83.9-94.5@1.071	1.457	0.0008-0.0021	0.0012-0.0025	0.2348-0.2354	0.2344-0.2350
	4.5	VK45DE	44°23'-44°67'	45°15'-45°45'	65-74@0.961	1.331	0.0008-0.0018	0.0012-0.0022	0.2351-0.2354	0.2347-0.2350

67162-FX35-C05

CRANKSHAFT AND CONNECTING ROD SPECIFICATIONS

All measurements are given in inches.

Year	Engine Displacement Liters	Engine ID	Crankshaft				Connecting Rod		
			Main Brg. Journal Dia.	Main Brg. Oil Clearance	Shaft End-play	Thrust on No.	Journal Diameter	Oil Clearance	Side Clearance
2003	3.5	VQ35DE	①	0.0014-0.0018	0.0039-0.0098	3	2.1654-2.1659	0.0013-0.0023	0.0079-0.0138
	4.5	VK45DE	②	③	0.0039-0.0098	3	2.1654-2.1659	0.0008-0.0018	0.0079-0.0138
2004	3.5	VQ35DE	①	0.0014-0.0018	0.0039-0.0098	3	2.1654-2.1659	0.0013-0.0023	0.0079-0.0138
	4.5	VK45DE	②	③	0.0039-0.0098	3	2.1654-2.1659	0.0008-0.0018	0.0079-0.0138
2005	3.5	VQ35DE	①	0.0014-0.0018	0.0039-0.0098	3	2.1654-2.1659	0.0013-0.0023	0.0079-0.0138
	4.5	VK45DE	②	③	0.0039-0.0098	3	2.1654-2.1659	0.0008-0.0018	0.0079-0.0138

① - Depends on the grade of the crankshaft, as listed below:

Grade No. A	59.975 - 59.974 (2.3612 - 2.3612)	Grade No. N	59.963 - 59.962 (2.3607 - 2.3607)
Grade No. B	59.974 - 59.973 (2.3612 - 2.3611)	Grade No. P	59.962 - 59.961 (2.3607 - 2.3607)
Grade No. C	59.973 - 59.972 (2.3611 - 2.3611)	Grade No. R	59.961 - 59.960 (2.3607 - 2.3606)
Grade No. D	59.972 - 59.971 (2.3611 - 2.3611)	Grade No. S	59.960 - 59.959 (2.3606 - 2.3606)
Grade No. E	59.971 - 59.970 (2.3611 - 2.3610)	Grade No. T	59.959 - 59.958 (2.3606 - 2.3605)
Grade No. F	59.970 - 59.969 (2.3610 - 2.3610)	Grade No. U	59.958 - 59.957 (2.3605 - 2.3605)
Grade No. G	59.969 - 59.968 (2.3610 - 2.3609)	Grade No. V	59.957 - 59.956 (2.3605 - 2.3605)
Grade No. H	59.968 - 59.967 (2.3609 - 2.3609)	Grade No. W	59.956 - 59.955 (2.3605 - 2.3604)
Grade No. J	59.967 - 59.966 (2.3609 - 2.3609)	Grade No. X	59.955 - 59.954 (2.3604 - 2.3604)
Grade No. K	59.966 - 59.965 (2.3609 - 2.3608)	Grade No. Y	59.954 - 59.953 (2.3604 - 2.3603)
Grade No. L	59.965 - 59.964 (2.3608 - 2.3608)	Grade No. 4	59.953 - 59.952 (2.3603 - 2.3603)
Grade No. M	59.964 - 59.963 (2.3608 - 2.3607)	Grade No. 7	59.952 - 59.951 (2.3603 - 2.3603)

② - Depends on the grade of the crankshaft, as listed below:

Journals 1 and 5 -   **Journals 2, 3 and 4 -**

Grade No. G	63.963 - 63.964 (2.5182 - 2.5183)	Grade No. A	63.963 - 63.964 (2.5182 - 2.5183)
Grade No. H	63.962 - 63.963 (2.5182 - 2.5182)	Grade No. B	63.962 - 63.963 (2.5182 - 2.5182)
Grade No. J	63.961 - 63.962 (2.5181 - 2.5182)	Grade No. C	63.961 - 63.962 (2.5181 - 2.5182)
Grade No. K	63.960 - 63.961 (2.5181 - 2.5181)	Grade No. D	63.960 - 63.961 (2.5181 - 2.5181)
Grade No. L	63.959 - 63.960 (2.5181 - 2.5181)	Grade No. E	63.959 - 63.960 (2.5181 - 2.5181)
Grade No. M	63.958 - 63.959 (2.5180 - 2.5181)	Grade No. F	63.958 - 63.959 (2.5180 - 2.5181)
Grade No. N	63.957 - 63.958 (2.5180 - 2.5180)	Grade No. G	63.957 - 63.958 (2.5180 - 2.5180)
Grade No. P	63.956 - 63.957 (2.5179 - 2.5180)	Grade No. H	63.956 - 63.957 (2.5179 - 2.5180)
Grade No. R	63.955 - 63.956 (2.5179 - 2.5179)	Grade No. J	63.955 - 63.956 (2.5179 - 2.5179)
Grade No. S	63.954 - 63.955 (2.5179 - 2.5179)	Grade No. K	63.954 - 63.955 (2.5179 - 2.5179)
Grade No. T	63.953 - 63.954 (2.5178 - 2.5179)	Grade No. L	63.953 - 63.954 (2.5178 - 2.5179)
Grade No. U	63.952 - 63.953 (2.5178 - 2.5178)	Grade No. M	63.952 - 63.953 (2.5178 - 2.5178)
Grade No. V	63.951 - 63.952 (2.5178 - 2.5178)	Grade No. N	63.951 - 63.952 (2.5178 - 2.5178)
Grade No. W	63.950 - 63.951 (2.5177 - 2.5178)	Grade No. P	63.950 - 63.951 (2.5177 - 2.5178)
Grade No. X	63.949 - 63.950 (2.5177 - 2.5177)	Grade No. R	63.949 - 63.950 (2.5177 - 2.5177)
Grade No. Y	63.948 - 63.949 (2.5176 - 2.5177)	Grade No. S	63.948 - 63.949 (2.5176 - 2.5177)
Grade No. 1	63.947 - 63.948 (2.5176 - 2.5176)	Grade No. T	63.947 - 63.948 (2.5176 - 2.5176)
Grade No. 2	63.946 - 63.947 (2.5176 - 2.5176)	Grade No. U	63.946 - 63.947 (2.5176 - 2.5176)
Grade No. 3	63.945 - 63.946 (2.5175 - 2.5176)	Grade No. V	63.945 - 63.946 (2.5175 - 2.5176)
Grade No. 4	63.944 - 63.945 (2.5175 - 2.5175)	Grade No. W	63.944 - 63.945 (2.5175 - 2.5175)
Grade No. 5	63.943 - 63.944 (2.5174 - 2.5175)	Grade No. X	63.943 - 63.944 (2.5174 - 2.5175)
Grade No. 6	63.942 - 63.943 (2.5174 - 2.5174)	Grade No. Y	63.942 - 63.943 (2.5174 - 2.5174)
Grade No. 7	63.941 - 63.942 (2.5174 - 2.5174)	Grade No. 1	63.941 - 63.942 (2.5174 - 2.5174)
Grade No. 9	63.940 - 63.941 (2.5173 - 2.5174)	Grade No. 2	63.940 - 63.941 (2.5173 - 2.5174)

③ - Journals 1 and 5: 0.00004-0.0004

    Journals 2, 3 & 4: 0.0003-0.0007

PISTON AND RING SPECIFICATIONS

All measurements are given in inches.

Year	Engine Displ. Liters	Engine ID	Piston Clearance	Ring Gap			Ring Side Clearance		
				Top Comp.	Bottom Comp.	Oil Control	Top Comp.	Bottom Comp.	Oil Control
2003	3.5	VQ35DE	0.004-0.0012	0.0091-0.0130	0.0130-0.0189	0.0079-0.0197	0.0018-0.0031	0.0012-0.0028	0.0026-0.0053
	4.5	VK45DE	0.004-0.0012	0.0087-0.0126	0.0087-0.0126	0.0079-0.0236	0.0018-0.0031	0.0012-0.0028	0.0026-0.0053
2004	3.5	VQ35DE	0.004-0.0012	0.0091-0.0130	0.0130-0.0189	0.0079-0.0197	0.0018-0.0031	0.0012-0.0028	0.0026-0.0053
	4.5	VK45DE	0.004-0.0012	0.0087-0.0126	0.0087-0.0126	0.0079-0.0236	0.0018-0.0031	0.0012-0.0028	0.0026-0.0053
2005	3.5	VQ35DE	0.004-0.0012	0.0091-0.0130	0.0130-0.0189	0.0079-0.0197	0.0018-0.0031	0.0012-0.0028	0.0026-0.0053
	4.5	VK45DE	0.004-0.0012	0.0087-0.0126	0.0087-0.0126	0.0079-0.0236	0.0018-0.0031	0.0012-0.0028	0.0026-0.0053

67162-FX35-C07

TORQUE SPECIFICATIONS
All readings in ft. lbs.

Year	Engine Displacement Liters	Engine ID	Cylinder Head Bolts	Main Bearing Bolts	Rod Bearing Bolts	Crankshaft Damper Bolts	Flywheel Bolts	Manifold		Spark Plugs	Oil Pan Drain Plug
								Intake	Exhaust		
2003	3.5	VQ35DE	①	②	③	④	65	⑤	22	18	25
	4.5	VK45DE	⑥	⑦	⑧	⑨	65	21	21	18	25
2004	3.5	VQ35DE	①	②	③	④	65	⑤	22	18	25
	4.5	VK45DE	⑥	⑦	⑧	⑨	65	21	21	18	25
2005	3.5	VQ35DE	①	②	③	④	65	⑤	22	18	25
	4.5	VK45DE	⑥	⑦	⑧	⑨	65	21	21	18	25

① Step 1: Tighten in sequence to 72 ft. lbs.

   Step 2: Completely loosen all in reverse sequence

   Step 3: Tighten in sequence to 29 ft. lbs.

   Step 4: Tighten 90 degrees.

   Step 5: Tighten another 90 degrees.

② Step 1: Tighten in sequence to 10 ft. lbs.

   Step 2: Tighten in sequence to 26 ft. lbs.

   Step 4: Tighten 90 degrees.

③ Step 1: 14 ft. lbs.

   Step 2: Plus 90 degrees

④ Step 1: 33 ft. lbs.

   Step 2: Plus 60 degrees

⑤ Upper manifold-to-lower mainfold bolts: 9 ft. lbs.

   Lower manifold-to-cylinder head bolts & nuts:

   Step 1: 5 ft. lbs.

   Step 2: 21 ft. lbs.

⑥ Step 1: Tighten in sequence to 72 ft. lbs.

   Step 2: Completely loosen all in reverse sequence

   Step 3: Tighten in sequence to 33 ft. lbs.

   Step 4: Tighten 60 degrees.

   Step 5: Tighten another 60 degrees.

⑦ Step 1: Tighten M12 bolts in sequence 1-10 to 29 ft. lbs.

   Step 2: Tighten M9 bolts in sequence 11-20 to 22 ft. lbs.

   Step 3: Tighten M12 bolts in sequence 1-10 another 40 degrees.

   Step 4: Tighten M9 bolts in sequence 11-20 another 30 degrees.

   Step 5: Tighten M10 bolts in sequence 21-30 to 36 ft. lbs.

⑧ Step 1: 11 ft. lbs.

   Step 2: Plus 60 degrees

⑨ Step 1: 69 ft. lbs.

   Step 2: Plus 90 degrees

WHEEL ALIGNMENT

Year	Model		Caster Range (+/-Deg.)	Caster Preferred Setting (Deg.)	Camber Range (+/-Deg.)	Camber Preferred Setting (Deg.)	Toe-in (in.)
2003	FX35/FX45	F	0.75	+3.78	0.75	-0.73	0.063+/-0.039
		R	—	—	0.50	-0.80	0.185+/-0.091
2004	FX35/FX45	F	0.75	+3.78	0.75	-0.73	0.063+/-0.039
		R	—	—	0.50	-0.80	0.185+/-0.091
2005	FX35/FX45	F	0.75	+3.78	0.75	-0.73	0.063+/-0.039
		R	—	—	0.50	-0.80	0.185+/-0.091

F: Front

R: Rear

67162-FX35-C09

TIRE, WHEEL AND BALL JOINT SPECIFICATIONS

Year	Model	OEM Tires Standard	OEM Tires Optional	Tire Pressures (psi) Front	Tire Pressures (psi) Rear	Wheel Size	Ball Joint Inspection	Lugnut Torque (ft. lbs.)
2003	FX35/FX45	P265/60R18	P265/50R20	32	32	—	①	80
2004	FX35/FX45	P265/60R18	P265/50R20	32	32	—	①	80
2005	FX35/FX45	P265/60R18	P265/50R20	32	32	—	①	80

OEM: Original Equipment Manufacturer

PSI: Pounds Per Square Inch

STD: Standard

OPT: Optional

① Replace if any measurable movement is found.

67162-FX35-C10

BRAKE SPECIFICATIONS

All measurements in inches unless noted

Year	Model		Brake Disc Original Thickness	Brake Disc Minimum Thickness	Brake Disc Maximum Runout	Minimum Lining Thickness	Brake Caliper Bracket Bolts (ft. lbs.)	Brake Caliper Mounting Bolts (ft. lbs.)
2003	FX35/FX45	F	1.102	1.024	0.0016	0.079	116	20
		R	0.630	0.551	0.0020	0.079	62	32
2004	FX35/FX45	F	1.102	1.024	0.0016	0.079	116	20
		R	0.630	0.551	0.0020	0.079	62	32
2005	FX35/FX45	F	1.102	1.024	0.0016	0.079	116	20
		R	0.630	0.551	0.0020	0.079	62	32

F: Front

R: Rear

67162-FX35-C11

SCHEDULED MAINTENANCE INTERVALS

INFINITI — FX35/FX45 — NORMAL MAINTENANCE SCHEDULE

TO BE SERVICED	TYPE OF SERVICE	VEHICLE MILEAGE INTERVAL (x1000)							
		7.5	15	23	30	37.5	45	52.5	60
Accessory drive belt	S/I								✓
Air cleaner filter	R				✓				✓
EVAP vapor lines	S/I				✓				✓
Fuel lines & connections	S/I				✓				✓
Engine coolant	R								✓
Engine oil	R	✓	✓	✓	✓	✓	✓	✓	✓
Engine oil filter	R	✓	✓	✓	✓	✓	✓	✓	✓
Spark plugs ①	R								
Brake lines & cables	S/I		✓		✓		✓		✓
Brake pads & rotors	S/I		✓		✓		✓		✓
AT fluid	S/I		✓		✓		✓		✓
Transfer case fluid	S/I		✓		✓		✓		✓
Steering system	S/I				✓				✓
CV-Joint boots & driveshafts	S/I		✓		✓		✓		✓
Exhaust system	S/I				✓				✓
In-cabin microfilter	R		✓		✓		✓		✓

INFINITI — FX35/FX45 — SEVERE MAINTENANCE SCHEDULE

TO BE SERVICED	TYPE OF SERVICE	VEHICLE MILEAGE INTERVAL (x1000)							
		3.75	7.5	11.25	15	17.5	22.5	26.25	30.00
Accessory drive belt	S/I				✓				✓
Air cleaner filter	R								✓
EVAP vapor lines	S/I								✓
Fuel lines & connections	S/I								✓
Engine coolant	R								✓
Engine oil	R	✓	✓	✓	✓	✓	✓	✓	✓
Engine oil filter	R	✓	✓	✓	✓	✓	✓	✓	✓
Spark plugs ①	R								
Brake lines & cables	S/I				✓				✓
Brake pads & rotors	S/I		✓		✓		✓		✓
AT fluid	S/I				✓				✓
Transfer case fluid	S/I				✓				✓
Steering system	S/I		✓		✓		✓		✓
CV-Joint boots & driveshafts	S/I		✓		✓		✓		✓
Exhaust system	S/I		✓		✓		✓		✓
In-cabin microfilter	R				✓		✓		

R: Replace   S/I: Service or Inspect

① Replace every 105,000 miles

PRECAUTIONS

Before servicing any vehicle, please be sure to read all of the following precautions, which deal with personal safety, prevention of component damage, and important points to take into consideration when servicing a motor vehicle:

• Never open, service or drain the radiator or cooling system when the engine is hot; serious burns can occur from the steam and hot coolant.

• Observe all applicable safety precautions when working around fuel. Whenever servicing the fuel system, always work in a well-ventilated area. Do not allow fuel spray or vapors to come in contact with a spark, open flame, or excessive heat (a hot drop light, for example). Keep a dry chemical fire extinguisher near the work area. Always keep fuel in a container specifically designed for fuel storage; also, always properly seal fuel containers to avoid the possibility of fire or explosion. Refer to the additional fuel system precautions later in this section.

• Fuel injection systems often remain pressurized, even after the engine has been turned **OFF**. The fuel system pressure must be relieved before disconnecting any fuel lines. Failure to do so may result in fire and/or personal injury.

• Brake fluid often contains polyglycol ethers and polyglycols. Avoid contact with the eyes and wash your hands thoroughly after handling brake fluid. If you do get brake fluid in your eyes, flush your eyes with clean, running water for 15 minutes. If eye irritation persists, or if you have taken brake fluid internally, IMMEDIATELY seek medical assistance.

• The EPA warns that prolonged contact with used engine oil may cause a number of skin disorders, including cancer. You should make every effort to minimize your exposure to used engine oil. Protective gloves should be worn when changing oil. Wash your hands and any other exposed skin areas as soon as possible after exposure to used engine oil. Soap and water, or waterless hand cleaner should be used.

• All new vehicles are now equipped with an air bag system. The system must be disabled before performing service on or around system components, steering column, instrument panel components, wiring and sensors. Failure to follow safety and disabling procedures could result in accidental air bag deployment, possible personal injury and unnecessary system repairs.

• Always wear safety goggles when working with, or around, the air bag system. When carrying a non-deployed air bag, be sure the bag and trim cover are pointed away from your body. When placing a non-deployed air bag on a work surface, always face the bag and trim cover upward, away from the surface. This will reduce the motion of the module if it is accidentally deployed. Refer to the additional air bag system precautions later in this section.

• NEVER disconnect the negative battery cable with the ignition **ON** or the engine running. Removing power from the computer control module with the ignition **ON** may destroy the module.

• Clean, high quality brake fluid from a sealed container is essential to the safe and

proper operation of the brake system. You should always buy the correct type of brake fluid for your vehicle. If the brake fluid becomes contaminated, completely flush the system with new fluid. Never reuse any brake fluid. Any brake fluid that is removed from the system should be discarded. Also, do not allow any brake fluid to come in contact with a painted surface; it will damage the paint.

• Never operate the engine without the proper amount and type of engine oil; doing so WILL result in severe engine damage.

• Timing belt maintenance is extremely important. Many models utilize an interference-type, non-freewheeling engine. If the timing belt breaks, the valves in the cylinder head may strike the pistons, causing potentially serious (also time-consuming and expensive) engine damage. Refer to the maintenance interval charts in the front of this manual for the recommended replacement interval for the timing belt, and to the timing belt section for belt replacement and inspection.

• Disconnecting the negative battery cable on some vehicles may interfere with the functions of the on-board computer system(s) and may require the computer to undergo a relearning process once the negative battery cable is reconnected.

• When servicing drum brakes, only disassemble and assemble one side at a time, leaving the remaining side intact for reference.

• Only an MVAC-trained, EPA-certified automotive technician should service the air conditioning system or its components.

ENGINE REPAIR

Ignition Coil

REMOVAL & INSTALLATION

1. Before servicing the vehicle, refer to the precautions in the beginning of this section.
2. Remove the engine cover.
3. Remove the air duct (for ignition coil of left bank side).
4. Move aside the wiring harness, wiring harness bracket, and hoses located above ignition coil.
5. Disconnect the wiring harness connector from the ignition coil.
6. Remove the ignition coil.

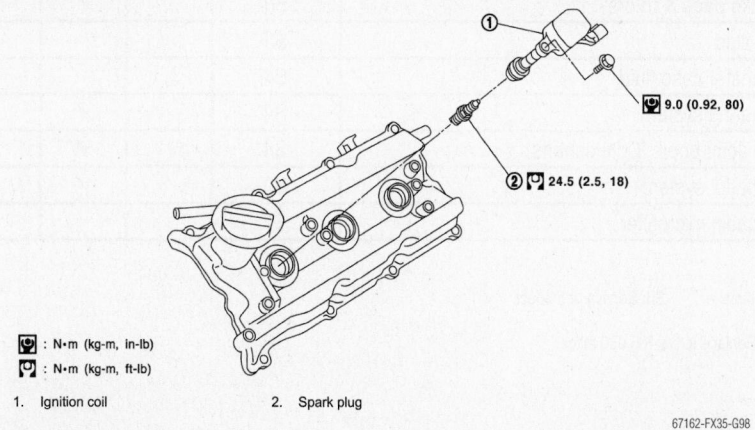

: N•m (kg-m, in-lb)
: N•m (kg-m, ft-lb)

1. Ignition coil
2. Spark plug

67162-FX35-G98

Exploded view of ignition coil mounting—VQ35DE Engine.

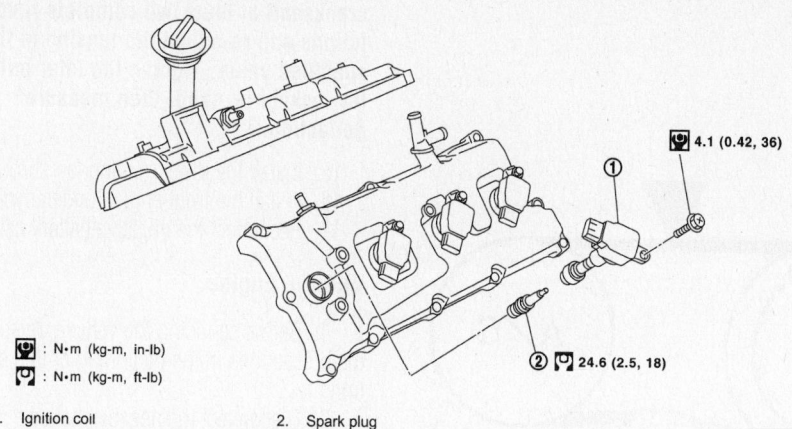

: N•m (kg-m, in-lb)

: N•m (kg-m, ft-lb)

1. Ignition coil

2. Spark plug

67162-FX35-G99

Exploded view of ignition coil mounting—VK45DE Engine.

❊❊ CAUTION

Do not subject the ignition coils to excessive shock or vibration.

To install:

7. Install the ignition coil on the engine.

8. Reconnect the wiring harness to the coil.

9. Reposition the wiring harness, bracket and hoses.

10. Install the air duct and the engine cover.

Alternator

REMOVAL

VQ35DE Engine

1. Before servicing the vehicle, refer to the precautions in the beginning of this section.

2. Disconnect the negative battery cable.

3. Remove the front engine undercover.

4. Remove the lower cooling fan shroud, as follows:

5. Remove the alternator/power steering belt, as follows:

a. Loosen idler pulley lock nut (A) and loosen belt tension by turning the adjusting bolt (B).

b. Remove the belt from the pulleys.

6. Remove the alternator mounting bolts.

7. Disconnect the alternator wiring connector.

8. Remove the water hose bracket, the oil pressure switch wiring harness clip (2WD models), and the oil pressure switch connector (2WD models).

9. Remove the alternator assembly, by lowering it out of the bottom of the engine compartment.

To install:

10. Reposition the alternator in place on the engine and tighten the mounting bolts. Tighten the long alternator bolt 48 ft. lbs. (65 Nm) and the short alternator bracket bolts to 21 ft. lbs. (28 Nm).

11. Tighten the B terminal nut carefully to 7 ft. lbs. (10 Nm).

12. Reattach the oil pressure switch connector, the oil pressure switch wiring harness clip, and the water hose bracket.

13. Reconnect the alternator wiring.

14. Install the accessory drive belt and adjust the belt tension, as follows:

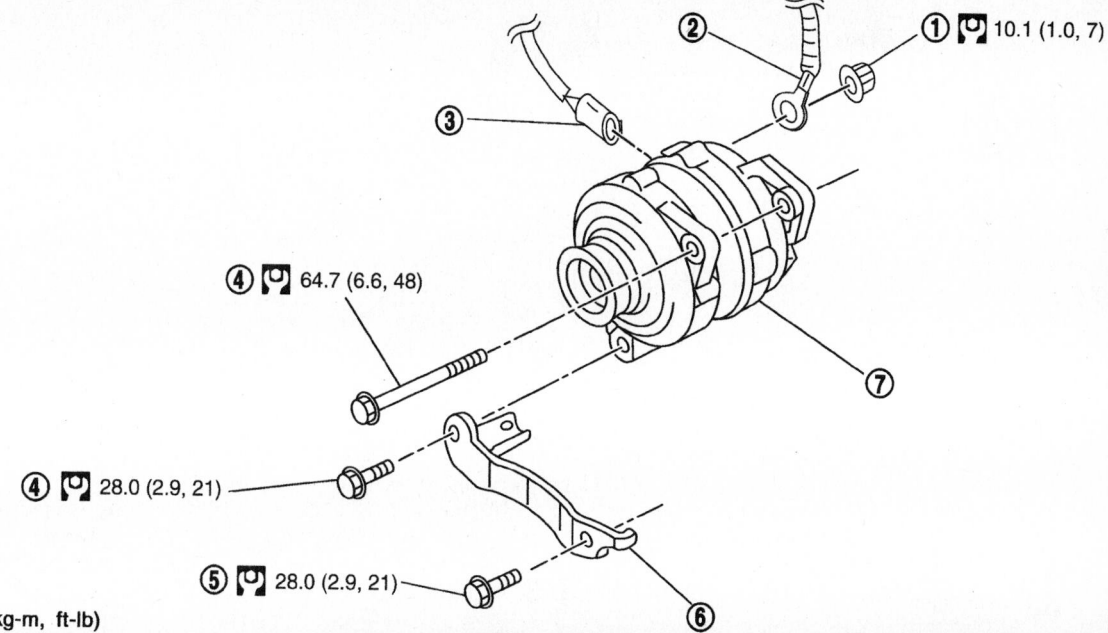

: N•m (kg-m, ft-lb)

1. B terminal nut	2. Alternator B terminal harness	3. Alternator connector
4. Alternator mounting bolt	5. Alternator stay mounting bolt	6. Alternator stay
7. Alternator		

Exploded view of alternator mounting—VQ35DE Engine.

67162-FX35-G100

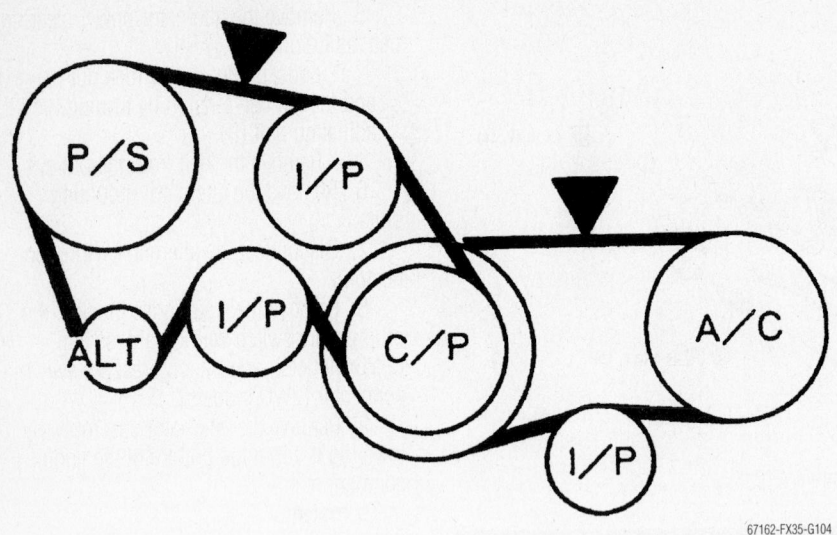

Accessory drive belt routing—VQ35DE Engine.

a. Inspect drive belt deflection or tension at a point on the belt midway between the Power Steering and Idler pulleys (power steering/alternator belt), and between the crankshaft and idler pulleys (A/C belt).

- Inspection should be done only when engine is cold, or after the engine has been off for at least 30 minutes.
- Measure the belt tension with tension gauge (SST: BT3373-F or equivalent) at points marked shown in the figure.
- When measuring deflection, apply 98 N (10 kg, 22 lb) at the marked point.
- Adjust the belts if deflection exceeds the limit or if the tension is not within specifications.

➡ When checking belt deflection or tension after installation, first adjust it to the specified value. Then, turn the crankshaft at least two complete revolutions and re-adjust the tension to the specified value. Tighten the idler pulley locknut by hand, then measure deflection.

15. Install the lower cooling fan shroud.
16. Install the front engine undercover.
17. Reconnect the negative battery cable.

VK45DE Engine

1. Before servicing the vehicle, refer to the precautions in the beginning of this section.

2. Disconnect the negative battery cable.

3. Remove the front engine undercover.

4. Remove the lower cooling fan shroud, as follows:

5. Remove the alternator/water pump/A/C belt as follows:

a. Remove air duct (inlet).

b. While securely holding the hexagonal part in the pulley center of the auto–tensioner with a wrench, move wrench handle in the direction of the arrow (loosening direction of the tensioner).

c. Under the above condition, insert a metal bar of approximately 6mm (0.24 in) in diameter through the holding boss to lock the auto–tensioner pulley arm in position.

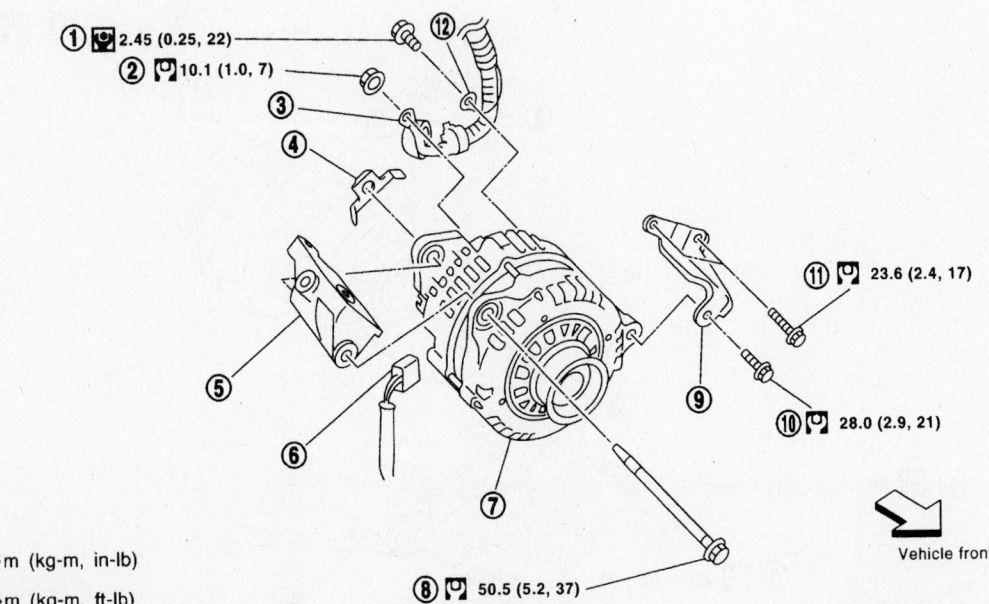

① 2.45 (0.25, 22)
② 10.1 (1.0, 7)
⑪ 23.6 (2.4, 17)
⑩ 28.0 (2.9, 21)
⑧ 50.5 (5.2, 37)

Vehicle front

: N•m (kg-m, in-lb)

: N•m (kg-m, ft-lb)

1. Alternator ground harness mounting bolt
2. B terminal nut
3. Alternator B terminal harness
4. Alternator Nut
5. Alternator bracket
6. Alternator connector
7. Alternator
8. Alternator mounting bolt
9. Alternator stay
10. Alternator mounting bolt
11. Alternator stay mounting bolt
12. Alternator ground harness

Exploded view of alternator mounting—VK45DE Engine.

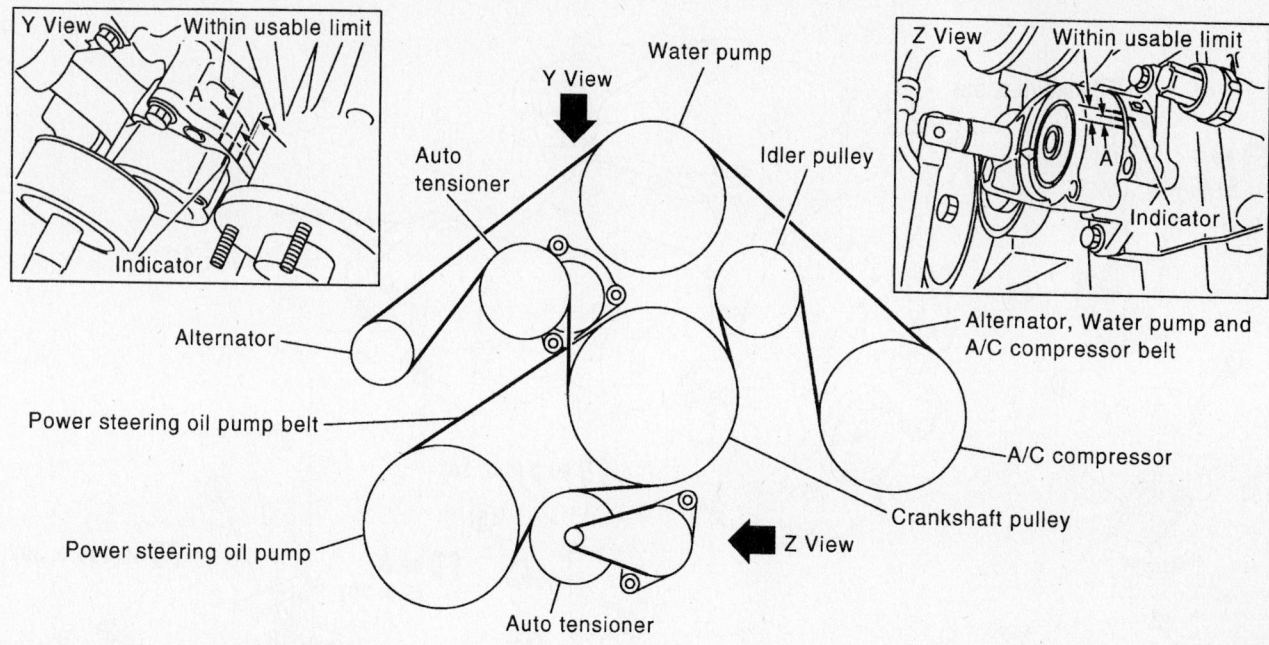

Accessory drive belt routing—VK45DE Engine.

67162-FX35-G103

➡**Leave the auto–tensioner pulley arm locked until the belt is reinstalled.**

6. Remove the alternator mounting bolts.

7. Disconnect the alternator wiring connector.

8. Remove the alternator assembly, by lowering it out of the bottom of the engine compartment.

To install:

9. Reposition the alternator in place on the engine and tighten the mounting bolts. Tighten the long alternator bolt 37 ft. lbs. (50 Nm), the alternator-to-bracket bolt to 21 ft. lbs. (28 Nm), and the bracket-to-engine bolts to 17 ft. lbs. (24 Nm).

10. Tighten the B terminal nut carefully to 7 ft. lbs. (10 Nm).

11. Reconnect the alternator wiring.

12. Install the alternator/water pump/A/C drive belt. The accessory drive belt uses an auto–tensioner; no manual tensioning is necessary.

13. Install the lower cooling fan shroud.

14. Install the front engine undercover.

15. Reconnect the negative battery cable.

Ignition Timing

ADJUSTMENT

Ignition timing is controlled by the electronic fuel injection system and is not adjustable. If ignition timing is incorrect, there could be an issue with the ECM,

Crankshaft Position Sensor or Camshaft Position Sensor.

Engine Assembly

REMOVAL & INSTALLATION

1. Before servicing the vehicle, refer to the precautions in the beginning of this section.

2. Situate vehicle on a flat and solid surface.

3. Place chocks at front and back of rear wheels.

4. For engines not equipped with engine slingers, attach proper slingers and bolts. Tighten the slinger bolts to 21 ft. lbs. (28 Nm).

✳✳ CAUTION

Always use the support point specified for lifting. Use either a 2-point lift type or a separate type lift as best you can. If a board-on type is used for unavoidable reasons, support at the rear axle jacking point with a transmission jack or similar tool before starting work. The rear axle jacking point needs to be supported since the center of gravity will shift rearward once the engine assembly is removed.

The engine and transmission assembly are removed from the vehicle with the sus-

pension member from the underside of the vehicle. After removal, separate the engine from the transmission.

5. Release the fuel system pressure.

6. Disconnect both battery terminals.

7. Remove the engine cover, battery cover and both front wheel/tire assemblies.

8. Remove the front and rear engine under covers and front cross bar.

9. Drain the engine coolant from the radiator.

10. On AWD models, remove the clips of the hood ledge cover and remove the hood ledge cover.

11. On AWD models, remove both wiper arms. Operate the wiper motor, and stop it at the auto stop position. Then, remove the washer tube from the washer tube joint. Remove the wiper arm mounting nuts and wiper arms from the vehicle.

12. Remove the air duct and air cleaner assembly.

13. Remove the radiator hoses.

14. Disconnect and plug the chassis-end of the heater hose.

15. Disconnect the chassis/left bank cylinder head ground wire.

16. Disconnect and tag all engine wiring harness connectors.

17. Disconnect the A/C tubing from the A/C compressor, and fasten it to the inside of the engine compartment with rope or strong cord.

18. Disconnect the two chassis ground cables.

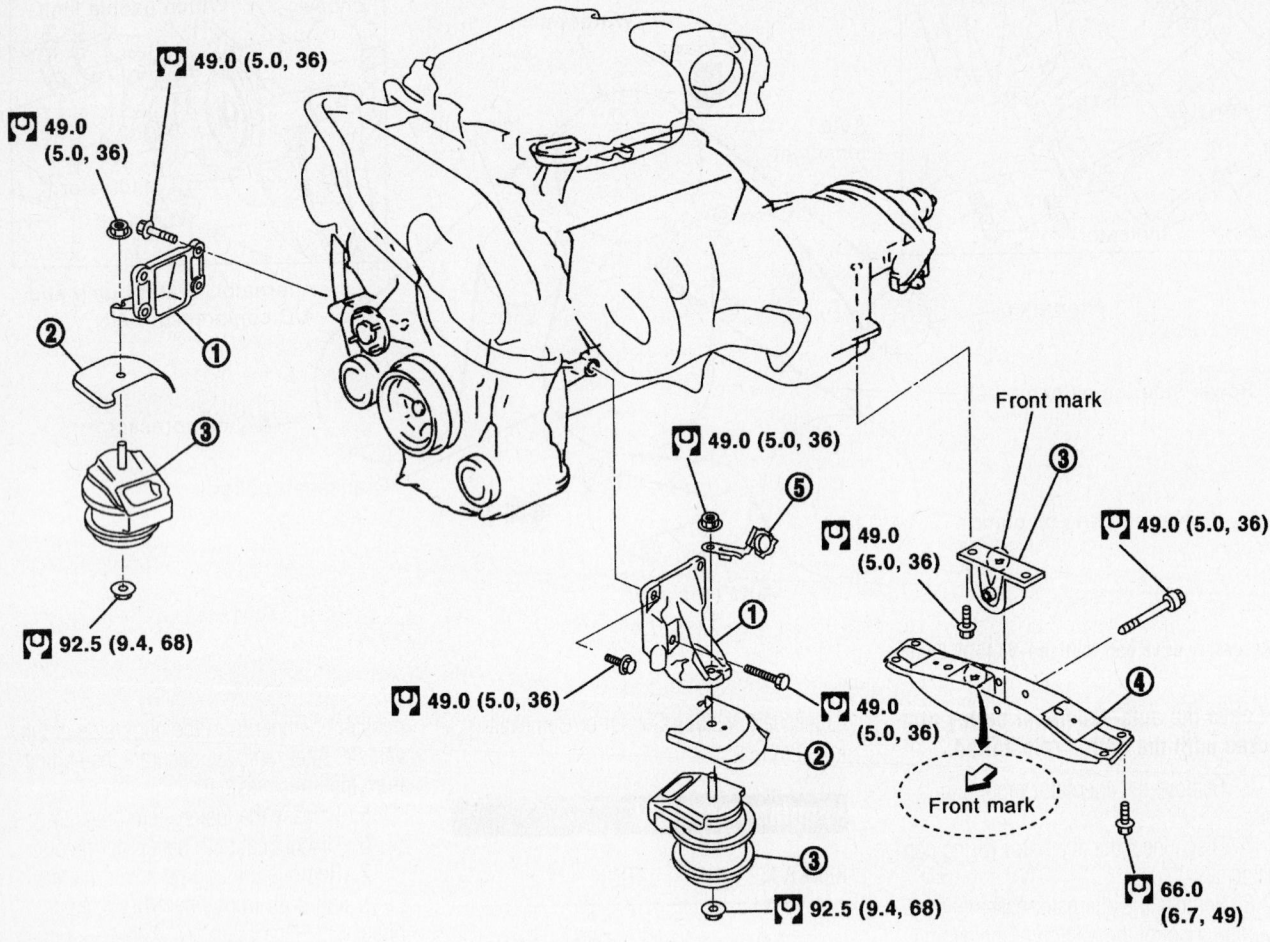

: N•m (kg-m, ft-lb)

1. Engine mounting bracket
2. Heat insulator
3. Engine mounting insulator
4. Rear member
5. Harness bracket

67162-FX35-G105

Exploded view of engine mounting—2WD VQ35DE Engine.

19. Disconnect the brake booster vacuum hose.

20. Disconnect and plug the fuel feed hose and EVAP hose.

21. Remove the power steering oil pump reservoir tank and tubing from the chassis, and fasten them onto the engine.

➡ **When securing the power steering tubing to the engine, situate it so that the open end is pointed up to avoid a fluid leak.**

22. Disconnect the engine compartment wiring harness connectors from the passenger compartment, as follows:

    a. Remove the passenger-side kick plate, dashboard side trim, and glove box.

    b. Disconnect the engine compartment wiring harness connectors from the ECM.

    c. Disengage the intermediate fixing point. Pull the engine compartment wiring harnesses through to the engine compartment and fasten them to the engine.

✳✳ CAUTION

Be careful not to damage the wiring harness when pulling it out of the passenger compartment. Also, cover the wiring harness connectors to prevent dirt or other material from contaminating the connector openings.

23. Remove the A/T fluid cooler hoses and power steering oil pump oil cooler hoses.

24. Remove the front exhaust pipe.

25. Disconnect the lower steering joint, and disengage the steering shaft.

26. Disconnect the drive shaft from the transmission.

27. Disconnect the shift control linkage from selector lever, then secure it onto the transmission so that it doesn't hang free.

28. Remove the rear plate cover from the upper oil pan. Remove the bolts holding the flexplate to the torque converter.

29. Remove the mounting bolts from the transmission to the lower rear side of the upper oil pan.

30. On 2WD models, remove the front stabilizer.

31. Detach the left-hand and right-hand side tie-rod ends from the steering knuckles.

32. Remove the lower ends of the left-hand and right-hand struts from the lower arms.

33. Disconnect the left-hand and right-

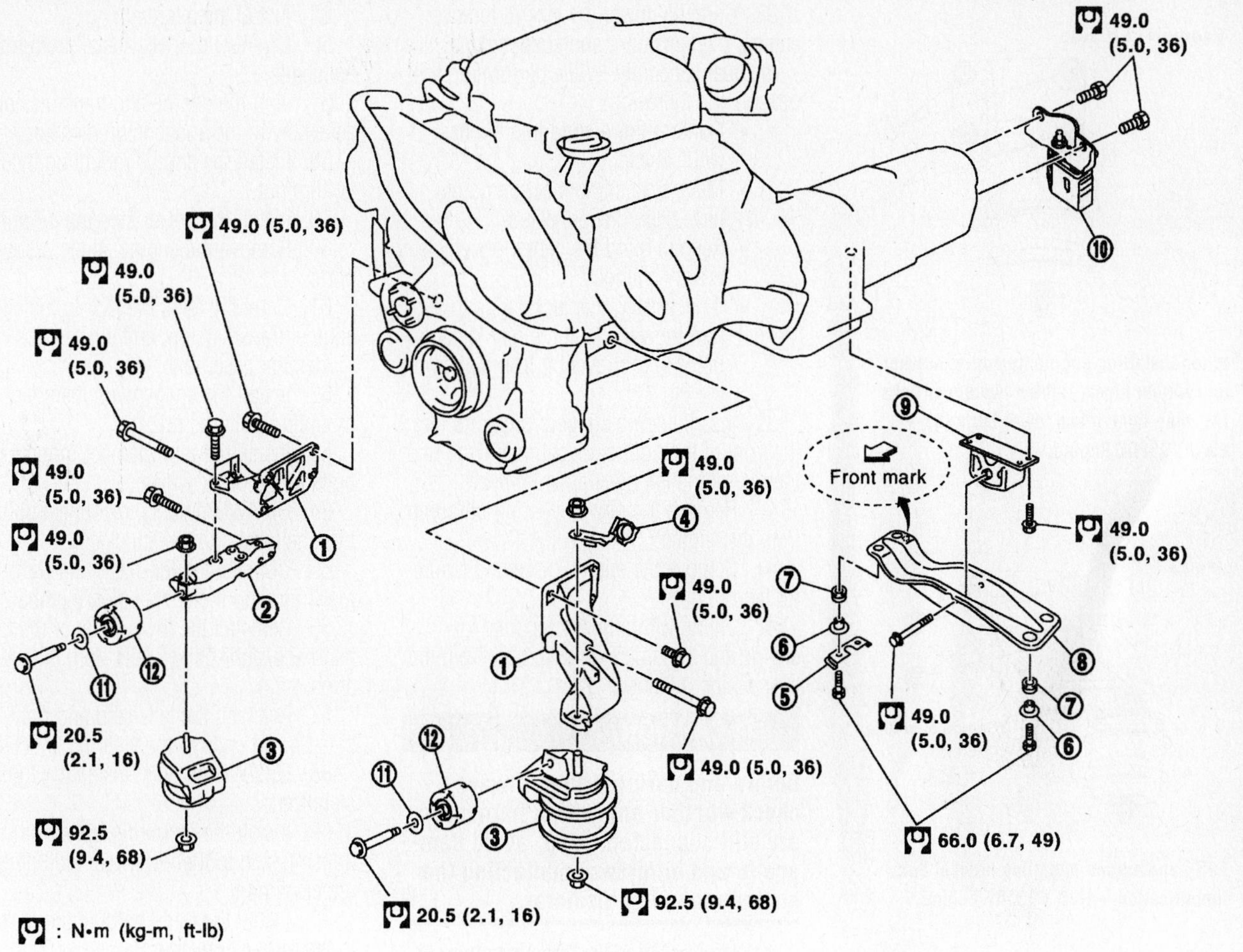

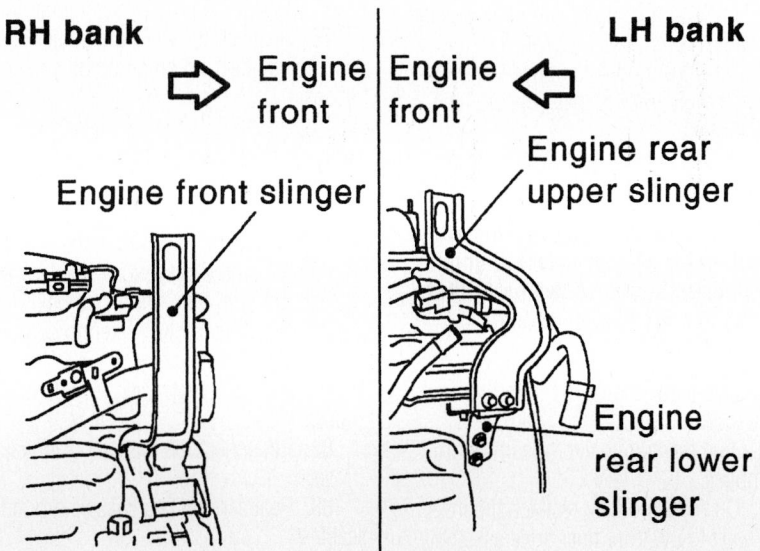

[Diagram torque values:]

49.0 (5.0, 36)
49.0 (5.0, 36)
49.0 (5.0, 36)
49.0 (5.0, 36)
49.0 (5.0, 36)
49.0 (5.0, 36)
49.0 (5.0, 36)
49.0 (5.0, 36)
49.0 (5.0, 36)
49.0 (5.0, 36)
49.0 (5.0, 36)
20.5 (2.1, 16)
92.5 (9.4, 68)
20.5 (2.1, 16)
92.5 (9.4, 68)
66.0 (6.7, 49)

Front mark

: N•m (kg-m, ft-lb)

1. Engine mounting bracket	2. Engine mounting bracket (Lower)	3. Engine mounting insulator FR
4. Harness bracket	5. Heat insulator	6. Caller
7. Rubber bush	8. Rear member	9. Engine mounting insulator RR
10. Dynamic damper	11. Washer	12. Dynamic damper

67162-FX35-G109

Exploded view of engine mounting—AWD VQ35DE Engine.

RH bank

⇨ Engine front

Engine front slinger

LH bank

Engine front ⇦

Engine rear upper slinger

Engine rear lower slinger

67162-FX35-G106A

Engine slinger installation positions—2WD VQ35DE Engine.

---

hand lower arms from the suspension member.

34. On AWD models, disconnect both front halfshafts from the knuckles.

35. Use a manual lift table caddy (commercial service tool) or equivalently rigid tool such as a transmission jack. Securely support the bottom of the suspension member and transmission.

✳✳ CAUTION

Put a piece of wood or something similar as the supporting surface, secure in a completely stable condition.

36. Remove the rear member mounting bolt.

37. Remove the suspension member mounting bolt and nut.

Example: LH side

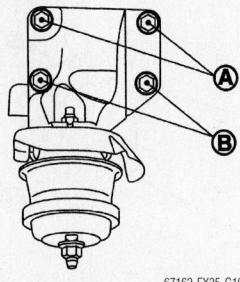

67162-FX35-G107

When installing engine mounting bracket on cylinder block, tighten two upper bolts (A), then tighten two lower bolts (B)—2WD VQ35DE Engine.

LH side

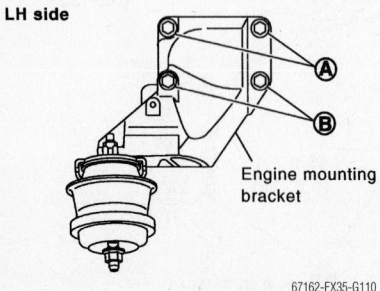

67162-FX35-G110

Left-hand engine mounting bracket bolt identification—AWD VQ35DE Engine.

RH side

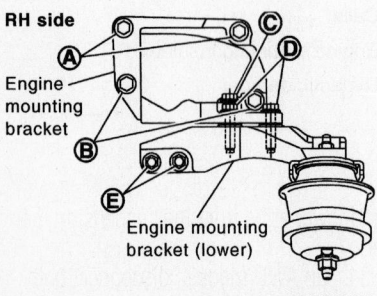

67162-FX35-G111

Right-hand engine mounting bracket bolt identification—AWD VQ35DE Engine.

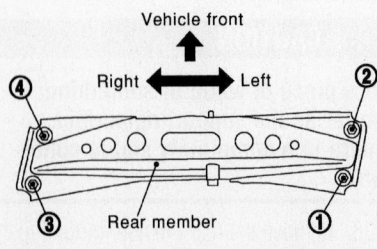

67162-FX35-G108

Rear member mounting bolt tightening sequence—2WD VQ35DE Engine.

38. Carefully lower the jack to remove engine, transmission, and suspension member assembly. While performing this, observe the following:
- Confirm there is no interference with vehicle.
- Make sure all connection points have been disconnected.
- Keep in mind the center of vehicle gravity changes.
- If necessary, use jackstands to support the vehicle at the rear jacking point(s) to prevent it from falling off the lift.

39. Install engine slingers/supports into the front of the right bank cylinder head and the rear of the left bank cylinder head.

40. Remove the power steering oil pump from the engine.

41. Remove the engine mounting insulator bottom nut.

42. Lifting with a hoist, separate the engine and transmission assembly from the suspension member.

✳✳ CAUTION

Before and during lifting, always check whether any wiring harnesses are still connected. Also, avoid damage to and oil/grease contacting the engine mounting insulator.

43. On AWD models, remove both front halfshafts.

44. Remove the alternator.

45. Remove the starter motor.

46. On AWD models, detach the front drive shaft from the front final drive assembly.

47. Separate the engine from the transmission assembly.

48. Remove the engine mounting insulator and bracket.

49. On AWD models, remove front drive assembly from oil pan (upper).

To install:

50. On AWD models, install the front final drive assembly onto the upper oil pan.

➡**When installing the engine mounting bracket on the cylinder block, tighten the two upper bolts (shown as A) first. Then tighten the two lower bolts (shown as B).**

51. Install the engine mounting insulator and bracket.

52. Join the engine and transmission assemblies.

53. On AWD models, reattach the front drive shaft to the front final drive assembly.

54. Install the starter motor.

55. Install the alternator.

56. On AWD models, install both front halfshafts.

57. Join the engine and transmission assembly to the suspension member.

58. Install the engine mounting insulator bottom nut.

59. Install the power steering oil pump.

60. Remove the engine slingers/supports.

61. Carefully raise the jack to install the engine, transmission, and suspension member assembly.

62. Install the suspension member mounting bolt and nut.

63. Install the rear member mounting bolts in the order shown.

64. On AWD models, reconnect both front halfshafts to the knuckles.

65. Connect the left-hand and right-hand lower arms to the suspension member.

66. Reinstall the lower ends of the left-hand and right-hand struts from the lower arms.

67. Reattach the left-hand and right-hand side tie-rod ends to the steering knuckles.

68. On 2WD models, install the front stabilizer.

69. Install the mounting bolts from the transmission to the lower rear side of the upper oil pan.

70. Install the bolts holding the flexplate to the torque converter.

71. Install the rear plate cover onto the upper oil pan.

72. Reconnect the shift control linkage to the selector lever.

73. Install the drive shaft to the transmission.

74. Connect the lower steering joint, and engage the steering shaft.

75. Install the front exhaust pipe.

76. Reattach the A/T fluid cooler hoses and power steering oil pump oil cooler hoses.

77. Route the engine compartment wiring harnesses through to the passenger compartment.

78. Connect the engine compartment wiring harness connectors to the ECM.

79. Install the passenger-side kick plate, dashboard side trim, and glove box.

80. Install the power steering oil pump reservoir tank and tubing.

81. Reconnect the fuel feed and EVAP hoses.

82. Reconnect the brake booster vacuum hose.

83. Reattach the two chassis ground cables.

84. Reconnect the A/C tubing to the A/C compressor.

85. Reconnect all engine wiring harness connectors.

86. Reconnect the chassis/left bank cylinder head ground wire.

87. Reconnect the heater hose.

88. Install the radiator hoses.

89. Install the air duct and air cleaner assembly.

90. On AWD models, install the wiper arms, as follows:

   a. Clean the pivot area, which will help reduce the possibility
of wiper arm looseness.

   b. Prior to wiper arm instillation, turn on the wiper switch to operate the wiper motor and then turn it "OFF" (auto stop).

   c. Push the wiper arm onto pivot shaft, paying attention to the blind spline.

   d. Attach the washer tube to the washer tube joint.

   e. Lift the blade up and then set it down onto the glass surface to set the blade center to clearance "L1" & "L2" immediately before tightening the nut.

   f. Operate the washer fluid pump. Turn on the wiper switch to operate the wiper motor and then turn it "OFF".

   g. Ensure that the wiper blades stop within the clearance "L1" & "L2".

   • Tighten the wiper arm nuts to 18 ft. lbs. (24 Nm).

   h. Install the cowl top seal rubber.

   i. Install the clips of the cowl top cover (right) and the cowl top cover (right).

   j. Install the clips, cap and bolts and cowl top cover (left).

   k. Install the washer hose from cowl top cover.

91. Install the top right-hand cowl cover.

92. Fill the engine cooling system.

93. Install the front crossbar and the front and rear lower engine covers.

94. Install the engine cover, the battery cover and both front wheel assemblies.

95. Connect both battery cables.

Water Pump

REMOVAL & INSTALLATION

VQ35DE Engine

✳✳ CAUTION

During service, be sure to prevent engine coolant from contacting the drive belt.

1. Before servicing the vehicle, refer to the precautions in the beginning of this section.

➡**Water pump cannot be disassembled and should be replaced as a unit.**

2. Remove the front engine undercover.

3. Remove the drive belts.

4. Drain the engine coolant from the radiator.

✳✳ WARNING

Make sure the engine is cold before draining the coolant.

5. Remove the air duct (inlet), power duct and air cleaner case assembly.

6. Remove the cylinder block drain plug (front side) of engine.

7. Remove the chain tensioner cover and water pump cover. Use a seal cutter (SST: KV10111100 (J37228) or equivalent) to separate the two mating surfaces.

✳✳ CAUTION

Be careful not to damage the mating surfaces.

8. Remove the timing chain tensioner (primary), as follows:

   a. Pull the lever down and release the plunger stopper tab.

➡**The plunger stopper tab can be pushed up to release (coaxial structure with lever).**

   b. Insert a stopper pin into the tensioner body hole to hold the lever and keep the plunger stopper tab released.

➡**An Allen wrench can be used for a stopper pin.**

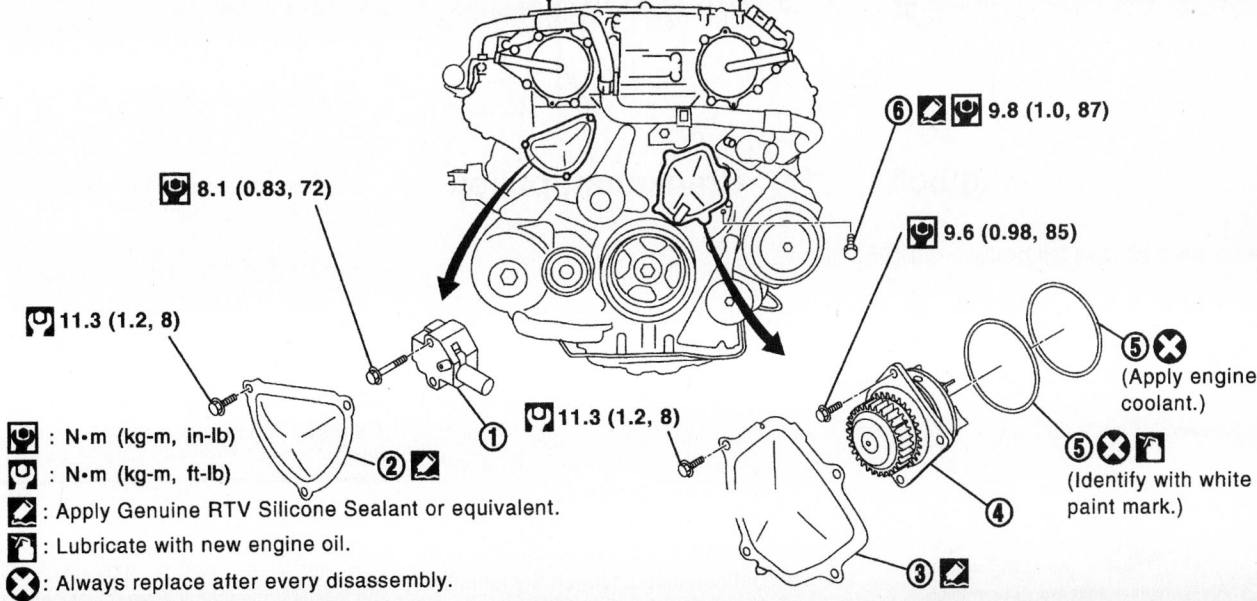

8.1 (0.83, 72)

11.3 (1.2, 8)

6 **9.8 (1.0, 87)**

9.6 (0.98, 85)

11.3 (1.2, 8)

5 ✕ (Apply engine coolant.)

5 ✕ (Identify with white paint mark.)

: N•m (kg-m, in-lb)

: N•m (kg-m, ft-lb)

: Apply Genuine RTV Silicone Sealant or equivalent.

: Lubricate with new engine oil.

✕ : Always replace after every disassembly.

1. Chain tensioner
2. Chain tensioner cover
3. Water pump cover
4. Water pump
5. O-rings
6. Cylinder block drain plug (front)

67162-FX35-G114

Exploded view of water pump and related components—VQ35DE Engine.

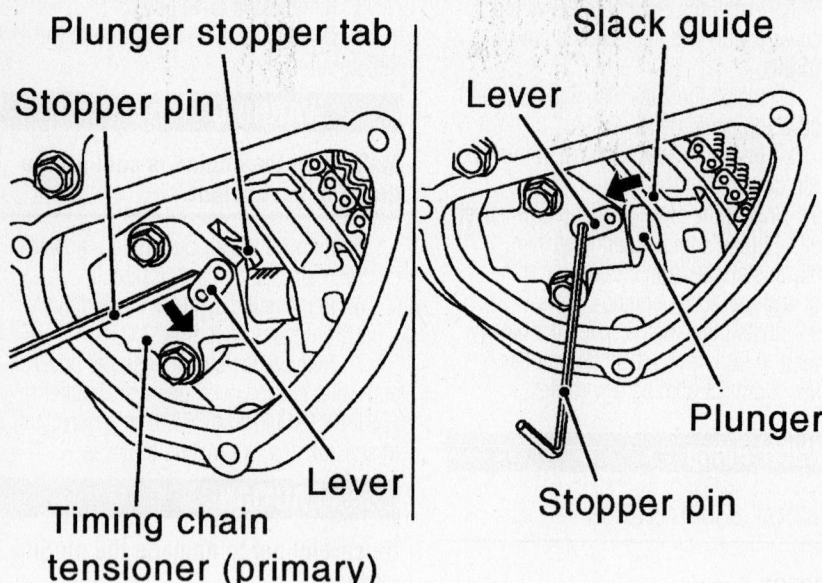

Primary timing chain tensioner details—VQ35DE Engine.

67162-FX35-G116

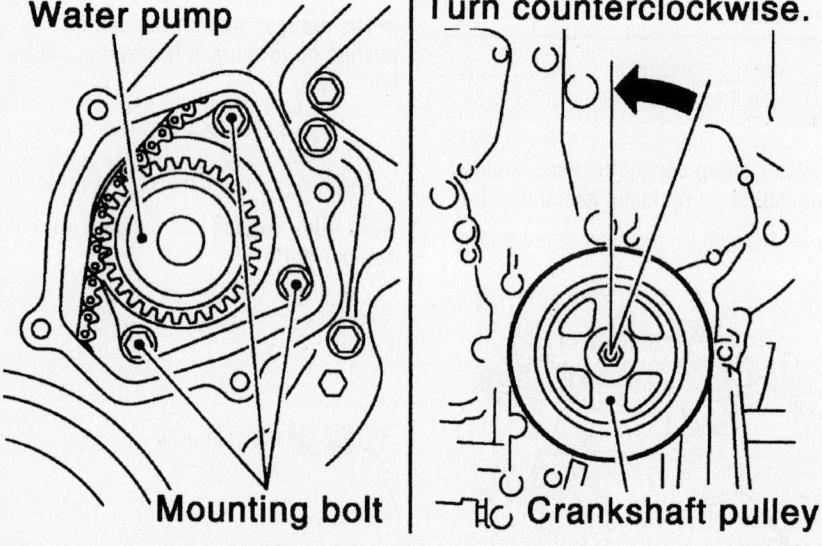

Water pump mounting bolt locations—VQ35DE Engine.

67162-FX35-G106

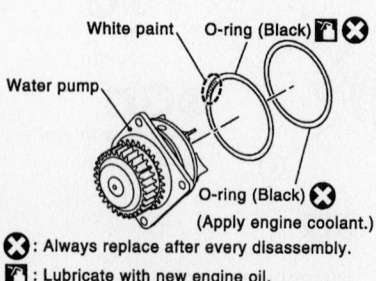

⊗ : Always replace after every disassembly.
▨ : Lubricate with new engine oil.

67162-FX35-G119

Water pump O-ring positioning for installation—VQ35DE Engine.

✽✽ CAUTION

Be careful not to drop mounting bolts inside the chain case.

9. Remove the three water pump fixing bolts. Secure a gap between the water pump gear and timing chain, by turning the crankshaft pulley counterclockwise until timing chain slack on the water pump sprocket is at its maximum.

10. Insert M8 bolts [pitch: 1.25 mm (0.049 in) length: approx. 50 mm (1.97 in)] into the water pump upper and lower mounting bolt holes until they contact the timing chain case. Then, alternately tighten each bolt for a half turn, and pull out water pump.

✽✽ CAUTION

Pull the water pump straight out while preventing the vane from contacting the socket in the installation area. Remove the water pump without allowing the sprocket to contact the timing chain.

11. Remove the M8 bolts and O-rings from the water pump.

✽✽ CAUTION

Do not disassemble the water pump.

12. Check for badly rusted or corroded water pump body assembly.
13. Check for rough operation due to excessive end play.
14. If any defects are found, replace the water pump.
   To install:
15. Install new O-rings on the water pump.
16. Apply engine oil and engine coolant to the water pump O-rings.
17. Position the O-ring with the white paint mark toward the engine front side.
18. Install the water pump.

✽✽ CAUTION

Do not allow the cylinder block to damage the O-rings during installation.

19. Make sure the timing chain and water pump sprocket are engaged properly.
20. Install the water pump by alternately and evenly tightening the mounting bolts.
21. Install the timing chain tensioner (primary), as follows:
   a. Remove all dust and foreign material completely from the backside of the

c. Insert the plunger into the tensioner body by pressing the timing chain slack guide.
   d. Keep the slack guide depressed and hold the plunger in by pushing the stopper pin deeper through the lever and into the tensioner body hole.
   e. Turn the crankshaft pulley approximately 20 degrees clockwise, so that the timing chain on the timing chain tensioner (primary) side is loose.
   f. Remove the mounting bolts and timing chain tensioner (primary).

chain tensioner and from the installation area of the rear timing chain case.

b. Turn the crankshaft pulley clockwise so that the timing chain on the timing chain tensioner (primary) side is loose.

c. Apply engine oil to the oil hole and tensioner, when installing the timing chain tensioner.

d. Install the timing chain tensioner (primary).

e. Remove the stopper pin.

22. Install the chain tensioner cover and water pump cover.

➡**Before installing, remove all traces of liquid gasket from the mating surface of the water pump cover and chain tensioner cover using a scraper. Also, remove traces of liquid gasket from the mating surface of the front timing chain case.**

23. Apply a continuous bead of liquid gasket 0.091-0.130 in. (2.3-3.3mm) thick to the mating surface of the chain tensioner cover and water pump cover with tube presser (SST: WS39930000 or equivalent).

➡**Use Genuine RTV Silicone Sealant or equivalent.**

24. Install the cylinder block drain plug (front side).

25. Apply thread sealant to the thread of cylinder block drain plug.

➡**Use Genuine Thread Sealant or equivalent.**

26. Install the air duct (inlet), power duct and air cleaner case assembly.

27. Fill the engine with coolant.

28. Install the drive belts.

29. Install the front engine undercover.

30. Check for leaks of engine coolant using radiator cap tester adapter (SST: EG17650301 (J33984-A) or equivalent) and radiator cap tester (commercial service tool).

31. Start the engine and let it idle for three minutes, then raise the engine RPM up to 3,000 rpm under no load to purge air from the high-pressure chamber of the chain tensioner. The engine may produce a rattling noise, which indicates that air remains in the chamber and is not a matter of concern.

32. With the engine idling, visually make sure that there are no leaks of engine coolant or A/T fluid.

VK45DE Engine

1. Before servicing the vehicle, refer to the precautions in the beginning of this section.

☀ CAUTION

When removing water pump, be careful not to get engine coolant on drive belt.

➡**Water pump can not be disassembled and should be replaced as a unit.**

2. Drain the engine coolant from the drain plugs on radiator and both sides of the engine block.

☀ WARNING

Make sure the engine is cold before draining the coolant.

3. Remove the engine front undercover.

4. Remove the air duct (inlet).

5. Remove the alternator, water pump and A/C compressor belt.

6. Remove the fan coupling with the cooling fan, and then the water pump pulley.

7. Remove the water pump.

➡**Engine coolant will leak from the cylinder block, so have a receptacle ready under vehicle.**

☀ CAUTION

Handle the water pump vane so that it does not contact any other parts.

8. Inspect the water pump after removal for the following:

- Significant dirt or rusting on water pump body and vane.
- Looseness in vane shaft, and that it turns smoothly when rotated by hand.

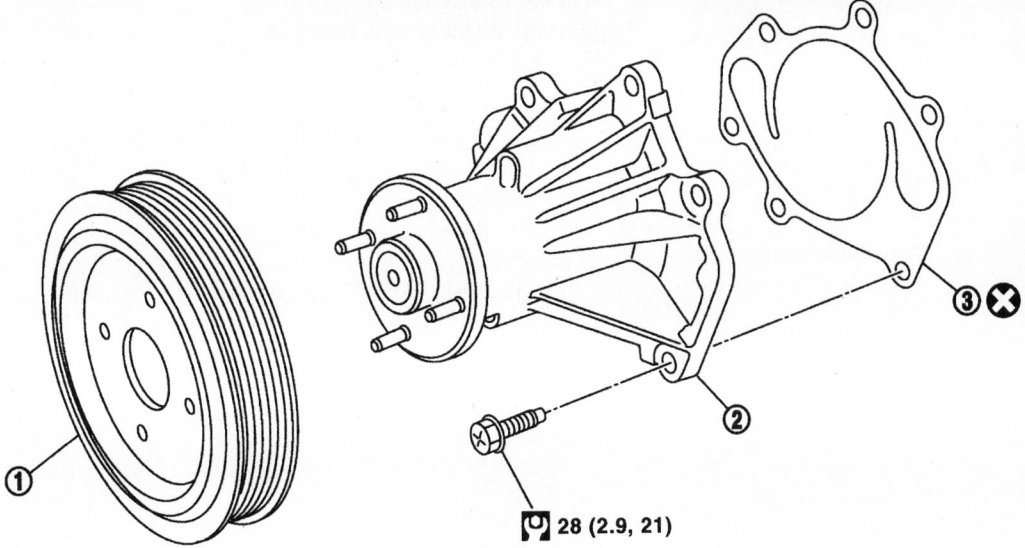

❌ : Always replace after every disassembly.

🔧 : N•m (kg-m, ft-lb)

28 (2.9, 21)

1. Water pump pulley   2. Water pump   3. Gasket

67162-FX35-G121

Exploded view of the water pump mounting—VK45DE Engine.

If anything is found, replace the water pump.

To install:

9. Install the water pump.

10. Install the water pump pulley, then the fan coupling with the cooling fan.

11. Install the alternator, water pump and A/C compressor belt.

12. Install the air duct (inlet).

13. Install the engine front undercover.

14. Tighten the drain plugs on the radiator and both sides of the engine block.

15. Fill the cooling system with coolant.

16. Check for leaks of engine coolant using a radiator cap tester adapter (SST: EG17650301 (J33984-A) or equivalent) and a radiator cap tester.

17. Start and warm up the engine. Visually make sure that there are no leaks of engine coolant.

Heater Core

REMOVAL & INSTALLATION

1. Before servicing the vehicle, refer to the precautions in the beginning of this section.

2. Use approved refrigerant collecting equipment (for HFC-134a) to discharge refrigerant.

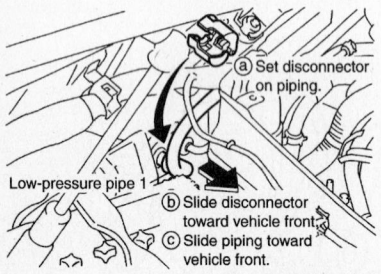

Heater core quick-connect coupling locations.

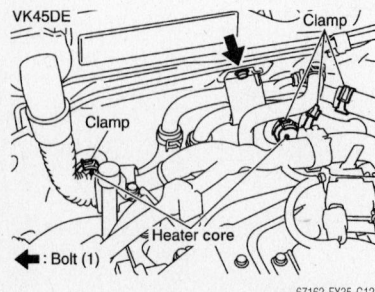

Heater hose clamp locations for heater core service.

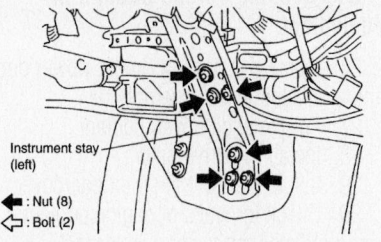

Instrument panel stay mounting bolt locations.

◆◆ WARNING

Make sure the engine is cold before draining the coolant.

3. Drain the coolant from the cooling system.

4. Remove the cowl top cover.

5. Remove the two high-pressure pipe mounting clips.

6. Remove the low-pressure flexible hose bracket mounting bolts.

7. Disconnect the evaporator-side one touch joint, as follows:

  a. Set a disconnector (High-pressure side: 92530-89908, Low pressure side: 92530-89916) on the A/C piping.

  b. Slide the disconnector toward the front of the vehicle until it clicks.

  c. Slide the A/C pipe toward the front of the vehicle front and disconnect it.

◆◆ CAUTION

Seal the connection opening of the pipe with a cap or vinyl tape to avoid exposure to atmosphere.

8. On VQ35DE engines, remove the electronic control throttle assembly.

9. Disconnect the two heater hoses from the heater core.

10. Remove the instrument panel assembly, as follows:

  a. Remove the front kicking plate on both sides of the vehicle.

  b. Remove dash side finisher plastic nuts, then remove the dash side finisher.

  c. Pull to the inside of the vehicle, disengage the metal clips and remove the front pillar garnish.

  d. To remove the A/T Select Lever Knob, pull down the knob cover. Remove the lock-pin of the select lever knob. Then, lift up the select lever knob and remove it.

  e. Insert a remover into the side between the gaps of the instrument clock finisher and pull back to your side.

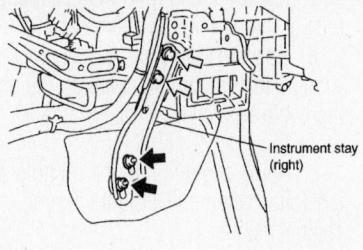

  f. Disconnect the clips and wiring harness connector, then remove instrument clock finisher.

  g. Insert a remover into the side between the gaps of the A/T console finisher and remove it by lifting the A/T console finisher.

  h. Disconnect the wiring harness connector.

  i. Remove the console finisher screws.

  j. Remove the console finishers.

  k. To remove the Center Console, remove the mounting screws, then remove the console sub-wiring harness.

◆◆ CAUTION

When removing console, be careful not to pull the wiring harness.

  l. Remove the Instrument Lower Cover by pulling down on the front instrument lower cover and disconnecting clips. Pull it horizontally, and remove it from the lower cover pawls.

  m. Remove the instrument passenger lower panel screws, disconnect the wiring harness connector, and remove the lower panel.

  n. Remove the Instrument Driver Lower Panel bolt and screws, detach the data link connector, pull to disengage the clip and pawl by removing panel in a horizontal direction. Then, disconnect the in-vehicle sensor and all electrical parts. Remove the grommet, and remove the hood lock cable.

  o. Remove the Steering Column Front Lower Cover screw, disengage the tab, and then remove the steering column front lower cover. Move the steering column telescopic to the rear most position, and move the steering column tilt to the top position.

  p. Remove the Steering Column Lower Cover screws, then disengage the tab and remove steering column lower cover.

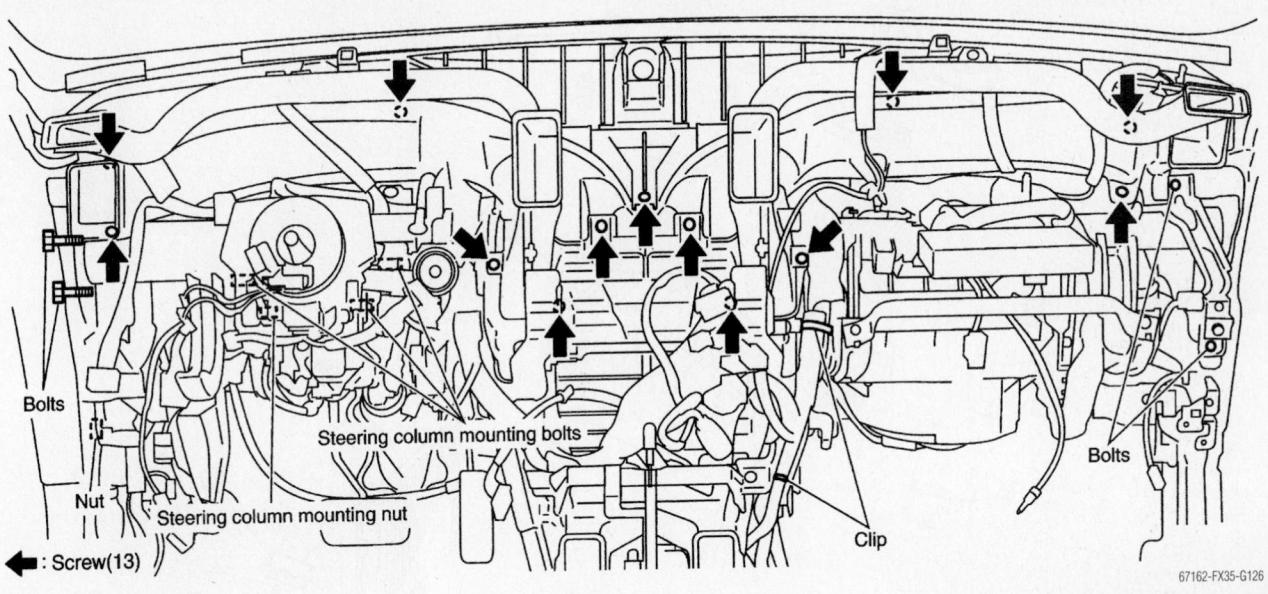

Instrument panel mounting bolt locations.

q. Remove the steering column upper cover.

r. Remove the wiper and washer switch

s. Remove the lighting and turn signal switch.

t. Pull the steering lock escutcheon back to your side, and remove it.

u. Remove the Combination Meter Assembly by removing the bolts and disconnecting the connector bracket. Remove the bolts and disconnect the wiring harness connector.

✴✴ CAUTION

To prevent it from damaged by interference with the combination meter assembly, protect the combination meter assembly with cloths.

v. Remove the instrument panel side panel screws, then pull the panels to the side, disconnect the clip and pawls, and remove the instrument side panels. Perform for both right-hand and left-hand panels.

w. To remove the Cluster Lid C, insert a pry tool into the gap between the instrument panel and pad, pull back towards yourself, and disconnect the metal clips. Then, disconnect the wiring harness connectors, and remove the cluster lid C.

✴✴ CAUTION

Cover surroundings with cloth to avid scratches or damages.

x. Remove the Display Unit and Audio Unit by removing the screws, disconnecting the wiring harness connector, and removing the display unit and audio unit.

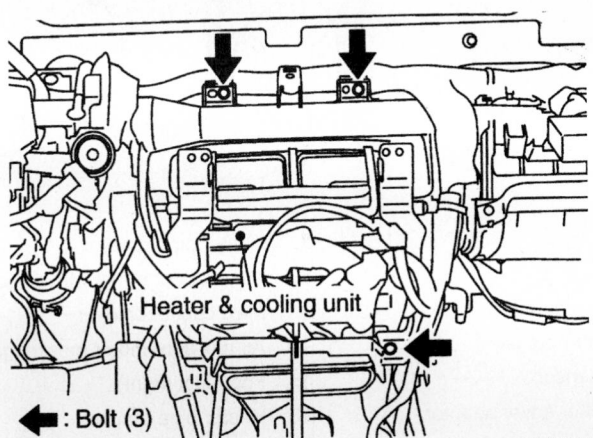

Heater core assembly mounting bolt locations.

✴✴ CAUTION

The Unit is heavy, so be careful not to pinch your fingers when working.

y. Insert a thin pry tool into the gaps between the front defroster grille and instrument panel and pad, lift the front defroster grille upward, and remove the front defroster grille. Perform for both right-hand and left-hand grilles.

z. Remove the Combination Meter Bracket bolts and remove the bracket from the vehicle.

aa. Once the mounting bolts of the wiring harness clip and steering column assembly are removed, pull the steering column assembly backward, and free the combination meter bracket from the instrument panel and pad.

bb. Remove the Side Ventilations by inserting a thin pry tool into the gaps between the instrument panel and pad, pull back to disconnect the metal retaining clips. Then, disconnect the door mirror switch wiring harness connectors, and remove the side ventilations.

cc. Remove the Instrument Panel and Pad by removing the bolts and screws. Then, remove the front passenger air bag module, disconnect the wiring harness connectors, and remove the instrument panel and pad from the passenger door opening.

11. Remove the blower unit, as follows:

a. Remove the ECM with bracket attached.

b. Disconnect the intake door motor

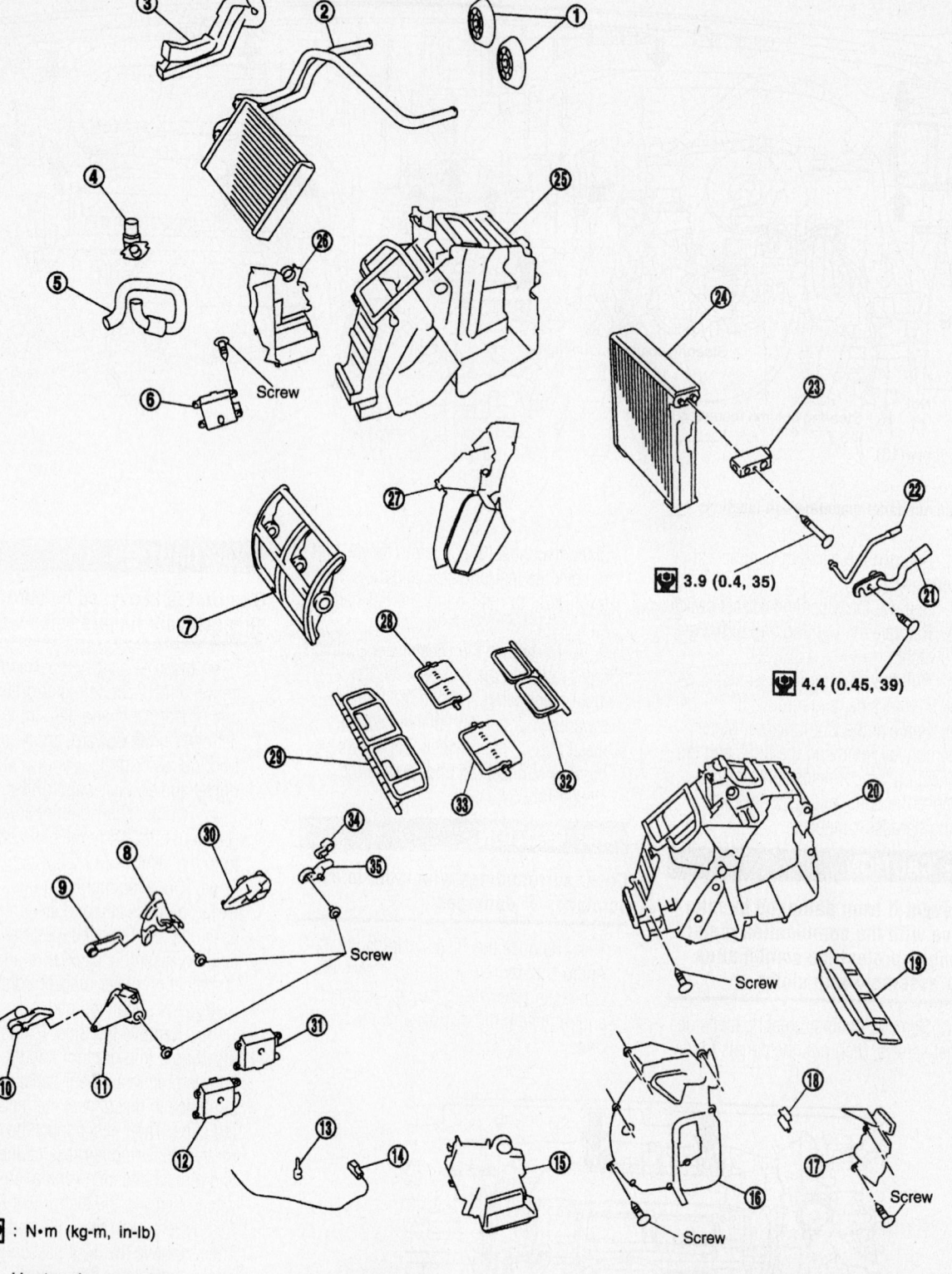

☒ : N•m (kg-m, in-lb)

1. Heater pipe grommet
2. Heater core
3. Heater pipe cover
4. Aspirator
5. Aspirator hose
6. Air mix door motor (driver side)
7. Air mix door (slide door)
8. Max. cool door link
9. Max. cool door lever
10. Ventilator door lever
11. Ventilator door link
12. Air mix door motor (passenger side)
13. Intake sensor bracket
14. Intake sensor
15. Foot duct (right)
16. Evaporator cover
17. Evaporator cover adaptor
18. Heater pipe bracket

Exploded view of the Heater & Cooling Unit, which contains the heater core.

67162-FX35-G127

connector and blower fan motor connector.

    c. Remove the wiring harness clip from the blower unit.

    d. Remove the mounting bolt and screws from the blower unit.

❊❊ CAUTION

Move blower unit rightward, and remove the locating pin and joint. Then, remove the blower unit downward.

    e. Remove the blower unit.

12. Remove the instrument stays (driver-side and passenger-side).

13. Remove the mounting bolts from the heater and cooling unit.

14. Disconnect the drain hose.

15. Remove the ventilator ducts, defroster nozzle and ducts.

16. Remove the steering member mounting bolts, nut and wiring harness clips.

17. Remove the steering member, then remove the heater and cooling unit.

To install:

18. Install the heater and cooling unit. Tighten the mounting bolts to 60 inch lbs. (7 Nm).

19. Install the steering member.

20. Install the steering member mounting bolts, nut and wiring harness clips. Tighten the steering member mounting bolts to 9 ft. lbs. (12 Nm).

21. Install the ventilator ducts, defroster nozzle and ducts.

22. Reconnect the drain hose.

23. Install the mounting bolts for the heater and cooling unit.

24. Install the instrument stays (driver-side and passenger-side).

❊❊ CAUTION

Make sure the locating pin and joint are securely inserted.

25. Install the blower unit, as follows:
    a. Install the blower unit.
    b. Install the mounting bolt and screws for the blower unit.
    c. Install the wiring harness clip for the blower unit.
    d. Reconnect the intake door motor connector and blower fan motor connector.
    e. Install the ECM with its bracket attached.

26. Install the instrument panel assembly, as follows:
    a. Install the front passenger air bag module and Instrument Panel and Pad.

    b. Install the Side Ventilations.
    c. Install the combination meter bracket and reinstall the steering column assembly.
    d. Install the front defroster grilles.
    e. Install the Display Unit and Audio Unit.
    f. Install the Cluster Lid C.
    g. Install the instrument side panels..
    h. Install the Combination Meter Assembly.
    i. Install the steering lock escutcheon.
    j. Install the lighting and turn signal switch.
    k. Install the wiper and washer switch
    l. Install the steering column upper cover.
    m. Install the Steering Column Lower Cover.
    n. Install the steering column front lower cover.
    o. Install the Instrument Driver Lower Panel.
    p. Reattach the data link connector.
    q. Reconnect the in-vehicle sensor and all electrical parts.
    r. Install the grommet and hood lock cable.
    s. Install the instrument passenger lower panel.
    t. Install the Instrument Lower Cover.
    u. Connect the center console sub-wiring harness.
    v. Install the console finishers.
    w. Install the console finisher screws.
    x. Reconnect the wiring harness connector.
    y. Install the A/T console finisher.
    z. Install instrument clock finisher.
    aa. Install the A/T Select Lever Knob.
    bb. Install the front pillar garnish.
    cc. Install the dash side finisher and plastic nuts.
    dd. Install the front kicking plate on both sides of the vehicle.

27. Reconnect the two heater hoses to the heater core.

28. On VQ35DE engines, install the electronic control throttle assembly.

❊❊ CAUTION

Replace the O-rings for A/C piping with new ones, then apply compressor oil to them when installing them.

29. Reconnect the evaporator-side one touch joint. The connection point for the female-side piping is thin. So, when inserting the male-side piping, take care not to deform the female-side piping. Slowly insert

it in the axial direction. Insert the one-touch joint connection point securely until it clicks. After the piping has been connected, pull on the male-side piping by hand to make sure the piping does not come off.

30. Install the low-pressure flexible hose bracket mounting bolts.

31. Install the two high-pressure pipe mounting clips.

32. Install the cowl top cover.

33. Fill the cooling system.

34. Recharge the vehicle A/C system and check for leaks.

Cylinder Head

REMOVAL & INSTALLATION

VQ35DE Engine

1. Before servicing the vehicle, refer to the precautions in the beginning of this section.

2. Remove the camshaft.

➡**It is also possible to perform the following steps 2 and 3 just before removing camshaft.**

3. Temporarily support the front suspension member to support the engine.

❊❊ CAUTION

Temporary support means that the engine is adequately stable although the weight supported by the hoist may be released.

➡**At the time of the start of this procedure the front suspension member is removed, and the cylinder head is hanged by the hoist with engine slinger installed.**

4. Release the hoist from hanging, then remove the engine slinger.

5. Remove the fuel tube and fuel injector assembly.

6. Remove the intake manifold.

7. Remove the exhaust manifold.

8. Remove the water inlet and thermostat housing.

9. Remove the water outlet and water piping.

10. Loosen the cylinder head bolts in the reverse of the tightening order using a cylinder head bolt wrench (commercial service tool).

11. Remove the cylinder head.

12. Remove the cylinder head gaskets.

13. Inspect the cylinder head bolt diameters. The cylinder head bolts are tightened by the plastic zone tightening method.

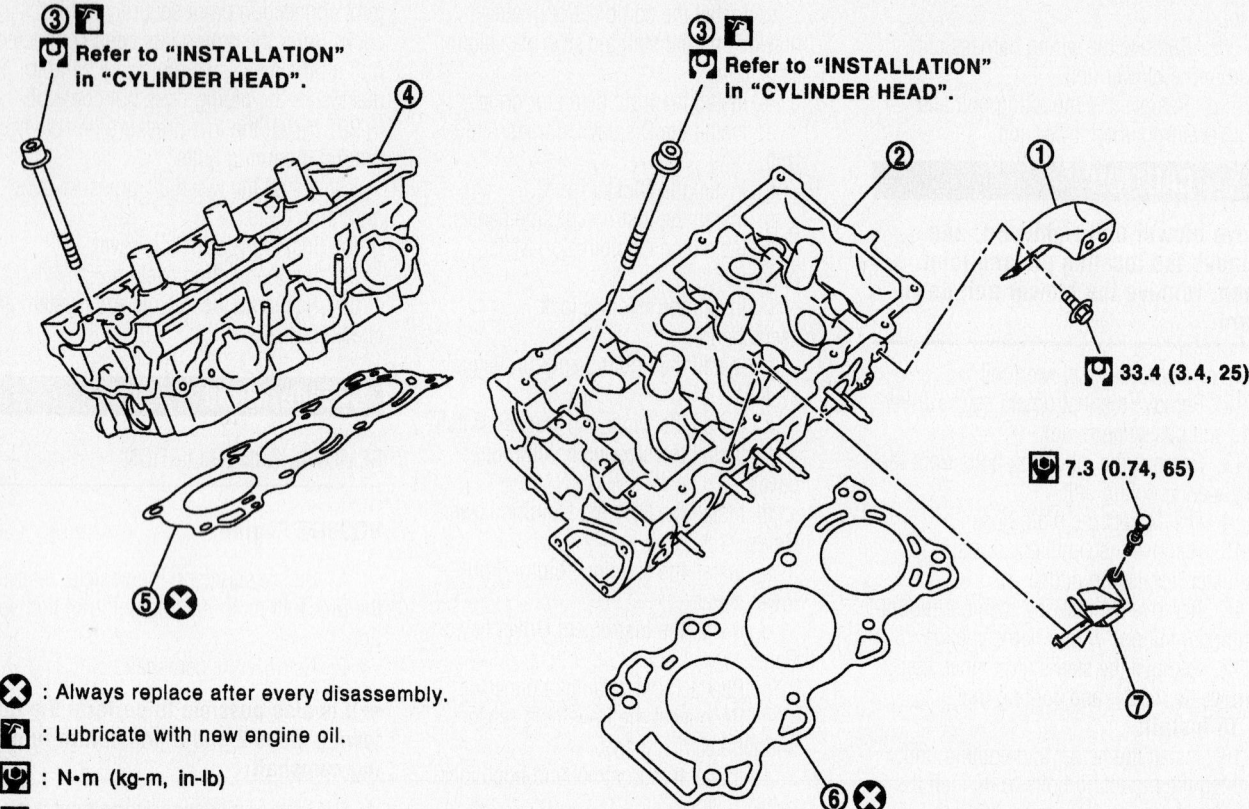

③ 🔧
📋 **Refer to "INSTALLATION" in "CYLINDER HEAD".**

③ 🔧
📋 **Refer to "INSTALLATION" in "CYLINDER HEAD".**

🔧 **33.4 (3.4, 25)**

🔧 **7.3 (0.74, 65)**

❌ : Always replace after every disassembly.

🔧 : Lubricate with new engine oil.

🔧 : N•m (kg-m, in-lb)

🔧 : N•m (kg-m, ft-lb)

1. Engine rear lower slinger
2. Cylinder head (left bank)
3. Cylinder head bolt
4. Cylinder head (right bank)
5. Cylinder head gasket (right bank)
6. Cylinder head gasket (left bank)
7. Oil level gauge guide

67162-FX35-G128

Exploded view of cylinder head mounting and related components—VQ35DE Engine.

Right bank

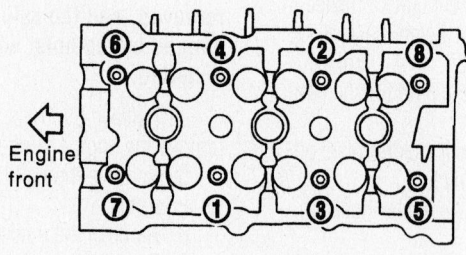

Left bank

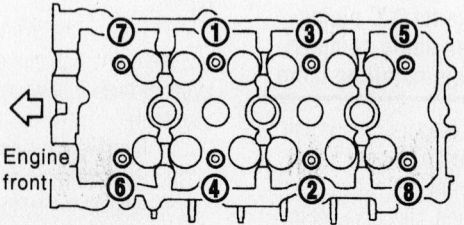

67162-FX35-G129

Cylinder head mounting bolt tightening sequence—VQ35DE Engine.

Whenever the size difference between d1 and d2 exceeds the limit, replace the bolt with a new one. The specification for d1-d2 for the cylinder head bolts is 0.0043 in. (0.11mm)

➡️**If the reduction of the outer diameter appears in a position other than at d2, use it as the d2 point value.**

14. Check the cylinder head for distortion. Using a scraper, wipe off oil, scale,

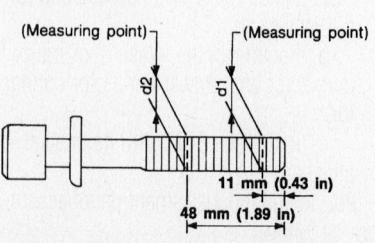

Cylinder head bolt

(Measuring point) — $d2$ | $d1$ — (Measuring point)

11 mm (0.43 in)

48 mm (1.89 in)

67162-FX35-G130

Cylinder head bolt inspection dimensions—VQ35DE Engine.

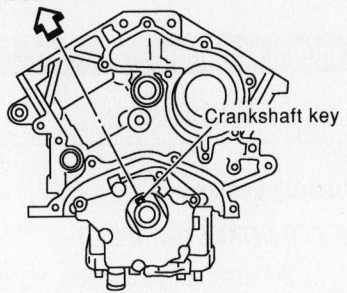

Crankshaft positioning for cylinder head installation—VQ35DE Engine.

gasket, sealant and carbon deposits from the surface of the cylinder head. At each of several locations on the bottom surface of the cylinder head, measure distortion in six directions. If cylinder head distortion exceeds the recommended limit of 0.004 in. (0.1mm), replace the cylinder head.

✳✳ CAUTION

Do not allow gasket fragments to enter engine oil or engine coolant passages.

To install:

15. Install a new cylinder head gasket.
16. Turn the crankshaft until No. 1 piston is set at TDC on the compression stroke. The crankshaft key should line up with the right bank cylinder center line.
17. Install the cylinder head.
18. Install and tighten the cylinder head bolts in the proper order as follows:
    a. Tighten all bolts to 72 ft. lbs. (98 Nm).
    b. Completely loosen all bolts in the reverse order of the tightening sequence.
    c. Retighten all bolts to 29 ft. lbs. (39 Nm)
    d. Turn all bolts "90" degrees clockwise (angle tightening).
    e. Turn all bolts "90" degrees clockwise again (angle tightening).

✳✳ CAUTION

Check and confirm the tightening angle by using angle wrench or equivalent, and cylinder head bolt wrench (commercial service tool). Avoid tightening the bolts "by eye."

19. After installing the cylinder head, measure the distance between the front end faces of the cylinder block and the cylinder head (left and right banks). If the measurement is outside the specified range, re-install the cylinder head.
20. Install the water outlet and water piping.
21. Install the water inlet and thermostat housing.
22. Install the exhaust manifold.
23. Install the intake manifold.
24. Install the fuel tube and fuel injector assembly.
25. Install the camshaft.

VK45DE Engine

1. Before servicing the vehicle, refer to the precautions in the beginning of this section.

➡**According to Nissan, cylinder head removal requires removal of the engine assembly from the vehicle.**

2. Remove the engine assembly from vehicle.
3. Remove the exhaust manifold.
4. Remove the camshaft.
5. Loosen the cylinder head bolts in the reverse of the tightening order using a cylinder head bolt wrench (commercial service tool).

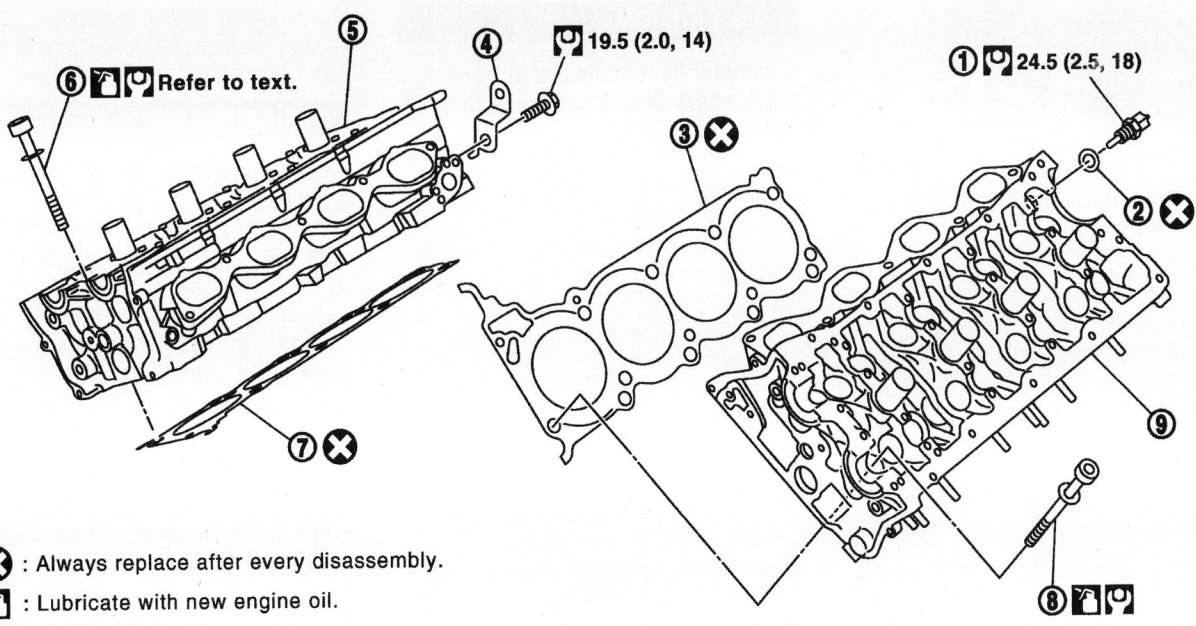

❌ : Always replace after every disassembly.

🛢 : Lubricate with new engine oil.

🔧 : N•m (kg-m, ft-lb)

1. Engine coolant temperature sensor	2. Washer	3. Cylinder head gasket (left bank)
4. Harness bracket	5. Cylinder head (right bank)	6. Cylinder head bolt
7. Cylinder head gasket (right bank)	8. Cylinder head bolt	9. Cylinder head (left bank)

Exploded view of the cylinder head mounting and related components—VK45DE Engine.

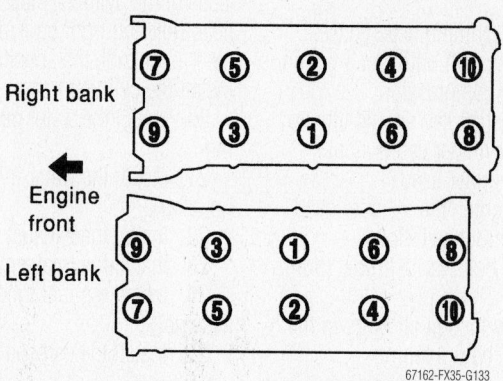

Cylinder head mounting bolt tightening sequence—VK45DE Engine.

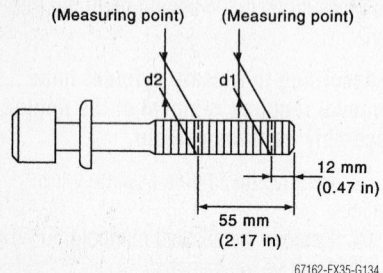

Cylinder head mounting bolt inspection dimensions—VK45DE Engine.

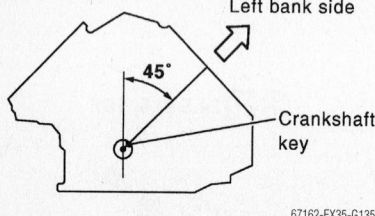

Crankshaft positioning for cylinder head installation—VK45DE Engine.

6. Remove the cylinder head.

7. Remove the cylinder head gaskets.

8. Inspect the cylinder head bolt diameters. The cylinder head bolts are tightened by the plastic zone tightening method. Whenever the size difference between d1 and d2 exceeds the limit, replace the bolt with a new one. The specification for d1-d2 for the cylinder head bolts is 0.0071 in. (0.18mm)

➡If the reduction of the outer diameter appears in a position other than at d2, use it as the d2 point value.

9. Check the cylinder head for distortion. Using a scraper, wipe off oil, scale, gasket, sealant and carbon deposits from the surface of the cylinder head. At each of several locations on the bottom surface of the cylinder head, measure distortion in six directions.

If cylinder head distortion exceeds the recommended limit of 0.004 in. (0.1mm), replace the cylinder head.

✳✳ CAUTION

Do not allow gasket fragments to enter engine oil or engine coolant passages.

To install:

10. Install a new cylinder head gasket.

11. Turn the crankshaft until No. 1 piston is set at TDC on the compression stroke. The crankshaft key should line up with the left bank cylinder center line.

12. Install the cylinder head.

✳✳ CAUTION

If cylinder head bolts are to be re-used, check their outer diameters before installation.

13. Install and tighten the cylinder head bolts in the proper order as follows:

   a. Tighten all bolts to 72 ft. lbs. (98 Nm).

   b. Completely loosen all bolts in the reverse order of the tightening sequence.

   c. Retighten all bolts to 33 ft. lbs. (44 Nm)

   d. Turn all bolts 60 degrees clockwise (angle tightening).

   e. Turn all bolts 60 degrees clockwise again (angle tightening).

✳✳ CAUTION

Check and confirm the tightening angle by using angle wrench or equivalent, and cylinder head bolt wrench (commercial service tool). Avoid tightening the bolts "by eye."

14. Install the camshaft.

15. Install the exhaust manifold.

16. Install the engine assembly into the vehicle.

Intake Manifold

REMOVAL & INSTALLATION

VQ35DE Engine

UPPER INTAKE MANIFOLD

1. Before servicing the vehicle, refer to the precautions in the beginning of this section.

The upper intake manifold is constructed of two halves: upper intake manifold collector and lower intake manifold collector. Together these two components create the upper intake manifold. The VQ35DE engine also uses a lower intake manifold plenum.

✳✳ WARNING

To avoid the danger of being scalded, never drain engine coolant when engine is hot.

➡The gasket for the intake manifold collector (upper) is secured together with the mounting bolt for the intake manifold collector (lower). Thus, even when only the gasket for upper side is replaced, the gasket for lower side must also be replaced.

2. Remove the engine cover.

3. Disconnect and plug the water hoses from intake manifold collector (upper).

✳✳ CAUTION

Do not spill engine coolant on the drive belts.

4. Remove the air cleaner case and air duct, as follows:

   a. Remove the air duct (inlet).

   b. Disconnect the mass air flow sensor wiring harness connector.

   c. Remove the air cleaner case/mass air flow sensor assembly and the air duct/resonator assembly disconnecting their joints.

➡Add marks as necessary for easier installation.

   d. Remove the mass air flow sensor from air cleaner case.

✳✳ CAUTION

Handle the mass air flow sensor with care. Do not expose it to harsh vibration or shock. Do not disassemble it or touch its sensor.

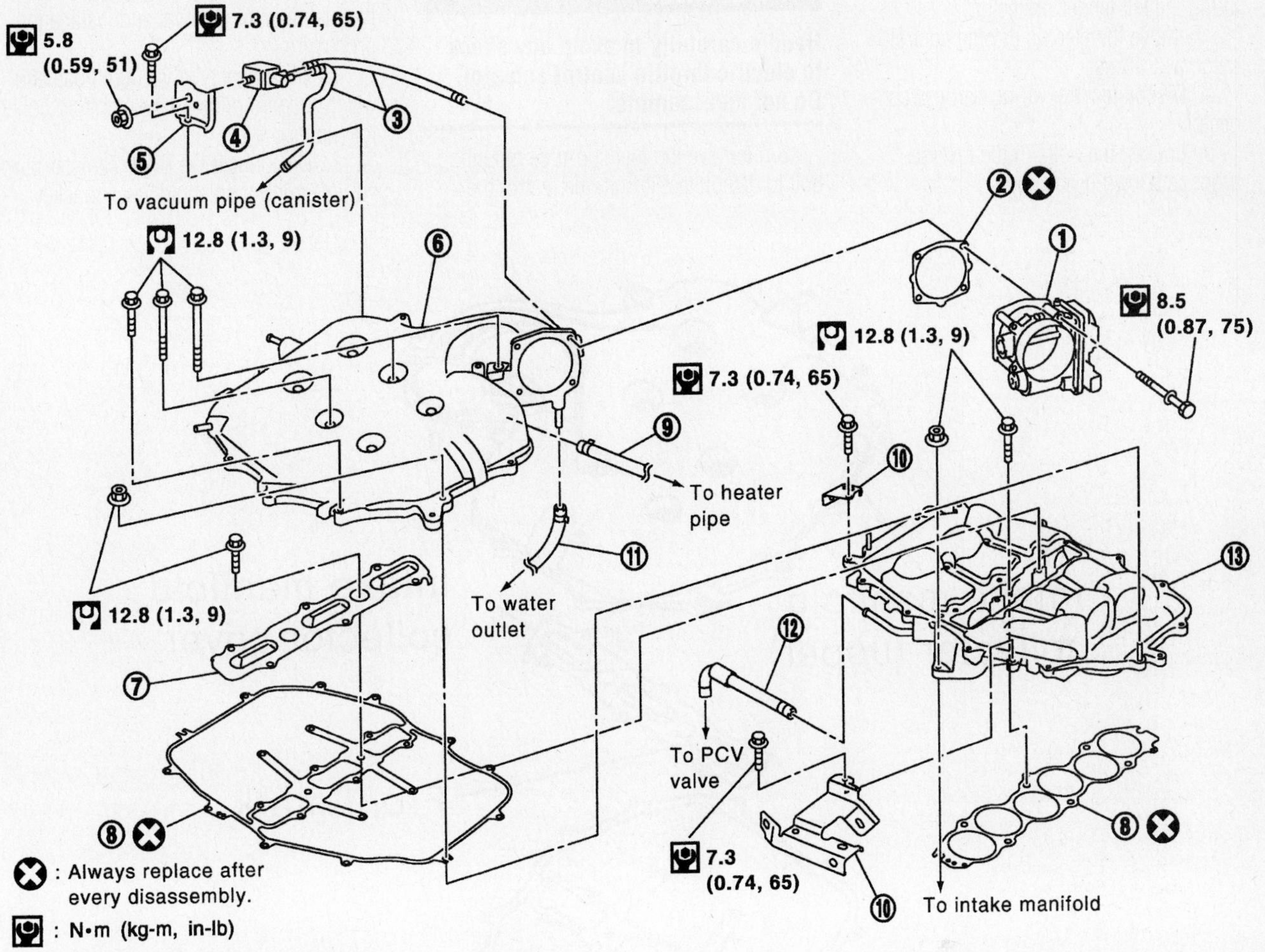

5.8 (0.59, 51)

7.3 (0.74, 65)

To vacuum pipe (canister)

12.8 (1.3, 9)

12.8 (1.3, 9)

7.3 (0.74, 65)

8.5 (0.87, 75)

To heater pipe

To water outlet

12.8 (1.3, 9)

To PCV valve

7.3 (0.74, 65)

To intake manifold

⊗ : Always replace after every disassembly.

🔧 : N•m (kg-m, in-lb)

1. Electric throttle control actuator
2. Gasket
3. Vacuum hose
4. EVAP canister purge volume control solenoid valve
5. Bracket
6. Intake manifold collector (upper)
7. Intake manifold collector cover
8. Gasket
9. Water hose
10. Bracket
11. Water hose
12. PCV hose
13. Intake manifold collector (lower)

67162-FX35-G136

Exploded view of the upper and lower halves of the upper intake manifold—VQ35DE Engine.

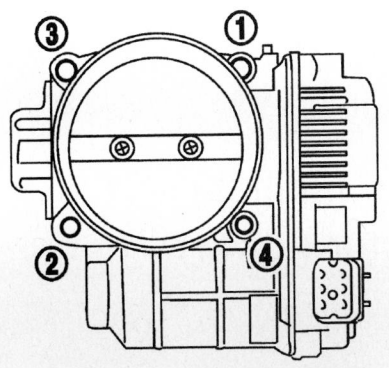

67162-FX35-G137

Throttle body mounting bolt tightening sequence—VQ35DE Engine.

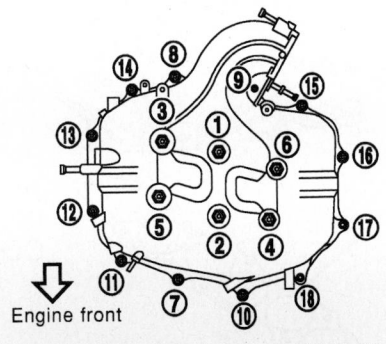

Engine front

67162-FX35-G138

Tightening sequence for the upper half of the upper intake manifold—VQ35DE Engine.

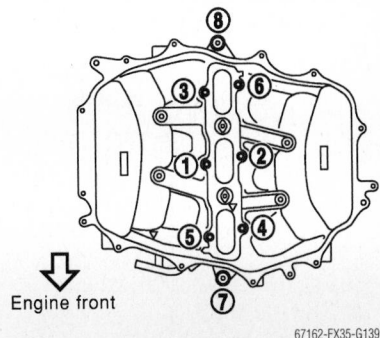

Engine front

67162-FX35-G139

Tightening sequence for the lower half of the upper intake manifold—VQ35DE Engine.

e. Remove the resonator in the fender, lifting the left fender protector.

5. Remove the electric throttle control actuator, as follows:

a. Disconnect the wiring harness connector.

b. Loosen the bolts in the reverse order as shown in the figure.

✳✳ CAUTION

Handle carefully to avoid any shock to electric throttle control actuator. Do not disassemble.

6. Remove the fuel sub-tube mounting bolt to disconnect it from the rear of the intake manifold collector (lower). Refer to the fuel injector removal and installation procedure.

7. Disconnect the vacuum hose and water hose from the intake manifold collector (upper).

8. Disconnect the EVAP canister purge volume control solenoid valve bracket

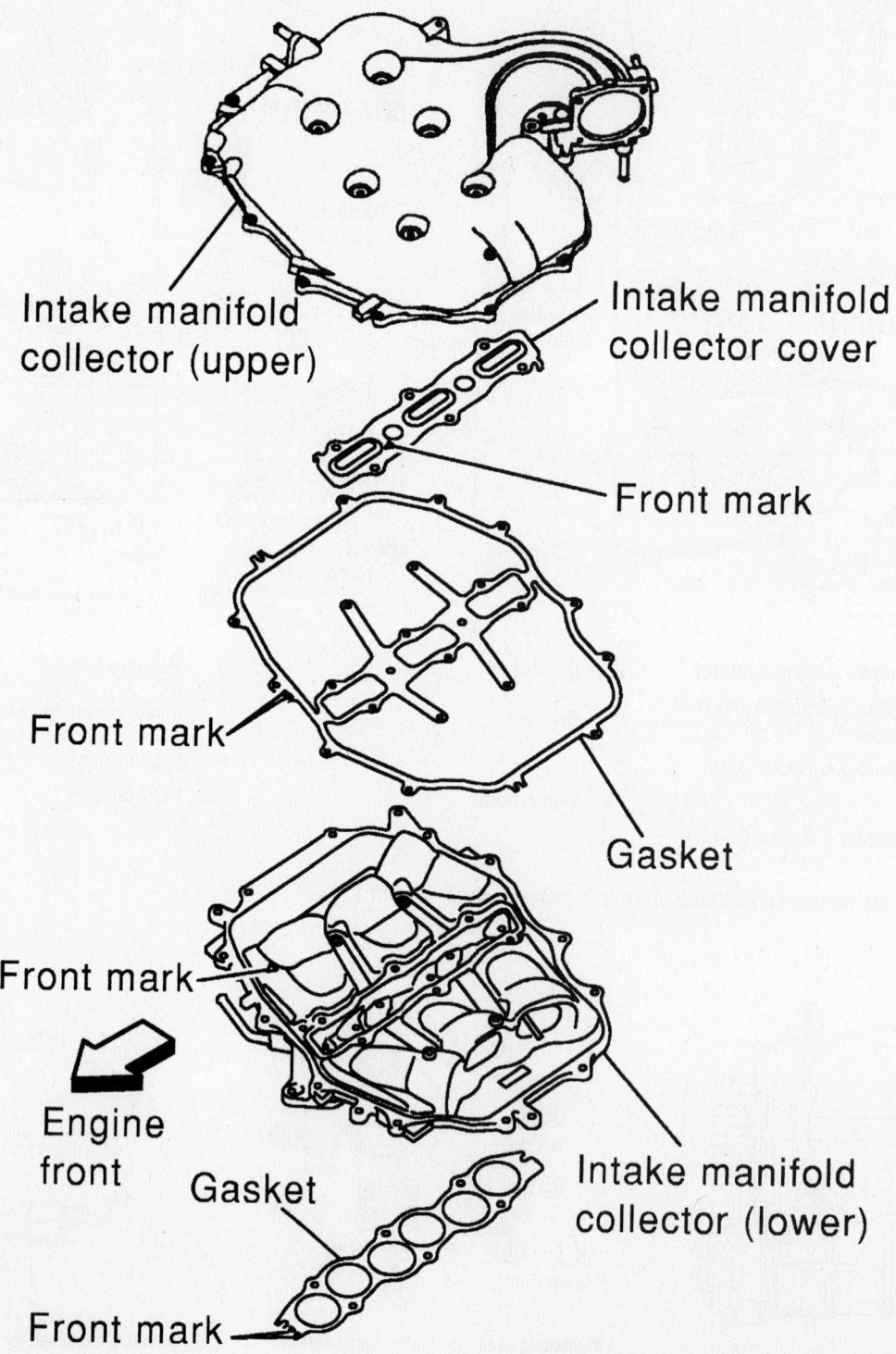

Intake manifold collector (upper)

Intake manifold collector cover

Front mark

Front mark

Gasket

Front mark

Engine front

Gasket

Intake manifold collector (lower)

Front mark

67162-FX35-G140

Upper intake manifold gasket positioning—VQ35DE Engine.

mounting bolt from the intake manifold collector (upper).

9. Loosen the bolts in the reverse order of the illustration to remove the intake manifold collector (upper).

10. Remove the PCV hose (between the intake manifold collector and the right-hand rocker cover).

11. Loosen the bolts in the reverse order of the illustration, and remove the intake manifold collector cover, gasket, intake manifold collector (lower) and gasket.

✳✳ CAUTION

Cover all engine openings to avoid entry of foreign materials.

12. Check the surface distortion of both the intake manifold collector (upper and lower) mating surfaces with a straightedge and feeler gauge. If it exceeds 0.004 in. (0.1mm), replace the intake manifold collector (upper and/or lower).

To install:

13. Install the intake manifold collector (lower). Tighten the mounting bolts in numerical order as shown in the figure.

➡**Tighten mounting bolts to secure gasket (lower), intake manifold collector (lower), gasket (upper), and intake manifold collector cover.**

14. Reconnect the PCV hose (between the intake manifold collector and the right-hand rocker cover).

15. Install the lower manifold. If the stud bolts were removed, install them and tighten them to the specified torque.

➡**The shank length from under the bolt head varies with bolt location. Install the bolts while referring to the numbers shown below and in the figure. (The bolt length does not include pilot portion.) Make sure to tighten them in numerical order as shown in the figure.**

- M6 bolt—length 0.98 in. (25mm): Positions 7, 8, 10, 11, 13, 14, 15, 16, and 18
- M6 bolt—length 1.77 in. (45mm): Positions 2, 4, 5
- M6 bolt—length 2.36 in. (60mm): Positions 1, 3, 6, and 9
- M6 nut: Positions 12, 17

16. Reattach the EVAP canister purge volume control solenoid valve bracket to the intake manifold collector (upper).

17. Reattach the vacuum hose and water hose to the intake manifold collector (upper).

18. Reattach the fuel sub-tube to the rear of the intake manifold collector (lower).

✳✳ CAUTION

Handle carefully to avoid any shock to electric throttle control actuator.

19. Install the electric throttle control actuator.

✳✳ CAUTION

Handle the mass air flow sensor with care. Do not expose it to harsh vibration or shock. Do not disassemble it or touch its sensor.

20. Install the mass air flow sensor in the air cleaner case.

21. Install the air cleaner case and air duct.

22. Reconnect the hoses to the intake manifold collector (upper).

23. Install the engine cover.

LOWER INTAKE MANIFOLD

1. Before servicing the vehicle, refer to the precautions in the beginning of this section.

2. Release the fuel pressure.

3. Remove the intake manifold collector (upper) and (lower). .

4. Remove the fuel tube and fuel injector assembly.

5. Loosen the mounting bolts and nuts in the reverse order of the illustration to remove the lower intake manifold.

6. Remove the intake manifold gaskets.

✳✳ CAUTION

Cover all engine openings to avoid entry of foreign materials.

7. Check the surface distortion of the intake manifold mating surface with a straightedge and feeler gauge. If it exceeds the limit of 0.04 in. (0.1mm), replace the lower intake manifold.

To install:

8. Install new lower intake manifold gaskets.

9. Install the lower intake manifold. If the stud bolts were removed, install them and tighten to 8 ft. lbs. (11 Nm). Tighten all mounting bolts and nuts to the specified

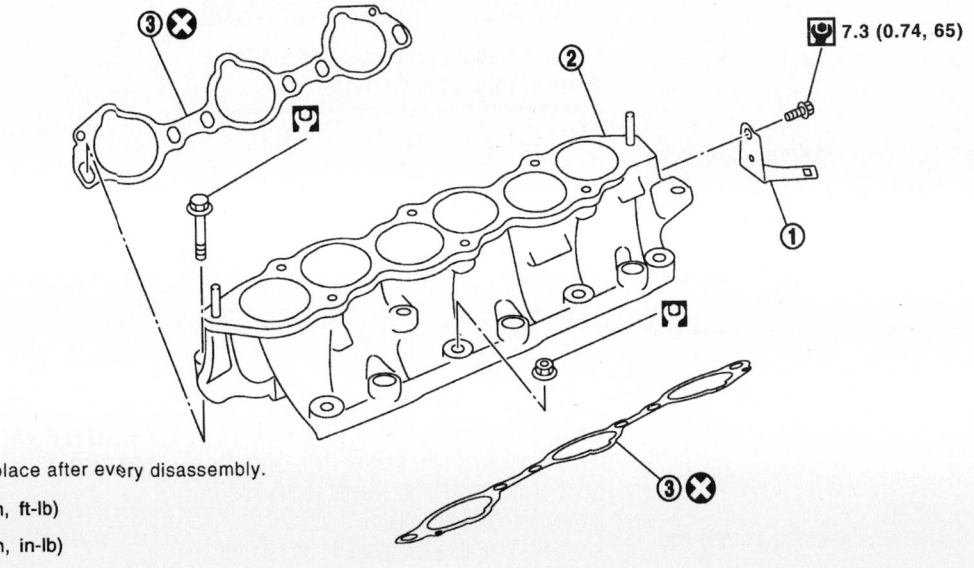

⊗ : Always replace after every disassembly.

🔧 : N•m (kg-m, ft-lb)

🔧 : N•m (kg-m, in-lb)

| 1. | Harness bracket | 2. | Intake manifold | 3. | Gasket |

67162-FX35-G141

Exploded view of the lower intake manifold—VQ35DE Engine.

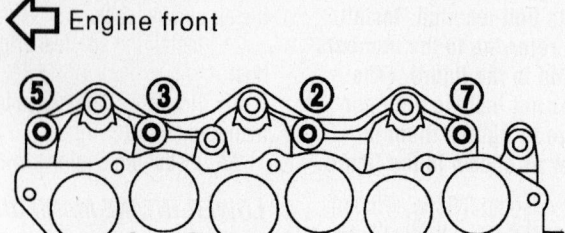

← Engine front

67162-FX35-G142

Lower intake manifold mounting bolt tightening sequence—VQ35DE Engine.

torque in two or more steps in numerical order shown in the figure, as follows:

   a. Step 1:5 ft. lbs. (7 Nm)

   b. Step 2—21 ft. lbs. (29 Nm)

10. The remainder of installation is the reverse of removal.

VK45DE Engine

1. Before servicing the vehicle, refer to the precautions in the beginning of this section.

✳✳ WARNING

To avoid the danger of being scalded, never drain the engine coolant when the engine is hot.

2. Remove the engine cover.

3. Release the fuel pressure.

4. Drain the engine coolant.

5. Remove the air duct (inlet), air cleaner case and mass air flow sensor assembly, air duct and resonator assembly.

6. Disconnect the fuel feed hose quick connector on the engine side, the fuel damper and fuel hose assembly. Refer to the fuel injector removal and installation procedure.

✳✳ CAUTION

While hoses are disconnected, plug them to prevent fuel from draining. Do not separate fuel damper and fuel hose.

7. Remove or disconnect the wiring harnesses, brackets, vacuum hose, vacuum gallery and PCV hose and tube from the intake manifold (upper).

8. Remove the electric throttle control actuator as follows:

   a. Disconnect the wiring harness connector.

   b. Loosen the mounting bolts diagonally.

✳✳ CAUTION

Handle carefully to avoid any shock to the electric throttle control actuator. Do not disassemble the electric throttle control actuator.

9. Disconnect the water hoses from the water gallery.

10. Remove the intake manifold adaptor and water gallery.

11. Loosen the bolts in the reverse order as shown in the figure to remove the intake manifold (upper).

12. Remove the vacuum tank from the intake manifold (lower).

13. Remove the fuel injector and fuel tube assembly. Refer to the fuel injector removal and installation procedure.

14. Loosen the bolts in the reverse order as shown in the figure to remove the intake manifold (lower).

15. Remove the intake manifold gaskets.

✳✳ CAUTION

Cover all engine openings to avoid entry of foreign materials.

16. Check the surface distortion of both the intake manifold (upper and lower) mating surfaces with a straightedge and feeler gauge. If it exceeds the limit of 0.004 in (0.1mm), replace the intake manifolds (lower and/or upper).

To install:

17. Install new intake manifold gaskets.

18. Install the intake manifold and tighten the mounting bolts in numerical order as shown in the figure. There are two types of mounting bolts—refer to the following for locating bolts:

- M8 bolts—length 3.54 in. (90mm): Positions 7, 8
- M8 bolts—length 1.38 in. (35mm): Positions except 7 and 8

19. Install the fuel injector and fuel tube assembly.

20. Install the vacuum tank onto the intake manifold (lower).

21. Install the intake manifold (upper). Tighten the mounting bolts in the numerical order shown in the figure. There are two types of mounting bolts—refer to the following for locating bolts:

- M8 bolts—length 3.15 in. (80mm): Positions 4, 5, 6, and 7
- M8 bolts—length 0.98 in. (25mm): Positions except 4, 5, 6, and 7

22. Install the intake manifold adaptor and water gallery.

23. Connect the water hoses to the water gallery.

✳✳ CAUTION

Handle carefully to avoid any shock to the electric throttle control actuator.

24. Install the electric throttle control actuator. Install the intake manifold adapter gasket and electric throttle control actuator gasket so that the three protrusions for installation do not face downward.

25. Tighten the mounting bolts of the electric throttle control actuator equally and diagonally in several steps.

26. Reconnect the wiring harnesses, brackets, vacuum hose, vacuum gallery and PCV hose and tube to the intake manifold (upper).

27. Connect the fuel feed hose quick connector on the engine side, the fuel damper and fuel hose assembly.

28. Install the air duct (inlet), air cleaner case and mass air flow sensor assembly, air duct and resonator assembly.

29. Fill the engine cooling system with engine coolant.

30. Install the engine cover.

31. Perform the following drivability adjustments:

32. Perform the "Throttle Valve Closed Position Learning" procedure (below) when the wiring harness connector of the electric throttle control actuator is disconnected, or perform the "Idle Air Volume Learning" and "Throttle Valve Closed Position Learning" procedures (below) when the electric throttle control actuator is replaced.

THROTTLE VALVE CLOSED POSITION LEARNING

1. Before servicing the vehicle, refer to the precautions in the beginning of this section.

The Throttle Valve Closed Position Learning procedure is an operation for the

ECM to relearn the fully closed position of the throttle valve by monitoring the throttle position sensor output signal. It must be performed each time the wiring harness connector of the electric throttle control actuator or ECM is disconnected.

2. Make sure that accelerator pedal is fully released.

3. Turn ignition switch ON.

4. Turn ignition switch OFF wait at least 10 seconds. Make sure that throttle valve moves during above 10 seconds by confirming the operating sound.

IDLE AIR VOLUME LEARNING

1. Before servicing the vehicle, refer to the precautions in the beginning of this section.

Idle Air Volume Learning is an operation to learn the idle air volume that keeps each engine within the specific range. It must be performed under any of the following conditions:

- Each time the electric throttle control actuator or ECM is replaced.
- Idle speed or ignition timing is out of specification.

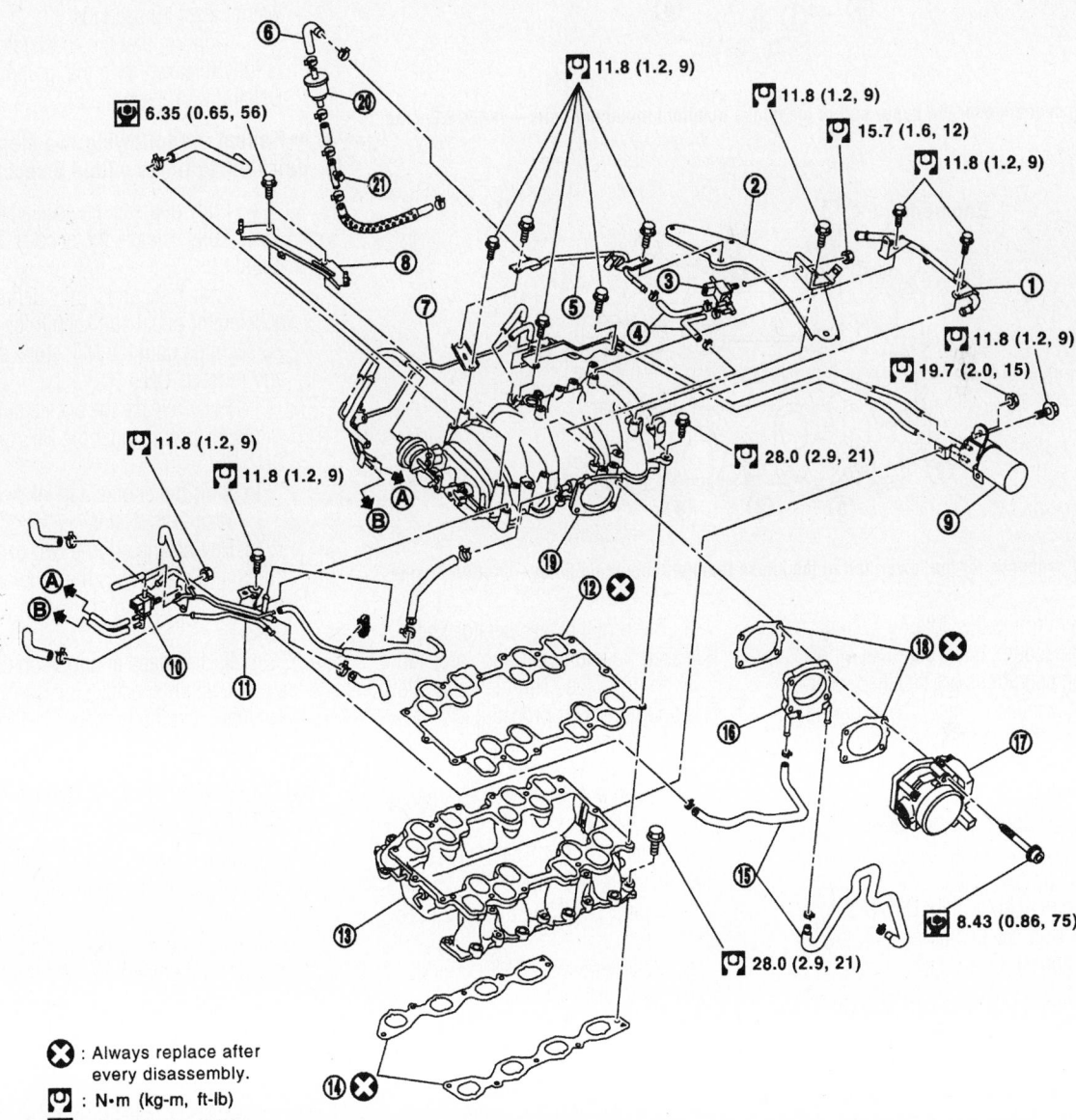

⊗ : Always replace after every disassembly.

🔧 : N·m (kg-m, ft-lb)

🔧 : N·m (kg-m, in-lb)

1.	PCV tube	2.	Engine cover rear bracket	3.	EVAP canister purge control solenoid valve
4.	EVAP hose	5.	EVAP tube	6.	EVAP hose
7.	Vacuum gallery	8.	Engine cover front bracket	9.	Vacuum tank
10.	VIAS control solenoid valve	11.	Water gallery	12.	Gasket
13.	Intake manifold (lower)	14.	Gasket	15.	Water hose
16.	Intake manifold adapter	17.	Electric throttle control actuator	18.	Gasket
19.	Intake manifold (upper)	20.	Resonator	21.	EVAP service port

67162-FX35-G143

Exploded view of the intake manifold mounting and related components— VK45DE Engine.

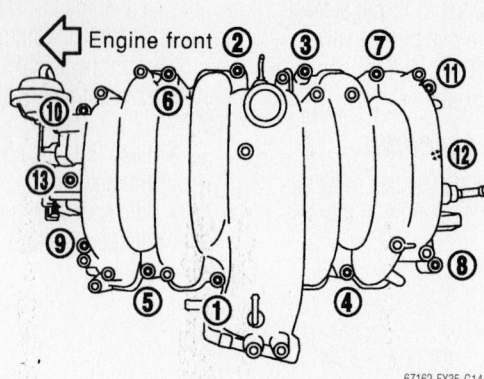

Tightening sequence for the upper half of the intake manifold mounting bolts—VK45DE Engine.

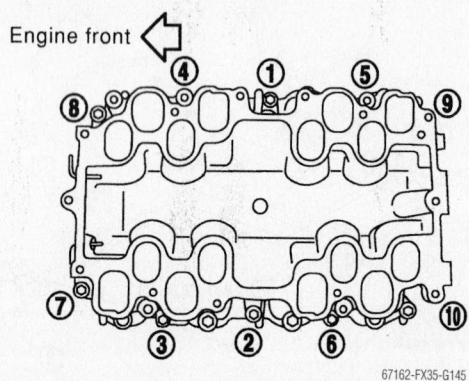

Tightening sequence for the lower half of the intake manifold mounting bolts—VK45DE Engine.

Before performing the "Idle Air Volume Learning" procedure, make sure that all of the following conditions are satisfied. Learning will be cancelled if any of the following conditions are missed for even a moment.

- Battery voltage: More than 12.9V (At idle)
- Engine coolant temperature: 70—100°C (158—212°F)
- PNP switch: ON
- Electric load switch: OFF (Air conditioner, headlamp, rear window defogger)

➡ **On vehicles equipped with daytime light systems, if the parking brake is applied before the engine is started, the headlamp will not be illuminated.**

- Steering wheel: Neutral (Straight-ahead position)
- Vehicle speed: Stopped
- Transmission: Warmed-up
- For models with CONSULT-II, drive vehicle until "FLUID TEMP SE" in "DATA MONITOR" mode of "A/T" system indicates less than 0.9V.
- For models without CONSULT-II, drive vehicle for 10 minutes.

2. If using the CONSULT-II tool, perform the following:

a. Perform the "Accelerator Pedal Released Position Learning" procedure.

b. Perform the "Throttle Valve Closed Position Learning" procedure.

c. Start the engine and warm it up to normal operating temperature.

d. Check that all items listed above are properly set.

e. Select "IDLE AIR VOL LEARN" in "WORK SUPPORT" mode.

f. Touch "START" and wait 20 seconds.

g. Make sure that "CMPLT" is displayed on CONSULT-II screen. If "CMPLT" is not displayed, the Idle Air Volume Learning procedure will not be carried out successfully.

h. Rev up the engine two or three times and make sure that idle speed and ignition timing are within specifications.

3. If NOT using the CONSULT-II tool, perform the following:

➡ **It is best to keep track of time accurately with a clock.**

➡ **It is impossible to switch the diagnostic mode when an accelerator pedal position sensor circuit has a malfunction.**

a. Perform the "Accelerator Pedal Released Position Learning" procedure.

b. Perform the "Throttle Valve Closed Position Learning" procedure.

c. Start the engine and warm it up to normal operating temperature.

d. Check that all items listed above are properly set.

e. Turn the ignition switch OFF and wait at least 10 seconds.

f. Confirm that the accelerator pedal is fully released, turn the ignition switch ON and wait 3 seconds.

➡ **Repeat the following two steps quickly five times within 5 seconds.**

g. Fully depress the accelerator pedal.

h. Fully release the accelerator pedal.

i. Wait 7 seconds, fully depress the accelerator pedal and keep it for approx. 20 seconds until the MIL stops blinking and remains ON.

j. Fully release the accelerator pedal within 3 seconds after the MIL turned ON.

k. Start the engine and let it idle.

l. Wait 20 seconds.

m. Rev up the engine two or three times and make sure that idle speed and ignition timing are within specifications.

n. If idle speed and ignition timing are not within specification, the Idle Air Volume Learning procedure will not be successful.

ACCELERATOR PEDAL RELEASED POSITION LEARNING

1. Before servicing the vehicle, refer to the precautions in the beginning of this section.

The "Accelerator Pedal Released Position Learning" procedure is an operation for the ECM to relearn the fully released position of the accelerator pedal by monitoring the accelerator pedal position sensor output signal. It must be performed each time the wiring harness connector of the accelerator pedal position sensor or ECM is disconnected.

2. Make sure that the accelerator pedal is fully released.

3. Turn the ignition switch ON and wait at least 2 seconds.

4. Turn the ignition switch OFF wait at least 10 seconds.

5. Turn the ignition switch ON and wait at least 2 seconds.

6. Turn the ignition switch OFF wait at least 10 seconds.

Exhaust Manifold

REMOVAL & INSTALLATION

VQ35DE Engine

1. Before servicing the vehicle, refer to the precautions in the beginning of this section.

※※ WARNING

Perform the work when the exhaust and cooling system have completely cooled down.

2. Remove the engine cover.
3. Remove the air cleaner case and air duct.

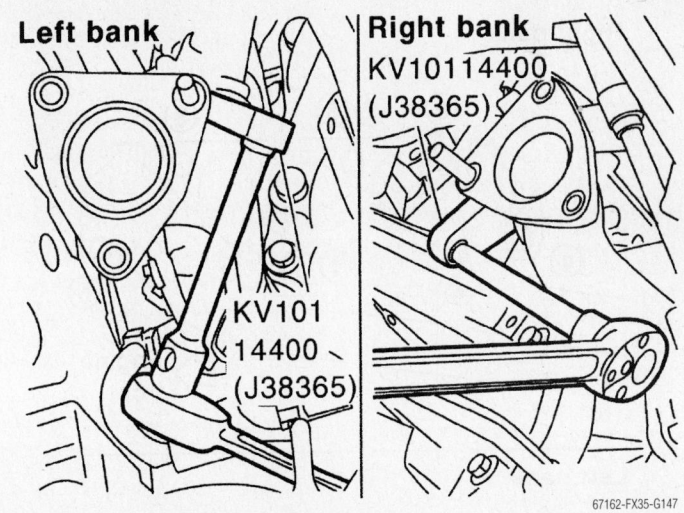

Left bank

Right bank
KV10114400 (J38365)

KV101 14400 (J38365)

67162-FX35-G147

Heated oxygen sensor locations—VQ35DE Engine.

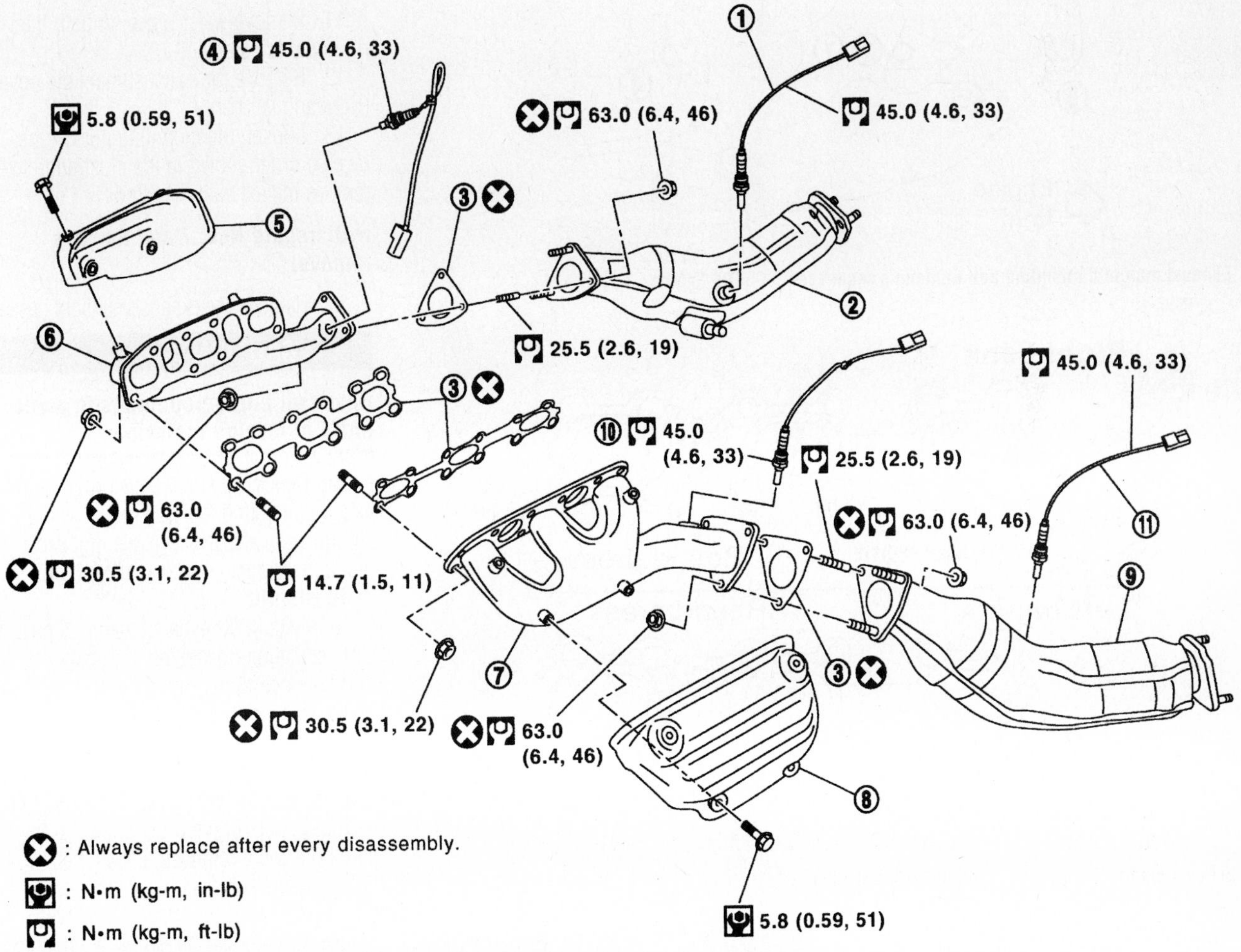

④ 🔧 45.0 (4.6, 33)

🔧 5.8 (0.59, 51)

❌ 🔧 63.0 (6.4, 46)

① 🔧 45.0 (4.6, 33)

③ ❌

🔧 25.5 (2.6, 19)

🔧 45.0 (4.6, 33)

⑩ 🔧 45.0 (4.6, 33)

🔧 25.5 (2.6, 19)

❌ 🔧 63.0 (6.4, 46)

③ ❌

❌ 🔧 63.0 (6.4, 46)

❌ 🔧 30.5 (3.1, 22)

🔧 14.7 (1.5, 11)

❌ 🔧 30.5 (3.1, 22)

❌ 🔧 63.0 (6.4, 46)

🔧 5.8 (0.59, 51)

❌ : Always replace after every disassembly.

🔧 : N•m (kg-m, in-lb)

🔧 : N•m (kg-m, ft-lb)

1. Heated oxygen sensor 2 (bank 1)
2. Three way catalyst (right bank)
3. Gasket
4. heated oxygen sensor 1 (bank 1)
5. Exhaust manifold cover (right bank)
6. Exhaust manifold (right bank)
7. Exhaust manifold (left bank)
8. Exhaust manifold cover (left bank)
9. Three way catalyst (left bank)
10. heated oxygen sensor 1 (bank 2)
11. Heated oxygen sensor 2 (bank 2)

67162-FX35-G146

Exploded view of the exhaust manifold mounting and related components—VQ35DE Engine.

Right bank

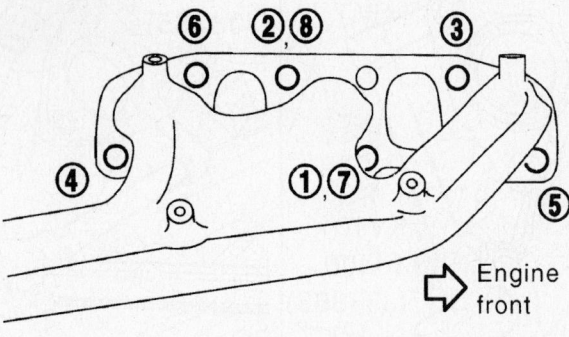

Left bank

Engine front

67162-FX35-G148

Exhaust manifold mounting bolt tightening sequence—VQ35DE Engine.

Right bank

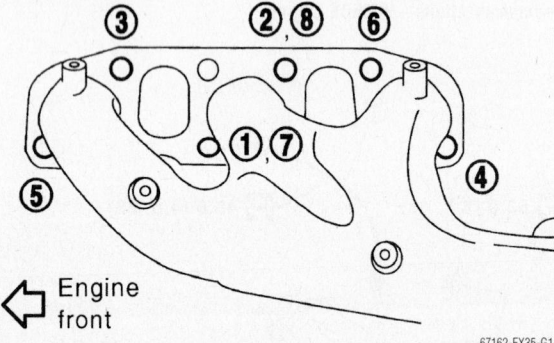

Engine front   Round press

Left bank   Round press

Engine front

67162-FX35-G149

Exhaust manifold gasket positioning—VQ35DE Engine.

4. Remove the front and rear engine undercover and front cross bar.

5. Disconnect the heated oxygen sensors 2 (bank 1 and bank 2) wiring harness connectors.

6. Using a heated oxygen sensor wrench , remove heated oxygen sensors 2 (bank 1 and bank 2).

❋❋ CAUTION

Be careful not to damage heated oxygen sensor. Discard any heated oxygen sensor which has been dropped from a height of more than 20 in. (0.5m) onto a hard surface such as a concrete floor; replace with a new sensor.

7. Remove the exhaust mounting bracket between the right/left catalytic converter and transmission.

8. Remove the three way catalyst (right and left bank).

9. Disconnect the heated oxygen sensor 1 (bank 1 and bank 2) wiring harness connectors and remove the wiring harness clip.

10. Using the heated oxygen sensor wrench , remove the heated oxygen sensor 1 (bank 1 and bank 2).

❋❋ CAUTION

Be careful not to damage heated oxygen sensor. Discard any heated oxygen sensor which has been dropped from a height of more than 20 in. (0.5m) onto a hard surface such as a concrete floor; replace with a new sensor.

11. Remove water pipes on both the right and left side.

12. Remove the exhaust manifold cover (right and left bank).

13. Loosen the mounting nuts in the reverse order shown in the illustration to remove the exhaust manifold.

➥Disregard Nos. 7 and No. 8 in removal.

14. Remove the exhaust manifold gaskets.

❋❋ CAUTION

Cover all engine openings to avoid entry of foreign materials.

15. Check the surface distortion of the exhaust manifold mating surface with a straightedge and feeler gauge. If it exceeds the limit, replace the exhaust manifold.

To install:

16. Install new exhaust manifold gaskets. On installation, locate the thick side of the port connecting part on the right-hand side (from your viewpoint). Locate the round press in the thick side of the port connecting part above the center level line of port.

17. Install the manifold and tighten the mounting nuts in the order shown. If the stud bolts were removed, install them and tighten them to the torque specified. Tighten nuts No. 1 and No. 2 in two steps. The numerical order No. 7 and No. 8 shows the second step.

18. Install the exhaust manifold cover (right and left bank).

19. Install the water pipes on both the right and left side.

20. Install the heated oxygen sensor 1 (bank 1 and bank 2).

21. Install the three way catalyst (right and left bank).

22. Install the exhaust mounting bracket between the right/left catalytic converter and transmission.

23. Install the heated oxygen sensors 2 (bank 1 and bank 2).

24. Reconnect the heated oxygen sensor wiring harness connectors.

25. Install the front and rear engine undercover and front cross bar.

26. Install the air cleaner case and air duct.

27. Install the engine cover.

VK45DE Engine

1. Before servicing the vehicle, refer to the precautions in the beginning of this section.

✳✳ WARNING

Perform the work, when the exhaust and cooling system have completely cooled down.

2. Remove the engine cover.

3. Remove the front and rear engine under covers.

4. Remove the air duct (inlet), air cleaner case and mass air flow sensor assembly, air duct and resonator assembly.

5. Remove the front cross bar. .

✳✳ CAUTION

Do not spill engine coolant on drive belts.

6. Drain the engine coolant from the radiator.

7. Remove the radiator.

8. Remove the drive belts.

9. Remove the heated oxygen sensors, as follows:

   a. Disconnect the wiring harness connector of each heated oxygen sensors.

   b. Remove the heated oxygen sensor 1 and 2 on both banks with a heated oxygen sensor wrench .

✳✳ CAUTION

Be careful not to damage heated oxygen sensor. Discard any heated oxygen sensor which has been dropped from a height of more than 20 in. (0.5m) onto a hard surface such as a concrete floor; replace with a new sensor.

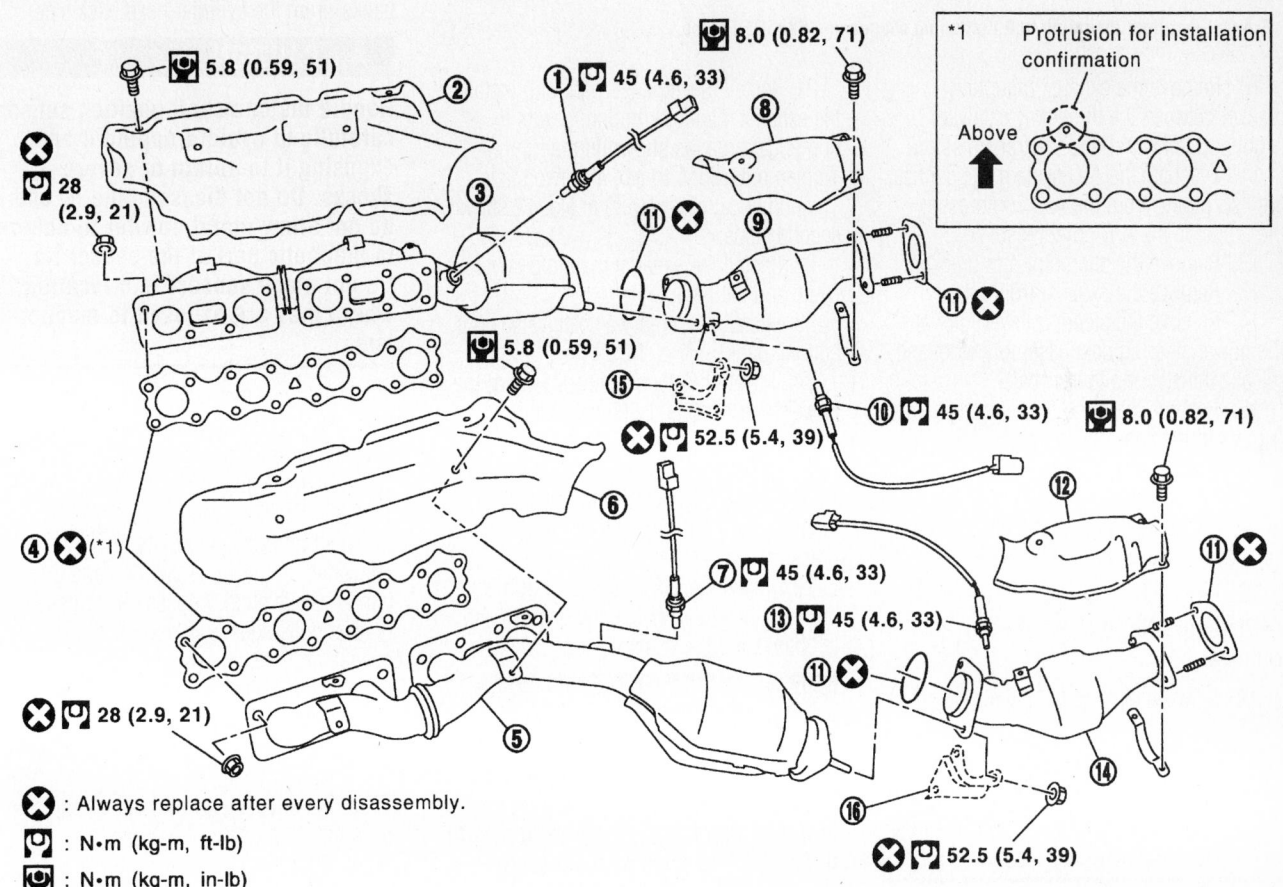

1. Heated oxygen sensor 1 (bank 2)
2. Exhaust manifold cover (right bank)
3. Exhaust manifold (right bank)
4. Gasket
5. Exhaust manifold (left bank)
6. Exhaust manifold cover (left bank)
7. Heated oxygen sensor 1 (bank 1)
8. Three way catalyst cover (right bank)
9. Three way catalyst (right bank)
10. Heated oxygen sensor 2 (bank 2)
11. Gasket
12. Three way catalyst cover (left bank)
13. Heated oxygen sensor 2 (bank 1)
14. Three way catalyst (left bank)
15. Mounting bracket
16. Mounting bracket

67162-FX35-G150

Exploded view of the exhaust manifold mounting and related components—VK45DE Engine.

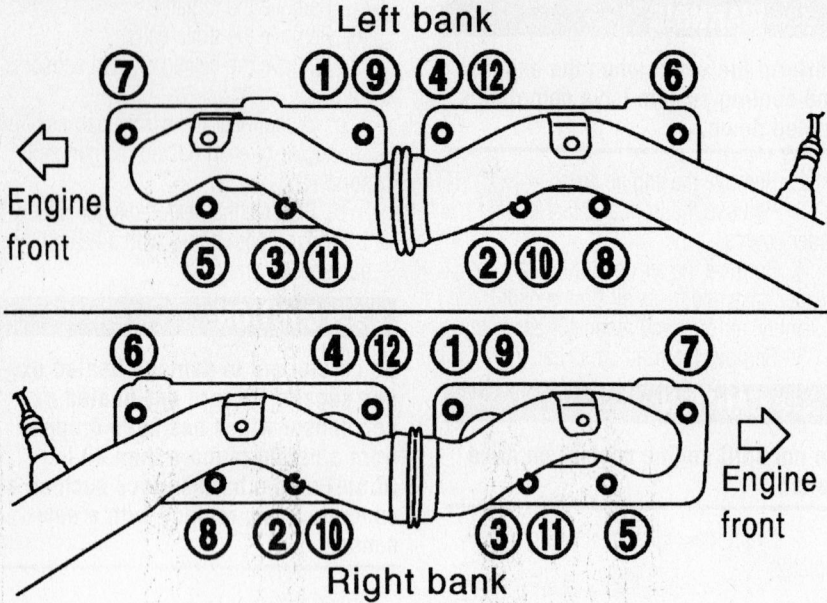

Left bank

Engine front

Right bank

Engine front

67162-FX35-G160

Exhaust manifold mounting bolt tightening sequence—VK45DE Engine.

10. Remove the exhaust mounting bracket between the three-way catalysts (right and left bank) and the transmission.

11. Evacuate the A/C system. Disconnect the A/C piping from the A/C compressor, then remove the A/C compressor.

12. Remove the alternator and bracket.

13. Remove the exhaust front tube.

14. Remove the steering lower joint at the power steering gear assembly side, and release the steering lower shaft.

15. Remove the three-way catalysts (right and left bank).

16. Remove the exhaust manifold covers. (right and left bank)

17. Loosen the nuts in the reverse order as shown in the figure to remove the exhaust manifold.

➡ **Disregard No. 9 to No. 12 in removal.**

18. Remove the exhaust manifold gaskets.

✲✲ CAUTION

Cover engine openings to avoid entry of foreign materials.

19. Check the surface distortion of each exhaust manifold flange mating surface with a straightedge and feeler gauge. If it exceeds the limit, replace exhaust manifold.

To install:

20. Install new exhaust manifold gaskets. Install each exhaust manifold gasket with its directional protrusion set upward. Refer to the illustration.

21. Install the exhaust manifold. Install the exhaust manifold mounting nuts in numerical order as shown in the figure. Tighten nuts No. 1 to No. 4 in two steps. The numerical order No. 9 to No. 12 shown second steps.

22. Install the exhaust manifold covers. (right and left bank)

23. Install the three-way catalysts (right and left bank).

24. Install the steering lower joint at the power steering gear assembly side.

25. Install the exhaust front tube.

26. Install the alternator and bracket.

27. Install the A/C compressor, and reconnect the A/C piping to the A/C compressor.

28. Install the exhaust mounting bracket between the three-way catalysts (right and left bank) and the transmission.

✲✲ CAUTION

Be careful not to damage heated oxygen sensor. Discard any heated oxygen sensor which has been dropped from a height of more than 20 in. (0.5m) onto a hard surface such as a concrete floor; replace with a new sensor.

29. Install the heated oxygen sensors, as follows:

30. Install the drive belts.

31. Install the radiator.

32. Install the front cross bar.

33. Install the air duct (inlet), air cleaner case and mass air flow sensor assembly, air duct and resonator assembly.

34. Install the front and rear engine under covers.

35. Install the engine cover.

36. Fill the engine cooling system with coolant.

37. Recharge the A/C system.

Camshaft and Valve Lifters

REMOVAL & INSTALLATION

VQ35DE Engine

1. Before servicing the vehicle, refer to the precautions in the beginning of this section.

2. Remove the front timing chain case, camshaft sprocket, timing chain and rear timing chain case.

3. If necessary, remove the camshaft position sensor (PHASE) (right and left banks) from the cylinder head back side.

✲✲ CAUTION

Handle the camshaft position sensor carefully to avoid dropping it and exposing it to abrupt or severe shocks. Do not disassemble it, and do not allow metal powder to adhere to magnetic part at the sensor tip. Do not place sensors in a location where they are exposed to magnetism.

4. Remove the intake valve timing control solenoid valve from the No.1 camshaft bracket.

5. Remove the intake and exhaust camshaft brackets. Mark the camshafts, camshaft brackets, and bolts so they are reinstalled in the same position and direction for installation. Loosen the camshaft bracket bolts equally in several steps in the reverse order as shown.

6. Remove the camshaft.

7. Remove the valve lifters. Identify the installation positions, and store them without mixing them up.

8. Remove the secondary timing chain tensioner from the cylinder head.

9. Remove the chain tensioner with its stopper pin attached.

➡ **The stopper pin was attached when the secondary timing chain was removed.**

10. Inspect camshaft run out, as follows:
   a. Put a V block on a precise flat bed, and support the No. 2 and No. 4 journals of the camshaft.

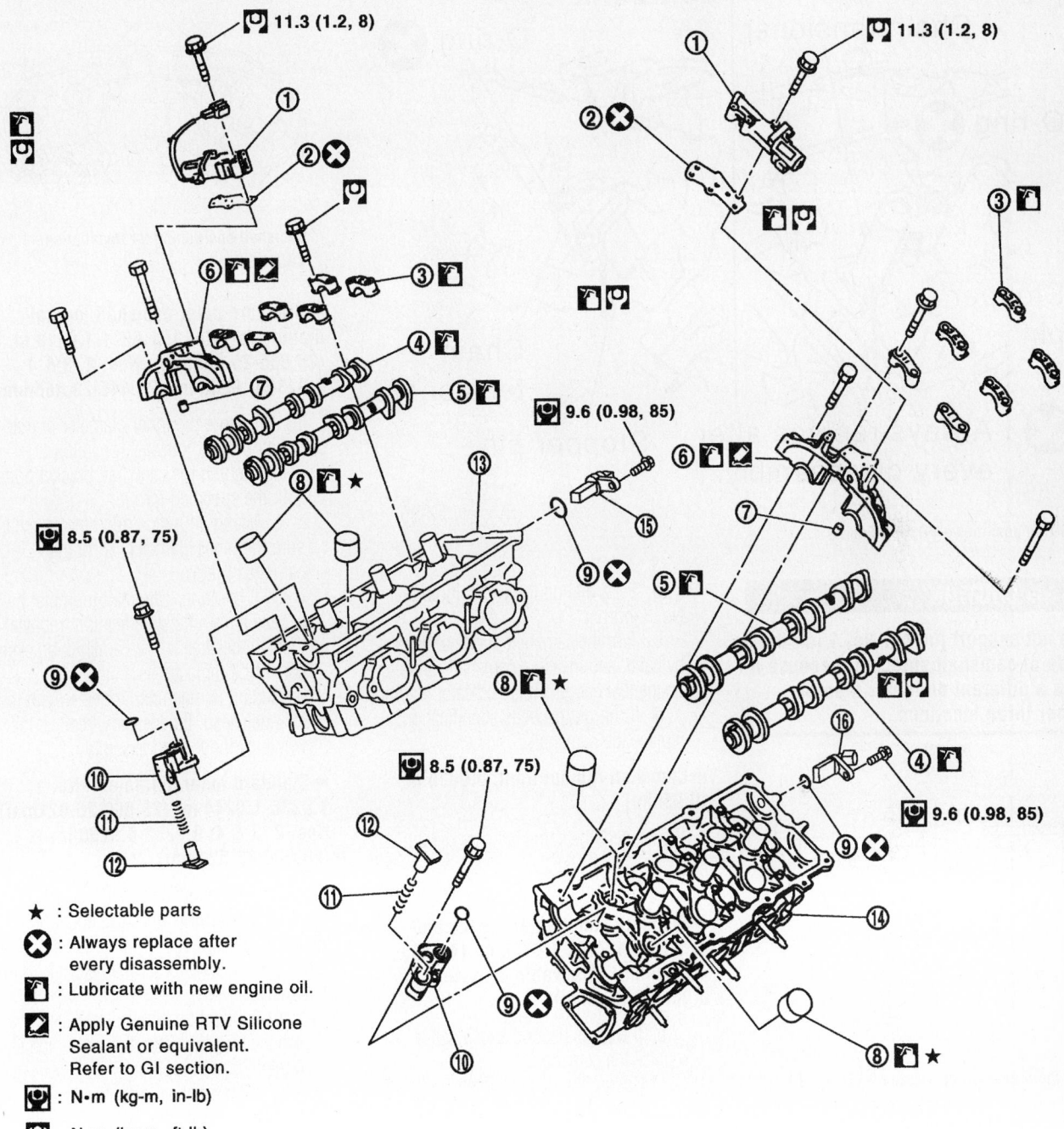

★ : Selectable parts

✖ : Always replace after
every disassembly.

🔧 : Lubricate with new engine oil.

✎ : Apply Genuine RTV Silicone
Sealant or equivalent.
Refer to GI section.

🔧 : N•m (kg-m, in-lb)

🔧 : N•m (kg-m, ft-lb)

1.	Intake valve timing control solenoid valve	2.	Gasket	3.	Camshaft bracket (No. 2 to No. 4)
4.	Camshaft (EXH)	5.	Camshaft (INT)	6.	Camshaft bracket (No. 1)
7.	Dowel pin	8.	Valve lifter	9.	O-ring
10.	Chain tensioner	11.	Spring	12.	Plunger
13.	Cylinder head (right bank)	14.	Cylinder head (left bank)	15.	Camshaft position sensor (PHASE) (right bank)
16.	Camshaft position sensor (PHASE) (left bank)				

Exploded view of camshaft and related components—VQ35DE Engine.

67162-FX35-G11

Right bank

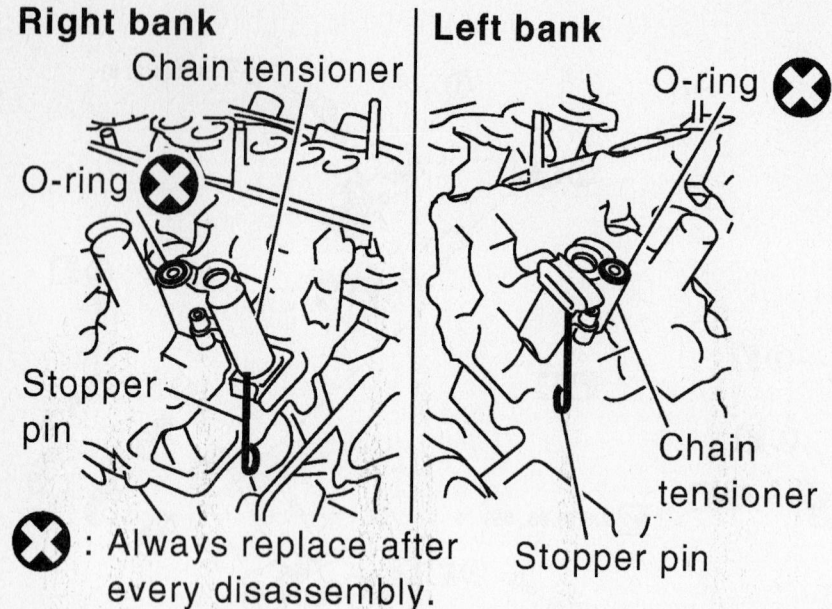

Chain tensioner

O-ring ⊗

Stopper pin

Left bank

O-ring ⊗

Chain tensioner

Stopper pin

⊗ : Always replace after every disassembly.

O-ring positions—VQ35DE Engine.

67162-FX35-G12

❊❊ CAUTION

Do not support journal No. 1 (on the side of camshaft sprocket) because it has a different diameter than the other three locations.

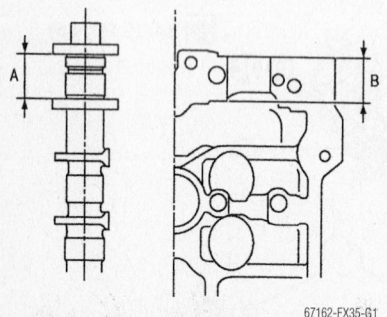

67162-FX35-G1

If the measurements A and B are out of range, replace the cylinder head or camshaft—VQ35DE Engine.

b. Set a dial gauge vertically to the No. 3 journal.

c. Turn the camshaft in one direction by hand, and measure camshaft run out on the dial gauge. (Total indicator reading)

d. If run out exceeds specification, replace the camshaft.

➡**Camshaft run out limit: 0.0020 in (0.05mm)**

11. Inspect the camshaft cam height, as follows:

a. Measure the camshaft cam height.

➡**Standard cam height (intake and exhaust): 1.7663-1.7738 in. (44.865-45.055mm). Allowable cam wear limit: 0.008 in. (0.2mm).**

b. If wear is beyond specification, replace the camshaft.

12. Measure the outer diameter of the camshaft journals.

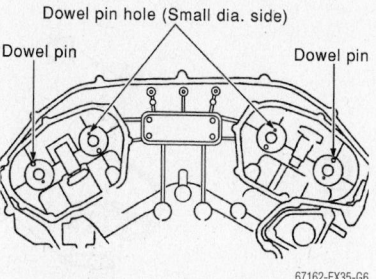

Dowel pin hole (Small dia. side)

Dowel pin   Dowel pin

67162-FX35-G6

Camshaft positioning for installation—VQ35DE Engine.

➡**Standard outer camshaft journal diameters—No. 1: 1.0211-1.0218 in. (25.935-25.955mm), Nos. 2, 3 & 4: 0.9230-0.9238 in. (23.445-23.465mm)**

13. Measure the inner diameter of the camshaft bracket journals, as follows:

a. Tighten the camshaft bracket bolts with the specified torque.

b. Using an inside micrometer, measure the inner diameter "A" of the camshaft bracket.

c. If the inner camshaft journals exceed allowable specification, replace the cylinder head. The camshaft brackets cannot be replaced as a single part, because it is machined together with the cylinder head. If necessary, replace the whole cylinder head assembly.

➡**Standard inner diameter—No. 1: 1.0236-1.0244 in. (26.000-26.021mm), Nos. 2, 3 & 4: 0.9252-0.9260 in. (23.500-23.521mm)**

14. Calculate the camshaft journal oil clearance, using the following equation: journal oil clearance = (inner diameter of camshaft bracket)−(outer diameter of camshaft journal).

a. Subtract the inner diameter of the camshaft bracket from the outer diameter of the camshaft journal.

b. If the oil clearance exceeds the allowable limit, replace either the cylin-

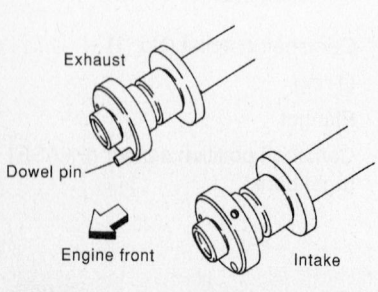

Exhaust

Dowel pin

Engine front   Intake

67162-FX35-G4

Exhaust and intake camshaft differences—VQ35DE Engine.

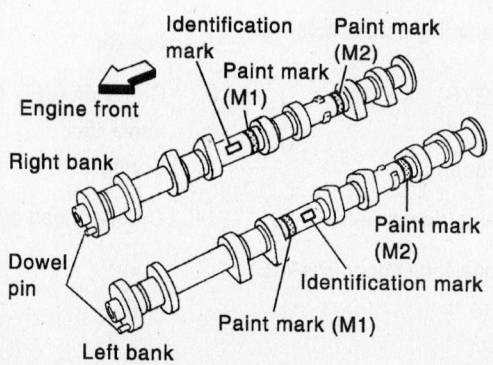

Identification mark   Paint mark (M2)

Paint mark (M1)

Engine front

Right bank

Dowel pin

Paint mark (M2)

Identification mark

Paint mark (M1)

Left bank

67162-FX35-G5

Camshaft identification mark locations—VQ35DE Engine.

Right camshaft brackets

Exhaust side

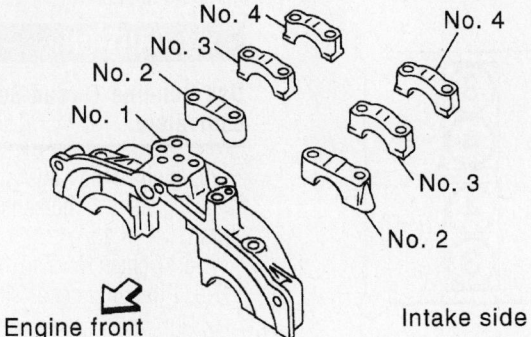

Left camshaft brackets

Intake side

Camshaft bracket positions—VQ35DE Engine.

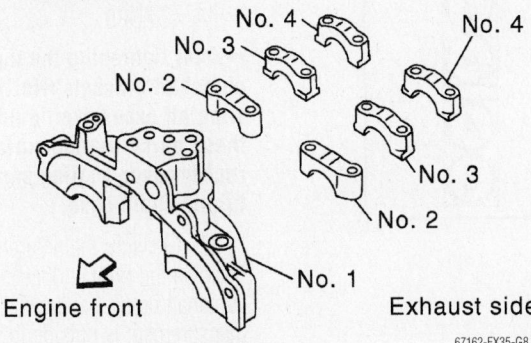

Camshaft bracket identification mark positions—VQ35DE Engine.

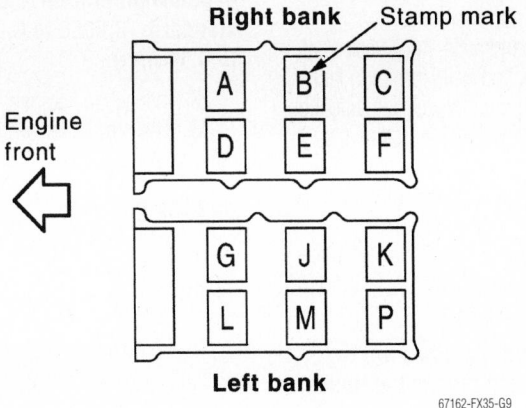

* : Remove the protruding sealant
from front face. (Remove the hardended
sealant from surface only.)

67162-FX35-G10

Liquid gasket positioning for camshaft bracket No. 1 installation—VQ35DE Engine

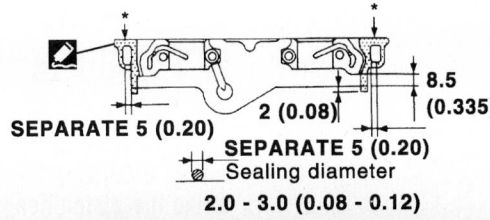

der head or camshaft or both, if necessary.

➡ **Standard oil clearance—No. 1: 0.0018-0.0034 in. (0.045-0.086mm), Nos. 2, 3 & 4: 0.0014-0.0030 in. (0.035-0.076mm), Maximum limit: 0.0059 in. (0.15mm)**

15. Inspect camshaft end-play, as follows:

a. Install a dial indicator in the thrust direction on the front end of the camshaft.

b. Measure the end play on the dial indicator when the camshaft is moved forward and backward (in direction to axis).

➡ **Camshaft end-play standard: 0.0045-0.0074 in. (0.115-0.188mm), limit: 0.0094 in. (0.24mm)**

c. Measure the following parts if the camshaft end-play exceeds specifications, and replace the camshaft and/or cylinder if the following measurements are out of specification:

- Dimension "A" for camshaft No. 1 journal—Standard: (1.0827-1.0846 in. (27.500-27.548mm)
- Dimension "B" for cylinder head No. 1 journal—Standard: 1.0772-1.0781 in. (27.360-27.385mm)

16. Inspect camshaft sprocket run out, as follows:

a. Put a V-block on precise flat table, and support the No. 2 and No. 4 journals of the camshaft.

➡ **Do not support journal No. 1 (on the side of camshaft sprocket) because it has a different diameter than the other three journals.**

b. Measure camshaft sprocket run out with a dial indicator. (Total indicator reading), and replace the camshaft sprocket it the run out measurement exceeds specification.

➡ **Camshaft sprocket run out limit : 0.0059 in. (0.15mm)**

17. Inspect the valve lifters, as follows:

a. Check the surface of valve lifter for wear or cracks. If any defects are found, replace the valve lifter.

18. Inspect valve lifter clearance, as follows:

a. Measure the outer diameter of each valve lifter.

➡ **Valve lifter outer diameter (Intake and exhaust): 1.3377-1.3381 in. (33.977-33.987mm)**

Right bank

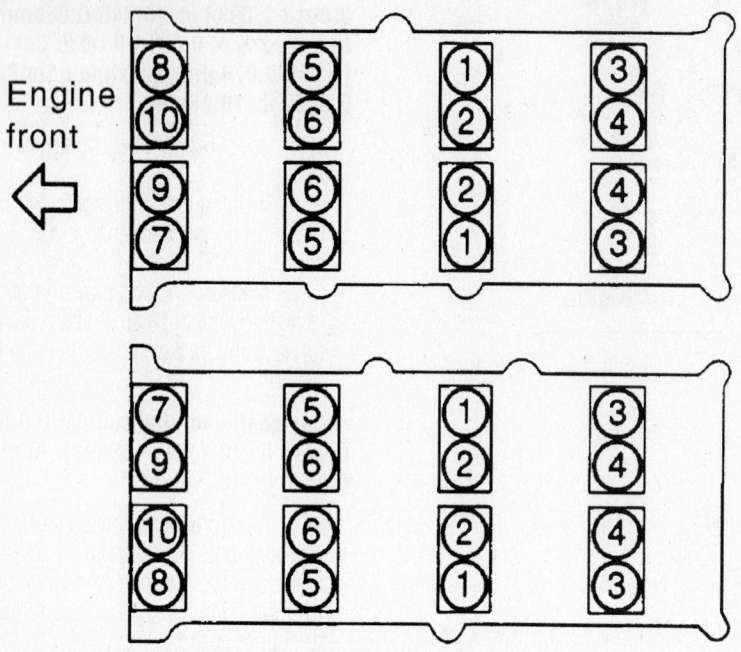

Engine front ⟵

Left bank

67162-FX35-G7

Camshaft bearing cap mounting bolt tightening sequence—VQ35DE Engine.

b. Using an inside micrometer, measure the diameter of valve lifter hole of the cylinder head.

➡**Valve lifter bore standard (Intake and exhaust): 1.3386—1.3392 in. (34.000-34.016mm)**

c. Calculate the valve lifter clearance by subtracting the valve lifter outer diameter from the valve lifter hole inner diameter. If result exceeds the standard, referring to each standard of valve lifter outer diameter and valve lifter hole diameter, replace either or both valve lifter and cylinder head.

➡**Valve lifter clearance standard (Intake and exhaust): 0.0005-0.0015 in. (0.013-0.039mm)**

To install:

19. Install the secondary chain tensioners on both sides of the cylinder head. Install the chain tensioner with its stopper pin attached. Install the tensioner with the sliding part facing downward on the right-hand cylinder head, and with the sliding part facing upward on the left-hand cylinder head.
20. Install new O-rings as shown.
21. Install the valve lifters. If using the

original valve lifters, make sure to install them in their original positions.

22. Install the camshafts. Install camshaft with dowel pin attached to its front end face on the exhaust side. Follow your identification marks made during removal, or follow the identification marks that are present on new camshafts for proper placement and direction. Install the camshaft so that the dowel pin hole and dowel pin on the front end face are positioned as shown in figure (No. 1 cylinder TDC on its compression stroke).

➡**Large and small pin holes are located on front end face of the intake camshaft, at intervals of 180 degrees. Position the small diameter side pin hole upward (in the cylinder head upper face direction).**

23. Install the camshaft brackets. Remove all foreign material completely from the camshaft bracket backside and from the cylinder head installation face. Install the camshaft brackets in their original positions and directions as shown in illustration. Install the No. 2 to 4 camshaft brackets aligning stamp marks as shown.

➡**There are no identification marks indicating left and right for No. 1 camshaft bracket.**

24. Apply sealant to the mating surface of No. 1 camshaft bracket (as shown) on the right and left banks.

※※ CAUTION

Use Genuine Thread Sealant or equivalent.

25. Tighten the camshaft brackets in the following steps, in numerical order as shown:
   a. Tighten No. 7 to 10, then tighten No. 1 to 6 in order as shown to 1 ft. lbs. (2 Nm)
   b. Tighten all bolts in numerical order to 4 ft. lbs. (6 Nm).
   c. Tighten all bolts in order to 8 ft. lbs. (10 Nm).

➡**After tightening the mounting bolts of camshaft brackets (No. 1), be sure to wipe off excessive liquid gasket from these parts: mating surface of the rocker cover, mating surface of the rear timing chain case**

26. Measure the difference in levels between the front end faces of the No. 1 camshaft bracket and cylinder head. If the measurement is outside the specified range, re-install the camshaft and camshaft brackets.

➡**Camshaft bracket and cylinder head standard: -0.0055 to 0.0055 in. (-0.14 to 0.14 mm)**

27. Inspect and adjust the valve clearance.
28. The remainder of installation is the reverse of removal.

VK45DE Engine

1. Before servicing the vehicle, refer to the precautions in the beginning of this section.
2. Remove the engine assembly from vehicle.
3. Remove the timing chain.
4. With the hexagonal part of the camshaft locked with a wrench, loosen the bolts securing the camshaft sprocket, then remove the camshaft sprocket.

※※ CAUTION

After removing timing chain, do not turn the crankshaft or camshaft separately, otherwise the valves will strike the piston head.

5. Remove the intake and exhaust camshaft brackets. Mark the camshafts, camshaft brackets and bolts so that they can be installed in their original positions and

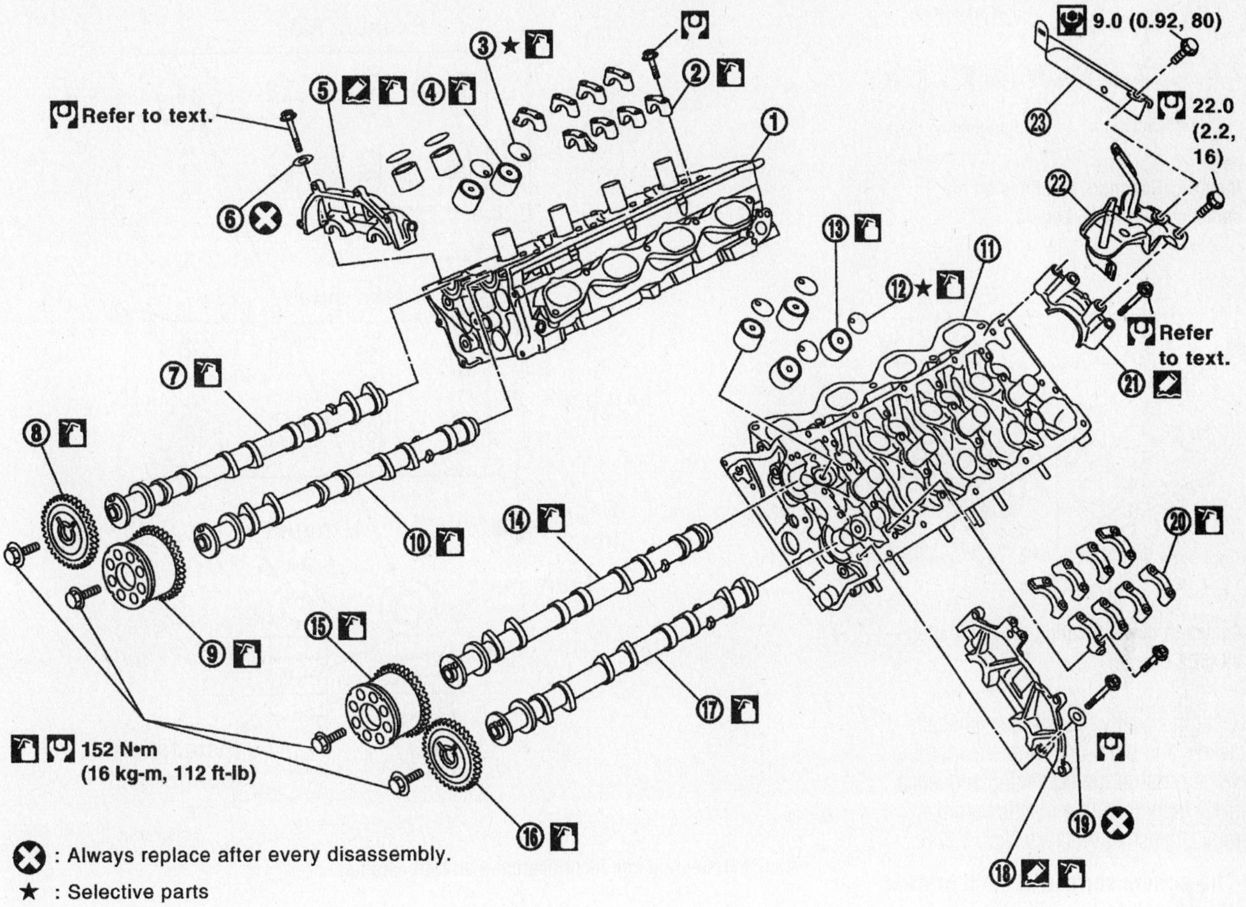

9.0 (0.92, 80)

22.0 (2.2, 16)

Refer to text.

Refer to text.

152 N•m
(16 kg-m, 112 ft-lb)

⊗ : Always replace after every disassembly.

★ : Selective parts

✐ : Apply Genuine RTV Silicone Sealant or equivalent.
Refer to GI section.

☖ : Lubricate with new engine oil.

1. Cylinder head (right bank)	2. Camshaft bracket (No. 2 to No. 5)	3. Adjusting shim
4. Valve lifter	5. Camshaft bracket (No. 1)	6. Seal washer
7. Camshaft (EXH)	8. Camshaft sprocket (EXH)	9. Camshaft sprocket (INT)
10. Camshaft (INT)	11. Cylinder head (left bank)	12. Adjusting shim
13. Valve lifter	14. Camshaft (INT)	15. Camshaft sprocket (INT)
16. Camshaft sprocket (EXH)	17. Camshaft (EXH)	18. Camshaft bracket (No. 1)
19. Seal washer	20. Camshaft bracket (No. 2 to No. 5)	21. Camshaft bracket (No. 6)
22. Harness bracket	23. Harness bracket	

67162-FX35-G13

Exploded view of the camshaft and related components—VK45DE Engine.

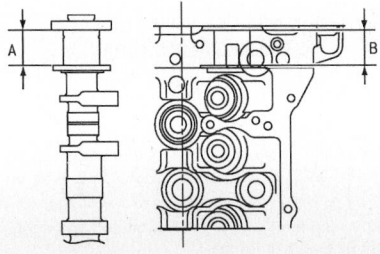

67162-FX35-G16

If the following measurements are out of
specifications, replace the camshaft
and/or cylinder head—VK45DE Engine.

Bank	INT/EXH	Identification rib	Paint marks		Identification mark
			M1	M2	
RH	INT	Yes	Blue	No	RH
	EXH	Yes	No	Orange	RH
LH	INT	No	Blue	No	LH
	EXH	No	No	Orange	LH

67162-FX35-G21

New camshaft identification marks—VK45DE Engine.

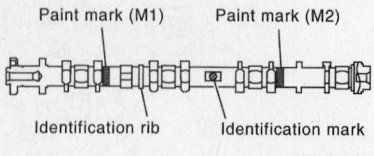

Identification mark locations on the camshaft—VK45DE Engine.

67162-FX35-G18

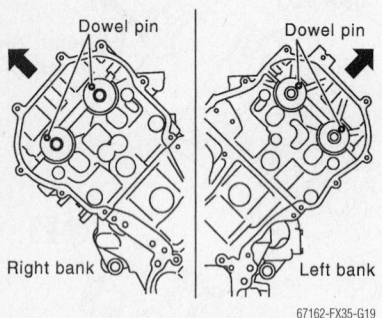

Camshaft dowel location for installation—VK45DE Engine.

67162-FX35-G19

direction. Equally loosen the camshaft brackets and bolts in several steps in the reverse order of the tightening sequence. Lightly tapping with a plastic hammer, remove camshaft brackets Nos. 1 & 6.

➡ **The bottom surface of each bracket will be stuck to the cylinder head because of liquid gasket.**

6. Remove the camshaft.

7. Remove the adjusting shims and valve lifters, if necessary.

8. Identify installation positions, and store all parts without mixing them up.

9. Inspect the camshaft run out, as follows:

a. Situate a V-block on a precise flat table, and support the camshaft by journals No. 2 and No. 5.

➡ **Do not support the camshaft by journal No. 1 (on the side of camshaft sprocket) since it has a different diameter from the other three locations.**

b. Set the dial indicator vertically to measure against journal No. 3.

c. Turn the camshaft in one direction by hand, and measure camshaft the run out on the dial indicator.

d. If run out exceeds the limit, replace the camshaft.

➡ **Camshaft run out limit: 0.0008 in. (0.02mm).**

10. Inspect the camshaft cam height, as follows:

a. Measure the camshaft cam height.

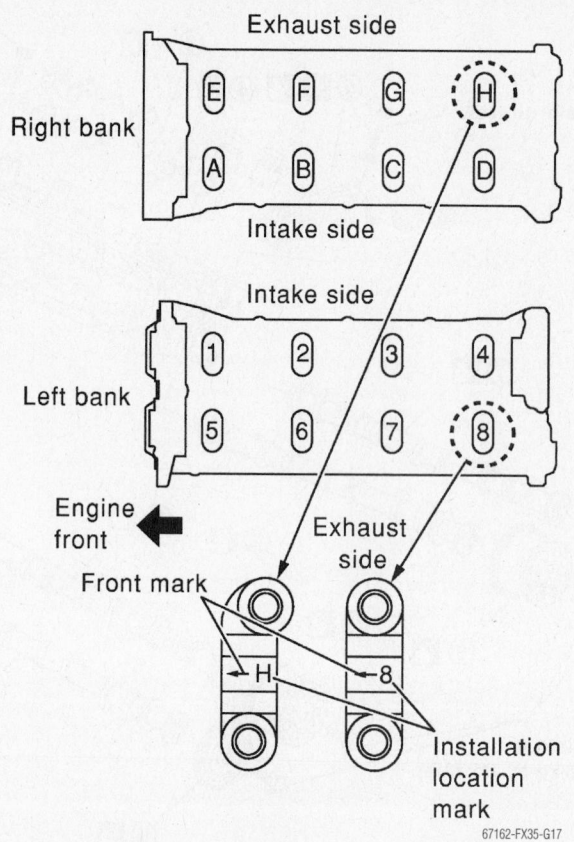

Camshaft bearing cap identification—VK45DE Engine.

67162-FX35-G17

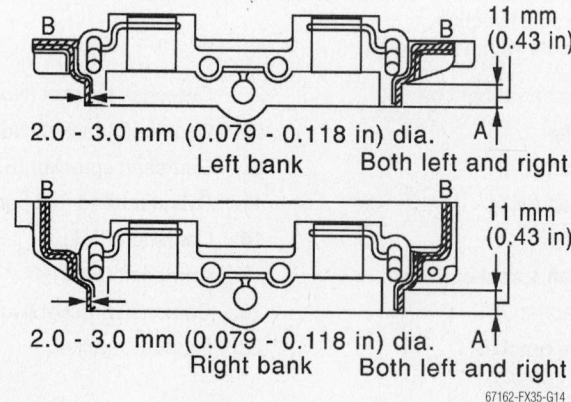

Liquid gasket positioning for No. 1 camshaft bracket installation—VK45DE Engine.

67162-FX35-G14

➡ **Standard cam height—intake: 1.7663-1.7738 in. (44.865-45.055mm), exhaust: 1.7293—1.7368 in. (43.925-44.115mm). Allowable cam wear limit: 0.008 in. (0.2mm).**

b. If wear is beyond specification, replace the camshaft.

11. Measure the outer diameter of the camshaft journals with a micrometer.

➡ **Standard outer camshaft journal diameters—No. 1: 1.0212-1.0218 in (25.938-25.955mm), Nos. 2, 3 & 4:**

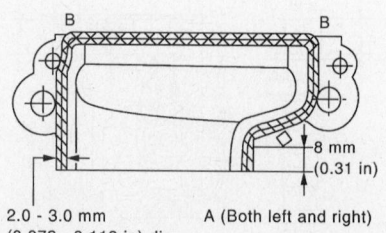

Liquid gasket positioning for No. 6 camshaft bracket installation—VK45DE Engine.

67162-FX35-G15

Right bank

Exhaust

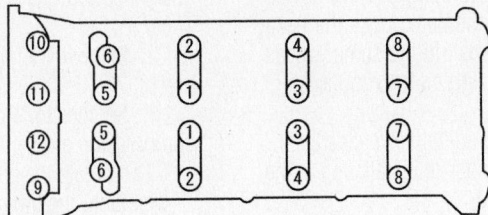

Engine front ← ——— Intake ———————

Left bank

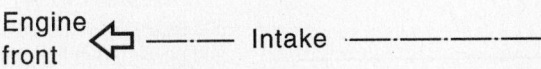

Exhaust

67162-FX35-G20

Tightening sequence for camshaft bearing caps—VK45DE Engine.

1.0218—1.0224 in. (25.953-25.970mm)

12. Measure the inner diameter of the camshaft bracket journals, as follows:

a. Tighten the camshaft bracket bolts with the specified torque.

b. Using an inside micrometer, measure the inner diameter of the camshaft bracket.

c. If the inner camshaft journals exceed allowable specification, replace the cylinder head. The camshaft brackets cannot be replaced as a single part, because it is machined together with the cylinder head. If necessary, replace the whole cylinder head assembly.

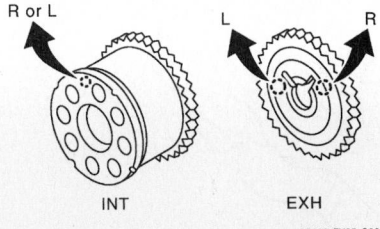

INT   EXH

67162-FX35-G22

Intake and Exhaust camshaft sprocket identification—VK45DE Engine.

➡**Standard inner diameter—1.0236-1.0244 in. (26.000-26.021mm)**

13. Calculate the camshaft journal oil clearance, using the following equation: journal oil clearance = (inner diameter of camshaft bracket)–(outer diameter of camshaft journal).

a. Subtract the inner diameter of the camshaft bracket from the outer diameter of the camshaft journal.

b. If the oil clearance exceeds the allowable limit, replace either the cylinder head or camshaft or both, if necessary.

➡**Standard oil clearance—No. 1: 0.0018-0.0033 in. (0.045-0.083mm), Nos. 2, 3 & 4: 0.0012-0.0027 in. (0.030-0.068mm)**

14. Inspect camshaft end-play, as follows:

a. Install a dial indicator in the thrust direction on the front end of the camshaft.

b. Measure the end play on the dial indicator when the camshaft is moved forward and backward (in direction to axis).

➡**Camshaft end-play standard: 0.0045-0.0074 in. (0.115-0.188mm).**

c. Measure the following parts if the camshaft end-play exceeds specifications, and replace the camshaft and/or cylinder if the following measurements are out of specification:

• Dimension "A" for camshaft No. 1 journal—1.2008-1.2027 in. (30.500-30.548mm)
• Dimension "B" for cylinder head No. 1 journal—1.1953-1.1963 in. (30.360-30.385mm)

15. Inspect camshaft sprocket run out, as follows:

a. Put a V-block on precise flat table, and support the No. 2 and No. 5 journals of the camshaft.

➡**Do not support journal No. 1 (on the side of camshaft sprocket) because it has a different diameter than the other three journals.**

b. Measure camshaft sprocket run out with a dial indicator. (Total indicator reading), and replace the camshaft sprocket it the run out measurement exceeds specification.

➡**Camshaft sprocket run out limit : 0.0059 in. (0.15mm)**

16. Inspect the valve lifters, as follows:

a. Check the surface of valve lifter for wear or cracks. If any defects are found, replace the valve lifter.

17. Inspect valve lifter clearance, as follows:

a. Measure the outer diameter of each valve lifter.

➡**Valve lifter outer diameter (Intake and exhaust): 1.3372-1.3376 in. (33.965-33.975mm)**

b. Using an inside micrometer, measure the diameter of valve lifter hole of the cylinder head.

➡**Valve lifter bore standard (Intake and exhaust): 1.3386—1.3392 in. (34.000-34.016mm)**

c. Calculate the valve lifter clearance by subtracting the valve lifter outer diameter from the valve lifter hole inner diameter. If the result exceeds the standard, referring to each valve lifter outer diameter and valve lifter hole diameter, replace either or both valve lifter and cylinder head.

➡**Valve lifter clearance standard (Intake and exhaust): 0.0010-0.0020 in. (0.025-0.51mm)**

To install:

18. Install valve lifters and adjusting shims if removed. Install them in their original positions.

19. Install the camshafts. Follow your identification marks made during removal, or follow the identification marks that are present on new camshafts for proper placement and direction. Install the camshafts so that the dowel pins on the front end faces are positioned as shown (No. 1 cylinder TDC on its compression stroke).

20. Install the camshaft brackets. Remove all foreign material completely from the camshaft bracket backsides and from the cylinder head installation faces. Install by referring to the installation location mark on the upper surface and front mark. Install so that the installation location mark can be correctly read when viewed from the side of the left exhaust bank.

21. Apply liquid gasket to the mating surface of the camshaft bracket (No. 1) as shown.

➡ **Use Genuine RTV Silicone Sealant or equivalent.**

✳✳ CAUTION

After installation, be sure to wipe off any excessive liquid gasket leaking from parts "A" and "B" (both on right and left sides). Remove completely any excess of liquid gasket inside the bracket.

22. Apply liquid gasket to the mating surface of camshaft bracket (No. 6) on left bank intake as shown.

23. Tighten the camshaft bracket bolts in the following steps, in the order shown.
   a. Tighten No. 9 to 12 to 1 ft. lbs. (2 Nm).
   b. Tighten No. 1 to 8 to 1 ft. lbs. (2 Nm).
   c. Tighten No. 13 to 14 to 1 ft. lbs. (2 Nm). (Left bank only)
   d. Tighten all bolts to 4 ft. lbs. (6 Nm).
   e. Tighten No. 1 to 12 to 8 ft. lbs. (10.5 Nm).
   f. Tighten No. 13 to 14 to 23 ft. lbs. (31 Nm). (Left bank only)

✳✳ CAUTION

After tightening mounting bolts of camshaft brackets, be sure to wipe off excessive liquid gasket from these parts: mating surface of the rocker cover, mating surface of the front cover

24. Install the camshaft sprockets. Make sure to position the sprockets as shown. Install the camshaft sprocket (EXH) by selectively using the groove of dowel pin according to the bank. Lock the hexagonal part of camshaft in the same way as for removal, and tighten the mounting bolts.

25. Check and adjust valve clearance.

26. The remainder of installation is the reverse of removal.

Valve Lash

ADJUSTMENT

VQ35DE Engine

1. Before servicing the vehicle, refer to the precautions in the beginning of this section.

Perform inspection after removal, installation or replacement of camshaft or valve-related parts, or if there is unusual engine conditions regarding valve clearance.

2. Remove the right and left rocker covers.

3. Set the No. 1 cylinder at TDC of its compression stroke, as follows:
   a. Rotate the crankshaft pulley clockwise until the timing mark (grooved line without color) is aligned with the timing indicator.
   b. Make sure the No. 1 cylinder intake and exhaust cam noses are facing inward and upward from the cylinder head, as shown. If they are not positioned as shown, rotate the crankshaft pulley 360 degrees clockwise (when viewed from the front).

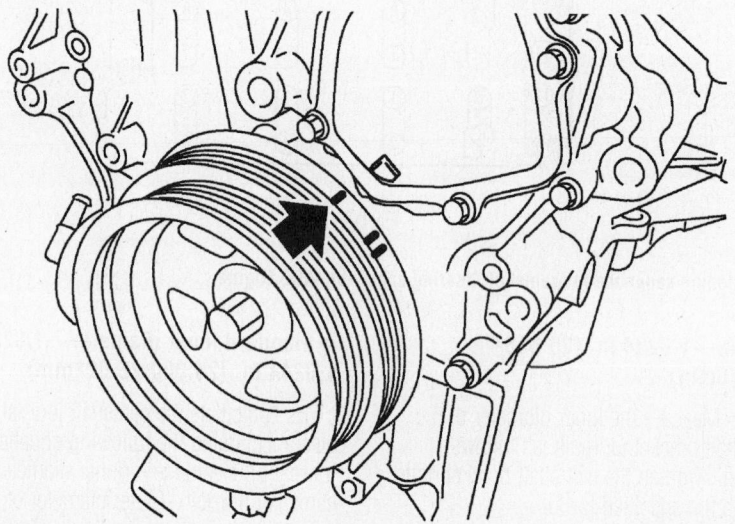

Position the crankshaft at TDC No. 1—VQ35DE Engine.

67162-FX35-G23

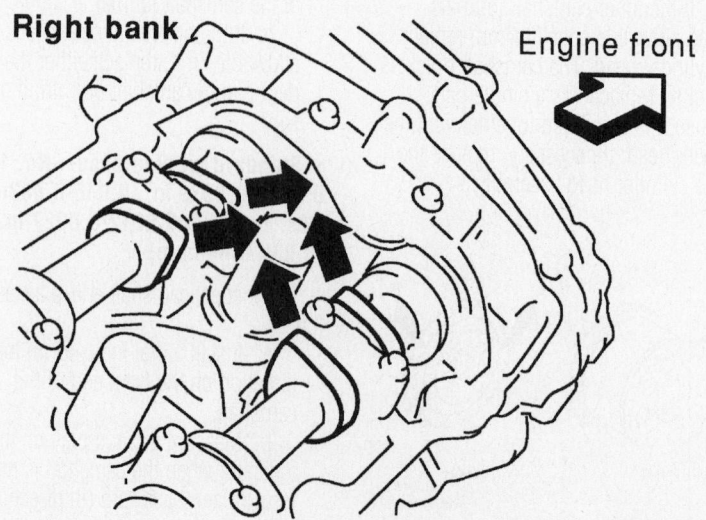

When at TDC No. 1 cylinder, the camshaft lobes should point as shown—VQ35DE Engine.

67162-FX35-G24

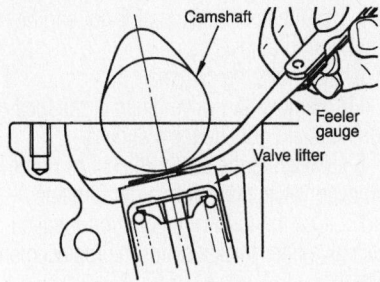

Measure the valve lash with the camshaft lobe positioned as shown—VQ35DE Engine.

4. Using a feeler gauge, measure the valve clearance for the following cylinders:
- Cylinder 1 Intake valves
- Cylinder 2 Exhaust valves
- Cylinder 3 Exhaust valves
- Cylinder 6 Intake valves

➡**Valve clearance standard—Cold Intake: 0.010-0.013 in. (0.26-0.34mm), Cold Exhaust: 0.011-0.015 in. (0.29-0.37mm), Hot* Intake: 0.012-0.016 in. (0.304-0.416mm), Hot* Exhaust: 0.012-0.016 in. (0.308-0.432mm) *Approximately 176 degrees F (80 degrees C)**

5. Rotate the crankshaft by 240 degrees clockwise (when viewed from front) to align the No. 3 cylinder at TDC of its compression stroke.

➡**Crankshaft pulley mounting bolt flange has a stamped line every 60 degrees. They can be used as a guide to rotation angle.**

6. Using a feeler gauge, measure the valve clearance for the following cylinders:
- Cylinder 2 Intake valves
- Cylinder 3 Intake valves
- Cylinder 4 Exhaust valves
- Cylinder 5 Exhaust valves

➡**Valve clearance standard—Cold Intake: 0.010-0.013 in. (0.26-0.34mm), Cold Exhaust: 0.011-0.015 in. (0.29-0.37mm), Hot* Intake: 0.012-0.016 in. (0.304-0.416mm), Hot* Exhaust: 0.012-0.016 in. (0.308-0.432mm) *Approximately 176 degrees F (80 degrees C)**

7. Rotate the crankshaft by 240 degrees clockwise (when viewed from front) to align the No. 5 cylinder at TDC of its compression stroke.

➡**Crankshaft pulley mounting bolt flange has a stamped line every 60 degrees. They can be used as a guide to rotation angle.**

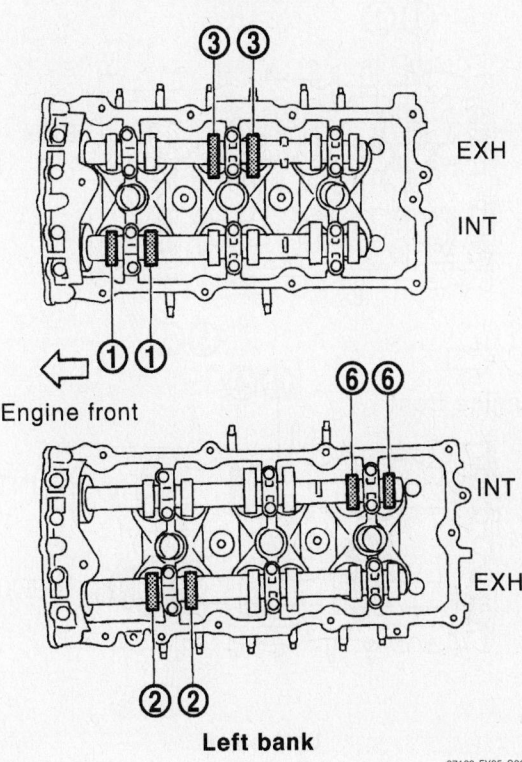

Measure the valve lash for the valves shown with the engine at No. 1 cylinder TDC—VQ35DE Engine

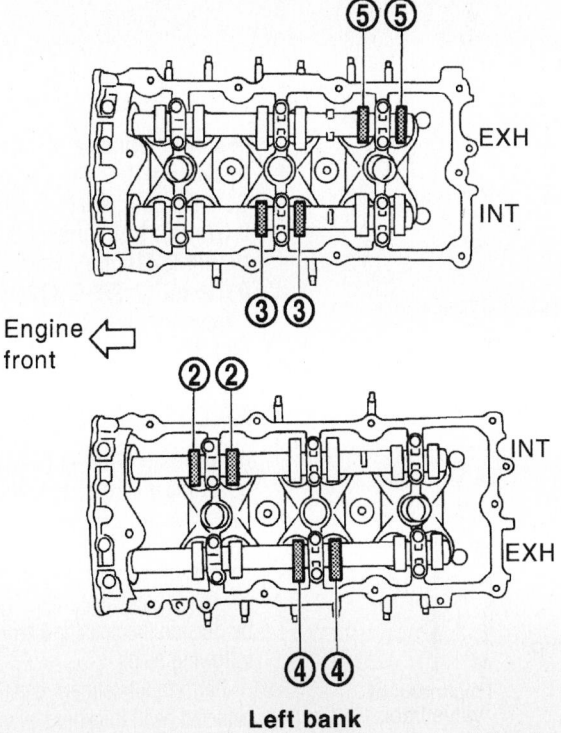

Measure the valve lash for the valves shown with the engine at No. 3 cylinder TDC—VQ35DE Engine

Right bank

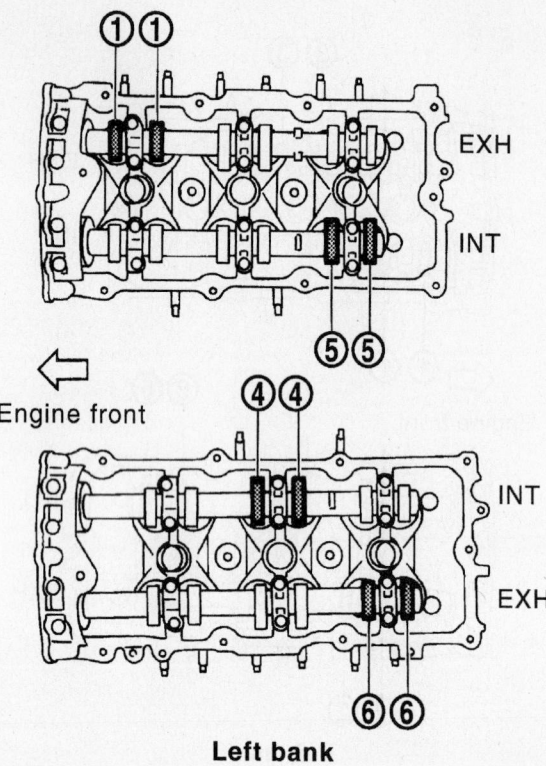

Engine front

Left bank

67162-FX35-G29

Measure the valve lash for the valves shown with the engine at No. 5 cylinder TDC—VQ35DE Engine

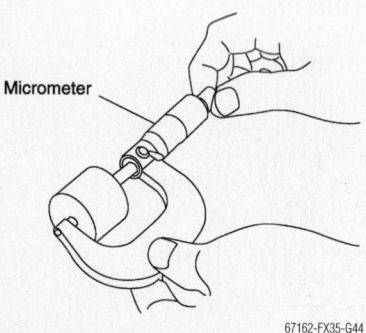

Micrometer

67162-FX35-G44

Measure the valve lifter height as shown—VQ35DE Engine.

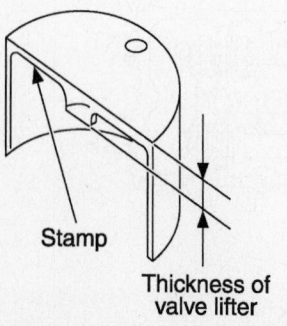

Stamp

Thickness of valve lifter

67162-FX35-G45

Valve lifter identification stamp location— VQ35DE Engine.

8. Using a feeler gauge, measure the valve clearance for the following cylinders:
- Cylinder 1 Exhaust valves
- Cylinder 4 Intake valves
- Cylinder 5 Intake valves
- Cylinder 6 Exhaust valves

➡ **Valve clearance standard—Cold Intake: 0.010-0.013 in. (0.26-0.34mm), Cold Exhaust: 0.011-0.015 in. (0.29-0.37mm), Hot* Intake: 0.012-0.016 in. (0.304-0.416mm), Hot* Exhaust: 0.012-0.016 in. (0.308-0.432mm) *Approximately 176 degrees F (80 degrees C)**

✳✳ CAUTION

If the inspection was carried out with cold engine, make sure values with fully warmed up engine are still within specifications.

9. For all valve lifters that are found to be outside the specified range, perform the following steps.

Perform adjustment depending on selected head thickness of valve lifter.

The specified valve lifter thickness is the dimension at normal temperatures. Ignore

dimensional differences caused by temperature. Use the specifications for hot engine condition to adjust.

10. Remove the camshaft.

11. Remove the valve lifters at the locations that are outside the standard.

12. Measure the center thickness of the removed valve lifters with a micrometer.

13. Use the following equation to calculate valve lifter thickness for the replacement lifters.

➡ **Valve lifter thickness calculation: thickness of replacement valve lifter = t1 + (C1–C2). t1 = Thickness of removed valve lifter, C1 = measured valve clearance, C2 = standard valve clearance.**

The thickness of a new valve lifter can be identified by stamp marks on the reverse side (inside the cylinder). Stamp mark 788U or 788R indicates 7.88 mm (0.3102 in) in thickness.

➡ **Two types of stamp marks are used for parallel setting and for manufacturer identification. Available thicknesses of valve lifters include 27 sizes covering a range of 0.3102-0.3307 in. (7.88-8.40mm) in steps of 0.0008 in. (0.02mm).**

14. Install the selected valve lifter(s).

15. Install the camshaft.

16. Manually turn crankshaft pulley a few turns.

17. Make sure the valve clearances for the cold engine are within specifications by referring to the specified values.

18. After completing the repair, check valve clearances again with the specifications for a warmed engine. Make sure the values are within specifications.

➡ **Valve clearance specifications:**

- Cold Intake (68 degrees F/20 degrees C): 0.010—0.013 in. (0.26—0.34mm)
- Cold Exhaust: 0.011—0.015 in. (0.29—0.37mm)
- Hot Intake (176 degrees F/80 degrees C): 0.012—0.016 in. (0.304—0.416mm)
- Hot Exhaust: 0.012—0.016 in. (0.308-0.432mm)

VK45DE Engine

1. Before servicing the vehicle, refer to the precautions in the beginning of this section.

Perform inspection after removal, installation or replacement of camshaft or valve-related parts, or if there is unusual engine conditions regarding valve clearance.

2. Warm up the engine. Then turn it OFF.

3. Remove the rocker covers (right and left bank).

4. Set the No. 1 cylinder at TDC of its compression stroke, as follows:

   a. Rotate the crankshaft pulley clockwise to align the TDC identification notch (without the paint mark) with the timing indicator on the front cover.

   b. Make sure that both intake and exhaust cam noses of the No. 1 cylinder (engine front side of the left engine bank) are located as shown. If not, turn the crankshaft one revolution (360 degrees) and align as shown.

5. Using a feeler gauge, measure the valve clearance for the following cylinders:

- Cylinder 1 Intake valves
- Cylinder 1 Exhaust valves
- Cylinder 2 Intake valves
- Cylinder 4 Intake valves
- Cylinder 5 Intake valves
- Cylinder 7 Exhaust valves
- Cylinder 8 Exhaust valves

➡ **Valve clearance standard—Cold* Intake: 0.010-0.013 in. (0.26-0.34mm), Cold* Exhaust: 0.011-0.015 in. (0.29-0.37mm), Hot Intake: 0.012-0.016 in. (0.304-0.416mm), Hot Exhaust: 0.012-0.016 in. (0.308-0.432mm) *Approximately 68 degrees F (20 degrees C)**

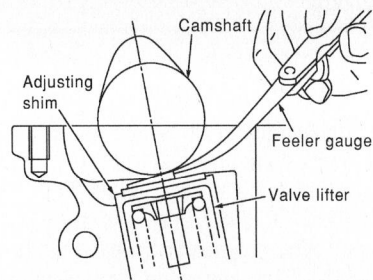

Checking valve clearance—VK45DE Engine.

67162-FX35-G32

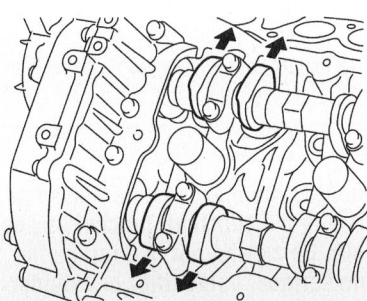

When at TDC No. 1 cylinder, the camshaft lobes should point as shown—VQ35DE Engine.

67162-FX35-G31

⬆ : Measurable at No. 1 cylinder compression TDC

⇧ : Measurable at No. 3 cylinder compression TDC

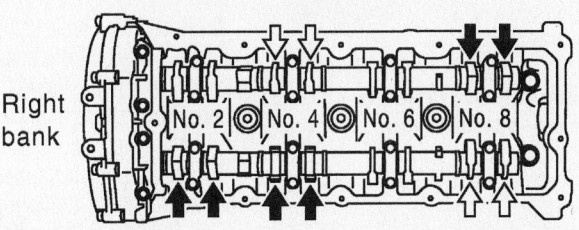

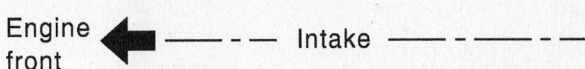

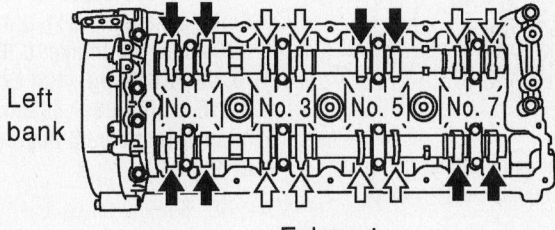

Valve lash inspection positions at No. 1 TDC and No. 3 TDC—VK45DE Engine.

67162-FX35-G33

✳✳ CAUTION

If inspection was carried out with cold engine, make sure values with fully warmed up engine are still within specifications. Valve clearance standard:

6. Rotate the crankshaft pulley clockwise (when viewed from engine front) by 270 degrees from the position of No. 1 cylinder compression TDC to align No. 3 cylinder at TDC of its compression stroke.

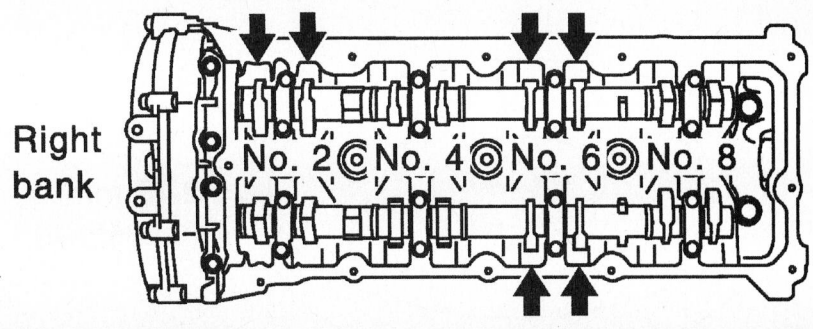

Valve lash inspection positions at No. 6 TDC—VK45DE Engine.

67162-FX35-G37

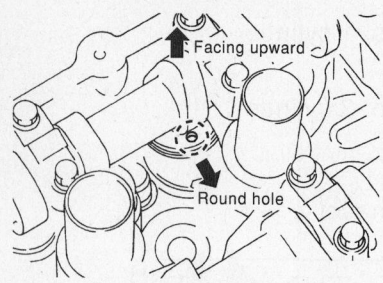

Position the camshaft lobe as shown for shim replacement—VK45DE Engine.

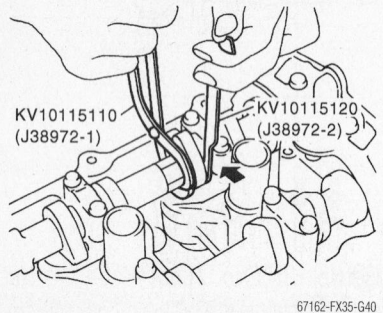

Use the tools shown to depress the valve spring for shim replacement—VK45DE Engine.

➡The crankshaft pulley mounting bolt flange has a angle mark every 90 degrees. They can be used as a guide to rotation angle.

7. Using a feeler gauge, measure the valve clearance for the following cylinders:

- Cylinder 3 Intake valves
- Cylinder 3 Exhaust valves
- Cylinder 4 Exhaust valves
- Cylinder 5 Exhaust valves
- Cylinder 7 Intake valves
- Cylinder 8 Intake valves

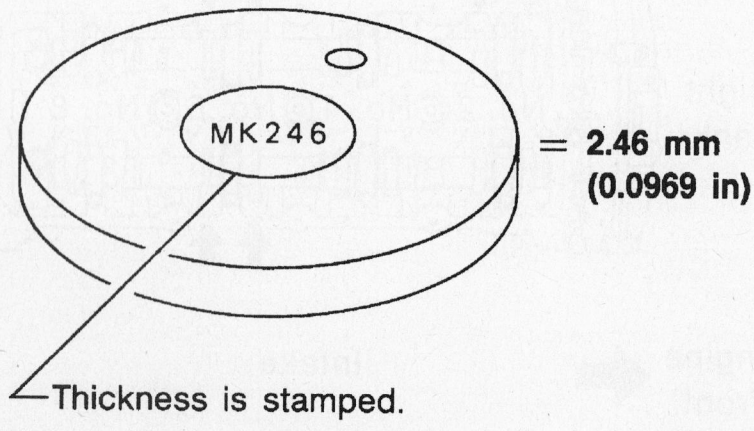

Thickness is stamped.

The shims are stamped with their thickness—VK45DE Engine.

➡Valve clearance standard—Cold* Intake: 0.010-0.013 in. (0.26-0.34mm), Cold* Exhaust: 0.011-0.015 in. (0.29-0.37mm), Hot Intake: 0.012-0.016 in. (0.304-0.416mm), Hot Exhaust: 0.012-0.016 in. (0.308-0.432mm) *Approximately 68 degrees F (20 degrees C)

8. Rotate the crankshaft pulley clockwise (when viewed from engine front) by 90 degrees from the position of No. 3 cylinder compression TDC to align No. 6 cylinder at TDC of its compression stroke.

9. Using a feeler gauge, measure the valve clearance for the following cylinders:

- Cylinder 2 Exhaust valves
- Cylinder 6 Exhaust valves
- Cylinder 6 Intake valves

➡Valve clearance standard—Cold* Intake: 0.010-0.013 in. (0.26-0.34mm), Cold* Exhaust: 0.011-0.015 in. (0.29-0.37mm), Hot Intake: 0.012-0.016 in. (0.304-0.416mm), Hot Exhaust: 0.012-0.016 in. (0.308-0.432mm) *Approximately 68 degrees F (20 degrees C)

10. For any valves that measure out of the standard allowable range, perform the following adjustment steps.

➡Adjust valve clearance while the engine is cold. After adjustment, make sure that the valve clearance is within the standard range while the engine is hot.

11. Thoroughly wipe off all engine oil around the adjusting shim.

12. Rotate the crankshaft to position the cam nose upward for the camshaft for the valve that must be adjusted.

13. Using small screwdriver or pick, turn the round hole of adjusting shim in the direction of the arrow (toward the center of the cylinder head).

14. For all valves except the exhaust side of No. 7 and No. 8 cylinders, install the lifter stopper SST: 10115120 (J38972-2) or equivalent, as follows:

a. Place the camshaft pliers around camshaft as shown in the figure.

b. Rotate the camshaft pliers so that the valve lifter is pushed down.

❊❊ CAUTION

Be careful not to damage the cam surface, valve lifter or cylinder head with the camshaft pliers.

c. Place the lifter stopper between the camshaft and the edge of the valve lifter to retain the valve lifter.

❊❊ CAUTION

The lifter stopper must be placed as close to the camshaft bracket as possible. Be careful not to damage the cam surface, valve lifter or cylinder head with the lifter stopper.

d. Remove camshaft pliers.

❊❊ CAUTION

The camshaft pliers should be removed by rotating it slowly because the lifter stopper hits and damages the journal portion when rotating the camshaft pliers quickly.

15. For the exhaust side of No. 7 and No. 8 cylinders, perform the following:

➡Exhaust side of No. 7 and No. 8 cylinders do not have space for installing with camshaft pliers SST: KV10115110 (J38972-1). Therefore, install lifter stopper SST: KV10115120 (J38972-2)] or equivalent according to the following instructions:

a. Rotate the crankshaft to press the cam nose onto the adjusting part of the valve lifter.

b. Place the lifter stopper between the camshaft and the edge of the valve lifter to retain the valve lifter.

❊❊ CAUTION

The lifter stopper must be placed as close to the camshaft bracket as possible. Be careful not to damage the cam surface, valve lifter or cylinder head with the lifter stopper.

c. Rotate the crankshaft slowly 180 degrees clockwise.

❄ CAUTION

Rotate the crankshaft slowly because the lifter stopper hits and damages the journal portion by rotating the crankshaft quickly.

16. Blow air into the round hole to separate the adjusting shim from the valve lifter.

❄ WARNING

When blowing, use goggles to protect your eye.

17. Remove the adjusting shim with a magnetic tool.
18. Use the following to calculate the adjusting shim thickness for replacement:

a. Use a micrometer to determine the thickness of the removed shim measured at the center.

b. Calculate the thickness of the new adjusting shim so that valve clearance falls within the specified acceptable range.

Valve lifter thickness calculation: $t = t_1 + (C_1 - C_2)$

t= Valve lifter thickness to be replaced
t1= Removed valve lifter thickness
C1= Measured valve clearance
C2= Standard valve clearance (Intake: 0.012 in./0.30mm, Exhaust: 0.013 in./0.33mm @ 68 degrees F/20 degrees C)

➡**Shims are available in 64 sizes from 0.0913-0.1161 in. (2.32-2.95mm) in steps of 0.0004 in. (0.01mm). And the thickness of new adjusting shims can be identified by stamp marks on the underside (inside the cylinder).**

19. Install the new adjusting shim using a suitable tool, with the surface on which the thickness is stamped facing down.
20. For all valve lifters except exhaust side of No. 7 and No. 8 cylinder, remove the lifter stopper as follows:

a. Perform the same procedure as described for removal using the camshaft pliers.

b. Remove the lifter stopper.

c. Remove the camshaft pliers.

21. For exhaust side of No. 7 and No. 8 cylinder valve lifters, rotate the crankshaft slowly 180 degrees clockwise, then remove the lifter stopper.
22. Manually rotate the crankshaft pulley a few turns.
23. Make sure that the valve clearance is within specifications for cold and hot settings.

➡**Valve clearance specifications:**

- Cold Intake (68 degrees F/20 degrees C): 0.010—0.013 in. (0.26—0.34mm)
- Cold Exhaust: 0.011—0.015 in. (0.29—0.37mm)
- Hot Intake (176 degrees F/80 degrees C): 0.012—0.016 in. (0.304—0.416mm)
- Hot Exhaust: 0.012—0.016 in. (0.308-0.432mm)

Oil Pan

REMOVAL & INSTALLATION

VQ35DE Engine

1. Before servicing the vehicle, refer to the precautions in the beginning of this section.

❄ WARNING

To avoid the danger of being scalded, never drain engine oil or engine coolant when the engine is hot.

➡**To remove only the lower oil pan, perform step 5 then skip to step 25.**

2. Remove the front tire.
3. Remove the hood assembly.
4. Remove the front and rear engine undercover.
5. Remove the front cross bar.
6. Drain the engine oil.
7. Drain the engine coolant.
8. Remove the engine cover.
9. Remove the air hose from the air duct to the mass air flow and the electric throttle control actuator side.
10. Remove the alternator and power steering pump and A/C compressor belt.
11. On AWD models, remove the front left and right halfshafts, and side shaft.
12. Remove the engine rear lower slinger, and install the engine rear slinger tool SST: 10006 31U00 (or equivalent) to hold the engine assembly in position. Tighten the engine rear slinger tool mounting bolts to 21 ft. lbs. (28 Nm).
13. Remove the front suspension member.
14. On AWD models, remove the engine mounting bracket, lower engine mounting bracket and insulator.
15. On AWD models, remove the front drive shaft.
16. On AWD models, remove the oil filter and oil filter bracket.
17. Remove the alternator stay.
18. Remove the starter motor.
19. Remove the alternator and power

steering pump and A/C compressor idler pulley and bracket assembly.

20. Disconnect the A/T fluid cooler hoses, and remove the oil cooler water pipe mounting bolt.
21. Disconnect the A/T fluid cooler tube.
22. On AWD models, remove the front final drive assembly.
23. Remove the crankshaft position sensor (POS).

❄ CAUTION

Handle the POS carefully to avoid dropping it and exposing it to abrupt shocks. Do not disassemble it, do not allow metal powder to adhere to magnetic part at the sensor tip, and do not place sensor in a location where it may be exposed to magnetism.

24. Remove the oil filter, as necessary.
25. Remove the oil cooler, as necessary.
26. Remove the oil pan (lower), as follows:

a. Loosen the mounting bolts in the reverse order of the tightening sequence.

b. Insert seal cutter SST: KV10111100 (J37228) or equivalent, between the upper oil pan and lower oil pan.

c. Slide the seal cutter by tapping on the side of the tool with a hammer.

d. Remove the lower oil pan.

❄ CAUTION

Be careful not to damage the mating surface. Do not use a flat-bladed screwdriver—this will damage the mating surfaces.

27. Remove the oil strainer.
28. Remove the transmission joint bolts which pass through the upper oil pan.
29. On 2WD models, remove the rear cover plate.
30. Loosen the upper oil pan bolts in the reverse order of the tightening sequence.
31. Insert seal cutter SST: KV10111100 (J37228) between the upper oil pan and cylinder block. Slide the seal cutter by tapping on the side of the tool with a hammer.
32. Remove the upper oil pan.

❄ CAUTION

Be careful not to damage the mating surface. Do not use a flat-bladed screwdriver—this will damage the mating surfaces.

33. Remove the O-rings from the bottom of the cylinder block and oil pump.

34. Remove the oil pan gaskets.
35. For AWD models, remove the axle pipe from the upper oil pan using a suitable drift, if necessary.
36. Clean the oil strainer, if necessary.

To install:
37. Install the upper oil pan, as follows:
   a. Use a scraper to remove the old liquid gasket from all mating surfaces. Remove old liquid gasket from the mating surface of the cylinder block, and the bolt holes and threads.

✳✳ CAUTION

Do not scratch or damage the mating surfaces when cleaning off old liquid gasket.

   b. Apply liquid gasket to the oil pan gaskets as shown.

➡**Use Genuine RTV Silicone Sealant or equivalent.**

   c. Install the new gasket. Align the

protrusion of the oil pan gasket with the notches of the front timing chain case and rear oil seal retainer.
   d. Install the oil pan gasket with the smaller arc to the front timing chain case side.
   e. Install new O-rings on the cylinder block and oil pump.
   f. Apply a continuous bead of liquid gasket with tube presser SST: WS39930000 (or equivalent) to the cylinder block mating surface of the

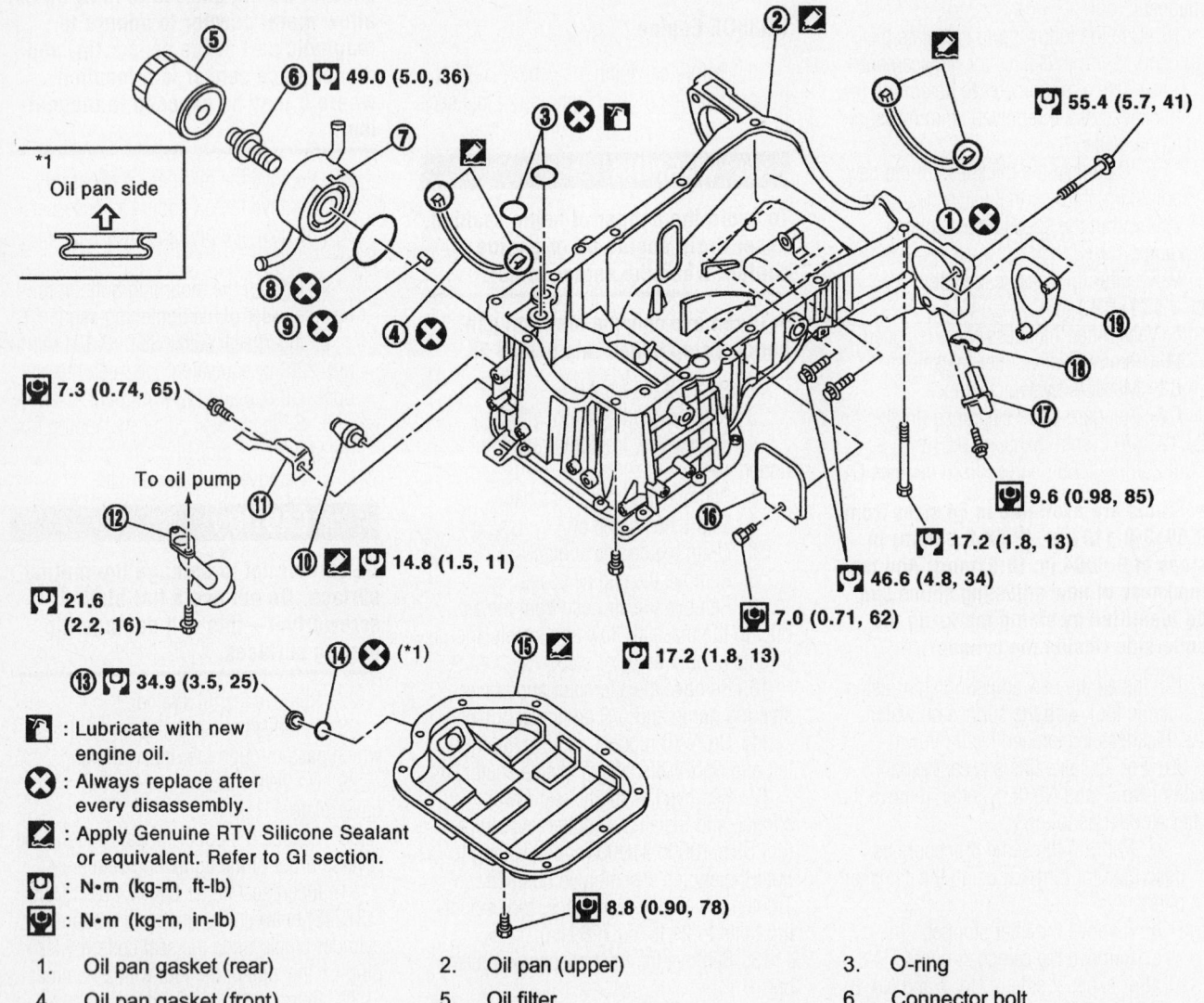

1. Oil pan gasket (rear)
4. Oil pan gasket (front)
7. Oil cooler
10. Oil pressure switch
13. Drain plug
16. Rear plate
19. Rear cover plate
2. Oil pan (upper)
5. Oil filter
8. O-ring
11. Bracket
14. Drain plug washer
17. Crankshaft position sensor (POS)
3. O-ring
6. Connector bolt
9. Relief valve
12. Oil strainer
15. Oil pan (lower)
18. Seal rubber

Exploded view of oil pan and related components—VQ35DE Engine 2WD.

upper oil pan to a limited portion as shown.

➡**Use Genuine RTV Silicone Sealant or equivalent.**

- For bolt holes with marks (5 locations), apply liquid gasket outside the holes.
- Apply a bead of 0.177-0.217 in. (4.5-5.5mm) in diameter to area "A".
- Installation should be done within 5 minutes after coating.

  g. Install the upper oil pan. Tighten the mounting bolts in the order shown.

There are two types of mounting bolts. Refer to the following for locating the bolt positions:

- M8 x 100 mm (3.97 in): positions 5, 7, 8 & 11
- M8 x 25 mm (0.98 in) : positions except 5, 7, 8 & 11

  h. Tighten the transmission joint bolts.

38. Install the oil strainer onto the oil pump.

39. Install the lower oil pan, as follows:

  a. Use a scraper to remove all old liq-

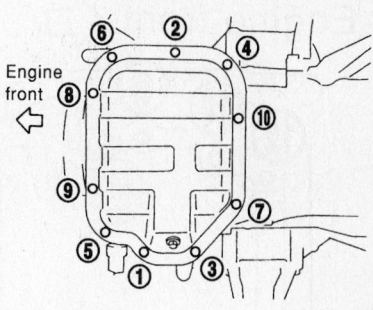

67162-FX35-G48

Lower oil pan mounting bolt tightening sequence—VQ35DE Engine.

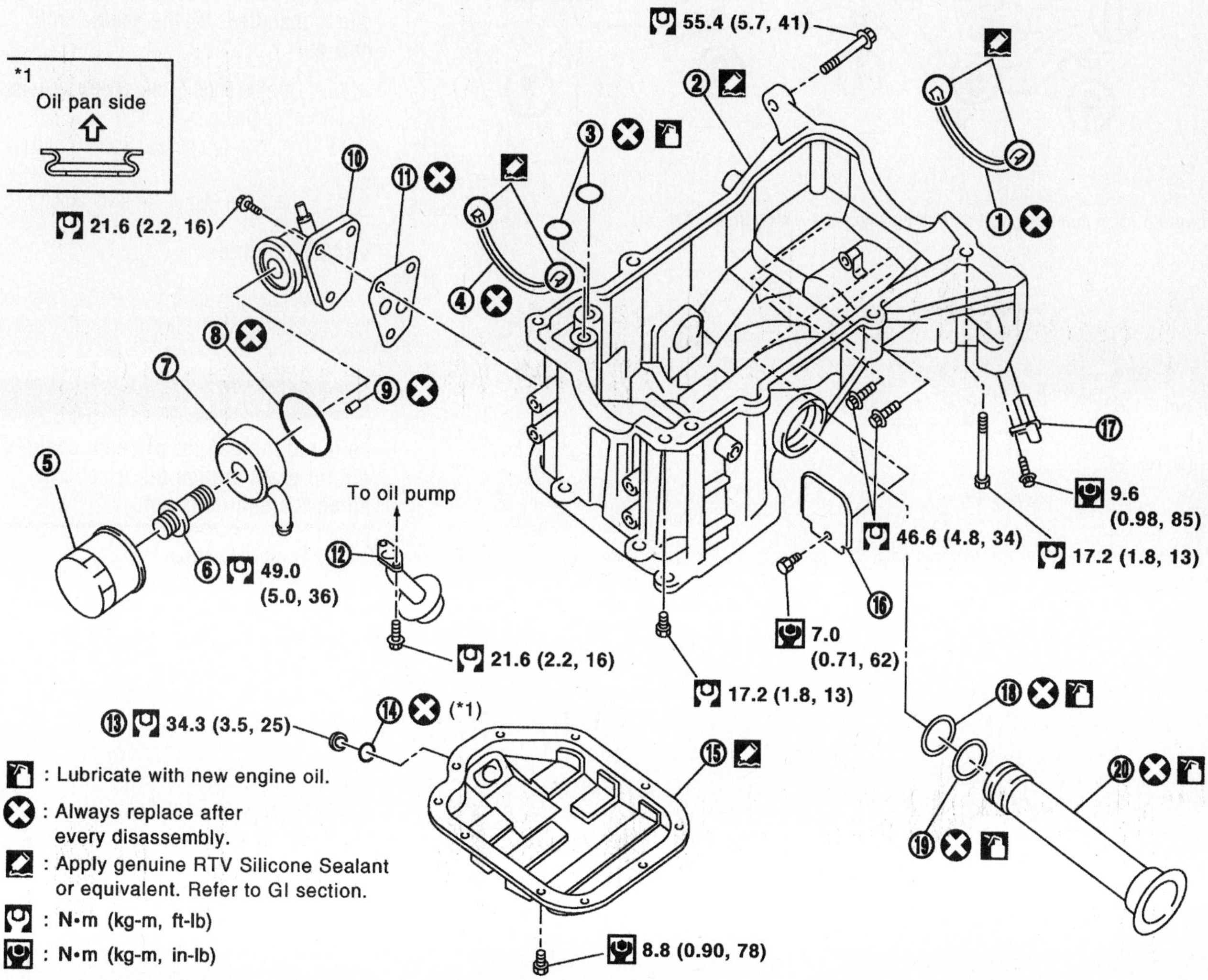

[Exploded view diagram with torque specifications:]

55.4 (5.7, 41)

21.6 (2.2, 16)

49.0 (5.0, 36)

21.6 (2.2, 16)

34.3 (3.5, 25) (*1)

*1 Oil pan side

To oil pump

9.6 (0.98, 85)

46.6 (4.8, 34)

17.2 (1.8, 13)

7.0 (0.71, 62)

17.2 (1.8, 13)

8.8 (0.90, 78)

: Lubricate with new engine oil.

: Always replace after every disassembly.

: Apply genuine RTV Silicone Sealant or equivalent. Refer to GI section.

: N•m (kg-m, ft-lb)

: N•m (kg-m, in-lb)

1. Oil pan gasket (rear)	2. Oil pan (upper)	3. O-ring
4. Oil pan gasket (front)	5. Oil filter	6. Connector bolt
7. Oil cooler	8. O-ring	9. Relief valve
10. Oil filter bracket	11. Oil filter bracket gasket	12. Oil strainer
13. Drain plug	14. Drain plug washer	15. Oil pan (lower)
16. Rear plate	17. Crankshaft position sensor (POS)	18. O-ring (small)
19. O-ring (large)	20. Axle pipe	

Exploded view of oil pan and related components—VQ35DE Engine AWD.

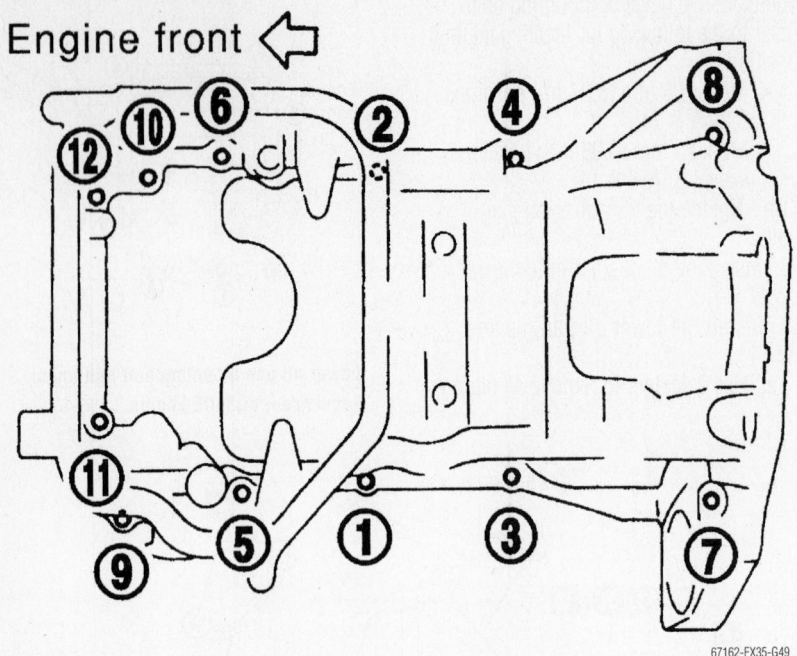

Engine front ⇐

Upper oil pan mounting bolt tightening sequence—VQ35DE Engine.

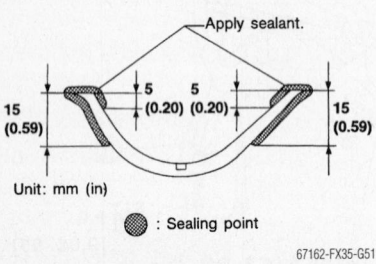

Apply sealant.

15 (0.59) 5 (0.20) 5 (0.20) 15 (0.59)

Unit: mm (in)

▨ : Sealing point

67162-FX35-G51

Sealant positioning for front oil pan seal installation—VQ35DE Engine.

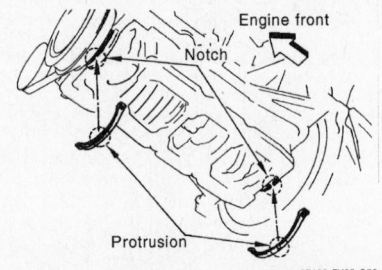

Engine front

Notch

Protrusion

67162-FX35-G52

Oil pan seal positioning—VQ35DE Engine.

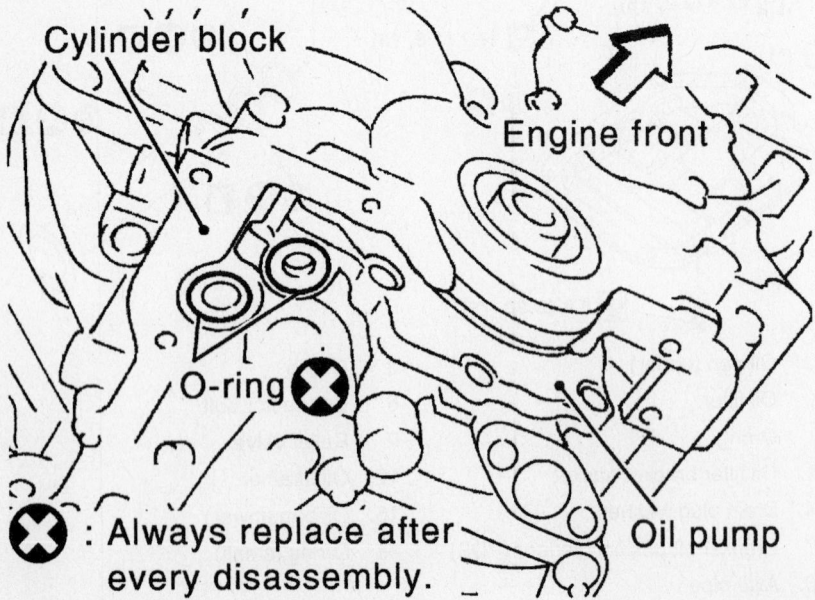

Cylinder block

Engine front

O-ring ✖

✖ : Always replace after every disassembly.

Oil pump

67162-FX35-G53

O-ring locations for oil pan service—VQ35DE Engine.

uid gasket from the mating surfaces. Also remove old liquid gasket from the mating surface of the upper oil pan.

b. Apply new liquid gasket thoroughly with tube presser SST: WS39930000 or equivalent.

➡ **Use Genuine RTV Silicone Sealant or equivalent. Attaching should be done within 5 minutes after coating.**

c. Tighten the mounting bolts in numerical order as shown.

40. Install the oil pan drain plug.

41. The remainder of installation is the reverse of removal.

➡ **Wait at least 30 minutes after the oil pan is installed, fill the engine with new oil.**

42. Start the engine, and check there is no leak of engine oil.

43. Stop the engine and wait 10 minutes.

44. Check the engine oil level again.

VK45DE Engine

1. Before servicing the vehicle, refer to the precautions in the beginning of this section.

❊❊ WARNING

To avoid the danger of being scalded, do not drain engine oil or coolant when the engine is hot.

2. Remove the front tire.

3. Remove the hood assembly.

4. Remove the engine cover.

5. Remove the front and rear engine under covers.

6. Drain the engine oil and engine coolant.

7. Remove the accessory drive belts.

8. Remove the auto tensioner for the power steering oil pump belt.

9. Remove the power steering oil pump with the piping connected, and temporarily secure it aside with ropes or equivalent.

10. Remove the A/C compressor with the piping connected, and temporarily secure it aside with ropes or equivalent.

11. Remove the A/C compressor fitting bolts, and install the A/C compressor temporarily on the vehicle side with ropes or equivalent.

12. Remove the wiring harness of the lower side of oil pan.

13. Remove the crankshaft position sensor (POS) from the transmission.

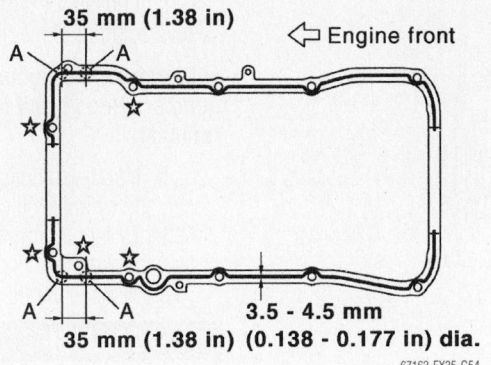

35 mm (1.38 in)

⇐ Engine front

A   A

☆   ☆

☆   ☆

A   A

3.5 - 4.5 mm
35 mm (1.38 in)   (0.138 - 0.177 in) dia.

67162-FX35-G54

Liquid gasket positioning for oil pan installation—VQ35DE Engine.

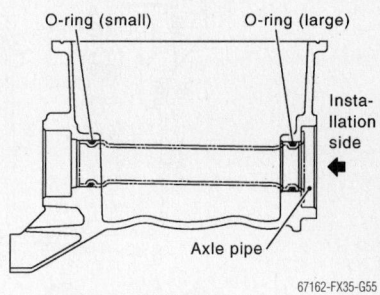

O-ring (small)   O-ring (large)

Installation side

Axle pipe

67162-FX35-G55

Cross-section view of the axle pipe installation—VQ35DE Engine AWD.

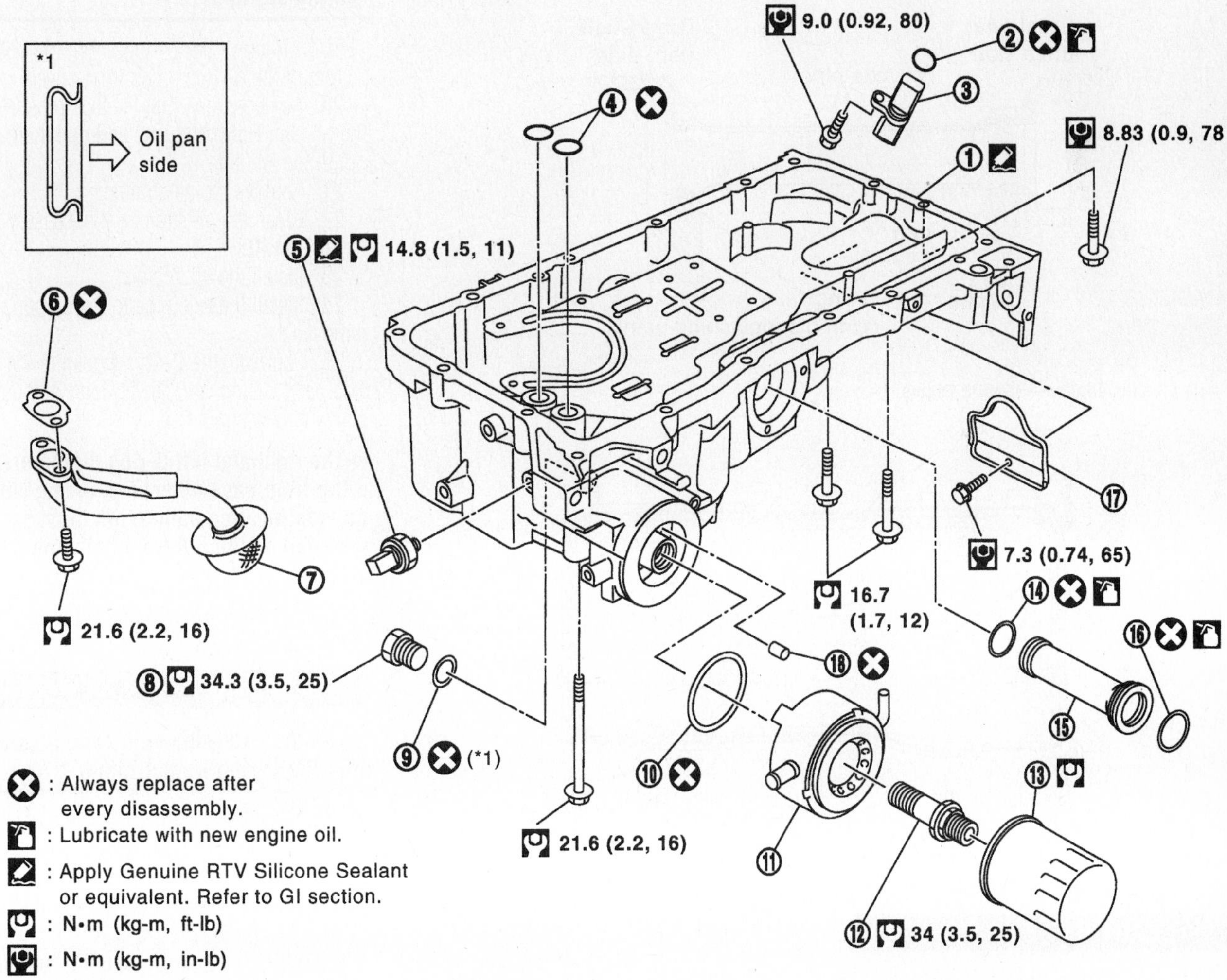

*1

→ Oil pan side

⑨ 9.0 (0.92, 80)   ② ✕

③

①

⑧ 8.83 (0.9, 78)

④ ✕

⑤ 14.8 (1.5, 11)

⑥ ✕

⑰

⑦

21.6 (2.2, 16)   16.7 (1.7, 12)

7.3 (0.74, 65)

⑧ 34.3 (3.5, 25)

⑭ ✕

⑯ ✕

⑮

⑨ ✕ (*1)

⑱ ✕

⑩ ✕

⑬

21.6 (2.2, 16)   ⑪

⑫ 34 (3.5, 25)

✕ : Always replace after
   every disassembly.

: Lubricate with new engine oil.

: Apply Genuine RTV Silicone Sealant
  or equivalent. Refer to GI section.

: N•m (kg-m, ft-lb)

: N•m (kg-m, in-lb)

1.	Oil pan	2.	O-ring	3.	Crankshaft position sensor (POS)
4.	O-ring	5.	Oil pressure switch	6.	Gasket
7.	Oil strainer	8.	Drain plug	9.	Drain plug washer
10.	O-ring	11.	Oil cooler	12.	Connector bolt
13.	Oil filter	14.	O-ring	15.	Axle pipe
16.	O-ring	17.	Rear plate cover	18.	Relief valve

67162-FX35-G56

Exploded view of the oil pan and related components—VK45DE Engine.

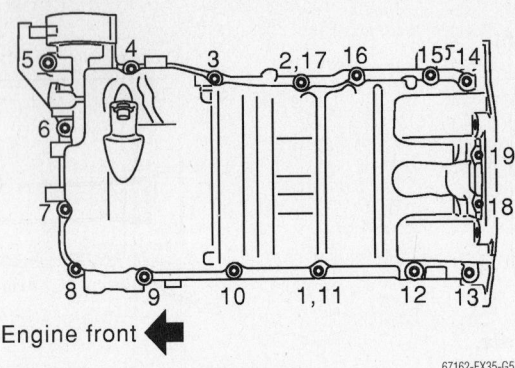

67162-FX35-G57

Oil pan mounting bolt tightening sequence—VK45DE Engine.

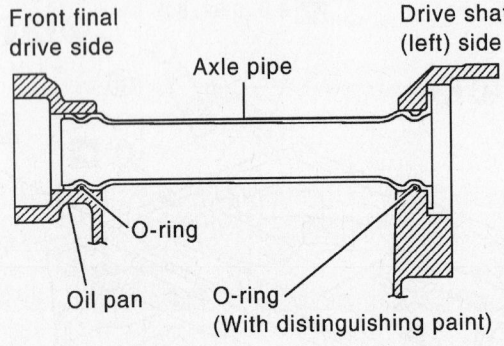

67162-FX35-G58

Axle pipe installation—VK45DE Engine.

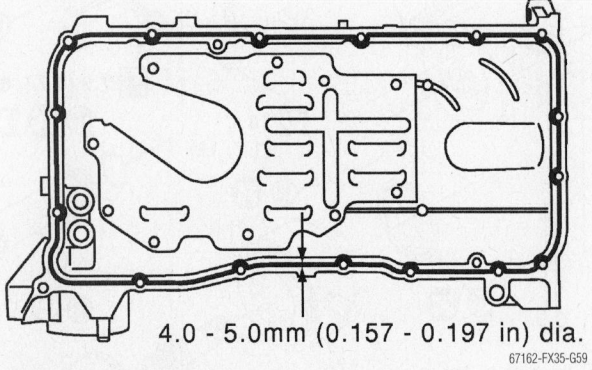

4.0 - 5.0mm (0.157 - 0.197 in) dia.

67162-FX35-G59

Liquid gasket positioning for oil pan installation—VK45DE Engine.

✳✳ CAUTION

Handle the POS carefully to avoid dropping it and exposing it to abrupt shocks. Do not disassemble it, do not allow metal powder to adhere to magnetic part at the sensor tip, and do not place sensor in a location where it may be exposed to magnetism.

14. Install the engine slinger to hold the engine assembly in a secure position.

Tighten the slinger mounting bolts to 25 ft. lbs. (33.5 Nm).

15. Remove the front suspension member.

16. Remove the front final drive assembly.

17. Remove the oil filter.

18. Disconnect the oil cooler water hoses, and remove the oil cooler water pipe and oil cooler.

19. Remove the oil pan as follows:

    a. Remove the rear plate cover.

    b. Remove the transmission joint bolts which pass through the oil pan.

    c. Loosen the oil pan bolts in the reverse order of the tightening sequence.

➡ **Disregard the tightening sequence numbers No. 11 and No. 17 during removal.**

    d. Insert seal cutter between the oil pan and cylinder block. Slide the seal cutter by tapping on the side of seal cutter with hammer.

    e. Remove the oil pan.

✳✳ CAUTION

Be careful not to damage the mating surfaces. Do not use a flat-bladed screwdriver—this will damage the mating surfaces.

    f. Remove the O-rings from the bottom of the oil pump and front cover.

20. As necessary, pull the axle pipe from the oil pan. Hold the pipes and pull them out to the left side.

21. Remove the oil strainer.

22. Clean the oil strainer, if necessary.

To install:

23. Install the oil strainer.

24. Install the axle pipe to the oil pan, if removed.

25. Lubricate the O-ring groove of the axle pipe, O-ring, and O-ring joint of the oil pan with new engine oil.

➡ **The right and left O-ring diameters differ from each other. The O-ring with an identification paint mark must be installed on the left front halfshaft side.**

26. Install the axle pipe to the oil pan on the left side.

✳✳ CAUTION

Insert the axle pipe with care to prevent the O-ring from sliding.

27. Install the oil pan as follows:

    a. Install new O-rings onto the oil pump and the side of the front cover.

    b. Apply a continuous bead of liquid gasket with tube presser SST: WS39930000, or equivalent, to the cylinder block mating surface of the oil pan as shown.

➡ **Use Genuine RTV Silicone Sealant or equivalent.**

✳✳ CAUTION

Installation should be done within 5 minutes after coating, otherwise the liquid gasket may not seal properly.

c. Install the oil pan and tighten the mounting bolts in order as shown. There are three types of mounting bolts. Refer to the following for locating bolts.

- M6 x 30 mm (1.18 in): positions 18 & 19
- M8 x 100 mm (3.94 in): positions 5 & 9
- M8 x 45 mm (1.77 in) : positions except 5, 9, 18 & 19

➡**Tighten bolts No. 1 and No. 2 in two steps—this is why they are shown as Nos. 1 & 11 and Nos. 2 & 17 in the illustration. Nos. 11 & 17 are the second steps for Nos. 1 & 2.**

d. Tighten the transmission joint bolts to 55 ft. lbs. (74 Nm).

e. Install the rear plate cover and tighten the mounting bolt to 65 inch lbs. (7.3 Nm).

28. Install the oil pan drain plug with a new drain plug washer. Tighten it to 25 ft. lbs. (34 Nm).

29. The remainder of installation is the reverse of the removal procedure.

➡**Wait at least 30 minutes after the oil pan is installed, to fill the engine with new oil.**

30. Add engine oil.

31. Start the engine, and check for leakage of engine oil.

32. Stop the engine and wait for 15 minutes.

33. Check the engine oil level again.

Oil Pump

REMOVAL & INSTALLATION

VQ35DE Engine

1. Before servicing the vehicle, refer to the precautions in the beginning of this section.

2. Remove the oil pan (lower and upper) and the oil strainer.

3. Remove the front timing chain case and the timing chain (primary).

4. Remove the oil pump assembly.

To install:

➡**For pump installation, align the crankshaft flat faces with the oil pump inner rotor flat faces.**

5. Installation is the reverse of the removal procedure.

6. After warming up the engine, check for engine oil leakage.

7. Check the engine oil level and add engine oil, as needed.

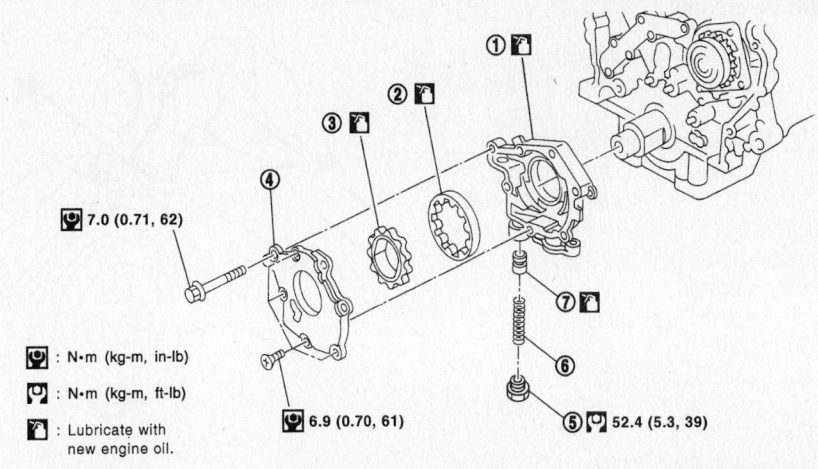

1. Oil pump body
2. Oil pump outer rotor
3. Oil pump inner rotor
4. Oil pump cover
5. Regulator valve plug
6. Regulator valve spring
7. Regulator valve

Exploded view of the oil pump assembly—VQ35DE Engine.

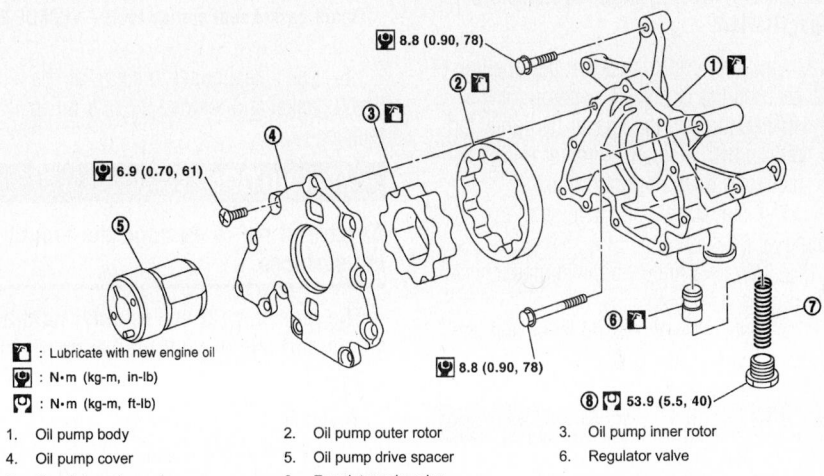

1. Oil pump body
2. Oil pump outer rotor
3. Oil pump inner rotor
4. Oil pump cover
5. Oil pump drive spacer
6. Regulator valve
7. Regulator valve spring
8. Regulator valve plug

Exploded view of the oil pump—VK45DE Engine.

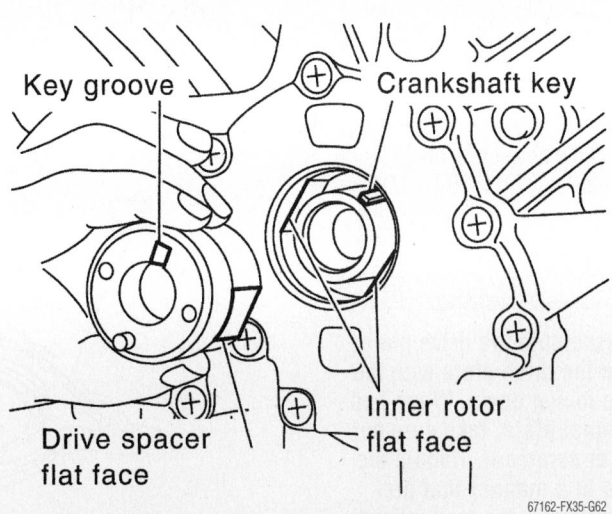

Oil pump drive spacer installation orientation—VK45DE Engine.

VK45DE Engine

1. Before servicing the vehicle, refer to the precautions in the beginning of this section.
2. Remove the front cover.
3. Remove the oil pump drive spacer.
4. Set bolts in the two bolt holes (M6 x 1.0mm/0.04 in) on the front surface. Using a suitable puller, pull oil pump drive spacer off the crankshaft.
5. Remove the oil pump.

To install:

6. Install the oil pump.
7. Install the oil pump drive spacer as follows:

   a. Insert the oil pump drive spacer so that the crankshaft key and flat surfaces of the oil pump inner rotor mesh properly.

➡ **If the positional relationship does not allow the insertion, rotate oil pump inner rotor with a finger to facilitate installation.**

   b. After confirming that the position of each part is in correct position for the spacer, force fit the spacer by lightly tapping it with a plastic hammer until it contacts and does not go further.

8. Installation is the reverse of the removal procedure.
9. After warming up the engine, check for engine oil leakage.
10. Check the engine oil level, and add more oil, if necessary.

Rear Main Seal

REMOVAL & INSTALLATION

VQ35DE Engine

1. Before servicing the vehicle, refer to the precautions in the beginning of this section.
2. Remove the upper oil pan.
3. Remove the transmission assembly.
4. Remove the drive plate. Install a ring gear stopper (SST: KV1011770 (J44716) or equivalent) on the crankshaft and remove the mounting bolts in a diagonal order.

❋❋ CAUTION

Do not disassemble the drive plate. Never place the drive plate with the signal plate facing down. When handling the signal plate, take care not to damage or scratch it. Handle the signal plate in a manner that prevents it from becoming magnetized.

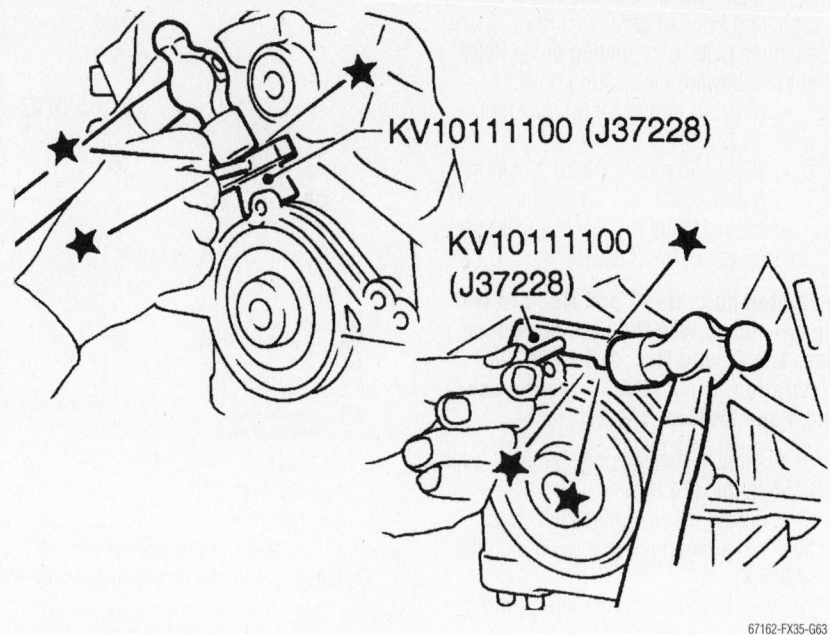

Liquid gasket seal cutting tools—VQ35DE Engine.

5. Use a seal cutter to cut away the old liquid gasket and remove the rear oil seal retainer.

❋❋ CAUTION

Be careful not to damage the mounting surfaces.

➡ **The rear oil seal and retainer form a single part and are handled as one assembly.**

To install:

6. Remove the old liquid gasket from the mating surface of the cylinder block and oil pan using a scraper.

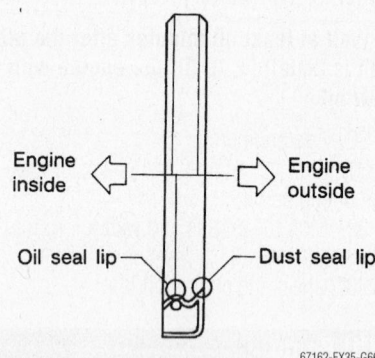

Rear main seal installation orientation—VK45DE Engine.

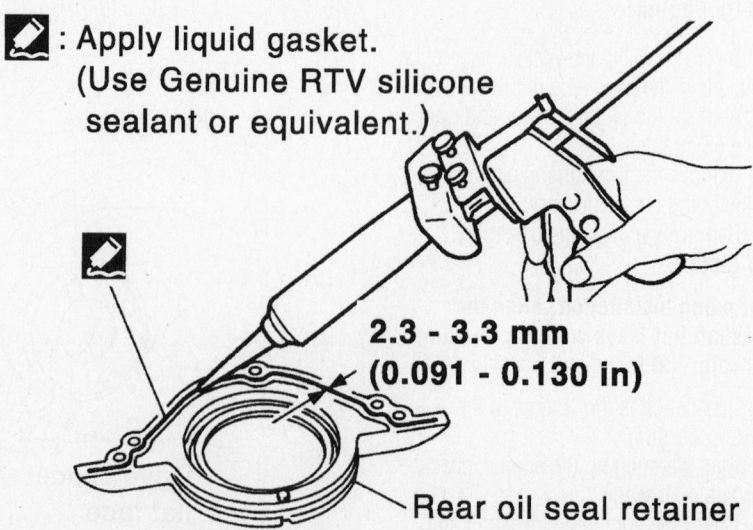

Rear main seal liquid gasket installation positioning—VQ35DE Engine.

7. Apply new engine oil to the oil and dust seal lips.

8. Apply liquid gasket to the rear oil seal retainer with tube presser SST: WS39930000 or equivalent.

➡**Use Genuine Thread Sealant or equivalent.**

❊❊ CAUTION

Installation should be done within 5 minutes after coating, otherwise the liquid gasket may not seal properly.

9. Install the rear oil seal retainer onto the cylinder block.

10. The remainder of installation is the reverse of the removal procedure.

VK45DE Engine

1. Before servicing the vehicle, refer to the precautions in the beginning of this section.

2. Remove the transmission and transfer assembly.

3. Remove the drive plate.

4. Remove the engine rear plate.

5. Remove the rear oil seal using a suitable tool.

❊❊ CAUTION

Be careful not to damage the crankshaft and oil seal retainer surface.

To install:

6. Apply engine oil to both the oil seal lip and dust seal lip.

7. Using a suitable drift, press fit the oil seal until the height of the oil seal is level with the mounting surface and so that each seal lip is oriented as shown.

❊❊ CAUTION

Press fit the seal straight and avoid causing burrs or tilting.

8. The remainder of installation is the reverse of the removal procedure.

Timing Chain, Sprockets, Front Cover and Seal

REMOVAL & INSTALLATION

VQ35DE Engine

WITH OIL PAN REMOVAL

1. Before servicing the vehicle, refer to the precautions in the beginning of this section.

This section describes procedures for removing/installing the front timing chain case and timing chain related parts, and the rear timing chain case, when the upper oil pan needs to be removed/installed for engine overhaul, etc.

When the upper oil pan needs to be removed or installed, or when rear timing chain case is removed or installed, remove the oil pans (upper and lower) first. Then remove the front timing chain case, timing chain related parts, and the rear timing chain case in this order, and install in the reverse order of removal.

2. Place the vehicle on a lift.

3. Remove the front tire.

4. Disconnect the negative battery terminal.

5. Remove the engine cover.

6. Remove the air cleaner case assembly.

7. Remove the front and rear engine under covers.

8. Drain the engine coolant from the radiator.

9. Drain the engine oil from the oil pan.

10. Remove the engine wiring harnesses.

11. Remove the upper and lower intake manifold collectors.

12. Remove the radiator cooling fan assembly.

13. Remove the A/C compressor from the bracket with its piping connected, and temporarily secure it aside.

14. Remove the power steering oil pump from the bracket with its piping connected, and temporarily secure it aside.

15. Remove the power steering oil pump bracket.

16. Remove the alternator.

17. Remove the water bypass hose, water hose clamp and idler pulley bracket from the front timing chain case.

18. Remove the upper and lower oil pan.

19. Remove the right and left intake valve timing control covers, by loosening the bolts in the reverse or the tightening sequence.

20. Use seal cutter SST: KV10111100 (J37228) or equivalent tool to cut the liquid gasket for removal.

❊❊ CAUTION

The shaft is internally joined to the intake camshaft sprocket center hole. During removal, keep the shaft horizontal until it is completely disconnected.

21. Remove the collared O-ring from the front timing chain case (left and right side).

22. Remove the right and left rocker covers.

23. Position the engine at compression TDC of No. 1 cylinder as follows:

a. Rotate the crankshaft pulley clockwise to align the timing mark (grooved line without color) with the timing indicator.

b. Make sure the intake and exhaust cam noses on the No. 1 cylinder (engine front side of right bank) are located so that they point inward and upward compared to the cylinder head.

c. If the cam lobes are not positioned pointing inward and upward, rotate the crankshaft one full revolution (360 degrees).

24. Remove the crankshaft pulley, as follows:

a. For 2WD models, remove the rear cover plate.

b. For AWD models, remove the starter motor.

c. Set the ring gear stopper to hold the crankshaft in position.

d. Loosen the crankshaft pulley bolt until the bolt seating surface is approximately 0.39 in. (10mm) from its original position.

❊❊ CAUTION

Do not completely remove the crankshaft pulley bolt, since it will be used as a supporting point for a suitable puller.

e. Place a suitable puller tab on the holes of the crankshaft pulley, and pull the crankshaft pulley through.

❊❊ CAUTION

Do not position the suitable puller tab on the outer edges of the crankshaft pulley since this can damage the internal damper.

25. Remove the front timing chain case, as follows:

a. Loosen the mounting bolts in the reverse order of the tightening sequence.

b. Insert the suitable tool into the notch at the top of the front timing chain case.

c. Pry off the case by moving the pry tool as shown. Use seal cutter SST: KV10111100 (J37228) or equivalent tool to cut the liquid gasket for removal.

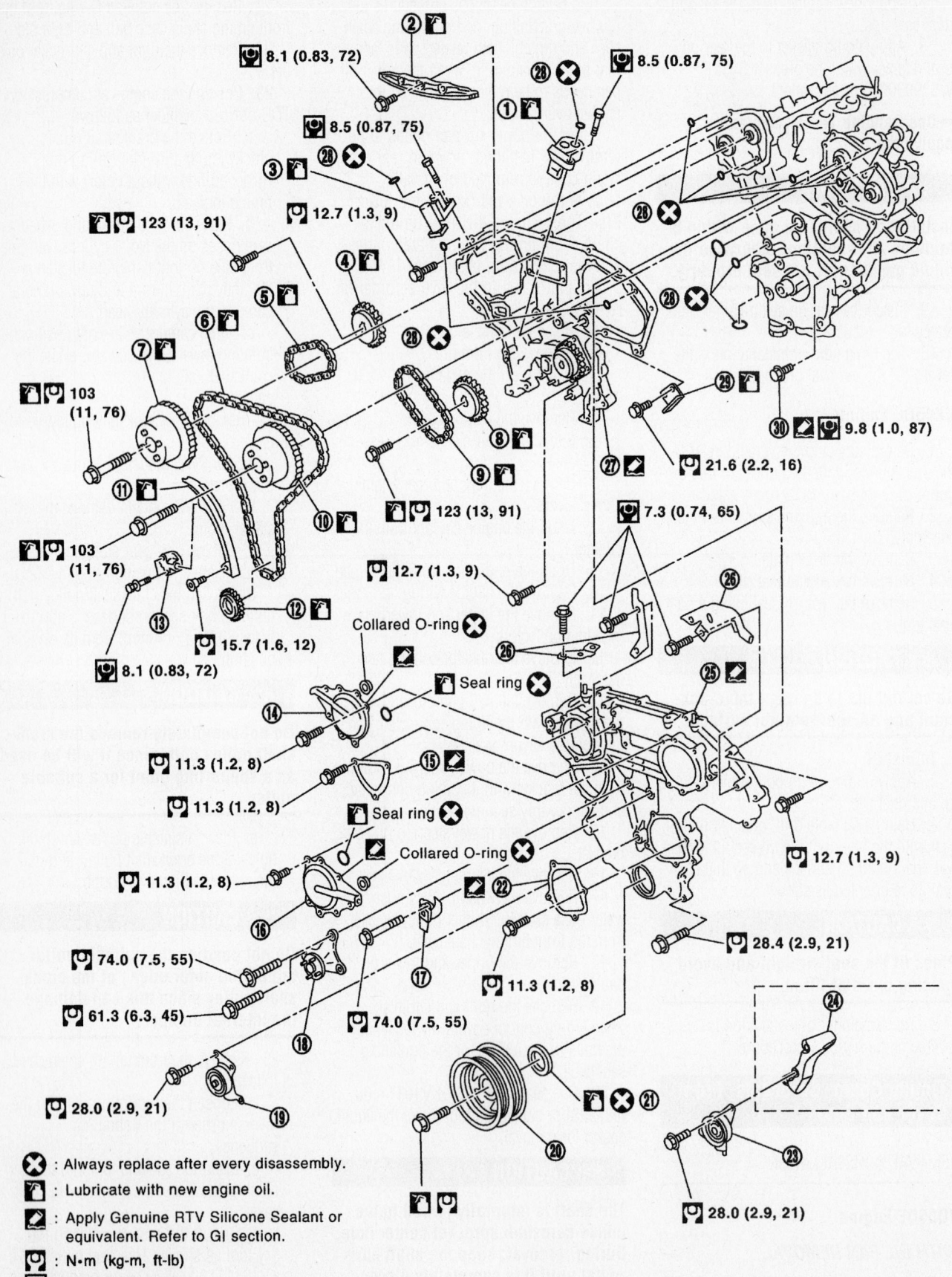

② 8.1 (0.83, 72)

8.5 (0.87, 75)

8.5 (0.87, 75)

③ 12.7 (1.3, 9)

123 (13, 91)

④

⑤

⑥

⑦ 103 (11, 76)

⑪

⑧

⑨

103 (11, 76)

⑬

⑫

15.7 (1.6, 12)

8.1 (0.83, 72)

123 (13, 91)

12.7 (1.3, 9)

9.8 (1.0, 87)

21.6 (2.2, 16)

7.3 (0.74, 65)

㉖

㉕

Collared O-ring

Seal ring

⑭

11.3 (1.2, 8)

11.3 (1.2, 8)

⑮

Seal ring

Collared O-ring

12.7 (1.3, 9)

11.3 (1.2, 8)

⑯

㉒

28.4 (2.9, 21)

74.0 (7.5, 55)

⑰

11.3 (1.2, 8)

㉔

61.3 (6.3, 45)

⑱

74.0 (7.5, 55)

28.0 (2.9, 21)

⑲

⑳

㉑

㉓

28.0 (2.9, 21)

✕ : Always replace after every disassembly.

🖐 : Lubricate with new engine oil.

✎ : Apply Genuine RTV Silicone Sealant or equivalent. Refer to GI section.

N•m (kg-m, ft-lb)

N•m (kg-m, in-lb)

67162-FX35-G67

Exploded view of timing chain and related components—VQ35DE Engine.

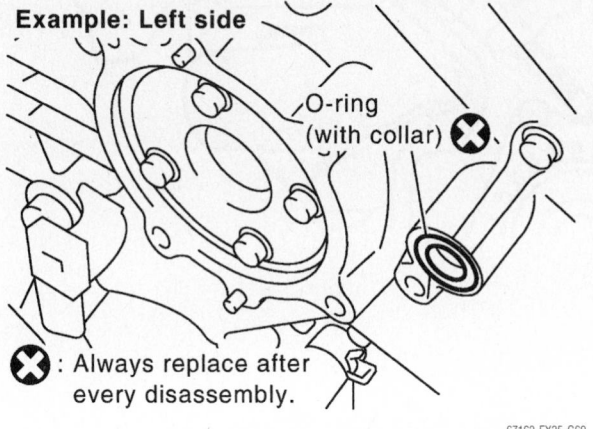

Right   **Left**

Dowel hole   Dowel hole

67162-FX35-G68

Tightening sequence for the timing control covers—VQ35DE Engine.

26. Remove the O-rings from the rear timing chain case.

27. Remove the water pump cover and chain tensioner cover from the front timing chain case. Use the seal cutter, or equivalent, to cut the liquid gasket for removal.

28. Remove the front oil seal from the front timing chain case using a suitable tool.

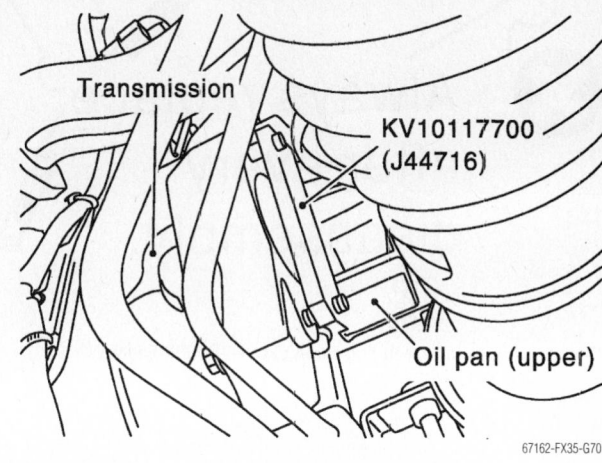

Example: Left side

O-ring (with collar) ✖

✖ : Always replace after every disassembly.

67162-FX35-G69

O-ring location in the front timing chain case—VQ35DE Engine.

Transmission

KV10117700 (J44716)

Oil pan (upper)

67162-FX35-G70

Ring gear stopper positioning—VQ35DE Engine.

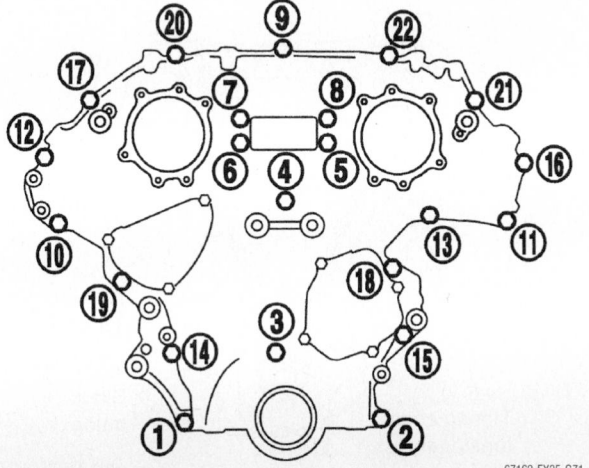

67162-FX35-G71

Tightening sequence for the front timing chain case mounting bolts—VQ35DE Engine.

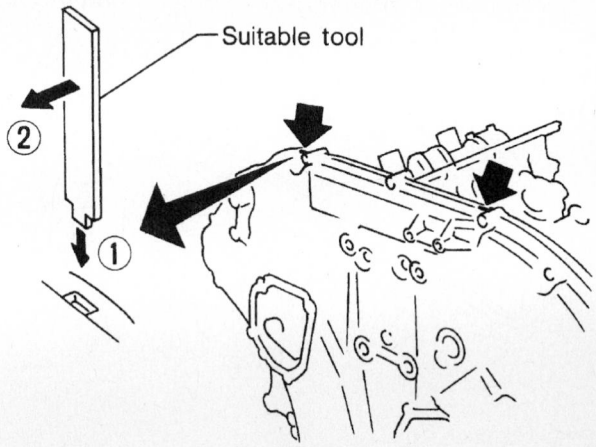

Suitable tool

67162-FX35-G72

Pry off the front timing chain case as shown—VQ35DE Engine.

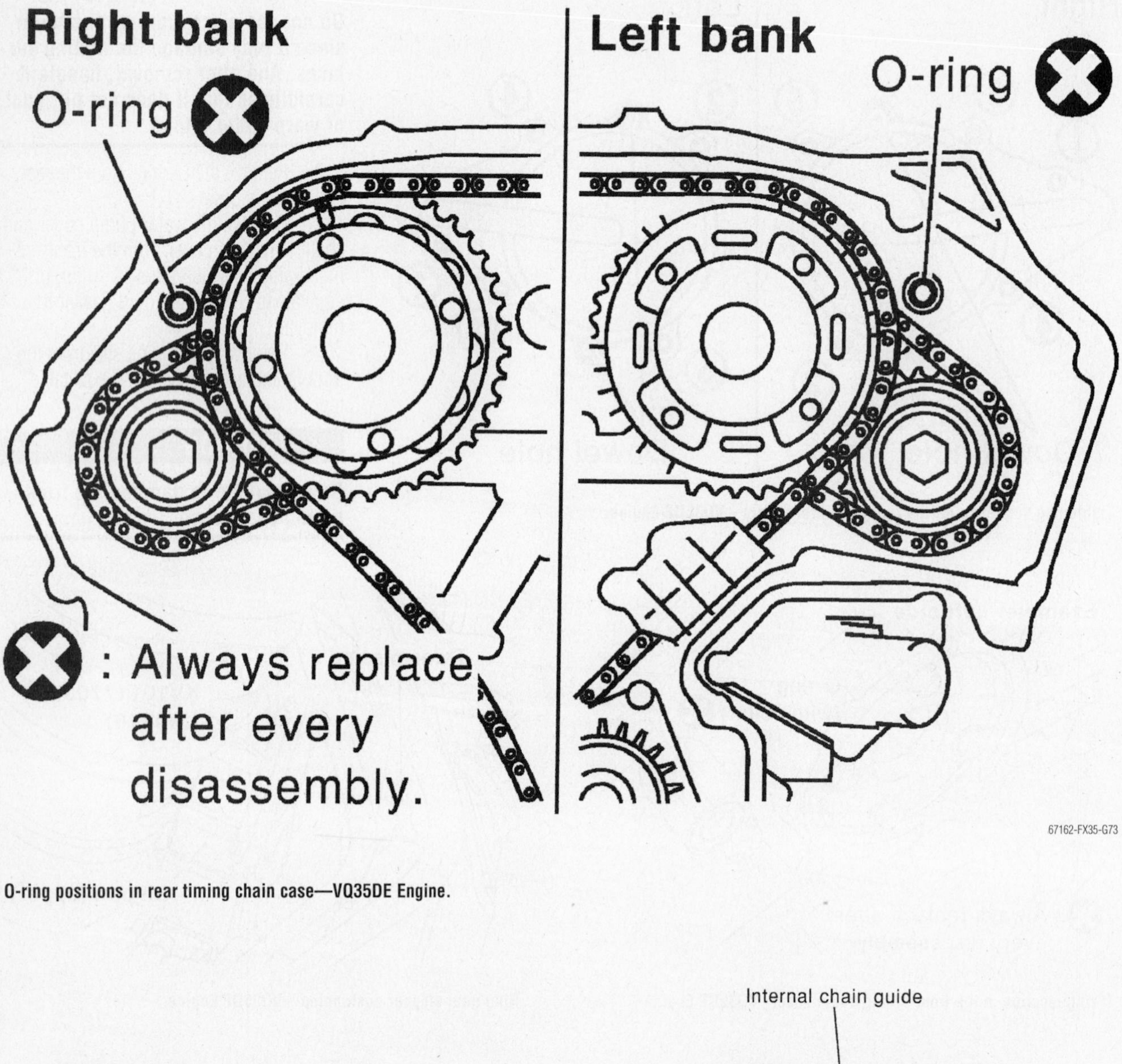

Right bank

O-ring ✖

Left bank

O-ring ✖

✖ : Always replace after every disassembly.

67162-FX35-G73

O-ring positions in rear timing chain case—VQ35DE Engine.

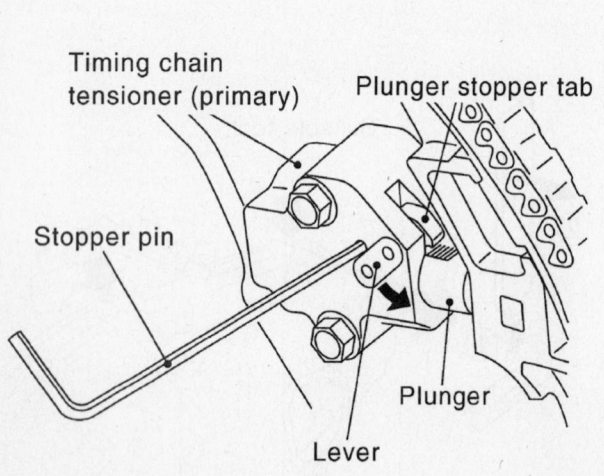

Timing chain tensioner (primary)

Plunger stopper tab

Stopper pin

Plunger

Lever

67162-FX35-G74

Timing chain tensioner detail—VQ35DE Engine.

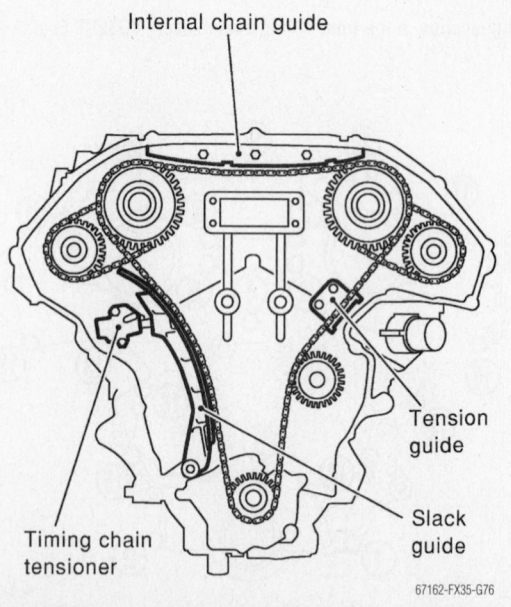

Internal chain guide

Tension guide

Slack guide

Timing chain tensioner

67162-FX35-G76

Internal chain guide, tension guide and slack guide positions—VQ35DE Engine.

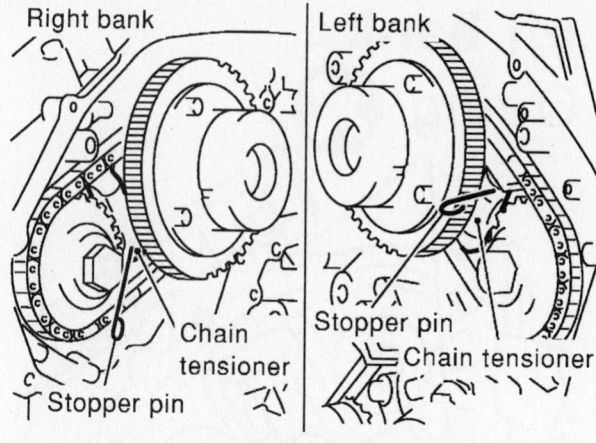

Secondary timing chain tensioner positions—VQ35DE Engine.

67162-FX35-G77

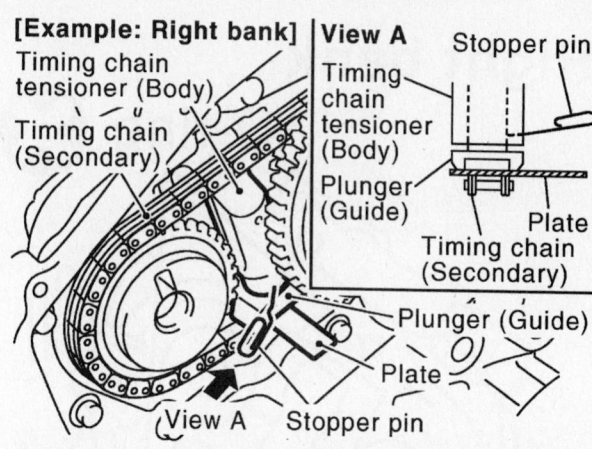

Secondary timing chain plunger setting for removal—VQ35DE Engine.

67162-FX35-G78

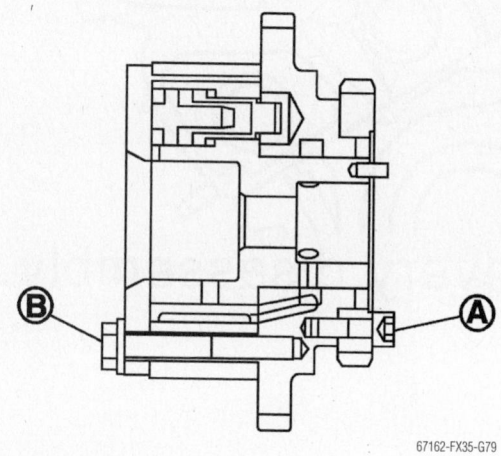

Cross-section of intake camshaft sprocket—Do NOT loosen bolts A or B—VQ35DE Engine.

67162-FX35-G79

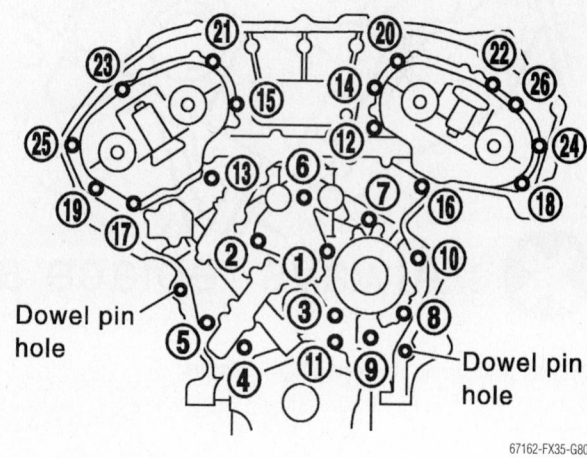

Tightening sequence for the rear timing chain case mounting bolts—VQ35DE Engine.

67162-FX35-G80

29. Remove the primary timing chain tensioner, as follows:

a. Pull the lever down and release the plunger stopper tab. The plunger stopper tab can be pushed up to release.

b. Insert a stopper pin into the tensioner body hole to hold the lever, and keep the tab released.

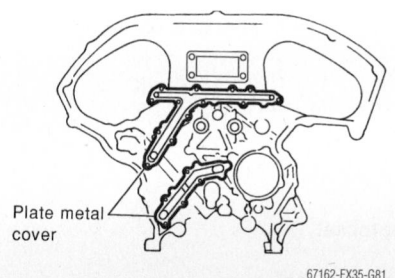

Plate metal cover

67162-FX35-G81

Do not remove these plate metal covers—VQ35DE Engine.

➡ **A 0.098 in. (2.5mm) Allen wrench can be used for a stopper pin.**

c. Insert the plunger into the tensioner body by pressing on the slack guide.

d. Keep the slack guide depressed and hold it by pushing the stopper pin through the lever hole and body hole.

e. Remove the mounting bolts and remove the primary timing chain tensioner.

30. Remove the internal chain guide, tension guide and slack guide.

➡ **The tension guide can be removed after removing the primary timing chain.**

31. Remove the primary timing chain, tension guide and crankshaft sprocket.

✳✳ **CAUTION**

After removing the timing chain, do not turn the crankshaft and camshaft

separately, or the valves will strike piston heads.

32. Remove the secondary timing chain and camshaft sprockets, as follows:

a. Attach a suitable stopper pin to the right and left secondary timing chain camshaft chain tensioners.

b. Remove the intake and exhaust camshaft sprocket bolts. Apply paint to the timing chain and camshaft sprockets for alignment during installation. Secure the hexagonal portion of the camshaft using an open-end wrench to hold the camshaft steady while loosening the mounting bolts.

c. Remove the secondary timing chain together with camshaft sprockets. Turn the camshaft slightly to create slack in the timing chain on the secondary timing chain tensioner side. Insert a 0.020 in. (0.5mm) thick metal or resin plate

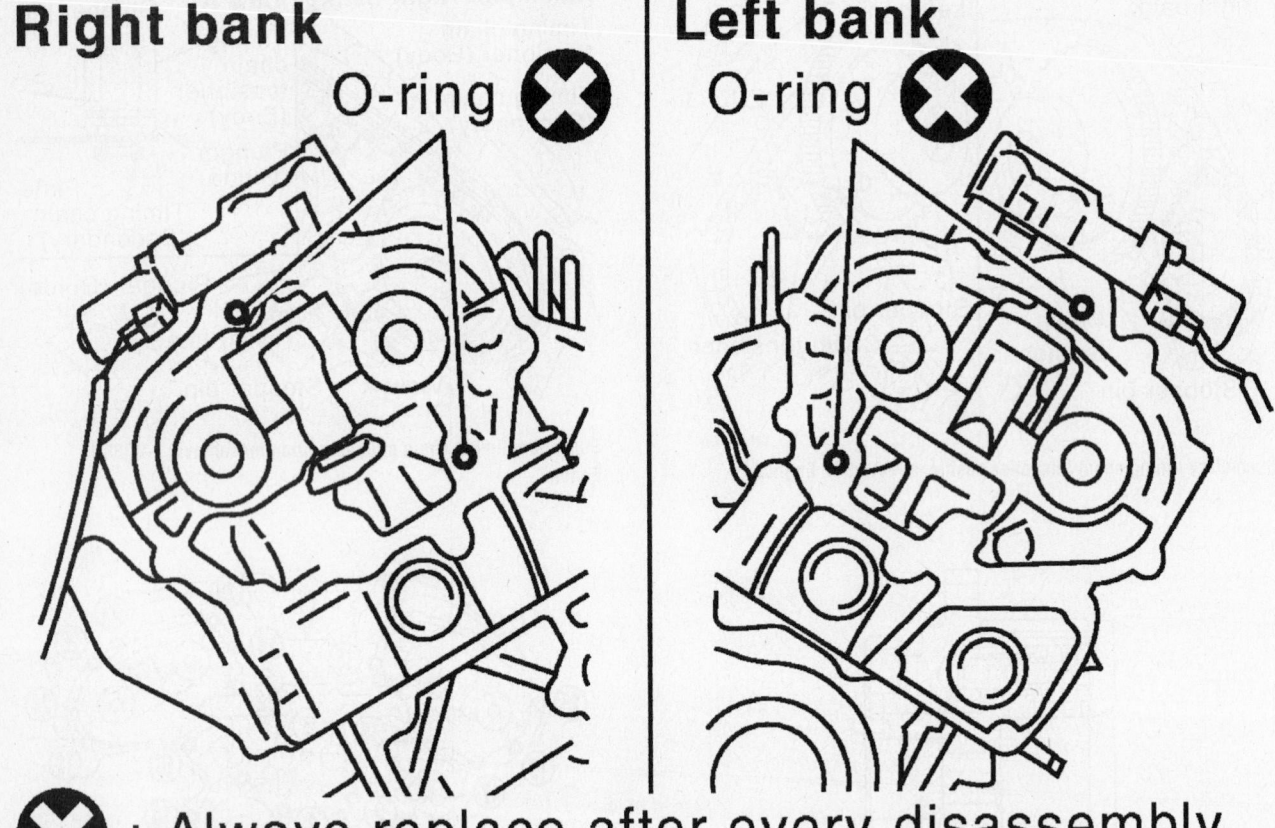

Right bank

O-ring ❌

Left bank

O-ring ❌

❌ : Always replace after every disassembly.

67162-FX35-G82

Cylinder head O-ring positions—VQ35DE Engine.

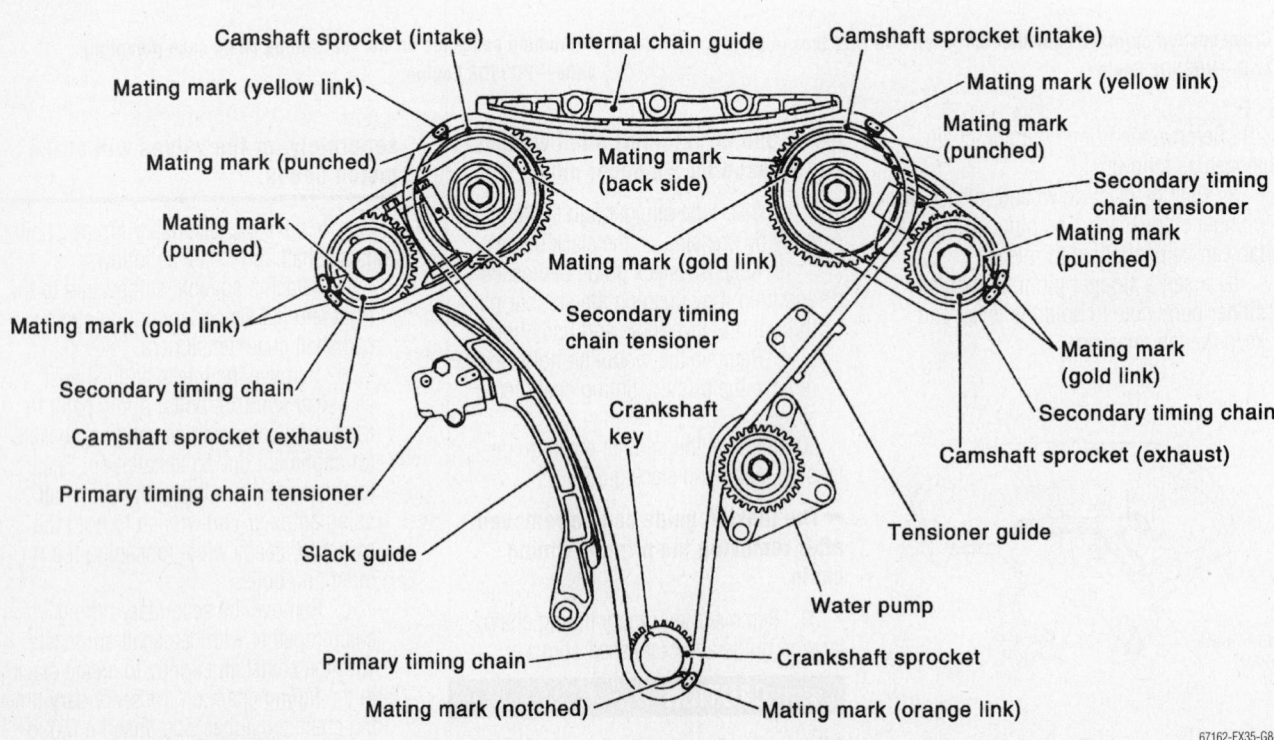

Camshaft sprocket (intake)

Internal chain guide

Camshaft sprocket (intake)

Mating mark (yellow link)

Mating mark (yellow link)

Mating mark (punched)

Mating mark (punched)

Mating mark (back side)

Secondary timing chain tensioner

Mating mark (punched)

Mating mark (punched)

Mating mark (gold link)

Mating mark (gold link)

Secondary timing chain tensioner

Mating mark (gold link)

Secondary timing chain

Camshaft sprocket (exhaust)

Crankshaft key

Secondary timing chain

Primary timing chain tensioner

Camshaft sprocket (exhaust)

Slack guide

Tensioner guide

Water pump

Primary timing chain

Crankshaft sprocket

Mating mark (notched)

Mating mark (orange link)

67162-FX35-G84

Alignment marks for timing chain installation—VQ35DE Engine.

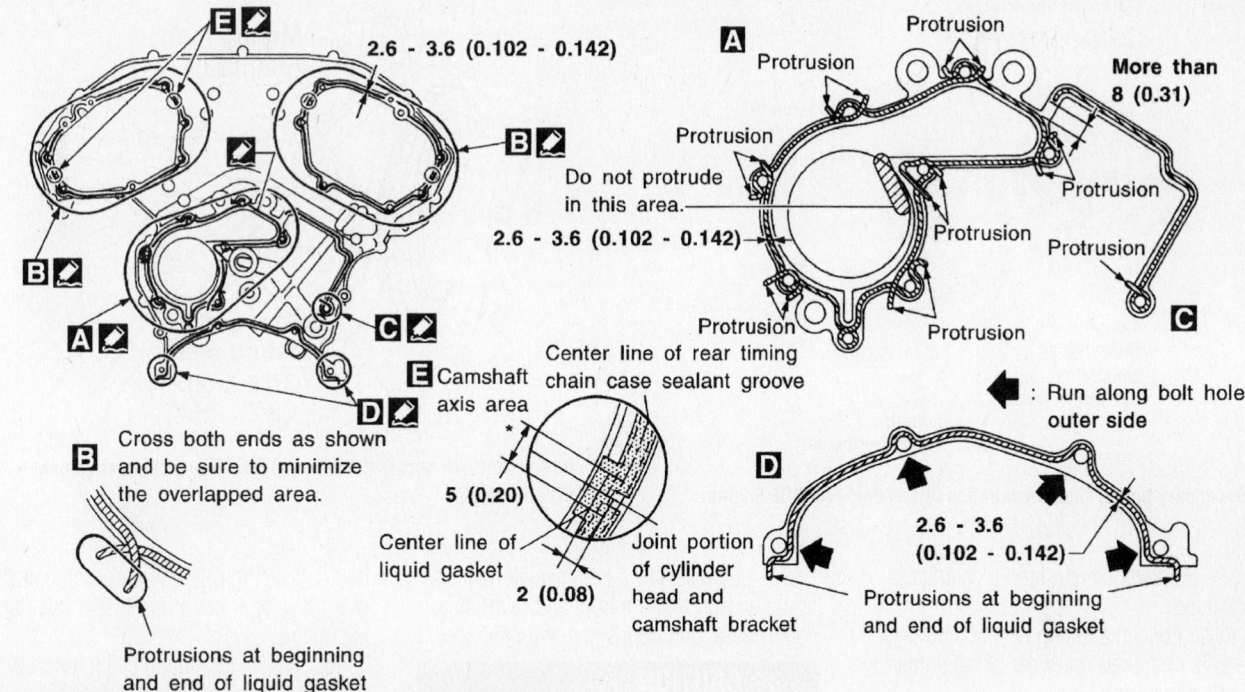

E Camshaft axis area

5 (0.20)

2 (0.08)

Center line of rear timing chain case sealant groove

Center line of liquid gasket

Joint portion of cylinder head and camshaft bracket

B Cross both ends as shown and be sure to minimize the overlapped area.

Protrusions at beginning and end of liquid gasket

D 2.6 - 3.6 (0.102 - 0.142)

Protrusions at beginning and end of liquid gasket

◀ : Run along bolt hole outer side

* : Apply liquid gasket to the chamfered surface between camshaft bracket and cylinder head.

: Apply liquid gasket. (Use Genuine RTV silicone sealant or equivalent. Refer to GI section.)

Unit: mm (in)

67162-FX35-G85

Liquid gasket positions for timing chain case installation—VQ35DE Engine.

between the timing chain and timing chain tensioner plunger guide. Remove the secondary timing chain together with the camshaft sprockets with the timing chain loose from the guide groove.

❊❊ CAUTION

Be careful of the plunger coming-off when removing the secondary timing

chain. This is because the plunger of the secondary timing chain tensioner moves during operation, which can result in the fixed stopper pin coming off.

➡Camshaft sprocket (INT) is a two-for-one assembly of the primary and secondary sprockets.

❊❊ CAUTION

When handling the camshaft sprocket (INT), be careful of the following:

- Handle it carefully to avoid any shock to the camshaft sprocket.
- Do not disassemble. (Do not loosen bolts "A" and "B" as shown.

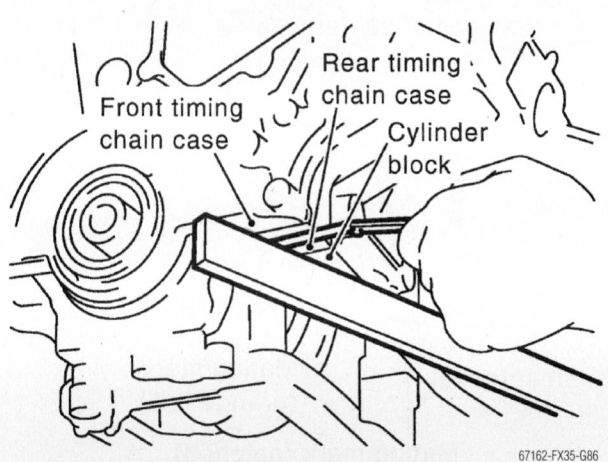

67162-FX35-G86

Check the surface height difference between the rear timing chain case and the cylinder block—VQ35DE Engine.

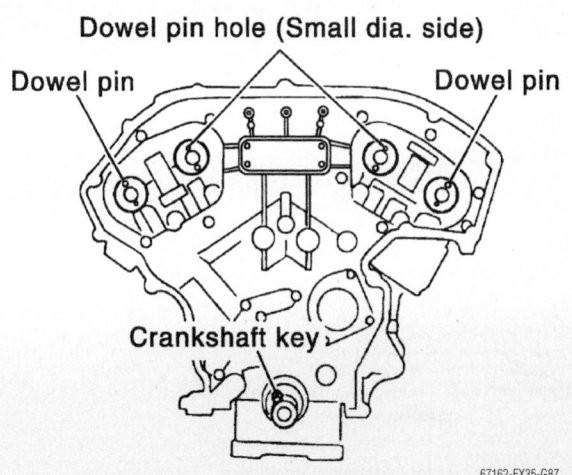

67162-FX35-G87

Camshaft positioning for timing chain installation—VQ35DE Engine.

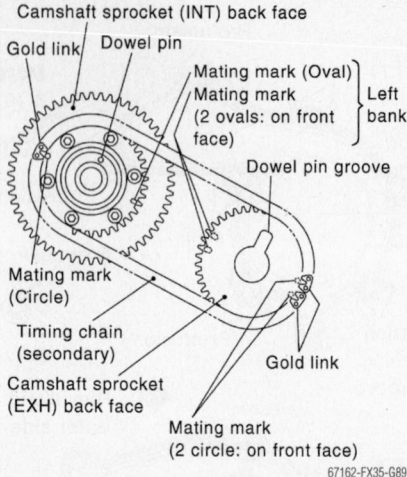

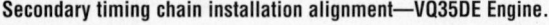

Secondary timing chain installation alignment—VQ35DE Engine.

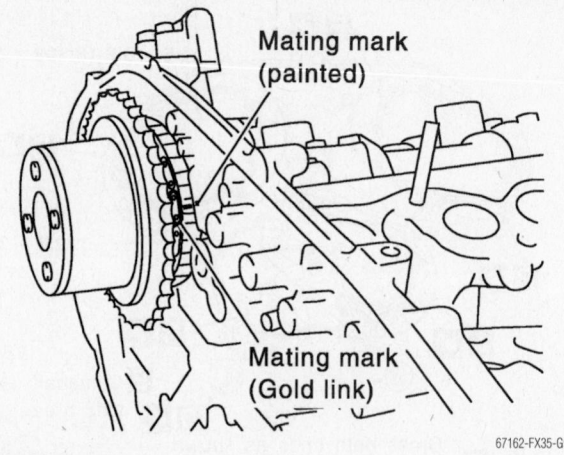

Intake camshaft sprocket and secondary timing chain alignment—VQ35DE Engine.

33. Remove the rear timing chain case, as follows:

    a. Loosen and remove the mounting bolts in the reverse order of the tightening sequence.

    b. Cut the sealant using seal cutter SST: KV10111100 (J37228), or equivalent, and remove the rear timing chain case.

✳✳ CAUTION

Do not remove plate metal cover of engine oil passage. After removing the chain case, do not apply any load on the case which could affect flatness.

34. Remove the O-rings from the cylinder head.

35. Remove the O-rings from the cylinder block.

36. If necessary, remove the secondary timing chain tensioners from the cylinder head, as follows:

    a. Remove the No. 1 camshaft brackets.

    b. Remove the secondary timing chain tensioners with the stopper pins attached.

37. Use a scraper to remove all traces of liquid gasket from the front and rear timing chain cases, and opposite mating surfaces.

✳✳ CAUTION

Be careful not to allow gasket fragments to enter the oil pan.

38. Remove the old liquid gasket from the bolt holes and threads.

39. Use a scraper to remove all traces of liquid gasket from the water pump cover, chain tensioner cover and intake valve timing control covers.

40. Inspect the timing chain for cracks and any excessive wear at link plates and roller links of the timing chain. Replace the timing chain, as necessary.

To install:

➡**The accompanying illustration shows the relationship between the mating**

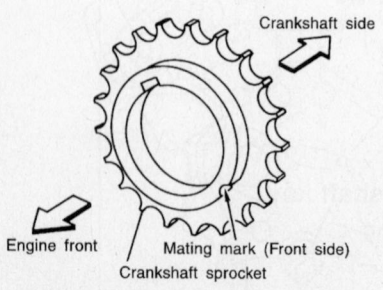

Crankshaft sprocket installation position—VQ35DE Engine.

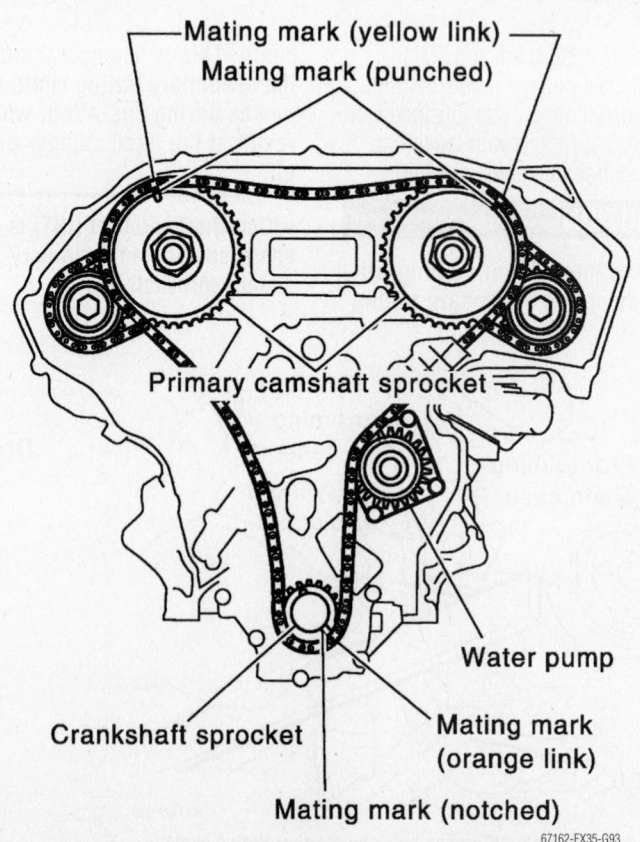

Primary and secondary timing chain installation positioning—VQ35DE Engine.

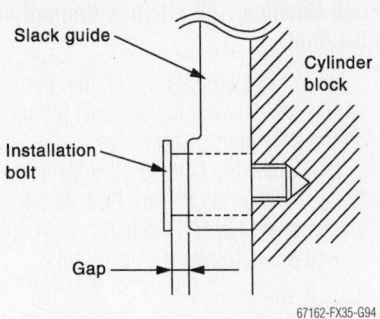

A gap between the slack guide bolt head and the slack guide is normal—VQ35DE Engine.

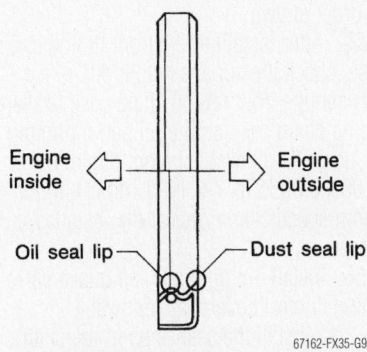

Crankshaft oil seal installation positioning—VQ35DE Engine.

mark on each timing chain and that on the corresponding sprocket, with the components installed.

41. If removed, install the secondary timing chain tensioners on the cylinder head, as follows:

   a. Install the chain tensioners with stopper pins attached and new O-rings.

   b. Install the No. 1 camshaft brackets.

Refer to the camshaft removal and installation procedure.

42. Install new O-rings onto the cylinder block.

43. Install new O-rings on the cylinder head.

44. Apply liquid gasket to the rear timing chain case backside, as shown, with tube presser SST: WS39930000 or equivalent.

❋❋ **CAUTION**

Use Genuine RTV Silicone Sealant or equivalent.

45. For "A" in the illustration, completely wipe out the liquid gasket extended on a portion touching at engine coolant.

46. Apply liquid gasket on the installation position of the water pump and cylinder head completely.

47. Align the rear timing chain case and water pump assembly with right and left dowel pins on cylinder block and install the case.

➡**Make sure the O-rings stay in place during installation on the cylinder block and cylinder head.**

48. Tighten the mounting bolts in the sequence shown. After all bolts are temporarily tightened, retighten them to 9 ft. lbs. (13 Nm) in the tightening sequence.

➡**There are two bolt lengths used for the timing chain case:**

   • 0.79 in. (20mm) length—Pos. 1, 2, 3, 6, 7, 8, 9, 10
   • 0.63 in. (16mm) length—Except Pos. 1, 2, 3, 6, 7, 8, 9, 10

➡**If RTV Silicone Sealant protrudes, wipe it off immediately.**

49. After installing the rear timing chain case, check the surface height difference between the rear timing chain case to the cylinder block at the oil pan mounting surface. If the difference is not within -0.0094-0.0055 in. (-0.24-0.14 mm), repeat the installation procedure.

50. Position the crankshaft so the No. 1 piston is set at TDC on the compression stroke. Make sure that the dowel pin hole, dowel pin and crankshaft key are located as shown.

➡**Though the camshaft does not stop at the position as shown in the figure, for the placement of the cam nose, it is generally accepted the camshaft is placed in the same direction of the figure and as follows:**

   • Camshaft dowel pin hole (intake side): At the cylinder head upper face side in each bank.
   • Camshaft dowel pin (exhaust side): At the cylinder head upper face side in each bank.
   • Crankshaft key: At the cylinder head side of the right bank.

❋❋ **CAUTION**

The hole on the small diameter side must be used for the intake side dowel pin hole. Do not misidentify (ignore the big diameter side).

51. Install the secondary timing chains and camshaft sprockets, as follows:

❋❋ **CAUTION**

The matching marks between the timing chain and sprockets slip easily. Confirm all matching mark positions repeatedly during the installation process.

   a. Push the plunger of the secondary chain tensioner and keep it pressed in with a stopper pin.

   b. Install the secondary timing chains and camshaft sprockets. Align the mating marks on the secondary timing chain (gold link) with the ones on the intake and exhaust camshaft sprockets (stamped), and install them.

➡**Ensure that the sprockets are properly positioned by heeding the following items:**

   • The mating marks for the intake camshaft sprocket are on the back side of the secondary camshaft sprocket.

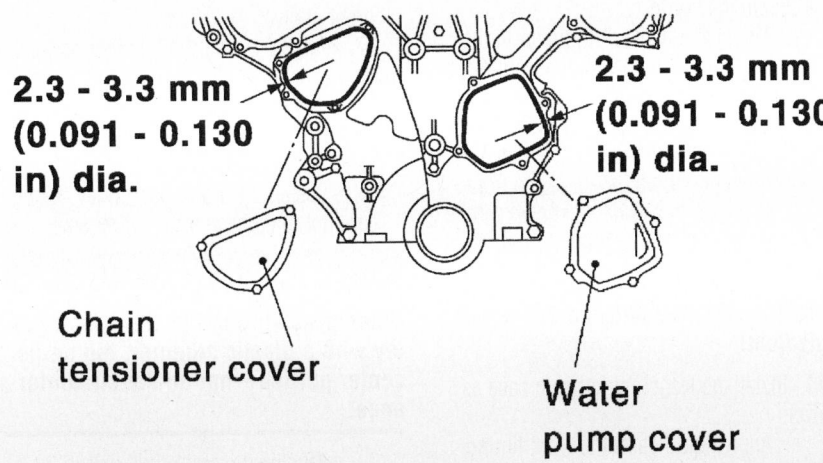

2.3 - 3.3 mm (0.091 - 0.130 in) dia.

2.3 - 3.3 mm (0.091 - 0.130 in) dia.

Chain tensioner cover

Water pump cover

Chain tensioner cover liquid gasket positioning—VQ35DE Engine.

- There are two types of mating marks, circle and oval types. They should be used for the right and left banks, respectively. For the right bank, use the circle type of mating mark, and for the left bank use the oval type of mating mark.
- Align the dowel pin and pin hole on the camshaft with the groove and dowel pin on the sprocket, and install them.
- On the intake side, align the pin hole on the small diameter side of the camshaft front end with the dowel pin on the back side of camshaft sprocket, and install them.
- On the exhaust side, align the dowel pin on the camshaft front end with the pin groove on the camshaft sprocket, and install them.
- In the case that positions of each mating mark and each dowel pin are not fit on the mating parts, make fine adjustments to the position holding the hexagonal portion of the camshaft with a wrench or equivalent.
- Mounting bolts for the camshaft sprockets must be tightened in the next step. Tightening them by hand is enough to prevent the dislocation of dowel pins.
- It may be difficult to visually check the dislocation of the mating marks during and after installation. To make the matching easier, make a mating mark on the top of the sprocket teeth and its extended line in beforehand with paint.

52. After confirming the mating marks are aligned, tighten the camshaft sprocket mounting bolts, while securing the camshaft using a wrench on a hexagonal portion of the camshaft.

53. Pull the stopper pins out from the secondary timing chain tensioners.

54. Install the primary timing chain, as follows:

➡**During alignment, be careful to prevent dislocation of the mating mark alignments of the secondary timing chains.**

a. Install the crankshaft sprocket, making sure the mating marks on the crankshaft sprocket face the front of the engine.

b. Install the primary timing chain, so that the mating mark (punched) on the camshaft sprocket is aligned with the yel-

low link on the timing chain, while the mating mark (notched) on the crankshaft sprocket is aligned with the orange one on the timing chain, as shown.

➡**If it is difficult to align mating marks of the primary timing chain with each sprocket, gradually turn the camshaft using a wrench on the hexagonal portion of the camshaft to align it with the mating marks.**

55. Install the internal chain guide and primary timing chain tensioner.

56. Install the slack guide.

✳✳ CAUTION

Do not over tighten the slack guide mounting bolts. It is normal for a gap to exist under the bolt seats when the mounting bolts are tightened to specification.

57. Remove all dirt and foreign materials completely from the back and mounting surfaces of the chain tensioner.

58. Install chain tensioner for slack guide. When installing the chain tensioner, push in the sleeve and keep it depressed with a stopper pin.

59. After chain tensioner installation, pull out the stopper pin by pressing in the slack guide.

60. Reconfirm that the mating marks on sprockets and timing chains have not slipped out of alignment.

61. Install new O-rings on the rear timing chain case.

62. Install the front oil seal on the front timing chain case, as follows:

a. Apply new engine oil to the oil seal edges.

b. Make sure the garter spring is in position and the seal lip is not inverted, and so that each seal lip is oriented as shown in the figure.

c. Using a suitable drift, press-fit the oil seal until it is flush with the front timing chain case end face.

63. Install the water pump cover and chain tensioner cover to the front timing chain case. Apply liquid gasket to the front timing chain case front side as shown with a tube presser SST: WS39930000 or equivalent.

➡**Use Genuine RTV Silicone Sealant or equivalent.**

64. Install the front timing chain case as follows:

a. Apply liquid gasket to front timing chain case back side as shown with tube presser SST: WS39930000 or equivalent.

➡**Use Genuine RTV Silicone Sealant or equivalent.**

b. Install the dowel pin on the rear timing chain case into the dowel pin hole on the front timing chain case.

c. Tighten the bolts to the specified torque in the order shown. Refer to the following for locating the bolts.
- 0.31 in. (8mm) diameter bolts (positions 1 & 2): 21 ft. lbs. (28 Nm)
- 0.24 in. (6mm) diameter bolts (except positions 1 & 2): 9 ft. lbs. (13 Nm)

d. After tightening, retighten them to the specified torque in the numerical order shown.

65. After installing the front timing chain case, check the surface height difference between the front timing chain case to rear timing chain case on the oil pan mounting surface. The allowable height difference is: -0.005-0.0055 in. (-0.14-0.14mm). If not within specification, repeat the installation procedure.

66. Install the right and left intake valve timing control covers, as follows:

a. Install the seal rings in the shaft grooves.

b. Apply liquid gasket to the intake valve timing control covers with tube presser SST: WS39930000 or equivalent.

➡**Use Genuine RTV Silicone Sealant or equivalent.**

c. Install the collared O-ring in the front cover engine oil hole (left and right sides).

d. Being careful not to move the seal ring from the installation groove, align the dowel pins on the chain case with the holes to install the intake valve timing control covers. Tighten the bolts in the order shown.

67. Install the crankshaft pulley, as follows:

a. Fix the crankshaft in position using ring gear stopper SST: KV10117700 (J44716) or equivalent.

b. Install the crankshaft pulley, taking care not to damage the front oil seal.

✳✳ CAUTION

When press-fitting the crankshaft pulley with a plastic hammer, tap on its center portion—not on the circumference.

c. Tighten the crankshaft bolt to 33 ft. lbs. (44 Nm).

d. Put a paint mark on the crankshaft pulley aligned with the angle mark on the crankshaft pulley bolt. Then, further tighten the bolt 60 degrees (equivalent to one graduation).

68. Rotate the crankshaft pulley in the normal direction (clockwise when viewed from front) to confirm that it turns smoothly.

69. The remainder of installation is the reverse of removal.

➡**If the hydraulic pressure inside the chain tensioner drops after removal/installation, the slack in the guide may generate a pounding noise during and just after engine start. However, this does not indicate a problem—the noise will stop after hydraulic pressure rises.**

70. Perform the following once installation is complete:

a. Before starting the engine, check the levels of the engine coolant, lubrications and working fluid. If less than required quantity, fill to the specified level.

b. Run the engine to check for unusual noise and vibration.

c. Warm up the engine thoroughly to make sure there is no leakage of engine coolant, engine oil and working fluid, fuel and exhaust gas.

d. Bleed air from passages in pipes and tubes of applicable lines, such as in the cooling system.

e. After cooling down the engine, again check the amounts of the engine coolant, engine oil and working fluid. Refill to the specified level, if necessary.

WITHOUT OIL PAN REMOVAL

1. Before servicing the vehicle, refer to the precautions in the beginning of this section.

This section describes the removal/installation procedure of the front timing chain case and timing chain related parts without removing the upper oil pan.

2. Position the vehicle onto a lift, or support it in a safe manner so that work can be performed on the underside of the vehicle.

3. Disconnect the negative battery terminal.

4. Remove the engine cover.

5. Remove the air cleaner case assembly.

6. Remove the front and rear engine undercover.

7. Drain the engine coolant from the radiator.

8. Drain the engine oil from oil pan.

9. Remove and label the engine wiring harnesses.

10. Remove the upper and lower intake manifold collectors.

11. Remove the power steering oil pump from the bracket with the piping connected, then temporarily secure it aside.

12. Remove the power steering oil pump bracket.

13. Remove the alternator.

14. Remove the water bypass hose, water hose clamp, idler pulley bracket, and accessory drive belt tensioner from the front timing chain case.

15. Remove the right and left intake valve timing control covers, by loosening the bolts in the reverse order as shown.

16. Use seal cutter SST: KV10111100 (J37228) or equivalent tool to cut the liquid gasket.

✳✳ CAUTION

A shaft is internally jointed with the intake camshaft sprocket center hole. When removing it, keep it horizontal until it is completely disconnected.

17. Remove the collared O-ring from the front timing chain case left and right sides.

18. Remove the right and left rocker covers.

➡**When the secondary timing chain is not removed/installed, the following step and associated sub-steps are not required.**

19. Position the engine with cylinder No. 1 on TDC of its compression stroke, as follows:

a. Rotate the crankshaft pulley clockwise to align the timing mark (grooved line without color) with the timing indicator.

b. Make sure the intake and exhaust camshaft lobes noses on the No. 1 cylinder (engine front side of right bank) are located as shown.

c. If not, turn the crankshaft one revolution (360 degrees) and align as shown.

➡**When only the primary timing chain is removed, the rocker cover does not need to be removed. To confirm that the No. 1 cylinder is at its compression TDC, remove the front timing chain case first. Then, check the mating marks on the camshaft sprockets.**

20. Remove the crankshaft pulley, as follows:

a. Remove the rear cover plate (2WD

models) or the starter motor (AWD models) and set install the ring gear stopper or equivalent as shown.

b. Loosen the crankshaft pulley bolt until the bolt seating surface is 0.39 in. (10mm) from its original position.

✳✳ CAUTION

Do not remove the crankshaft pulley bolt, since it will be used as a supporting point for the pulley puller.

c. Place a suitable puller tab in the holes of the crankshaft pulley, and remove the crankshaft pulley.

✳✳ CAUTION

Do not situate the puller tab on crankshaft pulley periphery, as this will damage the internal damper.

21. Remove the lower oil pan.

22. Loosen the two mounting bolts in the front of the upper oil pan in the reverse order shown.

23. Remove the front timing chain case, as follows:

a. Loosen the mounting bolts in the reverse order as shown.

b. Insert a suitable tool into the notch at the top of the front timing chain case, as shown.

c. Use seal cutter SST: KV10111100 (J37228) or an equivalent tool to cut the liquid gasket for removal.

✳✳ CAUTION

Use only an approved pry tool to remove the case.

d. Pry off the case.

e. After removal, handle the case carefully so it does not tilt, cant, or warp under a load.

24. Remove the O-rings from the rear timing chain case.

25. Remove the water pump cover and the chain tensioner cover from the front timing chain case.

26. Use seal cutter SST: KV10111100 (J37228) or an equivalent tool to cut the liquid gasket.

27. Remove front oil seal from front timing chain case using a suitable pry tool.

✳✳ CAUTION

Exercise care not to damage the front timing chain case.

28. Remove the timing chain and related parts.

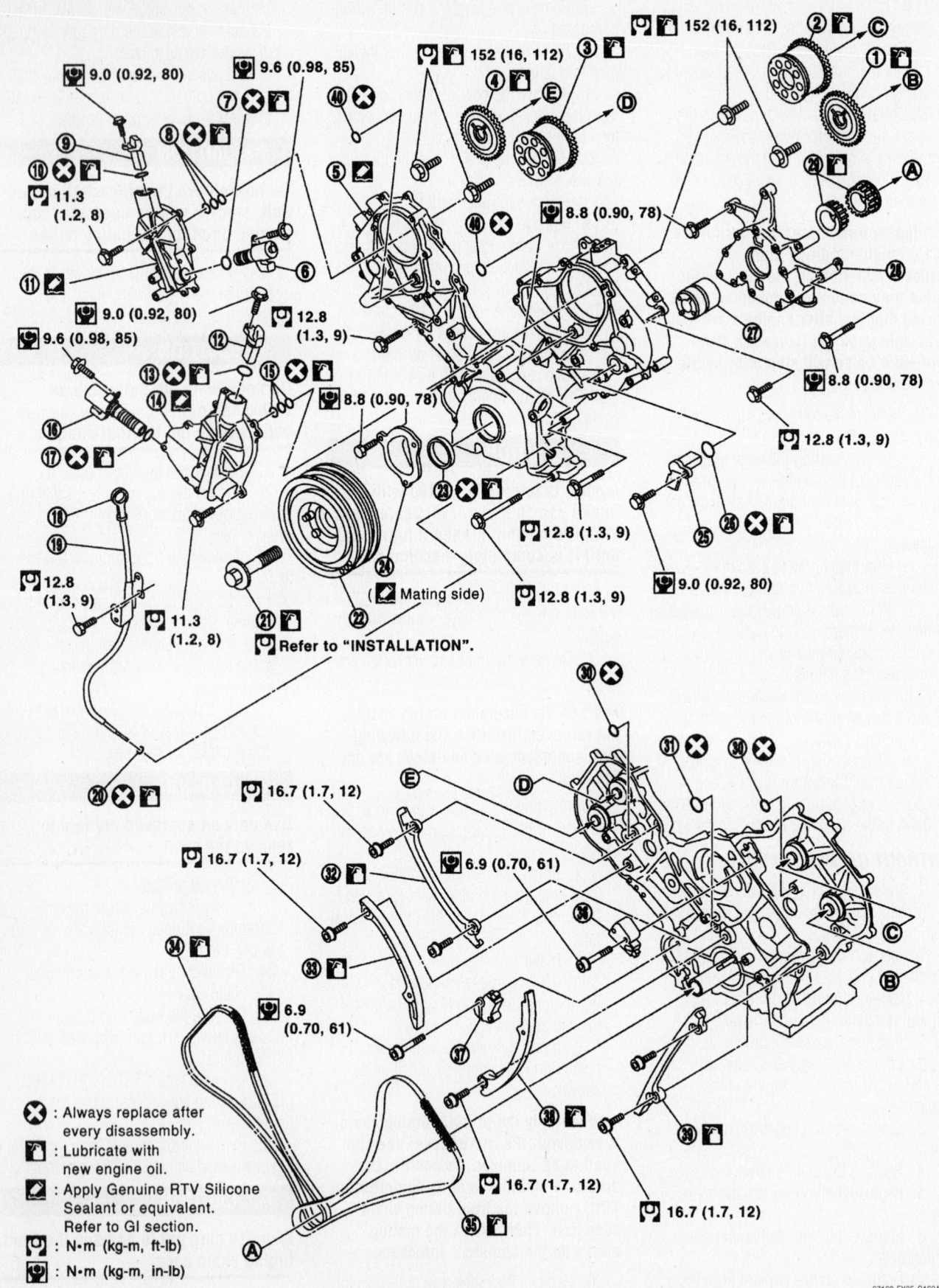

9.0 (0.92, 80)

9.6 (0.98, 85)

152 (16, 112)

152 (16, 112)

11.3 (1.2, 8)

9.0 (0.92, 80)

9.6 (0.98, 85)

12.8 (1.3, 9)

8.8 (0.90, 78)

8.8 (0.90, 78)

8.8 (0.90, 78)

12.8 (1.3, 9)

12.8 (1.3, 9)

12.8 (1.3, 9)

9.0 (0.92, 80)

11.3 (1.2, 8)

(🖊 Mating side)

🔧 Refer to "INSTALLATION".

16.7 (1.7, 12)

16.7 (1.7, 12)

6.9 (0.70, 61)

6.9 (0.70, 61)

16.7 (1.7, 12)

16.7 (1.7, 12)

❌ : Always replace after
    every disassembly.

🔧 : Lubricate with
    new engine oil.

🖊 : Apply Genuine RTV Silicone
    Sealant or equivalent.
    Refer to GI section.

🔧 : N•m (kg-m, ft-lb)

🔧 : N•m (kg-m, in-lb)

67162-FX35-G162A

Exploded view of the timing chain and related components—VK45DE Engine.

1. Camshaft sprocket (EXH)
2. Camshaft sprocket (INT)
3. Camshaft sprocket (INT)
4. Camshaft sprocket (EXH)
5. Front cover
6. Intake valve timing control solenoid valve (right bank)
7. O-ring
8. Seal ring
9. Intake valve timing control position sensor (right bank)
10. O-ring
11. Intake valve timing control cover (right bank)
12. Intake valve timing control position sensor (left bank)
13. O-ring
14. Intake valve timing control cover (left bank)
15. Seal ring
16. Intake valve timing control solenoid valve (left bank)
17. O-ring
18. Oil level gauge
19. Oil level gauge guide
20. O-ring
21. Crankshaft pulley bolt
22. Crankshaft pulley
23. Front oil seal
24. Chain tensioner cover
25. Camshaft position sensor (PHASE)
26. O-ring
27. Oil pump drive spacer
28. Oil pump assembly
29. Crankshaft sprocket
30. O-ring
31. O-ring
32. Timing chain tension guide (right bank)
33. Timing chain slack guide (right bank)
34. Timing chain (right bank)
35. Timing chain (left bank)
36. Chain tensioner (left bank)
37. Chain tensioner (right bank)
38. Timing chain slack guide (left bank)
39. Timing chain tension guide (right bank)
40. O-ring

67162-FX35-G162B

29. Use a scraper to remove all traces of old liquid gasket from the front and rear timing chain cases and upper oil pan, and liquid gasket mating surfaces. Remove the old liquid gasket from bolt holes and threads.

✳✳ CAUTION

Be careful not to allow gasket fragments to enter the oil pan.

30. Use a scraper to remove all traces of liquid gasket from the water pump cover, chain tensioner cover and intake valve timing control covers.

To install:

➡**Throughout the installation procedure, whenever liquid gasket is to be used make sure to use Genuine RTV Silicone Sealant or equivalent.**

31. Install the timing chain and related parts.
32. Hammer the dowel pins (left and left) into the front timing chain case up to a point close to the taper in order to shorten the protrusion length.
33. Install the front oil seal in the front timing chain case. During seal installation, heed the following:
- Apply new engine oil to the oil seal edges.
- Install the seal so that each seal lip is oriented as shown.

- Using a suitable drift, press-fit the oil seal until it becomes flush with front timing chain case end face.
- Make sure the retaining coil spring is in position and seal lip is not inverted.

34. Apply liquid gasket to the front timing chain case front side as shown with tube presser SST: WS39930000 or equivalent.
35. Install the water pump cover and chain tensioner cover to the front timing chain case.
36. Install the front timing chain case, as follows:
   a. Apply liquid gasket to the front timing chain case back side as shown with tube presser SST: WS39930000 or equivalent.
   b. Apply liquid gasket to the oil pan gasket as shown with tube presser SST: WS39930000 or equivalent.
   c. Install the new oil pan gasket, and heed the following:
   - Align the notch of the front timing chain case with the protrusion of the oil pan gasket.
   - Apply liquid gasket to the top surface of the upper oil pan as shown in figure with tube presser SST: WS39930000 or equivalent.
   d. Install new O-rings on the rear timing chain case.

✳✳ CAUTION

Be careful that oil pan gasket is in place.

   e. Assemble the front timing chain case by fitting the lower end of the front timing chain case tightly onto the top face of the upper oil pan. From the fitting point, make the entire front timing chain case contact the rear timing chain case completely. Then, while pressing the front timing chain case from its front and top as shown in figure, install the bolts and temporarily tighten them by hand. Hammer the dowel pin until the outer end becomes flush with the surface.
   f. After hand tightening the mounting bolts, retighten them to specified torque in numerical order shown.

➡**Refer to the following for locating the bolt positions and torque values:**

- 0.31 in. (8mm) diameter bolts— Positions 1 & 2: 21 ft. lbs. (28 Nm)
- 0.24 in. (6mm) diameter bolts— Exception positions 1 & 2: 9 ft. lbs. (13 Nm)

37. Install two mounting bolts in the front of the upper oil pan in numerical order shown to 13 ft. lbs. (17 Nm).
38. Install the lower oil pan.
39. Install the right and left intake valve timing control covers, as follows:

a. Install seal rings in the shaft grooves.

b. Apply liquid gasket to the intake valve timing control covers with tube presser SST: WS39930000 or equivalent.

c. Install the collared O-ring in the front timing chain case oil hole (left and right sides).

d. Being careful not to move the seal ring from the installation groove, align the dowel pins on the chain case with the holes in the intake valve timing control covers.

e. Tighten the bolts in the numerical order shown to 8 ft. lbs. (11 Nm).

40. Install the crankshaft pulley, as follows:

a. Fix the crankshaft in position using ring gear stopper SST: KV10117700 (J44716) or equivalent. b. Install the crankshaft pulley, taking care not to damage front oil seal. When press-fitting crankshaft pulley with a plastic hammer, tap on its center portion (not circumference).

c. Tighten the crankshaft pulley bolt to 33 ft. lbs. (44 Nm).

d. Put a paint mark on the crankshaft pulley aligned with the angle mark on the crankshaft pulley bolt. Then, further tighten the bolt by 60 degrees (equivalent to one graduation).

41. Rotate the crankshaft pulley in the normal direction (clockwise when viewed from the front of the engine) to confirm it turns smoothly.

42. The remainder of installation is the reverse of removal.

➡ **If hydraulic pressure inside the chain tensioner drops after removal/installation, slack in the guide may generate a pounding noise during and just after engine start. However, this is normal and the noise will stop after hydraulic pressure rises.**

43. Perform the following once installation is complete:

- Before starting the engine, check the levels of the engine coolant, lubrications and working fluids. If less than required quantity, fill to the specified level.
- Run the engine to check for unusual noise and vibration.
- Warm up the engine thoroughly to make sure there is no leakage of engine coolant, engine oil, working fluid, fuel and exhaust gas.
- Bleed air from passages in pipes and tubes of applicable lines, such as in the cooling system.

- After cooling down the engine, again check the amounts of engine coolant, engine oil and working fluid. Refill to specified level, if necessary.

VK45DE Engine

1. Before servicing the vehicle, refer to the precautions in the beginning of this section.

2. Remove the engine assembly from the vehicle.

3. Remove the drive belt auto tensioner and idler pulley, as follows:

a. Remove the front engine under-cover.

b. Remove the drive belts.

4. Remove the thermostat housing and hoses;

5. Remove the ignition coil;

6. Remove the rocker cover, as follows:

a. Remove the engine cover.

b. Remove the inlet air duct, air cleaner case and mass air flow sensor assembly, air duct and resonator assembly.

c. Move the wiring harness on the upper rocker cover and its peripheral aside.

d. Remove the wiring harness brackets from the No. 6 camshaft bracket.

e. Remove the electric throttle control actuator.

f. Remove the ignition coil.

g. Remove the PCV hose from the PCV valve.

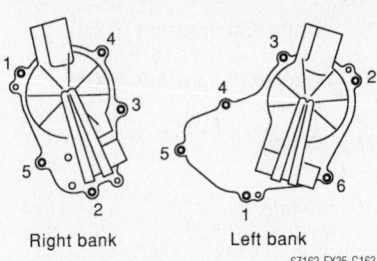

Intake valve timing control cover tightening sequence—VK45DE Engine.

h. Move the wiring harness on the upper rocker cover.

i. Remove the ignition coil.

j. Remove the PCV hose from the PCV valve.

k. Remove the grommets from the right and left cowl top panel. For the right side grommet, remove the battery, battery tray, then the grommet.

l. Loosen the rocker cover bolts in reverse order as shown.

✺✺ CAUTION

Do not hold the oil filler neck on the right bank so that it is not damaged.

➡**Loosen No. 10 bolt of the right bank and No. 10 and No. 12 bolts of the left bank from cowl top panel hole.**

m. Use a scraper to remove all traces of the liquid gasket from the cylinder head and camshaft bracket.

7. If necessary, remove the intake valve

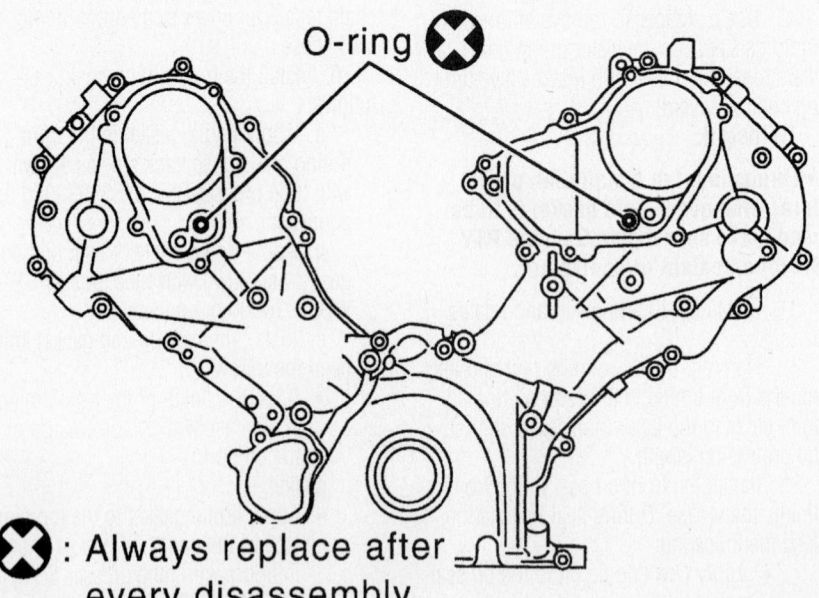

⊗ : Always replace after every disassembly.

Timing chain front cover O-ring locations—VK45DE Engine.

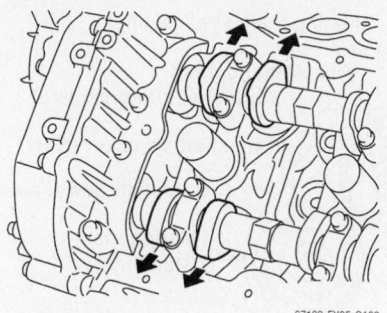

With the engine positioned at TDC on No. 1 cylinder, the camshaft lobes should be positioned as shown—VK45DE Engine.

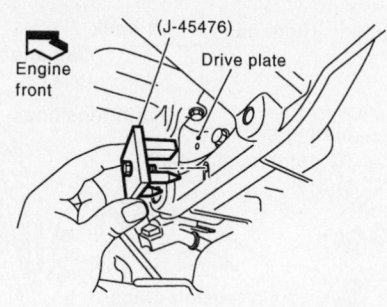

Use a ring gear holder to fix the crankshaft to loosen the crankshaft pulley bolt—VK45DE Engine.

timing control position sensor (right and left banks) and the camshaft position sensor (PHASE) from the intake valve timing control cover and front cover.

✳✳ CAUTION

Handle the sensors carefully to avoid dropping and/or shocking them. Also, do not disassemble them. Do not allow metal powder to adhere to the magnetic part at the sensor tip. Do not place sensors in a location where they are exposed to magnetism.

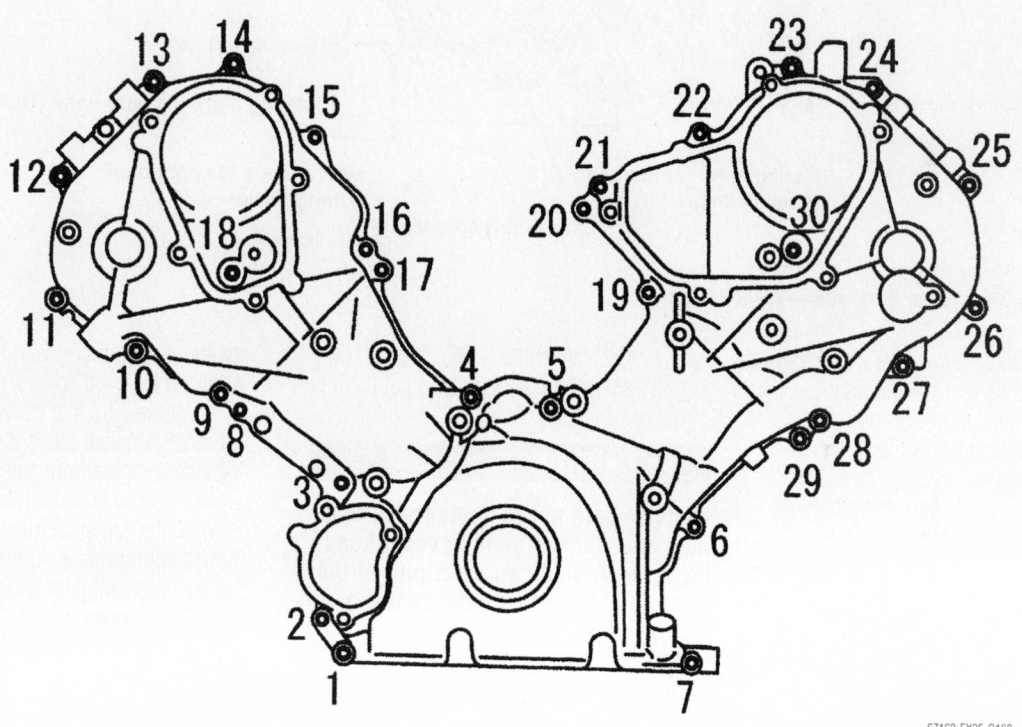

Front timing chain cover mounting bolt tightening sequence—VK45DE Engine.

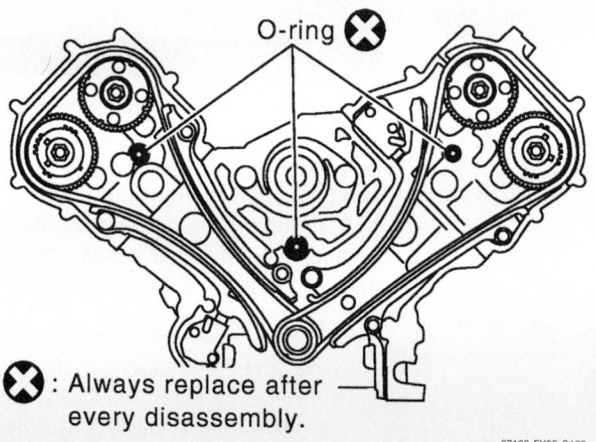

⊗ : Always replace after every disassembly.

O-ring positions on engine block and cylinder heads—VK45DE

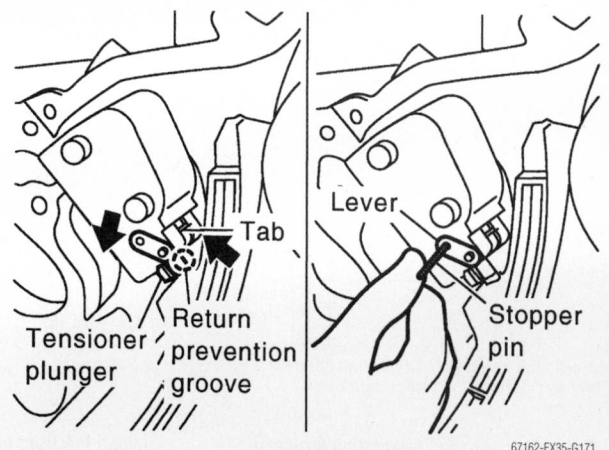

Timing chain tensioner details—VK45DE Engine.

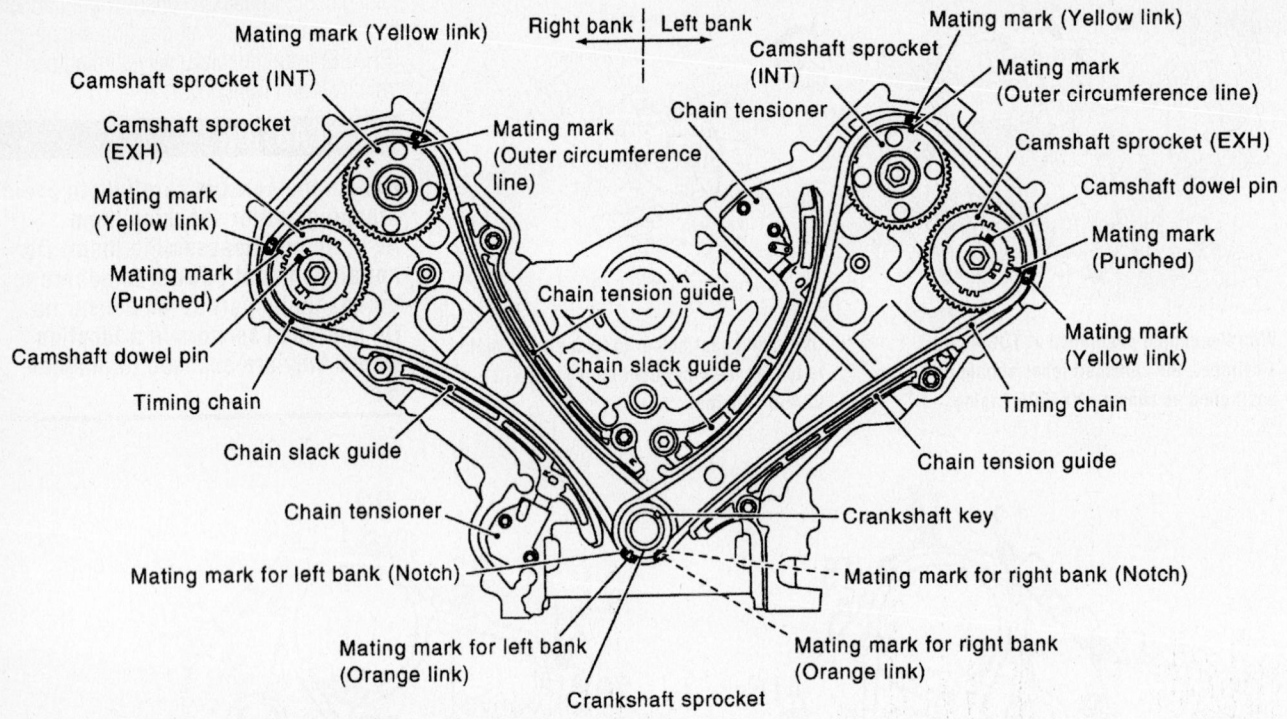

Timing chain positioning for installation—VK45DE Engine.

8. If necessary, remove the intake valve timing control solenoid valve from the intake valve timing control cover.

9. Remove the intake valve timing control cover, as follows:

a. Loosen and remove the mounting bolts in the reverse order shown.

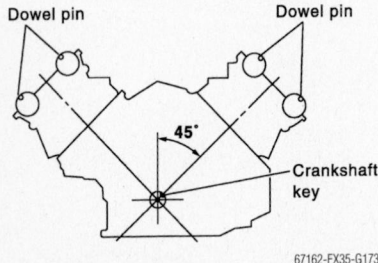

Crankshaft positioning for timing chain installation—VK45DE Engine.

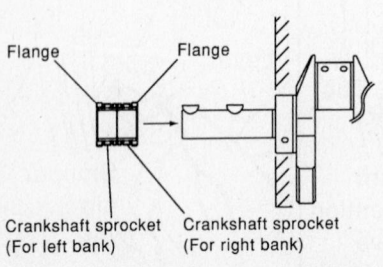

Crankshaft sprocket orientation for installation on the crankshaft—VK45DE Engine.

b. Use seal cutter SST: KV10111100 (J37228) or equivalent tool to cut the liquid gasket for removal.

✳✳ CAUTION

Exercise care not to damage the mating surfaces. Pull out the cover keeping it level, since an inner part of the cover is engaged with the center of the camshaft sprocket (INT).

10. Remove the O-rings from the front cover.

11. Position the crankshaft with No. 1 cylinder at TDC of its compression stroke, as follows:

a. Rotate the crankshaft pulley clockwise to align the TDC identification notch

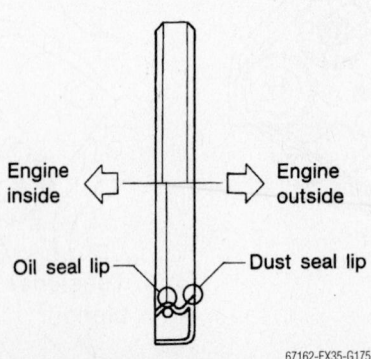

Install the front timing cover front seal as shown—VK45DE Engine.

(without paint mark) with the timing indicator on the front cover.

b. Make sure that both intake and exhaust camshaft lobes of the No. 1 cylinder (engine front side of left bank) are located as shown in the figure.

c. If the camshaft lobes are not positioned appropriately, turn the crankshaft pulley one revolution (360 degrees) and align it as shown.

12. Remove the crankshaft pulley, as follows:

a. Remove rear plate cover, and install the ring gear stopper or equivalent.

b. Loosen the crankshaft pulley bolt, and then pull the crankshaft pulley with both hands to remove it.

✳✳ CAUTION

Do not completely remove the crankshaft pulley bolt. Keep the loosened crankshaft pulley bolt in place to protect the removed crankshaft pulley from dropping. Do not remove the balance weight (inner hexagonal bolt) at the front of the crankshaft pulley.

13. Remove the oil pan and oil strainer.

14. Remove the front cover, as follows:

a. Loosen the mounting bolts in the reverse order shown.

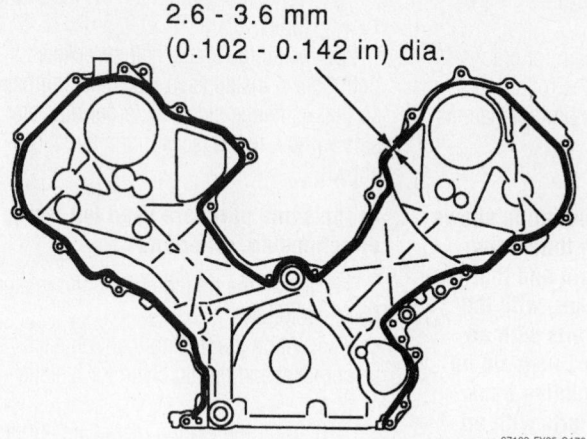

Liquid gasket positioning for timing chain cover installation—VK45DE Engine.

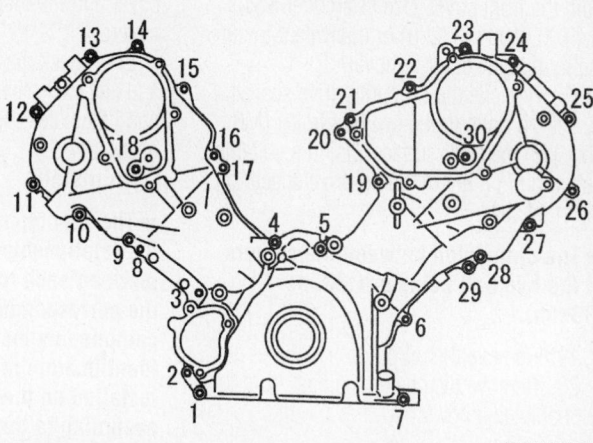

Front timing chain cover mounting bolt tightening sequence—VK45DE Engine.

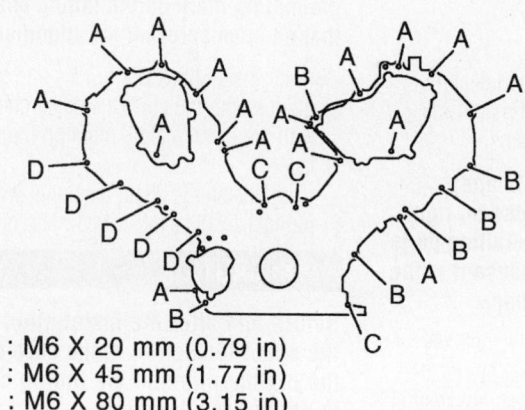

A : M6 X 20 mm (0.79 in)
B : M6 X 45 mm (1.77 in)
C : M6 X 80 mm (3.15 in)
D : M6 X 25 mm (0.98 in)

Front timing chain cover mounting bolt identification—VK45DE Engine.

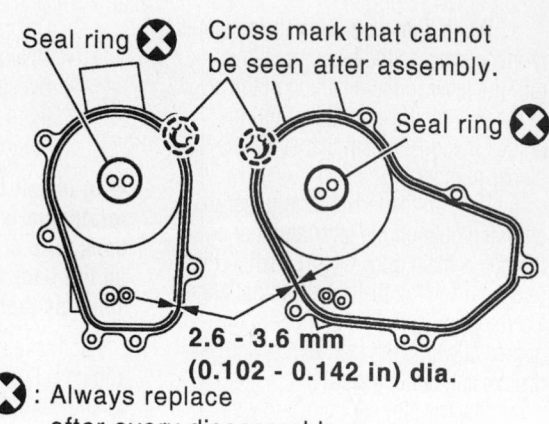

Seal ring ⊗   Cross mark that cannot
                be seen after assembly.
                            Seal ring ⊗

**2.6 - 3.6 mm
(0.102 - 0.142 in) dia.**

⊗ : Always replace
    after every disassembly.

Intake valve timing control cover liquid gasket positioning for installation—VK45DE Engine.

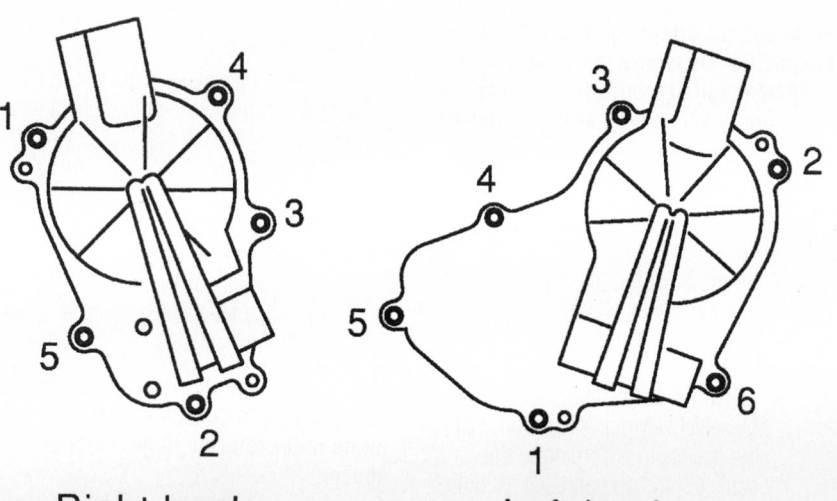

Right bank   **Left bank**

Intake valve timing control cover mounting bolt tightening sequence—VK45DE Engine.

b. Use seal cutter SST: KV10111100 (J37228) or equivalent to cut the liquid gasket.

✳ CAUTION

Exercise care not to damage mating surfaces. After removal, handle the front cover carefully so it does not tilt, cant, or warp under load.

15. Remove the front oil seal from the front cover using a suitable thin-bladed pry tool.

✳ CAUTION

Be careful not to damage the front cover.

16. Remove the O-rings from both cylinder heads and cylinder block.
17. Remove the chain tensioner cover

from the front cover. Use seal cutter SST: KV10111100 (J37228) or equivalent to cut the liquid gasket for removal.

18. Remove the oil pump drive spacer, by setting bolts in the two bolt holes (M6 pitch) on the front surface. Using a suitable puller, pull on the oil pump drive spacer off the crankshaft.

➡️**The dimension between the centers of the two bolt holes is 1.30 in. (33mm).**

19. Remove the oil pump.
20. Remove the chain tensioner from the left bank, as follows:

➡️**To remove timing chain and related parts, start with those on left bank. The procedure for removing parts on right bank is omitted because it is the same as that for the left bank.**

　　a. While lightly pressing the tensioner plunger, depress the tensioner tab in (or turn the lever in the direction of the arrow in the accompanying illustration) to unlock the mechanism that stops the tensioner plunger.

　　b. Push in the tensioner plunger to align the hole on the lever and that on the pump main body. If you push in the tensioner too far, the holes will not align. Therefore, push in the plunger to the degree at which the start of stopper groove and tab engages.

　　c. Insert a stopper pin (hard wire approximately 0.020 in./0.5mm thick or similar tool) to hold the plunger in position. With the plunger held, remove the chain tensioner.

21. Remove the chain tension guide and timing chain slack guide.
22. Remove the timing chain and crankshaft sprocket.

❈❈ CAUTION

After removing the timing chain, do not turn the crankshaft and camshaft separately, or the valves will strike the piston head.

23. With the hexagonal part of the camshaft held with a wrench, loosen the bolts securing the camshaft sprocket.
24. Perform the same procedure on the right side.
25. Use a scraper to remove all traces of old liquid gasket from the front cover and opposite mating surfaces. Remove the oil liquid gasket from the bolt holes and threads.
26. Use scraper to remove all trace of liquid gasket from the chain tensioner cover and the intake valve timing control covers.
27. Check the timing chain for cracks and any excessive wear at the roller links and link plates. Replace the timing chain as necessary.

To install:

➡️**The accompanying illustration shows the relationship between the mating mark on each timing chain and that on the corresponding sprocket, with the components installed. Parts with an identification mark (R or L) should be installed on the corresponding bank according to the mark. Parts with an identification mark include:**

- Camshaft sprocket (INT)
- Dowel pin groove of camshaft sprocket (EXH) (camshaft sprocket is same part both banks)
- Chain tension guide
- Chain slack guide
- Because of parallel manufacture, there are two types of mark (link colors) for timing chains.

➡️**To install the timing chain and related parts, start with those on right bank. The procedure for installing parts on left bank is omitted because it is the same as that for the right bank.**

28. Make sure that the crankshaft key and dowel pin of each camshaft are located as shown in the figure with the crankshaft at No. 1 cylinder at compression TDC. The positioning is as follows:
- Camshaft dowel pin—At the cylinder head upper face side in each bank
- Crankshaft key—At the cylinder head side of left bank

➡️**Though the camshaft may not stop at the position as shown in the figure, for the placement of camshaft lobe nose, it is generally accepted that the camshaft is placed for the same direction of the figure.**

29. Install the camshaft sprockets, while heeding the following items:
- Install the sprockets onto the correct side by checking with the identification mark on surface.
- Install the camshaft sprocket (EXH) by selectively using the groove of the dowel pin according to the bank. (Common part used for both banks.)
- Lock the hexagonal part of the camshaft in the same procedure as

removal, and tighten the mounting bolts.

30. Install the crankshaft sprockets for both banks. Install each crankshaft sprocket so that its flange side (the larger diameter side without teeth) faces in the direction shown.

➡️**The same parts are used but facing directions are different.**

31. Install the timing chains and related parts, as follows:
　　a. Align the mating mark on each sprocket and timing chain for installation.
　　b. After the mating marks are aligned, keep them aligned by holding them in position by hand.

➡️**Before installing the chain tensioner, it is possible to change the position of the mating mark on the timing chain for that on each sprocket for alignment.**

　　c. Install the slack guides and tension guides onto the correct side by checking with the identification mark on the surface.
　　d. Install the chain tensioner with the plunger fixed as described in its removal.

❈❈ CAUTION

Before and after the installation of the chain tensioner, make sure that the mating mark on the timing chain is not out of alignment.

　　e. After installing the chain tensioner, remove the stopper pin to release the tensioner. Make sure the tensioner is released.
　　f. To avoid chain-link skipping of timing chain teeth, do not move the crankshaft or camshafts until the front cover is installed.
32. Perform the same procedure on the right bank, installing the timing chain and related parts on the left side.
33. Install the oil pump.
34. Install the oil pump drive spacer, as follows:
　　a. Insert the oil pump drive spacer according to the directions of the crankshaft key and the two flat surfaces of the oil pump inner rotor.

➡️**If the positional relationship does not allow the insertion, rotate the oil pump inner rotor by finger to allow spacer.**

　　b. After confirming that the position of each part is in correct condition to allow for spacer, force fit the spacer by lightly tapping it with a plastic hammer until it contacts and does not go further.

35. Install the front oil seal in the front cover, as follows:

a. Apply new engine oil to both the oil seal lip and the dust seal lip.

b. Position the seal so that each seal is oriented as shown, then, using a suitable drift, press the front oil seal into the front cover until the face of the oil seal is flush with the mounting surface. The drift used to seat the oil seat should have an outside diameter of 2.20 in. (56mm) and inner diameter of 1.93 in. (49mm).

✳✳ CAUTION

Be careful not to scratch or make burrs on the circumference of the oil seal.

c. Make sure the garter spring is in position and the seal lips not inverted.

36. Install the chain tensioner cover on the front cover.

37. Apply a continuous bead of liquid gasket with tube presser SST: WS39930000 or equivalent to the front cover as shown.

➡**Use Genuine RTV Silicone Sealant or equivalent.**

38. Install the front cover, as follows:

a. Install new O-rings onto the right and left cylinder heads and cylinder block.

39. Apply a continuous bead of liquid gasket with tube presser SST: WS39930000 or equivalent to the front cover as shown.

➡**Use Genuine RTV Silicone Sealant or equivalent.**

a. Make sure again that the mating marks on the timing chain and that on each sprocket are aligned. Then, install the front cover.

✳✳ CAUTION

Be careful to avoid interference with the front end of the oil pump drive spacer. Such interference may damage the front oil seal.

b. Tighten the mounting bolts in numerical order shown. There are four type mounting bolts. The bolts are as follows:

- Position A—M6 x 20mm (0.79 in.)
- Position B—M6 x 45mm (1.77 in.)
- Position C—M6 x 80mm (3.15 in.)
- Position D—M6 x 25mm (0.98 in.)

c. After all bolts are tightened, retighten them in numerical order shown.

✳✳ CAUTION

Be sure to wipe off any excessive liquid gasket leaking onto the surface mating with the oil pan.

40. Install the intake valve timing control cover, as follows:

a. At the back of the intake valve timing control cover, install new seal rings (three for each bank) to the area to be inserted into the camshaft sprocket (INT).

✳✳ CAUTION

Do not spread the seal ring excessively to avoid breaks and deformation.

b. Install the new O-rings on the front cover.

c. Apply a continuous bead of liquid gasket with tube presser SST: WS3930000 or equivalent to the intake valve timing control covers as shown.

➡**Use Genuine RTV Silicone Sealant or equivalent.**

d. Tighten the mounting bolts in the numerical order shown.

41. Install the intake valve timing control position sensor, intake valve timing control solenoid valve and camshaft position sensor (PHASE) on the intake valve timing control cover and front cover, if removed. Be sure to tighten the bolts with flanges completely seated.

42. Install the oil pan and oil strainer.

43. Install the crankshaft pulley, as follows:

a. Hold the crankshaft with ring gear stopper SST: J-45476 or equivalent.

b. Install the crankshaft pulley, taking care not to damage the front oil seal. Install it according to the dowel pin of the oil pump drive spacer. Lightly tapping its center with plastic hammer, insert the pulley.

✳✳ CAUTION

Do not tap the pulley on the side surface where the belt is installed (outer circumference).

c. Apply engine oil onto the threaded parts of crankshaft pulley bolt and seating area.

d. Tighten the crankshaft pulley bolt to 69 ft. lbs. (93 Nm).

e. Mark the crankshaft pulley to align it with the angle mark on the crankshaft pulley bolt.

f. Further tighten the pulley bolt 90 degrees. (Angle tightening) Check the tightening angle by referencing it to the notches. The angle between two notches is 90 degrees.

44. Rotate the crankshaft pulley in the normal direction of rotation (clockwise when viewed from the front of the engine) to confirm it turns smoothly.

45. The remainder of installation is the reverse of removal.

➡**If hydraulic pressure inside the chain tensioner drops after removal/installation, slack in the guide may generate a pounding noise during and just after engine start. However, this is normal and the noise will stop after hydraulic pressure rises.**

46. For rocker arm cover installation, perform the following:

a. Apply liquid gasket to the joint part of the cylinder head and camshaft bracket as shown.

➡**The accompanying illustration shows an example of the left bank side (zoomed in shows camshaft bracket No. 1). Apply only to the camshaft bracket (No. 1) for the right bank side.**

b. Refer to "a" in the illustration to apply liquid gasket to the joint part of the camshaft bracket (both No. 1 and No. 6) and the cylinder head.

c. Refer to "b" in the illustration to apply liquid gasket in 90 degrees to "a".

➡**Use Genuine RTV Silicone Sealant or equivalent.**

d. Install the rocker cover. Check if the rocker cover gasket does not fall from the installation groove of the rocker cover.

e. Tighten the bolts in two separate steps in the numerical order shown, as follows:

- 1st Step: 18 inch lbs. (2 Nm)
- 2nd step: 73 inch lbs. (8.3 Nm)

✳✳ CAUTION

Do not hold the rocker cover by the oil filler neck (right bank) so that it won't be damaged.

➡**Tighten No. 10 bolt of the right bank and No. 10 and No. 12 bolts of the left bank from cowl top panel hole.**

47. Perform the following once installation is complete:

- Before starting the engine, check

the levels of engine coolant, lubrications and working fluid. If less than at the required level, fill them to the specified level.

- Run the engine to check for unusual noise and vibration.
- Warm up the engine thoroughly to make sure there is no leakage of engine coolant, engine oil, working fluid, fuel and exhaust gas.
- Bleed air from passages in pipes

and tubes of applicable lines, such as in the cooling system.

- After cooling down the engine, again check the amounts of engine coolant, engine oil and working fluid. Refill them to the specified level, if necessary.

Piston and Ring

POSITIONING

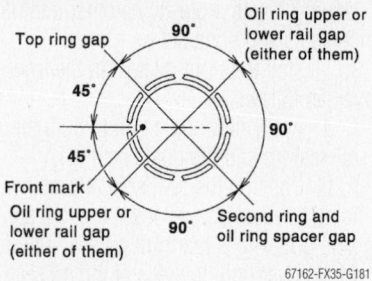

67162-FX35-G181

Piston ring gap positioning—VQ35DE and VK45DE Engines.

FUEL SYSTEM

Fuel System Service Precautions

Safety is the most important factor when performing not only fuel system maintenance but any type of maintenance. Failure to conduct maintenance and repairs in a safe manner may result in serious personal injury or death. Maintenance and testing of the vehicle's fuel system components can be accomplished safely and effectively by adhering to the following rules and guidelines.

- To avoid the possibility of fire and personal injury, always disconnect the negative battery cable unless the repair or test procedure requires that battery voltage be applied.
- Always relieve the fuel system pressure prior to disconnecting any fuel system component (injector, fuel rail, pressure regulator, etc.), fitting or fuel line connection. Exercise extreme caution whenever relieving fuel system pressure, to avoid exposing skin, face and eyes to fuel spray. Please be advised that fuel under pressure may penetrate the skin or any part of the body that it contacts.
- Always place a shop towel or cloth around the fitting or connection prior to loosening to absorb any excess fuel due to spillage. Ensure that all fuel spillage (should it occur) is quickly removed from engine surfaces. Ensure that all fuel soaked cloths or towels are deposited into a suitable waste container.
- Always keep a dry chemical (Class B) fire extinguisher near the work area.
- Do not allow fuel spray or fuel vapors to come into contact with a spark or open flame.
- Always use a back-up wrench when loosening and tightening fuel line connection fittings. This will prevent unnecessary stress and torsion to fuel line piping.
- Always replace worn fuel fitting O-rings with new. Do not substitute fuel hose or equivalent, where fuel pipe is installed.

Relieving Fuel System Pressure

WITH THE CONSULT-II TOOL

1. Before servicing the vehicle, refer to the precautions in the beginning of this section.
2. Turn the ignition switch ON.
3. Perform the "FUEL PRESSURE

FUEL PRESSURE RELEASE

FUEL PUMP WILL STOP BY TOUCHING START IN IDLING. CRANK A FEW TIMES AFTER ENGINE STALL.

67162-FX35-G112

Follow the directions on the CONSULT-II tool for fuel pressure release.

RELEASE" in "WORK SUPPORT" mode with the CONSULT-II.

4. Start the engine.
5. After engine stalls, crank it over two or three times to release all fuel pressure.
6. Turn the ignition switch OFF.

WITHOUT THE CONSULT-II TOOL

1. Before servicing the vehicle, refer to the precautions in the beginning of this section.
2. Remove the fuel pump fuse located in IPDM E/R.
3. Start the engine.
4. After the engine stalls, crank it over two or three times to release all fuel pressure.
5. Turn the ignition switch OFF.
6. Reinstall the fuel pump fuse after servicing the fuel system.

Fuel Filter

REMOVAL & INSTALLATION

The fuel filter is an integral component of the fuel pump. Please refer to the fuel pump removal and installation procedure.

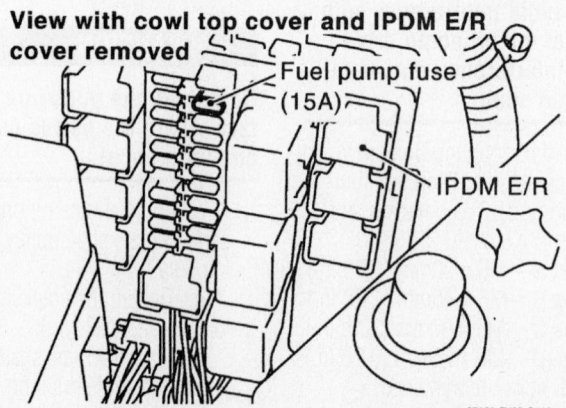

67162-FX35-G113

Fuel pump fuse location for fuel pressure release.

Fuel Pump

REMOVAL & INSTALLATION

1. Before servicing the vehicle, refer to the precautions in the beginning of this section.

2. Check the fuel level on the fuel gauge. If the fuel gauge indicates more than the level as shown in the figure (full or almost full), drain the fuel from the fuel tank until the gauge indicates a level as shown in the figure.

✳✳ CAUTION

Fuel will be spilled when removing the main and sub fuel level sensor units if level of the fuel in the tank is higher than the installation positions of the sensor units.

3. In the case that the fuel pump does not operate, perform the following procedure:

   a. Insert a hose of less than 1 in. (25mm) in diameter into the fuel filler tube through the fuel filler opening to draw fuel from the fuel filler tube.

   b. Disconnect the fuel filler hose from the fuel filler tube.

   c. Insert the fuel tube into the fuel tank through the fuel filler hose to draw the fuel from the fuel tank.

4. Release the fuel pressure from the fuel lines.

5. Open the fuel filler lid.

6. Open the filler cap and release the pressure inside the fuel tank.

7. Remove the rear seat cushion, as follows:

   a. Pull the lock at the front bottom of the seat cushion forward (1 for each side).

   b. Pull the seat cushion upward to release the retaining wire from the plastic hook.

   c. Pull the seat cushion forward to remove.

8. Lift up the floor carpet, then remove the inspection hole cover for the main and sub fuel level sensor units by turning the retaining clips clockwise by 90 degrees.

9. Disconnect the wiring harness connector and fuel feed tube.

10. Disconnect the fuel line quick connector, as follows:

   a. Hold the sides of connector, push in the tabs and pull out the tube.

➡**If the quick connector sticks to the tube of the main fuel level sensor unit, push and pull the quick connector several times until they start to move. Then, disconnect them by pulling.**

✳✳ CAUTION

When dealing with the fuel line quick connector, heed the following:

- The quick connector can be disconnected when the tabs are completely depressed. Do not twist it more than necessary.
- Do not use any tools to disconnected the quick connector.
- Keep the resin tube away from heat. Be especially careful when welding near the resin tube.
- Prevent acidic liquid such as battery electrolyte, etc. from getting on the resin tube.
- Do not bend or twist the resin tube during connection and disconnection.
- Do not remove the remaining retainer on the hard tube (or the equivalent) except when the resin tube or retainer is replaced.
- When the resin tube or hard tube (or the equivalent) is replaced, also replace the retainer with new one.
- To keep the connecting portion clean and to avoid damage and foreign materials, cover them completely with plastic bags or something similar.

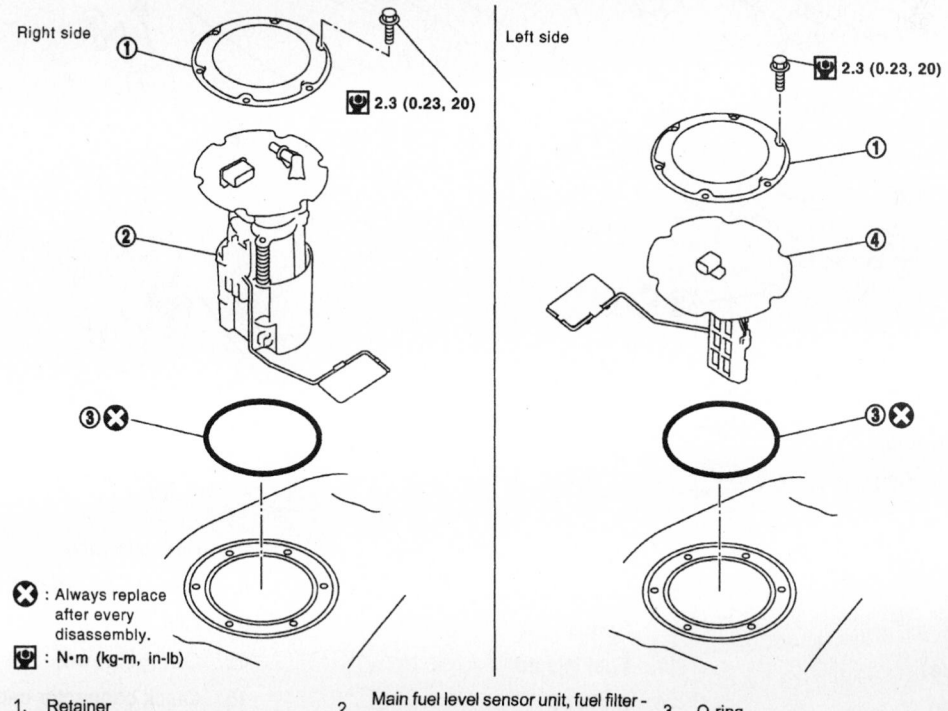

Right side Left side

2.3 (0.23, 20)

2.3 (0.23, 20)

❌ : Always replace after every disassembly.
🔧 : N•m (kg-m, in-lb)

1. Retainer
2. Main fuel level sensor unit, fuel filter - and fuel pump assembly
3. O-ring
4. Sub fuel level sensor unit

67162-FX35-G182

Exploded view of fuel pump and filter assembly and related components.

11. Remove the main fuel level sensor unit, fuel filter and fuel pump assembly, and sub fuel level sensor unit, as follows:

⁂ CAUTION

Make sure to not bend the float arm during removal, and avoid impacts, such as falling, when handling components.

    a. Remove the main fuel sensor unit retainer.

    b. Raise the main fuel level sensor unit, fuel filter and fuel pump assembly, and using snap-ring pliers, remove fuel hose connector.

⁂ CAUTION

Be careful not to damage the fuel hose connector by expanding them excessively.

    c. Removal of sub fuel level sensor unit:

    d. Remove the sub fuel level sensor unit retainer.

    e. Raise and release the sub fuel level sensor unit.

To install:

12. Installation is the reverse of removal.

Fuel Injectors

REMOVAL & INSTALLATION

VQ35DE Engine

    1. Before servicing the vehicle, refer to the precautions in the beginning of this section.

    2. Remove the engine cover.

    3. Release fuel pressure.

    4. Remove the fuel feed hose (with damper) from the fuel sub-tube.

➡ **There is no fuel return route.**

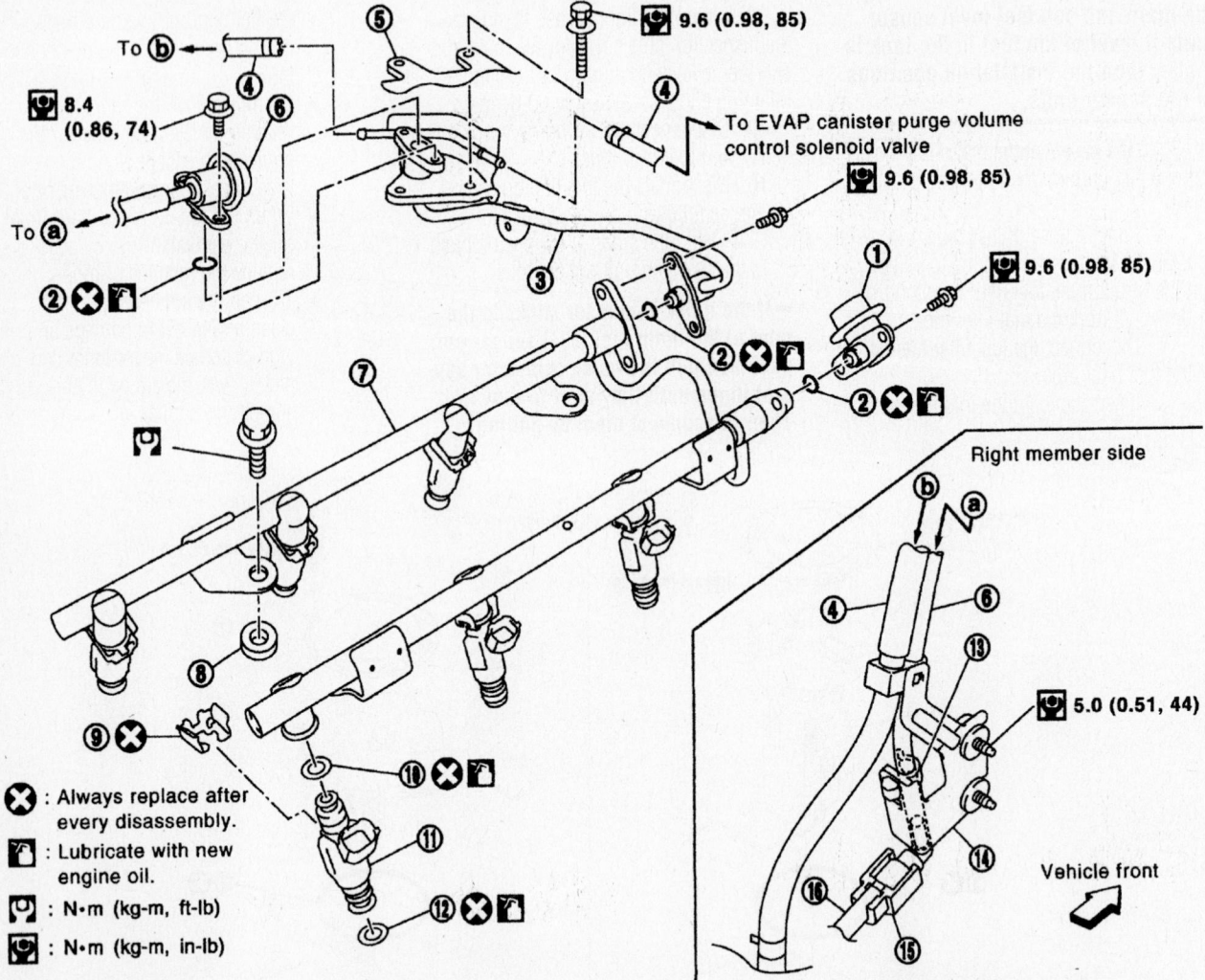

1.	Fuel damper	
4.	EVAP hose	
7.	Fuel tube	
10.	O-ring (blue)	
13.	Hose clamp	
16.	Centralized under-floor piping	
2.	O-ring	
5.	Intake manifold collector (lower)	
8.	Spacer	
11.	Fuel injector	
14.	Bracket	
3.	Fuel sub-tube	
6.	Fuel feed hose (with damper)	
9.	Clip	
12.	O-ring (brown)	
15.	Quick connector cap	

Exploded view of the fuel injector and rail assembly—VQ35DE Engine.

25. After installing the fuel tubes, make sure there is no fuel leakage at the connections in the following steps:

a. Apply fuel pressure to the fuel lines by turning the ignition switch "ON" (with the engine stopped). Then check for fuel leaks at all connections.

b. Start the engine and while holding it at a high rpm check for fuel leaks at all connections.

➡**Use mirrors to check hard-to-see connections.**

VK45DE Engine

1. Before servicing the vehicle, refer to the precautions in the beginning of this section.

✳✳ CAUTION

Do not remove or disassemble parts unless instructed as shown in the figure.

---

2. Remove the engine cover.
3. Release fuel pressure.
4. Disconnect the fuel feed hose on the engine side, as follows:

➡**Except to confirm whether or not there is a quick connector cap, perform the same procedure for the side of centralized under-floor piping as well.**

a. Remove the quick connector cap from the quick connector connection (engine side only).

✳✳ CAUTION

Disconnect the quick connector by using quick connector release tool SST: J-45488 or equivalent; not by picking out the retainer tabs. Inserting the quick connector release tool too hard will not disconnect the quick connector. Hold quick connector release where it contacts and goes no further. Do not pull with lateral force applied. O-ring inside quick connector may be damaged.

---

b. Disconnect the quick connector from the fuel feed damper. With the sleeve side of the quick connector release tool facing the quick connector,

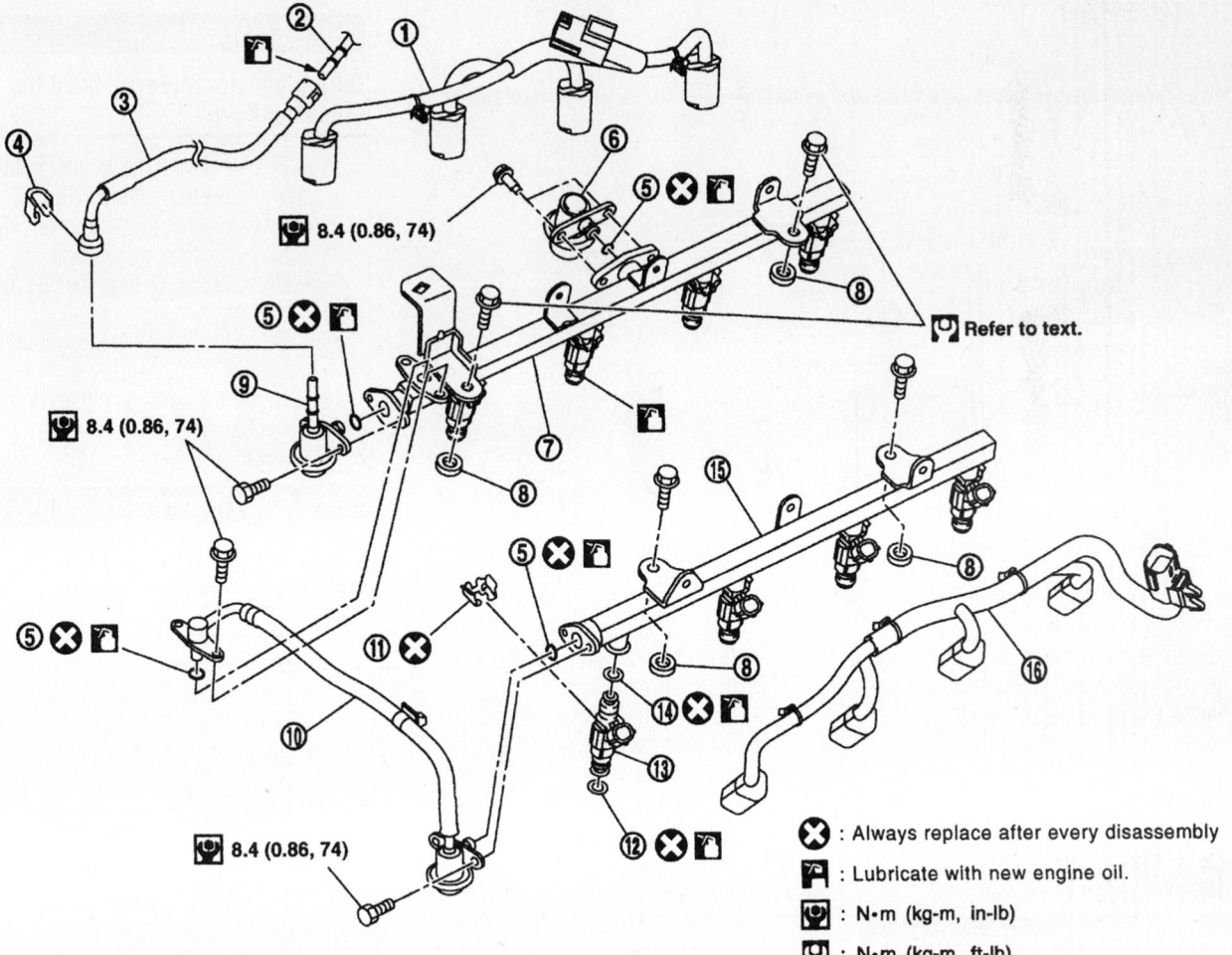

U Refer to text.

✕ : Always replace after every disassembly

⚑ : Lubricate with new engine oil.

🔧 : N·m (kg-m, in-lb)

🔧 : N·m (kg-m, ft-lb)

1.	Fuel injector sub harness (RH)	2.	Centralized under-floor piping	3.	Fuel feed hose
4.	Quick connector cap	5.	O-ring	6.	Fuel damper (RH)
7.	Fuel tube (RH)	8.	Spacer	9.	Fuel feed damper
10.	Fuel damper and fuel hose assembly	11.	Clip	12.	O-ring (Green)
13.	Fuel injector	14.	O-ring (Black)	15.	Fuel tube (LH)
16.	Fuel injector sub harness (LH)				

Exploded view of fuel rail and injector mounting and related components—VK45DE Engine.

67162-FX35-G189

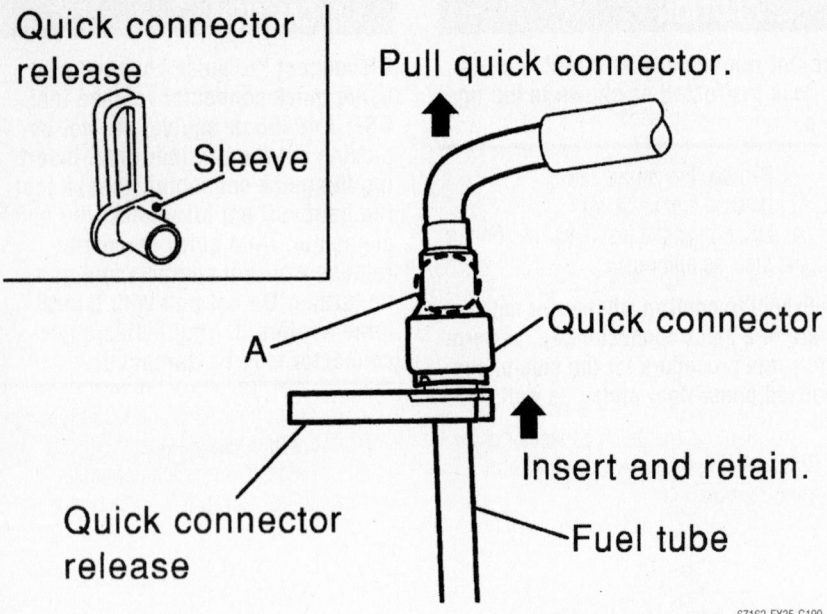

Quick connector release tool used to disconnect the fuel line quick connector—VK45DE Engine.

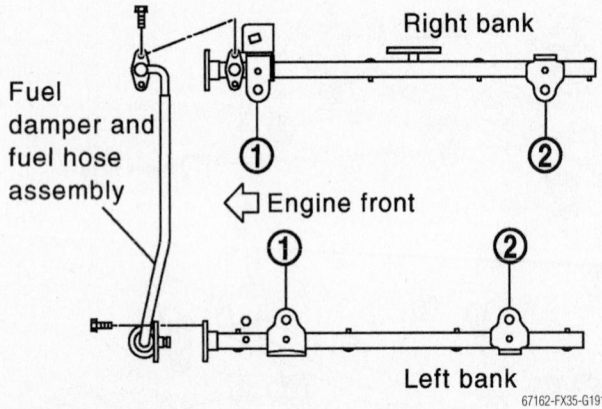

Fuel rail mounting bolt tightening sequence—VK45DE Engine.

install the quick connector release tool onto the fuel tube. Insert the quick connector release tool into the quick connector until the sleeve contacts and goes no further. Hold the quick connector release tool in that position. Draw and pull out the quick connector, holding position "A", straight from the fuel damper.

※ CAUTION

Heed the following when working with the fuel system tubing:

- Prepare a container and cloth beforehand since fuel will leak out.
- Avoid fire and sparks.
- Keep parts away from a heat source. Especially, be careful when welding is performed around them.

- Do not expose parts to battery electrolyte or other acids.
- Do not bend or twist the connection between the quick connector and the fuel feed hose during installation/removal.
- To keep the connecting portion clean and to avoid damage and foreign materials, cover them completely with plastic bags or something similar.

5. Disconnect the fuel damper and fuel hose assembly from the left-hand and right-hand fuel tubes.

※ CAUTION

While the hoses are disconnected, plug them to prevent fuel from leaking and do not separate the fuel damper and fuel hose.

6. Remove the upper intake manifold.
7. Disconnect the wiring harness connector from the fuel injector.
8. Loosen the mounting bolts in reverse order shown, and remove the fuel tube and fuel injector assembly.

※ CAUTION

Do not tilt the fuel injector assembly, or remaining fuel in the pipes may flow out from the pipes.

9. Remove the spacers on the lower intake manifold.
10. Remove the fuel injector from the fuel tube, as follows:
    a. Open and remove the clip.
    b. Remove the fuel injector from the fuel tube by pulling it straight out.

※ CAUTION

During injector removal, heed the following items:

- Be careful with the remaining fuel that leaks from the fuel tube.
- Be careful not to damage the injector nozzles during removal.
- Do not bump or drop the fuel injectors.
- Do not disassemble the fuel injectors.

11. Remove the right-hand fuel damper and fuel feed damper.
To install:

※ CAUTION

When handling all O-rings in this procedure, heed the following:

- Handle the O-ring with bare hands. Never wear gloves.
- Lubricate the O-ring with new engine oil.
- Do not clean the O-ring with solvent.
- Make sure that the O-ring and its mating part are free of foreign material.
- When installing the O-ring, be careful not to scratch it with a tool or fingernails.
- Be careful not to twist or stretch the O-ring. If the O-ring was stretched while it was being attached, do not insert it quickly into the fuel tube.
- Insert the O-ring straight into the fuel tube. Do not decenter or twist it.

12. Install the right-hand fuel damper and fuel feed damper, as follows:

a. Insert the right-hand fuel damper and fuel feed damper straight into the right-hand fuel tube.

b. Tighten the mounting bolts evenly in turn.

c. After tightening the mounting bolts, make sure that there is no gap between the flange and the right-hand fuel tube.

13. Install new O-rings on the fuel injector, paying attention to the fact that the upper and lower O-rings are different. Be careful not to confuse them. The fuel tube side O-ring is black, whereas the nozzle side O-ring is green.

14. Install the fuel injector onto the fuel tube, as follows:

a. Insert the clip into clip mounting groove on the fuel injector. Insert the clip so that lug "A" of the fuel injector matches notch "A" of the clip. Do not reuse clips; replace them with new ones.

➡**Be careful to keep the clip from interfering with the O-ring. If interference occurs, replace the O-ring with a new one.**

b. Insert the fuel injector into the fuel tube with the clip attached. Insert the fuel injector while matching it to the axial center. Insert the fuel injector so that lug "B" of the fuel tube matches notch "B" of the clip. Make sure that the fuel tube flange is securely fixed in the flange fixing groove on the clip.

c. Make sure that installation is complete by checking that the fuel injector

does not rotate or come off. Make sure that the protrusions of the fuel injectors are aligned with the cutouts of the clips after installation.

15. Install spacers on the lower intake manifold.

16. Install the fuel tube and fuel injector assembly onto the intake manifold.

✷✷ CAUTION

Be careful not to let the tip of the injector nozzle come in contact with other parts.

17. Tighten the mounting bolts in two steps in numerical order shown, as follows:
- 1st Step: 7 ft. lbs. (10 Nm)
- 2nd Step: 17 ft. lbs. (23.5 Nm)

18. Connect the fuel feed hose on the engine side, as follows

a. Make sure no foreign substances are deposited in and around the fuel tube and quick connector, and that they are not damaged.

b. Thinly apply new engine oil around the fuel tube from the tip end to the spool end.

c. Align the center to insert the quick connector straight into the fuel tube.

✷✷ CAUTION

Carefully align the center to avoid off-center or crooked insertion to prevent damage to the O-ring inside the quick connector.

d. Insert the fuel tube into the quick connector until the top spool is completely inside the quick connector, and the 2nd level spool exposes right below the quick connector. Hold position "A" when inserting fuel tube into the quick connector. Insert until you hear a "click" sound and actually feel the engagement. To avoid misidentification of engagement with a similar sound, be sure to perform the next step.

e. Pull the quick connector by hand holding position "A". Make sure it is completely engaged (connected) so that it does not disconnect from fuel tube.

f. Install the quick connector cap on quick connector connection. (on the engine side only).

g. Install the fuel feed hose to the hose clamps.

19. Perform the preceding step and associated sub steps for the side of the centralized under-floor piping as well.

20. The remainder of installation is the reverse of removal.

21. Perform the following once installation is complete:

a. Turn the ignition switch "ON" (with the engine stopped). With fuel pressure applied to the fuel piping, check for fuel leakage at all connection points.

➡**Use mirrors for checking fuel connections that are hard to see.**

23. Start the engine. With engine speed increased, check again for fuel leakage at all connection points.

DRIVE TRAIN

Automatic Transmission or Transaxle

REMOVAL & INSTALLATION

1. Before servicing the vehicle, refer to the precautions in the beginning of this section.

✷✷ CAUTION

When removing the automatic transmission assembly from the engine, first remove the crankshaft position sensor (POS) from the A/T assembly. Be careful not to damage the sensor edge.

2. Disconnect the negative battery terminal.
3. Remove the engine cover.
4. Remove the A/T fluid level gauge.
5. Remove the engine under cover.

6. Remove the exhaust front tube and center muffler.

7. Remove the drive shaft.

8. Detach the transmission shifter control rod by removing the snap pin retaining the control rod bracket to the control device assembly lever.

9. Remove the crankshaft position sensor (POS) from A/T assembly.

✷✷ CAUTION

When handling the POS, heed the following:

- Do not subject it to impact by dropping or hitting it.
- Do not disassemble it.
- Do not allow metal filings, etc., to get on the sensor's front edge magnetic area.
- Do not place in an area affected by magnetism.

10. Remove the starter motor, as follows:

a. Disconnect the negative battery terminal.

b. Remove the engine rear undercover.

c. Disconnect the S connector.

d. Remove the B terminal nut.

e. Remove the starter motor mounting bolts and wiring harness clip bracket.

f. Remove the starter motor toward the underside of the vehicle.

11. Disconnect the A/T fluid cooler tube from the transmission assembly.

12. Remove the dust cover from the converter housing part.

13. Turn the crankshaft, and remove the four tightening bolts for the driveplate and torque converter.

✷✷ CAUTION

When turning the crankshaft, turn it clockwise as viewed from the front of the engine.

14. For AWD models with the VQ35DE engine, remove the dynamic damper.

15. Support the transmission assembly with a transmission jack.

※※ CAUTION

When setting the transmission jack, be careful not to allow it to collide against the drain plug.

16. Remove the engine rear member.

17. Tilt the transmission slightly to keep the clearance between the body and the transmission, and then disconnect the air breather hose from the charging pipe.

18. Disconnect the A/T assembly connector.

19. Remove the A/T fluid charging pipe from the A/T assembly.

20. Plug up the openings such as the fluid charge pipe hole, etc.

21. Remove the bolts fixing the transmission assembly to the engine.

22. Secure the torque converter to prevent it from dropping.

23. Secure the transmission assembly to the jack.

24. For 2WD models, remove the transmission assembly from the vehicle with the transmission jack.

25. For AWD models, remove the transmission assembly with the transfer case from the vehicle with the transmission jack. Then remove the transfer mounting bolts and separate the transfer from the transmission.

To install:

26. After inserting the torque converter to the transmission, be sure to check dimension A to ensure it is within the reference value limit of 0.98 in. (25.0mm).

※※ CAUTION

When setting the transmission jack, be careful not to allow it to collide against the drain plug.

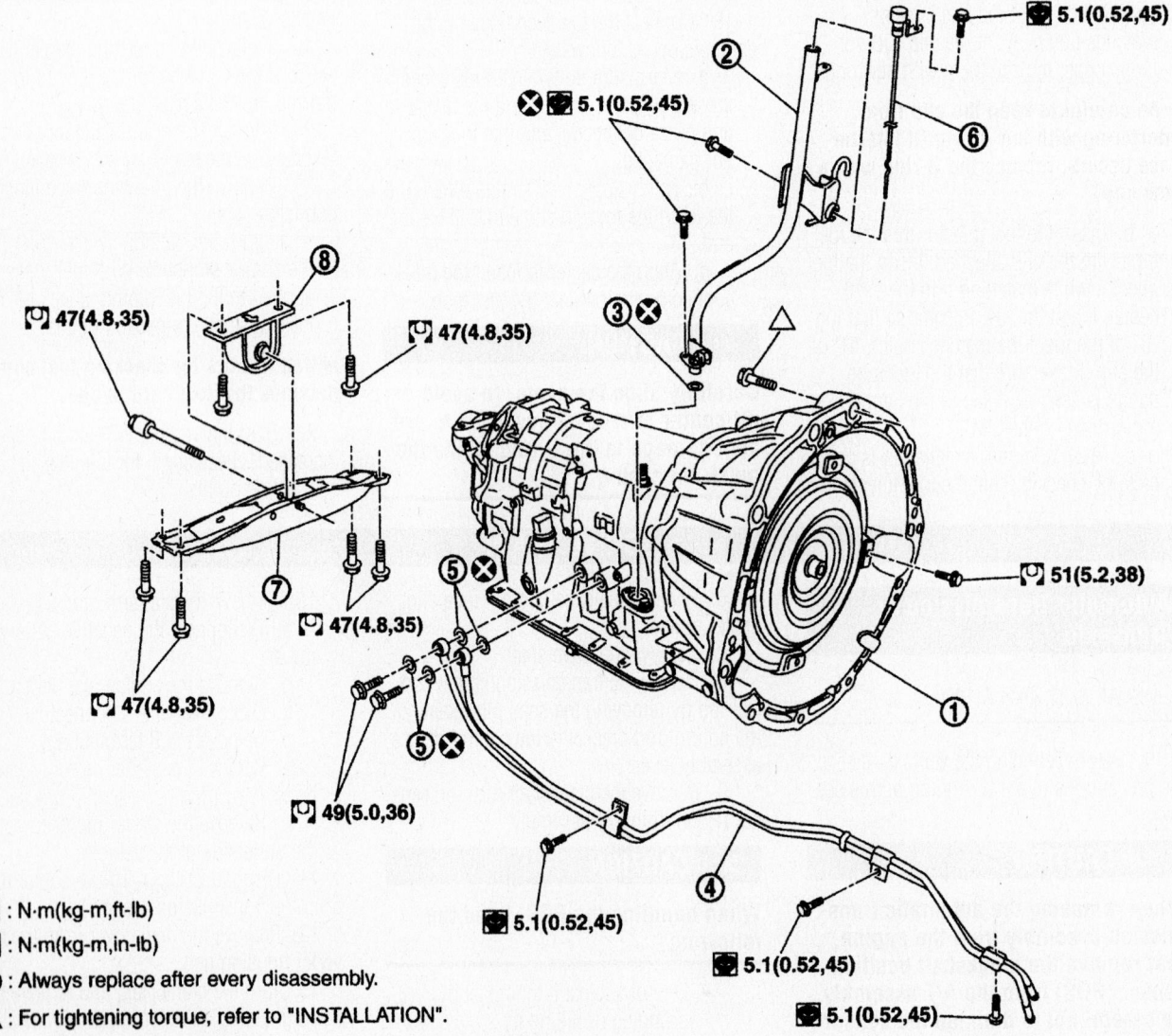

: N·m(kg-m,ft-lb)

: N·m(kg-m,in-lb)

: Always replace after every disassembly.

: For tightening torque, refer to "INSTALLATION".

1. Transmission assembly	2. A/T fluid charging pipe	3. O-ring
4. Fluid cooler tube	5. Copper washer	6. A/T fluid level gauge
7. Engine rear member	8. Insulator	

Exploded view of transmission mounting—2WD Models.

67162-FX35-G194

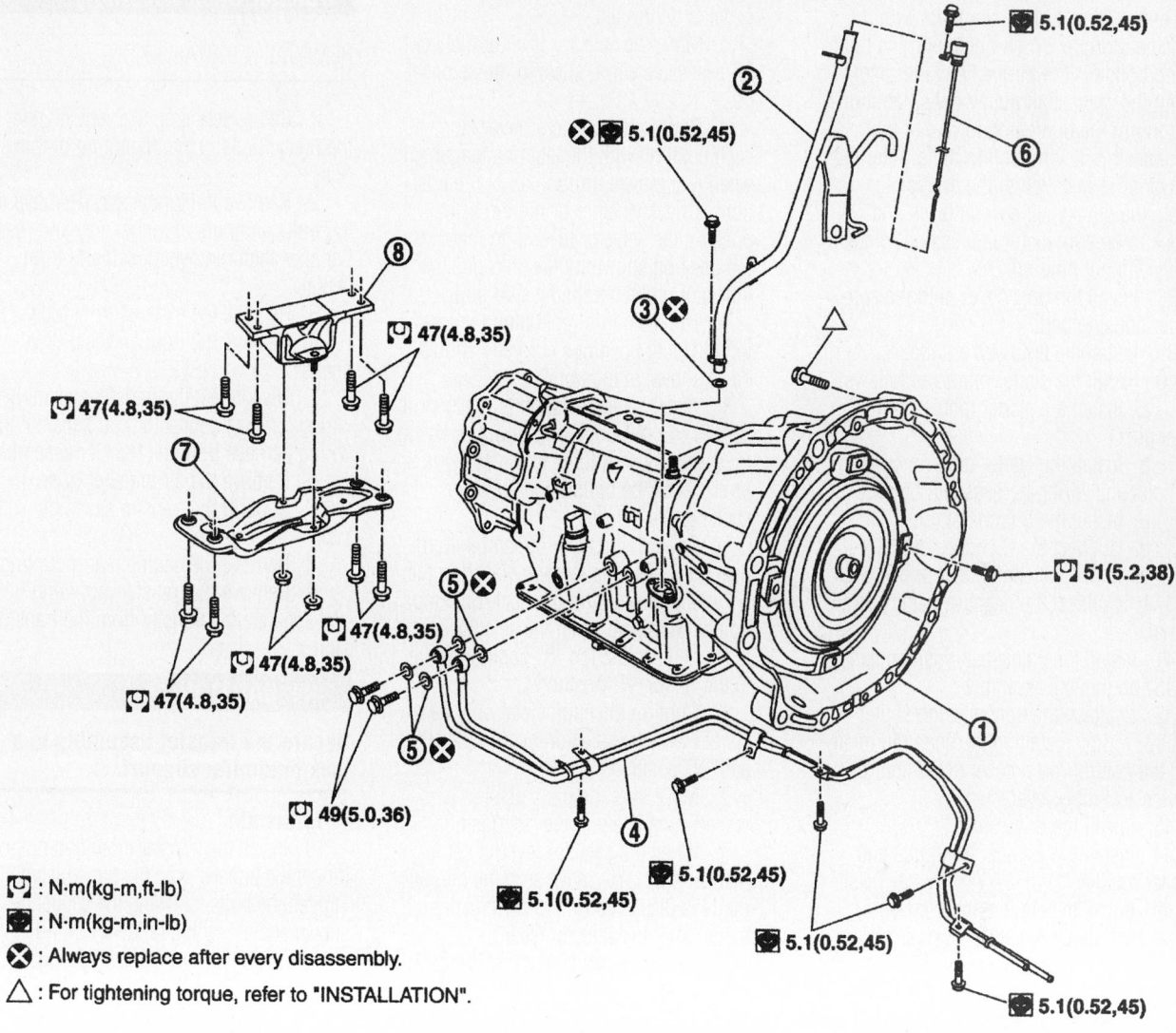

47(4.8,35)

5.1(0.52,45)

5.1(0.52,45)

51(5.2,38)

47(4.8,35)

47(4.8,35)

47(4.8,35)

47(4.8,35)

49(5.0,36)

5.1(0.52,45)

5.1(0.52,45)

5.1(0.52,45)

5.1(0.52,45)

5.1(0.52,45)

- : N·m(kg-m,ft-lb)
- : N·m(kg-m,in-lb)
- : Always replace after every disassembly.
- : For tightening torque, refer to "INSTALLATION".

1. Transmission assembly
2. A/T fluid charging pipe
3. O-ring
4. Fluid cooler tube
5. Copper washer
6. A/T fluid level gauge
7. Engine rear member
8. Insulator

67162-FX35-G195

Exploded view of transmission mounting—AWD Models.

27. For AWD models, install the transfer case to the transmission assembly.

28. Install the transmission assembly into the vehicle with the transmission jack.

29. Install the bolts fixing the transmission assembly to the engine. Identify the transmission-to-engine bolts in the accompanying illustration, as follows:

2WD Models:
- Bolt No. 1, 2.17 in. (55mm) long: 55 ft. lbs. (75 Nm)
- Bolt No. 2, 2.56 in. (65 mm) long: 55 ft. lbs. (75 Nm)
- Bolt No. 3, 2.20 in. (56mm) long: 41 ft. lbs. (55 Nm)
- Bolt No. 4, 1.38 in. (35mm) long: 35 ft. lbs. (47 Nm)

AWD Model/VQ35DE Engine:
- Bolt No. 1, 2.17 in. (55mm) long: 55 ft. lbs. (75 Nm)
- Bolt No. 2, 2.56 in. (65 mm) long: 55 ft. lbs. (75 Nm)
- Bolt No. 3, 1.38 in. (35mm) long: 35 ft. lbs. (47 Nm)
- Bolt No. 4, 2.56 in. (65 mm) long: 25 ft. lbs. (34 Nm)

AWD Model/VK45DE Engine:
- Bolt No. 1, 2.76 in. (70mm) long: 83 ft. lbs. (113 Nm)
- Bolt No. 2, 2.76 in. (70mm) long: 83 ft. lbs. (113 Nm)
- Bolt No. 3, 2.56 in. (65mm) long: 55 ft. lbs. (74 Nm)

30. Unplug the fluid holes.

31. Install the A/T fluid charging pipe onto the A/T assembly.

32. Connect the A/T assembly connector.

33. Install the air breather hose.

34. Install the engine rear member.

35. Carefully, remove the transmission jack.

❋❋ CAUTION

When turning the crankshaft, turn it clockwise as viewed from the front of the engine.

36. For AWD models with the VQ35DE engine, install the dynamic damper.

37. Align the positions of tightening bolts for the driveplate with those of the

torque converter, and temporarily tighten the bolts. Then, tighten the bolts with the specified torque. When tightening the tightening bolts for the torque converter after fixing the crankshaft pulley bolts, be sure to confirm the tightening torque of the crankshaft pulley mounting bolts. After the converter is installed to the driveplate, rotate the crankshaft several turns and check to be sure that transmission rotates freely without binding.

38. Install the dust cover onto the converter housing part.

39. Install the fluid cooler tube.

40. Install the starter motor, as follows:

    a. Install the starter motor on the engine.

    b. Install the starter motor mounting bolts and wiring harness clip bracket.

    c. Install the B terminal nut.

    d. Connect the S connector.

    e. Install the engine rear undercover.

    f. Connect the negative battery terminal.

41. Install the crankshaft position sensor (POS) on the A/T assembly.

42. Reattach the transmission shifter control rod by installing the snap pin retaining the control rod bracket to the control device assembly lever.

43. Install the drive shaft.

44. Install the exhaust front tube and center muffler.

45. Install the engine under cover.

46. Install the A/T fluid level gauge.

47. Install the engine cover.

48. Connect the negative battery terminal.

49. Check the adjustment of the A/T position, as follows:

    a. Loosen the nut of the control rod.

    b. Place the PNP switch and selector lever in the "P" position.

    c. While pressing the lower lever toward the rear of the vehicle (in P position direction), tighten the nut to the specified torque.

✳✳ CAUTION

Do not push the bracket.

50. Check the A/T position, as follows:

    a. Place the selector lever in "P" position, and turn ignition switch ON (engine OFF).

    b. Make sure the selector lever can be shifted to positions other than "P" when the brake pedal is depressed. Also, make sure the selector lever can be shifted

from "P" position only when the brake pedal is depressed.

    c. Move the selector lever and check for excessive effort, sticking, noise or rattle.

    d. Confirm the selector lever stops at each position with the feel of engagement when it is moved through all of the positions. Check whether or not the actual position the selector lever is in matches the position shown by the shift position indicator and the transmission body.

    e. The method of operating the lever to individual positions correctly should be as shown in the figure.

    f. When selector button is pressed in "P", "R", or "N" position without applying forward/backward force to the selector lever, check the button operation for sticking.

    g. Confirm the back-up lamps illuminate only when the lever is placed in the "R" position. Confirm the back-up lamps do not illuminate when the selector lever is pushed against the "R" position while in the "P" or "N" positions.

    h. Confirm the engine can only be started with the selector lever in the "P" and "N" positions.

    i. Make sure the transmission is locked completely in the "P" position.

    j. When the selector lever is set to manual shift gate, make sure the manual mode is displayed on the combination meter. Shift the selector lever to "+" and "-" sides, and make sure set shift position changes.

Transfer Case Assembly

REMOVAL & INSTALLATION

1. Before servicing the vehicle, refer to the precautions in the beginning of this section.

2. Remove the tunnel stay. Remove fixing bolts and nuts of tunnel stay and member stay, then remove those parts from vehicle.

3. Remove the exhaust front tube.

4. Remove the front and rear driveshafts.

5. Disconnect the transfer assembly wiring harness connector and separate the wiring harness from the transfer assembly.

6. Remove the air breather hose.

7. Support the transfer assembly with a jack.

8. Remove the engine rear mounting.

9. Remove the transfer mounting bolts and separate the transfer from the transmission.

✳✳ CAUTION

Secure the transfer assembly to a jack or similar support.

To install:

10. Install the transfer mounting bolts and mount the transfer onto the transmission. Tighten the bolts following the sequence shown and according to the following:

- Bolt No. 1, 2.95 in. (75mm): 27 ft. lbs. (37 Nm)

⊙ : Transfer to Transmission
⊗ : Transmission to transfer

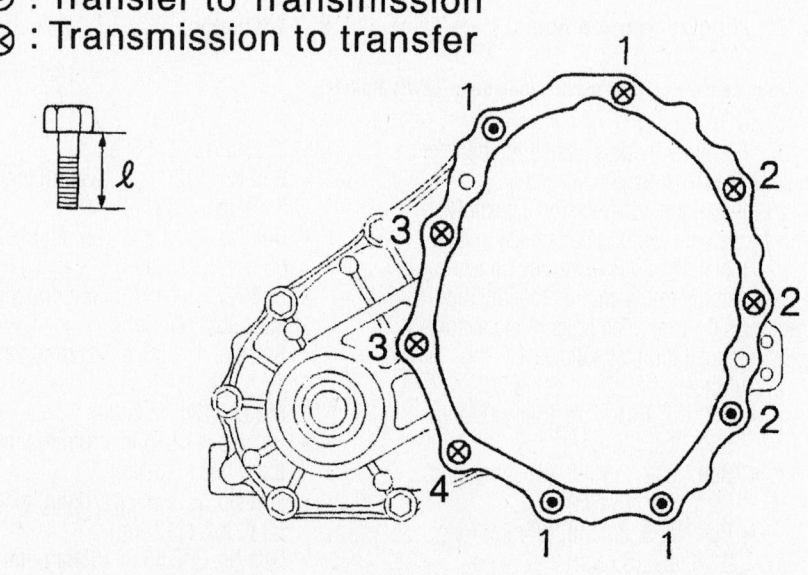

Transfer case-to-transmission mounting bolt tightening sequence.

67162-FX35-G196

- Bolt No. 2, 1.77 in. (45mm): 27 ft. lbs. (37 Nm)
- Bolt No. 3, 1.57 in. (40mm): 27 ft. lbs. (37 Nm)
- Bolt No. 4, 1.18 in. (30mm): 27 ft. lbs. (37 Nm)

11. Install the engine rear mounting.

12. Install the air breather hose.

13. Disconnect the transfer assembly wiring harness connector and separate the wiring harness from the transfer assembly.

14. Install the front and rear driveshafts.

15. Install the exhaust front tube.

16. Install the tunnel stay. Install fixing bolts and nuts of tunnel stay and member stay, then Install those parts from vehicle.

17. Check the fluid level and for fluid leakage.

Drive shaft

REMOVAL & INSTALLATION

Front

1. Before servicing the vehicle, refer to the precautions in the beginning of this section.

2. Remove the front and rear engine undercover.

3. Remove the front cross bar.

4. Remove the exhaust front tube bracket.

5. Disconnect the heated oxygen sensor wiring harness connector.

6. Remove the exhaust front tube mounting nuts.

7. Remove the right bank catalytic converter.

8. Remove the power steering piping mounting bolts.

9. Remove the power steering gear box fixing bolts to secure the working area for removal of the drive shaft.

> **❋❋ CAUTION**
>
> **Be careful not to damage the steering gear box piping during removal.**

10. Put matching marks on the propeller shaft flange and the companion flange on the front drive assembly.

> **❋❋ CAUTION**
>
> **For matching mark, use paint. Never damage the propeller shaft flange and companion flange on the front drive.**

11. Set a transmission jack under the transfer, remove the mounting bolts and rear engine mounting bracket. On the VK45DE engine, lower the transmission jack about 0.16-0.20 in. (40-50mm).

12. Remove the bolts and then remove the drive shaft from the front final drive and transfer.

13. As shown in the accompanying illustration, fix the yoke on one side and check the axial play of the joint. The standard is 0 in. (0mm), therefore if there is any play, replace the propeller shaft assembly.

14. Check the propeller shaft for bending and damage. If damage is detected, replace the drive shaft assembly.

To install:

15. Install the drive shaft into the transfer, then install the propeller shaft onto the front final drive flange while the matching marks are aligned.

16. Tighten the drive flange bolts to 29 ft. lbs. (39 Nm).

> **❋❋ CAUTION**
>
> **Do not reuse the bolts and nuts. Always replace them with new ones.**

17. The remainder of installation is the reverse of removal.

18. After installation, check the vibration

---

by driving the vehicle. If vibration is present, remove the drive shaft from the final drive companion flange and turn the propeller shaft 90, 180 or 270 degrees. Then, reinstall the drive shaft onto the companion flange.

19. Recheck vibration by driving the vehicle.

Rear

1. Before servicing the vehicle, refer to the precautions in the beginning of this section.

2. Move the A/T select lever to "N" position and release the parking brake.

3. Remove the tunnel stay.

4. Remove the center muffler.

5. Loosen the center bearing lower mounting bracket fixing nuts.

6. Put matching marks on the flange and the rear propeller shaft.

> **❋❋ CAUTION**
>
> **For matching marks, use paint. Never damage the propeller shaft flange and companion flange on the rear final drive.**

7. Remove the propeller shaft fixing bolts and nuts.

8. Remove the center bearing lower mounting bracket fixing nuts, remove the drive shaft from the vehicle.

9. Inspect drive shaft run out. Measure run out at the points shown below. If run out exceeds 0 in. (0mm), replace the drive shaft assembly.

2WD models:
- Position A: 7.56 in. (192mm)
- Position B: 7.48 in. (190mm)
- Position C: 7.28 in. (185mm)

2WD models:
- Position A: 6.38 in. (162mm)
- Position B: 9.65 in. (245mm)
- Position C: 7.28 in. (185mm)

To install:

➡ **During installation, refer to the following torque specifications:**

2WD models with VQ35DE engine:
- Center flange attaching bolts and nuts—70 ft. lbs. (95 Nm)
- First (front-most) flange-to-flange attaching nut—61 ft. lbs. (83 Nm)
- Center bearing lower mounting bracket nuts—32 ft. lbs. (44 Nm)
- Second (rear-most) rear-to-rear drive unit fasteners—42 ft. lbs. (57 Nm)

AWD models with VQ35DE engine:
- First (front-most) transmission-to-

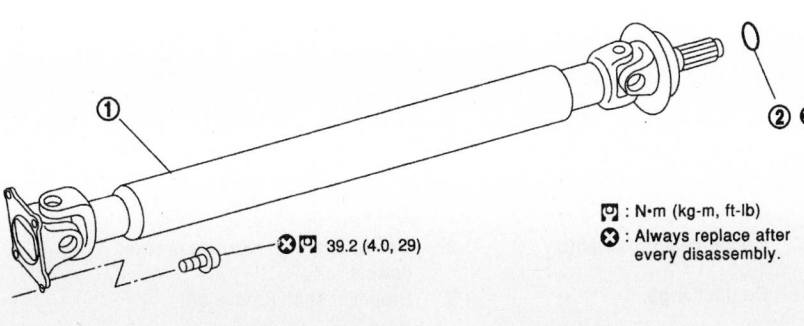

❋ 🔧 39.2 (4.0, 29)

🔧 : N·m (kg-m, ft-lb)
❌ : Always replace after every disassembly.

1. Propeller shaft assembly   2. O-ring

67162-FX35-G197

Front drive shaft—AWD Models.

VQ35DE 2WD MODEL
3S80A-1VL107

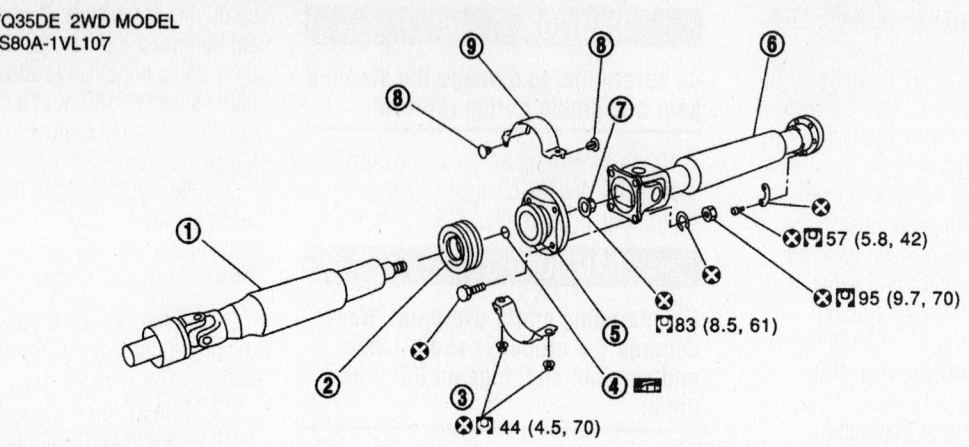

⊗ 🖳 57 (5.8, 42)
⊗ 🖳 95 (9.7, 70)
🖳 83 (8.5, 61)
⊗ 🖳 44 (4.5, 70)

VQ35DE AWD MODEL
3F80A-1VL107

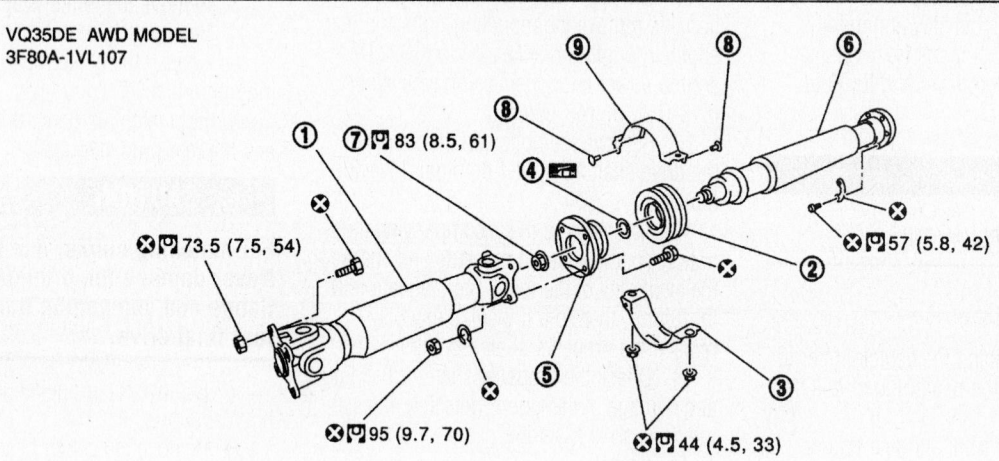

⑦ 🖳 83 (8.5, 61)
⊗ 🖳 73.5 (7.5, 54)
⊗ 🖳 57 (5.8, 42)
⊗ 🖳 95 (9.7, 70)
⊗ 🖳 44 (4.5, 33)

VK45DE AWD MODEL
3F80A-1VL107

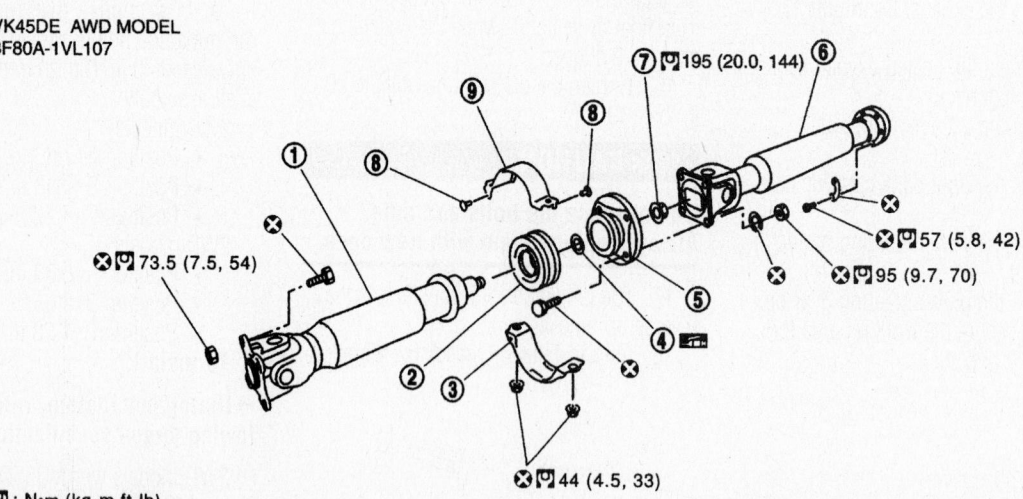

⑦ 🖳 195 (20.0, 144)
⊗ 🖳 73.5 (7.5, 54)
⊗ 🖳 57 (5.8, 42)
⊗ 🖳 95 (9.7, 70)
⊗ 🖳 44 (4.5, 33)

🖳 : N•m (kg-m, ft-lb)
🖳 : N•m (kg-m, in-lb)
⊗ : Always replase after every disassembly.
▣ : Apply multi-purpose grease.

1.	Propeller shaft (1st shaft)	2.	Center bearing assembly	3.	Center bearing mounting bracket (lower)
4.	Washer	5.	Center flange	6.	Propeller shaft (2nd shaft)
7.	Lock nut	8.	Clip	9.	Center bearing mounting bracket (upper)

67162-FX35-G198

Exploded view of the three rear drive shaft configurations.

transmission attaching nuts and bolts—54 ft. lbs. (74 Nm)

- Center flange-to-first drive shaft attaching bolts and nuts—70 ft. lbs. (95 Nm)
- Second (rear-most) flange-to-flange center attaching nut—61 ft. lbs. (83 Nm)
- Center bearing lower mounting bracket nuts—32 ft. lbs. (44 Nm)
- Second (rear-most) rear-to-rear drive unit fasteners—42 ft. lbs. (57 Nm)

AWD models with VK45DE engine:

- First (front-most) transmission-to-transmission attaching nuts and bolts—54 ft. lbs. (74 Nm)
- First (front-most) flange-to-flange center attaching nut—144 ft. lbs. (195 Nm)
- Center bearing lower mounting bracket nuts—32 ft. lbs. (44 Nm)
- Center flange-to-second drive shaft attaching bolts and nuts—70 ft. lbs. (95 Nm)
- Second (rear-most) rear-to-rear drive unit fasteners—42 ft. lbs. (57 Nm)

10. Install the propeller shaft onto the rear final drive companion flange while aligning the matching marks that were made during removal.

11. Adjust the position of the bearing cushion so as not to apply thrust play to the center bearing insulator.

12. Position the bearing cushion overlap as shown in the accompanying illustration.

13. Install the center bearing upper bracket with its arrow mark facing forward.

14. Tighten the center bearing upper mounting bracket fixing nuts to specified torque.

❋❋ CAUTION

Do not reuse the nuts. Always replace the nuts with a new ones.

15. If the companion flange has been removed, put a new alignment matching mark "C" on it. Then, reassemble using the following procedure:

➡ **Also, perform these steps when either of final drive and propeller shaft is replaced with a new one.**

   a. Erase the original mark C from companion flange with suitable solvent.
   b. Measure the companion flange vertical run out.
   c. Determine the position where maximum run out is read on the dial indica-

tor. Put a mark (shown by C in figure) on the flange perimeter corresponding to the maximum run out position.

16. If the propeller shaft or final drive has been replaced, connect the propeller shaft and final drive as follows:

➡ **Avoid damaging the rebro joint boot—protect it with a shop towel or equivalent.**

   a. Install the drive shaft while aligning its matching mark A with the mark C on the joint as close as possible.
   b. Tighten the joint bolts/nuts to the specified torque.

❋❋ CAUTION

Do not reuse the bolts, and washers. Always replace the them with new ones.

17. After installation, check driveline vibration by driving the vehicle. If vibration is present, remove the drive shaft from the final drive companion flange.

18. Turn the propeller shaft 60, 120, 180, 240 or 300 degrees and reinstall the drive shaft to the companion flange, then

check for vibration again by driving the vehicle on each angle position.

19. The remainder of installation is the reverse of removal.

Front Halfshaft

REMOVAL & INSTALLATION

Front

1. Before servicing the vehicle, refer to the precautions in the beginning of this section.

2. Remove the tire from vehicle.
3. Remove the undercover.
4. Remove the cotter pin. Then remove the locknut from the drive shaft.
5. Remove the wheel sensor wiring harness from the strut assembly.

❋❋ CAUTION

Do not pull on the wheel sensor wiring harness.

6. Remove the brake hose lockplate. Then remove the brake hose from the strut assembly.

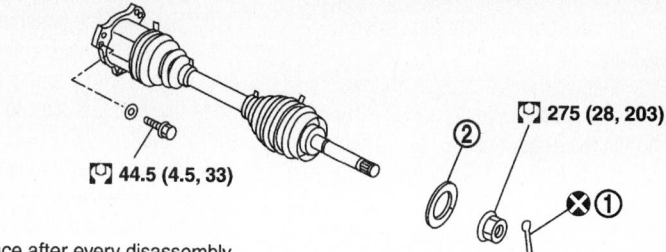

44.5 (4.5, 33)   275 (28, 203)

⊗ : Always replace after every disassembly
🔧 : N·m(kg-m,ft-lb)

1. Cotter pin   2. Washer

67162-FX35-G199

Exploded view of left-hand front halfshaft mounting.

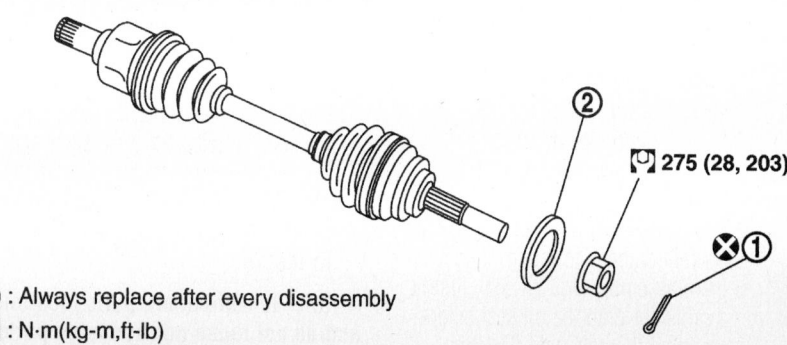

275 (28, 203)

⊗ : Always replace after every disassembly
🔧 : N·m(kg-m,ft-lb)

1. Cotter pin   2. Washer

67162-FX35-G200

Exploded view of right-hand front halfshaft mounting.

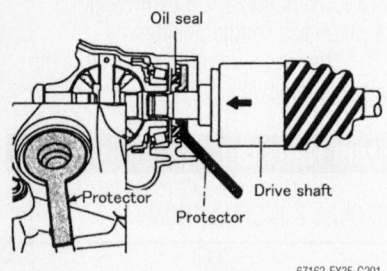

67162-FX35-G201

Use a protector to prevent damage to the side oil seal during installation.

7. Remove the mounting bolts and nuts between the strut assembly and the steering knuckle.

8. Separate the halfshaft from the steering knuckle.

❋❋ CAUTION

When removing the halfshaft, do not apply an excessive angle to the halfshaft joint. Also, be careful not to excessively extend the slide joint.

9. For the left-hand halfshaft, remove the fixing bolts of the front final drive side assembly halfshaft, then remove the halfshaft from the vehicle.

10. For the right-hand halfshaft, pry off the halfshaft from the front final drive assembly.

11. Inspect the halfshaft, as follows:

a. Move the joint up/down, left /right, and in the axial direction. Check for any rough movement or significant looseness.

b. Check the boot for cracks or other damage, and also for grease leakage.

c. If damage is found, disassemble the halfshaft and replace the defective part(s) with new one(s).

To install:

➡**Refer to component parts location and do not reuse non-reusable parts.**

12. Installation is the reverse of removal. During installation, tighten the fasteners to the following specifications:

- Halfshaft-to-hub locknut—203 ft. lbs. (275 Nm)
- For left-hand side, halfshaft-to-drive unit bolts—33 ft. lbs. (44.5 Nm)
- For right-hand side, in order to prevent damage to the front final drive assembly side oil seal, first fit a protector onto the oil seal before inserting the halfshaft. Slide the halfshaft into the slide joint and tap it with a hammer to install it securely.

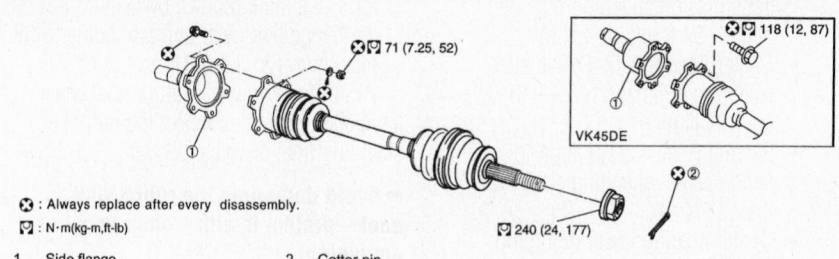

67162-FX35-G207

❊ : Always replace after every disassembly.

☑ : N·m(kg-m,ft-lb)

1. Side flange 2. Cotter pin

Exploded view of the rear halfshaft mounting.

❋❋ CAUTION

Be sure to check that circular clip is securely fastened.

13. Check the condition of the wheel sensor wiring harness

Rear

1. Before servicing the vehicle, refer to the precautions in the beginning of this section.

2. Remove the tire.

3. Remove the cotter pin. Then, remove the locknut from the outer end of the halfshaft.

4. Remove the fixing nuts and bolts between the side flange and halfshaft.

5. Separate the halfshaft from the wheel hub and bearing assembly by lightly tapping the end with a suitable hammer and wood block. If it is hard to separate, use a suitable puller.

6. Remove the halfshaft from the axle.

❋❋ CAUTION

When removing the halfshaft, do not apply an excessive angle to the halfshaft joint. Also be careful not to excessively extend the slide joint.

7. Inspect the halfshaft, as follows:

a. Move the joint up/down, left /right, and in the axial direction. Check for any rough movement or significant looseness.

b. Check the boot for cracks or other damage, and also for grease leakage.

c. If damage is found, disassemble the halfshaft and replace the defective part(s) with new one(s).

To install:

➡**Refer to component parts location and do not reuse non-reusable parts.**

8. Installation is the reverse of removal. During installation, tighten the fasteners to the following specifications:

- Halfshaft-to-hub locknut—177 ft. lbs. (240 Nm)
- Halfshaft-to-side flange bolts—52 ft. lbs. (71 Nm)

CV-Joints

OVERHAUL

Front

1. Before servicing the vehicle, refer to the precautions in the beginning of this section.

2. For the front final drive assembly side, perform the following:

a. Press the halfshaft in a vice.

❋❋ CAUTION

When retaining the shaft in a vice, always use copper or aluminum plates between the vise and shaft.

b. Remove the boot bands.

c. If the plug needs to be removed, move the boot toward the wheel side, and drive it out with a plastic hammer.

d. Put matching marks on the spider assembly and shaft.

❋❋ CAUTION

Use paint for the matching mark, but don't damage the spider assembly and halfshaft.

e. Remove the snap ring, then remove the spider assembly from the shaft.

f. Remove the boot from the shaft.

g. Remove the old grease on the slide joint assembly with paper towels.

3. For the wheel side, perform the following:

a. Place the halfshaft in a vice.

❋❋ CAUTION

When retaining the halfshaft in a vice, always use copper or aluminum plates between a vise and halfshaft.

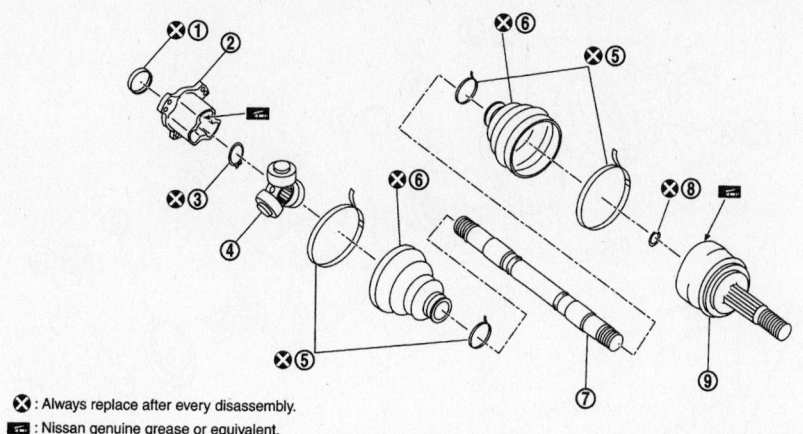

⊗ : Always replace after every disassembly.
▭ : Nissan genuine grease or equivalent.

1. Plug
2. Housing
3. Snap ring
4. Spider assembly
5. Boot band
6. Boot
7. Shaft
8. Circular clip
9. Joint sub-assembly

67162-FX35-G202

Exploded view of the left-hand halfshaft and CV-joint.

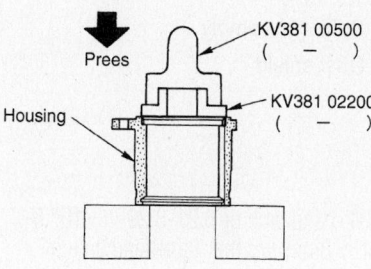

Prees — KV381 00500 (—)

Housing — KV381 02200 (—)

67162-FX35-G203

If the plug has been removed, use a press and tools to drive in a new one.

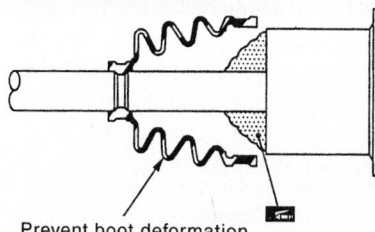

Prevent boot deformation

67162-FX35-G204

Install grease where indicated.

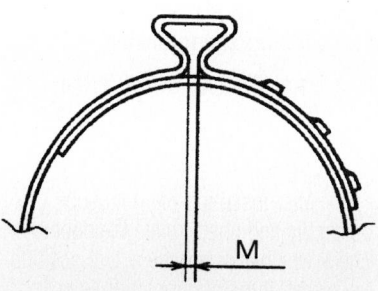

M

67162-FX35-G205

Install the band so that the gap M matches specification.

b. Remove the boot bands.

c. Remove the boot from the joint sub-assembly.

d. Screw a halfshaft puller or equivalent 1.18 in. (30 mm) or more into the threaded part of the joint sub-assembly. Pull the joint sub-assembly out of the halfshaft.

�֎ CAUTION

If the joint sub-assembly cannot be removed after five or more unsuccessful attempts, replace the shaft and joint subassembly as a set. Use a sliding hammer on the halfshaft and remove the halfshaft.

e. Remove the boot from the halfshaft.

f. Remove the circular clip from the halfshaft.

g. While rotating the ball cage, remove the old grease from the joint sub-assembly with paper towels.

4. Replace the halfshaft if there is any run out, cracking, or other damage.

5. Make sure there is no rough rotation or unusual axial looseness of the joint sub-assembly.

6. Make sure there is no foreign material inside joint sub-assembly.

7. Check joint sub-assembly for compression scar, cracks or fractures.

✖✖ CAUTION

If there are any irregular conditions of joint sub-assembly components, replace the entire joint sub-assembly.

8. Inspect the housing and spider assembly. If the roller or roller surface of the spider assembly has scratches or other wear, replace the housing and the spider assembly.

➡**The housing and spider assembly are components which are used as a set. To assemble:**

9. To assemble the front final drive assembly side, perform the following:

a. If the plug has been removed, use a drift to press in a new one.

➡**Discard the old plug, and replace it with a new one.**

b. Wind the serrated part of the shaft with tape. Install the boot band and boot on the shaft. Be careful not to damage the boot.

➡**Discard the old boot band and boot, and replace them with new ones.**

c. Remove the protective tape wound around the serrated part of the shaft.

d. Align the alignment marks, which were made when the spider assembly was removed. Install the spider assembly with the serration chamfer facing shaft.

e. Secure the spider assembly with a snap ring.

➡**Discard the old snap ring, and replace it with a new one.**

f. Apply Nissan genuine grease or equivalent to the spider assembly and sliding surface.

g. Install the housing onto the spider assembly. Apply Nissan genuine grease or equivalent to the housing.

h. Install the boot securely into the grooves (indicated by * marks) shown in the figure.

✖✖ CAUTION

If there is grease on the boot mounting surfaces (indicated by * marks) of the shaft and housing, the boot may come off. Remove all grease from surfaces.

i. Make sure the boot installation length "L" is 3.74-3.82 in. (95-97mm). Insert a flat-bladed pry tool or similar tool into the smaller side of the boot. Bleed air from the boot to prevent boot deformation.

✖✖ CAUTION

The boot may break if the boot installation length is less than the standard value. Take care not to touch the tip of the pry tool to the inside surface of the boot.

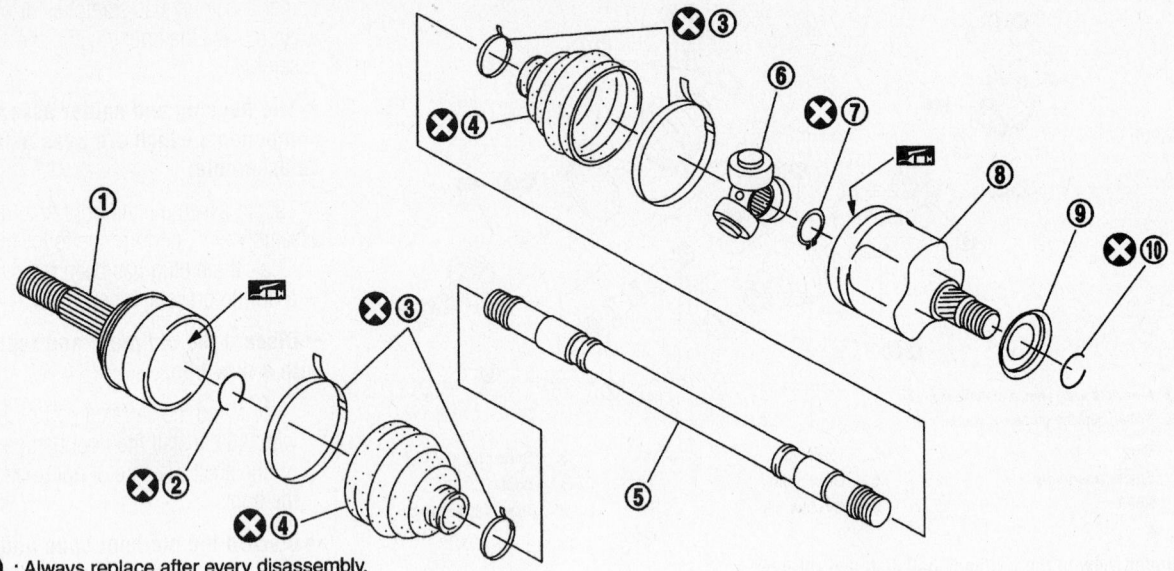

⊗ : Always replace after every disassembly.

▭ : Nissan genuine grease or equivalent.

1. Joint sub-assembly
2. Circular clip
3. Boot band
4. Boot
5. Shaft
6. Spider assembly
7. Snap ring
8. Housing
9. Dust shield
10. Circular clip

67162-FX35-G206

Exploded view of the right-hand halfshaft and CV-joint.

j. Install the new larger and smaller boot bands securely with a suitable tool.

➡**Discard the old boot bands, and replace them with new ones.**

k. Secure the boot band so that dimension "M" shown satisfies the following:
- Large diameter side: 0.118 in. (3.0mm)
- Small diameter side: 0.079 in. (2.0mm)

l. After installing the housing and shaft, rotate the boot to check whether or not the actual position is correct. If the boot position is not correct, secure the boot with new boot bands again.

10. Assemble the wheel side of the half-shaft as follows:

a. Insert grease (Nissan genuine grease or equivalent) into the joint sub-assembly serration hole until grease begins to ooze from the ball groove and serration hole. After inserting the grease, use a shop cloth to wipe off the old grease that has oozed out.

b. Wind the serrated part of the shaft with tape for protection of the boot, then install the boot band and boot on the shaft. Be careful not to damage the boot.

➡**Discard the old boot band and boot, and replace them with new ones.**

c. Remove the protective tape from the shaft.

d. Attach the circular clip to the shaft. At this time, the circular clip must fit securely into the shaft groove. Attach the nut to the joint sub-assembly. Use a wooden hammer to press-fit it in place.

➡**Discard the old circular clip, and replace it with a new one.**

e. Insert the specified amount of grease (Nissan genuine grease or equivalent) into the boot from the large end of the boot.

➡**Grease amount: 3.35-4.06 oz. (95–115 g) for the left-hand side or 3.99-4.34 oz. (113–123 g) for the right-hand side.**

f. Install the boot securely into the grooves (indicated by * marks) shown.

✳✳ CAUTION

If there is grease on the boot mounting surfaces (indicated by * marks) of the shaft and housing of the joint sub assembly, the boot may come off. Remove all grease from the surfaces.

g. Make sure the boot installation length "L" is 5.35 in. (136mm) for the

left-hand side or 6.21-6.29 in. (157.8-159.8mm) for the right-hand side or. Insert a flat-bladed pry tool or similar tool into the smaller side of boot. Bleed air from the boot to prevent boot deformation.

✳✳ CAUTION

The boot may brake if the boot installation length is less than the standard value. Be careful that the pry tool tip does not contact the inside surface of the boot.

h. Install the new larger and smaller boot bands securely with a suitable tool.

➡**Discard the old boot bands, and replace them with new ones. Secure the boot band so that dimension "M" shown satisfies the following:**
- Large diameter side: 0.118 in. (3.0mm)
- Small diameter side: 0.079 in. (2.0mm)

i. After installing the joint sub-assembly and shaft, rotate the boot to check whether or not the actual position is correct. If the boot position is not correct, secure the boot with new boot bands again.

Rear

1. Before servicing the vehicle, refer to the precautions in the beginning of this section.

2. Disassemble the final drive side, as follows:

   a. Press shaft in a vice.

✳✳ CAUTION

When retaining the halfshaft in a vice, always use copper or aluminum plates between a vise and halfshaft.

   b. Remove the boot bands.

   c. If the plug needs to be removed, move the boot toward the wheel side, and drive it out with a plastic hammer.

   d. Remove the stopper ring with a flat-bladed pry tool, and pull out the housing.

   e. Remove the snap-ring, then remove the ball cage/steel ball/inner race assembly from the shaft.

   f. Remove the boot from the shaft.

   g. Remove the old grease on the housing with paper towels.

3. Disassemble the wheel side, as follows:

   a. Place the shaft in a vice.

✳✳ CAUTION

When retaining the halfshaft in a vice, always use copper or aluminum plates between a vise and halfshaft.

   b. Remove the boot bands. Then remove the boot from the joint sub-assembly.

   c. Thread a halfshaft puller 30 mm (1.18 in) or more into threaded part of joint sub-assembly. Pull joint sub-assembly out of shaft.

✳✳ CAUTION

If the joint sub-assembly cannot be removed after five or more unsuccessful attempts, replace the shaft and joint subassembly as a set. Align the sliding hammer and halfshaft and remove them by pulling directly out.

   d. Remove the boot from shaft.

   e. Remove the circular clip from the shaft.

   f. While rotating the ball cage, remove the old grease on the joint sub-assembly with paper towels.

4. Replace the shaft if there is any run out, cracking, or other damage.

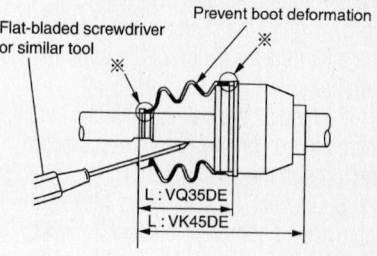

67162-FX35-G209

When installing the CV-joint boot make sure installation length L meets specification.

5. Inspect the joint sub-assembly. Make sure there is no rough rotation or unusual axial looseness. Make sure there is no foreign material inside the joint. Check the joint sub-assembly for compression scars, cracks, or fractures.

✳✳ CAUTION

If there are any irregular conditions of the joint sub-assembly components, replace the entire joint sub-assembly.

6. Inspect the sliding joint side housing. Make sure there are no compression scars, cracks, fractures or unusual wear on

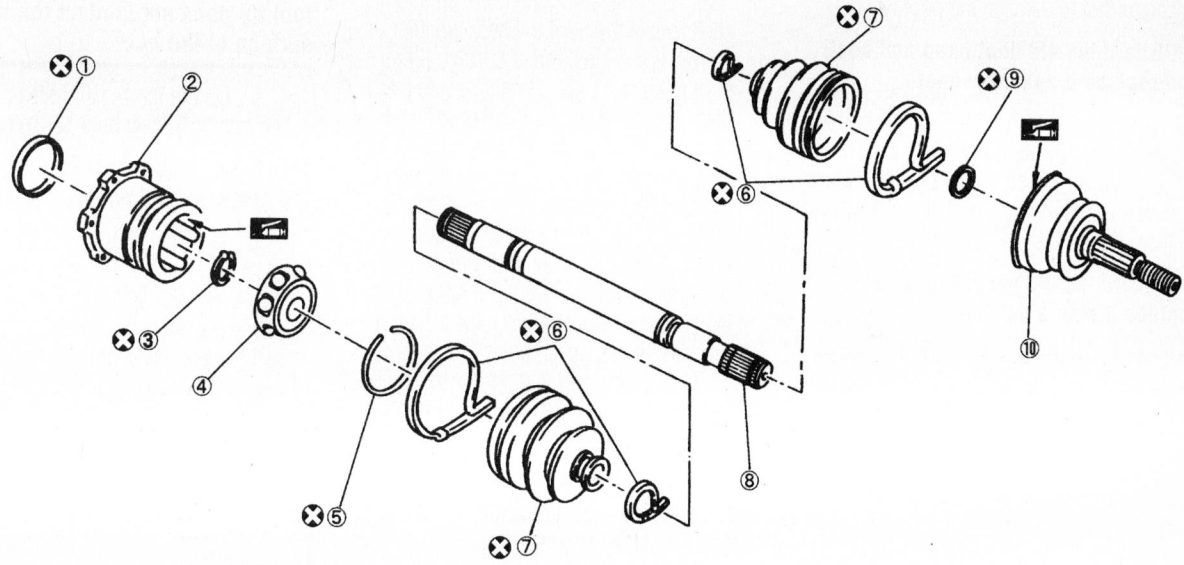

✖ : Always replace after every disassembly.

1.	Plug	2.	Housing	3.	Snap ring
4.	Ball cage/Steel ball/Inner race assembly	5.	Stopper ring	6.	Boot band
7.	Boot	8.	Shaft	9.	Circular clip
10.	Joint sub-assembly				

Exploded view of the rear halfshaft and CV-joint.

67162-FX35-G208

the ball rolling surface. Make sure there is no damage to the shaft screws. Make sure there is no deformation of the boot installation parts.

7. Inspect the ball cage. Make sure there are no compression scars, cracks, fractures of the sliding surface.

8. Inspect the steel balls. Make sure there are no compression scars, cracks, fractures or unusual wear.

9. Inspect the inner race. Check ball sliding surface for compression scars, cracks, or fractures. Make sure there is no damage to the serrated part.

※※ CAUTION

If there are any irregular conditions on the components, replace with a new set of housing, ball cage, steel ball and inner race. To assemble:

10. Assemble the final drive side as follows:

a. If the plug has been removed, use a drift to press in a new one.

➡ **Discard old plug, and replace with a new one.**

b. Wind the serrated part of the shaft with tape to protect the boot. Install the boot band and boot on the shaft. Be careful not to damage the boot.

➡ **Discard the old boot band and boot, and replace it with new ones.**

c. Remove the protective tape wound around the serrated part of shaft.

d. Install the ball cage/steel ball/inner race assembly onto the shaft, and secure them tightly with a snap-ring.

➡ **Discard the old snap ring, and replace it with a new one.**

e. Insert the proper amount of grease (NISSAN genuine grease or equivalent) into the housing (* point), and install it to shaft.
Grease amount:
- VK45DE models: 6.17-6.88 oz. (175–195 g)
- VQ35DE models: 4.37-4.73 oz. (124–134 g)

a. Install the stopper ring onto the housing.

b. After installation, pull the shaft to check engagement between the joint sub-assembly and the stopper ring.

c. Install the boot securely into the grooves (indicated by * marks) shown.

※※ CAUTION

If there is grease on the boot mounting surfaces (indicated by * marks) of the shaft and housing, the boot may come off. Remove all grease from the surfaces.

d. Make sure the boot installation length "L" is the length indicated. Insert a flat-bladed pry tool or similar tool into the smaller side of boot. Bleed air from the boot to prevent boot deformation. Boot installation length "L":
- VK45DE models: 5.82 in. (147.9mm)
- VQ35DE models: 3.697 in. (93.9mm)

※※ CAUTION

The boot may break if the boot installation length is less than the standard value. Take care not to touch the tip of the pry tool to the inside of the boot.

a. Secure the big and small ends of the boot with the new boot bands as shown.

➡ **Discard the old boot bands, and replace with new ones.**

b. After installing the housing and shaft, rotate the boot to check whether or not the actual position is correct. If the boot position is not correct, secure the boot with new boot band again.

11. Assemble the wheel side as follows:

a. Insert the proper amount of grease (NISSAN genuine grease or equivalent) into the joint sub-assembly serration hole until grease begins to ooze from the ball groove and the serration hole. Afterward, use a shop cloth to wipe off the old grease that has oozed out.

b. Wind the serrated part of the shaft with tape. Install the boot band and boot onto the shaft. Be careful not to damage the boot.

➡ **Discard the old boot band and boot, replace each with new ones.**

c. Remove the protective tape wound around the serrated part of the shaft.

➡ **Discard the old circular clip, and replace with a new one.**

d. Attach the circular clip to the shaft. At this time, the circular clip must fit securely into the shaft groove. Attach the nut to the joint sub-assembly. Use a wooden hammer to press-fit.

e. Insert the proper amount of grease (NISSAN genuine grease or equivalent) into the housing from the large end of the boot. Grease amount:
- VK45DE models: 4.93-5.64 in. (140–160 g)
- VQ35DE models: 3.03-3.39 in. (86–96 g)

a. Install the boot securely into grooves (indicated by * marks) shown.

※※ CAUTION

If there is grease on the boot mounting surfaces (indicated by * marks) of the shaft and the housing, the boot may come off. Remove all grease from the surfaces.

b. Make sure the boot installation length "L" is the length indicated below. Insert a flat-bladed pry tool or similar tool into the smaller side of the boot. Bleed air from the boot to prevent boot deformation. Boot installation length "L":
- VK45DE models: 5.57 in. (141.5mm)
- VQ35DE models: 3.82 in. (97mm)

※※ CAUTION

The boot may break if the boot installation length is less than the standard value. Be careful that the pry tool tip does not contact the inside surface of the boot.

a. Secure the big and small ends of the boot with new boot bands as shown.

➡ **Discard the old boot bands, and replace with new ones.**

b. After installing the joint sub-assembly and shaft, rotate the boot to check whether or not the actual position is correct. If the boot position is not correct, secure the boot with new boot bands again.

Front Differential Seals

REMOVAL & INSTALLATION

Front Seal

1. Before servicing the vehicle, refer to the precautions in the beginning of this section.

2. Remove the front drive shaft.

3. Put a matching mark on the end of the drive pinion corresponding to the B position matching mark on the final drive companion flange.

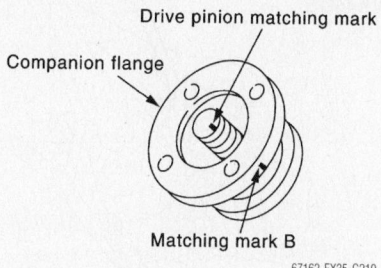

During installation the matching marks should align.

67162-FX35-G210

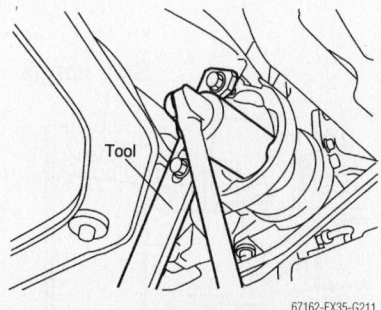

To remove the flange bolt, a spanner wrench will need to be used.

67162-FX35-G211

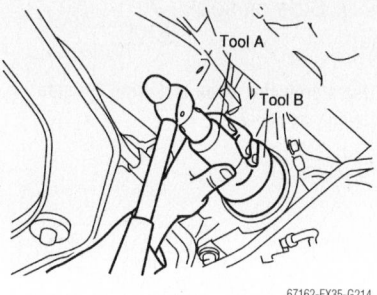

Drive the new seal into the differential housing as shown.

67162-FX35-G214

✳✳ CAUTION

Use paint for the matching mark; never damage drive pinion. The matching mark B on the final drive companion flange indicates the maximum vertical run out position.

4. Using the drive pinion flange wrench. Remove the drive pinion locknut with tool number: KV40104000 or equivalent.

5. Remove the companion flange using a puller (commercial service tool).

6. Remove the front oil seal using outer race puller ST33290001 (J34286) or equivalent.

To install:

7. Apply multi-purpose grease to the sealing lips of the oil seal. Drive the oil

---

seal into the differential case using special service tools ST33400001 (J26082) and KV38102510 or equivalents so that the oil seal is flush with the gear carrier end.

➡ **When installing the front oil seal, be careful not to get it inclined.**

8. Discard the old front oil seal. Always replace it with a new one.

9. Install the companion flange while aligning the matching mark of the drive pinion with the matching mark B of the companion flange.

10. Apply oil to the drive pinion threads and the seating surface of the drive pinion locknut.

11. Using the drive pinion flange wrench KV40104000 or equivalent. Install the drive pinion locknut with the tool.

✳✳ CAUTION

Do not reuse the drive pinion locknut. Always replace it with a new one.

12. Install the front drive shaft.

Side Seals

1. Before servicing the vehicle, refer to the precautions in the beginning of this section.

2. For the right side seal, remove the front ABS wheel sensor.

3. For the left side seal, remove the front drive shaft and side shaft assembly.

4. Remove the front halfshaft.

5. Remove the side oil seal using puller ST33290001 (J34286) or equivalent.

To install:

6. Apply multi-purpose grease to the sealing lips of the side oil seal.

7. Using drift ST33400001 (J26082) for the right side or ST33210000 for the left side or equivalent, press-fit the side oil seal so that its surface is flush with the end surface of the case.

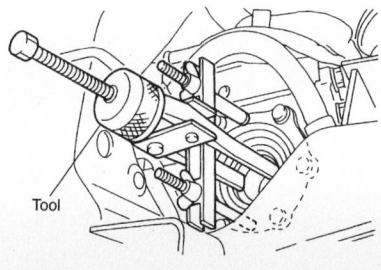

Remove the old side seal using a puller as shown.

67162-FX35-G215

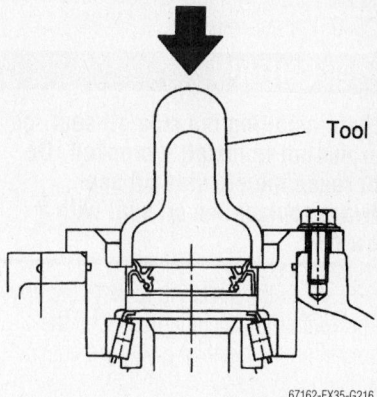

Tool

67162-FX35-G216

Drive the new seal into the housing as shown.

✳✳ CAUTION

When installing the side oil seal, be careful not to install it crooked. Do not reuse the old side oil seal. Always replace the oil seal with a new one.

8. Install the front halfshaft.

Rear Differential Seals

REMOVAL & INSTALLATION

Front Seal

1. Before servicing the vehicle, refer to the precautions in the beginning of this section.

2. Remove the rear drive shaft.

3. Put a matching mark on the end of the drive pinion corresponding to the C position matching mark on the final drive companion flange.

✳✳ CAUTION

Use paint for the matching mark; never damage drive pinion. The matching mark B on the final drive companion flange indicates the maximum vertical run out position.

4. Using the drive pinion flange wrench KV40104000 or equivalent, remove the drive pinion locknut.

5. Remove the companion flange using the puller.

6. Remove the front oil seal using the side bearing outer race puller ST33290001 (J34286) or equivalent.

To install:

7. Apply multi-purpose grease to sealing lips of oil seal. Press the front oil seal

into the carrier with tool ST30720000 (J25405) or equivalent.

❊❊ CAUTION

When installing the side oil seal, be careful not to install it crooked. Do not reuse the old side oil seal. Always replace the oil seal with a new one.

8. Align the matching mark of the drive pinion with the matching mark C of the companion flange, then install the companion flange.

9. Apply oil on the threaded part of the drive pinion and the seating surface of the drive pinion locknut.

10. Install the drive pinion nut with tool KV40104000 or equivalent.

❊❊ CAUTION

The drive pinion locknut is not reusable. Never reuse the drive pinion nut.

11. Install the rear drive shaft.

Side Seals

1. Before servicing the vehicle, refer to the precautions in the beginning of this section.

2. Remove the side flange, as follows:
   a. Remove ABS rear wheel sensor.
   b. Remove the halfshaft and axle assemblies.
   c. Install the axle stand onto the side flange, then pull out the side flange with a sliding hammer, as follows:
   - VQ35DE models—tools KV40104100 and ST36230000 (J25840-A) or equivalents
   - VK45DE models—tools KV40101000 and ST36230000 (J25840-A) or equivalents

3. Remove side oil seal using a flat-bladed pry tool.

To install:

4. Apply multi-purpose grease to the sealing lips of the side oil seal.

5. Using drift KV38100200 (J26233) or equivalent, press-fit the side oil seal so that its surface is flush with the end surface of the case.

❊❊ CAUTION

When installing the side oil seal be careful not to press it in crooked. Discard the old side oil seal. Always replace them with new ones.

6. Install the side flange, as follows:
   a. Attach protector KV38107900 (J39352) or equivalent to the side oil seal.
   b. After the side flange is inserted and the serrated part of the side gear has engaged the serrated part of the flange, remove the protector.
   c. Put a suitable drift on the center of the side flange, then drive it in until it clicks or snaps in place.

➠ **When installation is completed, the driving sound of the side flange turns into a sound which seems to affect the whole final drive.**

7. Confirm that the dimension of the side flange installation (Measurement A) in the illustration falls within the range of 12.83-12.91 in. (326-328mm).

8. Install the halfshaft and axle assembly.

9. Align the installation position of the ABS rear wheel sensor.

VQ35DE

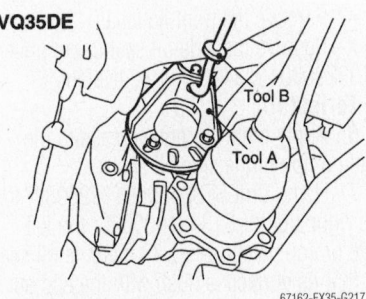

67162-FX35-G217

Remove the side seal using an adapter and puller—VQ35DE Engine.

VK45DE

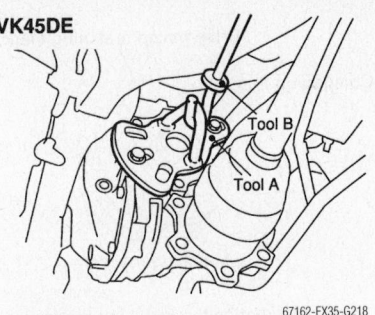

67162-FX35-G218

Remove the side seal using an adapter and puller—VK45DE Engine.

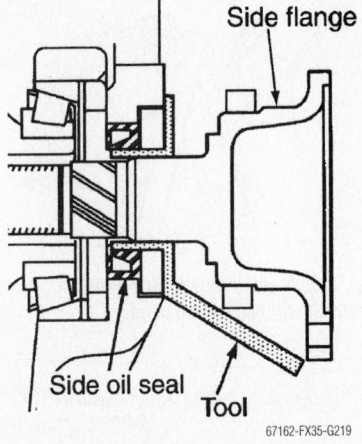

67162-FX35-G219

Use a protector when installing the side flange, as shown.

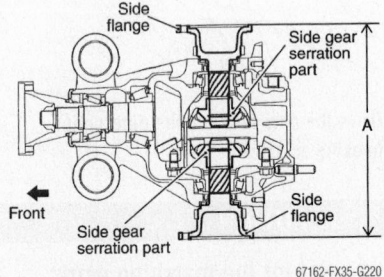

67162-FX35-G220

After installation, confirm that the side flange installed dimension A is approximately 12.83-12.91 in. (326-328mm).

STEERING AND SUSPENSION

Air Bag

❋❋ CAUTION

Some vehicles are equipped with an air bag system. The system must be disarmed before performing service on, or around, system components,

the steering column, instrument panel components, wiring and sensors. Failure to follow the safety precautions and the disarming procedure could result in accidental air bag deployment, possible injury and unnecessary system repairs.

PRECAUTIONS

Several precautions must be observed when handling the inflator module to avoid accidental deployment and possible personal injury.
- Never carry the inflator module by the

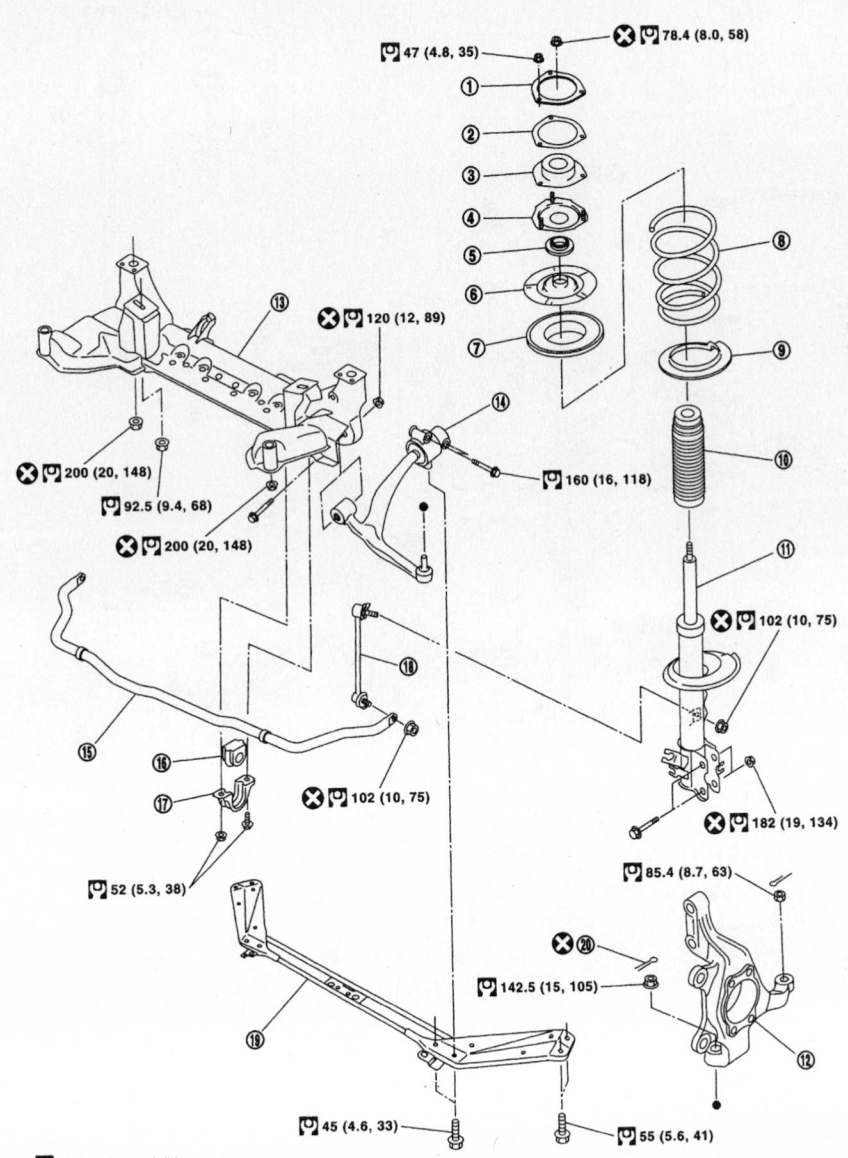

◻ : N•m (kg-m, ft-lb)

❌ : Always replace after every disassembly.

1.	Strut upper plate	2.	Strut spacer	3.	Mounting insulator
4.	Mounting insulator bracket	5.	Mounting bearing	6.	Spring upper seat
7.	Spring upper rubber seat	8.	Coil spring	9.	Spring lower rubber seat
10.	Bound bumper	11.	Strut	12.	Steering knuckle
13.	Front suspension member	14.	Transverse link	15.	Stabilizer bar
16.	Stabilizer bushing	17.	Stabilizer clamp	18.	Stabilizer connecting rod
19.	Front cross bar	20.	Cotter pin		

67162-FX35-G223

Exploded view of the front suspension components.

wires or connector on the underside of the module.

• When carrying a live inflator module, hold securely with both hands and ensure that the bag and trim cover are pointed away.

• Place the inflator module on a bench or other surface with the bag and trim cover facing up.

• With the inflator module on the bench, never place anything on or close to the

module which may be thrown in the event of an accidental deployment.

DISARMING THE SYSTEM

1. Before servicing the vehicle, refer to the precautions in the beginning of this section.
2. Turn the ignition switch to OFF.
3. Disconnect and isolate both battery cables.

4. Wait at least 3 minutes prior to servicing the SRS system.

ARMING THE SYSTEM

1. Before servicing the vehicle, refer to the precautions in the beginning of this section.
2. Reconnect the battery cables.

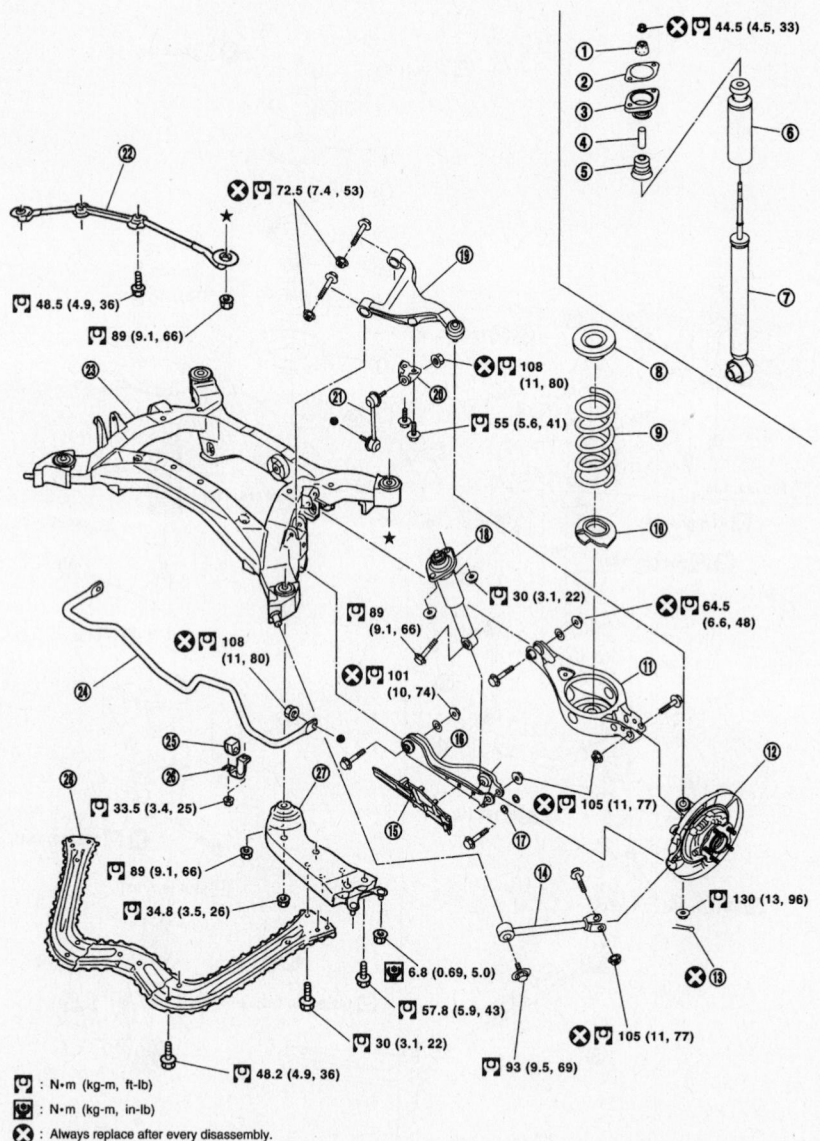

1. Bushing
2. Mounting seal
3. Mounting seal bracket
4. Distance tube
5. Bound bumper cover
6. Bound bumper
7. Shock absorber
8. Upper seat
9. Coil spring
10. Rubber seat
11. Rear lower link
12. Axle
13. Cotter pin
14. Radius rod
15. Front lower link protector
16. Front lower link
17. Stopper
18. Shock absorber assembly
19. Suspension arm
20. Stabilizer connecting rod mounting bracket
21. Stabilizer connecting rod
22. Rear pin stay
23. Rear suspension member
24. Stabilizer bar
25. Stabilizer bushing
26. Stabilizer clamp
27. Member stay
28. Tunnel stay

67162-FX35-G227

Exploded view of the rear suspension components.

Steering Angle Sensor Neutral Position

ADJUSTMENT

1. Before servicing the vehicle, refer to the precautions in the beginning of this section.

➡**After any of the following conditions, check the adjustment of the steering angle sensor neutral position before driving the vehicle:**

- Replacing ABS actuator and electric unit (control unit)
- Removing/Installing steering angle sensor
- Removing/Installing steering components
- Removing/Installing suspension components
- Adjusting wheel alignment

✳✳ CAUTION

To adjust the neutral position of steering angle sensor, make sure to use the CONSULT-II. (Adjustment cannot be done without CONSULT-II.)

2. Stop the vehicle with the front wheels in the straight-ahead position.
3. Connect the CONSULT-II and CONSULT-II CONVERTER to the data link connector on the vehicle, and turn the ignition switch ON (do not start engine).

✳✳ CAUTION

If the CONSULT-II is used with no connection of the CONSULT-II CONVERTER, malfunctions might be detected in self-diagnosis depending on the control unit which carry out CAN communication.

4. Touch "ABS", "WORK SUPPORT" and "ST ANGLE SENSOR ADJUSTMENT" on the CONSULT-II screen in this order.
5. Touch "START".

✳✳ CAUTION

Do not touch the steering wheel while adjusting the steering angle sensor.

6. After approximately 10 seconds, touch "END". (After approximately 60 seconds, it ends automatically.)
7. Turn the ignition switch OFF, then turn it ON again.

✳✳ CAUTION

Be sure to carry out above operation.

8. Run the vehicle with the front wheels in the straight-ahead position, then stop.
9. Select "DATA MONITOR", "SELECTION FROM MENU", and "STR ANGLE SIG" on the CONSULT-II screen. Then make sure "STR ANGLE SIG" is within 0 ±3.5 deg. If the value is more than specified, repeat steps 3 to 7.
10. Erase memory of the ABS actuator and electric unit (control unit) and the ECM.
11. Turn the ignition switch OFF.

Decel G-Sensor (AWD Models)

CALIBRATION

1. Before servicing the vehicle, refer to the precautions in the beginning of this section.

➡**After removing/installing or replacing the yaw rate/side/decel G-sensor, ABS actuator and electric unit (control unit), suspension components, or after adjusting the wheel alignment, make sure to calibrate the decel G-sensor before running the vehicle.**

✳✳ CAUTION

To calibrate the decel G-sensor, make sure to use the CONSULT-II. (Adjustment cannot be done without the CONSULT-II.)

2. Stop the vehicle with the front wheels in the straight-ahead position.

✳✳ CAUTION

The work should be done on a flat area when the vehicle is in the unloaded vehicle condition. Keep all tires inflated to the correct pressures. Adjust the tire pressure to the specified value.

3. Connect the CONSULT-II and CONSULT-II CONVERTER to data link connector on vehicle, and turn the ignition switch ON (do not start engine).

✳✳ CAUTION

If the CONSULT-II is used without the CONSULT-II CONVERTER, malfunctions might be detected in self-diagnosis, depending on control unit which carry out CAN communication.

4. Touch "ABS", "WORK SUPPORT" and "DECEL G-SEN CALIBRATION" on the CONSULT-II screen in this order.
5. Touch "START".
6. After approximately 10 seconds, touch "END". (After approximately 60 seconds, it ends automatically.)
7. Turn the ignition switch OFF, then turn it ON again.

✳✳ CAUTION

Be sure to carry out above operation.

8. Run the vehicle with the front wheels in the straight-ahead position, then stop.
9. Select "DATA MONITOR", "SELECTION FROM MENU", and "DECEL G-SEN" on the CONSULT-II screen. Then make sure "DECEL G-SEN" is within 0 ±0.08 G. If value is more than specified, repeat steps 3 to 7.
10. Erase the memory of the ABS actuator and electric unit (control unit) and ECM.
11. Turn the ignition switch OFF.

Power Steering Gear

REMOVAL & INSTALLATION

1. Before servicing the vehicle, refer to the precautions in the beginning of this section.

✳✳ CAUTION

The spiral cable may snap due to steering operation if the steering column is separated from the steering gear assembly. Therefore fix the steering wheel with a string to avoid turning too far.

2. Set the wheels in the straight-ahead position.
3. Remove the tires from vehicle.
4. Remove the undercover.
5. Confirm the slit of the lower joint fits with the projection on the rear cover cap, furthermore marking position on steering gear assembly nearly fits with the projection on rear cover cap.
6. Remove the cotter pin at steering outer socket, then loosen the mounting nut.
7. Use a ball joint remover to remove the steering outer socket from the steering knuckle. Be careful not to damage the ball joint boot.

✳✳ CAUTION

Temporarily tighten the mounting nut to prevent damage to the threads and to prevent the ball joint remover from coming off.

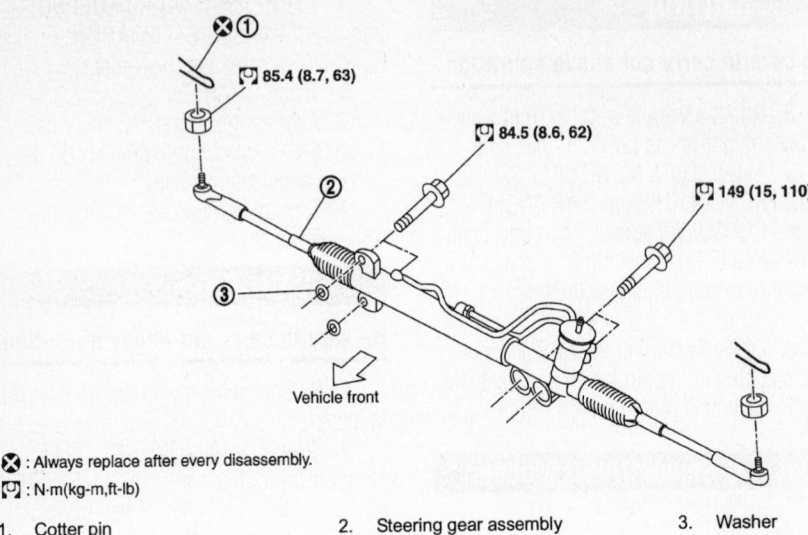

⊗ : Always replace after every disassembly.
🔧 : N·m(kg-m,ft-lb)

1. Cotter pin 2. Steering gear assembly 3. Washer

67162-FX35-G221

Exploded view of the power steering gear mounting.

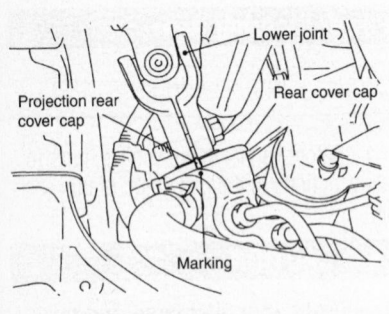

67162-FX35-G222

Steering gear projection position.

8. Remove the high pressure side and low pressure side oil pipes from the steering gear assembly, then drain the fluid from the pipes.

9. Remove the mounting bolt of the steering hydraulic pipe bracket from the steering gear assembly.

10. Remove the lower side mounting bolt of the lower joint.

11. Remove the mounting bolts of the steering gear assembly, then remove the steering gear assembly from the vehicle.

To install:

12. Installation is the reverse of removal procedure.

➡**Refer to component parts location and do not reuse non-reusable parts.**

13. After installation, check wheel alignment.

14. After adjusting wheel alignment, adjust the neutral position of the steering angle sensor.

15. When steering wheel is set in the straight ahead direction, confirm the slit of the lower joint fits with the projection on rear cover cap, and that the marking position on steering gear assembly nearly fits with the projection on rear cover cap.

16. Bleed all air from the power steering system, as follows:

Incomplete air bleeding causes the following. When this happens, bleed the system again:

- Generation of air bubbles in the reservoir tank.
- Generation of clicking noise in the oil pump.
- Excessive buzzing in the oil pump.

➡**When the vehicle is stationary or while the steering wheel is being turned slowly, some noise may be heard from the oil pump or gear. This noise is normal and does not affect any system.**

a. Stop the engine, then turn the steering wheel fully to the right and left several times.

b. Run the engine at idle speed. Turn the steering wheel fully to the right and then fully to the left, and keep hold for about three seconds. Then check whether any fluid leaks have occurred.

c. Repeat substep b several times at about three seconds intervals.

d. Check for air bubbles and cloudiness in the fluid.

e. If air bubbles and/or cloudiness don't fade, stop the engine, and stop air bleeding until the air bubbles and cloudiness fade.

f. Perform until all bubbles and cloudiness are gone.

g. Stop the engine and check the fluid level.

17. Check that the steering wheel turns smoothly when it is turned several times fully to the end of the left and right.

MacPherson Strut

REMOVAL & INSTALLATION

Front

1. Before servicing the vehicle, refer to the precautions in the beginning of this section.

2. Remove the tire from the vehicle.

3. Remove the brake hose lockplate. Then, remove the brake hose from the strut assembly.

4. Remove the wheel sensor wiring harness from the strut assembly.

5. Remove the stabilizer connecting rod upper nut, separate the stabilizer connecting rod and strut assembly.

6. Remove the attaching bolts and nuts between the strut assembly and the steering knuckle.

7. Remove the mounting nuts on the

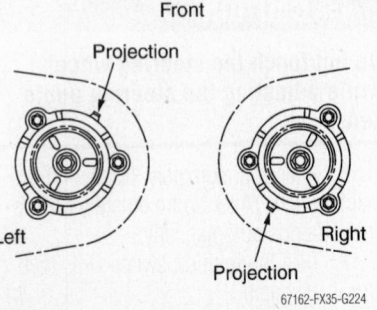

67162-FX35-G224

MacPherson strut projection positioning.

mounting insulator bracket, then remove the strut upper plate, strut spacer and the strut from the vehicle.

To install:

❋❋ CAUTION

Attach strut upper plate as shown in the accompanying illustration.

8. Install the strut upper plate, strut spacer and the strut onto the vehicle.

9. Install and tighten the mounting nuts on the mounting insulator bracket to 35 ft. lbs. (47 Nm).

10. Install and tighten the attaching bolts and nuts between the strut assembly and the steering knuckle to 134 ft. lbs. (182 Nm).

11. Reattach the stabilizer connecting rod to the strut and tighten the upper nut to 75 ft. lbs. (102 Nm).

12. Install the wheel sensor wiring harness onto the strut assembly.

13. Install the brake hose and the hose lockplate.

14. Install the tire from the vehicle.

➡**Refer to component parts location and do not reuse non-reusable parts.**

15. After installation, check wheel alignment.

16. After adjusting the wheel alignment, adjust the neutral position of steering angle sensor.

17. Double-check to ensure that the wheel sensor wiring harness is properly routed.

DISASSEMBLY

1. Before servicing the vehicle, refer to the precautions in the beginning of this section.

➡**Make sure the piston rod on the strut is not damaged when removing the components from the strut assembly.**

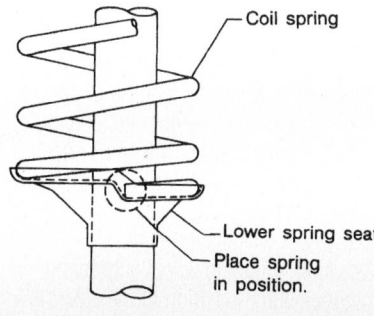

- Coil spring
- Lower spring seat
- Place spring in position.

67162-FX35-G225

Proper coil spring placement for installation.

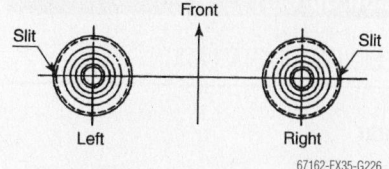

Front
Slit — Left ... Right — Slit

67162-FX35-G226

Upper spring seat positioning for installation.

2. Install the strut attachment SST: ST35652000 or equivalent to the strut and secure it in a vise.

❋❋ CAUTION

When installing the strut attachment to the strut, wrap a shop towel around the strut to protect it from damage.

3. Using a spring compressor (commercial service tool), compress the coil spring between the spring upper seat and spring lower seat (on the strut) until the coil spring is free.

❋❋ CAUTION

Be sure the spring compressor (commercial service tool) is securely attached to the coil spring before compressing the spring.

4. After making sure the coil spring is free between the spring upper seat and spring lower seat of the strut, remove the piston rod locknut.

5. Remove the mounting insulator, mounting insulator bracket, mounting bearing, spring upper seat, spring upper rubber seat, and bound bumper. Then, remove the coil spring and spring lower rubber seat from the strut.

6. Gradually release the spring compressor, and remove the coil spring.

❋❋ CAUTION

Loosen the spring compressor while making sure the coil spring attachment position does not move.

7. Remove the strut attachment from the strut.

8. Check the strut for deformation, cracks, damage, and replace if necessary.

9. Check the piston rod for damage, uneven wear or distortion, and replace if necessary.

10. Check the welded and sealed areas for oil leakage, and replace if necessary.

11. Check the mounting insulator for cracks and rubber parts for wear. Replace them if necessary.

12. Check the coil spring for cracks, wear or damage, and replace if necessary.

ASSEMBLY

1. Before servicing the vehicle, refer to the precautions in the beginning of this section.

➡**Make sure the piston rod on the strut is not damaged when attaching the components to the strut.**

2. Install the strut attachment to the strut and fix it in a vise.

❋❋ CAUTION

When installing strut attachment to strut, wrap a shop cloth around strut to protect it from damage.

3. Compress coil spring using a spring compressor (commercial service tool), and install it onto the strut.

❋❋ CAUTION

Face the tube side of the coil spring downward. Align the lower end to the spring rubber seat as shown. Be sure the spring compressor (commercial service tool) is securely attached to coil spring before compressing the coil spring.

4. Apply soapy water to the bound bumper and insert into the mounting insulator.

❋❋ CAUTION

Do not use machine oil.

5. Install the mounting insulator bracket, mounting bearing, bound bumper, spring upper seat, spring upper rubber seat and spring lower rubber seat.

➡**The installation position of spring upper seat is as shown.**

6. Fix the mounting insulator, then tighten the piston rod locknut to 58 ft. lbs. (78 Nm).

❋❋ CAUTION

Be sure not to deform the mounting insulator bracket.

7. Gradually the release spring compressor (commercial service tool), and remove the coil spring.

✳✳ CAUTION

Loosen the spring compressor while making sure coil spring attachment position does not move.

8. Remove the strut attachment from the strut.

Shock Absorber

REMOVAL & INSTALLATION

Rear

1. Before servicing the vehicle, refer to the precautions in the beginning of this section.

2. Remove the rear tire.

3. Position a jack or equivalent support under the rear lower link.

4. Remove the fixing bolt in the lower side of the shock absorber assembly.

5. Remove the attaching nuts in the upper side of the shock absorber assembly and remove the shock absorber assembly from the vehicle.

6. Check the shock absorber assembly for deformation, cracks, or damage, and replace if necessary.

7. Check the piston rod for damage, uneven wear, or distortion, and replace if necessary.

8. Check the welded and sealed areas for oil leakage, and replace if necessary.

To install:

9. Position the shock absorber assembly in the vehicle.

10. Install and tighten the attaching nuts in the upper side of the shock absorber assembly to 22 ft. lbs. (30 Nm).

11. Install the fixing bolt in the lower side of the shock absorber assembly, and tighten until snug.

12. Remove the jack or equivalent support from under the rear lower link.

13. Install the tire.

14. With the weight of the vehicle resting on the suspension (empty vehicle), tighten the shock absorber lower fixing bolt to 66 ft. lbs. (89 Nm).

➡**Refer to component parts location and do not reuse non-reusable parts.**

15. Check wheel alignment.

16. After adjusting wheel alignment, adjust the neutral position of the steering angle sensor.

Coil Spring

REMOVAL & INSTALLATION

Rear

Refer to the rear lower link procedure for the rear lower control arm.

Lower Ball Joint

REMOVAL & INSTALLATION

Front

The front lower control arm ball joints are not separately replaceable from the control arms themselves. If the joints are found to be defective, the entire assembly must be replaced.

Rear

The rear lower control arms (lower links) do not use replaceable bushings or ball joints. Replace the entire arm assembly if any damage is noticed, including cracks, dents or deformations.

Upper Control Arm

REMOVAL & INSTALLATION

Rear

1. Before servicing the vehicle, refer to the precautions in the beginning of this section.

2. Remove the rear tire.

3. Remove the stabilizer connecting rod mounting bracket from the suspension arm.

4. Remove the halfshaft from the vehicle.

5. Remove the cotter pin of the suspension arm ball joint, and loosen the nut.

6. Use a ball joint remover or suitable tool to remove the suspension arm from the axle. Be careful not to damage the ball joint boot.

✳✳ CAUTION

Temporarily tighten the mounting nut to prevent damage to the threads and to prevent the ball joint remover from coming off.

7. Remove the fixing nuts and bolts between the suspension arm and the rear suspension member.

8. Remove the suspension arm from vehicle.

9. Check the suspension arm and bushing for deformation, cracks, or damage. If any non-standard condition is found, replace it.

10. Check the boot of the ball joint for cracks, or damage, and also for grease leakage.

11. Manually move the ball stud to confirm it moves smoothly with no binding.

➡**Before measuring, move ball joint at least ten times by hand to check for smooth movement.**

12. Hook a spring scale at the cotter pin mounting hole. Confirm the spring scale measurement value is within 2.18-14.8 lbs. (9.7–66.0 N) when the ball joint stud begins moving. If it is outside the specified range, replace the suspension arm assembly.

13. Attach the mounting nut to the ball stud. Make sure the rotating torque is within 5-30 inch lbs. (0.5–3.4 Nm) with a preload gauge . If it is outside the specified range, replace the suspension arm assembly.

14. Move the tip of the ball joint in the axial direction to check for looseness. If it is outside the specified range of 0 in. (0mm), replace the suspension arm assembly.

To install:

15. Install the suspension arm onto the vehicle.

16. Install the fixing nuts and bolts between the suspension arm and the rear suspension member. Tighten them to 53 ft. lbs. (73 Nm).

17. Reattach the suspension arm ball joint to the axle. Tighten the ball joint nut to 96 ft. lbs. (130 Nm).

18. Install a new cotter pin.

19. Install the halfshaft.

20. Install the stabilizer connecting rod mounting bracket onto the suspension arm, and tighten the two mounting bolts to 41 ft. lbs. (55 Nm).

21. Install the rear tire.

➡**Refer to component parts location and do not reuse non-reusable parts.**

- Perform final tightening of the rear suspension member installation position (rubber bushing) under unladen conditions with the tires on level ground.

22. Check wheel alignment.

23. After adjusting wheel alignment, adjust the neutral position of the steering angle sensor.

Lower Control Arm

REMOVAL & INSTALLATION

Front

1. Before servicing the vehicle, refer to the precautions in the beginning of this section.
2. Remove the tire from vehicle.
3. Remove the undercover.
4. Remove the front cross bar.
5. Remove the cotter pin at the transverse link, then loosen the mounting nut.
6. Use a ball joint remover to remove the transverse link from the steering knuckle. Be careful not to damage the ball joint boot.

✳✳ CAUTION

Temporarily tighten the mounting nut to prevent damage to the threads and to prevent the ball joint remover from coming off.

7. Remove the mounting bolts which are at the back of the transverse link (mounting part with body), then separate the transverse link.
8. Remove the mounting bolts which are at the front of the transverse link (mounting part with the front suspension member), then separate the transverse link.
9. Remove the transverse link from the vehicle.
10. Check transverse link and bushing for deformation, cracks, or damage. If any non-standard condition is found, replace it.
11. Check the boot of the ball joint for cracks, or other damage, and also for grease leakage. If any non-standard condition is found, replace it.
12. Manually move ball stud to confirm it moves smoothly with no binding.

➡**Before measurement, move ball joint at least ten times by hand to check for smooth movement.**

13. Hook a spring scale onto the ball stud tip. Confirm that the spring scale measurement value is within specifications when the ball stud begins to move. If it is outside the specified range, replace the transverse link assembly.

➡**Swing torque specification: Less than 5-43 inch. lbs. (0.5–4.9 Nm), measure value of spring scale: less than 5-43 inch. lbs. (0.5–4.9 Nm).**

14. Attach the mounting nut onto the ball stud. Check that the rotating torque is within specifications with a preload gauge . If it is outside the specified range, replace transverse link assembly.

➡**Rotating torque specification: Less than 5-43 inch lbs. (0.5–4.9 Nm).**

15. Move the tip of ball joint in axial direction to check for looseness. If it is outside the specified range, replace transverse link assembly.

➡**Axial end play specification: 0.004 in. (0.1mm).**

To install:

16. Install the transverse link on the vehicle.
17. Install and tighten the mounting bolts at the front of the transverse link (mounting part with the front suspension member) to 89 ft. lbs. (120 Nm), then install and tighten the mounting bolts which are at the back of the transverse link (mounting part with body) to 118 ft. lbs. (160 Nm).
18. Reattach the transverse link to the steering knuckle, and tighten the ball joint nut to 105 ft. lbs. (142 Nm). Be careful not to damage the ball joint boot.
19. Install a new cotter pin.
20. Install the front cross bar. Tighten the inner two bolts on each end to 33 ft. lbs. (45 Nm) and the outer two bolts on each end to 41 ft. lbs. (55 Nm).
21. Install the undercover.
22. Install the tire.
23. Check the wheel alignment.
24. After adjusting wheel alignment, adjust the neutral position of the steering angle sensor.

Rear

FRONT LOWER LINK

1. Before servicing the vehicle, refer to the precautions in the beginning of this section.
2. Remove the rear tire.
3. Position a jack under the rear lower link for support.
4. Remove the front lower link protector.
5. Remove the shock absorber assembly from the vehicle.
6. Remove the mounting nut and bolt between the front lower link and the axle.
7. Remove the mounting nut and bolt between the front lower link and the rear suspension member.
8. Remove the front lower link from the vehicle.
9. Check the front lower link and bushing for any deformation, cracks, or damage. Replace it if necessary.

To install:

10. Position the front lower link on the vehicle.
11. Install and tighten the mounting nut and bolt between the front lower link and the rear suspension member to 74 ft. lbs. (101 Nm).
12. Install and tighten the mounting nut and bolt between the front lower link and the axle to 77 ft. lbs. (105 Nm).
13. Install the shock absorber assembly.
14. Install the front lower link protector.
15. Slowly lower the jack from under the rear lower link.
16. Install the rear tire.

✳✳ CAUTION

Perform final tightening of the rear suspension member and axle installation position (rubber bushing) under unladen conditions with the tires on level ground.

17. Check the wheel alignment.
18. After adjusting wheel alignment, adjust the neutral position of the steering angle sensor.

REAR LOWER LINK

1. Before servicing the vehicle, refer to the precautions in the beginning of this section.
2. Remove the rear tire.
3. Position a jack under the rear lower link for support.
4. Loosen the fixing bolt and nut of the rear lower link in the side of the suspension member, and then remove the fixing bolt and nut in the side of the axle.
5. Slowly lower the jack, then remove the upper seat, coil spring and rubber sheet from the rear lower link.
6. Remove the fixing bolt and nut in the side of the rear suspension member to remove the rear lower link.
7. Check rear the lower link, bushing and coil spring for deformation, cracks, and damage. Replace the rear lower link and coil spring, if necessary.

To install:

8. Position the rear lower link on the vehicle.
9. Install and tighten the fixing bolt and nut in the side of the rear suspension member to 48 ft. lbs. (65 Nm).

➡**Check that the upper seat is attached as shown. Insert the bracket into the upper seat with the setting three tabs of the upper seat to the projecting part of the bracket beforehand as shown.**

10. Position the upper seat, coil spring and rubber sheet in place, then slowly raise the jack under the rear lower link.

➡ **Match up the rubber seat indentions and rear lower link grooves. Also, make sure spring is not upside down. The top and bottom are indicated by paint color.**

11. Install and tighten the fixing bolt and nut in the side of the axle to 77 ft. lbs. (105 Nm).

12. Slowly lower the jack from under the rear lower link.

13. Install the rear tire.

※※ CAUTION

Perform final tightening of the rear suspension member and axle installation position (rubber bushing) under unladen conditions with the tires on level ground.

14. Check the wheel alignment.

15. After adjusting wheel alignment, adjust the neutral position of the steering angle sensor.

LOWER CONTROL ARM BUSHING REPLACEMENT

The front lower control arm bushings are not separately replaceable from the control arms themselves. If the bushings are found to be defective, the entire assembly must be replaced.

The rear lower control arms (lower links) do not use bushings or ball joints. Replace the entire arm assembly if any damage is noticed, including cracks, dents or deformations.

Wheel Hub and Knuckle

REMOVAL & INSTALLATION

Front

1. Before servicing the vehicle, refer to the precautions in the beginning of this section.

2. Remove the appropriate wheel.

3. Remove the brake caliper. Support it in a place where it will not interfere with work.

➡ **Avoid depressing brake pedal while brake caliper is removed.**

4. Remove the disc rotor.

5. Remove the wheel sensor from the wheel hub and bearing assembly.

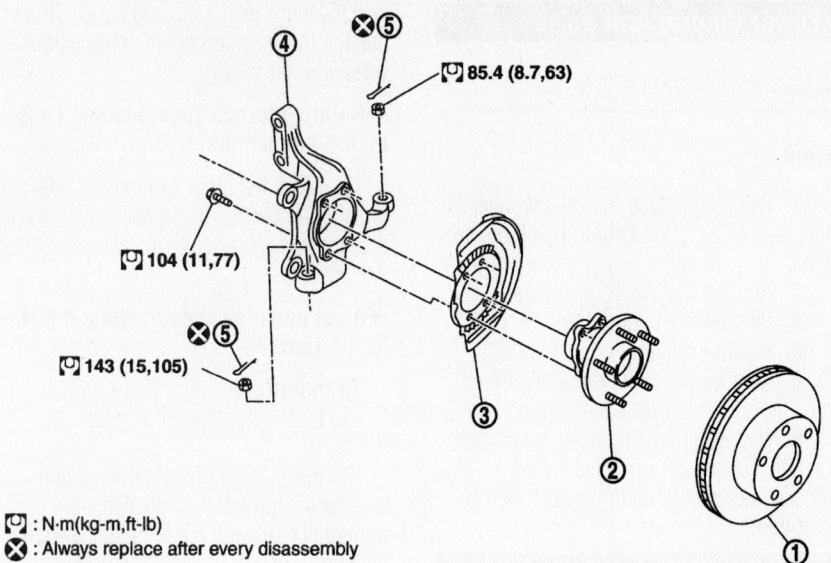

[🔧] : N·m(kg-m,ft-lb)
⊗ : Always replace after every disassembly

1. Disc rotor	2. Wheel hub and bearing assembly	3. Splash guard
4. Steering knuckle	5. Cotter pin	

67162-FX35-G228

Exploded view of front wheel hub mounting—2WD models.

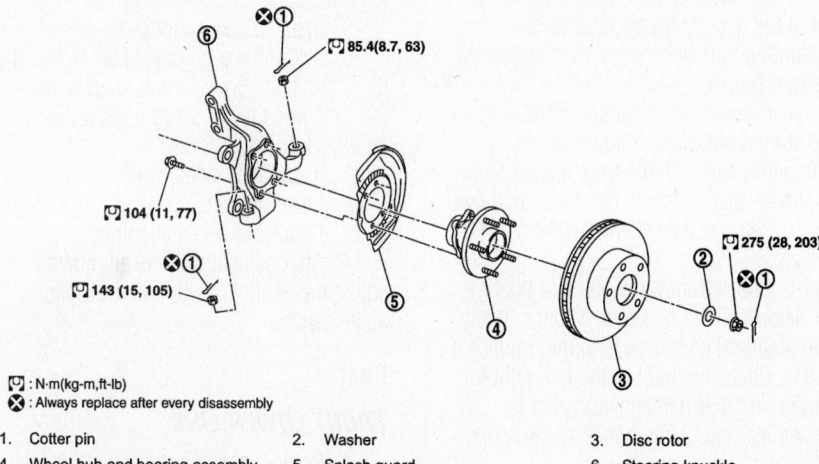

[🔧] : N·m(kg-m,ft-lb)
⊗ : Always replace after every disassembly

1. Cotter pin	2. Washer	3. Disc rotor
4. Wheel hub and bearing assembly	5. Splash guard	6. Steering knuckle

67162-FX35-G229

Exploded view of front wheel hub mounting—AWD models.

※※ CAUTION

Do not pull on wheel sensor wiring harness.

6. Remove the cotter pin from the steering outer socket, then loosen the mounting nut.

7. Use a ball joint remover to separate the steering outer socket from the steering knuckle. Be careful not to damage the ball joint boot.

※※ CAUTION

Temporarily tighten the mounting nut to prevent damage to the threads and

to prevent the ball joint remover from coming off.

8. Remove the cotter pin at the transverse link, then loosen the mounting nut.

9. Use a ball joint remover to separate the transverse link from the steering knuckle. Be careful not to damage ball joint boot.

※※ CAUTION

Temporarily tighten the mounting nut to prevent damage to the threads and to prevent the ball joint remover from coming off.

10. On AWD models, perform the following:

    a. Remove the cotter pin, then remove the lock nut from the halfshaft.

    b. Remove the steering knuckle from the halfshaft.

✳✳ CAUTION

When removing steering knuckle, do not apply an excessive angle to the halfshaft joint. Also be careful not to excessively extend the slide joint. Do not hang over the halfshaft without proper support.

11. Remove the mounting bolts and nuts between the strut assembly and the steering knuckle.

12. Remove the steering knuckle from the vehicle.

13. Remove the mounting bolts between the steering knuckle and the wheel hub/bearing assembly.

14. Remove the splash guard and wheel hub/bearing assembly from the steering knuckle.

15. Check for deformities, cracks and damage on all parts and replace if necessary.

16. Inspect the ball joint for boot breakage, axial looseness, and torque of transverse link and steering outer socket ball joint. Maximum of allowable axial end play is 0.002 in. (0.05mm) or less.

To install:

17. Install the splash guard and wheel hub/bearing assembly onto the steering knuckle.

18. Install the mounting bolts between the steering knuckle and the wheel hub/bearing assembly. Tighten the bolts to 77 ft. lbs. (104 Nm).

19. Install the steering knuckle on the vehicle.

20. Install the mounting bolts and nuts to the strut assembly and the steering knuckle. Tighten the bolts to 134 ft. lbs. (182 Nm).

21. On AWD models, perform the following:

    a. Install the steering knuckle onto the halfshaft.

    b. Install and tighten the lock nut on the halfshaft to 203 ft. lbs. (275 Nm). Install a new cotter pin.

22. Reattach the transverse link to the steering knuckle. Tighten the ball joint nut to 105 ft. lbs. (143 Nm).

23. Install a new cotter pin.

24. Reconnect the steering outer socket to the steering knuckle. Tighten the ball joint nut to 63 ft. lbs. (86 Nm).

25. Install a new cotter pin.

26. Install the wheel sensor onto the wheel hub and bearing assembly.

27. Install the disc rotor.

28. Install the brake caliper.

29. Install the wheel.

30. Check wheel alignment.

31. After adjusting wheel alignment, adjust the neutral position of the steering angle sensor.

32. Check the installation condition of the wheel sensor wiring harness.

Rear

1. Before servicing the vehicle, refer to the precautions in the beginning of this section.

2. Remove the rear wheel.

3. Remove the brake caliper. Hang it in a place where it will not interfere with work.

➡**Avoid depressing brake pedal while brake caliper is removed.**

4. Remove the disc rotor.

5. Remove the wheel sensor from the axle.

✳✳ CAUTION

Do not pull on the wheel sensor wiring harness.

6. Remove the cotter pin, then remove the lock nut from the halfshaft.

7. Separate the halfshaft from the wheel hub and bearing assembly by lightly tapping the end with a suitable hammer and block of wood. If it is hard to separate, use a suitable puller.

8. Remove the mounting bolts of the wheel hub/bearing assembly, then remove the wheel hub/bearing assembly from the axle.

9. Remove the parking brake cable and parking brake shoe from the brake backing plate.

10. Remove the mounting nuts of the anchor block, then remove the anchor block and backing plate from the axle.

11. Loosen mounting bolts and nuts of the front lower link, radius rod, and rear lower link on the side of the suspension member.

12. Set a jack under the rear lower link, then remove the mounting bolt in the front lower link side of the shock absorber.

13. Remove the bolt and nut in the axle side of the rear lower link, then remove the coil spring.

14. Remove the mounting bolts and nuts in the axle side of the front lower link and radius rod.

15. Remove the suspension arm and cotter pin at the axle, then loosen the mounting nut.

16. Use a ball joint remover (or equivalent) to remove the suspension arm from the axle. Be careful not to damage the ball joint boot.

✳✳ CAUTION

Temporarily tighten the mounting nut to prevent damage to the threads and to prevent the ball joint remover from coming off.

17. Remove the axle from the vehicle.

18. Inspect the Ball Joint for boot breakage, axial looseness, and torque of suspension arm ball joint. Maximum of allowable axial end play is 0.00 in. (0.0mm).

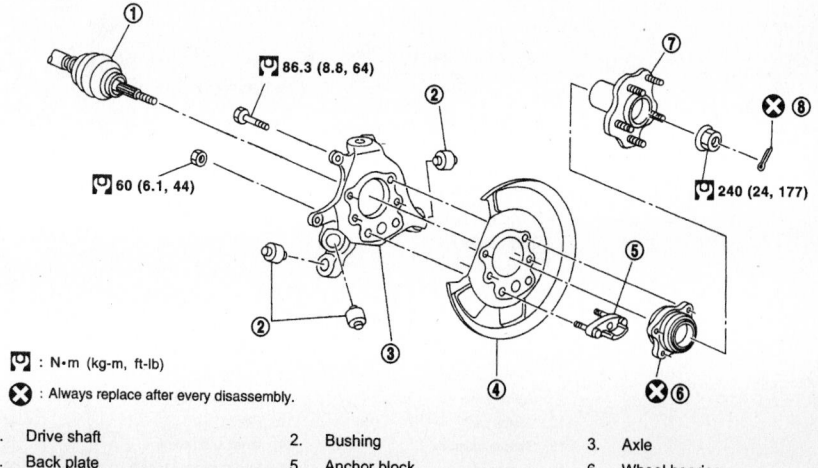

🔧 : N·m (kg-m, ft-lb)

✖ : Always replace after every disassembly.

1. Drive shaft   2. Bushing   3. Axle
4. Back plate   5. Anchor block   6. Wheel bearing
7. Wheel hub   8. Cotter pin

67162-FX35-G230

Exploded view of rear wheel hub mounting.

To install:

19. Install the axle in the vehicle.

20. Reattach the upper control arm to the axle.

21. Tighten the upper control arm mounting nut to 96 ft. lbs. (130 Nm), and install a new cotter pin at the axle.

22. Install the mounting bolts and nuts in the axle side of the front lower link and radius rod.

23. Install the coil spring, then install the bolt and nut in the axle side of the rear lower link.

24. Install the mounting bolt in the front lower link side of the shock absorber.

25. Tighten the mounting bolts and nuts of the front lower link, radius rod, and rear lower link on the side of the suspension member.

26. Install the anchor block and backing plate onto the axle.

27. Install and tighten the mounting nuts of the anchor block to 44 ft. lbs. (60 Nm).

28. Install the parking brake cable and parking brake shoe from the brake backing plate.

29. Install the wheel hub/bearing assembly onto the axle, then install the mounting bolts of the wheel hub/bearing assembly and tighten to 64 ft. lbs. (86 Nm).

30. Install the halfshaft to the wheel hub/bearing assembly.

31. Install and tighten the lock nut on the halfshaft to 177 ft. lbs. (240 Nm), then install a new cotter pin.

32. Install the wheel sensor on the axle.

33. Install the disc rotor.

34. Install the brake caliper.

35. Install the appropriate wheel.

✳✳ CAUTION

Perform final tightening of the rear suspension fasteners under unladen conditions with the tires on level ground.

36. Check the wheel alignment.

37. After adjusting wheel alignment, adjust the neutral position of the steering angle sensor.

BRAKES

Brake Caliper

REMOVAL & INSTALLATION

1. Before servicing the vehicle, refer to the precautions in the beginning of this section.

2. Remove the front wheel assemblies from the vehicle.

3. Drain the brake fluid.

4. Remove the union bolts and torque member bolts, and remove the brake caliper assembly from the vehicle.

5. If necessary, remove the disc rotor.

To install:

✳✳ CAUTION

Only use new "DOT3" brake fluid. Do not reuse drained brake fluid.

6. If removed, install the disc rotor.

7. Install the caliper assembly onto the vehicle. For front calipers, tighten the sliding pin bolts to 20 ft. lbs. (27 Nm) and the torque member bolts to 116 ft. lbs. (157 Nm). For rear calipers, tighten the sliding pin bolts to 32 ft. lbs. (43 Nm) and the torque member bolts to 62 ft. lbs. (84 Nm).

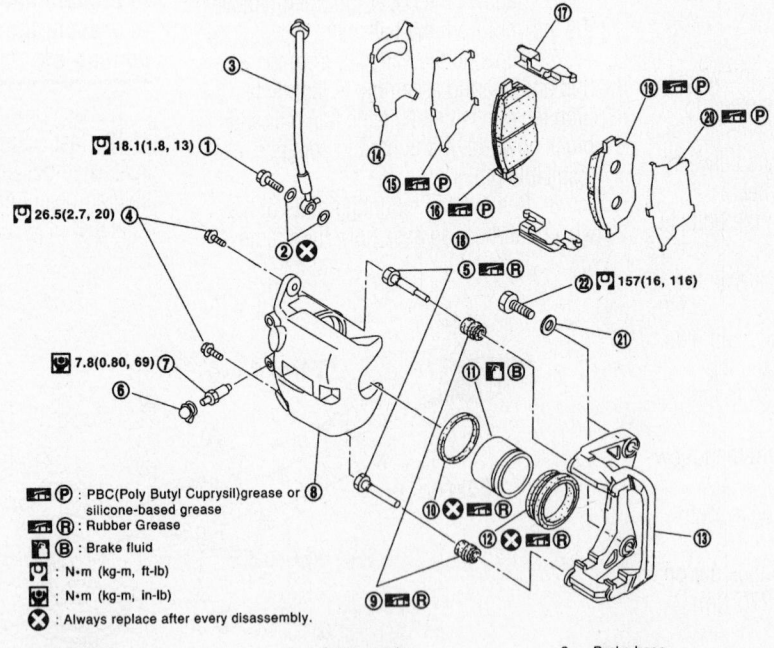

1. Union bolt	2. Copper washer	3. Brake hose
4. Sliding pin bolt	5. Sliding pin	6. Cap
7. Bleed valve	8. Cylinder body	9. Sliding pin boot
10. Piston seal	11. Piston	12. Piston boot
13. Torque member	14. Inner shim cover	15. Inner shim
16. Inner pad	17. Pad retainer (Upper)	18. Pad retainer (Lower)
19. Outer pad	20. Outer shim	21. Washer
22. Torque member bolt		

Exploded view of disc brake mounting and components—Front.

67162-FX35-G231

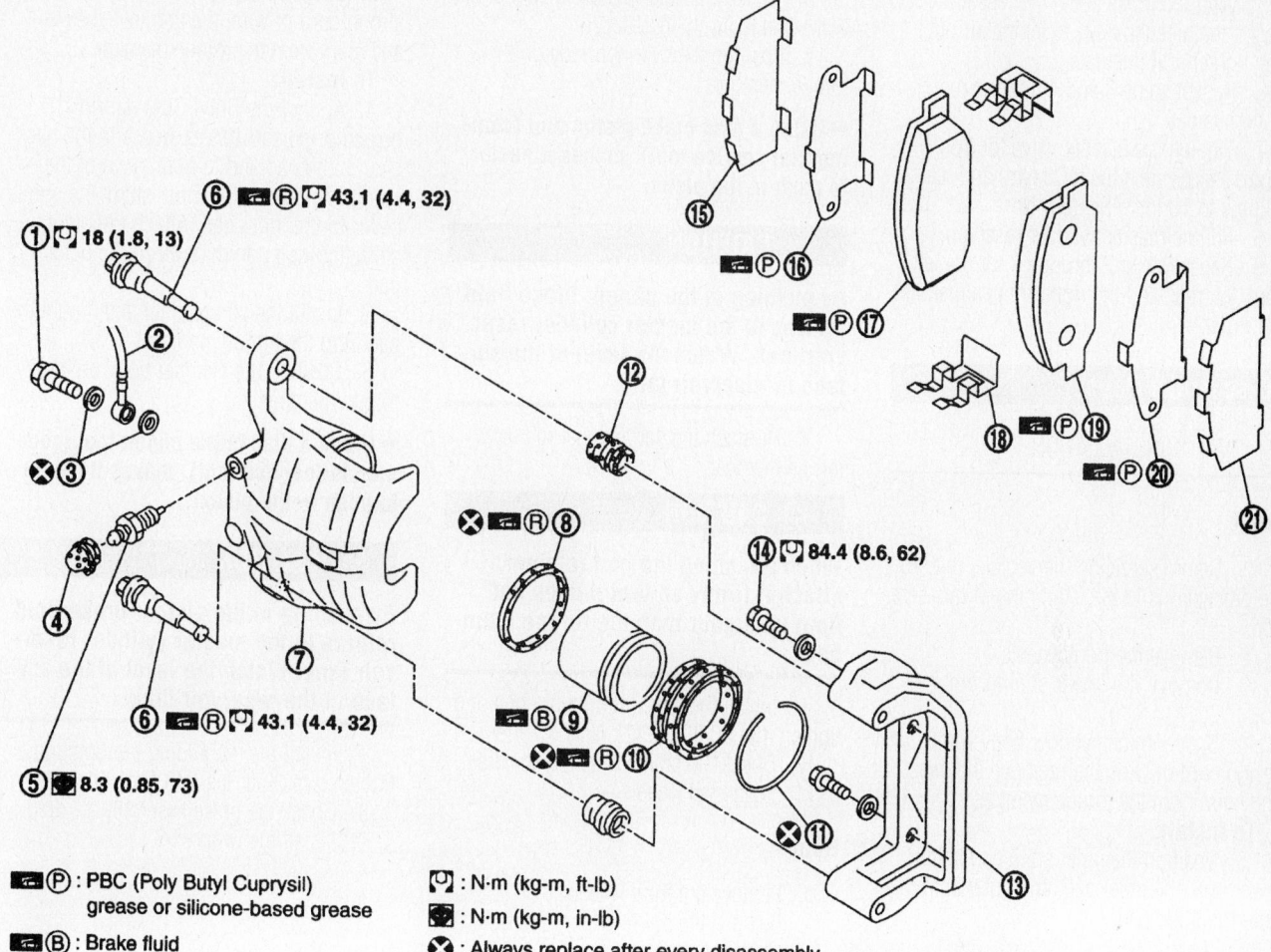

P : PBC (Poly Butyl Cuprysil)
grease or silicone-based grease

B : Brake fluid

R : Rubber grease

[] : N·m (kg-m, ft-lb)

[] : N·m (kg-m, in-lb)

X : Always replace after every disassembly.

1. Union bolt	2. Brake hose	3. Copper washer
4. Cap	5. Bleed valve	6. Sliding pin bolt
7. Cylinder body	8. Piston seal	9. Piston
10. Piston boot	11. Retaining ring	12. Sliding pin boot
13. Torque member	14. Torque member bolt	15. Inner shim cover
16. Inner shim	17. Inner pad	18. Pad retainer
19. Outer pad	20. Outer shim	21. Outer shim cover

67162-FX35-G232

Exploded view of disc brake mounting and components—Rear.

✳✳ CAUTION

When attaching the caliper assembly to the vehicle, wipe any oil off the knuckle spindle, washers and caliper assembly attachment surfaces.

8. Reattach the brake hose to the brake caliper assembly, and tighten the union bolt to 13 ft. lbs. (18 Nm).

✳✳ CAUTION

Do not reuse the old copper washers for the union bolt. Attach the brake

hose to the caliper assembly together only using the specified union bolt and washers.

9. Refill the brake system with new brake fluid, then bleed the system.
10. Install the wheels.

BRAKE SYSTEM BLEEDING

1. Before servicing the vehicle, refer to the precautions in the beginning of this section.

✳✳ CAUTION

While bleeding the brake system, pay attention to the master cylinder fluid level.

2. Turn the ignition switch OFF and disconnect the ABS actuator and electric unit (control unit) connector or negative battery cable.
3. Attach a vinyl tube to the right, rear bleeder valve.
4. Depress the brake pedal fully 4 or 5 times.

5. With the brake pedal depressed, loosen the bleeder valve to let the air out, then tighten it immediately.

6. Repeat steps 3 and 4 until no more air comes out.

7. Tighten the bleeder valve for front calipers to 69 inch lbs. (7.8 Nm) and rear calipers to 73 inch lbs. (8.3 Nm)

8. Fill the master cylinder reservoir.

9. Repeat steps 2 through 7 for the left front, left rear, and the right front calipers, in that order.

Disc Brake Pads

REMOVAL & INSTALLATION

Front

1. Before servicing the vehicle, refer to the precautions in the beginning of this section.

2. Remove the front wheels.

3. Remove the lower sliding pin bolt.

4. Suspend the cylinder body with strong cord or wire, then remove the pad and shim from the torque member.

To install:

5. Position the inner shim and shim cover onto the inner pad, and outer shim onto the outer pad.

6. Push the caliper piston in so that the brake pad is firmly installed.

7. Position the cylinder body on the torque member.

➡**Using a disc brake piston tool (commercial service tool), makes it easier to push in the piston.**

✳✳ CAUTION

By pushing in the piston, brake fluid returns to the master cylinder reservoir tank. Watch the level of the surface of reservoir tank.

8. Reattach the pad retainer to the torque member.

✳✳ CAUTION

When attaching the pad retainer, attach it firmly so that it does not float up higher than the torque member.

9. Install the lower sliding pin bolt and tighten it to 20 ft. lbs. (27 Nm).

10. Check the brake assembly for drag.

11. Install the wheels

Rear

1. Remove the front wheels.

2. Remove the upper sliding pin bolt.

3. Suspend the cylinder body with strong cord or wire, then remove the pad and shim from the torque member.

To install:

4. Apply Poly Butyl Cuprysil (PBC) grease or silicon-based grease to the backside of the pad and to both sides of the shim, then attach the inner shim and shim cover to the inner pad. Attach the outer shim and outer shim cover to the outer pad.

5. Install the pad retainer and mount the pad onto the torque member.

6. Position the cylinder body on the torque member.

➡**Using a disc brake piston tool (commercial service tool), makes it easier to push in the piston.**

✳✳ CAUTION

By pushing in the piston, brake fluid returns to the master cylinder reservoir tank. Watch the level of the surface of the reservoir tank.

7. Install the top sliding pin bolt and tighten it to 32 ft. lbs. (43 Nm).

8. Check the brake assembly for drag.

9. Install the wheels

INFINITI

G20 • G35 • I30 • I35 • M45 • Q45

SPECIFICATION AND MAINTENANCE CHARTS

ENGINE AND VEHICLE IDENTIFICATION

Code ①	Liters (cc)	Cu. In.	Cyl.	Fuel Sys.	Engine Type	Eng. Mfg.
SR20DE	2.0 (1998)	122	4	MFI	DOHC	Nissan
VQ30DE	3.0 (2988)	182	6	MFI	DOHC	Nissan
VQ35DE	3.5 (3498)	213	6	MFI	DOHC	Nissan
VH41DE	4.1 (4130)	252	8	MFI	DOHC	Nissan
VK45DE	4.5 (4494)	274	8	MFI	DOHC	Nissan

Code ②	Year
1	2001
2	2002
3	2003
4	2004
5	2005

MFI: Multi-port Fuel Injection

DOHC: Double Overhead Camshaft

① 4th digit of the Vehicle Identification Number (VIN)

② 10th digit of the Vehicle Identification Number (VIN)

67162-INFI-C01

GENERAL ENGINE SPECIFICATIONS

Year	Model	Engine Displacement Liters	Engine ID/VIN	Net Horsepower @ rpm	Net Torque @ rpm (ft. lbs.)	Bore x Stroke (in.)	Com-pression Ratio	Oil Pressure @ rpm
2001	G20	2.0	SR20DE	140@6400	132@4800	3.39x3.39	9.5:1	46-57@3200
	I30	3.0	VQ30DE	190@5600	205@4000	3.66x2.89	10.1:1	63-80@3000
	Q45	4.1	VH41DE	266@5600	278@4000	3.66x2.99	10.2:1	67-81@3000
2002	G20	2.0	SR20DE	140@6400	132@4800	3.39x3.39	9.5:1	46-57@3200
	G35	3.5	VQ35DE	260@6000	260@4800	3.76X3.20	10.3:1	43@2000
	I35	3.5	VQ35DE	255@5800	246@4400	3.76X3.20	10.3:1	43@2000
	Q45	4.5	VK45DE	340@6400	333@4000	3.66x3.25	10.5:1	43@3000
2003	G20	2.0	SR20DE	140@6400	132@4800	3.39x3.39	9.5:1	46-57@3200
	G35	3.5	VQ35DE	260@6000	260@4800	3.76X3.20	10.3:1	43@2000
	I35	3.5	VQ35DE	255@5800	246@4400	3.76X3.20	10.3:1	43@2000
	Q45	4.5	VK45DE	340@6400	333@4000	3.66x3.25	10.5:1	43@3000
2004	G35	3.5	VQ35DE	260@6000	260@4800	3.76X3.20	10.3:1	43@2000
	I35	3.5	VQ35DE	255@5800	246@4400	3.76X3.20	10.3:1	43@2000
	M45	4.5	VK45DE	340@6400	333@4000	3.66x3.25	10.5:1	43@3200
	Q45	4.5	VK45DE	340@6400	333@4000	3.66x3.25	10.5:1	43@3000

67162-INFI-C02

GASOLINE ENGINE TUNE-UP SPECIFICATIONS

Year	Engine Displacement Liters	Engine ID/VIN	Spark Plug Gap (in.)	Ignition Timing (deg.)		Fuel Pump (psi) ①	Idle Speed (rpm)		Valve Clearance	
				MT	AT		MT	AT	Intake	Exhaust
2001	2.0	SR20DE	0.031–0.035	15B	15B	34	800	800	HYD	HYD
	3.0	VQ30DE	0.039–0.043	15B	15B	34	625	700	HYD	HYD
	4.1	VH41DE	0.039–0.041	—	15B	34	—	650	HYD	HYD
2002	2.0	SR20DE	0.031–0.035	15B	15B	34	800	800	HYD	HYD
	3.5	VQ35DE	0.043	—	15B	34	—	600–700	HYD	HYD
	4.5	VK45DE	0.039–0.041	—	15B	34	—	650	HYD	HYD
2003	2.0	SR20DE	0.031–0.035	15B	15B	34	800	800	HYD	HYD
	3.5	VQ35DE	0.043	—	15B	34	—	600–700	HYD	HYD
	4.5	VK45DE	0.039–0.041	—	15B	34	—	650	HYD	HYD
2004	2.0	SR20DE	0.031–0.035	15B	15B	34	800	800	HYD	HYD
	3.5	VQ35DE	0.043	15B	15B	34	600–700	625–725	HYD	HYD
	4.5	VK45DE	0.043	—	12B	34	—	650	HYD	HYD

NOTE: The Vehicle Emission Control Information label often reflects specification changes made during production.

The label figures must be used if they differ from those in this chart.

B: Before top dead center

① 43 psi with regulator vacuum hose disconnected

67162-INFI-C03

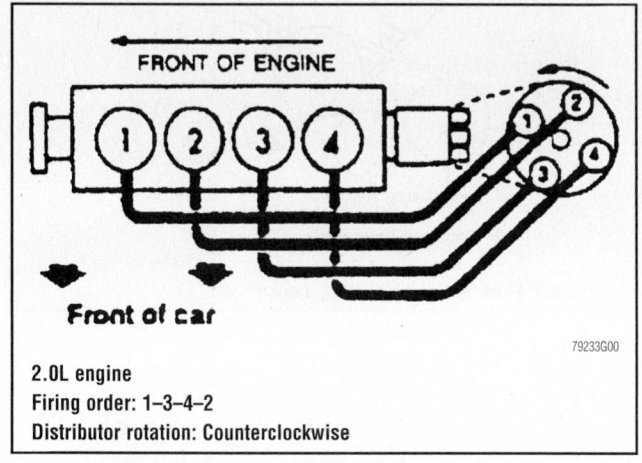

2.0L engine
Firing order: 1–3–4–2
Distributor rotation: Counterclockwise

79233G00

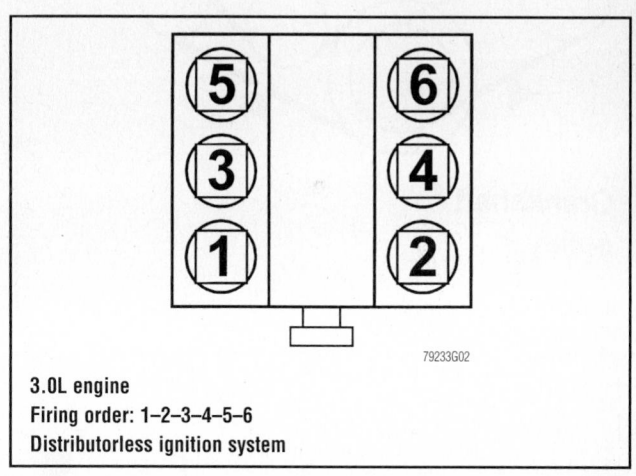

3.0L engine
Firing order: 1–2–3–4–5–6
Distributorless ignition system

79233G02

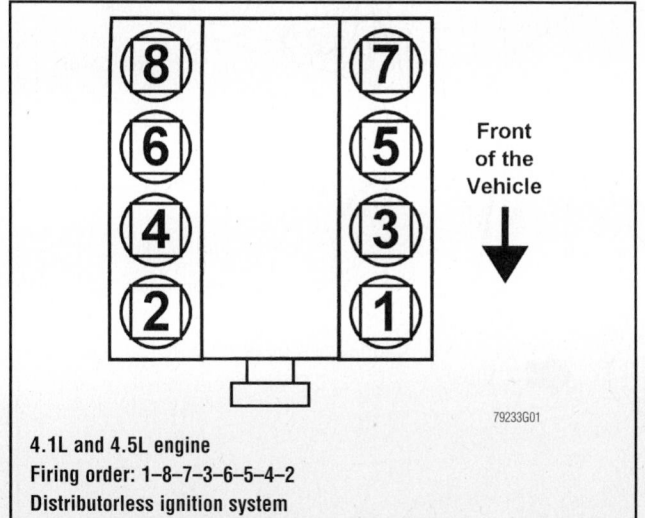

4.1L and 4.5L engine
Firing order: 1–8–7–3–6–5–4–2
Distributorless ignition system

79233G01

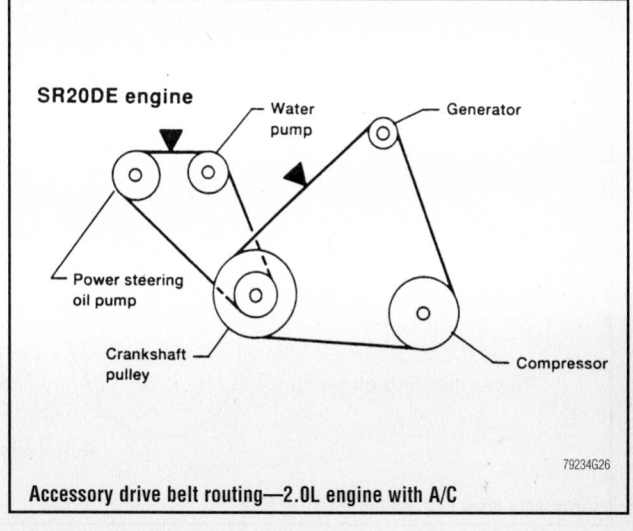

Accessory drive belt routing—2.0L engine with A/C

79234G26

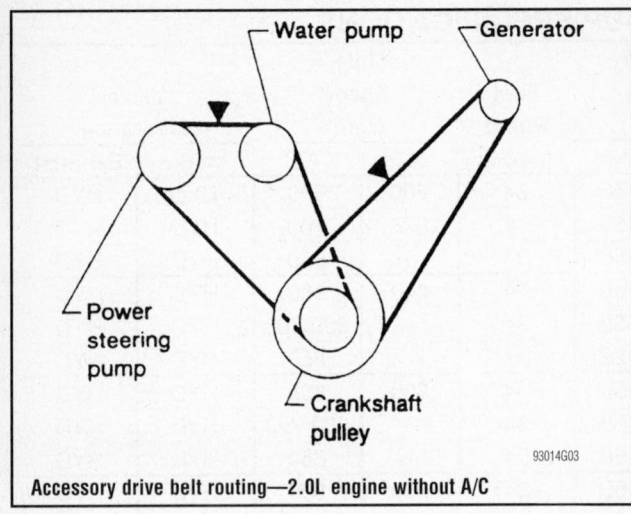

Accessory drive belt routing—2.0L engine without A/C

93014G03

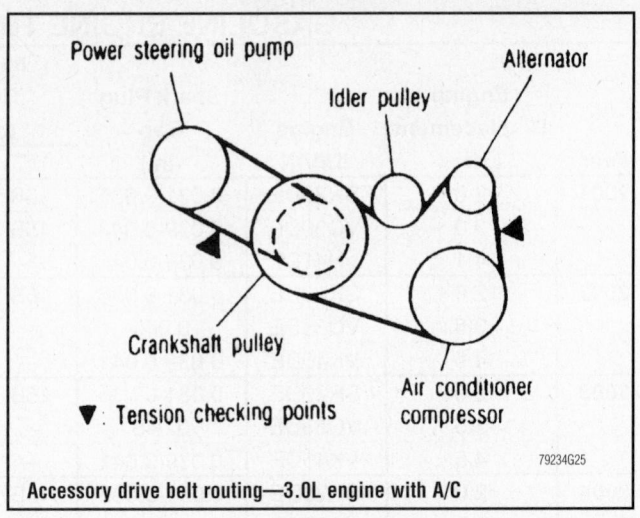

▼: Tension checking points

Accessory drive belt routing—3.0L engine with A/C

79234G25

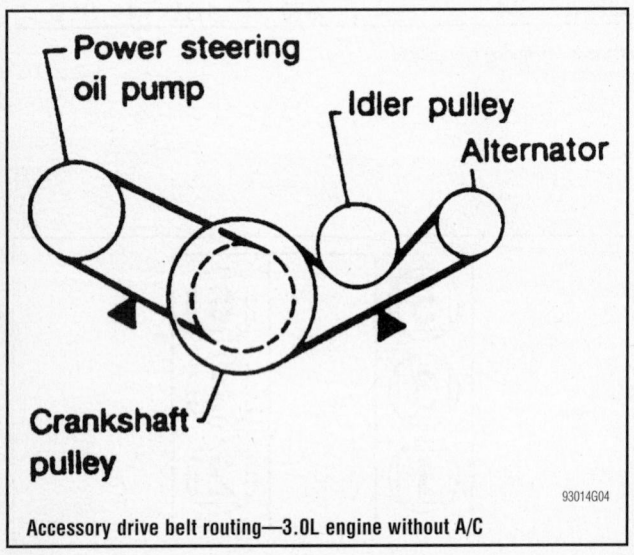

Accessory drive belt routing—3.0L engine without A/C

93014G04

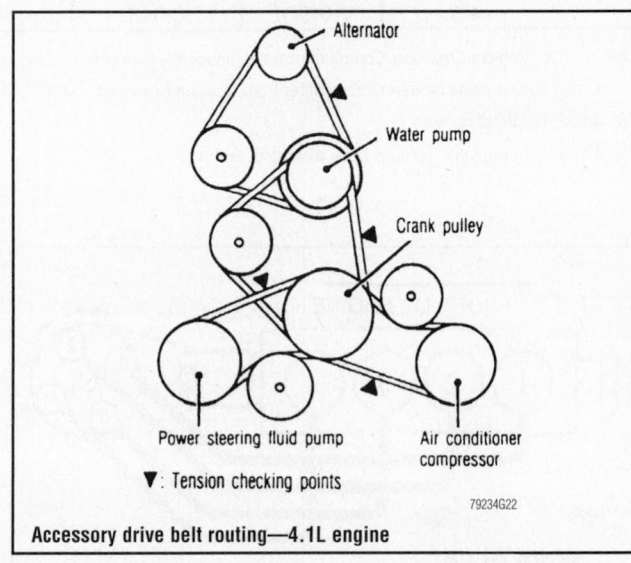

▼: Tension checking points

Accessory drive belt routing—4.1L engine

79234G22

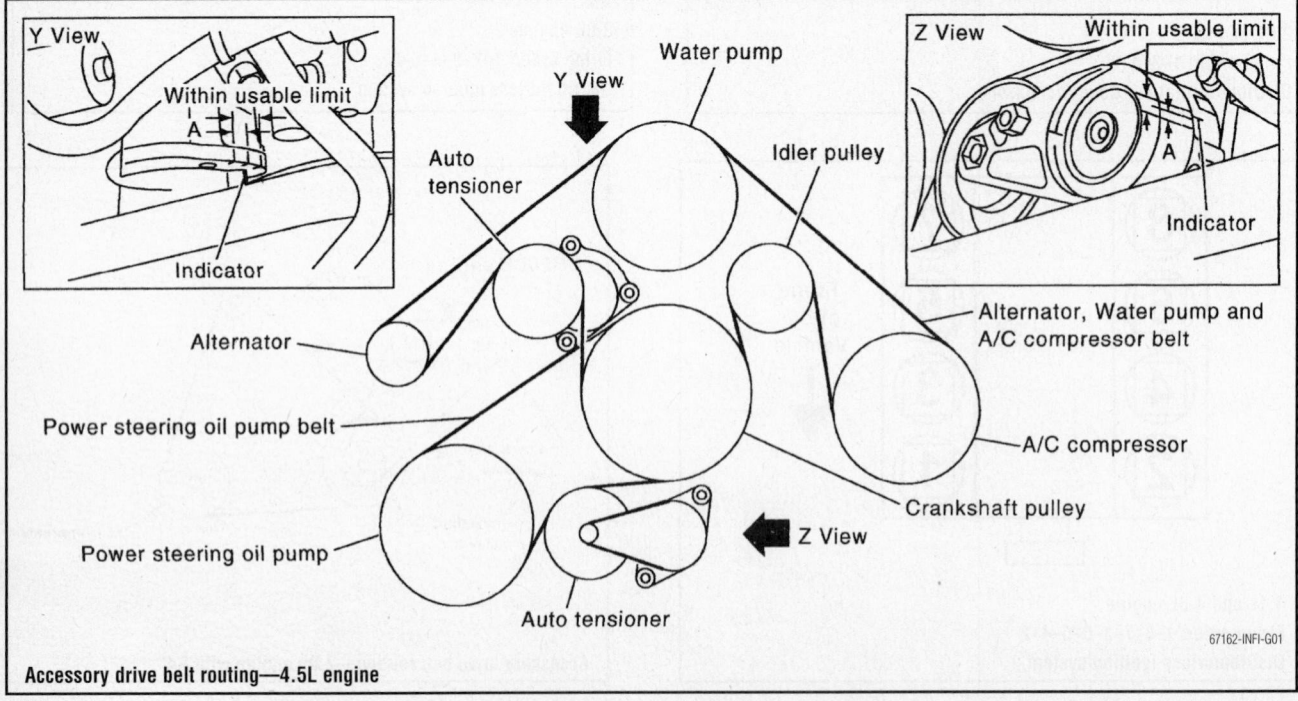

Accessory drive belt routing—4.5L engine

67162-INFI-G01

CAPACITIES

Year	Model	Engine Displacement Liters	Engine ID/VIN	Engine Oil with Filter	Transmission (pts.)		Drive Axle (pts.)	Fuel Tank (gal.)	Cooling System (qts.)
					5-Spd	Auto.			
2001	G20	2.0	SR20DE	3.62	①	14.8	—	15.9	6.50
	I30	3.0	VQ30DE	4.25	②	19.8	—	18.5	9.88
	Q45	4.1	VH41DE	5.63	—	22.2	2.75	21.1	13.00
2002	G20	2.0	SR20DE	3.62	②	14.8	—	15.9	6.50
	G35	3.5	VQ35DE	5.00	—	21.8	—	18.5	9.00
	I35	3.5	VQ35DE	5.00	—	21.8	—	18.5	9.00
	Q45	4.5	VK45DE	5.63	—	21.3	2.75	21.4	10.38
2003	G20	2.0	SR20DE	3.62	②	14.8	—	15.9	6.50
	G35	3.5	VQ35DE	5.00	—	21.8	—	18.5	9.00
	I35	3.5	VQ35DE	5.00	—	21.8	—	18.5	9.00
	Q45	4.5	VK45DE	5.63	—	21.3	2.75	21.4	10.38
2004	G35	3.5	VQ35DE	5.00	③	21.8	3.0	20.1	④
	I35	3.5	VQ35DE	5.00	—	18.0	—	18.5	9.00
	M45	4.5	VK45DE	5.91	—	21.7	3.20	21.4	10.38
	Q45	4.5	VK45DE	5.91	—	21.7	2.95	21.4	10.38

NOTE: All capacities are approximate. Add fluid gradually and check to be sure a proper fluid level is obtained.

① RS5F32A: 7.62 - 8.00
   RS5F32V: 7.86 - 8.25

② RSF50V: 9.13-9.50
   RSF50A: 9.50-10.13

③ 6-speed: 6.1 pts.

④ Man. Trans. 9.25
   Auto. Trans. 9.00

67162-INFI-C04

VALVE SPECIFICATIONS

Year	Engine Displacement Liters	Engine ID/VIN	Seat Angle (deg.)	Face Angle (deg.)	Spring Test Pressure (lbs. @ in.)	Spring Free Height (in.)	Stem-to-Guide Clearance (in.)		Stem Diameter (in.)	
							Intake	Exhaust	Intake	Exhaust
2001	2.0	SR20DE	45.25-45.75	44.85-45.10	127.9-144.3@1.181	1.943	0.0008-0.0021	0.0016-0.0029	0.2348-0.2354	0.2341-0.2346
	3.0	VQ30DE	45.25-45.75	NA	120.1@1.085	1.845	0.0008-0.0021	0.0016-0.0029	0.2348-0.2354	0.2341-0.2346
	4.1	VH41DE	45.25-45.75	44.85-45.10	120.4@1.055	1.946	0.0011-0.0020	0.0014-0.0020	0.2743-0.2744	0.3134-0.3136
2002	2.0	SR20DE	45.25-45.75	44.85-45.10	127.9-144.3@1.181	1.943	0.0008-0.0021	0.0016-0.0029	0.2348-0.2354	0.2341-0.2346
	3.5	VQ35DE	45.15-45.45	NA	91.5-103.2@1.094	1.796	0.010-0.013	0.011-0.015	0.2348-0.2354	0.2344-0.2350
	4.5	VK45DE	45.25-45.75	NA	65-74@1.331	2.0189-2.0268	0.010-0.013	0.011-0.015	0.2351-0.2354	0.2347-0.2350
2003	2.0	SR20DE	45.25-45.75	44.85-45.10	127.9-144.3@1.181	1.943	0.0008-0.0021	0.0016-0.0029	0.2348-0.2354	0.2341-0.2346
	3.5	VQ35DE	45.15-45.45	NA	91.5-103.2@1.094	1.796	0.010-0.013	0.011-0.015	0.2348-0.2354	0.2344-0.2350
	4.5	VK45DE	45.25-45.75	NA	65-74@1.331	2.0189-2.0268	0.010-0.013	0.011-0.015	0.2351-0.2354	0.2347-0.2350
2004	3.5	VQ35DE	45.15-45.45	NA	91.5-103.2@1.094	1.796	0.0008-0.0021	0.012-0.025	0.2348-0.2354	0.2344-0.2350
	4.5	VK45DE	45.15-45.45	NA	65-74@1.331	1.8247-1.8444	0.0008-0.0018	0.0012-0.0022	0.2351-0.2354	0.2347-0.2350

NA: Not Available

67162-INFI-C05

CRANKSHAFT AND CONNECTING ROD SPECIFICATIONS

All measurements are given in inches.

Year	Engine Displacement Liters	Engine ID/VIN	Crankshaft				Connecting Rod		
			Main Brg. Journal Dia.	Main Brg. Oil Clearance	Shaft End-play	Thrust on No.	Journal Diameter	Oil Clearance	Side Clearance
2001	2.0	SR20DE	2.1643-2.1646	0.0002-0.0009	0.0039-0.0102	3	1.8885-1.8887	0.0008-0.0018	0.0079-0.0138
	3.0	VQ30DE	2.3610-2.3612	0.0014-0.0021	0.0039-0.0098	3	1.7704-1.7706	0.0013-0.0023	0.0079-0.0138
	4.1	VH41DE	2.5181-2.5183	0.0002-0.0008	0.0039-0.0102	3	2.0460-2.0462	0.0008-0.0018	0.0079-0.0138
2002	2.0	SR20DE	2.1643-2.1646	0.0002-0.0009	0.0039-0.0102	3	1.8885-1.8887	0.0008-0.0018	0.0079-0.0138
	3.5	VQ35DE	2.3603-2.3612	0.0014-0.0021	0.0039-0.0098	3	1.7704-1.7706	0.0013-0.0023	0.0079-0.0138
	4.5	VK45DE	2.5173-2.5183	①	0.0039-0.0102	3	2.1654-2.1659	0.0008-0.0018	0.0079-0.0138
2003	2.0	SR20DE	2.1643-2.1646	0.0002-0.0009	0.0039-0.0102	3	1.8885-1.8887	0.0008-0.0018	0.0079-0.0138
	3.5	VQ35DE	2.3603-2.3612	0.0014-0.0021	0.0039-0.0098	3	1.7704-1.7706	0.0013-0.0023	0.0079-0.0138
	4.5	VK45DE	2.5173-2.5183	①	0.0039-0.0102	3	2.1654-2.1659	0.0008-0.0018	0.0079-0.0138
2004	3.5	VQ35DE	2.3603-2.3612	0.0014-0.0021	0.0039-0.0098	3	1.5890-1.5921	0.0013-0.0023	0.0079-0.0138
	4.5	VK45DE	2.5173-2.5183	①	0.0039-0.0980	3	2.0457-2.0462	0.0008-0.0018	0.0079-0.0138

① Nos. 1 and 5: 0.0004-0.0004 in.
Nos. 2, 3 and 4: 0.0003-0.0007

67162-INFI-C06

PISTON AND RING SPECIFICATIONS

All measurements are given in inches.

Year	Engine Displacement Liters	Engine ID/VIN	Piston Clearance	Ring Gap			Ring Side Clearance		
				Top Compression	Bottom Compression	Oil Control	Top Compression	Bottom Compression	Oil Control
2001	2.0	SR20DE	0.0004-0.0012	0.0079-0.0118	0.0138-0.0197	0.0079-0.0236	0.0018-0.0031	0.0012-0.0026	SNUG
	3.0	VQ30DE	0.0004-0.0012	0.0087-0.0161	0.0197-0.0291	0.0079-0.0272	0.0016-0.0031	0.0012-0.0028	SNUG
	4.1	VH41DE	0.0004-0.0012	0.0106-0.0181	0.0154-0.0248	0.0079-0.0272	0.0016-0.0031	0.0012-0.0028	SNUG
2002	2.0	SR20DE	0.0004-0.0012	0.0079-0.0118	0.0138-0.0197	0.0079-0.0236	0.0018-0.0031	0.0012-0.0026	SNUG
	3.5	VQ35DE	0.0004-0.0012	0.0091-0.0130	0.0130-0.0189	0.0079-0.0197	0.0018-0.0031	0.0012-0.0028	0.0026-0.0053
	4.5	VK45DE	0.0002-0.0007	0.0087-0.0126	0.0126-0.0185	0.0079-0.0236	0.0018-0.0039	0.0012-0.0028	0.0026-0.0053
2003	2.0	SR20DE	0.0004-0.0012	0.0079-0.0118	0.0138-0.0197	0.0079-0.0236	0.0018-0.0031	0.0012-0.0026	SNUG
	3.5	VQ35DE	0.0004-0.0012	0.0091-0.0130	0.0130-0.0189	0.0079-0.0197	0.0018-0.0031	0.0012-0.0028	0.0026-0.0053
	4.5	VK45DE	0.0002-0.0007	0.0087-0.0126	0.0126-0.0185	0.0079-0.0236	0.0018-0.0039	0.0012-0.0028	0.0026-0.0053
2004	3.5	VQ35DE	0.0004-0.0012	0.0091-0.0130	0.0130-0.0189	0.0079-0.0197	0.0018-0.0031	0.0012-0.0028	0.0026-0.0053
	4.5	VK45DE	0.0004-0.0012	0.0087-0.0126	0.0087-0.0126	0.0079-0.0236	0.0018-0.0031	0.0012-0.0028	0.0026-0.0053

67162-INFI-C07

TORQUE SPECIFICATIONS
All readings in ft. lbs.

Year	Engine Displacement Liters	Engine ID/VIN	Cylinder Head Bolts	Main Bearing Bolts	Rod Bearing Bolts	Crankshaft Damper Bolts	Flywheel Bolts	Manifold Intake	Manifold Exhaust	Spark Plugs	Oil Pan Drain Plug
2001	2.0	SR20DE	①	②	③	105-112	61-69	13-15	27-35	14-22	NA
	3.0	VQ30DE	④	⑤	⑥	⑦	61-69	20-23	21-24	14-22	NA
	4.1	VH41DE	⑧	⑨	⑩	260-275	61-69	12-15	20-23	14-22	NA
2002	2.0	SR20DE	①	②	③	105-112	61-69	13-15	27-35	14-22	NA
	3.5	VQ35DE	④	⑤	⑥	⑦	61-69	⑪	21-23	15-21	22-29
	4.5	VK45DE	⑫	⑬	③	⑭	62-68	18-23	19-22	15-21	22-28
2003	2.0	SR20DE	①	②	③	105-112	61-69	13-15	27-35	14-22	NA
	3.5	VQ35DE	④	⑤	⑥	⑦	61-69	⑪	21-23	15-21	22-29
	4.5	VK45DE	⑫	⑬	③	⑭	62-68	18-23	19-22	15-21	22-28
2004	3.5	VQ35DE	④	⑤	⑥	⑦	61-69	⑪	21-23	15-21	22-29
	4.5	VK45DE	⑫	⑬	③	⑭	62-68	21	19-22	18	22-28

NA: Not Available

① Step 1: 29 ft. lbs.
Step 2: 58 ft. lbs.
Step 3: Loosen bolts completely
Step 4: 25-33 ft. lbs.
Step 5: Tighten an additional 90-100 degrees
Step 6: Repeat Step 5.

② Step 1: 24-28 ft. lbs.
Step 2: Tighten an additional 45-50 degrees or 54-61 ft. lbs.

③ Step 1: 10-12 ft. lbs.
Step 2: Tighten an additional 60-65 degrees or 28-33 ft. lbs.

④ Step 1: 72 ft. lbs.
Step 2: Loosen bolts completely
Step 3: 25-33 ft. lbs.
Step 4: Tighten an additional 90-95 degrees
Step 5: Repeat Step 4

⑤ Step 1: Shift crankshaft to align the bearing beam
Step 2: Tighten all bolts to 24-28 ft. lbs.
Step 3: Tighten an additional 90-95 degrees

⑥ Step 1: Tighten to 15 ft. lbs.
Step 2: Tighten an additional 90-95 degrees

⑦ Step 1: 29-36 ft. lbs.
Step 2: Tighten an additional 60-66 degrees

⑧ Step 1: 22 ft. lbs.
Step 2: 69 ft. lbs.
Step 3: Loosen bolts completely
Step 4: 18-25 ft. lbs.
Step 5: Tighten an additional 90-95 degrees or 69-72 ft. lbs.

⑨ Step 1: Shift crankshaft back and forth to seat bearing caps
Step 2: Tighten inner cap bolts to 27-31 ft. lbs.
Step 3: Tighten outer cap bolts to 20-24 ft. lbs.
Step 4: Tighten No. 1-3, 5 inner cap bolts an additional 60 degrees
Step 5: Tighten No. 4 inner cap bolt and additional 35 degrees
Step 6: Tighten outer cap bolts an additional 35 degrees
Step 7: Tighten bearing cap side bolts to 34-38 ft. lbs.

⑩ Step 1: 10-12 ft. lbs.
Step 2: Tighten an additional 60-65 degrees or 43-48 ft. lbs.

⑪ Step 1: Tighten to 4-7 ft. lbs.
Step 2: Tighten to 20-23 ft. lbs.
Step 3: Tighten, again, to 20-23 ft. lbs.

⑫ Step 1: 72 ft. lbs.
Step 2: Loosen bolts completely
Step 3: 29-36 ft. lbs.
Step 4: Tighten an additional 60-65 degrees
Step 5: Repeat step 4

⑬ Step 1: M12 bolts: 27-31 ft. lbs., plus an additional 40-45 degrees
Step 2: M9 bolts: 20-23 ft. lbs., plus an additional 30-35 degrees
Step 3: M10 side bolts: 10-12 ft. lbs. exc. 2004, 2004 is 34-38 ft. lbs.

⑭ Step 1: 65-72 ft. lbs.
Step 2: Tighten an additional 90-96 degrees
Step 2: Tighten an additional 90-96 degrees

WHEEL ALIGNMENT

Year	Model		Caster Range (+/-Deg.)	Caster Preferred Setting (Deg.)	Camber Range (+/-Deg.)	Camber Preferred Setting (Deg.)	Toe-in (in.)
2001	G20	F	0.75	+1.92	0.75	0	0.04 +/- 0.04
		R	—	—	0.75	-1.03	0.16 +/- 0.16
	I30	F	0.75	+2.75	0.75	-0.25	0.04 +/- 0.04
		R	—	—	0.75	-1.00	0.04 +/- 0.15
	Q45	F	0.75	+6.42	0.75	-0.66	0.08 +/- 0.04
		R	—	—	0.50	-0.75	0.09 +/- 0.10
2002	G20	F	0.75	+1.92	0.75	0	0.04 +/- 0.04
		R	—	—	0.75	-1.03	0.16 +/- 0.16
	G35	F	0.75	+7.75	0.75	-0.08	0.04 +/- 0.04
		R	—	—	0.75	-1.00	0.04 +/- 0.15
	I35	F	0.75	+7.75	0.75	-0.08	0.04 +/- 0.04
		R	—	—	0.75	-1.00	0.04 +/- 0.15
	Q45	F	0.75	+6.42	0.75	-0.66	0.08 +/- 0.04
		R	—	—	0.50	-0.75	0.09 +/- 0.10
2003	G20	F	0.75	+1.92	0.75	0	0.04 +/- 0.04
		R	—	—	0.75	-1.03	0.16 +/- 0.16
	G35	F	0.75	+7.75	0.75	-0.08	0.04 +/- 0.04
		R	—	—	0.75	-1.00	0.04 +/- 0.15
	I35	F	0.75	+7.75	0.75	-0.08	0.04 +/- 0.04
		R	—	—	0.75	-1.00	0.04 +/- 0.15
	Q45	F	0.75	+6.42	0.75	-0.66	0.08 +/- 0.04
		R	—	—	0.50	-0.75	0.09 +/- 0.10
2004	G35 Coupe	F	0.75	①	0.75	-0.50	0.04 +/- 0.04
		R	—	—	0.75	-1.25	0.01 +/- 0.01
	G35 Sedan	F	0.75	+8.17	0.75	-0.08	0.04 +/- 0.04
		R	—	—	0.75	②	0.01 +/- 0.21
	I35	F	0.75	+2.75	0.75	-0.50	0.04 +/- 0.04
		R	—	—	0.75	-1.00	0.04 +/- 0.15
	M45	F	0.75	+6.58	0.75	-0.07	0.04 +/- 0.04
		R	—	—	0.75	-0.67	0.10 +/- 0.10
	Q45	F	0.75	+6.17	0.75	-0.75	0.05 +/- 0.05
		R	—	—	0.75	③	0.05 +/- 0.05

① 17 inch wheel: +8.17
  18 inch wheel: +8.00

② Auto. Trans.: -0.58
  Man. Trans.: -0.67

③ 17 inch wheel: -0.42
  18 inch wheel: -0.58

67162-INFI-C09

TIRE, WHEEL AND BALL JOINT SPECIFICATIONS

Year	Model	OEM Tires		Tire Pressures (psi)		Wheel Size	Ball Joint Inspection	Lug Nut (Ft. Lbs.)
		Standard	Optional	Front	Rear			
2001	G20	P195/65HR15	None	35	35	6-JJ	①	NA
	Q45	P215/60VR16	P225/50VR17	35	35	Std: 7-JJ Opt: 7.5-JJ	①	NA
	I30	P205/65R15	P215/60HR15	35	35	Std: 6-JJ Opt: 6.5-JJ	①	NA
2002	G20	P195/65HR15	None	35	35	6-JJ	①	NA
	G35	P205/65R16	P215/55R17	30	30	Std: 6.5-JJ Opt: 7-JJ	①	72-87
	I35	P205/65R16	P215/55R17	30	30	Std: 6.5-JJ Opt: 7-JJ	①	73-94
	Q45	P225/55VR17	P245/45VR18	33	33	Std: 7.5-JJ Opt: 7.5-JJ	①	72-87
2003	G20	P195/65HR15	None	35	35	6-JJ	①	NA
	G35	P205/65R16	P215/55R17	30	30	Std: 6.5-JJ Opt: 7-JJ	①	72-87
	I35	P205/65R16	P215/55R17	30	30	Std: 6.5-JJ Opt: 7-JJ	①	73-94
	Q45	P225/55VR17	P245/45VR18	33	33	Std: 7.5-JJ Opt: 7.5-JJ	①	72-87
2004	G35 Coupe	②	③	30	30	Std: 7.5-JJ Opt: 8-JJ	①	72-87
	G35 Sedan	P205/65R16	P215/55R17	30	30	Std: 6.5-JJ Opt: 7-JJ	①	72-87
	I35	P215/55R17	P225/50R17	30	30	Std: 6.5-JJ Opt: 7-JJ	①	73-94
	M45	P235/45R18	NA	33	33	Std: 7.5-JJ	①	72-87
	Q45	P225/55R17	P245/45R18	33	33	Std: 7.5-JJ Opt: 7.5-JJ	①	72-87

OEM: Original Equipment Manufacturer

PSI: Pounds Per Square Inch

STD: Standard

OPT: Optional

NA: Not Available

① Replace if any measurable movement is found.

② Front: P225/50R17, Rear: P235/50R17

③ Front: P225/45R18, Rear: P245/45R18

BRAKE SPECIFICATIONS
All measurements in inches unless noted

| Year | Model | Front Brake Disc | | | Rear Brake Disc | | | Minimum Lining Thickness | | Brake Caliper | |
		Original Thickness	Minimum Thickness	Maximum Run-out	Original Thickness	Minimum Thickness	Maximum Run-out	Front	Rear	Bracket Bolts (ft. lbs.)	Mounting Bolts (ft. lbs.)
2001	G20	0.870	0.787	0.003	0.350	0.310	0.003	0.079	0.059	①	16-23
	I30	0.870	0.787	0.003	0.350	0.315	0.003	0.079	0.059	①	16-23
	Q45	1.100	1.024	0.003	0.350	0.551	0.003	0.079	0.079	②	24-31
2002	G20	0.870	0.787	0.003	0.350	0.310	0.003	0.079	0.059	①	16-23
	G35	0.945	0.866	0.003	0.630	0.551	0.004	0.079	0.079	①	16-23
	I35	0.945	0.866	0.003	0.630	0.551	0.004	0.079	0.079	①	16-23
	Q45	1.100	1.020	0.003	0.630	0.550	0.004	0.079	0.079	②	24-31
2003	G20	0.870	0.787	0.003	0.350	0.310	0.003	0.079	0.059	①	16-23
	G35	0.945	0.866	0.003	0.630	0.551	0.004	0.079	0.079	①	16-23
	I35	0.945	0.866	0.003	0.630	0.551	0.004	0.079	0.079	①	16-23
	Q45	1.100	1.020	0.003	0.630	0.550	0.004	0.079	0.079	②	24-31
2004	G35	0.945	0.866	0.002	0.630	0.551	0.004	0.079	0.079	③	④
	I35	0.945	0.866	0.003	0.350	0.315	0.003	0.079	0.079	①	16-23
	M45	1.110	1.020	0.002	0.630	0.550	0.004	0.079	0.079	②	⑤
	Q45	1.100	1.020	0.002	0.630	0.550	0.004	0.079	0.079	②	⑤

① Front: 53-72
Rear: 28-38

② Front: 103-118
Rear: 28-38

③ Front: 113-114
Rear: 53-71

④ Front: 16-23
Rear: 53-71

⑤ Front: 16-23
Rear: 25-30

67162-INFI-C11

SCHEDULED MAINTENANCE INTERVALS
Infiniti—G20, G35, I30, I35, M45 & Q45

TO BE SERVICED	TYPE OF SERVICE	VEHICLE MILEAGE INTERVAL (x1000)												
		7.5	15	22.5	30	37.5	45	52.5	60	67.5	75	82.5	90	97.5
Engine oil & filter	R	✓	✓	✓	✓	✓	✓	✓	✓	✓	✓	✓	✓	✓
Automatic transaxle fluid	S/I		✓		✓		✓		✓		✓		✓	
Brake lines & cables	S/I		✓		✓		✓		✓		✓		✓	
Brake pads & discs	S/I		✓		✓		✓		✓		✓		✓	
Differential gear oil (Q45)	S/I		✓		✓		✓		✓		✓		✓	
Driveshaft boots (I30 & G20)	S/I		✓		✓		✓		✓		✓		✓	
Full-active suspension fluid (Q45) ①	S/I		✓		✓		✓		✓		✓		✓	
Manual transaxle oil (G20 & I30)	S/I		✓		✓		✓		✓		✓		✓	
Air cleaner filter	R				✓				✓				✓	
Exhaust system	S/I				✓				✓				✓	
Fuel lines	S/I				✓				✓				✓	
Steering gear linkage axle & suspension parts	S/I				✓				✓				✓	
SUPER HICAS linkage (Q45)	S/I				✓				✓				✓	
Vapor lines	S/I				✓				✓				✓	
Engine coolant	R								✓					
Spark plugs	R								✓					
Timing belt	R								✓					
Drive belts	S/I								✓					

R: Replace S/I: Service or Inspect

① Replace at 60,000 miles (if not previously replaced).

FREQUENT OPERATION MAINTENANCE (SEVERE SERVICE)

If a vehicle is operated under any of the following conditions it is considered severe service

- Extremely dusty areas.

- 50% or more of the vehicle operation is in 32°C (90°F) or higher temperatures, or constant operation in temperatures below 0°C (32°F).

- Prolonged idling (vehicle operation in stop and go traffic).

- Frequent short running periods (engine does not warm to normal operating temperatures).

- Police, taxi, delivery usage or trailer towing usage.

Oil & oil filter: change every 3750 miles.

Brake pads & discs: service or inspect every 7500 miles.

Driveshaft boots (G20 & I30): service or inspect every 7500 miles

Exhaust system: service or inspect every 7500 miles.

Steering gear, linkage, axle & suspension ball joints: service or inspect every 7500 miles.

Steering linkage, ball joints & front suspension ball joints: service or inspect every 7500 miles.

SUPER HICAS linkage (Q45): service or inspect every 7500 miles.

67162-INFI-C12

PRECAUTIONS

Before servicing any vehicle, please be sure to read all of the following precautions, which deal with personal safety, prevention of component damage, and important points to take into consideration when servicing a motor vehicle:

• Never open, service or drain the radiator or cooling system when the engine is hot; serious burns can occur from the steam and hot coolant.

• Observe all applicable safety precautions when working around fuel. Whenever servicing the fuel system, always work in a well-ventilated area. Do not allow fuel spray or vapors to come in contact with a spark, open flame, or excessive heat (a hot drop light, for example). Keep a dry chemical fire extinguisher near the work area. Always keep fuel in a container specifically designed for fuel storage; also, always properly seal fuel containers to avoid the possibility of fire or explosion. Refer to the additional fuel system precautions later in this section.

• Fuel injection systems often remain pressurized, even after the engine has been turned **OFF**. The fuel system pressure must be relieved before disconnecting any fuel lines. Failure to do so may result in fire and/or personal injury.

• Brake fluid often contains polyglycol ethers and polyglycols. Avoid contact with the eyes and wash your hands thoroughly after handling brake fluid. If you do get brake fluid in your eyes, flush your eyes

with clean, running water for 15 minutes. If eye irritation persists, or if you have taken brake fluid internally, IMMEDIATELY seek medical assistance.

• The EPA warns that prolonged contact with used engine oil may cause a number of skin disorders, including cancer! You should make every effort to minimize your exposure to used engine oil. Protective gloves should be worn when changing oil. Wash your hands and any other exposed skin areas as soon as possible after exposure to used engine oil. Soap and water, or waterless hand cleaner should be used.

• All new vehicles are now equipped with an air bag system. The system must be disabled before performing service on or around system components, steering column, instrument panel components, wiring and sensors. Failure to follow safety and disabling procedures could result in accidental air bag deployment, possible personal injury and unnecessary system repairs.

• Always wear safety goggles when working with, or around, the air bag system. When carrying a non-deployed air bag, be sure the bag and trim cover are pointed away from your body. When placing a non-deployed air bag on a work surface, always face the bag and trim cover upward, away from the surface. This will reduce the motion of the module if it is accidentally deployed. Refer to the additional air bag system precautions later in this section.

• Clean, high quality brake fluid from a sealed container is essential to the safe and proper operation of the brake system. You should always buy the correct type of brake fluid for your vehicle. If the brake fluid becomes contaminated, completely flush the system with new fluid. Never reuse any brake fluid. Any brake fluid that is removed from the system should be discarded. Also, do not allow any brake fluid to come in contact with a painted surface; it will damage the paint.

• Never operate the engine without the proper amount and type of engine oil; doing so WILL result in severe engine damage.

• Timing belt maintenance is extremely important! Many models utilize an interference-type, non-freewheeling engine. If the timing belt breaks, the valves in the cylinder head may strike the pistons, causing potentially serious (also time-consuming and expensive) engine damage. Refer to the maintenance interval charts in the front of this section for the recommended replacement interval for the timing belt.

• Disconnecting the negative battery cable on some vehicles may interfere with the functions of the on-board computer system(s) and may require the computer to undergo a relearning process once the negative battery cable is reconnected.

• When servicing drum brakes, only disassemble and assemble one side at a time, leaving the remaining side intact for reference.

ENGINE REPAIR

Distributor

REMOVAL

2.0L Engine

1. Before servicing the vehicle, refer to the precautions in the beginning of this section.
2. Remove or disconnect the following:
   • Negative battery cable
   • Splash shield, if equipped
   • Distributor connections but leave the ignition wires in place
   • Distributor cap hold-down screws and lift off the distributor cap with all ignition wires still connected
3. Matchmark the rotor to the distributor housing, and the distributor housing to the engine.

➡Do not crank the engine during this procedure. If the engine is cranked, the matchmark must be disregarded.

   • Hold-down bolt
   • Distributor from the engine

3.0L, 3.5L, 4.1L and 4.5L Engines

These engines are equipped with a distributorless ignition.

INSTALLATION

2.0L Engine

ENGINE NOT DISTURBED

1. If the engine was not disturbed, install or connect the following:
   • New distributor housing O-ring
   • Distributor in the engine so the rotor is aligned with the matchmark

on the housing and the housing is aligned with the matchmark on the engine. Be sure the distributor is fully seated and the distributor gear is fully engaged.
   • Snug the hold-down bolt
   • Distributor pick-up lead wires
   • Distributor cap and tighten the screws
   • Splash shield
   • Negative battery cable
2. Check and/or adjust the ignition timing and tighten the hold-down bolt to 10–12 ft. lbs. (14–16 Nm).

ENGINE DISTURBED

1. If the engine was disturbed (cranked or turned over with the distributor removed), install or connect the following:
   • New distributor housing O-ring
2. Position the engine so the No. 1 pis-

ton is at TDC of its compression stroke and the mark on the vibration damper is aligned with **0** on the timing indicator.

- Distributor in the engine so the rotor is aligned with the position of the No. 1 ignition wire on the distributor cap. Be sure the distributor is fully seated and that the distributor shaft is fully engaged.

➡ **There are distributor cap runners inside the cap on 2.0L engine. Be sure the rotor is pointing to where the No. 1 runner originates inside the cap.**

- Snug the hold-down bolt
- Distributor pick-up lead wires
- Distributor cap and tighten the screws
- Splash shield, if equipped
- Negative battery cable

3. Check and/or adjust the ignition timing and tighten the hold-down bolt to 10–12 ft. lbs. (14–16 Nm).

Alternator

REMOVAL

2.0L Engine

1. Before servicing the vehicle, refer to the precautions in the beginning of this section.
2. Remove or disconnect the following:
   - Negative battery cable
   - Drive belt
   - Alternator harness connector
   - Alternator bracket, if necessary
   - Alternator retainers
   - Alternator

3.0L Engine

1. Before servicing the vehicle, refer to the precautions in the beginning of this section.
2. Remove or disconnect the following:
   - Negative battery cable
   - Engine right-hand undercover
   - Right-hand side inspection cover
3. Loosen the belt idler pulley.
   - Drive belt
   - 4 A/C compressor mounting bolts
   - Cooling fan and fan shroud
4. Slide the A/C compressor forward.
   - Alternator harness connector
   - Upper and lower alternator bolts
   - Alternator

3.5L Engines

I35 MODELS

1. Before servicing the vehicle, refer to the precautions in the beginning of this section.

2. Remove or disconnect the following:
   - Negative battery cable
   - Right side engine undercover and side inspection cover
   - Radiator
   - Drive belt
   - Alternator and A/C compressor harness connectors
   - Upper and lower alternator bolts
   - Alternator

G35 MODELS

1. Before servicing the vehicle, refer to the precautions in the beginning of this section.
2. Remove or disconnect the following:
   - Negative battery cable
   - Engine undercover
   - Stabilizer bar clamps, then slide stabilizer downward
   - Loosen drive belts
   - Alternator electrical connector
   - Oil pressure switch harness connector
   - B terminal mounting nut
   - Upper and lower alternator mounting bolt
   - Both alternator bracket bolts
   - Alternator from the vehicle

4.1L Engine

1. Before servicing the vehicle, refer to the precautions in the beginning of this section.
2. Remove or disconnect the following:
   - Negative battery cable
   - Engine upper cover
   - Engine drive belt
   - Alternator electrical connector
   - Alternator

4.5L Engine

M45 MODELS

1. Before servicing the vehicle, refer to the precautions in the beginning of this section.

2. Remove or disconnect the following:
   - Negative battery cable
   - Air intake duct
   - Engine drive belt
   - Power steering hose clamp bolt
   - Alternator stay
   - Radiator reservoir tank
   - Ground harness bolt
   - Engine compartment harness clip
   - Alternator electrical connector
   - Alternator

Q45 MODELS

1. Before servicing the vehicle, refer to the precautions in the beginning of this section.

2. Remove or disconnect the following:
   - Battery
   - Air intake duct
   - Engine drive belt
   - Power steering hose clamp bolt
   - Alternator stay
   - Power steering reservoir tank
   - Ground harness bolt
   - Alternator electrical connector
   - Alternator

INSTALLATION

2.0L Engine

1. Install or connect the following:
   - Alternator
   - Alternator bracket, if removed
   - Alternator retainers as shown in the accompanying illustration

3.0L Engine

1. Install or connect the following:
   - Alternator
   - Upper and lower alternator bolts. Torque the upper bolt to 11–15 ft. lbs. (15–20 Nm) and the lower bolt to 32–38 ft. lbs. (44–52 Nm)
   - Alternator harness connector

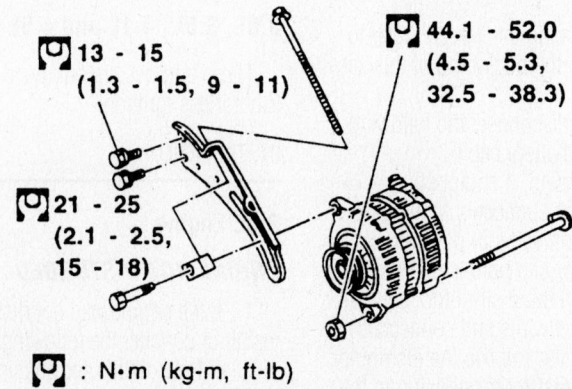

13 - 15
(1.3 - 1.5, 9 - 11)

44.1 - 52.0
(4.5 - 5.3, 32.5 - 38.3)

21 - 25
(2.1 - 2.5, 15 - 18)

: N·m (kg-m, ft-lb)

9307HG20

Alternator and bracket retainer locations and torque specifications—G20 models

2. Slide the A/C compressor rearward.
   - Fan shroud and cooling fan
   - 4 A/C compressor mounting bolts
   - Drive belt
3. Tighten the belt idler pulley.
   - Right-hand side inspection cover
   - Engine right-hand undercover
   - Negative battery cable

3.5L Engine

I35 MODELS

1. Install or connect the following:
   - Alternator
   - Upper and lower alternator bolts. Tighten the upper bolt to 12–15 ft. lbs. (16 –20 Nm) and the lower bolt to 32–38 ft. lbs. (44–52 Nm).
   - Alternator and A/C compressor harness connectors
   - Drive belt
   - Radiator
   - Right side engine undercover and side inspection cover
   - Negative battery cable

G35 MODELS

1. Install or connect the following:
   - Alternator into the vehicle
   - Alternator bracket bolts and torque to 18–23 ft. lbs. (24–31 Nm)
   - Upper and lower alternator mounting bolts and tighten to 45–51 ft. lbs. (60–70 Nm)
   - B terminal mounting nut. Torque the nut to 83–95 inch lbs. (9–11 Nm).
   - Oil pressure switch harness connector
   - Alternator electrical connector
   - Drive belts
   - Stabilizer into position and secure with the clamps
   - Engine undercover
   - Negative battery cable

4.1L Engine

1. Install or connect the following:
   - Alternator
   - Alternator bolts. Torque the alternator bolts marked **1**, (shown in the accompanying illustration) to 30–38 ft. lbs. (41–52 Nm) and the bolt marked **2**, (shown in the same illustration) to 15–20 ft. lbs. (21–26 Nm).
   - Alternator electrical connector
   - Engine drive belt
   - Engine upper cover
   - Negative battery cable

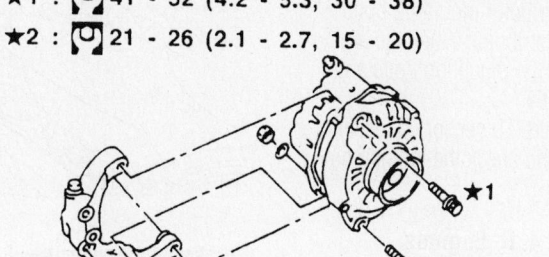

★1 : 41 - 52 (4.2 - 5.3, 30 - 38)
★2 : 21 - 26 (2.1 - 2.7, 15 - 20)

: N•m (kg-m, ft-lb)

9307HG01

Alternator bolt locations—4.1L Q45 models

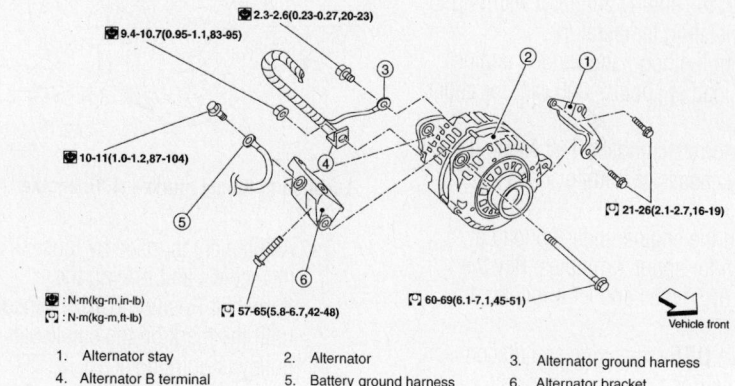

| : N•m(kg-m,in-lb)
| : N•m(kg-m,ft-lb)

1. Alternator stay
2. Alternator
3. Alternator ground harness
4. Alternator B terminal
5. Battery ground harness
6. Alternator bracket

67162-INFI-G02

Alternator mounting—4.5L engine

4.5L Engine

M45 AND Q45 MODELS

1. Installation is the reverse of the removal procedure.

Ignition Timing

ADJUSTMENT

2.0L Engine

➡ **The engine should be in good mechanical condition and all electrical connectors and vacuum hoses connected before making this adjustment.**

1. Before servicing the vehicle, refer to the precautions in the beginning of this section.
2. Start the engine and let it warm up to normal operating temperature.
3. Open the hood and run the engine under no load at about 2,000 rpm for about 2 minutes.
4. Perform Diagnostic Test Mode II and repair any causes of trouble codes as needed.

5. Run the engine under no load at 2,000 rpm for about 2 minutes. Rev the engine 2 or 3 times and let it idle for 1 minute.
6. Turn **OFF** the engine and disconnect the Throttle Position (TP) sensor connector. Connect a timing light to the No. 1 spark plug wire. Start the engine.

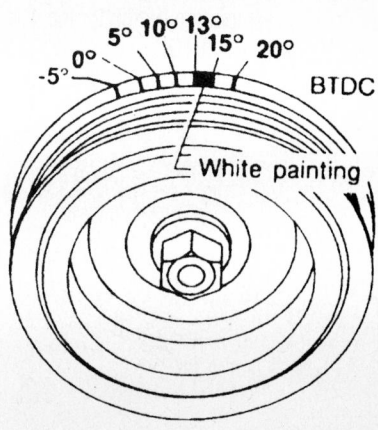

7923HG01

Crankshaft pulley and timing marks—2.0L engine

7. Adjust the timing to 13–17° BTDC by loosening the distributor mounting bolts and turning the distributor. When the timing is correct, tighten the mounting bolts and turn the engine **OFF**.

8. Reconnect the TP sensor connector. Start the engine and check the ignition timing again.

3.0L, 3.5L and 4.1L Engines

➡**The engine should be in good mechanical condition and all electrical connectors and vacuum hoses attached before making this adjustment.**

1. Before servicing the vehicle, refer to the precautions in the beginning of this section.

2. Start the engine and let it warm up to normal operating temperature.

3. Open the hood and run the engine under no load at about 2,000 rpm for about 2 minutes.

4. Perform Diagnostic Test Mode II and repair any causes of trouble codes as needed.

5. Run the engine under no load at 2,000 rpm for about 2 minutes. Rev the engine 2 or 3 times and let it idle for 1 minute.

6. Turn **OFF** the engine and disconnect the Throttle Position sensor connector. Remove the No. 1 ignition coil. Connect the coil to the spark plug using a spare piece of high-tension wire so you have a place to connect your timing light. Start the engine.

7. Run the engine under no load at 2,000 rpm for about 2 minutes. Rev the engine 2 or 3 times and let it idle.

8. Check the ignition timing and adjust if needed.

9. The correct ignition timing is as follows:

  a. Correct ignition timing for 3.0L and 3.5L engines is 8–12° BTDC.

  b. Correct ignition timing for the 4.1L engine is 13–17° BTDC.

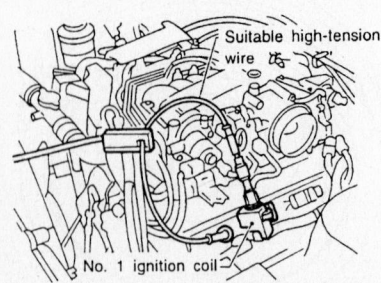

Connect the No. 1 ignition coil to the spark plug with the spare piece of high-tension wire—4.1L engine shown

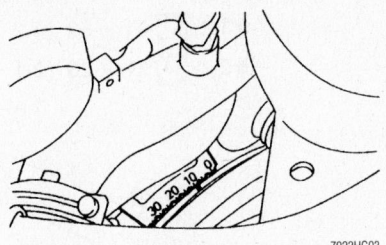

Location of timing marks—3.0L and 3.5L engines

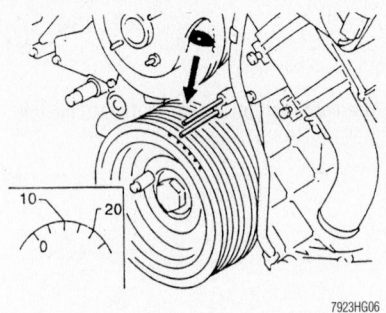

Location of timing marks—4.1L engine

- Adjustment is made by loosening the screws and turning the Camshaft Position (CMP) sensor until the mark on the crankshaft pulley is pointing at 10° BTDC. Tighten the mounting screws and confirm ignition timing has not changed.
- Turn the engine **OFF** and connect the TP sensor connector.

4.5L Engines

1. The ignition timing is controlled by the ECM and is not adjustable. With all electrical accessories off, wheels in the straight ahead position, the ignition timing is 7–17 degrees BTDC.

Engine Assembly

REMOVAL & INSTALLATION

2.0L Engine

1. Before servicing the vehicle, refer to the precautions in the beginning of this section.

2. Drain the coolant system.

3. Drain the engine oil.

4. Drain the transaxle fluid.

5. Release fuel system pressure and remove fuel line.

6. Remove or disconnect the following:
- Hood and hinges
- Negative battery cable
- Both front wheels
- Engine undercover
- Air cleaner assembly and duct
- Battery and battery tray
- All vacuum lines and wiring harness connectors
- Heater hoses
- Oil cooler lines, if equipped
- Power steering hoses
- Fuel lines
- Throttle cable
- Cruise control cable, if equipped
- A/T control cable, if equipped
- Cooling fans, radiator and recovery tank
- Drive shafts
- Front exhaust pipe
- Starter and intake manifold support
- Drive belts
- Alternator, A/C compressor, and the power steering pump from the engine and lay them aside. Do not disconnect the compressor or power steering pump lines.

7. Support the engine with a hoist and the transaxle with a suitable jack. Raise the engine and transaxle slightly and remove the center member.
- Engine mounting bolts from both sides and slowly lower the hoist and transaxle jack
- Engine and transaxle from beneath the vehicle

To install:

8. Install or connect the following:
- Center member bracket (manual transmission) on the engine, if removed. Ensure that all insulators are correctly positioned on the brackets. Torque the insulator through-bolts to 32–41 ft. lbs. (43–55 Nm).

9. If equipped with manual transaxle, ensure that the distance between the center of the insulator through-bolt and the center member is 2.28–2.36 in. (58–60mm). Torque the through-bolt to 46–58 ft. lbs. (62–78 Nm).
- Engine. Torque the center member-to-frame bolts to 57–72 ft. lbs. (77–98 Nm).
- Alternator, air conditioning compressor, and power steering pump
- Drive belts
- Starter and intake manifold support
- Front exhaust pipe
- Drive shafts
- Cooling fans, radiator and recovery tank
- A/T control cable, if equipped
- Cruise control cable, if equipped
- Throttle cable

- Fuel lines
- Power steering hoses
- Oil cooler lines, if equipped
- Heater hoses
- All vacuum lines and wiring harness connectors
- Engine undercover
- Front wheels
- Battery and battery tray
- Air cleaner assembly and duct
- Negative battery cable
- Hood and hinges

10. Fill and bleed the cooling system.
11. Fill the engine with clean oil.
12. Fill the transaxle to the proper level.
13. Start the vehicle, check for leaks and repair if necessary.

3.0L Engine

It is recommended the engine and transaxle be removed as a single unit. If necessary, the units may be separated after removal.

➡**The engine and transaxle assembly must be removed from the under side of the vehicle.**

1. Drain the cooling system.
2. Drain the engine oil.
3. Drain the transaxle fluid.
4. Properly relieve the fuel system pressure.
5. Before servicing the vehicle, refer to the precautions in the beginning of this section.
6. Remove or disconnect the following:
   - Negative battery cable
   - Hood
   - Both front wheels
   - Engine undercover
   - All necessary vacuum hoses, fuel hoses and electrical connections that would interfere with engine removal
   - Front exhaust tubes
   - Ball joints
   - Driveshafts
   - Radiator and fans
   - Drive belts
   - Alternator, compressor and power steering oil pump from the engine compartment
7. Set a suitable transmission jack under the transaxle. Hoist the engine with a suitable engine slinger.
   - Rear engine mounting
   - Control cable
   - Front engine mounting
   - Center member and slowly lower the transmission jack
8. Remove the engine and the transaxle assembly from the engine as shown in the accompanying illustration.

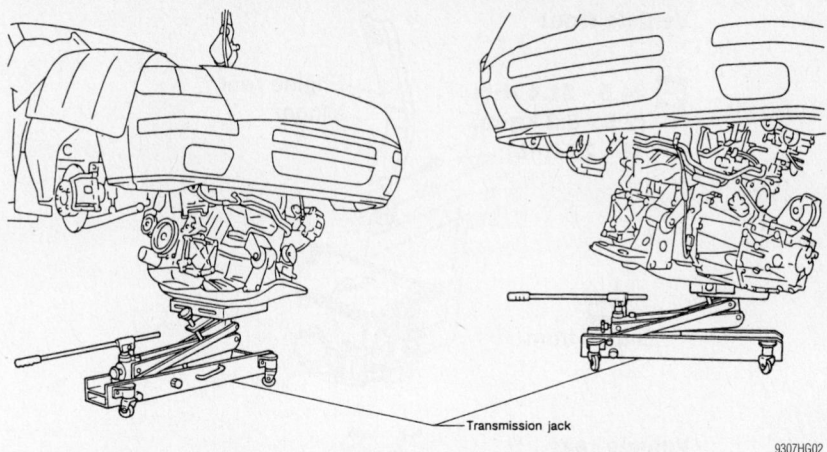

Transmission jack

9307HG02

Remove the engine and the transaxle as an assembly from beneath the vehicle using a transaxle jack and a slinger

To install:
9. Install or connect the following:
   - Center member bracket (manual transmission) on the engine, if removed. Ensure that all insulators are correctly positioned on the brackets. Torque the insulator through-bolts to 72 ft. lbs. (98 Nm).
   - Engine. Torque the center member-to-frame bolts to 57–72 ft. lbs. (77–98 Nm).
   - Front engine mount. Torque the bolts to 72 ft. lbs. (98 Nm).
   - Control cable
   - Rear engine mount. Torque the bolts to 72 ft. lbs. (98 Nm) and remove the transaxle jack
   - Alternator, air conditioning compressor, and power steering pump
   - Radiator and fans
   - Drive belts
   - Driveshafts
   - Ball joints
   - Front exhaust tubes
   - Vacuum hoses, electrical connectors, fuel hoses and wiring which was removed
   - Engine undercover
   - Front wheels
   - Battery and battery tray
   - Negative battery cable
   - Hood and hinges

10. Fill and bleed the cooling system.
11. Fill the engine with clean oil.
12. Fill the transaxle to the proper level.
13. Start the vehicle, check for leaks and repair if necessary.

3.5L Engines

I35 MODELS

It is recommended the engine and transaxle be removed as a single unit. If

need be, the units may be separated after removal.

➡**The engine and transaxle assembly must be removed from the underside of the vehicle.**

1. Before servicing the vehicle, refer to the precautions in the beginning of this section.
2. Release the fuel system pressure.
3. Drain the cooling system.
4. Drain the engine oil.
5. Drain the automatic transaxle, if equipped.
6. Remove or disconnect the following:
   - Negative battery cable
   - Hood
   - Engine undercover
   - All vacuum hoses, fuel lines, wires and connectors; tag before disconnecting
   - Front exhaust pipe from the manifold
   - Ball joints from the steering knuckle
   - Halfshafts
   - Radiator and fans
   - Drive belts
   - Alternator
   - A/C compressor. Position it aside with the lines attached. Do NOT disconnect the refrigerant lines.
   - Power steering pump and position aside with the lines attached. Do NOT disconnect the fluid lines.
7. Place a suitable jack under the transaxle. Install engine slingers and a suitable engine hoist. Raise the engine for access to the left side engine mount.
   - Left side engine mount
   - Control and support rods from the transaxle, manual transaxle only
   - Control cable from the transaxle, automatic transaxle only

Vehicle front

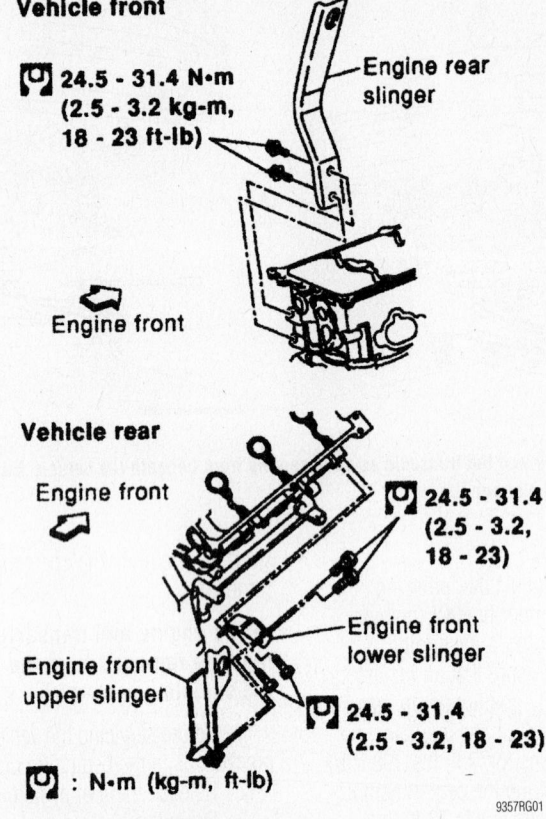

24.5 - 31.4 N•m
(2.5 - 3.2 kg-m,
18 - 23 ft-lb)

Engine rear
slinger

Engine front

Vehicle rear

Engine front

24.5 - 31.4
(2.5 - 3.2,
18 - 23)

Engine front
lower slinger

Engine front
upper slinger

24.5 - 31.4
(2.5 - 3.2, 18 - 23)

: N•m (kg-m, ft-lb)

9357RG01

Installation of engine slingers to lift the engine

- Right side engine mount
- Center member, then carefully and slowly lower the transmission jack

8. Lower the engine/transaxle assembly onto an engine stand.

➥**When lowering the engine out, guide it carefully to avoid hitting any other components.**

To install:

9. Installation is the reverse of the removal procedure, noting the following points:

a. Install the electronically controlled engine mount harness to the specifications shown in the accompanying figure.

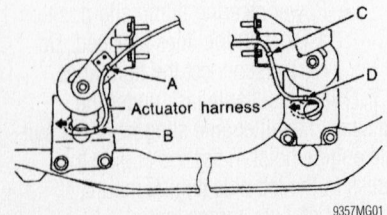

9357MG01

For the electronically controlled engine mount harness, the proper length from A to B is 6.69 in. (170mm) and from C to D is 5.12 in. (130mm)

b. Make sure to connect all vacuum hoses, lines, and electrical connectors as tagged during removal.

c. Fill the cooling system.

d. Fill the engine with clean oil.

e. Start the vehicle, check for leaks and repair if necessary.

G35 MODELS

1. Before servicing the vehicle, refer to the precautions in the beginning of this section.

2. Evacuate the A/C system.

3. Release the fuel system pressure.

4. Drain the engine oil.

5. Drain the cooling system.

6. Drain the transaxle fluid.

7. Remove or disconnect the following:

- Negative battery cable
- Hood
- Engine cover
- Battery cover
- Engine undercover
- Front wheels
- Wiper arm and cowl top cover
- Air cleaner assembly, including air ducts
- Fan, radiator shroud, radiator, reservoir tank and hoses
- Heater hose from the engine and plug to avoid leaks

- Ground wire from left hand cylinder head
- Positive battery cable from the vehicle and temporarily fasten it on the engine
- Battery
- Engine harness connector; tag before disconnecting
- A/C lines from the A/C compressor
- Body ground cables
- Brake booster vacuum hose
- Fuel feed and Evaporative Emission (EVAP) hoses. Plug the fuel lines to prevent fuel from leaking out.
- Power steering pump reservoir tank and pipes

8. Disconnect the connectors from the passenger compartment as follows:

a. Remove the passenger side kick plate, dashboard side trim and glove box.

b. From the engine compartment, detach the connectors from the Transmission Control Module (TCM), and Engine Control Module (ECM).

c. Unfasten the wire harnesses, pull the harnesses out into the engine compartment, then temporarily secure them to the engine. Cover all connectors with plastic or similar material to protect them.

9. Remove or disconnect the following:

- Front exhaust pipe from the manifold
- Steering lower joint and release the steering shaft
- Propeller shaft from the transmission
- Shift control linkage from the gear selector. Secure it temporarily to the transmission, so it doesn't drag or catch on any other components.
- Rear plate cover from upper oil pan
- Bolts securing the drive plate to the torque converter
- Bolts securing the transmission to the lower rear side of the oil pan
- Front stabilizer shaft
- Left and right tie rod ends from the steering knuckle
- Disconnect the lower strut from the lower control arms on the left and right sides
- Left and right control arms from the suspension crossmember

10. Position a suitable engine table or other suitable tool under the engine. Securely support the bottom of the suspension member and transmission.

- Rear member mounting bolt
- Suspension member mounting bolt and nut

11. Carefully lower the jack, or raise the

lift to remove the engine, transmission and suspension member assembly. Make sure that all lines, hoses, connectors etc. have been disconnected. If necessary, support the rear of the vehicle at the rear jacking point, as it's center of gravity has changed with the engine removed.

12. Remove and components and separate the engine and transmission as necessary.

To install:

13. Installation is the reverse of the removal procedure, noting the following points:

- Tighten all engine mounts and related components to the specification shown in the accompanying figure
- Tighten the rear member mounting bolts in sequence.
- Make sure to connect all vacuum hoses, lines, and electrical connectors as tagged during removal.
- Fill the cooling system.
- Fill the engine with clean oil.
- Fill the transmission to the proper level.
- Recharge the A/C system.

14. Start the vehicle, check for leaks and repair if necessary.

Vehicle front
Right
Left
④
②
③
Rear member
①
9357MG03

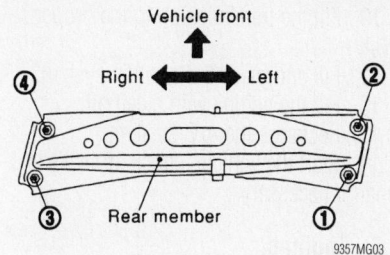

Rear member bolt torque sequence—G35 models

4.1L Engines

1. Before servicing the vehicle, refer to the precautions in the beginning of this section.
2. Evacuate the A/C system.
3. Release the fuel system pressure.
4. Drain the engine oil.
5. Drain the cooling system.
6. Drain the transmission fluid.
7. Remove or disconnect the following:
   - Negative battery cable
   - Hood
   - Engine undercover
   - Transmission
   - All necessary vacuum hoses, fuel hoses and electrical connections that would interfere with engine removal
   - Front exhaust tubes
   - Radiator and shroud
   - Drive belts

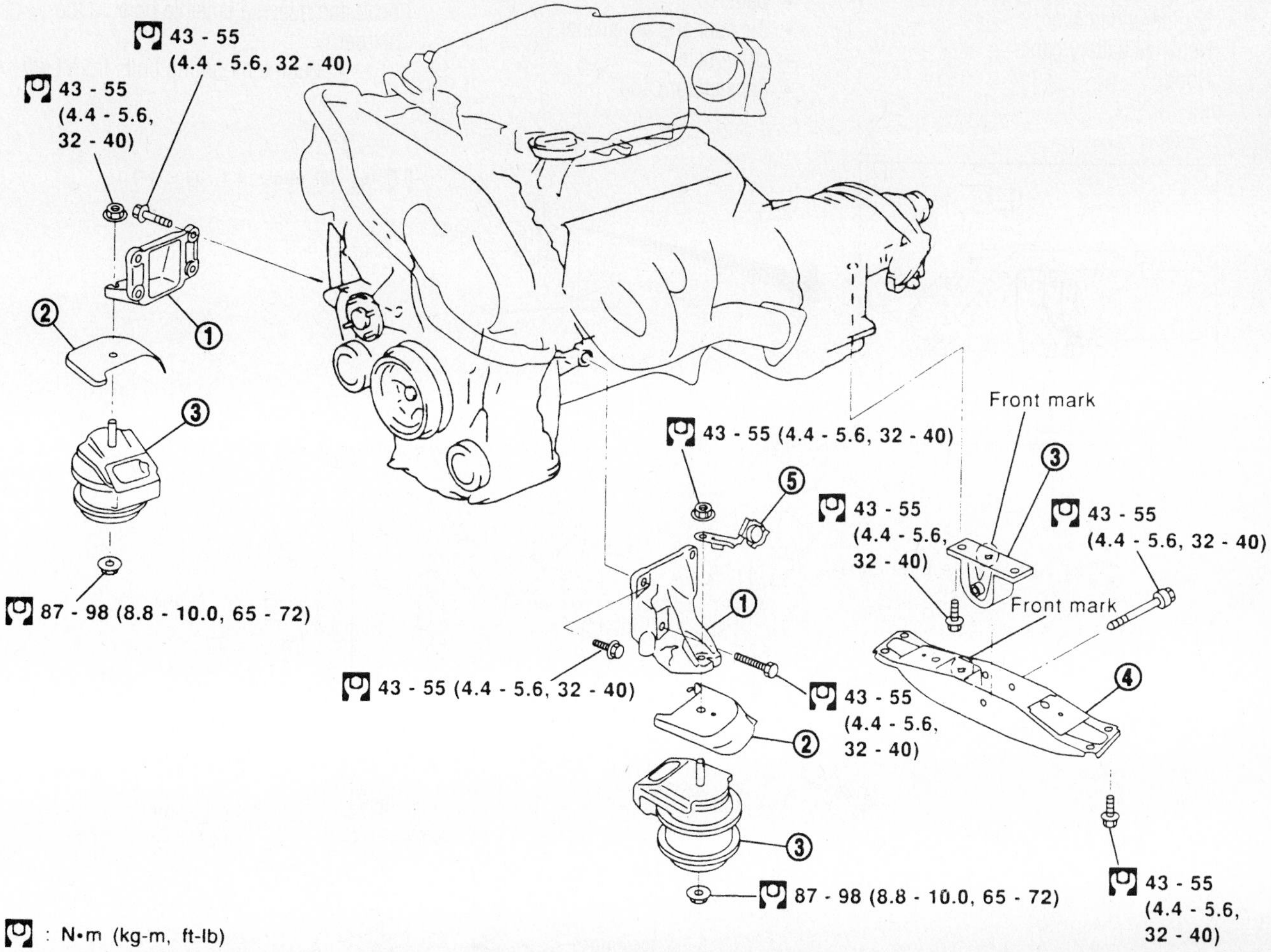

43 - 55 (4.4 - 5.6, 32 - 40)
43 - 55 (4.4 - 5.6, 32 - 40)
②
①
③
87 - 98 (8.8 - 10.0, 65 - 72)
43 - 55 (4.4 - 5.6, 32 - 40)
Front mark
43 - 55 (4.4 - 5.6, 32 - 40)
⑤
③
43 - 55 (4.4 - 5.6, 32 - 40)
43 - 55 (4.4 - 5.6, 32 - 40)
Front mark
①
④
43 - 55 (4.4 - 5.6, 32 - 40)
②
③
87 - 98 (8.8 - 10.0, 65 - 72)
43 - 55 (4.4 - 5.6, 32 - 40)
: N•m (kg-m, ft-lb)

1. Engine mounting bracket
2. Heat insulator
3. Insulator
4. Rear member
5. Harness bracket

9357MG02

Exploded view of the engine mounts and related components—G35 models

- Power steering oil pump from the engine compartment

8. Attach engine slingers to the cylinder head and attach a suitable hoist to the slinger
  - Engine mounting bolts from both sides and then slowly raise the engine
  - Engine from the engine compartment

To install:

9. Install or connect the following:
  - Engine. Torque the front mounting bolts to 41 ft. lbs. (55 Nm) and the rear mounting bolts to 21 ft. lbs. (28 Nm). Remove the engine supports
  - Power steering pump
  - Radiator and shroud
  - Drive belts
  - Front exhaust tubes
  - Vacuum hoses, electrical connectors, fuel hoses and wiring
  - Transmission
  - Engine undercover
  - Negative battery cable
  - Hood

10. Fill the transmission to the proper level.
11. Fill and bleed the cooling system.
12. Fill the engine with clean oil.
13. Recharge the A/C system.
14. Start the vehicle, check for leaks and repair if necessary.

4.5L Engines

1. Before servicing the vehicle, refer to the precautions in the beginning of this section.
2. Evacuate the A/C system.
3. Release the fuel system pressure.
4. Drain the engine oil.
5. Drain the cooling system.
6. Drain the transmission fluid.
7. Remove or disconnect the following:
  - Battery cables
  - Hood
  - Engine appearance cover
  - Engine undercover
  - Tower bar
  - Battery
  - Air duct and air cleaner
  - Drive belts
  - Accelerator cable

- Cooling fan and tubes
- Radiator hoses
- Radiator and shroud
- Engine harnesses from both sides
- Heater hoses
- Exhaust manifold cover
- Vacuum hoses to body
- Cooling reservoir tank
- A/C compressor
- Power steering reservoir and wire aside
- Power steering oil pump from the engine compartment
- Automatic transmission cooler line
- Front exhaust tubes
- Steering shaft lower joint
- Front cross bar
- Driveshaft
- Transmission
- All necessary vacuum hoses, fuel hoses and electrical connections that would interfere with engine removal

8. Attach engine slingers to the cylinder head and attach a suitable hoist to the slinger
  - Engine mounting bolts from both

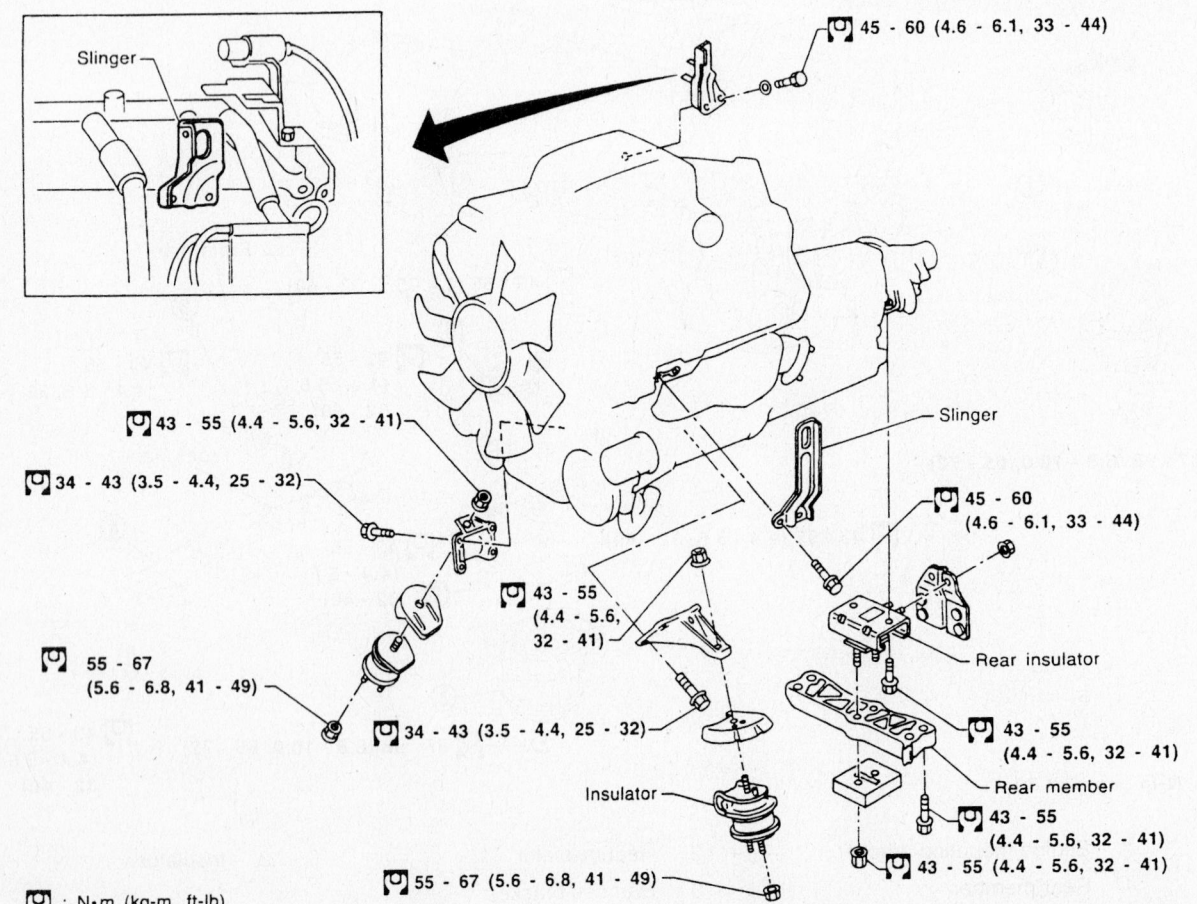

45 - 60 (4.6 - 6.1, 33 - 44)

43 - 55 (4.4 - 5.6, 32 - 41)

34 - 43 (3.5 - 4.4, 25 - 32)

55 - 67 (5.6 - 6.8, 41 - 49)

34 - 43 (3.5 - 4.4, 25 - 32)

43 - 55 (4.4 - 5.6, 32 - 41)

Insulator

55 - 67 (5.6 - 6.8, 41 - 49)

Slinger

45 - 60 (4.6 - 6.1, 33 - 44)

Rear insulator

43 - 55 (4.4 - 5.6, 32 - 41)

Rear member

43 - 55 (4.4 - 5.6, 32 - 41)

43 - 55 (4.4 - 5.6, 32 - 41)

: N•m (kg-m, ft-lb)

Engine mounting and torque specifications—4.1L engines

9307HG04

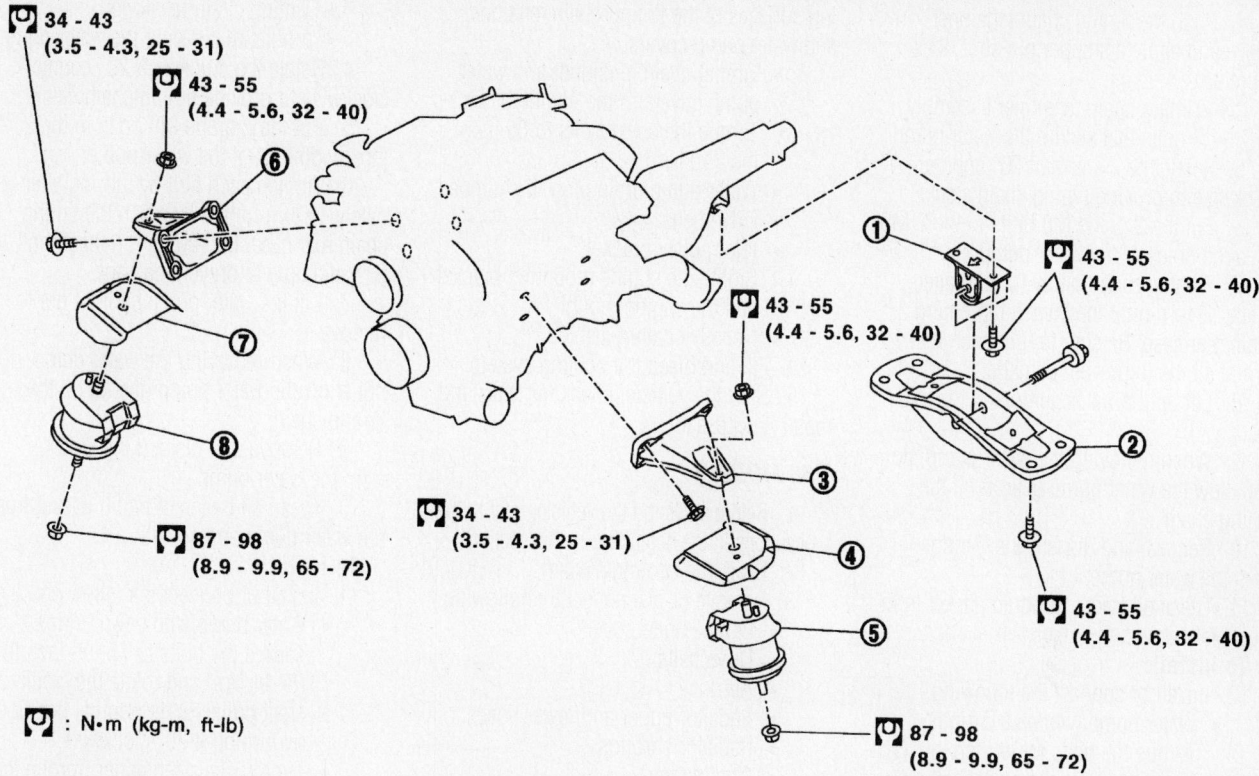

34 - 43
(3.5 - 4.3, 25 - 31)

43 - 55
(4.4 - 5.6, 32 - 40)

43 - 55
(4.4 - 5.6, 32 - 40)

43 - 55
(4.4 - 5.6, 32 - 40)

43 - 55
(4.4 - 5.6, 32 - 40)

87 - 98
(8.9 - 9.9, 65 - 72)

34 - 43
(3.5 - 4.3, 25 - 31)

87 - 98
(8.9 - 9.9, 65 - 72)

: N•m (kg-m, ft-lb)

1. Rear engine mounting insulator
4. Heat insulator
7. Heat insulator

2. Rear member
5. Left engine mounting insulator
8. Right engine mounting insulator

3. Left engine mounting bracket
6. Right engine mounting bracket

67162-INFI-G03

Engine mounting and torque specifications—4.5L engines M45 and Q45

sides and then slowly raise the engine
• Engine from the engine compartment

To install:

9. Installation is the reverse of the removal procedure. See the illustration for front cross bar and engine mounting torque specifications.

10. Fill the transmission to the proper level.

11. Fill and bleed the cooling system.

12. Fill the engine with clean oil.

13. Recharge the A/C system.

14. Start the vehicle, check for leaks and repair if necessary.

Water Pump

REMOVAL & INSTALLATION

2.0L Engine

1. Before servicing the vehicle, refer to the precautions in the beginning of this section.

2. Drain the coolant from the radiator

and engine block. The drain plug in the engine block is located at the left front of the cylinder block.

3. Remove or disconnect the following:
• Negative battery cable
• Right front wheel
• Engine side cover
• Drive belts
• Right front engine mount
• Water pump

To install:

4. Clean all mating surfaces and place a 2–3mm bead of liquid gasket on the water pump mating surface.

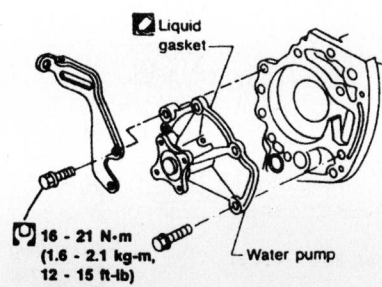

16 - 21 N•m
(1.6 - 2.1 kg-m,
12 - 15 ft-lb)

Liquid gasket

Water pump

7923HG08

Exploded view of the water pump mounting—2.0L engine

5. Install or connect the following:
• Water pump. Torque the bolts to 15 ft. lbs. (21 Nm).
• Drive belts
• Front engine mount
• Engine side cover
• Right front wheel
• Negative battery cable

6. Fill and bleed the cooling system.

7. Start the vehicle, check for leaks and repair if necessary.

3.0L and 3.5L Engines

EXCEPT G35 MODELS

1. Before servicing the vehicle, refer to the precautions in the beginning of this section.

2. Drain the cooling system.

3. Remove or disconnect the following:
• Negative battery cable
• Right side engine mount and bracket
• Drive belts
• Idler pulley bracket
• Water pump drain plug, if equipped
• Timing chain tensioner cover
• Water pump cover

4. Push the timing chain tensioner sleeve and apply a stopper pin so it does not return.

- Timing chain tensioner assembly
- 3 bolts that secure the water pump

5. Rotate the crankshaft 20° counter-clockwise to provide timing chain slack.

6. Put the 2 grade M8 bolts in the 2 M8 threaded holes of the water pump.

7. Tighten each bolt by turning alternately ½ turn until they reach the timing chain rear case. Be sure to turn each bolt ½ turn at a time to prevent damage.

8. Lift up the water pump and remove it.

9. When removing the water pump, do not allow the water pump gear to hit the timing chain.

10. Remove and discard the O-rings from the water pump.

11. Clean all traces of liquid gasket from the water pump and covers.

To install:

12. Install or connect the following:
- Water pump with new O-rings. Torque the bolts to 89 inch lbs. (10 Nm) and rotate the crankshaft pulley to its original position by turning it 20° clockwise.
- Timing chain tensioner. Torque the bolts to 89 inch lbs. (10 Nm). Remove the stopper pin from the timing chain tensioner.

13. Apply a continuous 0.091–0.130 in. (2–3mm) bead of liquid sealant to the mat-ing surfaces of the timing chain tensioner and water pump covers.

- Timing chain tensioner and water pump covers to the engine block. Torque the cover bolts to 89 inch lbs. (10 Nm).
- Water pump drain plug, if equipped
- Drive belts
- Idler pulley bracket
- Right side engine mounting bracket and the engine mount
- Negative battery cable

14. Fill and bleed the cooling system.

15. Start the vehicle, check for leaks and repair if necessary.

G35 MODELS

1. Before servicing the vehicle, refer to the precautions in the beginning of this section.

2. Drain the cooling system.

3. Remove or disconnect the following:
- Engine undercover
- Drive belts
- Air duct
- Radiator upper and lower hoses
- Radiator shrouds
- Cooling fan
- Water drain plug from the water pump side of the cylinder block
- Timing chain tensioner cover and water pump cover

✷✷ WARNING

Be careful not the drop the mounting bolts inside the chain case.

- Timing chain tensioner
- 3 bolts that secure the water pump

4. Rotate the crankshaft 20° counter-clockwise to provide timing chain slack.

5. Put the 2 grade M8 bolts in the 2 M8 threaded holes of the water pump.

6. Tighten each bolt by turning alternately ½ turn until they reach the timing chain rear case. Be sure to turn each bolt ½ turn at a time to prevent damage.

7. Lift the water pump straight out to remove it.

8. When removing the water pump, do not allow the water pump gear to hit the timing chain.

9. Remove and discard the O-rings from the water pump.

10. Clean all traces of liquid gasket from the water pump and covers.

To install:

11. Install or connect the following:
- Water pump with new O-rings. Torque the bolts to 75–95 inch lbs. (8–11 Nm) and rotate the crank-shaft pulley to its original position by turning it 20° clockwise.
- Timing chain tensioner. Torque the bolts to 89 inch lbs. (10 Nm). Remove the stopper pin from the timing chain tensioner.

12. Apply a continuous 0.091–0.130 in. (2–3mm) bead of liquid sealant to the mat-ing surfaces of the timing chain tensioner and water pump covers.

- Timing chain tensioner and water

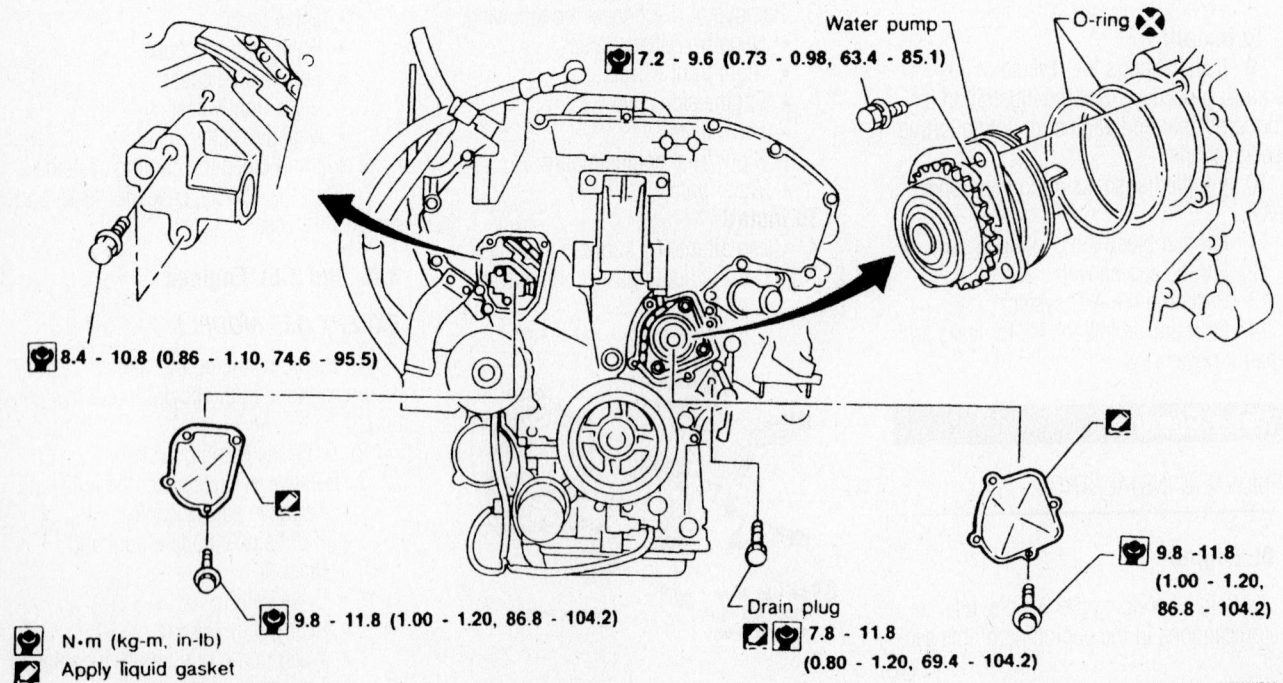

N·m (kg-m, in-lb)
Apply liquid gasket

Water pump: 7.2 - 9.6 (0.73 - 0.98, 63.4 - 85.1)
O-ring
8.4 - 10.8 (0.86 - 1.10, 74.6 - 95.5)
9.8 - 11.8 (1.00 - 1.20, 86.8 - 104.2)
Drain plug
7.8 - 11.8 (0.80 - 1.20, 69.4 - 104.2)
9.8 - 11.8 (1.00 - 1.20, 86.8 - 104.2)

Water pump and timing cover assembly—3.0L and 3.5L engines, except G35

7923HG09

SEC. 130•135•210

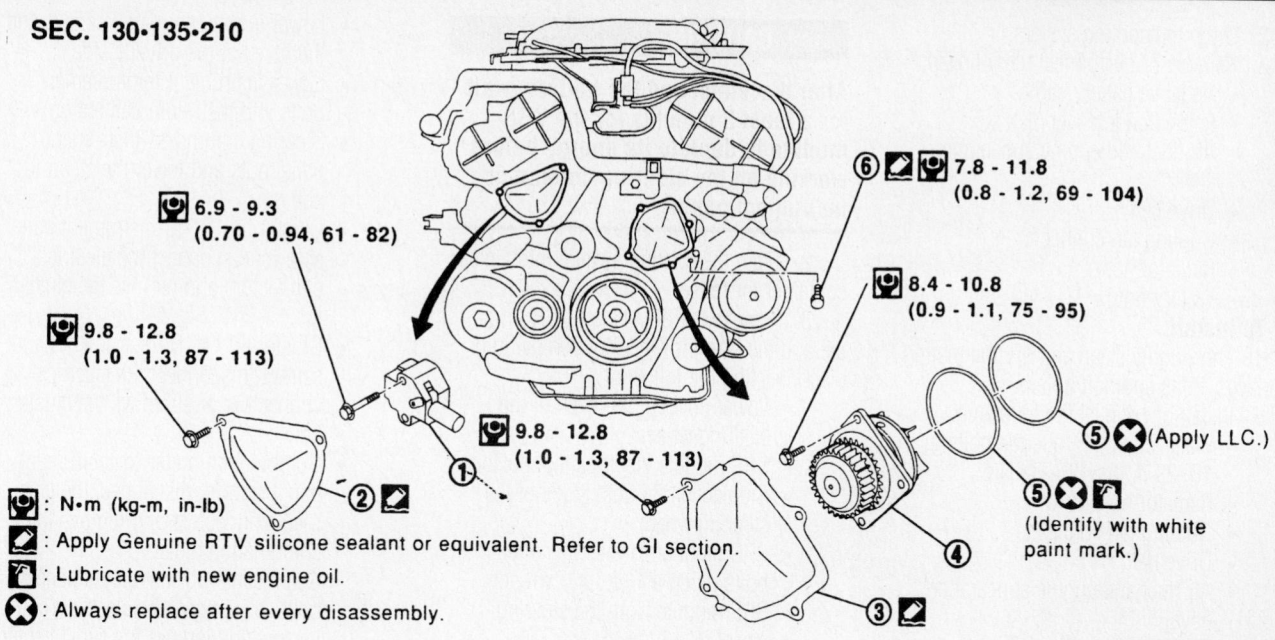

6.9 - 9.3
(0.70 - 0.94, 61 - 82)

9.8 - 12.8
(1.0 - 1.3, 87 - 113)

7.8 - 11.8
(0.8 - 1.2, 69 - 104)

8.4 - 10.8
(0.9 - 1.1, 75 - 95)

9.8 - 12.8
(1.0 - 1.3, 87 - 113)

(Apply LLC.)

(Identify with white
paint mark.)

: N•m (kg-m, in-lb)

: Apply Genuine RTV silicone sealant or equivalent. Refer to GI section.

: Lubricate with new engine oil.

: Always replace after every disassembly.

1. Chain tensioner
4. Water pump
2. Chain tensioner cover
5. O - rings
3. Water pump cover
6. Drain plug

9357MG04

Exploded view of the water pump mounting—G35 models

pump covers to the engine block. Torque the cover bolts to 89 inch lbs. (10 Nm).
- Water drain plug to the water pump side of the cylinder block
- Cooling fan
- Radiator shrouds
- Radiator upper and lower hoses
- Air duct
- Drive belts
- Engine undercover
- Negative battery cable

13. Fill and bleed the cooling system.
14. Start the vehicle, check for leaks and repair if necessary.

4.1L Engine

1. Before servicing the vehicle, refer to the precautions in the beginning of this section.
2. Drain the cooling system.
3. Remove or disconnect the following:
- Negative battery cable
- Loosen the drive belts
- Fan coupling and fan assembly
- Idler pulley bracket and drive belt
- Water pump

To install:

4. Thoroughly clean and dry the mating surfaces, bolts and bolt holes.
5. Apply liquid gasket to the water pump.
6. Install or connect the following:
- Water pump. Torque the bolts to 13 ft. lbs. (18 Nm).

7923HG11

Apply a continuous bead of RTV sealant to the mounting surface of the water pump assembly

- Idler pulley bracket. Torque the bolts to 89 inch lbs. (10 Nm).
- Drive belts
- Fan and coupling assembly
- Negative battery cable

7. Fill and bleed the cooling system.
8. Start the vehicle, check for leaks and repair if necessary.

4.5L Engines

1. Before servicing the vehicle, refer to the precautions in the beginning of this section.

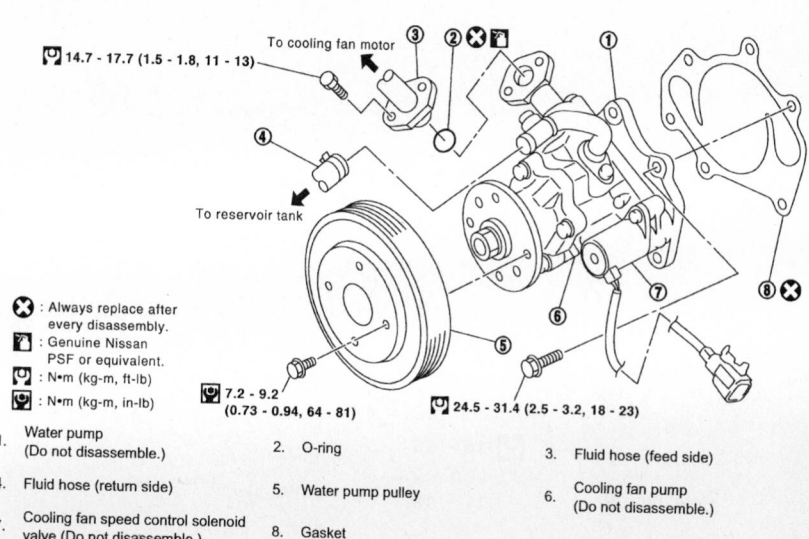

14.7 - 17.7 (1.5 - 1.8, 11 - 13)

To cooling fan motor

To reservoir tank

: Always replace after every disassembly.

: Genuine Nissan PSF or equivalent.

: N•m (kg-m, ft-lb)

: N•m (kg-m, in-lb)

7.2 - 9.2
(0.73 - 0.94, 64 - 81)

24.5 - 31.4 (2.5 - 3.2, 18 - 23)

1. Water pump
(Do not disassemble.)
2. O-ring
3. Fluid hose (feed side)

4. Fluid hose (return side)
5. Water pump pulley
6. Cooling fan pump
(Do not disassemble.)

7. Cooling fan speed control solenoid valve (Do not disassemble.)
8. Gasket

67162-INFI-G04

Exploded view of water pump—4.5L M45 and Q45 models

2. Drain the cooling system.

3. Remove or disconnect the following:
- Negative battery cable
- Engine undercover
- Air duct and engine appearance cover
- Drive belt
- Cooling fan connector
- Radiator hose
- Water pump

To install:

4. Thoroughly clean and dry the mating surfaces, bolts and bolt holes.

5. Install or connect the following:
- Water pump. Torque the bolts to 18–23 ft. lbs. (25–32 Nm).
- Radiator hose
- Cooling fan connector
- Drive belt
- Air duct and engine appearance cover
- Engine undercover
- Negative battery cable

6. Fill and bleed the cooling system.

7. Start the vehicle, check for leaks and repair if necessary.

Heater Core

REMOVAL AND INSTALLATION

G20

1. Disconnect both battery cables, the negative (–) cable first.

❊❊ CAUTION

After disconnecting the battery, wait for a least 3 minutes for the SRS module to deplete its energy before working on the steering column or instrument panel.

2. Drain the cooling system into a clean container for reuse.

3. Remove the driver's side SRS module and the steering wheel by removing or disconnecting the following:
- Lower cover at the base of the steering wheel
- SRS module electrical connector
- Side covers at both sides of the steering wheel
- SRS module-to-steering wheel bolts using a T50 Torx® wrench
- SRS module from the steering wheel
- Place the front wheel in the straight-ahead position
- Horn connector and remove the steering wheel nut
- Steering wheel from the steering column

4. Remove or disconnect the following:
- Dash side and floor trim
- Steering column cover screws and the covers
- Combination switch-to-steering column screws, disconnect the electrical harness connectors and remove the combination switch

- Lower instrument panel screws and the panel at the driver's side
- Lower instrument reinforcement bolts and the reinforcement
- Steering column-to-instrument panel nuts and lower the steering column
- Cluster lid "C"-to-instrument panel screws, disconnect the electrical connectors and remove the cluster lid
- Cluster lid "A"-to-instrument panel screws, disconnect the electrical connectors and remove the cluster lid
- Combination meter-to-instrument panel screws, disconnect the electrical connectors and remove the combination meter
- Audio center-to-instrument panel bolts, disconnect the electrical connectors and remove the audio center
- Air conditioning control unit-to-instrument panel screws, disconnect the electrical connectors and remove the air conditioning control unit
- Console finisher clips and the console finisher
- Console box assembly-to-instrument panel screws, disconnect the electrical connectors and remove the console box assembly
- Glove box assembly-to-instrument panel screws, disconnect the lamp socket and remove the glove box assembly

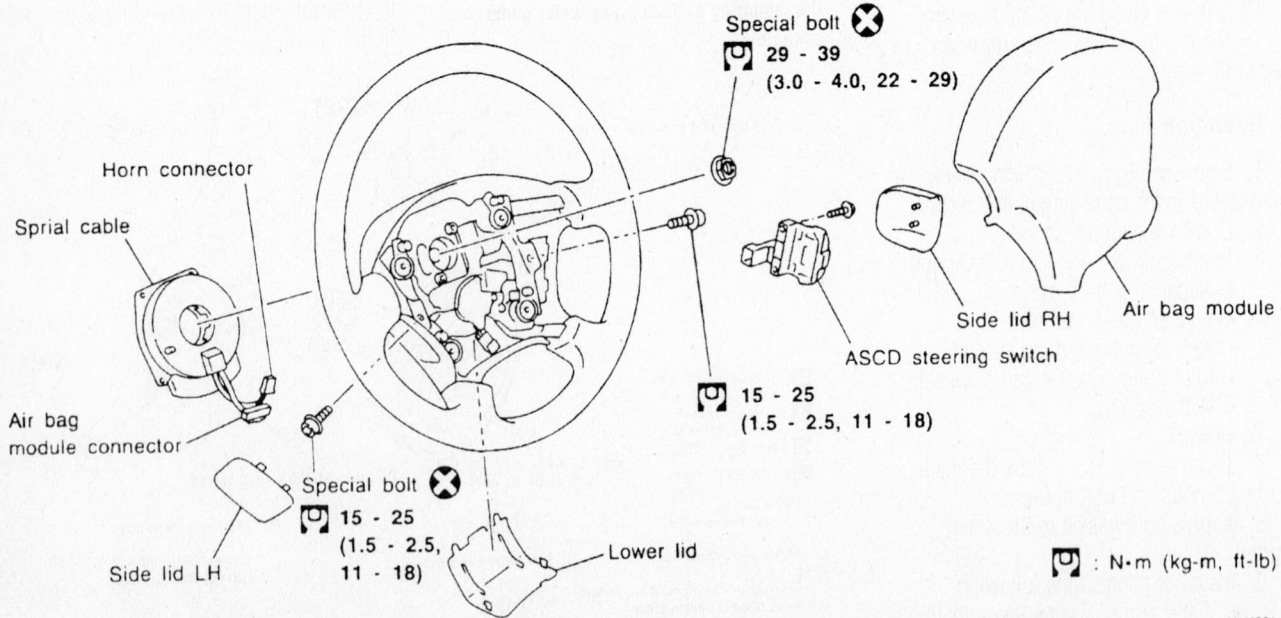

Horn connector
Sprial cable
Air bag module connector
Side lid LH
Special bolt ✖
15 - 25
(1.5 - 2.5, 11 - 18)
Lower lid
Special bolt ✖
29 - 39
(3.0 - 4.0, 22 - 29)
15 - 25
(1.5 - 2.5, 11 - 18)
ASCD steering switch
Side lid RH
Air bag module

⬚ : N•m (kg-m, ft-lb)

93112G04

Exploded view of the air bag module, the steering wheel and related components—G20

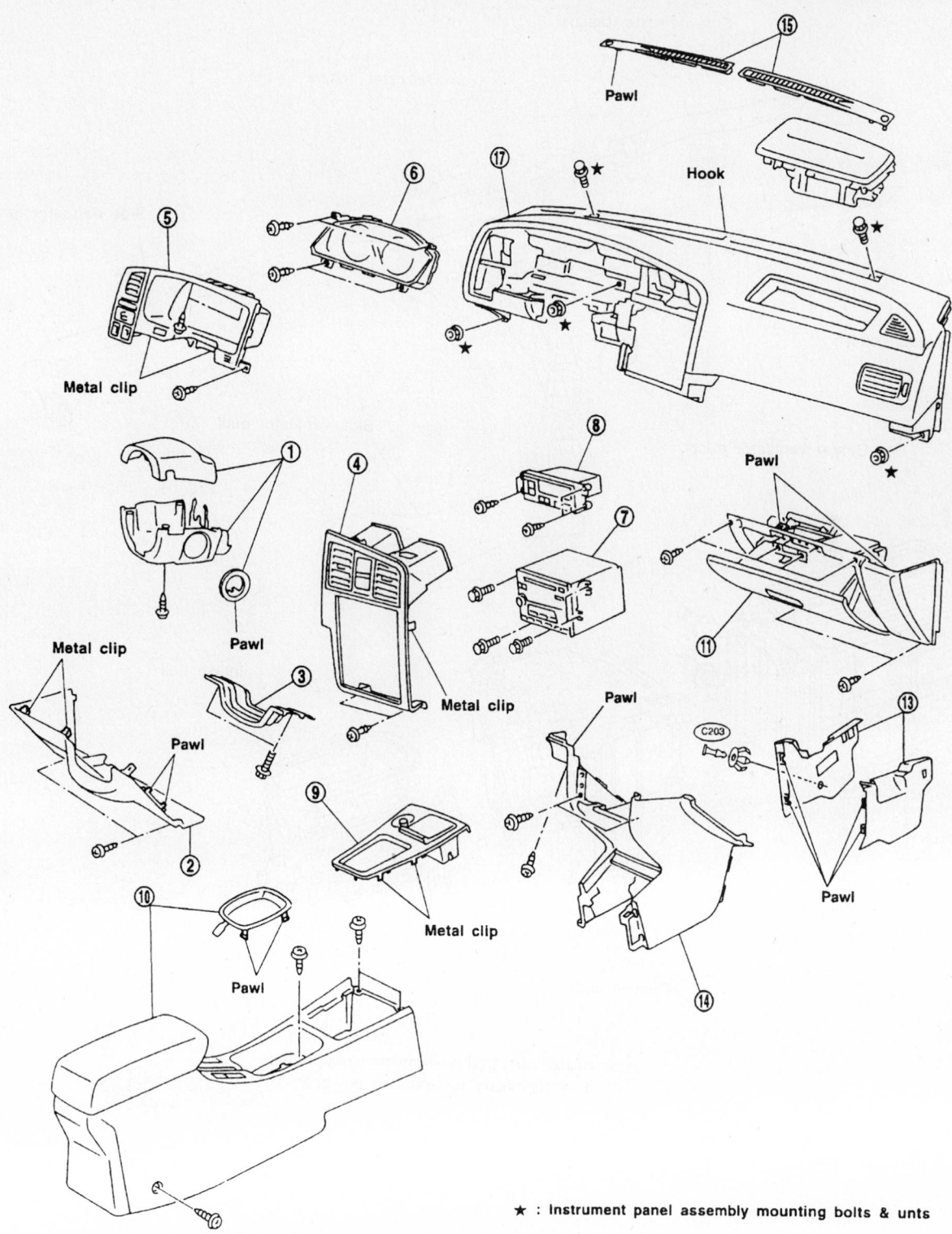

★ : Instrument panel assembly mounting bolts & unts

1. Steering column cover
   and combination switch
2. Instrument lower panel
   on driver side
3. Instrument lower reinforcement
4. Cluster lid C
5. Cluster lid A

6. Combination meter
7. Audio
8. A/C control unit
9. Console M/T or A/T finisher
10. Console box assembly
11. Glove box assembly

12. Passenger air bag module
13. Lower instrument cover
14. Lower instrument panel center
15. Defroster grille
16. Front pillar garnish
17. Instrument panel and pads

Exploded view of the instrument panel—G20

93112G05

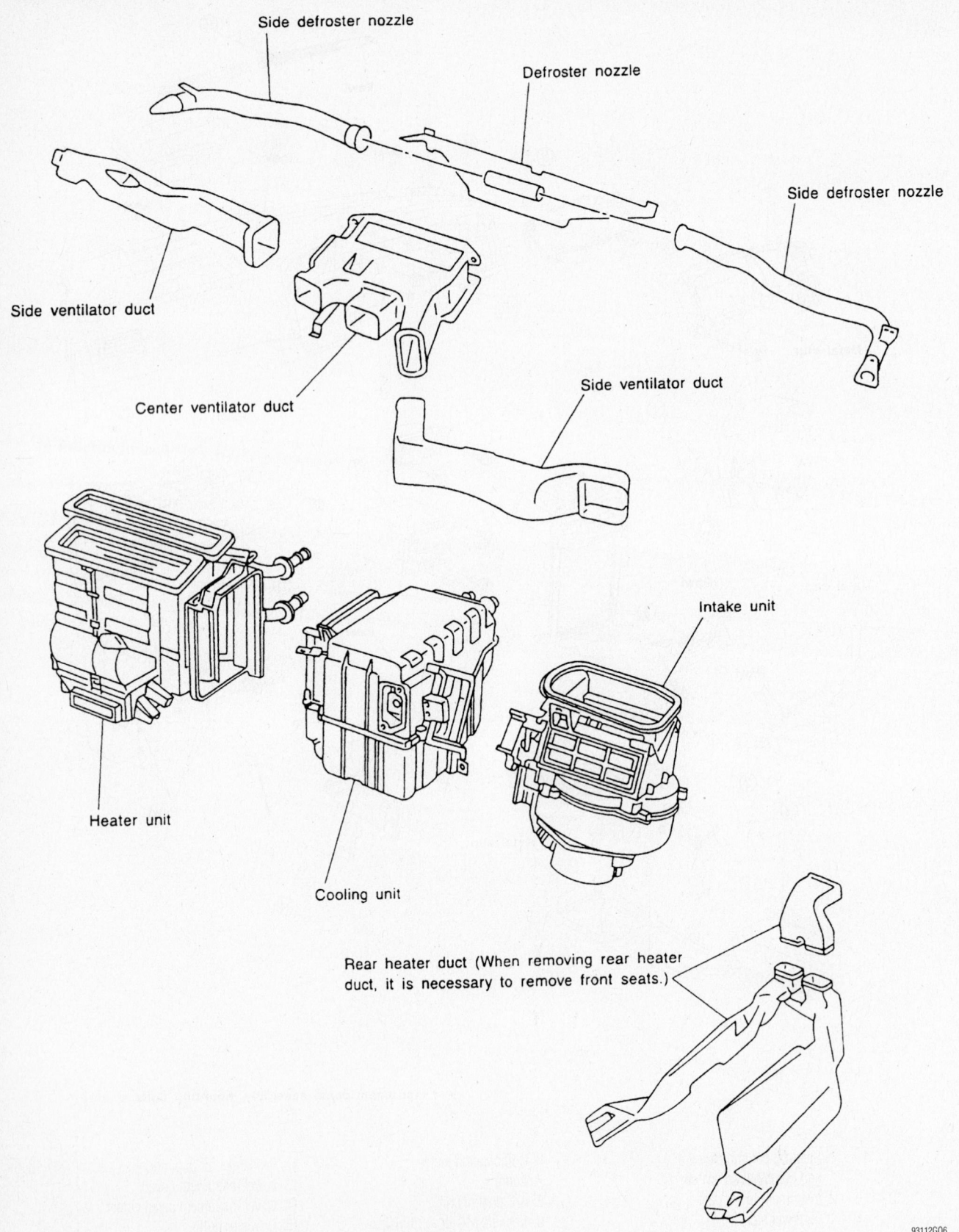

Side defroster nozzle

Defroster nozzle

Side defroster nozzle

Side ventilator duct

Center ventilator duct

Side ventilator duct

Heater unit

Intake unit

Cooling unit

Rear heater duct (When removing rear heater duct, it is necessary to remove front seats.)

93112G06

Exploded view of the heater housing, air conditioning housing, blower motor and ventilation ducts—G20

5. Remove the passenger's side SRS module by removing or disconnecting the following:
- Open the glove box and remove the SRS module cover
- SRS module electrical connector
- Glove box assembly
- SRS module-to-instrument panel bolts using a T50 Torx® wrench
- SRS module from the instrument panel

6. Remove or disconnect the following:
- Lower instrument cover clip and the cover
- Lower instrument panel center-to-instrument panel screws and the panel
- Connectors and remove the defroster grilles
- Front pillar garnish
- Instrument panel-to-chassis bolts/nuts and the instrument panel
- Heater hoses from the heater core
- Rear duct from the heater unit
- Air conditioning housing-to-heater housing fasteners
- Heater housing-to-chassis fasteners
- Heater unit from the vehicle

7. Disassemble and remove the heater core from the heater housing.

To install:

8. Assemble and install the heater core to the heater housing.

9. Install or connect the following:
- Heater unit to the vehicle
- Heater housing-to-chassis fasteners
- Air conditioning housing-to-heater housing fasteners
- Rear duct to the heater unit
- Heater hoses to the heater core
- Instrument panel and the instrument panel-to-chassis bolts/nuts
- Front pillar garnish
- Defroster grilles and connect the connectors
- Lower instrument panel center and the panel-to-instrument panel screws
- Lower instrument cover and the cover clip

10. Install the passenger's side SRS module by installing or connecting the following:
- SRS module to the instrument panel
- SRS module-to-instrument panel bolts and torque the bolts using a T50 Torx® wrench to 11–18 ft. lbs. (15–25 Nm)
- Glove box assembly

- SRS module electrical connector
- SRS module cover

11. Install or connect the following:
- Glove box assembly, connect the lamp socket and install the glove box assembly-to-instrument panel screws
- Console box assembly, connect the electrical connectors and install the console box assembly-to-instrument panel screws
- Console finisher and engage the clips
- Air conditioning control unit, connect the electrical connectors and Install the air conditioning control unit-to-instrument panel screws
- Audio center, connect the electrical connectors and Install the audio center-to-instrument panel bolts
- Combination meter, connect the electrical connectors and Install the combination meter-to-instrument panel screws
- Cluster lid "A", connect the electrical connectors and install the cluster lid-to-instrument panel screws
- Cluster lid "C", connect the electrical connectors and install the cluster lid-to-instrument panel screws
- Steering column and torque the steering column-to-instrument panel nuts to 11–14 ft. lbs. (15–19 Nm)
- Lower instrument reinforcement and the reinforcement bolts
- Lower instrument panel and the panel screws at the driver's side
- Combination switch and the combination switch-to-steering column screws and connect the electrical harness connectors
- Steering column covers and the cover screws
- Dash side and floor trim

12. Install the driver's side SRS module and the steering wheel by installing or connecting the following:
- Steering wheel to the steering column
- Steering wheel nut and torque to 22–29 ft. lbs. (29–39 Nm). Connect the horn connector
- SRS module to the steering wheel
- Torque the SRS module-to-steering wheel bolts to 11–18 ft. lbs. (15–25 Nm) using a T50 Torx® wrench
- Side covers at both sides of the steering wheel
- SRS module electrical connector

- Lower cover at the base of the steering wheel

13. Refill the cooling system.

14. Connect both battery cables, the negative (−) cable last.

15. Operate the engine to normal operating temperatures; then, check the climate control operation and check for leaks.

I30 and I35

1. Disconnect both battery cables, the negative (−) cable first.

2. Drain the cooling system into a clean container for reuse.

⁑ CAUTION

After disconnecting the battery, wait for a least 3 minutes for the SRS module to deplete its energy before working on the steering column or instrument panel.

3. Drain the cooling system into a clean container for reuse.

4. Remove the driver's side SRS module and the steering wheel by removing or disconnecting the following:
- Lower cover at the base of the steering wheel
- SRS module electrical connector
- Side covers at both sides of the steering wheel
- SRS module-to-steering wheel bolts using a T50 Torx® wrench
- SRS module from the steering wheel

⁑ CAUTION

Store the SRS module in a safe place with the front facing upward.

- Place the front wheel in the straight-ahead position
- Horn connector and remove the steering wheel nut
- Steering wheel from the steering column

5. Remove or disconnect the following:
- Instrument panel assembly
- Front seats
- Defroster ducts, the ventilator ducts and the floor ducts from the heater unit
- Vacuum hoses and electrical connectors leading to the heater unit
- Heater unit from the cooling unit. Take care not to damage the air conditioning tubes
- Heater unit attaching bolts. Remove the heater unit from the passenger compartment

6. Disassemble the heater unit and remove the heater core.

To install:

7. Install the heater core and assemble the heater unit.

8. Install or connect the following:
- Heater unit in the passenger compartment and tighten the attaching bolts securely
- All vacuum hoses and electrical connectors
- Defroster ducts, the ventilator ducts and the floor ducts to the heater unit

- Front seats
- Instrument panel assembly
- Heater hoses

9. Install the driver's side SRS module and the steering wheel by installing or connecting the following:
- Steering wheel to the steering column
- Steering wheel nut and torque to 22–29 ft. lbs. (29–39 Nm). Connect the horn connector
- SRS module to the steering wheel
- SRS module-to-steering wheel

bolts to 11–18 ft. lbs. (15–25 Nm) using a T50 Torx® wrench
- Side covers at both sides of the steering wheel
- SRS module electrical connector
- Lower cover at the base of the steering wheel

10. Refill the cooling system.

11. Connect both battery cables, the negative (–) cable last.

12. Operate the engine to normal operating temperatures; then, check the climate control operation and check for leaks.

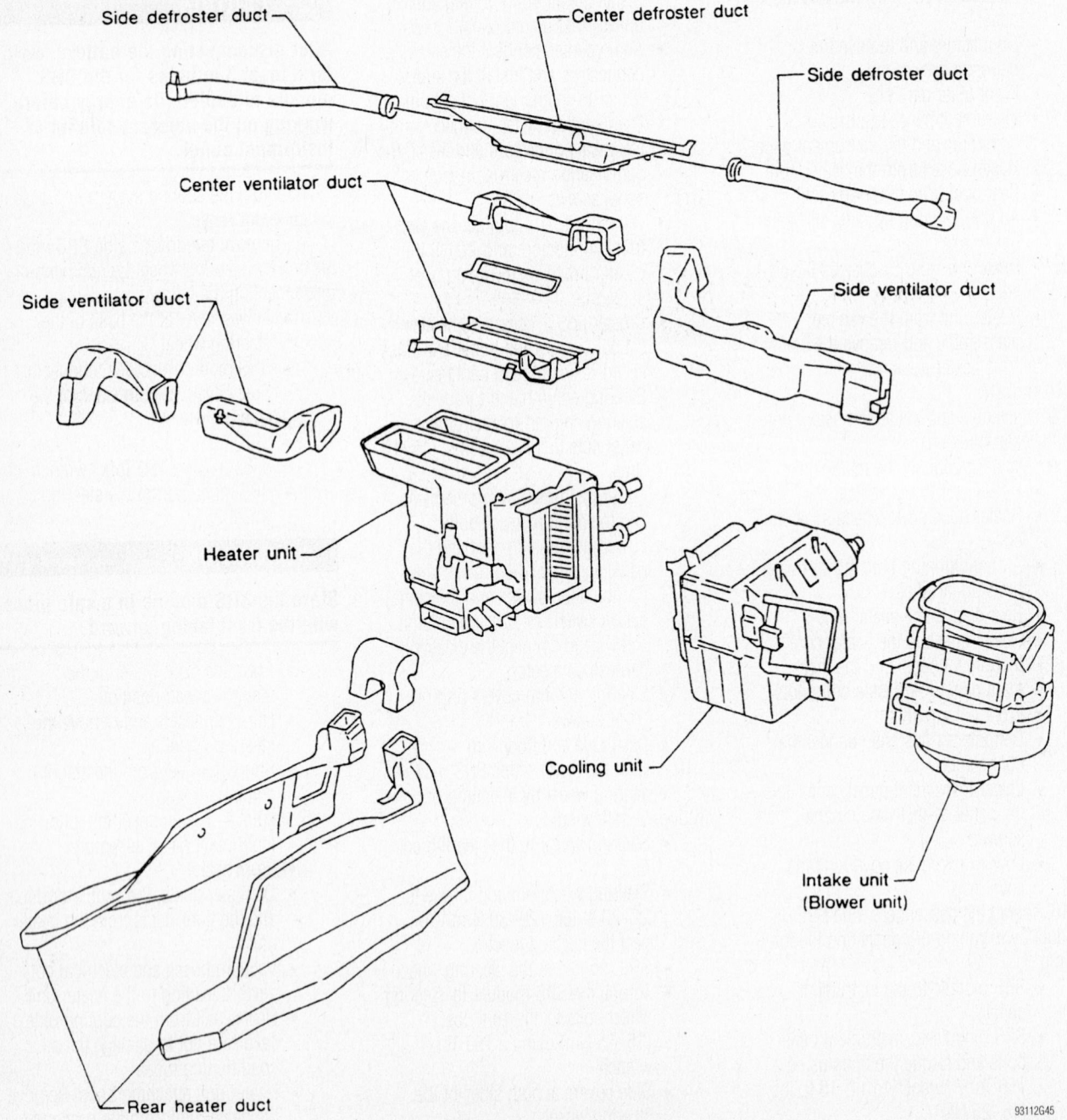

Side defroster duct — Center defroster duct — Side defroster duct

Center ventilator duct

Side ventilator duct — Side ventilator duct

Heater unit

Cooling unit

Intake unit (Blower unit)

Rear heater duct

93112G45

Exploded view of the heating-air conditioning system assemblies and related components—I30

M45

1. Disconnect both battery cables, the negative (–) cable first.

※※ CAUTION

After disconnecting the battery, wait for a least 3 minutes for the SRS module to deplete its energy before working on the steering column or instrument panel.

2. Recover the air conditioning refrigerant.
3. Drain the engine coolant.
4. Remove the engine appearance cover.
5. Remove the front suspension tower bar.
6. Remove or disconnect the following:
   • Heater hoses
   • Evaporator lines
   • Steering lock cover
   • Steering column upper and lower covers
   • Driver side lower instrument panel cover
   • Hood opening lever
   • Top instrument cluster cover
   • Combination meter
   • Inner instrument cluster panel
   • Instrument cluster
   • Clock
   • Automatic transmission shift lever cover
   • Cup holder
   • Ashtray and center lower panel
   • Console upper finish panel
   • Audio unit
   • Instrument panel lower cover
   • Glove box
   • Glove box cover
   • CD unit
   • Center ventilation grille
   • Console box assembly
   • Front defroster grille
   • Both A pillar trim panels
   • Ignition key lamp and instrument panel bracket
   • Instrument panel reinforcement bracket
   • Instrument panel and pad
   • Blower motor, intake door motor and amplifier connectors
   • Blower unit
   • Vehicle harness clips from steering member
   • Instrument panel stays at both sides
   • Defroster nozzle
   • Ventilator ducts
   • Heating and cooling unit

• Foot duct from heating and cooling unit
• Heater core from heating and cooling unit

To install:

7. Installation is the reverse of the removal procedure noting the following:
8. Refill the cooling system.
9. Connect both battery cables, the negative (–) cable last.
10. Operate the engine to normal operating temperatures; then, check the climate control operation and check for leaks.

Q45

2001–03 MODELS

1. Disconnect both battery cables, the negative (–) cable first.

※※ CAUTION

After disconnecting the battery, wait for a least 3 minutes for the SRS module to deplete its energy before working on the steering column or instrument panel.

2. Drain the cooling system into a clean container for reuse.
3. Remove the driver's side SRS module and the steering wheel by removing or disconnecting the following:
   • Lower cover at the base of the steering wheel
   • SRS module electrical connector
   • Side covers at both sides of the steering wheel
   • SRS module-to-steering wheel bolts using a T50 Torx® wrench
   • SRS module from the steering wheel
   • Place the front wheel in the straight-ahead position
   • Horn connector and remove the steering wheel nut
   • Steering wheel from the steering column
4. Remove or disconnect the following:
   • Dash side lower finishers
   • Steering column cover screws and the covers
   • Combination switch-to-steering column screws, disconnect the electrical harness connectors and remove the combination switch
   • Lower instrument panel screws/bolts and the panel. Disconnect the electrical harness connectors and the in-vehicle sensor at the driver's side
   • Lower instrument reinforcement bolts and the reinforcement

• Steering column-to-instrument panel nuts and lower the steering column
• Cluster lid "A"-to-instrument panel screws and remove the cluster lid
• Steering lock escutcheon screws and the escutcheon
• Cluster lid "D"
• Combination meter-to-instrument panel screws, disconnect the electrical connectors and remove the combination meter
• Glove box assembly-to-instrument panel screws, disconnect the lamp socket and remove the glove box assembly

5. Remove the passenger's side SRS module by removing or disconnecting the following:
   • Lower instrument panel cover
   • SRS module electrical connector
   • SRS module-to-instrument panel bolts using a T50 Torx® wrench
   • SRS module from the instrument panel

※※ CAUTION

Store the SRS module in a safe place with the front facing upward.

6. Remove or disconnect the following:
   • Center ventilation grille using a suitable prytool
   • Lock console
   • Card pocket assembly screws and the assembly
   • Console finisher clips and the console finisher
   • Audio center, the cluster lid "C" and the air conditioning control unit-to-instrument panel screws, disconnect the electrical connectors and remove the panel
   • Console box assembly-to-instrument panel screws, disconnect the electrical connectors and remove the console box assembly
   • Connectors and remove the defroster grilles
   • Connectors and remove the sunload sensor
   • Front pillar garnish
   • Instrument panel-to-chassis bolts/nuts and the instrument panel
   • Heater hoses from the heater core
   • Rear duct from the heater unit
   • Air conditioning housing-to-heater housing fasteners
   • Heater unit from the vehicle
7. Disassemble and remove the heater core from the heater housing.

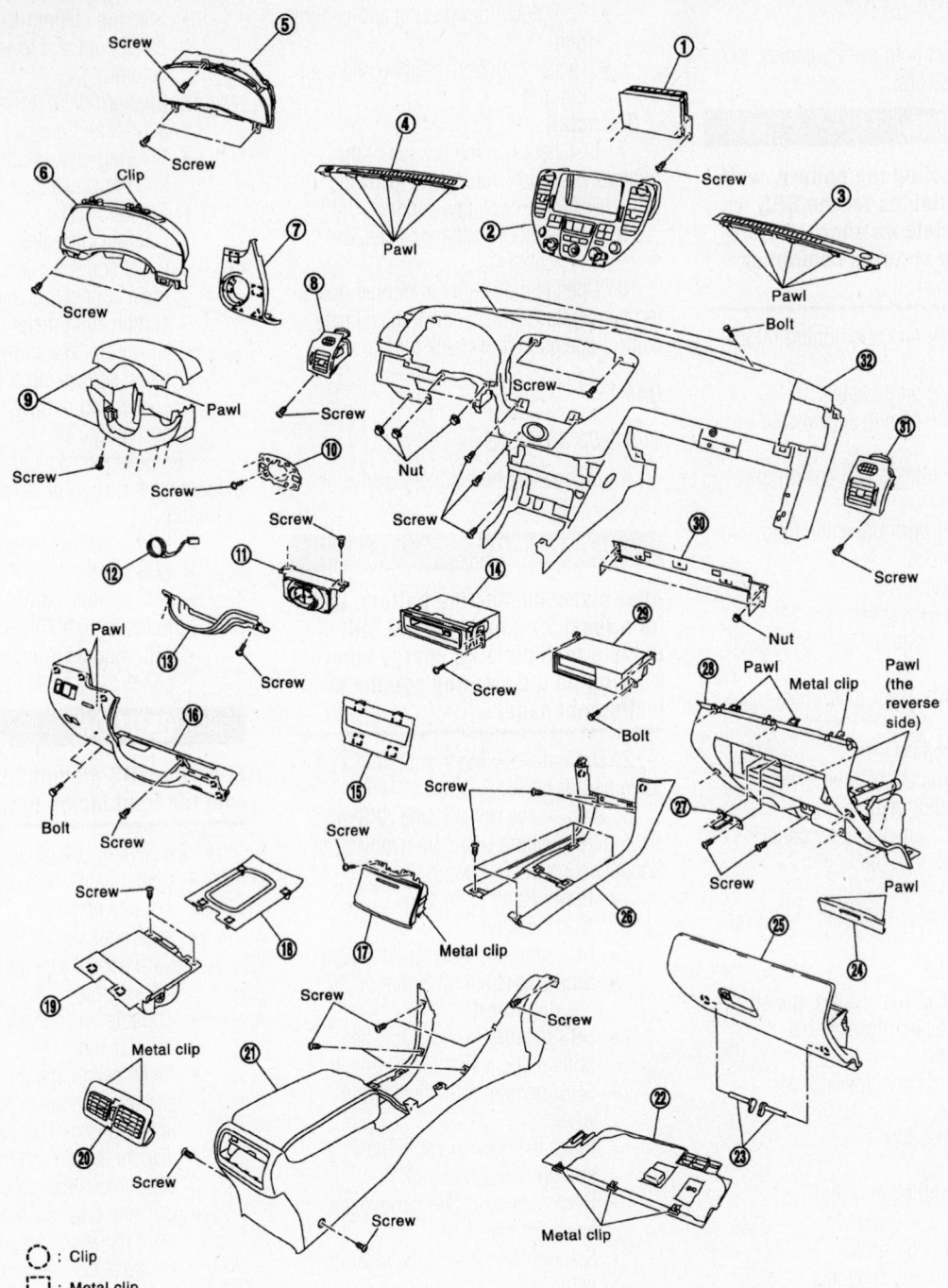

: Clip

: Metal clip

1. Display unit	2. Cluster lid C	3. Front defroster grille (RH)
4. Front defroster grille (LH)	5. Combination meter	6. Cluster lid A
7. Steering lock escutcheon	8. Side ventilation (LH)	9. Steering column cover
10. Instrument panel bracket	11. Clock	12. Ignition key lamp assembly
13. Knee protector lower	14. Audio unit	15. Cluster lid center lower
16. Instrument lower driver panel	17. Ashtray	18. A/T console finisher
19. Cup holder	20. Center ventilator grille	21. Console box assembly
22. Instrument lower cover	23. Glove box pin	24. Escutcheon glove box
25. Glove box assembly	26. Console upper finisher	27. Glove box striker
28. Glove box cover	29. CD auto changer	30. Instrument panel reinforcement bracket
31. Side ventilation (RH)	32. Instrument panel and pad	

Exploded view of the instrument panel assembly—M45

67162-INFI-G05

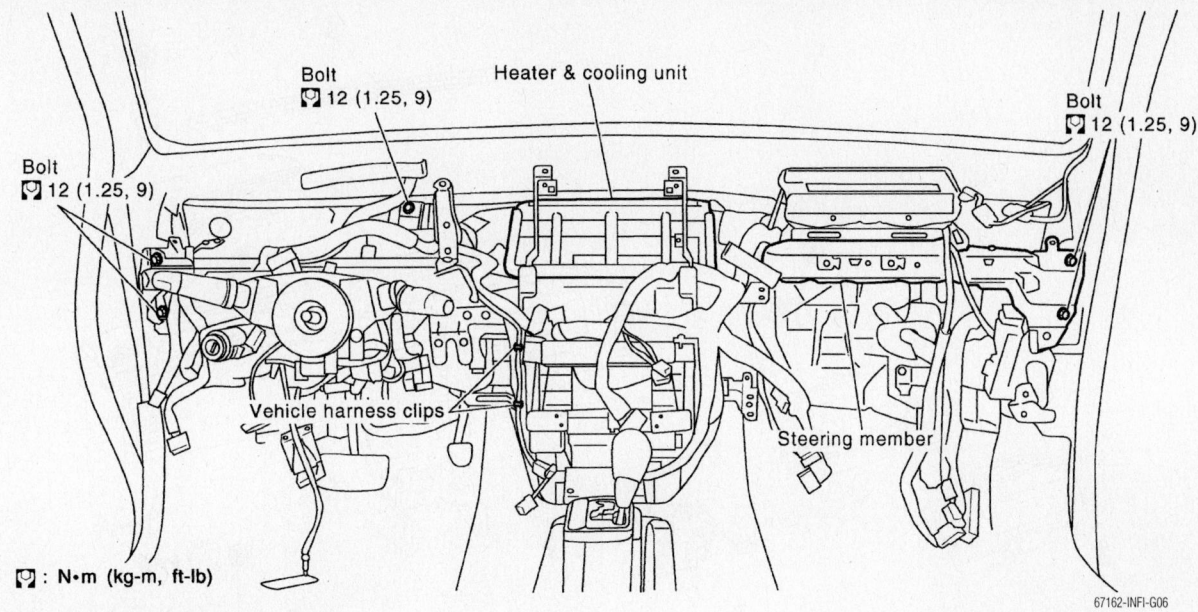

Heating and cooling unit mounting—M45 & Q45

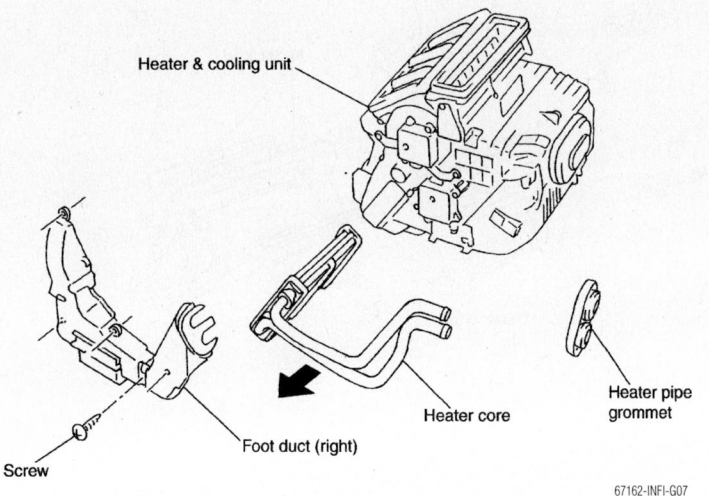

Heater core removal—M45

To install:

8. Assemble and install the heater core to the heater housing.

9. Install or connect the following:
   - Heater unit to the vehicle
   - Heater housing-to-chassis fasteners
   - Air conditioning housing-to-heater housing fasteners
   - Rear duct to the heater unit
   - Heater hoses to the heater core
   - Instrument panel and the instrument panel-to-chassis bolts/nuts
   - Front pillar garnish
   - Defroster grilles and connect the connectors
   - Lower instrument panel center and the panel-to-instrument panel screws

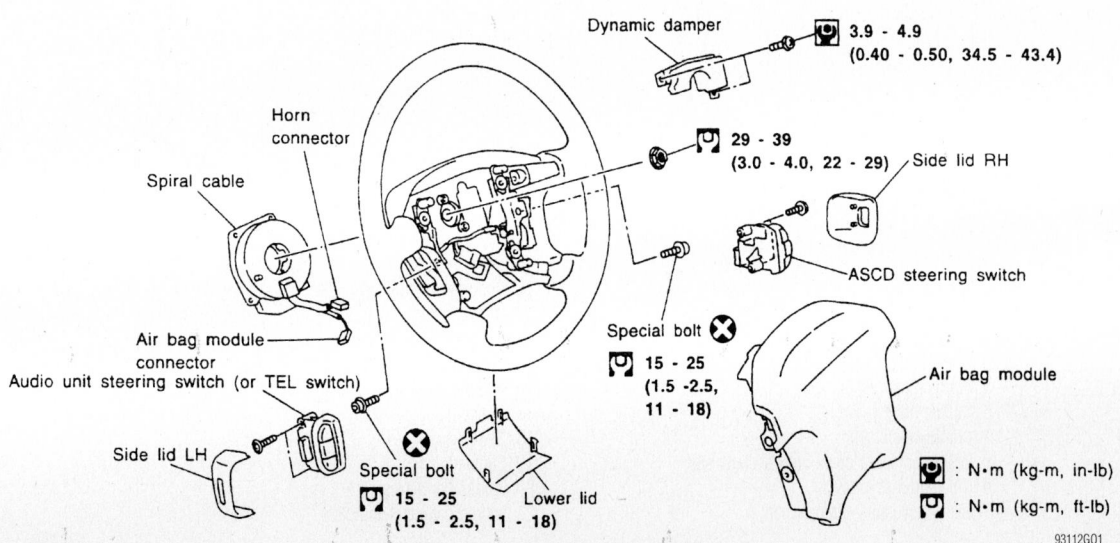

Exploded view of the air bag module, the steering wheel and related components—2001–03 Q45

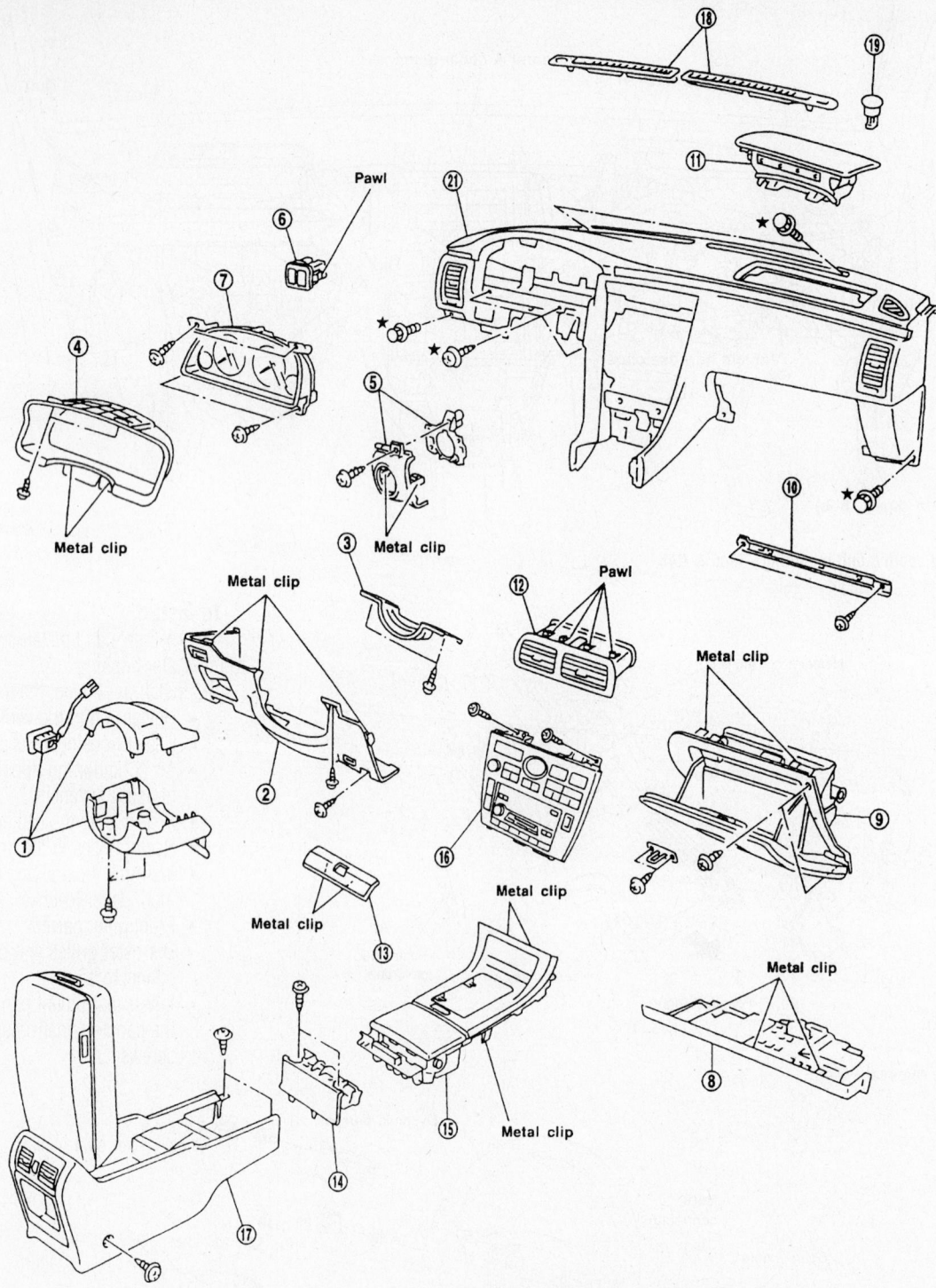

★ : Instrument panel assembly mounting bolts and screws

1. Steering column cover and combination switch
2. Instrument lower panel on driver side
3. Instrument lower reinforcement
4. Cluster lid A
5. Steering lock escutcheon
6. Cluster lid D
7. Combination meter
8. Instrument lower cover on passenger side
9. Glove box assembly
10. Instrument panel reinforcement

12. Center ventilation grille
13. Lock console
14. Card pocket assembly
15. Console A/T finisher
16. Audio, cluster lid C and A/C control unit
17. Console box assembly
18. Defroster grille
19. Sunload sensor
20. Front pillar garnish
21. Instrument panel and pads

93112G02

Exploded view of the instrument panel—2001–03 Q45

- Lower instrument cover and the cover clip
- Center ventilation grille
- Lock console
- Card pocket assembly and the assembly screws
- Console finisher and engage the clips

- Audio center, the cluster lid "C" and the air conditioning control unit panel, connect the electrical connectors and install the panel-to-instrument panel screws
- Console box assembly, connect the electrical connectors and install the

console box assembly-to-instrument panel screws
- Defroster grilles and connect the connector
- Sunload sensor and connectors

10. Install the passenger's side SRS module by installing or connecting the following:

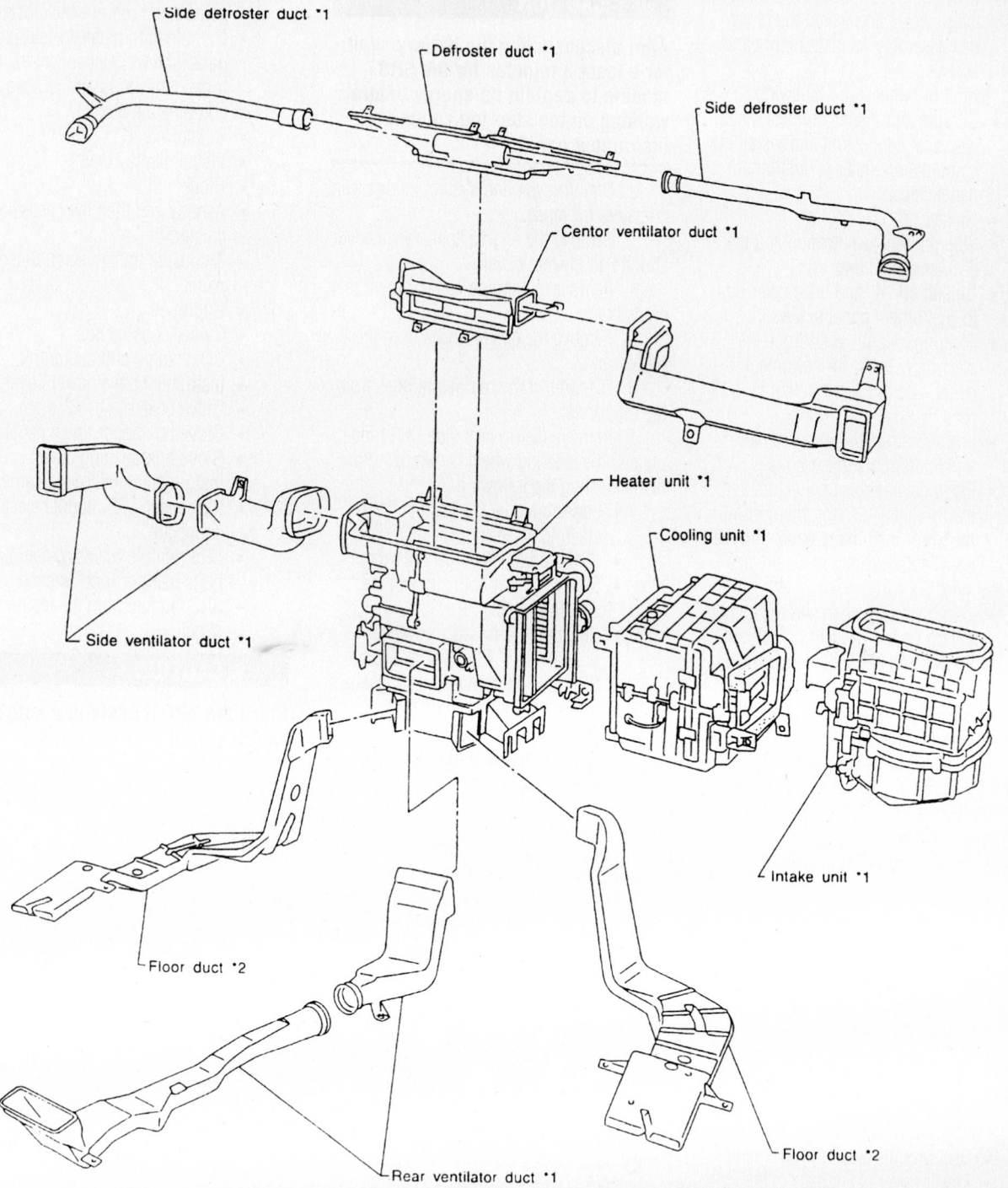

Side detroster duct *1

Defroster duct *1

Side defroster duct *1

Centor ventilator duct *1

Heater unit *1

Cooling unit *1

Side ventilator duct *1

Intake unit *1

Floor duct *2

Floor duct *2

Rear ventilator duct *1

*1 : For removal, it is necessary to remove instrument assembly.

*2 : For removal, it is necessary to remove front seat.

93112G03

Exploded view of the heater housing, air conditioning housing, blower motor and ventilation ducts—2001-03 Q45

- SRS module to the instrument panel
- SRS module-to-instrument panel bolts and torque the bolts to 11–18 ft. lbs. (15–25 Nm) using a T50 Torx® wrench
- SRS module electrical connector
- Lower instrument panel cover
- Glove box assembly, connect the lamp socket and install the glove box assembly-to-instrument panel screws

11. Install or connect the following:
- Combination meter, connect the electrical connectors and install the combination meter-to-instrument panel screws
- Cluster lid "D"
- Steering lock escutcheon and the escutcheon screws
- Cluster lid "A" and the cluster lid-to-instrument panel screws
- Steering column and torque the steering column-to-instrument panel nuts to 11–14 ft. lbs. (15–19 Nm)
- Lower instrument reinforcement and the reinforcement bolts.
- Electrical harness connectors and the in-vehicle sensor; then, install the lower instrument panel and the panel screws/bolts at the driver's side
- Combination switch and the combination switch-to-steering column screws and connect the electrical harness connectors
- Steering column covers and the cover screws
- Dash side lower finishers

12. Install the driver's side SRS module and the steering wheel by installing or connecting the following:
- Steering wheel to the steering column
- Steering wheel nut and torque to 22–29 ft. lbs. (29–39 Nm). Connect the horn connector
- SRS module to the steering wheel
- Torque the SRS module-to-steering wheel bolts to 11–18 ft. lbs. (15–25 Nm) using a T50 Torx® wrench
- Side covers at both sides of the steering wheel
- SRS module electrical connector
- Lower cover at the base of the steering wheel

13. Refill the cooling system.
14. Connect both battery cables, the negative (–) cable last.
15. Operate the engine to normal operat-

ing temperatures; then, check the climate control operation and check for leaks.

Q45

2004 MODELS

1. Disconnect both battery cables, the negative (–) cable first.

✳✳ CAUTION

After disconnecting the battery, wait for a least 3 minutes for the SRS module to deplete its energy before working on the steering column or instrument panel.

2. Drain the cooling system into a clean container for reuse.
3. Remove the engine appearance cover and the air cleaner cover.
4. Remove the engine compartment rear cross brace.
5. Remove the heater hoses from the heater core.
6. Disconnect the refrigerant lines from the evaporator.
7. Remove the driver's side SRS module and the steering wheel by removing or disconnecting the following:
- Lower cover at the base of the steering wheel
- SRS module electrical connector
- Side covers at both sides of the steering wheel
- SRS module-to-steering wheel bolts using a T30 Torx® wrench
- SRS module from the steering wheel
- Place the front wheel in the straight-ahead position
- Horn connector and remove the steering wheel nut

- Steering wheel from the steering column

8. Remove or disconnect the following:
- Steering column cover screws and the covers
- Steering lock escutcheon screws and the escutcheon
- Cluster lid "D"
- Cluster lid "A"-to-instrument panel screws and remove the cluster lid
- Combination meter-to-instrument panel screws, disconnect the electrical connectors and remove the combination meter
- Cluster lid "C"
- Visual display unit
- Clock
- Ashtray and then the center lower cluster lid
- Automatic transmission finish panel
- Cupholder
- Center console box
- CD changer and audio unit
- Instrument panel lower cover
- Glove box
- Glove box upper finish panel
- Glove box opening cover
- Instrument panel reinforcement
- Passenger SRS module electrical connector
- SRS module-to-instrument panel bolts using a Torx® wrench
- SRS module from the instrument panel

✳✳ CAUTION

Store the SRS module in a safe place with the front facing upward.

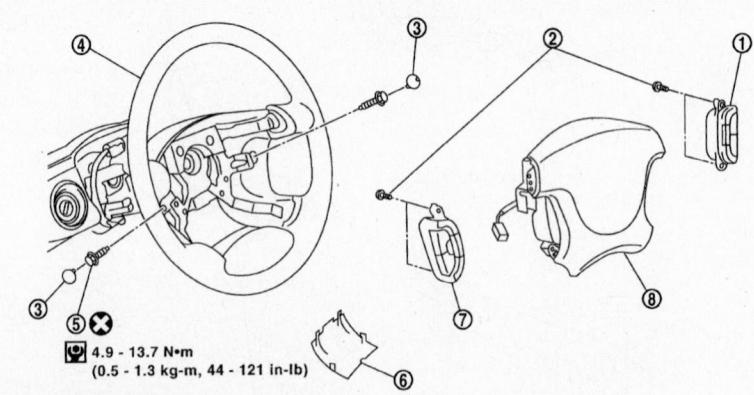

4.9 - 13.7 N•m
(0.5 - 1.3 kg-m, 44 - 121 in-lb)

❌ : Always replace after every disassembly.

1.	Audio steering switch	3.	Screw	3.	Side lid
4.	Steering wheel	5.	Special bolt	6.	Lower lid
7.	ASCD switch	8.	Driver air bag module		

67162-INFI-G08

Exploded view of the air bag module, the steering wheel and related components—2004 Q45

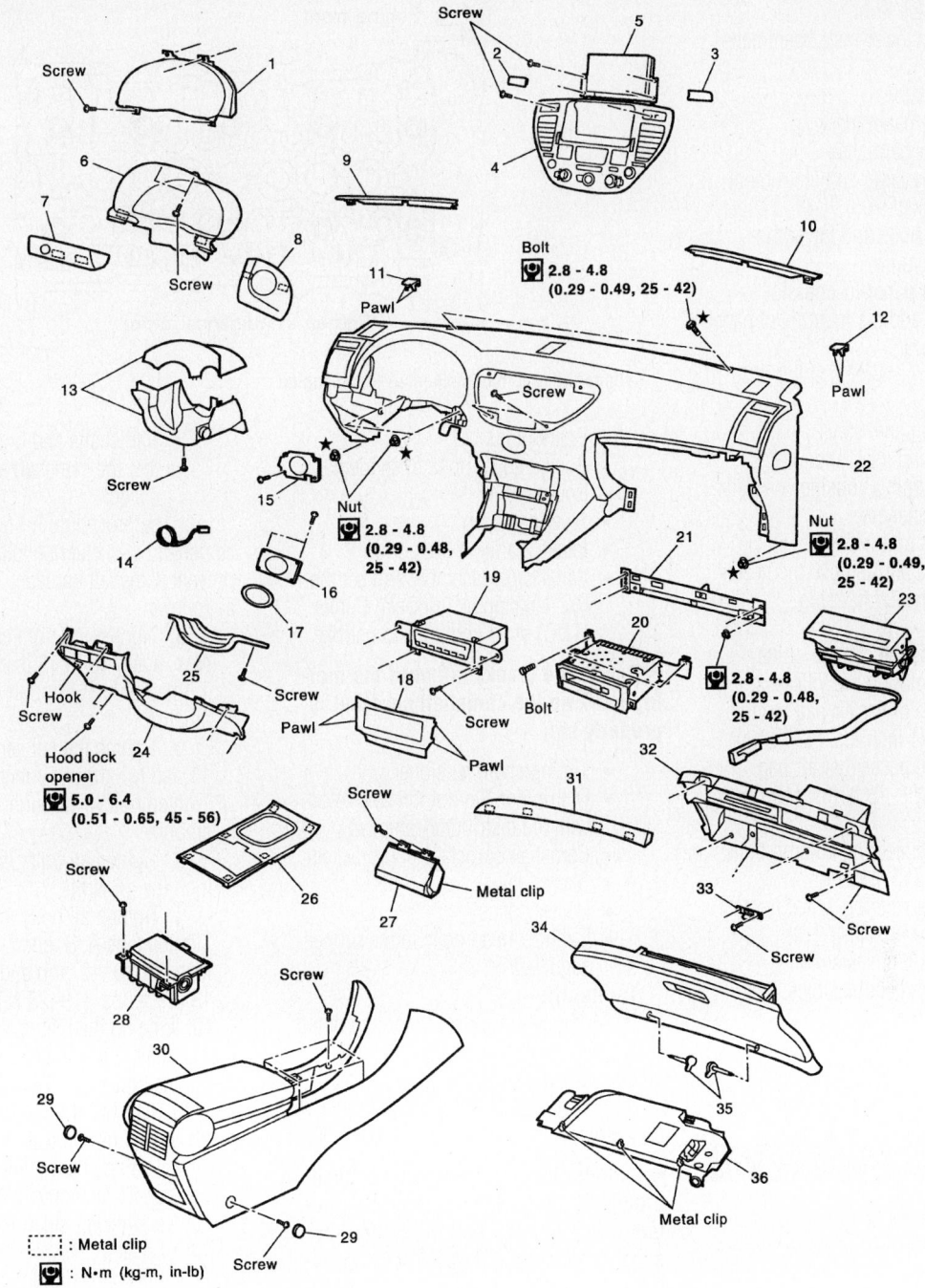

Screw

Screw

Screw

Screw

Screw

Screw

Bolt
2.8 - 4.8
(0.29 - 0.49, 25 - 42)

Pawl

Pawl

Screw

Nut
2.8 - 4.8
(0.29 - 0.48, 25 - 42)

Nut
2.8 - 4.8
(0.29 - 0.49, 25 - 42)

Screw

Screw

Pawl

Pawl

Bolt

2.8 - 4.8
(0.29 - 0.48, 25 - 42)

Hook
Screw

Hood lock
opener
5.0 - 6.4
(0.51 - 0.65, 45 - 56)

Screw

Screw

Screw

Metal clip

Screw

Screw

Screw

Bolt cap

Screw

Metal clip

Bolt cap

Screw

: Metal clip

: N•m (kg-m, in-lb)

1. Combination meter	2. Ventilation mask (LH)	3. Ventilation mask (RH)
4. Cluster lid C	5. Display unit	6. Cluster lid A
7. Cluster lid D	8. Steering lock escutcheon	9. Front defroster grille (LH)
10. Front defroster grille (RH)	11. Instrument panel mask (LH)	12. Instrument panel mask (RH)
13. Steering column cover	14. Ignition key lamp assembly	15. Instrument panel bracket
16. Clock	17. Cluster lid C lower	18. Cluster lid center lower
19. CD auto changer	20. Audio unit	21. Instrument panel bracket
22. Instrument panel and pad assembly	23. Front passenger air bag module	24. Instrument lower driver panel
25. Knee protector lower	26. A/T console finisher	27. Ashtray
28. Cup holder	29. Bolt cap	30. Console box assembly
31. Instrument finisher	32. Glove box cover	33. Glove box striker
34. Glove box assembly	35. Glove box pin	36. Instrument lower cover

67162-INFI-G09

Exploded view of the instrument panel—2004 Q45

- Hood release lever
- Driver side lower instrument panel cover
- Knee protector
- Instrument panel mask
- Front pillar garnishes
- Ignition key lamp and instrument panel bracket
- Remove 2 bolts and lower the steering column
- Instrument panel-to-chassis bolts/nuts and the instrument panel
- Blower motor
- Vehicle harness clips at left of A/C-heater unit
- Instrument panel stays
- Defroster and ventilation ducts
- Air conditioning housing-to-heater housing fasteners
- A/C-heater unit from the vehicle

9. Disassemble and remove the heater core from the A/C-heater housing.

To install:

10. Installation is the reverse of the removal procedure noting the following:

a. Tighten the driver air bag module screws to 44–121 inch lbs. (4–14 Nm).

b. Tighten the passenger air bag module bolts to 11–18 ft. lbs. (15–25 Nm).

c. Tighten the steering column bolts to 11–13 ft. lbs. (15–18 Nm).

d. Tighten the steering wheel nut to 23–28 ft. lbs. (30–39 Nm).

11. Refill the cooling system.

12. Connect both battery cables, the negative (−) cable last.

13. Operate the engine to normal operating temperatures; then, check the climate control operation and check for leaks.

Cylinder Head

REMOVAL & INSTALLATION

2.0L Engine

1. Before servicing the vehicle, refer to the precautions in the beginning of this section.

2. Relieve the fuel system pressure.

3. Drain the cooling system.

4. Remove or disconnect the following:

- Negative battery cable
- Right front wheel
- Engine side cover
- Radiator
- Air duct to intake manifold
- ASCD actuator
- Vacuum and fuel hoses
- Electrical connectors and wiring

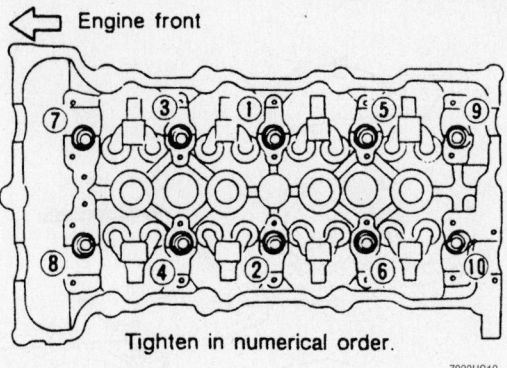

Tighten in numerical order.

7923HG12

Cylinder head torque sequence—2.0L engine

- Spark plugs
- Rocker cover bolts in sequence
- Rocker cover
- Steering pump
- Intake manifold supports
- Water pipe assembly and set the No. 1 piston at Top Dead Center (TDC) of its compression stroke

➡**Rotate the crankshaft until the mating mark on the camshaft sprocket is properly set.**

- Timing chain tensioner
- Distributor. Do not turn the rotor with the distributor removed.
- Camshaft sprockets and brackets
- Starter
- Heater hoses
- Cylinder head bolts in the proper sequence

To install:

5. Apply liquid gasket to the top of the chain cover where it meets the cylinder block before installing the head gasket.

6. Install the gasket and cylinder head on the block.

➡**Cylinder head bolts may be reused providing the dimension from the bottom of the head to the end of the bolt does not exceed 6.228 in. (158.2mm). If the dimension exceeds the specification, replace the cylinder head bolts.**

7. Tighten the cylinder head bolts as follows:

a. Tighten all bolts to 29 ft. lbs. (39 Nm) using the proper sequence.

b. Tighten all bolts to 58 ft. lbs. (78 Nm) using the proper sequence.

c. Loosen all bolts completely.

d. Tighten all bolts to 25–33 ft. lbs. (34–44 Nm) using the proper sequence.

e. Tighten all bolts 90–95°.

f. Tighten all bolts an additional 90–95°.

8. Install or connect the following:

- Heater hoses

- Camshafts and brackets. Ensure that the camshaft keys are at 12 o'clock.

9. The procedure for tightening camshaft bolts must be followed exactly to prevent camshaft damage. Tighten the bolts as follows:

a. Tighten the right camshaft bolts Nos. 9 and 10 to 18 inch lbs. (2 Nm). Tighten bolts 1 through 8 to the same amount.

b. Tighten the left camshaft bolts No. 11 and No. 12 to 18 inch lbs. (2 Nm). Tighten bolts 1 through 10 to the same amount.

c. Tighten all bolts in sequence to 54 inch lbs. (6 Nm).

d. Tighten all bolts in sequence again. Tighten type A, B, and D bolts to 78–102 inch lbs. (9–12 Nm) and type C bolts to 13–19 ft. lbs. (18–25 Nm).

10. Line up the mating marks on the timing chain and camshaft sprockets and install the sprockets. Tighten the sprocket bolts to 101–116 ft. lbs. (137–157 Nm).

11. Install or connect the following:

- Timing chain guide, distributor, chain tensioner, oil filter bracket and power steering oil pump bracket
- Intake manifold supports
- Rocker cover. Torque the bolts, in sequence, to 89 inch lbs. (10 Nm).
- Spark plugs, power steering pump, alternator, water pump pulley and drive belts, air duct to the intake manifold and the radiator
- Vacuum and fuel hoses and reconnect all electrical connections
- Engine undercover
- Right front wheel
- Negative battery cable

12. Fill and bleed the cooling system.

13. Start the vehicle, check for leaks and repair if necessary.

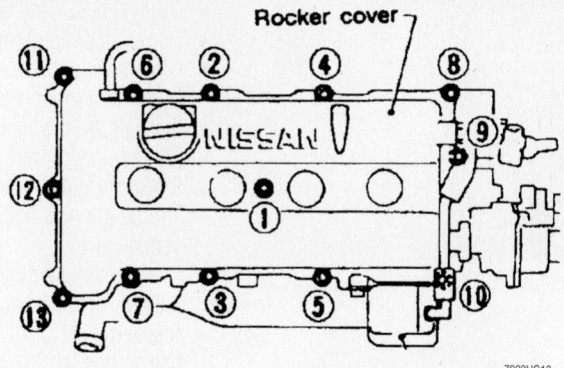

Tighten the rocker cover bolts according to the sequence shown—2.0L engine

3.0L Engine

1. Before servicing the vehicle, refer to the precautions in the beginning of this section.
2. Relieve the fuel system pressure.
3. Drain the engine oil.
4. Drain the cooling system.

➡**Before disconnecting any hoses or connectors, note the locations for reassembly.**

5. Remove or disconnect the following:
   • Negative battery cable
   • Right front wheel
   • Left side ornament cover

   • Air duct to intake manifold hose, collector hose, blow-by hose, and vacuum hoses
   • Fuel hoses and the harness connections
   • Evaporative emissions (EVAP) canister purge hose
   • Ignition coils from the spark plugs
   • Exhaust Gas Recirculation (EGR) tube
   • Right side intake manifold collector supports and remove the collector. Remove the manifold from the cylinder head.
   • Fuel tube assembly

   • Rocker arm covers
   • Engine undercover
   • Engine side cover
   • Power steering oil pump and belt
   • Camshaft Position (CMP) sensor (PHASE) and Crankshaft Position (CKP) sensors.

6. Set the No. 1 piston to Top Dead Center (TDC) of compression stroke by rotating the crankshaft.
   • Crankshaft pulley
   • Air compressor and bracket
   • Timing chain tensioner and slack side chain guide
   • Engine oil pan

7. Remove the camshaft sprockets first. Be sure to hold the flats of the camshafts while removing the sprocket bolts.

8. Loosen the camshaft bearing caps in several steps. The bearing caps MUST be loosened in sequence.

➡**Keep all bearing caps and camshafts in proper order for reinstallation.**

9. Loosen the cylinder head bolts in sequence.

To install:

10. Turn the crankshaft until the No. 1 piston is set 240° Before Top Dead Center (BTDC) on compression stroke.

11. Using new head gaskets, install the cylinder heads.

➡**If possible, replacement of the head bolts is suggested.**

12. If replacement of the head bolts is not possible, perform the following bolt measurement:
   a. Measure the diameter of the head bolt (11mm) from the bottom of the bolt.
   b. Measure the diameter of the head bolt (48mm) from the bottom of the bolt.
   c. Whenever the size difference between the 2 measurements exceeds 0.0043 in. (0.11mm) the head bolts must be replaced.

13. Lubricate the head bolt threads and the bolt seat surfaces with new engine oil.

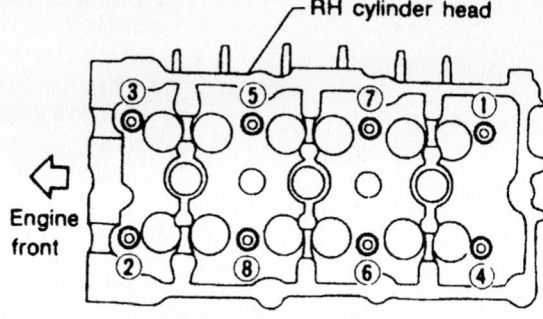

Right cylinder head bolt loosening sequence—3.0L engine

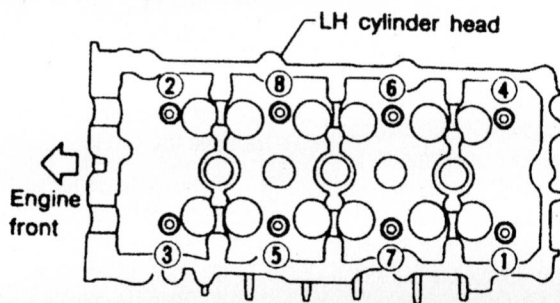

Left cylinder head bolt loosening sequence—3.0L engine

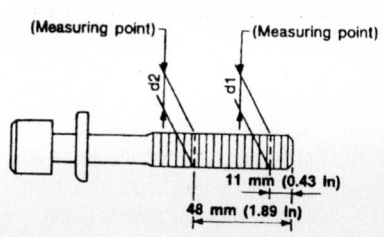

Measuring the cylinder head bolts—3.0L engine

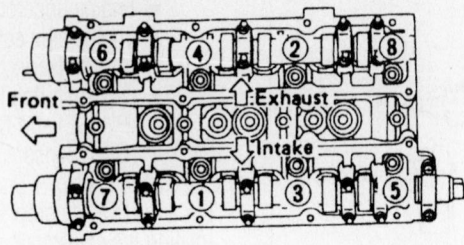

Right cylinder head

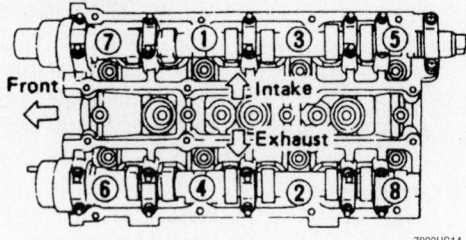

Left cylinder head

7923HG14

Cylinder head torque sequence—3.0L engine

14. Tighten the cylinder head bolts in sequence using the following steps:

   a. Step 1: All bolts in sequence to 72 ft. lbs. (98 Nm).

   b. Step 2: Completely loosen all bolts.

   c. Step 3: Tighten all bolts in sequence to 25–33 ft. lbs. (34–44 Nm).

   d. Step 4: Tighten all bolts in sequence an additional 90° clockwise.

   e. Step 5: Tighten all bolts in sequence an additional 90° clockwise.

15. Install or connect the following:

- Camshaft tensioners. Tighten the tensioner mounting bolts to 75–96 inch lbs. (8.4–10.8 Nm).

➡**The camshafts can be identified by the paint marks on the camshaft. The left cylinder head camshafts have a YELLOW paint mark and the right cylinder head camshafts have a WHITE paint mark.**

➡**When installing the camshafts, position the camshaft keys at the 12 o'clock position in respect to the cylinder head angle.**

- Camshafts and the bearing caps
- New O-rings to the front of the engine block
- Crankshaft sprocket with the mating mark facing out

16. Rotate the crankshaft clockwise and position the crankshaft to TDC of compression stroke and align the dowels of the camshaft sprockets to the 12 o'clock position.

- Lower chain guide on the dowel pin

with the front mark on the guide facing upward

- Timing chains and sprockets to the intake camshafts. Be sure to align the timing chain and sprocket mating marks.

17. Remove the left and right camshaft tensioner stopper pins.

18. Align the mating mark on the crankshaft with the matchmark (gold link) on the lower timing chain.

- Lower timing chain to the water pump sprocket

19. Working counterclockwise, install the lower timing chain camshaft sprockets. Be sure to align the sprocket marks with the blue links of the timing chain during installation.

20. Intake sprocket bolts and tighten to 88–95 ft. lbs. (119–128 Nm). Be sure to secure the camshafts while tightening the bolts

21. Timing chain guide, upper timing chain guide, lower timing chain tensioner and slack side timing chain guide

22. Timing cover evenly and gently. Be sure to align the dowel pin holes. Tighten the mounting bolts in sequence

- Front exhaust pipe and its support
- A/C compressor and bracket
- Crankshaft pulley to the crankshaft and install the mounting bolt

23. Torque the mounting bolt to 14–22 ft. lbs. (20–29 Nm). Torque the crankshaft bolt an additional 60–66° clockwise. This is approximately the angle from 1 hexagon bolt head corner to another.

- Ring gear cover plate

- CKP (PHASE) and CMP sensors
- Power steering pump assembly
- Drive belts and the idler pulley
- Right front inner wheel cover and the right front wheel
- Engine undercovers
- Intake manifold, using new gaskets. Torque the nuts and bolts in sequence.
- Intake manifold collector gasket with the arrow facing forward
- Intake manifold collector assembly and torque the mounting bolts to 16–18 ft. lbs. (22–25 Nm)
- Intake manifold collector support brackets
- EGR tube, using new gaskets and torque the mounting bolts to 15–20 ft. lbs. (21–26 Nm) in 2 progressive steps
- Spark plugs
- Ignition coils and torque the mounting bolts to 27–33 inch lbs. (3–4 Nm)
- Cylinder head cover ornament on the left side
- Water hoses to the cylinder head and intake manifold
- EVAP canister purge hoses
- Fuel hoses and wiring harness connections to the fuel rail
- Air duct-to-intake manifold hose, collector hose, blow-by hose, and vacuum hoses
- Negative battery cable

24. Fill the engine with clean oil.

25. Fill and bleed the cooling system.

26. Connect the negative battery cable.

27. Start the engine and run at 3000 rpm under no load to purge the air from the high pressure chamber. The engine may produce a rattling noise. This indicates that air still remains in the chamber and is not a matter of concern.

28. Inspect the vehicle for leaks and repair if necessary.

3.5L Engine

I35 MODELS

➡**For this procedure, you must remove the engine from the vehicle in order to remove the cylinder head.**

1. Before servicing the vehicle, refer to the precautions in the beginning of this section.

2. Relieve the fuel system pressure.

3. Drain the engine oil.

4. Drain the cooling system.

➡**Before detaching any hoses or connectors, note the locations for reassembly.**

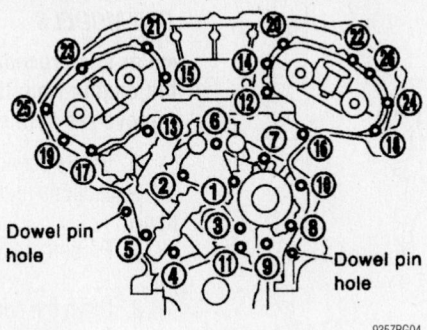

Loosen the rear timing chain case bolts in the reverse of this sequence—I35 with 3.5L engine

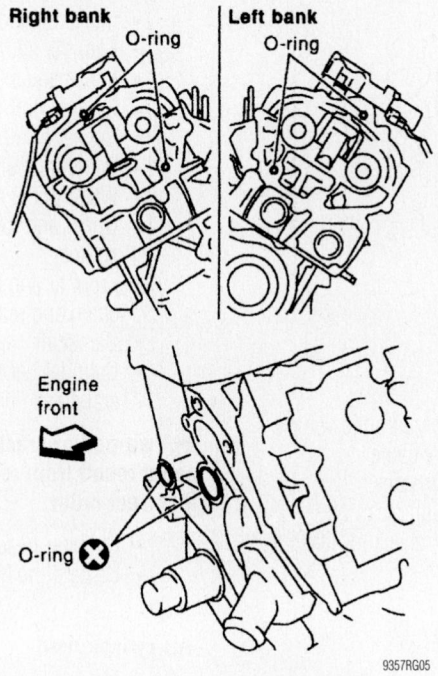

Remove the O-rings from the cylinder head and block— I35 with 3.5L engine

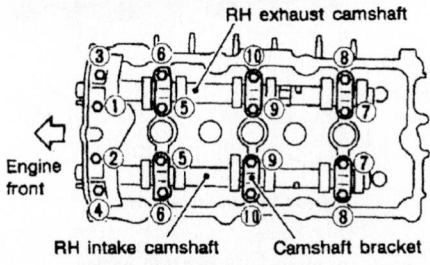

Loosen in numerical order.

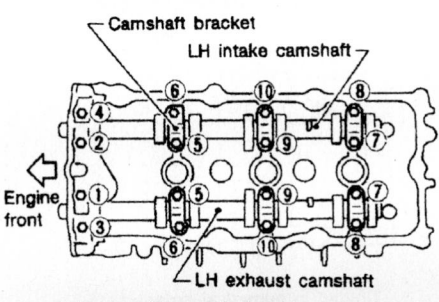

Loosen in numerical order.

Right and left camshaft bracket bolt loosening sequence—3.5L engine

5. Remove or disconnect the following:
- Negative battery cable
- Engine assembly
- Exhaust manifold

6. Place the engine on a suitable work-stand.
- Oil pan
- Timing chain
- Intake manifold
- Water outlet
- Rear timing chain case bolts, in the reverse of the sequence shown
- Rear timing chain case
- O-rings from the cylinder head and block
- Intake valve timing control solenoid valves

➡ **For installation purposes, matchmark the camshaft brackets before removing them.**

- Intake and exhaust camshafts and brackets. Loosen the bracket bolts in several steps, in the sequence shown.
- Right and left side cam chain tensioner from the cylinder head
- Cylinder head bolts. Loosen in several steps, in the sequence shown.

➡ **A warped or cracked cylinder head could result from removing the bolts in incorrect order.**

- Cylinder heads from the vehicle
- Discard the head gaskets

7. Remove all traces of liquid gasket from the timing chain case and from the water pump covers.

8. Remove all traces of liquid gasket from the engine block.

9. Inspect the timing chain for excessive wear or damage and replace as necessary.

To install:

10. Turn the crankshaft until the No. 1 piston is a Top Dead Center (TDC) on compression stroke. The crankshaft key should face toward the right bank.

11. Using new head gaskets, install the cylinder heads.

➡ **If possible, replacement of the head bolts is suggested.**

12. If replacement of the head bolts is not possible, perform the following bolt measurement:

a. Measure the diameter of the head bolt 0.43 in. (11mm) from the bottom of the bolt.

b. Measure the diameter of the head bolt 1.89 in. (48mm) from the bottom of the bolt.

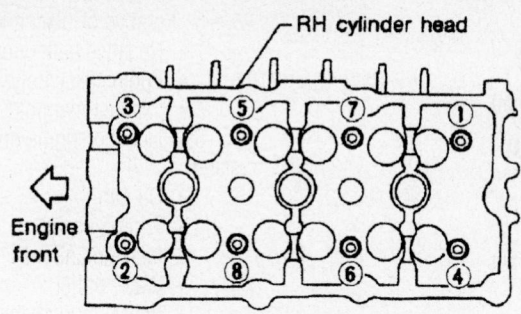

Loosen in numerical order.

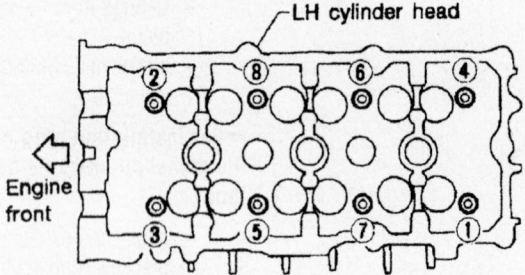

Loosen in numerical order.

Cylinder head bolt loosening sequence— I35 and G35 with 3.5L engine

c. Whenever the size difference between the 2 measurements exceeds 0.0043 in. (0.11mm) the head bolts must be replaced.

13. Install the cylinder head bolts and torque in sequence as follows:

a. Step 1: 72 ft. lbs. (98 Nm).

b. Step 2: Completely loosen all bolts.

c. Step 3: 26–32 ft. lbs. (34–44 Nm).

d. Step 4: plus 90–95 degrees clockwise.

e. Step 5: plus 90–95 degrees clockwise.

14. Install or connect the following:
- Camshafts and related components
- Intake valve timing control solenoid valves
- New O-rings to the front of the engine block and cylinder head

15. Apply sealant to the hatched portion of the of the rear timing chain case.

16. Align the rear timing chain case with the dowel pins and install onto the cylinder heads and engine block.

17. Torque the rear timing chain case mounting bolts in sequence to 105–121 inch lbs. (11.8–13.7 Nm).

18. Install or connect the following:
- Water outlet
- Intake manifold
- Timing chain
- Oil pan
- Exhaust manifold

- Engine assembly into the vehicle
- Negative battery cable

19. Fill the cooling system.

20. Fill the engine with clean oil.

21. Start the vehicle, check for leaks and repair if necessary.

G35 MODELS

➡For this procedure, you must remove the engine from the vehicle in order to remove the cylinder head.

1. Before servicing the vehicle, refer to the precautions in the beginning of this section.

2. Properly relieve the fuel system pressure.

3. Drain the cooling system.

4. Drain the engine oil.

5. Remove or disconnect the following:
- Engine and place on a suitable stand
- Intake manifold collector
- Fuel rail and injector assembly
- Intake and exhaust manifolds
- Ignition coil
- Rocker arm (valve) cover
- Water inlet and thermostat housing
- Water outlet and hoses
- Upper and lower oil pans and strainer
- Front timing chain case, timing chain and rear timing chain case
- Camshaft
- Cylinder head bolts. Loosen in several steps, in the sequence shown.

➡A warped or cracked cylinder head could result from removing the bolts in incorrect order.

- Cylinder heads from the vehicle
- Discard the head gaskets

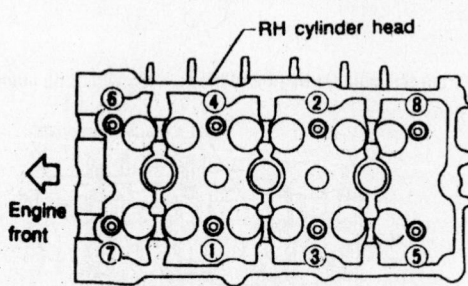

Tighten in numerical order.

Right cylinder head bolt torque sequence— I35 with 3.5L engine

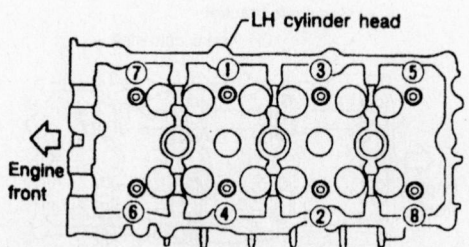

Tighten in numerical order.

Left cylinder head bolt torque sequence— I35 with 3.5L engine

6. Remove all traces of liquid gasket from the timing chain case and from the water pump covers.

7. Remove all traces of liquid gasket from the engine block.

8. Inspect the timing chain for excessive wear or damage and replace as necessary.

To install:

9. Turn the crankshaft until the No. 1 piston is a Top Dead Center (TDC) on compression stroke. The crankshaft key should face toward the right bank.

10. Using new head gaskets, install the cylinder heads.

➡**If possible, replacement of the head bolts is suggested.**

11. If replacement of the head bolts is not possible, perform the following bolt measurement:

a. Measure the diameter of the head bolt 0.43 in. (11mm) from the bottom of the bolt.

b. Measure the diameter of the head bolt 1.89 in. (48mm) from the bottom of the bolt.

c. Whenever the size difference between the 2 measurements exceeds 0.0043 in. (0.11mm) the head bolts must be replaced.

12. Install the cylinder head bolts and torque in sequence as follows:

a. Step 1: 72 ft. lbs. (98 Nm).

b. Step 2: Completely loosen all bolts, in the reverse of the tightening sequence.

c. Step 3: 26–32 ft. lbs. (34–44 Nm).

d. Step 4: plus 90–95 degrees clockwise.

e. Step 5: plus 90–95 degrees clockwise.

13. After installing the cylinder head, measure the distance between the front end faces of the cylinder block and cylinder head. If the specification does not fall within 0.555–0.587 in. (14.1–14.9mm), you must reinstall the cylinder head.

14. Install or connect the following:
- Camshaft
- Front timing chain case, timing chain and rear timing chain case
- Oil strainer and upper and lower oil pans
- Water outlet and hoses
- Thermostat housing and water inlet
- Rocker arm (valve) cover
- Ignition coil
- Intake and exhaust manifolds
- Fuel rail and injector assembly
- Intake manifold collector
- Engine into the vehicle
- Negative battery cable

15. Fill the cooling system.

16. Fill the engine with clean oil.

17. Start the vehicle, check for leaks and repair if necessary.

4.1L Engine

1. Before servicing the vehicle, refer to the precautions in the beginning of this section.

2. Properly relieve the fuel system pressure.

3. Drain the cooling system.

4. Drain the engine oil.

5. Remove or disconnect the following:
- Both battery cables
- Engine assembly from the vehicle and place it on a workstand
- Exhaust manifold
- Drain plugs on the sides of the engine and drain the coolant
- Intake manifold collector
- Ignition coil sub-harness
- Ignition coils
- Spark plugs
- Fuel rail with injectors. Do not disassemble the fuel hose.
- Intake manifold
- Rocker arm covers and bring the No. 1 piston to Top Dead Center (TDC) on the compression stroke
- Camshaft Position (CMP) sensor
- Right and left upper front covers. Be sure to remove the bolts in the correct sequence to prevent damage to the cover.
- Upper chain tensioners

6. Apply paint marks on the upper timing chains, camshaft and idler sprockets so they can be installed in their original positions.
- Camshaft sprocket
- Idler sprocket bolt
- Cylinder head sub-bolts. The bolts

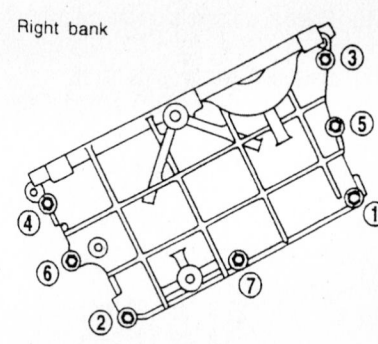

Right bank

7923HG19

Upper front cover bolt removal sequence—4.1L engine, right bank

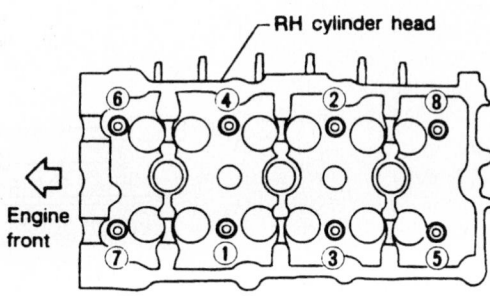

RH cylinder head

Engine front

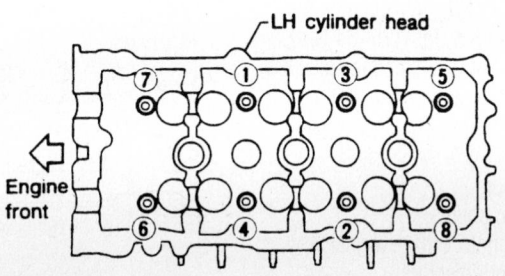

LH cylinder head

Engine front

9357MG05

Cylinder head bolt torque sequence— G35 with 3.5L engine

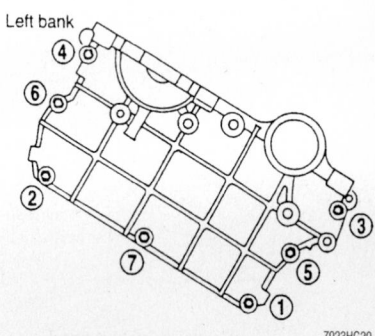

Left bank

7923HG20

Upper front cover bolt removal sequence—4.1L engine, left bank

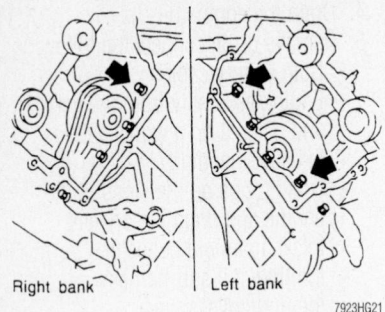

Be sure to install the long bolts in the positions indicated by the arrows—4.1L engine

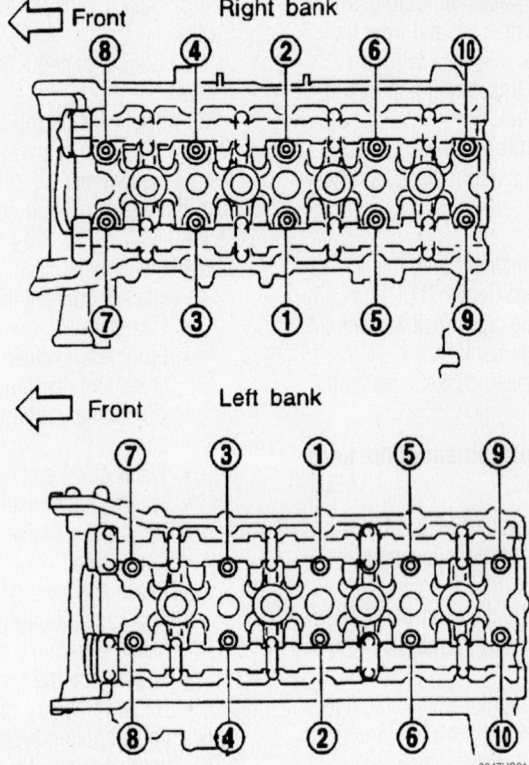

Cylinder head torque sequence—4.1L engine

are different lengths, note their differences so they can be installed in their original positions.

7. Loosen the cylinder head bolts gradually in the proper removal sequence, then remove the cylinder head.

To install:

8. Be sure all mating surfaces are clean before installation.

9. Check the cylinder head surface for warpage using a feeler gauge and a suitable straightedge. If the cylinder head is warped more than 0.004 in. (0.1mm), it must be resurfaced or replaced. The total amount machined from the head or head and block combined, cannot total more than 0.008 in. (0.2mm).

10. Place new gaskets on the cylinder block.

11. Be sure the knock pins on the camshafts are in the positions shown. Carefully place the cylinder heads on the engine. Do not damage the head gasket.

12. Lubricate the cylinder head bolt threats and seat surfaces with engine oil. Torque the bolts in sequence using the following sub-steps:

   a. Step 1: 22 ft. lbs. (29 Nm).
   b. Step 2: 69 ft. lbs. (93 Nm).
   c. Step 3: loosen the bolts completely.
   d. Step 4: 18–25 ft. lbs. (25–35 Nm).
   e. Step 5: Plus 90–95°.

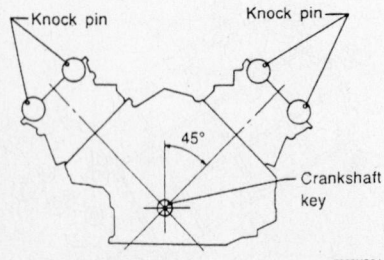

Be sure to position the camshaft knock pins as shown before installing the cylinder heads—4.1L engine

13. Install or connect the following:
- Cylinder head sub-bolts. Torque the bolts to 56–74 inch lbs. (6–8 Nm). Be sure the long bolts are returned to the positions shown.
- Idler and camshaft sprockets
- Chain tensioners

14. Using new gaskets, install the cylinder head covers. Be sure to apply RTV silicone sealant to the gasket arch and rubber plugs.

15. Torque the cylinder head cover bolts in sequence using the following sub-steps:
   a. Step 1: Nos. 1 through 16: 35–52 inch lbs. (4–6 Nm).
   b. Step 2: Nos. 1 through 16: 61–78 inch lbs. (7–9 Nm).
   c. Step 3: Nos. 1 and 2: 61–78 inch lbs. (7–9 Nm).

16. Install or connect the following:
- Intake valve timing control solenoid with a new O-ring. Torque the solenoid to 18–25 ft. lbs. (25–34 Nm).

17. Apply a bead of RTV silicone sealant to the upper front covers, then install them. Torque the bolts to 56–74 inch lbs. (6–8 Nm).
- CMP sensor
- Rocker arm covers
- Intake manifold with new gaskets
- Fuel rail and injectors
- Spark plugs and ignition coils

- Intake manifold collector
- Exhaust manifold with new gaskets
- Engine assembly to the vehicle
- Battery and cables

18. Fill and bleed the cooling system.

19. Fill the engine with clean oil.

20. Start the vehicle, check for leaks and repair if necessary.

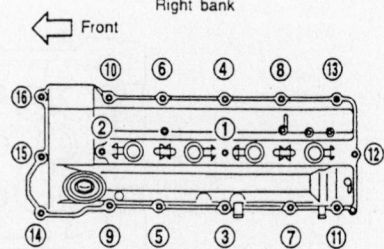

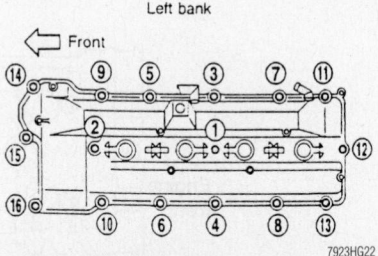

Cylinder head cover bolt tightening sequence—4.1L engine

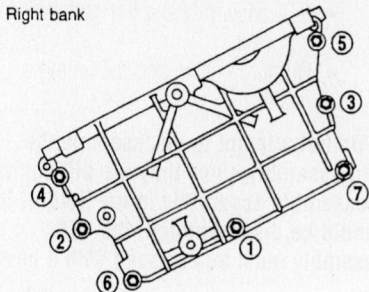

Right bank

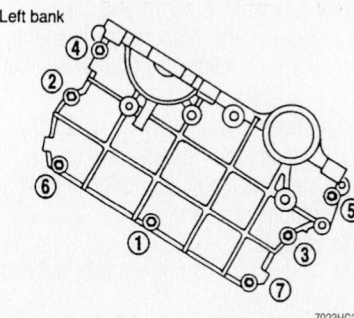

Left bank

7923HG23

Torque the upper front cover bolts in the sequence shown—4.1L engine

4.5L Engine

1. Before servicing the vehicle, refer to the precautions in the beginning of this section.

2. Properly relieve the fuel system pressure.

3. Drain the cooling system.

4. Drain the engine oil.

5. Remove or disconnect the following:
   - Both battery cables
   - Engine assembly from the vehicle and place it on a workstand
   - Drive belt and idler pulley
   - Thermostat housing
   - Oil pan and strainer
   - Upper and lower intake manifolds
   - Fuel rail and injector assembly
   - Ignition coils
   - Spark plugs
   - Valve cover
   - Timing chain
   - Camshafts

6. Loosen the cylinder head bolts gradually in reverse of the tightening sequence, then remove the cylinder head.

To install:

7. Be sure all mating surfaces are clean before installation.

8. Check the cylinder head surface for warpage using a feeler gauge and a suitable straightedge. If the cylinder head is warped more than 0.004 in. (0.1mm), it must be resurfaced or replaced. The total amount machined from the head or head and block combined, cannot total more than 0.008 in. (0.2mm).

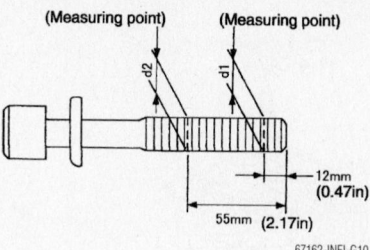

(Measuring point)   (Measuring point)

12mm (0.47in)
55mm (2.17in)

67162-INFI-G10

Checking the cylinder bolt threads for distortion—4.5L engine

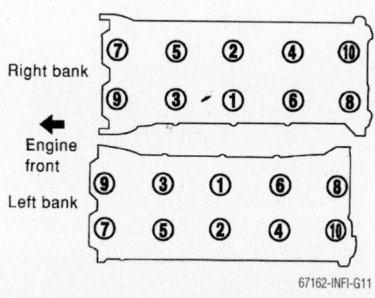

Right bank

Engine front

Left bank

67162-INFI-G11

Cylinder head bolt tightening sequence—4.5L engine

9. If the cylinder head bolts are to be reused, measure the outside diameter of the bolt threads for distortion. If the dimension between d1 and d2 exceeds 0.0071 inch (0.18 mm), replace the bolts.

10. Place new gaskets on the cylinder block.

11. Apply clean engine oil to the cylinder head bolt threads.

12. Install the cylinder head and tighten the bolts in sequence to 72 ft. lbs. (98 Nm).

13. Loosen all bolts in sequence to zero.

14. Tighten all bolts in sequence again to 33 ft. lbs. (44 Nm), then tighten them an additional 60 degrees, then an additional 60 degrees.

15. Install or connect the following:
   - Camshafts
   - Timing chain
   - Valve cover
   - Spark plugs
   - Ignition coils
   - Fuel rail and injector assembly
   - Upper and lower intake manifolds
   - Oil pan and strainer
   - Thermostat housing
   - Drive belt and idler pulley
   - Engine assembly
   - Both battery cables

16. Fill and bleed the cooling system.

17. Fill the engine with clean oil.

18. Start the vehicle, check for leaks and repair if necessary.

Rocker Arms/Shafts

REMOVAL & INSTALLATION

2.0L Engine

1. Before servicing the vehicle, refer to the precautions in the beginning of this section.

2. Relieve the fuel system pressure.

3. Drain the cooling system.

4. Remove or disconnect the following:

   - Negative battery cable
   - Right front wheel
   - Engine side cover
   - Radiator
   - Air duct to the intake manifold
   - Drive belts, water pump pulley, alternator and power steering pump
   - Vacuum hoses, fuel hoses and wiring harness connectors
   - Spark plugs, the Air Intake Valve (AIV) and resonator
   - Rocker cover and oil separator. Loosen rocker cover bolts, using 2 to 3 steps, in the opposite sequence of tightening.
   - Intake manifold supports
   - Oil filter bracket and power steering oil pump bracket

5. Set No. 1 piston at Top Dead Center (TDC) on the compression stroke by rotating the crankshaft.

   - Chain tensioner
   - Distributor. Do not turn the rotor with the distributor removed.
   - Timing chain guide, camshaft sprockets, camshafts, brackets, oil tubes and baffle plate. The camshaft bracket bolts must be loosened in sequence to prevent damage to the camshafts or the head.
   - Rocker arm assembly

To install:

6. Check the hydraulic lash adjusters to ensure they did not bleed down during disassembly by trying to compress them. If the lash adjuster can be compressed 0.04 in. (1mm), air has entered and it must be bleed.

➡**Air cannot be bled from the lash adjusters by running the engine.**

7. Clean the camshaft end bracket and coat with liquid gasket. Install the camshafts, camshaft brackets, oil tubes and baffle plate. Ensure the left camshaft key is at 12 o'clock and the right camshaft key is at 10 o'clock.

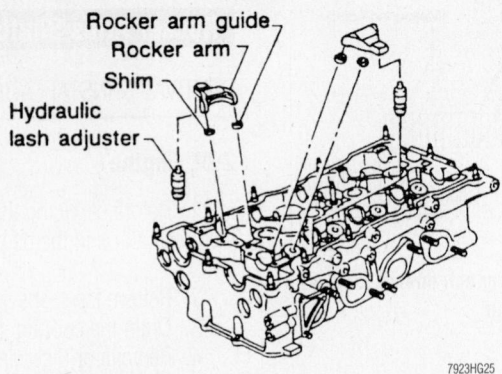

Exploded view of the rocker arms and related components—2.0L engine

➡The procedure for tightening camshaft bracket bolts must be followed exactly to prevent camshaft damage.

8. Line up the mating marks on the timing chain and camshaft sprockets and install the sprockets. Torque the sprocket bolts to 101–116 ft. lbs. (137–157 Nm).

9. Install or connect the following:
- Timing chain guide, distributor (ensure that rotor head is at 5 o'clock position) and chain tensioner
- Intake manifold supports. Clean the rocker cover and mating surfaces and apply a continuous bead of liquid gasket to the mating surface.
- Rocker cover and oil separator. Tighten the rocker cover bolts in sequence.
- Oil filter bracket and the power steering pump bracket
- Spark plugs, AIV valve and the resonator
- Fuel lines, vacuum hoses and wiring connectors
- Water pump pulley, alternator and power steering pump

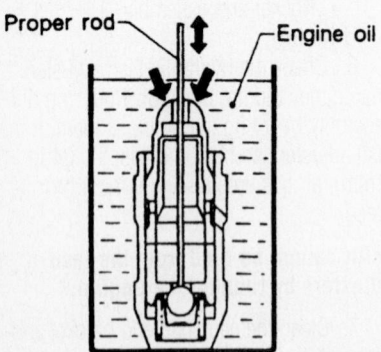

Submerge the lash adjuster in engine oil, lightly unseat the check ball with a thin rod and push on the plunger to release the air

- Drive belts
- Intake manifold air duct, engine side cover and right wheel
- Radiator
- Negative battery cable

10. Fill and bleed the cooling system.

11. Start the vehicle, check for leaks and repair if necessary.

3.0L, 3.5L and 4.5L Engines

➡The valves in the 3.0L, 3.5L and 4.5L engines are actuated directly by the camshaft. No rocker arms are used in these engines.

4.1 Engines

1. Before servicing the vehicle, refer to the precautions in the beginning of this section.

2. Remove or disconnect the following:
- Negative battery cable
- Engine and transmission assembly from the vehicle
- Suspension member and engine mounts from the engine
- Air compressor bracket
- Cooling fan with coupling and the engine gusset

3. Separate the engine from the transmission and mount the engine on a suitable workstand.
- Oil pan
- Crank angle sensor and the Valve Timing Control (VTC) solenoid
- Chain tensioners and the upper front covers
- Front timing chain cover

➡The timing chain will not be disengaged or dislocated from the crankshaft sprocket unless the front cover is removed. The cast portion of the front cover is located on the lower side of the crankshaft sprocket so the timing chain is not disengaged from the sprocket.

- VTC assembly and the camshaft sprocket
- Oil pump chain and the timing chains

➡Do not attempt to disassemble the VTC assembly since they are difficult to reassemble accurately in the field. If it should be disassembled, the VTC assembly must be replaced with a new one.

- Camshaft brackets and the camshafts. Mark the parts so they can be reinstalled in their original positions.
- Rocker arms. Be sure to identify each rocker arm so they can be reinstalled in their original positions.

To install:

4. Be sure all mating surfaces are clean before installation.

5. Install or connect the following:
- Rocker arms, camshafts and camshaft brackets on the right bank. Properly lubricate the rocker arms and camshafts prior to installation.
- VTC assembly and the exhaust cam sprocket on the right bank

6. Be sure the camshafts are still correctly positioned and the piston in the No. 1 cylinder is still at Top Dead Center (TDC).
- Timing chain on the right bank, aligning the mating marks on the chain with those on the crankshaft and camshaft sprockets
- Chain tensioner on the right bank

7. Turn the crankshaft approximately 120° clockwise from the point where the No. 1 piston is at TDC on the compression stroke. At this point, the valves on the left bank still remain closed.

8. Correctly position the camshafts and rocker arms for the left cylinder head. Properly lubricate the rocker arms and camshafts prior to installation. Install the VTC assembly and the exhaust cam sprocket.

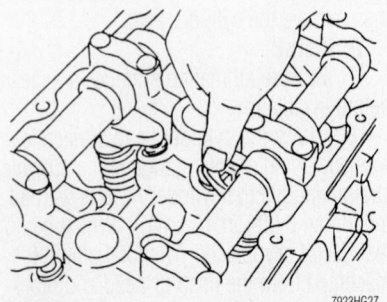

Press down on the lash adjuster to check for bleed-down—4.1L engines

9. Install the timing chain on the left bank, aligning the mating marks on the chain with those on the crankshaft and camshaft sprockets.

10. Install or connect the following:
- Oil pump chain and sprockets
- Oil pump chain guides. Place a 0.04 in. (1.0mm) feeler gauge between the upper chain guide and chain before assembling the chain guides. The force applied to the chain is equivalent to the upper chain guide weight.

11. Apply suitable sealer and install the front covers.
- Chain tensioner for the left bank

12. Apply suitable sealer to the rubber plugs and install them on the cylinder head.
- Crank angle sensor, VTC solenoid, rocker cover and crank pulley
- Transmission on the engine and install the engine assembly in the vehicle

Intake Manifold

REMOVAL & INSTALLATION

2.0L Engine

1. Before servicing the vehicle, refer to the precautions in the beginning of this section.

2. Properly relieve the fuel system pressure.

3. Drain the cooling system.

4. Remove or disconnect the following:
- Negative battery cable
- Fuel lines, vacuum hoses and electrical connectors
- Throttle linkage
- Intake manifold supports from the front and rear
- Intake manifold collector. Loosen the bolts in the sequence illustrated.
- Injector tube assembly

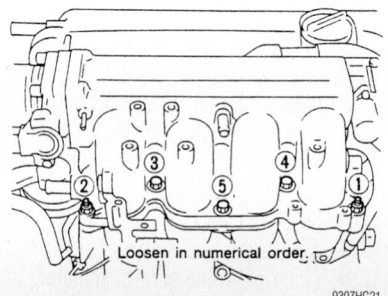

Intake manifold collector bolt loosening sequence—2.0L engine

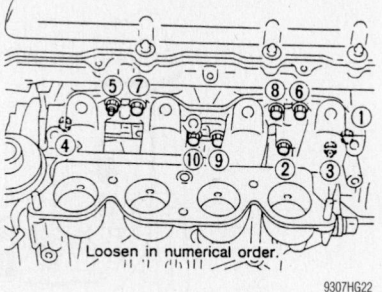

Intake manifold bolt loosening sequence—2.0L engine

- Power steering oil pump and the oil filter bracket
- Intake manifold-to-cylinder head bolts. Loosen the bolts in the sequence illustrated.
- Intake manifold and discard the gasket

To install:

5. Be sure all mating surfaces are clean prior to installation.

6. Fit a new gasket and the manifold into place. Start the support bolts to hold the manifold in place.

7. Install or connect the following:
- Intake manifold bolts. Torque the bolts in sequence to 13–15 ft. lbs. (18–21 Nm). Torque the bolts in 2 steps, starting at the center and working towards the ends.
- Injector tube assembly. Torque the bolts first to 84–96 inch lbs. (9–11 Nm), then to 15–20 ft. lbs. (21–26 Nm).
- Power steering oil pump and the oil filter bracket
- Intake manifold collector using a new gasket. Torque the bolts in sequence to 13–15 ft. lbs. (18–21 Nm).
- Fuel lines, vacuum hoses, electrical connectors and the throttle linkage
- Negative battery cable

8. Fill and bleed the cooling system.

9. Start the vehicle, check for leaks and repair if necessary.

3.0L and 3.5L Engines

EXCEPT G35 MODELS

1. Before servicing the vehicle, refer to the precautions in the beginning of this section.

2. Release the fuel system pressure.

3. Drain the cooling system.

4. Remove or disconnect the following:
- Negative battery cable
- Throttle body coolant hoses
- Electrical connectors from the Throttle Position (TP) sensor

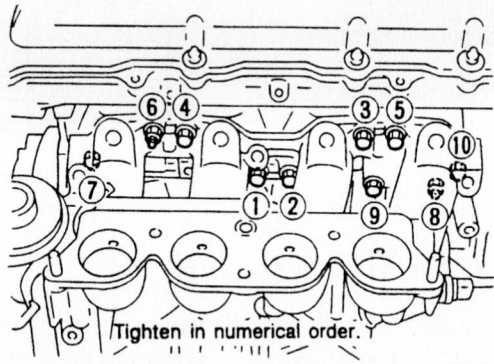

Lower intake manifold torque sequence—2.0L engine

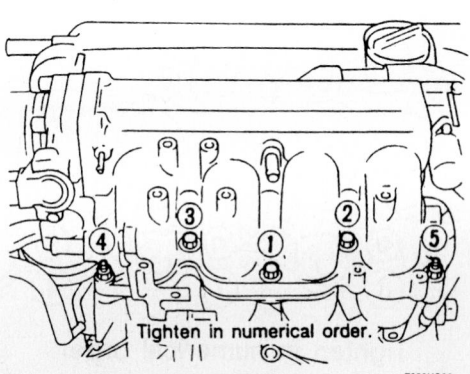

Upper intake manifold torque sequence—2.0L engine

- Hoses from the throttle body, the Exhaust Gas Recirculation (EGR) valve, intake manifold collector, Idle Air Control (IAC) valve, and the fuel pressure regulator
- Evaporative emissions (EVAP) canister purge hose and blow-by hose
- EGR guide tube
- Accelerator cable from the throttle body
- Intake manifold collector support brackets
- Right side electrical connectors from the ignition coils
- Electrical connector from the crank angle sensor and the power transistor, if necessary
- Intake manifold collector-to-intake manifold bolts/nuts and remove the intake manifold collector

5. Fuel injector assembly by removing or disconnecting the following:
- Electrical connectors from the fuel injectors
- Fuel lines from the fuel injector assembly
- Fuel rail-to-cylinder head bolts
- Fuel rail assembly from the engine
- Intake manifold bolts/nuts in the reverse sequence of the torque procedure

6. Remove the intake manifold from the engine and discard the gaskets
7. Clean all gasket mounting surfaces.

To install:

8. Using new gaskets, install the intake manifold to the engine.
9. Tighten the bolts/nuts in sequence as follows:
   a. Step 1: Tighten nuts and bolts to 44–86 inch lbs. (5–10 Nm).
   b. Step 2: Tighten nuts and bolts to 20–23 ft. lbs. (26–31 Nm).
   c. Step 3: Repeat step 2 at least 5 times until all nuts and bolts have a final torque of 20–23 ft. lbs. (26–31 Nm).

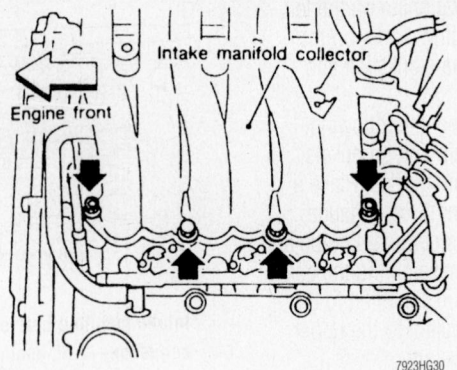

Upper intake manifold torque sequence—3.0L and 3.5L engine, except G35

10. Install the fuel injector assembly by installing or connecting the following:
- Fuel rail assembly to the engine

11. Install the fuel rail-to-cylinder head bolts and torque the bolts to 15–20 ft. lbs. (21–26 Nm) in the following sequence:
   a. Step 1: Tighten bolts in sequence to 84–96 inch lbs. (9–10 Nm).
   b. Step 2: Tighten the bolts in sequence to 15–20 ft. lbs. (21–26 Nm).
- Fuel lines to the fuel injector assembly
- Electrical connectors to the fuel injectors

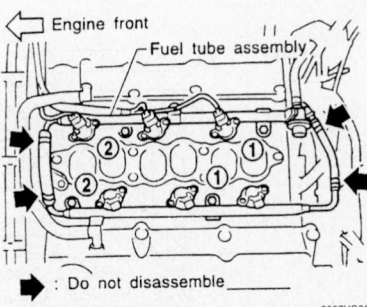

Tighten the fuel rail (tube) bolts in the sequence illustrated—3.0L and 3.5L engines, except G35

- Intake manifold collector using a new gasket, and torque the intake manifold collector-to-intake manifold bolts/nuts to 13–16 ft. lbs. (18–22 Nm) in the sequence illustrated
- Intake manifold collector supports. Torque the bolts to 14–18 ft. lbs. (20–25 Nm).
- Electrical connector to the crank angle sensor and the power transistor, if disconnected
- Electrical connectors to the ignition coils and torque the mounting bolts to 27–33 inch lbs. (3–4 Nm)
- Accelerator cable to the throttle body
- EGR guide tube. Torque the bolts to 15–20 ft. lbs. (21–26 Nm) in 2 progressive steps
- EVAP canister purge hose and blow-by hose
- Hoses to the throttle body, EGR valve, intake manifold collector, IAC valve, and the fuel pressure regulator
- Electrical connectors to the TP sensor
- Throttle body coolant hoses
- Negative battery cable

12. Fill and bleed the cooling system.
13. Start the vehicle, check for leaks and repair if necessary.

G35 MODELS

1. Before servicing the vehicle, refer to the precautions in the beginning of this section.
2. Release the fuel system pressure.
3. Drain the cooling system.
4. Remove or disconnect the following:
- Negative battery cable
- Engine cover
- Air cleaner and duct
- Electric throttle control actuator, loosening the bolts in a criss-cross pattern

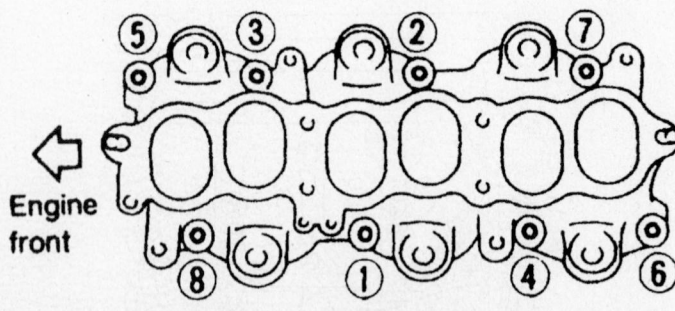

Tighten in numerical order.

Lower intake manifold torque sequence—3.0L and 3.5L engine, except G35

- Fuel sub-tube mounting bolt to disconnect it from the rear of the lower intake manifold (collector)
- Vacuum hose and water hose from the intake manifold collector
- Evaporative emissions (EVAP) canister purge volume control solenoid valve bracket mounting bolt from the intake manifold collector
- Intake manifold collector (upper) bolts in the sequence shown, then remove the collector
- Positive Crankcase Ventilation (PCV) hose between the intake manifold collector and right side rocker arm cover

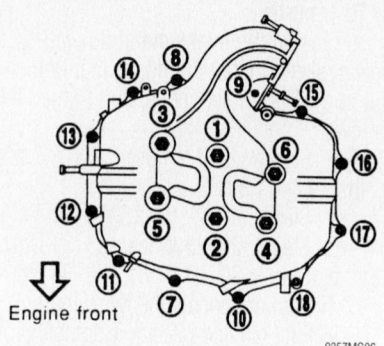

Intake manifold collector (upper) bolt loosening sequence—G35 models

9357MG06

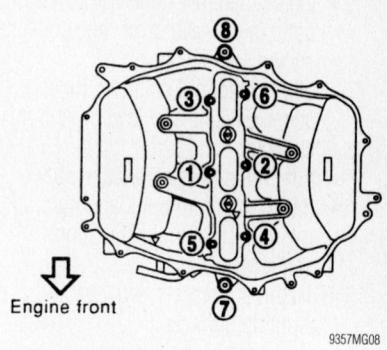

Intake manifold collector (lower) bolt tightening sequence. Use the reverse sequence for bolt removal—G35 models

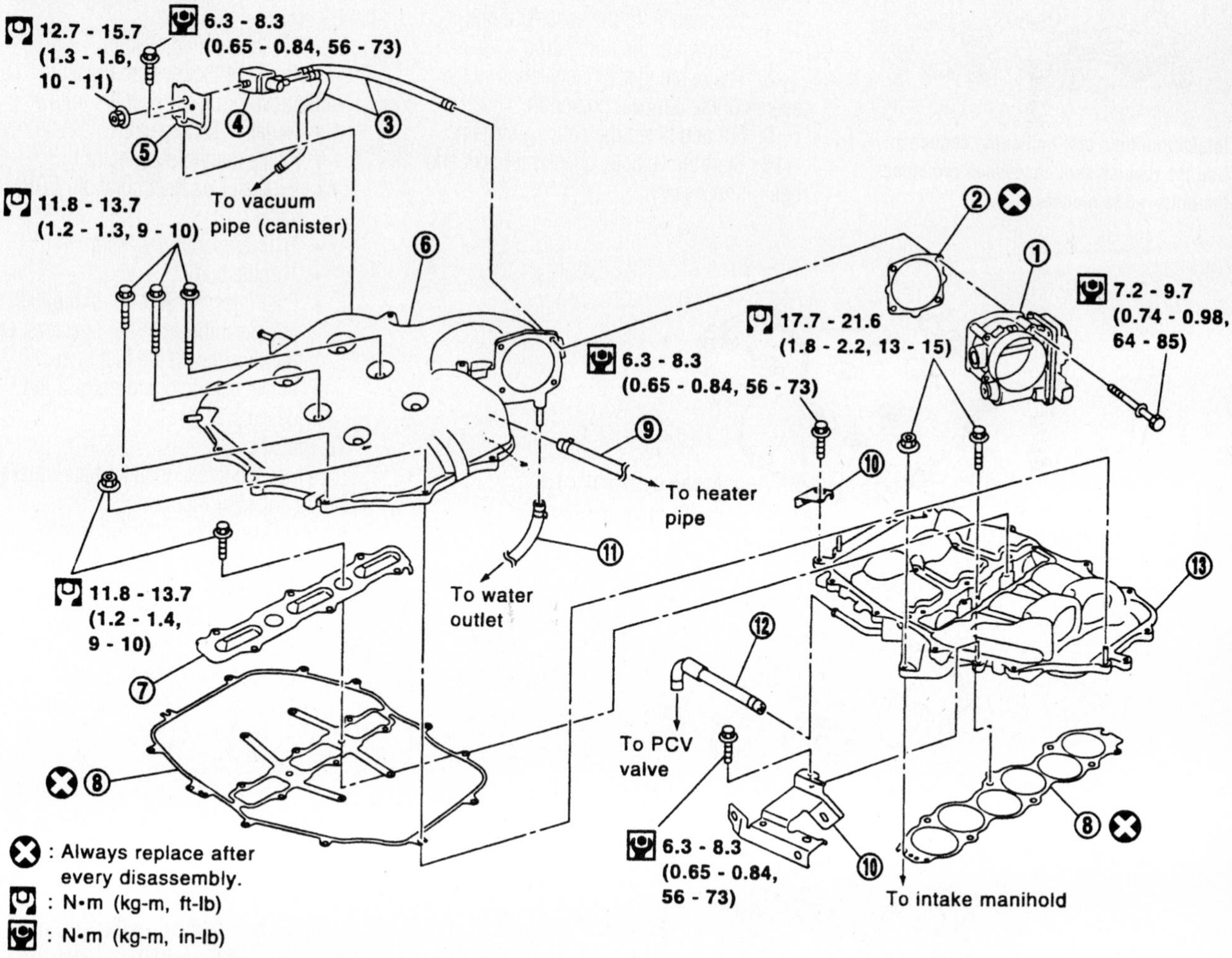

9357MG08

SEC. 140•147•163

12.7 - 15.7 (1.3 - 1.6, 10 - 11)

6.3 - 8.3 (0.65 - 0.84, 56 - 73)

11.8 - 13.7 (1.2 - 1.3, 9 - 10)

To vacuum pipe (canister)

17.7 - 21.6 (1.8 - 2.2, 13 - 15)

7.2 - 9.7 (0.74 - 0.98, 64 - 85)

6.3 - 8.3 (0.65 - 0.84, 56 - 73)

To heater pipe

11.8 - 13.7 (1.2 - 1.4, 9 - 10)

To water outlet

To PCV valve

6.3 - 8.3 (0.65 - 0.84, 56 - 73)

To intake manifold

✕ : Always replace after every disassembly.

▣ : N•m (kg-m, ft-lb)

▣ : N•m (kg-m, in-lb)

1. Electric throttle control actuator	2. Gasket	3. Vacuum hose
4. EVAP canister purge volume control solenoid valve	5. Bracket	6. Intake manifold collector (upper)
7. Intake manifold collector cover	8. Gasket	9. Water hose
10. Bracket	11. Water hose	12. PCV hose
13. Intake manifold collector (lower)		

Exploded view of the upper and lower intake manifold collectors and tightening specifications—G35 models

9357MG07

- Intake manifold collector (lower) bolts in the reverse of the sequence shown
- Intake manifold collector (lower) cover, gaskets and lower manifold collector
- Fuel rail and injector assembly
- Intake manifold bolts and nuts, in the reverse of the installation sequence

5. Discard all gaskets and clean all gasket mounting surfaces.

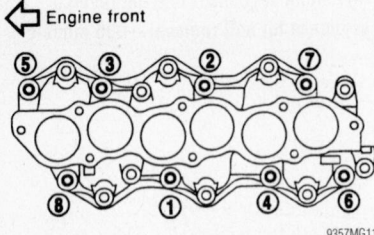

Intake manifold bolt tightening sequence. Use the reverse sequence when removing the bolts—G35 models

To install:

6. Install the intake manifold, with new gaskets. Install the intake manifold nut and bolts, in sequence, and tighten as follows:

  a. Stud bolts, if removed: 7.3–8.7 ft. lbs. (9.8–11.8 Nm).

  b. Step 1: 4–7 ft. lbs. (5–10 Nm).

  c. Step 2: 20–23 ft. lbs. (26–31 Nm).

  d. Step 3: 20–23 ft. lbs. (26–31 Nm).

7. Install or connect the following:

- Fuel rail and injector assembly
- New gaskets, lower manifold collector, and collector cover, as shown. Tighten the bolts, in sequence to 13–15 ft. lbs. (17.7–21.6 Nm).
- Intake manifold collector (upper). Torque the bolts and nuts, in sequence, to the specifications shown in the illustration.

8. The remainder of installation is the reverse of the removal procedure.

9. Fill and bleed the cooling system.

10. Start the vehicle, check for leaks and repair if necessary.

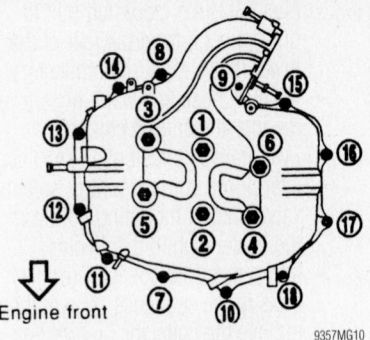

Intake manifold collector (upper) bolt tightening sequence—G35 models

4.1L Engine

1. Before servicing the vehicle, refer to the precautions in the beginning of this section.

2. Drain the cooling system.

3. Relieve the fuel system pressure.

4. Remove or disconnect the following:

- Negative battery cable
- Intake manifold collector
- Exhaust Gas Recirculation (EGR) valve
- EGR temperature sensor
- Throttle body
- Fuel injectors and lift up the fuel rail assembly with the injectors. Do not disconnect the fuel hose.
- Intake manifold and discard the gaskets

To install:

5. Clean the intake manifold and intake manifold collector mounting surfaces

6. Install or connect the following:

- Intake manifold using new gaskets. Torque the bolts in sequence to 13–17 ft. lbs. (18–21 Nm).
- Fuel tube assembly. Torque the bolts in sequence first to 84–96 inch lbs. (9–11 Nm), then to 15–20 ft. lbs. (21–26 Nm).
- Throttle body

7. Torque the throttle body bolts in the following steps:

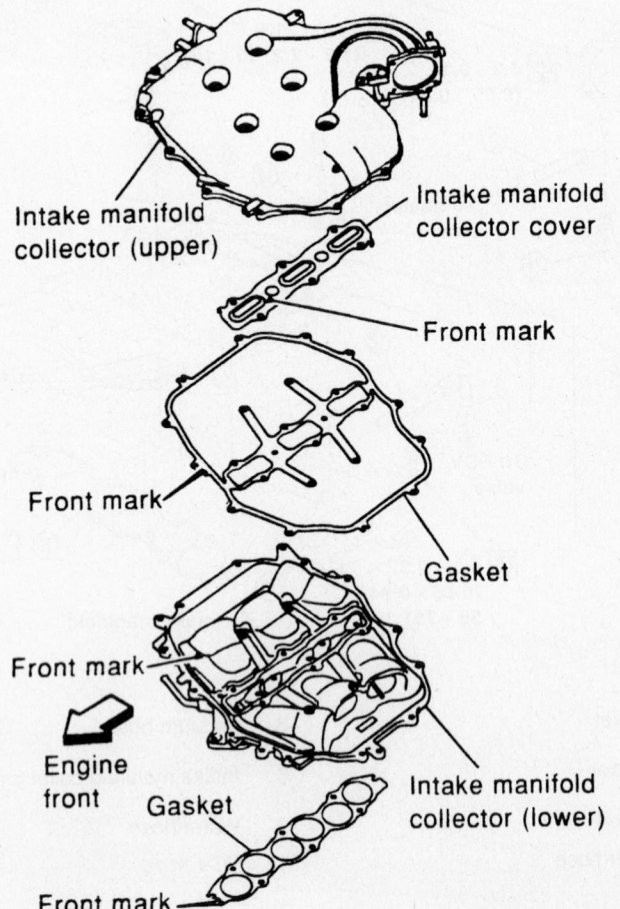

Exploded view of the intake manifold collector installation—G35 models

Intake manifold collector (upper)

Intake manifold collector cover

Front mark

Front mark

Gasket

Front mark

Engine front

Gasket

Intake manifold collector (lower)

Front mark

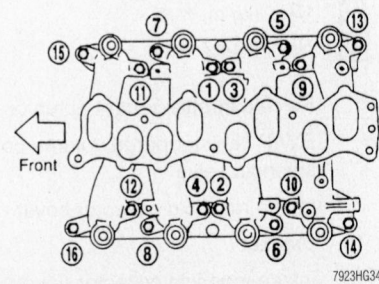

Lower intake manifold torque sequence—4.1L engine

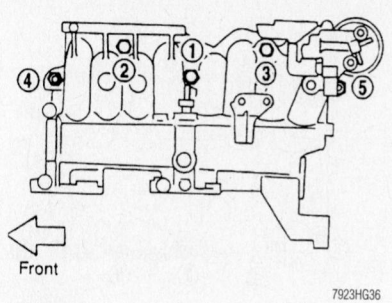

a. Step 1: bolts in sequence to 80–96 inch lbs. (9–11 Nm).

b. Step 2: Bolts in sequence to 13–16 ft. lbs. (18–22 Nm).

- EGR temperature sensor
- EGR valve
- Intake manifold collector using a new gasket. Torque the bolts in sequence to 13–16 ft. lbs. (18–22 Nm).
- Negative battery cable

8. Fill and bleed the cooling system.

9. Start the vehicle, check for leaks and repair if necessary.

Upper intake manifold torque sequence— 4.1L engine

7923HG36

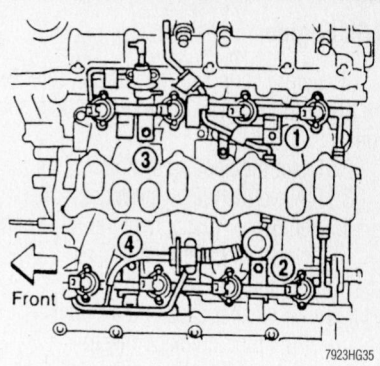

7923HG35

To prevent possible tube breakage, be sure to tighten the fuel tube mounting bolts according to the sequence shown

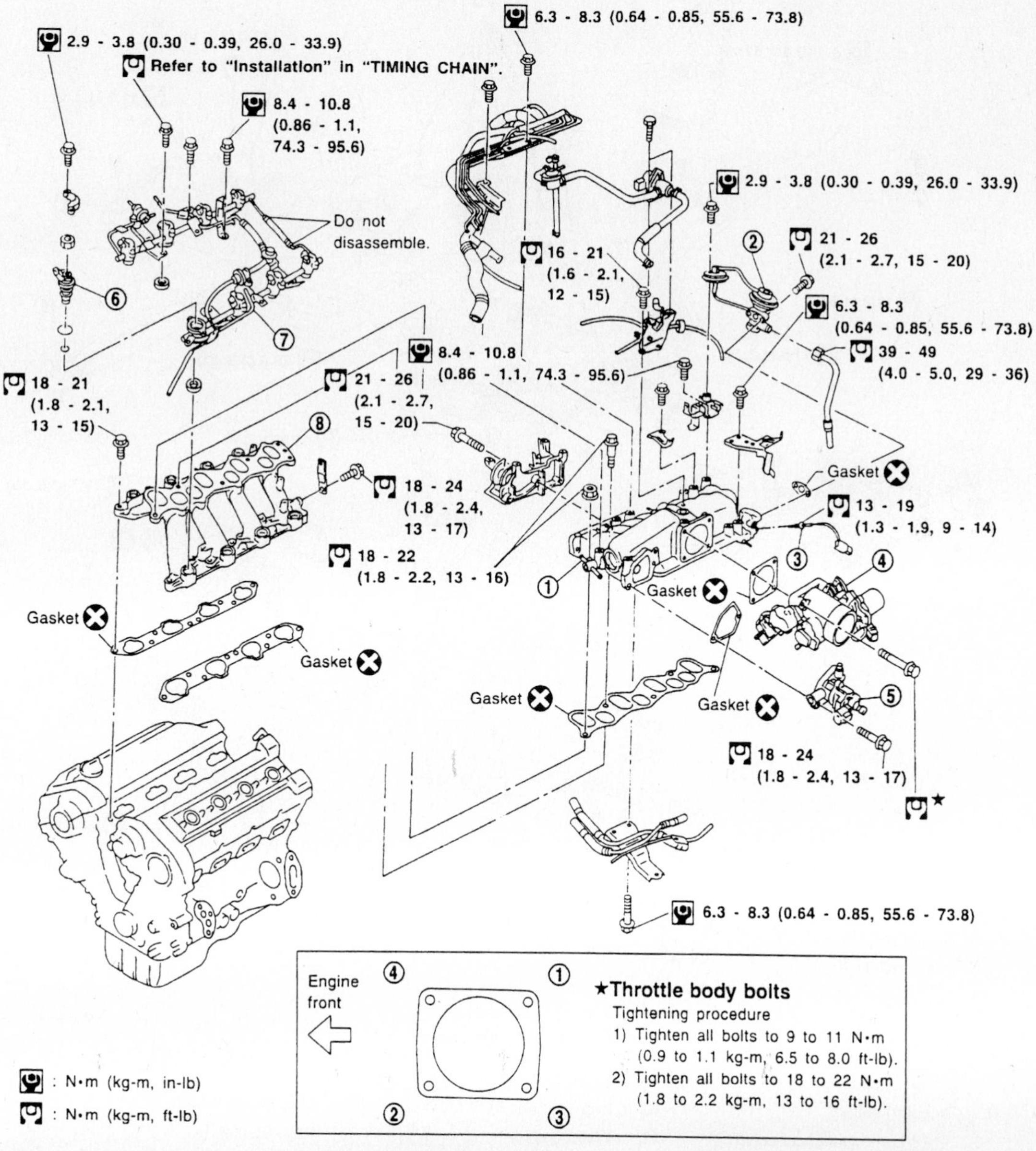

2.9 - 3.8 (0.30 - 0.39, 26.0 - 33.9)

Refer to "Installation" in "TIMING CHAIN".

8.4 - 10.8 (0.86 - 1.1, 74.3 - 95.6)

6.3 - 8.3 (0.64 - 0.85, 55.6 - 73.8)

Do not disassemble.

2.9 - 3.8 (0.30 - 0.39, 26.0 - 33.9)

21 - 26 (2.1 - 2.7, 15 - 20)

16 - 21 (1.6 - 2.1, 12 - 15)

6.3 - 8.3 (0.64 - 0.85, 55.6 - 73.8)

8.4 - 10.8 (0.86 - 1.1, 74.3 - 95.6)

39 - 49 (4.0 - 5.0, 29 - 36)

18 - 21 (1.8 - 2.1, 13 - 15)

21 - 26 (2.1 - 2.7, 15 - 20)

Gasket ⊗

18 - 24 (1.8 - 2.4, 13 - 17)

13 - 19 (1.3 - 1.9, 9 - 14)

18 - 22 (1.8 - 2.2, 13 - 16)

Gasket ⊗

Gasket ⊗

Gasket ⊗

Gasket ⊗

18 - 24 (1.8 - 2.4, 13 - 17)

★

6.3 - 8.3 (0.64 - 0.85, 55.6 - 73.8)

Engine front

★**Throttle body bolts**
Tightening procedure
1) Tighten all bolts to 9 to 11 N·m (0.9 to 1.1 kg-m, 6.5 to 8.0 ft-lb).
2) Tighten all bolts to 18 to 22 N·m (1.8 to 2.2 kg-m, 13 to 16 ft-lb).

: N·m (kg-m, in-lb)

: N·m (kg-m, ft-lb)

① Intake manifold collector   ④ Throttle body   ⑦ Fuel tube assembly
② EGR valve   ⑤ IACV-AAC valve   ⑧ Intake manifold
③ EGR temperature sensor   ⑥ Injector

9307HG07

Exploded view of the intake manifold assembly, related components and the throttle body torque sequence—4.1L engines

4.5L Engine

1. Before servicing the vehicle, refer to the precautions in the beginning of this section.
2. Drain the cooling system.
3. Relieve the fuel system pressure.
4. Remove or disconnect the following:
   - Negative battery cable
   - Engine appearance cover
   - Air cleaner case and air duct
   - Fuel tube quick connector
   - Accelerator cable
   - Wiring harnesses, bracket, vacuum

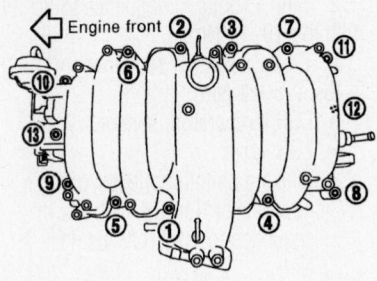

Lower intake manifold torque sequence—4.5L engine

67162-INFI-G12

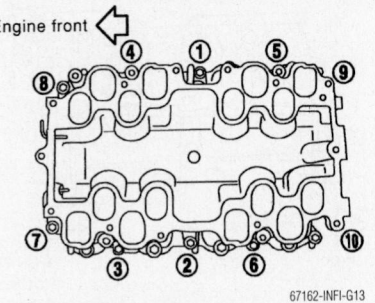

Upper intake manifold torque sequence—4.5L engine

67162-INFI-G13

Legend:

- ⊗ : Always replace after every disassembly.
- 🔧 : N·m (kg-m, ft-lb)
- 🔧 : N·m (kg-m, in-lb)

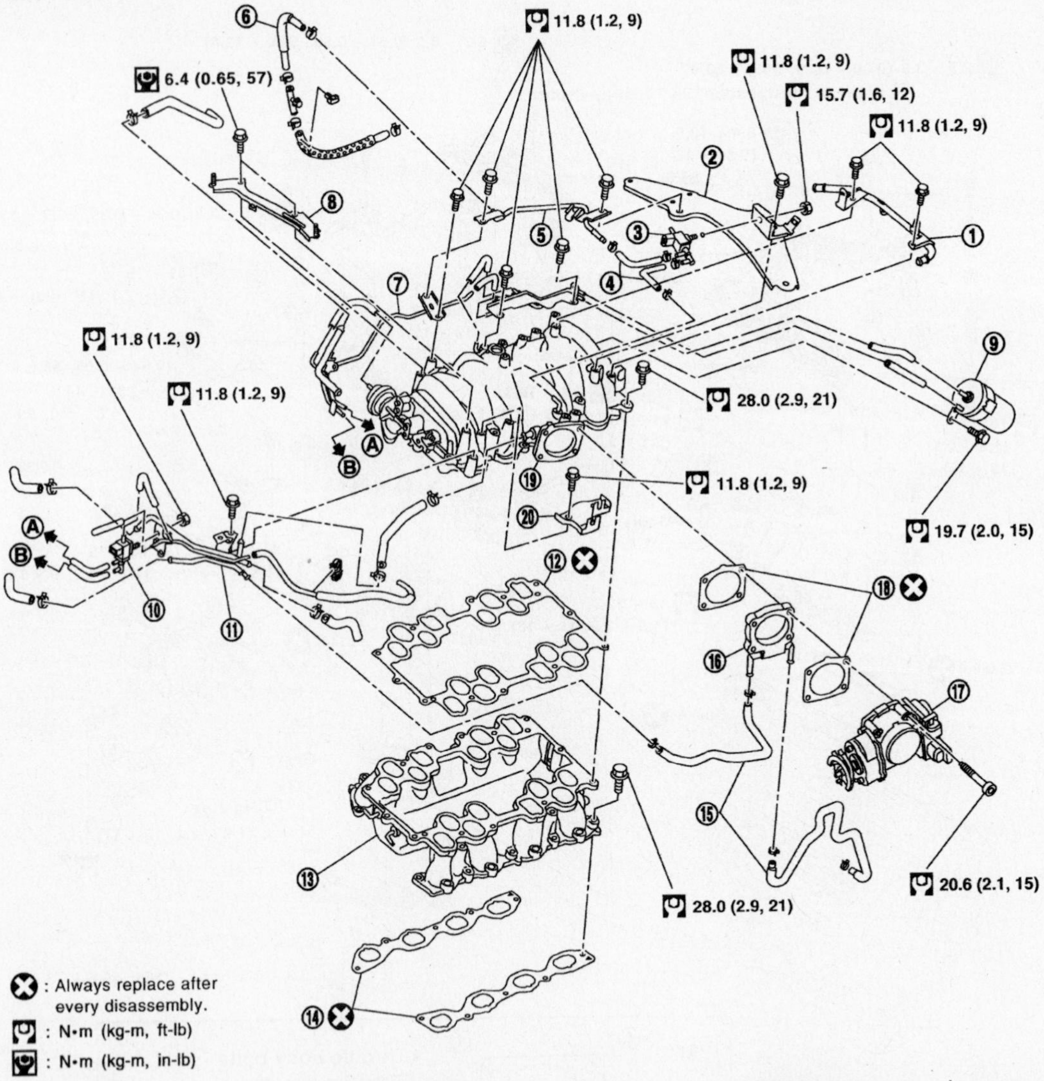

1.	PCV tube	2.	Engine cover rear bracket	3.	EVAP canister purge volume control solenoid valve
4.	EVAP hose	5.	EVAP tube	6.	EVAP hose
7.	Vacuum gallery	8.	Engine cover front bracket	9.	Vacuum tank
10.	VIAS control solenoid valve	11.	Water gallery	12.	Gasket
13.	Intake manifold lower	14.	Gasket	15.	Water hose
16.	Intake manifold adapter	17.	Electric throttle control actuator	18.	Gasket
19.	Intake manifold upper	20.	Bracket		

67162-INFI-G14

Exploded view of the intake manifold assembly—4.5L engine M45

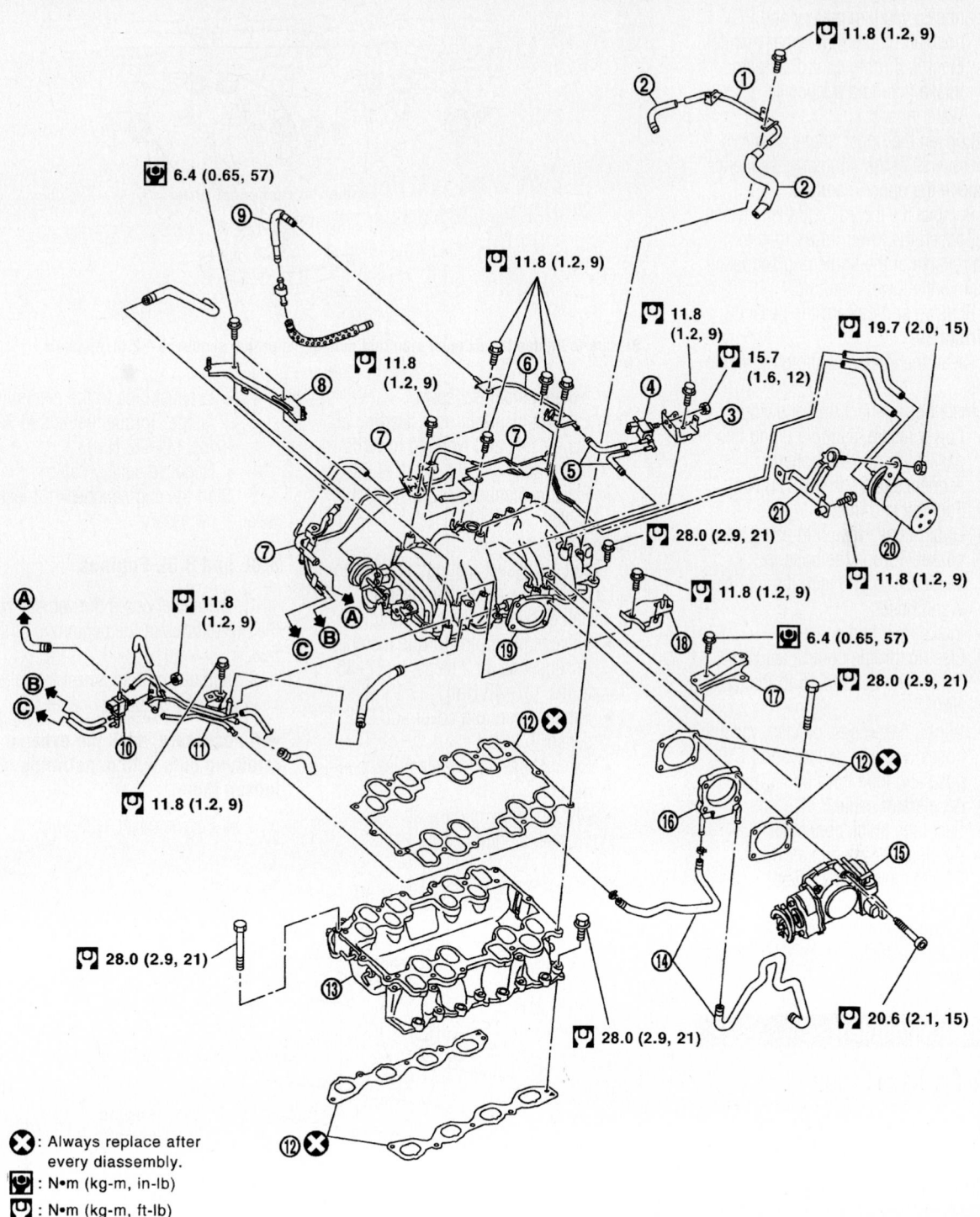

: Always replace after every diassembly.

: N•m (kg-m, in-lb)

: N•m (kg-m, ft-lb)

1. PCV tube	2. PCV hose	3. EVAP canister purge control sole-noid valve bracket
4. EVAP canister purge control solenoid valve	5. EVAP hose	6. EVAP tube
7. Vacuum gallery	8. Engine cover front bracket	9. EVAP hose
10. VIAS control solenoid valve	11. Water gallery	12. Gasket
13. Intake manifold lower	14. Water hose	15. Electric throttle control actuator
16. Intake manifold adapter	17. Engine cover rear bracket	18. Accelerator wire bracket
19. Intake manifold upper	20. Vacuum tank	21. Vacuum tank bracket

67162-INFI-G15

Exploded view of the intake manifold assembly—4.5L engine Q45

hoses, vacuum gallery and PCV hose and tube from upper manifold
- Electric throttle control actuator
- Intake manifold adapter
- Water hoses

5. Loosen the upper intake manifold bolts in reverse of the tightening sequence and remove the upper manifold.

6. Remove the fuel rail and fuel injectors.

7. Loosen the lower intake manifold bolts in reverse of the tightening sequence and remove the lower manifold.

8. Remove and discard the gaskets.

To install:

9. Clean the intake manifold mounting surfaces

10. Install or connect the following:
- Lower intake manifold using new gasket. Torque the bolts in sequence to 21 ft. lbs. (28 Nm).
- Fuel rail and injectors
- Upper intake manifold using new gasket. Torque the bolts in sequence to 106 inch lbs. (12 Nm).
- Water hoses
- Intake manifold adapter
- Electric throttle control actuator. Torque the bolts to 15 ft. lbs. (20 Nm).
- Wiring harnesses, bracket, vacuum hoses, vacuum gallery and PCV hose and tube from upper manifold
- Accelerator cable
- Fuel tube quick connector
- Air cleaner case and air duct
- Engine appearance cover
- Negative battery cable

11. Fill and bleed the cooling system.

12. Start the vehicle, check for leaks and repair if necessary.

Exhaust Manifold

REMOVAL & INSTALLATION

2.0L Engine

1. Before servicing the vehicle, refer to the precautions in the beginning of this section.

2. Remove or disconnect the following:
- Negative battery cable
- Undercover and dust covers, if equipped
- Exhaust pipe at the manifold flange
- Air Injection Valve (AIV), AIV tube, and the attaching bracket, if equipped
- Exhaust Gas Recirculation (EGR) sensor electrical connection and the sensor

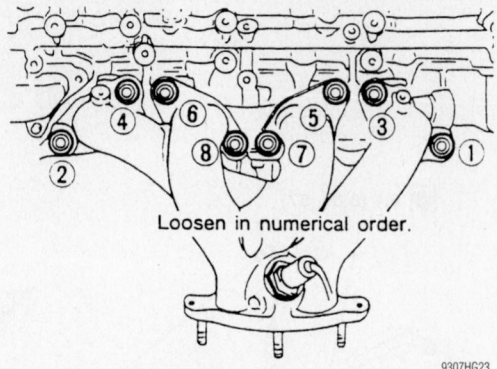

Be sure to tighten the exhaust manifold nuts in the proper sequence—2.0L engines

- Exhaust manifold cover
- Exhaust manifold nuts, starting at the outside and working towards the middle
- Exhaust manifold and discard the gasket

To install:

3. Clean the gasket mating surface and install a new exhaust manifold gasket.

4. Install or connect the following:
- Exhaust manifold. Torque the nuts in sequence, in 2 steps, to 27–35 ft. lbs. (37–48 Nm).
- Exhaust manifold cover and EGR sensor
- Exhaust gas sensor electrical connection
- AIV, AIV tube, and the attaching bracket, if equipped

- Exhaust pipe to the manifold flange. Torque the nuts to 30–35 ft. lbs. (41–48 Nm).
- Negative battery cable

5. Start the engine, check for leaks and repair if necessary.

3.0L and 3.5L Engines

1. Before servicing the vehicle, refer to the precautions in the beginning of this section.

2. Remove or disconnect the following:
- Negative battery cable

➡If necessary, soak the exhaust pipe retaining nuts with penetrating oil to loosen them.

- Engine cover, G35 only

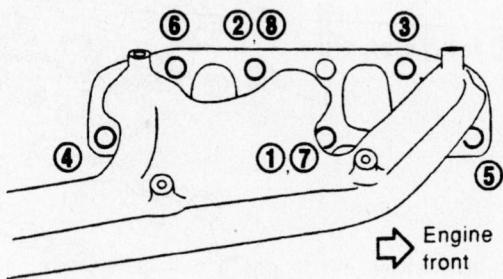

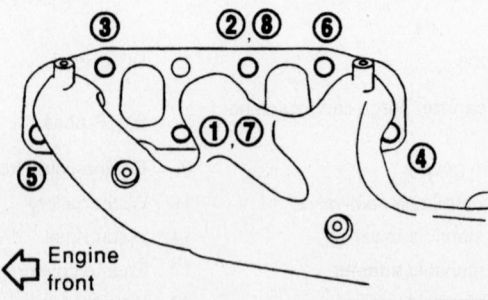

Exhaust manifold tightening sequence. Use the reverse of the sequence for removal—G35 shown

- Air cleaner assembly and duct, G35 only
- Front exhaust pipe from the exhaust manifolds
- Heated Oxygen Sensors (HO₂S) from the manifold, as necessary
- Protective covers from the manifolds
- Exhaust manifold-to-engine mounting nuts
- Manifold from the engine and discard the gaskets

To install:

3. Clean all gasket mounting surfaces.
4. Install or connect the following:

- Exhaust manifold with new gaskets. Torque the bolts, in sequence, in 2 steps to 21–23 ft. lbs. (28–32 Nm).
- Protective shields. Torque the bolts in 2 steps to 46–57 inch lbs. (5–6 Nm).
- HO₂S to the manifold, as necessary
- Exhaust manifolds to the exhaust pipes. Torque the bolts/nuts to 32–37 ft. lbs. (43–50 Nm) for all models except G35. For the G35, torque the nuts to 45–48 ft. lbs. (60–66 Nm).
- Air cleaner assembly and engine cover, G35 only

- Negative battery cable
5. Start the engine, check for exhaust leaks and repair if necessary.

4.1L Engines

1. Before servicing the vehicle, refer to the precautions in the beginning of this section.
2. Remove or disconnect the following:
- Negative battery cable
- Engine undercovers
- Exhaust pipe at the manifold flange
- Heat shield from the exhaust manifold, if equipped

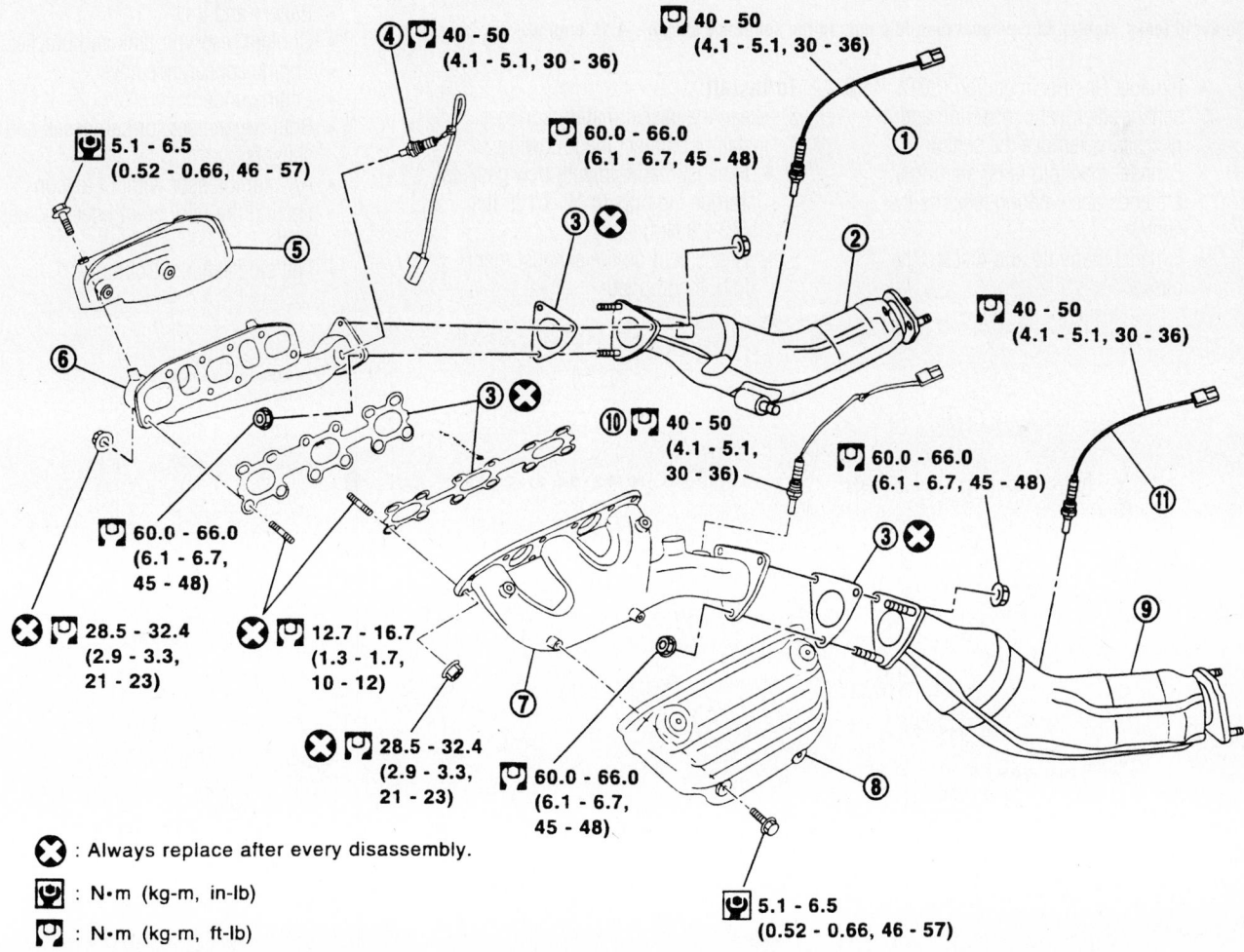

❌ : Always replace after every disassembly.

🔧 : N•m (kg-m, in-lb)

🔧 : N•m (kg-m, ft-lb)

1. Heated oxygen sensor 2 (rear) (bank1)	2. Three way catalyst (RH bank)	3. Gasket
4. Heated oxygen sensor 1 (front) (bank1)	5. Exhaust manifold cover (RH bank)	6. Exhaust manifold (RH bank)
7. Exhaust manifold (LH bank)	8. Exhaust manifold cover (LH bank)	9. Three way catalyst (LH bank)
10. Heated oxygen sensor 1 (front) (bank2)	11. Heated oxygen sensor 2 (rear) (bank2)	

Exploded view of the exhaust manifold and related components—G35 shown

9357MG13

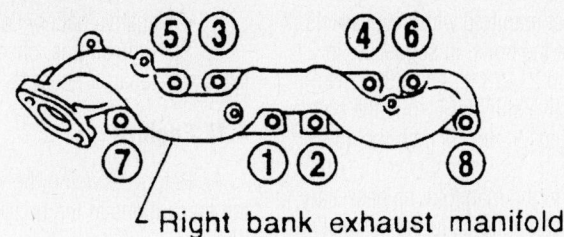

Right bank exhaust manifold

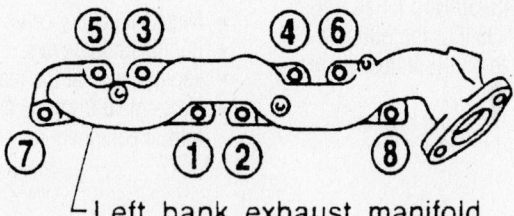

Left bank exhaust manifold

7923HG39

To avoid leaks, tighten the exhaust manifold nuts in the sequence shown—4.1L engine

- Exhaust Gas Recirculation (EGR) sensor electrical connection and if necessary, remove the sensor
- Exhaust manifold nuts, starting at the ends and working towards the center
- Exhaust manifold and discard the gasket

To install:
3. Clean the gasket mating surfaces.
4. Install or connect the following:
   - Exhaust manifold with new gaskets. Torque the nuts to 16–21 ft. lbs. (22–28 Nm).
   - Heat shield on the exhaust manifold, if equipped

- EGR sensor. Torque the fastener to 30–37 ft. lbs. (40–50 Nm), if removed.
- EGR sensor electrical connection
- Exhaust pipe to the manifold flange. Torque the nuts to 33–44 ft. lbs. (45–60 Nm).
- Negative battery cable
5. Start the engine, check for leaks and repair if necessary.

4.5L Engines

1. Before servicing the vehicle, refer to the precautions in the beginning of this section.
2. Remove or disconnect the following:
   - Both battery cables
   - Battery and tray
   - Coolant reservoir tank and bracket
   - Engine appearance cover
   - Engine undercovers
   - Both oxygen sensor harnesses and sensors
   - A/C compressor without disconnecting the refrigerant lines and set aside
   - Left side exhaust front tube

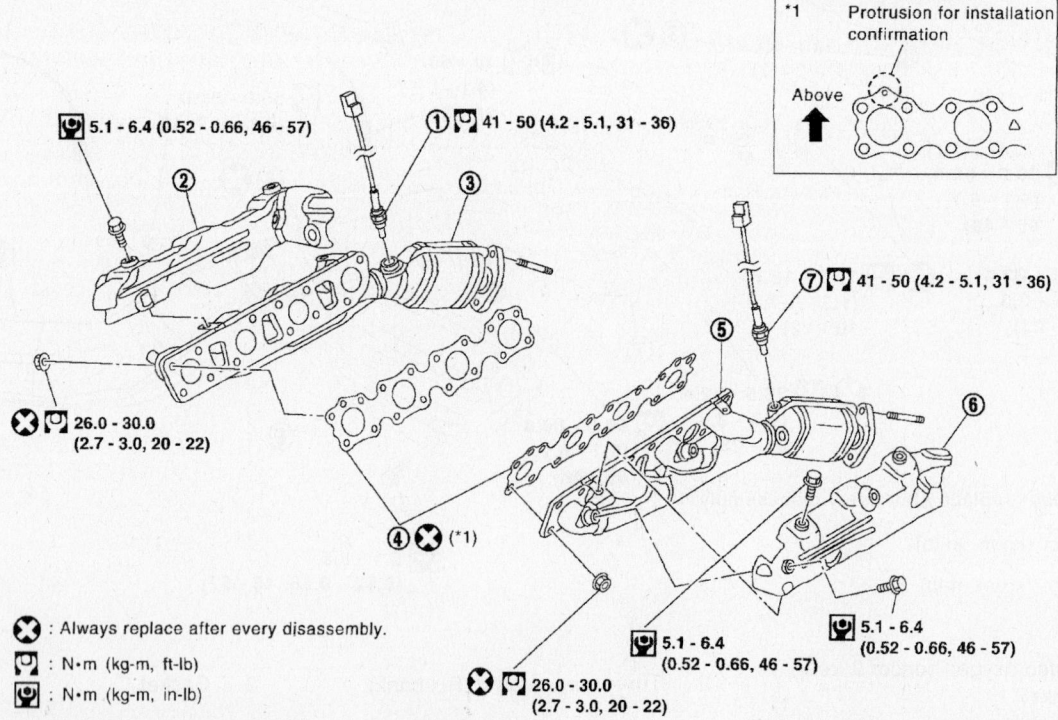

❌ : Always replace after every disassembly.

🔧 : N•m (kg-m, ft-lb)

🔧 : N•m (kg-m, in-lb)

1. Heated oxygen sensor 1 (bank 2) *
2. Exhaust manifold cover (right bank)
3. Exhaust manifold and three way catalyst (right bank)
4. Gasket
5. Exhaust manifold and three way catalyst (left bank)
6. Exhaust manifold cover (left bank)
7. Heated oxygen sensor 1 (bank1) *

67162-INFI-G16

Exploded view of the exhaust manifold assembly—4.5L engine

• Steering column lower joint

3. Use a jack and raise the bottom of the engine.

4. Remove the left engine mounting insulator and the engine mounting brackets.

5. Remove the left manifold heat shield.

6. Remove the left manifold mounting nuts in the reverse order of the tightening sequence and remove the manifold and catalyst.

7. Remove the right side front exhaust tube.

8. Remove the nuts from the bottom right engine mounting insulator and lift the right side of the engine up about 1 and a half inches.

9. Remove the starter, then remove the right engine mounting insulator and mounting brackets.

10. Remove the right manifold mounting nuts in the reverse order of the tightening sequence and remove the manifold and catalyst.

To install:

11. Clean the gasket mating surfaces.

12. Install or connect the following:
• Exhaust manifold with new gaskets. Torque the nuts to 20–22 ft. lbs. (26–30 Nm) in two steps.

13. Install the starter, then install the right engine mounting insulator and mounting brackets. Tighten the bracket bolts to 25–31 ft. lbs. (35–43 Nm). Tighten the upper insulator nut to 32–40 ft. lbs. (43–553 Nm) and the lower nut to 65–72 ft. lbs. (87–98 Nm).

14. Install the heat shields.

15. Install the right side front exhaust tube.

16. Install the left engine mounting insulator and the engine mounting brackets and tighten the fasteners to the same specification as the right side.

17. Connect the steering column lower

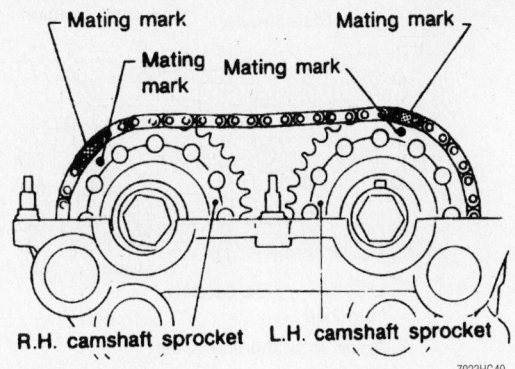

Be sure to align the marks on the sprockets with the marks on the chain—2.0L engine

joint and tighten the bolt to 20 ft. lbs. (27 Nm).

18. Install the left side front exhaust tube.

19. Install or connect the following:
• A/C compressor
• Both oxygen sensor harnesses and sensors
• Engine undercovers
• Engine appearance cover
• Coolant reservoir tank and bracket
• Battery and tray
• Both battery cables

20. Start the engine, check for leaks and repair if necessary.

Camshaft and Valve Lifters

REMOVAL & INSTALLATION

2.0L Engine

1. Before servicing the vehicle, refer to the precautions in the beginning of this section.

2. Relieve the fuel system pressure.

3. Remove or disconnect the following:
• Negative battery cable

• Rocker arm cover
• Oil separator

4. Rotate the crankshaft until the No. 1 piston is at Top Dead Center (TDC) on the compression stroke and the mating marks on the camshaft sprockets line up with the mating marks on the timing chain.
• Timing chain tensioner
• Distributor
• Timing chain guide
• Camshaft sprockets. Use a wrench to hold the camshaft while loosening the sprocket bolt.

5. Loosen the camshaft bearing cap bolts in sequence.
• Camshaft from the cylinder head

6. When removing the rocker arm, be careful not to drop the valve shims into the cylinder head. After removing the adjuster, set them upright or lay them down in a pan of clean engine oil. Do not lay them down on the bench or the oil will drain out and the adjuster will become air bound. Keep all of these parts in order so they can be installed in the same locations.

To install:

7. Install the adjusters, shims and rockers into their original locations.

8. Clean the left-hand camshaft end bearing cap and coat the mating surface with liquid gasket. Install the camshafts, bearing caps, oil tubes and baffle plate. Ensure the left camshaft key is at 12 o'clock and the right camshaft key is at 10 o'clock.

9. The procedure for tightening bearing cap bolts must be followed exactly to prevent camshaft damage. Torque bolts as follows:

a. Torque the right camshaft bolts 9 and 10 (in that order) to 17 inch lbs. (2 Nm), then bolts 1 through 8 (in that order) to the same specification.

b. Torque the left camshaft bolts 11 and 12 (in that order) to 17 inch lbs. (2 Nm), then bolts 1 through 10 (in that order) to the same specification.

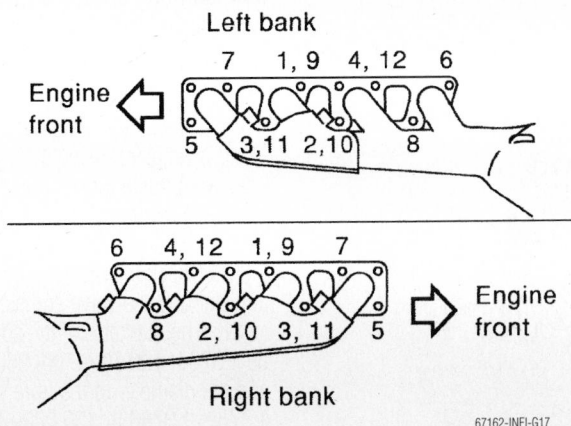

Exhaust manifold tightening sequence—4.5L engine

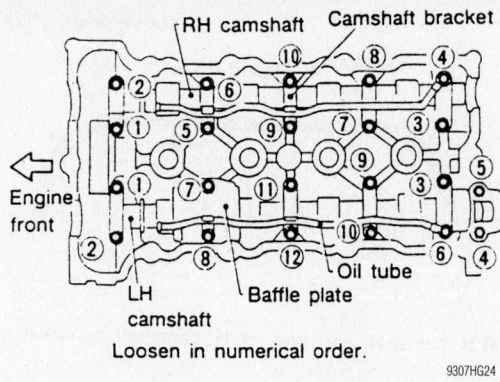

Camshaft bearing cap bolt loosening sequence—2.0L engine

c. Torque all bolts in sequence to 52 inch lbs. (6 Nm).

d. Torque all bolts again in sequence to 78–102 inch lbs. (9–12 Nm), then bolts 8 and 9 on the left camshaft to 13–19 ft. lbs. (18–25 Nm).

10. Line up the mating marks on the timing chain and camshaft sprockets and install the sprockets. Torque sprocket bolts to 101–116 ft. lbs. (137–157 Nm).

11. Install or connect the following:
- Timing chain guide and chain tensioner
- Distributor making certain that the rotor head is at the 5 o'clock position

12. Clean the rocker cover and mating surfaces and apply a continuous bead of liquid gasket to the mating surface.

13. Install the rocker cover and oil separator. Tighten the rocker cover bolts as follows:

a. Torque the nuts 1, 10, 11 and 8, in that order to 36 inch lbs. (4 Nm).

b. Torque the nuts 1 through 13 as indicated in the figure to 72–84 inch lbs. (8–10 Nm).

14. Connect the negative battery cable.

3.0L and 3.5L Engines

EXCEPT G35 MODELS

1. Before servicing the vehicle, refer to the precautions in the beginning of this section.

2. Relieve the fuel system pressure.

3. Drain the engine oil.

4. Drain the cooling system.

5. Remove or disconnect the following:
- Negative battery cable
- Left side rocker cover ornament

➡ **Before disconnecting any hoses or connectors, note the locations for reassembly.**

- Air duct to intake manifold hose, collector hose, blow-by hose, and vacuum hoses
- Fuel hoses and disconnect the harness connection
- Evaporative emissions (EVAP) canister purge hoses
- Water hoses from the cylinder head and intake manifold
- Ignition coils
- Spark plugs
- Exhaust Gas Recirculation (EGR) tube

- Intake manifold collector supports and the collector
- Fuel tube
- Intake manifold
- Rocker arm covers
- Engine undercovers
- Right front wheel
- Engine side covers
- Drive belts and idler pulley
- Power steering oil pump and belt
- Camshaft Position (CMP) sensor (PHASE) and Crankshaft Position (CKP) sensors (REF)/(POS)

6. Set the No. 1 piston to Top Dead center (TDC) of compression stroke by rotating the crankshaft.
- Ring gear cover access plate

7. Loosen the crankshaft pulley bolt while securing the ring gear so the crankshaft cannot rotate.

➡ **Use care not to damage the ring gear teeth.**

- Crankshaft pulley, using a suitable puller
- A/C compressor and bracket
- Front exhaust pipe and install engine slingers

8. Support the transaxle with jack.
- Right side engine mounting bracket
- Center crossmember assembly
- Oil pan bolts and oil pans
- Timing chain
- O-rings from the front of the engine block

9. Loosen the camshaft bearing caps in several steps. The bearing caps MUST be loosened in sequence.

➡ **Keep all bearing caps and camshafts in proper order for reinstallation.**

- Left-hand and right-hand camshaft tensioners from the cylinder head
- Camshafts from the cylinder heads

➡ **The valve adjusters have a replaceable shim on the top of the adjuster. Note the proper locations of each shim to adjuster and remove the shims from the adjusters.**

- Valve adjusting shim from the adjuster, using a magnet
- Adjuster assembly from the bore. Be sure to note the locations from where each adjuster came.

10. Check the diameter of the valve adjuster and the valve adjuster guide bore.

11. The diameter of the adjuster should be 1.3764–1.3770 in. (34.960–34.975mm) and the diameter of the bore should be 1.3780–1.3788 in. (35.000–35.021mm).
- All traces of liquid gasket from the

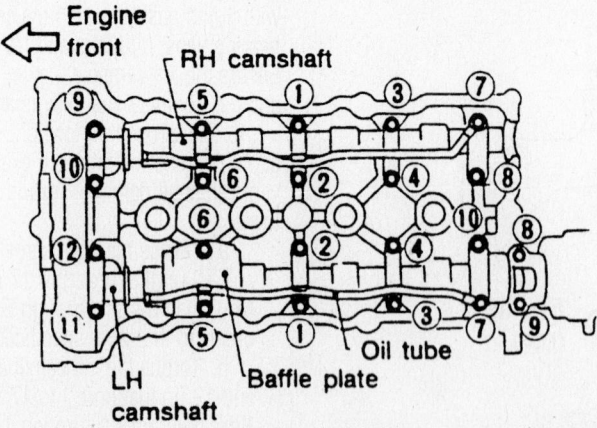

Camshaft bearing cap bolt torque sequence—2.0L engine

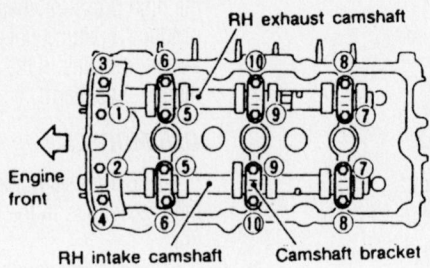

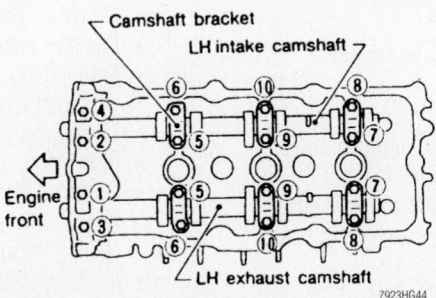

To avoid camshaft damage, loosen the bearing cap bolts in the sequence shown—3.0L and 3.5L engines, except G35

timing chain case and from the water pump covers
• All traces of liquid gasket from the engine block

12. Inspect the camshafts for excessive wear or damage and replace as necessary.

To install:

13. Lubricate the valve adjusters with clean engine oil and install the adjusters into the bore from which they were removed.

14. Lubricate the valve adjuster shims with clean engine oil and install the shims into the adjuster from which they were removed.

15. Turn the crankshaft clockwise until the No. 1 piston is set 240° before TDC on compression stroke.

16. Install or connect the following:
• Camshaft tensioners on both sides of the cylinder heads. Torque the tensioner mounting bolts to 75–96 inch lbs. (8–11 Nm).

➡**The camshafts can be identified by the paint marks on the camshaft. The left cylinder head camshafts have a YELLOW paint mark and the right cylinder head camshafts have a WHITE paint mark. When installing the camshafts, position the camshaft keys at the 12 o'clock position in respect to the cylinder head angle.**

• Exhaust and intake camshafts and install the bearing caps. Before

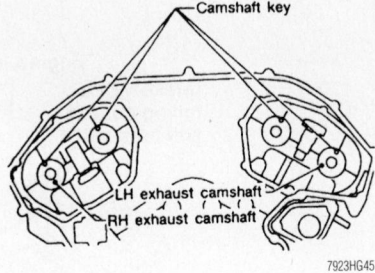

Positioning of the camshaft keys during installation—3.0L and 3.5L engines, except G35

installing the No. 1 bearing cap, apply liquid gasket to the corners of the cap.

17. Torque the camshaft bearing caps as follows:

a. Nos. 7 through 10 then, Nos. 1 through 6 to 17 inch lbs. (2 Nm).
b. All bolts in order to 52 inch lbs. (6 Nm).
c. Nos. 1 though 6 to 80–104 inch lbs. (9–11 Nm).
d. Nos. 7 though 10 to 74–91 inch lbs. (9–11 Nm).
• New O-rings to the front of the engine block

18. Apply sealant to the hatched portion of the of the rear timing chain case.

19. Align the rear timing chain case with the dowel pins and install onto the cylinder heads and engine block. Torque the rear timing chain case mounting bolts in sequence to 105–121 inch lbs. (11.8–13.7 Nm).
• Crankshaft sprocket with the mating mark facing out

20. Rotate the crankshaft clockwise and position the crankshaft to TDC of compression stroke and align the dowels of the camshaft sprockets to the 12 o'clock position in respect to the cylinder head
• Lower chain guide on the dowel pin with the front mark on the guide facing upward

21. On a workbench, align the marks on the intake and exhaust camshaft sprockets with the marks of the chain.
• Exhaust camshaft sprockets onto the dowel pin. Torque the bolts to 88–95 ft. lbs. (120–129 Nm). Be sure to secure the camshafts while tightening the bolts.
• Align and install the timing chains and sprockets to the camshafts
• Timing cover evenly and gently. Be sure to align the dowel pin holes

➡**Leave the bolts unattended for 30 minutes or more after tightening.**

22. Apply a 0.091–0.130 in. (2.3–3.3mm) continuous bead of liquid

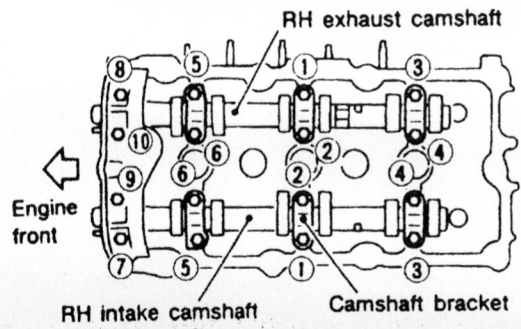

Tighten in numerical order.

Be sure to tighten the camshaft bearing cap bolts in the correct sequence—3.0L and 3.5L engines (except G35), right cylinder head

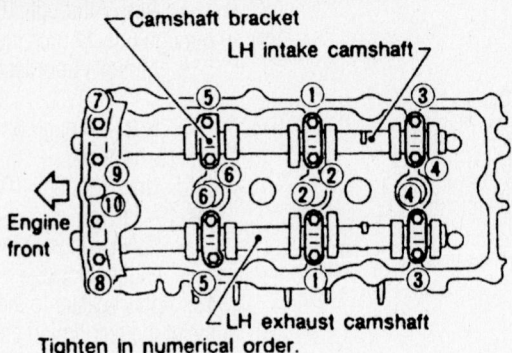

Tighten in numerical order.

7923HG47

Tighten the camshaft bearing cap bolts in the correct sequence—3.0L and 3.5L engines (except G35), left cylinder head

gasket to the water pump cover and install the cover. Tighten the bolts to 84–108 inch lbs. (10–13 Nm).

- Rocker covers

23. Apply sealant to the front and rear seal of the oil pan and install the oil pan.

- Center crossmember assembly
- Right side engine mounting bracket and mount assembly
- Engine slinger assembly
- Front exhaust pipe and its support
- Air conditioning compressor and bracket
- Crankshaft pulley to the crankshaft and install the mounting bolt

24. Torque the mounting bolt to 14–22 ft. lbs. (20–29 Nm). Torque the crankshaft bolt an additional 60–66° clockwise. This is approximately the angle from 1 hexagon bolt head corner to another.

- Ring gear cover plate
- CMP sensor (PHASE) and CKP sensors
- Power steering pump assembly, drive belts and the idler pulley
- Engine side cover
- Right front wheel
- Engine undercovers
- Intake manifold, using new gaskets
- Fuel tube assembly using new insulators. Torque the bolts in several steps to 15–20 ft. lbs. (21–26 Nm).
- Intake manifold collector gasket with the arrow facing forward
- Intake manifold collector assembly and support bracket
- EGR tube using new insulators. Torque the bolts in two steps to 15–20 ft. lbs. (21–26 Nm).
- Spark plugs and ignition coils. Torque the bolts to 27–33 inch lbs. (2.9–3.8 Nm).
- Water hoses to the cylinder head and intake manifold

- Fuel hoses and wiring harness connections to the fuel rail
- Air duct to intake manifold hose, collector hose, blow-by hose, and vacuum hoses
- Negative battery cable

25. Fill the engine with clean oil.
26. Fill and bleed the cooling system.
27. Start the engine and run at 3000 RPM under no load to purge the air from the high pressure chamber. The engine may produce a rattling noise. This indicates that air still remains in the chamber and is not a matter of concern.

G35 MODELS

1. Before servicing the vehicle, refer to the precautions in the beginning of this section.
2. Drain the engine oil and cooling system.
3. Relieve the fuel system pressure.
4. Remove or disconnect the following:

- Negative battery cable
- Front timing chain case
- Camshaft sprocket
- Timing chain
- Rear timing chain case
- Camshaft Position (CMP) sensors (PHASE) from the back side of the cylinder heads
- Intake valve timing control solenoid valves from the No. 1 camshaft bracket

➡**Before removal, matchmark the position of the camshafts, brackets and**

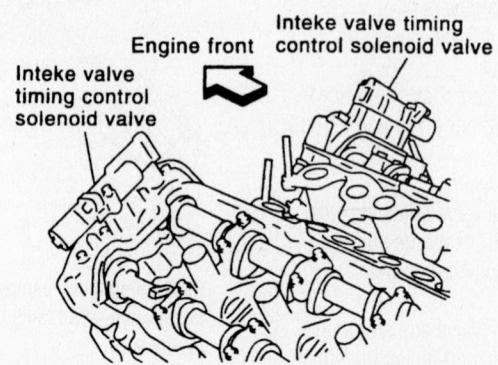

9357MG14

Location of the intake valve timing control solenoid valves—G35 models

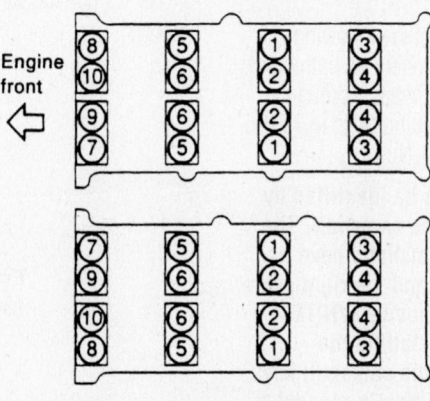

9357MG15

Intake and exhaust camshaft bracket bolts tightening sequence. Use the reverse sequence for removal—G35 models

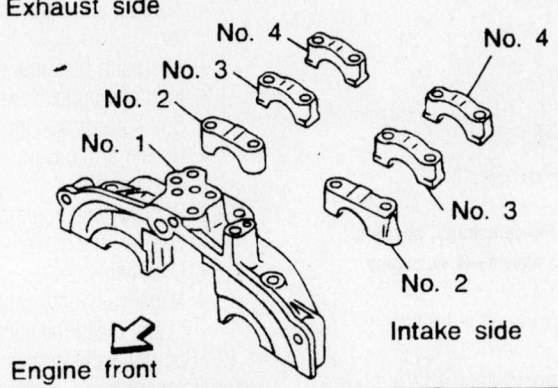

RH camshaft brackets

Exhaust side

Engine front

LH camshaft brackets

Intake side

Engine front

9357MG17

Camshaft identification and installation—G35 models

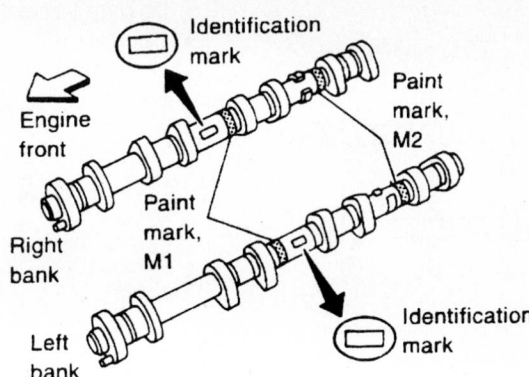

Identification mark

Engine front

Paint mark, M2

Right bank

Paint mark, M1

Left bank

Identification mark

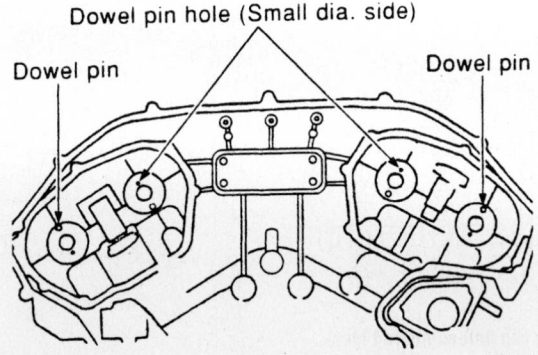

Dowel pin hole (Small dia. side)

Dowel pin

Dowel pin

9357MG16

Camshaft bracket installation—G35 models

bolts, so they are reinstalled in their original locations.

- Intake and exhaust camshaft brackets. Loosen the bolts in several stages, in reverse of the sequence shown.
- Camshafts
- Valve lifters if necessary, noting their installed positions

To install:

5. Install or connect the following:

- Valve lifters, in their original positions
- Camshafts with the dowel pin attached to its front end face on the exhaust side
- Camshaft brackets, as shown in the illustration

6. Torque the camshaft bracket bolts, in sequence, as follows:

   a. Step 1: Nos. 7–10, then Nos. 1–6 to 17 inch lbs. (2 Nm).

   b. Step 2: Nos. 1–10 to 52 inch lbs. (5.9 Nm).

   c. Step 3: Nos. 1–6 to 80–104 inch lbs. (9–12 Nm).

   d. Step 4: Nos. 7–10 to 74–91 inch lbs.

7. Measure the difference in levels between the front end faces of the No. 1 camshaft bracket and the cylinder head. If the measurement falls out of the range of -0.0055–0.0055 in. (-0.14–0.14mm), you must reinstall the camshaft and brackets.

8. Check and adjust the valve clearance.

- Intake valve timing control solenoid valves
- CMP sensors (PHASE)
- Rear timing chain case, timing chain, camshaft sprocket and front timing chain case
- Negative battery cable

9. Fill the engine with oil.

10. Fill and bleed the cooling system.

11. Start the vehicle, check for leaks and repair if necessary.

4.1L Engine

1. Before servicing the vehicle, refer to the precautions in the beginning of this section.

2. Drain the cooling system.

3. Relieve the fuel system pressure.

4. Remove or disconnect the following:

- Negative battery cable
- Ornament cover
- Undercover
- Radiator and cooling fan
- Inlet and outlet hoses
- Alternator belt and idler bracket

- Air duct and intake manifold collector
- Intake valve timing control solenoid
- Rocker arm covers
- Vacuum pipe
- Ignition coils and spark plugs

5. Turn the crankshaft to position the No. 1 piston at Top Dead Center TDC on compression.

- Camshaft Position (CMP) sensor
- Upper front covers

6. Paint alignment marks on the timing chain and camshaft sprockets.

- Upper chain tensioners
- Camshaft sprockets
- Camshaft bearing cap bolts in the proper sequence to prevent damage to the camshaft. Keep the caps in order so they can be installed in the correct locations.
- Camshafts, rocker arms and lash adjusters. The lash adjusters and rocker arms must be installed in their original positions. Keep them in order.

To install:

7. Install or connect the following:

- Lash adjusters and rocker arms

8. Lubricate the camshafts with engine oil and place them on the cylinder head with the knock pins facing away from the crankshaft.

9. Install the bearing caps in their origi-

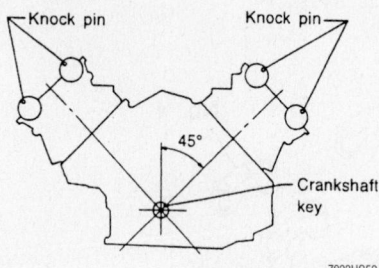

When installing the camshafts, position the knock pins as shown—4.1L engine

nal positions and torque the bolts as follows:

    a. Nos. 9 and 10, then 1 through 8: 17 inch lbs. (2 Nm).

    b. All bolts in order: 52 inch lbs. (6 Nm).

    c. All bolts in order: 96–121 inch lbs. (11–14 Nm).

10. Install or connect the following:

- Camshaft sprockets. Torque the intake sprocket bolt to 76–83 ft. lbs. (103–113 Nm) and the exhaust sprocket bolt to 12–15 ft. lbs. (16–21 Nm).
- Chain tensioner. Torque the bolts to 82–95 ft. lbs. (9–11 Nm).
- Upper chain covers
- Rocker covers using new gaskets
- Valve timing control solenoid valve using a new O-ring. Torque the

solenoid to 18–25 ft. lbs. (25–34 Nm).

- Air duct and intake manifold collector
- Alternator belt and idler bracket
- Inlet and outlet hoses
- Spark plugs and ignition coils
- Rocker arm covers
- Vacuum pipe
- Radiator and cooling fan
- Engine undercover
- Ornament cover
- Negative battery cable

11. Fill and bleed the cooling system.

12. Start the vehicle, check for leaks and repair if necessary.

4.5L Engine

1. Before servicing the vehicle, refer to the precautions in the beginning of this section.

2. Drain the cooling system.

3. Relieve the fuel system pressure.

4. Remove or disconnect the following:

- Engine
- Timing chain assemblies
- Camshaft sprockets while holding camshaft locked in place
- Camshaft bracket bolts in the reverse order of the tightening sequence to prevent damage to the camshaft. Keep the brackets in order so they can be installed in the correct locations.
- Camshafts, shims and lifters. Keep all parts in order so they can be installed in their original positions.

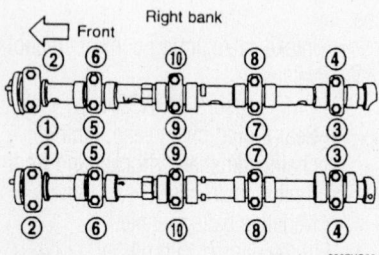

Camshaft bearing cap bolt numbered identification and loosening sequence—4.1L engine, right bank

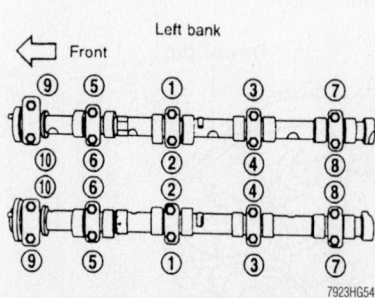

Camshaft bearing cap bolt numbered identification and tightening sequence—4.1L engine, right bank

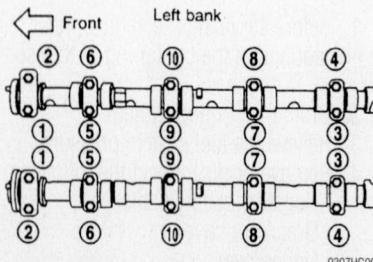

Camshaft bearing cap bolt numbered identification and loosening sequence—4.1L engine, left bank

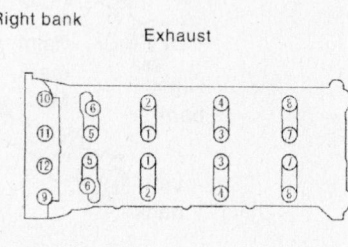

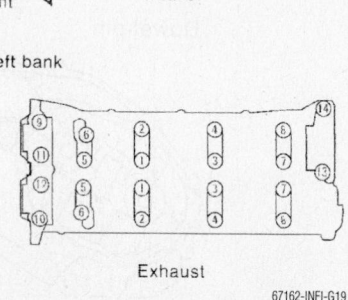

Camshaft bracket tightening sequence—4.5L engine

Camshaft bearing cap bolt numbered identification and tightening sequence—4.1L engine, left bank

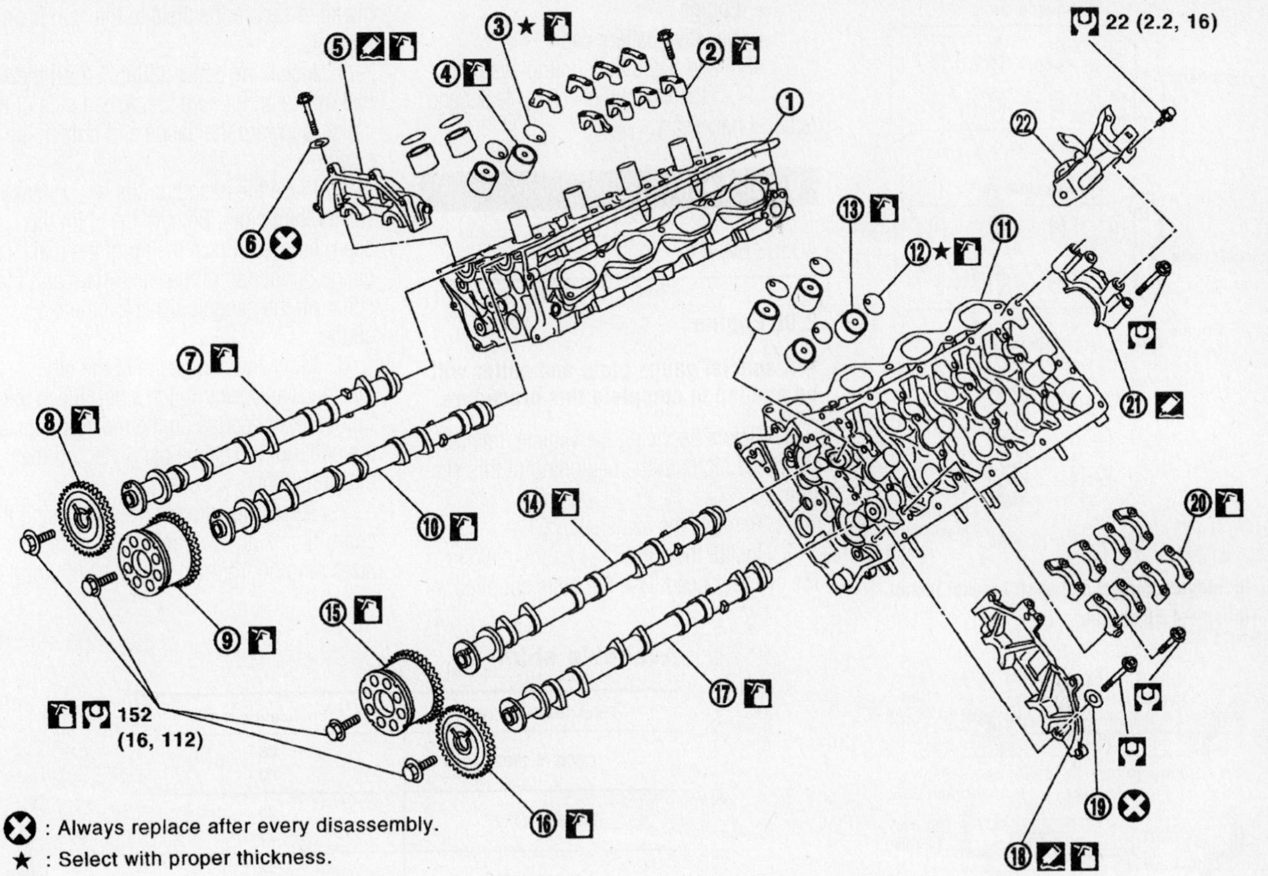

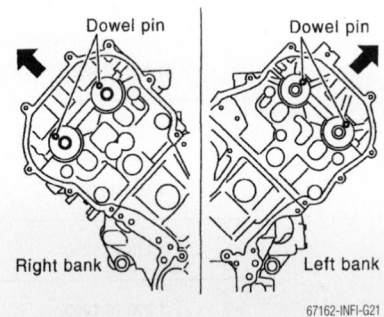

X : Always replace after every disassembly.

★ : Select with proper thickness.

✎ : Apply Genuine RTV Silicone Sealant or equivalent.

⬛ : Lubricate with new engine oil.

⬛ : N•m (kg-m, ft-lb)

1. Cylinder head (right bank)	2. Camshaft bracket (No. 2, 3, 4, 5)	3. Adjusting shim
4. Valve lifter	5. Camshaft bracket (No.1)	6. Washer
7. Camshaft (EXH)	8. Camshaft sprocket (EXH)	9. Camshaft sprocket (INT)
10. Camshaft (INT)	11. Cylinder head (left bank)	12. Adjusting shim
13. Valve lifter	14. Camshaft (INT)	15. Camshaft sprocket (INT)
16. Camshaft sprocket (EXH)	17. Camshaft (EXH)	18. Camshaft bracket (No.1)
19. Washer	20. Camshaft bracket (No. 2, 3, 4, 5)	21. Camshaft bracket (INT, No. 6)
22. Bracket		

67162-INFI-G18

Exploded view of camshaft assemblies—4.5L engine

To install:

5. Install or connect the following:
  • Valve lifters and shims
6. Identify the correct camshaft for proper placement as shown in the illustration.

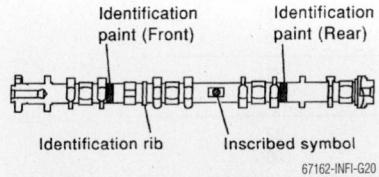

Identifying correct camshaft placement— 4.5L engine

67162-INFI-G20

Identifying correct camshaft dowel pin locations—4.5L engine

67162-INFI-G21

7. With no 1 cylinder at TDC, install the camshafts so the dowel pins are in the correct locations.

8. Install the camshaft brackets so the installation location can be correctly read when viewed from the side of the left exhaust bank.

9. Apply liquid gasket to camshaft brackets No. 1 and left side no. 6 in the locations shown.

10. Tighten the camshaft brackets in the sequence shown using the following steps:

  a. Tighten numbers 9 to 12 to 17 inch lbs. (1.96 Nm).

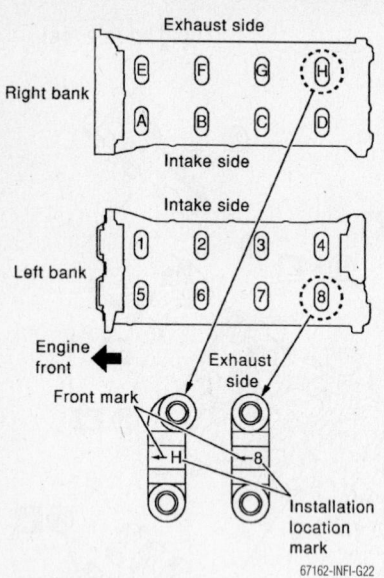

Identifying correct camshaft bracket installation—4.5L engine

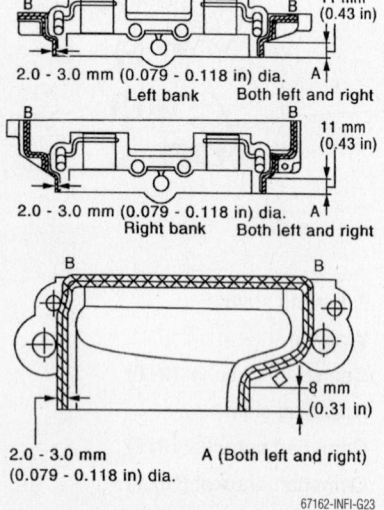

Identifying liquid gasket application areas to camshaft brackets—4.5L engine

b. Tighten numbers 1 to 8 to 17 inch lbs. (1.96 Nm).

c. Tighten numbers 13 and 14 to 17 inch lbs. (1.96 Nm).

d. Tighten all bolts in sequence to 52 inch lbs. (5.88 Nm).

e. Tighten numbers 1 to 12 to 92 inch lbs. (10.41 Nm).

f. Tighten numbers 13 and 14 to 23 ft. lbs. (31 Nm).

11. Install the camshaft sprockets and while holding the camshaft in place tighten the bolts to 112 ft. lbs. (152 Nm).

12. Adjust the valve clearances.

13. Install or connect the following:
- Timing chain assemblies
- Engine
- Negative battery cable

14. Fill and bleed the cooling system.

15. Start the vehicle, check for leaks and repair if necessary.

Valve Lash

ADJUSTMENT

2.0L Engine

→**A special gauge plate and collar will be needed to complete this procedure.**

1. Before servicing the vehicle, refer to the precautions in the beginning of this section.

2. Remove the camshafts.

3. Install the J38957–1 gauge plate to the cylinder head. Use the bolts supplied in the kit to secure the plate to the cam bearing journals.

4. Install the collar J38957–2 on the dial indicator. Be sure the dished side of the collar is toward the gauge and tighten the setscrew.

5. Place the gauge on the No. 1 intake valve (shim side). Be sure the shim has been removed. Place the tip of the dial gauge on the top of the valve stem and the collar on the gauge plate. Zero the dial gauge.

6. Move the dial gauge to the other intake valve (rocker guide side). Place the tip of the dial gauge on the rocker guide and the collar of the gauge plate. Record the measurement.

7. Select the correct size shim using the chart. Shims are available in 17 different sizes ranging from 0.1102 in. (2.800mm) to

Available shim

Thickness mm (in)	Identification mark
2.800 (0.1102)	28 00
2.825 (0.1112)	28 25
2.850 (0.1122)	28 50
2.875 (0.1132)	28 75
2.900 (0.1142)	29 00
2.925 (0.1152)	29 25
2.950 (0.1161)	29 50
2.975 (0.1171)	29 75
3.000 (0.1181)	30 00
3.025 (0.1191)	30 25
3.050 (0.1201)	30 50
3.075 (0.1211)	30 75
3.100 (0.1220)	31 00
3.125 (0.1230)	31 25
3.150 (0.1240)	31 50
3.175 (0.1250)	31 75
3.200 (0.1260)	32 00

Select the correct valve lash adjusting shim using the chart—2.0L engine

0.1260 in. (3.200mm) in increments of 0.001 in. (0.025mm).

3.0L, 3.5L and 4.1L Engines

➡**Check and adjust the valve clearances while the engine is cold and not running.**

1. Before servicing the vehicle, refer to the precautions in the beginning of this section.

2. Remove the intake manifold collector.

3. Remove the left and right rocker covers.

4. Remove the spark plugs.

5. Set the No. 1 cylinder at Top Dead Center (TDC) on its compression stroke. Align the pointer with the TDC mark on the crankshaft pulley. Check that the valve adjusters on the No. 1 cylinder are loose and valve adjusters on the No. 4 cylinder are tight. If not, turn the crankshaft 1 revolution (360°) and align the pointer with the TDC mark on the crankshaft pulley.

6. Check the following valves:
- Both No. 1 intake valves
- Both No. 2 exhaust valves
- Both No. 3 exhaust valves
- Both No. 6 intake valves

7. Using a feeler gauge, measure the clearance between the valve adjuster and the camshaft. Record any valve clearance measurements that are out of specification. Intake valve clearance (cold) is 0.010–0.013 in. (0.26–0.34mm) and exhaust valve clearance (cold) is 0.011–0.015 in. 0.29–0.37mm).

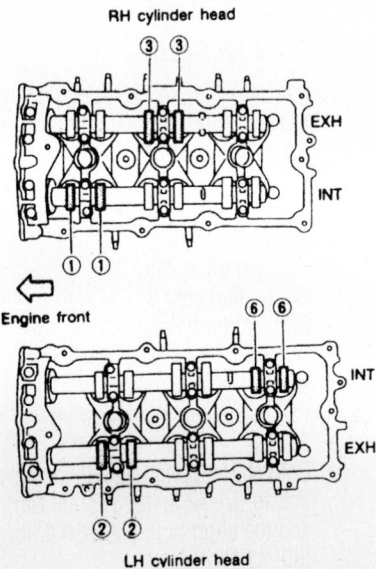

RH cylinder head

Engine front

LH cylinder head

7923HG57

Valve lash checking sequence at TDC of cylinder No. 1—3.0L engine

8. Turn the crankshaft 240° and set the No. 3 cylinder to TDC of its compression stroke.

9. Check the following valves:
- Both No. 2 intake valves
- Both No. 3 intake valves
- Both No. 4 exhaust valves
- Both No. 5 exhaust valves

10. Using a feeler gauge, measure the clearance between the valve adjuster and the camshaft. Record any valve clearance measurements that are out of specification. Intake valve clearance (cold) is 0.010–0.013 in. (0.26–0.34mm) and exhaust valve clearance (cold) is 0.011–0.015 in. (0.29–0.37mm).

11. Turn the crankshaft 240° and set the No. 5 cylinder to TDC of its compression stroke.

12. Check the following valves:
- Both No. 1 exhaust valves
- Both No. 4 intake valves
- Both No. 5 intake valves
- Both No. 6 exhaust valves

13. Using a feeler gauge, measure the clearance between the valve adjuster and the camshaft. Record any valve clearance measurements that are out of specification. Intake valve clearance (cold) is 0.010–0.013 in. (0.26–0.34mm) and exhaust valve clearance (cold) is 0.011–0.015 inches (0.29–0.37mm).

14. If all the valve clearances are within specification, install the cylinder head cover, spark plugs, and the intake manifold collector.

15. If an adjustment is necessary, adjust the valve clearance while engine is cold by removing the adjusting shim. The adjusting shim can be removed by using the following procedures:

a. Turn the crankshaft so the camshaft lobe of the valve to be adjusted is pointed straight up.

b. Turn the adjuster so the notch is pointed towards the center of the cylinder head; this will facilitate the shim removal process.

c. Using a depressor tool No. KV10115110 push down on the adjuster and insert a keeper tool on the edge of the adjuster to keep the adjuster in the depressed position.

d. Remove the depressor tool and remove the shim with a magnet.

➡**Compressed air can be blown into the hole of the adjuster to separate the adjusting shim from the adjuster.**

16. Determine the replacement adjusting shim size by using the following procedures and formula:

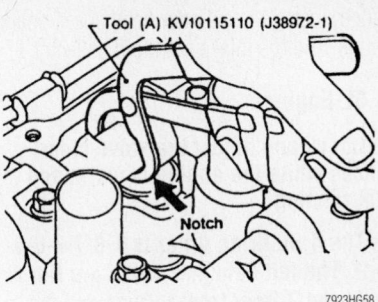

Tool (A) KV10115110 (J38972-1)

Notch

7923HG58

Install the depressor tool around the camshaft being careful not to damage the surfaces—3.0L engine shown

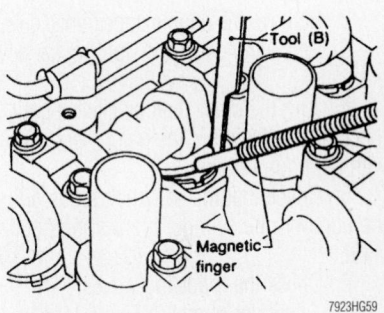

Tool (B)

Magnetic finger

7923HG59

Use a magnet to remove the shim from the adjuster. Sometimes a shot of compressed air can help lift the shim up—3.0L engine shown

a. Using a micrometer determine thickness of the removed shim.

b. Calculate the thickness of a new adjusting shim so valve clearance is within the specified values.
- R= thickness of the removed shim
- N= thickness of the new shim
- M= measured valve clearance
- Calculate the Intake Shim as follows: N = R + M—0.0118 in. (0.30mm)
- Calculate the Exhaust Shim as follows: N = R + M—0.0130 in. (0.33mm)

17. Shims are available in 64 sizes from 0.0913–0.1161 in. (2.32–2.95mm) in steps of 0.004 in. (0.01mm). The thickness is stamped on the shim; this side is always installed facing down. Select new shims with thickness as close as possible to calculated valve and install it in the adjuster.

18. Install the new shim onto the adjuster.

19. Depress the adjuster and remove the keeper tool. Remove the depressor tool and recheck the valve clearance. Repeat this procedure for any other valves requiring adjustment.

20. When all valve adjustments are fin-

ished, install the cylinder head cover, spark plugs, and the intake manifold collector.

4.5L Engines

➡**Check and adjust the valve clearances while the engine is warm and not running.**

➡**The 4.5L firing order is 1-8-7-3-6-5-4-2. The left bank has cylinders No. 1, 3, 5 and 7 from front to rear and the right bank has cylinders No. 2, 4, 6 and 8 from front to rear.**

1. Before servicing the vehicle, refer to the precautions in the beginning of this section.

2. Remove the engine appearance cover.

3. Remove the left and right rocker covers.

4. Turn the crankshaft clockwise until the TDC mark without the paint mark aligns with the timing pointer.

5. Check that the camshaft lobes on the number one cylinder are pointing outward.

6. Check the following valves:
- Cylinder numbers 1 and 2 intake valves
- Cylinder number 1 exhaust valves
- Cylinder numbers 4 and 5 intake valves
- Cylinder numbers 7 and 8 exhaust valves

7. Using a feeler gauge, measure the clearance between the valve lifter and the camshaft. Record any valve clearance measurements that are out of specification. Intake valve clearance (hot) is 0.012–0.016 in. (0.30–0.41mm) and exhaust valve clearance (hot) is 0.012–0.017 in. (0.30–0.43mm).

8. Turn the crankshaft 270° and set the No. 3 cylinder to TDC of its compression stroke.

9. Check the following valves:
- Cylinder numbers 3 and 4 exhaust valves
- Cylinder numbers 3 and 7 intake valves

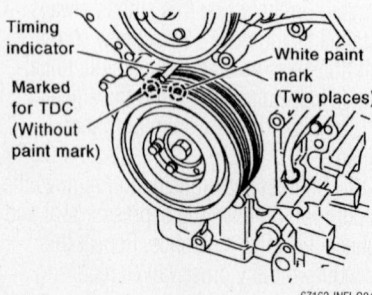

Identifying No. 1 cylinder TDC mark—4.5L engine

Timing indicator

White paint mark (Two places)

Marked for TDC (Without paint mark)

67162-INFI-G24

- Cylinder numbers 5 and 8 exhaust valves

10. Using a feeler gauge, measure the clearance between the valve lifter and the camshaft. Record any valve clearance measurements that are out of specification. Intake valve clearance (hot) is 0.012–0.016 in. (0.30–0.41mm) and exhaust valve clearance (hot) is 0.012–0.017 in. 0.30–0.43mm).

11. Turn the crankshaft 90° (360° from No. 1 TDC) and set the No. 6 cylinder to TDC of its compression stroke.

12. Check the following valves:
- Cylinder numbers 2 and 6 exhaust valves
- Cylinder number 6 intake valves

13. If all the valve clearances are within specification, install the cylinder head cover and the engine appearance cover.

14. If an adjustment is necessary, adjust the valve clearance while engine is cold by removing the adjusting shim. The adjusting shim can be removed by using the following procedures:

   a. Turn the crankshaft so the camshaft lobe of the valve to be adjusted is pointed straight up.

   b. Using an extra fine screwdriver, turn the round hole of the adjusting shim so it faces toward the center of the cylinder head.

   c. Using a depressor tool No. KV10115110 push down on the adjuster and insert a keeper tool on the edge of the adjuster to keep the adjuster in the depressed position.

   d. Remove the depressor tool and remove the shim with a magnet.

➡**Compressed air can be blown into the hole of the adjuster to separate the adjusting shim from the adjuster.**

15. Determine the replacement adjusting shim size by using the following procedures and formula:

   a. Using a micrometer determine thickness of the removed shim.

   b. Calculate the thickness of a new adjusting shim so valve clearance is within the specified values.
- R= thickness of the removed shim
- N= thickness of the new shim
- M= measured valve clearance
- Calculate the Intake Shim as follows: N = R + M—0.0118 in. (0.30mm)
- Calculate the Exhaust Shim as follows: N = R + M—0.0130 in. (0.33mm)

16. Shims are available in 64 sizes from 0.0913–0.1161 in. (2.32–2.95mm) in steps

of 0.004 in. (0.01mm). The thickness is stamped on the shim; this side is always installed facing down. Select new shims with thickness as close as possible to calculated valve and install it in the adjuster.

17. Install the new shim onto the adjuster.

18. Depress the adjuster and remove the keeper tool. Remove the depressor tool and recheck the valve clearance. Repeat this procedure for any other valves requiring adjustment.

19. When all valve adjustments are finished, install the cylinder head cover and the engine appearance cover.

Starter Motor

REMOVAL & INSTALLATION

2.0L Engine

1. Before servicing the vehicle, refer to the precautions in the beginning of this section.

2. Remove or disconnect the following:
- Negative battery cable
- Starter insulator
- Starter harness connector and cable
- Starter mounting bolt and nut
- Starter

To install:

3. Install or connect the following:
- Starter. Torque the bolts to 27 ft. lbs. (36 Nm).
- Starter harness connector and cable
- Starter insulator
- Negative battery cable

3.0L and 3.5L Engines

EXCEPT G35 MODELS

1. Before servicing the vehicle, refer to the precautions in the beginning of this section.

2. Remove or disconnect the following:
- Negative battery cable
- Air duct assembly
- Harness protector
- Starter harness
- Both starter bolts
- Starter

To install:

3. Install or connect the following:
- Starter
- Both starter bolts. Tighten the long bolt to 57–72 ft. lbs. (77–98 Nm) and the short bolt to 22–30 ft. lbs. (30–41 Nm).
- Starter harness
- Harness protector

- Air duct assembly
- Negative battery cable

G35 MODELS

1. Before servicing the vehicle, refer to the precautions in the beginning of this section.

2. Remove or disconnect the following:
   - Negative battery cable
   - Engine undercover
   - S and B terminals from the starter
   - Starter mounting bolts and harness bracket
   - Starter

To install:

3. Install or connect the following:
   - Starter motor. Tighten the mounting bolts to 37–45 ft. lbs. (49–62 Nm).
   - Terminals to the starter. Tighten the B terminal nut to 83–95 inch lbs. (9–10.8 Nm).
   - Engine undercover
   - Negative battery cable

4.1L Engine

1. Before servicing the vehicle, refer to the precautions in the beginning of this section.

2. Remove or disconnect the following:
   - Negative battery cable
   - Steering gear and linkage assembly
   - Harness connector
   - Starter retainers
   - Starter

To install:

3. Install or connect the following:
   - Starter
   - Starter retainers. Tighten the starter retainers to 30–37 ft. lbs. (40–50 Nm).
   - Harness connector
   - Steering gear and linkage assembly
   - Negative battery cable

4.5L Engine

1. Before servicing the vehicle, refer to the precautions in the beginning of this section.

2. Remove or disconnect the following:
   - Negative battery cable
   - Engine undercover
   - Harness connector
   - Starter retainers

3. Remove the engine mounting insulator bottom nut and raise the engine about 2 inches.

4. Remove the starter.

To install:

5. Install or connect the following:
   - Starter
   - Engine mounting insulator nut

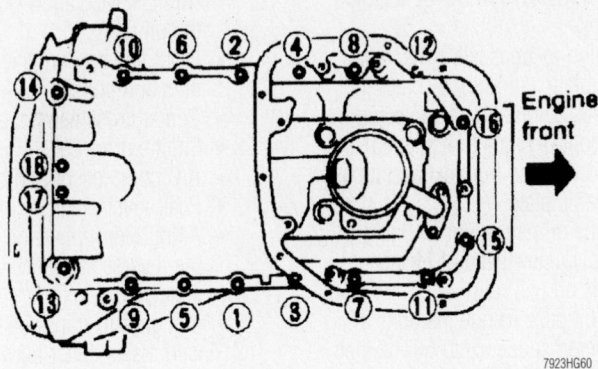

Tighten the aluminum oil pan mounting bolts in the sequence shown—2.0L engine

- Starter retainers. Tighten the starter retainers to 31–38 ft. lbs. (41–52 Nm).
- Harness connector
- Engine undercover
- Negative battery cable

Oil Pan

REMOVAL & INSTALLATION

2.0L Engine

1. Before servicing the vehicle, refer to the precautions in the beginning of this section.

2. Raise and support the vehicle safely.

3. Drain the engine oil.

4. Remove or disconnect the following:
   - Negative battery cable
   - Engine undercover
   - Steel oil pan bolts in the proper sequence
   - Steel oil pan. Insert tool KV10111100 between steel oil pan and aluminum oil pan to break the seal.
   - Front exhaust tube and support the transaxle with a suitable jack and

raise the engine with an engine hoist
   - Center crossmember
   - Transaxle shift control cable, if equipped with an automatic transaxle
   - A/C compressor bracket gussets and the rear cover plate
   - Aluminum oil pan bolts in sequence
   - Baffle plate
   - 2 engine-to-transaxle bolts and install them into vacant bolt holes on the oil pan. Tighten the bolts to release the oil pan from the cylinder block. Use tool KV10111100 to break the remaining seal.

To install:

5. Remove the 2 bolts previously installed in the oil pan.

6. Clean the oil pan rail of all liquid gasket and apply a new bead of 1/8 inch thickness to the oil pan rail.

7. Install or connect the following:
   - Aluminum oil pan.

8. Torque the bolts in the proper sequence as follows:
   a. Bolts 1 through 16 to 12–14 ft. lbs. (16–19 Nm).

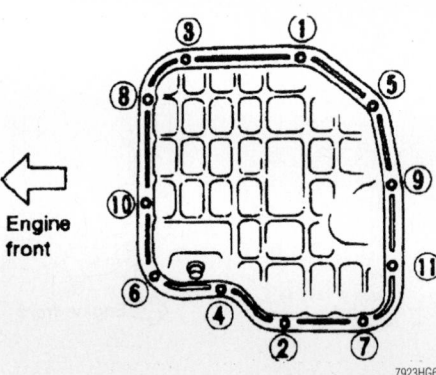

Be sure to tighten the steel oil pan mounting bolts in the proper order to prevent leakage—2.0L engine

b. Bolts 17 and 18 to 56–66 inch lbs. (6.5–7.5 Nm).
- 2 engine-to-transaxle bolts, rear cover plate, compressor bracket gussets, automatic transmission shift control cable (if equipped), center member, front exhaust tube and baffle plate

9. Clean the oil pan rail of all liquid gasket and apply a new bead of ⅛ inch thickness to the oil pan rail.
- Steel oil pan. Torque the bolts in numbered sequence to 56–66 inch lbs. (6–8 Nm). Wait 30 minutes before refilling engine case with oil.
- Negative battery cable

10. Fill the engine with clean oil.
- Start the vehicle, check for leaks and repair if necessary.

3.0L and 3.5L Engines

EXCEPT G35 MODELS

1. Before servicing the vehicle, refer to the precautions in the beginning of this section.

2. Drain the engine oil.

3. Remove or disconnect the following:
- Negative battery cable
- Engine undercover(s)
- Steel (lower) oil pan bolts in the reverse sequence of the torque sequence

4. Insert a seal cutter between the steel and aluminum oil pan.

5. Tapping the cutter with a hammer, slide it around the entire edge of the oil pan. Be careful not to damage the aluminum mating surface of the upper oil pan.
- Steel oil pan and the oil strainer
- Front exhaust pipe and its support

6. Hang the engine at the right and left side engine slingers with a suitable hoist.

7. Position a suitable jack under the transaxle.

8. Remove or disconnect the following:
- Crankshaft Position (CKP) sensors

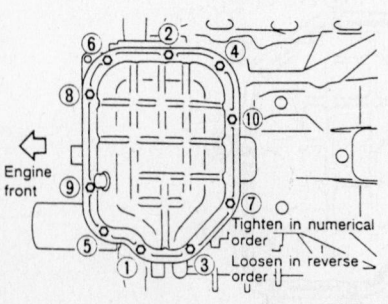

Steel (lower) oil pan loosening and tightening sequence—3.0L and 3.5L engines, except G35

9307HG10

(REFERENCE and POSITION) from the oil pan
- Front and rear engine mounting nuts and bolts
- Center crossmember assembly
- Engine drive belts
- A/C compressor and bracket
- Rear cover plate
- Aluminum (upper) oil pan bolts in the reverse sequence of the torque sequence
- 4 engine-to-transaxle bolts

9. Insert a seal cutter between the aluminum oil pan and the engine block.

10. Tapping the cutter with a hammer, slide it around the entire edge of the oil pan. Be careful not to damage the mating surfaces of the oil pan or engine block.
- Oil pan assembly
- O-rings from the cylinder block and oil pump body

To install:

11. Install or connect the following:
- Baffle plate to the oil pan. Torque the bolts to 22–27 inch lbs. (2–3 Nm).

12. Apply sealant to the front and rear seal of the oil pan.
- New O-rings to the cylinder block and the oil pump body

13. Apply a 0.177–0.217 in. (4.5–5.5mm) continuous bead of liquid

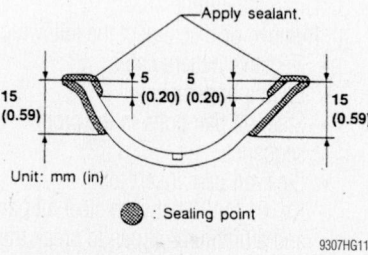

Apply sealant to the front and rear seal of the oil pan as shown—3.0L and 3.5L engines, except G35

9307HG11

gasket to the upper oil pan mating surface and install the oil pan. Torque the bolts in sequence to 12–14 ft. lbs. (16–19 Nm).
- Oil pan strainer. Torque the bolts to 12–14 ft. lbs. (16–19 Nm).
- Rear cover plate and the lower transaxle bolts
- A/C compressor and bracket
- Engine drive belts and adjust as necessary
- Center crossmember assembly
- Front and rear engine mounting nuts and bolts and remove the support jack and the engine hoist
- CKP sensors (REFERENCE and POSITION) to the oil pan. Torque the bolts to 75–96 inch lbs. (9–10 Nm).
- Front exhaust pipe and its support
- Oil strainer

14. Apply a 0.177–0.217 in. (4.5–5.5 mm) continuous bead of liquid gasket to the lower oil pan mating surface and install the oil pan. Tighten the mounting bolts in sequence to 57–66 inch lbs. (6–8 Nm)

➡ **Wait at least 30 minutes before refilling the engine oil.**

- Engine undercover(s)
- Negative battery cable

15. Fill the engine with clean oil.

16. Start the engine, check for leaks and repair if necessary.

G35 MODELS

1. Before servicing the vehicle, refer to the precautions in the beginning of this section.

2. Drain the engine oil and coolant.

3. Remove the hood.

4. Install a suitable engine slinger to secure the engine for crossmember removal.

5. Remove or disconnect the following:
- Front suspension crossmember
- Drive belts
- Alternator and starter
- Idler pulley and bracket
- Crankshaft Position (CKP) sensor
- Oil filter and oil cooler, as necessary
- Lower oil pan bolts in the reverse sequence of the torque sequence

6. Insert a seal cutter between the upper and lower oil pan. Tapping the cutter with a hammer, slide it around the entire edge of the oil pan. Be careful not to damage the aluminum mating surface of the upper oil pan.
- Transmission joint bolts which pierce upper oil pan
- Rear cover plate

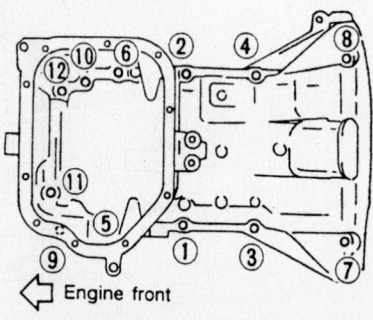

Aluminum oil pan torque sequence (loosen in reverse sequence)—3.0L and 3.5L engines, except G35

7923HG63

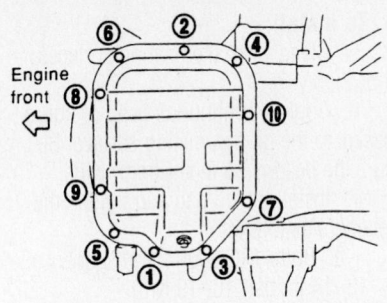

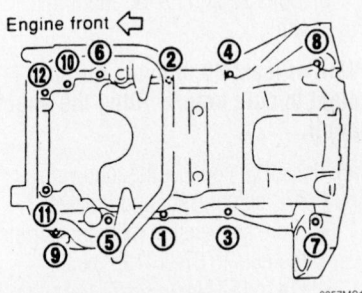

Lower and upper oil pan tightening sequence (use reverse for loosening)—G35 models

- Upper oil pan bolts in the reverse of the torque sequence

7. Insert a seal cutter between the steel and aluminum oil pan. Tapping the cutter with a hammer, slide it around the entire edge of the oil pan. Be careful not to damage the mating surfaces.
- Oil strainer
- O-rings and discard
8. Clean all gasket mating surfaces.

To install:

9. Install or connect the following:
- Oil strainer to the oil pump
- New O-rings to the cylinder block and oil pump side
- Oil pan gasket, applying RTV sealant as shown. Align the protrusion of the oil pan gasket with the notches of the front timing chain case and rear oil seal retainer.

10. Apply sealant as shown in the illustration. Install the upper oil pan and tighten the bolts, in sequence, to 12–13 ft. lbs. (15.7–18.6 Nm).

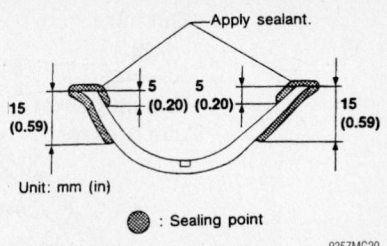

Proper sealant application for upper oil pan gasket—G35 models

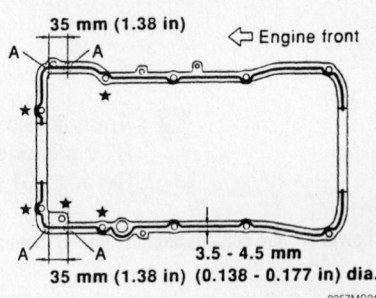

Upper oil pan sealant application—G35 models

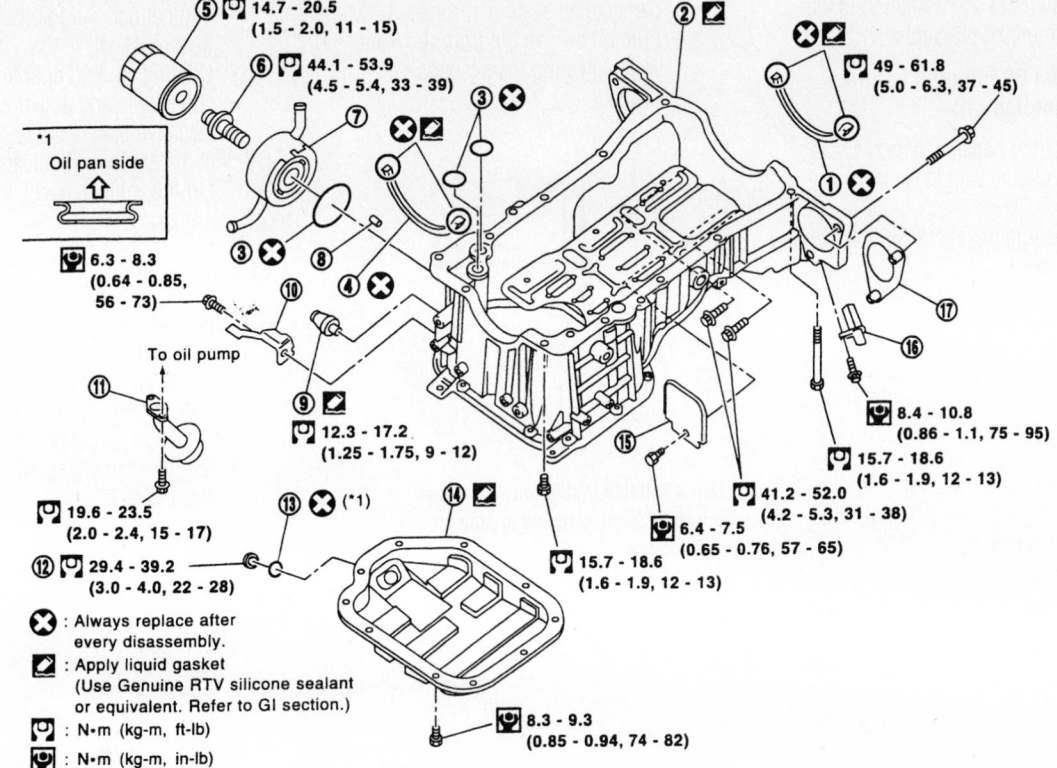

1. Oil pan gasket
2. Oil pan (upper)
3. O-ring
4. Oil pan gasket
5. Oil filter
6. Connector bolt
7. Oil cooler
8. Relief valve
9. Oil pressure switch
10. Bracket
11. Oil strainer
12. Drain plug
13. Drain plug washer
14. Oil pan (lower)
15. Rear plate
16. Crankshaft position sensor (POS)
17. Rear cover plate

Exploded view of the upper and lower oil pans and related components—G35 models

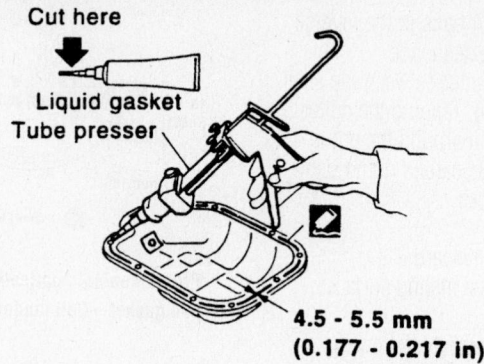

Cut here

Liquid gasket
Tube presser

4.5 - 5.5 mm
(0.177 - 0.217 in)

: Apply liquid gasket. (Use Genuine
RTV silicone sealant or equivalent.
Refer to GI section.)

9357MG22

Lower oil pan sealant application—G35 models

• Transmission joint bolts
11. For the lower oil pan, apply sealant
as shown in the illustrations, then install
and tighten the bolts, in sequence to 74–82
inch lbs. (8.3–9.3 Nm).
12. The remainder of installation is the
reverse of the removal procedure.

➡**Wait at least 30 minutes before
refilling the engine oil.**

13. Connect the negative battery cable
14. Fill the engine with clean oil and
coolant.
15. Start the engine, check for leaks and
repair if necessary.

4.1L Engine

1. Before servicing the vehicle, refer to
the precautions in the beginning of this section.
2. Drain the engine oil.
3. Attach an engine support fixture to
the engine so the right and left engine
mounts can be removed.
4. Remove or disconnect the following:
• Negative battery cable
• Drive belts
• Cooling fan and coupling
• Power steering oil pump
• Front stabilizer bar brackets from
the side members
• Right and left engine mounting
bolts
• Steering shaft lower joint
• Power steering tube bracket and
support the front suspension member
• Lower the front suspension member

• A/C compressor and bracket
• Oil pan mounting bolts, then insert
a tool into the notch on the oil pan
and break the seal between the pan
and engine block. Be careful not to
damage the sealing surface.
• Pull the oil pan out from the front
while lowering the suspension as
needed

Suitable tool

7923HG65

**Use a suitable tool to break the seal
between the oil pan and engine block—
4.1L engine**

To install:

5. Clean all gasket mating surfaces
thoroughly.
6. Apply a continuous bead of liquid
gasket to the oil pan mating surface. Be
sure the bead is ⅛ inch (3mm) wide.
7. Install the oil pan and tighten the
retainers as follows:
  a. Bolts 1 through 21 in sequence:
12–14 ft. lbs. (16–19 Nm).
  b. Bolts 22 and 23: 56–65 in lbs.
(6–7 Nm).

➡**Wait at least 30 minutes for the
sealant to cure before filling the engine
with oil.**

8. Install or connect the following:
• A/C compressor and bracket
• Front suspension member. Torque
the nuts to 87–101 ft. lbs.
(147–167 Nm).
• Power steering tube on the suspension member
• Lower steering shaft joint
• Stabilizer bar to the suspension
member. Torque the nuts to 35–46
ft. lbs. (47–62 Nm).
• Engine mounting bolts. Torque the
nuts to 41–49 ft. lbs. (55–67 Nm).
• Cooling fan and coupling
• Drive belts and adjust as required
• Negative battery cable
9. Fill the engine with clean oil.
10. Start the vehicle, check for leaks and
repair if necessary.

4.5L Engine

1. Before servicing the vehicle, refer to
the precautions in the beginning of this section.
2. Drain the engine oil.
3. Remove or disconnect the following:
• Negative battery cable
• Front wheels
• Hood
• Engine appearance cover

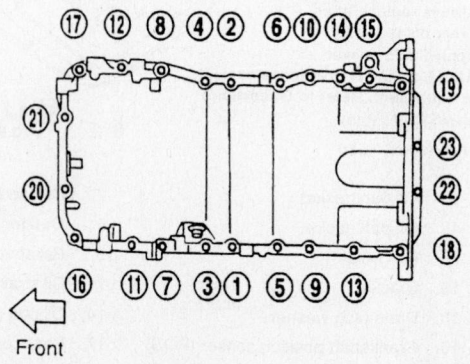

Front

7923HG66

Tighten the oil pan bolts in the sequence shown—4.1L engine

- Engine splash shield
- Drive belts
- Power steering pump belt tensioner
- Power steering oil pump and wire aside
- Oil filter
- A/C compressor and wire aside
- Wiring harnesses at oil pan
- Crankshaft Position Sensor (CKP)

4. Attach an engine lifting device and hold engine in place.

5. Remove the front lower crossmember.

6. Remove the oil pan rear plate cover.

7. Remove 4 transmission-to-oil pan bolts.

8. Remove oil pan bolts in reverse order of the tightening sequence.

9. Remove the oil pan and oil strainer.

To install:

10. Clean all gasket mating surfaces thoroughly.

11. Apply a continuous bead of liquid gasket to the oil pan mating surface. Be sure the bead is ⅛ inch (3mm) wide.

12. Install the oil pan and oil strainer. Tighten the bolts in sequence to 15–17 ft. lbs. (20–23 Nm).

13. Reverse the removal procedure to complete the installation. Tighten the front crossmember bolts to 80–93 ft. lbs. (108–123 Nm).

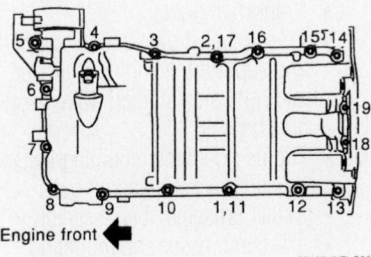

Oil pan bolt tightening sequence—4.5L engine

14. Fill the engine with clean oil.
15. Start the vehicle and check for leaks.

Oil Pump

REMOVAL & INSTALLATION

2.0L Engine

1. Before servicing the vehicle, refer to the precautions in the beginning of this section.

2. Relieve the fuel system pressure.

3. Drain the engine oil.

4. Remove or disconnect the following:
- Negative battery cable
- Drive belts
- Cylinder head with the intake and exhaust manifolds attached
- Oil pans
- Oil strainer and baffle plate
- Crankshaft pulley and the front cover assembly

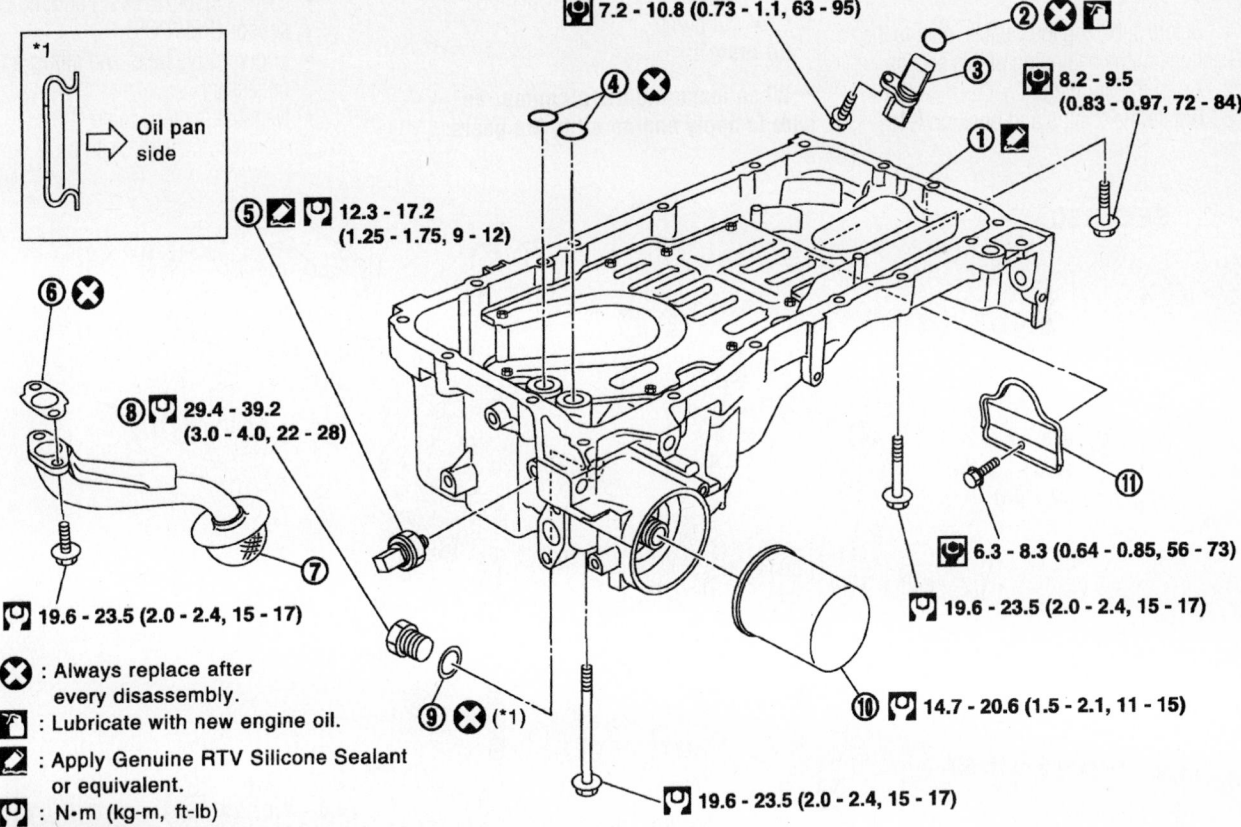

1. Oil pan
4. O-ring
7. Oil strainer
10. Oil filter

2. O-ring
5. Oil pressure switch
8. Oil pan drain plug
11. Rear plate cover

3. Crankshaft position sensor (POS)
6. Gasket
9. Washer

Exploded view of oil pan assembly—4.5L engine

67162-INFI-G25

- Oil pump from the inside of the front cover

To install:

5. Coat the oil pump gears with oil and fit the pump to the cover, using a new oil seal and O-ring.

6. Clean the mating surfaces of liquid gasket and apply a fresh bead of ⅛ inch (3mm) sealer to the sealing surface of the front cover.

7. Install or connect the following:
- Front cover assembly
- Crankshaft pulley
- Oil strainer, baffle plate, oil pans, cylinder head and drive belts
- Negative battery cable

8. Fill the engine with clean oil.

9. Start the vehicle, check for leaks and repair if necessary.

3.0L and 3.5L Engines

➡The oil pump bolts to the front of the engine block and is driven by the crankshaft. Removal of the timing cover and chains are necessary for oil pump service.

1. Before servicing the vehicle, refer to the precautions in the beginning of this section.

2. Drain the engine oil.

3. Rotate the engine and position it to Top Dead Center (TDC) compression stroke of cylinder No. 1.

4. Remove or disconnect the following:
- Negative battery cable
- Drive belts
- Camshaft Position (CMP) sensor (PHASE) and the Crankshaft Position (CKP) sensor (REF/POS)
- Right front wheel and inner fender cover
- Engine undercovers
- Crankshaft pulley
- Front exhaust pipe and its support and support the engine at the left and right side slingers with a suitable hoist
- Engine right side mounting insulator and bracket nuts and bolts
- Center crossmember assembly
- A/C compressor and mounting bracket
- Lower and upper oil pans
- Oil strainer from the oil pump
- Water pump cover and the front cover assembly
- Lower timing chain assembly
- Oil pump

To install:

➡When installing the oil pump, be sure to apply engine oil to the gears.

5. Install or connect the following:
- Oil pump. Torque the bolts to 57–66 inch lbs. (6.4–7.5 Nm) and the mounting screws to 53–69 inch lbs. (5.9–7.8 Nm).
- Lower timing chain assembly
- Front timing cover and water pump covers
- Oil strainer using a new gasket. Torque the bolts to 12–14 ft. lbs. (16–19 Nm).
- Upper and lower oil pans. Be sure to use new O-rings at the oil pump to upper oil pan mating surface.
- A/C compressor and mounting bracket
- Center crossmember assembly
- Engine right side mounting insulator and bracket and remove the engine support hoist
- Front exhaust pipe and its support
- Crankshaft pulley
- Engine undercovers and the right side inner fender cover
- Right front wheel
- CMP sensor (PHASE) and the CKP sensor (REF/POS)
- Engine drive belts and adjust as necessary
- Negative battery cable

6. Fill the engine with clean oil.

SEC. 150

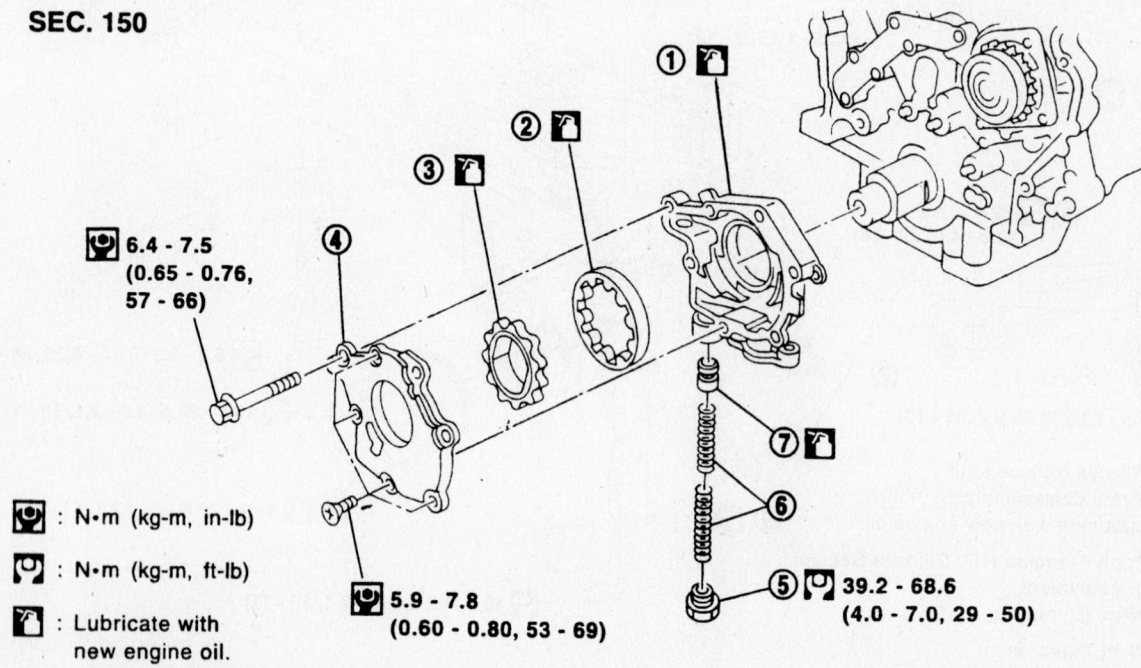

⌾ 6.4 - 7.5 (0.65 - 0.76, 57 - 66)

⌾ 5.9 - 7.8 (0.60 - 0.80, 53 - 69)

⌷ 39.2 - 68.6 (4.0 - 7.0, 29 - 50)

⌾ : N·m (kg-m, in-lb)

⌷ : N·m (kg-m, ft-lb)

▣ : Lubricate with new engine oil.

1. Oil pump body
2. Outer rotor
3. Inner rotor
4. Oil pump cover
5. Regulator valve
6. Spring
7. Regulator valve

Exploded view of the oil pump—G35 shown, other models similar

9357MG23

7. Start the engine, check the oil pressure, and check for oil leaks.

4.1L Engine

➡ **The oil pump is mounted in the cylinder block below the left bank and behind the left timing chain.**

1. Before servicing the vehicle, refer to the precautions in the beginning of this section.
2. Drain the engine oil.
3. Remove or disconnect the following:
   • Negative battery cable
   • Timing chains
   • Oil pump assembly from the front of the engine

To install:

4. Clean the oil pump mounting surface.
5. Install or connect the following:
   • Oil pump using a new gasket. Torque the short bolt to 56–66 ft. lbs. (6–8 Nm) and the long bolt to 12–14 ft. lbs. (16–19 Nm).
   • Timing chains
   • Negative battery cable
6. Fill the engine with clean oil.
7. Start the vehicle, check for leaks and repair if necessary.

4.5L Engine

➡ **The oil pump is mounted in the cylinder block below the left bank and behind the left timing chain.**

1. Before servicing the vehicle, refer to the precautions in the beginning of this section.
2. Drain the engine oil.
3. Remove or disconnect the following:
   • Negative battery cable
   • Timing chains
   • Oil pump assembly from the front of the engine

To install:

4. Clean the oil pump mounting surface.
5. Install or connect the following:
   • Oil pump using a new gasket. Torque the bolts to 78 inch lbs. (9 Nm).
   • Timing chains
   • Negative battery cable
6. Fill the engine with clean oil.
7. Start the vehicle, check for leaks and repair if necessary.

Rear Main Seal

REMOVAL & INSTALLATION

1. Before servicing the vehicle, refer to the precautions in the beginning of this section.

2. Remove or disconnect the following:
   • Transmission or transaxle
   • Drive plate from the crankshaft
3. Carefully pry the seal out of the retainer without damaging the crankshaft or the seal retainer.

To install:

4. Lubricate the seal with clean engine oil.
5. Install or connect the following:
   • Seal into the retainer using the appropriate seal driver
   • Driveplate and transmission or transaxle

Timing Chain, Sprockets, Front Cover and Seal

REMOVAL & INSTALLATION

2.0L Engine

1. Before servicing the vehicle, refer to the precautions in the beginning of this section.
2. Relieve the fuel system pressure.
3. Raise and support the vehicle safely.
4. Drain the cooling system.
5. Remove or disconnect the following:
   • Negative battery cable
   • Engine undercovers
   • Right front wheel
   • Engine side cover and lower the vehicle
   • Radiator
   • Intake manifold air duct
   • Drive belts, water pump pulley, alternator and power steering pump
   • Vacuum hoses, fuel hoses and wiring harness connectors
   • Spark plugs
   • Cylinder head cover and oil separator
   • Intake manifold supports
   • Oil filter bracket and the power steering oil pump bracket
6. Place the No. 1 piston at Top Dead Center (TDC) on the compression stroke.
   • Chain tensioner
   • Distributor. Do not turn the rotor while the distributor is removed.
   • Timing chain guide
   • Camshaft sprockets
   • Camshafts, camshaft brackets, oil tubes and baffle plate
   • Starter
   • Heater hoses and the water hoses from the cylinder head
   • Knock sensor harness connector
   • Cylinder head outside bolts
   • Cylinder head with the intake and

exhaust manifolds and raise and support the vehicle safely
   • Oil pans
   • Oil strainer and baffle plate
   • Crankshaft pulley and place a transmission jack under the main bearing beam and raise the engine slightly to take the weight off of the front engine mount
   • Front engine mount
   • Timing chain cover. Tap the seal out of the cover with a suitable seal driver.
   • Timing chain sprocket bolts
   • Timing chain guides, timing chain and sprockets

To install:

7. Be sure all sealing surfaces are clean and prepared for assembly.
8. Install or connect the following:
   • Crankshaft sprocket. Position the crankshaft so No. 1 piston is set at TDC (keyway at 12 o'clock, mating mark at 4 o'clock).
9. Fit the timing chain to crankshaft sprocket with the gold mating mark on the chain aligned with the mark on the sprocket. (The mating marks for the camshaft sprockets are silver).
   • Timing chain guides and hang the chain off the left (front) guide. If necessary, secure the chain so it does not disengage from the crankshaft sprocket during assembly.
   • New seal in the front cover and apply engine oil to the lip of the seal and apply a bead of liquid gasket to the front cover
   • Oil pump drive spacer and front cover. Torque the bolts evenly to 60 inch lbs. (6.7 Nm) and wipe away any excess liquid gasket.
   • Front engine mount
   • Crankshaft pulley and temporarily tighten the bolt to hold the sprocket in place. The timing mark should align with the TDC mark.
   • Oil strainer, baffle plate and oil pan
   • Cylinder head, camshafts, oil tubes and baffles. Position the left camshaft key at 12 o'clock and the right camshaft key at 10 o'clock.
   • Camshaft sprockets by lining up the mating marks on the timing chain with the mating marks on the camshaft sprockets. Torque the camshaft bolts to 101–116 ft. lbs. (137–157 Nm) and the crankshaft pulley bolt to 105–112 ft. lbs. (142–152 Nm).
   • Upper timing chain guide and dis-

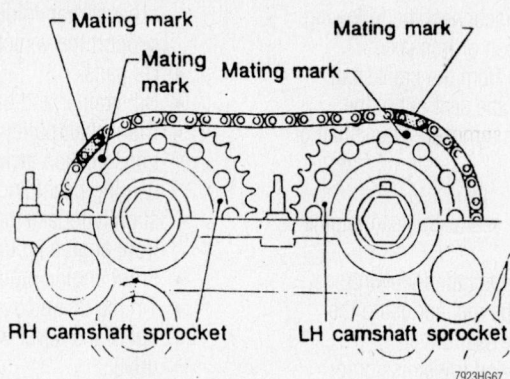

During disassembly, be sure to align the timing chain and camshaft sprocket mating marks—2.0L engine

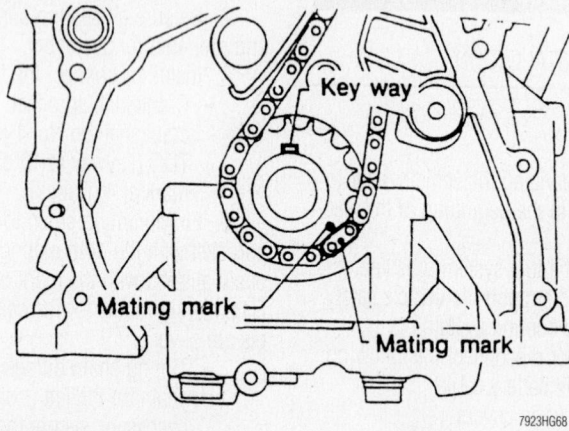

Crankshaft sprocket and timing chain alignment marks—2.0L engine

tributor. Ensure that the rotor is at the 5 o'clock position.

10. Before installing the chain tensioner, press the cam stopper down and the push in the sleeve until the hook can be engaged on the pin. When tensioner is bolted in position, the hook will release automatically. Ensure the arrow on the outside faces the front of the engine.

- Oil filter bracket and the power steering pump bracket
- Intake manifold supports
- Oil separator and the cylinder head cover
- Spark plugs
- Vacuum hoses, fuel hoses, and wiring harness connectors
- Alternator and power steering pump
- Water pump pulley
- Drive belts
- Radiator
- Engine undercovers
- Right front wheel
- Negative battery cable

11. Fill and bleed the cooling system.

12. Start the vehicle, check for leaks and repair if necessary.

4. Relieve the fuel system pressure.
5. Remove or disconnect the following:
- Negative battery cable
- Left side ornament cover
- Air duct to intake manifold hose, collector hose, blow-by hose, and vacuum hoses
- Fuel hoses and the harness connections
- Evaporative emissions (EVAP) canister purge hoses
- Water hoses from the cylinder head and intake manifold
- Ignition coils from the spark plugs
- Exhaust Gas Recirculation (EGR) tube
- Intake manifold collector supports and the collector
- Bolts that secure the fuel tube and the fuel tube from the vehicle
- Bolts that secure the intake manifold to the engine block and the manifold. Loosen the bolts in the reverse sequence of the tightening procedure.
- Left-hand and right-hand rocker covers from the cylinder head
- Engine undercovers
- Right front wheel and the engine side covers
- Drive belts and the idler pulley
- Power steering oil pump belt and the power steering oil pump assembly
- Camshaft Position (CMP) sensor (PHASE) and Crankshaft Position (CKP) sensors (REF)/(POS)

6. Set the No. 1 piston to Top Dead Center (TDC) of compression stroke by rotating the crankshaft.

3.0L Engine

1. Before servicing the vehicle, refer to the precautions in the beginning of this section.
2. Drain the engine oil.
3. Drain the cooling system.

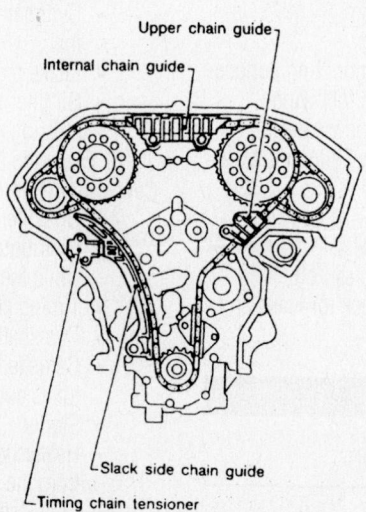

Timing chain tensioner and guide locations—3.0L engine

- Ring gear cover access plate

7. Loosen the crankshaft pulley bolt while securing the ring gear so the crankshaft cannot rotate.

➡ **Use care not to damage the ring gear teeth.**

- Crankshaft pulley using a suitable puller
- Air conditioning compressor and bracket
- Front exhaust pipe and its support

8. Hang the engine at the right and left side engine slingers with a suitable hoist.

9. Support the transaxle with jack.

- Right side engine mount and bracket
- Center crossmember assembly
- Steel oil pan and the oil strainer
- Aluminum oil pan
- Water pump cover
- Front timing chain case bolts in the proper sequence
- Timing chain case cover using the seal cutter being careful not to damage the sealing surfaces
- Internal timing chain guide and the upper chain guide
- Timing chain tensioner and slack side chain guide
- Left and right intake camshaft sprockets first. Be sure to hold the flats of the camshafts while removing the sprocket bolts.
- Lower timing chain assembly. Be sure to note the aligning marks of the chain before removal.

10. Insert a suitable stopper pin for the left and right camshaft tensioners.

- Left and right exhaust camshaft sprocket bolts. Be sure to hold the flats of the camshafts while removing the sprocket bolts.

- Upper timing chain assembly. Be sure to note the aligning marks of the chain before removal.
- Lower timing chain guide
- Crankshaft sprocket
- All traces of liquid gasket from the front timing chain case and from the water pump

11. Inspect the timing chain for excessive wear or damage and replace if necessary.

To install:

12. Install or connect the following:

- Crankshaft sprocket with the mating mark facing out

13. Position the crankshaft to TDC of compression stroke and align the dowels of the camshaft sprockets to the 12 o'clock position in respect to the cylinder head

- Lower timing chain guide. The front mark on the guide should face upwards.

14. On a workbench, align the marks on the intake and exhaust camshaft sprockets with the marks of the chain.

15. Put the exhaust camshaft sprockets onto the dowel pin and torque the mounting bolts to 88–95 ft. lbs. (119–128 Nm). Be sure to secure the camshafts while tightening the bolts.

- Timing chains and sprockets to the intake camshafts. Be sure to align the timing chain and sprocket mating marks.

16. Remove the left and right camshaft tensioner stopper pins.

17. Align the mating mark on the crankshaft with the matchmark (gold link) on the lower timing chain.

- Lower timing chain to the water pump sprocket

18. Working counterclockwise, install the lower timing chain camshaft sprockets.

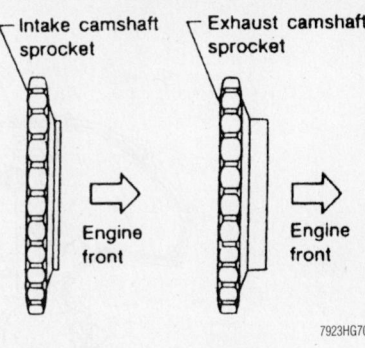

Identification of the intake and exhaust camshaft sprockets—3.0L engine

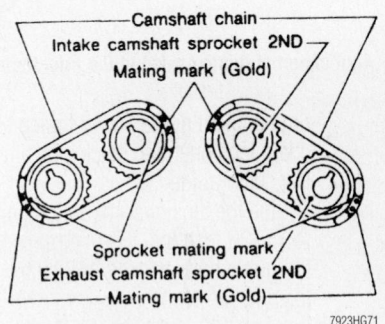

Upper timing chain alignment marks—3.0L engine

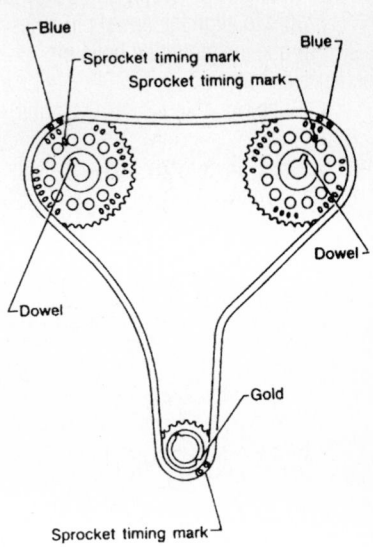

Lower timing chain alignment marks—3.0L engine

Be sure to align the sprocket marks with the blue links of the timing chain during installation.

- Intake sprocket bolts. Torque the bolts to 88–95 ft. lbs. (119–128 Nm). Be sure to secure the camshafts while tightening the bolts.
- Internal timing chain guide, upper

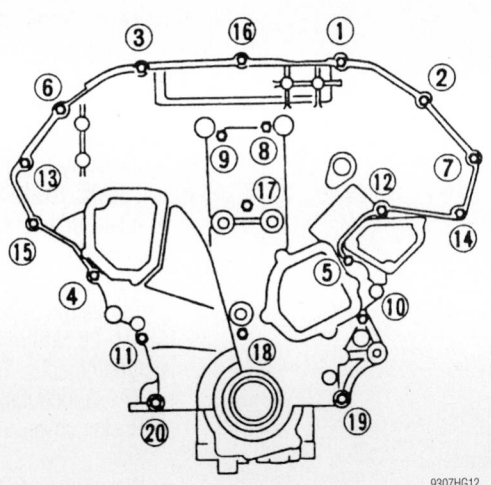

Front timing chain case bolt loosening sequence—3.0L engine

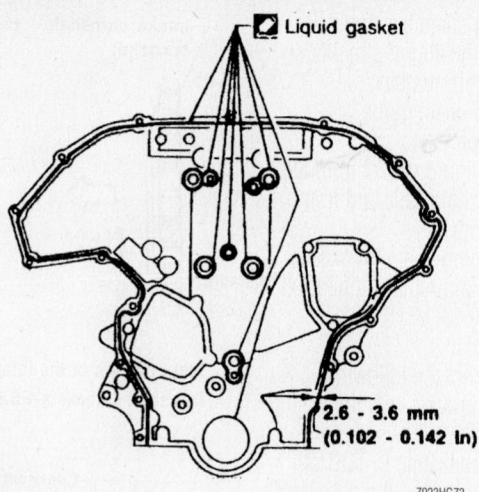

Application of liquid gasket to the front timing case—3.0L engine

timing chain guide, lower timing chain tensioner and slack side timing chain guide
- Torque the tensioner mounting bolt to 75–96 inch lbs. (8–11 Nm) and the guide bolts to 108–168 inch lbs. (13–19 Nm)
19. Apply a 0.102–0.142 in. (3–4mm) continuous bead of liquid gasket to all necessary areas on the front timing cover.
- Timing cover evenly and gently. Be sure to align the dowel pin holes.
20. Torque the mounting bolts in sequence as follows:
    a. Bolts No. 1 and 2: 19–23 ft. lbs. (26–31 Nm).
    b. Bolts No. 3 to 20: 105–121 inch lbs. (12–14 Nm).

➡ **Leave the bolts unattended for 30 minutes or more after tightening. This will allow the liquid gasket to cure sufficiently.**

21. Apply a 0.091–0.130 in. (2–3mm) continuous bead of liquid gasket to the water pump cover and install the cover. Torque the bolts to 84–108 inch lbs. (10–13 Nm).
22. Apply a 0.12 in. (3mm) continuous bead of liquid gasket to the rocker covers and install the covers. Torque the mounting bolts in sequence as follows:
    a. Bolts No. 1 to 10: 9–26 inch lbs. (1–3 Nm).
    b. Bolts No. 1 to 10: 52–69 inch lbs. (6–8 Nm).

23. Apply sealant to the front and rear seal of the oil pan.
24. Apply a 0.177–0.217 in. (4.5–5.5mm) continuous bead of liquid gasket to the upper oil pan mating surface and install the oil pan. Torque the bolts in sequence to 12–14 ft. lbs. (16–19 Nm).
25. Install or connect the following:
- Transaxle bolts that secure the oil pan
- Oil pan strainer. Torque the bolts to 12–14 ft. lbs. (16–19 Nm).
26. Apply a 0.177–0.217 in. (5–6mm) continuous bead of liquid gasket to the lower oil pan mating surface and install the oil pan. Torque the bolts in sequence to 57–66 inch lbs. (6–8 Nm)
- Center crossmember assembly
- Right side engine mounting bracket and mount assembly
27. Remove the engine slinger assembly.
- Front exhaust pipe and its support
- A/C compressor and bracket
- Crankshaft pulley to the crankshaft and the mounting bolt. Torque the mounting bolt to 14–22 ft. lbs. (20–29 Nm). Torque the crankshaft bolt an additional 60–66° clockwise. This is approximately the angle from 1 hexagon bolt head corner to another.
- Ring gear cover plate
- CMP sensor (PHASE) and CKP sensors (REF)/(POS)
- Power steering pump assembly
- Drive belts and the idler pulley
- Right front wheel
- Engine undercovers
28. Install the intake manifold using new gaskets. Torque the bolts in sequence and in 2 stages as follows:
    a. 44–86 inch lbs. (5–10 Nm).
    b. 16–18 ft. lbs. (22–25 Nm).
29. Using new insulators, install the fuel tube assembly. Torque the bolts to 15–20 ft. lbs. (21–26 Nm).
30. Install or connect the following:
- Intake manifold collector gasket with the arrow facing forward
- Intake manifold collector. Torque the bolts to 16–18 ft. lbs. (22–25 Nm).
- Intake manifold collector support brackets
- EGR tube using new gaskets. Torque the bolts to 15–20 ft. lbs. (21–26 Nm) in 2 progressive steps.
- Ignition coils. Torque the bolts to 27–33 inch lbs. (3–4 Nm).
- Rocker cover ornament on the left side
- Water hoses to the cylinder head and intake manifold

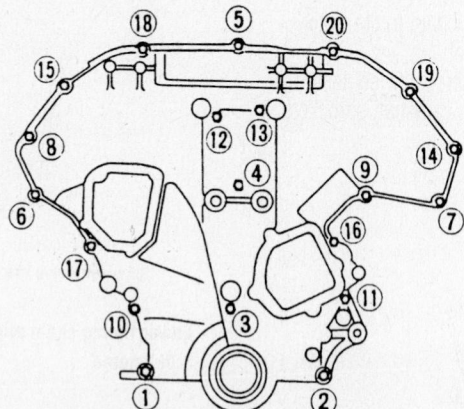

① - ② 8 mm dia. bolts
25.5 - 31.4 N·m
(2.6 - 3.2 kg-m, 18.8 - 23.1 ft-lb)
③ - ⑳ 6 mm dia. bolts
11.8 - 13.7 N·m
(1.2 - 1.4 kg-m, 8.7 - 10.1 ft-lb)

Front timing chain case bolt tightening sequence—3.0L engine

- EVAP canister purge hoses
- Fuel hoses and wiring harness connections to the fuel rail
- Air duct to intake manifold hose, collector hose, blow-by hose, and vacuum hoses
- Negative battery cable

31. Fill the engine with clean oil.

32. Fill and bleed the cooling system.

33. Start the engine and run at 3000 RPM under no load to purge the air from the high pressure chamber. The engine may produce a rattling noise. This indicates that air still remains in the chamber and is not a matter of concern.

3.5L Engines

1. Before servicing the vehicle, refer to the precautions in the beginning of this section.

2. Properly relieve the fuel system pressure.

3. Drain the engine oil and cooling system.

4. Remove or disconnect the following:
   - Negative battery cable
   - Right and left side rocker covers
   - Cooling fan and radiator
   - A/C compressor from bracket and position side with the lines attached
   - A/C compressor bracket

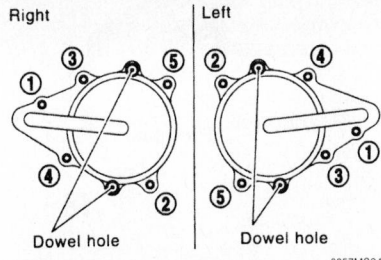

Intake valve timing control cover tightening sequence—3.5L engine

- Power steering pump from its bracket and position aside with the lines attached
- Power steering pump bracket
- Water bypass hose and cooling fan bracket from the front timing chain case
- Lower and upper oil pans
- Right and left side intake valve timing control covers. Loosen the bolts in the reverse of the tightening sequence. Use a seal cutter to cut the gasket.

5. Set the No. 1 piston to Top Dead Center (TDC) on its compression stroke. Align the timing mark (grooved line) on the crankshaft pulley with the timing indicator on the front cover.

6. Make sure the intake and exhaust cam nose for the No. 1 cylinder is positioned as shown in the illustration. If not, turn the crankshaft on revolution (360°) and align.

7. Remove or disconnect the following:
   - Crankshaft pulley using a suitable puller
   - Front timing chain case. Loosen the bolts in the reverse of the torque sequence.

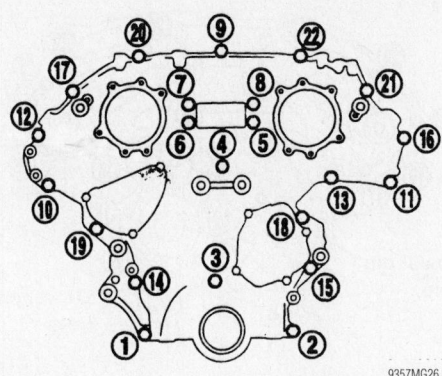

Front timing chain case torque sequence (loosen in reverse order)—3.5L engine

9357MG26

✱✱ WARNING

Do not use a screwdriver to pry the case off!

- Timing chain case. Insert the proper tool into the notch at the top of the case as shown. Pry the case off by levering the tool as shown. Use a seat cutter to cut the liquid gasket.
- Water pump cover and chain tensioner cover from the case
- Front oil seal from the case using a suitable prytool, being careful not the damage the case
- Internal chain guide, timing chain tensioner and slack guide. Remove the upper chain tensioner by pressing the tensioner in and inserting a 0.098 inch (2.5mm) diameter pin in the pin hole. Once secured remove the bolts and the tensioner.

✱✱ WARNING

After the timing chain is removed, do NOT turn the crankshaft and camshaft separately, or the valve will strike the pistons.

8. Remove the timing chain and camshaft sprocket, as follows:
   a. Attach a suitable stopper pin to the

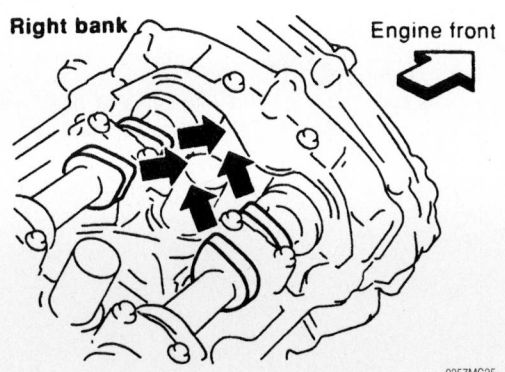

9357MG25

Proper orientation of the intake and exhaust cams—3.5L engine

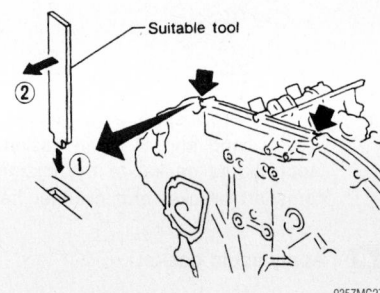

9357MG27

Front timing chain case removal, using a suitable tool—3.5L engine

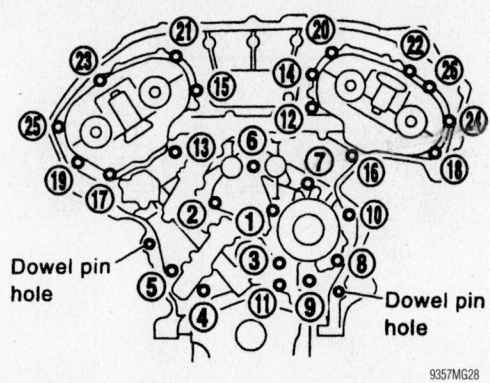

9357MG28

Rear timing chain case torque sequence (loosen in reverse order)—3.5L engines

right and left camshaft chain tensioners (for secondary timing chains).

b. Hold the hex part of the camshaft secure with a wrench, then remove the camshaft sprocket mounting bolts.

c. Remove the primary and secondary timing chains with the camshaft sprockets.

9. Remove or disconnect the following:
- Tension guide and crankshaft sprocket
- Rear timing chain case. Use the

reverse of the tightening sequence, then use a seal cutter to separate the gasket.

To install:

10. Install the rear timing chain case as follows:

a. Install new O-rings onto the cylinder block and head.

b. Apply suitable RTV sealant to the back side of the rear timing chain case as shown in the illustration.

c. Install the rear case and tighten the

bolts, in sequence, to 9–10 ft. lbs. (11.7–13.7 Nm). There are 2 bolts lengths: Bolts 1, 2, 3, 6–10 are 0.79 in. (20mm) long and the other bolts are 0.63 in. (16mm) long.

11. Set the No. 1 piston to Top Dead Center (TDC) on its compression stroke.

12. Install the crankshaft sprocket, making sure the mating marks on the sprocket face the front of the engine.

13. Push the plunger of the secondary chain tensioner and keep it pressed in with a stopper pin.

14. Install the secondary timing chains and camshaft sprockets, as follows:

a. Align the matchmarks on the secondary timing chain (gold link) with the stamped marks on the intake and exhaust sprockets, then install them.

➡**Matchmarks for the intake sprocket are on the back of the secondary sprocket. There are 2 kinds of marks, the right bank uses round marks and the left bank uses oval marks.**

b. Align the dowel pin and pin hole on the camshaft with the groove and

Rear timing chain case

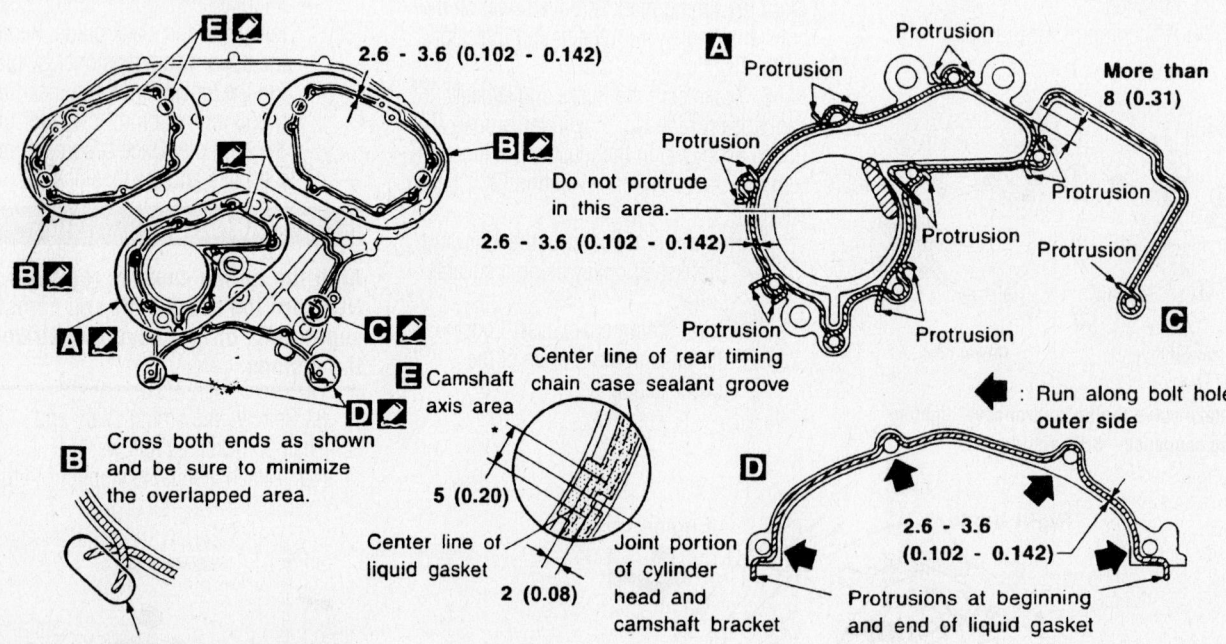

**: Apply liquid gasket

Unit: mm (in)

Rear timing chain case sealant application—3.5L engines

9357MG40

dowel pin on the sprocket, then install them.

   c. On the intake side, align the pin hold on the small diameter side of the camshaft front end with the dowel pin on the on the back side of the camshaft sprocket, and install them.

   d. On the exhaust side, align the dowel pin on the camshaft front end with the pin groove on the camshaft sprocket, then install them.

   e. Tighten the camshaft sprocket bolts hand-tight to prevent the dowel pins from dislocating.

15. Install the primary timing chain, as follows:

   a. Install the primary chain so the punched mating mark on the cam sprocket is aligned with the yellow link on the timing chain, while the notched mating mark on the crankshaft sprocket is aligned with the orange one on the timing chain.

   b. Use a wrench on the hex portion of the camshaft to secure it in place, tighten the camshaft sprocket mounting bolts to 73–78 ft. lbs. (98–108 Nm).

   c. Remove the stopper pins from secondary chain tensioners.

16. Install the internal chain guide, timing chain tensioner, tension guide and slack guide. Do not overtighten the slack guide mounting bolts. It's normal to have a gap under the bolt seats when the mounting bolts are properly tightened to 10–13 ft. lbs. (14–18 Nm).

17. Recheck that all matchmarks are still aligned.

18. Install or connect the following:

- New O-rings on the rear timing chain case
- New front oil seal in to the timing chain cover
- Water pump and chain tensioner covers to the front cover
- Suitable liquid gasket to the back side of the timing chain case
- Dowel pin on the rear timing chain case into the dowel pin hold on the front chain case
- Front timing chain case bolts, in sequence. Tighten the M6 bolts to 9–10 ft. lbs. (11.7–13.7 Nm) and the M8 bolts to 19–23 ft. lbs. (25–31 Nm).
- Right and left intake valve timing control cover
- Crankshaft pulley to the crankshaft and the mounting bolt. Torque the mounting bolt to 29–36 ft. lbs. (39–49 Nm). Torque the crankshaft bolt an additional 60–66° clock-

Example: Right bank side (Rear view)

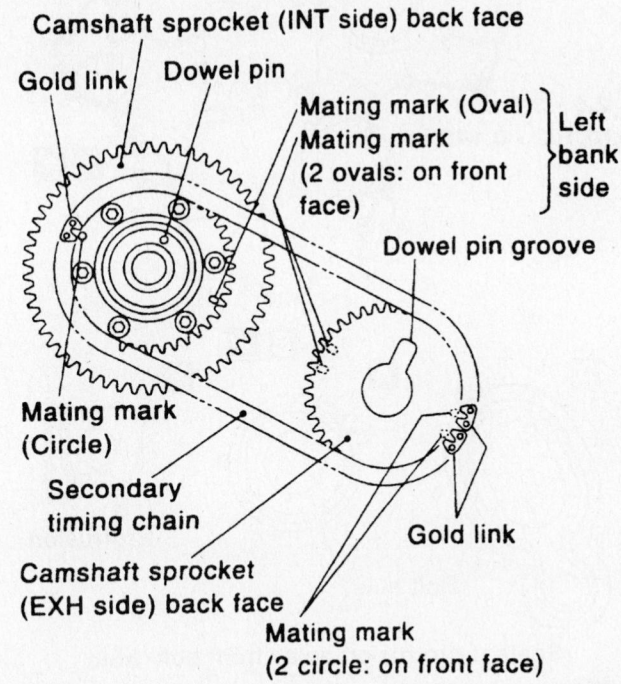

Camshaft sprocket alignment, right side bank shown—3.5L engine

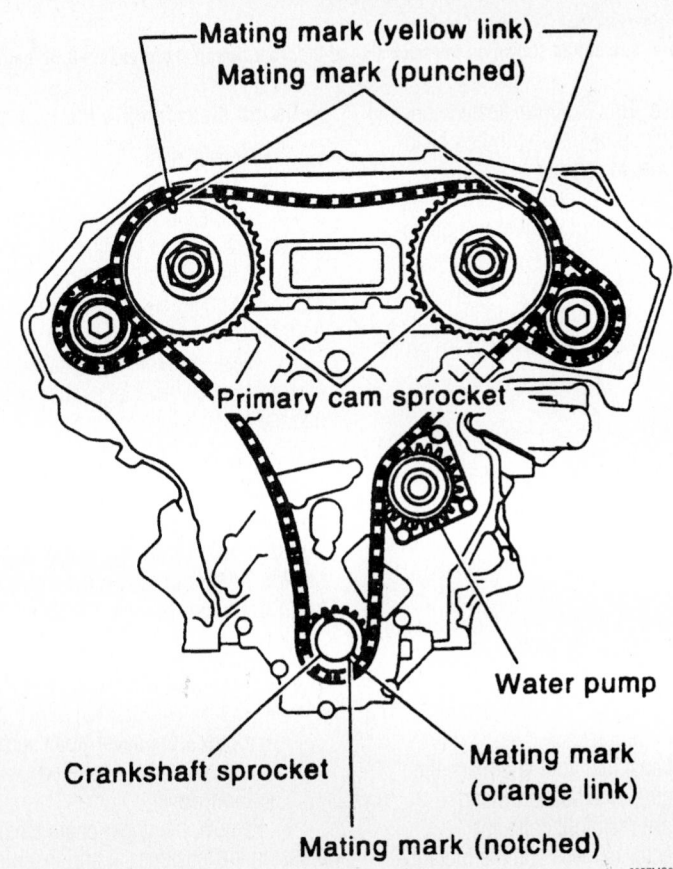

Proper alignment for the primary timing chain—3.5L engine

Front timing chain case

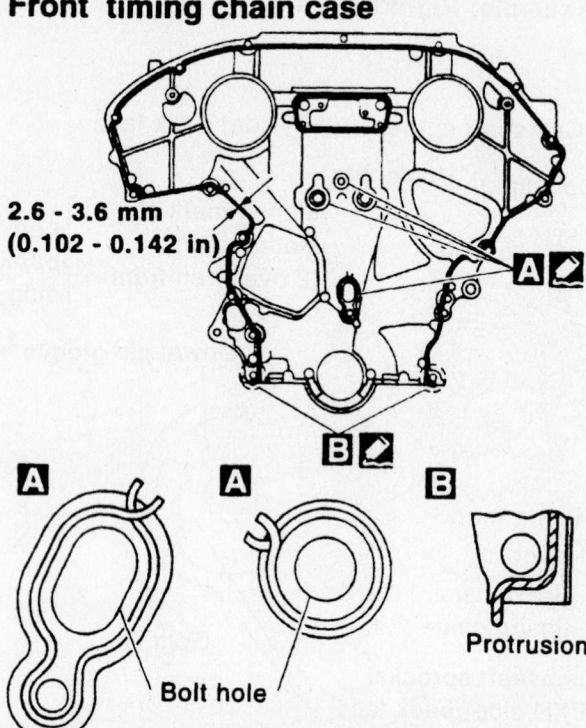

2.6 - 3.6 mm
(0.102 - 0.142 in)

A Bolt hole

B Protrusion

Sealant protrusion away from bolt hole

: Apply liquid gasket (Use Genuine RTV silicone sealant or equivalent. Refer to GI section.)

9357MG32

Apply proper sealant as shown on the back side of the front timing chain case—3.5L engine

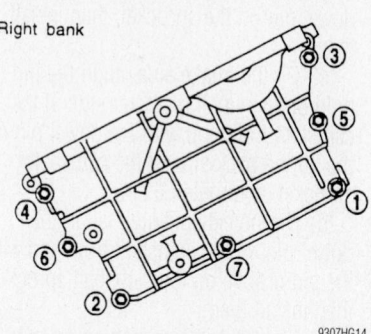

Right bank

9307HG14

Upper front cover (right bank) nut and bolt removal sequence—4.1L engine

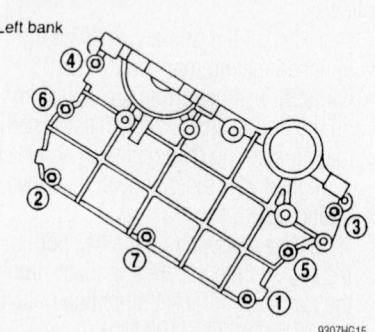

Left bank

9307HG15

Upper front cover (left bank) nut and bolt removal sequence—4.1L engine

wise. This is approximately the angle from 1 hexagon bolt head corner to another.

19. Installation of the remaining components is the reverse of the removal procedure.

20. Fill the engine with clean oil.

21. Fill and bleed the cooling system.

22. Start the engine and check for proper operation.

4.1L Engine

1. Before servicing the vehicle, refer to the precautions in the beginning of this section.

2. Properly relieve the fuel system pressure.

3. Remove or disconnect the following:
- Negative battery cable
- Engine from the vehicle
- Alternator
- A/C compressor
- Exhaust manifold and place the engine on a suitable stand
- Intake manifold collector
- Injector harness and the fuel tube assembly with the injector

➡**Do not disassemble the fuel hose.**

- Intake manifold
- Valve cover

4. Set the No. 1 piston to Top Dead Center (TDC) on its compression stroke. Align the timing mark (orange paint) on the crankshaft pulley with the timing indicator on the front cover.

5. Make sure the intake camshaft lobe for the No. 1 cylinder faces the intake port and the exhaust lobe faces the exhaust port.

➡**It is possible to check the camshaft positions by checking the notch positions on the camshaft when the No. 1 piston is at TDC of the compression stroke. At this position the cylinder head bolts can be removed.**

6. Remove or disconnect the following:
- Crankshaft pulley
- Crankshaft Position (CKP) sensor
- Upper front cover bolts and nuts in the sequence illustrated
- Front cover

7. Remove the upper chain tensioner by pressing the tensioner in and inserting a 0.04 inch (1mm) diameter pin in the pin

hole. Once secured remove the bolts and the tensioner

8. Matchmark the upper timing chain and camshaft sprockets to aid reassembly.
- Upper camshaft sprocket bolt while holding the hexagonal part of the camshaft with a wrench
- Camshaft sprockets

9. Remove the left and right tensioner covers from the front covers by pressing the tensioner in and inserting a 0.04 inch (1mm) diameter pin in the pin hole.
- Idler sprocket bolts
- Chain guide between the No. 1 camshaft bracket
- Cylinder head
- Upper timing chain

10. Matchmark the lower timing chain and idler sprocket to aid reassembly.
- Oil pan
- Front cover bolts in the proper sequence

11. Compress the lower chain tensioners and install a pin through the hole to secure it, then remove the tensioner.
- Oil pump drive chain
- Slack and chain guides. Be sure to note the locations of the bolts so they can be installed in their original positions.

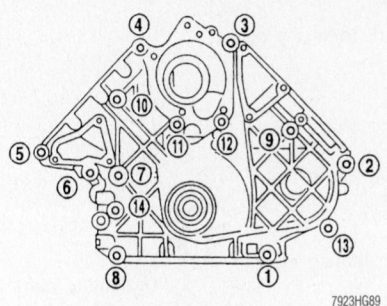

Remove the front cover bolts in the correct sequence—4.1L engine

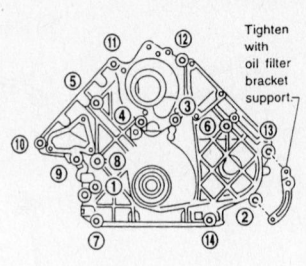

Position	Bolt dimensions	Tightening torque
①, ③, ⑤, ⑦	M6 x 45	6.3 - 8.3 N·m (0.64 - 0.85 kg-m, 55.6 - 73.8 in-lb)
②	M6 x 47	
⑥, ⑩	M6 x 65	
⑬	M6 x 67	
⑫	M6 x 84	
④, ⑧	M8 x 50	16 - 21 N·m (1.6 - 2.1 kg-m, 12 - 15 ft-lb)
⑨	M10 x 52	30 - 40 N·m (3.1 - 4.1 kg-m, 22 - 30 ft-lb)
⑭	M10 x 60	
⑪	M10 x 62	

Tighten with oil filter bracket support→

Front cover bolt location and torque specifications—4.1L engine

- Lower timing chains with the crankshaft sprockets

To install:

12. Be sure the crankshaft key is pointing toward the center of the left bank. This should be a 45° angle from the center.

13. Install or connect the following:
- Lower right bank timing chain by aligning the mark on the chain with the mark on the sprocket and installing the sprocket with chain on the crankshaft. Be sure the thick side of the sprocket faces the cylinder block to provide clearance between the block and chain.
- Slack and chain guides. Be sure to install the bolts in the correct locations. Torque the bolts to 10–14 ft. lbs. (13–19 Nm).
- Lower chain for the left bank in the same manner as the right one
- Left slack and chain guide. Torque the bolts to 10–14 ft. lbs. (13–19 Nm).
- Oil pump drive chain and sprockets. Torque the bolt on the driven gear to 22–30 ft. lbs. (30–40 Nm).
- Lower oil pump drive chain guide

14. Install the upper guide by installing the bolts loosely, then inserting a 0.04 in. (1mm) feeler gauge between the guide and

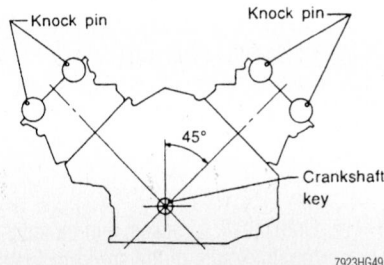

Before assembly, be sure to turn the crankshaft key towards the left cylinder head—4.1L engine

chain. Press on the guide lightly with the same force as the weight of the guide and torque the bolts to 56–74 inch lbs. (6–8 Nm).
- Chain tensioner with the pins installed using new gaskets. Torque the bolts to 82–95 inch lbs. (9–11 Nm).

15. Confirm that the timing marks on the crankshaft sprockets and chains are still aligned.
- Front cover. Be sure to install the bolts in the correct locations.

16. Apply engine oil to the idler shaft and install it on the idler sprocket.

17. Align the mark on the chain with the mark on the idler sprocket and install the sprocket.
- Place the upper chains on the idler sprockets. It is not necessary to align the mating marks at this time. The marks can be aligned after the cylinder head is installed.

18. Install the cylinder heads

19. Align the marks on the upper chains with the marks on the sprockets, then install the sprockets on the camshafts while keeping the marks aligned.
- Idler shaft bolts. Torque the bolts to 32–43 ft. lbs. (43–58 Nm).

20. Remove the lower chain tensioner pins.
- Chain guide between the No. 1 camshaft bracket

21. Align the upper timing chain mating marks with the marks on the sprockets and install the sprockets. Torque the intake sprocket bolt to 76–83 ft. lbs. (103–113 Nm) and the exhaust sprocket bolt to 12–15 ft. lbs. (16–21 Nm)

22. Compress the upper chain tensioners and install a pin through it to secure it in position.
- Tensioners. Torque the bolts to 82–95 inch lbs. (9–11 Nm).

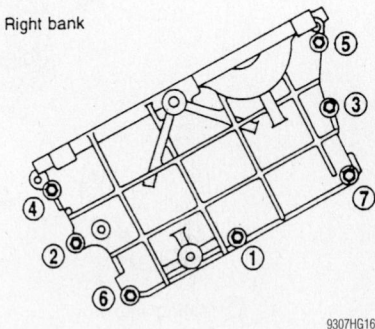

Upper front cover (right bank) nut and bolt tightening sequence—4.1L engine

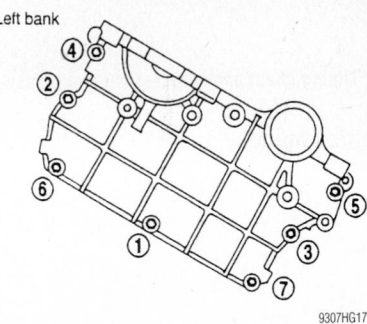

Upper front cover (left bank) nut and bolt tightening sequence—4.1L engine

23. Lubricate the timing chains and related parts with clean engine oil and install the upper covers. Torque the cover bolts in sequence.

24. Installation of the remaining components is the reverse of the removal procedure.

25. Fill the engine with clean oil.

26. Fill and bleed the cooling system.

27. Start the engine and check for proper operation.

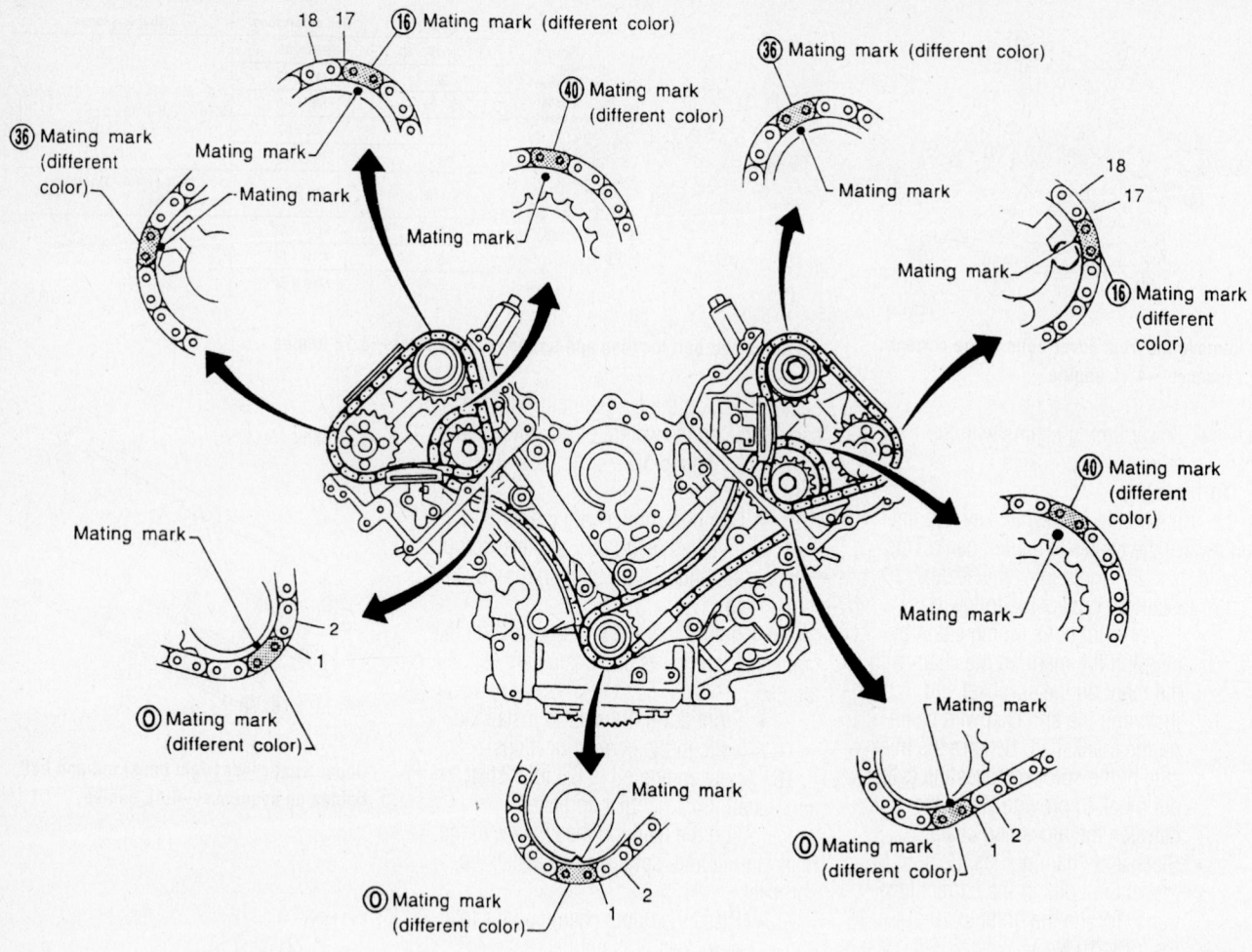

Timing chain and sprocket alignment marks—4.1L engine

7923HG48

4.5L Engine

1. Before servicing the vehicle, refer to the precautions in the beginning of this section.

2. Properly relieve the fuel system pressure.

3. Remove or disconnect the following:
- Negative battery cable
- Engine from the vehicle
- Drive belt tensioner and idler pulley
- Thermostat housing
- Ignition coils
- Cylinder head cover
- Intake valve timing control sensor from both sides
- Camshaft Position sensor (CMP)
- Intake valve timing control solenoid from both sides
- Intake valve timing control covers by loosening bolts in the reverse of the tightening sequence
- Front cover o-rings

4. Turn the crankshaft pulley clockwise until No. 1 cylinder is on the compression stroke. The TDC mark on the pulley without

the paint mark should align with the timing pointer.

5. Verify the correct TDC position by ensuring that the lobes of the No. 1 cylinder camshafts are pointing outward. If not rotate the crankshaft 360° until the lobes are correct.

6. Remove the oil pan rear plate cover and install a locking tool into the drive gear.

7. Remove the crankshaft pulley.

8. Remove the oil pan and strainer

9. Loosen and remove the front cover bolts in the reverse of the tightening sequence.

10. Remove the front cover and pry out the front cover oil seal.

11. Remove the 3 O-rings from the cylinder heads and block.

12. Remove the chain tensioner cover from the front cover.

13. Remove the oil pump drive spacer and oil pump.

14. Compress the left side chain tensioner and install a pin through the hole to secure it, then remove the tensioner.

15. Remove the chain tensioner, tension guide and slack guide.

16. Remove the left side timing chain.

17. While holding the camshaft in place, remove the left side camshaft sprockets.

18. Repeat the procedure on the right side to remove the chain tensioner, guides, timing chain and camshaft sprockets.

To install:

19. Be sure the crankshaft key is pointing toward the center of the left bank. This should be a 45° angle from the center. Check that the camshaft dowel pins are in the correct location as shown.

20. Install or connect the following:
- Camshaft sprockets by holding the camshaft with a wrench and tightening the sprocket bolts to 112 ft. lbs. (152 Nm).
- Crankshaft sprockets and be sure the thick side of the sprocket faces the cylinder block to provide clearance between the block and chain.
- Right bank timing chain by aligning

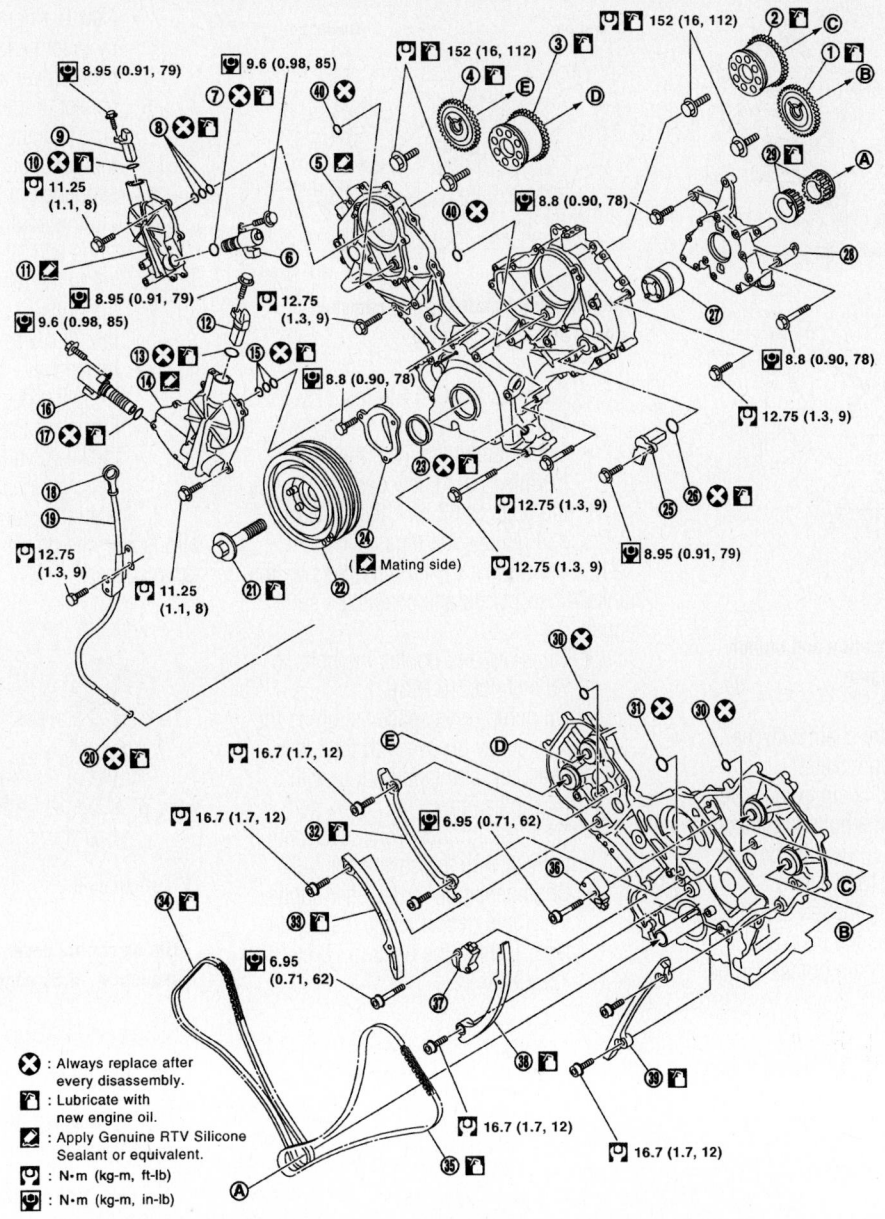

Exploded view of timing chain and front cover components—4.5L engine

1. Camshaft sprocket (left bank EXH)	2. Camshaft sprocket (left bank INT)	3. Camshaft sprocket (right bank INT)
4. Camshaft sprocket (right bank EXH)	5. Front cover	6. Intake valve timing control solenoid valve (right bank)
7. O-ring	8. Seal ring	9. Intake valve timing control position sensor (right bank)
10. O-ring	11. Intake valve timing control cover (right bank)	12. Intake valve timing control position sensor (left bank)
13. O-ring	14. Intake valve timing control cover (left bank)	15. Seal ring
16. Intake valve timing control solenoid valve (left bank)	17. O-ring	18. Oil level gauge
19. Oil level gauge guide	20. O-ring	21. Crankshaft pulley bolt
22. Crankshaft pulley	23. Front oil seal	24. Chain tensioner cover
25. Camshaft position sensor (PHASE)	26. O-ring	27. Oil pump drive spacer
28. Oil pump assembly	29. Crankshaft sprocket	30. O-ring
31. O-ring	32. Timing chain tension guide (right bank)	33. Timing chain slack guide (right bank)
34. Timing chain (right bank)	35. Timing chain (left bank)	36. Chain tensioner (left bank)
37. Chain tensioner (right bank)	38. Timing chain slack guide (left bank)	39. Timing chain tension guide (right bank)
40. O-ring		

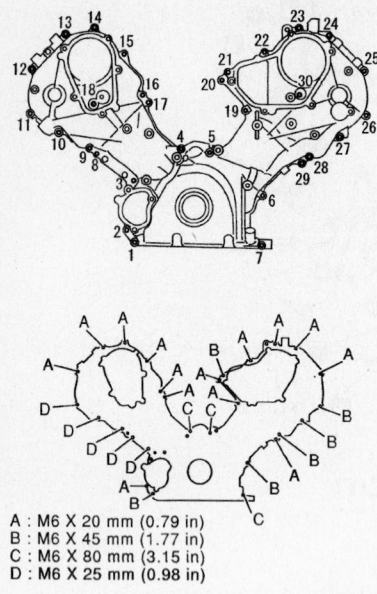

A : M6 X 20 mm (0.79 in)
B : M6 X 45 mm (1.77 in)
C : M6 X 80 mm (3.15 in)
D : M6 X 25 mm (0.98 in)

67162-INFI-G28

Front cover bolt identification and tightening sequence—4.5L engine

the marks on the chain with the marks on the sprockets

- Right slack and chain guides. Be sure to install the bolts in the correct locations. Torque the bolts to 10–14 ft. lbs. (13–19 Nm).
- Timing chain for the left bank in the same manner as the right one
- Left slack and chain guide. Torque

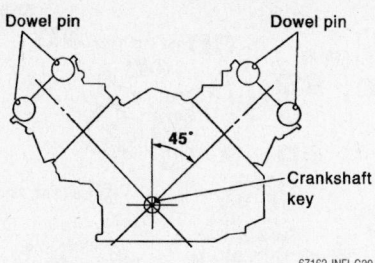

67162-INFI-G29

Aligning crankshaft key and camshaft dowel pins—4.5L engine

the bolts to 10–14 ft. lbs. (13–19 Nm).

- Chain tensioners with the pins installed using new gaskets. Torque the bolts to 62 inch lbs. (7 Nm) and remove the pins.

21. Confirm that the timing marks on the crankshaft sprockets and chains are still aligned.

- Oil pump and tighten the bolts to 78 inch lbs. (9 Nm).
- Oil pump drive spacer, aligning the spacer key groove with the crankshaft key, and tapping it in with a plastic hammer
- New seal into the front cover until it is flush with the cover face
- Chain tensioner cover after applying liquid gasket to the mating surface. Tighten the bolts to 78 inch lbs. (9 Nm).

- New O-rings into the block and cylinder heads
- Front cover after applying liquid gasket to the mating surface. Be sure to install the bolts in the correct locations. Tighten the bolts in sequence to 78 inch lbs. (9 Nm).
- Timing control cover using new O-rings and after applying liquid gasket to the mating surface. Tighten the bolts in sequence to 97 inch lbs. (11 Nm).
- Intake valve timing control sensor to both sides
- Camshaft Position sensor (CMP)
- Intake valve timing control solenoid to both sides

22. Install the crankshaft pulley so it aligns with the dowel pin of the oil pump drive spacer.

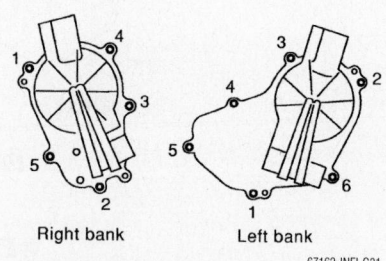

Right bank   Left bank

67162-INFI-G31

Timing control cover tightening sequence—4.5L engine

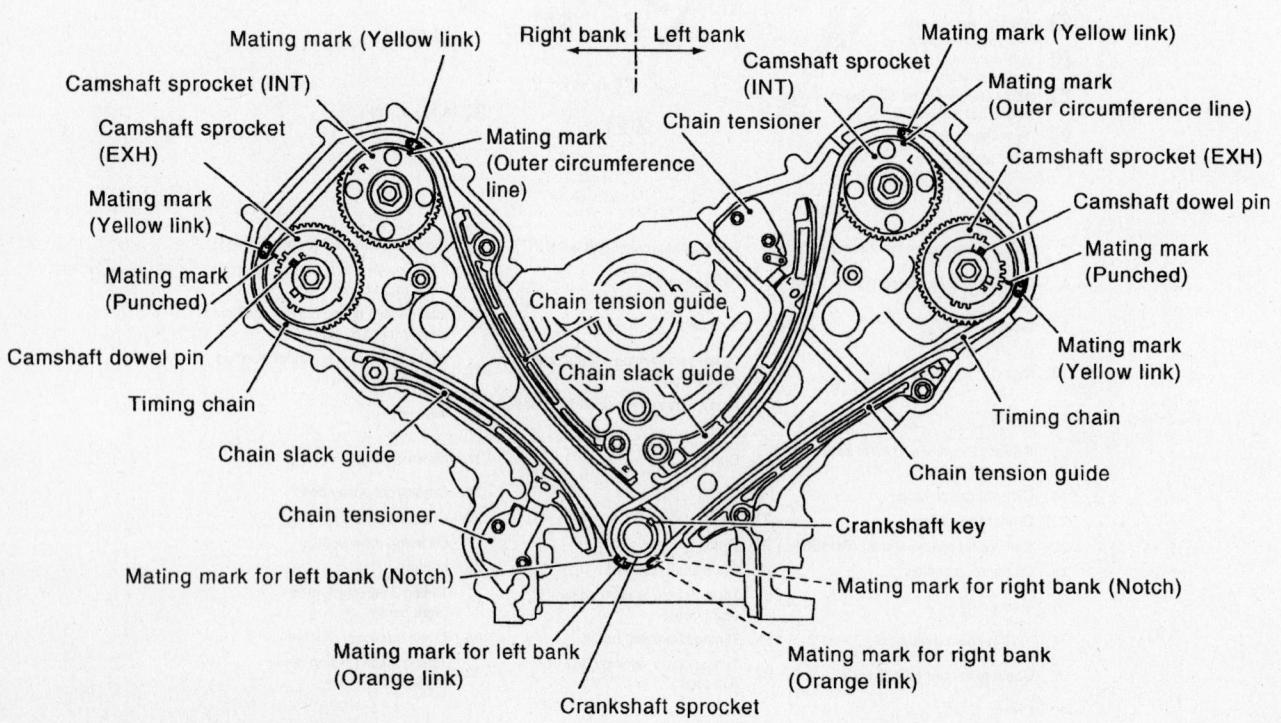

67162-INFI-G30

Timing chain and sprocket alignment marks—4.5L engine

23. Apply clean engine oil the crankshaft pulley bolt and tighten the bolt to 69 ft. lbs. (93 Nm), plus an additional 90˚.

24. Installation of the remaining components is the reverse of the removal procedure.

25. Fill the engine with clean oil.

26. Fill and bleed the cooling system.

27. Start the engine and check for proper operation.

Piston and Ring

POSITIONING

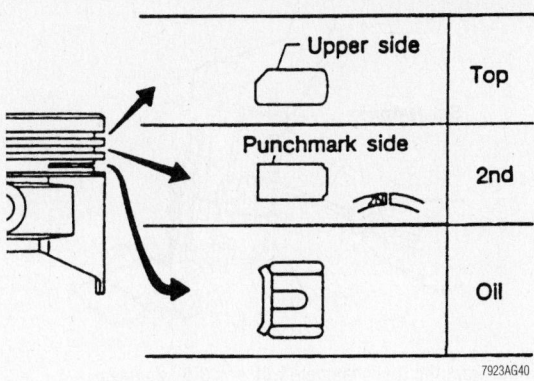

Piston ring positioning—Infiniti engines

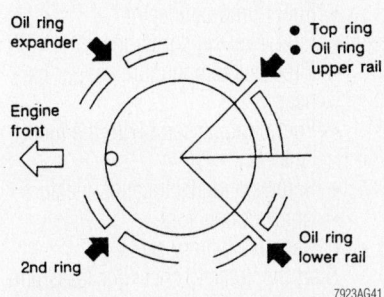

Piston ring end-gap spacing—Infiniti engines

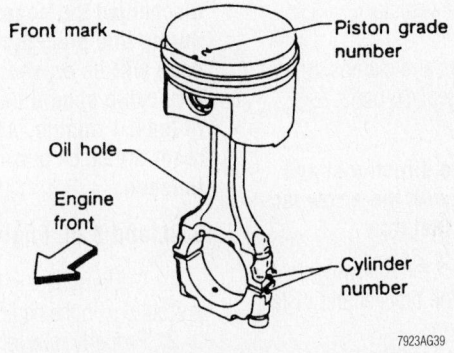

Piston/connecting rod assembly-to-engine orientation—Infiniti engines

FUEL SYSTEM

Fuel System Service Precautions

Safety is the most important factor when performing not only fuel system maintenance but any type of maintenance. Failure to conduct maintenance and repairs in a safe manner may result in serious personal injury or death. Maintenance and testing of the vehicle's fuel system components can be accomplished safely and effectively by adhering to the following rules and guidelines.

1. To avoid the possibility of fire and personal injury, always disconnect the negative battery cable unless the repair or test procedure requires that battery voltage be applied.

2. Always relieve the fuel system pressure prior to disconnecting any fuel system component (injector, fuel rail, pressure regulator, etc.), fitting or fuel line connection. Exercise extreme caution whenever relieving fuel system pressure, to avoid exposing skin, face and eyes to fuel spray. Please be advised that fuel under pressure may penetrate the skin or any part of the body that it contacts.

3. Always place a shop towel or cloth around the fitting or connection prior to loosening to absorb any excess fuel due to spillage. Ensure that all fuel spillage (should it occur) is quickly removed from engine surfaces. Ensure that all fuel soaked cloths or towels are deposited into a suitable waste container.

4. Always keep a dry chemical (Class B) fire extinguisher near the work area.

5. Do not allow fuel spray or fuel vapors to come into contact with a spark or open flame.

6. Always use a back-up wrench when loosening and tightening fuel line connection fittings. This will prevent unnecessary stress and torsion to fuel line piping. Always follow the proper torque specifications.

7. Always replace worn fuel fitting O-rings with new. Do not substitute fuel hose where fuel pipe is installed.

Fuel System Pressure

RELIEVING

1. Before servicing the vehicle, refer to the precautions in the beginning of this section.

2. Remove the fuel pump fuse.

3. Start the engine.

4. Allow the engine to run until it stalls.

5. After the engine stalls, crank the engine 2 or 3 times to release the remaining fuel pressure.

6. Turn the ignition switch **OFF**. Reinstall the fuel pump fuse into the fuse block.

➡**Do not crank the engine or turn the ignition switch ON after the fuel pump fuse has been reinstalled, or the fuel pressure will be reestablished.**

Fuel Filter

REMOVAL & INSTALLATION

Except 3.0L, 3.5L and 4.5L Engines

1. Before servicing the vehicle, refer to the precautions in the beginning of this section.

2. Relieve the fuel system pressure.

3. Remove or disconnect the following:
   • Negative battery cable
   • Fuel hoses from the fuel filter, located at the right side of the engine compartment

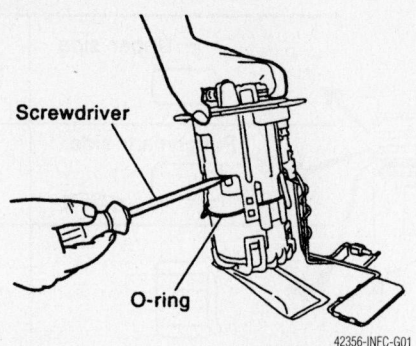

Screwdriver

O-ring

42356-INFC-G01

Remove the fuel filter from the fuel chamber—3.0L and 3.5L engines

- Filter mounting screws
- Filter from the vehicle

To install:

4. Inspect all hoses and clamps for damage of any type. Replace parts, as required.

➡ **The fuel filters are directional and should be installed with the arrow facing the direction of fuel flow.**

5. Install or connect the following:
- New filter in the bracket and install new hose clamp
- Negative battery cable

6. Start the vehicle, check for fuel leaks and repair if necessary.

➡ **On some vehicles, a code will be set and/or the check engine light will remain on after starting the vehicle. This is because a code was set for an open fuel pump circuit when the fuel pressure was released. If you did not disconnect the negative battery cable during this procedure, do it now so the code will be erased. The negative battery cable should be disconnected for at least 1 minute. Also, remember to reset the clock and radio stations when finished.**

3.0L and 3.5L Engines

1. Before servicing the vehicle, refer to the precautions in the beginning of this section.
2. Properly relieve fuel system pressure.
3. Remove or disconnect the following:
- Negative battery cable
- Rear seat bottom
- Inspection hole cover
- Electrical and quick connectors
- Six screws
- Fuel level sensor unit and fuel pump assembly

- Flange and snap fit portion of the fuel pump
- Fuel tank temperature sensor harness
- Fuel level sensor flange
- Fuel pump connector
- Quick connectors from the fuel level sensor
- Fuel level sensor from the chamber
- Fuel filter from the chamber

To install:

4. Install or connect the following:
- Fuel filter to the chamber
- Fuel level sensor to the chamber
- Quick connectors to the fuel level sensor
- Fuel pump connector
- Fuel level sensor flange
- Fuel tank temperature sensor harness
- Fuel pump assembly to the fuel tank
- Screws and electrical connectors
- Quick connectors
- Negative battery cable

5. Start the vehicle, check for leaks and repair if necessary.
- Inspection hole cover
- Rear seat bottom

4.5L Engine

1. Before servicing the vehicle, refer to the precautions in the beginning of this section.
2. Properly relieve fuel system pressure.

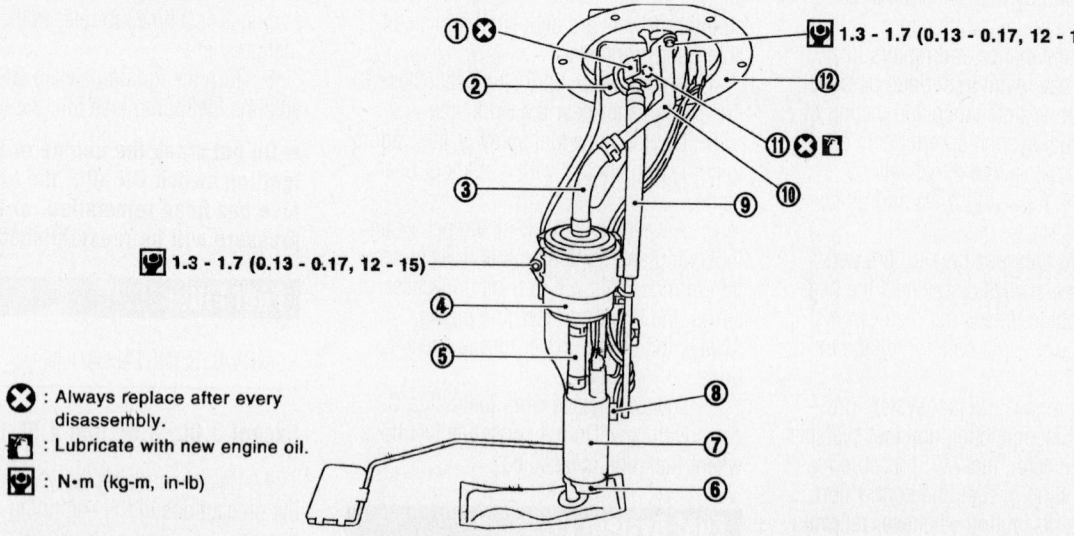

① ✖   ⊙ 1.3 - 1.7 (0.13 - 0.17, 12 - 15)

②

⑫

⑪ ✖ ⬛

③

⑨ ⑩

⊙ 1.3 - 1.7 (0.13 - 0.17, 12 - 15)

④

⑤

⑧

⑦

⑥

✖ : Always replace after every disassembly.

⬛ : Lubricate with new engine oil.

⊙ : N·m (kg-m, in-lb)

1. Clip	2. Pressure regulator	3. Fuel feed hose
4. Fuel filter	5. Fuel feed hose	6. Pump support rubber
7. Fuel pump	8. Temperature sensor	9. Fuel return hose
10. Pressure regulator housing	11. O-ring	12. Fuel level sensor unit

Fuel pump and fuel filter assembly—4.5L engine

67162-INFI-G32

3. Remove or disconnect the following:
- Negative battery cable
- Fuel filler cap
- Truck front finish panel
- Fuel pump connector
- Fuel feed line
- Fuel pump assembly
- Fuel filter

To install:

4. Installation is the reverse of the removal procedure.

5. Start the vehicle, check for leaks and repair if necessary.

Fuel Pump

REMOVAL & INSTALLATION

2.0L Engine

1. Before servicing the vehicle, refer to the precautions in the beginning of this section.
2. Release the fuel system pressure.
3. Remove or disconnect the following:
- Negative battery cable
- Rear seat back and bottom
- Inspection hole cover located beneath the rear seat
- Connectors and fuel tubes
- Six screws
- Fuel pump/gauge assembly and disconnect the tubes and connector
- Fuel pump by sliding it out on an angle

To install:

4. Install or connect the following:
- Fuel pump/gauge assembly
- All fuel lines and connectors
- Six screws
- Negative battery cable

5. Start the vehicle, check for leaks and repair if necessary.

6. Install the inspection cover.
7. Rear seat back and bottom.

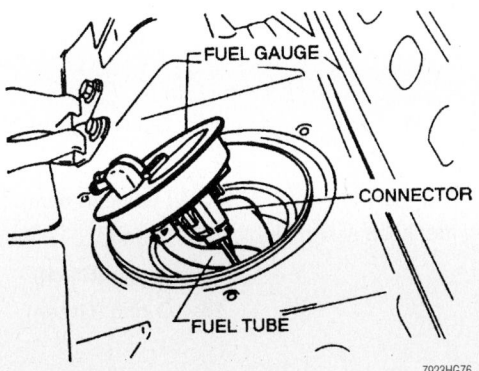

The fuel pump is located inside the tank—2.0L engine

3.0L and 3.5L Engines

1. Before servicing the vehicle, refer to the precautions in the beginning of this section.
2. Relieve the fuel system pressure
3. Remove or disconnect the following:
- Negative battery cable
- Rear seat bottom
- Inspection hole cover from under the rear seat
- Electrical and quick connectors
- Six screws
- Fuel level sensor/fuel pump assembly

➡ **If replacement of the fuel filter is required, proceed with the following steps.**

- Fuel level sensor/fuel pump assembly flange
- Fuel tank temperature sensor harness
- Fuel level sensor flange and raise the fuel level sensor
- Fuel pump electrical connector
- Quick connectors from the fuel level sensor
- Fuel level sensor from the assembly
- Snap fit connectors and remove the fuel filter from the fuel pump assembly

To install:

4. Install or connect the following:
- Fuel filter to the fuel pump assembly
- Fuel level sensor to the fuel pump assembly
- Fuel level sensor quick connectors
- Electrical connectors
- Fuel tank temperature sensor harness
- Fuel level sensor unit and fuel pump flanges. Make certain that they snap together.
- Fuel pump assembly to the fuel tank
- Six screws
- Quick and electrical connectors
- Negative battery cable

5. Start the vehicle, check for leaks and repair if necessary.
6. Install the inspection hole cover plate.
7. Install the rear seat bottom.

4.1L Engine

1. Before servicing the vehicle, refer to the precautions in the beginning of this section.
2. Relieve the fuel system pressure.
3. Remove or disconnect the following:
- Negative battery cable
- Trunk front finish panel
- Wiring harness connector and fuel tubes
- Fuel tank sender unit attaching bolts
- Fuel tank sender and discard the O-ring
- Fuel pump from the sender unit

To install:

4. Install or connect the following:
- New fuel pump on the sender unit assembly
- Sender unit in the fuel tank, using a new O-ring. Torque the bolts to 17–23 inch lbs. (2–2.5 Nm).
- Wiring harness connectors and fuel tubes
- Trunk room finish panel
- Negative battery cable

5. Start the engine, check for leaks and repair if necessary.

4.5L Engine

1. Before servicing the vehicle, refer to the precautions in the beginning of this section.
2. Properly relieve fuel system pressure.
3. Remove or disconnect the following:
- Negative battery cable
- Fuel filler cap
- Truck front finish panel
- Fuel pump connector
- Fuel feed line
- Fuel pump assembly
- Fuel pump

To install:

4. Installation is the reverse of the removal procedure.

5. Start the vehicle, check for leaks and repair if necessary.

Fuel Injector

REMOVAL & INSTALLATION

Except 4.5L Engines

1. Before servicing the vehicle, refer to the precautions in the beginning of this section.

2. Relieve the fuel system pressure
3. Remove or disconnect the following:
   - Negative battery cable
   - Engine cover, as necessary
   - Intake manifold collector
   - Vacuum hose from the fuel pressure regulator
   - Fuel hoses from fuel rail
   - Fuel rail bolts
   - Injector harness connectors
   - Injectors and the fuel rail as an assembly
   - Injector(s) from the fuel rail by pushing them out

4. Remove and discard the fuel injector O-rings

To install:

5. Lubricate the new O-rings with clean engine oil and install the O-rings on the injector(s).
6. Install or connect the following:
   - Fuel injectors to the fuel rail
   - Fuel rail and injectors as an assembly to the intake manifold
7. Tighten the fuel rail bolts in the following sequence;
   a. Step 1: 7–8 ft. lbs. (9–10 Nm).
   b. Step 2: 15–20 ft. lbs. (21–26 Nm).

- Fuel hoses to the fuel rail
- Vacuum hose to the fuel pressure regulator
- Intake manifold collector
- Engine cover, as necessary
- Negative battery cable

8. Start the vehicle, check for leaks and repair if necessary.

4.5L Engines

1. Before servicing the vehicle, refer to the precautions in the beginning of this section.
2. Relieve the fuel system pressure
3. Remove or disconnect the following:

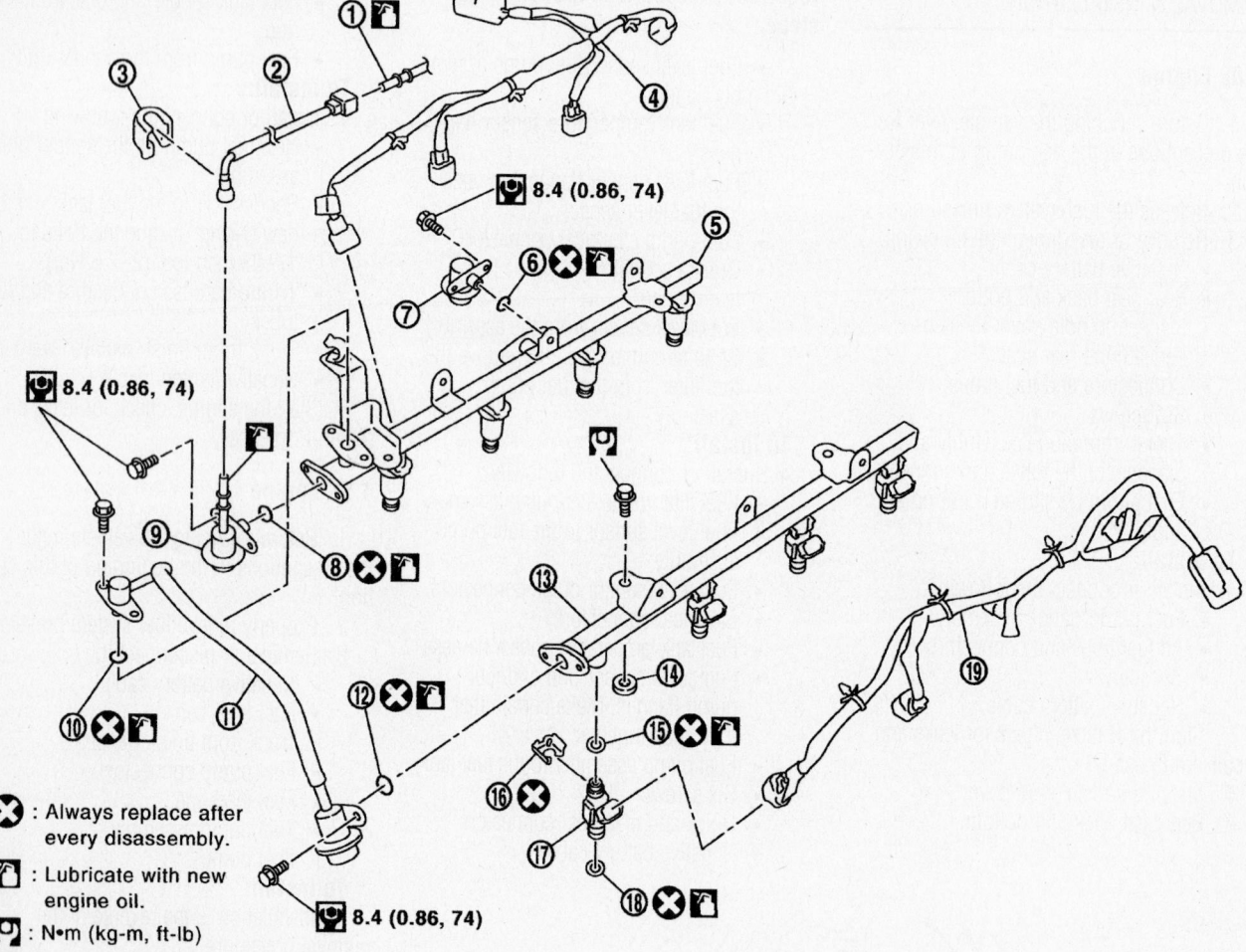

8.4 (0.86, 74)

8.4 (0.86, 74)

8.4 (0.86, 74)

❌ : Always replace after every disassembly.

🛢 : Lubricate with new engine oil.

🔧 : N•m (kg-m, ft-lb)

🔧 : N•m (kg-m, in-lb)

1.	Centralized underfloor piping	2.	Fuel feed hose	3.	Quick connector cap
4.	Fuel injector sub harness (right)	5.	Fuel tube (right)	6.	O-ring
7.	Fuel damper (right)	8.	O-ring	9.	Fuel feed damper
10.	O-ring	11.	Fuel damper and fuel hose assembly	12.	O-ring
13.	Fuel tube (left)	14.	Spacer	15.	O-ring (Black)
16.	Clip	17.	Fuel injector	18.	O-ring (Green)
19.	Fuel injector sub harness (left)				

Fuel rail and fuel injector assembly—4.5L engine

- Negative battery cable
- Engine appearance cover
- Fuel damper and hoses from fuel rail
- Fuel rail bolts
- Injector harness connectors
- Injectors and the fuel rail as an assembly

To install:

4. Lubricate the new O-rings with clean engine oil and install the O-rings on the injector(s). The Green O-ring goes on the bottom and the Black O-ring goes on the top of the injector.

5. Install or connect the following:
- Fuel injectors to the fuel rail
- Fuel rail and injectors as an assembly to the intake manifold

6. Tighten the fuel rail bolts in the following sequence;

a. Step 1: 7 ft. lbs. (10 Nm).
b. Step 2: 17 ft. lbs. (23 Nm).
- Fuel hoses to the fuel rail
- Fuel damper
- Engine appearance cover
- Negative battery cable

7. Start the vehicle, check for leaks and repair if necessary.

DRIVE TRAIN

Transaxle Assembly

REMOVAL & INSTALLATION

Manual

G20 MODELS

1. Before servicing the vehicle, refer to the precautions in the beginning of this section.

2. Drain the transaxle fluid.

3. Remove or disconnect the following:
- Negative battery cable
- Air cleaner and air duct assembly
- Clutch operating cylinder from the transaxle
- Back-up light switch, neutral switch and ground harness connectors
- Speedometer sensor
- Starter
- Air bleeder hose
- Shift control and support rods
- Front exhaust tube
- Halfshafts and support the engine with a suitable jack under the oil pan
- Rear and left engine mount

4. Raise the jack and remove the lower transaxle housing bolts. Lower jack and remove the upper housing bolts. Keep the bolts in order as they are different lengths and must be returned to the same position.

5. Lower the transaxle.

To install:

6. Raise the transaxle into place and install the attaching bolts. Torque the shortest bolt to 22–30 ft. lbs. (30–40 Nm) and the remaining bolts to 51–59 ft. lbs. (70–79 Nm).

7. Install or connect the following:
- Rear and left engine mounts
- Driveshafts
- Shift control rods, support rod, bleeder air hose and starter
- Air bleeder hose
- Starter
- Speedometer sensor
- Back-up light switch, neutral switch and ground harness connectors
- Clutch operating cylinder. Torque the bolts to 22–27 ft. lbs. (29–37 Nm).
- Negative battery cable

8. Fill the transaxle to the proper level.

9. Start the vehicle, check for leaks and repair if necessary.

I30 MODELS

1. Before servicing the vehicle, refer to the precautions in the beginning of this section.

2. Drain the transaxle fluid.

3. Remove or disconnect the following:
- Battery, battery bracket and tray
- Air cleaner assembly with the Mass Air Flow (MAF) sensor

- Clutch operating cylinder; do not disconnect the hydraulic line from the cylinder
- Clutch hose clamp
- Speedometer pinion and the neutral position switch connectors and the ground harness connectors
- Starter
- Back-up lamp switch and the neutral position switch
- Crankshaft Position (CKP) sensor (POS) from the transaxle front side
- Shifter control rod and the support rod bracket from the transaxle
- Both driveshafts from the transaxle assembly and support the transaxle

4. Support the engine of the transaxle by placing a jack under the oil pan. Be sure to use a block of wood between the oil pan and jack.
- Bolts that secure the center crossmember
- Left-hand engine mounts

➡The transaxle bolts are of different lengths, be sure to note the location of the bolts for reassembly.

- Transaxle bolts
- Transaxle from the vehicle by sliding the transaxle input shaft out of the clutch, lowering the rear of the transaxle, then lowering the transaxle from of the vehicle

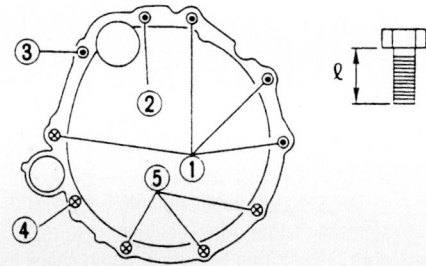

● M/T to engine
⊗ Engine to M/T

Bolt No.	Tightening torque N·m (kg-m, ft-lb)	"ℓ" mm (in)
①	70 - 79 (7.1 - 8.1, 51 - 59)	52 (2.05)
②	70 - 79 (7.1 - 8.1, 51 - 59)	65 (2.56)
③	70 - 79 (7.1 - 8.1, 51 - 59)	124 (4.88)
④	35.1 - 47.1 (3.58 - 4.80, 25.89 - 34.74)	40 (1.57)
⑤	35.1 - 47.1 (3.58 - 4.80, 25.89 - 34.74)	40 (1.57)

③ with starter
④ with support rod bracket

9307HG31

Manual transaxle-to-engine bolt torque sequence and specifications—I30 models

To install:

5. Install or connect the following:
- Transaxle assembly to the bell housing while aligning the output shaft of the transaxle with the clutch disc. Tighten the transaxle bolts to the specifications illustrated.
- Left-hand engine mount. Torque the through-bolt to 32–41 ft. lbs. (43–55 Nm).
- Center crossmember assembly. Torque the bolts to 57–72 ft. lbs. (77–98 Nm).
- Both driveshafts to the transaxle assembly
- Shifter control rod and the support rod bracket to the transaxle
- CKP sensor to the transaxle front side
- Back-up lamp switch and the neutral position switch
- Starter motor assembly to the transaxle
- Speedometer pinion and the ground harness connectors
- Clutch hose clamp
- Clutch operating cylinder. Torque the bolts to 22–30 ft. lbs. (30–40 Nm).
- Air cleaner assembly with the MAF sensor
- Battery tray, bracket and battery
- Positive and the negative battery cables

6. Fill the transaxle with proper amount and type of fluid.

7. Start the vehicle, check for leaks and repair if necessary.

Automatic

G20 MODELS

1. Before servicing the vehicle, refer to the precautions in the beginning of this section.

2. Drain the transaxle fluid.

3. Remove or disconnect the following:
- Battery and bracket
- Air duct assembly
- Transaxle solenoid harness connector, Park Neutral Position (PNP) switch connector and revolution switch connector
- Crankshaft Position (CKP) sensor from the transaxle
- Control cable and transaxle coolant lines
- Halfshafts, the intake manifold support bracket and the starter
- Upper bolts attaching transaxle to the engine

4. Support the engine with a suitable stand and use a suitable jack to support the transaxle.

➡**Bolts are of different lengths, note the locations that the bolts are removed from.**

- Center member
- Rear cover plate and the bolts securing the torque converter to the driveplate. Rotate the crankshaft to gain access to the bolts.
- Transaxle mounts
- Lower transaxle mounting bolts and lower the transaxle

To install:

5. Place a straightedge across the bell housing of the transaxle and measure the distance to the mounting bosses on the torque converter. The distance should be 0.626 in. (16mm). If not, the torque converter is not installed correctly.

6. Check the driveplate run-out with a dial indicator. Maximum allowable run-out is 0.008 in. (0.2mm).

7. Raise the transaxle into position and install the transaxle mounting bolts. Tighten the 50, 55, and 65mm long bolts to 51–59 ft. lbs. (70–79 Nm). Torque the 35 and 45 mm long bolts to 12–15 ft. lbs. (16–21 Nm).

8. Install or connect the following:
- Torque converter bolts. Torque the bolts to 33–43 ft. lbs. (44–59 Nm). Rotate the crankshaft to gain access to the bolts.
- Rear cover and center member
- Transaxle mounts
- Halfshafts, intake manifold support bracket and the starter
- Control cable and transaxle coolant lines
- CKP sensor to the transaxle
- Transaxle solenoid harness connector, PNP switch connector and revolution switch connector
- Air duct assembly
- Battery and bracket

9. Fill the transaxle with fluid.

10. Start the vehicle, check for leaks and repair if necessary.

I30 AND I35 MODELS

➡**The radio may contain a coded theft protection circuit. Always obtain the code number from the customer before disconnecting the battery.**

1. Before servicing the vehicle, refer to the precautions in the beginning of this section.

2. Drain the transaxle fluid.

3. Remove or disconnect the following:
- Battery and bracket
- Air cleaner and resonator
- Terminal cord assembly harness connector and Park Neutral Position (PNP) switch harness connector
- Revolution and Vehicle Speed Sensor (VSS) electrical connections
- Crankshaft Position (CKP) sensor from the transaxle
- Left-hand mounting bracket from transaxle and body
- Control cable from the transaxle
- Driveshafts
- Oil cooler pipes and cap pipes to avoid contamination
- Starter motor and place a jack under the oil pan to support the engine. Do not place the jack under the oil pan drain plug.
- Crossmember
- Rear cover plate and bolts attaching the torque converter to the drive plate

4. Support the transaxle with a jack.

5. Remove the transaxle-to-engine bolts and lower the transaxle using the jack.

To install:

6. Install or connect the following:
- Torque converter in the transmission. Be sure the torque converter is fully seated in the front pump assembly. The distance from the front edge of the transmission to the bolt hole of the torque converter should be 0.75 in. (19mm).
- Position the transmission to the engine and install a few bolts to hold the transmission in place. Do not fully tighten the bolts at this time.
- Torque converter-to-flexplate bolts. Torque the bolts to 33–43 ft. lbs. (44–59 Nm).

7. Secure the transmission to the engine. Torque the bolts in the sequence illustrated.

8. Install all remaining components in the reverse order of removal.

9. Fill the transmission with new fluid. Use the same amount of fluid that was drained before removal.

10. Connect negative battery cable and start the engine. Allow the engine to reach normal operating temperature and check the transmission fluid level. Add fluid as needed.

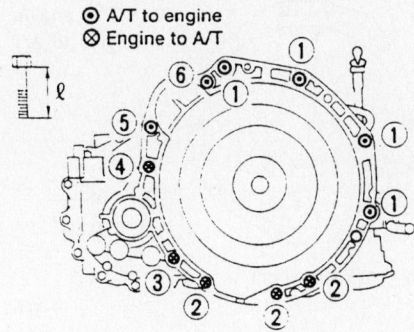

Bolt No.	Tightening torque N·m (kg-m, ft-lb)	ℓ mm (in)
①	39 - 49 (4.0 - 5.0, 29 - 36)	45 (1.77)
②	30 - 36 (3.1 - 3.7, 22 - 27)	30 (1.18)
③	30 - 36 (3.1 - 3.7, 22 - 27)	40 (1.57)
④	74 - 83 (7.5 - 8.5, 54 - 61)	45 (1.77)
⑤	30 - 36 (3.1 - 3.7, 22 - 27)	80 (3.15)
⑥	30 - 36 (3.1 - 3.7, 22 - 27)	65 (2.56)

9307HG19

Transaxle bolts tightening sequence and torque specifications–I30 and I35 models with automatic transaxle

Transmission Assembly

REMOVAL & INSTALLATION

Automatic

G35 MODELS

1. Before servicing the vehicle, refer to the precautions in the beginning of this section.

2. Remove or disconnect the following:
   • Negative battery cable

- Engine cover
- Positive battery cable and battery
- Exhaust pipe
- Propeller shaft
- Automatic Transmission (A/T) control rod and solenoid valve harness connector
- Crankshaft Position (CKP) sensor from the transmission
- Oil cooler and transmission fluid pipes, then plug the lines
- Air breather hose
- Starter motor

- Dust cover from converter housing

➡**When turning the crankshaft, turn it clockwise as viewed from the front of the engine.**

3. Turn the crankshaft, then remove the 4 tightening bolts for the drive plate and torque converter.

4. Support the transmission assembly with a suitable jack. Be careful not the let the jack hit the drain plug.

5. Remove or disconnect the following:
   • Rear member

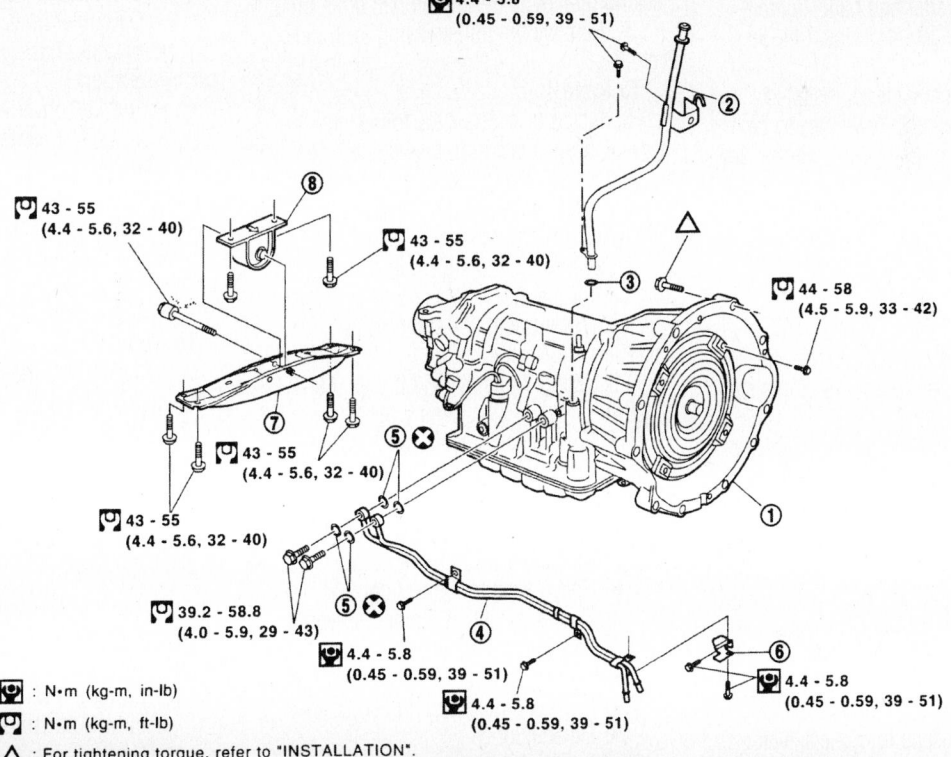

🔧 : N•m (kg-m, in-lb)

🔧 : N•m (kg-m, ft-lb)

△ : For tightening torque, refer to "INSTALLATION".

1. Transmission assembly
2. A/T fluid charging pipe
3. O-ring
4. Oil cooler tube
5. Copper washer
6. Bracket
7. Engine rear member
8. Insulator

9357MG33

Exploded view of the automatic transmission mounting—G35 models

Bolt No.	1	2	3
Number of bolts	1	5	2
Bolt length "ℓ" mm (in)	55 (2.17)	65 (2.56)	35 (1.38)
Tightening torque N·m (kg-m, ft-lb)	70 - 80 (7.2 - 8.1, 52 - 59)		41.2 - 52.0 (4.2 - 5.3, 31 - 38)

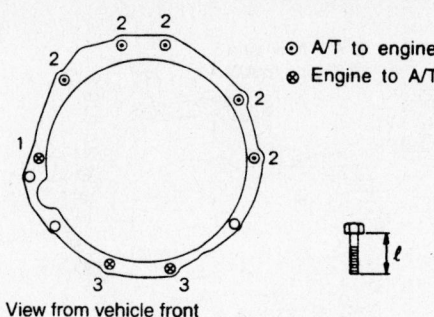

⊙ A/T to engine
⊗ Engine to A/T

View from vehicle front

9357MG34

Automatic transmission-to-engine bolt tightening specifications—G35 models

- Engine-to-transmission bolts

✳✳ WARNING

Before removal, secure the transmission to the jack and secure the torque converter to prevent it from falling.

- Transmission from the vehicle by carefully lowering it with the jack

To install:

6. Installation is the reverse of the removal procedure, noting the following:

   a. Tighten the transmission-to-engine bolts as shown in the illustration.

   b. Align the positions of the tightening bolts for the drive plate with those of the torque converter and hand-tighten. Then, tighten to 33–42 ft. lbs. (44–58 Nm).

   c. After the converter is installed, rotate the crankshaft a few times to be sure the transmission rotates without any binding

   d. Fill the transmission with fluid.

   e. Start the vehicle, check for leaks and repair if necessary.

2001–03 Q45 MODELS

1. Before servicing the vehicle, refer to the precautions in the beginning of this section.

2. Remove or disconnect the following:
   - Negative battery cable
   - Crankshaft Position (CKP) sensor
   - Rear Heated Oxygen Sensor (HO2S) connector
   - Exhaust tubes
   - Fluid charging pipe
   - Oil cooler lines. Plug fluid charging and oil cooler fittings after removing the lines.
   - Control linkage from the selector lever
   - Neutral safety switch and solenoid harness connectors
   - Speed sensor connection
   - Driveshaft (make matchmarks to ease in installation). Insert plug into rear seal opening to prevent loss of fluid.

3. Support the transmission safely.
   - Bolts securing the torque converter to the flexplate
   - Gussets securing the transmission to the engine
   - Bolts attaching the transmission to the engine

➡ **The bolts securing the transmission to the engine are of different lengths. Note the length of the bolts as they are removed.**

4. Support the engine safely. Avoid jacking directly under the oil pan drain plug.

5. Remove the transmission from the vehicle.

To install:

6. Install or connect the following:
   - Transmission in the vehicle and install the torque converter-to-flexplate bolts. Torque the bolts to 33–43 ft. lbs. (44–59 Nm).

7. Secure the transmission to the engine. Torque the bolts as follows:

   a. 70mm bolts: 80–87 ft. lbs. (108–118 Nm).

   b. 90mm bolts: 51–58 ft. lbs. (69–78 Nm).
   - Torque converter-to-drive plate bolts and tighten in 2 steps to 33–43 ft. lbs. (44–59 Nm)
   - Driveshaft, aligning the matchmarks made before removal
   - Speed sensor connection
   - Neutral safety switch and solenoid harness connectors
   - Control linkage to the selector lever
   - Fluid charging and oil cooler lines
   - Exhaust tubes

8. Lower the vehicle.

9. Fill the transmission with fluid.

10. Connect negative battery cable.

11. Start the vehicle, check for leaks and repair if necessary.

M45 AND 2004 Q45 MODELS

1. Before servicing the vehicle, refer to the precautions in the beginning of this section.

2. Remove or disconnect the following:
   - Negative battery cable
   - Engine splash shield
   - Front exhaust tube and center muffler
   - Driveshaft (make matchmarks to ease in installation. Insert plug into rear seal opening to prevent loss of fluid.
   - Transmission control rod
   - Transmission harness connectors
   - Crankshaft Position (CKP) sensor
   - Oil cooler lines
   - Breather hose
   - Starter
   - Torque converter cover
   - Torque converter bolts

3. Place a transmission jack under the transmission

4. Remove or disconnect the following:
   - Rear engine crossmember
   - Transmission-to-engine bolts
   - Transmission

To install:

5. Installation is the reverse of the removal procedure, noting the following:

   a. When installing the transmission, noting the correct bolt locations. The 65 mm bolts are the four bolts at the bottom. Tighten the 65 mm bolts to 55 ft. lbs. (74 Nm) and the 70 mm bolts to 84 ft. lbs. (114 Nm).

   b. Tighten the torque converter-to-drive plate bolts in 2 steps to 38 ft. lbs. (51 Nm)

   c. Lower the vehicle.

   d. Fill the transmission with fluid.

   e. Connect negative battery cable.

   f. Start the vehicle, check for leaks and repair if necessary.

Clutch

ADJUSTMENT

All models are equipped with a hydraulic clutch, which is self-adjusting.

REMOVAL & INSTALLATION

G20 Models

1. Before servicing the vehicle, refer to the precautions in the beginning of this section.
2. Raise and support the vehicle safely.
3. Remove or disconnect the following:
   - Negative battery cable
   - Transaxle
4. Insert alignment tool KV30101000 into the clutch disc hub and loosen the pressure plate bolts in small increments using a star-type pattern.
   - Pressure plate and clutch disc as an assembly
   - Release bearing by pulling the bearing retainers outward from the transaxle case
5. Inspect the clutch disc for surface wear. Measure from the friction surface to the top of the rivets. Wear limit is 0.012 in. (0.3mm). Replace clutch disc as necessary.
6. Inspect the contact surface of the flywheel for burns or discoloration. Check flywheel run-out. Maximum run-out is 0.0059 in. (0.15mm).
7. Using tools ST20050100 and ST20050010, check the pressure plate diaphragm springs. Measure from the pressure plate/flywheel mating surface to the top of the diaphragm spring. Height should be 1.201–1.280 in. (30.5–32.5mm). Replace pressure plate as necessary.
8. Inspect the release bearing for damage. Spin the bearing to see that it rolls freely.

To install:

9. Lightly lubricate the transaxle input shaft, input shaft collar, clutch lever assembly and the clutch release bearing with a lithium based grease.

➡**Keep clutch disc and all clutch components clean during installation. Do not allow grease to contact the clutch disc.**

10. Insert alignment tool KV30101000 into the clutch disc hub. Install the clutch disc and pressure plate on the tool and torque the pressure plate bolts to 16–22 ft. lbs. (22–29 Nm) in 2–3 steps using a criss-cross pattern. Remove the tool.
11. Install or connect the following:
    - Release bearing in the transaxle. Ensure that the bearing retainer clips are fully engaged.
    - Transaxle
    - Negative battery cable
12. If necessary, adjust clutch pedal height and free-play.

I30 Models

1. Before servicing the vehicle, refer to the precautions in the beginning of this section.
2. Drain the transaxle fluid.
3. Remove or disconnect the following:
   - Battery and battery bracket
   - Air cleaner and the air flow meter
4. Raise and safely support the vehicle so there is clearance to remove the transaxle from underneath. Securely support the engine via the oil pan using a cushioning wooden block and jack.
   - Transaxle from the engine and lower to the floor
5. Insert a clutch aligning bar or similar tool all the way into the clutch disc hub. This must be done so as to support the weight of the clutch disc during removal.

Mark the clutch assembly-to-flywheel relationship with paint or a center punch so the clutch assembly can be assembled in the same position from which it is removed.
   - Bolts in reverse order of tightening sequence, a turn at a time
   - Pressure plate and clutch disc
   - Release mechanism from the transaxle housing
6. Inspect the pressure plate for wear, scoring, etc., and resurface or replace, as necessary.
7. Measure the thickness of the clutch plate lining to the rivet heads; if the it is worn to a minimum of 0.012 in. (0.3mm), replace the clutch plate.
8. Inspect the release bearing and replace as necessary.
9. Using a dial indicator, mount it to the engine and inspect the flywheel run-out; if the run-out exceeds 0.0059 in. (0.15mm), replace it.

To install:

10. Apply a small amount of grease to the transaxle input shaft splines.
11. Install the disc on the splines and slide back and forth a few times. Remove the disc and remove excess grease on hub. Be sure no grease contacts the disc or pressure plate.
12. Apply lithium based molybdenum disulfide grease to the bearing sleeve inside groove, the contact point of the withdrawal lever and bearing sleeve, the contact surface of the lever ball pin and lever.
13. Install or connect the following:
    - Release mechanism and release bearing
    - Pressure plate and clutch disc, aligning it with a splined dummy shaft tool KV301010000
14. Torque the pressure plate bolts in sequence as follows:
    a. Step 1: torque in sequence to: 7–14 ft. lbs. (10–20 Nm).
    b. Step 2: torque in sequence to: 25–33 ft. lbs. (34–44 Nm).
15. Remove the dummy shaft.
16. Install or connect the following:
    - Transaxle in the correct position. Tighten the transaxle-to-engine bolts.
    - Rear and left-hand mounts
    - Speedometer cable
    - Electrical harness connector
    - Clutch release cylinder
    - Starter assembly
17. Securely support the transaxle and install the driveshafts.
18. Fill the transaxle with the required amount of approved fluid.
19. Install or connect the following:

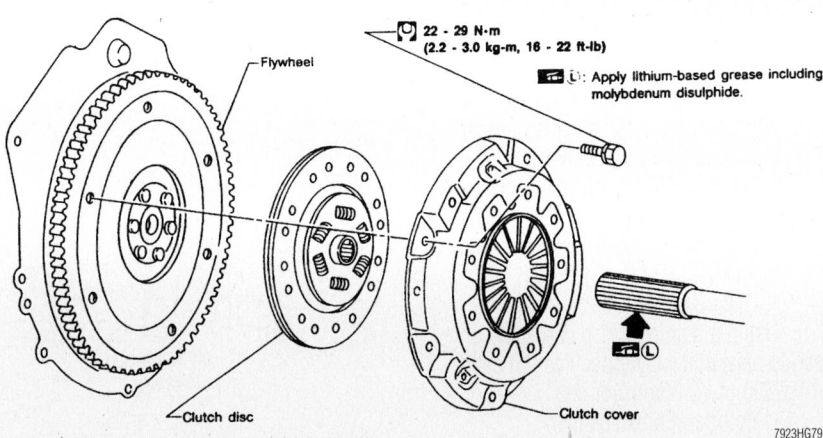

22 - 29 N·m
(2.2 - 3.0 kg-m, 16 - 22 ft-lb)

L: Apply lithium-based grease including molybdenum disulphide.

Flywheel

Clutch disc

Clutch cover

7923HG79

Clutch disc and pressure plate—G20 models

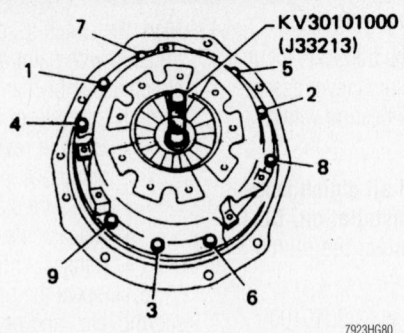

Tighten the pressure plate bolts according to the sequence shown—I30 models

- Air flow meter and the air cleaner
- Battery and battery bracket

20. Road test the vehicle for proper shift operation.

Hydraulic Clutch System

BLEEDING

G20 and I30 Models

Bleeding is required to remove air trapped in the hydraulic system. The bleed screw is located on the clutch slave (release) cylinder.

1. Before servicing the vehicle, refer to the precautions in the beginning of this section.
2. Remove the bleed screw dust cap.
3. Attach a transparent vinyl tube to the bleed screw, immersing the free end in a clean container of clean brake fluid.
4. Fill the clutch master cylinder with the proper fluid.
5. Slowly depress the clutch pedal all the way several times and hold it down.
6. Have an assistant open the bleeder valve about ¾ turn to release the air. Then, close the bleeder valve while the pedal is still depressed.
7. Repeat the above procedure until no more air bubbles are seen in the fluid container.
8. Remove the bleed tube.
9. Replace the dust cap and refill the master cylinder.
10. Bleed the clutch damper, if equipped.

Halfshaft

REMOVAL & INSTALLATION

G20 Models

FRONT

1. Before servicing the vehicle, refer to the precautions in the beginning of this section.

2. Raise and support the vehicle safely.
3. Remove or disconnect the following:
- Front wheel
- Wheel bearing locknut
- Brake caliper assembly and rotor. Using a piece of wire, position the caliper so that it is not supported by the brake line.
- Tie-rod from the ball joint
- Kingpin from the knuckle
- Halfshaft from the wheel hub/knuckle by lightly tapping it with a wood drift. Take care not to damage the CV-boots.
- Halfshaft from the transaxle by prying outward with a suitable tool at the transaxle case

4. On automatic transaxle models, remove the left halfshaft by tapping it out with a drift from the right side of the transaxle case. Take care not to damage the pinion mate shaft and side gear.

To install:

5. Drive a new oil seal into the transaxle. For the right side use tool KV38106800 along the inner circumference of the oil seal. For the left side use tool KV38106700.
6. Install or connect the following:
- Halfshaft into the transaxle. Ensure that the serrations are aligned. Remove the tool.
7. Push the halfshaft inward and install the circular clip in the groove of the side gear. After inserting the clip, pull outward on the flange of the slide joint to ensure the clip is properly meshed with the side gear. If it pulls out, the clip was not installed properly.
- Halfshaft into the wheel hub/knuckle. Torque the upper knuckle nut to 72–87 ft. lbs. (98–118 Nm) and wheel bearing locknut to 174–231 ft. lbs. (235–314 Nm).
8. Using a dial indicator, check wheel bearing axial end-play. Specification calls for 0.0020 in. (0.05mm) or less.
- Rotor and brake caliper
- Wheel

I30 and I35 Models

FRONT—RIGHT SIDE

1. Before servicing the vehicle, refer to the precautions in the beginning of this section.
2. Raise and support the front of the vehicle safely.
3. Remove or disconnect the following:
- Front wheel
- Anti-lock Brake System (ABS) wheel sensor and move it out of the way
- Brake hose from the strut
- Wheel bearing locknut
4. Matchmark and remove the bolts attaching the steering knuckle to the strut

➡ **Cover axle boots with waste cloth so as not to damage them when removing halfshaft.**

- Halfshaft from the knuckle by slightly tapping it
- Halfshaft from the transaxle using a suitable flat-bladed tool
- Circlip on the end of the halfshaft and discard circlip
- Seal from the transaxle

To install:

5. Install or connect the following:
- New seal into the transaxle and

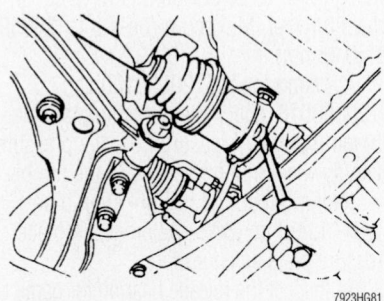

Separating the right halfshaft from the transaxle—I30 models

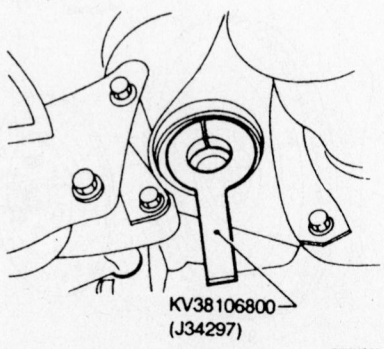

Right halfshaft alignment tool—I30 models

install a halfshaft alignment tool KV38106800 into the transaxle seal
- New circlip to the halfshaft, then insert the halfshaft into the transaxle

6. With the serration's aligned remove the alignment tool.

7. Push the halfshaft fully into the transaxle to seat the circlip. Try to pull the halfshaft from the transaxle by hand to verify that the circlip is properly seated.

- Halfshaft into the steering knuckle and install the hub locknut, do not tighten the hub nut at this time
- Steering knuckle to the strut
- Strut mounting bolts and align the matchmarks. Torque the bolts to 103–117 ft. lbs. (140–159 Nm).
- Brake hose to the strut
- ABS wheel sensor. Torque the attaching bolt to 13–17 ft. lbs. (18–24 Nm).
- Front wheels, lower the vehicle and torque the hub locknut to 174–231 ft. lbs. (235–314 Nm)

8. Check and/or adjust the wheel alignment as necessary.

FRONT—LEFT SIDE

1. Before servicing the vehicle, refer to the precautions in the beginning of this section.

2. Raise and support the front of the vehicle safely.

3. Remove or disconnect the following:
- Front wheel
- Anti-lock Brake System (ABS) wheel sensor and move it out of the way
- Brake hose from the strut
- Wheel bearing locknut
- Bolts attaching the steering knuckle to the strut. Matchmark the bolts prior to removal

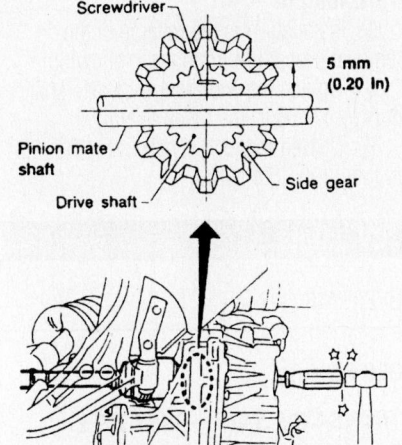

7923HG83

Separating the left halfshaft from an automatic transaxle—I30 models

➡ **Cover axle boots with waste cloth so as not to damage them when removing halfshaft.**

- Halfshaft from the knuckle by slightly tapping it
- Bolts attaching the support bearing to the support bearing bracket
- Halfshaft from the transaxle using a suitable prytool, if equipped with a manual transaxle

4. If equipped with a automatic transaxle perform the following:

a. Remove the right halfshaft from the vehicle.

b. Insert a flat-bladed tool into the transaxle where the right halfshaft was, place the end of the tool on the halfshaft and drive the left shaft from the pinion side gear.

- Support bearing bolts
- Halfshaft from the vehicle
- Circlip on the end of the halfshaft and discard circlip

- Seal from the transaxle

To install:

5. Install or connect the following:
- New seal into the transaxle and install a halfshaft alignment tool KV38106700 into the transaxle seal
- New circlip to the halfshaft, then insert the halfshaft into the transaxle

6. With the serration's aligned remove the alignment tool.

- Halfshaft fully into the transaxle to seat the circlip. Try to pull the halfshaft from the transaxle by hand to verify that the circlip is properly seated.
- Support bearing bolts and torque the bolts to 10–14 ft. lbs. (13–19 Nm)
- Halfshaft into the steering knuckle and install the hub locknut, do not tighten the hub nut at this time
- Steering knuckle to the strut
- Strut mounting bolts and align the matchmarks. Torque the bolts to 103–117 ft. lbs. (140–159 Nm).
- Brake hose to the strut
- ABS wheel sensor. Torque the attaching bolt to 13–17 ft. lbs. (18–24 Nm).
- Front wheels, lower the vehicle and torque hub locknut to 174–231 ft. lbs. (235–314 Nm)

7. Check and/or adjust the wheel alignment as necessary.

G35 and 2001–03 Q45 Models

REAR

✳✳ CAUTION

The amount of force need to loosen the rear wheel bearing nut is high enough to cause the vehicle to fall off the jack. Loosen and tighten this nut with the vehicle on the ground.

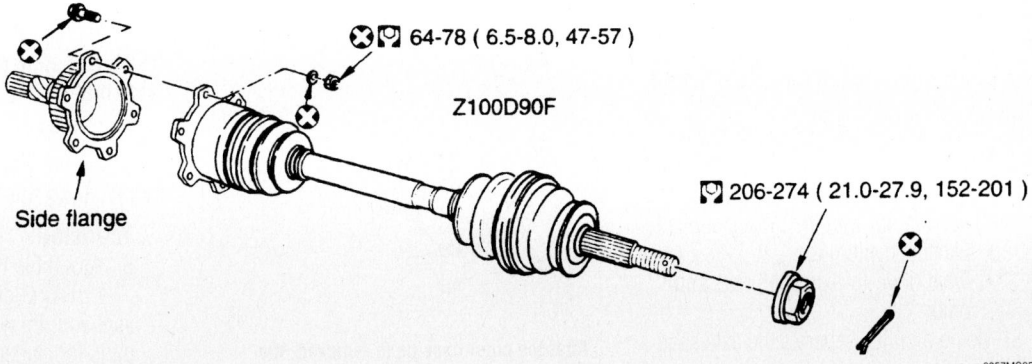

64-78 (6.5-8.0, 47-57)

Z100D90F

206-274 (21.0-27.9, 152-201)

Side flange

🔧 : N·m (kg-m, ft-lb)

9357MG35

Rear halfshaft removal and installation—G35 shown

1. Before servicing the vehicle, refer to the precautions in the beginning of this section.

2. Remove or disconnect the following:
- Rear wheel cotter pin, adjusting cap and insulator. Loosen the wheel bearing nut with the brakes applied and the vehicle sitting on the ground.

3. Raise the vehicle and support safely.
- Rear wheel
- Differential side flange bolts and nuts and separate shaft from the differential
- Wheel bearing locknut and washer from halfshaft
- Halfshaft by lightly tapping it with a copper hammer
- Halfshaft assembly from the vehicle

To install:

4. Install or connect the following:
- Halfshaft into wheel hub and install washer and wheel bearing locknut. Temporarily tighten the locknut.
- Halfshaft with the differential side flange. Install the nuts and bolts and tighten to 61–69 ft. lbs. (83–93 Nm) for the Q45 or to 47–57 ft. lbs. (64–78 Nm) for the G35.
- Wheels and lower the vehicle to the ground
- Torque the wheel bearing locknut with the brakes applied to 152–201 ft. lbs. (206–274 Nm)
- Insulator, adjusting cap and a new cotter pin

M45 and 2004 Q45 Models

REAR

❊❊ CAUTION

The amount of force need to loosen the rear wheel bearing nut is high enough to cause the vehicle to fall off the jack. Loosen and tighten this nut with the vehicle on the ground.

1. Before servicing the vehicle, refer to the precautions in the beginning of this section.

2. Remove or disconnect the following:
- Rear wheel cotter pin, adjusting cap and insulator. Loosen the wheel bearing nut with the brakes applied
- Exhaust center tube
- Final drive-to-axle shaft nuts and bolts

3. Using a puller, separate the axle shaft from the axle.

4. Remove the axle shaft.

To install:

5. Installation is the reverse of the removal procedure noting the following:

6. Tighten the axle flange-to-axle shaft nuts to 47–58 ft. lbs. (63–79 Nm).

7. Tighten the axle shaft nut to 152–202 ft. lbs. (206–274 Nm).

CV-Joints

OVERHAUL

G20 Models

TRANSAXLE SIDE—DS83 TYPE

1. Before servicing the vehicle, refer to the precautions in the beginning of this section.

2. Disassemble the joint as follows:
   a. Remove the boot bands.
   b. Matchmark the slide joint housing and inner race before separating the assembly.
   c. Using a suitable prytool, remove the stopper ring and pull out the slide joint.
   d. Matchmark the inner race and drive shaft.

3. Remove or disconnect the following:
- Snapring
- Ball cage, inner race and balls as a unit
- Boot

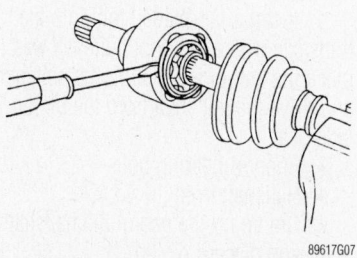

89617G07

The inner CV joint uses a large C-clip to retain the ball and cage assembly in the outer housing

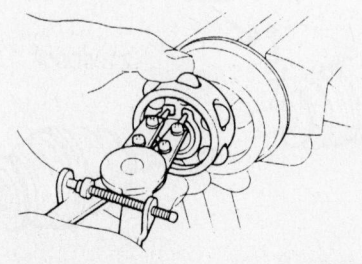

89617G08

After the outer housing is removed, the ball and cage assembly can slide from the shaft by removing the C-clip

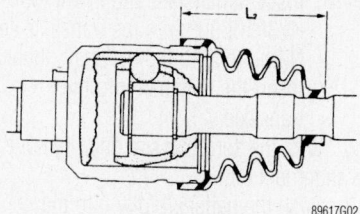

89617G02

Make sure to properly position the boot before tightening the boot clamps

➡**Cover the halfshaft serration's with tape, so as not to damage the boot.**

To install:

4. Assemble the joint as follows:
   a. Thoroughly clean all parts in solvent and dry with compressed air. Check parts for evidence of damage, and replace as necessary.
   b. Install the boot and new boot band on the halfshaft.
   c. Install a new inner snapring.
   d. Install the ball cage, inner race and balls as a unit. Confirm that the matchmarks are aligned.
   e. Install a new outer snapring.
   f. Pack the CV joint with 5.0–6.0 ounces of grease.
   g. Ensure that the boot is properly installed on the halfshaft groove.
   h. Set the boot so that it does not swell or deform when its length is 3.82–3.90 in. (97–99mm).
   i. Lock the new boot bands securely.

TRANSAXLE SIDE—TS83 TYPE

1. Before servicing the vehicle, refer to the precautions in the beginning of this section.

2. Disassemble the joint as follows:
   a. Remove the boot bands.
   b. Matchmark the slide joint housing and inner race before separating the assembly.
   c. Matchmark the spider assembly and driveshaft.
   d. Remove the snapring and the spider assembly.

➡**Do not disassemble the spider assembly.**

   e. Remove the boot.

➡**Cover the halfshaft serration's with tape, so as not to damage the boot.**

To install:

3. Assemble the joint as follows:
   a. Thoroughly clean all parts in solvent and dry with compressed air. Check parts for evidence of damage, and replace as necessary.

b. Install the boot and new boot band on the halfshaft.

c. Install the spider assembly making sure the matchmarks made during removal are properly aligned.

d. Install a new snapring.

e. Pack the joint with 4.5–5.11 ounces (124–145g) of grease.

f. Install the slide joint housing.

g. Ensure that the boot is properly installed on the halfshaft groove.

h. Set the boot so that it does not swell or deform when its length is 3.90 in. (99mm).

4. Lock the new boot bands securely.

WHEEL SIDE

➡ **The joint on the wheel side cannot be disassembled.**

1. Before servicing the vehicle, refer to the precautions in the beginning of this section.

2. Prior to separating the joint assembly, matchmark the halfshaft and joint assembly.

3. Separate the joint using a slide hammer.

4. Remove the boot bands.

To assemble:

5. Thoroughly clean all parts in solvent and dry with compressed air. Check parts for evidence of damage and replace as necessary.

➡ **Cover the halfshaft serration's with tape, so as not to damage the boot.**

6. Install the boot and small boot band on the halfshaft.

7. Set the joint assembly onto the halfshaft and align the matchmarks.

8. Attach the joint assembly to the halfshaft by lightly tapping the serrated end with a plastic hammer.

➡ **Using a metal hammer may damage the threads on the end of the joint.**

9. Pack the CV joint with 3.5–4.0 ounces of grease.

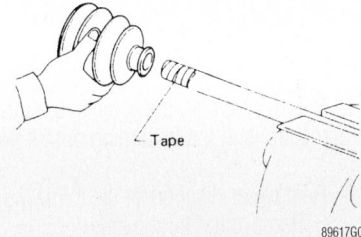

Use vinyl tape and wrap the end of the shaft to protect the boot during installation

89617G05

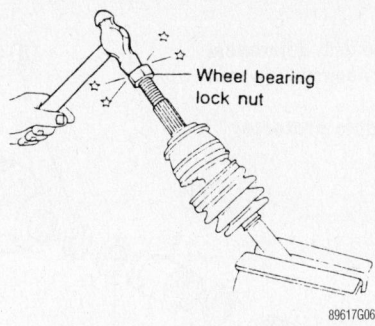

Wheel bearing lock nut

89617G06

Use an old nut to protect the threads when tapping the outer CV joint onto the shaft

10. Ensure that the boot is properly installed on the halfshaft groove.

11. Set the boot so that it does not swell or deform when its length is 3.327–3.406 in. (84.5–86.5mm).

12. Lock the new boot bands securely.

I30 and I35 Models

TRANSAXLE SIDE

1. Before servicing the vehicle, refer to the precautions in the beginning of this section.

2. Disassemble the joint as follows:

a. Remove the boot bands.

b. Matchmark the slide joint housing and inner race before separating the assembly.

c. Using a suitable prytool, remove the stopper ring and pull out the slide joint.

d. Matchmark the inner race and drive shaft.

e. Remove the snapring.

f. Remove the ball cage, inner race and balls as a unit.

g. Remove the boot.

➡ **Cover the halfshaft serration's with tape, so as not to damage the boot.**

To install:

3. Assemble the joint as follows:

a. Thoroughly clean all parts in solvent and dry with compressed air. Check parts for evidence of damage, and replace as necessary.

b. Install the boot and new boot band on the halfshaft.

4. Install a new inner snapring.

5. Install the ball cage, inner race and balls as a unit. Confirm that the matchmarks are aligned.

6. Install a new outer snapring.

7. Pack the CV joint with 5.8–6.17 ounces of grease.

8. Ensure that the boot is properly installed on the halfshaft groove.

9. Set the boot so that it does not swell or deform when its length is 3.82–3.90 in. (97–99mm).

10. Lock the new boot bands securely.

WHEEL SIDE

➡ **The joint on the wheel side cannot be disassembled.**

1. Before servicing the vehicle, refer to the precautions in the beginning of this section.

2. Prior to separating the joint assembly, matchmark the halfshaft and joint assembly.

3. Separate the joint using a slide hammer.

4. Remove the boot bands.

To assemble:

5. Thoroughly clean all parts in solvent and dry with compressed air. Check parts for evidence of damage and replace as necessary.

➡ **Cover the halfshaft serration's with tape, so as not to damage the boot.**

6. Install the boot and small boot band on the halfshaft.

7. Set the joint assembly onto the halfshaft and align the matchmarks.

8. Attach the joint assembly to the halfshaft by lightly tapping the serrated end with a plastic hammer.

➡ **Using a metal hammer may damage the threads on the end of the joint.**

9. Pack the CV joint with 4.7–5.11 ounces of grease.

10. Ensure that the boot is properly installed on the halfshaft groove.

11. Set the boot so that it does not swell or deform when its length is 3.78–3.86 in. (96–98mm).

12. Lock the new boot bands securely.

G35 and 2001–03 Q45 Models

TRANSMISSION SIDE

1. Before servicing the vehicle, refer to the precautions in the beginning of this section.

2. Remove or disconnect the following:

• Plug seal from the slide joint by gently tapping around the joint with a hammer

• Boot bands

3. Put matchmarks on the slide joint housing and halfshaft prior to separating the joint assembly.

4. Matchmark the spider assembly and driveshaft.

• Snapring and the spider assembly

Circular clip:
    Circular clips should be properly meshed with differential side gear (transaxle side) and with joint assembly (wheel side). Make sure they will not come out.
Be careful not to damage boots. Use suitable protector or cloth during removal and installation.

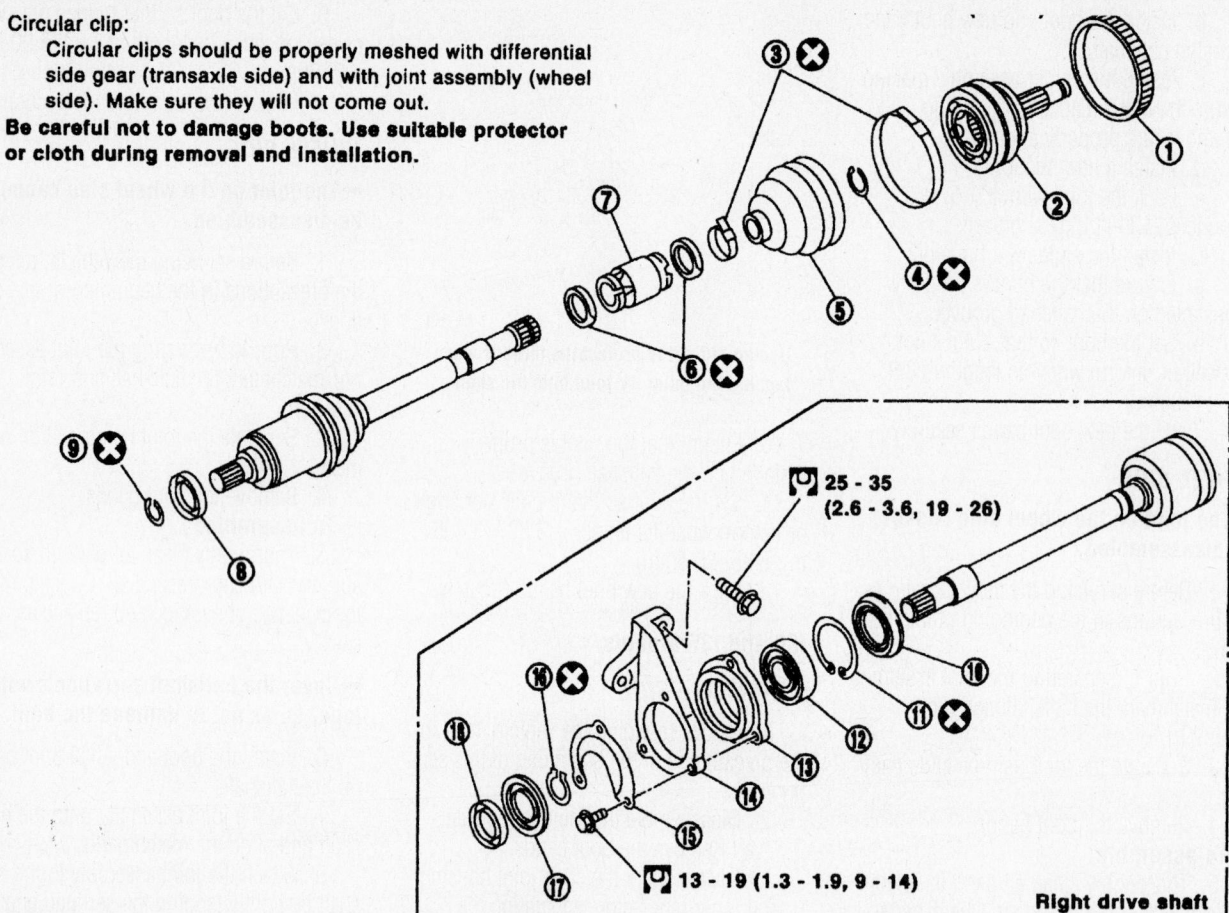

Transaxle side (SS86)

$\boxed{\text{N·m (kg-m, ft-lb)}}$

25 - 35 (2.6 - 3.6, 19 - 26)

13 - 19 (1.3 - 1.9, 9 - 14)

Right drive shaft

1. ABS sensor rotor	7. Dynamic damper	13. Support bearing retainer
2. Joint assembly	8. Dust shield	14. Support bearing bracket
3. Boot band	9. Circular clip	15. Shield heat plate
4. Circular clip	10. Support bearing dust shield	16. Snap ring
5. Boot	11. Snap ring	17. Support bearing dust shield
6. Dynamic damper band	12. Support bearing	18. Dust shield

9357MG36

Exploded view of the halfshafts—I35 shown, I30 similar

➡**Do not disassemble the spider assembly.**

• Slide joint housing and the boot

➡**Cover the halfshaft serration's with tape, so as not to damage the boot.**

To assemble:

5. Thoroughly clean all parts in solvent and dry with compressed air. Check parts for evidence of damage and replace as necessary.

6. Install or connect the following:
• Boot and small boot band on the halfshaft
• Joint housing onto halfshaft
• Spider assembly making sure the matchmarks are properly aligned

➡**The spider is press fit with the serration chamfer facing the shaft.**

• Snapring
• Coil spring, spring cap and new plug seal to the slide joint housing. Apply a suitable sealant to the plug seal prior to installation

➡**Hold the plug seal horizontally when pressing it into place. This will prevent the spring inside from falling down or tilting.**

7. Move the shaft in an axial direction to make sure that the spring is installed properly. If there is a drag or the spring is installed improperly, replace the plug seal with a new one.

8. Pack the halfshaft with 5.82–6.17 ounces (165–175g) of grease.

9. Ensure that the boot is properly installed on the halfshaft groove.

10. Set the boot so that it does not swell or deform when its length is 3.66–3.74 in. (93–95mm).

11. Lock the new boot bands securely.

WHEEL SIDE

1. Before servicing the vehicle, refer to the precautions in the beginning of this section.

2. Remove or disconnect the following:
  a. Remove the boot bands.
  b. Matchmark the housing with the shaft and halfshaft before separating the assembly.

c. Matchmark the spider assembly and halfshaft.

d. Remove the snapring and the spider assembly.

➡ **Do not disassemble the spider assembly.**

e. Remove the boot.

➡ **Cover the halfshaft serration's with tape, so as not to damage the boot.**

To install:

3. Install the joint as follows:

a. Thoroughly clean all parts in solvent and dry with compressed air. Check parts for evidence of damage, and replace as necessary.

b. Install the boot and new boot band on the halfshaft.

c. Install the spider assembly making sure the matchmarks made during removal are properly aligned.

➡ **The spider is press fit with the serration chamfer facing the shaft.**

d. Install a new snapring.

4. Pack the joint with 4–4.34 ounces (113–123g) of grease.

a. Install the slide joint housing and the snapring.

b. Ensure that the boot is properly installed on the halfshaft groove.

5. Set the boot so that it does not swell or deform when its length is 3.78–3.86 in. (96–98mm).

6. Lock the new boot bands securely.

M45 and 2004 Q45 Models

TRANSMISSION SIDE

1. Before servicing the vehicle, refer to the precautions in the beginning of this section.

2. Remove or disconnect the following:
- Plug seal from the slide joint by gently tapping around the joint with a hammer
- Boot bands

3. Put matchmarks on the slide joint housing and halfshaft prior to separating the joint assembly.

4. Matchmark the spider assembly and driveshaft.
- Snapring and the spider assembly

➡ **Do not disassemble the spider assembly.**
- Slide joint housing and the boot

➡ **Cover the halfshaft serration's with tape, so as not to damage the boot.**

To assemble:

5. Thoroughly clean all parts in solvent and dry with compressed air. Check parts for evidence of damage and replace as necessary.

a. Install the boot and new boot band on the halfshaft.

b. Install the spider assembly making sure the matchmarks made during removal are properly aligned.

➡ **The spider is press fit with the serration chamfer facing the shaft.**

c. Install a new snapring.

6. Pack the joint with 4.37–4.73 ounces (124–134g) of grease.

a. Install the slide joint housing and the snapring.

b. Ensure that the boot is properly installed on the halfshaft groove.

7. Set the boot so that it does not swell or deform when its length is 3.70 in. (94mm).

8. Lock the new boot bands securely.

WHEEL SIDE

1. Before servicing the vehicle, refer to the precautions in the beginning of this section.

2. Remove or disconnect the following:

a. Remove the boot bands.

b. Thread a slide hammer into the joint sub-assembly and pull the joint out of the shaft.

➡ **Do not disassemble the spider assembly.**

c. Remove the boot.

To install:

3. Install the joint as follows:

a. Thoroughly clean all parts in solvent and dry with compressed air. Check parts for evidence of damage, and replace as necessary.

b. Pack the joint with 3.03–3.39 ounces (86–96g) of grease.

c. Install the boot and new boot band on the halfshaft.

d. Install a new snapring.

e. Ensure that the boot is properly installed on the halfshaft groove.

4. Set the boot so that it does not swell or deform when its length is 3.82 in. (97mm).

5. Lock the new boot bands securely.

STEERING AND SUSPENSION

Air Bag

PRECAUTIONS

Several precautions must be observed when handling the inflator module to avoid accidental deployment and possible personal injury.

1. Never carry the inflator module by the wires or connector on the underside of the module.

2. When carrying a live inflator module, hold securely with both hands, and ensure that the bag and trim cover are pointed away.

3. Place the inflator module on a bench or other surface with the bag and trim cover facing up.

4. With the inflator module on the bench, never place anything on or close to the module that may be thrown in the event of an accidental deployment.

DISARMING

➡ **All Air Bag electrical wiring harnesses and connectors are covered with YELLOW outer insulation. Do not use electrical test equipment on any circuit related to the Air Bag sensors. When installing Air Bag components, always install with the arrow marks facing the front of the vehicle.**

1. Before servicing the vehicle, refer to the precautions in the beginning of this section.

2. Turn the ignition switch to the **OFF** position.

3. Disconnect both battery cables starting with the negative cable first and wait at least 10 minutes after the cables are disconnected. Be sure to insulate the battery terminal ends.

REARMING

1. Before servicing the vehicle, refer to the precautions in the beginning of this section.

2. Turn the ignition switch to the **OFF** position.

3. Connect both battery cables starting with the positive cable first.

➡ **The Air Bag or Air Bag system is equipped with a self-diagnostic operation. After turning the ignition key to the ON or START position, the AIR BAG warning lamp will illuminate for 7 seconds. After 7 seconds, the AIR BAG lamp will extinguish if no malfunction is detected. If the AIR BAG lamp does not extinguish after 7 seconds, check the Air Bag self-diagnostic system for a malfunction.**

Power Rack and Pinion Steering Gear

REMOVAL & INSTALLATION

G20 Models

1. Before servicing the vehicle, refer to the precautions in the beginning of this section.

✳✳ CAUTION

The air bag system must be disarmed before removing the steering wheel. Failure to do so may cause accidental deployment, property damage or personal injury.

2. Point the front tires straight ahead and lock the steering in this position.

✳✳ WARNING

Do not turn the steering wheel or column with the lower joint removed from the steering column or the spiral cable may be damaged.

3. Remove the steering wheel.

➡**The steering wheel must be removed before disconnecting the steering column lower joint to avoid damaging the Supplemental Restraint System (SRS) spiral cable.**

4. Raise and support the vehicle safely and remove the front wheels.

5. Remove or disconnect the following:
- Tie rod ends from the steering knuckles
- Carbon canister and properly support the engine
- Bolts attaching the engine mounts to the engine mounting center member
- Engine mounting center member
- Front stabilizer bar from the vehicle, if necessary

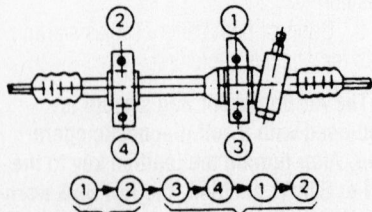

Temporary   Secure tightening
tightening

9307HG26

Exploded view of the steering gear assembly—G20 Models

- Nuts attaching the hole cover to the bulkhead

6. Move the hole cover aside and disconnect the lower joint from the rack and pinion. Matchmark the pinion shaft and the pinion housing to record the steering neutral position.
- Power steering fluid pipes from the rack and pinion
- Bolts attaching the mounting brackets and the rack and pinion from the vehicle

To install:

7. Install or connect the following:
- Rack and pinion in the vehicle
- Mounting brackets and tighten the mounting nuts and bolts in the proper sequence
- New O-rings to the power steering fluid pipes and connect them to the rack and pinion. Torque the low pressure line 20–29 ft. lbs. (27–39 Nm) and the high pressure line to 11–18 ft. lbs. (15–25 Nm).

8. Align the lower steering joint to the pinion shaft and install the joint onto the pinion shaft. Torque the bolt to 17–22 ft. lbs. (24–29 Nm).

9. Properly position the hole cover. Torque the nuts to 2.9–3.6 ft. lbs. (4–5 Nm).
- Front stabilizer
- Engine mounting center member and tighten the attaching bolts. Attach the engine mounts to the center member and tighten the bolts. Remove the support from the engine.
- Remaining components in the reverse order of removal

10. Torque the tie rod end nuts to 22–29 ft. lbs. (29–39 Nm), then install a new cotter pin.

11. Fill the power steering reservoir with fluid and bleed the air from the power steering system.

12. Check the vehicle front end alignment and adjust as necessary.

I30 Models

1. Before servicing the vehicle, refer to the precautions in the beginning of this section.

✳✳ CAUTION

The air bag system must be disarmed before removing the steering wheel. Failure to do so may cause accidental deployment, property damage or personal injury.

2. Point the front tires straight ahead and lock the steering in this position.

✳✳ WARNING

Do not turn the steering wheel or column with the lower joint removed from the steering column or the spiral cable may be damaged.

3. Remove the steering wheel.

➡**The steering wheel must be removed before disconnecting the steering column lower joint to avoid damaging the Supplemental Restraint System (SRS) spiral cable.**

4. Remove or disconnect the following:
- Both front wheels
- Tie rod ends from the steering knuckles
- Carbon canister from the vehicle and properly support the engine
- Bolts attaching the engine mounts to the engine mounting center member
- Engine mounting center member
- Front stabilizer bar from the vehicle, if necessary
- Nuts attaching the hole cover to the bulkhead

5. Move the hole cover aside and disconnect the lower joint from the rack and pinion. Matchmark the pinion shaft and the pinion housing to record the steering neutral position.
- Power steering fluid pipes from the rack and pinion
- Bolts attaching the mounting brackets and the rack and pinion from the vehicle

To install:

6. Install or connect the following:
- Rack and pinion in the vehicle
- Mounting brackets and tighten the mounting nuts and bolts in the proper sequence
- New O-rings to the power steering fluid pipes and connect them to the rack and pinion. Torque the low pressure line 20–29 ft. lbs. (27–39 Nm) and the high pressure line to 11–18 ft. lbs. (15–25 Nm).

7. Align the lower steering joint to the pinion shaft and install the joint onto the pinion shaft. Torque the bolt to 17–22 ft. lbs. (24–29 Nm).

8. Properly position the hole cover and install the attaching nuts. Torque the nuts to 2.9–3.6 ft. lbs. (4–5 Nm).
- Front stabilizer

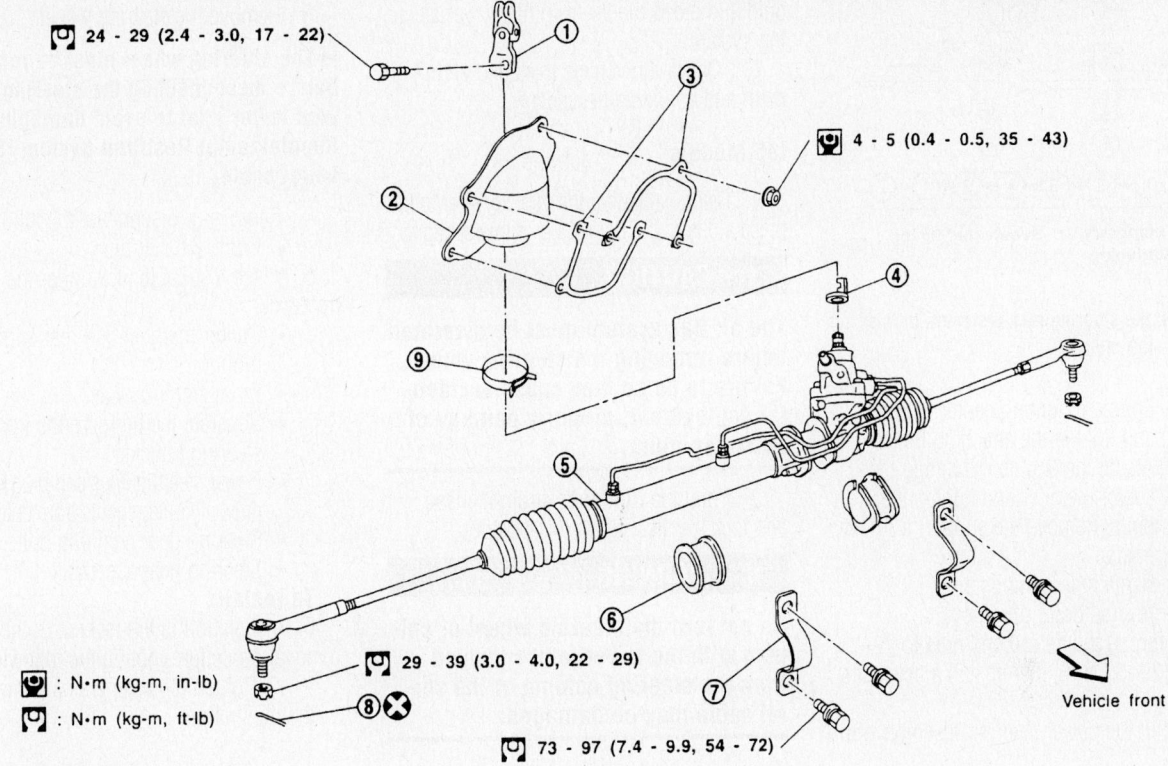

1. Lower joint
2. Hole cover
3. Insulator bracket
4. Rear cover cap
5. Gear and linkage assembly
6. Rack mounting insulator
7. Gear housing mounting bracket
8. Cotter pin
9. Clamp

24 - 29 (2.4 - 3.0, 17 - 22)

4 - 5 (0.4 - 0.5, 35 - 43)

29 - 39 (3.0 - 4.0, 22 - 29)

73 - 97 (7.4 - 9.9, 54 - 72)

: N•m (kg-m, in-lb)

: N•m (kg-m, ft-lb)

Vehicle front

9307HG25

Tighten the steering rack fasteners in this order—G20 Models

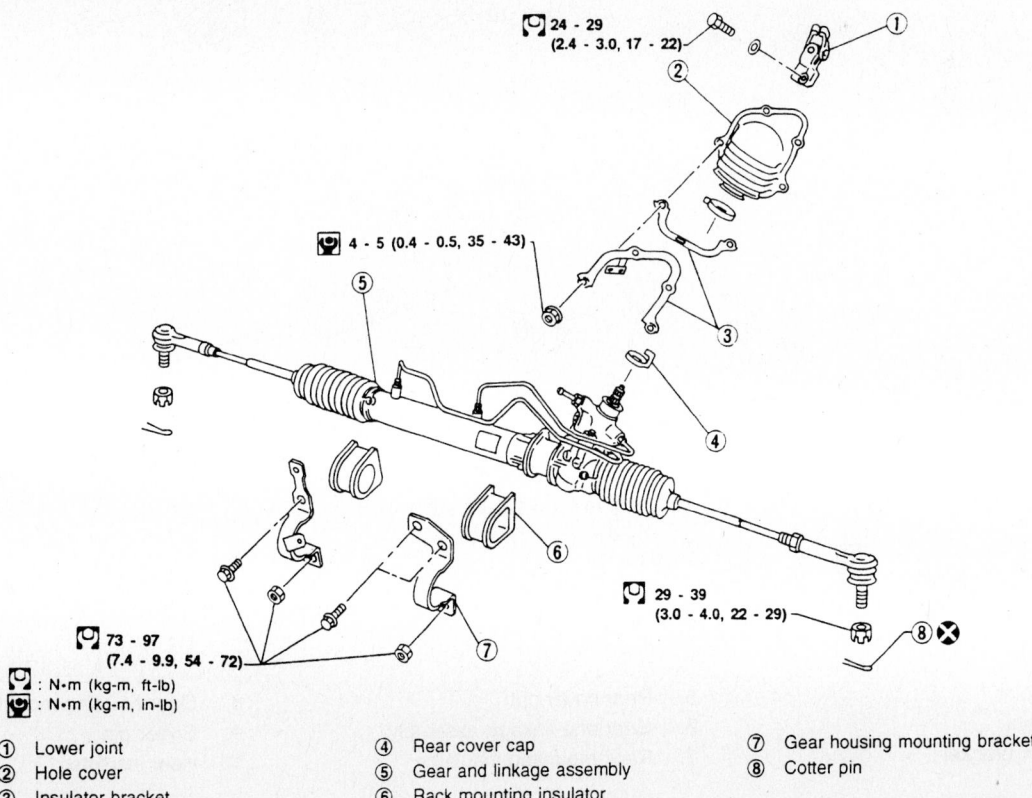

24 - 29
(2.4 - 3.0, 17 - 22)

4 - 5 (0.4 - 0.5, 35 - 43)

73 - 97
(7.4 - 9.9, 54 - 72)

29 - 39
(3.0 - 4.0, 22 - 29)

: N•m (kg-m, ft-lb)

: N•m (kg-m, in-lb)

1. Lower joint
2. Hole cover
3. Insulator bracket
4. Rear cover cap
5. Gear and linkage assembly
6. Rack mounting insulator
7. Gear housing mounting bracket
8. Cotter pin

9307HG32

Exploded view of the steering gear assembly—I30 models

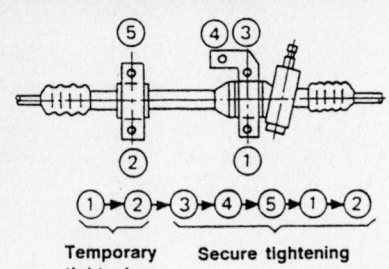

Temporary tightening   **Secure tightening**

9307HG33

Tighten the steering rack fasteners in this order—I30 Models

- Engine mounting center member and tighten the attaching bolts. Attach the engine mounts to the center member and tighten the bolts. Remove the support from the engine.
- Remaining components in the reverse order of removal

9. Torque the tie rod end nuts to 22–29 ft. lbs. (29–39 Nm), then install a new cotter pin.

10. Fill the power steering reservoir with fluid and bleed the air from the power steering system.

11. Check the vehicle front end alignment and adjust as necessary.

I35 Models

1. Before servicing the vehicle, refer to the precautions in the beginning of this section.

✳✳ CAUTION

The air bag system must be disarmed before removing the steering wheel. Failure to do so may cause accidental deployment, property damage or personal injury.

2. Point the front tires straight ahead and lock the steering in this position.

✳✳ WARNING

Do not turn the steering wheel or column with the lower joint removed from the steering column or the spiral cable may be damaged.

3. Remove the steering wheel.

➡**The steering wheel must be removed before disconnecting the steering column lower joint to avoid damaging the Supplemental Restraint System (SRS) spiral cable.**

4. Remove or disconnect the following:
- Front exhaust pipe

5. Place a suitable jack under the transaxle.
- Center member and rear engine mount
- Front stabilizer bar
- Separate the tie rod ends from the steering knuckle
- Power steering lines and plug them to prevent contaminants from entering
- Steering gear retaining bolts
- Steering gear assembly

To install:

6. Installation is the reverse of the removal procedure, noting the following:

a. Tighten all fasteners as shown in the illustration.

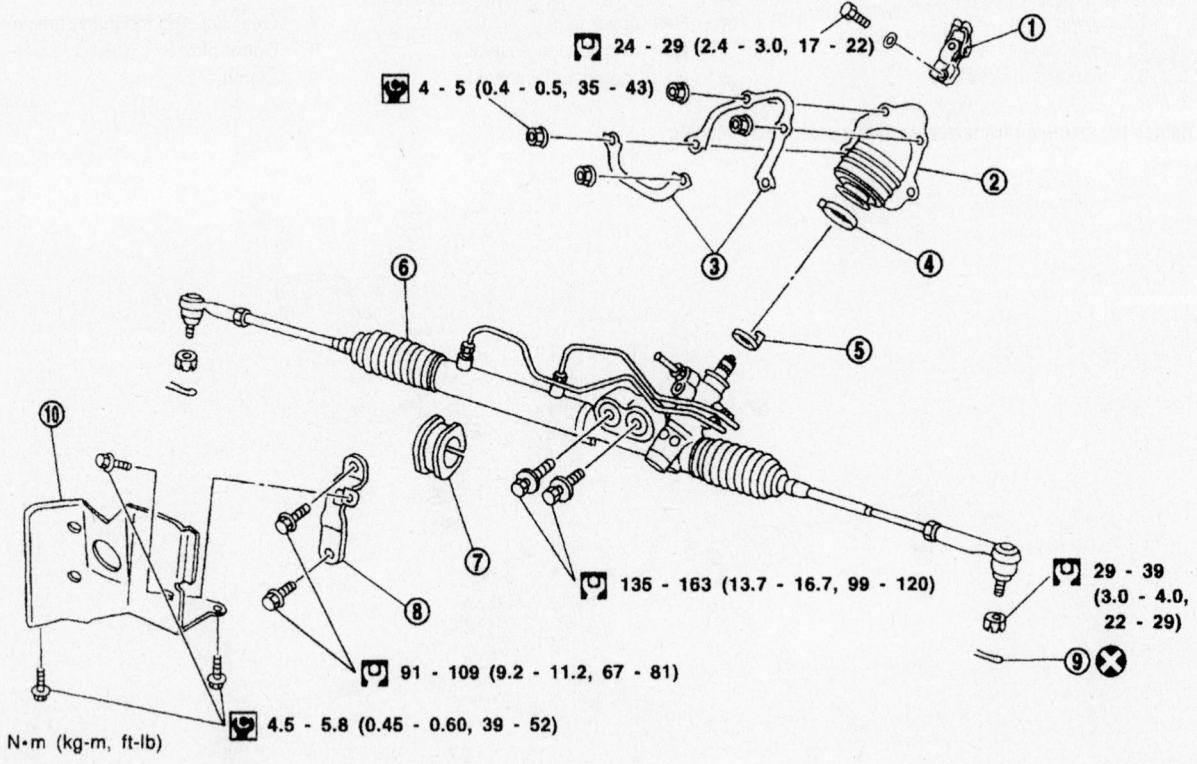

| : N•m (kg-m, ft-lb)

| : N•m (kg-m, in-lb)

1.	Lower joint	5.	Rear cover cap	8.	Gear housing mounting bracket
2.	Hole cover	6.	Gear and linkage assembly	9.	Cotter pin
3.	Insulator bracket	7.	Rack mounting insulator	10.	Heat insulator
4.	Clamp				

Exploded view of the power steering gear assembly—I35 models

9357MG37

b. Fill the power steering reservoir with fluid and bleed the air from the power steering system.

c. Check the vehicle front end alignment and adjust as necessary.

G35 Models

1. Before servicing the vehicle, refer to the precautions in the beginning of this section.

✳✳ WARNING

Do not turn the steering wheel or column with the steering gear is removed.

2. Drain the power steering fluid.
3. Remove or disconnect the following:
   - Both front wheels
   - Engine undercover
   - Tie rod ends from the steering knuckles
   - Pinch bolts from upper and lower sides
   - Power steering fluid pipes from the steering gear assembly
   - Bolt from rack mounting bracket insulator
   - Bolts attaching the mounting brackets and the steering gear from the vehicle

To install:

4. Installation is the reverse of the removal procedure, noting the following:

   a. Tighten all fasteners as shown in the illustration.

   b. Fill the power steering reservoir with fluid and bleed the air from the power steering system.

   c. Check the vehicle front end alignment and adjust as necessary.

M45 and 2004 Q45 Models

1. Before servicing the vehicle, refer to the precautions in the beginning of this section.

✳✳ WARNING

Do not turn the steering wheel or column with the steering gear is removed.

2. Remove or disconnect the following:
   - Both front wheels
   - Engine splash shield
   - Tie rod ends from the steering knuckles
   - Lower steering column joint bolt
   - Power steering fluid pipes from the rack and pinion
   - Bolts and the rack and pinion from the vehicle

To install:

3. Install or connect the following:

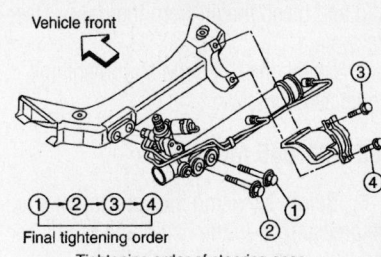

Vehicle front

① — ② — ③ — ④
Final tightening order

Tightening order of steering gear

67162-INFI-G34

Rack and pinion steering gear tightening sequence—4.5L engine M45 and 2004 Q45

- Rack and pinion in the vehicle.
- Mounting bolts. Torque the bolts in sequence. Tighten the mounting bracket bolts to 46–56 ft. lbs. (62–76 Nm) and the gear-to-frame bolts to 100–120 ft. lbs. (135–164 Nm).
- New O-rings to the power steering fluid pipes and connect them to the rack and pinion.

4. Align the lower steering joint to the pinion shaft and install the joint onto the pinion shaft. Install the bolt and torque to 18–21 ft. lbs. (24–29 Nm).

5. Torque the tie rod end nuts to 47–76 ft. lbs. (64–107 Nm) and install a new cotter pin.

6. Fill the power steering reservoir with

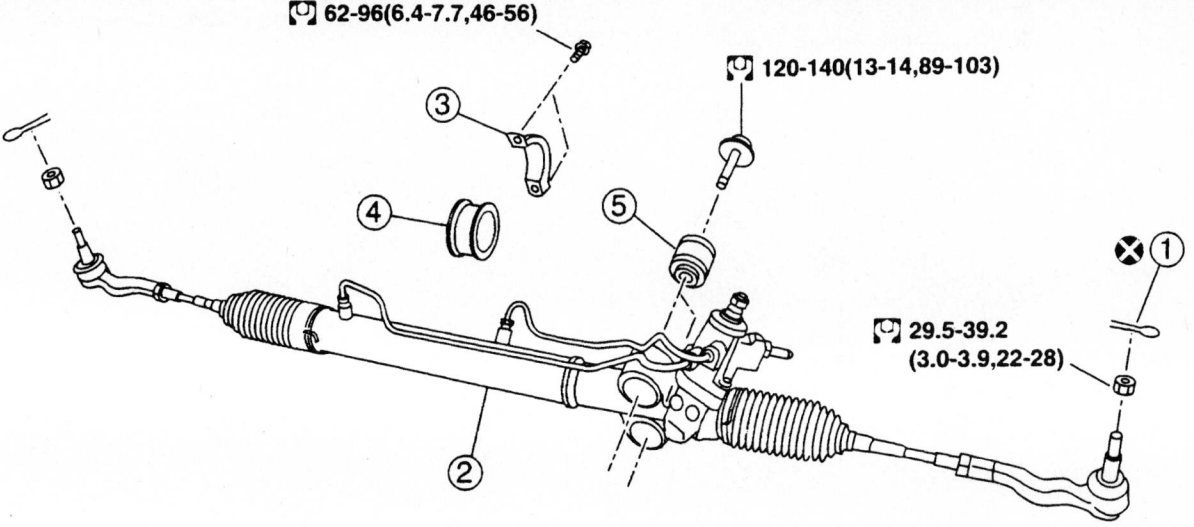

62-96(6.4-7.7,46-56)

120-140(13-14,89-103)

29.5-39.2 (3.0-3.9,22-28)

: N·m(kg-m,ft-lb)

| 1. | Cotter pin | 2. | Gear and linkage assembly | 3. | Rack mounting bracket |
| 4. | Rack mounting insulator | 5. | Rack mount housing insulator | | |

Exploded view of the power steering gear assembly—G35 models

9357MG38

fluid and bleed the air from the power steering system.

7. Check the front end alignment and adjust as necessary.

2001–03 Q45 Models

1. Before servicing the vehicle, refer to the precautions in the beginning of this section.

✳✳ WARNING

Do not turn the steering wheel or column with the steering gear is removed.

2. Remove or disconnect the following:
- Both front wheels
- Tie rod ends from the steering knuckles
- Carbon canister from the vehicle and properly support the engine
- Bolts attaching the engine mounts to the engine mounting center member
- Front stabilizer bar from the vehicle
- Lower joint bolts
- Power steering fluid pipes from the rack and pinion
- Bolts attaching the mounting brackets and the rack and pinion from the vehicle

To install:

3. Install or connect the following:
- Rack and pinion in the vehicle.
- Mounting brackets. Torque the bolts to 112–127 ft. lbs. (152–172 Nm).
- New O-rings to the power steering fluid pipes and connect them to the rack and pinion. Torque the low pressure line 30–33 ft. lbs. (40–44 Nm) and the high pressure line to 11–18 ft. lbs. (15–25 Nm).

4. Align the lower steering joint to the pinion shaft and install the joint onto the pinion shaft. Install the bolt and torque to 17–22 ft. lbs. (24–29 Nm).
- Front stabilizer, if removed
- Engine mounting center member and tighten the attaching bolts. Attach the engine mounts to the center member and tighten the bolts if removed. Remove the support from the engine.
- Remaining components in the reverse order of removal

5. Torque the tie rod end nuts to 47–80 ft. lbs. (64–108 Nm) and install a new cotter pin.

6. Fill the power steering reservoir with fluid and bleed the air from the power steering system.

7. Check the front end alignment and adjust as necessary.

Strut

REMOVAL & INSTALLATION

Front

G20 MODELS

1. Before servicing the vehicle, refer to the precautions in the beginning of this section.

2. Raise and support the vehicle safely.

3. Remove or disconnect the following:
- Strut mounting bolt at the lower suspension member and the 3 nuts inside the engine compartment. Do not remove the piston rod locknut
- Strut assembly and place in a suitable holding device

4. Using a prybar to hold the upper spring mount, loosen but do not remove the piston rod locknut.

5. Compress the spring with a spring compressor so the strut mounting insulator can be turned by hand.
- Piston rod locknut
- Coil spring from strut assembly

To install:

6. Inspect all components carefully for damage or wear. Replace as necessary.

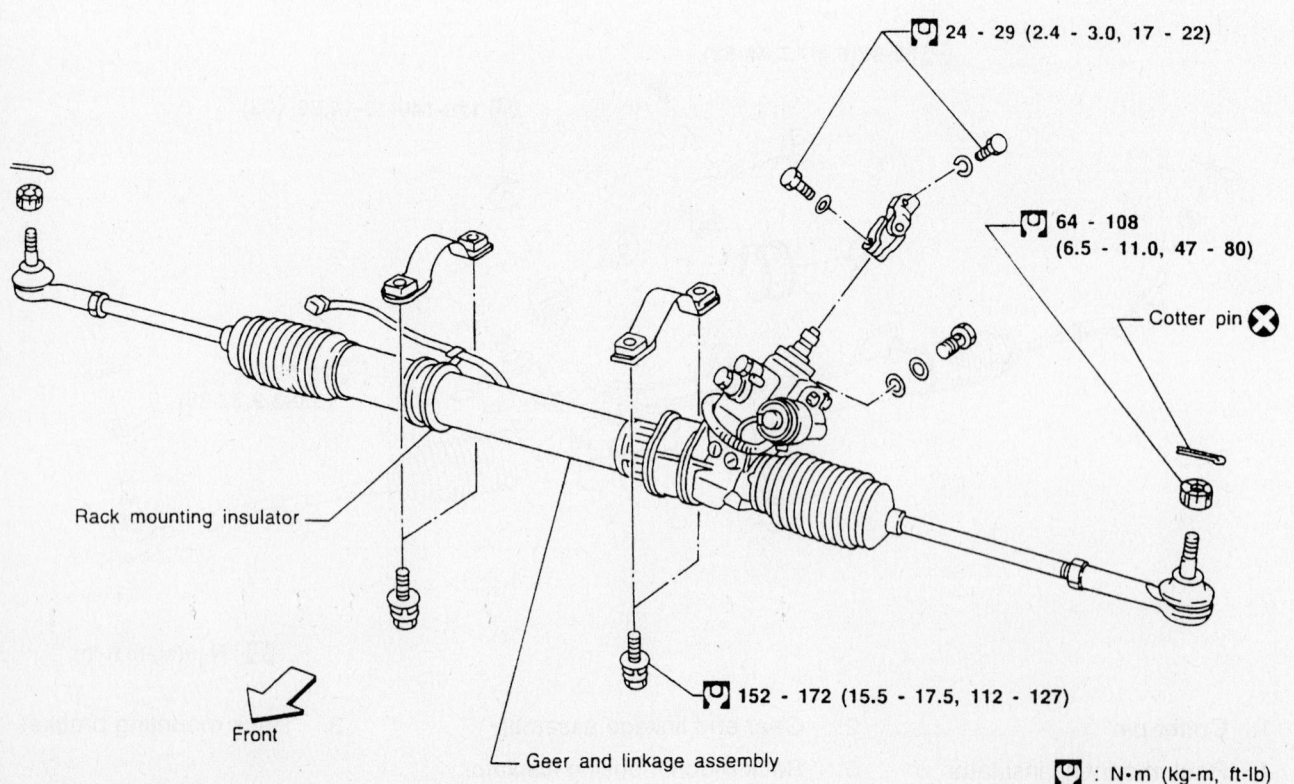

24 - 29 (2.4 - 3.0, 17 - 22)

64 - 108 (6.5 - 11.0, 47 - 80)

Cotter pin ⊗

Rack mounting insulator

Front

Geer and linkage assembly

152 - 172 (15.5 - 17.5, 112 - 127)

: N·m (kg-m, ft-lb)

9307HG34

Exploded view of the steering gear assembly—2001–03 Q45 Models

7. Install or connect the following:
- Compressed coil spring on the strut. Torque the locknut to 13–17 ft. lbs. (18–24 Nm).
- Strut. Ensure the bend in the lower strut bracket faces rearward on the left side and forward on the right side of the vehicle.
- Upper spring seat with the cutout facing the inside of the vehicle. Torque the upper mounting bolts to 31–40 ft. lbs. (42–54 Nm) and the lower through-bolt to 82–93 ft. lbs. (112–126 Nm). Final tightening must take place with the suspension loaded (vehicle at normal ride height).

I30, I35 AND G35 MODELS

1. Before servicing the vehicle, refer to the precautions in the beginning of this section.
2. Remove or disconnect the following:
- Wheel. Matchmark the position of the strut-to-steering knuckle location
- Brake hose from the strut
- Anti-lock Brake System (ABS) wheel sensor and move it out of the way
- Bolts attaching the steering knuckle transverse link to the strut. Matchmark the assembly prior to removing the bolts
- Strut attaching nuts while holding the strut from inside the engine compartment

❊❊ CAUTION

Do not remove the center locknut from the strut assembly until the strut is safely compressed.

- Strut from the vehicle
3. Place the strut assembly in a vise with the special holding tool ST35652000 or in a spring compressor.
4. Loosen the piston rod locknut.

❊❊ CAUTION

Do not remove the piston rod locknut, the spring is under tension and can cause serious personal injury.

5. Compress the spring with the spring compressor, then remove the piston rod locknut.

➡**Before removing the strut from the coil spring, note the positioning of the strut in relationship to the coil spring for reassembly.**

- Strut mounting insulator bracket, strut mounting bearing, upper spring seat, and the upper spring rubber seat
- Strut, leaving the coil spring compressed
- Piston boot and rebound bumper from the strut

To install:

6. Install or connect the following:
- Rebound bumper and the boot to the strut piston
- Strut into the coil spring, be sure the strut and spring are properly positioned
- Upper spring rubber seat, upper spring seat, strut mounting bearing, and the strut mounting insulator bracket. Be sure that the cutout on the upper spring seat is facing the outside of the vehicle.
- Piston rod locknut, then remove the spring compressor. Torque the piston rod locknut to 43–65 ft. lbs. (59–88 Nm) for I30 and I35 models, or to 40–47 ft. lbs. (54–65 Nm) for G35 models.
- Strut into the strut tower and install new attaching nuts. Torque the nuts to 29–40 ft. lbs. (39–54 Nm) for I30 models, to 32–38 ft. lbs. (43–51 Nm) for I35 models, or to 26–30 ft. lbs. (35–42 Nm) for G35 models.
- Bolts attaching the steering knuckle or transverse line to the strut and align the matchmarks. Torque the bolts to 103–117 ft. lbs. (140–159 Nm) for I30 models, to 130–139 ft. lbs. (176–189 Nm) for I35 models, or to 52–62 ft. lbs. (70–85 Nm) for G35 models.
- ABS wheel sensor. Torque the bolt to 13–17 ft. lbs. (18–24 Nm).
- Brake hose to the strut
- Front wheels and lower the vehicle
7. Check and/or adjust the wheel alignment as necessary.

M45 AND 2004 Q45 MODELS

1. Before servicing the vehicle, refer to the precautions in the beginning of this section.
2. Remove or disconnect the following:
- Front wheel
- Brake caliper and rotor
- Wheel speed sensor harness
- Brake hose
- Stabilizer connecting rod
- Steering outer socket from strut
- Lower strut end from control arm
- Suspension actuator assemblies on Q45

- Strut damper and bracket
- Strut tower bar and bracket
- Upper mounting insulator bolts
- Strut assembly
3. Secure the strut in a suitable holding fixture.
4. Loosen the piston rod locknut. Do not remove the locknut.
5. Compress the spring with the proper tool so the strut assembly mounting insulator can be turned by hand.
6. Remove the piston rod locknut. Remove the spring assembly, dust cover and rubber seat. Remove the strut insert.

To install:

7. Inspect the rubber parts for deterioration. If the rubber is pulling away from the metal, the mounting insulator should be replaced.
8. Fit the spring into the lower seat and compress the spring.
9. Install the rubber seat, upper seat and mounting insulator.
10. Install or connect the following:
- Piston rod locknut and torque to 44–54 ft. lbs. (59–47 Nm).
- Strut into place and torque the upper mounting nuts to 26–31 ft. lbs. (34–42 Nm).
- Strut tower bar and bracket and torque the bolts to 26–31 ft. lbs. (34–42 Nm).
- Strut damper and bracket and torque the bolts to 11–13 ft. lbs. (15–19 Nm).
- Lower strut end to control arm
- Steering outer socket to strut
- Brake hose
- Wheel speed sensor harness
- Brake caliper and rotor
- Front wheel

2001–03 Q45 MODELS WITH STANDARD SUSPENSION

1. Before servicing the vehicle, refer to the precautions in the beginning of this section.
2. Remove or disconnect the following:
- Front wheel
- Brake caliper and rotor
- Tie rod ball joint and lower ball joint with tool ST29020001
- Stabilizer connecting rod upper nut and separate the strut from the connecting rod
- Upper mounting insulator bolts
- Strut assembly
3. Secure the strut in a suitable holding fixture.
4. Loosen the piston rod locknut. Do not remove the locknut.
5. Compress the spring with the proper

tool so the strut assembly mounting insulator can be turned by hand.

6. Remove the piston rod locknut. Remove the spring assembly, dust cover and rubber seat. Remove the strut insert.

To install:

7. Inspect the rubber parts for deterioration. If the rubber is pulling away from the metal, the mounting insulator should be replaced.

8. Fit the spring into the lower seat, install the dust cover/bumper and upper seat and mounting insulator.

9. Install or connect the following:
- Piston rod locknut and torque to 13–17 ft. lbs. (17–23 Nm).
- Strut into place and torque the upper mounting nuts to 30–35 ft. lbs. (40–47 Nm).

10. Torque the lower mounting bolt to 80–94 ft. lbs. (108–128 Nm).
- Brake rotor and caliper
- Front wheel

2001–03 Q45 MODELS WITH ACTIVE SUSPENSION

➡The Nissan Consult or a scan tool that can issue commands to the control unit is required for bleeding the hydraulics in the Full Active Suspension system.

1. Before servicing the vehicle, refer to the precautions in the beginning of this section.

2. Relieve the hydraulic pressure as follows:
  a. Raise all 4 wheels off the ground and wait at least 3 minutes for the system to stabilize.
  b. Remove both front inner fenders and the rear pressure control unit cover.
  c. Loosen the locknut and slowly open the bypass valve on each pressure control unit. Open the valves all the way and leave them open until the job is finished.

3. Remove the flange joint from the top of the actuator.

4. Install 2, 15mm bolts into the actuator in the flange joint mounting bolt holes.

5. Insert a bar between the bolts and loosen the joint adapter. Do not remove it yet.

6. Remove or disconnect the following:
- Upper mount insulator nuts
- Hydraulic lines, then cap the lines to keep the system clean
- Lower actuator mounting nut and remove the assembly

7. Secure the actuator/spring assembly in a suitable holding fixture. Scribe align-

ment marks on the spring, upper mount insulator and actuator unit.

8. Compress the spring with the proper tool so the joint adapter can be turned by hand. Remove the joint adapter and lift off the mount insulator, spring, and any other components necessary.

To install:

9. If the actuator is being replaced, the rubber bumper should also be replaced. Fit the bumper, dust cover and rubber seat onto the actuator.

10. Fit the spring into the lower seat with the matchmarks aligned. Install the upper seat/mounting insulator with the marks aligned and start the joint adapter. The joint adapter will be tightened after installing the actuator assembly.

11. Fit the strut into place and torque the upper mounting nuts to 30–41 ft. lbs. (40–55 Nm).

12. Torque the lower mounting bolt to 76–94 ft. lbs. (103–128 Nm).

13. Torque the joint adapter to 63–72 ft. lbs. (85–98 Nm).

14. Install the flange adapter and torque the bolts to 11–13 ft. lbs. (15–18 Nm).

15. Close the bypass valves on the pressure control units.

16. Bleed the system as follows:
  a. With all 4 wheels about 2 in. (50mm) off the ground, run the engine for about 2 minutes.
  b. Connect the Consult scan tool and enter "WORK SUPPORT" mode. Select "4. AIR BLEEDING".
  c. Check the fluid level in the reservoir. It should be slightly overfilled.
  d. Touch "START" on the scan tool. The display will show a regular rise and fall in system pressure. When the pressure stabilizes, stop the engine.
  e. Connect a clear tube to the air bleeder at the actuator and place the other end in a container.

➡Do not allow the fluid to contact the body or the paint will be damaged.

  f. Open the bleeder and watch the fluid move through the tube. If there are still air bubbles in the fluid when the flow stops, check the fluid level, pressurize the system again and repeat the process.

Rear

G20 MODELS

1. Before servicing the vehicle, refer to the precautions in the beginning of this section.

2. Remove or disconnect the following:
- Wheels

- Brake calipers and suspend them with a piece of wire. Do not let them hang by the hose.

3. Using a transmission jack, raise the torsion beam slightly
- Strut lower mounting bolt

4. Open the trunk, remove the trim and remove the two nuts attaching the strut to the vehicle.
- Strut

5. Place the strut assembly in a vise with the special holding tool HT71780000 or in a spring compressor.

6. Loosen the piston rod locknut.

7. Compress the spring with the spring compressor then remove the piston rod locknut.

➡Before removing the strut from the coil spring, note the positioning of the strut in relationship to the coil spring for reassembly.

8. Remove or disconnect the following:
- Bushing, strut mounting bracket, and the upper spring seat rubber

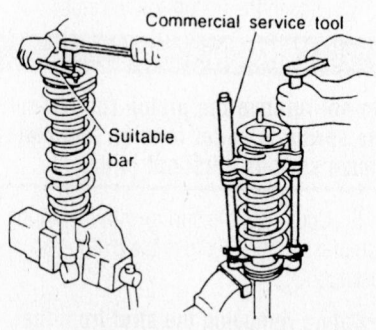

Be sure the spring is compressed before removing the piston locknut—G20 models

- Strut, leaving the coil spring compressed
- Bushing, bound bumper cover and the bound bumper

To install:

9. Install or connect the following:
- Bound bumper, bound bumper cover and the bushing
- Strut into the coil spring, make sure the strut and spring are properly positioned
- Upper spring seat rubber, strut mounting bracket, and the bushing. Make sure that the mounting bracket is properly positioned.
- Piston rod locknut then remove the spring compressor. Torque the locknut to 13–17 ft. lbs. (18–24 Nm).
- Strut with new attaching nuts. Torque the nuts to 14–16 ft. lbs. (19–22 Nm).
- Strut on the rear torsion beam. Torque the bolt to 80–94 ft. lbs. (108–127 Nm).

10. Remove the support from the rear torsion beam.

11. Install the rear wheels and lower the vehicle.

12. Check the vehicle's alignment and adjust as necessary.

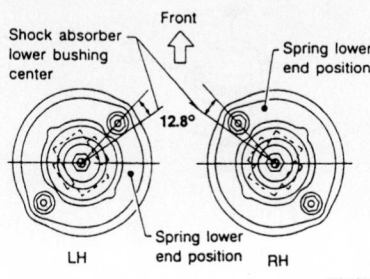

Align the spring seats as shown—G20 models

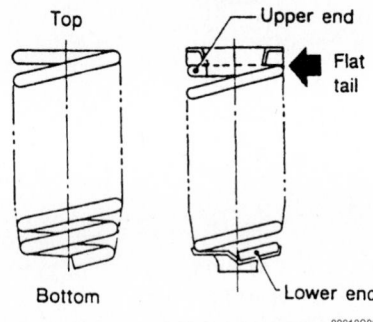

Make sure the springs are installed as shown—G20 models

I30 AND I35 MODELS

1. Before servicing the vehicle, refer to the precautions in the beginning of this section.

2. Remove or disconnect the following:
- Rear wheels

3. Support the rear torsion beam assembly with a jack.

> **✳✳ CAUTION**
>
> **Do not remove the center locknut from the strut assembly until the strut is safely compressed.**

- 2 nuts attaching the strut to the vehicle located inside the trunk
- Bolt attaching the strut to the rear torsion beam assembly and remove the strut

4. Place the strut assembly in a vise with the special holding tool HT71780000 or in a spring compressor.
- Piston rod locknut

> **✳✳ CAUTION**
>
> **Do not remove the piston rod locknut, the spring is under tension and can cause serious personal injury.**

5. Compress the spring with the spring compressor, then remove the piston rod locknut.

➡**Before removing the strut from the coil spring, note the positioning of the strut in relationship to the coil spring for reassembly.**

6. Remove or disconnect the following:
- Bushing, strut mounting bracket, and the upper spring seat rubber
- Strut, leaving the coil spring compressed
- Bushing, bound bumper cover, and the bound bumper

To install:

7. Install or connect the following:
- Bound bumper, bound bumper cover, and the bushing
- Strut into the coil spring, be sure the strut and spring are properly positioned
- Upper spring seat rubber, strut mounting bracket, and the bushing. Be sure that the mounting bracket is properly positioned.
- Piston rod locknut, then remove the spring compressor. Torque the locknut to 13–17 ft. lbs. (18–24 Nm)
- Strut with new attaching nuts. Torque the nuts to 12–16 ft. lbs. (16–22 Nm).

- Strut on the rear torsion beam. Torque the bolt to 80–94 ft. lbs. (108–127Nm).
- Support from the rear torsion beam
- Rear wheels and lower the vehicle

8. Check the vehicle's alignment and adjust as necessary.

2001–03 Q45 MODELS WITH STANDARD SUSPENSION

> **✳✳ CAUTION**
>
> **Do not remove piston rod locknut with the shock absorber on vehicle.**

1. Before servicing the vehicle, refer to the precautions in the beginning of this section.

2. Remove the upper strut mounting nuts.

3. Raise and safely support the vehicle and remove the lower mounting bolt. Remove coil spring/strut absorber assembly.

4. Place the assembly into a suitable holding fixture and matchmark the spring, strut and upper seat. Loosen but do not remove the piston rod locknut.

5. Install a spring compressor and compress the spring until the upper spring seat can be turned by hand.

6. Remove the locknut, spring seat components, spring, bushings and bumper.

To install:

7. Fit the bumper, spring, upper seat and other components onto the strut with the matchmarks aligned. The top of the spring is flat.

8. Install the locknut and torque it to 13–17 ft. lbs. (18–24 Nm) and remove the spring compressor.

9. Install strut assembly. Torque the upper shock mounting nuts to 12–14 ft. lbs. (16–19 Nm) and the lower bolt to 57–72 ft. lbs. (77–98 Nm).

2001–03 Q45 MODELS WITH FULL ACTIVE SUSPENSION

➡**The Nissan Consult or scan tool that can issue commands to the control unit is required for bleeding the hydraulics in the Full Active Suspension system.**

1. Before servicing the vehicle, refer to the precautions in the beginning of this section.

2. Relieve the hydraulic pressure as follows:
   a. Raise and safely support the vehicle with all 4 wheels off the ground and wait at least 3 minutes for the system to stabilize.

b. Remove both front inner fenders and the rear pressure control unit cover.

c. Loosen the locknut and slowly open the bypass valve on each pressure control unit. Do not open the bleeder valves.

d. Open the bypass valves all the way, and leave them open, until the job is finished.

3. Remove or disconnect the following:
- Upper mount insulator nuts
- Hydraulic lines. Cap the lines to keep the system clean
- Lower actuator mounting bolt and remove the actuator/spring assembly

4. Secure the actuator/spring assembly in a suitable holding fixture. Scribe alignment marks on the spring, upper mount insulator and actuator unit.

5. Compress the spring with the proper tool. Remove the piston rod locknut lift off the mount insulator, hose joint adapter, spring, and any other components necessary.

To install:

6. Fit the bumper and dust cover onto the actuator.

7. Fit the spring into the lower seat with the matchmarks aligned. Install the upper seat, mounting insulator and other components with the marks aligned.

8. Install the locknut and torque to 43–54 ft. lbs. (59–74 Nm).

9. Fit the assembly onto the vehicle. Torque the upper mounting nuts to 12–24 ft. lbs. (16–19 Nm) and the lower mounting bolt to 58–72 ft. lbs. (78–98 Nm).

10. Bleed the system as follows:

a. With all 4 wheels off the ground, run the engine for about 2 minutes.

b. Connect the scan tool and enter "WORK SUPPORT" mode. Select "4. AIR BLEEDING".

c. Check the fluid level in the reservoir and make it slightly overfilled.

d. Touch "START" on the scan tool. The display will show a regular rise and fall in system pressure that may last for several minutes. When the pressure stabilizes, stop the engine.

e. Connect a clear tube to the air bleeder at the actuator and place the other end in a container. Do not allow fluid to contact the body or the paint will be damaged.

f. Open the bleeder and watch the fluid move through the tube. If there are still air bubbles in the fluid when the flow stops, close the bleeder and check the fluid level. Pressurize the system again and repeat the bleeding process.

Shock Absorber

REMOVAL & INSTALLATION

Rear

G35 MODELS

1. Before servicing the vehicle, refer to the precautions in the beginning of this section.

2. Place a transmission jack under the rear axle to remove the fitting bolt and nut in the lower side of the shock absorber.

3. Remove or disconnect the following:
- Rear seat cushion, seat back and rear package shelf finish panel
- Fitting nut from the upper side of the shock absorber
- Shock absorber from the vehicle

4. Installation is the reverse of the

Figure shows rear left actuator.

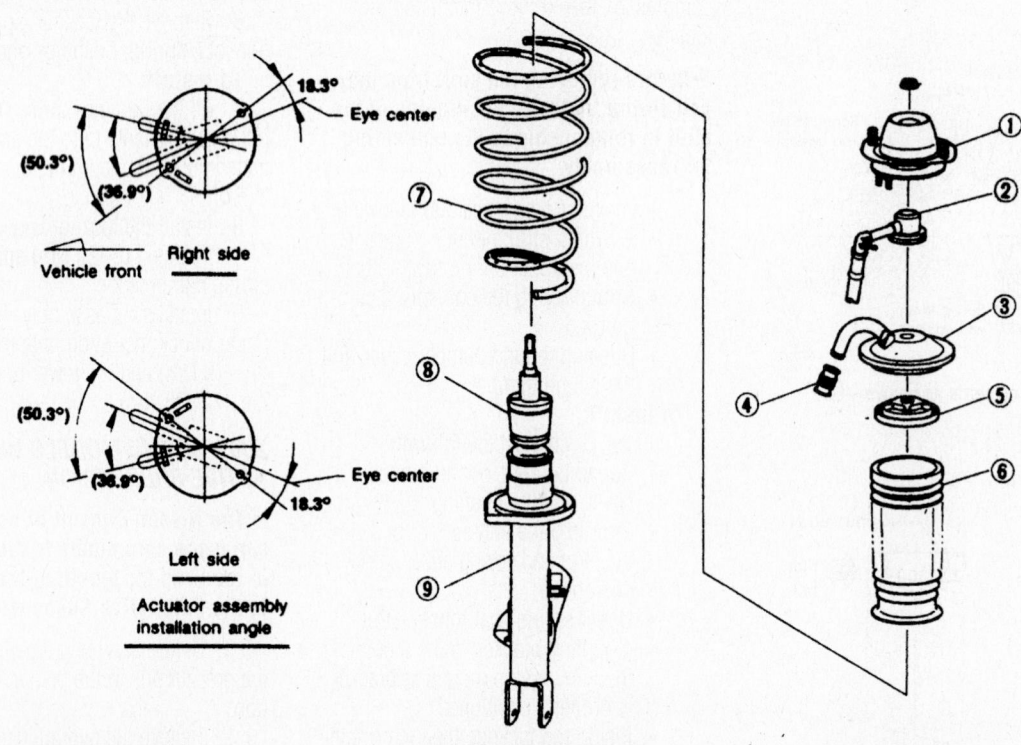

① Mount insulator
② Rear joint hose
③ Spring upper seat
④ Air tube connector
⑤ Bound bumper cover
⑥ Rear actuator dust cover
⑦ Coil spring
⑧ Bound bumper
⑨ Rear actuator

7923HG85

Rear actuator removal with Active Suspension—Q45 models

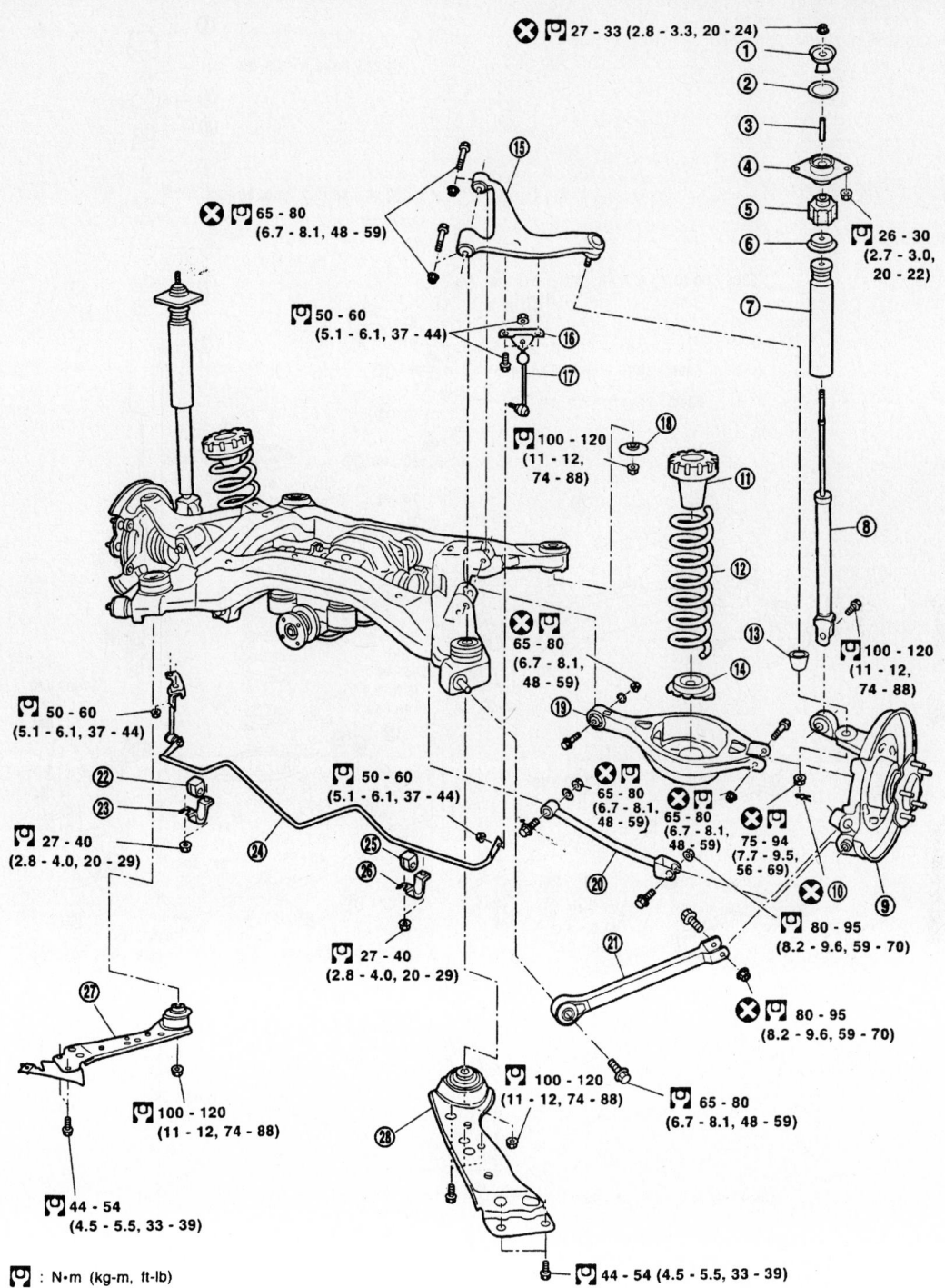

⊗ 🔧 27 - 33 (2.8 - 3.3, 20 - 24)

🔧 26 - 30 (2.7 - 3.0, 20 - 22)

⊗ 🔧 65 - 80 (6.7 - 8.1, 48 - 59)

🔧 50 - 60 (5.1 - 6.1, 37 - 44)

🔧 100 - 120 (11 - 12, 74 - 88)

🔧 50 - 60 (5.1 - 6.1, 37 - 44)

🔧 50 - 60 (5.1 - 6.1, 37 - 44)

⊗ 🔧 65 - 80 (6.7 - 8.1, 48 - 59)

🔧 100 - 120 (11 - 12, 74 - 88)

⊗ 🔧 65 - 80 (6.7 - 8.1, 48 - 59)

⊗ 🔧 65 - 80 (6.7 - 8.1, 48 - 59)

🔧 75 - 94 (7.7 - 9.5, 56 - 69)

⊗ 🔧 80 - 95 (8.2 - 9.6, 59 - 70)

🔧 27 - 40 (2.8 - 4.0, 20 - 29)

🔧 27 - 40 (2.8 - 4.0, 20 - 29)

⊗ 🔧 80 - 95 (8.2 - 9.6, 59 - 70)

🔧 100 - 120 (11 - 12, 74 - 88)

🔧 100 - 120 (11 - 12, 74 - 88)

🔧 65 - 80 (6.7 - 8.1, 48 - 59)

🔧 44 - 54 (4.5 - 5.5, 33 - 39)

🔧 44 - 54 (4.5 - 5.5, 33 - 39)

🔧 : N•m (kg-m, ft-lb)

1. Washer	2. Shock absorber mounting seal	3. Distance tube
4. Shock absorber mounting insulator	5. Bushing	6. Bound bumper cover
7. Bound bumper	8. Shock absorber	9. Axle assembly
10. Cotter pin	11. Upper seat	12. Coil spring
13. Ball seat	14. Rubber seat	15. Suspension arm
16. Connecting rod mounting bracket	17. Connecting rod	18. Mount stopper
19. Rear lower link	20. Front lower link	21. Radius rod
22. Bushing	23. Clamp	24. Stabilizer bar
25. Bushing	26. Clamp	27. Member stay
28. Member stay		

9357MG39

Exploded view of the rear suspension—G35 models

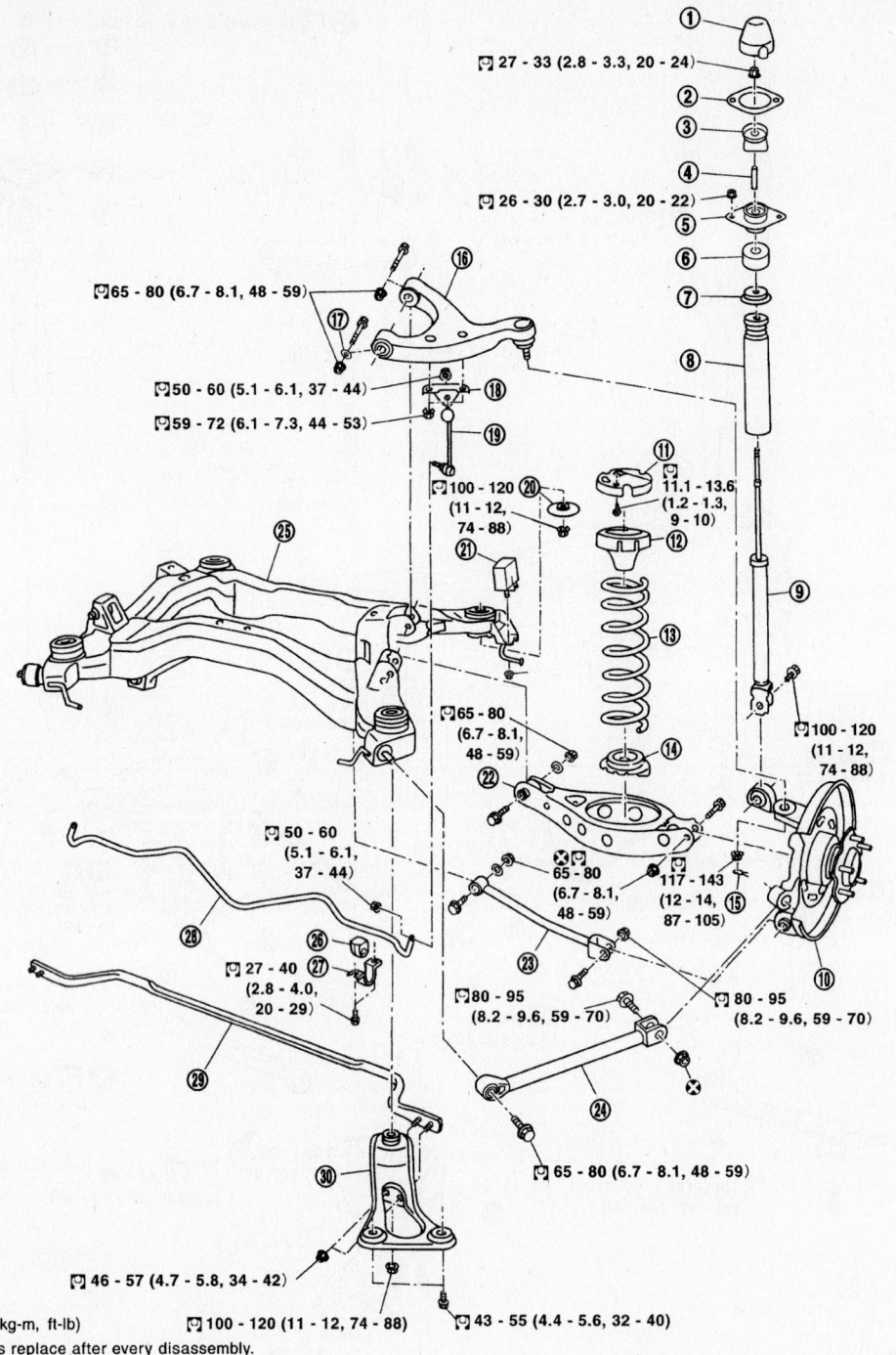

27 - 33 (2.8 - 3.3, 20 - 24)

26 - 30 (2.7 - 3.0, 20 - 22)

65 - 80 (6.7 - 8.1, 48 - 59)

50 - 60 (5.1 - 6.1, 37 - 44)

59 - 72 (6.1 - 7.3, 44 - 53)

100 - 120 (11 - 12, 74 - 88)

11.1 - 13.6 (1.2 - 1.3, 9 - 10)

65 - 80 (6.7 - 8.1, 48 - 59)

100 - 120 (11 - 12, 74 - 88)

50 - 60 (5.1 - 6.1, 37 - 44)

65 - 80 (6.7 - 8.1, 48 - 59)

117 - 143 (12 - 14, 87 - 105)

27 - 40 (2.8 - 4.0, 20 - 29)

80 - 95 (8.2 - 9.6, 59 - 70)

80 - 95 (8.2 - 9.6, 59 - 70)

65 - 80 (6.7 - 8.1, 48 - 59)

46 - 57 (4.7 - 5.8, 34 - 42)

100 - 120 (11 - 12, 74 - 88)

43 - 55 (4.4 - 5.6, 32 - 40)

: N•m (kg-m, ft-lb)

: Always replace after every disassembly.

1. Cap	2. Shock absorber mounting seal	3. Bushing
4. Distance tube	5. Shock absorber mounting bracket	6. Bushing
7. Bound bumper cover	8. Bound bumper	9. Shock absorber
10. Axle assembly	11. Bracket	12. Upper seat
13. Coil spring	14. Rubber seat	15. Cotter pin
16. Suspension arm	17. Stopper rubber	18. Stabilizer connecting rod mounting bracket
19. Stabilizer connecting rod	20. Mount stopper	21. Dynamic dumper
22. Rear lower link	23. Front lower link	24. Radius rod
25. Rear suspension member	26. Stabilizer bushing	27. Stabilizer clamp
28. Stabilizer bar	29. Cross bar	30. Member stay

Exploded view of the rear suspension—M45 models

67162-INFI-G35

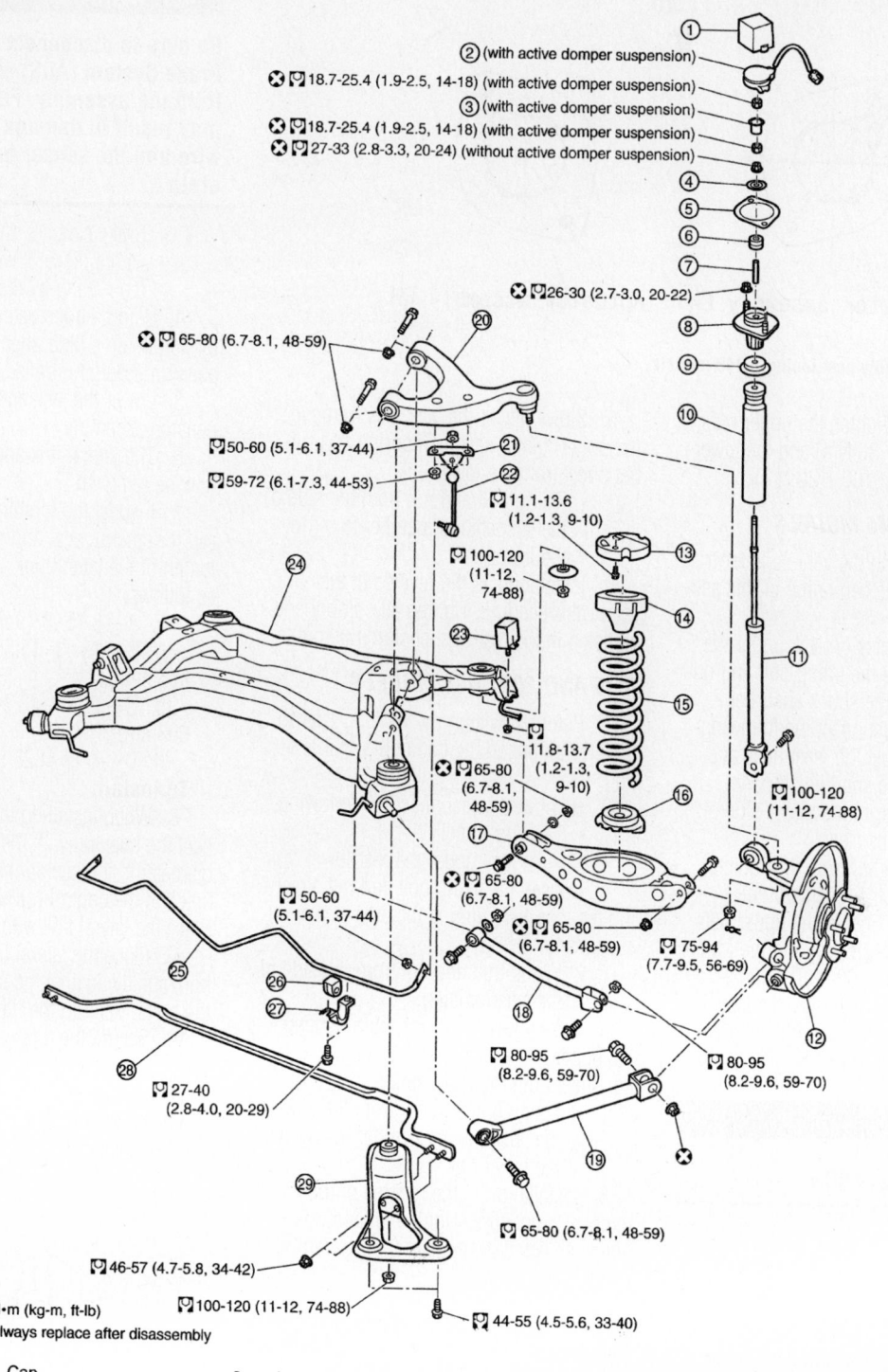

② (with active domper suspension)

❌ ☐18.7-25.4 (1.9-2.5, 14-18) (with active domper suspension)

③ (with active domper suspension)

❌ ☐18.7-25.4 (1.9-2.5, 14-18) (with active domper suspension)

❌ ☐27-33 (2.8-3.3, 20-24) (without active domper suspension)

❌ ☐26-30 (2.7-3.0, 20-22)

❌ ☐65-80 (6.7-8.1, 48-59)

☐50-60 (5.1-6.1, 37-44)

☐59-72 (6.1-7.3, 44-53)

☐11.1-13.6 (1.2-1.3, 9-10)

☐100-120 (11-12, 74-88)

☐11.8-13.7 (1.2-1.3, 9-10)

❌ ☐65-80 (6.7-8.1, 48-59)

☐100-120 (11-12, 74-88)

☐50-60 (5.1-6.1, 37-44)

❌ ☐65-80 (6.7-8.1, 48-59)

❌ ☐65-80 (6.7-8.1, 48-59)

☐75-94 (7.7-9.5, 56-69)

☐80-95 (8.2-9.6, 59-70)

☐80-95 (8.2-9.6, 59-70)

☐27-40 (2.8-4.0, 20-29)

☐65-80 (6.7-8.1, 48-59)

☐46-57 (4.7-5.8, 34-42)

☐ :N•m (kg-m, ft-lb)

❌ :Always replace after disassembly

☐100-120 (11-12, 74-88)

☐44-55 (4.5-5.6, 33-40)

1. Cap	2. Actuator assembly	3. Actuator plate
4. Washer	5. Shock absorber mounting seal	6. Bushing
7. Distance tube	8. Shock absorber mounting bracket	9. Bound bumper cover
10. Bound bumper	11. Shock absorber	12. Axle
13. Bracket	14. Upper seat	15. Coil spring
16. Rubber seat	17. Rear lower link	18. Front lower link
19. Radius rod	20. Suspension arm	21. Stabilizer connecting rod mounting bracket
22. Stabilizer connecting rod	23. Dynamic damper	24. Rear suspension member
25. Stabilizer bar	26. Stabilizer bushing	27. Stabilizer clamp
28. Cross bar	29. Member stay	

Exploded view of the rear suspension—Q45 models

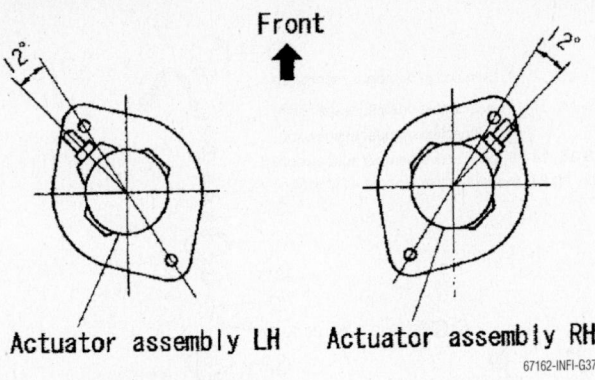

Rear actuator assembly positioning—Q45 models

removal procedure. Tighten the upper nuts to 20–22 ft. lbs. (26–30 Nm) and the lower bolt to 74–88 ft. lbs. (100–120 Nm).

M45 AND 2004 Q45 MODELS

1. Before servicing the vehicle, refer to the precautions in the beginning of this section.
2. Place a transmission jack under the rear axle and remove the fitting bolt and nut in the lower side of the shock absorber.
3. Remove or disconnect the following:
   - Rear seat cushion, seat back and rear package shelf finish panel
   - Fitting nut from the upper side of the shock absorber
   - Shock absorber from the vehicle
4. Installation is the reverse of the removal procedure. Tighten the upper nuts to 20–24 ft. lbs. (26–33 Nm) and the lower bolt to 74–88 ft. lbs. (100–120 Nm).
5. On Q45 models with active suspension, install the actuator assemblies in the correct position as shown.

Coil Spring

REMOVAL & INSTALLATION

Rear

G35 MODELS

For all models except the G35 and M45, refer to the Strut removal and installation procedure for coil spring replacement.

1. Before servicing the vehicle, refer to the precautions in the beginning of this section.
2. Remove or disconnect the following:
   - Tire and wheel
3. Place a jack under the rear lower link.
4. Loosen the fixing bolt and nut attaching the rear lower line to the side of the suspension member.
   - Fixing bolt and nut from the side of the axle housing

5. Slowly lower the jack, then remove the upper rubber seat, coil spring and rubber sheet from the lower link.
   - Fixing bolt and nut from the side of the suspension member to remove the lower link.
6. Installation is the reverse of the removal procedure. Tighten all components, as listed in the rear suspension illustration.

M45 AND 2004 Q45 MODELS

1. Before servicing the vehicle, refer to the precautions in the beginning of this section.
2. Remove or disconnect the following:
   - Tire and wheel
3. Place a jack under the rear lower link.
4. Loosen the fixing bolt and nut attaching the rear lower link to the side of the suspension member.
   - Fixing bolt and nut from the side of the axle housing
5. Slowly lower the jack, then remove the upper rubber seat, coil spring and rubber sheet from the lower link.
   - Fixing bolt and nut from the side of the suspension member to remove the lower link.
6. Installation is the reverse of the removal procedure. Tighten all components, as listed in the rear suspension illustration.

Torsion Bars

REMOVAL & INSTALLATION

G20 and I30 Models

1. Before servicing the vehicle, refer to the precautions in the beginning of this section.
2. Loosen the lug nuts.
3. Remove or disconnect the following:
   - Wheels

   - Brake calipers and suspend them with a piece of wire. Do not let them hang by the hose
4. Using a transmission jack, raise the torsion beam a little, then remove the suspension mounting bolts.
5. Lower the jack and remove the suspension assembly.
6. The lateral link and control rod can now be removed.
7. Inspect the torsion beam and control rod for cracks, wear and deformation. The length of the lateral link and control rod is as follows:
   a. A—8.15–8.19 in. (207–208mm).
   b. B—15.51–15.55 in. (394–395mm).
   c. C—23.66–23.74 in. (601–603mm).
   d. D—4.17–4.25 in. (106–108mm).
To install:
8. When installing the control rod, connect the bushing with the smaller inner diameter to the lateral link. Install the lateral link and the control rod on the torsion beam. Place the lateral link with the arrow topside.
9. Place the lateral link and control rod horizontally against the beam, and tighten the bolts. Refer to the illustration.
10. Secure the torsion beam to the vehicle. Make sure the lateral link is horizontal, then tighten the link to the chassis.
11. Attach the struts to the torsion beam and tighten the fasteners.

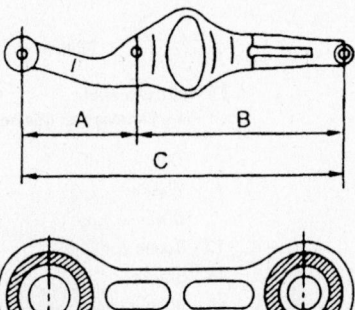
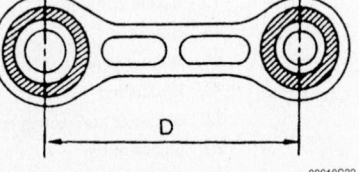

Measure the control rod and lateral links at these points—G20 and I30 models

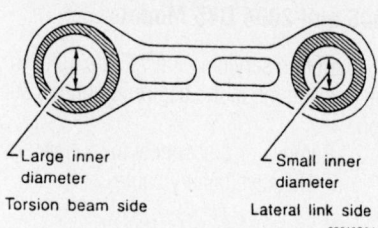

Be sure to install the control rod correctly—G20 and I30 models

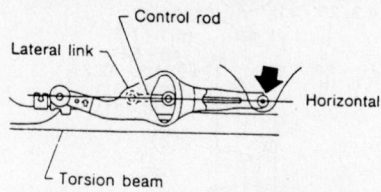

The lateral link must be in the horizontal position when tightening the bolts—G20 and I30 models

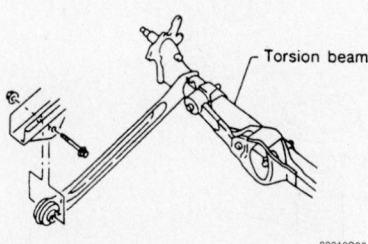

Tighten the torsion beam-to-chassis bolts with the suspension unloaded—G20 and I30 models

12. Tighten the torsion beam-to-chassis bolts.

13. Install the calipers, ABS sensor and wheels. Lower the vehicle to the ground. Final tighten the lug nuts.

Lower Ball Joint

REMOVAL AND INSTALLATION

The lower ball joint assembly is part of the lower control arm/transverse link. If replacement of the ball joint is required, the lower control arm needs to be replaced.

Lower Control Arms

REMOVAL AND INSTALLATION

G20 Models

1. Before servicing the vehicle, refer to the precautions in the beginning of this section.

2. Remove or disconnect the following:
   • Stabilizer bar

➡**Take note of paint mark and clamp position when removing stabilizer bar for correct reinstallation.**

3. Support the steering knuckle with a suitable jack and remove the lower ball joint nut. Separate the ball joint from the knuckle.
   • Bolts attaching the lower control arm to the chassis
   • Lower control arm

To install:

4. If the lower ball joint is worn or damaged, the lower control arm must be replaced. The ball joint is not serviceable separately.

5. Install or connect the following:
   • Lower control arm to the chassis with the attaching bolts and nut
   • Ball joint stud in the knuckle. Torque the nut to 52–64 ft. lbs. (71–86 Nm).
   • Stabilizer bar and wheel

6. Lower the vehicle.

➡**Final tightening must be done with the vehicle at normal ride height, tires on the ground and the chassis loaded.**

7. Torque front control arm bolts to 87–108 ft. lbs. (118–147 Nm) and rear gusset nut to 69–87 ft. lbs. (93–118 Nm).

I30 and I35 Models

1. Before servicing the vehicle, refer to the precautions in the beginning of this section.

2. Remove or disconnect the following:
   • Front wheels
   • Anti-lock Brake System (ABS) wheel sensor and move it out of the way
   • Wheel bearing locknut
   • Tie rod from the steering knuckle
   • Bolts attaching the strut to the steering knuckle. Matchmark prior to removal.
   • Halfshaft from the steering knuckle by lightly tapping the end of the shaft
   • Steering knuckle and the lower ball joint
   • Stabilizer bar from the lower control arm
   • Bolts attaching the link bushing pin to the chassis
   • Nut attaching the link to the control arm and remove the link
   • Bolts attaching the compression rod bushing clamp
   • Lower control arm/traverse link

To install:

3. Install or connect the following:
   • Lower control arm and the compression rod bushing clamp into the vehicle
   • Link bushing pin, if removed from the control arm

4. Tighten all bolts and nuts until they are snug enough to support the weight of the vehicle but not fully tight. The bolts should be tightened to specification with the vehicle on the floor.
   • Steering knuckle to the lower control arm and connect the ball joint. Torque the nut to 46–56 ft. lbs. (62–76 Nm).

➡**Always use a new nut when installing the ball joint to the control arm.**

   • Steering knuckle to the strut and to the halfshaft
   • Strut mounting bolts and align the matchmarks. Torque the bolts to 103–117 ft. lbs. (140–159 Nm).
   • Tie rod ball joint and tighten the nut to 46–54 ft. lbs. (63–73 Nm)
   • Wheel bearing locknut
   • ABS wheel sensor. Torque the attaching bolt to 13–17 ft. lbs. (18–24 Nm).
   • Front wheels

5. Lower the vehicle and torque the hub locknut to 174–231 ft. lbs. (235–314 Nm).

6. Torque the bolts attaching the compression rod bushing clamp and the link bushing pin, in the proper sequence to 87–108 ft. lbs. (118–147 Nm).

7. If the link bushing pin was removed from the control arm, torque the attaching nut to 87–108 ft. lbs. (118–147 Nm).

8. Tighten the sway bar attaching nut to 30–35 ft. lbs. (41–47 Nm).

9. Check the vehicle alignment.

G35 and 2001–03 Q45 Models

1. Before servicing the vehicle, refer to the precautions in the beginning of this section.

2. Remove or disconnect the following:
   • Negative battery cable
   • Front wheel
   • Engine undercover, if necessary
   • Nuts securing the tension rod to the transverse link (control arm)
   • Nut and separate the ball joint stud from the knuckle
   • Transverse link from the sub-frame

To install:

3. Install or connect the following:
   • Transverse link on the sub-frame. Temporarily install the bolt and nut.

- Tension rod on the transverse link. Torque the nuts to 87–94 ft. lbs. (118–127 Nm).
- Nut on the ball joint stud. Torque the nut to 71–88 ft. lbs. (96–120 Nm) for Q45 models or to 59–69 ft. lbs. (75–94 Nm) for G35 models.

- Front wheel and lower the vehicle to the floor
- Transverse link mounting bolt. Torque the bolt to 72–87 ft. lbs. (98–118 Nm).
- Engine undercover, if necessary
- Negative battery cable

M45 and 2004 Q45 Models

1. Before servicing the vehicle, refer to the precautions in the beginning of this section.
2. Remove or disconnect the following:
   - Negative battery cable

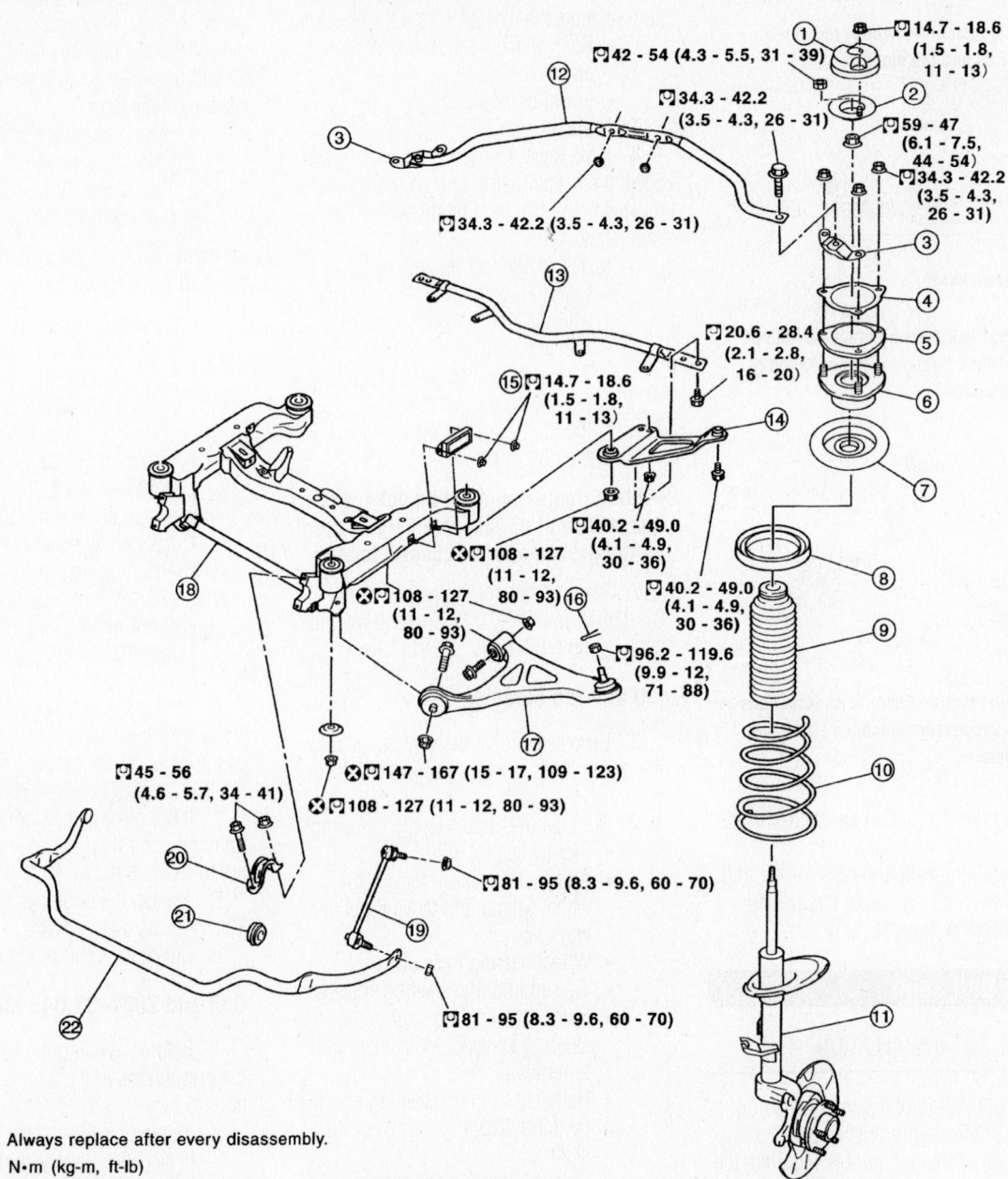

- ① : Always replace after every disassembly.
- ⌷ : N·m (kg-m, ft-lb)

1. Mass damper	2. Mass damper bracket	3. Tower bar bracket
4. Gasket	5. Strut mounting insulator bracket	6. Strut mounting insulator
7. Spring upper seat	8. Rubber seat	9. Bound bumper
10. Coil spring	11. Strut assembly	12 Tower bar
13. Front cross bar	14. Member stay	15. Dynamic dumper
16. Cotter pin	17. Suspension arm	18. Front suspension member
19. Stabilizer connecting rod	20. Stabilizer clamp	21. Stabilizer bushing
22. Stabilizer bar		

67162-INFI-G38

Exploded view of the front suspension—M45 Models

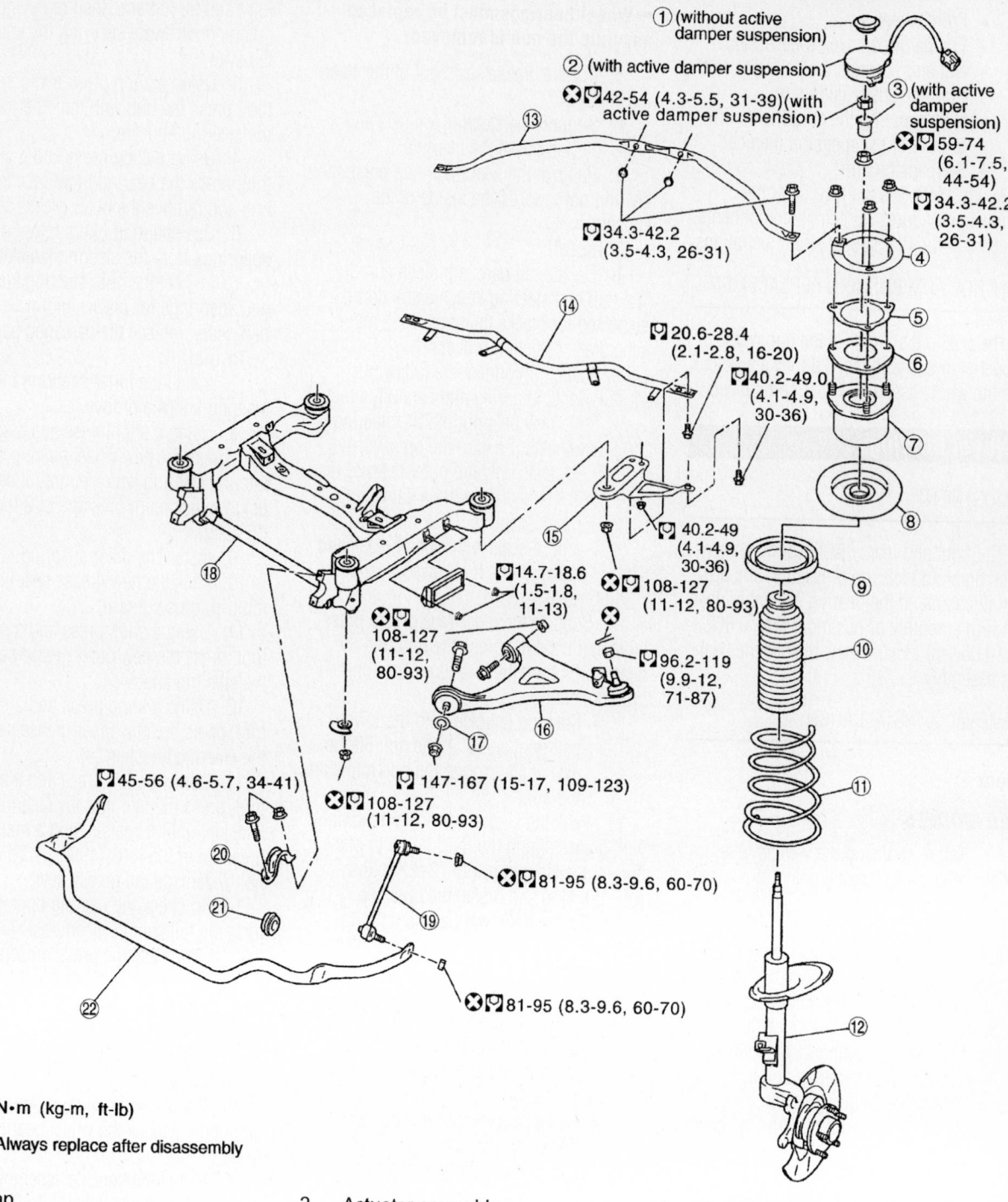

① (without active damper suspension)

② (with active damper suspension)

❌ 🔧42-54 (4.3-5.5, 31-39)(with active damper suspension)

③ (with active damper suspension)

❌ 🔧59-74 (6.1-7.5, 44-54)

🔧34.3-42.2 (3.5-4.3, 26-31)

🔧34.3-42.2 (3.5-4.3, 26-31)

🔧20.6-28.4 (2.1-2.8, 16-20)

🔧40.2-49.0 (4.1-4.9, 30-36)

🔧40.2-49 (4.1-4.9, 30-36)

🔧14.7-18.6 (1.5-1.8, 11-13)

❌ 🔧108-127 (11-12, 80-93)

❌ 🔧108-127 (11-12, 80-93)

🔧96.2-119 (9.9-12, 71-87)

🔧45-56 (4.6-5.7, 34-41)

❌ 🔧108-127 (11-12, 80-93)

🔧147-167 (15-17, 109-123)

❌ 🔧81-95 (8.3-9.6, 60-70)

❌ 🔧81-95 (8.3-9.6, 60-70)

🔧 : N•m (kg-m, ft-lb)

❌ : Always replace after disassembly

1. Cap	2. Actuator assembly	3. Actuator plate
4. Tower bar bracket	5. Strut mounting insulator bracket	6. Strut mounting insulator
7. Strut mounting bearing	8. Spring upper seat	9. Rubber seat
10. Bound bumper	11. Coil spring	12. Strut assembly
13. Tower bar	14. Front cross bar	15. Member stay
16. Suspension arm	17. Washer	18. Front suspension member
19. Stabilizer connecting rod	20. Stabilizer clamp	21. Stabilizer bushing
22. Stabilizer bar		

Exploded view of the front suspension—Q45 Models

67162-INFI-G39

- Front wheel
- Engine undercover, if necessary
- Nut and separate the ball joint stud from the suspension arm
- Nuts securing the suspension arm to the front suspension member
- Suspension arm

3. Installation is the reverse of the removal procedure. Tighten all components, as listed in the rear suspension illustrations.

CONTROL ARM BUSHING REPLACEMENT

The bushing are part of the transverse or suspension arm assembly, if they are defective the whole assembly must be replaced.

Wheel Bearings

ADJUSTMENT

The front and rear wheel bearing assemblies on all models are pressed in and are not adjustable. If the bearing assembly does not turn smoothly or has more than 0.002 in. (0.05mm) of axial play, replace the bearing assembly.

REMOVAL & INSTALLATION

Front

G20 MODELS

1. Before servicing the vehicle, refer to the precautions in the beginning of this section.
2. The axle nut torque is very high and should be loosened and tightened with the vehicle on the ground. Remove the cotter pin, adjusting cap and insulator and loosen the front axle nut.
3. Remove or disconnect the following:
- Brake caliper, carrier, and the rotor. Hang the caliper from the body with wire; do not let it hang by the brake hose.
- Cotter pin and nut and use a ball joint press to disconnect the tie rod end
- Cap and the upper king pin mounting nut and separate the kingpin from the third link

4. Hold a block of wood against the axle stub and strike it with a hammer to release it from the hub. Withdraw the axle from the hub and fold the steering knuckle down on the ball joint.
- Cotter pin and nut and use a ball joint press to disconnect the ball joint
- Steering knuckle

➡Wheel bearings must be replaced any time the hub is removed.

5. Pry the grease seals out of the steering knuckle.
6. Support the steering knuckle and press the hub out of the bearing.
7. Remove the snaprings and press the bearing out towards the inside of the knuckle.

To install:

8. Be sure all parts are clean and dry. The hub and steering knuckle should be inspected for cracks using dye or a magnetic crack detection process.
9. Install or connect the following:
- Inner snapring and carefully press the new bearing into the steering knuckle. Be sure the press tool contacts only the outer bearing race or the bearing will be damaged.
- Outer snapring. Pack the new grease seals with clean grease and install them. If removed, install the splash guard.

10. Support the inner race on the press table and carefully press the hub into the bearing. Be sure the hub turns smoothly in both directions.
- Steering knuckle onto the lower ball joint and start the nut. Fit the axle shaft through the hub and start the nut.

11. Pack the king pin bearing housing with grease and fit the third link into place. Torque the kingpin nut to 72–87 ft. lbs. (98–118 Nm) and install the dust cap.
12. Torque the lower ball joint nut to 52–64 ft. lbs. (71–86 Nm). Install a new cotter pin.
- Tie rod end. Torque the nut to 22–29 ft. lbs. (29–39 Nm). Torque as needed to install a new cotter pin but do not exceed 36 ft. lbs. (49 Nm).
- Brake caliper, carrier, rotor and the wheel

13. Lower the vehicle to the ground.
14. Torque the front axle nut to 174–231 ft. lbs. (235–314 Nm). Install the insulator, adjusting cap and cotter pin.

I30 AND I35 MODELS

➡Whenever the hub or bearing assembly is removed, the wheel bearing assembly must be replaced. Never reuse the old bearing assembly.

1. Before servicing the vehicle, refer to the precautions in the beginning of this section.
2. Remove the knuckle assembly from the vehicle by separating the ball

joint and tie rod end, then removing the retaining hardware securing the knuckle to the strut.
3. Using a shop press and a suitable tool, press the hub with the inner race from the steering knuckle.
4. Using a shop press and a suitable tool, press the bearing inner race from the hub and remove the outer grease seal.
5. Use snapring pliers to remove the snaprings from the steering knuckle.
6. Inspect the hub, steering knuckle and snaprings for cracks and/or wear; if necessary, replace the damaged part(s).

To install:

7. Install the inner snapring in the steering knuckle groove.
8. Using a shop press and a suitable tool, press the new wheel bearing assembly into the steering knuckle, until it seats, using a maximum pressure of 3 tons (2722kg).
9. Install the outer snapring.
10. Pack the new grease seal lips with multi-purpose grease.
11. Using a shop press and a suitable tool, press the new outer grease seal into the steering knuckle.
12. Using a shop press and a suitable tool, press the new inner grease seal into the steering knuckle.
13. Using a shop press and a suitable tool, press the hub into the steering knuckle, until it seats, using a maximum pressure of 5.5 tons (4990kg); be careful not to damage the grease seal.
14. To check the bearing operation, perform the following procedures:
a. Increase the press pressure to 3.5–5.0 tons (3175–4536kg).
b. Spin the steering knuckle, several turns, in both directions.
c. Be sure the wheel bearings operate smoothly.
15. If the wheel bearings do not operate smoothly, replace the wheel bearing assembly.
16. Install the knuckle assembly.
17. Install the halfshaft into the hub. Torque the locknut to 174–231 ft. lbs. (235–314 Nm) for I30 models, or to 188–245 ft. lbs. (255–333 Nm) for I35 models.
18. Install the wheel assembly and lower the vehicle.
19. Road test the vehicle and verify proper operation.

G35 AND 2001–03 Q45 MODELS

1. Before servicing the vehicle, refer to the precautions in the beginning of this section.

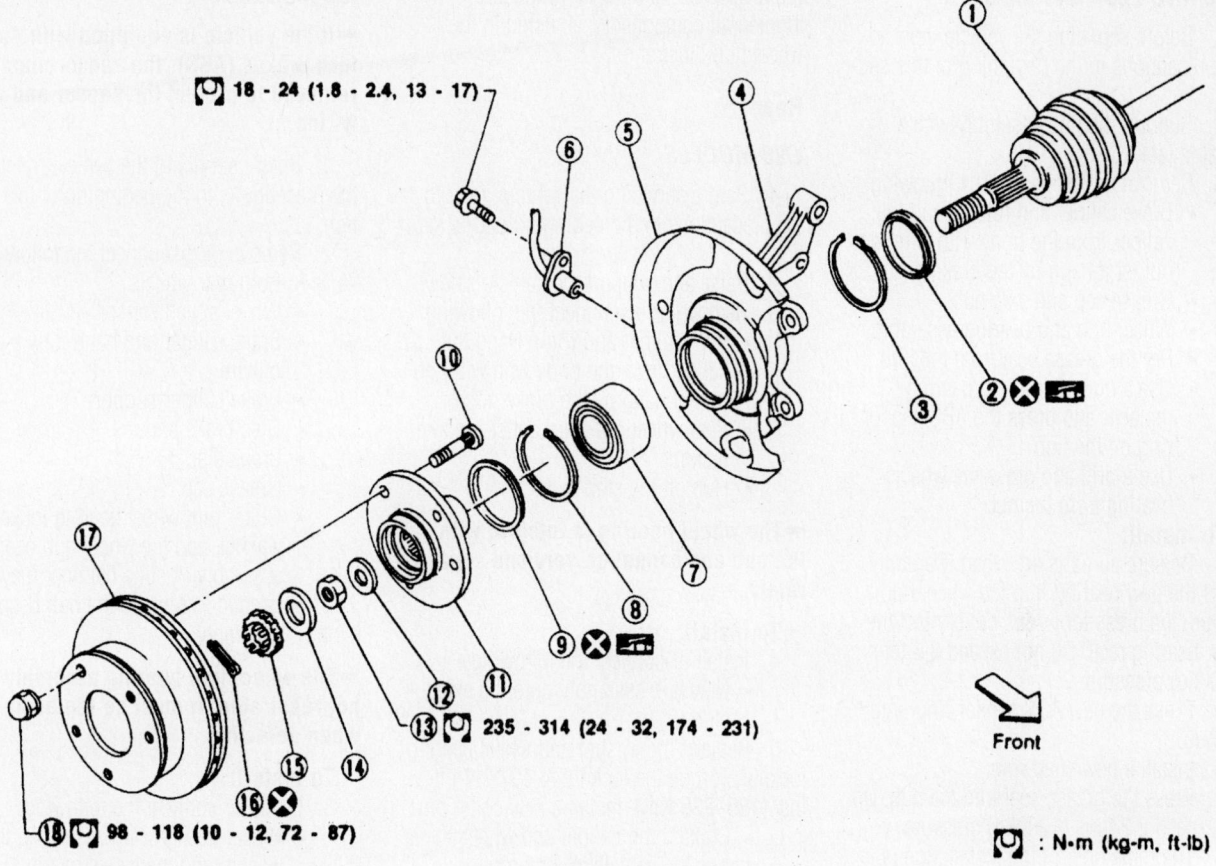

18 - 24 (1.8 - 2.4, 13 - 17)

235 - 314 (24 - 32, 174 - 231)

98 - 118 (10 - 12, 72 - 87)

Front

: N•m (kg-m, ft-lb)

①	Drive shaft	⑦	Wheel bearing assembly
②	Inner grease seal	⑧	Snap ring
③	Snap ring	⑨	Outer grease seal
④	Knuckle	⑩	Hub bolt
⑤	Baffle plate	⑪	Wheel hub
⑥	ABS sensor	⑫	Plain washer

⑬	Wheel bearing lock nut
⑭	Insulator
⑮	Adjusting cap
⑯	Cotter pin
⑰	Disc rotor
⑱	Wheel nut

7923HG87

Exploded view of the front knuckle assembly—I30 models

2. Support the hub assembly with a suitable jack.

3. Remove or disconnect the following:
- Brake caliper, carrier and rotor. Hang the caliper from the body with wire, do not let it hang by the brake hose.
- Cotter pins and nuts and use a ball joint press to disconnect the lower ball joint and tie rod end
- Kingpin lower mounting nut to remove the steering knuckle assembly

➡**Wheel bearings must be replaced any time the hub is removed.**

4. Use a vise or a wheel to hold the hub and remove the hub cap and nut from the back of the hub. Remove the wheel speed sensor rotor.

5. Use a press or large drift pin to press the hub out of the steering knuckle.

6. Remove the snapring and press the bearings and grease seal out of the steering knuckle.

To install:

7. Be sure all parts are clean. Carefully press the new bearing into the steering knuckle. Be sure the press tool contacts only the outer bearing race or the bearing will be damaged.

8. Install a new grease seal and the snapring. If removed, install the splash guard.

9. Lightly lubricate the lips of the seal with clean grease. Be careful not to grease the bearing or hub mating surfaces.

10. Carefully press the hub into the bearing. Support the inner race on the press table or the bearing will be damaged. Do not exceed 3.9 tons (3538kg) pressure.

11. Install or connect the following:
- Speed sensor rotor and nut on the hub and torque to 152–210 ft. lbs. (206–284 Nm). Stake the nut into place.

12. Lightly tap the cap into place and install the bolts.
- Steering knuckle to the king pin. Torque the nut to 108–137 ft. lbs. (88–108 Nm).
- Lower ball joint. Torque the nut to 65–80 ft. lbs. (88–108 Nm). Install a new cotter pin.
- Tie rod end. Torque the nut to 22–29 ft. lbs. (29–39 Nm). Torque as needed to install a new cotter pin but do not exceed 36 ft. lbs. (49 Nm).
- Rotor and the brake caliper
- Wheel and tire assembly

M45 AND 2004 Q45 MODELS

1. Before servicing the vehicle, refer to the precautions in the beginning of this section.

2. Support the hub assembly with a suitable jack.

3. Remove or disconnect the following:
   - Brake caliper and rotor. Hang the caliper from the body with wire, do not let it hang by the brake hose.
   - Grease cap and axle nut
   - Wheel hub and bearing assembly
   - Pry the grease seal from the hub
   - Use a puller, drift and bearing replacer and press the ABS sensor ring off the hub
   - Use a drift and press the wheel bearing from the hub

To install:

4. Be sure all parts are clean. Carefully press the new bearing into the wheel hub. Be sure the press tool does not contact the inner bearing race. Do not exceed 3.6 tons (3000kg) pressure.

5. Press the new ABS sensor ring onto the hub.

6. Install a new snap ring.

7. Press the grease seal into the hub. Do not exceed 2.2 tons (1000kg) pressure.

8. The remainder of the installation is the reverse of the removal procedure. Tighten all components, as listed in the wheel hub illustration.

Rear

G20 MODELS

1. Before servicing the vehicle, refer to the precautions in the beginning of this section.

2. Raise and support the vehicle safely.

3. Remove or disconnect the following:
   - Rear caliper and rotor. Hang the caliper from the body with wire, do not let hang by the brake hose.
   - Rear wheel hub cap, cotter pin and locknut
   - Hub off the stub axle

➡ **The wheel bearing is integral with the hub and cannot be serviced separately.**

To install:

4. Install or connect the following:
   - New hub assembly onto the axle stub

5. Replace the washer and wheel bearing locknut. Torque the locknut to 137–174 ft lbs. (186–235 Nm). Install a new cotter pin.
   - Brake rotor, caliper and wheel

6. Lower the vehicle to the ground.

I30 MODELS

➡ **If the vehicle is equipped with Anti-lock Brakes (ABS), the sensor must be removed to protect the sensor and its wiring.**

1. Before servicing the vehicle, refer to the precautions in the beginning of this section.

2. Remove or disconnect the following:
   - Both rear wheels
   - Wheel speed sensor
   - Brake caliper and hang it by a piece of wire
   - Brake caliper support
   - Disc brake pads
   - Brake disc
   - Grease cap
   - Cotter pin, wheel bearing locknut, washer, and the wheel hub bearing assembly. A slide hammer may be needed to remove the hub bearing assembly.

➡ **The wheel hub bearing assembly is not repairable; it must be replaced when defective.**

To install:

3. Install or connect the following:
   - Wheel hub bearing assembly, the washer and the wheel bearing lock-

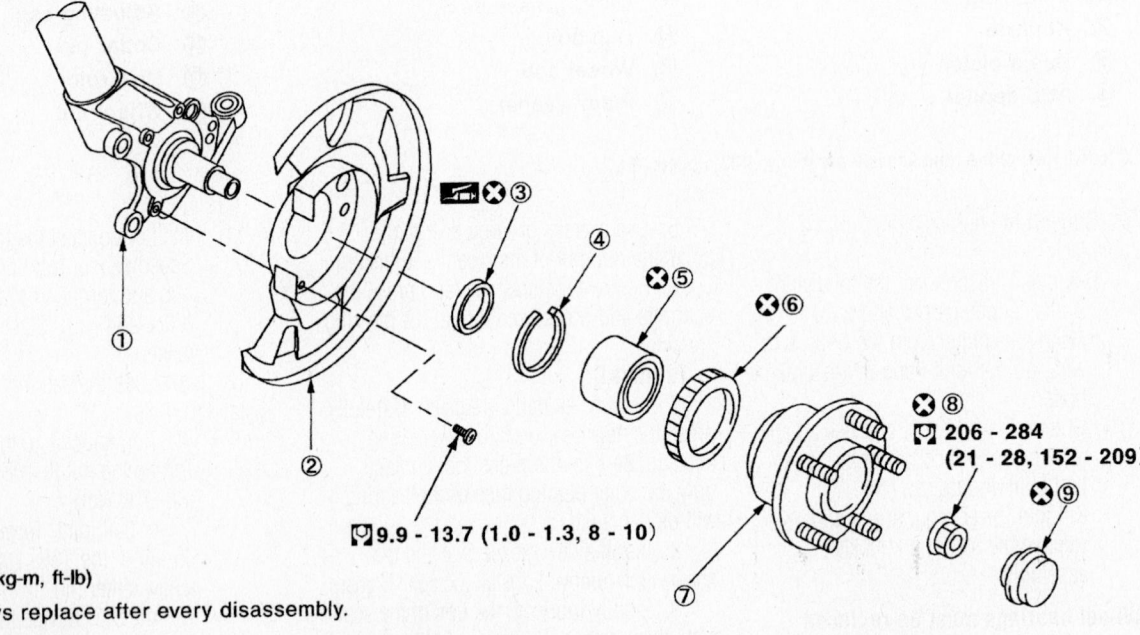

⌧ : N·m (kg-m, ft-lb)

✖ : Always replace after every disassembly.

1. Strut assembly
2. Splash guard
3. Grease seal
4. Snap ring
5. Wheel bearing
6. ABS sensor rotor
7. Wheel hub
8. Lock nut
9. Hub cap

67162-INFI-G40

Exploded view of the front hub and bearing—M45 Models

nut. Torque the wheel bearing locknut to 137–188 ft. lbs. (186–255 Nm).

4. Verify that the wheel bearings operate smoothly.

- New cotter pin into the spindle to hold the wheel bearing locknut

5. Install a dial micrometer to the rear wheel hub bearing assembly and check the axial end-play; it should be less than 0.0020 in. (0.05mm).

- Grease cap
- ABS wheel sensor and its wiring
- Brake assembly and the wheels

G35 AND 2001–03 Q45 MODELS

1. Before servicing the vehicle, refer to the precautions in the beginning of this section.

2. Remove or disconnect the following:

- Cotter pin and adjusting cap and loosen the wheel bearing nut. Carefully tap the end of the axle shaft or use a puller to loosen the shaft from the hub.
- Brake caliper and rotor. Do not let the caliper hang by the brake hose, support it with wire.
- Parking brake assembly
- Nuts and through-bolts to remove the axle housing from the suspension. If equipped with rear wheel

steering, use a ball joint press to separate the tie rod end.

- 4 bolts at the back and remove the bearing flange and hub from the bearing housing

3. Press the hub out of the bearing flange and use a puller to remove the bearing from the hub. If it is not damaged, the hub can be used again but the bearing and flange are supplied as a single unit.

To install:

➡**The wheel bearing and flange are supplied as an assembly.**

4. Place the hub on a press table and press the new bearing and flange onto the hub. Be sure the press tool contacts only the inner bearing race and take care not to damage the seal.

5. Install or connect the following:

- Bearing flange onto the axle housing. Torque the bolts to 58–72 ft. lbs. (78–98 Nm).
- Axle housing onto the lower ball joint, torque the nut to 58–69 ft. lbs. (78–93 Nm) and install a new cotter pin
- Torque the tie rod end nut to 33–44 ft. lbs. (45–60 Nm) and install a new cotter pin, if equipped with rear wheel steering
- Axle shaft into the hub and install

the bolts through the suspension bushings. Tighten the bolts temporarily, they will be tightened with the vehicle resting on the wheels.

- Brake components and apply the brake to hold the hub from turning
- Wheel bearing locknut and torque it to 152–203 ft. lbs. (206–275 Nm)
- Insulator and adjusting cap and a new cotter pin
- Wheel and lower the vehicle to the ground. Torque the suspension bushing bolts to 57–72 ft. lbs. (77–98 Nm).

M45 AND 2004 Q45 MODELS

1. Before servicing the vehicle, refer to the precautions in the beginning of this section.

2. Raise and support the vehicle.

3. Remove or disconnect the following:

- Rear tires
- Axle shaft nut
- Brake caliper and rotor. Hang the caliper from the body with wire, do not let it hang by the brake hose.
- Radius rod bolts and nuts
- Lower link nuts and bolts
- Coil spring
- Lower shock bolts
- Axle shaft
- Suspension arm ball joint

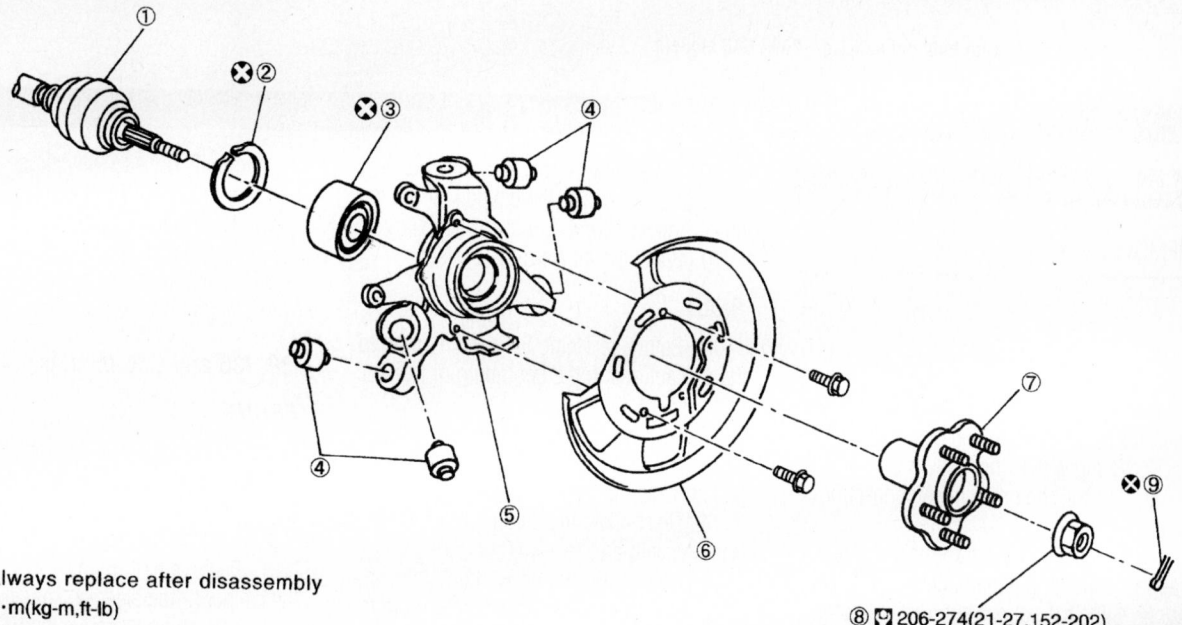

✖ : Always replace after disassembly

🔧 : N·m(kg-m,ft-lb)

⑧ 🔧 206-274(21-27,152-202)

1. Drive shaft	2. Snap ring	3. Wheel bearing
4. Bushing	5. Axle	6. Back plate
7. Wheel hub	8. Lock nut	9. Cotter pin

Exploded view of the rear hub and bearing—M45 Models

67162-INFI-G41

- Wheel hub and bearing assembly
- Use a puller, drift and bearing replacer and press the wheel bearing outer race off the hub
- Use a drift and remove the wheel bearing from the hub

To install:

4. Be sure all parts are clean. Carefully press the hub into the new bearing. Be sure the press tool does not contact the inner bearing race. Do not exceed 11,000 lbs. (5000kg) pressure.

5. Install backing plate and the hub and bearing assembly.

6. Reverse the remainder of the removal procedure to complete installation.

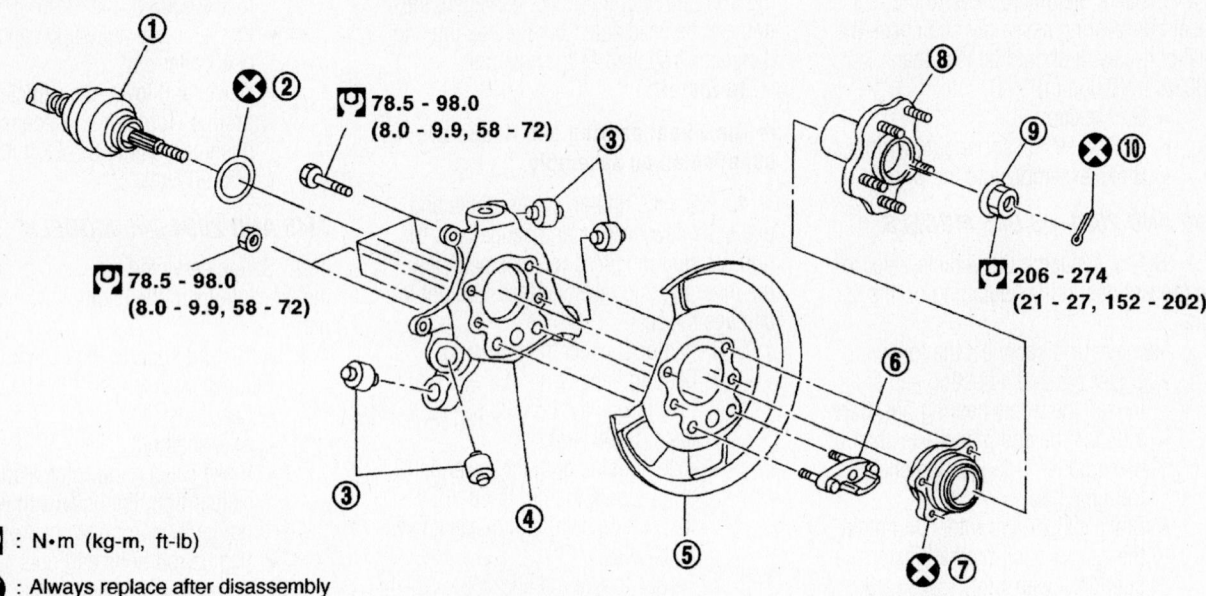

☒ : Always replace after disassembly

1. Drive shaft	2. Dust shield	3. Bushing
4. Axle	5. Back plate	6. Anchor block
7. Wheel bearing	8. Wheel hub	9. Lock nut
10. Cotter pin		

67162-INFI-G42

Exploded view of the rear hub and bearing—2004 Q45 Models

BRAKES

Brake Caliper

REMOVAL & INSTALLATION

G20 Models

FRONT

1. Before servicing the vehicle, refer to the precautions in the beginning of this section.
2. Remove the wheels.
3. Loosen the brake hose connecting bolt.
4. Remove the bolts connecting the caliper to the torque member.
5. Slide the caliper out from the rotor.
6. Remove the brake hose connecting bolt from the caliper.
7. Remove the caliper from the vehicle.

To install:

8. Fit the caliper onto the torque member and torque the bolts to 16–23 ft. lbs. (22–31 Nm).

9. Using new copper washers, connect the hydraulic hose to the caliper. Torque the union bolt to 12–14 ft. lbs. (17–19 Nm).
10. Bleed the air from the system.

REAR

1. Before servicing the vehicle, refer to the precautions in the beginning of this section.
2. Remove the rear wheels.
3. Remove the brake cable mounting bolt and lock spring.
4. Disconnect the parking brake cable from the caliper.
5. Disconnect the brake fluid hose.
6. Remove the torque member mounting bolts and remove the caliper assembly.

To install:

7. Fit the caliper onto the torque member and torque the bolts to 16–23 ft. lbs. (22–31 Nm).
8. Using new copper washers, connect

the hydraulic hose to the caliper. Torque the union bolt to 12–14 ft. lbs. (17–19 Nm).

9. Connect the parking brake cable to the rear caliper.
10. Bleed the brake system.
11. Install the rear wheels.

I30, I35 and G35 Models

FRONT

1. Before servicing the vehicle, refer to the precautions in the beginning of this section.
2. Remove the front wheels.
3. Remove both guide pin bolts securing the caliper to the steering knuckle.
4. Loosen and remove the brake hose connector from the caliper.
5. Remove the caliper assembly from the vehicle.

To install:

6. Using new copper washers, install the brake line to the brake caliper and torque

the connecting bolt to 12–14 ft. lbs. (17–20 Nm).

7. Install the caliper to the steering knuckle using the guide pins bolts.

8. Install the wheels and tighten the lug nuts to the proper specification.

9. Bleed the brake system and top off the master cylinder as necessary.

REAR

1. Before servicing the vehicle, refer to the precautions in the beginning of this section.

2. Remove the rear wheels.

3. Remove the parking brake cable stay fixing bolt and the lock spring.

4. Remove the brake fluid hose from the caliper.

5. Remove the guide pin bolts and remove the caliper.

To install:

6. Install the caliper body into position and torque the caliper-to-torque member pin bolts to 16–23 ft. lbs. (22–31 Nm).

7. Reconnect the brake fluid hose and tighten the flare nut to 12–14 ft. lbs. (17–20 Nm).

8. Install the lock spring and the parking brake stay attaching bolt.

9. Bleed the brake system and top off the master cylinder as necessary.

10. Install the wheels.

2001–03 Q45 Models

FRONT

1. Before servicing the vehicle, refer to the precautions in the beginning of this section.

2. Remove the wheels.

3. Remove the brake hose connecting bolt.

4. Remove the torque member mounting bolts and disconnect the brake fluid hose from the caliper.

5. Slide the caliper off of the rotor.

6. Remove the caliper from the vehicle.

To install:

7. Position the torque member on the knuckle assembly and install the mounting bolts. Torque the bolts to 118–137 ft. lbs. (160–186 Nm).

8. Using new copper washers, connect the hydraulic hose to the caliper. Torque the union bolt to 12–14 ft. lbs. (17–19 Nm).

9. Bleed the air from the system and fill the master cylinder with clean brake fluid.

REAR

1. Before servicing the vehicle, refer to the precautions in the beginning of this section.

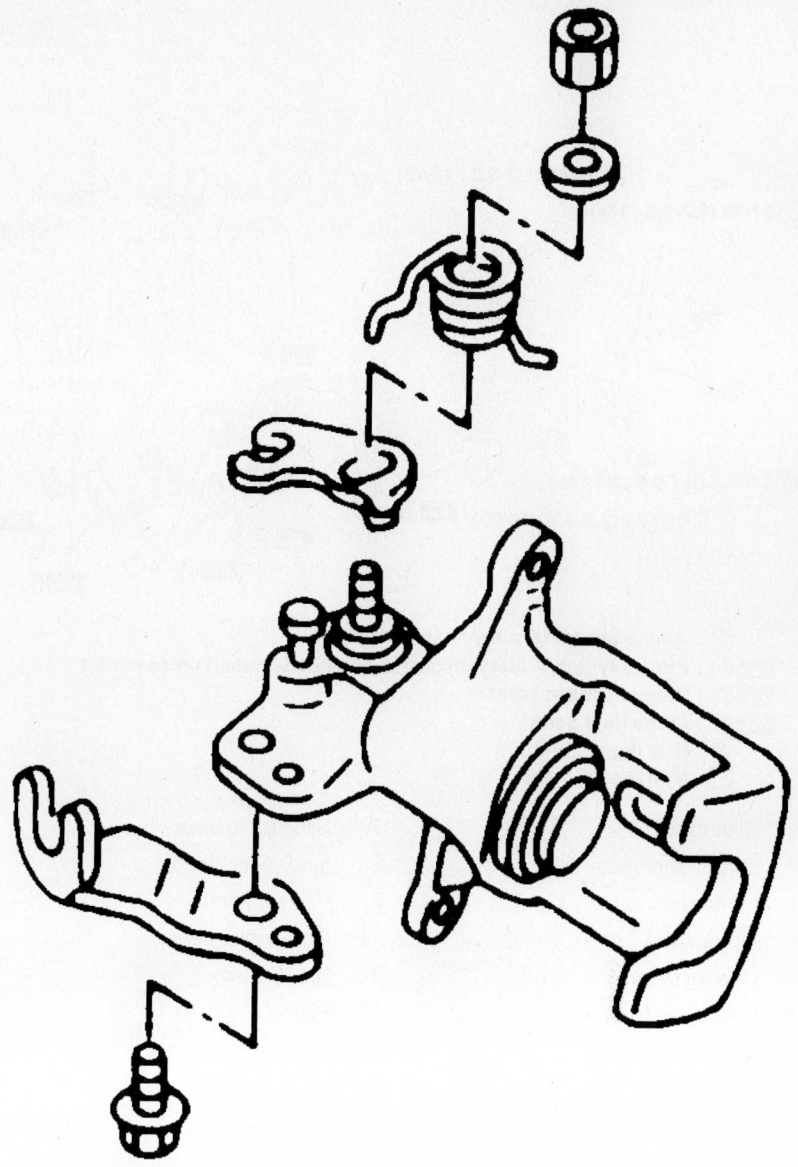

Rear caliper parking brake lever—G20 shown

93016G14

2. Remove the rear wheels.

3. Disconnect the brake fluid hose.

4. Remove the torque member mounting bolts and remove the caliper assembly.

To install:

5. Fit the caliper over the rotor and install the mounting bolts. Torque the bolts to 28–38 ft. lbs. (38–52 Nm).

6. Using new copper washers, connect the hydraulic hose to the caliper. Torque the union bolt to 12–14 ft. lbs. (17–19 Nm).

7. Bleed the brake system.

8. Install the rear wheels and lower the vehicle to the floor.

M45 and 2004 Q45 Models

FRONT

1. Before servicing the vehicle, refer to the precautions in the beginning of this section.

2. Remove the wheels.

3. Remove the brake hose connecting bolt.

4. Remove the torque member mounting bolts and disconnect the brake fluid hose from the caliper.

5. Slide the caliper off of the rotor.

6. Remove the caliper from the vehicle.

To install:

7. Position the torque member on the knuckle assembly and install the mounting

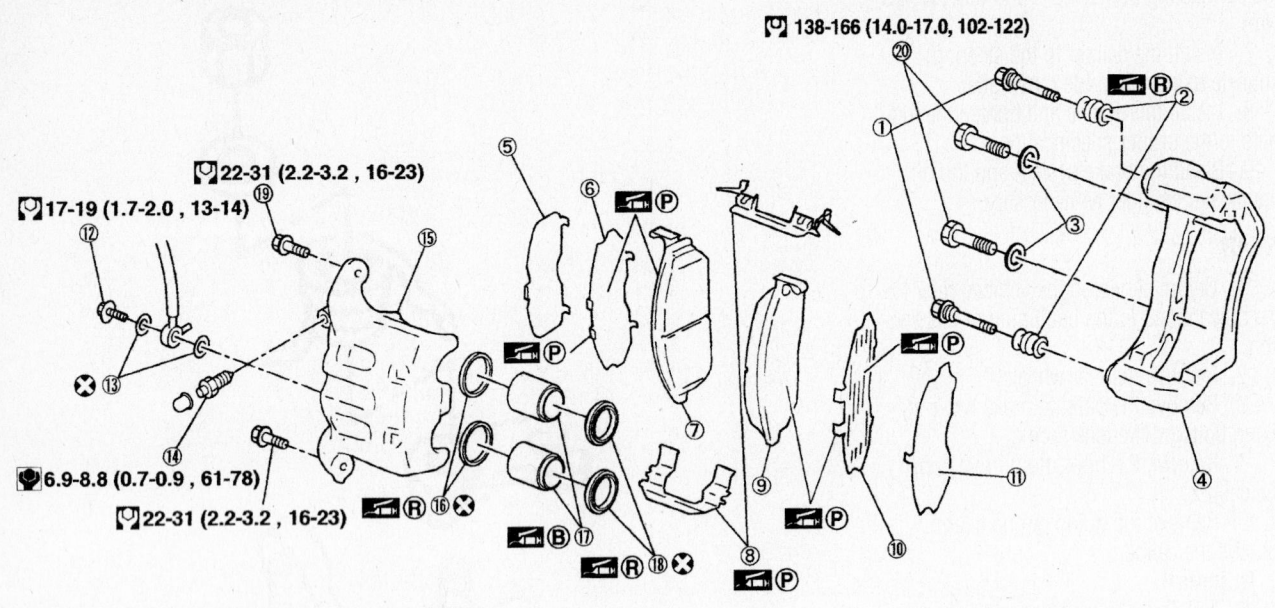

Ⓧ : Always replace after every disassembly
🔧Ⓟ : PBC(Poly Butyl Cuprysil) grease or silicone-based grease point
🔧Ⓡ : Rubber Grease point
🔧Ⓑ : Brake fluid point
🔧 : N·m (kg-m , in-lb)
🔧 : N·m (kg-m , ft-lb)

1.	Sliding pin	2.	Sliding pin boot	3.	Washer
4.	Torque member	5.	Inner shim cover	6.	inner shim
7.	Inner pad	8.	Pad retainer	9.	Outer pad
10.	Outer shim	11.	Outer shim cover	12.	Union bolt
13.	Copper washer	14.	Bleed valve	15.	Cylinder body
16.	Piston seal	17.	Piston	18.	Piston boot
19.	Sliding pin bolt	20.	Torque member bolt		

67162-INFI-G43

Exploded view of the front brake assembly—M45 and 2004 Q45 Models

bolts. Torque the bolts to 102–122 ft. lbs. (138–166 Nm).

8. Using new copper washers, connect the hydraulic hose to the caliper. Torque the union bolt to 12–14 ft. lbs. (17–19 Nm).

9. Bleed the air from the system and fill the master cylinder with clean brake fluid.

REAR

1. Before servicing the vehicle, refer to the precautions in the beginning of this section.

2. Remove the rear wheels.

3. Disconnect the brake fluid hose.

4. Remove the torque member mounting bolts and remove the caliper assembly.

To install:

5. Fit the caliper over the rotor and install the mounting bolts. Torque the bolts to 25–30 ft. lbs. (33–42 Nm).

6. Using new copper washers, connect the hydraulic hose to the caliper. Torque the union bolt to 12–14 ft. lbs. (17–19 Nm).

7. Bleed the brake system.

8. Install the rear wheels and lower the vehicle to the floor.

Disc Brake Pads

REMOVAL & INSTALLATION

G20 Models

FRONT

1. Before servicing the vehicle, refer to the precautions in the beginning of this section.

2. Remove the cap from the master cylinder reservoir and extract about ⅓ of

the brake fluid from the reservoir to prevent overflow when the caliper piston is compressed.

3. Remove the wheels.

4. Remove the lower pin bolt.

5. Pivot the caliper body upward and secure it with a length of wire. Remove the retainers and inner and outer shims and pads.

To install:

6. Place an old pad over the caliper piston. Use a C-clamp to compress the piston.

7. Install the new pads and shims and rotate caliper down onto rotor. Install the pin bolt and torque it to 16–23 ft. lbs. (22–31 Nm).

8. Install the wheels and lower the vehicle to the floor.

9. Check and then refill the master cylinder if needed.

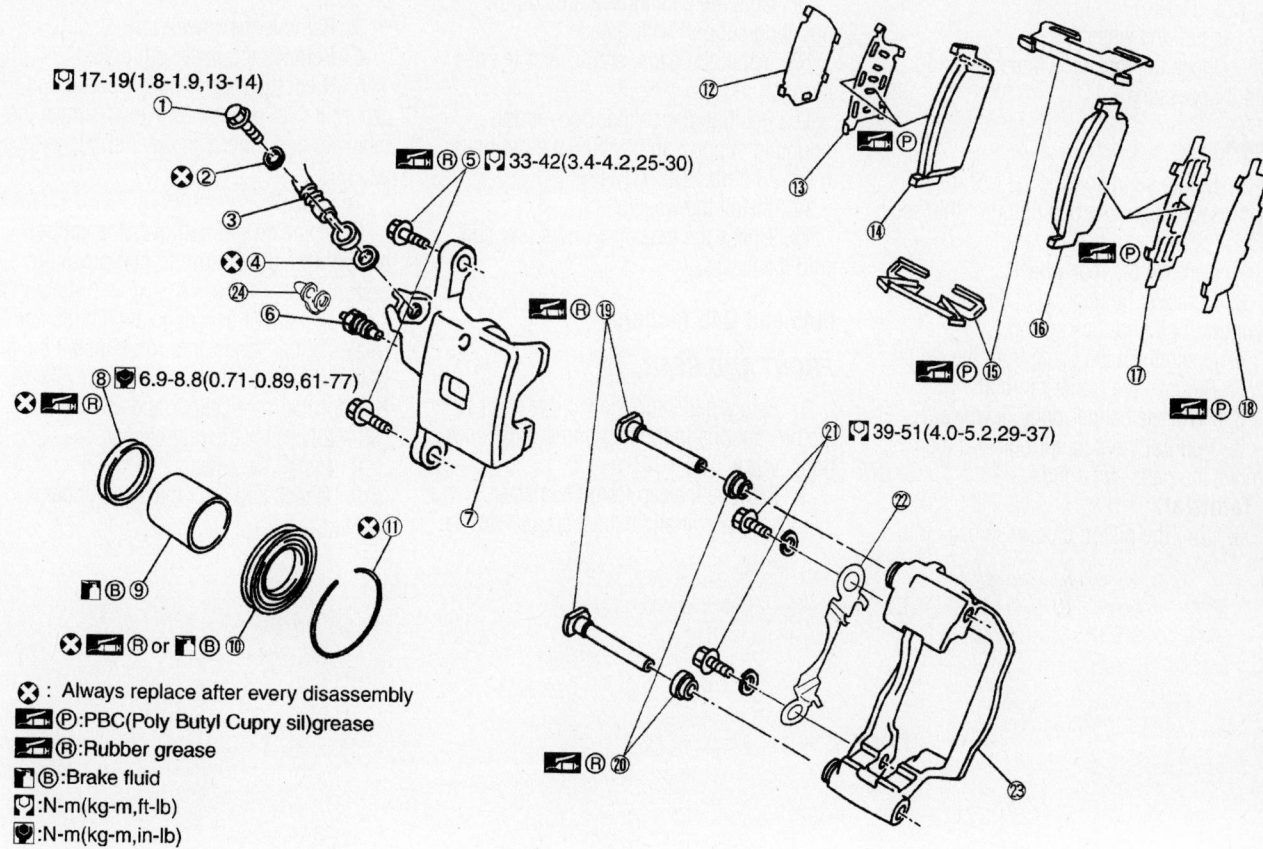

$\boxed{}$: Always replace after every disassembly

Ⓟ:PBC(Poly Butyl Cupry sil)grease

Ⓡ:Rubber grease

Ⓑ:Brake fluid

:N-m(kg-m,ft-lb)

:N-m(kg-m,in-lb)

1. Union bolt	2. Copper washer	3. Brake hose
4. Copper washer	5. Sliding pin bolt	6. bleed valve
7. Cylinder body	8. Piston seal	9. Piston
10. Piston boot	11. Retaining ring	12. Inner shim cover
13. Inner shim	14. Inner brake pad	15. Pad retainer
16. Outer brake pad	17. Outer shim	18. Outer shim cover
19. Sliding pin	20. Sliding pin boot	21. Torque member mounting bolts
22. Decrement shim (Not inserted in some vehicles.)	23. Torque member	24. Cap

Exploded view of the rear brake assembly—M45 and 2004 Q45 Models

67162-INFI-G44

REAR

1. Before servicing the vehicle, refer to the precautions in the beginning of this section.

2. Remove the cap from the master cylinder reservoir and extract about ⅓ of the brake fluid from the reservoir to prevent overflow when the caliper piston is compressed.

3. Remove the wheels.

4. Remove the brake cable mounting bracket bolt and lock spring.

5. Disconnect the parking brake cable.

6. Remove the lower pin bolt.

7. Pivot the caliper body upward and secure it with a length of wire. Remove the retainers and inner and outer shims and pads.

To install:

8. Push the piston into the cylinder body by turning the piston clockwise.

9. Install the new pads and shims and rotate the caliper down onto rotor. Install the pin bolt and torque it to 16–23 ft. lbs. (22–31 Nm).

10. Connect the parking brake cable and install the bracket.

11. Install the wheels.

12. Check and then refill the master cylinder if needed.

I30, I35 and G35 Models

FRONT

1. Before servicing the vehicle, refer to the precautions in the beginning of this section.

2. Remove the wheels.

3. Remove the bottom guide pin from the caliper and swing the caliper cylinder body upward. Support the caliper with a wire.

4. Remove the brake pad retainers and the pads.

To install:

5. Compress the piston of the disc brake caliper.

6. Install the brake pads and caliper assembly. Torque the guide pin to 16–23 ft. lbs. (22–31 Nm).

7. Install the wheels.

8. Check the master cylinder and add fluid if necessary.

REAR

1. Before servicing the vehicle, refer to the precautions in the beginning of this section.

2. Remove the rear wheels.

3. Remove the parking brake cable mounting bolt and lock spring.

4. Disconnect the cable from the caliper.

5. Remove the upper pin bolt.

6. Pivot the caliper body downward.

7. Pull out the pad springs and then remove the pads and shims.

To install:

8. Turn the piston clockwise back into the caliper body. Take care not to damage the piston boot.

9. Coat the pad contact area on the mounting support with grease.

10. Install the pads, shims, and the pad springs.

11. Position the caliper body in the mounting support and tighten the pin bolts to 16–23 ft. lbs. (22–31 Nm).

12. Install the wheels.

13. Check the master cylinder and add fluid if necessary.

M45 and Q45 Models

FRONT AND REAR

1. Before servicing the vehicle, refer to the precautions in the beginning of this section.

2. Remove the cap from the master cylinder reservoir and extract about ⅓ of the brake fluid from the reservoir to prevent overflow when the caliper piston is compressed.

3. Remove the wheels.

4. Remove the lower pin bolt.

5. Pivot the caliper body upward and secure it with a length of wire. Remove the retainers and inner and outer shims and pads.

To install:

6. Place an old pad over the caliper piston. Use a C-clamp to compress the piston.

7. Install the new pads and shims and rotate caliper down onto rotor. Install the pin bolt and torque it to 16–23 ft. lbs. (22–31 Nm) for the front caliper and 25–30 ft. lbs. (33–42 Nm) for the rear caliper.

8. Install the wheels.

9. Check and refill master cylinder if needed.

INFINITI

QX4

13

SPECIFICATION AND MAINTENANCE CHARTS

ENGINE AND VEHICLE IDENTIFICATION

Engine							Model Year	
Code ①	Liters (cc)	Cu. In.	Cyl.	Fuel Sys.	Engine	Eng. Mfg.	Code ②	Year
VG33E	3.3 (3277)	199.8	6	MFI	SOHC	Nissan	1	2001
VQ35DE	3.5 (3498)	213	6	MFI	DOHC	Nissan	2	2002
							3	2003
							4	2004
							5	2005

MFI: Multi-port Fuel Injection

SOHC: Single Overhead Camshaft

DOHC: Double Overhead Camshafts

① Located on the timing belt cover

② 10th digit of the Vehicle Identification Number (VIN)

67162-IQX4-C01

GENERAL ENGINE SPECIFICATIONS

Year	Model	Engine Displacement Liters	Engine ID	Net Horsepower @ rpm	Net Torque @ rpm (ft. lbs.)	Bore x Stroke (in.)	Com-pression Ratio	Oil Pressure @ rpm
2001	QX4	3.3	VG33E	170@4800	200@2800	3.60x3.27	8.9:1	60-65@2000
2002	QX4	3.5	VQ35DE	240@6000	265@3200	3.76X3.20	10.0:1	43@2000
2003	QX4	3.5	VQ35DE	240@6000	265@3200	3.76X3.20	10.0:1	43@2000

67162-IQX4-C02

ENGINE TUNE-UP SPECIFICATIONS

Year	Engine Displacement Liters	Engine ID	Spark Plug Gap (in.)	Ignition Timing	Fuel Pump (psi) ①	Idle Speed ②	Valve Clearance (in.) In.	Ex.
2001	3.3	VG33E	0.039-0.043	13-17B	34	700-800	HYD	HYD
2002	3.5	VQ35DE	0.044	15B	35	700-800	HYD	HYD
2003	3.5	VQ35DE	0.044	15B	35	700-800	HYD	HYD

NOTE: The Vehicle Emission Control Information label often reflects specification changes made during production. The label figures must be used if they differ from those in this chart.

B: Before top dead center

HYD: Hydraulic

① System pressure at idle with vacuum hose connected

　　Should increase to 43 psi when disconnected

② Automatic transmission in Neutral

67162-IQX4-C03

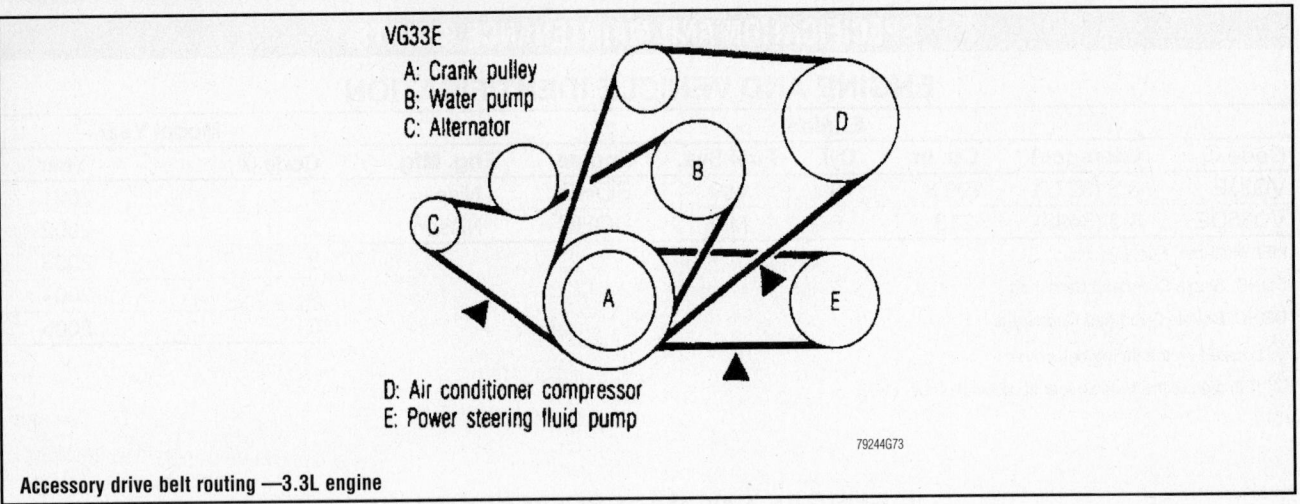

VG33E
A: Crank pulley
B: Water pump
C: Alternator

D: Air conditioner compressor
E: Power steering fluid pump

79244G73

Accessory drive belt routing —3.3L engine

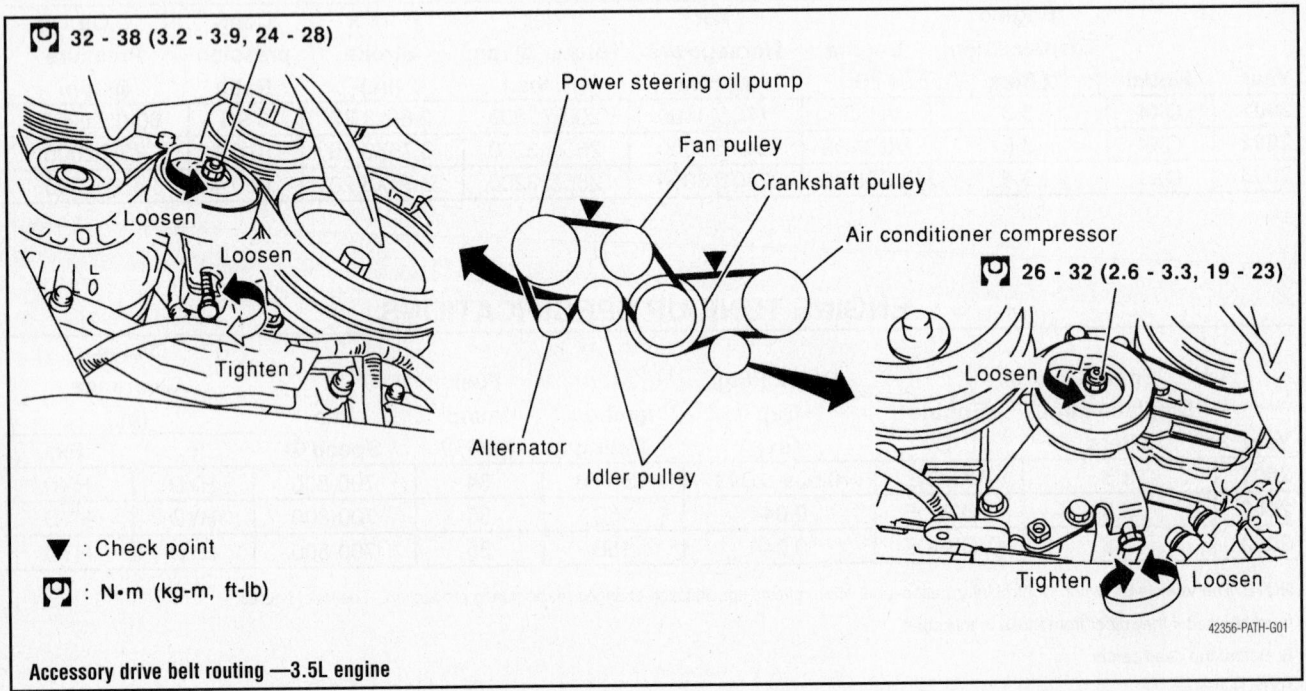

32 - 38 (3.2 - 3.9, 24 - 28)

Power steering oil pump

Fan pulley

Crankshaft pulley

Air conditioner compressor

Loosen

Loosen

Tighten

26 - 32 (2.6 - 3.3, 19 - 23)

Loosen

Alternator

Idler pulley

Tighten   Loosen

▼ : Check point

: N•m (kg-m, ft-lb)

42356-PATH-G01

Accessory drive belt routing —3.5L engine

CAPACITIES

Year	Model	Engine Displacement Liters	Engine ID	Engine Oil with Filter (qts.)	Transmission (pts.)	Transfer Case (pts.)	Drive Axle Front (pts.)	Drive Axle Rear (pts.)	Fuel Tank (gal.)	Cooling System (qts.)
2001	QX4	3.3	VG33E	3.8	18.0	5.3	4.4	5.9	21.1	11.25
2002	QX4	3.5	VQ35DE	5.25	18.0	5.3	3.9	5.9	21.1	9.75
2003	QX4	3.5	VQ35DE	5.25	18.0	5.3	3.9	5.9	21.1	9.75

NOTE: All capacities are approximate. Add fluid gradually and check to be sure a proper fluid level is obtained.

67162-IQX4-C04

VALVE SPECIFICATIONS

Year	Engine Displacement Liters	Engine ID	Seat Angle (deg.)	Face Angle (deg.)	Spring Test Pressure (lbs. @ in.)	Spring Installed Height (in.)	Stem-to-Guide Clearance (in.)		Stem Diameter (in.)	
							Intake	Exhaust	Intake	Exhaust
2001	3.3	VG33E	45	45.25-46.75	①	NA	0.0008-0.0021	0.0016-0.0029	0.2742-0.2748	0.3135-0.3138
2002	3.5	VQ35DE	45.15-45.45	45	45.4@1.457	1.457	0.0008-0.0021	0.0016-0.0029	0.2348-0.2354	0.2341-0.2346
2003	3.5	VQ35DE	45.15-45.45	45	45.4@1.457	1.457	0.0008-0.0021	0.0016-0.0029	0.2348-0.2354	0.2341-0.2346

NA: Not Available

① Inner: 57.3 @ 0.984

   Outer: 117.7 @ 1.181

67162-IQX4-C05

CRANKSHAFT AND CONNECTING ROD SPECIFICATIONS

All measurements are given in inches.

Year	Engine Displ. Liters	Engine ID	Crankshaft				Connecting Rod		
			Main Brg. Journal Dia.	Main Brg. Oil Clearance	Shaft End-play	Thrust on No.	Journal Diameter	Oil Clearance	Side Clearance
2001	3.3	VG33E	2.4790-2.4793	0.0011-0.0022	0.0020-0.0067	4	1.9967-1.9675	0.0006-0.0021	0.0079-0.0138
2002	3.5	VQ35DE	①	0.0014-0.0018	0.0118	4	②	0.0013-0.0023	0.0079-0.0138
2003	3.5	VQ35DE	①	0.0014-0.0018	0.0118	4	②	0.0013-0.0023	0.0079-0.0138

NA - Not Available

① There are 24 different grades, ranging from A (2.3612) to 7 (2.3603)

② Grade 0: 2.0460-2.0462

   Grade 1: 2.0457-2.0460

   Grade 2: 2.0445-2.0457

67162-IQX4-C06

PISTON AND RING SPECIFICATIONS

All measurements are given in inches.

Year	Engine Displacement Liters	Engine ID	Piston Clearance	Ring Gap			Ring Side Clearance		
				Top Comp.	Bottom Comp.	Oil Control	Top Comp.	Bottom Comp.	Oil Control
2001	3.3	VG33E	①	0.0083-0.0157	0.0197-0.0272	0.0079-0.0272	0.0009-0.0030	0.0012-0.0028	0.0006-0.0073
2002	3.5	VQ35DE	0.0004-0.0012	0.0091-0.0130	0.0130-0.0189	0.0079-0.0236	0.0016-0.0031	0.0012-0.0028	0.0006-0.0020
2003	3.5	VQ35DE	0.0004-0.0012	0.0091-0.0130	0.0130-0.0189	0.0079-0.0236	0.0016-0.0031	0.0012-0.0028	0.0006-0.0020

① Cylinders 1, 2, 6: 0.0010 - 0.0018 in.

   Cylinders 3 and 4: 0.0006 - 0.0010 in.

   Cylinder 5: 0.0012-0.0016 in.

67162-IQX4-C07

TORQUE SPECIFICATIONS
All readings in ft. lbs.

Year	Engine Displacement Liters	Engine ID	Cylinder Head Bolts	Main Bearing Bolts	Rod Bearing Bolts	Crankshaft Damper Bolts	Flywheel Bolts	Manifold Intake	Manifold Exhaust	Spark Plugs	Oil Pan Drain Plug
2001	3.3	VG33E	①	67-74	②	141-156	61-69	①	21-25	14-22	NA
2002	3.5	VQ35DE	③	④	⑤	⑥	61-69	⑦	21-24	14-22	22-28
2003	3.5	VQ35DE	③	④	⑤	⑥	61-69	⑦	21-24	14-22	22-28

NA: Information not available

① The cylinder heads and the lower intake manifold are installed together

Step 1: Tighten the cylinder head bolts to 22 ft. lbs.

Step 2: Tighten the cylinder head bolts to 43 ft. lbs.

Step 3: Loosen the cylinder head bolts completely

Step 4: Tighten the cylinder head bolts to 84 inch lbs.

Step 5: Tighten the intake manifold fasteners to 35 inch lbs.

Step 6: Tighten the intake manifold fasteners to 13 ft. lbs.

Step 7: Tighten the intake manifold fasteners to 12-14 ft. lbs.

Step 8: Loosen all intake manifold fasteners completely

Step 9: Tighten the cylinder head bolts to 22 ft. lbs.

Step 10: Tighten the cylinder head bolts 60-65 degrees

Step 11: Tighten the cylinder head sub-bolts to 80-105 inch lbs.

Step 12: Tighten the intake manifold fasteners to 35 inch lbs.

Step 13: Tighten the intake manifold fasteners to 78 inch lbs.

Step 14: Tighten the intake manifold fasteners to 70-84 inch lbs.

② 10-12 ft. lbs. plus 60-65 degrees or 28-33 ft. lbs.

③ Step 1: 72 ft. lbs.

Step 2: Loosen all bolts completely

Step 3: 25-33 ft. lbs.

Step 4: +90 degrees

Step 5: +90 degrees

④ Step 1: 24-28 ft. lbs.

Step 2: +90 degrees

⑤ Step 1: 15 ft. lbs.

Step 2: +90 degrees

⑥ 29-36 ft. lbs. +60-66 degrees

⑦ Step 1: 44-86 inch lbs.

Step 2: 20-23 ft. lbs.

67162-IQX4-C08

WHEEL ALIGNMENT

Year	Model	Caster Range (+/-Deg.)	Caster Preferred Setting (Deg.)	Camber Range (+/-Deg.)	Camber Preferred Setting (Deg.)	Toe-in (in.)
2001	QX4	0.75	+3.00	0.75	+0.10	0.08+/-0.04
2002	QX4 ①	0.75	+3.00	0.75	+0.17	0.08+/-0.04
2003	QX4 ①	0.75	+3.00	0.75	+0.17	0.08+/-0.04

① Assumes P245/65R17 tire

67162-IQX4-C09

TIRE, WHEEL AND BALL JOINT SPECIFICATIONS

Year	Model	OEM Tires		Tire Pressures (psi)		Wheel Size	Ball Joint Inspection	Lugnut Torque (ft. lbs.)
		Standard	Optional	Front	Rear			
2001	QX4	P245/70R16	None	35	35	7-JJ	①	87-108
2002	QX4	P245/70R16	P245/65R17	②	②	Std: 7J/Opt: 8J	③	87-108
2003	QX4	P245/65R17	None	②	②	Std: 7J/Opt: 8J	③	87-108

OEM: Original Equipment Manufacturer

PSI: Pounds Per Square Inch

STD: Standard

OPT: Optional

L: Lower

U: Upper

① Replace if any measurable movement is found.

② See placard on vehicle

③ Axial play

    Upper: 0

    Lower: 0.008 in.

67162-IQX4-C10

BRAKE SPECIFICATIONS
All measurements in inches unless noted

Year	Model	Brake Disc			Brake Drum Diameter			Minimum Lining Thickness		Brake Caliper	
		Original Thickness	Minimum Thickness	Maximum Runout	Original Inside Diameter	Max. Wear Limit	Maximum Machine Diameter	Front	Rear	Bracket Bolts (ft. lbs.)	Mounting Bolts (ft. lbs.)
2001	QX4	1.100	1.024	0.004	11.60	NA	11.67	0.079	0.059	53-72	24-31
2002	QX4	1.100	1.024	0.003	11.60	NA	11.67	0.079	0.059	127-134	24-31
2003	QX4	1.100	1.024	0.003	11.60	NA	11.67	0.079	0.059	127-134	24-31

NA: Not Available

67162-IQX4-C11

SCHEDULED MAINTENANCE INTERVALS
Infiniti—QX4

TO BE SERVICED	TYPE OF SERVICE	VEHICLE MILEAGE INTERVAL (x1000)															
		3.8	7.5	11	15	19	22.5	26	30	34	37.5	41	45	49	52.5	56	60
Engine oil & filter	R	✓	✓	✓	✓	✓	✓	✓	✓	✓	✓	✓	✓	✓	✓	✓	✓
Brake lines & cables	S/I				✓				✓				✓				✓
Brake pads, discs, drums & linings	I		✓		✓		✓		✓		✓		✓		✓		✓
Driveshaft boots & propeller shaft	L/I		✓		✓		✓		✓		✓		✓		✓		✓
Front wheel bearings (4x2)	I								✓								✓
Automatic & manual transmission, transfer & differential gear oil ①	I				✓				✓				✓				✓
LSD gear oil	R								✓								✓
Front wheel bearing grease (4x4)	R								✓								✓
Timing belt ②	R																
Air cleaner filter	R								✓								✓
Engine coolant ③	R																✓
Spark plugs	R	platinum tipped plugs every 105,000 miles															
Drive belt(s)	S/I								✓								✓
Cabin air filter	I/R		I		R		I		R		I		R		I		R
Exhaust system	I		✓		✓		✓		✓		✓		✓		✓		✓
Fuel lines	S/I								✓								✓
Steering gear (box) & linkage, axle & suspension parts	I		✓		✓		✓		✓		✓		✓		✓		✓
Vapor lines	S/I								✓								✓

R: Replace   S/I: Service or Inspect   L: Lubricate

① Differential (w/limited-slip differential) oil: replace oil every 30,000 miles.

② Timing belt: replace at 105,000 miles.

③ After 60,000, replace every 30,000

FREQUENT OPERATION MAINTENANCE (SEVERE SERVICE)

If a vehicle is operated under any of the following conditions it is considered severe service:

- Extremely dusty areas.

- 50% or more of the vehicle constant operation is in 32°C (90°F) or higher temperatures, or temperatures below 0°C (32°F).

- Prolonged idling (vehicle operation in stop and go traffic).

- Frequent short running periods (engine does not warm to normal operating temperatures).

- Police, taxi, delivery usage or trailer towing usage.

Oil & oil filter: replace every 3750 miles.

Brake pads, discs, drums & linings: service or inspect every 7500 miles.

Driveshaft boots & propeller shaft: service or inspect every 7500 miles.

Exhaust system: service or inspect every 7500 miles.

Steering gear (box) & linkage, (steering damper-4x4), axle & suspension parts: service or inspect every 7500 miles.

Steering linkage ball joints & front suspension ball joints: service or inspect every 7500 miles.

67162-IQX4-C12

PRECAUTIONS

Before servicing any vehicle, please be sure to read all of the following precautions, which deal with personal safety, prevention of component damage, and important points to take into consideration when servicing a motor vehicle:

• Never open, service or drain the radiator or cooling system when the engine is hot; serious burns can occur from the steam and hot coolant.

• Observe all applicable safety precautions when working around fuel. Whenever servicing the fuel system, always work in a well-ventilated area. Do not allow fuel spray or vapors to come in contact with a spark, open flame, or excessive heat (a hot drop light, for example). Keep a dry chemical fire extinguisher near the work area. Always keep fuel in a container specifically designed for fuel storage; also, always properly seal fuel containers to avoid the possibility of fire or explosion. Refer to the additional fuel system precautions later in this section.

• Fuel injection systems often remain pressurized, even after the engine has been turned **OFF**. The fuel system pressure must be relieved before disconnecting any fuel lines. Failure to do so may result in fire and/or personal injury.

• Brake fluid often contains polyglycol ethers and polyglycols. Avoid contact with the eyes and wash your hands thoroughly after handling brake fluid. If you do get brake fluid in your eyes, flush your eyes with clean, running water for 15 minutes. If eye irritation persists, or if you have taken brake fluid internally, IMMEDIATELY seek medical assistance.

• The EPA warns that prolonged contact with used engine oil may cause a number of skin disorders, including cancer! You should make every effort to minimize your exposure to used engine oil. Protective gloves should be worn when changing oil. Wash your hands and any other exposed skin areas as soon as possible after exposure to used engine oil. Soap and water, or waterless hand cleaner should be used.

• All new vehicles are now equipped with an air bag system. The system must be disabled before performing service on or around system components, steering column, instrument panel components, wiring and sensors. Failure to follow safety and disabling procedures could result in accidental air bag deployment, possible personal injury and unnecessary system repairs.

• Always wear safety goggles when working with, or around, the air bag system. When carrying a non-deployed air bag, be sure the bag and trim cover are pointed away from your body. When placing a non-deployed air bag on a work surface, always face the bag and trim cover upward, away from the surface. This will reduce the motion of the module if it is accidentally deployed. Refer to the additional air bag system precautions later in this section.

• Clean, high quality brake fluid from a sealed container is essential to the safe and proper operation of the brake system. You should always buy the correct type of brake fluid for your vehicle. If the brake fluid becomes contaminated, completely flush the system with new fluid. Never reuse any brake fluid. Any brake fluid that is removed from the system should be discarded. Also, do not allow any brake fluid to come in contact with a painted surface; it will damage the paint.

• Never operate the engine without the proper amount and type of engine oil; doing so WILL result in severe engine damage.

• Timing belt maintenance is extremely important! Many models utilize an interference-type, non-freewheeling engine. If the timing belt breaks, the valves in the cylinder head may strike the pistons, causing potentially serious (also time-consuming and expensive) engine damage. Refer to the maintenance interval charts in the front of this manual for the recommended replacement interval for the timing belt, and to the timing belt section for belt replacement and inspection.

• Disconnecting the negative battery cable on some vehicles may interfere with the functions of the on-board computer system(s) and may require the computer to undergo a relearning process once the negative battery cable is reconnected.

• When servicing drum brakes, only disassemble and assemble one side at a time, leaving the remaining side intact for reference.

ENGINE REPAIR

➡**Disconnecting the negative battery cable on some vehicles may interfere with the functions of the on board computer system. The computer may undergo a relearning process once the negative battery cable is reconnected.**

Distributor

REMOVAL

1. Before servicing the vehicle, refer to the precautions in the beginning of this section.
2. Remove or disconnect the following:
   • Negative battery cable
   • Distributor cap
   • Distributor wiring harness connector
3. Matchmark the rotor to the distributor housing and the distributor housing to the cylinder head.
4. Remove the distributor.

INSTALLATION

Timing Not Disturbed

1. Install or connect the following:
   • Distributor by aligning the matchmarks made during removal
   • Distributor wiring harness connector
   • Distributor cap
   • Negative battery cable
2. Check the ignition timing and adjust, as necessary.

Timing Disturbed

3.3L ENGINE

1. Set the engine to Top Dead Center (TDC) of the compression stroke for the No. 1 cylinder.
2. Align the index mark on the distributor shaft with the protrusion on the distributor housing.

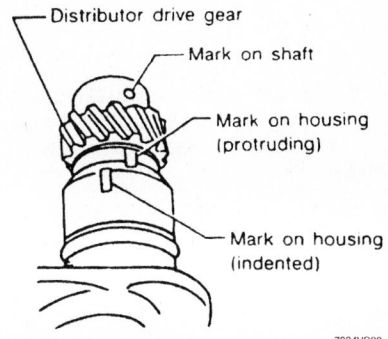

Distributor drive gear
Mark on shaft
Mark on housing (protruding)
Mark on housing (indented)

7924VG28

Distributor shaft alignment—3.3L engine

3. Install the distributor and check that the distributor rotor is aligned.
4. Install or connect the following:
   • Distributor cap
   • Distributor harness connector
5. Check the ignition timing and adjust, as necessary.

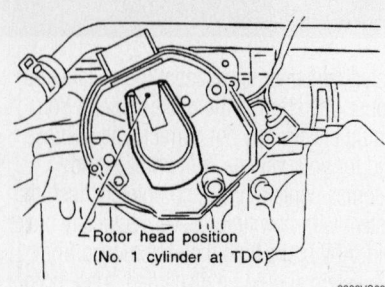

Distributor rotor alignment—3.3L engine

Rotor head position
(No. 1 cylinder at TDC)

9308VG03

Alternator

REMOVAL & INSTALLATOIN

3.3L and 3.5L Engines

1. Before servicing the vehicle, refer to the precautions in the beginning of this section.
2. Remove or disconnect the following:
   - Negative battery cable
   - Alternator harness connectors
   - Engine under cover
   - Alternator belt
   - Alternator

To install:

3. Install or connect the following:
   - Alternator
   - Alternator belt. Tighten the adjustment bolt to 12–14 ft. lbs. (16–19 Nm) and the pivot bolts to 16–22 ft. lbs. (22–30 Nm).
   - Engine under cover
   - Alternator harness connectors
   - Negative battery cable

Ignition Timing

ADJUSTMENT

➡️**Ignition timing is set with the engine at operating temperature, transmission in Neutral and all electrical accessories OFF.**

1. Before servicing the vehicle, refer to the precautions in the beginning of this section.
2. Attach a timing light to the No. 1 spark plug wire.
3. Start the engine and allow it to reach normal operating temperature.
4. Check that the idle speed is less than 1000 rpm.
5. Run the engine at 2000 rpm for 2 minutes.
6. Rev the engine to 3000 rpm 2–3 times and allow it to idle for 1 minute.

7. Check for the presence of Diagnostic Trouble Codes (DTC) and service as necessary.
8. Run the engine at 2000 rpm for 2 minutes.
9. Stop the engine and disconnect the Throttle Position (TP) sensor.
10. Start the engine and rev it to 3000 rpm 2–3 times and allow it to idle.
11. Set the base timing to 8–12 degrees Before Top Dead Center (BTDC) for 3.3L engines.
12. Tighten the distributor lockbolt to 83–113 inch lbs. (9–13 Nm).
13. Set the base idle speed to 700–800 rpm.
14. Stop the engine and connect the TP sensor.

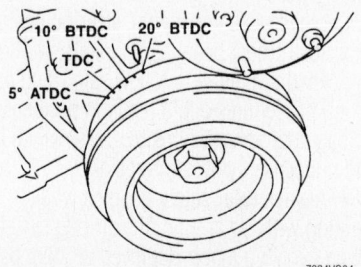

Timing indicator—3.3L engines

10° BTDC   20° BTDC
TDC
5° ATDC

7924VG04

Engine Assembly

REMOVAL & INSTALLATION

3.3L Engine

1. Before servicing the vehicle, refer to the precautions in the beginning of this section.
2. Drain the cooling system.
3. Relieve the fuel system pressure.
4. Recover the A/C refrigerant, if equipped.
5. Remove or disconnect the following:
   - Negative battery cable
   - Hood
   - Air cleaner assembly
   - Idle Air Control (IAC) valve and solenoid connectors
   - Throttle Position (TP) sensor and switch connectors
   - Engine Coolant Temperature (ECT) sensor connector
   - Manifold Absolute Pressure (MAP) sensor connector and vacuum line
   - Evaporative Emissions (EVAP) canister purge valve connector and vacuum line

   - Mass Air Flow (MAF) sensor connector
   - Brake booster vacuum line
   - Fuel lines
   - Exhaust Gas Recirculation (EGR) temperature sensor connector
   - Throttle cable
   - Accessory drive belts
   - Cooling fan and shroud
   - Radiator and hoses
   - Engine under cover
   - A/C compressor manifold
   - Power steering pump
   - Heated Oxygen (HO2S) sensor connectors
   - Exhaust front pipes
   - Crankshaft Position (CKP) sensor
   - Starter motor
   - Transmission
   - Left and right engine mounts
   - Engine

➡️**When removing the engine mounts, do not loosen the 4 mount cover nuts. The mount is fluid filled and will not function if the fluid leaks out.**

To install:

6. Install or connect the following:
   - Engine. Tighten the engine mount nuts to 43–58 ft. lbs. (59–78 Nm).
   - Transmission
   - Starter motor
   - CKP sensor
   - Exhaust front pipes
   - HO2S sensor connectors
   - Power steering pump
   - A/C compressor manifold
   - Engine under cover
   - Radiator and hoses
   - Cooling fan and shroud
   - Accessory drive belts
   - Throttle cable
   - EGR temperature sensor connector
   - Fuel lines
   - Brake booster vacuum line
   - MAF sensor connector
   - EVAP canister purge valve connector and vacuum line
   - MAP sensor connector and vacuum line
   - ECT sensor connector
   - TP sensor and switch connectors
   - IAC valve and solenoid connectors
   - Air cleaner assembly
   - Hood
   - Negative battery cable
7. Fill the cooling system.
8. Recharge the A/C system, if equipped.
9. Start the engine and check for leaks.

3.5L Engine

1. Release fuel pressure.
2. Remove engine hood and front RH and LH wheels.
3. Remove engine undercover and suspension member stay.
4. Drain coolant from radiator.
5. Remove the following parts.
   - Radiator shroud
   - Radiator
   - Cooling fan
   - Drive belts
   - Battery
   - Engine cover
   - Throttle wires
6. Air duct with air cleaner case.
7. Disconnect vacuum hoses, fuel hoses, heater hoses, EVAP canister hoses, harnesses, connectors and so on.
8. Remove air conditioner compressor from bracket, then put it aside holding with a suitable wire.
9. Remove power steering oil pump

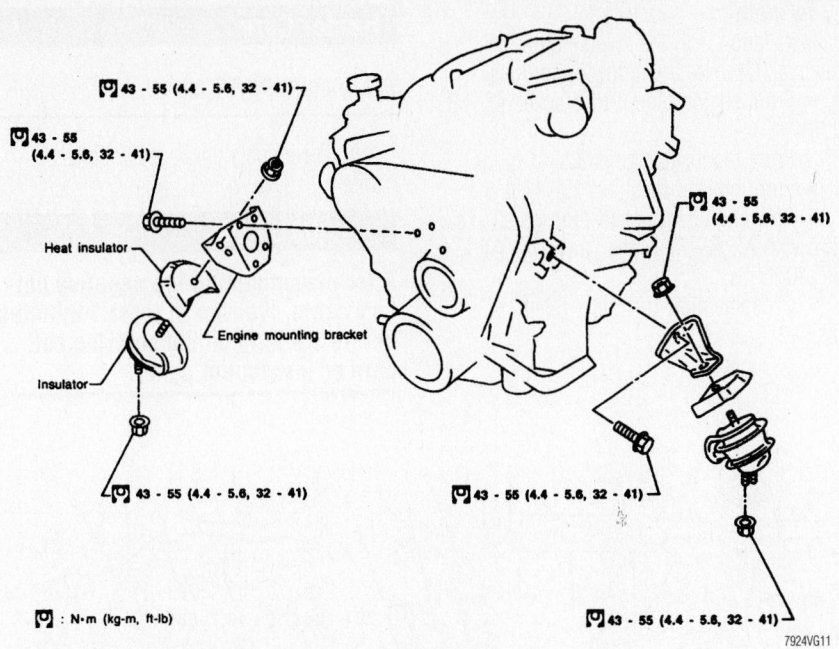

43 - 55 (4.4 - 5.6, 32 - 41)

43 - 55 (4.4 - 5.6, 32 - 41)

43 - 55 (4.4 - 5.6, 32 - 41)

Heat insulator

Engine mounting bracket

Insulator

43 - 55 (4.4 - 5.6, 32 - 41)

43 - 55 (4.4 - 5.6, 32 - 41)

43 - 55 (4.4 - 5.6, 32 - 41)

: N·m (kg-m, ft-lb)

Engine mounts and related components—3.3L engine

7924VG11

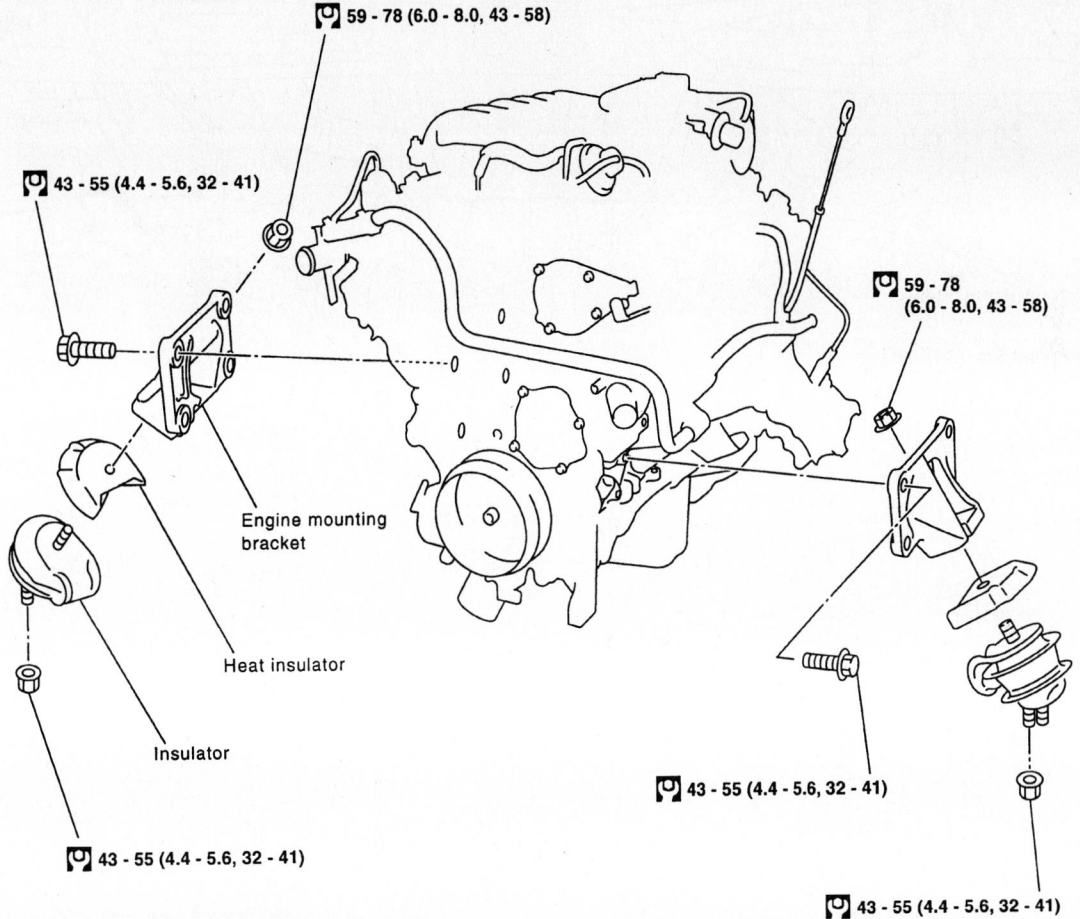

59 - 78 (6.0 - 8.0, 43 - 58)

43 - 55 (4.4 - 5.6, 32 - 41)

59 - 78 (6.0 - 8.0, 43 - 58)

Engine mounting bracket

Heat insulator

Insulator

43 - 55 (4.4 - 5.6, 32 - 41)

43 - 55 (4.4 - 5.6, 32 - 41)

43 - 55 (4.4 - 5.6, 32 - 41)

: N·m (kg-m, ft-lb)

9359VG01

Front engine mounting—3.5L engine

and reservoir tank with bracket, then put it aside holding with a suitable wire.

10. Remove alternator.

11. Remove exhaust front tube heat insulators, then remove rear heated oxygen sensors.

12. Remove exhaust front and rear tubes.

13. Remove transmission.

14. Remove TWC (manifold) heat insulators, then remove TWC (manifold).

15. Install engine slingers.

16. Hoist engine with engine slingers and remove front engine mounting nuts.

17. Remove engine from vehicle.

To install:

Installation is in the reverse order of removal. Observe the following torques:

• Front engine mount-to-bracket: 43–58 ft. lbs.

• Front mount-to-frame: 32–41 ft. lbs.

• Front bracket-to-block: 32–41 ft. lbs.

• Rear engine mount-to-bracket: all exc. 2wd with AT: 58–77 ft. lbs.; 2wd with AT: 32–40 ft. lbs.

• Crossmember-to-frame: 58–77 ft. lbs.

Heater Core

REMOVAL & INSTALLATION

1. Disconnect the negative battery cable.

✳✳ CAUTION

After disconnecting the negative battery cable, wait for at least 3 minutes before working on the steering column or instrument panel.

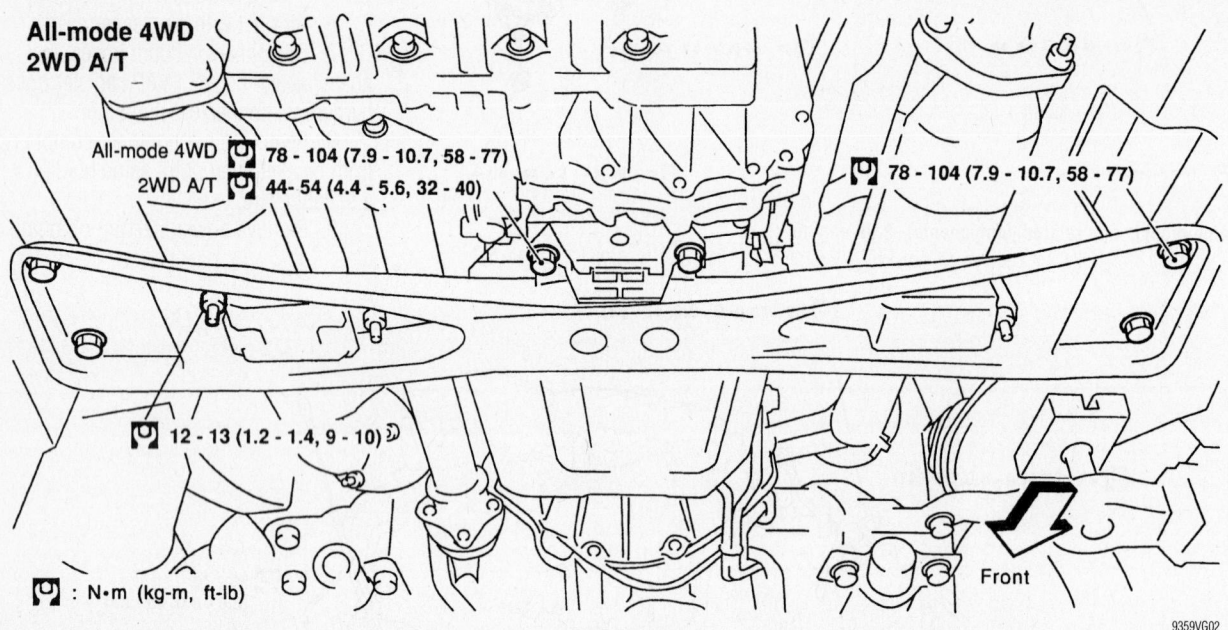

Rear engine mounting—3.5L engine

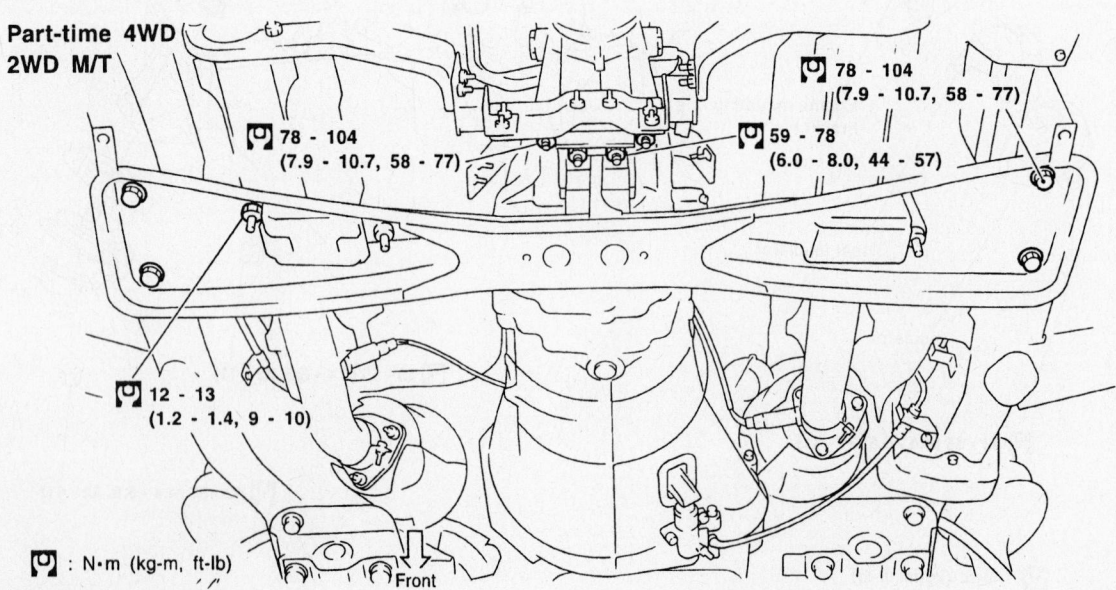

Rear engine mounting—3.5L engine

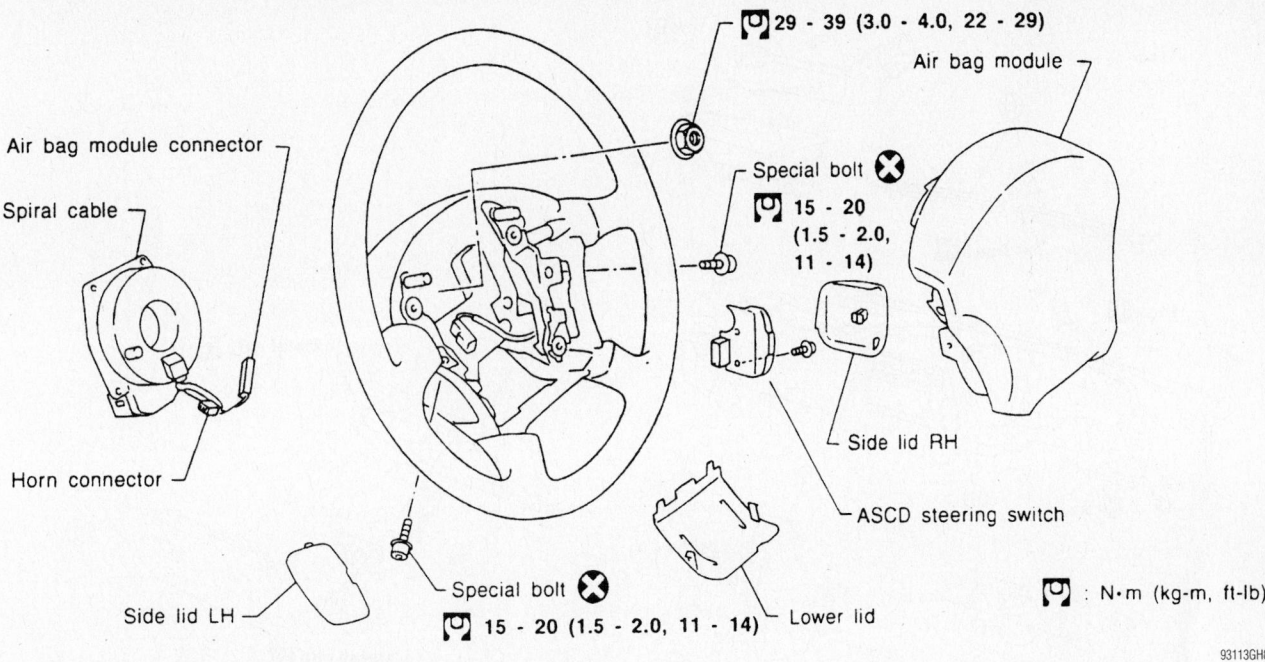

Exploded view of the driver's side air bag module and steering wheel

2. Drain the cooling system into a clean container for reuse.

3. Disconnect the heater hoses from the heater core.

4. Remove the driver's side air bag and steering wheel by performing the following procedure:

a. Place the front wheels in the straight-ahead position.

b. Remove the lower lid from the steering wheel and disconnect the air bag module connector.

c. Remove the side lids from both sides of the steering wheel.

d. Using the Tamper Resistant Torx® tool T50, remove the left and right Torx® bolts.

e. Carefully, remove the air bag module.

✳✳ CAUTION

Place the air bag module in safe place with the front facing upward.

f. Remove the steering wheel nut.

g. Using a steering wheel puller, press the steering wheel from the steering column.

5. Remove the passenger's side air bag by performing the following procedure:

a. Remove the glove box clips and disconnect the passenger's side air bag module connector.

b. Remove the lower panel screws; then, disconnect the harness connector and remove the air bag module bracket.

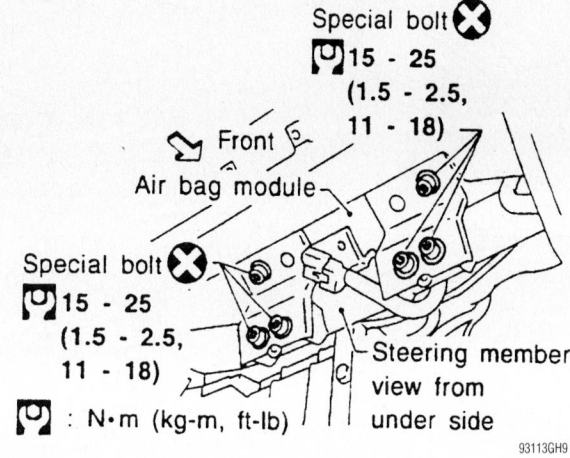

Exploded view of the passenger's side air bag module

c. Using the Tamper Resistant Torx® tool T50, remove the passenger's side air bag module bolts.

d. Carefully, remove the air bag module.

✳✳ CAUTION

Place the air bag module in safe place with the front facing upward.

6. Remove the instrument panel by performing the following procedure:

a. Remove the steering column cover and the combination switch.

b. Remove the instrument panel side lower finisher.

c. At the driver's side, remove the

lower panel screws, disconnect the electrical harness connectors and remove the panel.

d. Remove the cluster lid "A" screws and the cluster lid "A".

e. Remove the combination meter screws, disconnect the electrical harness connectors and remove the combination meter.

f. Remove the cluster lid "C" screws, disconnect the electrical harness connectors and remove the cluster lid "C".

g. Remove the audio assembly screws and the audio assembly.

h. Remove the air conditioning control unit screws, disconnect the electrical

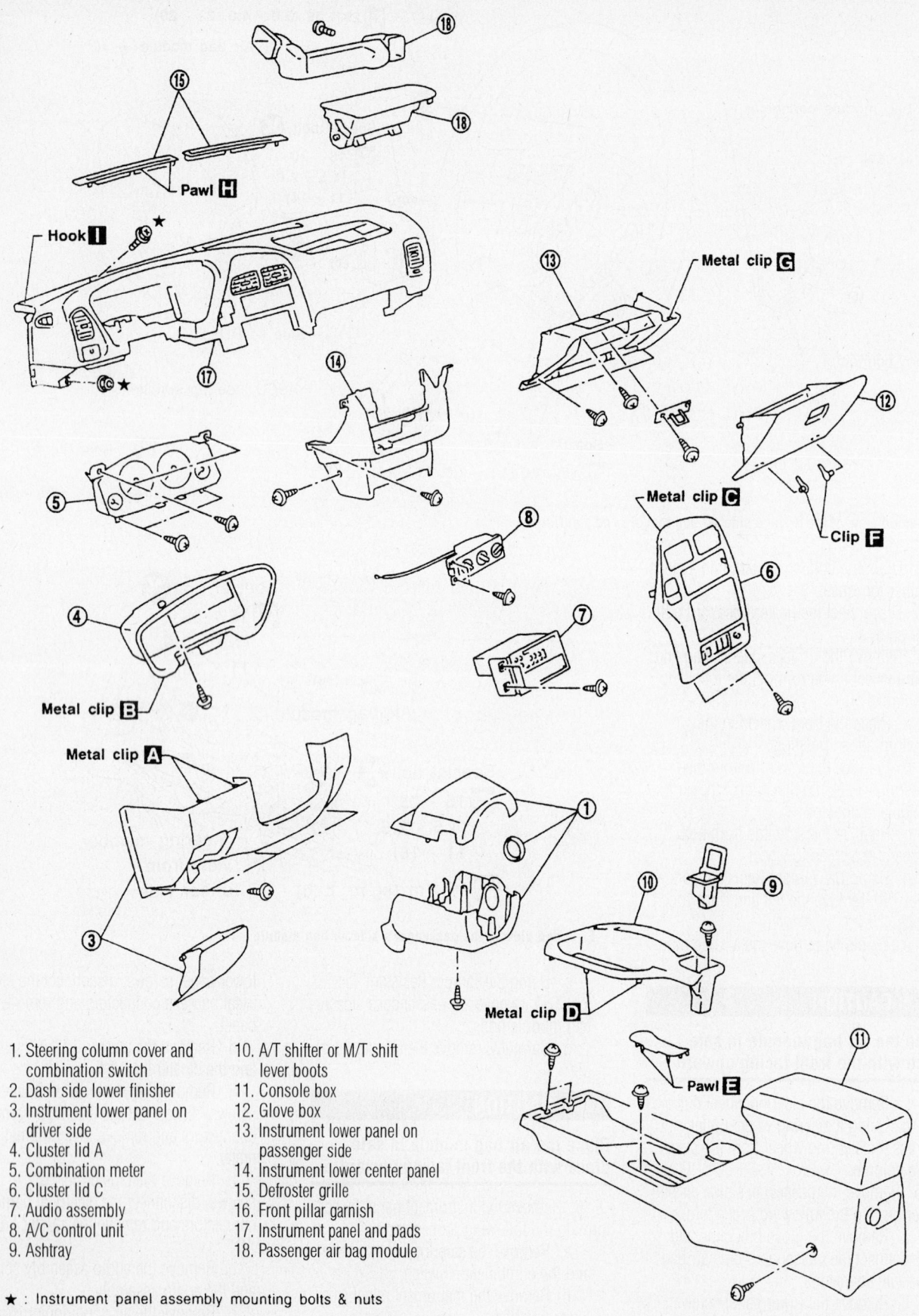

1. Steering column cover and combination switch
2. Dash side lower finisher
3. Instrument lower panel on driver side
4. Cluster lid A
5. Combination meter
6. Cluster lid C
7. Audio assembly
8. A/C control unit
9. Ashtray
10. A/T shifter or M/T shift lever boots
11. Console box
12. Glove box
13. Instrument lower panel on passenger side
14. Instrument lower center panel
15. Defroster grille
16. Front pillar garnish
17. Instrument panel and pads
18. Passenger air bag module

★ : Instrument panel assembly mounting bolts & nuts

Exploded view of the instrument panel and related accessories

93113GH0

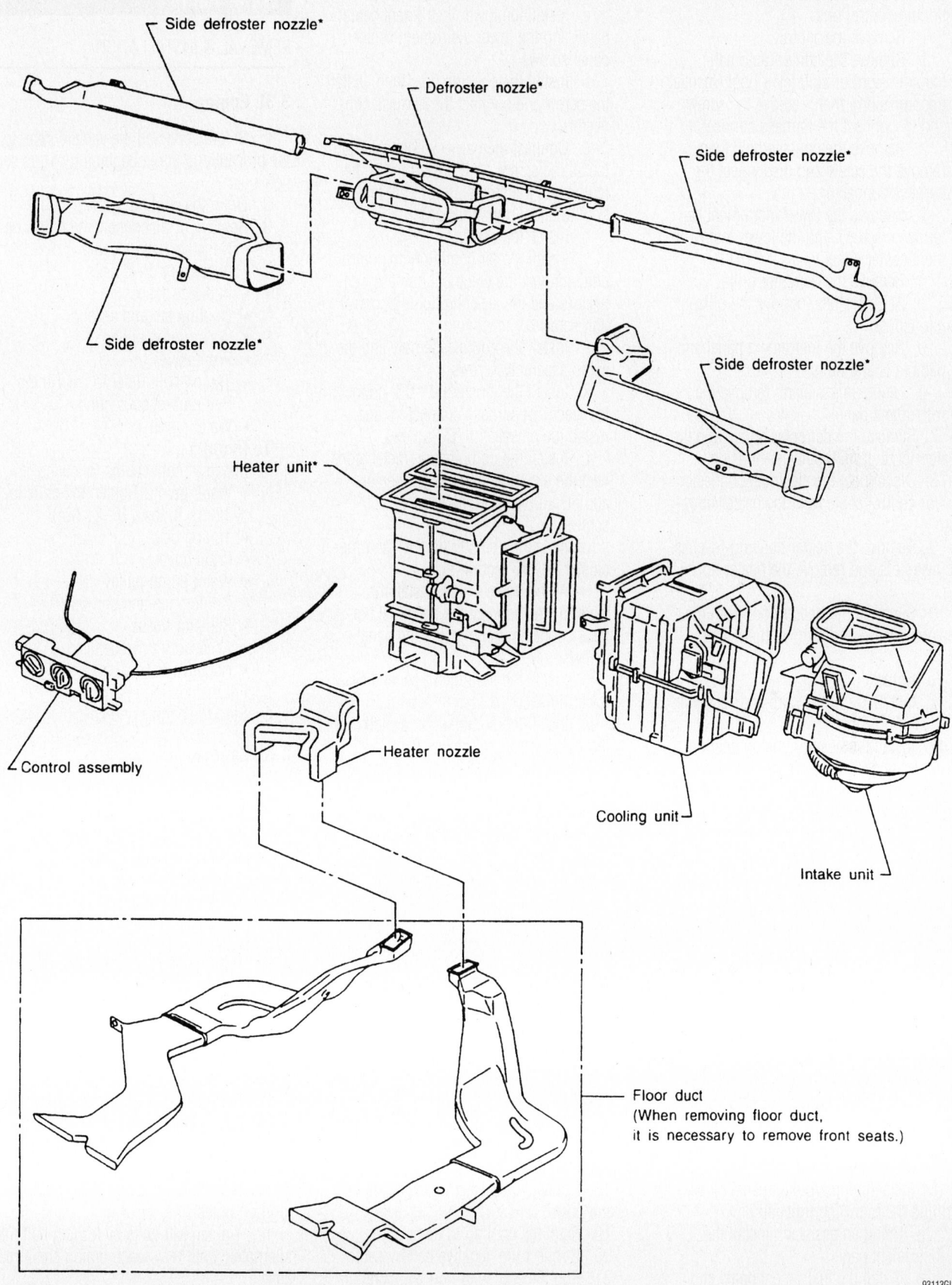

Side defroster nozzle*

Defroster nozzle*

Side defroster nozzle*

Side defroster nozzle*

Side defroster nozzle*

Heater unit*

Control assembly

Heater nozzle

Cooling unit

Intake unit

Floor duct
(When removing floor duct,
it is necessary to remove front seats.)

93113GI1

Exploded view of the heater housing, the evaporator housing, the ventilation dusts and related accessories

harness connectors and the air conditioning control unit.

i. Remove the ashtray.

j. Remove the shifter (automatic transmission) or shift lever boot (manual transmission); then, remove the screw and disconnect the harness connector.

k. Remove the console box; then, remove the screw and disconnect the harness connector.

l. Remove the lower instrument center panel screws and the lower instrument center panel.

m. Remove the defroster grille.

n. At both sides, remove the pillar garnishes.

o. Remove the instrument panel and pads nuts and bolts.

p. Using an assistant, remove the instrument panel.

7. Remove the defroster nozzle and the heater nozzle from the heater housing.

8. Disconnect the electrical connector and/or control cable from the heater housing.

9. Remove the heater housing-to-chassis fasteners and remove the heater housing.

10. Separate the heater core from the heater housing and remove the heater core.

To install:

11. Install the heater core and assemble the heater housing.

12. Install the heater housing and the heater housing-to-chassis fasteners.

13. Connect the electrical connector and/or control cable to the heater housing.

14. Install the defroster nozzle and the heater nozzle to the heater housing.

15. Install the passenger's side air bag by performing the following procedure:

a. Carefully, install the air bag module.

b. Using the Tamper Resistant Torx® tool T50, install the passenger's side air bag module bolts. Torque the bolts to 11–18 ft. lbs. (15–25 Nm).

c. Connect the harness connector and install the air bag module bracket; then, install the lower panel screws.

d. Connect the passenger's side air bag module connector and install the glove box clips.

16. Install the instrument panel by performing the following procedure:

a. Using an assistant, install the instrument panel.

b. Install the instrument panel and pads nuts and bolts.

c. At both sides, install the pillar garnishes.

d. Install the defroster grille.

e. Install the lower instrument center panel and the lower instrument center panel screws.

f. Install the console box; then, install the screw and connect the harness connector.

g. Connect the harness connector and install the screw; then, install the shifter (automatic transmission) or shift lever boot (manual transmission).

h. Install the ashtray.

i. Install the air conditioning control unit, connect the electrical harness connectors and the air conditioning control unit screws.

j. Install the audio assembly and the audio assembly screws.

k. Install the cluster lid **"C"**, connect the electrical harness connectors and install the cluster lid **"C"** screws.

l. Install the combination meter, connect the electrical harness connectors and install the combination meter screws.

m. Install the cluster lid **"A"** and the cluster lid **"A"** screws.

n. At the driver's side, install the lower panel, connect the electrical harness connectors and install the panel screws.

o. Install the instrument panel side lower finisher.

p. Install the combination switch and the steering column cover.

17. Install the driver's side air bag and steering wheel by performing the following procedure:

a. Install the steering wheel to the steering column.

b. Install the steering wheel nut. Torque the nut to 22–29 ft. lbs. (29–39 Nm).

c. Carefully, install the air bag module.

d. Using the Tamper Resistant Torx® tool T50, install the left and right Torx® bolts. Torque the bolts to 11–14 ft. lbs. (15–20 Nm).

e. Install the side lids to both sides of the steering wheel.

f. Connect the air bag module connector and install the lower lid to the steering wheel.

18. Connect the heater hoses to the heater core.

19. Refill the cooling system.

20. Connect the negative battery cable.

21. Run the engine to normal operating temperatures; then, check the climate control operation and check for leaks.

Water Pump

REMOVAL & INSTALLATION

3.3L Engine

1. Before servicing the vehicle, refer to the precautions in the beginning of this section.

2. Drain the cooling system.

3. Remove or disconnect the following:
- Negative battery cable
- Accessory drive belts
- Radiator hoses
- Cooling fan and shroud
- Water pump pulley
- Front cover
- Timing belt. Refer to the Timing Belt unit repair section.
- Water pump

To install:

4. Install or connect the following:
- Water pump. Tighten the bolts to 12–15 ft. lbs. (16–21 Nm).
- Timing belt
- Front cover
- Water pump pulley
- Cooling fan and shroud
- Radiator hoses
- Accessory drive belts
- Negative battery cable

5. Fill the cooling system.

6. Start the engine and check for leaks.

3.5L Engine

1. Remove undercover.

2. Remove suspension member stay.

3. Drain coolant from radiator.

4. Remove radiator shrouds.

5. Remove drive belts.

6. Remove cooling fan.

7. Remove water drain plug on water pump side of cylinder block.

8. Remove chain tensioner cover and water pump cover.

9. Pushing timing chain tensioner sleeve, apply a stopper pin so it does not return. Then remove the chain tensioner assembly.

10. Remove the 3 water pump fixing bolts. Secure a gap between water pump gear and timing chain, by turning crankshaft pulley 20° backwards.

11. Put M8 bolts to two water pump fixing bolt holes.

12. Tighten M8 bolts by turning half turn alternately until they reach timing chain rear case.

➡ **In order to prevent damages to water pump or timing chain rear case, do not**

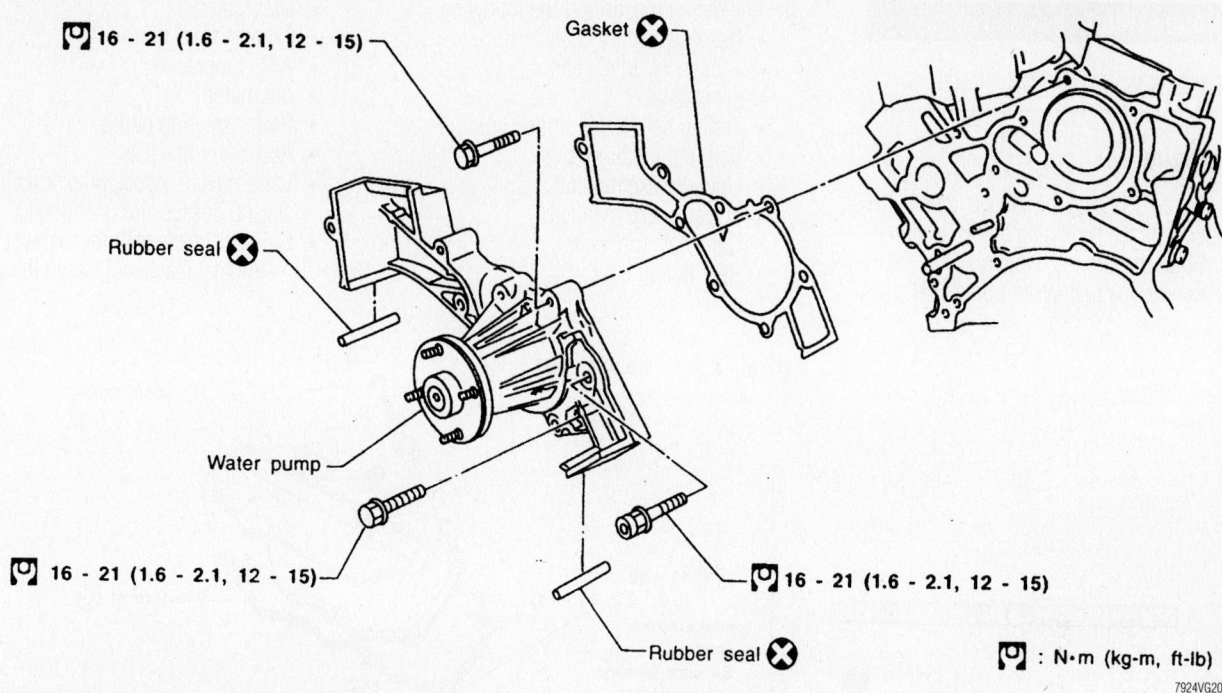

Exploded view of the water pump assembly—3.3L engine

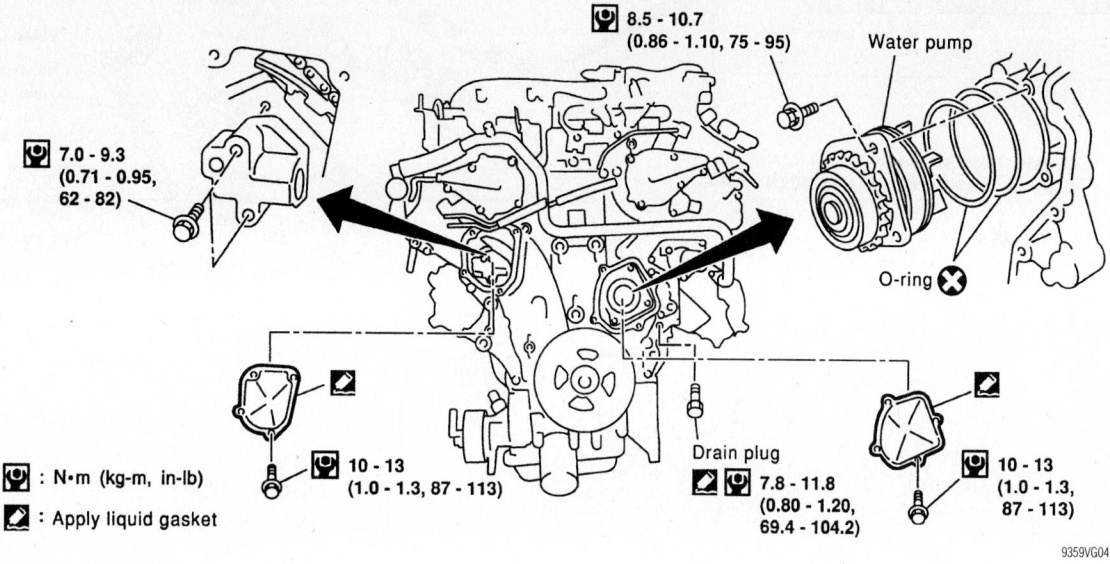

Exploded view of the water pump assembly—3.5L engine

tighten one bolt continuously. **Always turn each bolt half turn each time.**

13. Lift up water pump and remove it.

➡**When lifting up water pump, do not allow water pump gear to hit timing chain.**

To install:

14. Apply engine oil and coolant to O-rings as shown in the figure.

15. Install water pump.

➡**Do not allow cylinder block to nip O-rings when installing water pump.**

16. Before installing, remove all traces of liquid gasket from mating surface of water pump cover and chain tensioner cover using a scraper. Also remove traces of liquid gasket from mating surface of front cover.

17. Apply a continuous bead of liquid gasket to mating surface of chain tensioner cover and water pump cover. Use Genuine RTV silicone sealant or equivalent.

18. Return the crankshaft pulley to its original position by turning it 20° forward.

19. Install timing chain tensioner, then remove the stopper pin.

➡**When installing the timing chain tensioner, engine oil should be applied to the oil hole and tensioner.**

➡**After starting engine, let idle for three minutes, then rev engine up to 3,000 rpm under no load to purge air from the high-pressure chamber of the chain tensioners. The engine may produce a rattling noise. This indicates that air still remains in the chamber and is not a matter of concern.**

20. Reinstall any parts removed in reverse order of removal.

Cylinder Head

REMOVAL & INSTALLATION

3.3L Engine

1. Before servicing the vehicle, refer to the precautions in the beginning of this section.
2. Drain the cooling system.
3. Relieve the fuel system pressure.

4. Remove or disconnect the following:
   - Negative battery cable
   - Accessory drive belts
   - Front cover
   - Timing belt. Refer to the Timing Belt unit repair section.
   - Upper intake manifold
   - Lower intake manifold
   - Camshaft sprockets
   - Rear timing cover
   - Distributor
   - Exhaust front pipes
   - A/C compressor
   - Alternator
   - Power steering pump
   - Accessory brackets
   - Valve covers. Loosen the bolts in several passes and in sequence.
   - Cylinder heads with the exhaust manifolds attached. Loosen the

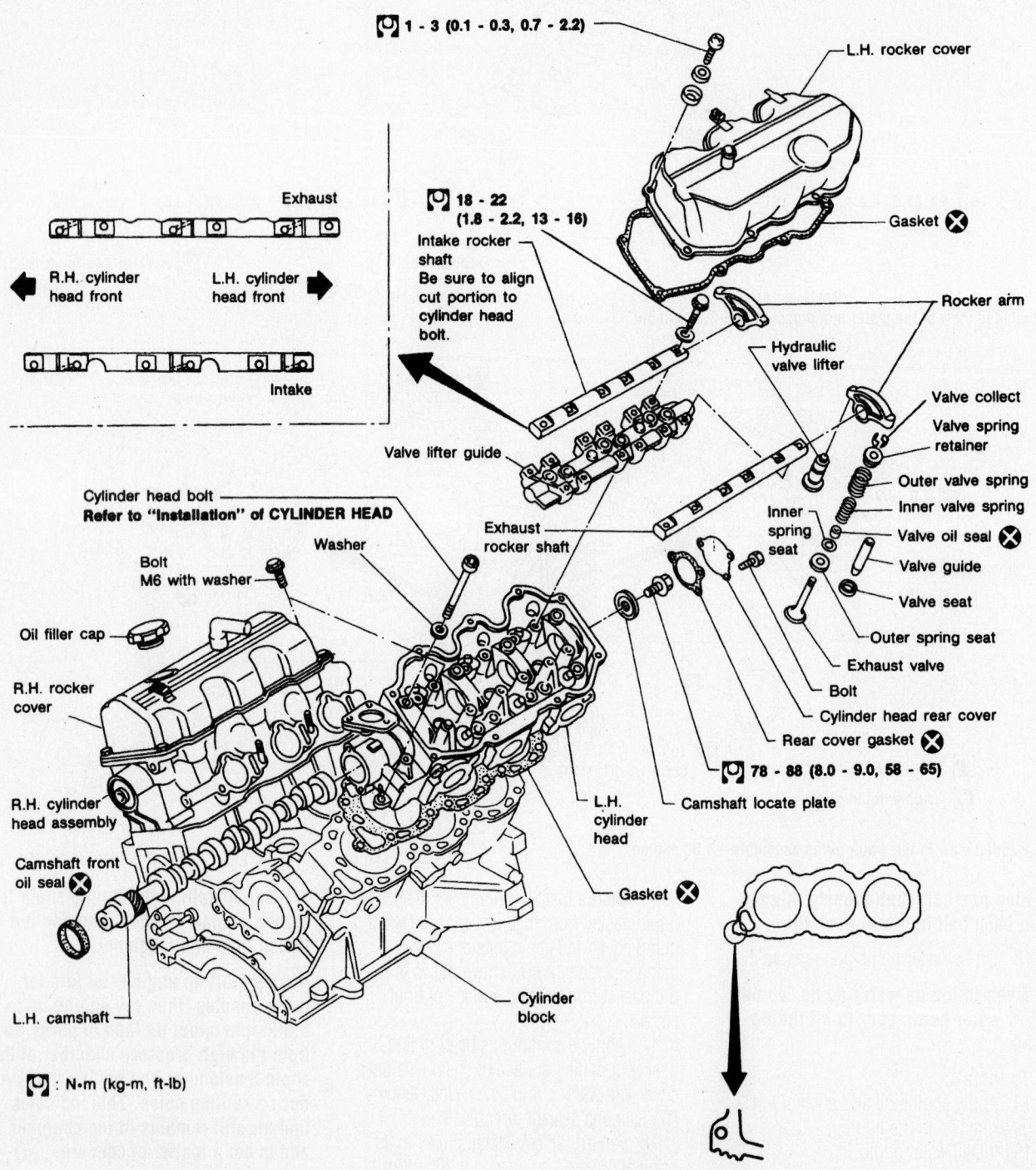

Exploded view of the cylinder head assembly—3.3L engine

7924VG25

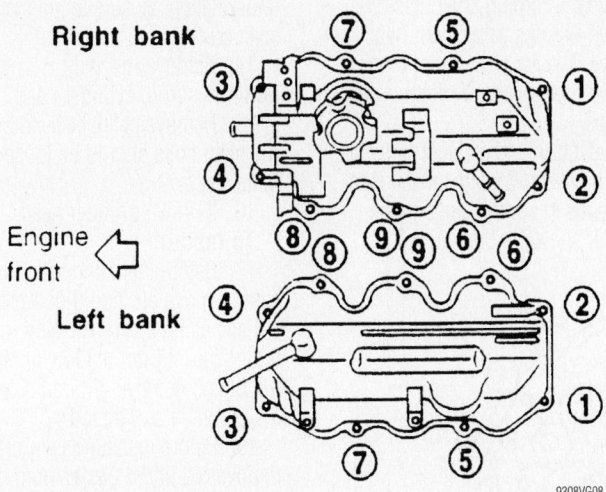

Valve cover bolt loosening sequence—3.3L engine

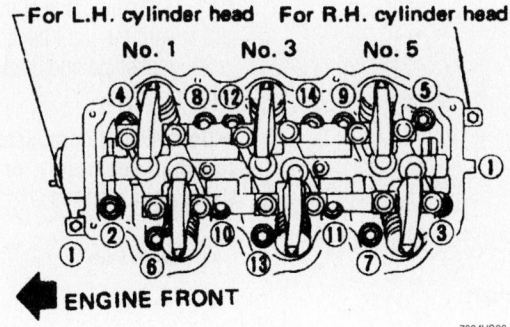

Cylinder head loosening sequence—3.3L engine

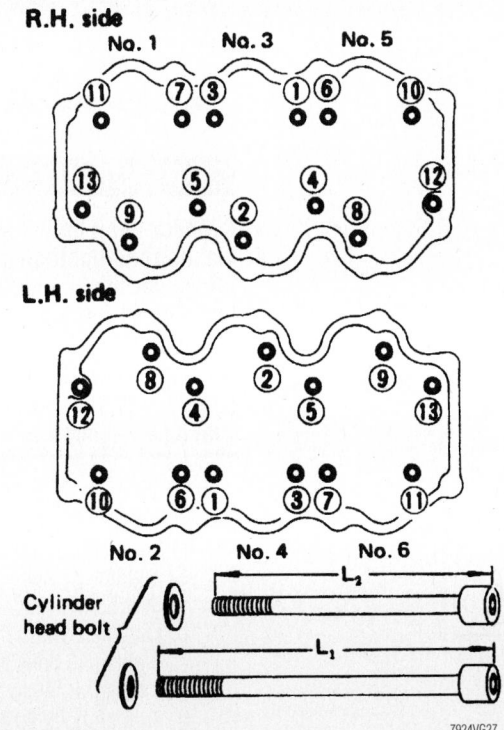

Cylinder head torque sequence—3.3L engine

bolts in several passes and in sequence.

➡ **The cylinder head bolts vary in length. Note the bolt locations for assembly.**

To install:

5. Install the cylinder heads and the lower intake manifold at the same time. Tighten the bolts in sequence as follows:

   a. Step 1: Tighten the cylinder head bolts to 22 ft. lbs. (29 Nm)

   b. Step 2: Tighten the cylinder head bolts to 43 ft. lbs. (59 Nm)

   c. Step 3: Loosen all cylinder head bolts completely

   d. Step 4: Tighten the cylinder head bolts to 84 inch lbs. (10 Nm)

   e. Step 5: Tighten the intake manifold fasteners to 35 inch lbs. (4 Nm)

   f. Step 6: Tighten the intake manifold fasteners to 13 ft. lbs. (18 Nm)

   g. Step 7: Tighten the intake manifold fasteners to 12–14 ft. lbs. (16–20 Nm)

   h. Step 8: Loosen all intake fasteners completely

   i. Step 9: Tighten the cylinder head bolts to 22 ft. lbs. (29 Nm)

   j. Step 10: Tighten the cylinder head bolts 60–65 degrees **OR** tighten to 40–47 ft. lbs. (54–64 Nm)

   k. Step 11: Tighten the cylinder head sub-bolts to 80–105 inch lbs. (9–12 Nm)

   l. Step 12: Tighten the intake manifold fasteners to 35 inch lbs. (4 Nm)

   m. Step 13: Tighten the intake manifold fasteners to 78 inch lbs. (9 Nm)

   n. Step 14: Tighten the intake manifold fasteners to 70–84 inch lbs. (6–7 Nm)

6. Install or connect the following:
   - Valve covers
   - Accessory brackets
   - Power steering pump
   - Alternator
   - A/C compressor
   - Exhaust front pipes
   - Distributor
   - Rear timing cover
   - Camshaft sprockets
   - Upper intake manifold
   - Timing belt
   - Front cover
   - Accessory drive belts
   - Negative battery cable

7. Fill the cooling system.
8. Start the engine and check for leaks.

3.5L Engine

1. Remove engine from vehicle.
2. Remove exhaust manifolds in reverse order of installation.

3. Place engine on a work stand.

4. Remove aluminum oil pan

5. Remove timing chain.

6. Remove intake manifold in reverse order of installation.

7. Remove water outlet.

8. Remove rear timing chain case bolts. Loosen in numerical order as shown in the figure.

9. Remove rear timing chain case.

10. Remove O-rings to cylinder head.

11. Remove O-rings to cylinder block.

12. Remove intake valve timing control solenoid valves.

13. Remove intake and exhaust camshafts and camshaft brackets. Equally loosen camshaft bracket bolts in several steps in the numerical order shown in the figure. For

reinstallation, be sure to put marks on camshaft bracket before removal.

14. Remove RH and LH camshaft chain tensioners from cylinder head.

15. Remove cylinder head bolts. Cylinder head bolts should be loosened in two or three steps.

16. Remove cylinder head.

To install:

17. Before installing rear timing chain case, remove old liquid gasket from mating surface using a scraper. Also remove old liquid gasket from mating surface of cylinder block. Remove old liquid gasket from the bolt hole and thread.

18. Before installing cam bracket, remove old liquid gasket from mating surface using a scraper.

19. Before installing the cylinder head gasket, be sure that No. 1 cylinder is at TDC. At this time, the crankshaft key should face toward the right bank.

20. Install cylinder heads with new gaskets.

➡ **Do not rotate crankshaft and camshaft separately, or valves will strike piston heads.**

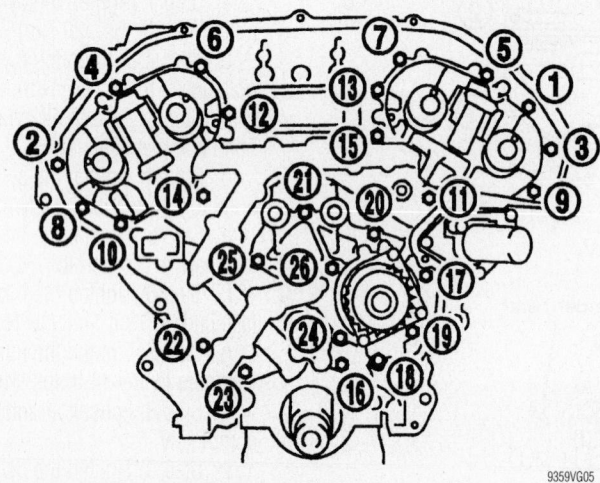

Rear timing case loosening sequence—3.5L engine

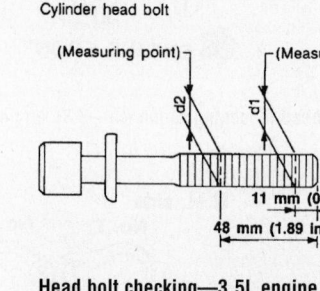

Head bolt checking—3.5L engine

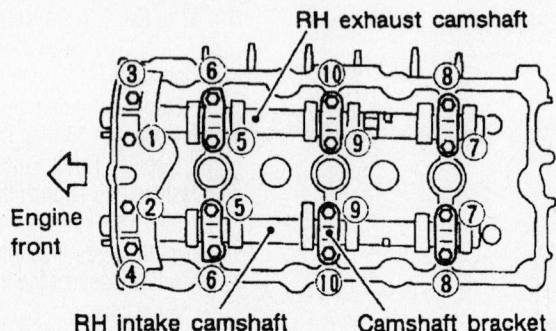

Right camshaft loosening sequence—3.5L engine

❋❋ CAUTION

Cylinder head bolts are tightened by plastic zone tightening method. Whenever the size difference between d1 and d2 exceeds the limit, replace them with new ones. Limit (d1 – d2): 0.0043 in. Lubricate threads and seat surfaces of the bolts with new engine oil.

21. Install cylinder head outside bolts Tighten in numerical order shown in the figure. Tightening procedure:

a. Tighten all bolts to 98 N·m (10 kg-m, 72 ft-lb).

b. Completely loosen all bolts.

c. Tighten all bolts to 34 to 44 N·m (3.5 to 4.5 kg-m, 25 to 33 ft-lb).

d. Turn all bolts 90 to 95 degrees clockwise.

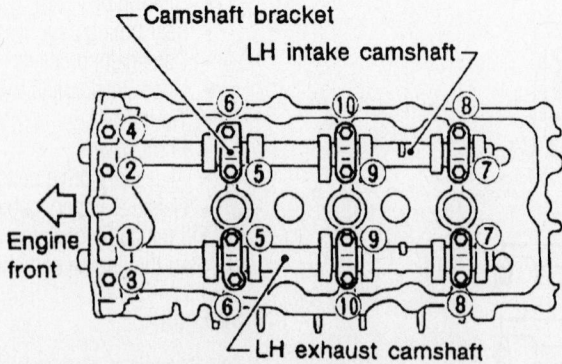

Left camshaft loosening sequence—3.5L engine

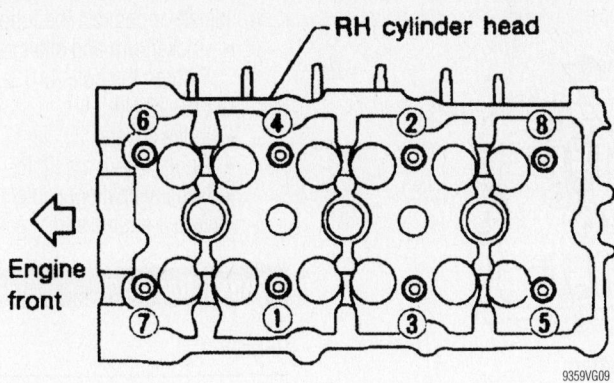

Right cylinder head bolt torque sequence—3.5L engine

LH cylinder head

Left cylinder head bolt torque sequence—3.5L engine

e. Turn all bolts 90 to 95 degrees clockwise.

22. Install camshaft chain tensioners on both sides of cylinder head.

23. Install exhaust and intake camshafts and camshaft brackets.

➡**Intake camshaft has a drill mark on camshaft sprocket mounting flange. Install it on the intake side. Position camshaft. RH exhaust camshaft dowel pin at about 10 o'clock; LH exhaust camshaft dowel pin at about 2 o'clock**

24. Before installing camshaft brackets, apply sealant to mating surface of No. 1 journal head. Use Genuine RTV silicone sealant or equivalent. Install camshaft brackets in their original positions. Align stamp mark as shown in the figure. If any part of valve assembly or camshaft is replaced, check valve clearance according to reference data. After completing assembly check valve clearance. Valve clearance (Cold):

- Intake 0.26—0.34 mm (0.010—0.013 in)
- Exhaust 0.29—0.37 mm (0.011—0.015 in)

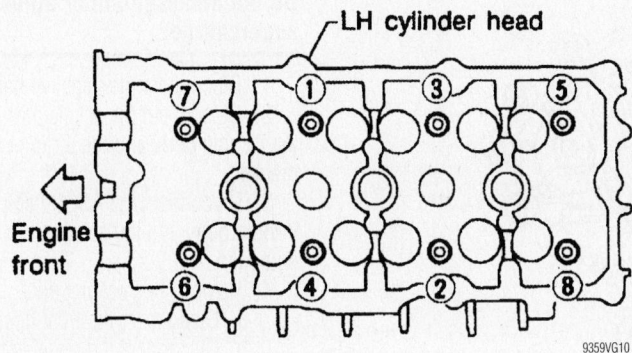

Camshaft identification—3.5L engine

● **Identification marks are present on camshafts.**

Bank	INT/EXH	ID mark	Drill mark	Paint mark	
				M1	M2
RH	INT	R3	Yes	Yes	No
	EXH	R3	No	No	Yes
LH	INT	L3	Yes	Yes	No
	EXH	L3	No	No	Yes

● Tighten the camshaft brackets in the following steps.

Step	Tightening torque	Tightening order
1	1.96 N·m (0.2 kg-m, 17 in-lb)	Tighten in the order of 7 to 10, then tighten 1 to 6.
2	5.88 N·m (0.6 kg-m, 52 in-lb)	Tighten in the numerical order.
3	9.02 - 11.8 N·m (0.92 - 1.20 kg-m, 79.9 - 104.2 in-lb)	Tighten in the order of 1 to 6.
	8.3 - 10.3 N·m (0.9 - 1.0 kg-m, 74 - 91 in-lb)	Tighten in the order of 7 to 10.

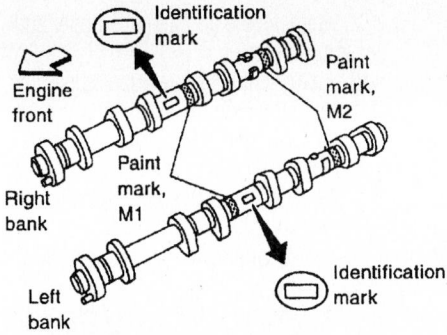

Right camshaft bolt torque sequence—3.5L engine

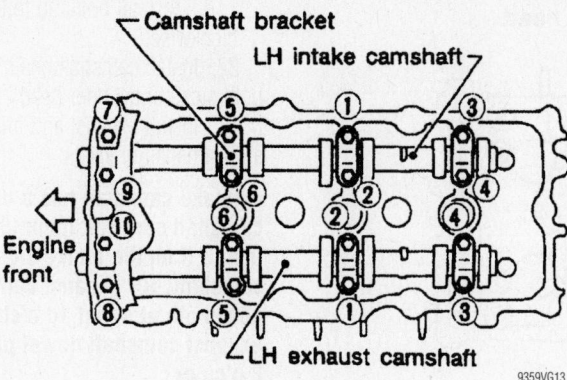

Left camshaft bolt torque sequence—3.5L engine

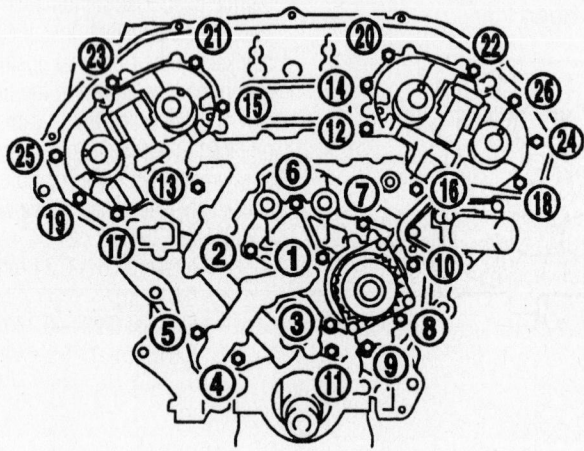

⊞ **12 - 13 N•m**
(1.2 - 1.4 kg-m, 9 - 10 ft-lb)

Rear timing case bolt torque sequence—3.5L engine

➡**Lubricate threads and seat surfaces of camshaft bracket bolts with new engine oil before installing them.**

25. Install intake valve timing control solenoid valves.

26. Install O-rings to cylinder block.

27. Install O-rings to cylinder head.

28. Apply sealant to the hatched portion of rear timing chain case. Apply continuous bead of liquid gasket to mating surface of rear timing chain case. Before installation, wipe off the protruding sealant.

29. Align rear timing chain case with dowel pins, then install on cylinder head and block.

30. Tighten rear chain case bolts.

a. Tighten bolts in numerical order shown in the figure.

b. Repeat above step a.

31. Reinstall all removed parts in reverse order of removal.

Rocker Arms/Shafts

REMOVAL & INSTALLATION

3.3L Engine

1. Before servicing the vehicle, refer to the precautions in the beginning of this section.

2. Remove or disconnect the following:
- Negative battery cable
- Upper intake manifold
- Valve covers
- Rocker arm and shaft assemblies
- Rocker arms from the shafts

➡**Keep all valvetrain components in order for assembly.**

To install:

3. Lubricate all contact points with clean engine oil and assemble the rocker arms to the shafts in their original positions.

4. Install or connect the following:
- Rocker arm and shaft assemblies. Tighten the bolts to 13–16 ft. lbs. (18–22 Nm).
- Valve covers
- Upper intake manifold
- Negative battery cable

5. Start the engine and check for leaks.

Supercharger

REMOVAL

❄❄ CAUTION

Do not disassemble or adjust the supercharger.

1. Disconnect the negative battery cable.

2. Disconnect the accelerator cable from the throttle body and the air inlet tube bracket.

3. Disconnect the ASCD cable from the throttle body and the air inlet tube bracket, if equipped.

4. Remove the air inlet duct

a. Disconnect the PCV hoses.

b. Disconnect the resonator hose.

5. Partially drain the cooling system.

6. Remove the supercharger pulley cover and the supercharger/air conditioning drive belt.

7. Remove the air inlet tube upper and lower supports.

8. Remove the air inlet tube bolts, nuts, and studs. Position the air inlet tube aside.

a. Disconnect the evaporative emission vacuum hose.

b. Disconnect the brake booster vacuum hose.

c. Disconnect the TPS sensor electrical connector.

d. Disconnect the TPS switch electrical connector.

9. Remove the supercharger bolts and the supercharger assembly.

a. Disconnect the boost control valve vacuum hose.

b. Disconnect the PCV hose.

INSPECTION

Supercharger Flange

1. Clean the mating surface of the supercharger flange.

2. Check the flange surface for any deformation and flatness.

Use a reliable straightedge and feeler gauge, or attach the supercharger flange to the intake collector mating flange, and

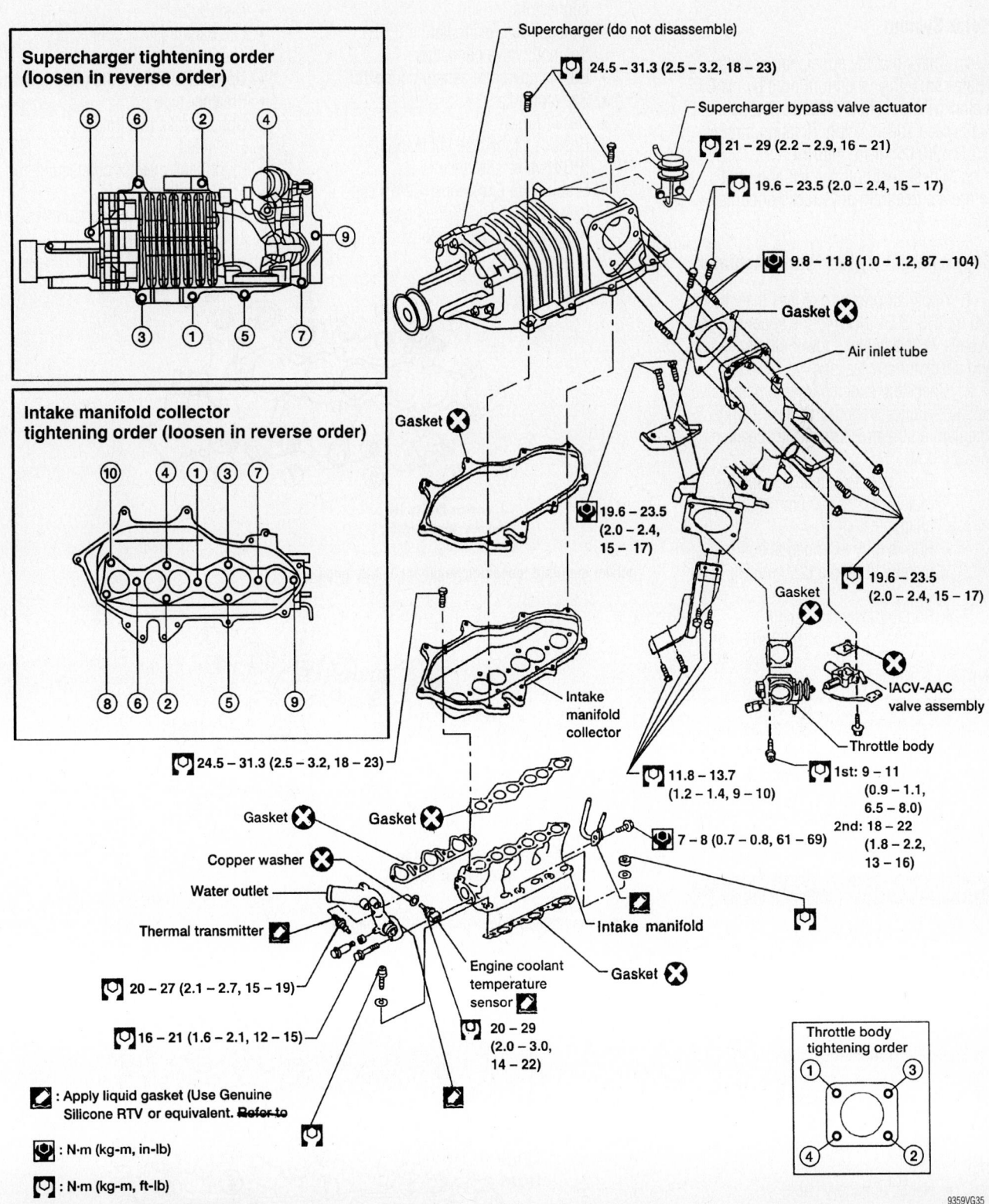

Supercharger tightening order (loosen in reverse order)

Intake manifold collector tightening order (loosen in reverse order)

Supercharger (do not disassemble)

24.5 – 31.3 (2.5 – 3.2, 18 – 23)

Supercharger bypass valve actuator

21 – 29 (2.2 – 2.9, 16 – 21)

19.6 – 23.5 (2.0 – 2.4, 15 – 17)

9.8 – 11.8 (1.0 – 1.2, 87 – 104)

Gasket

Air inlet tube

Gasket

19.6 – 23.5 (2.0 – 2.4, 15 – 17)

19.6 – 23.5 (2.0 – 2.4, 15 – 17)

Gasket

IACV-AAC valve assembly

Throttle body

1st: 9 – 11 (0.9 – 1.1, 6.5 – 8.0)
2nd: 18 – 22 (1.8 – 2.2, 13 – 16)

11.8 – 13.7 (1.2 – 1.4, 9 – 10)

7 – 8 (0.7 – 0.8, 61 – 69)

Intake manifold

24.5 – 31.3 (2.5 – 3.2, 18 – 23)

Gasket

Gasket

Copper washer

Water outlet

Thermal transmitter

Intake manifold

Gasket

Engine coolant temperature sensor

20 – 27 (2.1 – 2.7, 15 – 19)

16 – 21 (1.6 – 2.1, 12 – 15)

20 – 29 (2.0 – 3.0, 14 – 22)

: Apply liquid gasket (Use Genuine Silicone RTV or equivalent. Refer to

: N·m (kg-m, in-lb)

: N·m (kg-m, ft-lb)

Throttle body tightening order

Supercharger components

9359VG35

check that the flatness is within specification. Flange flatness limit: 0.12 mm (0.005 in).

Rotor System

1. Check that the supercharger pulley rotates smoothly when turning it by hand in a clockwise direction. Rotating torque must not exceed specification. Rotating torque: 0.5 N.m (0.05 kg-m, 4 in-lb).

2. Check that both the left and right rotors are free from any cracks or contamination.

Supercharger Bypass Valve Actuator

1. Apply air pressure of less than 12 kPa (90 mmHg, 3.54 inHg) to the supercharger bypass valve actuator's lower side hose port and check for any leakage.

2. Check the supercharger bypass valve actuator rod for smooth movement while maintaining the pressure at the specified levels below:

- Rod starts to extend at approximately: 12 Kpa (90 mmHg, 3.54 inHg)
- Rod is fully extended at approximately: 33.3 kPa (250 mmHg, 9.84 inHg)
- Rod full extended length: 20.83–22.71 mm (0.82–0.89 in)

INSTALLATION

To install the supercharger, follow the removal steps in reverse order. Replace all gaskets; make sure that all gasket surfaces are clean and undamaged. Follow all torque sequences for tightening. Refill the cooling system.

Intake Manifold

REMOVAL & INSTALLATION

3.3L and 3.5L Engines

1. Before servicing the vehicle, refer to the precautions in the beginning of this section.

2. Drain the cooling system.

3. Relieve the fuel system pressure.

4. Remove or disconnect the following:

- Negative battery cable
- Air intake duct
- Accelerator cable
- Cruise control cable
- Idle Air Control (IAC) valve connector

- Throttle Position (TP) sensor and switch connectors
- Ignition coil and power transistor connectors
- Exhaust Gas Recirculation (EGR) Solenoid valve connector
- EGR temperature sensor connector
- Radiator hoses
- Heater hoses
- Positive Crankcase Ventilation (PCV) valve and hose
- Evaporative Emissions (EVAP) canister vacuum and purge hoses
- Brake booster vacuum hose

- Fuel pressure regulator vacuum hose
- EGR tube
- Spark plug wires
- Distributor
- Left bank injector connectors
- Thermal transmitter
- Upper intake manifold ground cable
- Breather pipe
- Upper intake manifold
- Fuel lines
- Right bank injector connectors
- Fuel supply manifold
- Engine Coolant Temperature (ECT) sensor connector

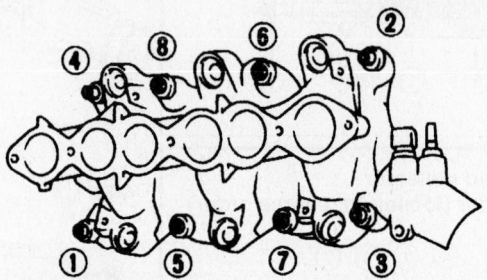

Loosen bolts in numerical order.

7924VG32

Intake manifold loosening sequence—3.3L engine

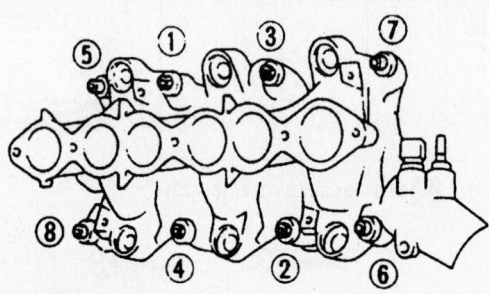

Tighten bolts in numerical order.

7924VG33

Intake manifold torque sequence—3.3L engine

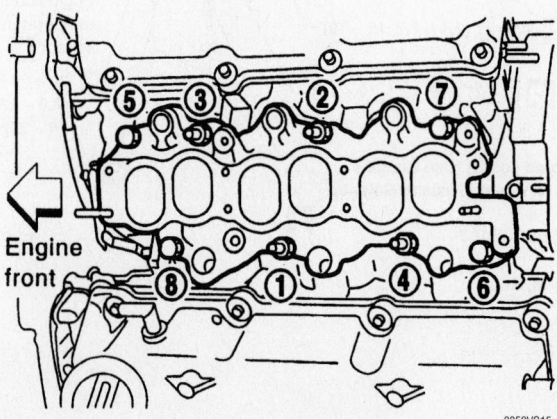

Engine front

9359VG15

Lower intake manifold torque sequence—3.5L engine

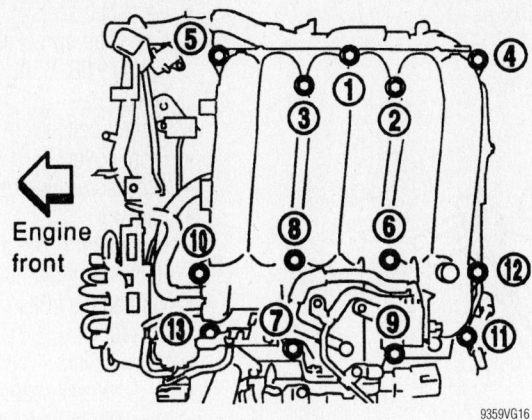

Upper intake manifold torque sequence—3.5L engine

- Lower intake manifold. Loosen the fasteners in the sequence shown.

To install:

5. Install the lower intake manifold with a new gasket.

6. For 3.3L engines, tighten the fasteners in sequence as follows:
   a. Step 1: 35 inch lbs. (4 Nm)
   b. Step 2: 78 inch lbs. (9 Nm)
   c. Step 3: 70–84 inch lbs. (8–10 Nm)

7. For 3.5L engines, tighten the fasteners in sequence as follows:
   a. Step 1: 86 inch lbs. (4 Nm)
   b. Step 2: 23 ft. lbs. (9 Nm)

8. Install or connect the following:
   - ECT sensor connector
   - Fuel supply manifold
   - Right bank injector connectors
   - Fuel lines
   - Upper intake manifold
   - Breather pipe
   - Upper intake manifold ground cable
   - Thermal transmitter
   - Left bank injector connectors
   - Distributor
   - Spark plug wires
   - EGR tube
   - Fuel pressure regulator vacuum hose
   - Brake booster vacuum hose
   - EVAP canister vacuum and purge hoses
   - PCV valve and hose
   - Heater hoses
   - Radiator hoses
   - EGR temperature sensor connector
   - EGR Solenoid valve connector
   - Ignition coil and power transistor connectors
   - TP sensor and switch connectors
   - IAC valve connector
   - Cruise control cable
   - Accelerator cable
   - Air intake duct
   - Negative battery cable

9. Fill the cooling system.
10. Start the engine and check for leaks.

Exhaust Manifold

REMOVAL & INSTALLATION

3.3L and 3.5L Engines

1. Before servicing the vehicle, refer to the precautions in the beginning of this section.

2. Remove or disconnect the following:
   - Negative battery cable
   - Exhaust manifold heat shields
   - Exhaust Gas Recirculation (EGR) tube
   - Heated Oxygen (HO$_2$S) sensor connectors
   - Exhaust front pipes
   - Exhaust manifolds with catalytic converters attached. Loosen the nuts in the reverse of the torque sequence.

To install:
3. Install or connect the following:
   - Exhaust manifolds with catalytic converters attached. Tighten the nuts in sequence to 21–25 ft. lbs. (28–33 Nm).
   - Exhaust front pipes. Tighten the bolts to 21–25 ft. lbs. (28–33 Nm).
   - Heated Oxygen (HO$_2$S) sensor connectors
   - EGR tube. Tighten the flange fittings to 29–36 ft. lbs. (39–49 Nm).
   - Exhaust manifold heat shields.

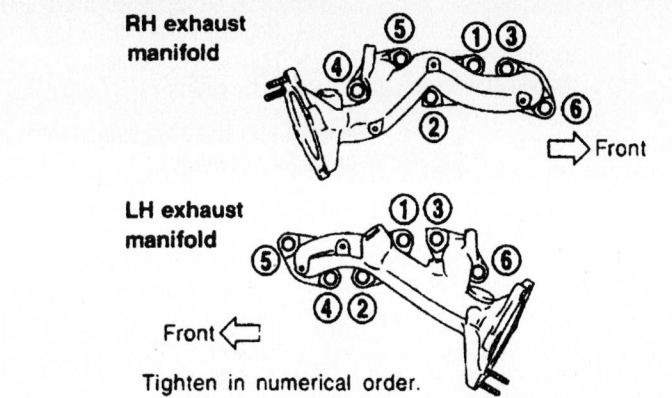

Exhaust manifold torque sequence—3.3L engine

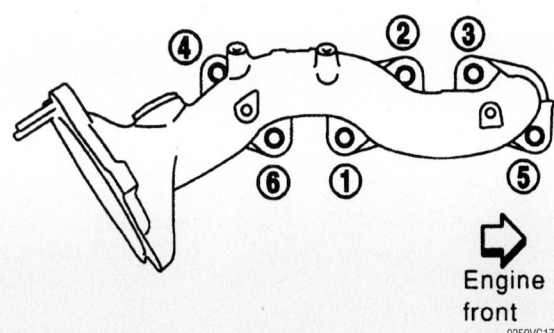

Right exhaust manifold torque sequence—3.5L engine

Left bank

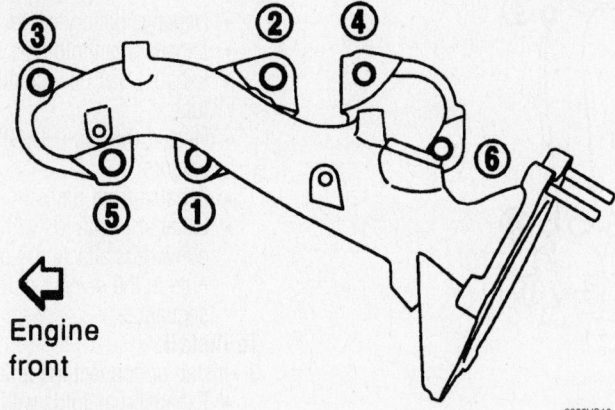

9359VG18

Exhaust manifold torque sequence—3.5L engine

Tighten the bolts to 84–96 inch lbs. (9–11 Nm)
- Negative battery cable
4. Start the engine and check for leaks.

Front Crankshaft Seal

REMOVAL & INSTALLATION

3.3L Engine

1. Before servicing the vehicle, refer to the precautions in the beginning of this section.
2. Drain the cooling system.
3. Remove or disconnect the following:
- Negative battery cable
- Accessory drive belts
- Radiator hoses
- Crankshaft pulley
- Front cover
- Timing belt. Refer to the Timing Belt unit repair section.
- Crankshaft timing sprocket
- Front crankshaft seal

To install:
4. Install or connect the following:
- Front crankshaft seal flush with the oil pump housing
- Crankshaft timing sprocket
- Timing belt
- Front cover. Tighten the bolts to 26–43 inch lbs. (3–5 Nm).
- Crankshaft pulley. Tighten the bolt to 141–156 ft. lbs. (191–211 Nm).
- Radiator hoses
- Accessory drive belts
- Negative battery cable
5. Fill the cooling system.
6. Start the engine and check for leaks.

Camshaft and Valve Lifters

REMOVAL & INSTALLATION

3.3L Engines

1. Before servicing the vehicle, refer to the precautions in the beginning of this section.
2. Drain the cooling system.
3. Remove or disconnect the following:
- Negative battery cable
- Upper intake manifold
- Valve covers

➡**Keep all valvetrain components in order for assembly.**

- Rocker arm and shaft assemblies
- Valve lifter guide and valve lifters. Attach a wire to the top of the lifters so that they will not drop from the lifter guide.
- Radiator
- Accessory drive belts
- Front cover
- Timing belt. Refer to the Timing Belt unit repair section.
- Camshaft sprockets
- Camshaft seals
- Rear timing cover
- Distributor
- Cylinder head rear covers
- Camshaft locating plates
- Camshafts

To install:
4. Install or connect the following:
- Camshafts
- Camshaft locating plates. Tighten the bolts to 58–65 ft. lbs. (78–88 Nm).
- Cylinder head rear covers
- Distributor
- Rear timing cover

- Camshaft seals
- Camshaft sprockets. Tighten the bolts to 58–65 ft. lbs. (78–88 Nm).
- Timing belt
- Front cover
- Accessory drive belts
- Radiator
- Valve lifter guide and valve lifters
- Rocker arm and shaft assemblies. Tighten the bolts to 13–16 ft. lbs. (18–22 Nm).
- Valve covers
- Upper intake manifold
- Negative battery cable
5. Fill the cooling system.
6. Start the engine and check for leaks.

3.5L Engines

See the Cylinder Head Removal and Installation procedure.

Valve Lash

ADJUSTMENT

3.3L Engines

These engines are equipped with hydraulic valve lifters that do not require periodic adjustment.

3.5L Engines

➡**Adjust valve clearance while engine is cold.**

1. Turn crankshaft, to position cam lobe on camshaft of valve that must be adjusted upward.
2. Thoroughly wipe off engine oil around adjusting shim using a rag.
3. Using an extra-fine screwdriver, turn the round hole of the adjusting shim in the direction of the arrow.
4. Place Tool (A) around camshaft as shown in figure.
   Before placing Tool (A), rotate notch toward center of cylinder head (See figure.), to simplify shim removal later.

✳✳ CAUTION

Be careful not to damage cam surface with Tool (A).

5. Rotate Tool (A) (See figure.) so that valve lifter is pushed down.
6. Place Tool (B) between camshaft and the edge of the valve lifter to retain valve lifter.

Tool (B) must be placed as close to camshaft bracket as possible. Be careful not to damage cam surface with Tool (B).

7. Remove Tool (A).

8. Blow air into the hole to separate adjusting shim from valve lifter.

9. Remove adjusting shim using a small screwdriver and a magnetic finger.

10. Determine replacement adjusting shim size following formula. Using a micrometer determine thickness of removed shim. Calculate thickness of new adjusting shim so valve clearance comes within specified values.

- R = Thickness of removed shim
- N = Thickness of new shim
- M = Measured valve clearance
- Intake: $N = R + [M - 0.30$ mm $(0.0118$ in$)]$
- Exhaust: $N = R + [M - 0.33$ mm $(0.0130$ in$)]$

Shims are available in 64 sizes from 2.32 mm (0.0913 in) to 2.95 mm (0.1161 in), in steps of 0.01 mm (0.0004 in). Select new shim with thickness as close as possible to calculated value.

11. Install new shim using a suitable tool. Install with the surface on which the thickness is stamped facing down.

12. Place Tool (A) as mentioned in steps 2 and 3.

13. Remove Tool (B).

14. Remove Tool (A).

15. Recheck valve clearance.
Valve clearance (Cold)
- Intake: 0.010—0.013
- Exhaust: 0.011—0.015

Starter Motor

REMOVAL & INSTALLATION

1. Before servicing the vehicle, refer to the precautions in the beginning of this section.

2. Remove or disconnect the following:
- Negative battery cable
- Engine under cover
- Starter harness connectors
- Starter motor

To install:

3. Install or connect the following:
- Starter motor. Tighten the bolts to 22–27 ft. lbs. (30–36 Nm) on the 3.3L; 37–45 ft. lbs. (61–69NM) on the 3.5L.
- Starter harness connectors

- Engine under cover
- Negative battery cable

Oil Pan

REMOVAL & INSTALLATION

3.3L Engine
2WD MODELS

1. Before servicing the vehicle, refer to the precautions in the beginning of this section.

2. Drain the engine oil.

3. Remove or disconnect the following:
- Negative battery cable
- Engine under cover
- Stabilizer bar
- Front crossmember
- Starter motor
- Transmission mount
- Left and right motor mounts
- Power steering gear

4. Raise and support the engine for clearance.

5. Remove or disconnect the following:
- Oil pan bolts in the sequence
- Oil pan

To install:

6. Apply a continuous bead of sealant 0.138–0.177 in. (3.5–4.5mm) to the oil pan mating surface.

7. Install or connect the following:
- Oil pan. Tighten the bolts in reverse of the removal sequence to 62 inch lbs. (7 Nm).
- Power steering gear
- Left and right motor mounts
- Transmission mount
- Starter motor
- Front crossmember
- Stabilizer bar
- Engine under cover
- Negative battery cable

➡Wait 30 minutes after installation of the oil pan to allow the sealant to cure before adding oil.

8. Fill the crankcase to the correct level.

9. Start the engine and check for leaks.

4WD MODELS

1. Before servicing the vehicle, refer to the precautions in the beginning of this section.

2. Drain the engine oil.

3. Remove or disconnect the following:
- Negative battery cable
- Engine under cover
- Stabilizer bar brackets
- Front driveshaft
- Axle halfshafts
- Front suspension crossmember
- Front differential and mounting bracket
- Starter motor
- Transmission mount
- Left and right motor mounts
- Power steering gear
- Relay rod

4. Raise and support the engine for clearance.

5. Remove or disconnect the following:
- Oil pan bolts in the sequence
- Oil pan

To install:

6. Apply a continuous bead of sealant 0.138–0.177 in. (3.5–4.5mm) to the oil pan mating surface.

7. Install or connect the following:
- Oil pan. Tighten the bolts in reverse of the removal sequence to 62 inch lbs. (7 Nm).
- Relay rod
- Power steering gear
- Left and right motor mounts
- Transmission mount
- Starter motor

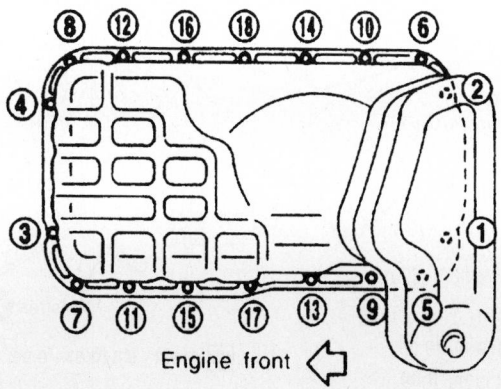

Oil pan bolt removal sequence—3.3L engine

7924VG42

- Front differential and mounting bracket
- Front suspension crossmember
- Axle halfshafts
- Front driveshaft
- Stabilizer bar brackets
- Engine under cover
- Negative battery cable

➡ **Wait 30 minutes after installation of the oil pan to allow the sealant to cure before adding oil.**

8. Fill the crankcase to the correct level.
9. Start the engine and check for leaks.

3.5L Engines

1. Remove front RH and LH wheels.
2. Remove battery.
3. Remove oil level gauge.
4. Remove engine undercover.
5. Remove suspension member stay.
6. Drain engine coolant from radiator drain plug.
7. Disconnect A/T oil cooler hoses. (A/T)
8. Drain engine oil.
9. Remove the crankshaft position sensors (REF and POS).
10. Remove drive belts and idler pulley with bracket.

11. Remove power steering oil pump, then put it aside holding with a suitable wire.
12. Remove alternator.
13. Install engine slingers.
14. Remove front propeller shaft. (4WD)
15. Remove exhaust front tube heat insulators, then remove rear heat oxygen sensors.
16. Remove exhaust front tube from both sides.
17. Remove front final drive. (4WD)
18. Remove starter motor.
19. Disconnect oil pressure switch harness connector.
20. Loosen and disconnect the bolts fixing the steering column assembly lower joint and the power steering gear.
21. Set a suitable transmission jack under the front suspension member and hoist engine with engine slingers.
22. Remove front engine mounting nuts from both sides.
23. Remove front suspension member bolts.
24. Lower the transmission jack carefully to secure clearance between the oil pan and suspension member.
25. Remove A/T oil cooler tube. (A/T)
26. Remove water hose and tube. (A/T)

27. Remove the four engine-to-transmission bolts.
28. Remove aluminum oil pan bolts in numerical order.
29. Remove aluminum oil pan.
   a. Insert tool between aluminum oil pan and cylinder block.

➡ **Be careful not to damage aluminum mating surface. I Do not insert screwdriver, or oil pan flange will be deformed.**

   b. Slide tool by tapping its side with a hammer.
30. Remove O-rings from cylinder block and oil pump body.
31. Remove front cover gasket and rear oil seal retainer gasket.

To install:

32. Before installing oil pan, remove old liquid gasket from mating surface using a scraper. Also remove old liquid gasket from mating surface of cylinder block. Remove old liquid gasket from the bolt hole and thread.
33. Apply sealant to front cover gasket and rear oil seal retainer gasket.
34. Install front cover gasket and rear oil seal retainer gasket.
35. Apply a continuous bead of liquid

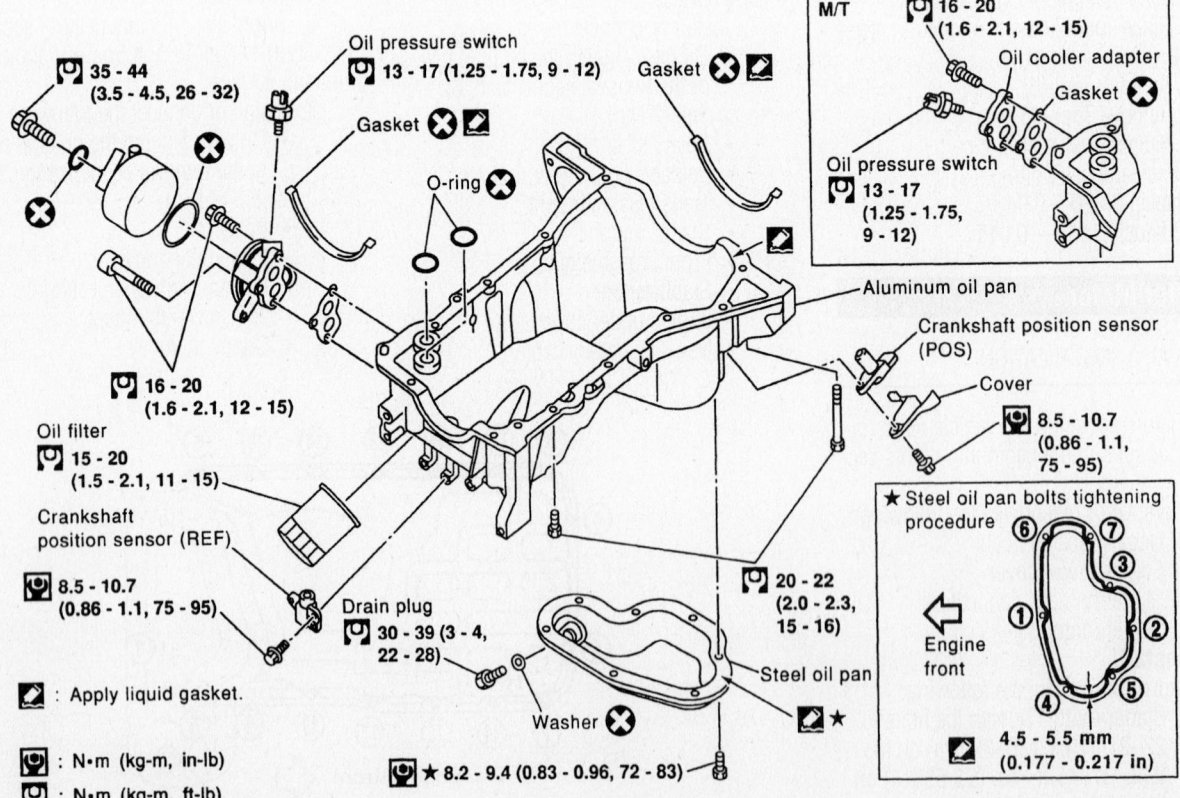

Oil pan exploded view—3.5L engine

9359VG19

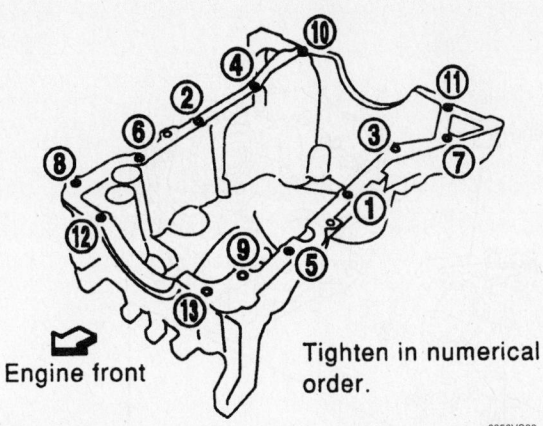

Engine front

Tighten in numerical order.

9359VG20

Oil pan bolt torque sequence—3.5L engine

gasket to mating surface of aluminum oil pan. Use RTV silicone sealant or equivalent.

36. Apply liquid gasket to inner sealing surface as shown in figure. Be sure liquid gasket is 4.0 to 5.0 mm (0.157 to 0.197 in) or 4.5 to 5.5 mm (0.177 to 0.217 in) wide. Attaching should be done within 5 minutes after coating.

37. Install O-rings, cylinder block and oil pump body.

38. Install aluminum oil pan. Tighten bolts in numerical order. Wait at least 30 minutes before refilling engine oil.

39. Install the four engine-to-transmission bolts.

40. Reinstall in the reverse order of removal.

Oil Pump

REMOVAL & INSTALLATION

3.3L Engine

1. Before servicing the vehicle, refer to the precautions in the beginning of this section.
2. Drain the engine oil.
3. Drain the cooling system.
4. Remove or disconnect the following:
   - Negative battery cable
   - Accessory drive belts
   - Radiator hoses
   - Crankshaft pulley

- Front cover
- Timing belt. Refer to the Timing Belt unit repair section.
- Crankshaft timing sprocket
- Oil pan
- Oil pump pickup tube
- Oil pump

To install:

5. Install or connect the following:
   - Oil pump. Tighten the large bolts to 16–22 ft. lbs. (22–29 Nm) and the small bolts to 55–74 inch lbs. (6–8 Nm).
   - Oil pump pickup tube. Tighten the flange bolts to 12 ft. lbs. (16 Nm) and the bracket bolt to 55–74 inch lbs. (6–8 Nm).
   - Oil pan
   - Crankshaft timing sprocket
   - Timing belt
   - Front cover
   - Crankshaft pulley
   - Radiator hoses
   - Accessory drive belts
   - Negative battery cable
6. Fill the cooling system.
7. Fill the crankcase to the correct level.
8. Start the engine and check for leaks.

3.5L Engine

1. Remove timing chain.
2. Remove oil pump assembly.
3. Reinstall any parts removed in reverse order of removal.

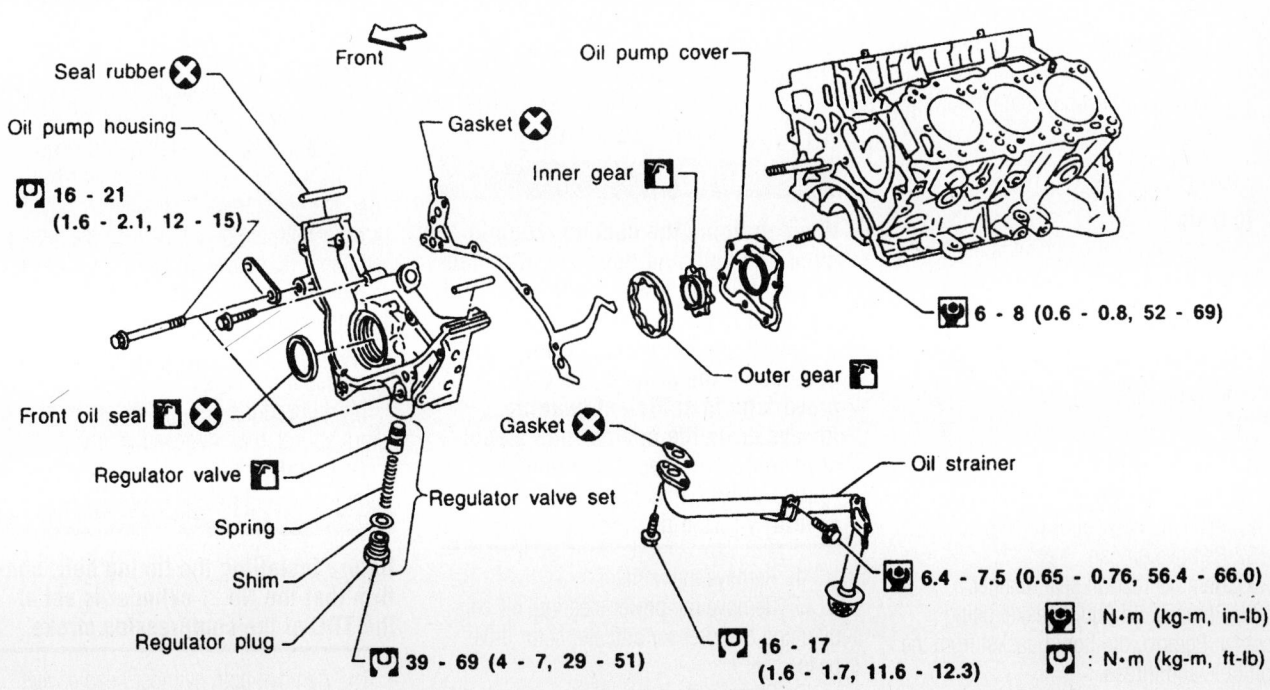

Oil pump assembly exploded view—3.3L engine

7924VG46

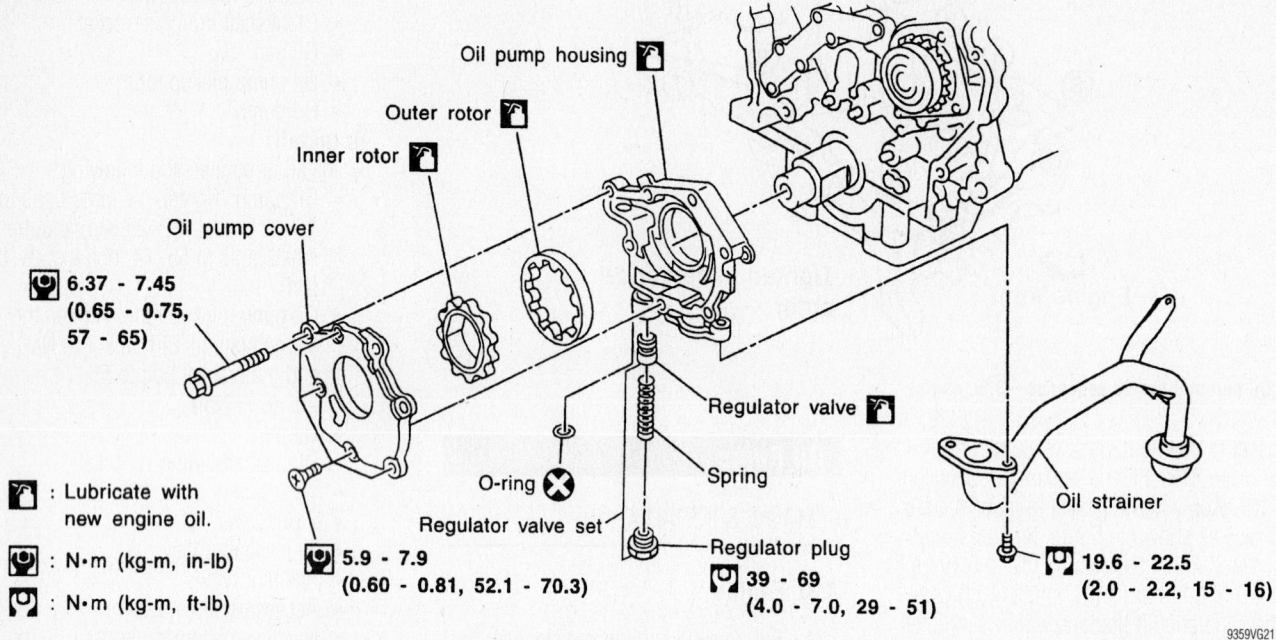

**6.37 - 7.45
(0.65 - 0.75,
57 - 65)**

Oil pump housing

Outer rotor

Inner rotor

Oil pump cover

Regulator valve

Spring

Oil strainer

O-ring

Regulator valve set

Regulator plug

: Lubricate with
new engine oil.

: N·m (kg-m, in-lb)

: N·m (kg-m, ft-lb)

**5.9 - 7.9
(0.60 - 0.81, 52.1 - 70.3)**

**39 - 69
(4.0 - 7.0, 29 - 51)**

**19.6 - 22.5
(2.0 - 2.2, 15 - 16)**

9359VG21

Oil pump assembly exploded view—3.5L engine

Rear Main Seal

REMOVAL & INSTALLATION

3.3L Engines

1. Before servicing the vehicle, refer to the precautions in the beginning of this section.
2. Remove or disconnect the following:
   - Transmission
   - Flywheel
   - Rear main seal

To install:

3. Install the seal so that it is flush with the retainer housing.
4. Install or connect the following:
   - Flywheel. Tighten the bolts to 61–69 ft. lbs. (83–93 Nm).
   - Transmission

3.5L Engine

1. Remove transmission.
2. Remove flywheel or drive plate.
3. Remove oil pan.
4. Remove rear oil seal retainer.
5. Remove old liquid gasket using scraper. Remove old liquid gasket from the bolt hole and thread.
6. Apply liquid gasket to rear oil seal retainer.

Timing Belt

REMOVAL & INSTALLATION

3.3L Engines

1. Remove the engine undercover.
2. Remove the radiator shroud, the fan and the pulleys.
3. Drain the coolant from the radiator and remove the water pump hose.

✳✳ CAUTION

When draining the coolant, keep in mind that cats and dogs are attracted by the ethylene glycol antifreeze, and are quite likely to drink any that is left in an uncovered container or in puddles on the ground. This will prove fatal in sufficient quantity. Always drain the coolant into a sealable container. Coolant should be reused unless it is contaminated or several years old.

4. Remove the radiator.
5. Remove the power steering, air conditioning compressor and alternator drive belts.
6. Remove the spark plugs.
7. Remove the distributor protector (dust shield).

8. Remove the air conditioning compressor drive belt idler pulley and bracket.
9. Remove the fresh air intake tube at the cylinder head cover.
10. Disconnect the radiator hose at the thermostat housing.
11. Remove the crankshaft pulley bolt, then pull off the pulley with a suitable puller.
12. Remove the bolts, then remove the front upper and lower timing belt covers.
13. Set the No. 1 piston at Top Dead Center (TDC) of its compression stroke. Align the punchmark on the left camshaft sprocket with the punchmark on the timing belt upper rear cover. Align the punchmark on the crankshaft sprocket with the notch on the oil pump housing. Temporarily install the crank pulley bolt so the crankshaft can be rotated if necessary.
14. Loosen the timing belt tensioner and return spring, then remove the timing belt.

To install:

✳✳ CAUTION

Before installing the timing belt, confirm that the No. 1 cylinder is set at the TDC of the compression stroke.

15. Remove both cylinder head covers and loosen all rocker arm shaft retaining bolts.

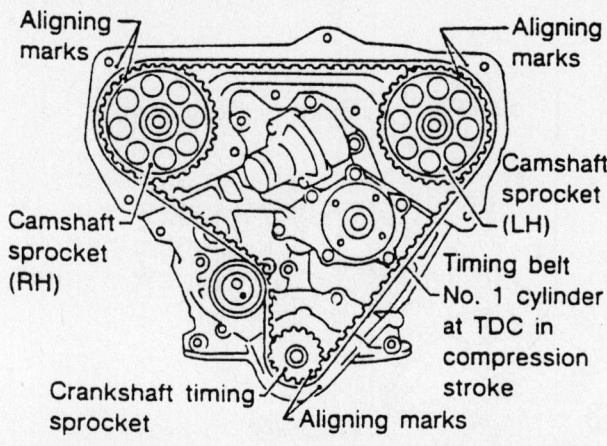

Timing belt alignment mark locations —3.0L and 3.3L engines

79245G35

➡**The rocker arm shaft bolts MUST be loosened so that the correct belt tension can be obtained.**

16. Install the tensioner and the return spring. Using a hexagon wrench, turn the tensioner clockwise and temporarily tighten the locknut.

17. Be sure that the timing belt is clean and free from oil or water.

18. When installing the timing belt, align the white lines on the belt with the punch-marks on the camshaft and crankshaft sprockets. Have the arrow on the timing belt pointing toward the front belt covers.

➡**A good way (although rather tedious!) to check for proper timing belt installation is to count the number of belt teeth between the timing marks. There are 133 teeth on the belt; there should be 40 teeth between the timing marks on the left and right side camshaft sprockets, and 43 teeth between the timing marks on the left side camshaft sprocket and the crankshaft sprocket.**

19. While keeping the tensioner steady, loosen the locknut with a hex wrench.

20. Turn the tensioner approximately 70 – 80 degrees clockwise with the wrench, then tighten the locknut.

✳✳ WARNING

If any binding is felt when adjusting the timing belt tension by turning the crankshaft, STOP turning the engine, because the pistons may be hitting the valves.

21. Turn the crankshaft in a clockwise direction several times, then **slowly** set the No. 1 piston to TDC of the compression stroke.

22. Apply 22 lbs. (10 kg) of pressure (push it in!) to the center span of the timing belt between the right side camshaft sprocket and the tensioner pulley, then loosen the tensioner locknut.

23. Using a 0.0138 in. (0.35mm) thick feeler gauge (the actual width of the blade **must** be ½ in. or 13mm!), turn the crankshaft clockwise (**slowly!**). The timing belt should move approximately 2½ teeth. Tighten the tensioner locknut, turn the crankshaft slightly and remove the feeler gauge.

24. Slowly rotate the crankshaft clockwise several more times, then set the No. 1 piston to TDC of the compression stroke.

25. Position the 2 timing covers on the block, then tighten the mounting bolts to 24 ft. lbs. (35 Nm).

26. Press the crankshaft pulley onto the shaft, then tighten the bolt to 90 – 98 ft. lbs. (123 – 132 Nm).

27. Connect the radiator hose to the thermostat housing.

28. Reconnect the fresh air intake tube at the cylinder head cover.

29. Install the air conditioning compressor drive belt idler pulley and bracket.

30. Install the distributor protector (dust shield).

31. Install the spark plugs.

32. Install the power steering, air conditioning compressor and alternator drive belts.

33. Install the radiator.

34. Reconnect the water pump hose and fill the engine with coolant. Install the fan shroud and pulleys.

35. Install the engine undercover.

36. Start the engine and check for any leaks.

Timing Chain, Sprockets, Front Cover and Seal

REMOVAL & INSTALLATION

3.5L Engine

1. Release fuel pressure.
2. Remove battery.
3. Remove radiator.
4. Drain engine oil.
5. Remove drive belts and idler pulley with brackets.
6. Remove cooling fan with bracket.
7. Remove engine cover.
8. Remove air duct with air cleaner case, collector, blow-by hose, vacuum hoses, fuel hoses, water hoses, wires, harnesses, connectors and so on.
9. Remove the air compressor, and tie it down using rope or the like to keep it from interfering.
10. Remove the power steering oil pump and reservoir tank. Tie them down using rope or the like to keep them from interfering.
11. Remove alternator.
12. Remove the following.

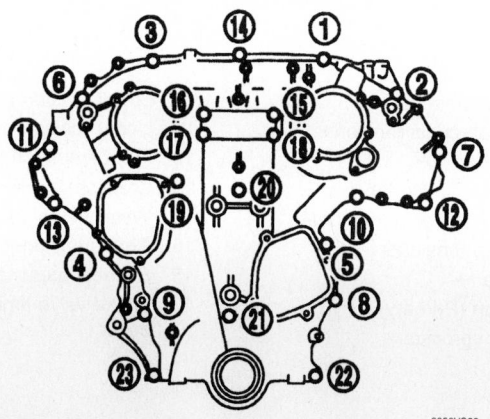

9359VG22

Rear timing case removal sequence—3.5L engine

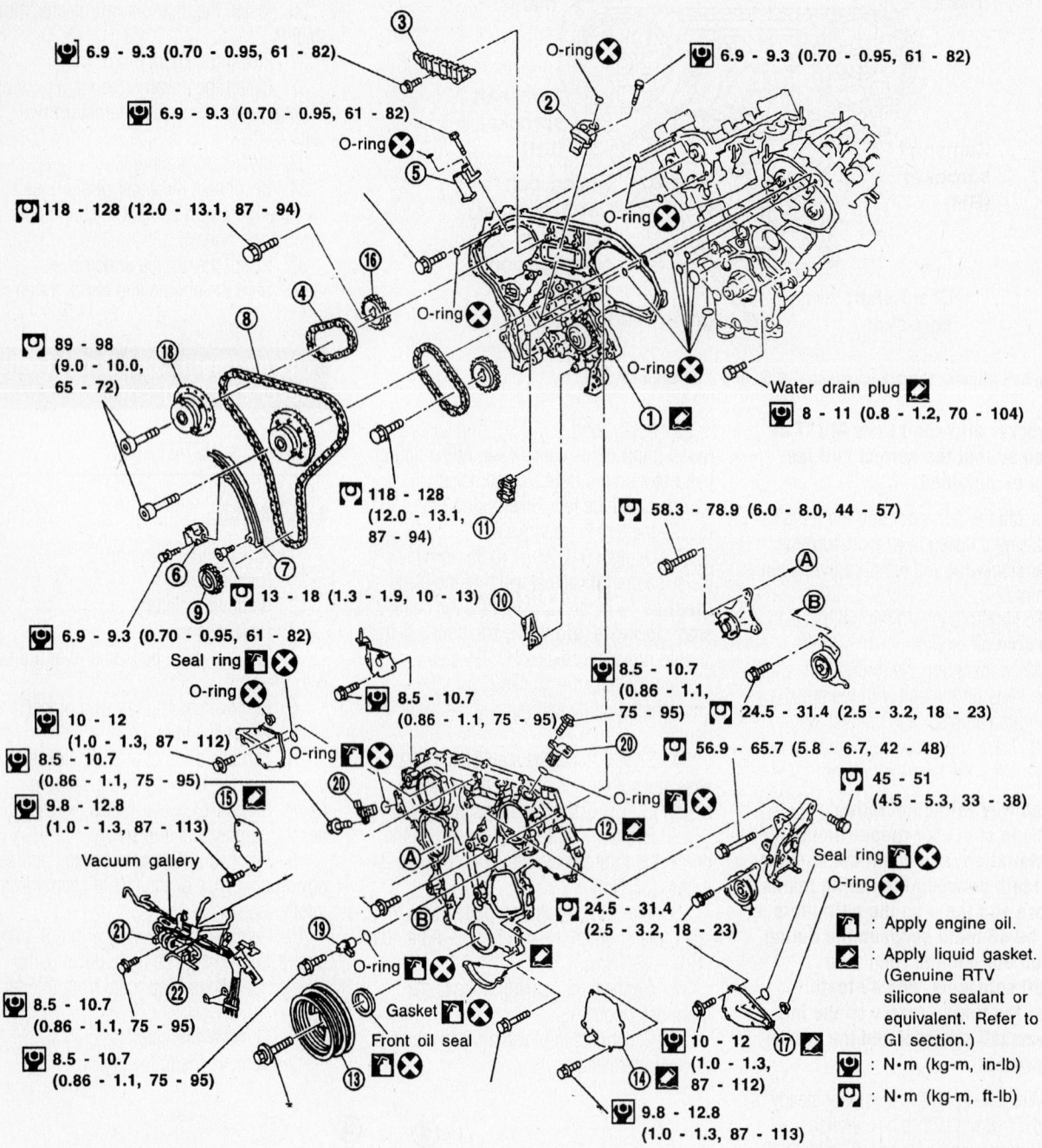

6.9 - 9.3 (0.70 - 0.95, 61 - 82)

O-ring

6.9 - 9.3 (0.70 - 0.95, 61 - 82)

6.9 - 9.3 (0.70 - 0.95, 61 - 82)

O-ring

118 - 128 (12.0 - 13.1, 87 - 94)

O-ring

O-ring

O-ring

Water drain plug

8 - 11 (0.8 - 1.2, 70 - 104)

89 - 98 (9.0 - 10.0, 65 - 72)

118 - 128 (12.0 - 13.1, 87 - 94)

58.3 - 78.9 (6.0 - 8.0, 44 - 57)

13 - 18 (1.3 - 1.9, 10 - 13)

6.9 - 9.3 (0.70 - 0.95, 61 - 82)

Seal ring

O-ring

8.5 - 10.7 (0.86 - 1.1, 75 - 95)

24.5 - 31.4 (2.5 - 3.2, 18 - 23)

10 - 12 (1.0 - 1.3, 87 - 112)

8.5 - 10.7 (0.86 - 1.1, 75 - 95)

8.5 - 10.7 (0.86 - 1.1, 75 - 95)

56.9 - 65.7 (5.8 - 6.7, 42 - 48)

9.8 - 12.8 (1.0 - 1.3, 87 - 113)

O-ring

45 - 51 (4.5 - 5.3, 33 - 38)

Vacuum gallery

O-ring

Seal ring

O-ring

: Apply engine oil.

: Apply liquid gasket. (Genuine RTV silicone sealant or equivalent. Refer to GI section.)

8.5 - 10.7 (0.86 - 1.1, 75 - 95)

O-ring

Gasket

10 - 12 (1.0 - 1.3, 87 - 112)

: N·m (kg-m, in-lb)

8.5 - 10.7 (0.86 - 1.1, 75 - 95)

Front oil seal

: N·m (kg-m, ft-lb)

24.5 - 31.4 (2.5 - 3.2, 18 - 23)

9.8 - 12.8 (1.0 - 1.3, 87 - 113)

1. Rear timing chain case
2. Left camshaft chain tensioner
3. Internal guide
4. Timing chain (Secondary)
5. Right camshaft chain tensioner
6. Timing chain tensioner
7. Slack guide
8. Timing chain (Primary)
9. Crankshaft sprocket

10. Lower tension guide
11. Upper tension guide
12. Front timing chain case
13. Crankshaft pulley
14. Water pump cover
15. Chain tensioner cover
16. Exhaust camshaft sprocket
17. Intake valve timing control valve cover

18. Intake camshaft sprocket
19. Camshaft position sensor (PHASE)
20. Intake valve timing control position sensor
21. Power valve actuator (A/T)
22. Swirl control valve control solenoid valve

9359VG30

Timing chain components—3.5L engine

Back side Primary
sprocket

Front Trigger teeth
section
(left bank only)

Secondary sprocket

9359VG23

Primary and secondary sprockets—3.5L engine

- Vacuum gallery
- Water bypass pipe
- Brackets

13. Remove camshaft position sensor (PHASE), intake valve timing control position sensors and crankshaft position sensor.

➡**Avoid impact such as dropping. Do not disassemble the components. Do not place them on areas where iron powder may adhere. Keep away from the objects susceptible to magnetism.**

14. Remove upper intake manifold collector in reverse order of installation.

15. Remove intake manifold collector support bolts.

16. Remove lower intake manifold collector in reverse order of installation.

17. Disconnect injector harness connectors.

18. Remove fuel tube assembly in reverse order of installation.

19. Remove ignition coils.

20. Remove RH and LH rocker covers from cylinder head.

21. Set No. 1 piston at TDC on the compression stroke by rotating crankshaft. Align pointer with TDC mark on crankshaft pulley. Check that intake and exhaust cam nose on No. 1 cylinder are installed as shown left. If not, turn the crankshaft one revolution (360°) and align as above.

22. Remove starter motor, and set ring gear stopper using the mounting bolt hole. Be careful not to damage the signal plate teeth.

23. Loosen the crankshaft pulley bolt.

24. Remove crankshaft pulley with a suitable puller.

25. Remove aluminum oil pan.

26. Temporarily install the suspension member bolts and engine mounting nuts.

27. Remove intake valve timing control valve covers. Loosen bolts in numerical order as shown in the figure. In the cover, the shaft is engaged with the center hole of the intake cam sprocket. Remove it straight out until the engagement comes off.

28. Remove front timing chain case bolts. Loosen bolts in numerical order as shown in the figure.

29. Remove front timing chain case. Do not scratch sealing surfaces.

30. Remove internal chain guide.

31. Remove upper tension guide.

32. Remove timing chain tensioner and slack guide. Remove timing chain tensioner. (Push piston and insert a suitable pin into pinhole.)

33. Attach a suitable stopper pin to RH and LH camshaft chain tensioners.

34. Remove intake and exhaust camshaft sprocket bolts. l Apply paint to timing chain and camshaft sprockets for alignment during installation. Secure the hexagonal head of the camshaft using a spanner
to loosen mounting bolts.

35. Remove primary and secondary timing chains along with the camshaft sprockets. Do not disassemble the intake camshaft

sprocket. Avoid damaging the signal mark protrusion area at the front of the left bank intake camshaft sprocket. Keep it away from magnetized objects.

36. Remove lower chain guide.

37. Remove crankshaft sprocket.

38. Use a scraper to remove all traces of liquid gasket from front timing chain case. Remove old liquid gasket from the bolt hole and thread.

39. Use a scraper to remove all traces of liquid gasket from intake valve timing control valve cover.

To install:

40. Position crankshaft so that No. 1 piston is set at TDC on compression stroke.

41. Install crankshaft sprocket on crankshaft. Make sure that mating marks on crankshaft sprocket face front of engine.

42. Install lower chain guide on dowel pin, with front mark on the guide facing upside.

43. Press and shrink the secondary chain tensioner sleeve, and fix it using stopper pins. Lubricate threads and seat surfaces of camshaft sprocket bolts with new engine oil.

44. Install secondary timing chain and sprocket to one of the banks (Right bank shown in the figure) as described below.

a. Align mating marks (golden links) on secondary timing chain with those (punched marks) on the intake and exhaust sprockets.

b. Align camshaft knock pins with the sprocket groove and hole. Because camshaft sprocket mounting bolts are tightened in step 7, perform manual tightening to the extent necessary to keep camshaft knock pin from dislocating. Matching marks of the intake sprocket are on the back side of the secondary sprockets. There are two types of the

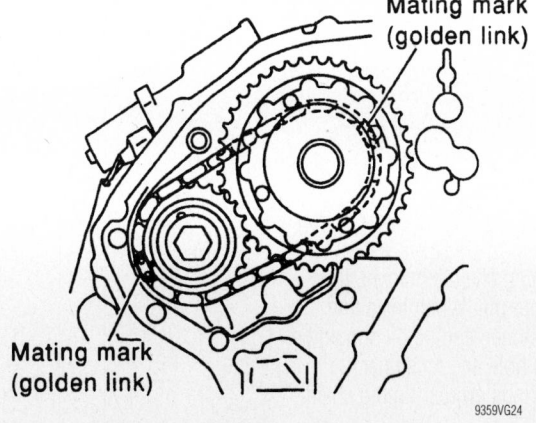

Mating mark
(golden link)

Mating mark
(golden link)

9359VG24

Secondary timing chain installed—3.5L engine

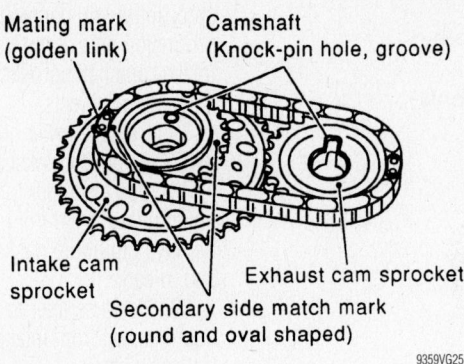

Intake sprocket mating marks—3.5L engine

marks; round and oval types, which should be used for right and left banks respectively.

- Right bank: Round
- Left bank: Oval

It may be difficult to visually check the dislocation of mating marks during and after installation. To make the matching easier, make a mating mark on the sprocket teeth in advance using paint.

45. Install secondary timing chain and sprocket to the other bank. Install primary timing chain at the same time. Installation of the secondary timing chain follows the procedure described in step 5.

46. Install primary timing chain so that mating mark (punched) on camshaft sprocket is aligned with that (dark blue link) on the timing chain, and mating mark (notched) on crankshaft sprocket is aligned with that on the timing chain, respectively.

47. When it is difficult to align mating marks of the primary timing chain with each sprocket, gradually turn the camshaft hexagonal head using a spanner so it is aligned with the mating mark.

48. During alignment, be careful to prevent dislocation of mating marks on the secondary timing chain.

49. After confirming the mating marks are aligned, tighten the camshaft sprocket mounting bolts. Secure the camshaft hexagonal head using a spanner to tighten mounting bolts.

50. Pull out the stopper pin from the secondary timing chain tensioner.

51. Install internal guide.

52. Install upper tension guide and slack guide.

53. Install timing chain tensioner, then remove the stopper pin. When installing the timing chain tensioner, engine oil should be applied to the oil hole and tensioner.

54. Install O-rings on rear timing chain case.

55. Apply liquid gasket to front timing case.

chain case. Before installation, wipe off the protruding sealant.

56. Install rear case pin into dowel pin hole on front timing chain case.

57. Tighten bolts to the specified torque in order shown in the figure. Leave the bolts unattended for 30 minutes or more after tightening.

58. Install intake valve timing control valve cover.

   a. Install O-rings at front timing chain case.

   b. Install seal ring at intake valve timing control valve covers.

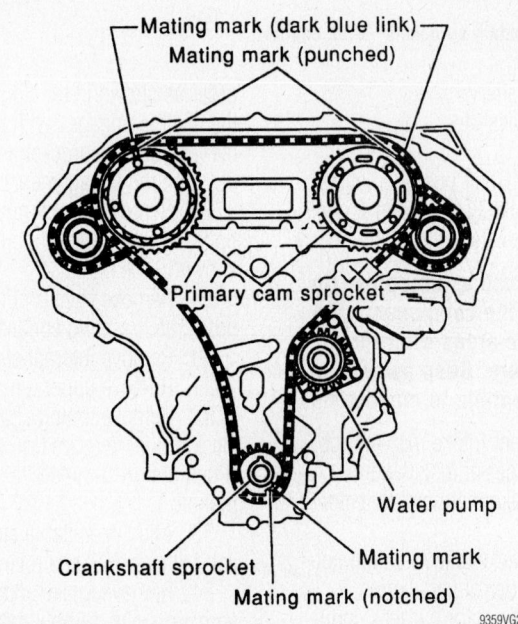

Primary timing chain installation—3.5L engine

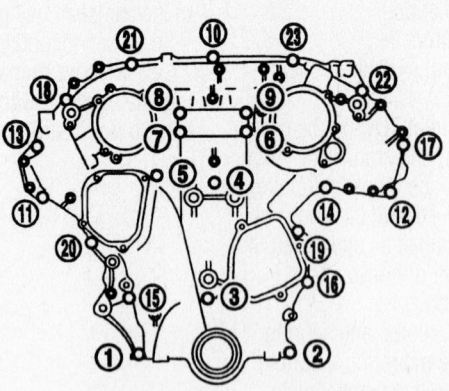

① - ② 8 mm dia. bolts
25.5 - 31.4 N·m
(2.6 - 3.2 kg-m, 18.8 - 23.1 ft-lb)
③ - ㉔ 6 mm dia. bolts
11.8 - 13.7 N·m
(1.2 - 1.4 kg-m, 8.7 - 10.1 ft-lb)

Rear timing case installation—3.5L engine

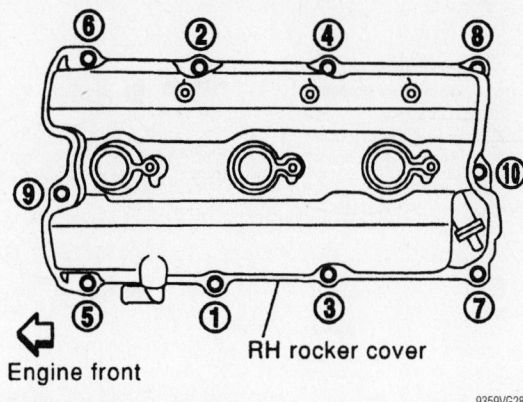

Right rocker cover installation—3.5L engine

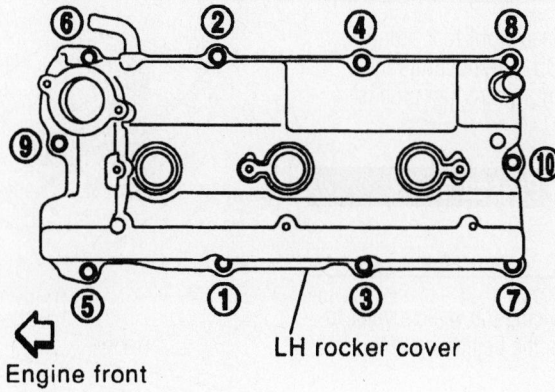

Left rocker cover installation—3.5L engine

Piston and Ring

POSITIONING

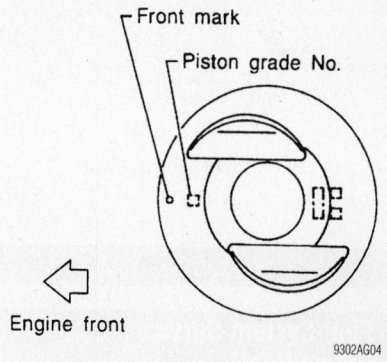

Piston ring positioning—3.3L engine

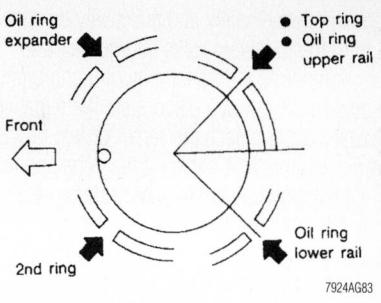

Piston ring end-gap spacing

c. Apply liquid gasket to intake valve timing control valve covers. Use RTV silicone sealant or equivalent. l Being careful not to move the seal ring from the installation groove, align the dowel pins on the chain case with the holes to install the intake valve timing control valve cover. Tighten in numerical order as shown in the figure.

59. Install RH and LH rocker covers. Rocker cover tightening procedure:

- Tighten in numerical order as shown in the figure.
- Tighten bolts 1 to 10 in that order to 6.9 to 8.8 N·m (0.7 to 0.9 kg-m, 61 to 78 in-lb).
- Then tighten bolts 1 to 10 as indicated in figure to 6.9 to 8.8 N·m (0.7 to 0.9 kg-m, 61 to 78 in-lb).

60. Hang engine using the right and left side engine slingers with a suitable hoist.

61. Set a suitable transmission jack under the suspension member.

62. Remove right and left side engine mounting nuts.

63. Remove right and left side suspension member bolts.

64. Install aluminum oil pan.

65. Set ring gear stopper using the mounting bolt hole. Be careful not to damage the signal plate teeth.

66. Install crankshaft pulley to crankshaft. Align pointer with TDC mark on crankshaft pulley.

67. Install crankshaft pulley bolt. Lubricate thread and seat surface of the bolt with new engine oil. Tighten to 39 to 49 N·m (4.0 to 5.0 kg-m, 29 to 36 ft-lb). Put a paint mark on the crankshaft pulley. Again tighten by turning 60° to 66°, about the angle from one hexagon bolt head corner to another.

68. Install camshaft position sensor (PHASE), crankshaft position sensors (REF)/(POS) and intake valve timing control position sensors.

69. Reinstall removed parts in the reverse order of removal. After starting engine, keep idling for three minutes. Then rev engine up to 3,000 rpm under no load to purge air from the high-pressure chamber of the chain tensioners. The engine may produce a rattling noise. This indicates that air still remains in the chamber and is not a matter of concern.

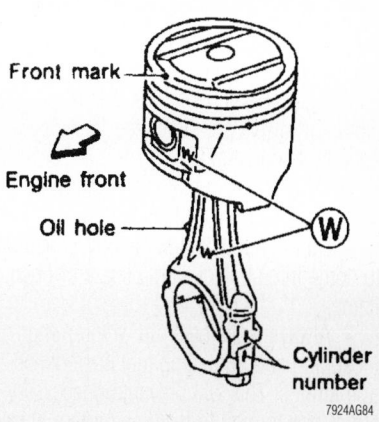

Piston and connecting rod positioning

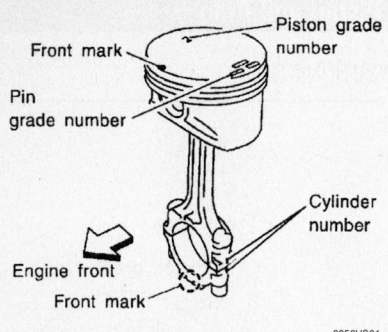

Piston and connecting rod positioning–3.5L

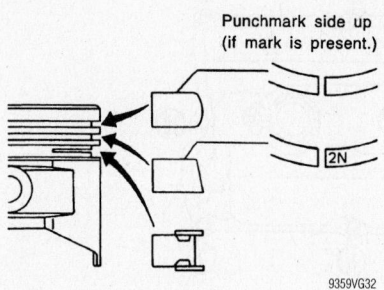

Piston ring positioning–3.5L

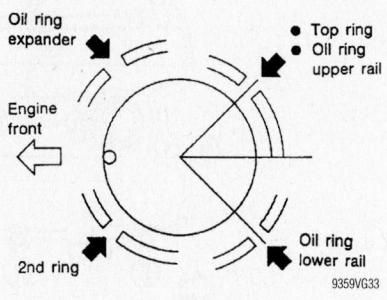

Piston ring positioning–3.5L

FUEL SYSTEM

Fuel System Service Precautions

Safety is the most important factor when performing not only fuel system maintenance but any type of maintenance. Failure to conduct maintenance and repairs in a safe manner may result in serious personal injury or death. Maintenance and testing of the vehicle's fuel system components can be accomplished safely and effectively by adhering to the following rules and guidelines.

• To avoid the possibility of fire and personal injury, always disconnect the negative battery cable unless the repair or test procedure requires that battery voltage be applied.

• Always relieve the fuel system pressure prior to disconnecting any fuel system component (injector, fuel rail, pressure regulator, etc.), fitting or fuel line connection. Exercise extreme caution whenever relieving fuel system pressure, to avoid exposing skin, face and eyes to fuel spray. Please be advised that fuel under pressure may penetrate the skin or any part of the body that it contacts.

• Always place a shop towel or cloth around the fitting or connection prior to loosening to absorb any excess fuel due to spillage. Ensure that all fuel spillage (should it occur) is quickly removed from engine surfaces. Ensure that all fuel soaked cloths or towels are deposited into a suitable waste container.

• Always keep a dry chemical (Class B) fire extinguisher near the work area.

• Do not allow fuel spray or fuel vapors to come into contact with a spark or open flame.

• Always use a back-up wrench when loosening and tightening fuel line connection fittings. This will prevent unnecessary stress and torsion to fuel line piping. Always follow the proper torque specifications.

• Always replace worn fuel fitting O-rings with new. Do not substitute fuel hose or equivalent, where fuel pipe is installed.

Fuel System Pressure

RELIEVING

1. Before servicing the vehicle, refer to the precautions in the beginning of this section.
2. Remove the fuel pump fuse from the panel.
3. Start the engine and allow it to run until it stalls. Crank the engine for a few seconds to relieve additional fuel pressure.
4. Disconnect the negative battery cable.
5. When repairs are complete, replace the fuel pump fuse and connect the negative battery cable.

Fuel Filter

REMOVAL & INSTALLATION

➡The fuel filter is located under the vehicle near the fuel tank.

1. Before servicing the vehicle, refer to the precautions in the beginning of this section.
2. Relieve the fuel system pressure.
3. Remove or disconnect the following:
   • Fuel filter shield, if equipped
   • Fuel lines
   • Fuel filter from the bracket

To install:
4. Install or connect the following:
   • Fuel filter to the bracket
   • Fuel lines
   • Fuel filter shield, if equipped
5. Start the engine and check for leaks.

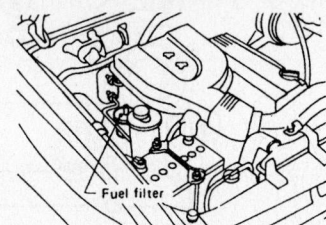

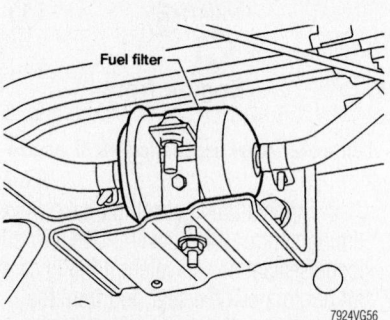

Typical fuel filter locations

Fuel Pump

REMOVAL & INSTALLATION

1. Before servicing the vehicle, refer to the precautions in the beginning of this section.
2. Relieve the fuel system pressure.
3. Remove or disconnect the following:
   • Negative battery cable
   • Access panel behind the rear seat
   • Fuel lines
   • Fuel pump and gauge harness connectors
   • Fuel gauge sender
   • Fuel pump

To install:
4. Install or connect the following:
   • Fuel pump
   • Fuel gauge sender. Tighten the screws to 17–23 inch lbs. (2.0–2.5 Nm).

- Fuel pump and gauge harness connectors
- Fuel lines
- Access panel
- Negative battery cable

5. Start the engine and check for leaks.

Fuel Injectors

REMOVAL & INSTALLATION

3.3L Engine

1. Before servicing the vehicle, refer to the precautions in the beginning of this section.
2. Drain the cooling system.
3. Relieve the fuel system pressure.
4. Remove or disconnect the following:
   - Negative battery cable
   - Air intake duct
   - Accelerator cable
   - Cruise control cable
   - Idle Air Control (IAC) valve connector
   - Throttle Position (TP) sensor and switch connectors
   - Ignition coil and power transistor connectors
   - Exhaust Gas Recirculation (EGR) Solenoid valve connector
   - EGR temperature sensor connector
   - Radiator hoses
   - Heater hoses
   - Positive Crankcase Ventilation (PCV) valve and hose
   - Evaporative Emissions (EVAP) canister vacuum and purge hoses
   - Brake booster vacuum hose
   - Fuel pressure regulator vacuum hose
   - EGR tube
   - Left bank injector connectors
   - Thermal transmitter
   - Upper intake manifold ground cable

- Breather pipe
- Upper intake manifold
- Fuel lines
- Right bank injector connectors
- Fuel supply manifold with the injectors attached
- Fuel injector caps
- Fuel injectors

To install:

→ **Use new insulators and O-ring seals for assembly.**

5. Install or connect the following:
   - Fuel injectors
   - Fuel injector caps. Tighten the screws to 26–34 inch lbs. (3–4 Nm).
   - Fuel supply manifold with the injectors attached. Tighten the bolts to 96–132 inch lbs. (11–15 Nm).
   - Right bank injector connectors
   - Fuel lines
   - Upper intake manifold
   - Breather pipe
   - Upper intake manifold ground cable
   - Thermal transmitter
   - Left bank injector connectors
   - EGR tube
   - Fuel pressure regulator vacuum hose
   - Brake booster vacuum hose
   - EVAP canister vacuum and purge hoses
   - PCV valve and hose
   - Heater hoses
   - Radiator hoses
   - EGR temperature sensor connector
   - EGR Solenoid valve connector
   - Ignition coil and power transistor connectors
   - TP sensor and switch connectors
   - IAC valve connector
   - Cruise control cable
   - Accelerator cable
   - Air intake duct
   - Negative battery cable

6. Fill the cooling system.
7. Start the engine and check for leaks.

3.5L

1. Release fuel pressure to zero.
2. Remove intake manifold collector.
3. Remove fuel tube assemblies in numerical sequence as shown in the figure at left.
4. Expand and remove clips securing fuel injectors.
5. Extract fuel injectors straight from fuel tubes.

→ **Be careful not to damage injector nozzles during removal. Do not bump or drop fuel injectors.**

6. Carefully install O-rings, including the one used with the pressure regulator. Lubricate O-rings with a smear of engine oil.

→ **Be careful not to damage O-rings with service tools, finger nails or clips. Do not expand or twist O-rings. Discard old clips; replace with new ones.**

7. Position clips in grooves on fuel injectors. Make sure that protrusions of fuel injectors are aligned with cutouts of clips after installation.
8. Align protrusions of fuel tubes with those of fuel injectors. Insert fuel injectors straight into fuel tubes.
9. After properly inserting fuel injectors, check to make sure that fuel tube protrusions are engaged with those of fuel injectors, and that flanges of fuel tubes are engaged with clips.
10. Tighten fuel tube assembly mounting nuts in numerical sequence (indicated in the figure at left) and in two stages. Tighten to:
    - Step 1: 84–96 inch lbs.
    - Step 2: 16–19 ft. lbs.
11. Install all parts removed in reverse order of removal.

DRIVE TRAIN

Automatic Transmission

REMOVAL & INSTALLATION

2 Wheel Drive

1. Before servicing the vehicle, refer to the precautions in the beginning of this section.
2. Remove or disconnect the following:
   - Negative battery cable
   - Crankshaft Position (CKP) sensor
   - Exhaust front pipes

- Exhaust rear pipes
- Transmission dipstick tube
- Transmission oil cooler lines
- Driveshaft
- Shift cable
- Transmission control harness connectors
- Vehicle Speed (VSS) sensor connector
- Starter motor
- Torque converter
- Transmission mount and crossmember. Support the transmission.

- Transmission flange bolts
- Transmission

→ **The transmission flange bolts vary in length. Note their positions for assembly.**

To install:
3. Install or connect the following:
   - Transmission. Tighten the large bolts to 29–36 ft. lbs. (39–49 Nm) and the small bolts to 22–29 ft. lbs. (29–39 Nm).
   - Transmission mount and cross-

member. Tighten the mount and crossmember fasteners to 30–38 ft. lbs. (41–52 Nm).
- Torque converter. Tighten the bolts to 33–43 ft. lbs. (44–59 Nm).
- Starter motor
- VSS sensor connector
- Transmission control harness connectors
- Shift cable
- Driveshaft
- Transmission oil cooler lines
- Transmission dipstick tube
- Exhaust rear pipes
- Exhaust front pipes
- CKP sensor
- Negative battery cable

4 Wheel Drive

1. Before servicing the vehicle, refer to the precautions in the beginning of this section.
2. Remove or disconnect the following:
   - Negative battery cable
   - Crankshaft Position (CKP) sensor
   - Exhaust front pipes
   - Exhaust rear pipes
   - Transmission dipstick tube
   - Transmission oil cooler lines
   - Front and rear driveshafts
   - Transfer case linkage

- Shift cable
- Transmission control harness connectors
- Vehicle Speed (VSS) sensor connector
- Starter motor
- Torque converter
- Transmission mount and crossmember. Support the transmission.
- Transmission flange bolts
- Transmission

➡The transmission flange bolts vary in length. Note their positions for assembly.

To install:
3. Install or connect the following:
   - Transmission. Tighten the large bolts to 29–36 ft. lbs. (39–49 Nm) and the small bolts to 22–29 ft. lbs. (29–39 Nm).
   - Transmission mount and crossmember. Tighten the mount and crossmember fasteners to 30–38 ft. lbs. (41–52 Nm).
   - Torque converter. Tighten the bolts to 33–43 ft. lbs. (44–59 Nm).
   - Starter motor
   - VSS sensor connector
   - Transmission control harness connectors
   - Shift cable

- Transfer case linkage
- Front and rear driveshafts
- Transmission oil cooler lines
- Transmission dipstick tube
- Exhaust rear pipes
- Exhaust front pipes
- CKP sensor
- Negative battery cable

Clutch

REMOVAL & INSTALLATION

1. Before servicing the vehicle, refer to the precautions in the beginning of this section.
2. Remove or disconnect the following:
   - Negative battery cable
   - Transmission
   - Pressure plate. Loosen the bolts evenly in ½ turn steps.
   - Clutch disc

To install:
3. Install or connect the following:
   - Clutch disc and pressure plate. Tighten the pressure plate bolts evenly in ½ turns to 16–22 ft. lbs. (22–29 Nm).
   - Transmission
   - Negative battery cable

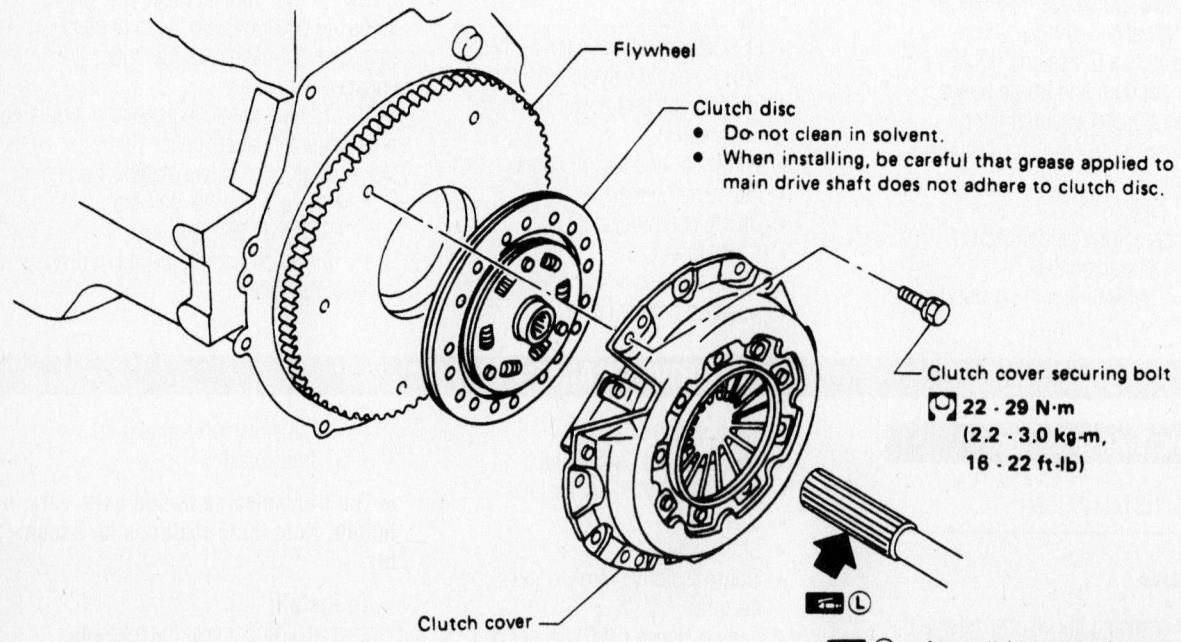

Flywheel

Clutch disc
- Do not clean in solvent.
- When installing, be careful that grease applied to main drive shaft does not adhere to clutch disc.

Clutch cover securing bolt
22 - 29 N·m
(2.2 - 3.0 kg-m,
16 - 22 ft-lb)

Clutch cover

(L): Apply lithium-based grease including molybdenum disulphide.

7924VG63

Exploded view of the pressure plate and clutch disc and related components—all models

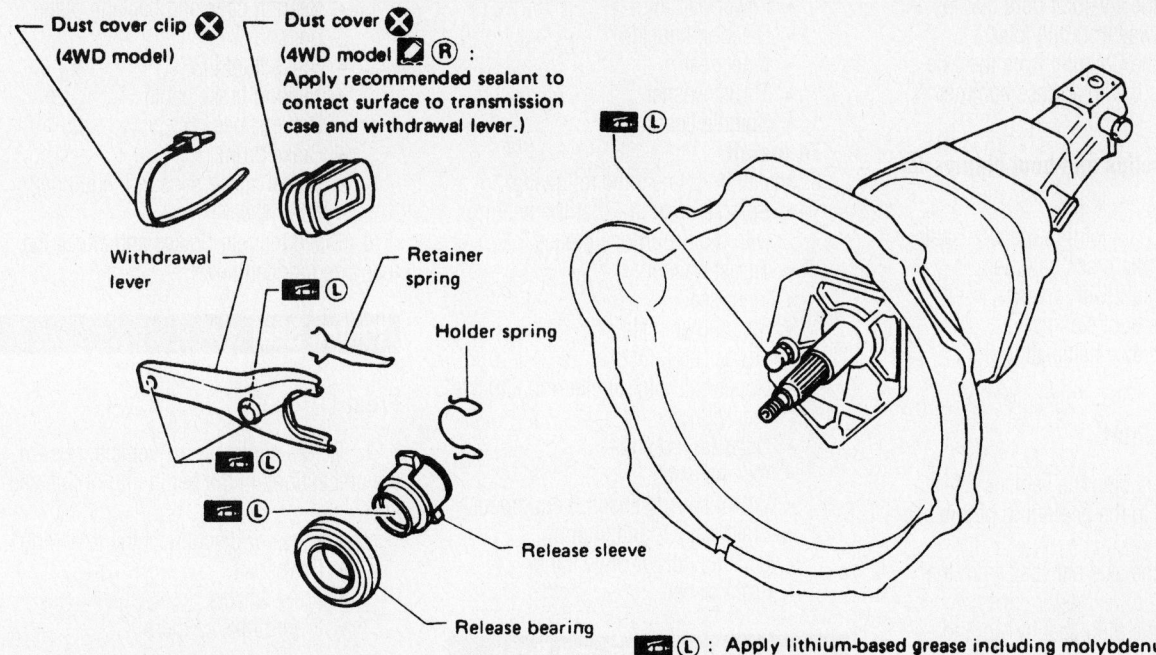

- Dust cover clip ⊗ (4WD model)
- Dust cover ⊗ (4WD model) ▨ Ⓡ : Apply recommended sealant to contact surface to transmission case and withdrawal lever.)
- Withdrawal lever
- Retainer spring
- Holder spring
- Release sleeve
- Release bearing
- ⊑ Ⓛ : Apply lithium-based grease including molybdenum disulphide

7924VG64

Clutch release mechanism exploded view—all models

Hydraulic Clutch System

BLEEDING

1. Before servicing the vehicle, refer to the precautions in the beginning of this section.
2. Fill the clutch master cylinder reservoir with fresh clean brake fluid.
3. Connect a clear plastic hose to the air bleeder.
4. Have an assistant pump the clutch pedal slowly several times and hold it depressed.
5. Open the slave cylinder bleeder screw and allow air to escape.
6. Close the bleeder screw before releasing the clutch pedal.
7. Repeat until all air is purged from the clutch hydraulic system.
8. Refill the reservoir to the full mark.

Transfer Case Assembly

REMOVAL & INSTALLATION

1. Before servicing the vehicle, refer to the precautions in the beginning of this section.
2. Remove or disconnect the following:
   - Negative battery cable
   - Front and rear driveshafts
   - Torsion bars and mounts
   - Rear torsion bar crossmember

- Exhaust front pipes
- Exhaust rear pipes
- Vehicle Speed (VSS) sensor connector
- Transfer case shift linkage
- Transfer case neutral switch connector
- 4 wheel drive switch connector
- Vent hose
- Transfer case flange bolts
- Transfer case

To install:

3. Install or connect the following:
   - Transfer case. Tighten the flange bolts to 23–30 ft. lbs. (31–41 Nm).
   - Vent hose
   - 4 wheel drive switch connector
   - Transfer case neutral switch connector
   - Transfer case shift linkage
   - VSS sensor connector
   - Exhaust rear pipes
   - Exhaust front pipes
   - Rear torsion bar crossmember
   - Torsion bars and mounts
   - Front and rear driveshafts
   - Negative battery cable

Halfshaft

REMOVAL & INSTALLATION

1. Before servicing the vehicle, refer to the precautions in the beginning of this section.

2. Remove or disconnect the following:
   - Front wheel
   - Wheel speed sensor, if equipped
   - Locking hub or drive flange
   - Snapring
   - Spindle washer
   - Thrust washer
   - Inner CV-joint bolts
   - Axle halfshaft. Separate the stub shaft from the spindle by tapping with a plastic hammer.

To install:

3. Install or connect the following:
   - Axle halfshaft. Guide the stub shaft into the spindle and tighten the inner CV-joint bolts to 25–33 ft. lbs. (34–44 Nm).
   - Thrust washer
   - Spindle washer
   - Snapring
   - Locking hub or drive flange
   - Wheel speed sensor, if equipped
   - Front wheel

CV-Joints

OVERHAUL

Outer CV-Joint

1. Before servicing the vehicle, refer to the precautions in the beginning of this section.
2. Remove the axle halfshaft from the vehicle.

3. Remove the CV-joint boot clamps and push the boot away from the joint.

4. Remove the CV-joint from the axle shaft by tapping it with a brass hammer.

To install:

➡ **Use new circlips and boot clamps for assembly.**

5. Install the CV-joint to the axle shaft by tapping it with a brass hammer.

6. Pack the joint with grease.

7. Install the boot clamps.

8. Install the axle halfshaft to the vehicle.

Inner Tri-Pot Joint

1. Before servicing the vehicle, refer to the precautions in the beginning of this section.

2. Remove the axle halfshaft from the vehicle.

3. Remove the plug seal by tapping around the joint housing flange with a brass hammer.

4. Remove or disconnect the following:
- CV-joint boot clamps
- Snapring
- Spider assembly
- CV-joint housing
- CV-joint boot

To install:

➡ **Use new snaprings and plug seals for assembly.**

5. Install or connect the following:
- CV-joint boot
- CV-joint housing
- Spider assembly
- Snapring. Pack the joint with grease.
- CV-joint boot clamps
- Plug seal

6. Install the axle halfshaft to the vehicle.

Spindle Bearings

REMOVAL, PACKING AND INSTALLATION

1. Before servicing the vehicle, refer to the precautions in the beginning of this section.

2. Remove or disconnect the following:
- Front wheel
- Locking hub or drive flange
- Brake caliper and support
- Wheel speed sensor, if equipped
- Axle halfshaft
- Outer tie rod ends
- Upper ball joint or steering knuckle bracket bolts

- Lower ball joint
- Steering knuckle
- Inner seal
- Thrust washer
- Spindle bearing

To install:

3. Install or connect the following:
- Spindle bearing. Coat the bearing with multi-purpose grease.
- Thrust washer
- Inner seal
- Steering knuckle
- Lower ball joint
- Upper ball joint or steering knuckle bracket bolts
- Outer tie rod ends
- Axle halfshaft
- Wheel speed sensor, if equipped
- Brake caliper and support
- Locking hub or drive flange
- Front wheel

Axle Shaft, Bearing and Seal

REMOVAL & INSTALLATION

1. Before servicing the vehicle, refer to the precautions in the beginning of this section.

2. Remove or disconnect the following:
- Rear wheel
- Wheel speed sensor, if equipped
- Brake drum
- Brake shoes
- Parking brake cable
- Brake fluid line
- Bearing cage and backing plate bolts
- Axle shaft assembly
- Axle seal
- Wheel speed sensor rotor, if equipped
- Lockwasher
- Bearing locknut
- Flat washer
- Wheel bearing
- Wheel bearing cage grease seal

To install:

➡ **Use new lockwashers, seals and bearings for assembly.**

3. Install or connect the following:
- Wheel bearing cage grease seal
- Wheel bearing
- Flat washer
- Bearing locknut
- Lockwasher
- Wheel speed sensor rotor, if equipped
- Axle seal
- Axle shaft assembly

- Bearing cage and backing plate bolts
- Brake fluid line
- Parking brake cable
- Brake shoes
- Brake drum
- Wheel speed sensor, if equipped
- Rear wheel

4. Bleed the rear brakes and check the rear axle lubricant level.

Pinion Seal

Front

1. Before servicing the vehicle, refer to the precautions in the beginning of this section.

2. Remove or disconnect the following:
- Driveshaft
- Front wheels
- Front brake calipers

➡ **The front brake calipers must be removed so that there is no additional drag when measuring pinion bearing preload.**

3. Use an inch lb. torque wrench and measure the amount of torque required to maintain pinion rotation through several revolutions.

4. Remove or disconnect the following:
- Pinion flange
- Oil seal

To install:

5. Install or connect the following:
- Pinion seal
- Pinion flange

6. Rotate the pinion flange occasionally while tightening the flange nut to make sure the pinion bearings seat correctly.

7. Take frequent bearing preload torque readings. Tighten the flange nut to achieve the preload torque readings originally recorded. Do not exceed 137–217 ft. lbs. (186–294 Nm) torque when tightening the pinion flange nut.

✳✳ CAUTION

If the bearing preload can not be achieved at the specified torque, remove the pinion bearing and install a new adjustment spacer.

8. Install or connect the following:
- Front brake calipers
- Front wheels
- Driveshaft. Tighten the fasteners to 29–33 ft. lbs. (39–44 Nm).

9. Fill the differential with gear lubricant and check for leaks.

Rear

2 WHEEL DRIVE

1. Before servicing the vehicle, refer to the precautions in the beginning of this section.

2. Remove or disconnect the following:
- Driveshaft
- Rear wheels
- Brake drums

➡**The rear brake drums must be removed so that there is no additional drag when measuring pinion bearing preload.**

3. Use an inch lb. torque wrench and measure the amount of torque required to maintain pinion rotation through several revolutions.

4. Remove or disconnect the following:
- Pinion flange
- Wheel speed sensor and rotor, if equipped
- Oil seal
- Pinion bearing
- Collapsible spacer

To install:

➡**Use a new collapsible spacer and wheel speed sensor rotor for assembly.**

5. Install or connect the following:
- Collapsible spacer
- Pinion bearing
- Pinion seal
- Pinion flange

6. Rotate the pinion flange occasionally while tightening the flange nut to make sure the pinion bearings seat correctly.

7. Take frequent bearing preload torque readings. Tighten the flange nut to achieve the preload torque readings originally recorded. Do not exceed 137–217 ft. lbs. (186–294 Nm) torque when tightening the pinion flange nut.

✳✳ CAUTION

Never loosen the pinion nut to reduce bearing preload. If it is necessary to reduce bearing preload, install a new collapsible spacer.

8. Install or connect the following:
- Brake drums
- Rear wheels
- Driveshaft. Tighten the fasteners to 58–65 ft. lbs. (78–88 Nm).

9. Fill the differential with gear lubricant and check for leaks.

4 WHEEL DRIVE

1. Before servicing the vehicle, refer to the precautions in the beginning of this section.

2. Remove or disconnect the following:
- Driveshaft
- Rear wheels
- Brake drums

➡**The rear brake drums must be removed so that there is no additional drag when measuring pinion bearing preload.**

3. Use an inch lb. torque wrench and measure the amount of torque required to maintain pinion rotation through several revolutions.

4. Remove or disconnect the following:
- Pinion flange
- Oil seal

To install:

5. Install or connect the following:
- Pinion seal
- Pinion flange

6. Rotate the pinion flange occasionally while tightening the flange nut to make sure the pinion bearings seat correctly.

7. Take frequent bearing preload torque readings. Tighten the flange nut to achieve the preload torque readings originally recorded. Do not exceed 137–217 ft. lbs. (186–294 Nm) torque when tightening the pinion flange nut.

✳✳ CAUTION

If the bearing preload can not be achieved at the specified torque, remove the pinion bearing and install a new adjustment spacer.

8. Install or connect the following:
- Brake drums
- Rear wheels
- Driveshaft. Tighten the fasteners to 58–65 ft. lbs. (78–88 Nm).

9. Fill the differential with gear lubricant and check for leaks.

STEERING AND SUSPENSION

Air Bag

✳✳ CAUTION

Some vehicles are equipped with an air bag system. The system must be disarmed before performing service on, or around, system components, the steering column, instrument panel components, wiring and sensors. Failure to follow the safety precautions and the disarming procedure could result in accidental air bag deployment, possible injury and unnecessary system repairs.

PRECAUTIONS

Several precautions must be observed when handling the inflator module to avoid accidental deployment and possible personal injury.

- Never carry the inflator module by the wires or connector on the underside of the module.
- When carrying a live inflator module, hold securely with both hands, and ensure that the bag and trim cover are pointed away.
- Place the inflator module on a bench or other surface with the bag and trim cover facing up.
- With the inflator module on the bench, never place anything on or close to the module which may be thrown in the event of an accidental deployment.

DISARMING

To disarm the **SRS** system turn the ignition switch to the **OFF** position. Then, disconnect both battery cables starting with the negative cable first and wait at least 3 minutes after the cables are disconnected.

To rearm the **SRS** system, turn the ignition switch to the **OFF** position. Connect both battery cables starting with the positive cable first.

Recirculating Ball Power Steering Gear

REMOVAL & INSTALLATION

1. Before servicing the vehicle, refer to the precautions in the beginning of this section.

2. Remove or disconnect the following:
- Pitman arm
- Steering column intermediate shaft
- Power steering hoses
- Steering gear

To install:

3. Install or connect the following:
- Steering gear. Tighten the bolts to 62–71 ft. lbs. (84–96 Nm).

- Power steering hoses. Tighten the banjo fittings to 29–38 ft. lbs. (39–51 Nm).
- Steering column intermediate shaft. Tighten the pinch bolt to 17–22 ft. lbs. (24–29 Nm).
- Pitman arm. Tighten the nut to 174–195 ft. lbs. (235–265 Nm).

4. Check the wheel alignment and adjust, as necessary.

Rack and Pinion Steering Gear

REMOVAL & INSTALLATION

1. Before servicing the vehicle, refer to the precautions in the beginning of this section.

2. Remove or disconnect the following:
- Front wheels
- Outer tie rod ends
- Steering shaft coupler
- Power steering hoses
- Steering gear

To install:

3. Install or connect the following:
- Steering gear. Tighten the bolts to 101 ft. lbs. (137 Nm).
- Power steering hoses. Tighten the fittings to 25 ft. lbs. (35 Nm).
- Steering shaft coupler. Tighten the bolt to 22 ft. lbs. (29 Nm).
- Outer tie rod ends. Tighten the nuts to 65 ft. lbs. (88 Nm).
- Front wheels

Strut

REMOVAL & INSTALLATION

Front

1. Before servicing the vehicle, refer to the precautions in the beginning of this section.
2. Remove or disconnect the following:
- Front wheel
- Stabilizer bar link
- Steering knuckle bracket bolts
- Upper strut mount nuts
- Strut

To install:

➡**Use new nuts and bolts for assembly.**

When installing rubber parts, final tightening must be carried out under unladen condition* with tires on ground. Fuel, radiator coolant and engine oil full.
Spare tire, jack, hand tools and mats in designated positions.

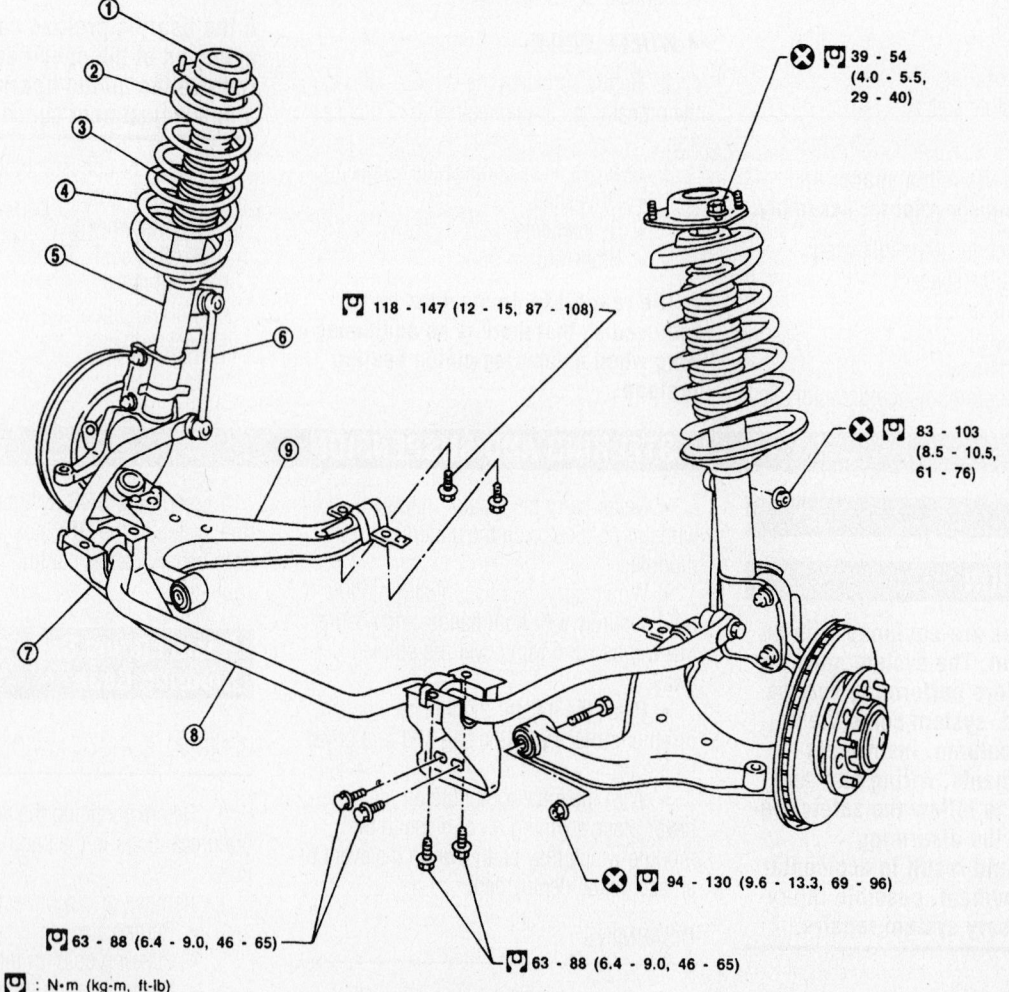

| N·m (kg-m, ft-lb)

① Strut mounting insulator
② Spring upper seat
③ Bound bumper
④ Coil spring
⑤ Strut assembly
⑥ Stabilizer connecting rod
⑦ Bracket
⑧ Stabilizer bar
⑨ Transverse link

Exploded view of the front suspension—2WD shown

7924VG66

3. Install or connect the following:
- Strut. Tighten the upper strut mount nuts to 29–40 ft. lbs. (39–54 Nm) and the knuckle bracket bolts to 111–122 ft. lbs. (151–165 Nm).
- Stabilizer bar link. Tighten the nut to 61–76 ft. lbs. (83–103 Nm).
- Front wheel

4. Check the wheel alignment and adjust, as necessary.

Shock Absorber

REMOVAL & INSTALLATION

Rear

1. Before servicing the vehicle, refer to the precautions in the beginning of this section.
2. Support the rear axle.
3. Remove or disconnect the following:
- Lower shock absorber bolt
- Upper shock absorber bolt
- Shock absorber

To install:

➡**Use new fasteners for assembly.**

4. Install the shock absorber and

tighten the bolts to 49–65 ft. lbs. (67–88 Nm).

Coil Spring

REMOVAL & INSTALLATION

Front

1. Before servicing the vehicle, refer to the precautions in the beginning of this section.
2. Remove the strut assembly.
3. Compress the coil spring and remove the piston rod nut.
4. Remove or disconnect the following:
- Upper strut mount
- Strut mount bracket
- Upper strut bearing
- Spring upper seat
- Coil spring

To install:

➡**Use new fasteners for assembly.**

5. Install or connect the following:
- Coil spring
- Spring upper seat
- Upper strut bearing
- Strut mount bracket
- Upper strut mount. Tighten the pis-

ton rod nut to 43–58 ft. lbs. (59–78 Nm).

6. Remove the spring compressor and install the strut assembly to the vehicle.
7. Check the wheel alignment and adjust, as necessary.

Rear

1. Before servicing the vehicle, refer to the precautions in the beginning of this section.
2. Support the vehicle at the frame.
3. Support the axle with a floor jack.
4. Remove or disconnect the following:
- Rear wheels
- Shock absorbers
- Stabilizer bar links
- Lateral control rod
- Coil springs

To install:

➡**Use new fasteners for assembly.**

5. Install or connect the following:
- Coil springs
- Lateral control rod. Tighten the nut to 80–94 ft. lbs. (108–127 Nm).
- Stabilizer bar links. Tighten the nuts to 30–35 ft. lbs. (41–47 Nm).
- Shock absorbers
- Rear wheels

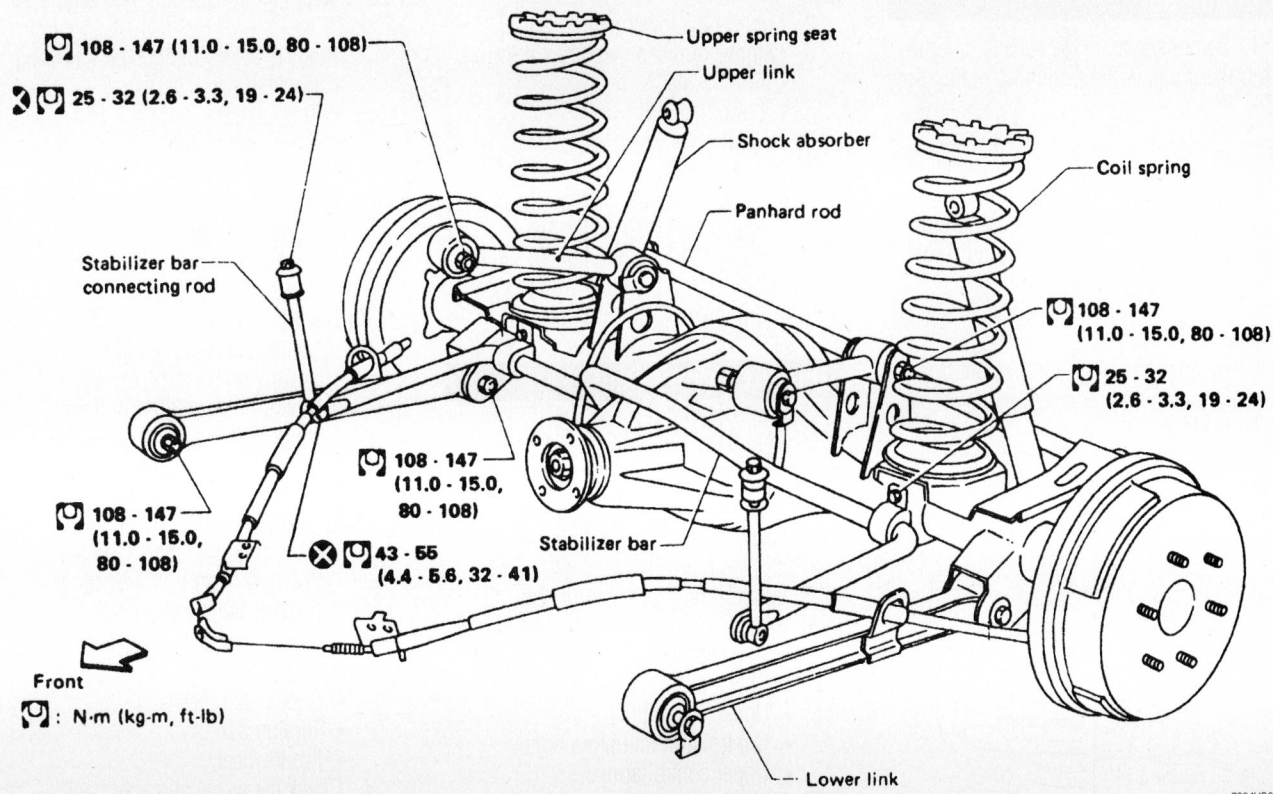

108 - 147 (11.0 - 15.0, 80 - 108)

25 - 32 (2.6 - 3.3, 19 - 24)

Upper spring seat

Upper link

Shock absorber

Coil spring

Panhard rod

Stabilizer bar connecting rod

108 - 147 (11.0 - 15.0, 80 - 108)

25 - 32 (2.6 - 3.3, 19 - 24)

108 - 147 (11.0 - 15.0, 80 - 108)

108 - 147 (11.0 - 15.0, 80 - 108)

43 - 55 (4.4 - 5.6, 32 - 41)

Stabilizer bar

Front

: N·m (kg-m, ft-lb)

Lower link

7924VG69

Rear suspension component identification

Leaf Springs

REMOVAL & INSTALLATION

1. Before servicing the vehicle, refer to the precautions in the beginning of this section.
2. Support the vehicle at the frame.
3. Support the axle with a floor jack.
4. Remove or disconnect the following:
   - Rear wheels
   - Shock absorbers
   - Axle U-bolts and spring pad
   - Spring shackle
   - Front mount bolt/pin
   - Leaf spring

To install:

➡**Use new fasteners for assembly.**

5. Install or connect the following:
   - Leaf spring. Tighten the front mount bolt to 86–108 ft. lbs. (117–147 Nm).
   - Spring shackle. Tighten the nuts to 58–72 ft. lbs. (78–98 Nm).
   - Axle U-bolts and spring pad. Tighten the nuts to 72–80 ft. lbs. (98–108 Nm).
   - Shock absorbers
   - Rear wheels

Torsion Bar

1. Before servicing the vehicle, refer to the precautions in the beginning of this section.
2. Move the dust cover, if equipped.
3. Matchmark the torsion bar to the control arm mount and the anchor arm.
4. Measure the adjustment bolt protrusion as shown and note the length (L) for assembly.
5. Loosen the adjustment bolt so that all tension is released.
6. Remove the torsion bar mount from the control arm and remove the torsion bar.

To install:

7. Align the matchmarks and install the torsion bar. Tighten the large mount nut to

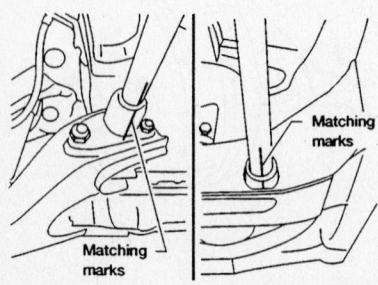

Torsion bar matchmarks

9308VG12

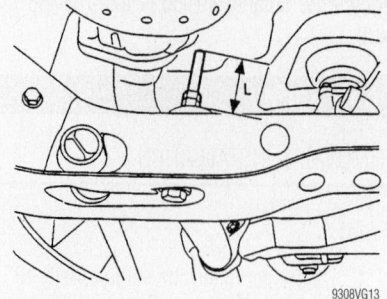

Adjustment bolt measurement (L)

9308VG13

66–87 ft. lbs. (89–118 Nm) and the small nut to 33–44 ft. lbs. (45–60 Nm).

8. Tighten the adjustment bolt to achieve the measurement (L) noted earlier. Tighten the locknut to 22–30 ft. lbs. (30–40 Nm).

9. If a new torsion bar is being installed, set length (L) as follows:
   - 2 wheel drive: 2.13 inches (54mm)
   - 4 wheel drive: 2.76 inches (70mm)

Lower Ball Joint

REMOVAL & INSTALLATION

1. Before servicing the vehicle, refer to the precautions in the beginning of this section.
2. Support the lower control arm.
3. Remove or disconnect the following:
   - Front wheel
   - Lower ball joint

To install:

4. Install or connect the following:
   - Lower ball joint. Tighten the control arm bolts to 76–94 ft. lbs. (103–127 Nm) and the stud nut to 87–123 ft. lbs. (118–167 Nm).
   - Front wheel

Upper Control Arm

REMOVAL & INSTALLATION

1. Before servicing the vehicle, refer to the precautions in the beginning of this section.
2. Support the lower control arm.
3. Remove or disconnect the following:
   - Front wheel
   - Shock absorber
   - Upper ball joint
   - Control arm mounting bolts
   - Upper control arm

To install:

4. Install or connect the following:
   - Upper control arm. Tighten the

mounting bolts to 72–87 ft. lbs. (98–118 Nm).
   - Upper ball joint. Tighten the nut to 58–108 ft. lbs. (78–147 Nm).
   - Shock absorber
   - Front wheel

5. Check the wheel alignment and adjust, as necessary.

CONTROL ARM BUSHING REPLACEMENT

1. Before servicing the vehicle, refer to the precautions in the beginning of this section.
2. Remove the control arm from the vehicle.
3. Remove the control arm bushing with a press.

To install:

4. Lubricate the control arm bushings with liquid soap.
5. Install the bushings with a press.
6. Install the control arm to the vehicle.
7. Check the wheel alignment and adjust, as necessary.

Lower Control Arm

REMOVAL & INSTALLATION

1. Before servicing the vehicle, refer to the precautions in the beginning of this section.
2. Remove or disconnect the following:
   - Front wheel
   - Torsion bar
   - Shock absorber
   - Stabilizer bar link
   - Axle halfshaft, if equipped
   - Lower ball joint
   - Control arm mounting bolts
   - Lower control arm

To install:

3. Install or connect the following:
   - Lower control arm. Tighten the mount bolts to 80–105 ft. lbs. (108–142 Nm) for 2001; 69–96 ft. lbs. (94–130 Nm) for 2002–03 models.
   - Lower ball joint. Tighten the nut to 87–141 ft. lbs. (118–191 Nm) for 2001; 87–123 ft. lbs. (118–167 Nm) for 2002–03 models.
   - Axle halfshaft, if equipped
   - Stabilizer bar link
   - Shock absorber
   - Torsion bar
   - Front wheel

4. Check the wheel alignment and adjust, as necessary.

CONTROL ARM BUSHING REPLACEMENT

1. Before servicing the vehicle, refer to the precautions in the beginning of this section.

2. Remove the control arm from the vehicle.

3. Remove the control arm bushing with a press.

To install:

4. Lubricate the control arm bushings with liquid soap.

5. Install the bushings with a press.

6. Install the control arm to the vehicle.

7. Check the wheel alignment and adjust, as necessary.

Wheel Bearings

ADJUSTMENT

2 Wheel Drive

➡**Use a new split pin for assembly.**

1. Before servicing the vehicle, refer to the precautions in the beginning of this section.

2. Remove or disconnect the following:
   - Dust cap
   - Split pin
   - Spindle nut cap

3. Tighten the spindle nut to 25–29 ft. lbs. (34–39 Nm).

4. Spin the hub several times to fully seat the bearings.

5. Retighten the spindle nut to 25–29 ft. lbs. (34–39 Nm).

6. Loosen the spindle nut 45–60 degrees and install the spindle nut cap and split pin.

7. Install the dust cap.

4 Wheel Drive

1. Before servicing the vehicle, refer to the precautions in the beginning of this section.

2. Remove or disconnect the following:
   - Locking hub or driveplate
   - Snapring
   - Spindle washer
   - Thrust washer
   - Lockwasher

3. Tighten the wheel bearing locknut to 58–72 ft. lbs. (78–98 Nm).

4. Loosen the locknut fully.

5. Tighten the wheel bearing locknut to 4–13 inch lbs. (0.5–1.5 Nm).

6. Spin the hub several times to fully seat the bearings.

7. Retighten the wheel bearing locknut to 4–13 inch lbs. (0.5–1.5 Nm).

8. Install or connect the following:
   - Lockwasher. Tighten the retaining screw to 10–16 inch lbs. (1–2 Nm).
   - Thrust washer
   - Spindle washer
   - Snapring
   - Locking hub or driveplate

REMOVAL & INSTALLATION

2 Wheel Drive

1. Before servicing the vehicle, refer to the precautions in the beginning of this section.

2. Remove or disconnect the following:
   - Front wheel
   - Brake caliper and support
   - Dust cap
   - Split pin
   - Spindle nut cap
   - Spindle nut
   - Bearing washer
   - Outer bearing
   - Hub and brake rotor assembly
   - Inner grease seal
   - Inner wheel bearing

To install:

3. Install or connect the following:
   - Inner wheel bearing

   - Inner grease seal
   - Hub and brake rotor assembly
   - Outer bearing
   - Bearing washer
   - Spindle nut. Adjust the wheel bearings.
   - Spindle nut cap
   - Split pin
   - Dust cap
   - Brake caliper and support
   - Front wheel

4 Wheel Drive

1. Before servicing the vehicle, refer to the precautions in the beginning of this section.

2. Remove or disconnect the following:
   - Front wheel
   - Brake caliper and support
   - Locking hub or driveplate
   - Snapring
   - Spindle washer
   - Thrust washer
   - Lockwasher
   - Wheel bearing locknut
   - Outer bearing
   - Hub and brake rotor assembly
   - Inner grease seal
   - Inner wheel bearing

To install:

3. Install or connect the following:
   - Inner wheel bearing
   - Inner wheel bearing
   - Inner grease seal
   - Hub and brake rotor assembly
   - Outer bearing
   - Wheel bearing locknut. Adjust the wheel bearings.
   - Lockwasher
   - Thrust washer
   - Spindle washer
   - Snapring
   - Locking hub or driveplate
   - Brake caliper and support
   - Front wheel

BRAKES

Brake Caliper

REMOVAL & INSTALLATION

1. Raise the vehicle and support safely.

2. Remove the appropriate tire and wheel assembly.

3. Remove the bolt attaching the brake

hose to the caliper. Plug the brake hose to prevent brake fluid loss.

4. Remove the caliper support mounting bolts and lift the caliper assembly from the knuckle.

To install

5. Position the caliper assembly onto the knuckle and install the bolts. Make sure the rotor fits between the brake pads. Torque the bolts to 24–31 ft. lbs.

6. Using new copper washers, connect the brake hose to the caliper. Torque the brake hose attaching bolt to 12–14 ft. lbs. (17–20 Nm).

7. Bleed the brake system.

8. Apply the brake pedal and inspect the system. Ensure proper operation and no leakage.

9. Install tire and wheel assembly. Lower the vehicle and road-test.

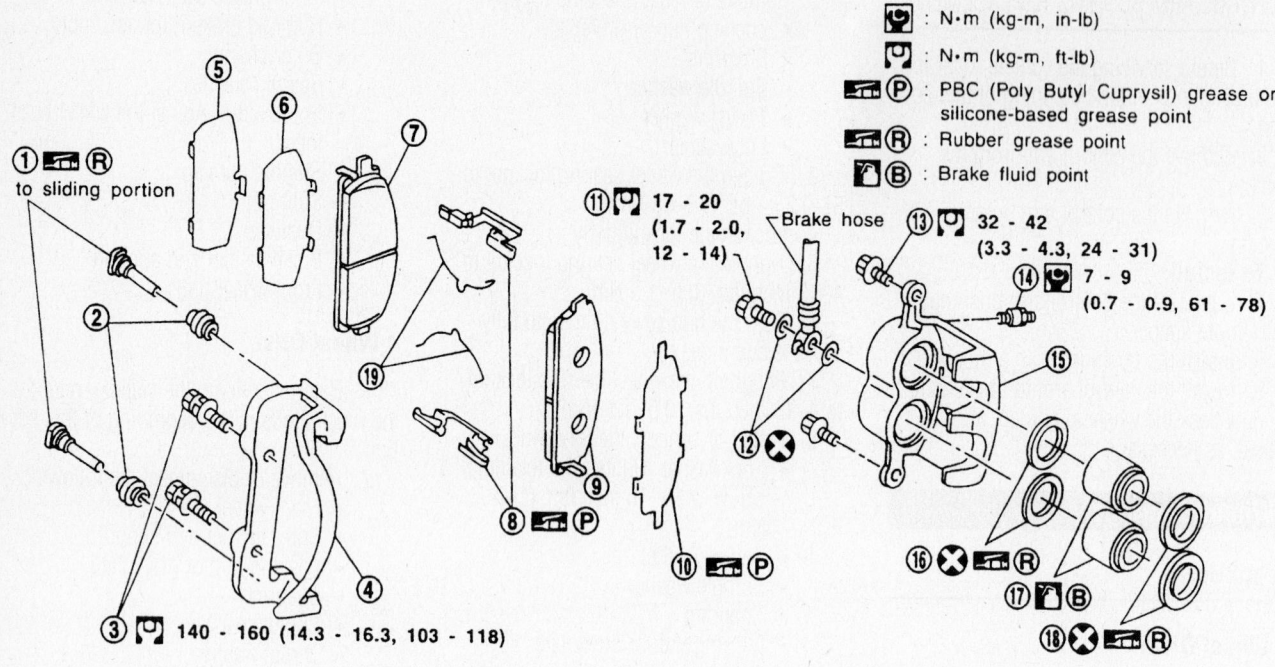

: N·m (kg-m, in-lb)

: N·m (kg-m, ft-lb)

(P) : PBC (Poly Butyl Cuprysil) grease or silicone-based grease point

(R) : Rubber grease point

(B) : Brake fluid point

① to sliding portion

⑪ 17 - 20 (1.7 - 2.0, 12 - 14)

Brake hose

⑬ 32 - 42 (3.3 - 4.3, 24 - 31)

⑭ 7 - 9 (0.7 - 0.9, 61 - 78)

③ 140 - 160 (14.3 - 16.3, 103 - 118)

① Main pin
② Pin boot
③ Torque member fixing bolt
④ Torque member
⑤ Shim cover
⑥ Inner shim
⑦ Inner pad

⑧ Pad retainer
⑨ Outer pad
⑩ Outer shim
⑪ Connecting bolt
⑫ Copper washer
⑬ Main pin bolt

⑭ Bleed valve
⑮ Cylinder body
⑯ Piston seal
⑰ Piston
⑱ Piston boot
⑲ Pad spring

93026G60

Exploded view of the dual piston caliper front brake components

Disc Brake Pads

REMOVAL & INSTALLATION

➡**Both the front and rear disc brake pads can be serviced using the same procedure.**

1. Using a syringe, siphon brake fluid from the reservoir, leaving reservoir approximately ½ full.
2. Raise and properly support the vehicle.
3. Remove the wheel assemblies.
4. Remove the lower pin bolt from the brake caliper.
5. Swivel the caliper up and away from the torque member. Tie the caliper to a suspension member so that it is out of the way.
6. Lift the 2 brake pads out of the torque member.
7. Remove the inner and outer shims. Remove the 2 pad retainers if they are not attached to the pads.

8. Check the pad thickness and replace the pads if they are less than 0.079 in. (2mm) thick.
To install:
9. Install the inner and outer shims into the torque member.
10. Install a pad retainer to the bottom of each pad.
11. Install the pads into the torque member.
12. Use a C-clamp or hammer handle and press the caliper piston(s) back into the housing.
13. Untie the caliper and swivel it back into position over the torque plate so that the dust boot is not pinched. Install the pin bolt and torque it to 24–31 ft. lbs.
14. Check the condition of the pin boot. Gently pull on it to expel any trapped air.
15. Install the wheel and lower the vehicle.
16. Pump the brakes until the pedal is firm and check the level of brake fluid. Road-test the vehicle.

Brake Drums

REMOVAL & INSTALLATION

1. Remove the hub cap and loosen the lug nuts.
2. Raise the rear of the vehicle and support it on jackstands.
3. Remove the lug nuts, tire and wheel.
4. Release the parking brake.
5. Pull the brake drum from the hub. If difficult to remove try the following:
   a. Strike the face of the drum with a plastic or rubber mallet. This will break free any rust that may develop between the drum and the hub.
   b. Install 2, M8x1.25mm bolts into the holes in the drum and gradually tighten them to pull the drum off the hub.
To install:
6. Install the brake drum to the hub.
7. Install the wheel.
8. Remove the jackstands and lower the vehicle.

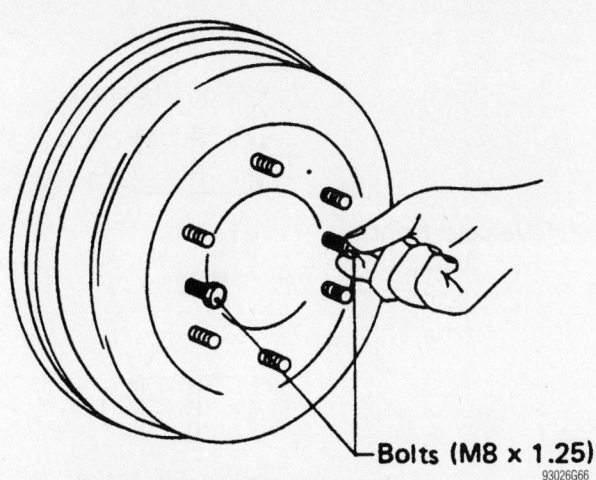

Install and tighten 2 bolts to remove a stubborn brake drum

9. Road-test the vehicle to ensure that the brakes are working properly.

Brake Shoes

REMOVAL & INSTALLATION

1. Release the parking brake.
2. Safely raise and support the vehicle.
3. Remove the rear wheel and drum.
4. Remove the hold-down pin retainers.
5. Remove the leading shoe and then the trailing shoe.
6. Remove the adjuster.
7. Disconnect the parking brake cable from the toggle lever on the rear shoe.

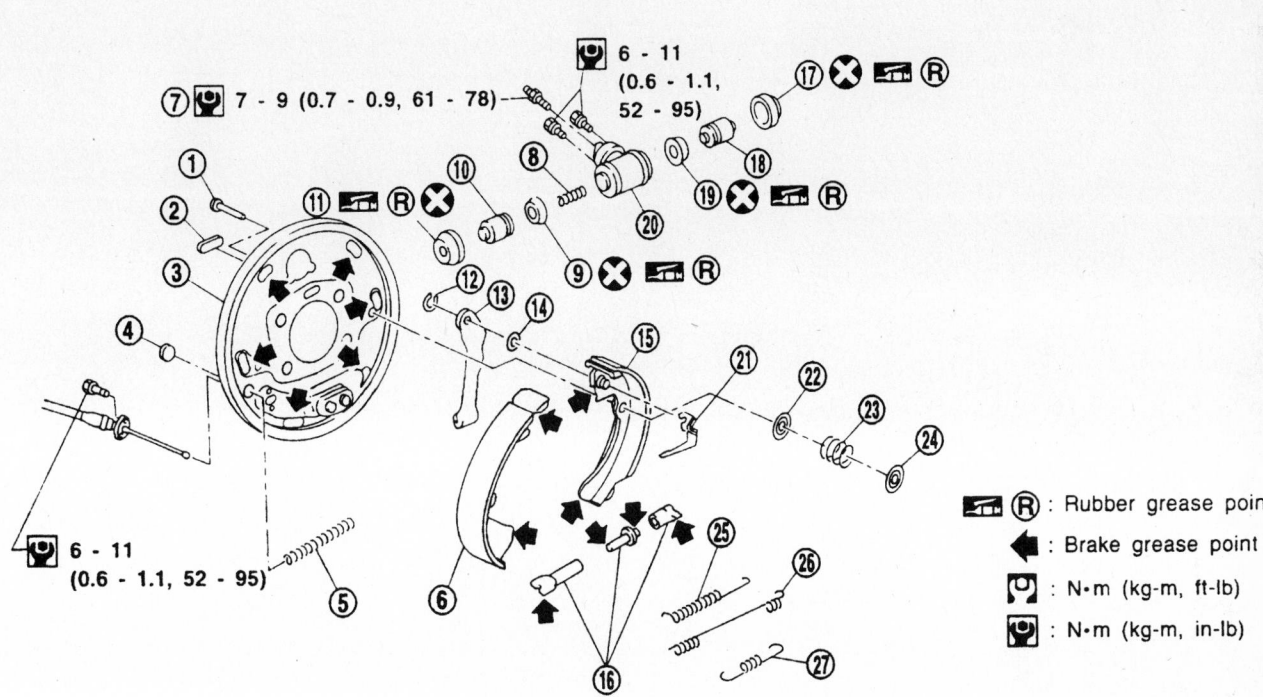

1. Shoe hold pin	10. Piston	19. Piston cup
2. Plug	11. Boot	20. Wheel cylinder
3. Back plate	12. Retainer ring	21. Adjuster lever
4. Check plug	13. Toggle lever	22. Spring seat
5. Spring	14. Wave washer	23. Shoe hold spring
6. Shoe (leading side)	15. Shoe (trailing side)	24. Retainer
7. Air bleeder	16. Adjuster	25. Adjuster spring
8. Spring	17. Boot	26. Return spring (upper)
9. Piston cup	18. Piston	27. Return spring (lower)

Drum brake assembly exploded view

To install:

8. Transfer the toggle lever to the new rear shoe.

9. Apply a small amount of brake grease to the tips of the shoes and the 6 pads on the backing plate that contact the brake shoe.

10. Shorten the adjuster by turning it.

11. Connect the parking brake cable to the toggle lever on the rear shoe.

12. Install the lower return spring to both shoes and install the shoes on the backing plate with the hold down pins and retainers.

13. Install the adjuster and the remaining springs. Pay attention to the direction of the adjuster assembly.

14. Inspect the complete assembly and install the brake drum.

15. Adjust the shoe to drum clearance.

16. Install the wheel assembly and lower the vehicle to the floor.

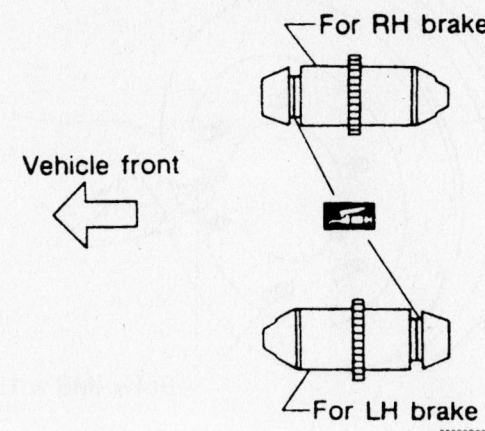

93026G63

Correct direction of brake shoe adjuster

INFINITI

QX56

SPECIFICATION AND MAINTENANCE CHARTS

ENGINE AND VEHICLE IDENTIFICATION

Engine								Model Year	
Code ①	Liters (cc)	Cu. In.	Cyl.	Fuel Sys.	Engine	Eng. Mfg.		Code ②	Year
VK56DE	5.6 (5552)	338.8	8	MFI	DOHC	Nissan		4	2004
								5	2005

MFI: Multi-port Fuel Injection

DOHC: Double Overhead Camshafts

① Located on the timing belt cover

② 10th digit of the Vehicle Identification Number (VIN)

67162-QX56-C01

GENERAL ENGINE SPECIFICATIONS

Year	Model	Engine Displacement Liters	Engine ID	Net Horsepower @ rpm	Net Torque @ rpm (ft. lbs.)	Bore x Stroke (in.)	Compression Ratio	Oil Pressure @ rpm
2004	QX56	5.6	VK56DE	315@4900	390@3600	3.86X3.62	9.8:1	43@2000

67162-QX56-C02

ENGINE TUNE-UP SPECIFICATIONS

Year	Engine Displacement Liters	Engine ID	Spark Plug Gap (in.)	Ignition Timing	Fuel Pump (psi) ①	Idle Speed ②	Valve Clearance (in.)	
							In.	Ex.
2004	5.6	VK56DE	0.043	15B	51	600-700	0.010-0.013	0.011-0.016

NOTE: The Vehicle Emission Control Information label often reflects specification changes made during production. The label figures must be used if they differ from those in this chart.

B: Before top dead center

① System pressure at idle with vacuum hose connected
Should increase to 43 psi when disconnected

② Automatic transmission in Neutral

67162-QX56-C03

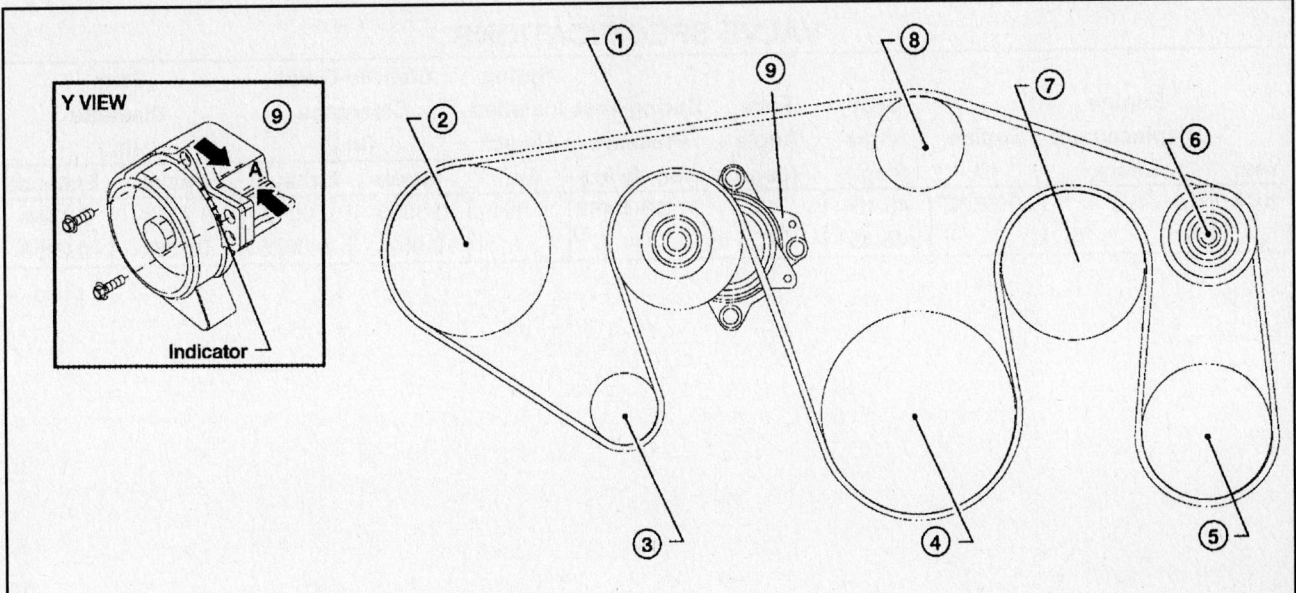

1. Drive Belt
4. Crankshaft Pulley
7. Cooling Fan Pulley

2. Power Steering Pump Pulley
5. A/C Compressor
8. Water Pump Pulley

3. Generator pulley
6. Idler Pulley
9. Drive Belt Tensioner

67162-QX56-G47

Accessory drive belt routing

CAPACITIES

Year	Model	Engine Displacement Liters	Engine ID	Engine Oil with Filter (qts.)	Transmission (pts.)	Transfer Case (pts.)	Drive Axle Front (pts.)	Rear (pts.)	Fuel Tank (gal.)	Cooling System (qts.)
2004	QX56	5.6	VK56DE	6.5	22.5	6.25	3.375	3.75	28.0	12.5

NOTE: All capacities are approximate. Add fluid gradually and check to be sure a proper fluid level is obtained.

67162-QX56-C04

VALVE SPECIFICATIONS

Year	Engine Displacement Liters	Engine ID	Seat Angle (deg.)	Face Angle (deg.)	Spring Test Pressure (lbs. @ in.)	Spring Installed Height (in.)	Stem-to-Guide Clearance (in.)		Stem Diameter (in.)	
							Intake	Exhaust	Intake	Exhaust
2004	5.6	VK56DE	45.15-45.45	45	37.0@1.457	1.991	0.0008-0.0021	0.0012-0.0025	0.2348-0.2354	0.2344-0.2350

67162-QX56-C05

CRANKSHAFT AND CONNECTING ROD SPECIFICATIONS

All measurements are given in inches.

Year	Engine Displ. Liters	Engine ID	Crankshaft				Connecting Rod		
			Main Brg. Journal Dia.	Main Brg. Oil Clearance	Shaft End-play	Thrust on No.	Journal Diameter	Oil Clearance	Side Clearance
2004	5.6	VK56DE	①	②	0.0118	3	③	0.0002-0.0007	0.0079-0.0157

NA - Not Available

① There are 17 different grades, ranging from 0 (2.483) to 78 (2.510)

② No. 1 and 5: 0.00004-0.0004

No. 2, 3 and 4: 0.0003-0.0007

③ Grade 0: 2.0441-2.0441

Grade 1: 2.0441-2.0442

Grade 2: 2.0442-2.0442

Grade 3: 2.0442-2.0443

Grade 4: 2.0443-2.0443

Grade 5: 2.0443-2.0443

Grade 6: 2.0443-2.0444

Grade 7: 2.0444-2.0444

Grade 8: 2.0444-2.0444

Grade 9: 2.0444-2.0445

Grade A: 2.0445-2.0445

Grade B: 2.0445-2.0446

Grade C: 2.0446-2.0446

67162-QX56-C06

PISTON AND RING SPECIFICATIONS

All measurements are given in inches.

Year	Engine Displacement Liters	Engine ID	Piston Clearance	Ring Gap			Ring Side Clearance		
				Top Comp.	Bottom Comp.	Oil Control	Top Comp.	Bottom Comp.	Oil Control
2004	5.6	VK56DE	0.0004-0.0012	0.0091-0.0110	0.0189-0.0217	0.0079-0.0197	0.0014-0.0033	0.0012-0.0028	0.0006-0.0073

67162-QX56-C07

TORQUE SPECIFICATIONS

All readings in ft. lbs.

Year	Engine Displacement Liters	Engine ID	Cylinder Head Bolts	Main Bearing Bolts	Rod Bearing Bolts	Crankshaft Damper Bolts	Flywheel Bolts	Manifold		Spark Plugs	Oil Pan Drain Plug
								Intake	Exhaust		
2004	5.6	VK56DE	①	②	③	④	65	6	25	18	25

NA: Information not available

① Step 1: 72 ft. lbs

   Step 2: Loosen all bolts completely

   Step 3: 33 ft. lbs.

   Step 4: +60 degrees

   Step 5: +60 degrees

② Step 1: Main Bolts to 29 ft. lbs.

   Step 2: Sub-bolts to 22 ft. lbs.

   Step 3: Main Bolts +40 degrees

   Step 4: Sub-Bolts +30 degrees

   Step 5: Side Bolts to 36 ft. lbs.

③ Step 1: 11 ft. lbs.

   Step 2: +90 degrees

④ Step 1: 65 ft. lbs.

   Step 2: +90 degrees

67162-QX56-C08

WHEEL ALIGNMENT

Year	Model	Caster		Camber		Toe-in (in.)
		Range (+/-Deg.)	Preferred Setting (Deg.)	Range (+/-Deg.)	Preferred Setting (Deg.)	
2004	QX56 ①	0.75	②	0.75	③	0.08+/-0.03

① Assumes P265/70R18 tire

② 4x2: +3.87

   4x4: +3.43

③ 4x2: -0.10

   4x4: +0.18

67162-QX56-C09

TIRE, WHEEL AND BALL JOINT SPECIFICATIONS

| Year | Model | OEM Tires | | Tire Pressures (psi) | | Wheel Size | Ball Joint Inspection | Lugnut Torque (ft. lbs.) |
		Standard	Optional	Front	Rear			
2004	QX56	P265/70R18	None	35	35	18	①	98

OEM: Original Equipment Manufacturer

PSI: Pounds Per Square Inch

① Axial play

  Upper: 0

67162-QX56-C10

BRAKE SPECIFICATIONS
All measurements in inches unless noted

| Year | Model | | Front Brake Disc | | | Minimum Lining Thickness | Brake Caliper | |
			Original Thickness	Minimum Thickness	Maximum Runout		Bracket Bolts (ft. lbs.)	Mounting Bolts (ft. lbs.)
2004	QX56	F	1.024	0.965	0.0016	0.039	155	32
		R	0.551	0.472	0.002	0.039	—	24

67162-QX56-C11

SCHEDULED MAINTENANCE INTERVALS
Infiniti—QX56

TO BE SERVICED	TYPE OF SERVICE	7.5	15	22.5	30	37.5	45	52.5	60
Engine oil & filter	R	✓	✓	✓	✓	✓	✓	✓	✓
Brake lines & cables	S/I		✓		✓		✓		✓
Brake pads and rotors	I	✓	✓	✓	✓	✓	✓	✓	✓
Driveshaft boots & propeller shaft (4x4)	L/I		✓		✓		✓		✓
Transmission, transfer & differential gear oil	I		✓		✓		✓		✓
Air cleaner filter	R					✓			✓
Engine coolant ①	R								✓
Spark plugs (Platinum)	R	Replace every 105,000 miles							
Drive belt(s) ②	S/I								✓
Cabin air filter	R		✓		✓		✓		✓
Exhaust system	I				✓				✓
Fuel lines	S/I				✓				✓
Fuel Filter ③									
Steering gear (box) & linkage, axle & suspension parts	I				✓				✓
Vapor lines	S/I				✓				✓

R: Replace   S/I: Service or Inspect   L: Lubricate

① Coolant: After 60,000 miles, inspect every 30,000 miles.

② Drive Belts: After 60,000 miles, inspect every 15,000 miles. Replace belts if damaged.

③ Fuel Filter: Maintenance free item.

FREQUENT OPERATION MAINTENANCE (SEVERE SERVICE)

If a vehicle is operated under any of the following conditions it is considered severe service:

- Extremely dusty areas.

- Rough, muddy, or salt spread roads.

- 50% or more of the vehicle constant operation is in 32°C (90°F) or higher temperatures, or temperatures below 0°C (32°F).

- Prolonged idling (vehicle operation in stop and go traffic).

- Frequent short running periods (engine does not warm to normal operating temperatures).

- Police, taxi, delivery usage or trailer towing usage.

Oil & oil filter: replace every 3750 miles.

Brake pads, discs, drums & linings: service or inspect every 7500 miles.

Driveshaft boots & propeller shaft: service or inspect every 7500 miles.

Exhaust system: service or inspect every 7500 miles.

Steering gear (box) & linkage, (steering damper-4x4), axle & suspension parts: service or inspect every 7500 miles.

Steering linkage ball joints & front suspension ball joints: service or inspect every 7500 miles.

67162-QX56-C12

Before servicing any vehicle, please be sure to read all of the following precautions, which deal with personal safety, prevention of component damage, and important points to take into consideration when servicing a motor vehicle:

• Never open, service or drain the radiator or cooling system when the engine is hot; serious burns can occur from the steam and hot coolant.

• Observe all applicable safety precautions when working around fuel. Whenever servicing the fuel system, always work in a well-ventilated area. Do not allow fuel spray or vapors to come in contact with a spark, open flame, or excessive heat (a hot drop light, for example). Keep a dry chemical fire extinguisher near the work area. Always keep fuel in a container specifically designed for fuel storage; also, always properly seal fuel containers to avoid the possibility of fire or explosion. Refer to the additional fuel system precautions later in this section.

• Fuel injection systems often remain pressurized, even after the engine has been turned **OFF**. The fuel system pressure must be relieved before disconnecting any fuel lines. Failure to do so may result in fire and/or personal injury.

• Brake fluid often contains polyglycol ethers and polyglycols. Avoid contact with the eyes and wash your hands thoroughly after handling brake fluid. If you do get brake fluid in your eyes, flush your eyes with clean, running water for 15 minutes. If eye irritation persists, or if you have taken brake fluid internally, IMMEDIATELY seek medical assistance.

• The EPA warns that prolonged contact with used engine oil may cause a number of skin disorders, including cancer. You should make every effort to minimize your exposure to used engine oil. Protective gloves should be worn when changing oil. Wash your hands and any other exposed skin areas as soon as possible after exposure to used engine oil. Soap and water, or waterless hand cleaner should be used.

• All new vehicles are now equipped with an air bag system, often referred to as a Supplemental Restraint System (SRS) or Supplemental Inflatable Restraint (SIR) system. The system must be disabled before performing service on or around system components, steering column, instrument panel components, wiring and sensors. Failure to follow safety and disabling procedures could result in accidental air bag deployment, possible personal injury, and unnecessary system repairs.

• Always wear safety goggles when working with, or around, the air bag system. When carrying a non-deployed air bag, be sure the bag and trim cover are pointed away from your body. When placing a non-deployed air bag on a work surface, always face the bag and trim cover upward, away from the surface. This will reduce the motion of the module if it is accidentally deployed. Refer to the additional air bag system precautions later in this section.

• Clean, high quality brake fluid from a sealed container is essential to the safe and proper operation of the brake system. You should always buy the correct type of brake fluid for your vehicle. If the brake fluid becomes contaminated, completely flush the system with new fluid. Never reuse any brake fluid. Any brake fluid that is removed from the system should be discarded. Also, do not allow any brake fluid to come in contact with a painted surface; it will damage the paint.

• Never operate the engine without the proper amount and type of engine oil; doing so will result in severe engine damage.

• Timing belt maintenance is extremely important. Many models utilize an interference-type, non-freewheeling engine. If the timing belt breaks, the valves in the cylinder head may strike the pistons, causing potentially serious (also time-consuming and expensive) engine damage. Refer to the maintenance interval charts in the front of this manual for the recommended replacement interval for the timing belt, and to the timing belt section for belt replacement and inspection.

• Disconnecting the negative battery cable on some vehicles may interfere with the functions of the on-board computer system(s) and may require the computer to undergo a relearning process once the negative battery cable is reconnected.

• When servicing drum brakes, only disassemble and assemble one side at a time, leaving the remaining side intact for reference.

Distributor

REMOVAL AND INSTALLATION

These models use a Distributorless Ignition System (DIS) controlled by the Powertrain Control Module (PCM).

Alternator

REMOVAL & INSTALLATION

1. Before servicing the vehicle, refer to the precautions in the beginning of this section.

2. Remove or disconnect the following:
• Negative battery cable
• Fan shroud
• Drive belt

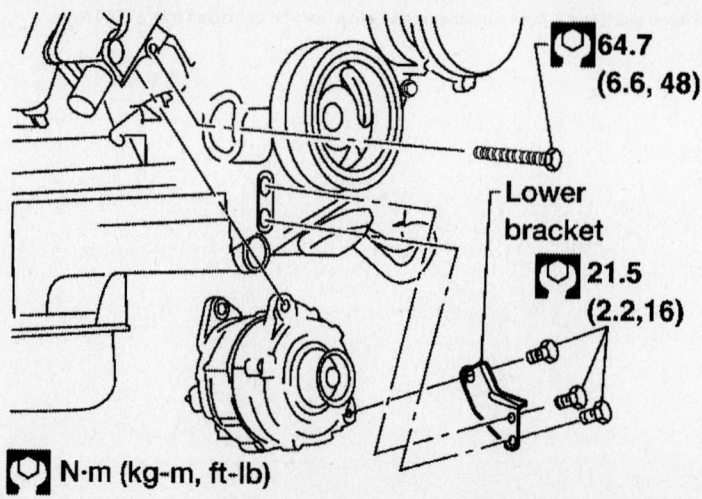

**64.7
(6.6, 48)**

Lower
bracket

**21.5
(2.2,16)**

N·m (kg-m, ft-lb)

67162-QX56-G41

Alternator mounting

- Lower alternator bracket
- Alternator upper bolt
- Alternator harness connectors
- Alternator

To install:

3. Install or connect the following:
- Alternator
- Alternator harness connectors
- Upper bolt, tighten to 48 ft. lbs. (65 Nm)
- Lower bracket, tighten to 16 ft. lbs (22 Nm)
- Drive belt
- Fan shroud
- Negative battery cable

Ignition Timing

ADJUSTMENT

Ignition timing is controlled by the ECM and manual adjustment is not possible.

Engine Assembly

REMOVAL & INSTALLATION

1. Before servicing the vehicle, refer to the precautions in the beginning of this section.
2. Drain the cooling system.
3. Partially drain the automatic transmission fluid.
4. Relieve the fuel system pressure.
5. Remove or disconnect the following:
- Hood
- Cowl extension
- Engine cover
- Air intake assembly
- Vacuum hose between vehicle and engine
- Radiator hoses
- Radiator
- Drive belts
- Engine fan

- Wiring harness
- ECM
- Power steering reservoir tank and oil pump
- A/C compressor
- Brake booster vacuum line
- EVAP line
- Fuel hose
- Heater hoses
- Automatic transmission dipstick tube assembly
- Automatic transmission

6. Install engine slings onto the left and right cylinder heads and tighten to 33 ft. lbs. (45 Nm).
7. Attach an engine hoist to slings and lift engine out of the vehicle

To install:

8. Lower engine into the vehicle
9. Install or connect the following:
- Automatic transmission
- Automatic transmission dipstick tube assembly

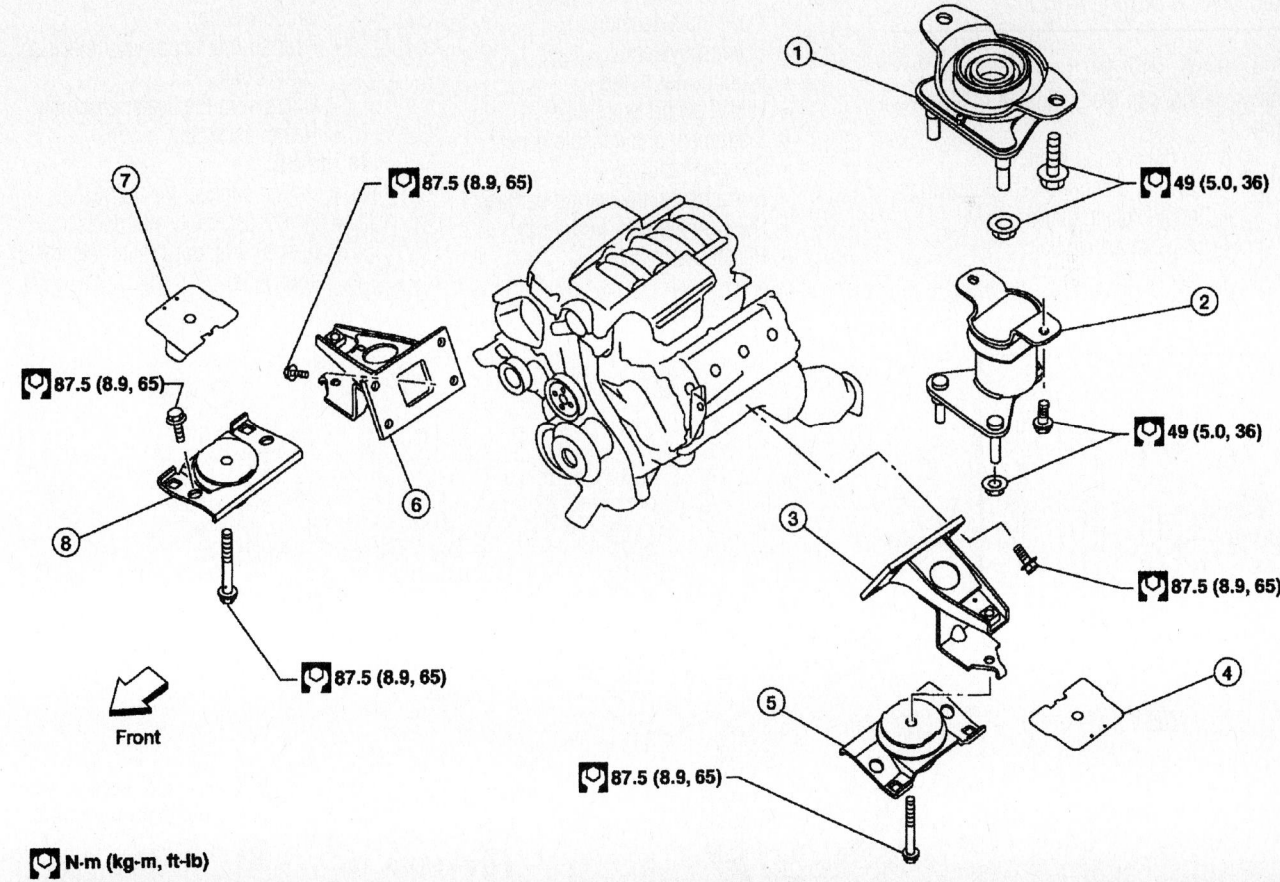

N·m (kg-m, ft-lb)

1. Rear engine mounting insulator 4x4
2. Rear engine mounting insulator 4x2
3. LH engine mounting bracket
4. LH Heat shield plate
5. LH engine mounting insulator
6. RH engine mounting bracket
7. RH Heat shield plate
8. RH engine mounting insulator

Engine mounts

67162-QX56-G40

- Heater hoses
- Fuel hose
- EVAP line
- Brake booster vacuum line
- A/C compressor
- Power steering reservoir tank and oil pump
- ECM
- Wiring harness
- Engine fan
- Drive belts
- Radiator and radiator hoses
- Vacuum hose between vehicle and engine
- Air intake assembly
- Engine cover
- Cowl extension
- Hood

10. Refill the automatic transmission fluid.
11. Refill the cooling system.
12. Start the engine and check for leaks.

Water Pump

REMOVAL & INSTALLATION

1. Before servicing the vehicle, refer to the precautions in the beginning of this section.
2. Drain the cooling system.
3. Remove or disconnect the following:
   - Engine splash guard
   - Accessory drive belt

➡**Leave tensioner pulley in its fixed position.**

- Water pump pulley
- Water pump

To install:
4. Install or connect the following:
   - Water pump with a new gasket. Tighten bolts to 18 ft. lbs. (25 Nm).
   - Water pump pulley and tighten bolts to 87 in. lbs. (10 Nm).
   - Accessory drive belt
   - Engine splash guard
5. Refill the cooling system.
6. Start the engine and check for leaks.

Heater Core

REMOVAL & INSTALLATION

Front Heater Assembly

1. Discharge the A/C system.
2. Drain the cooling system.
3. Remove or disconnect the following:

- Cowl top extension
- Exhaust system
- Front heater hoses
- High/Low pressure pipes
- Instrument and console panels
- Steering column
- Instrument panel wiring harness
- Steering member from body
- Heater assembly

To install:
4. Install or connect the following:
   - Heater assembly
   - Steering member
   - Instrument panel wiring harness
   - Steering column
   - Console panel
   - Instrument panel
   - High/Low pressure pipes
   - Heater hoses
   - Exhaust system
   - Cowl top extension
5. Refill the cooling system.
6. Recharge the A/C system.

Rear Heater Assembly

1. Discharge the A/C system.
2. Drain the cooling system.
3. Remove or disconnect the following:
   - Rear heater hoses
   - Rear A/C pipes
   - Rear right-hand interior trim panel
   - Rear blower motor electrical connection
   - Rear blower motor resistor electrical connection
   - Rear air mix door motor electrical connection
   - Ducts from the heater assembly
   - Heater assembly

To install:
4. Install or connect the following:
   - Heater assembly and ducts
   - Rear air mix door motor electrical connection

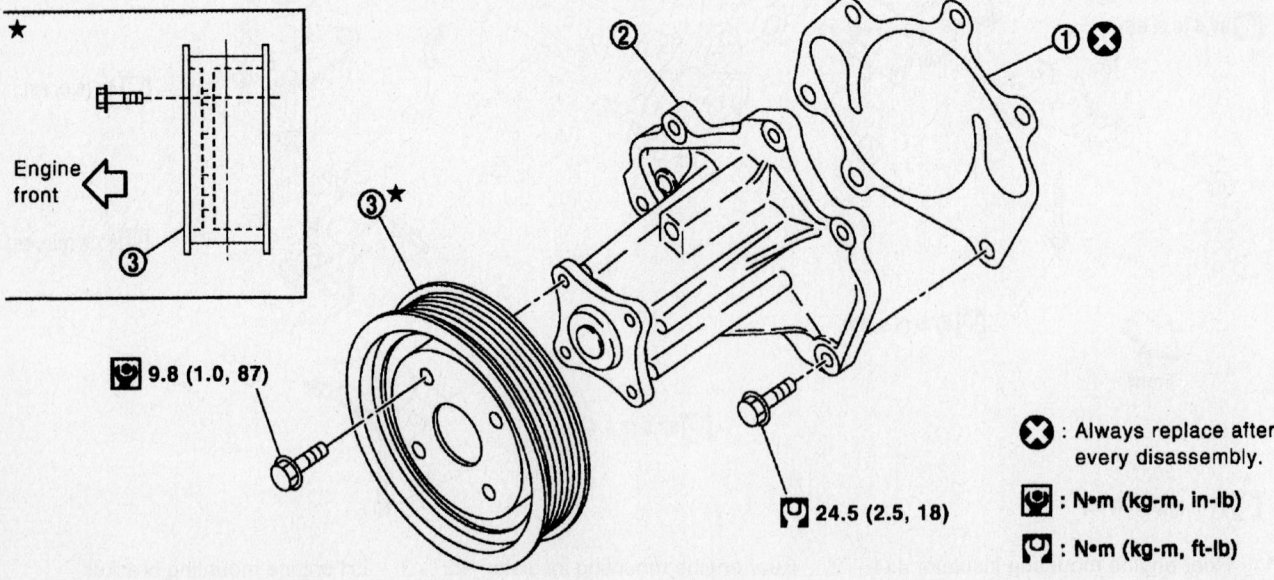

★

Engine front

⊡ 9.8 (1.0, 87)

⊡ 24.5 (2.5, 18)

❌ : Always replace after every disassembly.

⊡ : N•m (kg-m, in-lb)

⊡ : N•m (kg-m, ft-lb)

1. Gasket
2. Water pump
3. Water pump pulley

67162-QX56-G39

Water pump

- Rear blower motor resistor electrical connection
- Rear blower motor electrical connection
- Interior trim panel
- Rear A/C pipes
- Rear heater hoses
5. Refill the cooling system.
6. Recharge the A/C system.

Cylinder Head

REMOVAL & INSTALLATION

1. Remove or disconnect the following:
   - Engine assembly
   - Belt tensioner
   - Idler pulley
   - Thermostat housing and hose
   - Oil pan and strainer
   - Fuel tube and injector assembly
   - Intake manifold
   - Ignition coil
   - Rocker cover
   - Crankshaft pulley
   - Front engine cover
   - Oil pump
   - Timing chain
   - Camshaft sprockets
   - Camshaft
   - Cylinder head, removing bolts in reverse order shown in figure

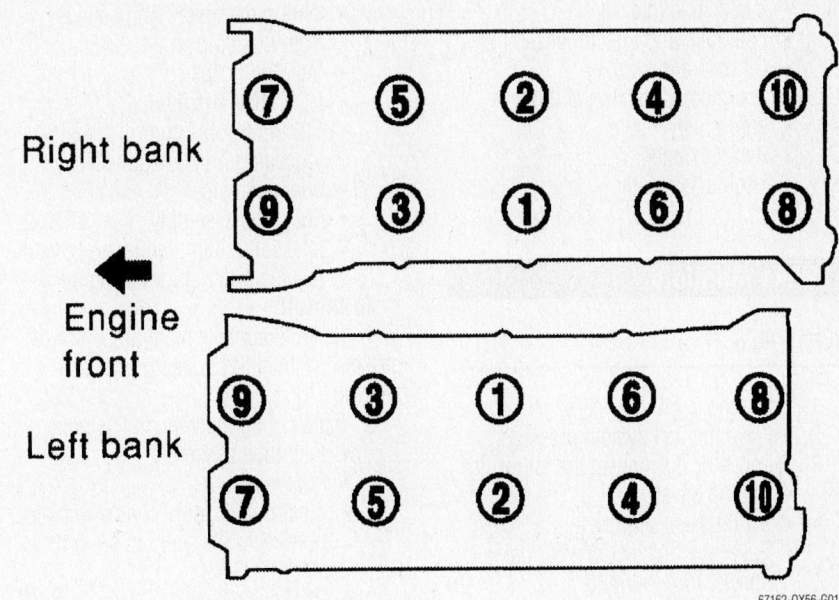

Right bank

← Engine front

Left bank

67162-QX56-G01

Cylinder head torque sequence

To install:
2. Install the cylinder head with a new gasket. Tighten the bolts in sequence as follows:
   a. Step 1: 72 ft. lbs. (98 Nm)
   b. Step 2: Loosen all bolts completely
   c. Step 3: 33 ft. lbs. (44 Nm)
   d. Step 4: Plus 60 degrees
   e. Step 5: Plus 60 degrees

3. Install or connect the following:
   - Camshaft
   - Camshaft sprockets
   - Timing chain
   - Oil pump
   - Front engine cover
   - Crankshaft pulley
   - Rocker cover
   - Ignition coil

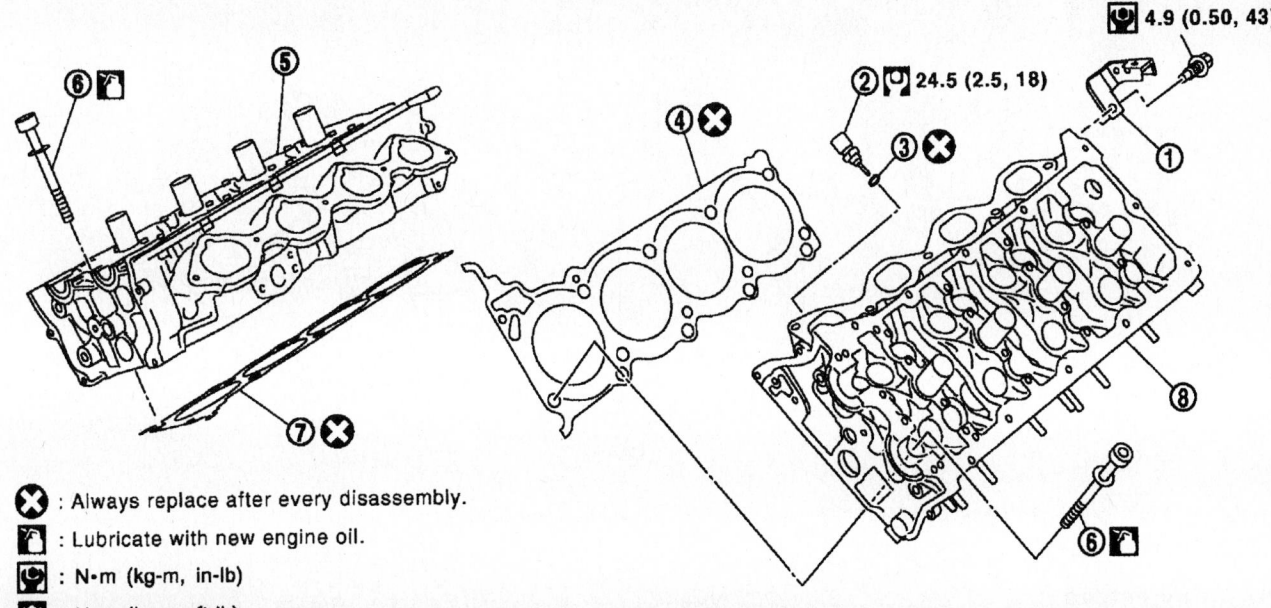

4.9 (0.50, 43)

2 24.5 (2.5, 18)

❌ : Always replace after every disassembly.

: Lubricate with new engine oil.

: N•m (kg-m, in-lb)

: N•m (kg-m, ft-lb)

1. Harness bracket	2. Engine coolant temperature sensor	3. Washer
4. Cylinder head gasket (left bank)	5. Cylinder head (right bank)	6. Cylinder head bolt
7. Cylinder head gasket (right bank)	8. Cylinder head (left bank)	

67162-QX56-G38

Cylinder heads and gaskets

- Intake manifold
- Fuel tube and injector assembly
- Oil pain and strainer
- Thermostat housing and hose
- Idler pulley
- Belt tensioner
- Engine assembly
4. Start the engine and check for leaks

Intake Manifold

REMOVAL & INSTALLATION

1. Drain the cooling system.
2. Relieve the fuel system pressure.
3. Remove or disconnect the following:
   - Engine cover
   - Air intake assembly
   - Fuel tube quick connector using special tool J-45488

- Wiring harnesses and brackets from manifold
- Vacuum hoses
- PCV hose and tube
- Electric throttle control actuator, loosening bolts diagonally
- Fuel injectors
- Fuel tube assembly
- Intake manifold, removing bolts in reverse order shown in figure

To install:

4. Install the intake manifold with new gaskets. Tighten the bolts in order as shown.
5. Install or connect the following:
   - Fuel tube assembly
   - Fuel injectors
   - Electronic throttle control actuator, tightening the bolts in several steps
   - PCV hose

- Vacuum hoses
- Wiring harnesses
6. Connect the fuel tube as follows:
   a. Apply a thin layer of engine oil on the tube from tip end to spool end.
   b. Insert tube into quick connector past the white identification mark
   c. Insert tube into quick connector until top spool is completely inside the connector and 2nd level spool is exposed right below the connector.
   d. Pull slightly on the quick connector to ensure it is fully engaged.
   e. Install quick connector cap on quick connector joint.
7. Install or connect the following:
   - Air intake assembly
   - Engine cover
8. Refill the cooling system.
9. Start engine and check for leaks.

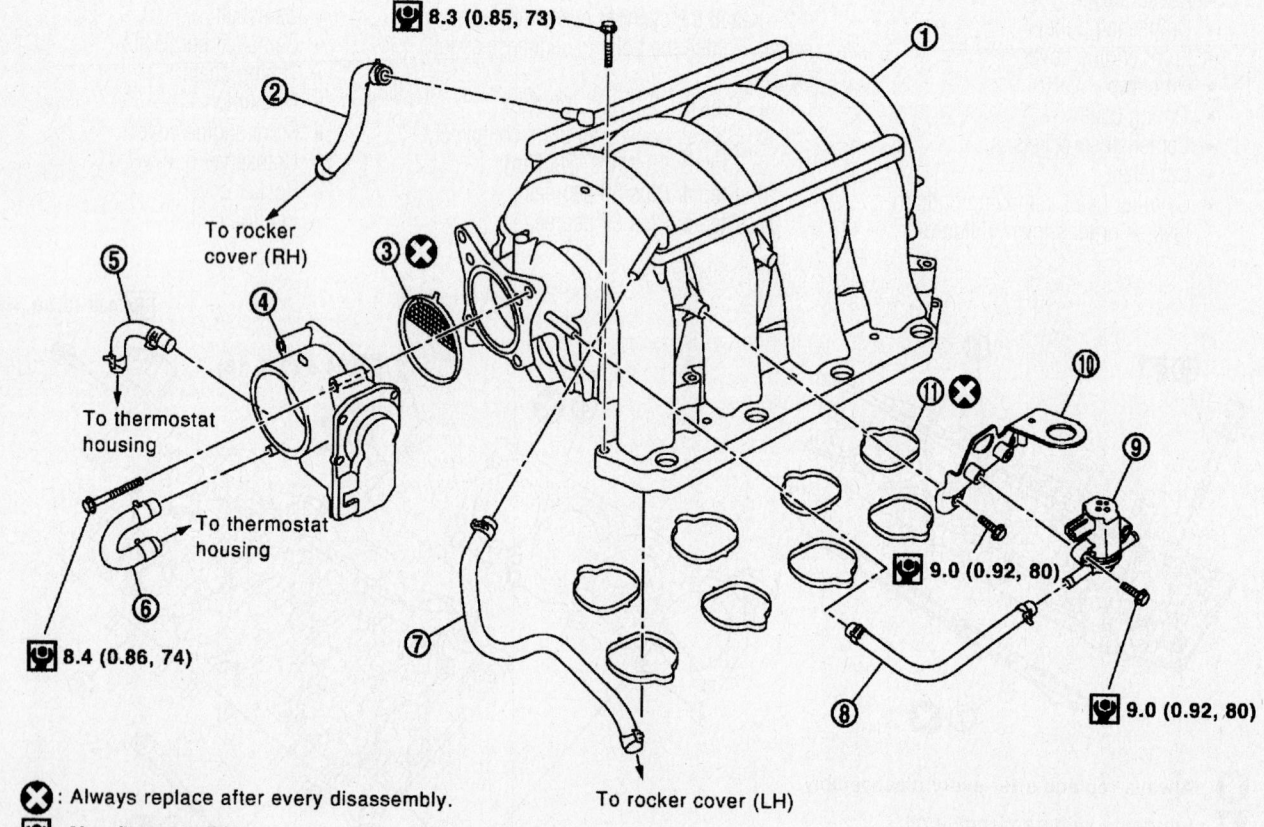

❽ : Always replace after every disassembly.

🔧 : N•m (kg-m, in-lb)

1.	Intake manifold	2.	PCV hose	3.	Gasket
4.	Electric throttle control actuator	5.	Water hose	6.	Water hose
7.	PCV hose	8.	EVAP hose	9.	EVAP canister purge control solenoid valve
10.	Bracket	11.	Gasket		

67162-QX56-G42

Intake manifold and related parts

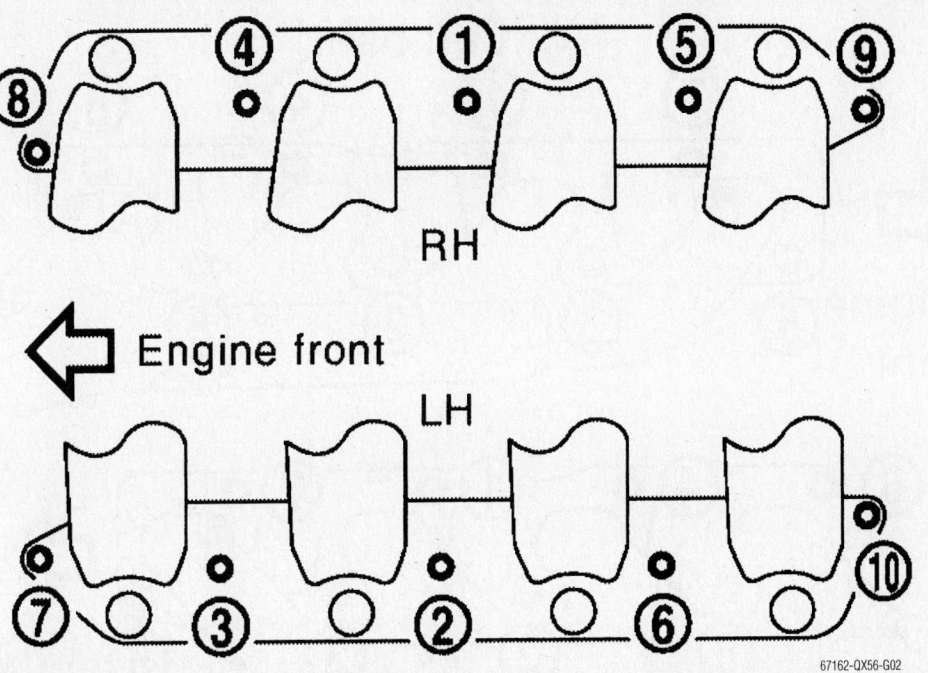

Intake manifold torque sequence

Exhaust Manifold

REMOVAL & INSTALLATION

1. Drain the cooling system.
2. Remove or disconnect the following:

- Air intake assembly
- Engine splash guard
- Radiator and radiator hoses
- Drive belts

3. Remove air fuel ratio sensors as follows:

   a. Remove engine cover.

   b. Remove wiring harness from each sensor

   c. Remove sensors, using special tool J-38356

4. Remove front cross bar
5. Remove left exhaust manifold as follows:

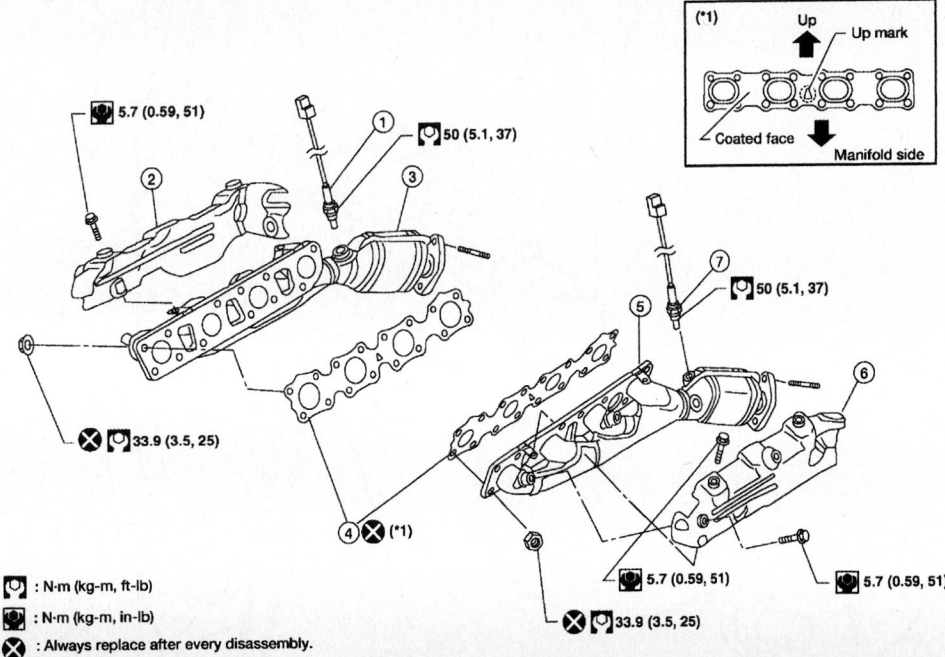

- : N·m (kg-m, ft-lb)
- : N·m (kg-m, in-lb)
- ⊗ : Always replace after every disassembly.

1. Air fuel ratio (A/F) sensor 1 (bank 2)
4. Gaskets
7. Air fuel ratio (A/F) sensor 1 (bank 1)
2 Exhaust manifold cover (bank 2)
5 Exhaust manifold (left bank 1)
3 Exhaust manifold (bank 2)
6 Exhaust manifold cover (bank 1)

67162-QX56-G43

Exhaust manifold and related parts

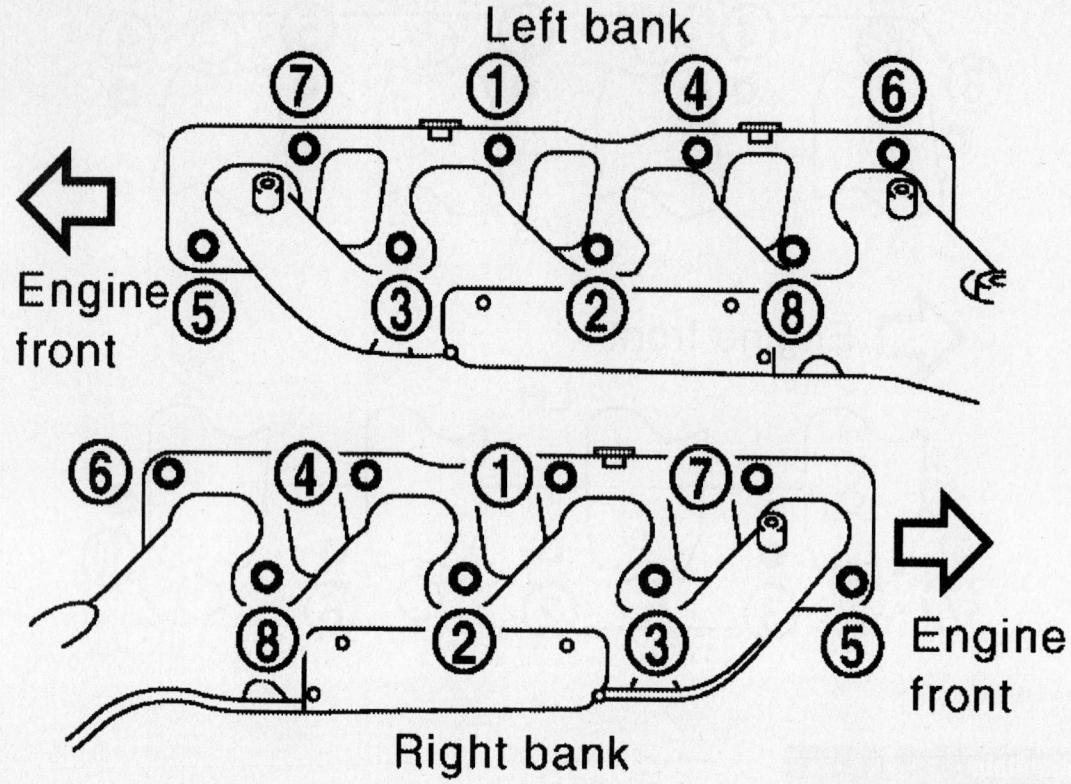

Left bank

⑦ ① ④ ⑥

← Engine front

⑤ ③ ② ⑧

⑥ ④ ① ⑦

Engine front →

⑧ ② ③ ⑤

Right bank

67162-QX56-G03

Exhaust manifold torque sequence

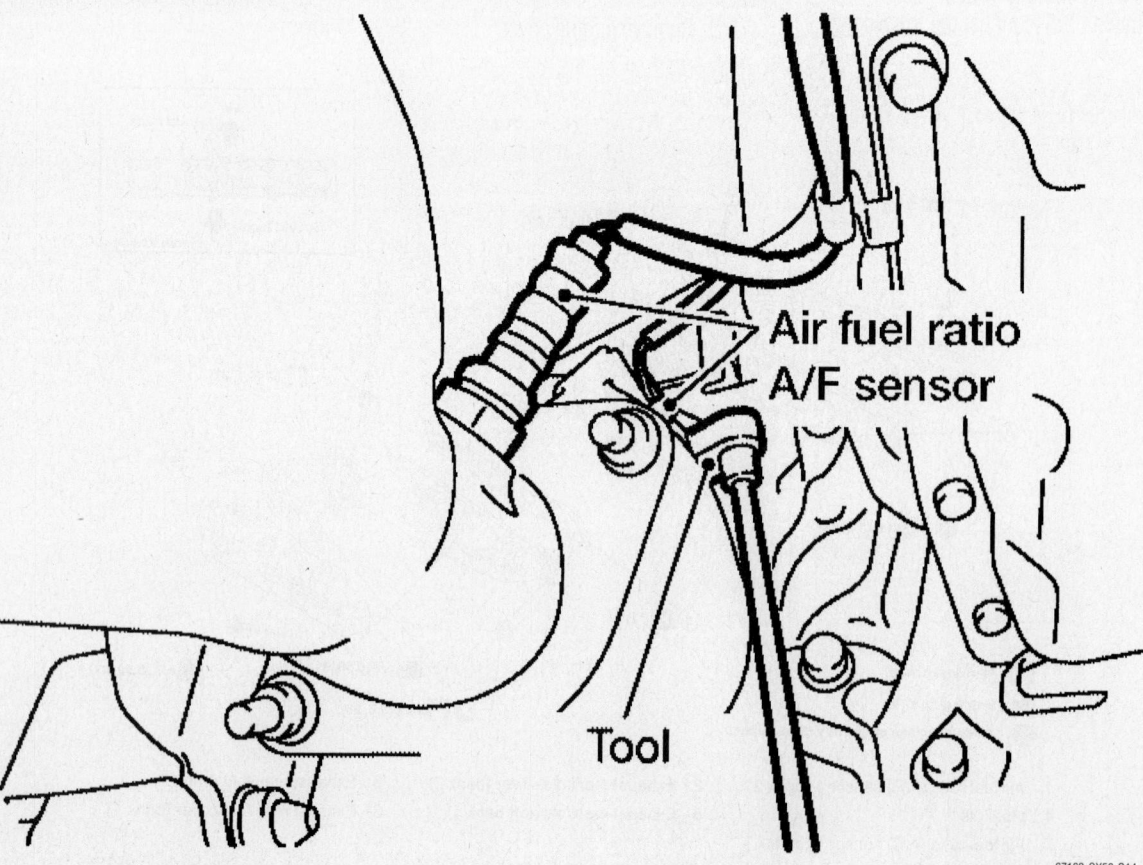

Air fuel ratio
A/F sensor

Tool

67162-QX56-G14

Removing Air-Fuel ratio sensors

a. Remove the exhaust front tube.

b. Remove the exhaust manifold cover.

c. Loosen nuts in reverse order shown in figure.

d. Remove studs from position 2, 4, 6, and 8 and remove manifold.

6. Remove right exhaust manifold as follows:

a. Remove the exhaust front tube.

b. Remove the oil level gauge guide.

c. Remove the exhaust manifold cover.

d. Loosen nuts in reverse order shown in figure.

e. Remove studs from position 2, 4, 6, and 8 and remove manifold.

To install:

7. Install or connect the following:
- Exhaust manifold gasket with triangle mark facing up and coated (gray) face toward exhaust manifold.
- Exhaust manifold, tightening the nuts as shown in figure
- Exhaust manifold cover
- Oil level gauge guide (right side only)
- Exhaust front tube
- Front cross bar
- Air fuel ratio sensors, with anti-seize lubricant
- Engine cover
- Drive belts
- Radiator and radiator hoses
- Engine splash guard
- Air intake assembly

8. Refill the cooling system
9. Start engine and check for leaks.

Camshaft and Valve Lifters

REMOVAL & INSTALLATION

1. Remove rocker cover.

2. Obtain compression Top Dead Center (TDC) of No. 1 cylinder.

3. Remove timing chain case cover.

4. Matchmark the timing chain, aligning with the camshaft sprocket marks.

5. Remove chain tensioner from left bank as follows:

Right bank
Exhaust

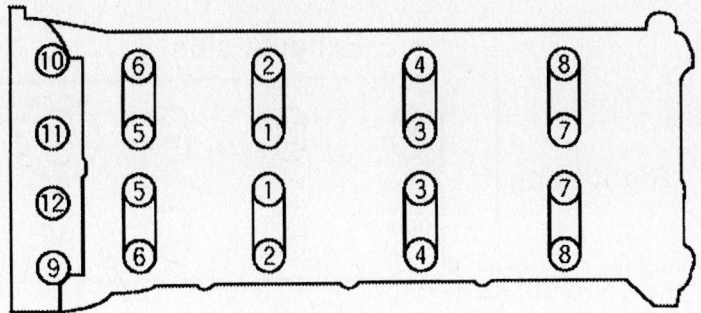

Engine front ⟸ ——— Intake ———

Left bank

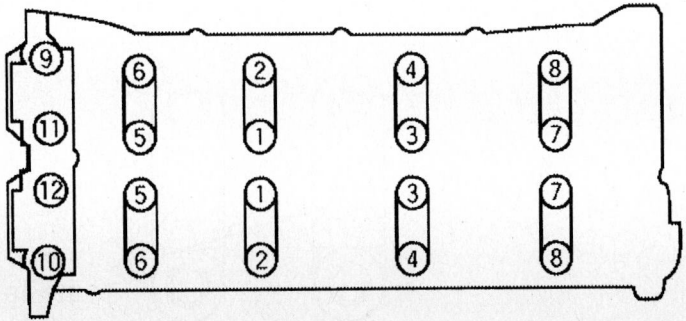

Exhaust

67162-QX56-G04

Camshaft torque sequence

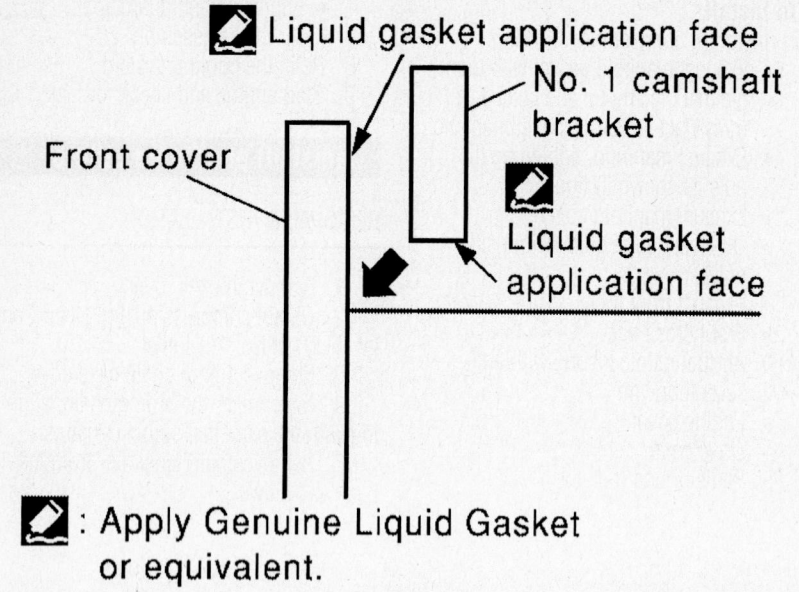

Liquid gasket application face

No. 1 camshaft bracket

Front cover

Liquid gasket application face

: Apply Genuine Liquid Gasket or equivalent.

67162-QX56-G05

Gasket application for camshaft bracket

a. Squeeze end clips and push plunger into tensioner body.

b. Secure plunger using stopper pin.

c. Remove chain tensioner.

6. Remove chain tensioner from right bank as follows:

a. Remove chain tensioner cover using special tool J-37228.

b. Squeeze end clips and push plunger into tensioner body.

c. Secure plunger using stopper pin.

d. Remove chain tensioner.

7. With camshaft locked with a wrench, loosen bolts to remove camshaft sprocket.

8. Remove front cover bolts.

9. Remove camshaft brackets, removing bolts in reverse order shown in figure.

10. Remove camshaft.

11. Remove valve lifters.

To install:

12. Install valve lifters.

13. Install camshaft, refer to table for correct placement.

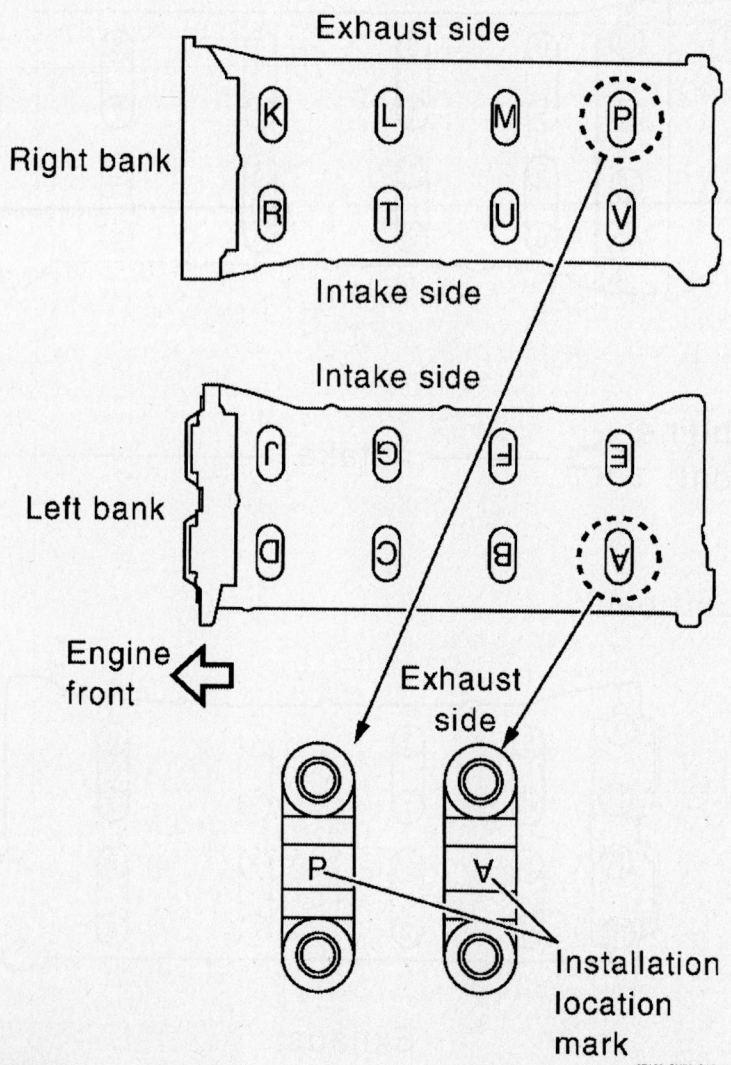

67162-QX56-G15

Camshaft bracket installation markings

Bank	INT EXH	Identification paint (front)	Identification paint (rear)	Identification rib
RH	INT	White	—	Yes.
	EXH	—	Light blue	Yes.
LH	INT	White	—	No.
	EXH	—	Light blue	No.

67162-QX56-G16

Camshaft installation markings

14. Install camshaft brackets as follows:
   a. Refer to location mark on upper surface of bracket.
   b. Installation mark should be correctly read when viewed from intake side.
15. Install camshaft bracket #1 as follows:
   a. Apply liquid gasket to bracket and backside of front cover as shown in figure.
   b. Carefully position and mount camshaft bracket #1.
   c. Temporarily tighten front cover bolts
16. Tighten fixing bolts for camshaft brackets as follows:
   a. Step 1: Bolts 9-12: 17 in. lbs. (1.9 Nm)
   b. Step 2: Bolts 1-8: 17 in. lbs. (1.9 Nm)
   c. Step 3: All bolts: 52 in. lbs. (5.9 Nm)
   d. Step 4: All bolts: 92 in. lbs. (10 Nm)

17. Tighten front cover bolts to 8 ft. lbs. (11 Nm)
18. Install camshaft sprocket as follows:
   a. Install camshaft sprocket aligning matchmarks with timing chain. Align camshaft sprocket key groove with dowel pin on camshaft front edge.
   b. Temporarily tighten bolts.
   c. Lock the camshaft with a wrench and tighten the bolts.
19. Install chain tensioner as shown:
   a. Install chain tensioner, compress plunger and hold with stopper pin.
   b. Tighten chain tensioner bolts to 61 in. lbs. (7 Nm)
   c. Remove stopper pin, release plunger and apply tension to timing chain.
   d. Install chain tensioner front cover (Right-hand bank only) and tighten bolts to 80 in. lbs. (9 Nm).
20. Install timing chain cover.
21. Install rocker cover.

Valve Lash

INSPECTION

1. Run engine to operating temperature.
2. Remove or disconnect the following:
   - Engine cover
   - Battery cover
   - Air intake assembly
   - Left and right rocker covers
3. Turn the crankshaft pulley clockwise to Top Dead Center (TDC) identification notch with timing indicator.
4. Ensure that both the intake and exhaust cam noses of the No. 1 cylinder face outside.
5. Measure the valve clearances at locations marked 'x' shown in figure.
6. Turn the crankshaft pulley clockwise 270 degrees from the position of No. 1 cylinder compression to obtain No. 3 cylinder compression TDC.
7. Measure the valve clearances at locations marked 'x' shown in next figure.

Measuring position (RH bank)		No.2 CYL	No.4 CYL	No.6 CYL	No. 8 CYL
No. 1 cylinder at TDC	EXH				×
	INT	×	×		
Measuring position (LH bank)		No.1 CYL	No. 3 CYL	No. 5 CYL	No. 7 CYL
No. 1 cylinder at TDC	INT	×		×	
	EXH	×			×

67162-QX56-G06

Locations to measure clearance with No. 1 cylinder at TDC

Measuring position (RH bank)		No.2 CYL	No.4 CYL	No.6 CYL	No. 8 CYL
No. 3 cylinder at TDC	EXH		×		
	INT				×
Measuring position (LH bank)		No.1 CYL	No. 3 CYL	No. 5 CYL	No. 7 CYL
No. 3 cylinder at TDC	INT		×		×
	EXH		×	×	

67162-QX56-G07

Locations to measure clearance with No. 3 cylinder at TDC

8. Turn crankshaft pulley clockwise 90 degrees and measure the intake and exhaust valve clearance of No. 6 cylinder and exhaust valve clearance of No. 2 cylinder.

ADJUSTMENT

1. Remove camshaft and valve lifter(s) out of specification.
2. Install replacement valve lifter(s).
3. Install the camshaft.

4. Manually turn the crankshaft pulley several turns.
5. Recheck valve clearances with engine at operating temperature.

Oil Pan

REMOVAL & INSTALLATION

1. Before servicing the vehicle, refer to the precautions in the beginning of this section.

2. Remove engine assembly.
3. Remove lower oil pan, loosening bolts in reverse order shown in figure.
4. Remove oil strainer from upper oil pan.
5. Gently pry and remove upper oil pan from engine block.
To install:
6. Apply liquid gasket to upper oil pan mating surfaces.
7. Install new O-rings to oil pump and front cover side.

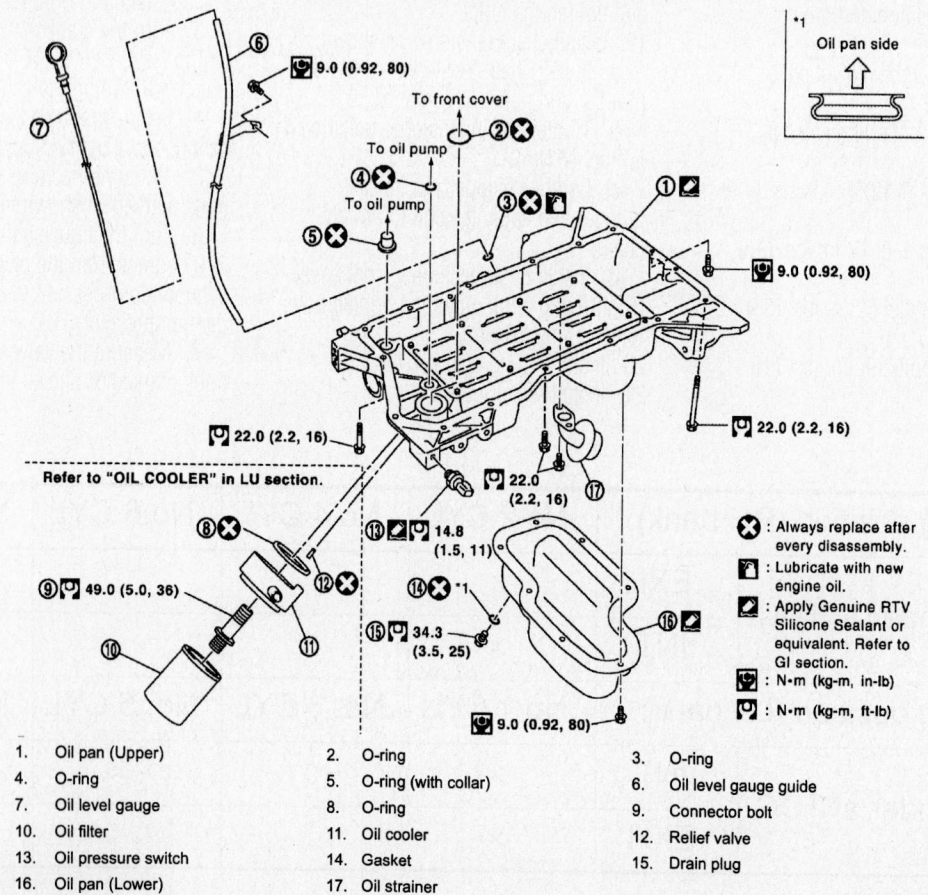

1.	Oil pan (Upper)	2.	O-ring	3.	O-ring
4.	O-ring	5.	O-ring (with collar)	6.	Oil level gauge guide
7.	Oil level gauge	8.	O-ring	9.	Connector bolt
10.	Oil filter	11.	Oil cooler	12.	Relief valve
13.	Oil pressure switch	14.	Gasket	15.	Drain plug
16.	Oil pan (Lower)	17.	Oil strainer		

67162-QX56-G44

Oil pan and related parts

8
6
3
2
1
4
Engine front
5
7
67162-QX56-G08

Lower oil pan torque sequence

8. Tighten upper oil pan bolts in following numerical order:
   a. No. 15, 16
   b. No. 1, 3, 5, 7, 11, 13
   c. No. 2, 4, 6, 8, 10, 14
   d. No. 9, 12
9. Install or connect the following:
   - Rear plate cover
   - Oil strainer to upper oil pan
   - Lower oil pan, tightening bolts in order shown in figure

Oil Pump

REMOVAL & INSTALLATION

1. Remove or disconnect the following:
   - Timing chain cover
   - Oil pump drive spacer
   - Oil pump

To install:

2. Install or connect the following:
   - Oil pump
   - Oil pump drive spacer
   - Timing chain cover

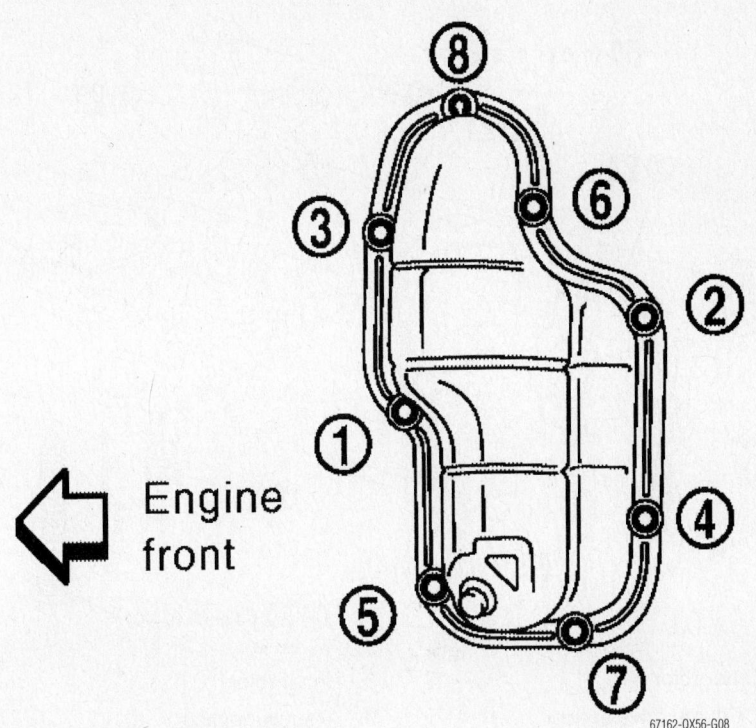

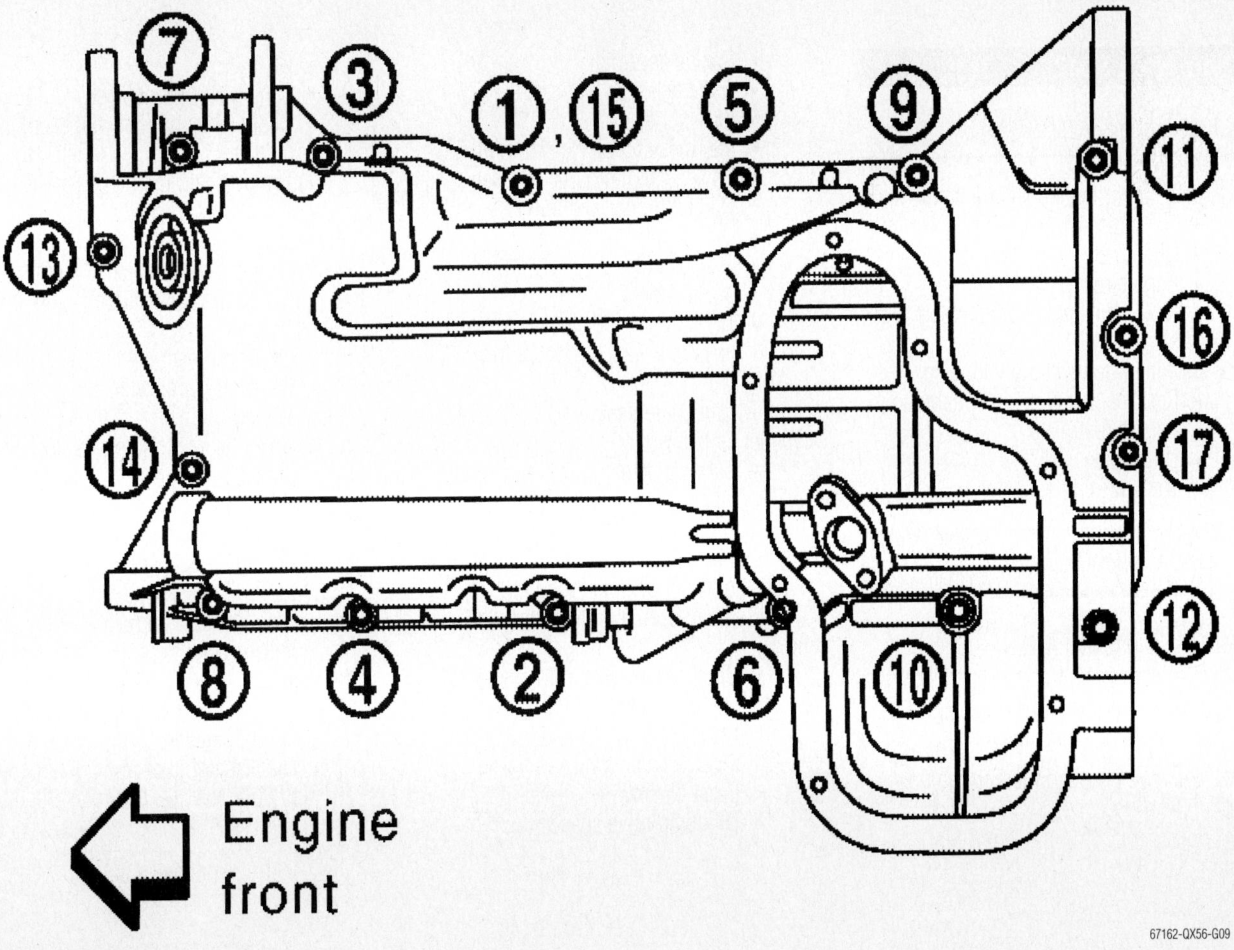

7
3
1, 15
5
9
11
13
16
14
17
12
8
4
2
6
10
Engine front
67162-QX56-G09

Upper oil pan bolt identification

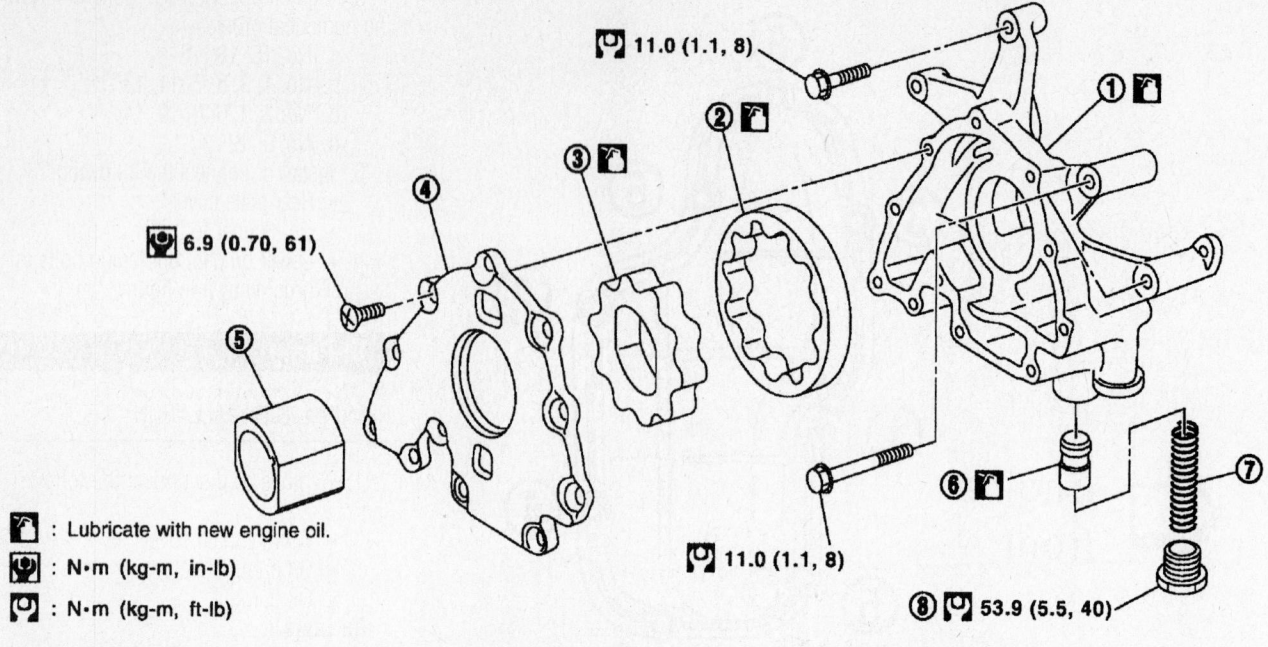

6.9 (0.70, 61)

11.0 (1.1, 8)

11.0 (1.1, 8)

8 **53.9 (5.5, 40)**

: Lubricate with new engine oil.

: N·m (kg-m, in-lb)

: N·m (kg-m, ft-lb)

1. Oil pump body
4. Oil pump cover
7. Regulator spring

2. Outer rotor
5. Oil pump drive spacer
8. Regulator plug

3. Inner rotor
6. Regulator valve

67162-QX56-G37

Oil pump exploded view

Rear Main Seal

REMOVAL & INSTALLATION

1. Remove or disconnect the following:
   - Transmission assembly
   - Drive plate
   - Engine rear plate
   - Rear main seal using suitable tool

To install:
2. Install or connect the following:
   - Rear main seal using suitable tool
   - Engine rear plate
   - Drive plate
   - Transmission assembly

Timing Chain, Sprockets, Front Cover and Seal

REMOVAL & INSTALLATION

➡**Left side timing chain must be removed before right side.**

1. Remove or disconnect the following:

- Engine assembly
- Drive belt auto tensioner
- Idler pulley
- Thermostat housing and water hose
- Power steering pump bracket
- Oil pan (upper and lower)
- Oil strainer
- Ignition coil
- Rocker cover
- Timing chain case cover, loosening bolts in reverse order shown in figure

2. Obtain compression TDC of No. 1 cylinder as follows:

   a. Turn crankshaft pulley to align the TDC identification notch with timing indicator on front cover.

   b. Ensure intake and exhaust cam lobes of No. 1 cylinder point outside.

3. Remove or disconnect the following:

- Crankshaft pulley from crankshaft using a suitable puller
- Front cover, loosening bolts in reverse order shown in figure
- Front oil seal
- Oil pump drive spacer

- Oil pump
- Timing chain tensioner
- Chain tension guide and slack guide
- Timing chain
- Camshaft sprocket

To install:
4. Ensure that the crankshaft key and dowel pin of each camshaft are facing the same direction.

5. Install or connect the following:

- Camshaft sprockets
- Timing chain
- Chain tension guide and slack guide
- Oil pump
- Oil pump drive spacer
- Front oil seal, using suitable tool
- Front cover, using new O-rings and tighten bolts in order shown in figure
- Chain case cover, and tighten bolts in order shown in figure
- Crankshaft pulley and tighten bolt to 69 ft. lbs. (93 Nm) plus 90 degrees

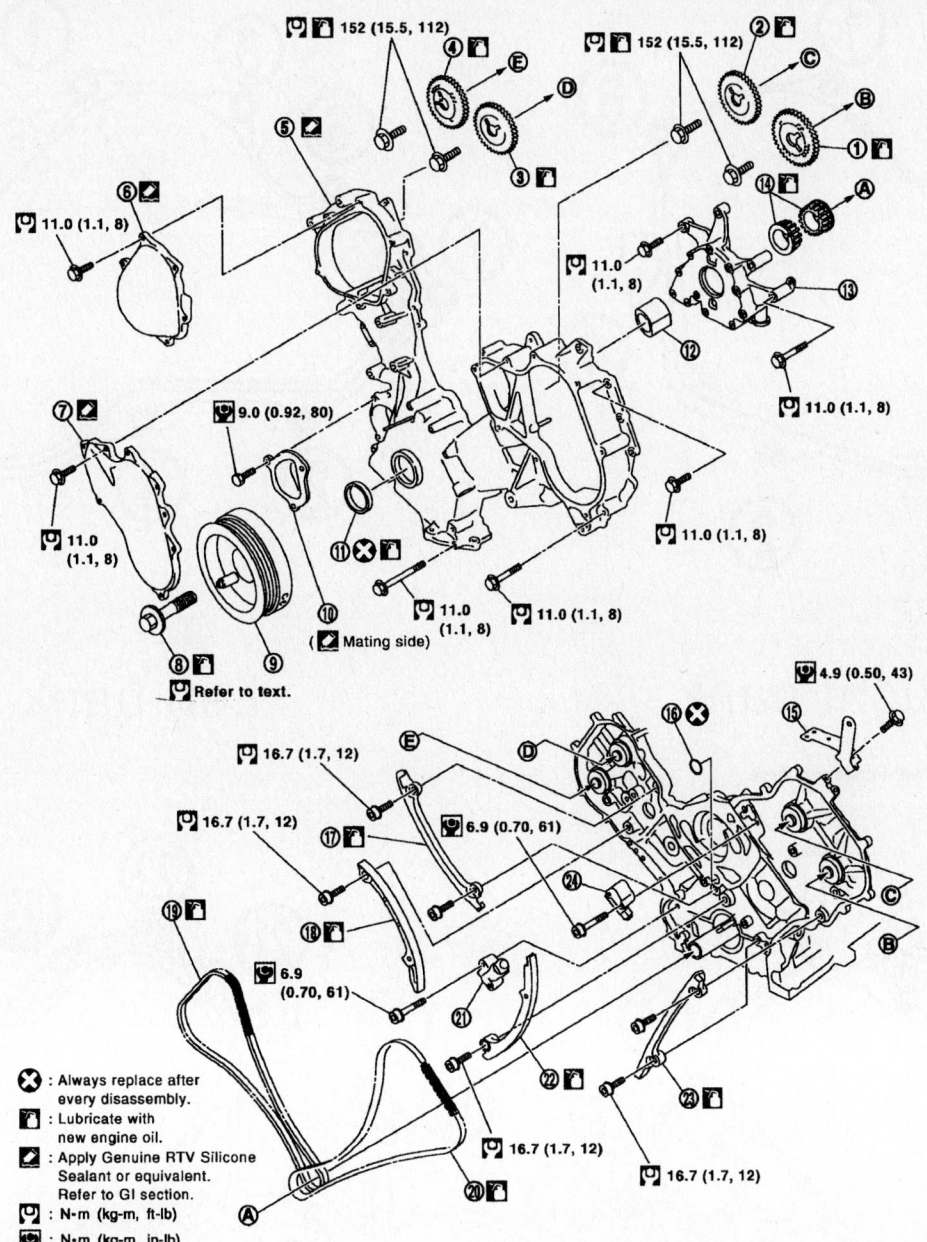

152 (15.5, 112) 152 (15.5, 112)

11.0 (1.1, 8)

11.0 (1.1, 8)

9.0 (0.92, 80)

11.0 (1.1, 8)

7

11.0 (1.1, 8)

11.0 (1.1, 8)

11.0 (1.1, 8)

11.0 (1.1, 8)

(Mating side)

Refer to text.

4.9 (0.50, 43)

16.7 (1.7, 12)

16.7 (1.7, 12)

6.9 (0.70, 61)

6.9 (0.70, 61)

16.7 (1.7, 12)

16.7 (1.7, 12)

: Always replace after every disassembly.
: Lubricate with new engine oil.
: Apply Genuine RTV Silicone Sealant or equivalent. Refer to GI section.
: N·m (kg-m, ft-lb)
: N·m (kg-m, in-lb)

1. Camshaft sprocket (left bank EXH)	2. Camshaft sprocket (left bank INT)	3. Camshaft sprocket (right bank INT)
4. Camshaft sprocket (right bank EXH)	5. Front cover	6. Chain case cover (right bank)
7. Chain case cover (left bank)	8. Crankshaft pulley bolt	9. Crankshaft pulley
10. Chain tensioner cover	11. Front oil seal	12. Oil pump drive spacer
13. Oil pump assembly	14. Crankshaft sprocket	15. Bracket
16. O-ring	17. Timing chain tension guide (right bank)	18. Timing chain slack guide (right bank)
19. Timing chain (right bank)	20. Timing chain (left bank)	21. Chain tensioner (right bank)
22. Timing chain slack guide (left bank)	23. Timing chain tension guide (left bank)	24. Chain tensioner (left bank)

67162-QX56-G45

Timing chain and related parts

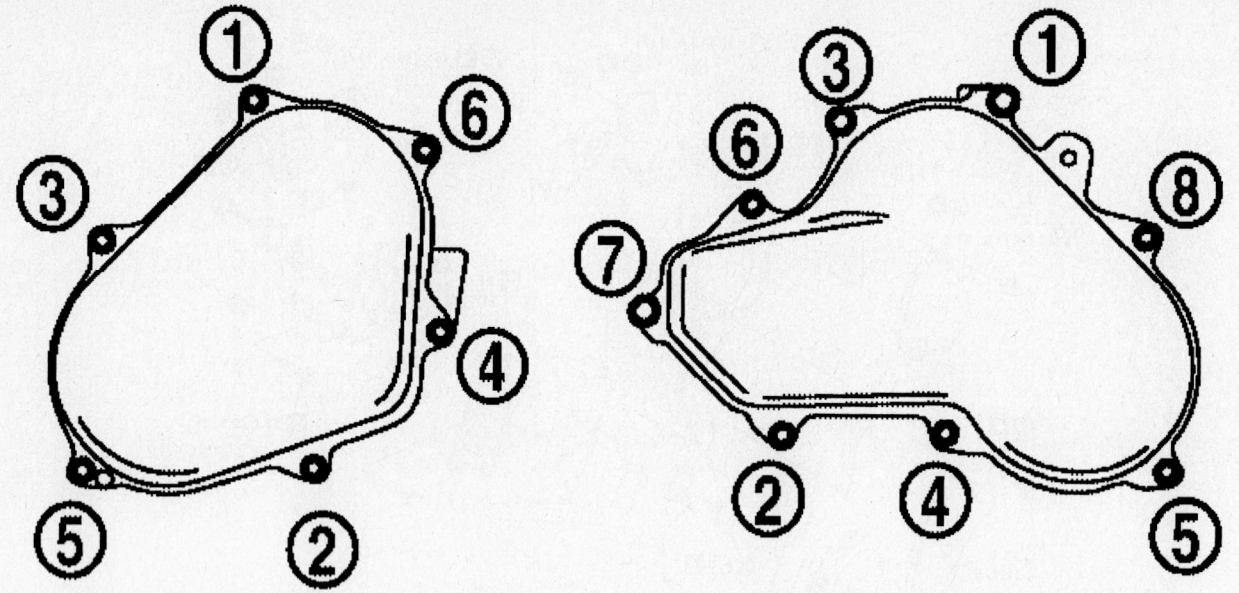

Right bank   Left bank

67162-QX56-G10

Timing chain case cover torque sequence

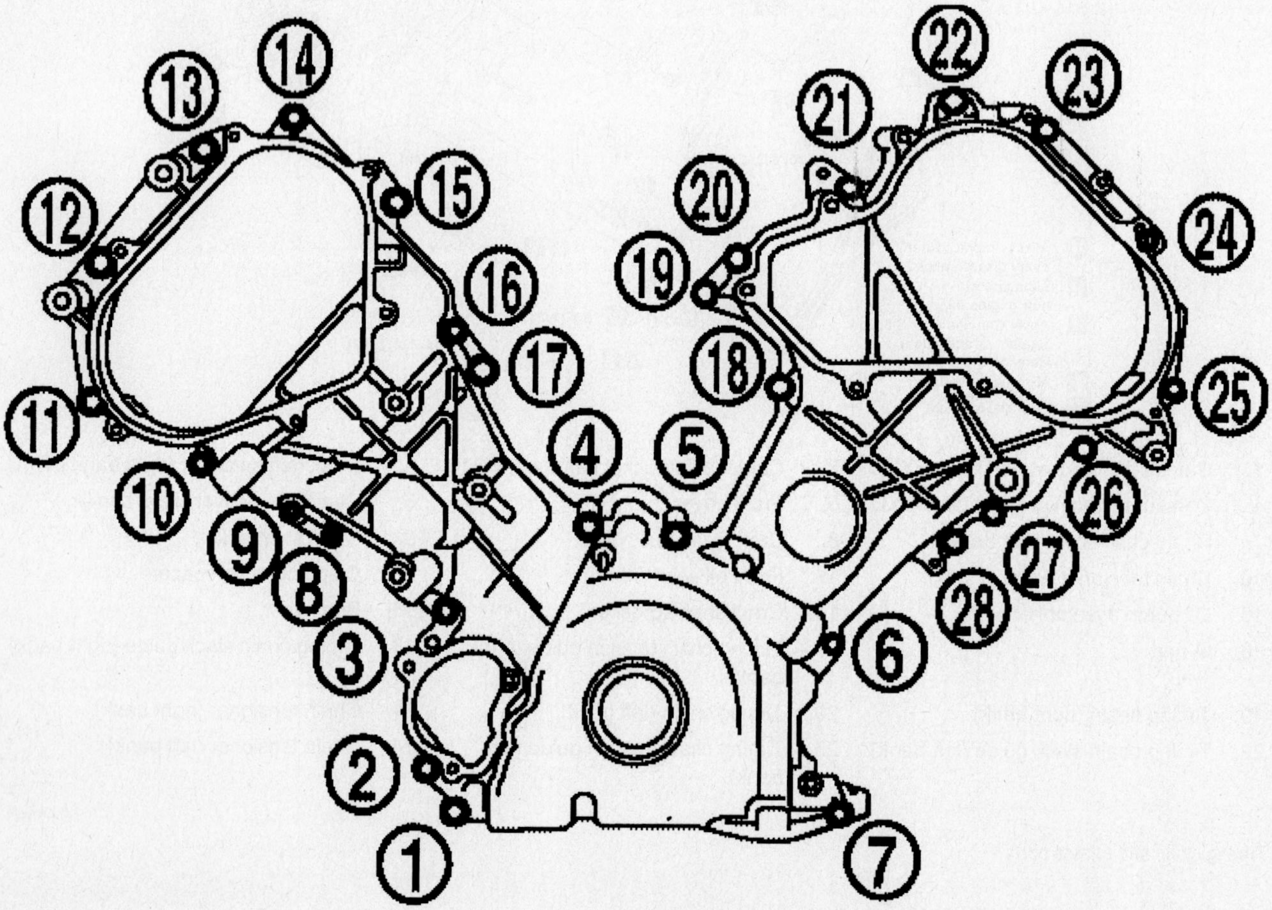

Front cover torque sequence

67162-QX56-G11

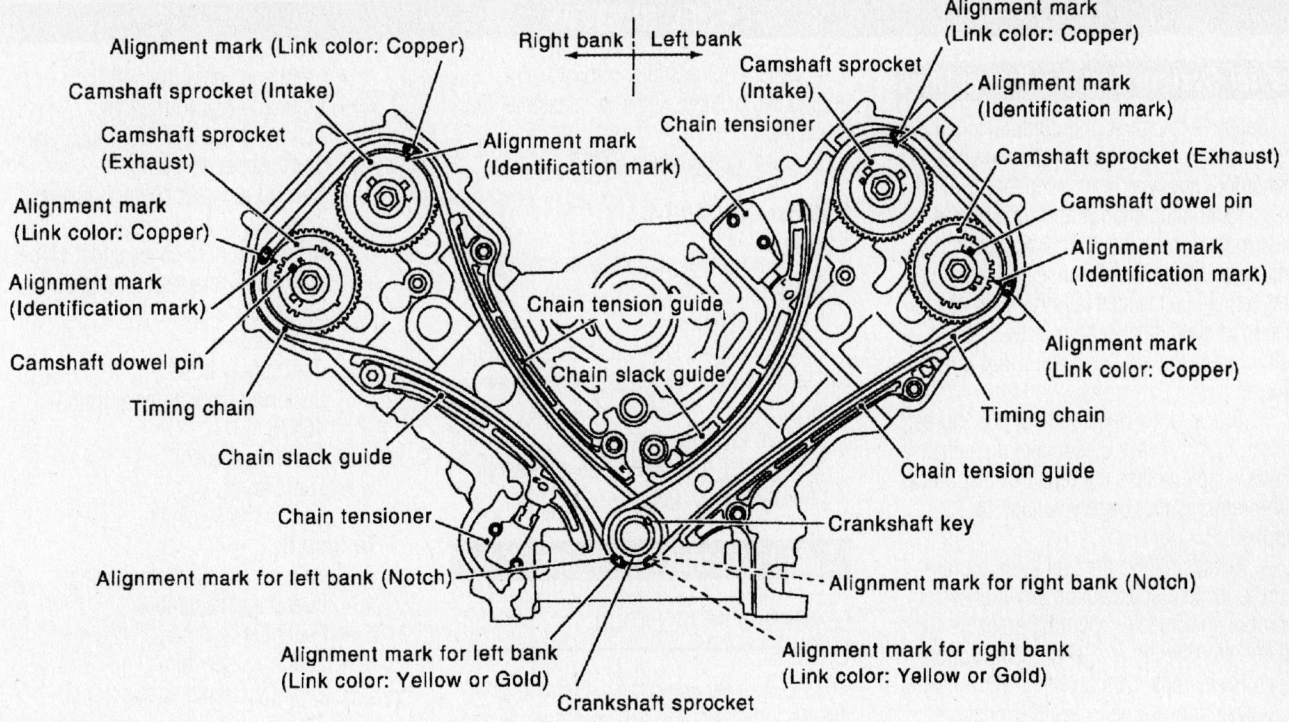

Timing mark alignment

- Ignition coil
- Oil strainer
- Lower and upper oil pan
- Power steering pump bracket
- Thermostat housing and water hose
- Idler pulley
- Drive belt auto tensioner
- Engine assembly

Piston and Ring

POSITIONING

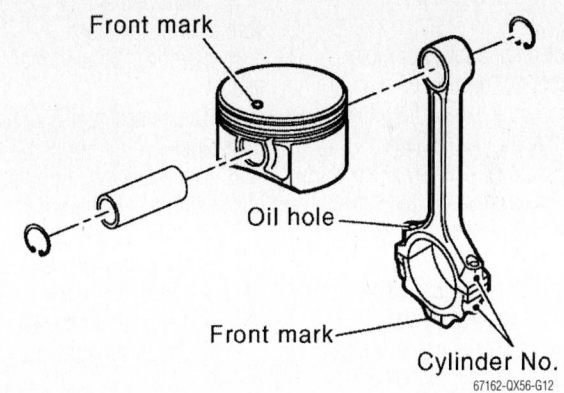

Piston and rod positioning and identification

FUEL SYSTEM

Fuel System Service Precautions

Safety is the most important factor when performing not only fuel system maintenance but any type of maintenance. Failure to conduct maintenance and repairs in a safe manner may result in serious personal injury or death. Maintenance and testing of the vehicle fuel system components can be accomplished safely and effectively by adhering to the following rules and guidelines.

• To avoid the possibility of fire and personal injury, always disconnect the negative battery cable unless the repair or test procedure requires that battery voltage be applied.

• Always relieve the fuel system pressure prior to disconnecting any fuel system component (injector, fuel rail, pressure regulator, etc.), fitting or fuel line connection. Exercise extreme caution whenever relieving fuel system pressure to avoid exposing skin, face and eyes to fuel spray. Please be advised that fuel under pressure may penetrate the skin or any part of the body that it contacts.

• Always place a shop towel or cloth around the fitting or connection prior to loosening to absorb any excess fuel due to spillage. Ensure that all fuel spillage (should it occur) is quickly removed from engine surfaces. Ensure that all fuel soaked cloths or towels are deposited into a suitable waste container.

• Always keep a dry chemical (Class B) fire extinguisher near the work area.

• Do not allow fuel spray or fuel vapors to come into contact with a spark or open flame.

• Always use a back-up wrench when loosening and tightening fuel line connection fittings. This will prevent unnecessary stress and torsion to fuel line piping. Always follow the proper tighten specifications.

• Always replace worn fuel fitting O-rings with new. Do not substitute fuel hose or equivalent, where fuel pipe is installed.

Relieving Fuel System Pressure

With CONSULT-II

1. Turn ignition switch **ON**.
2. Perform "FUEL PRESSURE RELEASE" in "WORK SUPPORT" mode with CONSULT-II.
3. Start engine.

4. After engine stalls, turn over the engine two or three times to release all fuel pressure.
5. Turn ignition switch **OFF**.

Without CONSULT-II

1. Remove fuel pump fuse located in IPDM E/R.
2. Start engine.
3. After engine stalls, turn over engine two or three times to release all fuel pressure.
4. Turn ignition switch **OFF**.
5. Reinstall fuel pump fuse after servicing fuel system.

Fuel Filter and Fuel Pump

REMOVAL & INSTALLATION

1. Before servicing the vehicle, refer to the precautions in the beginning of this section.
2. Relieve the fuel system pressure.
3. Remove fuel filler cap to release pressure from inside tank.
4. Remove left hand rear inner fender liner.
5. Disconnect fuel filler hose from fuel filler pipe.
6. Drain fuel tank through the fuel filler hose using a suitable hose.
7. Remove or disconnect the following:
   • Second row left hand seat
   • Third row seat
   • Second and third row seat belt buckles mounted on floor

• Left hand center pillar trim
• Left hand rear trim panel
• Left hand rear side door kick plate and weather stripping
• Second row rear center console and base, if equipped
• Inspection hole cover under carpet by turning retainers 90 degrees
• Electrical connectors
• EVAP hose
• Fuel supply hose
• Lock ring using special tool J-46536
• Fuel level sensor
• Fuel filter
• Fuel pump assembly

To install:
8. Install or connect the following:
   • Fuel pump assembly
   • Fuel filter
   • Fuel level sensor
   • Lock ring using special tool J-46536
   • Fuel supply hose
   • EVAP hose
   • Electrical connectors
   • Inspection hole cover
   • Second row rear center console and base, if equipped
   • Left hand rear side door kick plate and weather stripping
   • Left hand rear trim panel
   • Left hand center pillar trim
   • Second and third row seat belt buckles
   • Third row seat
   • Second row left hand seat
   • Fuel filler hose to fuel filler pipe

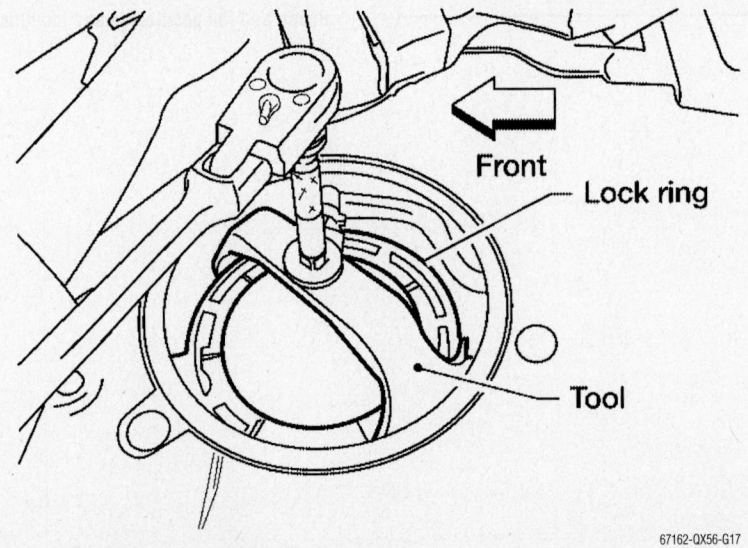

Front
Lock ring
Tool

67162-QX56-G17

Removing fuel assembly lock ring

- Left hand rear inner fender liner
9. Start the engine and check for leaks.

Fuel Injectors

REMOVAL & INSTALLATION

1. Remove engine cover
2. Relieve fuel system pressure.

3. Remove or disconnect the following:
- Negative battery cable
- Fuel injector harness connectors
- Fuel hose assembly from right and left fuel rails
- Fuel injectors with fuel rail as an assembly
- Fuel injector from fuel rail

To install:
4. Install or connect the following:
- New clip onto the fuel injector

- Fuel injector to fuel rail
- Fuel injectors and fuel rail as an assembly to the intake manifold. Torque the bolts in 2 steps, first to 9 ft. lbs. (12.8 Nm); then to 18 ft. lbs. (24.5 Nm).
- Fuel hose assembly
- Fuel injector harness connectors
- Negative battery cable
- Engine cover
5. Start engine and check for leaks.

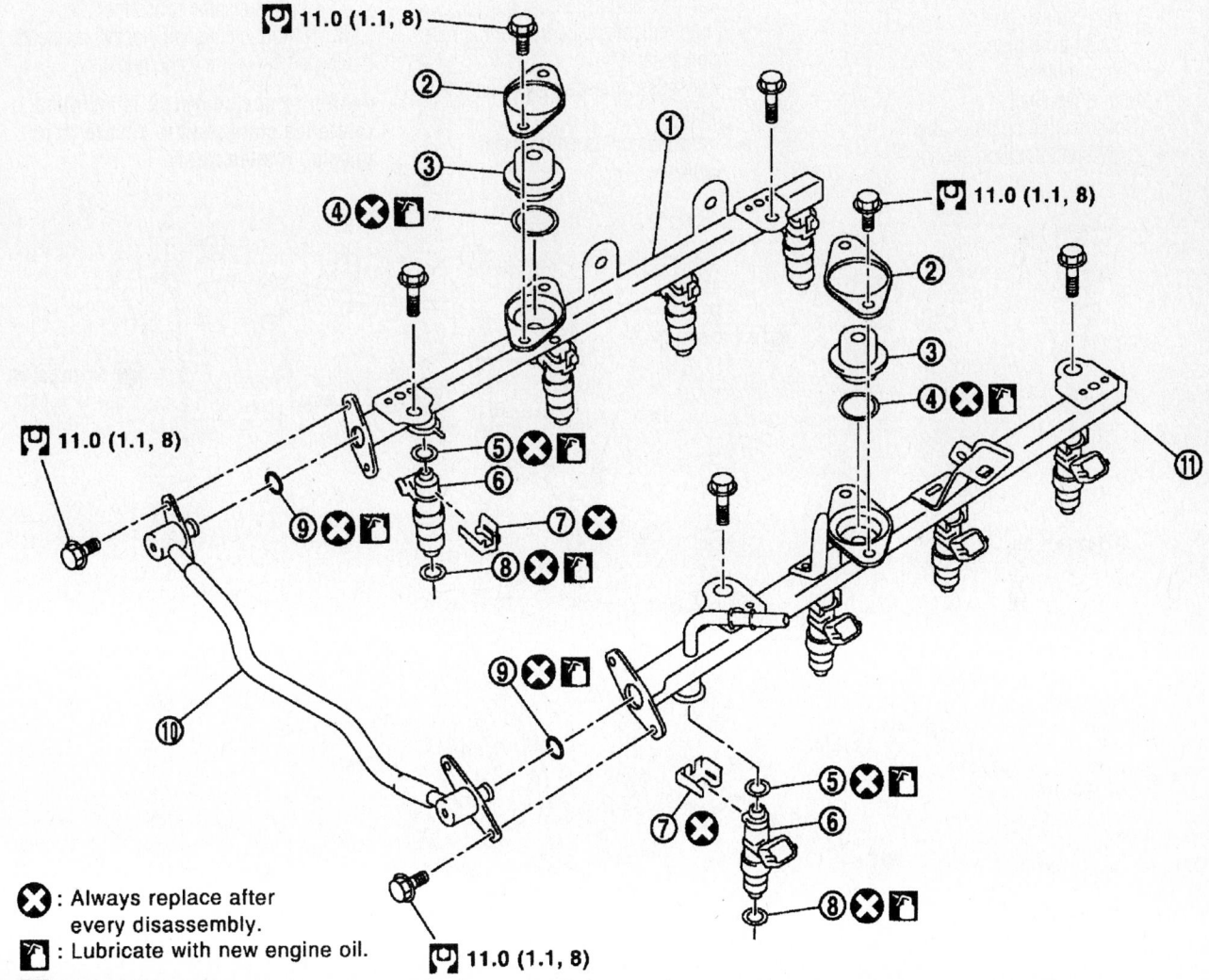

⊗ : Always replace after every disassembly.

▣ : Lubricate with new engine oil.

▣ : N•m (kg-m, ft-lb)

1. Fuel tube (right bank)	2. Cap	3. Fuel damper
4. O-ring	5. O-ring (Blue)	6. Fuel injector
7. Clip	8. O-ring (Brown)	9. O-ring
10. Fuel hose assembly	11. Fuel tube (left bank)	

67162-QX56-G36

Fuel injectors and related parts

DRIVE TRAIN

Transmission Assembly

REMOVAL & INSTALLATION

2-Wheel Drive

1. Remove or disconnect the following:
   - Negative battery cable
   - Engine cover
   - Transmission fluid indicator gauge
   - Engine splash guard
   - Exhaust front pipe
   - Center muffler
   - Rear drive shaft
   - Transmission control cable
   - Crankshaft position sensor
   - Fluid cooler tube
   - Dust cover from converter housing
2. Turning crankshaft clockwise, remove the four tightening bolts for drive plate and torque converter
3. Support the transmission with a suitable jack.
4. Remove or disconnect the following:
   - Transmission cross member
   - Air breather hose
   - Transmission assembly connector
   - Fluid indicator tube from transmission assembly
   - Transmission assembly to engine bolts
   - Transmission assembly from vehicle

To install:

5. Install or connect the following:
   - Transmission assembly into vehicle
   - Transmission assembly to engine bolts tightening to 83 ft. lbs. (113 Nm)
   - Fluid indicator tube to transmission assembly
   - Transmission assembly connector
   - Air breather hose
   - Transmission cross member
6. Turning crankshaft clockwise, install the torque converter to drive plate.

➡**After torque converter is installed, rotate the crankshaft to ensure transmission rotates freely.**

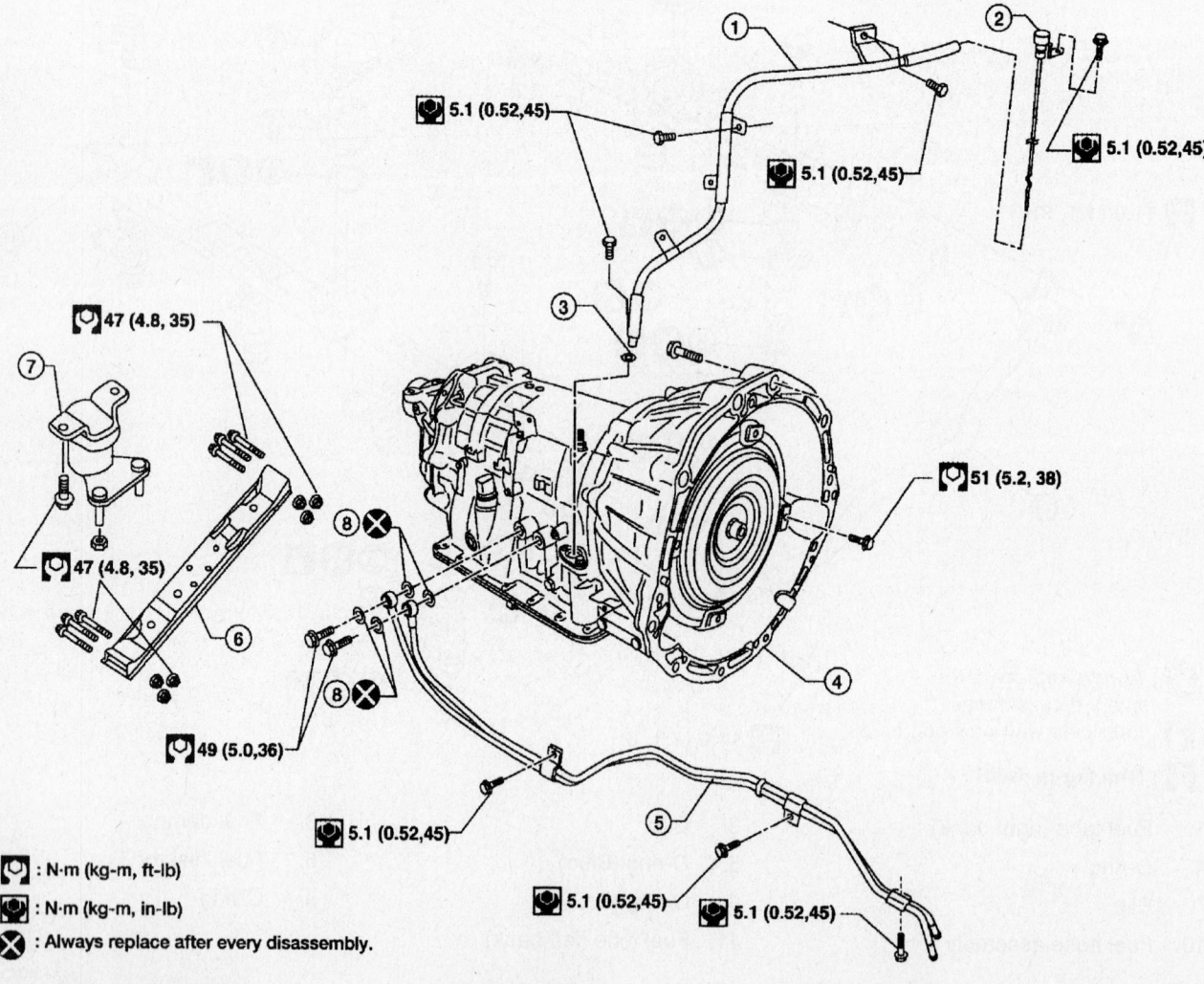

5.1 (0.52,45)

5.1 (0.52,45)

5.1 (0.52,45)

47 (4.8, 35)

47 (4.8, 35)

51 (5.2, 38)

49 (5.0,36)

5.1 (0.52,45)

5.1 (0.52,45)

5.1 (0.52,45)

: N·m (kg-m, ft-lb)

: N·m (kg-m, in-lb)

⊗ : Always replace after every disassembly.

1. A/T fluid indicator pipe
2. A/T fluid indicator
3. O-ring
4. Transmission assembly
5. Fluid cooler tube
6. A/T cross member
7. Insulator
8. Copper washers

Transmission, with 2WD

67162-QX56-G35

7. Install or connect the following:
- Dust cover for converter housing
- Fluid cooler tube
- Crankshaft position sensor
- Transmission control cable
- Rear drive shaft
- Center muffler
- Exhaust front pipe
- Engine splash guard
- Transmission fluid indicator gauge
- Engine cover
- Negative battery cable

8. Start engine and check for leaks.

4-Wheel Drive

1. Remove or disconnect the following:
- Negative battery cable
- Engine cover
- Transmission fluid indicator gauge
- Engine splash guard
- Exhaust front pipe
- Center muffler
- Drive shaft
- Transmission control cable
- Crankshaft position sensor
- Fluid cooler tube
- Dust housing for torque converter

2. Turning the crankshaft clockwise, remove the four tightening bolts for drive plate and torque converter.

3. Support the transmission assembly with a suitable jack.

4. Remove transmission cross member.

5. Tilt the transmission slightly to keep clearance between the body and the transmission assembly, then disconnect the air breather hose.

6. Remove or disconnect the following:
- Transmission assembly connector and transfer case connector

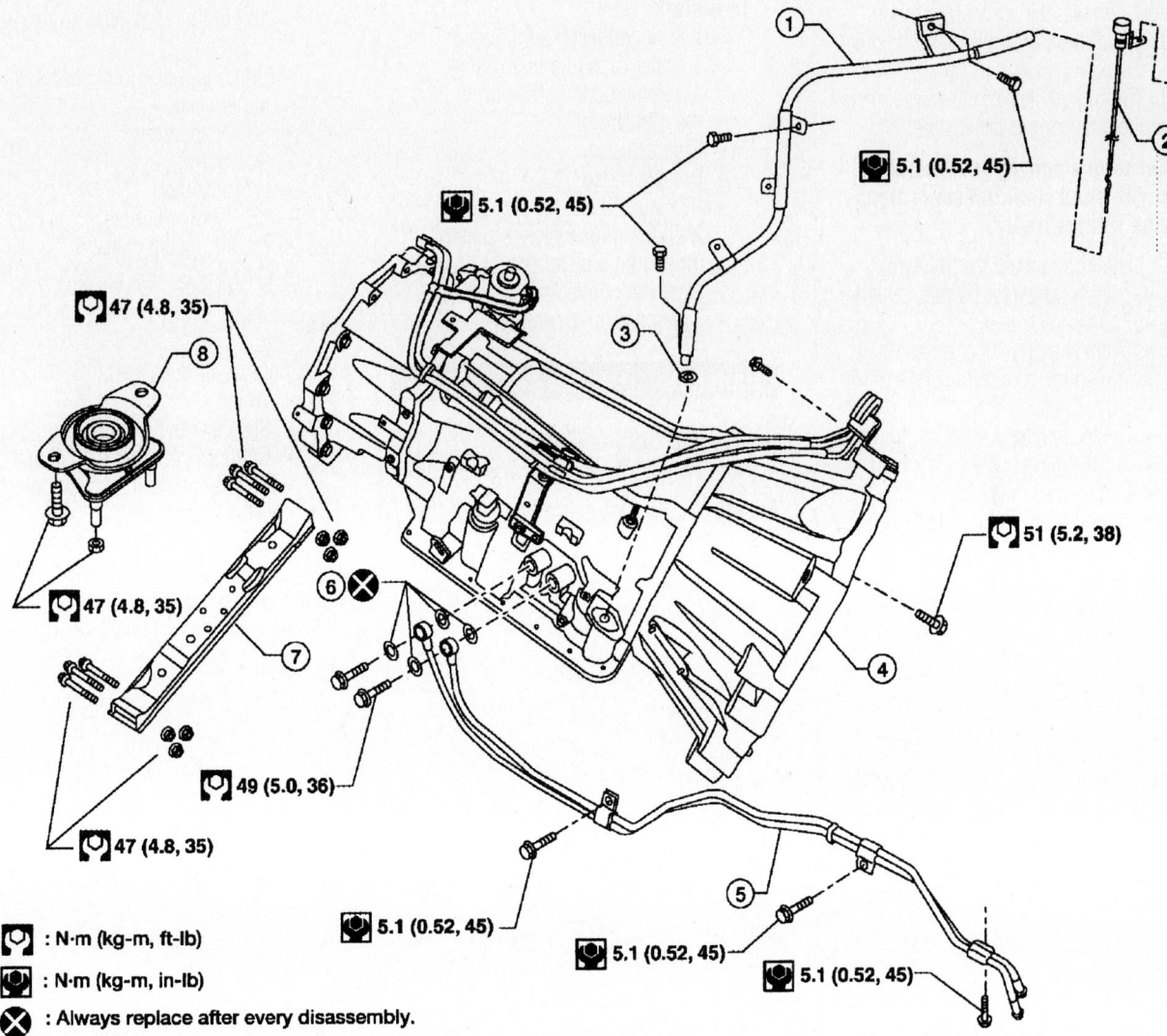

47 (4.8, 35)

47 (4.8, 35)

47 (4.8, 35)

49 (5.0, 36)

5.1 (0.52, 45)

5.1 (0.52, 45)

5.1 (0.52, 45)

5.1 (0.52, 45)

5.1 (0.52, 45)

51 (5.2, 38)

: N·m (kg-m, ft-lb)

: N·m (kg-m, in-lb)

: Always replace after every disassembly.

1. A/T fluid indicator pipe
2. A/T fluid indicator
3. O-ring
4. Transmission assembly
5. Fluid cooler tube
6. Copper washer
7. A/T cross member
8. Insulator

67162-QX56-G34

Transmission, with 4WD

- Fluid indicator pipe
- Transmission assembly to engine bolts
- Transmission assembly, with transfer case attached, from vehicle
- Transmission assembly from transfer case

To install:

7. Install or connect the following:
- Transfer case to transmission assembly
- Transmission assembly into vehicle
- Transmission assembly to engine bolts tightening to 83 ft. lbs. (113 Nm)

8. With the transmission slightly tilted to allow clearance between body and transmission, connect the air breather hose.

9. Install the transmission cross member.

10. Turning crankshaft clockwise, install the torque converter to drive plate.

➡**After torque converter is installed, rotate the crankshaft to ensure transmission rotates freely.**

11. Install or connect the following:
- Dust housing for torque converter
- Fluid cooler tube
- Crankshaft position sensor
- Transmission control cable
- Drive shaft
- Center muffler
- Front exhaust pipe
- Engine splash guard
- Transmission fluid indicator gauge
- Engine cover
- Negative battery cable

12. Start engine and check for leaks.

Transfer Case Assembly

REMOVAL & INSTALLATION

1. Remove or disconnect the following:
- Transmission assembly
- ATP switch, neutral 4LO switch, wait detection switch, transfer motor and transfer control device electrical connectors
- Breather hoses
- Shift actuator from the extension housing
- Transfer case to transmission assembly bolts
- Transfer case assembly

To install:

2. Install or connect the following:
- Transfer case to transmission assembly bolts tightening to 26 ft. lbs. (36 Nm)
- Shift actuator
- Breather hoses
- ATP switch, neutral 4LO switch, wait detection switch, transfer motor and transfer control device electrical connectors
- Transmission assembly

Halfshaft

REMOVAL & INSTALLATION

Front

1. Remove or disconnect the following:
- Wheel
- Engine splash guard

- ABS sensor harness on knuckle
- Brake caliper
- Coil spring
- Shock absorber

2. Separate upper ball link joint stud from steering knuckle using special tool J-24319-01.

3. Remove or disconnect the following:
- Cotter pin and drive shaft nut
- Half shaft mounting bolts
- Halfshaft

To install:

4. Install or connect the following:
- Half shaft
- Half shaft mounting bolts and tighten to 54 ft. lbs. (74 Nm)
- Half shaft nut and tighten nut to 101 ft. lbs. (137 Nm) and replace cotter pin
- Upper ball link joint stud to steering knuckle
- Shock absorber
- Coil spring
- Brake caliper
- ABS sensor harness
- Engine splash guard
- Wheel

Rear

1. Remove or disconnect the following:
- Wheel
- Stabilizer bar clamp
- Cotter pin and drive shaft nut
- Bolts from the inside flange of the drive shaft

2. Separate the drive shaft from the wheel hub by lightly tapping the end with suitable hammer and wood block.

3. Remove the half shaft.

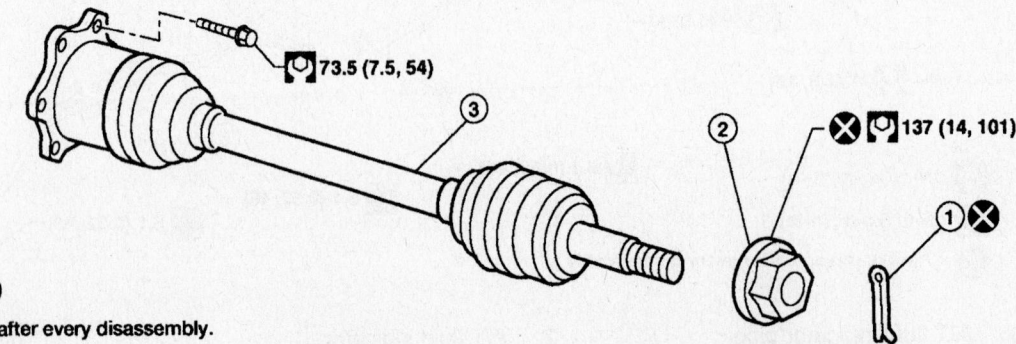

🔧 73.5 (7.5, 54)

❌ 🔧 137 (14, 101)

🔧 : N·m (kg-m, ft-lb)

❌ : Always replace after every disassembly.

1. Cotter pin
2. Drive shaft nut
3. Drive shaft

67162-QX56-G33

Front halfshaft

✳ CAUTION

Do not excessively extend the slide joint.

To install:

4. Install or connect the following:
- Half shaft
- Bolts for the inside flange and tighten to 87 ft. lbs. (118 Nm)
- Drive shaft nut and tighten nut to 101 ft. lbs. (137 Nm) and replace cotter pin

- Stabilizer bar clamp
- Wheel

CV-Joints

OVERHAUL

Inner

1. Remove the halfshaft from the vehicle.
2. Mount halfshaft in a vise.
3. Remove the dust boot bands.

4. Remove the stopper ring with a flat-bladed screwdriver or suitable tool.
5. Remove the snap ring.
6. Disassemble the cage, ball and inner race assembly and dust boot for cleaning and inspection.

To install:

➡**Discard old dust boot, dust boot bands and snap ring and use new ones for assembly.**

7. Wrap the serrated part of the half-shaft with tape.

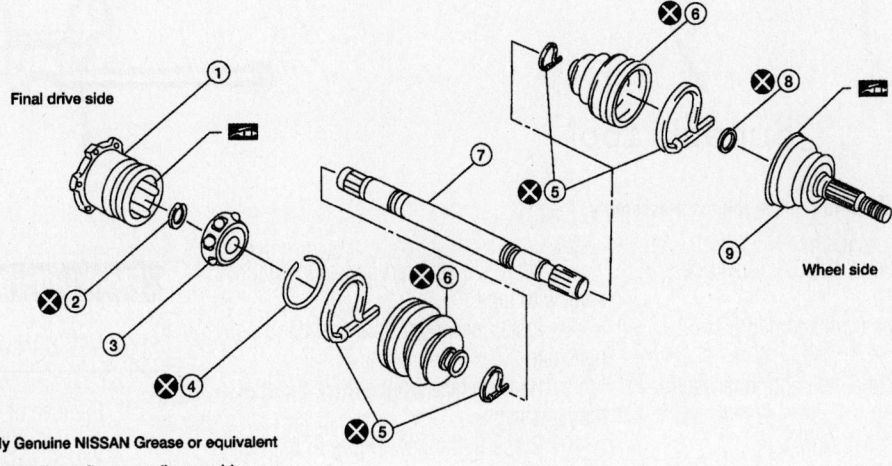

⊞ : Apply Genuine NISSAN Grease or equivalent

❌ : Always replace after every disassembly.

1. Sliding joint housing	2. Snap ring	3. Ball cage, steel ball, inner race assembly
4. Stopper ring	5. Boot band	6. Boot
7. Drive shaft	8. Circlip	9. Joint sub-assembly

67162-QX56-G31

Front halfshaft exploded view

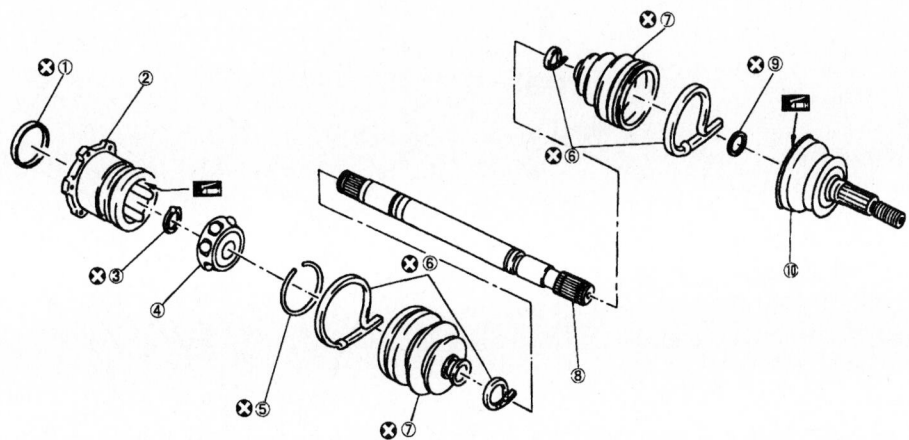

❌ : Always replace after every disassembly.

1. Plug	2. Housing	3. Snap ring
4. Ball cage, steel ball, liner race assembly	5. Stopper ring	6. Boot band
7. Boot	8. Shaft	9. Circlip
10. Joint sub-assembly		

67162-QX56-G32

Rear halfshaft exploded view

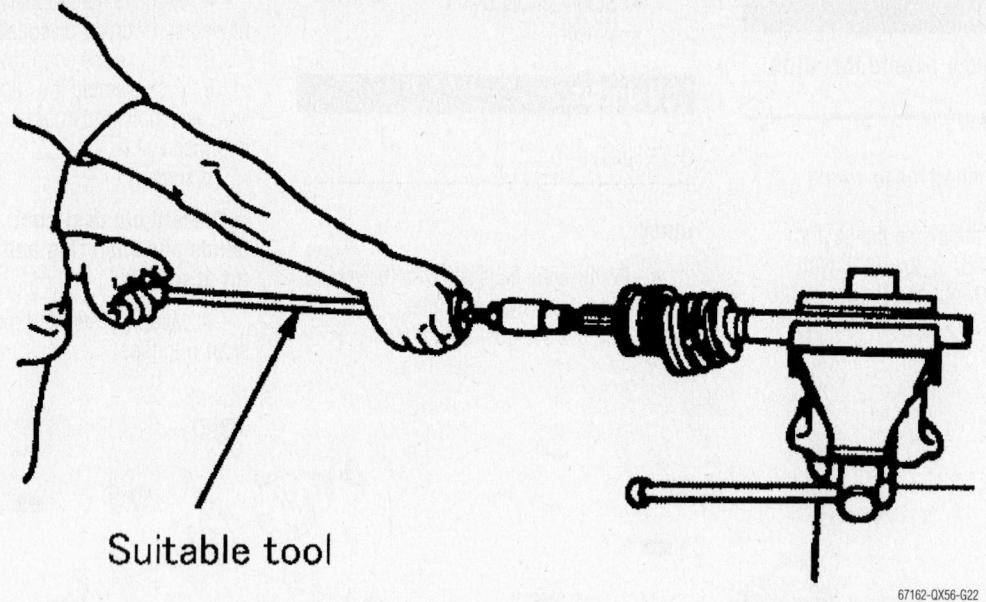

Suitable tool

Using a suitable puller to remove joint sub-assembly.

8. Install new dust boot and band onto halfshaft.

9. Remove tape from serrated part of halfshaft.

10. Install the cage, ball and inner race assembly.

11. Install new snap ring.

12. Insert 4.50-5.3 oz of genuine NISSAN grease or equivalent onto the housing and install onto halfshaft.

13. Install the stopper ring onto the housing.

14. Install the dust boot into the grooves on joint sub-assembly.

15. Secure the big and small ends of the dust boot using new boot bands.

Outer

1. Remove the halfshaft from the vehicle.

2. Mount halfshaft in a vise.

3. Remove the dust boot bands and dust boot from joint sub-assembly.

4. Insert a suitable puller into the threaded part of the halfshaft. Pull the joint sub-assembly off of the halfshaft as shown in figure.

5. Remove dust boot and circlip from halfshaft for cleaning and inspection.

To install:

➡Discard old dust boot, dust boot bands and circlip and use new ones for assembly.

6. Insert genuine NISSAN grease or equivalent into the joint sub-assembly until grease oozes from the ball groove and serration hole.

7. Wrap the serrated part of the halfshaft with tape.

8. Install new dust boot and band onto halfshaft.

9. Remove tape from serrated part of the halfshaft.

10. Press-fit the new circlip to the halfshaft.

11. Insert 5.1-5.8 oz of genuine NISSAN grease or equivalent into the joint sub-assembly and large end of boot.

12. Install the dust boot into the grooves on the joint sub-assembly.

13. Secure the big and small ends of the dust boot using new boot bands.

Front Differential Pinion Seal

REMOVAL & INSTALLATION

1. Remove or disconnect the following:
   - Front drive shaft
   - Halfshafts

2. Measure and record the pinion bearing preload using special tool J-25765-A.

3. Loosen the pinion nut while holding the companion flange using special tool J-44195.

4. Remove the companion flange using a suitable tool.

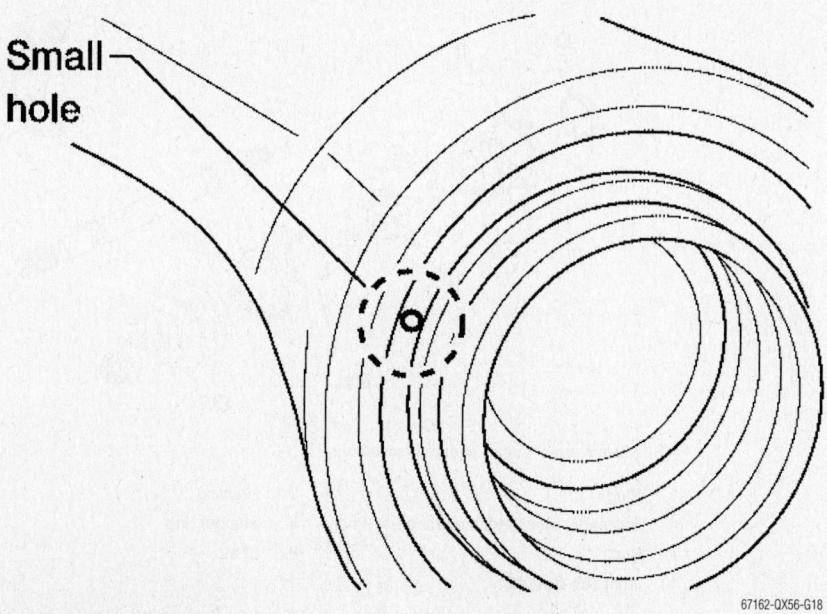

Small hole

Small hole in casing

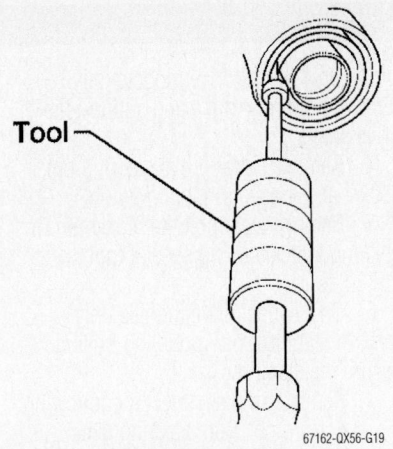

67162-QX56-G19

Removing the pinion seal

5. Using a punch or drill, place a small hole in the case.

6. Remove the seal using special tool SP8P.

To install:

7. Press front seal into carrier using a suitable tool.

8. Install companion flange and new pinion nut. Tighten pinion nut until there is no end play and until recorded pinion bearing preload is met plus an additional 5 inch lbs. (0.5 Nm).

9. Install or connect the following:
- Halfshafts
- Front drive shaft

Rear Differential Pinion Seal

REMOVAL & INSTALLATION

1. Remove the rear drive shaft.

2. Measure and record the total pre-load.

3. Matchmark the drive pinion to position 'B' on the companion flange.

4. Remove the drive pinion nut using suitable tool.

5. Remove the companion flange using suitable tool.

6. Remove the rear pinion seal using special tool J-34286.

To install:

7. Press the rear pinion seal into the carrier using suitable tool.

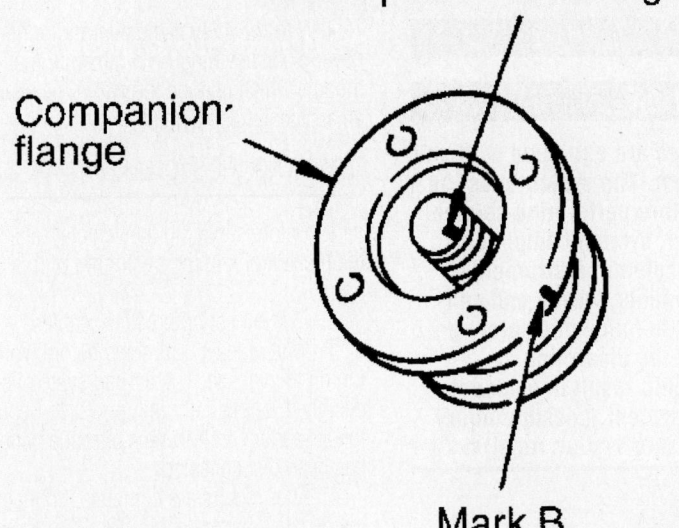

67162-QX56-G20

Companion flange marking

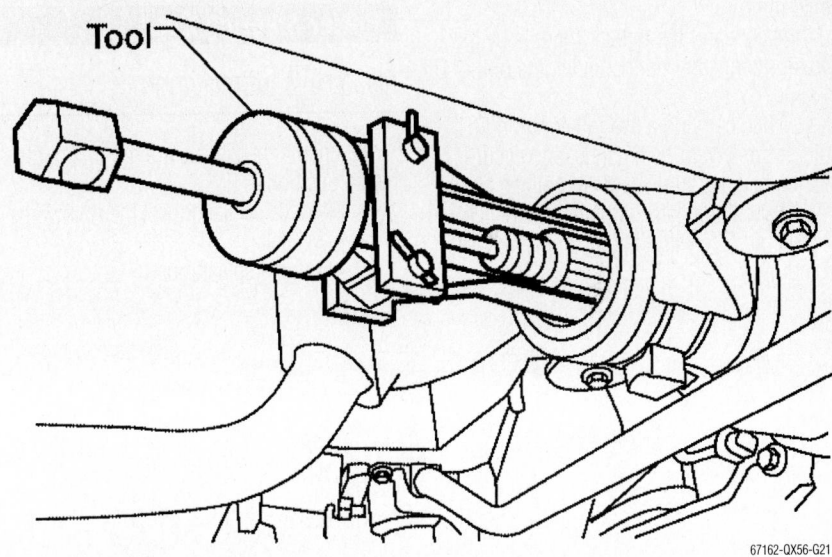

67162-QX56-G21

Removing the pinion seal

8. Align the matchmark on the companion flange to the drive pinion and install the companion flange.

9. Lubricate the drive pinion threads and seating surfaces of the drive pinion nut with grease.

10. Using a new drive pinion nut, tighten to 124-274 ft. lbs. (167-372 Nm).

➡**Final torque is determined when adjusting total preload using special tool J-25765-A.**

11. Install rear drive shaft.

STEERING AND SUSPENSION

Air Bag

✳✳ CAUTION

Some vehicles are equipped with an air bag system. The system must be disarmed before performing service on, or around, system components, the steering column, instrument panel components, wiring and sensors. Failure to follow the safety precautions and the disarming procedure could result in accidental air bag deployment, possible injury and unnecessary system repairs.

PRECAUTIONS

Several precautions must be observed when handling the inflator module to avoid accidental deployment and possible personal injury.

• Never carry the inflator module by the wires or connector on the underside of the module.

• When carrying a live inflator module, hold securely with both hands, and ensure that the bag and trim cover are pointed away.

• Place the inflator module on a bench or other surface with the bag and trim cover facing up.

• With the inflator module on the bench, never place anything on or close to the module which may be thrown in the event of an accidental deployment.

DISARMING THE SYSTEM

1. Before servicing the vehicle, refer to the precautions in the beginning of this section.

2. Disconnect both battery cables.

3. Wait at least 3 minutes before working on the vehicle. The air bag system is designed to retain enough power to deploy the air bag for a short time after the battery has been disconnected.

4. After repairs are complete, connect the negative battery cable. Turn the ignition switch to the **ON** position and check the air bag warning light blinks for proper operation.

Power Steering Gear

REMOVAL & INSTALLATION

1. Ensure the wheels are in the straight-ahead position.

2. Remove or disconnect the following:
   • Wheels
   • Engine splash guard

3. On 4-wheel drive models only, remove front final drive and support the drive shafts.

4. Remove cotter pin at steering outer socket and loosen mounting nut.

5. Remove steering outer socket from steering knuckle using special tool J-25730-A.

6. On 2-wheel drive models only, remove stabilizer bar mounting bolts and secure the stabilizer bar.

7. Remove or disconnect the following:
   • Oil pipes from steering gear assembly
   • Lower joint mounting bolt from lower shaft
   • Mounting bolts and nuts from steering gear assembly
   • Steering gear assembly

To install:

8. Install or connect the following:
   • Steering gear assembly, tighten nuts to 140 ft. lbs. (190 Nm)
   • Lower joint mounting bolt
   • Oil pipes to steering gear assembly
   • Stabilizer bar, 2 wheel-drive models only
   • Steering outer socket to steering knuckle, tighten nut to 63 ft. lbs. (86 Nm)

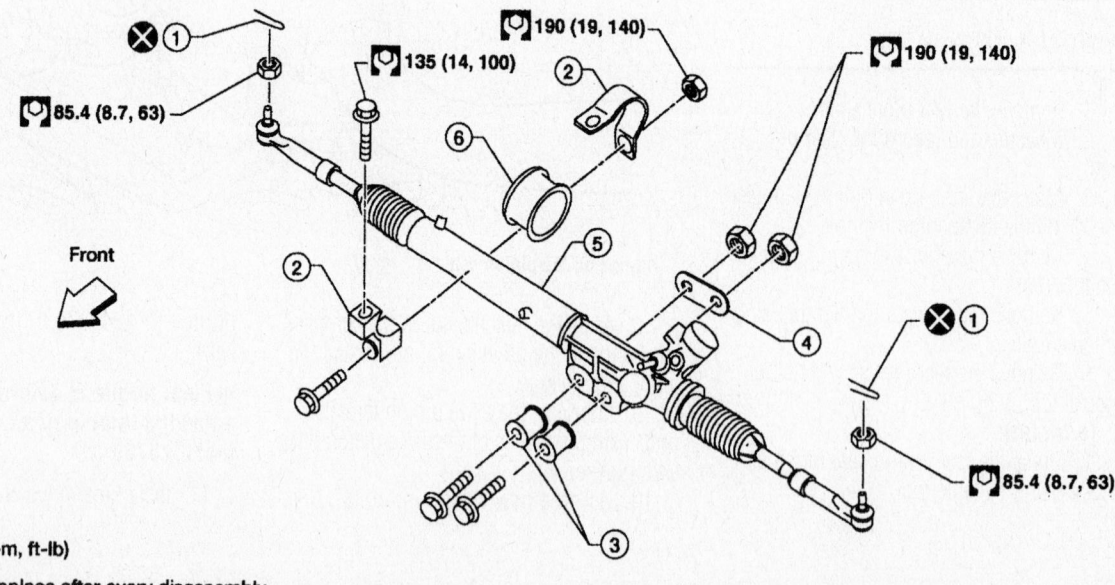

🔧 : N•m (kg-m, ft-lb)

❌ : Always replace after every disassembly.

1. Cotter pin	2. Mounting bracket	3. Bushing
4. Washer	5. Steering gear assembly	6. Mounting insulator

Steering gear

67162-QX56-G30

- Front final drive, 4-wheel drive models only
- Engine splash guard
- Wheels

9. Check the wheel alignment and adjust as necessary.

Shock Absorber

REMOVAL & INSTALLATION

Front

1. Remove or disconnect the following:
   - Wheel
   - Lower shock absorber bolt
   - Upper shock absorber bolts
   - Coil spring and shock absorber assembly
2. Secure the shock absorber in a vice and loosen (without removing) the piston rod lock nut.
3. Install a spring compressor and tighten until the shock absorber mounting insulator can be turned by hand.
4. Remove piston rod lock nut and remove shock absorber.

To install:

5. Install upper mounting insulator in line with the lower shock absorber mount and step in shock absorber lower seat as shown in figure.
6. Tighten the new piston rod lock nut to 40 ft. lbs. (54 Nm).
7. Install or connect the following:
   - Coil spring and shock absorber assembly
   - Upper shock absorber bolts and tighten to 22 ft. lbs (30 Nm)
   - Lower shock absorber bolt and tighten to 99 ft. lbs. (134 Nm)
   - Wheel
8. Check wheel alignment and adjust as necessary.

Rear

1. Remove the rear wheel.
2. Release the air pressure from the rear load leveling air suspension system using the CONSULT-II "EXHAUST SOLENOID" active test.
3. Remove or disconnect the following:
   - Rear fender protector
   - Rear load leveling air suspension hose from the shock absorber
   - Shock absorber upper and lower end bolts
   - Shock absorber

To install:

4. Install or connect the following:

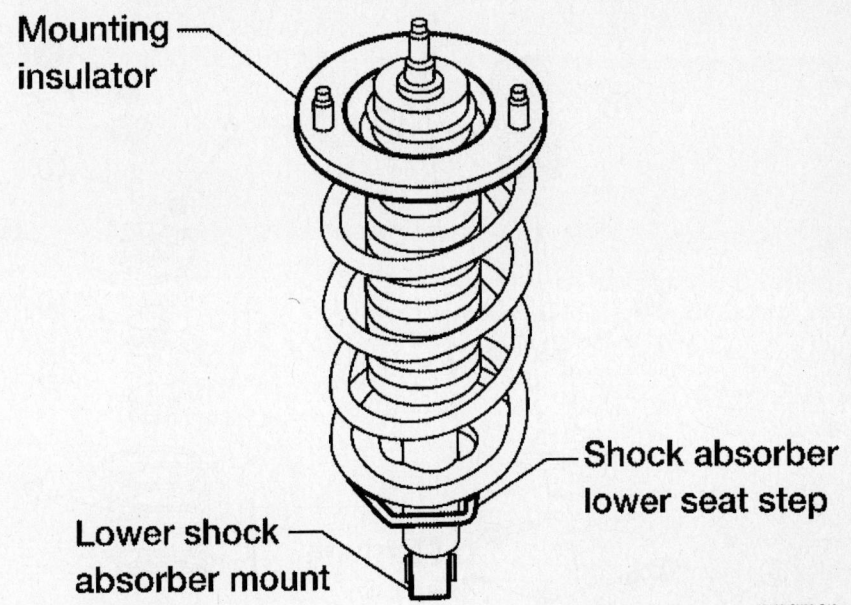

67162-QX56-G13

Shock absorber installation

- Shock absorber and tighten end bolts to 129 ft. lbs. (175 Nm)
- Rear load leveling air suspension hose
- Rear fender protector
- Rear wheel

Coil Spring

REMOVAL & INSTALLATION

Front

1. Remove or disconnect the following:
   - Wheel
   - Lower shock absorber bolt
   - Upper shock absorber bolts
   - Coil spring and shock absorber assembly
2. Secure the shock absorber in a vice and loosen (without removing) the piston rod lock nut.
3. Install a spring compressor and tighten until the shock absorber mounting insulator can be turned by hand.
4. Remove piston rod lock nut and remove shock absorber from the coil spring.

To install:

5. Install upper mounting insulator in line with the lower shock absorber mount and step in shock absorber lower seat as shown in figure.
6. Tighten the new piston rod lock nut to 40 ft. lbs. (54 Nm).
7. Install or connect the following:
   - Coil spring and shock absorber assembly

- Upper shock absorber bolts and tighten to 22 ft. lbs (30 Nm)
- Lower shock absorber bolt and tighten to 99 ft. lbs. (134 Nm)
- Wheel

8. Check wheel alignment and adjust as necessary.

Rear

1. Remove the rear wheel.
2. Release the air pressure from the rear load leveling air suspension system using the CONSULT-II "EXHAUST SOLENOID" active test.
3. Remove the height sensor arm bracket bolt from the left-hand rear lower link.
4. Place a suitable jack under the rear lower link and relieve the coil spring tension.
5. Loosen the rear lower link adjusting bolt and nut connected to the rear suspension member.
6. Remove the rear lower link bolt and nut from the knuckle.
7. Slowly lower the jack to relieve the coil spring tension.
8. Remove the coil spring.

To install:

9. Install or connect the following:
   - Coil spring

➡**When installing the rubber seats for the coil spring, ensure the embossed arrows point outward towards the wheel.**

- Rear lower link bolt to knuckle and tighten nut to 70 ft. lbs. (95 Nm)

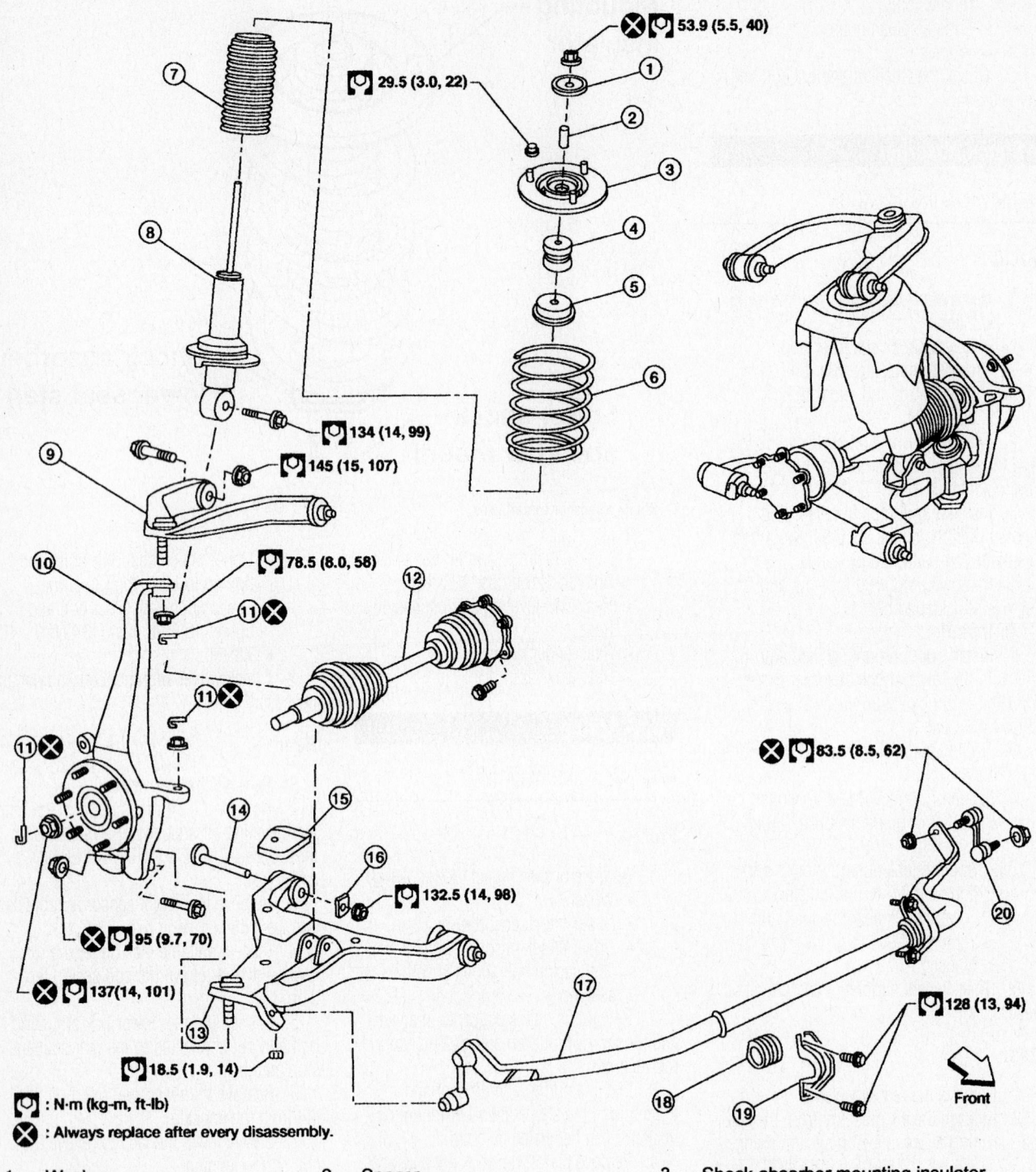

⊗ ◎ 53.9 (5.5, 40)

◎ 29.5 (3.0, 22)

◎ 134 (14, 99)

◎ 145 (15, 107)

◎ 78.5 (8.0, 58)

◎ 83.5 (8.5, 62)

⊗ ◎ 95 (9.7, 70)

◎ 132.5 (14, 98)

⊗ ◎ 137(14, 101)

◎ 18.5 (1.9, 14)

◎ 128 (13, 94)

Front

◎ : N·m (kg-m, ft-lb)

⊗ : Always replace after every disassembly.

1.	Washer	2.	Spacer	3.	Shock absorber mounting insulator
4.	Shock absorber bushing	5.	Upper seat	6.	Coil spring
7.	Dust cover	8.	Shock absorber	9.	Upper link
10.	Steering knuckle	11.	Cotter pin	12.	Drive shaft
13.	Lower link	14.	Cam bolt	15.	Jounce bumper
16.	Cam washer	17.	Stabilizer bar	18.	Stabilizer bar bushing
19.	Stabilizer bar mounting bracket	20.	Connecting rod		

67162-QX56-G29

Front suspension components

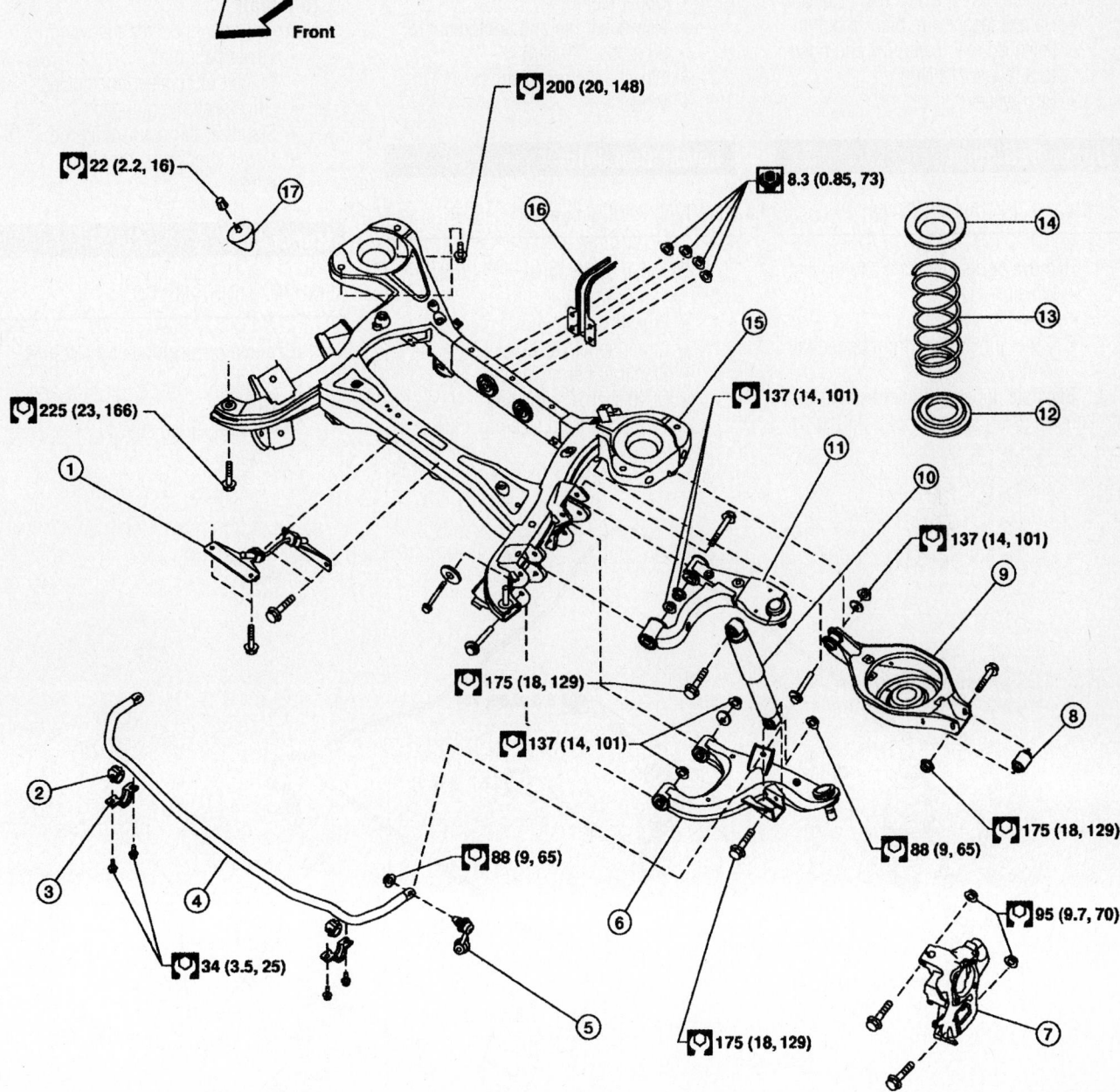

Front

200 (20, 148)

22 (2.2, 16)

8.3 (0.85, 73)

225 (23, 166)

137 (14, 101)

137 (14, 101)

175 (18, 129)

137 (14, 101)

88 (9, 65)

88 (9, 65)

175 (18, 129)

34 (3.5, 25)

95 (9.7, 70)

175 (18, 129)

N·m (kg-m, in-lb)

N·m (kg-m, ft-lb)

1.	Seat belt latch anchor	2.	Stabilizer bar bushing	3.	Stabilizer bar clamp
4.	Stabilizer bar	5.	Connecting rod	6.	Front lower link
7.	Knuckle	8.	Bushing	9.	Rear lower link
10.	Shock absorber	11.	Suspension arm	12.	Lower rubber seat
13.	Coil spring	14.	Upper rubber seat	15.	Rear suspension member
16.	Spare tire bracket	17.	Bound bumper		

67162-QX56-G28

Standard rear suspension

- Rear lower link adjusting bolt to rear suspension member and tighten nut to 101 ft. lbs. (137 Nm)
- Height sensor arm bracket bolt to left-head rear lower link and tighten to 9 ft. lbs. (12 Nm)
- Rear wheel

Upper Ball Joint

REMOVAL & INSTALLATION

1. Remove or disconnect the following:
   - Wheel
   - Wheel opening shield
   - Cotter pin and nut from upper ball joint
2. Separate upper ball joint from steering knuckle using special tool J-24319-01

To install:

3. Install or connect the following:
   - Upper ball joint
   - New cotter pin and tighten nut to 58 ft. lbs. (79 Nm)
   - Wheel opening shield
   - Wheel

Lower Ball Joint

REMOVAL & INSTALLATION

1. Remove or disconnect the following:
   - Wheel
   - Lower shock absorber bolt
   - Stabilizer bar connecting rod
   - Drive shaft, if equipped
   - Pinch bolt from steering knuckle

2. Separate lower ball joint from steering knuckle

To install:

3. Install or connect the following:
   - Lower ball joint
   - Pinch bolt to steering knuckle
   - Drive shaft, if equipped
   - Stabilizer bar connecting rod
   - Lower shock absorber bolt
   - Wheel

Upper Link

REMOVAL & INSTALLATION

1. Remove or disconnect the following:
   - Wheel
   - Wheel opening shield

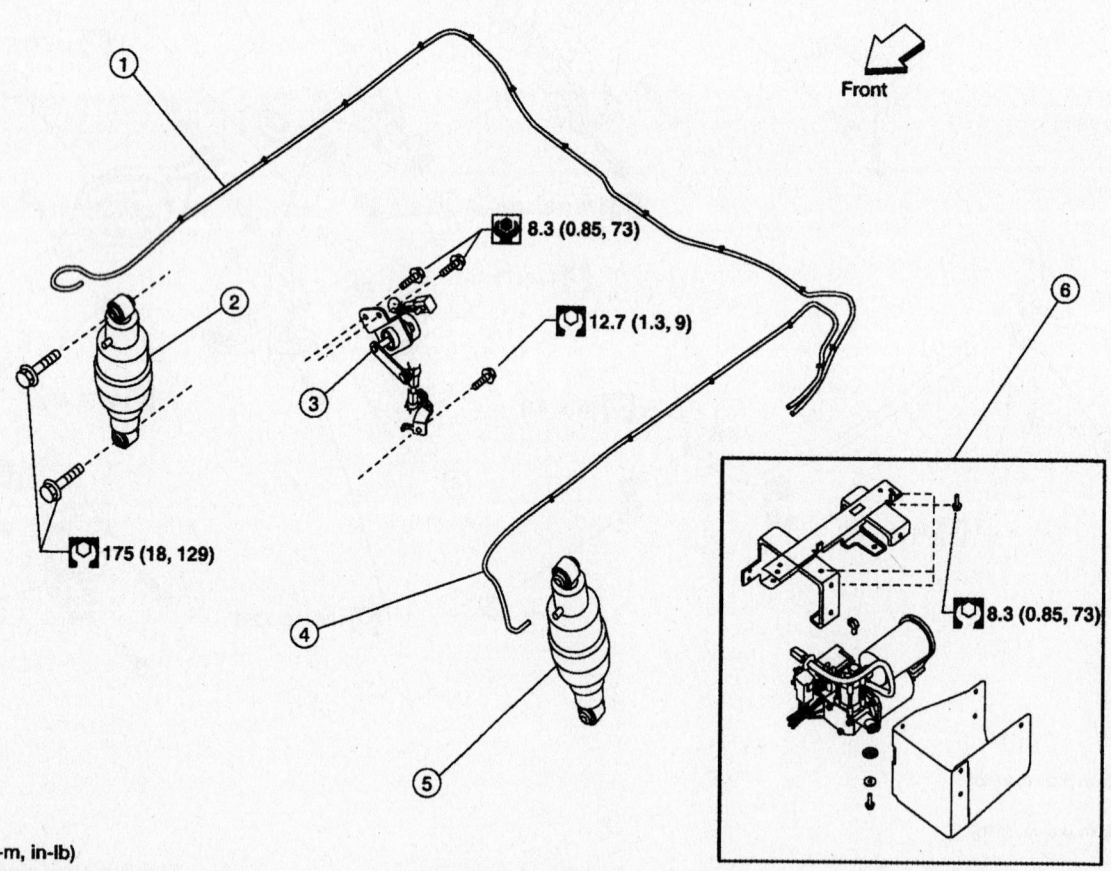

8.3 (0.85, 73)

12.7 (1.3, 9)

175 (18, 129)

8.3 (0.85, 73)

Front

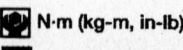 N·m (kg-m, in-lb)

N·m (kg-m, ft-lb)

1. Rear load leveling air suspension hose, RH	2. Shock absorber, RH	3. Height sensor
4. Rear load leveling air suspension hose, LH	5. Shock absorber, LH	6. Rear load leveling air suspension compressor assembly

Rear load leveling air suspension system

67162-QX56-G27

- Cotter pin and nut from upper ball joint
2. Separate upper ball joint stud from steering knuckle using special tool J-24319-01.
3. Remove the following:
   - Upper link mounting bolts
   - Upper link
To install:
4. Install or connect the following:
   - Upper link and tighten bolts to 107 ft. lbs. (145 Nm)
   - Upper ball joint with new cotter pin and tighten nut to 58 ft. lbs. (79 Nm)
   - Wheel opening shield
   - Wheel

Lower Link

REMOVAL AND & INSTALLATION

1. Remove or disconnect the following:
   - Wheel
   - Lower shock absorber bolt
   - Stabilizer bar connecting rod

- Drive shaft, if equipped
- Pinch bolt from steering knuckle
2. Separate lower ball joint from steering knuckle.
3. Remove the following:
   - Lower link adjusting bolts
   - Lower link
To install:
4. Install or connect the following:
   - Lower link and tighten adjusting bolts to 98 ft. lbs. (133 Nm)
   - Lower ball joint
   - Pinch bolt
   - Drive shaft, if equipped
   - Stabilizer bar connected rod
   - Lower shock absorber bolt
   - Wheel

Wheel Bearings

REMOVAL & INSTALLATION

Front

1. Remove or disconnect the following:

- Wheel
- Engine splash guard
- Brake caliper without disconnecting the hydraulic lines, and reposition aside with wire
2. Matchmark the brake rotor to the wheel hub and remove the brake rotor.
3. Remove or disconnect the following:
   - Cotter pin and lock nut from drive shaft
   - Drive shaft from wheel hub and bearing assembly
   - ABS sensor
   - Wheel hub and bearing assembly bolts
   - Wheel hub and bearing assembly
To install:
4. Install or connect the following:
   - Wheel hub and bearing assembly, using new bolts and tighten to 155 ft. lbs. (210 Nm)
   - ABS sensor
   - Drive shaft to wheel hub and bearing assembly
   - Cotter pin and lock nut and tighten to 101 ft. lbs. (137 Nm)

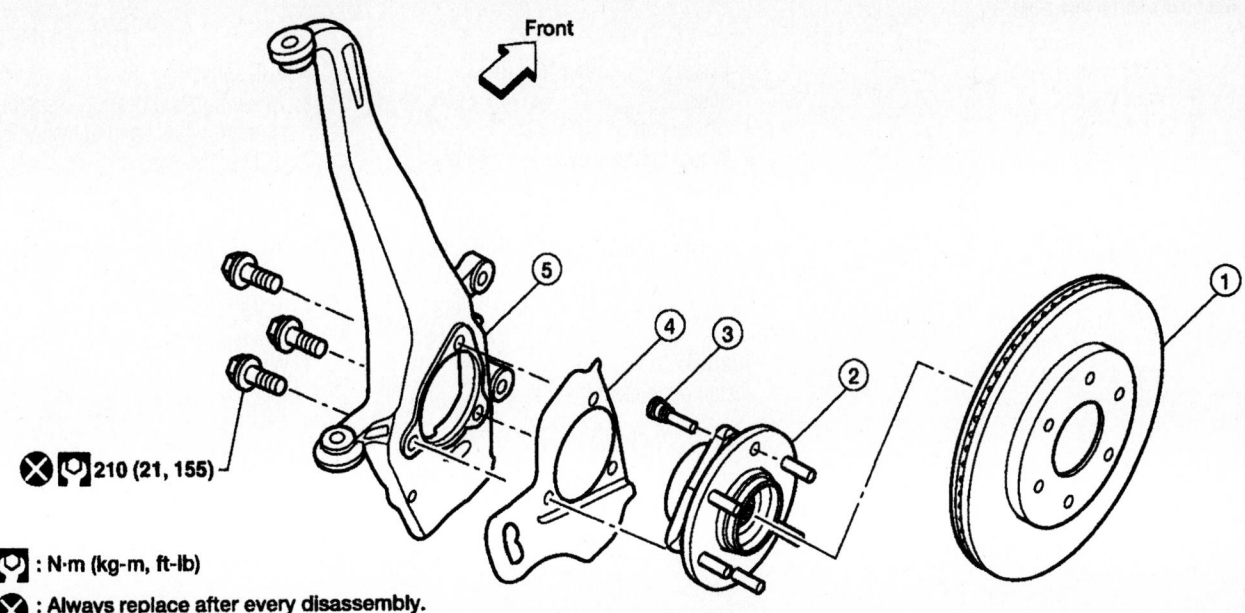

Front

210 (21, 155)

: N·m (kg-m, ft-lb)

: Always replace after every disassembly.

1. Disc rotor
4. Splash guard
2. Wheel hub and bearing assembly
5. Steering knuckle
3. Wheel stud

Front hub and related parts

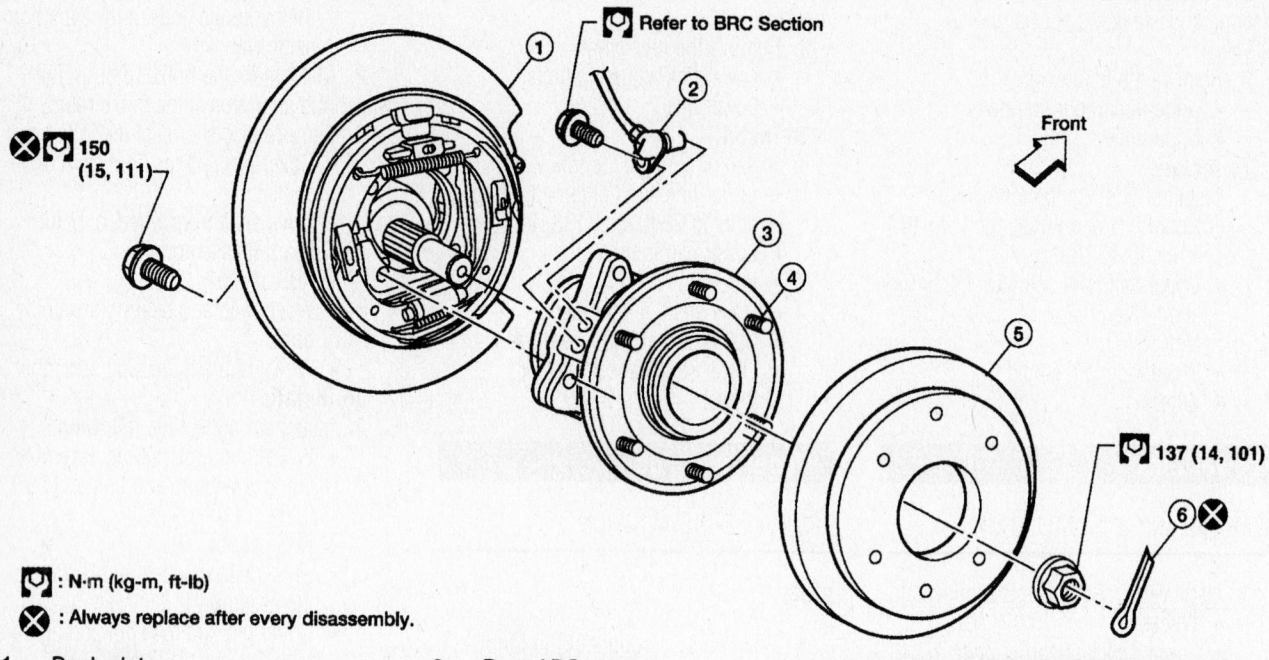

Refer to BRC Section

Front

⊗ 🔧 150
(15, 111)

🔧 137 (14, 101)

🔧 : N·m (kg-m, ft-lb)

⊗ : Always replace after every disassembly.

1. Back plate
4. Wheel stud

2. Rear ABS sensor
5. Rear disc rotor

3. Wheel hub and bearing assembly
6. Cotter pin

67162-QX56-G25

Rear hub and related parts

- Brake rotor
- Brake caliper
- Engine splash guard
- Wheel

Rear

1. Remove or disconnect the following:
   - Wheel
   - Brake caliper without disconnecting the hydraulic lines, and reposition aside with wire
   - Brake rotor
   - Cotter pin and nut from drive shaft
   - Drive shaft
   - Wheel hub and bearing assembly bolts

2. Pulling out the wheel hub and bearing assembly slightly, remove the ABS sensor.

3. Remove the wheel hub and bearing assembly.

To install:

4. Install or connect the following:
   - ABS sensor
   - Wheel hub and bearing assembly, using new bolts and tighten to 111 ft. lbs. (150 Nm)
   - Drive shaft
   - Lock nut and tighten to 101 ft. lbs. (137 Nm) and new cotter pin
   - Brake rotor
   - Brake caliper
   - Wheel

BRAKES

Brake Caliper

REMOVAL & INSTALLATION

Front

1. Drain brake fluid as necessary.
2. Remove or disconnect the following:
   - Wheel
   - Union bolt
   - Caliper-to-torque member slide

pins, or remove the caliper and torque member as an assembly.
   - Brake caliper

To install:

3. Install or connect the following:
   - Brake caliper, tighten torque member bolts to 155 ft. lbs. (210 Nm); the caliper slide pins to 32 ft. lbs. (44 Nm)
   - Union bolt and tighten to 13 ft. lbs. (18 Nm)
4. Fill the master cylinder and bleed the brake system.
5. Install the wheels.

Rear

1. Drain brake fluid as necessary.
2. Remove or disconnect the following:
   - Wheel
   - Union bolt
   - Mounting bolts
   - Brake caliper assembly

To install:

3. Install or connect the following:
   - Brake caliper assembly and tighten mounting bolts to 24 ft. lbs. (32 Nm)

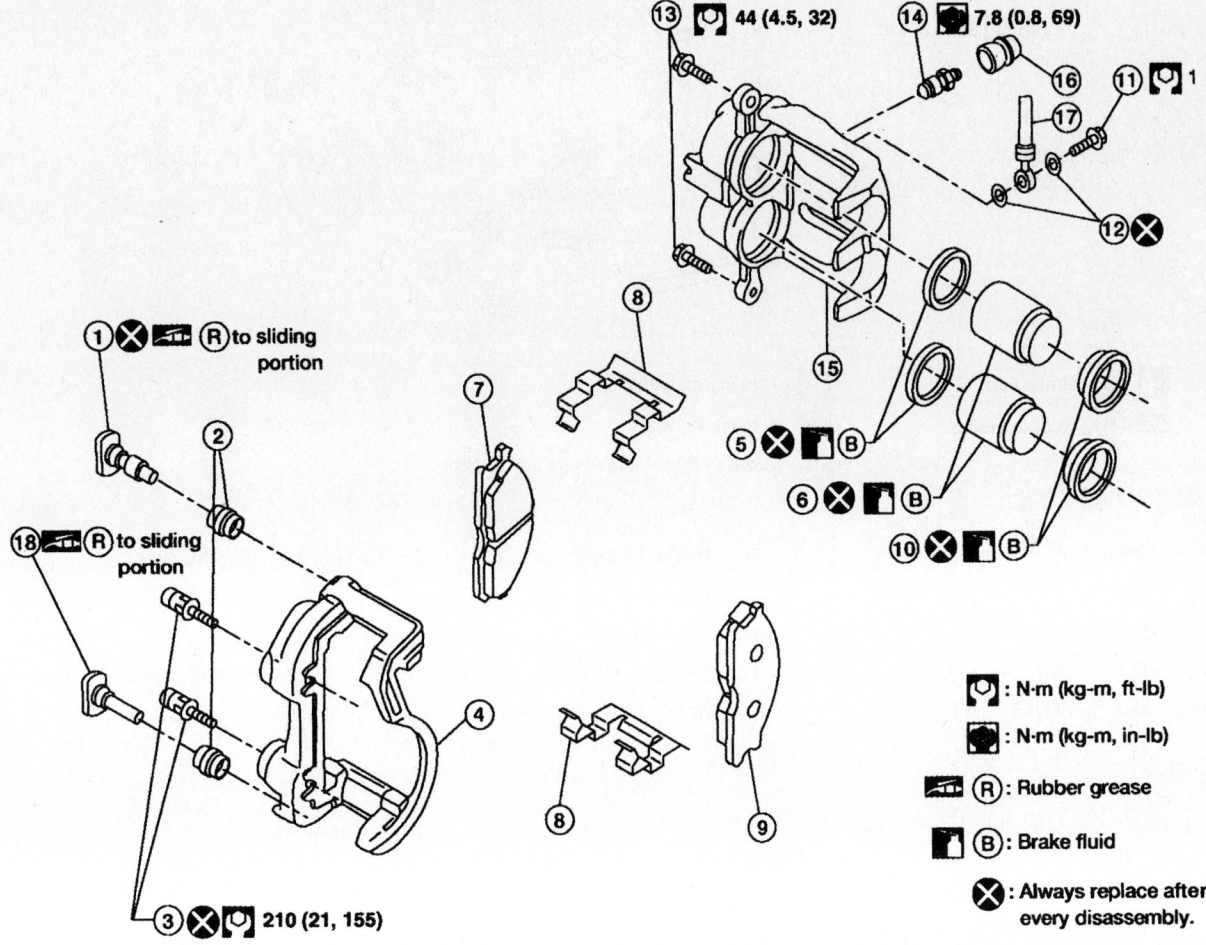

1. Upper sliding pin	2. Sliding pin boot	3. Torque member bolt
4. Torque member	5. Piston seal	6. Piston
7. Inner pad	8. Pad retainer	9. Outer pad
10. Piston boot	11. Union bolt	12. Copper washer
13. Sliding pin bolt	14. Bleed valve	15. Brake caliper
16. Cap	17. Brake hose	18. Lower sliding pin

67162-QX56-G24

Front brakes

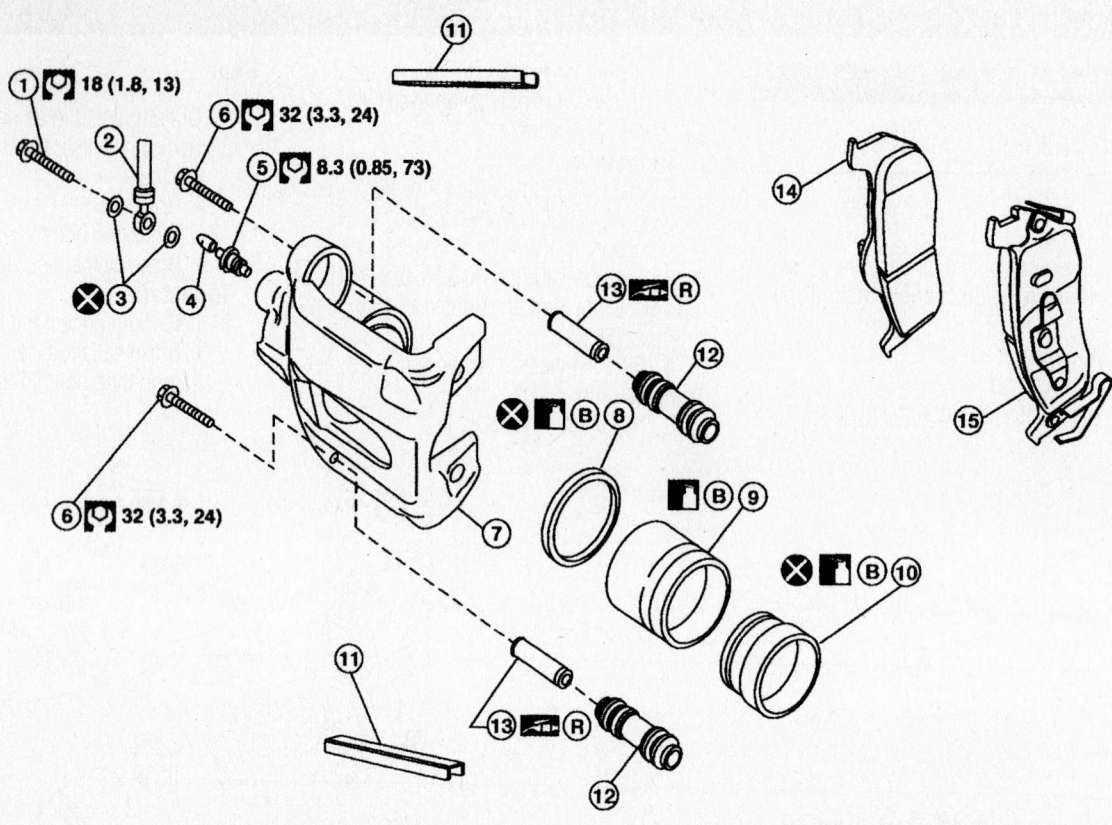

Legend:

- [B] : Brake fluid
- [R] : Rubber grease
- [N·m] : N·m (kg-m, ft-lb)
- [N·m] : N·m (kg-m, in-lb)
- ⊗ : Always replace after every disassembly.

1.	Union bolt	2.	Brake hose	3.	Copper washer
4.	Cap	5.	Bleed valve	6.	Mounting bolt
7.	Brake caliper	8.	Piston seal	9.	Piston
10.	Piston boot	11.	Knuckle slide	12.	Sliding sleeve boot
13.	Sliding sleeve	14.	Inner pad	15.	Outer pad

67162-QX56-G23

Rear brakes

- Union bolt and tighten to 13 ft. lbs. (18 Nm)
4. Fill the master cylinder and bleed the brake system.
5. Install the wheels.

Disc Brake Pads

REMOVAL AND INSTALLATION

Front

1. Remove the wheel.
2. Remove lower sliding pin bolt.

3. Suspend brake caliper with a remove and remove brake pad and shim from torque member.

To install:

4. Push pistons in so that the pad is firmly installed, using a suitable tool.
5. Mount the brake caliper to torque member.
6. Attach pad retainer to torque member.
7. Lubricate lower sliding pin bolt with a thin layer of silicone grease and install. Torque to 32 ft. lbs. (44 Nm).
8. Install the wheel.

Rear

1. Remove the wheel.
2. Remove mounting bolt from the top mount.
3. Swing brake caliper open and remove the brake pads.

To install:

4. Push pistons in so that the pad is firmly installed, using a suitable tool.
5. Install pads to the brake caliper.
6. Install top mounting bolt and tighten to 24 ft. lbs. (32 Nm).
7. Install the wheel.

ISUZU

Ascender

SPECIFICATION AND MAINTENANCE CHARTS

ENGINE AND VEHICLE IDENTIFICATION

Code ①	Liters (cc)	Cu. In.	Cyl.	Fuel Sys.	Engine Type	Eng. Mfg.
S	4.2 (4200)	256	6	MFI	DOHC	CPC
P	5.3 (5326)	325	8	SFI	OHV	CPC

Code ②	Year
Y	2001
1	2002
2	2003
3	2004
4	2005

CPC: Chevrolet/Pontiac/Canada

MFI: Multi-port Fuel Injection

① 8th position of VIN

② 10th position of VIN

67162-ASCE-C01

GENERAL ENGINE SPECIFICATIONS
All measurements are given in inches.

Year	Model	Engine Displacement Liters (VIN)	Net Horsepower @ rpm	Net Torque B rpm (ft. lbs.)	Bore x Stroke (in.)	Compression Ratio	Oil Pressure @ rpm
2003-04	Ascender	4.2 (S)	275@6000	275@3600	3.66x4.02	10.0:1	12@1200
		5.3 (P)	290@5300	325@4000	3.78x3.62	9.45:1	6@1000

67162-ASCE-C02

GASOLINE ENGINE TUNE-UP SPECIFICATIONS

Year	Engine Displacement Liters (VIN)	Spark Plugs Gap (in.)	Ignition Timing (deg.) MT	Ignition Timing (deg.) AT	Fuel Pump (psi)	Idle Speed (rpm) MT	Idle Speed (rpm) AT	Valve Clearance In.	Valve Clearance Ex.
2003-04	4.2 (S)	0.050	—	①	50-57 ②	—	③	HYD	HYD
	5.3 (P)	0.060	④	④	55-62 ②	—	625	HYD	HYD

NOTE: The Vehicle Emission Control Information label often reflects specification changes made during production. The label figures must be used if they differ from those in this chart.

HYD: Hydraulic

① Distributorless ignition, cannot be adjusted

② With key ON and engine OFF

③ Idle speed is maintained by the PCM

④ Distributorless ignition, cannot be adjusted

67162-ASCE-C03

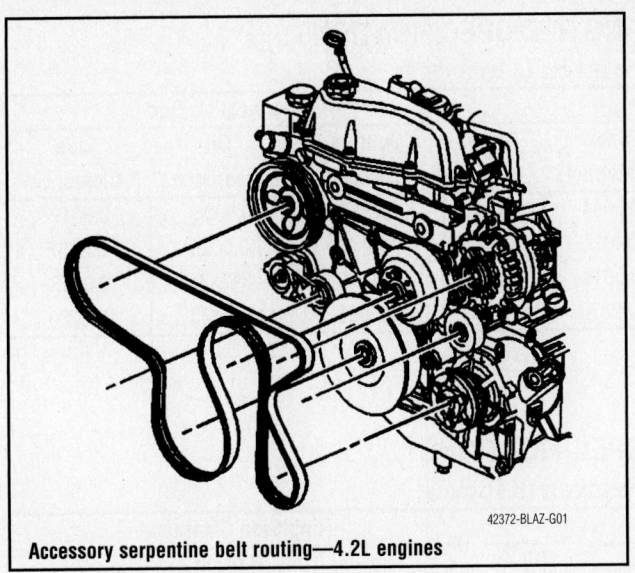

Accessory serpentine belt routing—4.2L engines

42372-BLAZ-G01

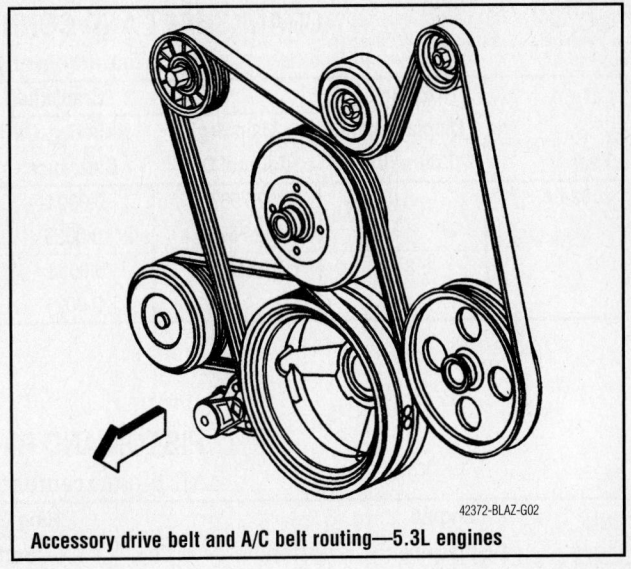

Accessory drive belt and A/C belt routing—5.3L engines

42372-BLAZ-G02

CAPACITIES

Year	Model	Engine Displacement Liters (VIN)	Engine Oil with Filter (qts.)	Transmission (pts.) 5-Spd	Transmission (pts.) Auto.	Transfer Case (pts.)	Drive Axle Front (pts.)	Drive Axle Rear (pts.)	Fuel Tank (gal.)	Cooling System (qts.)
2003-04	Ascnder	4.2 (S)	7.0	—	10.0	2.6	2.6	4.0	25.0	13.9
		5.3 (P)	6.0	—	10.0	4.0	2.6	4.0	25.0	14.9

NOTE: All capacities are approximate. Add fluid gradually and check to be sure a proper fluid level is obtained.

67162-ASCE-C04

VALVE SPECIFICATIONS

Year	Engine Displacement Liters (VIN)	Seat Angle (deg.)	Face Angle (deg.)	Spring Test Pressure (lbs. @ in.)	Spring Installed Height (in.)	Stem-to-Guide Clearance (in.) Intake	Stem-to-Guide Clearance (in.) Exhaust	Stem Diameter (in.) Intake	Stem Diameter (in.) Exhaust
2003-04	4.2 (S)	NA	NA	130-142@1.26	NA	0.0011-0.0025	0.0015-0.0030	NA	NA
	5.3 (P)	46	45	220@1.32	1.80	0.0010-0.0026	0.0010-0.0026	0.313-0.314	0.313-0.314

NA: Not Available

67162-ASCE-C05

CRANKSHAFT AND CONNECTING ROD SPECIFICATIONS
All measurements are given in inches.

Year	Engine Displacement Liters (VIN)	Crankshaft				Connecting Rod		
		Main Brg. Journal Dia.	Main Brg. Oil Clearance	Shaft End-play	Thrust on No.	Journal Diameter	Oil Clearance	Side Clearance
2003-04	4.2 (S)	2.7567-2.7574	0.0004-0.0025	0.0044-0.0153	4	2.2337-2.2342	0.0008-0.0025	0.0019-0.0137
	5.3 (P)	2.558	0.0008-0.0021	0.0015-0.0078	4	2.0991-2.0999	0.0009-0.0003	0.0043-0.0200

67162-ASCE-C06

PISTON AND RING SPECIFICATIONS
All measurements are given in inches.

Year	Engine Displacement Liters (VIN)	Piston Clearance	Ring Gap			Ring Side Clearance		
			Top Compression	Bottom Compression	Oil Control	Top Compression	Bottom Compression	Oil Control
2003-04	4.2 (S)	0.0004-0.0017	0.0079-0.0157	0.0118-0.0197	0.0098-0.0299	0.0017-0.0037	0.0017-0.0037	0.0023-0.0085
	5.3 (P)	-0.0014 0.0006	0.009-0.017	0.017-0.027	0.007-0.029	0.0015-0.0033	0.0015-0.0031	0.0005-0.0078

67162-ASCE-C07

TORQUE SPECIFICATIONS
All readings in ft. lbs.

Year	Engine Displacement Liters (VIN)	Cylinder Head Bolts	Main Bearing Bolts	Rod Bearing Bolts	Crankshaft Damper Bolts	Flywheel Bolts	Manifold		Spark Plugs	Oil Pan Drain Plug
							Intake *	Exhaust		
2003-04	4.2 (S)	①	②	③	④	⑤	12	⑥	13	19
	5.3 (P)	⑦	⑧	⑨	⑩	⑪	⑫	⑬	11	19

* NOTE: Applies to Lower Manifold only.

① Cylinder head bolts (14)
    1st pass: 22 ft. lbs.
    2nd pass: Plus 155 degrees
    2 short end bolts: 62 inch. lbs.
    2 short end bolts: plus 60 degrees
    1 long end bolt: 62 inch lbs.
    1 long end bolt: plus 120 degrees

② 18 ft. lbs., plus 155 depress

③ 18 ft. lbs., plus 110 degrees

④ 111 ft. lbs., plus 180 degrees

⑤ 18 ft. lbs., plus 50 degrees

⑥ 1st pass: 18 ft. lbs.
    2nd pass: 18 ft. lbs.
    3rd pass: 18 ft. lbs.

⑦ 1st pass: 22 ft. lbs.
    2nd pass: Plus 90 degrees
    Final pass:
    Except medium bolts at front and rear of head: Plus 90 degrees
    Medium bolts at front and rear of head: Plus 50 degrees

⑧ Inner bolts:
    1st pass: 15 ft. lbs.
    Final pass: Plus 80 degrees
    Outer bolts:
    1st pass: 15 ft. lbs.
    Final pass: Plus 51 degrees

⑨ 15 ft. lbs. plus 75 degrees

⑩ Installation pass: 240 ft. lbs. (discard bolt)
    First pass: 37 ft. lbs. (new bolt)
    Final pass: Plus 140 degrees

⑪ 1st pass: 15 ft. lbs.
    2nd pass: 37 ft. lbs.

⑫ 1st pass: 44 inch lbs.
    2nd pass: 89 inch lbs.

⑬ 1st pass: 11 ft. lbs.
    2nd pass: 18 ft. lbs.

67162-ASCE-C08

WHEEL ALIGNMENT

Year	Model		Caster Range (+/-Deg.)	Caster Preferred Setting (Deg.)	Camber Range (+/-Deg.)	Camber Preferred Setting (Deg.)	Toe-in (in.)
2003-04	Ascender	2WD	1.00	+3.50	1.00	-0.5	-0.5+/-0.5
		4WD	1.00	+3.50	1.00	-0.5	-0.5+/-0.5

67162-ASCE-C09

TIRE, WHEEL AND BALL JOINT SPECIFICATIONS

Year	Model	OEM Tires Standard	OEM Tires Optional	Tire Pressures (psi) Front	Tire Pressures (psi) Rear	Wheel Size	Ball Joint Inspection	Lug Nut
2002	Ascender	P245/65SR17	NA	36	36	17x7	L ①	103

NA: Not Available

OEM: Original Equipment Manufacturer

PSI: Pounds Per Square Inch

STD: Standard

OPT: Optional

L: Lower

U: Upper

① Do not lift truck. Inspect the boss into which the grease fitting is threaded. Replace if the boss is flush or receded below the surface of the ball joint

67162-ASCE-C10

BRAKE SPECIFICATIONS

All measurements in inches unless noted

Year	Model		Brake Disc Original Thickness	Brake Disc Minimum Thickness	Brake Disc Maximum Runout	Brake Drum Diameter Original Inside Diameter	Brake Drum Diameter Max. Wear Limit	Brake Drum Diameter Maximum Machine Diameter	Minimum Lining Thickness	Brake Caliper Bracket Bolts (ft. lbs.)	Brake Caliper Mounting Bolts (ft. lbs.)
2003-04	Ascender	F	1.030	0.965	0.003	—	—	—	0.030	52	①
		R	0.787	0.728	0.004	9.50	9.59	9.56	0.030	NA	—

NA: Not Available

① 2WD: 38 ft. lbs.

4WD: 77 ft. lbs.

67162-ASCE-C11

SCHEDULED MAINTENANCE INTERVALS
ISUZU—ASCENDER

TO BE SERVICED	TYPE OF SERVICE	VEHICLE MILEAGE INTERVAL (x1000)															
		7.5	15	22.5	30	37.5	45	52.5	60	67.5	75	82.5	90	97.5	105	112.5	120
Accessory drive belt	S/I								✓								✓
Air cleaner filter	R				✓				✓				✓				✓
Automatic transmission fluid	R	Every 50,000 miles															
Brake system ①	S/I	✓	✓	✓	✓	✓	✓	✓	✓	✓	✓	✓	✓	✓	✓	✓	✓
Chassis & suspension grease points	L	✓	✓	✓	✓	✓	✓	✓	✓	✓	✓	✓	✓	✓	✓	✓	✓
CV-joint boots & axle seals	S/I	✓	✓	✓	✓	✓	✓	✓	✓	✓	✓	✓	✓	✓	✓	✓	✓
Engine coolant system ②	S/I	Every 150,000 miles															
Engine oil & filter	R	✓	✓	✓	✓	✓	✓	✓	✓	✓	✓	✓	✓	✓	✓	✓	✓
Front wheel bearings	S/I & L				✓				✓				✓				✓
Fuel filter	R				✓				✓				✓				✓
Fuel tank, cap & lines	S/I								✓								✓
PCV valve	S/I	Every 100,000 miles															
Rear/front axle fluid level	S/I	✓	✓	✓	✓	✓	✓	✓	✓	✓	✓	✓	✓	✓	✓	✓	✓
Rotate tires	S/I	✓	✓	✓	✓	✓	✓	✓	✓	✓	✓	✓	✓	✓	✓	✓	✓
Spark plug wires	S/I	Every 100,000 miles															
Spark plugs	R	Every 100,000 miles															

R: Replace   S/I: Inspect and service, if necessary   L: Lubricate

① This should be performed when the tires are removed for rotation.

② Drain, flush and refill the cooling system, inspect the system hoses, and clean the radiator and condenser.

③ 2-wheel drive models only.

FREQUENT OPERATION MAINTENANCE (SEVERE SERVICE)

If a vehicle is operated under any of the following conditions it is considered severe service:

- Towing a trailer or using a camper or car-top carrier.

- Repeated short trips of less than 5 miles in temperatures below freezing, or trips of less than 10 miles in any temperature.

- Extensive idling or low-speed driving for long distances as in heavy commercial use, such as delivery, taxi or police cars.

- Operating on rough, muddy or salt-covered roads.

- Operating on unpaved or dusty roads.

- Driving in extremely hot (over 90°) conditions.

Engine oil & filter: replace every 3000 miles or 3 months, whichever occurs first.

Chassis and suspension grease points: lubricate every 3000 miles.

Rear/front axle fluid level: inspect every 3000 miles.

Rotate the tires ever 6000 miles.

Brake system components: inspect ever 6000 miles.

Front wheel bearings (2-wheel drive only): clean, inspect and repack every 15,000 miles.

Air cleaner filter: inspect every 15,000 miles.

Automatic transmission fluid & filter: replace every 15,000 miles.

67162-ASCE-C12

PRECAUTIONS

Before servicing any vehicle, please be sure to read all of the following precautions, which deal with personal safety, prevention of component damage, and important points to take into consideration when servicing a motor vehicle:

• Never open, service or drain the radiator or cooling system when the engine is hot; serious burns can occur from the steam and hot coolant.

• Observe all applicable safety precautions when working around fuel. Whenever servicing the fuel system, always work in a well-ventilated area. Do not allow fuel spray or vapors to come in contact with a spark, open flame, or excessive heat (a hot drop light, for example). Keep a dry chemical fire extinguisher near the work area. Always keep fuel in a container specifically designed for fuel storage; also, always properly seal fuel containers to avoid the possibility of fire or explosion. Refer to the additional fuel system precautions later in this section.

• Fuel injection systems often remain pressurized, even after the engine has been turned **OFF**. The fuel system pressure must be relieved before disconnecting any fuel lines. Failure to do so may result in fire and/or personal injury.

• Brake fluid often contains polyglycol ethers and polyglycols. Avoid contact with the eyes and wash your hands thoroughly after handling brake fluid. If you do get brake fluid in your eyes, flush your eyes with clean, running water for 15 minutes. If eye irritation persists, or if you have taken brake fluid internally, IMMEDIATELY seek medical assistance.

• The EPA warns that prolonged contact with used engine oil may cause a number of skin disorders, including cancer! You should make every effort to minimize your exposure to used engine oil. Protective gloves should be worn when changing oil. Wash your hands and any other exposed skin areas as soon as possible after exposure to used engine oil. Soap and water, or waterless hand cleaner should be used.

• All new vehicles are now equipped with an air bag system. The system must be disabled before performing service on or around system components, steering column, instrument panel components, wiring and sensors. Failure to follow safety and disabling procedures could result in accidental air bag deployment, possible personal injury and unnecessary system repairs.

• Always wear safety goggles when working with, or around, the air bag system. When carrying a non-deployed air bag, be sure the bag and trim cover are pointed away from your body. When placing a non-deployed air bag on a work surface, always face the bag and trim cover upward, away from the surface. This will reduce the motion of the module if it is accidentally deployed. Refer to the additional air bag system precautions later in this section.

• Clean, high quality brake fluid from a sealed container is essential to the safe and proper operation of the brake system. You should always buy the correct type of brake fluid for your vehicle. If the brake fluid becomes contaminated, completely flush the system with new fluid. Never reuse any brake fluid. Any brake fluid that is removed from the system should be discarded. Also, do not allow any brake fluid to come in contact with a painted surface; it will damage the paint.

• Never operate the engine without the proper amount and type of engine oil; doing so WILL result in severe engine damage.

• Timing belt maintenance is extremely important! Many models utilize an interference-type, non-freewheeling engine. If the timing belt breaks, the valves in the cylinder head may strike the pistons, causing potentially serious (also time-consuming and expensive) engine damage.

• Disconnecting the negative battery cable on some vehicles may interfere with the functions of the on-board computer system(s) and may require the computer to undergo a relearning process once the negative battery cable is reconnected.

• When servicing drum brakes, only disassemble and assemble one side at a time, leaving the remaining side intact for reference.

ENGINE REPAIR

Alternator

REMOVAL

4.2L Engine

1. Before servicing the vehicle, refer to the precautions in the beginning of this section.
2. Remove or disconnect the following:
   • Negative battery cable
   • Accessory belt
   • Positive battery cable nut from the generator
   • A/C line mounting bracket bolt at the engine lift hook
   • Right engine lift hook bolts
   • Engine lift hook
   • Mounting bolts
   • Alternator

5.3L Engine

1. Before servicing the vehicle, refer to the precautions in the beginning of this section.
2. Remove or disconnect the following:
   • Negative battery cable
   • Accessory belt
   • Electrical connector
   • Terminal stud nut, after sliding boot down
   • Alternator cable
   • Mounting bolts
   • Alternator

INSTALLATION

4.2L Engine

1. Install or connect the following:
   • Alternator and loosely install the mounting blots

• Tighten the alternator mounting bolts to 37 ft. lbs. (50 Nm)
• Positive battery cable and secure with the nut; tighten the nut to 80 inch lbs. (9 Nm)
• Engine lift hook and bolts; tighten the bolts to 37 ft. lbs. (50 Nm)
• A/C line bracket to the lift hook, then tighten the retaining bolt to 89 inch lbs. (10 Nm)
• Accessory belt
• Negative battery cable

5.3L Engine

1. Install or connect the following:
   • Alternator and loosely install the mounting bolts
   • Tighten the bolts to 37 ft. lbs. (50 Nm)
   • Alternator cable

- Terminal stud nut and tighten to 80 inch lbs. (9 Nm)
- Boot back over terminal stud
- Electrical connector
- Accessory belt
- Negative battery cable

Ignition Timing

ADJUSTMENT

The ignition timing is preset and cannot be adjusted.

Engine Assembly

REMOVAL & INSTALLATION

4.2L Engine

1. Before servicing the vehicle, refer to the precautions in the beginning of this section.
2. Drain the engine cooling system

➥Keep the oil drain plug removed during the engine removal and installation.

3. Drain the engine oil. Install a suitable plug into the oil pan to prevent oil leakage during the remainder of the procedure.
4. Using the proper equipment, discharge and recover the refrigerant from the A/C system, if equipped.
5. Remove or disconnect the following:
- Hood
- Negative battery cable
- Fuel system pressure
- Air cleaner assembly
- Throttle body
- Manifold Absolute Pressure (MAP) sensor
- Windshield washer solvent container
- Air intake baffle
- Grille
- Headlight housing
- Radiator support brace
- Hood latch
- A/C lines from the condenser
- Transmission cooler lines from the engine, not the radiator
6. Remove the cooling fan and shroud, tilting the radiator forward, and the cooling fan and shroud rearward for clearance.
- Accessory belt
- Power steering pump bolts; position the pump aside
- Heater hoses from the heater core

- Transmission filler tube bracket nut from the Air Injector Reactor (AIR) adapter
- AIR adapter
7. Install a suitable lift hook to the AIR adapter
8. Remove or disconnect the following:
- Oxygen (O_2) sensor connector
- A/C line from the accumulator
- Front axle actuator electrical connector
- Camshaft phaser actuator valve electrical connector
- Transmission cooler lines from the clips on the right side of the engine block
- Ignition coil harness connectors
- Harness retainer from the clips
- Power brake hose from the booster
- Powertrain Control Module (PCM)
- Fuel lines from the fuel pressure regulator. Cap the lines to avoid excessive fuel leakage.
- All harnesses from the engine harness bracket
- Engine harness bracket bolt and bracket
- Starter electrical connections
- A/C pressure sensor and clutch electrical connector
- Alternator electrical connector and battery lead
- Knock Sensor (KS), Crankshaft Position (CKP) and Camshaft Position (CMP) sensor electrical connectors
- 4 ground on the left side of the block
9. Raise and safely support the vehicle.
- Left and right side driveshafts
- Propeller shaft from the front axle pinion yoke
- Engine protection shield
- Exhaust pipe from the exhaust manifold. Slide the exhaust pipe backward slightly.
- Fuel tank shield, if equipped
- Torque converter access cover and bolts
10. Place a jack on the transmission fluid pan for support.
11. Remove the transmission support.
12. Lower the transmission enough to reach the top bell housing bolts.
13. Remove the top 4 bell housing bolts, there may be 2 harness clips that will need to be removed in order to have access to 2 of the top bolts.
14. Raise the transmission.
15. Reinstall the transmission support using only 2 through bolts.

16. Remove or disconnect the following:
- Remaining bell housing bolts (11 total)
- Left and right engine lower mount nuts
- Oil level sensor electrical connector
- Oil pressure switch electrical connector
17. Carefully lower the vehicle.
18. Remove the left, then the right upper engine mount nut.
19. Install a suitable engine hoist.
20. Raise the engine out of the compartment slowly, keeping the transmission supported.
21. Remove both engine mounts for clearance.
22. Continue raising the engine out of the vehicle.
23. Place the engine on a suitable engine stand.
To install:
24. Remove the engine from the engine stand.
25. Slowly install the engine into the engine compartment, aligning the engine mounts with the brackets.
26. When the engine mounts are aligned, install the engine mounts, putting the mount up through the engine mount brackets before inserting into the chassis mount brackets.
27. Lower the engine onto the mounts and install the upper engine mounting nuts. Tighten the nuts to 51 ft. lbs. (71 Nm).
28. Remove the engine hoist.
29. Lay the radiator into the radiator support, but do not install the radiator completely.
30. Raise and safely support the vehicle.
31. Install the lower bell housing bolts, except the top four.
32. Remove the 2 through bolts secure the transmission support, then lower the transmission.
33. Install the top 4 bell housing bolts and tighten all 11 bolts to 37 ft. lbs. (50 Nm).
34. Raise the transmission.
35. Install or connect the following:
- Transmission support
- 3 torque converter bolts and tighten to 44 ft. lbs. (60 Nm)
- Torque converter bolt cover
- Fuel tank shield, if equipped
- Engine protection shield
- Propeller shaft to the front axle pinion yoke
- Exhaust pipe to the manifold and tighten the bolts to 37 ft. lbs. (50 Nm)

- Oil level switch and oil pressure sender electrical connectors
- Oil pan drain plug and tighten to 19 ft. lbs. (26 Nm)
- Lower radiator hose
- Left and right wheel driveshafts

36. Lower the vehicle.

- 4 grounds on the left side of the block
- CMP, CKP and knock sensor electrical connectors
- Alternator and starter electrical connectors and battery leads. Torque the nuts to 80 inch lbs. (9 Nm).
- Fuel lines at the fuel pressure regulator
- Engine harness bracket and bolt. Torque the bolt to 37 ft. lbs. (50 Nm).
- Front differential vent hose, to the engine harness bracket
- PCM
- Power brake hose to the booster
- Harness retainer to its original location
- Ignition coil harness connectors
- Transmission cooler lines to clips on right side of engine block
- Camshaft phase actuator valve electrical connector
- Front axle actuator electrical connector
- A/C line at the accumulator

37. Remove the lift hook.
38. Install or connect the following:

- AIR adapter and secure with the studs. Tighten to 18 ft. lbs. (25 Nm).
- Transmission filler tube bracket to AIR adapter stud and secure the bracket with the nut. Torque the nut to 89 inch lbs. (10 Nm).
- Heater hoses to the heater core
- Power steering pump and tighten the bolts to 18 ft. lbs. (25 Nm).

39. The remainder of installation is the reverse of removal, but please note the following important steps:
40. Connect the negative battery cable
41. Check all powertrain fluid levels and add, as necessary.
42. Refill the engine crankcase.
43. Refill the engine cooling system.
44. Perform the CKP System Variation Learn Procedure, as follows:

  a. Install a suitable scan tool and check for Diagnostic Trouble Codes (DTCs). If any DTCs, other than P1336 are set, resolve those codes first, before proceeding with this procedure.

  b. With the scan tool, select the crankshaft position variation learn procedure.

  c. Observe the fuel cut-off for the 4.2L engine.

  d. The scan tool will instruct you to perform certain steps, make sure you follow all directions given by the scan tool exactly.

  e. Enable the crankshaft position system variation learn procedure.

➡ **While the learn procedure is in progress, release the throttle immediately when the engine started to decelerate. The engine control is returned to the operator and the engine responds to throttle position after the learn procedure is complete.**

  f. Slowly increase the engine speed to the RPM that you observed.

  g. Immediately release the throttle when fuel cut-out is reached.

  h. The scan tool displays: Learn Status: Learned this ignition. If the scan tool does NOT display this message and not other DTCs set, you must perform further troubleshooting.

  i. Turn the ignition **OFF** for 30 seconds after the learn procedure has been completed successfully.

45. Start and run the engine, then check for leaks.

5.3L Engine

1. Before servicing the vehicle, refer to the precautions in the beginning of this section.
2. Drain the engine cooling system

3. Drain the engine oil.
4. Remove and recover the refrigerant, if equipped with A/C.
5. Remove or disconnect the following:

- Negative battery cable
- Hood
- Radiator
- Radiator support brace
- Front axle, if 4WD
- Drive shafts
- Intake manifold
- Oil pressure sensor connector
- Oxygen (O_2) sensor connector
- Camshaft Position (CMP) sensor connector
- A/C compressor hose
- Rear auxiliary A/C compressor pipe fitting
- Rear auxiliary A/C compressor pipe nut and bolt. Tie the pipe out of the way.
- Engine Coolant Temperature (ECT) sensor
- Ground terminal bolt
- Retaining clips from the brackets
- A/C pressure switch electrical connector
- Retaining clip from the cylinder head
- Ground terminal bolts
- Starter
- Battery cable channel bolt
- Battery cable channel from the oil pan
- A/C compressor electrical connector

6. Collect all branches of the engine wiring harness, then position the harness out of the way.

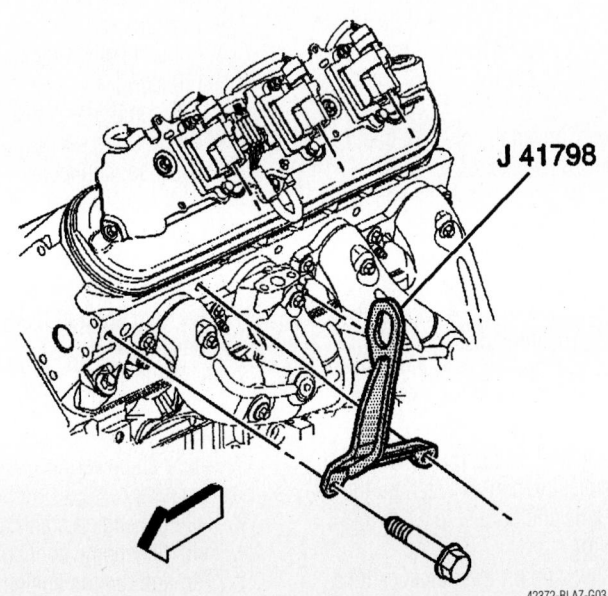

J 41798

42372-BLAZ-G03

If necessary, remove ignition coil(s) to install the engine lifting brackets—5.3L engine

- Alternator cable from the alternator
- Alternator bracket bolts, then position the bracket and alternator assembly aside
- Inlet and outlet hoses from the water outlet, using J 38185 to move the hose clamps
- Auxiliary heater inlet and outlet hose/pipe assembly from the heater water shutoff valve pipes
- Auxiliary heater inlet and outlet hoses/pipes from the water pump, using Hose Clamp Pliers J 38185
- Remove ignition coils, if necessary, to install Engine Lifting Brackets J 41798 to the cylinder heads

7. Install Engine Lifting Brackets J 41798 to the cylinder heads. Tighten the M8 bolts to 18 ft. lbs. (25 Nm) and the M10 bolts to 37 ft. lbs. (50 Nm).

- Catalytic converter
- 3 frame engine mount bracket bolts from the right and left sides
- Torque converter bolts
- Transmission oil level dipstick tube nut and tube
- Transmission bolt and stud on the right side
- Lower transmission bolt/studs
- 3 upper transmission bolts/studs

8. Install a suitable engine hoist to the engine lifting brackets.

9. Place a floor jack under the transmission for support.

10. Separate the engine from the transmission.

11. Remove the engine from the vehicle and place on a suitable engine stand.

12. Install Converter Holding Strap J 21366 to the transmission to hold the torque converter.

To install:

13. Remove Converter Holding Strap J 21366 from the transmission.

14. Attach the engine to a hoist and remove it from the engine stand

15. Install or connect the following:

- Engine into the vehicle. Match the transmission up to the engine, then remove the floor jack.
- 3 upper transmission bolts/studs and tighten to 37 ft. lbs. (50 Nm)
- Lower transmission bolts/studs and tighten to 37 ft. lbs. (50 Nm)
- Transmission bolt and stud on the right side and tighten to 37 ft. lbs. (50 Nm)
- Transmission oil level dipstick tube and nut. Torque to 89 inch lbs. (10 Nm).

- Torque converter bolts and tighten to 44 ft. lbs. (60 Nm)
- 3 frame engine mount bracket bolts to both the right and left sides. Torque the bolts to 37 ft. lbs. (50 Nm).
- Catalytic converter

16. Remove the engine lifting brackets from the cylinder heads

- Ignition coils, if removed, and tighten the bolts to 71 inch lbs. (8 Nm)
- Auxiliary heater inlet and outlet hoses
- Auxiliary heater inlet and outlet hose/pipe assembly to the heater water shutoff valve pipes
- Outlet and inlet hoses to the water outlet
- Bracket and alternator assembly. Tighten the bolts to 37 ft. lbs. (50 Nm).
- Cable to the alternator
- Position the engine wiring harness back over the engine
- A/C compressor electrical connector
- Battery cable channel to the oil pan and secure with the bolt. Torque to 106 inch lbs. (12 Nm).
- Starter
- Ground terminal bolts and tighten to 18 ft. lbs. (25 Nm)
- Retaining clip to the cylinder head
- A/C pressure switch electrical connector
- Retaining clips to the brackets
- Ground terminal bolt and tighten to 18 ft. lbs. (25 Nm)
- ECT sensor connector
- Rear auxiliary A/C compressor pipe nut and bolt. Torque to 15 ft. lbs. (20 Nm).
- A/C compressor hose
- Oil pressure sensor connector
- O_2 sensor connector
- CMP sensor connector
- Intake manifold
- Drive shafts
- Front axle, if removed
- Radiator support brace
- Radiator

17. Recharge the A/C system

- Negative battery cable
- Hood

18. Check all powertrain fluid levels and add, as necessary.

19. Refill the engine crankcase.

20. Refill the engine cooling system.

21. Start and run the engine, then check for leaks.

Water Pump

REMOVAL & INSTALLATION

1. Before servicing the vehicle, refer to the precautions in the beginning of this section.

2. Disconnect the negative battery cable.

3. Drain the engine cooling system.

4. For 5.3L engines, loosen the air cleaner outlet duct clamps at the throttle body and Mass Airflow/Intake Air Temperature (MAP/IAT) sensor. Remove the bolt and air cleaner outlet duct.

5. Relieve the belt tension and remove the accessory drive belts or the serpentine drive belt, as applicable.

6. Remove or disconnect the following:

- Upper fan shroud
- Fan or fan and clutch assembly, as applicable
- Water pump pulley; use a suitable tool to hold the pulley while removing the bolts
- Coolant hose(s) from the water pump

➡ **For the hoses on some engines, removal may be easier if the hose is left attached until the pump is free from the block. Once the pump is removed from the engine, the pump may be pulled (giving a better grip and greater leverage) from the tight hose connection.**

- Water pump retainers
- Water pump from the engine

✳✳ WARNING

Note the positions of all retainers as some engines will utilize different length fasteners in different locations and/or bolts and studs in different locations.

To install:

7. Clean the gasket mounting surfaces.

➡ **The water pumps on some of the engines covered may have been installed using sealer only, no gasket, at the factory. If a gasket is supplied with the replacement part, it should be used. Otherwise, a 1/8 in. (3mm) bead of RTV sealer should be used around the sealing surface of the pump.**

8. Apply sealant to the water pump retainer threads.

9. Install or connect the following:

- Water pump using a new gasket. Tighten the water pump retainers to

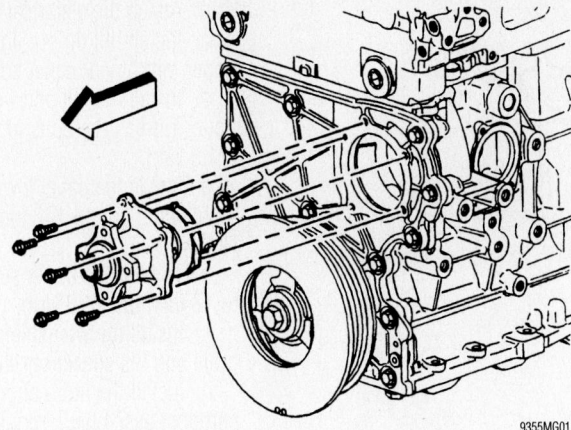

9355MG01

Exploded view of the water pump assembly mounting—4.2L engine

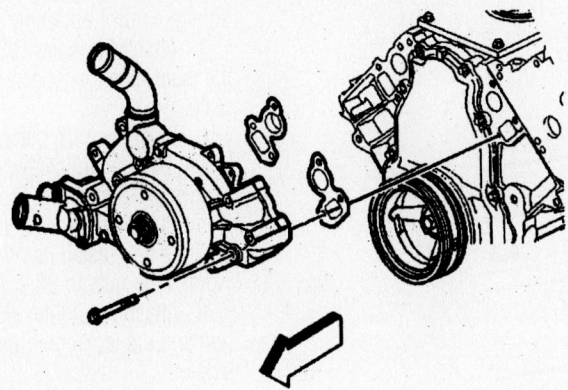

42372-BLAZ-G04

Exploded view of the water pump assembly mounting—5.3L engine

89 inch lbs. (10 Nm) for 4.2L engines. For 5.3L engines, tighten the bolts to 11 ft. lbs. (15 Nm), then to 22 ft. lbs. (30 Nm).
- Coolant hose(s)
- Water pump pulley. Tighten the pulley bolts to 18 ft. lbs. (25 Nm).
- Fan or fan and clutch assembly
- Serpentine drive belt (if equipped) by positioning the belt over the pulleys and carefully allow the tensioner back into contact with the belt.
- V-belts (if equipped) and adjust the tension
- Upper fan shroud
- Negative battery cable

10. Refill the engine cooling system.
11. Run the engine and check for leaks.

Heater Core

REMOVAL & INSTALLATION

1. Before servicing the vehicle, refer to the precautions in the beginning of this section.
2. Drain the engine cooling system.

3. Remove or disconnect the following:
- Negative battery cable
- Heater hoses from the heater core
4. Remove the instrument panel as follows:
   a. Disable the air bag system.
   b. Set the parking brake and block the wheels.
   c. Disconnect the parking brake release cable from the parking brake lever.
   d. Unfasten the screws that retain the DLC instrument panel left side sound insulator. Feed the DLC through the hole in the sound insulator.
   e. Unfasten the right side sound insulator panel screws and remove the panel.
   f. Unfasten the screws that attach the instrument panel left side sound insulator to the knee bolster and cowl panel.
   g. Unfasten the nut that attaches the left side sound insulator to the accelerator pedal bracket.
   h. Unplug the remote control door lock receiver module electrical connector.
   i. Remove the door lock receiver module from the left side sound insulator. Remove the left side sound insulator.

   j. Unfasten the screws that attach the instrument panel center sound insulator to the knee bolster, instrument panel, heater assembly and floor duct.
   k. Remove the center sound insulator.
   l. Unfasten the screws that attach the courtesy lamp to the knee bolster.
   m. Unfasten the screws that attach the knee bolster to the instrument panel.
   n. Disconnect the lap cooler duct from the knee bolster.
   o. Unplug the lighter electrical connection and remove the knee bolster.
   p. Unfasten the steering column-to-instrument panel nuts and lower the column.
   q. Unfasten the screws that attach the instrument panel accessory trim plate to the instrument panel.
   r. Remove the trim plate and unplug all necessary electrical connection.
   s. Remove the heater and/or air conditioning control assembly.
   t. Remove the radio and the storage compartment assembly (if equipped).
   u. If necessary, remove the instrument cluster.
   v. Unfasten the left and right instrument panel pivot bolts and the panel lower support bolt.
   w. Unfasten the speaker grilles retaining screws and remove the speaker grilles.
   x. Remove the windshield defroster grille using a flat-bladed prytool. Start at one end of the grille and work your way down the grille.
   y. Unfasten the 4 instrument panel upper support screws.
   z. Tag and unplug all necessary electrical connections.
   aa. Remove the instrument panel from the vehicle.
5. Remove or disconnect the following:
- Air inlet assembly, if equipped
- Vacuum hoses
- Heater assembly studs, from inside the engine compartment
- Blower motor resistor
- Stud from inside the heater case assembly; the stud is located behind the blower motor resistor
- Heater assembly-to-chassis screws
- Heater assembly from the vehicle
- Access cover screws and cover from the heater assembly
- Heater core from the heater case assembly

To install:
6. Install or connect the following:
- Heater core to the heater case assembly

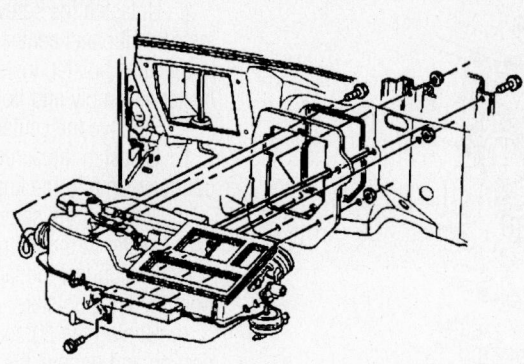

View of the heater case assembly

93113G77

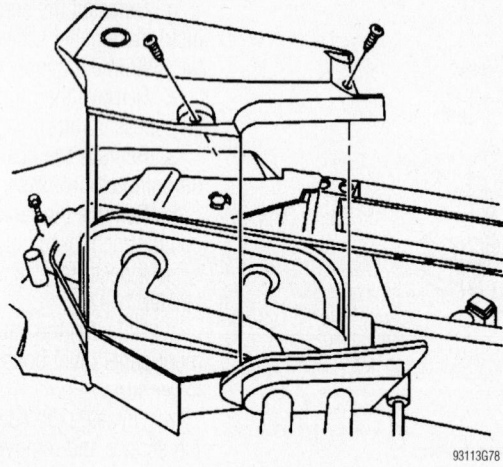

View of the heater case cover

93113G78

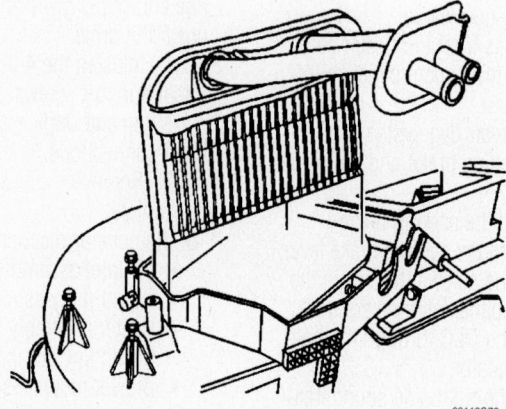

View of the heater core

93113G79

- Access cover to the heater assembly and the cover screws
- Heater assembly to the vehicle
- Heater assembly-to-chassis screws and torque them to 40 inch lbs. (4.5 Nm)
- Stud, working from inside the heater case assembly; the stud is located behind the blower motor resistor
- Blower motor resistor
- Heater assembly studs, working inside the engine compartment, torque them to 17 inch lbs. (1.9 Nm)
- Vacuum hoses
- Air inlet assembly, if equipped

7. Install the instrument panel as follows:

a. Rest the instrument panel on the lower pivot studs.

b. Attach the electrical connections.

c. Install but do not tighten the 4 upper instrument panel support screws.

d. Install the left and right panel pivot bolts. Tighten the bolts to 102 inch lbs. (11.5 Nm).

e. Install the panel lower support bolt. Tighten the bolt to 102 inch lbs. (11.5 Nm).

f. Tighten the upper support screws to 17 inch lbs. (1.9 Nm).

g. Install the windshield defroster grille and the speaker grilles.

h. Install the radio and storage compartment assembly (if equipped).

i. If removed, install the instrument cluster.

j. Install the heater and/or air conditioning control assembly.

k. Attach the electrical connections to the instrument panel accessory trim plate.

l. Place the trim plate in position and install its retaining screws. Tighten the screws to 17 inch lbs. (1.9 Nm).

m. Place the steering column into position and install its retaining nuts. Tighten the nuts to 22 ft. lbs. (30 Nm).

n. Attach the lighter electrical connection and the lap cooler duct to the knee bolster.

o. Place the knee bolster into position and install its retaining screws. Tighten the Torx® head screws to 80 inch lbs. (9 Nm) and the hex head screws to 17 inch lbs. (1.9 Nm).

p. Place the courtesy lamp in position and install its screws. Tighten the screws to 17 inch lbs. (1.9 Nm).

q. Place the instrument panel center sound insulator in position. Install the screws that attach the center sound insulator to the knee bolster, instrument panel and the floor duct. Tighten the screws to 17 inch lbs. (1.9 Nm).

r. Install the screw that attaches the center sound insulator to the heater assembly. Tighten the screw to 13 inch lbs. (1.5 Nm).

s. Install the remote control door lock receiver module to the instrument panel left side sound insulator.

t. Attach the door lock receiver electrical connection.

u. Install the nut that attaches the left side sound insulator to the accelerator pedal bracket. Tighten the nut to 35 inch lbs. (4 Nm).

v. Install the screw that attaches the left side sound insulator to cowl panel. Tighten the screw to 13 inch lbs. (1.5 Nm).

w. Install the screws that attach the left side sound insulator to knee bolster. Tighten the screw to 17 inch lbs. (1.9 Nm).

x. Feed the DLC through the hole in the sound insulator, place the DLC in position and install its retaining screws. Tighten the screws to 21 inch lbs. (2.4 Nm).

y. Install the right side sound insulator and tighten the screws

z. Connect the parking brake release cable to the lever.

aa. Enable the air bag system.

8. Install the heater hoses to the heater core.

9. Refill the cooling system.

10. Connect the negative battery cable.

11. Run the engine to normal operating temperatures; then, check the climate control operation and check for leaks.

Cylinder Head

REMOVAL & INSTALLATION

4.2L Engine

1. Before servicing the vehicle, refer to the precautions in the beginning of this section.

2. Disconnect the negative battery cable.

3. Drain the engine cooling system.

4. Remove or disconnect the following:

- Camshaft cover
- Exhaust manifold
- Front cover
- Cylinder head access hole plugs
- Timing chain tensioner shoe bolt and shoe
- Timing chain tensioner guide bolts and guide
- Timing chain and sprockets

5. Unfasten the cylinder head bolts by loosening them in the reverse of the torque sequence, then carefully remove the cylinder head.

6. Remove the cylinder head gasket.

To install:

7. Carefully clean and inspect the cylinder head and the gasket mounting surfaces.

➥The gasket surfaces on both the head and block must be clean of any foreign matter and free of nicks or heavy scratches. The cylinder bolt threads in the block and thread on the bolts must be cleaned (dirt will affect the bolt torque).

➥**DO NOT apply sealer to composition steel-asbestos gaskets.**

✳✳ WARNING

Make sure the number 1 cylinder is at Top Dead Center (TDC).

8. If using a steel only gasket, apply a thin and even coat of sealer to both sides of the gaskets.

9. Place a new gasket over the dowel pins with the bead or the words "This Side Up" facing upwards (as applicable), then carefully lower the cylinder head into position over the gasket and dowels.

10. Apply a coating of 12345493 or equivalent sealer to the threads of the cylinder head bolts, then thread the bolts into position until finger-tight.

11. Tighten the cylinder head bolts in sequence as follows:

a. Tighten the long bolts (1-14), in sequence, to 22 ft. lbs. (30 Nm).

b. Tighten the long bolts, in sequence, an additional 155 degrees.

c. Tighten the 2 short end bolt to 62 inch lbs. (7 Nm).

d. Tighten the 2 short end bolts an additional 60 degrees.

e. Tighten the 1 long end bolts to 62 inch lbs. (7 Nm).

f. Tighten the 1 long end bolt an additional 120 degrees.

12. Install or connect the following:

- Cylinder head access hole plugs and tighten to 44 inch lbs. (5 Nm)
- Timing chain and sprockets
- Front cover
- Camshaft cover
- Exhaust manifold

- Negative battery cable

13. Properly refill the engine cooling system.

14. Run the engine to check for leaks.

5.3L Engine

LEFT SIDE

1. Before servicing the vehicle, refer to the precautions in the beginning of this section.

2. Drain the engine cooling system.

3. Remove or disconnect the following:

- Negative battery cable
- Alternator bracket
- Coolant air bleed pipe
- Left exhaust manifold
- Pushrods
- Auxiliary A/C bracket bolt, if equipped
- Cylinder head bolts 1, 2 and 3. Discard the bolts
- Cylinder head
- Cylinder head gasket and discard

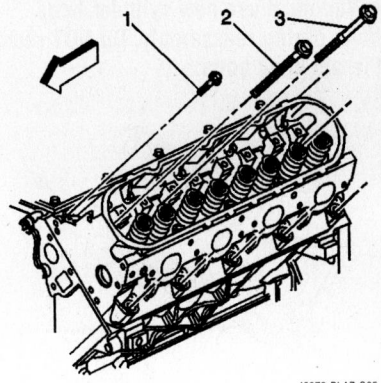

42372-BLAZ-G05

Remove and discard cylinder head bolts 1, 2 and 3—5.3L engine

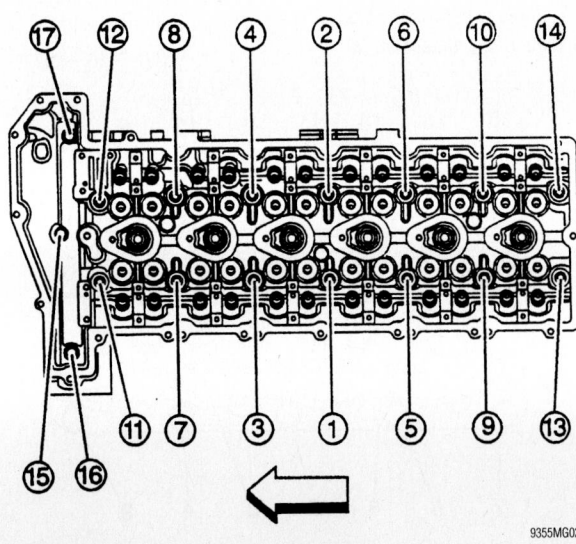

9355MG02

Cylinder head bolt tightening sequence—4.2L engine

To install:

4. Carefully clean and inspect the cylinder head and the gasket mounting surfaces.

➡The gasket surfaces on both the head and block must be clean of any foreign matter and free of nicks or heavy scratches. The cylinder bolt threads in the block and thread on the bolts must be cleaned (dirt will affect the bolt torque).

➡DO NOT apply any type sealer to the cylinder head gasket, unless otherwise specified.

5. Check the cylinder head locating pins for proper installation, location (a) 0.327 in. (8.3mm), as shown.

6. Place a new gasket over the dowel pins. When installed properly, the word "FRONT" on the left side, the tab on the gasket should be left of center or closer to the front of the engine.

7. Install or connect the following:
- Cylinder head

➡You must use new cylinder head bolts during reassembly. Do NOT reuse the old head bolts.

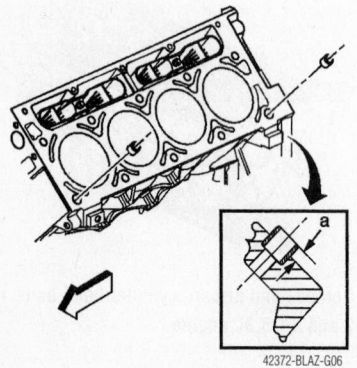

42372-BLAZ-G06

Make sure the cylinder head locating pins are properly installed, see dimension (a)

- NEW cylinder head bolts 1, 2 and 3.

8. Tighten the cylinder head bolts in sequence as follows:

a. Tighten the M11 bolts to 22 ft. lbs. (30 Nm).

b. Tighten the M11 an additional 90 degrees.

c. Tighten M11 bolts 1–8, an additional 90 degrees and M11 bolts 9 and 10 an additional 50 degrees.

d. Tighten the M8 bolts (11–15) to 22 ft. lbs. (30 Nm). Tighten all the bolts beginning with the center bolt and working outward, alternating sides

9. Install or connect the following:
- Auxiliary A/C bracket, if equipped. Torque the bolt to 15 ft. lbs. (20 Nm).
- Pushrods
- Left exhaust manifold
- Coolant air bleed pipe
- Alternator bracket

10. Properly refill the engine cooling system.

11. Run the engine to check for leaks.

RIGHT SIDE

1. Before servicing the vehicle, refer to the precautions in the beginning of this section.

2. Drain the engine cooling system.

3. Remove or disconnect the following:
- Negative battery cable
- Oil level dipstick
- Coolant air bleed pipe
- Right exhaust manifold
- Pushrods
- Auxiliary A/C bracket nut, if equipped
- Cylinder head bolts 1, 2 and 3. Discard the bolts
- Cylinder head
- Cylinder head gasket and discard

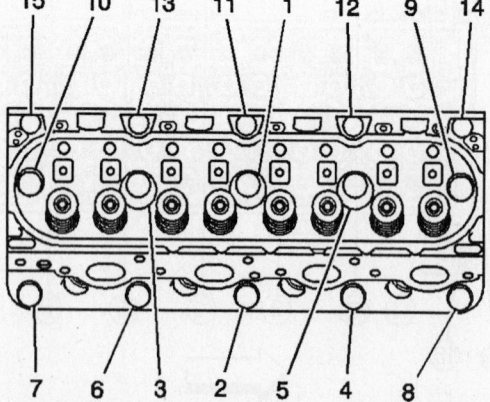

67162-ASCE-G01

Cylinder head bolt torque sequence—5.3L engine

To install:

4. Carefully clean and inspect the cylinder head and the gasket mounting surfaces.

➡The gasket surfaces on both the head and block must be clean of any foreign matter and free of nicks or heavy scratches. The cylinder bolt threads in the block and thread on the bolts must be cleaned (dirt will affect the bolt torque).

➡DO NOT apply any type sealer to the cylinder head gasket, unless otherwise specified.

5. Check the cylinder head locating pins for proper installation, location (a) 0.327 in. (8.3mm), as shown.

6. Place a new gasket over the dowel pins. When installed properly, the word "FRONT" on the right side, the tab on the gasket should be right of center or closer.

7. Install or connect the following:
- Cylinder head

➡You must use new cylinder head bolts during reassembly. Do NOT reuse the old head bolts.

- NEW cylinder head bolts 1, 2 and 3.

8. Tighten the cylinder head bolts in sequence as follows:

a. Tighten the M11 bolts to 22 ft. lbs. (30 Nm).

b. Tighten the M11 an additional 90 degrees.

c. Tighten M11 bolts 1–8, an additional 90 degrees and M11 bolts 9 and 10 an additional 50 degrees.

d. Tighten the M8 bolts (11–15) to 22 ft. lbs. (30 Nm). Tighten all the bolts beginning with the center bolt and working outward, alternating sides

9. Install or connect the following:
- Auxiliary A/C bracket, if equipped. Torque the nut to 15 ft. lbs. (20 Nm).
- Pushrods
- Right exhaust manifold
- Coolant air bleed pipe
- Oil level dipstick

10. Properly refill the engine cooling system.

11. Run the engine to check for leaks.

Rocker Arms/Shafts

REMOVAL & INSTALLATION

4.2L Engine

1. Before servicing the vehicle, refer to the precautions in the beginning of this section.

2. Remove or disconnect the following:
- Camshaft cover

➡**Make sure to place the camshaft caps in a rack to keep them in order, so they may be installed in their original locations.**

- Camshaft cap bolts and caps
- Camshafts

➡**If valve train components, such as the rocker arms or lash adjusters, are to be reused, they must be tagged or arranged to insure installation in their original locations.**

- Rocker arms
- Valve lash adjusters

To install:

3. Lubricate and fill the valve lash adjusters and the rocker arm roller with engine oil.

4. Install or connect the following
- Valve lash adjusters, in their original locations
- Rocker arm rollers in their original positions
- Camshafts
- Camshaft cap bolts
- Camshaft cover

5.3L Engine

1. Before servicing the vehicle, refer to the precautions in the beginning of this section.

2. Remove or disconnect the following:
- Rocker arm cover(s)

➡**If valve train components, such as the rocker arms, pushrods or pivot supports, are to be reused, they must be tagged or arranged to insure installation in their original locations.**

- Rocker arm bolts
- Rocker arms
- Rocker arm pivot support
- Pushrods

To install:

3. Inspect and replace components if worn or damaged.

4. Coat the bearing surfaces of the rocker arms, pushrods and the flange of the rocker arm bolts with clean engine oil.

5. Install or connect the following:
- Rocker arm pivot support
- Pushrods making sure they seat properly in the lifter sockets

➡**Make sure the pushrods are seated properly to the ends of the rocker arms, but do not tighten the bolts yet.**

- Rocker arms and bolts

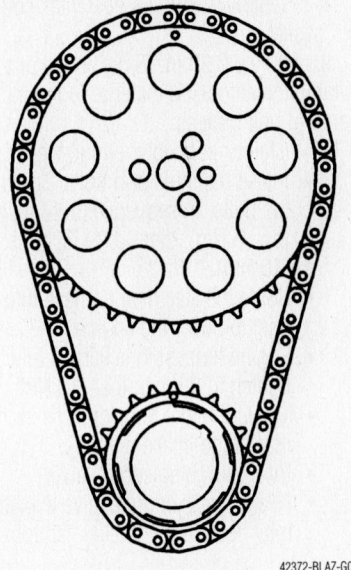

42372-BLAZ-G07

View of the crankshaft key with the No. 1 piston at TDC—5.3L engine

6. Rotate the crankshaft until the No. 1 piston is at Top Dead Center (TDC) of the compressor stroke. In this position, the cylinder No. 1 rocker arms will be off lobe lift, and the crankshaft sprocket key will be at the 1:30 position.

➡**The engine firing order is: 1–8–7–2–6–5–4–3. Cylinders 1, 3, 5 and 7 are the left bank. Cylinders 2, 4, 6 and 8 are the right bank.**

7. With the engine in the No. 1 firing position, tighten the following rocker arm bolts:

a. Tighten cylinders 1, 2, 7 and 8 exhaust valve rocker arm bolts to 22 ft. lbs. (30 Nm).

b. Tighten cylinder 1, 3, 4 and 5 intake valve rocker arm bolts to 22 ft. lbs. (30 Nm).

8. Rotate the crankshaft 360 degrees, then tighten the following rocker arm bolts:

a. Tighten cylinders 3, 4, 5 and 6 exhaust valve rocker arm bolts to 22 ft. lbs. (30 Nm).

b. Tighten cylinder 2, 6, 7 and 8 intake valve rocker arm bolts to 22 ft. lbs. (30 Nm).

9. Install the rocker arm cover(s).

Intake Manifold

REMOVAL & INSTALLATION

4.2L Engine

1. Before servicing the vehicle, refer to the precautions in the beginning of this section.

2. Properly relieve the fuel system pressure.

3. Disconnect the negative battery cable.

4. Drain the engine cooling system.

5. Remove or disconnect the following:
- Throttle body
- Powertrain Control Module (PCM)
- All electrical harnesses from the engine harness bracket
- Front differential vent hose from the bracket clip
- Engine harness bracket bolt and bracket
- Manifold Absolute Pressure (MAP) sensor connector
- Crankcase ventilation hose
- Brake hose from the booster
- Alternator
- Intake manifold bolts and manifold.
- Manifold gasket

To install:

6. Clean the gasket mounting surfaces. Be sure to inspect the manifold for warpage and/or cracks. If necessary, replace it.

7. Properly position a new intake manifold gasket.

8. Install or connect the following:
- Intake manifold and bolts. Torque the bolts to 16 ft. lbs. (22 Nm).
- Alternator
- Brake hose to the booster
- Crankcase ventilation hose, lubricating the inner diameter first with 12345884, or equivalent lubricant
- MAP sensor electrical connector
- Engine harness bracket. Tighten the retaining bolt to 37 ft. lbs. (50 Nm).
- Front differential vent hose to the engine harness bracket clip
- All harnesses to their original locations onto the engine harness bracket
- PCM
- Throttle body
- Negative battery cable

9. Refill the engine cooling system.

5.3L Engine

➡**The intake manifold, throttle body, fuel rail and injectors can be removed as an assembly. If you are not servicing these components individually, remove the intake manifold as a complete assembly.**

1. Before servicing the vehicle, refer to the precautions in the beginning of this section.

2. Properly relieve the fuel system pressure.

3. Disconnect the negative battery cable.

4. Drain the engine cooling system.

5. Remove or disconnect the following:
- Air cleaner outlet duct
- A/C compressor pressure switch electrical connector
- Harness clip from the cylinder head and fuel raid
- Mass Airflow/Intake Air Temperature sensor connector

6. Disconnect the electrical connectors from the following:

a. Main coil

b. Electronic Throttle Control (ETC)

c. Fuel injectors. Matchmark the connectors, pull the Connector Position Assurance (CPA) retainer up 1 click. Push the tab on the connector in, then detach the injector connector.
- Alternator connector
- Evaporative emission (EVAP) purge solenoid electrical connector
- Knock Sensor (KS) electrical connector
- Main coil
- Fuel injector electrical connector
- Electrical harness clips from the fuel rail
- KS harness electrical connector from the intake manifold
- Positive Crankcase Ventilation (PCV) valve hose and valve
- Heater water shutoff valve actuator inlet hose from the intake manifold
- EVAP purge solenoid vent tube
- Vacuum brake booster hose from the rear of the intake manifold
- Upper engine wire harness retainer nut. Position the wire harness aside.
- Intake manifold bolts
- Intake manifold and gaskets. Discard the gaskets.

To install:

7. Clean the gasket mounting surfaces. Be sure to inspect the manifold for warpage and/or cracks. If necessary, replace it.

8. Properly position a new intake manifold gasket.

9. Apply a 0.20 in. (5mm) band of a suitable threadlocking material to the intake manifold bolt threads.

10. Install or connect the following:
- Intake manifold and bolts. Torque the bolts, in sequence to 44 inch lbs. (5 Nm), then to 89 inch lbs. (10 Nm).
- Route the electrical harness into position over the engine.
- Engine harness bracket nut and tighten to 89 inch lbs. (10 Nm)
- Vacuum brake booster hose to the rear of the intake manifold
- EVAP purge solenoid valve
- Heater water shutoff valve actuator inlet hose to the intake manifold
- PCV valve and hose
- EVAP purge solenoid, KS, MAP sensor, main coil & fuel injector electrical connectors
- Harness clips to the fuel rail
- Alternator electrical connector
- Main coil, ETC, fuel injector electrical connectors
- Electrical harness clips to the fuel rail
- A/C compressor pressure switch electrical connector
- Harness clip to the cylinder head
- Mass Airflow/Intake Air Temperature sensor connector
- Air cleaner outlet duct
- Fuel fill cap
- Negative battery cable

11. Refill the engine cooling system.

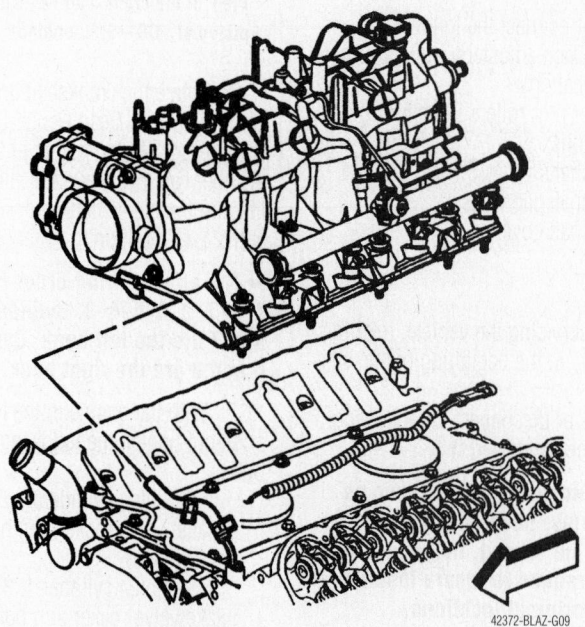

Exploded view of the intake manifold—5.3L engine

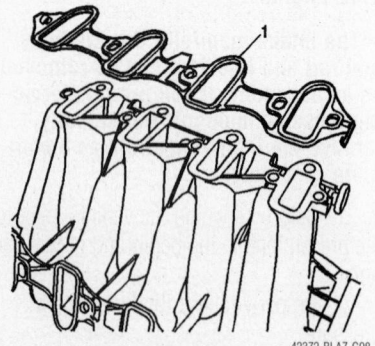

Make sure to use NEW intake manifold gaskets (1)—5.3L engine

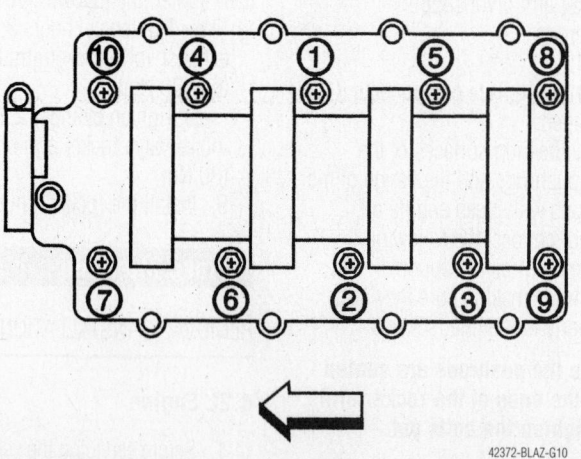

Intake manifold bolt tightening sequence—5.3L engine

Exhaust Manifold

REMOVAL & INSTALLATION

4.2L Engine

1. Before servicing the vehicle, refer to the precautions in the beginning of this section.

2. Remove or disconnect the following:
- Negative battery cable

➡ **It will be easier if the vehicle is only supported to a height where underhood access is still possible, the vehicle may be left in position for the entire procedure. If the vehicle is raised too high for underhood access, it will have to lowered, raised and lowered again during the procedure.**

- Air cleaner resonator outlet duct
- Transmission filler tube stud nut from the Air Injector Reactor (AIR) adapter and move the tube aside
- Oil level indicator tube
- Oxygen (O_2) sensor from the exhaust manifold
- 4 manifold heat shield nuts and shield
- Exhaust pipe bolts from the exhaust manifold
- Exhaust manifold bolts, and manifold
- Old gaskets and discard

To install:

3. Using a putty knife, clean the gasket mounting surfaces. Inspect the exhaust manifold for distortion, cracks or damage; replace if necessary.

4. Apply a threadlock such as GM 12345493 to the threads of the manifold retainers prior to installation.

5. Install or connect the following:
- Exhaust manifold to the cylinder using a new gasket, then tighten the bolts, in 3 passes, in sequence, to 18 ft. lbs. (25 Nm)
- Heat shield studs, if necessary, and tighten to 89 inch lbs. (10 Nm)
- O_2 sensor
- Exhaust manifold heat shield

➡ **Apply a suitable anti-seize compound to the exhaust manifold heat shield nuts prior to installation.**

- Heat shield nuts and tighten to 44 inch lbs. (5 Nm)
- Exhaust pipe to the manifold with seal and retaining nuts. Tighten the nuts to 37 ft. lbs. (50 Nm).
- Oil level indicator tube
- Transmission filler tube back onto the AIR adapter block stud and secure with the nut. Tighten the bracket nut to 89 inch lbs. (10 Nm).
- Air cleaner resonator outlet duct
- Negative battery cable.

5.3L Engine

1. Before servicing the vehicle, refer to the precautions in the beginning of this section.

2. Remove or disconnect the following:
- Negative battery cable
- Spark plug wires from the spark plugs. Don't disconnect the wires from the ignition coil unless necessary for clearance.
- Exhaust manifold bolts, manifold and gasket. Discard the gasket.

- Heat shield bolt and shield, if necessary

To install:

3. Apply a 0.2 inch (5mm) bead of threadlock GM P/N 12345493, or equivalent to the threads of the exhaust manifold bolts. Do NOT apply sealer to the first 3 threads of the bolts.

4. Install or connect the following:
- New exhaust manifold gasket
- Exhaust manifold
- Exhaust manifold bolts. Tighten in two passes. First to 11 ft. lbs. (15 Nm), then to 18 ft. lbs. (25 Nm) starting with the center bolts and working outward.

5. Bend over the exposed edge of the gasket at the rear of the cylinder head using a flat punch or equivalent tool.
- Heat shield and bolts, if removed. Torque the bolts to 80 inch lbs. (9 Nm).
- Spark plug wires to the spark plugs
- Negative battery cable

Camshaft and Valve Lifters

REMOVAL & INSTALLATION

4.2L Engine

1. Before servicing the vehicle, refer to the precautions in the beginning of this section.

2. Disconnect the negative battery cable.

3. Discharge and recover the refrigerant from the air conditioning system, using the proper equipment.

4. Remove or disconnect the following:
- Intake manifold
- A/C line from the oil level indicator tube
- A/C line from the accumulator
- A/C bracket bolt from the engine lift hook
- Engine lift bracket
- Ignition control module electrical connectors
- Ignition control module bolts and module

✳✳ WARNING

Be careful not to damage the clips that hold the harness housing in place.

- Engine electrical harness housing from the camshaft cover
- Fuel injection harness electrical connector

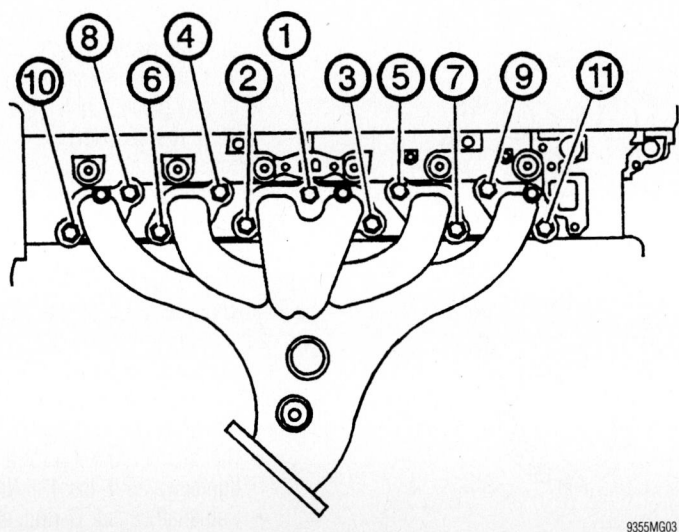

9355MG03

Exhaust manifold bolt tightening sequence—4.2L engine

- Camshaft cover bolts and cover
- Exhaust and intake sprocket bolts

5. Install a suitable sprocket holding tool onto the cylinder head and adjust the horizontal bolts into the camshaft sprockets to maintain timing chain tension and avoid disturbing the timing chain components.

6. Carefully move the sprockets with the timing chain off of the camshafts.

➡**Make sure to place the camshaft caps in a rack to keep them in order, so they may be installed in their original locations.**

7. Remove or disconnect the following:
- Camshaft cap bolts and caps
- Camshafts

To install:

8. Coat the camshaft journals with engine oil.
- Camshafts, in their original position
- Camshaft caps, in their original locations. Tighten the bolts to 106 inch lbs. (12 Nm).

9. Carefully place the camshaft sprockets back onto the camshafts and remove the holding tool.

10. Install or connect the following:
- Intake camshaft sprocket washer and bolt and the exhaust camshaft actuator bolt. Tighten the intake camshaft sprocket bolt to 22 ft. lbs. (30 Nm), plus an additional 135 degrees and the exhaust camshaft actuator bolt to 18 ft. lbs. (25 Nm), plus an additional 135 degrees.
- New camshaft cover seal
- New rubber ignition control module seals
- Camshaft cover and bolts. Tighten the bolts to 89 inch lbs. (10 Nm).
- Ignition control module. Tighten the bolts to 89 inch lbs. (10 Nm).
- Ignition control module electrical connectors
- Fuel injector electrical connectors
- Engine electrical harness housing
- A/C line bracket to the oil level indicator tube stud and secure with the nut. Tighten the nut to 62 inch lbs. (7 Nm).
- Engine lift bracket and secure the lift hook with the bolts. Tighten the bolts to 37 ft. lbs. (50 Nm).
- A/C line bracket to the engine lift bracket. Tighten the bolt to 89 inch lbs. (10 Nm).
- Intake manifold

11. Using the proper equipment, recharge the A/C system.

5.3L Engine

1. Before servicing the vehicle, refer to the precautions in the beginning of this section.

2. Disconnect the negative battery cable.

3. Discharge and recover the refrigerant from the air conditioning system, using the proper equipment.

4. Remove or disconnect the following:
- Condenser
- Cylinder head and gasket
- Valve lifter guide bolts
- Valve lifters and guide

➡**If the lifters are stuck in the bores due to built up deposits, use Valve Lifter Remover tool No. J 3049-A or equivalent to remove the lifters**

- Valve lifters from the guide

➡**Make sure to keep the lifters in order as you are removing them. They must be installed in their original locations.**

5. Clean and inspect the lifters for damage.

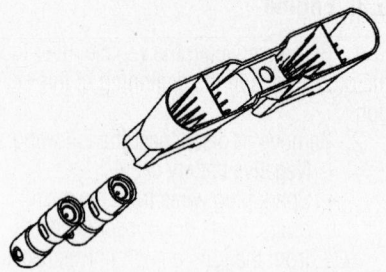

42372-BLAZ-G11

Remove the lifters from the guides, making sure to keep them in order—5.3L engine

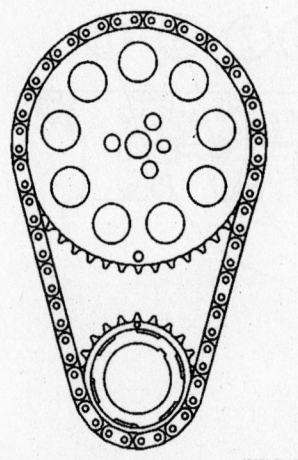

42372-BLAZ-G12

Make sure the crankshaft and camshaft timing marks are aligned

- Camshaft sensor bolt and sensor

6. Rotate the crankshaft until the timing marks on the crankshaft and camshaft sprockets are aligned.
- Camshaft sprocket bolts

Do NOT turn the crankshaft after the timing chain has been removed to avoid damaging the pistons or valves!

- Camshaft sprocket and reposition the timing chain
- Camshaft retaining bolts and retainer
- Camshaft by installing three M8-1.25 x 4.0 in. (M8-1.25 x 1.00mm) bolts in the front of the camshaft to act as a handle; then, remove the camshaft while turning slightly from side to side, as necessary. Remove the bolts from the camshaft.

➡**Take care not to damage the camshaft bearings when removing the camshaft.**

7. Clean and inspect the camshaft and bearings.

To install:

➡**If the camshaft must be replaced, you must also replace the lifters.**

8. Lubricate the camshaft journals with clean engine oil.

9. Install or connect the following:
- Three bolts used during removal into the bolt hold in the front of the camshaft
- Camshaft carefully into the engine block, using the bolts as a handle. Remove the bolts.
- Camshaft retainer and bolts. Make sure the retaining plate is installed with the sealing gasket facing the engine block. Tighten the bolts to 18 ft. lbs. (25 Nm).

10. Properly locate the camshaft sprocket locating pin with the cam sprocket alignment hole. The sprocket teeth and timing chain must mesh. The camshaft and crankshaft sprocket alignment marks MUST be aligned properly. Locate the camshaft sprocket alignment mark in the 6 o'clock position. It may be necessary to rotate the camshaft or crankshaft to align the marks.
- Camshaft sprocket and timing chain
- Camshaft sprocket bolts and tighten to 26 ft. lbs. (35 Nm)
- Camshaft sensor O-ring, after making sure it is not damaged and lubricating it with clean engine oil

- Camshaft sensor and bolt. Torque the bolt to 18 ft. lbs. (25 Nm).
11. Lubricate the valve lifters and engine block lifter bores with clean engine oil.
12. Install or connect the following:
   - Lifters into the lifter guides. Align the area on top of the lifter with the flat area in the lifter guide bore. Push the lifter completely into the guide bore.
   - Valve lifters and guide to the engine block
   - Valve lifter guide bolt and tighten to 106 inch lbs. (12 Nm)
   - Cylinder head and gasket
   - Condenser
13. Using the proper equipment, recharge the A/C system.

Valve Lash

ADJUSTMENT

The engines covered in this section do not require a periodic valve lash adjustment.

Starter Motor

REMOVAL & INSTALLATION

4.2L Engine

1. Before servicing the vehicle, refer to the precautions in the beginning of this section.
2. Disconnect the negative battery cable
3. Raise and safely support the vehicle.
4. Remove the left front tire and wheel assembly.
5. Working in the left fender area, disconnect the positive battery lead from the solenoid.
6. Remove or disconnect the following:
   - Starter mount bolt and nut
   - Starter motor

To install:
7. Install or connect the following:
   - Starter motor
   - Starter mounting bolt and nut. Tighten to 37 ft. lbs. (50 Nm).
   - Positive battery cable to the starter. Tighten the nut to 80 inch lbs. (9 Nm).
   - Left front tire and wheel assembly
8. Carefully lower the vehicle, then connect the negative battery cable.

5.3L Engine

1. Before servicing the vehicle, refer to the precautions in the beginning of this section.

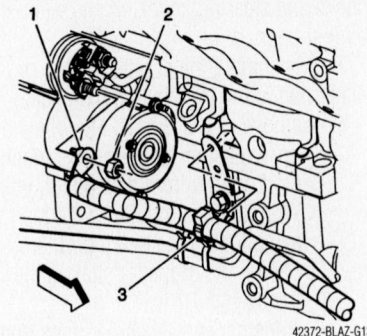

View of the starter, positive cable (1) and starter lead nut —5.3L engine

2. Remove or disconnect the following:
   - Negative battery cable
   - Catalytic converter
   - Engine shield bolts and shield
   - Right transmission cover bolt
   - Starter bolts
   - Transmission cover and shield, after repositioning the starter
3. Position the starter down, with the terminals facing toward the front of the vehicle.
   - Starter solenoid nut
   - Starter lead from the solenoid stud
   - Starter lead nut
   - Positive cable from the starter stud
   - Starter

To install:
4. Install or connect the following:
   - Starter in the vehicle. Position the starter down , with the terminals facing toward the front of the vehicle.
   - Positive cable to the starter stud.
   - Starter lead nut and tighten to 80 inch lbs. (9 Nm)
   - Starter solenoid lead to the stud
   - Starter solenoid nut and tighten to 30 inch lbs. (3.4 Nm)
   - Install the shield and transmission cover, after repositioning the starter
5. Slide the starter rearward.
   - Starter bolts and tighten to 37 ft. lbs. (50 Nm)
   - Right transmission cover bolt and tighten to 80 inch lbs. (9 Nm)
   - Catalytic converter
   - Negative battery cable
6. Start the vehicle to check for proper operation.

Oil Pan

REMOVAL & INSTALLATION

4.2L Engine

1. Before servicing the vehicle, refer to the precautions in the beginning of this section.

2. Disconnect the negative battery cable.
3. Remove or disconnect the following:
   - A/C compressor bottom bolts and loosen the top bolts
   - Oil dipstick and tube
4. Raise and safely support the vehicle.
5. Drain the engine crankcase oil.
6. Remove or disconnect the following:
   - Left and right front tire and wheel assemblies
   - Engine protection shield mounting bolts and shield
   - Front steering gear crossmember
   - Left and right driveshafts
   - Front drive axle clutch fork assembly
   - Prop shaft from the front axle pinion yoke
   - Unclip the transmission cooler lines from the engine block
   - Front differential bolts and position the differential aside
   - 4 transmission bell housing-to-oil pan bolts
   - Remaining oil pan bolts
   - Oil pan, by placing 2 oil pan bolts in the jack screws on the oil pan and tighten evenly to release the oil pan from the engine

To install:
7. Clean the gasket mounting surfaces.

➡**The alignment between the rear of the oil pan and the rear of the block is critical. When the oil pan is installed it could be inadvertently shifted front or back a small amount which could cause a transmission alignment problem. The back to the oil pan needs to be flush with the engine block.**

8. Apply a 0.12 in. (3mm) bead of sealant to engine block, rather than the oil pan.

➡**The oil pan MUST be installed within 10 minutes of applying the sealant to the engine block.**

9. Install or connect the following:
   - Oil pan, maneuvering it to clear the oil pump and screen assembly

➡**After the bolts are installed, before tightening them to specifications, check the oil pan alignment. Use a straight edge on the back to the block and the oil pan transmission mounting surface.**

   - Oil pan bolts; tighten the side bolts to 18 ft. lbs. (25 Nm) and the end bolts to 89 inch lbs. (10 Nm)
   - Transmission bell housing-to-oil

pan bolts and tighten to 35 ft. lbs. (47 Nm)

- A/C compressor bottom bolts. Tighten to 37 ft. lbs. (50 Nm)
- Front differential bolts and tighten to 63 ft. lbs. (85 Nm)
- Front drive axle and clutch fork assembly
- Transmission cooler lines to block
- Prop shaft to front differential
- Steering gear crossmember
- Left and right driveshaft
- Oil pan drain plug. Tighten to 19 ft. lbs. (26 Nm)
- Engine protection shield. Tighten the bolts to 18 ft. lbs. (25 Nm)
- Left and right front wheel and tire assemblies

10. Carefully lower the vehicle.
11. Refill the crankcase with fresh oil. Start the engine, establish normal operating temperatures and check for leaks.

5.3L Engine

1. Before servicing the vehicle, refer to the precautions in the beginning of this section.
2. Disconnect the negative battery cable.
3. Drain the engine crankcase oil and differential oil.
4. Remove or disconnect the following:
   - Oil level dipstick
   - Front shock upper retaining nuts
   - Tires and wheels
   - Engine shield bolts and shield
   - Power steering gear
   - Left and right Antilock Brake System (ABS) wiring harnesses from the retainers
   - Wheel Speed Sensor (WSS) electrical connectors
   - Brake hose retaining bolts from the frame
   - Sway bar link pins from the lower control arm on both sides
5. Place an adjustable jackstand under the lower control arm.
   - Upper ball joint pinch bolt and nut
   - Upper control arm from the upper ball joint
6. Lower and remove the jackstand, letting the suspension hang.
   - Left driveshaft
   - Right driveshaft from the intermediate shaft bearing only. Do not remove the driveshaft from the steering knuckle. Position the driveshaft aside.
7. Using wire or hooks, secure the front shock modules to the frame. Do NOT let the

shocks and steering knuckle hang without being supported.

8. Matchmark the position of the propeller shaft to the front axle pinion yoke.
9. Remove or disconnect the following:
   - Yoke retainer bolt and yoke retainers from the front axle pinion yoke. Wrap the bearing caps with tap to avoid losing the bearing rollers. Secure the propeller shaft to the frame.
   - Transmission oil cooler lines from the retainer
   - Transmission oil cooler line retaining bracket bolt and bracket
   - Inner axle shaft
   - Starter
   - Flywheel inspection cover from the left side of the transmission
   - Battery cable channel bolt from the front of the oil pan
   - Battery cable channel from the oil pan
   - Loosen the 2 upper A/C compressor bracket bolts
   - 2 lower A/C compressor bracket bolts
   - Front differential attachment bolts. Secure the front differential to the frame.
   - 2 lower bellhousing bolts
   - Oil pan bolts
   - Oil pan by tilting the rear of the oil pan down to clear the transmission, pull the oil pan rearward past the front wire harness, then lower the oil pan clear of the vehicle

➡ **The oil pan gasket is reusable if it is not damaged.**

10. Drill out the oil pan gasket retaining rivets, if necessary. Remove the gaskets. Discard the rivets. Inspect the gasket, if it is damaged, it must be replaced.

To install:

➡ **The proper alignment of the oil pan is very important. The rear bolt hold location of the oil pan provide mounting points for the transmission bellhousing. To ensure the rigidity of the powertrain and correct transmission alignment, make sure that the rear of the block and rear of the oil pan NEVER protrude beyond the engine block and transmission bellhousing plane.**

➡ **If replacing the oil pan gasket, it is not necessary to rivet the NEW gasket to the pan.**

11. Apply a 0.20 in. (5mm) bead of sealant 0.80 in. (20mm) long to the engine

block. Apply the sealant directly onto the tabs of the front cover gasket that protrudes into the oil pan surface.

12. Apply a 0.20 in. (5mm) bead of sealant 0.80 in. (20mm) long to the engine block. Apply the sealant directly onto the tabs of the rear cover gasket that protrudes into the oil pan surface.

13. Pre-assemble the oil pan gasket and bolts to the pan. Install the gasket onto the pan. Install the oil pan bolts to the pan and through the gasket.

14. Install the oil pan, oil pan gasket and bolts to the engine block as an assembly.

15. Hand-start the bolts into the engine block snug-tight. Do not fully tighten yet.

16. Install the 2 lower bellhousing bolts and tighten to 37 ft. lbs. (50 Nm).

17. Tighten the 2 rear oil pan-to-rear cover bolts to 106 inch lbs. (12 Nm) and the remaining oil pan bolts to 18 ft. lbs. (25 Nm).

18. Release the differential from the frame and install to the oil pan. Install and tighten the bolts to 63 ft. lbs. (85 Nm).

19. Install or connect the following:

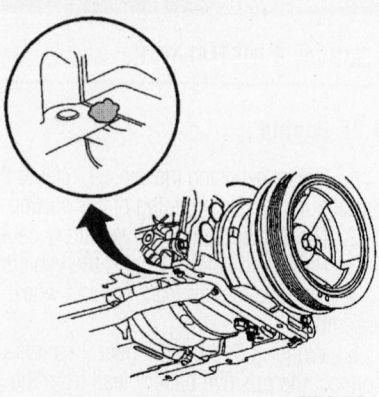

42372-BLAZ-G14

Proper sealant application to the front cover gasket

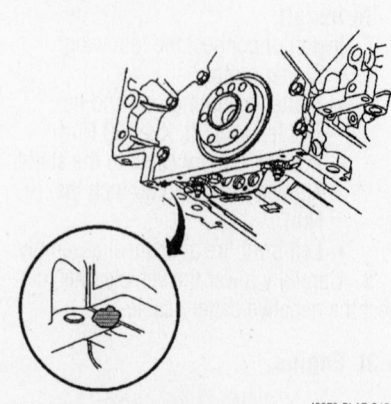

42372-BLAZ-G15

Proper sealant application to the rear cover gasket

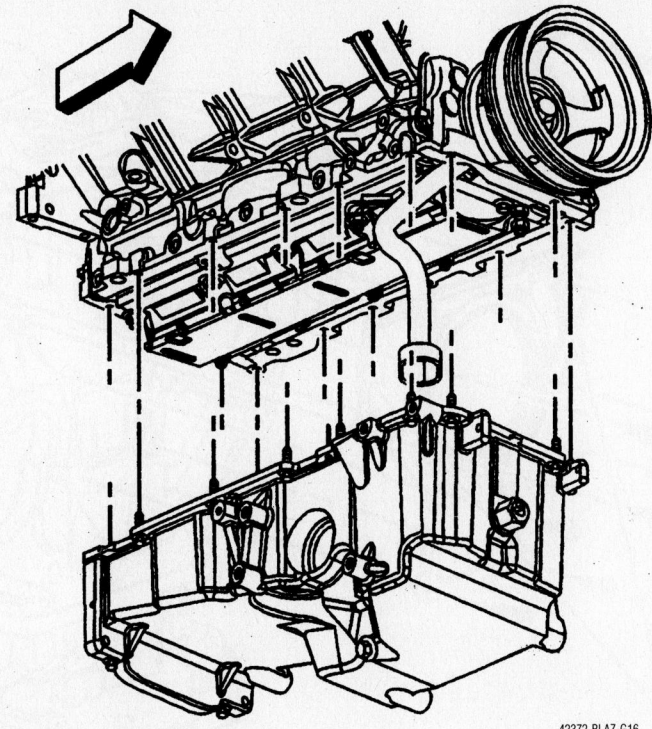

Oil pan mounting—5.3 engine

- 2 lower A/C compressor bracket bolts. Tighten the lower and upper compressor bolts to 37 ft. lbs. (50 Nm).
- Battery cable channel to the oil pan
- Battery cable channel bolt and tighten to 106 inch lbs. (12 Nm)
- Flywheel inspection cover to the left side of the transmission
- Starter
- Inner axle shaft
- Transmission oil cooler line retaining bracket and bolt. Torque the bolt to 80 inch lbs. (9 Nm).
- Transmission oil cooler lines to the retainer

20. Unhook the right driveshaft from the frame.
- Left and right driveshafts

21. Unsecure the shocks from the frame. Put adjustable jackstand under the lower control arm. Using the jackstand, raise the lower control arm and knuckle assembly in order to connect the upper ball joint to the upper control arm.
- Upper ball joint pinch nut and bolt and tighten to 30 ft. lbs. (40 Nm). Remove the jackstand.
- Sway bar link pins to the lower control arm on both sides
- Steering gear

22. Unsecure the prop shaft from the frame. Align the matchmarks on the prop shaft to the marks on the front axle pinion yoke.

- Propeller shaft to the front axle pinion yoke
- Yoke retainers and yoke retainer bolts to the front axle pinion yoke. Torque the bolts to 15 ft. lbs. (20 Nm).
- Brake hose retaining bolts to the frame and tighten to 18 ft. lbs. (25 Nm).
- WSS electrical connectors
- Left and right ABS wiring harnesses to the retainers
- Differential with oil
- Engine shield and bolts. Tighten the bolts to 18 ft. lbs. (25 Nm).
- Tires and wheels

23. Fill the engine with oil. Fill the power steering system with fluid.
- Upper shock nuts and tighten to 74 ft. lbs. (100 Nm).
- Oil dipstick
- Negative battery cable

Oil Pump

REMOVAL & INSTALLATION

4.2L Engine

1. Before servicing the vehicle, refer to the precautions in the beginning of this section.

2. Remove or disconnect the following:
- Engine front cover

- Oil pump cover bolts
- Oil pump cover. Mark the inner and outer gears in relation to the pump housing.
- Inner and outer pump gears
- Oil pump pressure relief valve plug
- Oil pump pressure relief valve and spring

To install:

3. Install or connect the following:
- Oil pump pressure relief valve and spring
- Oil pump pressure relief valve plug. Tighten to 10 ft. lbs. (14 Nm).
- Oil pump outer and inner gears, as marked during removal
- Oil pump cover and bolts. Tighten the bolts to 89 inch lbs. (10 Nm).
- Front cover

5.3L Engine

1. Before servicing the vehicle, refer to the precautions in the beginning of this section.

2. Remove or disconnect the following:
- Oil pan
- Engine front cover
- Oil pump screen bolt and nuts
- Oil pump screen with O-ring seal
- O-ring seal from the pump screen. Discard the O-ring seal.
- Remaining crankshaft oil deflector nuts
- Crankshaft oil deflector
- Oil pump bolts
- Oil pump

➡**Do not let any dirt or debris into the oil pump or cap end.**

- Clean and inspect the oil pump.

To install:

3. Align the splined surfaces of the crankshaft sprocket and the oil pump drive gear and install the oil pump.

4. Install or connect the following:
- Oil pump onto the crankshaft

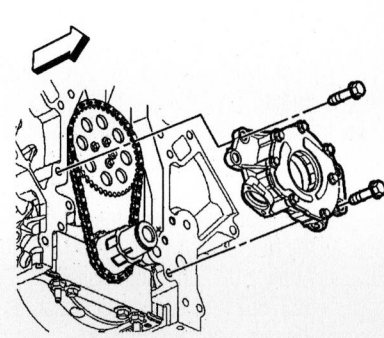

Exploded view of the oil pump mounting—5.3L engine

sprocket until the pump housing contacts the face of the engine block

- Oil pump bolts and tighten to 18 ft. lbs. (25 Nm)
- Crankshaft oil deflector and nuts until snug
- New oil pump screen O-ring seal into the oil pump screen, after lubricating with clean engine oil

➡**Push the oil pump screen tube completely into the oil pump prior to tightening the bolt. Do not let the bolt pull the tube into the pump.**

5. Align the oil pump screen mounting brackets with the correct crankshaft bearing cap studs.

- Oil pump screen
- Oil pump screen bolts and nuts. Tighten the bolts to 106 inch lbs. (12 Nm) and the nuts to 18 ft. lbs. (25 Nm).
- Engine front cover
- Oil pan

Rear Main Seal

REMOVAL & INSTALLATION

4.2L Engine

Please note that the transmission assembly must be removed to perform this procedure.

1. Before servicing the vehicle, refer to the precautions in the beginning of this section.

2. Remove or disconnect the following:

- Negative battery cable
- Transmission
- Flywheel
- Crankshaft rear main seal housing bolts. Install 2 bolts into the jackscrew holes to release the cover from the block
- Crankshaft and rear main seal housing
- Rear main seal from the crankshaft snout

To install:

3. Install or connect the following:

- Rear main seal, using a suitable seal installation tool, then remove the tool
- Apply a 0.12 in. (3mm) bead of 12378521, or equivalent sealant to the rear mail seal housing
- Suitable cover alignment pins into the block

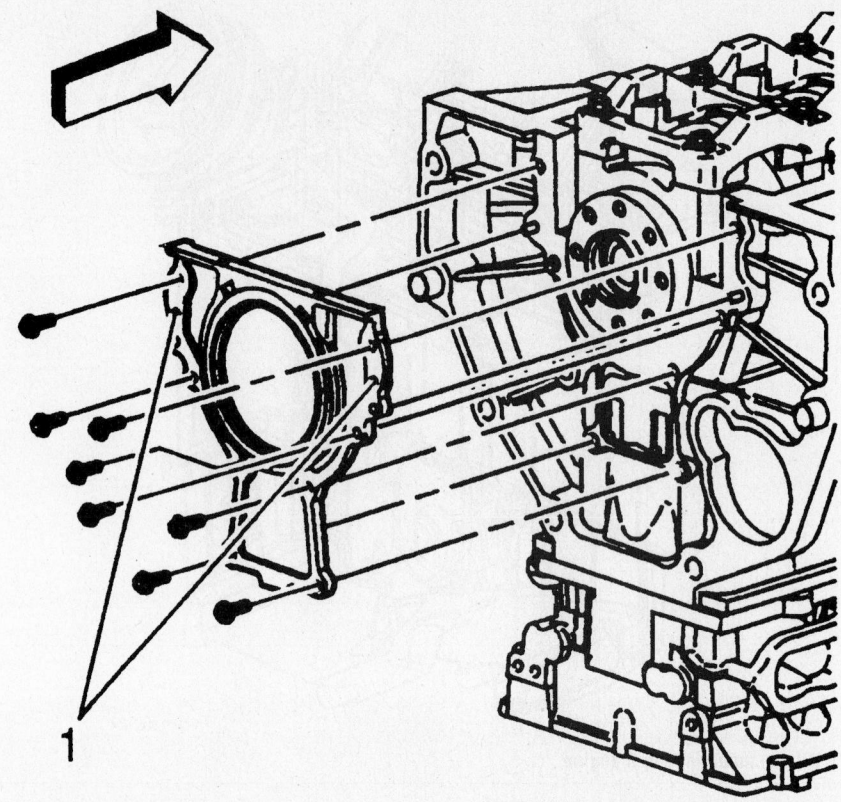

Install 2 bolts into the jackscrew holes (1) to push the cover off of the block—4.2L engine

9355MG04

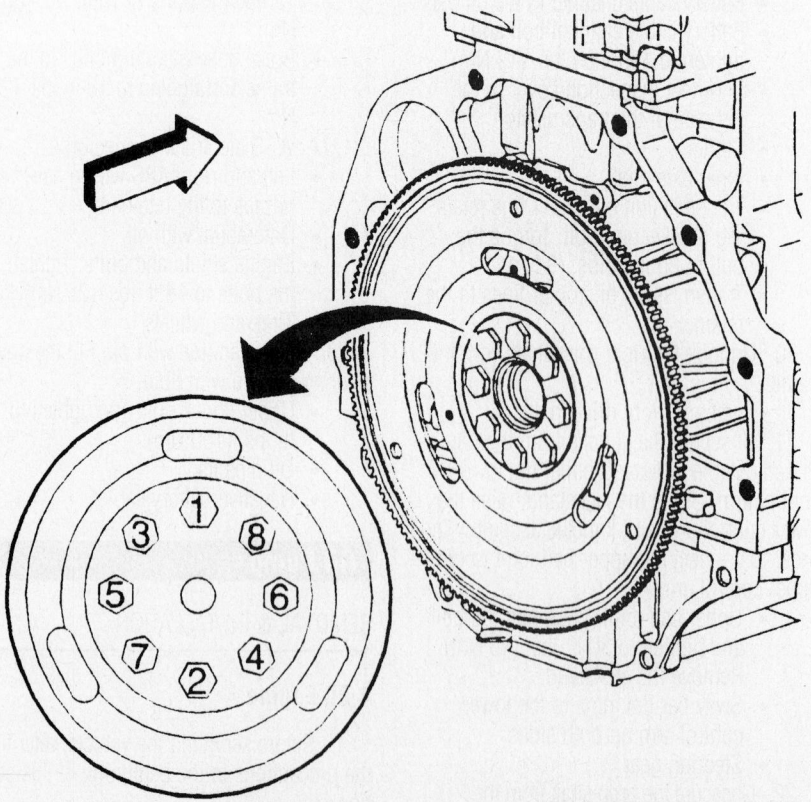

Flywheel bolt tightening sequence—4.2L engine

9355MG05

➡ **When you install a new seal, make sure to use the plastic installation sleeve supplies with the new seal. The sleeve should come off and be discarded after the seal is installed.**

4. Slide the crankshaft rear main seal housing over the alignment pins and crankshaft.

5. Install the crankshaft rear main seal housing bolts, except the 2 in place of the guide pins.

6. Remove the guide pins.

7. Install or connect the following:
- Remaining 2 crankshaft rear main seal housing bolts and tighten to 89 inch lbs. (10 Nm). Wipe off any excess sealant.
- Flywheel and secure with the mounting bolts. Tighten, in sequence, to 18 ft. lbs. (25 Nm), plus an additional 50 degrees.
- Transmission

5.3L Engine

Please note that the transmission assembly must be removed to perform this procedure.

1. Before servicing the vehicle, refer to the precautions in the beginning of this section.

2. Remove or disconnect the following:
- Negative battery cable
- Transmission
- Flywheel
- Crankshaft rear main oil seal from the rear cover

To install:

➡ **The flywheel spacer (if applicable) must be removed prior to oil seal installation. Do not lubricate the oil seal Inside Diameter (ID) or crankshaft surface. Never reuse the rear main seal. Once it is removed, it must be replaced with a new seal.**

3. Lubricate the Outside Diameter (OD) of the rear main seal and the rear cover oil seal bore with clean engine oil. Do NOT let oil contact the seal surface or the crankshaft surface.

4. Install or connect the following:
- Crankshaft Rear Oil Seal Installer Tool No. J 41479 tapered cone and bolts onto the rear of the crankshaft. Tighten the bolts until just snug, being careful not to overtighten.
- Rear oil seal onto the tapered cone until the tool contacts the oil seal

5. Align the oil seal into the tool, Rotate

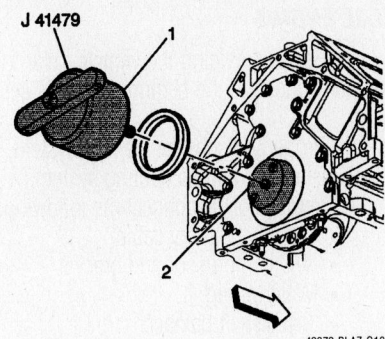

J 41479

View of the rear main seal installation— 5.3L engine

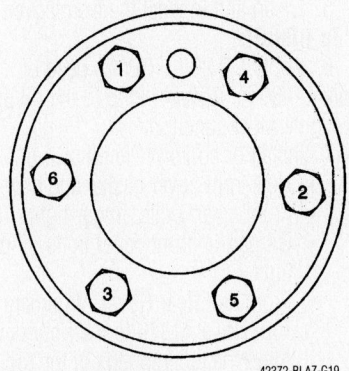

42372-BLAZ-G19

Flywheel bolt tightening sequence—5.3L engine

the handle of the tool clockwise until the seal enters the rear cover and bottoms into the cover bore. Remove the tool.
- Flywheel and secure with the mounting bolts.

6. Tighten the flywheel mounting bolts, in sequence, as follows:
a. 1st pass: 15 ft. lbs. (20 Nm)
b. 2nd pass: 37 ft. lbs. (50 Nm)
c. Final pass: 74 ft. lbs. (100 Nm)
- Transmission
- Negative battery cable

7. Start the engine and verify no oil leaks.

Timing Chain, Sprockets, Front Cover and Seal

REMOVAL & INSTALLATION

Front Cover and Seal

4.2L ENGINE

1. Before servicing the vehicle, refer to the precautions in the beginning of this section.

2. Remove or disconnect the following:
- Negative battery cable
- Drain the engine cooling system.
- Cooling fan and shroud

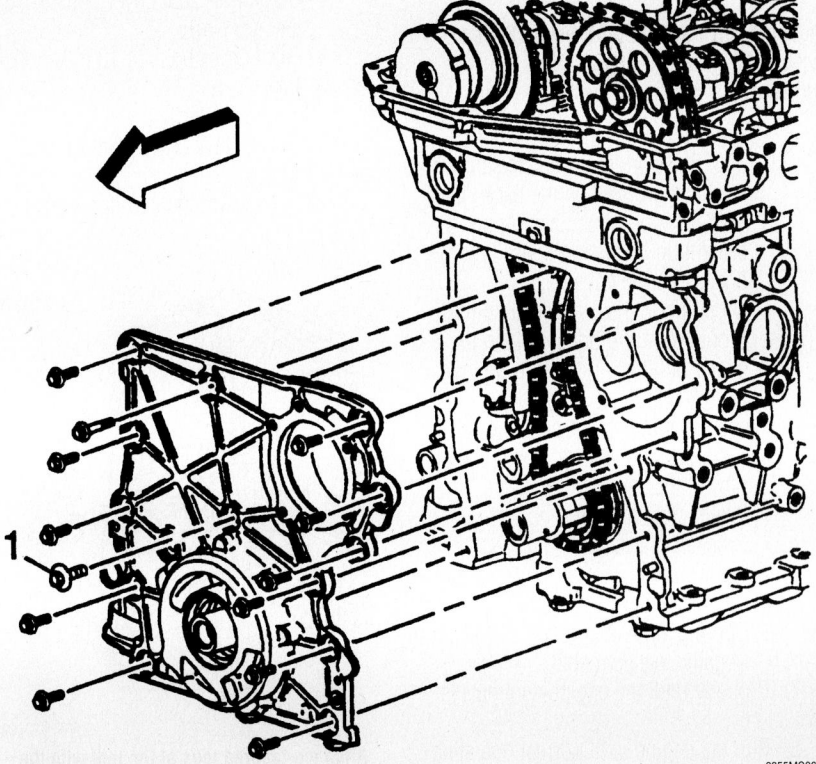

9355MG06

Place 2 front cover bolts in the jackscrew holes on the cover and tighten to push the cover off of the engine—4.2L engine

- Accessory belt
- Water pump
- Crankshaft balancer

❋❋ WARNING

When removing the seal, be careful not to damage the front cover or crankshaft.

- Seal from the front cover, using a suitable prytool in the slots provided
- Power steering pump

3. Raise and safely support the vehicle.
- Oil pan, then carefully lower the vehicle
- 7mm center bolt
- Remaining front cover bolts. Place two of the front cover bolts in the jackscrew holes on the front cover and tighten the bolts evenly to release the front cover from the engine.
- 2 bolts from the front cover
- Oil pump

To install:

4. Clean the gasket mating surfaces of the engine and cover of all remaining gasket or sealer material. Be careful not to score or damage the surfaces.

5. Install or connect the following:
- Suitable cover alignment pins, onto the engine

➡**The front cover MUST be installed within 10 minutes of applying the sealant.**

- Apply a 0.12 in. (3mm) beat of 12378521 or equivalent sealant to the trace grooves on the back side of the engine front cover. Apply sealant on the inside 3 bolt hole bosses on the cover also.
- Oil pump to the crankshaft splines
- Front cover and bolts, tighten the center bolt last. Tighten to 89 inch lbs. (10 Nm).

6. Remove the alignment pins and raise and safely support the vehicle. Install the oil pan, then lower the vehicle.
- Power steering pump
- Crankshaft balancer
- Water pump
- Accessory belt
- Cooling fan and shroud
- Negative battery cable

7. Properly refill the engine cooling system.

8. Run the engine until normal operating temperature has been reached, then check for leaks.

5.3L ENGINE

1. Before servicing the vehicle, refer to the precautions in the beginning of this section.

2. Properly discharge the A/C system.

3. Drain the engine cooling system.

4. Remove or disconnect the following:
- Negative battery cable
- A/C compressor and bracket
- Water pump
- Crankshaft balancer
- Oil pan-to-front cover bolts
- Front cover bolts
- Front cover and gasket. Discard the gasket.

5. Clean and inspect the front cover.

To install:

6. Apply a 0.20 in. (5mm) bead of sealant 0.80 in. (20mm) long to the oil pan-to-engine block junction.

7. Install or connect the following:
- New front cover gasket and cover
- Front cover bolts, finger-tight
- Oil pan-to-front cover bolts, finger-tight
- Front and Rear Cover Alignment Tool No. J 41476 to the front cover. Align the tapered legs of the tool with the machined alignment surfaces on the front cover
- Crankshaft balancer bolt, finger-tight
- Oil pan-to-front cover bolts to 18 ft. lbs. (25 Nm)
- Front cover bolts to 18 ft. lbs. (25 Nm)

8. Remove the tool.

9. Install a NEW crankshaft front oil seal as follows:
   a. Remove the radiator for access.

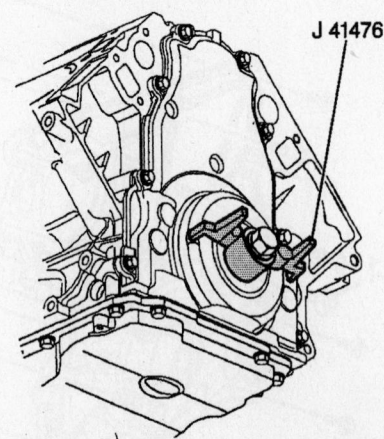

Align the tapered legs of the tool with the machined alignment surfaces on the front cover—5.3L engine

42372-BLAZ-G20

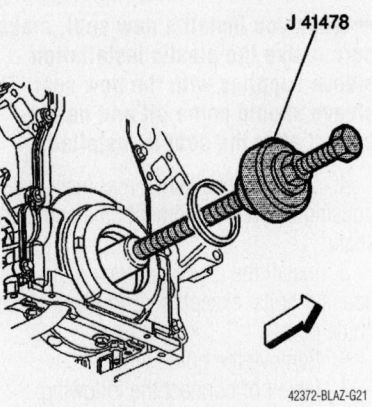

J 41478

Front cover seal installation using the proper tool—5.3L engine

42372-BLAZ-G21

   b. Remove the crankshaft balancer.
   c. Remove the crankshaft oil seal.
   d. Lubricate the outer edge ONLY of the NEW crankshaft oil seal with clean engine oil.
   e. Install the crankshaft front oil seal into the Crankshaft Front Seal Installation Tool No. J 41478 guide.
   f. Install the J 41478 threaded rod (with nut, washer, guide and oil seal) into the end of the crankshaft.
   g. Use J 41478 to install the oil seal into the cover bore. Use a wrench and hold the hex on the installer bolt. Use a second wrench to rotate the installer nut clockwise until the seal bottoms in the cover bore. Remove the tool.
   h. Check the seal for proper installation. It should be installed evenly and completely into the front cover bore.
   i. Install the crankshaft balancer. Tighten the bolt to 37 ft. lbs. (50 Nm), plus an additional 140 degrees using a torque angle meter.
   j. Install the radiator.

10. Install or connect the following:
- Water pump
- A/C compressor and bracket
- Cooling system with coolant
- Negative battery cable

11. Properly recharge the A/C system

Timing Chain and Sprockets

4.2L ENGINE

➡**The following procedure requires the use of the Crankshaft Holding tool No. J-44221 and a suitable torque angle meter.**

1. Before servicing the vehicle, refer to the precautions in the beginning of this section.

2. Remove or disconnect the following:
- Camshaft cover

- Timing chain (front) cover
- Tension on the timing chain by moving the tensioner shoe in. Place a tee into the tension to hold the shoe in place.
- Top chain guide bolts and guide
- Exhaust camshaft position actuator bolt and actuator
- Intake camshaft sprocket bolt and sprocket
- Timing chain
- Crankshaft sprocket
- Cylinder head access hole plugs
- Timing chain tensioner shoe bolt and shoe
- Timing chain tensioner guide bolts and guide
- Timing chain tensioner bolts and tensioner

To install:

➡**Every seventh link of the timing chain is darkened to help in aligning the timing marks.**

3. Install or connect the following:
- Timing chain tensioner and bolts. Tighten to 18 ft. lbs. (25 Nm).
- Timing chain guide and bolts. Tighten to 89 inch lbs. (10 Nm).
- Timing chain tensioner shoe and bolt. Tighten to 19 ft. lbs. (26 Nm).

- Cylinder head access hole plugs and tighten to 44 inch lbs. (5 Nm)
- Crankshaft Holding tool No. J-44221, or equivalent with the camshaft flats up and the No. 1 cylinder at Top Dead Center (TDC)
- Crankshaft sprocket
- Intake camshaft sprocket into the timing chain

4. Align the dark link of the timing chain with the timing mark on the intake camshaft sprocket. Feed the timing chain down through the opening in the head.
- Timing chain onto the crankshaft sprocket. Align the dark link of the timing chain with the timing mark on the crankshaft sprocket.

➡**It may be necessary to remove the crankshaft holding tool to rotate and hold the camshaft hex to align the pin to the camshaft sprocket**

- Intake camshaft sprocket onto the intake camshaft
- Intake camshaft washer and bolt
- Exhaust camshaft actuator into the timing chain. Align the dark link of the timing chain with the timing mark on the exhaust camshaft actuator.

➡**It may be necessary to remove the crankshaft holding tool to rotate and hold the camshaft hex to align the pin to the camshaft sprocket**

- Exhaust camshaft actuator onto the exhaust camshaft

➡**Rotate the camshaft actuator clockwise relative to the camshaft prior to tightening the bolt.**

5. Rotate the camshaft actuator clockwise (as seen from the front of the vehicle).

※※ **WARNING**

The camshaft actuator must be fully advanced during installation. Engine damage may occur if the camshaft actuator is not fully advanced.

6. Install the exhaust camshaft actuator bolt and tighten to 18 ft. lbs. (25 Nm), plus an additional 135 degrees, using a torque angle meter.
7. Tighten the intake camshaft sprocket bolt to 22 ft. lbs. (30 Nm), plus an additional 135 degrees, using a torque angle meter.
8. Remove the tee from the timing chain tensioner to regain tension on the timing chain.
9. Remove the crankshaft holding tool. The dark lines on the timing chain should be aligned with the marks on the sprockets.
10. Install or connect the following:
- Top chain guide
- Suitable threadlock to the top chain guide bolt threads, then install and tighten to 89 inch lbs. (10 Nm)
- Engine front cover
- Camshaft cover

5.3L ENGINE

1. Before servicing the vehicle, refer to the precautions in the beginning of this section.

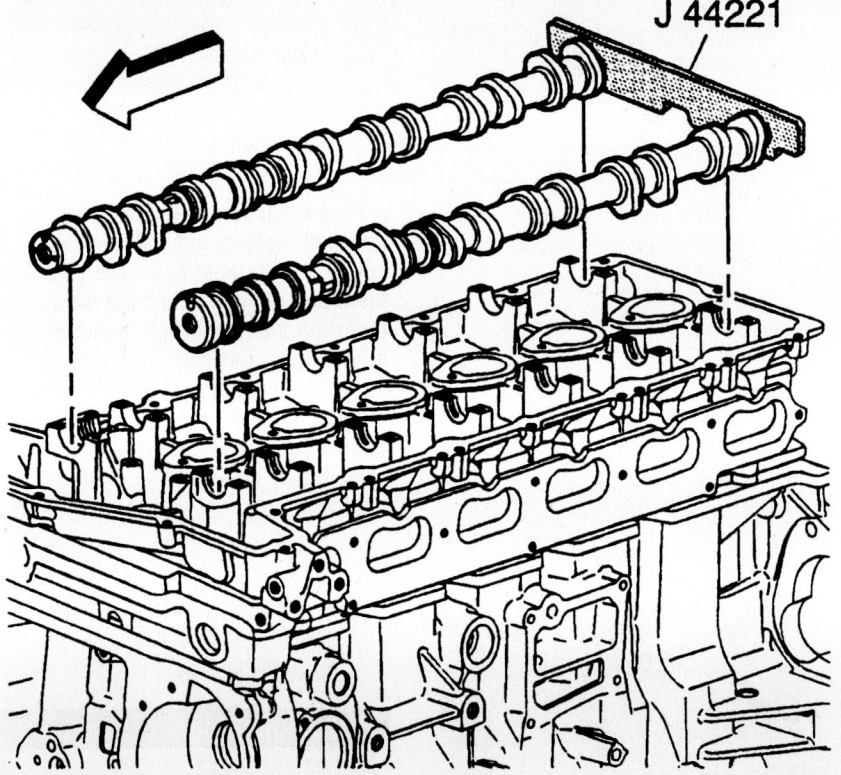

J 44221

Proper installation of the crankshaft holding tool with the No. 1 cylinder at TDC—4.2L engine

9355MG07

SEE MANUAL

DELPHI
25178506

Rotate the camshaft actuator clockwise—4.2L engine

9355MG08

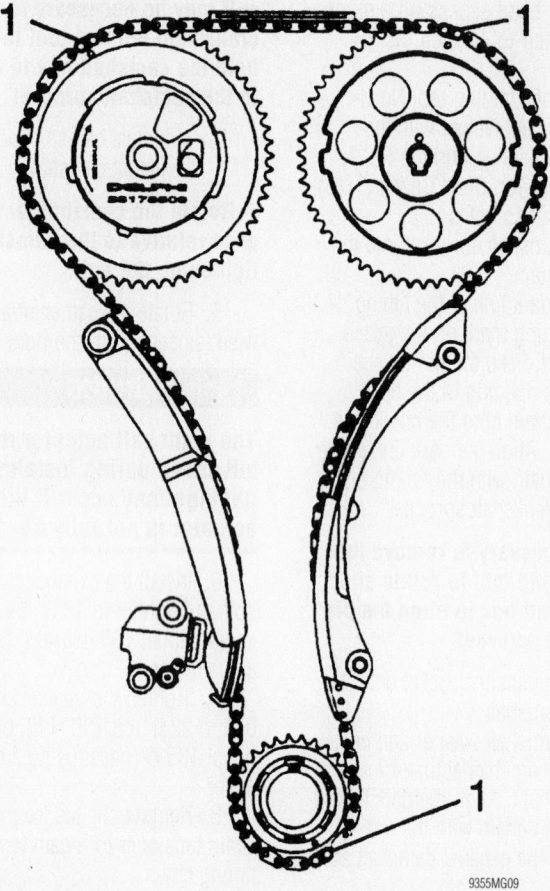

The dark lines on the timing chain should be aligned with the marks on the sprockets—4.2L engine

2. Remove the oil pump.
3. Rotate the crankshaft until the timing marks on the crankshaft and the camshaft sprockets are aligned.

✳✳ WARNING

Do NOT turn the crankshaft after the timing chain has been removed to prevent damage to the pistons and valves.

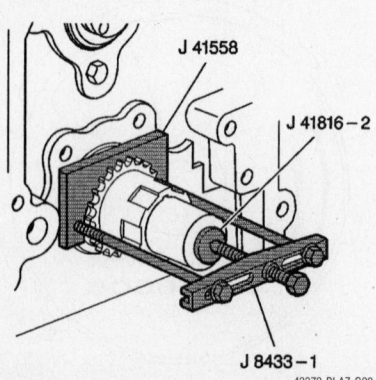

Use the proper tools to remove the crankshaft sprocket—5.3L engine

4. Remove or disconnect the following:
   • Camshaft sprocket bolts
   • Camshaft sprocket and timing chain
   • Crankshaft sprocket using Pulley Puller No. J 8433, Crankshaft End Protector Tool No. J 41816-2 and Crankshaft Sprocket Removal Tool No. J 41558

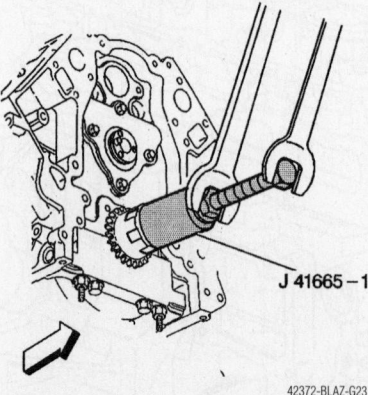

Crankshaft sprocket installation—5.3L engine

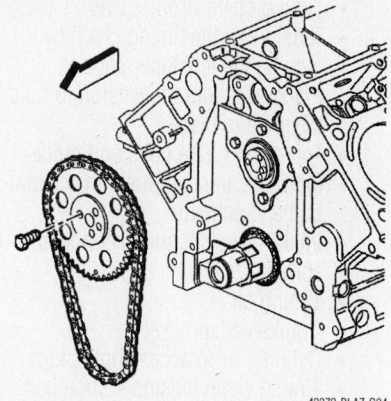

Proper alignment of the timing marks for timing chain installation—5.3L engine

   • Crankshaft sprocket key, if necessary
5. Clean and inspect the timing chain and sprockets.
   To install:
6. Install or connect the following:
   • Key into the crankshaft keyway, if removed. Tap the key into the keyway until both ends of the key bottom into the crankshaft.
   • Crankshaft sprocket onto the front of the crankshaft. Align the crankshaft key with the sprocket keyway.
   • Crankshaft sprocket using Sprocket Installation Tool No. J 41665. Install the sprocket onto the crankshaft until fully seated against the crankshaft flange. Rotate the crankshaft sprocket until the alignment mark is in the 12 o'clock position.

➡**Properly locate the camshaft sprocket locating pin with the cam sprocket alignment hole. The sprocket teeth and timing chain must mesh. The camshaft and crankshaft sprocket alignment marks MUST be aligned properly. Locate the camshaft sprocket alignment mark in the 6 o'clock position. It may be necessary to rotate the camshaft or crankshaft to align the marks.**

   • Camshaft sprocket and timing chain
   • Camshaft sprocket bolts and tighten to 26 ft. lbs. (35 Nm)
   • Oil pump

Piston and Ring

POSITIONING

9355MG10

Piston ring positioning—4.2L engine

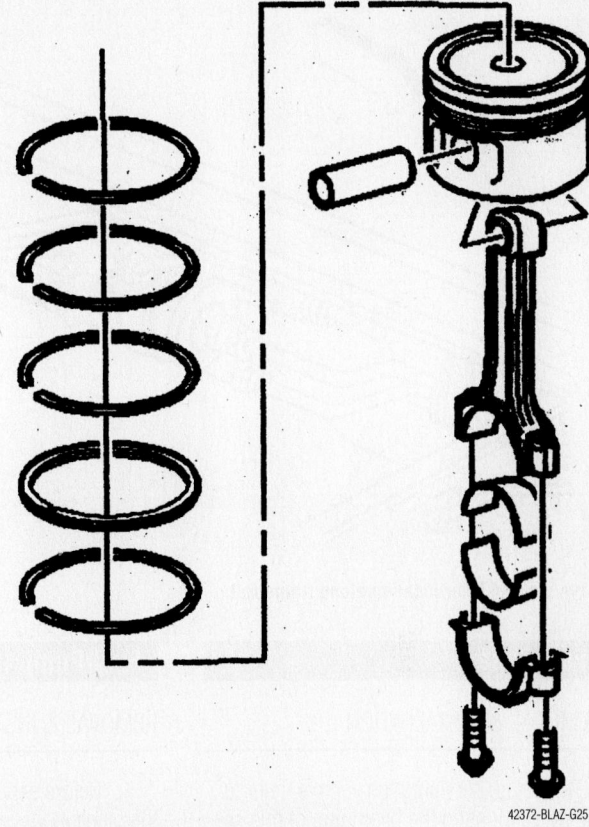

42372-BLAZ-G25

Piston ring positioning—5.3L engine

FUEL SYSTEM

Fuel System Service Precautions

Safety is the most important factor when performing not only fuel system maintenance but also any type of maintenance. Failure to conduct maintenance and repairs in a safe manner may result in serious personal injury or death. Maintenance and testing of the vehicle's fuel system components can be accomplished safely and effectively by adhering to the following rules and guidelines.

• To avoid the possibility of fire and personal injury, always disconnect the negative battery cable unless the repair or test procedure requires that battery voltage be applied.

• Always relieve the fuel system pressure prior to disconnecting any fuel system component (injector, fuel rail, pressure regulator, etc.), fitting or fuel line connection. Exercise extreme caution whenever relieving fuel system pressure, to avoid exposing skin, face and eyes to fuel spray. Please be advised that fuel under pressure may penetrate the skin or any part of the body that it contacts.

• Always place a shop towel or cloth around the fitting or connection prior to loosening to absorb any excess fuel due to spillage. Ensure that all fuel spillage (should it occur) is quickly removed from engine surfaces. Ensure that all fuel soaked cloths or towels are deposited into a suitable waste container.

• Always keep a dry chemical (Class B) fire extinguisher near the work area.

• Do not allow fuel spray or fuel vapors to come into contact with a spark or open flame.

• Always use a back-up wrench when loosening and tightening fuel line connection fittings. This will prevent unnecessary stress and torsion to fuel line piping. Always follow the proper torque specifications.

• Always replace worn fuel fitting O-rings with new. Do not substitute fuel hose or equivalent where fuel pipe is installed.

Fuel System Pressure

RELIEVING

The fuel systems operate under high fuel pressures. It is very important that the pressure be properly relieved prior to servicing the system or any of its components.

1. Before servicing the vehicle, refer to the precautions in the beginning of this section.

✹✹ WARNING

Do not perform this procedure for more than 2 minutes to avoid damaging the catalytic converter.

2. Loosen the fuel filler cap to release the fuel tank pressure.

3. Remove the fuel pump relay from the junction block.

4. Crank the engine, allowing it to start and stall.

5. Crank the engine for an additional 3 seconds to relieve any remaining fuel pressure.

6. Disconnect the negative battery cable to avoid repressurizing the fuel system.

7. Install the fuel pump relay in the junction block.

8. Tighten the fuel filler cap.

9. After you are finished working on the fuel system, connect the negative battery cable.

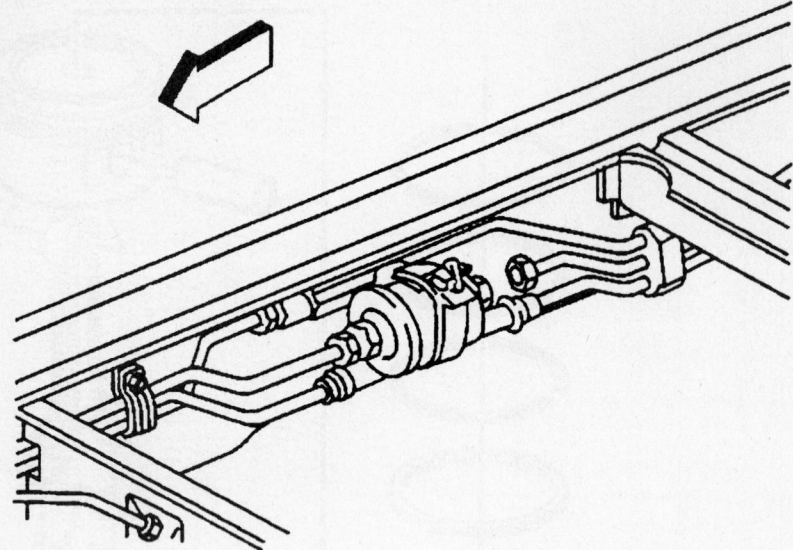

Typical fuel filter location along frame rail

7924JG43

Fuel Filter

REMOVAL & INSTALLATION

1. Before servicing the vehicle, refer to the precautions in the beginning of this section.
2. Properly relieve the fuel system pressure.
3. Remove or disconnect the following:
   - Negative battery cable and fuel filler cap, if not already done
4. Raise and support the vehicle.
   - Fuel tank shield, if equipped
   - Quick connect fittings from the filter
   - Filter feed nut and the clamp bolt
   - Filter and the clamp from the vehicle

To install:
5. Install or connect the following:
   - Filter and clamp with the directional arrow facing away from the fuel tank, towards the throttle body

➡️The filter has an arrow (fuel flow direction) on the side of the case, be sure to install it correctly in the system, the with arrow facing away from the fuel tank.

   - Tighten the fuel feed nut
   - Tighten the filter clamp assembly bolt
   - Fuel quick disconnect fittings to the filter
   - Fuel tank shield, if equipped
   - Fuel filler cap
   - Negative battery cable
6. Start the engine and check for leaks.

Fuel Pump

REMOVAL & INSTALLATION

1. Before servicing the vehicle, refer to the precautions in the beginning of this section.
2. Properly relieve the fuel system pressure.
3. Drain the fuel tank.
4. Support the fuel tank.
5. Remove or disconnect the following:
   - Negative battery cable
   - Filler neck from the tank
   - Frame brace

View of the in-tank fuel pump assembly

7924JG21

- Shield from tank and tank straps
- Fuel lines and vapor hose from pump
- Electrical connection from fuel pump
- Fuel tank
- Fuel pump/sending unit assembly by turning the locking ring (located on top of the fuel tank) counterclockwise using a spanner wrench
- Fuel pump from the fuel lever sending device

To install:
6. Install or connect the following:
   - Fuel pump in tank with new seal around opening

➡️The fuel pump strainer must be in a horizontal position when the fuel sender is installed in the tank. When installing the sender assembly, make sure that the fuel pump strainer does not block full travel of the float arm.

   - Tank and connect fuel lines and vapor hose
   - Tank and torque the strap fasteners to 24 ft. lbs. (32 Nm).
   - Shield
   - Fuel filler neck and clamp
   - Frame brace and tighten to 33 ft. lbs. (45 Nm)
   - Negative battery cable
7. Refill the tank.
8. Run the engine and check for leaks.

Fuel Injector

REMOVAL & INSTALLATION

4.2L Engine

1. Before servicing the vehicle, refer to the precautions in the beginning of this section.
2. Relieve the fuel system pressure. Refer to the fuel system relief procedure in this section.
3. Remove or disconnect the following:
   - Negative battery cable, if not done already
   - Intake manifold

➡️Clean the fuel rail assembly with a suitable spray cleaner before proceeding. Never soak the fuel rail in a cleaning solvent.

   - Fuel pressure regulator vacuum line
   - Fuel feed and return pipes
   - Fuel injector in-line electrical connector

- Fuel rail attaching bolts and fuel rail
- Fuel injector harness connector from the fuel injectors
- Injector retaining clip
- Injector from the fuel rail
- Retainer clip and O-ring seals from each end of the injector and discard

To install:

→**Each injector is calibrated. When replacing the fuel injectors, be sure to replace it with the correct injector.**

4. Lubricate the new injector O-ring seats with engine oil.

5. Install or connect the following:
- O-rings on the injector
- New retainer clip on the injector

6. Push the fuel injector into the fuel rail socket, making sure the connector faces outward. The retainer clip locks to a flange on the fuel rail injector socket.
- Fuel rail assembly. Tighten the bolts to 89 inch lbs. (10 Nm).
- Fuel feed and return lines to the rail
- Fuel injector electrical connectors
- Fuel pressure regulator vacuum line
- Intake manifold
- Negative battery cable

7. Turn the ignition **ON** for 2 seconds and then turn it **OFF** for 10 seconds. Again turn the ignition **ON** and check for leaks.

5.3L Engine

1. Before servicing the vehicle, refer to the precautions in the beginning of this section.

2. Relieve the fuel system pressure. Refer to the fuel system relief procedure in this section.

3. Remove or disconnect the following:
- Negative battery cable, if not done already
- A/C compressor pressure switch electrical connector
- Wire harness from the clip on the cylinder head
- Mass Airflow/Intake Air Temperature (MAF/IAT) sensor connector
- Alternator electrical connector
- Right side electrical connectors from the coil main electrical harness, Electronic Throttle Control (ETC) and fuel injectors.

4. To detach the injector connector: Matchmark the connectors, pull the Connector Position Assurance (CPA) retainer up 1 click. Push the tab on the connector in, then detach the injector connector.

- Alternator connector
- Electrical harness from the clips on the ignition coil bracket
- Evaporative emission (EVAP) purge solenoid electrical connector
- Knock Sensor (KS) electrical connector
- Manifold Absolute Pressure (MAP) electrical connector
- Main coil
- Fuel injector electrical connector (right side)
- Electrical harness from the clips on the ignition coil bracket
- Upper engine wire harness retainer nut. Position the wire harness aside.
- Fuel feed and return pipes from the rail
- Fuel pressure regulator vacuum line
- Fuel rail bolts
- Fuel rail, after cleaning with a spray-type cleaner

❊❊ WARNING

Be very careful when removing the fuel rail and injectors not to damage the connector terminals or injector spray tips

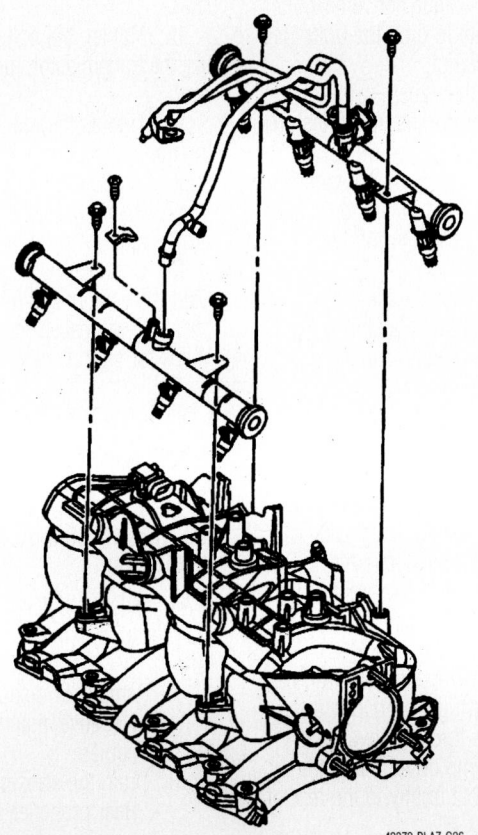

42372-BLAZ-G26

Exploded view of the fuel rail mounting—5.3L engine

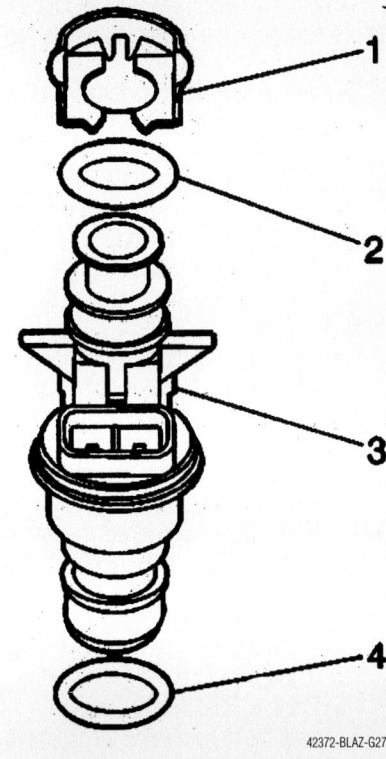

42372-BLAZ-G27

Exploded view of the fuel injector (3), retainer (1) and O-ring seals (2, 4)—5.3L engine

- Fuel injector from the fuel rail
- Fuel injector retainer clip and discard
- Fuel injector lower O-ring seals and discard

To install:

5. Install or connect the following:
  - New O-ring seals on the injectors, after lubricating with clean engine oil
  - New retainer clip on the injector
  - Fuel injector by pushing it into the fuel rail socket
  - Fuel rail
  - Apply 0.20 inch (5mm) band of threadlock to the threads of the fuel rail bolts
  - Fuel rail bolts and tighten to 89 inch lbs. (10 Nm)
  - Fuel pressure regulator vacuum line
  - Fuel feel and return pipes
  - Route the upper electrical harness into position over the engine.
  - Engine harness bracket nut and tighten to 89 inch lbs. (10 Nm)
  - PCV valve and hose
  - EVAP purge solenoid, KS, MAP sensor, main coil & fuel injector electrical connectors
  - Harness to the clips on the ignition coil bracket
  - Main coil, ETC, fuel injector electrical connectors
  - Harness to the clips on the ignition coil bracket
  - Alternator electrical connector
  - MAF/IAT sensor connector
  - Wire harness to the clip on the cylinder head
  - A/C compressor switch electrical connector
  - Air cleaner outlet duct
  - Fuel fill cap
  - Negative battery cable

6. Refill the engine cooling system.

DRIVE TRAIN

Automatic Transmission Assembly

REMOVAL & INSTALLATION

LL8 Transmission

➡ **This procedure requires the use of a Converter Holding Strap tool No. J 21366 to secure the torque converter to the transmission during removal and installation.**

1. Before servicing the vehicle, refer to the precautions in the beginning of this section.
2. Remove or disconnect the following:
  - Negative, then the positive battery cables
  - Fill tube nut, located on the right side of the engine
3. Drain the transmission fluid.
  - Rear propeller shaft
4. Support the transmission with a transmission jack.
  - Nuts securing the transmission mount to the transmission support
  - Evaporative emission (EVAP) canister from its mounting bracket on the left inside of the frame to get access to the transmission support bolts. Do not disconnect the canister lines.
  - Fuel tank shield
  - Transmission support
  - Transmission mount bolts and mount
  - Front exhaust pipe assembly
5. Lower the transmission for access to the top and sides of the transmission.
  - Transfer case, if equipped
  - Range selector cable end from the transmission range selector lever ball stud and bracket
  - Transmission heat shield, transmission vent hose park/neutral position switch connector, and main connector from the transmission
  - Bolt that secures the fuel line bracket to the left side of the transmission
  - Torque converter access plug
6. Matchmark the flywheel and torque converter orientation for reassembly.
  - Flywheel-to-torque converter bolts. Be careful not to drop the bolts into the bell housing!
  - Disconnect the transmission oil cooler lines from the transmission. Plug the transmission oil cooler lines connectors in the transmission case.
7. Install a safety chain around the transmission.
  - Bolt that secures the fuel line bracket to the bell housing
  - Bolts that secure the coolant pipe to the bell housing
  - Remaining nuts, studs and/or bolts that secure the transmission to the engine
8. Install Converter Holding Strap tool No. J 21366 onto the transmission bell housing to hold the torque converter.
9. Pull the transmission straight back and remove it from the vehicle.

To install:

Installation is the reverse of removal, but please note the following important steps.

10. Make sure the torque converter is fully seated in the pump drive. If not, the transmission will not fit tightly to the rear of the engine block.
11. Raise the transmission into position and remove the torque converter holding strap. Carefully slide the transmission forward until the dowel pins are engaged while lining up the marks on the flywheel made during removal.
12. The torque converter should be flush with the flywheel and turn freely by hand.
13. Install the transmission-to-engine nuts, studs and or bolts. Tighten the studs and/or bolts to 37 ft. lbs. (50 Nm).
14. Tighten the bolts securing the heat shield to the transmission to 13 ft. lbs. (17 Nm).
15. Tighten the bolts and washers securing the transmission mount to 18 ft. lbs. (25 Nm).
16. Tighten the nut and washer securing the transmission mount to the transmission support to 35 ft. lbs. (46 Nm).
17. Refill the transmission with the proper amount and type of fluid.
18. Connect the negative battery cable. Start the vehicle and allow to warm while checking for leaks. Road test the vehicle to check for shift quality.

LM4 Transmission

➡ **This procedure requires the use of a Converter Holding Strap tool No. J 21366 to secure the torque converter to the transmission during removal and installation.**

1. Before servicing the vehicle, refer to the precautions in the beginning of this section.
2. Remove or disconnect the following:
  - Negative, then the positive battery cables
3. Drain the transmission fluid.
  - Rear propeller shaft
4. Support the transmission with a transmission jack.

- Nuts securing the transmission mount to the transmission support and remove the support
- Transmission mount bolts and mount
- Front exhaust pipe assembly

5. Lower the transmission for access to the top and sides of the transmission.

- Transfer case, if equipped
- Range selector cable end from the transmission range selector lever ball stud and bracket
- Transmission heat shield, transmission vent hose park/neutral position switch connector, and main connector from the transmission
- Transmission harness from its retainers
- Bolt that secures the fuel line bracket to the left side of the transmission
- Torque converter access plug

6. Matchmark the flywheel and torque converter orientation for reassembly.

- Flywheel-to-torque converter bolts. Be careful not to drop the bolts into the bell housing!
- Disconnect the transmission oil cooler lines from the transmission. Plug the transmission oil cooler lines connectors in the transmission case.

7. Install a safety chain around the transmission.

- Bolt that attaches the fuel tube to the bell housing and remove the tube
- Remaining nuts, studs and/or bolts that secure the transmission to the engine

8. Install Converter Holding Strap tool No. J 21366 onto the transmission bell housing to hold the torque converter.

9. Pull the transmission straight back and remove it from the vehicle.

To install:

Installation is the reverse of removal, but please note the following important steps.

10. Make sure the torque converter is fully seated in the pump drive. If not, the transmission will not fit tightly to the rear of the engine block.

11. Raise the transmission into position and remove the torque converter holding strap. Carefully slide the transmission forward until the dowel pins are engaged while lining up the marks on the flywheel made during removal.

12. The torque converter should be flush with the flywheel and turn freely by hand.

13. Install the transmission-to-engine

nuts, studs and or bolts. Tighten the studs and/or bolts to 37 ft. lbs. (50 Nm).

14. Tighten the bolts securing the heat shield to the transmission to 13 ft. lbs. (17 Nm).

15. Tighten the bolts and washers securing the transmission mount to 18 ft. lbs. (25 Nm).

16. Tighten the nut and washer securing the transmission mount to the transmission support to 35 ft. lbs. (46 Nm).

17. Refill the transmission with the proper amount and type of fluid.

18. Connect the negative battery cable. Start the vehicle and allow to warm while checking for leaks. Road test the vehicle to check for shift quality.

Transfer Case Assembly

REMOVAL & INSTALLATION

1. Before servicing the vehicle, refer to the precautions in the beginning of this section.

2. Disconnect the negative battery cable.

3. Raise and support the vehicle. Drain the transfer case.

4. Remove or disconnect the following:

- Fuel tank shield mounting bolts and shield
- Front and rear propeller shaft. Matchmark the shafts prior to removal.
- Fuel lines from the retainer
- Electrical harness from the retainers on the right and left sides
- Speed sensor electrical connectors
- Motor/encoder electrical connector
- Transfer case wiring harness
- Vent hose

5. Install a transmission jack to support the transfer case.

- Transfer case mounting bolts
- Transfer case from the vehicle
- Transfer case gasket and discard if damaged

To install:

6. Install or connect the following:

➡**You must replace the transfer case gasket if it is damaged. Never use silicone sealant in place of, or with the transfer case gasket.**

- Transfer case, using a new gasket if necessary
- Transfer case mounting bolts and tighten to 35 ft. lbs. (47 Nm)

7. Remove the transmission jack.

8. The remainder of installation is the reverse of removal.

9. Refill the transfer case.

Halfshaft

REMOVAL & INSTALLATION

1. Before servicing the vehicle, refer to the precautions in the beginning of this section.

2. Remove or disconnect the following:

- Front wheel

➡**Place a drift through the caliper into the edge of the rotor to keep the rotor from turning when the nut is removed**

- Wheel center cap, if equipped
- Halfshaft nut and discard. A new nut must be used for installation.
- Drift from the rotor
- Brake caliper and support it with a piece of wire to avoid damaging the brake hose
- Brake rotor

3. To remove the steering knuckle, remove or disconnect the following:

- Wheel hub and bearing
- Outer tie rod retaining nut
- Outer tie rod end from the steering knuckle using a puller
- Brake hose bracket retaining bolts
- Brake hose bracket
- Anti-lock Brake System (ABS) wheel speed sensor wiring harness bracket, if necessary
- Upper control arm-to-steering knuckle pinch bolt and nut
- Upper control arm from the steering knuckle
- Lower ball joint retaining nut
- Steering knuckle from the control arm using a puller
- Steering knuckle

4. Remove the left side halfshaft from differential carrier, or right halfshaft from the clutch fork housing as follows:

a. Place a brass drift against the tripot housing.

b. Use a hammer to strike the drift outward from the case, striking hard enough to overcome the snaping tension holding the halfshaft.

5. Pull the halfshaft straight out of the differential carrier or clutch fork housing.

To install:

6. Install the halfshaft as follows:

a. With both hands on the tripot housing, align the splines on the shaft with the differential carrier assembly (left) or clutch fork housing (right).

b. Center the halfshaft into the differential carrier or clutch fork housing assembly seal.

c. Firmly push the shaft straight into

the differential carrier or clutch fork housing assembly until the snapring is properly seated.

7. To install the steering knuckle, install or connect the following:

- Steering knuckle to the lower control arm
- Lower ball joint retaining nut and tighten to 81 ft. lbs. (110 Nm)
- Upper control arm to the steering knuckle
- Upper control arm pinch bolt and nut and tighten to 30 ft. lbs. (40 Nm)
- ABS wheel speed sensor harness bracket
- Brake hose bracket. Tighten the bolts to 7 ft. lbs. (10 Nm).
- Outer tie rod to the steering knuckle and tighten the nut to 33 ft. lbs. (45 Nm)
- Hub and bearing

8. Install or connect the following:

- New halfshaft nut and tighten to 103 ft. lbs. (140 Nm)
- Wheel

9. Lower the vehicle. Adjust the front toe.

CV-Joints

OVERHAUL

Outer CV-Joint

1. Before servicing the vehicle, refer to the precautions in the beginning of this section.

2. Remove or disconnect the following:

- Front wheel
- Halfshaft and position it in a vise
- Large CV-joint boot clamp and discard it
- Small CV-joint boot clamp and discard it
- CV-joint boot and slide it back on the shaft
- Outer race from the halfshaft, by spreading the outer race-to-halfshaft retaining ring, using Snapring Pliers J-8059
- Retaining ring from the halfshaft and discard it
- CV-joint boot from the halfshaft and discard it, if damaged

3. Disassemble the chrome alloy balls from the CV-joint cage as follows:

a. Position a brass drift against the CV-joint cage and tap it with a hammer to tilt the cage.

b. Remove the 1st chrome alloy ball from the cage.

c. Tilt the cage in the opposite direction.

d. Remove the opposite chrome alloy ball.

e. Repeat the procedure until all 6 balls are removed.

4. Disassemble the CV-joint cage and inner race as follows:

a. Pivot the cage and race 90 degrees to the center line of the outer race.

b. Align the cage windows with outer race lands.

c. Remove the cage from the outer race.

d. Rotate the inner race upward and remove it from the cage.

5. Thoroughly clean and inspect all parts.

To install:

6. Lubricate the parts with a light coat of grease.

7. Assemble the CV-joint cage and inner race, as follows:

a. Rotate the inner race 90 degrees to the cage centerline.

b. Align the cage windows with inner race lands.

c. Insert the inner race into the cage by rotating the inner race downward.

d. Insert the cage/inner race into the outer race.

8. Assemble the chrome alloy balls into the CV-joint cage, as follows:

a. Position a brass drift against the CV-joint cage and tap it with a hammer to tilt the cage.

b. Insert the 1st chrome alloy ball into the cage.

c. Tilt the cage in the opposite direction.

d. Insert the opposite chrome alloy ball.

e. Repeat the procedure until all 6 balls are inserted.

9. Install ½ kit grease into the CV-joint.

10. Install or connect the following:

- Small ring clamp on the CV boot
- New retaining ring on the halfshaft
- Large ring clamp on the CV boot
- Outer race assembly onto the halfshaft until the ring engages the halfshaft groove

11. Slide the small end of the CV-joint boot/clamp into place, with the seal lip in the halfshaft groove

➡**Make sure the boot lies flat against the halfshaft.**

12. Using the Crimp tool J-35910, a torque wrench and a breaker bar, crimp the small CV-joint boot clamp to 100 ft. lbs. (136 Nm).

13. Check the clamp gap dimension; if it is not 0.085 in. (2.15mm), continue tightening the clamp until it is.

14. Install ½ kit grease into the CV-joint boot.

15. Measure approximately 0.687 in. (17.5mm) up from the bottom edge of the outer CV-joint assembly.

16. Slide the large end of the CV boot/clamp into place, with the seal lip in place over the outer race.

➡**Make sure the boot lies flat against the outer race.**

17. Using the Crimp tool J-35910, a torque wrench and a breaker bar, crimp the large CV-joint boot clamp to 130 ft. lbs. (176 Nm).

18. Check the clamp gap dimension; if it is not 0.102 in. (2.60mm), continue tightening the clamp until it is.

19. Install the halfshaft and the front wheel.

Inner (Tri-Pot) Joint

1. Before servicing the vehicle, refer to the precautions in the beginning of this section.

2. Remove or disconnect the following:

- Front wheel
- Halfshaft and place it in a vise
- Snapring from the stub shaft and discard it
- Small CV-joint boot clamp, cut and discard it
- Large CV-joint boot clamp, cut and discard it
- CV-joint boot by sliding it away from the tri-pot joint

3. Install a Stub Shaft Removal tool J-38868-A to the stub shaft snapring groove.

4. Using a slide hammer puller, press the stub shaft from the tri-pot housing.

5. Remove or disconnect the following:

- Tri-pot housing from the tri-pot spider
- Inboard spacer ring slide it rearward on the shaft using Snapring Pliers tool J-8059
- Outboard retaining ring using Snapring Pliers tool J-8059 and discard it
- Tri-pot joint spider assembly
- Inboard spacer ring and discard it
- CV-joint boot
- Trilobal tri-pot bushing from the housing

6. Thoroughly clean and inspect all parts.

To install:

7. Install or connect the following:

- New snapring onto the stub shaft

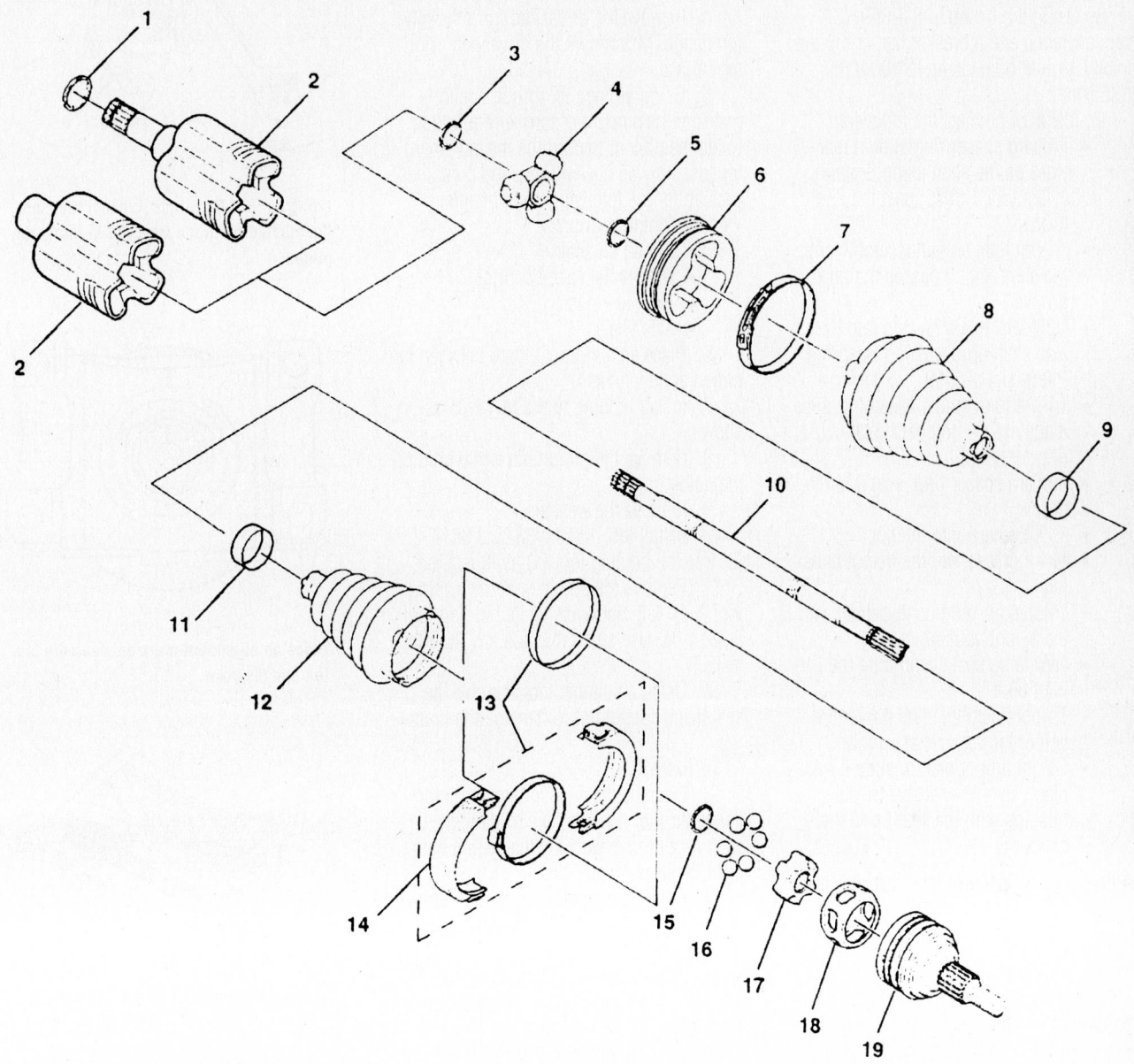

(1) Differential Shaft Ring
(2) Tripot Housing Assembly
(3) Spacer Ring
(4) Tripot Joint Spider Assembly
(5) Spacer Ring
(6) Tripot Bushing
(7) Boot Retaining Clamp
(8) Tripot Joint Boot
(9) Halfshaft Swage Ring
(10) Halfshaft Bar

(11) Halfshaft Swage Ring
(12) CV Joint Boot
(13) Swage Ring
(14) Clamp Protector
(15) Race Retaining Ring
(16) Ball
(17) CV Joint Inner Race
(18) CV Joint Cage
(19) CV Joint Outer Race

9308JG09

Exploded view of the CV-Joint Assembly

- Small boot clamp
- CV-joint boot

8. Using the Crimp tool J-35910, a torque wrench and a breaker bar, crimp the small CV-joint boot clamp to 100 ft. lbs. (136 Nm).

9. Install or connect the following:

- Inboard spacer ring slide it rearward on the shaft using Snapring Pliers tool J-8059, past the 2nd groove
- Tri-pot joint spider assembly onto the shaft until it passes the 2nd groove
- Outboard retaining ring into the axle shaft groove using Snapring Pliers tool J-8059
- Tri-pot joint spider assembly, slide it against the outboard retaining ring
- Inboard spacer ring, seat it in the groove
- ½ kit grease into the boot
- ½ kit grease into the tri-pot housing
- Trilobal tip-pot bushing flush with the tri-pot housing face
- New large seal clamp onto the CV-joint boot
- Tri-pot housing, slide it over the tri-pot joint spider assembly
- CV-joint boot/clamp, slide it into place, over the trilobal tri-pot bushing with the seal lip in the groove

➡ **Make sure the boot lies flat against the trilobal bushing.**

10. Position the CV-joint boot so it measures 4.9 in. (125mm).

11. Using the Crimp tool J-35566, latch the large CV-joint boot clamp.

12. Install the halfshaft and the front wheel.

Axle Shaft, Bearing and Seal

REMOVAL & INSTALLATION

For the Axle Shaft, Bearing and Seal, Removal and Installation, please refer to Wheel Bearing procedure located in the section.

Pinion Seal

REMOVAL & INSTALLATION

1. Before servicing the vehicle, refer to the precautions in the beginning of this section.

2. Remove the wheels.

3. Remove the rear calipers and rotors.

4. Remove the driveshaft from the pinion flange. Matchmark the driveshaft prior to removal.

5. Using an inch lb. torque wrench, measure the amount of torque required to rotate the pinion. Record the measurement for assembly as this will give the combined preload for the following components:

- Pinion bearings
- Pinion oil seal
- Differential case bearings
- Axle bearings
- Axle seals

6. Place an alignment mark between the pinion and the yoke.

7. Install holding tool J 8614-01 as shown.

8. Remove the pinion nut while holding the holding tool.

9. Remove the washer.

10. Install tool J 8614-3 (2), J 8614-2 (3) into the holding tool (1) as illustrated.

11. Remove the pinion yoke by turning tool J 8614-3 clockwise while holding tool J 8614-01. Use a pan to catch any fluid that leaks.

12. Using a suitable tool, remove the seal being careful not to damage the housing.

To install:

13. Examine the seal surface of pinion flange for tool marks, nicks or damage, such as a groove worn by the seal. If damaged, replace flange.

14. Examine the carrier bore and remove any burrs that might cause leaks around the O.D. of the seal.

15. Apply GM seal lubricant 1050169 to the outside diameter of the pinion flange and sealing lip of new seal.

16. Install or connect the following:

- New pinion oil seal using a seal installer tool such as J 33782 for the 8 inch axle or tool J 38694 for the 8.6 inch axle.
- Pinion flange and tighten nut to the same position as marked earlier. Tighten the nut a little at a time and turn the pinion flange several times after each tightening in order to set the rollers.

17. Measure the torque necessary to turn the pinion and compare this to the reading taken during removal. Tighten the nut additionally, as necessary to achieve the same preload as measured earlier. The rotating torque should be 3–5 inch lbs. (0.40–0.57 nm).

18. Install the driveshaft assembly to the pinion flange.

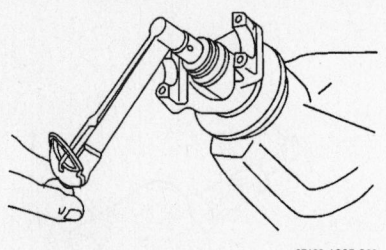

67162-ASCE-G02

Using an inch lb. torque wrench, measure the amount of torque required to rotate the pinion

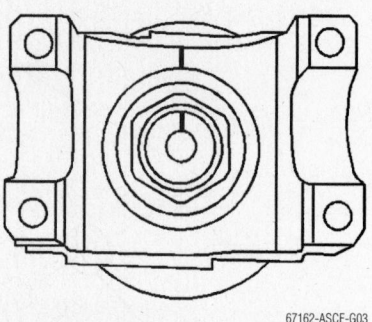

67162-ASCE-G03

Place an alignment mark between the pinion and the yoke

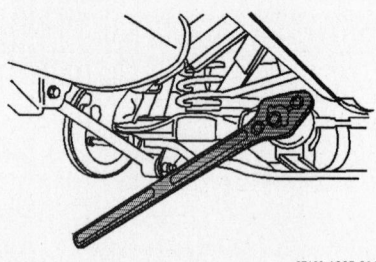

67162-ASCE-G04

Install holding tool J 8614-01 as shown

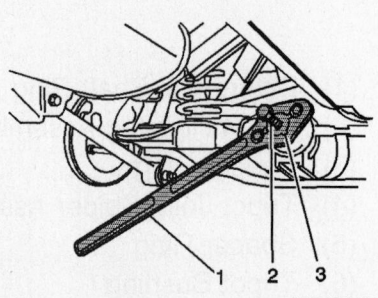

67162-ASCE-G05

Install tool J 8614-3 (2), J 8614-2 (3) into the holding tool (1) as shown

➡ **The original matchmarks MUST be aligned to assure proper shaft balance and prevent vibration.**

19. Install the rotors and calipers.

20. Install the wheels

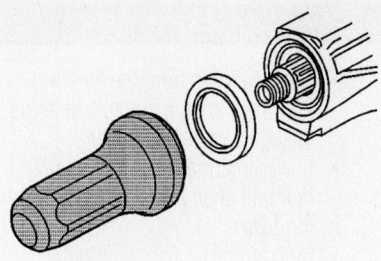

67162-ASCE-G06

Use the appropriately sized installation tool to drive the new seal into position.

➡**If fluid was lost from the differential housing during this procedure, be sure to check and add additional fluid, as necessary.**

Axle Housing

REMOVAL & INSTALLATION

1. Before servicing the vehicle, refer to the precautions in the beginning of this section.

2. Support the rear axle housing. If a floor jack is being used, take care when removing the U-bolts to keep the axle from suddenly dislodging.

3. Remove or disconnect the following:
- Rear wheels and drums for clearance and to remove some weight from the axle housing
- Axle vibration dampener, if equipped
- Rear driveshaft from the pinion flange. Either remove the shaft completely from the vehicle or support it aside from the undercarriage using safety wire, but DO NOT allow the shaft to hang from the slip joint.
- Shock absorber-to-axle housing retainers, then swing the shock absorbers away from the axle housing
- Brake lines from the axle housing clips and the backing plates (wheel cylinders)

➡**When disconnecting the brake lines from the wheel cylinders, immediately plug or cap the lines to prevent system contamination or excessive fluid loss.**

- Speed sensor connectors at the junction block, if applicable
- Parking brake cable(s)
- Axle housing-to-spring U-bolt nuts, washers, U-bolts and the anchor plates
- Vent hose from the top of the axle housing
- Axle with the help of an assistant by moving it to clear the leaf spring

To install:

4. With the help of an assistant, carefully position the rear axle into the vehicle.

5. Install or connect the following:
- Vent hose to the axle housing

6. Be sure the housing is properly positioned on the leaf spring, then loosely install the U-bolts, anchor plates, washers and nuts.
- Tighten the U-bolt nuts in a cross pattern to 18 ft. lbs. (25 Nm) to made sure everything is evenly seated. Then tighten the nuts in steps to 74 ft. lbs. (100 Nm).
- Brakes lines secure them to the axle housing
- Parking brake cable(s), if removed
- Speed sensor connectors to the junction block, if equipped
- Driveshaft assembly
- Shock absorbers to the lower mounts, then tighten the mount nuts
- Axle vibration dampener, if equipped
- Brake drums and the tire/wheel assemblies

7. Bleed the hydraulic brake system.

8. Check the fluid level in the rear axle assembly and add, as necessary. Make sure the vehicle is level when checking and adding fluid.

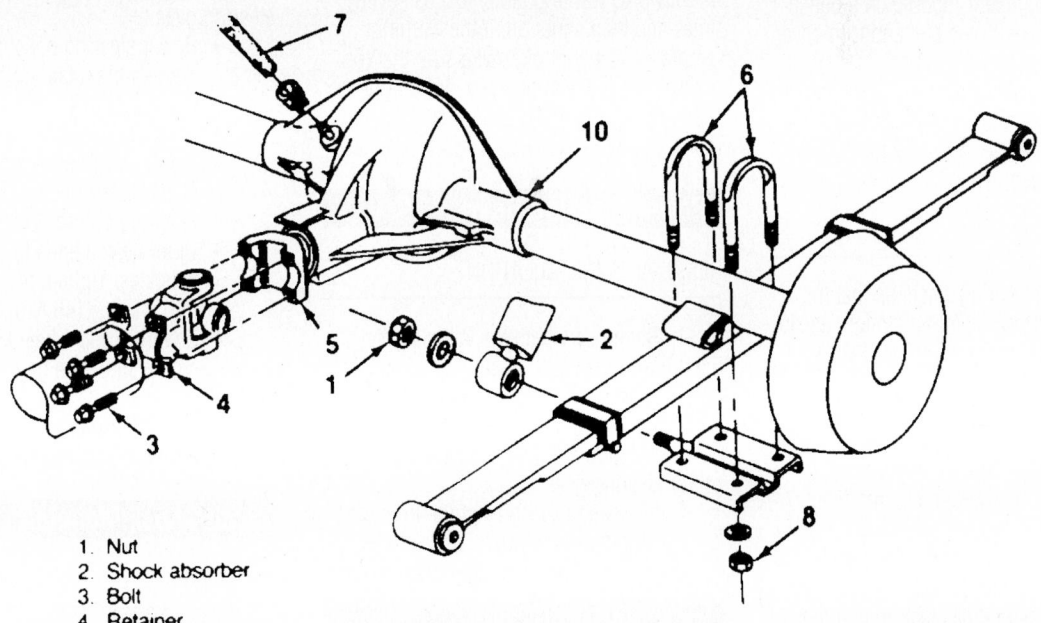

1. Nut
2. Shock absorber
3. Bolt
4. Retainer
5. Pinion flange
6. U-bolts
7. Vent hose
8. Nut
10. Axle housing

Exploded view of the rear axle mounting

88457G85

STEERING AND SUSPENSION

Air Bag

☼ CAUTION

Some vehicles are equipped with an air bag system, also known as the Supplemental Inflatable Restraint (SIR) system. The system must be disabled before performing service on or around system components, steering column, instrument panel components, wiring and sensors. Failure to follow safety and disabling procedures could result in accidental air bag deployment, possible personal injury and unnecessary system repairs.

PRECAUTIONS

Several precautions must be observed when handling the inflator module to avoid accidental deployment and possible personal injury.

• Never carry the inflator module by the wires or connector on the underside of the module.

• When carrying a live inflator module, hold securely with both hands, and ensure that the bag and trim cover are pointed away.

• Place the inflator module on a bench or other surface with the bag and trim cover facing up.

• With the inflator module on the bench, never place anything on or close to the module, that may be thrown in the event of an accidental deployment.

DISARMING

1. Turn the steering wheel so that the vehicle's wheels are pointing straight ahead.
2. Turn the ignition switch to **LOCK**, remove the key, then disconnect the negative battery cable.
3. Remove the AIR BAG or SIR fuse from the fuse block.
4. Remove the steering column filler panel or knee bolster.
5. Unplug the Connector Position Assurance (CPA) and yellow two way connector at the base of the steering column.
6. Open the glove compartment door, lift the stop and let the door fully open.
7. Remove the Connector Position Assurance (CPA) from the passenger yellow two way connector located behind the glove box.

8. Unplug the yellow two way connector located behind the glove box.
9. Connect the negative battery cable.

➡ **With the AIR BAG fuse removed, the battery cable connected and the ignition in the ON position, the AIR BAG warning lamp will be ON. This is normal and does not indicate a system malfunction.**

ARMING

1. Disconnect the negative battery cable.
2. Attach the yellow two way connector located behind the glove box.
3. Install the Connector Position Assurance (CPA) to the passenger yellow two way connector located behind the glove box.
4. Close the glove compartment door.
5. Turn the ignition switch to **LOCK**, then remove the key.
6. Attach the two way connector at the base of the steering column and the Connector Position Assurance (CPA).
7. Install the steering column filler panel or knee bolster.
8. Install the AIR BAG fuse to the fuse block.
9. Connect the negative battery cable.
10. From the passenger seat, turn the ignition switch to **RUN** and make sure that the AIR BAG warning lamp flashes seven times and then shuts off. If the warning lamp does not shut off, make sure that the wiring is properly connected. If the light remains on, take the vehicle to a reputable repair facility for service.

Power Steering Gear

REMOVAL & INSTALLATION

1. Before servicing the vehicle, refer to the precautions in the beginning of this section.
2. Raise and support the vehicle.
3. Position a fluid catch pan under the power steering gear.
4. Remove or disconnect the following:
   • Front tire and wheel assemblies
   • Outer tie rod retaining nuts

☼ WARNING

Do not try to separate a steering linkage joint by driving a wedge between the joint and the attached part. Doing this can cause seal damage and premature failure of the part.

• Outer tie rods from the steering knuckles using a suitable steering linkage and tie rod puller
• Lower intermediate shaft retaining bolt and shaft from the power steering gear
• Steering gear crossmember
• Feed and return fluid hoses from the steering gear. Immediately cap or plug all openings to prevent system contamination or excessive fluid loss.
5. Support the power steering gear.
• Power steering gear mounting bolts, then remove the gear from the vehicle.
6. Loosen the outer tie rod jam nuts, then remove the outer tie rods from the inner tie rods and discard the jam nut.
To install:
7. Lubricate the inner tie rod threads with a suitable lubricant before installing the outer tie rod.
8. Install or connect the following:
   • New jam nuts to the outer tie rods
   • Outer tie rods to the inner tie rods
   • Power steering gear to the vehicle. Tighten the retaining bolts to 81 ft. lbs. (110 Nm).
9. Remove the support from the power steering gear.
   • Power steering hose(s) to the gear. Tighten the retaining bolt to 9 ft. lbs. (12 Nm).
   • Steering gear crossmember
   • Lower intermediate shaft to the power steering gear. Tighten the retaining bolt to 30 ft. lbs. (40 Nm).
   • Outer tie rod ends to the steering knuckles. Tighten the retaining nuts to 33 ft. lbs. (45 Nm).
   • Front tire and wheel assemblies
10. Remove the drain pan, then lower the vehicle.
11. Bleed the power steering system and adjust the front toe as necessary.

Strut/Shock Module

REMOVAL & INSTALLATION

Front

➡ **In these models a "shock module", similar to a strut was used on these vehicles. This procedure requires the use of a suitable steering linkage and tie rod puller.**

1. Before servicing the vehicle, refer to the precautions in the beginning of this section.

2. Remove or disconnect the following:

- Shock module upper retaining nuts
- Tire and wheel
- Shock module-to-lower control arm retaining nut
- Shock module yoke from the lower control arm using a suitable puller such as J 24319-B
- Shock module from the shock tower and lower control arm

To install:

3. Install or connect the following:

- Shock module to the shock tower and lower control arm
- Shock module yoke to the lower control arm
- Shock module upper retaining nuts and tighten to 33 ft. lbs. (45 Nm)
- Shock module-to-lower control arm retaining nut and tighten to 81 ft. lbs. (110 Nm)
- Tire and wheel

Shock Absorbers

REMOVAL & INSTALLATION

Rear

1. Before servicing the vehicle, refer to the precautions in the beginning of this section.

2. Properly support the rear axle assembly.

3. Remove or disconnect the following:

- Automatic level control air lines from the shock absorber, if equipped
- Shock absorber-to-frame retainer(s) at the top of the shock
- Shock-to-axle retainer(s) at the bottom of the shock
- Shock absorber

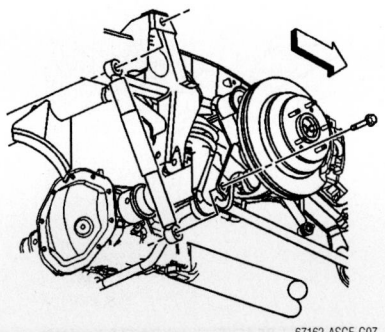

67162-ASCE-G07

Rear shock absorber mounting

To install:

4. Install the shock in the vehicle and loosely install the upper mounting fasteners to retain it

5. Align the lower-end of the shock absorber with the axle mounting, then loosely install the retainers.

6. Tighten the upper and lower shock retainers to 59 ft. lbs. (80 Nm).

7. If equipped, attach the automatic level control air lines to the shock absorber.

Coil Springs

REMOVAL & INSTALLATION

Front

➡**This procedure requires the use of a suitable spring compressor.**

1. Before servicing the vehicle, refer to the precautions in the beginning of this section.

2. Remove or disconnect the following:

- Wheel
- Shock module
- Shock module yoke-to-shock absorber pinch bolt and nut

3. Spread the shock module yoke at the pinch bolt using a suitable flat-bladed tool.

- Shock module yoke from the shock absorber

4. Install pieces of heater hose or equivalent material to the shock module spring where the spring compressor contacts the lower part of the spring.

5. Install the shock module into the spring compressor.

➡**The spring is compressed when the shock absorber moves freely.**

6. Turn the spring compressor forcing screw until the coil spring is compressed.

7. Remove or disconnect the following:

- Shock absorber upper retaining nut
- Shock absorber from the shock module

8. Loosen the compressor forcing screw until the upper mounting plate and coil spring can be removed.

- Upper mounting plate and coil spring from the spring compressor

To install:

9. Install or connect the following:

- Coil spring and upper mounting plate to the spring compressor

10. Turn the compressor forcing screw until the coil spring is compressed.

- Shock absorber to the shock module. Tighten the retaining nut to 33 ft. lbs. (45 Nm)

11. Remove the shock module from the spring compressor. Remove the pieces of heater hose from the spring..

- Shock module yoke to the shock absorber
- Shock module yoke-to-shock pinch bolt and nut and tighten to 52 ft. lbs. (70 Nm)
- Shock module to the vehicle
- Tire and wheel

12. Lower the vehicle

Rear

1. Before servicing the vehicle, refer to the precautions in the beginning of this section.

2. Raise and support the vehicle.

3. Support the rear axle.

4. Remove the shock absorber lower mounting bolts.

5. Lower the rear axle, then remove the coil springs.

To install:

6. Install the coil springs, then raise the rear axle.

7. Install the shock absorber lower mounting bolts and tighten to 59 ft. lbs. (80 Nm).

8. Remove the rear axle support.

9. Lower the vehicle.

Torsion Bar

REMOVAL & INSTALLATION

1. Before servicing the vehicle, refer to the precautions in the beginning of this section.

➡**The following procedure requires the use of the Torsion Bar Unloader tool J-36202.**

2. Remove or disconnect the following:

- Transmission shield, if equipped
- Torsion bar unloader tool to relax the tension on the torsion bar adjusting arm screw; record the number of turns necessary to properly install the tool. Remove the adjusting screw and the unloader tool.
- Lower link mount nut from one side
- Torsion bars by disengaging them

➡**Note the direction of the forward end and side of the torsion bar being removed**

- Lower link nut from the opposite side
- Lower link mount, upper link mount nut

- Upper link mount
- Torsion bar from the frame

To install:

3. Install or connect the following:
   - Torsion bar and support
   - Upper link mount. Torque the nut to 48 ft. lbs. (68 Nm).
4. Place a jack under the torsion bar to release tension.
5. Install or connect the following:
   - Lower link mount bushing and nut. Torque the nut to 37 ft. lbs. (50 Nm).
   - Torsion bar unloader tool. Tighten the tool against the adjusting arm the same number turns recorded earlier and remove the tool. This loads the torsion bars.
   - Transmission shield, if removed

Upper Ball Joints

REMOVAL & INSTALLATION

➡ **This procedure requires the use of the following special tools: J 9519-E Lower Ball Joint Remover and Installer, J 21474-01 Control Arm Bushing Set and J 45117 Ball Joint Installation Spacer.**

1. On 4WD vehicles, remove the wheel center cap and drive axle nut.
2. Raise and support the vehicle.
3. Remove or disconnect the following:
   - Tire and wheel
   - Wheel hub and bearing, if necessary
   - Outer tie rod retaining nut
   - Out tie rod from the steering knuckle using a suitable puller
   - Brake hose bracket retaining bolts and bracket
   - Upper control arm-to-steering knuckle pinch bolt and nut
   - Upper control arm from the steering knuckle
   - Lower ball joint retaining nut
   - Steering knuckle from the lower control arm using a suitable ball joint removal tool
   - Steering knuckle from the vehicle
   - Upper ball joint retaining clip
   - Upper ball joint from the steering knuckle using Lower Ball Joint Removal and Installer tool No. J 9519-E

To install:

4. Install or connect the following:
   - Upper ball joint to the steering knuckle using J 9519-E, J 21474-01 and J 45117

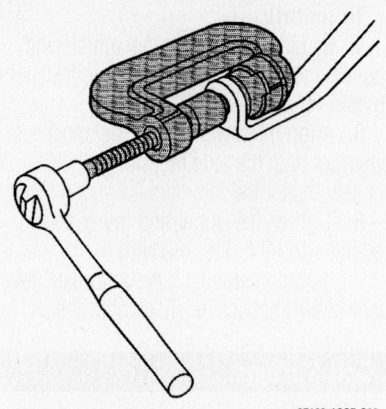

Remove the upper ball joint from the steering knuckle using tool No. J 9519-E

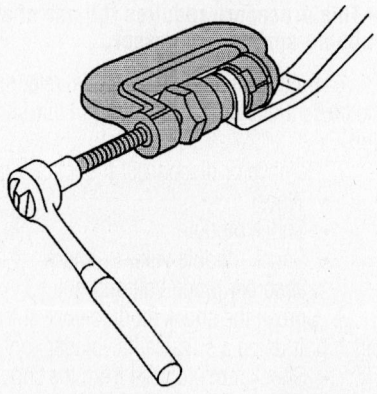

Install the upper ball joint to the steering knuckle using J 9519-E, J 21474-01 and J 45117

- Upper ball joint retaining clip
- Steering knuckle to the lower control arm
- Lower ball joint retaining nut and tighten to 81 ft. lbs. (110 Nm)
- Upper control arm to the steering knuckle
- Upper control arm pinch bolt and nut and tighten to 30 ft. lbs. (41 Nm)
- Brake hose bracket to the steering knuckle
- Brake hose bracket retaining nuts and tighten to 7 ft. lbs. (10 Nm)
- Outer tie rod to the steering knuckle
- Outer tie rod retaining nut and tighten to 33 ft. lbs. (45 Nm)
- Wheel hub and bearing, if removed
- Tire and wheel
5. Lower the vehicle
   - Drive axle nut, if 4WD, and tighten to 103 ft. lbs. (140 Nm)
   - Wheel enter cap, if removed
6. Check the front wheel alignment.

Lower Ball Joints

REMOVAL & INSTALLATION

➡ **This procedure requires the use of the following special tools: J 9519-E Lower Ball Joint Remover and Installer, J 34874 Booster Seal Remover/ Installer, J 41435 Ball Joint Installer, J 45105-1 Ball Joint Flaring Adapter and J 45105-2 Receiver.**

1. On 4WD vehicles, remove the wheel center cap and drive axle nut.
2. Raise and support the vehicle.
3. Remove or disconnect the following:
   - Tire and wheel
   - Wheel hub and bearing, if necessary
   - Outer tie rod retaining nut
   - Out tie rod from the steering knuckle using a suitable puller
   - Brake hose bracket retaining bolts and bracket
   - Upper control arm-to-steering knuckle pinch bolt and nut

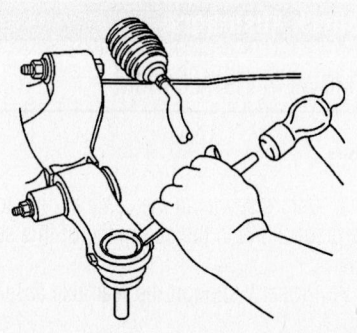

Remove the lower ball joint flange with a chisel

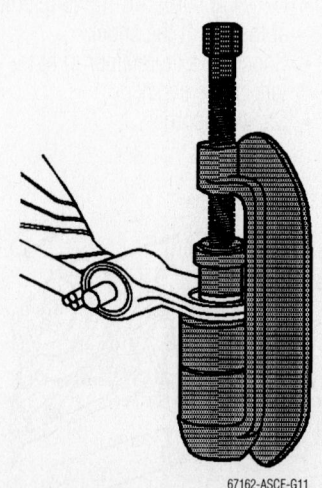

Use tools J 9519-E and J 34874 to remove the lower ball joint

- Upper control arm from the steering knuckle
- Lower ball joint retaining nut
- Steering knuckle from the lower control arm using a suitable ball joint removal tool
- Steering knuckle from the vehicle
- Lower ball joint flange with a chisel

4. Install tools J 9519-E and J 34874 to the lower ball joint, then use those tools to remove the lower ball joint from the lower control arm.

To install:

5. Install or connect the following:
- Lower ball joint to the lower control arm, using tools J 9519-E, J 41435 and J 45105-2

6. Remove the tools from the lower control arm.
- Tools J 9519-E and J 45105-1 to the lower ball joint

7. Flare the lower ball joint flange with J 9519-E and J 45105-1, then remove the tools from the lower ball joint.
- Steering knuckle to the lower control arm
- Lower ball joint retaining nut and tighten to 81 ft. lbs. (110 Nm)
- Upper control arm to the steering knuckle
- Upper control arm pinch bolt and nut and tighten to 30 ft. lbs. (41 Nm)
- Brake hose bracket to the steering knuckle
- Brake hose bracket retaining nuts and tighten to 7 ft. lbs. (10 Nm)
- Outer tie rod to the steering knuckle
- Outer tie rod retaining nut and tighten to 33 ft. lbs. (45 Nm)
- Wheel hub and bearing, if removed
- Tire and wheel

8. Lower the vehicle

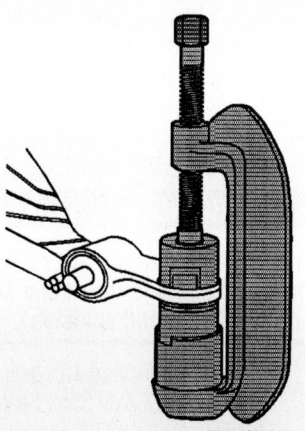

Use tools J 9519-E and J 45105-1 to install the lower ball joint

- Drive axle nut, if 4WD, and tighten to 103 ft. lbs. (140 Nm)
- Wheel enter cap, if removed

9. Check the front wheel alignment.

Upper Control Arm

REMOVAL & INSTALLATION

1. Before servicing the vehicle, refer to the precautions in the beginning of this section.
2. Remove or disconnect the following:
- Tire and wheel assembly
- Upper ball joint-to-upper control arm pinch bolt and nut
- Upper control arm from the knuckle
- Anti-lock Brake System (ABS) wheel speed sensor wiring harness
- Upper control arm mounting bolts
- Upper control arm

To install:

3. Install or connect the following:
- Upper control arm and tighten the bolts to 111 ft. lbs. (150 Nm)
- ABS wheel speed sensor wiring harness
- Upper control arm to the steering knuckle
- Upper ball joint-to-upper control arm pinch bolt and nut and tighten to 30 ft. lbs. (40 Nm)
- Tire and wheel

4. Check the front wheel alignment.

Lower Control Arm

REMOVAL & INSTALLATION

1. Before servicing the vehicle, refer to the precautions in the beginning of this section.
2. Raise the vehicle.
3. Remove or disconnect the following:
- Tire and wheel
- Tie rod from the steering knuckle
- Stabilizer shaft link lower nut and remove the shaft and washer from the control arm
- Shock module lower mounting bolt and disconnect the module from the lower control arm using puller J 24319-B
- Lower control arm-to-bracket bolts. Note the direction of the bolts prior to removal.
- Lower ball joint from the steering knuckle.

➡On 4WD models, do not disengage the axle shaft from the transmission.

- Pivot the lower control arm out and down to disconnect it from the lower bracket and remove the control arm.

To install:

4. Install or connect the following:
- Lower control arm to the knuckle. Pivot the arm out and up to engage the arm to the lower bracket.
- Lower control arm–to–lower bracket bolts. Make sure the arm is parallel to the control arm bracket when installing and tightening of the bolts and nuts to ensure correct alignment of the bushings. Tighten the nuts to 81 ft. lbs. (110 Nm).
- Shock module yoke to the lower control arm and install the mounting nut.

➡There is a washer between the stabilizer shaft and the lower control arm which is made of hardened steel with a felt inner liner. Make sure if replacing this washer it is replaced with the identical washer only.

- Stabilizer link and washer. Tighten the nut to 74 ft. lbs. (100 Nm).
- Outer tie rod to the knuckle and tighten the nut to 33 ft. lbs. (40 Nm).
- Wheel

5. Check the wheel alignment.

Wheel Bearings

ADJUSTMENT

The wheel bearings on these vehicles are not adjustable. If the bearings become loose or make noise, they must be replaced.

REMOVAL & INSTALLATION

Front

REAR WHEEL DRIVE

1. Before servicing the vehicle, refer to the precautions in the beginning of this section.
2. Raise and support the vehicle.
3. Remove or disconnect the following:
- Tire and wheel
- Caliper, leaving the fluid lines connected
- Brake rotor
- Wheel speed sensor
- Wheel hub and bearing-to-steering knuckle bolts and hub and bearing

➡Lay the hub and bearing on the wheel studs on the outboard side. This will avoid damaging the bearing seal.

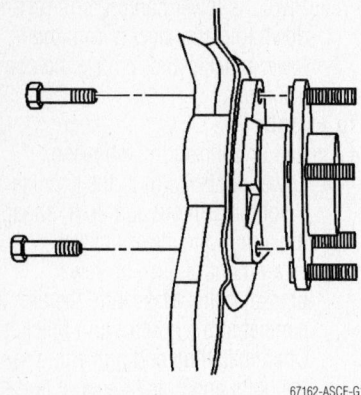

67162-ASCE-G13

Front wheel bearing and hub mounting—rear wheel drive models

- Splash shield from the steering knuckle

To install:

4. Install or connect the following:
   - Splash shield to the steering knuckle, making sure it's properly aligned
   - Hub and bearing to the steering knuckle, aligning the threaded holes
   - Hub and bearing bolts and tighten to 77 ft. lbs. (105 Nm)
   - Wheel speed sensor. Tighten the bolt to 13 ft. lbs. (18 Nm).
   - Rotor and brake caliper
   - Tire and wheel
5. Lower the vehicle

FOUR WHEEL DRIVE

1. Before servicing the vehicle, refer to the precautions in the beginning of this section.
2. Remove the wheel center cap, if equipped, and the drive axle nut and washer
3. Raise and support the vehicle.
4. Remove or disconnect the following:
   - Tire and wheel
   - Caliper, leaving the fluid lines connected
   - Brake rotor
   - Halfshaft from the hub and bearing. Place a brass drift against the outer edge of the halfshaft to protect the shaft threads. Use a hammer to sharply strike the brass drift, but to do not remove the halfshaft at this time.
   - Wheel speed sensor
   - Wheel hub and bearing-to-steering knuckle bolts and hub and bearing

➡️ **Lay the hub and bearing on the wheel studs on the outboard side. This will avoid damaging the bearing seal.**

- Splash shield from the steering knuckle
- Seal from the hub and bearing

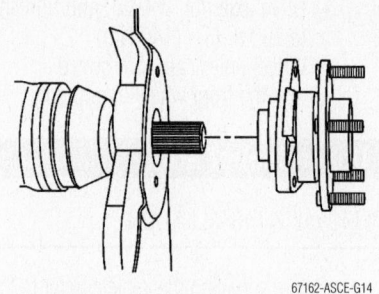

67162-ASCE-G14

Front wheel bearing and hub mounting—four wheel drive models

To install:

5. Install or connect the following:
   - Wheel hub and bearing seal
   - Splash shield to the steering knuckle, making sure it's properly aligned
   - Hub and bearing to the steering knuckle, aligning the threaded holes
   - Hub and bearing bolts and tighten to 77 ft. lbs. (105 Nm)
   - Wheel speed sensor. Tighten the bolt to 13 ft. lbs. (18 Nm).
   - Rotor and brake caliper
   - Tire and wheel
6. Lower the vehicle
7. Install the drive axle nut and tighten to 103 ft. lbs. (140 Nm), then install the center cap.

Rear

A new pinion shaft lockbolt should be installed whenever either of the axle shafts is removed.

The axle shaft and seal may be removed and replaced without disturbing the bearing or seal but it is highly recommended to replace the seals when removing the axle shaft.

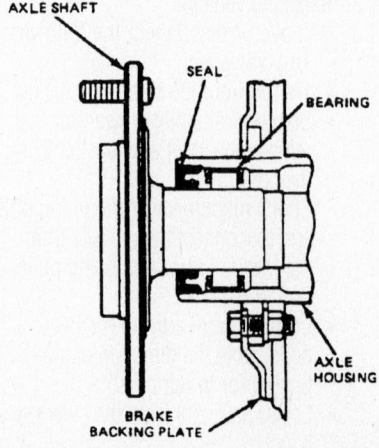

7924JG55

Cross-sectional view of the rear axle, bearing and seal assembly

1. Before servicing the vehicle, refer to the precautions in the beginning of this section.
2. Remove or disconnect the following:
   - Rear wheels
   - Brake drums
3. Using a wire brush, clean the dirt/rust from around the rear axle cover.
4. Drain the fluid.
5. Remove or disconnect the following:
   - Rear pinion shaft lockbolt and the pinion shaft
   - C-lock from the button end of the axle shaft by pushing the axle shaft inward
   - Axle shaft from the axle housing

➡️ **Be careful not to damage the oil seal.**

✳✳ WARNING

If equipped with an Anti-Lock Brake System (ABS), be careful not to damage the reflector ring on the axle shaft or the speed sensor bolted to the backing plate, immediately adjacent to the shaft.

6. Remove or disconnect the following:
   - Oil seal by prying the it from the end of the rear axle housing

✳✳ WARNING

DO NOT damage the housing oil seal surface.

- Wheel bearing using the GM Slide Hammer tool J-2619, the GM Adapter tool J-2619-4 and the GM Axle Bearing Puller tool J-22813-01

To install:

7. Clean and inspect the components for excessive wear or damage and replace them, if necessary.
8. Install or connect the following:
   - New or reused bearing, coated with gear lubricant, using the Axle Shaft Bearing Installer tool J-34974 to drive the bearing in until it bottoms against the seat

✳✳ WARNING

Be sure the bearing installer does not contact and damage the speed sensor on ABS equipped vehicles.

- New seal lubricated with gear oil using the GM Axle Shaft Seal Installer tool J-33782 to seat it in the housing until it is flush with the axle tube

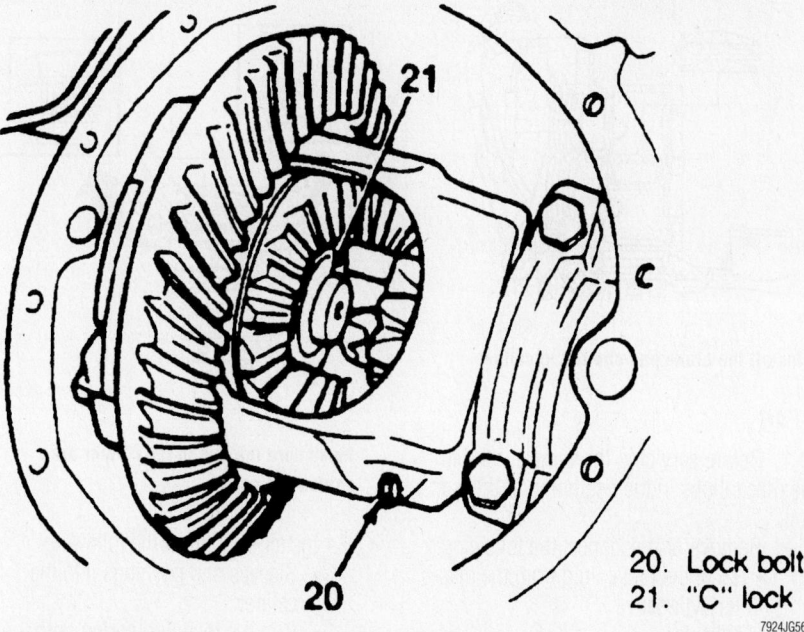

20. Lock bolt
21. "C" lock

7924JG56

Pinion shaft lockbolt and axle C-lock locations, inside the differential

➡ **Be sure the seal installer does not contact and damage the speed sensor on ABS equipped vehicles.**

- Axle shaft into the housing by engaging the splines
- C-lock retainer on the axle shaft button end

✳✳ WARNING

BE CAREFUL not to damage the wheel bearing seal.

- Axle shaft by pulling it outward to seat the C-lock retainer in the counterbore of the side gears
- Pinion shaft through the case and the pinions. Tighten the new lock-bolt to 27 ft. lbs. (36 Nm).
- New rear axle cover gasket
- Housing cover
- Brake drums
- Wheels
9. Refill the housing.

BRAKES

Brake Caliper

REMOVAL AND INSTALLATION

FRONT

1. Before servicing the vehicle, refer to the precautions in the beginning of this section.
2. Remove or disconnect the following:
   - ⅔ of the brake fluid from the master cylinder reservoir
   - Tire and wheel assembly
   - Caliper fluid line, plug the line and discard the copper washers
   - Bolts retaining the caliper to the rotor
   - Caliper from the rotor

To install:
3. Clean and lubricate the sleeves and bushings with silicon grease.
4. Install or connect the following:
   - Caliper in position over the rotor
   - Mounting bolts and tighten to 38 ft. lbs. (51 Nm)
   - Fluid lines to the caliper using new washers and tighten to 33 ft. lbs. (45 Nm)
   - Wheel and tire assembly
5. Refill the master cylinder to the correct level. Bleed the brake system if the fluid lines were disconnected from the caliper.

REAR

1. Before servicing the vehicle, refer to the precautions in the beginning of this section.

2. Raise and safely support the vehicle.
3. Remove or disconnect the following:
   - Rear wheels
   - Brake hose and cap the line. Discard the copper washers.
   - Retainers from caliper and remove caliper

To install:
4. Coat the caliper guide pin with a high temperature silicone brake lubricant. Make sure the lubricant does not get on the pads.
5. Install or connect the following:
   - Caliper over rotor, and onto mounts
   - Retainers, and tighten to 23 ft. lbs. or (31 Nm)
   - Brake hose with new washers, and tighten to 33 ft. lbs. (44 Nm)
6. Bleed brake system.
7. Install tires.
8. Refill the master cylinder and pump pedal to attain full brake pedal before Road-testing the vehicle.

Disc Brake Pads

REMOVAL AND INSTALLATION

FRONT

1. Before servicing the vehicle, refer to the precautions in the beginning of this section.
2. Remove or disconnect the following:

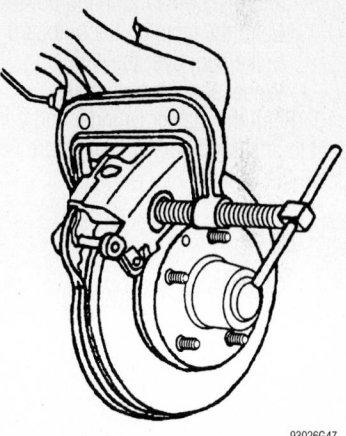

93026G47

Compressing the caliper piston with a C-clamp

   - ⅔ of the brake fluid from the master cylinder
3. Place a C-clamp around the outer pad and caliper; tighten the C-clamp until the piston is fully compressed in the caliper.
   - Brake pads
   - Inboard pad and retaining spring from the caliper
   - Outboard pad from the caliper
   - Sleeves and bushings

To install:
4. Clean and lubricate the sleeves and bushing with silicone lubricant and install them in the caliper.
5. Install the pad retaining clips onto the caliper mounting bracket.

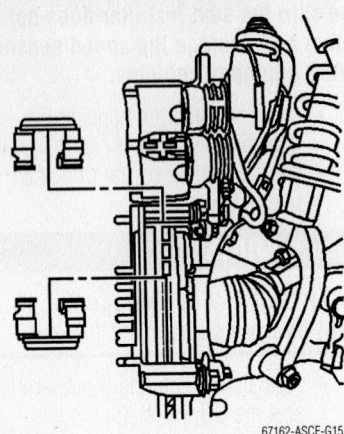

67162-ASCE-G15

Install the pad retaining clips onto the caliper mounting bracket

6. Install the brake pad clip to the caliper.

7. Make sure the clip in the caliper are seated correctly.

8. Install the inboard pad in the caliper.

9. Install or connect the following:
- Outboard pad into the caliper
- Caliper in position over the rotor and install the mounting bolts. Bend the tabs, on the outboard brake pad, over the caliper.
- Wheel and tire assemblies

10. Refill the master cylinder and pump pedal to attain full brake pedal before Road-testing the vehicle.

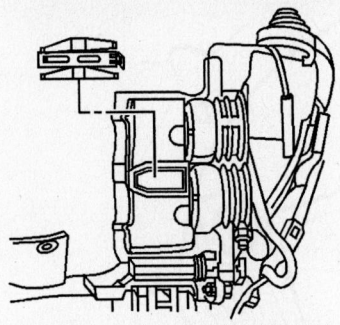

67162-ASCE-G16

Install the brake pad clip to the caliper

REAR

1. Before servicing the vehicle, refer to the precautions in the beginning of this section.

2. Remove or disconnect the following:
- ⅔ of the brake fluid from the master cylinder
- Wheels

3. Place a C-clamp around the outer pad and caliper; tighten the C-clamp until the piston is fully compressed in the caliper.
- Top caliper retainer, and rotate caliper away from rotor
- Inboard pad and retaining spring from the caliper
- Outboard pad from the caliper

To install:

4. Clean and lubricate the sleeves and bushing with silicone lubricant

67162-ASCE-G17

Make sure the clip in the caliper are seated correctly

5. Install or connect the following:
- Sleeves and bushings into the caliper
- Clip the retaining spring onto the inboard pad and install the pad in the caliper
- Outboard pad into the caliper
- Caliper in position over the rotor and install the mounting bolts and tighten to 23 ft. lbs. (31 Nm).
- Wheel and tire assemblies

6. Refill the master cylinder and pump pedal to attain full brake pedal before Road-testing the vehicle.

SPECIFICATION AND MAINTENANCE CHARTS

ENGINE AND VEHICLE IDENTIFICATION

			Engine						Model Year	
Code	Liters (cc)	Cu. In.	Cyl.	Fuel Sys.	Engine Type	Eng. Mfg.		Code ①	Year	
X	3.5 (3494)	213	6	MFI	DOHC	Isuzu		1	2001	

Code ①	Year
1	2001
2	2002
3	2003
4	2004
5	2005

NA: Not available

MFI: Multi-port Fuel Injection

DOHC: Double Overhead Camshaft

① 10th position of VIN

67162-AXIO-C01

GENERAL ENGINE SPECIFICATIONS

Year	Model	Engine Displacement Liters	Engine Series VIN)	Net Horsepower @ rpm	Net Torque @ rpm (ft. lbs.)	Bore x Stroke (in.)	Com-pression Ratio	Oil Pressure @ rpm
2002	Axiom	3.5	X	230@5400	230@3000	3.68x3.35	9.1:1	60-80@3000
2003	Axiom	3.5	X	230@5400	230@3000	3.68x3.35	9.1:1	60-80@3000

MFI: Multiport fuel injection

67162-AXIO-C02

ENGINE TUNE-UP SPECIFICATIONS

Year	Engine Displacement Liters	Engine VIN	Spark Plug Gap (in.)	Ignition Timing (deg.) MT	Ignition Timing (deg.) AT	Fuel Pump (psi)	Idle Speed (rpm) MT	Idle Speed (rpm) AT	Valve Clearance In.	Valve Clearance Ex.
2002	3.5	X	0.040	①	①	48-55	750	750	0.009-0.013	0.010-0.014
2003	3.5	X	0.040	①	①	48-55	750	750	0.009-0.013	0.010-0.014

NOTE: The Vehicle Emission Control Information label often reflects specification changes made during production.

The label figures must be used if they differ from those in this chart.

B: Before top dead center

HYD: Hydraulic

① Controlled by the PCM

67162-AXIO-C03

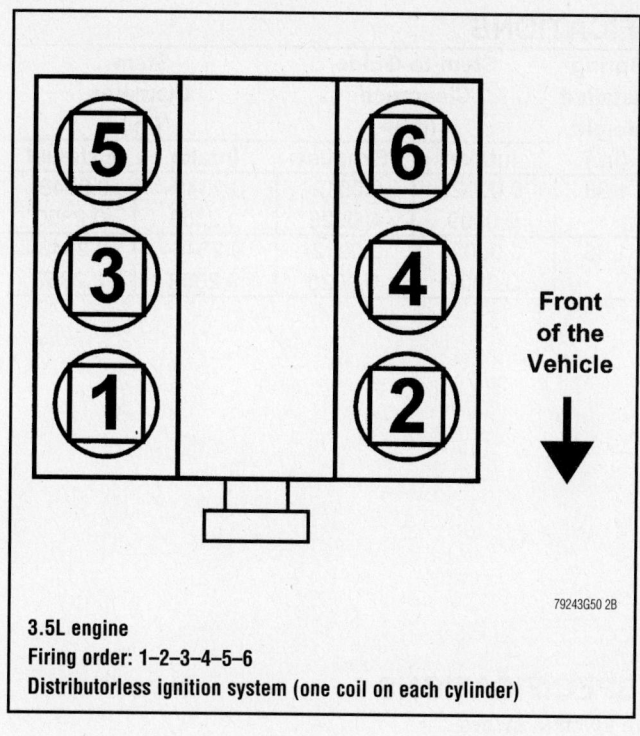

3.5L engine
Firing order: 1–2–3–4–5–6
Distributorless ignition system (one coil on each cylinder)

79243G50 2B

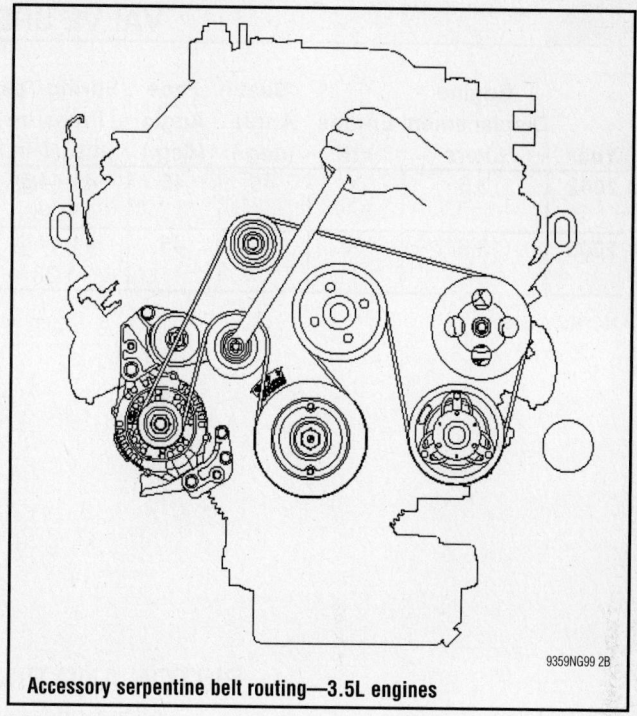

Accessory serpentine belt routing—3.5L engines

9359NG99 2B

CAPACITIES

Year	Model	Engine Displacement Liters	Engine VIN	Oil with Filter (qts.)	Transmission (pts.) Man.	Transmission (pts.) Auto.	Transfer Case (pts.)	Drive Axle Front (pts.)	Drive Axle Rear (pts.)	Fuel Tank (gal.)	Cooling System (qts.)
2002	Axiom	3.5	X	5.0	—	18.2	2.8	2.66	3.64	19.5	11.7
2003	Axiom	3.5	X	5.0	—	18.2	2.8	2.66	3.64	19.5	11.7

NOTE: All capacities are approximate. Add fluid gradually and check to ensure a proper level has been reached.

67162-AXIO-C04

CRANKSHAFT AND CONNECTING ROD SPECIFICATIONS
All measurements are given in inches.

Year	Engine Displacement Liters	Engine VIN	Main Brg. Journal Dia.	Main Brg. Oil Clearance	Shaft End-play	Thrust on No.	Journal Diameter	Oil Clearance	Side Clearance
2002	3.5	X	2.5165-2.5170	0.0007-0.0017	0.0024-0.0094	3	2.1229-2.1235	0.0010-0.0023	0.0050-0.0150
2003	3.5	X	2.5165-2.5170	0.0007-0.0017	0.0024-0.0094	3	2.1229-2.1235	0.0010-0.0023	0.0050-0.0150

67162-AXIO-C05

VALVE SPECIFICATIONS

Year	Engine Displacement Liters	Engine VIN	Seat Angle (deg.)	Face Angle (deg.)	Spring Test Pressure (lbs. @ in.)	Spring Installed Height (in.)	Stem-to-Guide Clearance (in.)		Stem Diameter (in.)	
							Intake	Exhaust	Intake	Exhaust
2002	3.5	X	45	45	41-44@ 1.38	1.38	0.0002- 0.0009	0.0012- 0.0025	0.2346- 0.2353	0.2343- 0.2350
2003	3.5	X	45	45	41-44@ 1.38	1.38	0.0002- 0.0009	0.0012- 0.0025	0.2346- 0.2353	0.2343- 0.2350

NA: Not Available

67162-AXIO-C06

PISTON AND RING SPECIFICATIONS
All measurements are given in inches.

Year	Engine Displacement Liters	Engine VIN	Piston Clearance	Ring Gap			Ring Side Clearance		
				Top Compression	Bottom Compression	Oil Control	Top Compression	Bottom Compressior	Oil Control
2002	3.5	X	NA	0.0118- 0.0157	0.0177- 0.0236	0.006- 0.018	0.0006- 0.0015	0.0006- 0.0015	NA
2003	3.5	X	NA	0.0118- 0.0157	0.0177- 0.0236	0.006- 0.018	0.0006- 0.0015	0.0006- 0.0015	NA

NA: Not Available

67162-AXIO-C07

TORQUE SPECIFICATIONS
All readings in ft. lbs.

Year	Engine Displacement Liters	Engine VIN	Cylinder Head Bolts	Main Bearing Bolts	Rod Bearing Bolts	Crankshaft Damper Bolts	Flywheel Bolts	Manifold Intake	Exhaust	Spark Plugs	Oil Drain Plug
2002	3.5	X	①	②	40	123	40	18	38	13	58
2003	3.5	X	①	②	40	123	40	18	38	13	58

① Step 1: 21 ft. lbs.
　Step 2: 47 ft. lbs.
② Step 1: 22 ft. lbs.
　Step 2: Plus 55-65 degrees
　Step 3: Crankcase side bolts to 29 ft. lbs.

67162-AXIO-C08

WHEEL ALIGNMENT

Year	Model	Caster		Camber		Toe-in (in.)
		Range (+/-Deg.)	Preferred Setting (Deg.)	Range (+/-Deg.)	Preferred Setting (Deg.)	
2002	Axiom	1.00	+2.50	0.50	0	0+/-0.08
2003	Axiom	1.00	+2.50	0.50	0	0+/-0.08

67162-AXIO-C09

TIRE, WHEEL AND BALL JOINT SPECIFICATIONS

Year	Model	OEM Tires		Tire Pressures (psi)		Wheel Size	Ball Joint Inspection	Wheel Lug Torque (ft.lbs.)
		Standard	Optional	Front	Rear			
2002	Axiom	P235/65R17	none	26	26	7JJ	U: 4-28 ① L: 4-55	87
2003	Axiom	P235/65R17	none	26	26	7JJ	U: 4-28 ① L: 4-55	87

L: Lower

U: Upper

NA: Not available

NS: Not specified by manufacturer

① Torque required in inch lbs. to rotate ball joint when removed from the knuckle

67162-AXIO-C10

BRAKE SPECIFICATIONS

All measurements in inches unless noted

Year	Model		Brake Disc				Brake Drum Diameter			Minimum Lining Thickness	Brake Caliper	
			Original Thickness	Machine Thickness	Minimum Thickness	Maximum Runout	Original Inside Diameter	Max. Wear Limit	Maximum Machine Diameter		Bracket Bolts (ft. lbs.)	Mounting Bolts (ft. lbs.)
2002	Axiom	F	1.020	0.983	0.969	0.005	—	—	—	0.039	115	54
		R	0.710	0.668	0.654	0.005	11.6	11.67	NA	0.039	76	32
2003	Axiom	F	1.020	0.983	0.969	0.005	—	—	—	0.039	115	54
		R	0.710	0.668	0.654	0.005	11.6	11.67	NA	0.039	76	32

NA: Not Available

67162-AXIO-C11

SCHEDULED MAINTENANCE INTERVALS
Isuzu—Axiom

TO BE SERVICED	TYPE OF SERVICE	VEHICLE MILEAGE INTERVAL (x1000)															
		7.5	15	22.5	30	37.5	45	52.5	60	67.5	75	82.5	90	97.5	105	112.5	120
Accelerator linkage ①	L	✓	✓	✓	✓	✓	✓	✓	✓	✓	✓	✓	✓	✓	✓	✓	✓
Accessory drive belts ②	S/I				✓				✓				✓				✓
Air cleaner filter	R				✓				✓				✓				✓
Auto cruise control linkage & hose ③	S/I		✓		✓		✓		✓		✓		✓		✓		✓
Automatic transmission fluid level ③	S/I	✓		✓		✓		✓		✓		✓		✓		✓	
Battery fluid level ③	S/I	✓	✓	✓	✓	✓	✓	✓	✓	✓	✓	✓	✓	✓	✓	✓	✓
Body and chassis ①	L	✓	✓	✓	✓	✓	✓	✓	✓	✓	✓	✓	✓	✓	✓	✓	✓
Brake fluid level ③	S/I	✓	✓	✓	✓	✓	✓	✓	✓	✓	✓	✓	✓	✓	✓	✓	✓
Brake lines & hoses ③	S/I	✓	✓	✓	✓	✓	✓	✓	✓	✓	✓	✓	✓	✓	✓	✓	✓
Brake pedal play ③	S/I		✓		✓		✓		✓		✓		✓		✓		✓
Clutch fluid level ③	S/I	✓	✓	✓	✓	✓	✓	✓	✓	✓	✓	✓	✓	✓	✓	✓	✓
Clutch lines & hose ③	S/I				✓				✓				✓				✓
Clutch pedal free-play ③	S/I		✓		✓		✓		✓		✓		✓		✓		✓
Clutch pedal spring, bushing and clevis pin ①	S/I		✓		✓		✓		✓		✓		✓		✓		✓
Cooling and heating system hoses ③	S/I		✓		✓		✓		✓		✓		✓		✓		✓
Driveshaft flange torque ③	S/I	✓		✓		✓		✓		✓		✓		✓		✓	
Drum and disc brakes ③	S/I		✓		✓		✓		✓		✓		✓		✓		✓
Engine coolant	R				✓								✓				
Engine coolant level ③	S/I	✓	✓	✓	✓	✓	✓	✓	✓	✓	✓	✓	✓	✓	✓	✓	✓
Engine oil & filter ③	R	✓	✓	✓	✓	✓	✓	✓	✓	✓	✓	✓	✓	✓	✓	✓	✓
Exhaust system ③	S/I	✓	✓	✓	✓	✓	✓	✓	✓	✓	✓	✓	✓	✓	✓	✓	✓
Front and rear axle lubricant	R		✓						✓					✓			✓
Front and rear driveshafts ①	S/I	✓	✓		✓		✓		✓		✓		✓		✓		✓
Front wheel bearings	S/I & L				✓				✓					✓			✓
Fuel lines & tank cap ③	S/I								✓								✓
Inspect for fluid leaks ③	S/I	✓	✓	✓	✓	✓	✓	✓	✓	✓	✓	✓	✓	✓	✓	✓	✓
Key lock cylinder ③	L		✓		✓		✓		✓		✓		✓		✓		✓
Manual transmission and transfer case fluid ④	R		✓		✓				✓				✓				✓
Parking brake system ③	S/I		✓		✓		✓		✓		✓		✓		✓		✓
Power steering fluid	R				✓				✓				✓				✓
Radiator core and A/C condenser	S/I & C								✓								✓
Rotate tires	S/I	✓	✓	✓	✓	✓	✓	✓	✓	✓	✓	✓	✓	✓	✓	✓	✓
Shift-on-the-fly system gear fluid ③	S/I		✓		✓		✓		✓		✓		✓		✓		✓
Spark plug wires ⑤	S/I								✓								✓

67162-AXIO-C12

SCHEDULED MAINTENANCE INTERVALS
Isuzu—Axiom

TO BE SERVICED	TYPE OF SERVICE	VEHICLE MILEAGE INTERVAL (x1000)															
		7.5	15	22.5	30	37.5	45	52.5	60	67.5	75	82.5	90	97.5	105	112.5	120
Spark plugs	R	Every 100,000 miles															
Starter safety switch ③	S/I	✓	✓	✓	✓	✓	✓	✓	✓	✓	✓	✓	✓	✓	✓	✓	✓
Steering operation ③	S/I	✓	✓	✓	✓	✓	✓	✓	✓	✓	✓	✓	✓	✓	✓	✓	✓
Suspension & steering ③	S/I	✓	✓	✓	✓	✓	✓	✓	✓	✓	✓	✓	✓	✓	✓	✓	✓
Throttle linkage ③	S/I		✓		✓		✓		✓		✓		✓		✓		✓
Timing belt	R										✓						
Tires and wheels ③	S/I	✓	✓	✓	✓	✓	✓	✓	✓	✓	✓	✓	✓	✓	✓	✓	✓
Valve clearance ④	A							✓									✓

R: Replace   S/I: Service or Inspect   L: Lubricate   A: Adjust   C: Clean

① Perform this at the mileage indicated or every 6 months, whichever occurs first.

② Perform this at the mileage indicated or every 24 months, whichever occurs first.

③ Perform this at the mileage indicated or every 12 months, whichever occurs first.

④ 3.2L V6 engine.

⑤ 2.2L 4 cyl. engine.

FREQUENT OPERATION MAINTENANCE (SEVERE SERVICE)

If a vehicle is operated under any of the following conditions it is considered severe service:

- Towing a trailer or using a camper or car-top carrier.

- Repeated short trips of less than 5 miles in temperatures below freezing.

- Extensive idling or low-speed driving for long distances as in heavy commercial use, such as delivery, taxi or police cars.

- Operating on rough, muddy or salt-covered roads.

- Operating on unpaved or dusty roads.

Air cleaner element: replace every 15,000 miles

Engine oil and filter: replace every 3000 miles or 3 months, whichever occurs first.

Automatic transmission fluid: replace every 20,000 miles.

Rear axle lubricant: replace every 15,000 miles.

67162-AXIO-C13

PRECAUTIONS

Before servicing any vehicle, please be sure to read all of the following precautions, which deal with personal safety, prevention of component damage and important points to take into consideration when servicing a motor vehicle:

• Never open, service or drain the radiator or cooling system when the engine is hot; serious burns can occur from the steam and hot coolant.

• Observe all applicable safety precautions when working around fuel. Whenever servicing the fuel system, always work in a well-ventilated area. Do not allow fuel spray or vapors to come in contact with a spark, open flame, or excessive heat (a hot drop light, for example). Keep a dry chemical fire extinguisher near the work area. Always keep fuel in a container specifically designed for fuel storage; also, always properly seal fuel containers to avoid the possibility of fire or explosion. Refer to the additional fuel system precautions later in this section.

• Fuel injection systems often remain pressurized, even after the engine has been turned OFF. The fuel system pressure must be relieved before disconnecting any fuel lines. Failure to do so may result in fire and/or personal injury.

• Brake fluid often contains polyglycol ethers and polyglycols. Avoid contact with the eyes and wash your hands thoroughly after handling brake fluid. If you do get brake fluid in your eyes, flush your eyes with clean, running water for 15 minutes. If eye irritation persists, or if you have taken brake fluid internally, seek medical assistance IMMEDIATELY.

• The EPA warns that prolonged contact with used engine oil may cause a number of skin disorders, including cancer. You should make every effort to minimize your exposure to used engine oil. Protective gloves should be worn when changing oil. Wash your hands and any other exposed skin areas as soon as possible after exposure to used engine oil. Soap and water, or waterless hand cleaner should be used.

• All new vehicles are now equipped with an air bag system, often referred to as a Supplemental Restraint System (SRS) or Supplemental Inflatable Restraint (SIR) system. The system must be disabled before performing service on or around system components, steering column, instrument panel components, wiring and sensors. Failure to follow safety and disabling procedures could result in accidental air bag deployment, possible personal injury and unnecessary system repairs.

• Always wear safety goggles when working with, or around, the air bag system. When carrying a non-deployed air bag, be sure the bag and trim cover are pointed away from your body. When placing a non-deployed air bag on a work surface, always face the bag and trim cover upward, away from the surface. This will reduce the motion of the module if it is accidentally deployed. Refer to the additional air bag system precautions later in this section.

• Clean, high quality brake fluid from a sealed container is essential to the safe and proper operation of the brake system. You should always buy the correct type of brake fluid for your vehicle. If the brake fluid becomes contaminated, completely flush the system with new fluid. Never reuse any brake fluid. Any brake fluid that is removed from the system should be discarded. Also, do not allow any brake fluid to come in contact with a painted surface; it will damage the paint.

• Never operate the engine without the proper amount and type of engine oil; doing so WILL result in severe engine damage.

• Timing belt maintenance is extremely important. Many models utilize an interference-type, non-freewheeling engine. If the timing belt breaks, the valves in the cylinder head may strike the pistons, causing potentially serious (also time-consuming and expensive) engine damage. Refer to the maintenance interval charts in the front of this manual for the recommended replacement interval for the timing belt and to the timing belt section for belt replacement and inspection.

• Disconnecting the negative battery cable on some vehicles may interfere with the functions of the on-board computer system(s) and may require the computer to undergo a relearning process once the negative battery cable is reconnected.

• When servicing drum brakes, only disassemble and assemble one side at a time, leaving the remaining side intact for reference.

• Only an MVAC-trained, EPA-certified automotive technician should service the A/C system or its components.

ENGINE REPAIR

➡ Disconnecting the negative battery cable on some vehicles may interfere with the functions of the on board computer system. The computer may undergo a relearning process once the negative battery cable is reconnected.

Distributor

REMOVAL

These engines are equipped with a Distributorless Ignition System (DIS).

Alternator

REMOVAL

1. Before servicing the vehicle, refer to the precautions in the beginning of this section.
2. Remove or disconnect the following:
   • Negative battery cable
   • Accessory drive belt
   • Alternator wiring connectors
   • Alternator

INSTALLATION

1. Before servicing the vehicle, refer to the precautions in the beginning of this section.

2. Install or connect the following:
   • Alternator. Tighten the 10mm bolts to 30 ft. lbs. (41 Nm) and the 8mm bolts to 15 ft. lbs. (21 Nm).
   • Alternator wiring connectors
   • Accessory drive belt
   • Negative battery cable

Ignition Timing

ADJUSTMENT

These engines are equipped with a Distributorless Ignition System (DIS). No adjustment is possible.

Engine Assembly

REMOVAL & INSTALLATION

1. Before servicing the vehicle, refer to the precautions in the beginning of this section.
2. Drain the cooling system.
3. Remove or disconnect the following:
   - Battery
   - Hood
   - Air cleaner assembly
   - Canister vacuum line
   - Brake booster vacuum line
   - Engine wiring harness connectors
   - Transmission harness connectors and bracket
   - Engine ground cable
   - Bonding cable connectors
   - Starter harness connector
   - Alternator harness connector
   - Coolant reservoir tank hose
   - Radiator hoses
   - Upper fan shroud
   - Cooling fan
   - Accessory drive belt
   - Power steering pump
   - A/C compressor
   - Heated Oxygen (HO$_2$S) sensor connectors
   - Left and right exhaust front pipes
   - Flywheel dust cover
   - Heater hoses
   - Fuel lines
   - Transmission. Refer to the transmission procedure in this section.
   - Accelerator cable
   - Cruise control cable
   - Left and right engine mounts
   - Engine

To install:

4. Install or connect the following:
   - Engine
   - Left and right engine mounts. Tighten the bolts to 30 ft. lbs. (41 Nm).
   - Transmission
   - Flywheel dust cover
   - Fuel lines
   - Left and right exhaust front pipes
   - HO$_2$S sensor connectors
   - A/C compressor
   - Power steering pump
   - Accessory drive belt
   - Cooling fan. Tighten the nuts to 16 ft. lbs. (22 Nm).
   - Radiator
   - Upper fan shroud
   - Heater hoses
   - Radiator hoses
   - Coolant reservoir tank hose

- Alternator harness connector
- Starter harness connector
- Engine ground cable
- Transmission harness connectors and bracket
- Engine wiring harness connectors
- Brake booster vacuum line
- Canister vacuum line
- Cruise control cable
- Accelerator cable
- Air cleaner assembly
- Hood
- Battery

5. Fill the cooling system. Check all fluid levels and adjust as necessary.
6. Start the engine and check for leaks.

Water Pump

REMOVAL & INSTALLATION

1. Before servicing the vehicle, refer to the precautions in the beginning of this section.

2. Drain the cooling system.
3. Remove or disconnect the following:
   - Negative battery cable
   - Upper radiator hose
   - Timing belt. Refer to the Timing Belt Unit Repair Section.
   - Idler pulley
   - Water pump

To install:

➡**Apply Loctite® 262 to bolt number 3 prior to installation.**

4. Install or connect the following:
   - Water pump. Tighten the bolts in two passes, in sequence, to 13 ft. lbs. (18 Nm) for 3.2L engines or to 18 ft. lbs. (25 Nm) for 3.5L engines.
   - Idler pulley
   - Timing belt
   - Upper radiator hose
   - Negative battery cable
5. Fill the cooling system.
6. Start the engine and check for leaks.

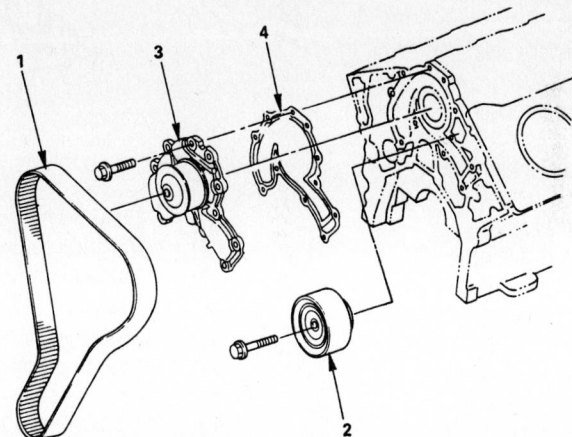

1. Timing belt
2. Idle pulley
3. Water pump assembly
4. Gasket

7924BG41

Exploded view of the water pump mounting

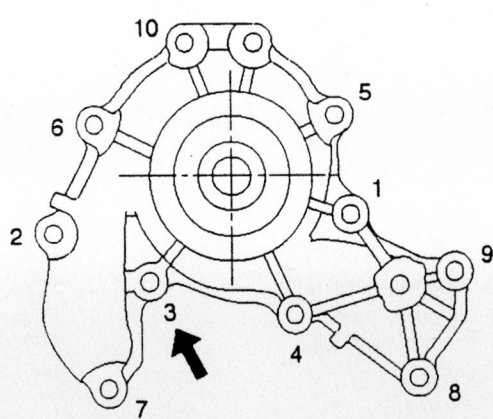

9302BG01

Water pump torque sequence. Apply LOCTITE® 262 to bolt number 3 (arrow)—3.5L engine

Heater Core

REMOVAL & INSTALLATION

1. Before servicing the vehicle, refer to the precautions in the beginning of this section.
2. Disconnect the battery ground.
3. Drain the coolant.
4. Discharge and recover the refrigerant.
5. Remove the instrument panel knee pads.
6. Remove the upper cluster and connections.
7. Remove the center cluster assembly.
8. Disconnect the cigarette lighter, ash tray light, and hazard switch connectors.
9. Remove the display unit and radio.
10. Remove the front and rear consoles.
11. Remove the dash side trim panels.
12. Remove the glove box.
13. Remove the driver side lower trim panel.
14. Remove the meter cluster assembly.
15. Remove the driver side knee bolster.
16. Remove the instrument panel as follows:

   a. Disconnect the six driver's side harness connectors.

   b. Disconnect the three passenger's side harness connectors.

   c. Disconnect the two center harness connectors.

   d. Disconnect the antenna cable.

   e. Disconnect the ground cable bolts.

   f. Disconnect the passenger side air bag.

   g. Unbolt the fuse box.

   h. Remove the three nuts at the base of the windshield.

   i. Remove the six retaining bolts and one nut.

   j. Remove the passenger air bag.

   k. Remove the vent duct assembly.

   l. Remove the passenger side lower bracket.

   m. Remove the glove box side reinforcements.

   n. Remove the instrument panel upper reinforcement.

   o. Remove the instrument panel center reinforcement.

   p. Remove the instrument panel harness assembly.

   q. Remove the instrument panel stays.

   r. Remove the instrument panel.

17. Remove the instrument panel with the suspension control unit.
18. Remove the cross beam.

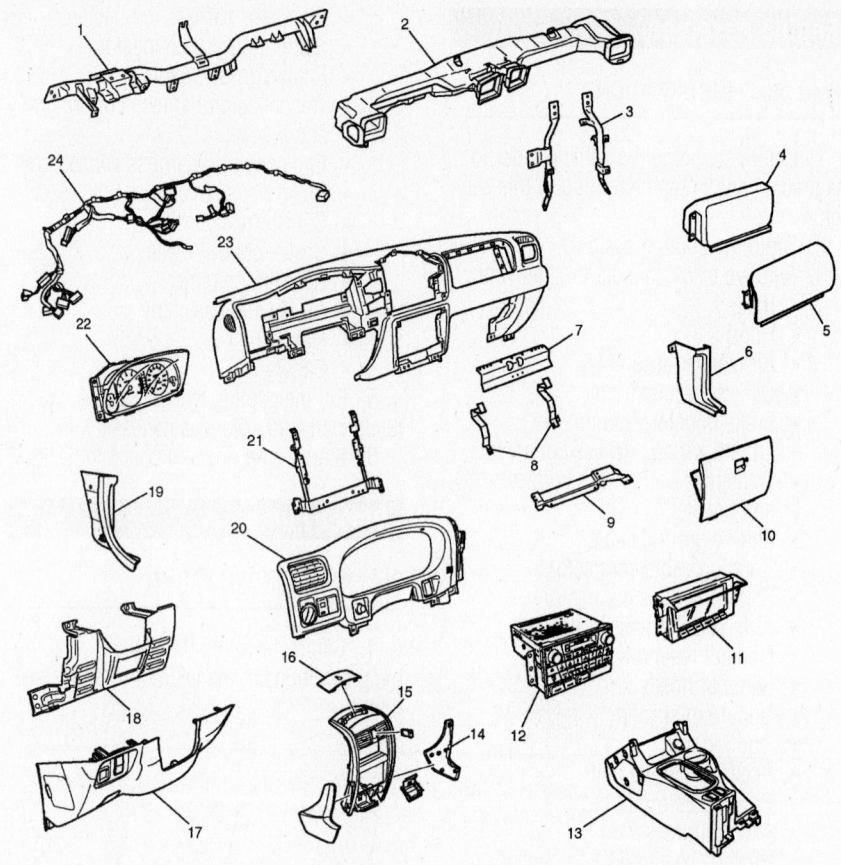

(1)	Cross Beam
(2)	Vent Duct Assembly
(3)	Instrument Panel Stay
(4)	Passenger Air Bag
(5)	Passenger Air Bag Cover
(6)	Dash Side Trim Panel (RH)
(7)	Instrument Upper Reinforcement
(8)	Glove Box Side Reinforcement
(9)	Passenger Lower Bracket
(10)	Glove Box
(11)	Display Unit
(12)	Audio Kit
(13)	Front Console Assembly
(14)	Knee Pad
(15)	Center Cluster Assembly
(16)	Cluster Upper Cover
(17)	Instrument Panel Driver Lower Cover Assembly
(18)	Driver Knee Bolster Assembly
(19)	Dash Side Trim Panel (LH)
(20)	Meter Cluster Assembly
(21)	Instrument Panel Center Reinforcement
(22)	Meter Assembly
(23)	Instrument Panel Assembly
(24)	Instrument Harness Assembly

Exploded view of the instrument panel mounting

67162-AXIO-G01

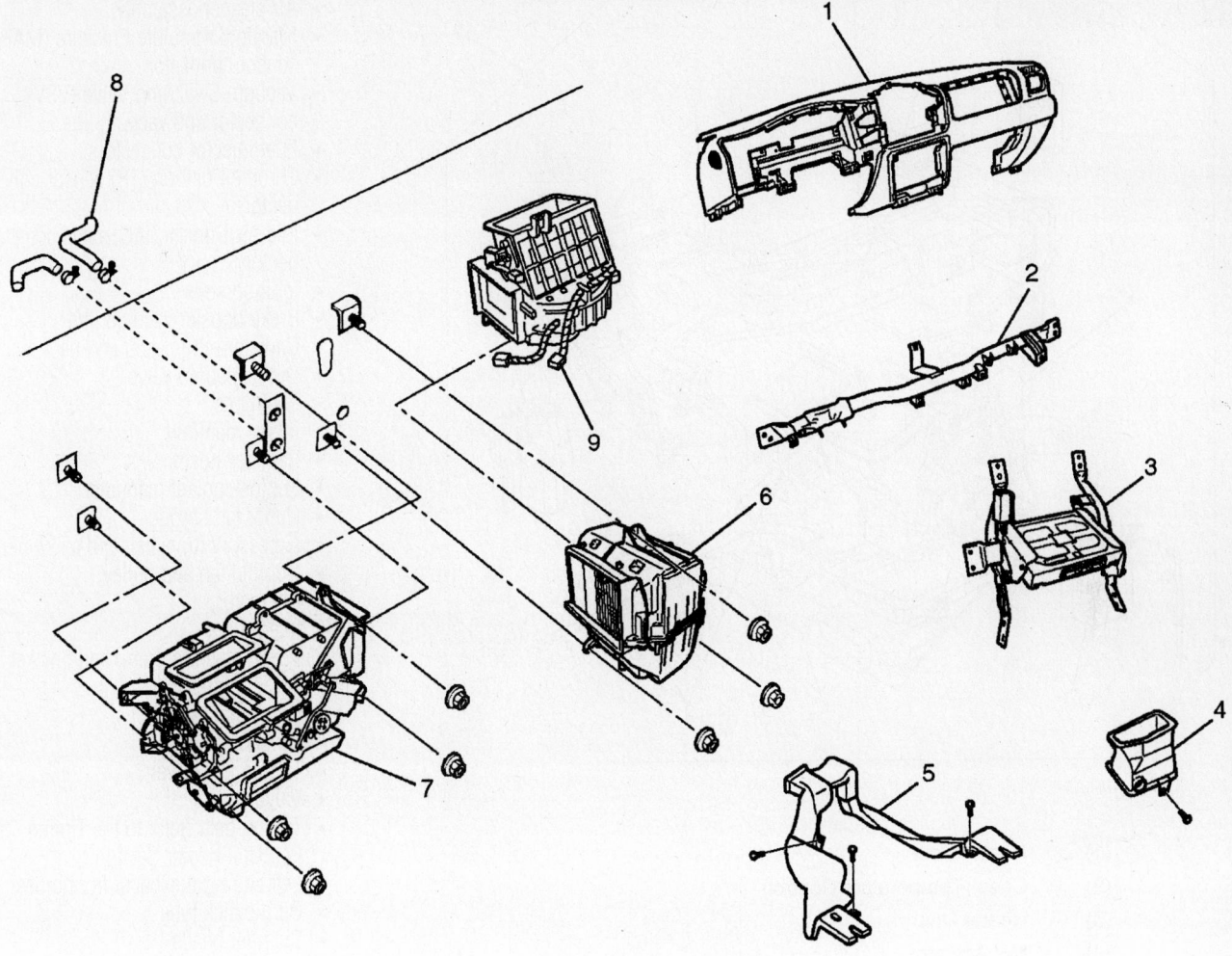

(1)   Instrument Panel Assembly

(2)   Cross Beam Assembly

(3)   Instrument Panel Bracket w/Suspension Control Unit

(4)   Ventilation Lower Duct

(5)   Rear Heater Duct

(6)   Evaporator Assembly

(7)   Heater Unit Assembly

(8)   Heater Hose

(9)   Power Transistor Connector

67162-AXIO-G02

Exploded view of the HVAC mounting

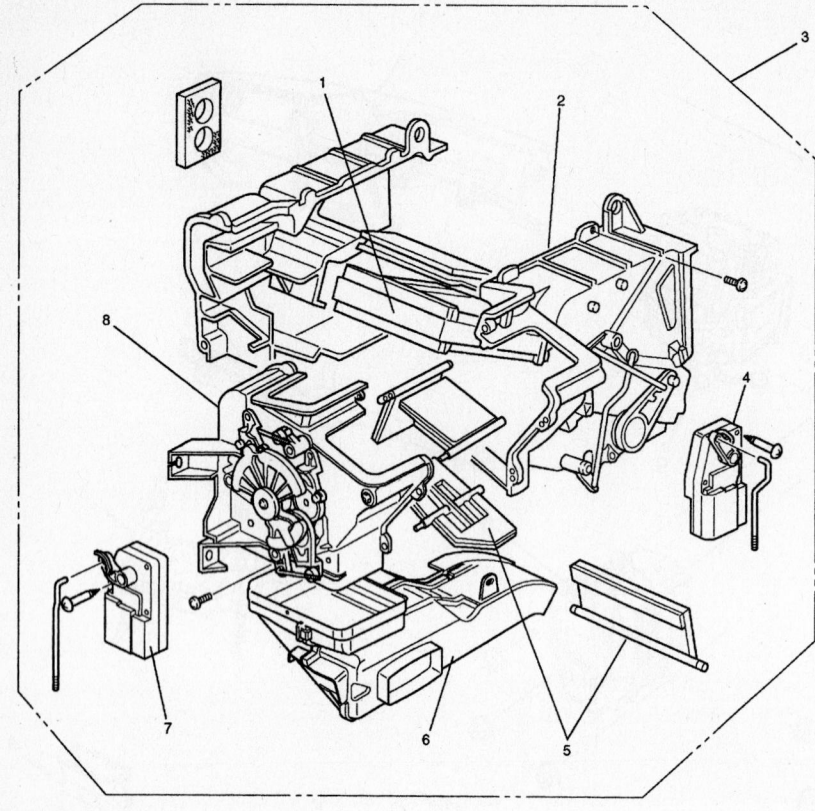

(1)   Heater Core

(2)   Case (Temperature Control)

(3)   Heater Unit

(4)   Mix Actuator

(5)   Mode Door

(6)   Duct

(7)   Mode Actuator

(8)   Case (Mode Control)

67162-AXIO-G03

Exploded view of the heater case

19. Disconnect the power transistor connector.

20. Remove the evaporator assembly.

21. Remove the lower duct.

22. Remove the rear heater duct.

23. Remove the heater unit assembly.

24. Remove the duct, mix actuator, and mode actuator.

25. Remove the mode control case.

26. Separate the heater case and remove the heater core.

To install:

27. Installation is the reverse of the removal procedure.

Cylinder Head

REMOVAL & INSTALLATION

1. Before servicing the vehicle, refer to the precautions in the beginning of this section.

2. Drain the cooling system.

3. Remove or disconnect the following:
- Negative battery cable
- Hood
- Engine cover
- Mass Air Flow (MAF) sensor connector
- Intake Air Temperature (IAT) sensor connector
- Positive Crankcase Ventilation (PCV) valve and hose

- Air cleaner assembly
- Manifold Absolute Pressure (MAP) sensor connector
- Vacuum Switching Valve (VSV) connector and vacuum line
- Fuel injector connectors
- Throttle Position (TP) sensor connector
- Idle Air Control (IAC) valve connector
- Ignition coils
- Brake booster vacuum line
- Canister purge vacuum line
- Duty solenoid valve
- Fuel lines
- Intake manifold
- Radiator hoses
- Engine coolant manifold
- Upper fan shroud
- Accessory drive belt and tensioner
- Cooling fan and pulley
- Alternator
- Idler pulley
- Power steering pump and bracket
- A/C compressor
- Crankshaft pulley
- Oil cooler hoses
- Timing belt cover
- Valve covers
- Timing belt. Refer to the Timing Belt Unit Repair Section.
- Left and right exhaust front pipes
- Oil dipstick tube
- Cylinder heads

To install:

➡**Use new head bolts when installing the cylinder head. Do not apply oil to the head bolt threads.**

➡**The left and right cylinder head gaskets are not interchangeable.**

4. Install the cylinder heads with new gaskets. Tighten the bolts to 22 ft. lbs. in sequence, then to 47 ft. lbs. (64 Nm) in sequence.

5. Install or connect the following:
- Oil dipstick tube
- Left and right exhaust front pipes
- Timing belt
- Valve covers
- Timing belt cover
- Oil cooler hoses
- Crankshaft pulley. Tighten the pulley bolt to 123 ft. lbs. (167 Nm).
- A/C compressor
- Power steering pump and bracket. Tighten the bolts to 34 ft. lbs. (46 Nm).
- Idler pulley
- Alternator
- Cooling fan and pulley

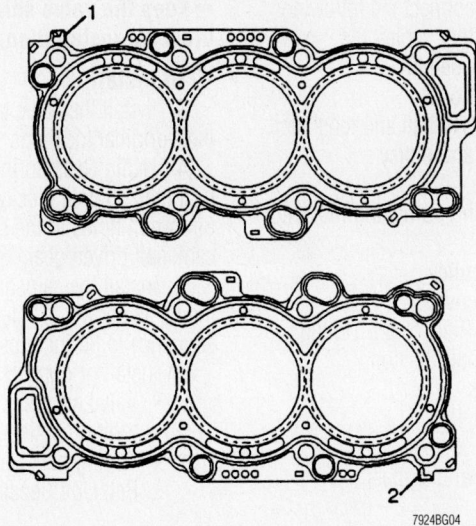

Right (1) and left (2) head gasket identification mark locations—3.5L engine

7924BG04

- Accessory drive belt and tensioner
- Upper fan shroud
- Engine coolant manifold
- Radiator hoses
- Intake manifold
- Fuel lines
- Duty solenoid valve
- Canister purge vacuum line
- Brake booster vacuum line
- Ignition coils
- IAC valve connector
- TP sensor connector
- Fuel injector connectors
- VSV connector and vacuum line
- MAP sensor connector
- Air cleaner assembly
- PCV valve and hose
- IAT sensor connector
- MAF sensor connector
- Engine cover
- Hood
- Negative battery cable
6. Fill the cooling system.
7. Start the engine. Check for leaks and proper operation.

Rocker Arms/Shafts

REMOVAL & INSTALLATION

➡These engines are not equipped with rocker arms. The camshaft lobes act directly on the valve shims.

Intake Manifold

REMOVAL & INSTALLATION

1. Before servicing the vehicle, refer to the precautions in the beginning of this section.

2. Remove or disconnect the following:
- Negative battery cable
- Engine cover
- Air cleaner assembly
- Accelerator cable
- Cruise control cable
- Brake booster vacuum line
- Manifold Absolute Pressure (MAP) sensor connector
- Idle Air Control (IAC) valve connector
- Throttle Position (TP) sensor connector
- Canister purge solenoid connector
- Electronic Vacuum Sensing Valve (EVSV) connector and vacuum line
- Exhaust Gas Recirculation (EGR) valve
- Positive Crankcase Ventilation (PCV) valve and hose
- Pressure regulator vacuum line
- Ventilation hose
- Throttle body
- Fuel lines
- Fuel injector connectors
- Intake manifold

To install:
3. Install or connect the following:
- Intake manifold. Tighten the fasteners to 18 ft. lbs. (25 Nm).
- Fuel injector connectors
- Fuel lines
- Throttle body. Tighten the bolts to 88 inch lbs. (10 Nm).
- Ventilation hose
- Pressure regulator vacuum line
- PCV valve and hose
- EGR valve
- EVSV connector and vacuum line
- Canister purge solenoid connector
- TP sensor connector

- IAC valve connector
- MAP sensor connector
- Brake booster vacuum line
- Cruise control cable
- Accelerator cable
- Air cleaner assembly
- Engine cover
- Negative battery cable
4. Start the engine and check for proper operation.

Exhaust Manifold

REMOVAL & INSTALLATION

1. Before servicing the vehicle, refer to the precautions in the beginning of this section.
2. Remove or disconnect the following:
- Negative battery cable
- Air cleaner assembly
- Heated Oxygen (HO$_2$S) sensor connectors
- Right torsion bar
- Exhaust Gas Recirculation (EGR) pipe and bracket
- Left and right exhaust front pipes
- Heat shields
- Accessory drive belt
- A/C compressor and bracket
- Exhaust manifolds

To install:
3. Install or connect the following:
- Exhaust manifolds. Tighten the bolts to 38 ft. lbs. (52 Nm).
- A/C compressor and bracket
- Accessory drive belt
- Heat shields
- Left and right exhaust front pipes
- EGR pipe and bracket
- Right torsion bar
- HO$_2$S sensor connectors
- Air cleaner assembly
- Negative battery cable
4. Start the engine and check for leaks.

Front Crankshaft Seal

REMOVAL & INSTALLATION

1. Before servicing the vehicle, refer to the precautions in the beginning of this section.
2. Remove or disconnect the following:
- Negative battery cable
- Air cleaner assembly
- Upper fan shroud
- Accessory drive belt and tensioner
- Cooling fan and pulley
- Idler pulley

- Power steering pump and move it aside
- Crankshaft pulley
- Timing belt cover
- Timing belt. Refer to the Timing Belt Unit Repair Section.
- Crankshaft timing sprocket
- Oil seal

To install:

3. Install or connect the following:
- Oil seal so that it is flush with the oil pump housing
- Crankshaft timing sprocket
- Timing belt
- Timing belt cover
- Crankshaft pulley. Tighten the bolt to 123 ft. lbs. (167 Nm).
- Power steering pump
- Idler pulley
- Cooling fan and pulley
- Accessory drive belt and tensioner
- Upper fan shroud
- Air cleaner assembly
- Negative battery cable

4. Start the engine and check for leaks.

Camshaft and Valve Lifters

REMOVAL & INSTALLATION

1. Before servicing the vehicle, refer to the precautions in the beginning of this section.

2. Remove or disconnect the following:
- Negative battery cable
- Air cleaner assembly
- Upper fan shroud
- Accessory drive belt and tensioner
- Cooling fan and pulley
- Idler pulley
- Power steering pump and move it aside
- Crankshaft pulley
- Timing belt cover
- Timing belt. Refer to the Timing Belt Unit Repair Section.
- Ignition coils
- Valve covers
- Camshafts
- Valve shims and tappets

➡ **Keep the valve shims and tappets in order for installation.**

To install:

3. Install the valve tappets and shims in their original locations.

4. Using Gear Spring Lever J-42686, turn the sub gear clockwise to align the 5mm bolt holes in the sub gear and the camshaft driven gear. Tighten the 5mm bolt.

5. Install the camshafts. Align the timing marks as shown. Tighten the bolts in sequence to 89 inch lbs. (10 Nm).

6. Install or connect the following:
- Valve covers
- Ignition coils
- Timing belt. Refer to the Timing Belt Unit Repair Section.

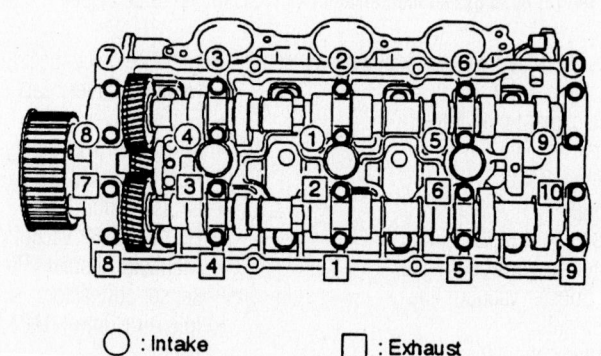

○ : Intake   □ : Exhaust

7924BG12

Camshaft retaining bracket tightening sequence—3.5L engine

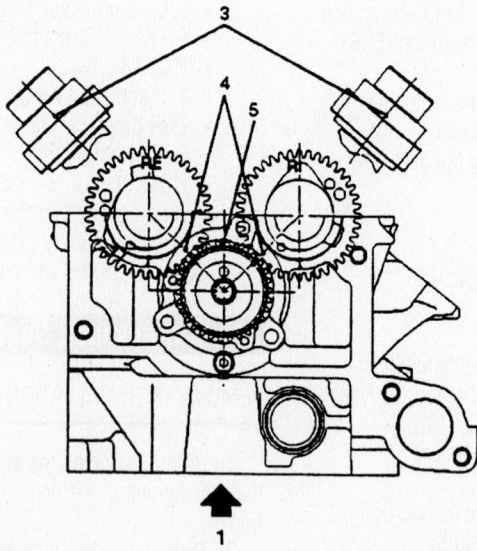

1	**Right Bank**
2	**Left Bank**

3	**Alignment Mark on Camshaft Drive Gear**
4	**Alignment Mark on Camshaft**
5	**Alignment Mark on Retainer**

7924BG11

Camshaft alignment marks for the left and right cylinder heads—3.5L engine

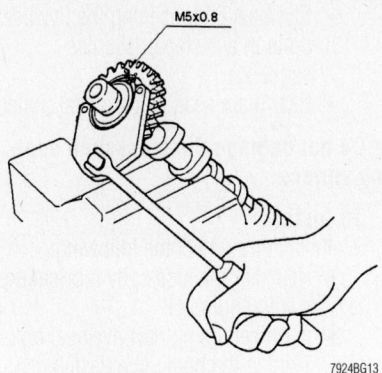

Aligning the sub gear with the Gear Spring Lever J-42686—3.5L engine

- Timing belt cover
- Crankshaft pulley
- Power steering pump
- Idler pulley
- Cooling fan and pulley
- Accessory drive belt and tensioner
- Upper fan shroud
- Air cleaner assembly
- Negative battery cable

Valve Lash

ADJUSTMENT

➡ **Measure valve clearance with the engine cold.**

1. Before servicing the vehicle, refer to the precautions in the beginning of this section.

2. Remove the valve covers.

3. Check the valve clearance with the camshafts positioned as shown. Intake valve clearance should be 0.0091–0.0130 inches. Exhaust valve clearance should be 0.0098–0.0138 inches.

4. If adjustment is required, replace the shims as follows:

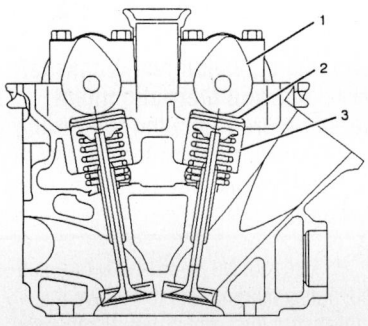

Cross section of the 3.5L cylinder head. Note the position of the camshaft lobe (1), adjustment shim (2) and the tappet (3)

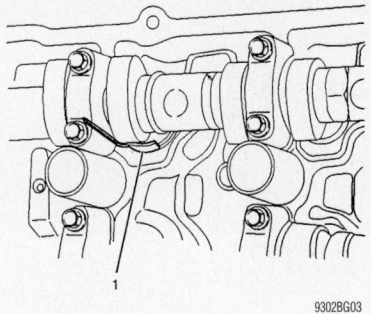

Valve clearance adjusting tool J–42689 (1)

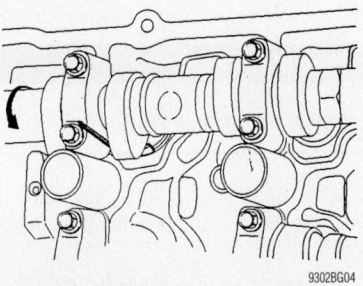

Using the valve clearance adjusting tool to hold the tappet for shim replacement

   a. Step 1: Position special tool J–42689 on the edge of the tappet.

   b. Step 2: Rotate the crankshaft until the maximum lift portion of the camshaft lobe contacts the upper edge of the special tool and presses the tappet down to create enough clearance between the adjustment shim and the camshaft for the shim to be removed.

   c. Step 3: Replace shims as necessary to achieve correct valve clearance.

   d. Step 4: Repeat for each valve to be adjusted.

5. Replace the valve covers. Tighten the bolts to 80 inch lbs. (9 Nm).

Starter Motor

REMOVAL & INSTALLATION

1. Before servicing the vehicle, refer to the precautions in the beginning of this section.

2. Remove or disconnect the following:
- Negative battery cable
- Heated Oxygen (HO$_2$S) sensor connectors
- Exhaust front pipe
- Heat shield
- Starter wiring connectors
- Starter motor

To install:

3. Install or connect the following:

- Starter motor. Tighten the bolts to 30 ft. lbs. (40 Nm).
- Starter wiring connectors
- Heat shield
- Exhaust front pipe
- Heated Oxygen (HO$_2$S) sensor connectors
- Negative battery cable

Oil Pan

REMOVAL & INSTALLATION

1. Before servicing the vehicle, refer to the precautions in the beginning of this section.

2. Drain the engine oil.

3. Remove or disconnect the following:

- Negative battery cable
- Front wheels
- Oil level dipstick
- Stone guard
- Radiator under fan shroud
- Shift-on-the-fly from the axle
- Suspension crossmember

4. If equipped with 4 wheel drive, unbolt and lower the front axle housing assembly for clearance.

5. Remove or disconnect the following:

- Steering gear
- Starter
- Oil pan
- Lower crankcase

To install:

6. Apply a bead of silicone sealant to the crankcase flange and install the crankcase. Tighten the fasteners in sequence to 89 inch lbs. (10 Nm).

7. Apply a bead of silicone sealant to the oil pan flange and install the oil pan. Tighten the fasteners to 89 inch lbs. (10 Nm).

8. Install or connect the following:
- Starter. Torque to 30 ft. lbs. (40 Nm)

9. If equipped, raise the axle housing assembly into position. Tighten the axle case bolts to 61 ft. lbs. (82 Nm) and the mounting bolts to 112 ft. lbs. (152 Nm).

10. Install or connect the following:
- Steering gear
- Suspension crossmember. Tighten the bolts to 58 ft. lbs. (78 Nm).
- Radiator under fan shroud
- Stone guard
- Oil level dipstick
- Front wheels
- Negative battery cable

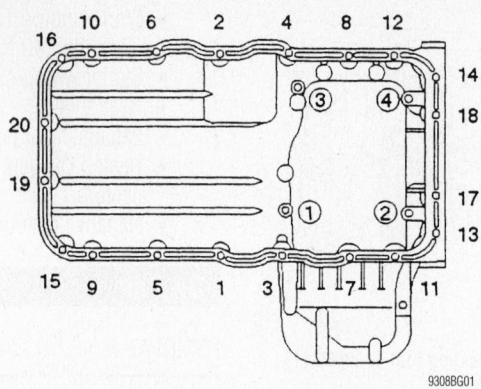

Lower crankcase torque sequence—3.5L engine

11. Fill the crankcase with engine oil.
12. Start the engine and check for leaks.

Oil Pump

REMOVAL & INSTALLATION

1. Before servicing the vehicle, refer to the precautions in the beginning of this section.
2. Remove or disconnect the following:
   - Timing belt
   - Oil pan
   - Oil pick-up tube
   - Oil filter adapter
   - Oil pump

To install:
3. Apply silicone sealant to the oil pump mounting surface and install the oil pump. Tighten the bolts in sequence to 18 ft. lbs. (25 Nm).
4. Install or connect the following:
   - Oil filter adapter
   - Oil pickup tube
   - Oil pan
   - Timing belt

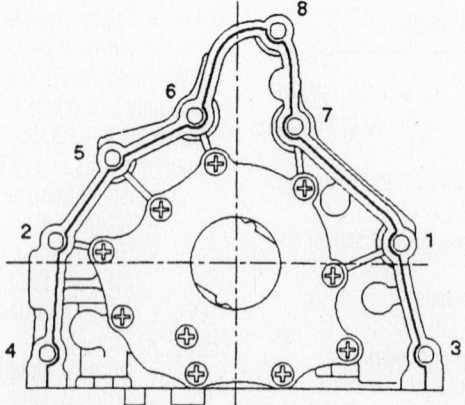

Oil pump torque sequence

Rear Main Seal

REMOVAL & INSTALLATION

1. Before servicing the vehicle, refer to the precautions in the beginning of this section.
2. Remove or disconnect the following:
   - Negative battery cable
   - Transmission
   - Clutch assembly, if equipped with a manual transmission

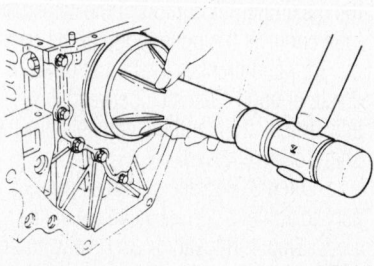

Installing a one-piece rear crankshaft oil seal

- Flywheel by loosening the flywheel bolts in a 2-step crisscross sequence
- Rear main seal, using a seal puller

➡**Do not damage the crankshaft sealing surface.**

To install:
3. Install or connect the following:
   - New rear main seal, by lubricating it with engine oil
   - Flywheel, using new flywheel bolts. Tighten the bolts, in a 2-step crisscross pattern, to 40 ft. lbs. (54 Nm).
   - Clutch assembly, if removed
   - Transmission
   - Negative battery cable
4. Check the oil and refill as necessary.

Timing Belt and Cover

REMOVAL & INSTALLATION

1. Disconnect the negative battery cable.
2. Remove the air cleaner assembly and intake air duct.
3. Remove the upper fan shroud from the radiator.
4. Remove the 4 nuts retaining the cooling fan assembly. Remove the cooling fan from the fan pulley.
5. Loosen and remove the drive belts.
6. Remove the upper timing belt covers.
7. Remove the fan pulley assembly.
8. Rotate the crankshaft to align the camshaft timing marks with the pointer dots on the back covers. Verify that the pointer on the crankshaft aligns with the mark on the lower timing cover.

➡**When the timing marks are aligned, the No. 2 piston is at Top Dead Center (TDC) compression.**

✳✳ WARNING

Align the camshaft and crankshaft sprockets with their alignment marks before removing the timing belt. Failure to align the belt and sprocket marks may result in valve damage.

9. Use tool No. J-8614-01, or a suitable pulley holding tool to remove the crankshaft pulley center bolt. Remove the crankshaft pulley.
10. If present, disconnect the 2 oil cooler hose bracket bolts on the timing cover.

Move the oil cooler hoses and bracket off of the lower timing cover.

11. Remove the lower timing belt cover.

12. Remove the pusher assembly (tensioner) from below the belt tensioner pulley. The pusher rod must always face upward to prevent oil leakage. Depress the pusher rod, and insert a wire pin into the hole to keep the pusher rod retracted.

13. Remove the timing belt.

14. Inspect the water pump and replace it if there is any doubt about its condition.

15. Repair any oil or coolant leaks before installing a new timing belt. If the timing belt has been contaminated with oil or coolant, or is damaged, it must be replaced.

To install:

16. Verify that the sprocket timing marks are still aligned and that the groove and the keyway on the crankshaft timing sprocket align with the mark on the oil pump. The white pointers on the camshaft timing sprockets should align with the dots on the front plate.

17. Install the timing belt. Use clips to secure the belt onto each sprocket until the installation is complete. Align the dotted marks on the timing belt with the timing mark opposite the groove on the crankshaft sprocket.

➡The arrows on the timing belt must follow the belt's direction of rotation. The manufacturer's trademark on the belt's spine should be readable left-to-right when the belt is installed.

18. Align the white line on the timing belt with the alignment mark on the right bank camshaft timing pulley. Secure the belt with a clip.

✷✷ WARNING

If any binding is felt when adjusting the timing belt tension by turning the crankshaft, STOP turning the engine, because the pistons may be hitting the valves.

19. Rotate the crankshaft counterclockwise to remove the slack between the crankshaft sprocket and the right camshaft timing belt sprocket.

20. Install the belt around the water pump pulley.

21. Install the belt on the idler pulley.

22. Align the white alignment mark on the timing belt with the alignment mark on the left bank camshaft timing belt sprocket.

23. Install the crankshaft pulley and tighten the center bolt by hand. Rotate the crankshaft pulley clockwise to give slack

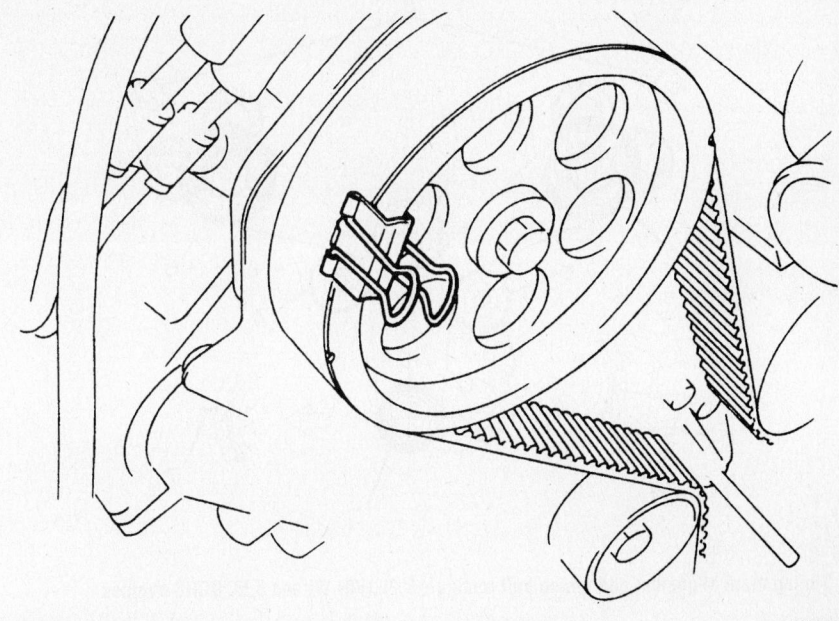

Using a double clip to hold the belt in place—3.2L and 3.5L engines

79245G08

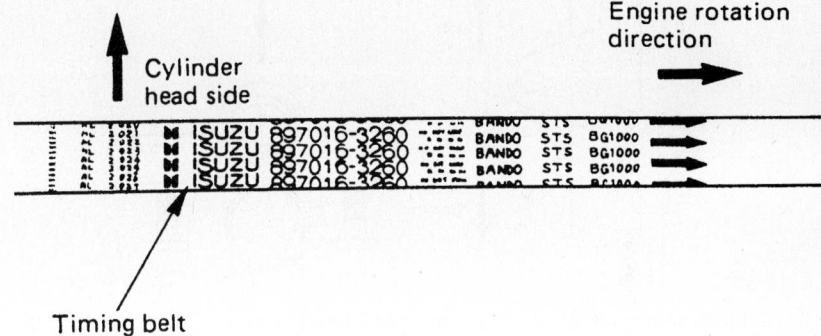

For maximum timing belt life, install the belt as shown—3.2L and 3.5L engines

79245G07

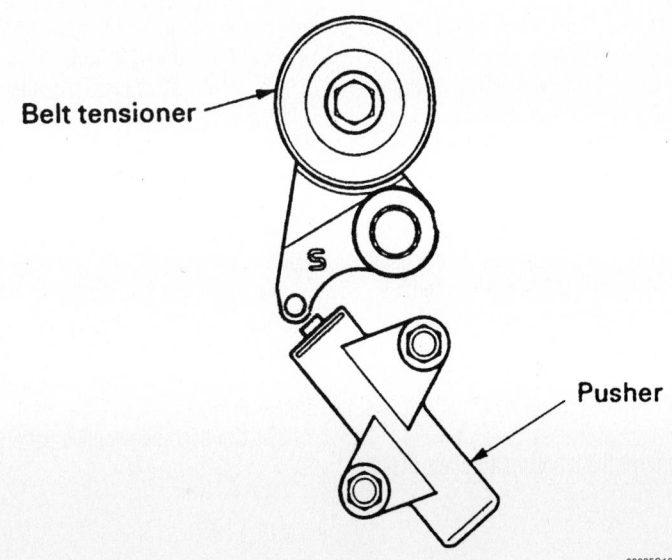

View of timing belt tensioner and pusher—3.2L and 3.5L engines

93025G10

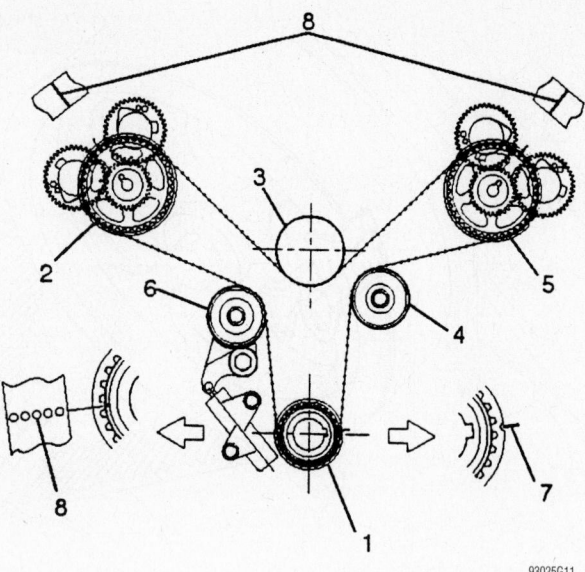

Timing mark alignment and timing belt routing—3.2L (VIN W) and 3.5L DOHC engines

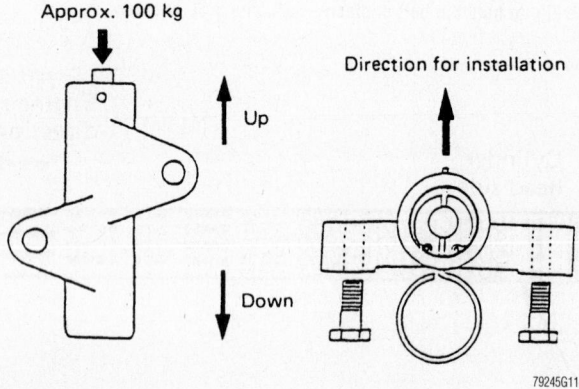

Approx. 100 kg

Up

Down

Direction for installation

View of timing belt tensioner pusher—3.2L and 3.5L engines

Piston and Ring

POSITIONING

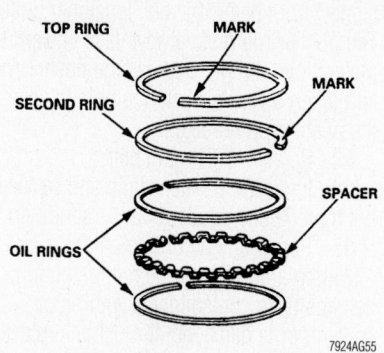

TOP RING

MARK

SECOND RING

MARK

OIL RINGS

SPACER

Piston ring positioning and top mark locations—3.5L engine

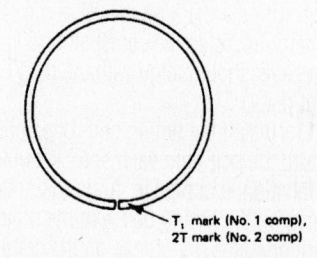

T_1 mark (No. 1 comp), 2T mark (No. 2 comp)

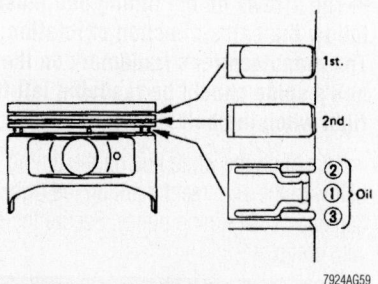

1st.

2nd.

② ① ③ Oil

Piston ring positioning—3.2L and 3.5L engines

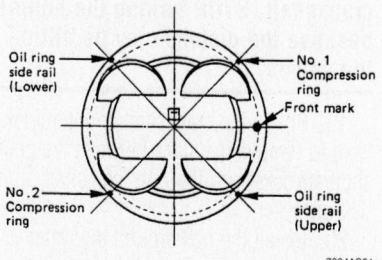

Oil ring side rail (Lower)

No. 1 Compression ring

Front mark

No.2 Compression ring

Oil ring side rail (Upper)

Piston ring end-gap spacing—3.2L and 3.5L engines

between the crankshaft timing belt pulley and the right bank camshaft timing belt pulley.

24. Insert a 1.4mm piece of wire through the hole in the pusher to hold the rod in. Install the pusher assembly while pushing the tension pulley toward the belt.

25. Pull the pin out from the pusher to release the rod.

26. Remove the clamps from the sprockets. Rotate the crankshaft pulley clockwise 2 turns. Measure the rod protrusion to ensure it is between 0.16 – 0.24 in. (4 – 6mm).

27. If the tensioner pulley bracket pivot bolt was removed, tighten it to 31 ft. lbs. (42 Nm).

28. Tighten the pusher bolts to 14 ft. lbs. (19 Nm).

29. Remove the crankshaft pulley. Install the lower and upper timing belt covers and tighten their bolts to 12 ft. lbs. (17 Nm).

30. Fit the oil cooler hose onto the timing cover and tighten its mounting bracket bolts to 16 ft. lbs. (22 Nm).

31. Install the crankshaft pulley and tighten the pulley bolt to 123 ft. lbs. (167 Nm).

32. Install fan pulley assembly and tighten the bolts to 16 ft. lbs. (22 Nm).

33. Install and adjust the accessory drive belts.

34. Install the cooling fan assembly and tighten the bolts to 72 inch lbs. (8 Nm).

35. Install the upper fan shroud.

36. Install the air cleaner assembly and intake air duct.

37. Connect the negative battery cable.

FUEL SYSTEM

Fuel System Service Precautions

Safety is the most important factor when performing not only fuel system maintenance but any type of maintenance. Failure to conduct maintenance and repairs in a safe manner may result in serious personal injury or death. Maintenance and testing of the vehicle's fuel system components can be accomplished safely and effectively by adhering to the following rules and guidelines:

• To avoid the possibility of fire and personal injury, always disconnect the negative battery cable unless the repair or test procedure requires that battery voltage be applied.

• Always relieve the fuel system pressure prior to disconnecting any fuel system component (injector, fuel rail, pressure regulator, etc.), fitting or fuel line connection. Exercise extreme caution whenever relieving fuel system pressure, to avoid exposing skin, face and eyes to fuel spray. Please be advised that fuel under pressure may penetrate the skin or any part of the body that it contacts.

• Always place a shop towel or cloth around the fitting or connection prior to loosening to absorb any excess fuel due to spillage. Ensure that all fuel spillage (should it occur) is quickly removed from engine surfaces. Ensure that all fuel soaked cloths or towels are deposited into a suitable waste container.

• Always keep a dry chemical (Class B) fire extinguisher near the work area.

• Do not allow fuel spray or fuel vapors to come into contact with a spark or open flame.

• Always use a backup wrench when loosening and tightening fuel line connection fittings. This will prevent unnecessary stress and torsion to fuel line piping. Always follow the proper tightening specifications.

• Always replace worn fuel fitting O-rings with new. Do not substitute fuel hose or equivalent, where fuel pipe is installed.

Fuel System Pressure

RELIEVING

1. Before servicing the vehicle, refer to the precautions in the beginning of this section.

2. Remove the fuel filler cap.
3. Remove the fuel pump relay from the underhood relay box.
4. Start the engine and let it run until it stalls, then crank the engine for an additional 30 seconds.
5. Turn the ignition switch to the **OFF** position and remove the key. Disconnect the negative battery cable.
6. When service is completed, install the fuel pump relay and connect the negative battery cable.

Fuel Filter

REMOVAL & INSTALLATION

1. Before servicing the vehicle, refer to the precautions in the beginning of this section.
2. Relieve the fuel system pressure.
3. Remove or disconnect the following:
    • Fuel lines from the fuel filter
    • Fuel filter

To install:
4. Install or connect the following:
    • Fuel filter and tighten the bracket bolt. Note the fuel flow directional arrow.
    • Fuel lines to the fuel filter

• Negative battery cable
5. Start the engine and inspect the fuel filter connections for leaks.

Fuel Pump

REMOVAL & INSTALLATION

1. Before servicing the vehicle, refer to the precautions in the beginning of this section.
2. Relieve fuel system pressure.
3. Drain the fuel tank.
4. Remove or disconnect the following:
    • Negative battery cable
    • Fuel filler and vent hoses
    • Fuel tank skid plate
    • Fuel tank wiring connectors
    • Fuel supply and return lines
    • Fuel tank
    • Fuel pump assembly

To install:
5. Install or connect the following:
    • Fuel pump assembly
    • Fuel tank. Tighten the bolts to 27 ft. lbs. (36 Nm).
    • Fuel supply and return lines
    • Fuel tank wiring connectors
    • Fuel tank skid plate
    • Fuel filler and vent hoses

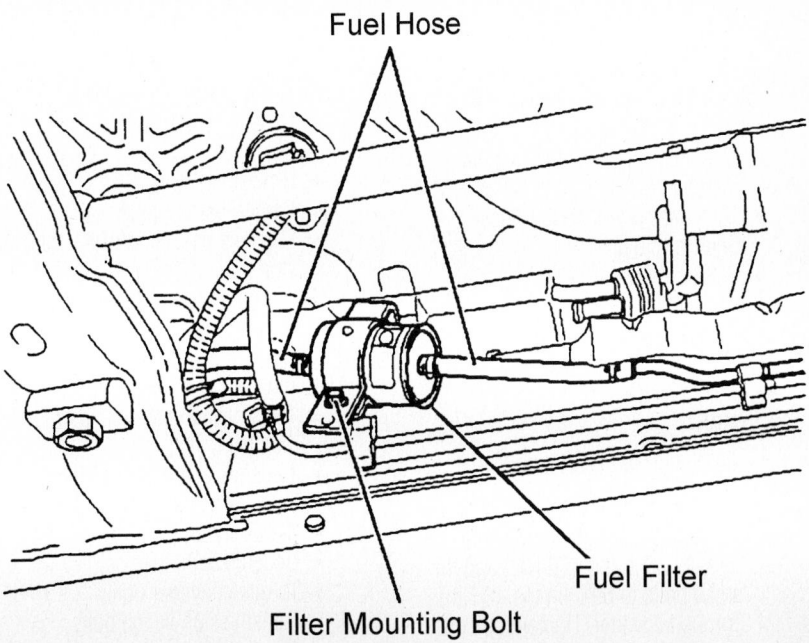

Fuel Hose

Fuel Filter

Filter Mounting Bolt

7924NG25

Fuel filter mounting location under the vehicle

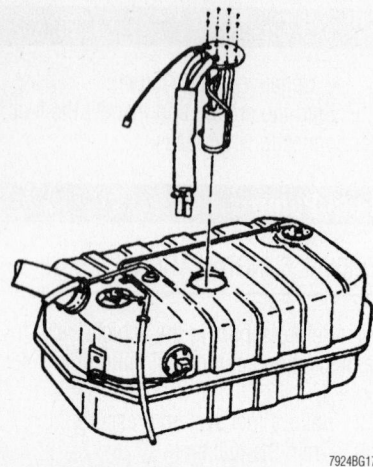

Fuel pump assembly mounting

7924BG17

- Negative battery cable
6. Start the engine and check for leaks.

Fuel Injector

REMOVAL & INSTALLATION

1. Before servicing the vehicle, refer to the precautions in the beginning of this section.
2. Relieve fuel system pressure.
3. Remove or disconnect the following:
   - Negative battery cable
   - Engine cover
   - Fuel injector wiring connectors
   - Fuel lines
   - Fuel supply manifold with injectors attached
   - Clips
   - Injectors from the supply manifold

To install:
4. Install or connect the following:
   - New O-rings on the fuel injectors
   - Fuel injectors
   - Fuel supply manifold with injectors attached. Tighten the bolts to 60 inch lbs. (6.5 Nm).
   - Fuel lines
   - Fuel injector wiring connectors
   - Engine cover
   - Negative battery cable
5. Start the engine and check for leaks.

DRIVE TRAIN

Automatic Transmission Assembly

REMOVAL & INSTALLATION

2-Wheel Drive

1. Before servicing the vehicle, refer to the precautions in the beginning of this section.
2. Remove or disconnect the following:
   - Negative battery cable
   - Driveshaft
   - Fuel line bracket from the transmission
   - Wiring harness heat shield
   - Transmission harness connectors
3. Support the transmission with a jack.
4. Remove or disconnect the following:
   - Transmission mount and crossmember
   - Transmission oil cooler lines
   - Selector cable
   - Starter motor
   - Undercovers
   - Torque converter bolts
   - Transmission flange bolts
   - Transmission

To install:

➡ **Use new torque converter bolts.**

5. Install or connect the following:
   - Transmission. Tighten the large bolts to 56 ft. lbs. (76 Nm) and the small bolts to 69 inch lbs. (8 Nm).
   - Torque converter. Tighten the bolts to 40 ft. lbs. (54 Nm).
   - Under covers
   - Starter motor. Tighten the bolts to 30 ft. lbs. (40 Nm).

- Selector cable
- Transmission oil cooler lines
- Crossmember. Tighten the bolts to 85 ft. lbs. (116 Nm).
- Transmission mount. Tighten the bolts to 37 ft. lbs. (50 Nm).
- Transmission harness connectors
- Wiring harness heat shield
- Fuel line bracket
- Driveshaft. Tighten the flange bolts to 46 ft. lbs. (63 Nm).
- Negative battery cable

4-Wheel Drive

1. Before servicing the vehicle, refer to the precautions in the beginning of this section.
2. Remove or disconnect the following:
   - Negative battery cable
   - Skid plate
   - Driveshafts
   - Center exhaust pipe
   - Fuel line bracket from the transmission
   - Wiring harness heat shield
   - Transmission harness connectors
3. Support the transmission with a jack.
4. Remove or disconnect the following:
   - Transmission mount and crossmember
   - Transmission oil cooler lines
   - Selector cable
   - Starter motor
   - Undercovers
   - Torque converter bolts
   - Transmission flange bolts
   - Transmission

To install:

➡ **Use new torque converter bolts.**

5. Install or connect the following:
   - Transmission. Tighten the large bolts to 56 ft. lbs. (76 Nm) and the small bolts to 69 inch lbs. (8 Nm).
   - Torque converter. Tighten the bolts to 40 ft. lbs. (54 Nm).
   - Under covers
   - Starter motor. Tighten the bolts to 30 ft. lbs. (40 Nm).
   - Selector cable
   - Transmission oil cooler lines
   - Crossmember. Tighten the bolts to 85 ft. lbs. (116 Nm).
   - Transmission mount. Tighten the bolts to 37 ft. lbs. (50 Nm).
   - Transmission harness connectors
   - Wiring harness heat shield
   - Fuel line bracket
   - Center exhaust pipe
   - Driveshafts. Tighten the flange bolts to 46 ft. lbs. (63 Nm).
   - Skid plate
   - Negative battery cable

Transfer Case Assembly

REMOVAL & INSTALLATION

1. Before servicing the vehicle, refer to the precautions in the beginning of this section.
2. Remove or disconnect the following:
   - Negative battery cable
   - Transfer case skid plate
   - Front and rear driveshafts
   - Breather hose
   - Center exhaust pipe
   - Vehicle Speed (VSS) sensor connector
   - Harness connector
3. Support the transfer case

4. Remove or disconnect the following:
   - Transfer case flange fasteners
   - Transfer case

To install:

5. Install or connect the following:
   - Transfer case. Tighten the flange fasteners to 34 ft. lbs. (46 Nm).
   - Harness connector

- Breather hose
- VSS sensor connector
- Center exhaust pipe
- Front and rear driveshafts. Tighten the bolts to 46 ft. lbs. (63 Nm).
- Transfer case skid plate. Tighten the bolts to 27 ft. lbs. (37 Nm).
- Negative battery cable

REMOVAL & INSTALLATION

1. Before servicing the vehicle, refer to the precautions in the beginning of this section.

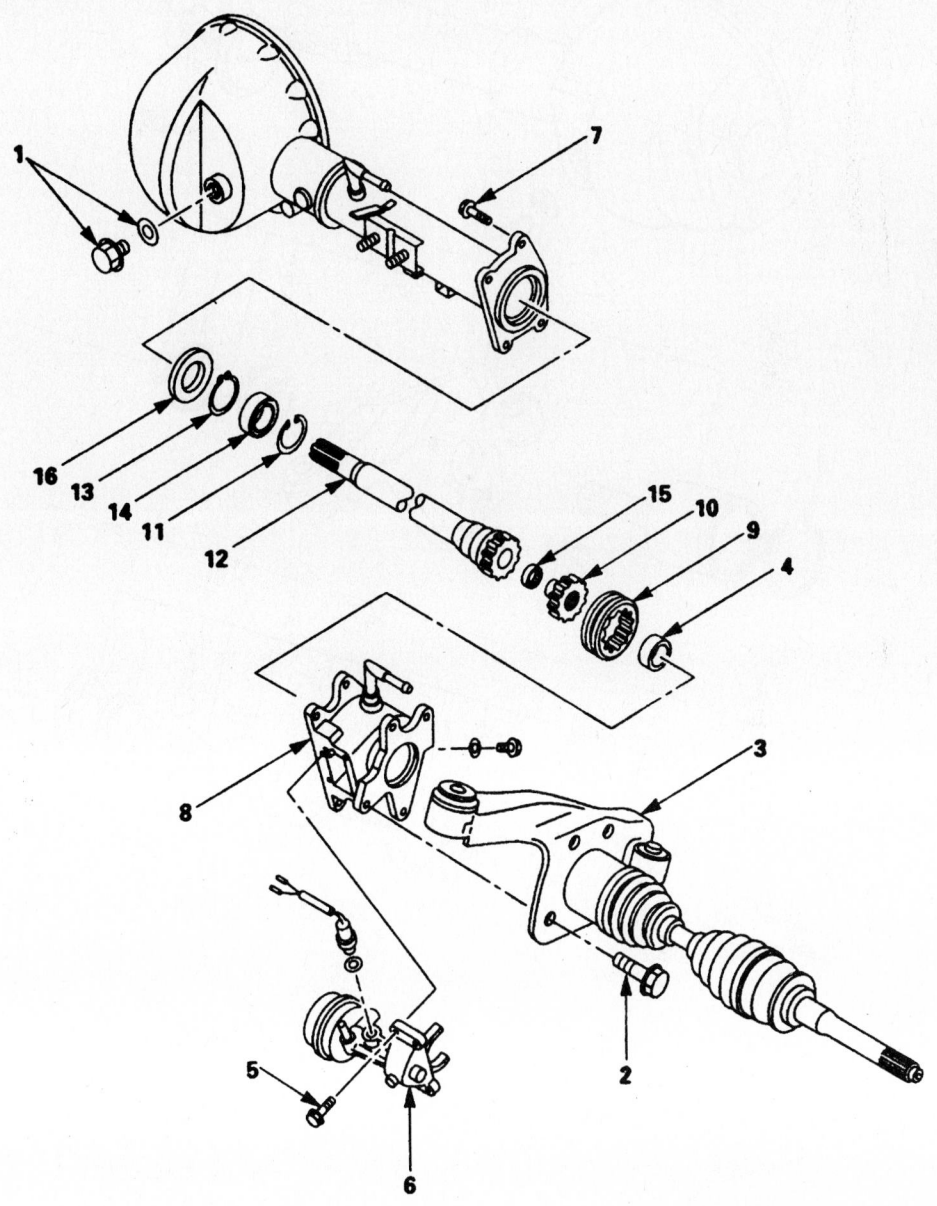

1. Filler plug
2. Bolt
3. Front axle drive shaft (LH side)
4. Spacer
5. Bolt
6. Actuator assembly
7. Bolt
8. Housing
9. Sleeve
10. Clutch gear
11. Snap ring
12. Inner shaft
13. Snap ring
14. Inner shaft bearing
15. Needle bearing
16. Oil seal

7924BG26

Exploded view of the left halfshaft, axle shaft and axle disconnect

2. Remove or disconnect the following:
- Negative battery cable
- Front wheel
- Radiator skid plate
- Transfer case skid plate

- Brake calipers and mounting bracket
- Brake rotor
- Wheel speed sensor
- Steering knuckle

3. Support the axle housing with a jack. Unbolt the axle mounting bracket and remove the halfshaft/bracket assembly.

To install:

4. Install or connect the following:

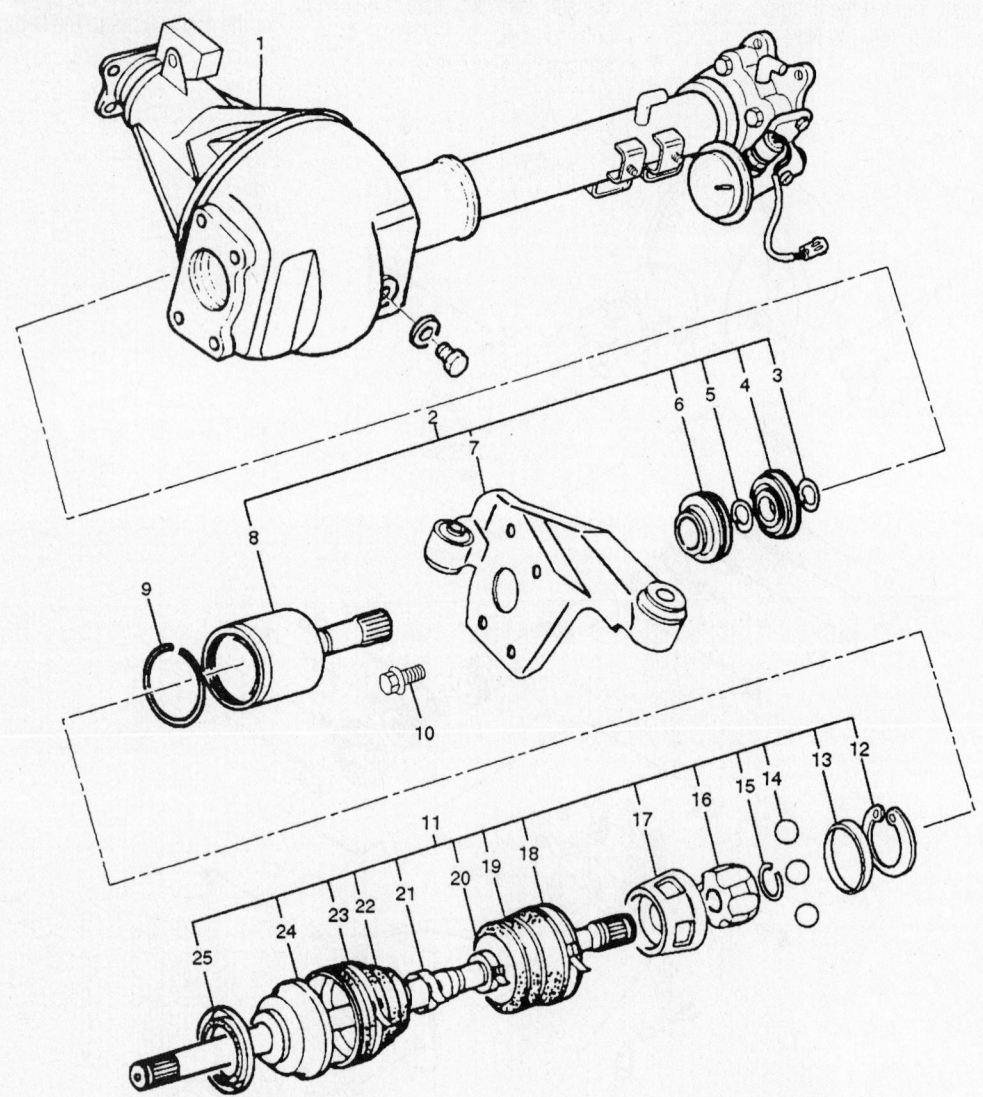

1 Axle Case and Differential
2 DOJ Case Assembly
3 Snap Ring
4 Bearing
5 Snap Ring
6 Oil Seal
7 Bracket
8 DOJ Case
9 Circlip
10 Bolt
11 Drive Shaft Joint Assembly
12 Snap Ring
13 Spacer
14 Ball
15 Snap Ring
16 Ball Retainer
17 Ball Guide
18 Band
19 Bellows
20 Band
21 Band
22 Bellows
23 Band
24 BJ Shaft
25 Dust Seal

9308BG03

Exploded view of the right halfshaft and mounting bracket

- Axle/bracket assembly. Tighten the bracket flange bolts to 85 ft. lbs. (116 Nm) and the bracket mounting bolts to 112 ft. lbs. (152 Nm).
- Steering knuckle
- Wheel speed sensor
- Brake rotor
- Brake caliper and mounting bracket. Tighten the bracket bolts to 115 ft. lbs. (155 Nm).
- Transfer case skid plate. Tighten the bolts to 27 ft. lbs. (37 Nm).
- Radiator skid plate. Tighten the bolts to 58 ft. lbs. (78 Nm).
- Front wheel
- Negative battery cable

5. Check the wheel alignment and adjust as necessary.

CV-Joints

OVERHAUL

Outer CV-Joint

The outer CV-joint is serviced with the axle shaft as an assembly. The outer CV-joint boot can be serviced by removing the inner CV-joint.

Inner CV-Joint

1. Before servicing the vehicle, refer to the precautions in the beginning of this section.
2. Remove or disconnect the following:
- Halfshaft from the vehicle
- Snapring and bearing
- Snapring and oil seal
- Mounting bracket
- CV-joint boot
- Circlip and inner joint housing
- Snapring and spacer
- Inner joint balls

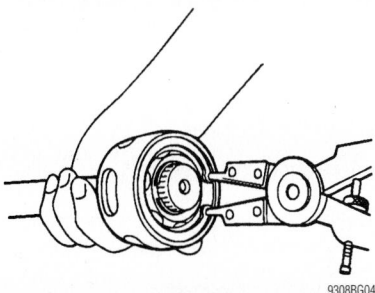

CV-joint spacer snapring—Inner CV-Joint

- Snapring and inner CV-joint

To install:

3. Install or connect the following:
- Inner CV-joint and snapring
- Inner joint balls
- Spacer and snapring
- Inner joint housing and circlip. Add 150 grams CV-joint grease.
- CV-joint boot
- Mounting bracket
- Oil seal and snapring
- Bearing and snapring

4. Install the halfshaft and mounting bracket to the vehicle.

5. Check the wheel alignment and adjust as necessary.

Rear Axle Shaft, Bearing and Seal

REMOVAL & INSTALLATION

1. Before servicing the vehicle, refer to the precautions in the beginning of this section.
2. Remove or disconnect the following:
- Rear wheel
- Disc brake caliper and bracket
- Disc brake rotor
- Wheel speed sensor bracket
- Parking brake cable and bracket
- Parking brake shoes
- Axle shaft
- Snapring and discard it
- Bearing, press it off the axle shaft with the bearing holder and oil seal

To install:

3. Install or connect the following:
- New oil seal into the bearing housing
- Bearing housing onto the axle shaft
- Bearing, press it onto the axle shaft
- New snapring
- Axle shaft. Use new lockwashers and tighten the bearing holder nuts to 54 ft. lbs. (74 Nm).
- Parking brake shoes
- Parking brake cable and bracket
- Wheel speed sensor bracket
- Disc brake rotor
- Disc brake caliper and bracket. Tighten the bracket bolts to 76 ft. lbs. (103 Nm).
- Rear wheel

4. Check the rear axle oil level and adjust as necessary.

Pinion Seal

REMOVAL & INSTALLATION

1. Before servicing the vehicle, refer to the precautions in the beginning of this section.
2. Remove or disconnect the following:
- Driveshaft
- Wheels
- Brake calipers and pads

➡ **The brake calipers and pads must be removed so that there is no additional drag when measuring pinion bearing preload.**

3. Use an inch lb. torque wrench and measure and record the amount of torque required to maintain pinion rotation through several revolutions.

4. Remove or disconnect the following:
- Pinion flange
- Pinion seal
- Pinion bearing
- Collapsible spacer

To install:

➡ **Use a new collapsible spacer and flange nut for assembly.**

5. Install or connect the following:
- Collapsible spacer
- Pinion bearing
- Pinion seal
- Pinion flange

6. Rotate the pinion flange occasionally while tightening the flange nut to make sure the pinion bearings seat correctly.

7. Take frequent bearing preload torque readings. Tighten the flange nut to achieve the preload torque readings originally recorded.

✳✳ CAUTION

Never loosen the pinion nut to reduce bearing preload. If it is necessary to reduce bearing preload, install a new collapsible spacer and pinion nut.

8. Install or connect the following:
- Driveshaft
- Brake calipers and pads
- Wheels

9. Fill the differential with gear lubricant and check for leaks.

STEERING AND SUSPENSION

Air Bag

❄❄ CAUTION

Some vehicles are equipped with an air bag system. The system must be disarmed before performing service on, or around, system components, the steering column, instrument panel components, wiring and sensors. Failure to follow the safety precautions and the disarming procedure could result in accidental air bag deployment, possible injury and unnecessary system repairs.

PRECAUTIONS

Several precautions must be observed when handling the inflator module to avoid accidental deployment and possible personal injury.

• Never carry the inflator module by the wires or connector on the underside of the module.

• When carrying a live inflator module, hold securely with both hands, and ensure that the bag and trim cover are pointed away from you.

• Place the inflator module on a bench or other surface with the bag and trim cover facing up.

• With the inflator module on the bench, never place anything on or close to the module which may be thrown in the event of an accidental deployment.

DISARMING

1. Before servicing the vehicle, refer to the precautions in the beginning of this section.

2. Turn the ignition switch to the **LOCK** position. Remove the key.

3. Disconnect the negative battery cable. Wait 1 minute before working around the air bags.

4. Disconnect the yellow 2-pin connector at the base of the steering column.

5. Disconnect the yellow 2-pin connector behind the glove box assembly.

6. When repairs are completed, connect the yellow 2-pin connectors.

7. Connect the negative battery cable.

8. Turn the ignition to the **ON** position, but don't start the engine. The AIR BAG warning light should turn ON and flash ON and OFF 7 times, and then turn OFF. This light sequence indicates that the SRS system is functioning normally. If the AIR BAG light doesn't come ON, or stays ON longer than 7 seconds, the system must be diagnosed.

Rack and Pinion Steering Gear

REMOVAL AND INSTALLATION

2-Wheel Drive

1. Before servicing the vehicle, refer to the precautions in the beginning of this section.

2. Disable the air bag system.

3. Remove or disconnect the following:
   • Stone guard
   • Tie rod from knuckle
   • Power steering pressure and return lines
   • Steering gear

To install:

4. Install or connect the following:
   • Steering gear. Torque the mounting bolts to 85 ft. lbs. (116 Nm)
   • Fluid lines. Torque the fitting to 18 ft. lbs. (25 Nm)
   • Tie rod ends. 87 ft. lbs. (118 Nm)
   • Stone guard

5. Fill and bleed the system.

4-Wheel Drive

1. Before servicing the vehicle, refer to the precautions in the beginning of this section.

2. Disable the air bag system.

3. Remove or disconnect the following:
   • Stone guard
   • Tie rod from knuckle
   • Power steering pressure and return lines
   • Torsion bar
   • Lower control arm frame side bolt
   • Crossmember bolts
   • Steering gear with crossmember
   • Steering gear

To install:

4. Install or connect the following:
   • Steering gear. Torque the mounting bolts to 85 ft. lbs. (116 Nm)
   • Crossmember Torque the bolts to 128 ft. lbs. (173 Nm)
   • Lower control arm bolt
   • Torsion bar
   • Fluid lines. Torque the fitting to 18 ft. lbs. (25 Nm)
   • Tie rod ends. 87 ft. lbs. (118 Nm)
   • Stone guard

5. Fill and bleed the system.

Shock Absorber

REMOVAL & INSTALLATION

Front

1. Before servicing the vehicle, refer to the precautions in the beginning of this section.

2. Support the lower control arm with a jackstand.

3. Remove or disconnect the following:
   • Front wheels
   • Upper shock retaining nut and rubber bushing
   • ISC actuator and bracket
   • Suspension bump stops
   • Shock absorber

To install:

4. Install or connect the following:
   • Shock absorber. Tighten the lower bolt to 69 ft. lbs. (93 Nm).
   • ISC actuator and bracket
   • Bump stop. Tighten the bolts to 30 ft. lbs. (41 Nm).
   • Upper shock retaining nut and rubber bushing. Tighten the nut to 14 ft. lbs. (20 Nm).
   • Front wheels

Rear

1. Before servicing the vehicle, refer to the precautions in the beginning of this section.

2. Support the rear axle with jackstands.

3. Remove the rear shock absorbers.

To install:

4. Install the rear shock absorbers. Tighten the upper bolt to 14 ft. lbs. (20 Nm). Tighten the lower bolt to 58 ft. lbs. (78 Nm).

➡**With ISC, do not use any grease on or near the bushings.**

5. Remove the jackstands.

Coil Spring

REMOVAL & INSTALLATION

1. Before servicing the vehicle, refer to the precautions in the beginning of this section.

2. Support the vehicle under the frame.

3. Support the rear axle with a jack.

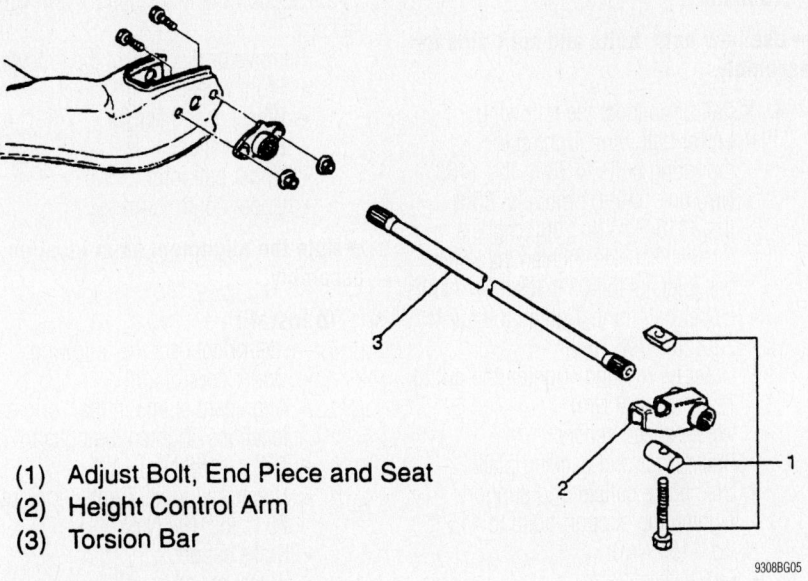

(1) Adjust Bolt, End Piece and Seat
(2) Height Control Arm
(3) Torsion Bar

9308BG05

Exploded view of the torsion bar assembly

4. Remove or disconnect the following:
   - Rear wheels
   - Breather hose
   - Upper link fixing bolt, left side only
   - Stabilizer bar links
   - Shock absorbers

5. Lower the rear axle with the jack to release the coil spring tension. Remove the coil springs and insulators.

To install:

6. Place the coil springs on the axle assembly and the insulators on top of the springs.

7. Raise the axle assembly into position.

8. Install or connect the following:

- Shock absorbers. Torque to 58 ft. lbs. (78 Nm)
- Stabilizer bar links. Tighten the nuts to 23 ft. lbs. (31 Nm).
- Upper link. Torque to 101 ft. lbs. (137 Nm)
- Rear wheels

Torsion Bar

REMOVAL & INSTALLATION

1. Before servicing the vehicle, refer to the precautions in the beginning of this section.

2. Matchmark the adjusting bolt and end piece, then remove the bolt, end piece, and seat.

3. Matchmark the height control arm to the torsion bar, then remove the height control arm.

4. Matchmark the torsion bar to the lower control arm, then remove the torsion bar.

To install:

5. Apply grease to the torsion bar splines.

6. Apply grease to the contact points of the height control arm, adjusting bolt end piece and seat.

7. Align the matchmarks and install the torsion bar.

8. Align the matchmarks and install the height control arm.

9. Install the adjusting bolt, seat and end piece.

10. Tighten the adjusting bolt to align the matchmarks.

Upper Ball Joint

REMOVAL & INSTALLATION

1. Before servicing the vehicle, refer to the precautions in the beginning of this section.

2. Support the lower control arm with a floor jack.

3. Remove or disconnect the following:
   - Front wheel
   - Wheel speed sensor
   - Upper ball joint

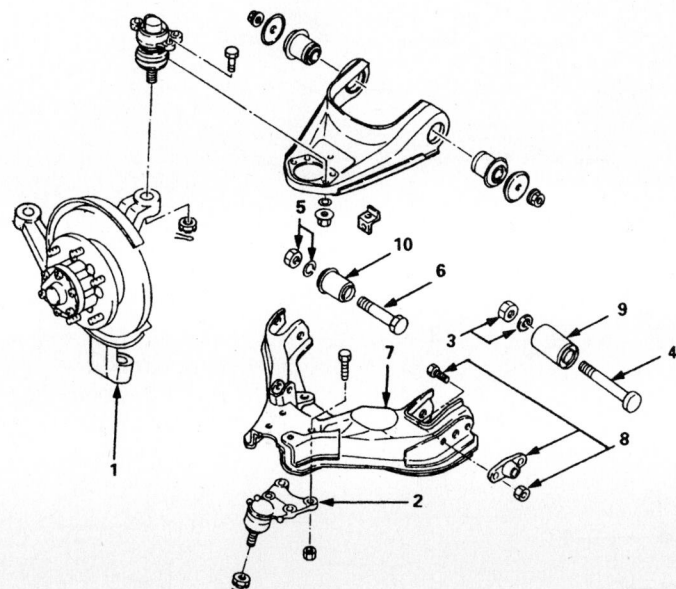

1. Knuckle
2. Lower end
3. Nut and washer, rear
4. Bolt, rear
5. Nut and washer, front
6. Bolt, front
7. Lower control arm assembly
8. Torsion bar arm bracket
9. Bushing, rear
10. Bushing, front

7924BG34

Exploded view of the control arm and ball joint components

To install:

➡ **Use new nuts, bolts and split pins for assembly.**

4. Install or connect the following:
  - Upper ball joint. Tighten the mounting bolts to 42 ft. lbs. (57 Nm) and the nut to 72 ft. lbs. (96 Nm).
  - Wheel speed sensor
  - Front wheel

Lower Ball Joint

REMOVAL & INSTALLATION

1. Before servicing the vehicle, refer to the precautions in the beginning of this section.

2. Support the lower control arm with a jackstand.

3. Remove or disconnect the following:
  - Front wheel
  - Disc brake caliper and support
  - Brake rotor and backing plate
  - Wheel speed sensor
  - Outer tie rod end
  - Upper ball joint
  - Steering knuckle
  - Lower ball joint

To install:

➡ **Use new nuts, bolts and split pins for assembly.**

4. Install or connect the following:
  - Lower ball joint. Tighten the mounting bolts to 76 ft. lbs. (103 Nm) on 1999–01 models; 85 ft. lbs. (116 Nm) on 2002 models.
  - Steering knuckle. Tighten the lower ball joint nut to 108 ft. lbs. (147 Nm).
  - Upper ball joint. Tighten the nut to 72 ft. lbs. (96 Nm).
  - Outer tie rod end. Tighten the nut to 72 ft. lbs. (98 Nm).
  - Wheel speed sensor
  - Brake rotor and backing plate
  - Disc brake caliper and support. Tighten the support bolts to 115 ft. lbs. (155 Nm).
  - Front wheel

5. Check the wheel alignment and adjust as necessary.

Upper Control Arm

REMOVAL & INSTALLATION

1. Before servicing the vehicle, refer to the precautions in the beginning of this section.

2. Support the lower control arm with a jackstand.

3. Remove or disconnect the following:
  - Front wheel
  - Wheel speed sensor
  - Brake caliper
  - Upper ball joint
  - Upper control arm

➡ **Note the alignment shim location for assembly.**

To install:

4. Install or connect the following:
  - Upper control arm
  - Alignment shims in their original locations. Tighten the bolts to 112 ft. lbs. (152 Nm).
  - Upper ball joint. Tighten the nut to 72 ft. lbs. (98 Nm).
  - Brake caliper
  - Wheel speed sensor
  - Front wheel

5. Check the wheel alignment and adjust as necessary.

CONTROL ARM BUSHING REPLACEMENT

1. Before servicing the vehicle, refer to the precautions in the beginning of this section.

2. Remove the upper control arm.

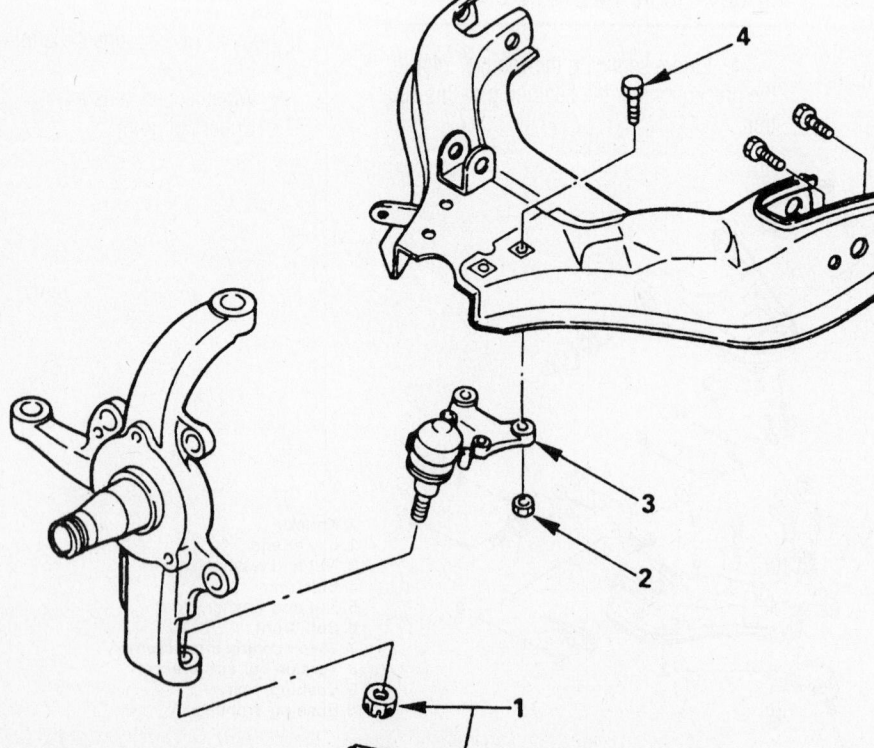

1. Nut and cotter pin
2. Nut
3. Lower ball joint
4. Bolt

Exploded view of the lower ball joint mounting and related components

7924BG35

3. Remove the nuts and washers from the fulcrum pin.

4. Press the bushings out of the control arm.

To install:

5. Press the bushings into the control arm.

6. Install the fulcrum pin washers and nuts.

7. Install the upper control arm.

8. Raise the suspension so that there is 0.79 inches (20mm) between the bump stop and the lower control arm. Tighten the fulcrum pin nuts to 80 ft. lbs. (108 Nm).

9. Check the wheel alignment and adjust as necessary.

Lower Control Arm

REMOVAL & INSTALLATION

1. Before servicing the vehicle, refer to the precautions in the beginning of this section.

2. Remove or disconnect the following:
   - Front wheel
   - Torsion bar
   - Lower ball joint
   - Stabilizer bar link
   - Shock absorber
   - Lower control arm

To install:

3. Install or connect the following:
   - Lower control arm
   - Shock absorber
   - Stabilizer bar link
   - Lower ball joint
   - Torsion bar
   - Front wheel

4. Raise the suspension so that there is 0.79 inches (20mm) between the bump stop and the lower control arm. For 1999–01, tighten the front control arm bolt to 116 ft. lbs. (157 Nm) and tighten the rear control arm bolt to 145 ft. lbs. (196 Nm). For 2002, tighten the rear nut to 174 ft. lbs. (235 Nm); the rear nut to 137 ft. lbs. (186 Nm).

5. Check the wheel alignment and adjust as necessary.

CONTROL ARM BUSHING REPLACEMENT

1. Before servicing the vehicle, refer to the precautions in the beginning of this section.

2. Remove or disconnect the following:
   - Lower control arm
   - Bushings, press them from the control arm, using Remover/Installer J-36833 for the front bushing and Remover/Installer J-36834 for the rear bushing.

To install:

3. Install or connect the following:
   - New bushings
   - Lower control arm

Wheel Bearings

ADJUSTMENT

2- and 4-Wheel Drive

1. Before servicing the vehicle, refer to the precautions in the beginning of this section.

2. Remove or disconnect the following:
   - Front wheel
   - Brake caliper and pads
   - Hub dust cap
   - Snapring and shim
   - Hub flange
   - Lockscrew and washer

3. Tighten the hub nut to 22 ft. lbs. (29 Nm) to seat the bearings and then fully loosen the nut.

4. Tighten the hub nut to achieve a bearing preload of 2.6–4.0 lbs. (1.2–1.8 Kg) for used bearings. If the bearings were replaced, set the preload to 4.4–5.5 lbs. (2.0–2.5 Kg).

5. Install or connect the following:
   - Lockwasher and screw

   - Hub flange
   - Shim and snapring
   - Hub dust cap. Tighten the bolts to 43 ft. lbs. (59 Nm).
   - Brake caliper and pads
   - Front wheel

REMOVAL & INSTALLATION

1. Before servicing the vehicle, refer to the precautions in the beginning of this section.

2. Remove or disconnect the following:
   - Front wheel
   - Brake caliper and support
   - Hub dust cap
   - Snapring and shim
   - Hub flange
   - Lockscrew and washer
   - Hub nut
   - Brake rotor and hub assembly
   - Wheel speed sensor ring
   - Outer bearing
   - Grease seal
   - Inner bearing

To install:

3. Clean and inspect the bearings. Replace if necessary.

4. Apply clean wheel bearing grease to the inner and outer bearings.

5. Apply grease in the hub.

6. Install the wheel bearings into the hub along with a new grease seal.

7. Install or connect the following:
   - Wheel speed sensor ring. Tighten the bolts to 13 ft. lbs. (18 Nm).
   - Brake rotor and hub assembly
   - Hub nut. Set the bearing preload.
   - Lockscrew and washer
   - Hub flange
   - Snapring and shim
   - Hub dust cap
   - Brake caliper and support. Tighten the support bolts to 115 ft. lbs. (155 Nm).
   - Front wheel

BRAKES

Brake Caliper

REMOVAL AND INSTALLATION

Front

1. Raise and safely support the vehicle. Remove the front wheels.

2. Remove some brake fluid from the master cylinder reservoir.

3. Disconnect and plug the brake fluid line from the caliper.

4. Remove the brake caliper mounting bolt and guide bolt and remove the caliper from the mount. The brackets can be remove for additional work space.

5. Remove the brake pads and clips from the caliper. Inspect the brake pads for wear; replace them, if necessary.

To install:

6. Install the brake pads and clips onto the caliper.

7. If the caliper bracket was removed, tighten the bolts to 103–126 ft. lbs. (139–171 Nm).

8. Install the caliper on the mounting bracket. Torque the caliper-to-mounting bracket bolts to:

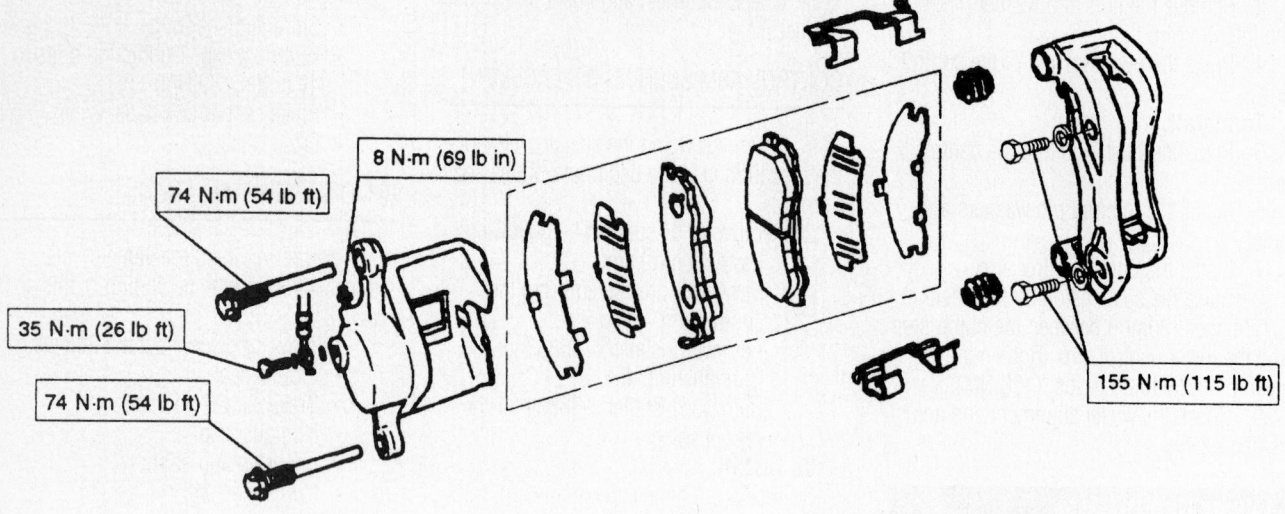

74 N·m (54 lb ft)

8 N·m (69 lb in)

35 N·m (26 lb ft)

74 N·m (54 lb ft)

155 N·m (115 lb ft)

93026G54

Exploded view of the front caliper components

- 4-cylinder models: 20–27 ft. lbs. (27–37 Nm)
- 6-cylinder models: 54 ft. lbs. (74 Nm)

9. Connect the fluid line to the caliper using new washers. Torque the brake line banjo fitting to 26 ft. lbs. (35 Nm).

✳ WARNING

Be sure the hook end of the flexible brake line is positioned in the anti-rotation cavity.

10. Refill the master cylinder reservoir and bleed the brake system.

11. Install the front wheels and lower the vehicle.

Rear

1. Raise and safely support the vehicle. Remove the rear wheels.

2. Remove some fluid from the master cylinder reservoir.

3. Disconnect and plug the brake fluid line from the caliper.

➡**Discard the parking brake cable mounting pin after removal.**

4. If equipped with caliper actuated parking brakes; remove the mounting pin from the parking brake cable and disconnect the parking cable from the disc caliper.

5. Remove the brake caliper mounting bolt and guide bolt and remove the caliper from the mount.

6. Remove the brake pads and clips from the caliper. Inspect the brake pads for wear; replace them, if necessary.

To install:

7. Install the brake pads and clips onto the caliper.

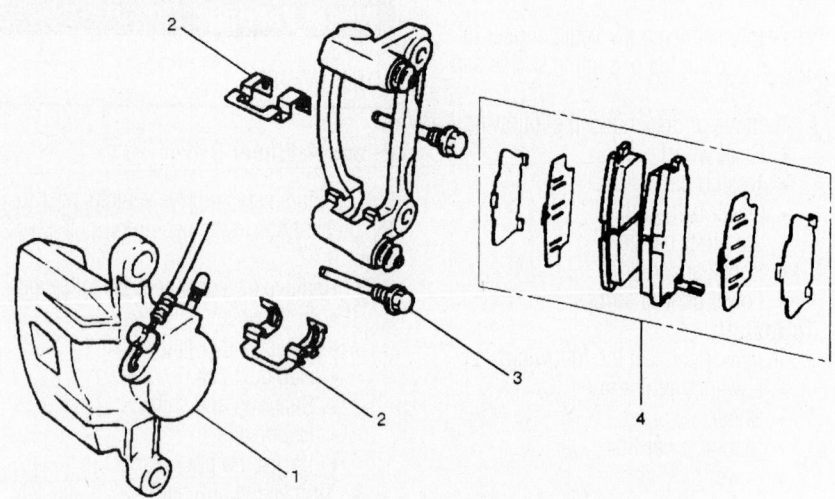

(1) Caliper Assembly
(2) Clip
(3) Lock Bolt
(4) Pad Assembly

93026G55

Exploded view of the rear caliper components

8. If the mounting bracket was removed, tighten the bolts to 69–84 ft. lbs. (93–114 Nm).

9. Install the caliper on the mounting bracket. Torque the caliper-to-mounting bracket bolts to 12–17 ft. lbs. (16–24 Nm), or 32 ft. lbs. (43 Nm) on vehicles with shoe-type parking brakes.

10. Connect the parking brake cable to the caliper and install a new mounting pin.

11. Connect the fluid line to the caliper using new washers. Torque the brake line banjo fitting to 26 ft. lbs. (35 Nm).

12. Refill the master cylinder reservoir and bleed the brake system.

13. Install the rear wheels and lower the vehicle.

Disc Brake Pads

REMOVAL AND INSTALLATION

Front

Most disc brake pads are equipped with wear indicators. If a squealing noise occurs from the brakes while driving, check the pad wear indicator plate. If there is evidence of the indicator plate contacting the brake disc, the brake pad should be replaced.

1. Remove ½ of the volume of brake fluid from the master cylinder to prevent overflow when the caliper piston is compressed.

2. Raise and safely support the vehicle.

3. Remove the wheel and tire assemblies.

4. Remove the brake caliper without disconnecting the brake line. Support the caliper with a length of wire. Do not let the caliper hang from the brake hose.

➡**On some disc brake systems it is not necessary to remove the caliper when installing new brake pads. Remove the lower slide bolt and rotate the caliper upward to remove the pads.**

5. Remove the brake pads and shims. Inspect the brake rotor and machine or replace as necessary. Check the minimum thickness (specification is cast into the rotor) before machining.

To install:

6. Use a suitable tool to push the caliper piston into its bore.

7. Apply a thin coat of grease to the rear face of the brake pad and install the shim. Install the brake pads.

8. Install the calipers. Lubricate the caliper bolts and boots. If equipped with a 4-cylinder engine, tighten the caliper mounting bolts to 24 ft. lbs. (33 Nm). If equipped with a 6-cylinder engine, tighten the caliper mounting bolts to 54 ft. lbs. (74 Nm).

9. Install the wheel and tire assemblies and lower the vehicle.

10. Apply the brakes several times to seat the pads before moving the vehicle. Check the fluid in the master cylinder and add as necessary.

Rear

1. Raise and safely support the vehicle. Remove the rear wheels.

2. Remove the brake caliper mounting bolts and remove the caliper without disconnecting the brake fluid line. Support the caliper so it does not hang on the brake line.

3. Remove the brake pads and retaining clips from the caliper.

4. If equipped with caliper-activated parking brakes; use tool J-37617 or equivalent to rotate the piston clockwise until it retracts into the bore. Align the notches of the piston face so the centerline of the notches is perpendicular to the centerline of the mounting bosses.

To install:

5. Install the new brake pads and clips in the caliper and install the caliper in the mounting bracket.

6. Tighten the caliper mounting bolts to 12–17 ft. lbs. (16–24 Nm), or 32 ft. lbs. (43 Nm) on vehicles with shoe-type parking brakes.

7. Install the rear wheels. Check the brake fluid level.

8. Pump the brake pedal until pressure is felt before moving the vehicle.

Brake Drums

REMOVAL AND INSTALLATION

1. Raise and safely support the vehicle. Release the parking brake.

2. Remove the rear wheels.

3. Use chalk to mark the brake drum to one of the wheel studs as an index mark for reinstallation.

4. Remove the retaining screw that holds the brake drum to the axle flange.

5. Pull the brake drum from the axle flange.

To install:

6. Align the index mark and install the brake drum to the axle flange.

7. Install the retaining screw to secure the brake drum to the axle flange.

8. Install the rear wheels.

Rear Brake Shoes

REMOVAL AND INSTALLATION

1. Raise and safely support the vehicle.

2. Remove the rear wheels.

3. Remove the brake drums.

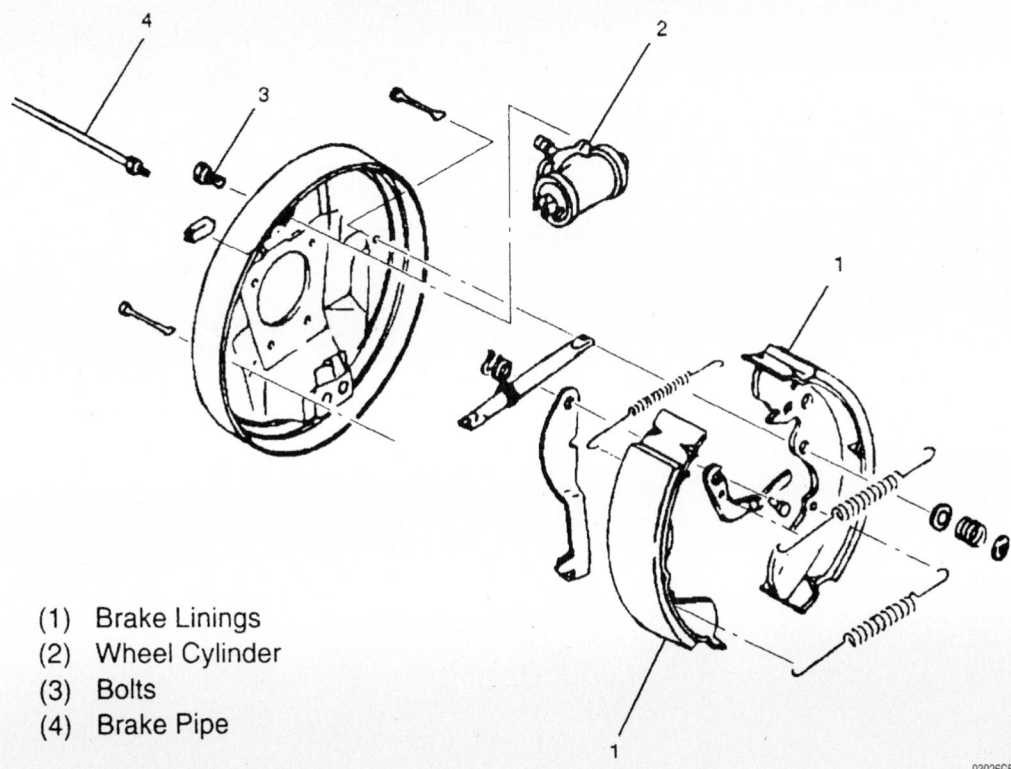

(1) Brake Linings
(2) Wheel Cylinder
(3) Bolts
(4) Brake Pipe

93026G56

Exploded view of the rear drum brakes

4. Remove the brake return springs.

5. Remove the leading shoe holding pin and spring, and then the leading shoe.

6. Remove the self-adjuster and the adjuster lever.

7. Remove the trailing shoe holding pin and spring.

8. Disconnect the parking brake cable from the trailing shoe and remove the trailing shoe. Remove the parking brake lever from the trailing shoe.

To install:

9. Attach the parking brake lever to the trailing shoe.

10. Connect the parking brake cable to the parking brake lever.

11. Apply a thin coat of high temperature grease to the shoe contact points on the brake backing plate (locations A and C in the accompanying illustration), piston contact surface (B), and self-adjuster (D).

12. Position the trailing shoe on the backing plate and install the hold-down pin, spring, and retainer. Don't stretch the return spring when fitting the shoes onto the backing plate.

13. Connect the upper return spring and the leading shoe to the trailing shoe and position the leading brake shoe on the backing plate.

14. Install the adjuster assembly and the hold-down pin, spring, and retainer.

15. Use a brake spring tool to install the lower return spring.

16. Install the self-adjuster lever and adjuster spring.

17. Adjust the shoe-to-drum clearance to 0.0098–0.0157 in. (0.25–0.40mm) and install the brake drum.

18. Check the brake drum for scoring or other wear. Machine or replace as necessary. Check the maximum brake drum diameter specification when machining.

19. Install the rear wheels. Lower the vehicle.

20. Road-test the vehicle.

ISUZU AND HONDA

Amigo • Passport • Rodeo • Rodeo Sport • Trooper • VehiCROSS

17

SPECIFICATIONS AND MAINTENANCE CHARTS

ENGINE AND VEHICLE IDENTIFICATION

	Engine								Model Year	
Code	Liters (cc)	Cu. In.	Cyl.	Fuel Sys.	Engine Type	Eng. Mfg.		Code ①		Year
D	2.2 (2198)	134	4	MFI	DOHC	Isuzu		1		2001
W	3.2 (3165)	193	6	MFI	DOHC	Isuzu		2		2002
X	3.5 (3494)	213	6	MFI	DOHC	Isuzu		3		2003

NA: Not available

MFI: Multi-port Fuel Injection

DOHC: Double Overhead Camshaft

① 10th position of VIN

67162-ISUZ-C01

GENERAL ENGINE SPECIFICATIONS

Year	Model	Engine Displacement Liter	Engine Series VIN	Net Horsepower @ rpm	Net Torque @ rpm (ft. lbs.)	Bore x Stroke (in.)	Compression Ratio	Oil Pressure @ rpm
2001	Rodeo	2.2	D	130@5200	144@4000	3.39x3.72	10.0:1	22@800
		3.2	W	205@5400	214@3000	3.68x3.03	9.1:1	60-80@3000
	Rodeo Sport	2.2	D	130@5200	144@4000	3.39x3.72	10.0:1	22@800
		3.2	W	205@5400	214@3000	3.68x3.03	9.1:1	60-80@3000
	Passport	3.2	W	205@5400	214@3000	3.68x3.03	9.1:1	57-80@3000
	Trooper	3.5	X	215@5400	230@3000	3.68x3.35	9.1:1	60-80@3000
	VehiCROSS	3.5	X	215@5400	230@3000	3.68x3.35	9.1:1	60-80@3000
2002	Rodeo	2.2	D	130@5200	144@4000	3.39x3.72	10.0:1	22@800
		3.2	W	205@5400	214@3000	3.68x3.03	9.1:1	60-80@3000
	Rodeo Sport	2.2	D	130@5200	144@4000	3.39x3.72	10.0:1	22@800
		3.2	W	205@5400	214@3000	3.68x3.03	9.1:1	60-80@3000
	Passport	3.2	W	205@5400	214@3000	3.68x3.03	9.1:1	57-80@3000
	Trooper	3.5	X	215@5400	230@3000	3.68x3.35	9.1:1	60-80@3000
2003	Rodeo	2.2	D	130@5200	144@4000	3.39x3.72	10.0:1	22@800
		3.2	W	205@5400	214@3000	3.68x3.03	9.1:1	60-80@3000
	Rodeo Sport	2.2	D	130@5200	144@4000	3.39x3.72	10.0:1	22@800
		3.2	W	205@5400	214@3000	3.68x3.03	9.1:1	60-80@3000

MFI: Multiport fuel injection

67162-ISUZ-C02

ENGINE TUNE-UP SPECIFICATIONS

Year	Engine Displacement Liters	Engine VIN	Spark Plug Gap (in.)	Ignition Timing (deg.) MT	Ignition Timing (deg.) AT	Fuel Pump (psi)	Idle Speed (rpm) MT	Idle Speed (rpm) AT	Valve Clearance In.	Valve Clearance Ex.
2001	2.2	D	0.040	①	①	41-55	800	800	HYD	HYD
	3.2	W	0.040	①	①	48-55	750	750	0.009-0.013	0.010-0.014
	3.5	X	0.040	①	①	48-55	750	750	0.009-0.013	0.010-0.014
2002	2.2	D	0.040	①	①	41-55	800	800	HYD	HYD
	3.2	W	0.040	①	①	48-55	750	750	0.009-0.013	0.010-0.014
	3.5	X	0.040	①	①	48-55	750	750	0.009-0.013	0.010-0.014
2003	2.2	D	0.040	①	①	41-55	800	800	HYD	HYD
	3.2	W	0.040	①	①	48-55	750	750	0.009-0.013	0.010-0.014

NOTE: The Vehicle Emission Control Information label figures must be used if they differ from those in this chart.

B: Before top dead center

HYD: Hydraulic

① Controlled by the PCM

67162-ISUZ-C03

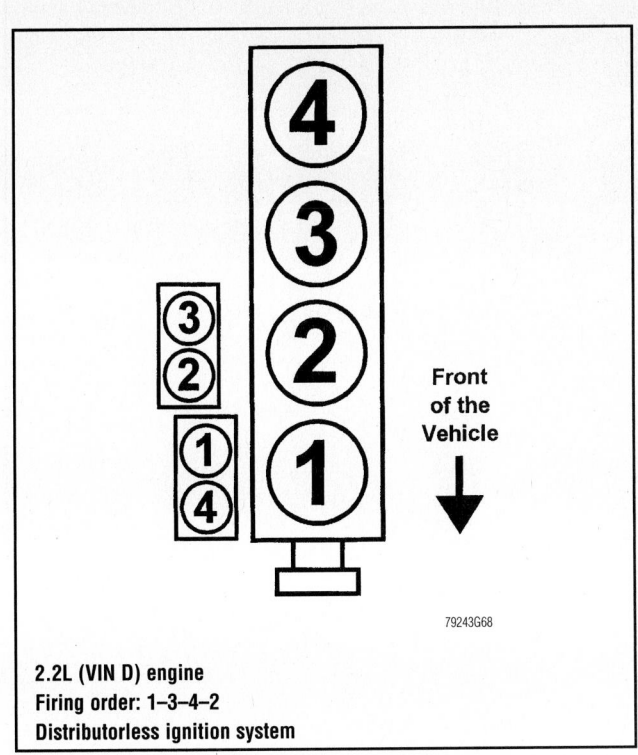

2.2L (VIN D) engine
Firing order: 1–3–4–2
Distributorless ignition system

79243G68

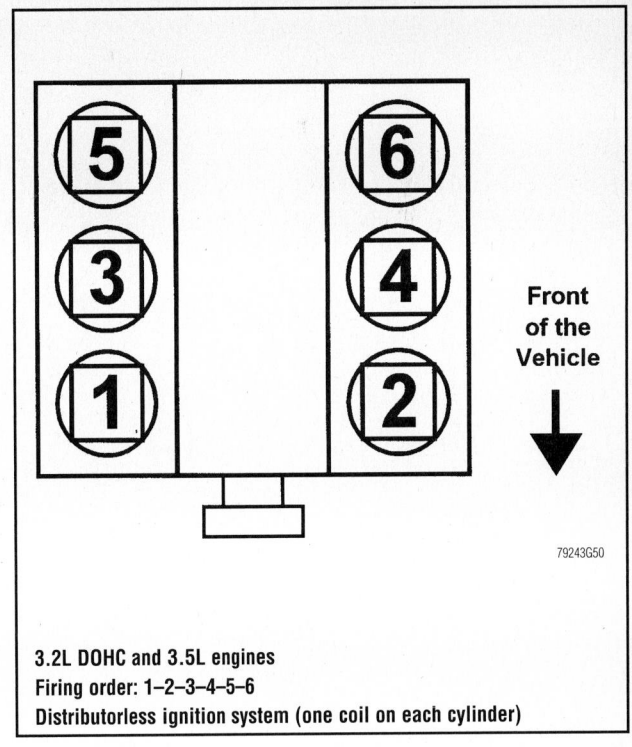

3.2L DOHC and 3.5L engines
Firing order: 1–2–3–4–5–6
Distributorless ignition system (one coil on each cylinder)

79243G50

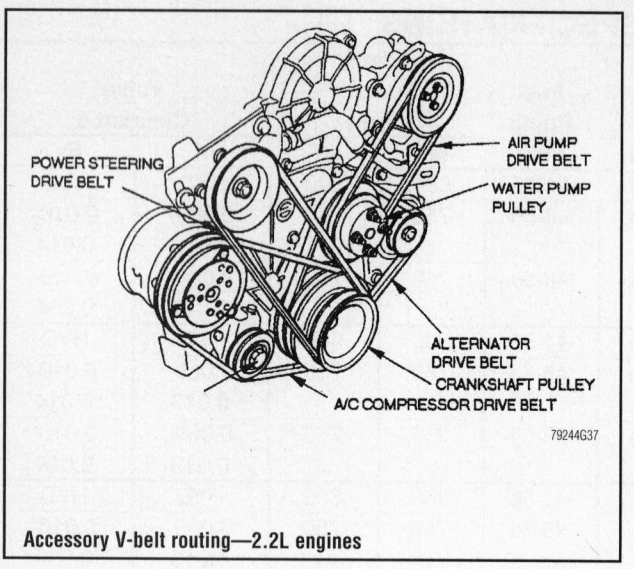

POWER STEERING DRIVE BELT

AIR PUMP DRIVE BELT

WATER PUMP PULLEY

ALTERNATOR DRIVE BELT

CRANKSHAFT PULLEY

A/C COMPRESSOR DRIVE BELT

79244G37

Accessory V-belt routing—2.2L engines

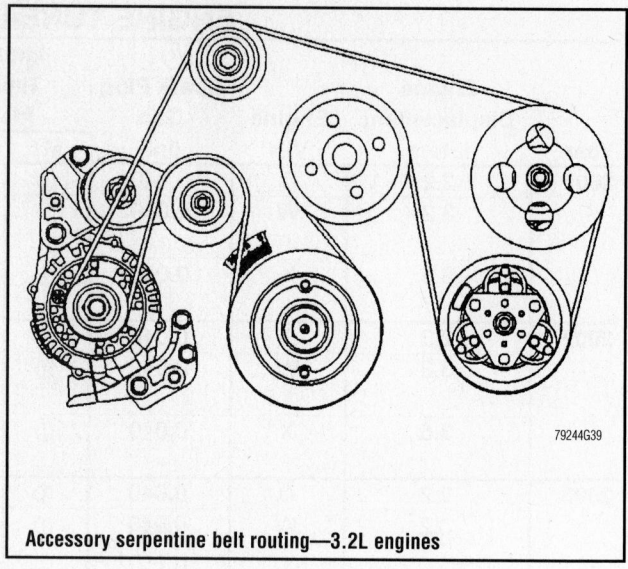

79244G39

Accessory serpentine belt routing—3.2L engines

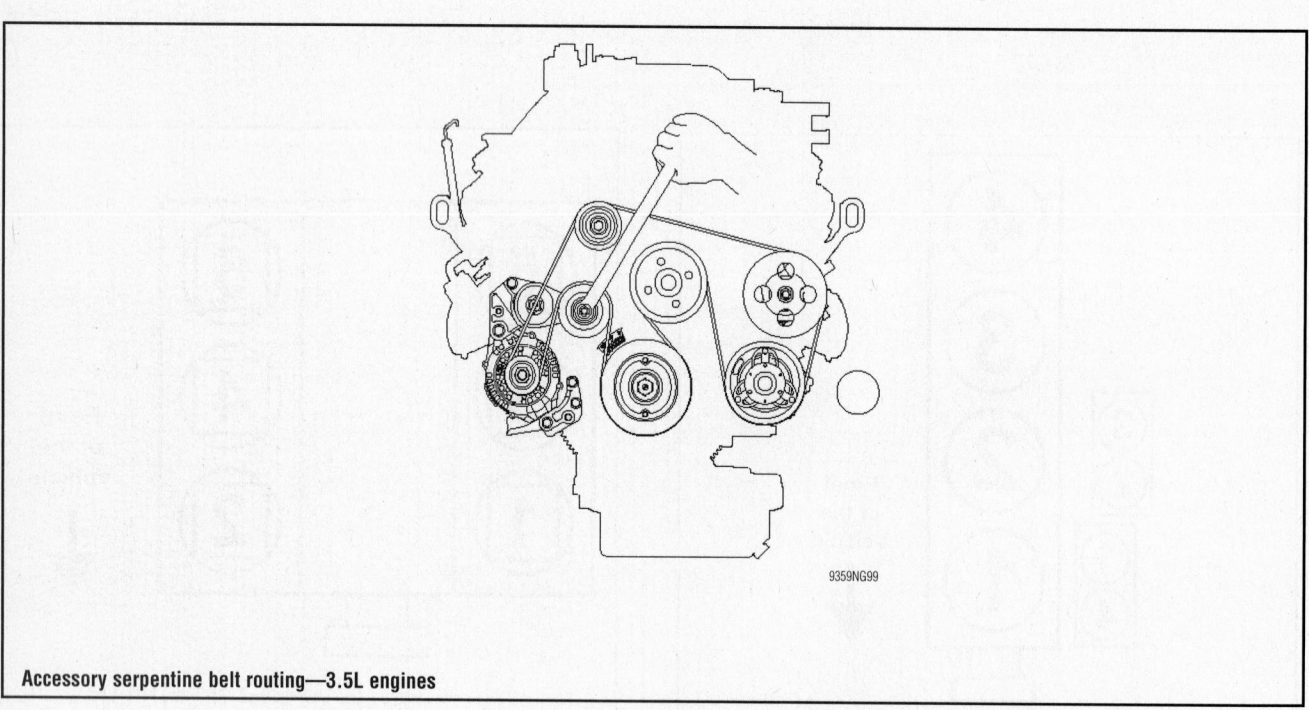

9359NG99

Accessory serpentine belt routing—3.5L engines

CAPACITIES

Year	Model	Engine Displacement Liters	Engine VIN	Oil with Filter (qts.)	Engine Transmission (pts.) Man.	Auto.	Transfer Case (pts.)	Drive Axle Front (pts.)	Rear (pts.)	Fuel Tank (gal.)	Cooling System (qts.)
2001	Rodeo Sport	2.2	D	4.8	6.2	—	3.0	2.2	3.7	17.7	7.3
		3.2	W	5.0	6.2	18.2	3.0	2.6	3.7	17.7	11.2
	Rodeo	2.2	D	4.8	4.5	—	3.0	2.2	3.7	20.0	7.3
		3.2	W	5.0	6.2	18.2	3.0	2.6	3.7	20.0	11.6
	Passport	3.2	W	5.0	6.2	18.2	3.0	2.6	3.74	19.5	11.6
	Trooper	3.5	X	5.0	5.8	18.2	3.0①	3.0	6.4	22.5	②
	VehiCROSS	3.5	X	5.0	—	18.2	4.0	3.0	4.6	22.5	7.4
2002	Rodeo Sport	2.2	D	4.8	6.2	—	3.0	2.2	3.7	17.7	7.3
		3.2	W	5.0	6.2	18.2	3.0	2.6	3.7	17.7	11.2
	Rodeo	2.2	D	4.8	6.2	—	3.0	2.2	3.7	19.5	7.3
		3.2	W	5.0	6.2	18.2	3.0	2.6	3.7	19.5	11.6
	Passport	3.2	W	5.0	6.2	18.2	3.0	2.6	3.74	19.5	11.6
	Trooper	3.5	X	5.0	5.8	18.2	3.0①	3.0	6.4	22.5	②
2003	Rodeo Sport	2.2	D	4.8	6.2	—	2.7	2.2	3.7	15.6	7.3
		3.2	W	5.0	6.2	18.2	2.7	2.6	3.7	15.6	11.2
	Rodeo	2.2	D	4.8	6.2	—	2.7	2.2	3.7	19.5	7.3
		3.2	W	5.0	6.2	18.2	2.7	2.6	3.7	19.5	11.2

NOTE: All capacities are approximate. Add fluid gradually and check to ensure a proper level has been reached.

① 4.0 pts. if equipped with Torque On Demand

② A/T: 7.4 qts.

M/T: 7.0 qts.

67162-ISUZ-C04

VALVE SPECIFICATIONS

Year	Engine Displacement Liters	Engine VIN	Seat Angle (deg.)	Face Angle (deg.)	Spring Test Pressure (lbs. @ in.)	Spring Installed Height (in.)	Stem-to-Guide Clearance (in.) Intake	Exhaust	Stem Diameter (in.) Intake	Exhaust
2001	2.2	D	NA	NA	NA	NA	0.0012-0.0022	0.0016-0.0026	NA	NA
	3.2	W	45	45	41-44@1.38	1.38	0.0002-0.0009	0.0012-0.0025	0.2346-0.2353	0.2343-0.2350
	3.5	X	45	45	41-44@1.38	1.38	0.0002-0.0009	0.0012-0.0025	0.2346-0.2353	0.2343-0.2350
2002	2.2	D	NA	NA	NA	NA	0.0012-0.0022	0.0016-0.0026	NA	NA
	3.2	W	45	45	41-44@1.38	1.38	0.0002-0.0009	0.0012-0.0025	0.2346-0.2353	0.2343-0.2350
	3.5	X	45	45	41-44@1.38	1.38	0.0002-0.0009	0.0012-0.0025	0.2346-0.2353	0.2343-0.2350
2003	2.2	D	NA	NA	NA	NA	0.0012-0.0022	0.0016-0.0026	NA	NA
	3.2	W	45	45	41-44@1.38	1.38	0.0002-0.0009	0.0012-0.0025	0.2346-0.2353	0.2343-0.2350

NA: Not Available

67162-ISUZ-C05

CRANKSHAFT AND CONNECTING ROD SPECIFICATIONS

All measurements are given in inches.

| Year | Engine Displacement Liters | Engine VIN | Crankshaft | | | | Connecting Rod | | |
			Main Brg. Journal Dia.	Main Brg. Oil Clearance	Shaft End-play	Thrust on No.	Journal Diameter	Oil Clearance	Side Clearance
2001	2.2	D	2.2590-2.2610	0.0007-0.0016	0.0004-0.0008	2	1.9090-1.9100	0.0002-0.0012	0.0050-0.0150
	3.2	W	2.5165-2.5170	0.0007-0.0017	0.0024-0.0094	3	2.1229-2.1235	0.0010-0.0023	0.0050-0.0150
	3.5	X	2.5165-2.5170	0.0007-0.0017	0.0024-0.0094	3	2.1229-2.1235	0.0010-0.0023	0.0050-0.0150
2002	2.2	D	2.2590-2.2610	0.0007-0.0016	0.0004-0.0008	2	1.9090-1.9100	0.0002-0.0012	0.0050-0.0150
	3.2	W	2.5165-2.5170	0.0007-0.0017	0.0024-0.0094	3	2.1229-2.1235	0.0010-0.0023	0.0050-0.0150
	3.5	X	2.5165-2.5170	0.0007-0.0017	0.0024-0.0094	3	2.1229-2.1235	0.0010-0.0023	0.0050-0.0150
2003	2.2	D	2.2590-2.2610	0.0007-0.0016	0.0004-0.0008	2	1.9090-1.9100	0.0002-0.0012	0.0050-0.0150
	3.2	W	2.5165-2.5170	0.0007-0.0017	0.0024-0.0094	3	2.1229-2.1235	0.0010-0.0023	0.0050-0.0150

67162-ISUZ-C06

PISTON AND RING SPECIFICATIONS

All measurements are given in inches.

| Year | Engine Displacement Liters | Engine VIN | Piston Clearance | Ring Gap | | | Ring Side Clearance | | |
				Top Compression	Bottom Compression	Oil Control	Top Compression	Bottom Compression	Oil Control
2001	2.2	D	NA	0.0118-0.0195	0.0118-0.0195	0.016-0.055	0.0008-0.0546	0.0008-0.0546	NA
	3.2	W	NA	0.0118-0.0157	0.0177-0.0236	0.0060-0.018	0.0006-0.002	0.0006-0.0015	NA
	3.5	X	NA	0.0118-0.0157	0.0177-0.0236	0.006-0.018	0.0006-0.0015	0.0006-0.0015	NA
2002	2.2	D	NA	0.0118-0.0195	0.0118-0.0195	0.016-0.055	0.0008-0.0546	0.0008-0.0546	NA
	3.2	W	NA	0.0118-0.0157	0.0177-0.0236	0.0060-0.018	0.0006-0.002	0.0006-0.0015	NA
	3.5	X	NA	0.0118-0.0157	0.0177-0.0236	0.006-0.018	0.0006-0.0015	0.0006-0.0015	NA
2003	2.2	D	NA	0.0118-0.0195	0.0118-0.0195	0.016-0.055	0.0008-0.0546	0.0008-0.0546	NA
	3.2	W	NA	0.0118-0.0157	0.0177-0.0236	0.0060-0.018	0.0006-0.002	0.0006-0.0015	NA

NA: Not Available

67162-ISUZ-C07

SCHEDULED MAINTENANCE INTERVALS
Isuzu—Amigo, Rodeo, Rodeo Sport, Trooper & vehiCROSS; Honda—Passport

TO BE SERVICED	TYPE OF SERVICE	VEHICLE MILEAGE INTERVAL (x1000)															
		7.5	15	22.5	30	37.5	45	52.5	60	67.5	75	82.5	90	97.5	105	112.5	120
Tires and wheels ③	S/I	✓	✓	✓	✓	✓	✓	✓	✓	✓	✓	✓	✓	✓	✓	✓	✓
Valve clearance ④	A								✓								✓

R: Replace   S/I: Service or Inspect   L: Lubricate   A: Adjust   C: Clean

① Perform this at the mileage indicated or every 6 months, whichever occurs first.

② Perform this at the mileage indicated or every 24 months, whichever occurs first.

③ Perform this at the mileage indicated or every 12 months, whichever occurs first.

④ 3.2L V6 engine.

⑤ 2.2L 4 cyl. engine.

FREQUENT OPERATION MAINTENANCE (SEVERE SERVICE)

 If a vehicle is operated under any of the following conditions it is considered severe service:

- Towing a trailer or using a camper or car-top carrier.

- Repeated short trips of less than 5 miles in temperatures below freezing.

- Extensive idling or low-speed driving for long distances as in heavy commercial use, such as delivery, taxi or police cars.

- Operating on rough, muddy or salt-covered roads.

- Operating on unpaved or dusty roads.

Air cleaner element: replace every 15,000 miles

Engine oil and filter: replace every 3000 miles or 3 months, whichever occurs first.

Automatic transmission fluid: replace every 20,000 miles.

Rear axle lubricant: replace every 15,000 miles.

67162-ISUZ-C13

PRECAUTIONS

Before servicing any vehicle, please be sure to read all of the following precautions, which deal with personal safety, prevention of component damage and important points to take into consideration when servicing a motor vehicle:

• Never open, service or drain the radiator or cooling system when the engine is hot; serious burns can occur from the steam and hot coolant.

• Observe all applicable safety precautions when working around fuel. Whenever servicing the fuel system, always work in a well-ventilated area. Do not allow fuel spray or vapors to come in contact with a spark, open flame, or excessive heat (a hot drop light, for example). Keep a dry chemical fire extinguisher near the work area. Always keep fuel in a container specifically designed for fuel storage; also, always properly seal fuel containers to avoid the possibility of fire or explosion. Refer to the additional fuel system precautions later in this section.

• Fuel injection systems often remain pressurized, even after the engine has been turned **OFF**. The fuel system pressure must be relieved before disconnecting any fuel lines. Failure to do so may result in fire and/or personal injury.

• Brake fluid often contains polyglycol ethers and polyglycols. Avoid contact with the eyes and wash your hands thoroughly after handling brake fluid. If you do get brake fluid in your eyes, flush your eyes with clean, running water for 15 minutes. If eye irritation persists, or if you have taken brake fluid internally, seek medical assistance IMMEDIATELY.

• The EPA warns that prolonged contact with used engine oil may cause a number of skin disorders, including cancer. You should make every effort to minimize your exposure to used engine oil. Protective gloves should be worn when changing oil. Wash your hands and any other exposed skin areas as soon as possible after exposure to used engine oil. Soap and water, or waterless hand cleaner should be used.

• All new vehicles are now equipped with an air bag system, often referred to as a Supplemental Restraint System (SRS) or Supplemental Inflatable Restraint (SIR) system. The system must be disabled before performing service on or around system components, steering column, instrument panel components, wiring and sensors. Failure to follow safety and disabling procedures could result in accidental air bag deployment, possible personal injury and unnecessary system repairs.

• Always wear safety goggles when working with, or around, the air bag system. When carrying a non-deployed air bag, be sure the bag and trim cover are pointed away from your body. When placing a non-deployed air bag on a work surface, always face the bag and trim cover upward, away from the surface. This will reduce the motion of the module if it is accidentally deployed. Refer to the additional air bag system precautions later in this section.

• Clean, high quality brake fluid from a sealed container is essential to the safe and proper operation of the brake system. You should always buy the correct type of brake fluid for your vehicle. If the brake fluid becomes contaminated, completely flush the system with new fluid. Never reuse any brake fluid. Any brake fluid that is removed from the system should be discarded. Also, do not allow any brake fluid to come in contact with a painted surface; it will damage the paint.

• Never operate the engine without the proper amount and type of engine oil; doing so WILL result in severe engine damage.

• Timing belt maintenance is extremely important. Many models utilize an interference-type, non-freewheeling engine. If the timing belt breaks, the valves in the cylinder head may strike the pistons, causing potentially serious (also time-consuming and expensive) engine damage. Refer to the maintenance interval charts in the front of this manual for the recommended replacement interval for the timing belt and to the timing belt section for belt replacement and inspection.

• Disconnecting the negative battery cable on some vehicles may interfere with the functions of the on-board computer system(s) and may require the computer to undergo a relearning process once the negative battery cable is reconnected.

• When servicing drum brakes, only disassemble and assemble one side at a time, leaving the remaining side intact for reference.

• Only an MVAC-trained, EPA-certified automotive technician should service the A/C system or its components.

ENGINE REPAIR

➡**Disconnecting the negative battery cable on some vehicles may interfere with the functions of the on board computer system. The computer may undergo a relearning process once the negative battery cable is reconnected.**

Distributor

REMOVAL

These engines are equipped with a Distributorless Ignition System (DIS).

Alternator

REMOVAL

2.2L Engine

1. Before servicing the vehicle, refer to the precautions in the beginning of this section.
2. Remove or disconnect the following:
   • Negative battery cable
   • Accessory drive belt
   • Alternator harness connectors
   • Alternator

3.2L Engine

1. Before servicing the vehicle, refer to the precautions in the beginning of this section.
2. Remove or disconnect the following:
   • Negative battery cable
   • Accessory drive belt
   • Alternator harness connectors
   • Alternator

3.5L Engine

1. Before servicing the vehicle, refer to the precautions in the beginning of this section.

2. Remove or disconnect the following:
- Negative battery cable
- Accessory drive belt
- Alternator wiring connectors
- Alternator

INSTALLATION

2.2L Engine

Install or connect the following:
- Alternator. Tighten the long bolt to 26 ft. lbs. (35 Nm) and the short bolt to 15 ft. lbs. (20 Nm).
  - Alternator harness connectors
  - Accessory drive belt
  - Negative battery cable

3.2L Engine

Install or connect the following:
- Alternator. Tighten the 10mm bolt to 30 ft. lbs. (41 Nm) and the 8mm bolt to 15 ft. lbs. 21 Nm).
  - Alternator harness connectors
  - Accessory drive belt
  - Negative battery cable

3.5L Engine

1. Before servicing the vehicle, refer to the precautions in the beginning of this section.
2. Install or connect the following:
   - Alternator. Tighten the 10mm bolts to 30 ft. lbs. (41 Nm) and the 8mm bolts to 15 ft. lbs. (21 Nm).
   - Alternator wiring connectors
   - Accessory drive belt
   - Negative battery cable

Ignition Timing

ADJUSTMENT

These engines are equipped with a Distributorless Ignition System (DIS). No adjustment is possible.

Engine Assembly

REMOVAL & INSTALLATION

2.2L engines

1. Before servicing the vehicle, refer to the precautions in the beginning of this section.
2. Drain the cooling system.
3. Relieve the fuel system pressure.
4. Remove or disconnect the following:
   - Battery

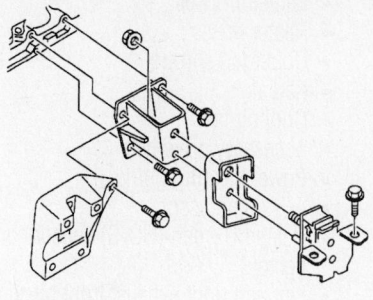

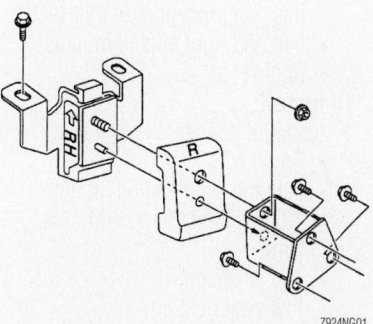

7924NG01

Left and right motor mounts—2.2L engine

- Hood
- Accessory drive belt
- Accelerator cable
- Air intake assembly
- Engine wiring harness connectors at left rear of the engine compartment
- Brake booster vacuum line
- Engine ground cables
- Clutch fluid line bracket and slave cylinder
- Fuel lines and bracket
- Exhaust front pipe
- Transmission
- A/C compressor
- Power steering pump
- Chassis harness connectors at right rear of the engine compartment
- Frame ground cable
- Radiator hoses
- Heater hoses
- Cooling fan connector
- Cooling fan and shroud
- Radiator
- Left and right engine mounts
5. Lift the engine from the vehicle.
To install:
6. Position the engine in the engine compartment.
7. Install or connect the following:
   - Left and right engine mounts. Tighten the fasteners to 30 ft. lbs. (41 Nm).
   - Radiator

- Cooling fan and shroud
- Cooling fan connector
- Heater hoses
- Radiator hoses
- Frame ground cable
- Chassis harness connectors at right rear of the engine compartment
- Power steering pump
- A/C compressor
- Transmission
- Exhaust front pipe
- Fuel lines and bracket
- Clutch fluid line bracket and slave cylinder
- Engine ground cables
- Brake booster vacuum line
- Engine wiring harness connectors at left rear of the engine compartment
- Air intake assembly
- Accelerator cable
- Accessory drive belts
- Hood
- Battery
8. Fill the cooling system.
9. Start the engine and check for leaks.

3.2L Engines

1. Before servicing the vehicle, refer to the precautions in the beginning of this section.
2. Drain the cooling system.
3. Relieve the fuel system pressure.
4. Remove or disconnect the following:
   - Battery
   - Hood
   - Accelerator cable
   - Cruise control cable
   - Air intake assembly
   - Canister vacuum hose
   - Brake booster vacuum hose
   - Engine wiring harness connectors
   - Front axle harness connector, if equipped
   - Transmission harness connector and bracket
   - Frame ground cable
   - Firewall ground cable
   - Starter harness connectors
   - Alternator harness connectors
   - Coolant overflow reservoir hose
   - Radiator hoses
   - Cooling fan and shroud
   - Accessory drive belt
   - Power steering pump
   - A/C compressor
   - Heated Oxygen (HO_2S) sensor connectors
   - Exhaust front pipes
   - Heater hoses
   - Fuel lines

- Transmission
- Left and right engine mounts

5. Lift the engine from the vehicle.

To install:

6. Lower the engine into the vehicle.

7. Install or connect the following:
- Left and right engine mounts. Tighten the bolts to 30 ft. lbs. (41 Nm) and the nuts to 37 ft. lbs. (50 Nm).
- Transmission
- Fuel lines
- Heater hoses
- Exhaust front pipes
- Heated Oxygen (HO2S) sensor connectors
- A/C compressor
- Power steering pump
- Accessory drive belt
- Cooling fan and shroud
- Radiator hoses
- Coolant overflow reservoir hose
- Alternator harness connectors
- Starter harness connectors
- Firewall ground cable
- Frame ground cable
- Transmission harness connector and bracket
- Front axle harness connector, if equipped
- Engine wiring harness connectors
- Brake booster vacuum hose
- Canister vacuum hose
- Air intake assembly
- Cruise control cable
- Accelerator cable
- Hood
- Battery

8. Fill the cooling system.

9. Start the engine and check for leaks.

3.5L Engine

1. Before servicing the vehicle, refer to the precautions in the beginning of this section.

2. Drain the cooling system.

3. Remove or disconnect the following:
- Battery
- Hood
- Air cleaner assembly
- Accelerator cable
- Cruise control cable
- Canister vacuum line
- Brake booster vacuum line
- Engine wiring harness connectors
- Transmission harness connectors and bracket
- Engine ground cable
- Starter harness connector
- Alternator harness connector
- Coolant reservoir tank hose

- Radiator hoses
- Heater hoses
- Upper fan shroud
- Radiator
- Cooling fan
- Accessory drive belt
- Power steering pump
- A/C compressor
- Heated Oxygen (HO2S) sensor connectors
- Left and right exhaust front pipes
- Fuel lines
- Flywheel dust cover
- Transmission. Refer to the transmission procedure in this section.
- Left and right engine mounts
- Engine

To install:

4. Install or connect the following:
- Engine
- Left and right engine mounts. Tighten the bolts to 30 ft. lbs. (41 Nm).
- Transmission
- Flywheel dust cover
- Fuel lines
- Left and right exhaust front pipes
- HO2S sensor connectors
- A/C compressor
- Power steering pump
- Accessory drive belt
- Cooling fan. Tighten the nuts to 16 ft. lbs. (22 Nm).
- Radiator
- Upper fan shroud
- Heater hoses
- Radiator hoses
- Coolant reservoir tank hose
- Alternator harness connector
- Starter harness connector

- Engine ground cable
- Transmission harness connectors and bracket
- Engine wiring harness connectors
- Brake booster vacuum line
- Canister vacuum line
- Cruise control cable
- Accelerator cable
- Air cleaner assembly
- Hood
- Battery

5. Fill the cooling system. Check all fluid levels and adjust as necessary.

6. Start the engine and check for leaks.

Water Pump

REMOVAL & INSTALLATION

2.2L Engine

1. Before servicing the vehicle, refer to the precautions in the beginning of this section.

2. Drain the cooling system.

3. Remove or disconnect the following:
- Negative battery cable
- Radiator hose
- Accessory drive belt
- Front cover
- Timing belt. Refer to the Timing Belt unit repair section.
- Water pump

To install:

4. Install a new O-ring and coat the water pump sealing surface with silicone grease.

5. Install or connect the following:
- Water pump. Tighten the bolts to 18 ft. lbs. (25 Nm).

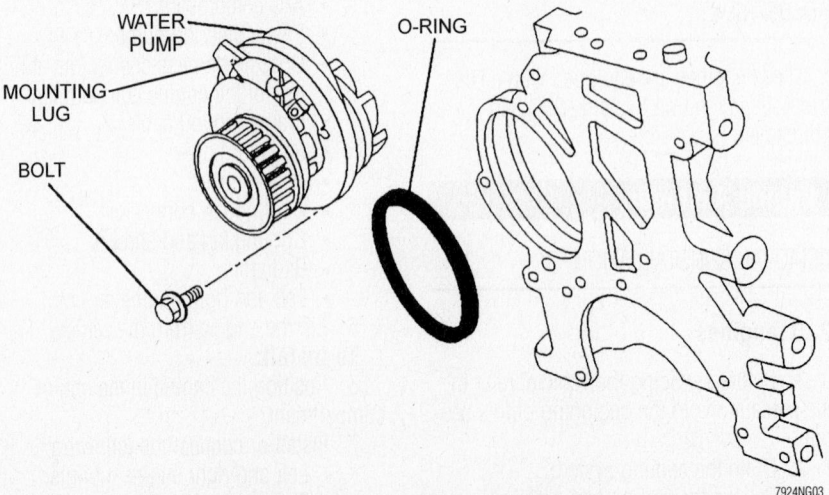

Exploded view of the water pump mounting, showing the location of the mounting lug—2.2L engine

7924NG03

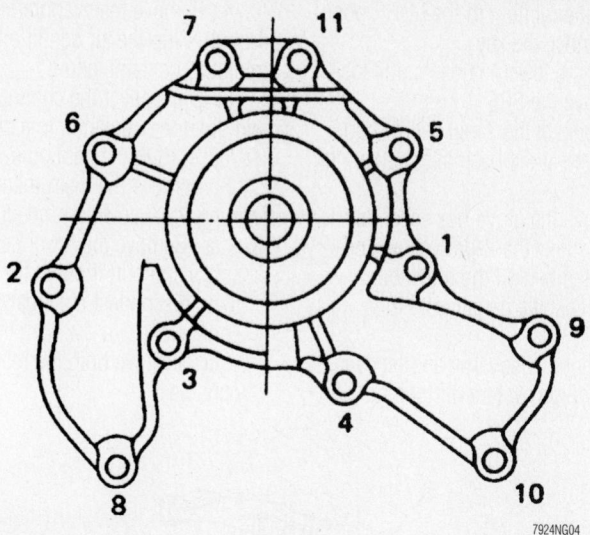

Water pump bolt tightening sequence—3.2L engines

7924NG04

- Timing belt
- Front cover
- Accessory drive belt
- Radiator hose
- Negative battery cable
6. Fill the cooling system.
7. Start the engine and check for leaks.

3.2L Engines

1. Before servicing the vehicle, refer to the precautions in the beginning of this section.
2. Drain the cooling system.
3. Remove or disconnect the following:
   - Negative battery cable
   - Radiator hose
   - Accessory drive belt
   - Front cover
   - Timing belt. Refer to the Timing Belt unit repair section.
   - Timing belt idler pulley
   - Water pump

To install:

4. Install or connect the following:
   - Water pump. Tighten the bolts in sequence to 18 ft. lbs. (25 Nm).
   - Timing belt idler pulley. Tighten the bolt to 38 ft. lbs. (52 Nm).
   - Timing belt
   - Front cover
   - Accessory drive belt
   - Radiator hose
   - Negative battery cable
5. Fill the cooling system.
6. Start the engine and check for leaks.

3.5L Engine

1. Before servicing the vehicle, refer to the precautions in the beginning of this section.

2. Drain the cooling system.
3. Remove or disconnect the following:
   - Negative battery cable
   - Upper radiator hose
   - Timing belt. Refer to the Timing Belt Unit Repair Section.
   - Idler pulley
   - Water pump

To install:

➡ Apply Loctite® 262 to bolt number 3 prior to installation.

4. Install or connect the following:
   - Water pump. Tighten the bolts in two passes, in sequence, to 13 ft. lbs. (18 Nm) for 3.2L engines or to 18 ft. lbs. (25 Nm) for 3.5L engines.
   - Idler pulley

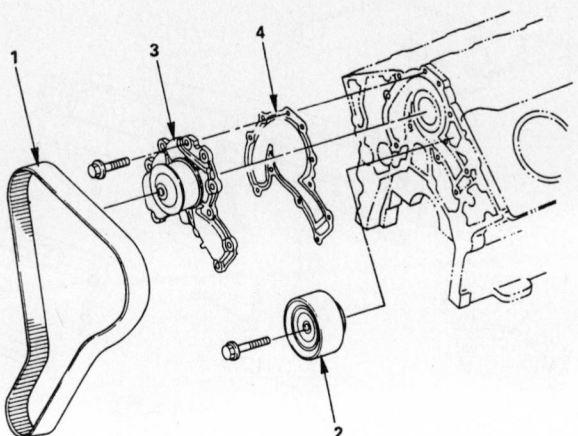

1. Timing belt
2. Idle pulley
3. Water pump assembly
4. Gasket

7924BG41

Exploded view of the water pump mounting

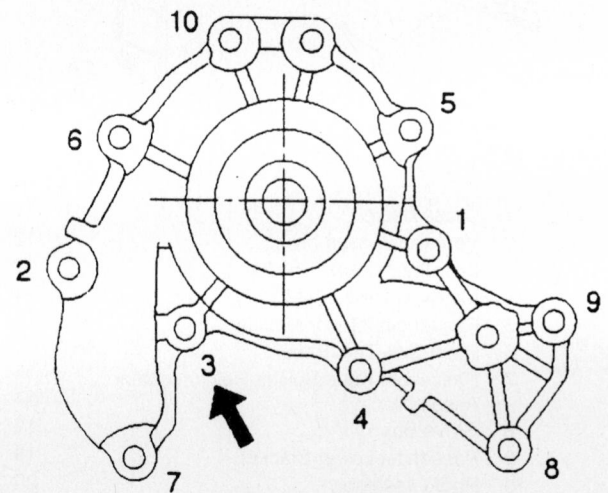

9302BG01

Water pump torque sequence. Apply LOCTITE® 262 to bolt number 3 (arrow)—3.5L engine

- Timing belt
- Upper radiator hose
- Negative battery cable
5. Fill the cooling system.
6. Start the engine and check for leaks.

Heater Core

REMOVAL & INSTALLATION

Amigo

1. If equipped with an air bag, perform the following procedure:

a. Turn the ignition to the LOCK position and remove the key.

b. From the lower left dash side fuse block, remove the SRS-1 fuse.

c. Disconnect the 2-pin yellow connector located at the base of the steering column.

d. Remove the glove box assembly.

e. Disconnect the 2-pin yellow connector located behind the glove box.

2. Disconnect the negative battery cable.

3. If equipped, discharge and recover the air conditioning system refrigerant.

4. Remove the evaporator lines at the firewall. Plug the air conditioning lines to minimize contamination.

5. Disconnect the cooling system hoses and drain the coolant into a clean container for reuse. Plug the cooling system hoses.

6. Remove the instrument panel by performing the following procedure:

a. Remove the lower center cover screw and pull it out at the clip positions; then, disconnect the cigarette lighter connector.

b. Remove both the rear and front console.

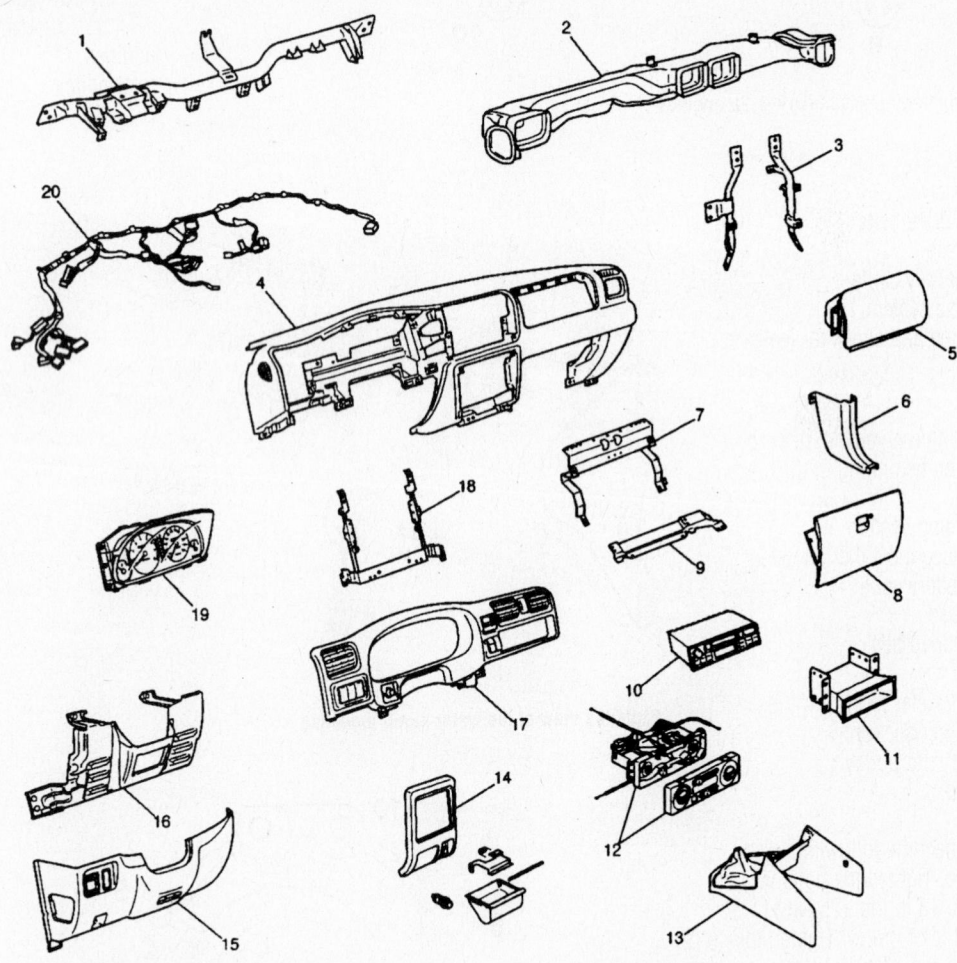

1	Cross Beam	11	Audio Sub Box
2	Vent Duct Assembly	12	Control Lever Assembly
3	Instrument Panel Bracket	13	Front Console Assembly
4	Instrument Panel Assembly	14	Lower Center Cover
5	Passenger Inflator Module	15	Instrument Panel Driver Lower Cover Assembly
6	Dash Side Trim Panel		
7	Passenger Knee Bolster Reinforcement Assembly	16	Driver Knee Bolster Assembly
8	Glove Box	17	Meter Cluster Assembly
9	Passenger Lower Bracket	18	Instrument Panel Center Reinforcement
10	Radio Assembly	19	Meter Assembly
		20	Instrument Harness Assembly

93113GB8

Exploded view of the instrument panel—Isuzu Amigo

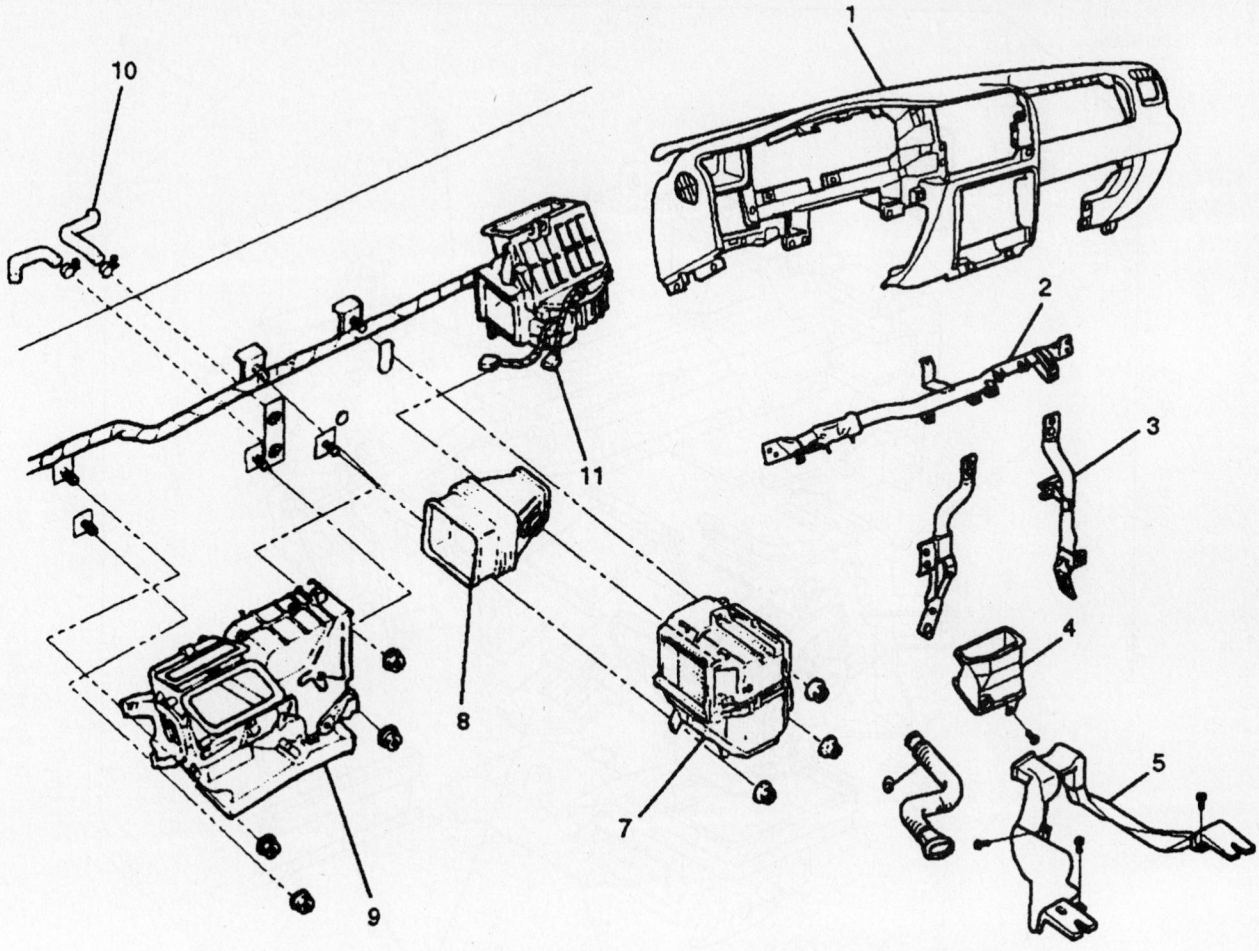

1	**Instrument Panel Assembly**	**6**	**Driver Lap Vent Duct**
2	**Cross Beam Assembly**	**7**	**Evaporator Assembly (A/C only)**
3	**Instrument Panel Bracket**	**8**	**Duct**
4	**Ventilation Lower Duct**	**9**	**Heater Unit Assembly**
5	**Rear Heater Duct**	**10**	**Heater Hose**
		11	**Resistor Connector**

93113GB9

View of the heater and air conditioning housing assemblies and related components—Isuzu Amigo

c. Remove the dash side trim panel sill plates and the panels.

d. Remove the 2 glove box screws and the glove box.

e. Remove the 2 hood release screws, the 6 instrument panel driver's lower cover assembly screws and the cover assembly.

f. Remove 5 instrument cluster screws, the 2 clips; then, disconnect the 8 switch connectors and remove the instrument cluster assembly.

g. Remove the 6 driver's knee bolster assembly bolts and screws and the knee bolster assembly.

h. Remove the 4 control lever assembly bolts; then, disconnect the 3 control cables (unit side) and the 3 harness connectors.

i. Remove the 4 radio/audio sub box assembly screws and the radio/audio sub box assembly.

j. Disconnect or remove the following instrument panel harness connectors or items:

- The 6 driver's side connectors
- The 3 passenger's side connectors
- The 2 center connectors
- Passenger's inflator module connector

- Radio antenna cable plug
- Ground cable bolt on the left dash side panel
- The 8 instrument panel-to-chassis bolts and the 3 nuts.

k. Remove the instrument panel assembly.

7. Remove the instrument panel bracket by performing the following procedure:

a. Remove the 2 passenger's inflator module bolts and 4 nuts.

b. Remove the 4 meter assembly screws; then, disconnect the meter wiring harness connectors and remove the meter assembly.

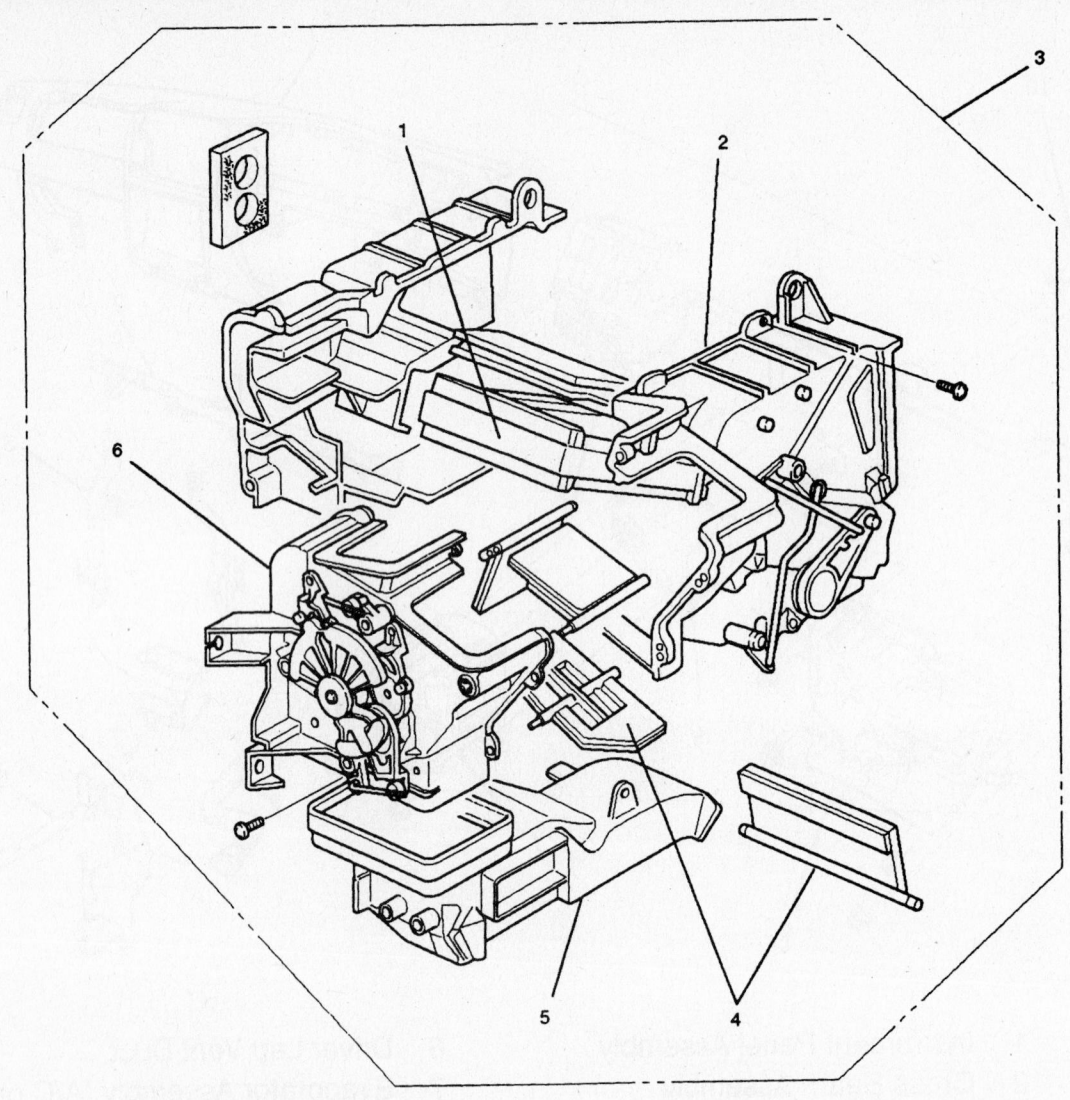

1	Heater Core	4	Mode Door
2	Case (Temperature Control)	5	Duct
3	Heater Unit	6	Case (Mode Control)

93113GB0

Exploded view of the heater housing assembly—Isuzu Amigo

c. Remove the 5 vent duct assembly screws and the assembly.

d. Remove the 3 lower passenger bracket screws and the bracket.

e. Remove the 9 passenger knee bolster reinforcement screws and the reinforcement.

f. Remove the 6 instrument panel center reinforcement screws and the reinforcement.

g. Remove the instrument panel wiring harness assembly clips and the wiring harness.

h. Remove the 2 instrument panel bracket nuts and 2 bolts for each bracket; then, remove the bracket(s).

8. Remove the 5 cross beam assembly nuts, 2 bolts and the 6 lower bolts; then, remove the crossbeam.

9. Disconnect the resistor wiring connector.

10. Remove the duct from the heater assembly.

11. If equipped with air conditioning, remove the evaporator assembly.

12. Remove the driver's lap vent.

13. Remove the lower ventilation duct.

14. Remove the footrest, the carpet, the 3 clips and the rear heater duct.

15. Remove the heater assembly.

16. Remove the mode control case-to-temperature control case screws and

remove the mode control case; do not remove the link unit.

17. Remove the temperature control case screws and separate the cases.

18. Remove the heater core from the case.

To install:

19. Install the heater core to the case.

20. Assemble the temperature control cases and install the case screws.

21. Install the mode control case and the mode control case-to-temperature control case screws.

22. Install the heater assembly.

23. Install the rear heater duct, the footrest, the carpet, and the 3 clips.

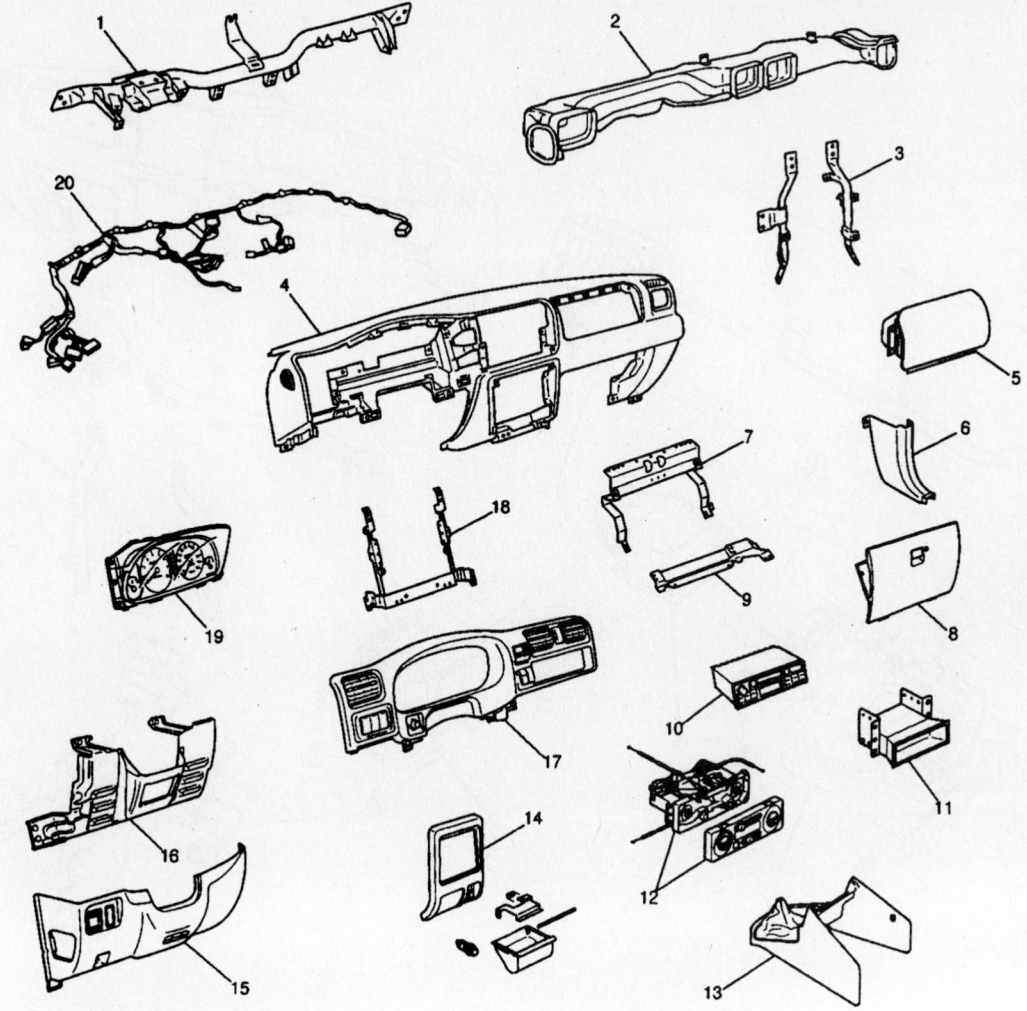

1	Cross Beam	11	Audio Sub Box
2	Vent Duct Assembly	12	Control Lever Assembly
3	Instrument Panel Bracket	13	Front Console Assembly
4	Instrument Panel Assembly	14	Lower Center Cover
5	Passenger Inflator Module	15	Instrument Panel Driver Lower Cover Assembly
6	Dash Side Trim Panel	16	Driver Knee Bolster Assembly
7	Passenger Knee Bolster Reinforcement Assembly	17	Meter Cluster Assembly
8	Glove Box	18	Instrument Panel Center Reinforcement
9	Passenger Lower Bracket	19	Meter Assembly
10	Radio Assembly	20	Instrument Harness Assembly

93113GB8

Exploded view of the instrument panel—Isuzu Rodeo

24. Install the lower ventilation duct.
25. Install the driver's lap vent.
26. If equipped with air conditioning, install the evaporator assembly.
27. Install the duct to the heater assembly.
28. Connect the resistor wiring connector.
29. Install the crossbeam, the 5 crossbeam assembly nuts, 2 bolts and the 6 lower bolts.
30. Install the instrument panel bracket by performing the following procedure:

a. Install the instrument panel bracket and the 2 nuts and 2 bolts for each bracket.
b. Install the instrument panel wiring harness assembly and the wiring harness clips.
c. Install the instrument panel center reinforcement and the 6 reinforcement screws.
d. Install the passenger knee bolster reinforcement and the 9 reinforcement screws.

e. Install the lower passenger bracket and the 3 bracket screws.
f. Install the vent duct assembly and the 5 vent duct assembly screws.
g. Install the meter assembly and the 4 meter assembly screws; then, connect the meter wiring harness connectors.
h. Install the 2 passenger's inflator module bolts and 4 nuts.
31. Install the instrument panel by performing the following procedure:
a. Install the instrument panel assembly.

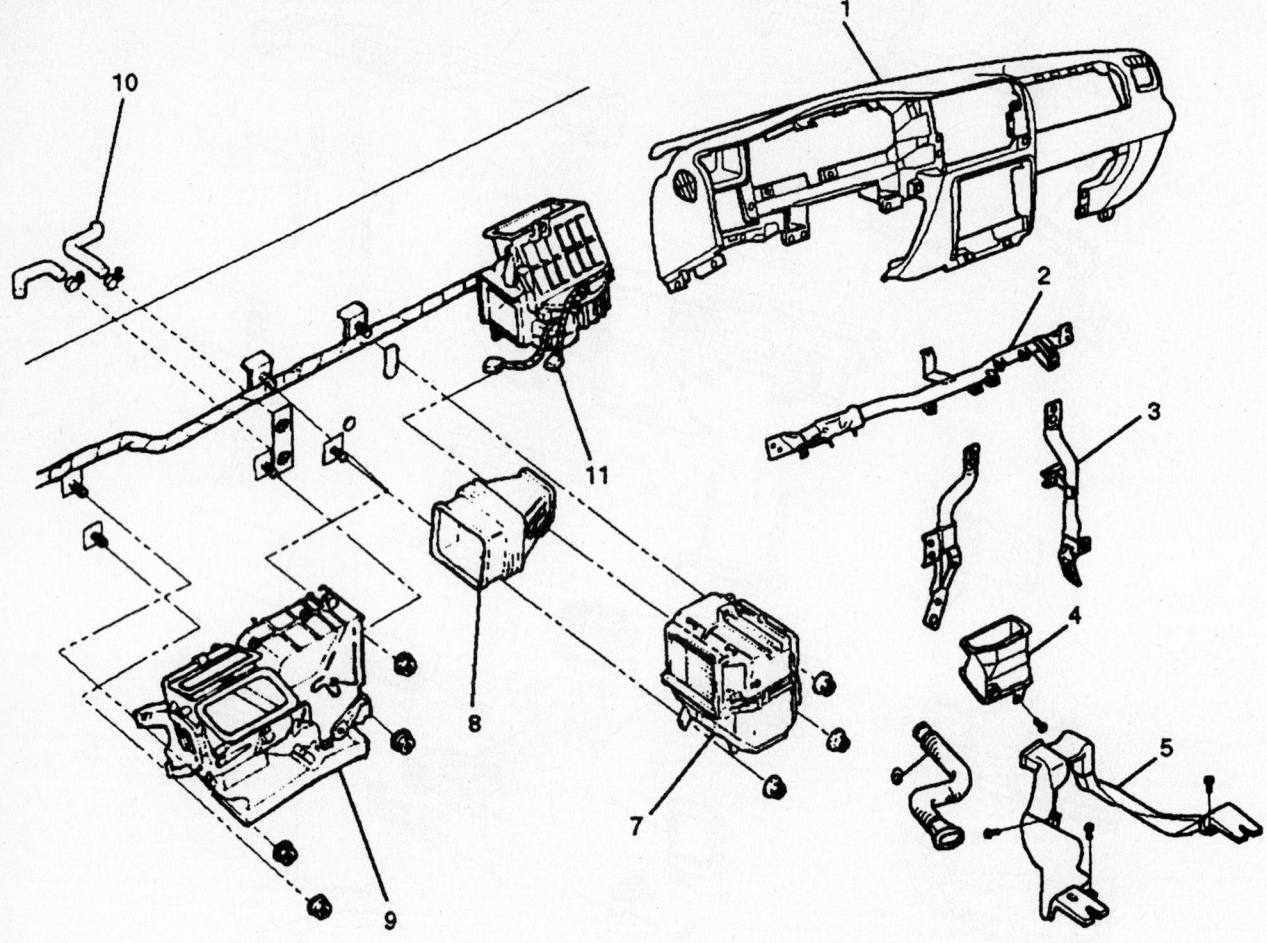

1	Instrument Panel Assembly	6	Driver Lap Vent Duct
2	Cross Beam Assembly	7	Evaporator Assembly (A/C only)
3	Instrument Panel Bracket	8	Duct
4	Ventilation Lower Duct	9	Heater Unit Assembly
5	Rear Heater Duct	10	Heater Hose
		11	Resistor Connector

93113GB9

View of the heater and air conditioning housing assemblies and related components—Isuzu Rodeo

b. Connect or install the following instrument panel harness connectors or items:
- The 6 driver's side connectors
- The 3 passenger's side connectors
- The 2 center connectors
- Passenger's inflator module connector
- Radio antenna cable plug
- Ground cable bolt on the left dash side panel
- The 8 instrument panel-to-chassis bolts and the 3 nuts.

c. Install the radio/audio sub box assembly and the 4 radio/audio sub box assembly screws.

d. Connect the 3 control cables (unit side) and the 3 harness connectors. Install the 4 control lever assembly bolts.

e. Install the knee bolster assembly and the 6 driver's knee bolster assembly bolts and screws.

f. Install the instrument cluster assembly. Connect the 8 switch connectors. Install 5 instrument cluster screws and the 2 clips.

g. Install the 2 hood release screws, the 6 instrument panel driver's lower cover assembly screws and the cover assembly.

h. Install the glove box and the 2 glove box screws.

i. Install the dash side trim panel sill plates and the panels.

j. Install both the rear and front console.

k. Connect the cigarette lighter connector and install the lower center cover screw.

32. Connect the cooling system hoses.

33. Refill the cooling system.

34. Install the evaporator lines at the firewall.

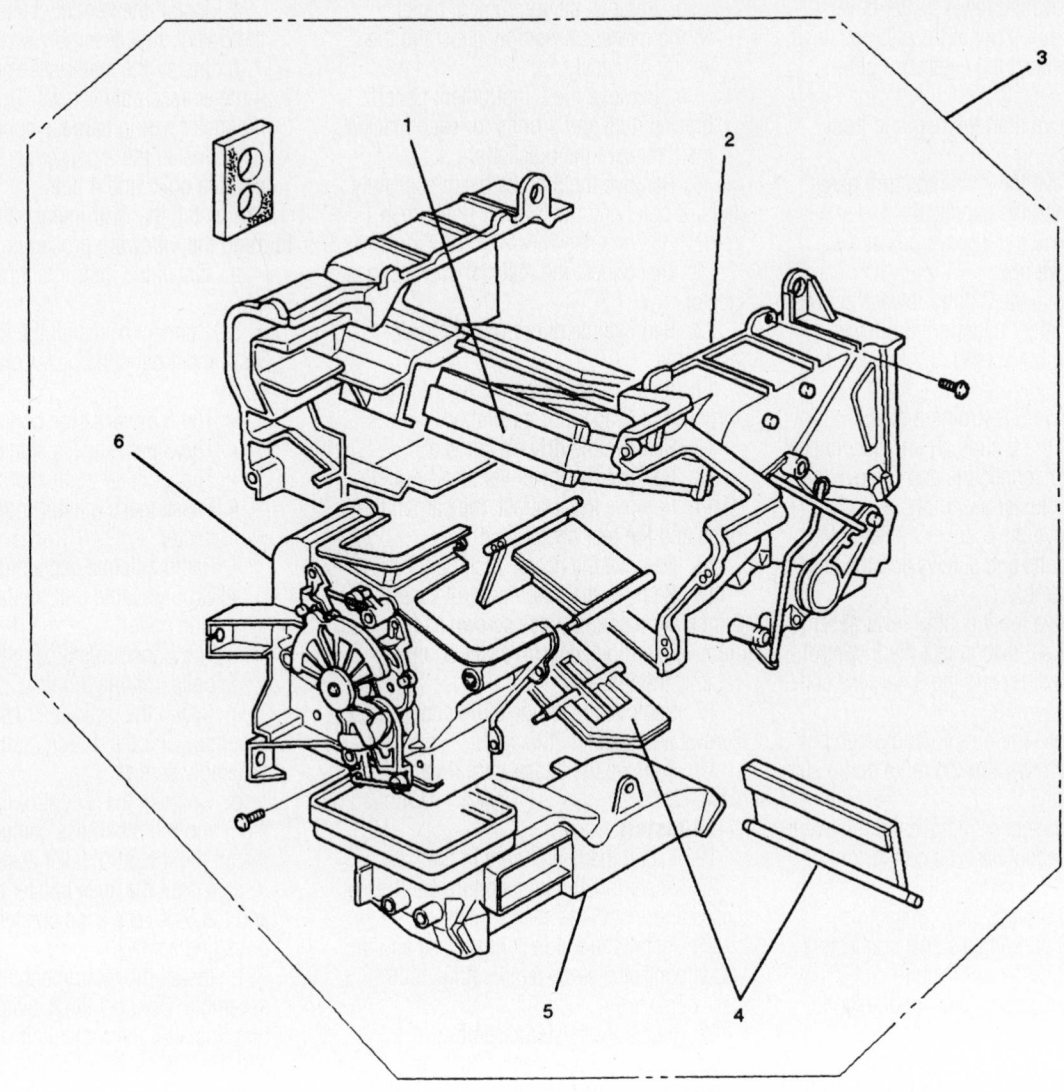

1	Heater Core	4	Mode Door
2	Case (Temperature Control)	5	Duct
3	Heater Unit	6	Case (Mode Control)

93113GB0

Exploded view of the heater housing assembly—Isuzu Rodeo

35. If equipped, evacuate and charge the air conditioning system.

36. Connect the negative battery cable.

37. If equipped with an air bag, perform the following procedure:

    a. Turn the ignition to the LOCK position and remove the key.

    b. Connect the 2-pin yellow connector located behind the glove box.

    c. Install the glove box assembly.

    d. Connect the 2-pin yellow connector located at the base of the steering column.

    e. At the lower left dash side fuse block, install the SRS-1 fuse.

    f. Turn the ignition switch to ON and verify that the AIR BAG warning light flashes 7 times and turns OFF.

38. Run the engine to normal operating temperatures; then, check the climate control operation and check for leaks.

Rodeo

1. If equipped with an air bag, perform the following procedure:

    a. Turn the ignition to the LOCK position and remove the key.

    b. From the lower left dash side fuse block, remove the SRS-1 fuse.

    c. Disconnect the 2-pin yellow connector located at the base of the steering column.

    d. Remove the glove box assembly.

    e. Disconnect the 2-pin yellow connector located behind the glove box.

2. Disconnect the negative battery cable.

3. If equipped, discharge and recover the air conditioning system refrigerant.

4. Remove the evaporator lines at the firewall. Plug the air conditioning lines to minimize contamination.

5. Disconnect the cooling system hoses and drain the coolant into a clean container for reuse. Plug the cooling system hoses.

6. Remove the instrument panel by performing the following procedure:

a. Remove the lower center cover screw and pull it out at the clip positions; then, disconnect the cigarette lighter connector.

b. Remove both the rear and front console.

c. Remove the dash side trim panel sill plates and the panels.

d. Remove the 2 glove box screws and the glove box.

e. Remove the 2 hood release screws, the 6 instrument panel driver's lower cover assembly screws and the cover assembly.

f. Remove 5 instrument cluster screws and the 2 clips. Then, disconnect the 8 switch connectors and remove the instrument cluster assembly.

g. Remove the 6 driver's knee bolster assembly bolts and screws and the knee bolster assembly.

h. Remove the 4 control lever assembly bolts; then, disconnect the 3 control cables (unit side) and the 3 harness connectors.

i. Remove the 4 radio/audio sub box assembly screws and the radio/audio sub box assembly.

j. Disconnect or remove the following instrument panel harness connectors or items:
- The 6 driver's side connectors
- The 3 passenger's side connectors
- The 2 center connectors
- Passenger's inflator module connector
- Radio antenna cable plug
- Ground cable bolt on the left dash side panel
- The 8 instrument panel-to-chassis bolts and the 3 nuts.

k. Remove the instrument panel assembly.

7. Remove the instrument panel bracket by performing the following procedure:

a. Remove the 2 passenger's inflator module bolts and 4 nuts.

b. Remove the 4 meter assembly screws. Then, disconnect the meter wiring harness connectors and remove the meter assembly.

c. Remove the 5 vent duct assembly screws and the assembly.

d. Remove the 3 lower passenger bracket screws and the bracket.

e. Remove the 9 passenger knee bolster reinforcement screws and the reinforcement.

f. Remove the 6 instrument panel center reinforcement screws and the reinforcement.

g. Remove the instrument panel wiring harness assembly clips and the wiring harness.

h. Remove the 2 instrument panel bracket nuts and 2 bolts for each bracket; then, remove the bracket(s).

8. Remove the 5 cross beam assembly nuts, 2 bolts and the 6 lower bolts; then, remove the crossbeam.

9. Disconnect the resistor wiring connector.

10. Remove the duct from the heater assembly.

11. If equipped with air conditioning, remove the evaporator assembly.

12. Remove the driver's lap vent.

13. Remove the lower ventilation duct.

14. Remove the footrest, the carpet, the 3 clips and the rear heater duct.

15. Remove the heater assembly.

16. Remove the mode control case-to-temperature control case screws and remove the mode control case; do not remove the link unit.

17. Remove the temperature control case screws and separate the cases.

18. Remove the heater core from the case.

To install:

19. Install the heater core to the case.

20. Assemble the temperature control cases and install the case screws.

21. Install the mode control case and the mode control case-to-temperature control case screws.

22. Install the heater assembly.

23. Install the rear heater duct, the footrest, the carpet, and the 3 clips.

24. Install the lower ventilation duct.

25. Install the driver's lap vent.

26. If equipped with air conditioning, install the evaporator assembly.

27. Install the duct to the heater assembly.

28. Connect the resistor wiring connector.

29. Install the crossbeam, the 5 cross-beam assembly nuts, 2 bolts and the 6 lower bolts.

30. Install the instrument panel bracket by performing the following procedure:

a. Install the instrument panel bracket and the 2 nuts and 2 bolts for each bracket.

b. Install the instrument panel wiring harness assembly and the wiring harness clips.

c. Install the instrument panel center reinforcement and the 6 reinforcement screws.

d. Install the passenger knee bolster reinforcement and the 9 reinforcement screws.

e. Install the lower passenger bracket and the 3 bracket screws.

f. Install the vent duct assembly and the 5 vent duct assembly screws.

g. Install the meter assembly and the 4 meter assembly screws. Then, connect the meter wiring harness connectors.

h. Install the 2 passenger's inflator module bolts and 4 nuts.

31. Install the instrument panel by performing the following procedure:

a. Install the instrument panel assembly.

b. Connect or install the following instrument panel harness connectors or items:
- The 6 driver's side connectors
- The 3 passenger's side connectors
- The 2 center connectors
- Passenger's inflator module connector
- Radio antenna cable plug
- Ground cable bolt on the left dash side panel
- The 8 instrument panel-to-chassis bolts and the 3 nuts.

c. Install the radio/audio sub box assembly and the 4 radio/audio sub box assembly screws.

d. Connect the 3 control cables (unit side) and the 3 harness connectors. Install the 4 control lever assembly bolts.

e. Install the knee bolster assembly and the 6 driver's knee bolster assembly bolts and screws.

f. Install the instrument cluster assembly. Connect the 8 switch connectors. Install 5 instrument cluster screws and the 2 clips.

g. Install the 2 hood release screws, the 6 instrument panel driver's lower cover assembly screws and the cover assembly.

h. Install the glove box and the 2 glove box screws.

i. Install the dash side trim panel sill plates and the panels.

j. Install both the rear and front console.

k. Connect the cigarette lighter connector and install the lower center cover screw.

32. Connect the cooling system hoses.

33. Refill the cooling system.

34. Install the evaporator lines at the firewall.

35. If equipped, evacuate and charge the air conditioning system.

36. Connect the negative battery cable.

37. If equipped with an air bag, perform the following procedure:

a. Turn the ignition to the LOCK position and remove the key.

b. Connect the 2-pin yellow connector located behind the glove box.

c. Install the glove box assembly.

d. Connect the 2-pin yellow connector located at the base of the steering column.

e. At the lower left dash side fuse block, install the SRS-1 fuse.

f. Turn the ignition switch to ON and verify that the AIR BAG warning light flashes 7 times and turns OFF.

38. Run the engine to normal operating temperatures; then, check the climate control operation and check for leaks.

Trooper

✳✳ CAUTION

The vehicle is equipped with a driver's side and a passenger's side air bag. Before starting service procedures on components, especially under the instrument panel and/or near the steering column, disable the air bag systems. There is sufficient voltage in the system to cause a

deployment for up to 15 seconds after the battery has been disconnected, the ignition turned OFF or fuse C-21 is removed from the fuse panel.

1. If equipped with an air bag, perform the following procedures:

a. Disconnect the negative battery cable, then disconnect the positive battery cable.

b. Disconnect the yellow 2-pin con-

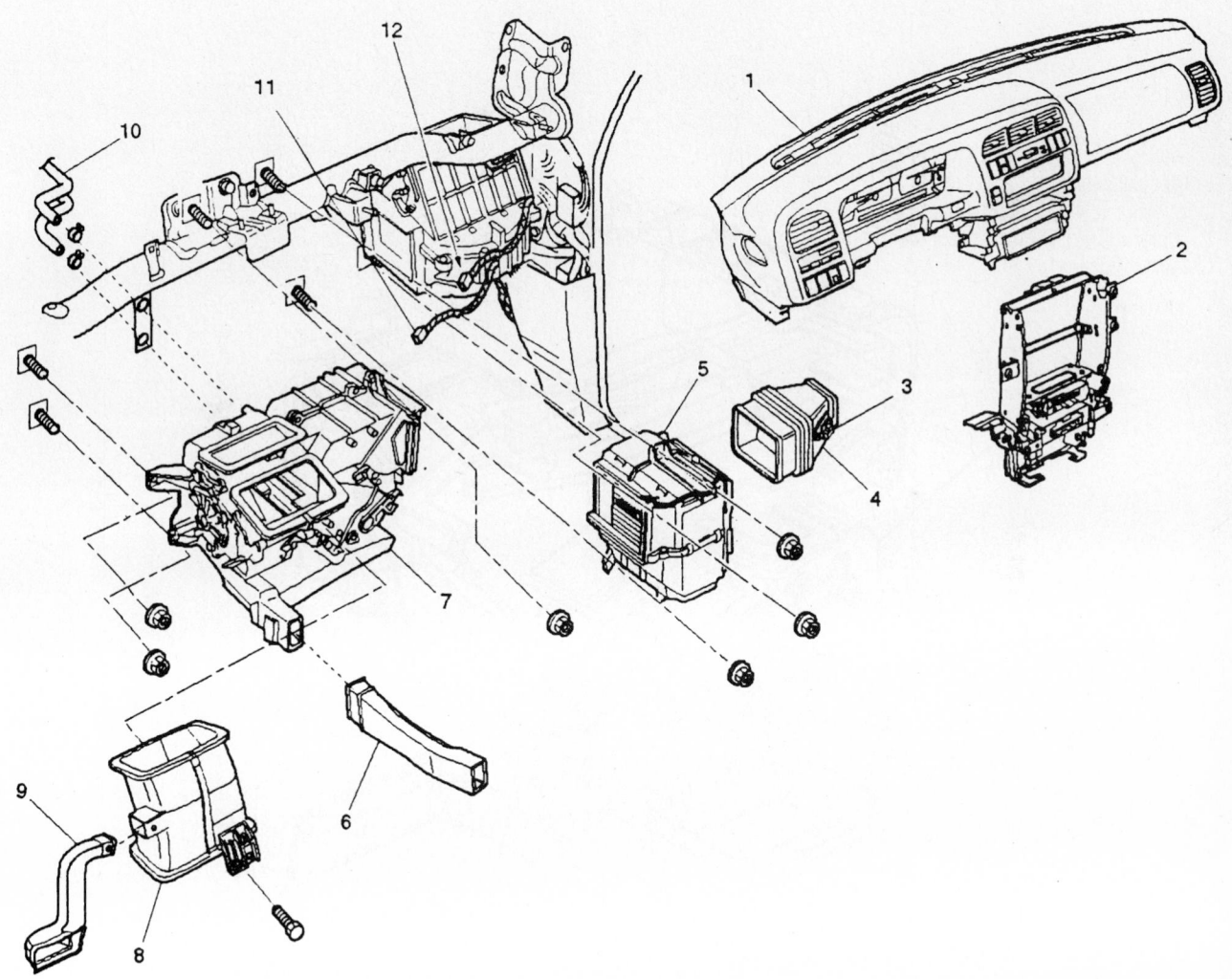

(1) Instrument Panel Assembly
(2) Instrument Panel Center Bracket
(3) Resistor
(4) Duct
(5) Evaporator Assembly (A/C only)
(6) Rear Heater Duct
(7) Heater Unit Assembly
(8) Center Ventilation Lower Duct
(9) Driver Lap Vent Nozzle
(10) Water Hose
(11) Electro Thermo Connector (With A/C)
(12) Resistor Connector

93113G01

Exploded view of the heater unit and related components—Isuzu Trooper

nector located at the base of the steering column.

c. Remove the glove box and disconnect the yellow 2-pin connector located behind the glove box.

2. Disconnect the negative battery cable.

3. Drain the cooling system.

4. If equipped with air conditioning, discharge and recover the refrigerant.

5. Remove the instrument panel assembly by performing the following procedure:

a. At the front console assembly, disconnect the switch connectors; then,

remove the console-to-chassis screws and the console.

b. At the lower cluster assembly, remove the cluster-to-instrument panel screws, disconnect the cigarette lighter and light connectors and remove the lower cluster.

c. Remove the glove box and the instrument panel lower cover and the passenger knee bolster reinforcement.

d. At the left side, remove the instrument panel lower cover and the knee bolster assembly.

e. At the top of the instrument panel,

pry the 8 claws on the front side toward you, raise the defroster grille and remove it.

f. At the SRS adjust bracket and cross beam under the passenger air bag module, remove the 2 fixing bolts and remove the instrument panel assembly.

g. Disconnect the air conditioning control cables from the unit.

h. Remove the instrument panel harness connectors (5 on the driver's side and 3 on the passenger's side), the passenger air bag module connector, the radio antenna plug and the center bracket ground cable bolt.

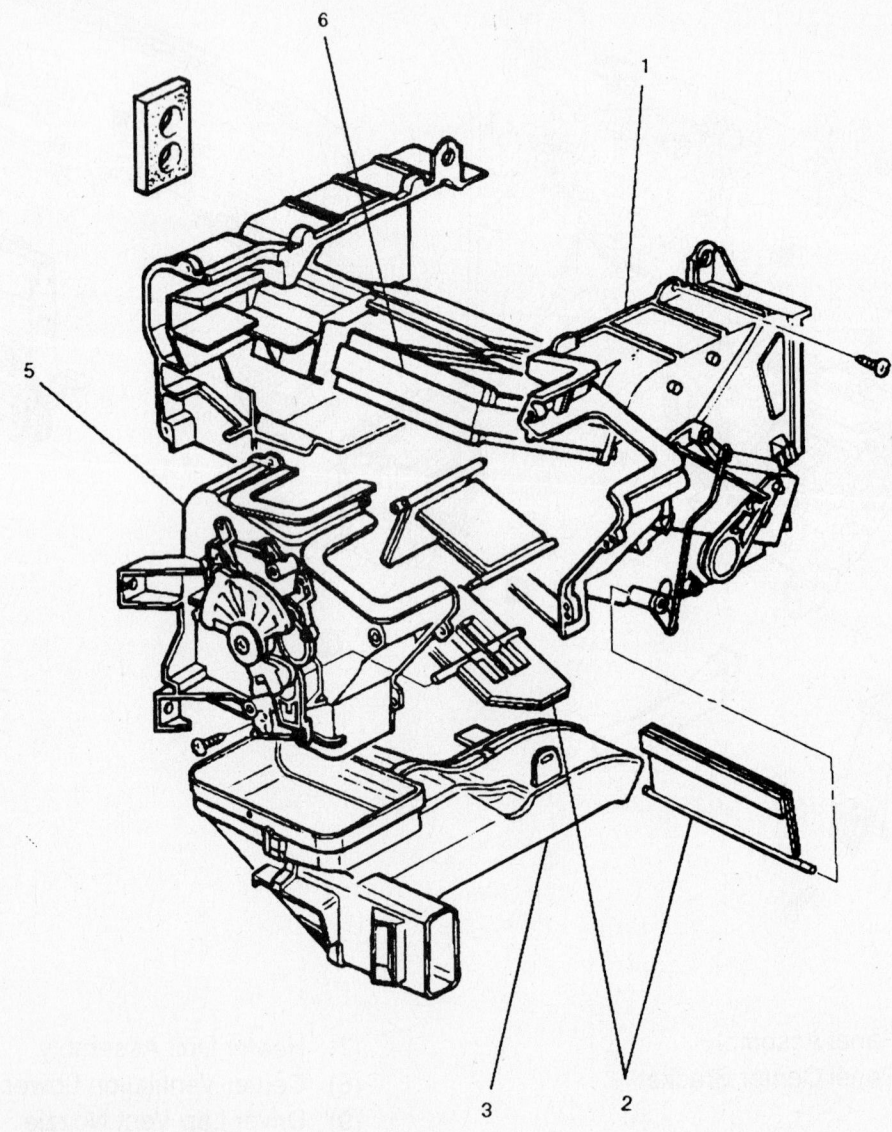

(1) Case (Temperature Control)
(2) Mode Door
(3) Duct
(5) Case (Mode Control)
(6) Heater Core

93113G02

Exploded view of the heater unit—Isuzu Trooper

i. Remove the passenger's air bag module nuts, disconnect the connectors and remove the module.

j. Remove the instrument panel cluster assembly screws, disconnect the switch connectors and the instrument panel assembly.

6. Disconnect the heater hoses from the heater unit.

7. Disconnect the heater resistor connector and the electro thermo connector (if equipped with air conditioning).

8. Remove the heater duct.

9. If equipped with air conditioning, remove the evaporator assembly by performing the following procedure:

a. Disconnect the drain hose.

b. Using a backup wrench, disconnect the refrigerant lines from the evaporator.

c. Plug or cap the refrigerant lines.

d. Remove the evaporator assembly.

10. Remove the instrument panel center bracket (crossbeam assembly) by performing the following procedure:

a. Remove the side support bracket bolts and brackets from both sides of the vehicle.

b. Remove the crossbeam center bracket nuts, disconnect the electrical connectors and the center bracket.

11. Remove the rear heater duct and heater assembly.

12. Disassemble the heater unit assembly by performing the following procedure:

a. Remove the lower air duct; do not remove the link unit.

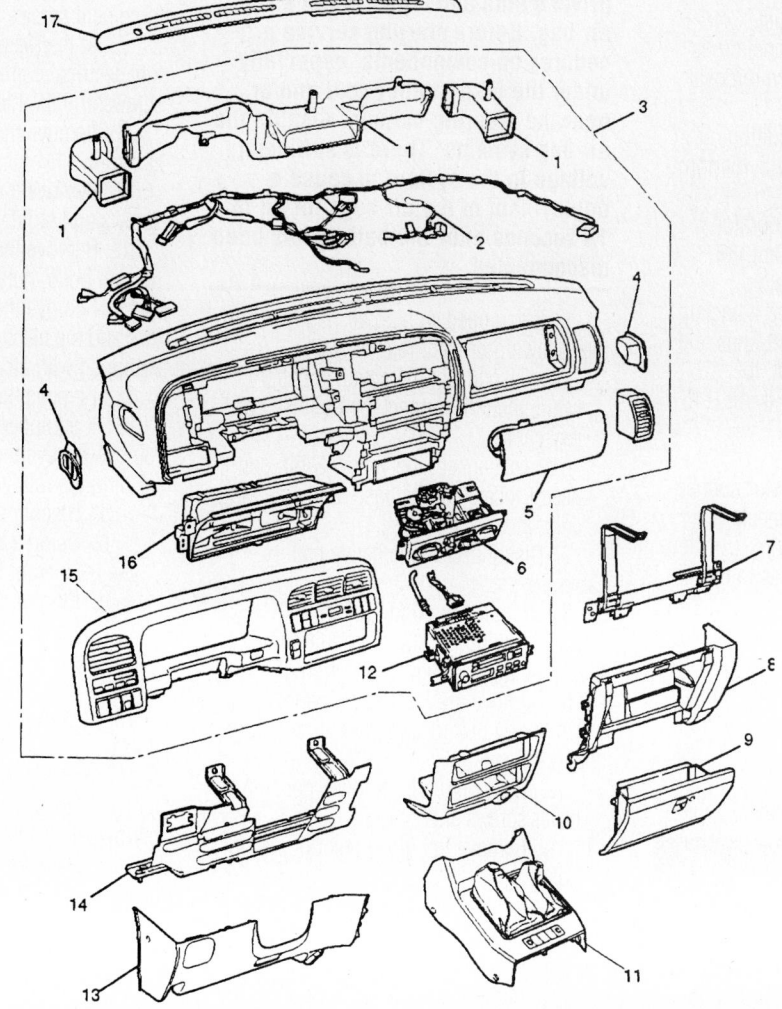

(1) Vent Duct Assembly	(9) Glove Box
(2) Instrument Harness Assembly	(10) Lower Cluster Assembly
(3) Instrument Panel Assembly	(11) Front Console Assembly
(4) Side Defroster Grille	(12) Radio Assembly
(5) Passenger Inflator Module	(13) Instrument Panel Driver Lower Cover Assembly
(6) Control Lever Assembly	(14) Driver Knee Bolster Assembly
(7) Passenger Knee Bolster Reinforcement Assembly	(15) Instrument Panel Cluster Assembly
(8) Instrument Panel Passenger Lower Cover Assembly	(16) Meter Assembly
	(17) Front Defroster Grille

93113G03

Exploded view of the instrument panel and accessories—Isuzu Trooper

b. Remove the temperature control case screws and lift the case from the heater unit.

c. Remove the heater core.

To install:

13. Assemble the heater unit assembly by performing the following procedure:

a. Install the heater core into the heater unit.

b. Install the temperature control case onto the unit and secure with screws.

c. Install the lower air duct.

14. Install the heater unit assembly into the vehicle.

15. Install the rear heater duct.

16. Install the instrument panel cross beam assembly by reversing the removal procedures.

17. If equipped with air conditioning, install the evaporator assembly by performing the following procedures:

a. If installing a new evaporator assembly, add 1.7 fl. oz. (50mL) of refrigerant oil to the evaporator.

b. Using new O-rings and a backup wrench, install the refrigerant lines and torque the outlet line to 18 ft. lbs. (25 Nm) and the inlet line to 11 ft. lbs. (15 Nm).

18. Install the heater duct.

19. Connect the heater resistor connector and the electro-thermo connector (if equipped with air conditioning).

20. Connect the heater hoses to the heater unit.

21. Install the instrument panel assembly by reversing the removal procedures.

22. If equipped with air conditioning, evacuate and charge and leak-test the system.

23. Refill the cooling system.

24. Connect the negative battery cable.

✲✲ CAUTION

Never use an air bag assembly from another vehicle and/or different model year. Starting in 1999, the air bag assemblies are equipped with identification colors on the bar code label as follows: YELLOW for the driver's air bag assembly, WHITE for the passenger's air bag assembly.

25. Enable the air bags by performing the following procedure:

a. Connect the passenger's side air bag yellow 2-pin connector.

b. Install the glove box.

c. At the base of the steering column, connect the yellow 2-pin connector.

d. Install the air bag fuse C-21 (if

removed) or connect the negative battery cable.

e. Turn the ignition switch ON and verify that the AIR BAG warning light flashes 7 times and then turns OFF.

26. Run the engine to normal operating temperatures and check for leaks. Check the systems for correct operation.

VehiCROSS

✲✲ CAUTION

The vehicle is equipped with a driver's side and a passenger's side air bag. Before starting service procedures on components, especially under the instrument panel and/or near the steering column, disable the air bag systems. There is sufficient voltage in the system to cause a deployment of the air bags for up to 15 seconds after the battery has been disconnected.

1. If equipped with an air bag, perform the following procedures:

a. Disconnect the negative battery cable, then disconnect the positive battery cable.

b. Disconnect the yellow 3-pin connector located at the base of the steering column.

2. Disconnect the negative battery cable.

3. Drain the cooling system.

4. If equipped with air conditioning, discharge and recover the refrigerant.

5. Remove the instrument panel assembly by performing the following procedure:

a. Remove the front lower console cover screws and cover.

b. Remove the glove box door and box.

c. At the passenger's side, remove the instrument panel lower cover screws and panel.

d. At the driver's side, disconnect the accelerator cable from the pedal and remove the instrument panel lower cover screws and panel.

e. Remove the lower cluster.

f. At the meter cluster assembly, disconnect the switch connectors, then remove the screws, clips and the meter assembly.

g. At the driver's side, disconnect the data link connector, then remove the bolts and the knee bolster.

h. At the lower cluster cover, remove the cover-to-instrument panel screws,

disconnect the cigarette lighter and remove the lower cover.

i. Disconnect the air conditioning control cables from the unit.

j. Remove the instrument panel cluster assembly screws, disconnect the switch connectors and the instrument panel assembly.

k. Remove the instrument harness connectors, the radio antenna plug.

l. Remove the side defroster grille. Remove the instrument panel nuts, bolts and screws and the instrument panel.

6. Remove the passenger's air bag reinforcement screws and the reinforcement.

7. At the meter assembly, disconnect the electrical connector, then remove the screws and the meter assembly.

8. Remove the radio and the vent duct assembly.

9. Remove the passenger's knee bolster screws and bolster.

10. Remove the instrument panel center bracket bolts, nuts and bracket.

11. Remove the heater resistor connectors and the electro thermo connector (if equipped with air conditioning).

12. Remove the blower motor assembly.

13. If equipped with air conditioning, remove the evaporator assembly by performing the following procedure:

a. Disconnect the drain hose.

b. Using a backup wrench, disconnect the refrigerant lines from the evaporator.

c. Plug or cap the refrigerant lines.

d. Remove the evaporator assembly.

14. Remove the driver's side lap vent duct.

15. Remove the center and lower vent ducts.

16. Remove the heater assembly.

17. Disassemble the heater unit assembly by performing the following procedure:

a. Remove the lower air duct; do not remove the link unit.

b. Remove the temperature control case screws and lift the case from the heater unit.

c. Remove the heater core.

To install:

18. Assemble the heater unit assembly by performing the following procedure:

a. Install the heater core into the heater unit.

b. Install the temperature control case onto the unit and secure with screws.

c. Install the lower air duct.

19. Install the heater unit assembly into the vehicle.

20. Install the center and lower vent ducts.

21. Install the driver's side lap vent duct.

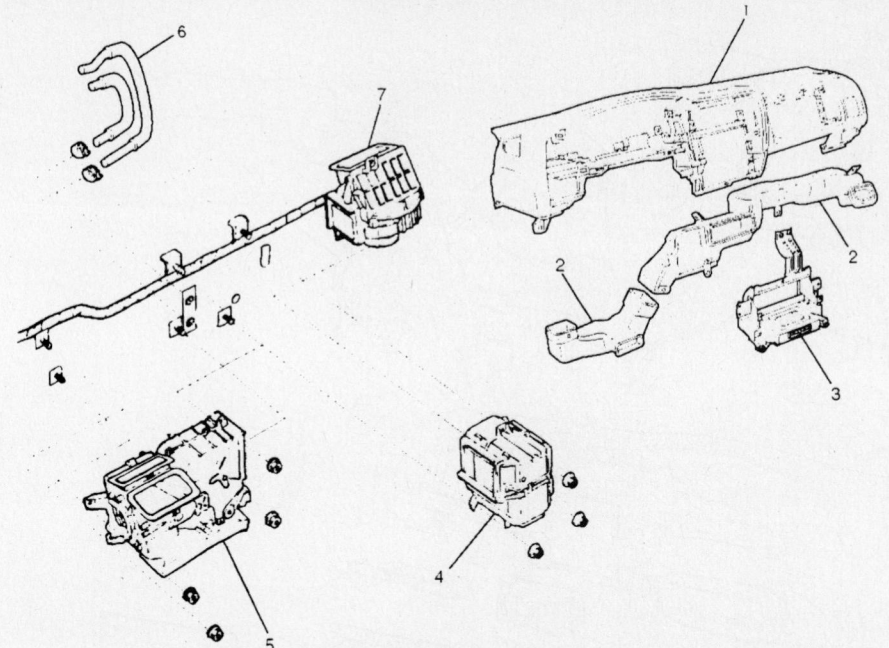

1 Instrument Panel Assembly

2 Center & Lower Vent Duct

3 Instrument Panel Center Bracket

4 Evaporator Assembly

5 Heater Unit Assembly

6 Heater Hose

7 Blower Unit

93113G04

Exploded view of the heater unit and related components—Isuzu VehiCROSS

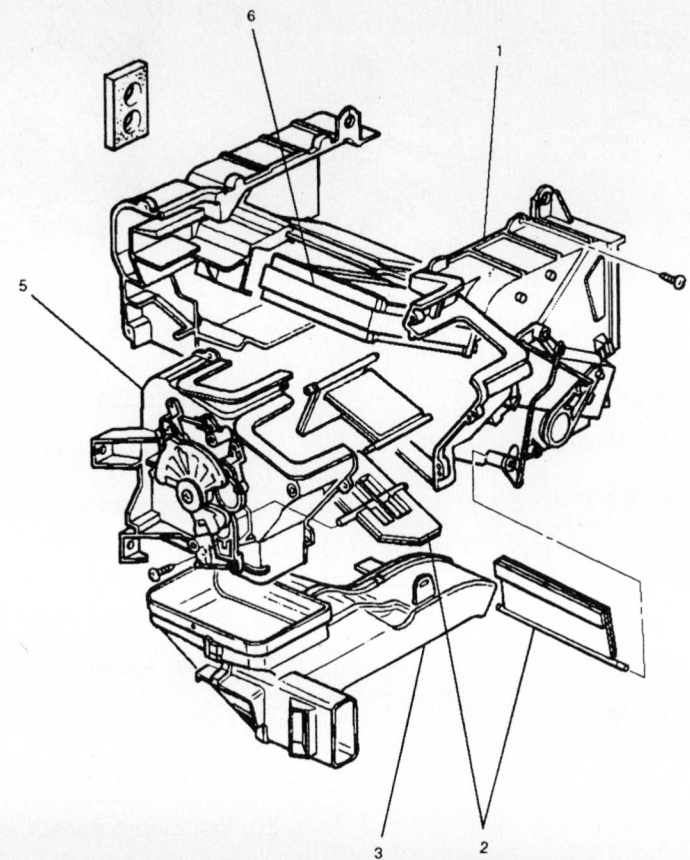

(1) Case (Temperature Control)

(2) Mode Door

(3) Duct

(5) Case (Mode Control)

(6) Heater Core

93113G02

Exploded view of the heater unit—Isuzu VehiCROSS

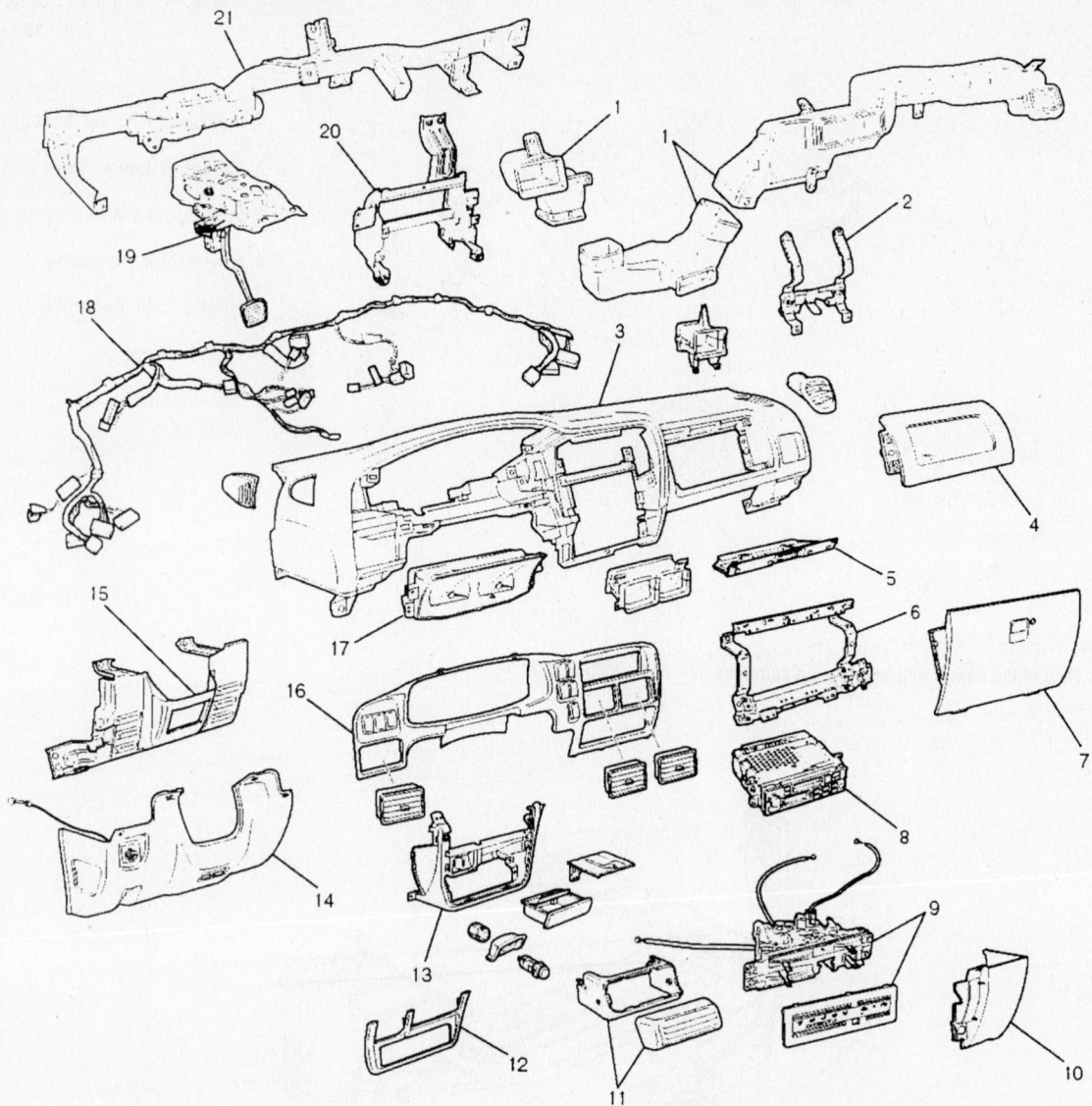

1. Vent Duct Assembly

2. Passenger Air Bag Reinforcement

3. Instrument Panel Assembly

4. Passenger Air Bag Assembly

5. Glove Box Cover

6. Passenger Knee Bolster

7. Glove Box Assembly

8. Radio Assembly

9. Air Conditioner Control Lever Assembly

10. Instrument Panel Passenger Lower Cover

11. Front Lower Console Cover

12. Lower Cluster

13. Instrument Panel Lower Center Cover

14. Instrument Panel Driver Lower Cover

15. Driver Knee Bolster

16. Meter Cluster Assembly

17. Meter Assembly

18. Instrument Harness Assembly

19. Brake Pedal & Bracket Assembly

20. Instrument Panel Center Bracket

21. Cross Beam Assembly

93113G05

Exploded view of the instrument panel and accessories—Isuzu VehiCROSS

22. Install the instrument panel cross beam assembly by reversing the removal procedures.

23. If equipped with air conditioning, install the evaporator assembly by performing the following procedures:

a. If installing a new evaporator assembly, add 1.7 fl. oz. (50mL) of refrigerant oil to the evaporator.

b. Using new O-rings and a backup wrench, install the refrigerant lines and torque the outlet line to 18 ft. lbs. (25 Nm) and the inlet line to 11 ft. lbs. (15 Nm).

24. Install the blower motor.

25. Connect the heater resistor connectors and the electro-thermo connector (if equipped with air conditioning).

26. Connect the heater hoses to the heater unit.

27. Install the instrument panel assembly by reversing the removal procedures.

28. If equipped with air conditioning, evacuate and recharge the system.

29. Refill the cooling system.

30. Connect the negative battery cable.

✳✳ CAUTION

Never use an air bag assembly from another vehicle and/or different model year.

31. Enable the air bag by performing the following procedure:

a. At the base of the steering column, connect the yellow 3-pin connector.

b. Connect the negative battery cable.

c. Turn the ignition switch ON and verify that the AIR BAG warning light flashes 7 times and then turns OFF.

32. Run the engine to normal operating temperatures and check for leaks. Check the systems for correct operation.

Passport

1. If equipped with an air bag, perform the following procedure:

a. Turn the ignition to the LOCK position and remove the key.

b. From the lower left dash side fuse block, remove the SRS-1 fuse.

c. Disconnect the 2-pin yellow connector located at the base of the steering column.

d. Remove the glove box assembly.

e. Disconnect the 2-pin yellow connector located behind the glove box.

2. Disconnect the negative battery cable.

3. If equipped, discharge and recover the air conditioning system refrigerant.

4. Remove the evaporator lines at the firewall. Plug the air conditioning lines to minimize contamination.

5. Disconnect the cooling system hoses and drain the coolant into a clean container for reuse. Plug the cooling system hoses.

6. Remove the instrument panel by performing the following procedure:

a. Remove the lower center cover screw and pull it out at the clip positions; then, disconnect the cigarette lighter connector.

b. Remove both the rear and front console.

c. Remove the dash side trim panel sill plates and the panels.

d. Remove the 2 glove box screws and the glove box.

e. Remove the 2 hood release screws, the 6 instrument panel driver's lower cover assembly screws and the cover assembly.

f. Remove the 5 instrument cluster screws and the 2 clips. Disconnect the 8 switch connectors and remove the instrument cluster assembly.

g. Remove the 6 driver's knee bolster assembly bolts and screws and the knee bolster assembly.

h. Remove the 4 control lever assembly bolts; then, disconnect the 3 control cables (unit side) and the 3 harness connectors.

i. Remove the 4 radio/audio sub box assembly screws and the radio/audio sub box assembly.

j. Disconnect or remove the following instrument panel harness connectors or items:

- The 6 driver's side connectors
- The 3 passenger's side connectors
- The 2 center connectors
- Passenger's inflator module connector
- Radio antenna cable plug
- Ground cable bolt on the left dash side panel
- The 8 instrument panel-to-chassis bolts and the 3 nuts.

k. Remove the instrument panel assembly.

7. Remove the instrument panel bracket by performing the following procedure:

a. Remove the 2 passenger's inflator module bolts and 4 nuts.

b. Remove the 4 meter assembly screws. Then, disconnect the meter wiring harness connectors and remove the meter assembly.

c. Remove the 5 vent duct assembly screws and the assembly.

d. Remove the 3 lower passenger's bracket screws and the bracket.

e. Remove the 9 passenger's knee bolster reinforcement screws and the reinforcement.

f. Remove the 6 instrument panel center reinforcement screws and the reinforcement.

g. Remove the instrument panel wiring harness assembly clips and the wiring harness.

h. Remove the 2 instrument panel bracket nuts and 2 bolts for each bracket; then, remove the bracket(s).

8. Remove the 5 cross beam assembly nuts, 2 bolts and the 6 lower bolts; then, remove the crossbeam.

9. Disconnect the resistor wiring connector.

10. Remove the duct from the heater assembly.

11. If equipped with air conditioning, remove the evaporator assembly.

12. Remove the driver's lap vent.

13. Remove the lower ventilation duct.

14. Remove the footrest, the carpet, the 3 clips and the rear heater duct.

15. Remove the heater assembly.

16. Remove the mode control case-to-temperature control case screws and remove the mode control case; do not remove the link unit.

17. Remove the temperature control case screws and separate the cases.

18. Remove the heater core from the case.

To install:

19. Install the heater core to the case.

20. Assemble the temperature control cases and install the case screws.

21. Install the mode control case and the mode control case-to-temperature control case screws.

22. Install the heater assembly.

23. Install the rear heater duct, the footrest, the carpet, and the 3 clips.

24. Install the lower ventilation duct.

25. Install the driver's lap vent.

26. If equipped with air conditioning, install the evaporator assembly.

27. Install the duct to the heater assembly.

28. Connect the resistor wiring connector.

29. Install the crossbeam, the 5 cross beam assembly nuts, 2 bolts and the 6 lower bolts.

30. Install the instrument panel bracket by performing the following procedure:

a. Install the instrument panel bracket and the 2 nuts and 2 bolts for each bracket.

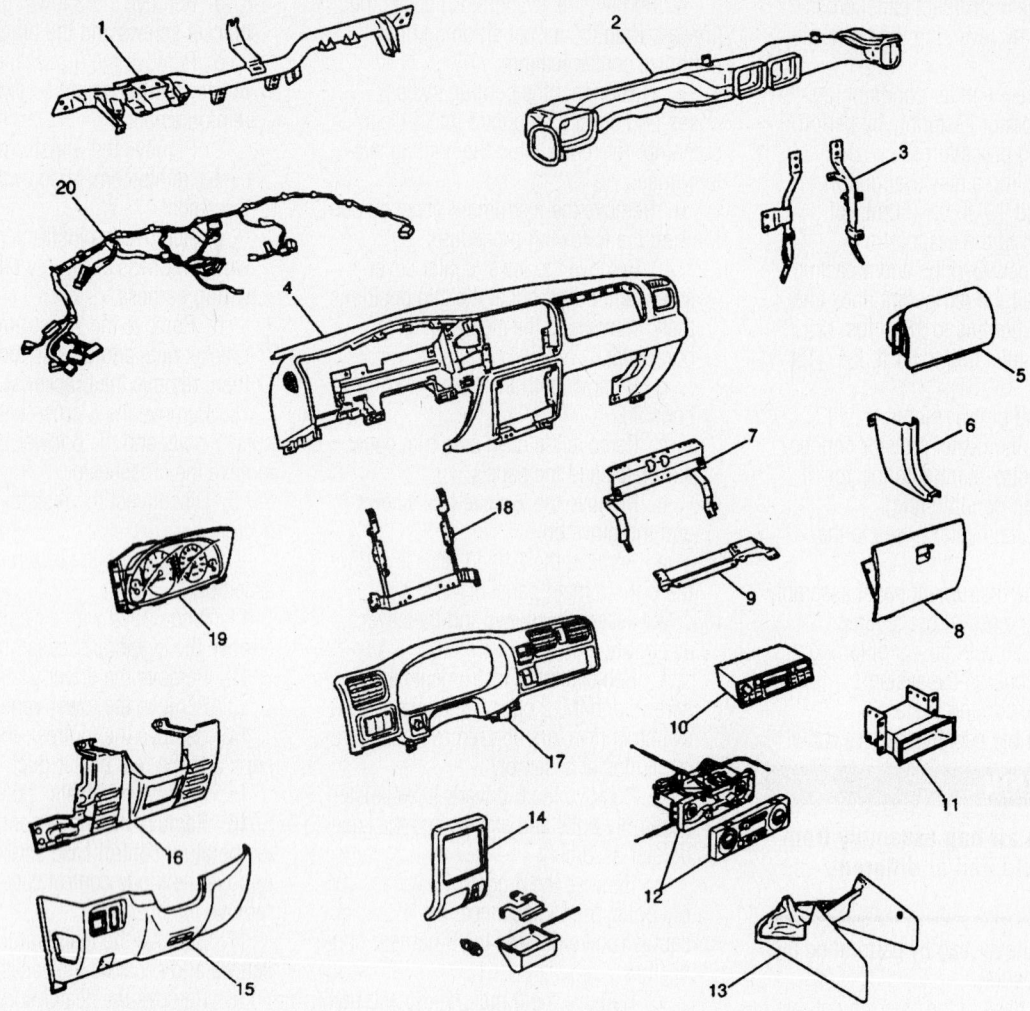

1 Cross Beam	11 Audio Sub Box
2 Vent Duct Assembly	12 Control Lever Assembly
3 Instrument Panel Bracket	13 Front Console Assembly
4 Instrument Panel Assembly	14 Lower Center Cover
5 Passenger Inflator Module	15 Instrument Panel Driver Lower Cover Assembly
6 Dash Side Trim Panel	16 Driver Knee Bolster Assembly
7 Passenger Knee Bolster Reinforcement Assembly	17 Meter Cluster Assembly
8 Glove Box	18 Instrument Panel Center Reinforcement
9 Passenger Lower Bracket	19 Meter Assembly
10 Radio Assembly	20 Instrument Harness Assembly

93113GB8

Exploded view of the instrument panel—Honda Passport

b. Install the instrument panel wiring harness assembly and the wiring harness clips.

c. Install the instrument panel center reinforcement and the 6 reinforcement screws.

d. Install the passenger knee bolster reinforcement and the 9 reinforcement screws.

e. Install the lower passenger bracket and the 3 bracket screws.

f. Install the vent duct assembly and the 5 vent duct assembly screws.

g. Install the meter assembly and the 4 meter assembly screws; then, connect the meter wiring harness connectors.

h. Install the 2 passenger's inflator module bolts and 4 nuts.

31. Install the instrument panel by performing the following procedure:

a. Install the instrument panel assembly.

b. Connect or install the following instrument panel harness connectors or items:

- The 6 driver's side connectors
- The 3 passenger's side connectors
- The 2 center connectors
- Passenger's inflator module connector
- Radio antenna cable plug
- Ground cable bolt on the left dash side panel
- The 8 instrument panel-to-chassis bolts and the 3 nuts.

c. Install the radio/audio sub box

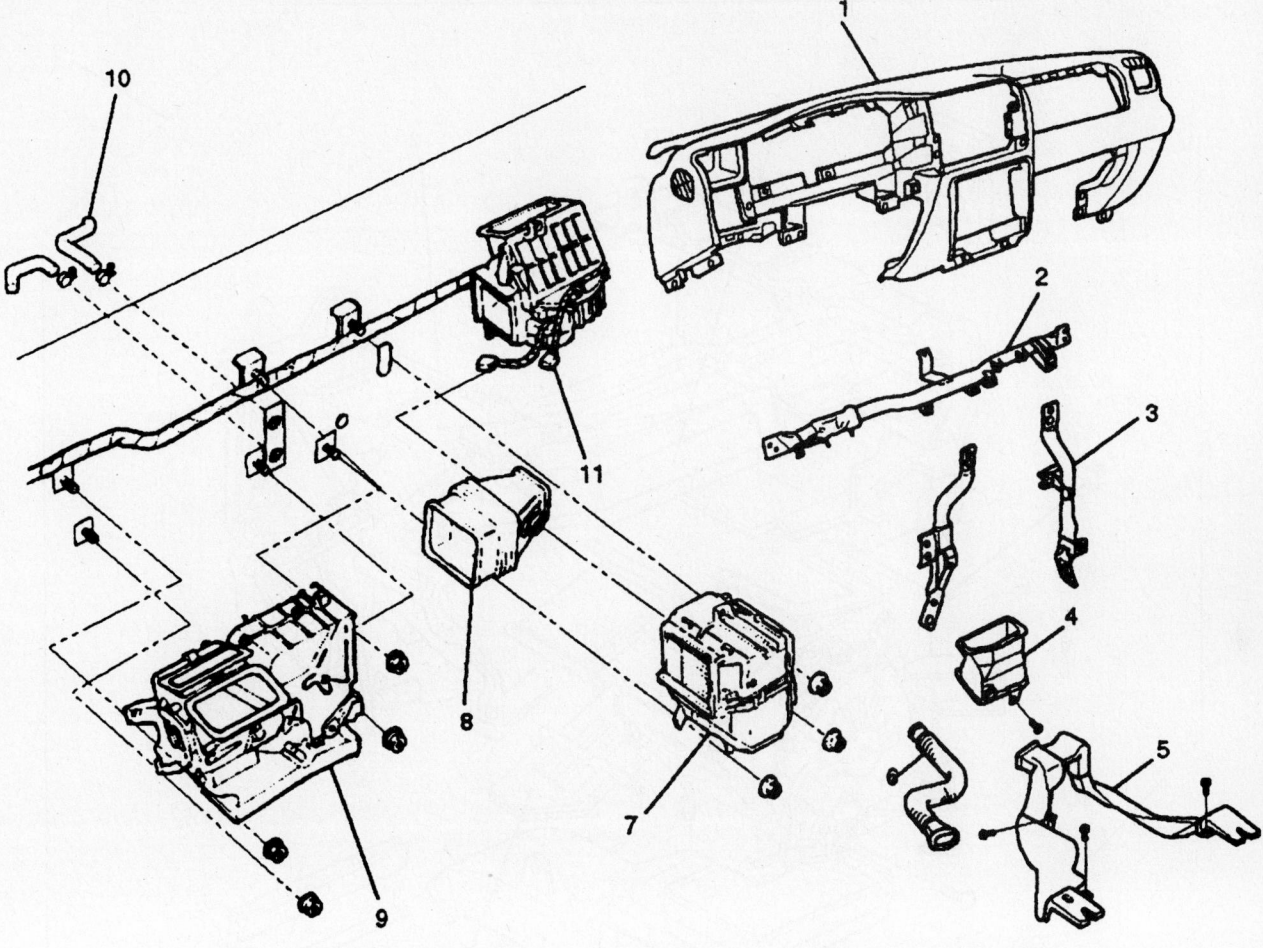

1	Instrument Panel Assembly	6	Driver Lap Vent Duct
2	Cross Beam Assembly	7	Evaporator Assembly (A/C only)
3	Instrument Panel Bracket	8	Duct
4	Ventilation Lower Duct	9	Heater Unit Assembly
5	Rear Heater Duct	10	Heater Hose
		11	Resistor Connector

93113GB9

View of the heater and air conditioning housing assemblies and related components—Honda Passport

assembly and the 4 radio/audio sub box assembly screws.

d. Connect the 3 control cables (unit side) and the 3 harness connectors. Install the 4 control lever assembly bolts.

e. Install the knee bolster assembly and the 6 driver's knee bolster assembly bolts and screws.

f. Install the instrument cluster assembly. Connect the 8 switch connectors. Install 5 instrument cluster screws and the 2 clips.

g. Install the 2 hood release screws,

the 6 instrument panel driver's lower cover assembly screws and the cover assembly.

h. Install the glove box and the 2 glove box screws.

i. Install the dash side trim panel sill plates and the panels.

j. Install both the rear and front console.

k. Connect the cigarette lighter connector and install the lower center cover screw.

32. Connect the cooling system hoses.

33. Refill the cooling system.

34. Install the evaporator lines at the firewall.

35. If equipped, evacuate and charge the air conditioning system.

36. Connect the negative battery cable.

37. If equipped with an air bag, perform the following procedure:

a. Turn the ignition to the LOCK position and remove the key.

b. Connect the 2-pin yellow connector located behind the glove box.

c. Install the glove box assembly.

d. Connect the 2-pin yellow connec-

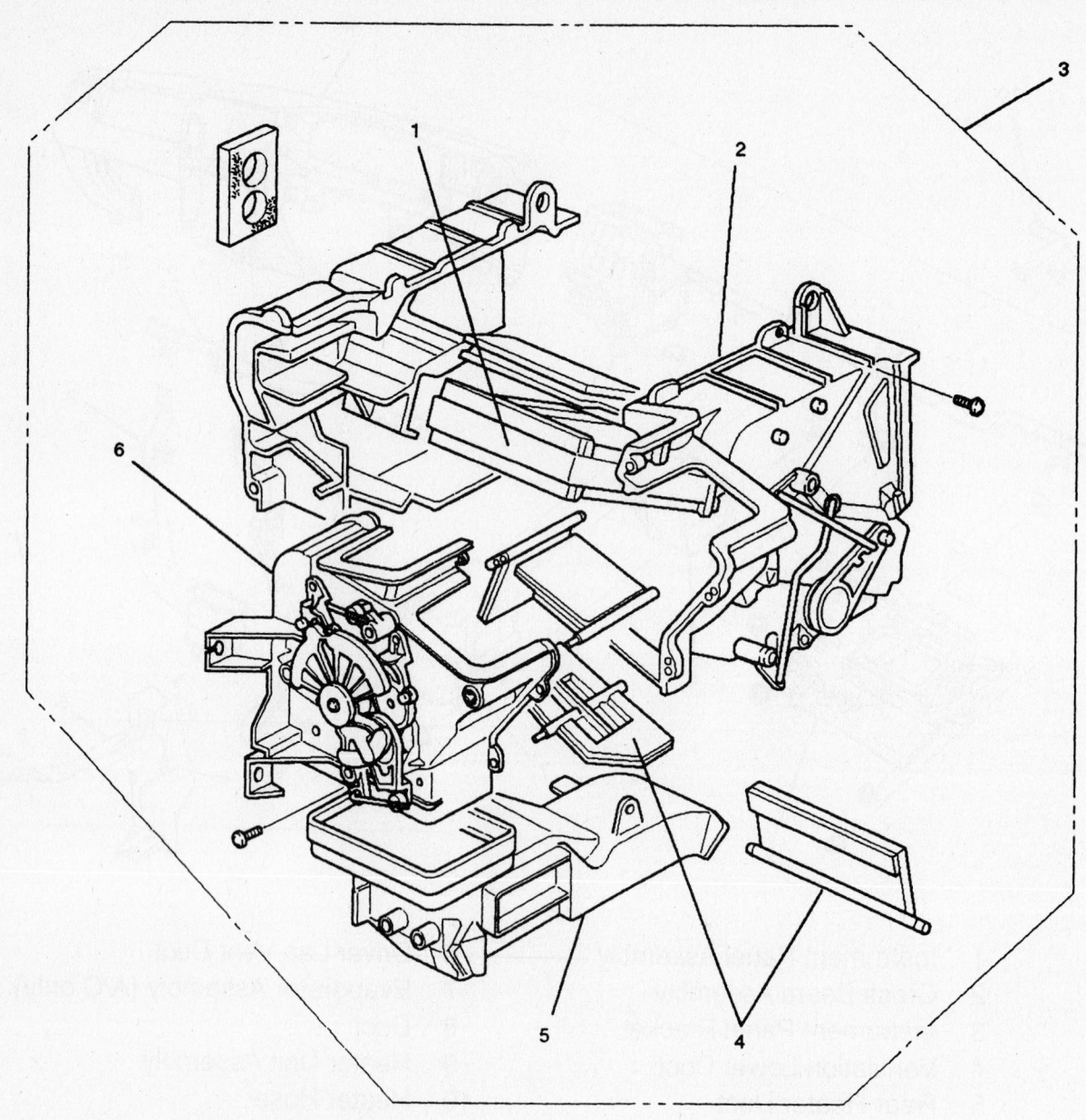

1 Heater Core
2 Case (Temperature Control)
3 Heater Unit
4 Mode Door
5 Duct
6 Case (Mode Control)

93113GB0

Exploded view of the heater housing assembly—Honda Passport

tor located at the base of the steering column.

e. At the lower left dash side fuse block, install the SRS-1 fuse.

f. Turn the ignition switch to ON and verify that the AIR BAG warning light flashes 7 times and turns OFF.

38. Run the engine to normal operating temperatures; then, check the climate control operation and check for leaks.

Cylinder Head

REMOVAL & INSTALLATION

2.2L Engine

1. Before servicing the vehicle, refer to the precautions in the beginning of this section.

2. Drain the cooling system.

3. Relieve the fuel system pressure.

4. Remove or disconnect the following:
   - Negative battery cable
   - Intake Air Temperature (IAT) sensor connector
   - Positive Crankcase Ventilation (PCV) valve and hose
   - Air intake assembly
   - Upper radiator hose
   - Accessory drive belt
   - Exhaust front pipe

- Alternator and brackets
- Crankshaft Position (CKP) sensor connector
- Knock sensor connector
- Heater hoses
- Water bypass hose
- Fuel lines
- Evaporative Emissions (EVAP) valve connector
- Canister hose
- Intake manifold
- Engine wiring harness connectors at left rear of the engine compartment
- Power steering pump pressure switch connector
- Front cover
- Spark plugs and wires
- Camshaft Position (CMP) sensor
- Valve cover
- Timing belt. Refer to the Timing Belt unit repair section.
- Timing belt idler pulleys
- Timing belt rear cover
- Oil pressure switch connector
- Camshafts
- Cylinder head. Remove the bolts in reverse of the tightening sequence.

To install:

➥**Use new cylinder head bolts for assembly.**

5. Install the cylinder head with a new gasket. Tighten the bolts in sequence as follows:
   a. Step 1: 18 ft. lbs. (25 Nm)
   b. Step 2: Plus 90 degrees
   c. Step 3: Plus 90 degrees
   d. Step 4: Plus 90 degrees
6. Install or connect the following:
   - Camshafts
   - Oil pressure switch connector
   - Timing belt rear cover
   - Timing belt idler pulleys. Tighten the bolts to 18 ft. lbs. (25 Nm).
   - Timing belt
   - Valve cover
   - CMP sensor
   - Spark plugs and wires
   - Front cover
   - Power steering pump pressure switch connector
   - Engine wiring harness connectors at left rear of the engine compartment
   - Intake manifold
   - Canister hose
   - EVAP valve connector
   - Fuel lines
   - Water bypass hose
   - Heater hoses
   - Knock sensor connector

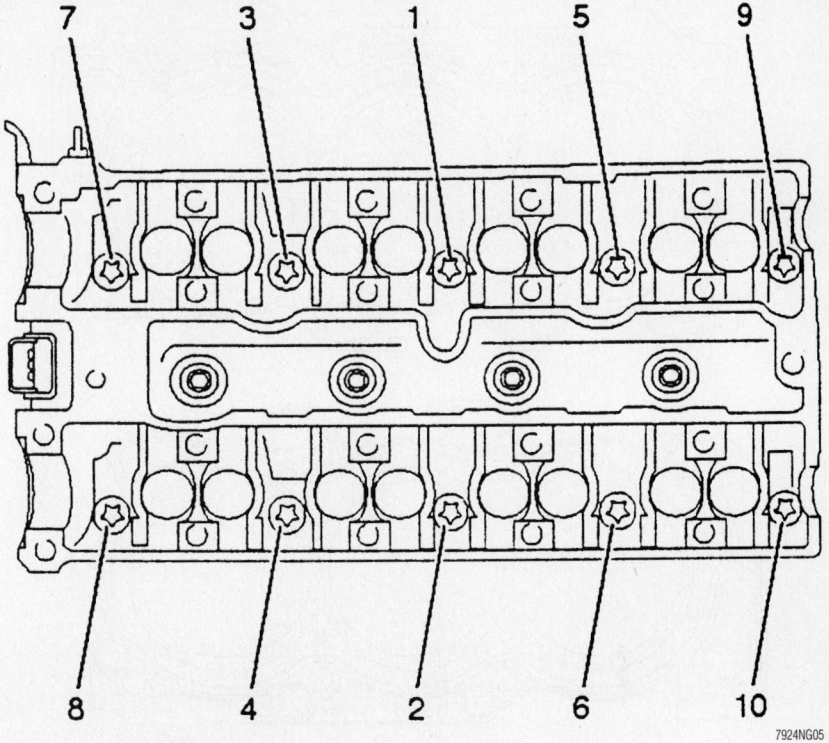

Cylinder head torque sequence—2.2L (VIN D) engine

- CKP sensor connector
- Alternator and brackets
- Exhaust front pipe
- Accessory drive belt
- Upper radiator hose
- Air intake assembly
- PCV valve and hose
- IAT sensor connector
- Negative battery cable
7. Fill the cooling system.
8. Start the engine and check for leaks.

3.2L Engines

1. Before servicing the vehicle, refer to the precautions in the beginning of this section.
2. Drain the cooling system.
3. Relieve the fuel system pressure.
4. Remove or disconnect the following:
   - Negative battery cable
   - Hood
   - Engine cover
   - Mass Air Flow (MAF) sensor connector
   - Intake Air Temperature (IAT) sensor connector
   - Positive Crankcase Ventilation (PCV) valve and hose
   - Air cleaner assembly
   - Manifold Absolute Pressure (MAP) sensor connector
   - Vacuum Switching Valve (VSV) connector and vacuum line

- Fuel injector connectors
- Throttle Position (TP) sensor connector
- Idle Air Control (IAC) valve connector
- Ignition coils
- Brake booster vacuum line
- Canister purge vacuum line
- Duty solenoid valve
- Fuel lines
- Intake manifold
- Radiator hoses
- Engine coolant manifold
- Upper fan shroud
- Accessory drive belt and tensioner
- Cooling fan and pulley
- Alternator
- Idler pulley
- Power steering pump and bracket
- A/C compressor
- Crankshaft pulley
- Oil cooler hoses
- Timing belt cover
- Valve covers
- Timing belt. Refer to the Timing Belt Unit Repair Section.
- Left and right exhaust front pipes
- Oil dipstick tube
- Cylinder heads

To install:

➥**Use new head bolts when installing the cylinder head.**

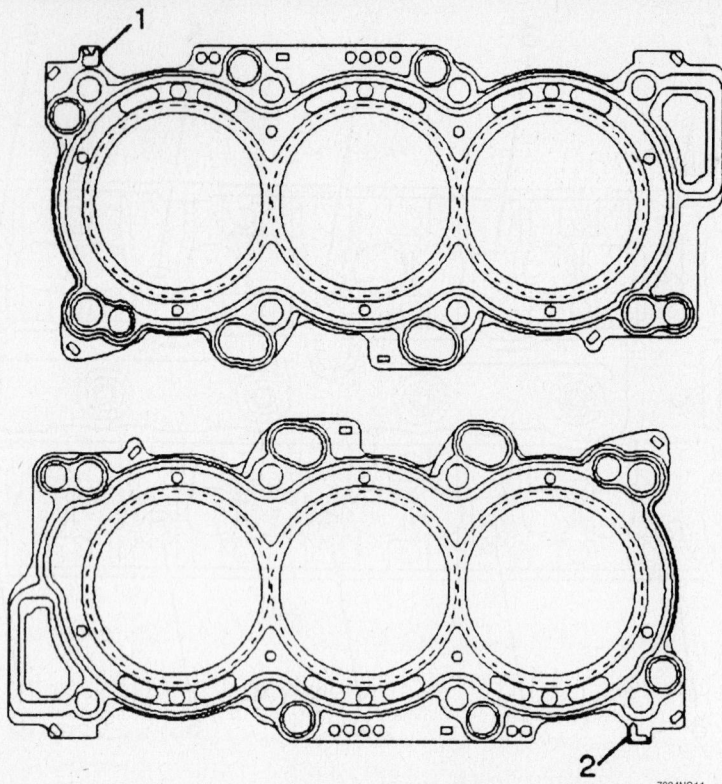

Right (1) and left (2) head gasket identification mark locations—3.2L DOHC engine

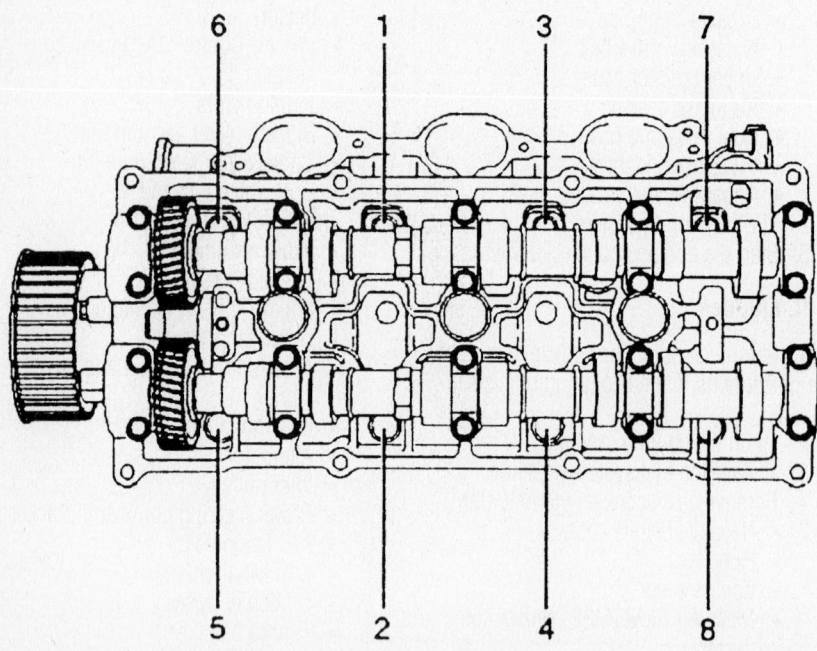

Cylinder head torque sequence—3.2L DOHC and 3.5L engines

➡**The left and right cylinder head gaskets are not interchangeable.**

5. Install the cylinder heads with new gaskets. Tighten the bolts in sequence as follows:

  a. Step 1: 21 ft. lbs. (29 Nm)

  b. Step 2: 47 ft. lbs. (64 Nm)

6. Install or connect the following:
- Oil dipstick tube
- Left and right exhaust front pipes
- Timing belt
- Valve covers
- Timing belt cover
- Oil cooler hoses
- Crankshaft pulley. Tighten the pulley bolt to 123 ft. lbs. (167 Nm).
- A/C compressor
- Power steering pump and bracket. Tighten the bolts to 34 ft. lbs. (46 Nm).
- Idler pulley
- Alternator
- Cooling fan and pulley
- Accessory drive belt and tensioner
- Upper fan shroud
- Engine coolant manifold
- Radiator hoses
- Intake manifold
- Fuel lines
- Duty solenoid valve
- Canister purge vacuum line
- Brake booster vacuum line
- Ignition coils
- IAC valve connector
- TP sensor connector
- Fuel injector connectors
- VSV connector and vacuum line
- MAP sensor connector
- Air cleaner assembly
- PCV valve and hose
- IAT sensor connector
- MAF sensor connector
- Engine cover
- Hood
- Negative battery cable

7. Fill the cooling system.

8. Start the engine and check for leaks.

3.5L Engine

1. Before servicing the vehicle, refer to the precautions in the beginning of this section.

2. Drain the cooling system.

3. Remove or disconnect the following:
- Negative battery cable
- Hood
- Engine cover
- Mass Air Flow (MAF) sensor connector
- Intake Air Temperature (IAT) sensor connector
- Positive Crankcase Ventilation (PCV) valve and hose
- Air cleaner assembly
- Manifold Absolute Pressure (MAP) sensor connector
- Vacuum Switching Valve (VSV) connector and vacuum line
- Fuel injector connectors
- Throttle Position (TP) sensor connector
- Idle Air Control (IAC) valve connector

- Ignition coils
- Brake booster vacuum line
- Canister purge vacuum line
- Duty solenoid valve
- Fuel lines
- Intake manifold
- Radiator hoses
- Engine coolant manifold
- Upper fan shroud
- Accessory drive belt and tensioner
- Cooling fan and pulley
- Alternator
- Idler pulley
- Power steering pump and bracket
- A/C compressor
- Crankshaft pulley
- Oil cooler hoses
- Timing belt cover
- Valve covers
- Timing belt. Refer to the Timing Belt Unit Repair Section.
- Left and right exhaust front pipes
- Oil dipstick tube
- Cylinder heads

To install:

➡**Use new head bolts when installing the cylinder head. Do not apply oil to the head bolt threads.**

➡**The left and right cylinder head gaskets are not interchangeable.**

4. Install the cylinder heads with new gaskets. Tighten the bolts to 47 ft. lbs. (64 Nm).

5. Install or connect the following:
- Oil dipstick tube
- Left and right exhaust front pipes
- Timing belt
- Valve covers
- Timing belt cover
- Oil cooler hoses
- Crankshaft pulley. Tighten the pulley bolt to 123 ft. lbs. (167 Nm).
- A/C compressor
- Power steering pump and bracket. Tighten the bolts to 34 ft. lbs. (46 Nm).
- Idler pulley
- Alternator
- Cooling fan and pulley
- Accessory drive belt and tensioner
- Upper fan shroud
- Engine coolant manifold
- Radiator hoses
- Intake manifold
- Fuel lines
- Duty solenoid valve
- Canister purge vacuum line
- Brake booster vacuum line
- Ignition coils
- IAC valve connector
- TP sensor connector

- Fuel injector connectors
- VSV connector and vacuum line
- MAP sensor connector
- Air cleaner assembly
- PCV valve and hose
- IAT sensor connector
- MAF sensor connector
- Engine cover
- Hood
- Negative battery cable

6. Fill the cooling system.
7. Start the engine. Check for leaks and proper operation.

Rocker Arms/Shafts

REMOVAL & INSTALLATION

➡**These engines are not equipped with rocker arms. The camshaft lobes act directly on the valve shims.**

Intake Manifold

REMOVAL & INSTALLATION

2.2L Engine

1. Before servicing the vehicle, refer to the precautions in the beginning of this section.
2. Drain the cooling system.
3. Relieve the fuel system pressure.
4. Remove or disconnect the following:
- Negative battery cable
- Accessory drive belt
- Positive Crankcase Ventilation (PCV) valve and hose
- Air intake duct
- Throttle body water hoses
- Throttle Position (TP) sensor connector
- Idle Air Control (IAC) valve connector
- Fuel lines
- Fuel injector connectors
- Fuel pressure regulator vacuum line
- Fuel supply manifold
- Accelerator cable
- Alternator and brackets
- Water pipe
- Intake manifold bracket
- Ignition coil and bracket
- Brake booster vacuum line
- Intake manifold

To install:

5. Install or connect the following:
- Intake manifold. Use a new gasket and tighten the bolts to 16 ft. lbs. (22 Nm).

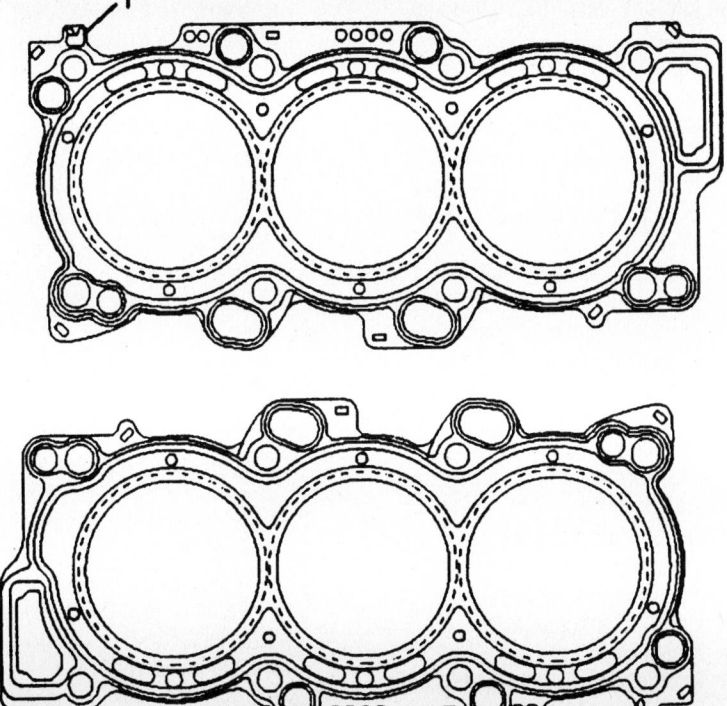

Right (1) and left (2) head gasket identification mark locations—3.5L engine

7924BG04

- Brake booster vacuum line
- Ignition coil and bracket
- Intake manifold bracket. Tighten the bolts to 16 ft. lbs. (22 Nm).
- Water pipe
- Alternator and brackets. Tighten the short bolts to 14 ft. lbs. (20 Nm) and the long bolts to 25 ft. lbs. (35 Nm).
- Accelerator cable
- Fuel supply manifold
- Fuel pressure regulator vacuum line
- Fuel injector connectors
- Fuel lines
- IAC valve connector
- TP sensor connector
- Throttle body water hoses
- Air intake duct
- PCV valve and hose
- Accessory drive belt
- Negative battery cable
6. Fill the cooling system.
7. Start the engine and check for leaks.

3.2L Engine

1. Before servicing the vehicle, refer to the precautions in the beginning of this section.
2. Remove or disconnect the following:
- Negative battery cable
- Engine cover
- Air cleaner assembly
- Accelerator cable
- Cruise control cable
- Brake booster vacuum line
- Manifold Absolute Pressure (MAP) sensor connector
- Idle Air Control (IAC) valve connector
- Throttle Position (TP) sensor connector
- Canister purge solenoid connector
- Electronic Vacuum Sensing Valve (EVSV) connector and vacuum line
- Exhaust Gas Recirculation (EGR) valve
- Positive Crankcase Ventilation (PCV) valve and hose
- Pressure regulator vacuum line
- Ventilation hose
- Throttle body
- Fuel lines
- Fuel injector connectors
- Intake manifold

To install:
3. Install or connect the following:
- Intake manifold. Tighten the fasteners to 18 ft. lbs. (25 Nm).
- Fuel injector connectors
- Fuel lines

- Throttle body. Tighten the bolts to 88 inch lbs. (10 Nm).
- Ventilation hose
- Pressure regulator vacuum line
- PCV valve and hose
- EGR valve
- EVSV connector and vacuum line
- Canister purge solenoid connector
- TP sensor connector
- IAC valve connector
- MAP sensor connector
- Brake booster vacuum line
- Cruise control cable
- Accelerator cable
- Air cleaner assembly
- Engine cover
- Negative battery cable
4. Start the engine and check for proper operation.

3.5L Engine

1. Before servicing the vehicle, refer to the precautions in the beginning of this section.
2. Remove or disconnect the following:
- Negative battery cable
- Engine cover
- Air cleaner assembly
- Accelerator cable
- Cruise control cable
- Brake booster vacuum line
- Manifold Absolute Pressure (MAP) sensor connector
- Idle Air Control (IAC) valve connector
- Throttle Position (TP) sensor connector
- Canister purge solenoid connector
- Electronic Vacuum Sensing Valve (EVSV) connector and vacuum line
- Exhaust Gas Recirculation (EGR) valve
- Positive Crankcase Ventilation (PCV) valve and hose
- Pressure regulator vacuum line
- Ventilation hose
- Throttle body
- Fuel lines
- Fuel injector connectors
- Intake manifold

To install:
3. Install or connect the following:
- Intake manifold. Tighten the fasteners to 18 ft. lbs. (25 Nm).
- Fuel injector connectors
- Fuel lines
- Throttle body. Tighten the bolts to 88 inch lbs. (10 Nm).
- Ventilation hose
- Pressure regulator vacuum line
- PCV valve and hose

- EGR valve
- EVSV connector and vacuum line
- Canister purge solenoid connector
- TP sensor connector
- IAC valve connector
- MAP sensor connector
- Brake booster vacuum line
- Cruise control cable
- Accelerator cable
- Air cleaner assembly
- Engine cover
- Negative battery cable
4. Start the engine and check for proper operation.

Exhaust Manifold

REMOVAL & INSTALLATION

2.2L Engine

1. Before servicing the vehicle, refer to the precautions in the beginning of this section.
2. Remove or disconnect the following:
- Negative battery cable
- Air intake duct
- Exhaust front pipe
- Exhaust manifold heat shield
- Exhaust manifold

To install:
3. Install the exhaust manifold. Tighten the nuts in sequence as follows:
   a. Step 1: 10 ft. lbs. (14 Nm)
   b. Step 2: 14 ft. lbs. (20 Nm)
   c. Step 3: 14 ft. lbs. (20 Nm)
4. Install or connect the following:
- Exhaust manifold heat shield. Tighten the bolts to 71 inch lbs. (8 Nm).
- Exhaust front pipe. Tighten the bolts to 18 ft. lbs. (25 Nm).
- Air intake duct
- Negative battery cable
5. Start the engine and check for leaks.

3.2L Engines

1. Before servicing the vehicle, refer to the precautions in the beginning of this section.
2. Remove or disconnect the following:
- Negative battery cable
- Air cleaner assembly
- Heated Oxygen (HO2S) sensor connectors
- Right torsion bar
- Exhaust Gas Recirculation (EGR) pipe and bracket
- Left and right exhaust front pipes
- Heat shields
- Accessory drive belt

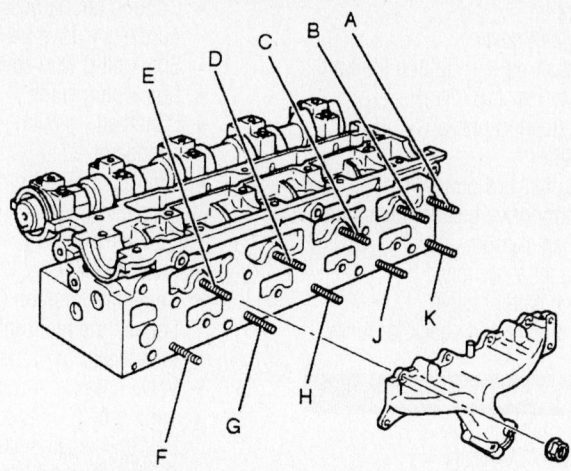

- **Tightening sequence:**
  - **Step1:** J G H B D C J G B D
  - **Step2:** A B C D E F G H J K
  - **Step3:** A B C D E F G H J K
- **Tightening torque:**
  - **Step1:** 14 N·m (10 lb ft)
  - **Step2:** 20 N·m (14 lb ft)
  - **Step3:** 20 N·m (14 lb ft)

7924NG17

Exhaust manifold torque sequence—2.2L engine

- A/C compressor and bracket
- Exhaust manifolds

To install:
3. Install or connect the following:
  - Exhaust manifolds. Tighten the bolts to 42 ft. lbs. (57 Nm).
  - A/C compressor and bracket
  - Accessory drive belt
  - Heat shields
  - Left and right exhaust front pipes
  - EGR pipe and bracket
  - Right torsion bar
  - HO$_2$S sensor connectors
  - Air cleaner assembly
  - Negative battery cable
4. Start the engine and check for leaks.

3.5L Engine

1. Before servicing the vehicle, refer to the precautions in the beginning of this section.
2. Remove or disconnect the following:
  - Negative battery cable
  - Air cleaner assembly
  - Heated Oxygen (HO$_2$S) sensor connectors
  - Right torsion bar
  - Exhaust Gas Recirculation (EGR) pipe and bracket
  - Left and right exhaust front pipes

- Heat shields
- Accessory drive belt
- A/C compressor and bracket
- Exhaust manifolds

To install:
3. Install or connect the following:
  - Exhaust manifolds. Tighten the bolts to 38 ft. lbs. (52 Nm).
  - A/C compressor and bracket
  - Accessory drive belt
  - Heat shields
  - Left and right exhaust front pipes
  - EGR pipe and bracket
  - Right torsion bar
  - HO$_2$S sensor connectors
  - Air cleaner assembly
  - Negative battery cable
4. Start the engine and check for leaks.

Front Crankshaft Seal

REMOVAL & INSTALLATION

2.2L Engines

1. Before servicing the vehicle, refer to the precautions in the beginning of this section.
2. Remove or disconnect the following:
  - Negative battery cable

- Accessory drive belts
- Cooling fan
- A/C belt tensioner, if equipped
- Water pump pulley
- Power steering pump
- Crankshaft pulley
- Front cover
- Timing belt. Refer to the Timing Belt unit repair section.
- Crankshaft timing sprocket
- Rear timing cover
- Crankshaft oil seal

To install:
3. Install or connect the following:
  - Crankshaft oil seal
  - Rear timing cover
  - Crankshaft timing sprocket. Tighten the bolt to 94 ft. lbs. (130 Nm) plus 45 degrees.
  - Timing belt. Refer to the Timing Belt unit repair section.
  - Front cover
  - Crankshaft pulley
  - Power steering pump
  - Water pump pulley
  - A/C belt tensioner, if equipped
  - Cooling fan
  - Accessory drive belts
  - Negative battery cable
4. Start the engine and check for leaks.

3.2L Engines

1. Before servicing the vehicle, refer to the precautions in the beginning of this section.
2. Remove or disconnect the following:
  - Negative battery cable
  - Air cleaner assembly
  - Upper fan shroud
  - Accessory drive belt and tensioner
  - Cooling fan and pulley
  - Idler pulley
  - Power steering pump
  - Crankshaft pulley
  - Timing belt cover
  - Timing belt. Refer to the Timing Belt Unit Repair Section.
  - Crankshaft timing sprocket
  - Oil seal

To install:
3. Install or connect the following:
  - Oil seal so that it is flush with the oil pump housing
  - Crankshaft timing sprocket
  - Timing belt
  - Timing belt cover
  - Crankshaft pulley. Tighten the bolt to 123 ft. lbs. (167 Nm).
  - Power steering pump
  - Idler pulley
  - Cooling fan and pulley

- Accessory drive belt and tensioner
- Upper fan shroud
- Air cleaner assembly
- Negative battery cable
4. Start the engine and check for leaks.

3.5L Engine

1. Before servicing the vehicle, refer to the precautions in the beginning of this section.
2. Remove or disconnect the following:
- Negative battery cable
- Air cleaner assembly
- Upper fan shroud
- Accessory drive belt and tensioner
- Cooling fan and pulley
- Idler pulley
- Power steering pump and move it aside
- Crankshaft pulley
- Timing belt cover
- Timing belt. Refer to the Timing Belt Unit Repair Section.
- Crankshaft timing sprocket
- Oil seal

To install:
3. Install or connect the following:
- Oil seal so that it is flush with the oil pump housing
- Crankshaft timing sprocket

- Timing belt
- Timing belt cover
- Crankshaft pulley. Tighten the bolt to 123 ft. lbs. (167 Nm).
- Power steering pump
- Idler pulley
- Cooling fan and pulley
- Accessory drive belt and tensioner
- Upper fan shroud
- Air cleaner assembly
- Negative battery cable
4. Start the engine and check for leaks.

Camshaft and Valve Lifters

REMOVAL & INSTALLATION

2.2L Engine

1. Before servicing the vehicle, refer to the precautions in the beginning of this section.
2. Remove or disconnect the following:
- Negative battery cable
- Positive Crankcase Ventilation (PCV) valve and hose
- Air intake duct and bracket
- Ground cables
- Engine wiring harness connectors at left rear of the engine compartment

- Cooling fan harness connector
- Accessory drive belt
- Spark plug wire cover
- Spark plug wires
- Camshaft Position (CMP) sensor connector
- Crankshaft Position (CKP) sensor connector
- Crankshaft pulley
- Front cover
- Camshaft Position (CMP) sensor. Loosen the rear timing cover bolt for access.
- Valve cover
- Timing belt. Refer to the Timing Belt unit repair section.
- Camshaft sprockets
- Camshaft bearing caps
- Camshaft seals
- Camshafts
- Hydraulic tappets

➡**Keep all valvetrain components in order for assembly.**

To install:
3. Install or connect the following:
- Hydraulic tappets in their original locations
- Camshafts
- Camshaft bearing caps. Tighten the

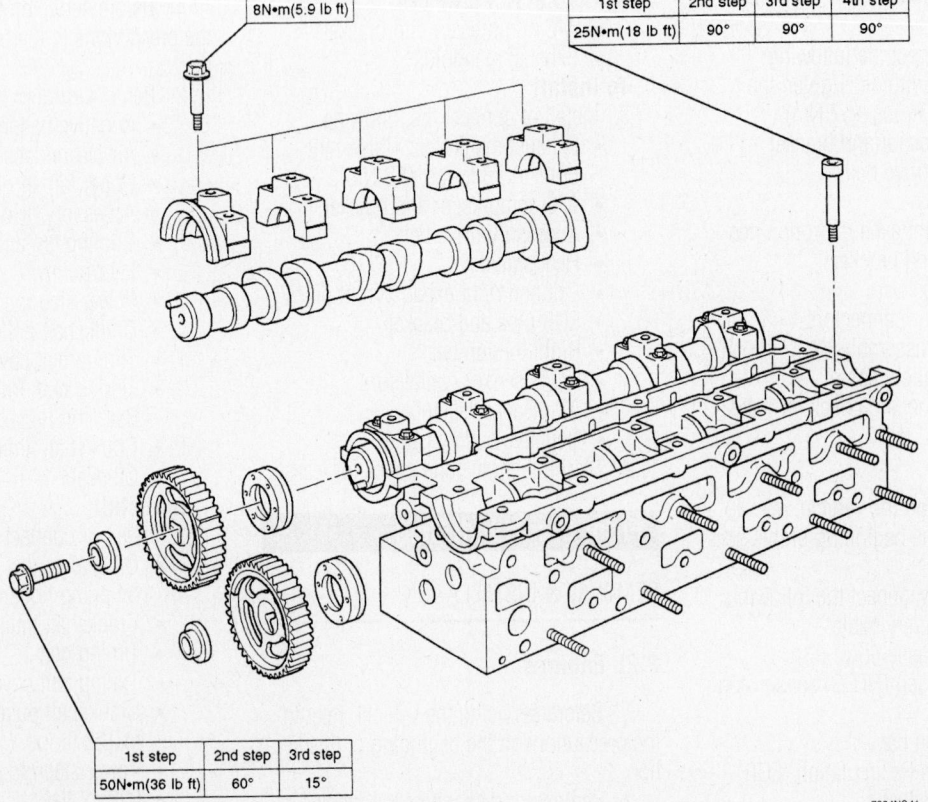

1st step	2nd step	3rd step	4th step
25N•m(18 lb ft)	90°	90°	90°

8N•m(5.9 lb ft)

1st step	2nd step	3rd step
50N•m(36 lb ft)	60°	15°

7924NG41

Exploded view of the cylinder head and camshaft components—2.2L engine

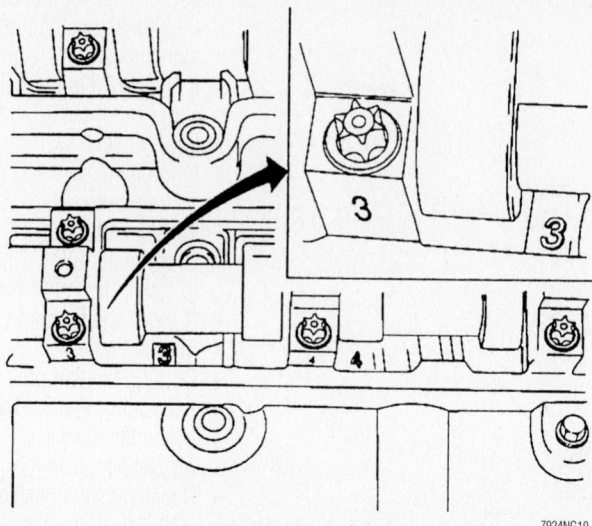

Camshaft bearing cap identification locations—2.2L engine

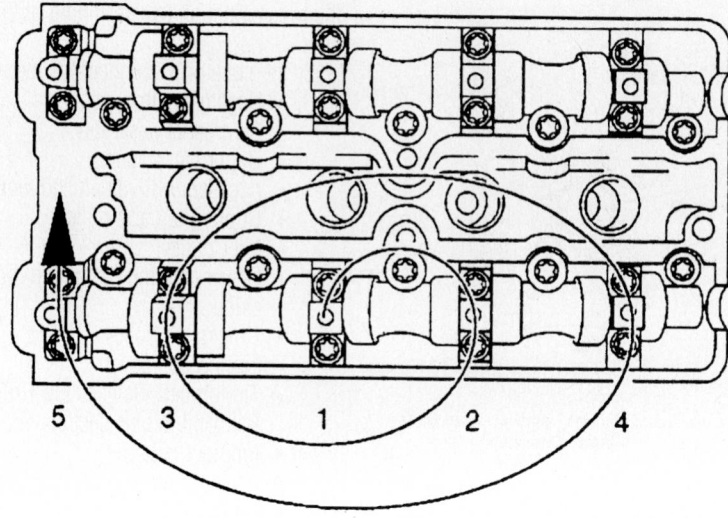

Camshaft bearing cap tightening sequence—2.2L engine

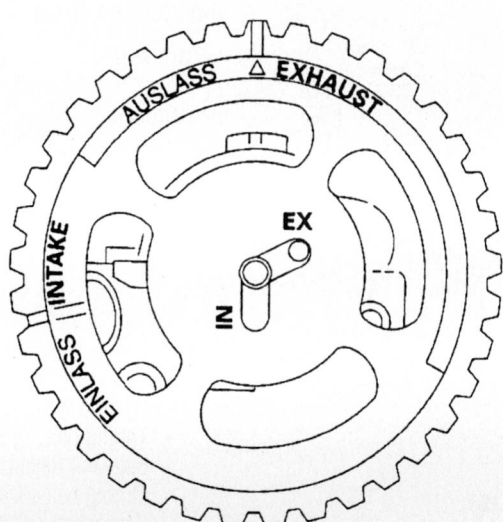

Guide pin location for the exhaust cam gear—2.2L engine

bolts in sequence to 71 inch lbs. (8 Nm).

- Camshaft seals
- Camshaft sprockets

4. Tighten the camshaft sprocket bolts as follows:

    a. Step 1: 36 ft. lbs. (50 Nm)
    b. Step 2: Plus 60 degrees
    c. Step 3: Plus 15 degrees

5. Install or connect the following:

- Timing belt
- Valve cover
- CMP sensor. Loosen the rear timing cover bolt for access.
- Front cover
- Crankshaft pulley. Tighten the bolts to 14 ft. lbs. (20 Nm).
- CKP sensor connector
- CMP sensor connector
- Spark plug wires
- Spark plug wire cover
- Accessory drive belt
- Cooling fan harness connector
- Engine wiring harness connectors at left rear of the engine compartment
- Ground cables
- Air intake duct and bracket
- PCV valve and hose
- Negative battery cable

3.2L Engine

1. Before servicing the vehicle, refer to the precautions in the beginning of this section.

2. Remove or disconnect the following:

- Negative battery cable
- Air cleaner assembly
- Upper fan shroud
- Accessory drive belt and tensioner
- Cooling fan and pulley
- Idler pulley
- Power steering pump and move it aside
- Crankshaft pulley
- Timing belt cover
- Timing belt. Refer to the Timing Belt Unit Repair Section.
- Ignition coils
- Valve covers
- Camshafts
- Valve shims and tappets

➡**Keep the valve shims and tappets in order for installation.**

To install:

3. Install the valve tappets and shims in their original locations.

4. Using Gear Spring Lever J-42686, turn the sub gear clockwise to align the 5mm bolt holes in the sub gear and the camshaft driven gear. Tighten the 5mm bolt.

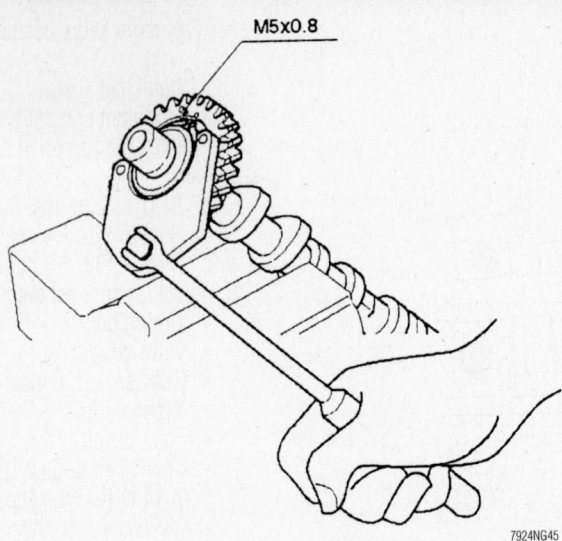

M5x0.8

7924NG45

Aligning the sub gear with the Gear Spring Lever J-42686—3.2L DOHC engine

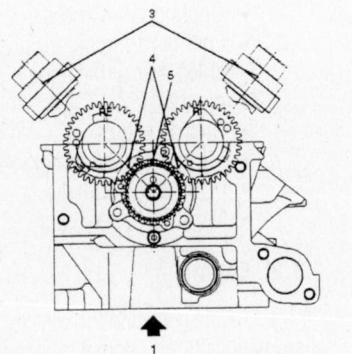 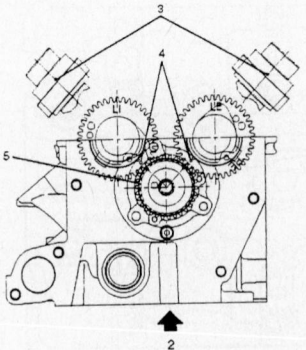

Legend
(1) Right Bank
(2) Left Bank

(3) Alignment Mark on Camshaft Drive Gear
(4) Alignment Mark on Camshaft
(5) Alignment Mark on Retainer

7924NG46

Camshaft alignment marks for the left and right cylinder heads—3.2L DOHC engine

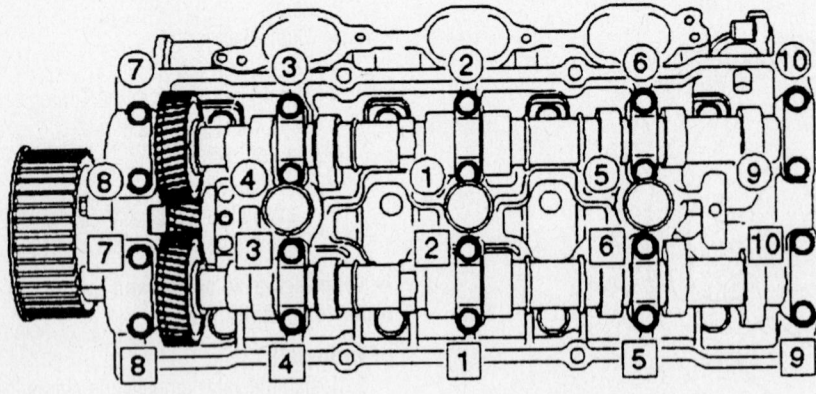

◯ : **Intake**   ▢ : **Exhaust**

7924NG47

Camshaft retaining bracket tightening sequence—3.2L DOHC engine

5. Install or connect the following:
- Camshafts by aligning the timing marks as shown. Tighten the bolts in sequence to 89 inch lbs. (10 Nm).
- Valve covers
- Ignition coils
- Timing belt
- Timing belt cover
- Crankshaft pulley. Tighten the bolt to 123 ft. lbs. (167 Nm).
- Power steering pump
- Idler pulley
- Cooling fan and pulley
- Accessory drive belt and tensioner
- Upper fan shroud
- Air cleaner assembly
- Negative battery cable

3.5L Engine

1. Before servicing the vehicle, refer to the precautions in the beginning of this section.

2. Remove or disconnect the following:
- Negative battery cable
- Air cleaner assembly
- Upper fan shroud
- Accessory drive belt and tensioner
- Cooling fan and pulley
- Idler pulley
- Power steering pump and move it aside
- Crankshaft pulley
- Timing belt cover
- Timing belt. Refer to the Timing Belt Unit Repair Section.
- Ignition coils
- Valve covers
- Camshafts
- Valve shims and tappets

➡ **Keep the valve shims and tappets in order for installation.**

To install:

3. Install the valve tappets and shims in their original locations.

4. Using Gear Spring Lever J-42686, turn the sub gear clockwise to align the 5mm bolt holes in the sub gear and the camshaft driven gear. Tighten the 5mm bolt.

5. Install the camshafts. Align the timing marks as shown. Tighten the bolts in sequence to 89 inch lbs. (10 Nm).

6. Install or connect the following:
- Valve covers
- Ignition coils
- Timing belt. Refer to the Timing Belt Unit Repair Section.
- Timing belt cover
- Crankshaft pulley
- Power steering pump

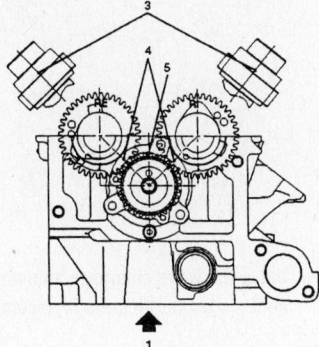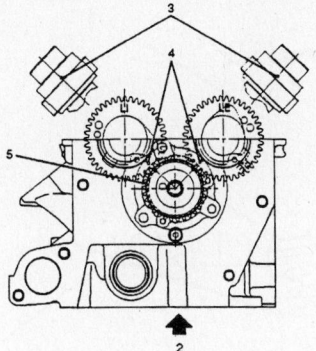

1	Right Bank	3	Alignment Mark on Camshaft Drive Gear
2	Left Bank	4	Alignment Mark on Camshaft
		5	Alignment Mark on Retainer

7924BG11

Camshaft alignment marks for the left and right cylinder heads—3.5L engine

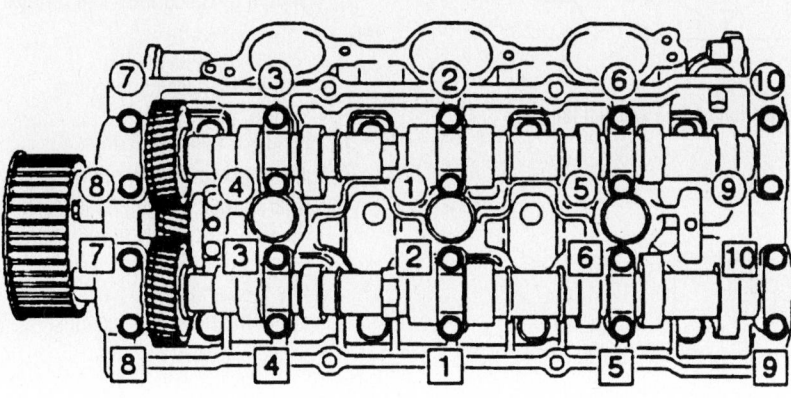

○ : Intake   □ : Exhaust

7924BG12

Camshaft retaining bracket tightening sequence—3.5L engine

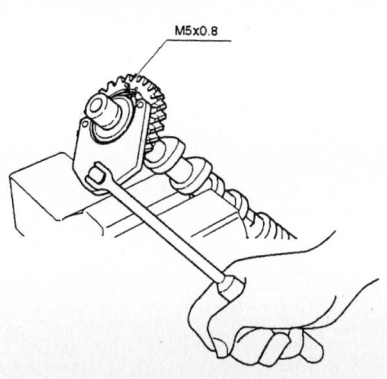

7924BG13

Aligning the sub gear with the Gear Spring Lever J-42686—3.5L engine

- Idler pulley
- Cooling fan and pulley
- Accessory drive belt and tensioner
- Upper fan shroud
- Air cleaner assembly
- Negative battery cable

Valve Lash

ADJUSTMENT

2.2L Engine

The 2.2L DOHC engine is equipped with hydraulic lash adjusters. No valve adjustment is necessary.

3.2L Engines

➡**Measure valve clearance with the engine cold.**

1. Before servicing the vehicle, refer to the precautions in the beginning of this section.

2. Remove the valve covers.

3. Check the valve clearance with the camshafts positioned as shown. Intake valve clearance should be 0.0091–0.0130 in. (0.2311–0.3302mm). Exhaust valve clearance should be 0.0098–0.0138 in. (0.2489–0.3505mm).

4. If adjustment is required, replace the shims as follows:

a. Step 1: Position special tool J-42689 on the edge of the tappet.

b. Step 2: Rotate the crankshaft until the maximum lift portion of the camshaft lobe contacts the upper edge of the special tool and presses the tappet down to create enough clearance between the adjustment shim and the camshaft for the shim to be removed.

c. Step 3: Replace shims as necessary to achieve correct valve clearance.

d. Step 4: Repeat for each valve to be adjusted.

5. Replace the valve covers. Tighten the bolts to 80 inch lbs. (9 Nm).

3.5L Engine

➡**Measure valve clearance with the engine cold.**

1. Before servicing the vehicle, refer to the precautions in the beginning of this section.

2. Remove the valve covers.

3. Check the valve clearance with the camshafts positioned as shown. Intake valve clearance should be 0.0091–0.0130 inches. Exhaust valve clearance should be 0.0098–0.0138 inches.

4. If adjustment is required, replace the shims as follows:

a. Step 1: Position special tool J–42689 on the edge of the tappet.

b. Step 2: Rotate the crankshaft until the maximum lift portion of the camshaft lobe contacts the upper edge of the special tool and presses the tappet down to create enough clearance between the adjustment shim and the camshaft for the shim to be removed.

c. Step 3: Replace shims as necessary to achieve correct valve clearance.

d. Step 4: Repeat for each valve to be adjusted.

5. Replace the valve covers. Tighten the bolts to 80 inch lbs. (9 Nm).

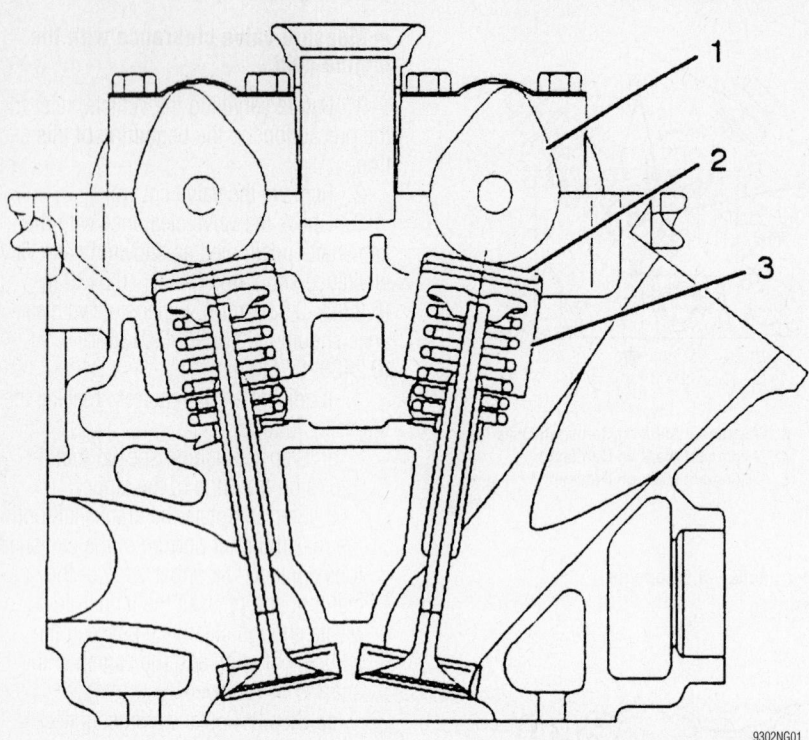

Cross section of the 3.2L DOHC cylinder head. Note the position of the camshaft lobe (1), adjustment shim (2) and the tappet (3)

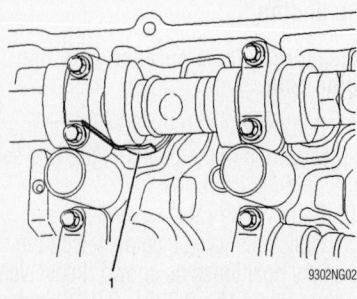

Insert special tool J-42689 (1) and use the camshaft to press the tappet down—3.2L DOHC engine

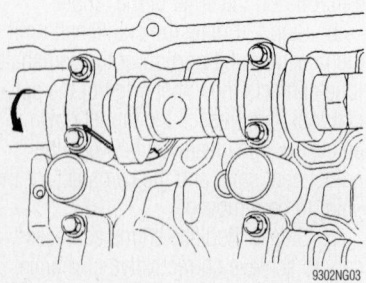

Rotate the camshaft to depress the valve with special tool J-42689—3.2L DOHC engine

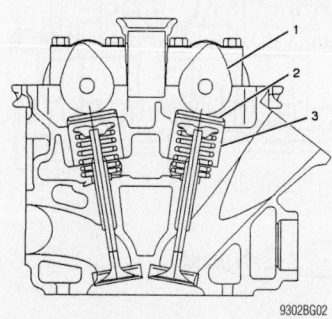

Cross section of the 3.5L cylinder head. Note the position of the camshaft lobe (1), adjustment shim (2) and the tappet (3)

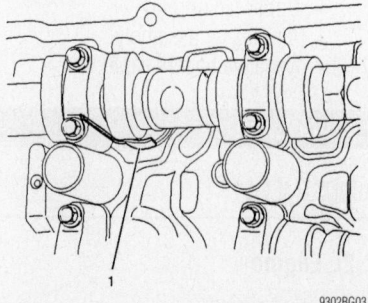

Valve clearance adjusting tool J–42689 (1)

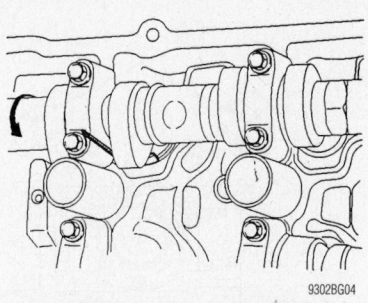

Using the valve clearance adjusting tool to hold the tappet for shim replacement

Starter Motor

REMOVAL & INSTALLATION

2.2L Engines

1. Before servicing the vehicle, refer to the precautions in the beginning of this section.

2. Remove or disconnect the following:

- Negative battery cable
- Starter harness connections
- Starter motor

To install:

3. Install or connect the following:

- Starter motor. Tighten the fasteners to 18 ft. lbs. (25 Nm) for the 2.2L engine or to 30 ft. lbs. (40 Nm) for the 2.6L engine.
- Starter harness connections
- Negative battery cable

3.2L Engines

1. Before servicing the vehicle, refer to the precautions in the beginning of this section.

2. Remove or disconnect the following:

- Negative battery cable
- Heated Oxygen (HO$_2$S) sensor connectors
- Exhaust front pipe
- Heat shield
- Starter wiring connectors
- Starter motor

To install:

3. Install or connect the following:

- Starter motor. Tighten the bolts to 30 ft. lbs. (40 Nm).
- Starter wiring connectors
- Heat shield
- Exhaust front pipe
- Heated Oxygen (HO$_2$S) sensor connectors
- Negative battery cable

3.5L Engine

1. Before servicing the vehicle, refer to the precautions in the beginning of this section.

2. Remove or disconnect the following:
   • Negative battery cable
   • Heated Oxygen (HO2S) sensor connectors
   • Exhaust front pipe
   • Heat shield
   • Starter wiring connectors
   • Starter motor

To install:

3. Install or connect the following:
   • Starter motor. Tighten the bolts to 30 ft. lbs. (40 Nm).
   • Starter wiring connectors
   • Heat shield
   • Exhaust front pipe
   • Heated Oxygen (HO2S) sensor connectors
   • Negative battery cable

Oil Pan

REMOVAL & INSTALLATION

2.2L Engines

1. Before servicing the vehicle, refer to the precautions in the beginning of this section.

2. Drain the engine oil.

3. Remove or disconnect the following:
   • Flywheel dust cover
   • Left and right engine mounts. Raise the engine for access.
   • Oil pan
   • Oil pan support, for 2.2L engine

To install:

4. Perform the following:

   a. Install the oil pan support and tighten the bolts to 14 ft. lbs. (20 Nm).

   b. Install the oil pan and tighten the bolts to 70 inch lbs. (8 Nm) plus 30 degrees.

5. Install or connect the following:
   • Left and right engine mounts. Tighten the nuts to 41 ft. lbs. (55 Nm).
   • Flywheel dust cover

6. Fill the crankcase to the correct level.

7. Start the engine and check for leaks.

3.2L Engines

1. Before servicing the vehicle, refer to the precautions in the beginning of this section.

2. Drain the engine oil.

3. Remove or disconnect the following:
   • Negative battery cable

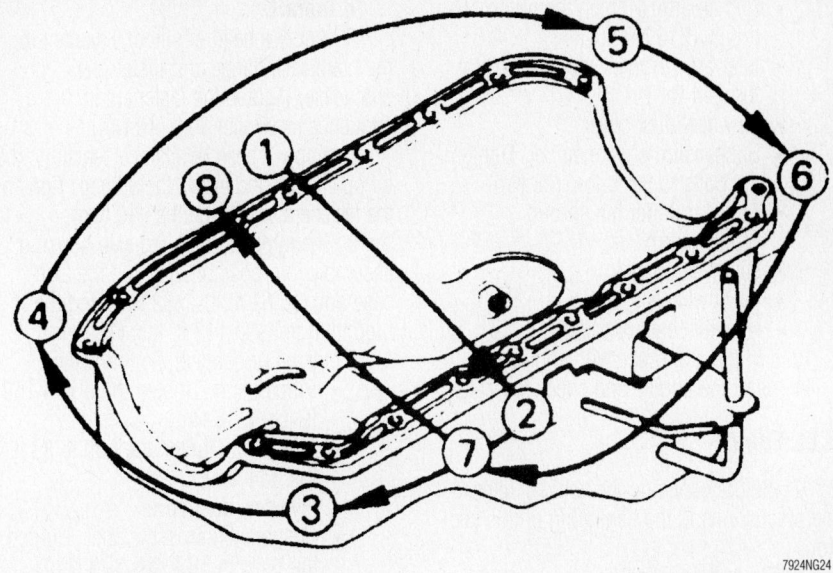

Oil pan bolt tightening sequence—2.2L engines

   • Front wheels
   • Oil level dipstick
   • Stone guard
   • Radiator under fan shroud
   • Suspension crossmember
   • Flywheel dust cover
   • Pitman arm
   • Idler arm

4. If equipped with 4 wheel drive, unbolt and lower the front axle housing assembly for clearance.

5. Remove or disconnect the following:
   • Oil pan
   • Lower crankcase

To install:

6. Apply a bead of silicone sealant to the crankcase flange and install the crankcase. Tighten the fasteners in sequence to 89 inch lbs. (10 Nm).

7. Apply a bead of silicone sealant to the oil pan flange and install the oil pan. Tighten the fasteners to 89 inch lbs. (10 Nm).

8. If equipped, raise the axle housing assembly into position. Tighten the axle case bolts to 61 ft. lbs. (82 Nm) and the mounting bolts to 112 ft. lbs. (152 Nm).

9. Install or connect the following:

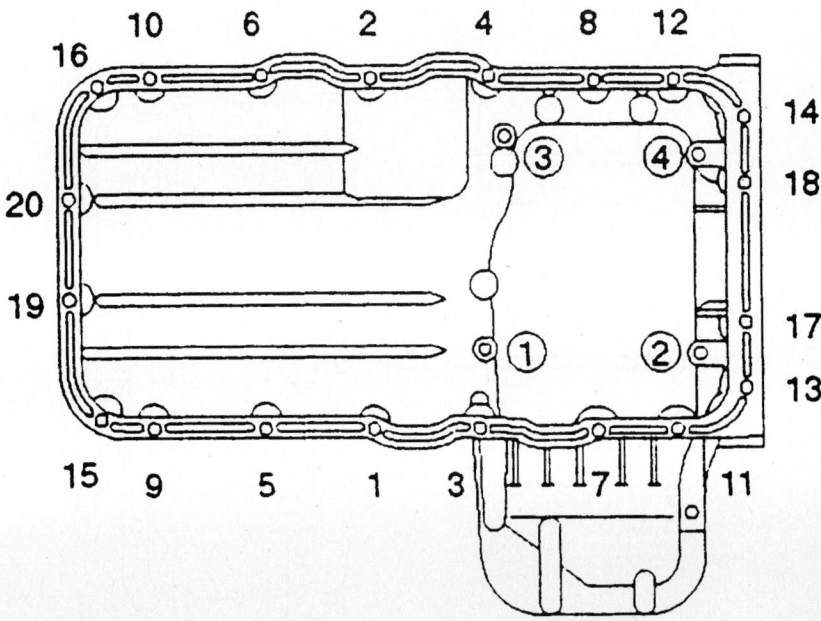

Lower crankcase torque sequence—3.2L DOHC engine

- Pitman arm. Tighten the nut to 159 ft. lbs. (216 Nm).
- Idler arm. Tighten the bolt to 33 ft. lbs. (44 Nm).
- Flywheel dust cover
- Suspension crossmember. Tighten the bolts to 58 ft. lbs. (78 Nm).
- Radiator under fan shroud
- Stone guard
- Oil level dipstick
- Front wheels
- Negative battery cable

10. Fill the crankcase with engine oil.
11. Start the engine and check for leaks.

3.5L Engine

1. Before servicing the vehicle, refer to the precautions in the beginning of this section.
2. Drain the engine oil.
3. Remove or disconnect the following:
   - Negative battery cable
   - Front wheels
   - Oil level dipstick
   - Stone guard
   - Radiator under fan shroud
   - Suspension crossmember
   - Flywheel dust cover
   - Pitman arm
   - Idler arm
4. If equipped with 4 wheel drive, unbolt and lower the front axle housing assembly for clearance.
5. Remove or disconnect the following:
   - Oil pan
   - Lower crankcase

Lower crankcase torque sequence—3.5L engine

To install:

6. Apply a bead of silicone sealant to the crankcase flange and install the crankcase. Tighten the fasteners in sequence to 89 inch lbs. (10 Nm).
7. Apply a bead of silicone sealant to the oil pan flange and install the oil pan. Tighten the fasteners to 89 inch lbs. (10 Nm).
8. If equipped, raise the axle housing assembly into position. Tighten the axle case bolts to 61 ft. lbs. (82 Nm) and the mounting bolts to 112 ft. lbs. (152 Nm).
9. Install or connect the following:
   - Pitman arm. Tighten the nut to 159 ft. lbs. (216 Nm).
   - Idler arm. Tighten the bolt to 33 ft. lbs. (44 Nm).
   - Flywheel dust cover
   - Suspension crossmember. Tighten the bolts to 58 ft. lbs. (78 Nm).
   - Radiator under fan shroud
   - Stone guard
   - Oil level dipstick
   - Front wheels
   - Negative battery cable
10. Fill the crankcase with engine oil.
11. Start the engine and check for leaks.

Oil Pump

REMOVAL & INSTALLATION

2.2L Engine

1. Before servicing the vehicle, refer to the precautions in the beginning of this section.

2. Drain the engine oil.
3. Remove or disconnect the following:
   - Negative battery cable
   - Accessory drive belts
   - Cooling fan
   - A/C belt tensioner, if equipped
   - Water pump pulley
   - Power steering pump
   - Crankshaft pulley
   - Front cover
   - Timing belt. Refer to the Timing Belt unit repair section.
   - Crankshaft timing sprocket
   - Rear timing cover
   - Crankshaft oil seal
   - Oil pan
   - Oil pump pickup tube
   - Oil pump

To install:

4. Install or connect the following:
   - Oil pump. Use a new gasket and tighten the bolts to 53 inch lbs. (6 Nm).
   - Oil pump pickup tube. Tighten the bolts to 70 inch lbs. (8 Nm).
   - Oil pan
   - Crankshaft oil seal
   - Rear timing cover
   - Crankshaft timing sprocket. Tighten the bolt to 94 ft. lbs. (130 Nm) plus 45 degrees.
   - Timing belt. Refer to the Timing Belt unit repair section.
   - Front cover
   - Crankshaft pulley. Tighten the bolts to 14 ft. lbs. (20 Nm).
   - Power steering pump
   - Water pump pulley
   - A/C belt tensioner, if equipped
   - Cooling fan
   - Accessory drive belts
   - Negative battery cable
5. Fill the crankcase to the correct level.
6. Start the engine and check for leaks.

3.2L Engine

1. Before servicing the vehicle, refer to the precautions in the beginning of this section.
2. Remove or disconnect the following:
   - Timing belt
   - Oil pan
   - Oil pickup tube
   - Oil filter adapter
   - Oil pump

To install:

3. Apply silicone sealant to the oil pump mounting surface and install the oil pump. Tighten the bolts in sequence to 18 ft. lbs. (25 Nm).
4. Install or connect the following:

9308BG01

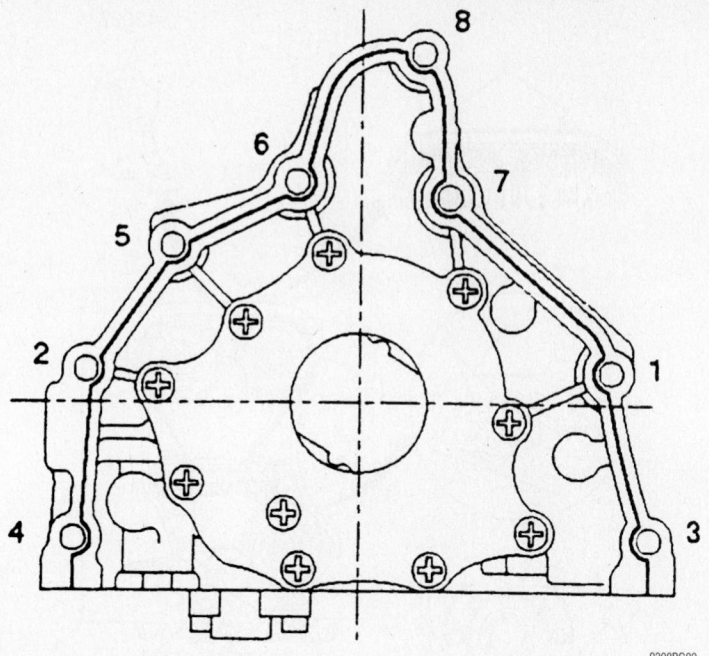

Oil pump torque sequence–3.2L engine

- Oil filter adapter
- Oil pickup tube
- Oil pan
- Timing belt

5. Fill the crankcase to the correct level.
6. Start the engine and check for leaks.

3.5L Engine

1. Before servicing the vehicle, refer to the precautions in the beginning of this section.

2. Remove or disconnect the following:
   - Timing belt
   - Oil pan
   - Oil pick-up tube
   - Oil filter adapter
   - Oil pump

To install:

3. Apply silicone sealant to the oil pump mounting surface and install the oil pump. Tighten the bolts in sequence to 18 ft. lbs. (25 Nm).

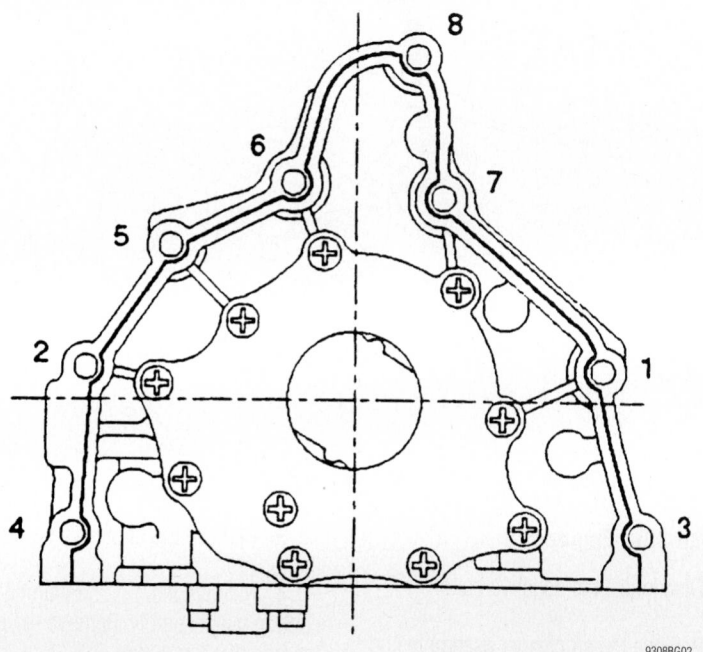

Oil pump torque sequence

4. Install or connect the following:
   - Oil filter adapter
   - Oil pickup tube
   - Oil pan
   - Timing belt

Rear Main Seal

REMOVAL & INSTALLATION

1. Before servicing the vehicle, refer to the precautions in the beginning of this section.
2. Remove or disconnect the following:
   - Negative battery cable
   - Transmission
   - Clutch assembly, if equipped with a manual transmission

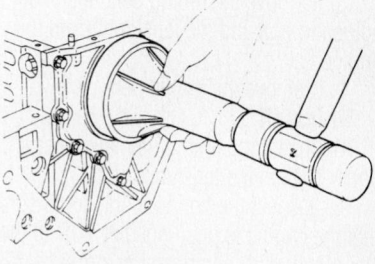

Installing a one-piece rear crankshaft oil seal

- Flywheel by loosening the flywheel bolts in a 2-step crisscross sequence
- Rear main seal, using a seal puller

➡**Do not damage the crankshaft sealing surface.**

To install:
3. Install or connect the following:
   - New rear main seal, by lubricating it with engine oil
   - Flywheel, using new flywheel bolts. Tighten the bolts, in a 2-step crisscross pattern, to 40 ft. lbs. (54 Nm).
   - Clutch assembly, if removed
   - Transmission
   - Negative battery cable
4. Check the oil and refill as necessary.

Timing Belt and Cover

REMOVAL & INSTALLATION

2.2L Engine

1. Disconnect the negative battery cable.

2. Using a box-end wrench on the drive belt adjuster, turn the adjuster clockwise and remove the drive belt.

3. From the left rear of the engine compartment, disconnect the 3 electrical connectors from the chassis harness.

4. Remove the crankshaft pulley-to-crankshaft bolts and remove the pulley.

5. From the front of the engine, remove the nut and the engine harness cover.

6. Remove the timing belt cover.

7. Rotate the crankshaft to position the timing marks at Top Dead Center (TDC) of the No. 1 cylinder's compression stroke.

➡ **Mark the rotational direction of the timing belt for reinstallation purposes.**

8. Remove the timing belt tensioner adjusting bolt and the tensioner from the engine.

9. Remove the timing belt.

To install:

10. Install the timing belt tensioner and finger-tighten the tensioner bolt.

11. Inspect the timing marks to be sure that the engine is positioned at TDC of the No. 1 cylinder's compression stroke.

12. Using tool J-43037, or equivalent, place it between the intake and exhaust sprockets to prevent the camshaft gear from moving during the timing belt installation.

13. Install the timing belt.

14. Position the timing belt to ensure that the tension side of the belt is taut and move the timing belt tension adjusting lever clockwise until the tensioner pointer is flowing.

15. If installing a used timing belt (used over 60 min. from new), the pointer should be positioned approximately 0.16 in. (4mm) to the left of the "V" notch when viewed from the front of the engine.

16. If installing a new timing belt, the pointer should be positioned at the center of the "V" notch when viewed from the front of the engine.

17. Torque the timing belt tensioner adjusting bolt to 18 ft. lbs. (25 Nm).

18. Install the timing belt front cover and torque the bolts to 53 inch lbs. (6 Nm).

19. Install the engine harness connectors.

20. Install the crankshaft pulley and toque the pulley-to-crankshaft bolts to 14 ft. lbs. (20 Nm).

21. Move the drive belt tensioner to the loose side and install the drive belt to its normal position.

22. Connect the negative battery cable.

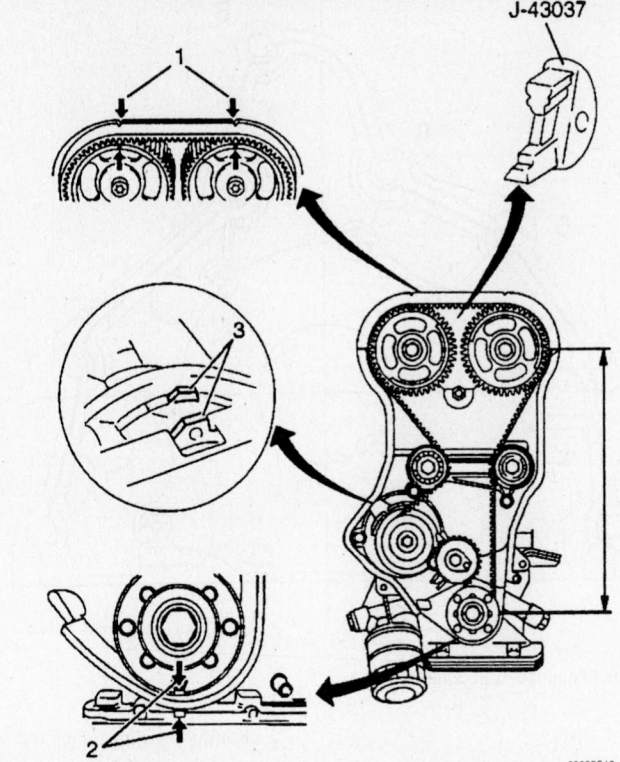

Aligning the timing marks and installing the timing belt —2.2L engine

93025G12

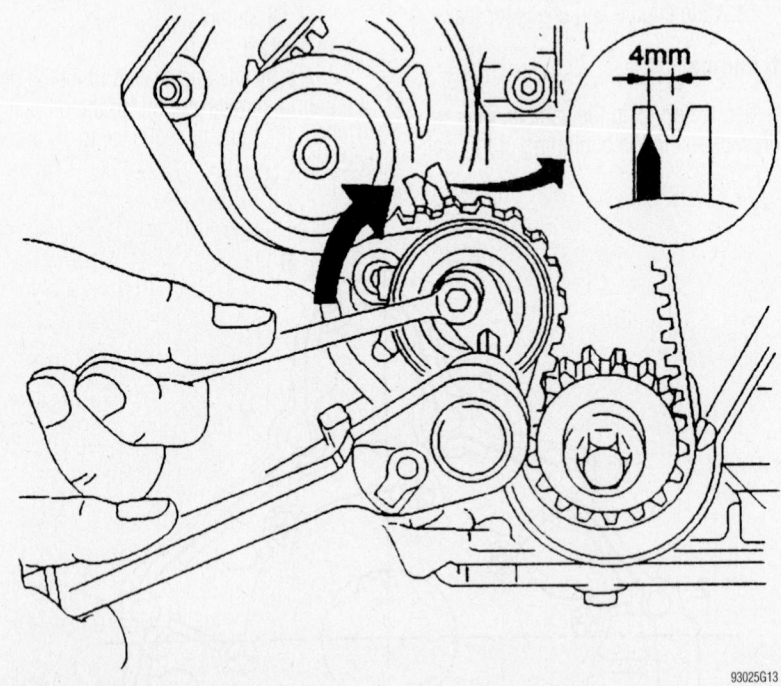

Tensioning the timing belt for a used timing belt —2.2L engine

93025G13

3.2L and 3.5L Engines

1. Disconnect the negative battery cable.

2. Remove the air cleaner assembly and intake air duct.

3. Remove the upper fan shroud from the radiator.

4. Remove the 4 nuts retaining the cooling fan assembly. Remove the cooling fan from the fan pulley.

5. Loosen and remove the drive belts.

6. Remove the upper timing belt covers.

7. Remove the fan pulley assembly.

8. Rotate the crankshaft to align the camshaft timing marks with the pointer dots on the back covers. Verify that the pointer on the crankshaft aligns with the mark on the lower timing cover.

➡ **When the timing marks are aligned, the No. 2 piston is at Top Dead Center (TDC) compression.**

✳✳ WARNING

Align the camshaft and crankshaft sprockets with their alignment marks before removing the timing belt. Failure to align the belt and sprocket marks may result in valve damage.

9. Use tool No. J-8614-01, or a suitable pulley holding tool to remove the crankshaft pulley center bolt. Remove the crankshaft pulley.

10. If present, disconnect the 2 oil cooler hose bracket bolts on the timing cover. Move the oil cooler hoses and bracket off of the lower timing cover.

11. Remove the lower timing belt cover.

12. Remove the pusher assembly (tensioner) from below the belt tensioner pulley. The pusher rod must always face upward to prevent oil leakage. Depress the pusher rod, and insert a wire pin into the hole to keep the pusher rod retracted.

13. Remove the timing belt.

14. Inspect the water pump and replace it if there is any doubt about its condition.

15. Repair any oil or coolant leaks before installing a new timing belt. If the timing belt has been contaminated with oil or coolant, or is damaged, it must be replaced.

To install:

16. Verify that the sprocket timing marks are still aligned and that the groove and the keyway on the crankshaft timing sprocket align with the mark on the oil pump. The white pointers on the camshaft timing sprockets should align with the dots on the front plate.

17. Install the timing belt. Use clips to secure the belt onto each sprocket until the installation is complete. Align the dotted marks on the timing belt with the timing mark opposite the groove on the crankshaft sprocket.

➡ **The arrows on the timing belt must follow the belt's direction of rotation. The manufacturer's trademark on the belt's spine should be readable left-to-right when the belt is installed.**

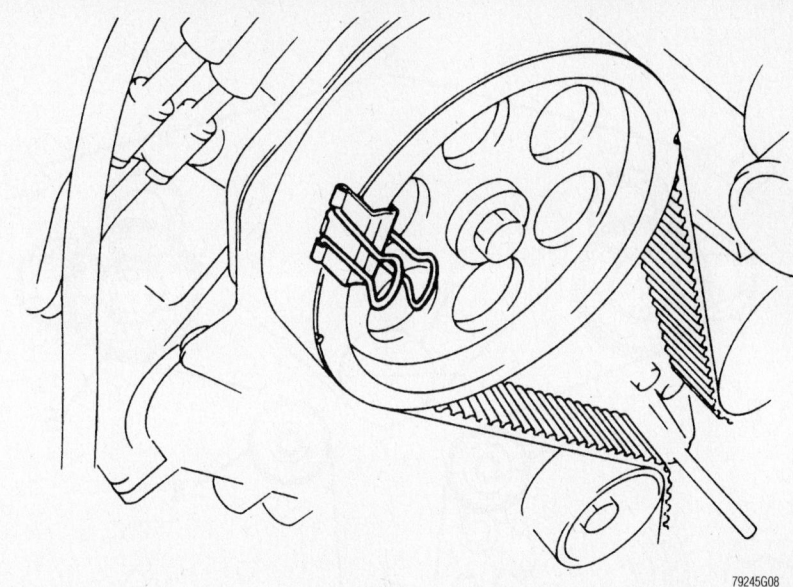

Using a double clip to hold the belt in place—3.2L and 3.5L engines

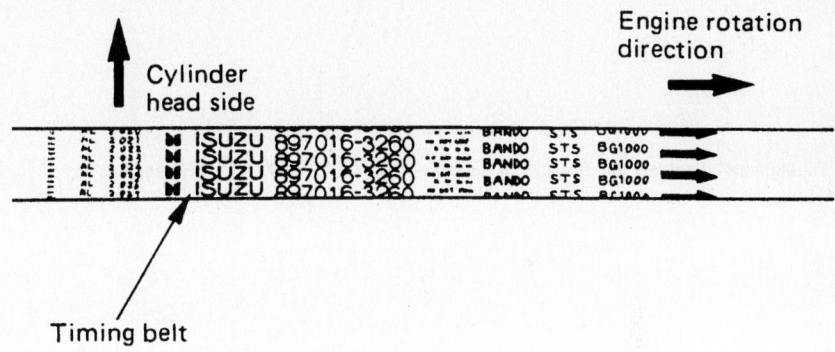

For maximum timing belt life, install the belt as shown—3.2L and 3.5L engines

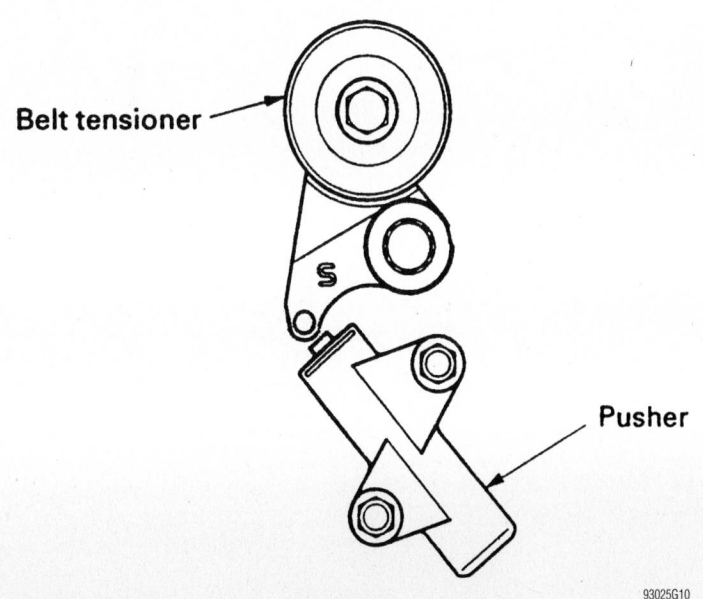

View of timing belt tensioner and pusher—3.2L and 3.5L engines

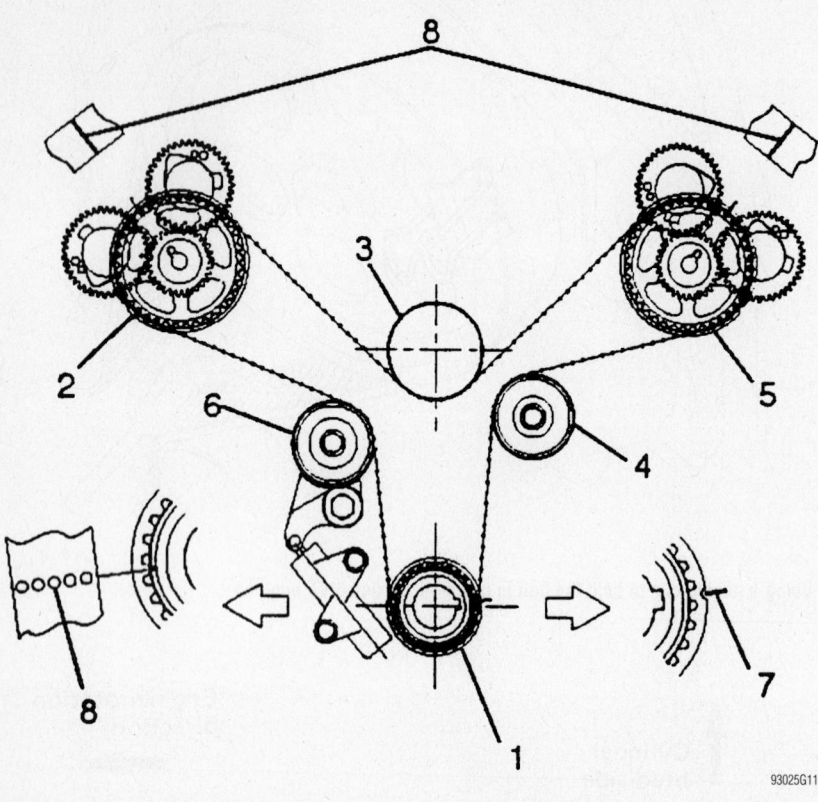

Timing mark alignment and timing belt routing—3.2L (VIN W) and 3.5L DOHC engines

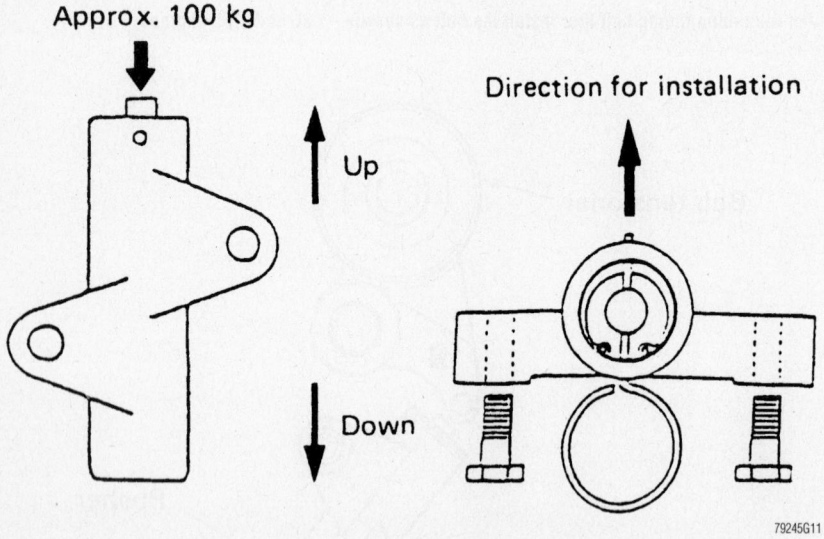

View of timing belt tensioner pusher—3.2L and 3.5L engines

18. Align the white line on the timing belt with the alignment mark on the right bank camshaft timing pulley. Secure the belt with a clip.

✳✳ WARNING

If any binding is felt when adjusting the timing belt tension by turning the crankshaft, STOP turning the engine, because the pistons may be hitting the valves.

19. Rotate the crankshaft counterclockwise to remove the slack between the crankshaft sprocket and the right camshaft timing belt sprocket.

20. Install the belt around the water pump pulley.

21. Install the belt on the idler pulley.

22. Align the white alignment mark on the timing belt with the alignment mark on the left bank camshaft timing belt sprocket.

23. Install the crankshaft pulley and tighten the center bolt by hand. Rotate the crankshaft pulley clockwise to give slack between the crankshaft timing belt pulley and the right bank camshaft timing belt pulley.

24. Insert a 1.4mm piece of wire through the hole in the pusher to hold the rod in. Install the pusher assembly while pushing the tension pulley toward the belt.

25. Pull the pin out from the pusher to release the rod.

26. Remove the clamps from the sprockets. Rotate the crankshaft pulley clockwise 2 turns. Measure the rod protrusion to ensure it is between 0.16 – 0.24 in. (4 – 6mm).

27. If the tensioner pulley bracket pivot bolt was removed, tighten it to 31 ft. lbs. (42 Nm).

28. Tighten the pusher bolts to 14 ft. lbs. (19 Nm).

29. Remove the crankshaft pulley. Install the lower and upper timing belt covers and tighten their bolts to 12 ft. lbs. (17 Nm).

30. Fit the oil cooler hose onto the timing cover and tighten its mounting bracket bolts to 16 ft. lbs. (22 Nm).

31. Install the crankshaft pulley and tighten the pulley bolt to 123 ft. lbs. (167 Nm).

32. Install fan pulley assembly and tighten the bolts to 16 ft. lbs. (22 Nm).

33. Install and adjust the accessory drive belts.

34. Install the cooling fan assembly and tighten the bolts to 72 inch lbs. (8 Nm).

35. Install the upper fan shroud.

36. Install the air cleaner assembly and intake air duct.

37. Connect the negative battery cable.

Piston and Ring

POSITIONING

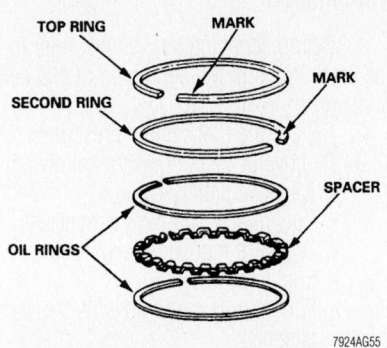

Piston ring positioning and top mark locations—all engines

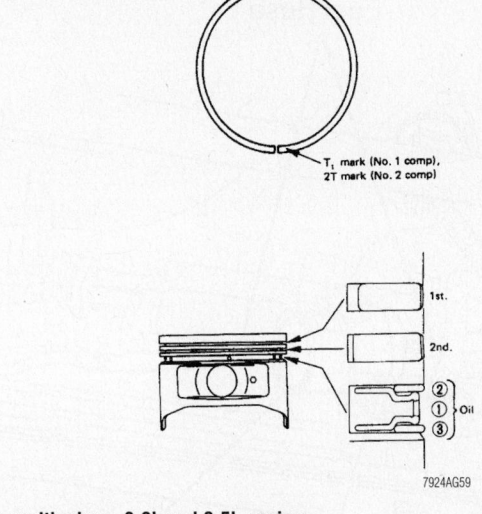

Piston ring positioning—3.2L and 3.5L engines

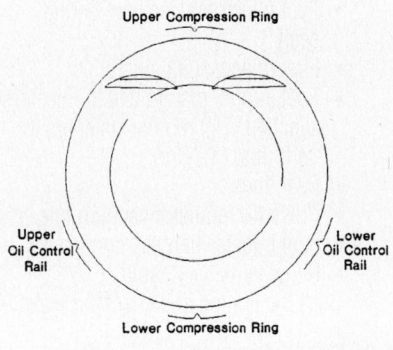

Piston ring end-gap spacing—2.2L engine

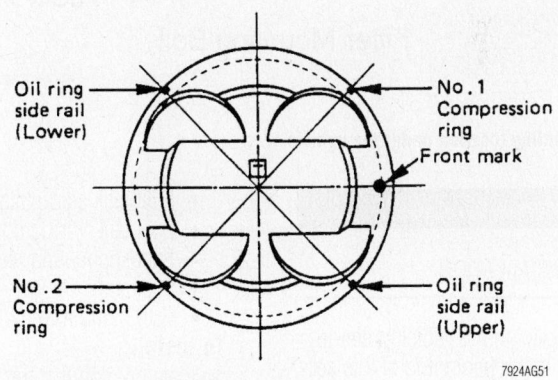

Piston ring end-gap spacing—3.2L and 3.5L engines

FUEL SYSTEM

Fuel System Service Precautions

Safety is the most important factor when performing not only fuel system maintenance but any type of maintenance. Failure to conduct maintenance and repairs in a safe manner may result in serious personal injury or death. Maintenance and testing of the vehicle's fuel system components can be accomplished safely and effectively by adhering to the following rules and guidelines:

• To avoid the possibility of fire and personal injury, always disconnect the negative battery cable unless the repair or test procedure requires that battery voltage be applied.

• Always relieve the fuel system pressure prior to disconnecting any fuel system component (injector, fuel rail, pressure regulator, etc.), fitting or fuel line connection. Exercise extreme caution whenever relieving fuel system pressure, to avoid exposing skin, face and eyes to fuel spray. Please be advised that fuel under pressure may penetrate the skin or any part of the body that it contacts.

• Always place a shop towel or cloth around the fitting or connection prior to loosening to absorb any excess fuel due to spillage. Ensure that all fuel spillage (should it occur) is quickly removed from engine surfaces. Ensure that all fuel soaked cloths or towels are deposited into a suitable waste container.

• Always keep a dry chemical (Class B) fire extinguisher near the work area.

• Do not allow fuel spray or fuel vapors to come into contact with a spark or open flame.

• Always use a backup wrench when loosening and tightening fuel line connection fittings. This will prevent unnecessary stress and torsion to fuel line piping. Always follow the proper tightening specifications.

• Always replace worn fuel fitting O-rings with new. Do not substitute fuel hose or equivalent, where fuel pipe is installed.

Fuel System Pressure

RELIEVING

1. Before servicing the vehicle, refer to the precautions in the beginning of this section.
2. Remove the fuel filler cap.
3. Remove the fuel pump relay from the underhood relay box.
4. Start the engine and let it run until it stalls, then crank the engine for an additional 30 seconds.
5. Turn the ignition switch to the **OFF** position and remove the key. Disconnect the negative battery cable.
6. When service is completed, install the fuel pump relay and connect the negative battery cable.

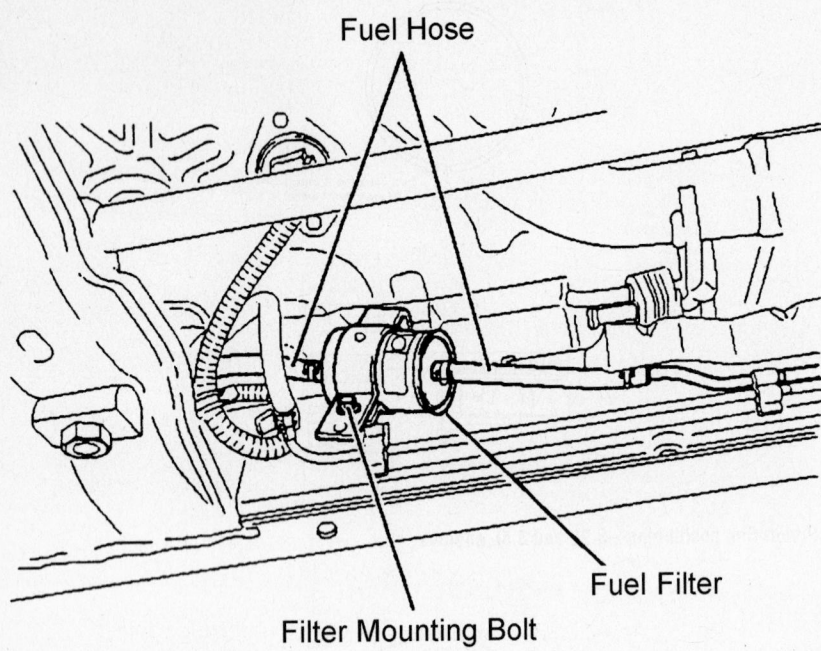

Fuel Hose

Fuel Filter

Filter Mounting Bolt

7924NG25

Fuel filter mounting location under the vehicle

Fuel Filter

REMOVAL & INSTALLATION

1. Before servicing the vehicle, refer to the precautions in the beginning of this section.
2. Relieve the fuel system pressure.
3. Remove or disconnect the following:
   - Fuel lines from the fuel filter
   - Fuel filter

To install:

4. Install or connect the following:
   - Fuel filter and tighten the bracket bolt. Note the fuel flow directional arrow.
   - Fuel lines to the fuel filter
   - Negative battery cable
5. Start the engine and inspect the fuel filter connections for leaks.

Fuel Pump

REMOVAL & INSTALLATION

1. Before servicing the vehicle, refer to the precautions in the beginning of this section.
2. Relieve fuel system pressure.
3. Drain the fuel tank.
4. Remove or disconnect the following:
   - Negative battery cable
   - Fuel filler and vent hoses

- Fuel tank skid plate
- Fuel tank wiring connectors
- Fuel supply and return lines
- Fuel tank
- Fuel pump assembly

To install:

5. Install or connect the following:
   - Fuel pump assembly
   - Fuel tank. Tighten the bolts to 27 ft. lbs. (36 Nm).
   - Fuel supply and return lines
   - Fuel tank wiring connectors
   - Fuel tank skid plate
   - Fuel filler and vent hoses
   - Negative battery cable
6. Start the engine and check for leaks.

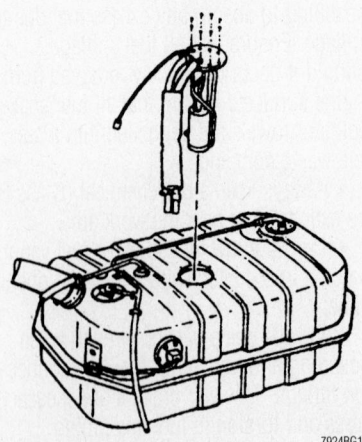

7924BG17

Fuel pump assembly mounting

Fuel Injector

REMOVAL & INSTALLATION

2.2L Engine

1. Before servicing the vehicle, refer to the precautions in the beginning of this section.
2. Relieve the fuel system pressure.
3. Remove or disconnect the following:
   - Negative battery cable
   - Fuel injector harness connectors
   - Pressure regulator vacuum line
   - Fuel lines
   - Fuel supply manifold with injectors attached
   - Fuel injector retaining clips
   - Fuel injectors

To install:

4. Install or connect the following:
   - Fuel injectors. Use new O-ring seals.
   - Fuel injector retaining clips
   - Fuel supply manifold with injectors attached. Tighten the fasteners to 14 ft. lbs. (19 Nm).
   - Fuel lines
   - Pressure regulator vacuum line
   - Fuel injector harness connectors
   - Negative battery cable
5. Start the engine and check for leaks.

3.2L Engines

1. Before servicing the vehicle, refer to the precautions in the beginning of this section.
2. Relieve fuel system pressure.
3. Remove or disconnect the following:
   - Negative battery cable
   - Engine cover
   - Fuel injector wiring connectors
   - Fuel lines
   - Fuel supply manifold with injectors attached
   - Fuel injector retaining clips
   - Fuel injectors

To install:

4. Install or connect the following:
   - Fuel injectors. Use new O-ring seals.
   - Fuel injector retaining clips
   - Fuel supply manifold with injectors attached. Tighten the bolts to 60 inch lbs. (6.5 Nm).
   - Fuel lines
   - Fuel injector wiring connectors
   - Engine cover
   - Negative battery cable
5. Start the engine and check for leaks.

3.5L Engine

1. Before servicing the vehicle, refer to the precautions in the beginning of this section.
2. Relieve fuel system pressure.
3. Remove or disconnect the following:
   - Negative battery cable
   - Engine cover
   - Fuel injector wiring connectors
   - Fuel lines
   - Fuel supply manifold with injectors attached
   - Clips
   - Injectors from the supply manifold

To install:
4. Install or connect the following:
   - New O-rings on the fuel injectors
   - Fuel injectors
   - Fuel supply manifold with injectors attached. Tighten the bolts to 60 inch lbs. (6.5 Nm).
   - Fuel lines
   - Fuel injector wiring connectors
   - Engine cover
   - Negative battery cable
5. Start the engine and check for leaks.

DRIVE TRAIN

Transmission Assembly

REMOVAL & INSTALLATION

Manual Transmissions

2 WHEEL DRIVE

➡**The transmission flange bolts vary in length. Note their locations for assembly.**

1. Before servicing the vehicle, refer to the precautions in the beginning of this section.
2. Remove the hood.
3. Install a support fixture to the engine lifting eyes.
4. Remove or disconnect the following:
   - Negative battery cable
   - Shift lever knob
   - Rear console assembly
   - Grommet assembly
   - Shift lever
   - Clutch slave cylinder and hose bracket
   - Driveshaft
   - Fuel line heat shield
   - Vehicle Speed (VSS) sensor connector
   - Reverse light switch connector
   - Flywheel under cover
   - Transmission mount and crossmember
   - Transmission flange bolts
   - Transmission

To install:
5. Install or connect the following:
   - Transmission. Tighten the large flange bolts to 52 ft. lbs. (71 Nm) and the small bolts to 30 ft. lbs. (41 Nm).

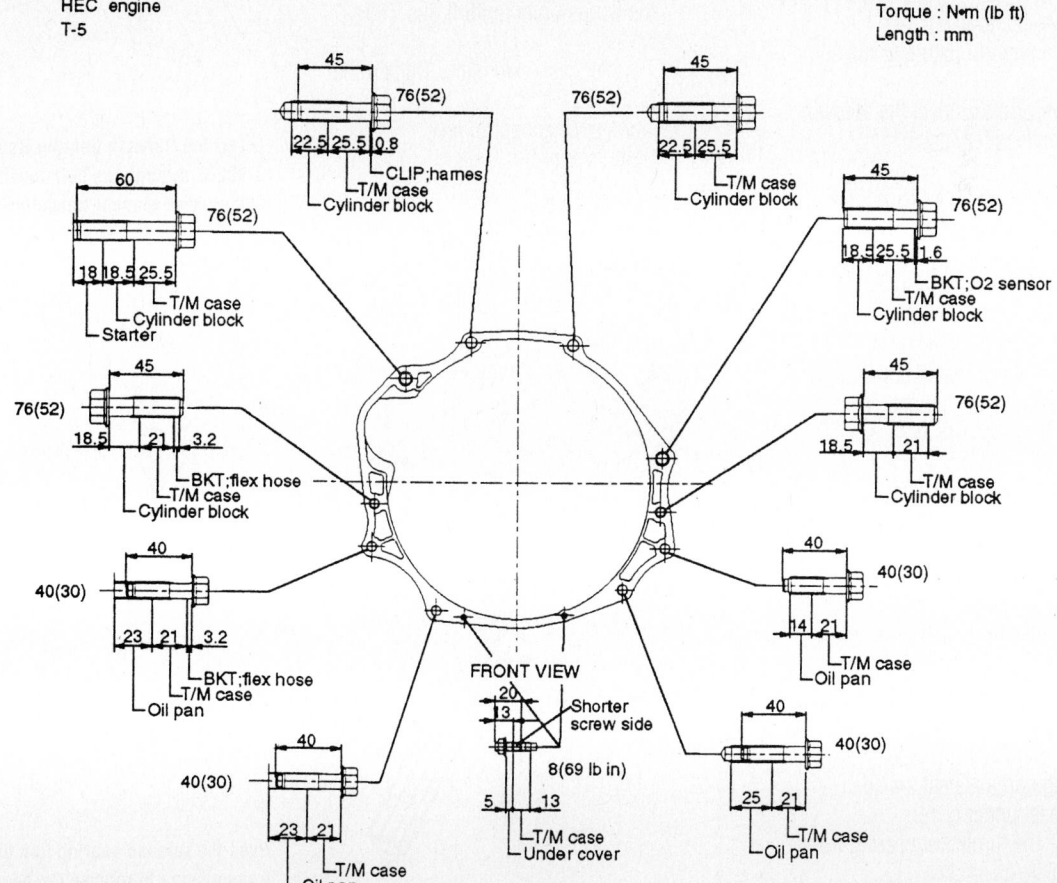

Transmission flange bolt identification and torque—2 wheel drive transmission

- Crossmember. Tighten the bolts to 56 ft. lbs. (76 Nm).
- Transmission mount. Tighten the fasteners to 30 ft. lbs. (41 Nm).
- Flywheel under cover. Tighten the bolts to 69 inch lbs. (8 Nm).
- Reverse light switch connector
- Vehicle Speed (VSS) sensor connector
- Fuel line heat shield
- Driveshaft. Tighten the bolts to 37 ft. lbs. (50 Nm).
- Clutch slave cylinder and hose bracket
- Shift lever
- Grommet assembly
- Rear console assembly
- Shift lever knob
- Negative battery cable

6. Remove the engine support fixture and install the hood.

4 WHEEL DRIVE

➡The transmission flange bolts vary in length. Note their locations for assembly.

1. Before servicing the vehicle, refer to the precautions in the beginning of this section.
2. Remove the hood.
3. Install a support fixture to the engine lifting eyes.
4. Remove or disconnect the following:
- Negative battery cable
- Shift lever knob
- Console assembly
- Grommet assembly
- Shift lever
- Transfer case control lever
- Transfer case skid plate
- Front and rear driveshafts
- Reverse lamp switch connector
- Indicator switch connectors
- Vehicle Speed (VSS) sensor connector
- 4WD actuator connector
- Transmission harness clamps
- Fuel pipe bracket
- Clutch slave cylinder and heat shield
- Transmission mount and crossmember
- Heated Oxygen (HO2S) sensor connectors
- Right exhaust front pipe
- Wiring harness heat shield
- Flywheel under cover

5. Release the throw out bearing from the pressure plate as shown.
6. Remove the trasmision flange bolts and remove the transmission.

To install:

7. Install the transmission. Tighten the large bolts to 56 ft. lbs. (76 Nm) and the small bolts to 52 inch lbs. (6 Nm).
8. Apply 13–18 lbs. (59–78 N) of force to the clutch fork to engage the throw out bearing to the pressure plate.
9. Install or connect the following:
- Flywheel under cover
- Wiring harness heat shield
- Right exhaust front pipe. Tighten the manifold flange fasteners to 49 ft. lbs. (67 Nm) and the exhaust flange bolts to 32 ft. lbs. (43 Nm).
- HO2S sensor connectors
- Crossmember. Tighten the bolts to 37 ft. lbs. (50 Nm).
- Transmission mount. Tighten the bolts to 30 ft. lbs. (41 Nm).
- Clutch slave cylinder and heat shield
- Fuel pipe bracket
- Transmission harness clamps
- 4WD actuator connector
- VSS sensor connector
- Indicator switch connectors
- Reverse lamp switch connector
- Front and rear driveshafts. Tighten the flange bolts to 46 ft. lbs. (63 Nm).
- Transfer case skid plate. Tighten the bolts to 27 ft. lbs. (37 Nm).
- Transfer case control lever
- Shift lever

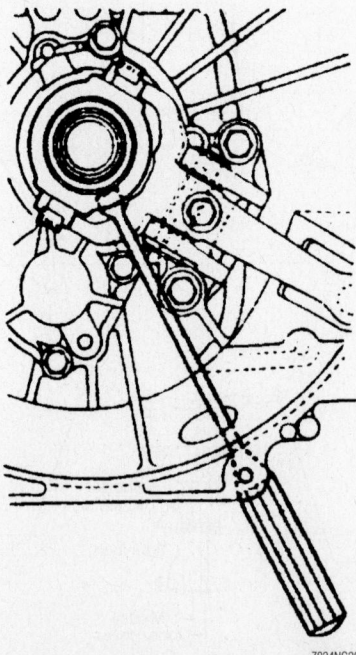

Insert the tool between the wedge collar and the releose bearing

- Grommet assembly
- Console assembly
- Shift lever knob
- Negative battery cable

10. Remove the engine support fixture and install the hood.

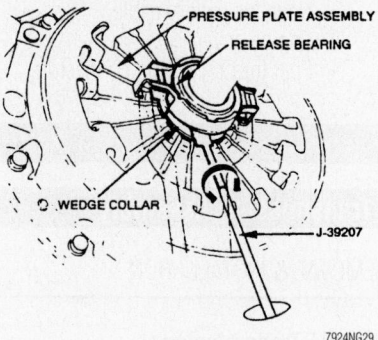

Turn the remover to separate the release bearing

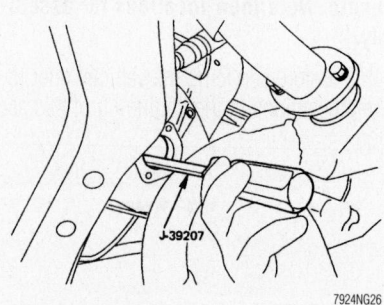

Insert the Release Bearing Remover tool J-39207 through the bell housing—4 wheel drive manual transmission

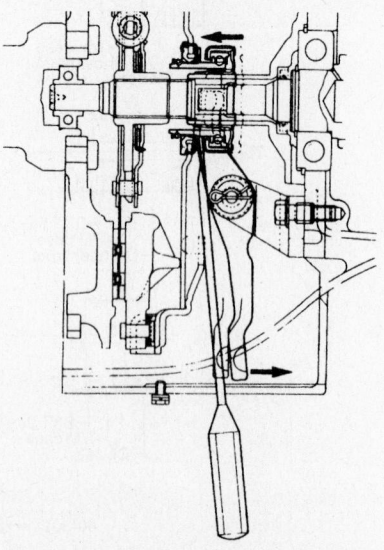

Push the release bearing fork toward the transmission to release the bearing from the pressure plate—4 wheel drive manual transmission

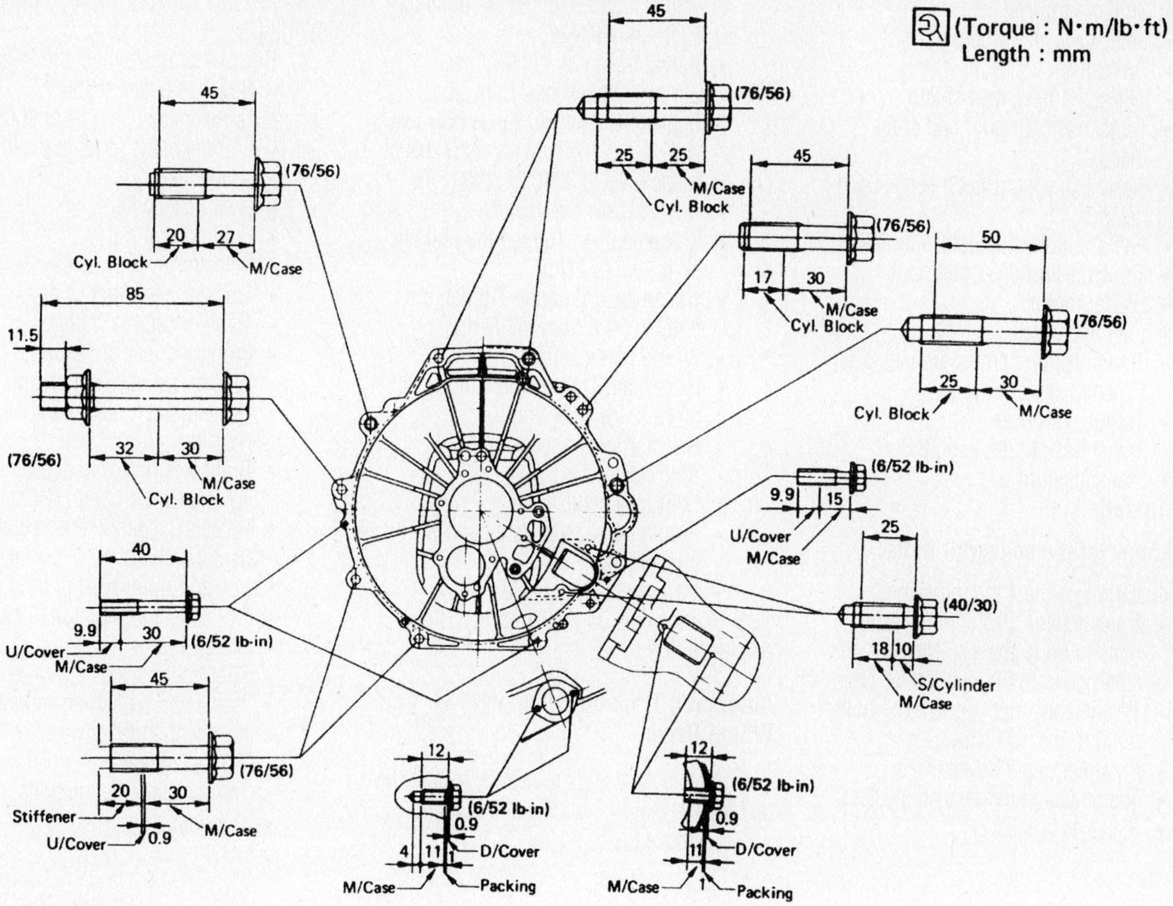

(Torque : N·m/lb·ft)
Length : mm

45

(76/56)

25 | 25
M/Case
Cyl. Block

45

(76/56)

17 | 30
M/Case
Cyl. Block

50

(76/56)

25 | 30
Cyl. Block | M/Case

45

(76/56)

20 | 27
Cyl. Block | M/Case

85

11.5

(76/56)

32 | 30
M/Case
Cyl. Block

(6/52 lb-in)

9.9 | 15
U/Cover
M/Case

25

(40/30)

18 | 10
S/Cylinder
M/Case

40

9.9 | 30 | (6/52 lb-in)
U/Cover
M/Case

45

(76/56)

20 | 30
Stiffener
U/Cover | 0.9 | M/Case

12

(6/52 lb-in)
0.9
4 | 11 | D/Cover
M/Case | Packing

12

(6/52 lb-in)
0.9
11 | D/Cover
M/Case | 1 | Packing

7924NG30

Transmission mounting bolt identification and torque specifications—V6 engine with 4 wheel drive transmission shown

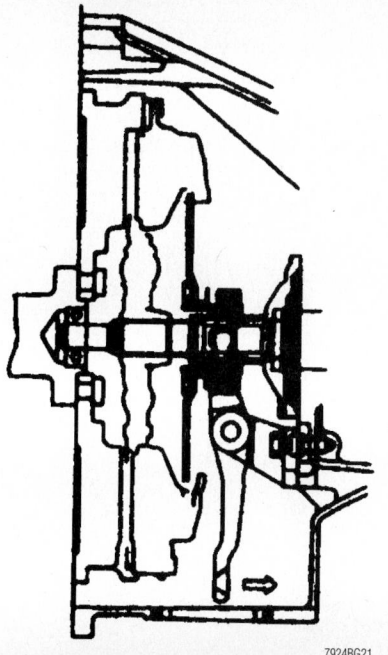

7924BG21

Push the release bearing fork toward the transmission to engage the release bearing with the pressure plate

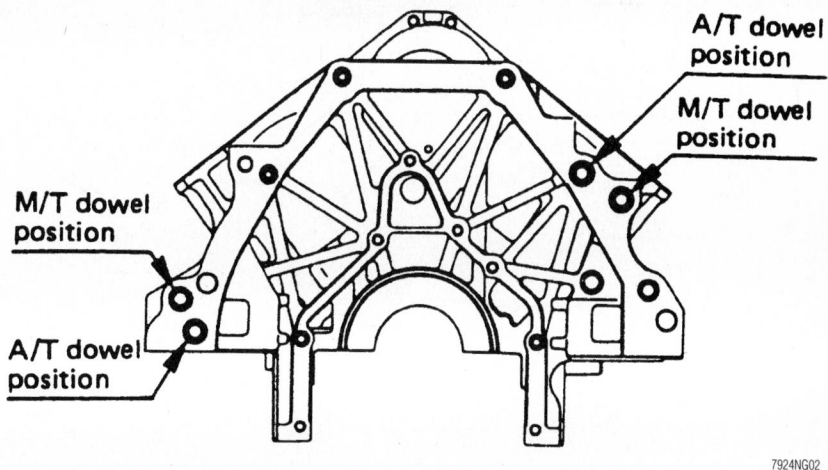

A/T dowel position

M/T dowel position

M/T dowel position

A/T dowel position

7924NG02

Dowel pin locations for automatic and manual transmissions—3.2L engine

Automatic Transmissions with 2-Wheel Drive

1. Before servicing the vehicle, refer to the precautions in the beginning of this section.
2. Remove the hood.
3. Install a support fixture to the engine lifting eyes.
4. Remove or disconnect the following:
   • Negative battery cable
   • Front console assembly and wiring connectors
   • Shift lock cable

- Shift control rod
- Selector lever assembly
- Driveshaft
- Wiring harness heat shield
- Transmission mount and cross-member
- Heated Oxygen (HO2S) sensor connectors
- Left and right exhaust front pipes
- Transmission oil cooler lines
- Starter motor
- Fuel line bracket
- Transmission harness connectors
- Flywheel under covers
- Torque converter
- Transmission flange bolts
- Transmission

To install:

➡ **Use new torque converter bolts.**

5. Install or connect the following:
- Transmission. Tighten the large bolts to 56 ft. lbs. (76 Nm) and the small bolts to 69 inch lbs. (8 Nm).
- Torque converter. Tighten the bolts to 40 ft. lbs. (54 Nm).
- Flywheel under covers
- Transmission harness connectors
- Fuel line bracket

- Starter motor. Tighten the bolts to 30 ft. lbs. (40 Nm).
- Transmission oil cooler lines
- Left and right exhaust front pipes. Tighten the manifold flange fasteners to 49 ft. lbs. (67 Nm) and the exhaust flange bolts to 32 ft. lbs. (43 Nm).
- HO2S sensor connectors
- Crossmember. Tighten the bolts to 37 ft. lbs. (50 Nm).
- Transmission mount. Tighten the bolts to 30 ft. lbs. (41 Nm).
- Wiring harness heat shield
- Driveshaft. Tighten the flange bolts to 46 ft. lbs. (63 Nm).
- Selector lever assembly
- Shift control rod
- Shift lock cable
- Front console assembly and wiring connectors
- Negative battery cable

6. Remove the engine support fixture and install the hood.

Automatic Transmissions with 4-Wheel Drive

1. Before servicing the vehicle, refer to the precautions in the beginning of this section.
2. Remove the hood.

3. Install a support fixture to the engine lifting eyes.
4. Remove or disconnect the following:
- Negative battery cable
- Transfer case shift lever knob
- Front console assembly and wiring connectors
- Shift lock cable
- Shift control rod
- Selector lever assembly
- Transfer case shift lever
- Transfer case skid plate
- Front and rear driveshafts
- Wiring harness heat shield
- Transmission mount and cross-member
- Right torsion bar, if equipped with Torque On Demand (TOD) system
- Heated Oxygen (HO2S) sensor connectors
- Left and right exhaust front pipes
- Transmission oil cooler lines
- Starter motor
- Fuel line bracket
- Transmission harness connectors
- Flywheel under covers
- Torque converter
- Transmission flange bolts
- Transmission

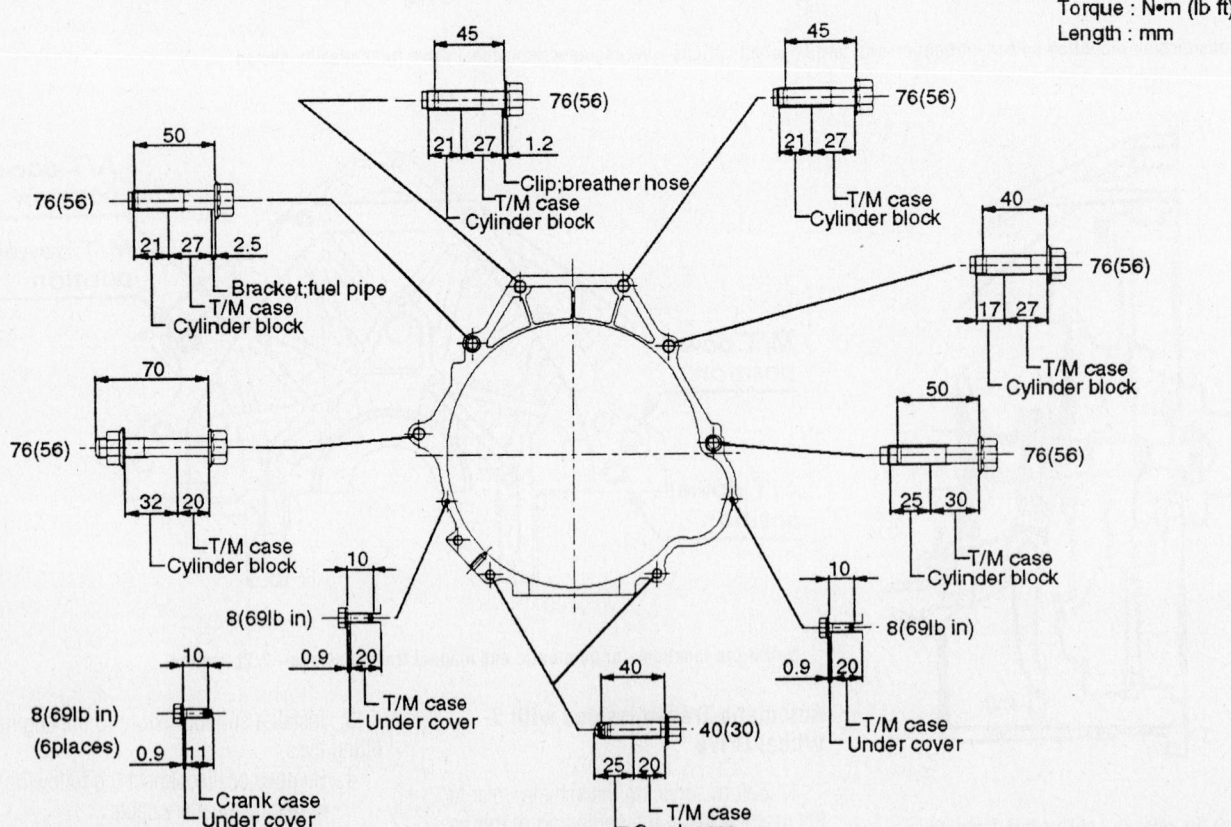

Torque : N•m (lb ft)
Length : mm

Automatic transmission mounting bolt locations and torque specifications

9308NG03

To install:

➡ **Use new torque converter bolts.**

5. Install or connect the following:
- Transmission. Tighten the large bolts to 56 ft. lbs. (76 Nm) and the small bolts to 30 ft. lbs. (40 Nm).
- Torque converter. Tighten the bolts to 40 ft. lbs. (54 Nm).
- Flywheel under covers
- Transmission harness connectors
- Fuel line bracket
- Starter motor. Tighten the bolts to 30 ft. lbs. (40 Nm).
- Transmission oil cooler lines
- Left and right exhaust front pipes. Tighten the manifold flange fasteners to 49 ft. lbs. (67 Nm) and the exhaust flange bolts to 32 ft. lbs. (43 Nm).
- HO2S sensor connectors
- Right torsion bar, if removed
- Crossmember. Tighten the bolts to 37 ft. lbs. (50 Nm).
- Transmission mount. Tighten the bolts to 30 ft. lbs. (41 Nm).
- Wiring harness heat shield
- Front and rear driveshafts. Tighten the flange bolts to 46 ft. lbs. (63 Nm).
- Transfer case skid plate. Tighten the bolts to 27 ft. lbs. (37 Nm).
- Transfer case shift lever
- Selector lever assembly
- Shift control rod
- Shift lock cable
- Front console assembly and wiring connectors
- Transfer case shift lever knob
- Negative battery cable

6. Remove the engine support fixture and install the hood.

Clutch

ADJUSTMENTS

➡ **This vehicle is equipped with a hydraulic clutch linkage. No adjustment is necessary.**

REMOVAL & INSTALLATION

1. Before servicing the vehicle, refer to the precautions in the beginning of this section.
2. Remove the transmission.
3. Loosen the pressure plate mounting bolts in a 2-step crisscross sequence until the spring tension is relieved.
4. Remove the pressure plate and the clutch disc.

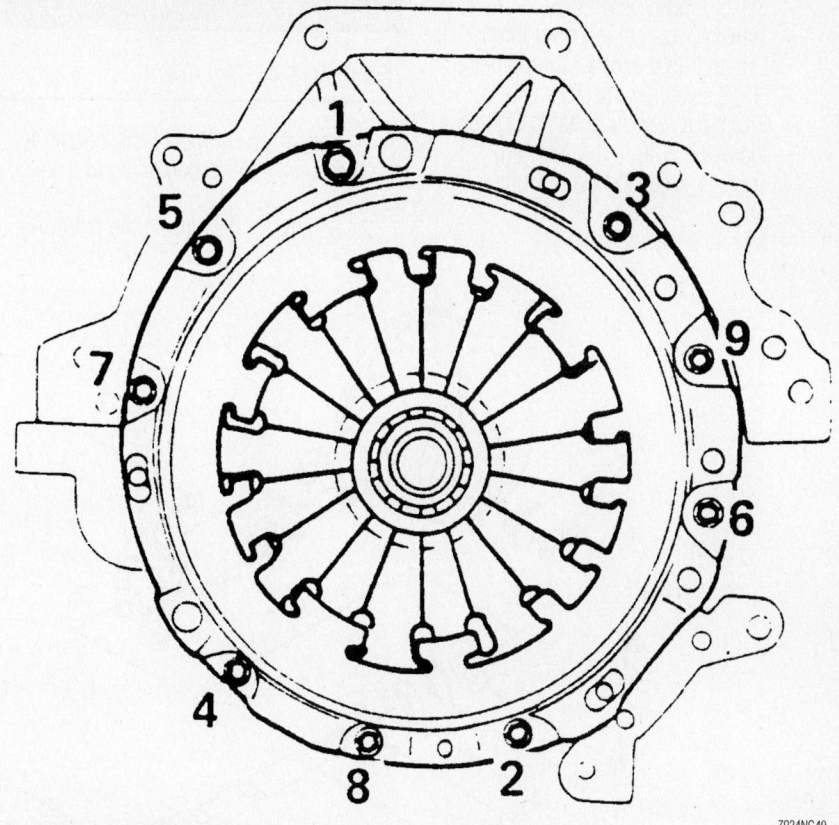

Pressure plate tightening sequence

To install:

5. Install a new wedge collar and wire snapring into the pressure plate.
6. Using a clutch alignment tool, assemble the clutch disc and pressure plate onto the flywheel.
7. Tighten the pressure plate bolts in sequence and in two passes to 13 ft. lbs. (8 Nm).
8. Install the transmission.
9. Road test the vehicle and check for proper clutch operation.

Hydraulic Clutch System

BLEEDING

1. Before servicing the vehicle, refer to the precautions in the beginning of this section.
2. Have an assistant pump the clutch pedal slowly several times and hold it depressed.
3. Open the slave cylinder bleeder screw and allow air to escape.
4. Close the bleeder screw before releasing the clutch pedal.
5. Repeat until all air is purged from the clutch hydraulic system.
6. Refill the reservoir to the full mark.

Transfer Case Assembly

REMOVAL & INSTALLATION

1. Before servicing the vehicle, refer to the precautions in the beginning of this section.
2. Remove or disconnect the following:
- Negative battery cable
- Transfer case skid plate
- Front and rear driveshafts
- Heated Oxygen (HO2S) sensor connectors
- Left and right exhaust front pipes
- Transfer case control lever knob
- Selector lever assembly
- Transfer case control lever
- Vehicle Speed (VSS) sensor connector
- 4 wheel drive switch connector
- 4 wheel drive actuator connector
- Transfer case flange fasteners
- Transfer case

To install:
3. Install or connect the following:
- Transfer case. Tighten the flange fasteners to 34 ft. lbs. (46 Nm).
- 4 wheel drive actuator connector
- 4 wheel drive switch connector
- VSS sensor connector
- Transfer case control lever

7924NG49

- Selector lever assembly
- Transfer case control lever knob
- Left and right exhaust front pipes
- HO2S sensor connectors
- Front and rear driveshafts. Tighten the bolts to 46 ft. lbs. (63 Nm).
- Transfer case skid plate. Tighten the bolts to 27 ft. lbs. (37 Nm).
- Negative battery cable

Halfshaft

REMOVAL & INSTALLATION

1. Before servicing the vehicle, refer to the precautions in the beginning of this section.
2. Remove or disconnect the following:

- Negative battery cable
- Front wheel
- Radiator skid plate
- Transfer case skid plate
- Brake calipers and mounting bracket
- Brake rotor
- Wheel speed sensor
- Steering knuckle

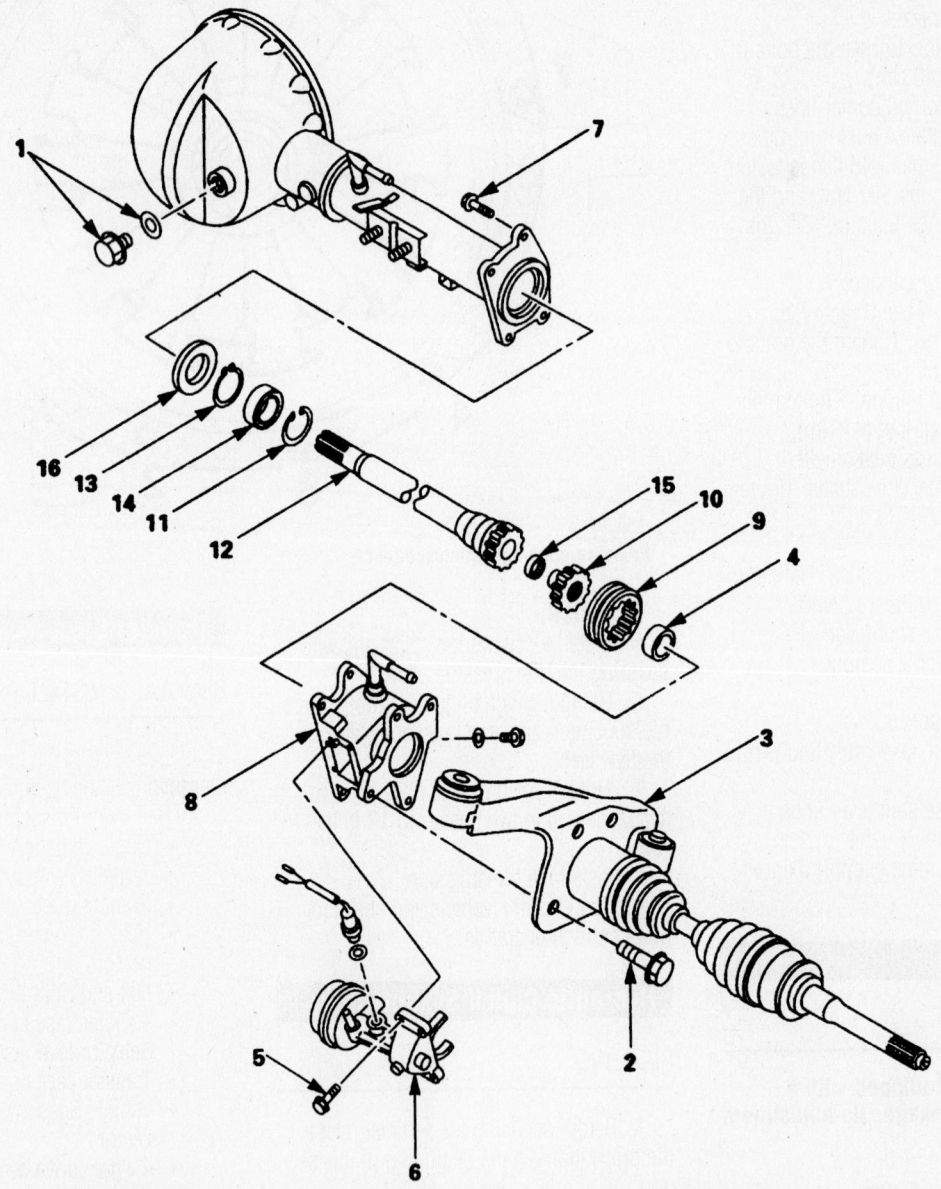

1. Filler plug	9. Sleeve
2. Bolt	10. Clutch gear
3. Front axle drive shaft (LH side)	11. Snap ring
4. Spacer	12. Inner shaft
5. Bolt	13. Snap ring
6. Actuator assembly	14. Inner shaft bearing
7. Bolt	15. Needle bearing
8. Housing	16. Oil seal

7924BG26

Exploded view of the left halfshaft, axle shaft and axle disconnect

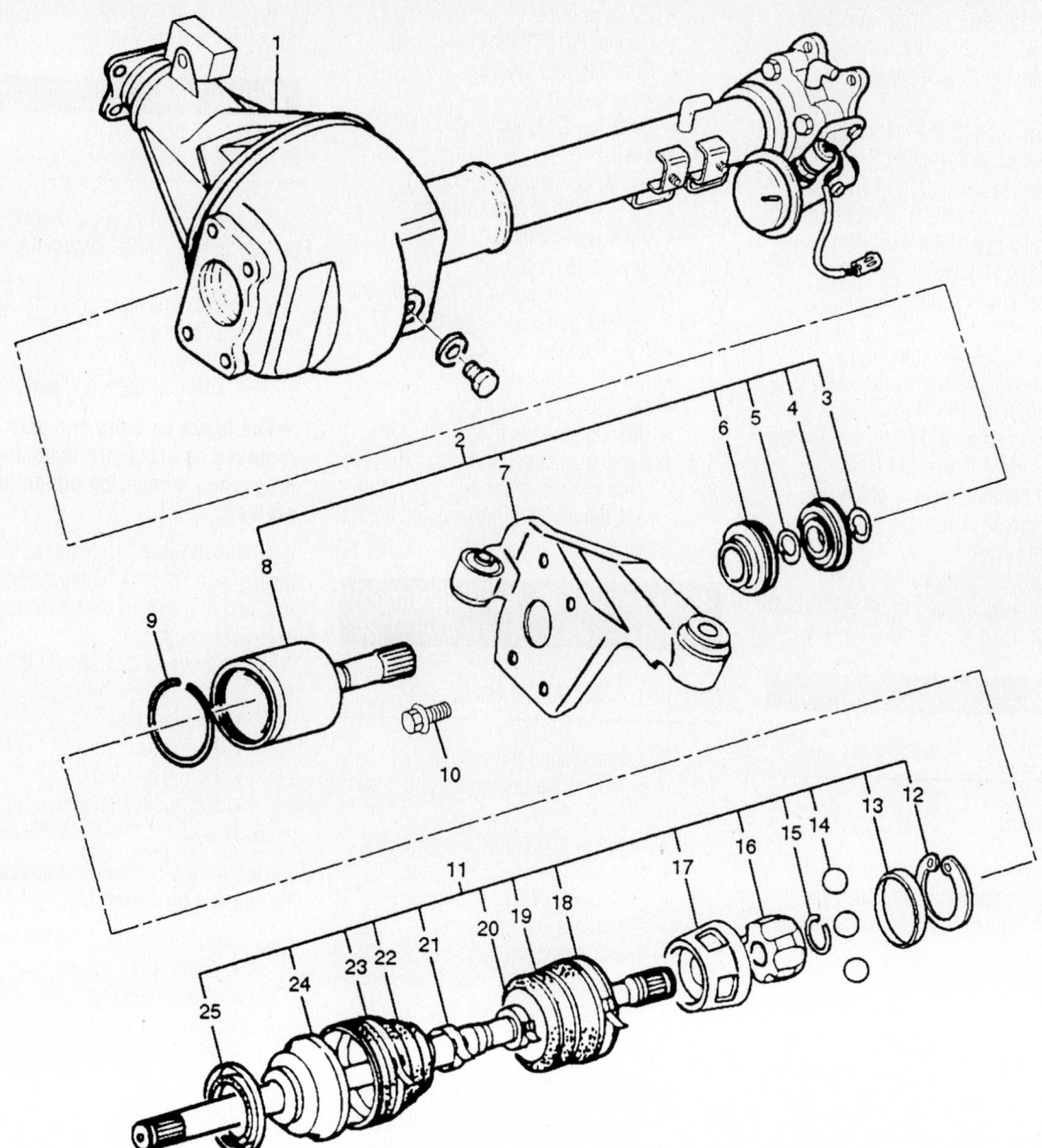

1 Axle Case and Differential
2 DOJ Case Assembly
3 Snap Ring
4 Bearing
5 Snap Ring
6 Oil Seal
7 Bracket
8 DOJ Case
9 Circlip
10 Bolt
11 Drive Shaft Joint Assembly
12 Snap Ring

13 Spacer
14 Ball
15 Snap Ring
16 Ball Retainer
17 Ball Guide
18 Band
19 Bellows
20 Band
21 Band
22 Bellows
23 Band
24 BJ Shaft
25 Dust Seal

9308BG03

Exploded view of the right halfshaft and mounting bracket

3. Support the axle housing with a jack. Unbolt the axle mounting bracket and remove the halfshaft/bracket assembly.

To install:

4. Install or connect the following:
- Axle/bracket assembly. Tighten the bracket flange bolts to 85 ft. lbs. (116 Nm) and the bracket mounting bolts to 112 ft. lbs. (152 Nm).
- Steering knuckle
- Wheel speed sensor
- Brake rotor
- Brake caliper and mounting bracket. Tighten the bracket bolts to 115 ft. lbs. (155 Nm).
- Transfer case skid plate. Tighten the bolts to 27 ft. lbs. (37 Nm).
- Radiator skid plate. Tighten the bolts to 58 ft. lbs. (78 Nm).
- Front wheel
- Negative battery cable

5. Check the wheel alignment and adjust as necessary.

CV-Joints

OVERHAUL

Outer CV-Joint

The outer CV-joint is serviced with the axle shaft as an assembly. The outer CV-joint boot can be serviced by removing the inner CV-joint.

Inner CV-Joint

1. Before servicing the vehicle, refer to the precautions in the beginning of this section.

2. Remove or disconnect the following:
- Halfshaft from the vehicle
- Snapring and bearing
- Snapring and oil seal
- Mounting bracket

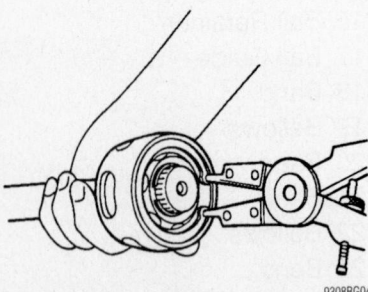

9308BG04

CV-joint spacer snapring—Inner CV-Joint

- CV-joint boot
- Circlip and inner joint housing
- Snapring and spacer
- Inner joint balls
- Snapring and inner CV-joint

To install:

3. Install or connect the following:
- Inner CV-joint and snapring
- Inner joint balls
- Spacer and snapring
- Inner joint housing and circlip. Add 150 grams CV-joint grease.
- CV-joint boot
- Mounting bracket
- Oil seal and snapring
- Bearing and snapring

4. Install the halfshaft and mounting bracket to the vehicle.

5. Check the wheel alignment and adjust as necessary.

Rear Axle Shaft, Bearing and Seal

REMOVAL & INSTALLATION

1. Before servicing the vehicle, refer to the precautions in the beginning of this section.

2. Remove or disconnect the following:
- Rear wheel
- Disc brake caliper and bracket
- Disc brake rotor
- Wheel speed sensor bracket
- Parking brake cable and bracket
- Parking brake shoes
- Axle shaft
- Snapring and discard it
- Bearing, press it off the axle shaft with the bearing holder and oil seal

To install:

3. Install or connect the following:
- New oil seal into the bearing housing
- Bearing housing onto the axle shaft
- Bearing, press it onto the axle shaft
- New snapring
- Axle shaft. Use new lockwashers and tighten the bearing holder nuts to 54 ft. lbs. (74 Nm).
- Parking brake shoes
- Parking brake cable and bracket
- Wheel speed sensor bracket
- Disc brake rotor
- Disc brake caliper and bracket. Tighten the bracket bolts to 76 ft. lbs. (103 Nm).
- Rear wheel

4. Check the rear axle oil level and adjust as necessary.

Pinion Seal

REMOVAL & INSTALLATION

1. Before servicing the vehicle, refer to the precautions in the beginning of this section.

2. Remove or disconnect the following:
- Driveshaft
- Wheels
- Brake calipers and pads

➡ **The brake calipers and pads must be removed so that there is no additional drag when measuring pinion bearing preload.**

3. Use an inch lb. torque wrench and measure and record the amount of torque required to maintain pinion rotation through several revolutions.

4. Remove or disconnect the following:
- Pinion flange
- Pinion seal
- Pinion bearing
- Collapsible spacer

To install:

➡ **Use a new collapsible spacer and flange nut for assembly.**

5. Install or connect the following:
- Collapsible spacer
- Pinion bearing
- Pinion seal
- Pinion flange

6. Rotate the pinion flange occasionally while tightening the flange nut to make sure the pinion bearings seat correctly.

7. Take frequent bearing preload torque readings. Tighten the flange nut to achieve the preload torque readings originally recorded.

✳✳ CAUTION

Never loosen the pinion nut to reduce bearing preload. If it is necessary to reduce bearing preload, install a new collapsible spacer and pinion nut.

8. Install or connect the following:
- Driveshaft
- Brake calipers and pads
- Wheels

9. Fill the differential with gear lubricant and check for leaks.

STEERING AND SUSPENSION

Air Bag

✳✳ CAUTION

Some vehicles are equipped with an air bag system. The system must be disarmed before performing service on, or around, system components, the steering column, instrument panel components, wiring and sensors. Failure to follow the safety precautions and the disarming procedure could result in accidental air bag deployment, possible injury and unnecessary system repairs.

PRECAUTIONS

Several precautions must be observed when handling the inflator module to avoid accidental deployment and possible personal injury.

• Never carry the inflator module by the wires or connector on the underside of the module.

• When carrying a live inflator module, hold securely with both hands, and ensure that the bag and trim cover are pointed away from you.

• Place the inflator module on a bench or other surface with the bag and trim cover facing up.

• With the inflator module on the bench, never place anything on or close to the module which may be thrown in the event of an accidental deployment.

DISARMING

1. Before servicing the vehicle, refer to the precautions in the beginning of this section.
2. Turn the ignition switch to the **LOCK** position. Remove the key.
3. Disconnect the negative battery cable. Wait 1 minute before working around the air bags.
4. Disconnect the yellow 2-pin connector at the base of the steering column.
5. Disconnect the yellow 2-pin connector behind the glove box assembly.
6. When repairs are completed, connect the yellow 2-pin connectors.
7. Connect the negative battery cable.
8. Turn the ignition to the **ON** position, but don't start the engine. The AIR BAG warning light should turn on and flash on and off 7 times, and then turn off. This light sequence indicates that the SRS system is functioning normally. If the AIR BAG light doesn't come on, or stays on longer than 7 seconds, the system must be diagnosed.

Recirculating Ball Steering Gear

REMOVAL & INSTALLATION

1. Before servicing the vehicle, refer to the precautions in the beginning of this section.
2. Disable the air bag system.
3. Remove or disconnect the following:
   • Skid plates
   • Lower fan shroud
   • Stabilizer bar
   • Power steering pressure and return lines
   • Pitman arm
   • Steering column intermediate shaft
   • Steering gear

To install:
4. Install or connect the following:
   • Steering gear. Tighten the bolts to 33 ft. lbs. (44 Nm).
   • Steering column intermediate shaft. Tighten the pinch bolt to 18 ft. lbs. (25 Nm).
   • Pitman arm. Tighten the nut to 159 ft. lbs. (216 Nm).
   • Power steering pressure and return lines. Tighten the fittings to 33 ft. lbs. (44 Nm).
   • Stabilizer bar
   • Lower fan shroud
   • Skid plates
5. Fill the power steering fluid reservoir.
6. Check the wheel alignment and adjust as necessary.

Rack and Pinion Steering Gear

REMOVAL & INSTALLATION

2-Wheel Drive

1. Before servicing the vehicle, refer to the precautions in the beginning of this section.
2. Disable the air bag system.
3. Remove or disconnect the following:
   • Stone guard
   • Tie rod from knuckle
   • Power steering pressure and return lines
   • Steering gear

To install:
4. Install or connect the following:
   • Steering gear. Torque the mounting bolts to 85 ft. lbs. (116 Nm)
   • Fluid lines. Torque the fitting to 18 ft. lbs. (25 Nm)
   • Tie rod ends. 87 ft. lbs. (118 Nm)
   • Stone guard
5. Fill and bleed the system.

4-Wheel Drive

1. Before servicing the vehicle, refer to the precautions in the beginning of this section.
2. Disable the air bag system.
3. Remove or disconnect the following:
   • Stone guard
   • Tie rod from knuckle
   • Power steering pressure and return lines
   • Torsion bar
   • Lower control arm frame side bolt
   • Crossmember bolts
   • Steering gear with crossmember
   • Steering gear

To install:
4. Install or connect the following:
   • Steering gear. Torque the mounting bolts to 85 ft. lbs. (116 Nm)
   • Crossmember Torque the bolts to 128 ft. lbs. (173 Nm)
   • Lower control arm bolt
   • Torsion bar
   • Fluid lines. Torque the fitting to 18 ft. lbs. (25 Nm)
   • Tie rod ends. 87 ft. lbs. (118 Nm)
   • Stone guard
5. Fill and bleed the system.

Shock Absorber

REMOVAL & INSTALLATION

Front

1. Before servicing the vehicle, refer to the precautions in the beginning of this section.
2. Support the lower control arm with a jackstand.
3. Remove or disconnect the following:
   • Front wheels
   • Upper shock retaining nut and rubber bushing
   • Suspension bump stops
   • Shock absorber

To install:
4. Install or connect the following:
   • Shock absorber. Tighten the lower bolt to 60–61 ft. lbs. (82–84 Nm) for 2001–02; 69 ft. lbs. (93 Nm) for 2003.

- Bump stop. Tighten the bolts to 30 ft. lbs. (41 Nm).
- Upper shock retaining nut and rubber bushing. Tighten the nut to 14–15 ft. lbs. (19–20 Nm).
- Front wheels

Rear

1. Before servicing the vehicle, refer to the precautions in the beginning of this section.
2. Support the rear axle with jackstands.
3. Remove the rear shock absorbers.

To install:

4. Install the rear shock absorbers. Tighten the upper bolt to 14 ft. lbs. (20 Nm) for Rodeo and Rodeo Sport; 58 ft. lbs. (78 Nm) for Trooper. Tighten the lower bolt to 58 ft. lbs. (78 Nm) for Rodeo and Rodeo Sport; 70 ft. lbs. (95 Nm) for Trooper.
5. Remove the jackstands.

Coil Spring

REMOVAL & INSTALLATION

1. Before servicing the vehicle, refer to the precautions in the beginning of this section.
2. Support the vehicle under the frame.
3. Support the rear axle with a jack.
4. Remove or disconnect the following:
   - Rear wheels
   - Stabilizer bar links
   - Parking brake cable brackets
   - Shock absorbers
5. Lower the rear axle with the jack to release the coil spring tension. Remove the coil springs and insulators.

To install:

6. Place the coil springs on the axle assembly and the insulators on top of the springs.
7. Raise the axle assembly into position.
8. Install or connect the following:
   - Shock absorbers
   - Parking brake cable brackets
   - Stabilizer bar links. Tighten the nuts to 37 ft. lbs. (50 Nm).
   - Rear wheels

Torsion Bar

REMOVAL & INSTALLATION

1. Before servicing the vehicle, refer to the precautions in the beginning of this section.
2. Matchmark the adjusting bolt and end piece, then remove the bolt, end piece, and seat.

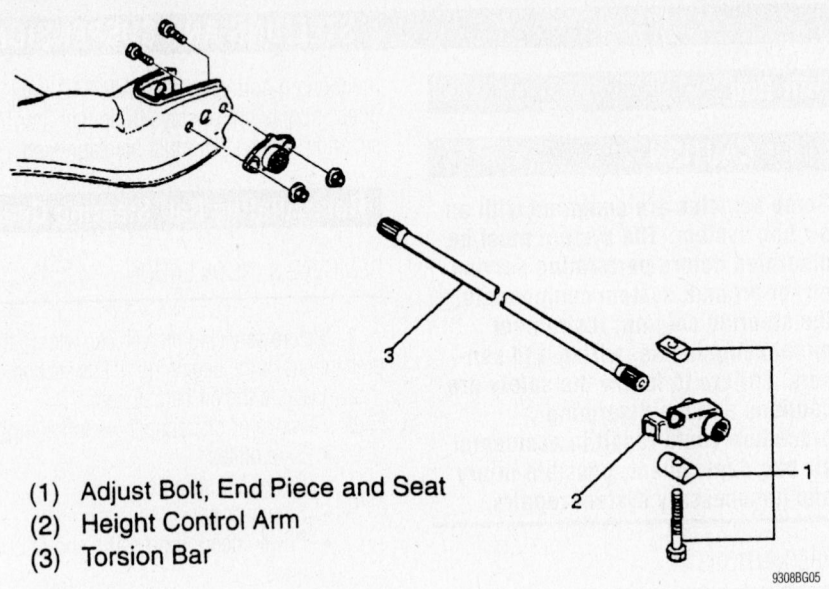

(1) Adjust Bolt, End Piece and Seat
(2) Height Control Arm
(3) Torsion Bar

Exploded view of the torsion bar assembly

3. Matchmark the height control arm to the torsion bar, then remove the height control arm.
4. Matchmark the torsion bar to the lower control arm, then remove the torsion bar.

To install:

5. Apply grease to the torsion bar splines.
6. Apply grease to the contact points of the height control arm, adjusting bolt end piece and seat.
7. Align the matchmarks and install the torsion bar.
8. Align the matchmarks and install the height control arm.
9. Install the adjusting bolt, seat and end piece.

10. Tighten the adjusting bolt to align the matchmarks.

Upper Ball Joint

REMOVAL & INSTALLATION

1. Before servicing the vehicle, refer to the precautions in the beginning of this section.
2. Support the lower control arm with a floor jack.
3. Remove or disconnect the following:

- Front wheel
- Wheel speed sensor
- Upper ball joint

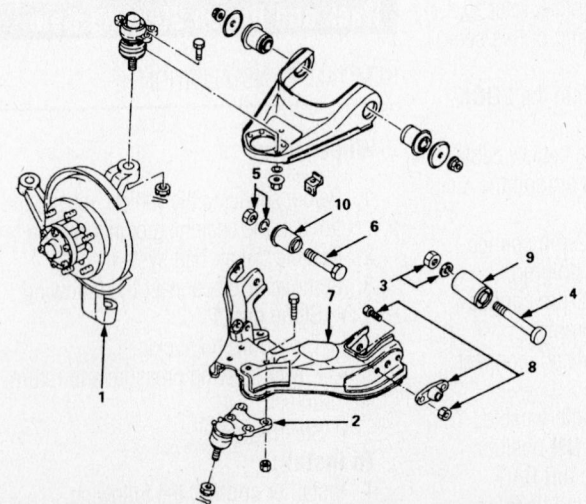

1. Knuckle
2. Lower end
3. Nut and washer, rear
4. Bolt, rear
5. Nut and washer, front
6. Bolt, front
7. Lower control arm assembly
8. Torsion bar arm bracket
9. Bushing, rear
10. Bushing, front

Exploded view of the control arm and ball joint components

To install:

➡**Use new nuts, bolts and split pins for assembly.**

4. Install or connect the following:
   • Upper ball joint. Tighten the mounting bolts to 42 ft. lbs. (57 Nm) and the nut to 72 ft. lbs. (96 Nm).
   • Wheel speed sensor
   • Front wheel

Lower Ball Joint

REMOVAL & INSTALLATION

1. Before servicing the vehicle, refer to the precautions in the beginning of this section.
2. Support the lower control arm with a jackstand.
3. Remove or disconnect the following:
   • Front wheel
   • Disc brake caliper and support
   • Brake rotor and backing plate
   • Wheel speed sensor
   • Outer tie rod end

   • Upper ball joint
   • Steering knuckle
   • Lower ball joint

To install:

➡**Use new nuts, bolts and split pins for assembly.**

4. Install or connect the following:
   • Lower ball joint. Tighten the mounting bolts to 76 ft. lbs. (103 Nm) on 1999–01 models; 85 ft. lbs. (116 Nm) on 2002–03 models.
   • Steering knuckle. Tighten the lower ball joint nut to 108 ft. lbs. (147 Nm).
   • Upper ball joint. Tighten the nut to 72 ft. lbs. (96 Nm).
   • Outer tie rod end. Tighten the nut to 72 ft. lbs. (98 Nm).
   • Wheel speed sensor
   • Brake rotor and backing plate
   • Disc brake caliper and support. Tighten the support bolts to 115 ft. lbs. (155 Nm).
   • Front wheel
5. Check the wheel alignment and adjust as necessary.

Upper Control Arm

REMOVAL & INSTALLATION

1. Before servicing the vehicle, refer to the precautions in the beginning of this section.
2. Support the lower control arm with a jackstand.
3. Remove or disconnect the following:
   • Front wheel
   • Wheel speed sensor
   • Brake caliper
   • Upper ball joint
   • Upper control arm

➡**Note the alignment shim location for assembly.**

To install:
4. Install or connect the following:
   • Upper control arm
   • Alignment shims in their original locations. Tighten the bolts to 112 ft. lbs. (152 Nm).
   • Upper ball joint. Tighten the nut to 72 ft. lbs. (98 Nm).
   • Brake caliper

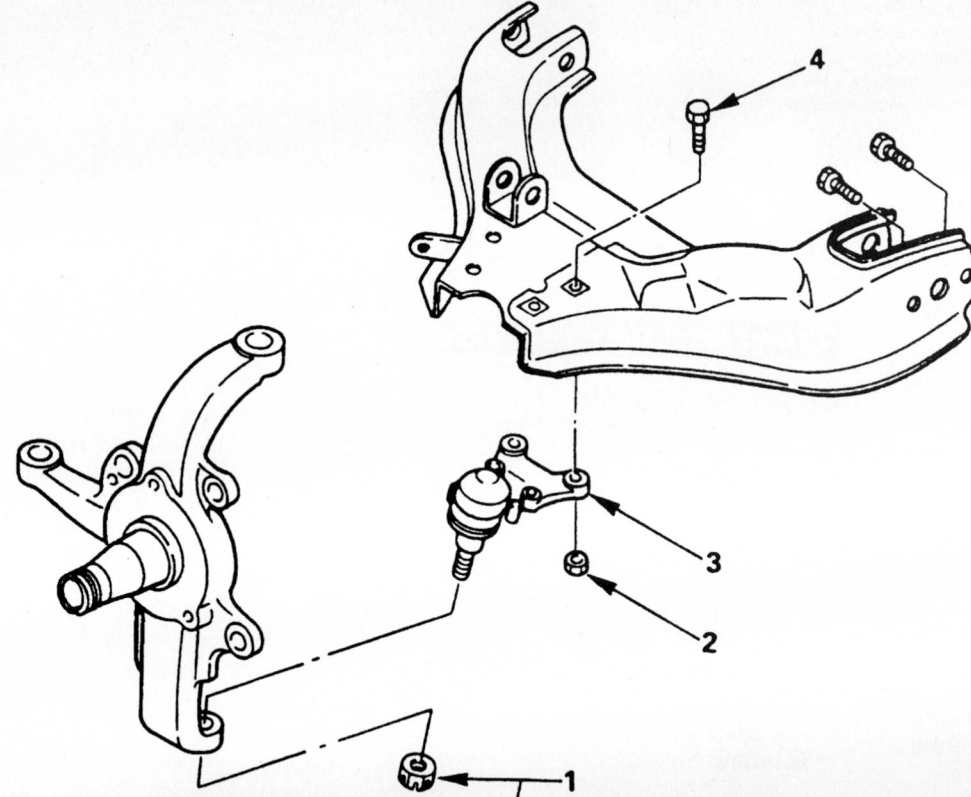

1. Nut and cotter pin
2. Nut
3. Lower ball joint
4. Bolt

7924BG35

Exploded view of the lower ball joint mounting and related components

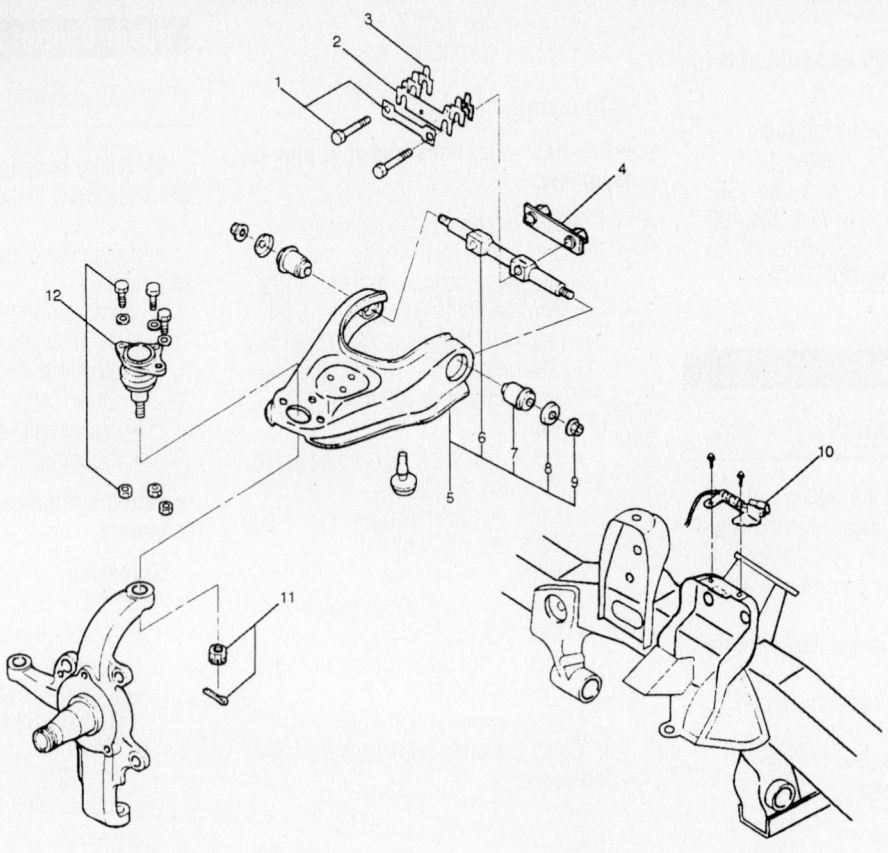

1	Bolt and Plate		7	Bushing
2	Camber Shims		8	Plate
3	Caster Shims		9	Nut
4	Nut Assembly		10	Speed Sensor Cable
5	Upper Control Arm Assembly		11	Nut and Cotter Pin
6	Fulcrum Pin		12	Upper Ball Joint

9308BG06

Upper control arm and related parts

- Wheel speed sensor
- Front wheel

5. Check the wheel alignment and adjust as necessary.

CONTROL ARM BUSHING REPLACEMENT

1. Before servicing the vehicle, refer to the precautions in the beginning of this section.
2. Remove the upper control arm.
3. Remove the nuts and washers from the fulcrum pin.
4. Press the bushings out of the control arm.

To install:

5. Press the bushings into the control arm.
6. Install the fulcrum pin washers and nuts.
7. Install the upper control arm.
8. Raise the suspension so that there is 0.79 inches (20mm) between the bump stop and the lower control arm. Tighten the fulcrum pin nuts to 80 ft. lbs. (108 Nm).

9. Check the wheel alignment and adjust as necessary.

Lower Control Arm

REMOVAL & INSTALLATION

1. Before servicing the vehicle, refer to the precautions in the beginning of this section.
2. Remove or disconnect the following:
- Front wheel
- Torsion bar
- Lower ball joint
- Stabilizer bar link
- Shock absorber
- Lower control arm

To install:

3. Install or connect the following:
- Lower control arm
- Shock absorber
- Stabilizer bar link
- Lower ball joint

- Torsion bar
- Front wheel

4. Raise the suspension so that there is 0.79 inches (20mm) between the bump stop and the lower control arm. For 1999–01, tighten the front control arm bolt to 116 ft. lbs. (157 Nm) and tighten the rear control arm bolt to 145 ft. lbs. (196 Nm). For 2002–03, tighten the rear nut to 174 ft. lbs. (235 Nm); the rear nut to 137 ft. lbs. (186 Nm).

5. Check the wheel alignment and adjust as necessary.

CONTROL ARM BUSHING REPLACEMENT

1. Before servicing the vehicle, refer to the precautions in the beginning of this section.
2. Remove or disconnect the following:
- Lower control arm
- Bushings, press them from the control arm, using Remover/Installer J-36833 for the front bushing and

Remover/Installer J-36834 for the rear bushing.

To install:

3. Install or connect the following:
- New bushings
- Lower control arm

Wheel Bearings

ADJUSTMENT

Trooper

1. Before servicing the vehicle, refer to the precautions in the beginning of this section.

2. Remove or disconnect the following:
- Front wheel
- Brake caliper and pads
- Hub dust cap
- Snapring and shim
- Hub flange
- Lockscrew and washer

3. Tighten the hub nut to 22 ft. lbs. (29 Nm) to seat the bearings and then fully loosen the nut.

4. Tighten the hub nut to achieve a bearing preload of 2.6–4.0 lbs. (1.2–1.8 Kg) for used bearings. If the bearings were replaced, set the preload to 4.4–5.5 lbs. (2.0–2.5 Kg).

5. Install or connect the following:
- Lockwasher and screw
- Hub flange
- Shim and snapring
- Hub dust cap. Tighten the bolts to 43 ft. lbs. (59 Nm).
- Brake caliper and pads
- Front wheel

Amigo, Passport, Rodeo and Rodeo Sport

The front wheel bearings are not adjustable.

REMOVAL & INSTALLATION

Trooper

1. Before servicing the vehicle, refer to the precautions in the beginning of this section.

2. Remove or disconnect the following:
- Front wheel
- Brake caliper and support
- Hub dust cap
- Snapring and shim
- Hub flange
- Lockscrew and washer
- Hub nut
- Brake rotor and hub assembly
- Wheel speed sensor ring
- Outer bearing
- Grease seal
- Inner bearing

To install:

3. Clean and inspect the bearings. Replace if necessary.

4. Apply clean wheel bearing grease to the inner and outer bearings.

5. Apply grease in the hub.

6. Install the wheel bearings into the hub along with a new grease seal.

7. Install or connect the following:
- Wheel speed sensor ring. Tighten the bolts to 13 ft. lbs. (18 Nm).
- Brake rotor and hub assembly
- Hub nut. Set the bearing preload.
- Lockscrew and washer
- Hub flange

- Snapring and shim
- Hub dust cap
- Brake caliper and support. Tighten the support bolts to 115 ft. lbs. (155 Nm).
- Front wheel

Amigo, Passport, Rodeo and Rodeo Sport

2-WHEEL DRIVE

1. Raise and support the front end.

2. Remove the caliper from the anchor plate and suspend it out of the way.

3. Remove the rotor.

4. Remove the bolts attaching the hub to the knuckle.

5. Installation is the reverse of removal. Torque the hub bolts to 76 ft. lbs. (103 Nm).

4-WHEEL DRIVE

1. Place the transfer case in 2WD.

2. Raise and support the front end.

3. Remove the caliper from the anchor plate and suspend it out of the way.

4. Remove the rotor.

5. Loosen the caulking around the front driveshaft nuts. Remove the nuts and discard them.

6. Remove the 4 hub retaining bolts.

➡**If the hub is difficult to remove, install 2 long bolt in the bolt holes and strike them to loosen the hub.**

7. Installation is the reverse of removal. Torque the hub bolts to 76 ft. lbs. (103 Nm). Use new driveshaft nuts, torquing them to 181 ft. lbs. (245 Nm). Apply new caulking. The caulk should be free of cracks when applied.

BRAKES

Brake Caliper

REMOVAL & INSTALLATION

Trooper

FRONT

1. Raise and safely support the vehicle.

2. Remove some brake fluid from the reservoir.

3. Remove the front wheels.

4. Disconnect the brake fluid line from the caliper. Plug the line to prevent fluid loss.

5. Loosen the brake caliper mounting bolt and guide bolt. Remove the caliper from the mount.

6. Remove the brake pads and clips

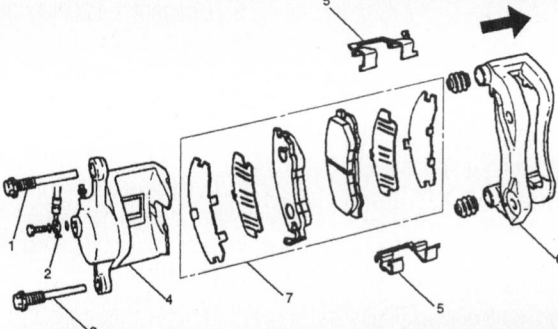

(1)	Guide Bolt	(4)	Caliper Assembly
(2)	Brake Flexible Hose	(5)	Clip
(3)	Lock Bolt	(6)	Support Bracket with Pad Assembly
		(7)	Pad Assembly

93026G02

Front caliper assembly—Trooper

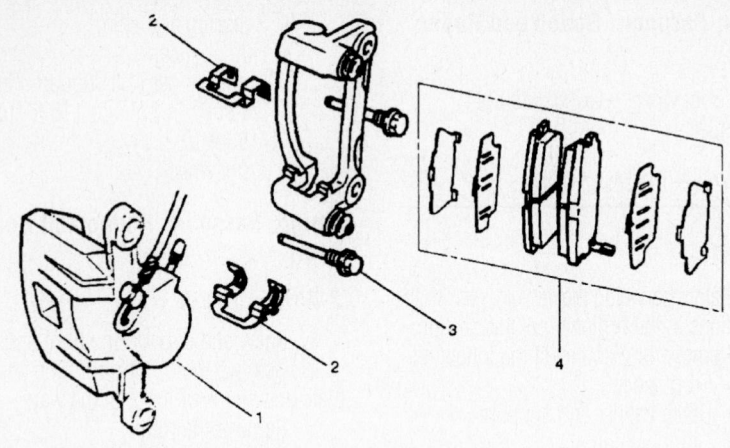

(1)	Caliper Assembly	(3)	Lock Bolt
(2)	Clip	(4)	Pad Assembly

93026G01

Rear caliper assembly—Trooper

from the caliper. Inspect the brake pads for wear and replace them if necessary.

To install:

7. Fill the brake caliper with clean brake fluid and connect the fluid line to the caliper using new washers. Tighten the brake line banjo fitting to 26 ft. lbs. (35 Nm). Install the brake pads and clips onto the caliper.

8. Install the caliper onto the mounting bracket. Lubricate the caliper bolts and their boots. Then, install the caliper mounting bolts and tighten them to 54 ft. lbs. (74 Nm).

9. Refill and bleed the brake system.

10. Install the front wheels and lower the vehicle.

REAR

1. Raise and safely support the vehicle.

2. Remove some brake fluid from the reservoir.

3. Remove the rear wheels.

4. Disconnect the brake fluid line from the caliper. Plug the line to prevent fluid loss.

5. Loosen the brake caliper mounting bolt and guide bolt. Remove the caliper from the mount bracket.

6. Remove the brake pads and clips from the caliper. Inspect the brake pads for wear; replace them if necessary.

7. If necessary for servicing, unbolt the caliper mounting bracket from the backing plate.

To install:

8. If removed, install the caliper mounting bracket and tighten its bolts to 76 ft. lbs. (103 Nm).

9. Fill the brake caliper with clean

brake fluid and connect the fluid line to the caliper using new washers. Tighten the brake line banjo fitting to 26 ft. lbs. (35 Nm). Install the brake pads and clips onto the caliper.

10. Install the caliper on the mounting bracket. Lubricate the caliper bolts and their boots. Then, install the caliper mounting bolts. Tighten them to 32 ft. lbs. (44 Nm).

11. Refill and bleed the brake system.

12. Install the rear wheels and lower the vehicle.

Amigo, Passport and Rodeo

FRONT

1. Raise and safely support the vehicle. Remove the front wheels.

2. Remove some brake fluid from the master cylinder reservoir.

3. Disconnect and plug the brake fluid line from the caliper.

4. Remove the brake caliper mounting bolt and guide bolt and remove the caliper from the mount. The brackets can be remove for additional work space.

5. Remove the brake pads and clips from the caliper. Inspect the brake pads for wear; replace them, if necessary.

To install:

6. Install the brake pads and clips onto the caliper.

7. If the caliper bracket was removed, tighten the bolts to 103–126 ft. lbs. (139–171 Nm).

8. Install the caliper on the mounting bracket. Torque the caliper-to-mounting bracket bolts to:

2001–02:

- 4-cylinder models: 20–27 ft. lbs. (27–37 Nm)
- 6-cylinder models: 54 ft. lbs. (74 Nm)

2003:

- 33 ft. lbs. (45 Nm)

9. Connect the fluid line to the caliper using new washers. Torque the brake line banjo fitting to 26 ft. lbs. (35 Nm).

✲✲ WARNING

Be sure the hook end of the flexible brake line is positioned in the anti-rotation cavity.

10. Refill the master cylinder reservoir and bleed the brake system.

11. Install the front wheels and lower the vehicle.

REAR

1. Raise and safely support the vehicle. Remove the rear wheels.

2. Remove some fluid from the master cylinder reservoir.

3. Disconnect and plug the brake fluid line from the caliper.

➡ **Discard the parking brake cable mounting pin after removal.**

4. If equipped with caliper actuated parking brakes; remove the mounting pin from the parking brake cable and disconnect the parking cable from the disc caliper.

5. Remove the brake caliper mounting bolt and guide bolt and remove the caliper from the mount.

6. Remove the brake pads and clips from the caliper. Inspect the brake pads for wear; replace them, if necessary.

To install:

7. Install the brake pads and clips onto the caliper.

8. If the mounting bracket was removed, tighten the bolts to 69–84 ft. lbs. (93–114 Nm).

9. Install the caliper on the mounting bracket. Torque the caliper-to-mounting bracket bolts to 12–17 ft. lbs. (16–24 Nm), or 32 ft. lbs. (43 Nm) on vehicles with shoe-type parking brakes.

10. Connect the parking brake cable to the caliper and install a new mounting pin.

11. Connect the fluid line to the caliper using new washers. Torque the brake line banjo fitting to 26 ft. lbs. (35 Nm).

12. Refill the master cylinder reservoir and bleed the brake system.

13. Install the rear wheels and lower the vehicle.

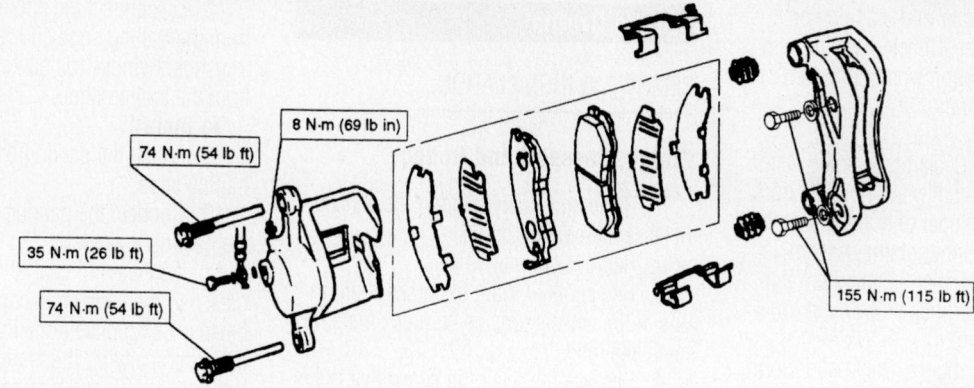

8 N·m (69 lb in)

74 N·m (54 lb ft)

35 N·m (26 lb ft)

74 N·m (54 lb ft)

155 N·m (115 lb ft)

93026G54

Exploded view of the front caliper components—2001 Rodeo

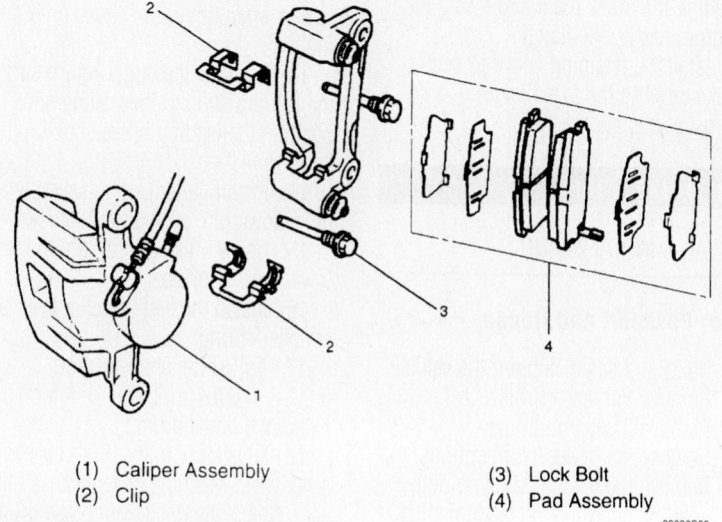

(1) Caliper Assembly
(2) Clip
(3) Lock Bolt
(4) Pad Assembly

93026G55

Exploded view of the rear caliper components—Rodeo

Disc Brake Pads

REMOVAL & INSTALLATION

Amigo, Passport and Rodeo

FRONT

Most disc brake pads are equipped with wear indicators. If a squealing noise occurs from the brakes while driving, check the pad wear indicator plate. If there is evidence of the indicator plate contacting the brake disc, the brake pad should be replaced.

1. Remove ½ of the volume of brake fluid from the master cylinder to prevent overflow when the caliper piston is compressed.

2. Raise and safely support the vehicle.

3. Remove the wheel and tire assemblies.

4. Remove the brake caliper without disconnecting the brake line. Support the caliper with a length of wire. Do not let the caliper hang from the brake hose.

➡ On some disc brake systems it is not necessary to remove the caliper when installing new brake pads. Remove the lower slide bolt and rotate the caliper upward to remove the pads.

5. Remove the brake pads and shims. Inspect the brake rotor and machine or replace as necessary. Check the minimum thickness (specification is cast into the rotor) before machining.

To install:

6. Use a suitable tool to push the caliper piston into its bore.

7. Apply a thin coat of grease to the rear face of the brake pad and install the shim. Install the brake pads.

8. Install the calipers. Lubricate the caliper bolts and boots.

9. Install the wheel and tire assemblies and lower the vehicle.

10. Apply the brakes several times to seat the pads before moving the vehicle. Check the fluid in the master cylinder and add as necessary.

REAR

1. Raise and safely support the vehicle. Remove the rear wheels.

2. Remove the brake caliper mounting bolts and remove the caliper without disconnecting the brake fluid line. Support the caliper so it does not hang on the brake line.

3. Remove the brake pads and retaining clips from the caliper.

4. If equipped with caliper-activated parking brakes; use tool J-37617 or equivalent to rotate the piston clockwise until it retracts into the bore. Align the notches of the piston face so the centerline of the notches is perpendicular to the centerline of the mounting bosses.

To install:

5. Install the new brake pads and clips in the caliper and install the caliper in the mounting bracket.

6. Tighten the caliper mounting bolts to 12–17 ft. lbs. (16–24 Nm), or 32 ft. lbs. (43 Nm) on vehicles with shoe-type parking brakes.

7. Install the rear wheels. Check the brake fluid level.

8. Pump the brake pedal until pressure is felt before moving the vehicle.

Trooper

FRONT

1. Remove about ½ of the brake fluid from the master cylinder reservoir to prevent overflow when the caliper piston is compressed.

2. Raise and safely support the vehicle.

3. Remove the front wheels.

4. Remove the brake caliper from the caliper bracket without disconnecting the brake line. Support the caliper with a length of wire. Do not let the caliper hang from the brake hose.

5. Remove the brake pads and shims.

Inspect the brake rotor and machine or replace as necessary. Check the minimum thickness (specification is cast into the rotor) before machining.

To install:

6. Use a large C-clamp or brake piston tool to push the caliper piston into its bore.

7. Apply a thin coat of brake grease to both sides of both inner shims. Assemble the pads and shims, then install them into the caliper. The wear indicator on the inner pad must face down.

8. Install the calipers. Clean and lubricate the caliper mounting bolts and lubricate the mounting bolt boots. Install the mounting bolts and tighten them to 54 ft. lbs. (74 Nm).

9. Install the front wheels and lower the vehicle.

10. Apply the brakes several times to seat the pads before moving the vehicle. Check the fluid level in the master cylinder reservoir and add as necessary.

REAR

1. Use a vacuum pump to remove some brake fluid from the master cylinder reservoir to prevent overflow when the caliper piston is compressed.

2. Raise and safely support the vehicle.

3. Remove the rear wheels.

4. Remove the brake caliper from the caliper bracket without disconnecting the brake line. Support the caliper with a length of wire. Do not let the caliper hang from the brake hose.

5. Remove the brake pads and shims. Inspect the brake rotor and machine or replace as necessary. Check the minimum thickness (specification is cast into the rotor) before machining.

To install:

6. Use a large C-clamp or brake piston tool to push the caliper piston into its bore.

7. Apply a thin coat of brake grease to both sides of both inner shims. Assemble the pads and shims, then install them into the caliper. The wear indicator on the inner pad must face down.

8. Install the calipers. Clean and lubricate the caliper mounting bolts and lubricate the mounting bolt boots. Install the mounting bolts and tighten them to 32 ft. lbs. (44 Nm).

9. Install the rear wheels and lower the vehicle.

10. Apply the brakes several times to seat the pads before moving the vehicle. Check the fluid level in the master cylinder reservoir and add as necessary.

Brake Drums

REMOVAL & INSTALLATION

Amigo, Passport and Rodeo

1. Raise and safely support the vehicle. Release the parking brake.

2. Remove the rear wheels.

3. Use chalk to mark the brake drum to one of the wheel studs as an index mark for reinstallation.

4. Remove the retaining screw that holds the brake drum to the axle flange.

5. Pull the brake drum from the axle flange.

To install:

6. Align the index mark and install the brake drum to the axle flange.

7. Install the retaining screw to secure the brake drum to the axle flange.

8. Install the rear wheels.

Rear Brake Shoes

REMOVAL & INSTALLATION

Amigo, Passport and Rodeo

1. Raise and safely support the vehicle.

2. Remove the rear wheels.

3. Remove the brake drums.

4. Remove the brake return springs.

5. Remove the leading shoe holding pin and spring, and then the leading shoe.

6. Remove the self-adjuster and the adjuster lever.

7. Remove the trailing shoe holding pin and spring.

8. Disconnect the parking brake cable from the trailing shoe and remove the trailing shoe. Remove the parking brake lever from the trailing shoe.

To install:

9. Attach the parking brake lever to the trailing shoe.

10. Connect the parking brake cable to the parking brake lever.

11. Apply a thin coat of high temperature grease to the shoe contact points on the brake backing plate (locations A and C in the accompanying illustration), piston contact surface (B), and self-adjuster (D).

12. Position the trailing shoe on the backing plate and install the hold-down pin, spring, and retainer. Don't stretch the return spring when fitting the shoes onto the backing plate.

13. Connect the upper return spring and the leading shoe to the trailing shoe and position the leading brake shoe on the backing plate.

14. Install the adjuster assembly and the hold-down pin, spring, and retainer.

15. Use a brake spring tool to install the lower return spring.

16. Install the self-adjuster lever and adjuster spring.

17. Adjust the shoe-to-drum clearance to 0.0098–0.0157 in. (0.25–0.40mm) and install the brake drum.

18. Check the brake drum for scoring or other wear. Machine or replace as necessary. Check the maximum brake drum diameter specification when machining.

19. Install the rear wheels. Lower the vehicle.

20. Road-test the vehicle.

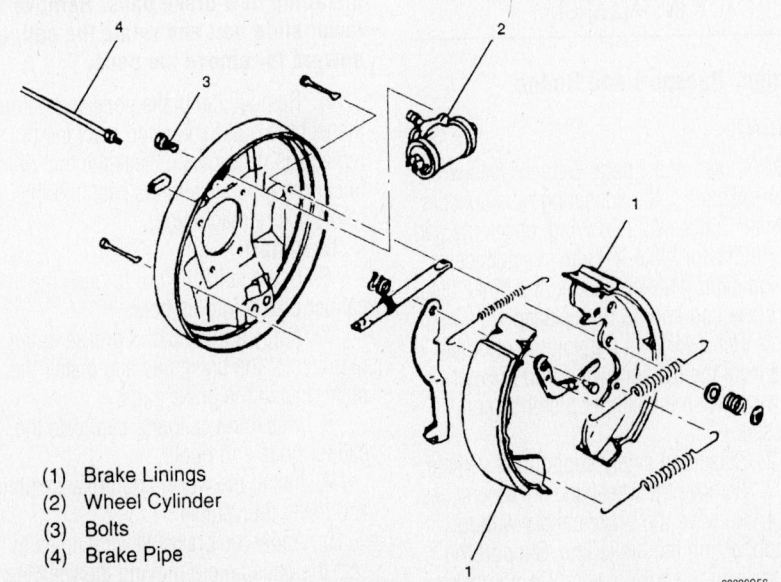

(1) Brake Linings
(2) Wheel Cylinder
(3) Bolts
(4) Brake Pipe

93026G56

Exploded view of the rear drum brakes—Amigo and Rodeo

SPECIFICATION AND MAINTENANCE CHARTS

ENGINE AND VEHICLE IDENTIFICATION

Engine								Model Year	
Code ①	Liters (cc)	Cu. In.	Cyl.	Fuel Sys.	Engine Type	Eng. Mfg.		Code ②	Year
1	1.8 (1793)	109	4	EGI	DOHC	KIA		1	2001
3	1.5 (1493)	91	4	EGI	DOHC	KIA		2	2002
5	1.6 (1594)	97	4	EGI	DOHC	KIA		3	2003
2	2.0 (1975)	120.5	4	EGI	DOHC	KIA		4	2004
S	2.4 (2351)	144	4	EGI	DOHC	KIA		5	2005
8	2.7 (2656)	163	6	EGI	DOHC	KIA			

EGI: Electronic Gasoline Injection

DOHC: Double Overhead Camshafts

① 8th digit of VIN

② 10th digit of VIN

67162-KIAC-C01

GENERAL ENGINE SPECIFICATIONS

Year	Model	Engine Displacement Liters	Engine VIN	Net Horsepower @ rpm	Net Torque @ rpm (ft. lbs.)	Bore x Stroke (in.)	Compression Ratio	Oil Pressure @ rpm
2001	Rio	1.5	3	96@5800	98@4500	NA	9.5:1	43-57@3000
	Sephia	1.8	1	125@6000	120@4500	3.19x3.43	9.5:1	43-57@3000
	Optima	2.4	S	149@6000	159@4500	NA	10.0:1	43-57@3000
		2.7	8	178@6000	181@4000	NA	10.0:1	43-57@3000
2002	Rio	1.5	3	96@5800	98@4500	NA	9.5:1	43-57@3000
	Spectra	1.8	1	125@6000	120@4500	3.19x3.43	9.5:1	43-57@3000
	Optima	2.4	S	149@6000	159@4500	NA	10.0:1	43-57@3000
		2.7	8	178@6000	181@4000	NA	10.0:1	43-57@3000
2003	Rio	1.6	5	104@5800	104@4700	NA	10.0:1	43-57@3000
	Spectra	1.8	1	125@6000	120@4500	3.19x3.43	9.5:1	43-57@3000
	Optima	2.4	S	149@6000	159@4500	NA	10.0:1	43-57@3000
		2.7	8	178@6000	181@4000	NA	10.0:1	43-57@3000
2004	Rio	1.6	5	104@5800	104@4700	NA	10.0:1	43-57@3000
	Spectra	1.8	1	125@6000	120@4500	3.19x3.43	9.5:1	43-57@3000
		2.0	2	138@6000	136@4500	3.23x8.68	10.1:1	43-57@3000
	Optima	2.4	S	149@6000	159@4500	NA	10.0:1	43-57@3000
		2.7	8	178@6000	181@4000	NA	10.0:1	43-57@3000

NA: Not available

67162-KIAC-C02

ENGINE TUNE-UP SPECIFICATIONS

Year	Engine Displacement Liters	Engine VIN	Spark Plug Gap (in.)	Ignition Timing (deg.)		Fuel Pump (psi)	Idle Speed (rpm)		Valve Clearance	
				MT	AT		MT	AT	Intake	Exhaust
2001	1.5	3	0.039-0.043	1-11B	1-11B	65-94	700-800	700-800	HYD	HYD
	1.8	1	0.039-0.043	3-13B	3-13B	47-48	750-850	750-850	HYD	HYD
	2.4	S	0.039-0.043	3-7B	3-7B	46-49	700-900	700-900	HYD	HYD
	2.7	8	0.039-0.043	—	7-17B	46-49	—	600-800	HYD	HYD
2002	1.5	3	0.039-0.043	1-11B	1-11B	65-94	700-800	700-800	HYD	HYD
	1.8	1	0.039-0.043	3-13B	3-13B	47-48	750-850	750-850	HYD	HYD
	2.4	S	0.039-0.043	3-7B	3-7B	46-49	700-900	700-900	HYD	HYD
	2.7	8	0.039-0.043	—	7-17B	46-49	—	600-800	HYD	HYD
2003	1.6	5	0.039-0.043	1-11B	1-11B	65-94	700-800	700-800	HYD	HYD
	1.8	1	0.039-0.043	3-13B	3-13B	47-48	750-850	750-850	HYD	HYD
	2.4	S	0.039-0.043	3-7B	3-7B	46-49	700-900	700-900	HYD	HYD
	2.7	8	0.039-0.043	—	7-17B	46-49	—	600-800	HYD	HYD
2004	1.6	5	0.039-0.043	1-11B	1-11B	65-94	700-800	700-800	HYD	HYD
	1.8	1	0.039-0.043	3-13B	3-13B	47-48	750-850	750-850	HYD	HYD
	2.0	2	0.039-0.043	3-7B	3-7B	49.8	750-850	750-850	HYD	HYD
	2.4	S	0.039-0.043	3-7B	3-7B	46-49	700-900	700-900	HYD	HYD
	2.7	8	0.039-0.043	—	7-17B	46-49	—	600-800	HYD	HYD

NOTE: The Vehicle Emission Control Information label often reflects specification changes made during production.

The label figures must be used if they differ from those in this chart.

B: Before top dead center

HYD: Hydraulic

67162-KIAC-C03

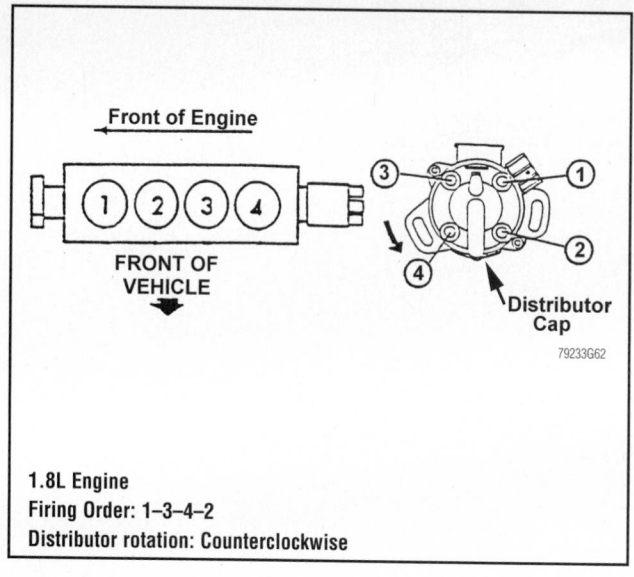

1.8L Engine
Firing Order: 1–3–4–2
Distributor rotation: Counterclockwise

79233G62

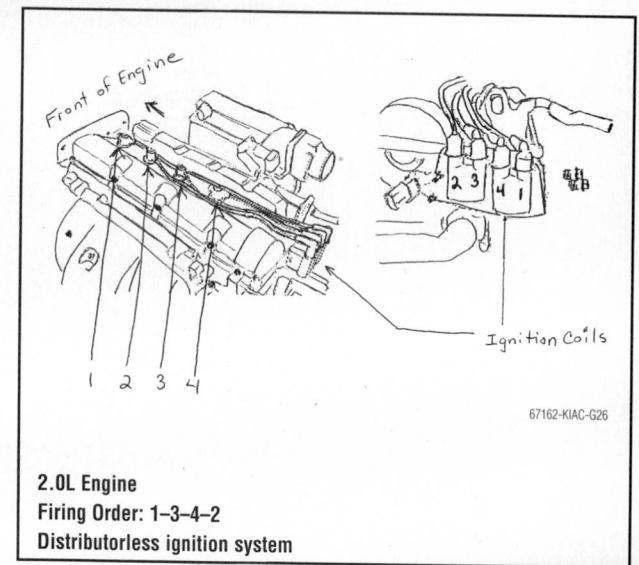

67162-KIAC-G26

2.0L Engine
Firing Order: 1–3–4–2
Distributorless ignition system

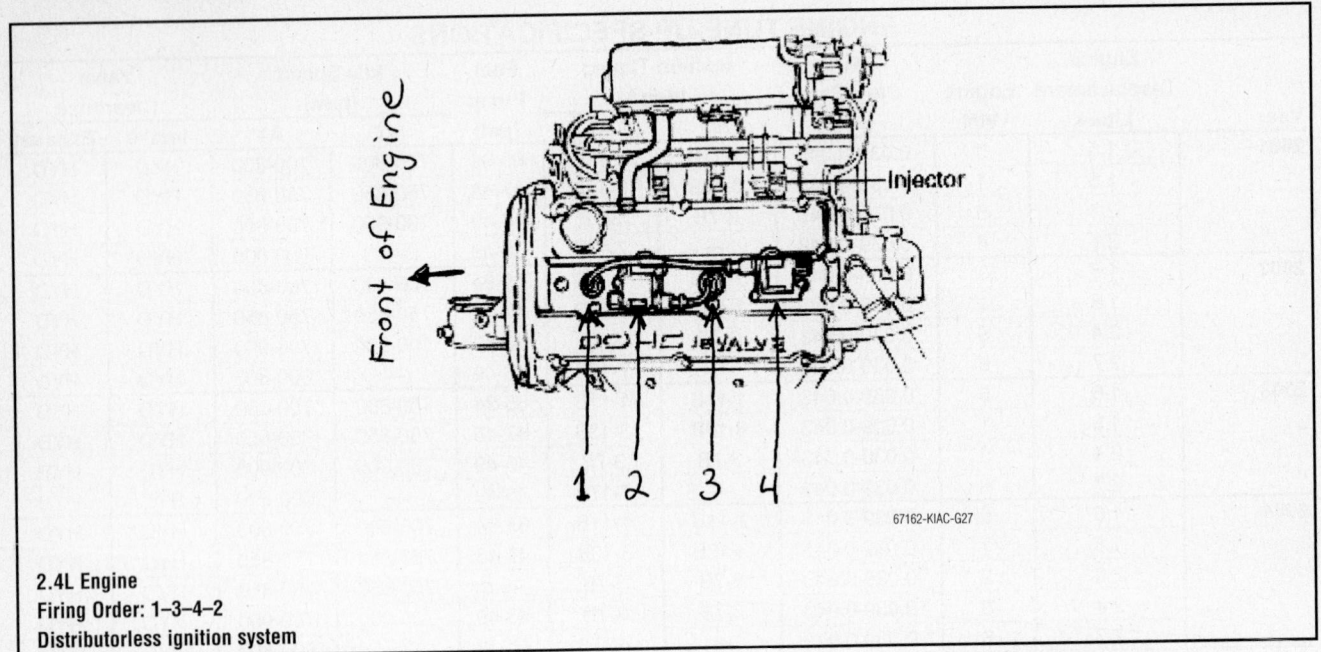

Front of Engine

Injector

DOHC 16 VALVE

1 2 3 4

67162-KIAC-G27

2.4L Engine
Firing Order: 1–3–4–2
Distributorless ignition system

67162-KIAC-G28

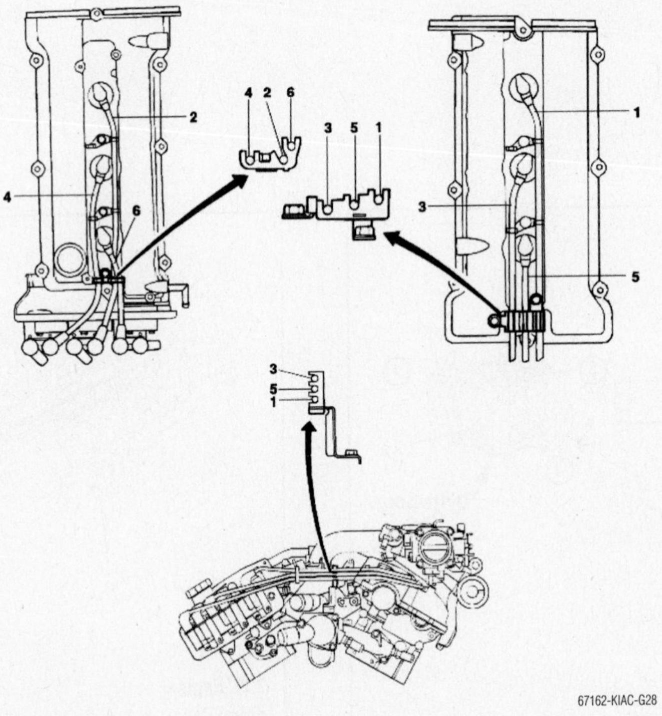

2.7L Engine
Firing Order: 1–2–3–4–5–6
Distributorless ignition system

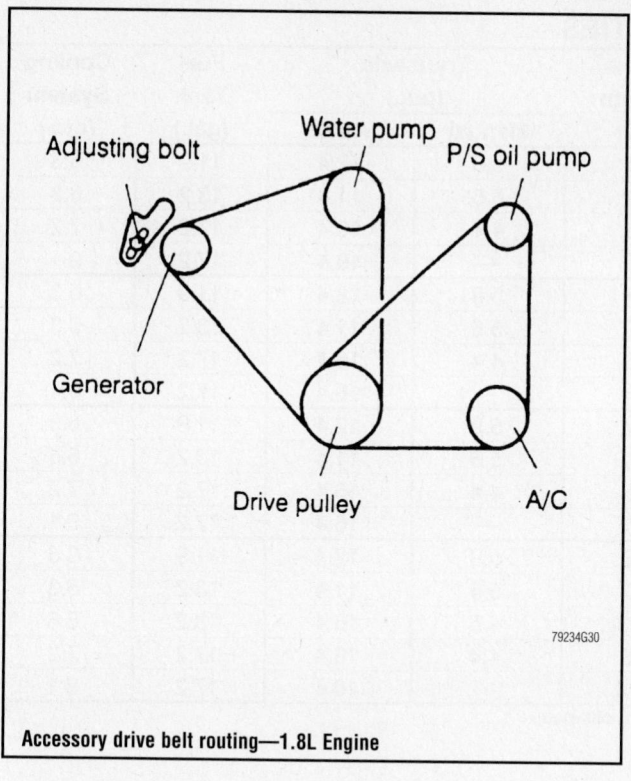

Accessory drive belt routing—1.8L Engine

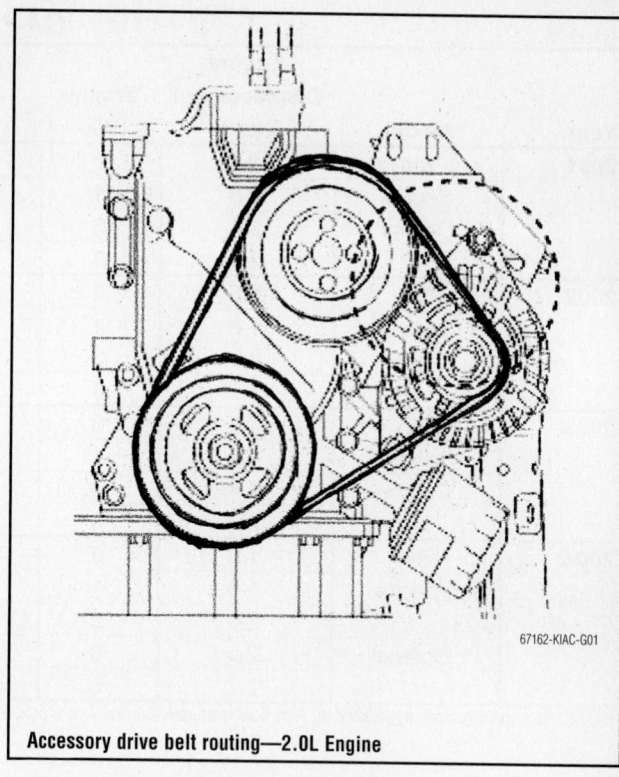

Accessory drive belt routing—2.0L Engine

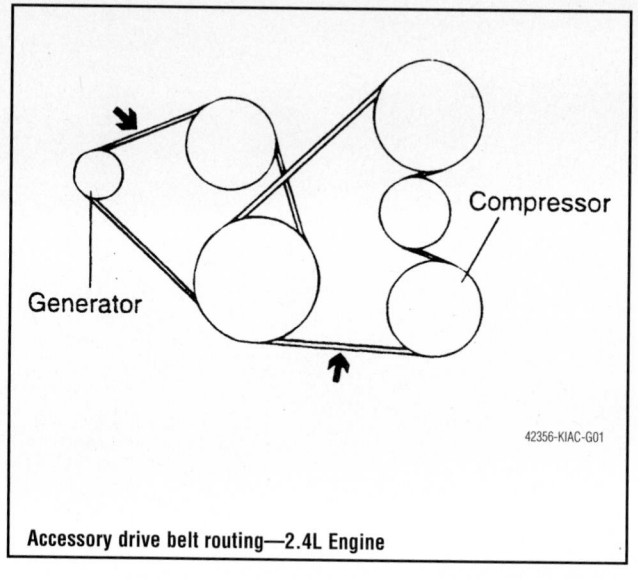

Accessory drive belt routing—2.4L Engine

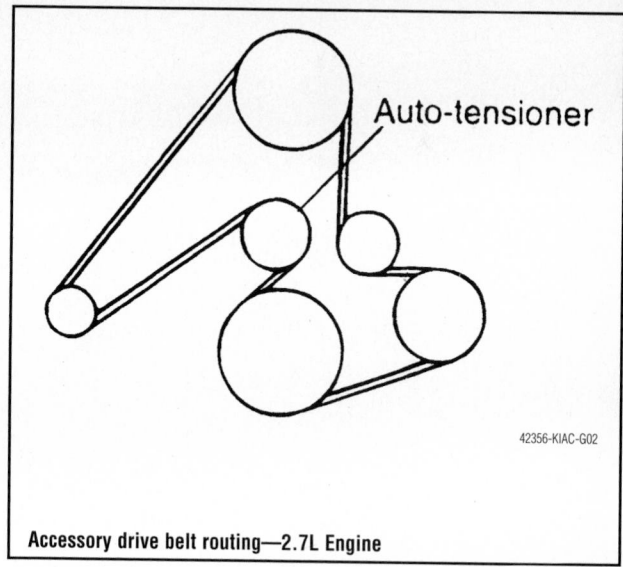

Accessory drive belt routing—2.7L Engine

CAPACITIES

Year	Model	Engine Displacement Liters	Engine VIN	Engine Oil with Filter	Transaxle (pts.)		Fuel Tank (gal.)	Cooling System (qts.)
					Manual	Auto.		
2001	Rio	1.5	3	3.4	5.8	12.4	11.9	6.3
	Sephia	1.8	1	4.0	5.6	11.4	13.2	6.3
	Optima	2.4	S	4.5	4.4	16.4	17.2	7.2
		2.7	8	4.75	—	16.4	17.2	9.1
2002	Rio	1.5	3	3.4	5.8	12.4	11.9	6.3
	Spectra	1.8	1	4.0	5.6	11.4	13.2	6.3
	Optima	2.4	S	4.5	4.4	16.4	17.2	7.2
		2.7	8	4.75	—	16.4	17.2	9.1
2003	Rio	1.6	5	3.2	5.9	12.4	11.9	6.3
	Spectra	1.8	1	4.0	5.6	11.4	13.2	6.3
	Optima	2.4	S	4.5	4.4	16.4	17.2	7.2
		2.7	8	4.75	—	16.4	17.2	9.1
2004	Rio	1.6	5	3.2	5.9	12.4	11.9	6.3
	Spectra	1.8	1	4.0	5.6	11.4	13.2	6.3
		2.0	2	4.23	4.5	16.4	13.2	8.6
	Optima	2.4	S	4.5	4.4	16.4	17.2	7.2
		2.7	8	4.75	—	16.4	17.2	9.1

NOTE: All capacities are approximate. Add fluid gradually and ensure a proper level is obtained.

67162-KIAC-C04

VALVE SPECIFICATIONS

Year	Engine Displacement Liters	Engine VIN	Seat Angle (deg.)	Face Angle (deg.)	Maximum out of Square (in.)	Spring Free Length (in.)	Stem-to-Guide Clearance (in.)		Stem Diameter (in.)	
							Intake	Exhaust	Intake	Exhaust
2001	1.5	3	45	45	0.433	1.653	0.0007-0.0019	0.0019-0.0033	0.2131-0.2137	0.2117-0.2125
	1.8	1	45	45	0.0638	1.840	①	②	0.2350-0.2356	0.2348-0.2354
	2.4	S	44-44.5	45-45.5	0.0638	1.804	0.0008-0.0019	0.0020-0.0033	0.2585-0.2591	0.2571-0.2579
	2.7	8	45	45-45.5	0.0638	1.673	0.0009-0.0020	0.0014-0.0026	0.2348-0.2354	0.2343-0.2348
2002	1.5	3	45	45	0.433	1.653	0.0007-0.0019	0.0019-0.0033	0.2131-0.2137	0.2117-0.2125
	1.8	1	45	45	0.0638	1.840	①	②	0.2350-0.2356	0.2348-0.2354
	2.4	S	44-44.5	45-45.5	0.0638	1.804	0.0008-0.0019	0.0020-0.0033	0.2585-0.2591	0.2571-0.2579
	2.7	8	45	45-45.5	0.0638	1.673	0.0009-0.0020	0.0014-0.0026	0.2348-0.2354	0.2343-0.2348
2003	1.6	5	45	45	0.0402	1.535	0.0008-0.0020	0.0020-0.0033	0.2152-0.2157	0.2138-0.2146
	1.8	1	45	45	0.0638	1.840	①	②	0.2350-0.2356	0.2348-0.2354
	2.4	S	44-44.5	45-45.5	0.0638	1.804	0.0008-0.0019	0.0020-0.0033	0.2585-0.2591	0.2571-0.2579
	2.7	8	45	45-45.5	0.0638	1.673	0.0009-0.0020	0.0014-0.0026	0.2348-0.2354	0.2343-0.2348
2004	1.6	5	45	45	0.0402	1.535	0.0008-0.0020	0.0020-0.0033	0.2152-0.2157	0.2138-0.2146
	1.8	1	45	45	0.0638	1.840	①	②	0.2350-0.2356	0.2348-0.2354
	2.0	2	45-45.5	45-45.5	0.0638	1.923	0.0008-0.0020	0.0014-0.0026	0.2348-0.2354	0.2343-0.2348
	2.4	S	44-44.5	45-45.5	0.0638	1.804	0.0008-0.0019	0.0020-0.0033	0.2585-0.2591	0.2571-0.2579
	2.7	8	45	45-45.5	0.0638	1.673	0.0009-0.0020	0.0014-0.0026	0.2348-0.2354	0.2343-0.2348

NA: Not Available

① Standard range: 0.0010-0.0023 in.
  Maximum value: 0.0080 in.

② Standard range: 0.0012-0.0025 in.
  Maximum value: 0.0080 in.

CRANKSHAFT AND CONNECTING ROD SPECIFICATIONS
All measurements are given in inches.

Year	Engine Displacement Liters	Engine VIN	Crankshaft				Connecting Rod		
			Main Brg. Journal Dia.	Main Brg. Oil Clearance	Shaft End-play	Thrust on No.	Journal Diameter	Oil Clearance	Side Clearance
2001	1.5	3	1.9365-1.9562	0.0007-0.0014	0.0032-0.0111	3	1.7693-1.7700	0.0008-0.0019	0.0044-0.0103
	1.8	1	2.1629-2.1636	0.0010-0.0017	0.0032-0.0111	3	1.7693-1.7700	0.0008-0.0019	0.0044-0.0019
	2.4	S	2.2434-2.2442	①	0.0020-0.0098	3	1.7709-1.7717	0.0008-0.0020	0.0040-0.0098
	2.7	8	2.4402-2.4409	0.0002-0.0009	0.0028-0.0098	3	1.8891-1.8898	0.0007-0.0014	0.0039-0.0098
2002	1.5	3	1.9365-1.9562	0.0007-0.0014	0.0032-0.0111	3	1.7693-1.7700	0.0008-0.0019	0.0044-0.0103
	1.8	1	2.1629-2.1636	0.0010-0.0017	0.0032-0.0111	3	1.7693-1.7700	0.0008-0.0019	0.0044-0.0019
	2.4	S	2.2434-2.2442	①	0.0020-0.0098	3	1.7709-1.7717	0.0008-0.0020	0.0040-0.0098
	2.7	8	2.4402-2.4409	0.0002-0.0009	0.0028-0.0098	3	1.8891-1.8898	0.0007-0.0014	0.0039-0.0098
2003	1.6	5	1.9365-1.9562	0.0007-0.0014	0.0032-0.0111	3	1.7693-1.7700	0.0008-0.0019	0.0044-0.0103
	1.8	1	2.1629-2.1636	0.0010-0.0017	0.0032-0.0111	3	1.7693-1.7700	0.0008-0.0019	0.0044-0.0019
	2.4	S	2.2434-2.2442	①	0.0020-0.0098	3	1.7709-1.7717	0.0008-0.0020	0.0040-0.0098
	2.7	8	2.4402-2.4409	0.0002-0.0009	0.0028-0.0098	3	1.8891-1.8898	0.0007-0.0014	0.0039-0.0098
2004	1.6	5	1.9365-1.9562	0.0007-0.0014	0.0032-0.0111	3	1.7693-1.7700	0.0008-0.0019	0.0044-0.0103
	1.8	1	2.1629-2.1636	0.0010-0.0017	0.0032-0.0111	3	1.7693-1.7700	0.0008-0.0019	0.0044-0.0019
	2.0	2	2.2418-2.2426	0.0011-0.0019	0.0024-0.0102	3	1.8898-1.8905	0.0009-0.0017	0.0039-0.0098
	2.4	S	2.2434-2.2442	①	0.0020-0.0098	3	1.7709-1.7717	0.0008-0.0020	0.0040-0.0098
	2.7	8	2.4402-2.4409	0.0002-0.0009	0.0028-0.0098	3	1.8891-1.8898	0.0007-0.0014	0.0039-0.0098

① Journal Nos. 1, 2, 4 & 5: 0.0007-0.0014 in.
 Journal Nos. 3: 0.0009-0.0016 in.

67162-KIAC-C06

PISTON AND RING SPECIFICATIONS

All measurements are given in inches.

Year	Engine Displacement Liters	Engine VIN	Piston Clearance	Ring Gap			Ring Side Clearance		
				Top Compression	Bottom Compression	Oil Control	Top Compression	Bottom Compression	Oil Control
2001	1.5	3	0.0015-0.0021	0.006-0.011	0.0016-0.0021	0.008-0.027	0.0020-0.0030	0.0010-0.0030	SNUG
	1.8	1	0.0015-0.0021	0.006-0.011	0.012-0.018	0.008-0.027	0.0020-0.0030	0.0010-0.0030	SNUG
	2.4	S	0.0008-0.0120	0.0098-0.0138	0.0157-0.0216	0.0039-0.0157	0.0012-0.0028	0.0008-0.0024	SNUG
	2.7	8	0.0004-0.0012	0.0079-0.0138	0.0146-0.0205	0.0079-0.0276	0.0016-0.0031	0.0012-0.0028	SNUG
2002	1.5	3	0.0015-0.0021	0.006-0.011	0.0016-0.0021	0.008-0.027	0.0020-0.0030	0.0010-0.0030	SNUG
	1.8	1	0.0015-0.0021	0.006-0.011	0.012-0.018	0.008-0.027	0.0020-0.0030	0.0010-0.0030	SNUG
	2.4	S	0.0008-0.0120	0.0098-0.0138	0.0157-0.0216	0.0039-0.0157	0.0012-0.0028	0.0008-0.0024	SNUG
	2.7	8	0.0004-0.0012	0.0079-0.0138	0.0146-0.0205	0.0079-0.0276	0.0016-0.0031	0.0012-0.0028	SNUG
2003	1.6	5	0.0015-0.0021	0.006-0.011	0.0016-0.0021	0.008-0.027	0.0012-0.0028	0.0012-0.0028	SNUG
	1.8	1	0.0015-0.0021	0.006-0.011	0.012-0.018	0.008-0.027	0.0020-0.0030	0.0010-0.0030	SNUG
	2.4	S	0.0008-0.0120	0.0098-0.0138	0.0157-0.0216	0.0039-0.0157	0.0012-0.0028	0.0008-0.0024	SNUG
	2.7	8	0.0004-0.0012	0.0079-0.0138	0.0146-0.0205	0.0079-0.0276	0.0016-0.0031	0.0012-0.0028	SNUG
2004	1.6	5	0.0015-0.0021	0.006-0.011	0.0016-0.0021	0.008-0.027	0.0012-0.0028	0.0012-0.0028	SNUG
	1.8	1	0.0015-0.0021	0.006-0.011	0.012-0.018	0.008-0.027	0.0020-0.0030	0.0010-0.0030	SNUG
	2.0	2	0.0008-0.0016	0.0091-0.0150	0.0130-0.0189	0.0079-0.0236	0.0016-0.0031	0.0012-0.0028	0.0024-0.0059
	2.4	S	0.0008-0.0120	0.0098-0.0138	0.0157-0.0216	0.0039-0.0157	0.0012-0.0028	0.0008-0.0024	SNUG
	2.7	8	0.0004-0.0012	0.0079-0.0138	0.0146-0.0205	0.0079-0.0276	0.0016-0.0031	0.0012-0.0028	SNUG

67162-KIAC-C07

TORQUE SPECIFICATIONS
All readings in ft. lbs.

Year	Engine Displacement Liters	Engine VIN	Cylinder Head Bolts	Main Bearing Bolts	Rod Bearing Bolts	Crankshaft Damper Bolts	Flywheel Bolts	Manifold Intake	Manifold Exhaust	Spark Plugs	Oil Pan Drain Plug
2001	1.5	3	①	40-43	22-24	28-38	71-76	11-14	11-14	11-17	22-30
	1.8	1	②	③	35-37	9-13 ④	71-76	14-19	28-34	11-17	29-32.5
	2.4	S	⑤	⑥	⑦	80-94	94-101	⑧	⑨	15-21	26-32
	2.7	8	⑩	⑪	⑫	130-138	53-55	14-15	12-22	15-21	26-32
2002	1.5	3	①	40-43	22-24	28-38	71-76	11-14	11-14	11-17	22-30
	1.8	1	②	③	35-37	9-13 ④	71-76	14-19	28-34	11-17	29-32.5
	2.4	S	⑤	⑥	⑦	80-94	94-101	⑧	⑨	15-21	26-32
	2.7	8	⑩	⑪	⑫	130-138	53-55	14-15	12-22	15-21	26-32
2003	1.6	5	①	40-43	22-24	28-38	71-76	11-14	11-14	11-17	22-30
	1.8	1	②	③	35-37	9-13 ④	71-76	14-19	28-34	11-17	29-32.5
	2.4	S	⑤	⑥	⑦	80-94	94-101	⑧	⑨	15-21	26-32
	2.7	8	⑩	⑪	⑫	130-138	53-55	14-15	12-22	15-21	26-32
2004	1.6	5	①	40-43	22-24	28-38	71-76	11-14	11-14	11-17	22-30
	1.8	1	②	③	35-37	9-13 ④	71-76	14-19	28-34	11-17	29-32.5
	2.0	2	⑬	⑭	36.2-38.3	123-130.2	86.8-94	13-16.5	31-40	11-17	29-32.5
	2.4	S	⑤	⑥	⑦	80-94	94-101	⑧	⑨	15-21	26-32
	2.7	8	⑩	⑪	⑫	130-138	53-55	14-15	12-22	15-21	26-32

① Step 1: 36 inch lbs.
   Step 2: Loosen fully
   Step 3: 18 ft. lbs.
   Step 4: Tighten 90 degrees
   Step 5: Tighten 60 degrees

② Step 1: 36 ft. lbs.
   Step 2: Loosen fully
   Step 3: 29 ft. lbs.
   Step 4: Tighten 90 degrees
   Step 5: Additional 90 degrees

③ Step 1: 29 ft. lbs.
   Step 2: Loosen fully
   Step 3: 14.5 ft. lbs.
   Step 4: Tighten 90 degrees
   Step 5: Tighten 60 degrees

④ Crankshaft pulley

⑤ Step 1: 14 ft. lbs.
   Step 2: Loosen fully
   Step 3: 14 ft. lbs.
   Step 4: Tighten 90 degrees
   Step 5: Tighten 90 degrees

⑥ 18 ft. lbs., plus 90-94 degrees

⑦ 13-16 ft. lbs., plus 90-94 degrees

⑧ M10 bolts: 13-18 ft. lbs.
   M8 bolts: 11-14 ft. lbs.
   Nut: 22-30 ft. lbs.

⑨ M10 bolts: 25-40 ft. lbs.
   M8 bolts: 18-22 ft. lbs.

⑩ Step 1: 18 ft. lbs., plus 60-65 degrees
   Step 2: Tighten 45-49 degrees
   Step 3: 26-32 ft. lbs.

⑪ M10 bolts: 19.5-23.9 ft. lbs., plus 90-95 degrees
   M8 bolts: 9.6-14 ft. lbs., plus 90-95 degrees

⑫ 11.6-14 ft. lbs., plus 90-94 degrees

⑬ M10 bolts: 17-19.5 ft. lbs., plus 60-65 degrees, plus additional 60-65 degrees
   M12 bolts: 20.5-23 ft. lbs., plus 60-65 degrees, plus additional 60-65 degrees

⑭ 20.3-23.1 ft. lbs., plus 60-64 degrees

67162-KIAC-C08

WHEEL ALIGNMENT

Year	Model		Caster Range (+/-Deg.)	Caster Preferred Setting (Deg.)	Camber Range (+/-Deg.)	Camber Preferred Setting (Deg.)	Toe-in (in.)
2001	Rio	F	0.45	+1.41	0.45	+0.6	0.12 +/- 0.12
		R	—	—	0.18	-0.50	0.20 +/- 0.24
	Sephia	F	0.75	+2.45	0.50	0	0.11 +/- 0.12
		R	—	—	0.50	-0.52	0.07 +/- 0.12
	Optima	F	1.00	+3.15	0.30	0	0.11 +/- 0.12
		R	—	—	0.30	-0.30	0.07 +/- 0.12
2002	Rio	F	0.45	+1.41	0.45	+0.6	0.12 +/- 0.12
		R	—	—	0.18	-0.50	0.20 +/- 0.24
	Spectra	F	0.75	+2.45	0.50	0	0.11 +/- 0.12
		R	—	—	0.50	-0.52	0.07 +/- 0.12
	Optima	F	1.00	+3.15	0.30	0	0.11 +/- 0.12
		R	—	—	0.30	-0.30	0.07 +/- 0.12
2003	Rio	F	0.45	+1.41	0.45	+0.6	0.12 +/- 0.12
		R	—	—	0.18	-0.50	0.20 +/- 0.24
	Spectra	F	0.75	+2.45	0.50	0	0.11 +/- 0.12
		R	—	—	0.50	-0.52	0.07 +/- 0.12
	Optima	F	1.00	+3.15	0.30	0	0.11 +/- 0.12
		R	—	—	0.30	-0.30	0.07 +/- 0.12
2004	Rio	F	0.45	+1.41	0.45	+0.6	0.12 +/- 0.12
		R	—	—	0.18	-0.50	0.20 +/- 0.24
	Spectra	F	0.75	+2.45	0.50	0	0.11 +/- 0.12
		R	—	—	0.50	-0.52	0.07 +/- 0.12
	Optima	F	1.00	+3.15	0.30	0	0.11 +/- 0.12
		R	—	—	0.30	-0.30	0.07 +/- 0.12

67162-KIAC-C09

TIRE, WHEEL AND BALL JOINT SPECIFICATIONS

| Year | Model | OEM Tires | | Tire Pressures (psi) | | Wheel Size | Ball Joint Inspection | Wheel Lug Torque (ft. lbs.) |
		Standard	Optional	Front	Rear			
2001	Rio	P155/80SR13	P175/70R13	①	①	Std: 5-JJ Opt: 5-JJ	②	65-87
	Sephia	P175/70SR13	P185/60HR14	29	29	Std: 5-JJ Opt: 6-JJ	②	65-87
	Optima	P175/70R14	P205/60R15	30	30	Std: 5.5-JJ Opt: 6-JJ	②	65-87
2002	Rio	P155/80SR13	P175/70R13	①	①	Std: 5-JJ Opt: 5-JJ	②	65-87
	Spectra	P175/70SR13	P185/60HR14	29	29	Std: 5-JJ Opt: 6-JJ	②	65-87
	Optima	P175/70R14	P205/60R15	30	30	Std: 5.5-JJ Opt: 6-JJ	②	65-87
2003	Rio	P155/80SR13	P175/70R13	①	①	Std: 5-JJ Opt: 5-JJ	②	65-87
	Spectra	P175/70SR13	P185/60HR14	29	29	Std: 5-JJ Opt: 6-JJ	②	65-87
	Optima	P175/70R14	P205/60R15	30	30	Std: 5.5-JJ Opt: 6-JJ	②	65-87
2004	Rio	P155/80SR13	P175/70R13	①	①	Std: 5-JJ Opt: 5-JJ	②	65-87
	Spectra	P175/70SR13	P185/60HR14	29	29	Std: 5-JJ Opt: 6-JJ	②	65-87
	Optima	P175/70R14	P205/60R15	30	30	Std: 5.5-JJ Opt: 6-JJ	②	65-87

OEM: Original Equipment Manufacturer

PSI: Pounds Per Square Inch

STD: Standard

OPT: Optional

① Standard front and rear tire pressure: 32 psi

   Optional front and rear tire pressure: 29 psi

② Replace if any measurable movement is found.

67162-KIAC-C10

BRAKE SPECIFICATIONS
All measurements in inches unless noted

Year	Model		Brake Disc			Brake Drum			Minimum Lining Thickness	Brake Caliper	
			Original Thickness	Minimum Thickness	Maximum Run-out	Original Inside Diameter	Max. Wear Limit	Maximum Machine Diameter		Bracket Bolts (ft. lbs.)	Mounting Bolts (ft. lbs.)
2001	Rio	F	0.870	0.710	0.0040	—	—	—	0.080	33-49	19-21
		R	—	—	—	7.87	7.91	7.91	0.079	33-49	22-29
	Sephia	F	0.940	0.710	0.0040	—	—	—	0.080	33-49	19-21
		R	0.400	0.320	0.0039	7.87	7.91	7.91	0.079	33-49	22-29
	Optima	F	0.965	0.880	0.0012	—	—	—	0.079	51-63	16-24
		R	0.390	0.320	0.0039	9.00	9.08	9.08	0.079	51-63	16-24
2002	Rio	F	0.870	0.710	0.0040	—	—	—	0.080	33-49	19-21
		R	—	—	—	7.87	7.91	7.91	0.079	33-49	22-29
	Spectra	F	0.940	0.710	0.0040	—	—	—	0.080	33-49	19-21
		R	0.400	0.320	0.0039	7.87	7.91	7.91	0.079	33-49	22-29
	Optima	F	0.965	0.880	0.0012	—	—	—	0.079	51-63	16-24
		R	0.390	0.320	0.0039	9.00	9.08	9.08	0.079	51-63	16-24
2003	Rio	F	0.870	0.710	0.0040	—	—	—	0.080	33-49	19-21
		R	—	—	—	7.87	7.91	7.91	0.079	33-49	22-29
	Spectra	F	0.940	0.710	0.0040	—	—	—	0.080	33-49	19-21
		R	0.400	0.320	0.0039	7.87	7.91	7.91	0.079	33-49	22-29
	Optima	F	0.965	0.880	0.0012	—	—	—	0.079	51-63	16-24
		R	0.390	0.320	0.0039	9.00	9.08	9.08	0.079	51-63	16-24
2004	Rio	F	0.870	0.710	0.0040	—	—	—	0.080	33-49	19-21
		R	—	—	—	7.87	7.91	7.91	0.079	33-49	22-29
	Spectra	F	0.940	0.710	0.0040	—	—	—	0.080	33-49	19-21
		R	0.400	0.320	0.0039	7.87	7.91	7.91	0.079	33-49	22-29
	Optima	F	0.965	0.880	0.0012	—	—	—	0.079	51-63	16-24
		R	0.390	0.320	0.0039	9.00	9.08	9.08	0.079	51-63	16-24

F: Front
R: Rear

67162-KIAC-C11

SCHEDULED MAINTENANCE INTERVALS
Kia—Sephia, Spectra, Rio & Optima

TO BE SERVICED	TYPE OF SERVICE	VEHICLE MILEAGE INTERVAL (x1000)																
		7.5	15	22.5	30	37.5	45	52.5	60	67.5	75	82.5	90	97.5	105	112.5	120	127
Accessory drive belts	S/I				✓				✓				✓				✓	✓
Air cleaner element	R				✓				✓				✓				✓	✓
Air conditioner system	S/I	Inspect the system operation and refrigerant amount annually.																
Brake lines, hoses and connections	S/I				✓				✓				✓				✓	✓
Chassis and body fasteners	T				✓				✓				✓				✓	✓
Clutch pedal height, freeplay and operation	S/I				✓				✓				✓				✓	✓
Cooling system hoses and coolant level	S/I				✓				✓				✓				✓	✓
CV-joint boots	S/I				✓				✓				✓				✓	✓
Engine coolant	R				✓				✓				✓				✓	✓
Engine oil and filter	R	✓	✓	✓	✓	✓	✓	✓	✓	✓	✓	✓	✓	✓	✓	✓	✓	✓
Exhaust system heat shields	S/I				✓				✓				✓				✓	✓
Front and rear brakes	S/I				✓				✓				✓				✓	✓
Front ball joints S/I	S/I				✓				✓				✓				✓	✓
Fuel filter	R								✓								✓	✓
Fuel lines and hoses	S/I				✓				✓				✓				✓	✓
Idle speed	A				✓				✓				✓				✓	✓
Locks and hinges	L	✓	✓	✓	✓	✓	✓	✓	✓	✓	✓	✓	✓	✓	✓	✓	✓	✓
Spark plugs	R				✓				✓				✓				✓	✓
Steering operation and linkage	S/I				✓				✓				✓				✓	✓
Timing belt (California models)	R														✓			
Timing belt (California models)	S/I								✓				✓					
Timing belt (except California models)	R								✓								✓	✓

R: Replace   S/I: Inspect and service, if needed   L: Lubricate   A: Adjust   T: Tighten

FREQUENT OPERATION MAINTENANCE (SEVERE SERVICE)

If a vehicle is operated under any of the following conditions it is considered severe service

- Towing a trailer or using a camper or car-top carrier.

- Repeated short trips of less than 5 miles in temperatures below freezing, or trips of less than 10 miles in any temperature.

- Extensive idling or low-speed driving for long distances as in heavy commercial use, such as delivery, taxi or police cars.

- Operating on rough, muddy or salt-covered roads.

- Operating on unpaved or dusty roads.

- Driving in extremely hot (over 90°F) conditions.

Engine oil and filter: replace every 5000 miles or 5 months, whichever occurs first.

Air cleaner element: inspect ever 15,000 miles or 15 months and replace every 30,000 miles or 30 months, whichever occurs first.

Fuel system hoses (California models only): replace every 105,000 miles.

Emission system hoses (non-California models): inspect every 55,000 miles or 55 months, whichever occurs first.

Emission system hoses (California models): inspect every 60,000 miles or 60 months, whichever occurs first.

Front and rear disc brakes: inspect every 15,000 miles or 15 months, whichever occurs first.

Chassis and body fasteners: tighten every 15,000 miles or 15 months, whichever occurs first.

Locks and hinges: lubricate every 5000 miles or 5 months, whichever occurs first.

PRECAUTIONS

Before servicing any vehicle, please be sure to read all of the following precautions, which deal with personal safety, prevention of component damage, and important points to take into consideration when servicing a motor vehicle:

• Never open, service or drain the radiator or cooling system when the engine is hot; serious burns can occur from the steam and hot coolant.

• Observe all applicable safety precautions when working around fuel. Whenever servicing the fuel system, always work in a well-ventilated area. Do not allow fuel spray or vapors to come in contact with a spark, open flame, or excessive heat (a hot drop light, for example). Keep a dry chemical fire extinguisher near the work area. Always keep fuel in a container specifically designed for fuel storage; also, always properly seal fuel containers to avoid the possibility of fire or explosion. Refer to the additional fuel system precautions later in this section.

• Fuel injection systems often remain pressurized, even after the engine has been turned **OFF**. The fuel system pressure must be relieved before disconnecting any fuel lines. Failure to do so may result in fire and/or personal injury.

• Brake fluid often contains polyglycol ethers and polyglycols. Avoid contact with the eyes and wash your hands thoroughly after handling brake fluid. If you do get brake fluid in your eyes, flush your eyes with clean, running water for 15 minutes. If

eye irritation persists, or if you have taken brake fluid internally, IMMEDIATELY seek medical assistance.

• The EPA warns that prolonged contact with used engine oil may cause a number of skin disorders, including cancer! You should make every effort to minimize your exposure to used engine oil. Protective gloves should be worn when changing oil. Wash your hands and any other exposed skin areas as soon as possible after exposure to used engine oil. Soap and water, or waterless hand cleaner should be used.

• All new vehicles are now equipped with an air bag system. The system must be disabled before performing service on or around system components, steering column, instrument panel components, wiring and sensors. Failure to follow safety and disabling procedures could result in accidental air bag deployment, possible personal injury and unnecessary system repairs.

• Always wear safety goggles when working with, or around, the air bag system. When carrying a non-deployed air bag, be sure the bag and trim cover are pointed away from your body. When placing a non-deployed air bag on a work surface, always face the bag and trim cover upward, away from the surface. This will reduce the motion of the module if it is accidentally deployed. Refer to the additional air bag system precautions later in this section.

• Clean, high quality brake fluid from a sealed container is essential to the safe and

proper operation of the brake system. You should always buy the correct type of brake fluid for your vehicle. If the brake fluid becomes contaminated, completely flush the system with new fluid. Never reuse any brake fluid. Any brake fluid that is removed from the system should be discarded. Also, do not allow any brake fluid to come in contact with a painted surface; it will damage the paint.

• Never operate the engine without the proper amount and type of engine oil; doing so WILL result in severe engine damage.

• Timing belt maintenance is extremely important! Many models utilize an interference-type, non-freewheeling engine. If the timing belt breaks, the valves in the cylinder head may strike the pistons, causing potentially serious (also time-consuming and expensive) engine damage. Refer to the maintenance interval charts in the front of this section for the recommended replacement interval for the timing belt.

• Disconnecting the negative battery cable on some vehicles may interfere with the functions of the on-board computer system(s) and may require the computer to undergo a relearning process once the negative battery cable is reconnected.

• When servicing drum brakes, only dissemble and assemble one side at a time, leaving the remaining side intact for reference.

• Only an MVAC-trained, EPA-certified automotive technician should service the air conditioning system or its components.

ENGINE REPAIR

Alternator

REMOVAL

1. Before servicing the vehicle, refer to the precautions at the beginning of this section.

2. Disconnect the negative battery cable.

3. Remove the front air intake inlet pipe bolts, 1.8L engine only.

4. Remove the top hose from the air intake inlet pipe, 1.8L engine only.

5. Remove the air intake inlet pipe clamp and the pipe, 1.8L engine only.

6. Remove the power steering pump, 2.0L engine only.

7. Remove the power steering pump bracket, 2.0L engine only.

8. Remove the alternator **B** terminal cover cap.

9. Disconnect the alternator electrical connections.

10. Loosen, but do not remove the alternator pivot bolt and the tensioner mounting bolt.

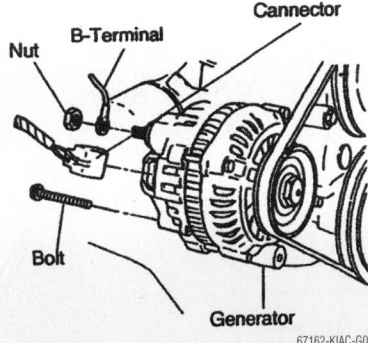

Alternator mounting—1.5L and 1.6L engine shown

67162-KIAC-G02

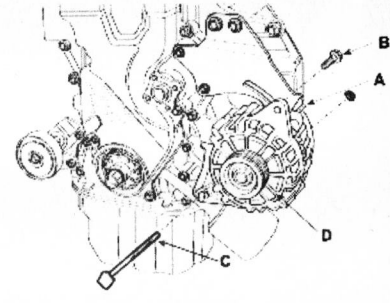

A - adjustment bolt
B - tension mounting bolt
C - pivot bolt
D - alternator

67162-KIAC-G03

Alternator mounting—2.0L engine shown

2.4L DOHC ENGINE

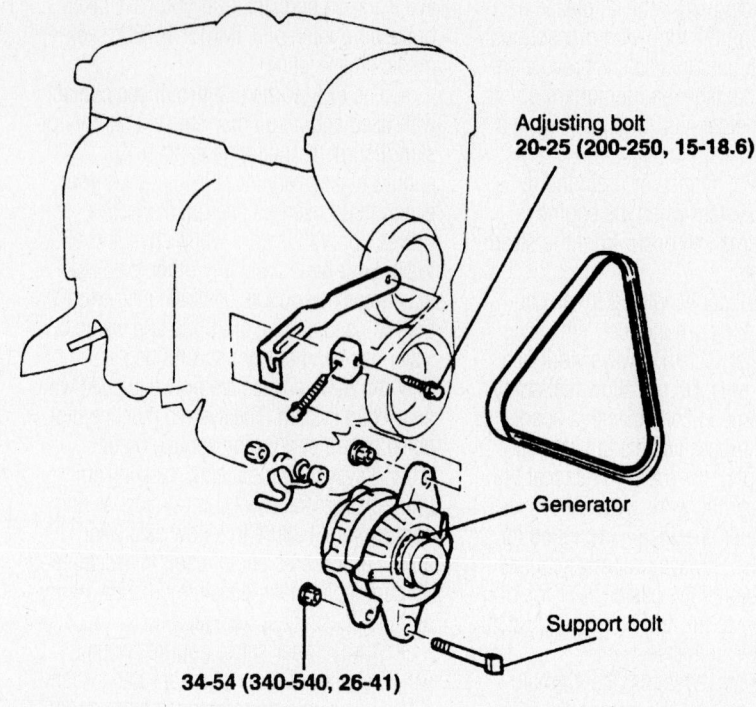

Adjusting bolt
20-25 (200-250, 15-18.6)

Generator

Support bolt

34-54 (340-540, 26-41)

V6 ENGINE

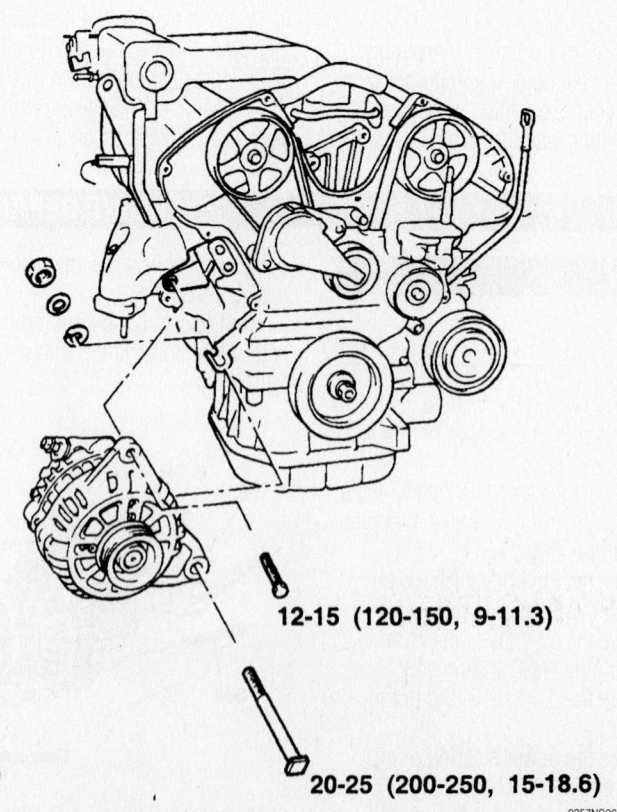

12-15 (120-150, 9-11.3)

20-25 (200-250, 15-18.6)

TORQUE : N·m (kg·cm, lb·ft)

9357NG26

Alternator mounting—2.4L and 2.7L engine

11. Remove the drive belt(s); relieve tension on the belt by rotating the adjustment bolt.

12. Remove the alternator tensioner mounting bolt and the belt tensioner.

13. Remove the alternator pivot bolt.

14. Loosen the bolt at the base of the adjusting bracket and rotate the bracket up.

15. Remove the alternator.

INSTALLATION

1. Before servicing the vehicle, refer to the precautions at the beginning of this section.

2. Install the alternator.

3. Install the alternator pivot bolt and hand-tighten at this time.

4. Rotate the adjusting bracket into position, place the belt tensioner into position. Hand-tighten the mounting bolt.

5. Install the drive belt.

6. Adjust the belt tension by rotating the adjustment bolt.

7. The belt deflection for 1.5L, 1.6L, 2.4L and 2.7L engines, is as follows:

    a. New belt: 0.22–0.28 in.
(5.5–7mm).

    b. Used belt: 0.24–0.28 in. (6–7mm).

8. The belt deflection for the 2.0L engine, is as follows:

    a. New belt: 0.22–0.31 in.
(5.5–8mm).

    b. Used belt: 0.33–0.45 in.
(8.5–11.5mm).

9. The belt deflection for the 1.8L engine, is as follows:

    a. New belt: 0.23–0.31 in. (6–8mm).

    b. Used belt: 0.28–0.35 in. (7–9mm).

10. Tighten the tensioner bolt to 14–19 ft. lbs. (19–26 Nm) and the pivot bolt to 28–38 ft. lbs. (38–51 Nm).

11. Connect the alternator electrical connections.

12. Install the alternator **B** terminal cover cap.

13. Install the power steering pump and adjust the power steering belt tension, 2.0L engine only.

14. Install the air intake inlet pipe and fasten the clamp, 1.8L engine only.

15. Install the top hose to the air intake inlet pipe, 1.8L engine only.

16. Install the front air intake inlet pipe bolts, 1.8L engine only.

17. Connect the negative battery cable.

Ignition Timing

This vehicle is equipped with a Distributorless Ignition System (DIS). No adjustment is necessary or possible.

Engine Assembly

REMOVAL & INSTALLATION

1.5L and 1.6L Engine

1. Before servicing the vehicle, refer to the precautions at the beginning of this section.

2. Properly relieve the fuel system pressure.

3. Drain and recycle the engine coolant.

4. Disconnect the battery cables.

5. Remove the battery and tray.

6. Remove the fresh air intake duct.

7. Remove the upper and lower radiator hoses.

8. Disconnect the accelerator cable.

9. Disconnect the fuel hose from the fuel rail.

10. Remove the heater hose.

11. Disconnect the brake vacuum hose and purge control hose from the dynamic chamber.

12. Disconnect the injector connectors.

13. Disconnect the electrical connectors, tag before removing.

14. Remove the transaxle linkage, if equipped with an automatic transaxle.

15. Remove the manual transaxle linkage and extension bar, if equipped.

16. Remove the clutch release cylinder and pipe, if equipped.

17. Disconnect the transaxle range switch connector, solenoid valve connector and fluid cooler hose, if equipped with an automatic transaxle.

18. Disconnect the power steering pump hose.

19. Disconnect the B and S-terminal connectors from starter.

20. Disconnect the alternator B-terminal connector.

21. Remove the 4 A/C compressor mounting bolts, but do not disconnect the fluid line. Position the compressor aside.

22. Remove the front wheels.

23. Disconnect the front exhaust pipe from the manifold.

24. Remove the right and left tie rod ends from the steering knuckle.

25. Remove the bolt and nut, then separate the right and left lower control arm from the steering knuckle.

26. Remove the 2 bolts and nuts from the damper, then separate the damper from the knuckle.

27. Remove the halfshafts from the transaxle, by carefully prying them.

28. Support the engine with a suitable hoist.

29. Remove the 4 nuts and bolts from the engine mounting member.

30. Remove the 2 bolts from the No. 1 engine mounting bracket.

31. Remove the 4 bolts from the No. 2 engine mounting bracket.

32. Remove the 2 nuts from the No. 3 engine mounting rubber.

33. Remove the engine and transaxle assembly by lifting it out of the engine compartment as a unit.

To install:

34. Installation is the reverse of the removal procedure. Note the following important steps.

35. When possible, leave the engine mounting nuts/bolts loose (hand tight) until all mounts are aligned and bolted. This may help in aligning the engine/transaxle assembly in the vehicle.

36. Tighten the engine mount bolts/nuts as follows:

- No. 3 engine mounting rubber (insulator) bolts: 49–68 ft. lbs. (68–93 Nm)
- No. 2 engine mounting nuts: 49–68 ft. lbs. (68–93 Nm)
- No. 2 engine mounting bolts: 28–38 ft. lbs. (38–51 Nm)
- Engine mounting member nuts: 28–38 ft. lbs. (39–52 Nm)
- Engine mounting member bolts: 47–66 ft. lbs. (65–91 Nm)

37. Install new circlips on the inner CV-joint stub shafts, if equipped, intermediate shaft. Grease the shaft splines before installing the halfshaft/intermediate shaft into the transaxle.

38. Always install new gaskets and/or O-rings. Use new self-locking nuts, especially on the exhaust.

39. Fill the engine and the transaxle with the proper types and quantities of oil. Fill the cooling system.

40. Connect the negative battery cale, start the engine and check for leaks. Check all fluid levels.

1.8L Engine

1. Before servicing the vehicle, refer to the precautions at the beginning of this section.

2. Properly relieve the fuel system pressure.

3. Drain and recycle the engine coolant.

4. Disconnect the battery cables.

5. Remove the battery and tray.

6. Remove the Data Link Connector (DLC) from the Mass Air Flow (MAF) sensor bracket.

7. Disconnect the Intake Air Tempera-

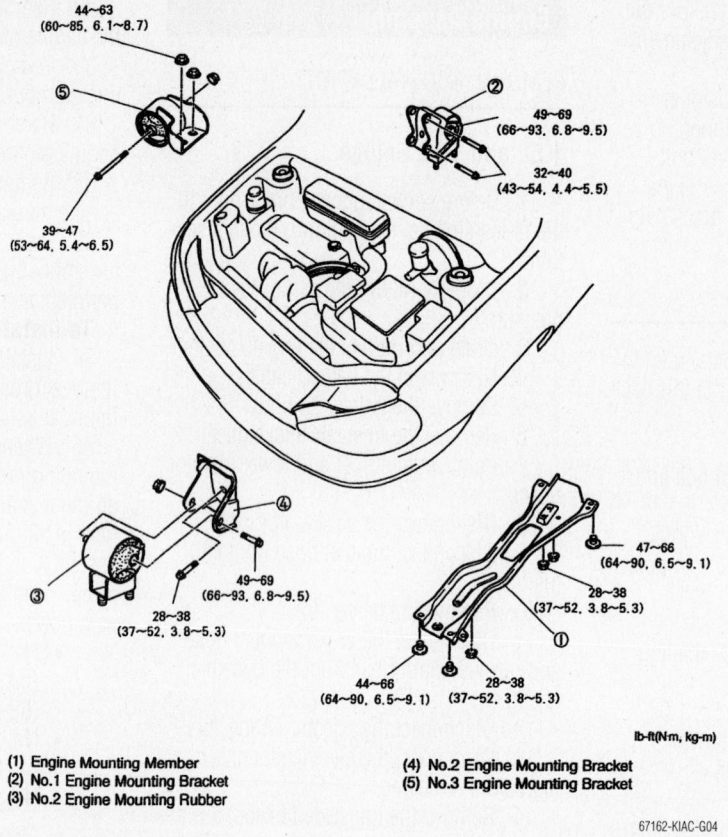

44~63
(60~85, 6.1~8.7)

49~69
(66~93, 6.8~9.5)

32~40
(43~54, 4.4~5.5)

39~47
(53~64, 5.4~6.5)

47~66
(64~90, 6.5~9.1)

28~38
(37~52, 3.8~5.3)

49~69
(66~93, 6.8~9.5)

28~38
(37~52, 3.8~5.3)

44~66
(64~90, 6.5~9.1)

28~38
(37~52, 3.8~5.3)

lb-ft(N·m, kg-m)

(1) Engine Mounting Member
(2) No.1 Engine Mounting Bracket
(3) No.2 Engine Mounting Rubber
(4) No.2 Engine Mounting Bracket
(5) No.3 Engine Mounting Bracket

67162-KIAC-G04

Exploded view of the engine mount and brackets—1.5L and 1.6L engine

ture (IAT) and Mass Air Flow (MAF) sensor connectors.

8. Remove the air intake hose from the throttle body.

9. Remove the ventilation hose and fresh air duct.

10. Disconnect the accelerator and, if equipped, cruise control cables.

11. Remove the air cleaner assembly.

12. Disconnect the brake booster and purge control vacuum hoses from the intake manifold.

13. Remove the upper and lower radiator hoses.

14. Disconnect the fuel hose from the fuel injector rail.

15. Remove the heater hoses.

16. Disconnect the Idle Air Control (IAC) and Throttle Position (TP) sensor connectors.

17. Disconnect the fuel injector electrical connectors.

18. Disconnect the starter and alternator electrical connectors.

19. Disconnect the engine ground strap.

20. Remove the left and right splash shields.

21. Remove the exhaust manifold heat shield and 3 power steering pump bracket bolts.

22. Remove the air conditioning com-pressor mounting bolts and position it aside leaving the hoses attached.

23. Disconnect the ground strap from the top of the transaxle.

24. Remove the No. 4 engine mount.

25. Disconnect the input/turbine speed sensor connector, if equipped with an auto-matic transaxle.

26. Remove the back-up light switch, if equipped with a manual transaxle.

27. Remove the Vehicle Speed Sensor (VSS).

28. Remove the U-clip from the selector cable and the nut and washer from the transaxle linkage, if equipped with an auto-matic transaxle.

29. Remove the linkage and extension bar, then the clutch release cylinder and hydraulic hose, if equipped with a manual transaxle.

30. Disconnect the transaxle range switch and solenoid valve connectors, then the 2 transaxle oil cooler hoses, if equipped with an automatic transaxle.

31. Remove the front wheels.

32. Disconnect the Oxygen (O_2) sensor electrical connectors.

33. Disconnect the front exhaust pipe.

34. Remove the halfshafts.

35. Properly support the engine/transaxle assembly.

36. Remove the two No. 2 engine mount-to-mounting member bolts.

37. Remove the one bolt from the No. 3 engine mount.

38. Remove the three No. 1 engine mounting bolts, then carefully lift the engine/transaxle assembly from the vehicle.

To install:

39. Installation is the reverse of the removal procedure. Note the following important steps.

40. When possible, leave the engine mounting nuts/bolts loose (hand tight) until all mounts are aligned and bolted. This may help in aligning the engine/transaxle assembly in the vehicle.

41. Tighten the engine mount bolts/nuts as follows:

- No. 1 mounting bolts: 50–70 ft. lbs. (67–93 Nm)
- No. 3 mounting bolts: 63–86 ft. lbs. (85–116 Nm)
- No. 2 mounting nuts: 28–38 ft. lbs. (38–51 Nm)
- No. 4 mounting bolts: 47–66 ft. lbs. (64–89 Nm)
- No. 4 mounting nuts: 49–68 ft. lbs. (68–93 Nm)

42. Install new circlips on the inner CV-joint stub shafts, if equipped, intermediate shaft. Grease the shaft splines before

installing the halfshaft/intermediate shaft into the transaxle.

43. Always install new gaskets and/or O-rings. Use new self-locking nuts, especially on the exhaust.

44. Fill the engine and the transaxle with the proper types and quantities of oil. Fill the cooling system.

45. Connect the negative battery cable, start the engine and check for leaks. Check all fluid levels.

2.0L Engine

1. Disconnect the battery terminals and remove the heat shield.

2. Remove the battery and battery tray.

3. Remove the engine cover.

4. Drain the engine coolant. Remove the radiator cap to speed draining.

5. Remove the intake air hose and air cleaner assembly.

6. Disconnect the AFS (Air Flow Sensor) connector.

7. Disconnect the breather hose from intake air hose.

8. Remove the intake air hose and air cleaner upper cover.

9. Remove the upper radiator hose and lower radiator hose.

10. Disconnect the ATF (Automatic Transaxle Fluid) cooler hose.

11. Remove the radiator.

12. Remove the heater hose.

13. Disconnect the accelerator cable.

14. Remove the engine wire harness connectors and wire harness clamps from the cylinder head and the intake manifold.

15. Disconnect the OCV (Oil Control Valve) connector.

16. Disconnect the oil temperature sensor connector.

17. Disconnect the ECT (Engine Coolant Temperature) sensor connector.

18. Disconnect the ignition coil connector.

19. Disconnect the TPS (Throttle Position Sensor) connector.

20. Disconnect the ISA (Idle Speed Actuator) connector.

21. Disconnect the air conditioner compressor switch.

22. Disconnect the CMP (Camshaft Position Sensor) connector.

23. Disconnect the four fuel injector connectors.

24. Disconnect the knock sensor connector.

25. Disconnect the ground cables from the intake manifold and vehicle's body.

26. Disconnect the front heated oxygen sensor connector.

27. Disconnect the CKP (Crankshaft Position Sensor) connector.

28. Disconnect the oil pressure switch connector.

29. Disconnect the PCSV (Purge Control Solenoid Valve) connector.

30. Disconnect the transaxle wire harness connectors and control cable from transaxle (A/T).

31. Disconnect the transaxle range switch connector.

32. Disconnect the solenoid valve connector.

33. Disconnect the input shaft speed sensor connector.

34. Disconnect the output shaft speed sensor connector.

35. Disconnect the Vehicle Speed Sensor (VSS) connector.

36. Disconnect the transaxle ground cable.

37. Disconnect the control cable nut from transaxle range switch.

38. Disconnect the control cable.

39. Disconnect the fuel inlet hose of the delivery pipe side.

40. Disconnect the hose of the Purge Control Solenoid Valve (PCSV) side.

41. Remove the brake booster vacuum hose.

42. Remove the power steering pump drive belt.

43. Remove the power steering pump and use a wire to secure the pump to the vehicle so that it is out of the way.

44. Remove the front wheel (RH, LH).

45. Remove the bolts and RH side cover.

46. Remove the air conditioner compressor drive belt.

47. Remove the air conditioner compressor and fix the compressor to vehicle with a wire.

48. Disconnect the rear oxygen sensor connector and remove the clamp.

49. Disconnect the starter motor connector and "B" terminal connection.

50. Remove the mounting clip of the starter cable.

51. Remove the alternator connector and "B" terminal connection.

52. Remove the front muffler.

53. Remove the drive shaft.

54. Remove the front roll stopper.

55. Remove the rear roll stopper.

56. Install the engine hanger lift to the engine and transaxle assembly.

57. Remove the bolt, nuts and engine mounting support bracket.

58. Remove the transaxle mounting bracket.

59. Using engine hanger lift, remove the engine and transaxle assembly on vehicle.

To install:

60. Installation is the reverse of the removal procedure. Note the following important steps.

61. When possible, leave the engine mounting nuts/bolts loose (hand tight) until all mounts are aligned and bolted. This may help in aligning the engine/transaxle assembly in the vehicle.

62. Tighten the engine mount bolts/nuts as follows:

- Front roll stopper bolt and nut: 36–47 ft. lbs. (49–63 Nm)
- Rear roll stopper bolt and nut: 36–47 ft. lbs. (49–63 Nm)
- Engine mount support bracket bolt (B) and nut (C): 36–47 ft. lbs. (49–63 Nm)

67162-KIAC-G05

Front roll stopper—2.0L engine

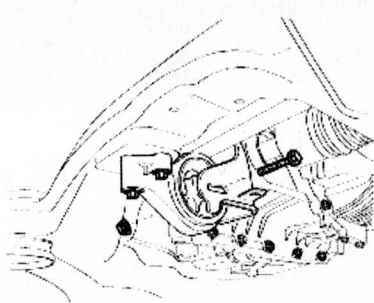

67162-KIAC-G06

Rear roll stopper—2.0L engine

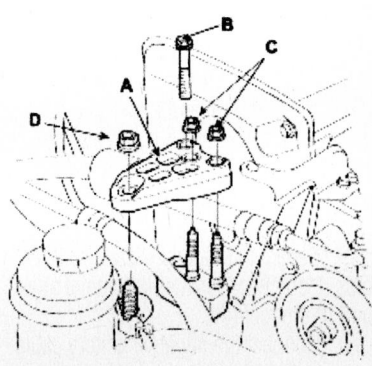

67162-KIAC-G07

Engine mount support bracket—2.0L engine

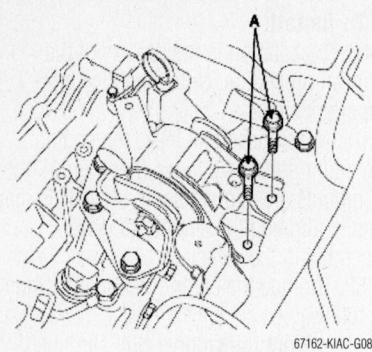

Transaxle mounting bracket bolts—2.0L engine

67162-KIAC-G08

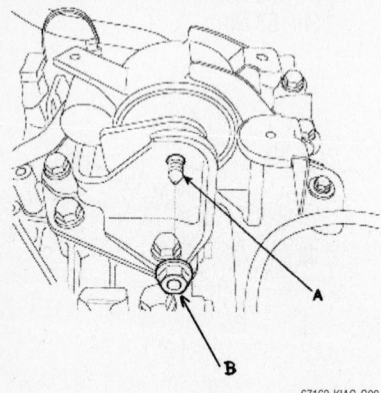

Transaxle mounting bracket bolt and nut—2.0L engine

67162-KIAC-G09

- Engine mount support bracket nut (D): 51–68 ft. lbs. (69–93 Nm)
- Transaxle mounting bracket bolts (A): 29–36 ft. lbs. (39–49 Nm)
- Transaxle mounting bracket nut (B): 65–79 ft. lbs. (88–108 Nm)

63. Always install new gaskets and/or O-rings. Use new self-locking nuts, especially on the exhaust.

64. Fill the engine and the transaxle with the proper types and quantities of oil. Fill the cooling system.

65. Adjust the shift cable.

66. Adjust the throttle cable.

67. Bleed the cooling system.

68. After, assemble the fuel line, turn on the ignition switch (do not operate the starter) so that the fuel pump runs for approximately two seconds and fuel line pressurizes.

69. Repeat this operation two or three times, then check for fuel leakage at any point in the fuel line.

70. Connect the negative battery cable, start the engine and check for leaks.

71. Bleed the cooling system.

72. Check all fluid levels.

2.4L Engine

1. Before servicing the vehicle, refer to the precautions at the beginning of this section.

2. Properly relieve the fuel system pressure.

3. Drain and recycle the engine coolant, transaxle fluid and engine oil.

4. Remove the battery.

5. Remove the air cleaner.

6. Disconnect the back-up lamp and engine harness connectors.

7. Disconnect the select control valve connector, if equipped with a manual transaxle.

8. Disconnect the alternator harness and oil pressure gauge wiring.

9. Disconnect the transaxle oil cooler hoses, if equipped with an automatic transaxle.

10. Remove the upper and lower radiator hoses from the engine.

11. Disconnect the engine ground.

12. Disconnect the brake booster vacuum hose.

13. Disconnect the main fuel line and return and vapor hoses from the engine side.

14. Disconnect the inlet and outlet heater hoses from the engine.

15. Disconnect the accelerator cable from the engine.

16. Disconnect the clutch cable, shift control rod and extension rod if equipped with a manual transaxle.

17. Disconnect the control cable from the transaxle, if equipped with an automatic transaxle.

18. Disconnect the Vehicle Speed Sensor (VSS) connector from the transaxle.

19. Disconnect the column shaft from the steering gear box.

20. Remove the engine and transaxle mounting insulators.

21. Remove the wheel.

22. Remove the caliper. Unbolt and support with a piece of wire. Do not disconnect the brake fluid line.

23. Remove the lower arm and fork and separate the upper arm and knuckle.

24. Disconnect the front exhaust pipe from the manifold. Use wire to hang the exhaust pipe from the bottom of the vehicle.

25. Place a suitable supporter under the sub-frame. Make sure all hoses, vacuum lines and connectors are detached from the engine.

26. Loosen the sub-frame bolts, then slowly lift the vehicle. The engine/transaxle, sub-frame, steering gear and halfshafts are removed as an assembly.

27. Remove the halfshafts, raise the engine/transaxle assembly with a hoist, and separate the engine and transaxle from the sub-frame.

28. To separate the engine/transaxle from the sub-frame, remove the engine mounting bracket and transaxle mounting bracket, then lift up the engine hoist.

To install:

29. To place the engine and transaxle assembly on the sub-frame, install the front roll stopper and rear roll stopper among the sub-frame and engine/transaxle assembly.

30. Installation is the reverse of the removal procedure, noting the following steps:

   a. Fill the engine and the transaxle with the proper types and quantities of oil.

   b. Fill the cooling system.

   c. Start the engine and check for leaks. Check all fluid levels.

2.7L Engine

1. Before servicing the vehicle, refer to the precautions at the beginning of this section.

2. Properly relieve the fuel system pressure.

3. Drain and recycle the engine coolant and engine oil.

4. Remove the battery and engine cover.

5. Remove the air cleaner.

6. Disconnect the engine harness connectors.

7. Remove the alternator harness, oil pressure switch and oil pressure sender connectors.

8. Disconnect the transaxle oil cooler hoses.

9. Remove the upper and lower radiator hoses from the engine.

10. Remove the radiator.

11. Disconnect the spark plug wires.

12. Disconnect the engine ground.

13. Disconnect the brake booster vacuum hose.

14. Disconnect the main fuel line and return and vapor hoses from the engine side.

15. Remove the inlet and outlet heater hoses from the engine.

16. Disconnect the accelerator and cruise control cables from the engine.

17. Disconnect the control cable from the transaxle.

18. Remove the speedometer cable from the transaxle.

19. Remove the power steering pump hose.

20. Remove the oil pan shield.

21. Disconnect the front exhaust pipe

from the manifold. Use a wire to support the exhaust pipe from the bottom of the vehicle.

22. Remove the lower control arm ball joint bolts and upper arm bolt from the steering knuckle.

23. Remove the halfshafts from the transaxle. Plug the openings in the transaxle case and discard the halfshaft circlips.

24. Attach a cable or chain to the engine and use a chain hoist to lift the engine enough to pull the cable tight.

25. Remove the front and rear roll stoppers.

26. Disconnect the connector from the starter motor harness.

27. Remove the engine mount bolts.

28. Remove the bolts and nuts that fasten the engine mount bracket to the body.

29. Slowly raise the engine and transaxle assembly and temporarily hold in the raised position. Make sure that all hoses, cables, vacuum lines and connectors are detached from the engine.

30. Remove the transaxle mounting bracket bolts.

31. Remove the left mount insulator bolt.

32. Remove the engine and transaxle assembly. While directing the transaxle side down, lift the engine and transaxle up and out of the vehicle.

To install:

33. Installation is the reverse of the removal procedure, noting the following steps:

   a. Fill the engine and the transaxle with the proper types and quantities of oil.

   b. Fill the cooling system.

   c. Start the engine and check for leaks. Check all fluid levels.

Water Pump

REMOVAL & INSTALLATION

1.5L and 1.6L Engine

1. Before servicing the vehicle, refer to the precautions at the beginning of this section.

2. Disconnect the negative battery cable.

3. Drain and recycle the engine coolant.

4. Remove the drive belt.

5. Remove the timing belt.

➡**The power steering pump must be removed to access the water inlet pipe. Do not disconnect the pump fluid lines.**

6. Remove the power steering pump and position aside.

7. Remove the water inlet pipe and gasket.

8. Remove the water bypass pipe and O-ring.

9. Remove the water pump bolts, pump and gasket.

10. Clean all gasket mating surfaces.

To install:

11. Install the water pump, using a new gasket and tighten the mounting bolts to 14–19 ft. lbs. (19–26 Nm)

12. Install the water pump bypass pump, with a new O-ring.

13. Install the water inlet pipe, with a new O-ring.

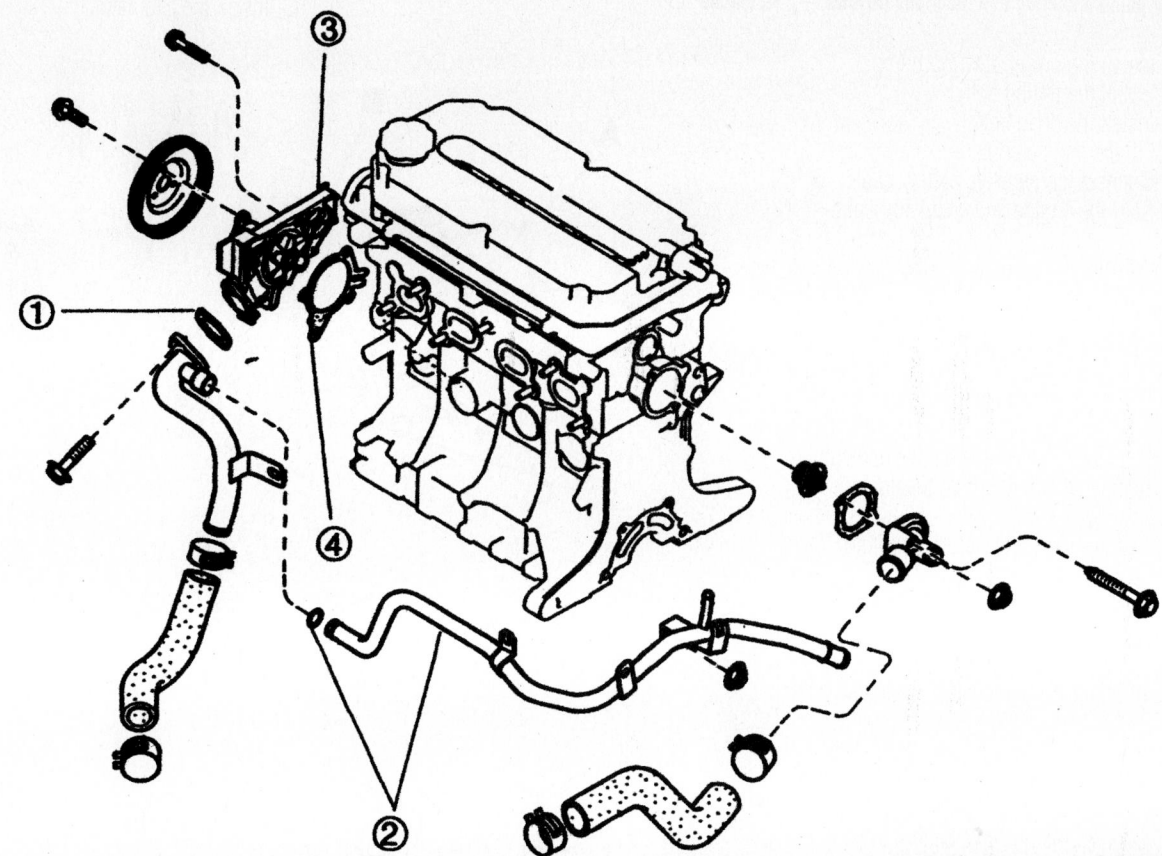

(1) Water inlet pipe and gasket
(2) Water bypass pipe and O-ring

(3) Water pump assembly
(4) Water pump gasket

Water pump mounting—1.5L and 1.6L engine

67162-KIAC-G10

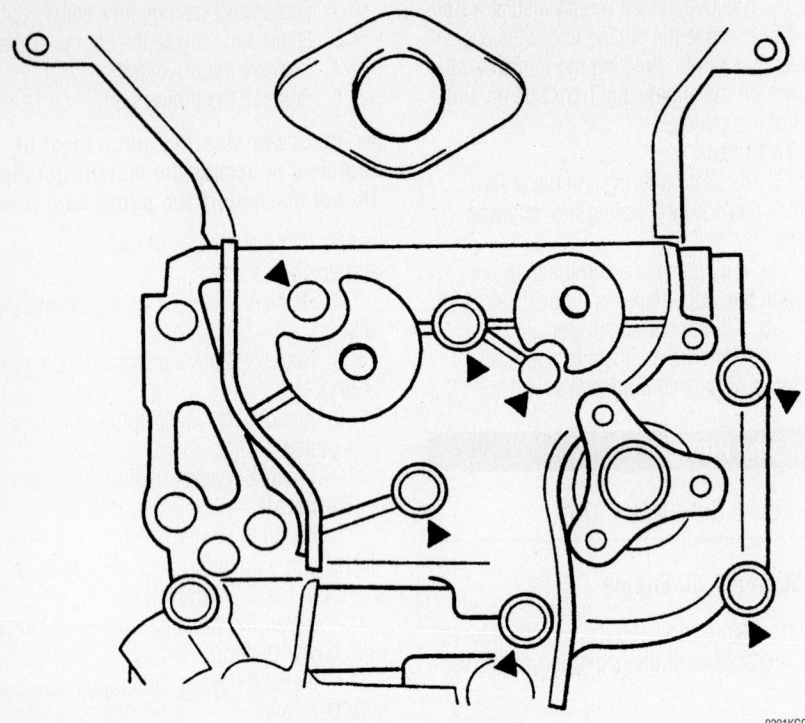

Water pump mounting bolt locations (arrows)—1.8L engine

9301KG01

14. Install the power steering pump.
15. Install the timing belt.
16. Install the drive belt.
17. Fill the engine coolant.
18. Connect the negative battery cable.
19. Start the engine and check for leaks.

1.8L Engine

1. Before servicing the vehicle, refer to the precautions at the beginning of this section.
2. Disconnect the negative battery cable.
3. Drain and recycle the engine coolant.
4. Remove the timing belt, tensioner and idler pulleys.
5. Remove the water pump mounting bolts, then the pump.
6. Clean all gasket mating surfaces.

To install:

7. Install the water pump, using a new gasket and tighten the mounting bolts to 14–19 ft. lbs. (19–26 Nm).
8. Install the tensioner and idler pulleys
9. Install the timing belt
10. Fill the engine coolant.
11. Connect the negative battery cable.
12. Start the engine and check for leaks.

2.0L Engine

1. Before servicing the vehicle, refer to the precautions at the beginning of this section.

2. Disconnect the negative battery cable.
3. Drain the engine coolant.
4. Remove the drive belts.
5. Remove the timing belt.
6. Remove the timing belt idler.
7. Remove the power steering pump and use a wire to secure the pump to the vehicle so that it is out of the way.
8. Remove the bolts and power steering pump bracket.
9. Remove the alternator.
10. Remove the 2 bolts (D) and alternator brace (A).
11. Remove the 3 bolts (C) and remove the water pump (B) and gasket.

To install:

12. Install the water pump.
13. Install the water pump (B) and a new gasket with the 3 bolts (C). Tighten the bolts to 8.5–10.5 ft. lbs. (11.5–14.5 Nm)
14. Install the alternator brace (A) with the 2 bolts (D). Tighten the bolts to 14.5–19.5 ft. lbs. (19.5–26.5 Nm)
15. Install the power steering pump bracket and bolts.
16. Install the alternator.

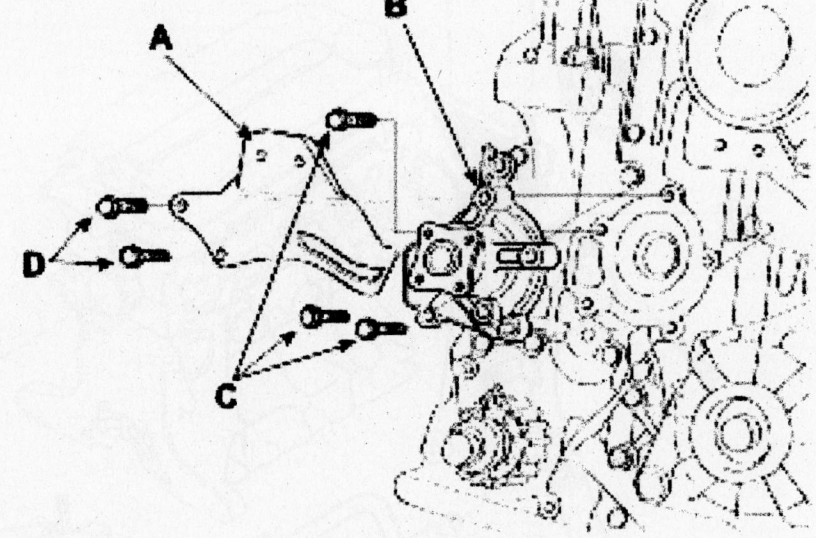

A. Alternator brace
B. Water pump
C. Water pump bolts
D. Alternator bracket bolts

Exploded view of the water pump mounting—2.0L engine

67162-KIAC-G11

17. Install the power steering pump.

18. Install the timing belt idler.

19. Install the timing belt.

20. Install the water pump pulley.

21. Install the drive belts.

22. Tighten the water pump pulley bolts. Tighten the bolts to 5.5–7 ft. lbs. (7.5–10 Nm)

23. Fill with engine coolant.

24. Connect the negative battery cable.

25. Start engine and check for leaks.

26. Recheck engine coolant level.

2.4L and 2.7L Engines

1. Before servicing the vehicle, refer to the precautions at the beginning of this section.

2. Disconnect the negative battery cable.

3. Drain and recycle the engine coolant.

4. Disconnect the radiator outlet hose and coolant bypass hose from the water pump.

5. Remove the drive belt.

6. Remove the water pump pulley.

7. Remove the timing belt covers and timing belt tensioner.

8. Remove the water pump mounting bolts.

9. Remove the alternator brace.

10. Remove the water pump from the cylinder block.

11. Clean all gasket mating surfaces.

To install:

12. Install the water pump, using a new gasket and tighten the mounting bolts to 14–20 ft. lbs. (19–27 Nm) for 2.4L engines and 11–16 ft. lbs. (15–22 Nm) for 2.7L engines.

13. Install the timing belt tensioner and timing belt.

14. Install the timing belt covers.

15. Install the coolant pump pulley and drive belt.

16. Connect the radiator outlet hose and coolant bypass hose to the water pump.

17. Fill the engine coolant.

18. Connect the negative battery cable.

19. Start the engine and check for leaks.

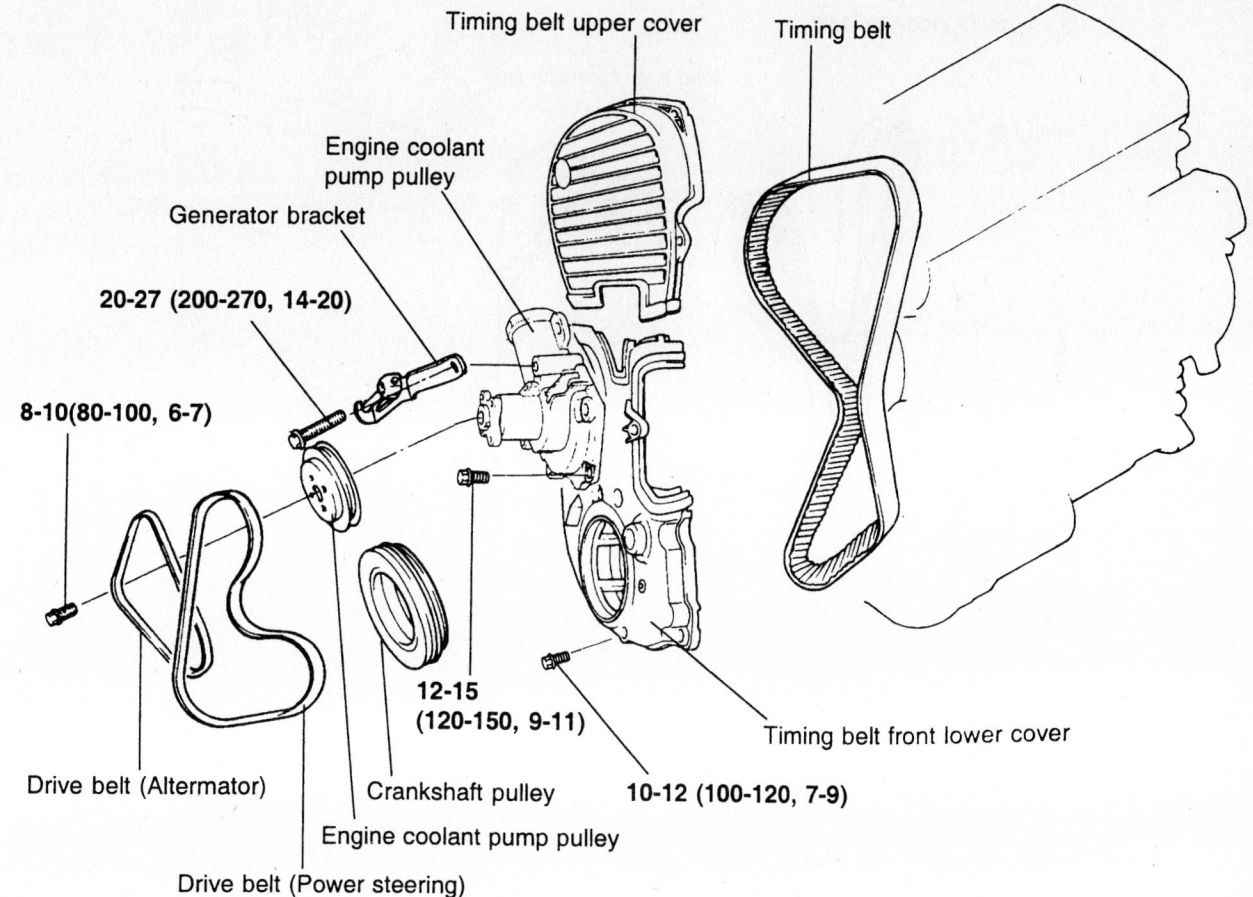

TORQUE : Nm (kg.cm, lb.ft)

9357NG27

Exploded view of the water pump mounting—2.4L engine

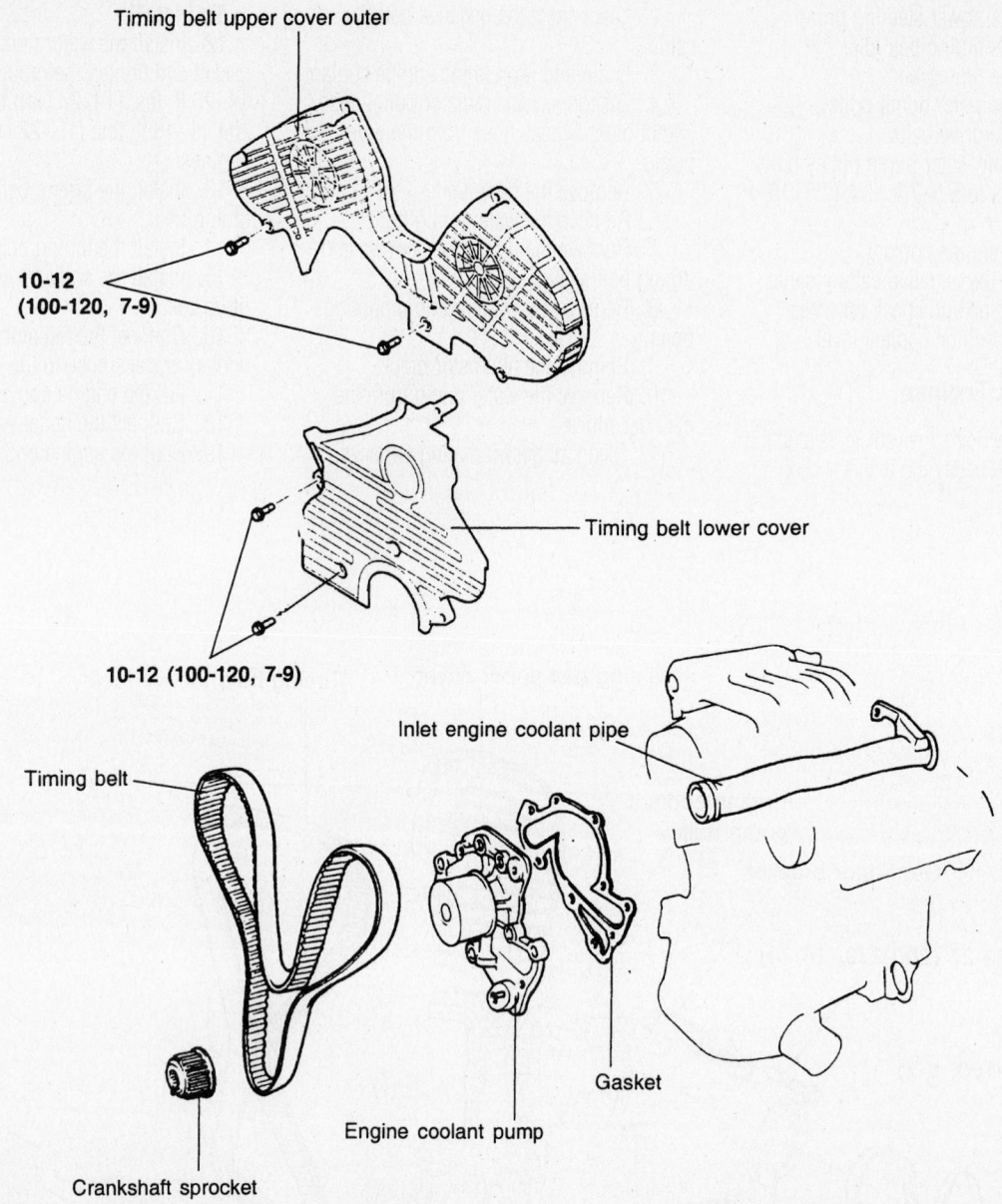

Timing belt upper cover outer

10-12
(100-120, 7-9)

Timing belt lower cover

10-12 (100-120, 7-9)

Inlet engine coolant pipe

Timing belt

Gasket

Engine coolant pump

Crankshaft sprocket

TORQUE : Nm (kg.cm, lb.ft)

9357NG28

Exploded view of the water pump mounting—2.7L engine

Heater Core

REMOVAL & INSTALLATION

1. Before servicing the vehicle, refer to the precautions at the beginning of this section.
2. Disconnect the negative battery cable.

✷✷ CAUTION

After disconnecting the negative battery cable, wait for at least 10 minutes for the SRS module to deplete its stored energy.

3. Position the front wheels in the straight-ahead position.
4. Remove the 4 steering wheel-to-air bag module bolts.
5. Remove the air bag module and disconnect the electrical connector.
6. Remove the steering wheel-to-steering column nut.
7. Remove the steering wheel from the steering column.
8. Remove the 2 screws and the glove box at the bottom of the glove box.
9. Remove the passenger side cover and pull the connector from the "T" bar side bracket.

10. Remove the 4 air bag module-to-instrument panel bolts.
11. Remove the air bag module, disconnect the electrical connector and remove the air bag module.
12. Drain the cooling system into a clean container for reuse.
13. Discharge and recover the air conditioning system refrigerant.
14. Remove the center panel trim.
15. Remove the console.
16. Remove the side cover.
17. Remove the lower left side cover.
18. Remove the steering column-to-

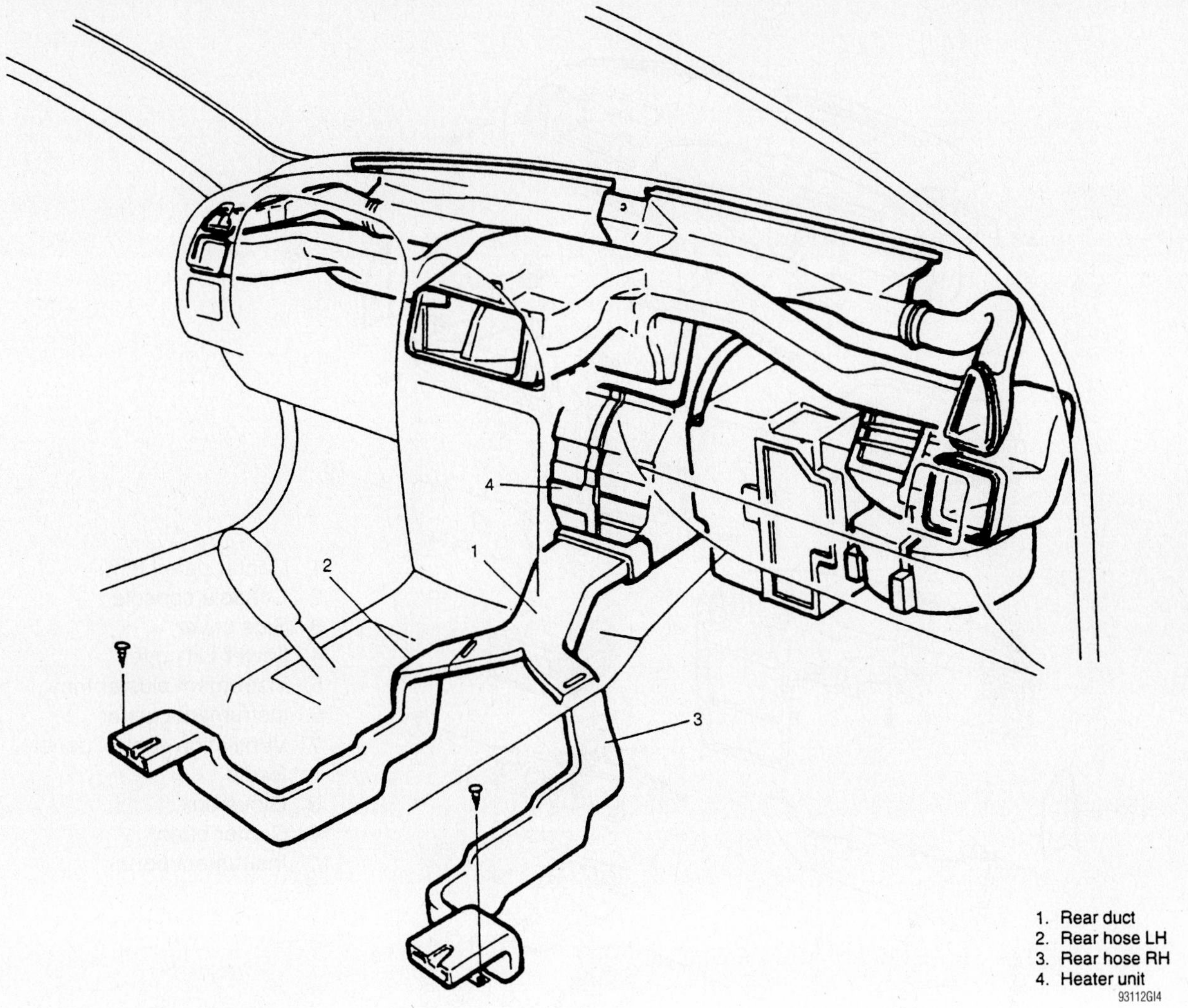

1. Rear duct
2. Rear hose LH
3. Rear hose RH
4. Heater unit

93112GI4

Exploded view of the steering wheel and air bag module assembly—Sephia

instrument panel bolts and lower the steering column.

19. Remove the instrument cluster trim.

20. Remove the instrument cluster and disconnect the electrical connectors.

21. Remove the ventilation control panel and disconnect the electrical connectors.

22. Remove the radio and disconnect antenna and electrical connector.

23. Remove the center panel.

24. Disconnect the electrical harness connectors.

25. Remove the instrument panel-to-chassis bolts and remove the instrument panel.

26. Disconnect the control cable from the heater housing.

27. Disconnect the heater hoses from the heater core.

28. Remove the heater housing.

29. Remove the heater core from the heater housing.

To install:

30. Install the heater core to the heater housing.

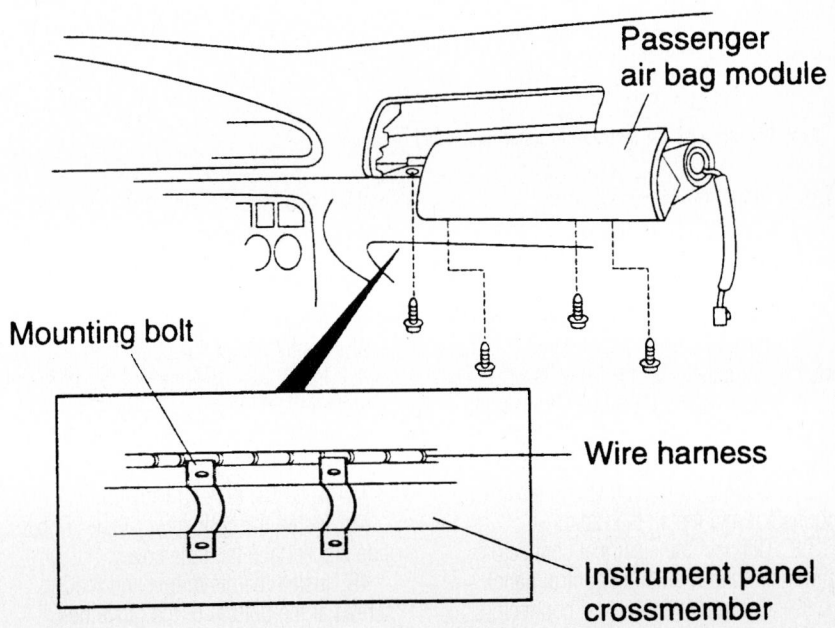

Passenger air bag module

Mounting bolt

Wire harness

Instrument panel crossmember

93112GJ5

Exploded view of the passenger's side air bag module assembly—Sephia

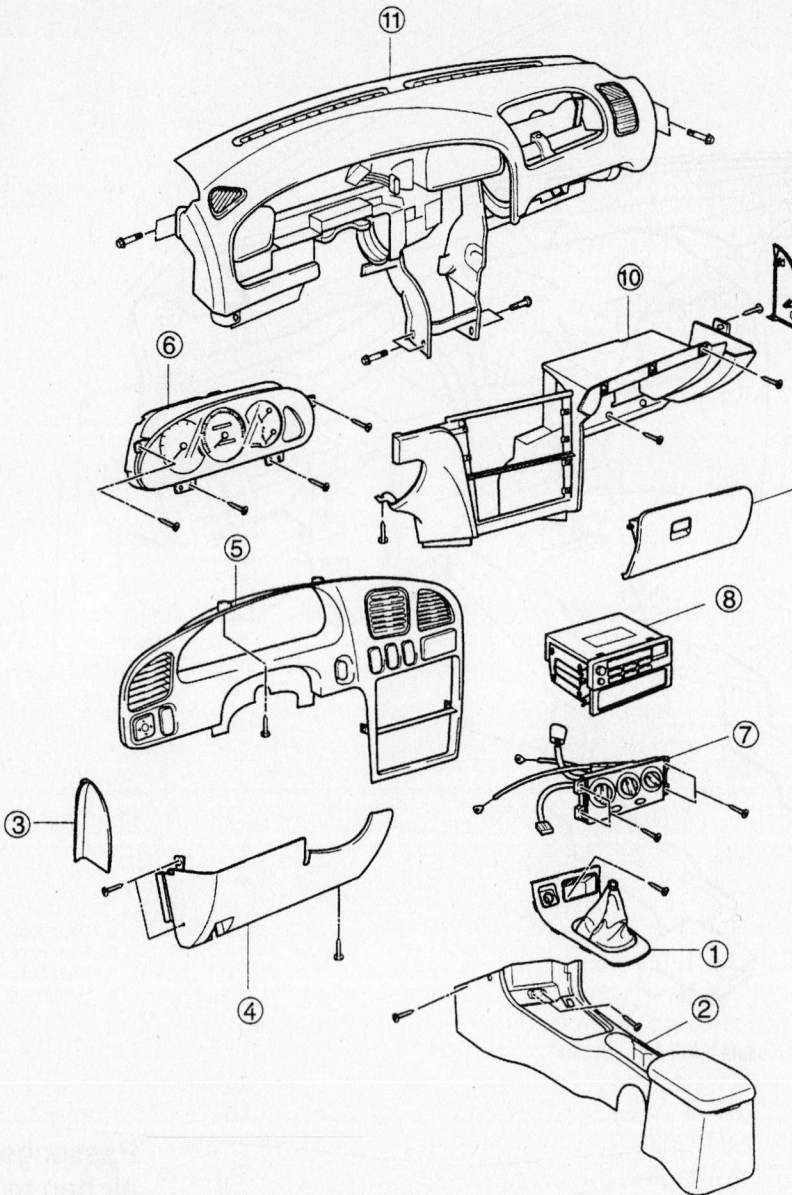

1 Center panel trim
2 Console console
3 Side cover
4 Lower LH cover
5 Instrument cluster trim
6 Instrument cluster
7 Ventilation control panel
8 Radio
9 Glove box
10 Center panel
11 Instrument panel

93112GJ6

Exploded view of the instrument panel assembly—Sephia

31. Install the heater housing.
32. Connect the heater hoses to the heater core.
33. Connect the control cable to the heater housing.
34. Install the instrument panel and the instrument panel-to-chassis bolts.
35. Connect the electrical harness connectors.
36. Install the center panel.
37. Connect the antenna and electrical connector and install the radio.
38. Connect the electrical connectors and install the ventilation control panel.
39. Connect the electrical connectors and install the instrument cluster.
40. Install the instrument cluster trim.

41. Install the steering column-to-instrument panel bolts.
42. Install the lower left side cover.
43. Install the side cover.
44. Install the console.
45. Install the center panel trim.
46. Install the passenger's side air bag module and connect the electrical connector.
47. Install the 4 air bag module-to-instrument panel bolts and torque the bolts to 18–32 ft. lbs. (24–43 Nm).
48. Install the connector to the "T" bar side bracket and the side cover.
49. Install the glove box and the 2 screws at the bottom of the glove box.
50. Install the steering wheel to the steering column.

51. Install the steering wheel-to-steering column nut and torque the nut to 33 ft. lbs. (45 Nm).
52. Install the driver's side air bag module and connect the electrical connector.
53. Install the 4 steering wheel-to-air bag module bolts and torque to 72–106 inch lbs. (8–12 Nm).
54. Refill the cooling system.
55. Connect the negative battery cable.
56. Evacuate, charge and leak test the air conditioning system.
57. Operate the engine to normal operating temperatures; then, check the climate control operation and check for leaks.

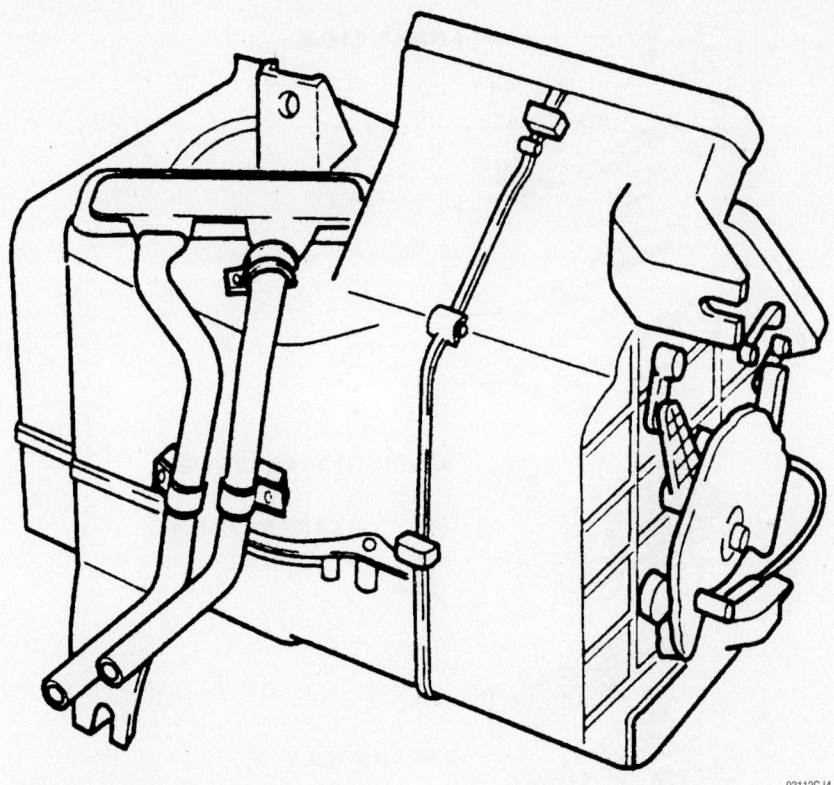

View of the heater housing—Sephia

Cylinder Head

REMOVAL & INSTALLATION

1.5L and 1.6L Engines

1. Before servicing the vehicle, refer to the precautions at the beginning of this section.
2. Disconnect the negative battery cable.
3. Properly relieve the fuel system pressure.
4. Drain and recycle the engine coolant.
5. Drain and recycle the engine oil.
6. Remove the upper radiator hose.
7. Remove the breather hose from the between the air cleaner and rocker arm (valve) cover.
8. Remove the air intake hose.
9. Disconnect the vacuum hose, fuel hose and coolant hose.
10. Disconnect the spark plug wires, tag before disconnecting.
11. Remove the ignition coil.
12. Remove the power steering pump and bracket and position aside. Do not disconnect the fluid lines.
13. Remove the intake manifold.
14. Remove the heat shield from the exhaust manifold.

15. Remove the exhaust manifold.
16. Remove the surge tank.
17. Remove the water pump and crankshaft pulleys.
18. Remove the timing belt cover, belt tensioner and timing belt.
19. Remove the cylinder head cover and cam carrier assembly.
20. Remove the cylinder head bolts, in several steps in the proper sequence.
21. Remove the cylinder head and gasket. Discard the gasket.

To install:

22. Thoroughly, clean the cylinder head and the block contact surfaces. Examine the head gasket and check the cylinder head for cracks. Check the cylinder head for warpage using a feeler gauge and straightedge. The maximum allowable distortion is 0.006 in. (0.15mm).
23. Clean the cylinder head bolts and the threads in the block. Be sure the bolts turn freely in the block.
24. Install a new head gasket on the engine block.
25. Install the cylinder head.
26. Install the cylinder head bolts.
27. Tighten the head bolts in the following step using the proper sequence:
   a. Step 1: Tighten to 36 ft. lbs. (49 Nm).
   b. Step 2: Loosen the bolts in the reverse order shown.
   c. Step 3: Tighten to 18 ft. lbs. (25 Nm).
   d. Step 4: Tighten 90° (¼ turn).
   e. Step 5: Tighten 90° (¼ turn).
28. Install the timing belt tensioner pulley and timing belt. Make sure all timing marks are aligned.
29. Install the cylinder head cover. Tighten the bolts to 3.6–6.5 ft. lbs. (5–9 Nm).
30. Install the timing belt cover.
31. Install the intake manifold, with a new gasket. Torque the bolts to 11–14 ft. lbs. (15–20 Nm).
32. Install the exhaust manifold with a new gasket. Torque the bolts to 11–14 ft. lbs. (15–20 Nm).
33. Install the exhaust manifold heat shield.
34. Install the surge tank and tighten the bolts to 11–14 ft. lbs. (15–20 Nm).

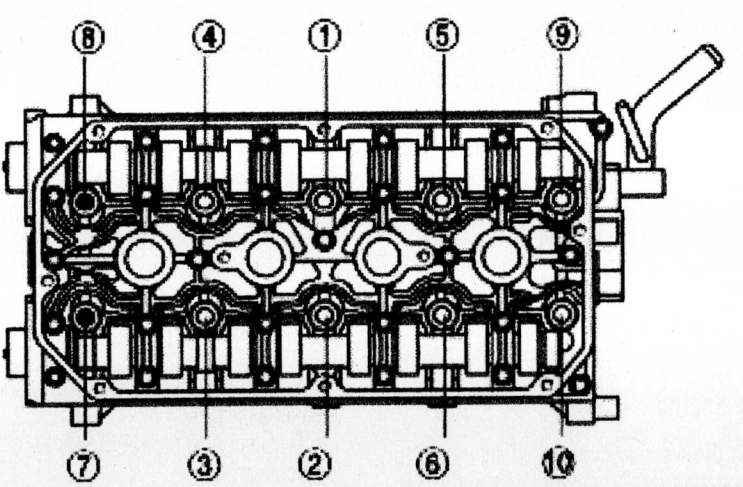

Cylinder head bolt tightening (loosen the reverse order)—1.5L and 1.6L engines

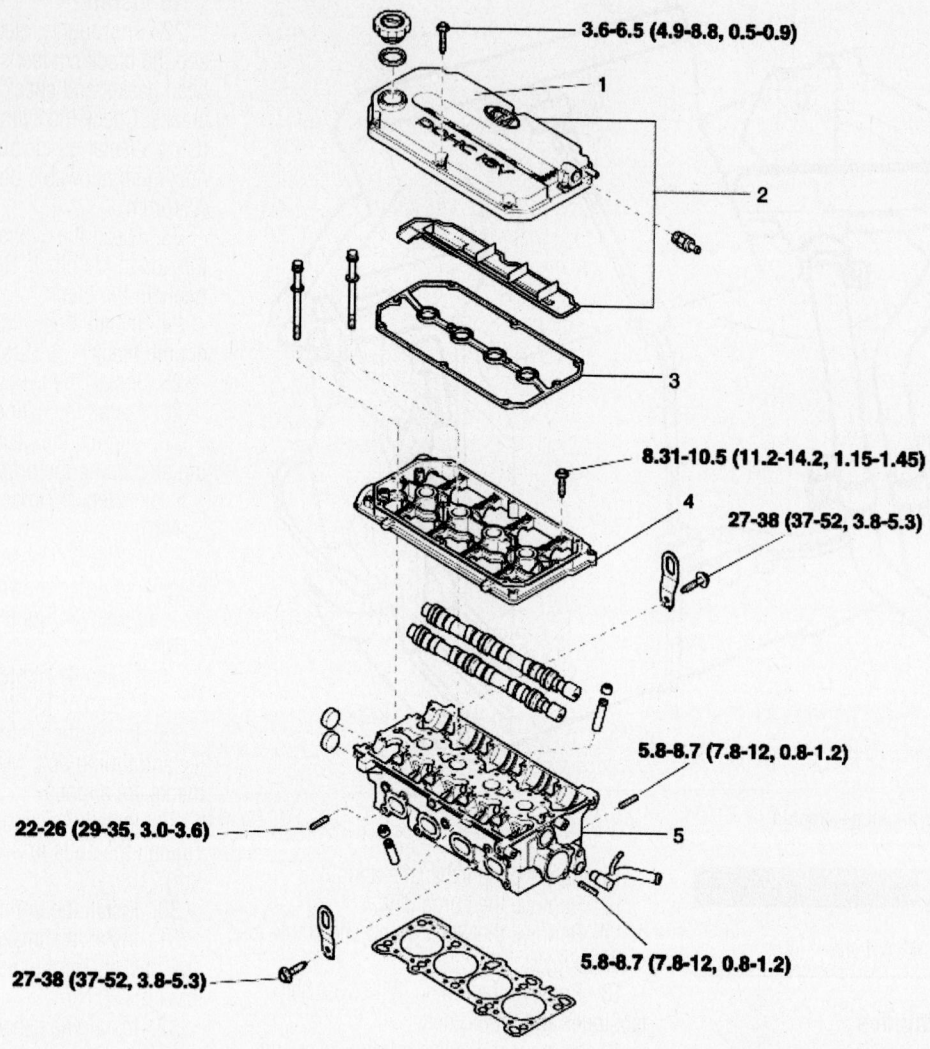

3.6-6.5 (4.9-8.8, 0.5-0.9)

1

2

3

8.31-10.5 (11.2-14.2, 1.15-1.45)

4

27-38 (37-52, 3.8-5.3)

5.8-8.7 (7.8-12, 0.8-1.2)

22-26 (29-35, 3.0-3.6)

5

27-38 (37-52, 3.8-5.3)

5.8-8.7 (7.8-12, 0.8-1.2)

TORQUE : lb-ft (N·m, kg·m)

1. Cylinder head cover
2. Cylinder head cover
3. Cylinder head cover gasket
4. Cam carrier assembly
5. Cylinder head

67162-KIAC-G13

Exploded view of the cylinder head and related components—1.5L and 1.6L engines

35. Install the power steering pump and bracket.

36. Install the ignition coil and connect the spark plug wires.

37. Install the air intake hose.

38. Connect the vacuum hose, fuel hose and water hose.

39. Install the breather hose.

40. Fill the engine oil and coolant.

41. Start the vehicle and check for leaks.

2.0L Engine

1. Before servicing the vehicle, refer to the precautions at the beginning of this section.

2. Disconnect the negative battery cable.

3. Turn the crankshaft pulley so that the No. 1 piston is at top dead center.

4. Disconnect the terminals from battery and remove the heat shield.

5. Remove the engine cover.

6. Drain the engine coolant.

7. Remove the radiator cap to speed draining.

8. Remove the intake air hose and air cleaner assembly.

9. Disconnect the AFS (Air Flow Sensor) connector.

10. Disconnect the breather hose from intake air hose.

11. Remove the intake air hose and air cleaner upper cover.

12. Remove the air cleaner element.

13. Remove the bolts and air cleaner lower cover.

14. Remove the upper radiator hose and lower radiator hose.

15. Remove the heater hoses.

16. Disconnect the accelerator cable by loosening the lock-nut, then slip the cable end out of the throttle linkage.

17. Remove the engine wire harness connectors and wire harness clamps from cylinder head and the intake manifold.

18. Disconnect the OCV (Oil Control Valve) connector.

19. Disconnect the oil temperature sensor connector.

20. Disconnect the ECT (Engine Coolant Temperature) sensor connector.

21. Disconnect the ignition coil connector.

22. Disconnect the TPS (Throttle Position Sensor) connector.

23. Disconnect the ISA (Idle Speed Actuator) connector.

24. Disconnect the CMP (Camshaft Position Sensor) connector.

25. Disconnect the four injector connectors.

26. Disconnect the knock sensor connector.

27. Disconnect the ground cables from the intake manifold and vehicle's body.

28. Disconnect the air conditioner compressor switch connector.

29. Disconnect the front heated oxygen sensor connector.

30. Disconnect the CKP (Crankshaft Position Sensor) connector.

31. Disconnect the oil pressure switch connector.

32. Disconnect the PCSV (Purge Control Solenoid Valve) connector.

33. Disconnect the fuel inlet hose of the delivery pipe side.

34. Disconnect the hose of the PCSV (Purge Control Solenoid Valve) side.

35. Remove the brake booster vacuum hose.

36. Remove the power steering pump drive belt.

37. Remove the power steering pump and use a wire to secure the pump to the vehicle so that it is out of the way.

38. Remove the bolts and power steering pump bracket.

39. Remove the spark plug cables.

40. Remove the exhaust manifold.

41. Remove the intake manifold.

42. Remove the timing belt.

43. Remove the PCV (Positive Crankcase Ventilation) hose.

44. Remove the cylinder head cover.

45. Remove the camshaft sprocket.

46. Insert a stopper pin or other device into timing chain auto tensioner and remove the auto tensioner.

47. Remove the camshaft bearing caps and camshafts.

48. Remove the OCV (Oil Control Valve) and filter.

49. Disconnect the water hose from water pipe.

50. Using 8mm and 10mm hexagon wrench, uniformly loosen and remove the 10 cylinder head bolts, in several passes, in the sequence shown.

51. Remove the 10 cylinder head bolts and plate washers.

52. Lift the cylinder head from the dowels on the cylinder block and place the cylinder head on wooden blocks on a bench.

To install:

53. Install the cylinder head gasket onto the cylinder block. Be careful of the installation direction.

54. Place the cylinder head onto the block carefully in order to prevent damaging the gasket. If the gasket is damaged, fluid leakage could occur.

55. Apply a light coat if engine oil on the threads and under the heads of the cylinder head bolts. Using an 8mm and 10mm hexagon wrench, install and tighten the 10 cylinder head bolts and plate washers, in several passes, in the the proper sequence:

   a. Step 1: Tighten the M10 bolts to: 17–19.5 ft. lbs. (23–26.5 Nm)

   b. Step 2: Tighten 60–65°.

   c. Step 3: Tighten an additional 60–65°.

   d. Step 4: Tighten the M12 bolts to: 20.5–23 ft. lbs. (27.5–31 Nm)

   e. Step 5: Tighten 60–65°.

   f. Step 6: Tighten an additional 60–65°.

56. Install the OCV filter along with a new filter gasket. Tighten the filter to 30–37 ft. lbs. (40–50 Nm).

57. Install the OCV. Tighten the bolt to 7–8.5 ft. lbs. (10–11.5 Nm).

58. Install the camshafts.

59. Align the camshaft timing chain with the intake timing chain sprocket and exhaust timing chain sprocket as shown.

60. Install the camshaft and bearing caps. Tighten the bearing cap bolts to 10.5 ft. lbs. (14 Nm).

61. Install the timing chain auto tentioner. Tighten the bolts to 6 ft. lbs. (8.5 Nm).

62. Remove the auto tentioner stopper pin.

63. Check and adjust valve clearance.

64. Using the SST (09221—21000), install the camshaft bearing oil seal.

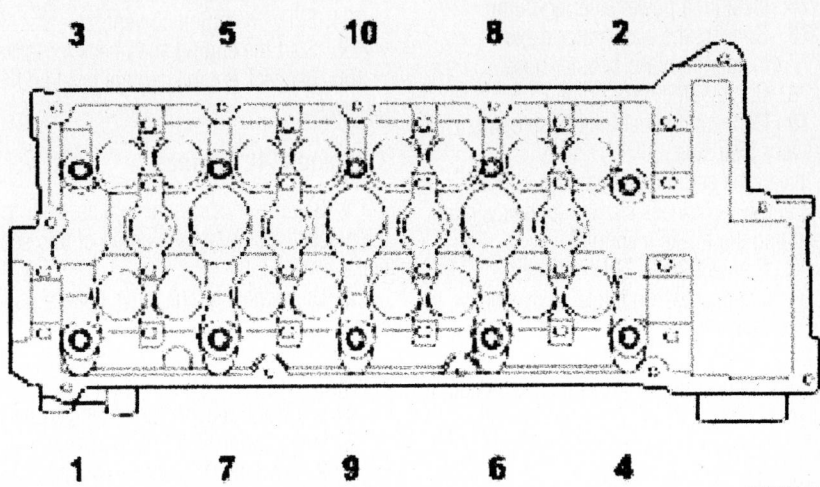

Cylinder head bolt loosening sequence—2.0L engine

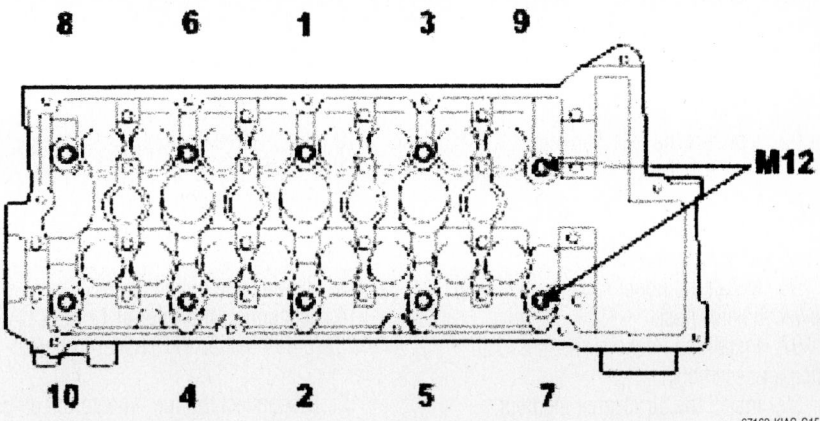

Cylinder head bolt tightening sequence—2.0L engine

65. Install the camshaft sprocket.
66. Install the cylinder head cover.
67. Install the cylinder head cover gasket in the groove of the cylinder head cover. Before installing the cylinder head cover gasket, thoroughly clean the cylinder head cover and the groove. When installing, make sure the cylinder head cover gasket is seated securely in the corners of the recesses with no gap. Apply liquid gasket to the head cover gasket at the corners of the recess.
68. Install the cylinder head cover with the 12 bolts. Tighten all cylinder head cover bolts temporarily to half of the standard torque, the re-tighten completely using the order specified in the illustration.
69. Install the PCV(Positive Crankcase Ventilation) hose.
70. Install the timing belt.
71. Install the intake manifold.
72. Install the exhaust manifold.
73. Install the spark plug cables.
74. Install the power steering pump bracket and bolts.
75. Install the power steering pump.
76. Connect the accelerator cable.
77. Install the brake booster hose.
78. Connect the hose of the PCSV side.
79. Connect the fuel inlet hose of the delivery pipe side.
Install the engine wire harness connectors and wire harness clamps to the cylinder head and the intake manifold.
80. Connect the PCSV connector.
81. Connect the front heated oxygen sensor connector.
82. Connect the CKP connector.
83. Connect the oil pressure switch connector.
84. Connect the air conditioner compressor switch connector.
85. Connect the ground cables to intake manifold and vehicle's body.
86. Connect the knock sensor connector.
87. Connect the fuel injector connectors.
88. Connect the CMP connector.
89. Connect the ISA connector.
90. Connect the TPS connector.
91. Connect the ignition coil connector.
92. Connect the ECT connector.
93. Connect the oil temperature sensor connector.
94. Connect the OCV connector.
95. Install the heater hose.
96. Install the upper radiator hose and lower radiator hose.
97. Install the intake air hose and air cleaner assembly.
98. Install the air cleaner element.
99. Install the intake air hose and air cleaner upper cover.

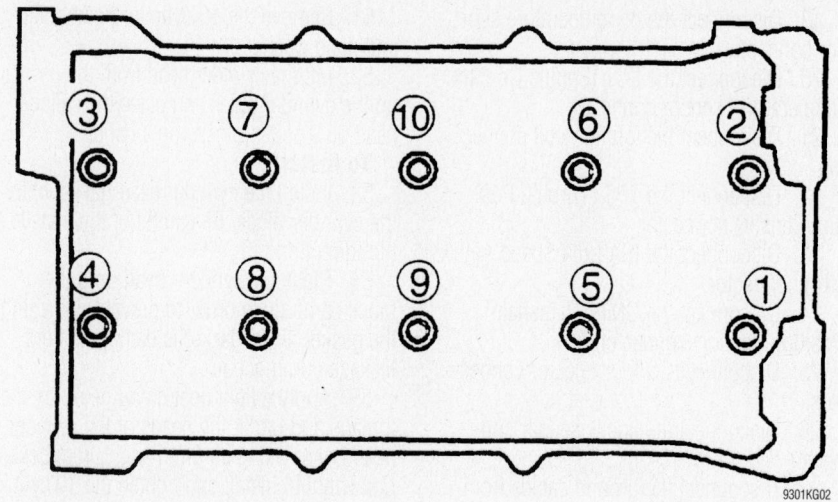

Cylinder head bolt removal sequence—1.8L engine

100. Install the breather hose to intake air hose.
101. Connect the AFS connector.
102. Install the engine cover.
103. Install the heat shield and connect the battery terminals to the battery.
104. Fill with engine coolant.
105. Start the engine and check for leaks.
106. Recheck engine coolant level and oil level.

1.8L and 2.4L Engines

1. Before servicing the vehicle, refer to the precautions at the beginning of this section.
2. Disconnect the negative battery cable.
3. Properly relieve the fuel system pressure.
4. Drain and recycle the engine coolant.
5. Drain and recycle the engine oil.
6. Remove the Positive Crankcase Ventilation (PCV) and crankcase ventilation hoses.
7. Disconnect the accelerator and, if equipped, the cruise control cables.
8. Disconnect the air intake hose from the throttle body.
9. Disconnect the brake vacuum hose and the purge control vacuum hose.
10. Remove the ventilation hose and fresh air duct.
11. Remove the upper radiator hose.
12. Disconnect the fuel hose from the fuel injector rail.
13. Remove the heater hoses.
14. Disconnect the Idle Air Control (IAC) and Throttle Position (TP) sensor connectors.
15. Disconnect the fuel injector electrical connectors.
16. Remove the engine ground strap.

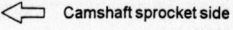

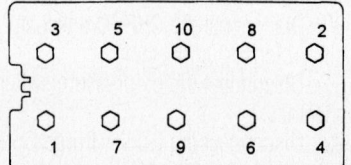

Cylinder head bolt removal sequence— 2.4L engine

17. Remove the exhaust manifold heat shield.
18. Remove the front exhaust pipe from the manifold.
19. Remove the exhaust manifold.
20. Remove the coolant bypass pipe from the cylinder head.
21. Remove the intake manifold support bracket.
22. Remove the camshaft and Hydraulic Lash Adjusters (HLA's).
23. Remove the cylinder head bolts in several steps, in the order illustrated.
24. Remove the cylinder head bolts, then the cylinder head.
To install:
25. Thoroughly, clean the cylinder head and the block contact surfaces. Examine the head gasket and check the cylinder head for

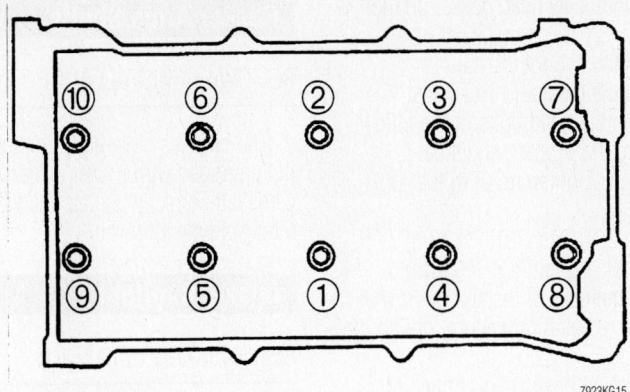

Cylinder head torque sequence—1.8L engine

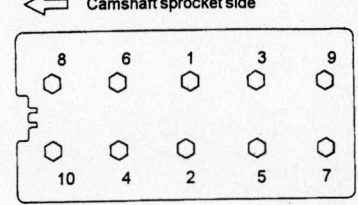

Cylinder head torque sequence—2.4L engine

cracks. Check the cylinder head for warpage using a feeler gauge and straightedge. The maximum allowable distortion is 0.006 in. (0.15mm).

26. Clean the cylinder head bolts and the threads in the block. Be sure the bolts turn freely in the block.

27. Install a new head gasket on the engine block.

28. Install the cylinder head.

29. Install the cylinder head bolts.

30. For 1.8L engines, tighten the head bolts in the following step using the proper sequence:

   a. Step 1: Tighten to 36 ft. lbs. (49 Nm).

   b. Step 2: Loosen the bolts in the reverse order shown.

   c. Step 3: Tighten to 29 ft. lbs. (39 Nm).

   d. Step 4: Tighten 90° (¼ turn).

   e. Step 5: Tighten 90° (¼ turn).

31. For 2.4L engines, tighten the head bolts in the following step using the proper sequence:

   a. Step 1: Tighten to 14 ft. lbs. (20 Nm).

   b. Step 2: Tighten 90° (¼ turn).

   c. Step 3: Loosen the bolts in the reverse order shown.

   d. Step 4: Tighten to 14 ft. lbs. (20 Nm).

   e. Step 5: Tighten 90° (¼ turn).

   f. Step 6: Tighten 90° (¼ turn).

32. Install the camshaft and HLA's.

33. Install the intake manifold support bracket and tighten the mounting bolts to 28–38 ft. lbs. (37–52 Nm).

34. Install the coolant bypass pipe and tighten the mounting bolt to 66–86 ft. lbs. (89–117 Nm).

35. Install the exhaust manifold, tighten the manifold-to-cylinder head mounting nuts to 28–34 ft. lbs. (38–46 Nm) and the manifold-to-exhaust pipe mounting nuts to 16–21 ft. lbs. (22–28 Nm).

36. Install the exhaust manifold heat shield and tighten the mounting bolts to 13–22 ft. lbs. (19–30 Nm).

37. The remaining steps of the installation procedure is the reverse of the removal, while keeping in mind the following:

   a. Attach all hoses and connectors.

   b. Fill the engine oil and coolant.

   c. Start the vehicle and check for leaks.

2.7L Engine

1. Before servicing the vehicle, refer to the precautions at the beginning of this section.

2. Disconnect the negative battery cable.

3. Properly relieve the fuel system pressure.

4. Drain and recycle the engine coolant.

5. Drain and recycle the engine oil.

6. Remove the upper radiator hose.

7. Remove the breather hose and air intake hose.

8. Remove the vacuum hose, fuel hose and coolant hose.

9. Remove the intake manifold.

10. Remove the spark plug wires from the spark plugs.

11. Remove the ignition coil.

12. Remove the upper and lower timing belt covers.

13. Remove the timing belt and camshaft sprockets.

14. Remove the heat protector and exhaust manifold.

15. Remove the water pump pulley and head cover.

16. Remove the intake and exhaust camshafts.

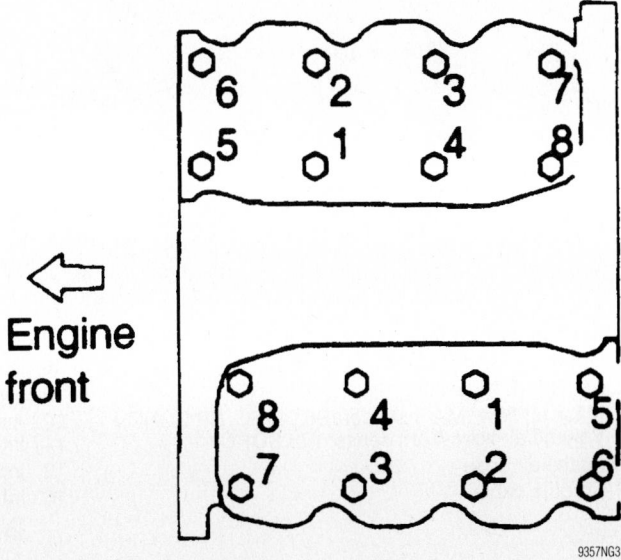

Cylinder head tightening sequence (use the reverse for removal)—2.7L engine

17. Remove the cylinder head bolts, loosening in the reverse of the torque sequence, in several steps.

18. Remove the cylinder head.

19. Remove the gaskets and discard.

To install:

20. Thoroughly, clean the cylinder head and the block contact surfaces. Examine the head gasket and check the cylinder head for cracks. Check the cylinder head for warpage using a feeler gauge and straightedge. The maximum allowable distortion is 0.006 in. (0.15mm).

21. Clean the cylinder head bolts and the threads in the block. Be sure the bolts turn freely in the block.

22. Install the new head gasket on the engine block.

23. Install the cylinder head.

24. Install the cylinder head bolts.

25. Tighten the head bolts in the following step using the proper sequence:

   a. Step 1: Tighten to 18 ft. lbs. (25 Nm).

   b. Step 2: Tighten 60° (⅙ turn).

   c. Step 3: Tighten 45° (⅛ turn).

26. The remainder of installation is the reverse or the removal procedure, noting the following steps:

   a. Attach all hoses and connectors.

   b. Fill the engine oil and coolant.

   c. Start the vehicle and check for leaks.

Rocker Arms/Shafts

REMOVAL & INSTALLATION

The engines covered in this section are not equipped with rocker arms/shafts. The camshafts directly actuate the valve through a bucket type follower.

Intake Manifold

REMOVAL & INSTALLATION

1. Before servicing the vehicle, refer to the precautions at the beginning of this section.

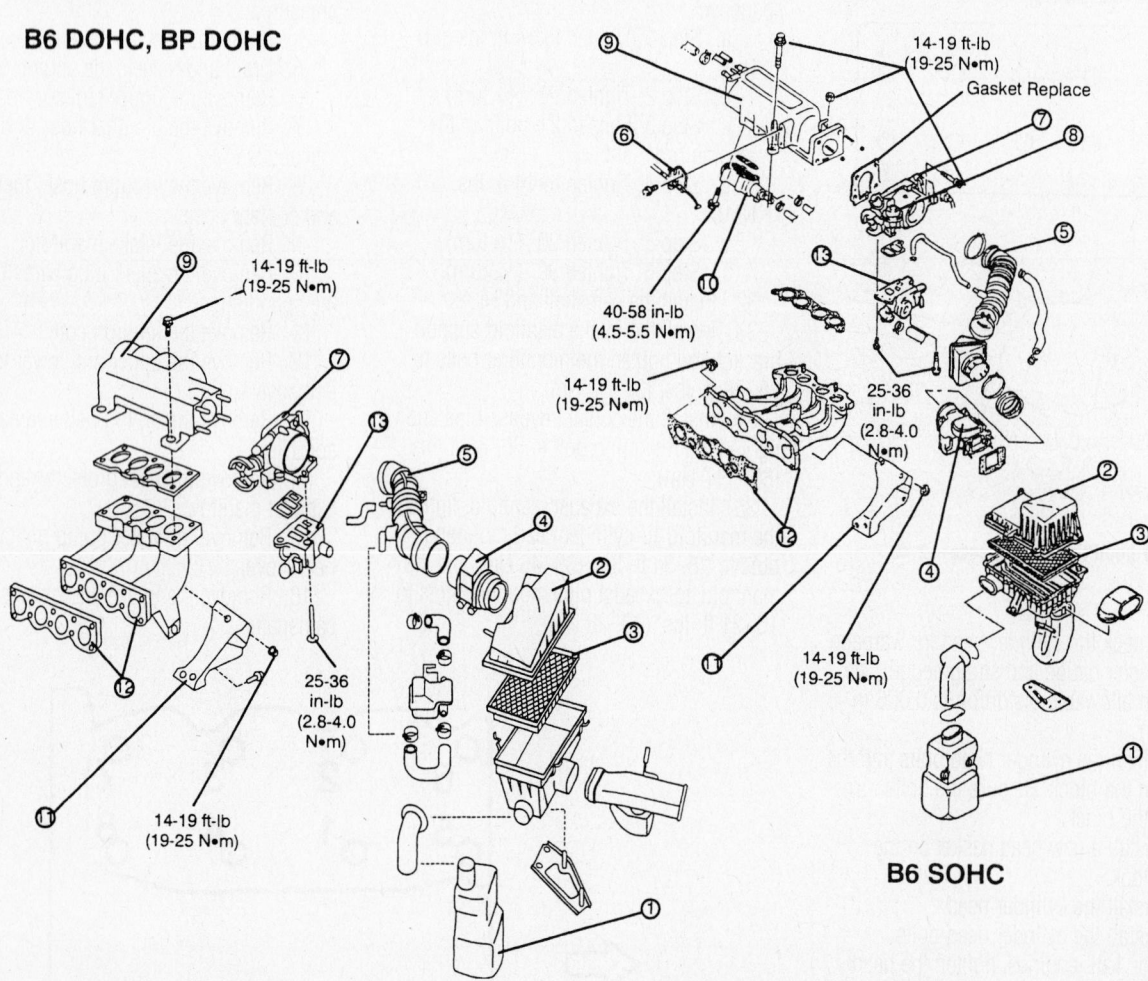

B6 DOHC, BP DOHC

14-19 ft-lb (19-25 N•m)

Gasket Replace

40-58 in-lb (4.5-5.5 N•m)

14-19 ft-lb (19-25 N•m)

25-36 in-lb (2.8-4.0 N•m)

14-19 ft-lb (19-25 N•m)

14-19 ft-lb (19-25 N•m)

25-36 in-lb (2.8-4.0 N•m)

14-19 ft-lb (19-25 N•m)

B6 SOHC

1. Resonance chamber
2. Upper air filter housing
3. Air filter (B6 SOHC)
4. Mass air flow (MAF) sensor (B6 DOHC, BP DOHC)/ Volume air flow (VAF) sensor (B6 SOHC)
5. Intake air hose
6. Throttle cable
7. Throttle body
8. Dashpot (B6 SOHC)
9. Dynamic chamber
10. Air valve (B6 SOHC)
11. Intake manifold support bracket
12. Intake manifold and gasket (Replace)
13. Idle air control valve (B6 SOHC)/ Bypass air control (BAC) valve (B6 DOHC, BP DOHC)

7923KG17

Exploded view of the intake manifold assembly—1.8L engine shown

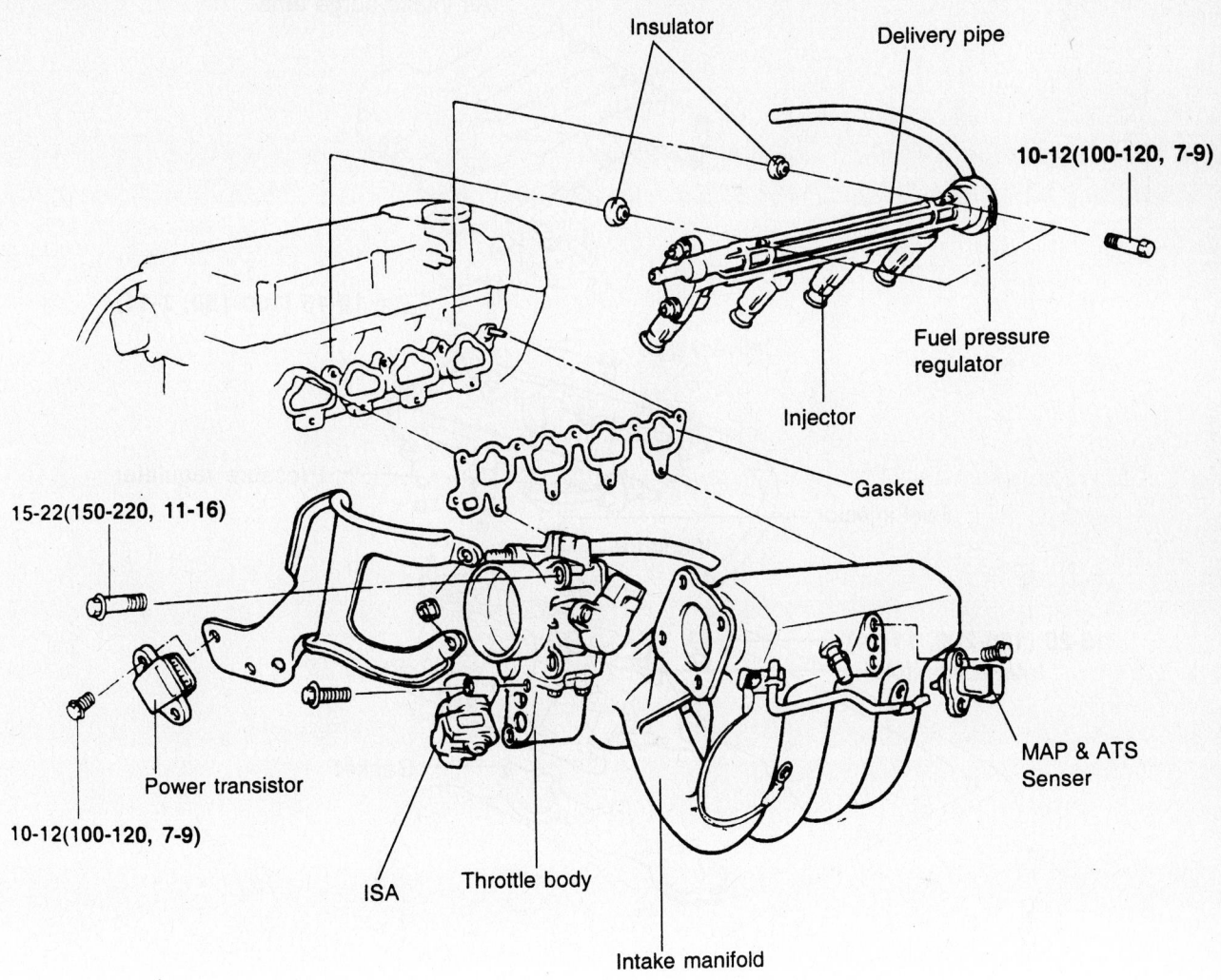

Insulator

Delivery pipe

10-12(100-120, 7-9)

Fuel pressure regulator

Injector

Gasket

15-22(150-220, 11-16)

MAP & ATS Senser

Power transistor

10-12(100-120, 7-9)

ISA

Throttle body

Intake manifold

TORQUE : Nm (kg.cm, lb.ft)

9357NG32

Exploded view of the intake manifold—2.4L engine

2. Properly relieve the fuel system pressure. Disconnect the negative battery cable and drain the cooling system.

3. Remove the air intake hose from the throttle body.

4. Remove the air intake hose and air cleaner assembly, if necessary.

5. Remove the air intake surge tank, if necessary.

6. Disconnect the accelerator cable.

7. Disconnect the fuel lines. Plug the lines to avoid contamination.

8. Disconnect all necessary vacuum hoses and electrical connectors.

9. Remove the coolant hoses.

10. Remove the Exhaust Gas Recirculation (EGR) tube, if equipped.

11. Remove the air valve, if equipped.

12. Remove the fuel rail attaching bolts.

13. Remove the fuel rail and injectors as an assembly.

14. Remove the intake manifold support bracket.

15. Remove the bolt retaining the dipstick tube bracket to the intake manifold, if necessary.

16. Remove the intake manifold-to-cylinder bolts/nuts and the intake manifold assembly.

17. Remove the throttle body, if necessary and separate the intake manifold upper and lower halves.

To install:

18. Clean all gasket mating surfaces.

19. Install the upper and lower intake manifolds using a new gasket, if separated. Tighten the nuts/bolts to 19 ft. lbs. (25 Nm).

20. Install the throttle body using a new gasket, if removed. Tighten the retaining nuts/bolts to 19 ft. lbs. (25 Nm).

21. Install the intake manifold assembly to the cylinder head using a new gasket. Tighten the nuts/bolts to 11–14 ft. lbs. (15–20 Nm) for 1.5L, 1.6L and 2.7L engines, or 13–16.5 ft. lbs. (18–22.5 Nm) for 2.0L engine, or to 19 ft. lbs. (25 Nm) for 1.8L and 2.4L engines.

➡**Tighten the bolts in the center of the manifold first and work outward toward the ends.**

22. Install the bolt retaining the dipstick tube to the intake manifold, if equipped.

23. Install the intake manifold bracket. Tighten the attaching nuts/bolts to 19 ft. lbs. (25 Nm).

24. Install the EGR tube, if equipped.

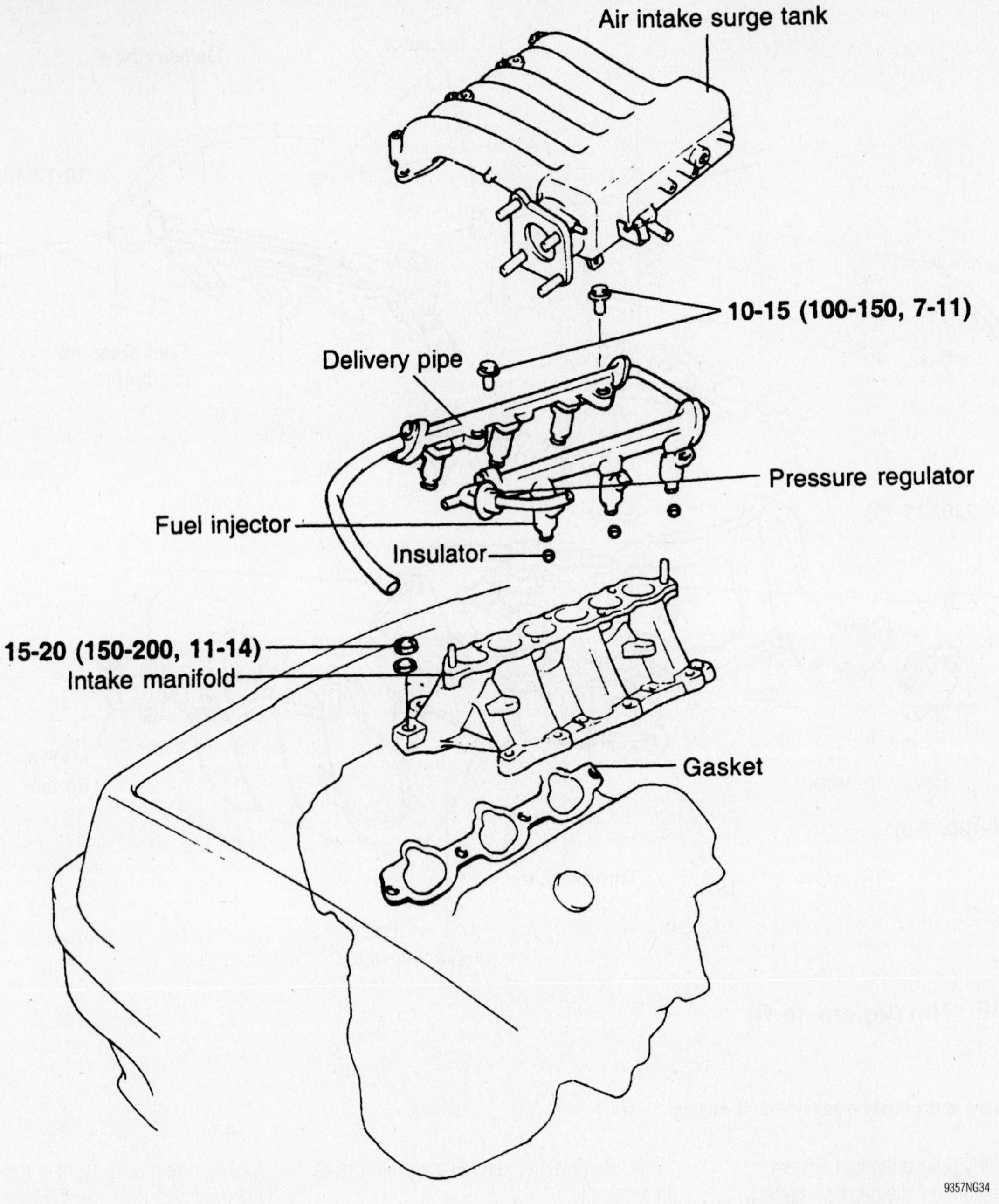

Air intake surge tank

10-15 (100-150, 7-11)

Delivery pipe

Pressure regulator

Fuel injector

Insulator

15-20 (150-200, 11-14)
Intake manifold

Gasket

9357NG34

Exploded view of the intake manifold—2.7L engine

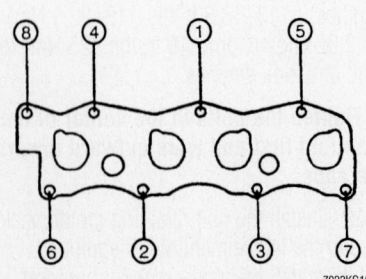

7923KG18

Intake manifold torque sequence—1.8L engine

25. Connect the coolant and vacuum hoses, electrical connectors and fuel lines.

26. Connect the accelerator cable.

27. Install the air intake surge tank, if removed.

28. Install the air cleaner assembly, if removed.

29. Install the air intake tube to the throttle body.

30. Connect the negative battery cable.

31. Fill the cooling system.

32. Start the engine and bring to normal operating temperature. Check for leaks. Check the idle speed.

Exhaust Manifold

REMOVAL & INSTALLATION

1. Before servicing the vehicle, refer to the precautions at the beginning of this section.

2. Disconnect the negative battery cable.

3. Remove the air cleaner.

4. Remove the air hose.

5. Remove the water bypass pipe bolt.

6. Remove the exhaust manifold heat shield bolts and the heat shield.

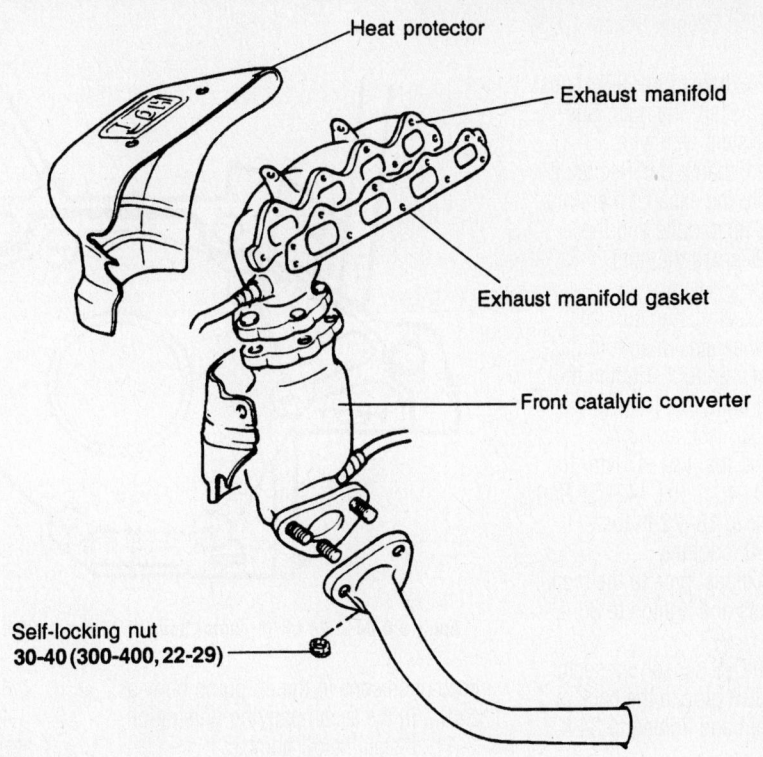

Heat protector

Exhaust manifold

Exhaust manifold gasket

Front catalytic converter

Self-locking nut
30-40 (300-400, 22-29)

TORQUE : Nm (kg.cm, lb.ft)

9357NG33

Exploded view of the exhaust manifold—2.4L engine

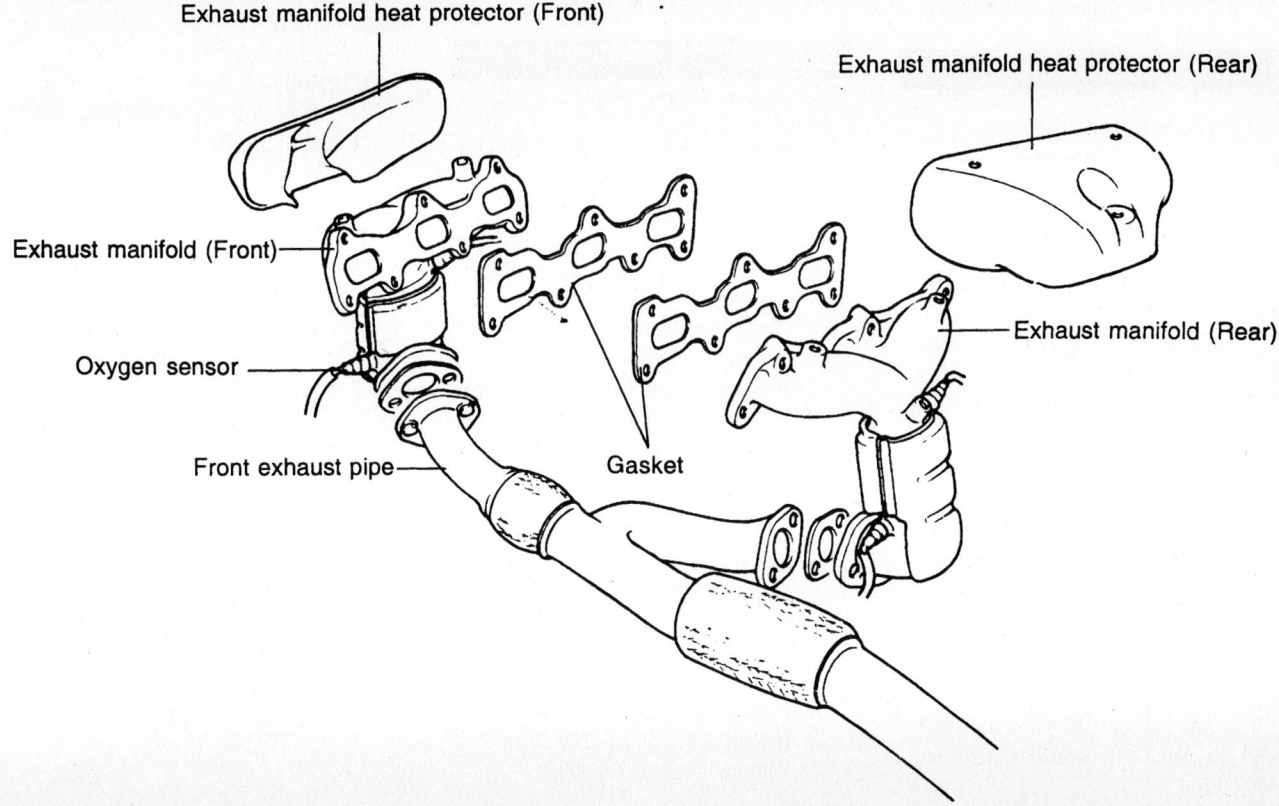

Exhaust manifold heat protector (Front)

Exhaust manifold heat protector (Rear)

Exhaust manifold (Front)

Oxygen sensor

Exhaust manifold (Rear)

Front exhaust pipe

Gasket

9357NG35

Exploded view of the exhaust manifolds—2.7L engine

7. Disconnect the Oxygen Sensor (O$_2$S) electrical connector.

8. Remove the exhaust pipe-to-exhaust manifold nuts and discard the nuts. Suspend the exhaust system with wire.

9. Remove the Exhaust Gas Recirculation (EGR) pipe from the exhaust manifold.

10. Remove the nuts, bolts and the exhaust manifold. Discard the nuts.

To install:

11. Clean all gasket mating surfaces.

12. Install a new exhaust manifold gasket and the exhaust manifold. Tighten the mounting nuts and bolts to 11–14 ft. lbs. (15–20 Nm) for 1.5L, 1.6L and 2.7L engines, to 29–34 ft. lbs. (39–47 Nm) for 1.8L engines, to 31–40 ft. lbs. (42–54 Nm) for 2.0L engines, or to 18–22 ft. lbs. (25–30 Nm) for 2.4L engines.

13. Install the exhaust pipe to the manifold. Install new nuts and tighten to 38 ft. lbs. (52 Nm).

14. Connect the O$_2$S sensor connector.

15. Install the EGR pipe to the back of the exhaust manifold and tighten to 34 ft. lbs. (47 Nm).

16. Install the heat shield and tighten the bolts to 88 inch lbs. (10 Nm).

17. Install the water bypass pipe bolt, and tighten to 48–65 ft. lbs. (64–89 Nm).

18. Install the air hose.

19. Install the air cleaner.

20. Connect the negative battery cable.

Front Crankshaft Seal

REMOVAL & INSTALLATION

1. Before servicing the vehicle, refer to the precautions at the beginning of this section.

2. Disconnect the negative battery cable.

3. Remove the timing belt covers and belt.

4. Remove the timing belt pulley using a puller.

5. Remove the oil pump bolts and the pump.

6. Wrap a suitable prytool with a rag and work the old seal from the oil pump housing.

To install:

7. Lubricate the seal lip with clean engine oil and push the seal slightly in by hand.

8. Install the seal using a seal installer. Install the seal until it is flush with the oil pump body.

9. Install a new O-ring seal.

10. Apply a 0.04–0.08 in. (1–2mm)

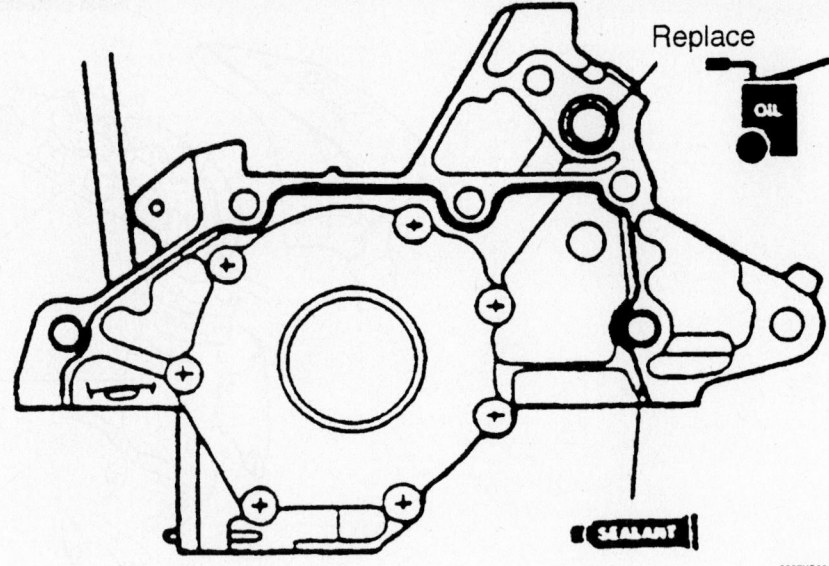

Apply a 0.04–0.08 in. (1–2mm) bead of silicone to the oil pump body

9307KG08

bead of silicone to the oil pump body as shown in the accompanying illustration.

11. Install the oil pump.

12. Install the timing belt pulley. Align the keyway groove on the pulley to the keyway on the crankshaft.

13. Install the woodruff key with the tapered side towards the oil pump body.

14. Install the remaining components in the reverse order of removal.

Camshaft

REMOVAL & INSTALLATION

1.5L and 1.6L Engines

1. Before servicing the vehicle, refer to the precautions at the beginning of this section.

2. Disconnect the negative battery cable.

3. Remove the breather hose and Positive Crankcase Ventilation (PCV) hose.

4. Remove the water pump and crankshaft pulleys.

5. Remove the timing belt cover.

6. Loosen the timing bolt tensioner pulley and temporarily secure it.

7. Remove the timing belt from the camshaft sprocket.

8. Remove the center cover bolts and cover.

9. Remove the ignition coil.

10. Remove the cylinder head cover.

11. Remove the camshaft pulley (sprocket).

12. Remove the cam carrier assembly and timing belt.

13. Remove the camshafts.

14. Remove the camshaft oil seal using a seal removal tool.

➡ **The Hydraulic Lash Adjusters (HLA's) must be installed in the location from which they were removed.**

15. Mark the HLA's to identify their original positions.

16. Remove the HLA's from the cylinder head using a magnet, and store upside-down in a oil-filled container

To install:

17. Apply a coat of the clean engine oil to the sides of the HLA's.

18. Install the HLA's into the cylinder head bore. Check that the HLA moves freely in its bore.

19. Lubricate the camshaft journals and lobes with clean engine oil.

20. Install the camshafts in the cylinder head making sure that the camshaft dowel pins point straight up.

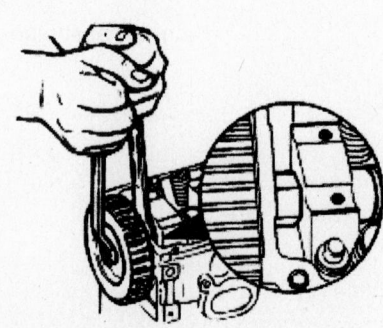

67162-KIAC-G16

Remove the camshaft pulley (sprocket)— 1.5L and 1.6L engines

21. Install the cam carrier assembly. Tighten the bolts, in several steps, to 8–11 ft. lbs. (11–15 Nm).

22. Install the camshaft oil seal, using a suitable seal installer.

23. Install the camshaft sprockets onto the camshaft. Be sure to align the **I** mark with the intake camshaft dowel pin and the **E** mark with the exhaust camshaft dowel pin, then tighten the retaining bolt to 36–44 ft. lbs. (49–61 Nm).

24. The remainder of installation is the reverse of the removal procedure.

25. Check the engine fluid levels, start the vehicle and check for leaks.

1.8L Engine

1. Before servicing the vehicle, refer to the precautions at the beginning of this section.

2. Disconnect the negative battery cable.

3. Remove the ignition coils and high-tension cords.

4. Remove the Positive Crankcase Ventilation (PCV) valve and ventilation hoses.

5. Position the engine so that No. 1 cylinder is at Top Dead Center (TDC).

6. Remove the timing belt.

7. Remove the cylinder head cover.

8. Remove the Camshaft Position (CMP) sensor.

9. Remove the camshaft sprockets.

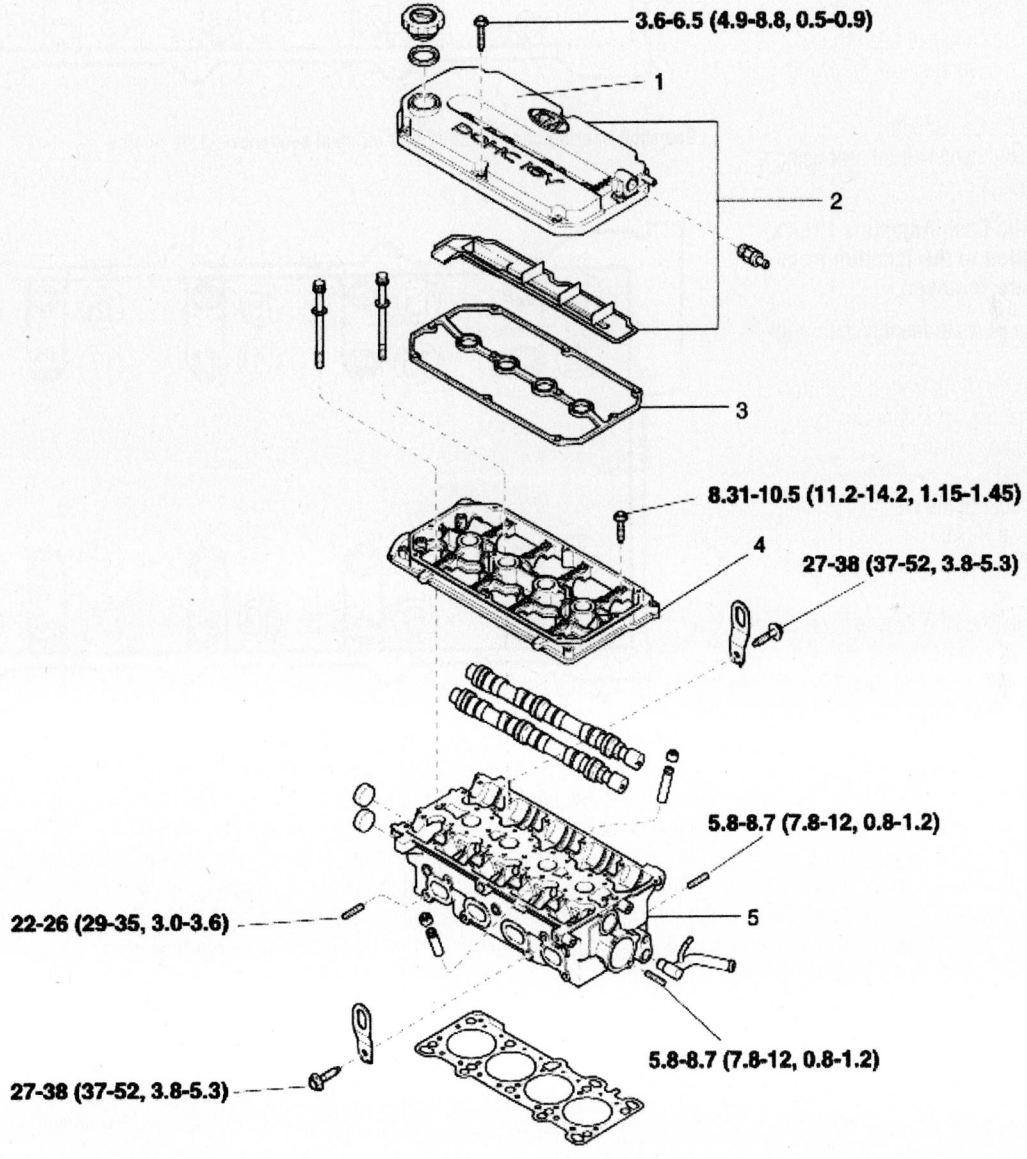

3.6-6.5 (4.9-8.8, 0.5-0.9)

8.31-10.5 (11.2-14.2, 1.15-1.45)

27-38 (37-52, 3.8-5.3)

5.8-8.7 (7.8-12, 0.8-1.2)

22-26 (29-35, 3.0-3.6)

5.8-8.7 (7.8-12, 0.8-1.2)

27-38 (37-52, 3.8-5.3)

TORQUE : lb·ft (N·m, kg·m)

1. Cylinder head cover
2. Cylinder head cover
3. Cylinder head cover gasket
4. Cam carrier assembly
5. Cylinder head

67162-KIAC-G17

Exploded view of the camshafts and related components—1.5L and 1.6L engines

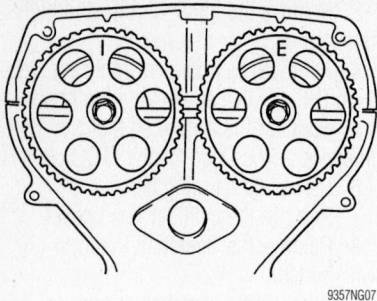

Proper alignment, prior to camshaft installation—1.8L engine

10. Loosen the camshaft bearing cap retaining bolts in several steps, following the proper sequence.

11. Remove the camshafts.

12. Remove the camshaft oil seal using a seal removal tool.

➡**The Hydraulic Lash Adjusters (HLA's) must be installed in the location from which they were removed.**

13. Mark the HLA's to identify their original positions.

14. Remove the HLA's from the cylinder head using a magnet, and store upside-down in a oil-filled container

To install:

15. Apply a coat of the clean engine oil to the sides of the HLA's.

16. Install the HLA's into the cylinder head bore.

17. Check that the HLA moves freely in its bore.

18. Lubricate the camshaft journals and lobes with clean engine oil.

19. Install the camshafts in the cylinder head making sure that the camshaft dowel pins point straight up.

20. Install the camshaft caps in their original positions. Loosely install the cap bolts.

21. Install the camshaft cap bolts in 5–6 steps to 13–20 ft. lbs. (18–27 Nm) in the proper sequence.

22. Oil the lip of the new camshaft oil seal and, using a seal installer. Drive the seal into the cylinder head until it is flush with the edge of the camshaft bearing cap.

23. Install the camshaft sprockets onto the camshaft. Be sure to align the **I** mark with the intake camshaft dowel pin and the **E** mark with the exhaust camshaft dowel pin, then tighten the retaining bolt to 36–44 ft. lbs. (49–61 Nm).

24. Install the CMP sensor.

25. Install the cylinder head cover.

26. Install the timing belt.

27. Install the PCV and ventilation hoses.

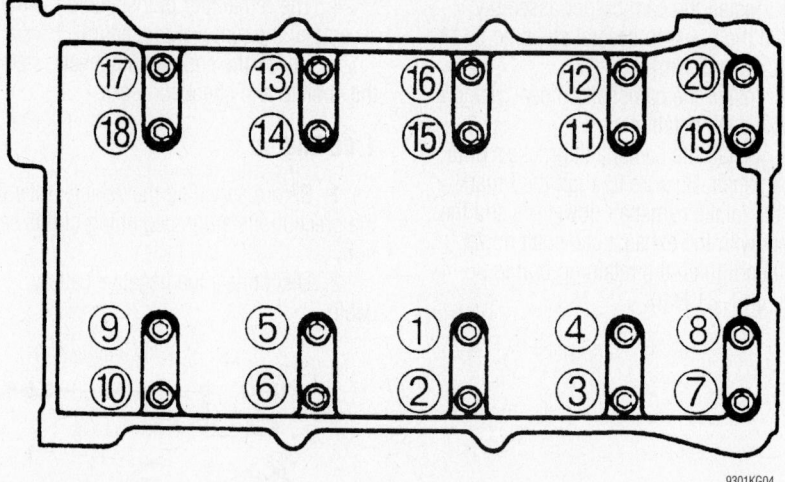

Camshaft bearing cap mounting bolt removal sequence—1.8L engine

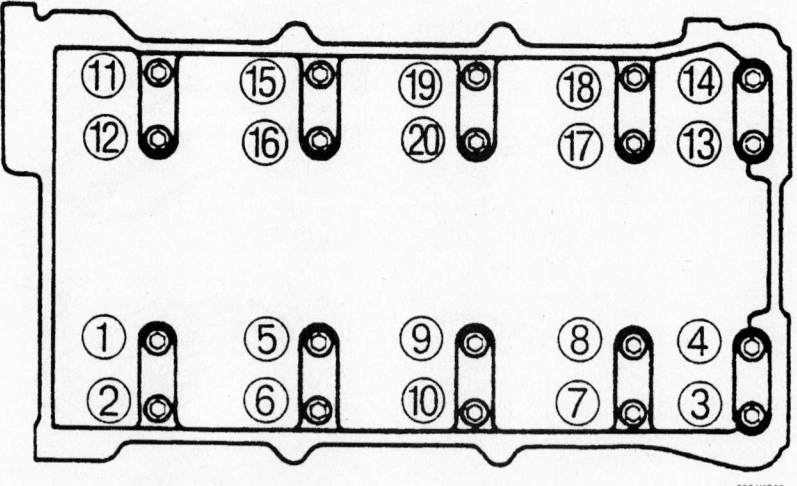

Camshaft bearing cap mounting bolt tightening sequence—1.8L engine

28. Install the ignition coils and high-tension cords.

29. Connect the negative battery cable.

30. Check the engine fluid levels, start the vehicle and check for leaks.

2.0L Engine

1. Before servicing the vehicle, refer to the precautions at the beginning of this section.

2. Disconnect the negative battery cable.

3. Turn the crankshaft pulley so that the No. 1 piston is at top dead center.

4. Disconnect the terminals from battery and remove the heat shield.

5. Remove the engine cover.

6. Drain the engine coolant.

7. Remove the radiator cap to speed draining.

8. Remove the intake air hose and air cleaner assembly.

9. Disconnect the AFS (Air Flow Sensor) connector.

10. Disconnect the breather hose from intake air hose.

11. Remove the intake air hose and air cleaner upper cover.

12. Remove the air cleaner element.

13. Remove the bolts and air cleaner lower cover.

14. Remove the upper radiator hose and lower radiator hose.

15. Remove the heater hoses.

16. Disconnect the accelerator cable by loosening the lock-nut, then slip the cable end out of the throttle linkage.

17. Remove the engine wire harness connectors and wire harness clamps from cylinder head and the intake manifold.

18. Disconnect the OCV (Oil Control Valve) connector.

19. Disconnect the oil temperature sensor connector.

20. Disconnect the ECT (Engine Coolant Temperature) sensor connector.

21. Disconnect the ignition coil connector.

22. Disconnect the TPS (Throttle Position Sensor) connector.

23. Disconnect the ISA (Idle Speed Actuator) connector.

24. Disconnect the CMP (Camshaft Position Sensor) connector.

25. Disconnect the four injector connectors.

26. Disconnect the knock sensor connector.

27. Disconnect the ground cables from the intake manifold and vehicle's body.

28. Disconnect the air conditioner compressor switch connector.

29. Disconnect the front heated oxygen sensor connector.

30. Disconnect the CKP (Crankshaft Position Sensor) connector.

31. Disconnect the oil pressure switch connector.

32. Disconnect the PCSV (Purge Control Solenoid Valve) connector.

33. Disconnect the fuel inlet hose of the delivery pipe side.

34. Disconnect the hose of the PCSV (Purge Control Solenoid Valve) side.

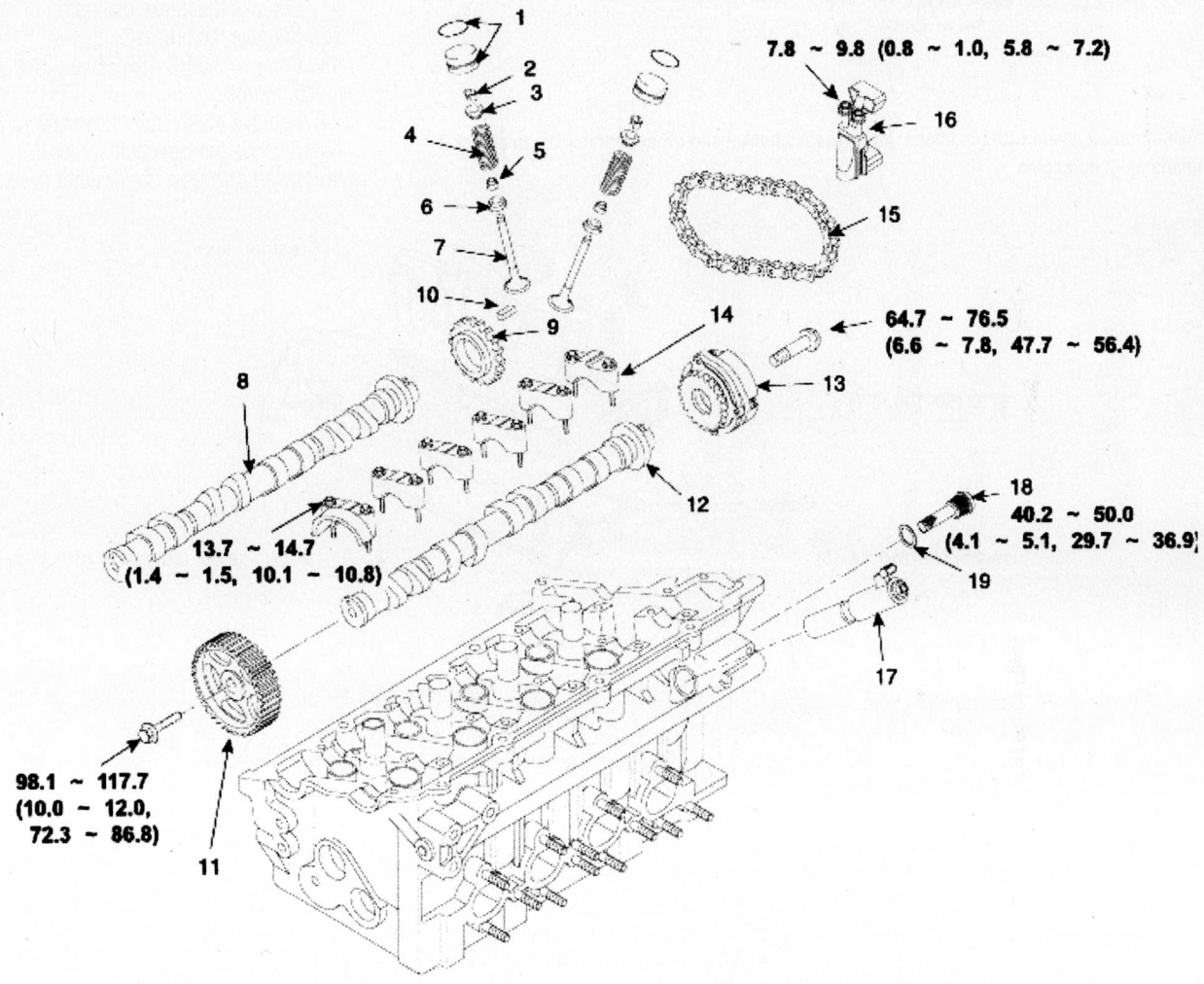

TORQUE : N-m (kg-m, lb-ft)

1. MLA (Mechanical Lash Adjuster)
2. Retainer lock
3. Retainer
4. Valve spring
5. Stem seal
6. Spring seat
7. Valve
8. Intake camshaft
9. Chain sprocket
10. Key
11. Camshaft sprocket
12. Exhaust camshaft
13. CVVT (Continuously Variable Valve Timing) assembly
14. Camshaft bearing cap
15. Timing chain
16. Auto tentioner
17. OCV (Oil Control Valve)
18. OCV (Oil Control Valve) filter
19. Washer

Exploded view of the camshafts and related components—2.0L engine

67162-KIAC-G18

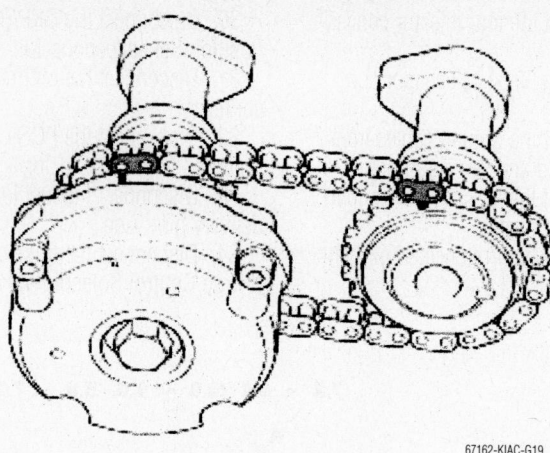

67162-KIAC-G19

Camshaft timing chain with the intake timing chain sprocket and exhaust timing chain sprocket alignment—2.0L engine

35. Remove the brake booster vacuum hose.

36. Remove the power steering pump drive belt.

37. Remove the power steering pump and use a wire to secure the pump to the vehicle so that it is out of the way.

38. Remove the bolts and power steering pump bracket.

39. Remove the spark plug cables.

40. Remove the exhaust manifold.

41. Remove the intake manifold.

42. Remove the timing belt.

43. Remove the PCV (Positive Crankcase Ventilation) hose.

44. Remove the cylinder head cover.

45. Remove the camshaft sprocket.

46. Insert a stopper pin or other device

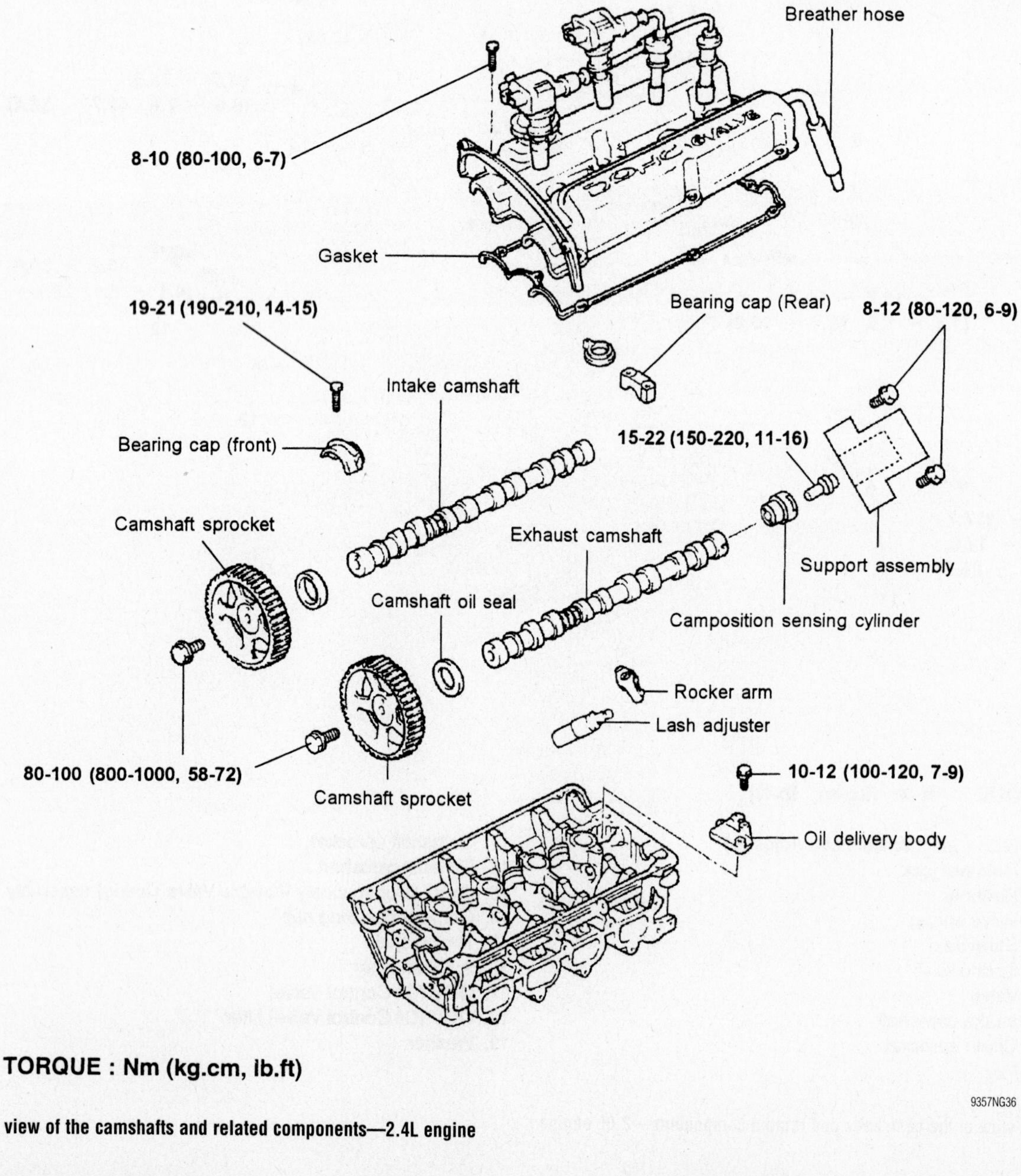

Breather hose

8-10 (80-100, 6-7)

Gasket

19-21 (190-210, 14-15)

Bearing cap (Rear)

8-12 (80-120, 6-9)

Intake camshaft

Bearing cap (front)

15-22 (150-220, 11-16)

Camshaft sprocket

Camshaft oil seal

Exhaust camshaft

Support assembly

Camposition sensing cylinder

Rocker arm

Lash adjuster

80-100 (800-1000, 58-72)

Camshaft sprocket

10-12 (100-120, 7-9)

Oil delivery body

TORQUE : Nm (kg.cm, lb.ft)

9357NG36

Exploded view of the camshafts and related components—2.4L engine

into timing chain auto tensioner and remove the auto tensioner.

47. Remove the camshaft bearing caps and camshafts.

To install:

48. Install the camshafts.

49. Align the camshaft timing chain with the intake timing chain sprocket and exhaust timing chain sprocket as shown.

50. Install the camshaft and bearing caps. Tighten the bearing cap bolts to 10.5 ft. lbs. (14 Nm).

51. Install the timing chain auto tentioner. Tighten the bolts to 6 ft. lbs. (8.5 Nm).

52. Remove the auto tentioner stopper pin.

53. Check and adjust valve clearance.

54. Using the SST (09221—21000), install the camshaft bearing oil seal.

55. Install the camshaft sprocket.

56. Install the cylinder head cover.

57. Install the cylinder head cover gasket in the groove of the cylinder head cover. Before installing the cylinder head cover gasket, thoroughly clean the cylinder head cover and the groove. When installing, make sure the cylinder head cover gasket is seated securely in the corners of the recesses with no gap. Apply liquid gasket to the head cover gasket at the corners of the recess.

58. Install the cylinder head cover gasket with the 12 bolts. Tighten all cylinder head cover bolts temporarily to half of the standard torque, the re-tighten completely using the order specified in the illustration.

59. Install the PCV(Positive Crankcase Ventilation) hose.

60. Install the timing belt.

61. Install the intake manifold.

62. Install the exhaust manifold.

63. Install the spark plug cables.

64. Install the power steering pump bracket and bolts.

65. Install the power steering pump.

66. Connect the accelerator cable.

67. Install the brake booster hose.

68. Connect the hose of the PCSV side.

69. Connect the fuel inlet hose of the delivery pipe side.

Install the engine wire harness connectors and wire harness clamps to the cylinder head and the intake manifold.

70. Connect the PCSV connector.

71. Connect the front heated oxygen sensor connector.

72. Connect the CKP connector.

73. Connect the oil pressure switch connector.

74. Connect the air conditioner compressor switch connector.

75. Connect the ground cables to intake manifold and vehicle's body.

76. Connect the knock sensor connector.

77. Connect the fuel injector connectors.

78. Connect the CMP connector.

79. Connect the ISA connector.

80. Connect the TPS connector.

81. Connect the ignition coil connector.

82. Connect the ECT connector.

83. Connect the oil temperature sensor connector.

84. Connect the OCV connector.

85. Install the heater hose.

86. Install the upper radiator hose and lower radiator hose.

87. Install the intake air hose and air cleaner assembly.

88. Install the air cleaner element.

89. Install the intake air hose and air cleaner upper cover.

90. Install the breather hose to intake air hose.

91. Connect the AFS connector.

92. Install the engine cover.

93. Install the heat shield and connect the battery terminals to the battery.

94. Fill with engine coolant.

95. Start the engine and check for leaks.

96. Recheck engine coolant level and oil level.

2.4L Engine

1. Before servicing the vehicle, refer to the precautions at the beginning of this section.

2. Drain the engine coolant.

3. Disconnect the negative battery cable.

4. Remove the breather hose from between the air cleaner and rocker arm cover.

5. Remove the air cleaner.

6. Remove the timing belt cover.

7. Remove the rocker arm cover and crank angle sensor.

8. Remove the camshaft sprocket bolts and sprockets.

➡**Keep all valvetrain components in order as you remove them. The components must be installed in their original locations.**

9. Remove the bearing cap bolts, bearing caps, camshafts, rocker arms and valve adjusters.

To install:

10. Lubricate the camshaft journals and lobes with clean engine oil.

11. Install the camshafts on the cylinder head. The intake camshaft has a slit on its rear side to drive the Crankshaft Position (CKP) sensor.

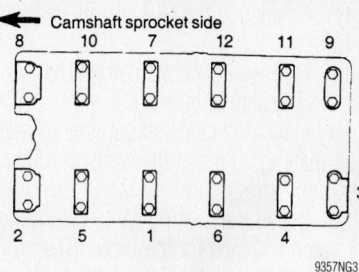

Camshaft bearing cap tightening sequence—2.4L engine

12. Install the camshaft bearing caps. Make sure the cam can be easily turned by hand, then remove the bearing caps and install the rocker arms.

13. Install the camshafts and bearings caps. Torque the bearing caps, in sequence, to 14–15 ft. lbs. (19–21 Nm).

14. Install the camshaft oil seal.

15. Install the camshaft sprockets. Torque the bolts to 58–72 ft. lbs. (80–100 Nm).

16. Install the rocker cover.

17. Install the remaining components in the reverse of the removal procedure.

18. Fill the cooling system. Check the engine fluid levels, start the vehicle and check for leaks.

2.7L Engine

1. Before servicing the vehicle, refer to the precautions at the beginning of this section.

2. Disconnect the negative battery cable.

3. Remove the engine cover and intake manifold.

4. Remove the breather hose and engine harness.

5. Remove the water pump pulley, crankshaft pulley, idler pulley and tensioner pulley.

6. Remove the timing belt cover.

7. Remove the timing belt tensioner pulley, loosen and secure temporarily.

8. Remove the timing belt from the camshaft sprocket.

9. Remove the spark plug cables.

10. Remove the cylinder head cover.

11. Remove the camshaft sprocket.

12. Remove the camshaft bearing cap bolts.

13. Remove the camshafts.

14. Remove the HLA's from the cylinder head using a magnet, and store upside-down in a oil-filled container

To install:

15. Apply a coat of the clean engine oil to the sides of the HLA's.

16. Install the HLA's into the cylinder head bore.

17. Check that the HLA moves freely in its bore.

18. Lubricate the camshaft journals and lobes with clean engine oil.

19. Install the camshafts in the cylinder head making sure that the camshaft dowel pins point straight up.

20. Install the bearing caps. Check the marks on the caps for the Intake (I) and Exhaust (E) markings. Torque the bolts, working from the center outward to 10–11 ft. lbs. (14–15 Nm).

21. Install the remaining components in the reverse of the removal procedure.

22. Check the engine fluid levels, start the vehicle and check for leaks.

Valve Lash

ADJUSTMENT

The valve lash on all engines is kept in adjustment hydraulically. No adjustment is necessary or possible.

Starter Motor

REMOVAL & INSTALLATION

1.5L and 1.6L Engines

1. Before servicing the vehicle, refer to the precautions at the beginning of this section.

2. Disconnect the negative battery cable.

3. Remove the 4 upper intake manifold stay bracket bolts.

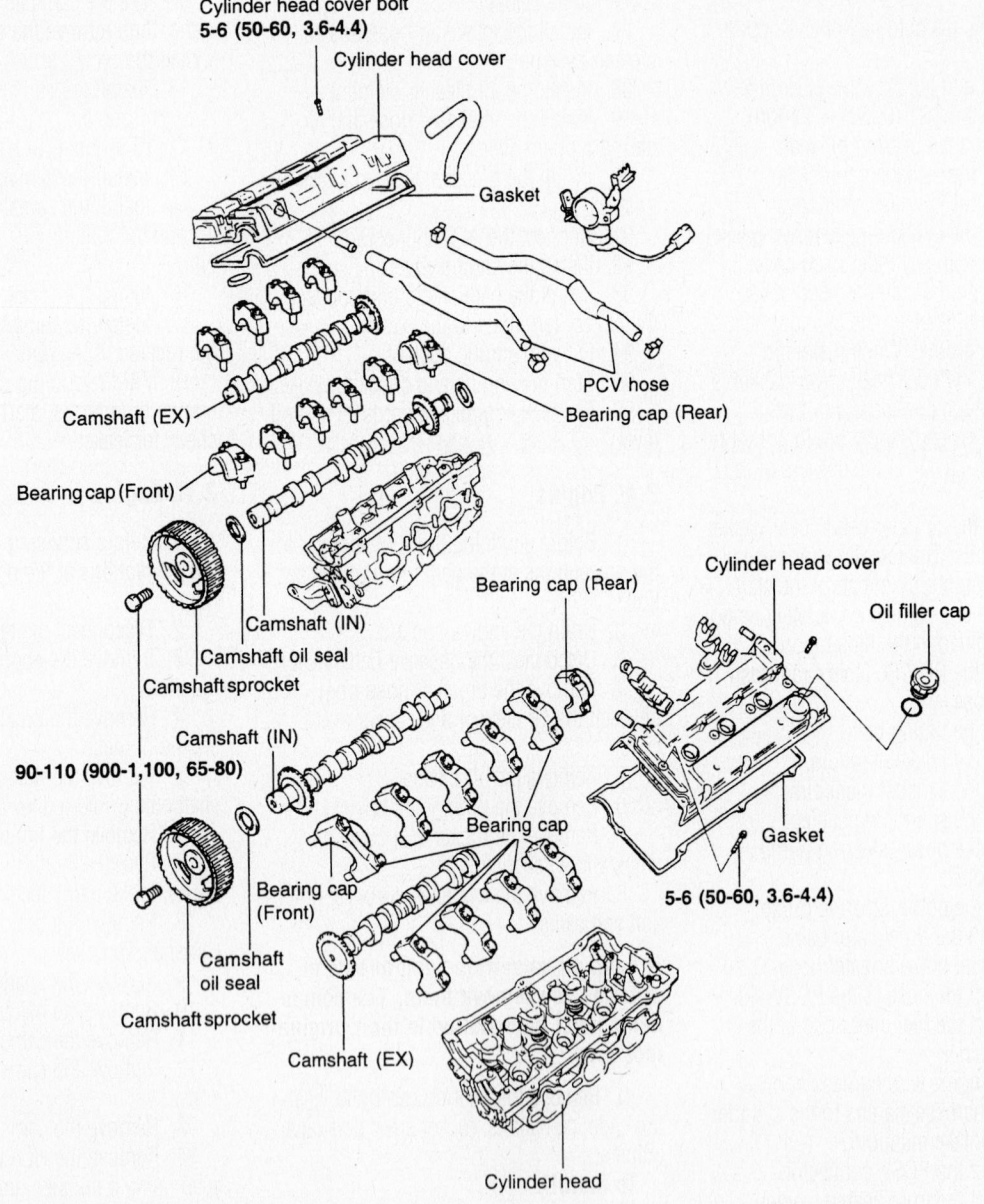

Cylinder head cover bolt
5-6 (50-60, 3.6-4.4)
Cylinder head cover
Gasket
PCV hose
Bearing cap (Rear)
Camshaft (EX)
Bearing cap (Front)
Camshaft (IN)
Camshaft oil seal
Camshaft sprocket
90-110 (900-1,100, 65-80)
Camshaft (IN)
Bearing cap (Front)
Camshaft oil seal
Camshaft sprocket
Camshaft (EX)
Bearing cap
Bearing cap (Rear)
Cylinder head cover
Oil filler cap
Gasket
5-6 (50-60, 3.6-4.4)
Cylinder head

TORQUE : Nm (kg.cm, lb.ft)

9357NG38

Exploded view of the camshafts and related components—2.7L engine

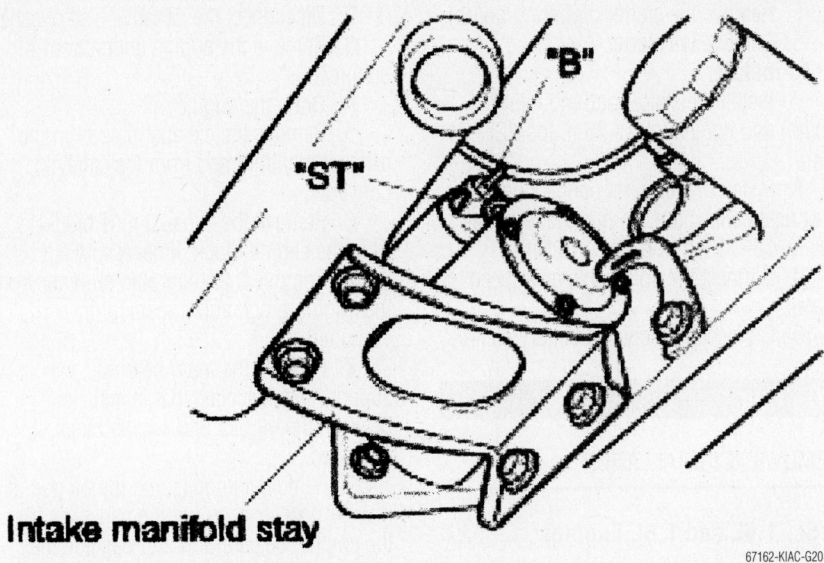

Intake manifold stay

67162-KIAC-G20

Starter mounting—1.5L and 1.6L engines

4. Disconnect the starter electrical connections.

5. Remove the upper starter bolts.

6. Remove the lower starter mounting bolt(s) and the starter.

To install:

7. Install the starter.

8. Install the lower starter bolt and tighten to 27–38 ft. lbs. (37–52 Nm).

9. Connect the starter electrical connections.

10. Install the intake manifold support bracket and finger-tighten the 3 mounting bolts. Tighten the lower bolt to 27–38 ft. lbs. (37–52 Nm).

11. Install the 4 upper intake manifold stay bracket bolts and tighten to 27–38 ft. lbs. (37–52 Nm).

12. Install the upper starter bolts and tighten to 27–38 ft. lbs. (37–52 Nm).

13. Connect the negative battery cable.

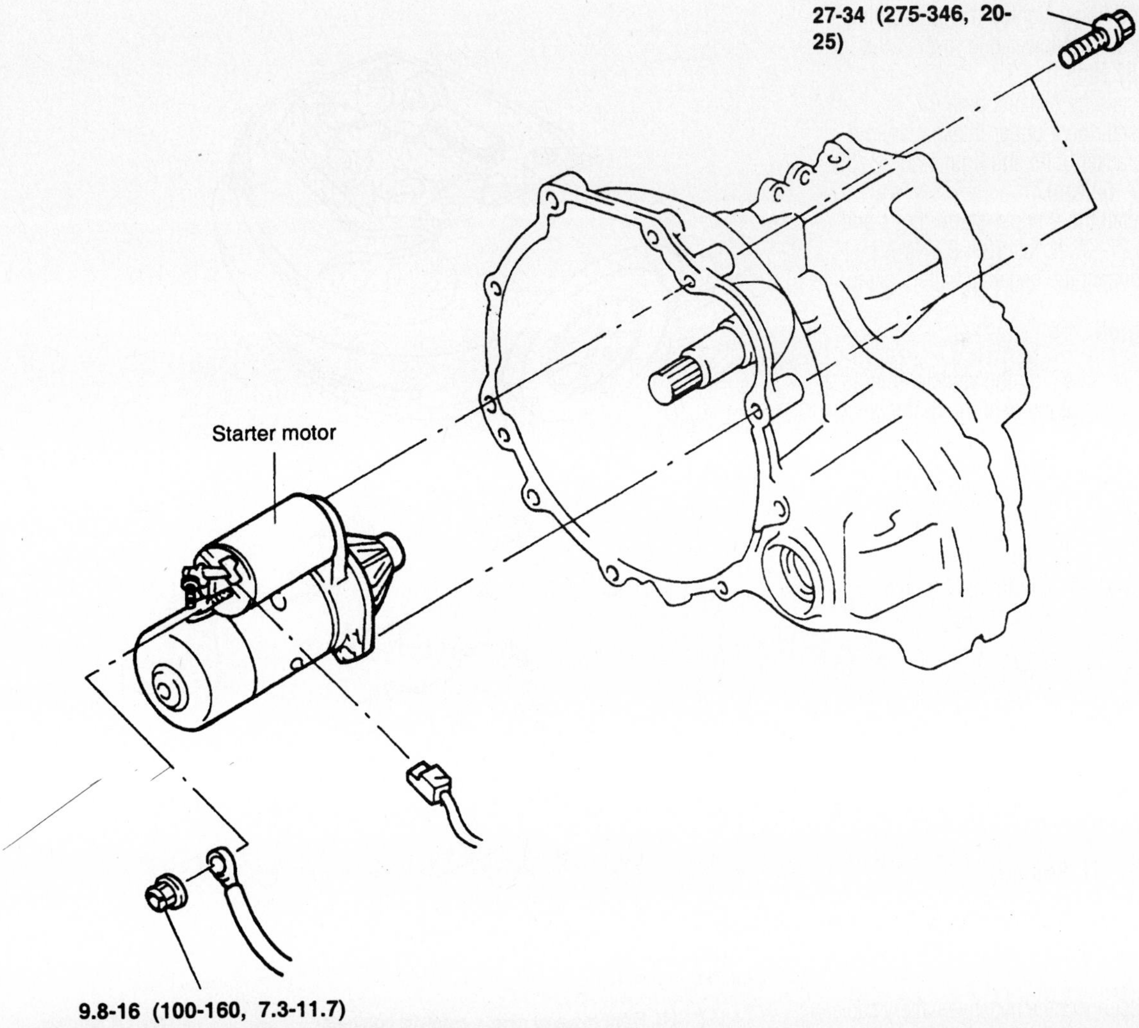

27-34 (275-346, 20-25)

Starter motor

9.8-16 (100-160, 7.3-11.7)

TORQUE : N·m (kg·cm, lb·ft)

9357NG39

Exploded view of the starter mounting—2.4L and 2.4L engines

1.8L Engine

1. Before servicing the vehicle, refer to the precautions at the beginning of this section.

2. Disconnect the negative battery cable.

3. Remove the 2 upper intake manifold support bracket bolts.

4. Disconnect the starter electrical connections.

5. Remove the 2 upper starter bolts.

6. Remove the exhaust pipe.

7. Remove the lower intake manifold support bracket bolt and the bracket.

8. Remove the lower starter bolt and the starter.

To install:

9. Install the starter.

10. Install the lower starter bolt and tighten to 27–38 ft. lbs. (37–52 Nm).

11. Connect the starter electrical connections.

12. Install the intake manifold support bracket and finger-tighten the 3 mounting bolts. Tighten the lower bolt to 27–38 ft. lbs. (37–52 Nm).

13. Install the exhaust pipe.

14. Install the 2 upper intake manifold support bracket bolts and tighten to 27–38 ft. lbs. (37–52 Nm).

15. Install the 2 upper starter bolts and tighten to 27–38 ft. lbs. (37–52 Nm).

16. Connect the negative battery cable.

2.0L Engine

1. Before servicing the vehicle, refer to the precautions at the beginning of this section.

2. Disconnect the negative battery cable.

3. Disconnect the starter cable from the B terminal on the solenoid, then disconnect the connector from the S terminal.

4. Remove the 2 bolts holding the starter, then remove the starter.

To install:

5. Install the starter.

6. Install the starter bolts and tighten to 27–38 ft. lbs. (37–52 Nm).

7. Connect the starter electrical connections.

8. Connect the negative battery cable.

2.4L and 2.7L Engines

1. Before servicing the vehicle, refer to the precautions at the beginning of this section.

2. Disconnect the negative battery cable.

3. Disconnect the speed meter and shift cable.

4. Disconnect the starter connector and terminal connection.

5. Remove the starter mounting bolt(s).

6. Remove the starter.

To install:

7. Install the starter and mounting bolt(s) and tighten to 20–25 ft. lbs. (27–34 Nm).

8. Install the terminal connection and electrical connection. Torque the terminal nut to 7.3–11.7 ft. lbs. (9.8–16 Nm).

9. Connect the shift cable and speed meter.

10. Connect the negative battery cable.

Oil Pan

REMOVAL & INSTALLATION

1.5L, 1.6L and 1.8L Engines

1. Before servicing the vehicle, refer to the precautions at the beginning of this section.

2. Disconnect the negative battery cable.

3. Remove the engine undercover, if equipped.

4. Drain the engine oil.

5. Remove the exhaust pipe from the exhaust manifold and from the catalytic converter.

6. Remove the exhaust pipe bracket from the engine block, if necessary.

7. Remove the integrated stiffener from the engine block and transaxle, if equipped.

8. Remove the main bearing support/stiffener plate that is installed between the oil pan and engine block, if equipped.

9. Remove the bolts and the oil pan. It may be necessary to pry the pan away from the engine; be careful not to damage the gasket contact surfaces.

10. Remove the oil strainer, if necessary.

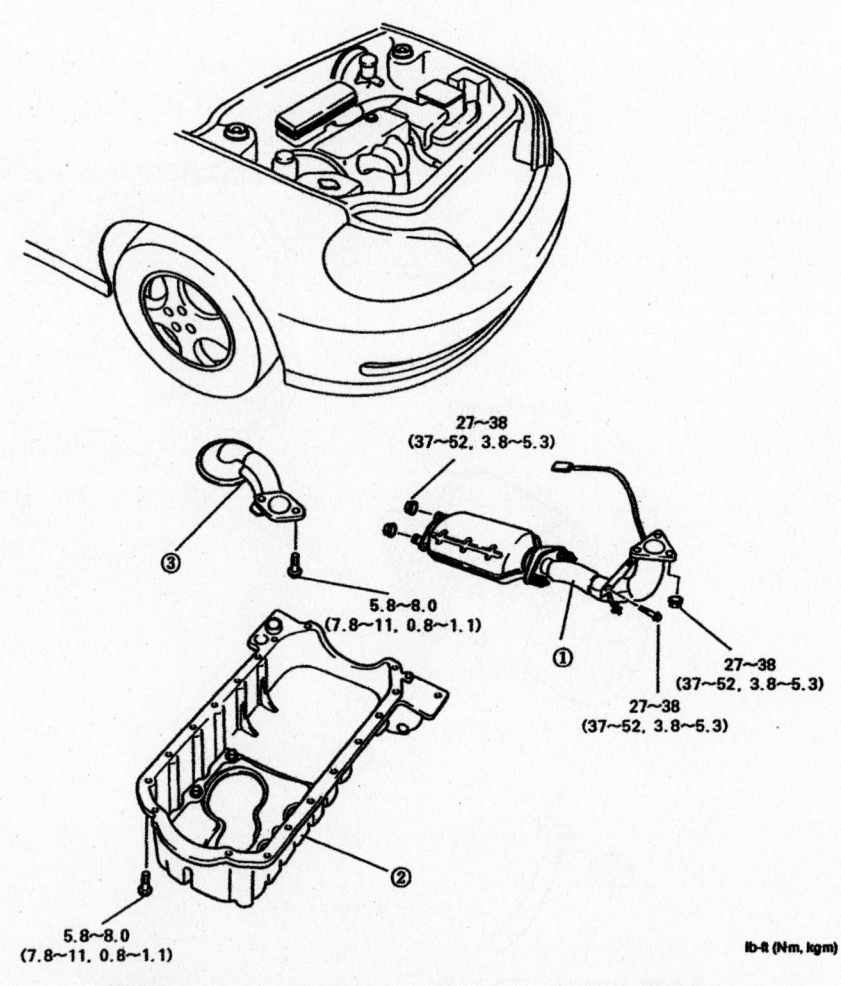

27~38
(37~52, 3.8~5.3)

5.8~8.0
(7.8~11, 0.8~1.1)

27~38
(37~52, 3.8~5.3)

27~38
(37~52, 3.8~5.3)

5.8~8.0
(7.8~11, 0.8~1.1)

lb-ft (Nm, kgm)

(1) Front exhaust pipe & catalytic converter
(2) Oil pan
(3) Oil strainer

67162-KIAC-G21

Exploded view of the oil pan and related components—1.5L and 1.6L engines

Tightening torque:

Ⓐ Ⓑ Ⓒ Ⓓ : 5.8~8.0 lb-ft
(7.8~10.8 N•m, 0.8~1.1 kg-m)

Ⓔ : 28~38 lb-ft (38~51 N•m, 3.8~5.3 kg-m)

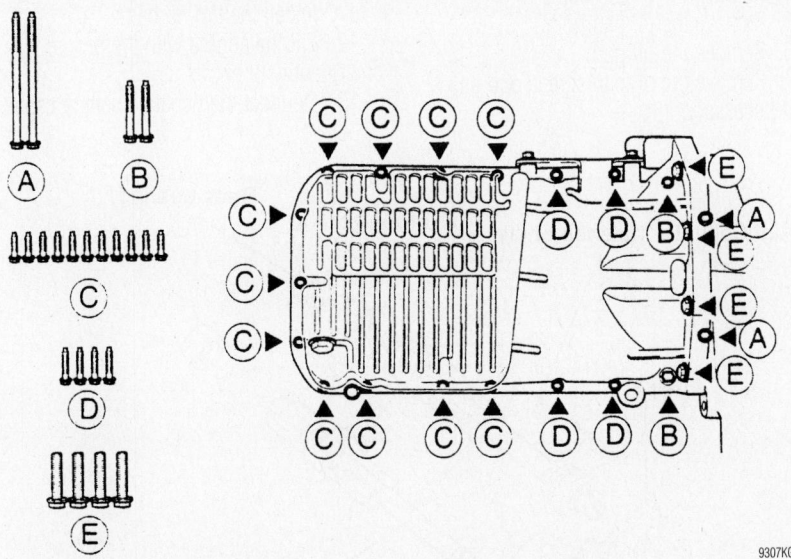

9307KG14

Oil pan mounting bolt locations and torque specifications—1.8L engine

To install:

11. Clean all oil, dirt, old gasket material and sealer from the oil pan, support/stiffener plate, oil pan bolts and all gasket mating surfaces. If removed, clean the oil strainer.

12. If equipped with the main bearing support/stiffener plate, run a bead of silicone sealer around the perimeter of the plate, going inside the bolt holes. Install the plate and tighten the bolts.

✳✳ **WARNING**

Be sure all old sealer is removed from the bolts prior to installation. Installing a bolt coated with old sealer could result in cracking of the bolt holes.

13. Install the oil strainer using a new gasket, if removed.

14. If used, apply silicone sealer to new rubber end gaskets and press them into place on the engine.

15. Apply a bead of silicone to the perimeter of the oil pan, going around the inside of the bolt holes.

16. Install the pan to the engine and the oil pan bolts finger-tight.

17. Tighten the oil pan bolts to the spec-

ifications shown in the accompanying illustrations.

✳✳ **WARNING**

Be sure all old sealer is removed from the bolts prior to installation. Installing a bolt coated with old sealer could result in cracking of the bolt holes.

18. Install the integrated stiffener to the engine block and transaxle, if removed. Tighten the bolts to 38 ft. lbs. (52 Nm).

19. Install the transverse member, if removed. Tighten the bolts to 93 ft. lbs. (126 Nm).

20. Install the front exhaust pipe bracket, if equipped.

21. Install the front exhaust pipe, using new gaskets. Tighten the exhaust manifold flange nuts to 27–38 ft. lbs. (37–52 Nm) for 1.5L and 1.6L engines, or 34 ft. lbs. (46 Nm) for 1.8L engines.

22. Install the oil pan drain plug using a new gasket. Tighten the drain plug to 30 ft. lbs. (41 Nm).

23. Install the engine undercover.

24. Fill the engine with the proper type and quantity of oil.

25. Connect the negative battery cable. Start the engine and bring to normal operating temperature. Check for leaks.

2.0L Engine

1. Before servicing the vehicle, refer to the precautions at the beginning of this section.

2. Disconnect the negative battery cable.

3. Drain the engine oil.

4. Disconnect the rear oxygen sensor connector.

5. Remove the front muffler.

6. Remove the exhaust manifold.

7. Remove the front muffler bracket.

8. Remove the oil pan.

To install:

9. Install the oil pan.

10. Using a razor blade and gasket scraper, remove all the old packing material from the gasket surfaces.

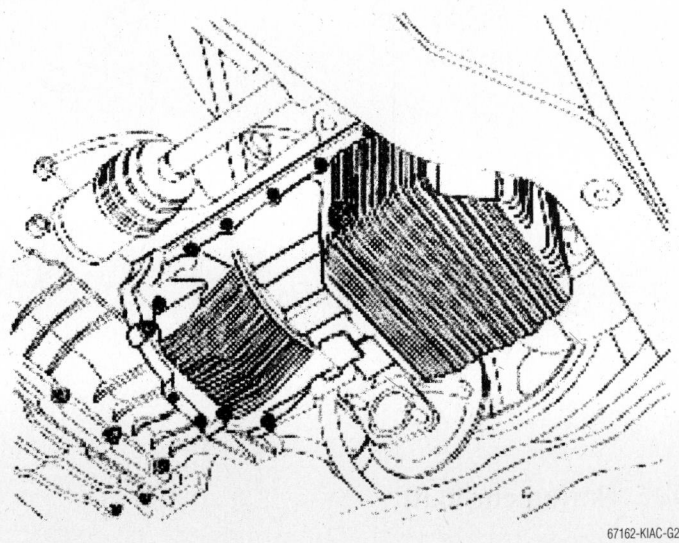

67162-KIAC-G22

Oil pan—2.0L engine

11. Apply liquid gasket as an even bead, centered between the edges of the mating surface.

12. Install the oil pan with the bolts. Uniformly tighten the bolts in several passes to 7–8.5 ft. lbs. (10–12 Nm).

13. Install the front muffler bracket.

14. Install the exhaust manifold.

15. Install the front muffler.

16. Connect the rear oxygen sensor connector.

17. Fill with engine oil

18. Connect the negative battery cable.

Start the engine and bring to normal operating temperature. Check for leaks.

2.4L and 2.7L Engines

1. Before servicing the vehicle, refer to the precautions at the beginning of this section.

2. Drain the engine oil.

3. Disconnect the negative battery cable.

4. Remove the timing belt.

5. Remove the oil pan bolts.

6. Remove the oil pan.

7. Remove the oil pan screen and gasket, if necessary.

To install:

8. Install the oil pump screen. Torque the bolts to 11–16 ft. lbs. (15–22 Nm).

9. Install a 0.16 in. (44mm) bead of sealant into the groove of the oil pan flange. Install the oil pan within 15 minutes of applying the sealant.

10. Install the oil pan and tighten the bolts to 7–9 ft. lbs. (10–12 Nm).

11. Install the timing belt.

12. Fill the engine with the proper type and quantity of oil.

13. Connect the negative battery cable.

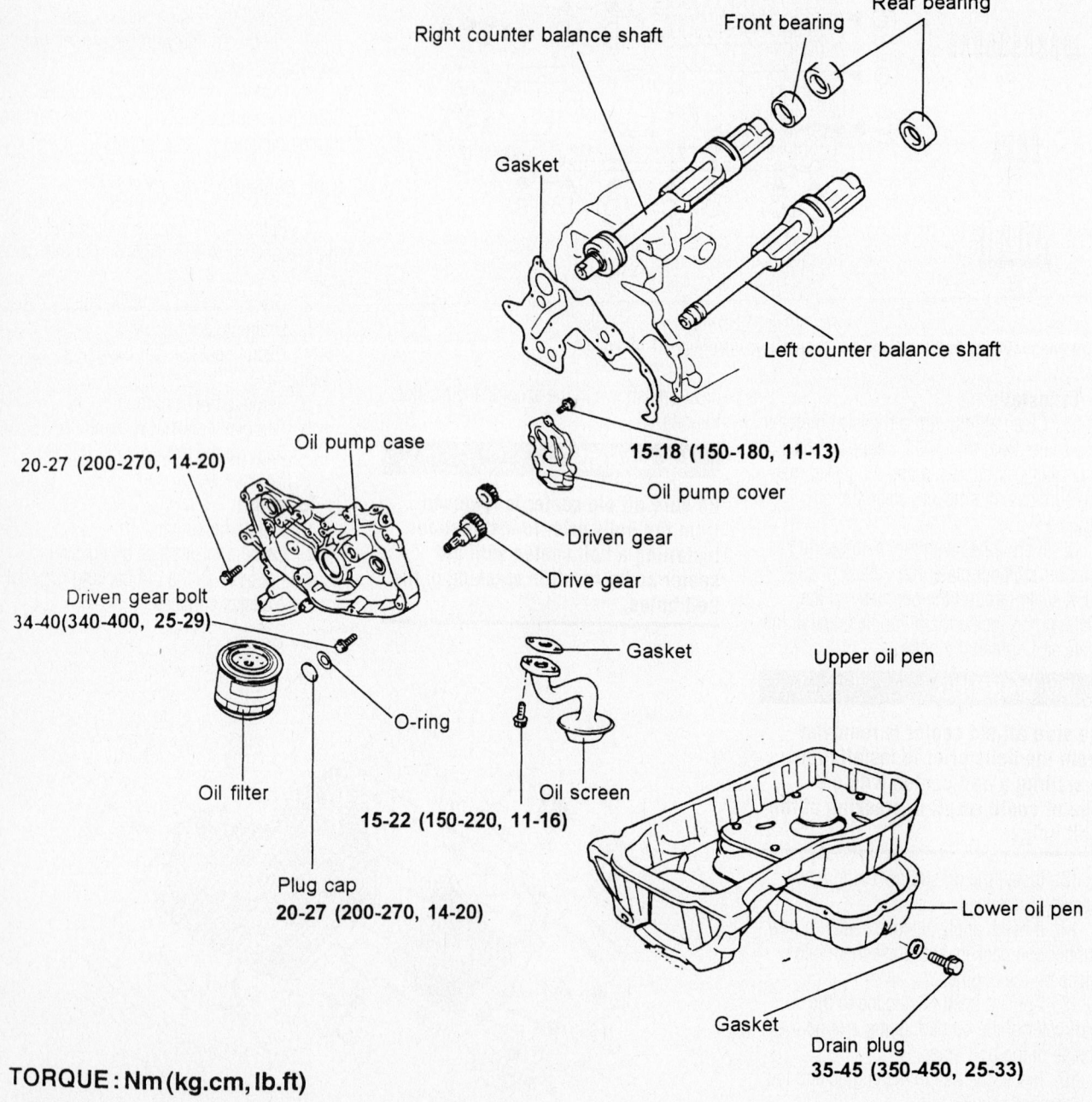

Right counter balance shaft

Front bearing

Rear bearing

Gasket

Left counter balance shaft

20-27 (200-270, 14-20)

Oil pump case

15-18 (150-180, 11-13)

Oil pump cover

Driven gear

Drive gear

Driven gear bolt
34-40(340-400, 25-29)

Oil filter

O-ring

Gasket

Oil screen

15-22 (150-220, 11-16)

Plug cap
20-27 (200-270, 14-20)

Upper oil pen

Lower oil pen

Gasket

Drain plug
35-45 (350-450, 25-33)

TORQUE : Nm (kg.cm, lb.ft)

Exploded view of the oil pan, pump and related components—2.4L engine

Fuel Pump

REMOVAL & INSTALLATION

1. Before servicing the vehicle, refer to the precautions at the beginning of this section.
2. Relieve the fuel system pressure.
3. Disconnect the negative battery cable.
4. Remove the rear seat cushion from the vehicle.
5. Clean any dirt that has accumulated around the fuel pump cover so it will not enter the tank during pump removal and installation.
6. Remove the fuel pump cover.
7. Disconnect the fuel gauge connector, hoses, and the gauge.
8. Disconnect the fuel pump electrical connector.

9. Remove the fuel pump from the bracket assembly.
10. Remove the seal ring and discard.
To install:
11. Clean the fuel pump mounting flange, fuel tank mounting surface and seal ring groove.
12. Apply a light coating of grease on a new seal ring to hold it in place during assembly.
13. Install the seal ring.
14. Install the fuel pump to the bracket assembly carefully to ensure the filter is not damaged. Be sure the seal ring remains in the groove.
15. Hold the pump assembly in place, and pull the fuel pump down so that it is tight against the bracket.
16. Connect the fuel pump electrical connector.
17. Connect the fuel gauge, hoses, and gauge connector.

18. Install the fuel pump cover.
19. Install the rear seat cushion.
20. Connect the negative battery cable.
21. Start the engine and check for proper system operation and for fuel leaks.

Fuel Injector

REMOVAL & INSTALLATION

1. Before servicing the vehicle, refer to the precautions at the beginning of this section.
2. Relieve the fuel system pressure.
3. Disconnect the negative battery cable.
4. Disconnect the injector electrical connectors.
5. Disconnect the fuel line from the fuel rail.

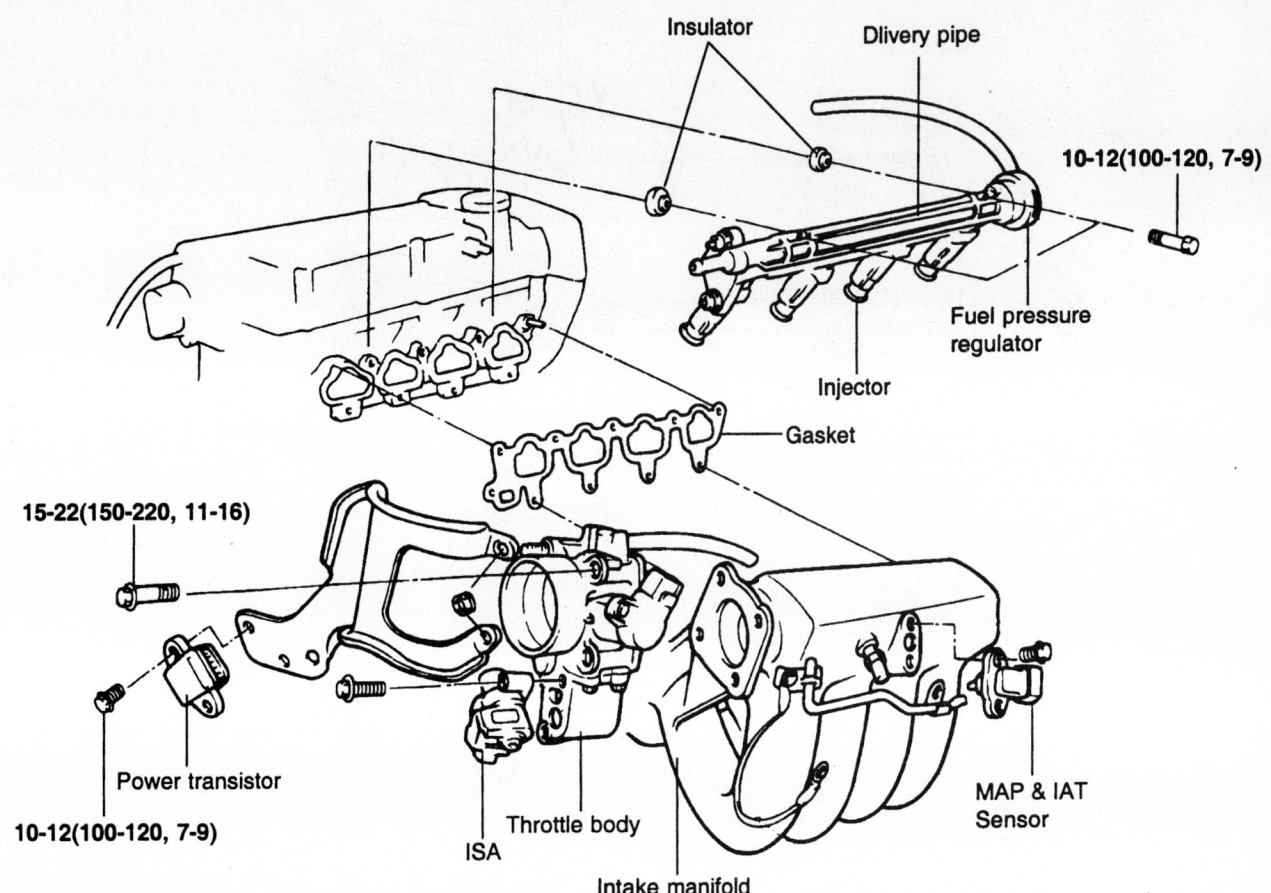

TORQUE : N·m (kg·cm, lb·ft)

9357NG44

Exploded view of the fuel rail and related components—2.4L engine

6. Remove the accelerator cable bracket and the cable, if necessary.

7. Remove the fuel rail retainers and the fuel rail.

8. Remove the injector retaining clips and the injectors.

9. Remove the injector O-rings and discard.

To install:

10. Apply a small amount of clean engine oil to the new O-rings and install them.

11. Install the injectors to the fuel rail and the retaining clips.

12. Install the fuel rail and the fuel rail retainers.

13. Install the accelerator cable and bracket, if removed.

14. Connect the fuel line to the fuel rail.

15. Connect the injector electrical connectors.

16. Connect the negative battery cable.

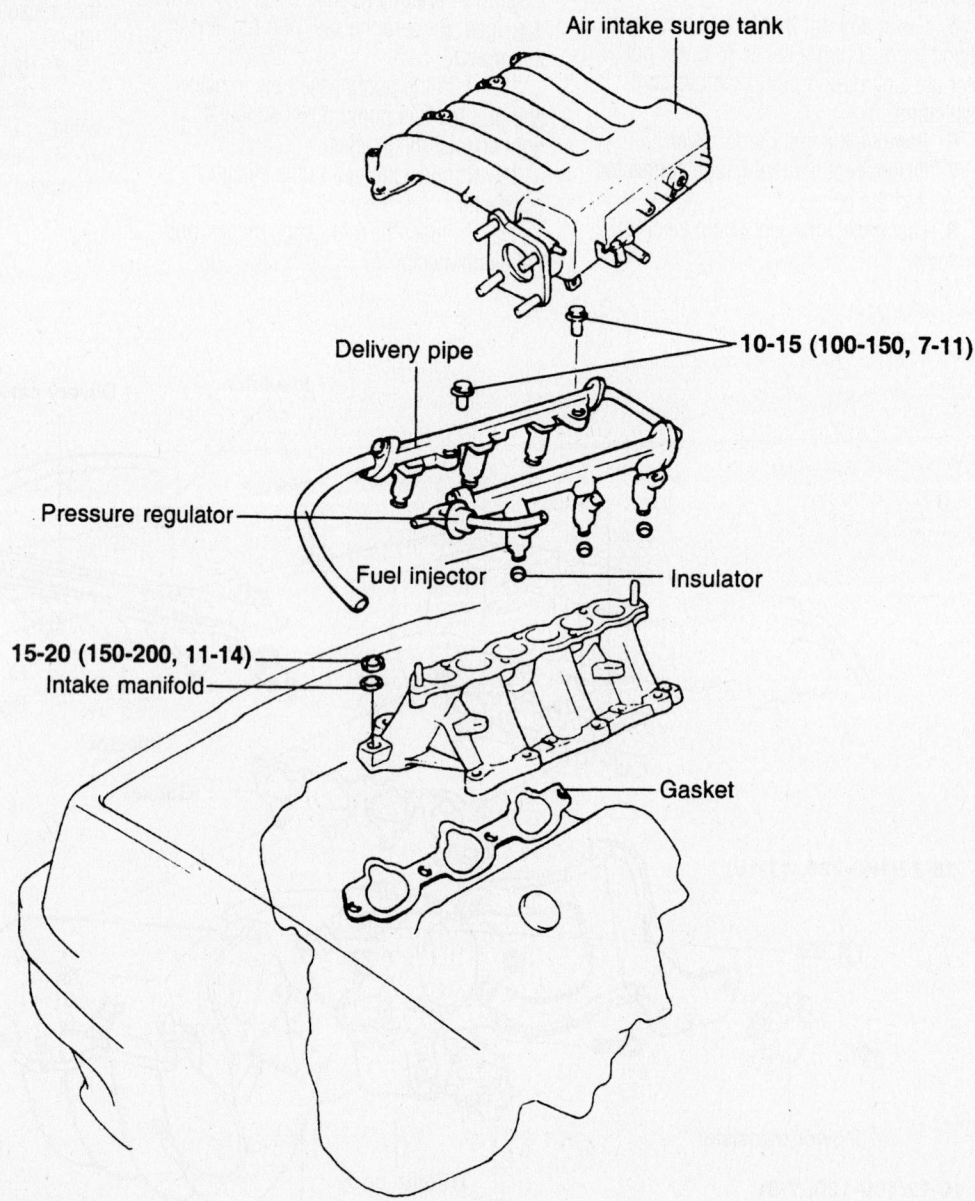

Air intake surge tank

Delivery pipe

10-15 (100-150, 7-11)

Pressure regulator

Fuel injector

Insulator

15-20 (150-200, 11-14)

Intake manifold

Gasket

TORQUE : N·m (kg·cm, lb·ft)

9357NG45

Exploded view of the fuel rail and related components—2.7L engine

DRIVE TRAIN

Transaxle Assembly

REMOVAL & INSTALLATION

Manual

RIO

1. Before servicing the vehicle, refer to the precautions at the beginning of this section.

2. Drain the transaxle oil.

3. Disconnect the negative, then positive battery cables.

4. Remove the coolant reservoir tank. Position it aside for access to the battery bracket.

5. Remove the battery bracket.

6. Remove the battery, battery tray and fixing holder.

7. Remove the fresh air duct and air cleaner assembly.

8. Remove the hose from the intake manifold.

9. Remove the Manifold Absolute Pressure (MAP) sensor connector.

10. Remove the air temperature sensor connector.

11. Remove the back-up switch connector from the connector bracket case.

12. Remove the Vehicle Speed Sensor (VSS) connector from the right side of the transaxle.

13. Remove the ground strap bolt and strap.

14. Remove the Crankshaft Position (CKP) sensor connector.

15. Remove the release lever, position the lever and clutch line aside.

16. Remove the upper starter mounting bolt.

17. Remove the 3 clutch housing bolts.

18. Support the engine with a suitable support bar.

19. Remove the front wheels.

20. Remove the splash shields.

21. Remove the engine mounting member. The member is secured with 2 bolts and 4 nuts.

22. Remove the extension bar and shift control rod.

23. Remove the bolts from the No. 1 and No. 2 engine mounts and the mounts. There are 2 bolts for the No. 1 mount and 3 bolts for the No. 2 mount.

24. Remove the tension rod mounting nuts.

25. Remove the stabilizer bar and control link.

26. Separate the tie rod ends from the steering knuckle.

27. Remove the ball joint bolt and nut from the control arm, then separate the ball joint from the control arm.

28. Remove the halfshaft from the transaxle. Support the halfshaft with wire to prevent it from hanging unsupported.

29. Remove the starter lower mounting bolt and starter.

30. Remove the 3 lower clutch housing bolts, support the transaxle with a jack, then remove the 2 remaining bolts.

31. Carefully separate the transaxle from the engine and lower it.

To install:

32. Raise the transaxle into position and seat it against the back of the engine. Install the transaxle-to-engine bolts and tighten as follows:

    a. Upper bolts (1–4): 47–66 ft. lbs. (64–89 Nm).

    b. Lower bolts (5–8): 28–38 ft. lbs. (37–51 Nm).

33. Remove the jack from the transaxle.

34. Install the starter and lower mounting bolt.

35. Install a new clip on the driveshaft.

36. Install the driveshaft into the transaxle with the opening of the clip pointing upward.

37. Install the lower ball joint into the steering knuckle. Torque the nut to 43–50 ft. lbs. (54–68 Nm).

38. Install the tie rod end to the steering knuckle. Torque the nut to 22–33 ft. lbs. (30–44 Nm). Install a new cotter pin.

39. Install the stabilizer bar and control link. Torque the retainers to 32–45 ft. lbs. (43–61 Nm).

40. Install the tension rod mounting nut and tighten to 87–108 ft. lbs. (118–147 Nm).

41. Install the No. 2 engine mount. Tighten the bolt to 32–40 ft. lbs. (43–54 Nm) and the nut to 49–69 ft. lbs. (67–93 Nm).

42. Install the No. 1 engine mount and tighten to 32–40 ft. lbs. (43–54 ft. lbs.).

43. Install the extension bar and shift control rod.

44. Install the transaxle pan drain plug.

45. Install the engine mounting member-to-chassis bolts and nuts. There are 2 bolt and 4 nuts. Torque the engine mounting member bolts and nuts to 47–66 ft. lbs. (64–89 Nm) and the No. 1 and 2 engine mount nuts to 27–38 ft. lbs. (37–52 Nm).

46. Install the splash shields.

47. Install the front wheels.

48. Remove the engine support bar.

49. Install the 3 clutch housing bolts and torque to 66–86 ft. lbs. (89–116 Nm).

50. Install the upper starter mounting bolts and tighten to 27–38 ft. lbs. (37–52 Nm).

51. Install the release lever.

52. Install the CKP sensor electrical connector.

53. Install the ground strap and bolt.

54. Install the VSS and back-up switch connectors.

55. Install the air cleaner assembly.

56. Install the air temperature and MAF sensor connectors.

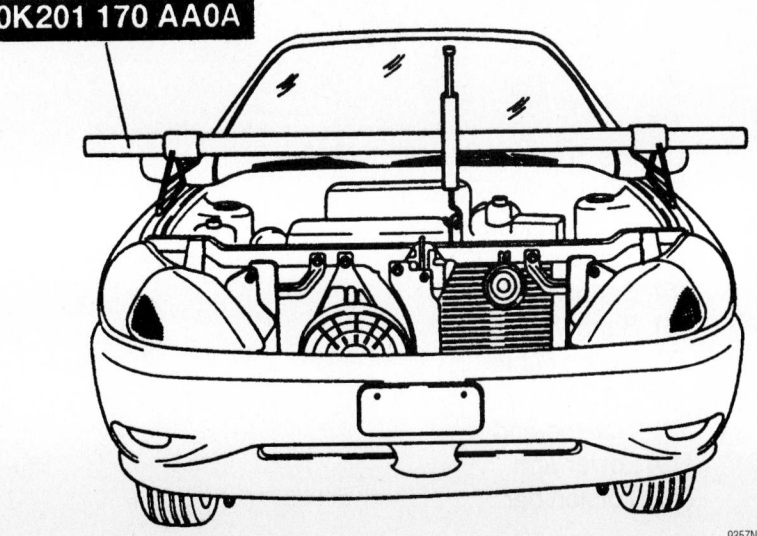

OK201 170 AA0A

9357NG13

Use the proper tool to support the engine—Rio

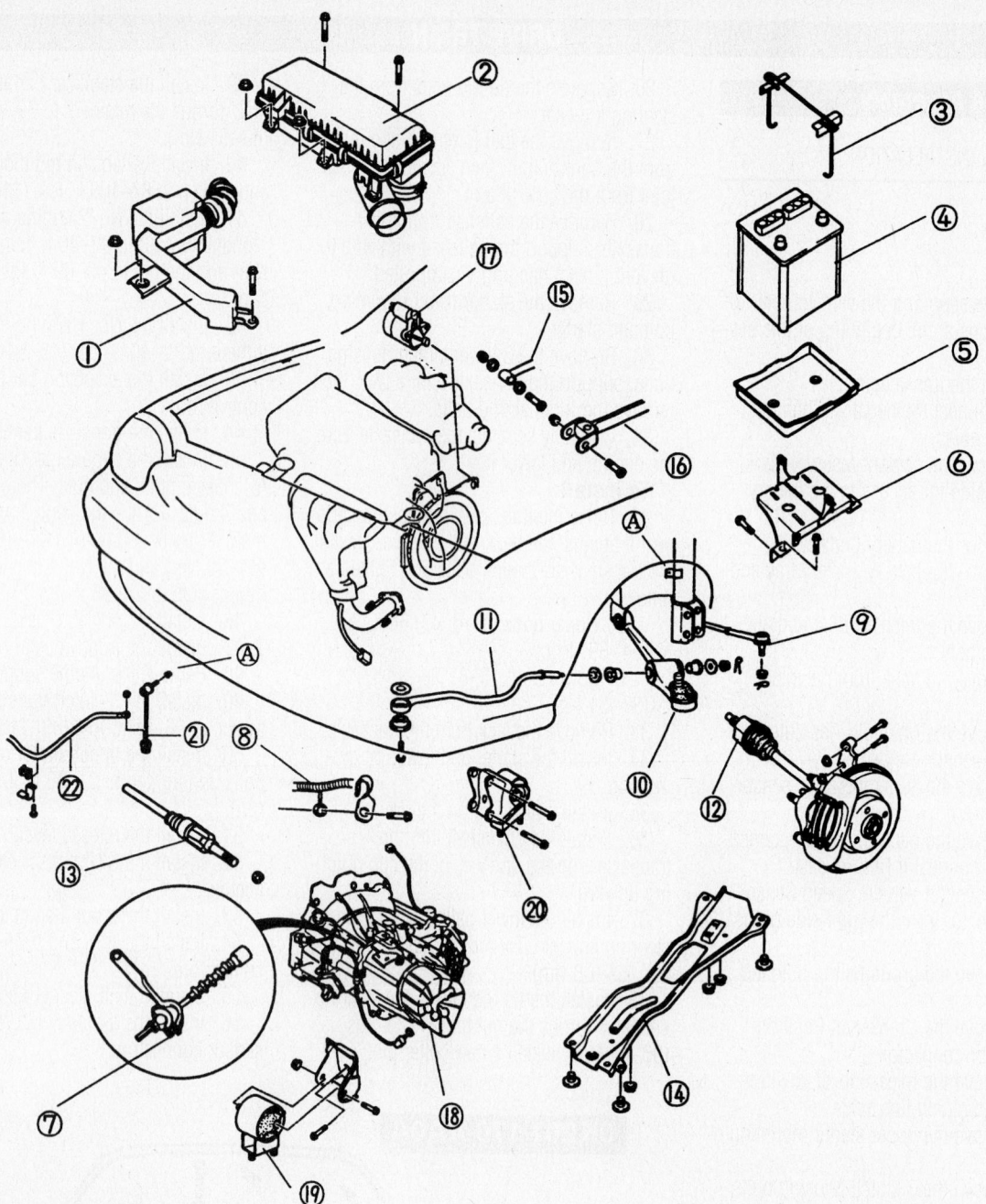

(1) Fresh air duct
(2) Air cleaner assembly
(3) Battery fixing holder
(4) Battery
(5) Battery tray
(6) Battery bracket
(7) Clutch cable
(8) Ground
(9) Tire-rod end
(10) Lower arm
(11) Torsion bar

(12) Driveshaft (Right side)
(13) Driverhaft (Left side)
(14) Engine mounting member
(15) Shift control rod
(16) Extension bar
(17) Starter
(18) Manual transaxle
(19) Engine mounting NO.2
(20) Engine mounting NO.1
(21) Shabilizer control link
(22) Stabilizer bar

9357NG12

Exploded view of the manual transaxle mounting and related components—Rio

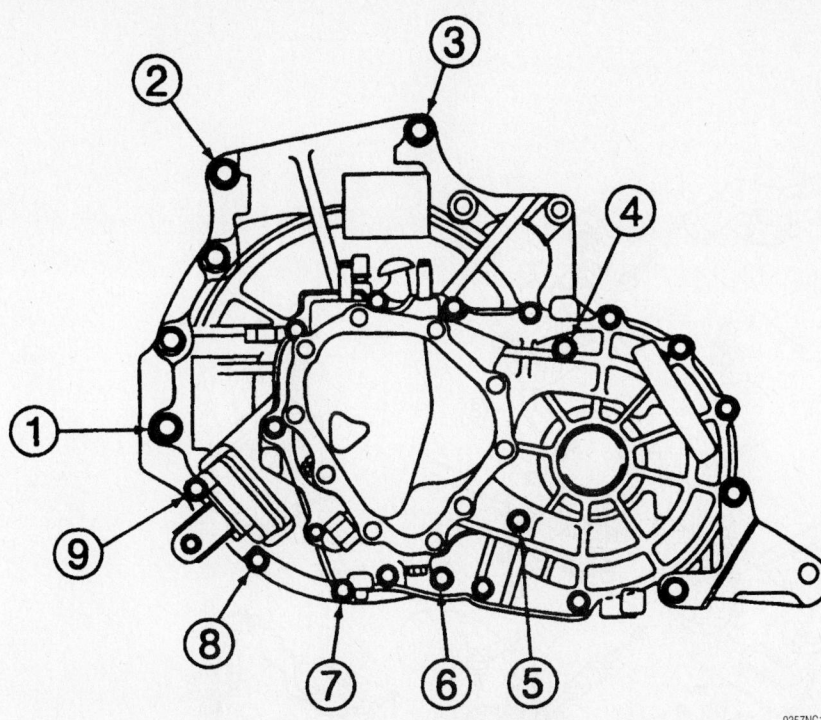

Location of the manual transaxle-to-engine mounting bolts—Rio

9357NG14

57. Install the fresh air duct.

58. Install the hose from the intake manifold.

59. Install the coolant reservoir tank. Position it aside for access to the battery bracket.

60. Install the battery fixing holder, battery tray and battery.

61. Install the battery bracket.

62. Connect the Positive, then negative battery cables.

63. Fill the transaxle with fluid.

64. Check for proper clutch operation.

SEPHIA AND SPECTRA

1. Before servicing the vehicle, refer to the precautions at the beginning of this section.

2. Drain the transaxle oil.

3. Remove the battery box and disconnect the battery.

4. Remove the air cleaner assembly.

5. Remove the battery carrier.

6. Remove the back-up light switch and the bracket.

7. Disconnect the ground strap from the top of the transaxle.

8. Disconnect the neutral switch connector and the vehicle speed sensor (VSS) connector.

9. Remove the wire harness bracket and ground cable.

10. Remove the Crankshaft Position (CKP) sensor.

11. Remove the wheels.

12. Remove the splash shield.

13. Remove the transverse member.

14. Remove the extension bar and the change control rod.

15. Remove the tie-rod ends.

16. Remove the stabilizer control link.

17. Remove the halfshaft and the joint shaft.

18. Remove the intake manifold bracket.

19. Remove the starter.

20. Remove the front exhaust pipe.

21. Support the engine and remove the engine mounting member.

22. Remove the rear engine/transaxle mount.

23. Remove the front engine/transaxle mount.

24. Remove the clutch release cylinder.

25. Remove the side engine/transaxle mount.

26. Support the transaxle on a jack

27. Remove the transaxle mounting bolts.

28. Remove the transaxle.

To install:

29. Install the transaxle into position and install the mounting bolts. Tighten to 28–38 ft. lbs. (37–52 Nm).

30. Remove the support jack.

31. Install the side mount. Tighten the body side nuts and bolts to 32–44 ft. lbs. (44–60 Nm). Tighten the transaxle side nuts to 50–68 ft. lbs. (67–93 Nm).

32. Install the clutch release cylinder.

33. Install the front mount, loosely tighten the mount nut and bolt.

34. Install the rear mount, align and set all bolts, then tighten to 50–68 ft. lbs. (67–93 Nm).

35. Install the engine mounting member. Tighten the 4 outer nuts and bolts to 50–65 ft. lbs. (67–89 Nm) and the 2 remaining nuts to 28–38 ft. lbs. (38–51 Nm).

36. Tighten the front mount nut and bolt to 50–68 ft. lbs. (67–93 Nm).

37. Install the starter.

38. Install the manifold bracket and front exhaust pipe.

39. Install the joint shaft and the half-shaft.

40. Install the stabilizer control link.

41. Install the tie-rod ends.

42. Install the control rod and extension bar.

43. Install the transverse member.

44. Install the splash shield.

45. Install the wheels.

46. Install the harness bracket.

47. Install the CKP sensor.

48. Install the VSS and the neutral switch connectors.

49. Install the back-up light switch connector bracket and the switch.

50. Install the No. 4 engine mount.

51. Install the battery carrier.

52. Install the air cleaner assembly.

53. Install the battery and battery box.

54. Fill the transaxle with fluid.

55. Check for proper clutch operation.

OPTIMA

1. Before servicing the vehicle, refer to the precautions at the beginning of this section.

2. Drain the transaxle oil.

3. Disconnect the negative battery cable.

4. Remove the air duct.

5. Remove the air cleaner and air flow assembly.

6. Remove the back-up light switch connector.

7. Remove the clutch tube and clip.

8. Remove the clutch release cylinder.

9. Disconnect the speedometer cable.

10. Disconnect the select cable and shift cable.

11. Remove the starter mounting bolts.

12. Remove the upper transaxle mounting bolts.

13. Install engine hooks and support the engine with a suitable support bar.

14. Remove the transaxle mounting bracket and insulator.

15. Remove the front wheels.

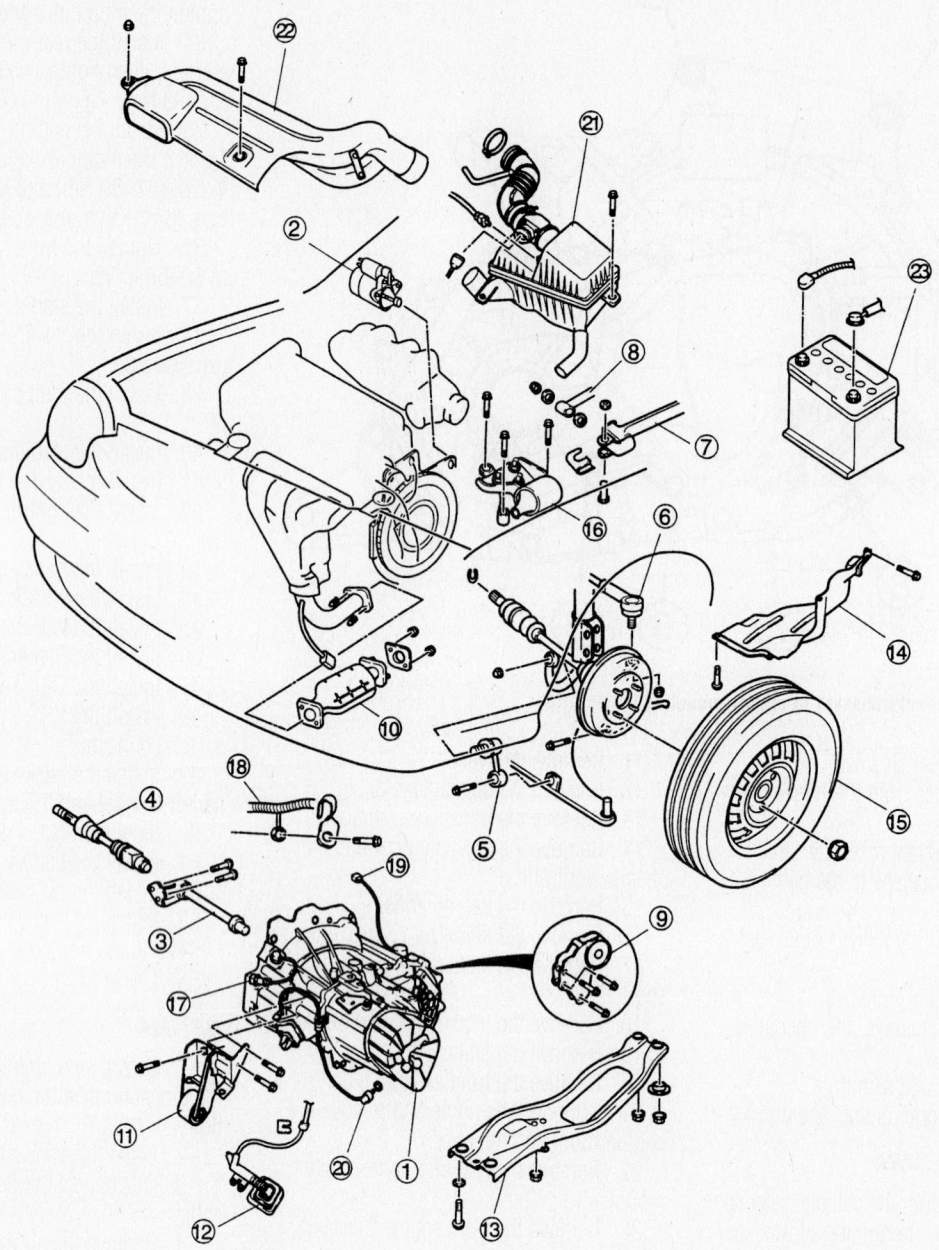

(1) Transaxle
(2) Starter
(3) Joint shaft
(4) Driveshaft
(5) Stabilizer control link
(6) Tie-rod end
(7) Change control rod
(8) Extension bar

(9) Engine mount No. 1
(10) Catalytic converter
(11) Engine mount No. 2
(12) Clutch release cylinder
(13) Engine mount member
(14) Splash shield
(15) Front wheel and tire
(16) Engine mount No. 4

(17) Crankshaft position sensor
(18) Ground
(19) Vehicle speed sensor connector
(20) Back-up switch connector
(21) Air cleaner assembly
(22) Fresh air duct
(23) Battery

9301KG07

Exploded view of the manual transaxle assembly mounting and related components—Sephia

16. Remove the steering gear box U-joint bolt and return tube mounting bolts.

17. Remove the muffler.

18. Remove the sub-frame mounting bolts and sub-frame.

19. Remove the transaxle front and rear mounting bracket.

20. Remove the transaxle side mounting bolts. Support the transaxle with a jack.

21. Remove the transaxle from the vehicle.

22. Installation is the reverse of the removal procedure. Refer to the illustration for tightening specifications.

23. Fill the transaxle with fluid.

24. Check for proper clutch operation.

Automatic

RIO

1. Before servicing the vehicle, refer to the precautions at the beginning of this section.

2. Drain the transaxle fluid.

3. Disconnect the negative, then positive battery cables.

4. Remove the fresh air duct.

5. Disconnect the input/turbine speed sensor, solenoid and transaxle range connector.

6. Remove the ground strap bolt and ground strap from the top of the transaxle.

7. Disconnect the Vehicle Speed Sensor (VSS) connector from the right side of the transaxle.

8. Remove the U-clip connecting the selector cable to the linkage.

9. Remove the nut and washer from the transaxle linkage.

10. Disconnect the Crankshaft Position (CKP) and Oxygen (O$_2$) sensor connectors.

11. Disconnect the 2 transaxle fluid cooler hoses.

12. Remove the top 2 upper converter housing bolts.

13. Support the engine with a suitable support bar.

14. Remove the front wheels.

15. Remove the splash shield.

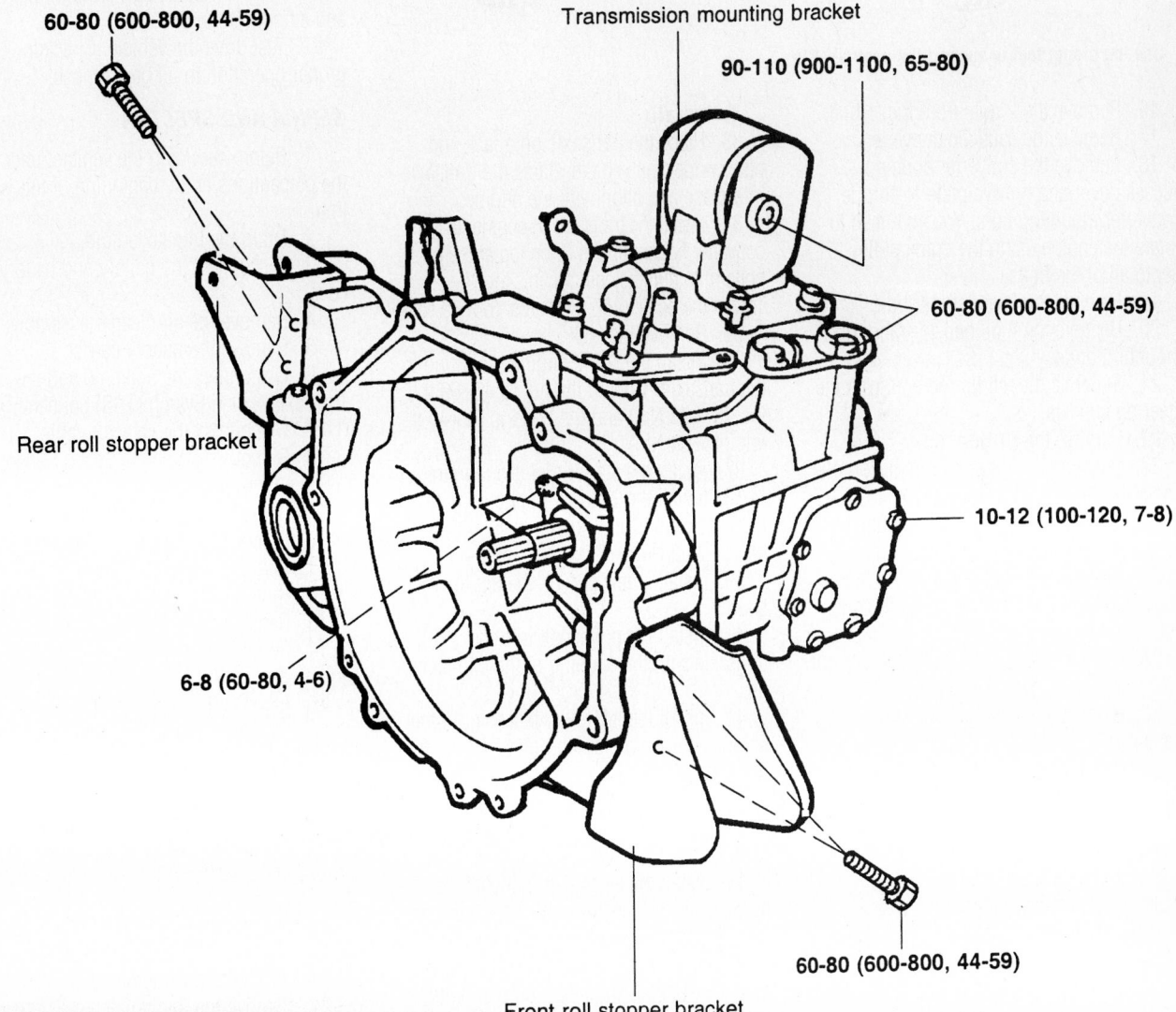

60-80 (600-800, 44-59)

Transmission mounting bracket

90-110 (900-1100, 65-80)

60-80 (600-800, 44-59)

Rear roll stopper bracket

10-12 (100-120, 7-8)

6-8 (60-80, 4-6)

60-80 (600-800, 44-59)

Front roll stopper bracket

TORQUE : Nm (kg·cm, lb·ft)

9357NG46

Manual transaxle mounting and tightening specifications—Optima

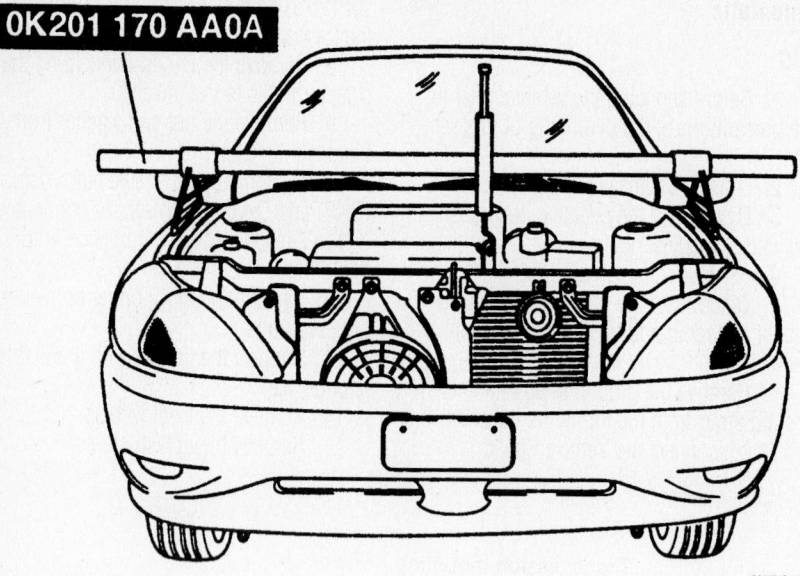

0K201 170 AA0A

Use the proper tool to support the engine—Rio

9357NG13

16. Remove the 2 nuts from the U-bolt.

17. Remove the catalytic converter.

18. Remove the converter housing access cover and 4 drive plate-to-torque converter mounting nuts. You will need to rotate the engine using the crank pulley to get to all of the bolts.

19. Remove the lower starter bolt.

20. Remove the 4 oil pan-to-transaxle mounting bolts.

21. Separate the left tie rod end from the steering knuckle.

22. Remove the tension rod.

23. Separate the lower ball joint from the control arm.

24. Remove the No. 1 and 2 mounting nuts from the engine mounting member.

25. Remove the engine mounting member-to-chassis bolts and nuts and the member.

26. Remove the stabilizer bar and control link.

27. Remove the left halfshaft from the transaxle.

28. Install special tool 0K201 270 014 to prevent the side gear from becoming misaligned.

29. Remove the No. 2 engine mount from the transaxle.

30. Support and secure the transaxle with a suitable jack.

31. Remove the 2 remaining converter housing bolts from the front and rear sides of the transaxle

32. Remove the transaxle. Slowly lower the drivetrain, letting the transaxle tilt toward the ground. Carefully separate the transaxle from the engine and pull the unit out through the wheel well.

To install:

33. Place the transaxle on a jack and place under the vehicle. Raise the transaxle into place and align with the engine.

34. Install the transaxle to engine, using 4 converter housing bolts (2 on top and 2 on bottom) to pull the components together. Torque the bolts to 47–66 ft. lbs. (64–89 Nm).

35. Remove the jack.

36. Install the No. 2 engine mounting to the transaxle. Tighten the nut to 49–69 ft. lbs. (67–93 Nm) and the bolts to 32–40 ft. lbs. (43–54 Nm).

37. Install the starter and ground strap.

38. Install the 3 No. 1 engine mounting-to-chassis bolts and tighten to 32–46 ft. lbs. (43–52 Nm).

39. Install a new clip on the left driveshaft.

40. Install the driveshaft into the transaxle with the opening of the clip facing up.

41. Install the left lower ball joint to the spindle. Torque the pinch bolt to 40–50 ft. lbs. (54–68 Nm).

42. Install the stabilizer bar and control link and tighten to 32–45 ft. lbs. (43–61 Nm).

43. Install the tension rod. Install the front part of the tension rod to the body after inserting the bushing into the tensioner rod. Tighten to 87–108 ft. lbs. (118–147 Nm).

44. Install the engine-to-oil pan bolts and access cover. Torque the bolts to 27–38 ft. lbs. (37–52 Nm).

45. Install the drive plate-to-torque converter mounting nuts and torque to 25–36 ft. lbs. (34–49 Nm). Rotate the engine with the crank pulley to get to all of the nuts.

46. Install the engine mounting member-to-chassis nuts and bolts and tighten to 48–65 ft. lbs. (64–89 Nm).

47. Install the 2 nuts to the No. 1 and 2 engine mount-to-mounting member and tighten to 28–38 ft. lbs. (38–51 Nm).

48. Install the catalytic converter and U-clip. Torque to 27–38 ft. lbs. (37–52 Nm).

49. Install the splash shield.

50. Install the front wheels.

51. Connect the transaxle fluid hoses onto the cooler.

52. Remove the engine support bar.

53. Install the CKP and O_2 sensor connectors.

54. Install the solenoid valve connector.

55. Install the remaining components in the reverse of the removal procedure.

56. Fill the transaxle with the proper type and amount of fluid.

57. Test drive the vehicle. Check for proper operation in all gear ranges.

SEPHIA AND SPECTRA

1. Before servicing the vehicle, refer to the precautions at the beginning of this section.

2. Drain the transaxle fluid.

3. Disconnect the battery and battery cover.

4. Remove the air cleaner assembly.

5. Remove the battery carrier.

6. Disconnect the solenoid and the Transaxle Range Switch (TRS) connectors.

7. Disconnect the selector cable.

8. Remove the Vehicle Speed Sensor (VSS) connector.

9. Remove the harness bracket.

10. Disconnect the throttle cable.

11. Remove the front wheels.

12. Remove the splash shields.

13. Remove the front exhaust pipe.

14. Remove the transverse member, if equipped.

15. Remove the tie-rod ends.

16. Remove the stabilizer control links.

17. Remove the lower arm by removing the cinch bolt from the lower arm ball joints. Pry the lower arm out of the knuckle.

18. Support the engine.

19. Remove the engine mounting member.

20. Remove the left and right halfshaft. Install Differential Side Gear holder K49A-4208-AT to hold the side gears.

21. Remove the joint shaft.

22. Remove the intake manifold bracket.

23. Remove the starter.

24. Remove the front engine/transaxle mount.

25. Remove the rear engine/transaxle mount.

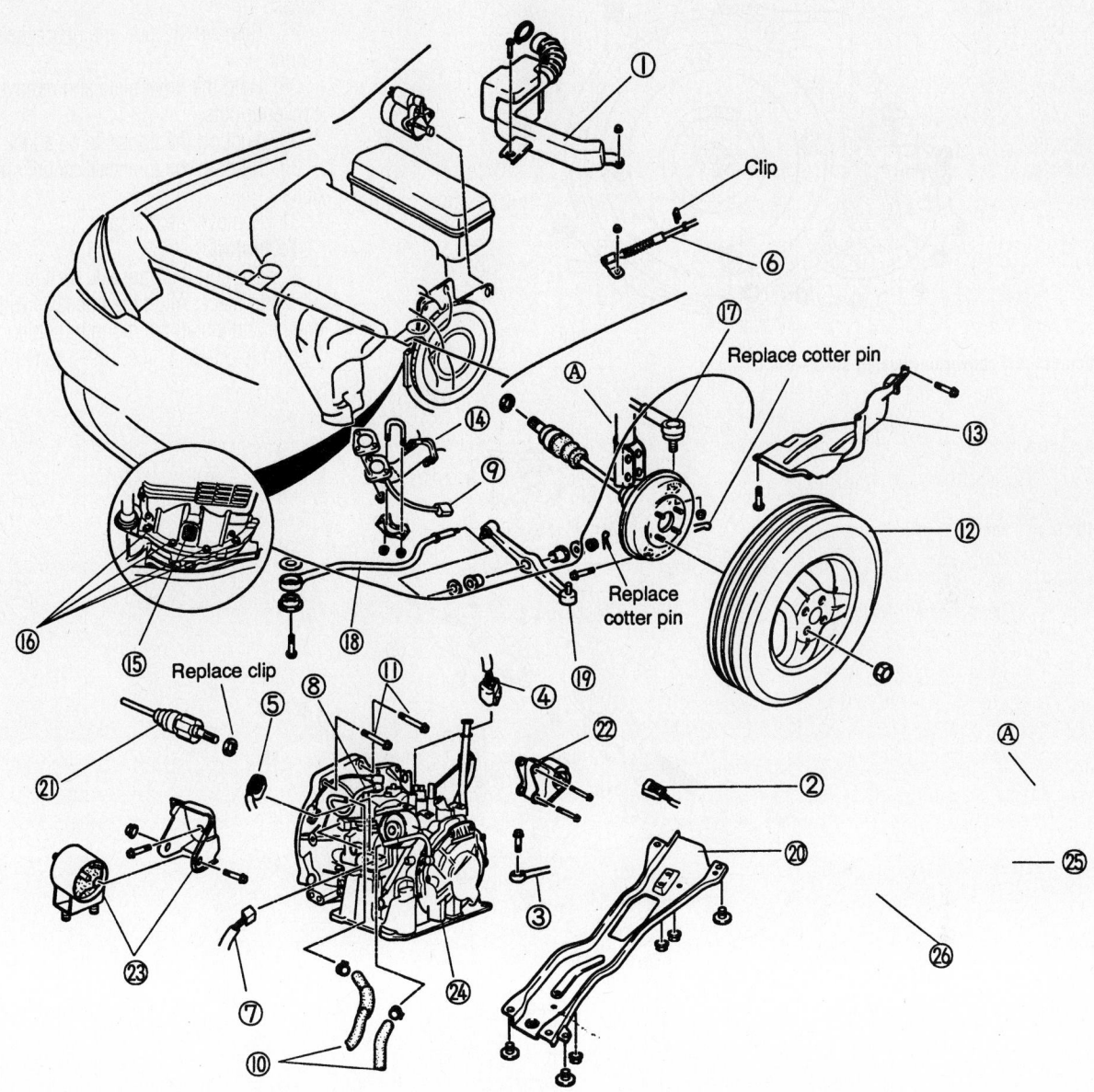

Exploded view of the automatic transaxle mounting—Rio

(1) Air cleaner assembly
(2) Input/turbine speed sensor connector
(3) Ground strap bolt
(4) Vehicle speed sensor connector
(5) Transaxle range switch connector
(6) Selector cable
(7) Solenoid valve connector
(8) Crankshaft position connector
(9) Oxygen sensor connector
(10) ATF cooler hose
(11) Upper converter housing bolts
(12) Wheel and tire
(13) Splash shield

(14) Catalytic converter
(15) Access cover
(16) Engine oil pan-to-transaxle mounting bolt
(17) Tie-rod end
(18) Tension rod
(19) Ball joint
(20) Engine mounting member
(21) Driveshaft
(22) No.1 engine mounting
(23) No.2 engine mounting
(24) Auto transaxle
(25) Stabilizer control link
(24) Stabilizer bar

9357NG15

Location of the 4 converter housing bolts—Rio

26. Remove the inner and outer oil hoses.

27. Remove the side engine/transaxle mount.

28. Hold the drive plate and remove the converter nuts.

29. Support the transaxle on a jack.

30. Remove the transaxle mounting bolts.

31. Remove the transaxle.

To install:

32. Support the transaxle on a jack and lift it into place. Align the transaxle with the engine and install the mounting bolts. Tighten to 65–86 ft. lbs. (89–116 Nm).

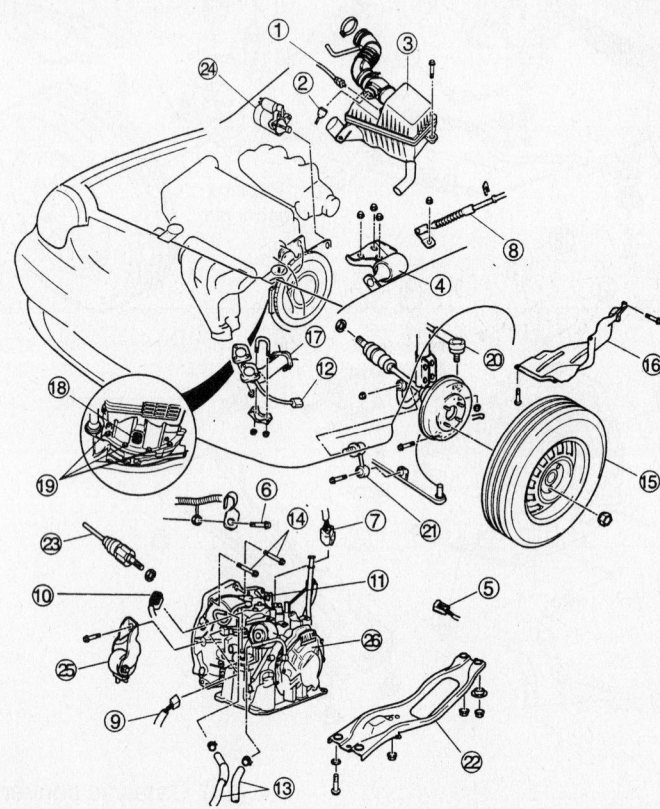

(1) Air temperature sensor connector
(2) MAF sensor connector
(3) Air cleaner assembly
(4) No. 4 mounting
(5) Input/turbine speed sensor connector
(6) Ground strap bolt
(7) Vehicle speed sensor connector
(8) Selector cable
(9) Transaxle range switch connector
(10) Solenoid valve connector
(11) Crankshaft position connector
(12) Oxygen sensor connector
(13) ATF cooler hose
(14) Upper converter housing bolts
(15) Wheel and tire

(16) Gravel shield
(17) Catalytic converter
(18) Converter housing
(19) Engine oil pan-to-transaxle mounting bolt
(20) Tie-rod end
(21) Stabilizer control link
(22) Engine mounting member
(23) Driveshaft
(24) Starter
(25) No. 2 engine mounting
(26) Auto transaxle

Exploded view of the automatic transaxle assembly mounting and related components—Sephia

9301KG08

33. Hold the driveplate and install the torque converter mount nuts. 26–36 ft. lbs. (35–49 Nm).

34. Install the side engine/transaxle mount. Loosely tighten the nuts of the transaxle side. Tighten the nuts and bolts of the body side to 32–44 ft. lbs. (44–60 Nm). Tighten the nuts of the transaxle side to 50–68 ft. lbs. (67–93 Nm).

35. Install the inner and outer oil hoses.

36. Install the rear engine/transaxle mount. Tighten the bolts to 50–68 ft. lbs. (67–93 Nm).

37. Install the front engine/transaxle mount. Tighten the mount bracket to the transaxle to 28–38 ft. lbs. (38–51 Nm). Loosely tighten the nuts and bolts of the engine mount rubber, then tighten to 50–68 ft. lbs. (67–98 Nm).

38. Install the starter.

39. Install the manifold bracket.

40. Install the joint shaft into the transaxle.

41. Install the joint shaft to the cylinder block and tighten the bolts in sequence (counterclockwise). Tighten to 32–46 ft. lbs. (42–62 Nm).

42. Install the halfshafts, be sure that the shafts are properly installed and do not pull out.

43. Install the engine mounting member, the mounting nuts/bolts and tighten the nuts/bolts at the far corners to 48–65 ft. lbs. (64–89 Nm), tighten the remaining 2 nuts to 28–38 ft. lbs. (38–51 Nm).

44. Install the front exhaust pipe.

45. Install the lower arm to the knuckle.

46. Install the stabilizer control link.

47. Install the tie-rod ends.

48. Install the transverse member, if removed.

49. Install the splash shields.

50. Install the wheels.

51. Connect the throttle cable.

52. Install the harness bracket.

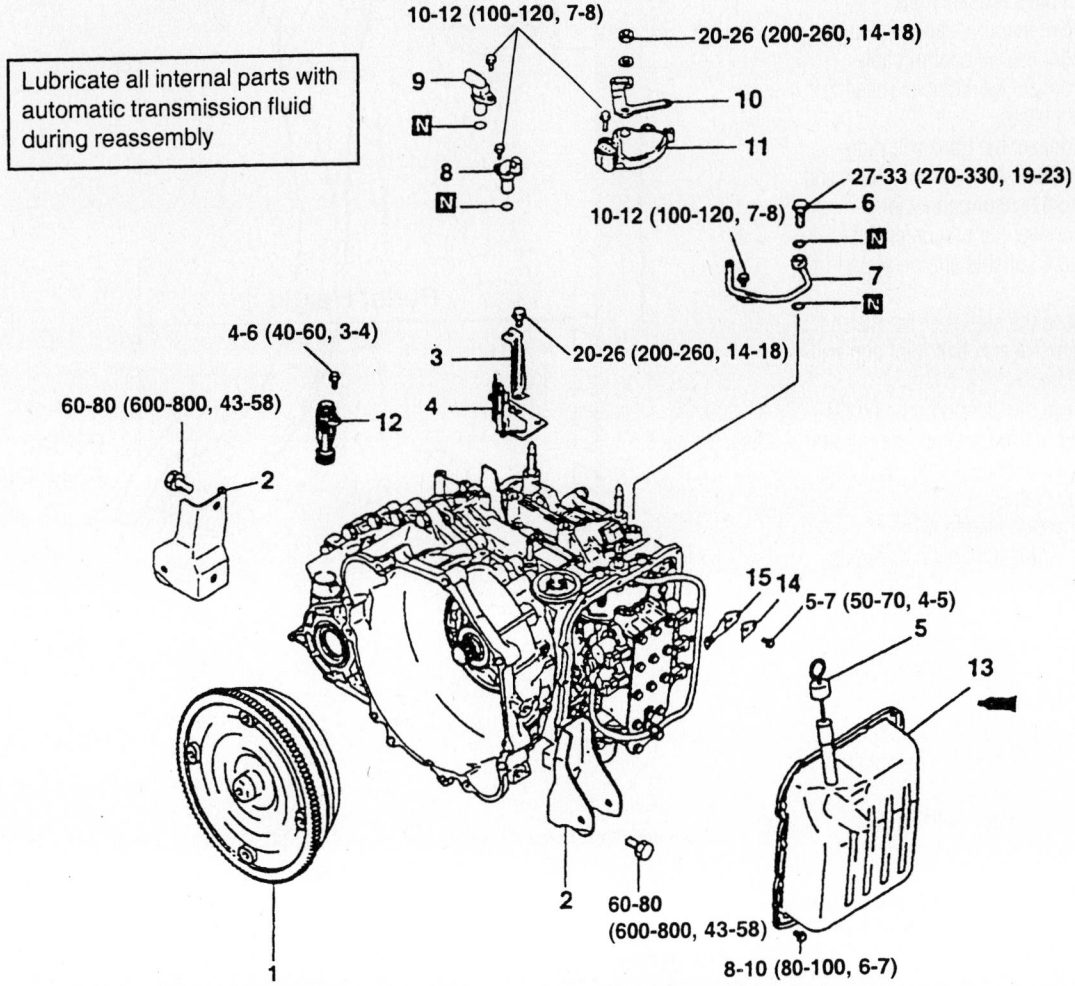

Lubricate all internal parts with automatic transmission fluid during reassembly

1. Torque converter
2. Roll stopper bracket
3. Harness bracket
4. Control cable support bracket
5. Oil level gauge
6. Eye bolt
7. Oil cooler feed tube
8. Input shaft speed sensor
9. Output shaft speed sensor
10. Manual control lever
11. Transaxle range switch
12. Vehicle speed sensor
13. Valve body cover
14. Manual control shaft detent spring
15. Manual control shaft detent

TORQUE : Nm (kg·cm, lb·ft)

Automatic transaxle mounting and tightening specifications—Optima

9357NG47

53. Connect the VSS connector.
54. Connect the selector cable.
55. Connect the TRS connector.
56. Connect the solenoid connector.
57. Install the battery carrier.
58. Install the air cleaner assembly.
59. Install the battery and battery cover.
60. Fill the transaxle with the proper type and amount of fluid.
61. Test drive the vehicle. Check for proper operation in all gear ranges.

OPTIMA

1. Before servicing the vehicle, refer to the precautions at the beginning of this section.
2. Drain the transaxle fluid.
3. Remove the air cleaner assembly.
4. Disconnect the control cable.
5. Disconnect the Vehicle Speed Sensor (VSS) connector.
6. Disconnect the transaxle range switch connector, solenoid connector and oil temperature sensor connector.
7. Disconnect the oil cooler hose.
8. Install a suitable engine support to the engine.
9. Remove the stabilizer bar, tie rod end, lower control arm ball joint and half-shafts.
10. Remove the steering gear box U-joint bolt and return tube mounting bolts.
11. Remove the sub-frame mounting bolts and sub-frame.
12. Remove the starter.
13. Remove the automatic transaxle mounting bolts.
14. Remove the engine-to-transaxle bolts. Place a jack under the transaxle.
15. Remove the transaxle from the vehicle.
16. Installation is the reverse of the removal procedure. Refer to the illustration for tightening specifications.
17. Fill the transaxle with fluid.
18. Check for proper clutch operation.

Clutch

ADJUSTMENTS

Pedal Height

1. Before servicing the vehicle, refer to the precautions at the beginning of this section.
2. Measure the distance from the upper surface of the pedal pad to the carpet.
3. The distance should be as follows:
   a. Sephia, Spectra and Optima: 7.83–8.15 in. (199–207mm).

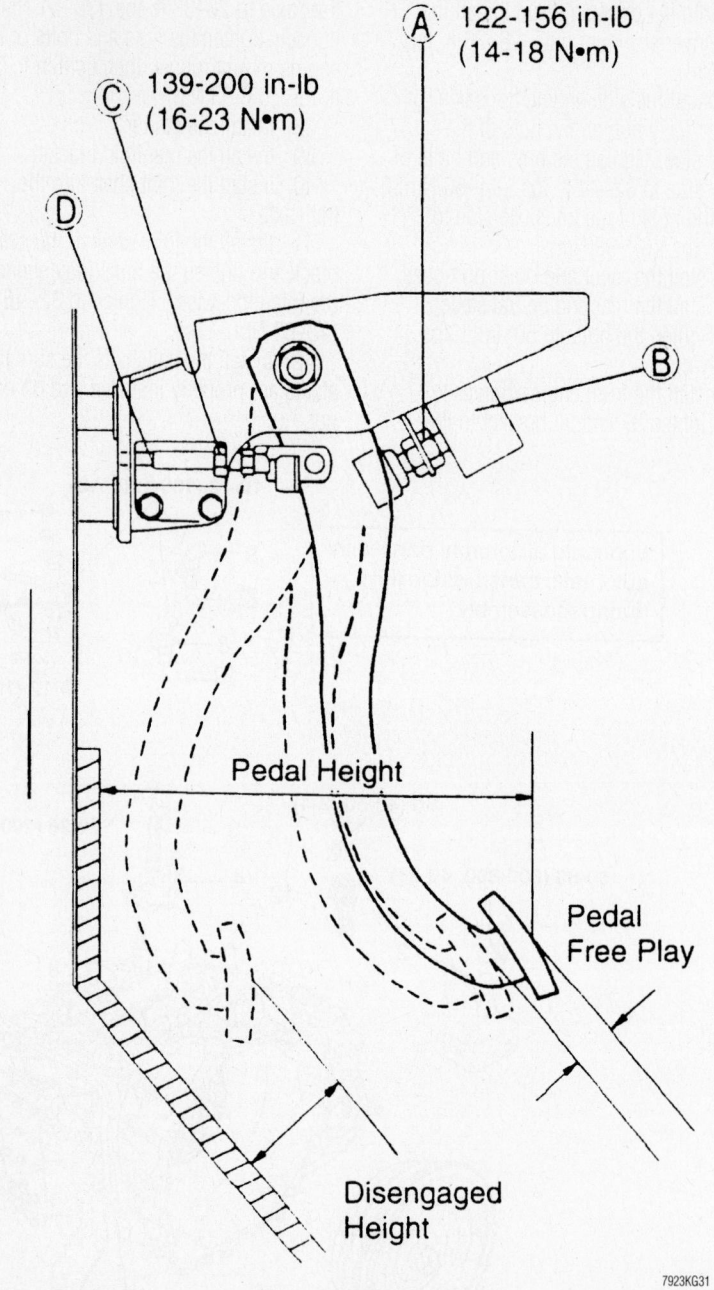

Clutch pedal measurement and adjustment points. (A) and (B) are for adjusting the pedal height, while (C) and (D) are for the free-play adjustment

   b. Rio: 7.67 in. (195mm).
4. If the distance is not as specified, loosen the locknut on the stopper bolt or switch.
5. Turn the switch or bolt until the distance is correct, then tighten the locknut.

Free-Play

1. Before servicing the vehicle, refer to the precautions at the beginning of this section.
2. Depress the clutch pedal by hand until resistance is felt. The free-play should be as follows:

   a. Sephia and Spectra: 0.12–0.20 in. (3.0–5.0mm).
   b. Rio: 0.35–0.59 in. (9–15mm).
   c. Optima: 0.2–0.5 in. (6–13mm).
3. If the free-play is not correct, loosen the clutch master cylinder pushrod locknut and turn the pushrod to adjust.

REMOVAL & INSTALLATION

1. Before servicing the vehicle, refer to the precautions at the beginning of this section.

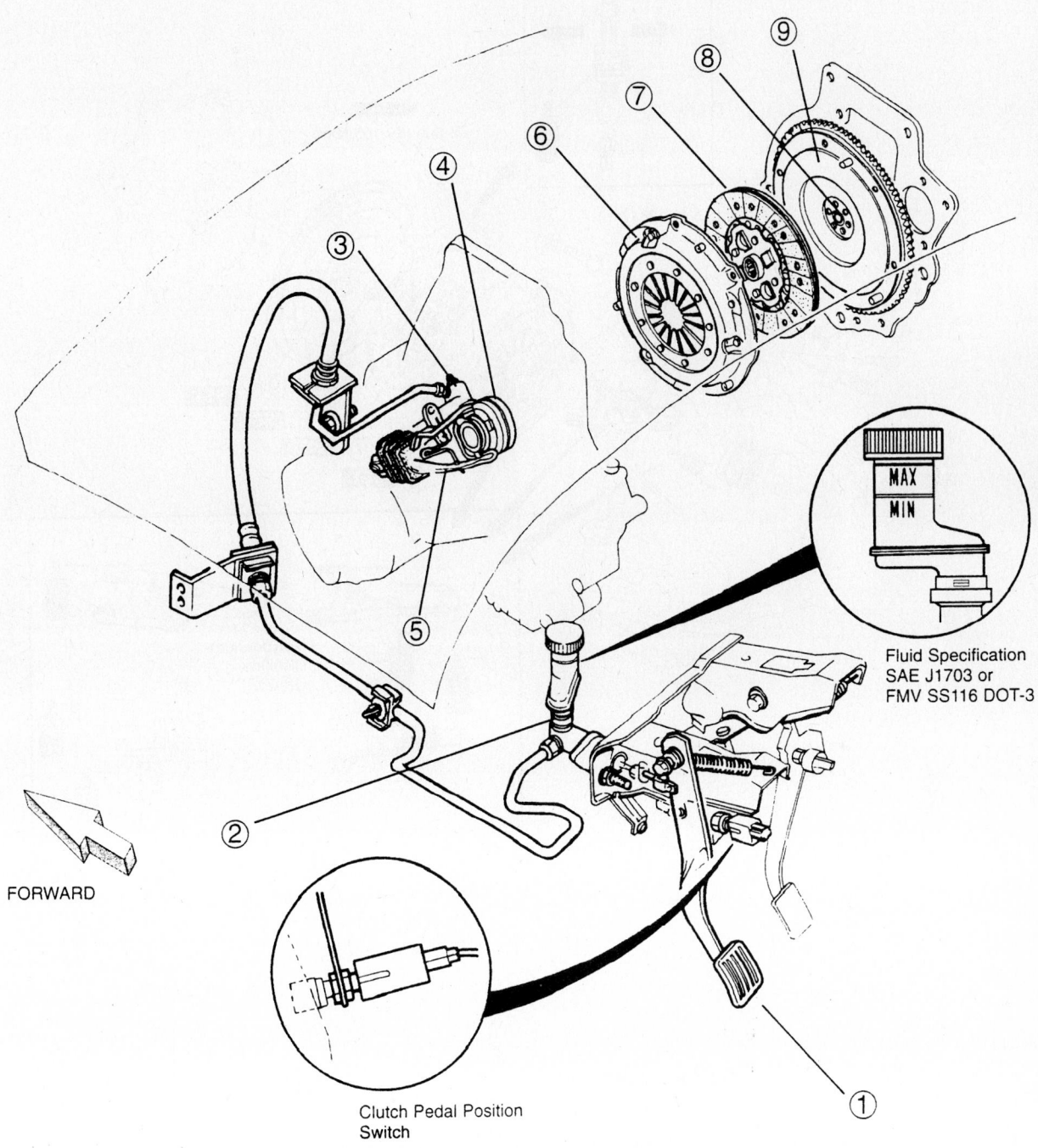

FORWARD

MAX
MIN

Fluid Specification
SAE J1703 or
FMV SS116 DOT-3

Clutch Pedal Position
Switch

1. Clutch pedal
2. Clutch master cylinder
3. Clutch release cylinder
4. Release bearing
5. Clutch release fork

6. Clutch cover
7. Clutch disc
8. Pilot bearing
9. Flywheel

7923KG32

Structural view of the hydraulic clutch system

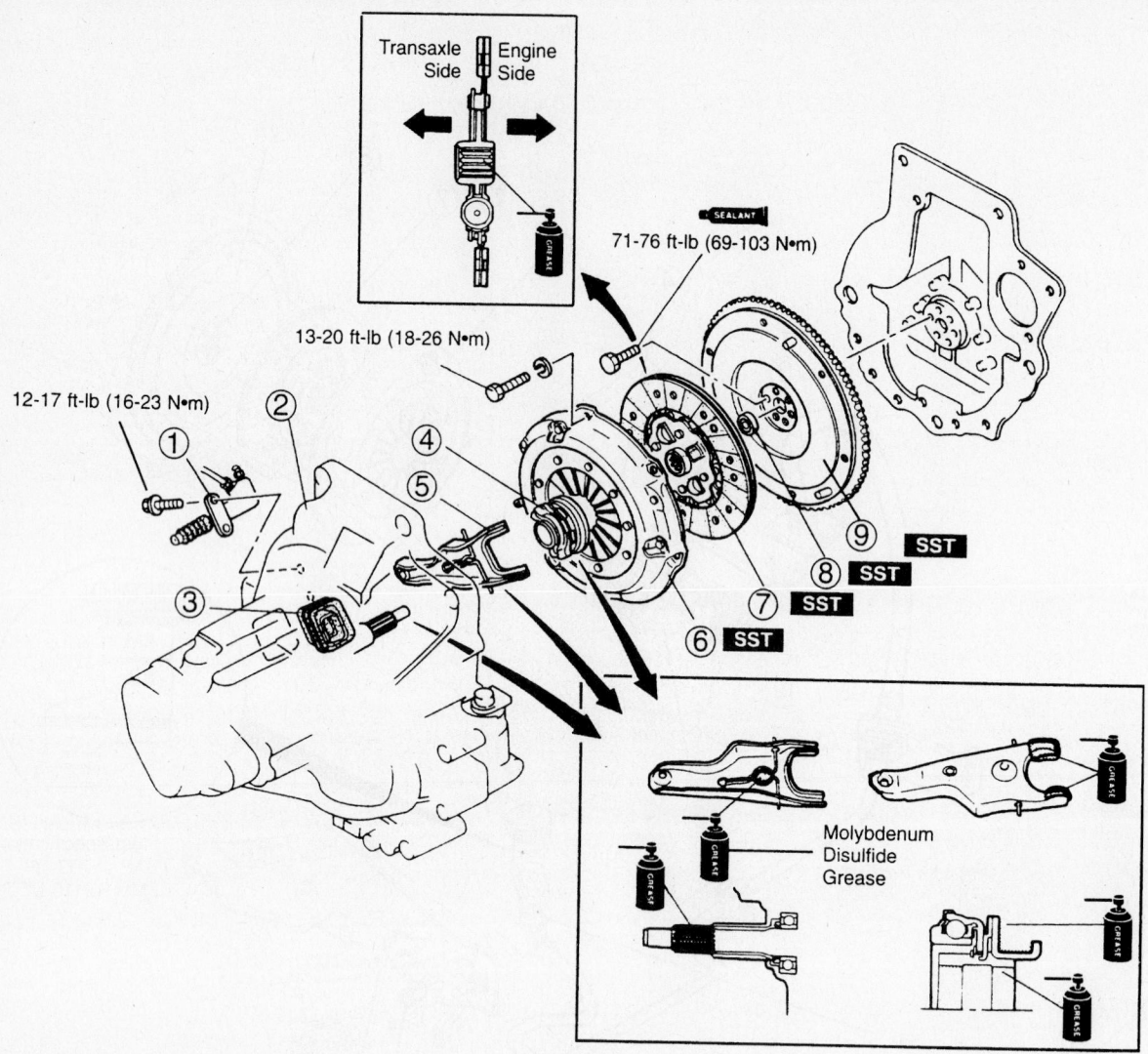

1. Clutch release cylinder
2. Transaxle housing
3. Boot
4. Release bearing
5. Clutch release fork

6. Clutch cover
7. Clutch disc
8. Pilot bearing
9. Flywheel

7923KG33

Exploded view of the clutch assembly

2. Disconnect the negative battery cable.

3. Remove the transaxle.

4. Gradually loosen the clutch pressure plate bolts, in a crisscross pattern. Support the pressure plate and remove the bolts. Remove the pressure plate and clutch disc.

5. Inspect the pilot bearing. If it is worn or damaged and does not turn easily by hand, remove it using a puller/slide hammer.

6. Check the flywheel surface for scoring, cracks or burning and machine or replace, as necessary.

7. Install a flywheel holder to keep the flywheel from turning. Loosen the flywheel bolts evenly and gradually in a crisscross pattern. Remove the flywheel.

8. Inspect the clutch release bearing for wear. Replace it if it sticks or does not turn easily.

9. Inspect the release fork for wear or damage and replace as necessary.

To install:

10. Lubricate the release fork fingers and pivot with molybdenum grease and install in the release fork boot.

11. Install the clutch release bearing on the release fork.

12. Install a new pilot bearing in the flywheel, if removed.

13. Be sure the flywheel mounting surface and the crankshaft or eccentric shaft mounting surfaces are clean. Remove any old sealant from the flywheel bolt hole threads and the flywheel bolts.

14. Install the flywheel.

15. Apply sealant to the flywheel bolt threads and install them hand-tight. Install the flywheel holding tool. Tighten the bolts, in a crisscross pattern, to 71–76 ft. lbs. (96–103 Nm).

16. Apply a small amount of molybdenum grease to the clutch disc splines and

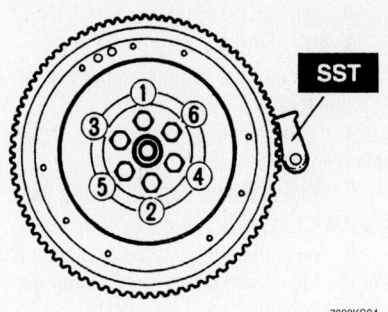

Flywheel tightening sequence

7923KG34

install the clutch disc on the flywheel, spring side toward the transaxle. Install a suitable alignment tool in the pilot bearing to position the clutch disc.

17. Install the clutch pressure plate, aligning the dowel holes with the flywheel dowels.

18. Install the pressure plate bolts and gradually tighten, in a crisscross pattern to 20 ft. lbs. (26 Nm). Remove the alignment tool.

19. Install the transaxle.

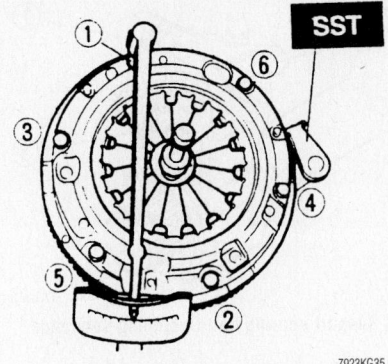

Pressure plate tightening sequence

7923KG35

Hydraulic Clutch System

BLEEDING

1. Before servicing the vehicle, refer to the precautions at the beginning of this section.

2. If necessary, remove the gravel shield from the drivers side.

3. Remove the rubber cap from the bleeder screw on the release cylinder.

4. Place a bleeder tube over the end of the bleeder screw.

5. Submerge the other end of the tube in a jar half filled with hydraulic brake fluid.

6. Slowly pump the clutch pedal fully and allow it to return slowly, several times.

7. While pressing the clutch pedal to the floor, loosen the bleeder screw until the fluid starts to run out. Then, close the bleeder screw. Keep repeating this Step, while watching the hydraulic fluid in the jar. As soon as the air bubbles disappear, close the bleeder screw.

8. During the bleeding procedure the reservoir must be kept at least ¾ full.

Halfshaft

REMOVAL & INSTALLATION

1. Before servicing the vehicle, refer to the precautions at the beginning of this section.

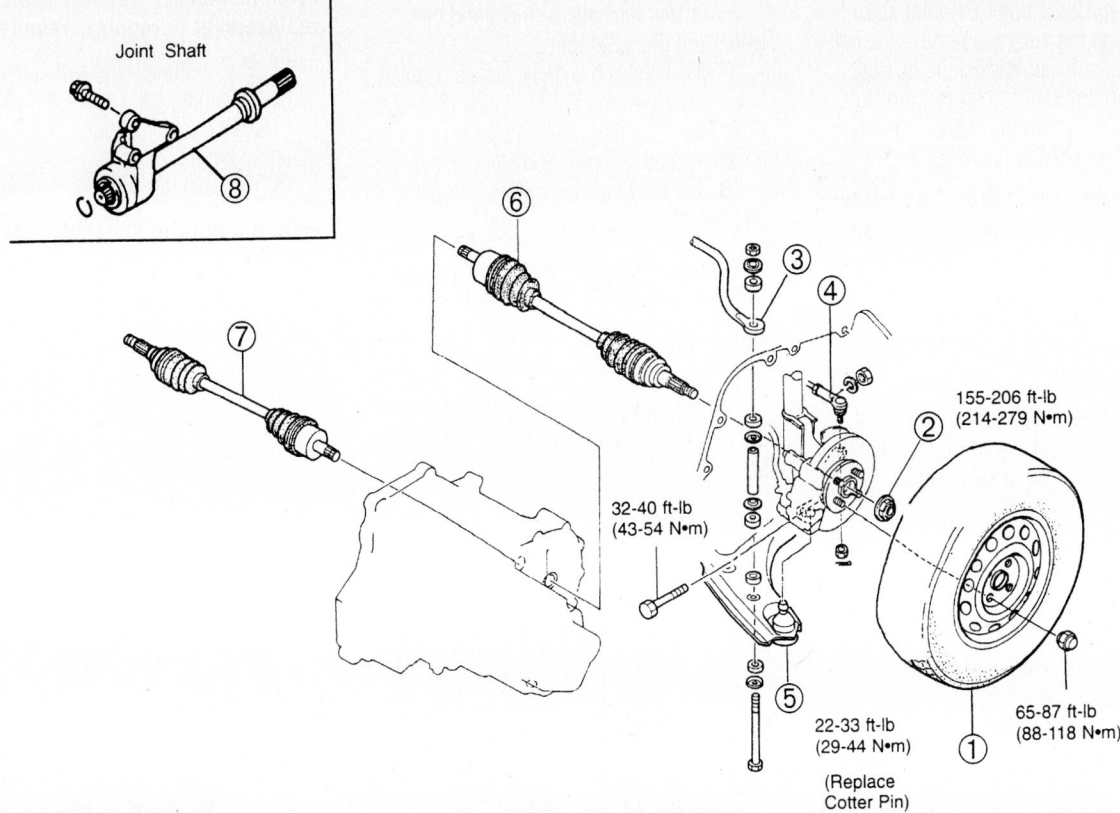

1. Wheel and tire
2. Locknut
3. Stabilizer bar
4. Tie-rod end
5. Ball joint
6. Left driveshaft
7. Right driveshaft
8. Joint shaft

7923KG36

Exploded view of the halfshafts and related components

2. Remove the wheel and tire assemblies.

3. Remove the splash shield, if equipped.

4. Drain the transaxle.

5. Raise the staked portion of the hub locknut with a hammer and chisel. Lock the hub by applying the brakes and loosen the nut.

6. Separate the stabilizer bar from the lower control arm.

7. Remove the cotter pin and nut from the tie rod end ball stud.

8. Remove the tie rod end from the knuckle, using a suitable tool.

9. Remove the lower ball joint pinch bolt and nut.

10. Separate the ball joint from the knuckle, using a prybar on the control arm.

11. Position a prybar between the inner CV-joint and transaxle case. Carefully pry the halfshaft from the transaxle being careful not damage the oil seal. If equipped with a right side intermediate shaft, insert the prybar between the halfshaft and intermediate shaft and tap on the bar to uncouple them.

12. Remove the hub locknut and discard.

13. Pull outward on the hub/knuckle assembly, push the outer CV-joint stub shaft through the hub, and remove the halfshaft. If the halfshaft is stuck in the hub, install the old hub nut to protect the stub shaft threads. Tap on the nut, using only a soft mallet, to remove the halfshaft.

➡**Install Differential Side Gear holder K49A-4208-AT, into the transaxle after removing the halfshaft, to keep the differential side gear in position. If the gear becomes out of position the differential may have to be removed to realign the gear.**

14. Remove the intermediate shaft, if necessary, by removing the support bearing bolts and pulling the shaft from the transaxle.

To install:

15. Install a new circlip on the end of the intermediate shaft, with the end gap facing upward, if removed.

16. Install the intermediate shaft in the transaxle, being careful not to damage the oil seals.

17. Install the support bearing bolts and tighten, in sequence, to 45 ft. lbs. (61 Nm).

18. Install a new circlip on the end of the halfshaft, with the end gap facing upward.

19. Install the halfshaft into the transaxle, being careful not to damage the oil seal.

20. Install the halfshaft into the intermediate shaft, if equipped.

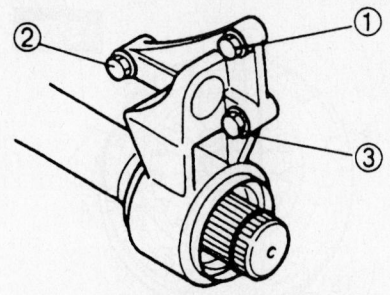

7923KG37

Support bearing bolt tightening sequence

21. Install the other end of the halfshaft through the hub. Loosely install a new locknut.

22. Install the lower ball joint into the knuckle.

23. Install the pinch bolt and nut and tighten to 40 ft. lbs. (54 Nm).

24. Install the tie rod end to the steering knuckle and tighten the nut to 42 ft. lbs. (57 Nm). Install a new cotter pin. Tighten the nut, if necessary, to align the ball stud hole with the nut castellation.

25. Install the stabilizer bar to the lower control arm.

26. Install the splash shield and the wheel and tire assemblies.

27. Lock the hub with the brakes. Tighten the new hub nut to 155–206 ft. lbs. (214–279 Nm). After tightening, stake the locknut using a hammer and dull bladed chisel.

28. Fill the transaxle with the proper type and quantity of fluid.

CV-Joints

OVERHAUL

1. Before servicing the vehicle, refer to the precautions at the beginning of this section.

2. Remove the halfshaft.

3. Pry up the locking clip of the transaxle side boot retaining band with a suitable tool.

4. Using a pair of pliers, remove the retaining band.

5. Slide the boot away to access the CV-joint.

6. Matchmark the CV-joint housing and the shaft to ensure proper positioning during assembly.

7. Remove the outer ring.

8. Matchmark the shaft and tripod assembly to ensure proper positioning during assembly.

9. Remove the snapring.

➡**Be careful not to damage the needle bearings.**

10. Using a brass drift and a hammer, drive the tripod joint from the shaft.

➡**Cover the halfshaft serration's with tape, so as not to damage the boot.**

11. Slide the boot off the shaft.

➡**It is not necessary to remove the dynamic damper from the shaft unless replacement or repair is required.**

12. Pry up the locking clip of the dynamic damper retaining band with a suitable tool.

13. Using a pair of pliers, remove the retaining band.

14. Remove the dynamic damper.

➡**Do not remove the outer boot from the shaft unless it is necessary.**

15. Pry up the locking clip of the outer boots small and large retaining bands with a suitable tool.

16. Using a pair of pliers, remove the retaining bands.

17. Cover the halfshaft serration's with tape and remove the boot.

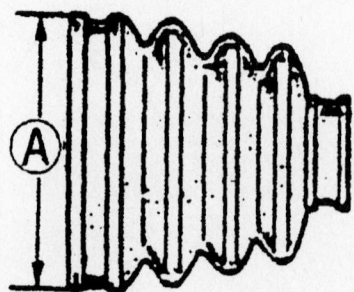

Transaxle End

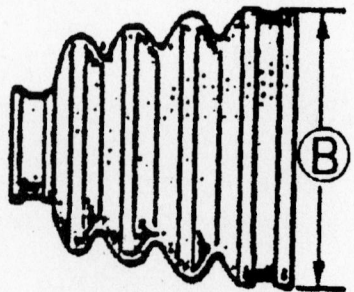

Wheel End

9307KG12

Measure the boots as shown to ensure the larger boot is placed on the wheel side of the halfshaft.

Length of driveshaft   in (mm)

Item	Right side		Left side	
	MTX	ATX	MTX	ATX
TED	24.93 (633.3)	25.54 (648.7)	25.15 (638.9)	25.16 (639)

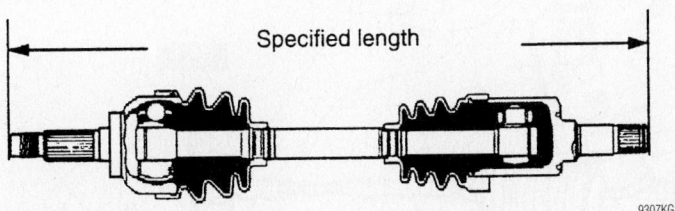

Specified length

9307KG13

Make sure the driveshaft's specified length is correct

To install:

➡**The boot on the wheel side of the driveshaft is larger than the boot on the differential side.**

18. Cover the halfshaft serration's with tape and install the inner boot.

➡**The bands should be installed so that their pointed ends initially point in the forward direction of rotation.**

19. Install the dynamic damper and band. Fold the band back by pulling on the end of it with a pair of pliers, then lock the end of the band by bending the locking clip.

20. Install the outer boot.

21. Align the marks made during removal and install the tripod joint using a brass drift and a hammer.

22. Install the snapring.

23. Apply the grease supplied with the joint rebuild kit to the tripod joint, outer ring and the boot.

24. Install the outer ring.

25. If the outer boot was removed, fill it with the correct amount of grease as follows:

    a. Transaxle side: 4.94 oz. (140g).
    b. Wheel side: 4.58 oz. (130g).

26. Make sure the boots are not damaged, then carefully up the small end of the boots to release any trapped air.

27. Measure the length of the driveshaft to ensure it is the correct length. Refer to the accompanying illustration for the driveshaft measuring points and the correct specifications.

28. Install the boot retaining bands. Fold the bands back by pulling on the ends with a pair of pliers, then lock the end of the bands by bending the locking clips.

29. Install the halfshaft.

STEERING AND SUSPENSION

Air Bag

❈❈ CAUTION

Some vehicles are equipped with an air bag system. The system must be disabled before performing service on or around system components, steering column, instrument panel components, wiring and sensors. Failure to follow safety and disabling procedures could result in accidental air bag deployment, possible personal injury and unnecessary system repairs.

PRECAUTIONS

Several precautions must be observed when handling the inflator module to avoid accidental deployment and possible personal injury.

• Never carry the inflator module by the wires or connector on the underside of the module.

• When carrying a live inflator module, hold securely with both hands, and ensure that the bag and trim cover are pointed away.

• Place the inflator module on a bench or other surface with the bag and trim cover facing up.

• With the inflator module on the bench, never place anything on or close to the module which may be thrown in the event of an accidental deployment.

• An air bag is an explosive device. Handle with extreme caution.

• Always disconnect the battery and the air bag connector before removing the steering wheel or beginning work on the air bag system.

• Air bag components must not be repaired or opened. Always use new parts, including the wiring harness.

• Always place a removed air bag unit with the horn pad facing up. Put it in a safe place where it will not be disturbed.

• The air bag unit must not be exposed to grease, fluids, or cleaning agents.

• The air bag unit must not be exposed to temperatures above 194° F (90° C) at any time. Even the heat of a soldering iron can damage or ignite the charge.

• Storage and transport of air bags is subject to rules governing explosive devices and should be done only in the original package.

• Failure to follow proper safety precautions may result in personal injury through accidental firing of the air bag, or through failure of the air bag in an accident.

DISARMING

1. Before servicing the vehicle, refer to the precautions at the beginning of this section.

2. Turn the ignition switch to the **LOCK** position.

3. Disconnect the negative battery cable.

4. Wait 10 minutes for the battery back-up power to discharge.

ARMING

1. Before servicing the vehicle, refer to the precautions at the beginning of this section.

2. Connect the negative battery cable.

3. Turn the ignition switch **ON**.

4. Verify that the air bag indicator illuminates for 4–8 seconds, then goes off.

Manual Rack and Pinion Steering Gear

REMOVAL & INSTALLATION

Sephia

1. Before servicing the vehicle, refer to the precautions at the beginning of this section.

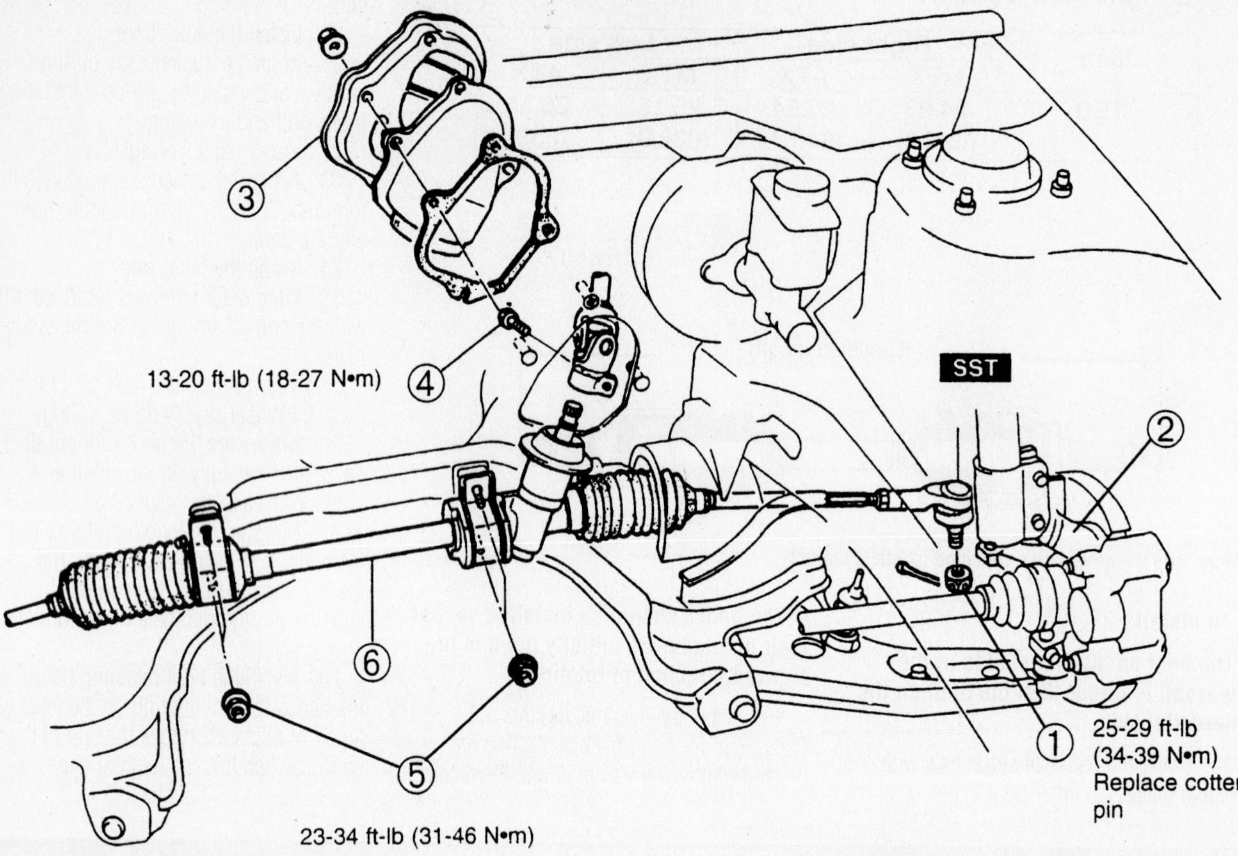

13-20 ft-lb (18-27 N•m)

SST

25-29 ft-lb
(34-39 N•m)
Replace cotter
pin

23-34 ft-lb (31-46 N•m)

1. Tie-rod end nut
2. Steering knuckle
3. Bulkhead sealing cover
4. Pinch bolts
5. Steering rack nuts
6. Steering rack

7923KG38

Exploded view of the manual steering gear assembly mounting

2. Disconnect the negative battery cable.

3. Remove the front wheels.

4. Remove the cotter pins from both steering tie rod ends and the nuts.

5. Remove the tie rod out of the knuckle arm, using a suitable tool.

6. Remove the set plate from the firewall.

7. Remove the fixing bolt from the steering shaft to steering gear pinion shaft.

8. Remove the steering shaft from the steering gear.

9. Remove the steering gear mounting nuts.

10. Move the steering gear to the right of the vehicle.

To install:

11. Install the steering gear to the vehicle.

12. Install the mounting nuts in the order

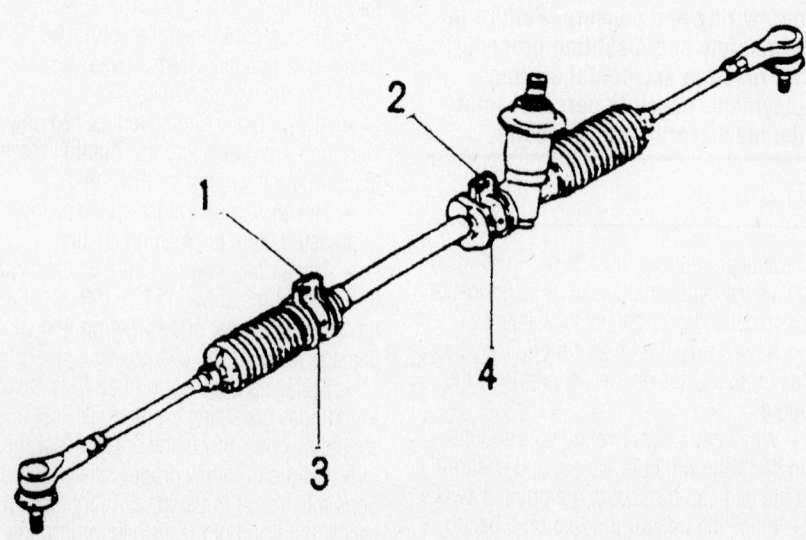

Rack mounting tightening sequence

7923KG39

shown. Tighten the nuts to 23–34 ft. lbs. (31–46 Nm).

13. Install the steering shaft to the steering gear pinion shaft. Tighten the bolt/nut to 13–20 ft. lbs. (18–27Nm).

14. Install the set plate to the firewall.

15. Install the tie rod ends to the knuckle arm and tighten the nuts to 25–29 ft. lbs. (34–39 Nm)

16. Install new cotter pins.

17. Install the wheels.

18. Connect the negative battery cable.

19. Check the front end alignment.

Power Rack and Pinion Steering Gear

REMOVAL & INSTALLATION

1. Before servicing the vehicle, refer to the precautions at the beginning of this section.

2. Disconnect the negative battery cable.

3. Remove the front wheels.

4. Remove the cotter pins from both steering tie rod ends and the nuts.

5. Remove the tie rod out of the knuckle arm, using a suitable tool.

6. Remove the catalytic converter, if necessary for access.

7. Disconnect the pressure line and return pipe from the steering gear.

8. Remove the set plate from the firewall, if necessary.

9. Remove the fixing bolt from the steering shaft to steering gear pinion shaft.

10. Separate the steering shaft from the steering gear.

11. Remove the manual transaxle shifter linkage, if necessary.

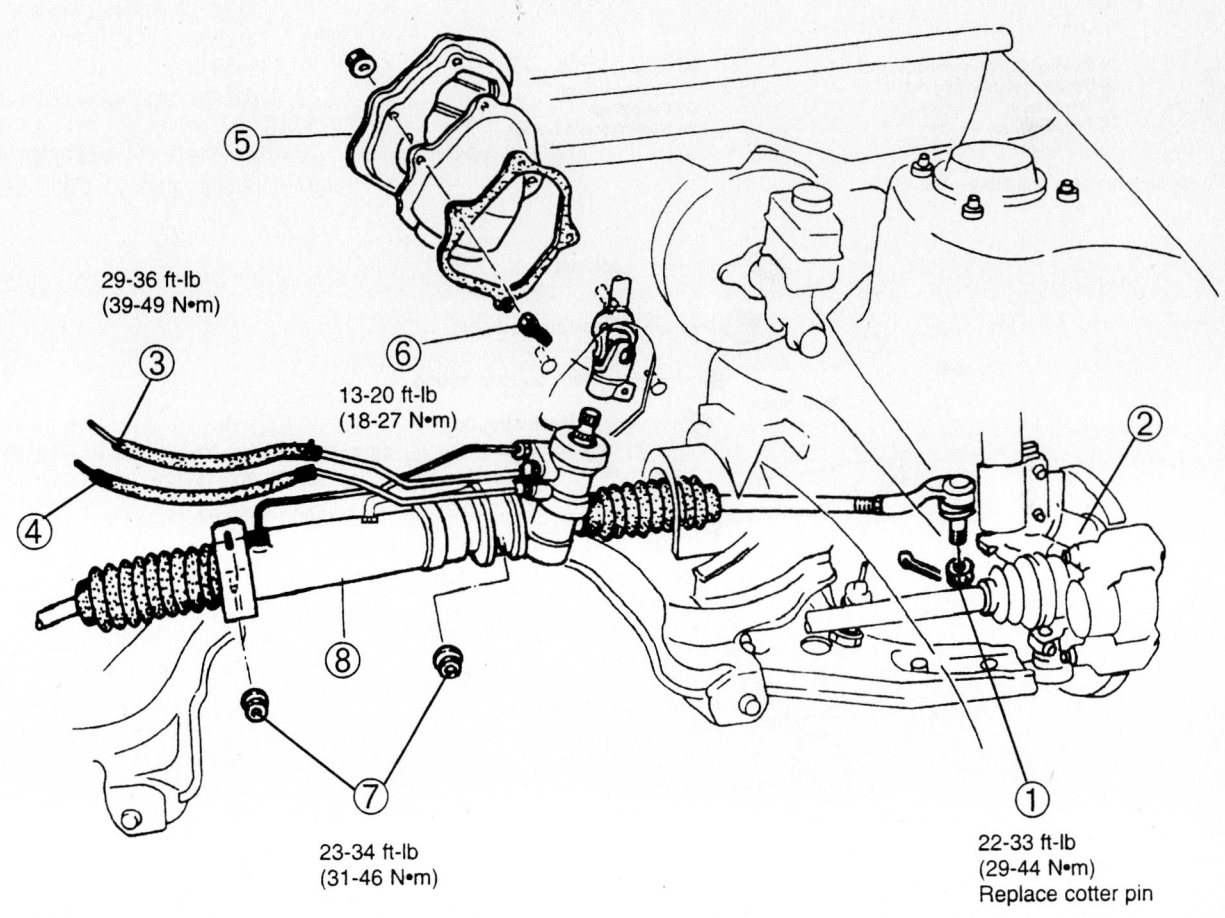

29-36 ft-lb
(39-49 N•m)

13-20 ft-lb
(18-27 N•m)

23-34 ft-lb
(31-46 N•m)

22-33 ft-lb
(29-44 N•m)
Replace cotter pin

1. Tie-rod end nut
2. Steering knuckle
3. High-pressure line
4. Return line

5. Sealing cover
6. Pinch bolt
7. Steering rack nuts
8. Steering rack and linkage

7923KG40

Exploded view of the power steering gear assembly mounting—Sephia

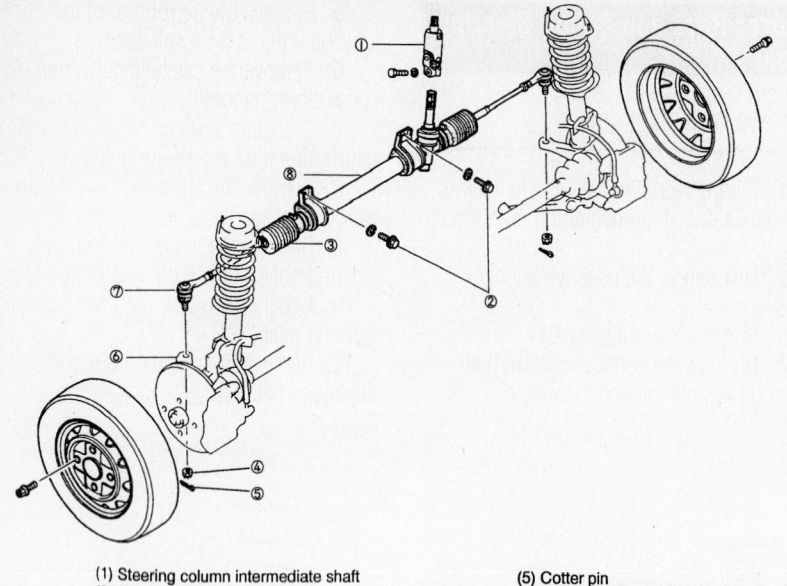

Power steering gear mounting—Rio

(1) Steering column intermediate shaft
(2) Steering gear mounting bolt
(3) Tie-rod boot
(4) Tie-rod nut

(5) Cotter pin
(6) Steering knuckle
(7) Tie-rod end
(8) Steering gear & Linkage

9357NG52

12. Remove the steering gear mounting nuts.

13. Remove the steering gear to the right of the vehicle.

To install:

14. Install the steering gear to the vehicle and the mounting nuts/bolts. Tighten the nuts to 23–34 ft. lbs. (31–46 Nm) for the Sephia and Spectra and 27–38 ft. lbs. (37–52 Nm) for Rio and Optima.

15. Install the steering shaft to the steering gear pinion shaft. Tighten the bolt/nut to the specified torque.

16. Install the manual transaxle shift linkage, if disconnected.

17. Install the set plate to the firewall, if removed.

18. Connect the pressure line and return hose to the steering gear. Torque to 29–43 ft. lbs. (39–59 Nm).

19. Install the catalytic converter, if removed.

20. Install the tie rod ends to the knuckle arm. Tighten the nuts to 22–33 ft. lbs.

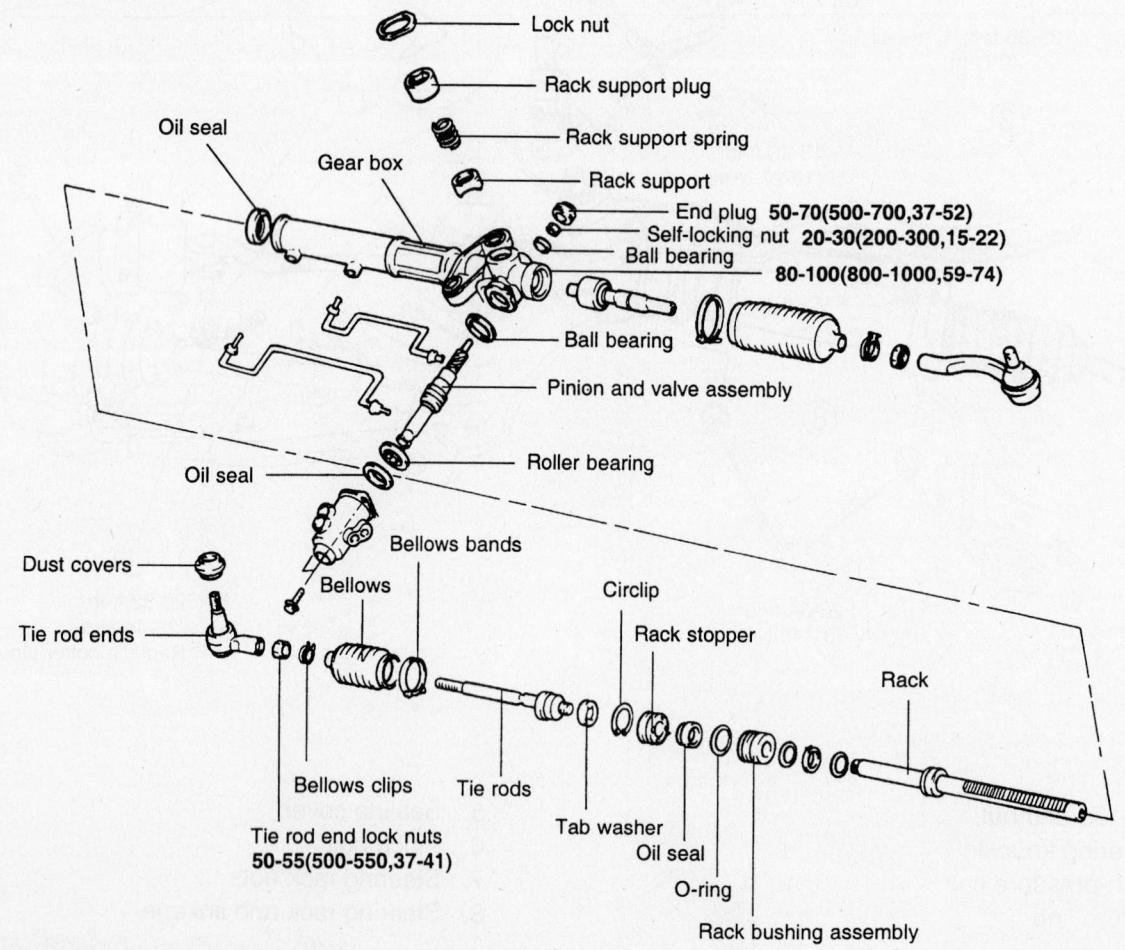

TORQUE : Nm (kg·cm, lb·ft)

Exploded view of the power steering gear—Optima

9357NG48

(29–44 Nm) for Sephia and Spectra and to 27–38 ft. lbs. (37–52 Nm) for Rio and Optima.

21. Install new cotter pins.
22. Install the wheels.
23. Connect the negative battery cable.
24. Check and add power steering fluid, bleed the system, and check the front end alignment.

Strut

REMOVAL & INSTALLATION

Front

1. Before servicing the vehicle, refer to the precautions at the beginning of this section.

2. Remove the wheel and tire assembly.
3. Support the lower control arm with a jack.
4. Remove the bolts or clips attaching the brake hose and/or Anti-lock Brake System (ABS) sensor harness to the strut.
5. Remove the stabilizer control link from the bracket mounted to the strut, Rio only.

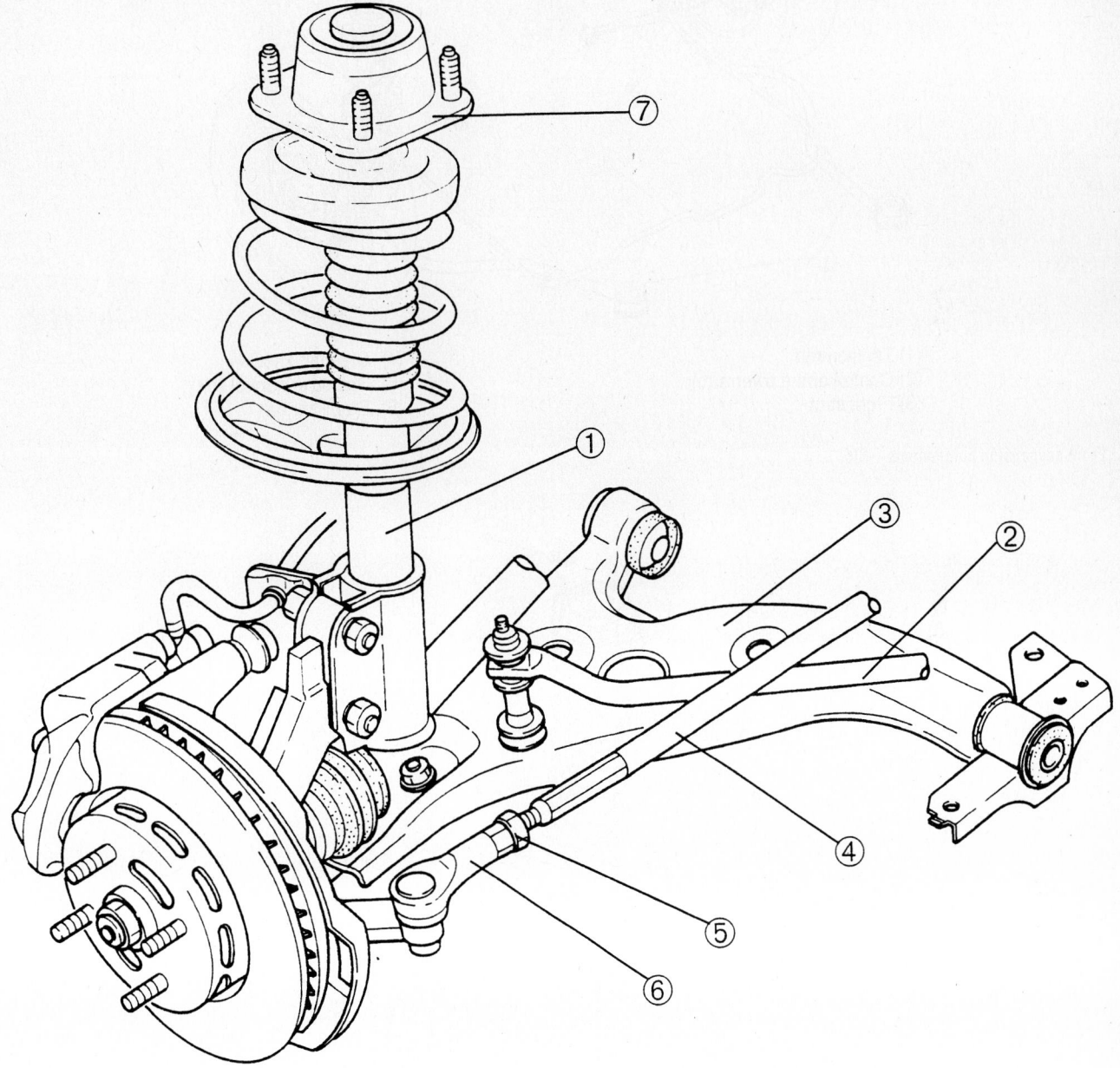

1. Front strut
2. Front stabilizer
3. Lower control arm
4. Tie-rod
5. Jam nut
6. Tie-rod end
7. Mounting block

7923KG41

Front suspension component identification

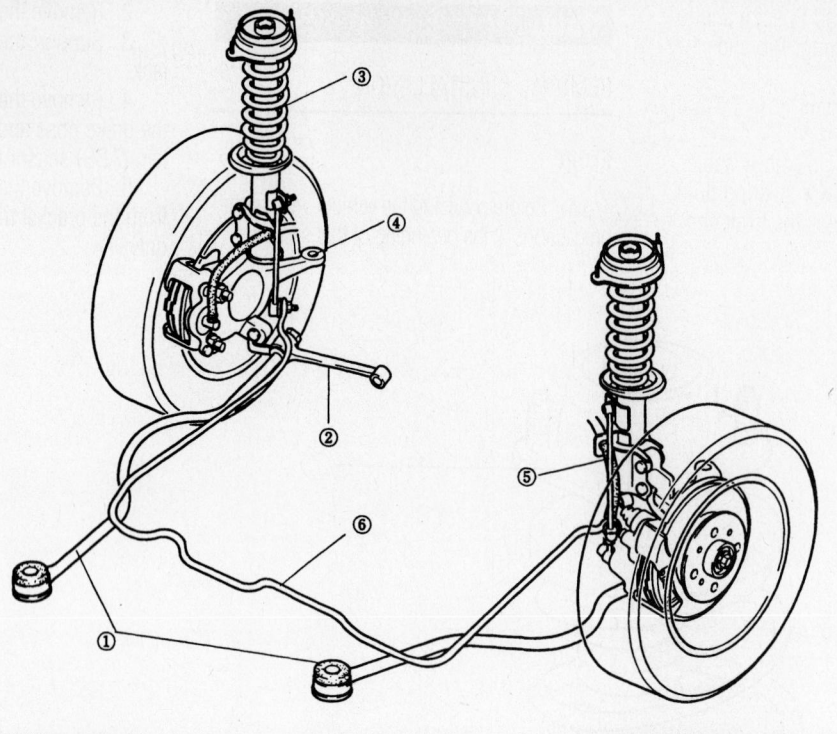

(1) Tension rod
(2) Control arm (Lower arm)
(3) Front strut

(4) Steering knuckle
(5) Stabilizer control link
(6) Stabilizer bar

9357NG17

Front suspension components—Rio

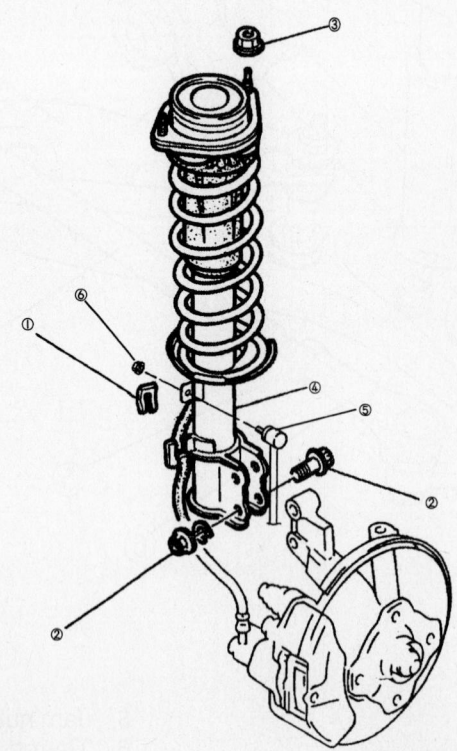

(1) Hose clip
(2) Bolt and nut
(3) Mounting block nut

(4) MacPherson strut
(5) Stabilizer control link
(6) Stabilizer control link nut

9357NG18

Exploded view of the front strut mounting—Rio

13. Install the coil spring and compress the coil spring with the spring compressor.

14. Install the rubber seat, the spring upper seat, the bearing and the mounting block. Be sure that the spring upper seat

notched portion is facing inward and tighten the piston rod upper nut.

15. Remove the spring compressor from the strut. Secure the upper mounting block in the vise. Tighten the nut to 41–50 ft. lbs.

(55–68 Nm) for the front strut and 47–59 ft. lbs. (64–80 Nm) for the rear strut.

16. Be sure that the spring is well seated in the upper seats.

17. Install the strut to the vehicle.

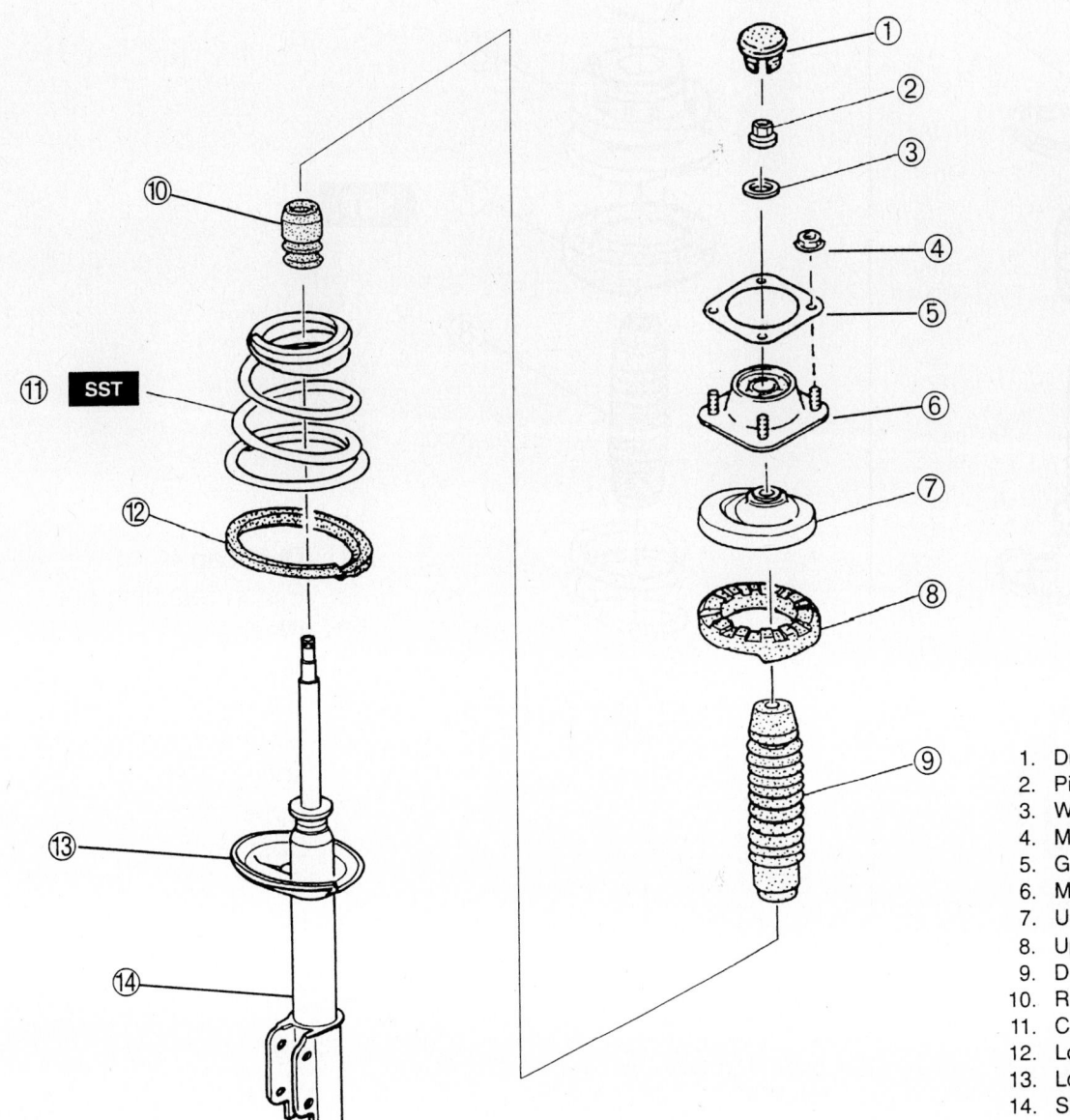

1. Dust cap
2. Piston retaining nut
3. Washer
4. Mounting nut
5. Gasket
6. Mounting block
7. Upper spring seat
8. Upper spring isolator
9. Dust boot
10. Rebound stopper
11. Coil spring
12. Lower spring isolator
13. Lower spring seat
14. Shock absorber

7932KG43

Exploded view of the front strut assembly

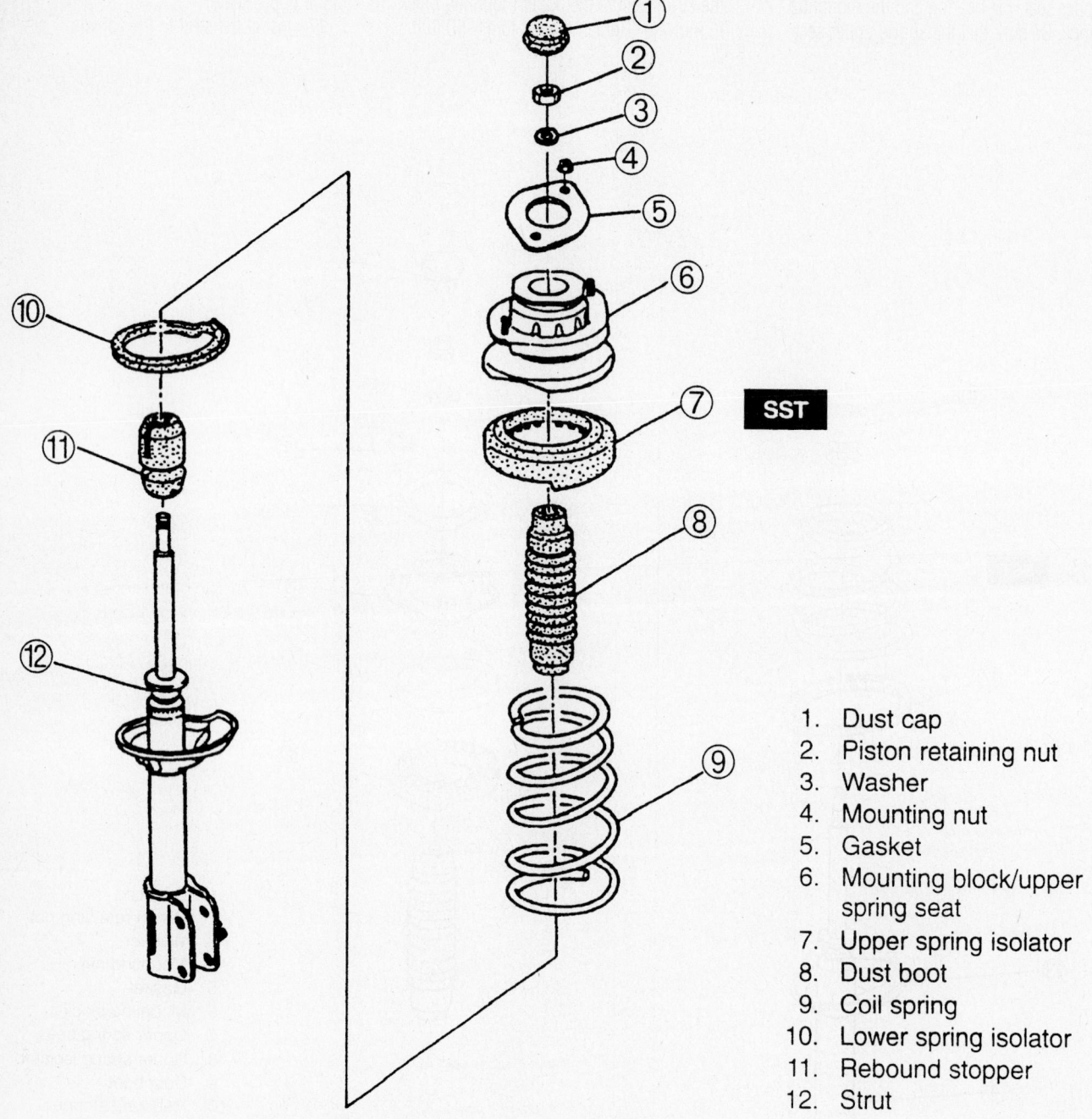

1. Dust cap
2. Piston retaining nut
3. Washer
4. Mounting nut
5. Gasket
6. Mounting block/upper spring seat
7. Upper spring isolator
8. Dust boot
9. Coil spring
10. Lower spring isolator
11. Rebound stopper
12. Strut

7923KG44

Exploded view of the rear strut assembly

Lower Ball Joint

REMOVAL & INSTALLATION

1. Before servicing the vehicle, refer to the precautions at the beginning of this section.

2. Remove the wheel and tire assembly.

3. Remove the ball joint stud pinch bolt and nut from the steering knuckle.

4. Remove the ball joint from the knuckle, using a prytool.

5. Remove the bolt, nut and the ball joint from the lower control arm.

To install:

6. Installation is the reverse of the removal procedure. Tighten the ball joint-to-lower control arm bolt and nut to 86 ft. lbs. (117 Nm). Tighten the ball joint pinch bolt and nut to 43 ft. lbs. (59 Nm). Check the front wheel alignment.

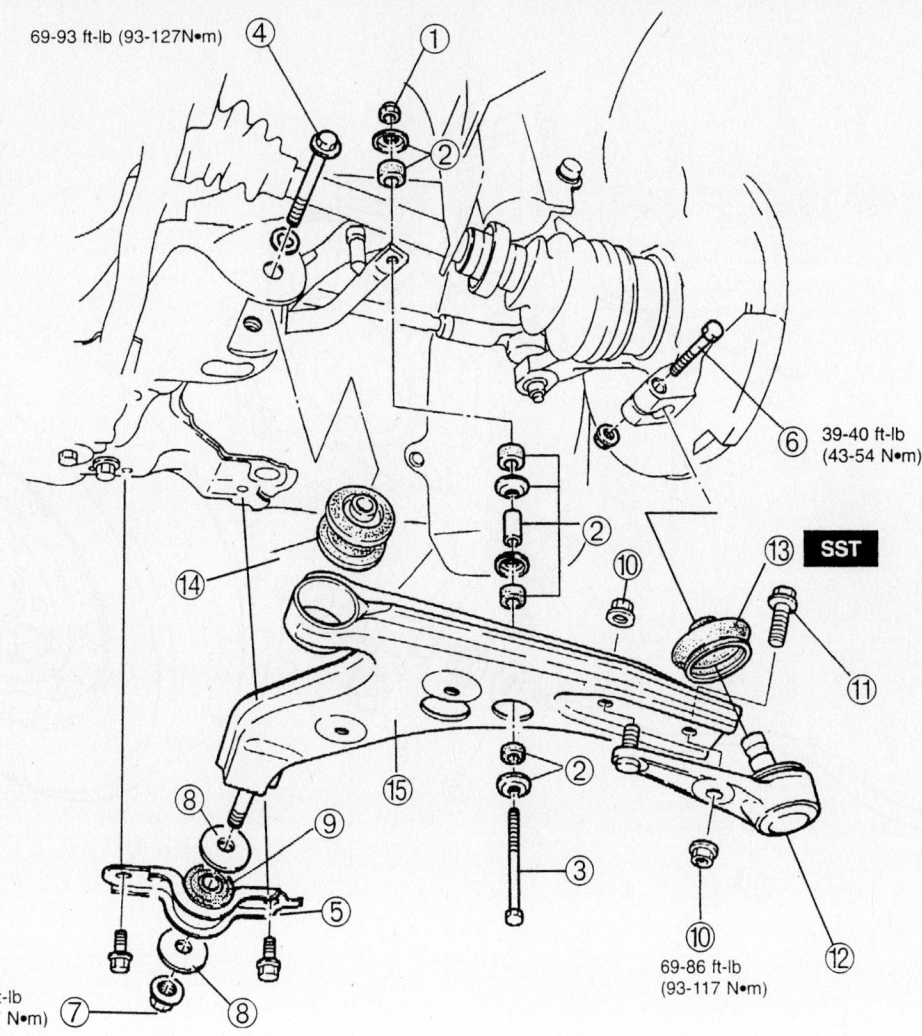

69-93 ft-lb (93-127N•m)

39-40 ft-lb (43-54 N•m)

SST

69-86 ft-lb (93-117 N•m)

69-86 ft-lb (93-117 N•m)

1. Stabilizer retaining nut
2. Stabilizer hardware - spacer, retainers, bushings
3. Stabilizer bolt
4. Pivot bolt
5. Mounting bolts
6. Pinch bolt,
7. Retaining nut
8. Washers
9. Control arm bushing - rear
10. Ball joint mounting nuts
11. Ball joint mounting bolt
12. Ball joint
13. Ball joint dust boot (Replace)
14. Control arm bushing - front
15. Control arm

7923KG45

Exploded view of the lower control arm with replaceable ball joint

Lower Control Arm

REMOVAL & INSTALLATION

Rio

1. Before servicing the vehicle, refer to the precautions at the beginning of this section.
2. Remove the front wheel.
3. Remove the tension rod attaching bolt from the frame bracket.
4. Remove the tensioner rod bushing cotter pin and nut from the rear of the lower arm.
5. Remove the washer and bushing.
6. Remove the tension rod from the lower arm.
7. Remove the front bushing and washer.
8. Remove the lower arm, by lowering it and prying the ball joint from the steering knuckle after loosening the bolt and nut.
9. Remove the lower arm after loosening the lower arm attaching bolt from the frame bracket.

To install:

10. Raise the inner end of the lower arm into the pivot bracket on the frame, and loosely tighten the arm frame bracket pivot to hold in place.
11. Install the lower control arm ball joint in the clamp bore of the knuckle. Tighten the bolt and nut to 40–50 ft. lbs. (54–68 Nm).
12. Torque the lower arm frame bracket pivot bolt to 40–50 ft. lbs. (54–68 Nm).
13. Install the tension rod bolt into the lower arm, tighten washer and bushing and with a new cotter pin.
14. Install the bushing and insulator on the tensioner rod.
15. Tighten the tension rod bolt to 87–108 ft. lbs. (118–147 Nm).

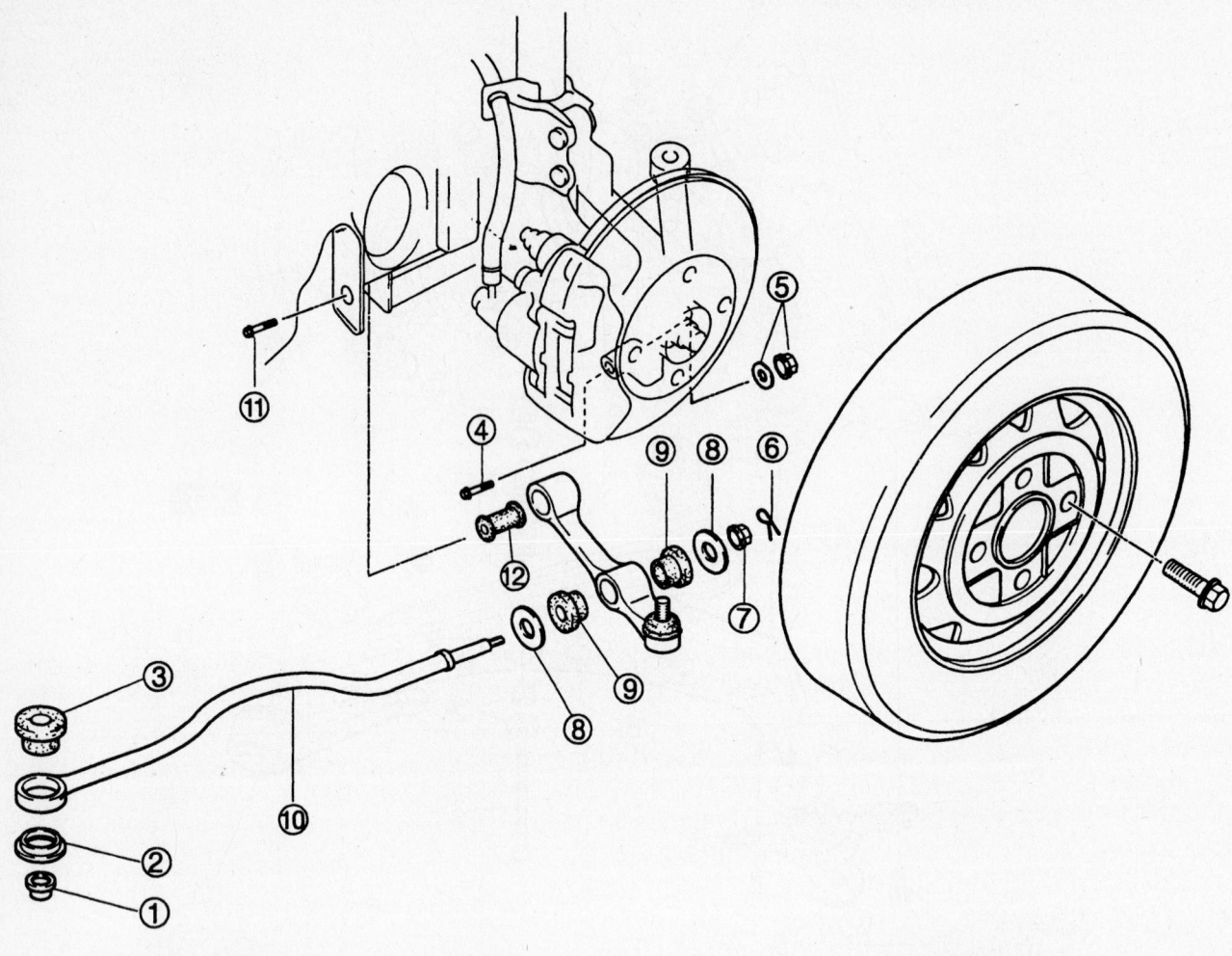

(1) Nut
(2) Stopper bushing
(3) Rubber
(4) Bolt
(5) Nut & Washer
(6) Cotter pin

(7) Nut
(8) Washer
(9) Bushing
(10) Tension rod
(11) Bolt
(12) Bushing

9357NG20

Exploded view of the lower control arm and tension rod—Rio

Sephia and Spectra

1. Before servicing the vehicle, refer to the precautions at the beginning of this section.

2. Remove the front wheel.

3. Remove the stabilizer control link nut from the bracket on the lower control arm.

4. Remove the pivot bolt.

5. Remove the lower control arm ball joint bolt and nut from the steering knuckle.

6. Remove the 3 mounting bolts.

7. Remove the retaining nut, washer and the rear control arm bushing.

8. Remove the 2 ball joint mounting nuts and bolts from the lower control arm.

9. Place the ball joint in a vise and use a chisel top carefully remove the dust boot from the ball joint.

10. Remove the control arm front bushing.

11. Remove the control arm.

To install:

12. Apply a suitable general purpose grease to the new tire dust boot and press the boot onto the ball joint using a suitable tool.

13. Install the ball joint onto the control arm. Tighten the mounting nuts and bolts to 86 ft. lbs. (117 Nm).

14. Install the rear control arm bushing

with 2 washers and tighten the nut to 86 ft. lbs. (117 Nm).

15. Install the 3 mounting bolts. Tighten the long mount bolt 86 ft. lbs. (117 Nm) and the short mounting bolts to 50 ft. lbs. (68 Nm).

16. Install the ball joint into the knuckle and tighten the pinch bolt and nut to 40 ft. lbs. (54 Nm).

17. Install the front control arm bushing.

18. Install the pivot bolt and tighten to 86 ft. lbs. (117 Nm).

19. Install the stabilizer control link nut to the bracket on the control arm and tighten the nut to 45 ft. lbs. (61 Nm).

20. Install the wheels.

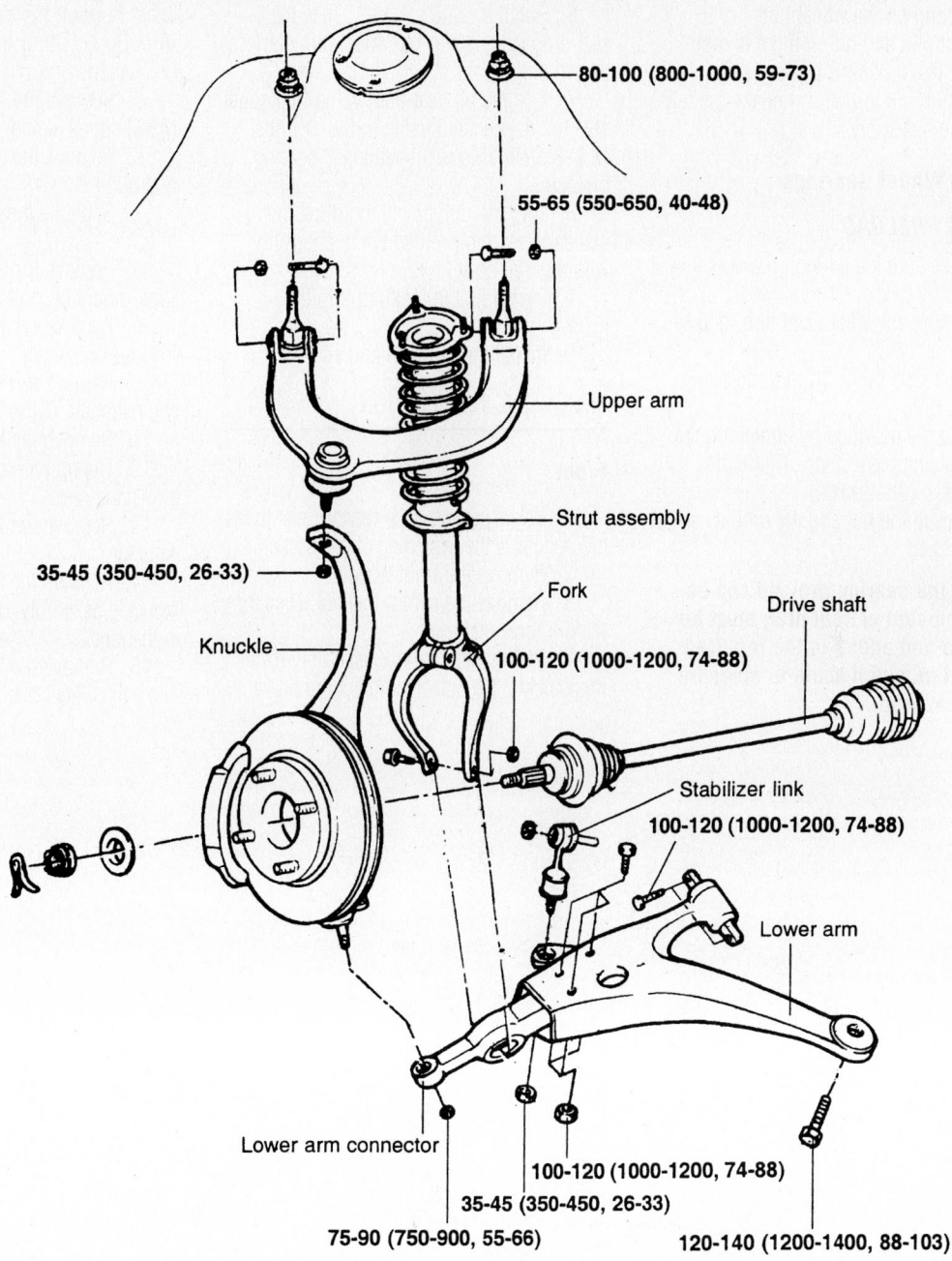

80-100 (800-1000, 59-73)

55-65 (550-650, 40-48)

Upper arm

Strut assembly

Fork

Drive shaft

35-45 (350-450, 26-33)

Knuckle

100-120 (1000-1200, 74-88)

Stabilizer link

100-120 (1000-1200, 74-88)

Lower arm

Lower arm connector

100-120 (1000-1200, 74-88)

35-45 (350-450, 26-33)

75-90 (750-900, 55-66)

120-140 (1200-1400, 88-103)

TORQUE : Nm (kg·cm, lb·ft)

9357NG51

Exploded view of the lower control arm—Optima

Optima

1. Before servicing the vehicle, refer to the precautions at the beginning of this section.
2. Remove the front wheel.
3. Remove the ball joint nut. Loosen, but do not remove it.
4. Separate the ball joint from the lower control arm.
5. Remove the ball joint.
6. Remove the bolt connecting the fork to the lower control arm.

7. Remove the stabilizer link from the lower control arm.
8. Remove the 2 bolts from the lower control arm bushing.
9. Remove the lower control arm bushing bolt.
10. Remove the steering gear box.
11. Remove the stabilizer bar.
12. Installation is the reverse of removal procedure. Refer to the illustration for tightening specifications.

Wheel Bearings

ADJUSTMENT

Except Rio Rear Wheel Bearings

The wheel bearings on these vehicles are not adjustable. To check if the bearing requires service, remove the wheel and tire assembly, brake caliper and disc brake rotor. Install a dial indicator with the indicator foot resting on the wheel hub. Try to move the hub in and out. If there is more than 0.002 in. (0.05mm) bearing play, check the wheel hub nut torque or replace the hub and bearing assembly.

Rio Rear Wheel Bearings

BEARING PRELOAD

1. Make sure the parking brake is fully released.
2. Remove the wheel and hub (grease) cap.
3. Rotate the brake drum to be sure there is no drag.
4. Seat the bearings by tightening the nut after raising the nut tap. Tighten to 18–22 ft. lbs. (25–29 Nm).
5. Loosen the nut slightly until it can be turned by hand.

➡ **Before the bearing preload can be set, the amount of seal drag must be measured and added to the required preload. Use a pull scale to measure the oil seal drag.**

6. Pull the scale squarely. Take the oil seal drag value when the wheel hub starters to turn and record it.
7. Add the oil seal drag value in the last step to the specified value to 0.6–1.9 lbs. (2.6–8.5 N). This is the standard bearing preload.
8. Turn the nut slowly to adjust the standard bearing preload while checking with the pull scale.
9. Firmly fix the lock nut into the groove.
10. Install the hub cap and wheel.

REMOVAL & INSTALLATION

Front

1. Before servicing the vehicle, refer to the precautions at the beginning of this section.
2. Remove the front wheels.
3. Remove the center locknut. Discard the old locknut.
4. Remove the caliper assembly from the knuckle. Do not disconnect the brake

lines. Support the caliper with a piece of wire. Do not allow the caliper to hang by the hose at any time. Remove the brake disc.
5. Remove the Anti-lock Brake System (ABS) speed sensor, if equipped.
6. Remove the tie rod end cotter pin and nut.
7. Separate the tie rod end from the knuckle assembly.
8. Remove the outer lower arm-to-ball joint mounting bolt and nut.
9. Remove the lower arm from the knuckle assembly.
10. Remove the knuckle assembly free of the halfshaft, using a plastic mallet.
11. Remove the knuckle assembly.
12. Clamp the knuckle in a vise with protected jaws.
13. Remove the inner oil seal from the knuckle.
14. Remove the front wheel hub from the knuckle assembly, using an appropriate hub-puller.
15. Remove the bearing inner race from the front wheel hub.

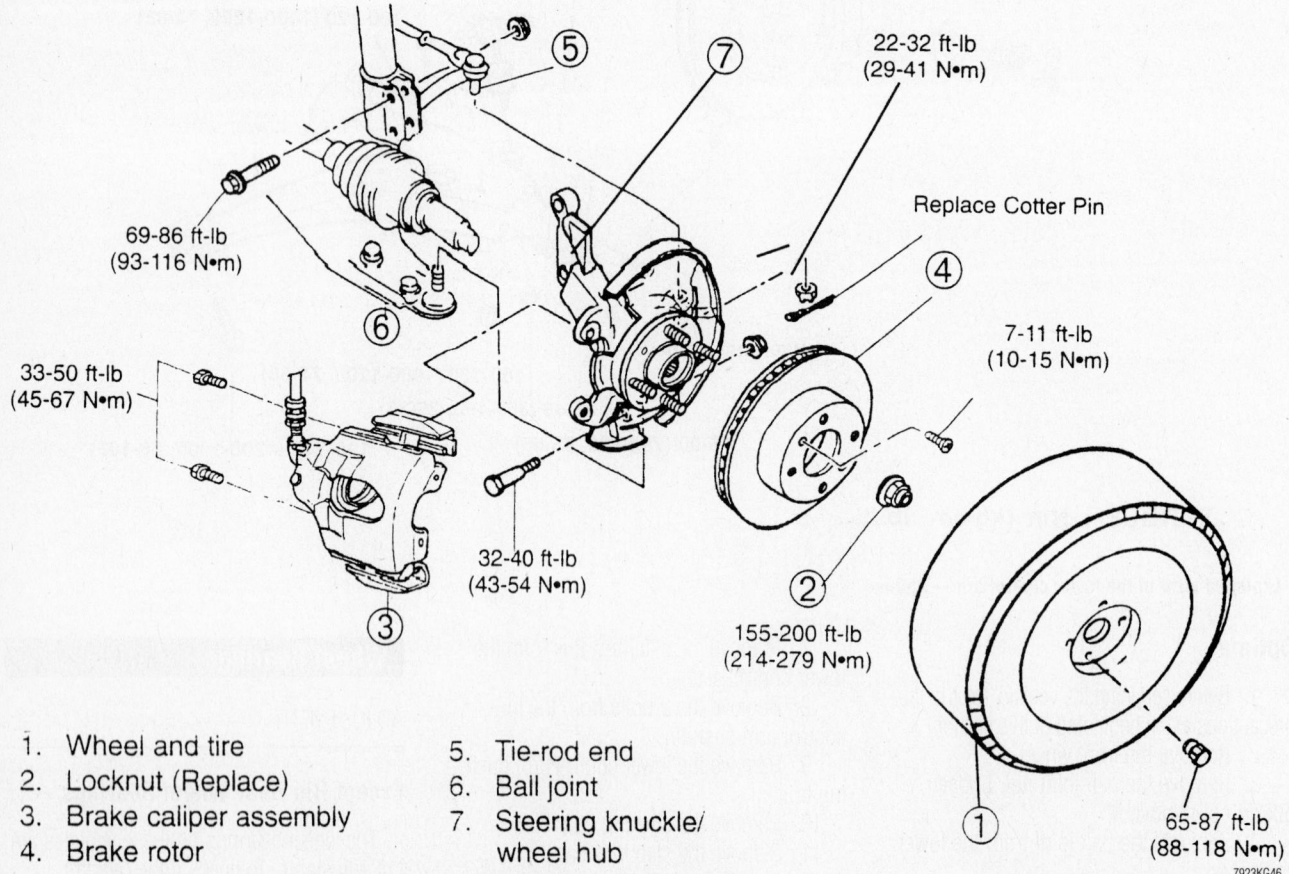

1. Wheel and tire
2. Locknut (Replace)
3. Brake caliper assembly
4. Brake rotor
5. Tie-rod end
6. Ball joint
7. Steering knuckle/wheel hub

22-32 ft-lb (29-41 N•m)
Replace Cotter Pin
69-86 ft-lb (93-116 N•m)
7-11 ft-lb (10-15 N•m)
33-50 ft-lb (45-67 N•m)
32-40 ft-lb (43-54 N•m)
155-200 ft-lb (214-279 N•m)
65-87 ft-lb (88-118 N•m)

7923KG46

Exploded view of the front steering knuckle and related components—Sephia and Spectra

➡**If the bearing inner race still remains on the hub assembly, grind a section of the bearing inner race until about 0.02 in. (0.50mm) remains. Remove with a chisel.**

16. Remove the retaining ring from within the knuckle.

17. Remove the front wheel bearing from the knuckle, using a wheel bearing removal tool to press it out.

18. Clean and inspect all parts but do not wash or clean the wheel bearing.

To install:

19. Install the new wheel bearing into the knuckle assembly, using the press tools.

20. Install the wheel bearing retaining ring.

21. Install the front wheel hub, using a press and the correct bearing driver.

22. Install a new oil seal using the appropriate seal driver and a hammer. Tap the oil seal in evenly until the special tool contacts the steering knuckle. Coat the lip of the oil seal with grease.

23. Install the bearing/hub and knuckle assembly in place. Loosely tighten the knuckle to shock absorber bolt.

24. Install the lower arm ball joint to the knuckle. Tighten the nut to 32–40 ft. lbs. (43–54 Nm) for Sephia and Spectra, to

40–50 ft. lbs. (54–68 Nm) for Rio or to 74–88 ft. lbs. (100–120 Nm) for Optima.

25. Install the halfshaft to the knuckle assembly.

26. Install the wheel speed sensor, if equipped with ABS and tighten the bolts to 12–17 ft. lbs. (16–23 Nm)

27. Install the tie rod end to the knuckle and tighten the nut to 22–32 ft. lbs. (29–41 Nm) for Rio, Sephia and Spectra or to 18–25 ft. lbs. (24–34 Nm) for Optima.

28. Install a New cotter pin.

29. Install a new wheel hub locknut. Tighten the nut to 155–200 ft. lbs. (214–279 Nm) for Sephia and Spectra, to

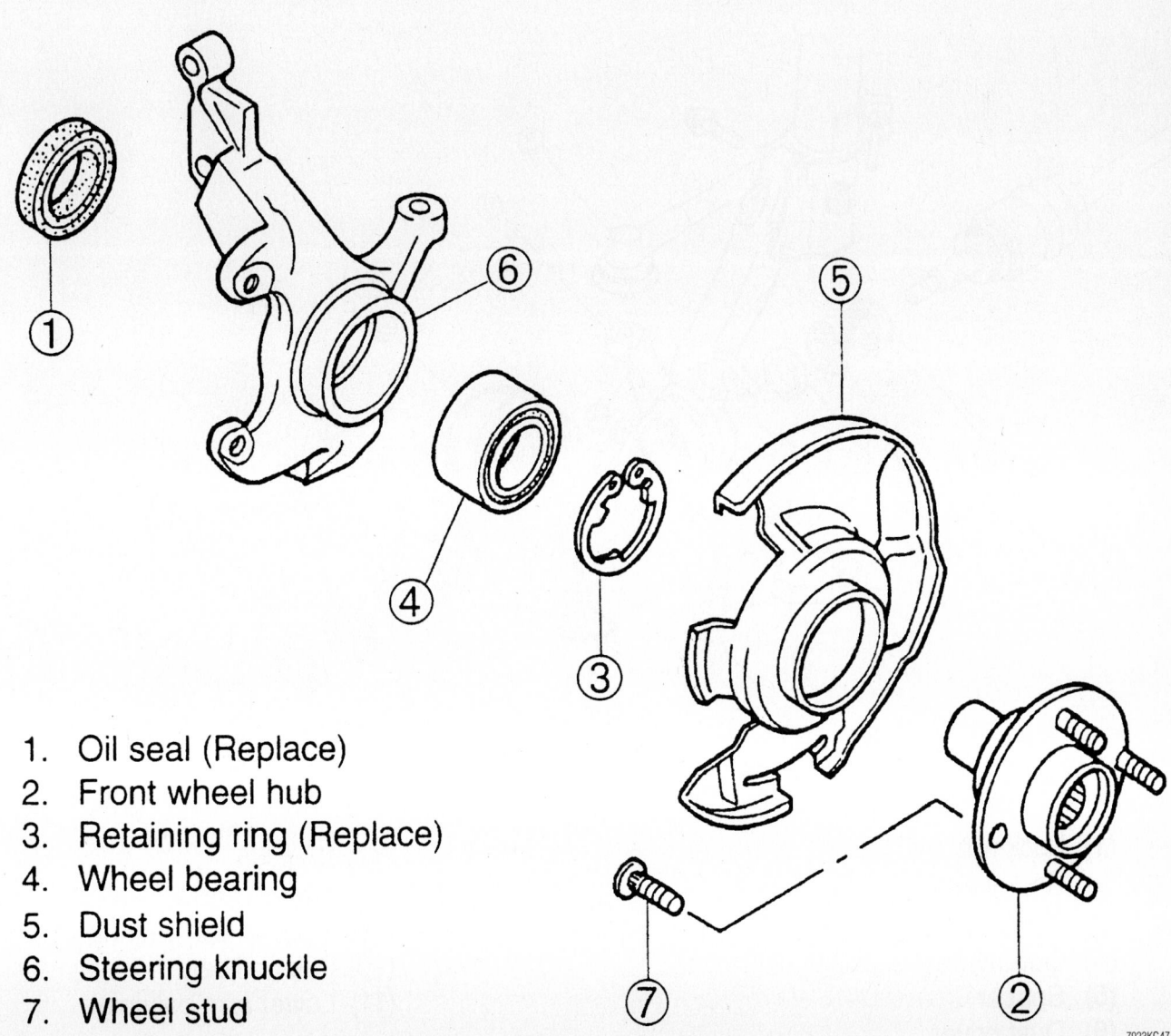

1. Oil seal (Replace)
2. Front wheel hub
3. Retaining ring (Replace)
4. Wheel bearing
5. Dust shield
6. Steering knuckle
7. Wheel stud

7923KG47

Exploded view of the front hub and bearing assembly—Sephia and Spectra

116–174 ft. lbs. (157–235 Nm) for Rio or to 148–192 ft. lbs. (200–260 Nm) for Optima.

30. Check the end-play of the wheel bearing by installing a dial indicator against the wheel hub and tire to move the brake disc back and forth. There should be no more than 0.002 in. (0.05mm) of free-play present.

31. Stake the locknut into place by bending it into the groove.

32. Install the brake caliper(s) and tighten the bolts to 33–50 ft. lbs. (45–67 Nm)

33. Install the front wheels and lower the vehicle.

34. With the vehicle lowered check all of the bolts and retighten as necessary.

35. Inspect the front end alignment and adjust as is necessary.

Rear

RIO

1. Before servicing the vehicle, refer to the precautions at the beginning of this section.

2. Remove the rear wheels.

3. Remove the hub cap.

4. Use a small chisel to raise the staked edge of the hub lock nut.

5. Remove the drum, washer and bearings as an assembly.

6. Remove the Anti-lock Brake System (ABS) sensor rotor using a suitable puller.

7. Remove the bearing oil seal and discard.

➡ **If the bearings will be reused, they must be matchmarked so they can be installed in their original positions.**

8. Remove the inner bearings from the bearing hub.

9. Remove the inner and outer bearing outer races using special tool 0K68A 173 002.

To install:

10. Install the outer bearing race using a hammer. Tap the race in until it is fully seated in the hub.

11. Install the inner bearing in the bearing hub.

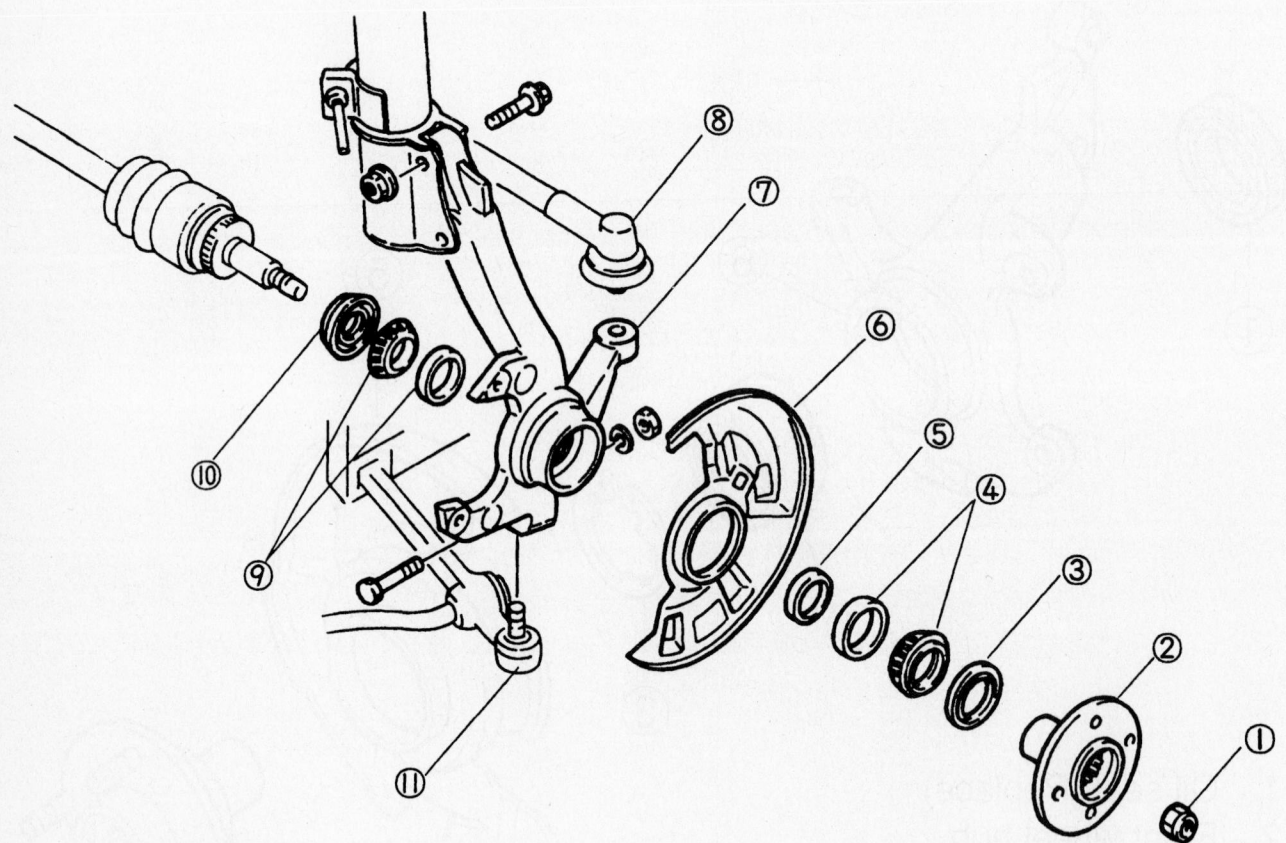

(1) Lock nut
(2) Wheel hub
(3) Outer oil seal
(4) Outer wheel bearing
(5) Spacer
(6) Dust cover
(7) Knuckle
(8) Tie-rod end
(9) Inner wheel bearing
(10) Inner oil seal
(11) Lower arm ball-joint

9357NG21

Exploded view of the front hub and bearing assembly—Rio

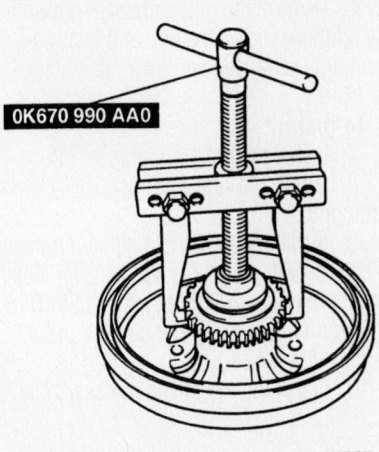

Removing the ABS rotor, using a puller—Rio

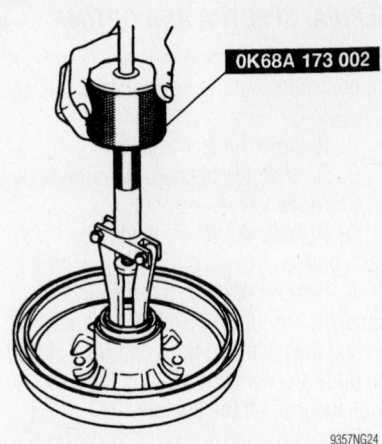

View of the special tool need to remove the inner and outer bearing outer races—Rio

12. Install the oil seal after lubricating it with lithium grease, using a hammer and seal installation tool.

13. Install the ABS sensor rotor using a flat plate to press it into place.

14. Refer to the accompanying illustration and completely fill in the shaded area with lithium grease.

15. Install the brake drum bearings and hub on the spindle. Keep the drum centered on the spindle to prevent damage to the oil seal and spindle threads.

16. Install the outer bearing and washer.

17. Install a new hub lock nut.

18. Adjust the bearing preload to 5.6–8.6 ft. lbs. (0.63–0.98 Nm).

19. Install the hub lock nut to the groove firmly.

20. Install the hub cap.

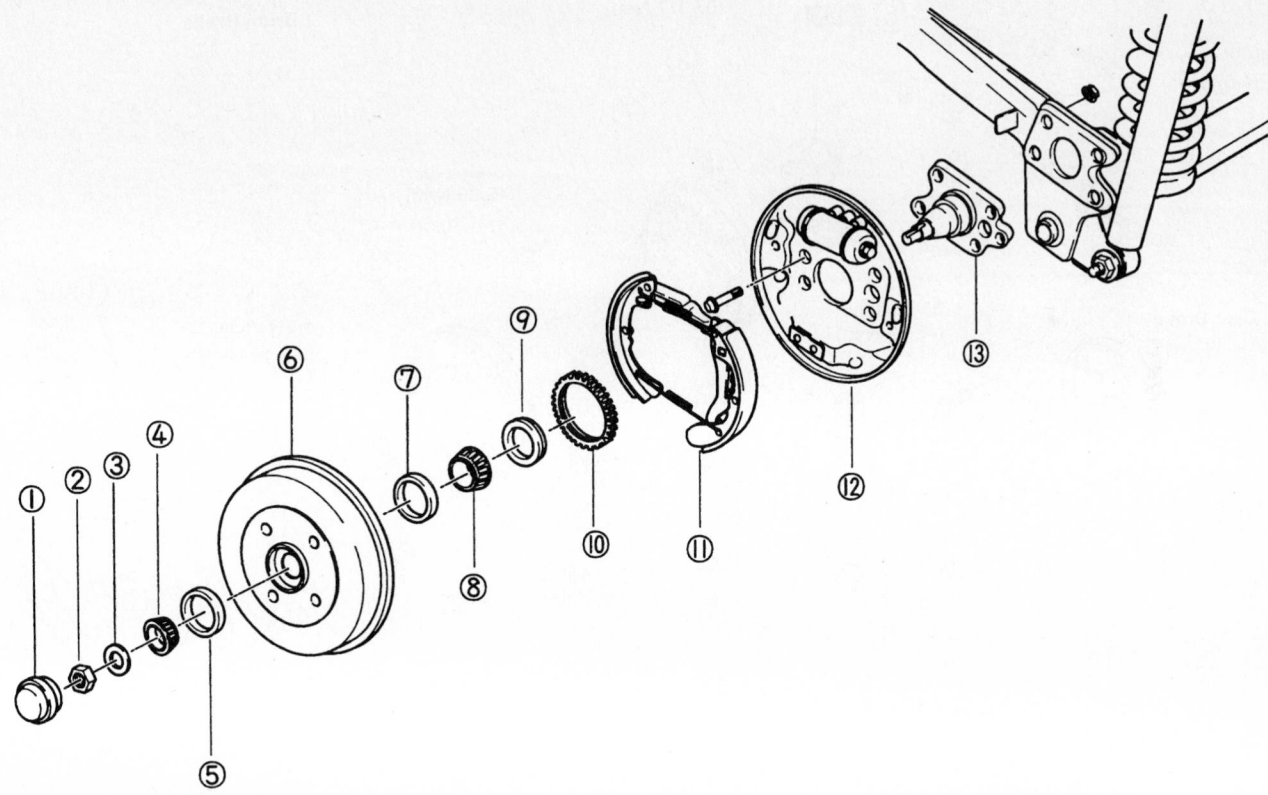

(1) Hub cap
(2) Lock nut
(3) Washer
(4) Outer bearing
(5) Outer bearing outer race
(6) Brake drum
(7) Inner bearing outer race

(8) Inner bearing
(9) Oil seal
(10) Sensor rotor(for ABS)
(11) Brake assembly
(12) Back plate
(13) Spindle

Exploded view of the rear wheel bearings and related components—Rio

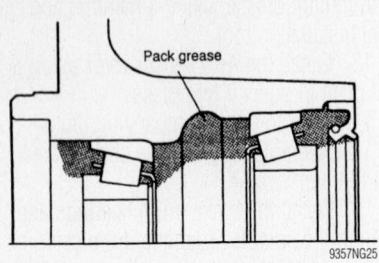

Fill in the shaded areas with lithium grease—Rio

SEPHIA, SPECTRA AND OPTIMA

1. Before servicing the vehicle, refer to the precautions at the beginning of this section.

2. Remove the rear wheels.

3. Remove the hubcap. Hold the brake to remove the center axle nut.

4. Remove the drum, if equipped with drum brakes.

5. Remove the disc brake caliper, if equipped with disc brakes without disconnecting the hydraulic hose and hang it from the body. Do not let it hang by the hose. Slide the disc off the spindle.

6. Remove the hub and bearing assembly off the spindle. The hub and bearing cannot be separated and must be replaced as 1 piece.

To install:

7. Install the hub and drum or rotor.

8. Install the brake caliper, if equipped.

9. Install a new spindle nut and tighten to 131–173 ft. lbs. (177–235 Nm) for disk brakes and 155–200 ft. lbs. (209–279 Nm) for drum brakes. Stake the nut into place. Replace the hubcap.

10. Install the wheel and tire assembly.

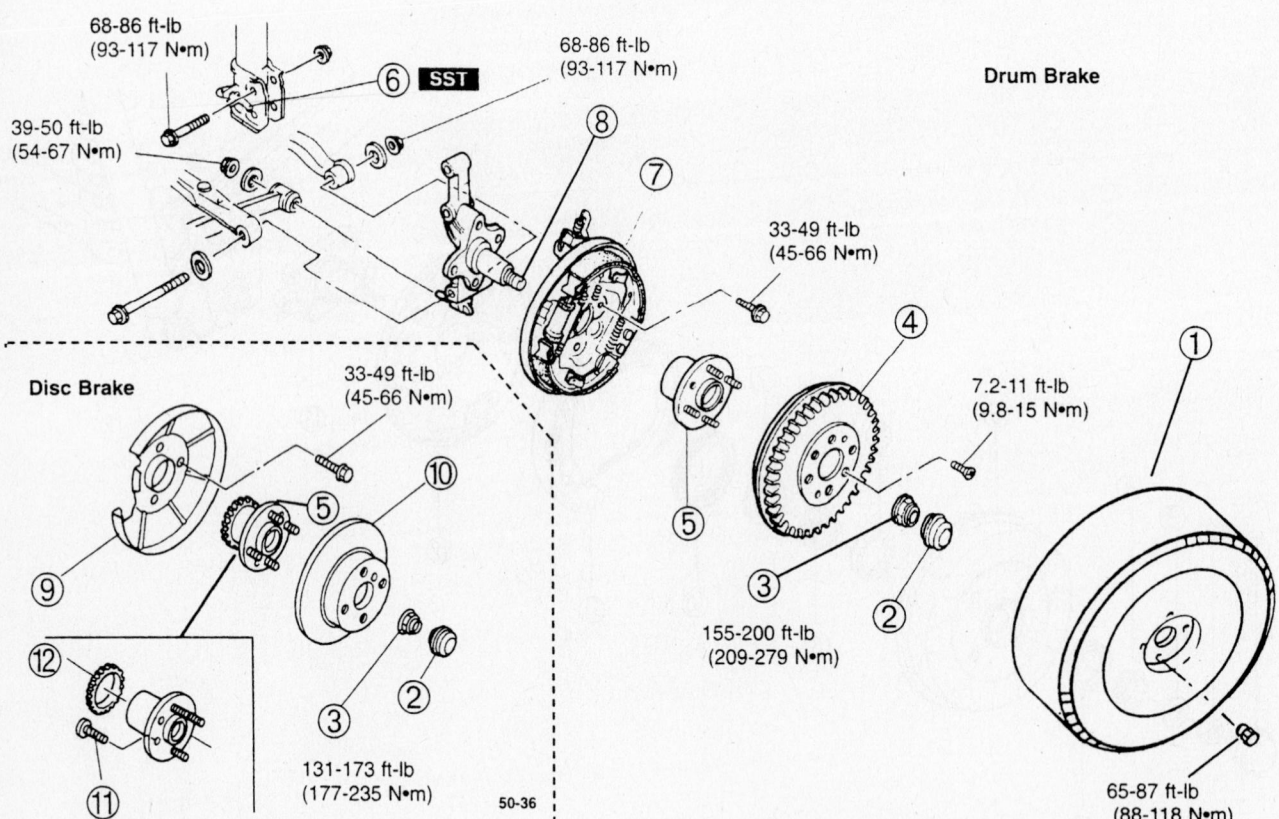

1. Wheel and tire
2. Dust cap
3. Locknut (Replace)
4. Brake drum (or disc)
5. Hub with bearing assembly
6. Brake line
7. Rear brake assembly (drum or disc)
8. Spindle
9. Dust cover
10. Brake rotor
11. Hub bolt
12. ABS sensor rotor

Exploded view of the rear axle assembly—Sephia and Spectra

BRAKES

Brake Caliper

REMOVAL & INSTALLATION

Front

1. Before servicing the vehicle, refer to the precautions at the beginning of this section.
2. Remove the wheels.
3. Remove the brake hose from the caliper.

4. Remove the caliper guide pin bolts and caliper.

To install:

5. Install the caliper and tighten the guide pin bolts to 19–21 ft. lbs. (26–28 Nm)
6. Install the brake hose and tighten to 9–13 ft. lbs. (13–18 Nm)
7. Bleed the brakes and fill the master cylinder with clean brake fluid.
8. Install the wheels.

Rear

1. Before servicing the vehicle, refer to the precautions at the beginning of this section.
2. Remove the wheels.
3. Remove the parking brake cable and clip.
4. Remove the brake hose banjo bolt.
5. Remove the caliper lock bolts and the caliper.

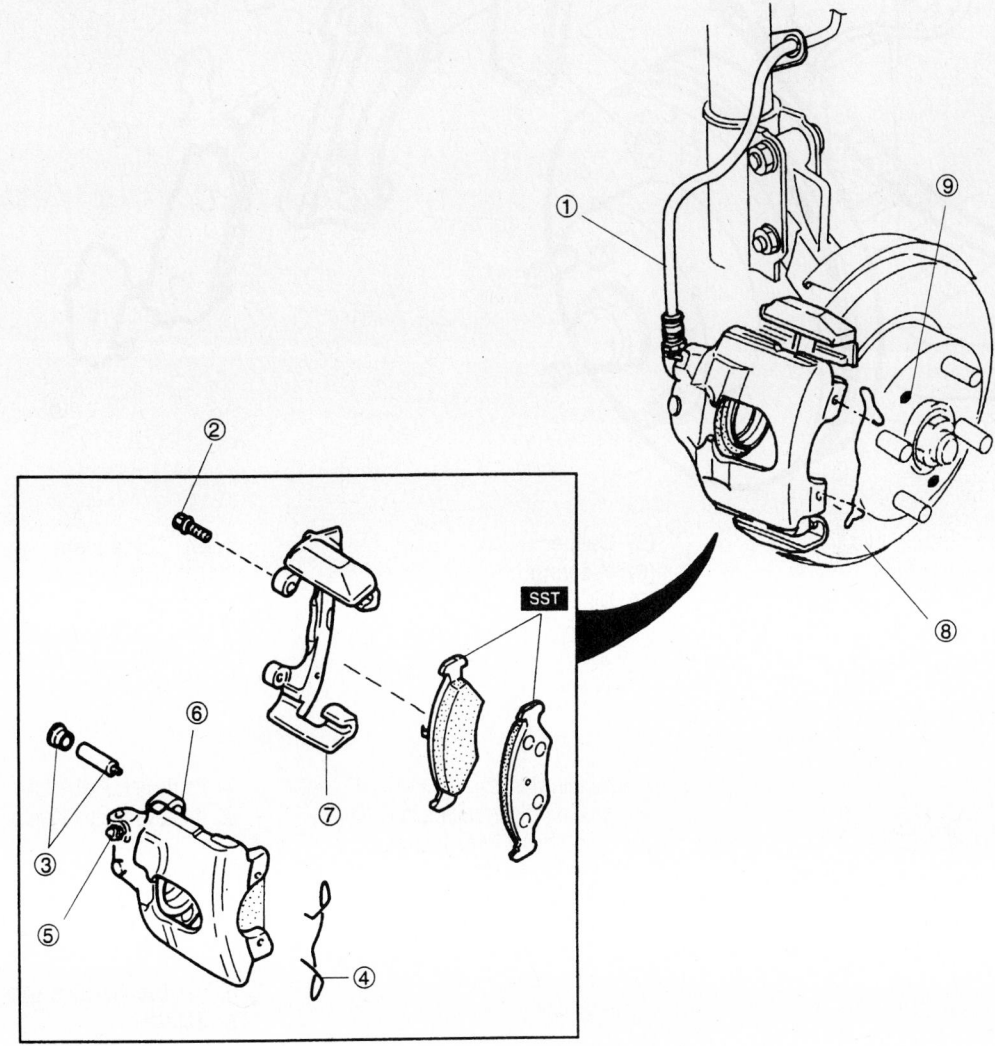

(1) Flexible hose
(2) Bolt
(3) Cap, bolt (square head) and bushing
(4) Spring
(5) Cap and bleeder screw
(6) Caliper
(7) Supporting plate
(8) Brake rotor (Disc)
(9) Mounting screws

Exploded view of the front brakes

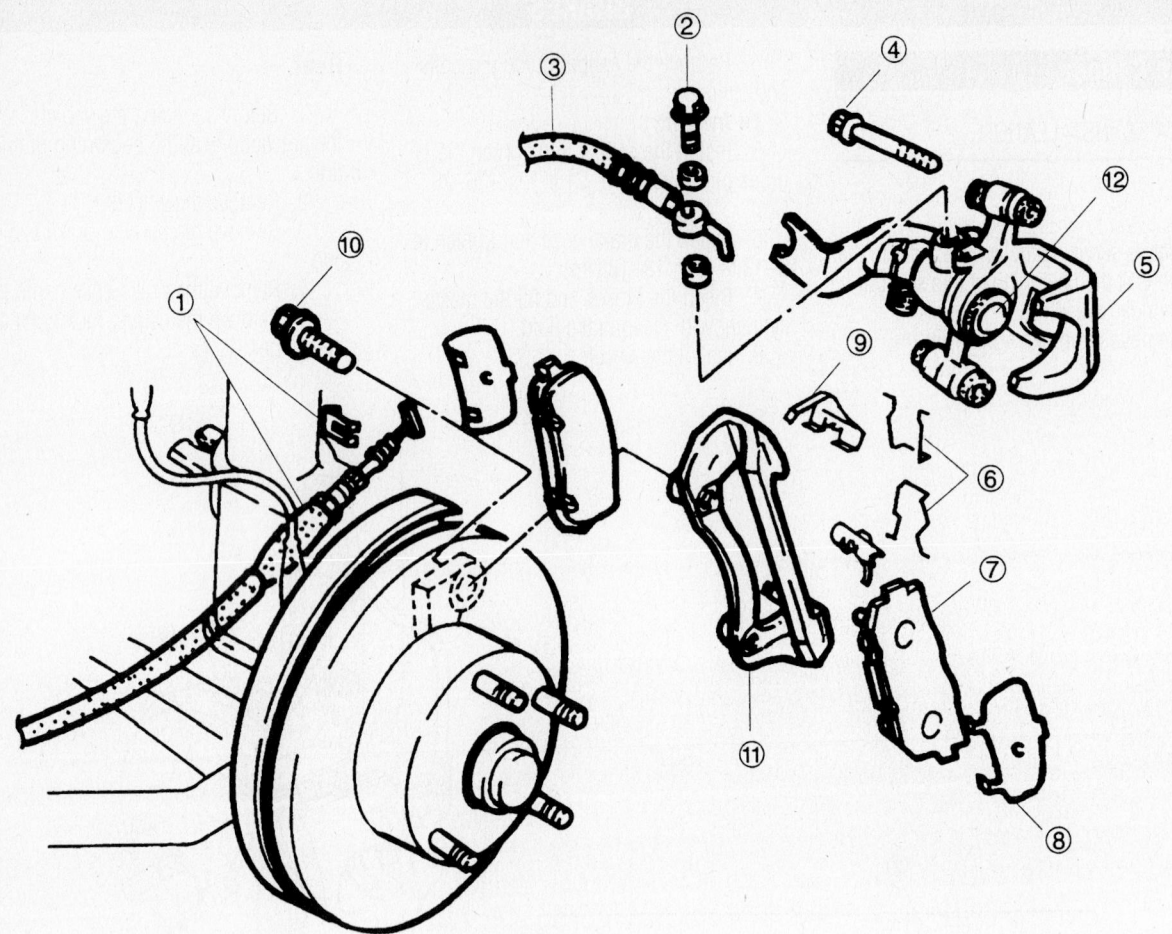

(1) Parking cable, clip
(2) Connecting bolt
(3) Brake hose
(4) Lock bolt

(5) Caliper
(6) V-spring
(7) Disc pad
(8) Shim

(9) Guide plate
(10) Bolt
(11) Mounting support
(12) Caliper piston

93016G16

Exploded view of the rear disc brakes

To install:

6. Install the caliper and tighten the lock bolts to 22–29 ft. lbs. (29–39 Nm)

7. Install the brake hose and tighten the banjo bolt to 16–22 ft. lbs. (22–29 Nm)

8. Install the park brake cable and clip.

9. Bleed the brakes and fill the master cylinder with clean brake fluid.

10. Install the wheels.

Disc Brake Pads

REMOVAL & INSTALLATION

Front

1. Before servicing the vehicle, refer to the precautions at the beginning of this section.

2. Remove the wheels.

3. Remove the caliper guide pin bolts and lift the caliper away from the rotor.

4. Remove the brake pads and retainer spring from the caliper.

To install:

5. Compress the caliper piston into the bore.

6. Install the brake pads and retainer spring to the caliper.

7. Install the caliper in position on the caliper support bracket.

8. Install the guide pin bolts.

9. Install the wheels.

Rear

1. Before servicing the vehicle, refer to the precautions at the beginning of this section.

2. Remove the wheels.

3. Remove the parking brake cable and clip.

4. Remove the caliper lock bolts and remove the caliper.

5. Remove the V-springs from the brake pads.

6. Remove the brake pads and shims.

To install:

7. Compress the caliper piston into the bore by rotating the piston with special tool OK9A4 263 001.

8. Install the new pads and shims.

9. Install the V-springs.

10. Install the caliper and tighten the lock bolts to 22–29 ft. lbs. (29–39 Nm)

11. Install the park brake cable and clip.

12. Install the wheels.

Brake Drums

REMOVAL & INSTALLATION

1. Before servicing the vehicle, refer to the precautions at the beginning of this section.
2. Remove the wheels.
3. Remove the retaining screws and brake drum. Two 8mm x 1.25 bolts can be used to press the drum from the hub.

To install:
4. Install the brake drum to the hub.
5. Install the wheels.

Brake Shoes

REMOVAL & INSTALLATION

1. Before servicing the vehicle, refer to the precautions at the beginning of this section.

2. Remove the wheels.
3. Remove the retaining screws and brake drum. Two 8mm x 1.25 bolts can be used to press the drum from the hub.
4. Remove the top return spring.
5. Remove the shoe retainer springs and pins.
6. Remove the adjuster spring and anti-rattle spring from the operating lever assembly.

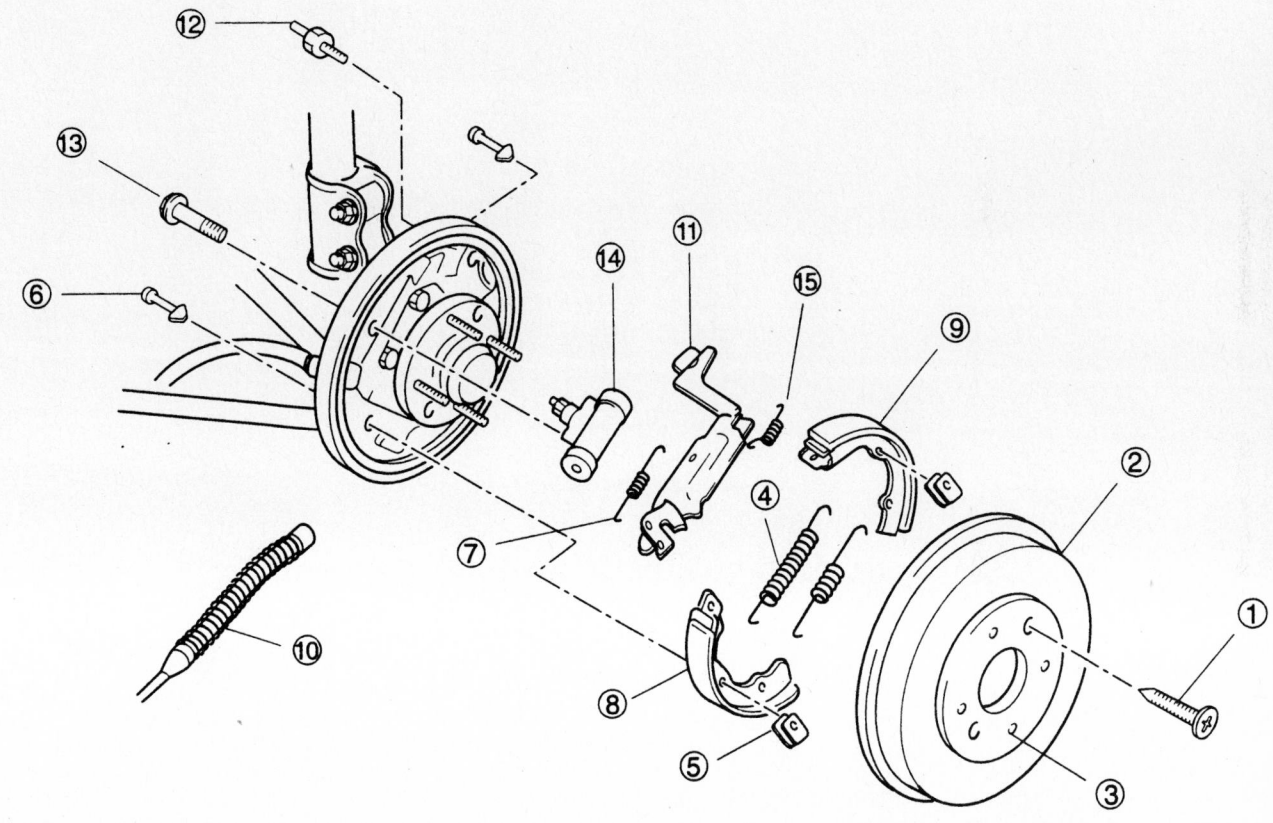

(1) Mounting screws
(2) Brake drum
(3) Drum pulling threads
(4) Return springs
(5) Spring clips
(6) Hold down pins
(7) Adjuster spring
(8) Brake shoe-leading
(9) Brake shoe-trailing
(10) Parking brake cable
(11) Operating lever assembly
(12) Brake line
(13) Bolts
(14) Wheel cylinder assembly
(15) Anti-rattle spring

93016G17

Exploded view of the rear drum brakes

7. Remove the bottom return spring.

8. Remove the parking brake cable from the rear shoe and remove the brake shoes.

To install:

9. Position the operating lever assembly above the hub.

10. Install the parking brake cable to the rear shoe.

11. Fit the operating lever assembly between the front and rear shoes and install the bottom return spring.

12. Install the shoe retainer pins and springs.

13. Install the top return spring.

14. Install the adjuster spring and the anti-rattle spring to the operating lever assembly.

15. Install the brake drum.

16. Install the wheels and adjust the brakes.

SPECIFICATIONS AND MAINTENANCE CHARTS

ENGINE AND VEHICLE IDENTIFICATION

Code ①	Liters (cc)	Cu. In.	Cyl.	Fuel Sys.	Engine Type	Eng. Mfg.
1	3.5 (3497)	213	6	EGI	DOHC	KIA

EGI: Electronic Gasoline Injection

DOHC: Double Overhead Camshafts

① 8th digit of VIN

② 10th digit of VIN

Model Year Code ②	Year
2	2002
3	2003
4	2004
5	2005

67162-SED0-C01

GENERAL ENGINE SPECIFICATIONS

Year	Model	Engine Displacement Liters	Engine VIN	Net Horsepower @ rpm	Net Torque @ rpm (ft. lbs.)	Bore x Stroke (in.)	Compression Ratio	Oil Pressure @ rpm
2002	Sedona	3.5	1	195@5500	218@3500	3.66x3.88	10:01	43-57@3000
2003	Sedona	3.5	1	195@5500	218@3500	3.66x3.88	10:01	43-57@3000
2004	Sedona	3.5	1	195@5500	218@3500	3.66x3.88	10:01	43-57@3000

67162-SED0-C02

ENGINE TUNE-UP SPECIFICATIONS

Year	Engine Displacement Liters	Engine VIN	Spark Plug Gap (in.)	Ignition Timing (deg.) MT	Ignition Timing (deg.) AT	Fuel Pump (psi)	Idle Speed (rpm) MT	Idle Speed (rpm) AT	Valve Clearance Intake	Valve Clearance Exhaust
2002	3.5	1	0.039-0.043	—	①	46-49	—	600-800	HYD	HYD
2003	3.5	1	0.039-0.043	—	①	46-49	—	600-800	HYD	HYD
2004	3.5	1	0.039-0.043	—	①	46-49	—	600-800	HYD	HYD

NOTE: The Vehicle Emission Control Information label often reflects specification changes made during production.

The label figures must be used if they differ from those in this chart

B: Before top dead center

HYD: Hydraulic

① Computer controled, no adjustment possible

67162-SED0-C03

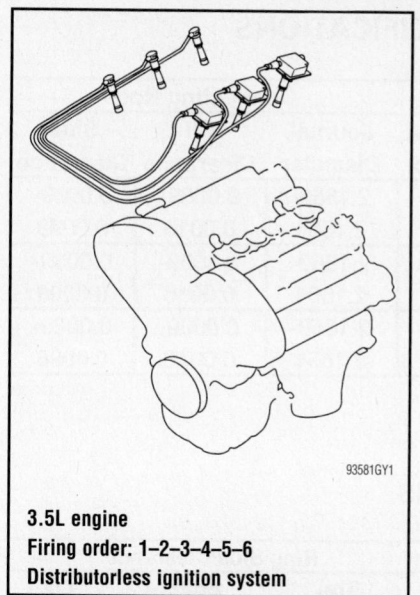

3.5L engine
Firing order: 1–2–3–4–5–6
Distributorless ignition system

93581GY1

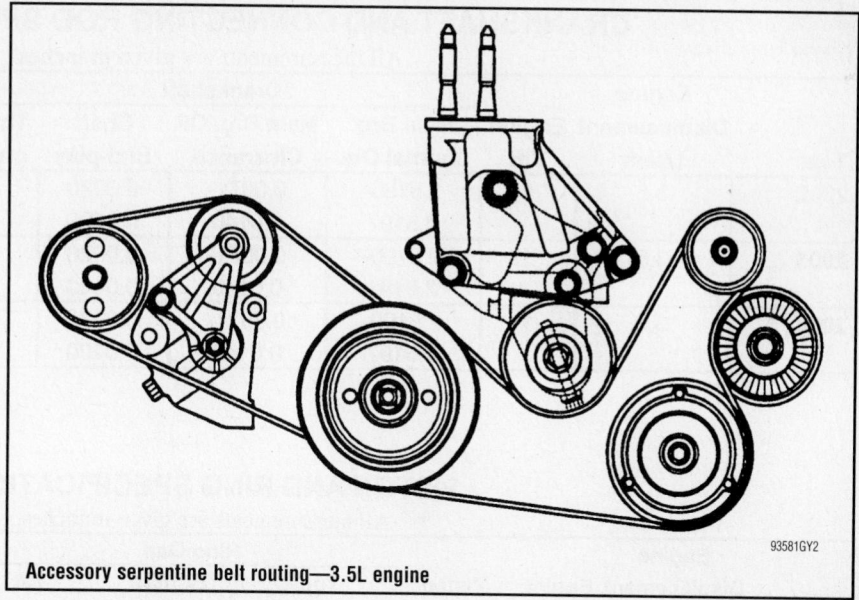

Accessory serpentine belt routing—3.5L engine

93581GY2

CAPACITIES

Year	Model	Engine Displacement Liters	Engine VIN	Engine Oil with Filter	Transaxle (pts.) Manual	Transaxle (pts.) Auto.	Fuel Tank (gal.)	Cooling System (qts.)
2002	Sedona	3.5	1	4.3	—	8.5	19.8	8.2
2003	Sedona	3.5	1	4.3	—	8.5	19.8	8.2
2004	Sedona	3.5	1	4.3	—	8.5	19.8	8.2

NOTE: All capacities are approximate. Add fluid gradually and ensure a proper level is obtained.

67162-SEDO-C04

VALVE SPECIFICATIONS

Year	Engine Displacement Liters	Engine VIN	Seat Angle (deg.)	Face Angle (deg.)	Maximum out of Square (degrees)	Spring Free Length (in.)	Stem-to-Guide Clearance (in.) Intake	Stem-to-Guide Clearance (in.) Exhaust	Stem Diameter (in.) Intake	Stem Diameter (in.) Exhaust
2002	3.5	1	45	45	2	1.8268	0.0009–0.0020	0.0020–0.0039	0.258–0.259	0.257–0.258
2003	3.5	1	45	45	2	1.8268	0.0009–0.0020	0.0020–0.0039	0.258–0.259	0.257–0.258
2004	3.5	1	45	45	2	1.8268	0.0009–0.0020	0.0020–0.0039	0.258–0.259	0.257–0.258

NA: Not Available

67162-SEDO-C05

CRANKSHAFT AND CONNECTING ROD SPECIFICATIONS

All measurements are given in inches.

Year	Engine Displacement Liters	Engine VIN	Crankshaft Main Brg. Journal Dia.	Crankshaft Main Brg. Oil Clearance	Crankshaft Shaft End-play	Crankshaft Thrust on No.	Connecting Rod Journal Diameter	Connecting Rod Oil Clearance	Connecting Rod Side Clearance
2002	3.5	1	2.5190-2.5197	0.0071-0.0140	0.0020-0.0100	3	2.1650-2.1654	0.0009-0.0016	0.0039-0.0098
2003	3.5	1	2.5190-2.5197	0.0071-0.0140	0.0020-0.0100	3	2.1650-2.1654	0.0009-0.0016	0.0039-0.0098
2004	3.5	1	2.5190-2.5197	0.0071-0.0140	0.0020-0.0100	3	2.1650-2.1654	0.0009-0.0016	0.0039-0.0098

67162-SEDO-C06

PISTON AND RING SPECIFICATIONS

All measurements are given in inches.

Year	Engine Displacement Liters	Engine VIN	Piston Clearance	Ring Gap Top Compression	Ring Gap Bottom Compression	Ring Gap Oil Control	Ring Side Clearance Top Compression	Ring Side Clearance Bottom Compression	Ring Side Clearance Oil Control
2002	3.5	1	0.0012-0.0020	0.0079-0.0118	0.0177-0.0236	0.0079-0.0276	0.0016-0.0315	0.0008-0.0024	SNUG
2003	3.5	1	0.0012-0.0020	0.0079-0.0118	0.0177-0.0236	0.0079-0.0276	0.0016-0.0315	0.0008-0.0024	SNUG
2004	3.5	1	0.0012-0.0020	0.0079-0.0118	0.0177-0.0236	0.0079-0.0276	0.0016-0.0315	0.0008-0.0024	SNUG

67162-SEDO-C07

TORQUE SPECIFICATIONS

All readings in ft. lbs.

Year	Engine Displacement Liters	Engine VIN	Cylinder Head Bolts	Main Bearing Bolts	Rod Bearing Bolts	Crankshaft Damper Bolts	Flywheel Bolts	Manifold Intake	Manifold Exhaust	Spark Plugs	Oil Pan Drain Plug
2002	3.5	1	77-85	52-59	①	130-138	54-57	②	30-37	15-22	26-30
2003	3.5	1	77-85	52-59	①	130-138	54-57	②	30-37	15-22	26-30
2004	3.5	1	77-85	52-59	①	130-138	54-57	②	30-37	15-22	26-30

① Step 1: 24-27 ft. lbs.
   Step 2: Plus 90-94 degrees

② Upper: 15-17 ft. lbs.
   Lower: 15-22 ft. lbs.

67162-SEDO-C08

WHEEL ALIGNMENT

Year	Model		Caster Range (+/-Deg.)	Caster Preferred Setting (Deg.)	Camber Range (+/-Deg.)	Camber Preferred Setting (Deg.)	Toe-in (in.)
2002	Sedona	F	0.50	+1.88	0.50	0.51	-0.04
		R	—	—	—	—	—
2003	Sedona	F	0.50	+1.88	0.50	0.51	-0.04
		R	—	—	—	—	—
2004	Sedona	F	0.50	+1.88	0.50	0.51	-0.04
		R	—	—	—	—	—

67162-SEDO-C09

TIRE, WHEEL AND BALL JOINT SPECIFICATIONS

Year	Model	OEM Tires Standard	OEM Tires Optional	Tire Pressures (psi) Front	Tire Pressures (psi) Rear	Wheel Size	Ball Joint Inspection	Wheel Lug Torque (ft. lbs.)
2002	Sedona	P215/70R15	—	35	35	6JJ	①	65-79
2003	Sedona	P215/70R15	—	35	35	6JJ	①	65-79
2004	Sedona	P215/70R15	—	35	35	6JJ	①	65-79

OEM: Original Equipment Manufacturer

PSI: Pounds Per Square Inch

STD: Standard

OPT: Optional

① Replace if any measurable movement is found.

67162-SEDO-C10

BRAKE SPECIFICATIONS

All measurements in inches unless noted

Year	Model		Brake Disc Original Thickness	Brake Disc Minimum Thickness	Brake Disc Maximum Run-out	Brake Drum Original Inside Diameter	Brake Drum Max. Wear Limit	Brake Drum Maximum Machine Diameter	Minimum Lining Thickness	Brake Caliper Bracket Bolts (ft. lbs.)	Brake Caliper Mounting Bolts (ft. lbs.)
2002	Sedona	F	1.020	0.940	0.0020	—	—	—	0.100	7.2-9.4	18-26
		R	—	—	—	10.00	10.05	10.05	0.040	—	—
2003	Sedona	F	1.020	0.940	0.0020	—	—	—	0.100	7.2-9.4	18-26
		R	—	—	—	10.00	10.05	10.05	0.040	—	—
2004	Sedona	F	1.020	0.940	0.0020	—	—	—	0.100	7.2-9.4	18-26
		R	—	—	—	10.00	10.05	10.05	0.040	—	—

F: Front

R: Rear

67162-SEDO-C11

SCHEDULED MAINTENANCE INTERVALS
Kia—Sedona

TO BE SERVICED	TYPE OF SERVICE	7.5	15	22.5	30	37.5	45	52.5	60	67.5	75	82.5	90	97.5	105	112.5	120
								VEHICLE MILEAGE INTERVAL (x1000)									
Accessory drive belts	S/I			✓			✓			✓			✓			✓	
Air cleaner element	I/R		✓		✓		✓		✓		✓		✓		✓		✓
Air conditioner system	S/I	Inspect the system operation and refrigerant amount annually.															
Brake lines, hoses and connections	S/I		✓		✓		✓		✓		✓		✓		✓		✓
Chassis and body fasteners	T	✓	✓	✓	✓	✓	✓	✓	✓	✓	✓	✓	✓	✓	✓	✓	✓
Cooling system hoses and coolant level	S/I		✓		✓		✓		✓		✓		✓		✓		✓
CV-joint boots	S/I				✓				✓				✓				✓
Engine coolant	R				✓				✓				✓				✓
Engine oil and filter	R	✓	✓	✓	✓	✓	✓	✓	✓	✓	✓	✓	✓	✓	✓	✓	✓
Exhaust system heat shields	S/I				✓				✓				✓				✓
Front and rear brakes	S/I				✓				✓				✓				✓
Front ball joints S/I	S/I				✓				✓								✓
Fuel filter	R								✓								✓
Fuel lines and hoses	S/I				✓				✓				✓				✓
Idle speed	A				✓				✓								✓
Locks and hinges	L	✓	✓	✓	✓	✓	✓	✓	✓	✓	✓	✓	✓	✓	✓	✓	✓
Spark plugs	R				✓				✓				✓				✓
Steering operation and linkage	S/I				✓				✓				✓				✓
Timing belt	R								✓								✓

R: Replace   S/I: Inspect and service, if needed   L: Lubricate   A: Adjust   T: Tighten

FREQUENT OPERATION MAINTENANCE (SEVERE SERVICE)

If a vehicle is operated under any of the following conditions it is considered severe service

- Towing a trailer or using a camper or car-top carrier.
- Repeated short trips of less than 5 miles in temperatures below freezing, or trips of less than 10 miles in any temperature.
- Extensive idling or low-speed driving for long distances as in heavy commercial use, such as delivery, taxi or police cars.
- Operating on rough, muddy or salt-covered roads.
- Operating on unpaved or dusty roads.
- Driving in extremely hot (over 90°F) conditions.

Engine oil and filter: replace every 5000 miles or 5 months, whichever occurs first.

Air cleaner element: inspect ever 15,000 miles or 15 months and replace every 30,000 miles or 30 months, whichever occurs first.

Fuel system hoses (California models only): replace every 105,000 miles.

Emission system hoses (non-California models): inspect every 55,000 miles or 55 months, whichever occurs first.

Emission system hoses (California models): inspect every 60,000 miles or 60 months, whichever occurs first.

Front and rear disc brakes: inspect every 15,000 miles or 15 months, whichever occurs first.

Chassis and body fasteners: tighten every 15,000 miles or 15 months, whichever occurs first.

Locks and hinges: lubricate every 5000 miles or 5 months, whichever occurs first.

67162-SEDO-C12

PRECAUTIONS

Before servicing any vehicle, please be sure to read all of the following precautions, which deal with personal safety, prevention of component damage, and important points to take into consideration when servicing a motor vehicle:

• Never open, service or drain the radiator or cooling system when the engine is hot; serious burns can occur from the steam and hot coolant.

• Observe all applicable safety precautions when working around fuel. Whenever servicing the fuel system, always work in a well-ventilated area. Do not allow fuel spray or vapors to come in contact with a spark, open flame, or excessive heat (a hot drop light, for example). Keep a dry chemical fire extinguisher near the work area. Always keep fuel in a container specifically designed for fuel storage; also, always properly seal fuel containers to avoid the possibility of fire or explosion. Refer to the additional fuel system precautions later in this section.

• Fuel injection systems often remain pressurized, even after the engine has been turned **OFF**. The fuel system pressure must be relieved before disconnecting any fuel lines. Failure to do so may result in fire and/or personal injury.

• Brake fluid often contains polyglycol ethers and polyglycols. Avoid contact with the eyes and wash your hands thoroughly after handling brake fluid. If you do get brake fluid in your eyes, flush your eyes with clean, running water for 15 minutes. If eye irritation persists, or if you have taken brake fluid internally, IMMEDIATELY seek medical assistance.

• The EPA warns that prolonged contact with used engine oil may cause a number of skin disorders, including cancer. You should make every effort to minimize your exposure to used engine oil. Protective gloves should be worn when changing oil. Wash your hands and any other exposed skin areas as soon as possible after exposure to used engine oil. Soap and water, or waterless hand cleaner should be used.

• All new vehicles are now equipped with an air bag system, often referred to as a Supplemental Restraint System (SRS) or Supplemental Inflatable Restraint (SIR) system. The system must be disabled before performing service on or around system components, steering column, instrument panel components, wiring and sensors. Failure to follow safety and disabling procedures could result in accidental air bag deployment, possible personal injury and unnecessary system repairs.

• Always wear safety goggles when working with, or around, the air bag system. When carrying a non-deployed air bag, be sure the bag and trim cover are pointed away from your body. When placing a non-deployed air bag on a work surface, always face the bag and trim cover upward, away from the surface. This will reduce the motion of the module if it is accidentally deployed. Refer to the additional air bag system precautions later in this section.

• Clean, high quality brake fluid from a sealed container is essential to the safe and proper operation of the brake system. You should always buy the correct type of brake fluid for your vehicle. If the brake fluid becomes contaminated, completely flush the system with new fluid. Never reuse any brake fluid. Any brake fluid that is removed from the system should be discarded. Also, do not allow any brake fluid to come in contact with a painted surface; it will damage the paint.

• Never operate the engine without the proper amount and type of engine oil; doing so WILL result in severe engine damage.

• Timing belt maintenance is extremely important. Many models utilize an interference-type, non-freewheeling engine. If the timing belt breaks, the valves in the cylinder head may strike the pistons, causing potentially serious (also time-consuming and expensive) engine damage. Refer to the maintenance interval charts for the recommended replacement interval for the timing belt.

• Disconnecting the negative battery cable on some vehicles may interfere with the functions of the on-board computer system(s) and may require the computer to undergo a relearning process once the negative battery cable is reconnected.

• When servicing drum brakes, only disassemble and assemble one side at a time, leaving the remaining side intact for reference.

• Only an MVAC-trained, EPA-certified automotive technician should service the air conditioning system or its components.

ENGINE REPAIR

Alternator

REMOVAL

1. Before servicing the vehicle, refer to the precautions at the beginning of this section.
2. Disconnect the negative battery cable.
3. Remove the accessory drive belt.
4. Remove the alternator mounting bolts.
5. Remove the alternator.
To install:
6. Install the alternator.
7. Connect the alternator electrical connectors. Tighten the battery terminal connector nut to 60 inch lbs. (7 Nm).
8. Install the accessory drive belt.
9. Connect the negative battery cable.

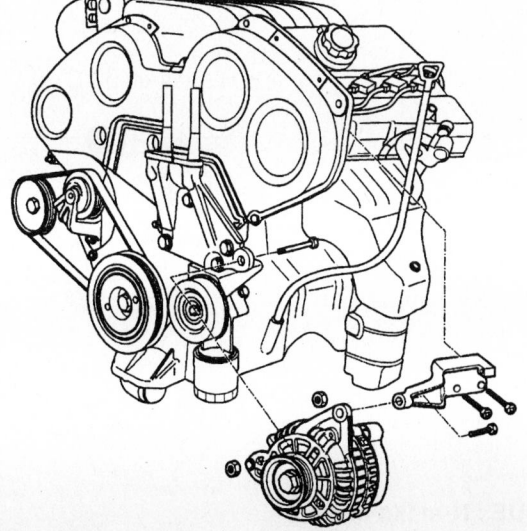

9358HG01

Alternator mounting exploded view

Ignition Timing

This vehicle is equipped with a Distributorless Ignition System (DIS). No adjustment is necessary or possible.

Engine Assembly

REMOVAL & INSTALLATION

1. Before servicing the vehicle, refer to the precautions at the beginning of this section.
2. Drain the cooling system.
3. Drain the transaxle fluid.
4. Relieve the fuel system pressure.
5. Remove the battery.
6. Remove the engine cover.
7. Remove the air cleaner assembly.
8. Disconnect the engine wiring harness connectors.
9. Disconnect the alternator wiring harness connectors.
10. Disconnect the oil pressure switch connector.
11. Disconnect the oil pressure sensor connector.
12. Disconnect the starter motor wiring harness connectors.
13. Disconnect the transaxle oil cooler hose.
14. Remove the upper and lower radiator hoses.
15. Remove the radiator.
16. Remove the engine ground cable.
17. Disconnect the brake booster vacuum hose.
18. Disconnect the EVAP canister hose.
19. Disconnect the fuel lines.
20. Remove the heater hoses.
21. Disconnect the throttle and cruise control cables.
22. Disconnect the transaxle shift cable.
23. Disconnect the power steering pump hose.
24. Remove the oil pan shield.

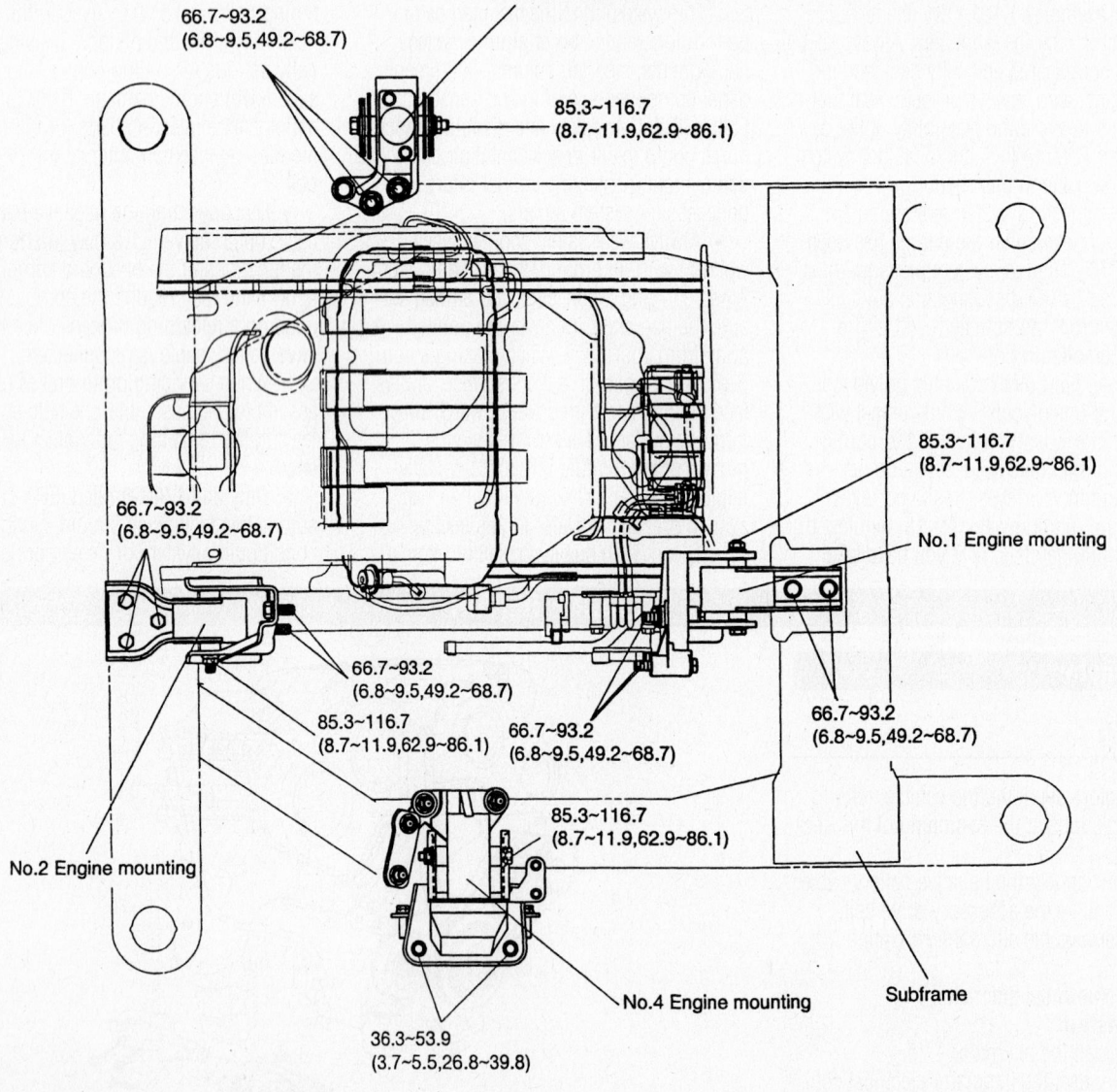

No.3 Engine mounting

66.7~93.2
(6.8~9.5,49.2~68.7)

85.3~116.7
(8.7~11.9,62.9~86.1)

85.3~116.7
(8.7~11.9,62.9~86.1)

No.1 Engine mounting

66.7~93.2
(6.8~9.5,49.2~68.7)

66.7~93.2
(6.8~9.5,49.2~68.7)

85.3~116.7
(8.7~11.9,62.9~86.1)

66.7~93.2
(6.8~9.5,49.2~68.7)

66.7~93.2
(6.8~9.5,49.2~68.7)

No.2 Engine mounting

85.3~116.7
(8.7~11.9,62.9~86.1)

No.4 Engine mounting

Subframe

36.3~53.9
(3.7~5.5,26.8~39.8)

TORQUE : N·m (kg·m, lb·ft)

9358HG02

Engine mount locations and torque specifications

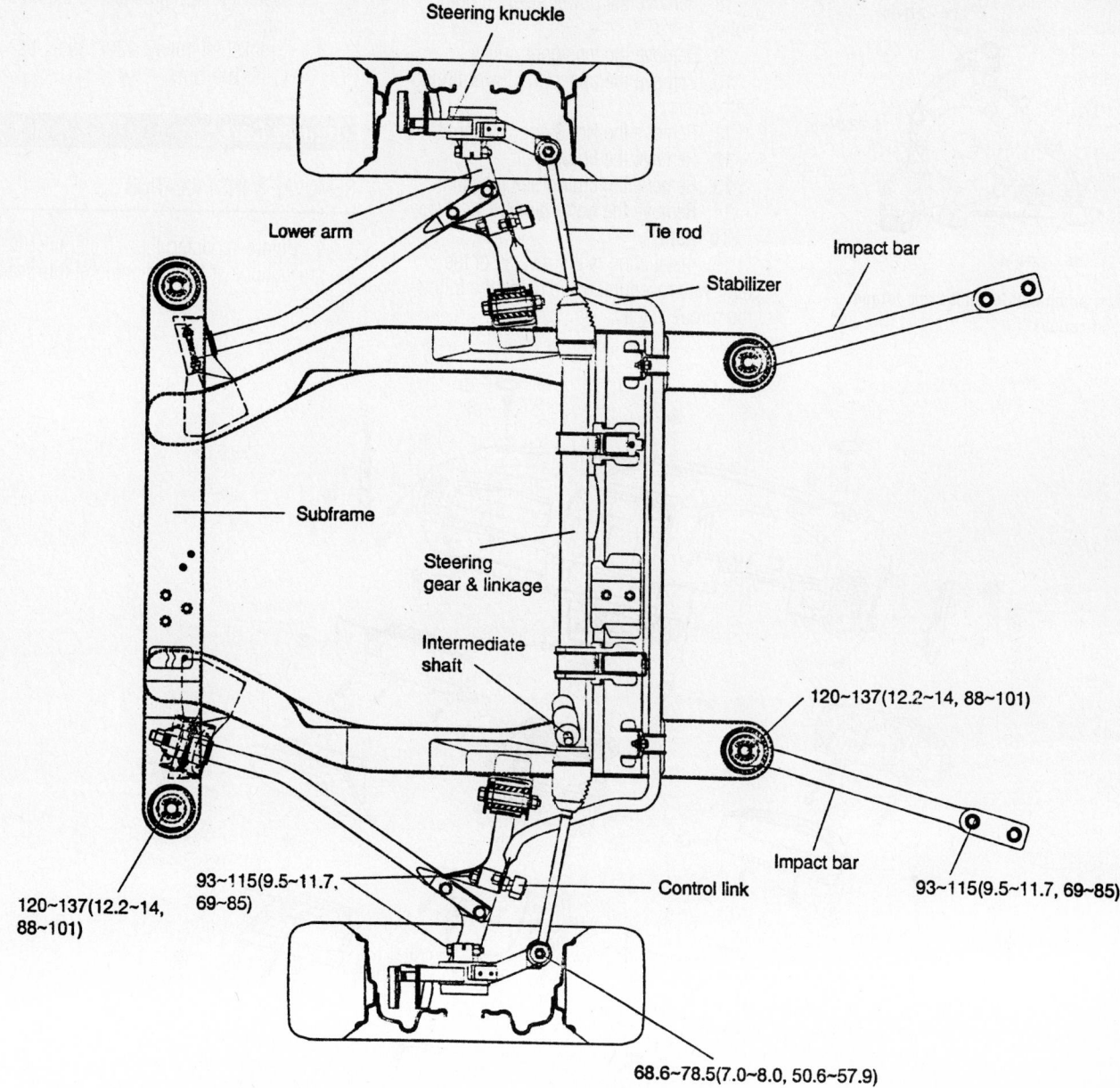

Steering knuckle

Lower arm

Tie rod

Impact bar

Stabilizer

Subframe

Steering gear & linkage

Intermediate shaft

120~137(12.2~14, 88~101)

Impact bar

93~115(9.5~11.7, 69~85)

Control link

93~115(9.5~11.7, 69~85)

120~137(12.2~14, 88~101)

68.6~78.5(7.0~8.0, 50.6~57.9)

TORQUE : N•m (kg•m, lb•ft)

9358HG03

Front subframe bolt locations and torque specifications

25. Remove the exhaust front pipe.
26. Remove the outer tie rod ends.
27. Remove the sway bar links.
28. Remove the lower ball joints.
29. Remove the axle halfshafts.
30. Remove the intermediate shaft bolt.
31. Remove the No. 3 and 4 engine mounts.
32. Support the front subframe with a suitable powertrain jack.
33. Remove the subframe bolts.
34. Remove the impact bar bolts.
35. Lower the powertrain and subframe away from the vehicle.

To install:

36. Installation is the reverse of the removal procedure, while using the following torque values:
- Subframe bolts: 88-101 ft. lbs. (120-137 Nm)
- Impact bar bolts: 69-85 ft. lbs. (93-115 Nm)
- Engine mount bolts: 49-69 ft. lbs. (67-93 Nm)
- Engine mount through-bolts: 63-86 ft. lbs. (85-117 Nm)
- Tie rod ends: 51-58 ft. lbs. (69-79 Nm)

Water Pump

REMOVAL & INSTALLATION

1. Before servicing the vehicle, refer to the precautions at the beginning of this section.
2. Drain the cooling system.
3. Disconnect the negative battery cable.
4. Remove the engine cover.
5. Remove the accessory drive belt.
6. Remove the idler pulley.
7. Remove the crankshaft pulley.

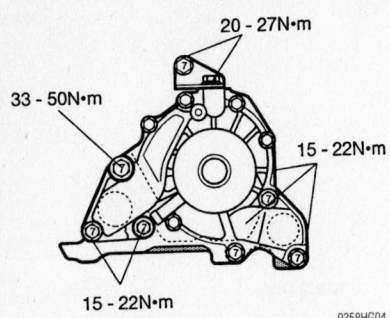

Water pump bolt locations and torque specifications

20 - 27N·m

33 - 50N·m

15 - 22N·m

15 - 22N·m

9358HG04

8. Remove the power steering pump pulley.

9. Remove the tensioner pulley.

10. Remove the upper and lower timing belt covers.

11. Remove the No. 3 engine mount.

12. Remove the timing belt.

13. Remove the timing belt tensioner.

14. Remove the water pump.

To install:

15. Installation is the reverse of the removal procedure, while using the following torque values:

- Water pump bolts: refer to the illustration
- Crankshaft pulley: 130-138 ft. lbs. (180-190 Nm)

Heater Core

REMOVAL & INSTALLATION

1. Before servicing the vehicle, refer to the precautions at the beginning of this section.

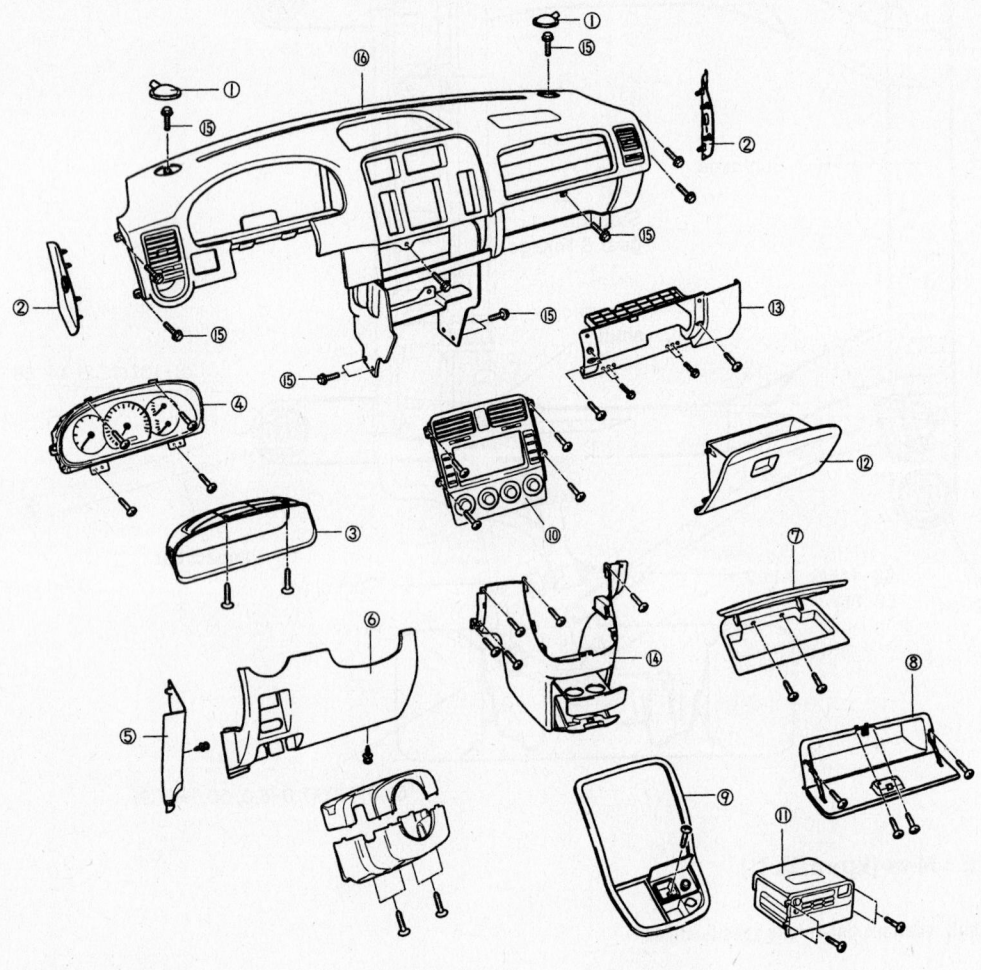

(1) Speaker assembly

(2) Side cover

(3) Instrument cluster trim

(4) Instrument cluster

(5) A-pillar lower trim

(6) Lower LH panel

(7) Center upper tray

(8) Multi box

(9) Audio panel

(10) Heater control panel

(11) Audio

(12) Glove box

(13) Lower RH panel

(14) Center console

(15) Mounting bolt

(16) Instrument panel

9358HG29

Instrument panel exploded view

2. Drain the cooling system.
3. Disconnect the negative battery cable.
4. Remove the heater hoses.
5. Remove the driver air bag module.
6. Remove the steering wheel.
7. Remove the turn signal assembly.
8. Remove the upper and lower steering column covers.
9. Remove the A-pillar upper trim.
10. Remove the instrument cluster trim.
11. Remove the instrument cluster.
12. Remove the A-pillar lower trim.
13. Remove the hood release handle.
14. Remove the left lower trim panel.
15. Remove the multi box.
16. Remove the audio panel.
17. Remove the ventilation control panel.
18. Remove the radio.
19. Remove the glove box.
20. Remove the center console.
21. Remove the audio system speakers and mounting bolts from the top of the instrument panel.
22. Remove the side covers.
23. Remove the T-bar mounting bolts.
24. Remove the bottom mounting bolts.
25. Remove the instrument panel wiring harness connectors.
26. Remove the instrument panel.

27. Remove the heater unit.
To install:
28. Install the heater unit.
29. Install the instrument panel.
30. Connect the instrument panel wiring harness connectors.
31. Install the bottom mounting bolts.
32. Install the T-bar mounting bolts.
33. Install the side covers.
34. Install the upper mounting bolts and audio system speakers.
35. Install the center console.
36. Install the glove box.
37. Install the radio.
38. Install the ventilation control panel.
39. Install the audio panel.
40. Install the multi box.
41. Install the left lower trim panel.
42. Install the hood release handle.
43. Install the A-pillar lower trim.
44. Install the instrument cluster.
45. Install the instrument cluster trim.
46. Install the A-pillar upper trim.
47. Install the upper and lower steering column covers.
48. Install the turn signal assembly.
49. Install the steering wheel.
50. Install the driver air bag module.
51. Install the heater hoses.

52. Connect the negative battery cable.
53. Fill the cooling system.
54. Start the engine and check for leaks and proper heater operation.

Cylinder Head

REMOVAL & INSTALLATION

1. Before servicing the vehicle, refer to the precautions at the beginning of this section.
2. Drain the cooling system.
3. Relieve the fuel system pressure.
4. Disconnect the negative battery cable.
5. Remove the engine cover.
6. Remove the timing belt.
7. Remove the intake manifold.
8. Remove the exhaust manifolds.
9. Remove the cylinder head covers.
10. Remove the camshafts.
11. Remove the rocker arms and lash adjusters.

➡**Keep all valvetrain components in order for assembly.**

12. Remove the cylinder head bolts.
13. Remove the cylinder heads.

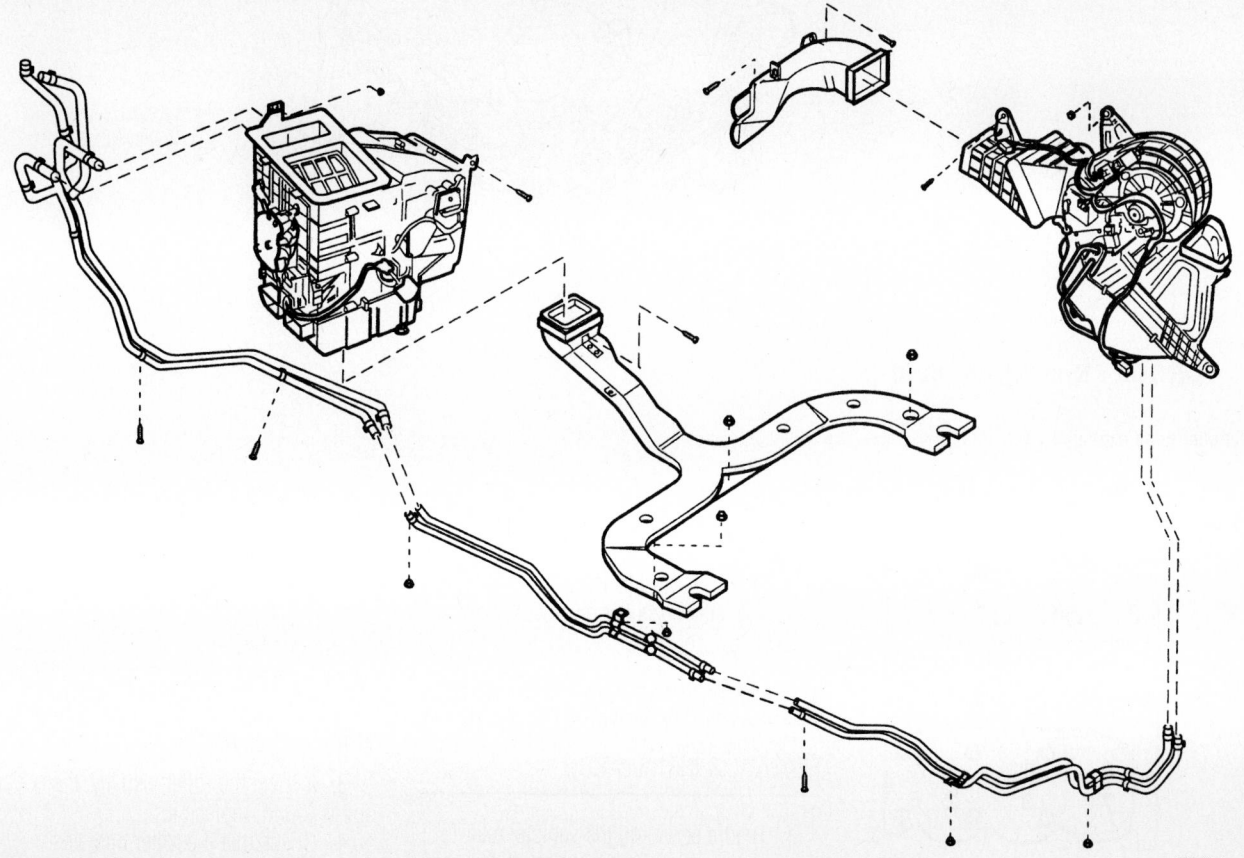

9358HG30

Heater unit mounting exploded view

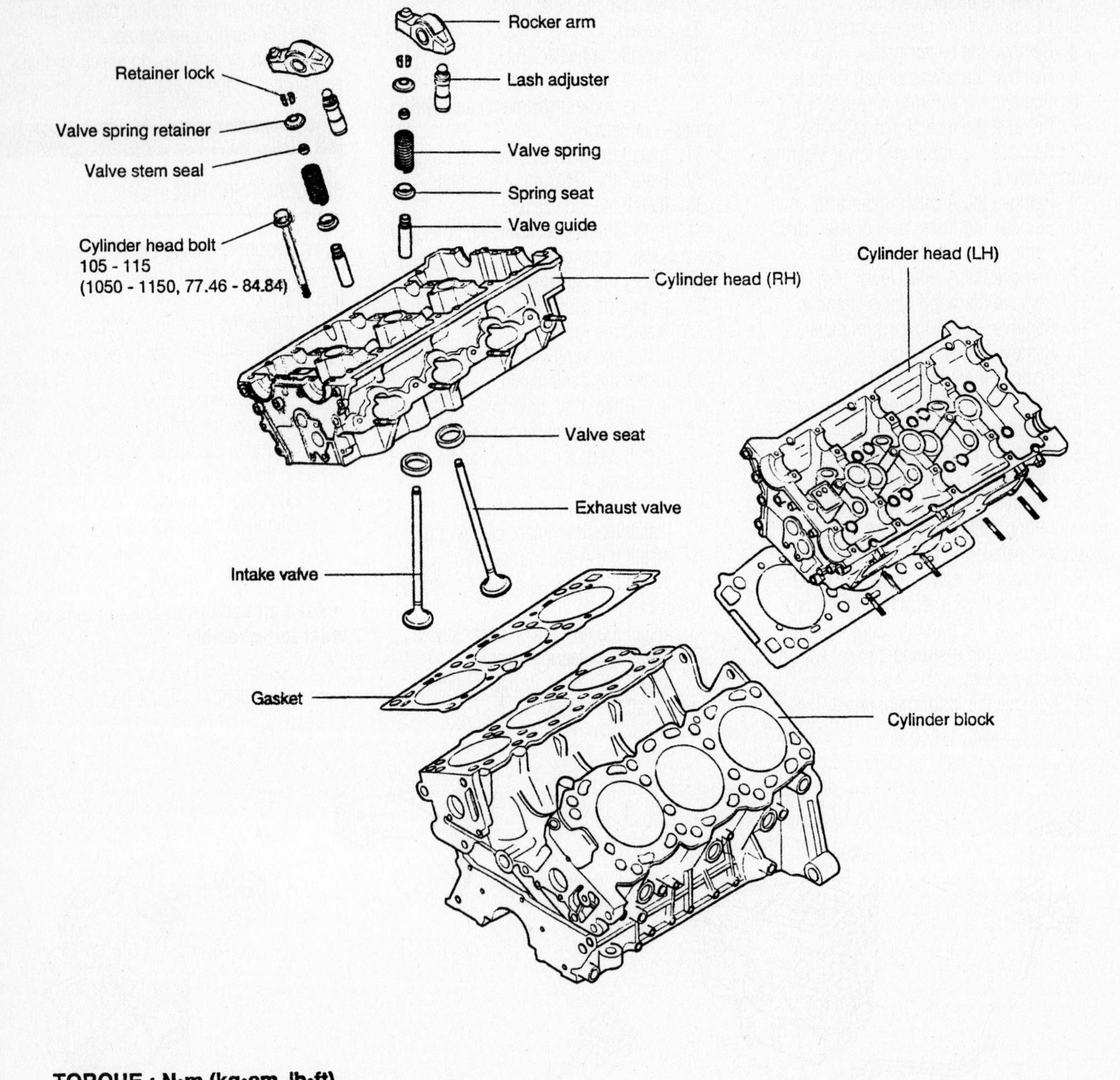

Rocker arm

Retainer lock

Lash adjuster

Valve spring retainer

Valve stem seal

Valve spring

Spring seat

Cylinder head bolt
105 - 115
(1050 - 1150, 77.46 - 84.84)

Valve guide

Cylinder head (RH)

Cylinder head (LH)

Valve seat

Exhaust valve

Intake valve

Gasket

Cylinder block

TORQUE : N•m (kg•cm, lb•ft)

9358HG05

Cylinder head exploded view

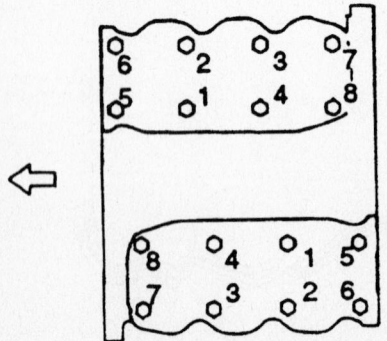

9358HG06

Cylinder head bolt torque sequence

To install:

14. Install the cylinder heads with new gaskets. Tighten the bolts in sequence to 78-85 ft. lbs. (105-115 Nm).

15. Installation is the reverse of the removal procedure.

Rocker Arms

REMOVAL & INSTALLATION

1. Before servicing the vehicle, refer to the precautions at the beginning of this section.

2. Remove or disconnect the following:
- Negative battery cable
- Engine cover
- Timing belt
- Cylinder head covers
- Camshafts
- Rocker arms and lash adjusters

➡**Keep all valvetrain components in order for installation.**

3. Inspect the roller visually. If any damage is found, replace it.

4. Check that the roller operates smoothly. If there is excessive clearance, replace it.

To install:

5. Installation is the reverse of the removal procedure.

Intake Manifold

REMOVAL & INSTALLATION

1. Before servicing the vehicle, refer to the precautions at the beginning of this section.

2. Drain the cooling system.

3. Relieve the fuel system pressure.

4. Remove or disconnect the following:
- Negative battery cable
- Engine cover
- Upper radiator hose
- Positive Crankcase Ventilation (PCV) hose
- Brake booster vacuum hose
- Surge tank stays
- Fuel lines
- Upper intake manifold (surge tank)
- Fuel injector harness connectors
- Fuel supply manifold with injectors attached
- Engine Coolant Temperature (ECT) sensor connector
- Coolant temperature gauge sensor connector
- Thermostat
- Lower intake manifold

To install:

5. Installation is the reverse of the removal procedure, while using the following torque values:
- Lower intake manifold nuts: 15-22 ft. lbs. (20-30 Nm)
- Upper intake manifold bolts: 11-15 ft. lbs. (15-20 Nm)

Exhaust Manifold

REMOVAL & INSTALLATION

1. Before servicing the vehicle, refer to the precautions at the beginning of this section.

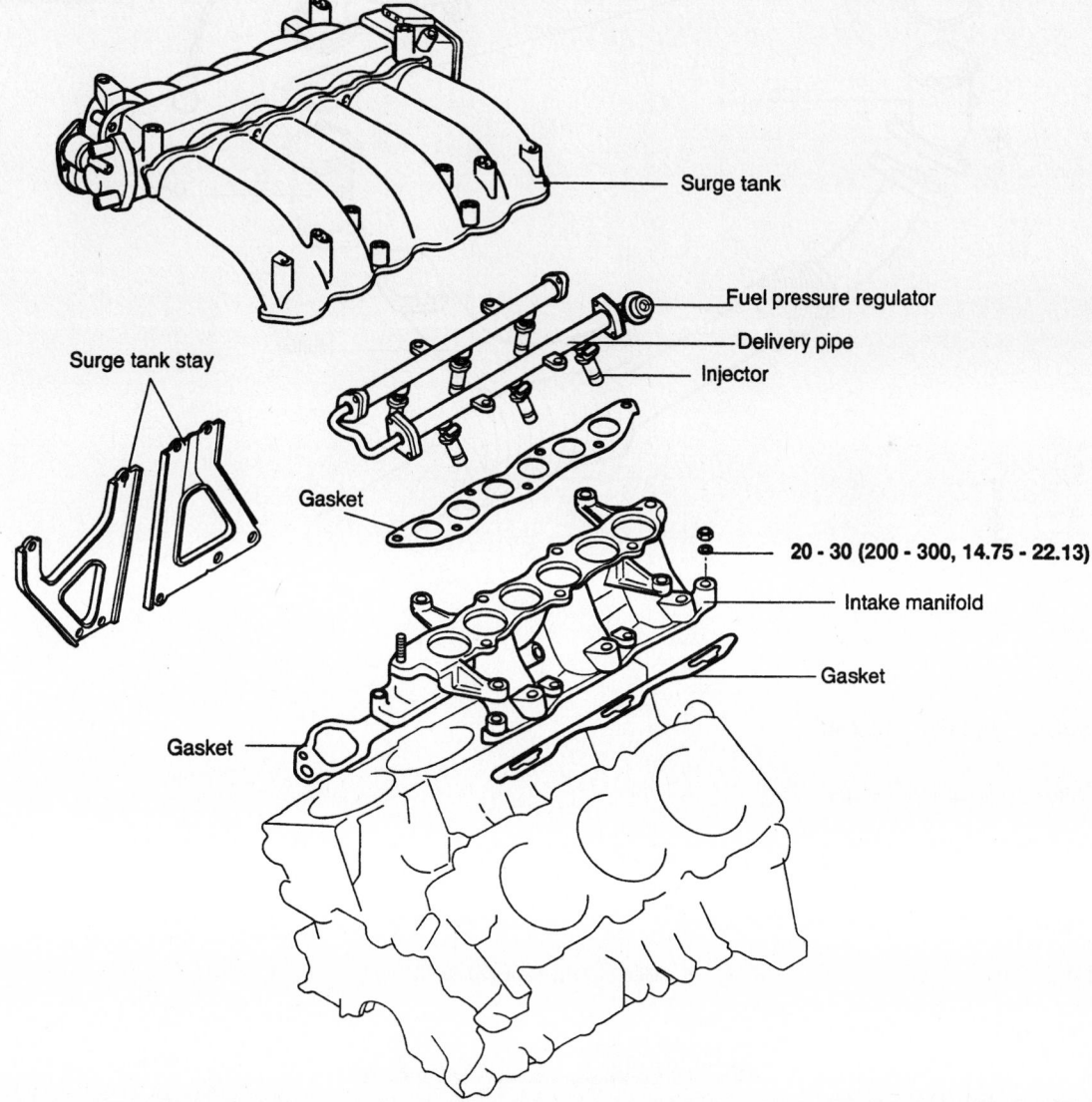

Surge tank

Fuel pressure regulator

Delivery pipe

Injector

Surge tank stay

Gasket

20 - 30 (200 - 300, 14.75 - 22.13)

Intake manifold

Gasket

Gasket

TORQUE : N•m (kg•cm, lb•ft)

9358HG07

Intake manifold exploded view

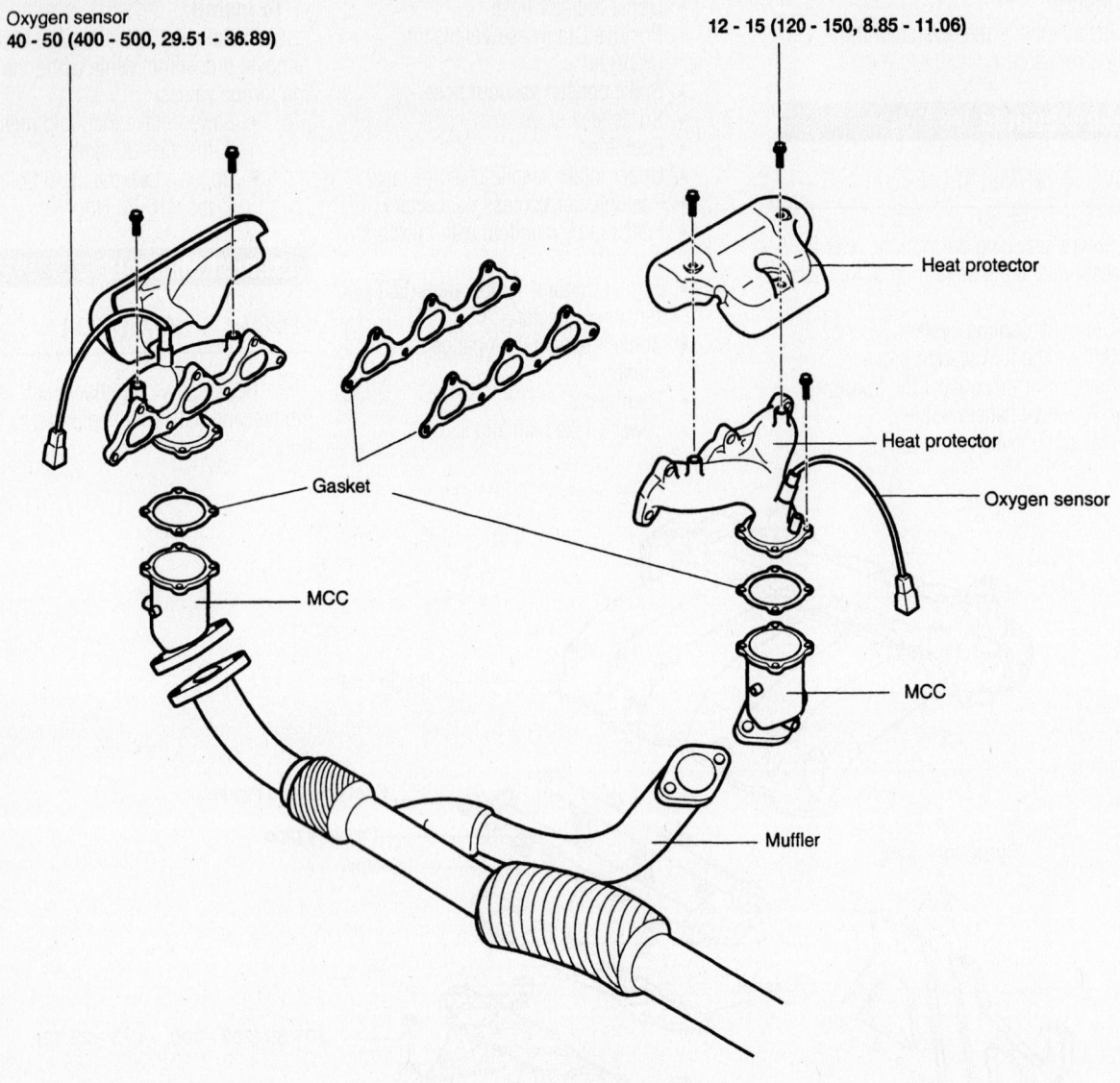

Oxygen sensor
40 - 50 (400 - 500, 29.51 - 36.89)

12 - 15 (120 - 150, 8.85 - 11.06)

Heat protector

Heat protector

Oxygen sensor

Gasket

MCC

MCC

Muffler

9358HG08

TORQUE : N•m (kg•cm, lb•ft)

Exhaust manifold mounting exploded view

2. Remove or disconnect the following:

- Heated Oxygen (HO2S) sensor connectors
- Exhaust Y pipe
- Exhaust manifold heat shields
- Exhaust manifolds

To install:

➡ **Use only new gaskets and nuts for assembly.**

3. Install or connect the following:

- Exhaust manifolds with new gaskets. Tighten the fasteners to 30-37 ft. lbs. (40-50 Nm)

- Exhaust manifold heat shields. Tighten the bolts to 106-132 inch lbs. (12-15 Nm)
- Exhaust Y pipe
- Heated Oxygen (HO2S) sensor connectors

Front Crankshaft Seal

REMOVAL & INSTALLATION

1. Before servicing the vehicle, refer to the precautions at the beginning of this section.

2. Remove or disconnect the following:

- Negative battery cable
- Engine cover
- Accessory drive belts
- Idler pulley
- Crankshaft pulley
- Power steering pump pulley
- Belt tensioner pulley
- Upper and lower timing belt covers
- No. 3 engine mount
- Timing belt
- Timing belt crankshaft sprocket
- Crankshaft Position (CKP) sensor tone ring
- Front crankshaft seal

To install:

3. Install or connect the following:
- Front crankshaft seal. Use Seal Driver 09214-33000 or similar.
- CKP sensor tone ring
- Timing belt crankshaft sprocket
- Timing belt
- No. 3 engine mount
- Upper and lower timing belt covers
- Belt tensioner pulley
- Power steering pump pulley
- Crankshaft pulley
- Idler pulley

- Accessory drive belts
- Engine cover
- Negative battery cable

Camshaft and Valve Lifters

REMOVAL & INSTALLATION

1. Before servicing the vehicle, refer to the precautions at the beginning of this section.
2. Remove or disconnect the following:
- Negative battery cable

- Engine cover
- Valve covers
- Accessory drive belts
- Idler pulley
- Crankshaft pulley
- Power steering pump pulley
- Belt tensioner pulley
- Upper and lower timing belt covers
- No. 3 engine mount
- Timing belt
- Camshaft sprockets

➡ **Keep all valvetrain components in order for assembly.**

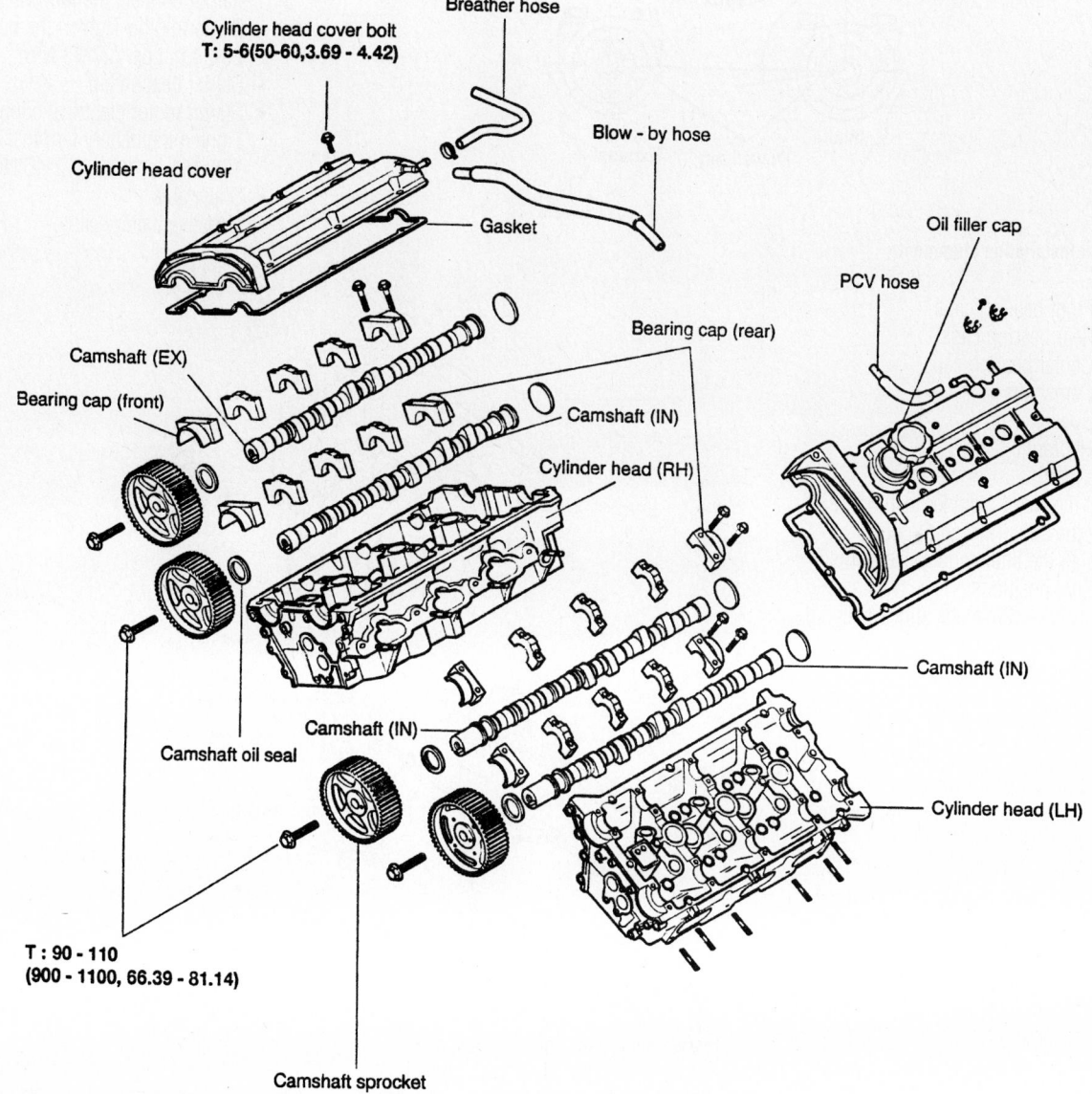

Cylinder head cover bolt
T: 5-6(50-60,3.69 - 4.42)

Breather hose

Blow - by hose

Cylinder head cover

Gasket

Oil filler cap

PCV hose

Bearing cap (rear)

Camshaft (EX)

Camshaft (IN)

Bearing cap (front)

Cylinder head (RH)

Camshaft (IN)

Camshaft oil seal

Camshaft (IN)

Cylinder head (LH)

T : 90 - 110
(900 - 1100, 66.39 - 81.14)

Camshaft sprocket

TORQUE : N•m (kg•cm, lb•ft)

9358HG09

Camshaft mounting exploded view

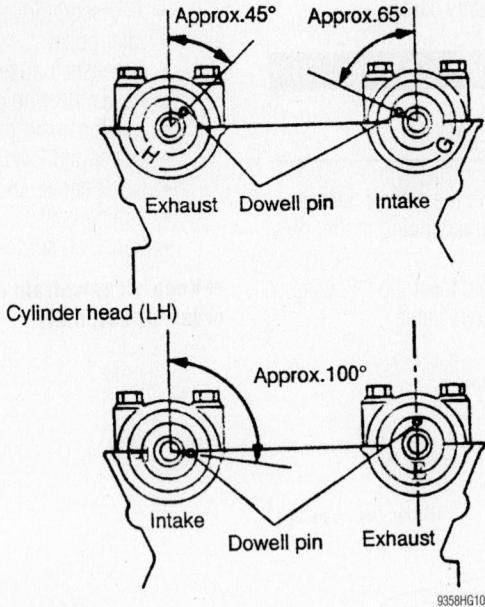

Cylinder head (RH)

Approx.45°   Approx.65°

Exhaust   Dowell pin   Intake

Cylinder head (LH)

Approx.100°

Intake   Dowell pin   Exhaust

9358HG10

Camshaft installation alignment

- Front bearing caps
- Rear bearing caps
- Center bearing caps
- Camshafts
- Rocker arms
- Hydraulic lifters

To install:

3. Set the No. 1 cylinder to Top Dead Center of the compression stroke.

4. Install the lifters and rocker arms in their original positions.

5. Install the camshafts aligned according to the illustration.

6. Install the bearing caps. Tighten the bolts evenly in several passes to the following torque specifications:
- Front and rear bearing caps: 19-21 ft. Lbs. (14-15 Nm)
- Center bearing caps: 88-106 inch lbs. (10-12 Nm)

7. Install or connect the following:
- Camshaft sprockets. Tighten the bolts to 67-81 ft. Lbs. (90-110 Nm).
- Timing belt
- No. 3 engine mount
- Upper and lower timing belt covers
- Belt tensioner pulley
- Power steering pump pulley
- Crankshaft pulley
- Idler pulley
- Accessory drive belts
- Valve covers. Tighten the bolts to 44-53 inch lbs. (5-6 Nm).
- Engine cover
- Negative battery cable

Starter Motor

REMOVAL & INSTALLATION

1. Before servicing the vehicle, refer to the precautions at the beginning of this section.

2. Remove or disconnect the following:
- Negative battery cable
- Shift cable
- Starter motor electrical connectors
- Starter heat shield
- Starter motor

To install:

3. Install or connect the following:
- Starter motor. Tighten the bolts to 20-24 ft. Lbs. (27-33 Nm).
- Starter heat shield
- Starter motor electrical connectors. Tighten the battery terminal nut to 106-141 inch lbs. (12-16 Nm).
- Shift cable
- Negative battery cable

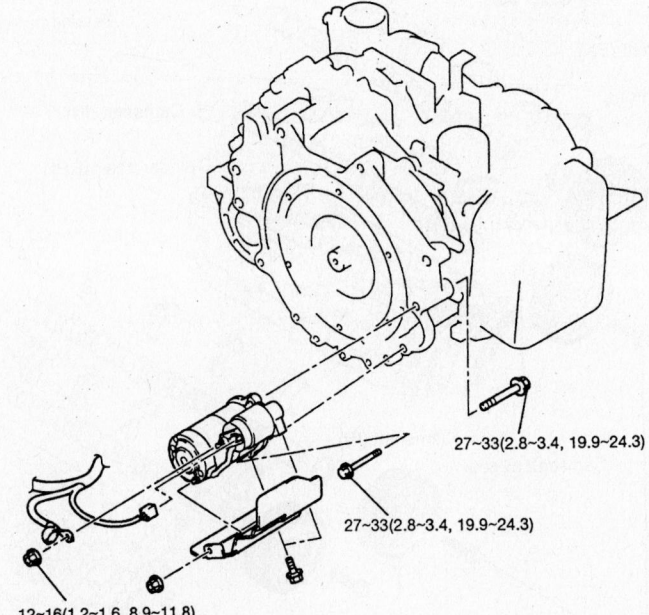

27~33(2.8~3.4, 19.9~24.3)

27~33(2.8~3.4, 19.9~24.3)

12~16(1.2~1.6, 8.9~11.8)

N•m(kg•m, lb•ft)

9358HG11

Starter motor mounting exploded view

Valve Lash

ADJUSTMENT

This vehicle is equipped with hydraulic valve lifters. No adjustment is necessary.

Oil Pan

REMOVAL & INSTALLATION

1. Before servicing the vehicle, refer to the precautions at the beginning of this section.

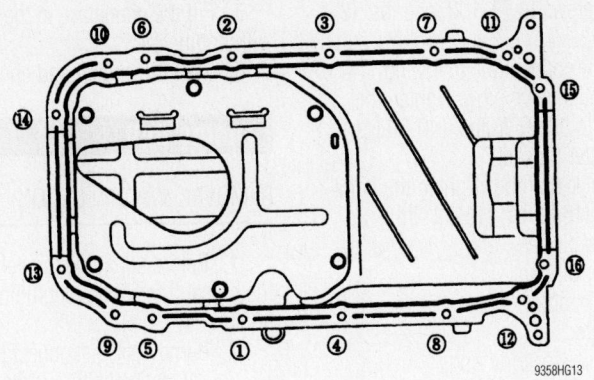

Upper oil pan torque sequence

9358HG13

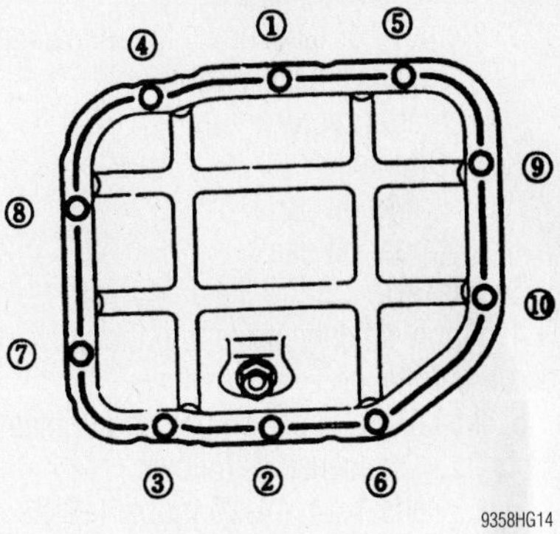

Lower oil pan torque sequence

9358HG14

Oil pump cover

Oil pump outer rotor

Oil pump inner rotor

Oil pump case

**Oil pressure switch
8-15 (80-150, 5.90-11.06)**

Oil filter

**12 - 15
(120 - 150, 8.85 - 11.06)**

Relief valve plunger

Relief valve spring

Sensing blade

Relief valve plug
40 - 50 (400 - 500, 29.51 - 36.88)

Crankshaft sprocket

Upper baffle plate

10 - 12 (100 - 120, 7.38 - 8.85)

Gasket

Oil screen

Lower baffle plate

Upper oil pan

Lower oil pan

Drain plug
30 - 45 (300 - 450, 22.13 - 33.19)

10 - 12 (100 - 120, 7.38 - 8.85)

TORQUE : N•m (kg•cm, lb•ft)

9358HG12

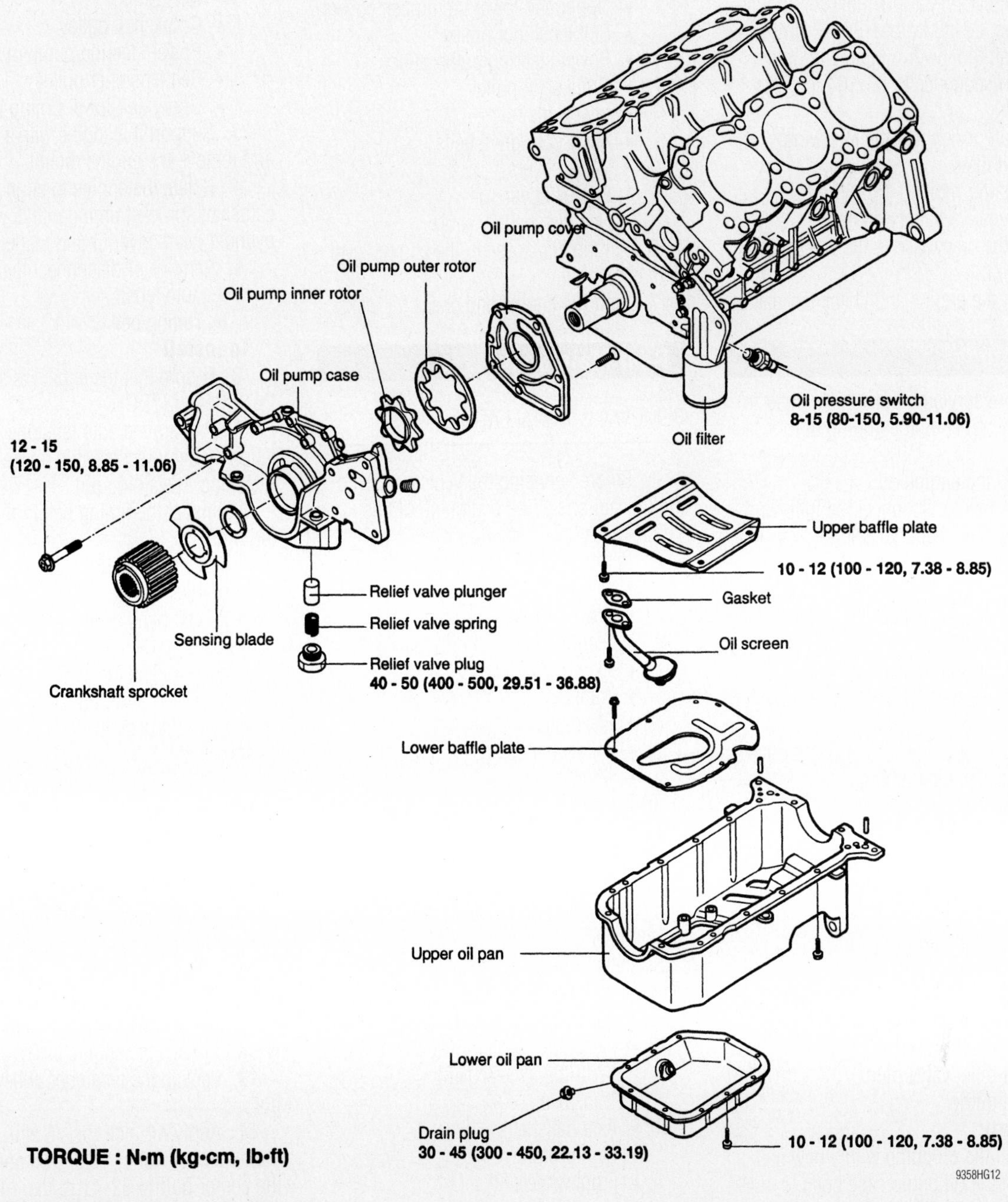

Oil pump and oil pan mounting exploded view

2. Drain the engine oil.

3. Remove or disconnect the following:
- Negative battery cable
- Starter motor
- Oil filter
- Lower oil pan
- Upper oil pan

To install:

4. Apply silicone sealant to the grove of the oil pan flange.

5. Install the upper oil pan and tighten the bolts in sequence as follows:

    a. Bolts 1-14: 14-20 ft. lbs. (19-28 Nm)

    b. Bolts 15 and 16: 44-62 inch lbs. (5-7 Nm)

    c. Upper oil pan-to-transaxle mounting bolts: 22-33 ft. lbs. (30-42 Nm)

6. Install the lower oil pan and tighten the bolts in sequence to 86-104 inch lbs. (10-12 Nm).

7. Install or connect the following:
- Oil filter
- Starter motor
- Negative battery cable

8. Fill the crankcase to the correct level with engine oil.

9. Start the engine and check for leaks.

Oil Pump

1. Before servicing the vehicle, refer to the precautions at the beginning of this section.

2. Drain the engine oil.

3. Remove or disconnect the following:
- Negative battery cable
- Engine cover
- Valve covers
- Accessory drive belts
- Idler pulley
- Crankshaft pulley
- Power steering pump pulley
- Belt tensioner pulley
- Upper and lower timing belt covers
- No. 3 engine mount
- Timing belt
- Crankshaft sprocket
- Crankshaft Position (CKP) sensor tone ring
- Oil filter
- Starter motor
- Lower oil pan
- Upper oil pan
- Oil pick-up tube
- Oil filter bracket
- Oil relief valve plug
- Oil pump

To install:

4. Install the oil pump with a new gasket. Tighten the oil pump case bolts to 106-133 inch lbs. (12-15 Nm). Tighten the oil

pump cover screws to 71-106 inch lbs. (8-12 Nm).

5. Install or connect the following:
- Oil relief valve plug. Tighten the plug to 30-37 ft. lbs. (40-50 Nm).
- Oil filter bracket
- Oil pick-up tube. Tighten the bolts to 11-16 ft. lbs. (15-22 Nm).
- Upper oil pan
- Lower oil pan
- Starter motor
- Oil filter
- CKP sensor tone ring
- Crankshaft sprocket
- Timing belt
- No. 3 engine mount
- Upper and lower timing belt covers
- Belt tensioner pulley
- Power steering pump pulley
- Crankshaft pulley
- Idler pulley
- Accessory drive belts
- Valve covers
- Engine cover
- Negative battery cable

6. Fill the crankcase to the correct level with engine oil.

7. Start the engine and check for leaks.

Rear Main Seal

REMOVAL & INSTALLATION

1. Before servicing the vehicle, refer to the precautions at the beginning of this section.

2. Remove or disconnect the following:
- Negative battery cable
- Front wheels
- Starter motor
- Axle halfshafts
- Transaxle
- Flexplate
- Oil seal housing
- Oil seal

To install:

3. Install the oil seal to the seal housing using special tool 09231-33000 or similar seal driver.

4. Apply silicone sealant to the oil seal housing flange.

5. Install the seal housing and tighten the bolts to 94-106 inch lbs. (10-12 Nm).

6. Install or connect the following:
- Flexplate. Tighten the bolts to 54-57 ft. lbs. (73-77 Nm)
- Transaxle
- Axle halfshafts
- Starter motor
- Front wheels
- Negative battery cable

7. Fill the crankcase to the correct level with engine oil.

8. Start the engine and check for leaks.

Timing Belt

REMOVAL & INSTALLATION

1. Before servicing the vehicle, refer to the precautions at the beginning of this section.

2. Remove or disconnect the following:
- Negative battery cable
- Engine cover
- Accessory drive belts
- Idler pulley
- Crankshaft pulley
- Power steering pump pulley
- Belt tensioner pulley
- Upper and lower timing belt covers

3. Support the engine with a floor jack and remove the engine mount.

4. Rotate the engine to align the camshaft sprocket timing marks with the cylinder head cover timing marks.

5. Remove or disconnect the following:
- Auto tensioner
- Timing belt

To install:

6. Ensure that the engine is set to Top Dead Center (TDC).

7. Prepare the auto tensioner for installation by compressing it in a vise and installing a retaining pin.

8. Install the timing belt in the following order:

    a. Crankshaft sprocket

    b. Idler pulley

    c. Left bank exhaust camshaft sprocket

    d. Left bank intake camshaft sprocket

    e. Water pump pulley

    f. Right bank intake camshaft sprocket

    g. Right bank exhaust camshaft sprocket

    h. Tensioner pulley

9. Install the auto tensioner. Do not remove the retaining pin at this time.

10. Check that the crankshaft and camshaft timing marks are aligned correctly.

11. Rotate the crankshaft ¼ turn **Counterclockwise**.

12. Rotate the crankshaft ¼ turn **Clockwise** to return the engine to TDC.

13. Loosen the tensioner pulley center bolt.

14. Apply 44 inch lbs. (5 Nm) torque to the tensioner pulley as shown and tighten the center bolt to 32-41 ft. lbs. (43-55 Nm).

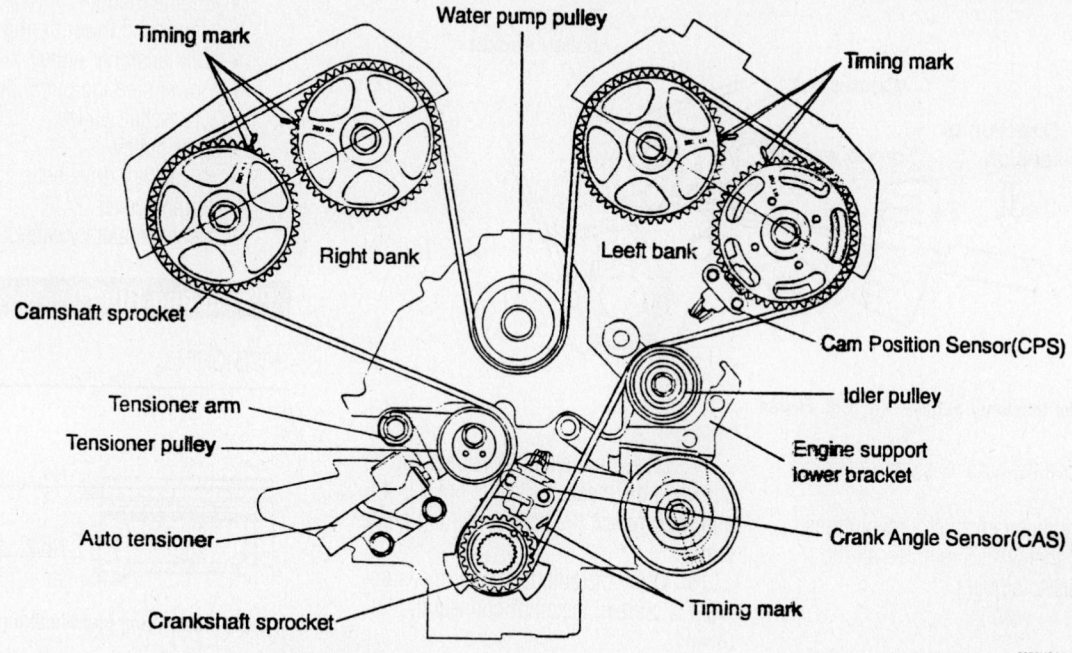

Timing belt routing and timing marks—Kia 3.5L Engine

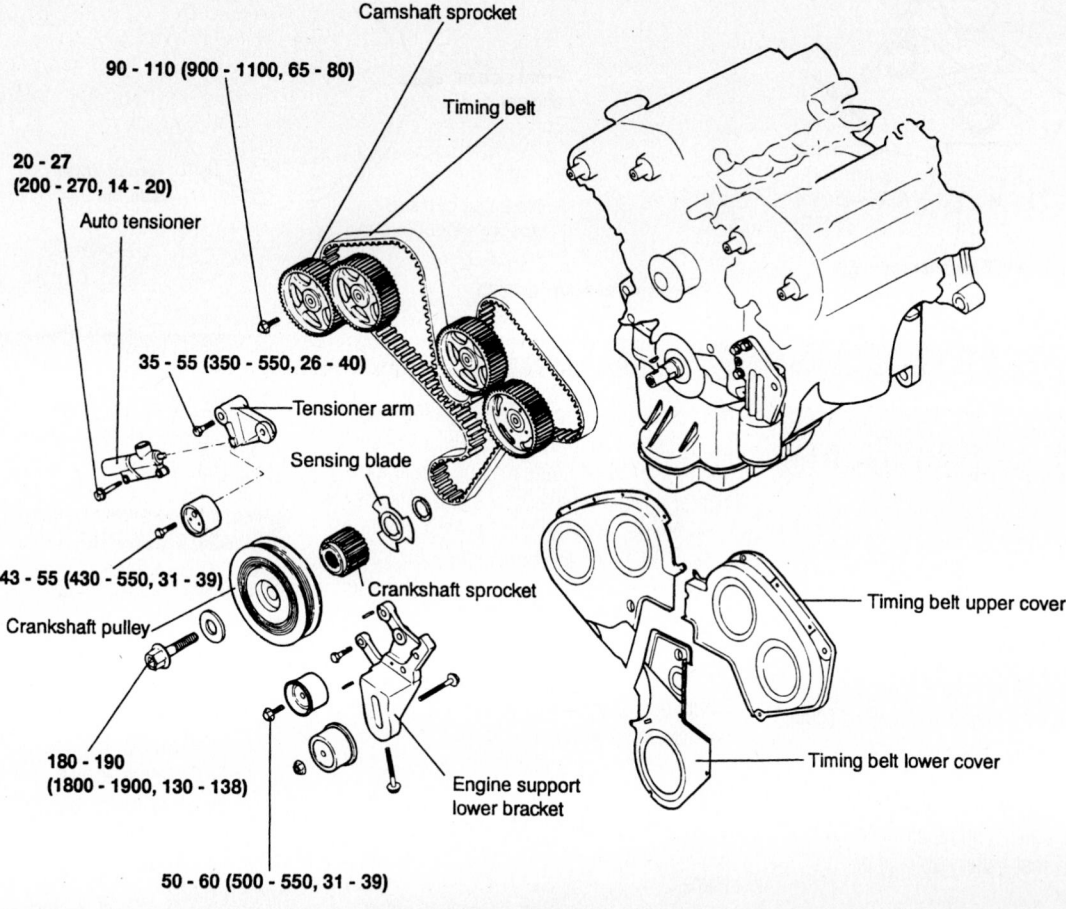

TORQUE : N·m (kg·cm, lb·ft)

Timing belt exploded view—Kia 3.5L Engine

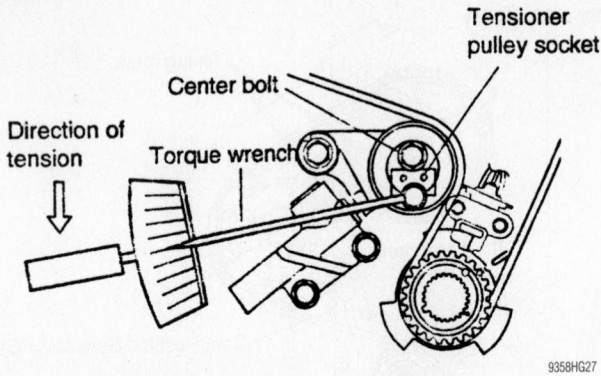

Adjusting the tensioner pulley—Kia 3.5L Engine

15. Remove the auto tensioner retaining pin.

16. Rotate the crankshaft 2 revolutions **Clockwise**, then wait 5 minutes for the auto tensioner to adjust.

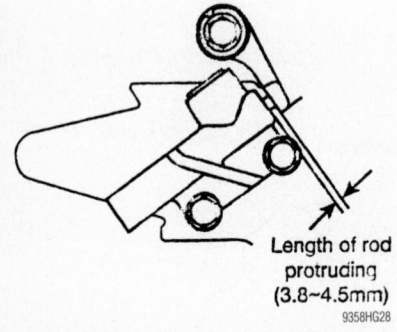

Measuring the auto tensioner rod—Kia 3.5L Engine

17. Measure the auto tensioner rod as shown. If the measurement is not 3.8-4.5 mm, then repeat the belt tensioning procedure.

18. When the auto tensioner measurement is correct, install or connect the following:

- Engine mount
- Upper and lower timing belt covers
- Belt tensioner pulley
- Power steering pump pulley
- Crankshaft pulley
- Idler pulley
- Accessory drive belts
- Engine cover
- Negative battery cable

Piston and Ring

POSITIONING

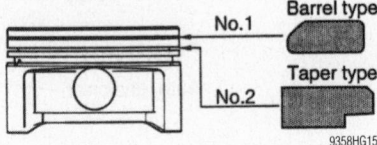

Compression ring identification

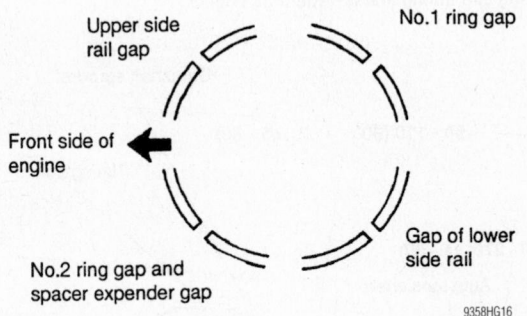

Piston ring end gap spacing

FUEL SYSTEM

Fuel System Service Precautions

Safety is the most important factor when performing not only fuel system maintenance, but any type of maintenance. Failure to conduct maintenance and repairs in a safe manner may result in serious personal injury or death. Maintenance and testing of the vehicle's fuel system components can be accomplished safely and effectively by adhering to the following rules and guidelines.

- To avoid the possibility of fire and personal injury, always disconnect the negative battery cable unless the repair or test procedure requires that battery voltage be applied.

- Always relieve the fuel system pressure prior to disconnecting any fuel system component (injector, fuel rail, pressure regulator, etc.), fitting or fuel line connection. Exercise extreme caution whenever relieving fuel system pressure, to avoid exposing skin, face and eyes to fuel spray. Please be advised that fuel under pressure may penetrate the skin or any part of the body that it contacts.

- Always place a shop towel or cloth around the fitting or connection prior to loosening to absorb any excess fuel due to spillage. Ensure that all fuel spillage (should it occur) is quickly removed from engine surfaces. Ensure that all fuel soaked cloths or towels are deposited into a suitable waste container.

- Always keep a dry chemical (Class B) fire extinguisher near the work area.

- Do not allow fuel spray or fuel vapors to come into contact with a spark or open flame.

- Always use a back-up wrench when loosening and tightening fuel line connection fittings. This will prevent unnecessary stress and torsion to fuel line piping. Always follow the proper torque specifications.

- Always replace worn fuel fitting O-rings with new. Do not substitute fuel hose or equivalent where fuel pipe is installed.

Fuel System Pressure

RELIEVING

1. Before servicing the vehicle, refer to the precautions at the beginning of this section.

2. Remove the rear seat.

3. Remove the access cover

4. Disconnect the fuel pump electrical connector.

5. Start the engine and allow it to run until it stalls.

6. Turn the ignition switch to the **OFF** position.

7. Restore the electrical connections after fuel system repairs are completed.

Fuel Filter

REMOVAL & INSTALLATION

The fuel filter is located inside the fuel tank and is replaced with the fuel pump as an assembly.

Fuel Pump

REMOVAL & INSTALLATION

1. Before servicing the vehicle, refer to the precautions at the beginning of this section.
2. Remove or disconnect the following:
   - Rear seat
   - Access cover
   - Fuel pump electrical connector
3. Relieve the fuel system pressure.
   - Fuel lines
   - Fuel pump and filter assembly

To install:
4. Install or connect the following:
   - Fuel pump and filter assembly
   - Fuel lines
   - Fuel pump electrical connector
   - Access cover
   - Rear seat
5. Start the engine and check for leaks.

Fuel Injector

REMOVAL & INSTALLATION

1. Before servicing the vehicle, refer to the precautions at the beginning of this section.
2. Relieve the fuel system pressure.
3. Remove or disconnect the following:
   - Negative battery cable
   - Engine cover
   - Positive Crankcase Ventilation (PCV) hose
   - Brake booster vacuum hose
   - Surge tank stays
   - Fuel lines
   - Upper intake manifold (surge tank)
   - Fuel injector harness connectors
   - Fuel supply manifold with injectors attached

To install:
4. Replace all injector seals.
5. Install or connect the following:
   - Fuel supply manifold with injectors attached
   - Fuel injector harness connectors
   - Upper intake manifold (surge tank)
   - Fuel lines
   - Surge tank stays
   - Brake booster vacuum hose
   - Positive Crankcase Ventilation (PCV) hose
   - Engine cover
   - Negative battery cable
6. Start the engine and check for leaks.

DRIVE TRAIN

Transaxle Assembly

REMOVAL & INSTALLATION

1. Before servicing the vehicle, refer to the precautions at the beginning of this section.
2. Drain the cooling system.
3. Drain the transaxle fluid.
4. Remove or disconnect the following:
   - Battery and tray
   - Engine wiring harness
   - Transaxle wiring harness
   - Air filter assembly
   - Shift cable
   - Transaxle cooler lines
   - Radiator hoses
   - Radiator
   - Heater hoses
   - Steering intermediate shaft
   - Power steering pressure and return lines from the steering gear
   - Engine upper roll stopper and bracket
   - Front wheels
   - Heated Oxygen (HO2S) sensor connector
   - Front muffler
   - Outer tie rod ends
   - Lower ball joints
   - Axle halfshafts
   - Steering tube mounting bolt
5. Support the engine from below.
   - Front subframe
   - Starter motor
   - Transaxle housing cover
6. Support the transaxle with a transmission jack.
   - Torque converter bolts
   - Transaxle flange bolts
   - Transaxle

To install:
7. Install or connect the following:
   - Transaxle. Tighten the flange bolts to 29-38 ft. lbs. (42-54 Nm)
   - Torque converter bolts
   - Transaxle housing cover
   - Starter motor
   - Front subframe
   - Steering tube mounting bolt
   - Axle halfshafts
   - Lower ball joints
   - Outer tie rod ends
   - Front muffler
   - Heated Oxygen (HO2S) sensor connector
   - Front wheels
   - Engine upper roll stopper and bracket
   - Power steering pressure and return lines from the steering gear
   - Steering intermediate shaft
   - Heater hoses
   - Radiator
   - Radiator hoses
   - Transaxle cooler lines
   - Shift cable
   - Air filter assembly
   - Transaxle wiring harness
   - Engine wiring harness
   - Battery and tray
8. Fill the transaxle to the correct level with the proper transmission fluid.
9. Fill the cooling system.
10. Start the engine and check for leaks.

Halfshaft

REMOVAL & INSTALLATION

1. Before servicing the vehicle, refer to the precautions at the beginning of this section.
2. Drain the transaxle fluid.
3. Remove or disconnect the following:
   - Front wheels
   - Hub retaining nuts
   - Sway bar links
   - Outer tie rod ends
   - Lower ball joints
4. Using a prybar, remove the left halfshaft from the transaxle. Separate the right halfshaft from the intermediate shaft.
5. Remove the axle halfshafts from the front hubs. It may be necessary to tap the stub shaft with a brass hammer to remove the axles.

To install:

➡**Use new circlips for assembly.**

6. Install the left halfshaft to the transaxle.
7. Install the right halfshaft to the intermediate shaft.
8. Install the halfshaft stub shafts to the wheel hubs.

9. Install or connect the following:
- Lower ball joints. Tighten the pinch bolts to 69-85 ft. lbs. (93-115 Nm).
- Outer tie rod ends. Tighten the nuts to 43-58 ft. lbs. (59-78 Nm).
- Sway bar links. Tighten nuts to 69-85 ft. lbs. (93-115 Nm).

- Hub retaining nuts. Tighten the hub nuts to 177-199 ft. lbs. (240-270 Nm).
- Front wheels. Tighten the lug nuts to 65-79 ft. lbs. (88-108 Nm).

10. Fill the transaxle to the correct level with the proper fluid.

11. Start the engine and check for leaks.

CV Joint

OVERHAUL

Outer CV Joint

The outer CV joint is serviced only with the axle halfshaft as an assembly. The outer

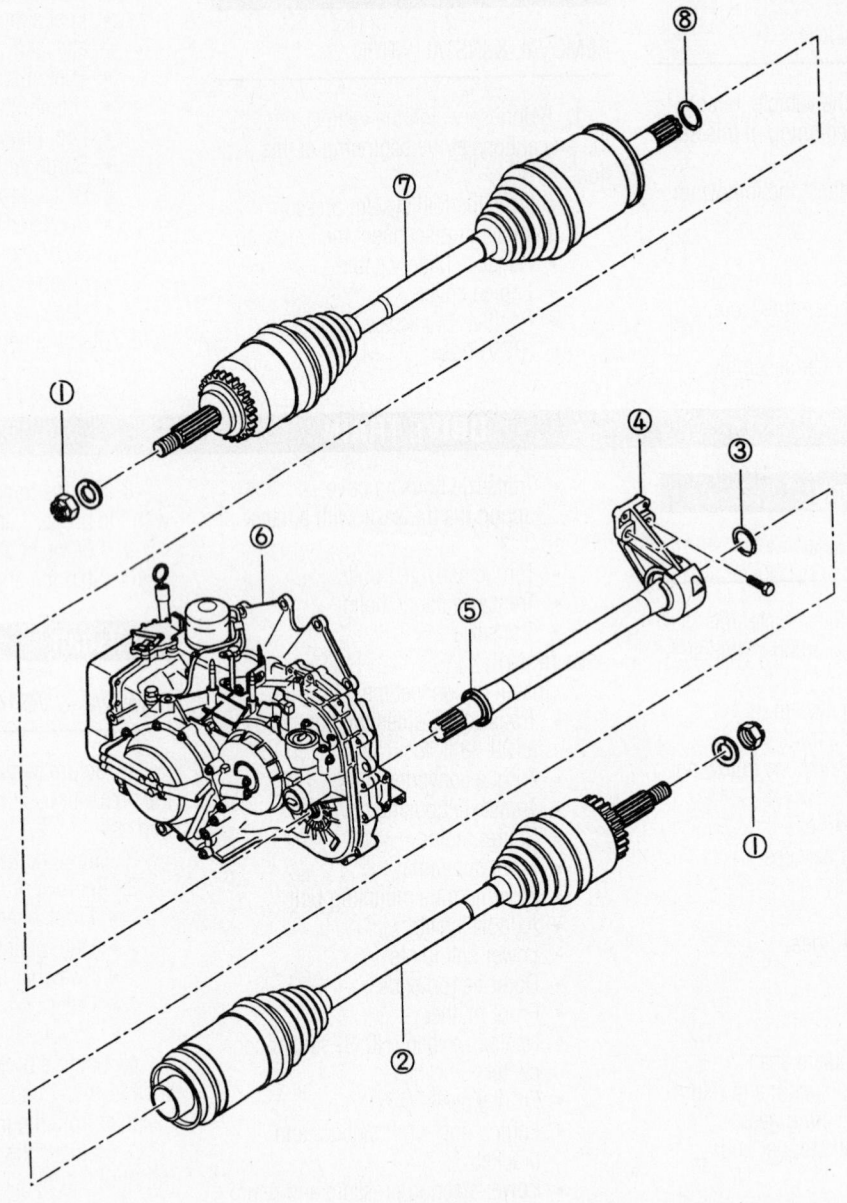

(1) Wheel nut
(2) Driveshaft(RH)
(3) Circlip
(4) Center bearing bracket

(5) constant velocity shaft
(6) Automatic transaxle
(7) Driveshaft(LH)
(8) Circle pin

9358HG17

Halfshaft mounting exploded view

CV joint boot can be serviced by first removing the inner Tripod joint.

Inner Tripod Joint

1. Before servicing the vehicle, refer to the precautions at the beginning of this section.
2. Remove the axle halfshaft from the vehicle and place it in a vise.
3. Remove or disconnect the following:

- Tripod joint boot clamps. Slide the boot on the axle shaft to expose the joint.
- Circle pin
- Tripod joint housing
- Tripod joint snapring
- Tripod joint
- Inner Tripod joint boot

To install:

4. Install or connect the following:
- Inner Tripod joint boot

- Tripod joint
- Tripod joint snapring
- Tripod joint housing. Fill with 7.5 ounces of CV Joint grease.
- Circle pin
- Tripod joint boot clamps. Pull on the clamp end with pliers and fold the locking tabs over to lock the clamp in place.

5. Install the axle halfshaft to the vehicle.

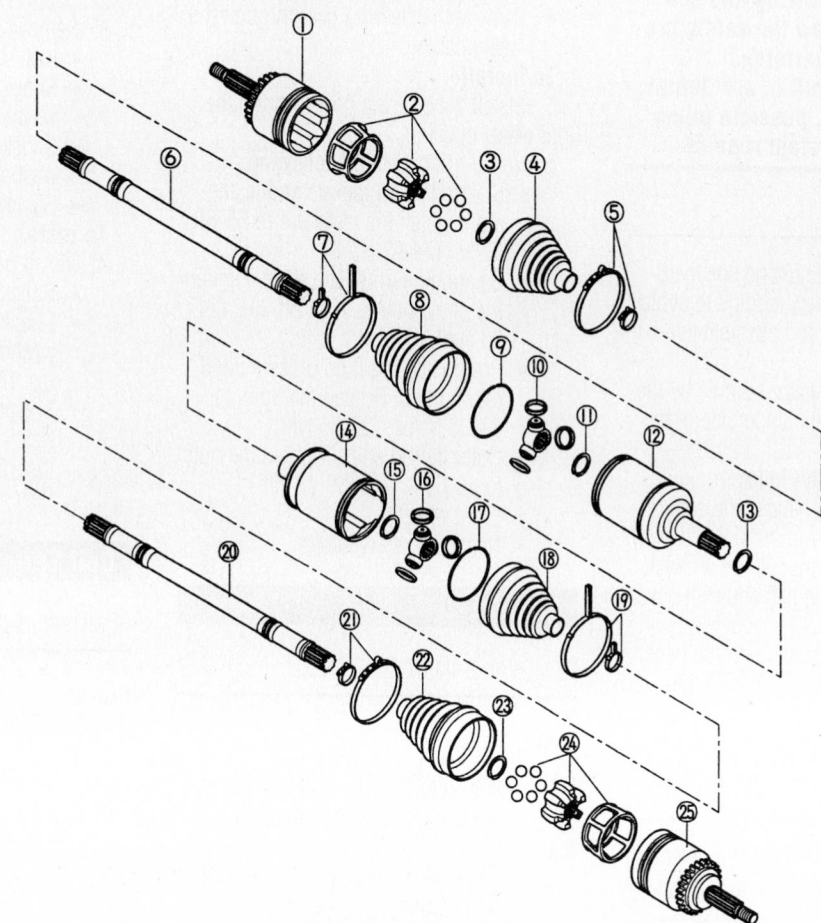

Driveshaft(LH)	(10) Spider assembly	(18) UTJ boot
(1) BJ assembly	(11) Snap ring	(19) UTJ boot band
(2) BJ inner race and ball	(12) UTJ assembly	(20) Driveshaft(RH)
(3) Snap ring	(13) Circlip	(21) BJ boot band
(4) BJ boot		(22) BJ boot
(5) BJ boot band	**Driveshaft(RH)**	(23) Snap ring
(6) Driveshaft(LH)	(14) UTJ assembly	(24) BJ inner race and ball
(7) UTJ boot band	(15) Snap ring	(25) BJ assembly
(8) UTJ boot	(16) Spider assembly	
(9) Circle pin	(17) Circle pin	

✏ Caution
   a) *Install a protective material in a vice, and secure joint in the vice.*
   b) *Keep dust or foreign material from joint during procedure.*
   c) *Do not disassemble wheel side ball joint.*
   d) *Do not wash joint unless disassembling it.*

9358HG18

Inner and outer CV Joint exploded view and parts identification

STEERING AND SUSPENSION

Air Bag

✳✳ CAUTION

Some vehicles are equipped with an air bag system. The system must be disarmed before performing service on, or around, system components, the steering column, instrument panel components, wiring and sensors. Failure to follow the safety precautions and the disarming procedure could result in accidental air bag deployment, possible injury and unnecessary system repairs.

PRECAUTIONS

Several precautions must be observed when handling the inflator module to avoid accidental deployment and possible personal injury.

• Never carry the inflator module by the wires or connector on the underside of the module.

• When carrying a live inflator module, hold securely with both hands, and ensure that the bag and trim cover are pointed away.

• Place the inflator module on a bench or other surface with the bag and trim cover facing up.

• With the inflator module on the bench, never place anything on or close to the module which may be thrown in the event of an accidental deployment.

DISARMING

1. Before servicing the vehicle, refer to the precautions at the beginning of this section.
2. Position the vehicle with the front wheels in a straight-ahead position.
3. Disconnect both battery cables.
4. Wait at least 1 minute for the air bag back-up power supply to deplete its stored energy before continuing.
5. Proceed with the repair.
6. Reconnect both battery cables once the repair is complete.

Rack and Pinion Steering Gear

REMOVAL & INSTALLATION

1. Before servicing the vehicle, refer to the precautions at the beginning of this section.

2. Drain the power steering fluid.
3. Remove or disconnect the following:
   • Negative battery cable
   • Front wheels
   • Outer tie rod ends
   • Power steering fluid pressure and return lines
   • Steering intermediate shaft
   • Steering rack brackets and fasteners
4. Remove the steering gear through the right wheel opening.

To install:
5. Install the steering gear through the right wheel opening.
6. Install or connect the following:
   • Steering rack brackets and fasteners. Tighten the fasteners to 55-69 ft. lbs. (74-93 Nm).
   • Steering intermediate shaft. Tighten the pinch bolt to 16-20 ft. lbs. (21-26 Nm).
   • Power steering fluid pressure and return lines. Tighten the bolts to 17-26 ft. lbs. (24-35 Nm).
   • Outer tie rod ends. Tighten the nuts to 43-58 ft. lbs. (59-78 Nm).
   • Front wheels
   • Negative battery cable

Strut

REMOVAL & INSTALLATION

Front

1. Before servicing the vehicle, refer to the precautions at the beginning of this section.
2. Remove or disconnect the following:
   • Front wheel
   • Brake hose from the bracket
   • Wheel speed sensor cable
   • Steering knuckle mounting bolts
   • Upper strut mount nuts
3. Remove the strut from the vehicle.

To install:
4. Install the strut to the vehicle.
5. Install or connect the following:
   • Upper strut mount nuts and tighten them to 33-46 ft. lbs. (46-62 Nm).
   • Steering knuckle mounting bolts and tighten them to 88-101 ft. lbs. (119-137 Nm).
   • Wheel speed sensor cable
   • Brake hose to the bracket
   • Front wheel
6. Check the front end alignment and adjust as necessary.

Shock Absorber

REMOVAL & INSTALLATION

Rear

1. Before servicing the vehicle, refer to the precautions at the beginning of this section.
2. Support the rear axle on jack stands.
3. Remove or disconnect the following:
   • Rear wheel
   • Shock absorber upper nut and washer
   • Shock absorber lower nut and washer
   • Shock absorber

To install:
4. Install or connect the following:
   • Shock absorber
   • Shock absorber lower nut and washer. Tighten the nut to 55-69 ft. lbs. (74-93 Nm).
   • Shock absorber upper nut and washer. Tighten the nut to 41-47 ft. lbs. (55-64 Nm).
   • Rear wheel

Coil Spring

REMOVAL & INSTALLATION

Front

1. Before servicing the vehicle, refer to the precautions at the beginning of this section.
2. Remove the strut from the vehicle and attach Service Tool 0K2A1 341 001 or other suitable spring compressor.
3. Compress the coil spring.
4. Remove or disconnect the following:
   • Upper mount retaining nut
   • Mounting block
   • Upper spring seat
   • Upper spring isolator
   • Coil spring
   • Dust boot
   • Bump stopper
   • Lower spring seat

To install:
5. Install or connect the following:
   • Lower spring seat
   • Bump stopper
   • Dust boot
   • Coil spring
   • Upper spring isolator
   • Upper spring seat

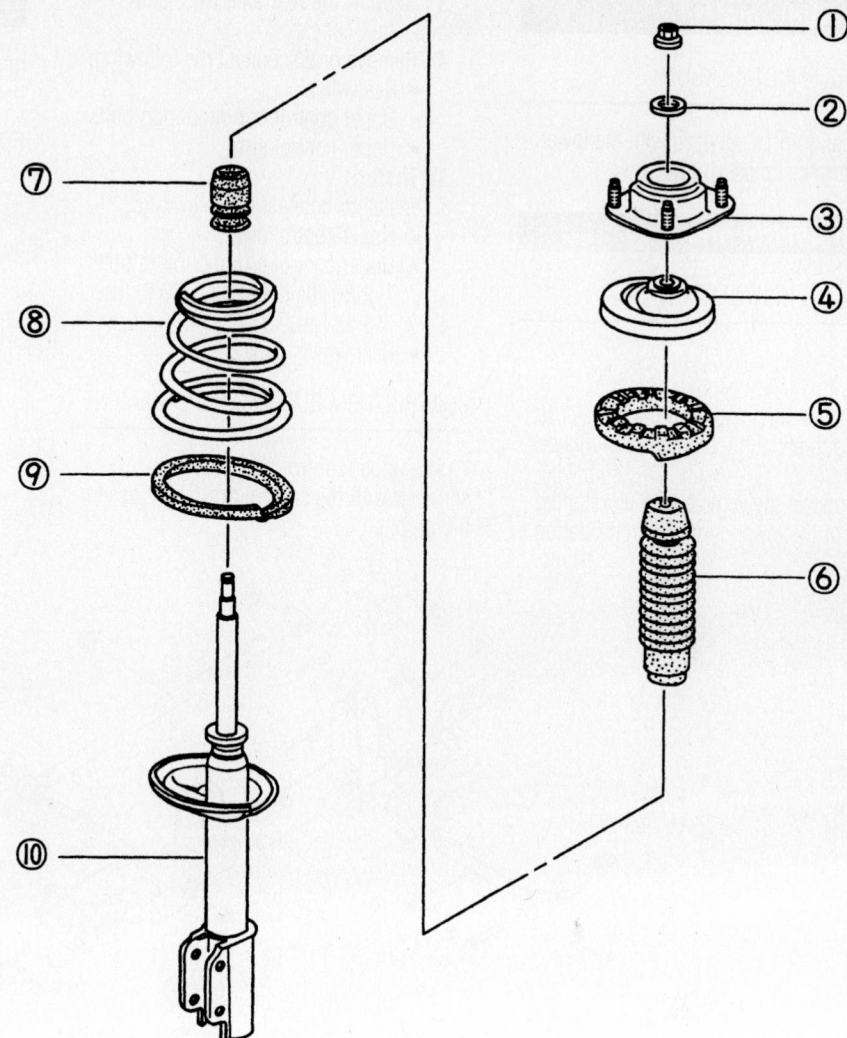

(1) **Piston retaining nut**
(2) **Washer**
(3) **Mounting block**
(4) **Upper spring seat**
(5) **Upper spring isolator**

(6) **Dust boot**
(7) **Bump stopper**
(8) **Coil spring**
(9) **Lower spring isolator**
(10) **Shock absorber**

9358HG19

Front strut and coil spring exploded view

- Mounting block
- Upper mount retaining nut and tighten it to 88-101 ft. lbs. (120-137 Nm).
6. Install the strut to the vehicle.
7. Check the front end alignment and adjust as necessary.

Rear

1. Before servicing the vehicle, refer to the precautions at the beginning of this section.
2. Support the vehicle with jackstands forward of the lower control arm mounting.
3. Support the rear axle with jackstands.
4. Support the lower control arm with a floor jack.
5. Remove or disconnect the following:
   - Rear wheel
   - Sway bar link
   - Parking brake cable
   - Lower control arm bolts
6. Carefully lower the floor jack and remove the lower control arm, coil spring, and spring seats from the vehicle.

To install:
7. Install or connect the following:
   - Spring seats
   - Coil spring
   - Lower control arm. Tighten the bolts to 87-101 ft. lbs. (118-137 Nm).
   - Parking brake cable. Tighten the bracket bolts to 14-19 ft. lbs. (16-23 Nm).
   - Sway bar link. Tighten the locknut to 17-20 ft. lbs. (24-28 Nm).
   - Rear wheel

Lower Ball Joint

REMOVAL & INSTALLATION

The ball joint is serviced with the lower control arm as an assembly.

Upper Control Arm

REMOVAL & INSTALLATION

Rear

1. Before servicing the vehicle, refer to the precautions at the beginning of this section.
2. Support the vehicle with jackstands forward of the lower control arm mounting.

3. Support the rear axle with jackstands.
4. Remove or disconnect the following:
   - Rear wheel
   - Upper control arm mounting bolts
   - Upper control arm

To install:

5. Install or connect the following:
   - Upper control arm
   - Upper control arm mounting bolts. Tighten the bolts to 55-69 ft. lbs. (74-93 Nm).
   - Rear wheel

CONTROL ARM BUSHING REPLACEMENT

The upper control arm bushings are serviced with the control arm as an assembly.

Lower Control Arm

REMOVAL & INSTALLATION

Front

1. Before servicing the vehicle, refer to the precautions at the beginning of this section.
2. Remove or disconnect the following:
   - Front wheel
   - Sway bar link
   - Tension rod bolts
   - Lower ball joint
   - Lower control arm subframe bolt
   - Lower control arm

To install:

3. Install or connect the following:

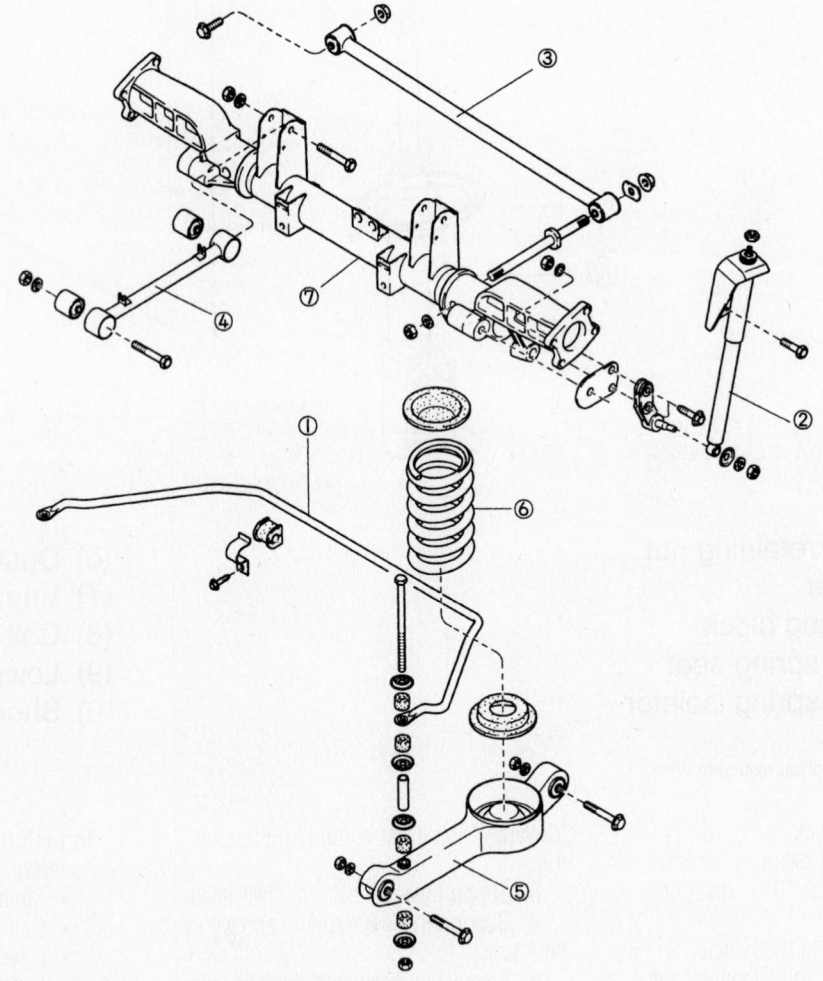

(1) Stabilizer bar
(2) Shock absorber
(3) Panhard rod
(4) Upper arm assembly
(5) Lower arm assembly
(6) Coil spring
(7) Rear casing

9358HG20

Rear suspension exploded view

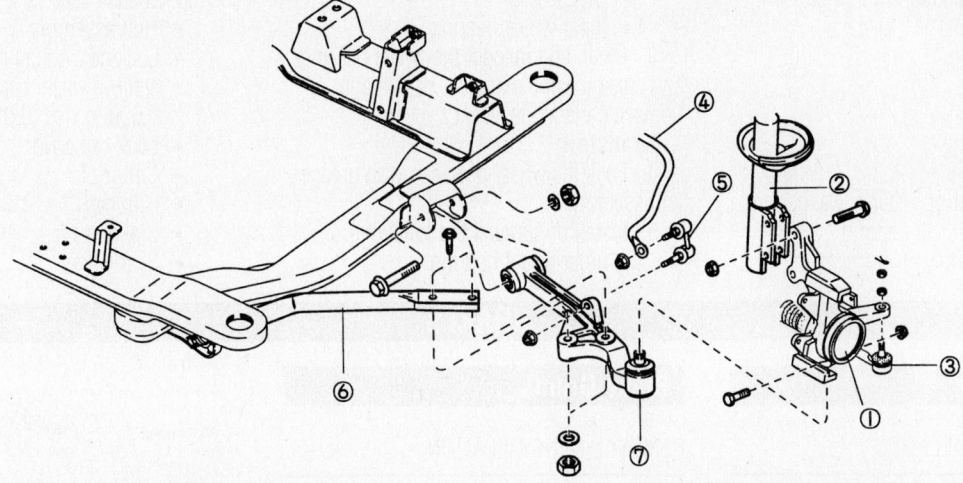

(1) Knuckle
(2) Shock absorber
(3) Tie rod end
(4) Stabilizer bar

(5) Stabilizer control link
(6) Tension rod
(7) Lower arm

9358HG21

Front suspension exploded view

- Lower control arm. Tighten the subframe bolt to 88-101 ft. lbs. (120-137 Nm).
- Lower ball joint. Tighten the pinch bolt to 69-85 ft. lbs. (93-115 Nm).
- Tension rod bolts and tighten them to 88-101 ft. lbs. (120-137 Nm)
- Sway bar link. Tighten the nut to 69-85 ft. lbs. (93-115 Nm).
- Front wheel

4. Check the front end alignment and adjust as necessary.

Rear

1. Before servicing the vehicle, refer to the precautions at the beginning of this section.
2. Support the vehicle with jackstands forward of the lower control arm mounting.
3. Support the rear axle with jackstands.
4. Support the lower control arm with a floor jack.
5. Remove or disconnect the following:
   - Rear wheel
   - Sway bar link
   - Parking brake cable
   - Lower control arm bolts
6. Carefully lower the floor jack and remove the lower control arm, coil spring, and spring seats from the vehicle.

To install:

7. Install or connect the following:
   - Spring seats
   - Coil spring
   - Lower control arm. Tighten the bolts to 87-101 ft. lbs. (118-137 Nm).

- Parking brake cable. Tighten the bracket bolts to 14-19 ft. lbs. (16-23 Nm).
- Sway bar link. Tighten the locknut to 17-20 ft. lbs. (24-28 Nm).
- Rear wheel

CONTROL ARM BUSHING REPLACEMENT

The lower control arm bushings are serviced with the control arm as an assembly.

Wheel Bearings

ADJUSTMENT

Front

The front wheel bearings are not adjustable. Replace the wheel bearings if play exceeds 0.002 inches (0.05 mm).

Rear

1. Before servicing the vehicle, refer to the precautions at the beginning of this section.
2. Remove or disconnect the following:

   - Rear wheel
   - Brake drum
   - Hub cover
   - Cotter pin
   - Lock nut cover

3. Adjust the locknut to achieve wheel bearing play of 0.001-0.006 inches (0.025-0.152 mm).

REMOVAL & INSTALLATION

Front

➡**The front wheel bearings are serviced with the steering knuckle as an assembly.**

1. Before servicing the vehicle, refer to the precautions at the beginning of this section.
2. Remove or disconnect the following:
   - Front wheel
   - Brake caliper and rotor
   - Wheel speed sensor harness
   - Hub nut
   - Lower ball joint
   - Strut mounting bolts
   - Steering knuckle

To install:

3. Install or connect the following:
   - Steering knuckle
   - Strut mounting bolts and tighten them to 88-101 ft. lbs. (119-137 Nm).
   - Lower ball joint. Tighten the pinch bolt to 69-85 ft. lbs. (93-115 Nm).
   - Hub nut and tighten it to 177-199 ft. lbs. (240-270 Nm).
   - Wheel speed sensor harness
   - Brake caliper and rotor
   - Front wheel

4. Check the alignment and adjust as necessary.

Rear

1. Before servicing the vehicle, refer to the precautions at the beginning of this section.

2. Remove or disconnect the following:
- Rear wheel
- Brake drum
- Hub cap
- Cotter pin
- Lock nut cover
- Lock nut
- Wheel bearing retainer washer and the outer wheel bearing
- Hub assembly

- Grease seal
- Inner wheel bearing

3. Clean and inspect the wheel bearings and races for unusual wear or damage. Replace parts as necessary.

To install:

4. Pack the wheel bearings with grease for assembly.

5. Install or connect the following:
- Inner wheel bearing

- Grease seal
- Hub assembly
- Lock nut. Adjust the locknut to achieve wheel bearing play of 0.001-0.006 inches (0.025-0.152 mm).
- Lock nut cover
- Cotter pin
- Hub cap
- Brake drum
- Rear wheel

BRAKES

Brake Caliper

REMOVAL & INSTALLATION

1. Before servicing the vehicle, refer to the precautions at the beginning of this section.
2. Remove or disconnect the following:
- Front wheel
- Brake fluid hose
- Caliper mounting bolts
- Brake caliper

To install:

3. Install or connect the following:
- Brake caliper. Tighten the mounting bolts to 18–26 ft. lbs. (24–35 Nm).
- Brake fluid hose
- Front wheel
4. Bleed the brake system.
5. Before attempting to move the vehicle, pump the brake pedal to seat the pads against the rotors. Make sure the vehicle has a firm brake pedal. Check the level of the brake fluid and add DOT 3 or 4 brake fluid if necessary.

Disc Brake Pads

REMOVAL & INSTALLATION

1. Before servicing the vehicle, refer to the precautions at the beginning of this section.
2. Remove or disconnect the following:
- Front wheel
- Brake hose from the support bracket
- Outer brake pad retaining clip
- Brake caliper
- Inner and outer brake pads

To install:

3. Compress the caliper piston into the caliper bore.
4. Install or connect the following:
- Inner and outer brake pads
- Brake caliper. Tighten the mounting bolts to 18–26 ft. lbs. (24–35 Nm).
- Brake hose to the support bracket
- Front wheel

Brake Drums

REMOVAL & INSTALLATION

1. Before servicing the vehicle, refer to the precautions at the beginning of this section.
2. Remove or disconnect the following:
- Rear wheel
- Brake drum retaining screws
- Brake drum

To install:

3. Install or connect the following:
- Brake drum
- Brake drum retaining screws
- Rear wheel

Brake Shoes

REMOVAL & INSTALLATION

1. Before servicing the vehicle, refer to the precautions at the beginning of this section.
2. Remove or disconnect the following:
- Rear wheels
- Brake drum
- Brake adjuster spring
- Adjuster lever
- Strut spring and retracting spring
- Upper return spring
- Hold-down spring and pins
- Leading (primary) brake shoe
- Trailing (secondary) brake shoe
- Parking brake cable

To install:

3. Transfer the parking brake lever to the new trailing (secondary) brake shoe.
4. Install or connect the following:
- Parking brake cable
- Trailing (secondary) brake shoe
- Leading (primary) brake shoe
- Hold-down spring and pins
- Upper return spring
- Strut spring and retracting spring
- Adjuster lever
- Brake adjuster spring
- Brake drum
- Rear wheels

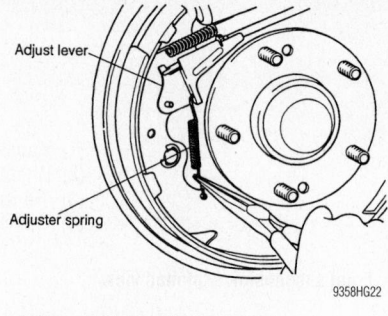

Brake adjuster spring removal

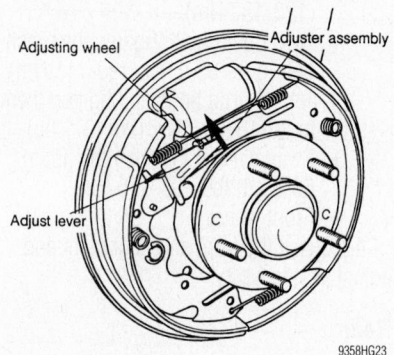

Brake adjuster assembly

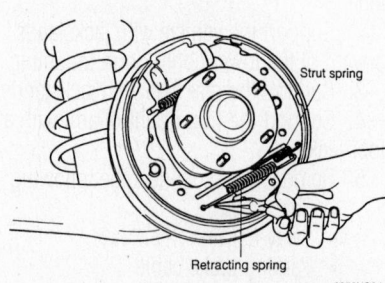

Brake strut spring and retracting spring

5. Work the parking brake control several times to complete the brake shoe adjustment and to check the parking brake adjustment as well.
6. Pump the brake pedal several times to assure a good pedal.
7. Road test the vehicle and check for proper brake system operation.

KIA

Sorento

20

SPECIFICATION AND MAINTENANCE CHARTS

ENGINE AND VEHICLE IDENTIFICATION

Engine							Model Year	
Code ①	Liters (cc)	Cu. In.	Cyl.	Fuel Sys.	Engine Type	Eng. Mfg.	Code ②	Year
3	3.5 (3497)	213	6	EGI	DOHC	KIA	3	2003
							4	2004
							5	2005

EGI: Electronic Gasoline Injection

DOHC: Double Overhead Camshafts

① 8th digit of VIN

② 10th digit of VIN

67162-SORE-C01

GENERAL ENGINE SPECIFICATIONS

Year	Model	Engine Displacement Liters	Engine VIN	Net Horsepower @ rpm	Net Torque @ rpm (ft. lbs.)	Bore x Stroke (in.)	Com-pression Ratio	Oil Pressure @ rpm
2003	Sorento	3.5	3	192@5500	217@3000	3.66x3.38	10:01	43-57@3000
2004	Sorento	3.5	3	192@5500	217@3000	3.66x3.38	10:01	43-57@3000

67162-SORE-C02

ENGINE TUNE-UP SPECIFICATIONS

Year	Engine Displacement Liters	Engine VIN	Spark Plug Gap (in.)	Ignition Timing (deg.) MT	Ignition Timing (deg.) AT	Fuel Pump (psi)	Idle Speed (rpm) MT	Idle Speed (rpm) AT	Valve Clearance Intake	Valve Clearance Exhaust
2003	3.5	3	0.039-0.043	—	①	46-49	—	②	HYD	HYD
2004	3.5	3	0.039-0.043	①	①	46-49	②	②	HYD	HYD

NOTE: The Vehicle Emission Control Information label often reflects specification changes made during production.

The label figures must be used if they differ from those in this chart

B: Before top dead center

HYD: Hydraulic

① Computer controlled, no adjustment possible

② N/P range with A/C OFF: 700-900 rpm

N/P range with A/C ON: 800-1000 rpm

D range: 650-850 rpm

67162-SORE-C03

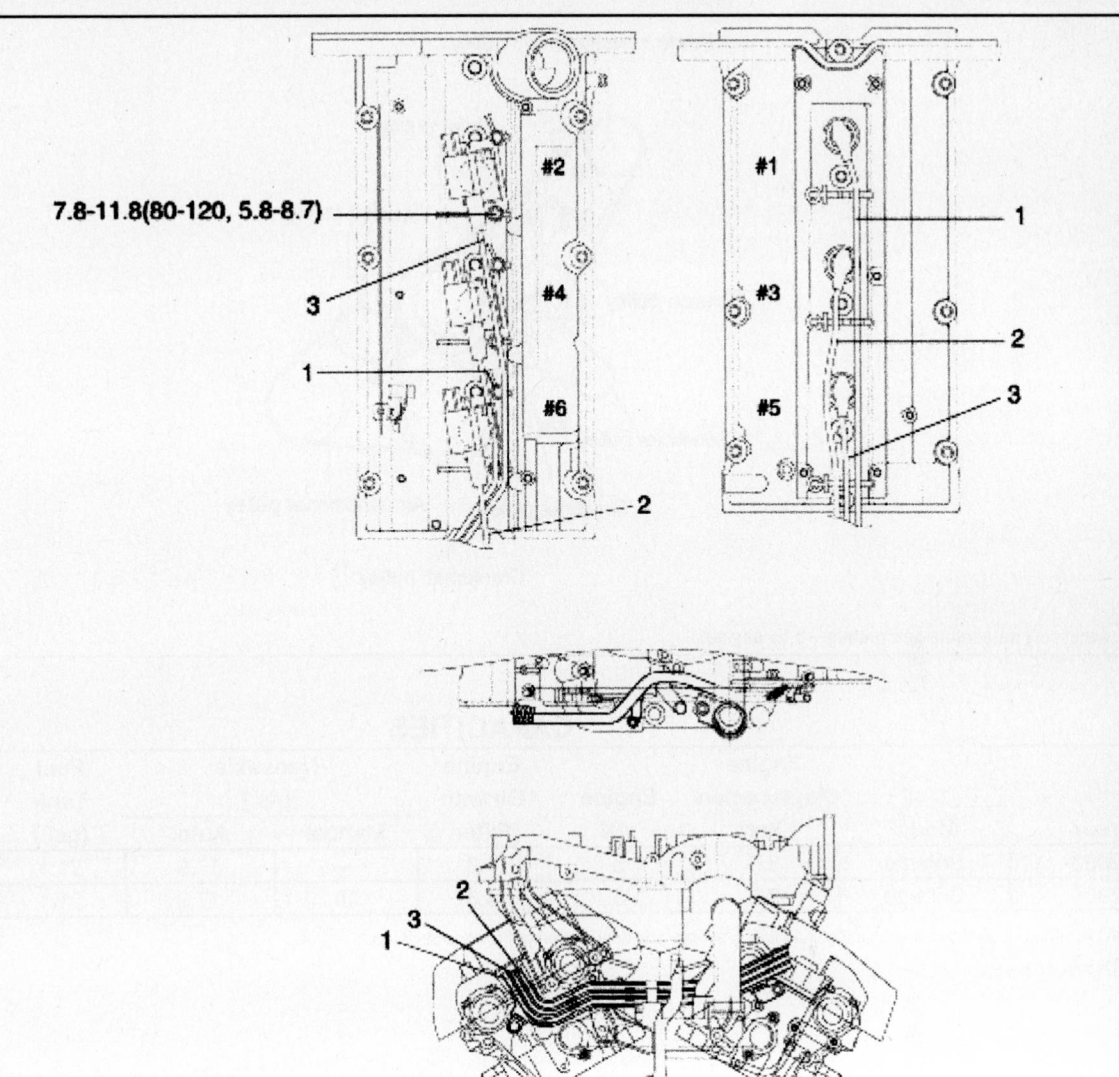

7.8-11.8(80-120, 5.8-8.7)

#2 #4 #6 #1 #3 #5

TORQUE : N·m (kg·cm, lb·ft)

1. No.1 Spark plug cable
2. No.3 Spark plug cable
3. No.5 Spark plug cable

67162-SORE-G01

3.5L engine
Firing order: 1–2–3–4–5–6
Distributorless ignition system

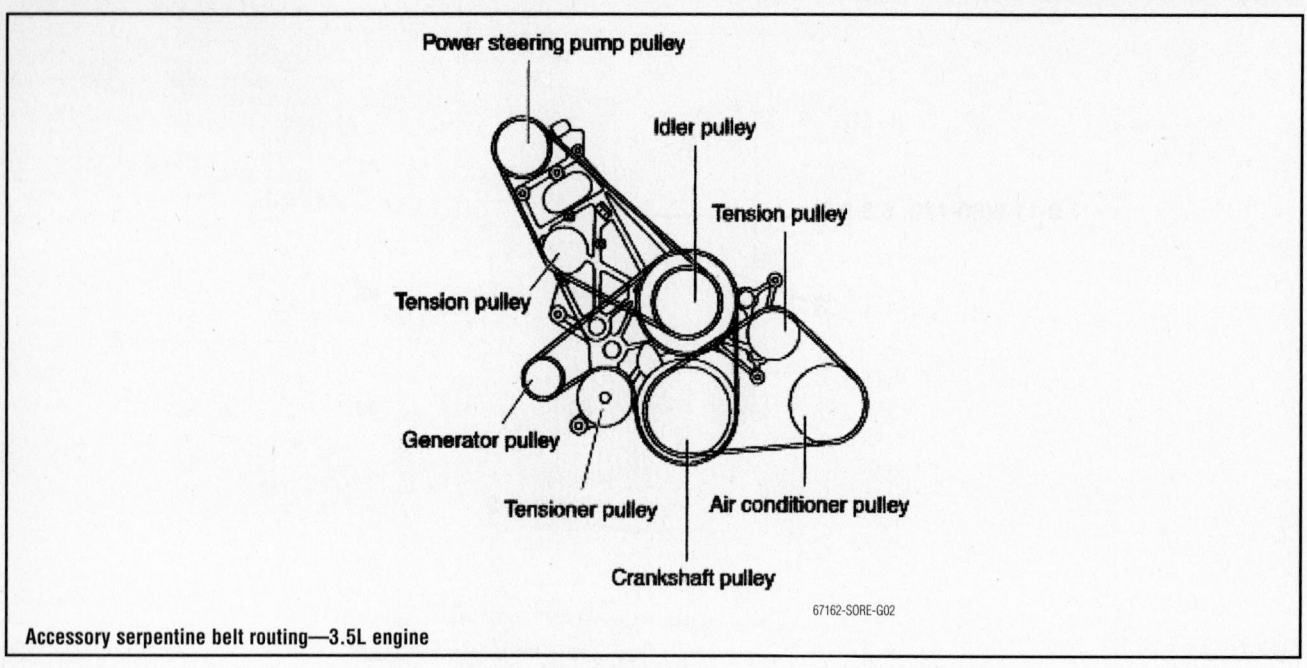

Accessory serpentine belt routing—3.5L engine

67162-SORE-G02

CAPACITIES

Year	Model	Engine Displacement Liters	Engine VIN	Engine Oil with Filter	Transaxle (pts.) Manual	Transaxle (pts.) Auto.	Fuel Tank (gal.)	Cooling System (qts.)
2003	Sorento	3.5	3	4.3	—	17.9	21.1	11.6
2004	Sorento	3.5	3	4.3	①	17.9	21.1	11.6

NOTE: All capacities are approximate. Add fluid gradually and ensure a proper level is obtained.

① 2WD: 5.6 pts.
  4WD: 4.75 pts.

67162-SORE-C04

VALVE SPECIFICATIONS

Year	Engine Displacement Liters	Engine VIN	Seat Angle (deg.)	Face Angle (deg.)	Maximum out of Square (degrees)	Spring Free Length (in.)	Stem-to-Guide Clearance (in.) Intake	Stem-to-Guide Clearance (in.) Exhaust	Stem Diameter (in.) Intake	Stem Diameter (in.) Exhaust
2003	3.5	3	45	45-45.5	2	1.8268	0.0009-0.0020	0.0020-0.0039	0.258-0.259	0.257-0.258
2004	3.5	3	45	45-45.5	2	1.8268	0.0009-0.0020	0.0020-0.0033	0.258-0.259	0.257-0.258

NA: Not Available

67162-SORE-C05

CRANKSHAFT AND CONNECTING ROD SPECIFICATIONS
All measurements are given in inches.

Year	Engine Displacement Liters	Engine VIN	Crankshaft				Connecting Rod		
			Main Brg. Journal Dia.	Main Brg. Oil Clearance	Shaft End-play	Thrust on No.	Journal Diameter	Oil Clearance	Side Clearance
2003	3.5	3	2.5190-2.5197	0.0007-0.0014	0.002-0.0098	3	2.1650-2.1653	0.0009-0.0016	0.0039-0.0098
2004	3.5	3	2.5190-2.5197	0.0007-0.0014	0.002-0.0098	3	2.1650-2.1653	0.0009-0.0016	0.0039-0.0098

67162-SORE-C06

PISTON AND RING SPECIFICATIONS
All measurements are given in inches.

Year	Engine Displacement Liters	Engine VIN	Piston Clearance	Ring Gap			Ring Side Clearance		
				Top Compression	Bottom Compression	Oil Control	Top Compression	Bottom Compression	Oil Control
2003	3.5	3	0.0012-0.0020	0.0079-0.0118	0.0177-0.0236	0.0079-0.0276	0.0016-0.0031	0.0008-0.0024	SNUG
2004	3.5	3	0.0012-0.0020	0.0079-0.0118	0.0177-0.0236	0.0079-0.0276	0.0016-0.0031	0.0008-0.0024	SNUG

67162-SORE-C07

TORQUE SPECIFICATIONS
All readings in ft. lbs.

Year	Engine Displacement Liters	Engine VIN	Cylinder Head Bolts	Main Bearing Bolts	Rod Bearing Bolts	Crankshaft Damper Bolts	Flywheel Bolts	Manifold		Spark Plugs	Oil Pan Drain Plug
								Intake	Exhaust		
2003	3.5	3	75-82	51-58	①	130-138	53-55	14-17	29-32	15-22	26-32
2004	3.5	3	75-82	51-58	①	130-138	53-55	14-17	29-32	15-22	26-32

① Step 1: 26 ft. lbs.
　 Step 2: Plus 92 degrees

67162-SORE-C08

WHEEL ALIGNMENT

Year	Model		Caster Range (+/-Deg.)	Caster Preferred Setting (Deg.)	Camber Range (+/-Deg.)	Camber Preferred Setting (Deg.)	Toe-in (in.)
2003	Sorento	F	0.50	3.30	0.50	0.39	0.1024
		R	—	—	—	—	—
2004	Sorento	F	0.50	3.30	0.50	0.39	0.1024
		R	—	—	—	—	—

67162-SORE-C09

TIRE, WHEEL AND BALL JOINT SPECIFICATIONS

Year	Model	OEM Tires Standard	OEM Tires Optional	Tire Pressures (psi) Front	Tire Pressures (psi) Rear	Wheel Size	Ball Joint Inspection	Wheel Lug Torque (ft. lbs.)
2003	Sorento	P245/70R16	—	30	30	7JJ	①	65-86
2004	Sorento	P245/70R16	—	30	30	7JJ	①	65-86

OEM: Original Equipment Manufacturer

PSI: Pounds Per Square Inch

STD: Standard

OPT: Optional

① Replace if any measurable movement is found.

67162-SORE-C10

BRAKE SPECIFICATIONS

All measurements in inches unless noted

Year	Model		Brake Disc Original Thickness	Brake Disc Minimum Thickness	Brake Disc Maximum Run-out	Brake Drum Original Inside Diameter	Brake Drum Max. Wear Limit	Brake Drum Maximum Machine Diameter	Minimum Lining Thickness	Brake Caliper Bracket Bolts (ft. lbs.)	Brake Caliper Mounting Bolts (ft. lbs.)
2003	Sorento	F	1.100	1.020	0.0012	—	—	—	0.079	7.2-9.4	16-24
		R	0.787	0.724	—	—	—	—	0.079	16-23	47-54
2004	Sorento	F	1.100	1.020	0.0012	—	—	—	0.079	7.2-9.4	16-24
		R	0.787	0.724	—	—	—	—	0.079	16-23	47-54

F: Front

R: Rear

67162-SORE-C11

SCHEDULED MAINTENANCE INTERVALS
Kia—Sorento

TO BE SERVICED	TYPE OF SERVICE	VEHICLE MILEAGE INTERVAL (x1000)															
		7.5	15	22.5	30	37.5	45	52.5	60	67.5	75	82.5	90	97.5	105	112.5	120
Accessory drive belts	S/I				✓				✓				✓				✓
Air cleaner element	I/R		✓		✓		✓		✓		✓		✓		✓		✓
Air conditioner system	S/I	Inspect the system operation and refrigerant amount annually.															
Brake lines, hoses and connections	S/I		✓		✓		✓		✓		✓		✓		✓		✓
Chassis and body fasteners	T				✓				✓				✓				✓
Cooling system hoses and coolant level	S/I				✓				✓				✓				✓
CV-joint boots	S/I		✓		✓		✓		✓		✓		✓		✓		✓
Engine coolant	R				✓				✓				✓				✓
Engine oil and filter	R	✓	✓	✓	✓	✓	✓	✓	✓	✓	✓	✓	✓	✓	✓	✓	✓
Exhaust system heat shields	S/I				✓				✓				✓				✓
Front and rear brakes	S/I		✓		✓		✓		✓		✓		✓		✓		✓
Front ball joints S/I	S/I				✓				✓				✓				✓
Fuel filter	R								✓								✓
Fuel lines and hoses	S/I				✓				✓				✓				✓
Rear differential fluid	S/I	✓	✓	✓	✓	✓	✓	✓	✓	✓	✓	✓	✓	✓	✓	✓	✓
Front differential fluid (if equipped)	S/I	✓	✓	✓	✓	✓	✓	✓	✓	✓	✓	✓	✓	✓	✓	✓	✓
Automatic transmission fluid	S/I		✓		✓		✓		✓		✓		✓		✓		✓
Manual transmission fluid (if equipped)	S/I	✓	✓	✓	✓	✓	✓	✓	✓	✓	✓	✓	✓	✓	✓	✓	✓
Transfer case oil (if equipped)	S/I	✓	✓	✓	✓	✓	✓	✓	✓	✓	✓	✓	✓	✓	✓	✓	✓
Ignition wires	S/I								✓								✓
Emission hoses	S/I				✓				✓				✓				✓
Driveshaft U-joints	L		✓		✓		✓		✓		✓		✓		✓		✓
Brake fluid	S/I		✓		✓		✓		✓		✓		✓		✓		✓
Clutch fluid (if equipped)	S/I		✓		✓		✓		✓		✓		✓		✓		✓
Idle speed	A				✓				✓				✓				✓
Locks and hinges	L	✓	✓	✓	✓	✓	✓	✓	✓	✓	✓	✓	✓	✓	✓	✓	✓
Spark plugs	R								✓								✓
Steering operation and linkage	S/I				✓				✓				✓				✓
Timing belt	R								✓								✓

R: Replace   S/I: Inspect and service, if needed   L: Lubricate   A: Adjust   T: Tighten

FREQUENT OPERATION MAINTENANCE (SEVERE SERVICE)

If a vehicle is operated under any of the following conditions it is considered severe service

- Towing a trailer or using a camper or car-top carrier.
- Repeated short trips of less than 5 miles in temperatures below freezing, or trips of less than 10 miles in any temperature.
- Extensive idling or low-speed driving for long distances as in heavy commercial use, such as delivery, taxi or police cars.
- Operating on rough, muddy or salt-covered roads.
- Operating on unpaved or dusty roads.
- Driving in extremely hot (over 90°F) conditions.

Engine oil and filter: replace every 3000 miles or 3 months, whichever occurs first.

Air cleaner element: inspect every 10,000 miles or 10 months and replace every 20,000 miles or 20 months, whichever occurs first.

Front differential fluid (if equipped): inspect every 5000 miles and replace every 15,000 miles.

Transfer case fluid (if equipped): inspect every 5000 miles.

Manual transmission fluid (if equipped): inspect every 5000 miles and replace every 60,000 miles.

Emission system hoses: inspect every 30,000 miles or 30 months, whichever occurs first.

Front and rear disc brakes: inspect every 15,000 miles or 15 months, whichever occurs first.

Chassis and body fasteners: tighten every 15,000 miles or 15 months, whichever occurs first.

Locks and hinges: lubricate every 5000 miles or 5 months, whichever occurs first.

PRECAUTIONS

Before servicing any vehicle, please be sure to read all of the following precautions, which deal with personal safety, prevention of component damage, and important points to take into consideration when servicing a motor vehicle:

• Never open, service or drain the radiator or cooling system when the engine is hot; serious burns can occur from the steam and hot coolant.

• Observe all applicable safety precautions when working around fuel. Whenever servicing the fuel system, always work in a well-ventilated area. Do not allow fuel spray or vapors to come in contact with a spark, open flame, or excessive heat (a hot drop light, for example). Keep a dry chemical fire extinguisher near the work area. Always keep fuel in a container specifically designed for fuel storage; also, always properly seal fuel containers to avoid the possibility of fire or explosion. Refer to the additional fuel system precautions later in this section.

• Fuel injection systems often remain pressurized, even after the engine has been turned **OFF**. The fuel system pressure must be relieved before disconnecting any fuel lines. Failure to do so may result in fire and/or personal injury.

• Brake fluid often contains polyglycol ethers and polyglycols. Avoid contact with the eyes and wash your hands thoroughly after handling brake fluid. If you do get brake fluid in your eyes, flush your eyes with clean, running water for 15 minutes. If eye irritation persists, or if you have taken brake fluid internally, IMMEDIATELY seek medical assistance.

• The EPA warns that prolonged contact with used engine oil may cause a number of skin disorders, including cancer. You should make every effort to minimize your exposure to used engine oil. Protective gloves should be worn when changing oil. Wash your hands and any other exposed skin areas as soon as possible after exposure to used engine oil. Soap and water, or waterless hand cleaner should be used.

• All new vehicles are now equipped with an air bag system, often referred to as a Supplemental Restraint System (SRS) or Supplemental Inflatable Restraint (SIR) system. The system must be disabled before performing service on or around system components, steering column, instrument panel components, wiring and sensors. Failure to follow safety and disabling procedures could result in accidental air bag deployment, possible personal injury and unnecessary system repairs.

• Always wear safety goggles when working with, or around, the air bag system. When carrying a non-deployed air bag, be sure the bag and trim cover are pointed away from your body. When placing a non-deployed air bag on a work surface, always face the bag and trim cover upward, away from the surface. This will reduce the motion of the module if it is accidentally deployed. Refer to the additional air bag system precautions later in this section.

• Clean, high quality brake fluid from a sealed container is essential to the safe and proper operation of the brake system. You should always buy the correct type of brake fluid for your vehicle. If the brake fluid becomes contaminated, completely flush the system with new fluid. Never reuse any brake fluid. Any brake fluid that is removed from the system should be discarded. Also, do not allow any brake fluid to come in contact with a painted surface; it will damage the paint.

• Never operate the engine without the proper amount and type of engine oil; doing so WILL result in severe engine damage.

• Timing belt maintenance is extremely important. Many models utilize an interference-type, non-freewheeling engine. If the timing belt breaks, the valves in the cylinder head may strike the pistons, causing potentially serious (also time-consuming and expensive) engine damage. Refer to the maintenance interval charts for the recommended replacement interval for the timing belt.

• Disconnecting the negative battery cable on some vehicles may interfere with the functions of the on-board computer system(s) and may require the computer to undergo a relearning process once the negative battery cable is reconnected.

• When servicing drum brakes, only disassemble and assemble one side at a time, leaving the remaining side intact for reference.

• Only an MVAC-trained, EPA-certified automotive technician should service the air conditioning system or its components.

ENGINE REPAIR

Alternator

REMOVAL & INSTALLATION

1. Before servicing the vehicle, refer to the precautions at the beginning of this section.
2. Disconnect the negative battery cable.
3. Remove the accessory drive belt.
4. Disconnect the wiring to the alternator.
5. Remove the alternator mounting bolts.
6. Remove the alternator.
To install:
7. Install the alternator and insert a support bolt (Do not insert a nut this time).

8. After pushing the alternator forward, check to see if any spacers are required (a spacer thickness: 0.198 mm) between the alternator front leg and front case.
9. If necessary, insert spacer(s), then insert and tighten nut securely.
10. Connect the alternator electrical connectors. Tighten the battery terminal connector nut to 60 inch lbs. (7 Nm).
11. Install the accessory drive belt.
12. Connect the negative battery cable.

Ignition Timing

This vehicle is equipped with a Distributorless Ignition System (DIS). No adjustment is necessary or possible.

Engine Assembly

REMOVAL & INSTALLATION

1. Before servicing the vehicle, refer to the precautions at the beginning of this section.
2. Remove the battery and air cleaner assembly.
3. Remove the hood from the vehicle.
4. Drain the cooling system.
5. Drain the engine oil.
6. Drain the transmission fluid.
7. Relieve fuel system pressure.
8. Remove the upper and lower radiator hoses.
9. Remove the heater hoses.
10. Remove the radiator.

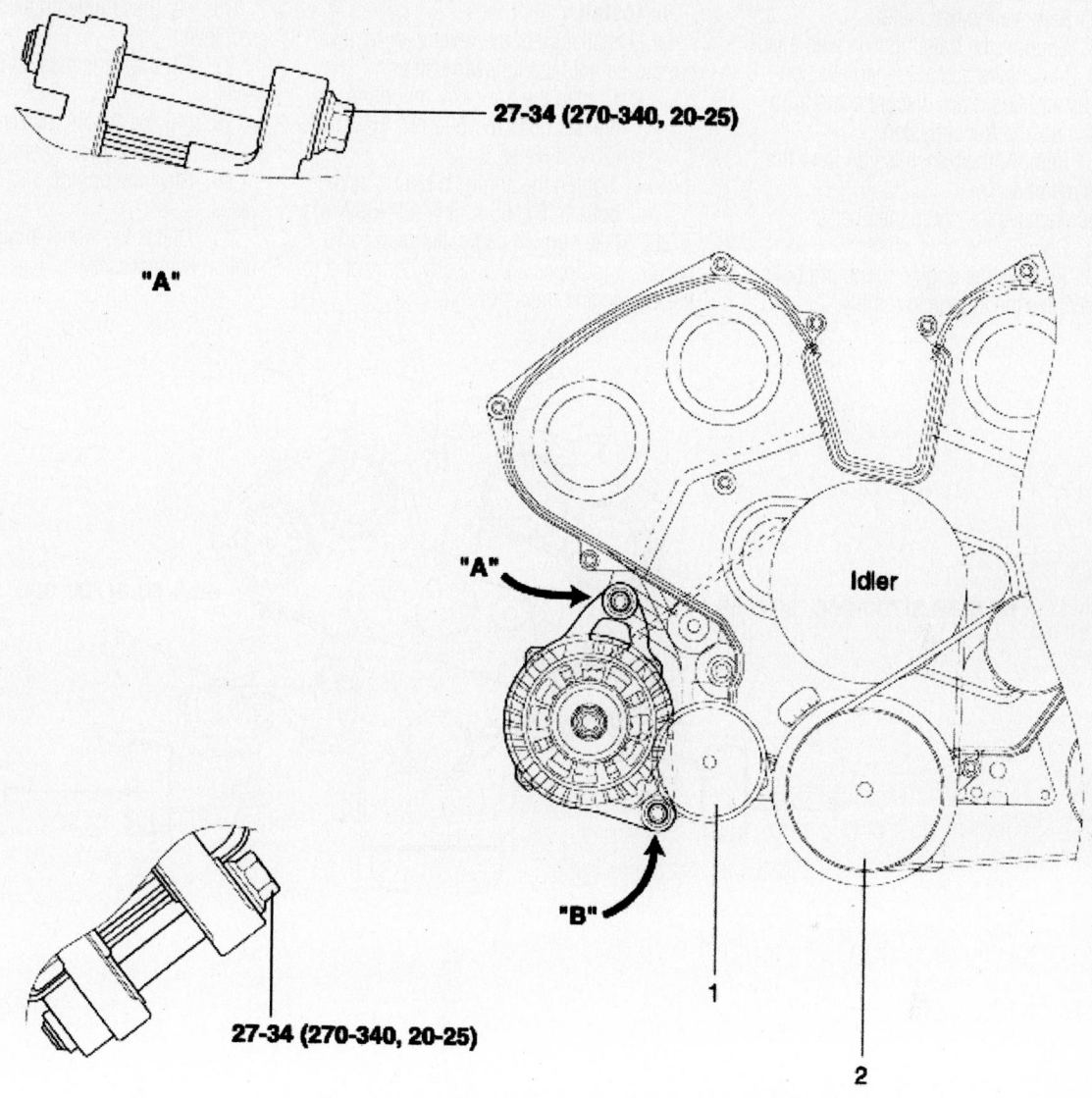

27-34 (270-340, 20-25)

"A"

"A" → Idler

"B"

1

2

27-34 (270-340, 20-25)

"B"

TORQUE : N-m (kg-cm, lb-ft)

1. Generator tensioner
2. Crankshaft pulley

67162-SORE-G03

Alternator mounting exploded view

11. Remove the breather hose and air-intake hose.

12. Disconnect the throttle and cruise control cables.

13. Disconnect the oil pressure switch connector.

14. Disconnect the oil pressure sensor connector.

15. Remove the engine ground cable.

16. Disconnect the EVAP canister hose.

17. Disconnect the brake booster vacuum hose.

18. Disconnect the fuel lines.

19. Remove the accessory drive belts.

20. Disconnect the power steering hoses from the pump.

21. Disconnect the alternator wiring harness connectors.

22. If equipped with an automatic transmission, disconnect the oil cooler lines.

23. Disconnect the starter motor wiring harness connectors.

24. Remove the starter motor.

25. Remove the exhaust front pipes from the exhaust manifolds.

26. Disconnect the transmission shift cable or clutch cable, if equipped.

27. Remove the halfshafts.

28. Support the transmission with a jack.

29. Make sure all cable, harness connectors and hoses are disconnected from the engine and transmission.

30. Remove the transmission from the cross member.

31. Remove the cross member.

32. Remove the drive shaft.

33. Remove the engine mounting bolts and remove the engine assembly.

To install:

34. Installation is the reverse of removal but please note the following steps:
- Tighten the transaxle mounting bracket bolts to 15–21 ft. lbs. (20–28 Nm)
- Tighten the engine mount bracket bolts to 51–65 ft. lbs. (69–88 Nm)

35. Make sure all cable, harness connectors and hoses are properly connected to the engine and transmission.

36. Fill the engine crankcase to the correct level.

37. Fill the transmission to the correct level.

38. Fill the cooling system.

39. Fill the power steering system.

40. Start the engine and check for leaks.

41. Check the wheel alignment and adjust as necessary.

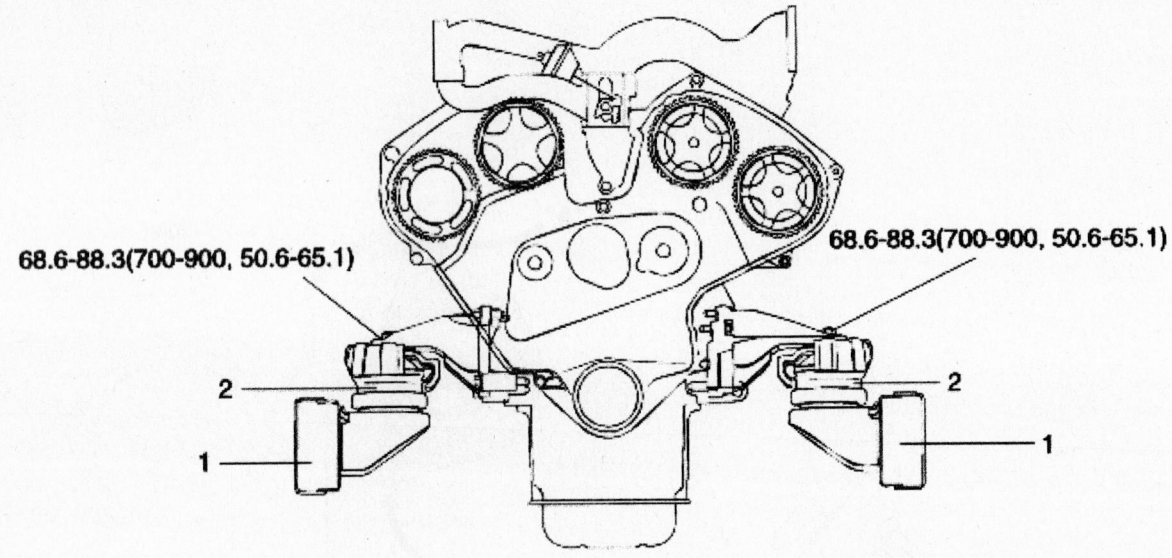

68.6-88.3(700-900, 50.6-65.1)   68.6-88.3(700-900, 50.6-65.1)

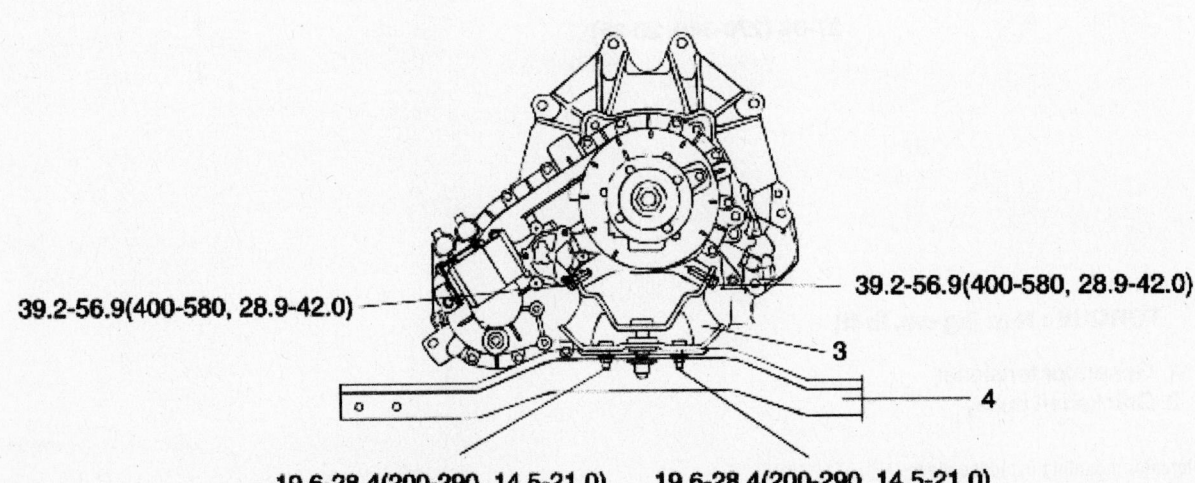

39.2-56.9(400-580, 28.9-42.0)   39.2-56.9(400-580, 28.9-42.0)

19.6-28.4(200-290, 14.5-21.0)   19.6-28.4(200-290, 14.5-21.0)

TORQUE : N·m (kg·cm, lb·ft)

1. Frame
2. Engine mounting
3. Transmission mounting
4. Cross member

67162-SORE-G04

Engine mount locations and torque specifications

Water Pump

REMOVAL & INSTALLATION

1. Before servicing the vehicle, refer to the precautions at the beginning of this section.
2. Drain the cooling system.
3. Disconnect the negative battery cable.
4. Remove the drive belt and the water pump pulley.
5. Remove the timing belt cover, timing belt, auto tensioner and idler pulley.
6. Remove the water outlet fitting, the thermostat case and the water pump fitting.
7. Remove the water pump mounting bolts.
8. Remove the water pump.

To install:

9. Clean the gasket surfaces of the water pump body and the cylinder block.
10. Install the new water pump gasket and pump assembly. Tighten the bolts to 12–16 ft. lbs. (17–22 Nm).

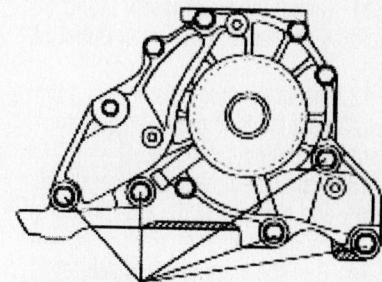

Water pump mounting bolt

67162-SORE-G06

Water pump bolt locations

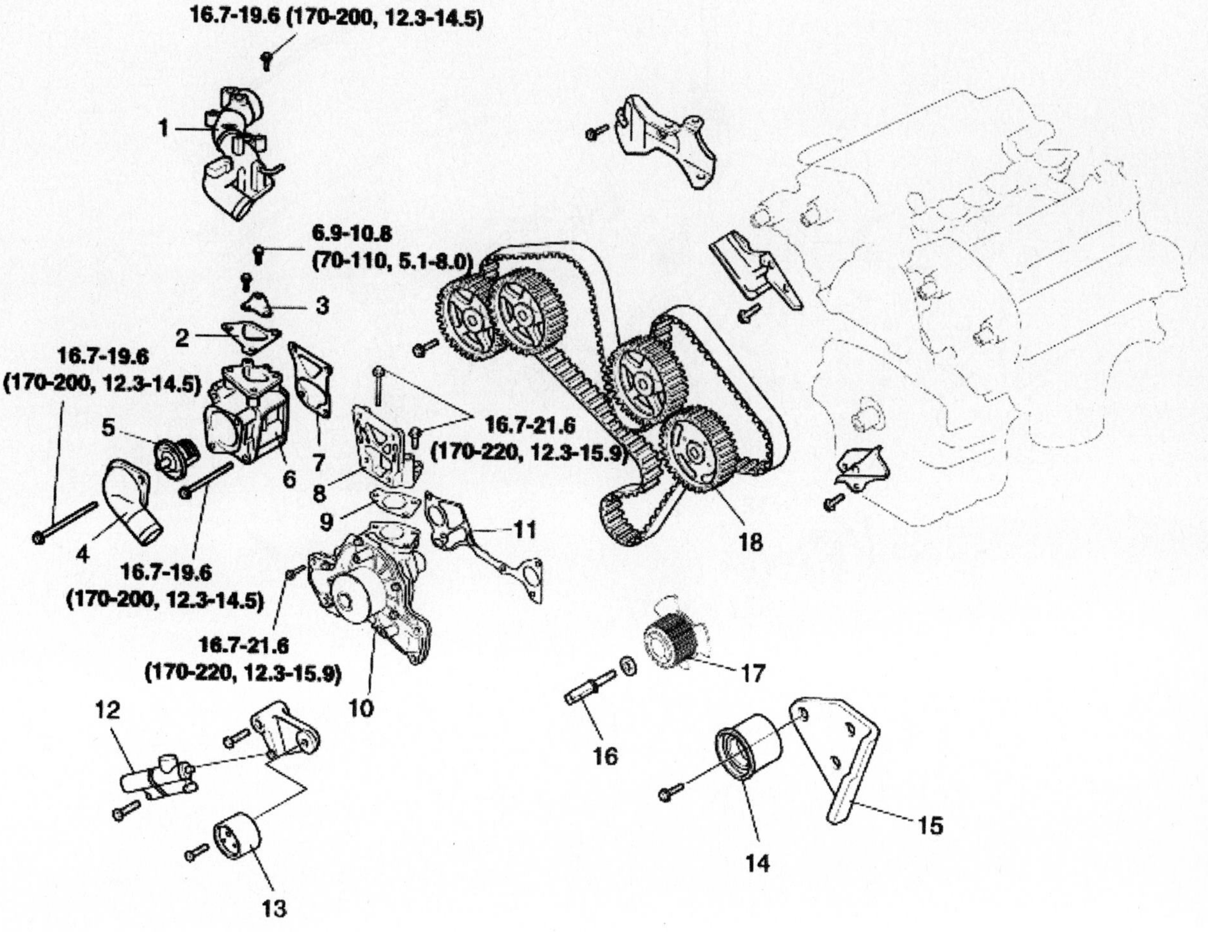

16.7-19.6 (170-200, 12.3-14.5)
1
6.9-10.8
(70-110, 5.1-8.0)
3
2
16.7-19.6
(170-200, 12.3-14.5)
5
16.7-21.6
(170-220, 12.3-15.9)
7
6
8
9
11
4
18
16.7-19.6
(170-200, 12.3-14.5)
16.7-21.6
(170-220, 12.3-15.9)
17
12
10
16
15
14
13

TORQUE : N·m (kg·cm, lb·ft)

1. Water outlet fitting
2. Gasket
3. Thermostat case bracket
4. Water inlet fitting
5. Thermostat
6. Thermostat case
7. Gasket
8. Water pump fitting
9. Gasket
10. Water pump
11. Gasket
12. Auto tensioner
13. Tension bearing
14. Idler
15. Idler bracket
16. Damper pulley bolt
17. Crankshaft sprocket
18. Camshaft sprocket

67162-SORE-G05

Water pump and timing belt assembly—exploded view

11. Install the water pump fitting, the thermostat case and the water outlet fitting.

12. Install the auto tensioner and timing belt. Adjust the timing belt tension, then install the timing belt cover.

13. Install the drive belt, water pump pulley and then adjust the auto tensioner.

14. Refill the coolant.

15. Run the engine and check for leaks.

Heater Core

REMOVAL & INSTALLATION

1. Before servicing the vehicle, refer to the precautions at the beginning of this section.

2. Drain the cooling system.

3. Disconnect the negative battery cable.

4. Remove the heater hoses.

5. Remove the A pillar trim.

6. Remove the cowl side trim.

7. Remove the front door scuff trim.

8. Remove the shroud cover.

9. Remove the left side cover.

10. Remove the hood release cable.

11. Remove the lower panel screws (4), disconnect electrical connectors and remove the lower instrument panel.

12. Remove the driver air bag module.

13. Remove the steering wheel.

14. Remove the steering shroud cover.

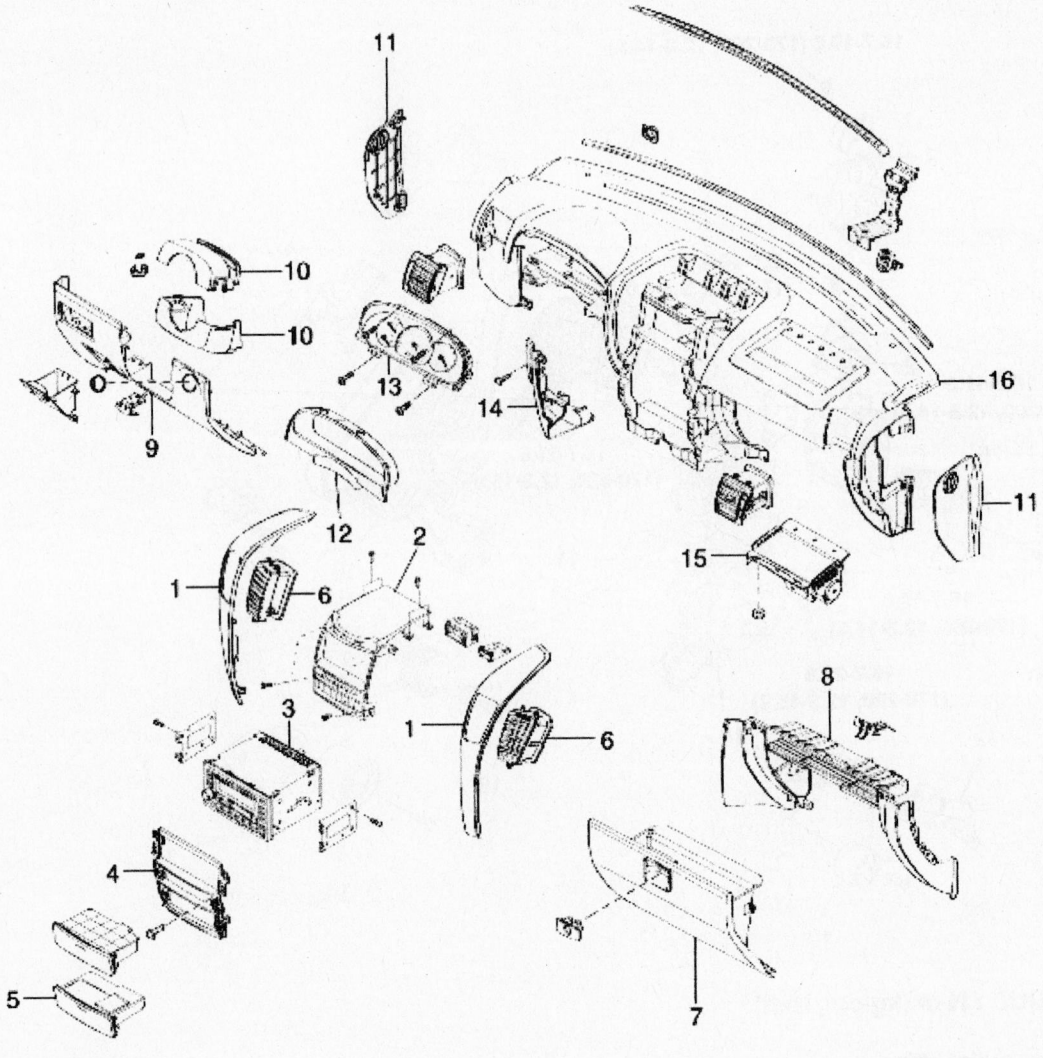

1. Center facia side panel assembly
2. Audio facia panel
3. Audio assembly
4. Center facia panel assembly
5. Ashtray
6. Center air-vent assembly
7. Glove box assembly
8. Lower panel assembly (RH)
9. Lower panel assembly (LH)
10. Steering column shroud
11. Crash pad side cover
12. Cluster facia panel assembly
13. Instrument cluster assembly
14. Center lower panel assembly
15. Passenger airbag assembly
16. Instrument panel assembly

67162-SORE-G07

Instrument panel exploded view

15. Remove the lower instrument panel and steering shaft.

16. Remove the fascia panel screws (2) and remove the instrument cluster fascia panel.

17. Remove the cluster screws (4), disconnect electrical connector and remove the instrument cluster assembly.

18. Open ashtray, the remove instrument center fascia side panel.

19. Remove audio fascia panel screws (8), disconnect the electrical connector and remove audio fascia panel.

20. Remove the radio mounting screws (4), disconnect the electrical connector, antenna cable and remove the radio unit.

21. Remove the screws (4), disconnect the electrical connector and remove heater controller.

22. Remove the screws (6), disconnect the electrical connector and remove center fascia panel.

23. Remove the screws (4), then remove the left center lower panel.

24. Remove the glove box.

25. Remove the right side cover, instrument right lower panel screws (10), disconnect the electrical connector and remove the instrument right lower panel.

26. Disconnect the electrical connector and remove the main instrument panel assembly.

27. Remove crossbar side mounting bolts (4).

28. Remove crossbar lower mounting bolts (4).

29. Remove crossbar bracket bolts (2).

30. Remove the heater unit.

To install:

31. Install the heater unit.

32. Install the crossbar and tighten the

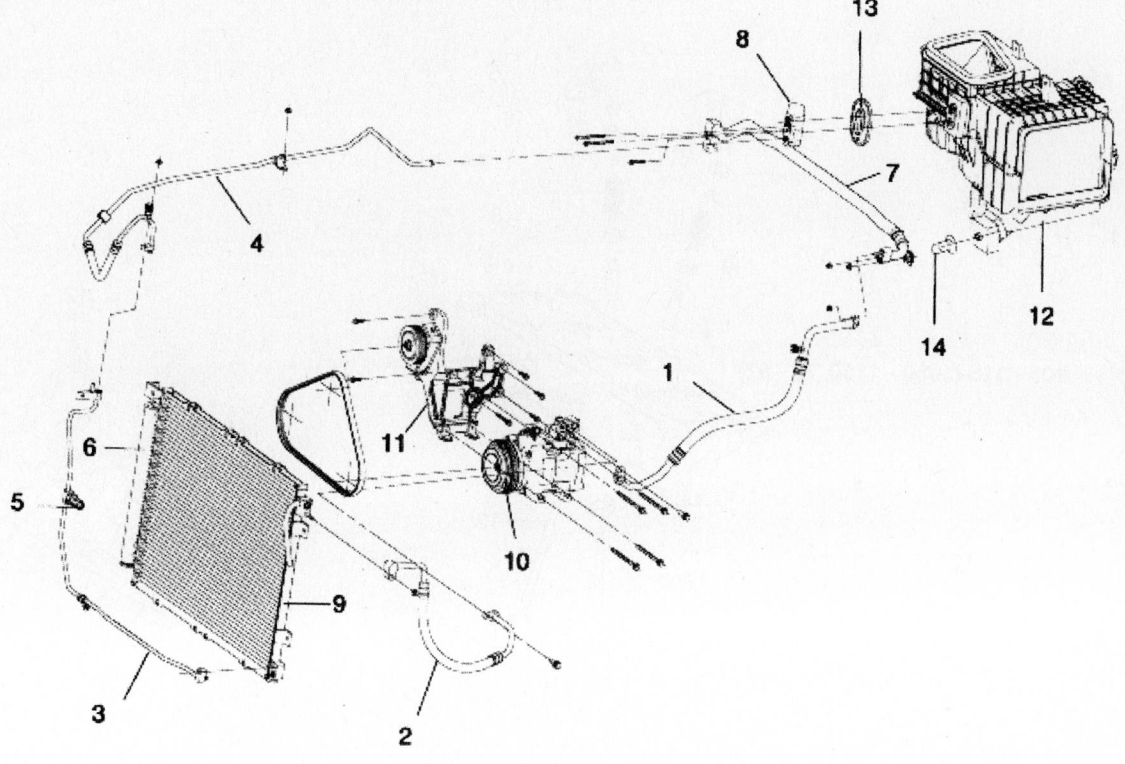

1. Suction hose
2. Discharge hose
3. Liquid pipe, A
4. Liquid pipe, B
5. Triple pressure switch
6. Receiver drier
7. Suction pipe
8. Expansion valve
9. Condenser
10. Compressor
11. Compressor mounting bracket
12. Blower & Evaporator
13. Evaporator pipe seal
14. Drain hose

Heater unit mounting exploded view

67162-SORE-G08

mounting bolts to 12–19 ft. lbs. (17–26 Nm).

33. Install the main instrument panel assembly and reconnect the wiring harness connector.

34. Install the instrument right lower panel, reconnect the wiring harness connector and install the right side cover.

35. Install the glove box.
36. Install the left center lower panel.
37. Install the center fascia panel.
38. Install the heater controller.
39. Install the radio.
40. Install the audio fascia panel.
41. Install the instrument center fascia side panel.

42. Install the instrument cluster assembly.

43. Install the instrument cluster fascia panel.

44. Install the steering shaft and lower instrument panel.

45. Install the steering shroud cover.
46. Install the steering wheel.
47. Install the driver air bag module.
48. Install the lower instrument panel.
49. Install the hood release cable.
50. Install the left side cover.
51. Install the shroud cover.
52. Install the front door scuff trim.
53. Install the cowl side trim.
54. Install the A pillar trim.

55. Install the heater hoses.
56. Connect the negative battery cable.
57. Fill the cooling system.
58. Start the engine and check for leaks and proper heater operation.

Cylinder Head

REMOVAL & INSTALLATION

1. Before servicing the vehicle, refer to the precautions at the beginning of this section.
2. Drain the cooling system.
3. Relieve the fuel system pressure.

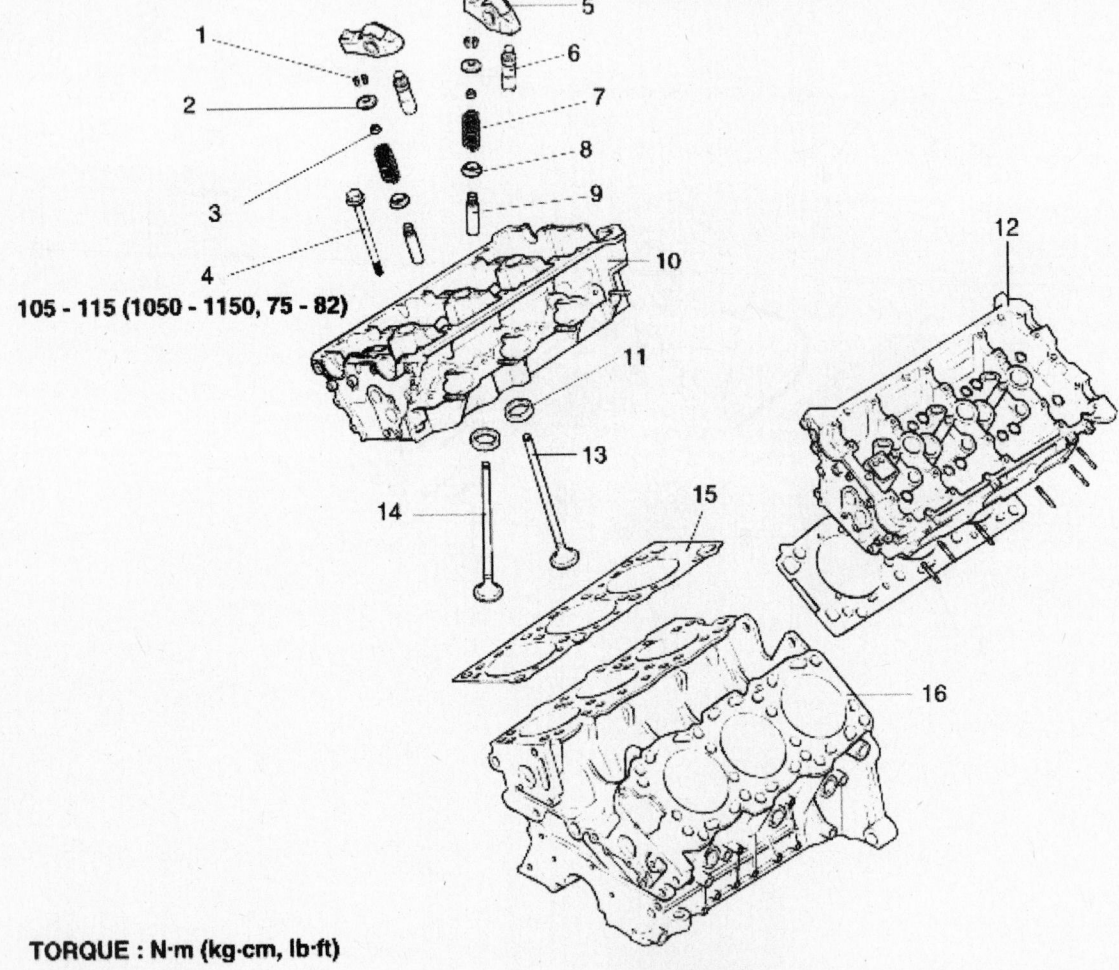

105 - 115 (1050 - 1150, 75 - 82)

TORQUE : N·m (kg·cm, lb·ft)

1. Retainer lock
2. Valve spring retainer
3. Valve stem seal
4. Cylinder head bolt
5. Rocker arm
6. Lash adjuster
7. Vlve spring
8. Spring sheet

9. Valve guide
10. Cylinder head (RH)
11. Exhaust valve seat ring
12. Cylinder head (LH)
13. Exhaust vlave
14. Intake valve
15. Gasket
16. Cylinder block

Cylinder head and rocker arm exploded view

67162-SORE-G09

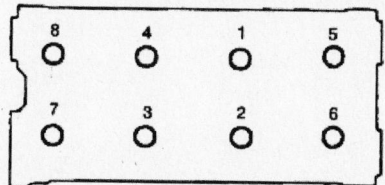

Cylinder head bolt torque sequence

67162-SORE-G10

4. Disconnect the negative battery cable.

5. Disconnect the upper radiator hose.

6. Remove the breather hose and air-intake hose.

7. Remove the vacuum hose, fuel hose and coolant hose.

8. Remove the intake manifold.

9. Remove the cables from the spark plugs. The cables should be removed holding the boot portion.

10. Remove the ignition coil.

11. Remove the upper and lower timing belt cover.

12. Remove the timing belt and camshaft sprockets.

13. Remove the heat protector and exhaust manifold assembly.

14. Remove the coolant pump pulley and head cover.

15. Remove the intake and exhaust camshaft.

➡ **Keep all valvetrain components in order for assembly.**

16. Remove the cylinder head assembly. The cylinder head bolts should be removed using the 12 mm socket, in two or three steps. Clean the gasket pieces from the cylinder block top surface and cylinder bottom surface.

To install:

17. Install the cylinder heads with new gaskets. Tighten the bolts in sequence to 75–82 ft. lbs. (105–115 Nm).

18. Installation is the reverse of the removal procedure.

Rocker Arms

REMOVAL & INSTALLATION

1. Before servicing the vehicle, refer to the precautions at the beginning of this section.

2. Drain the cooling system.

3. Relieve the fuel system pressure.

4. Disconnect the negative battery cable.

5. Disconnect the upper radiator hose.

6. Remove the breather hose and air-intake hose.

7. Remove the vacuum hose, fuel hose and coolant hose.

8. Remove the intake manifold.

9. Remove the cables from the spark plugs. The cables should be removed holding the boot portion.

10. Remove the ignition coil.

11. Remove the upper and lower timing belt cover.

12. Remove the timing belt and camshaft sprockets.

13. Remove the heat protector and exhaust manifold assembly.

14. Remove the coolant pump pulley and head cover.

15. Remove the camshafts.

16. Remove the rocker arms and lash adjusters.

➡ **Keep all valvetrain components in order for installation.**

17. Inspect the roller visually. If any damage is found, replace it.

18. Check that the roller operates smoothly. If there is excessive clearance, replace it.

To install:

19. Installation is the reverse of removal procedure.

Intake Manifold

REMOVAL & INSTALLATION

1. Before servicing the vehicle, refer to the precautions at the beginning of this section.

2. Drain the cooling system.

3. Relieve the fuel system pressure.

4. Disconnect the negative battery cable.

5. Remove the air intake hose connected to the throttle body.

6. Remove the accelerator and cruise control cables.

7. Remove the engine coolant hose and throttle body.

8. Remove the PCV hose and brake booster vacuum hoses.

9. Disconnect the vacuum hose connections.

10. Remove the surge tank stay.

11. Bleed off the pressure in the fuel pipe line to prevent the fuel from spilling.

12. Disconnect the connector from high pressure hose.

13. Disconnect the surge tank and gasket.

14. Disconnect the fuel injector harness connector.

15. Remove the delivery pipe with the fuel injector and the pressure regulator.

16. Disconnect the wiring harness of the coolant sensor assembly.

17. Remove the intake manifold.

To install:

18. Installation is the reverse of the removal procedure, while using the following torque values:

- Lower intake manifold nuts, in sequence to: 15-17 ft. lbs. (20-23 Nm)
- Surge tank stay bolts: 11-14 ft. lbs. (15-20 Nm)

Exhaust Manifold

REMOVAL & INSTALLATION

1. Before servicing the vehicle, refer to the precautions at the beginning of this section.

2. Disconnect the negative battery cable.

3. Disconnect the Heated Oxygen (HO2S) sensor connectors.

4. Separate the exhaust Y pipe from the exhaust manifold(s).

5. Remove the heat protector.

6. Remove the exhaust manifold.

7. Remove the exhaust manifold gasket.

To install:

➡ **Use only new gaskets and nuts for assembly.**

8. Install the exhaust manifolds with new gaskets. Tighten the fasteners to 20–24 ft. lbs. (27–32 Nm).

9. Install the exhaust manifold heat protectors. Tighten the bolts to 9–11 ft. lbs. (12-15 Nm).

10. Install the exhaust Y pipe.

11. Connect the Heated Oxygen (HO2S) sensor connectors.

12. Connect the negative battery cable.

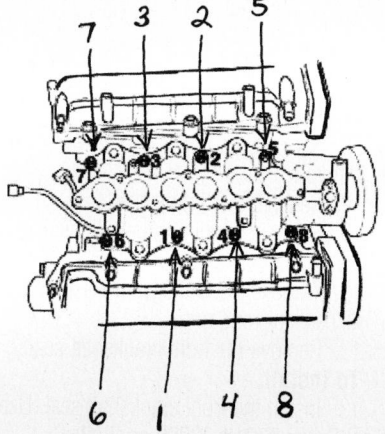

67162-SORE-G11

Intake manifold torque sequence

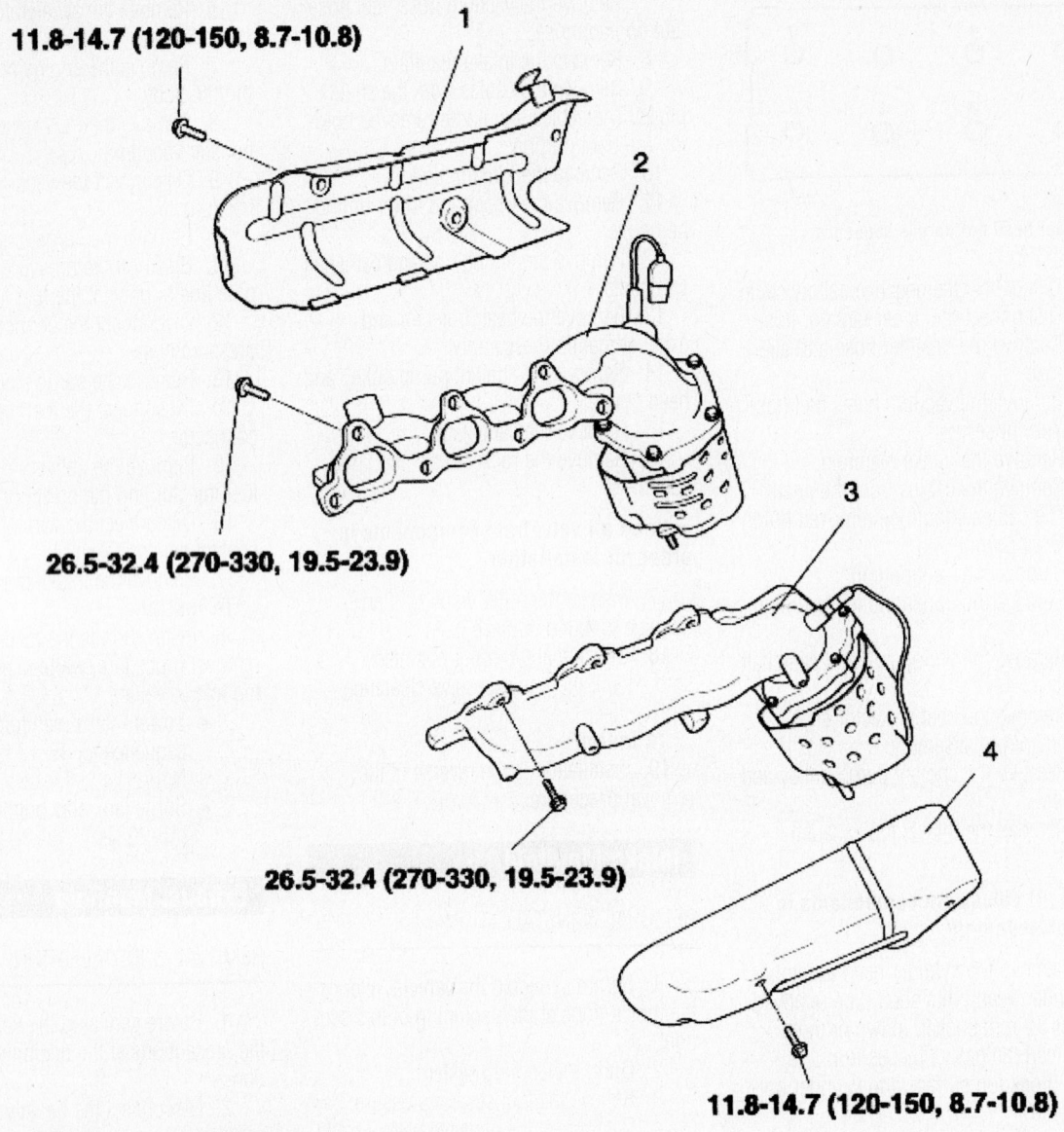

11.8-14.7 (120-150, 8.7-10.8)

26.5-32.4 (270-330, 19.5-23.9)

26.5-32.4 (270-330, 19.5-23.9)

11.8-14.7 (120-150, 8.7-10.8)

TORQUE : N·m (kg·cm, lb·ft)

1. Heat protector
2. Exhaust manifold
3. Exhaust manifold
4. Heat protector

67162-SORE-G12

Exhaust manifold mounting exploded view

Front Crankshaft Seal

REMOVAL & INSTALLATION

1. Before servicing the vehicle, refer to the precautions at the beginning of this section.
2. Disconnect the negative battery cable.
3. Remove the accessory drive belts.
4. Remove the idler pulley.
5. Remove the crankshaft pulley.
6. Remove the power steering pump pulley.
7. Remove the belt tensioner pulley.
8. Remove the upper and lower timing belt covers.
9. Remove the timing belt.
10. Remove the timing belt crankshaft sprocket.
11. Remove the Crankshaft Position (CKP) sensor tone ring.
12. Remove the front crankshaft seal.

To install:

13. Install the front crankshaft seal. Use Seal Driver 09214-33000 or similar.
14. Install the CKP sensor tone ring.
15. Install the timing belt crankshaft sprocket.
16. Install the timing belt.
17. Install the upper and lower timing belt covers.
18. Install the belt tensioner pulley.
19. Install the power steering pump pulley.
20. Install the crankshaft pulley.
21. Install the idler pulley.
22. Install the accessory drive belts.
23. Connect the negative battery cable.

Camshaft and Valve Lifters

REMOVAL & INSTALLATION

1. Before servicing the vehicle, refer to the precautions at the beginning of this section.
2. Disconnect the negative battery cable.
3. Remove intake manifold.
4. Disconnect the breather hose and the engine harness.
5. Remove the power steering pulley, air conditioner pulley, crankshaft pulley, idler pulley and tensioner pulley.
6. Remove the timing belt cover.
7. Loosen the auto tensioner.
8. Remove the timing belt from the camshaft sprocket.
9. Remove the spark plug cables.
10. Loosen the cylinder head cover bolts and then remove it.
11. Remove the camshaft sprockets.

➡ Keep all valvetrain components in order for assembly.

12. Remove the camshaft bearing caps.
13. Remove the camshafts.
14. Remove the rocker arms.
15. Remove the hydraulic lifters.

To install:

16. Set the No. 1 cylinder to Top Dead Center of the compression stroke.
17. Install the lifters and rocker arms in their original positions.

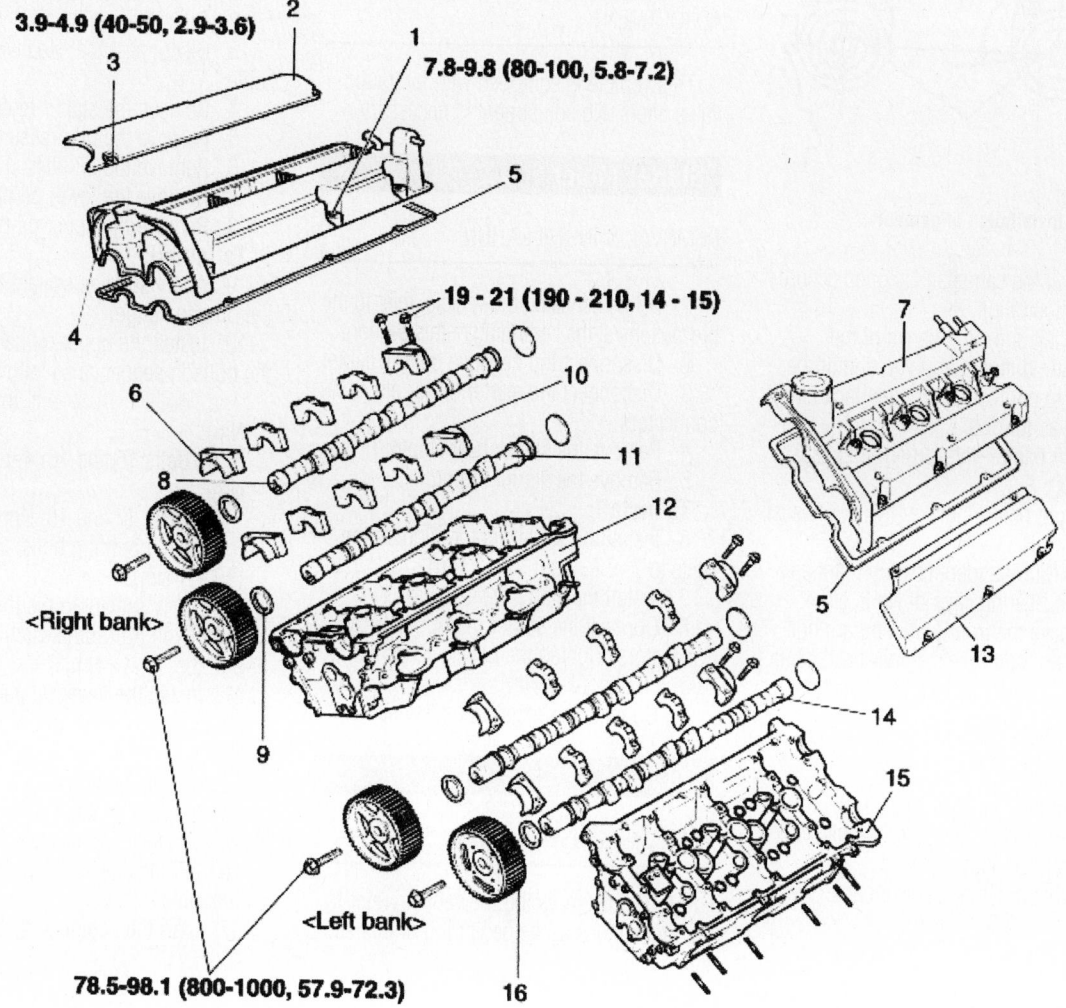

3.9-4.9 (40-50, 2.9-3.6)

7.8-9.8 (80-100, 5.8-7.2)

19 - 21 (190 - 210, 14 - 15)

<Right bank>

<Left bank>

78.5-98.1 (800-1000, 57.9-72.3)

TORQUE : N·m (kg·cm, lb·ft)

1. Cylinder head cover bolt
2. Center cover
3. Center cover bolt
4. Cylinder head cover
5. Gasket
6. Bearing cap (Front)
7. Cylinder head cover
8. Camshaft (EX)

9. Camshaft oil seal
10. Bearing cap (Rear)
11. Camshaft (IN)
12. Cylinder head (RH)
13. Center cover
14. Camshaft (EX)
15. Cylinder head (LH)
16. Camshaft sprocket

Camshaft mounting exploded view

67162-SORE-G13

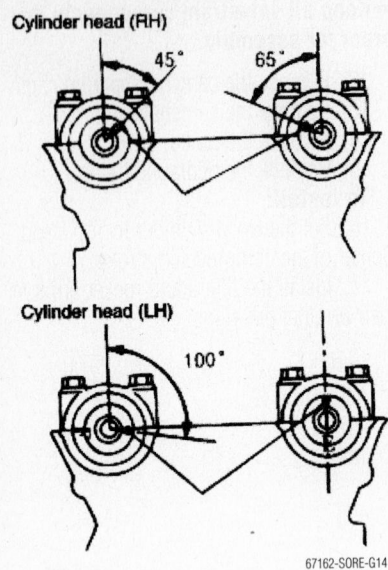

Camshaft installation alignment

18. Install the camshafts aligned according to the illustration.

19. The left and right banks of the camshafts are different and you should be careful not to confuse them. Identification signals are as follows:

- Left bank—Intake (IN): I; Exhaust (Ex): E
- Right bank—Intake (IN): J; Exhaust (Ex): H

20. Confirm the identification mark and the number. Bearing caps of No.3, No.4, and No.5 have the front mark and arrange the front mark upon the cylinder head while installing the bearing caps. Identification marks are as follows:

- Intake: I
- Exhaust: E

21. Install the bearing caps. Tighten the bolts evenly in 2 or 3 steps to the following torque specifications:

- Front and rear bearing caps: 14–15 ft. lbs. (19–21 Nm)
- Center bearing caps: 7–9 ft. lbs. (10–12 Nm)

22. Install the gasket to the cylinder head cover correctly.

23. Clean sealing surface on camshaft cap.

24. Apply sealant to the sealing surface on cylinder head cover and camshaft cap.

25. Install the cylinder head cover to the cylinder head.

26. Install the camshaft sprockets. Tighten the bolts to 58–72 ft. lbs. (79–98 Nm).

27. Install the timing belt.

28. Install the upper and lower timing belt covers.

29. Install the belt tensioner pulley.

30. Install the power steering pump pulley.

31. Install the crankshaft pulley.

32. Install the idler pulley.

33. Install the accessory drive belts.

34. Install the intake manifold.

35. Connect the breather hose and the engine harness.

36. Connect the spark plug cables.

37. Connect the negative battery cable.

Valve Lash

ADJUSTMENT

This vehicle is equipped with hydraulic valve lifters. No adjustment is necessary.

Starter Motor

REMOVAL & INSTALLATION

1. Before servicing the vehicle, refer to the precautions at the beginning of this section.

2. Disconnect the negative battery cable.

3. Disconnect the starter motor electrical connectors.

4. Remove the starter heat shield.

5. Remove the starter motor.

To install:

6. Install the starter motor. Tighten the bolts to 7–9 ft. lbs. (10–12 Nm).

7. Install the starter heat shield.

8. Connect the starter motor electrical connectors. Tighten the battery terminal nut to 3–4 ft. lbs. (4–6 Nm).

9. Connect the negative battery cable.

Oil Pan

REMOVAL & INSTALLATION

1. Before servicing the vehicle, refer to the precautions at the beginning of this section.

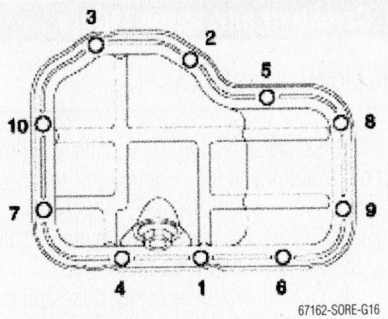

Lower oil pan torque sequence

2. Drain the engine oil.

3. Disconnect the negative battery cable.

4. Remove the starter motor.

5. Remove the oil pressure switch.

6. Remove the oil filter.

7. Remove the lower oil pan.

8. Remove the upper oil pan.

To install:

9. Apply silicone sealant to the grove of the oil pan flange.

10. Install the upper oil pan and tighten the bolts in sequence as follows:

   a. Bolts 1-14: 7–9 ft. lbs. (10–12 Nm).

   b. Bolts 15 and 16: 4–5 ft. lbs. (5–7 Nm).

   c. Bolts 17 and 18: Upper oil pan-to-transaxle mounting bolts: 22–30 ft. lbs. (29–41 Nm).

11. Apply sealant to the threads, then install the oil pressure switch and tighten to 6–9 ft. lbs. (8–12 Nm).

12. Install the lower oil pan and tighten the bolts in sequence to 7–9 ft. lbs. (10–12 Nm).

13. Install the oil filter.

14. Install the starter motor.

15. Connect the negative battery cable.

16. Fill the crankcase to the correct level with engine oil.

17. Start the engine and check for leaks.

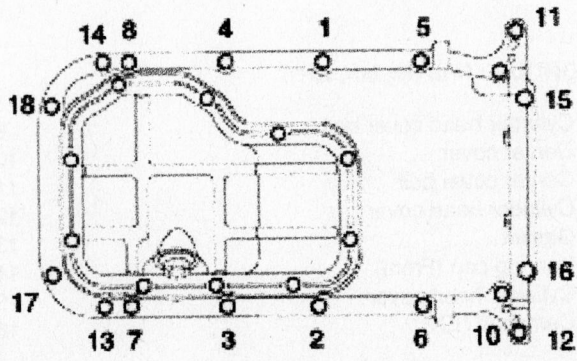

Upper oil pan torque sequence

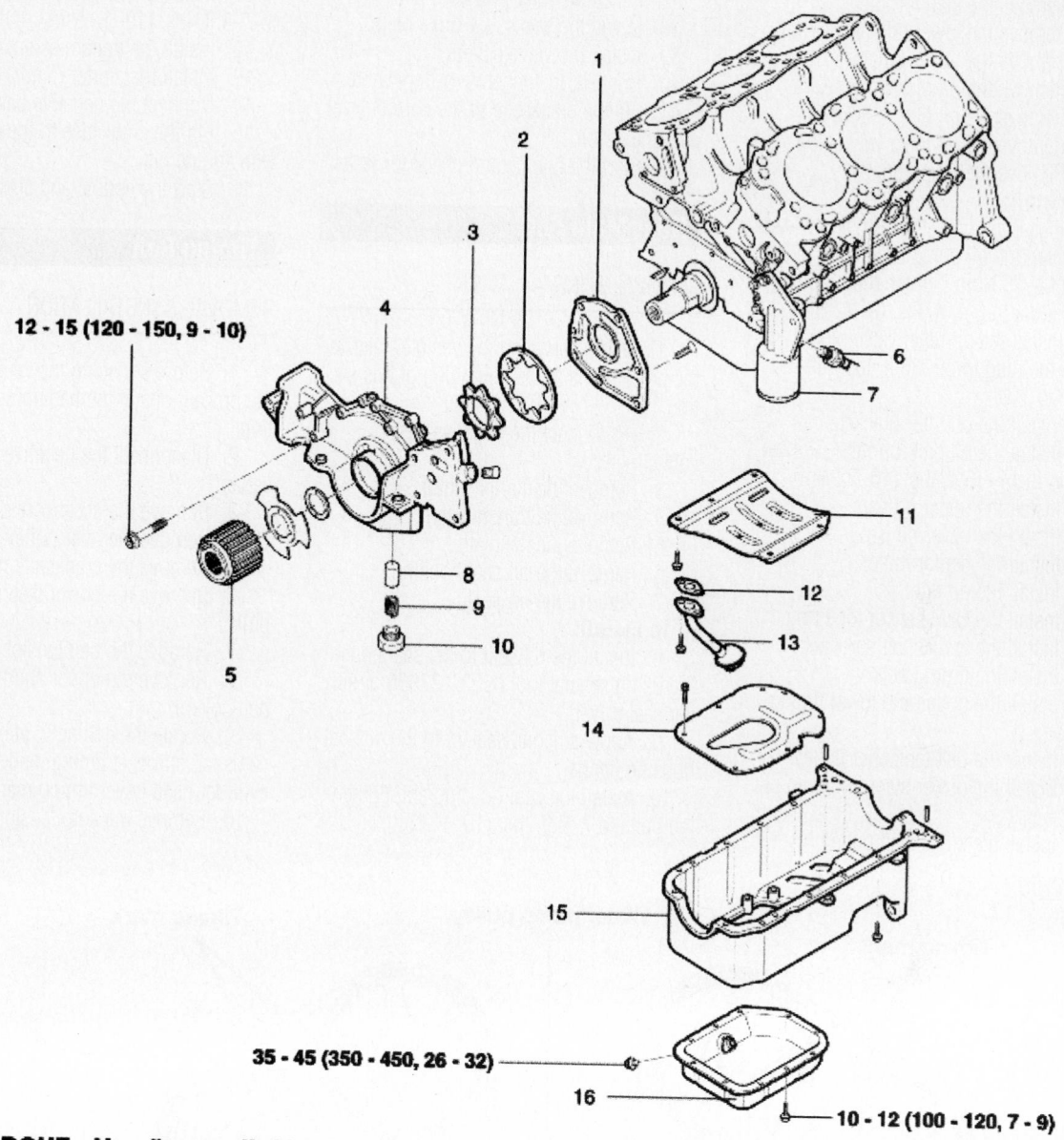

12 - 15 (120 - 150, 9 - 10)

35 - 45 (350 - 450, 26 - 32)

10 - 12 (100 - 120, 7 - 9)

67162-SORE-G17

TORQUE : N·m (kg·cm, lb·ft)

1. Oil pump cover
2. Oil pump outer rotor
3. Oil pump inner rotor
4. Oil pump case
5. Crankshaft sprocket
6. Oil pressure switch
7. Oil pump
8. Relief valve planger

9. Relief valve spring
10. Relief valve plug
11. Upper baffle plate
12. Gasket
13. Oil screen
14. Lower baffle plate
15. Upper oil pan
16. Lower oil pan

Oil pump and oil pan mounting exploded view

Oil Pump

REMOVAL & INSTALLATION

1. Before servicing the vehicle, refer to the precautions at the beginning of this section.

2. Drain the engine oil.
3. Disconnect the negative battery cable.
4. Remove the valve covers.
5. Remove the accessory drive belts.
6. Remove the idler pulley.
7. Remove the crankshaft pulley.
8. Remove the power steering pump pulley.

9. Remove the belt tensioner pulley.
10. Remove the upper and lower timing belt covers.
11. Remove the timing belt.
12. Remove the crankshaft sprocket.
13. Remove the Crankshaft Position (CKP) sensor tone ring.
14. Remove the oil filter.

15. Remove the starter motor.
16. Remove the lower oil pan.
17. Remove the upper oil pan.
18. Remove the oil pick-up tube.
19. Remove the oil filter bracket.
20. Remove the oil relief valve plug.
21. Remove the oil pump.

To install:

22. Install the oil pump with a new gasket. Tighten the oil pump case bolts to 9–10 ft. lbs. (12–15 Nm). Tighten the oil pump cover screws to 6–9 ft. lbs. (8–12 Nm).

23. Install the oil relief valve plug. Tighten the plug to 29–36 ft. lbs. (40–50 Nm).

24. Install the oil filter bracket.

25. Install the oil pick-up tube. Tighten the bolts to 11–15 ft. lbs. (15–22 Nm).

26. Install the upper oil pan.
27. Install the lower oil pan.
28. Install the starter motor.
29. Install the oil filter.
30. Install the CKP sensor tone ring.
31. Install the crankshaft sprocket.
32. Install the timing belt.
33. Install the upper and lower timing belt covers.
34. Install the belt tensioner pulley.
35. Install the power steering pump pulley.
36. Install the crankshaft pulley.

37. Install the idler pulley.
38. Install the accessory drive belts.
39. Install the valve covers.
40. Connect the negative battery cable.
41. Fill the crankcase to the correct level with engine oil.
42. Start the engine and check for leaks.

Rear Main Seal

REMOVAL & INSTALLATION

1. Before servicing the vehicle, refer to the precautions at the beginning of this section.

2. Disconnect the negative battery cable.

3. Remove the starter motor.
4. Remove the transmission.
5. Remove the flexplate.
6. Remove the oil seal housing.
7. Remove the oil seal.

To install:

8. Install the oil seal to the seal housing using special tool 09231-33000 or similar seal driver.

9. Apply silicone sealant to the oil seal housing flange.

10. Install the seal housing and tighten the bolts to 7–9 ft. lbs. (10–12 Nm).

11. Install the flexplate. Tighten the bolts to 7–9 ft. lbs. (10–12 Nm).

12. Install the transmission.
13. Install the starter motor.
14. Connect the negative battery cable.
15. Fill the crankcase to the correct level with engine oil.
16. Start the engine and check for leaks.

Timing Belt

REMOVAL & INSTALLATION

1. Before servicing the vehicle, refer to the precautions at the beginning of this section.

2. Disconnect the negative battery cable.

3. Remove the accessory drive belts.
4. Remove the idler pulley.
5. Remove the crankshaft pulley.
6. Remove the power steering pump pulley.
7. Remove the belt tensioner pulley.
8. Remove the upper and lower timing belt covers.

9. Rotate the engine to align the camshaft sprocket timing marks with the cylinder head cover timing marks.

10. Remove the auto tensioner.

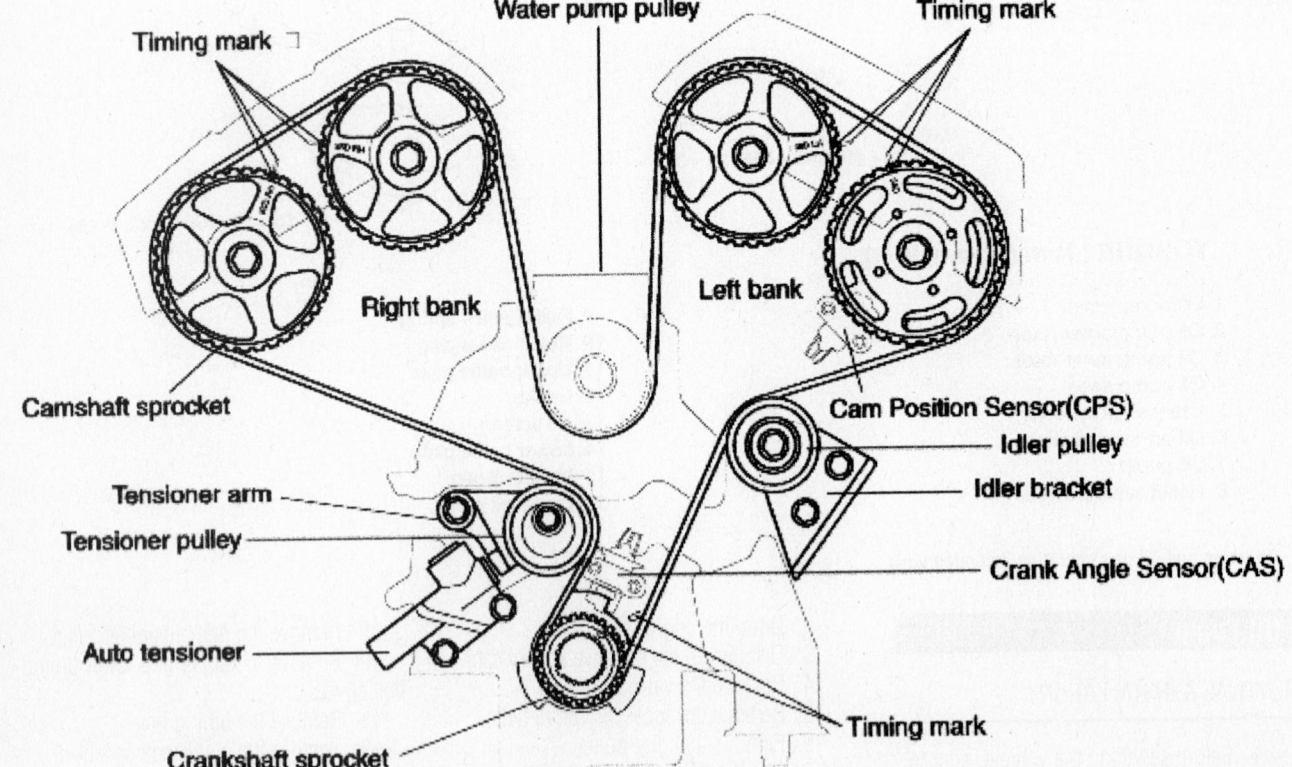

Timing belt routing and timing marks—Kia 3.5L Engine

67162-SORE-G18

11. Remove the timing belt.

To install:

12. Ensure that the engine is set to Top Dead Center (TDC).

13. Prepare the auto tensioner for installation by compressing it in a vise and installing a retaining pin.

14. Install the timing belt in the following order:

a. Crankshaft sprocket
b. Idler pulley
c. Left bank exhaust camshaft sprocket
d. Left bank intake camshaft sprocket
e. Water pump pulley
f. Right bank intake camshaft sprocket
g. Right bank exhaust camshaft sprocket
h. Tensioner pulley

15. Install the auto tensioner. Do not remove the retaining pin at this time.

16. Check that the crankshaft and camshaft timing marks are aligned correctly.

17. Rotate the crankshaft ¼ turn **Counterclockwise**.

18. Rotate the crankshaft ¼ turn **Clockwise** to return the engine to TDC.

19. Loosen the tensioner pulley center bolt.

20. Apply 44 inch lbs. (5 Nm) torque to the tensioner pulley and tighten the center bolt to 31–40 ft. lbs. (43–55 Nm).

21. Remove the auto tensioner retaining pin.

22. Rotate the crankshaft 2 revolutions **Clockwise**, then wait 5 minutes for the auto tensioner to adjust.

23. Measure the auto tensioner rod as

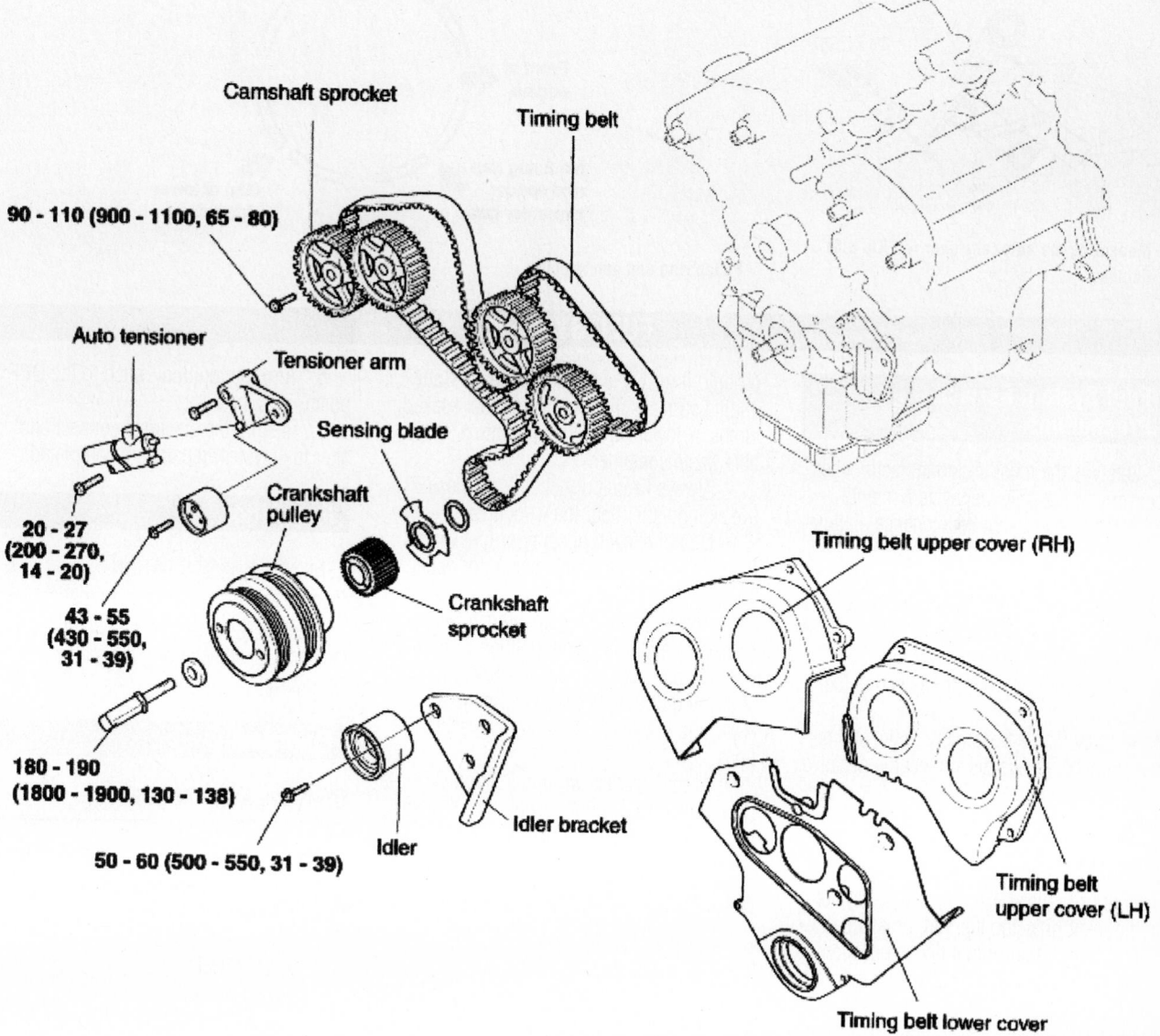

TORQUE : N·m (kg·cm, lb·ft)

Timing belt exploded view-Kia 3.5L Engine

67162-SORE-G19

shown. If the measurement is not 3.8–4.5 mm, then repeat the belt tensioning procedure.

24. After the auto tensioner measurement is correct, install the upper and lower timing belt covers.

25. Install the belt tensioner pulley.

26. Install the power steering pump pulley.

27. Install the crankshaft pulley.

28. Install the idler pulley.

29. Install the accessory drive belts.

30. Connect the negative battery cable.

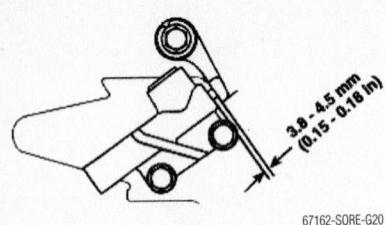

67162-SORE-G20

Measuring the auto tensioner rod-Kia 3.5L Engine

Piston and Ring

POSITIONING

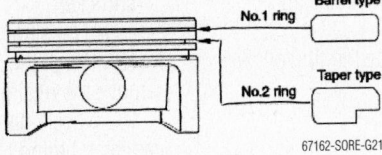

67162-SORE-G21

Compression ring identification

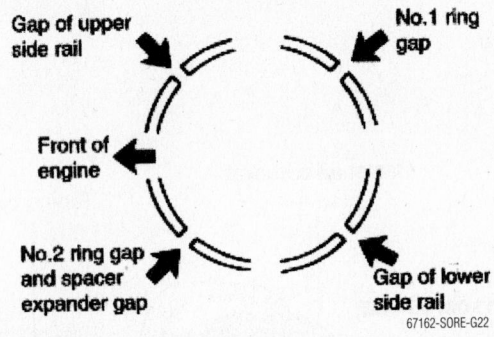

67162-SORE-G22

Piston ring end gap spacing

FUEL SYSTEM

Fuel System Service Precautions

Safety is the most important factor when performing not only fuel system maintenance, but any type of maintenance. Failure to conduct maintenance and repairs in a safe manner may result in serious personal injury or death. Maintenance and testing of the vehicle's fuel system components can be accomplished safely and effectively by adhering to the following rules and guidelines.

• To avoid the possibility of fire and personal injury, always disconnect the negative battery cable unless the repair or test procedure requires that battery voltage be applied.

• Always relieve the fuel system pressure prior to disconnecting any fuel system component (injector, fuel rail, pressure regulator, etc.), fitting or fuel line connection. Exercise extreme caution whenever relieving fuel system pressure, to avoid exposing skin, face and eyes to fuel spray. Please be advised that fuel under pressure may penetrate the skin or any part of the body that it contacts.

• Always place a shop towel or cloth around the fitting or connection prior to loosening to absorb any excess fuel due to spillage. Ensure that all fuel spillage

(should it occur) is quickly removed from engine surfaces. Ensure that all fuel soaked cloths or towels are deposited into a suitable waste container.

• Always keep a dry chemical (Class B) fire extinguisher near the work area.

• Do not allow fuel spray or fuel vapors to come into contact with a spark or open flame.

• Always use a back-up wrench when loosening and tightening fuel line connection fittings. This will prevent unnecessary stress and torsion to fuel line piping. Always follow the proper torque specifications.

• Always replace worn fuel fitting O-rings with new. Do not substitute fuel hose or equivalent where fuel pipe is installed.

Fuel System Pressure

RELIEVING

1. Before servicing the vehicle, refer to the precautions at the beginning of this section.

2. Remove the center floor carpet.

3. Remove the access cover.

4. Disconnect the fuel pump electrical connector.

5. Start the engine and allow it to run until it stalls.

6. Turn the ignition switch to the **OFF** position.

7. Restore the electrical connections after fuel system repairs are completed.

Fuel Filter

REMOVAL & INSTALLATION

The fuel filter is located inside the fuel tank and is replaced with the fuel pump as an assembly.

Fuel Pump

REMOVAL & INSTALLATION

1. Before servicing the vehicle, refer to the precautions at the beginning of this section.

2. Remove or disconnect the following:
   • Rear seat
   • Access cover
   • Fuel pump electrical connector

3. Relieve the fuel system pressure.
   • Fuel lines
   • Fuel pump and filter assembly

To install:

4. Install or connect the following:
   • Fuel pump and filter assembly
   • Fuel lines

- Fuel pump electrical connector
- Access cover
- Rear seat

5. Start the engine and check for leaks.

Fuel Injector

REMOVAL & INSTALLATION

1. Before servicing the vehicle, refer to the precautions at the beginning of this section.

2. Relieve the fuel system pressure.

3. Remove or disconnect the following:
- Negative battery cable
- Positive Crankcase Ventilation (PCV) hose
- Brake booster vacuum hose
- Surge tank stays
- Fuel lines
- Upper intake manifold (surge tank)
- Fuel injector harness connectors
- Fuel supply manifold with injectors attached

To install:

4. Replace all injector seals.

5. Install or connect the following:
- Fuel supply manifold with injectors attached
- Fuel injector harness connectors
- Upper intake manifold (surge tank)
- Fuel lines
- Surge tank stays
- Brake booster vacuum hose
- Positive Crankcase Ventilation (PCV) hose
- Negative battery cable

6. Start the engine and check for leaks.

DRIVE TRAIN

Transmission Assembly

REMOVAL & INSTALLATION

Manual

1. Before servicing the vehicle, refer to the precautions at the beginning of this section.

2. Disconnect the negative battery cable.

3. Remove the knob and the control lever.

4. Raise the vehicle.

5. Remove the transmission under cover.

6. Remove the clutch release cylinder.

7. Remove the front propeller shaft (4WD vehicle).

8. Remove the front muffler and the heater protector.

9. Remove the transfer case connector (4WD vehicle).

10. Remove the rear propeller shaft.

11. Support the transmission by the jack.

12. Remove the rear crossmember.

13. Remove the transmission with transfer case (4WD vehicle).

To install:

14. Install the transmission and transfer case (if equipped). Tighten the flange bolts to 31–40 ft. lbs. (43–55 Nm).

15. Install the rear crossmember.

16. Install the rear propeller shaft.

17. Connect the transfer case connector (4WD).

18. Install the front muffler and the heater protector.

19. Install the front propeller shaft (4WD).

20. Install the clutch release cylinder.

21. Install the transmission under cover.

22. Install the control lever and shift knob.

23. Connect the negative battery cable.

24. Start the engine and check for proper operation.

Automatic

1. Before servicing the vehicle, refer to the precautions at the beginning of this section.

2. Disconnect the negative battery cable.

3. Drain the automatic transmission fluid.

4. Remove the control cable.

5. Remove the under cover.

6. Remove the front propeller shaft (4WD).

7. Remove the front muffler and the heater protector.

8. Remove the transfer case connector (4WD).

9. Remove the rear propeller shaft.

10. Remove the oil cooler pipe.

11. Remove the speed sensor connector.

12. Remove the back-up lamp switch connector.

13. Remove the starter motor.

14. Remove the transmission mounting bolt.

15. Support the transmission on the jack.

16. Remove the transmission with transfer case (4WD vehicle).

To install:

17. Install the transmission and transfer case (if equipped). Tighten the flange bolts to 29–38 ft. lbs. (42–54 Nm).

18. Install the starter motor.

19. Connect the back-up lamp switch connector.

20. Connect the speed sensor connector.

21. Install the oil cooler pipe.

22. Install the rear propeller shaft.

23. Connect the transfer case connector (4WD)

24. Install the front muffler and the heater protector.

25. Install the front propeller shaft (4WD).

26. Install the under cover.

27. Install the control cable.

28. Connect the negative battery cable.

29. Fill the transaxle to the correct level with the proper transmission fluid.

30. Start the engine and check for leaks.

Clutch

ADJUSTMENT

Clutch Pedal Height and Free-Play

1. Measure the clutch pedal height (From the face of the pedal pad to the floorboard) and the clutch pedal free-play (measured at the face of the pedal pad). The standard value is as follows:

   a. (A) 7.3–13.9mm (0.29–0.55 in)
   b. (A') 163.8mm (6.45 in)

2. If the clutch pedal free-play is not within the standard value range, adjust as follows :

   a. Turn and adjust the bolt, then secure it by tightening the lock nut.

   b. Turn the push rod to coincide with the standard value and then secure the push rod with the lock nut.

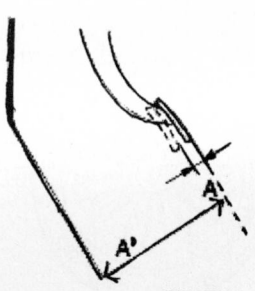

67162-SORE-G23

Clutch pedal height (A') and free-play (A) adjustment points

➡**When adjusting the clutch pedal height or the clutch pedal clevis pin play, be careful not to push the push rod toward the master cylinder.**

3. After completing the adjustments, check that the clutch pedal free play (measured at the face of the pedal pad) falls within the standard value range of 6–13mm (0.2–0.5 in.).

4. If the clutch pedal free play and the distance between the clutch pedal and the floor board when the clutch is disengaged do not meet the standard values, the cause may be either air in the hydraulic system or a faulty master cylinder clutch. Bleed the system or disassemble and inspect the master cylinder or clutch.

REMOVAL & INSTALLATION

1. Before servicing the vehicle, refer to the precautions at the beginning of this section.

2. Remove the transmission.

3. Insert the special tool (09411-43000) in the clutch disc to prevent the disc from shifting.

4. Loosen the bolts which attach the clutch cover to the flywheel in a star pattern. Loosen the bolts in succession, one or two turns at a time, to avoid bending the cover.

5. Remove the pressure plate bolts and remove the clutch plate and disc.

To install:

6. Apply multipurpose grease to the spline of the disc.

7. Install the clutch disc assembly to the flywheel using a centering tool.

8. Install the clutch cover assembly to the flywheel and temporarily tighten the bolts one or two steps at a time in a star pattern. Tighten the clutch cover bolts in sequence to 11–16 ft. lbs. (15–22 Nm) and remove the centering tool.

9. Check the release bearing condition and lubricate or replace as necessary.

10. Install the transmission.

Hydraulic Clutch System

BLEEDING

1. Before servicing the vehicle, refer to the precautions at the beginning of this section.

2. With an assistant in the vehicle, raise and safely support the vehicle.

3. Have your assistant pump the clutch pedal three times and hold the pedal to the floor.

4. Open the bleeder valve on the clutch slave cylinder until the air is purged from the cylinder.

5. Tighten the bleeder valve.

6. Have your assistant release the clutch pedal.

7. Fill the clutch master cylinder if below minimum.

8. Repeat Steps 2 through 6 until no air exits from the bleeder valve.

9. Lower the vehicle.

10. Fill the clutch master cylinder fluid reservoir.

Transfer Case Assembly

REMOVAL & INSTALLATION

1. Before servicing the vehicle, refer to the precautions at the beginning of this section.

2. Drain the transfer case fluid.

3. Disconnect the negative battery cable.

4. Remove the front and rear console mounting screws. Slide the console forward to clear the parking brake handle and set aside. Open the shift boot.

5. Loosen the transfer case shift lever locknut and remove the lever knob.

6. Pull the console up to access the wiring connector(s). Unplug the connector(s) and remove the console.

7. Shift the transfer lever to the 4L position.

8. Remove the cover plate.

9. Remove the retaining bolts from the transfer case and lift the shifter lever assembly straight out and properly support the transmission.

10. Matchmark the driveshafts at the flanges and remove the driveshafts.

11. Remove the crossmember bolts and support the transmission on the jack.

12. Disconnect the 4WD light switch connector.

13. Remove the transfer case mounting bolts.

14. Remove the crossmember.

15. Separate the transfer case from the transmission by striking the transfer case with a plastic mallet at the seal area.

16. Lower the transfer case from the vehicle.

To install:

17. Install the transfer case in position with a new gasket. Torque the bolts to 32 ft. lbs. (44 Nm).

18. Install the crossmember.

19. Connect the 4WD light switch connector.

20. Align the matchmarks on the driveshafts to the flanges. Torque the bolts to 27 ft. lbs. (36 Nm) and remove the transmission support

21. Install the retaining bolts to the transfer case and install the shift lever assembly.

22. Remove the transmission support jack.

23. Install the cover plate.

24. Install the floor console unit and connect or install the following:
- Switch wiring connector(s)
- Front console
- Lever knob
- Front console and tie the shift boot draw strings
- Slide the console over the parking brake handle

25. Connect the negative battery cable

26. Fill the transfer case to the proper level

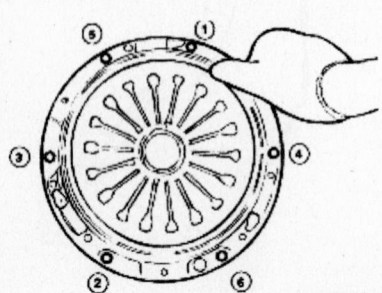

67162-SORE-G24

Clutch cover torque sequence

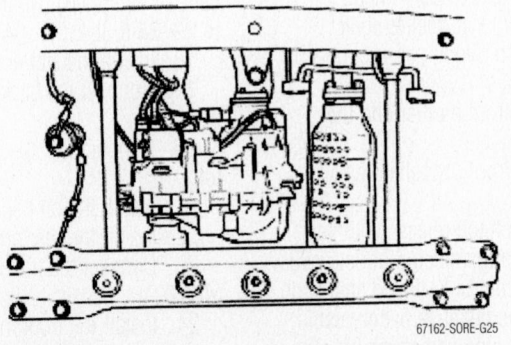

67162-SORE-G25

Transfer case and crossmember location

27. Start the vehicle and check for leaks, repair if necessary.

Halfshaft

REMOVAL & INSTALLATION

1. Before servicing the vehicle, refer to the precautions at the beginning of this section.
2. Remove the front wheels.
3. Remove the lock nut from front hub.
4. Remove the upper control arm link lock bolt, spring washer and nut.
5. Remove tie rod end cotter pin and using a ball joint puller, remove tie rod end from steering knuckle.
6. Mark drive shaft for identical installation position.
7. Using tool, pry the drive shaft from the differential housing. Separate the right halfshaft from the intermediate shaft.
8. Remove the axle halfshafts from the front knuckle. It may be necessary to tap the stub shaft with a brass hammer to remove the axles.

To install:

9. Coincide the joining mark between the drive shaft and the differential and insert the shaft.
10. Install the knuckle assembly. Tighten the ball joint castle nut to 51–57 ft. lbs. (70–80 Nm).
11. Install the upper arm link lock bolt and tighten to 32–39 ft. lbs. (44–55 Nm).
12. Tighten the lock nut to 177–198 ft. lbs. (245–275 Nm) and then caulk the flange of lock nut on the end of drive shaft.
13. Install the front wheels.

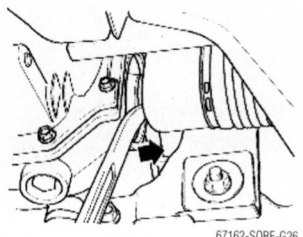

67162-SORE-G26

Pry the drive shaft from the differential housing

Axle Shaft, Bearing and Seal

REMOVAL & INSTALLATION

Rear

1. Before servicing the vehicle, refer to the precautions at the beginning of this section.

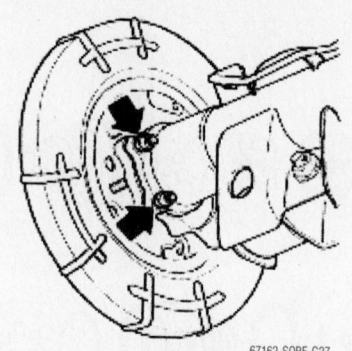

67162-SORE-G27

Rear axle shaft mounting bolt locations

2. Disconnect the negative battery cable.
3. Remove the rear wheels.
4. Remove the disc brake and parking brake assembly.
5. Remove the parking brake cable and wheel speed sensor cable.
6. Remove the rear axle shaft mounting bolts.
7. Remove the rear axle shaft.
8. Remove the bearing collar and bearing from the axle.
9. Using a slide hammer, remove the oil seal.

To install:

10. Apply grease to the oil seal lip and using the appropriate seal driver, install the new axle seal into the differential.
11. Install the new wheel bearing and retainer collar to the rear axle shaft.
12. Install the rear axle shaft. Torque the axle shaft mounting bolts to 32–44 ft. lbs. (43–60 Nm).
13. Install the wheel speed sensor and parking brake cables.
14. Install the disc brake and parking brake assembly and the rear wheels.
15. Adjust the parking brake lever.
16. Connect the negative battery cable.

Pinion Seal

REMOVAL & INSTALLATION

1. Before servicing the vehicle, refer to the precautions at the beginning of this section.
2. Drain the gear oil.
3. Disconnect the negative battery cable.
4. Remove the wheels.
5. Remove the brake drums (if removing the rear pinion seal).
6. Remove the driveshaft.
7. Remove the drive pinion.
8. Remove the pinion seal

To install:

9. Install a new pinion seal lightly coated with clean gear oil.

10. Install the drive pinion.
11. Rotate the pinion flange occasionally while tightening the flange nut and make certain that the pinion bearings are seated properly.
12. Install the driveshaft after aligning the matchmarks.
13. Install the brake drums.
14. Install the wheels.
15. Connect the negative battery cable.
16. Fill the gear oil to the proper level.
17. Start the vehicle and check for leaks, repair if necessary.

CV Joint

OVERHAUL

Outer CV Joint (B.J.)

The outer CV joint (B.J.) is serviced only with the axle halfshaft as an assembly. The outer CV joint (B.J.) boot can be serviced by first removing the inner Tripod joint (T.S.J.).

Inner Tripod Joint (T.S.J.)

1. Before servicing the vehicle, refer to the precautions at the beginning of this section.
2. Remove the axle halfshaft from the vehicle and place it in a vise.
3. Remove the T.S.J boot band and pull the boot away from T.S.J outer race.
4. Remove the circlip using a screwdriver.
5. Remove the drive shaft from the T.S.J outer race.
6. Remove the snap ring and disassemble the inner race and ball from the shaft.
7. Remove the B.J boot band and pull out the T.S.J boot and the B.J boot.

➡ **If the boot is reused, wrap a tape around the drive shaft splines to protect the boot.**

To install:

8. Wrap tape around the drive shaft spline (T.S.J side) to avoid boot damage.
9. Install the outer and inner joint boots.
10. Install the inner race and ball to the shaft, then install the snap ring.
11. Install the drive shaft into the T.S.J outer race and install the circlip.
12. Apply CV joint grease to the drive shaft and position the boots over each joint.
13. Install the boot clamps. Pull on the clamp end with pliers and fold the locking tabs over to lock the clamp in place.
14. Install the axle halfshaft to the vehicle.

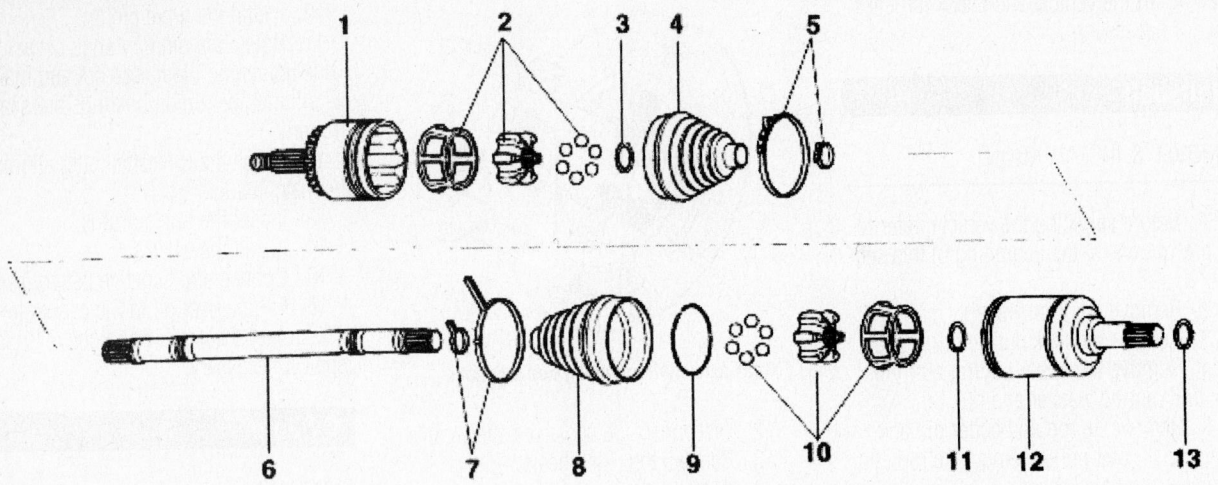

1. B.J assembly
2. B.J inner race and ball
3. Snap ring
4. B.J boot
5. B.J boot band
6. Drive shaft

7. T.S.J boot band
8. T.S.J boot
9. Circlip
10. T.S.J inner race and ball
11. Snap ring
12. T.S.J assembly
13. Clip

67162-SORE-G28

Inner and outer CV Joint exploded view and parts identification

STEERING AND SUSPENSION

Air Bag

✳✳ CAUTION

Some vehicles are equipped with an air bag system. The system must be disarmed before performing service on, or around, system components, the steering column, instrument panel components, wiring and sensors. Failure to follow the safety precautions and the disarming procedure could result in accidental air bag deployment, possible injury and unnecessary system repairs.

PRECAUTIONS

Several precautions must be observed when handling the inflator module to avoid accidental deployment and possible personal injury.

• Never carry the inflator module by the wires or connector on the underside of the module.

• When carrying a live inflator module, hold securely with both hands, and ensure that the bag and trim cover are pointed away.

• Place the inflator module on a bench or other surface with the bag and trim cover facing up.

• With the inflator module on the bench, never place anything on or close to the module which may be thrown in the event of an accidental deployment.

DISARMING

1. Before servicing the vehicle, refer to the precautions at the beginning of this section.

2. Position the vehicle with the front wheels in a straight-ahead position.

3. Disconnect the negative battery cable and keep the cable secure from touching the battery.

4. Remove the ignition key from the vehicle.

5. Wait at least 1 minute for the air bag back-up power supply to deplete its stored energy before continuing.

6. Proceed with the repair.

7. Reconnect the negative battery cable once the repair is complete.

Rack and Pinion Steering Gear

REMOVAL & INSTALLATION

1. Before servicing the vehicle, refer to the precautions at the beginning of this section.

2. Drain the power steering fluid.

3. Disconnect the negative battery cable.

4. Remove the front wheels.

5. Disconnect the power steering fluid pressure and return lines.

6. Remove the joint assembly connecting bolt and separate the steering intermediate shaft from the steering gear box.

7. Separate the outer tie rod ends from the steering knuckles.

8. Remove the steering gear box mounting bolts and remove the steering gear box assembly together with mounting rubber.

To install:

9. Install the steering gear into position on the vehicle.

10. Install the steering gear mounting rubber, brackets and fasteners. Tighten the

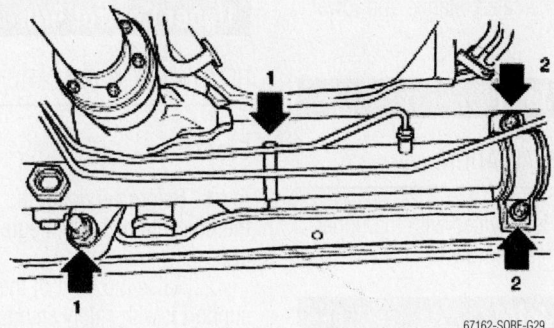

Power steering gear fastener (1) and mounting bracket fastener (2) locations

67162-SORE-G29

mounting bracket fasteners (2) to 54–68 ft. lbs. (75–95 Nm). Tighten the steering gear fasteners (1) to 88–114 ft. lbs. (122–158 Nm).

11. Connect the intermediate shaft to the steering gear. Tighten the pinch bolt to 15–19 ft. lbs. (22–27 Nm).

12. Connect the outer tie rod ends to the steering knuckles. Tighten the nuts to 50–57 ft. lbs. (70–80 Nm).

13. Connect the power steering fluid pressure and return lines. Tighten the bolts to 23–34 ft. lbs. (32–48 Nm).

14. Install the front wheels

15. Connect the negative battery cable.

16. Fill the power steering system with the correct amount of power steering fluid.

Strut

REMOVAL & INSTALLATION

Front

1. Before servicing the vehicle, refer to the precautions at the beginning of this section.

2. Loosen battery cable and mounting bolt and then remove battery.

3. Remove three shock absorber mounting block nuts from the mounting block.

4. Raise the front of the vehicle and support it with safety stands.

5. Remove the front wheels.

6. Remove the bolt on the steering knuckle side that secures the upper arm ball joint.

7. Remove the brake hose bracket and then the remove the upper arm bolts and nuts.

8. Remove the strut lower mounting nut.

9. Remove the strut from the vehicle.

To install:

10. After making sure identification mark on the spring seat. Position the strut assembly into the upper mounting block.

11. Install the mounting nuts by 3-4 threads only.

12. Insure the front of the vehicle is raised and supported with safety stands.

13. Tighten the lower nut of the strut assembly to 88–101 ft. lbs. (122–140 Nm).

14. Position the upper arm to the frame brackets, insert the bolts and hand tighten the nuts.

15. Install the upper arm ball joint into the top of the steering knuckle and tighten the side bolt and nut to 31–39 ft. lbs. (44–55 Nm).

16. Tighten the upper arm bolts and nuts to 54–68 ft. lbs. (76–95 Nm) and then install brake hose brackets.

17. Install the front wheels.

18. Lower the vehicle.

19. Tighten the mounting block nuts to 31–39 ft. lbs. (44–55 Nm).

20. Install the battery mounting bracket and the battery.

21. Check the front end alignment and adjust as necessary.

Shock Absorber

REMOVAL & INSTALLATION

Rear

1. Before servicing the vehicle, refer to the precautions at the beginning of this section.

2. Raise the rear of the vehicle and support it with safety stands.

3. Remove the rear wheels.

4. Raise the rear axle housing to facilitate removal of the shock absorbers.

5. Remove stabilizer link upper mounting nut.

6. Remove the rear shock absorber lower nut and washer. Remove the shock absorber upper bolt, and then remove the shock absorber.

To install:

7. Install the shock absorber and upper nut. Tighten the nut to 88–101 ft. lbs. (122–140 Nm).

8. Install the shock absorber lower bolt and tighten the bolt to 88–101 ft. lbs. (122–140 Nm).

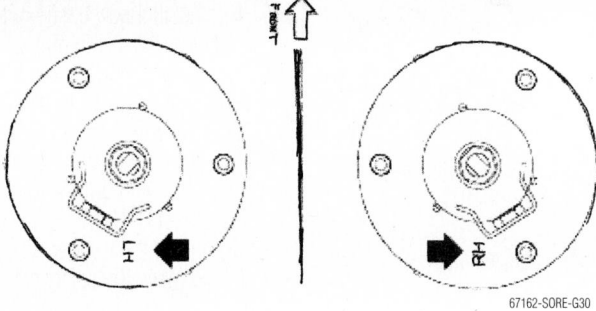

Correct strut mount placement

67162-SORE-G30

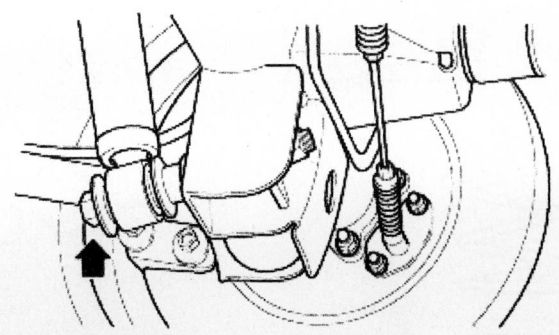

Lower shock absorber mounting bolt

67162-SORE-G31

9. Install the stabilizer link upper mounting nut.

10. Install the rear wheels.

11. Remove the safety stands and lower the vehicle.

Coil Spring

REMOVAL & INSTALLATION

Rear

1. Before servicing the vehicle, refer to the precautions at the beginning of this section.

2. Raise the rear of the vehicle and support it with safety stands.

3. Remove the rear wheels.

4. Raise the rear axle housing to facilitate removal of the shock absorbers.

5. Remove stabilizer link upper mounting nut.

6. Remove the rear shock absorber lower nut and washer. Remove the shock absorber upper bolt, and then remove the shock absorber.

7. Lower the rear axle housing slowly to facilitate removal of the coil spring.

8. Remove the upper rubber seat.

To install:

9. Position the upper rubber seat to the coil spring. Align the spring end with the groove of the spring pad and fix the spring and the spring pad by adhering the 3 parts with tape.

10. Slowly raise the rear axle housing while installing the coil spring.

11. Install the shock absorber and upper nut. Tighten the nut to 88–101 ft. lbs. (122–140 Nm).

12. Install the shock absorber lower bolt and tighten the bolt to 88–101 ft. lbs. (122–140 Nm).

13. Install the stabilizer link upper mounting nut.

14. Install the rear wheels.

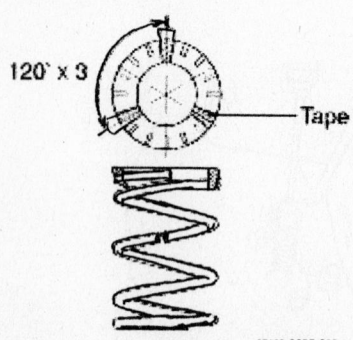

Proper placement of coil spring upper rubber seat

15. Remove the safety stands and lower the vehicle.

Upper Ball Joint

REMOVAL & INSTALLATION

The ball joint is serviced with the upper control arm as an assembly.

Lower Ball Joint

REMOVAL & INSTALLATION

1. Before servicing the vehicle, refer to the precautions at the beginning of this section.

2. Raise the front of the vehicle and support it with safety stands.

3. Remove the front wheels.

4. Remove the lower nut of the control link of the stabilizer bar.

5. Remove the lower nut of the strut.

6. Remove the bolts and nuts that joins the lower arm and lower arm ball joint.

7. Remove the cotter pin and castle nut from the lower arm ball joint.

8. Separate the lower arm ball joint from the steering knuckle.

To install:

9. Install the lower arm ball joint to the steering knuckle. Tighten the ball joint-to-control arm fasteners to 116–145 ft. lbs. (157–196 Nm).

10. Install a new castle nut and cotter pin through the castle nut.

11. Install the lower nut of the strut assembly and tighten to 88–101 ft. lbs. (122–140 Nm).

12. Install the lower nut of control link of stabilizer bar and tighten the nut to 68–84 ft. lbs. (95–117 Nm).

13. Install the front wheels.

14. Remove the safety stands and lower the vehicle.

15. Check the front end alignment and adjust as necessary.

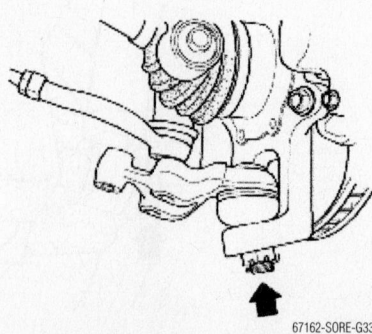

Lower ball joint, castle nut and cotter pin

Upper Control Arm

REMOVAL & INSTALLATION

Front

1. Before servicing the vehicle, refer to the precautions at the beginning of this section.

2. Raise the front of the vehicle and support it with safety stands.

3. Remove the front wheels.

4. Remove the bolt on the steering knuckle side that secures the upper arm ball link.

5. Remove the brake hose bracket and then remove the upper arm bolts and nuts.

6. Remove the upper control arm from the vehicle.

To install:

7. Raise the front of the vehicle and support it with safety stands.

8. Position the upper arm to the frame brackets, insert the bolts and hand tighten the nuts.

9. Install the upper arm ball joint into the top of the steering knuckle and tighten the side bolt and nut to 31–39 ft. lbs. (44–55 Nm).

10. Tighten the upper arm bolts and nuts to 54–68 ft. lbs. (76–95 Nm) and then install brake hose brackets.

11. Install the front wheels.

12. Check the front end alignment and adjust as necessary.

Rear

1. Before servicing the vehicle, refer to the precautions at the beginning of this section.

2. Raise the rear of the vehicle and support it with safety stands.

3. Remove the rear wheels.

4. Raise the rear axle housing to facilitate removal of the upper arm.

5. Remove shock absorber lower bolt.

6. Loosen the upper arm bolts and remove the upper arm.

7. Inspect the upper arm for bends, cracks and/or other damage. Inspect the upper arm bushings for wear and/or deterioration.

To install:

8. Install the upper arm and the bolts. Tighten the bolts to 88–101 ft. lbs. (122–140 Nm).

9. Install shock absorber lower bolt and tighten to 88–101 ft. lbs. (122–140 Nm).

10. Lower the rear axle housing.

11. Install the rear wheels.

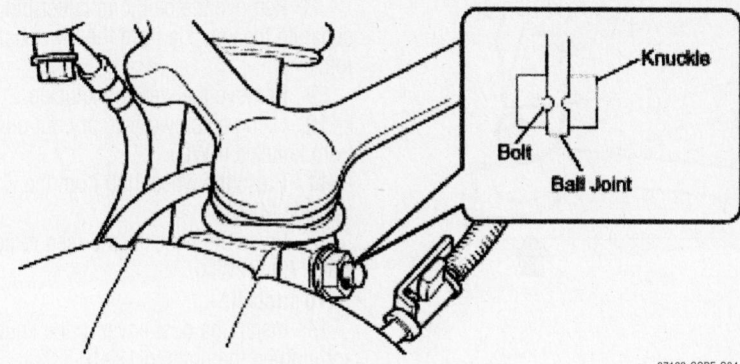

Installation of upper ball joint to steering knuckle

67162-SORE-G34

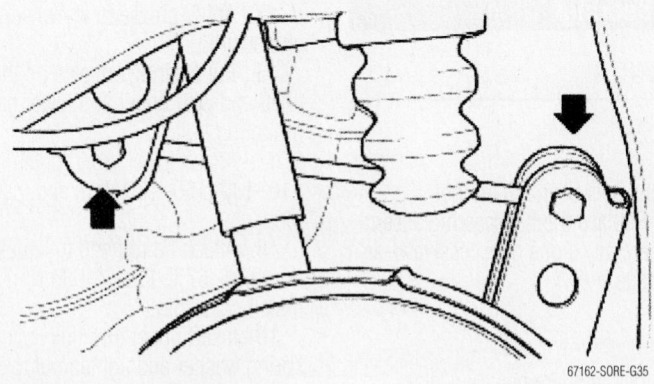

Rear upper control arm bolt locations

67162-SORE-G35

12. Remove the safety stands and lower the vehicle.

CONTROL ARM BUSHING REPLACEMENT

1. Secure the upper control arm in a suitable vise.

2. Using a standard bearing press, remove the old bushing.

3. Install the new bushing and then press it into the upper arm with a standard bearing press.

4. Apply lubricant to the new bushings to facilitate insertion into the upper control arm.

Lower Control Arm

REMOVAL & INSTALLATION

Front

1. Before servicing the vehicle, refer to the precautions at the beginning of this section.

2. Raise the front of the vehicle and support it with safety stands.

3. Remove the front wheels.

4. Remove the lower nut of the control link of the stabilizer bar.

5. Remove the lower nut of the strut.

6. Remove the bolts and nuts that joins the lower arm and lower arm ball joint.

7. Remove the cotter pin and castle nut from the lower arm ball joint.

8. Separate the lower arm ball joint from the steering knuckle.

9. Remove the steering gear mounting bolts and nuts.

10. Remove the spindle from the front frame crossmember brackets during raising the steering gear box by using a suitable bar.

➡Before loosening the nuts of the spindles, make note of the numerical setting and mark the location on the frame bracket and plate so it can be re-installed to the same setting and location.

11. Remove the lower control arm.

To install:

12. Install the lower arm ball joint to the steering knuckle. Tighten the ball joint-to-control arm fasteners to 116–145 ft. lbs. (157–196 Nm).

13. Install a new cotter pin through the castle nut.

14. Position the lower arm to the front frame crossmember brackets and then position the spindle up to the steering gear box by using a suitable bar.

15. Install the lower arm spindles and tighten the fasteners to 159–181 ft. lbs. (216–245 Nm).

16. Install the lower nut of the strut assembly and tighten to 88–101 ft. lbs. (122–140 Nm).

17. Install the lower nut of control link of stabilizer bar and tighten the nut to 68–84 ft. lbs. (95–117 Nm).

18. Install the front wheels.

19. Remove the safety stands and lower the vehicle.

20. Check the front end alignment and adjust as necessary.

Rear

1. Before servicing the vehicle, refer to the precautions at the beginning of this section.

2. Raise the rear of the vehicle and support it with safety stands.

3. Remove the rear wheels.

4. Raise the rear axle housing to facilitate removal of the lower arm.

5. Remove the shock absorber lower bolt.

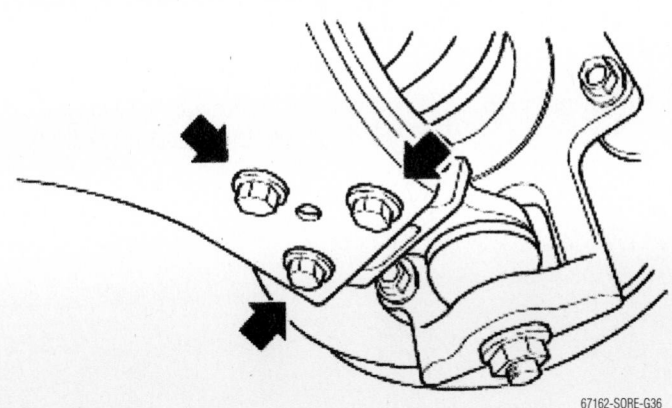

67162-SORE-G36

Lower ball joint-to-control arm fastener locations

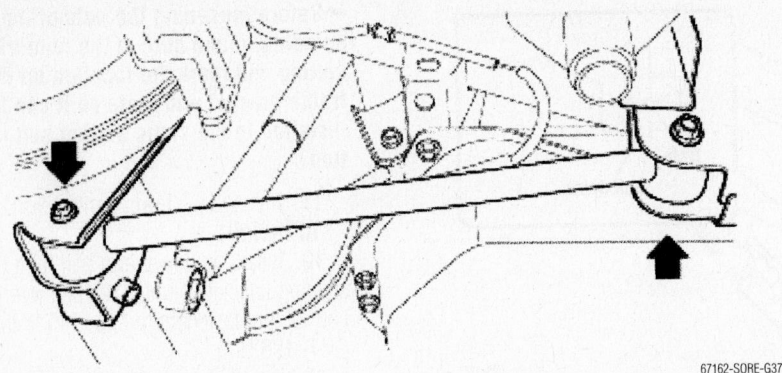

67162-SORE-G37

Rear lower control arm bolt locations

6. Remove the wheel speed sensor cable from rear lower arm.

7. Loosen the lower arm bolts and remove the lower arm.

8. Inspect the lower arm for bends, cracks and/or other damage. Inspect the lower arm bushings for wear and/or deterioration.

To install:

9. Install the lower arm and the bolts. Tighten the bolts to 101–116 ft. lbs. (137–157 Nm).

10. Install wheel speed sensor cable to the rear lower arm.

11. Install shock absorber lower bolt. Tighten the bolt to 88–101 ft. lbs. (122–140 Nm).

12. Lower the rear axle housing.

13. Install the rear wheels.

14. Remove the safety stands and lower the vehicle.

CONTROL ARM BUSHING REPLACEMENT

1. Secure the lower control arm in a suitable vise.

2. Using a standard bearing press, remove the old bushing.

3. Install the new bushing and then press it into the lower arm with a standard bearing press.

4. Apply lubricant to the new bushings to facilitate insertion into the lower control arm.

Wheel Bearings

ADJUSTMENT

Front

The front wheel bearings are not adjustable. Replace the hub/bearing assembly if there are any signs of excessive wear and/or damage.

Rear

The rear bearings are not adjustable. Replace the rear bearings if play exceeds 0.002 inches (0.05 mm).

REMOVAL & INSTALLATION

Front

➡ **The front wheel bearings are serviced with the steering knuckle as an assembly.**

1. Before servicing the vehicle, refer to the precautions at the beginning of this section.

2. Remove the front wheel.

3. Remove the brake caliper and rotor.

4. Remove the wheel speed sensor harness.

5. Remove the hub nut.

6. Separate the knuckle from the upper control arm.

7. Remove the tie rod end cotter pin and separate the knuckle from the tie rod end.

8. Remove the ball joint cotter pin, then separate the knuckle from the lower ball joint.

9. Remove the steering knuckle.

10. Using a screwdriver, pry out oil seal from knuckle (4WD).

11. Press the wheel hub from the knuckle (4WD).

12. Press the knuckle and then remove wheel hub (2WD).

To install:

13. Install the dust cover to the knuckle and tighten the bolts to 12–16 ft. lbs. (16–23 Nm).

14. Install new oil seal and then install the wheel hub to the knuckle by pressing.

15. Apply grease to the wheel bearing and seal lip.

16. Put steering knuckle on the drive shaft end with upper and lower ball joints in mounting holes.

17. Attach lower arm, tighten lock nut to 116–130 (160–180 Nm), and install cotter pin.

18. Attach tie rod end to knuckle, tighten nut to 51–57 ft. lbs. (70–80 Nm), and install cotter pin.

19. Insert upper arm link lock bolt with spring washer and tighten nut to 32–39 ft. lbs. (44–55 Nm).

20. Install the chamfer of plain washer toward the bearing (2WD).

21. Screw lock nut up against wheel hub assembly and using a lock nut wrench, tighten nut to tightening torque to set bearing preload. Use spring scale to measure:

- Bearing preload: 10 inch lbs.
- Tightening torque: 178–198 ft. lbs. (245–275 Nm)

22. Caulk the flange of lock nut on the end of drive shaft.

23. Install the wheel speed sensor harness.

24. Install the brake caliper and rotor.

25. Install the front wheel.

26. Check the alignment and adjust as necessary.

Rear

Refer to Axle Shaft, Bearing and Seal in the DRIVE TRAIN Section of this chapter.

BRAKES

Brake Caliper

REMOVAL & INSTALLATION

Front

1. Before servicing the vehicle, refer to the precautions at the beginning of this section.
2. Remove the front wheel.
3. Disconnect the brake fluid hose.
4. Remove the caliper mounting bolts.
5. Remove the brake caliper.

To install:

6. Install the brake caliper. Tighten the mounting bolts to 16–24 ft. lbs. (22–32 Nm).
7. Install the brake fluid hose. Tighten the hose fitting to 12–14 ft. lbs. (17–20 Nm).
8. Bleed the brake system.
9. Install the front wheel.
10. Before attempting to move the vehicle, pump the brake pedal to seat the pads against the rotors. Make sure the vehicle has a firm brake pedal. Check the level of the brake fluid and add DOT 3 or 4 brake fluid if necessary.

Rear

1. Before servicing the vehicle, refer to the precautions at the beginning of this section.
2. Remove the rear wheel.
3. Disconnect the brake fluid hose.
4. Remove the caliper guide bolts.
5. Remove the brake caliper.

To install:

6. Install the brake caliper. Tighten the guide bolts to 16–23 ft. lbs. (22–32 Nm).

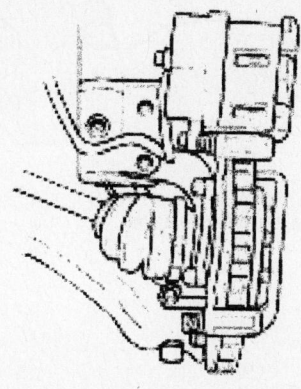

Front disc brake pad components

7. Install the brake fluid hose. Tighten the hose fitting to 12–14 ft. lbs. (17–20 Nm).
8. Bleed the brake system.
9. Install the rear wheel.
10. Before attempting to move the vehicle, pump the brake pedal to seat the pads against the rotors. Make sure the vehicle has a firm brake pedal. Check the level of the brake fluid and add DOT 3 or 4 brake fluid if necessary.

Disc Brake Pads

REMOVAL & INSTALLATION

Front

1. Before servicing the vehicle, refer to the precautions at the beginning of this section.

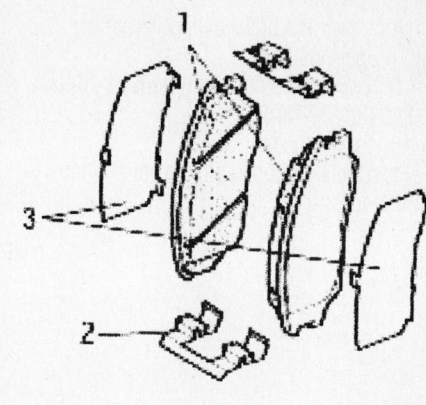

67162-SORE-G39

2. Remove the front wheel.
3. Remove the guide pin, lift the caliper assembly up and suspend it with a wire.
4. Remove the following parts from the caliper support:
   - Pad and wear sensor assembly
   - Pad spring
   - Outer shim

To install:

5. Compress the caliper piston into the caliper bore.
6. Install the pad clips.
7. Install the inner and outer pads on each pad clip.
8. Lower the brake caliper carefully so as not to damage the boot.
9. Tighten the guide pin bolt to 16–24 ft. lbs. (22–32 Nm).
10. Install the front wheel.

Rear

1. Before servicing the vehicle, refer to the precautions at the beginning of this section.
2. Remove the rear wheel.
3. Remove the guide pin bolts, lift the caliper assembly up and suspend it with a wire.
4. Before replacing the brake pads, drain brake fluid from the master cylinder reservoir until it remains half full.
5. Remove the brake pads by turning the piston in the housing assembly using special tool 09581-11000 to compress the piston.
6. Remove the inner and outer pads from the caliper.

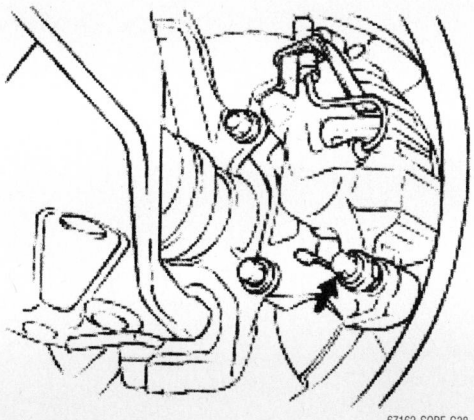

67162-SORE-G38

Front brake caliper mounting bolt locations

To install:

7. Install the inner and outer brake pads, engaging the clips securely onto the caliper assembly.

8. Lower the brake caliper assembly into proper position.

9. Tighten the guide pin bolts to 16–23 ft. lbs. (22–32 Nm).

10. Install the rear wheel.

11. Check the brake fluid level and top off, if necessary.

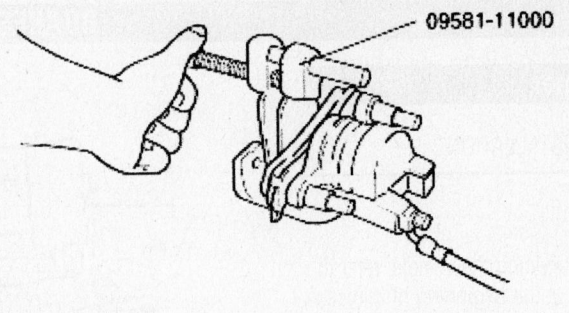

09581-11000

67162-SORE-G40

Rear disc brake pad removal

KIA

Sportage

21

SPECIFICATION AND MAINTENANCE CHARTS

ENGINE AND VEHICLE IDENTIFICATION

Engine							Model Year	
Code ①	Liters (cc)	Cu. In.	Cyl.	Fuel Sys.	Engine Type	Eng. Mfg.	Code ②	Year
3	2.0 (1998)	122	4	MFI	DOHC	KIA	1	2001
							2	2002
							3	2003
							4	2004
							5	2005

MFI: Multi-port Fuel Injection

DOHC: Double Overhead Camshafts

① 8th digit of VIN

② 10th digit of VIN

67162-KSPO-C01

GENERAL ENGINE SPECIFICATIONS

Year	Model	Engine Displacement Liters	Engine VIN	Net Horsepower @ rpm	Net Torque @ rpm (ft. lbs.)	Bore x Stroke (in.)	Compression Ratio	Oil Pressure @ rpm
2001	Sportage	2.0	3	130@5500	127@4000	3.39x3.39	9.2:1	43-57 ①
2002	Sportage	2.0	3	130@5500	127@4000	3.39x3.39	9.2:1	43-57 ①

MFI: Multi-port Fuel Injection

① The manufacturer does not provide an engine speed specification for oil pump pressure.

67162-KSPO-C02

ENGINE TUNE-UP SPECIFICATIONS

Year	Engine Displacement Liters	Engine VIN	Spark Plug Gap (in.)	Ignition Timing (deg.)		Fuel Pump (psi)	Idle Speed (rpm)		Valve Clearance	
				MT	AT		MT	AT	Intake	Exhaust
2001	2.0	3	0.039-0.043	4B	4B	38	750-850	750-850	HYD.	HYD.
2002	2.0	3	0.039-0.043	4B	4B	38	750-850	750-850	HYD.	HYD.

B: Before Top Dead Center

HYD: Hydraulic lash adjusters

67162-KSPO-C03

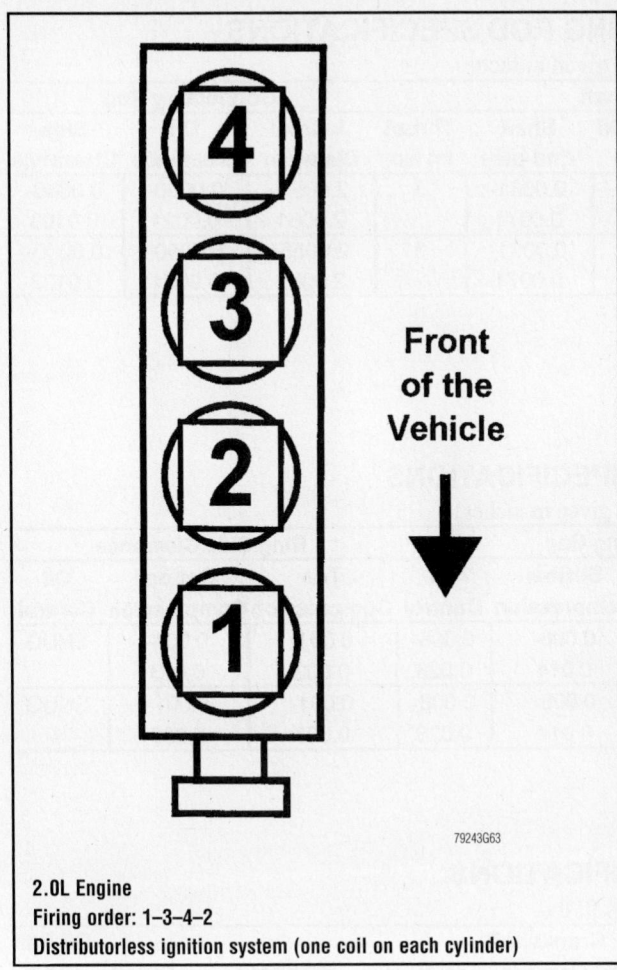

2.0L Engine
Firing order: 1–3–4–2
Distributorless ignition system (one coil on each cylinder)

Front of the Vehicle

79243G63

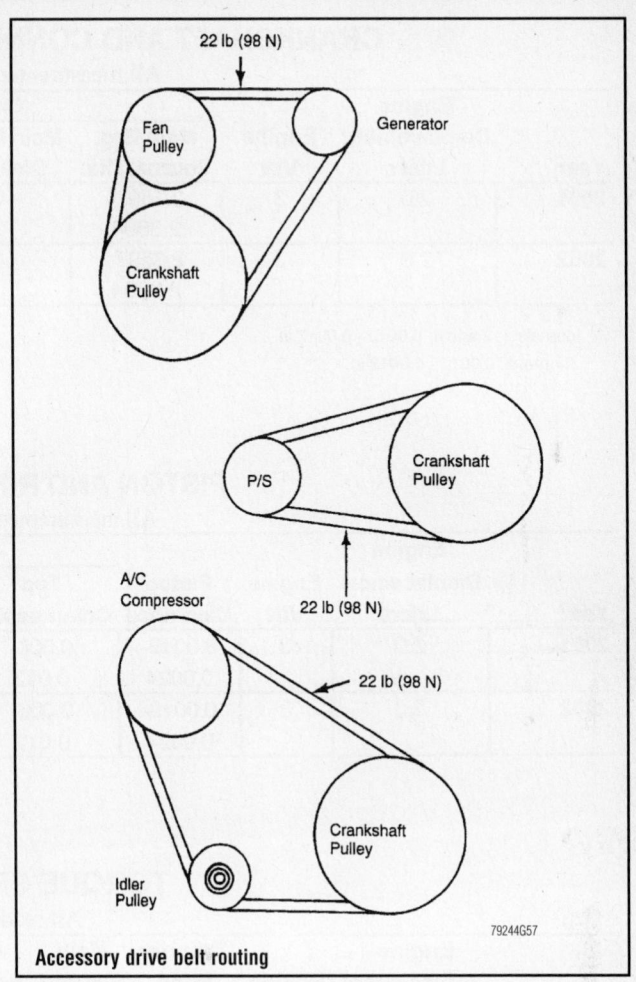

Accessory drive belt routing

79244G57

CAPACITIES

Year	Model	Engine Displacement Liters	Engine VIN	Engine Oil with Filter (qts.)	Transmission (pts.) Manual	Transmission (pts.) Auto.	Transfer Case (pts.)	Drive Axle Front (pts.)	Drive Axle Rear (pts.)	Fuel Tank (gal.)	Cooling System (qts.)
2001	Sportage	2.0	3	4.4	2.6	5.4	2.8	2.6	2.6	15.8	7.9
2002	Sportage	2.0	3	4.4	2.6	5.4	2.8	2.6	2.6	15.8	7.9

NOTE: All capacities are approximate. Add fluid gradually and check to be sure a proper fluid level is obtained.

67162-KSPO-C04

VALVE SPECIFICATIONS

Year	Engine Displacement Liters	Engine VIN	Seat Angle (deg.)	Face Angle (deg.)	Spring Test Pressure (lbs. @ in.)	Spring Installed Height (in.)	Stem-to-Guide Clearance (in.) Intake	Stem-to-Guide Clearance (in.) Exhaust	Stem Diameter (in.) Intake	Stem Diameter (in.) Exhaust
2001	2.0	3	45	45	①	①	0.0010-0.0024	0.0012-0.0026	0.2350-0.2356	0.2348-0.2354
2002	2.0	3	45	45	①	①	0.0010-0.0024	0.0012-0.0026	0.2350-0.2356	0.2348-0.2354

① Spring test pressure or installed height not provided by the manufacturer.
  Valve Spring Free Length:
  Outer spring: 1.524-1.539 in.
  Inner spring: 1.484-1.496 in.

67162-KSPO-C05

CRANKSHAFT AND CONNECTING ROD SPECIFICATIONS

All measurements are given in inches.

Year	Engine Displacement Liters	Engine VIN	Crankshaft				Connecting Rod		
			Main Brg. Journal Dia.	Main Brg. Oil Clearance	Shaft End-play	Thrust on No.	Journal Diameter	Oil Clearance	Side Clearance
2001	2.0	3	2.3597-2.3604	①	0.0031-0.0071	3	2.0055-2.0061	0.0090-0.0021	0.0040-0.0103
2002	2.0	3	2.3597-2.3604	①	0.0031-0.0071	3	2.0055-2.0061	0.0090-0.0021	0.0040-0.0103

① Journals 1, 2 and 4: 0.0010 - 0.0017 in.
　Journal 3: 0.0012 - 0.0019 in.

67162-KSPO-C06

PISTON AND RING SPECIFICATIONS

All measurements are given in inches.

Year	Engine Displacement Liters	Engine VIN	Piston Clearance	Ring Gap			Ring Side Clearance		
				Top Compression	Bottom Compression	Oil Control	Top Compression	Bottom Compression	Oil Control
2001	2.0	3	0.0019-0.0024	0.006-0.012	0.008-0.014	0.008-0.028	0.001-0.003	0.001-0.003	SNUG
2002	2.0	3	0.0019-0.0024	0.006-0.012	0.008-0.014	0.008-0.028	0.001-0.003	0.001-0.003	SNUG

67162-KSPO-C07

TORQUE SPECIFICATIONS

All readings in ft. lbs.

Year	Engine Displacement Liters	Engine VIN	Cylinder Head Bolts	Main Bearing Bolts	Rod Bearing Bolts	Crankshaft Damper Bolts	Flywheel Bolts	Manifold		Spark Plugs	Oil Pan Drain Plug
								Intake	Exhaust		
2001	2.0	3	62	63	50	11	73	16	31	11-17	26
2002	2.0	3	62	63	50	11	73	16	31	11-17	26

67162-KSPO-C08

WHEEL ALIGNMENT

Year	Model		Caster		Camber		Toe-in (in.)
			Range (+/-Deg.)	Preferred Setting (Deg.)	Range (+/-Deg.)	Preferred Setting (Deg.)	
2001	Sportage	F	0.75	+3.58	0.75	+0.44	0.10+/-0.10
		R	—	—	—	0	0
2002	Sportage	F	0.75	+3.58	0.75	+0.44	0.10+/-0.10
		R	—	—	—	0	0

67162-KSPO-C09

TIRE, WHEEL AND BALL JOINT SPECIFICATIONS

Year	Model	OEM Tires		Tire Pressures (psi)		Wheel Size	Ball Joint Inspection	Lug Nut Torque (ft. lbs.)
		Standard	Optional	Front	Rear			
2001	Sportage	P205/75R15	none	26	26	6-JJ	①	73
2002	Sportage	P205/75R15	none	26	26	6-JJ	①	73

OEM: Original Equipment Manufacturer

PSI: Pounds Per Square Inch

STD: Standard

OPT: Optional

① Replace if any measurable movement is found.

67162-KSPO-C10

BRAKE SPECIFICATIONS

All measurements in inches unless noted

Year	Model	Brake Disc			Brake Drum			Minimum Lining Thickness		Brake Caliper Mounting Bolts (ft. lbs.)
		Original Thickness	Minimum Thickness	Maximum Run-out	Original Inside Diameter	Max. Wear Limit	Maximum Machine Diameter	Front	Rear	
2001	Sportage	0.940	0.880	0.004	NA	9.89	NA	0.080	0.060	72
2002	Sportage	0.940	0.880	0.004	NA	9.89	NA	0.080	0.060	72

NA: Not Available

67162-KSPO-C11

SCHEDULED MAINTENANCE INTERVALS
Kia—Sportage

TO BE SERVICED	TYPE OF SERVICE	7.5	15	22.5	30	37.5	45	52.5	60	67.5	75	82.5	90	97.5	105	112.5	120
Accessory drive belt	S/I				✓				✓				✓				✓
Air cleaner filter	R				✓				✓				✓				✓
Automatic transmission fluid	R				✓		✓		✓		✓		✓		✓		✓
Ball joints	S/I				✓				✓				✓				✓
Brake lines & connections	S/I				✓				✓				✓				✓
Chassis/body fasteners	S/I				✓				✓				✓				✓
Cooling system	S/I				✓				✓				✓				✓
CV-joint boots	S/I		✓		✓		✓		✓		✓		✓		✓		✓
Disc brakes	S/I		✓		✓		✓		✓		✓		✓		✓		✓
Driveshaft U-joints	L		✓		✓		✓		✓		✓		✓		✓		✓
Drum brakes	S/I				✓				✓				✓				✓
Emission hoses & tubes	S/I								✓								✓
Emission hoses & tubes (Cal)	R															✓	
Engine coolant	R				✓				✓				✓				✓
Engine oil & filter	R	✓	✓	✓	✓	✓	✓	✓	✓	✓	✓	✓	✓	✓	✓	✓	✓
Exhaust system heat shields	S/I				✓				✓				✓				✓
Front differential fluid	R				✓				✓				✓				✓
	S/I	✓	✓	✓	✓	✓	✓	✓	✓	✓	✓	✓	✓	✓	✓	✓	✓
Fuel filter	R				✓				✓				✓				✓
Fuel lines & hoses	S/I				✓				✓				✓				✓
Idle speed	S/I				✓				✓				✓				✓
Locks & hinges	L	✓	✓	✓	✓	✓	✓	✓	✓	✓	✓	✓	✓	✓	✓	✓	✓
Manual transmission fluid	R				✓				✓				✓				✓
PCV valve	S/I								✓								✓
Rear differential fluid	R				✓				✓				✓				✓
	S/I	✓	✓	✓	✓	✓	✓	✓	✓	✓	✓	✓	✓	✓	✓	✓	✓
Spark plug wires	S/I								✓								✓
Spark plugs	R				✓				✓				✓				✓
Steering operation & linkage	S/I				✓				✓				✓				✓
Timing belt	R														✓		
	S/I								✓				✓				
Timing belt (non-California)	R								✓								✓
Transfer case fluid	R				✓				✓				✓				✓
Transfer case fluid	S/I		✓		✓		✓		✓		✓		✓		✓		✓
Transmission fluid	S/I	✓	✓	✓	✓	✓	✓	✓	✓	✓	✓	✓	✓	✓	✓	✓	✓

R: Replace   S/I: Inspect and service, if needed   L: Lubricate

FREQUENT OPERATION MAINTENANCE (SEVERE SERVICE)

If a vehicle is operated under any of the following conditions it is considered severe service:

- Towing a trailer or using a camper or car-top carrier.
- Repeated short trips of less than 5 miles in temperatures below freezing, or trips of less than 10 miles in any temperature.
- Extensive idling or low-speed driving for long distances as in heavy commercial use, such as delivery, taxi or police cars.
- Operating on rough, muddy or salt-covered roads.
- Operating on unpaved or dusty roads.
- Driving in extremely hot (over 90°) conditions.

Engine oil & filter: replace every 5000 miles or 5 months, whichever occurs first.

Air cleaner filter: inspect and replace if necessary, every 15,000 miles or 15 months, whichever occurs first.

Transfer case fluid: inspect the level every 5000 miles or 5 months, and replace every 15,000 miles or 15 months, whichever occurs first.

Transmission fluid: inspect the level every 5000 miles or 5 months, and replace every 15,000 miles or 15 months, whichever occurs first.

Front differential fluid: inspect the level every 5000 miles or 5 months, and replace every 15,000 miles or 15 months, whichever occurs first.

Rear differential fluid: inspect the level every 5000 miles or 5 months, and replace every 15,000 miles or 15 months, whichever occurs first.

PRECAUTIONS

Before servicing any vehicle, please be sure to read all of the following precautions, which deal with personal safety, prevention of component damage, and important points to take into consideration when servicing a motor vehicle:

• Never open, service or drain the radiator or cooling system when the engine is hot; serious burns can occur from the steam and hot coolant.

• Observe all applicable safety precautions when working around fuel. Whenever servicing the fuel system, always work in a well-ventilated area. Do not allow fuel spray or vapors to come in contact with a spark, open flame, or excessive heat (a hot drop light, for example). Keep a dry chemical fire extinguisher near the work area. Always keep fuel in a container specifically designed for fuel storage; also, always properly seal fuel containers to avoid the possibility of fire or explosion. Refer to the additional fuel system precautions later in this section.

• Fuel injection systems often remain pressurized, even after the engine has been turned **OFF**. The fuel system pressure must be relieved before disconnecting any fuel lines. Failure to do so may result in fire and/or personal injury.

• Brake fluid often contains polyglycol ethers and polyglycols. Avoid contact with the eyes and wash your hands thoroughly after handling brake fluid. If you do get brake fluid in your eyes, flush your eyes with clean, running water for 15 minutes. If eye irritation persists, or if you have taken

brake fluid internally, IMMEDIATELY seek medical assistance.

• The EPA warns that prolonged contact with used engine oil may cause a number of skin disorders, including cancer! You should make every effort to minimize your exposure to used engine oil. Protective gloves should be worn when changing oil. Wash your hands and any other exposed skin areas as soon as possible after exposure to used engine oil. Soap and water, or waterless hand cleaner should be used.

• All new vehicles are now equipped with an air bag system, often referred to as a Supplemental Restraint System (SRS) or Supplemental Inflatable Restraint (SIR) system. The system must be disabled before performing service on or around system components, steering column, instrument panel components, wiring and sensors. Failure to follow safety and disabling procedures could result in accidental air bag deployment, possible personal injury and unnecessary system repairs.

• Always wear safety goggles when working with, or around, the air bag system. When carrying a non-deployed air bag, be sure the bag and trim cover are pointed away from your body. When placing a non-deployed air bag on a work surface, always face the bag and trim cover upward, away from the surface. This will reduce the motion of the module if it is accidentally deployed. Refer to the additional air bag system precautions later in this section.

• Clean, high quality brake fluid from a

sealed container is essential to the safe and proper operation of the brake system. You should always buy the correct type of brake fluid for your vehicle. If the brake fluid becomes contaminated, completely flush the system with new fluid. Never reuse any brake fluid. Any brake fluid that is removed from the system should be discarded. Also, do not allow any brake fluid to come in contact with a painted surface; it will damage the paint.

• Never operate the engine without the proper amount and type of engine oil; doing so WILL result in severe engine damage.

• Timing belt maintenance is extremely important! Many models utilize an interference-type, non-freewheeling engine. If the timing belt breaks, the valves in the cylinder head may strike the pistons, causing potentially serious (also time-consuming and expensive) engine damage. Refer to the maintenance interval charts in the front of this section for the recommended replacement interval for the timing belt.

• Disconnecting the negative battery cable on some vehicles may interfere with the functions of the on-board computer system(s) and may require the computer to undergo a relearning process once the negative battery cable is reconnected.

• When servicing drum brakes, only disassemble and assemble one side at a time, leaving the remaining side intact for reference.

• Only an MVAC-trained, EPA-certified automotive technician should service the A/C system or its components.

ENGINE REPAIR

Alternator

REMOVAL

1. Before servicing the vehicle, refer to the precautions at the beginning of this section.
2. Remove or disconnect the following:
   • Negative battery cable
   • Air cleaner inlet pipe front bolts
   • Top hose from the resonance chamber
   • Air cleaner inlet pipe
   • Alternator electrical connectors
   • Loosen the pivot and tensioner mounting bolts, do mot remove them
   • Drive belt from the alternator pulley

• Drive belt tensioner
• Alternator pivot bolt
• Loosen the bolt at the base of the adjusting bracket and rotate the bracket up
• Alternator

To install:
3. Install or connect the following:
   • Alternator
   • Pivot bolt and hand tighten
   • Rotate the bracket down on top of the alternator
   • Belt tensioner on the adjustment bracket
   • Tensioner mounting bolt and hand tighten it
   • Drive belt
   • Torque the tensioner bolt to 19 ft. lbs. (26 Nm).
   • Torque the pivot bolt to 38 ft. lbs. (51 Nm).

• Alternator electrical connectors
• Air cleaner inlet pipe and tighten the clamp
• Hose to the resonance chamber
• Air inlet pipe bolts and tighten
• Negative battery cable

Ignition Timing

ADJUSTMENT

The 2.0L engine in the Sportage is equipped with a distributorless ignition system. The ignition timing is controlled by the Powertrain Control Module (PCM) through the input of engine control system sensors. The ignition timing is set at 4 degrees BTDC for vehicles equipped with manual or automatic transmissions. The ignition timing cannot be adjusted.

Engine Assembly

REMOVAL & INSTALLATION

1. Before servicing the vehicle, refer to the precautions at the beginning of this section.

2. Properly relieve the fuel system pressure.

3. Drain the cooling system.
4. Drain the engine oil.
5. Drain the transmission fluid.
6. Remove or disconnect the following:
   - Both battery cables
   - Windshield washer hose from the hood
   - Hood
   - 2 air duct mounting bolts from the top of the radiator. Loosen the clamp at the air intake housing and remove the duct
   - Accelerator cable by pulling the throttle back and rotating the cable until it aligns with the slot in the pulley
   - Transmission control cable
   - Resonance chamber mounting bolt, chamber bolt and air silencer
   - Idle Air Control (IAC) air hose,

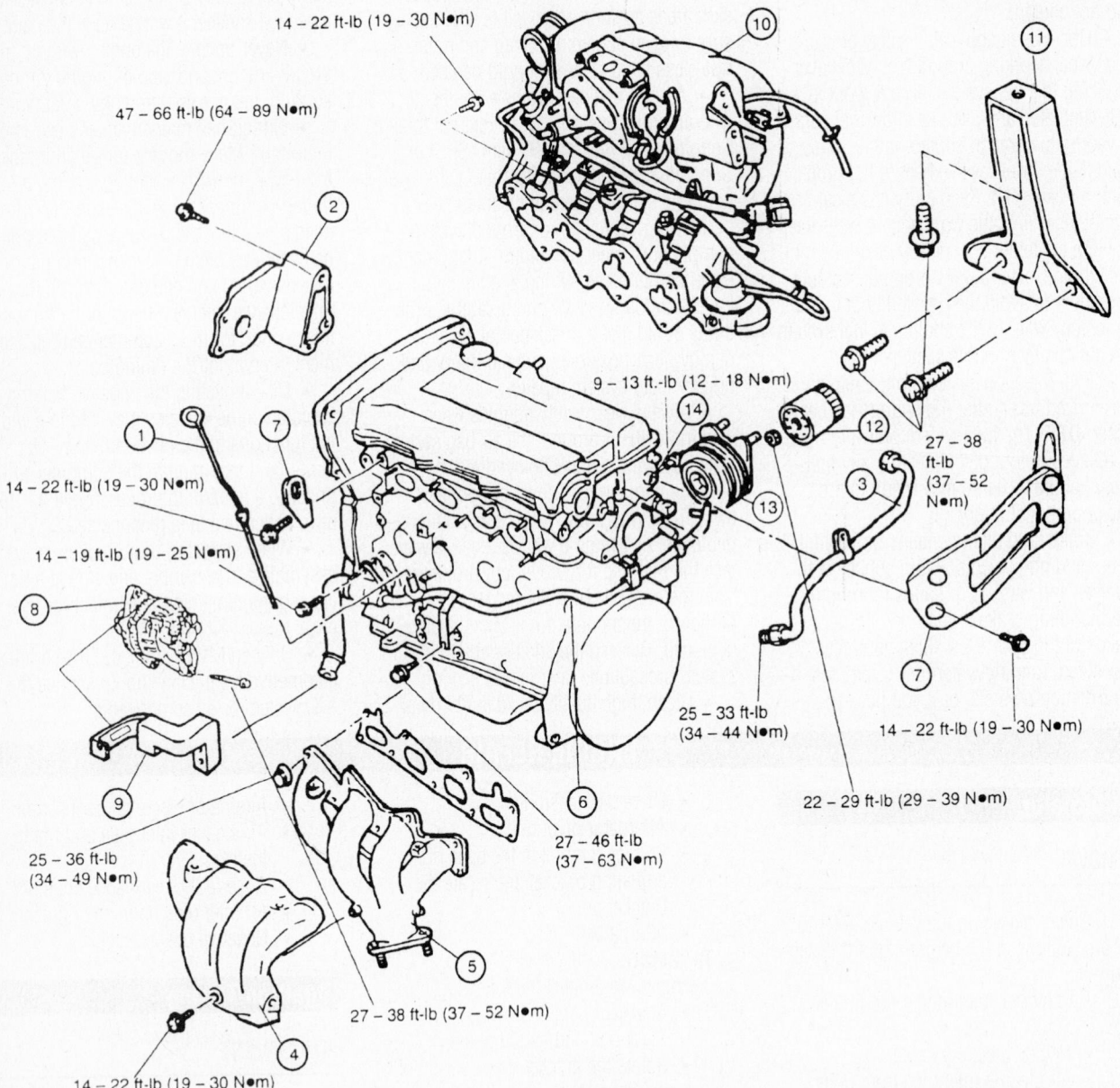

14 – 22 ft-lb (19 – 30 N•m)

47 – 66 ft-lb (64 – 89 N•m)

9 – 13 ft.-lb (12 – 18 N•m)

14 – 22 ft-lb (19 – 30 N•m)

14 – 19 ft-lb (19 – 25 N•m)

25 – 33 ft-lb (34 – 44 N•m)

27 – 38 ft-lb (37 – 52 N•m)

14 – 22 ft-lb (19 – 30 N•m)

22 – 29 ft-lb (29 – 39 N•m)

25 – 36 ft-lb (34 – 49 N•m)

27 – 46 ft-lb (37 – 63 N•m)

27 – 38 ft-lb (37 – 52 N•m)

14 – 22 ft-lb (19 – 30 N•m)

1. Oil Level Gauge
2. Thermo-Modulated Fan Bracket
3. EGR Pipe
4. Exhaust Manifold Heat Shield
5. Exhaust Manifold
6. Coolant Inlet Pipe and Bypass Pipe
7. Engine Hanger
8. Generator
9. Generator Strap and Bracket
10. Intake Manifold Assembly
11. Intake Manifold Support Bracket
12. Oil Filter
13. Oil Cooler
14. Oil Pressure Switch

7924QG36

Exploded view of some peripheral engine component mountings

breather hose and vacuum line from the air intake tube
- Manifold Air Flow (MAF) sensor connector
- Loosen the air inlet hose clamp from the MAF sensor
- 3 bolts from the air intake tube to the throttle body
- Air intake hose and tube as an assembly
- Upper radiator hose
- Clutch fan nuts
- Cooling fan shroud bolts
- Fan and shroud at the same time
- Alternator drive belt
- Fan pulley
- Alternator electrical connectors
- Exhaust Gas Recirculation (EGR) solenoid valve connector on the intake manifold in front of the dynamic chamber
- Both heater hoses from the pipes
- Engine-to-body ground wire from the intake manifold and the harness bracket
- Brake booster vacuum hose from the dynamic chamber
- Fuel lines and fuel pressure regulator from the rear of the dynamic chamber
- Vacuum hose from the bottom of the EGR valve
- Purge solenoid valve vacuum hose from dynamic chamber
- Vacuum hoses from the top of the charcoal canister and slide the charcoal canister up and out of the bracket
- Lower radiator hose
- Cooling lines from the radiator, if equipped with an automatic transmission
- Radiator and raise and safely support the vehicle
- Lower splash panel
- Drive belt
- A/C pulley assembly
- A/C compressor mounting bolts and move the A/C compressor out of the way
- Power steering drive belt
- Intake manifold support bracket
- Starter wiring harness
- Starter
- Front exhaust pipe from the exhaust manifold
- Bracket bolt from the front exhaust pipe
- Exhaust-to-clutch (manual transmission) or converter (automatic transmission) housing bolts and the bracket

- Clutch housing (manual transmission) or converter (automatic transmission) housing bolts.
- Drive plate-to-torque converter bolts, if equipped

7. Support the transmission from underneath the vehicle.

8. Connect the engine hoist to the engine assembly.
- Left and right side engine mounting bolts

9. Lift the engine up and forward slightly to provide access to three electrical connectors on the rear of the cylinder head.
- Electrical connectors from the Camshaft Position (CMP) sensor, coil and condenser on the rear of the cylinder head
- Engine from the vehicle

To install:

10. Lower the engine enough to connect the three electrical connectors to the CMP sensor, coil and condenser on the rear of the cylinder head.

11. Position the engine to the transmission. Install the transmission bolts and tighten the bolts. Torque according to bolt size:
- 14mm bolts to 80 ft. lbs. (108 Nm)
- 10mm bolts to 28 ft. lbs. (38 Nm)
- 6mm bolts to 60 inch lbs. (7 Nm)
- Right and left side engine mounting bolts. Torque the bolt s to 38 ft. lbs. (52 Nm).

12. Disconnect the engine hoist from the engine assembly.

13. Raise and safely support the vehicle.
- Drive plate-to-torque converter bolts, if equipped.
- Connect the front exhaust pipe to the exhaust manifold. Torque the flange bolts to 24 ft. lbs. (31 Nm).
- Front exhaust pipe. Torque the bolts to 20 ft. lbs. (27 Nm).
- Starter
- Connect the starter wiring harness
- Intake manifold support bracket bolts and the bracket.

14. Install the power steering pump lock and mounting bolts. Install the power steering drive belt. Torque the bolts to 30 ft. lbs. (42 Nm).
- A/C compressor mounting bolts. Torque the bolts to 18 ft. lbs. (24 Nm).
- A/C belt pulley assembly and drive belt. Install the two A/C idler pulley bracket bolts and torque to 24 ft. lbs. (32 Nm).
- Lower splash panel
- Radiator

- Cooling lines to the radiator, if equipped
- Lower radiator hose
- Slide the charcoal canister in the bracket
- Vacuum hoses to the top of the charcoal canister
- Engine-to-body ground wire to the intake manifold and the harness bracket
- Brake booster vacuum hose to the dynamic chamber
- Fuel lines and fuel pressure regulator to the rear of the dynamic chamber
- Vacuum hose to the bottom of the EGR valve
- Purge solenoid valve vacuum hose to dynamic chamber
- EGR solenoid valve connector on the intake manifold in front of the dynamic chamber
- Heater hoses
- Electrical terminal connectors to the alternator
- Fan pulley
- Alternator drive belt. Torque the adjusting bolt 16 ft. lbs. (22 Nm) and the mounting bolt to 32 ft. lbs. (45 Nm).
- Fan and shroud as an assembly. Torque the five cooling fan shroud bolts to 72 inch lbs. (8 Nm)
- Clutch fan nuts. Torque the nuts to 27 ft. lbs. (37 Nm).
- Upper radiator hose
- Air intake hose and tube as an assembly
- Air intake tube to the throttle body and tighten the air inlet hose clamp to the MAF sensor
- MAF sensor electrical connector
- Resonance chamber mounting bolt, chamber bolt and air silencer
- IAC air hose, breather hose and vacuum line to the air intake tube
- Accelerator cable by pulling the throttle back and rotating the cable until it aligns with the slot in the pulley
- Transmission control cable
- Air duct mounting bolts to the top of the radiator. Torque the clamp at the air intake housing
- Hood
- Windshield washer hose to the hood

15. Fill the engine with clean engine oil.
16. Connect the battery cables.
17. Fill the cooling system.
18. Fill the transmission fluid.
19. Recharge the A/C system.

20. Start the vehicle. Check for leaks, repair if necessary.

21. Road test the vehicle to check engine performance.

Water Pump

REMOVAL & INSTALLATION

1. Before servicing the vehicle, refer to the precautions at the beginning of this section.

2. Drain the cooling system.

3. Remove or disconnect the following:
   - Negative battery cable
   - Lower splash shield
   - Upper and lower radiator hoses
   - Coolant reservoir tank hose
   - Fresh air duct
   - Fan and shroud
   - Loosen the alternator mounting and adjusting bolts
   - Alternator belt
   - Fan pulley and bracket
   - Upper and lower timing belt covers and turn the crankshaft until No. 1 cylinder is at Top Dead Center (TDC).
   - Loosen the tensioner lockbolt and pry the tensioner away from the belt
   - Timing belt
   - Loosen the tensioner bolt
   - Water pump
   - Tensioners from the water pump

To install:

4. Clean the surface of any old gasket material.

5. Install or connect the following:
   - Tensioners on the water pump
   - Water pump and gasket. Torque the bolts to 19 ft. lbs. 25 Nm).

- Timing belt
- Loosen the tensioner lockbolt and allow the tensioner to rest against the belt. Torque the tensioner lockbolt to 32 ft. lbs. (43 Nm).
- Upper and lower timing belt covers
- Fan bracket assembly and fan pulley
- Drive belt
- Cooling fan and shroud
- Position the radiator and torque the bracket bolts to 89 inch lbs. (10 Nm).
- Torque the shroud bolts to 89 inch lbs. (10 Nm) and torque the alternator adjusting and mounting bolts
- Position the fresh air duct over the radiator and tighten the retaining bolt 89 inch lbs. (10 Nm)
- Radiator hoses and tighten the clamps
- Lower splash shield
- Negative battery cable

6. Fill the cooling system.

7. Start the vehicle and bring the engine to operating temperature. Check for leaks and repair if necessary.

Heater Core

REMOVAL & INSTALLATION

1. Disconnect the negative battery cable.

✱✱ CAUTION

After disconnecting the negative battery cable, wait for at least 10 minutes for the air bag module to deplete its stored energy.

2. Remove the driver's side air bag and steering wheel by performing the following procedure:

   a. Position the front wheels in the straight-ahead position.

   b. Remove the 4 steering wheel-to-air bag module bolts.

   c. Carefully, lift the air bag module and disconnect the electrical connector.

✱✱ CAUTION

Place the air bag module in a safe location with the front facing upward.

   d. Remove the steering wheel-to-steering column nut.

➡ **If may be necessary to mark the steering wheel to steering column alignment.**

   e. Using a steering wheel puller, press the steering wheel from the steering column.

3. Drain the cooling system into a clean container for reuse.

4. Discharge and recover the air conditioning system refrigerant.

5. Remove the instrument panel by performing the following procedure:

   a. Remove both the rear and front consoles.

   b. Remove the knee bolster assembly.

   c. Remove the "T" bar section.

   d. Remove the relay bracket.

   e. Remove the turn signal assembly and the upper/lower steering column covers.

   f. Remove the hood release handle lockscrew, the hood release handle and the cable assembly nut.

   g. Remove the left side front pillar trim and the lower left side cover.

   h. Remove the 2 left side of the "T" bar-to-chassis bolts.

   i. At the left side of the instrument panel, remove the 3 instrument panel-to-chassis bolts.

   j. Remove the ashtray.

   k. Remove the center panel trim.

   l. Remove the ventilation control panel.

   m. At the center of the windshield next to the windshield, remove the cap and the mounting bolt.

   n. Remove the right side front pillar trim and the lower right side cover.

   o. Remove the 2 right side of the "T" bar-to-chassis bolts.

   p. At the right side of the instrument panel, remove the 3 instrument panel-to-chassis bolts.

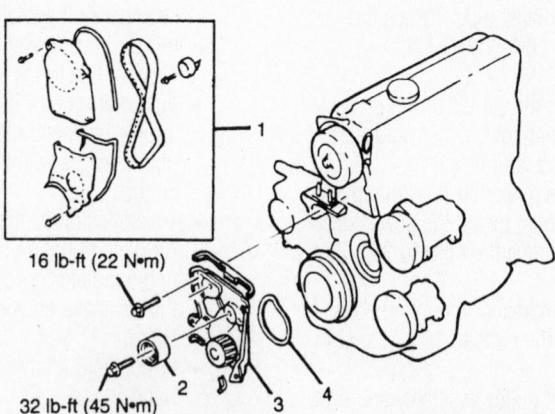

16 lb-ft (22 N•m)

32 lb-ft (45 N•m)

1	TIMING BELT COVERS, GASKETS AND TIMING BELT	3	COOLANT PUMP
2	IDLER PULLEY	4	GASKET

7924QG01

Exploded view of the water pump mounting

q. Remove the steering column-to-instrument panel bolts and lower the steering column.

r. Disconnect the instrument panel electrical connectors.

s. Remove the instrument panel.

6. Remove the blower/evaporator housing by performing the following procedure:

a. Disconnect the air conditioning refrigerant lines from the evaporator core and discard the gaskets. Plug the openings to prevent contamination.

b. Disconnect the fresh air control cable from the blower/evaporator housing inlet duct.

c. Disconnect the 5 connectors from the bottom of the blower/evaporator housing.

d. Move the carpeting from the bulkhead to gain access to the hole cover plate.

e. Remove the 4 hole cover plate nuts and the plate.

f. Remove the 2 upper blower/evaporator housing bolts.

g. Remove the 2 lower blower/evaporator housing-to-bulkhead nuts.

h. Remove the blower/evaporator housing.

7. Disconnect the heater hoses from the heater core.

8. Remove the temperature control cable from the heater housing.

9. Remove the 2 lower heater housing nuts and the upper heater housing-to-bulkhead nut.

10. Remove the heater housing.

11. Disassemble the heater housing by performing the following procedure:

a. Remove the seal from the heater core tube connections.

b. Remove the vent seal.

c. Remove the 2 wiring harness-to-heater servo screws.

d. Remove the 8 heater housing clips located on the servo side (left side).

e. Remove the left side of the heater housing.

f. Remove the 6 heater housing assembly clips.

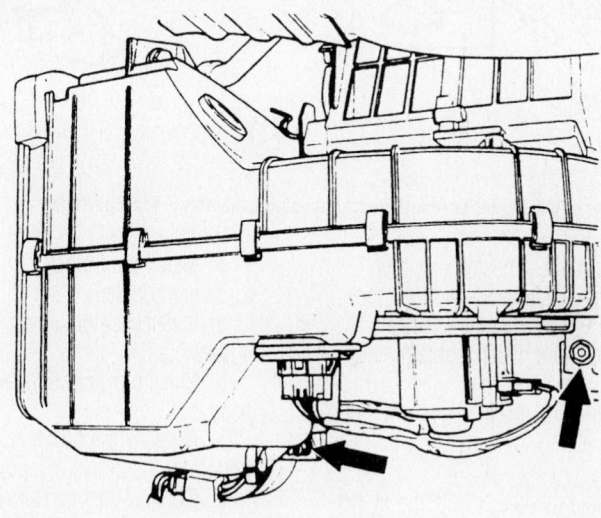

93113GI5

View of the blower/evaporator housing assembly—Kia Sportage

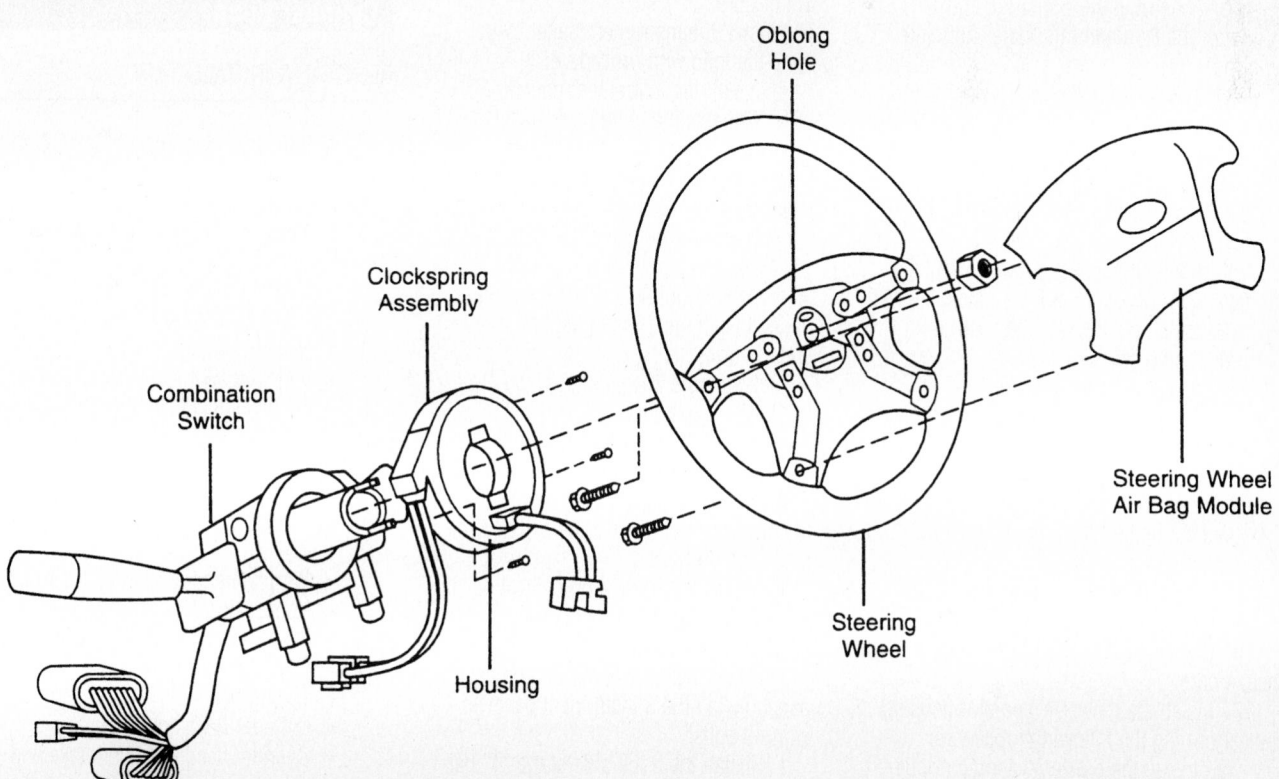

93113GI4

Exploded view of the steering wheel and air bag module assembly—Kia Sportage

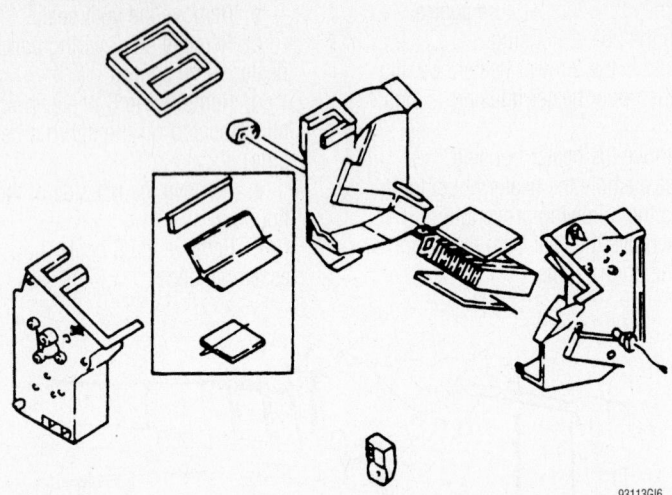

93113GI6

Exploded view of the heater core and heater housing assembly—Kia Sportage

g. Remove the 4 heater core tube mounting bracket screws and the bracket.

h. Remove the 8 remaining heater housing clips and disassemble the housings.

i. Remove the heater core from the housing.

To install:

12. Assemble the heater housing by performing the following procedure:

a. Install the heater core to the housing.

b. Assemble the housings and install the 8 remaining heater housing clips.

c. Install the heater core tube mounting bracket and the 4 bracket screws.

d. Install the 6 heater housing assembly clips.

e. Install the left side of the heater housing.

f. Install the 8 heater housing clips located on the servo side (left side).

g. Install the 2 wiring harness-to-heater servo screws.

h. Install the vent seal.

i. Install the seal to the heater core tube connections.

13. Install the heater housing.

14. Install the 2 lower heater housing nuts and the upper heater housing-to-bulkhead nut.

15. Install the temperature control cable to the heater housing.

16. Connect the heater hoses to the heater core.

17. Install the blower/evaporator housing by performing the following procedure:

a. Install the blower/evaporator housing.

b. Install the 2 lower blower/evaporator housing-to-bulkhead nuts.

c. Install the 2 upper blower/evaporator housing bolts.

d. Install the hole cover plate and the 4 plate nuts.

e. Move the carpeting over the bulkhead.

f. Connect the 5 connectors to the bottom of the blower/evaporator housing.

g. Connect the fresh air control cable to the blower/evaporator housing inlet duct.

h. Using new gaskets, connect the air conditioning refrigerant lines to the evaporator core.

18. Install the instrument panel by performing the following procedure:

a. Install the instrument panel.

b. Connect the instrument panel electrical connectors.

c. Install the steering column and lower the steering column-to-instrument panel bolts. Torque the bolts to 15 ft. lbs. (20 Nm).

d. At the right side of the instrument panel, install the 3 instrument panel-to-chassis bolts.

e. Install the 2 right side of the "T" bar-to-chassis bolts.

f. Install the right side front pillar trim and the lower right side cover.

g. At the center of the windshield next to the windshield, install the cap and the mounting bolt.

h. Install the ventilation control panel.

i. Install the center panel trim.

j. Install the ashtray.

k. At the left side of the instrument panel, install the 3 instrument panel-to-chassis bolts.

l. Install the 2 left side of the "T" bar-to-chassis bolts.

m. Install the lower left side cover and the left side front pillar trim.

n. Install the cable assembly nut, the hood release handle and the hood release handle lockscrew.

o. Install the turn signal assembly and the upper/lower steering column covers.

p. Install the relay bracket.

q. Install the "T" bar section.

r. Install the knee bolster assembly.

s. Install both the rear and front consoles.

19. Evacuate and charge the air conditioning system refrigerant.

20. Refill the cooling system.

21. Install the driver's side air bag and steering wheel by performing the following procedure:

a. Install the steering wheel to the steering column.

b. Install the steering wheel-to-steering column nut and torque the nut to 33 ft. lbs. (45 Nm).

c. Carefully, install the air bag module and connect the electrical connector.

d. Install the 4 steering wheel-to-air bag module bolts and torque to 72–106 inch lbs. (8–12 Nm).

22. Connect the negative battery cable.

23. Run the engine to normal operating temperatures; then, check the climate control operation and check for leaks.

Cylinder Head

REMOVAL & INSTALLATION

1. Before servicing the vehicle, refer to the precautions at the beginning of this section.

2. Properly relieve the fuel system pressure.

3. Drain the cooling system.

4. Remove or disconnect the following:
- Negative battery cable
- Brake booster vacuum hose from the dynamic chamber
- Fuel line from the pressure regulator and the return line from the rear of the dynamic chamber
- Ground wire from the intake manifold
- Purge solenoid valve vacuum hose
- Upper radiator hose
- Intake manifold support bracket
- Converter flange inlet pipe
- Timing belt
- Cylinder head cover
- Cylinder head with the intake and exhaust manifolds attached
- 3 wire harness connectors at the rear of the cylinder head

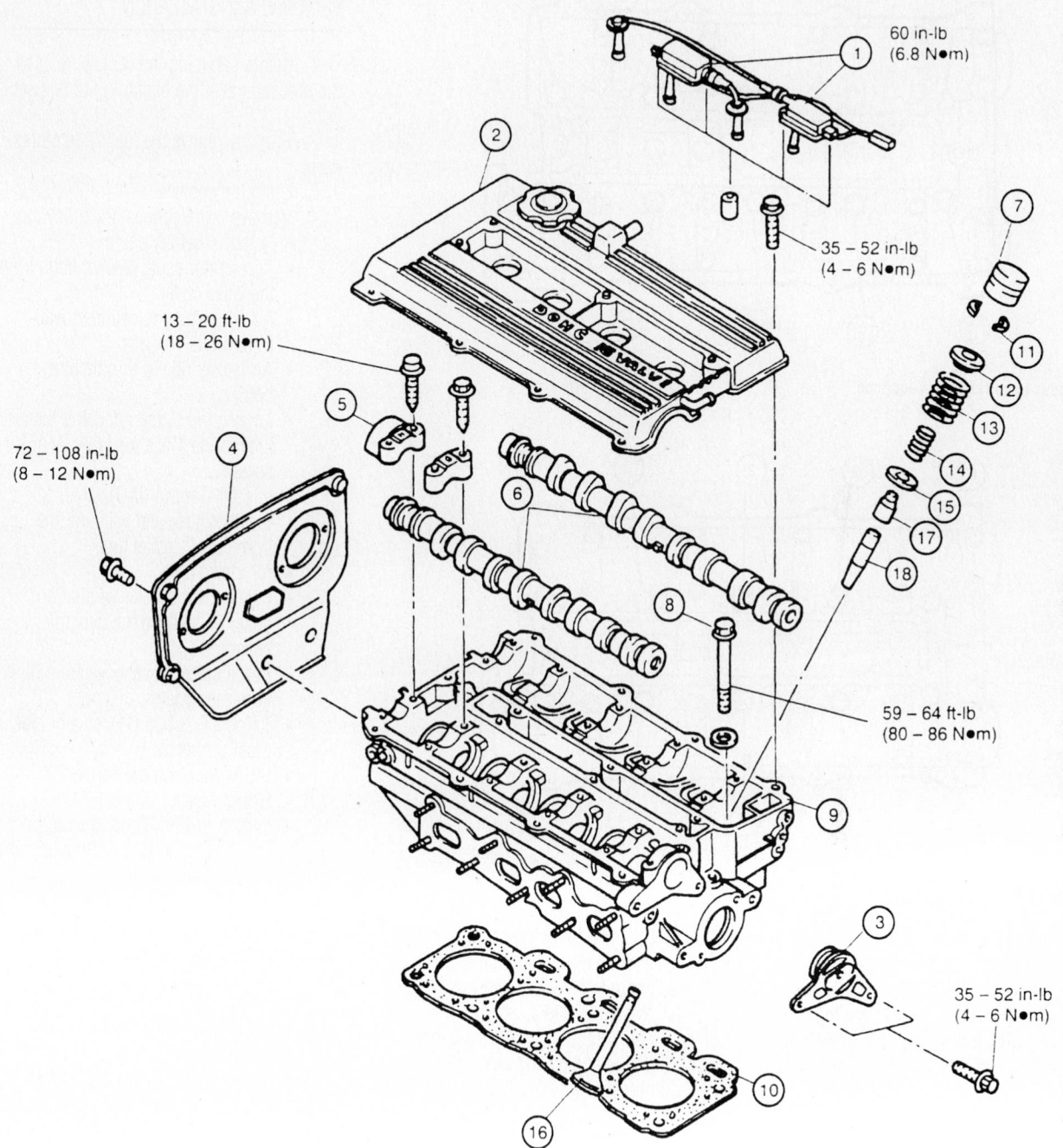

60 in-lb
(6.8 N•m)

35 – 52 in-lb
(4 – 6 N•m)

13 – 20 ft-lb
(18 – 26 N•m)

72 – 108 in-lb
(8 – 12 N•m)

59 – 64 ft-lb
(80 – 86 N•m)

35 – 52 in-lb
(4 – 6 N•m)

1. Ignition Coils and High Tension
   Leads
2. Cylinder Head Cover
3. Camshaft Position Sensor
4. Seal Plate
5. Camshaft Caps
6. Camshafts

7. Hydraulic Lash Adjuster
8. Cylinder Head Bolt
9. Cylinder Head
10. Cylinder Head Gasket
11. Valve Locks
12. Upper Spring Seat

13. Outer Valve Spring
14. Inner Valve Spring
15. Lower Spring Seat
16. Valve
17. Valve Stem Seal
18. Valve Guide

7924QG37

Exploded view of the cylinder head assembly

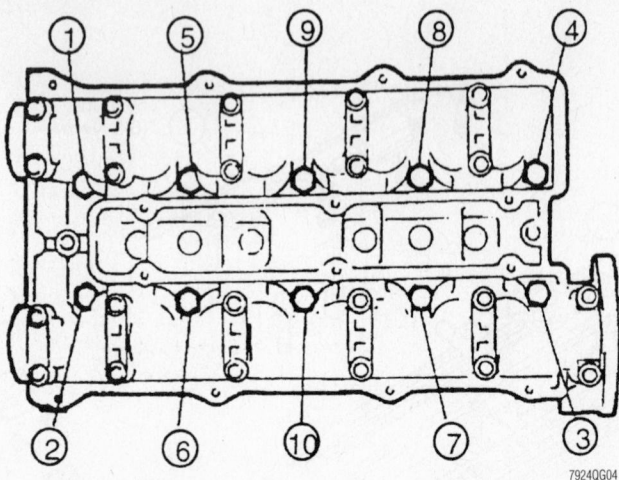

Cylinder head removal sequence

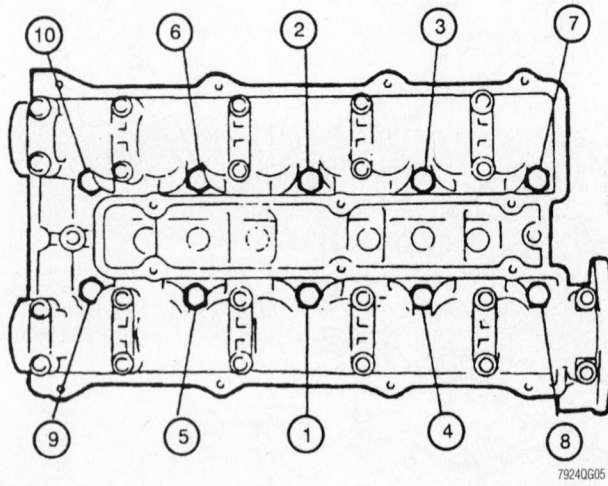

Cylinder head torque sequence—Kia Sportage

Intake Manifold

REMOVAL & INSTALLATION

1. Before servicing the vehicle, refer to the precautions at the beginning of this section.

2. Properly relieve the fuel system pressure.

3. Drain the coolant.

4. Remove or disconnect the following:
- Negative battery cable
- Accelerator cable bracket bolts from the valve cover
- Air intake tube to cylinder head cover bolts
- Air intake tube to throttle body bolts
- Loosen the clamp attaching the air tube to the Mass Air Flow (MAF) sensor
- Idle Air Control (IAC) valve, breather hose and vacuum line from the air intake tube
- Air intake tube
- Positive Crankcase Ventilation hose (PCV) from the dynamic chamber
- Purge solenoid valve vacuum hose from the dynamic chamber
- Throttle Position (TP) sensor electrical connector
- IAC valve electrical connector
- Heater hoses
- Engine-to-body ground strap from the intake manifold and the harness bracket below it
- Brake booster vacuum line
- Vacuum hose from the fuel pressure regulator hose
- Dynamic chamber support bracket bolts
- Fuel injector electrical connectors
- Fuel pressure and return lines
- Intake manifold support bracket

To install:

5. Place the new head gasket on the engine block.

6. Install or connect the following:
- Cylinder head with the intake and exhaust manifolds attached
- 3 wiring connectors at the rear of the cylinder head
- Cylinder head bolts in the proper sequence. Torque the bolts to 64 ft. lbs. (87 Nm).
- Cylinder head cover
- Timing belt
- Converter inlet pipe flange nuts. Torque the nuts to 24 ft. lbs. (33 Nm).
- Upper radiator hose
- Vacuum hose from the intake manifold to the charcoal canister
- Purge solenoid vacuum hose
- Ground wire and harness bracket to the intake manifold. Torque the bolts to 18 ft. lbs. (25 Nm).

- Fuel line to the pressure regulator and the return line to the fuel rail
- Brake booster vacuum hose
- Negative battery cable

7. Properly fill the cooling system.

8. Start the engine and check for leaks, repair if necessary.

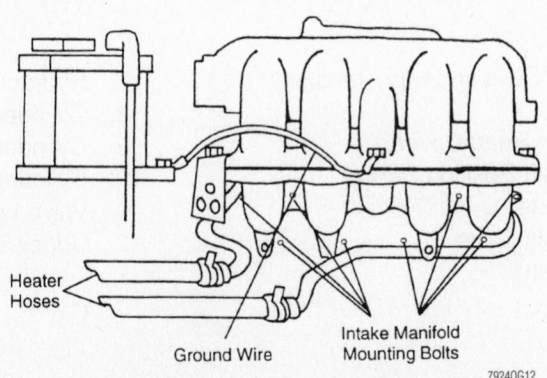

Intake manifold mounting bolt locations. Be sure to connect the ground cable

- Oil filter
- Intake manifold bolts
- Bypass pipe from the heater hose
- Intake manifold and discard the gasket

To install:

5. Install or connect the following:
- Intake manifold with a new gasket to the cylinder head
- Heater hose to the bypass pipe
- Bolts and nuts attaching the intake manifold to the cylinder head. Torque the bolts to 14–22 ft. lbs. (19–30 Nm).
- New oil filter
- Intake manifold support bracket. Torque the bolts to 27–38 ft. lbs. (37–52 Nm)
- Fuel lines
- Fuel injector electrical connectors

- Engine to body ground wire. Torque the bolt to 18 ft. lbs. (25 Nm).
- Dynamic chamber support bracket. Torque the bolts to 18 ft. lbs. (25 Nm).
- Purge solenoid valve vacuum hose to the dynamic chamber
- Vacuum hose to the fuel pressure regulator
- Coolant hoses to the throttle body
- IAC valve electrical connector
- Brake booster vacuum hose
- Heater hoses
- TP sensor electrical connector
- Air intake tube and hose to the throttle body. Torque the bolts to 16 ft. lbs. (22 Nm).
- PCV hose to the dynamic chamber
- Accelerator cable to the throttle body pulley

- Accelerator cable bracket. Torque the bolts to 10 ft. lbs. (15 Nm).
- Air intake hose to the MAF sensor
- Resonance chamber
- IAC hose, breather hose and vacuum line to the intake manifold
- Negative battery cable

6. Properly fill the cooling system.
7. Start the engine and check for leaks, repair if necessary.

Exhaust Manifold

REMOVAL & INSTALLATION

1. Before servicing the vehicle, refer to the precautions at the beginning of this section.
2. Remove or disconnect the following:

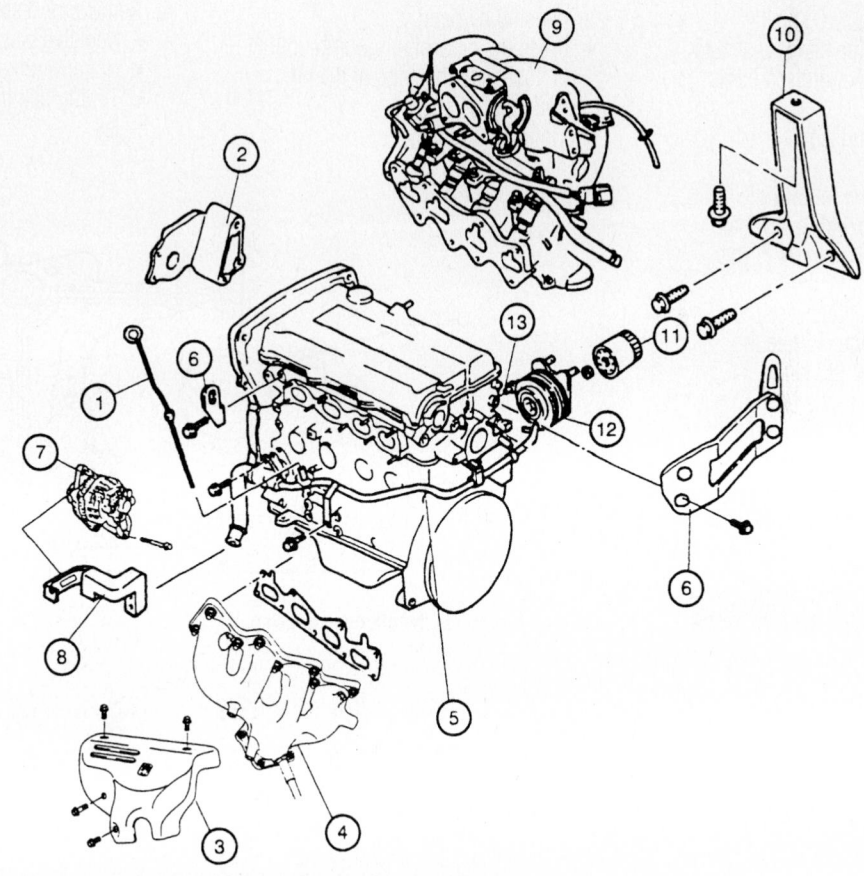

1. Oil Level Gauge
2. Thermo-Modulated Fan Bracket
3. Exhaust Manifold Heat Shield
4. Exhaust Manifold
5. Coolant Inlet Pipe and Bypass Pipe
6. Engine Hanger
7. Generator
8. Generator Strap and Bracket
9. Intake Manifold Assembly
10. Intake Support Bracket
11. Oil Filter
12. Oil Cooler
13. Oil Pressure Switch

Exploded view of the intake, exhaust manifold and related components

9308QG01

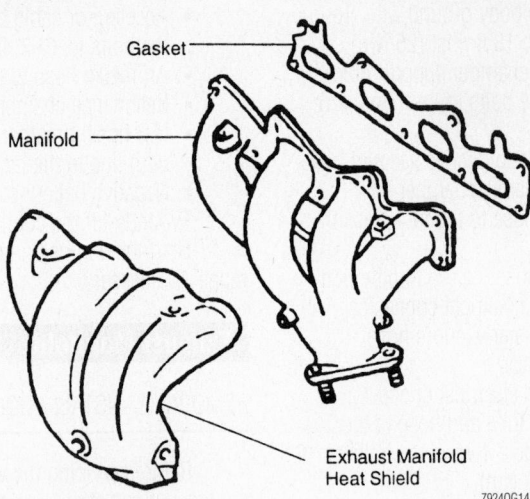

Exploded view of the exhaust manifold assembly

- Negative battery cable
- Air intake hose
- Exhaust manifold heat shield
- Converter inlet pipe flange locknuts
- Exhaust manifold and discard the gasket

3. Clean the mating surfaces.

To install:

4. Install or connect the following:
- Exhaust manifold with a new gasket. Torque the bolts to 31 ft. lbs. (42 Nm).
- New flange gasket and install the converter inlet pipe. Torque the bolts to 24 ft. lbs. (33 Nm).
- Exhaust manifold heat shield. Torque the bolts to 18 ft. lbs. (25 Nm).
- Air intake hose
- Negative battery cable

5. Start the vehicle and check for leaks, repair if necessary.

Front Crankshaft Seal

REMOVAL & INSTALLATION

1. Before servicing the vehicle, refer to the precautions at the beginning of this section.

2. Remove or disconnect the following:
- Negative battery cable
- Engine under cover
- Timing belt
- Timing belt pulley lock bolt
- Timing belt pulley
- Pulley woodruff key
- Oil seal by carefully cutting it out of the oil pump housing

To install:

3. Lubricate the lip of the new seal with clean engine oil.

4. Install or connect the following:
- New oil seal into the oil pump housing by hand
- Press the oil seal into pump until it is flush with the edge of the oil pump body
- Timing belt pulley

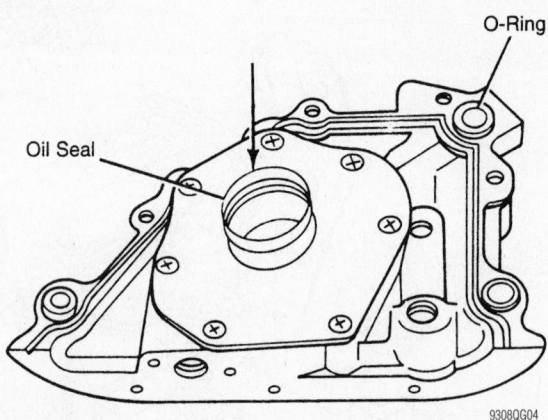

Install the oil seal into the oil pump housing

- Pulley woodruff key
- Pulley lock bolt. Torque the bolt to 18 ft. lbs. (25 Nm).
- Timing belt
- Engine under cover. Torque the bolts to 18 ft. lbs. (25 Nm).
- Negative battery cable

5. Start the engine and check for leaks, repair if necessary.

Camshaft

REMOVAL & INSTALLATION

1. Before servicing the vehicle, refer to the precautions at the beginning of this section.

2. Properly relieve the fuel system pressure.

3. Drain the coolant into a suitable container.

4. Remove or disconnect the following:
- Negative battery cable
- Upper timing belt cover
- Timing belt from the camshaft pulley

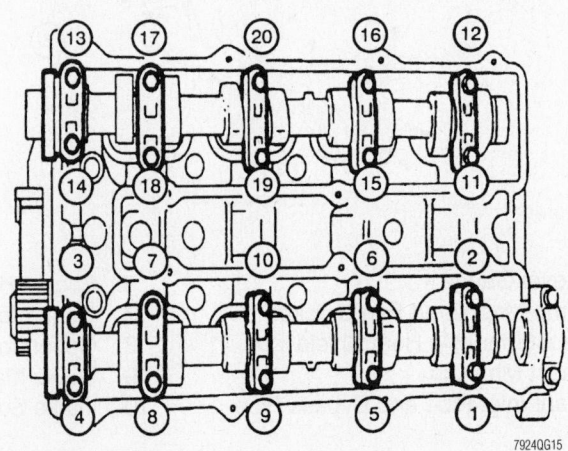

Camshaft cap bolt removal sequence

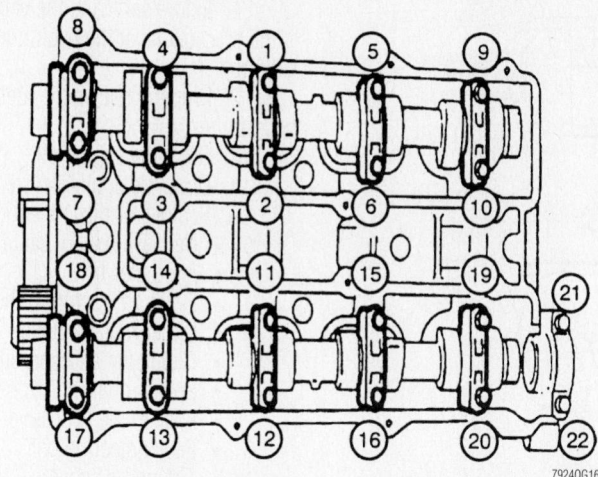

Camshaft journal bolt tightening sequence

- Camshaft pulleys
- Camshaft cap bolts in the proper sequence
- Camshaft caps
- Camshafts

To install:

5. Install or connect the following:
- Camshafts into the cylinder head. The exhaust camshaft has a steel dowel pin at the rear for the camshaft position sensor
- Clean engine oil to the journals and bearings
- Camshaft oil seal
- Silicone sealant to the front camshaft cap and the camshaft position sensor mounting cap
- Camshaft caps in the proper sequence. Torque the bolts in three steps to 20 ft. lbs. (26 Nm).
- Camshaft pulleys
- Timing belt
- Timing belt cover
- Negative battery cable

Valve Lash

ADJUSTMENT

The DOHC engine uses Hydraulic Lash Adjusters (HLA's), which automatically maintain the proper amount of valve lash. Therefore, the DOHC engine does not need manual valve lash adjustment.

Starter

REMOVAL & INSTALLATION

1. Before servicing the vehicle, refer to the precautions at the beginning of this section.

2. Drain the engine oil.
3. Remove or disconnect the following:
- Negative battery cable
- Intake manifold bracket upper bolts
- Clutch release cylinder and move it aside, if equipped
- Intake manifold bracket lower bolts and remove the bracket
- Starter from the clutch, manual transmissions only
- Starter from the torque converter, automatic transmissions only
- Starter electrical connectors
- Move the transmission wire harness aside
- Starter

To install:

4. Install or connect the following:
- Starter to the engine well
- Starter electrical connectors
- Lower intake manifold bracket and install the upper bolts
- Starter into position. When aligned

properly torque the bolts to 40 ft. lbs. (54 Nm).
- Torque the intake manifold bracket bolts to 40 ft. lbs. (54 Nm).
- Properly position the clutch release cylinder, if equipped. Torque the bolts to 40 ft. lbs. (54 Nm).
- Torque the intake manifold bracket upper bolts to 40 ft. lbs. (54 Nm).
- Negative battery cable

Oil Pan

REMOVAL & INSTALLATION

1. Before servicing the vehicle, refer to the precautions at the beginning of this section.
2. Drain the engine oil.
3. Remove or disconnect the following:
- Negative battery cable
- 2 Top intake manifold bracket bolts
- Front 3 axle housing mounting bolts, 4WD only
- Left front bushing from the axle housing mount and lower the front axle housing
- Both gusset plates from the engine
- Transmission under cover
- Engine under cover
- Oil pan mounting bolts and using a scrapper tool separate the oil pan
- Oil pan
- Oil strainer assembly
- Oil baffle

To install:

4. Clean the engine block, oil pan and baffle pan surfaces of any gasket material.
5. Apply a continuous bead of Loctite Ultra Blue 587® silicone sealant around the baffle pan.
6. Install or connect the following:

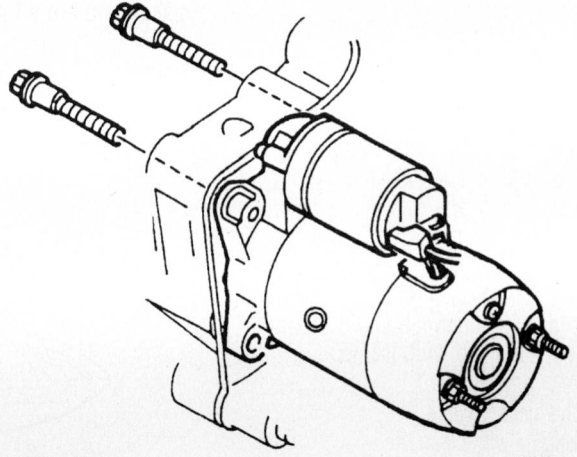

Exploded view of the starter

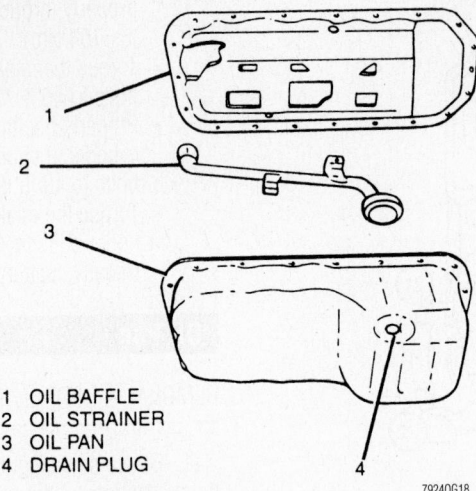

1 OIL BAFFLE
2 OIL STRAINER
3 OIL PAN
4 DRAIN PLUG

7924QG18

Exploded view of the oil pan assembly mounting

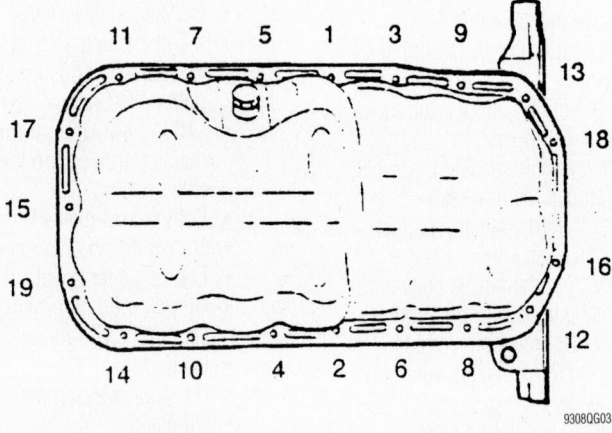

9308QG03

Tighten the oil pan bolts in sequence

- Oil baffle. Torque the bolt to 84 inch lbs. (9.5 Nm).
- Oil strainer. Torque the bolts to 84 inch lbs. (9.5 Nm).

7. Apply a continuous bead of Loctite Ultra Blue 587® silicone sealant around the oil pan.

- Oil pan. Torque the bolts to 84 inch lbs. (9.5 Nm).
- Transmission under cover. Torque the bolts to 84 inch lbs. (9.5 Nm).
- Gusset plates to the engine. Torque the bolts to 33 ft. lbs. (45 Nm).
- Engine under cover
- Front axle housing into position. When properly aligned, torque the bolts to 48 ft. lbs. (65 Nm).
- Intake manifold bracket bolts. Torque the bolts to 34 ft. lbs. (65 Nm).
- Negative battery cable

8. Fill the engine with clean oil.

9. Start the vehicle and check for leaks, repair if necessary.

Oil Pump

REMOVAL & INSTALLATION

➡The oil pump is externally-mounted, but still requires the removal of the oil pan to disconnect the oil pump strainer.

1. Before servicing the vehicle, refer to the precautions at the beginning of this section.

2. Properly relieve the fuel system pressure.

3. Drain the engine oil.

4. Drain the cooling system.

5. Remove or disconnect the following:
- Negative battery cable
- Alternator belt
- Fresh air duct from the radiator
- Upper radiator hose
- Clutch fan and shroud
- Splash guard
- Loosen the A/C drive belt
- Power steering belt
- Timing belt covers
- Lower timing belt pulley and lock bolt and place a support under the front axle
- Axle attaching bolts and lower the axle enough to gain access to the oil pan
- Transmission under cover
- Oil pan
- Oil pump

To install:

6. Clean the engine block, oil pan and baffle pan surfaces of any gasket material.

7. Apply a continuous bead of silicone sealant around the oil pump.

➡Do not allow sealant to get in the oil passages when applying sealant to the contact surface.

8. Install or connect the following:
- New O-ring and mount the oil pump to the engine. Torque the "A" bolts to 16 ft. lbs. (22 Nm) and the "B" bolts to 33 ft. lbs. (45 Nm).

9. Remove the upper and lower A/C compressor mounting bolts.

10. Loosen the A/C compressor bracket.
- Power steering pump bracket and hand tighten the bolts

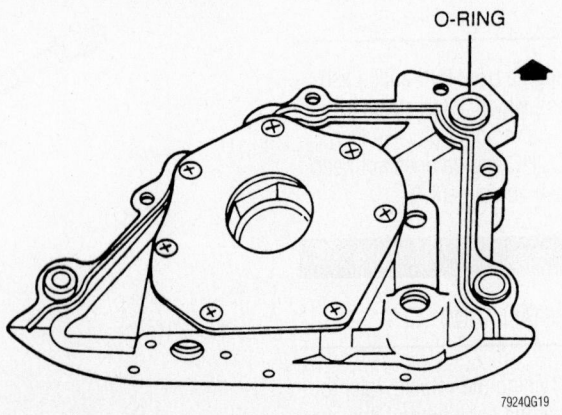

O-RING

7924QG19

Be sure the oil pump O-ring is in the proper location prior to installation

- A/C compressor bracket
- A/C compressor. Torque the mounting bolts to 17 ft. lbs. (23 Nm).
- Torque the power steering pump bracket bolts to 24 ft. lbs. (33 Nm).
- Power steering pump. Torque the bolts to 43 ft. lbs. (58 Nm).
- Timing belt gear on the crankshaft. Torque the large crank bolt to 119 ft. lbs. (162 Nm).
- Oil baffle after applying sealant to the mating surface. Torque the bolt to 84 inch lbs. (9.5 Nm).
- Oil strainer. Torque the bolts to 84 inch lbs. (9.5 Nm).
- Oil pan
- Transmission under cover. Torque the bolts to 84 inch lbs. (9.5 Nm).
- Both gusset plates. Torque the bolts to 33 ft. lbs. (45 Nm).
- Raise the front axle into position. When aligned properly, torque the bolts to 123 ft. lbs. (167 Nm).
- Timing belt and cover
- Alternator belt
- A/C and power steering belt and adjust as needed
- Splash shield
- Upper radiator hose
- Clutch fan and shroud as an assembly
- Air duct to the top of the radiator
- Engine under cover. Torque the bolts to 18 ft. lbs. (25 Nm).
- Negative battery cable

11. Properly fill the cooling system.
12. Fill the engine with clean oil.
13. Start the engine, check for leaks, and repair if necessary.

Rear Main Seal

REMOVAL & INSTALLATION

1. Before servicing the vehicle, refer to the precautions at the beginning of this section.
2. Drain the transmission fluid.
3. Remove or disconnect the following:
- Negative battery cable
- Transmission
- Clutch cover and disc, if equipped
- Flywheel, if equipped
- Rear cover
- Rear main oil seal

To install:
4. Coat the new seal with clean oil and press the seal into the cover.
5. Install or connect the following:
- Rear cover

- Flywheel onto the crankshaft. While holding the flywheel torque the bolts in sequence:
  a. Step 1: 30 ft. lbs. (41 Nm).
  b. Step 2: 60 ft. lbs. (81 Nm).
  c. Step 3: 73 ft. lbs. (99 Nm).
- Clutch disc and cover. Torque the bolts to 16 ft. lbs. (22 Nm).
- Transmission
- Negative battery cable

6. Fill the transmission to the proper level.
7. Start the vehicle and check for leaks, repair if necessary.

Timing Belt

REMOVAL & INSTALLATION

1. Disconnect the negative battery cable.
2. Properly relieve the fuel system pressure.
3. Remove the alternator drive belt.
4. Remove the fresh air duct from the top of the radiator.
5. Remove the upper radiator hose.
6. Remove the 4 attaching nuts to the clutch fan.

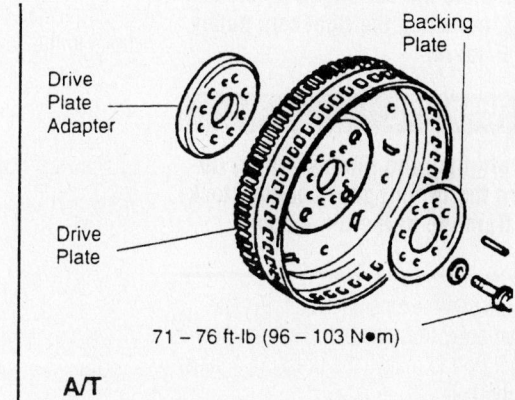

Drive Plate Adapter

Drive Plate

Backing Plate

71 – 76 ft-lb (96 – 103 N•m)

A/T

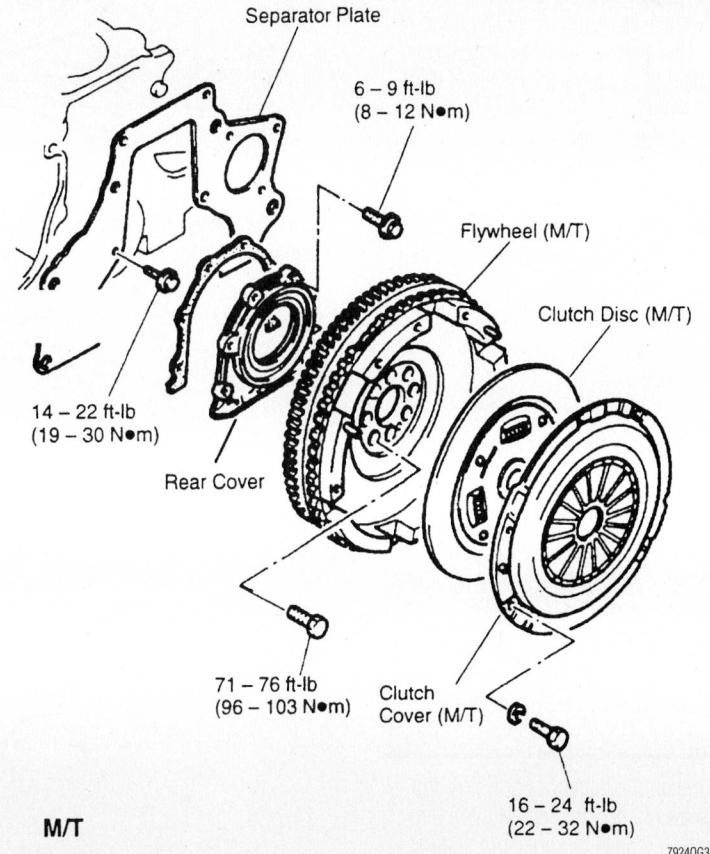

Separator Plate

6 – 9 ft-lb (8 – 12 N•m)

Flywheel (M/T)

Clutch Disc (M/T)

14 – 22 ft-lb (19 – 30 N•m)

Rear Cover

71 – 76 ft-lb (96 – 103 N•m)

Clutch Cover (M/T)

16 – 24 ft-lb (22 – 32 N•m)

M/T

79240G38

Exploded view of the rear main seal and related components

7. Remove the 5 fan shroud bolts. Remove the fan and shroud as an assembly.

8. Remove the 4 splash guard mounting bolts and the splash guard.

9. Loosen the lockbolts and loosen the air conditioning drive belt.

10. Loosen the power steering lock and mounting bolt. Remove the power steering belt.

11. Remove the 5 upper timing belt cover bolts and remove the cover.

12. Remove the 2 lower timing belt cover bolts and remove the cover.

13. Align the timing marks.

➡ When aligning the cam pulleys with the seal plate marks, align the left cam pulley I mark and the right cam pulley on the E mark.

✳✳ WARNING

When aligning the timing marks, do not turn the timing gear counterclockwise. Damage to the engine will occur.

14. Loosen the tensioner bolt. Pry the tensioner away from the belt. Tighten the tensioner bolt to relieve the pressure against the timing belt.

15. Remove the timing belt.

16. Remove the camshaft pulley attaching bolts. Use a driver placed through one of the holes in the pulley to prevent it from moving when the attaching bolt is removed. Remove and mark the pulleys.

17. Remove the lower timing belt pulley and locking bolt.

To install:

18. Install the camshaft pulleys. Tighten the bolts to 35 – 48 ft. lbs. (47 – 65 Nm).

19. Install the lower timing belt pulley and locking bolt. Tighten the bolt to 120 ft. lbs. (162 Nm).

20. If necessary, align the timing marks.

➡ When aligning the cam pulleys with the seal plate marks, align the left cam pulley "I" mark and the right cam pulley on the "E" mark.

✳✳ WARNING

When aligning the timing marks, do not turn the timing gear counterclockwise. Damage to the engine will occur.

21. Loosen the tensioner bolt. Pry the tensioner away from the belt. Tighten tensioner bolt to relieve the pressure against the timing belt.

22. Install the timing belt.

✳✳ WARNING

If any binding is felt when adjusting the timing belt tension by turning the crankshaft, STOP turning the engine, because the pistons may be hitting the valves.

23. Loosen the tensioner bolt and allow the tensioner to tighten the timing belt. Tighten the tensioner bolt 27 – 38 ft. lbs. (37 – 52 Nm).

24. Check the timing belt deflection. If there is more than 0.30 – 0.33 in. (7.5 – 8.5mm) replace the tensioner spring.

25. Install the 2 lower timing belt cover bolts to the cover.

26. Install the 5 upper timing belt cover bolts to the cover.

27. Install and adjust the air conditioning and power steering drive belts.

28. Install the splash guard.

29. Install and tighten the alternator belt.

30. Install the upper radiator hose.

31. Install the fan and shroud as an assembly.

32. Install the 4 attaching nuts to the clutch fan.

33. Install the 5 fan shroud bolts.

34. Install the fresh air duct to the top of the radiator.

35. Properly fill the cooling system.

36. Connect the negative battery cable.

37. Start the engine and check for leaks.

38. Road test the vehicle.

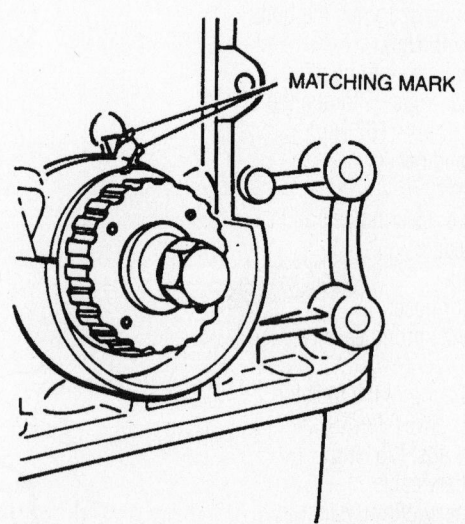

Align the crankshaft marks before removing the timing belt — KIA Sportage 2.0L (DOHC) engine

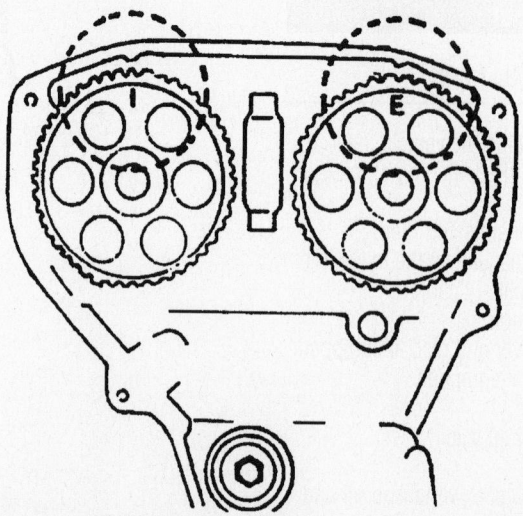

Proper alignment of the intake and exhaust camshaft pulley timing marks — KIA Sportage 2.0L (DOHC) engine

Piston and Ring

POSITIONING

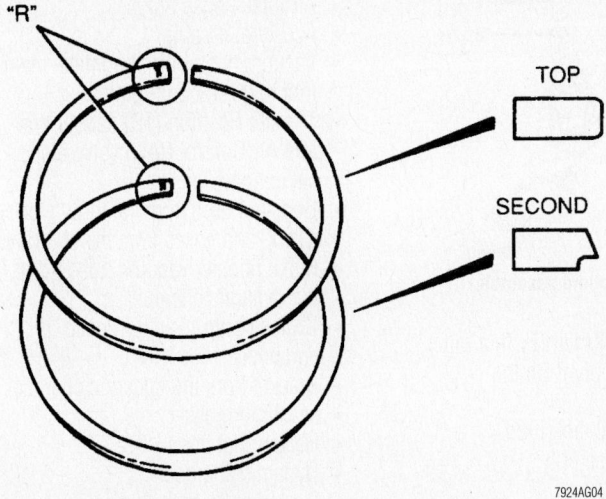

Kia 2.0L engine—compression ring positioning mark locations

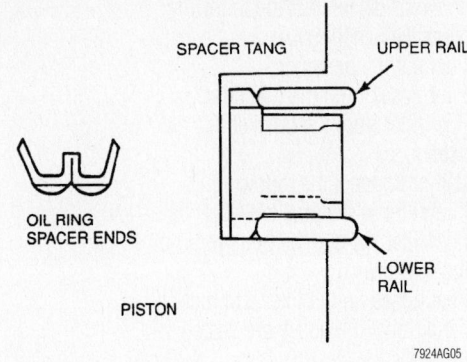

Kia 2.0L engine—oil control ring rail and spacer positioning

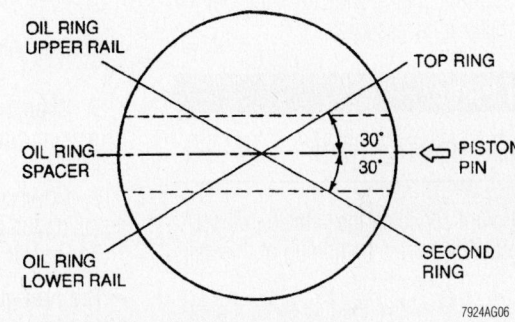

Kia 2.0L engine—piston ring end-gap spacing

FUEL SYSTEM

Fuel System Service Precautions

Safety is the most important factor when performing not only fuel system maintenance but any type of maintenance. Failure to conduct maintenance and repairs in a safe manner may result in serious personal injury or death. Maintenance and testing of the vehicle's fuel system components can be accomplished safely and effectively by adhering to the following rules and guidelines.

• To avoid the possibility of fire and personal injury, always disconnect the negative battery cable unless the repair or test procedure requires that battery voltage be applied.

• Always relieve the fuel system pressure prior to disconnecting any fuel system component (injector, fuel rail, pressure regulator, etc.), fitting or fuel line connection. Exercise extreme caution whenever relieving fuel system pressure to avoid exposing skin, face and eyes to fuel spray. Please be advised that fuel under pressure may penetrate the skin or any part of the body that it contacts.

• Always place a shop towel or cloth around the fitting or connection prior to loosening to absorb any excess fuel due to spillage. Ensure that all fuel spillage (should it occur) is quickly removed from engine surfaces. Ensure that all fuel soaked cloths or towels are deposited into a suitable waste container.

• Always keep a dry chemical (Class B) fire extinguisher near the work area.

• Do not allow fuel spray or fuel vapors to come into contact with a spark or open flame.

• Always use a back-up wrench when loosening and tightening fuel line connection fittings. This will prevent unnecessary stress and torsion to fuel line piping.

• Always replace worn fuel fitting O-rings with new. Do not substitute fuel hose or equivalent where fuel pipe is installed.

Fuel System Pressure

RELIEVING

1. Before servicing the vehicle, refer to the precautions at the beginning of this section.

2. Disconnect the fuel pump harness connector located behind the rear seat.

3. Start the engine and allow the engine to run out of fuel.

4. Once the engine has stalled, turn the key to the **OFF** position and connect the electrical connector.

5. Disconnect the negative battery cable so pressure cannot build up until work has been completed.

Fuel Filter

REMOVAL & INSTALLATION

1. Before servicing the vehicle, refer to the precautions at the beginning of this section.

2. Properly relieve the fuel system pressure.

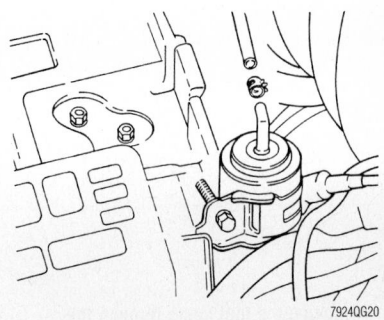

Fuel filter underhood mounting location

3. Remove or disconnect the following:
- Negative battery cable
- Fuel pump connector
- Fuel hoses from the fuel filter
- Fuel filter from the bracket

To install:

4. Install or connect the following:
- Fuel filter in the bracket
- Fuel filter. Torque the bolts to 95 ft. lbs. (129 Nm).
- Fuel hoses on the filter and make certain that the hoses are seated properly
- Fuel pump connector
- Negative battery cable

5. Start the engine and check for fuel leaks, repair if necessary.

Fuel Pump

REMOVAL & INSTALLATION

1. Before servicing the vehicle, refer to the precautions at the beginning of this section.

2. Properly relieve the fuel system pressure.

3. Release the catch for the back seat and tilt the seat out of the way.

4. Move the carpet behind the seat that covers the fuel pump access panel.

5. Remove or disconnect the following:
- Negative battery cable
- Fuel pump electrical connectors
- Bolt securing the ground wire
- Fuel pump access panel
- Hose clamps connecting the fuel hoses to the fuel pump
- Hoses from the fuel pump
- Screws securing the fuel pump to the fuel tank
- Gradually lift the fuel pump from the tank
- Plastic retaining bracket from the fuel pump assembly
- Fuel hose from the fuel pump
- Fuel tank pressure sensor

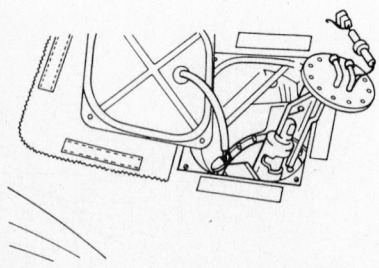

7924QG21

Removing the fuel pump through the access panel

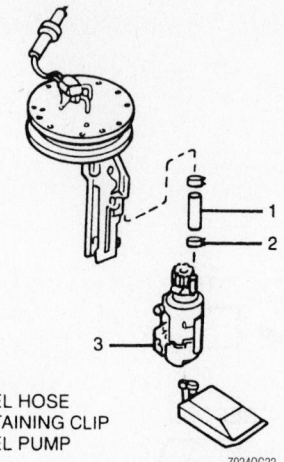

1 FUEL HOSE
2 RETAINING CLIP
3 FUEL PUMP

7924QG22

Exploded view of the fuel pump assembly

6. Wrap the fuel pump assembly in a rag before removing the assembly from the vehicle.

7. Cover or seal the fuel tank until installing the fuel pump assembly.

To install:

➡ **The fuel pump is part of the assembly and is replaced as a complete unit.**

8. Install or connect the following:
- Plastic mounting bracket to the fuel pump and secure the bracket to the pump with 4 screws
- Fuel hose to the pump and secure with a new clamp
- Fuel pump into the access port on top of the fuel tank
- Twist the fuel pump as necessary to properly position it in the fuel tank
- 8 screws around the top surface of the fuel pump
- Fuel hoses to the pump and secure with clamps
- Fuel tank pressure sensor
- Fuel pressure sensor electrical connector
- Ground wire and fuel pump electrical connector
- 2 halves of the fuel pump electrical connector
- Access plate and reposition the carpet at seat
- Negative battery cable

Fuel Injector

REMOVAL & INSTALLATION

1. Before servicing the vehicle, refer to the precautions at the beginning of this section.

2. Properly relieve the fuel system pressure.

3. Drain the cooling system.

4. Remove or disconnect the following:
- Negative battery cable
- Air intake hose assembly
- Breather hoses from the air intake duct
- Mass Air Flow (MAF) sensor electrical connector
- Air intake hose bracket
- Accelerator cable
- Vacuum hose from the intake manifold to the vacuum pipe
- Throttle Position (TP) sensor and Idle Air Control (IAC) valve electrical connectors
- Coolant hoses from the throttle body
- Clamp and hoses from the IAC valve
- Brake booster vacuum hose from the dynamic chamber
- Cruise control vacuum hose, if equipped
- Bracket from the dynamic chamber
- Manifold bracket
- Heater inlet hose
- Dynamic chamber
- Fuel hose from the pressure regulator
- Fuel injector rail clips
- Fuel rail
- Fuel injector insulators
- Pressure regulator
- Fuel injectors

To install:

5. Install or connect the following:
- Fuel injectors to the fuel rail
- New insulators
- New injector clips
- Fuel rail
- Clamps and air hose to the fuel rail
- Fuel hose to the pressure regulator
- Dynamic chamber with a new gasket. Torque the bolts to 16 ft. lbs. (22 Nm).
- IAC valve bracket bottom bolt
- IAC valve and TP sensor electrical connectors
- Heater inlet hose
- Cruise control hose, if equipped
- Vacuum hose to the pressure regulator
- Air hose and clamp to the air rail
- IAC valve vacuum hose
- Manifold bracket bolts
- Coolant hoses to the throttle body
- Vacuum hose to the vacuum pipe
- MAF sensor bracket
- MAF sensor electrical connector
- Accelerator cable
- Breather hoses
- Negative battery cable

6. Fill the cooling system.

7. Start the vehicle and check for leaks, repair if necessary.

DRIVE TRAIN

Transmission Assembly

REMOVAL & INSTALLATION

Manual

➡ **The removal of the manual transmission is virtually the same for 4WD and 2WD vehicles.**

1. Before servicing the vehicle, refer to the precautions at the beginning of this section.
2. Drain the transmission fluid.
3. Drain the transfer case.
4. Remove or disconnect the following:
   - Negative battery cable
   - Rear portion of the center console
   - Shift lever and transfer lever knobs
   - Slide the boot cover over the shifters and remove the center console
   - Shift lever
   - Transfer lever
   - Front driveshaft by removing the bolts at the front differential and the bolts at the transfer case, if equipped
   - Bolts from the rear differential flange and the center support. Pull the driveshaft out of the tail shaft housing, if equipped with a 4 x 4
   - Bolts from the rear differential flange and center support. Pull the driveshaft out of the tail shaft housing, if equipped with a 4 x 2
   - Back-up light electrical connector and the Vehicle Speed Sensor (VSS) electrical connectors and move the wire harness aside
   - Crankshaft Position (CKP) sensor from the transmission housing
   - Clutch release cylinder and move it aside
   - Front exhaust pipe bracket
   - Front lower transmission housing bolts
   - Transfer case side mount, if equipped and properly support the transmission
   - Transmission crossmember
   - transmission mount bolts
   - Starter from the front housing
   - Transmission and transfer case, if equipped

To install:

5. Install or connect the following:
   - Transmission into position at the rear of the engine
   - Wire harness along the right side of the transmission and route the VSS wire over the transmission to the rear of the control rod extension
   - Transmission to engine 14mm mounting bolts. Torque the bolts to 80 ft. lbs. (108 Nm).
   - Transmission to engine 10mm mounting bolts. Torque the bolts to 29 ft. lbs. (39 Nm).
   - Transmission to engine 6mm mounting bolts. Torque the bolts to 5 ft. lbs. (7 Nm).
   - Exhaust pipe to the bracket. Torque the bolts to 20 ft. lbs. (27 Nm).
   - Starter and ground wire. Torque the bolts to 29 ft. lbs. (39 Nm).
   - Transmission crossmember mount. Torque the bolts to 80 ft. lbs. (108 Nm).
   - Crossmember to the chassis. Torque the bolts to 32 ft. lbs. (44 Nm).
   - CKP sensor. Torque the bolt to 5 ft. lbs. (7 Nm).
   - 4WD indicator switch connector
   - Back-up light electrical connector
   - VSS electrical connector
   - Clutch release cylinder. Torque the bolts to 29 ft. lbs. (39 Nm).
   - Driveshaft to the rear differential flange, 4 x 4 only
   - Forward end of the driveshaft into the extension housing and attach the center support to the chassis
   - Shaft to the rear differential flange. Torque the bolts to 27 ft. lbs. (36 Nm).
   - Front driveshaft at the transfer case and install the bolts to the front differential, 4 x 4 only
   - Transfer case side mount, if equipped. Torque the bolts to 38 ft. lbs. (52 Nm).
   - Transfer case side mount to the chassis. Torque the bolts to 38 ft. lbs. (52 Nm).
   - Shift lever assembly. Torque the shift lever bracket bolts to 18 ft. lbs. (25 Nm).
   - Transfer lever assembly, if equipped. Torque the bolts to 18 ft. lbs. (25 Nm).
   - Dust cover plate over the shifter lever handles. Torque the bolts to 15 ft. lbs. (20 Nm).
   - Front console
   - Shifter lever knobs
   - Negative battery cable
   - Fill the transfer case.
6. Fill the transmission assembly.
7. Start the vehicle and check for leaks, repair if necessary.

Automatic

1. Before servicing the vehicle, refer to the precautions at the beginning of this section.
2. Drain the transmission fluid.
3. Drain the cooling system.
4. Remove or disconnect the following:
   - Negative battery cable
   - Automatic transmission control cable from the throttle body
5. Slide the front seats forward and remove the 2 rear shift console mounting screws and set the parking brake.
   - Rear console and slide the front seats rearward
   - Front console mounting screws and untie the shifter boot draw strings
   - Loosen the transfer case lock nut and remove the transfer case shifter lever knob
   - Power/economy switch electrical connector
   - Front console and shift the transfer lever to the 4L position
   - Transfer shift lever cover plate

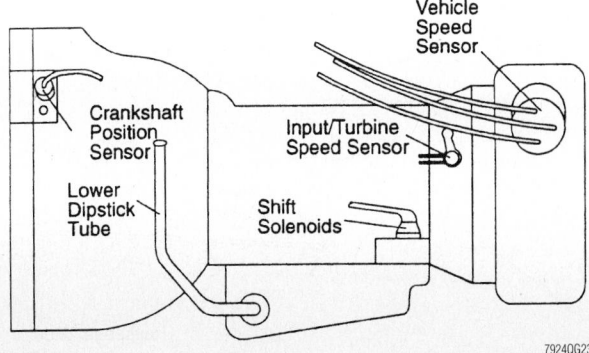

7924QG23

Automatic transmission wiring connections

- Transfer case shift lever assembly and place the transmission in the park position
- Selector lever nuts
- Split pin from the shift selector lever
- Shift selector rod and spring washers
- Electrical connectors from the base of the selector lever

➤**There is a fifth wire connection (Park Position). This wire is hard-wired, do not disconnect it.**

- Selector lever assembly
- Control cable from the throttle linkage and slide the cable from the bellcrank
- Upper dipstick tube from the lower dipstick tube
- Input/Turbine speed sensor from the top rear of the transmission
- Shift solenoids from the lower left side of the transmission
- Vehicle Speed Sensor (VSS) from the center of the transfer case
- Input speed sensor electrical connectors
- Matchmark the front and rear driveshafts. Remove the attaching bolts from both flanges and remove the driveshafts
- Oil cooler pipes at the transmission
- Starter
- Transfer case mounting bolts
- Transfer case nuts from the crossmember and support the transmission
- Crossmember
- Transmission to engine mounting bolts
- Front lower splash shield

- Left side lower gusset
- Torque converter inspection cover
- Torque converter to drive plate bolts
- Slide the transmission away from the engine and lower the transmission slightly
- Crankshaft Position (CKP) sensor attaching bolt and sensor
- 4WD, 4WD LOW indicator switches and the VSS from the transfer case

6. Lower the transmission. Be sure all wiring is clear and disconnected. Be sure the throttle cables come out without binding or attaching to anything.

To install:

7. Install or connect the following:
- Transmission into position to attach the sensor and indicator switch wiring. Be sure the throttle cables are guided into the engine compartment without binding or attaching to anything.
- 4WD, 4WD LOW indicator switches and the VSS to the transfer case
- Input/turbine speed sensor and CKP sensor. Torque the bolt to 12 ft. lbs. (16 Nm).
- Raise the transmission and position it to the engine. Install the upper housing bolts. Torque the bolts in the following sequence:

8. 10mm bolts to 38–60 ft. lbs. (57–81 Nm).

9. 12mm bolts to 51–65 ft. lbs. (69–88 Nm).

- Exhaust hanger and bracket. Torque the bolts to 38–60 ft. lbs. (57–81 Nm).
- Oil cooler pipe to the lower engine mount. Torque the bolt to 60 ft. lbs. (81 Nm).

- Crossmember. Torque the bolts to 23–34 ft. lbs. (31–46 Nm).
- Two transfer case mounting bolts located at the right center of the transfer case. Torque the bolts to 23–34 ft. lbs. (31–46 Nm).
- Four transfer case nuts to the crossmember. Torque the nuts to 23–34 ft. lbs. (31–46 Nm).
- Torque converter-to-drive plate bolts. Torque the bolts to 12–20 ft. lbs. (16–27 Nm).
- Torque converter inspection cover. Torque the bolts to 41–62 inch lbs. (5–7 Nm).
- Front splash guard. Torque the bolts to 41–62 inch lbs. (5–7 Nm).
- Starter. Torque bolts to 27–40 ft. lbs. (37–54 Nm).
- Left side lower gusset. Torque the bolts to 38–60 ft. lbs. (57–81 Nm).
- Right side lower gusset in 3 steps:

a. Install the bottom mounting bolts, but do not tighten.

b. Install the top bolt to the intake manifold support bracket and manifold. Tighten to 27–40 ft. lbs. (37–54 Nm).

c. Secure the manifold intake bracket by tightening the two attaching bolts to 38–60 ft. lbs. (57–81 Nm).

- Oil cooler pipes at the transmission. Torque the lines to 42–62 inch lbs. (5–7 Nm).
- Two oil cooling tube clamps to the lines
- Driveshafts with the matchmarks aligned. Torque the bolts to the differential flanges to 20–22 ft. lbs. (27–30 Nm) and the transfer case flange bolts to 36–43 ft. lbs. (49–59 Nm).
- Undercover splash shield. Torque the bolts to 42–62 inch lbs. (5–7 Nm).
- Upper dipstick tube to the lower tube and lower the vehicle

10. Provide automatic transmission control cable slack by gently pulling the cable to the left until the cable pin has rotated sufficiently to line the automatic transmission control cable up with the slot in the rear of the throttle body bellcrank.

11. Slide the automatic transmission control cable and cable pin into the bellcrank.

12. Tighten the locknut.
- Throttle kickdown cable to the mounting bracket
- Automatic transmission control cable to the throttle body
- Air silencer
- Shifter lever assembly to the trans-

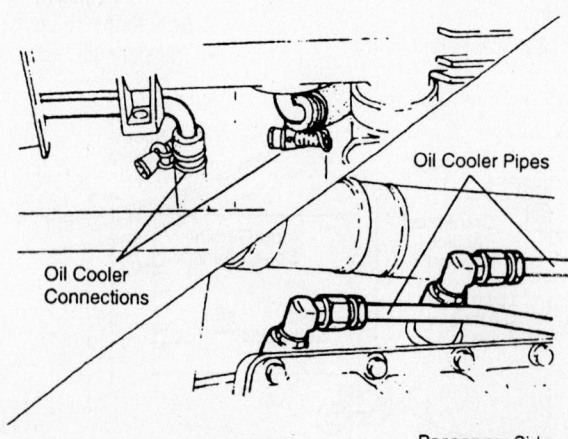

Radiator Side

Oil Cooler Connections

Oil Cooler Pipes

Passenger Side

7924QG24

Exploded view of the oil cooler pipe connections

fer case. Torque the bolts to 72–102 inch lbs. (22–28 Nm).
- Four wiring connectors under the shift selector lever
- Shift selector lever, do not exert any force when installing the shift selector lever
- Four shift selector lever nuts. Tighten to 72–102 inch lbs. (22–28 Nm).
- Shift rod and washers to the selector lever and install the split pin to the shift selector lever
- Power/Economy switch wiring connector
- Rear console
- Front console and tie the shift boot draw strings
- Negative battery cable

13. Fill the transmission to the proper level.

14. Fill the cooling system.

15. Start the vehicle, check for leaks, and repair if necessary.

Clutch

ADJUSTMENT

Clutch Pedal Height

1. Before servicing the vehicle, refer to the precautions at the beginning of this section.

2. Pull back the carpet to measure the distance from the firewall to the top of the pedal. The standard height is 9.84 in. (250 mm).

3. If adjustment is required, loosen the locknut and turn the stopper bolt.

4. After adjustment is made tighten the locknut to 12 ft. lbs. (16 Nm).

Clutch Pedal Free-Play

1. Before servicing the vehicle, refer to the precautions at the beginning of this section.

2. Depress the clutch pedal gently by hand and measure the amount of free-play (distance the pedal travels before resistance is felt). The proper amount of free-play is 0.5 in. (12.7mm). If the free-play is not within the proper specifications, continue with the procedure.

3. Measure from the floor pan to the middle point of the clutch pedal when the pedal is in the fully released position. The proper clutch pedal height is 7.25 in. (184mm).

4. If the pedal height is incorrect, loosen locknut (A) and turn the pedal adjusting bolt (B) until the proper height is achieved, then retighten the locknut.

5. Remeasure the free-play. If it is still out of specification, loosen the clutch pushrod locknut (C) and turn the pushrod (D) until the proper free-play is achieved. Tighten locknut (C) securely.

REMOVAL & INSTALLATION

1. Before servicing the vehicle, refer to the precautions at the beginning of this section.

2. Remove or disconnect the following:
- Transmission
- Pressure plate bolts and remove the clutch plate and disc

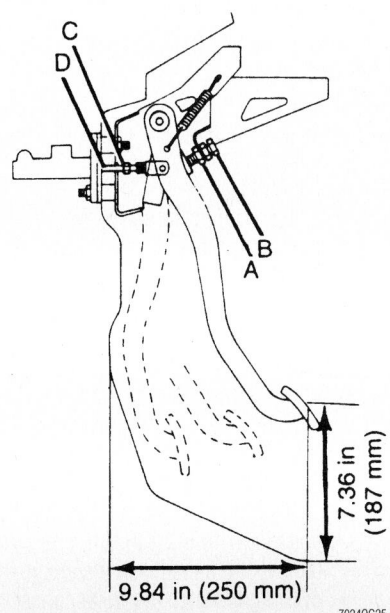

Clutch pedal height and free-play adjustment points

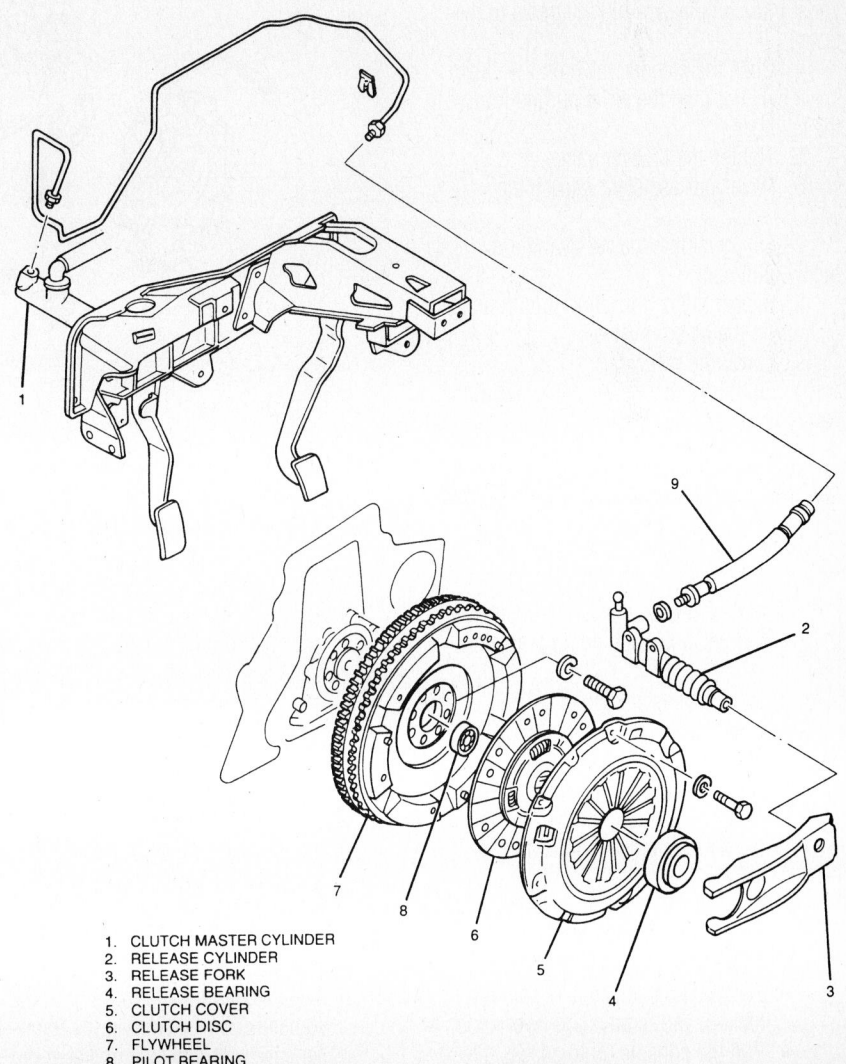

1. CLUTCH MASTER CYLINDER
2. RELEASE CYLINDER
3. RELEASE FORK
4. RELEASE BEARING
5. CLUTCH COVER
6. CLUTCH DISC
7. FLYWHEEL
8. PILOT BEARING
9. FLEXIBLE HOSE

Exploded view of the clutch assembly

To install:

3. Install or connect the following:
    - Clutch disc and plate using a centering tool
    - Clutch cover. Torque the bolts to 73 ft. lbs. (99 Nm) and remove the centering tool
    - Check the release bearing condition and lubricate or replace as necessary
    - Transmission

Hydraulic Clutch System

BLEEDING

1. Before servicing the vehicle, refer to the precautions at the beginning of this section.
2. With an assistant in the vehicle, raise and safely support the vehicle.
3. Have your assistant pump the clutch pedal three times and hold the pedal to the floor.
4. Open the bleeder valve on the clutch slave cylinder until the air is purged from the cylinder.
5. Tighten the bleeder valve.
6. Have your assistant release the clutch pedal.
7. Fill the clutch master cylinder if below minimum.
8. Repeat Steps 2 through 6 until no air exits from the bleeder valve.
9. Lower the vehicle.
10. Fill the clutch master cylinder fluid reservoir.

Transfer Case Assembly

REMOVAL & INSTALLATION

1. Before servicing the vehicle, refer to the precautions at the beginning of this section.
2. Drain the transfer case.
3. Remove or disconnect the following:
    - Negative battery cable
    - Two rear console mounting screws. Slide the console forward to clear the parking brake handle and set aside
    - Three mounting screws from the front console. Untie the shift boot draw strings and open the boot
    - Loosen the transfer case shift lever locknut and remove the lever knob
    - Pull the console up to access the Power/Economy switch wiring connector. Unplug the connector and remove the console

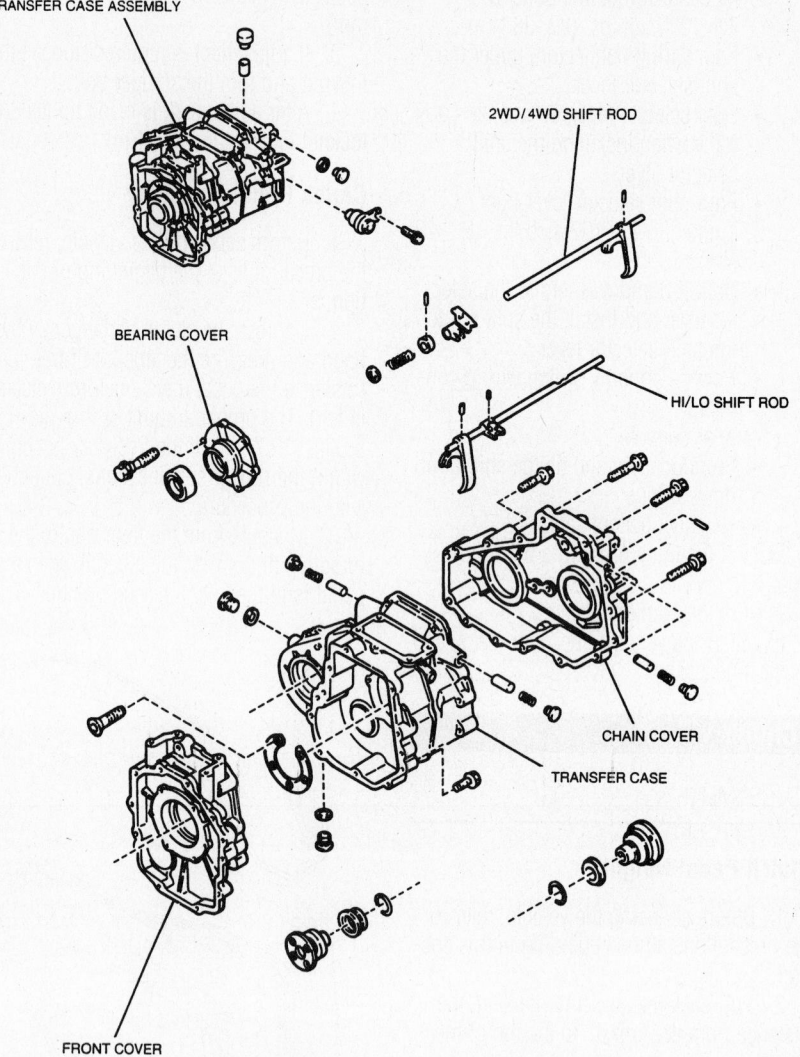

TRANSFER CASE ASSEMBLY

2WD/4WD SHIFT ROD

BEARING COVER

HI/LO SHIFT ROD

CHAIN COVER

TRANSFER CASE

FRONT COVER

9308QG05

Exploded view of the transfer case assembly

4. Shift the transfer lever to the 4L position.
    - Cover plate
    - Retaining bolts from the transfer case and lift the shifter lever assembly straight out and properly support the transmission
    - Matchmark the driveshafts at the flanges and remove the driveshafts
    - Crossmember bolts
    - 4WD light switch connector
    - Transfer case mounting bolts located at the right center of the transfer case
    - Transfer case nuts from the crossmember
    - Separate the transfer case from the transmission by striking the transfer case with a plastic mallet at the seal area
5. Lower the transfer case from the vehicle.

To install:

6. Install or connect the following:
    - Transfer case in position with a new gasket. Torque the bolts to 32 ft. lbs. (44 Nm).
    - Crossmember. Torque bolts to 32 ft. lbs. (44 Nm).
    - Transfer case mounting bolts located at the right center of the transfer case. Torque the bolts to 38 ft. lbs. (52 Nm).
    - Four transfer case nuts to the crossmember. Torque the nuts to 15 ft. lbs. (20 Nm).
    - Align the matchmarks on the driveshafts to the flanges. Torque the bolts to 27 ft. lbs. (36 Nm) and remove the transmission support
    - 4WD light switch electrical connector
    - VSS electrical connector
    - Shifter lever assembly. Torque

retaining bolts to 20 ft. lbs. (27 Nm).

- Cover plate. Torque the bolts to 15 ft. lbs. (20 Nm).
- Power/Economy switch wiring connector
- Front console
- Lever knobs
- Front console and tie the shift boot draw strings
- Slide the console over the parking brake handle
- Rear console
- Negative battery cable

7. Fill the transfer case to the proper level

8. Start the vehicle and check for leaks, repair if necessary.

Halfshaft

REMOVAL & INSTALLATION

1. Before servicing the vehicle, refer to the precautions at the beginning of this section.

2. Remove or disconnect the following:
- Negative battery cable
- Both front wheels
- Six free wheel hub bolts and remove the hub
- Snapring and spacer from the hub
- Carefully remove the fixed cam assembly
- Caliper from the brake rotor
- Upper control arm link lockbolt, spring washer and nut
- Tie rod end from the steering knuckle
- Loosen the drop link lower locknut
- Loosen the four upper drop link locknuts
- Spread open the drop link fork with a rubber mallet
- Matchmark the halfshaft and differential.
- Carefully pry the halfshaft from the differential
- Halfshaft

To install:

3. Install or connect the following:
- Halfshaft with the matchmarks aligned with the differential
- Torque the upper and lower drop link nuts to 36 ft. lbs. (49 Nm).
- Tie rod end to the steering knuckle. Torque the locknut to 27 ft. lbs. (36 Nm) and install a new cotter pin
- Upper control arm link lockbolt, spring washer and nut. Torque the bolt to 36 ft. lbs. (49 Nm).
- Fixed cam assembly
- Snapring and spacer in the hub
- Wheel hub and the six free wheel hub bolts. Torque the bolts in two passes, 14 ft. lbs. (17 Nm), then to 23 ft. lbs. (31 Nm).
- Front wheel assemblies
- Negative battery cable

Locking Hubs

REMOVAL & INSTALLATION

1. Before servicing the vehicle, refer to the precautions at the beginning of this section.

2. Remove or disconnect the following:
- Negative battery cable
- Wheel assembly
- Six free wheel hub bolts and remove the hub
- Snapring and spacer from the hub
- Carefully remove the fixed cam assembly

To install:

3. Install or connect the following:
- Fixed cam assembly
- Snapring and spacer in the hub

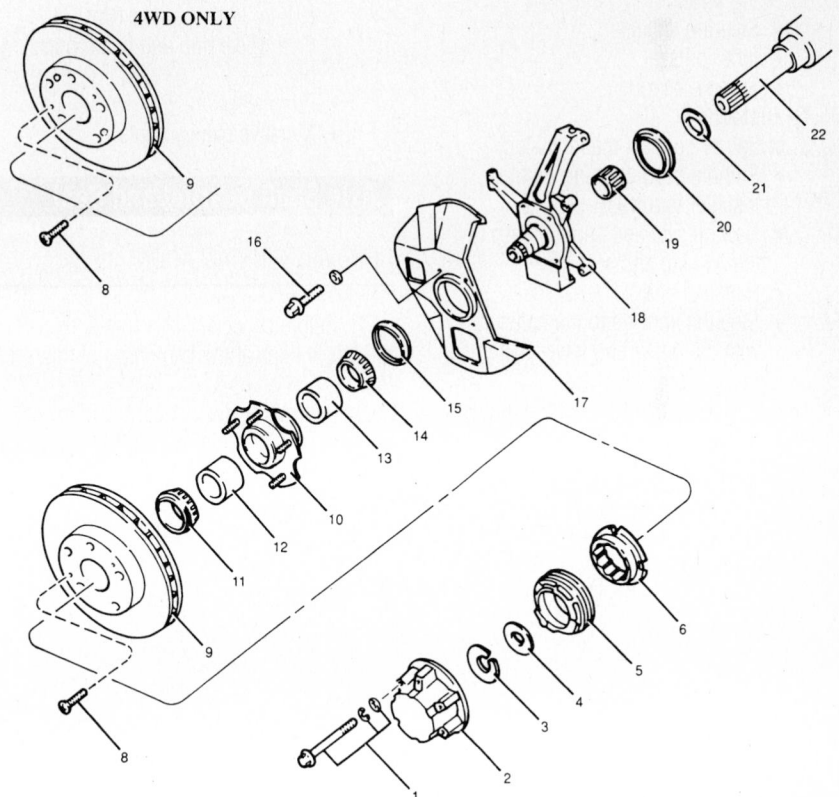

4WD ONLY

1	BOLT/WASHER	9	ROTOR	16	BOLT & SPRING WASHER
2	FREE WHEEL HUB BODY	10	WHEEL HUB	17	DUST COVER
3	SNAP RING	11	INNER BEARING INNER RACE	18	KNUCKLE
4	SPACER	12	INNER BEARING OUTER RACE	19	NEEDLE BEARING
5	FIXED CAM ASSEMBLY	13	OUTER BEARING OUTER RACE	20	OIL SEAL
6	LOCK NUT	14	OUTER BEARING INNER RACE	21	SPACER
8	SCREW	15	OIL SEAL	22	DRIVE SHAFT (LH)

7924QG28

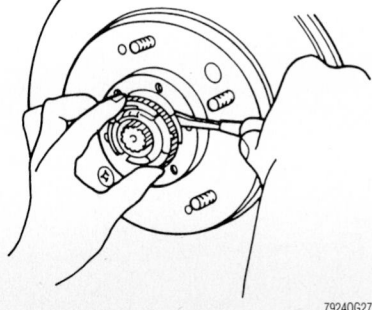

7924QG27

Removing the 4WD fixed cam assembly

Exploded view of the 4WD locking hub assembly

- Wheel hub and the six free wheel hub bolts. Torque the bolts in two passes, 14 ft. lbs. (17 Nm), then to 23 ft. lbs. (31 Nm).
- Wheel assembly
- Negative battery cable

Spindle Bearings

REMOVAL & INSTALLATION

1. Before servicing the vehicle, refer to the precautions at the beginning of this section.
2. Remove or disconnect the following:
   - Negative battery cable
   - Wheel assembly
   - Free wheel hub and bearing
   - Upper tie rod end from the steering knuckle
   - Lower control arm from the steering knuckle
   - Steering knuckle
   - Inner oil seal
   - Spindle bearing

To install:

3. Install or connect the following:
   - Spindle bearing using Bearing Installer tool K95B-5011-A
   - New oil seal and apply grease to the bearing and seal lip
   - Halfshaft end
   - Steering knuckle to the halfshaft with the upper and lower ball joints in the mounting holes
   - Lower control arm. Torque the lock

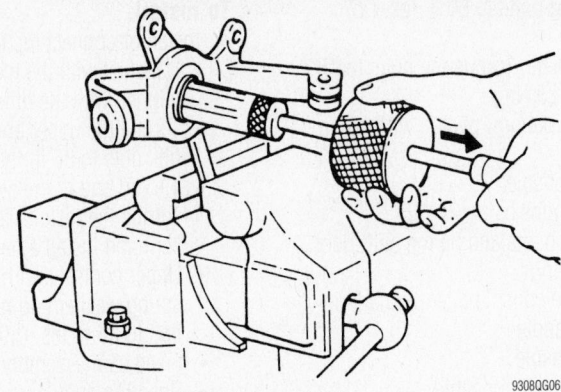

9308QG06

Remove the spindle bearing from the steering knuckle

nut to 110 ft. lbs. (148 Nm) and install a cotter pin
- Tie rod end to the steering knuckle. Torque the nut to 27 ft. lbs. (36 Nm) and install a cotter pin
- Upper control arm. Torque the lock bolt to 36 ft. lbs. (49 Nm).
- Free wheel hub and bearing assembly
- Front wheel
- Negative battery cable

Axle Shaft Bearing and Seal

REMOVAL & INSTALLATION

1. Before servicing the vehicle, refer to the precautions at the beginning of this section.
2. Remove or disconnect the following:

- Negative battery cable
- Wheel assembly
- Rear wheel
- Brake drum
- wheel speed sensor, if equipped
- Bearing retaining nuts
- Axle shaft and bearing

To install:

3. Install or connect the following:
   - Oil seal retainer, oil seal and wheel bearing to the axle shaft. Torque the new lock nut to 220 ft. lbs. (300 Nm).

➡**Turn the right side halfshaft lock nut clockwise and the left side halfshaft lock nut counterclockwise.**

- Axle shaft assembly into the axle housing. Torque the nuts to 74 ft. lbs. (100 Nm).

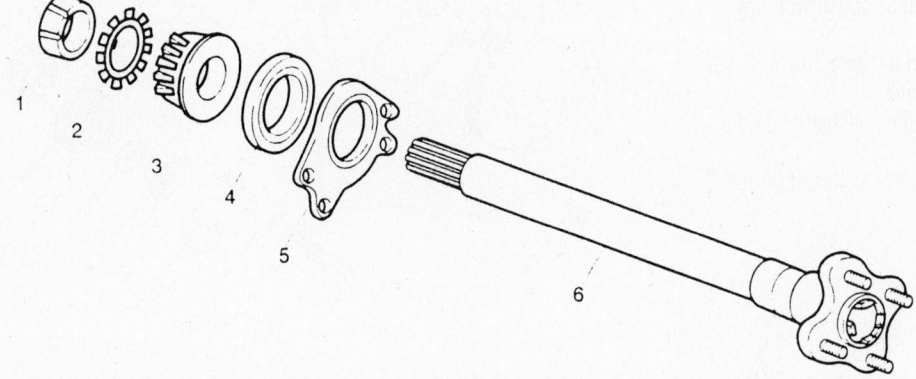

1. Lock Nut
2. Lock Washer
3. Bearing
4. Oil Seal
5. Oil Seal Retainer
6. Axle Shaft

9308QG07

Exploded view of the axle shaft, bearing and seal

- Wheel speed sensor, if equipped
- Brake drum
- Rear wheel assembly
- Negative battery cable

4. Check the fluid level and top off if necessary.

Pinion Seal

REMOVAL & INSTALLATION

1. Before servicing the vehicle, refer to the precautions at the beginning of this section.
2. Drain the gear oil.
3. Remove or disconnect the following:

- Negative battery cable
- Wheel assemblies
- Brake drums
- driveshaft

➡ **Use an inch lb. (Nm) torque wrench, measure and record the amount of torque required to maintain rotation of the pinion.**

- Pinion flange
- Pinion seal

To install:

4. Install or connect the following:
- New pinion seal lightly coated with clean gear oil
- Pinion flange
5. Rotate the pinion flange occasionally

while tightening the flange nut and make certain that the pinion bearings are seated properly.

6. Take several bearing preload torque readings. Tighten the flange nut to achieve the preload torque reading. The maximum torque reading should not exceed 14 inch lbs. (1.6 Nm).

- Driveshaft after aligning the match-marks
- Brake drums
- Wheel assemblies
- Negative battery cable

7. Fill the gear oil to the proper level.
8. Start the vehicle and check for leaks, repair if necessary.

STEERING AND SUSPENSION

Air Bag

PRECAUTIONS

Several precautions must be observed when handling the inflator module to avoid accidental deployment and possible personal injury.

1. Never carry the inflator module by the wires or connector on the underside of the module.
2. When carrying a live inflator module, hold securely with both hands, and ensure that the bag and trim cover are pointed away from you.
3. Place the inflator module on a bench or other surface with the bag and trim cover facing up.
4. With the inflator module on the bench, never place anything on or close to the module which may be thrown in the event of an accidental deployment.

DISARMING

1. Before servicing the vehicle, refer to the precautions at the beginning of this section.
2. Turn the ignition switch to the **LOCK** position.
3. Disconnect the negative battery cable.
4. Wait 10 minutes for the back-up power to discharge.

ARMING

Assuming the system components (air bag control module, sensors, air bag, etc.) are installed correctly and are in good working order, the system is armed whenever the battery positive and negative battery cables are connected.

If you have disarmed the air bag system for any reason, to rearm, be sure no one is in the vehicle (as an added safety measure), then connect the negative battery cable.

Power Steering Gear

ADJUSTMENT

1. Before servicing the vehicle, refer to the precautions at the beginning of this section.
2. Place the steering gear in a vise with protective jaws.
3. Place a torque wrench on the Pitman arm end of the shaft.
4. Loosen the locknut on the adjusting bolt.
5. Slowly turn the adjusting bolt to until the breakaway torque is 65 ft. lbs. (88 Nm).
6. Hold the adjusting bolt in position and tighten the locknut to 25 ft. lbs. (34 Nm).

REMOVAL & INSTALLATION

1. Before servicing the vehicle, refer to the precautions at the beginning of this section.
2. Center the steering wheel.
3. Drain the power steering fluid.
4. Remove or disconnect the following:
- Negative battery cable
- Left front wheel
- Pitman arm-to-centerlink attaching nut
- Separate the Pitman arm from the centerlink with a ball joint puller
- Power steering hoses
- Coolant recovery tank and the power steering reserve tank
- Set bolt from the intermediate shaft
- Intermediate shaft from the steering gear
- Steering gear-to-frame bolts
- Steering gear

To install:

5. Install or connect the following:

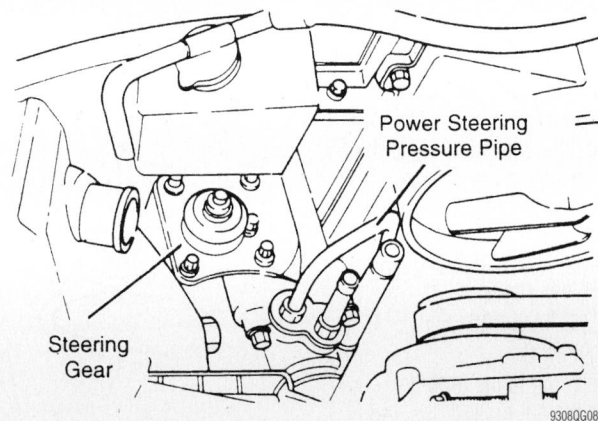

Exploded view of the steering gear

Power Steering Pressure Pipe

Steering Gear

9308QG08

- Position the steering gear to the frame. Torque the bolts to 159 ft. lbs. (215 Nm).
- Pitman arm to the centerlink. Torque the bolt to 36 ft. lbs. (49 Nm) and install a new cotter pin
- Intermediate shaft to the steering gear shaft. Torque the set bolt o 25 ft. lbs. (34 Nm).
- Power steering hoses/lines. Torque the fasteners to 29 ft. lbs. (34 Nm).
- Power steering reserve tank over the bracket and press until full engagement is reached
- Coolant recovery tank
- Left front wheel
- Negative battery cable

6. Fill the power steering fluid to the proper level and bleed the system.

7. Start the vehicle and check for leaks, repair if necessary.

8. Road test the vehicle to check that the steering wheel is straight.

Shock Absorber

REMOVAL & INSTALLATION

Front

The front shock absorber and coil spring are removed as a single unit.

1. Before servicing the vehicle, refer to the precautions at the beginning of this section.

2. Remove or disconnect the following:
- Negative battery cable
- Both front wheels
- Upper shock absorber mounting block nuts
- Stabilizer bar
- Drop link nut and allow the drop link to remain in place
- Both halves of the front fork
- Drop link
- Shock absorber and coil spring as an assembly

To install:

3. Install or connect the following:
- Coil spring to the shock absorber and position the assembly to the upper mounting block
- Upper mounting block nuts and hand tighten them
- Both front forks . Torque the bolts to 36 ft. lbs. (48 Nm).
- Drop link. Torque the nut to 145 ft. lbs. (197 Nm).
- Stabilizer bar to the drop link. Torque the nut to 36 ft. lbs. (48 Nm).

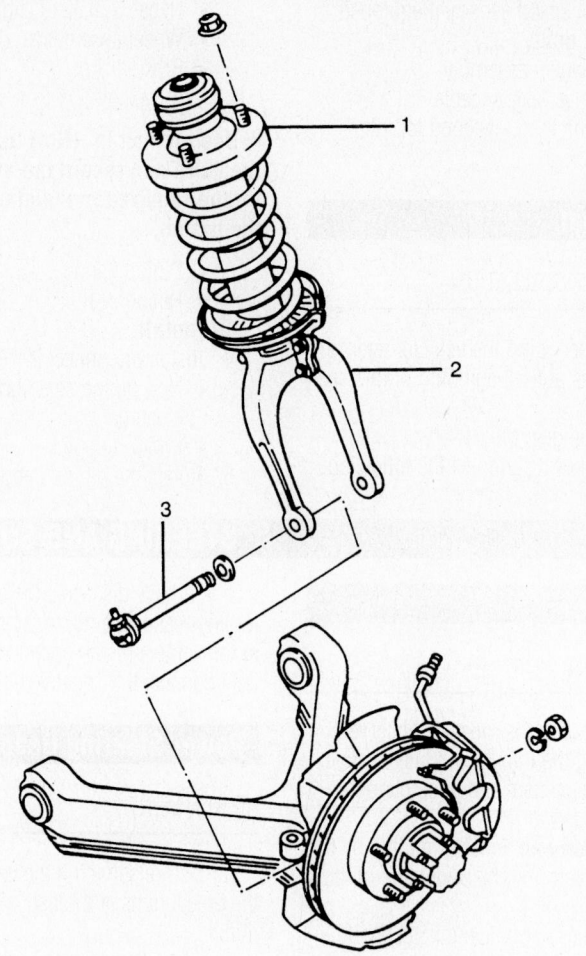

1 Front Shock Absorber & Coil Spring Assembly
2 Front Fork
3 Drop Link

9308QG09

Exploded view of the front shock absorber assembly

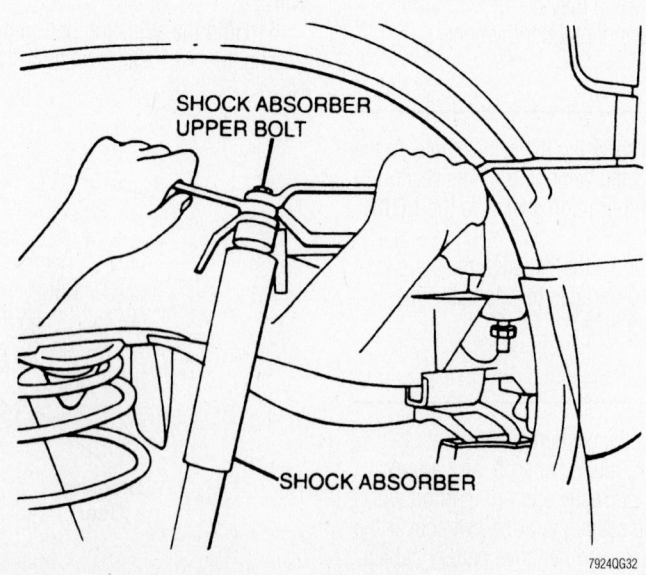

SHOCK ABSORBER UPPER BOLT

SHOCK ABSORBER

7924QG32

Upper shock absorber mounting nut

- Front wheels and torque the mounting block nuts to 18 ft. lbs. (25 Nm).
- Negative battery cable

Rear

1. Before servicing the vehicle, refer to the precautions at the beginning of this section.

2. Remove or disconnect the following:
- Negative battery cable
- Rear wheels
- Raise the rear axle with a floor jack to relax the shock absorbers and support the rear axle when the shock absorber is removed.
- Rear safety nut, upper nut and washer
- Upper rubber plate
- Lower bolt from the shock absorber
- Lower the rear axle housing
- Shock absorber

3. Remove the lower mounting bolt and remove the shock absorber.

To install:

4. Install or connect the following:
- Bottom washer and rubber cushion on the top of the shock absorber. Position the shock absorber on the vehicle
- Lower bolt
- Rubber cushion, washer and nut. Tighten to 53 ft. lbs. (72 Nm)

- Safety nut. Torque the lower bolt to 62 ft. lbs. (84 Nm).
- Rear wheels
- Negative battery cable

Coil Spring

REMOVAL & INSTALLATION

Front

1. Before servicing the vehicle, refer to the precautions at the beginning of this section.

2. Remove or disconnect the following:
- Negative battery cable
- Shock absorber and coil spring as an assembly
- Place the shock absorber in a vice
- Loosen the pivot rod nut several turns
- While still secured in a vise compress the coil spring
- Piston rod nut and disassemble the coil spring as needed

To install:

3. Install or connect the following:
- Bottom portion of the shock absorber in a vice and compress the coil spring
- End of the coil spring to the rubber seat and install the spring
- Assemble the dust boot and lower

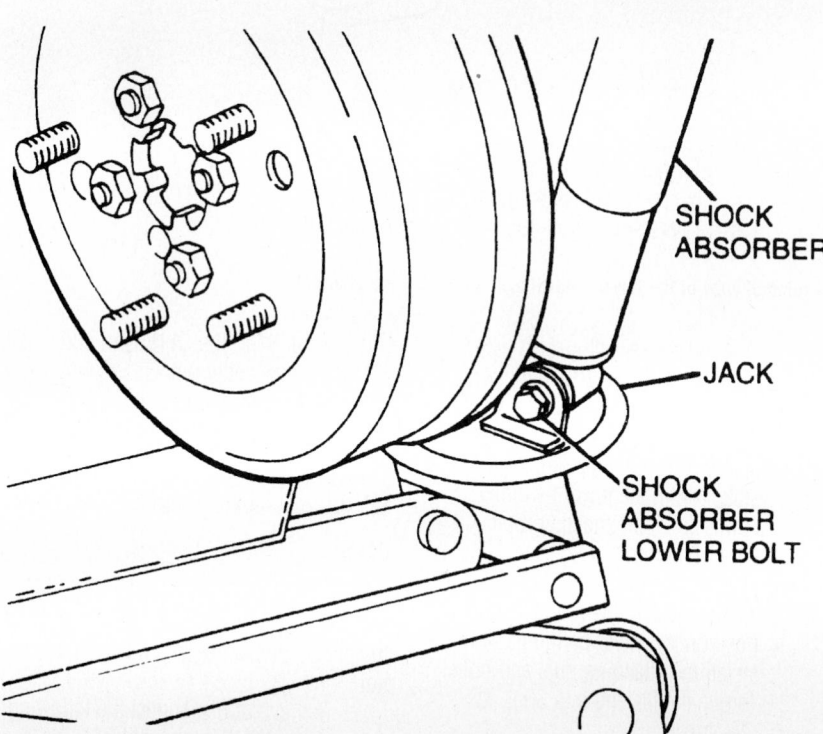

Lower shock absorber mounting bolt

7924QG33

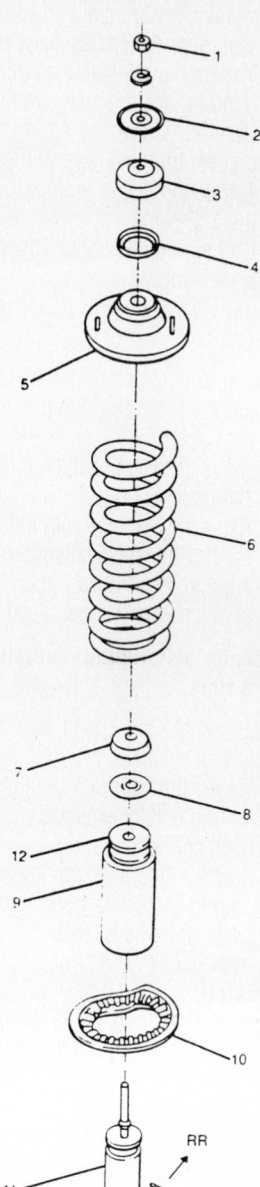

1. Nut
2. Upper retainer
3. Upper insulator
4. Centering washer
5. Spring seat
6. Coil spring
7. Lower insulator
8. Lower retainer
9. Dust boot
10. Rubber seat
11. Shock absorber
12. Front jounce stop

9308QG10

Exploded view of the front shock absorber and coil spring assembly

retainer, lower insulator, spring seat, boss, center washer, upper insulator and install the coil spring
- Hand tighten the piston rod nut.
- Carefully loosen the spring compressor tool and remove the tool
- Torque the piston rod nut to 31 ft. lbs. (42 Nm).
- Coil spring and shock absorber as an assembly

Rear

1. Before servicing the vehicle, refer to the precautions at the beginning of this section.
2. Remove or disconnect the following:
   - Both rear wheels
   - Raise the rear axle with a floor jack to relax the shock absorbers and support the rear axle when the shock absorber is removed.

➡**For easier installation, complete one side at a time.**

- Lower mounting bolt, then the shock absorber
- Lower the floor jack until the coil spring is fully expanded
- Coil spring
- Inspect the upper and lower rubber spring seats and jounce stop for wear or damage, replace if necessary

To install:

3. Install or connect the following:
   - Position the spring in the upper and lower saddles
   - Raise the floor jack and connect the lower shock absorber bolt. Torque the bolt to 62 ft. lbs. (84 Nm).
   - Rear wheels

Upper Ball Joint

REMOVAL & INSTALLATION

The upper ball joint is an integral part of the upper control arm. If the ball joint is worn, replacement of the upper control arm is necessary.

Lower Ball Joint

REMOVAL & INSTALLATION

1. Before servicing the vehicle, refer to the precautions at the beginning of this section.
2. Remove or disconnect the following:
   - Negative battery cable

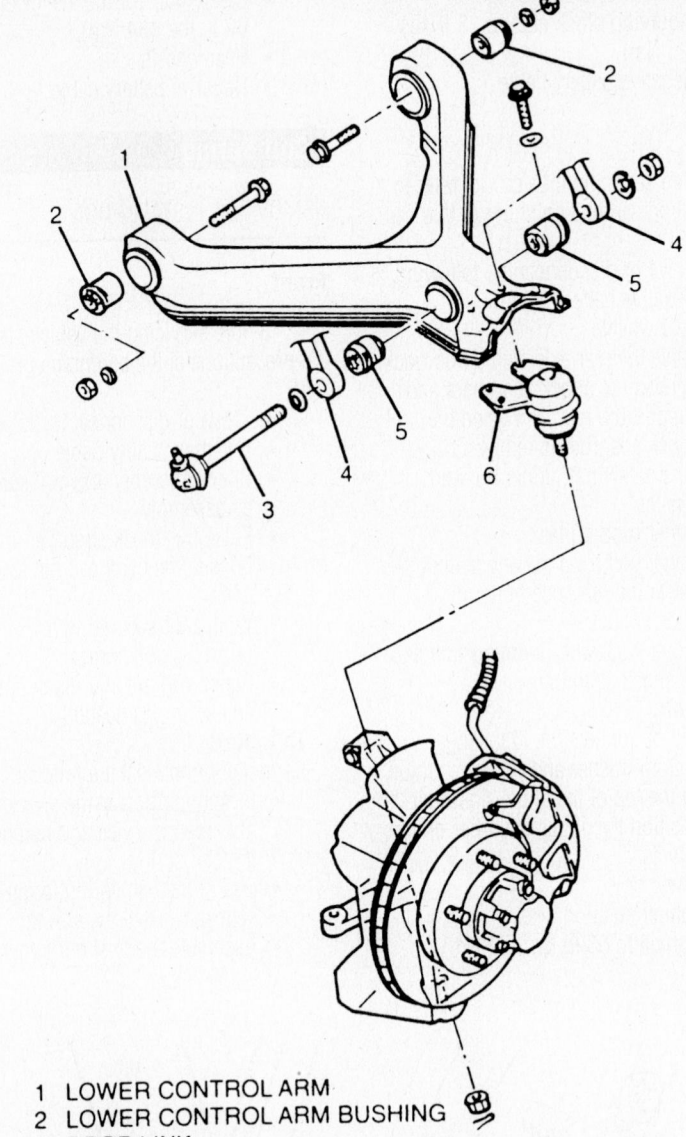

1. LOWER CONTROL ARM
2. LOWER CONTROL ARM BUSHING
3. DROP LINK
4. FRONT FORK
5. DROP LINK BUSHING
6. LOWER CONTROL ARM BALL JOINT

7924QG34

Exploded view of the lower control arm and ball joint assembly

- Front wheel assembly
- Cotter pin and lower ball joint nut
- Separate the lower ball joint from the spindle with a puller tool by prying down on the spindle to separate it from the lower ball joint
- Lower ball joint attaching bolts
- Lower ball joint

To install:

3. Install or connect the following:
   - Position the lower ball joint and install the attaching nuts and bolts. Torque the fasteners to 36 ft. lbs. (48 Nm).
   - Pry down and guide the spindle onto the lower ball joint. Torque the nut 87 ft. lbs. (118 Nm) and install a new cotter pin
   - Front wheel assembly.
   - Negative battery cable

Upper Control Arm

REMOVAL & INSTALLATION

1. Before servicing the vehicle, refer to the precautions at the beginning of this section.
2. Remove or disconnect the following:
   - Negative battery cable
   - Front wheel assembly

• Bolt securing the upper ball joint to the steering knuckle

➡**Note the matchmark setting on the upper control arm mounting bolts before removal.**

  • Upper control arm mounting bolts
  • Upper control arm from the vehicle.

To install:

3. Install or connect the following:
  • Position the upper control arms in the frame mounting. Hand tighten the bolts
  • Position the ball joint in the spindle. Torque the through bolt to 36 ft. lbs. (48 Nm).

➡**Be sure the slot in the ball joint aligns with the through-bolt during installation.**

  • Align the upper control arm bolts to the previous settings. Torque bolts to 62 ft. lbs. (108 Nm).

• Front wheel assembly
4. Check and adjust the alignment, if necessary.

UPPER CONTROL ARM BUSHING REPLACEMENT

1. Before servicing the vehicle, refer to the precautions at the beginning of this section.

2. Remove or disconnect the following:
  • Upper control arm assembly
  • Secure the control arm in a vise
  • Using a standard press, remove the bushing

To install:

3. Install or connect the following:
  • Lubricate the new bushing and press it into the upper control arm
  • Upper control arm to the vehicle
4. Check the wheel alignment and adjust if necessary.

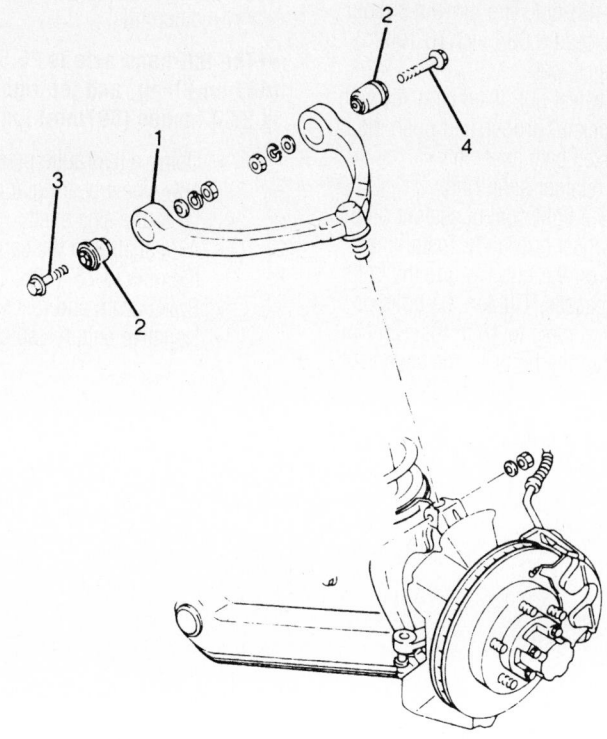

1 UPPER CONTROL ARM
2 UPPER CONTROL ARM BUSHING
3 FRONT SPINDLE
4 REAR SPINDLE

7924QG31

Exploded view of the upper control arm assembly

Lower Control Arm

REMOVAL & INSTALLATION

1. Before servicing the vehicle, refer to the precautions at the beginning of this section.
2. Remove or disconnect the following:
  • Negative battery cable
  • Front wheel assembly
  • Stabilizer bar
  • Driveshaft from the differential and steering knuckle
  • Shock absorber and coil spring from the front half of the fork
  • Cotter pin and castle nut from the lower control arm ball joint
  • Lower control arm ball joint from the steering knuckle
  • Lower control arm bushing bolts from the front frame crossmember brackets
  • Lower control arm

To install:

3. Install or connect the following:
  • Lower control arm to the front frame crossmember brackets and hand tighten the bushing bolts
  • Lower control arm ball joint and bolt to the steering knuckle. Torque the bolt to 87 ft. lbs. (118 Nm).
  • New cotter pin and castle nut
  • Torque the lower control arm bushing bolts to 206 ft. lbs. (280 Nm).
  • Front fork halves to the shock absorber and coil spring and position the lower portion over the lower control arm drop link holes
  • Drop link. Torque the nut to 145 ft. lbs. (197 Nm).
  • Driveshaft to the front differential and steering knuckle
  • Stabilizer bar
  • Front wheel assembly
  • Negative battery cable
4. Check and adjust the front wheel alignment if necessary.

LOWER CONTROL ARM BUSHING REPLACEMENT

1. Before servicing the vehicle, refer to the precautions at the beginning of this section.
2. Remove or disconnect the following:
  • Lower control arm assembly
  • Secure the control arm in a vise
  • Using a standard press, remove the bushing

To install:

3. Install or connect the following:

- Lubricate the new bushing and press it into the lower control arm
- lower control arm to the vehicle

4. Check the wheel alignment and adjust if necessary.

Wheel Bearings

ADJUSTMENT

Front

1. Remove or disconnect the following:
2. Before servicing the vehicle, refer to the precautions at the beginning of this section.
   - Negative battery cable
   - Front wheel
   - Brake caliper from the rotor and hang it out of the way
3. Attach a dial indicator to the axle hub and measure the bearing play
4. If the play exceeds.004 inch (.10mm), check and adjust locknut torque. The bolts should be tightened to 23 ft. lbs. (31 Nm).

Rear

The rear wheel bearings are not adjustable.

REMOVAL & INSTALLATION

Front

1. Before servicing the vehicle, refer to the precautions at the beginning of this section.
2. Remove or disconnect the following:
   - Negative battery cable
   - Front wheel

- Free wheel hub body
- Brake caliper
- Brake rotor
- Wheel bearing and using a screw driver, pry out the oil seal
- Inner and outer bearings
- Using a drift punch, remove the inner and outer bearing race

To install:

3. Pack the new bearings with grease.
4. Install or connect the following:
   - Inner bearing and race and a new oil seal
   - Outer bearing and race and secure the dust cover with four screws
   - Apply grease to the new bearings and the lip of the oil seal
   - Hub assembly in the steering knuckle
   - Screw the locknut against the hub assembly until there is 10 inch lbs. (1.3 Nm) of preload on the hub
   - Brake rotor and retaining screws
   - Attach a run-out gauge to check the rotor run-out. The run-out should not exceed 0.004 inch (0.10mm)
   - Brake caliper
   - Free wheel hub fixed cam key with the locknut groove and push the on the fixed cam assembly
   - Axle retainer snap ring
   - Apply a light coat of sealant on the free wheel hub body. Install the body on the hub. Torque the bolts in 2 passes. Tighten the bolts on the first pass to 18 ft. lbs. (25 Nm). Tighten the bolts on the second pass to 23 ft. lbs. (31 Nm).

- Negative battery cable

Rear

1. Before servicing the vehicle, refer to the precautions at the beginning of this section.
2. Remove or disconnect the following:
   - Negative battery cable
   - Rear wheels
   - Brake drum
   - Oil seal retainer flange

➡ **The axle shafts are different from side-to-side, mark the to ensure they are returned to the proper side.**

- Using a slide hammer, remove the axle shaft assembly
- Using a hydraulic press, remove the bearing collar and bearing from the axle
- oil seal from the differential

To install:

3. Install or connect the following:
   - Using the appropriate seal driver, install the new axle seal into the differential

➡ **The left-hand axle is 25.5 inches (647mm) long, and the right-hand axle is 27.4 inches (697mm) long.**

- Using a hydraulic press, install the new wheel bearing and retainer collar to the axle shaft
- Axle shaft into the carrier. Torque the nuts to 75 ft. lbs. (100 Nm).
- Brake drum and rear wheels
- Negative battery cable

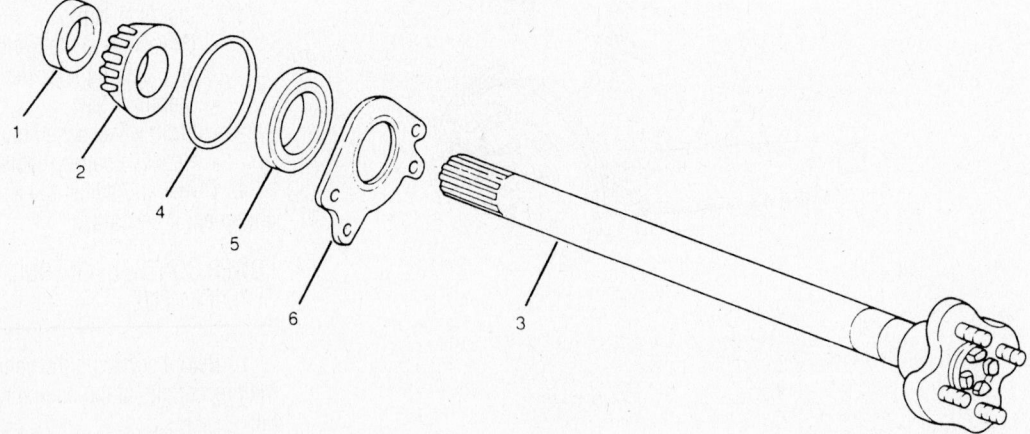

1 Bearing Collar
2 Bearing
3 Axle Shaft
4 Rib Ring
5 Oil Seal
6 Oil Seal Retainer

7924QG35

Rear axle bearing component identification

BRAKES

Brake Caliper

REMOVAL & INSTALLATION

Sportage

1. Raise and safely support the vehicle.
2. Remove the front wheels.
3. Remove the 2 caliper bolts and lift the caliper from the disc.
4. Disconnect the brake fluid flex line

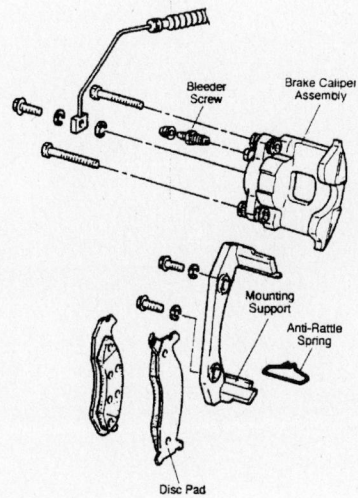

Exploded view of the front disc brake assembly—Sportage

by removing the retaining bolt if the caliper is to be replaced.

To install:

5. Seat the caliper piston using a C-clamp.
6. Connect the brake fluid flex line bolt to the caliper. Tighten the bolt to 17 ft. lbs. (23.5 Nm).
7. Position the caliper over the disc assembly. Install the caliper bolts. Tighten to 72 ft. lbs. (98 Nm).
8. Install the front wheels. Tighten the lug nuts 77 ft. lbs. (99 Nm).
9. Bleed the hydraulic system if the flex hoses were removed.
10. Lower the vehicle.

Disc Brake Pads

REMOVAL & INSTALLATION

1. Raise and safely support the vehicle.
2. Remove the front wheels.
3. Remove the 2 caliper bolts and lift the caliper from the disc.
4. Slide the disc pads off the caliper bracket.

To install:

5. Clean the caliper bracket contact surface with a wire brush and lightly coat with assembly lube.
6. Position the disc pads.

7. Position the caliper over the disc assembly. Install the caliper bolts. Tighten to 72 ft. lbs. (98 Nm).
8. Install the front wheels. Tighten the lug nuts 77 ft. lbs. (99 Nm).
9. Bleed the hydraulic system if the flex hoses were removed.
10. Lower the vehicle.

Brake Drums

REMOVAL & INSTALLATION

1. Raise and safely support the vehicle.
2. Remove the rear wheels.
3. Apply the parking brake.
4. Remove the 4 attaching nuts.
5. Release the parking brake and remove the brake drum.

Installation is the reverse of the removal procedure.

Brake Shoes

REMOVAL & INSTALLATION

1. Raise and safely support the vehicle.
2. Remove the rear wheels.
3. Apply the parking brake.
4. Remove the 4 attaching nuts.
5. Release the parking brake and remove the brake drum.

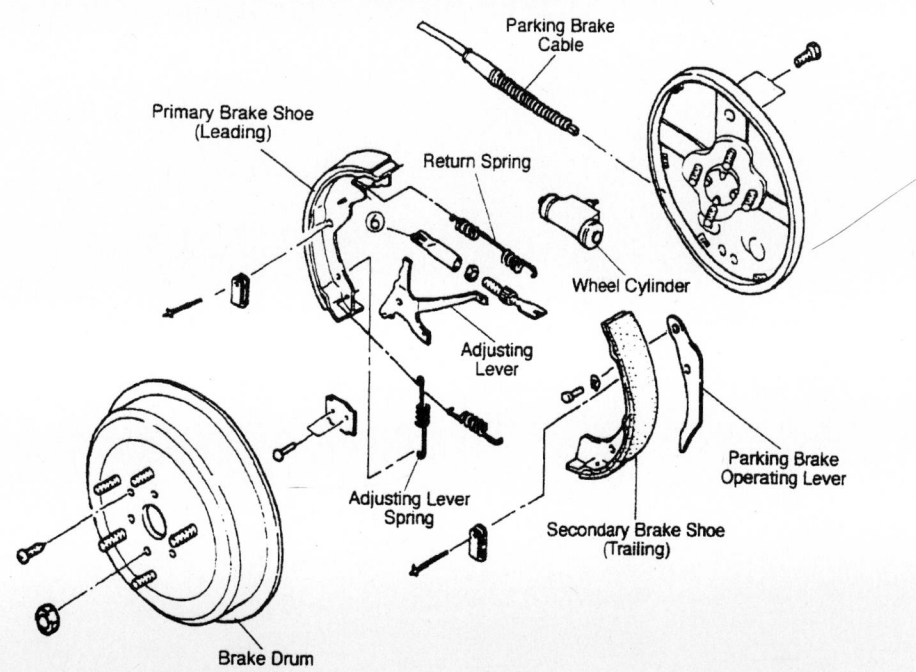

Exploded view of the rear drum brake assembly—Sportage

6. Remove the spring from the secondary shoe to the adjusting lever.

7. Remove the adjusting lever.

8. Remove the return spring above the star adjusting wheel.

9. Turn the starwheel clockwise to relieve tension on the brake shoes.

10. Remove the starwheel.

11. Remove the hold-down pin clips.

12. Remove the primary shoe.

13. Disconnect the C-clip and pin attaching the parking brake lever to the secondary brake shoe.

To install:

14. Lubricate the backing plate contact points.

15. Connect the parking brake lever to the secondary brake shoe.

16. Attach the primary and secondary brake shoes to the backing plate.

17. Install the starwheel.

18. Install the return spring above the starwheel.

19. Install the adjusting lever.

20. Install the spring from the secondary shoe to the adjusting lever. Be sure the lever contacts the starwheel.

21. Install the brake drum and retaining nuts.

22. Adjust the rear brakes through the slot in the rear of the backing plate.

23. Install the wheels. Tighten the wheel lugs to 77 ft. lbs. (99 Nm).

24. Lower the vehicle.

LEXUS

ES 300 • ES 330 • IS 300 • GS 300 • GS 430 • LS 430

22

SPECIFICATION AND MAINTENANCE CHARTS

ENGINE AND VEHICLE IDENTIFICATION

Code ①	Liters (cc)	Cu. In.	Cyl.	Fuel Sys.	Engine Type	Eng. Mfg.	Code ②	Year
		Engine					Model Year	
1MZ-FE	3.0 (2995)	183	6	SFI	DOHC	Toyota	1	2001
1UZ-FE	4.0 (3969)	242	8	SFI	DOHC	Toyota	2	2002
2JZ-GE	3.0 (2997)	183	6	SFI	DOHC	Toyota	3	2003
3UZ-FE	4.3 (4293)	262	8	SFI	DOHC	Toyota	4	2004
3MZ-FE	3.3 (3311)	202	6	SFI	DOHC	Toyota	5	2005

SFI: Sequential Multi-port Fuel Injection

DOHC: Double Overhead Camshaft

① Located on the timing belt cover

② 10th digit of the VIN

67162-LEXU-C01

GENERAL ENGINE SPECIFICATIONS

All measurements are given in inches.

Year	Model	Engine Displacement Liters (cc)	Engine ID	Fuel System Type	Net Horsepower @ rpm	Net Torque @ rpm (ft. lbs.)	Bore x Stroke (in.)	Compression Ratio	Oil Pressure @ rpm
2001	ES 300	3.0 (2995)	1MZ-FE	SFI	210@5200	220@4400	3.44x3.27	10.5:1	43-78@3000
	GS 300	3.0 (2997)	2JZ-GE	SFI	220@5800	220@3800	3.39x3.39	10.0:1	47-84@3000
	GS 430	4.3 (4264)	3UZ-FE	SFI	300@5300	325@4500	3.58x3.25	10.4:1	43-85@3000
	IS 300	3.0 (2997)	2JZ-GE	SFI	215@5800	218@3800	3.39x3.39	10.0:1	47-84@3000
	LS 430	4.3 (4264)	3UZ-FE	SFI	290@5300	320@4500	3.58x3.25	10.4:1	43-85@3000
2002	ES 300	3.0 (2995)	1MZ-FE	SFI	210@5200	220@4400	3.44x3.27	10.5:1	43-78@3000
	GS 300	3.0 (2997)	2JZ-GE	SFI	220@5800	220@3800	3.39x3.39	10.0:1	47-84@3000
	GS 430	4.3 (4264)	3UZ-FE	SFI	300@5300	325@4500	3.58x3.25	10.4:1	43-85@3000
	IS 300	3.0 (2997)	2JZ-GE	SFI	215@5800	218@3800	3.39x3.39	10.0:1	47-84@3000
	LS 430	4.3 (4264)	3UZ-FE	SFI	290@5300	320@4500	3.58x3.25	10.4:1	43-85@3000
2003	ES 300	3.0 (2995)	1MZ-FE	SFI	210@5200	220@4400	3.44x3.27	10.5:1	43-78@3000
	GS 300	3.0 (2997)	2JZ-GE	SFI	220@5800	220@3800	3.39x3.39	10.0:1	47-84@3000
	GS 430	4.3 (4264)	3UZ-FE	SFI	300@5300	325@4500	3.58x3.25	10.4:1	43-85@3000
	IS 300	3.0 (2997)	2JZ-GE	SFI	215@5800	218@3800	3.39x3.39	10.0:1	47-84@3000
	LS 430	4.3 (4264)	3UZ-FE	SFI	290@5300	320@4500	3.58x3.25	10.4:1	43-85@3000
2004	ES 330	3.3 (3311)	3MZ-FE	SFI	225@5600	240@3600	NA	10.8:1	36-78@3000
	GS 300	3.0 (2997)	2JZ-GE	SFI	220@5800	220@3800	3.39x3.39	10.0:1	47-84@3000
	GS 430	4.3 (4264)	3UZ-FE	SFI	300@5300	325@4500	3.58x3.25	10.4:1	43-85@3000
	IS 300	3.0 (2997)	2JZ-GE	SFI	215@5800	218@3800	3.39x3.39	10.0:1	47-84@3000
	LS 430	4.3 (4264)	3UZ-FE	SFI	290@5300	320@4500	3.58x3.25	10.4:1	43-85@3000
	SC 430	4.3 (4293)	3UZ-FE	SFI	300@5600	325@3400	3.58x3.25	10.4:1	43-85@3000

NA: Information not available

67162-LEXU-C02

ENGINE TUNE-UP SPECIFICATIONS

Year	Engine Displacement Liters (cc)	Engine ID	Spark Plug Gap (in.)	Ignition Timing (deg.)	Fuel Pump (psi)	Idle Speed (rpm)	Valve Clearance Intake	Valve Clearance Exhaust
2001	3.0 (2995)	1MZ-FE	0.043	8-12B ①	44-50	650-750	0.006-0.010	0.010-0.014
	3.0 (2997)	2JZ-GE	0.043	8-12B ②	44-50	650-750	0.006-0.010	0.010-0.014
	4.3 (4264)	3UZ-FE	0.043	8-12B ③	44-50	700-800	0.006-0.010	0.010-0.014
2002	3.0 (2995)	1MZ-FE	0.039-0.043	8-12B ①	44-50	650-750	0.006-0.010	0.010-0.014
	3.0 (2997)	2JZ-GE	0.039-0.043	8-12B ②	44-50	650-750	0.006-0.010	0.010-0.014
	4.3 (4264)	3UZ-FE	0.043	8-12B ③	44-50	700-800	0.006-0.010	0.010-0.014
2003	3.0 (2995)	1MZ-FE	0.039-0.043	8-12B ①	44-50	650-750	0.006-0.010	0.010-0.014
	3.0 (2997)	2JZ-GE	0.039-0.043	8-12B ②	44-50	650-750	0.006-0.010	0.010-0.014
	4.3 (4264)	3UZ-FE	0.043	8-12B ③	44-50	700-800	0.006-0.010	0.010-0.014
2004	3.0 (2997)	2JZ-GE	0.039-0.043	8-12B ①	44-50	650-750	0.006-0.010	0.010-0.014
	3.3 (3311)	3MZ-FE	0.043-0.047	8-12B ③	44-50	650-750	0.006-0.010	0.010-0.014
	4.3 (4293)	3UZ-FE	0.043	8-12B ③	44-50	700-800	0.010-0.024	0.012-0.026

NOTE: The Vehicle Emission Control Information label often reflects specification changes made during production.

The label figures must be used if they differ from those in this chart.

B: Before top dead center

① Terminals TE1 and E1 of check connector must be connected

② Terminals TC and E1 of check connector must be connected

③ LS 430: Terminals TC and CG of check connector must be connected

　GS 430/300: Terminals TC and E1 of check connector must be connected

　ES 330: Terminals TC and CG of check connector must be connected

67162-LEXU-C03

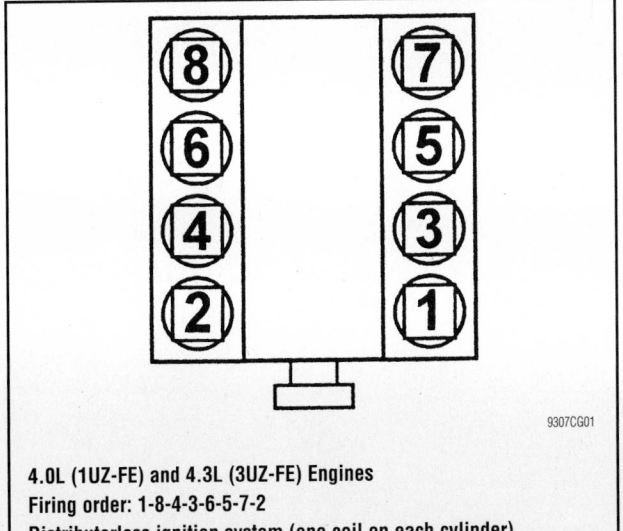

4.0L (1UZ-FE) and 4.3L (3UZ-FE) Engines
Firing order: 1-8-4-3-6-5-7-2
Distributorless ignition system (one coil on each cylinder)

9307CG01

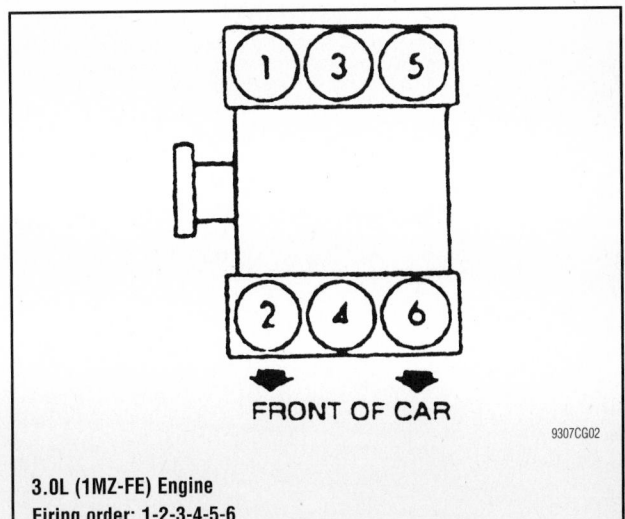

FRONT OF CAR

3.0L (1MZ-FE) Engine
Firing order: 1-2-3-4-5-6
Distributorless ignition system (one coil per cylinder)

9307CG02

3.0L (2JZ-GE) Engine
Firing order: 1-5-3-6-2-4
Distributorless ignition system (one coil on each cylinder)

9307CG04

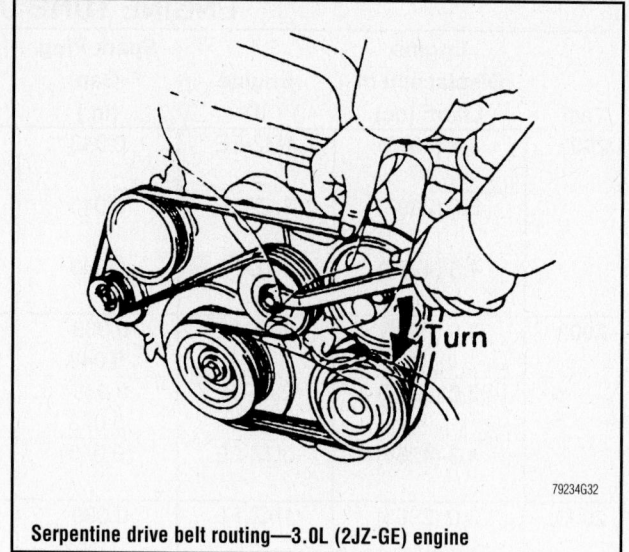

Serpentine drive belt routing—3.0L (2JZ-GE) engine

79234G32

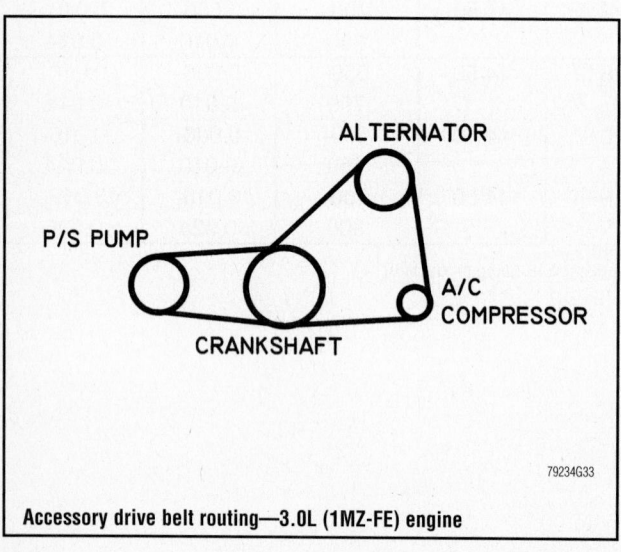

Accessory drive belt routing—3.0L (1MZ-FE) engine

79234G33

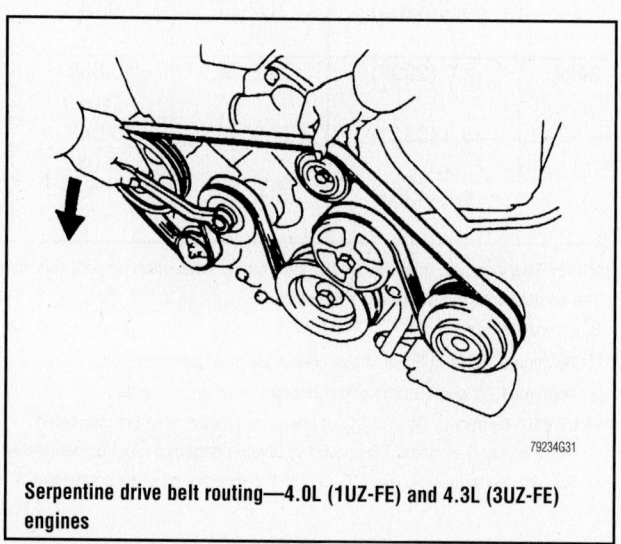

Serpentine drive belt routing—4.0L (1UZ-FE) and 4.3L (3UZ-FE) engines

79234G31

CAPACITIES

Year	Model	Engine Displacement Liters (cc)	Engine ID	Engine Oil with Filter	Transmission (pts.)①	Drive Axle (pts.)	Fuel Tank (gal.)	Cooling System (qts.)
2001	ES 300	3.0 (2995)	1MZ-FE	5.0	8.2	—	18.5	9.7
	GS 300	3.0 (2997)	2JZ-GE	6.0	4.0	2.8	19.8	8.1
	GS 430	4.3 (4264)	3UZ-FE	5.5	4.0	2.8	19.8	9.5
	IS 300	3.0 (2997)	2JZ-GE	6.0	4.0	2.8	17.5	7.9
	LS 430	4.3 (4264)	3UZ-FE	6.5	4.0	2.8	22.2	10.4
2002	ES 300	3.0 (2995)	1MZ-FE	5.0	8.2	—	18.5	9.7
	GS 300	3.0 (2997)	2JZ-GE	5.7	4.0	2.8	19.8	8.1
	GS 430	4.3 (4264)	3UZ-FE	5.5	4.0	2.8	19.8	9.5
	IS 300	3.0 (2997)	2JZ-GE	6.0	4.0	2.8	17.5	7.9
	LS 430	4.3 (4264)	3UZ-FE	6.5	4.0	2.8	22.2	10.4
2003	ES 300	3.0 (2995)	1MZ-FE	5.0	8.2	—	18.5	9.7
	GS 300	3.0 (2997)	2JZ-GE	5.7	4.0	2.8	19.8	8.1
	GS 430	4.3 (4264)	3UZ-FE	5.5	4.0	2.8	19.8	9.5
	IS 300	3.0 (2997)	2JZ-GE	6.0	4.0	2.8	17.5	7.9
	LS 430	4.3 (4264)	3UZ-FE	6.5	4.0	2.8	22.2	10.4
2004	ES 330	3.3 (3311)	3MZ-FE	—	—	—	18.5	—
	GS 300	3.0 (2997)	2JZ-GE	5.7	4.0	2.8	19.8	8.1
	GS 430	4.3 (4264)	3UZ-FE	5.5	4.0	2.8	19.8	9.5
	IS 300	3.0 (2997)	2JZ-GE	6.0	4.0	2.8	17.5	7.9
	LS 430	4.3 (4293)	3UZ-FE	—	—	—	19.8	—

NOTE: All capacities are approximate. Add fluid gradually and check to be sure a proper fluid level is obtained.

① Specification is for transmission drain and refill, not overhaul.

67162-LEXU-C04

VALVE SPECIFICATIONS

Year	Engine Displacement Liters (cc)	Engine ID	Seat Angle (deg.)	Face Angle (deg.)	Spring Test Pressure (lbs. @ in.)	Spring Free-Length (in.)	Stem-to-Guide Clearance (in.)		Stem Diameter (in.)	
							Intake	Exhaust	Intake	Exhaust
2001	3.0 (2995)	1MZ-FE	NA	44.5	41.9-46.3@ 1.331	1.791	0.0010-0.0024	0.0012-0.0026	0.2154-0.2159	0.2152-0.2157
	3.0 (2997)	2JZ-GE	NA	44.5	41.9-46.3@ 1.358	①	0.0010-0.0024	0.0012-0.0026	0.2350-0.2356	0.2348-0.2354
	4.3 (4264)	3UZ-FE	45	44.5	45.9-50.7@ 1.3795	2.130	0.0010-0.0024	0.0012-0.0026	0.2154-0.2159	0.2152-0.2157
2002	3.0 (2995)	1MZ-FE	NA	44.5	41.9-46.3@ 1.331	1.791	0.0010-0.0024	0.0012-0.0026	0.2154-0.2159	0.2152-0.2157
	3.0 (2997)	2JZ-GE	NA	44.5	41.9-46.3@ 1.358	①	0.0010-0.0024	0.0012-0.0026	0.2350-0.2356	0.2348-0.2354
	4.3 (4264)	3UZ-FE	45	44.5	45.9-50.7@ 1.3795	2.130	0.0010-0.0024	0.0012-0.0026	0.2154-0.2159	0.2152-0.2157
2003	3.0 (2995)	1MZ-FE	NA	44.5	41.9-46.3@ 1.331	1.791	0.0010-0.0024	0.0012-0.0026	0.2154-0.2159	0.2152-0.2157
	3.0 (2997)	2JZ-GE	NA	44.5	41.9-46.3@ 1.358	①	0.0010-0.0024	0.0012-0.0026	0.2350-0.2356	0.2348-0.2354
	4.3 (4264)	3UZ-FE	45	44.5	45.9-50.7@ 1.3795	2.130	0.0010-0.0024	0.0012-0.0026	0.2154-0.2159	0.2152-0.2157
2004	3.0 (2997)	2JZ-GE	45	44.5	41.9-46.3@ 1.358	①	0.0010-0.0024	0.0012-0.0026	0.2350-0.2356	0.2348-0.2354
	3.3 (3311)	3MZ-FE	NA	44.5	41.9-46.3@ 1.331	NA	0.0010-0.0024	0.0012-0.0026	0.2154-0.2159	0.2152-0.2157
	4.3 (4293)	3UZ-FE	45	44.5	45.9-50.7@ 1.3795	2.131	0.0010-0.0024	0.0012-0.0026	0.2154-0.2159	0.2152-0.2157

NA: Information not available

① Pink: 1.7209
  Yellow: 1.7362

67162-LEXU-C05

CRANKSHAFT AND CONNECTING ROD SPECIFICATIONS
All measurements are given in inches.

Year	Engine Displacement Liters (cc)	Engine ID	Crankshaft				Connecting Rod		
			Main Brg. Journal Dia.	Main Brg. Oil Clearance	Shaft End-play	Thrust on No.	Journal Diameter	Oil Clearance	Side Clearance
2001	3.0 (2995)	1MZ-FE	2.4011	①	0.0016-0.0095	2	2.0863-2.0866	0.0015-0.0025	0.0059-0.0118
	3.0 (2997)	2JZ-GE	2.4403-2.4409	0.0010-0.0016	0.0008-0.0087	4	2.0465-2.0472	0.0009-0.0016	0.0098-0.0158
	4.3 (4264)	3UZ-FE	2.6373-2.6378	②	0.0008-0.0087	3	2.0465-2.0472	0.0008-0.0019	0.0063-0.0138
2002	3.0 (2995)	1MZ-FE	2.4011	①	0.0016-0.0095	2	2.0863-2.0866	0.0015-0.0025	0.0059-0.0118
	3.0 (2997)	2JZ-GE	2.4403-2.4409	0.0010-0.0016	0.0008-0.0087	4	2.0465-2.0472	0.0009-0.0016	0.0098-0.0158
	4.3 (4264)	3UZ-FE	2.6373-2.6378	②	0.0008-0.0087	3	2.0465-2.0472	0.0008-0.0019	0.0063-0.0138
2003	3.0 (2995)	1MZ-FE	2.4011	①	0.0016-0.0095	2	2.0863-2.0866	0.0015-0.0025	0.0059-0.0118
	3.0 (2997)	2JZ-GE	2.4403-2.4409	0.0010-0.0016	0.0008-0.0087	4	2.0465-2.0472	0.0009-0.0016	0.0098-0.0158
	4.3 (4264)	3UZ-FE	2.6373-2.6378	②	0.0008-0.0087	3	2.0465-2.0472	0.0008-0.0019	0.0063-0.0138
2004	3.0 (2997)	2JZ-GE	2.4403-2.4409	0.0010-0.0016	0.0008-0.0087	4	2.0465-2.0472	0.0009-0.0016	0.0098-0.0158
	3.3 (3311)	3MZ-FE	2.4403-2.4409	①	0.0016-0.0094	4	2.0863-2.0866	0.0015-0.0026	0.0059-0.0118
	4.3 (4293)	3UZ-FE	2.6373-2.6378	②	0.0008-0.0087	3	2.0465-2.0472	0.0008-0.0019	0.0063-0.0138

① Journal No. 1 and 4: 0.0006 - 0.0013 inch
   Journal No. 2 and 3: 0.0010 - 0.0018 inch

② Journal No. 1 and 5: 0.0007 - 0.0013 inch
   Remaining journals: 0.0011 - 0.0018 inch

PISTON AND RING SPECIFICATIONS

All measurements are given in inches.

Year	Engine Displacement Liters (cc)	Engine ID	Piston Clearance	Ring Gap			Ring Side Clearance		
				Top Compression	Bottom Compression	Oil Control	Top Compression	Bottom Compression	Oil Control
2001	3.0 (2995)	1MZ-FE	0.0033-0.0042	0.0098-0.0138	0.0138-0.0177	0.0059-0.0157	0.0008-0.0028	0.0008-0.0024	SNUG
	3.0 (2997)	2JZ-GE	0.0014-0.0027	0.0118-0.0185	0.0138-0.0205	0.0051-0.0177	0.0004-0.0028	0.0012-0.0028	SNUG
	4.3 (4264)	3UZ-FE	0.0033-0.0041	0.0118-0.0157	0.0157-0.0197	0.0059-0.0157	0.0012-0.0031	0.0008-0.0024	SNUG
2002	3.0 (2995)	1MZ-FE	0.0033-0.0042	0.0098-0.0138	0.0138-0.0177	0.0059-0.0157	0.0008-0.0028	0.0008-0.0024	SNUG
	3.0 (2997)	2JZ-GE	0.0022-0.0031	0.0118-0.0185	0.0138-0.0205	0.0051-0.0177	0.0004-0.0028	0.0012-0.0028	SNUG
	4.3 (4264)	3UZ-FE	0.0031-0.0041	0.0118-0.0197	0.0157-0.0236	0.0059-0.0197	0.0012-0.0031	0.0008-0.0024	SNUG
2003	3.0 (2995)	1MZ-FE	0.0033-0.0042	0.0098-0.0138	0.0138-0.0177	0.0059-0.0157	0.0008-0.0028	0.0008-0.0024	SNUG
	3.0 (2997)	2JZ-GE	0.0022-0.0031	0.0118-0.0185	0.0138-0.0205	0.0051-0.0177	0.0004-0.0028	0.0012-0.0028	SNUG
	4.3 (4264)	3UZ-FE	0.0031-0.0041	0.0118-0.0197	0.0157-0.0236	0.0059-0.0197	0.0012-0.0031	0.0008-0.0024	SNUG
2004	3.0 (2997)	2JZ-GE	0.0022-0.0031	0.0118-0.0185	0.0138-0.0205	0.0051-0.0177	0.0004-0.0028	0.0012-0.0028	SNUG
	3.3 (3311)	3MZ-FE	0.0013-0.0023	0.0118-0.0157	0.0197-0.0236	0.0059-0.0157	0.0012-0.0031	0.0008-0.0024	SNUG
	4.3 (4293)	3UZ-FE	0.0031-0.0040	0.0118-0.0197	0.0157-0.0236	0.0059-0.0197	0.0012-0.0031	0.0008-0.0024	SNUG

67162-LEXU-C07

TORQUE SPECIFICATIONS
All readings in ft. lbs.

Year	Engine Displacement Liters (cc)	Engine ID	Cylinder Head Bolts	Main Bearing Bolts	Rod Bearing Bolts	Crankshaft Damper Bolts	Flywheel Bolts	Manifold Intake	Manifold Exhaust	Spark Plugs	Lug Nuts
2001	3.0 (2995)	1MZ-FE	①	②	③	159	61	11	36	13	76
	3.0 (2997)	2JZ-GE	④	⑤	⑥	243	61	21	30	13	76
	4.3 (4264)	3UZ-FE	⑦	⑧	⑨	181	⑩	13	32	13	76
2002	3.0 (2995)	1MZ-FE	①	②	⑦	159	61	11	36	13	76
	3.0 (2997)	2JZ-GE	④	⑤	⑥	243	61	21	30	13	76
	4.3 (4264)	3UZ-FE	⑨	⑧	⑦	181	⑩	13	32	13	76
2003	3.0 (2995)	1MZ-FE	①	②	⑦	159	61	11	36	13	76
	3.0 (2997)	2JZ-GE	④	⑤	⑥	243	61	21	30	13	76
	4.3 (4264)	3UZ-FE	⑨	⑧	⑦	181	⑩	13	32	13	76
2004	3.0 (2997)	2JZ-GE	④	⑤	⑥	243	61	21	30	13	76
	3.3 (3311)	3MZ-FE	①	②	③	162	61	11	36	18	76
	4.3 (4293)	3UZ-FE	⑨	⑧	⑦	181	⑩	13	32	13	76

① Head bolt:
Step 1: 40 ft. lbs.
Step 2: Plus 90 degrees
Recessed head bolt: 13 ft. lbs.

② 6-point bolts: 20 ft. lbs.
12-point bolts:
Step 1: 16 ft. lbs.
Step 2: Plus an additional 90 degrees

③ Step 1: 18 ft. lbs.
Step 2: Plus 90 degrees

④ Step 1: 25 ft. lbs.
Step 2: plus 90 degrees
Step 3: plus 90 degrees

⑤ Step 1: 33 ft. lbs.
Step 2: plus 90 degrees

⑥ Step 1: 22 ft. lbs.
Step 2: plus 90 degrees

⑦ Step 1: 29 ft. lbs.
Step 2: plus 90 degrees

⑧ Nuts:
Step 1: 20 ft. lbs.
Step 2: Plus 90 degrees
Bolts: 36 ft. lbs.

⑨ Step 1: 44 ft. lbs.
Step 2: Plus 90 degrees

⑩ Step 1: 36 ft. lbs.
Step 2: Plus 90 degrees

67162-LEXU-C08

WHEEL ALIGNMENT

Year	Model		Caster Range (+/-Deg.)	Caster Preferred Setting (Deg.)	Camber Range (+/-Deg.)	Camber Preferred Setting (Deg.)	Toe-in (in.)	Steering Axis Inclination (Deg.)
2001	ES 300	F	0.75	+2.30	0.75	-0.62	0 +/- 0.08	13.06
		R	—	—	0.75	-0.80	0.16 +/- 0.08	—
	GS 300	F	0.50	+7.55	0.50	-0.27	0.06 +/- 0.08	8.83
		R	—	—	0.50	-0.78	0.06 +/- 0.08	—
	GS 430	F	0.50	+7.55	0.50	-0.27	0.06 +/- 0.08	8.83
		R	—	—	0.50	-0.78	0.06 +/- 0.08	—
	IS 300	F	0.50	+6.12	0.50	-0.50	0.04 +/- 0.08	9.42
		R	—	—	0.50	-0.07	0.08 +/- 0.08	
	LS 430 ①	F	0.75	+6.75	0.75	-0.08	0.04 +/- 0.08	9.00
		R	—	—	0.75	-1.00	0.12 +/- 0.08	—
	LS 430 ②	F	0.75	+7.25	0.75	-0.25	0.04 +/- 0.08	9.25
		R	—	—	0.75	-1.55	0.12 +/- 0.08	—
2002	ES 300	F	0.75	+2.77	0.75	-0.72	0 +/- 0.08	11.45
		R	—	—	0.75	-1.38	0.16 +/- 0.08	—
	GS 300	F	0.50	+7.55	0.50	-0.27	0.06 +/- 0.08	8.83
		R	—	—	0.50	-0.78	0.06 +/- 0.08	—
	GS 430	F	0.50	+7.55	0.50	-0.27	0.06 +/- 0.08	8.83
		R	—	—	0.50	-0.78	0.06 +/- 0.08	—
	IS 300	F	0.50	+6.12	0.50	-0.50	0.04 +/- 0.08	9.42
		R	—	—	0.50	-0.07	0.08 +/- 0.08	
	LS 430 ①	F	0.75	+6.75	0.75	-0.08	0.04 +/- 0.08	9.00
		R	—	—	0.75	-1.00	0.12 +/- 0.08	—
	LS 430 ②	F	0.75	+7.25	0.75	-0.25	0.04 +/- 0.08	9.25
		R	—	—	0.75	-1.55	0.12 +/- 0.08	—
2003	ES 300	F	0.75	+2.77	0.75	-0.72	0 +/- 0.08	11.45
		R	—	—	0.75	-1.38	0.16 +/- 0.08	—
	GS 300	F	0.50	+7.55	0.50	-0.27	0.06 +/- 0.08	8.83
		R	—	—	0.50	-0.78	0.06 +/- 0.08	—
	GS 430	F	0.50	+7.55	0.50	-0.27	0.06 +/- 0.08	8.83
		R	—	—	0.50	-0.78	0.06 +/- 0.08	—
	IS 300	F	0.50	+6.12	0.50	-0.50	0.04 +/- 0.08	9.42
		R	—	—	0.50	-0.07	0.08 +/- 0.08	
	LS 430 ①	F	0.75	+6.75	0.75	-0.08	0.04 +/- 0.08	9.00
		R	—	—	0.75	-1.00	0.12 +/- 0.08	—
	LS 430 ②	F	0.75	+7.25	0.75	-0.25	0.04 +/- 0.08	9.25
		R	—	—	0.75	-1.55	0.12 +/- 0.08	—
2004	ES 330	F	0.75	+2.77	0.75	-0.72	0 +/- 0.08	11.45
		R	—	—	0.75	-1.38	0.16 +/- 0.08	—
	GS 300	F	0.50	+7.55	0.50	-0.27	0.06 +/- 0.08	8.87
		R	—	—	0.50	-0.78	0.06 +/- 0.08	—
	GS 430	F	0.50	+7.55	0.50	-0.27	0.06 +/- 0.08	8.87
		R	—	—	0.50	-0.78	0.06 +/- 0.08	—
	IS 300	F	0.50	+6.12	0.50	-0.50	0.04 +/- 0.08	9.42
		R	—	—	0.50	-0.07	0.08 +/- 0.08	
	LS 430 ①	F	0.75	+6.75	0.75	-0.08	0.04 +/- 0.08	9.00
		R	—	—	0.75	-1.00	0.12 +/- 0.08	—

67162-LEXU-C09

WHEEL ALIGNMENT

Year	Model		Caster		Camber		Toe-in (in.)	Steering Axis Inclination (Deg.)
			Range (+/-Deg.)	Preferred Setting (Deg.)	Range (+/-Deg.)	Preferred Setting (Deg.)		
2004 cont.	LS 430 ②	F	0.75	+7.25	0.75	-0.25	0.04 +/- 0.08	9.25
		R	—	—	0.75	-1.55	0.12 +/- 0.08	—
	SC430	F	0.75	+7.92	0.75	-0.58	0.06 +/- 0.08	9.16
		R	—	—	0.50	-1.17	0.06 +/- 0.08	—

① Except with air suspension

② With air suspension

67162-LEXU-C10

TIRE, WHEEL AND BALL JOINT SPECIFICATIONS

Year	Model	OEM Tires		Tire Pressures (psi)		Wheel Size	Ball Joint Inspection
		Standard	Optional	Front	Rear		
2001	GS 430	225/55VR16	235/45ZR17	Std: 32 / Opt: 33	Std: 32 / Opt: 33	Std: 7.5-JJ / Opt: 8-JJ	U: 9-30 in. ①
	GS 300	P215/60VR16	225/55VR16	30	30	7.5-JJ	U: 9-30 in. ①
	LS 430	P225/60VR16		29	29	7-JJ	U: 9-30 in. ①
			P225/55R17 95H	32	35		
			225/55R17 97W	35	36		
	IS 300	P205/55R16 89V	215/45ZR17	33	33	NA	U: 9-30 in. ①
	ES 300	P205/65VR15	None	26	26	6-JJ	U: 9-30 in. ①
2002	GS 430	225/55VR16	235/45ZR17	Std: 32 / Opt: 33	Std: 32 / Opt: 33	Std: 7.5-JJ / Opt: 8-JJ	U: 9-30 in. ①
	GS 300	P215/60VR16	225/55VR16	30	30	7.5-JJ	U: 9-30 in. ①
	LS 430	P225/60VR16		29	29	7-JJ	U: 9-30 in. ①
			P225/55R17 95H	32	35		
			225/55R17 97W	35	36		
	IS 300	P205/55R16 89V	215/45ZR17	33	33	NA	U: 9-30 in. ①
	ES 300	P205/65VR15	None	26	26	6-JJ	U: 9-30 in. ①
2003	GS 430	225/55VR16	235/45ZR17	Std: 32 / Opt: 33	Std: 32 / Opt: 33	Std: 7.5-JJ / Opt: 8-JJ	U: 9-30 in. ①
	GS 300	P215/60VR16	225/55VR16	30	30	7.5-JJ	U: 9-30 in. ①
	LS 430	P225/60VR16		29	29	7-JJ	U: 9-30 in. ①
			P225/55R17 95H	32	35		
			225/55R17 97W	35	36		
	IS 300	P205/55R16 89V	215/45ZR17	33	33	NA	U: 9-30 in. ①
	ES 300	P205/65VR15	None	26	26	6-JJ	U: 9-30 in. ①
2004	GS 430	225/55VR16	235/45ZR17	Std: 32 / Opt: 33	Std: 32 / Opt: 33	Std: 7.5-JJ / Opt: 8-JJ	U: 9-30 in. ①
	GS 300	P215/60VR16	225/55VR16	30	30	7.5-JJ	U: 9-30 in. ①
	LS 430	P225/60VR16		29	29	7-JJ	U: 9-30 in. ①
			P225/55R17 95H	32	35		
			225/55R17 97W	35	36		
	SC 430	P245/40ZR18	None	33	33	NA	U: 9-26 in. ①
	IS 300	P205/55R16 89V	215/45ZR17	33	33	NA	U: 9-30 in. ①
	ES 330	P215/60VR16	None	29	29	NA	U: 9-30 in. ①

NA: Not Available

OEM: Original Equipment Manufacturer

PSI: Pounds Per Square Inch

STD: Standard

OPT: Optional

L: Lower

U: Upper

① Torque required in inch lbs. to rotate ball joint when removed from the knuckle

BRAKE SPECIFICATIONS
All measurements in inches unless noted

| Year | Model | Front Brake Disc | | | Rear Brake Disc | | | Minimum Lining Thickness | Brake Caliper | |
		Original Thickness	Minimum Thickness	Maximum Run-out	Original Thickness	Minimum Thickness	Maximum Run-out		Bracket Bolts (ft. lbs.)	Mounting Bolts (ft. lbs.)
2001	ES 300	1.102	1.024	0.0020	0.394	0.354	0.0059	0.0390	①	②
	GS 300	1.260	1.181	0.0020	0.472	0.413	0.0020	0.0390	③	25
	GS 430	1.260	1.181	0.0020	0.472	0.413	0.0020	0.0390	③	25
	IS 300	1.260	1.181	0.0020	0.472	0.413	0.0020	0.0390	③	25
	LS 430	1.181	1.102	0.0020	0.630	0.571	0.0020	0.0390	—	④
2002	ES 300	1.102	1.024	0.0020	0.472	0.413	0.0059	0.0390	①	②
	GS 300	1.260	1.181	0.0020	0.472	0.413	0.0020	0.0390	③	25
	GS 430	1.260	1.181	0.0020	0.472	0.413	0.0020	0.0390	③	25
	IS 300	1.260	1.181	0.0020	0.472	0.413	0.0020	0.0390	③	25
	LS 430	1.181	1.102	0.0020	0.630	0.571	0.0020	0.0390	—	④
2003	ES 300	1.102	1.024	0.0020	0.472	0.413	0.0059	0.0390	①	②
	GS 300	1.260	1.181	0.0020	0.472	0.413	0.0020	0.0390	③	25
	GS 430	1.260	1.181	0.0020	0.472	0.413	0.0020	0.0390	③	25
	IS 300	1.260	1.181	0.0020	0.472	0.413	0.0020	0.0390	③	25
	LS 430	1.181	1.102	0.0020	0.630	0.571	0.0020	0.0390	—	④
2004	ES 330	1.102	1.024	0.0020	0.472	0.413	0.0059	0.0390	①	②
	GS 300	1.260	1.181	0.0020	0.472	0.413	0.0020	0.0390	③	25
	GS 430	1.260	1.181	0.0020	0.472	0.413	0.0020	0.0390	③	25
	IS 300	1.260	1.181	0.0020	0.472	0.413	0.0020	0.0390	③	25
	LS 430	1.181	1.102	0.0020	0.630	0.571	0.0020	0.0390	③	④
	SC 430	1.260	1.181	0.0020	0.472	0.413	0.0020	0.0390	③	④

① Front: 79 ft. lbs.
   Rear: 34 ft. lbs.

② Front: 25 ft. lbs.
   Rear: 14 ft. lbs.

③ Front: 87 ft. lbs.
   Rear: 77 ft. lbs.

④ Front: 81 ft. lbs.
   Rear: 58 ft. lbs.

67162-LEXU-C12

SCHEDULED MAINTENANCE INTERVALS
Lexus—ES300, ES330, IS300, GS300, GS400, GS430, SC300, SC400, SC430, LS400 & LS430

TO BE SERVICED	TYPE OF SERVICE	VEHICLE MILEAGE INTERVAL (x1000)												
		7.5	15	22.5	30	37.5	45	52.5	60	67.5	75	82.5	90	97.5
Engine oil & filter	R	✓	✓	✓	✓	✓	✓	✓	✓	✓	✓	✓	✓	✓
Air conditioning filter (LS 400) ①	S/I	✓	✓	✓	✓	✓	✓	✓	✓	✓	✓	✓	✓	✓
Automatic transaxle fluid & filter	S/I		✓		✓		✓		✓		✓		✓	
Ball joints & dust covers	S/I		✓		✓		✓		✓		✓		✓	
Bolts & nuts on chassis & body	S/I		✓		✓		✓		✓		✓		✓	
Brake fluid ②	S/I		✓		✓		✓		✓		✓		✓	
Brake line pipes & hoses	S/I		✓		✓		✓		✓		✓		✓	
Brake linings & drums	S/I		✓		✓		✓		✓		✓		✓	
Brake pads & discs (front & rear)	S/I		✓		✓		✓		✓		✓		✓	
Differential oil	S/I		✓		✓		✓		✓		✓		✓	
Driveshaft boots (ES 300)	S/I		✓		✓		✓		✓		✓		✓	
Steering gear housing oil	S/I		✓		✓		✓		✓		✓		✓	
Steering linkage	S/I		✓		✓		✓		✓		✓		✓	
Air filter	R				✓				✓				✓	
Exhaust pipes & mountings	S/I				✓				✓				✓	
Fuel lines & connections	S/I				✓				✓				✓	
Engine coolant	R						✓					✓		
Fuel tank cap gasket	R								✓					
Spark plugs	R								✓					
Charcoal canister	S/I								✓					
Drive belts	S/I								✓					
Valve clearance	S/I								✓					

R: Replace   S/I: Service or Inspect

① Replace at 15,000 miles.

② Replace at 30,000 miles (unless previously replaced).

FREQUENT OPERATION MAINTENANCE (SEVERE SERVICE)

If a vehicle is operated under any of the following conditions it is considered severe service

- Extremely dusty areas.

- 50% or more of the vehicle operation is in 32°C (90°F) or higher temperatures, or constant operation in temperatures below 0°C (32°F).

- Prolonged idling (vehicle operation in stop and go traffic).

- Frequent short running periods (engine does not warm to normal operating temperatures).

- Police, taxi, delivery usage or trailer towing usage.

Oil & oil filter: change every 3750 miles.

Ball joints & dust covers: service or inspect every 7500 miles.

Bolts & nuts on chassis & body: service or inspect every 7500 miles.

Brake linings & drums: service or inspect every 7500 miles.

Brake pads & discs (front & rear): service or inspect every 7500 miles.

Driveshaft boots (ES 300): service or inspect every 7500 miles.

Brake linings & drums: service or inspect every 7500 miles.

Steering linkage: service or inspect every 7500 miles.

Air filter: service or inspect every 15,000 miles.

Automatic transmission fluid & filter: replace every 15,000 miles.

Differential oil: replace every 15,000 miles.

Exhaust pipes & mountings: service or inspect every 15,000 miles.

Drive belts: service or inspect at 60,000 miles & every 7500 miles thereafter.

PRECAUTIONS

Before servicing any vehicle, please be sure to read all of the following precautions that deal with personal safety, prevention of component damage, and important points to take into consideration when servicing a motor vehicle:

• Never open, service or drain the radiator or cooling system when the engine is hot; serious burns can occur from the steam and hot coolant.

• Observe all applicable safety precautions when working around fuel. Whenever servicing the fuel system, always work in a well-ventilated area. Do NOT allow fuel spray or vapors to come in contact with a spark, open flame or excessive heat (a hot drop light, for example). Keep a dry chemical fire extinguisher near the work area. Always keep fuel in a container specifically designed for fuel storage; also, always properly seal fuel containers to avoid the possibility of fire or explosion. Refer to the additional fuel system precautions later in this section.

• Fuel injection systems often remain pressurized, even after the engine has been turned **OFF**. The fuel system pressure must be relieved before disconnecting any fuel lines. Failure to do so may result in fire and/or personal injury.

• Brake fluid often contains Polyglycol ethers and P-Polyglycols. Avoid contact with the eyes and wash your hands thoroughly after handling brake fluid. If you do get brake fluid in your eyes, flush your eyes with clean, running water for 15 minutes. If eye irritation persists, or if you have taken

brake fluid internally, IMMEDIATELY seek medical assistance.

• The EPA warns that prolonged contact with used engine oil may cause a number of skin disorders, including cancer. You should make every effort to minimize your exposure to used engine oil. Protective gloves should be worn when changing oil. Wash your hands and any other exposed skin areas as soon as possible after exposure to used engine oil. Soap and water, or waterless hand cleaner should be used.

• All new vehicles are now equipped with an air bag system. The system must be disabled before performing service on or around system components, steering column, instrument panel components, wiring and sensors. Failure to follow safety and disabling procedures could result in accidental air bag deployment, possible personal injury and unnecessary system repairs.

• Always wear safety goggles when working with, or around, the air bag system. When carrying a non-deployed air bag, be sure the bag and trim cover are pointed away from your body. When placing a non-deployed air bag on a work surface, always face the bag and trim cover upward, away from the surface. This will reduce the motion of the module if it is accidentally deployed. Refer to the additional air bag system precautions later in this section.

• Clean, high quality brake fluid from a sealed container is essential to the safe and proper operation of the brake system. You should always buy the correct type of brake

fluid for your vehicle. If the brake fluid becomes contaminated, completely flush the system with new fluid. Never reuse any brake fluid. Any brake fluid that is removed from the system should be discarded. Also, do not allow any brake fluid to come in contact with a painted surface; it will damage the paint.

• Never operate the engine without the proper amount and type of engine oil; doing so WILL result in severe engine damage.

• Timing belt maintenance is extremely important. Many models utilize an interference-type, non-freewheeling engine. If the timing belt breaks, the valves in the cylinder head may strike the pistons, causing potentially serious (also time-consuming and expensive) engine damage. Refer to the maintenance interval charts in the front of this section for the recommended replacement interval for the timing belt, and to the timing belt procedure for belt replacement and inspection.

• Disconnecting the negative battery cable on some vehicles may interfere with the functions of the on-board computer system(s) and may require the computer to undergo a relearning process once the negative battery cable is reconnected.

• When servicing drum brakes, only disassemble and assemble one side at a time, leaving the remaining side intact for reference.

• Only an MVAC-trained, EPA-certified automotive technician should service the air conditioning system or its components.

ENGINE REPAIR

Alternator

REMOVAL

3.0L (1MZ-FE) Engine

1. Before servicing the vehicle, refer to the precautions in the beginning of this section.
2. Remove or disconnect the following:
   • Negative battery cable
   • Accessory drive belt
   • Alternator harness connectors
   • Alternator

3.0L (2JZ-GE) Engine

1. Before servicing the vehicle, refer to the precautions in the beginning of this section.

2. Remove or disconnect the following:
   • Negative battery cable
   • Engine under cover
   • Accessory drive belt
   • Alternator connector
   • Cap and nut
   • Alternator wire
   • Alternator wire clamp from the wire clip on the alternator
   • Bolt and pipe clamp
   • 2 automatic transmission oil cooler pipes from the alternator
   • Bolt, nut, pipe bracket and alternator

4.0L (1UZ-FE) and 4.3L (3UZ-FE) Engines

1. Before servicing the vehicle, refer to the precautions in the beginning of this section.

2. Remove or disconnect the following, as equipped:
   • Negative battery cable
   • Air cleaner inlet
   • Accessory drive belt
   • Oil pan protector
   • Engine under cover
   • Power steering pump pulley
   • Alternator harness connectors
   • Power steering oil hose from oil pan
   • Heated Oxygen (HO2S) sensor wiring
   • Alternator

3.3L, (3MZ-FE) Engines

1. Before servicing the vehicle, refer to the precautions in the beginning of this section.
2. Remove or disconnect the following:

- Remove two wire harness clamps.
- Disconnect the alternator connector.
- Open the terminal clamp.
- Remove the nut, then disconnect the alternator wire.

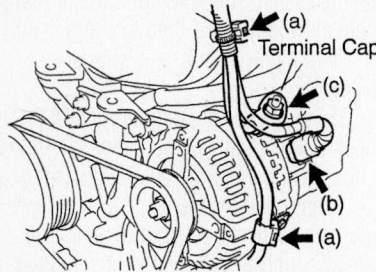

Removing wire harness clamps—3.3L (3MZ-FE)

67162-LEXU-G01

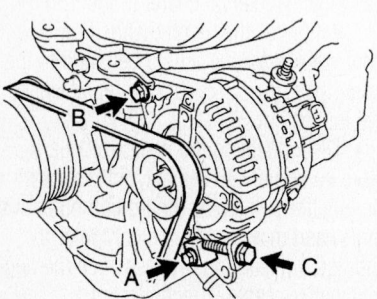

Loosening belt tension bolt—3.3L (3MZ-FE)

67162-LEXU-G02

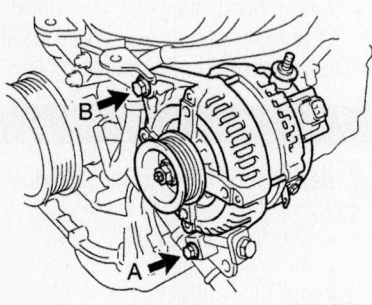

Removing alternator bolts—3.3L (3MZ-FE)

67162-LEXU-G03

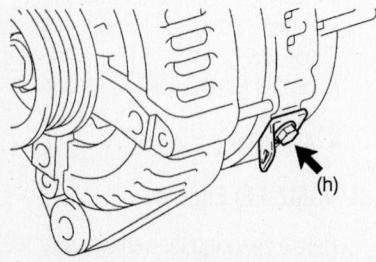

Removing wiring harness clamp from alternator—3.3L (3MZ-FE)

67162-LEXU-G04

- Loosen mounting bolts to lessen tension of the alternator belt
- Remove mounting bolts
- Remove the wiring harness clamp bolt.

INSTALLATION

3.0L (1MZ-FE) Engine

1. Install or connect the following:
   - Alternator
   - Alternator harness connectors
   - Accessory drive belt. Tighten the adjusting lock bolt to 13 ft. lbs. (18 Nm) and the pivot bolt to 41 ft. lbs. (56 Nm).
   - Negative battery cable

3.0L (2JZ-GE) Engine

1. Install or connect the following:
   - Bolt, nut, pipe bracket and alternator. Tighten the fasteners to 30 ft. lbs. (40 Nm).
   - 2 automatic transmission oil cooler pipes to the alternator
   - Bolt and pipe clamp
   - Alternator wire clamp to the wire clip on the alternator
   - Alternator wire
   - Cap and nut
   - Alternator connector
   - Accessory drive belt
   - Engine under cover
   - Negative battery cable

4.0L (1UZ-FE) and 4.3L (3UZ-FE) Engines

1. Install or connect the following, as equipped:
   - Alternator. Tighten the fasteners to 29 ft. lbs. (39 Nm).
   - HO2S sensor wiring
   - Power steering pump oil hose to oil pan
   - Alternator harness connectors
   - Power steering pump pulley
   - Engine under cover
   - Oil pan protector
   - Accessory drive belt
   - Air cleaner inlet
   - Negative battery cable

4.3L, 2004 (3UZ-FE) Engines

- The 2 clamps from cord clips on alternator
- The nut, bolt, stay and alternator, torque to 29 ft lbs. (39Nm)
- PS oil hose to #1 oil pan bolt
- The alternator wire connector from the clamp on cord clip

- Alternator wire from the clamp on the cord clip
- The rubber cap and nut and alternator wire
- The alternator connector
- PS pulley or pump, torque to 32 ft. lbs (43Nm)
- Air cleaner inlet
- Negative battery cable
- Drive belt
- Engine undercover (if necessary)

3.3L, (3MZ-FE) Engines

1. Install or connect the following:
   - Wiring harness clamp bracket with bolt
   - Temporarily install mounting bolts
   - Adjust belt with tensioner
   - Tighten mounting bolts to lower mounting bolt 13ft.lbs. (18 Nm) upper 43 ft. lbs.(58 Nm)
   - Alternator wire nut, tighten to 7 ft. lbs. (9.8 Nm)
   - Alternator wire connector
   - Wiring harness clamp
   - Battery negative cable

Ignition Timing

ADJUSTMENT

The engines covered in this section are equipped with a Distributorless Ignition System (DIS). No timing adjustments are possible.

Engine Assembly

REMOVAL & INSTALLATION

ES 300

1. Before servicing the vehicle, refer to the precautions in the beginning of this section.
2. Release the fuel pressure.
3. Drain the engine coolant and engine oil.
4. Remove or disconnect the following:
   - Battery and tray
   - Hood
   - Accelerator cable and the throttle cable
   - Air cleaner cover
   - Volume air flow meter and air cleaner duct as an assembly
   - Cruise control actuator, if equipped
   - Radiator
   - Engine relay box and 2 bolts

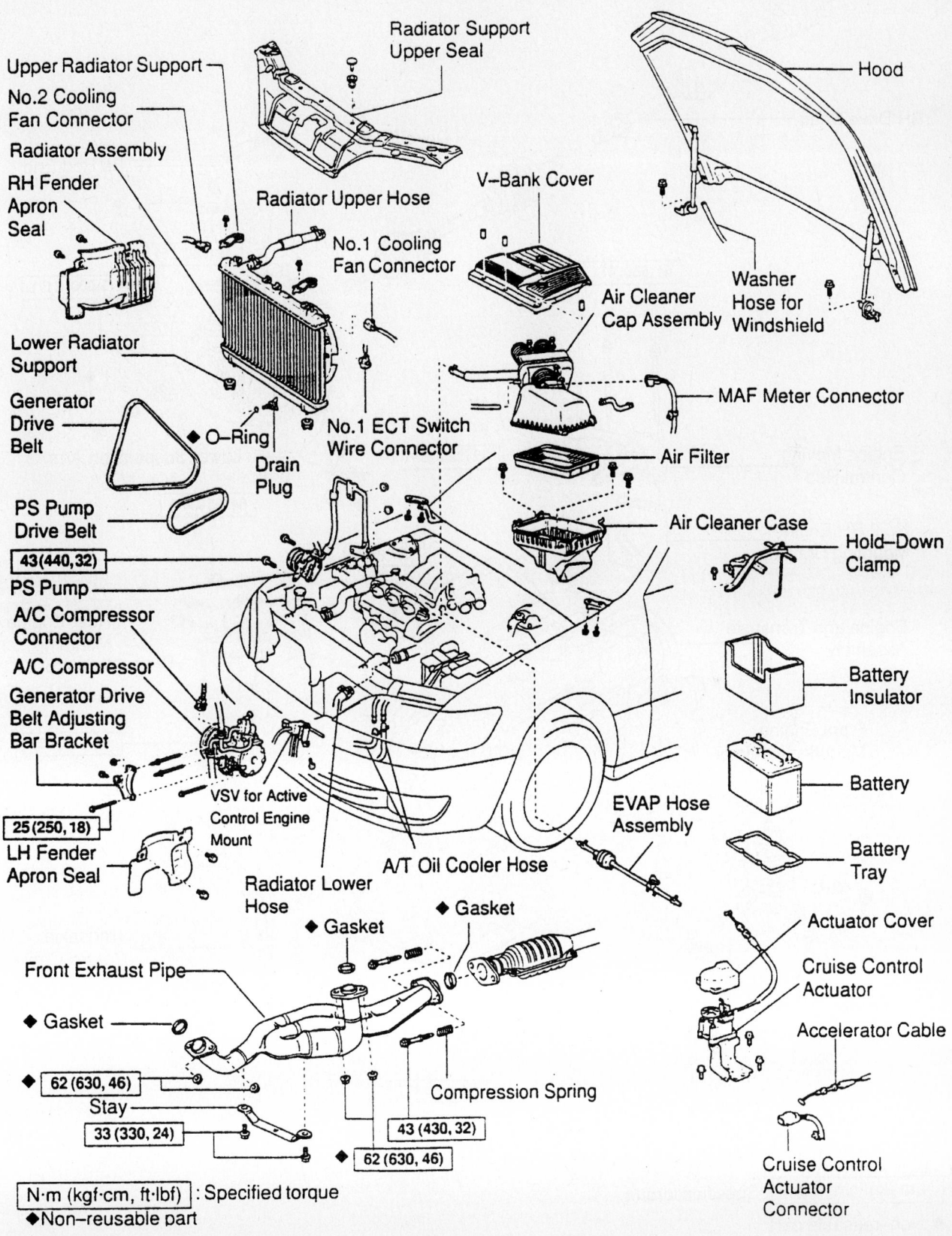

Upper Radiator Support

No.2 Cooling Fan Connector

Radiator Assembly

RH Fender Apron Seal

Radiator Support Upper Seal

Radiator Upper Hose

No.1 Cooling Fan Connector

V–Bank Cover

Hood

Air Cleaner Cap Assembly

Washer Hose for Windshield

MAF Meter Connector

Lower Radiator Support

Generator Drive Belt

◆ O–Ring

Drain Plug

No.1 ECT Switch Wire Connector

Air Filter

Air Cleaner Case

PS Pump Drive Belt

43(440,32)

PS Pump

A/C Compressor Connector

A/C Compressor

Generator Drive Belt Adjusting Bar Bracket

25(250,18)

LH Fender Apron Seal

VSV for Active Control Engine Mount

Radiator Lower Hose

◆ Gasket

◆ Gasket

A/T Oil Cooler Hose

EVAP Hose Assembly

Hold–Down Clamp

Battery Insulator

Battery

Battery Tray

Actuator Cover

Cruise Control Actuator

Accelerator Cable

Cruise Control Actuator Connector

Front Exhaust Pipe

◆ Gasket

◆ **62(630,46)**

Stay

33(330,24)

◆ **62(630,46)**

Compression Spring

43(430,32)

N·m (kgf·cm, ft·lbf) : Specified torque

◆Non–reusable part

Exploded view of the engine removal and related components—ES 300

9301LG01

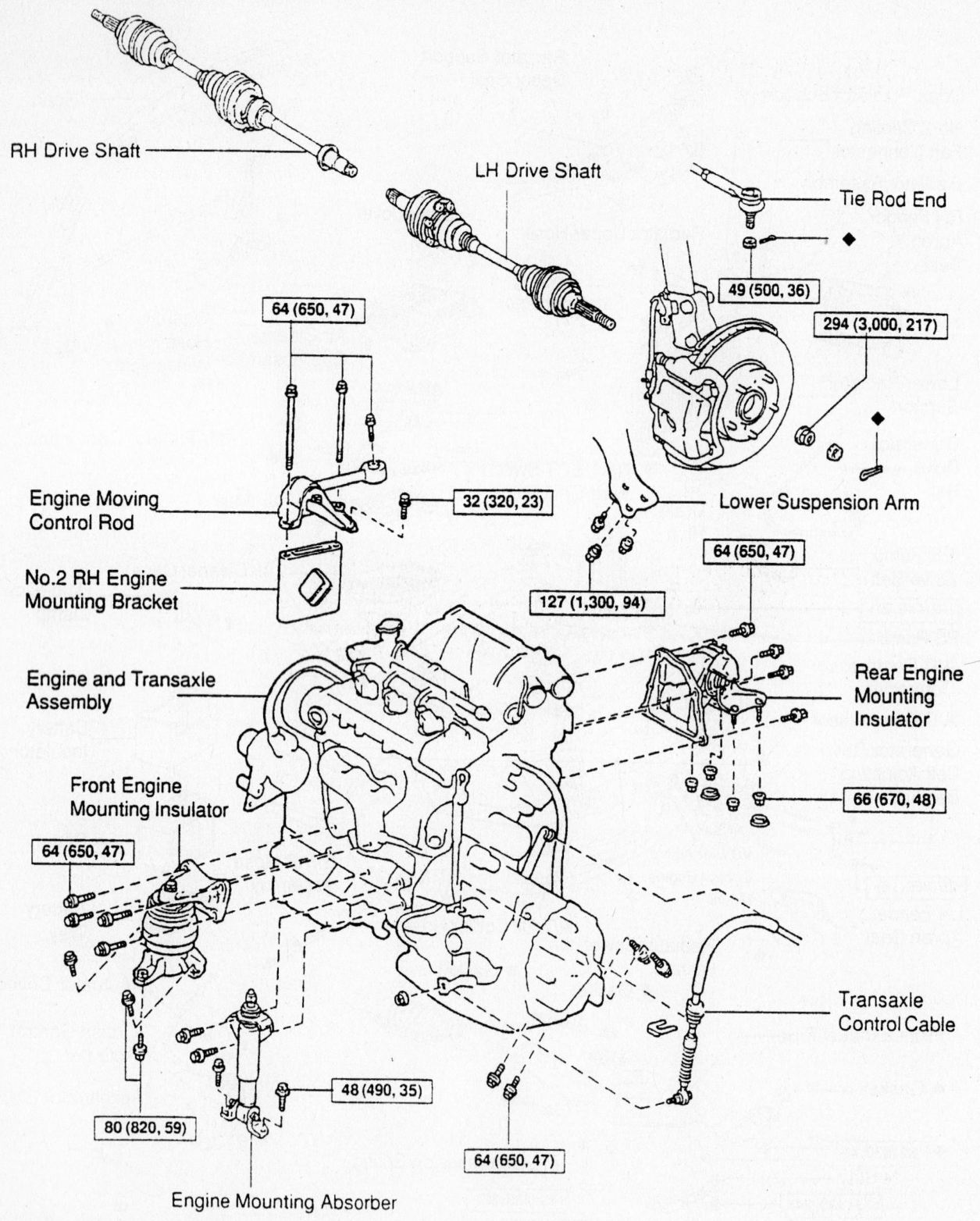

RH Drive Shaft

LH Drive Shaft

Tie Rod End

49 (500, 36)

294 (3,000, 217)

64 (650, 47)

Engine Moving Control Rod

32 (320, 23)

Lower Suspension Arm

No.2 RH Engine Mounting Bracket

64 (650, 47)

127 (1,300, 94)

Engine and Transaxle Assembly

Rear Engine Mounting Insulator

Front Engine Mounting Insulator

66 (670, 48)

64 (650, 47)

Transaxle Control Cable

48 (490, 35)

80 (820, 59)

64 (650, 47)

Engine Mounting Absorber

N·m (kgf·cm, ft·lbf) : Specified torque

◆ Non–reusable part

9301LG02

Exploded view of the engine removal and related components (cont.)—ES 300

- 5 connections from the relay box
- 2 igniter connectors
- Left fender apron connector
- Noise filter connector
- 2 ground straps
- Engine wiring harness from the engine.
- Vacuum hoses from the following connections: intake air control valve vacuum tank, charcoal canister, brake booster vacuum hose from the intake chamber
- 2 heater hoses from the bulkhead
- Fuel feed and return lines
- Control cable from the transaxle
- Wiring harness from the PCM and route it through the bulkhead.
- Air conditioning compressor from the engine without disconnecting the lines and position it out of the way
- Front exhaust pipe
- Halfshafts
- 2 power steering air hoses from the engine
- Hydraulic cooling fan pressure hose
- Power steering pump without disconnecting the lines and position it out of the way
- Right and left lower engine mounts from the body
- Engine mounting shock absorber
- 3 front engine mounting bolts from the body

5. Attach a lifting device to the engine.
6. Remove or disconnect the following:
- Coolant reservoir tank.
- Right engine mounting bracket
- Engine moving control rod and right No. 2 engine mounting bracket
- Engine and transaxle as an assembly

❊❊ WARNING

Be careful not to hit the power steering or PNP switches.

- Engine mounting insulator below the oil filter
- Right rear engine-mounting insulator
- Front exhaust pipe stay

7. Label and detach the following connectors:
- Overdrive solenoid
- PNP switch
- Speedometer
- Starter terminal
- Speed sensor.

- 2 wire clamps from the transaxle
- Oil dipstick and guide
- Starter
- Flywheel housing cover

8. Turn the crankshaft pulley to gain access to the 8 torque converter bolts. Secure the crankshaft and remove them as they become accessible.
9. Install or connect the following:
- 2 exhaust manifold stays and plate
- 2 bolts attaching the transaxle to the oil pan
- 6 transaxle mounting bolts
- Transaxle

To install:

10. Position the transaxle to the engine. Tighten the transaxle mounting bolts: 47 ft. lbs. (64 Nm).
11. Install or connect the following:
- Bolts that attach the transaxle to the oil pan bolts and tighten them to 34 ft. lbs. (46 Nm)
- Exhaust manifold support. Tighten the bolts to 14 ft. lbs. (20 Nm).
- Bolts that attach the flywheel to the torque converter. Coat the threads with a locking compound. Rotate the engine and tighten the bolts alternately to 30 ft. lbs. (41 Nm).
- Starter to the engine
- Flywheel cover and tighten the bolts to 13 ft. lbs. (18 Nm)
- Dipstick and tube with a new O-ring

12. Attach the clamps and following connectors:
- Overdrive solenoid
- PNP switch
- Speedometer
- Starter terminal
- Speed sensor

13. Install or connect the following:
- Exhaust pipe's stay. Tighten the bolts to 15 ft. lbs. (21 Nm).
- Right rear insulator. Tighten the bolts to 47 ft. lbs. (64 Nm).
- Front engine mounting insulator. Tighten the bolts to 47 ft. lbs. (64 Nm).

14. Lower the engine and transaxle into the engine compartment. Tilt the transaxle downward and clear the left mount.
15. Keep the engine level and align the right and left engine mounts.
16. Install or connect the following:
- Engine mounting bracket and moving control rod. Tighten bolts to 47 ft. lbs. (64 Nm).
- Right engine stay. Tighten bolts to 23 ft. lbs. (32 Nm).
- Ground straps
- Coolant reservoir

- Front engine mounting insulator to the body. Tighten bolts to 59 ft. lbs. (81 Nm).
- Engine mounting shock absorber. Tighten bolts to 35 ft. lbs. (48 Nm).
- Right engine mount. Tighten bolts to 48 ft. lbs. (66 Nm).
- Left engine mount. Tighten bolts to 47 ft. lbs. (64 Nm).
- Engine lifting device
- Power steering pump and belt
- Hydraulic cooling fan pressure hose. Tighten the fitting to 33 ft. lbs. (44 Nm).
- Power steering air tube and hoses
- Halfshafts
- Front exhaust pipe with new gaskets
- Air conditioning compressor. Tighten bolts to 18 ft. lbs. (25 Nm).
- Harness to the Powertrain Control Module and assemble the instrument panel
- Control cable to the transaxle
- Fuel lines and tighten the fittings to 22 ft. lbs. (30 Nm)
- Heater hoses

17. Connect the vacuum hoses to the following connections:
- Intake air control valve vacuum tank
- Charcoal canister
- Air intake chamber from the brake booster

18. Install or connect the following:
- Engine wiring harness to the engine
- Engine relay box
- 2 bolts and attach the following connectors:
- 5 connections from the relay box
- 2 igniter connectors
- Left fender apron connector
- Noise filter connector
- 2 ground straps
- Radiator
- Cruise control actuator, if equipped
- Air cleaner cover
- Volume airflow meter and air cleaner duct assembly
- Throttle cable
- Accelerator cable

19. Fill the engine to the proper level with the recommended grade of oil.
20. Align the matchmarks and install the hood.
21. Fill the engine to the proper level with coolant.
22. Bleed the cooling system.
23. Install the battery and tray.
24. Check and/or adjust the ignition timing.

25. Start the engine and check for leaks.
26. Road test the vehicle.
27. Recheck the engine oil and coolant levels.

GS 300 & GS 330

1. Before servicing the vehicle, refer to the precautions in the beginning of this section.
2. Release the fuel pressure.
3. Drain the fuel from the tank.
4. Remove or disconnect the following:
   - Negative battery cable. Wait at least 90 seconds before performing any other work.
   - Hood insulator pad and the hood
   - Engine undercover, then drain the engine coolant and oil
   - Front suspension member brace, if equipped
   - Engine cover, if equipped
   - Accelerator cable, cruise control actuator cable and the automatic transmission throttle control cable from the throttle body
   - Air cleaner assembly
   - Volume air flow meter (or MAF meter) and the air intake hose
   - Air cleaner duct
   - Accelerator cable from engine
   - Drive belt
   - Radiator
5. Label and detach the following wires and electrical connectors, as equipped:
   - Ground strap from floor
   - Fuel inlet hose
   - Igniter
   - Ground strap from dash panel
   - Oxygen sensor connectors
   - Ignition coil
   - Wiring harness from the wire clamp and coolant tank
   - Alternator wiring
   - Ground strap from engine block and from left engine mount
   - Starter
   - DLC1 connector
6. Remove or disconnect the following:
   - Fuel lines from the intake and return lines
   - Power steering pump without disconnecting the lines and position it aside
   - Air conditioning compressor without disconnecting the air conditioning lines and position it aside
   - Brake booster vacuum hose
   - Evaporative Emissions (EVAP) hose
   - Heater hoses from the firewall
   - Heater valve and engine wire from the firewall

- Electrical harness from the PCM and route it through the firewall
- Connectors from ECM box
- Oxygen (O_2) sensor (if equipped) from the front exhaust pipe
- Front exhaust pipes and heat insulator

7. Remove the rear center floor crossmember brace.
8. Disconnect the power steering pump from it mounting and disconnect the A/C compressor from its mounting, without disconnecting hoses from either component. Position these out of the way.
9. Remove the transmission shift control rod.
10. Remove the driveshaft.
11. Disconnect the power steering gear housing from the steering column, then remove the gear housing assembly from the mounting brackets and suspend it securely.
12. Support the transmission with a jack. Use a piece of wood to prevent damage to the transmission oil pan.
13. Attach a lifting device to the engine.
14. Remove or disconnect the following:
   - Rear transmission crossmember
   - 2 hole plugs in the front crossmember
   - 2 nuts holding the engine insulators to the front crossmember
15. Slowly and carefully remove the engine and transmission from the engine compartment as an assembly.

To install:
16. Position the engine assembly into the vehicle.
17. Install or connect the following:
   - Stud bolts for the front engine mount into their bores in the front engine crossmember. Temporarily install the 2 nuts.
18. Remove the engine hoist
19. Install or connect the following:
   - Temporarily, the rear engine support with the 4 nuts
   - The 4 support bolts and tighten them to 19 ft. lbs. (25 Nm). Tighten the nuts to 10 ft. lbs. (13 Nm).
   - Front engine crossmember to mount nuts to 54 ft. lbs. (74 Nm) and the hole plugs
   - Driveshaft
20. Shift the transmission control shift rod into **N** (neutral) by shifting the lever all the way back and returning it 2 notches. Connect the shift rod to the lever and tighten it to 108 inch lbs. (13 Nm).
21. Install the power steering gear to its mountings. Torque the bolts to 48 ft. lbs. (65 Nm). Install and tighten the gear-to-

steering column yoke bolt to 26 ft. lbs. (35 Nm). Connect the power steering switch connector.
22. Install or connect the following:
   - Transmission control rod; torque nuts to 9 ft. lbs. (13 Nm)
   - Rear center floor crossmember brace and tighten the bolts to 108 inch lbs. (13 Nm)
   - Exhaust pipe heat insulator
   - Front exhaust pipes
   - Oxygen sensor connectors, if equipped
   - Engine wiring harness to the PCM
   - PCM and its cover
   - The lower portion of the passenger side instrument panel, the vent, the carpet and the scuff panel
   - Heater water valve and engine wire to the cowl panel
   - Heater hoses
   - EVAP hose
   - Brake booster hose
   - The air conditioning compressor and connectors; tighten the Torx® bolt to 19 ft. lbs. (26 Nm). Tighten the nut and bolts to 38 ft. lbs. (52 Nm).
   - Power steering pump and connector; tighten the upper and lower P/S bracket bolts to 43 ft. lbs. (58 Nm) and the side bracket bolt to 38 ft. lbs. (52 Nm)
   - Power steering pump rear stay; torque bolts to 29 ft. lbs. (39 Nm)
   - Wiring to EMC box
   - Fuel lines with new gaskets and tighten the union bolts to 22 ft. lbs. (29 Nm)
   - Igniter
   - Ignition coil
   - Wiring harness from the wire clamp and coolant tank
   - Alternator and ground strap from the left engine mount
   - Starter
   - Radiator
   - Drive belt
   - Air cleaner
   - Volume air flow meter and intake air connector pipe as an assembly
   - Air cleaner duct
   - Accelerator, cruise control and the automatic transmission throttle control cables
   - Fuel
   - Engine oil
   - Coolant
   - Negative battery cable
23. Start the engine and check for leaks.
24. Check the automatic transmission fluid level.

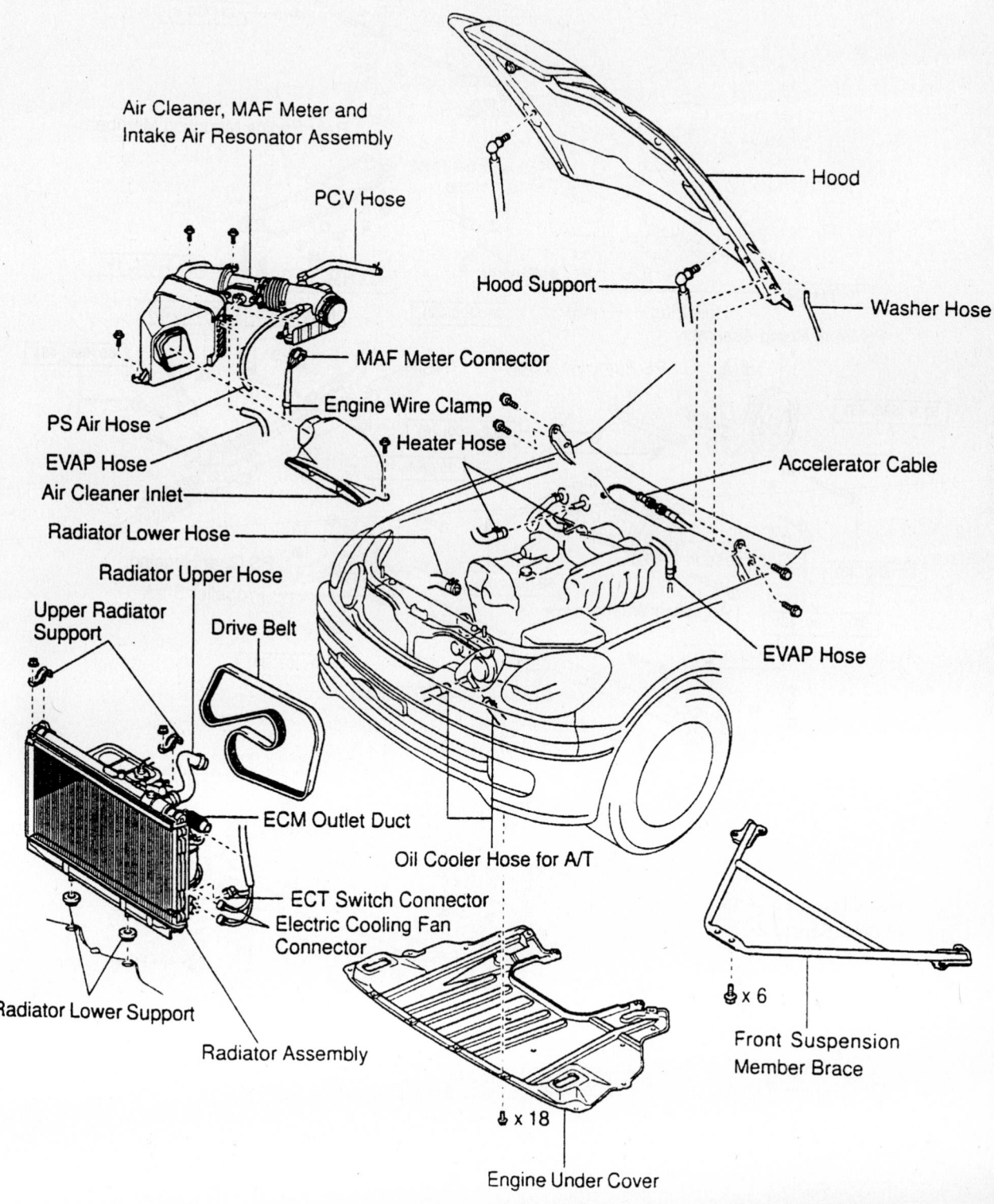

Air Cleaner, MAF Meter and Intake Air Resonator Assembly

PCV Hose

Hood

Hood Support

Washer Hose

MAF Meter Connector

Engine Wire Clamp

PS Air Hose

Heater Hose

Accelerator Cable

EVAP Hose

Air Cleaner Inlet

Radiator Lower Hose

EVAP Hose

Radiator Upper Hose

Upper Radiator Support

Drive Belt

ECM Outlet Duct

Oil Cooler Hose for A/T

ECT Switch Connector
Electric Cooling Fan Connector

Radiator Lower Support

☗ x 6

Radiator Assembly

Front Suspension Member Brace

☗ x 18

Engine Under Cover

9301LG03

Exploded view of the related engine compartment components for 2JZ-GE engine removal—GS 300

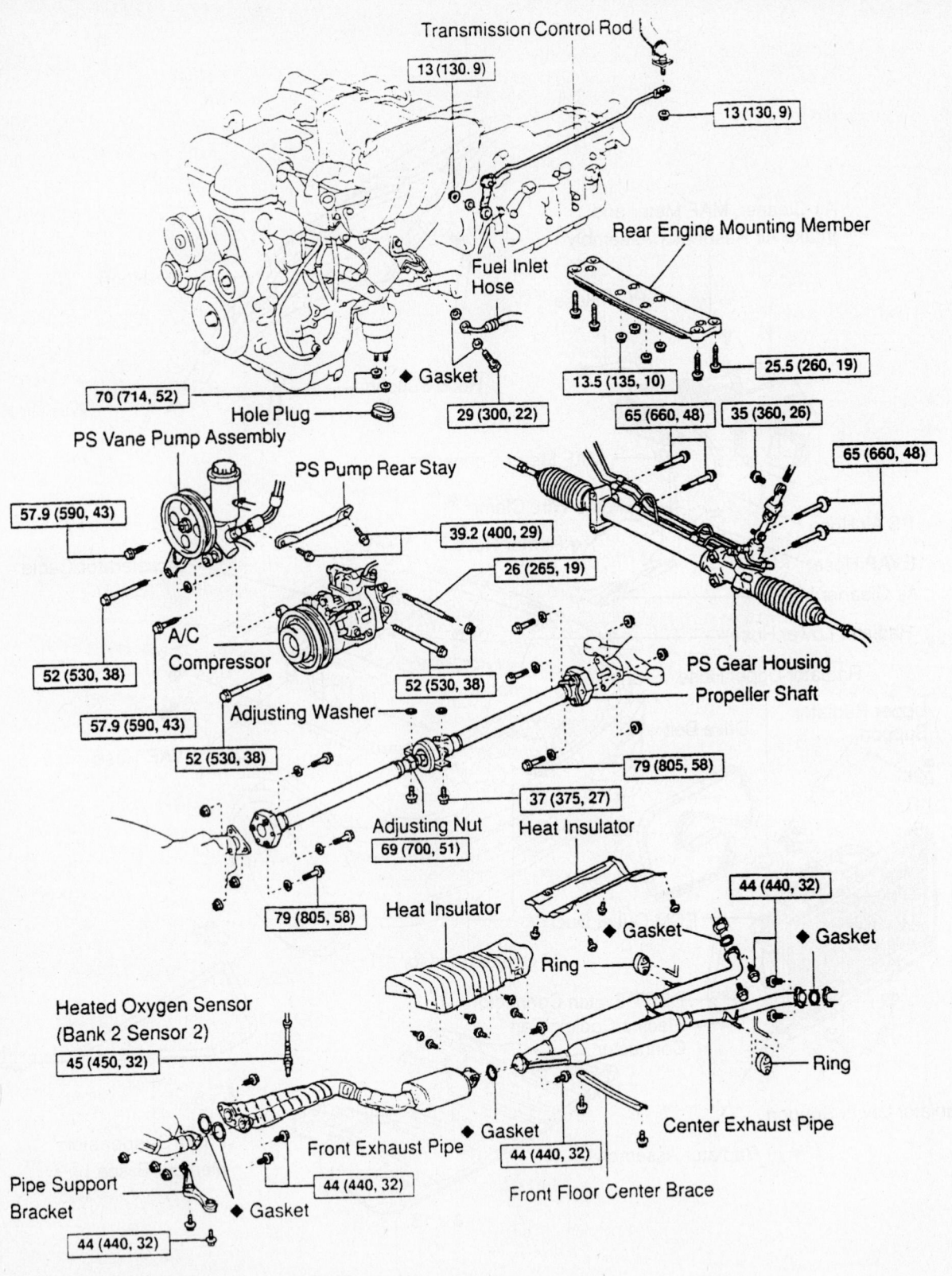

Transmission Control Rod

13 (130. 9)

13 (130, 9)

Rear Engine Mounting Member

Fuel Inlet Hose

◆ Gasket

70 (714, 52)

Hole Plug

29 (300, 22)

13.5 (135, 10)

25.5 (260, 19)

65 (660, 48)

35 (360, 26)

65 (660, 48)

PS Vane Pump Assembly

57.9 (590, 43)

PS Pump Rear Stay

39.2 (400, 29)

26 (265, 19)

52 (530, 38)

A/C Compressor

52 (530, 38)

57.9 (590, 43)

Adjusting Washer

52 (530, 38)

PS Gear Housing

Propeller Shaft

79 (805, 58)

37 (375, 27)

Adjusting Nut

69 (700, 51)

Heat Insulator

79 (805, 58)

Heat Insulator

44 (440, 32)

◆ Gasket

◆ Gasket

Ring

Heated Oxygen Sensor
(Bank 2 Sensor 2)

45 (450, 32)

Ring

Center Exhaust Pipe

◆ Gasket

Front Exhaust Pipe

44 (440, 32)

44 (440, 32)

Front Floor Center Brace

Pipe Support Bracket

◆ Gasket

44 (440, 32)

N·m (kgf·cm, ft·lbf) : Specified torque

◆ Non-reusable part

9301LG04

Exploded view of the related underbody components for 2JZ-GE engine removal (cont.)—GS 300

25. Check and/or adjust the ignition timing.

26. If equipped, install the front suspension brace and tighten the mounting bolts to 43 ft. lbs. (58 Nm).

27. Install the engine cover, hood and the hood insulator pad.

28. Refill and check all fluids.

29. Road test the vehicle.

30. Recheck all fluids.

IS 300

1. Before servicing the vehicle, refer to the precautions in the beginning of this section.

2. Drain the cooling system.

3. Relieve the fuel system pressure.

4. Drain the engine oil.

5. Remove or disconnect the following:

- Negative battery cable
- Engine under cover
- Air cleaner inlet
- Brake booster vacuum hose
- Radiator hoses
- Mass Air Flow (MAF) sensor connector
- Positive Crankcase Ventilation (PCV) hose
- Air intake resonator
- Accelerator cable
- Accessory drive belt
- Front subframe brace
- Floor ground strap
- Starter motor wiring harness
- Fuel inlet hose support
- Dash panel ground strap
- Heater hoses
- Evaporative Emissions (EVAP) canister hose
- Heated Oxygen (HO2S) sensor connectors
- Alternator wiring harness and clamp
- Cylinder block ground cable bracket
- Igniter connector
- Data link connector and harness clamps
- Powertrain Control Module (PCM) harness connectors
- Power steering pump
- A/C compressor
- Drive shaft
- Transmission control rod
- Exhaust front pipe
- Exhaust center pipe
- Stabilizer bar
- Front shock absorbers
- Lower ball joints from the steering knuckles

- Transmission mount crossmember. Support the powertrain from below.
- Steering intermediate shaft
- Front subframe

6. Lower the engine, transmission and subframe away from the vehicle.

To install:

7. Installation is the reverse of the removal procedure, while using the following torque values:

- Transmission flange bolts: 53 ft. lbs. (72 Nm)
- Transmission flange-to-oil pan bolts: 27 ft. lbs. (37 Nm)
- Torque converter bolts: 74 ft. lbs. (100 Nm)
- Left and right motor mount nuts: 52 ft. lbs. (70 Nm)
- Front subframe bolts: 52 ft. lbs. (70 Nm)
- Transmission mount crossmember: 19 ft. lbs. (26 Nm)
- Lower ball joint bolts: 181 ft. lbs. (245 Nm)
- Lower shock absorber bolts: 47 ft. lbs. (64 Nm)
- Stabilizer bar bolts: 13 ft. lbs. (18 Nm)
- Stabilizer bar nuts: 36 ft. lbs. (49 Nm)
- Steering intermediate shaft pinch bolt: 26 ft. lbs. (35 Nm)
- Transmission control rod nuts: 108 inch lbs. (13 Nm)

LS 430 (Except 2004)

1. Before servicing the vehicle, refer to the precautions in the beginning of this section.

2. Relieve the fuel system pressure.

3. Drain the engine coolant and engine oil

4. Remove or disconnect the following:
- The battery clamp cover
- Battery cables
- Battery
- Hood
- The oil pan protector
- Air cleaner inlet
- Air cleaner and intake air connector assembly
- Drive belt, fan clutch and fan pulley
- Accelerator, cruise control actuator and automatic transmission throttle cables from the throttle body
- Radiator
- Engine oil level sensor connector
- Alternator connector and wire
- Engine wire clamp from the bracket on the alternator
- 2 igniter connectors

- Engine wire clamp from the igniter bracket
- Ground strap from the right-hand engine mounting bracket
- Ground strap from under the left-hand fender apron
- Engine wire clamp from the cowl panel
- Radiator reservoir hose from the water bypass pipe
- Brake booster vacuum hose from the air intake chamber
- Heater hose from the heater water valve and water bypass pipe
- Fuel inlet hose from the fuel inlet pipe
- Fuel return hose to the return pipe
- Power steering air hose from the air intake chamber
- 2 power steering hoses from the clamp on the right-hand No. 3 timing belt cover
- Evaporative Emission (EVAP) hose from the pipe (from the charcoal canister).
- Engine wire from the cabin as follows:
- Undercover from under the glove compartment
- Glove compartment
- 3 Powertrain Control Module (PCM) connectors
- 2 cowl wire connectors from the connector on the bracket
- Wire clamp from the bracket
- Grommet from the cowl panel, then pull the engine wire out.
- Power steering oil cooler pipe from the oil pan
- Heated Oxygen (HO2) sensors
- Front exhaust pipe
- 2 catalytic converters
- Center exhaust pipe
- Heat insulator from the rear side of the front exhaust pipe
- Front center floor and rear center floor crossmember braces
- Driveshaft
- Air conditioning compressor without disconnecting the air conditioning lines
- Power steering pump
- Heat insulators for the front side of the front exhaust pipe
- Heater water valve from the cowl panel by removing the 2 nuts

5. Attach the engine chain hoist to the engine hangers.

6. Remove or disconnect the following:
- Engine mounting insulators from the engine suspension crossmember by removing the 2 nuts

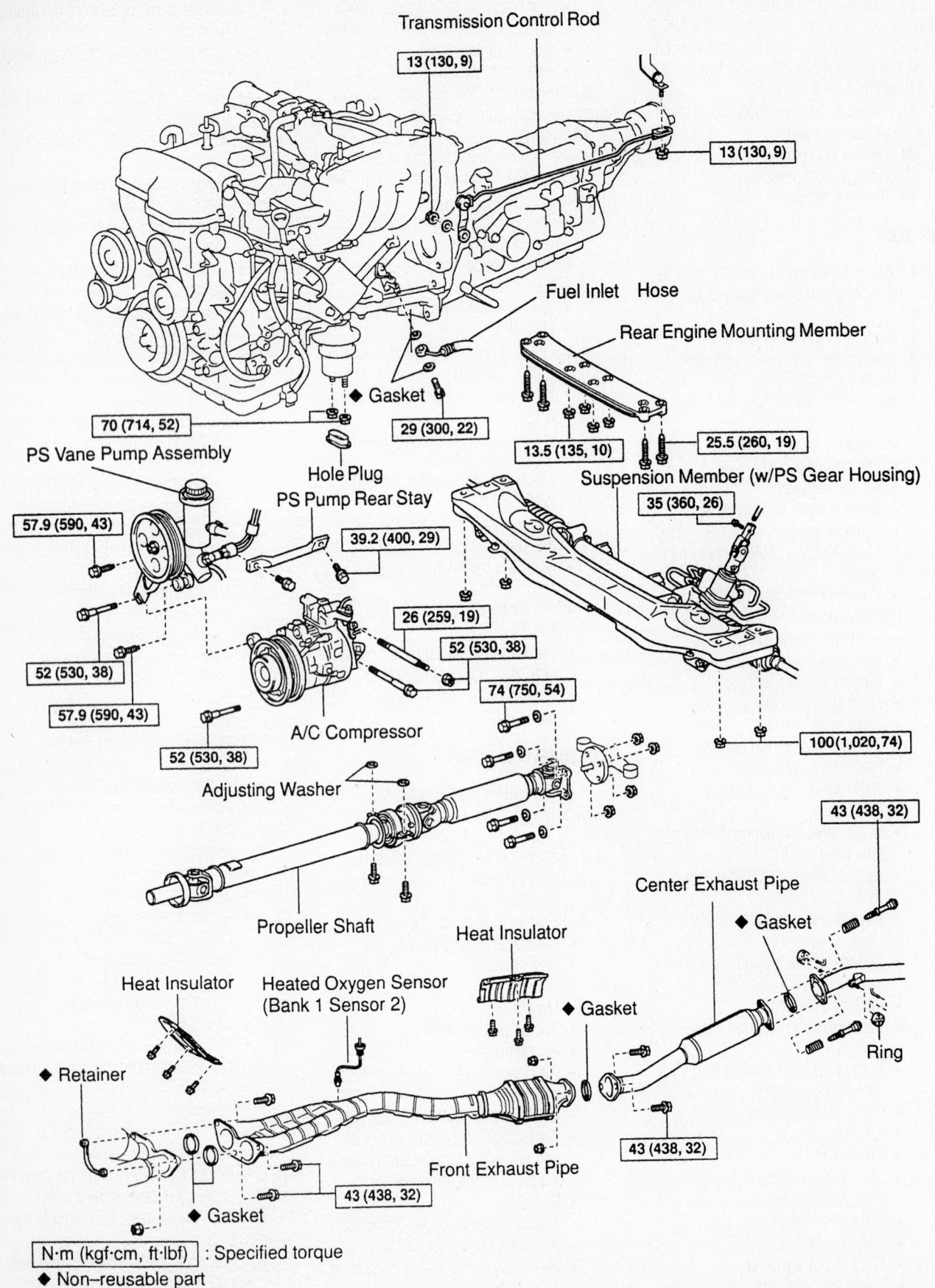

Transmission Control Rod

13 (130, 9)

13 (130, 9)

Fuel Inlet Hose

Rear Engine Mounting Member

◆ Gasket

29 (300, 22)

70 (714, 52)

Hole Plug

13.5 (135, 10)

25.5 (260, 19)

Suspension Member (w/PS Gear Housing)

PS Vane Pump Assembly

PS Pump Rear Stay

35 (360, 26)

57.9 (590, 43)

39.2 (400, 29)

26 (259, 19)

52 (530, 38)

52 (530, 38)

57.9 (590, 43)

74 (750, 54)

52 (530, 38)

A/C Compressor

100 (1,020, 74)

Adjusting Washer

43 (438, 32)

Center Exhaust Pipe

◆ Gasket

Propeller Shaft

Heat Insulator

Heat Insulator

◆ Gasket

Heat Insulator

Heated Oxygen Sensor
(Bank 1 Sensor 2)

Ring

◆ Retainer

43 (438, 32)

Front Exhaust Pipe

43 (438, 32)

43 (438, 32)

◆ Gasket

N·m (kgf·cm, ft·lbf) : Specified torque

◆ Non−reusable part

Exploded view of the engine mounting—IS 300

9347LG03

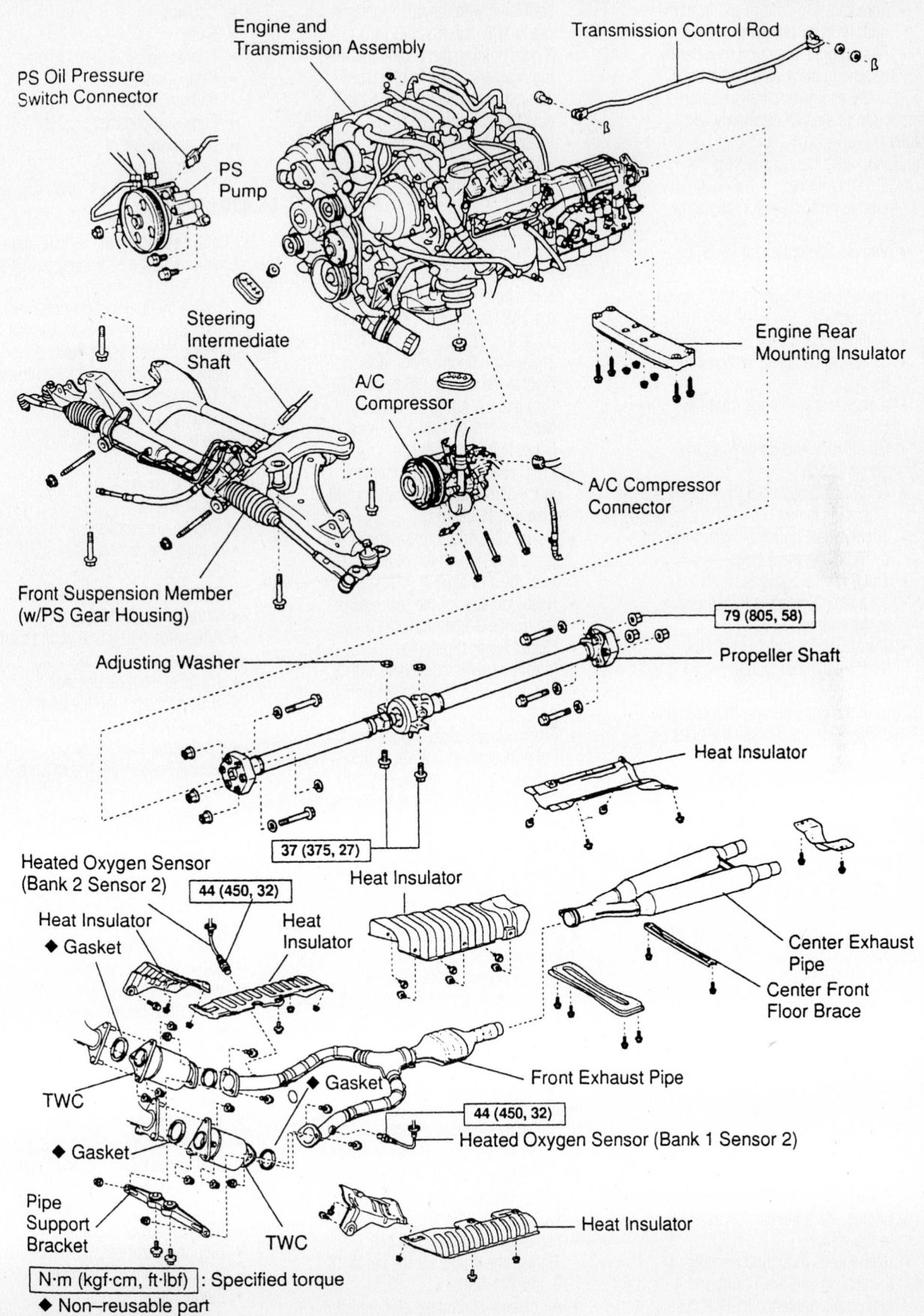

Engine and Transmission Assembly

Transmission Control Rod

PS Oil Pressure Switch Connector

PS Pump

Steering Intermediate Shaft

A/C Compressor

Engine Rear Mounting Insulator

A/C Compressor Connector

Front Suspension Member (w/PS Gear Housing)

Adjusting Washer

Propeller Shaft

79 (805, 58)

Heat Insulator

37 (375, 27)

Heated Oxygen Sensor (Bank 2 Sensor 2)

44 (450, 32)

Heat Insulator

Heat Insulator

Heat Insulator

Center Exhaust Pipe

Center Front Floor Brace

◆ Gasket

TWC

◆ Gasket

◆ Gasket

Front Exhaust Pipe

44 (450, 32)

Heated Oxygen Sensor (Bank 1 Sensor 2)

Pipe Support Bracket

TWC

Heat Insulator

N·m (kgf·cm, ft·lbf) : Specified torque

◆ Non–reusable part

9347LG04

Exploded view of the engine removal and related components—LS 430

- Transmission control rod from the shift lever by removing the nut
- Rear engine mounting member by removing the 4 nuts and 4 bolts

7. Lift the engine and transmission assembly out of the vehicle slowly and carefully.

8. Disconnect the engine from the transmission as follows:

9. Remove or disconnect the following:

- Vehicle Speed Sensor (VSS) connector
- Park/Neutral Position (PNP) switch connector
- Solenoid connector
- Direct clutch speed sensor connector
- 4 engine wire clamps from the brackets
- Oil dipstick and guide from the transmission
- Oil cooler pipes from the transmission and clamps
- The flywheel housing undercover by removing the 2 bolts
- The 6 torque converter bolts
- 10 bolts holding the transmission to the engine
- Transmission together with the torque converter clutch

To install:

10. Install the transmission to the engine and install the 10 bolts. Tighten the bolts as follows:

- 14mm: 27 ft. lbs. (37 Nm)
- 17mm: 53 ft. lbs. (72 Nm)

11. Install or connect the following:

- Torque converter clutch bolts. Apply adhesive to 2 or 3 threads of the bolt end. Tighten the bolts to 30 ft. lbs. (41 Nm).
- Flywheel housing undercover with the 2 bolts. Tighten bolts to 14 ft. lbs. (19 Nm).
- Oil cooler pipe for the transmission
- Dipstick guide and dipstick for the transmission
- Engine wire to the transmission
- VSS connector
- PNP switch connector
- Solenoid connector
- Direct clutch speed sensor connector
- 4 wire clamps to the brackets
- Engine and transmission assembly to the vehicle
- Rear engine mounting member to the vehicle and the 4 bolts and 4 nuts; bolts tightened to 19 ft. lbs. (25 Nm), nuts to 10 ft. lbs. (14 Nm)
- The transmission control rod to the

shift lever with the nut. Tighten the nut to 108 inch lbs. (13 Nm).
- 2 nuts holding the engine mounting brackets to the front suspension crossmember. Tighten the 2 nuts to 52 ft. lbs. (70 Nm).
- Heater water valve to the cowl panel with the 2 nuts

12. Remove the engine hoist.

13. Install or connect the following:

- Heat insulators for the front side of the front exhaust pipe
- Power steering pump with the nut and 3 bolts-tighten the nut to 32 ft. lbs. (43 Nm); tighten the bolts to 29 ft. lbs. (39 Nm)
- The air conditioning compressor. Tighten the bolts to 36 ft. lbs. (49 Nm) and the nut to 22 ft. lbs. (29 Nm).
- Driveshaft to the vehicle
- Front center floor crossmember brace and tighten the bolts to 108 inch lbs. (13 Nm)
- Rear center floor crossmember brace and tighten the bolts to 108 inch lbs. (13 Nm)
- Heat insulator for the rear side of the front exhaust pipe
- Center exhaust pipe
- 2 front catalytic converters with 3 new nuts each. Tighten the nuts to 46 ft. lbs. (62 Nm).
- Front exhaust pipe. 4 bolts holding the pipe support bracket to the transmission: 32 ft. lbs. (44 Nm).
- HO_2 sensors. Tighten the sensors to 33 ft. lbs. (44 Nm).
- Power steering oil cooler pipe
- Engine wire harness to the passenger's compartment
- 3 PCM connectors
- 2 engine wire connectors to the connector on the bracket
- Engine wire clamp to bracket
- Glove compartment and the dash undercover
- All of the engine assembly connectors, wires, straps, clamps and hoses
- Radiator
- Accelerator and cruise control cables to the throttle body
- Throttle control cable to the throttle body, if equipped with automatic transmission
- Fan pulley
- Fan clutch and the drive belt. Tighten the 4 nuts for the fan to 16 ft. lbs. (21 Nm).
- Air cleaner and intake air connector assembly
- Air cleaner inlet

- Coolant
- Battery
- Engine oil
- Battery cables
- Battery cover
- Engine undercover
- Oil pan protector
- Hood

LS 430 (2004)

1. Before servicing the vehicle, refer to the precautions in the beginning of this section.

2. Relieve the fuel system pressure

3. Drain the engine coolant, engine oil and automatic transmission fluid

4. Remove or disconnect the following:

- The battery clamp cover
- Battery cables
- Battery
- Hood assembly
- V-bank covers
- Front wheels
- Engine undercovers
- Radiator assembly
- Fuel pipe No 2
- Fan and alternator drive belts
- Wiring from ECM box
- Alternator wiring and wiring clamps
- Ground strap from alternator stay
- PS hose from oil pan bolt
- Ground strap from the body
- PS air hose
- Fuel vapor feed hose
- Heater hoses from heater core
- PS oil reservoir
- PS pump
- PS air hoses
- PS oil pressure switch connector
- A/C compressor. Lay assembly to the side and support. Do NOT discharge system
- Front center floor brace
- Exhaust pipe assembly
- w/Catalyst converter assembly
- Front floor heat insulator
- Parking brake cable heat insulator
- Drive shaft w/center support bearing
- Floor shift gear shifting rod
- Steering shaft with yoke
- Front disc brake calipers. Lay to the side and support. Do NOT disconnect fluid lines
- RF & LF upper suspension arms
- RF & LF shock absorbers
- Height control sensors
- Rack and Pinion assembly
- 2 hole plugs
- 4 nuts on engine mounting insulators

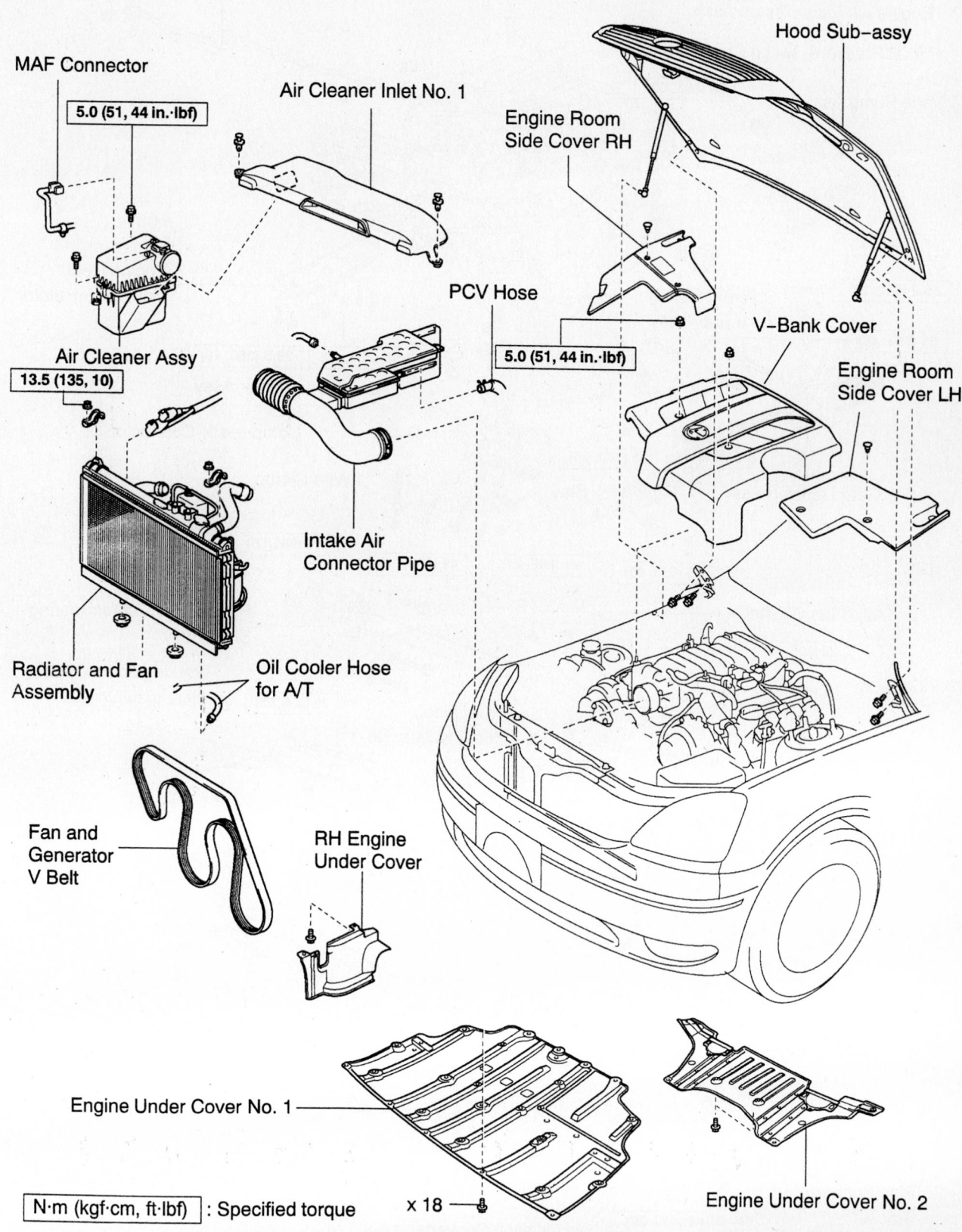

MAF Connector

5.0 (51, 44 in.·lbf)

Air Cleaner Inlet No. 1

Engine Room
Side Cover RH

Hood Sub–assy

PCV Hose

5.0 (51, 44 in.·lbf)

V-Bank Cover

Engine Room
Side Cover LH

Air Cleaner Assy

13.5 (135, 10)

Intake Air
Connector Pipe

Radiator and Fan
Assembly

Oil Cooler Hose
for A/T

Fan and
Generator
V Belt

RH Engine
Under Cover

Engine Under Cover No. 1

Engine Under Cover No. 2

N·m (kgf·cm, ft·lbf) : Specified torque

x 18

67162-LEXU-G120

Exploded view underhood covers requiring removal—2004 LS 430

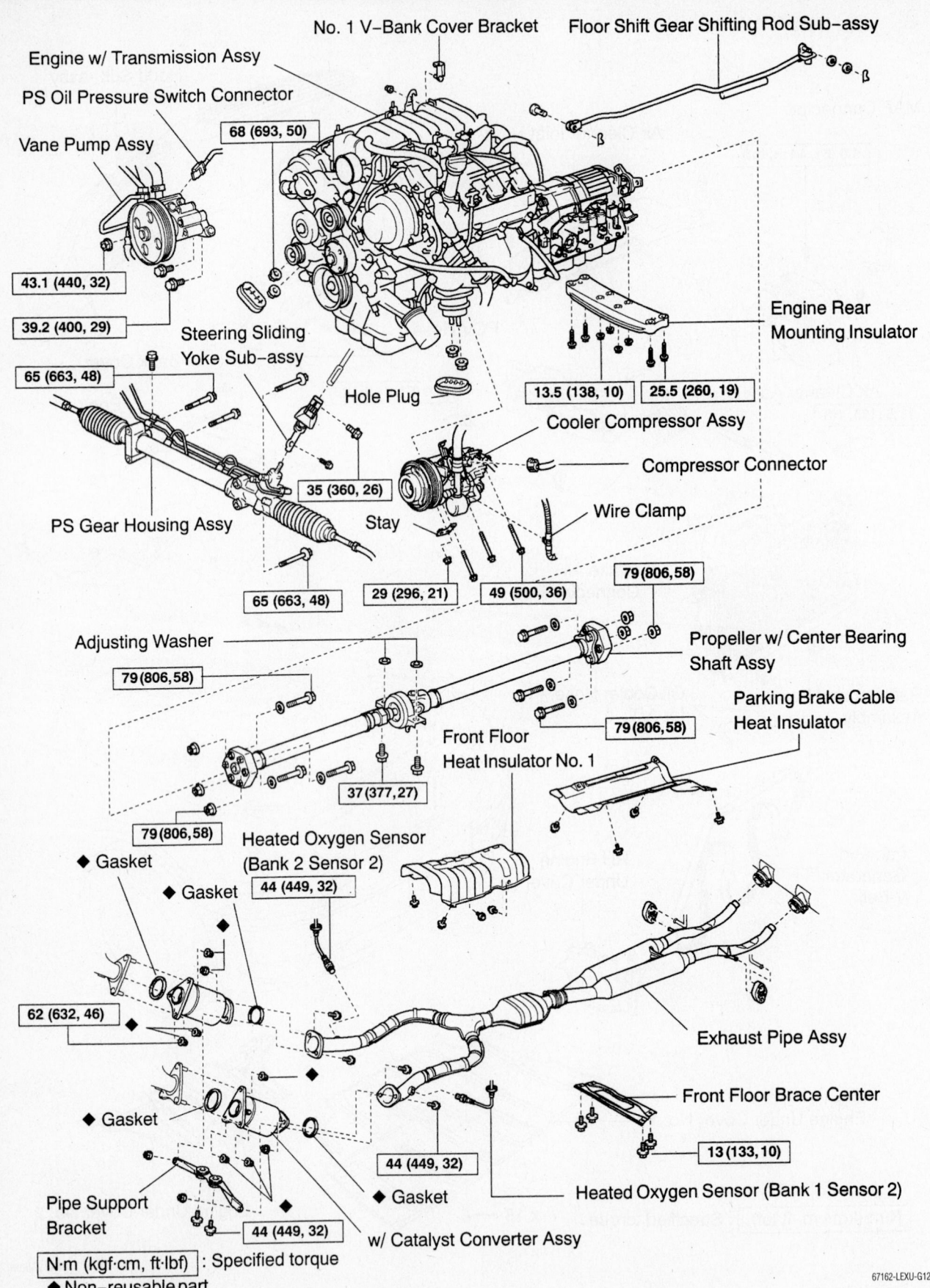

No. 1 V-Bank Cover Bracket

Floor Shift Gear Shifting Rod Sub-assy

Engine w/ Transmission Assy

PS Oil Pressure Switch Connector

Vane Pump Assy

68 (693, 50)

43.1 (440, 32)

39.2 (400, 29)

Steering Sliding Yoke Sub-assy

65 (663, 48)

Hole Plug

Engine Rear Mounting Insulator

13.5 (138, 10)   25.5 (260, 19)

Cooler Compressor Assy

Compressor Connector

35 (360, 26)

PS Gear Housing Assy

Stay

Wire Clamp

65 (663, 48)   29 (296, 21)   49 (500, 36)   79 (806, 58)

Adjusting Washer

79 (806, 58)

Propeller w/ Center Bearing Shaft Assy

Parking Brake Cable Heat Insulator

79 (806, 58)

Front Floor Heat Insulator No. 1

37 (377, 27)

79 (806, 58)

Heated Oxygen Sensor (Bank 2 Sensor 2)

◆ Gasket

44 (449, 32)

◆ Gasket

62 (632, 46)

◆ Gasket

Exhaust Pipe Assy

Front Floor Brace Center

44 (449, 32)

13 (133, 10)

Heated Oxygen Sensor (Bank 1 Sensor 2)

Pipe Support Bracket

44 (449, 32)

w/ Catalyst Converter Assy

◆ Gasket

N·m (kgf·cm, ft·lbf) : Specified torque

◆ Non-reusable part

Exploded view engine attached components—2004 LS 430

67162-LEXU-G122

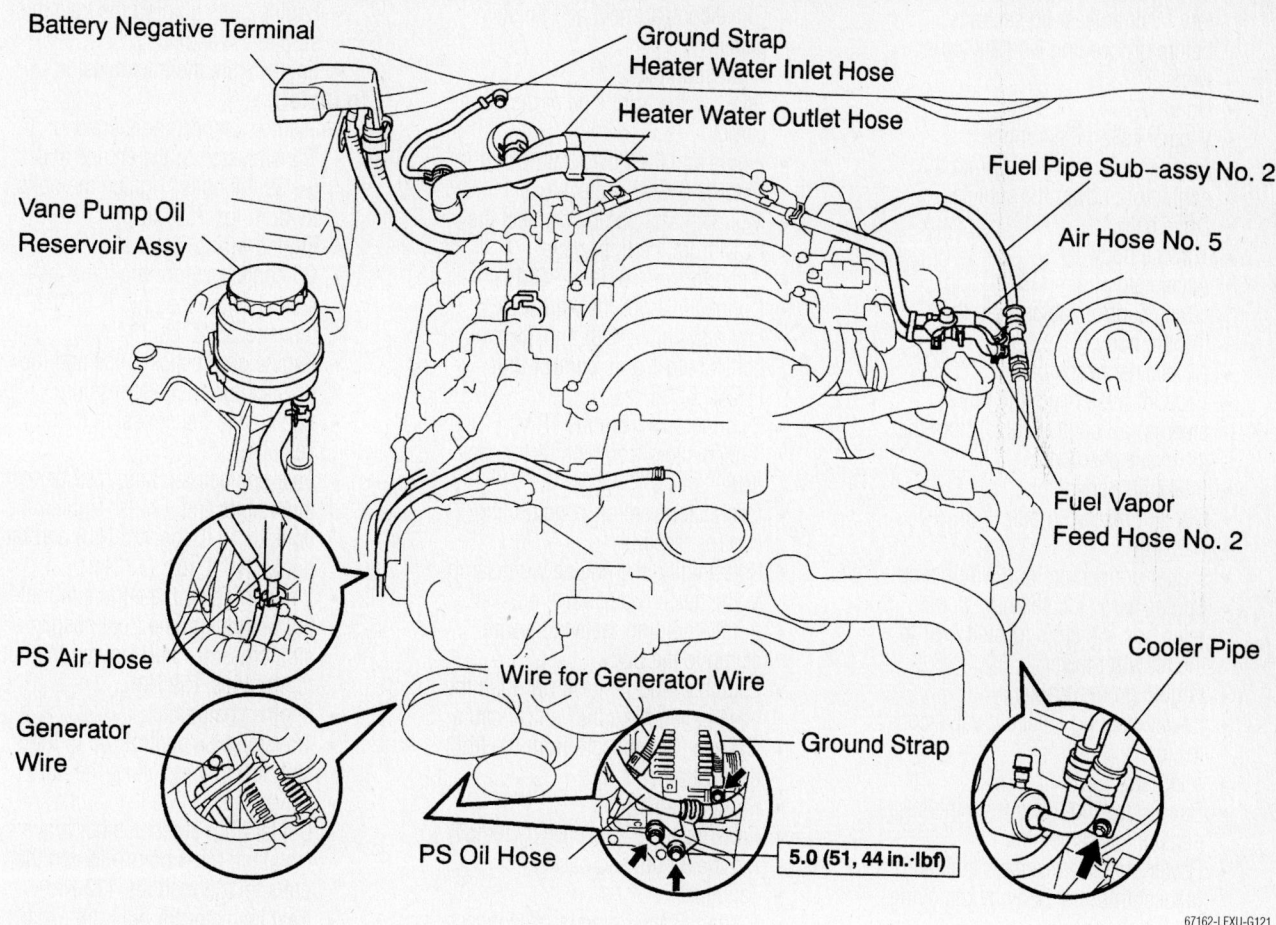

Battery Negative Terminal

Ground Strap
Heater Water Inlet Hose

Heater Water Outlet Hose

Vane Pump Oil
Reservoir Assy

Fuel Pipe Sub–assy No. 2

Air Hose No. 5

Fuel Vapor
Feed Hose No. 2

PS Air Hose

Generator
Wire

Wire for Generator Wire

Ground Strap

Cooler Pipe

PS Oil Hose

5.0 (51, 44 in.·lbf)

67162-LEXU-G121

Exploded view of engine component removed during engine removal—2004 LS 430

- 4 bolts and 4 nuts on rear engine mounting member

5. Attach engine hoist and gently remove the engine and transmission assembly

6. Disconnect transmission from the engine

To install:

7. Install the transmission to the engine and install the 10 bolts. Tighten the bolts as follows:
  - 14mm: 27 ft. lbs. (37 Nm)
  - 17mm: 53 ft. lbs. (72 Nm)

8. Install or connect the following:
  - Torque converter clutch bolts. Apply adhesive to 2 or 3 threads of the bolt end. Tighten the bolts to 30 ft. lbs. (41 Nm).
  - Flywheel housing undercover with the 2 bolts. Tighten bolts to 14 ft. lbs. (19 Nm).
  - Oil cooler pipe for the transmission
  - Dipstick guide and dipstick for the transmission
  - Engine wire to the transmission
  - VSS connector
  - PNP switch connector

- Solenoid connector
- Direct clutch speed sensor connector
- 4 wire clamps to the brackets
- Engine and transmission assembly to the vehicle
- Rack and Pinion assembly. Tighten to 48 ft. lbs. (65 Nm)
- Height control sensor link
- Front shock absorbers
- Upper suspension arms
- Front disc brake calipers
- Steering shaft and yoke
- Floor gear shift rod
- Drive shaft and center bearing assembly
- Front brake cable heat insulator
- Front floor heat insulator
- Catalyst converter
- Exhaust pipe assembly
- Front floor center brace
- A/C compressor
- PS pump assembly
- PS pump reservoir
- Heater hoses to heater core
- Fuel vapor feed hose
- Air hose No 5

- Wiring to ECM, alternator, wiring clamp, ground wire on alternator stay, PS hose to oil pan bolt, body ground strap

9. reinstall remaining components in reverse order of removal.

10. Add engine oil, automatic transmission fluid and coolant

11. Add power steering fluid and bleed system

12. Start engine and check for fluid, fuel and exhaust leaks

13. Set ignition timing and engine speed

14. Inspect and adjust front end alignment

15. Check speed sensor signal

GS 430 (Except 2004)

1. Before servicing the vehicle, refer to the precautions in the beginning of this section.

2. Relieve the fuel pressure from the fuel lines.

3. Drain the engine coolant from the cooling system.

4. Remove or disconnect the following:
  - Battery cables and remove the bat-

tery. Wait at least 90 seconds before proceeding with any other work.
- Hood
- V-bank cover, if equipped
- Engine undercover and drain the engine oil. Lower the vehicle.
- Drive belt
- Throttle body
- Accelerator, transmission and cruise control cables from the throttle body.
- Air cleaner assembly
- Vacuum hose (from the power steering air control valve) from the air intake chamber.
- Intake air connector
- Coolant reservoir tank
- Radiator
- Igniter connectors and wire clamp
- Engine wires located next to the relay box, which is located next to the left strut tower
- Engine ground cable
- Power steering solenoid valve connector
- Alternator
- Power steering tubes from the suspension crossmember
- Power steering reservoir tank and bracket from the body by removing the 3 bolts
- Power steering pump by removing the pump mounting bolts and nut. Do NOT disconnect the power steering lines and place the pump off to the side.
- Air conditioning compressor from the engine. Do NOT remove the compressor pressure lines.
- Heater water hose from the water bypass hose
- Heater water hose from the heater water valve
- Brake booster hose from the union on the air intake chamber
- Vacuum hose from the Vacuum Switching Valve (VSV) for the heater water valve from the air intake chamber
- Ground strap from the bracket on the body
- Fuel inlet hose from fuel tube
- Charcoal canister from the engine
- Engine wire from the cabin as follows:
- Passenger's side lower instrument panel undercover
- 4 screws to the lower instrument panel finish panel and glove compartment door assembly

- Glove compartment and finish panel.
- Right scuff plate
- Take out the front side of the floor carpet
- 2 nuts and the Powertrain Control Module (PCM) protector
- Mounting nut and disconnect the PCM from the floor panel.
- 2 connectors from the PCM
- Connector from the Anti-lock brake system (ABS) and Traction control electronic control unit (TRAC ECU)
- 2 connectors from the TRAC ECU
- 4 connectors from connector cassette
- Connector from air conditioning control assembly
- Bolt holding the engine wire clamp to the heater water valve bracket.
- 2 bolts holding the engine wire clamp to the body.
- Engine wiring harness (through the cowl panel) from the vehicle cabin
- Oxygen (O_2) sensors from the front exhaust pipe
- Front exhaust pipe
- Front catalytic converter by removing the 3 nuts and gasket
- Tailpipes
- Center exhaust pipe by disconnecting the 2 hooks
- Heat insulator by removing the 4 nuts
- Center floor crossmember brace by removing the 4 bolts
- Driveshaft from the vehicle using the proper tools (2 of tool SST 09922–10010), loosen the adjusting nut on the driveshaft. Place matchmarks on the transmission flange and the flexible coupling.
- The transmission control rod from the shift lever by removing the nut

5. Attach the engine chain hoist to the engine hangers.
6. Remove or disconnect the following:
- 2 nuts holding the engine mounting insulators to the front suspension crossmember.
- 4 bolts, 4 nuts and the rear engine mounting member
- Ground strap to the rear mounting member
- Engine out of the vehicle
- Oil dipstick guide and dipstick for transmission
- Oil cooler pipes for the transmission
- All the engine wiring

- Engine bolts holding the transmission to the engine
- Engine from the transmission

To install:
7. Install or connect the following:
- Transmission to the engine and install the bolts. Tighten the bolts to 42 ft. lbs. (57 Nm).
- Engine wiring
- Oil cooler pipe for the transmission. Tighten the unions on the pipes to 25 ft. lbs. (34 Nm).
- Engine oil dipstick guide and the dipstick for the transmission
- Engine and transmission to the vehicle
- Rear engine mounting member with the 4 bolts and 4 nuts. Tighten the bolts to 19 ft. lbs. (25 Nm) and the nuts to 10 ft. lbs. (13 Nm).
- 2 nuts holding the engine mounting brackets to the front suspension crossmember. Tighten the nuts to 43 ft. lbs. (59 Nm).
- Engine chain hoist
- Transmission control rod to the shift lever by installing the nut
- Driveshaft
- Center floor crossmember brace by installing the 4 bolts. Tighten the bolts to 108 inch lbs. (13 Nm).
- Heat insulator for the front exhaust pipe by installing the 4 bolts
- Center exhaust pipe by installing the 2 hooks
- Tailpipe and tighten the 2 bolts to 14 ft. lbs. (19 Nm)
- Front catalytic converter and tighten the nuts to 46 ft. lbs. (62 Nm)

8. Torque the front exhaust retainers as follows:
- 4 bolts and nuts holding the catalytic converter to the front exhaust pipe to 32 ft. lbs. (43 Nm)
- 2 bolts and nuts holding the front exhaust pipe to the center exhaust pipe to 32 ft. lbs. (43 Nm)
- 4 bolts holding the pipe support bracket to the transmission to 32 ft. lbs. (43 Nm)

9. Install or reconnect the following:
- O_2 sensors to the front exhaust and tighten the sensors to 33 ft. lbs. (44 Nm)
- Engine wiring harness in through the cowl panel
- Engine wire retainer with the 3 bolts
- Reattach the connectors under the dash panel
- PCM with the nut
- PCM protector with the 2 nuts

- Floor carpet
- Scuff plate
- Lower instrument panel finish panel and glove compartment door assembly with the 4 screws
- Instrument panel undercover with the 2 screws
- Charcoal canister
- All hoses and grounds
- Air conditioning compressor with the nut and 3 bolts. Tighten the bolts to 36 ft. lbs. (49 Nm) and the nut to 22 ft. lbs. (29 Nm).
- Power steering pump with the nut and 3 bolts. Tighten the bolts to 29 ft. lbs. (39 Nm) and the nut to 32 ft. lbs. (43 Nm).
- Power steering reservoir tank and bracket with the 3 bolts
- Power steering tubes with the clamp and bolt
- Alternator, tighten the nut and bolt to 27 ft. lbs. (37 Nm)
- Power steering solenoid valve connector
- Engine wire connectors
- Theft deterrent horn connector

- Ground cable to the body from the engine
- Igniter connectors
- Yellow taped connector to the igniter on the rear side
- Radiator assembly
- The reservoir tank and the inlet pipe to the fan shroud and tighten the 4 bolts to 43 inch lbs. (5 Nm)
- The 2 hydraulic lines for the fan motor and tighten the bolts to 47 ft. lbs. (64 Nm)
- Upper and lower radiator hoses to the radiator
- 2 oil cooler hoses for the transmission to the radiator
- Coolant tank
- Intake air connector
- Vacuum hose (from the power steering air control valve) to the air intake chamber
- Air cleaner
- Accelerator, transmission throttle control and the cruise control actuator cables to the engine
- Throttle cover and hose clamp with the cap nut and 2 bolts

- Evaporative emission control (EVAP) hose to the hose clamp
- Drive belt to the engine
- Battery to the engine compartment and attach the electrical connectors
- Engine coolant
- V-bank cover if it was removed
- Engine undercover
- Hood.

10. Fill the engine oil and check the transmission oil.

11. Start the engine, bleed the cooling system, and check for leaks.

2004 GS 430

1. Before servicing the vehicle, refer to the precautions in the beginning of this section.

2. Disconnect the negative battery cable. Wait at least 90 seconds before proceeding with any other work. This provides time to disarm the airbag system.

3. Remove the hood.

4. Relieve the fuel pressure from the fuel lines.

5. Remove the engine under cover.

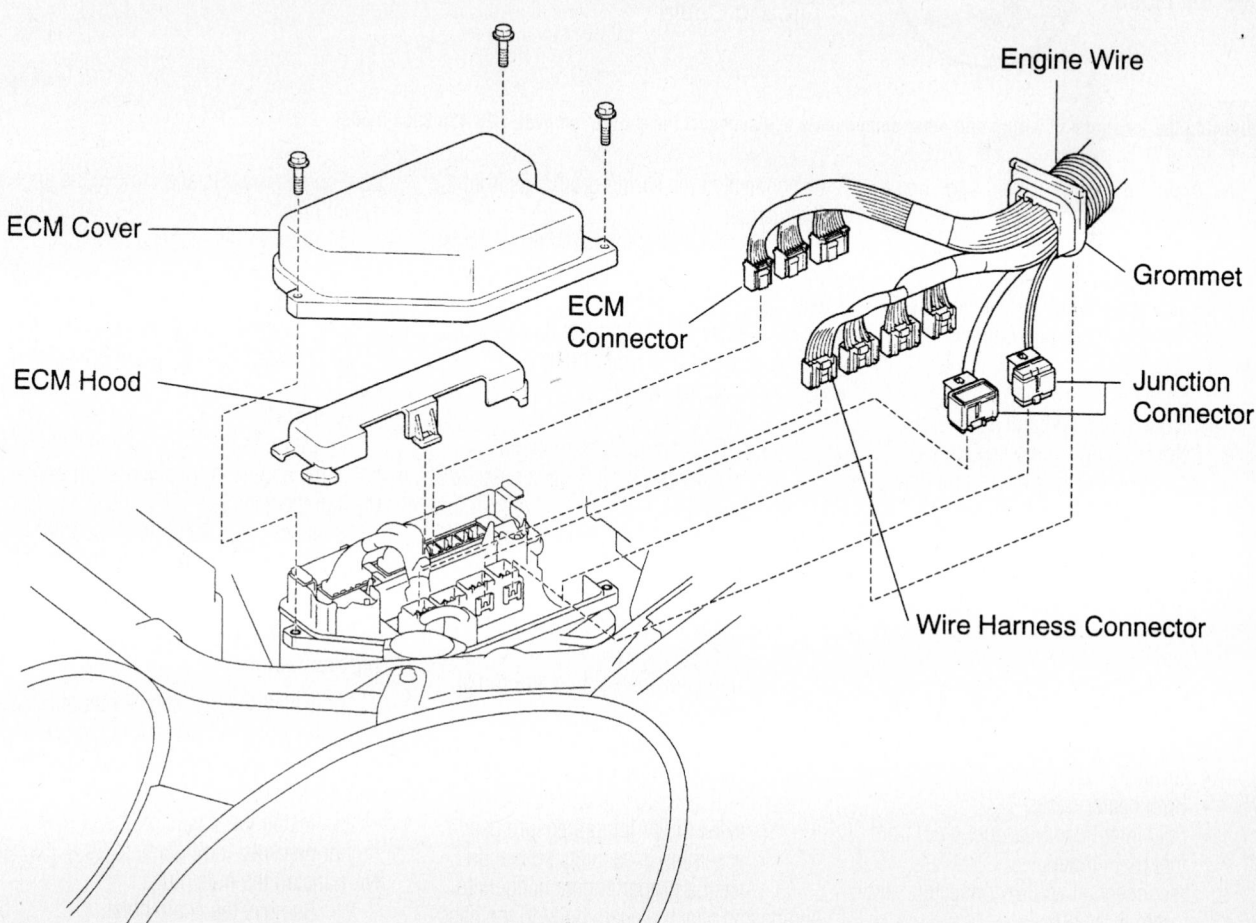

Disconnecting the wiring from the ECM box—GS 430 (2004 model)

67162-LEXU-G91

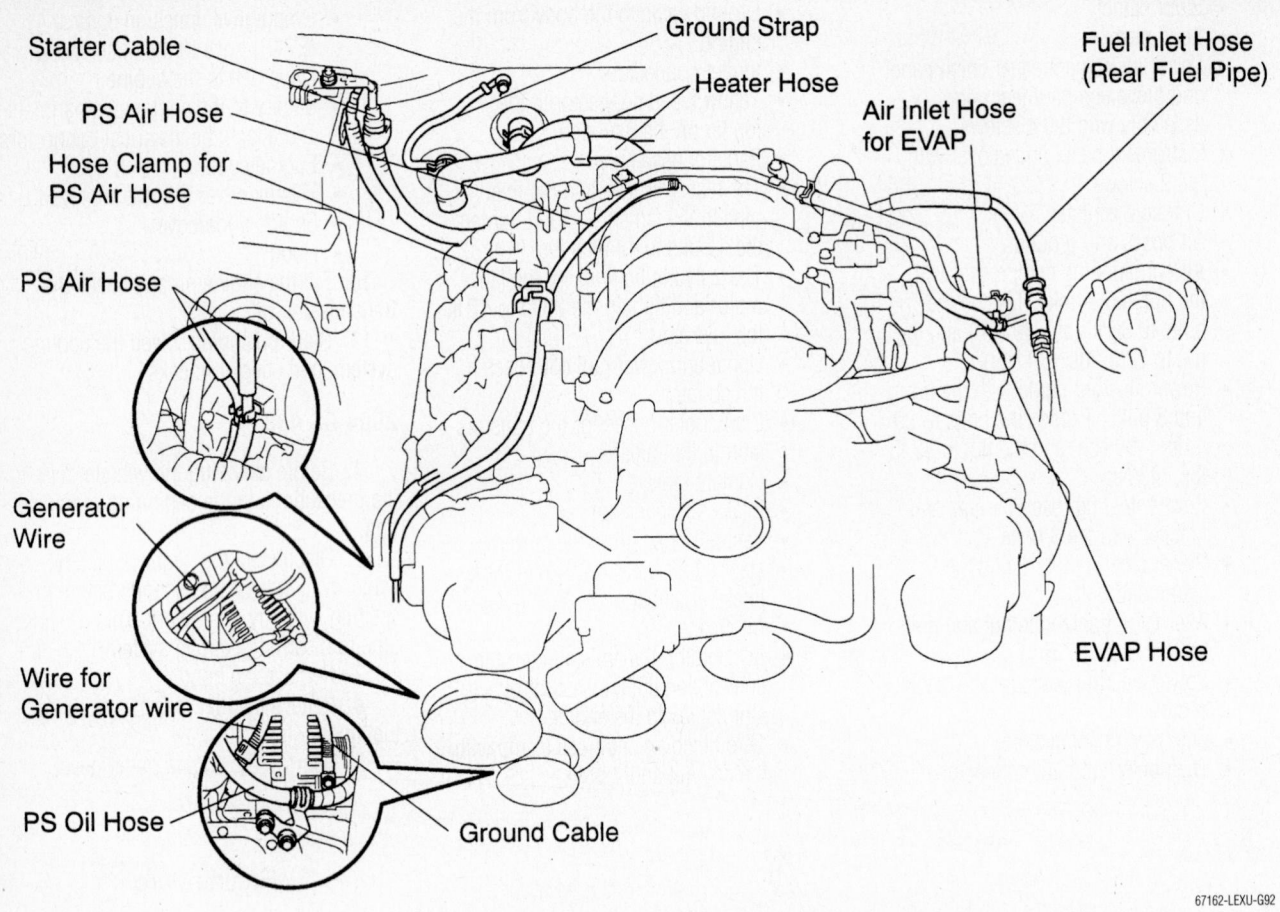

Showing the locations of wiring and other components to disconnect for engine removal—GS 430 2004 model

6. Drain the engine coolant and engine oil.

7. Remove or disconnect the following:
- V-bank cover, if equipped
- Air cleaner inlet, air cleaner assembly, and intake air connector pipe
- Radiator
- Serpentine drive belt
- Front suspension member brace.

8. Before disconnecting any wiring, connectors, cables or hoses, be sure they are properly marked for proper reinstallation. As items are marked, remove the following:
- Alternator wiring
- Power steering oil hose from oil pan
- Power steering hoses and clamp from timing belt cover
- Starter cable from battery
- Ground strap from body
- Both heater hoses
- Fuel inlet hose (rear fuel pipe) from fuel main tube

9. Disconnect all wiring connectors and grommet from ECM box

10. Disconnect the Heated Oxygen Sensors (HEGO) from the exhaust pipe, after disconnecting the wiring grommets from the floor panel.

11. Remove the front and center exhaust pipes and then remove the catalytic converters.

12. Remove or disconnect the following:
- Center front floor brace
- Heat insulators
- Driveshaft
- A/C compressor from engine mounting (Do NOT disconnect A/C hoses)
- Power steering pump oil pressure switch connector
- Power steering pump from mounting (Do NOT disconnect power steering hoses)
- Transmission shift control rod at both ends
- Power steering rack and pinion gear housing from its body mountings (Do NOT disconnect from axles or CV joints); ensure gear assembly is securely suspended

13. Attach a proper engine hangers to engine and attach a chain hoist to the hangers.

14. Remove the hole plugs to access the front engine mounts and remove the engine mount nuts.

15. Remove the rear engine mount crossmember.

16. Carefully lift the engine and transmission assembly out of the vehicle. Ensure assembly clears all wiring, hoses and components.

To install:

17. Using the chain hoist, carefully reposition the engine and transmission assembly into the vehicle.

18. Insert the stud bolts of the front engine mounting brackets into the holes of the front suspension crossmember.

➡**Ensure the engine is kept level during reattachment.**

19. Install the rear engine mount crossmember (with "V8" mark facing forward) and torque the bolts to 19 ft. lbs. (26 Nm) and the nuts to 10 ft. lbs. (14 Nm).

20. Install the 4 nuts on the front engine mount brackets and tighten to 50 ft. lbs. (68 Nm). Install the hole plugs.

21. Remove the chain hoist.

22. Reinstall the V-bank cover bracket to the engine hanger, with a nut.

23. Install or connect the following:
- Heat insulators to front side of front exhaust pipe
- Power steering gear housing to mounting; torque nuts to 48 ft. lbs. (65 Nm)
- Power steering gear oil tube
- Steering column yoke bolt to power steering gear; torque to 26 ft. lbs. (25 Nm)
- Transmission control rod
- A/C compressor to mounting; torque bolt to 36 ft. lbs. (49 Nm) and nut to 21 ft. lbs. (29 Nm)
- A/C compressor wiring
- Power steering pump to mounting; torque bolts to 29 ft. lbs. (39 Nm) and nut to 21 ft. lbs. (29 Nm)
- Power steering oil pressure switch connector
- Driveshaft
- Front center floor brace
- Heat insulators
- Catalytic converters; torque new nuts to 46 ft. lbs. (62 Nm)
- Front and center exhaust pipes
- Oxygen sensors; tighten to 32 ft. lbs. (44 Nm)

➡ **Before installing oxygen sensors, twist sensor wiring 3-1/2 turns in counterclockwise direction; hold it in this position, then install the sensors. This method prevents sensor wires from being twisted after installation.**

- Wiring connectors and grommet to ECM box
- All wiring, connectors, straps and hoses to original positions as marked
- Front suspension member brace; torque bolts to 43 ft. lbs. (58 Nm)
- Serpentine drive belt
- Radiator
- Intake air connector, air cleaner assembly and air cleaner inlet

24. Refill engine and cooling system with proper fluids and amounts.

25. Install the hood, then road test the vehicle. Recheck fluid levels after road test.

SC 430

1. Before servicing the vehicle, refer to the precautions in the beginning of this section.
2. Relieve the fuel pressure from the fuel lines.

3. Drain the engine coolant from the cooling system.
4. Remove or disconnect the following:

- Battery negative cable
- Hood sub-assembly
- V-Bank cover
- Air cleaner inlet
- Air cleaner assembly and connector pipe
- Engine undercover and drain the oil
- Drain Automatic Transmission fluid
- Remove radiator
- Fuel sub-pipe assembly
- V-belts
- Engine wire from ECU
- Alternator wires
- PS hose from # 1 oil pan bolt
- Ground strap from the body
- PS Air hose
- Fuel vapor feed hose
- Heater inlet and outlet water hose
- PS pump reservoir
- RH engine under cover
- PS oil switch connector
- PS pump from the engine
- A/C Compressor electrical connector and clamp

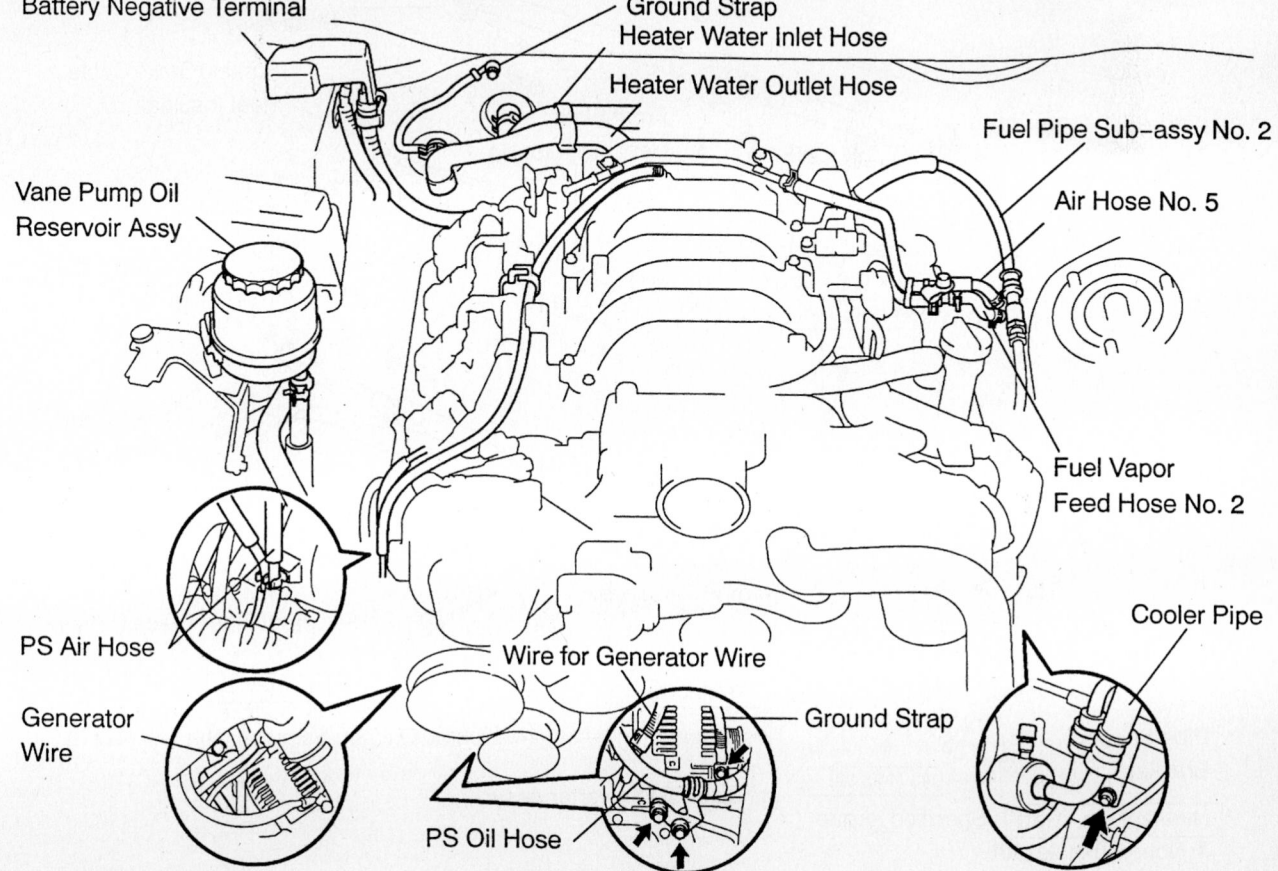

Engine component locations—SC 430

67162-LEXU-G05

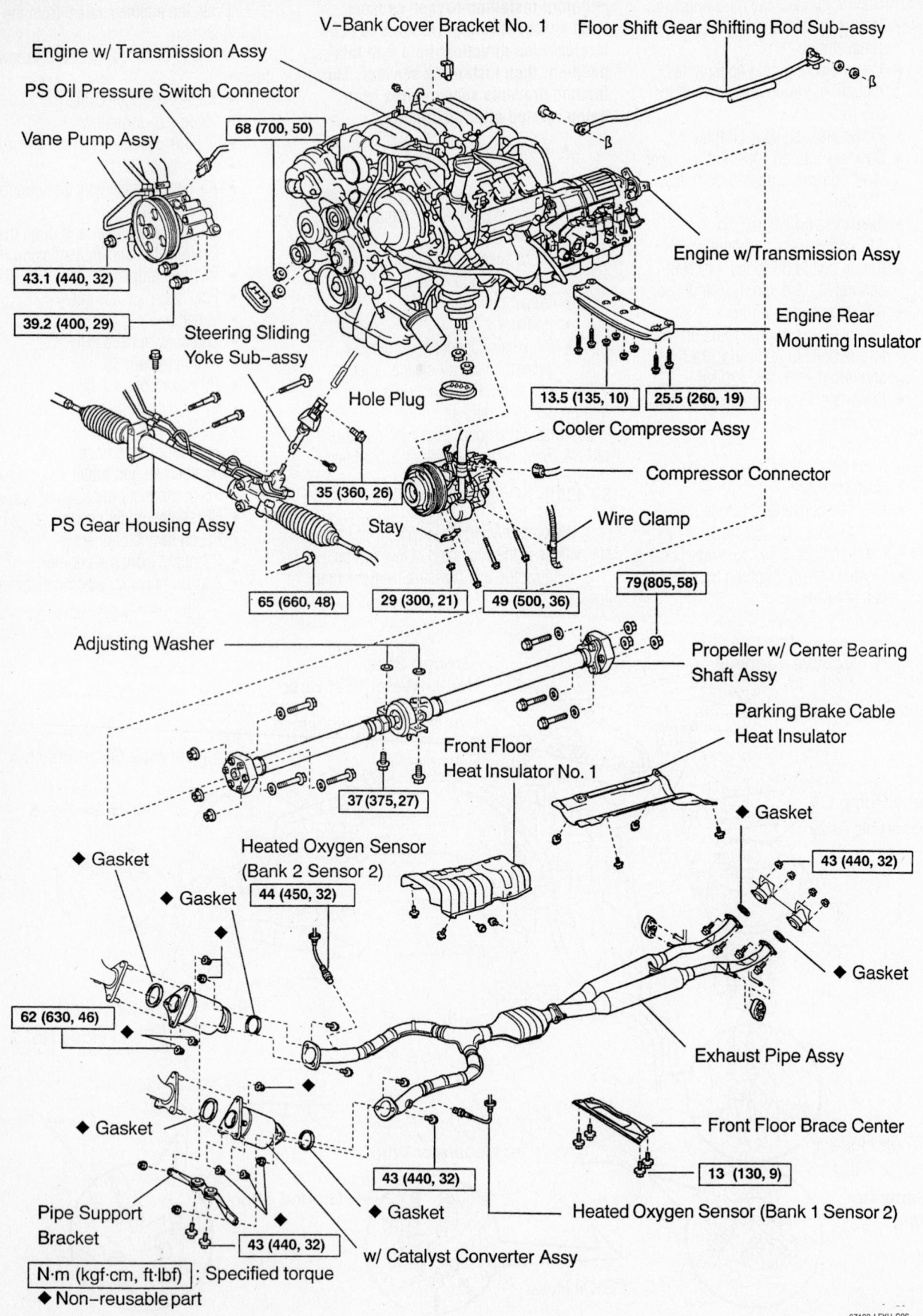

PS Oil Pressure Switch Connector

Engine w/ Transmission Assy

V-Bank Cover Bracket No. 1

Floor Shift Gear Shifting Rod Sub-assy

Vane Pump Assy

68 (700, 50)

43.1 (440, 32)

39.2 (400, 29)

Engine w/Transmission Assy

Engine Rear Mounting Insulator

Steering Sliding Yoke Sub-assy

Hole Plug

13.5 (135, 10)   25.5 (260, 19)

Cooler Compressor Assy

35 (360, 26)

Compressor Connector

PS Gear Housing Assy

Stay

Wire Clamp

65 (660, 48)   29 (300, 21)   49 (500, 36)   79 (805, 58)

Adjusting Washer

Propeller w/ Center Bearing Shaft Assy

Parking Brake Cable Heat Insulator

Front Floor Heat Insulator No. 1

37 (375, 27)

Gasket

Heated Oxygen Sensor (Bank 2 Sensor 2)

43 (440, 32)

44 (450, 32)

Gasket

Gasket

Gasket

62 (630, 46)

Exhaust Pipe Assy

Gasket

Front Floor Brace Center

43 (440, 32)

13 (130, 9)

Pipe Support Bracket

43 (440, 32)

Gasket

Heated Oxygen Sensor (Bank 1 Sensor 2)

w/ Catalyst Converter Assy

N·m (kgf·cm, ft·lbf) : Specified torque

◆ Non-reusable part

Engine removal components, exploded view—(SC 430)

67162-LEXU-G06

- A/C compressor from the engine (Do NOT disconnect hoses)
- Support compressor assembly
- Front floor brace assembly
- Exhaust pipe assembly
- Front suspension member brace sub-assembly
- Catalytic converter assembly
- Front floor heat insulator
- Parking brake cable heat insulator
- Drive shaft W/Center bearing assembly
- Steering sliding yoke assembly
- The bolts and disconnect the 2 PS oil tubes from the front frame.
- The 4 bolts securing the PS gear housing from the front frame
- Suspend the PS gear housing securely
- V-bank cover bracket from number engine hanger
- Install engine chain hoist to the engine hangers and support engine
- Remove 2 hole plugs
- 4 nuts holding the engine mounting insulator to the front suspension cross member
- 4 nuts and bolts from the rear engine mounting member
- Remove engine slowly and carefully. Make sure engine is clear of all wiring, hoses and cables.

To install:

5. Lower engine and transmission assembly onto the engine compartment
6. Install or connect the following:
   - front engine mounting bolts in front suspension crossmember
   - Rear engine mounting member with 4 bolts and nuts. Torque to: bolts 19 ft. lbs (25.5 Nm), Nuts 10 ft. lbs (13.5 Nm)
   - 4 nuts holding engine mounting brackets to front crossmember. Torque to: 50 ft. lbs. (68 Nm)
   - Front V-bank cover bracket to engine hanger with nut
   - Sliding PS gear housing yoke to intermediate shaft
   - PS gear housing with 4 bolts. Torque to 48 ft. lbs (65 Nm)
   - PS oil tube with bolts
   - Steering sliding yoke sub-assembly with 2 bolts. Torque to 26 ft. lbs (35 Nm)
   - Floor gear shift rod sub-assembly
   - Drive shaft W/Center bearing shaft assembly
   - Parking brake cable heat insulator
   - Front floor heat insulator
   - Catalytic converter assembly

- Front suspension member sub-assembly. Torque to 43 ft. lbs. (58Nm)
- Exhaust pipe assembly
- Front floor brace center
- A/C compressor, stay and wire bracket with 3 bolts and nuts. Torque bolts to 36 ft. lbs (49 Nm) and nuts to 21 ft. lbs. (29 Nm)
- A/C wiring connector and clamps
- PS pump with 2 bolts and nuts. Alternately tighten the bolts and nuts: bolts to 29 ft. lbs. (39 Nm) and nuts to 32 ft. lbs. (43 Nm)
- PS oil pressure switch connector
- PS air hoses to the PS pump
- PS pump reservoir assembly
- Heater inlet and outlet water hoses
- Vapor feed hose
- Upper PS air hose
- Engine wire from ECM box
- Alternator wires and clamp
- Ground cable to alternator
- PS hose to #1 oil pan bolt
- Ground strap to body
- V-belts
- Air cleaner assembly and connector pipes
- V-bank cover
- Hood sub-assembly
- Battery negative cable
- Automatic transmission fluid
- Engine coolant
- Engine oil
- Start engine
- Inspect for coolant, oil and transmission fluid leaks
- Adjust engine settings
- Test drive vehicle

ES 330

1. Before servicing the vehicle, refer to the precautions in the beginning of this section.
2. Relieve the fuel pressure from the fuel lines.
3. Drain the engine coolant from the cooling system.
4. Remove or disconnect the following:
   - Front wheels
   - Engine under covers
   - Front fender apron seal
   - Drain engine oil
   - Drain automatic transmission fluid
   - V-bank cover sub-assembly
   - Radiator lower air deflector
   - Battery cables and remove battery
   - Battery tray
   - Air cleaner assembly, brackets and inlets
   - Intake air resonator sub-assembly

- A/C compressor V-belt
- Alternator wiring and alternator
- Engine stabilizing control rod
- Engine mounting stay
- Alternator bracket
- Alternator adjusting bar
- A/C compressor (Do NOT disconnect hoses)
- Transmission control cable assembly
- Union to check valve hose
- Fuel vapor feed hose
- Fuel pipe sub-assembly
- Heater inlet and outlet hoses
- Radiator inlet and outlet hoses
- Oil cooler inlet and outlet hoses
- Steering gear outlet return tube
- Glove compartment door assembly
- Engine wire from ECU and junction box
- Engine harness from engine compartment junction block
- Front exhaust pipe support bracket
- Rear exhaust pipe support bracket
- Front exhaust pipe assembly
- Front stabilizer link assembly
- Rear stabilizer link assembly
- Both front axel hub nuts
- Both front speed sensors
- Separate both outer tie rod ends
- Separate front lower suspension arm sub assembly, both sides
- Left and Right drive axels
- Separate steering intermediate shaft assembly
- Attach engine hoist
- 4 bolts and 2 nuts from RH & LH frame side rail plate
- 4 bolts and 2 nuts from RH & LH front suspension member
- Carefully remove engine assembly

To install:

5. Lower engine and transmission assembly onto the engine compartment
6. Install or connect the following:
   - RH & LH side rail plates; torque large bolt 63 ft. lbs. (85 Nm) and small bolt and nuts 24 ft lbs (32 Nm)
   - RH & LH front suspension member brace. Torque large bolt to 63 ft. lbs (85 Nm) and small bolt and nuts 24 ft. lbs. (32 Nm)
   - Steering intermediate shaft assembly
   - LH & RH axel shaft assemblies
   - LH & RH lower suspension arm sub-assemblies
   - LH & RH tie rod assemblies
   - LH & RH speed sensors
   - LH & RH front axle hub nuts. Torque to 217 ft. lbs. (294 Nm)

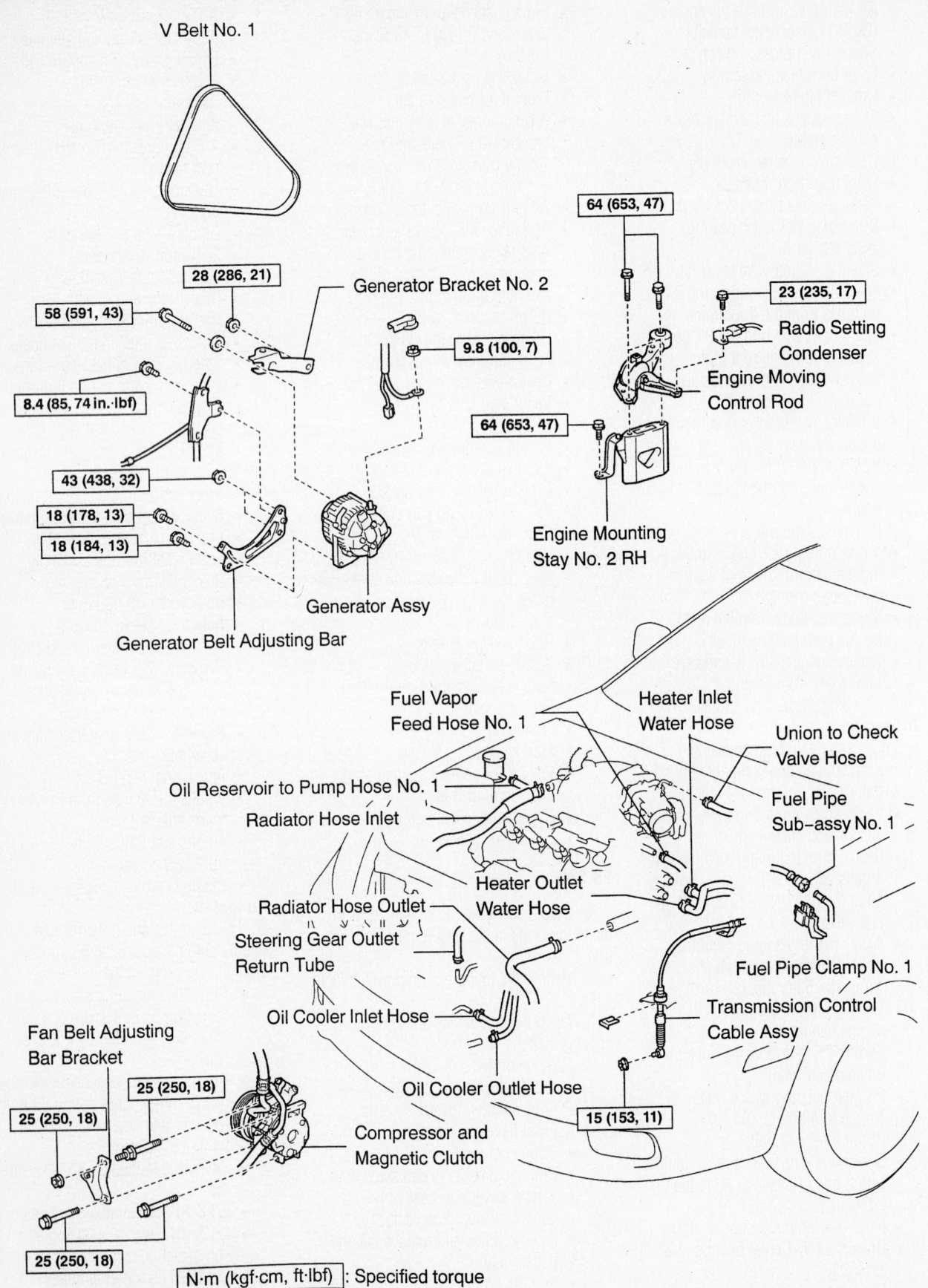

V Belt No. 1

28 (286, 21)

Generator Bracket No. 2

58 (591, 43)

9.8 (100, 7)

8.4 (85, 74 in.·lbf)

43 (438, 32)

18 (178, 13)

18 (184, 13)

Generator Assy

Generator Belt Adjusting Bar

64 (653, 47)

23 (235, 17)

Radio Setting Condenser

Engine Moving Control Rod

64 (653, 47)

Engine Mounting Stay No. 2 RH

Fuel Vapor Feed Hose No. 1

Heater Inlet Water Hose

Union to Check Valve Hose

Oil Reservoir to Pump Hose No. 1

Radiator Hose Inlet

Fuel Pipe Sub-assy No. 1

Radiator Hose Outlet

Heater Outlet Water Hose

Steering Gear Outlet Return Tube

Oil Cooler Inlet Hose

Fuel Pipe Clamp No. 1

Transmission Control Cable Assy

Fan Belt Adjusting Bar Bracket

25 (250, 18)

25 (250, 18)

Oil Cooler Outlet Hose

15 (153, 11)

Compressor and Magnetic Clutch

25 (250, 18)

N·m (kgf·cm, ft·lbf) : Specified torque

67162-LEXU-G07

Engine compartment accessory locations—ES 330

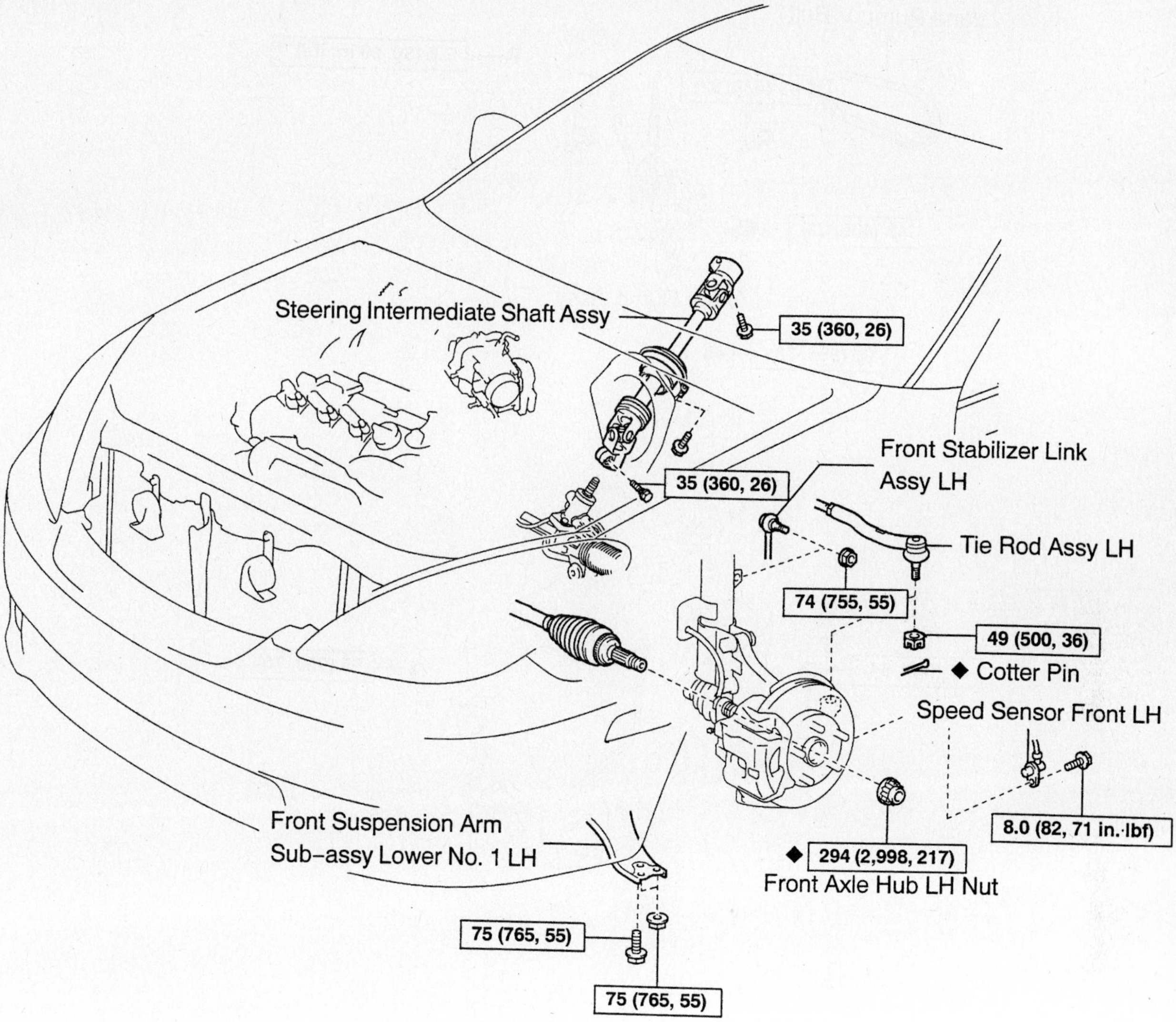

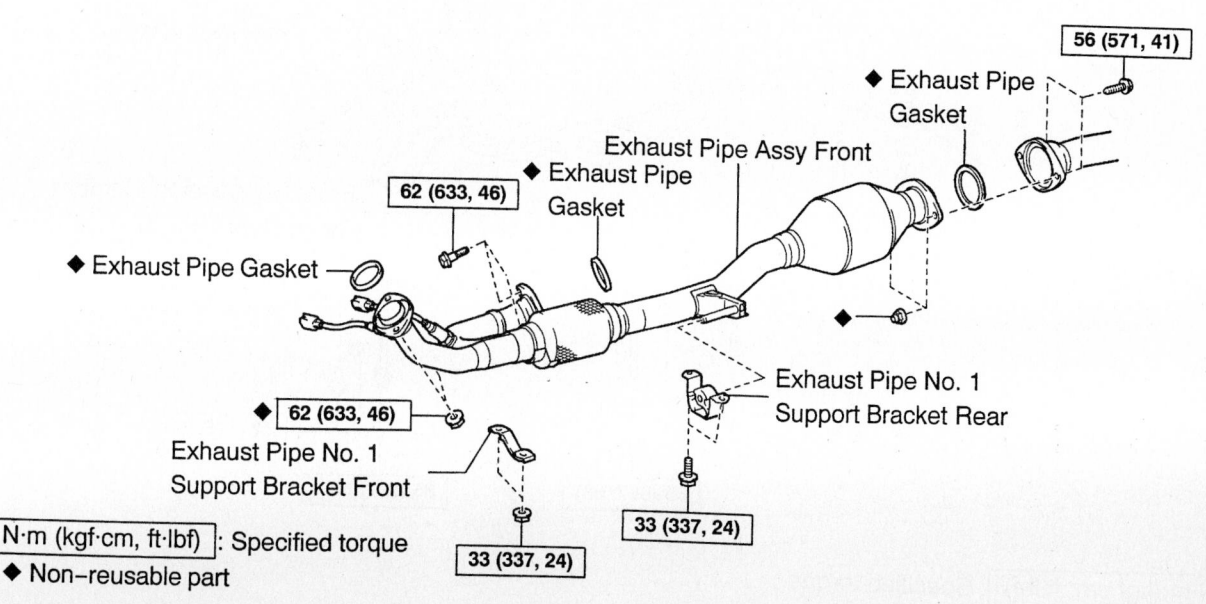

Steering Intermediate Shaft Assy

35 (360, 26)

35 (360, 26)

Front Stabilizer Link Assy LH

Tie Rod Assy LH

74 (755, 55)

49 (500, 36)

◆ Cotter Pin

Speed Sensor Front LH

8.0 (82, 71 in.·lbf)

Front Suspension Arm Sub–assy Lower No. 1 LH

◆ 294 (2,998, 217)
Front Axle Hub LH Nut

75 (765, 55)

75 (765, 55)

56 (571, 41)

◆ Exhaust Pipe Gasket

Exhaust Pipe Assy Front

62 (633, 46)

◆ Exhaust Pipe Gasket

◆ Exhaust Pipe Gasket

Exhaust Pipe No. 1 Support Bracket Rear

◆ 62 (633, 46)

Exhaust Pipe No. 1 Support Bracket Front

33 (337, 24)

N·m (kgf·cm, ft·lbf) : Specified torque

◆ Non–reusable part

33 (337, 24)

Steering and exhaust components removal—ES 330

67162-LEXU-G08

67162-LEXU-G09

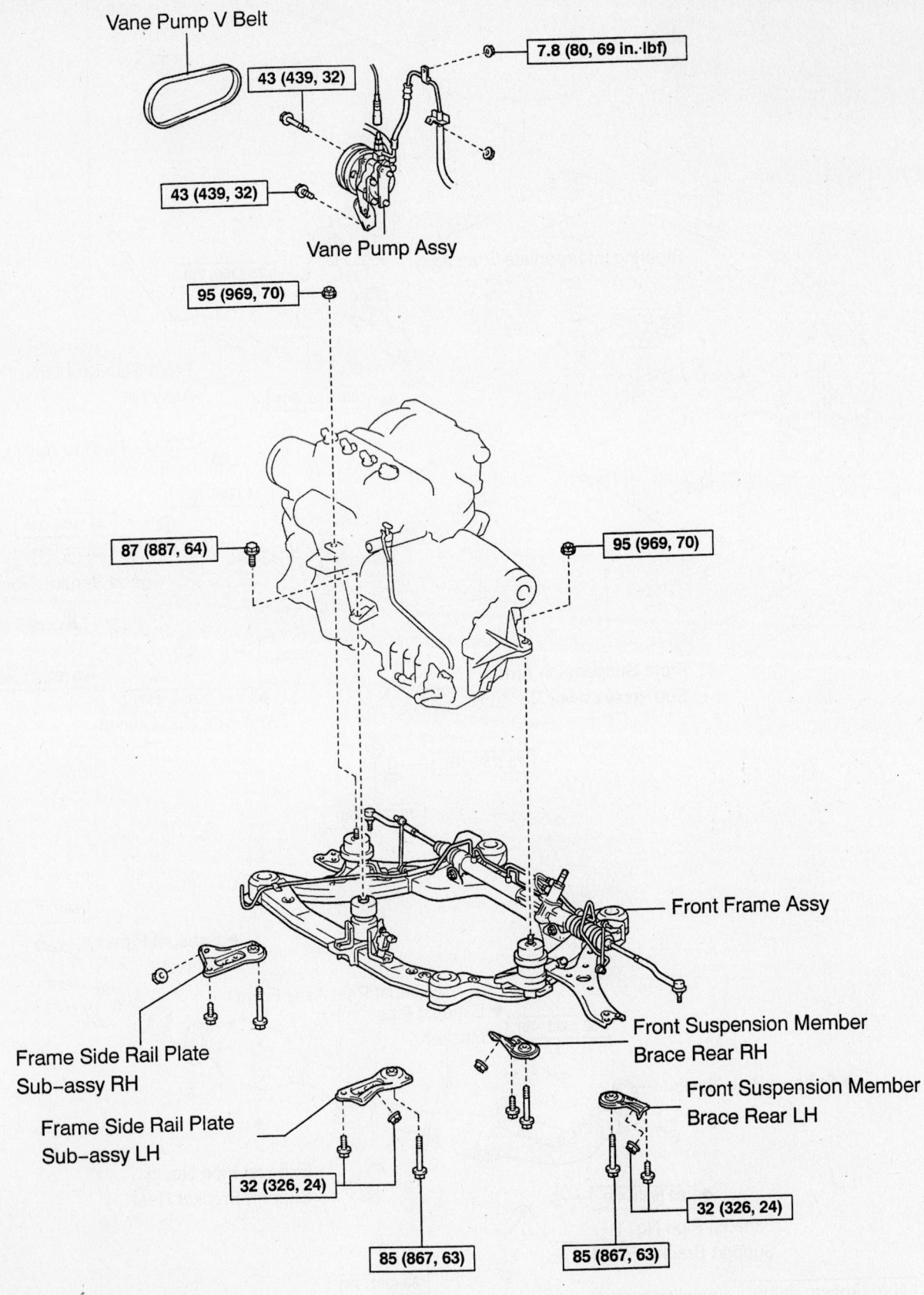

Vane Pump V Belt

43 (439, 32)

7.8 (80, 69 in.·lbf)

43 (439, 32)

Vane Pump Assy

95 (969, 70)

87 (887, 64)

95 (969, 70)

Front Frame Assy

Frame Side Rail Plate
Sub–assy RH

Frame Side Rail Plate
Sub–assy LH

Front Suspension Member
Brace Rear RH

Front Suspension Member
Brace Rear LH

32 (326, 24)

32 (326, 24)

85 (867, 63)

85 (867, 63)

N·m (kgf·cm, ft·lbf) : Specified torque

Sub frame component removal—ES 330

- LH & RH stabilizer link assemblies
- Front exhaust pipe assembly
- Rear exhaust support bracket
- Front exhaust support bracket
- Fuel pipe sub-assembly
- Transmission control cable assembly
- A/C compressor assembly
- Alternator belt adjusting bar and bracket
- RH engine mounting stay
- Engine stabilizing control rod
- Alternator assembly
- A/C compressor V-belt
- Inspect drive belt deflector and tensioner
- Intake air resonator. Torque to 44 inch lbs. (5 Nm)
- Air cleaner assembly, brackets and inlets
- Verify vacuum hose connections
- Add automatic transmission fluid
- Add engine oil
- Add power steering fluid
- Inspect automatic transaxle fluid
- Check for oil, coolant, fuel and exhaust leaks
- Adjust front wheel alignment
- Adjust ignition timing and engine idle speed
- Inspect CO/HC
- Check ABS sensor signal
- System initialization

Water Pump

REMOVAL & INSTALLATION

3.0L (1MZ-FE) Engine

1. Before servicing the vehicle, refer to the precautions in the beginning of this section.
2. Remove or disconnect the following:
   - Negative battery terminal
   - Engine coolant
   - Timing belt
   - Right and left camshaft pulleys
   - No. 2 idler pulley by removing the bolt
   - 3 clamps and engine wire from the rear timing belt cover
   - 6 bolts holding the rear timing belt cover to the engine block
   - 4 bolts and 2 nuts to the water pump
   - Water pump
   - All the old packing (sealant) and gasket material from the water pump and clean the mounting surfaces.

- All gasket material from the upper inner timing belt cover

To install:

3. Check that the water pump turns smoothly. Also, check the air hole for coolant leakage.
4. Using a new gasket, apply liquid sealer to the gasket, water pump and engine block.
5. Install or connect the following:
   - Gasket and pump to the engine and install the 4 bolts and 2 nuts. Tighten the nuts and bolts to 53 inch lbs. (6 Nm).
   - Rear timing belt cover and tighten the 6 bolts to 74 inch lbs. (9 Nm)
   - Engine wire with the 3 clamps to the rear timing belt cover
   - No. 2 idler pulley with the bolt. Tighten the bolt to 32 ft. lbs. (43 Nm). After tightening the bolt, be sure the idler pulley moves smoothly.
   - Right-hand camshaft pulley, with the flange side **outward**. Be sure to align the knock pinhole on the camshaft pulley with the knock pin on the camshaft. Camshaft bolt to 65 ft. lbs. (88 Nm).
   - Left-hand camshaft pulley with the flange side **inward**. Be sure to align the knock pin hole on the camshaft pulley with the knock pin on the camshaft. Camshaft bolt to 94 ft. lbs. (125 Nm).
   - Timing belt
   - Engine coolant
   - Negative battery cable to the battery and start the engine.

3.0L (2JZ-GE) Engine

1. Before servicing the vehicle, refer to the precautions in the beginning of this section.
2. Disconnect the negative battery cable. Wait at least 90 seconds before performing any work.
3. Drain the engine coolant.
4. Remove or disconnect the following:
   - Radiator assembly
   - Air cleaner
   - Mass Air Flow (MAF) meter
   - Intake air connector pipe assembly
   - Serpentine drive belt
   - Water pump pulley
   - Timing belt
   - Idler pulley
   - Water bypass outlet and the No. 1 water bypass pipe
   - Water inlet and the thermostat
   - Alternator (position aside)

- Bolt and engine wire bracket
- Bolt and clamp bracket for the Crankshaft Sensor (CKP) connector
- Nuts and the No. 2 water bypass pipe from the water pump
- 6 bolts and the water pump and gasket
- Drain hose and the O-ring from the cylinder block

To install:

5. Install or connect the following:
   - New O-ring to the cylinder block
   - Drain hose
   - New gasket to the water pump
   - Water pump to the water bypass pipe. Do NOT install the nut yet.
   - Water pump with the 2 bolts (A) and the 4 bolts (B)

➡ **Hand-tighten the bolts (A) first. Tighten all 6 bolts to 15 ft. lbs. (21 Nm).**

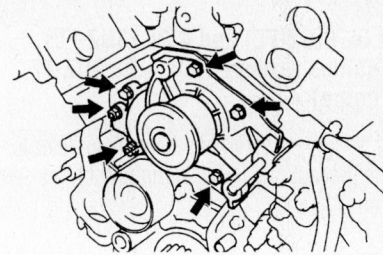

Water pump mounting bolts—3.0L (1MZ-FE) engine

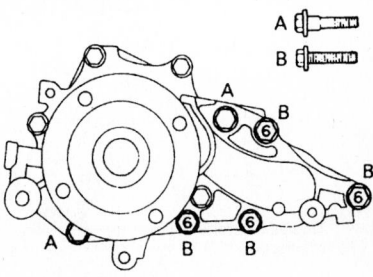

Water pump mounting bolt locations—3.0L (2JZ-GE) engine

Be sure to use new O-rings when installing the water bypass pipe—3.0L (2JZ-GE) engine

- 2 nuts holding the No. 2 water bypass pipe to the water pump. Tighten the nuts to 15 ft. lbs. (21 Nm).
- Clamp bracket for the CKP sensor connector
- Engine wire bracket
- Alternator
- New O-rings to the No. 1 water bypass pipe.
- New O-ring and the water bypass outlet with the 2 bolts and tighten them to 78 inch lbs. (9 Nm)
- Thermostat and the water inlet
- Idler pulley
- Timing belt
- Serpentine drive belt
- Air cleaner, the MAF meter and the intake air connector pipe assembly
- Radiator assembly
- Negative battery cable
- Coolant. Start the engine, check for leaks and bleed the cooling system.

4.0L (1UZ-FE) and 4.3L (3UZ-FE) Engines (except 2004 4.3L 3UZ-FE Engine)

1. Before servicing the vehicle, refer to the precautions in the beginning of this section.

2. Disconnect the negative battery cable.
3. Drain the cooling system.
4. Remove or disconnect the following:
   - Timing belt
   - No. 2 idler pulley
   - Throttle body
   - Bypass hose(s) from the water inlet housing
   - 2 bolts holding the water inlet housing to water pump
   - Water inlet housing and discard the gasket
   - Mounting bolts, studs and the nut to the water pump
   - Water pump by carefully prying between the pump and the cylinder block
   - The old gasket and clean all mounting surfaces

To install:

5. Install new seal packing to the water pump groove and a new O-ring to the water bypass pipe end.
6. Install or connect the following:
   - Water pump to the water bypass pipe end
   - Water pump and tighten the mounting bolts and nut to 13 ft. lbs. (18 Nm)
   - Sealant to the groove of the water inlet housing

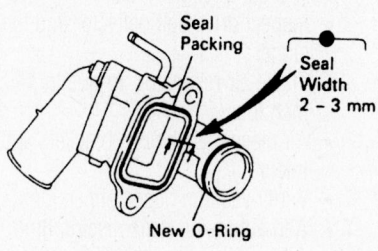

Apply silicone sealant to the water pump as shown—4.0L engine

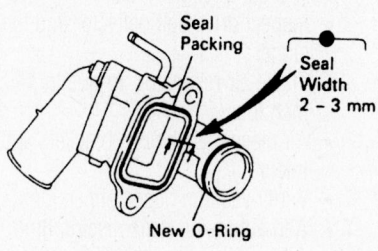

Apply sealant to the water inlet housing as shown—4.0L engine

- A new O-ring to the water inlet housing
- Water inlet housing end into the water pump hole
- Water inlet and housing assembly with the 2 bolts. Alternately tighten the bolts to 13 ft. lbs. (18 Nm).
- Bypass hose(s) to the water inlet housing
- Throttle body. Tighten the mounting bolts to 13 ft. lbs. (18 Nm).
- No. 2 idler pulley
- Timing belt
- Coolant
- Negative battery cable

7. Fill the cooling system.
8. Start the engine and check for leaks.

4.3L (3UZ-FE) Engine 2004

1. Before servicing the vehicle, refer to the precautions in the beginning of this section.
2. Disconnect the negative battery cable.
3. Drain the cooling system.
4. Remove or disconnect the following:
   - Engine under cover
   - Timing belt
   - Water inlet housing
   - Timing belt idler sub assembly
   - Water pump

To install:

5. Install or connect the following:
   - New gasket, water pump with 5 bolts and 2 stud bolts and nuts. Tighten bolt to 15 ft lbs (21 Nm) and stud nuts to 13 ft. lbs (18 Nm)

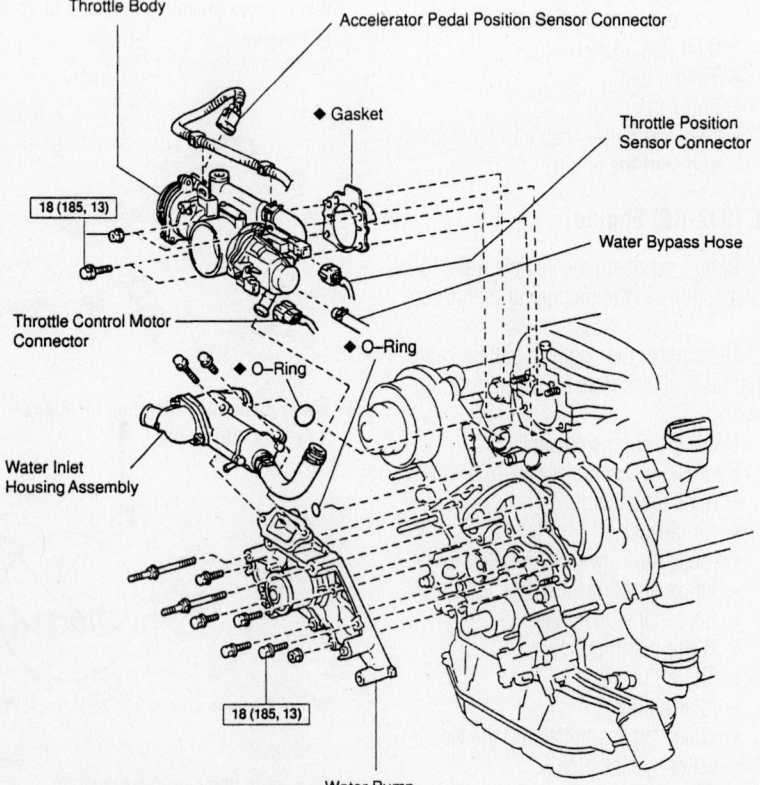

Exploded view of the water pump mounting and related components—4.0L engine shown

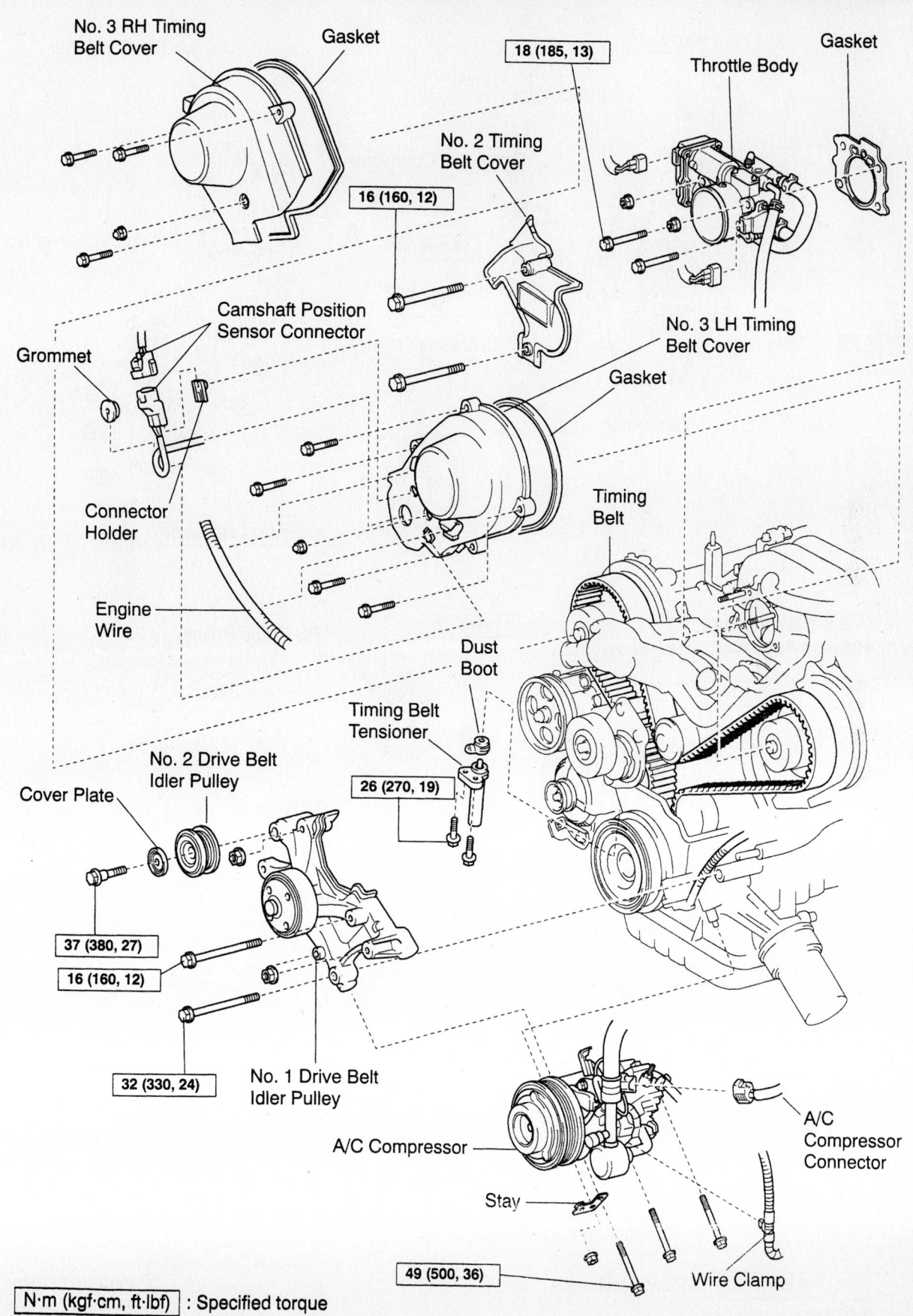

No. 3 RH Timing Belt Cover

Gasket

18 (185, 13)

Throttle Body

Gasket

No. 2 Timing Belt Cover

16 (160, 12)

No. 3 LH Timing Belt Cover

Camshaft Position Sensor Connector

Grommet

Connector Holder

Gasket

Engine Wire

Timing Belt

Dust Boot

Timing Belt Tensioner

26 (270, 19)

No. 2 Drive Belt Idler Pulley

Cover Plate

37 (380, 27)

16 (160, 12)

32 (330, 24)

No. 1 Drive Belt Idler Pulley

A/C Compressor

Stay

A/C Compressor Connector

49 (500, 36)

Wire Clamp

N·m (kgf·cm, ft·lbf) : Specified torque

67162-LEXU-G200

Exploded view of components that need to be removed to access the water pump—2004 (3UZ-FE) Engine

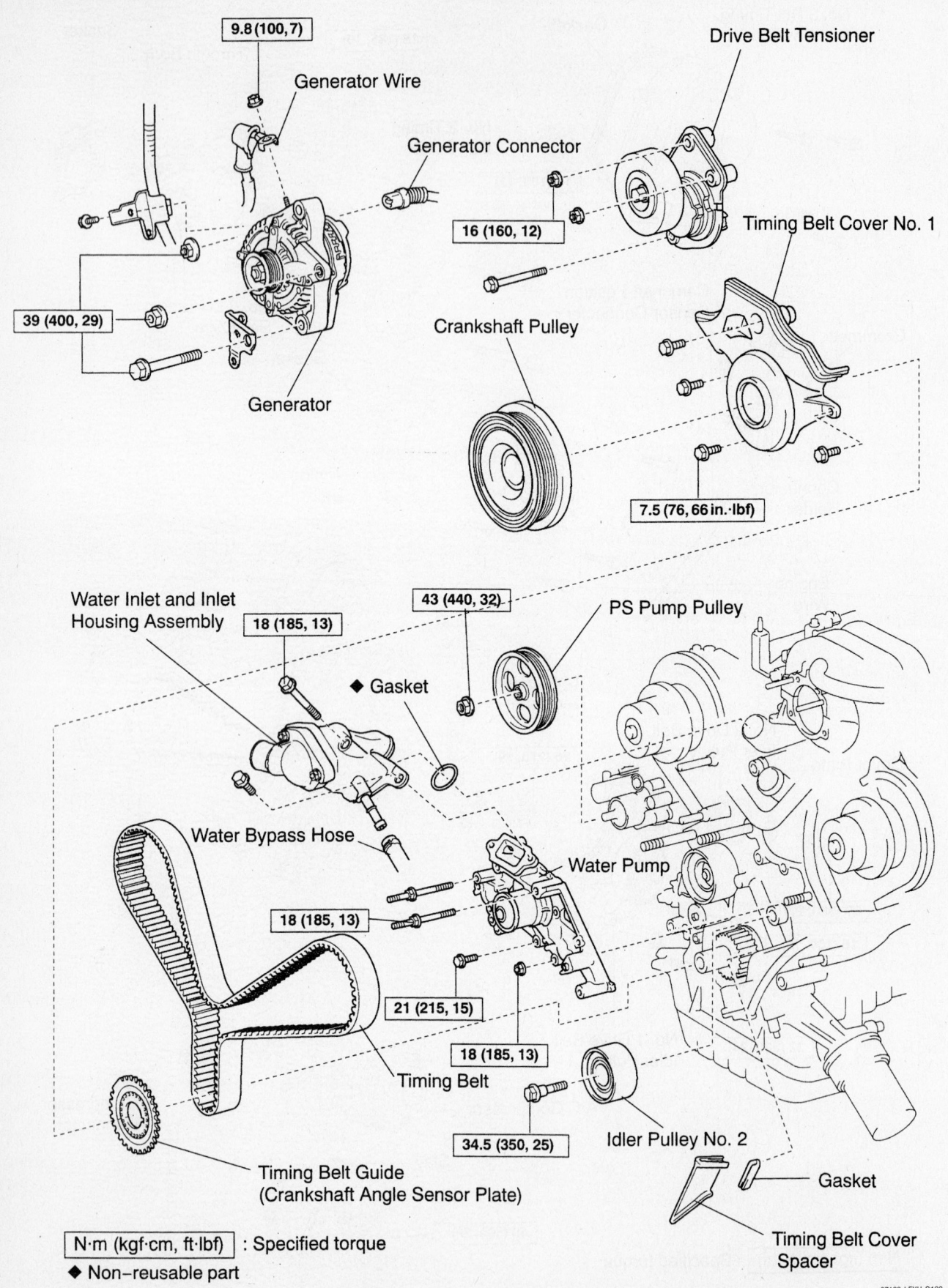

9.8 (100, 7)

Generator Wire

Generator Connector

Drive Belt Tensioner

16 (160, 12)

Timing Belt Cover No. 1

39 (400, 29)

Generator

Crankshaft Pulley

7.5 (76, 66 in.·lbf)

Water Inlet and Inlet
Housing Assembly

18 (185, 13)

43 (440, 32)

PS Pump Pulley

◆ Gasket

Water Bypass Hose

Water Pump

18 (185, 13)

21 (215, 15)

18 (185, 13)

Timing Belt

34.5 (350, 25)

Idler Pulley No. 2

Timing Belt Guide
(Crankshaft Angle Sensor Plate)

Gasket

Timing Belt Cover
Spacer

N·m (kgf·cm, ft·lbf) : Specified torque

◆ Non-reusable part

67162-LEXU-G123

Exploded view of water pump assembly—GS430 & LS 430

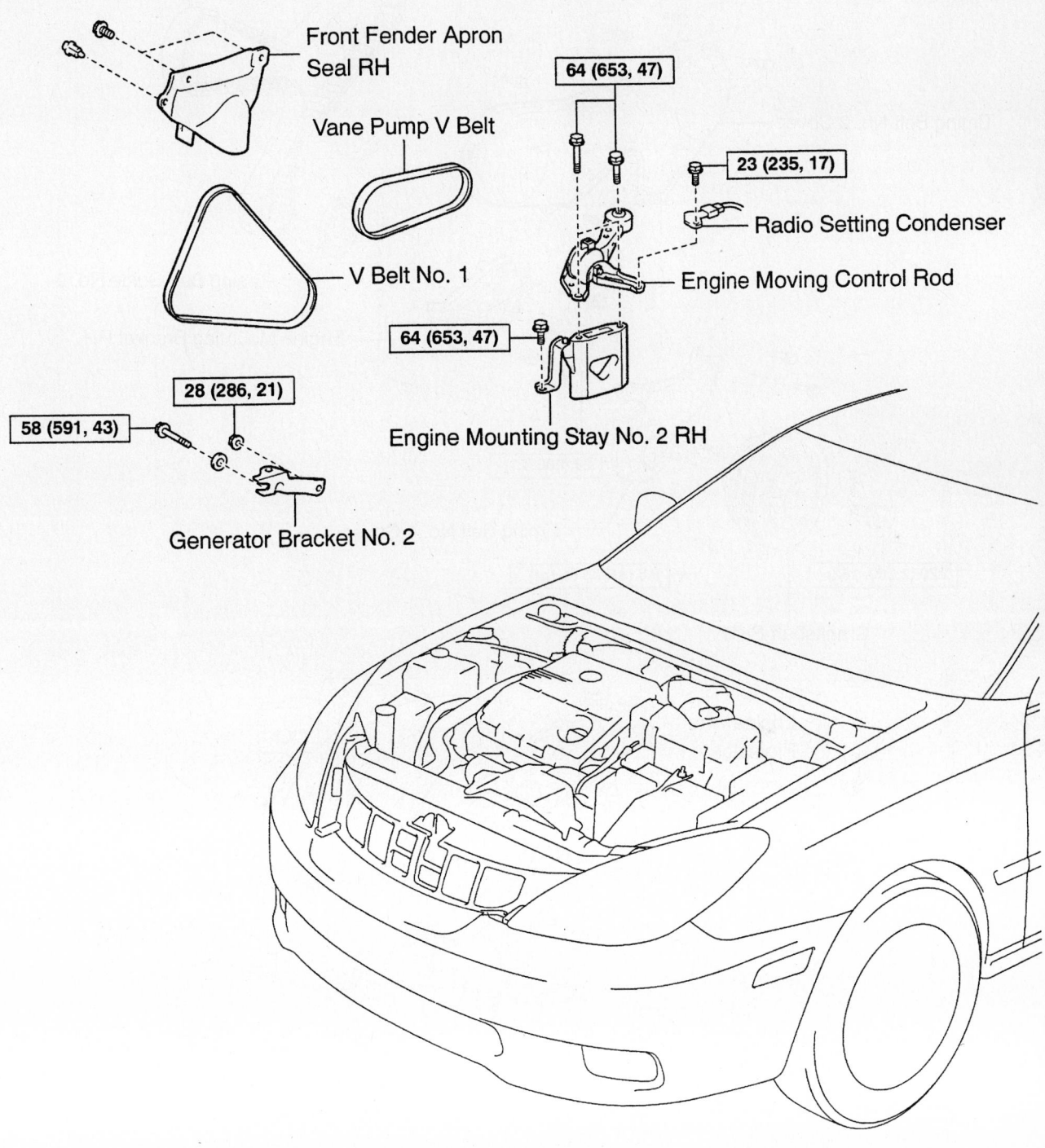

Front Fender Apron Seal RH

Vane Pump V Belt

V Belt No. 1

64 (653, 47)

23 (235, 17)

Radio Setting Condenser

Engine Moving Control Rod

64 (653, 47)

Engine Mounting Stay No. 2 RH

28 (286, 21)

58 (591, 43)

Generator Bracket No. 2

N·m (kgf·cm, ft·lbf) : Specified torque

67162-LEXU-G10

Drive belt and mount locations—3.3L (3MZ-FE)

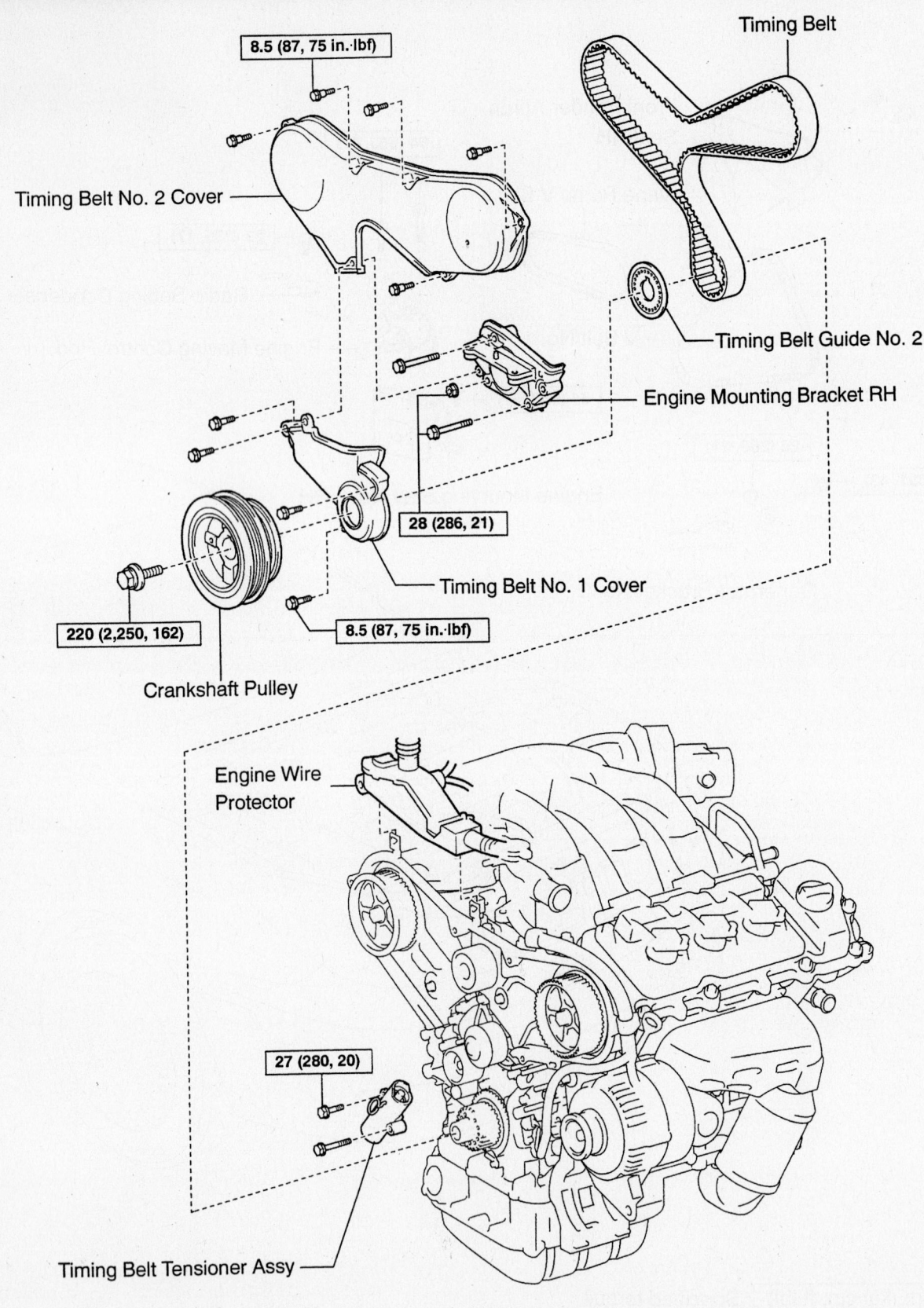

Timing Belt

8.5 (87, 75 in.·lbf)

Timing Belt No. 2 Cover

Timing Belt Guide No. 2

Engine Mounting Bracket RH

28 (286, 21)

Timing Belt No. 1 Cover

220 (2,250, 162)

8.5 (87, 75 in.·lbf)

Crankshaft Pulley

Engine Wire Protector

27 (280, 20)

Timing Belt Tensioner Assy

N·m (kgf·cm, ft·lbf) : Specified torque

67162-LEXU-G11

Timing cover and belt components—3.3L (3MZ-FE)

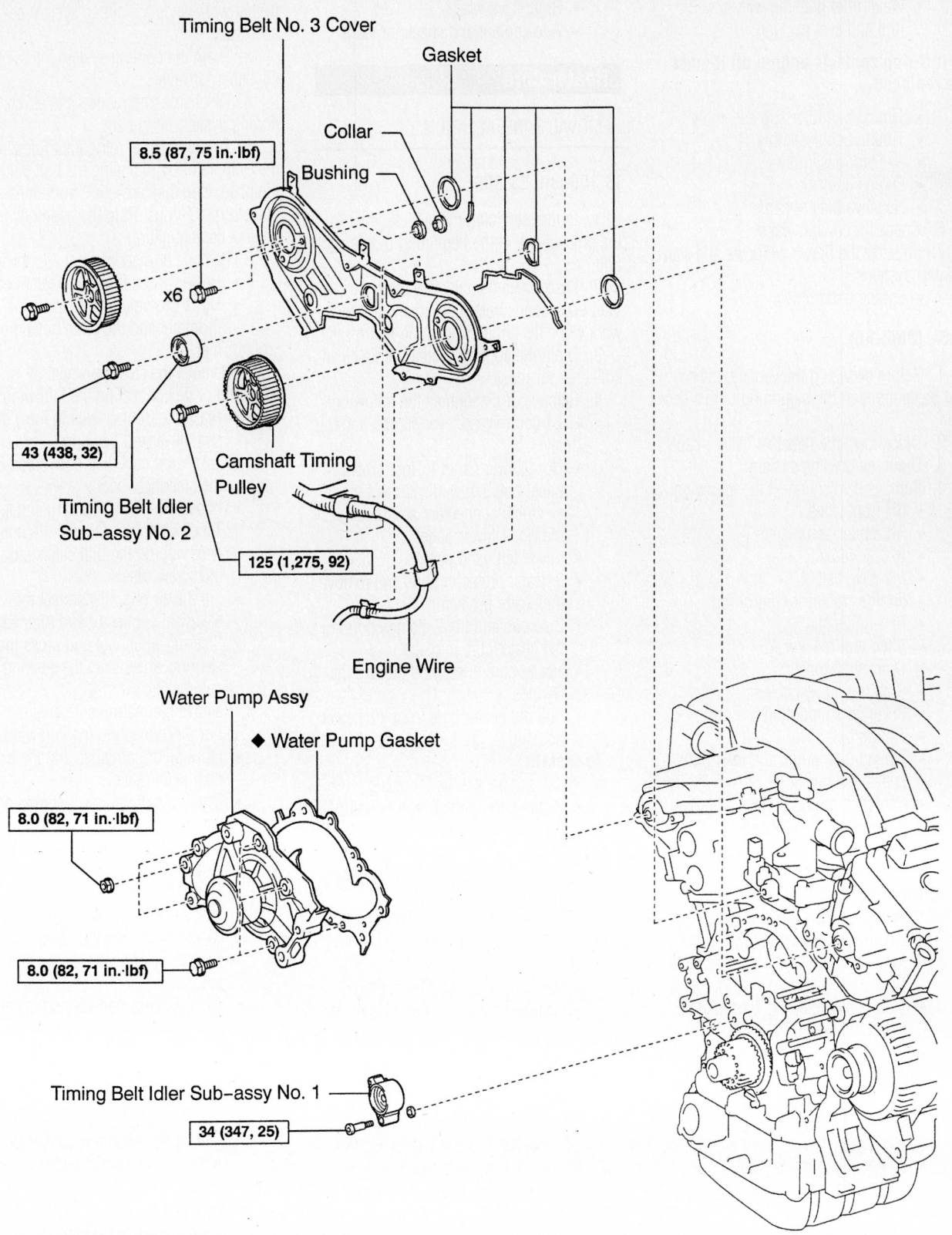

Timing Belt No. 3 Cover

Gasket

Collar

8.5 (87, 75 in.·lbf)

Bushing

x6

43 (438, 32)

Camshaft Timing Pulley

Timing Belt Idler Sub-assy No. 2

125 (1,275, 92)

Engine Wire

Water Pump Assy

◆ Water Pump Gasket

8.0 (82, 71 in.·lbf)

8.0 (82, 71 in.·lbf)

Timing Belt Idler Sub-assy No. 1

34 (347, 25)

N·m (kgf·cm, ft·lbf) : Specified torque

◆ Non-reusable part

67162-LEXU-G12

Camshaft pulleys, rear cover and pump components—3.3L (3MZ-FE)

- Water inlet housing with new O-ring and seal packing

➡**If O-ring contacts engine oil it must be replaced**

- Timing belt idler sub assembly
- Timing belt assembly
- Radiator assembly
- Engine coolant
- Negative battery cable

6. Check for coolant leaks
7. Initialize the power windows and seat control systems

- Engine under cover

3.3L (3MZ-FE)

1. Before servicing the vehicle, refer to the precautions in the beginning of this section.
2. Disconnect the negative battery cable.
3. Drain the cooling system.
4. Remove or disconnect the following:

- RH front wheel
- RH fender apron seal
- A/C drive belt
- PS drive belt
- Engine stabilizer control rod
- RH engine stay
- Alternator bracket #2
- Crankshaft pulley
- Both timing belt covers
- RH engine mounting bracket
- Timing belt cover 1 and 2
- Timing belt, guide and idler pulley sub-assembly #1
- Camshaft timing pulleys and idler pulley sub-assembly #2
- Timing belt cover #3
- Water pump assembly

To install:
5. Install or connect the following:

- Water pump assembly with new gasket. Torque to 71 inch lbs. (8.0 Nm)
- Timing belt idler #1. Torque to 25 ft. lbs (24 Nm)
- Timing belt cover #3
- Camshaft timing pulleys
- Timing belt idler sub-assembly
- Timing belt, tensioner assembly and guide
- Engine mounting bracket
- Upper and lower timing belts covers
- Crankshaft pulley
- Alternator bracket
- Engine mounting stay
- Engine stabilizing control rod
- PS pump
- A/C drive belt
- Inspect drive belt tension

- Right front wheel
- Add coolant and check for leaks

Heater Core

REMOVAL & INSTALLATION

IS 300 and ES 300

1. Before servicing the vehicle, refer to the precautions in the beginning of this section.
2. Disconnect the negative battery cable. Wait 90 seconds before doing any further work while the airbag system de-energizes.
3. Drain the cooling system into a clean container for reuse.
4. Remove or disconnect the following:

- 2 hood release lever screws and the lever
- No. 1 lower panel-to-instrument panel bolt/screw, disconnect the electrical connectors and remove the No. 1 lower panel
- Lower left hand panel
- 3 heater protector clips and remove the heater protector
- 2 screws and the 2 clamps holding the heater core in place
- Heater core hoses and discard the O-rings
- Pull out heater core from the heater housing

To install:
5. Install or connect the following:

- Heater core to the heater housing
- Heater core hoses using new O-rings
- 2 clamps and the 2 screws holding the heater core in place
- Heater protector and connect the 3 heater protector clips
- Lower left hand panel
- No. 1 lower panel, connect the electrical connectors and install the No. 1 lower panel-to-instrument panel bolt/screw
- Hood release lever and the 2 lever screws

6. Refill the cooling system.
7. Connect the negative battery cable.
8. Operate the engine to normal operating temperatures; then, check the climate control operation and check for leaks.

GS 300, GS 330 and GS 430

1. Before servicing the vehicle, refer to the precautions in the beginning of this section.
2. Disconnect the negative battery cable. Wait 90 seconds before doing any

further work while the airbag system de-energizes.
3. Drain the cooling system into a clean container for reuse.
4. Discharge and recover the air conditioning system refrigerant.
5. Disconnect the refrigerant lines from the evaporator by removing the bolt, sliding the plate, then disconnecting both lines and discard the O-rings. Plug the openings to prevent contamination.
6. Remove or disconnect the following:

- Heater hoses from the heater core
- No. 1 grommet, the heater pipe grommet and the drain hose grommet

7. Remove the steering wheel by removing or disconnecting the following:

- Place the front wheels in the straight-ahead position.
- Torx® bolt covers at both sides of the steering wheel
- Loosen the Torx® bolts (using a Torx® wrench), until the circumference ring on the bolt catches on the screw case
- Lift the air bag, disconnect the electrical connector and remove it
- Steering wheel nut and press the steering wheel from the steering column

8. Remove the instrument panel by removing or disconnecting the following:

- Front pillar garnishes and the front door scuff plates
- Steering column cover screws and the covers
- Electrical connectors and remove the combination switch
- End pad
- 2 No. 1 undercover-to-instrument panel screws and the undercover
- 2 hood lock release screws and the release
- 4 No. 1 safety pad-to-instrument panel bolts, screw and the safety pad
- Parking brake handle and the No. 1 switch hole base
- 4 steering column-to-instrument panel nuts, disconnect the spring from the brake pedal and remove the steering column
- Instrument cluster finish panel using a suitable prytool
- 4 instrument cluster-to-instrument panel screws, disconnect the electrical connectors and remove the instrument cluster
- No. 2 undercover using a suitable prytool
- Plate and disconnect the air bag

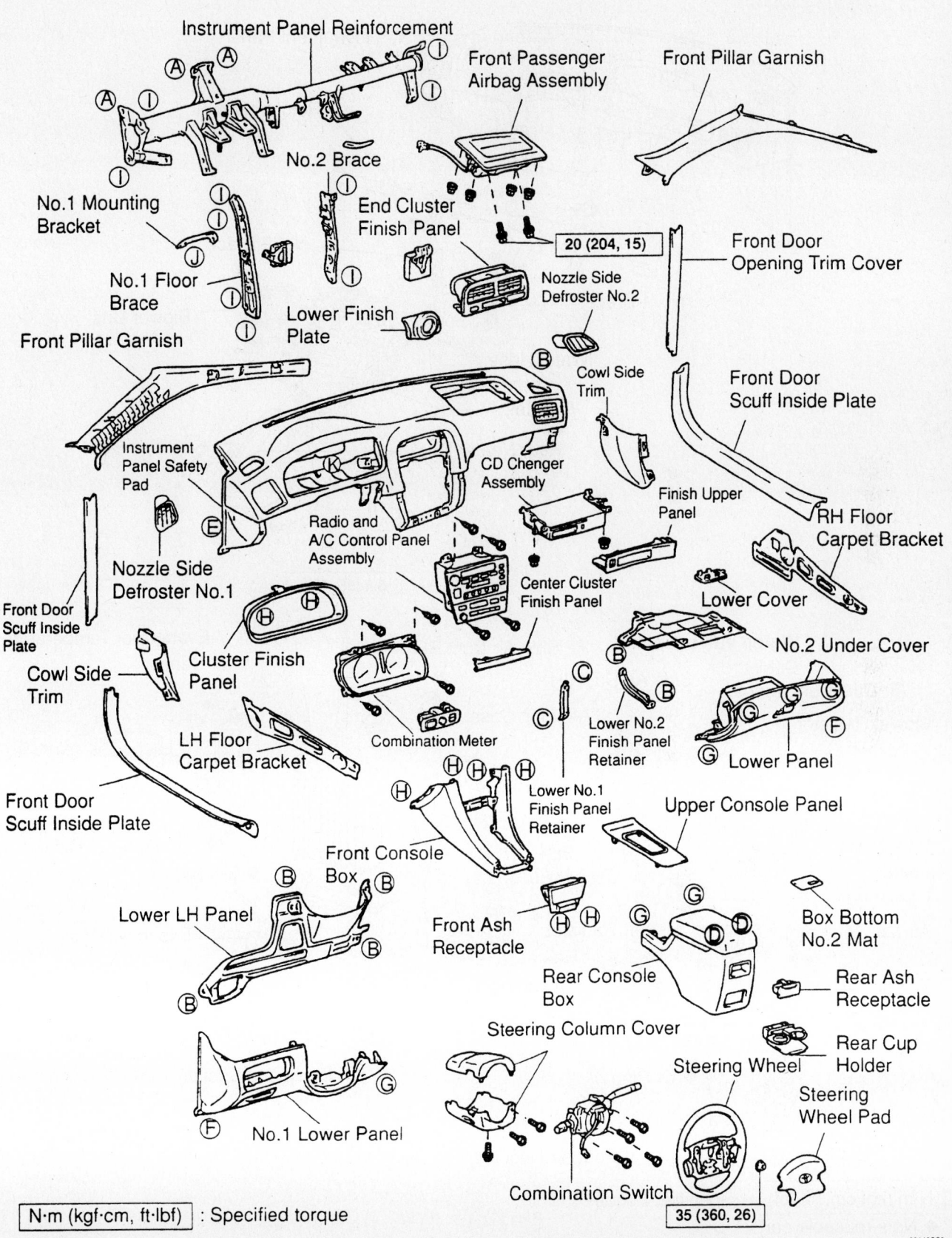

Instrument Panel Reinforcement

Front Passenger Airbag Assembly

Front Pillar Garnish

No.2 Brace

No.1 Mounting Bracket

End Cluster Finish Panel

20 (204, 15)

No.1 Floor Brace

Nozzle Side Defroster No.2

Front Door Opening Trim Cover

Lower Finish Plate

Cowl Side Trim

Front Door Scuff Inside Plate

Front Pillar Garnish

Instrument Panel Safety Pad

CD Chenger Assembly

Front Door Scuff Inside Plate

Nozzle Side Defroster No.1

Radio and A/C Control Panel Assembly

Finish Upper Panel

RH Floor Carpet Bracket

Cowl Side Trim

Cluster Finish Panel

Center Cluster Finish Panel

Lower Cover

No.2 Under Cover

Front Door Scuff Inside Plate

LH Floor Carpet Bracket

Combination Meter

Lower No.2 Finish Panel Retainer

Lower Panel

Lower No.1 Finish Panel Retainer

Upper Console Panel

Front Console Box

Lower LH Panel

Front Ash Receptacle

Box Bottom No.2 Mat

Rear Console Box

Rear Ash Receptacle

Steering Column Cover

Rear Cup Holder

Steering Wheel

Steering Wheel Pad

No.1 Lower Panel

Combination Switch

35 (360, 26)

N·m (kgf·cm, ft·lbf) : Specified torque

93112GS3

Exploded view of the instrument assembly—ES 300

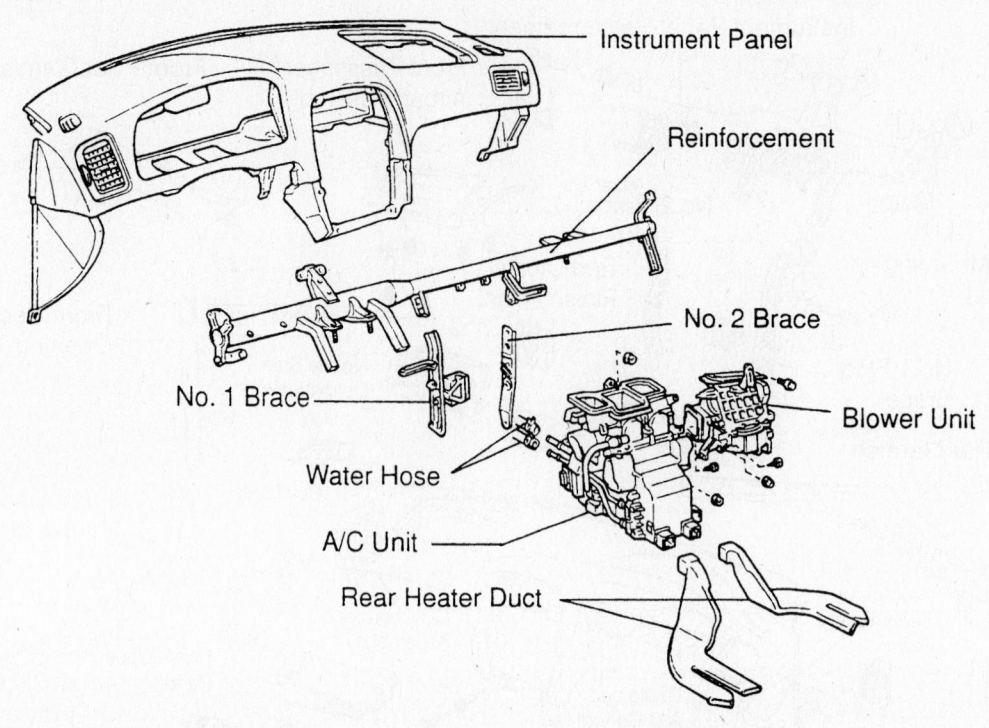

Instrument Panel

Reinforcement

No. 2 Brace

No. 1 Brace

Blower Unit

Water Hose

A/C Unit

Rear Heater Duct

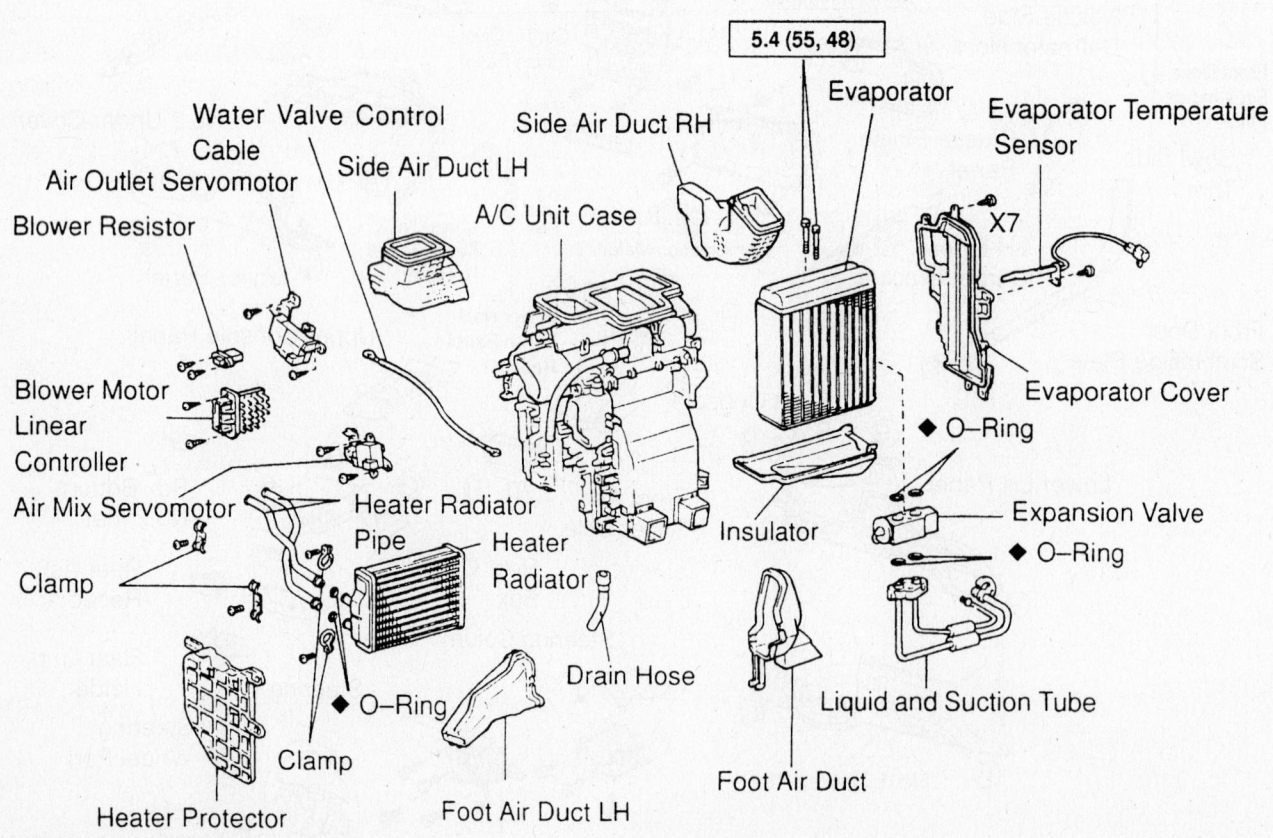

5.4 (55, 48)

Water Valve Control Cable

Side Air Duct RH

Evaporator

Evaporator Temperature Sensor

Air Outlet Servomotor

Side Air Duct LH

Blower Resistor

A/C Unit Case

X7

Blower Motor Linear Controller

Evaporator Cover

◆ O-Ring

Air Mix Servomotor

Expansion Valve

Clamp

◆ O-Ring

Heater Radiator Pipe

Heater Radiator

Insulator

Liquid and Suction Tube

Drain Hose

◆ O-Ring

Clamp

Foot Air Duct

Heater Protector

Foot Air Duct LH

N·m (kgf·cm, in.·lbf) : Specified torque

◆ Non-reusable part

93112GS4

Exploded view of the heater core, heater/air conditioning housing and related components—ES 300

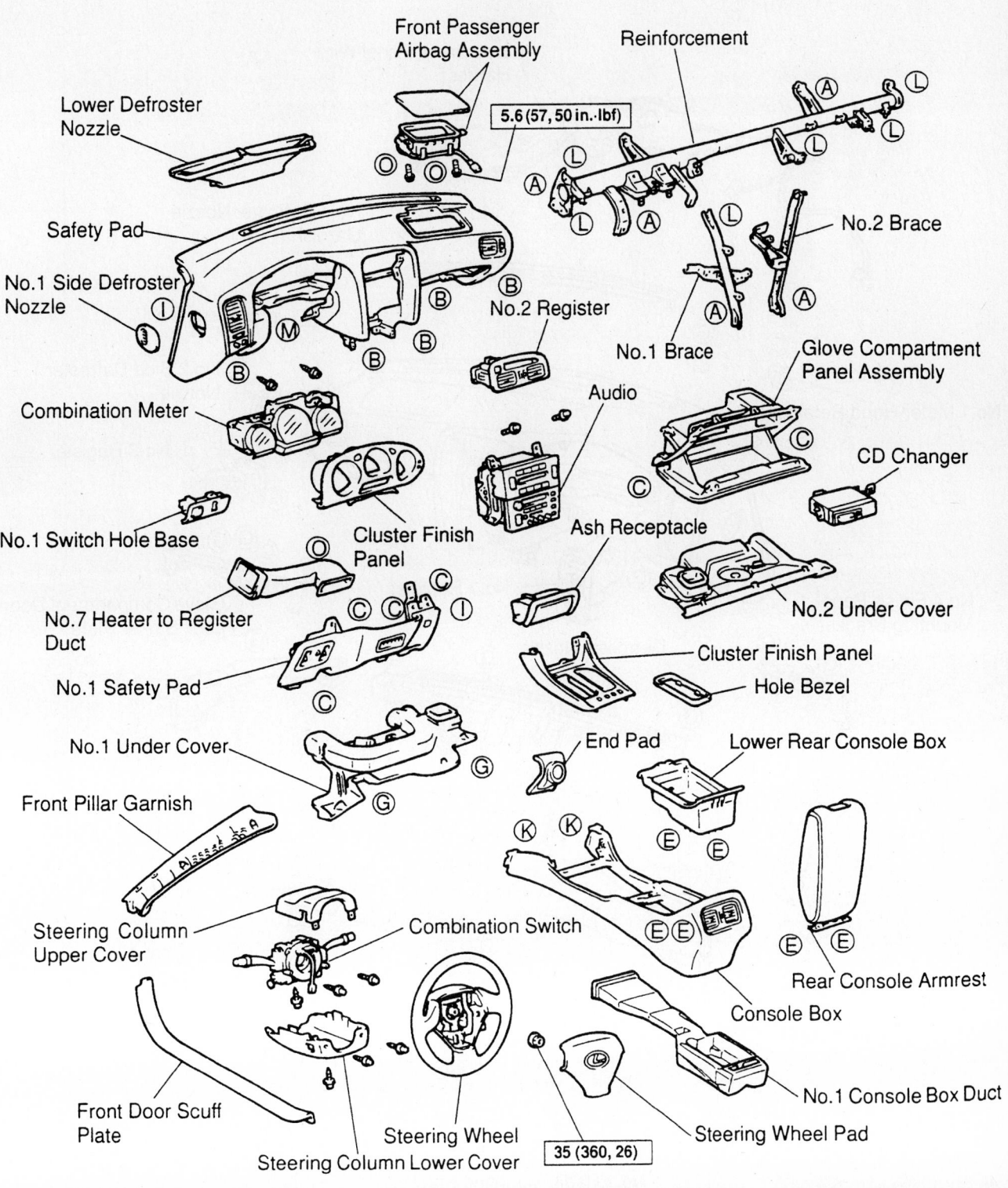

Front Passenger Airbag Assembly

Reinforcement

5.6 (57, 50 in.·lbf)

Lower Defroster Nozzle

Safety Pad

No.1 Side Defroster Nozzle

No.2 Register

No.2 Brace

No.1 Brace

Glove Compartment Panel Assembly

Combination Meter

Audio

CD Changer

No.1 Switch Hole Base

Cluster Finish Panel

Ash Receptacle

No.2 Under Cover

No.7 Heater to Register Duct

Cluster Finish Panel

Hole Bezel

No.1 Safety Pad

No.1 Under Cover

End Pad

Lower Rear Console Box

Front Pillar Garnish

Rear Console Armrest

Steering Column Upper Cover

Combination Switch

Console Box

Front Door Scuff Plate

No.1 Console Box Duct

Steering Wheel

Steering Wheel Pad

Steering Column Lower Cover

35 (360, 26)

N·m (kgf·cm, ft·lbf) : Specified torque

Exploded view of the instrument panel and related components—GS 300, GS 330 and GS 430

93112GT4

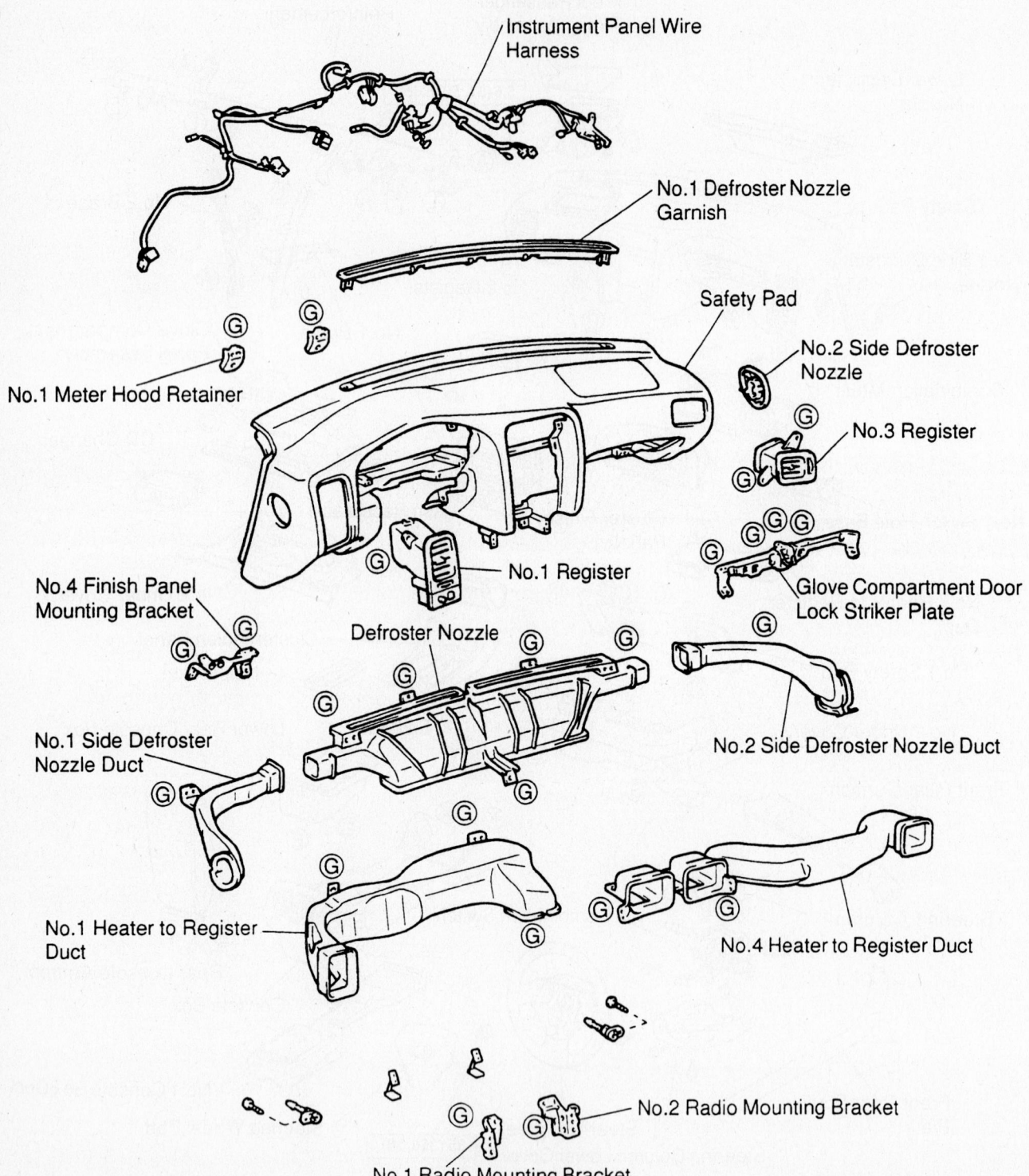

Instrument Panel Wire Harness

No.1 Defroster Nozzle Garnish

Safety Pad

No.2 Side Defroster Nozzle

No.1 Meter Hood Retainer

No.3 Register

No.1 Register

No.4 Finish Panel Mounting Bracket

Glove Compartment Door Lock Striker Plate

Defroster Nozzle

No.1 Side Defroster Nozzle Duct

No.2 Side Defroster Nozzle Duct

No.1 Heater to Register Duct

No.4 Heater to Register Duct

No.2 Radio Mounting Bracket

No.1 Radio Mounting Bracket

93112GT5

Exploded view of the instrument panel, ventilation ducts and related components—GS 300, GS 330 and GS 430

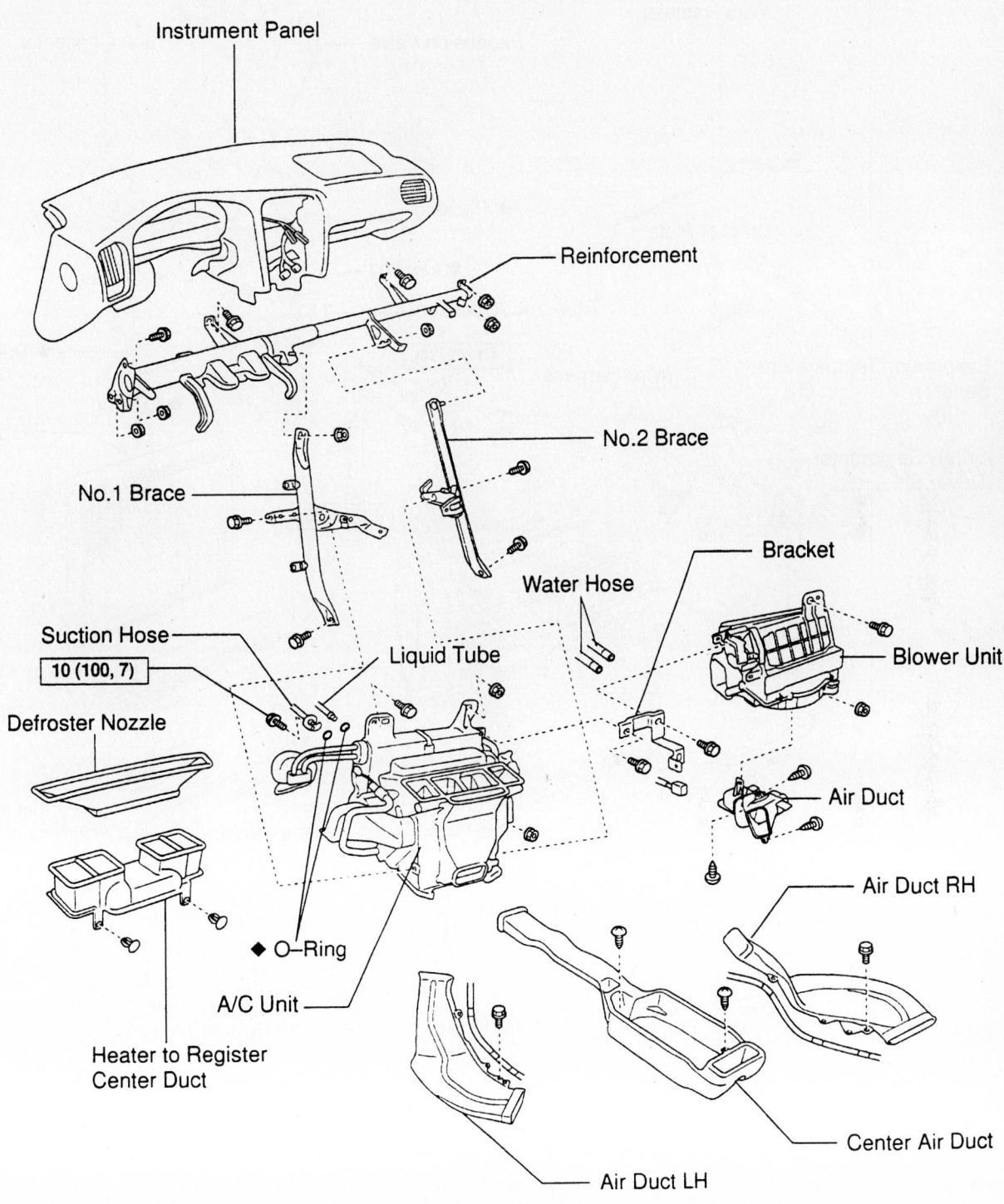

Instrument Panel

Reinforcement

No.2 Brace

No.1 Brace

Bracket

Water Hose

Suction Hose
10 (100, 7)

Liquid Tube

Blower Unit

Defroster Nozzle

◆ O–Ring

Air Duct

A/C Unit

Air Duct RH

Heater to Register
Center Duct

Air Duct LH

Center Air Duct

N·m (kgf·cm, ft·lbf) : Specified torque

◆ Non–reusable part

93112GT6

Exploded view of the heater/air conditioning housing and related components—GS 300, GS 330 and GS 430

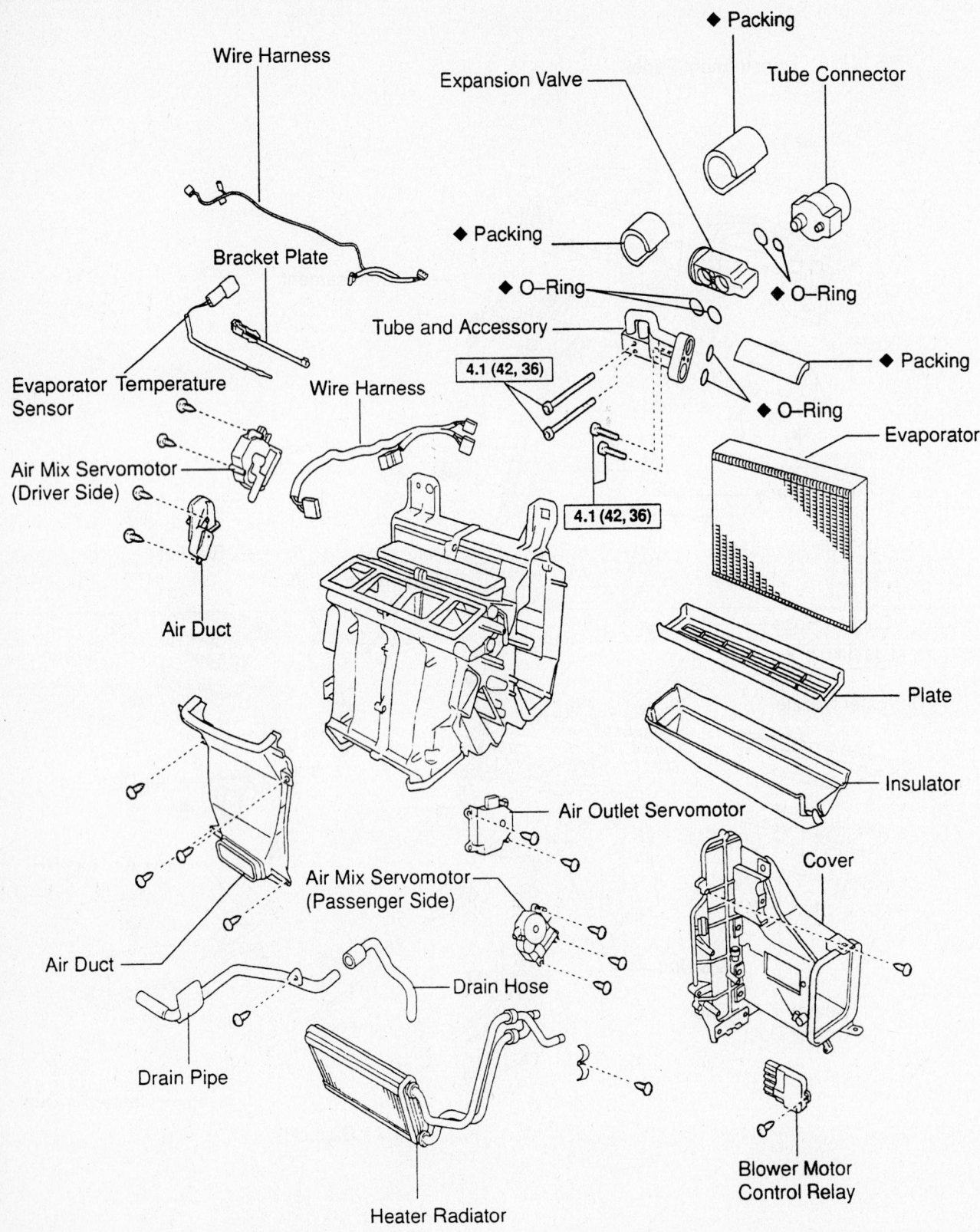

Packing

Tube Connector

Wire Harness

Expansion Valve

◆ Packing

Bracket Plate

◆ O–Ring

◆ O–Ring

Tube and Accessory

◆ Packing

Evaporator Temperature Sensor

4.1 (42, 36)

◆ O–Ring

Wire Harness

Air Mix Servomotor (Driver Side)

Evaporator

4.1 (42, 36)

Air Duct

Plate

Insulator

Air Outlet Servomotor

Cover

Air Mix Servomotor (Passenger Side)

Air Duct

Drain Hose

Drain Pipe

Heater Radiator

Blower Motor Control Relay

N·m (kgf·cm, in.·lbf) : Specified torque

◆ Non–reusable part

93112GT7

Exploded view of the heater core, heater housing and related components—GS 300, GS 330 and GS 430

electrical connector inside the glove box

- Glove box-to-instrument panel 2 bolts, 3 screws, and the glove box
- 3 CD changer-to-instrument panel nuts, disconnect the electrical connectors and remove the CD changer
- Ashtray
- No. 2 register using a suitable prytool and disconnect the connector
- Audio unit-to-instrument panel 2 bolts, 2 screws and the audio unit
- Cluster finish panel using a suitable prytool, disconnect the connectors and remove the panel.
- Console box carpet
- Lower rear console box
- Rear console armrest
- No. 3 console box mounting bracket
- Console box
- No. 1 console box duct
- No. 7 heater-to-register
- 5 instrument panel-to-chassis bolts, the nut, the screw and the instrument panel
- No. 1 and No. 2 brace
- Reinforcement-to-chassis 5 nuts, 4 bolts and the reinforcement
- Ventilation nozzles from the heater/air conditioning housing

9. Remove the blower unit by removing or disconnecting the following:
- Connector clamp
- 3 air duct-to-blower housing screws and the air duct
- Electrical connector bracket, the wiring harness clamps and the wiring harness
- Blower housing connectors
- 2 blower housing-to-bracket bolts and the bracket
- Blower housing-to-chassis bolt, screw, nut and the blower housing

10. Remove or disconnect the following:
- 2 center air duct-to-heater/air conditioning housing screws and the air duct
- Move the floor carpet rearward
- Wiring harness clamps
- 2 air duct bolts and the ducts at both sides

11. Remove the heater/air conditioning housing by removing or disconnecting the following:
- Electrical connector
- Wiring harness set nut
- Wiring harness clamp
- Heater/air conditioning housing 2 nuts and bolt
- Heater/air conditioning housing

12. Remove the heater core-to-heater/air conditioning housing clamp screw and the clamp.

13. Pull the heater core from the heater/air conditioning housing.

To install:

14. Install the heater core to the heater/air conditioning housing.

15. Install the heater core clamp and the clamp-to-heater/air conditioning housing screw.

16. Install the heater/air conditioning housing by installing or connecting the following:
- Heater/air conditioning housing
- Heater/air conditioning housing 2 nuts and bolt
- Wiring harness clamp
- Wiring harness set nut
- Electrical connector

17. Install or connect the following:
- Air duct and the 2 duct bolts on both sides
- Wiring harness clamps
- Move the floor carpet forward
- Center air duct and the 2 air duct-to-heater/air conditioning housing screws

18. Install the blower unit by installing or connecting the following:
- Blower housing and the blower housing-to-chassis bolt, screw and nut
- Blower housing and the 2 bracket-to-bracket bolts
- Blower housing connectors
- Electrical connector bracket, the wiring harness clamps and the wiring harness
- Air duct and the 3 air duct-to-blower housing screws
- Connector clamp

19. Install the instrument by installing or connecting the following:
- Ventilation nozzles to the heater/air conditioning housing
- Reinforcement and the reinforcement-to-chassis 5 nuts and 4 bolts
- No. 1 and No. 2 brace
- Instrument panel and the 5 instrument panel-to-chassis bolts, the nut and the screw

20. Install or connect the following:
- No. 7 heater-to-register
- No. 1 console box duct
- Console box
- No. 3 console box mounting bracket
- Rear console armrest
- Lower rear console box
- Console box carpet
- Connectors and install the cluster finish panel

- Audio unit and the audio unit-to-instrument panel 2 bolts and 2 screws
- Connector and install the No. 2 register
- Ashtray
- CD changer, connect the electrical connectors and install the 3 CD changer-to-instrument panel nuts
- Glove box and the glove box-to-instrument panel 2 bolts and 3 screws
- Air bag electrical connector and install the plate inside the glove box
- No. 2 undercover
- Instrument cluster, connect the electrical connectors and install the 4 instrument cluster-to-instrument panel screws
- Instrument cluster finish panel
- Steering column, connect the spring to the brake pedal and install the 4 steering column-to-instrument panel nuts
- Parking brake handle and the No. 1 switch hole base
- No. 1 safety pad and the 4 safety pad-to-instrument panel bolts and screw
- Hood lock release and the 2 release screws
- No. 1 undercover and the 2 undercover-to-instrument panel screws
- End pad
- Combination switch and connect the electrical connectors
- Steering column covers and the cover screws
- Front pillar garnishes and the front door scuff plates

21. Install the steering wheel by installing or connecting the following:
- Steering wheel and torque the steering wheel nut to 26 ft. lbs. (35 Nm)
- Electrical connector and install the air bag
- Tighten the Torx® bolts to 80 inch lbs. (9.0 Nm) using a Torx® wrench
- Torx® bolt covers at both sides of the steering wheel
- No. 1 grommet, the heater pipe grommet and the drain hose grommet
- Heater hoses to the heater core
- Refrigerant lines (using new O-rings) and the refrigerant lines-to-evaporator bolt

22. Refill the cooling system.

23. Connect the negative battery cable.

24. Evacuate, charge and leak test the air conditioning system refrigerant.

25. Operate the engine to normal operating temperatures; then, check the climate control operation and check for leaks.

LS 430

1. Before servicing the vehicle, refer to the precautions in the beginning of this section.

2. Disconnect the negative battery cable. Wait 90 seconds before doing any further work while the airbag system de-energizes.

3. Disconnect the negative battery cable.

4. Drain the cooling system into a clean container for reuse.

5. Remove or disconnect the following:
- Undercover and the No. 1 safety pad-to-instrument panel screws and the panel at the driver's side
- No. 2 heater-to-register duct
- Heater core-to-heater housing screw and clamp
- Heater hoses from the heater core
- Heater core from the heater housing
- Discard the O-rings

To install:

6. Install or connect the following:
- New O-rings to the heater core
- Heater core to the heater housing
- Heater hoses to the heater core
- Heater core clamp and the heater core-to-heater housing screw
- No. 2 heater-to-register duct
- Undercover and the No. 1 safety pad-to-instrument panel and the panel screws at the driver's side

7. Refill the cooling system.

8. Connect the negative battery cable.

9. Operate the engine to normal operating temperatures; then, check the climate control operation and check for leaks.

SC 430

➡**Removal of the heater core requires removal of the entire heater air conditioning assembly.**

1. Before servicing the vehicle, refer to the precautions in the beginning of this section.

2. Drain the cooling system into a clean container for reuse.

3. Discharge and recover the air conditioning system refrigerant.

4. Remove or disconnect the following:
- A/C Suction and pressure hose bolt and plate at firewall

➡**Do Not use pry tools to separate**

➡**Cap the end to prevent system contamination**

- Heater hoses from the heater core

5. Set CD changer to ship mode setting using following procedure
- Remove all CDs
- Simultaneously press "Seek Up" and "Disc" while turning the ignition switch to "Acc"

➡**When mode setting is complete "Ship" appears on the display**

6. Remove or disconnect the following:
- Disconnect the negative battery cable. Wait 90 seconds before doing any further work while the airbag system de-energizes.
- Center front wheels and steering wheel
- Lower steering wheel cover
- Switch & volume case
- Steering pad switch modulator
- Horn button assembly
- Steering wheel assembly
- RH & LH door scuff plates
- RH & LH door opening trim covers
- Instrument panel safety pad. LH side under steering column
- Instrument cluster finish panel
- Upper and lower steering column covers
- Turn signal switch assembly
- Combination meter assembly
- Upper/Rear shift console garnish panel
- Shift console panel sub-assembly upper
- Air conditioning control assembly
- Instrument cluster radio finish panel assembly lower center
- Instrument panel safety pad garnish sub-assembly
- Instrument panel garnish sub-assembly No. 2
- Instrument cluster finish panel sub-assembly center
- Instrument panel under cover sub-assembly No. 2
- Console glove box carpet
- Console glove box panel No. 3
- Console glove box assembly
- Instrument panel brace sub-assembly No. 2
- Instrument panel brace sub-assembly No. 1
- RH & LH front pillar garnish
- Instrument side panel
- Instrument panel safety pad sub-assembly
- 3 bolts and separate air bag sensor assembly center
- Heater to register duct No. 2

7. Air Conditioning unit assembly
- Wire harness clamps and discon-

nect wire harness from air conditioning unit assembly
- 2 screws, 3 nuts, bolt and air conditioning unit

8. Remove or disconnect the following:
- Remove heater bracket

9. Air conditioning evaporator assembly
- Connector and clamp for the blower assembly
- Release claw and disconnect the connector from the bracket
- 2 screws and release the claw or the evaporator assembly
- evaporator assembly

10. Remove or disconnect the following:
- Cooler thermistor hose
- Air duct No.1, 2, 3 and 4

11. Remove heater core unit sub-assembly
- Screw and clamp
- Heater core assembly from A/C assembly

To install:

12. Install or connect the following:

13. Heater core unit sub assembly
- Heater core into A/C assembly
- Screw and clamp

14. Install or connect the following:
- Air duct No 1,2, 3 and 4
- Cooler thermistor hose

15. Install Air Conditioning Unit Assembly
- 2 screws and 3 nuts and air conditioning unit
- Bolt to air conditioning unit assembly. Torque: 87 inch lbs. (9.8 Nm)
- Clamps to the air conditioner unit assembly

16. Install or connect the following:
- Heater to register duct No. 2
- Air bag sensor assembly center
- Instrument panel safety pad sub-assembly
- Instrument side panel
- RH & LH front pillar garnish
- Instrument panel brace sub-assembly No. 1
- Instrument panel brace sub-assembly No. 2
- Console glove box assembly, panel and box carpet
- Instrument panel under cover sub-assembly No. 2
- Instrument cluster finish panel sub-assembly center
- Instrument panel garnish sub-assembly No. 2
- Instrument panel safety pad garnish sub-assembly
- Instrument cluster radio finish assembly lower center

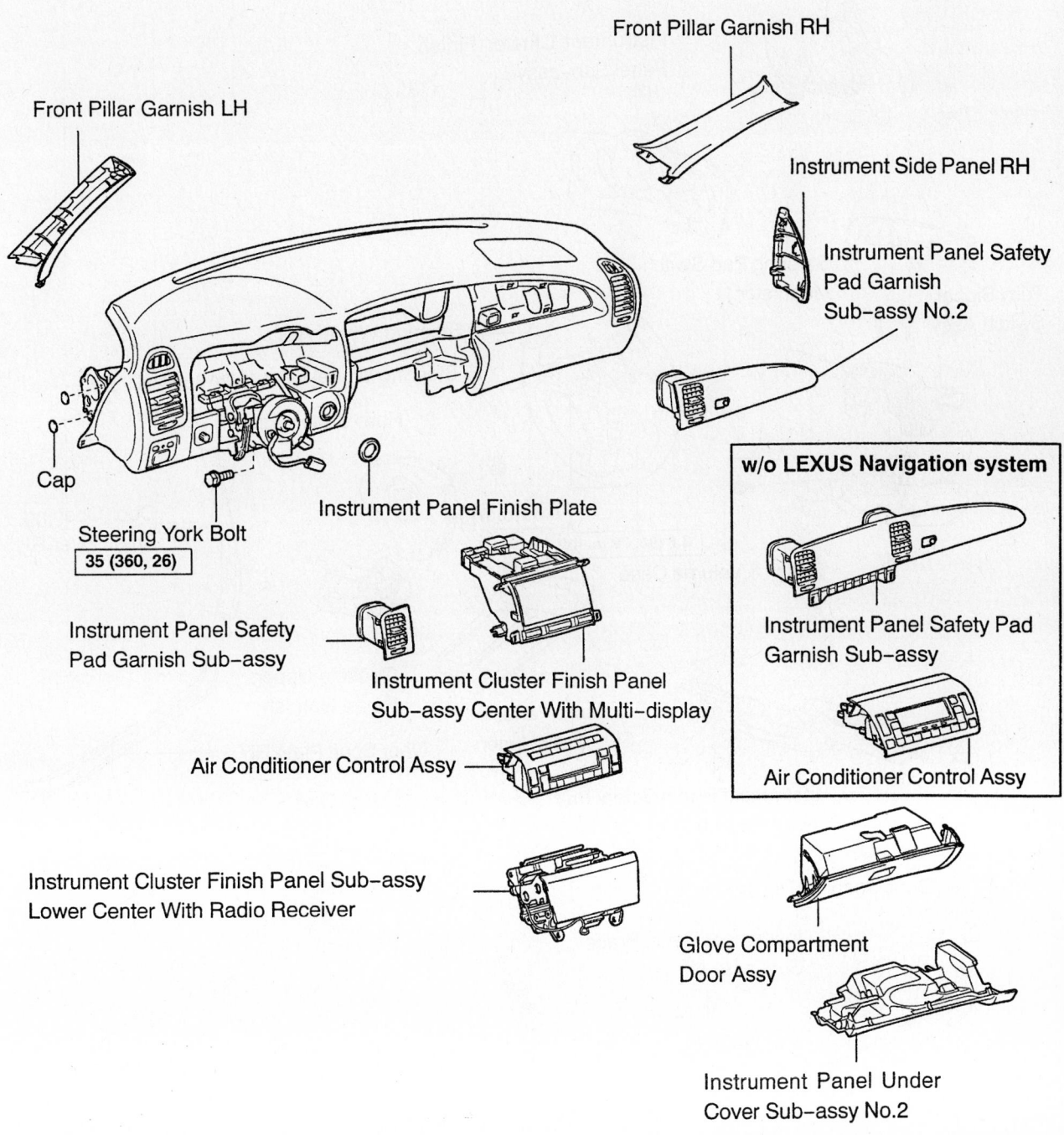

Front Pillar Garnish RH

Front Pillar Garnish LH

Instrument Side Panel RH

Instrument Panel Safety Pad Garnish Sub–assy No.2

Cap

Steering York Bolt
35 (360, 26)

Instrument Panel Finish Plate

w/o LEXUS Navigation system

Instrument Panel Safety Pad Garnish Sub–assy

Instrument Panel Safety Pad Garnish Sub–assy

Instrument Cluster Finish Panel Sub–assy Center With Multi–display

Air Conditioner Control Assy

Air Conditioner Control Assy

Instrument Cluster Finish Panel Sub–assy Lower Center With Radio Receiver

Glove Compartment Door Assy

Instrument Panel Under Cover Sub–assy No.2

N·m (kgf·cm, ft·lbf) : Specified torque

67162-LEXU-G13

Exploded view of the dash components—SC 430

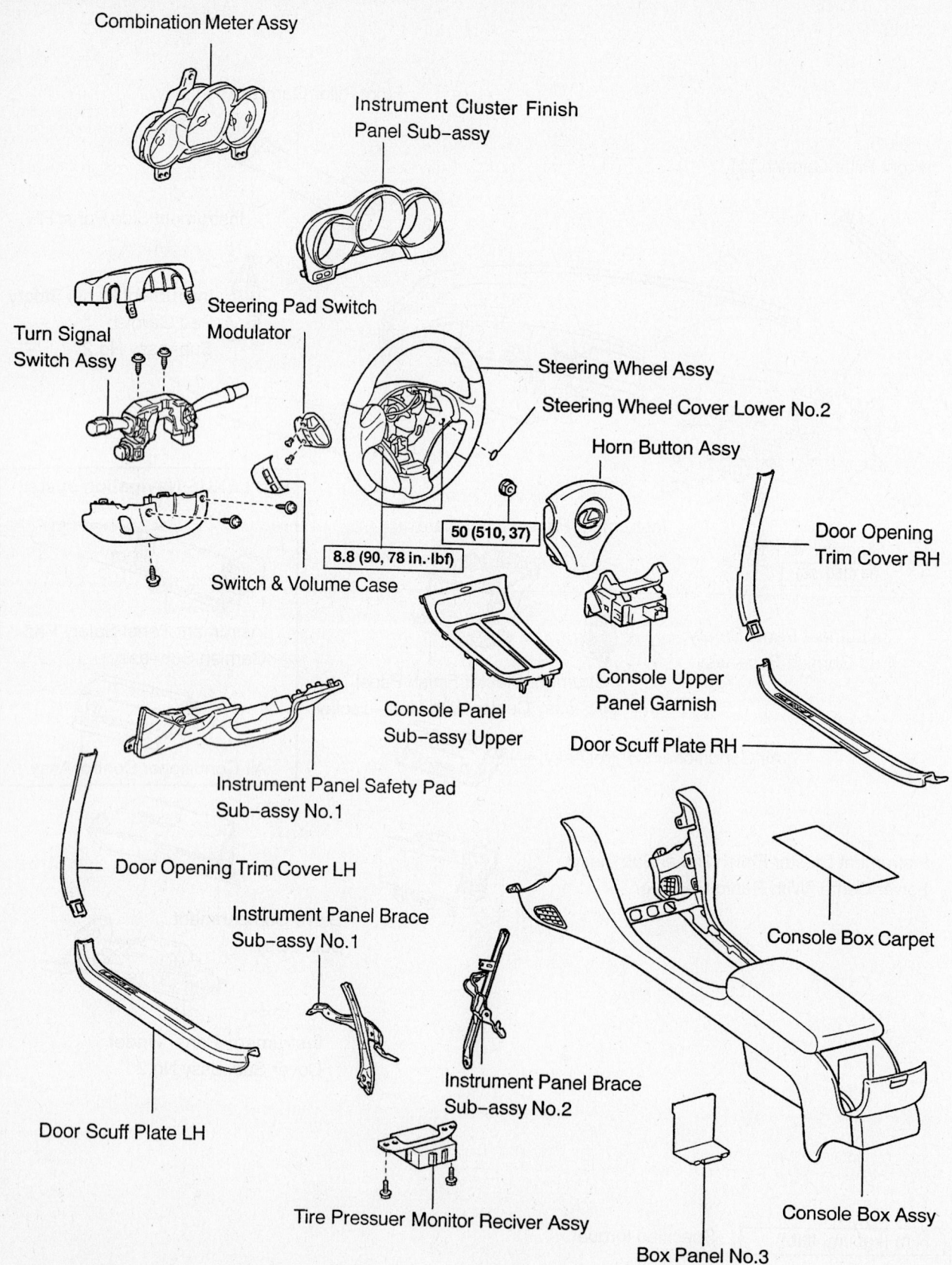

Combination Meter Assy

Instrument Cluster Finish Panel Sub-assy

Steering Pad Switch Modulator

Turn Signal Switch Assy

Steering Wheel Assy

Steering Wheel Cover Lower No.2

Horn Button Assy

50 (510, 37)

8.8 (90, 78 in.·lbf)

Switch & Volume Case

Door Opening Trim Cover RH

Console Upper Panel Garnish

Console Panel Sub-assy Upper

Door Scuff Plate RH

Instrument Panel Safety Pad Sub-assy No.1

Door Opening Trim Cover LH

Console Box Carpet

Instrument Panel Brace Sub-assy No.1

Door Scuff Plate LH

Instrument Panel Brace Sub-assy No.2

Tire Pressuer Monitor Reciver Assy

Box Panel No.3

Console Box Assy

N·m (kgf·cm, ft·lbf) : Specified torque

Exploded view gauge cluster, steering wheel and console—SC 430

67162-LEXU-G14

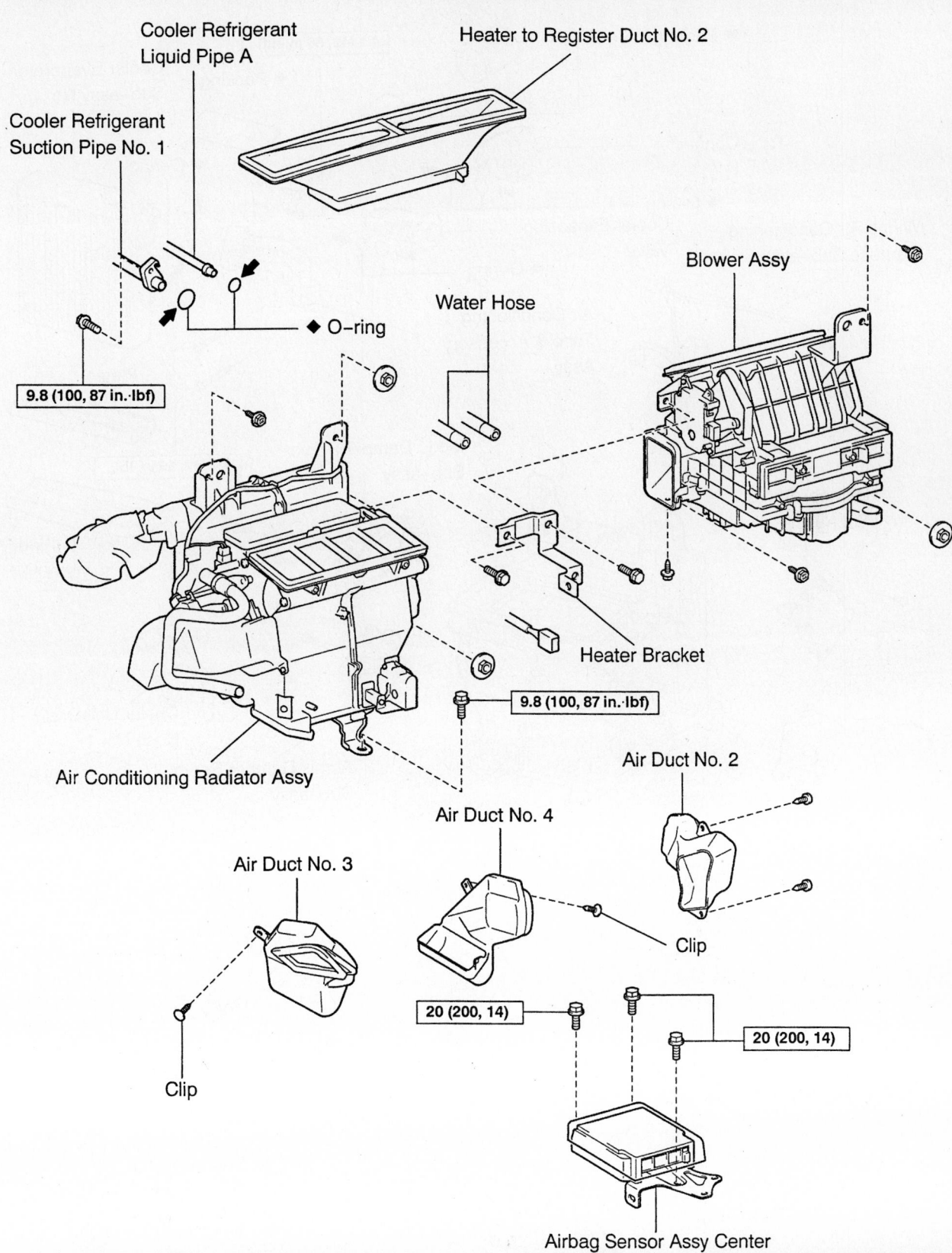

Cooler Refrigerant Liquid Pipe A

Heater to Register Duct No. 2

Cooler Refrigerant Suction Pipe No. 1

Blower Assy

Water Hose

◆ O-ring

9.8 (100, 87 in.·lbf)

Heater Bracket

Air Conditioning Radiator Assy

9.8 (100, 87 in.·lbf)

Air Duct No. 2

Air Duct No. 4

Air Duct No. 3

Clip

Clip

20 (200, 14)

20 (200, 14)

Airbag Sensor Assy Center

N·m (kgf·cm, ft·lbf) : Specified torque

◆ Non-resable part

◀ Compressor oil ND-OIL 8 or equivalent

67162-LEXU-G15

View of the Heater & Air Conditioning assembly—SC 430

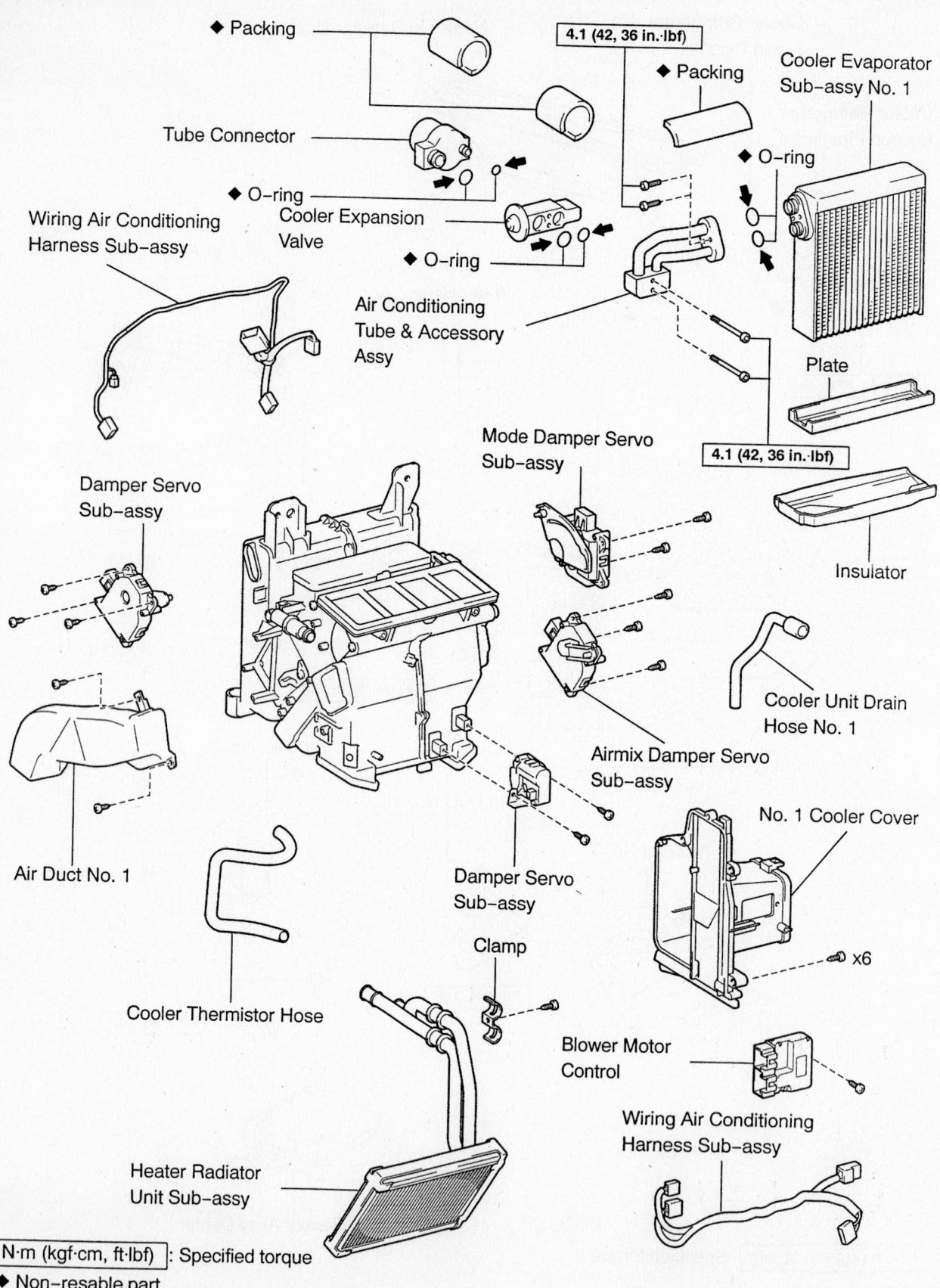

◆ Packing

Tube Connector

Wiring Air Conditioning Harness Sub-assy

◆ O-ring

Cooler Expansion Valve

◆ O-ring

Air Conditioning Tube & Accessory Assy

4.1 (42, 36 in.·lbf)

◆ Packing

Cooler Evaporator Sub-assy No. 1

◆ O-ring

Plate

4.1 (42, 36 in.·lbf)

Insulator

Mode Damper Servo Sub-assy

Damper Servo Sub-assy

Cooler Unit Drain Hose No. 1

Airmix Damper Servo Sub-assy

Air Duct No. 1

No. 1 Cooler Cover

x6

Damper Servo Sub-assy

Clamp

Cooler Thermistor Hose

Blower Motor Control

Wiring Air Conditioning Harness Sub-assy

Heater Radiator Unit Sub-assy

N·m (kgf·cm, ft·lbf) : Specified torque

◆ Non-resable part

◄ Compressor oil ND-OIL 8 or equivalent

Exploded view of Heater & Air Conditioning assembly—SC 430

- Air conditioning control assembly
- Shift console panel sub-assembly upper
- Upper/Rear shift console garnish panel
- Combination meter assembly
- Turn signal switch assembly
- Upper and lower steering column covers
- Instrument cluster finish panel
- Instrument panel safety pad. LH side under steering wheel
- RH & LH door opening trim covers
- RH & LH door scuff plates
- Steering wheel assembly
- Horn button assembly
- Steering pad switch modulator
- Switch and volume case
- Lower steering wheel cover
- Heater core hoses
- A/C suction and pressure hoses, attach with bolt and plate

➥**Lubricate O-rings with compressor oil**

- Negative battery cable
- Fill cooling system with coolant
- Evacuate and recharge A/C system
- Warm up engine and inspect for coolant leaks

ES 330

➥**Removal of the heater core requires removal of the entire heater air conditioning assembly.**

1. Before servicing the vehicle, refer to the precautions in the beginning of this section.

2. Drain the cooling system into a clean container for reuse.

3. Set CD changer to ship mode setting using following procedure
- Remove all CDs
- Simultaneously press "Seek Up" and "Disc" while turning the ignition switch to "Acc"

➥**When mode setting is complete "Ship" appears on the display**

4. Disconnect the negative battery cable. Wait 90 seconds before doing any further work while the airbag system de-energizes.

5. Discharge and recover the air conditioning system refrigerant.

6. Disconnect A/C suction hose (No. 1) and Liquid pipe (A)
   a. Install SST to piping clamp
   b. Push down SST and release the clamp lock

※※ WARNING

Be careful not to deform the tube, when pushing the SST

➥**Cap the open fittings immediately to prevent system contamination**

7. Remove or disconnect the following:
- Heater core hoses

8. Disassemble the dash components as follows
- RH & LH door scuff plates
- Instrument panel sub-assembly
- Lower LH instrument panel sub-assembly
- Lower steering column cover
- Steering column cover

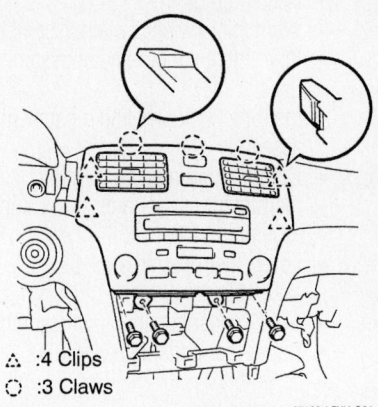

△ :4 Clips
○ :3 Claws

67162-LEXU-G21

Exploded view center finish panel—ES 330

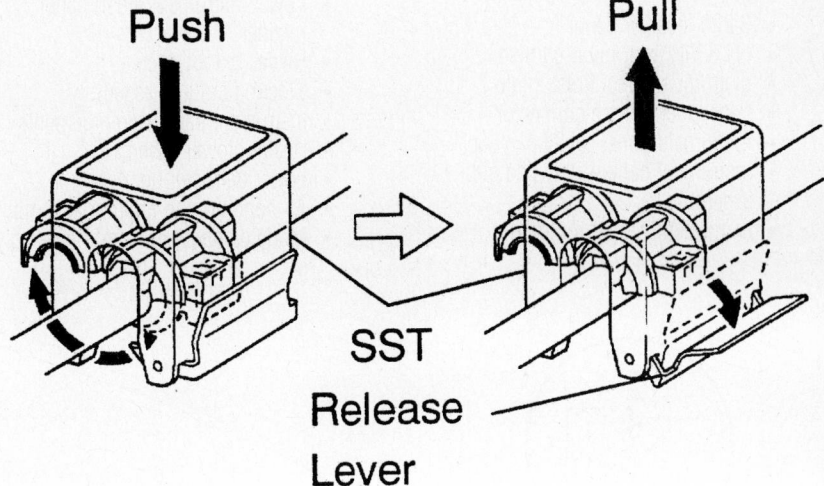

Push   Pull

SST Release Lever

67162-LEXU-G19

Using SST tool—ES 330

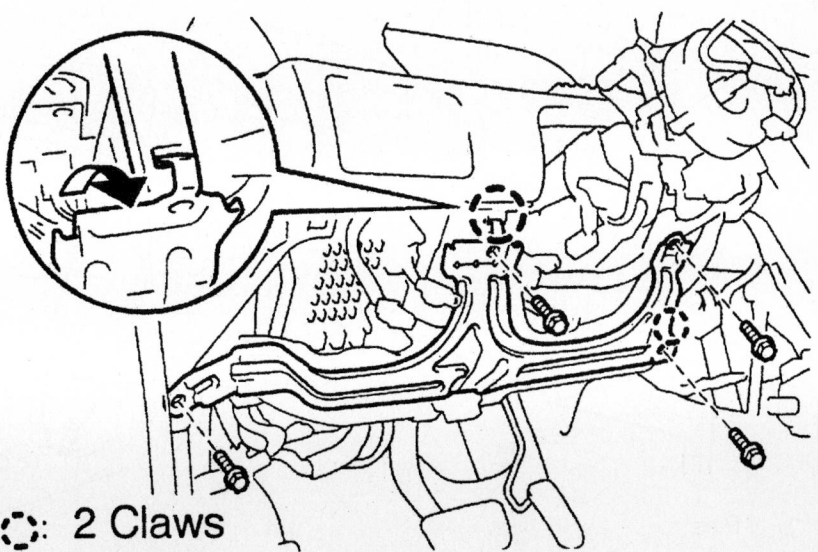

⬚ : 2 Claws

67162-LEXU-G20

Exploded view lower instrument panel sub-assembly—ES 330

- Headlamp dimmer switch assembly
- Windshield wiper switch assembly
- Instrument panel register assemblies
- Instrument cluster finish panel sub-assembly
- Combination meter assembly
- Instrument panel under cover sub-assembly
- Lower instrument panel sub-assembly
- Upper console panel sub-assembly
- Upper rear console panel sub-assembly
- Console box
- Instrument panel LH & RH end panel
- Air conditioning control assembly
- Instrument panel center cluster finish panel
- Radio receiver panel
- LH & RH front pillar garnish
- Instrument panel finish plate
- Passenger air bag connector
- Instrument panel safety pad cap
- instrument panel safety pad sub-assembly
- Instrument cluster molding

- Switch hole base
- Glove box lamp assembly
- Automatic light control sensor
- Side defroster nozzle ducts
- Defroster nozzle assembly
- Heater register ducts
- Left door instrument panel air bag assembly

❋❋ WARNING

Follow air bag removal procedures

9. Remove or disconnect the following:
   - Rear air ducts
   - Console box duct
   - Floor shift parking lock cable assembly
   - Headlamp leveling ECU assembly
   - Instrument panel brace assemblies
   - Lower instrument finish panel retainer
   - Heater to foot ducts
   - Steering column assembly
   - Instrument panel reinforcements
   - Heater blower assembly
   - Lower defroster nozzles
   - Air conditioning radiator assembly
   - Mode damper servo sub-assembly

- Airmix damper servo sub-assembly
- Heater radiator (core) sub-assembly

To install:

10. Install or connect the following:
11. Heater core unit sub assembly
    - Heater core into A/C assembly
    - Screw and clamp
12. Install or connect the following:
    - Air conditioning radiator assembly
    - Lower defroster nozzles
    - Heater blower assembly
    - Instrument panel reinforcements
    - Steering column assembly
    - Heater to foot ducts
    - Lower instrument finish panel retainer
    - Instrument panel brace assemblies
    - Headlamp leveling ECU assembly
    - Floor shift parking lock cable assembly
    - Console box duct
    - Rear air ducts
13. Reassemble the dash components as follows:
    - Left door instrument panel air bag assembly
    - Heater register ducts

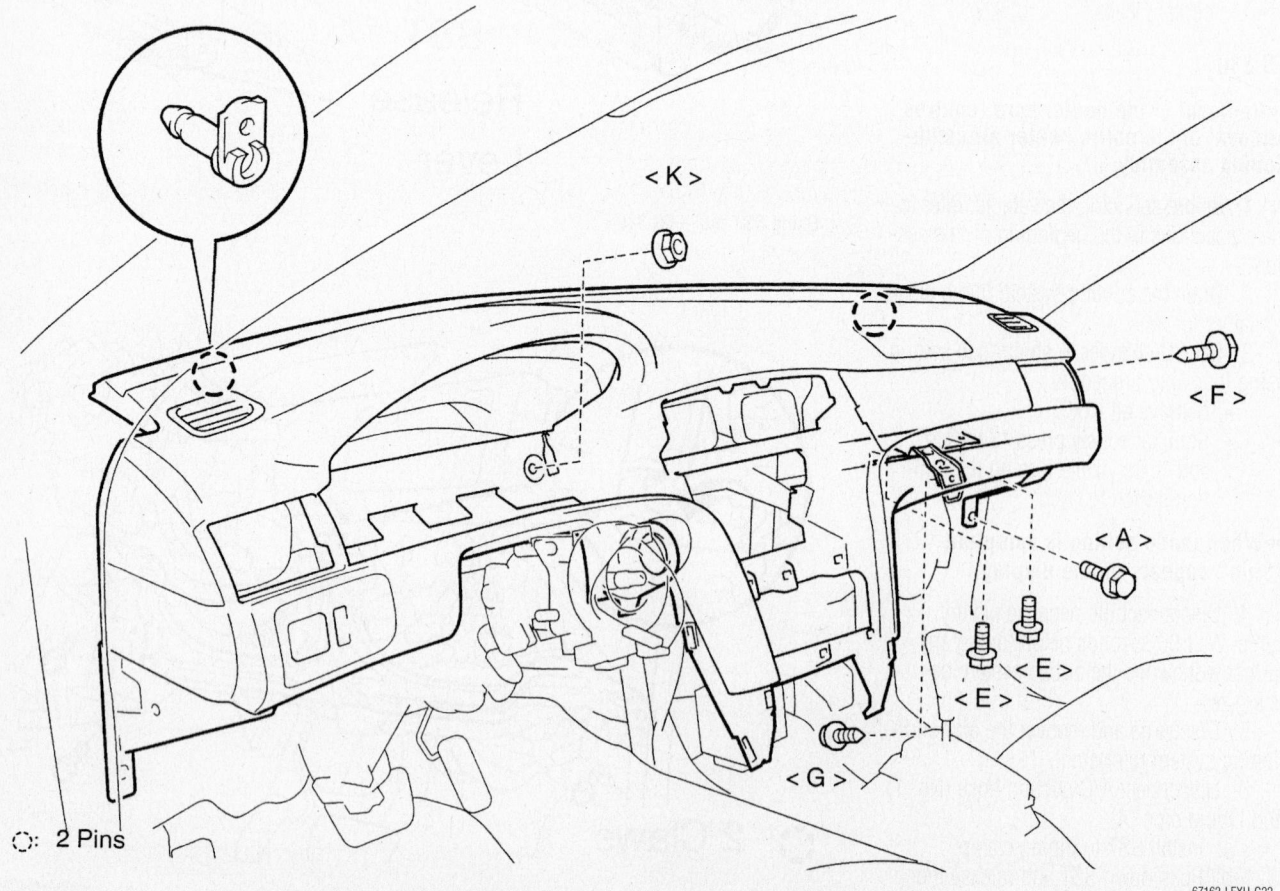

< K >

< F >

< A >

< E >

< E >

< G >

◌ : 2 Pins

Exploded view of instrument panel safety pad sub-assembly—ES 330

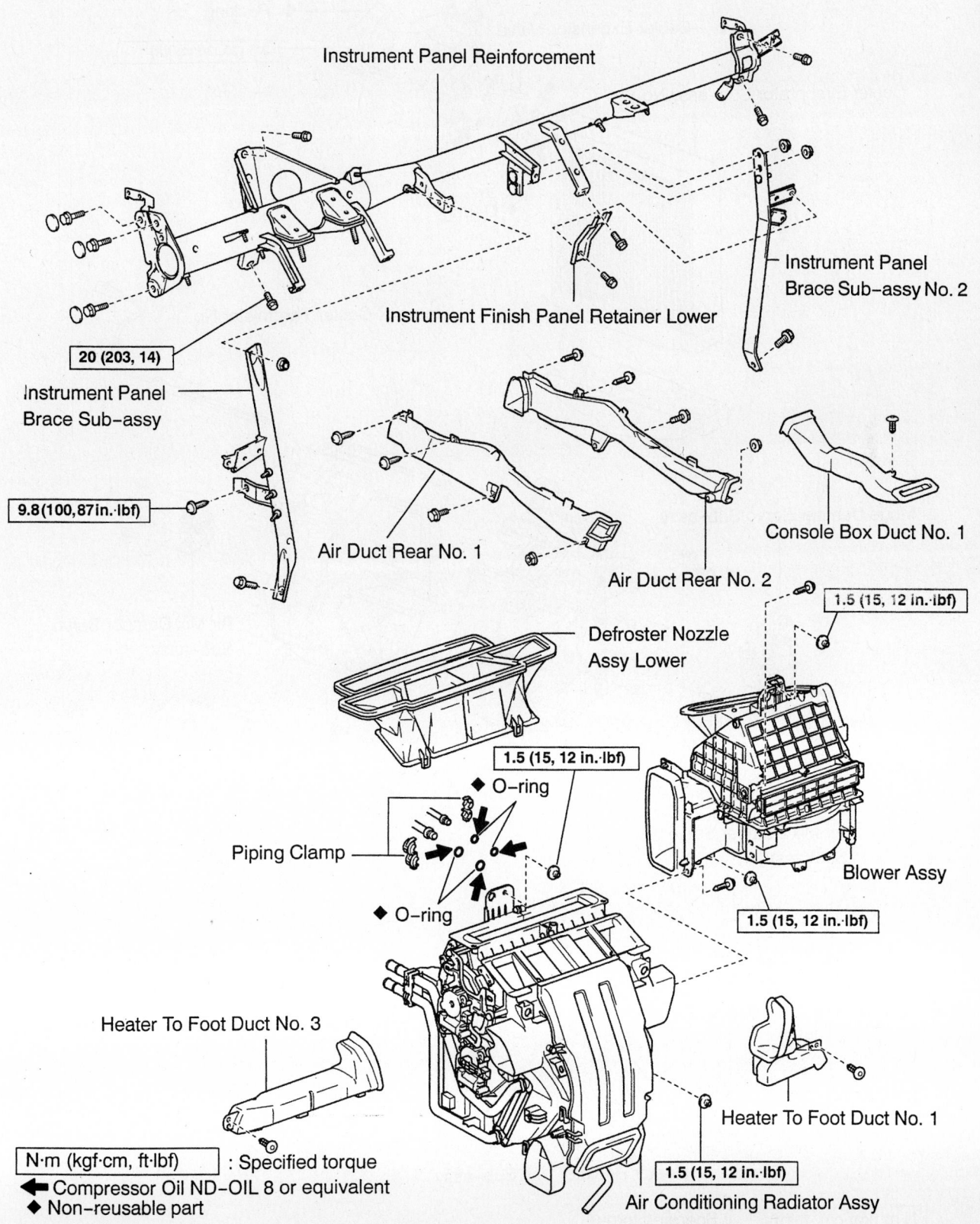

Instrument Panel Reinforcement

Instrument Panel
Brace Sub-assy No. 2

Instrument Finish Panel Retainer Lower

20 (203, 14)

Instrument Panel
Brace Sub-assy

9.8 (100, 87 in.·lbf)

Air Duct Rear No. 1

Air Duct Rear No. 2

Console Box Duct No. 1

1.5 (15, 12 in.·lbf)

Defroster Nozzle
Assy Lower

1.5 (15, 12 in.·lbf)

◆ O-ring

Piping Clamp

Blower Assy

◆ O-ring

1.5 (15, 12 in.·lbf)

Heater To Foot Duct No. 3

Heater To Foot Duct No. 1

N·m (kgf·cm, ft·lbf) : Specified torque

1.5 (15, 12 in.·lbf)

◀ Compressor Oil ND-OIL 8 or equivalent
◆ Non-reusable part

Air Conditioning Radiator Assy

67162-LEXU-G23

Exploded view Air conditioning sub-assembly—ES 330

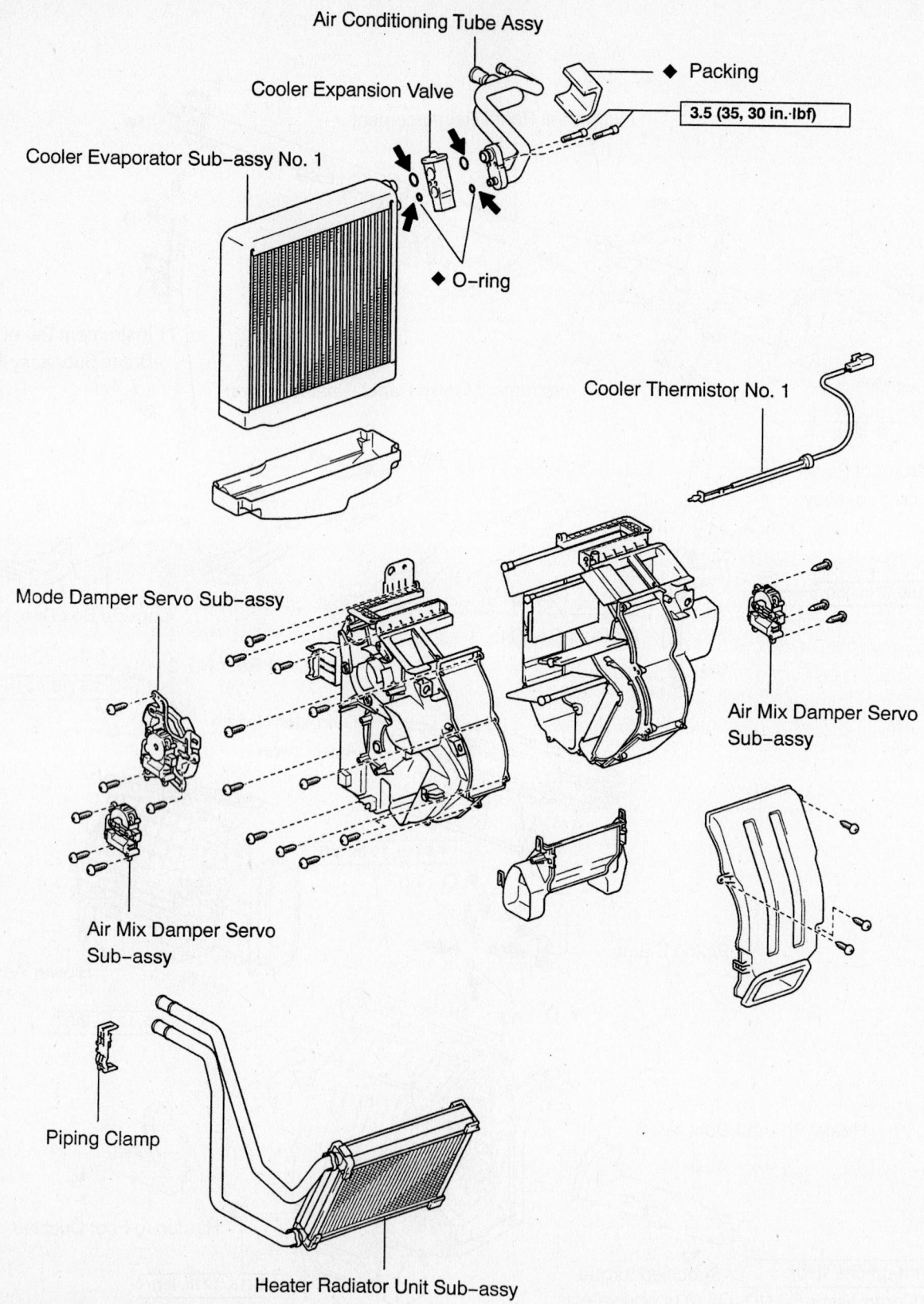

Air Conditioning Tube Assy

◆ Packing

3.5 (35, 30 in.·lbf)

Cooler Expansion Valve

Cooler Evaporator Sub-assy No. 1

◆ O-ring

Cooler Thermistor No. 1

Mode Damper Servo Sub-assy

Air Mix Damper Servo Sub-assy

Air Mix Damper Servo Sub-assy

Piping Clamp

Heater Radiator Unit Sub-assy

N·m (kgf·cm, ft·lbf) : Specified torque

◀ Compressor Oil ND-OIL 8 or equivalent

◆ Non-reusable part

Exploded view of the heater case assembly—ES330

- Defroster nozzle assembly
- Side defroster nozzle ducts
- Automatic light control sensor
- Glove box lamp assembly
- Instrument cluster molding
- Instrument panel safety pad sub-assembly
- Passenger air bag connector
- Instrument panel finish plate
- LH & RH front pillar garnish
- Radio receiver panel
- Instrument panel center cluster finish panel
- Air conditioning control assembly
- Instrument panel LH & RH end panel
- Console box
- Upper rear console panel sub-assembly
- Upper console panel sub-assembly
- Lower instrument panel sub-assembly
- Instrument panel under cover sub-assembly
- Combination meter assembly
- Instrument cluster finish pale sub-assembly
- Instrument panel register assemblies
- Windshield wiper switch assembly
- Headlamp dimmer switch assembly
- Steering column cover
- Lower steering column cover
- Lower LH instrument panel sub-assembly
- Instrument panel sub-assembly
- RH & LH door scuff plates

14. Install or connect the following:
- Heater core hoses
- A/C suction and pressure hoses, attach with bolt and plate

➡ **Lubricate O-rings with compressor oil**

- Negative battery cable
- Fill cooling system with coolant
- Evacuate and recharge A/C system
- Warm up engine and inspect for coolant leaks

Cylinder Head

REMOVAL & INSTALLATION

3.0L (1MZ-FE) Engine

1. Before servicing the vehicle, refer to the precautions in the beginning of this section.
2. Drain the cooling system.
3. Relieve the fuel system pressure.

4. Remove or disconnect the following:
- Negative battery cable
- Accelerator and the throttle cables
- Air cleaner cover, air flow meter and the air duct
- Cruise control actuator and bracket, if equipped
- 2 engine ground straps
- Right engine mounting support
- Radiator hoses
- 2 heater hoses

5. Plug the fuel feed and return lines from the fuel rail assembly.

6. Plug the pressure hose from the hydraulic motor

7. Remove or disconnect the following:
- V-bank cover
- Fuel pressure control Vacuum Switching Valve (VSV)
- Fuel pressure regulator
- Cylinder head rear plate
- Intake air control valve VSV
- Exhaust Gas Recirculation (EGR) vacuum modulator
- EGR valve
- Intake Air Control (IAC) valve
- Fuel pressure regulator
- EGR VSV
- 2 nuts and the emission control valve set
- Brake booster vacuum hose
- Positive Crankcase Ventilation (PCV) hose
- Intake air control valve vacuum hose
- Data link connector from the mounting bracket
- 2 ground straps from the intake chamber
- Hydraulic motor pressure hose from the intake chamber
- Right Oxygen (O$_2$) sensor connector from the power steering pressure tube
- 2 nuts and the power steering pressure tube from the intake chamber
- 2 power steering air hoses
- Engine hanger and the intake chamber support
- EGR pipe and gaskets
- Throttle Pressure (TP) sensor connector
- Idle Air Control (IAC) valve connector
- EGR gas temperature connector
- air conditioning idle up connector
- 2 vacuum hoses from the Thermal Vacuum Valve (TVV)
- Vacuum hose from the cylinder head rear plate
- Vacuum hose from the charcoal canister

- Air assist hose and the 2 water bypass hoses
- Air intake chamber
- Left engine wiring harness and position it out of the way
- Wiring harness from the rear of the engine
- Right engine wiring harness and position it out of the way
- Ignition coils and the spark plugs
- Timing belt
- Camshaft pulleys and the timing belt rear cover
- Cylinder head rear plate
- Water inlet pipe
- Air assist hose and vacuum hose
- Intake manifold and fuel rail assembly
- Water outlet
- EGR pipe from the right exhaust manifold
- Exhaust manifolds
- Dipstick assembly and the power steering pump bracket
- Valve covers and the Camshaft Position (CMP) sensor
- Camshafts

8. Be sure the engine is at or near ambient temperature and remove the 2 (one on each head) 8mm recessed hex bolts. Loosen and remove the 8 head bolts evenly, in 3 passes, in the reverse order of the tightening sequence. Carefully lift the head from the engine; if it is necessary to pry the head loose, take great care not to damage the mating surfaces. Place the head on wood blocks in a clean work area.

✳✳ WARNING

If the cylinder head bolts are loosened out of sequence, warpage or cracking could result.

9. Remove the cylinder head gasket. With a gasket scraper, carefully remove all the old gasket material from the cylinder head and engine block surfaces.

To install:

10. Place the new cylinder head gasket onto the cylinder block. Place the cylinder head onto the gasket.

11. Coat the threads of the 8 cylinder head bolts (12-sided) with clean engine oil and install the bolts into the cylinder head. Uniformly tighten the bolts in sequence in 3 steps to an ultimate tighten of 40 ft. lbs. (54 Nm). If any of the bolts does not meet the torque, replace it.

12. Mark the forward edge of each bolt with paint, then retighten each bolt, in proper sequence, an additional 90 degrees.

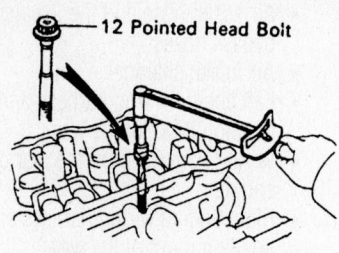

12 Pointed Head Bolt

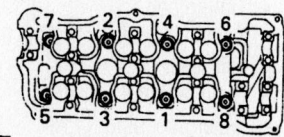

Front ←

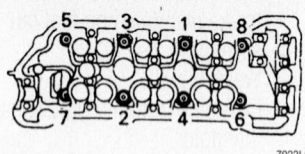

7923LG13

Cylinder head torque sequence—3.0L (1MZ-FE) engine

Check that each painted mark is now at a 90 degrees angle to the front. The paint mark should have been applied to the bolt in the 9 o'clock position and should now be in the 12 o'clock position.

13. Coat the threads of the 2 remaining 8mm bolts with engine oil and install them. Tighten to 13 ft. lbs. (18 Nm).

14. Install or connect the following:
- Camshafts and adjust the valves

→**Apply sealant to the cylinder heads where the camshaft supports meet the cylinder heads.**

- Cylinder head covers. Use new gaskets.
- Dipstick and power steering pump bracket
- Exhaust manifolds. Tighten the nuts to 36 ft. lbs. (49 Nm).
- EGR pipe to the right exhaust manifold
- Water outlet
- Intake manifold and the fuel rail assembly. Tighten the intake manifold nuts and bolts to 11 ft. lbs. (15 Nm).
- Air assist hose and the 2 water bypass hoses
- Water inlet pipe and the cylinder head rear plate
- Timing belt rear cover and the camshaft pulleys
- Timing belt
- Spark plugs and the ignition coils
- Right engine wiring harness
- Wiring harness to the rear of the engine

- Left engine wiring harness
- Air intake chamber
- EGR pipe. Use new gaskets
- 2 TVV vacuum hoses
- Vacuum hose to the rear cylinder head plate
- Charcoal canister vacuum hose
- TP sensor connector
- IAC valve connector
- EGR gas temperature connector
- Air conditioning idle up connector
- Engine hanger and the intake chamber support
- 2 power steering air hoses
- Power steering pressure tube to the intake chamber
- O_2 sensor connector to the pressure tube
- 2 ground straps to the intake chamber
- Data link connector to the bracket
- Power brake booster vacuum hose
- PCV hose
- IAC valve vacuum hose
- Emission control valve set and related vacuum hoses and connectors
- V-bank cover
- Pressure hose to the hydraulic motor
- Fuel lines to the fuel rail assembly
- Heater and radiator hoses
- Right engine mounting support
- 2 engine ground straps
- Cruise control actuator and bracket
- Air cleaner, air flow meter and air duct assembly
- Accelerator and the throttle cables
- Negative battery cable

15. Fill the cooling system to the proper level with coolant.

16. Start the engine and check for leaks. Bleed the air from the cooling system.

17. Adjust the ignition timing.

18. Road test the vehicle and check for unusual noise, shock, slippage, correct shift points and smooth operation.

19. Recheck the coolant and engine oil levels.

3.0L (2JZ-GE) Engine (Except 2004)

1. Before servicing the vehicle, refer to the precautions in the beginning of this section.

2. Disconnect the negative battery cable. Wait at least 90 seconds before performing any other work.

3. Relieve the fuel pressure from the fuel lines.

4. Remove or disconnect the following:

- Coolant
- Undercovers
- Accelerator, throttle control (automatic transmission only) and cruise control cables from the throttle body
- Cleaner duct
- Air cleaner, airflow meter and the intake air pipe
- Drive belt, the fan and fluid coupling and the water pump pulley
- No. 2 front exhaust pipe
- Exhaust manifold cover
- 2 Heated Oxygen (HO_2) sensor connector(s)
- Exhaust manifolds and gaskets by removing the 8 bolts
- Water bypass outlet and the No. 1 water bypass pipe
- Power steering air hose from the No. 4 timing belt cover
- Power steering hose from the air intake chamber
- 2 bolts and the vane pump from the pump bracket
- 2 bolts, the pump rear stay. Put aside the vane pump and suspend it.
- Fuel return hose from the fuel return pipe. Plug the hose end.
- Fuel return hose from the oil dipstick guide
- Bolt and bracket
- Engine wire from the intake manifold stay
- Throttle body and intake air connector assembly
- Bolt, pull out the oil dipstick guide with the dipstick and remove the O-ring from the dipstick guide
- Transmission dipstick and guide, if equipped with automatic transmission
- Connector from the No. 2 vacuum pipe
- Exhaust Gas Recirculation (EGR) gas temperature sensor wiring harness
- 2 nuts and the vacuum pipe from the air intake chamber and intake manifold
- No. 2 vacuum pipe and Vacuum Switching Valve (VSV) assembly
- Nuts and the vacuum tank from the intake manifold
- VSV connector and hoses
- Vacuum hose (from the air intake chamber) from port B of the vacuum tank
- Vacuum hose (from actuator) from the VSV
- Vacuum control valve set

- Data Link Connector (DLC1) bracket and VSV assembly
- Vacuum hose from the brake booster union and the Evaporative Emission (EVAP) hose from the No. 2 vacuum pipe
- Bolt holding the engine wire protector to the air intake chamber
- 5 bolts, nut, air intake chamber and gasket
- No. 3 (top) timing belt cover by removing the oil filler cap and the 6 bolts using a 5mm hexagon wrench.
- 4 bolts, using a 5mm hexagon wrench, and the rear cylinder head cover
- Spark plugs
- Drive belt tensioner by removing the 3 bolts

5. Set the engine to Top Dead Center (TDC)/compression for cylinder No. 1 piston

6. Remove or disconnect the following:
- Timing belt tensioner and dust boot. Remove the timing belt from the camshaft pulleys. Support the belt so that it remains in contact with the crankshaft pulley.
- Wire clamp from the bracket.
- HO$_2$ and the Crankshaft Position (CKP) sensors.
- 2 ground straps from the intake manifold.
- Engine Coolant Temperature (ECT) sender gauge
- Knock Sensor (KS)
- Oil pressure switch
- Oil level sensor
- air conditioning compressor
- 6 injector electrical connectors
- 3 nuts and the engine wire protector from the intake manifold
- Water bypass hose from the clamp on the oil filter bracket
- Water outlet, 2 nuts, and the bolt with the water bypass hose
- 2 bolts and the intake manifold stay
- Fuel pressure pulsation damper
- Clamp bolt from the intake manifold
- Union bolt and gaskets
- Fuel inlet pipe
- 6 bolts, 2 nuts, the intake manifold
- Delivery pipe assembly and gasket
- Cylinder head covers (valve covers)
- Camshaft timing pulleys
- Rear (No. 4) timing belt cover
- Camshafts
- Cylinder head bolts in several passes and in the reverse order of the tightening sequence

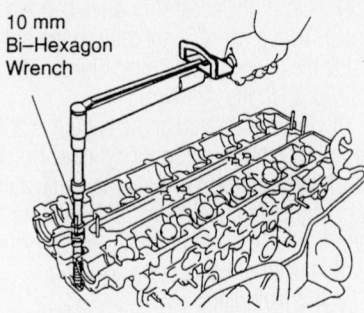

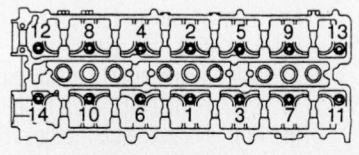

Cylinder head torque sequence—3.0L (2JZ-GE) engine

- Head from the engine

7. Clean the head and block of all gasket material.

To install:

8. Install or connect the following:
- New gasket and cylinder head on the block
- Head bolts, lightly coated with engine oil and plate washers. Uniformly tighten the head bolts in several passes, in sequence to 26 ft. lbs. (35 Nm). Following the correct order, tighten each bolt an additional 90 degrees. Again following the correct order, tighten the bolts another 90 degrees of rotation.

⁂ WARNING

Correct bolt torque must be achieved in 3 steps; do not attempt to shorten the procedure by combining the two 90 degree steps.

- Camshafts. Coat the thrust portions of each with engine oil.
- No. 3 and No. 7 bearing caps in place. Coat the bolt threads with oil, then uniformly and alternately tighten them temporarily.
- New oil seals, coated with multipurpose grease, over the camshafts
- Seal packing to the No. 1 bearing cap
- Remaining bearing caps in their proper locations. Coat the threads of each bolt with clean oil, then tighten them, in several passes, in the correct sequence, to 14 ft. lbs. (20 Nm).

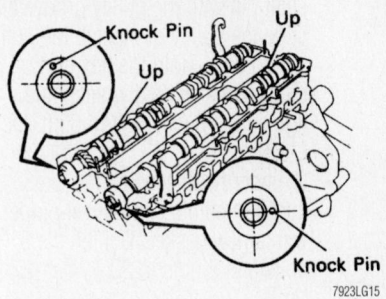

Position the knock pins as shown when installing the camshafts—3.0L (2JZ-GE) engine

- The 2 oil seals in as far as it will go

9. Rotate each camshaft until the forward straight (knock) pin is straight up. Loosen the exhaust Nos. 1, 2 and 6 bearing cap bolts until they can be turned by hand; retighten the bolts, in several passes, to 14 ft. lbs. (20 Nm). Loosen the intake Nos. 1, 2 and 5 bearing cap bolts and retighten the bolts, in several passes, to 14 ft. lbs. (20 Nm).

10. Turn each camshaft ⅓ of a revolution (120 degrees). Loosen the exhaust Nos. 4 and 7 bearing cap bolts; retighten the bolts, in several passes, to 14 ft. lbs. (20 Nm). Loosen the intake Nos. 4 and 6 bearing cap bolts; retighten the bolts, in several passes, to 14 ft. lbs. (20 Nm).

11. Turn each camshaft an additional ⅓ of a revolution, loosen the exhaust bearing cap bolts Nos. 3 and 5, then retighten the bolts, in several passes, to 14 ft. lbs. (20 Nm). Loosen the intake bearing cap bolts Nos. 3 and 7, then retighten the bolts, in several passes, to 14 ft. lbs. (20 Nm).

12. Check and adjust the valve clearance.

13. Install or connect the following:
- Rear (No. 4) timing belt cover. Tighten the bolts to 78 inch lbs. (9 Nm).
- Camshaft timing pulleys. Align the

Apply sealant to the areas indicated on the cylinder head before installing the cover—3.0L (2JZ-GE) engine

shaft pin with the pulley groove and slide the pulley on. Install the bolt temporarily. Hold the hex portion of the camshaft with a wrench and tighten the pulley bolt to 59 ft. lbs. (79 Nm).

- Cylinder head covers
- Intake manifold and delivery pipe with a new gasket. Tighten the 6 bolts and 2 nuts to 20 ft. lbs. (27 Nm).
- Fuel inlet pipe to the fuel rail. Tighten the union bolt to 30 ft. lbs. (42 Nm).
- Clamp bolt to the intake manifold
- Fuel pressure pulsation damper
- Intake manifold stay and tighten the bolts to 29 ft. lbs. (39 Nm)
- Water outlet and the bypass hose. Tighten the bolts to 15 ft. lbs. (21 Nm).
- Engine wiring harness. Secure the wiring in all clamps and retainers.
- Wiring leads to the proper sender, sensor or switch
- Injector leads

14. Compress the timing belt tensioner in a vise and retain the pin with a 1.5mm hex wrench. Install the dust boot onto the tensioner.

15. Install the tensioner. Alternately tighten the bolts to 20 ft. lbs. (26 Nm). Remove the hex wrench with a pair of pliers, allowing the tensioner to be applied to the timing belt.

16. Turn the crankshaft 2 full revolutions clockwise. Check that all timing marks align as before. If the marks (cam and crankshaft) do not align, remove the timing belt and reinstall it.

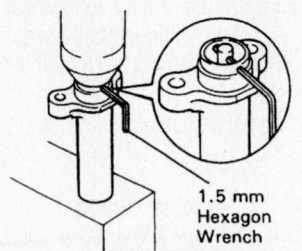

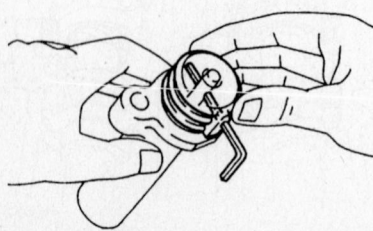

7923LG16

Compressing the timing belt tensioner—3.0L (2JZ-GE) engine

17. Install the accessory drive belt tensioner. Take great care not to drop the bolts inside the lower timing cover. Tighten the bolts to 15 ft. lbs. (21 Nm).

18. Double check that the engine is still set to TDC/compression for cylinder No. 1. Check the alignment of both the crank and camshaft timing marks. Install the timing belt.

19. Install or connect the following:

- Spark plugs
- Wiring to the spark plugs
- No. 3 timing belt cover
- Cylinder head rear cover
- Air intake chamber with a new gasket. Tighten the bolts to 20 ft. lbs. (27 Nm). Install the bolt to hold the engine wire protector to the air intake chamber.
- Vacuum hose to the brake booster union and the EVAP hose to the No. 2 vacuum pipe.
- DLC connector and bracket and VSV connector
- Vacuum control set
- No. 2 vacuum pipe assembly and connect the hoses. Tighten the nuts to 20 ft. lbs. (27 Nm).
- EGR gas temperature sensor. Tighten it to 14 ft. lbs. (20 Nm).
- Vacuum hoses
- Dipstick tubes. Always use a new O-ring on each tube.
- Intake chamber supports and tighten the bolts to 13 ft. lbs. (18 Nm). The supports are marked **F** and **R** for the front and rear positions.
- Throttle body and intake air connector assembly
- Engine wire bracket
- Fuel return hose
- Vane pump to the pump bracket
- Power steering air hose to the No. 4 timing belt cover and intake chamber.
- Water bypass outlet and the bypass pipe. Always use new O-rings.
- Exhaust manifolds with new gaskets. Tighten the bolts to 29 ft. lbs. (39 Nm).
- O_2 sensor leads
- Front exhaust pipe. Tighten the bolts to 46 ft. lbs. (62 Nm).
- Manifold cover
- Water pump pulley
- Fan and coupling and the drive belt. Tighten the 4 nuts to 12 ft. lbs. (16 Nm).
- Air cleaner, airflow meter and the intake air connector pipe
- Air cleaner duct

- Control and accelerator cables to the throttle body
- Coolant
- Negative battery cable. Start the engine and check for leaks.
- Engine undercovers

3.0L (2JZ-GE) Engine (2004)

1. Before servicing the vehicle, refer to the precautions in the beginning of this section.

2. Disconnect the negative battery cable. Wait at least 90 seconds before performing any other work.

3. Relieve the fuel pressure from the fuel lines.

4. Remove or disconnect the following:

- Undercover(s)
- Coolant
- Upper radiator hose from water outlet
- Engine cover
- Air cleaner inlet, air cleaner assembly, MAF meter, and intake air resonator assembly
- Serpentine drive belt
- Power steering pump from mounting (Do NOT disconnect hoses)
- Oxygen sensor wiring and grommet for bank 2, No. 2 sensor
- Front exhaust pipe from exhaust manifold
- Exhaust manifold
- Vacuum control valve set and No. 2 vacuum pipe from lower side of engine
- No. 3 timing belt cover
- Ignition coils and high-tension wires
- Spark plugs

5. Properly mark all engine wiring before disconnecting as needed for access to cylinder head, then disconnect the following:

- Ground strap from cylinder head
- 2 water bypass hoses from cylinder head and oil filter bracket
- Bank 2, No. 1 and bank 1, No. 1 oxygen sensor connectors and wire clamp
- Alternator wiring
- Wiring from water pump clamp
- 2 ground terminals from intake manifold
- 2 engine wire clamps from No. 1 oil pipe and clamp on intake manifold
- ECT, knock, oil pressure switch, and oil level sensor connectors
- Starter connector
- Injector connectors

- Camshaft timing oil control valve connector
- Camshaft position sensor connector
- Wiring protector from No. 2 cylinder head cover
- Engine wiring protector from intake manifold

6. Remove the fuel pressure pulsation damper.

7. Remove the intake manifold assembly, including stays.

8. Remove both cylinder head covers and gaskets.

9. While holding the timing belt so its position to timing marks does not change, remove the timing belt from the camshaft pulleys.

10. Remove both camshaft pulleys (exhaust pulley first, while holding the hex portion of the camshaft in order to loosen the pulley bolt).

➡ **See CAMSHAFT & VALVE LIFTERS for detailed removal procedure, if needed.**

11. Remove the No. 4 timing belt cover.

12. Uniformly remove the No. 3 camshaft bearing cap bolts from both camshafts. Use a screwdriver to pry out the No. 1 and No. 3 camshaft bearing caps and oil seals. Then, evenly loosen all remaining 12 camshaft bearing cap bolts, in the order shown. Remove the No. 2 camshaft bearing caps and remove the camshafts.

13. Loosen and remove all 14 cylinder head bolts in the sequence shown, using several passes during loosening sequence.

14. Lift the cylinder head from the block dowels. Disconnect the heater hose from the union. Remove the cylinder head from the

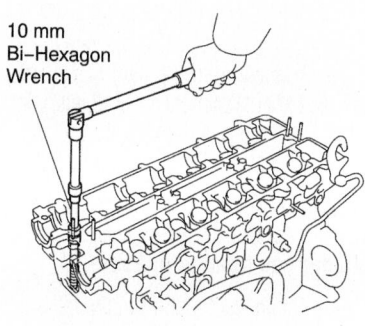

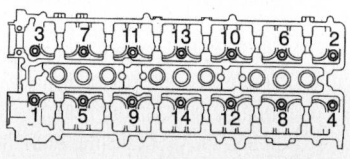

Showing the cylinder head bolt removal sequence—3.0L (2JZ-GE) engine (2004 model)

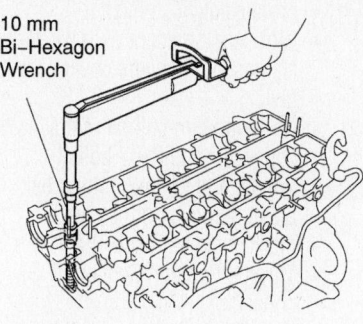

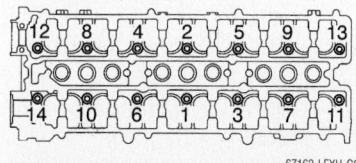

Showing the cylinder head bolt tightening sequence—3.0L (2JZ-GE) engine (2004 model)

vehicle and place on wooden blocks on the workbench.

To install:

15. With a new gasket, position the cylinder head onto the block.

16. Apply a light coat of engine oil to the cylinder head bolt threads and install the washers.

17. Install the cylinder head bolts, following the bolt tightening sequence shown. Torque bolts to 26 ft. lbs. (35 Nm).

18. Mark the front of each cylinder head bolt with paint, then retighten each bolt, in the same sequence, an additional 90°. The paint mark should now face 90° to the side.

19. Install the camshafts. See CAMSHAFTS & VALVE LIFTERS section.

20. Check and adjust the valve lash.

21. Install the No. 4 timing belt cover.

22. Install both camshaft pulleys. Torque the exhaust camshaft pulley bolt to 60 ft. lbs. (81 Nm).

23. Install the timing belt back onto the camshaft pulleys, making sure the timing marks align properly.

24. Install both cylinder head covers, with new gaskets.

25. Install the intake manifold assembly.

26. Install the fuel pressure pulsation damper to the lower side of the engine.

27. Reconnect all engine wiring and hoses in reverse order of removal. Ensure components are reinstalled in original positions, as marked.

28. Install or reconnect the following:
- Spark plugs
- Ignition coils and wires
- No. 3 timing belt cover
- Vacuum control valve set and No. 2

vacuum pipe; torque nuts to 15 ft. lbs. (21 Nm)
- Air intake chamber
- Oil dipstick and guide
- Throttle body and intake air connector
- Water bypass outlet and No. 1 water bypass pipe
- Exhaust manifold, with new gaskets; torque nuts to 30 ft. lbs. (40 Nm)
- Front exhaust pipe to manifold; torque bolts and nuts to 32 ft. lbs. (44 Nm)
- Reconnect wire grommet and wiring for bank 2, No. 2 oxygen sensor
- Power steering pump to mounting; torque bolts to 43 ft. lbs. (58 Nm)
- Serpentine drive belt
- Air cleaner, MAF meter and intake air resonator
- Upper radiator hose
- Engine cover

29. Refill the coolant and engine oil. Start the engine and check for leaks or abnormal conditions. Perform and road test. Then, recheck for leaks and recheck fluid levels.

4.0L (1UZ-FE) and 4.3L (3UZ-FE) Engines (Except 2004)

1. Before servicing the vehicle, refer to the precautions in the beginning of this section.

2. Relieve the fuel system pressure.

3. Remove or disconnect the following:
- Negative battery cable. Wait at least 90 seconds before performing any other work.
- Oil pan protector
- Engine undercover
- Coolant
- Battery clamp cover
- Air cleaner inlet
- V bank cover by removing the bolt and 2 cap nuts
- Air cleaner and intake air connector assembly
- Drive belt, fluid coupling and the fan pulley. The drive belt tension may be slackened by turning the tensioner counterclockwise. The pulley bolt for the drive belt tensioner has a left-handed thread.
- Radiator
- Right-hand No. 3 timing belt cover
- Left-hand No. 3 timing belt cover
- Drive belt idler pulley by removing the pulley bolt and cover plate
- Right-hand No. 2 timing belt cover

- Left-hand No. 2 timing belt cover
- No. 1 ignition coil
- Air conditioning compressor from the engine
- Fan bracket by removing the 2 bolts and 2 nuts

4. Set the engine to Top Dead Center (TDC) on cylinder No. 1.

❊❊ WARNING

Since the thrust clearance of the camshaft is small, the camshaft must be kept level while it is being removed. If the camshaft is not kept level, the portion of the cylinder head receiving the shaft thrust may crack or be damaged, causing the camshaft to seize or break.

5. Turn the crankshaft pulley approximately 50 degrees clockwise and put the timing mark of the crankshaft pulley in line with the centers of the crankshaft pulley bolt and the idler pulley bolt.

❊❊ WARNING

If the timing belt is disengaged, having the crankshaft pulley at the wrong angle can cause the piston head and valve head to come into contact with each other when you remove the camshaft timing pulley. Always set the crankshaft pulley at the correct angle before removing the timing belt.

6. If the timing belt is to be reused, turn the crank pulley slowly; check that the 3 installation marks are present on the belt. If the marks are not present, make new installation marks before removing the belt. The marks should align with the timing marks on each camshaft pulley and the crank pulley.

7. Remove the timing belt tensioner. Alternately loosen the 2 bolts; remove the bolts, the tensioner and the dust protector.

8. Loosen the tension between the left side and the right side timing pulleys by slightly turning the left side camshaft clockwise.

9. Remove or disconnect the following:

- Timing belt from the camshaft timing pulleys. Using the proper tool, remove the bolt and the camshaft timing pulleys.
- Power steering pump from the engine. Do NOT disconnect the hoses or lines from the power steering pump. Support the power steering pump with a piece of wire. Do NOT allow the pump to hang.

- Front catalytic converter
- High tension spark plug wires, wire clamps and the wire cover assembly
- No. 2 ignition coil by removing the connector and the 2 bolts
- 2 bolts and the rear timing belt plate. Remove both plates
- Intake chamber assembly
- Throttle Position Sensor (TPS) connector
- With TRAC system, sub TP sensor connector
- With TRAC system, sub TP connector
- Idle Air Control (IAC) valve connector
- Exhaust Gas Recirculation (EGR) valve connector
- Vacuum Switching Valve (VSV) connector for fuel pressure control
- VSV connector for Evaporative Emissions (EVAP) system
- EGR gas temperature sensor connector
- Brake booster vacuum hose from the union on the air intake chamber
- Positive Crankcase Ventilation (PCV) hose from the PCV valve on the left-hand cylinder head
- Water bypass hose (from the EGR valve) from the rear water bypass joint
- Water bypass hose (from the throttle body) from the rear water bypass joint
- Vacuum hose (from the VSV for fuel pressure control) from the fuel pressure regulator
- EVAP hose (from charcoal canister) from the VSV for EVAP
- Heater hose from the water bypass pipe
- Fuel inlet hose from the delivery pipe
- Fuel return hose from the fuel return pipe
- Engine wire from the delivery pipes and rear water bypass joint
- Fuel hose from the fuel pressure regulator
- 2 bolts and fuel return pipe from the intake manifold
- 8 injector connectors
- 6 bolts, 4 nuts, the intake manifold assembly and the 2 gaskets
- Water inlet and inlet housing
- Front water bypass joint
- Rear water bypass joint and No. 1 EGR pipe assembly
- Oil dipstick and guide for the automatic transmission

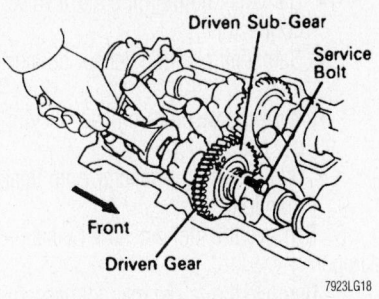

Securing the exhaust camshaft on the right cylinder head—4.0L (1UZ-FE) and 4.3L (3UZ-FE) Engines

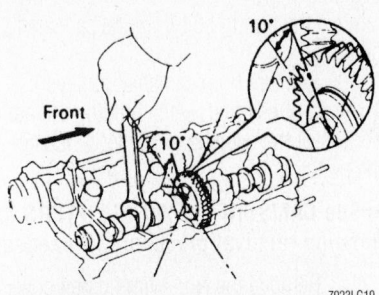

Turning the exhaust camshaft 10 degrees on the right cylinder head—4.0L (1UZ-FE) and 4.3L (3UZ-FE) Engines

- Oil dipstick and guide for the engine
- Engine hangers
- Right and left cylinder head covers by removing the 8 bolts, seal washers and gaskets
- Semi-circular plug, if necessary
- Exhaust camshaft from the right side cylinder head. See the camshaft procedure for tightening sequence.
- Intake camshaft from the right side cylinder head. See the camshaft procedure for tightening sequence.
- Exhaust camshaft of the left side cylinder head. See the camshaft procedure for tightening sequence.

➡**When removing the camshaft, be sure the torsional spring force of the subgear has been eliminated.**

- Intake camshaft from the left side cylinder head. See the camshaft procedure for tightening sequence.
- Main Heated Oxygen (HO$_2$) sensor
- Bolt and HO$_2$ the ground cable from the right cylinder head
- Bolt and the ground strap from the left cylinder head
- Bolt and the engine wire protector from the left-hand cylinder head

- 2 bolts, seal washers, bearing cap and the camshaft housing plug from the right-hand cylinder head
- 10 cylinder head bolts and plate washers to each cylinder head. Loosen the bolts in the reverse order of the tightening sequence. Lift the heads from the dowels on the block with the exhaust manifolds attached. Place the heads on blocks of wood on the workbench.

➡**Do NOT drop anything in the opening in the front of the right side cylinder head. The opening leads through the block and into the oil pan. If anything falls into the opening the oil pan will have to be removed in order to retrieve it.**

⁕⁕ WARNING

If necessary to pry the head loose, take great care not to damage the contact surfaces of the head or block.

- 2 bolts, seal washers, bearing cap and camshaft housing plug from the right-hand cylinder head
- Right exhaust manifold from the cylinder head by removing the heat insulator, 8 nuts and the gasket
- Left exhaust manifold from the cylinder head by removing the heat insulator, 8 nuts and the gasket

To install:

10. Install or connect the following:
    - Right exhaust manifold. The new gasket must be installed with the white marks facing the manifold side. Tighten the bolts to 33 ft. lbs. (44 Nm). Install the right O_2 sensor.
    - Left exhaust manifold. The new gasket must be installed with the white marks facing the manifold side. Tighten the bolts to 33 ft. lbs. (44 Nm). Install the left O_2 sensor.
    - The 2 new cylinder head gaskets in position on the engine block. Each gasket has a painted mark denoting the rear of the gasket. The gasket for the right bank has a white mark and the gasket for the left bank has a yellow mark. Double check the gasket position and placement.
11. For 4.0L engines, install the cylinder heads and tighten the bolts in sequence as follows:
    a. Step 1: 29 ft. lbs. (39 Nm)
    b. Step 2: Plus 90 degrees
12. For 4.3L engines, install the cylinder

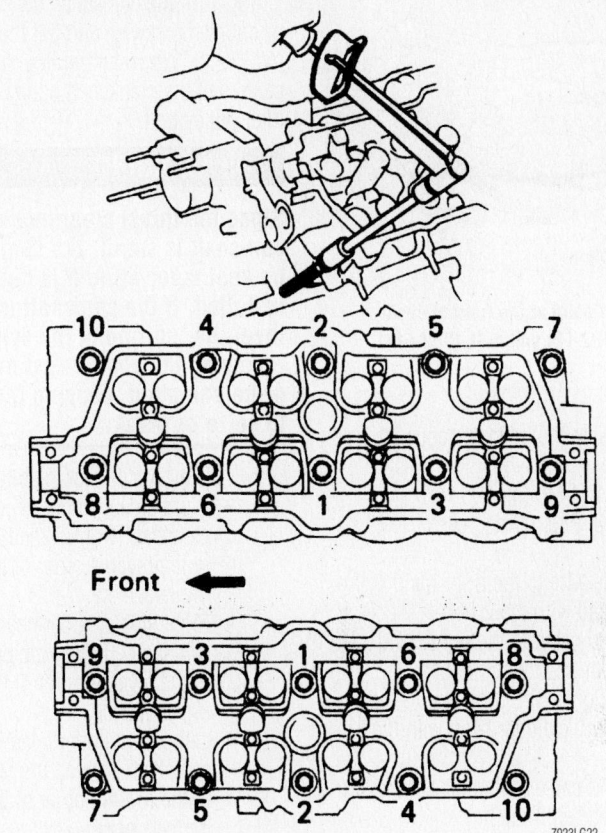

Cylinder head torque sequence—4.0L (1UZ-FE) engine

RH Bank

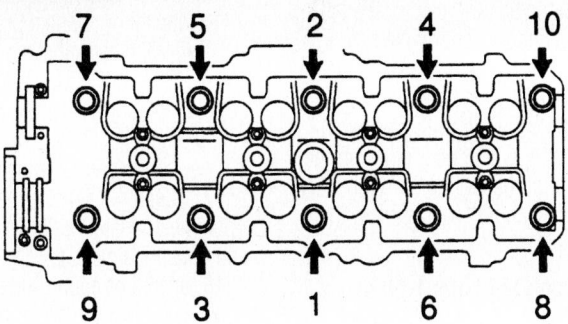

LH Bank

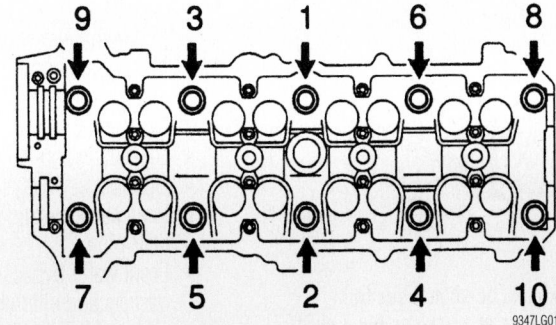

Cylinder head torque sequence—4.3L (3UZ-FE) engine

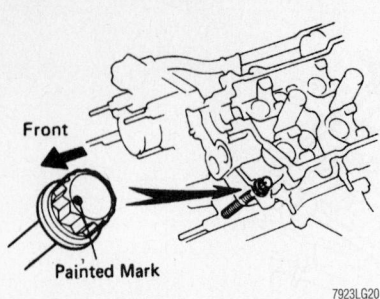

Apply a dot of paint at the front of each bolt—4.0L (1UZ-FE) and 4.3L (3UZ-FE) Engines

heads and tighten the bolts in sequence as follows:

   a. Step 1: 44 ft. lbs. (60 Nm)
   b. Step 2: Plus 90 degrees

- HO_2.
- Engine wire to the right-hand cylinder head with the 2 bolt
- Ground cable to the right-hand cylinder head with the bolt
- The engine wire protector to the left-hand cylinder head with the bolt
- Ground cable to the left-hand cylinder head with the bolt
- Old packing and apply new seal packing to the bearing caps
- Bearing cap on the right side cylinder head, marked **I1**, in position with the arrow mark facing the rear. Install the bearing cap on the left side cylinder head, marked **I6**, in position with the arrow mark facing the front.
- Bearing cap bolts with new washers. Apply a light coat of oil on the threads of the cap bolts. Alternately tighten each bolt to 12 ft. lbs. (16 Nm).

➡**Use silver colored bolts 1.50 in. (38mm) in length.**

- Camshaft housing plugs on the cylinder heads. Be sure to face the cupped side forward.

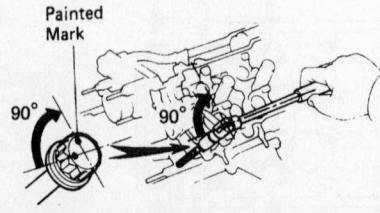

The paint mark must be 90 degrees from the starting point—4.0L (1UZ-FE) and 4.3L (3UZ-FE) Engines

13. Turn the crankshaft pulley clockwise or counterclockwise and put the timing mark of the crankshaft pulley in line with the centers of the crankshaft pulley bolt and the idler pulley bolt.

✷✷ WARNING

Since the thrust clearance of the camshaft is small, the camshaft must be kept level while it is being installed. If the camshaft is not kept level, the portion of the cylinder head receiving the shaft thrust may crack or be damaged, causing the camshaft to seize or break.

14. Install or connect the following:
- Right side cylinder head intake camshaft. Tighten the bracket bolt in the reverse order of the loosening sequence.
- Right side cylinder head exhaust camshaft. Tighten the bracket bolt in the reverse order of the loosening sequence.
- Left side cylinder head intake camshaft. Tighten the bracket bolt in the reverse order of the loosening sequence.
- Left side cylinder head exhaust camshaft. Tighten the bracket bolt in the reverse order of the loosening sequence.

15. Check and adjust the valve clearance.
16. Install or connect the following:
- Camshaft oil seals with the proper tool (SST 09223-46011). Be sure to apply MP grease to the new oil seal lip.
- Semi-circular plugs, if removed

17. Clean the cylinder head covers. Apply new sealant in the correct locations and install the gaskets.

18. Install or connect the following:
- Right cylinder head cover and bolts. Tighten the bolts to 52 inch lbs. (6 Nm).
- Left cylinder head cover and bolts. Tighten the bolts to 52 inch lbs. (6 Nm).
- Engine hanger with the 2 bolts. Install both engine hangers. Tighten the bolts to 27 ft. lbs. (37 Nm).
- Oil dipstick guide for the engine
- Oil dipstick for the transmission
- Rear water bypass joint and No. 1 EGR pipe
- Front water bypass joints. Install 2 gaskets and alternately tighten the nuts to 13 ft. lbs. (18 Nm).

- Water inlet and inlet housing, alternately tighten the bolts to 13 ft. lbs. (18 Nm)
- Delivery pipe and intake manifold
- Return pipe with 2 new gaskets. Tighten the union bolt to 26 ft. lbs. (35 Nm).
- Engine wire to the delivery pipes and rear water bypass joint
- Fuel return hose to the fuel return pipe
- Fuel inlet hose to the left-hand delivery pipe
- Fuel hose to the fuel pressure regulator
- Air intake chamber assembly
- Brake booster vacuum hose to the union on the air intake chamber
- PCV hose to the PCV valve on the left-hand cylinder head
- Water bypass hose (from EGR valve) to the rear water bypass joint
- Water bypass hose (from throttle body) to the rear water bypass joint
- Vacuum hose (from VSV for fuel pressure control) to the fuel pressure regulator
- EVAP hose (from charcoal canister) from the VSV for EVAP
- TPS connector
- With TRAC system, sub TPS connector
- With TRAC system, sub throttle actuator connector
- IAC valve connector
- EGR valve connector
- EGR gas temperature sensor connector
- VSV connector for fuel pressure control
- VSV connector for EVAP
- Accelerator bracket with the 2 bolts
- Accelerator, automatic transmission throttle control and the cruise control actuator cable
- Spark plug wires and clamps to the right and left cylinder head cover
- Belt rear plates by installing the bolts. Tighten the bolts to 66 inch lbs. (8 Nm).
- No. 2 ignition coil
- A new gasket to the exhaust manifold and install the catalytic converters. Tighten the 3 nuts to each converter to 46 ft. lbs. (62 Nm).
- Front exhaust pipe, tighten the bolts and nuts to 32 ft. lbs. (44 Nm). Tighten the 4 bolts holding the pipe support bracket to the transmission. Tighten the bolts to 32 ft. lbs. (44 Nm).
- Power steering pump with the nut

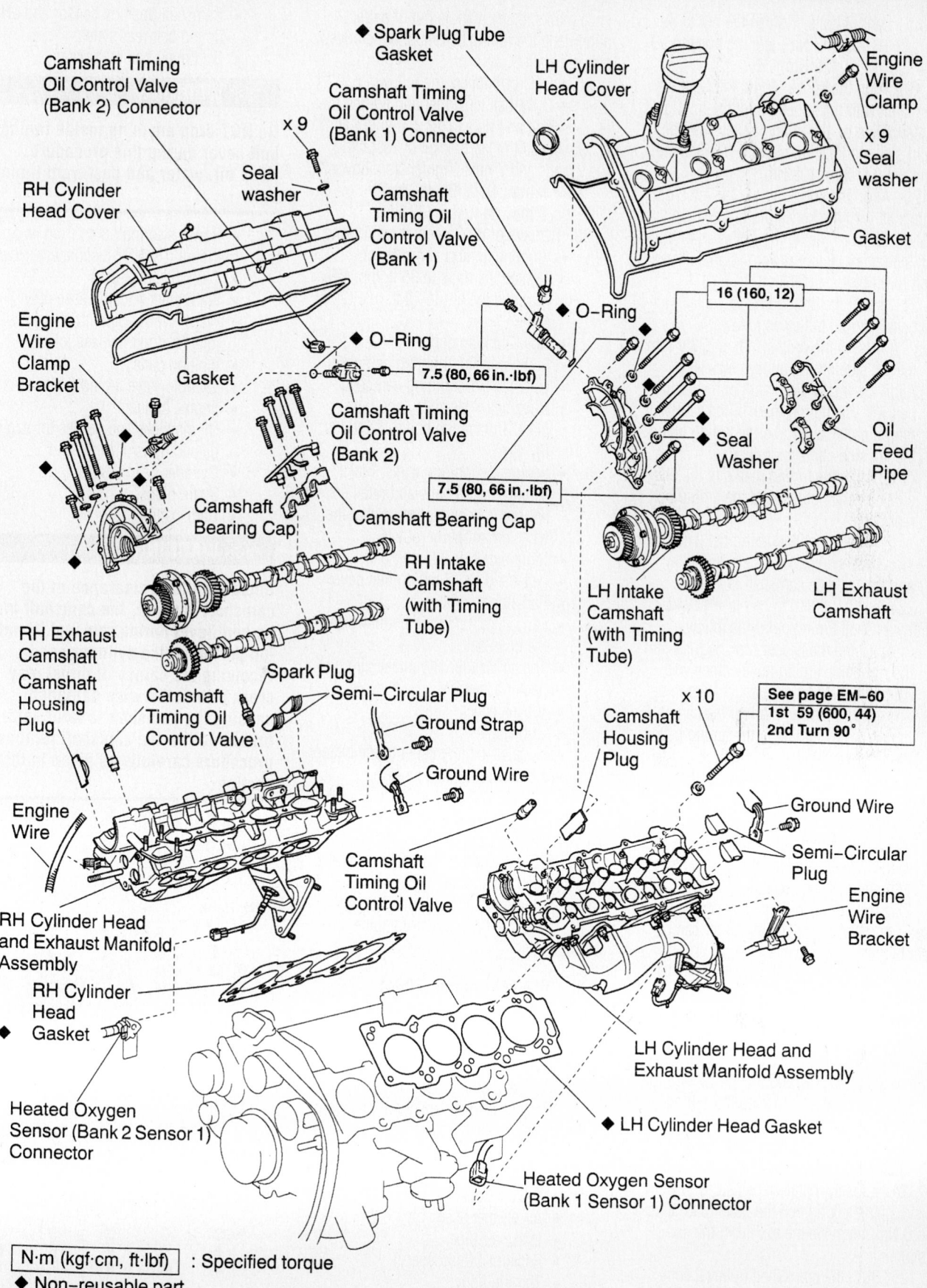

Camshaft Timing Oil Control Valve (Bank 2) Connector

◆ Spark Plug Tube Gasket

Camshaft Timing Oil Control Valve (Bank 1) Connector

LH Cylinder Head Cover

Engine Wire Clamp

x 9 Seal washer

x 9

Seal washer

RH Cylinder Head Cover

Camshaft Timing Oil Control Valve (Bank 1)

Engine Wire Clamp Bracket

Gasket

Gasket

◆ O-Ring

16 (160, 12)

7.5 (80, 66 in.·lbf)

◆ O-Ring

Camshaft Timing Oil Control Valve (Bank 2)

7.5 (80, 66 in.·lbf)

Seal Washer

Oil Feed Pipe

Camshaft Bearing Cap

Camshaft Bearing Cap

RH Intake Camshaft (with Timing Tube)

LH Intake Camshaft (with Timing Tube)

LH Exhaust Camshaft

RH Exhaust Camshaft

Camshaft Housing Plug

Camshaft Timing Oil Control Valve

Spark Plug

Semi-Circular Plug

Ground Strap

Ground Wire

x 10

Camshaft Housing Plug

See page EM-60
1st 59 (600, 44)
2nd Turn 90°

Ground Wire

Semi-Circular Plug

Engine Wire

Camshaft Timing Oil Control Valve

Engine Wire Bracket

RH Cylinder Head and Exhaust Manifold Assembly

RH Cylinder Head Gasket ◆

LH Cylinder Head and Exhaust Manifold Assembly

◆ LH Cylinder Head Gasket

Heated Oxygen Sensor (Bank 2 Sensor 1) Connector

Heated Oxygen Sensor (Bank 1 Sensor 1) Connector

N·m (kgf·cm, ft·lbf) : Specified torque
◆ Non-reusable part

67162-LEXU-G95

Exploded view of the cylinder heads and related components—4.3L (3UZ-FE) Engine (2004 Model)

and 3 bolts. Tighten the nut to 32 ft. lbs. (43 Nm) and the bolts to 29 ft. lbs. (39 Nm).

19. Align the knock pin on the right side camshaft with the knock pin of the timing pulley. Slide on the timing pulley with the right side mark facing forward. Tighten the bolt to 80 ft. lbs. (108 Nm).

20. Align the knock pin on the left side camshaft with the knock pin of the timing pulley. Slide on the timing pulley with the left side mark facing forward. Tighten the bolt to 80 ft. lbs. (108 Nm).

21. Install the timing belt to the left side camshaft timing pulley as follows:

    a. Using the proper tool, slightly turn the left side timing pulley clockwise. Align the installation mark of the timing belt with the timing mark of the camshaft timing pulley and hang the timing belt on the left side camshaft pulley.

    b. Align the timing marks of the left side camshaft pulley and the timing belt rear plate.

    c. Check that the timing belt has tension between crankshaft timing pulley and the left side camshaft pulley.

22. Install the timing belt to the right side camshaft timing pulley as follows:

    a. Using the proper tool, slightly turn the right side timing pulley clockwise. Align the installation mark of the timing belt with the timing mark of the camshaft timing pulley and hang the timing belt on the right side camshaft pulley.

    b. Align the timing marks of the right side camshaft pulley and the timing belt rear plate.

    c. Check that the timing belt has tension between crankshaft timing pulley and the right side camshaft pulley.

23. The timing belt tensioner must be set prior to installation. The tensioner can be set by:

    a. Place a plate washer between the tensioner and a block. Using a press, press in the pushrod using 220–225 lbs. of pressure.

    b. Align the holes of the pushrod and housing, pass a 0.05 inch (1.27mm) rod through the holes to keep the setting position of the pushrod.

    c. Release the press and install the dust boot to the tensioner.

24. Loosely install the tensioner. Evenly and alternately tighten the bolts to 20 ft. lbs. (26 Nm). Remove the tool from the tensioner.

25. Turn the crankshaft pulley 2 complete revolutions from TDC to TDC. Always turn the crankshaft clockwise. Check that all belt and pulley marks align with their refer-

ence marks. If any mark is out of perfect alignment, the timing belt must be removed and reinstalled.

26. Install or connect the following:

- Drive belt tensioner and tighten the bolt and nuts to 12 ft. lbs. (16 Nm).

27. Install the fan bracket by installing the 2 bolts and 2 nuts. Tighten as follows:

    a. 12mm: 12 ft. lbs. (16 Nm)

    b. 14mm: 24 ft. lbs. (32 Nm)

28. Remove or disconnect the following:

- Air conditioning compressor. Tighten the bolts to 36 ft. lbs. (49 Nm) and the nut to 22 ft. lbs. (29 Nm).
- No. 1 ignition coil
- Right side No. 2 timing belt cover
- Left side No. 2 timing belt cover
- Drive belt idler pulley and cover plate. Tighten the bolt to 27 ft. lbs. (37 Nm).
- Secure the ignition wires. Make certain that all clips and retainers are securely engaged and that the wires are properly routed.
- Right side No. 3 timing belt
- Left-hand No. 3 timing belt cover
- Radiator assembly
- Fan pulley, fan, fluid coupling and the drive belt
- The air cleaner and intake air connector assembly
- V bank cover
- Coolant
- Negative battery cable to the battery
- Air cleaner inlet
- Battery clamp cover
- Engine undercover
- Oil pan protector

29. Start the engine and check for leaks

30. Bleed the cooling system and recheck the engine coolant level.

31. Make all the necessary engine adjustments.

4.3L (3UZ-FE) Engine (2004)

1. Before servicing the vehicle, refer to the precautions in the beginning of this section.

2. Relieve the fuel system pressure.

3. Remove or disconnect the following, as applicable to each engine:

- Engine under cover
- Drain engine coolant
- V-bank cover
- Air cleaner assembly
- Intake air pipe
- Radiator (if necessary)
- Throttle body
- Upper and lower intake manifold assembly

- Camshaft position sensor and LH timing belt rear plates
- RH timing belt rear plates

✳✳ WARNING

Do NOT drop anything inside timing belt cover during this procedure. Keep oil, water and dust from timing belt.

- Power steering pump from engine mount (Do NOT disconnect hoses)
- Catalytic converters
- Water inlet housing assembly
- Water bypass pipe, front bypass joint, and rear bypass joint
- Ignition coils
- Variable valve timing (VVT) sensors
- Engine hangers
- Oil dipsticks and guides for engine oil and transmission fluid
- Cylinder head covers
- Spark plugs
- Camshafts

✳✳ WARNING

Since the thrust clearance of the camshaft is small, the camshaft must be kept level during removal. If not, the portion of the cylinder head receiving the camshaft thrust may crack or be otherwise damaged, causing the camshaft to later seize or break. Follow the camshaft removal procedure carefully as given in this section.

- Both oxygen sensor connectors
- Ground wire from LH cylinder head

RH Bank

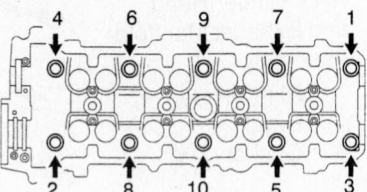

LH Bank

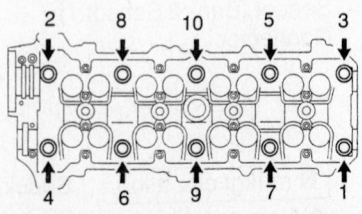

67162-LEXU-G96

Cylinder head bolt loosening sequence— 4.3L (3UZ-FE) Engine (2004 Model)

- Engine wire bracket for oxygen sensor on LH cylinder head

4. Uniformly loosen the 10 cylinder head bolts on each cylinder head, in several passes, following loosening sequence as shown.

✳✳ WARNING

Use care so that no bolts or washers are dropped into the recesses or enclosed portions of the cylinder head or block.

5. Carefully lift the cylinder head from the locating dowels on the engine block. Place cylinder heads on wooden blocks on the workbench.

6. If necessary, exhaust manifolds may be removed from the cylinder heads at this time.

To install:

7. If removed, install exhaust manifolds to cylinder heads, with new gaskets. Ensure the white mark on the gasket is facing the manifold side.

8. Install and tighten the exhaust manifold bolts, in an alternating pattern, to a final torque of 32 ft. lbs. (44 Nm).

9. Install the heat shields.

10. With a new cylinder head gasket in place, carefully position the cylinder head onto the engine block locating dowels.

➡ **The cylinder head gaskets have a "3R" marks for the RIGHT cylinder head, and a "3L" mark for the LEFT cylinder head.**

➡ **If any cylinder head bolt appears stretched or damaged, replace it. If a bolt will not reach final torque setting, replace it.**

11. Apply a light coat of oil to the cylinder head bolt threads. Install the washers and insert the cylinder head bolts into position.

12. In several passes, following the tightening sequence shown, tighten the cylinder head bolts to a final torque of 44 ft. lbs. (59 Nm).

13. Once the bolts reach this setting, then place a white paint mark on the front of each bolt head. Using the torque wrench, turn each bolt, in the sequence shown, an additional 90°, using the paint mark as a reference.

14. Install or reconnect the following:
- Engine wiring and ground straps
- Oxygen sensor wire bracket on LH cylinder head
- Spark plugs
- Camshafts

15. Inspect and adjust the valve lash.

16. Install or reconnect the following:
- Cylinder head covers, with new gaskets
- Engine hangers
- VVT sensors
- Engine and transmission dipsticks and tubes
- Ignition coils
- Water bypass joints and pipe; torque nuts and bolt to 13 ft. lbs. (18 Nm)
- Water inlet housing assembly
- Catalytic converters, with new gaskets and new nuts; torque nuts to 46 ft. lbs. (62 Nm)
- Power steering pump to mounting; torque bolts to 29 ft. lbs. (39 Nm) and nut to 32 ft. lbs. (43 Nm)
- LH timing belt rear plates and camshaft position sensor; torque bolts to 66 inch lbs. (7.5 Nm)
- RH right belt rear plates; torque bolts to 66 inch lbs. (7.5 Nm)
- Camshaft pulleys
- Timing belt to camshaft pulleys
- Upper and lower intake manifold assembly, with new gaskets; torque bolts, in alternating pattern, to 13 ft. lbs. (18 Nm)
- Throttle body
- Radiator
- Intake air connector and air cleaner assembly

17. Refill the engine cooling system. Start the engine and check for leaks and proper operation.

18. Recheck the engine oil level.

19. Install the V-bank cover and the engine under cover.

3.3L (3MZ-FE) Engine

1. Before servicing the vehicle, refer to the precautions in the beginning of this section.

2. Relieve the fuel system pressure.

3. Remove or disconnect the following:
- Negative battery cable. Wait at least 90 seconds before performing any other work
- Oil pan protector
- Engine undercover
- Coolant
- Battery clamp cover
- Air cleaner inlet
- Lower radiator air deflector
- RF wheel
- V bank cover by removing the bolt and 2 cap nuts
- Air cleaner and intake air connector assembly
- Emission control valve
- Air intake surge tank
- Drive belt, fluid coupling and the fan pulley. The drive belt tension may be slackened by turning the tensioner counterclockwise. The pulley bolt for the drive belt tensioner has a left-handed thread.
- PS pump drive belt
- Radiator

4. Remove intake manifold

5. Remove timing belt

6. Remove PS pump assembly

7. Front and rear exhaust pipe and brackets

8. Remove camshafts

9. Remove LH or RH cylinder head assemblies

10. Remove or disconnect the following:
- The VVT Sensor connector
- Camshaft timing oil valve connector
- Engine wire harness clamp
- 8 cylinder head bolts uniformly in the sequence

✳✳ WARNING

Head warpage or cracking could result from removing the bolts in an incorrect order.

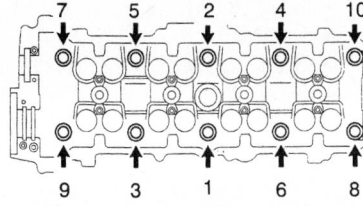

RH Bank

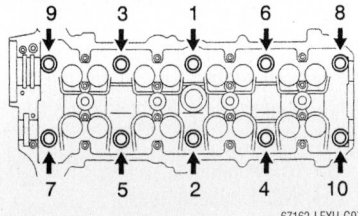

LH Bank

67162-LEXU-G97

Cylinder head bolt tightening sequence— 4.3L (3UZ-FE) Engine (2004 Model)

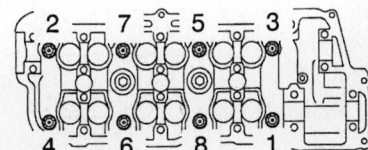

◀ Front

67162-LEXU-G31

Cylinder head bolt removal sequence— 3.3L (3MZ-FE) Engine

11. Inspect the cylinder head set bolts. Ensure they match the following:
  - Outside diameter is .3524 to .3563 in. (8.95 to 9.05mm)
  - Minimum diameter is .3445 in. (8.75mm)

12. If diameter is lees than minimum, replace the bolts.

To install:

13. Install new head gasket with R mark upward

14. Install cylinder head assembly
  - Apply light oil to the cylinder head bolts
  - Install plate washers on the cylinder head bolts

➡**Cylinder head bolts are tightened in two successive steps. Install and tighten 8 cylinder head bolts in required sequence.**

  a. Tighten to 40 ft. lbs. Torque 40 ft. lbs (54 Nm)
  b. Mark the front side of each head bolt with paint.
  c. Retighten cylinder head bolts 90 degrees in the same sequence
  d. Check that each painted mark is now at a 90 degree angle to the front

15. Install or connect the following:
  - Wiring harness clamp
  - Camshaft timing oil valve connector
  - Camshaft assemblies
  - Valve cover assemblies

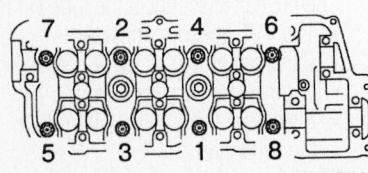

◀ Front

Cylinder head bolt tightening sequence— 3.3L (3MZ-FE) Engine

67162-LEXU-G32

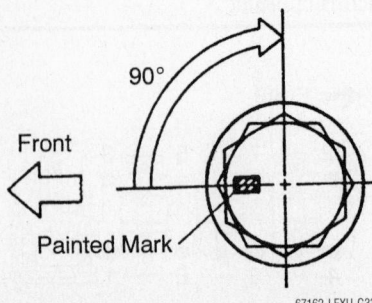

90°

Front

Painted Mark

67162-LEXU-G33

Head bolt marking procedure—3.3L (3MZ-FE) Engine

  - Exhaust manifold assemblies and support brackets. Torque to 36 ft. lbs. (49 Nm)
  - Exhaust manifold heat insulators
  - PS pump assembly
  - Timing belt inner cover
  - Camshaft timing pulleys
  - Timing belt and idler assemblies
  - RH engine mounting bracket
  - Timing belt covers
  - Alternator bracket
  - Engine mounting stay No 2
  - Engine stabilizer rod
  - PS drive belt
  - A/C compressor drive belt
  - Water outlet
  - Intake manifold assembly
  - Intake air surge tank
  - Emission control valve set
  - Air cleaner assembly
  - Vacuum hoses
  - V-bank cover sub-assembly
  - Front suspension upper brace
  - RF wheel
  - Install engine oil
  - Installed engine coolant

16. Inspect for fuel leaks
17. Inspect for oil leaks
18. Inspect for exhaust leaks
19. Adjust engine timing and idle speed
20. Run system initialization

Intake Manifold

REMOVAL & INSTALLATION

3.0L (1MZ-FE) Engine

1. Before servicing the vehicle, refer to the precautions in the beginning of this section.
2. Remove or disconnect the following:
  - Negative battery cable
  - Coolant
  - Throttle/accelerator cable from the throttle body
  - Air cleaner hose at the air intake chamber and remove it
  - All lines and hoses. Tag them for installation.
  - Idle Speed Control (ISC) valve and the throttle body
  - Exhaust Gas Recirculation (EGR) valve and vacuum modulator
  - Cylinder head rear plate
  - Intake chamber stays, any wires, then, the air intake chamber
  - Fuel injection delivery pipe and the injectors
  - Water outlet and the bypass outlet
  - 2 bolts and the No. 2 idler pulley bracket stay

  - 8 bolts and 4 nuts, then lift out the intake manifold

To install:

3. Thoroughly clean the intake manifold and cylinder head surfaces. Using a machinist's straight edge and a feeler gauge, check the surface of the intake manifold for warpage. If the warpage is greater than 0.0039 in. (0.10mm), replace the intake manifold.

4. Place new gaskets onto the intake manifold and position the intake manifold between the cylinder heads. Tighten the nuts and bolts to 13 ft. lbs. (18 Nm). Tighten the No. 2 pulley bracket bolts to 13 ft. lbs. (18 Nm).

5. Install or connect the following:
  - Water bypass outlet and tighten the bolts to 74 inch lbs. (8.3 Nm). Tighten the water outlet to 74 inch lbs. (8 Nm).
  - Injectors and delivery pipe
  - Air intake chamber and tighten the 2 bolts and 2 nuts to 32 ft. lbs. (43 Nm); use an 8mm hex wrench
  - Chamber stays and tighten the mounting bolts to 29 ft. lbs. (39 Nm)
  - Remaining components. Tighten the emission control valve set to 73 inch lbs. (8 Nm).
  - All hoses
  - Accelerator cable, if equipped with automatic transaxle
  - Coolant
  - Negative battery cable

3.0L (2JZ-GE) Engine (Except 2004)

1. Before servicing the vehicle, refer to the precautions in the beginning of this section.
2. Remove or disconnect the following:
  - Negative battery cable
  - Coolant
  - Spark plug wires at the spark plugs
  - Spark plugs
  - Radiator
  - Water pump pulley
  - Timing belt
  - No. 2 front exhaust pipe
  - 2 Oxygen (O_2) sensor leads
  - 4 nuts, and the manifold heat shield
  - Exhaust manifolds
  - Water bypass outlet and the No. 1 bypass pipe. Remove the 3 O-rings.
  - Water outlet
  - No. 1 bypass hose
  - Vacuum Control Valve (VCV) set and the No. 2 vacuum pipe

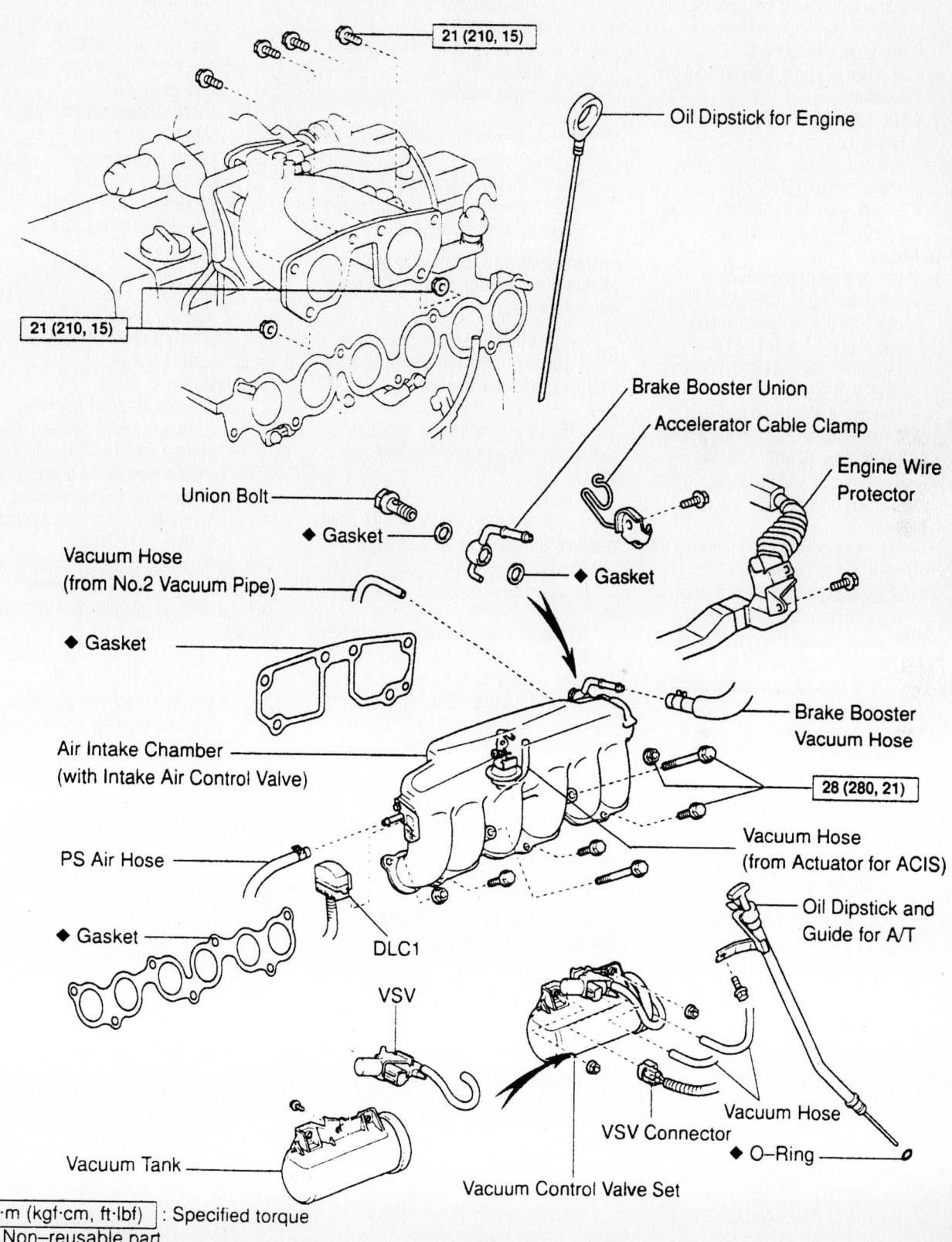

21 (210, 15)

Oil Dipstick for Engine

21 (210, 15)

Brake Booster Union

Accelerator Cable Clamp

Engine Wire Protector

Union Bolt

◆ Gasket

◆ Gasket

Vacuum Hose
(from No.2 Vacuum Pipe)

◆ Gasket

Brake Booster Vacuum Hose

Air Intake Chamber
(with Intake Air Control Valve)

28 (280, 21)

Vacuum Hose
(from Actuator for ACIS)

PS Air Hose

Oil Dipstick and
Guide for A/T

◆ Gasket

DLC1

VSV

Vacuum Hose

VSV Connector

◆ O-Ring

Vacuum Tank

Vacuum Control Valve Set

N·m (kgf·cm, ft·lbf) : Specified torque
◆ Non-reusable part

9301LG09

Exploded view of the intake manifold mounting and related components—3.0L (2JZ-GE) engine

- Fuel return hose from the oil dipstick guide; remove the mounting bolt and pull the guide and dipstick from the pan. Plug the hole.
- Air intake chamber
- Fuel delivery pipe, then pull out the injectors
- No. 1 and 2 fuel pipes
- Engine harness from the intake manifold
- Intake manifold stay
- Loosen the 6 bolts and 2 nuts, then lift out the intake manifold

To install:

3. Install or connect the following:
   - Install the intake manifold, with a new gasket, and tighten the bolts and nuts to 15 ft. lbs. (21 Nm)
   - Mounting stay and tighten the bolts to 29 ft. lbs. (39 Nm)
   - Engine harness to the manifold
   - 2 fuel pipes and tighten the bolts to 78 inch lbs. (9 Nm)
   - Delivery pipe and injectors. Tighten the pipe bolts to 15 ft. lbs. (21 Nm).
   - Air intake chamber and tighten it to 15 ft. lbs. (21 Nm)
   - 2 stays and tighten them to 13 ft. lbs. (18 Nm); The No. 1 stay is marked with an **F** and the No. 2 stay is marked with an **R**.
   - The oil dipstick and guide, using a new O-ring
   - VCV set and the vacuum pipe. Tighten the set mounting bolts to 15 ft. lbs. (21 Nm).
   - Water bypass outlet and the pipe, tighten the bolts to 78 inch lbs. (9 Nm)
   - Exhaust manifolds. Tighten the bolts to 29 ft. lbs. (39 Nm).
   - Heat shield and tighten it to 13 ft. lbs. (18 Nm)
   - No. 2 front pipe
   - Timing belt
   - Radiator and water pump pulley
   - Plug wires to the plugs
   - Coolant
   - Negative battery cable

3.0L (2JZ-GE) Engine (2004)

1. Before servicing the vehicle, refer to the precautions in the beginning of this section.
2. Remove or disconnect the following:
   - Negative battery cable
   - Engine under cover
   - Engine coolant
   - Engine cover

- Air cleaner assembly
- Throttle body assembly
- Water bypass pipe running through intake manifold
- Engine and transmission dipsticks and tubes
- Air intake chamber
- Drive belt
- Vacuum control valve set
- Power steering pump from mounting (Do NOT disconnect hoses)
- Engine wiring, and hoses, as need for access to intake manifold

➡ **Mark each wire, connector or hose, as it is removed, for proper reinstallation reference.**

   - Fuel pressure pulsation damper
3. Disconnect the starter wire from the intake manifold stay. Remove the manifold stay.
4. Remove 7 bolts and 2 nuts and remove the intake manifold and delivery pipe and gasket.

To install:

5. Position a new gasket and the intake manifold and delivery pipe assembly into position.
6. Install the 7 bolts and 2 nuts; torque to 21 ft. lbs. (28 Nm).
7. Insert the water bypass hose between the No. 2 and No. 3 intake ports of the manifold and delivery pipe.
8. Install the manifold stay and torque the bolts to 30 ft. lbs. (40 Nm).
9. Reconnect the starter wire to the manifold stay.
10. Reinstall the remaining components in reverse of the removal procedure.

4.0L (1UZ-FE) and 4.3L (3UZ-FE) Engines (Except 2004)

1. Before servicing the vehicle, refer to the precautions in the beginning of this section.
2. Properly relieve the fuel system pressure.
3. Remove or disconnect the following:
   - Negative battery cable
   - Coolant
   - Accelerator cable
   - Throttle and accelerator pedal position electrical connectors
   - Throttle motor electrical connector
   - Vacuum Switching Valve (VSV) electrical connectors
   - Injector electrical connectors
   - Noise filter electrical connector
   - Brake booster vacuum hose from the intake manifold

- Positive Crankcase Ventilation (PCV) hose from the left-hand valve cover
- Evaporative Emission (EVAP) hoses and label for installation
- Power steering air hose from the intake manifold
- Coolant hoses from the throttle body
- 2 EVAP-to-intake manifold mounting bolts
- Accelerator cable bracket
- 3 V-bank cover brackets
- EVAP VSV
- Fuel supply and return lines
- 6 bolts and 4 nuts, then the intake manifold assembly

4. Clean the gasket mating surfaces of old gasket and sealant.

To install:

5. Install or connect the following:
   - 2 intake manifold gaskets on the cylinder heads with the white painted mark facing upward.
   - Intake manifold assembly and uniformly, tighten the mounting bolts to 13 ft. lbs. (18 Nm).
6. The remaining steps of the installation procedure are the reverse of the removal, while keeping in mind the following items:
   - Tighten the V-bank cover brackets to 66 inch lbs. (8 Nm)
   - Tighten the accelerator cable bracket mounting bolts to 13 ft. lbs. (18 Nm)
   - Tighten the EVAP VSV to 13 ft. lbs. (18 Nm)
   - All remaining components
   - Coolant
   - Negative battery cable

4.3L (3UZ-FE) Engines (2004)

1. Before servicing the vehicle, refer to the precautions in the beginning of this section.
2. Properly relieve the fuel system pressure.
3. Remove or disconnect the following:
   - Negative battery cable
   - Engine under cover
   - Coolant
   - V-bank cover
   - Air cleaner assembly and connectors
   - Throttle body assembly
4. Disconnect the fuel inlet hose (rear fuel pipe) from the fuel main tube.
5. Remove and disconnect the following:

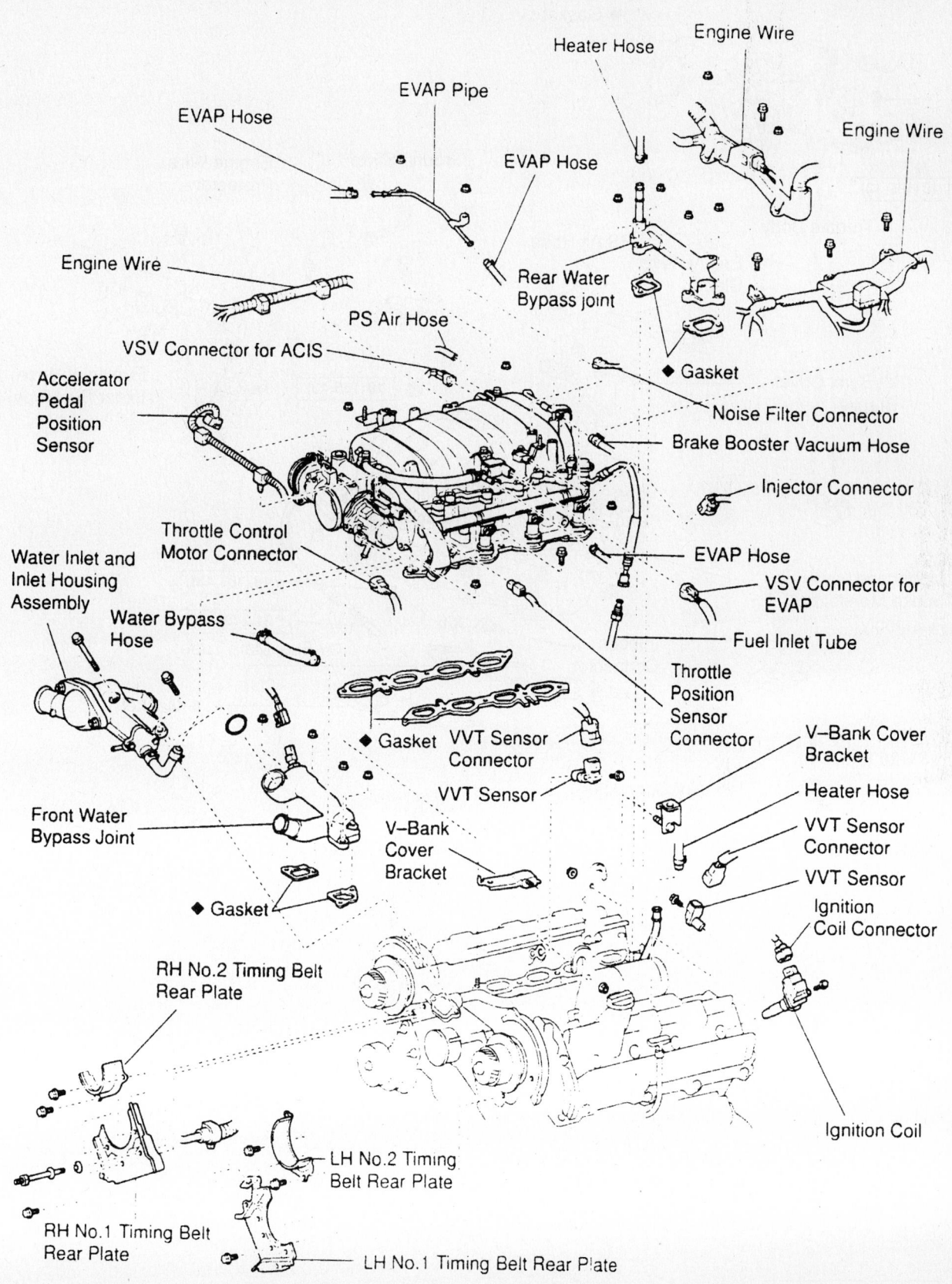

Heater Hose

Engine Wire

Engine Wire

EVAP Pipe

EVAP Hose

EVAP Hose

Rear Water Bypass joint

◆ Gasket

Engine Wire

PS Air Hose

Noise Filter Connector

VSV Connector for ACIS

Brake Booster Vacuum Hose

Accelerator Pedal Position Sensor

Injector Connector

EVAP Hose

VSV Connector for EVAP

Water Inlet and Inlet Housing Assembly

Throttle Control Motor Connector

Fuel Inlet Tube

Water Bypass Hose

◆ Gasket VVT Sensor Connector

Throttle Position Sensor Connector

V–Bank Cover Bracket

VVT Sensor

Heater Hose

Front Water Bypass Joint

V–Bank Cover Bracket

VVT Sensor Connector

◆ Gasket

VVT Sensor

Ignition Coil Connector

RH No.2 Timing Belt Rear Plate

LH No.2 Timing Belt Rear Plate

Ignition Coil

RH No.1 Timing Belt Rear Plate

LH No.1 Timing Belt Rear Plate

◆ Non–reusable part

Exploded view of the intake manifold mounting and related components—4.0L (1UZ-FE) and 4.3L (3UZ-FE) Engines

9301LG10

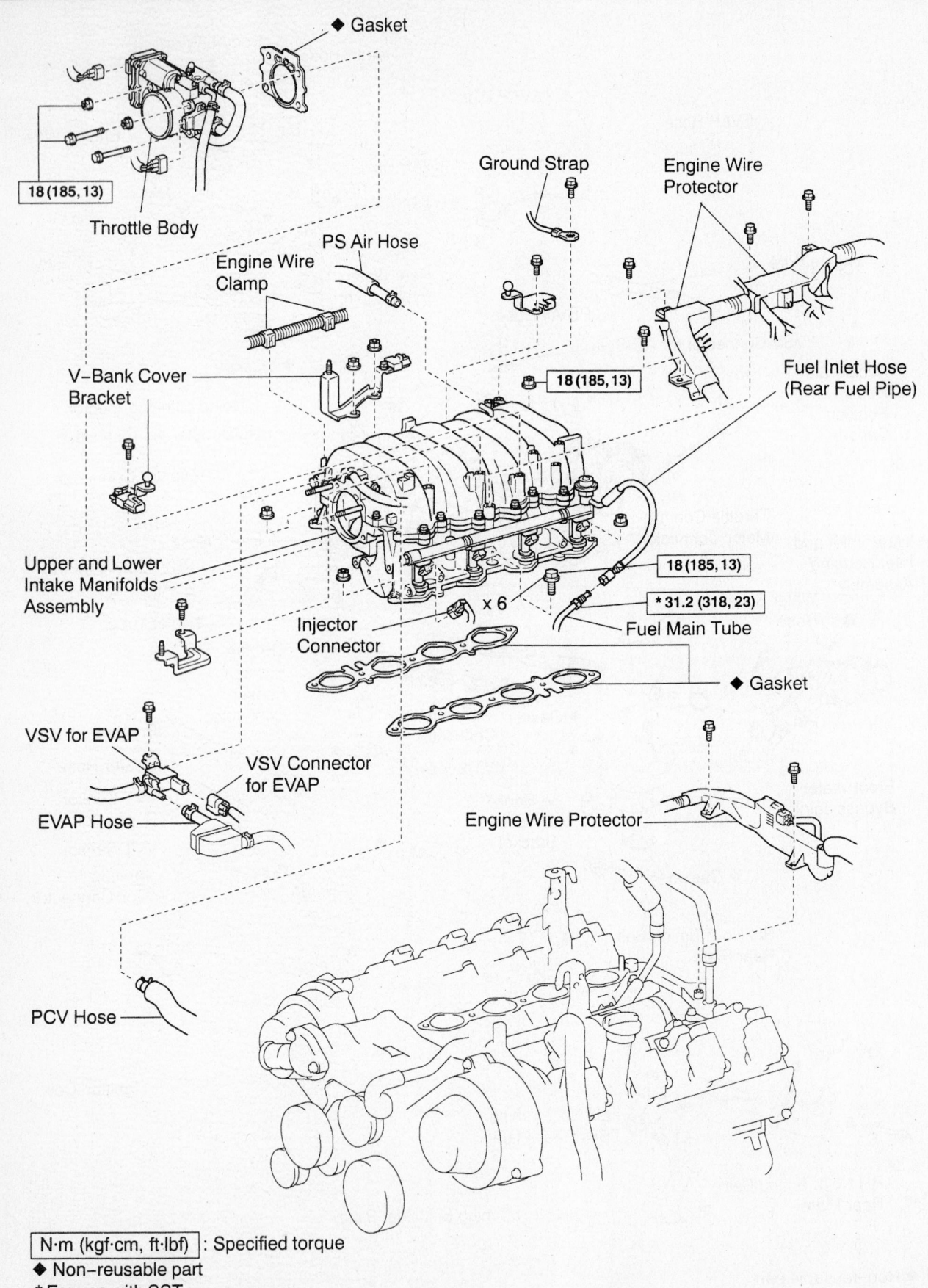

◆ Gasket

Ground Strap

Engine Wire Protector

Throttle Body

PS Air Hose

18 (185, 13)

Engine Wire Clamp

V–Bank Cover Bracket

Fuel Inlet Hose (Rear Fuel Pipe)

18 (185, 13)

Upper and Lower Intake Manifolds Assembly

Injector Connector

x 6

18 (185, 13)

* 31.2 (318, 23)

Fuel Main Tube

◆ Gasket

VSV for EVAP

VSV Connector for EVAP

EVAP Hose

Engine Wire Protector

PCV Hose

N·m (kgf·cm, ft·lbf) : Specified torque
◆ Non–reusable part
* For use with SST

67162-LEXU-S98

Exploded view of the intake manifold mounting and related components—4.3L (3UZ-FE) Engines (2004)

- VSV connector for EVAP
- EVAP hose from VSV
- VSV from upper intake manifold
- 4 V-bank cover brackets
- Engine wiring protector (LH side) from the upper intake manifold and camshaft bearing cap
- 2 wire clamps (RH side) from the brackets on the delivery pipe
- Engine wire protector (rear side) from the rear water bypass joint and the RH cylinder head
- VSV connector for the ACIS
- 8 injector connectors

6. Remove the 6 bolts and 4 nuts and remove the upper and lower intake manifold assembly.

7. If necessary, the upper intake manifold can be disassembled from the lower intake manifold by removing or disconnecting the following:

- Vacuum hose for VSV from air control valve actuator
- Vacuum tank hose from lower intake manifold
- Vacuum hose (VSV for ACIS) from clamp
- Wire clamp from lower intake manifold
- Vacuum tank and VSV assembly from ACIS
- Air control valve actuator
- 15 bolts and 5 nuts to remove upper intake manifold from lower intake manifold

To install:

➡**Always be sure to use new gaskets at each component mounting.**

8. Reassemble the upper and lower intake manifold in reverse of disassembly procedure given. Torque the upper intake manifold-to-lower manifold bolts and nuts to 13 ft. lbs. (18 Nm).

9. Install the vacuum tank and VSV assembly. Torque the nuts to 13 ft. lbs. (18 Nm).

10. Reconnect all of the wiring connectors, clamps and vacuum hoses in reverse of the removal procedure.

11. With new gaskets on the cylinder heads (white marks facing outward), position the upper and lower intake manifold assembly into position. Install the 6 bolts and 4 nuts and torque them to 13 ft. lbs. (18 Nm).

12. Reinstall and reconnect all remaining components in reverse of the removal procedure.

13. Refill the cooling system. Start the engine and check for leaks and proper operation.

3.3L (3MZ-FE) Engine

1. Before servicing the vehicle, refer to the precautions in the beginning of this section.

2. Relieve the fuel system pressure.

3. Remove or disconnect the following:

- Negative battery cable. Wait at least 90 seconds before performing any other work
- Oil pan protector
- Engine undercover
- Coolant
- Battery clamp cover
- Air cleaner inlet
- Lower radiator air deflector
- RF wheel
- V bank cover by removing the bolt and 2 cap nuts
- Air cleaner and intake air connector assembly
- Emission control valve
- Air intake surge tank
- Engine stabilizer control rod
- Fuel pipe
- Heater inlet water hose
- Battery ground cable form the engine
- 6 fuel injector connectors
- Remove intake bolts following the proper sequence as shown.
- Water outlet
- Radiator reserve tank pipe
- Upper radiator hose
- Engine coolant temp sensor
- Water outlet housing
- Knock sensor
- Gently remove the intake manifold

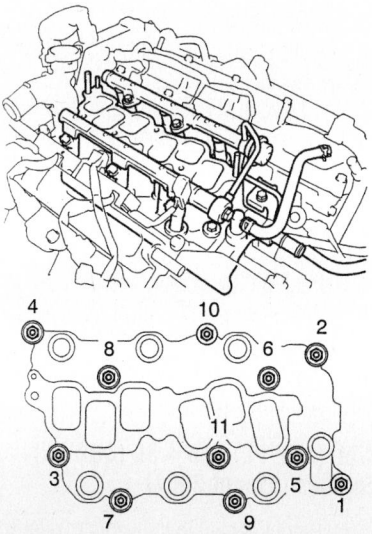

Intake bolt removal sequence—3.3L (3MZ-FE) Engine

67162-LEXU-G25

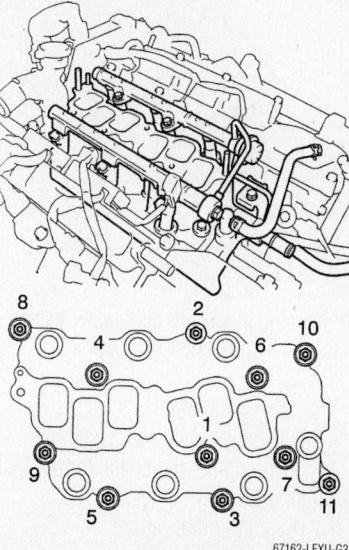

Intake manifold tightening sequence— 3.3L (3MZ-FE) Engine

67162-LEXU-G34

To install:

4. Install or connect the following:
- Two knock sensors and wiring connectors. Torque to: 14 ft. lbs (20 Nm)

5. Install water outlet

a. Install 2 new gaskets to the cylinder heads

b. Install water outlet with water bypass hose

c. Tighten 2 bolts, 2 nuts and 2 washers to 11 ft. lbs. (15 Nm)

d. Connect engine temperature connector

e. Connect radiator reserve tank connector

f. Radiator hose

6. Install Intake Manifold

7. Install or connect the following:
- 9 bolts, 2 nuts and 2 washers; torque to 11 ft. lbs. (15 Nm)

➡**Using several steps, tighten the bolts and nuts uniformly in sequence**

- Retighten 9 bolts and 2 nuts to 11 ft. lbs. (15 Nm)
- Fuel pipe making sure the connector is seated until it clicks.

8. Install or connect the following:
- Engine stabilizing control rod; torque bolt "A" to 17 ft. lbs. (23 Nm) and bolt "B" to 47 ft. lbs. (64 Nm)
- Intake air surge tank
- Emission control valve set
- Air Cleaner assembly
- Vacuum hoses
- Battery negative wire
- Front center suspension upper brace

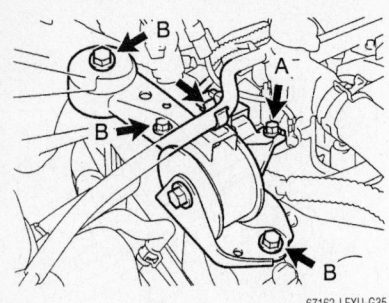

Tightening procedure for Engine Stabilizing Rod —3.3L (3MZ-FE) Engine

67162-LEXU-G35

- Engine coolant
- V-bank cover sub-assembly
9. Run engine and check for coolant and fuel leaks
10. System initialization

Exhaust Manifold

REMOVAL & INSTALLATION

3.0L (1MZ-FE) Engine

1. Before servicing the vehicle, refer to the precautions in the beginning of this section.
2. Remove or disconnect the following:
   - Negative battery cable
   - Engine undercovers
   - 2 front exhaust pipe stay bolts
   - Front pipe from the center pipe
   - 3 nuts and the front pipe
   - Oxygen (O_2) sensor at the right side manifold
   - 3 mounting nuts and lift off the outside heat insulator
   - 6 nuts and lift off the right side manifold and gasket
   - Left side heat insulator
   - 6 nuts and lift off the left side manifold and gaskets

To install:
3. Install or connect the following:
   - Right manifold with a new gasket. Tighten the nuts to 29 ft. lbs. (39 Nm).
   - Outer insulator
   - Left manifold. Use a new gasket. Tighten the nuts to 29 ft. lbs. (39 Nm).
   - Outer insulator
   - Front exhaust pipe and tighten the manifold-to-pipe nuts to 46 ft. lbs. (62 Nm). Tighten the pipe-to-converter nuts to 32 ft. lbs. (43 Nm).
   - O_2 sensor
   - Undercovers
   - Battery cable

3.0L (2JZ-GE) Engine (Except 2004)

1. Before servicing the vehicle, refer to the precautions in the beginning of this section.
2. Remove or disconnect the following:
   - Negative battery cable
   - Engine undercovers
   - No. 2 front exhaust pipe bolts and disconnect it from the front exhaust pipe. Loosen the 4 nuts, then remove the front pipe.
   - Both Oxygen (O_2) sensors at the manifold
   - 4 mounting nuts and lift off the outside heat insulator
   - 4 nuts and disconnect the manifolds from the pipe. Loosen the mounting bolts and remove the 2 manifolds and the gasket.

To install:
3. Install or connect the following:
   - Manifolds with a new gasket. Tighten the nuts to 30 ft. lbs. (40 Nm).
   - Outer insulator. Tighten the nuts to 13 ft. lbs. (18 Nm).
   - No. 2 front pipe. Tighten the nuts to 46 ft. lbs. (62 Nm).
   - Front exhaust pipe. Tighten the bolts/nuts to 32 ft. lbs. (43 Nm).
   - O_2 sensors
   - Undercovers
   - Battery cable

3.0L (2JZ-GE) Engine (2004)

1. Before servicing the vehicle, refer to the precautions in the beginning of this section.
2. Remove or disconnect the following:
   - Negative battery cable
   - Engine cover
   - Air cleaner assembly, as needed for access
   - Oxygen sensor wiring connectors
   - Front exhaust pipe support bracket
   - Front exhaust pipe from manifold
   - Exhaust manifold (8 nuts accessible with a deep-socket wrench)

To install:
3. Position the exhaust manifold in place, with new gaskets. Torque the nuts, in an alternating pattern, to 30 ft. lbs. (40 Nm).
4. Reinstall the remaining components in reverse of the removal procedure.

4.0L (1UZ-FE) and 4.3L (3UZ-FE) Engines (Except 2004)

1. Before servicing the vehicle, refer to the precautions in the beginning of this section.
2. Remove or disconnect the following:
   - Negative battery cable
   - Coolant
   - Camshaft timing pulleys
   - Cooling fan hydraulic pump, on the SC 400
   - Accelerator, throttle control and cruise control actuator cables
   - High tension cord cover and the right side ignition coil
   - Water inlet housing mounting bolts
   - Water bypass hose from the Idle Speed Control (ISC) valve
   - Water inlet and inlet housing assemblies
   - O-ring from the water inlet housing
   - Exhaust Gas Recirculation (EGR) pipe
   - Vacuum Switching Valve (VSV) connector
   - Vacuum pipe hose
   - EGR water bypass pipe
   - Fuel pressure VSV
   - EGR vacuum hoses and the EGR VSV
   - Water bypass pipe hose from the ISC valve
   - Water bypass joint hose
   - Vacuum pipe hoses
   - EGR gas temperature sensor
   - EGR valve adapter
   - Fuel pressure regulator vacuum hose
   - Air intake chamber vacuum hose
   - Vacuum hose from the Evaporative Emission (EVAP) BVSV
   - Mounting bolts, hoses and the vacuum pipe
   - ISC valve
   - Throttle body sensor connectors and the water bypass pipe from the rear water bypass joint
   - Positive Crankcase Ventilation (PCV) valve hose
   - Throttle body and gasket
   - Accelerator cable bracket and the brake booster vacuum union and hose
   - Cold start injector connector and the cold start injector tube from the right side delivery pipe, if equipped
   - Check connector from the intake chamber and remove the mounting nuts and bolts
   - Air intake chamber and the cold start injector, tube and wire assembly, if equipped
   - Engine wire from the intake manifold and from the right side cylinder head
   - Heater hoses

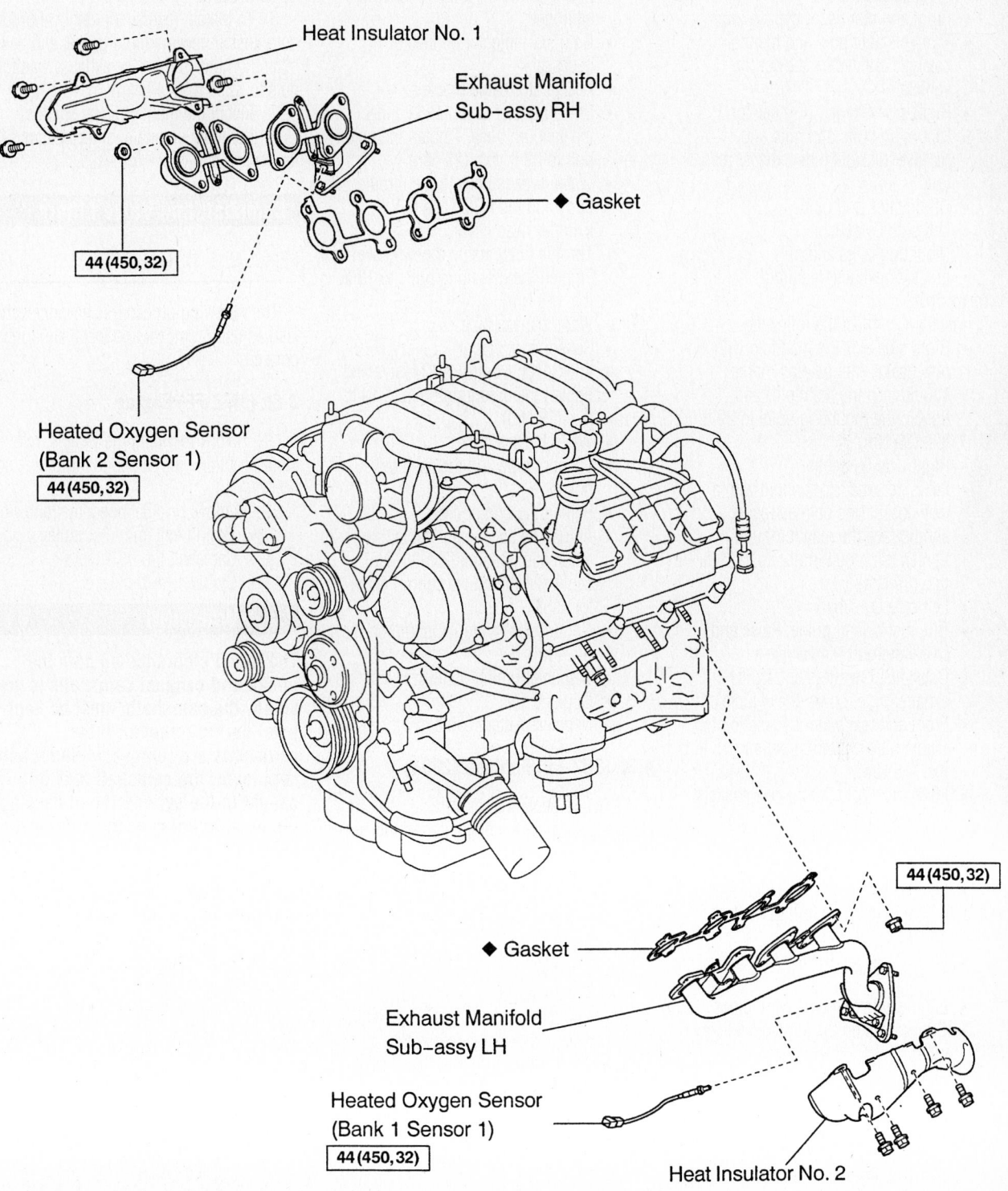

Heat Insulator No. 1

Exhaust Manifold
Sub-assy RH

◆ Gasket

44 (450, 32)

Heated Oxygen Sensor
(Bank 2 Sensor 1)
44 (450, 32)

44 (450, 32)

◆ Gasket

Exhaust Manifold
Sub-assy LH

Heated Oxygen Sensor
(Bank 1 Sensor 1)
44 (450, 32)

Heat Insulator No. 2

N·m (kgf·cm, ft·lbf) : Specified torque

◆ Non-reusable part

67162-LEXU-G17

Exhaust manifolds—2004 3UZ-FE

- Delivery pipes and the fuel injectors
- Mounting bolts and nuts. Lift up the intake manifold
- Front and rear water bypass joint
- Front exhaust pipe and the main catalytic converters. Lower the vehicle.
- Right side Oxygen (O2) sensor
- Mounting bolts and nuts and remove the right side exhaust manifold.
- Oil dipstick and guide
- Left side O2 sensor
- Mounting bolts and nuts
- Left side exhaust manifold

To install:

3. Install or connect the following:
- Right side exhaust manifold with a new gasket (the painted marks should face the manifold) and tighten the mounting bolts to 29 ft. lbs. (39 Nm).
- Right side O2 sensor
- Left side exhaust manifold with a new gasket (the painted marks should face the manifold) and tighten the mounting bolts to 29 ft. lbs. (39 Nm)
- Left side O2 sensor
- Oil dipstick and guide. Raise and safely support the vehicle.
- Catalytic converters and front exhaust pipe. Lower the vehicle.
- Front and rear water bypass joints. Tighten the mounting bolts to 13 ft. lbs. (18 Nm).
- Intake manifold, using new gaskets. Tighten the mounting nuts and bolts to 13 ft. lbs. (18 Nm).
- Delivery pipes and fuel injectors
- Fuel return pipe with new gaskets. Tighten the union bolt to 26 ft. lbs. (35 Nm).
- Fuel hoses and the injector connectors
- Engine wire to the delivery pipes
- Connectors on the left side delivery pipe
- ECT sensor
- Cold start injector time switch
- Water temperature sender gauge connectors
- Heater hoses and engine wire bracket
- Engine wire to the bracket.
- Cold start injector, tube and wire assembly. Tighten the mounting bolts to 69 inch lbs. (8 Nm), if equipped.
- Air intake chamber with new gaskets and tighten the mounting bolts to 13 ft. lbs. (18 Nm).

- Cold start injector tube to the right side delivery pipe and tighten the union bolt to 11 ft. lbs. (15 Nm), if equipped.
- Cold start injector connector, if necessary
- Accelerator cable bracket
- Brake booster union and connect the vacuum hose. Tighten the union bolt to 22 ft. lbs. (29 Nm).
- Water bypass hose to the throttle body and the PCV hose to the cylinder head cover.
- Throttle body, using a new gasket. Tighten the mounting bolts to 13 ft. lbs. (18 Nm).
- Water bypass pipe
- Sensor connectors
- Idle Speed Control (ISC) valve and tighten the mounting bolts to 13 ft. lbs. (18 Nm).
- Water bypass hose.
- Vacuum pipe and the assorted hoses
- Remaining components.
- Adjust the accelerator cable, the automatic transmission throttle cable and the cruise control actuator cable.
- Cooling fan hydraulic pump on the SC 400
- Camshaft timing pulleys
- Coolant
- Negative battery cable

4.3L (3UZ-FE) Engines (2004)

1. Before servicing the vehicle, refer to the precautions in the beginning of this section.
2. Remove or disconnect the following:

- Negative battery cable
- Engine under cover
- Coolant
- V-bank cover
- Air cleaner assembly (if needed for access to exhaust manifold)

3. Remove cylinder heads. See CYLINDER HEADS in this section.
4. Remove 4 bolts and heat shield from exhaust manifold.

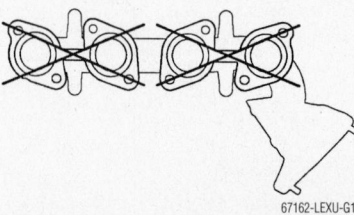

67162-LEXU-G18
Measuring exhaust manifold warpage

5. Remove 8 nuts and remove exhaust manifold and gasket.

To install:

6. To install, reverse the removal procedure. Install new manifold gasket and new retaining nuts. Torque the exhaust manifold nuts to 32 ft. lbs. (44 Nm).
7. Install the manifold heat shields.
8. Refill the engine cooling system. Start the engine and check for leaks.

Camshaft and Valve Lifters

REMOVAL & INSTALLATION

The following procedures have the valve lash adjuster removal and installation incorporated.

3.0L (1MZ-FE) Engine

1. Before servicing the vehicle, refer to the precautions in the beginning of this section.
2. Remove or disconnect the following:
- Timing belt and idler pulley
- Camshaft timing pulleys
- Cylinder head covers

✲✲ WARNING

The thrust clearance on both the intake and exhaust camshafts is very small, the camshafts must be kept level during removal. If the camshafts are removed without being kept level, the camshaft may be caught in the cylinder head causing the head to break or the camshaft to seize.

3. To remove the exhaust and intake camshafts from the right side cylinder head:
   a. Turn the camshaft with a wrench until the 2 pointed marks drive and driven gears are aligned. (The right camshaft gears have 2 marks apiece; the

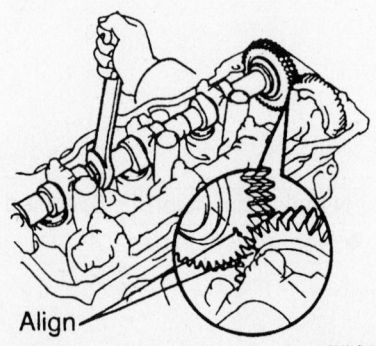

7923LG40
Aligning the right side camshaft timing marks—3.0L (1MZ-FE) engine

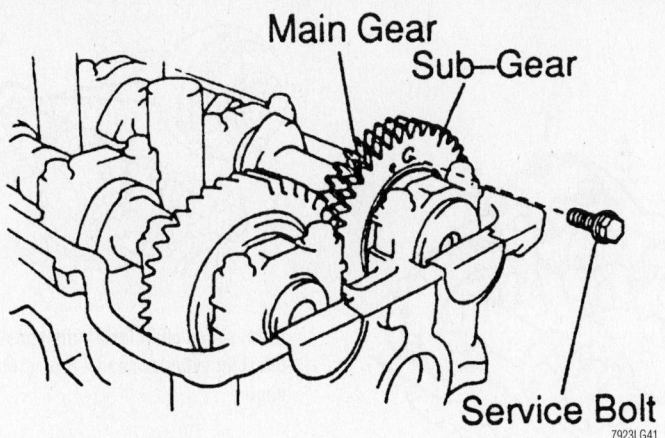

Securing the subgear and driven gear, right side—3.0L (1MZ-FE) engine

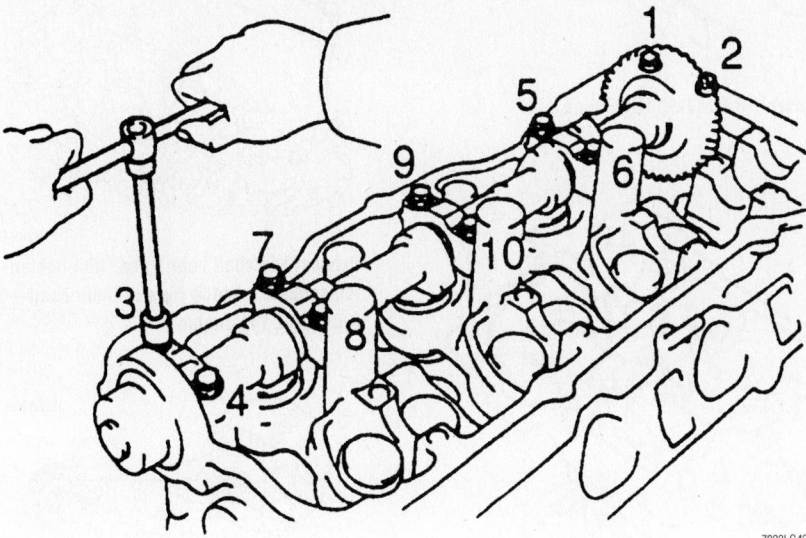

Right exhaust camshaft bearing loosening sequence—3.0L (1MZ-FE) engine

Right intake camshaft bearing loosening sequence—3.0L (1MZ-FE) engine

left side camshaft gears have 1 mark each.)

b. Secure the exhaust camshaft sub-gear to the main gear using a service bolt. A bolt 0.63–0.79 in. (16–20mm) long with a 6mm thread diameter and a 1mm pitch is recommended. When removing the exhaust camshaft be sure the subgear is not loaded; all the force must be eliminated.

c. Uniformly loosen and remove the exhaust camshaft bearing cap bolts in several passes and in the proper sequence. Remove the 8 bearing cap bolts and remove the caps, keeping them in the correct order.

d. Remove the exhaust camshaft from the engine.

e. Uniformly loosen and remove the 10 bearing cap bolts in several passes, in the proper sequence. Remove the bearing caps, keeping them in order, remove the oil seal, then lift out the intake camshaft.

4. To remove the exhaust and intake camshafts from the left side cylinder head:

a. Turn the camshaft with a wrench until the pointed marks on the drive and driven gears are aligned. (The right camshaft gears have 2 marks apiece; the left side camshaft gears have 1 mark each.)

b. Secure the exhaust camshaft sub-gear to the main gear using a service

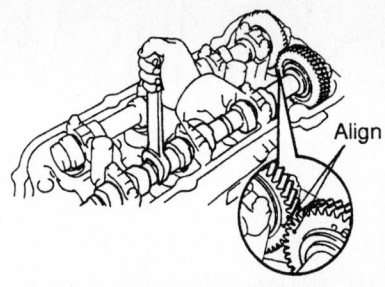

Aligning the left side camshaft timing marks—3.0L (1MZ-FE) engine

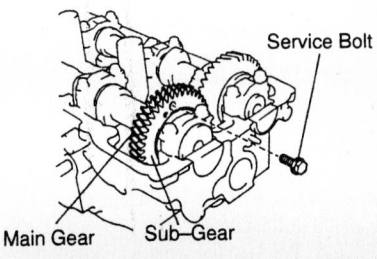

Securing the subgear and driven gear, left side—3.0L (1MZ-FE) engine

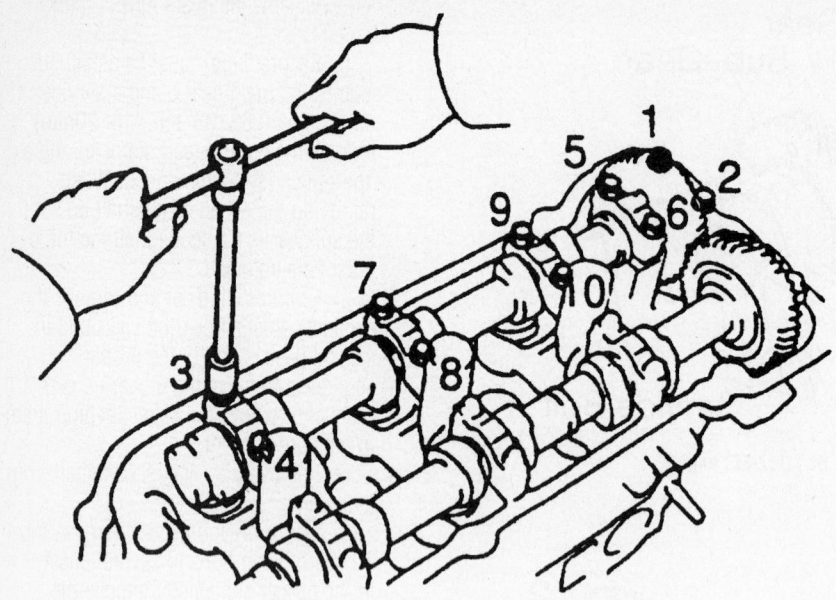

Left intake camshaft bearing cap bolt loosening sequence—3.0L (1MZ-FE) engine

7923LG46

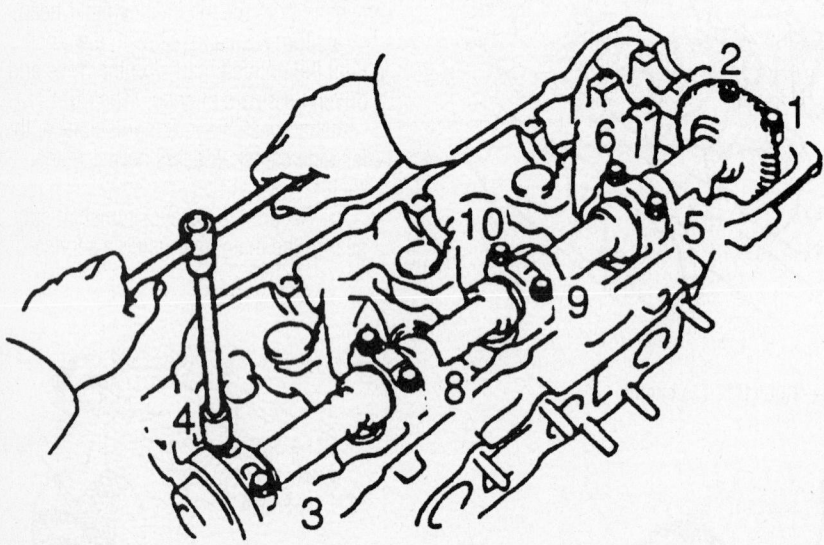

Left exhaust camshaft bearing cap bolt loosening sequence—3.0L (1MZ-FE) engine

7923LG47

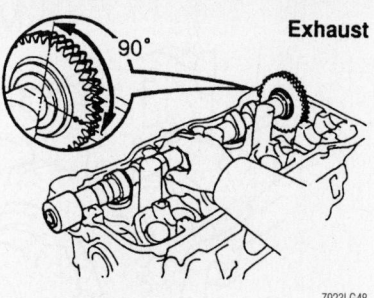

7923LG48

Exhaust camshaft installation position on the right cylinder head—3.0L (1MZ-FE) engine

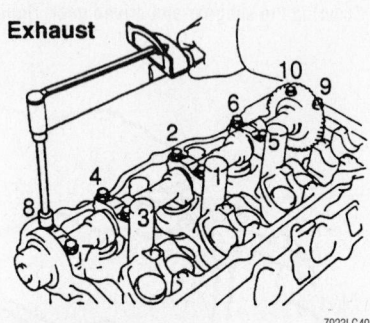

7923LG49

Exhaust camshaft bearing cap bolt tightening sequence on the right cylinder head—3.0L (1MZ-FE) engine

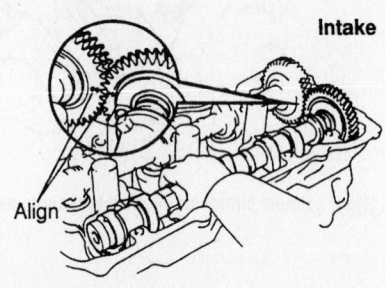

7923LG50

Intake camshaft installation position on the right cylinder head—3.0L (1MZ-FE) engine

bolt. A bolt 0.63–0.79 in. (16–20mm) long with a 6mm thread diameter and a 1mm pitch is recommended. When removing the exhaust camshaft be sure the subgear is not loaded; all the force must be eliminated.

c. Uniformly loosen and remove the exhaust camshaft bearing cap bolts in several passes and in the proper sequence. Remove the 8 bearing cap bolts and remove the caps. Keep the caps in the correct order.

d. Remove the exhaust camshaft from the engine.

e. Uniformly loosen and remove the

10 bearing cap bolts in several passes, in the proper sequence. Remove the bearing caps, keeping them in order, remove the oil seal, then lift out the intake camshaft.

5. Remove the valve lash adjuster shims and hydraulic lash adjusters. Identify each lash adjuster and shim as it is removed so it can be reinstalled in the same position. If the lash adjusters are to be reused, store them upside down in a sealed container.

To install:

6. Install or connect the following:
- Valve lash adjusters and shims.

Check valve clearance and replace the shims as necessary.

➡**Before installing the camshafts in either cylinder head, apply multi-purpose grease to the thrust portions of each camshaft.**

7. To install the right camshafts:
a. Position the intake camshaft on the head so that the alignment marks are at a 90° angle from vertical. The mark should be at the 3 o'clock position.

b. Apply sealant to the No. 1 bearing cap.

c. Apply a light coat of clean engine oil to the bolt threads and under the bolt

Intake

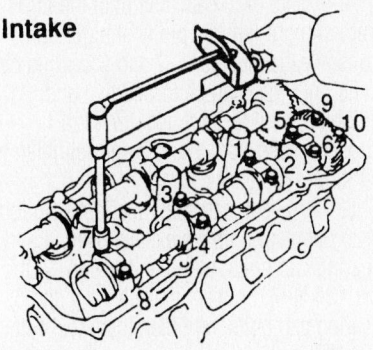

7923LG51

Intake camshaft bearing cap bolt tightening sequence on the right cylinder head—3.0L (1MZ-FE) engine

Exhaust

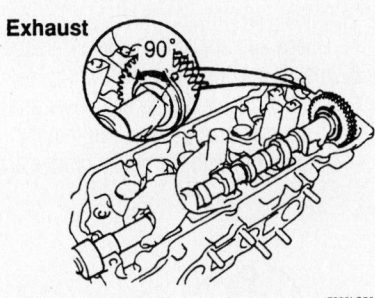

7923LG52

Exhaust camshaft installation position on the left cylinder head—3.0L (1MZ-FE) engine

Exhaust

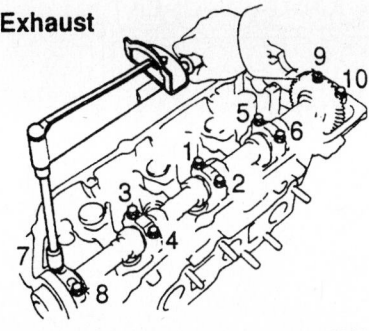

7923LG53

Exhaust camshaft bearing cap bolt tightening sequence on the right cylinder head—3.0L (1MZ-FE) engine

head. Install the bearing caps to their proper position. Tighten the bolts evenly and in several passes in the reverse order of loosening to 12 ft. lbs. (16 Nm) in the proper sequence.

d. Position the exhaust camshaft on the head so that the alignment marks are at a 90°angle from vertical. The mark should be at the 9 o'clock position and must align with the marks on the other gear.

e. Apply a light coat of clean engine

Intake

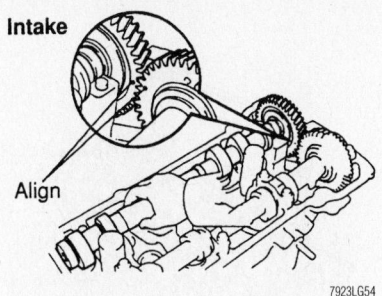

7923LG54

Intake camshaft installation position on the left cylinder head—3.0L (1MZ-FE) engine

Intake

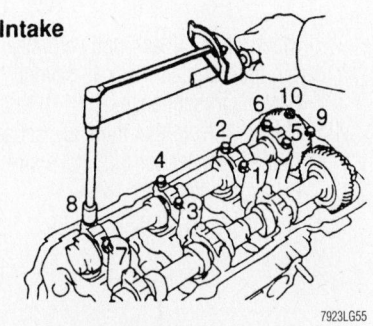

7923LG55

Intake camshaft bearing cap bolt tightening sequence on the right cylinder head—3.0L (1MZ-FE) engine

oil to the bolt threads and under the bolt head. Install the bearing caps to their proper position. Tighten the bolts evenly and in several passes in the reverse order of loosening to 12 ft. lbs. (16 Nm) in the proper sequence.

f. Remove the service bolt.

8. To install the left camshafts:

a. Position the intake camshaft on the head so that the alignment mark is at a 90 degree angle from vertical. The mark should be at the 9 o'clock position.

b. Apply sealant to the No. 1 bearing cap.

c. Apply a light coat of clean engine oil to the bolt threads and under the bolt head. Install the bearing caps to their proper position. Tighten the bolts evenly and in several passes to 12 ft. lbs. (16 Nm) in the proper sequence.

d. Position the exhaust camshaft on the head so that the alignment marks are at a 90-degree angle from vertical. The mark should be at the 3 o'clock position and must align with the marks on the other gear.

e. Apply a light coat of clean engine oil to the bolt threads and under the bolt head. Install the bearing caps to their proper position. Tighten the bolts evenly

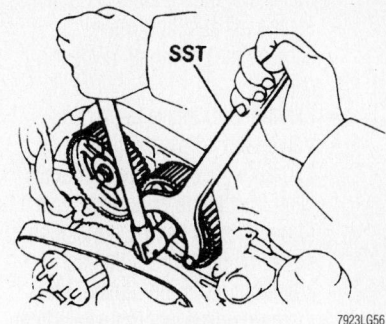

7923LG56

Removing the camshaft sprockets—3.0L (2JZ-GE) engine

and in several passes to 12 ft. lbs. (16 Nm) in the proper sequence.

f. Remove the service bolt.

9. Apply multi-purpose grease to new camshaft oil seals. Install the seals.

10. Install or connect the following:
- No. 3 (rear) timing belt cover
- Camshaft timing gears
- Idler pulley, timing belt and covers

11. Check and adjust the valve clearance.

12. Install the cylinder head (valve) covers.

3.0L (2JZ-GE) Engine (Except 2004)

1. Before servicing the vehicle, refer to the precautions in the beginning of this section.

2. Remove or disconnect the following:
- Negative battery cable from the battery
- Timing belt from the engine
- Cylinder head covers
- Camshaft sprocket
- 4 bolts and lift out the No. 4 (inner) timing belt cover.
- No. 1 camshaft bearing cap bolts. These are the bolts directly behind the sprockets.
- Bearing caps
- Remaining bearing cap bolts. Note

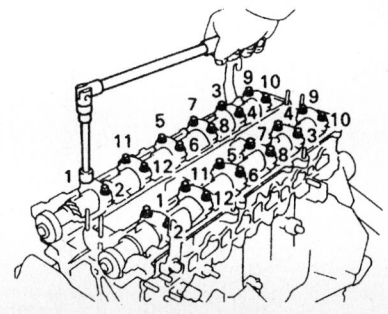

7923LG57

Camshaft bearing cap bolt removal sequence—3.0L (2JZ-GE) engine

that there are separate sequences for the exhaust and intake camshafts. Lift off all 12 bearing caps.

- Exhaust and intake camshafts
- Valve lash adjuster shims and hydraulic lash adjusters. Identify each lash adjuster and shim as it is removed so it can be reinstalled in the same position. If the lash adjusters are to be reused, store them upside down in a sealed container.

To install:
3. Install or connect the following:
   - Valve lash adjusters and shims. Check valve clearance and replace the shims as necessary.
   - Camshafts. Coat the thrust portions of each with engine oil, then position them in the cylinder head with the cam lobes and the knock pins in the correct position.
   - No. 3 and No. 7 bearing caps in place, coat the bolt threads with oil, then tighten them temporarily
   - New oil seals, coated with multi-purpose grease, over the camshafts
   - No. 1 bearing caps, then apply some sealant. Install the bolts.
   - All remaining bearing caps. Coat

Camshaft bearing cap bolt tightening sequence—3.0L (2JZ-GE) engine

the threads of each bolt with clean oil, then tighten them, in several passes, in sequence, to 14 ft. lbs. (20 Nm). Note that there are separate sequences for the intake and exhaust sides.

- Oil seal in as far as it will go

4. Rotate each camshaft until the forward straight (knock) pin is straight up. Loosen exhaust Nos. 1, 2 and 6 bearing cap bolts until they can be turned by hand; retighten them to 14 ft. lbs. (20 Nm). Loosen intake Nos. 1 and 2 and retighten to 14 ft. lbs. (20 Nm).

5. Turn each camshaft 1/3 of a revolution (120 degrees). Loosen exhaust Nos. 4 and 7 bearing cap bolts; retighten them to 14 ft. lbs. (20 Nm). Loosen intake Nos. 4 and 6 bearing cap bolts; retighten them to 14 ft. lbs. (20 Nm).

6. Turn each camshaft an additional 1/3 of a revolution, loosen exhaust bearing cap bolts Nos. 3 and 5, then retighten them to 14 ft. lbs. (20 Nm). Loosen intake bearing cap bolts Nos. 3 and 7, then retighten them to 14 ft. lbs. (20 Nm).

7. Check and adjust the valve clearance.
8. Install or connect the following:
   - No. 4 inside timing belt cover and the camshaft pulleys. Align the shaft pin with the pulley groove and

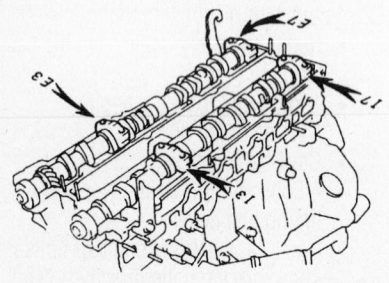

Installing No. 3 and 7 bearing caps—3.0L (2JZ-GE) engine

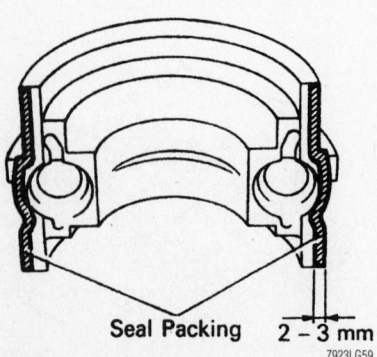

Applying sealant to the No. 1 bearing cap—3.0L (2JZ-GE) engine

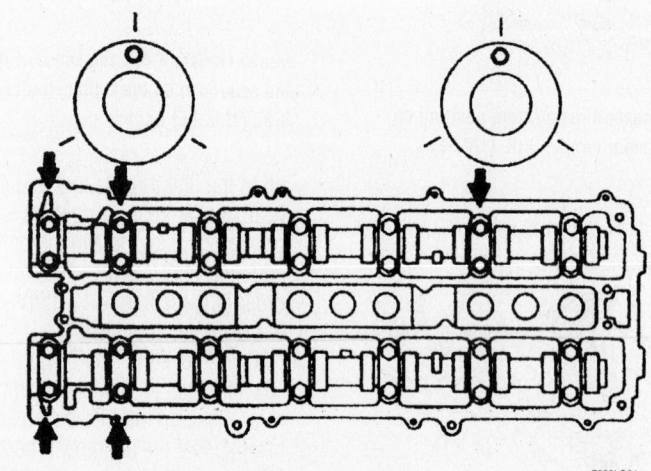

Tightening the camshafts (Step 1)—3.0L (2JZ-GE) engine

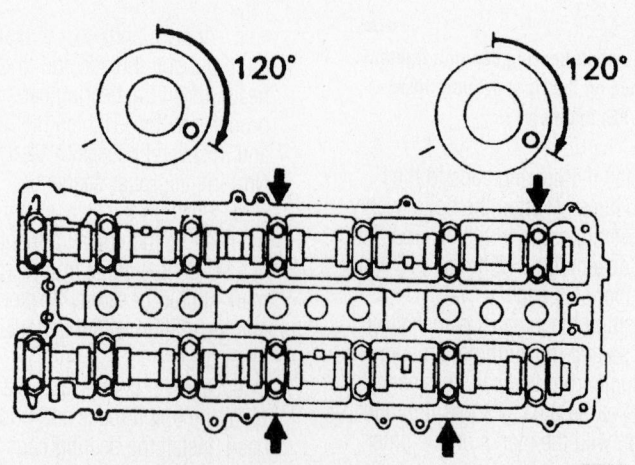

Tightening the camshafts (Step 2)—3.0L (2JZ-GE) engine

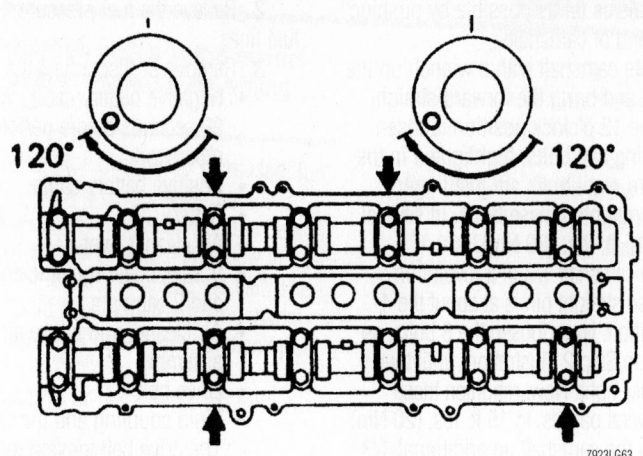

Tightening the camshafts (Step 3)—3.0L (2JZ-GE) engine

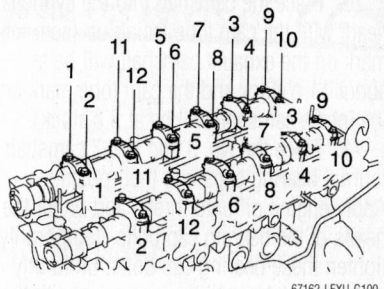

Camshaft bearing cap bolt loosening sequence—3.0L (2JZ-GE) Engine (2004 Model)

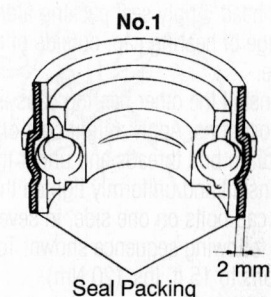

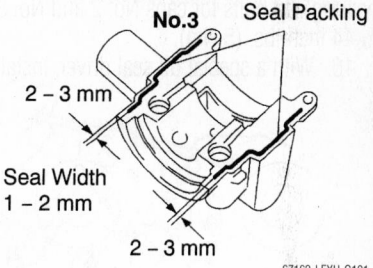

Installing packing onto camshaft bearing caps—3.0L (2JZ-GE) Engine (2004 Model)

slide the pulley on. Install the bolt temporarily. Hold the hex portion of the camshaft with a wrench; tighten the pulley bolt to 59 ft. lbs. (79 Nm).
- Cylinder head covers
- Timing belt to the engine
- Negative battery cable to the battery

9. Check and/or adjust the ignition timing as necessary

3.0L (2JZ-GE) Engines (2004)

1. Before servicing the vehicle, refer to the precautions in the beginning of this section.

2. Remove or disconnect the following:
- Negative battery cable from the battery
- Timing belt from the engine
- Cylinder head covers

3. Make reference marks on the timing belt to the crankshaft and idler pulley timing marks.

4. Remove the timing belt. See TIMING BELT.

> **⁂ WARNING**
>
> **Do NOT allow anything to drop into the lower part of the timing belt cover. Do NOT allow the timing belt to come in contact with any oil, water or dust.**

5. Remove both the exhaust and intake camshaft pulleys. See procedure under TIMING BELT.

> **⁂ WARNING**
>
> **Since the thrust clearance of the camshaft is small, the camshaft must be kept level during removal. If not, the portion of the cylinder head receiv-**

ing the camshaft thrust may crack or be otherwise damaged, causing the camshaft to later seize or break. Follow the camshaft removal procedure carefully as given in this section.

6. With a 5mm hex wrench, remove the No. 3 camshaft bearing cap bolts. Then, uniformly loosen and remove the 4 camshaft bearing cap bolts, located just rear of the camshaft pulley locations.

7. Carefully pry out the No. 1 and No. 3 camshaft bearing caps and oil seals.

8. Uniformly loosen and remove the 12 camshaft bearing cap bolts, in several passes, and following the sequence illustrated. Remove the 6 No. 2 bearing caps and the camshafts from the engine.

To install:

9. Apply engine oil to the thrust portion of the camshaft.

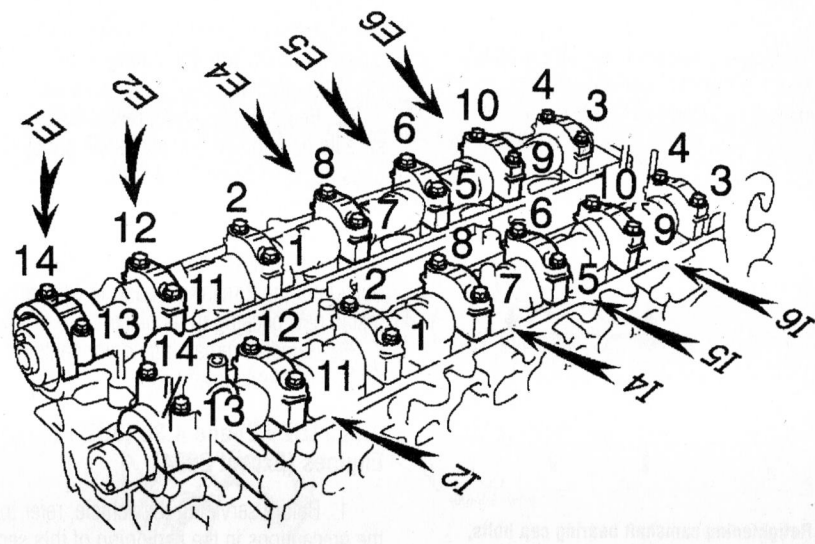

Camshaft bearing cap bolt tightening sequence—3.0L (2JZ-GE) Engine (2004 Model)

i. Remove the bearing caps, oil seal and the intake camshaft.

9. To remove the exhaust camshaft of the left side cylinder head:

a. Position the service bolt hole of the drive subgear to the upright position. Secure the camshaft subgear to drive gear with a service bolt.

➡ **When removing the camshaft, be sure the torsional spring force of the subgear has been eliminated.**

b. Set the timing mark (2 dots) on the camshaft drive gear at approximately 15 degrees, by turning the camshaft with the proper tool.

c. Alternately loosen and remove the bearing cap bolts holding the intake camshaft side of the oil feed pipe to the cylinder head.

d. Uniformly loosen (in several passes) and remove the bearing cap bolts in the proper sequence.

e. Remove the oil feed pipe and the bearing caps. Remove the camshaft.

10. To remove the intake camshaft from the left side cylinder head:

a. Set the timing mark (single dot) of the camshaft drive gear at approximately 60 degrees, by turning the camshaft with the proper tool.

b. Uniformly loosen (in several passes) and remove the bearing cap bolts, in sequence.

c. Remove the bearing caps, oil seal and the intake camshaft.

d. Remove the valve lash adjuster shims and hydraulic lash adjusters. Identify each lash adjuster and shim as it is removed so it can be reinstalled in the same position. If the lash adjusters are to be reused, store them upside down in a sealed container.

To install:

11. Install or connect the following:
- Valve lash adjusters and shims. Check the valve clearance and replace the shims as necessary.
- New seal packing to the bearing caps
- Bearing cap on the right side cylinder head, marked **I1**, in position with the arrow mark facing the rear. Install the bearing cap on the left

side cylinder head, marked **I6**, in position with the arrow mark facing the front.
- Oil on the threads of the cap bolts. Install the bearing cap bolts with new washers and tighten to 12 ft. lbs. (16 Nm).

12. To install the right side cylinder head intake camshaft:

a. Apply grease to the thrust portion of the camshaft.

b. Place the intake camshaft at a 45 degree angle of the timing mark (single dot) on the cylinder head.

c. Remove any old packing and apply new seal packing to the bearing cap marked **I6** and install the front bearing cap, marked **I6** with the arrow facing rearward.

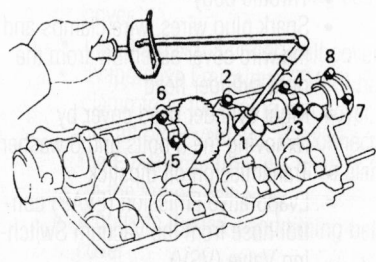

Right side intake camshaft bracket bolt tightening sequence—4.0L (1UZ-FE) and 4.3L (3UZ-FE) Engines (Except 2004)

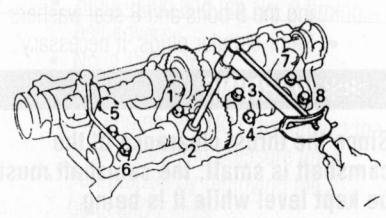

Right side exhaust camshaft bracket bolt tightening sequence—4.0L (1UZ-FE) and 4.3L (3UZ-FE) Engines (Except 2004)

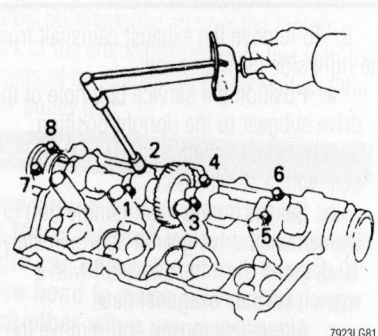

Left side intake camshaft bracket bolt tightening sequence—4.0L (1UZ-FE) and 4.3L (3UZ-FE) Engines (Except 2004)

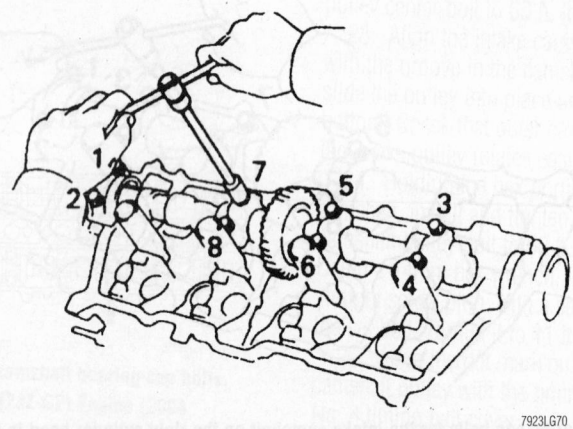

Bolt removal sequence for the intake camshaft on the left cylinder head—4.0L (1UZ-FE) and 4.3L (3UZ-FE) Engines (Except 2004)

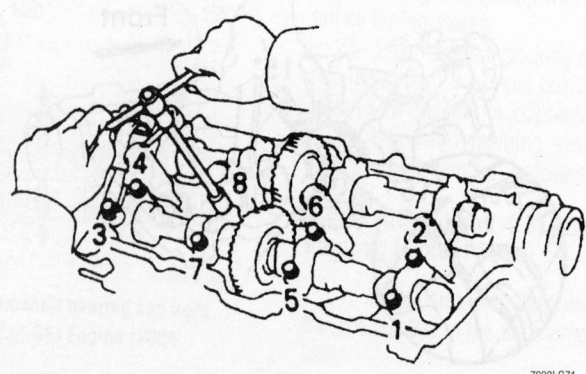

Bolt removal sequence for the exhaust camshaft on the left cylinder head—4.0L (1UZ-FE) and 4.3L (3UZ-FE) Engines (Except 2004)

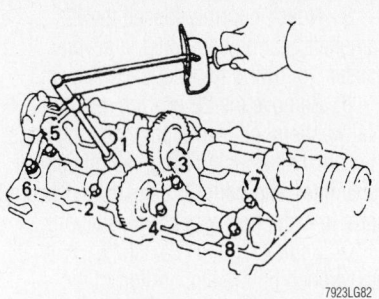

7923LG82

Left side exhaust camshaft bracket bolt tightening sequence—4.0L (1UZ-FE) and 4.3L (3UZ-FE) Engines (Except 2004)

d. Align the arrows at the front and rear of the cylinder head with the bearing cap.

e. Install the remaining bearing caps in the proper sequence with the arrow mark facing rearward. Install the oil feed pipe and the mounting bolts.

f. Uniformly tighten the bearing cap bolts in the proper sequence to 12 ft. lbs. (16 Nm).

13. To install the right side cylinder head exhaust camshaft:

a. Set the timing mark (single dot) on the camshaft drive gear at a 10 degree angle by turning the intake camshaft with the proper tool.

b. Apply grease to the thrust portion of the camshaft.

c. Align the timing marks (single dots) on the camshaft drive and driven gears.

d. Place the exhaust camshaft in the cylinder head. Install the rear bearing cap with the arrow mark facing rearward.

e. Align the arrow marks at the front and rear of the cylinder head with the mark on the bearing cap. Apply a light coat of oil on the threads of the bearing cap bolts.

f. Uniformly tighten the bearing cap bolts in the proper sequence to 12 ft. lbs. (16 Nm).

g. Bring the service bolt installed upward by turning the camshaft with the proper tool. Remove the service bolt.

14. To install the left side cylinder head intake camshaft:

a. Apply grease to the thrust portion of the camshaft.

b. Place the intake camshaft with the timing mark (single dot) at a 60 degree angle on the cylinder head.

c. Remove any old packing and apply new seal packing to the bearing cap marked **I6** and install the front bearing cap, marked **I1** with the arrow facing rearward.

d. Align the arrows at the front and rear of the cylinder head with the bearing cap. Apply a light coat of oil on the threads of the bearing cap bolts.

e. Install the remaining bearing caps in the proper sequence with the arrow mark facing rearward. Install the oil feed pipe and the mounting bolts.

f. Uniformly tighten the bearing cap bolts in the proper sequence to 12 ft. lbs. (16 Nm).

15. Install the left side cylinder head exhaust camshaft by:

a. Set the timing mark (2 dots) on the camshaft drive gear at a 15 degree angle by turning the intake camshaft with the proper tool.

b. Apply grease to the thrust portion of the camshaft.

c. Align the timing marks (2 dots each) on the camshaft drive and driven gears.

d. Place the exhaust camshaft on the cylinder head. Install the rear bearing cap with the arrow mark facing rearward.

e. Align the arrow marks at the front and rear of the cylinder head with the mark on the bearing cap. Apply a light coat of oil on the threads of the bearing cap bolts.

f. Uniformly tighten the bearing cap bolts in the proper sequence to 12 ft. lbs. (16 Nm).

16. Bring the service bolt installed upward by turning the camshaft with the proper tool. Remove the service bolt.

17. Install or connect the following:

• Camshaft oil seals. Be sure to apply Multi-Purpose (MP) grease to the new oil seal lip.
• Semi-circular plugs to the cylinder heads
• Left cylinder head cover and bolts. Tighten the bolts to 52 inch lbs. (6 Nm).
• Spark plug wires and clamps to the left cylinder head cover
• Engine wire clamp to the wire bracket on the delivery pipe
• EVAP hose to the VSV
• Transmission oil dipstick
• Right cylinder head cover and bolts. Tighten the bolts to 52 inch lbs. (6 Nm).
• Spark plug wires and clamps to the right cylinder head cover
• Throttle body to the air intake chamber. Install the 2 bolts and 2 nuts and tighten to 13 ft. lbs. (18 Nm).
• Timing belt rear plates by installing the bolts. Tighten the bolts to 66 inch lbs. (8 Nm).

• No. 2 ignition coil

18. Align the knock pin on the right side camshaft with the knock pin of the timing pulley. Slide on the timing pulley with the right side mark facing forward. Tighten the bolt to 80 ft. lbs. (108 Nm).

19. Align the knock pin on the left side camshaft with the knock pin of the timing pulley. Slide on the timing pulley with the left side mark facing forward. Tighten the bolt to 80 ft. lbs. (108 Nm).

20. Turn the crankshaft pulley and align its groove with the **0** timing mark on the timing belt cover.

21. Turn each camshaft timing pulley and align the timing marks of the pulley with the timing belt rear plate.

22. Attach the timing belt to the left side camshaft timing pulley:

23. Using the proper tool, slightly turn the left side timing pulley clockwise. Align the installation mark of the timing belt with the timing mark of the camshaft timing pulley and hang the timing belt on the left side camshaft pulley.

24. Align the timing marks of the left side camshaft pulley and the timing belt rear plate.

25. Check that the timing belt has tension between crankshaft timing pulley and the left side camshaft pulley.

26. Install the timing belt to the right side camshaft timing pulley:

27. Using the proper tool, slightly turn the right side timing pulley clockwise. Align the installation mark of the timing belt with the timing mark of the camshaft timing pulley and hang the timing belt on the right side camshaft pulley.

28. Align the timing marks of the right side camshaft pulley and the timing belt rear plate.

29. Check that the timing belt has tension between crankshaft timing pulley and the right side camshaft pulley.

30. The timing belt tensioner must be set prior to installation. The tensioner can be set by:

a. Place a plate washer between the tensioner and a block. Using a press, press in the pushrod using 220–225 lbs. (100–102 kg.) of pressure.

b. Align the holes of the pushrod and housing, pass a 0.05 in. (1.27mm) rod through the holes to keep the setting position of the pushrod.

c. Release the press and install the dust boot to the tensioner.

d. Loosely install the tensioner. Evenly and alternately tighten the bolts to 20 ft. lbs. (26 Nm). Remove the tool from the tensioner.

e. Turn the crankshaft pulley 2 complete revolutions from TDC to TDC. Always turn the crankshaft clockwise. Check that all belt and pulley marks align with their reference marks. If any mark is out of perfect alignment, the timing belt must be removed and reinstalled.

31. Install or connect the following:
- Remaining components
- V bank cover
- Coolant
- Battery and battery tray
- Battery cables, positive cable first
- Air cleaner inlet
- Battery clamp cover
- Engine undercover
- Oil pan protector

4.3L (3UZ-FE) Engines (2004)

1. Before servicing the vehicle, refer to the precautions in the beginning of this section.

2. Relieve the fuel pressure from the fuel lines.

3. Remove or disconnect the following:
- Engine under cover
- Drain engine coolant and engine oil
- V-bank cover
- Air cleaner assembly
- Intake air pipe
- Radiator
- Throttle body
- Upper and lower intake manifold assembly
- Timing belt from camshaft pulleys
- Camshaft position sensor and LH timing belt rear plates
- RH timing belt rear plates

> **✳✳ WARNING**
>
> **Do NOT drop anything inside timing belt cover during this procedure. Keep oil, water and dust from timing belt.**

- Power steering pump from engine mount (Do NOT disconnect hoses)
- Catalytic converters
- Water inlet housing assembly
- Water bypass pipe, front bypass joint, and rear bypass joint
- Ignition coils
- Variable valve timing (VVT) sensors
- Engine hangers
- Oil dipsticks and guides for engine oil and transmission fluid
- Cylinder head covers
- Spark plugs
- Semi-circular plugs and camshaft housing plugs
- Camshaft timing oil control valve

> **✳✳ WARNING**
>
> **Since the thrust clearance of the camshaft is small, the camshaft must be kept level during removal steps. If it is not kept level, the portion of the cylinder head receiving the shaft thrust may crack or be damaged, causing the camshaft to later seize or break. Follow the procedure carefully to avoid this damage.**

4. Check the crankshaft pulley position and ensure the timing mark of the pulley is aligned with the centers of the crankshaft pulley bolt and the No. 2 timing belt idler pulley bolt.

> **✳✳ WARNING**
>
> **Having the crankshaft pulley at the wrong angle can cause the piston head and valve head to come into contact with each other during camshaft removal. Always set the crankshaft pulley at the described angle.**

5. Using a special wrench, rotate the camshaft timing tube from left to right about 2-3 times, within only a 25°range of movement. Use a waste cloth to collect oil from the camshaft timing oil control valve installation hole.

6. Remove the LH camshafts first. With a hex wrench on the hex portion of the camshaft, rotate so that a 6mm service bolt can be inserted into the bolt hole in the rear face of the camshaft pulley into order to secure the camshaft in place. The bolt should be about 0.63-0.79 in. (16-20mm) long.

7. Align the timing mark (2 dots) of the camshaft drive gear by turning the camshaft with a hex wrench until the timing mark aligns.

8. Now, uniformly loosen the 22 camshaft bearing cap bolts, in several passes, following the sequence shown.

9. Remove the 22 bearing cap bolts, 4 seal washers, oil feed pipe, 9 bearing caps, the camshaft housing plug, the oil control valve filter, and both LH camshafts. Keep all parts in order for proper reinstallation.

10. Remove the RH camshafts. With a hex wrench on the hex portion of the camshaft, rotate so that a 6mm service bolt can be inserted into the bolt hole in the rear face of the camshaft pulley into order to secure the camshaft in place. The bolt should be about 0.63-0.79 in. (16-20mm) long.

11. Align the timing mark (1 dot) of the camshaft main gear about 10° angle by turning the camshaft with a hex wrench until the timing mark aligns.

12. Now, uniformly loosen the 22 camshaft bearing cap bolts, in several passes, following the sequence shown.

13. Remove the 22 bearing cap bolts, 4 seal washers, oil feed pipe, 9 bearing caps, the camshaft housing plug, the oil control valve filter, and both LH camshafts. Keep all parts in order for proper reinstallation.

To install:

> **✳✳ WARNING**
>
> **Since the thrust clearance of the camshaft is small, the camshaft must be kept level during removal steps. If it is not kept level, the portion of the cylinder head receiving the shaft thrust may crack or be damaged, causing the camshaft to later seize or break. Follow the procedure carefully to avoid this damage.**

14. Ensure the crankshaft pulley is in position so that its timing mark is in line with the centers of the pulley bolt and idler pulley bolt.

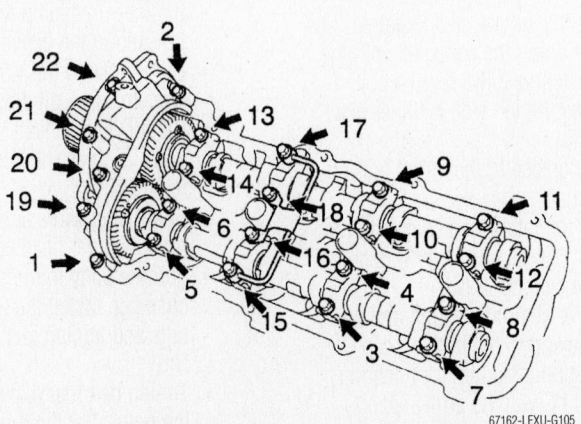

67162-LEXU-G105

LH camshaft bearing cap bolt loosening sequence—4.3L (3UZ-FE) Engine (2004 Model)

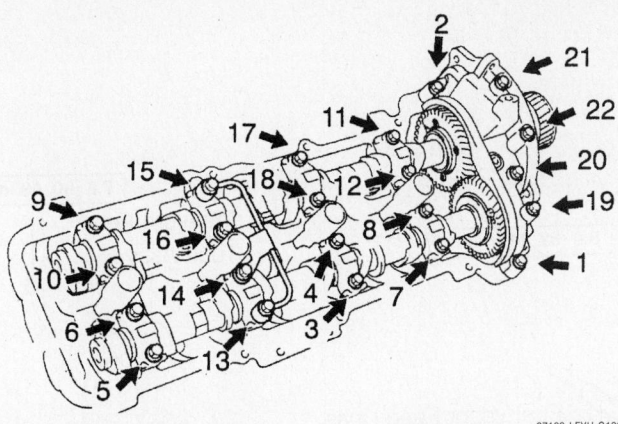

RH camshaft bearing cap bolt loosening sequence—4.3L (3UZ-FE) Engine (2004 Model)

✳✲ WARNING

Having the crankshaft pulley at the wrong angle can cause the piston head and valve head to come into contact with each other during camshaft removal. Always set the crankshaft pulley at the described angle.

15. Apply grease to the thrust portion of the LH intake and exhaust camshafts. Align the timing marks (2 dots) of the camshaft

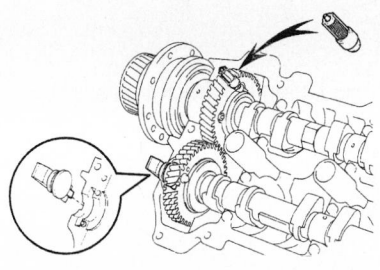

Installing the camshaft housing plug and oil control valve filter for the LH camshaft—4.3L (3UZ-FE) Engine (2004 Model)

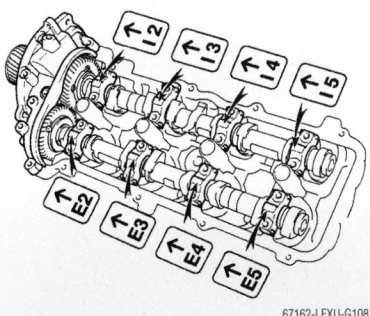

Installing the camshaft bearing caps in sequence on the LH camshaft—4.3L (3UZ-FE) Engine (2004 Model)

drive and driven main gears. Place the camshafts into the LH cylinder head.

16. Apply new seal packing material around the opening of the camshaft housing plug. Install the camshaft housing plug and

the oil control valve filter into the cylinder head, as shown.

17. Remove any old packing material, then install new packing material around the mounting edge (not in the grooves) of the front bearing cap.

18. Position the front bearing cap in place. This will determine the thrust portion of the camshaft.

   a. Install the other bearing caps, in sequence shown, with the arrow marks facing forward.

   b. Push the camshaft oil seal into place by pushing from the front of the engine. Install a new seal washer to the front bearing cap bolts.

   c. Apply a light coat of oil to bearing cap bolt threads and under the heads of the bearing cap bolts "D" and "E", as shown. Do NOT apply engine oil under the heads of bearing cap bolts "A", "B" and "C".

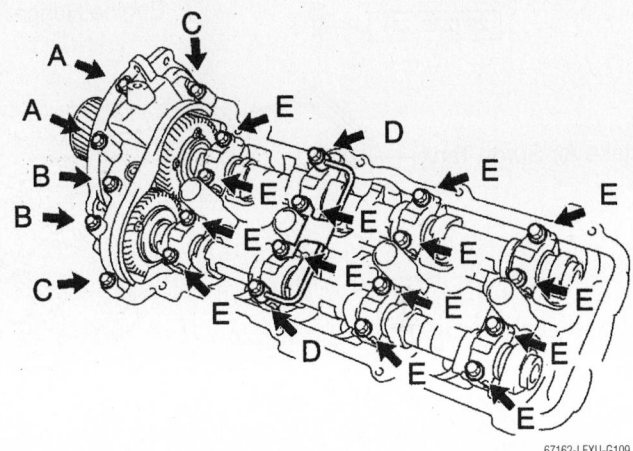

Identifying the locations of camshaft bearing caps on the LH camshaft—4.3L (3UZ-FE) Engine (2004 Model)

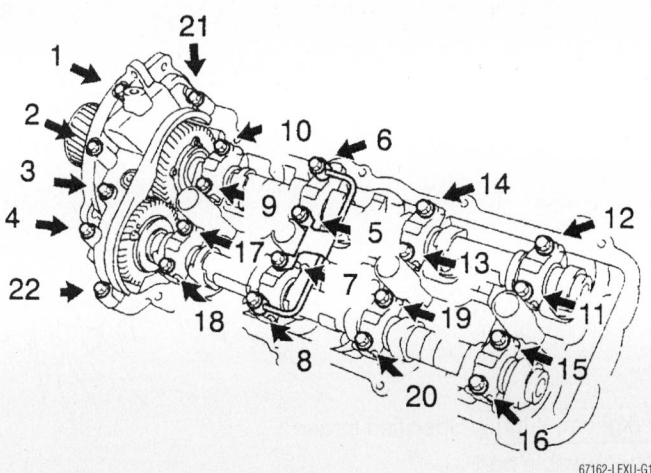

Identifying the locations of camshaft bearing caps on the LH camshaft—4.3L (3UZ-FE) Engine (2004 Model)

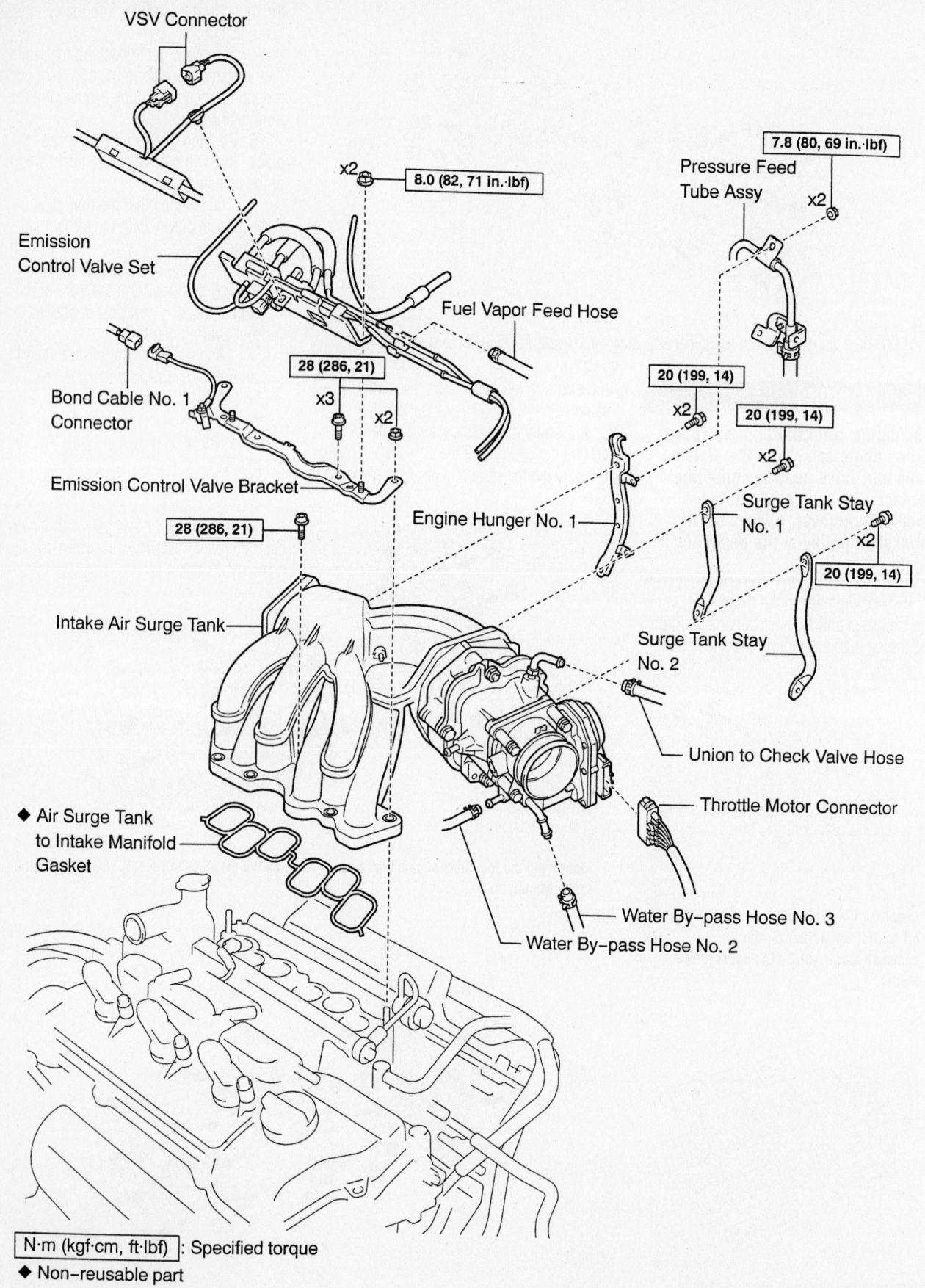

VSV Connector

7.8 (80, 69 in.·lbf)

Pressure Feed
Tube Assy

x2

8.0 (82, 71 in.·lbf)

x2

Emission
Control Valve Set

Fuel Vapor Feed Hose

20 (199, 14)

Bond Cable No. 1
Connector

28 (286, 21)

x2

x3

x2

20 (199, 14)

Emission Control Valve Bracket

x2

Engine Hunger No. 1

Surge Tank Stay
No. 1

28 (286, 21)

x2

20 (199, 14)

Intake Air Surge Tank

Surge Tank Stay
No. 2

◆ Air Surge Tank
to Intake Manifold
Gasket

Union to Check Valve Hose

Throttle Motor Connector

Water By-pass Hose No. 3

Water By-pass Hose No. 2

N·m (kgf·cm, ft·lbf) : Specified torque

◆ Non-reusable part

67162-LEXU-G36

Exploded view Intake Manifold—3.3L (3MZ-FE) Engine

d. Bolt lengths vary for each bearing cap. Refer to the illustration for each of the following bolts:
- Bolt "A" with seal washer is 3.70 in. (94mm)
- Bolt "B" with seal washer is 2.83 in. (72mm)
- Bolt "C" is 0.98 in. (25mm)
- Bolt "D" is 2.05 in. (52mm)
- Bolt "E" is 1.50 in. (38mm)

19. Install the oil feed pipe and the 22 bearing cap bolts in their respective locations. Uniformly tighten the 22 bearing cap bolts, in several passes, following the sequence as shown. Torque bolt "C" to a final torque of 66 inch lbs. (7.5 Nm). Torque all other bearing cap bolts to a final torque of 12 ft. lbs. (16 Nm).

20. Remove the service bolt installed in the rear face of the gear.

21. Repeat this entire procedure for the RH camshafts. Use the illustrations given for the LH camshafts, as the sequences are the same on the RH camshafts.

22. Check and adjust valve lash. See VALVE LASH.

23. Install or reconnect the following:
- Camshaft timing oil control valve.
- Semi-circular plugs in rear ends of each cylinder head
- Cylinder head covers, with new gaskets and packing material
- Engine hangers
- Variable valve timing sensors
- Oil dipsticks and tubes
- All remaining components in reverse of removal procedure.

3.3 L (3MZ-FE) Engine

1. Before servicing the vehicle, refer to the precautions in the beginning of this section.

2. Remove or disconnect the following:
- Negative battery cable. Wait at least 90 seconds before performing any other work
- RH front wheel
- Front suspension upper brace center
- V-bank cover sub-assembly
- Air cleaner assembly
- Emission control valve set
- Intake air surge tank
- Ignition coil assembly
- Cylinder head valve covers
- Front fender apron seal
- A/C Compressor and PS drive belts
- Engine stabilizing rod
- Engine mounting stay
- Alternator bracket
- Crankshaft pulley

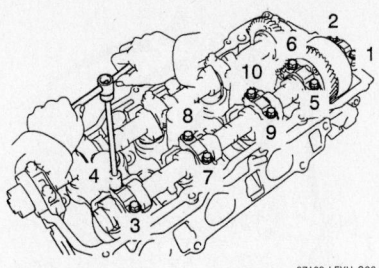

67162-LEXU-G26

Aligning timing marks—3.3L (3MZ-FE) Engine

- Timing belt covers
- Timing belt, tensioners, idlers and guides
- Camshaft timing pulleys

➡**Align the camshaft pulleys so that they can be returned to the original locations when reassembling.**

- Inner timing belt cover

✲✲ WARNING

Since the thrust clearance of the camshaft is small, the camshaft must be kept level while it is being removed. If the camshaft is not kept level, the portion of the cylinder head receiving the shaft thrust may crack or be damaged, causing the camshaft to seize or break.

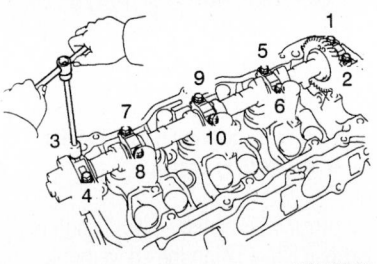

67162-LEXU-G27

Camshaft bolt removal sequence Camshaft No.1—3.3L (3MZ-FE) Engine

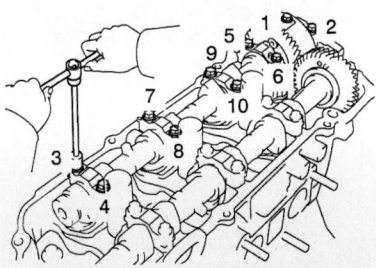

67162-LEXU-G28

Camshaft bolt removal sequence Camshaft No.2 —3.3L (3MZ-FE) Engine

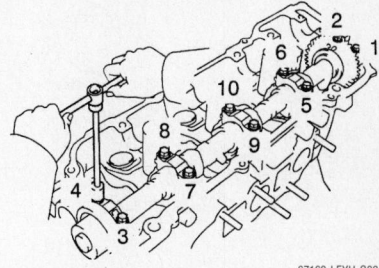

67162-LEXU-G30

Camshaft bolt removal sequence Camshaft No.4 —3.3L (3MZ-FE) Engine

3. Remove the camshaft using the following procedures

4. RH bank camshaft No.1 & LH bank camshaft No.2

a. Align the (2 dot marks) of the camshaft drive and driven gear by turning the camshaft with a wrench.

b. Secure the exhaust camshaft sub gear to the main gear with service bolt. Torque to 48 inch lbs. (5.4 Nm).

➡**When removing the camshaft, make certain that the torsional spring force of the sub gear has been eliminated by installation of the service bolt.**

c. Using several steps, loosen the 10 bearing cap bolts uniformly in the sequence shown in the illustration. Remove the 5 bearing caps and camshaft

✲✲ WARNING

Do Not pry out camshaft

✲✲ WARNING

Do Not damage contact surface of the cylinder head that receives the shaft thrust.

5. LH bank camshaft No. 3 & No. 4

a. Using several steps, loosen the 10 bearing cap bolts uniformly in the sequence shown in the illustration. Remove the 5 bearing caps and camshaft

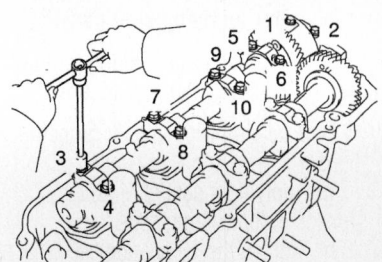

67162-LEXU-G37

Camshaft bolt removal sequence Camshaft No. 3—3.3L (3MZ-FE) Engine

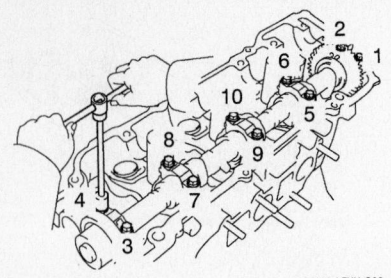

Camshaft bolt removal sequence Camshaft No. 4—3.3L (3MZ-FE) Engine

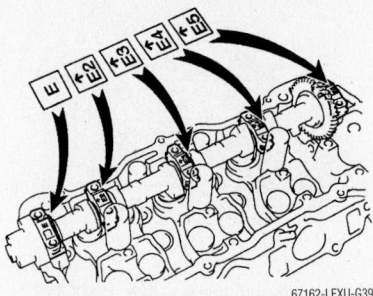

67162-LEXU-G39

Bolt torque procedure RH camshaft No. 2—3.3L (3MZ-FE) Engine

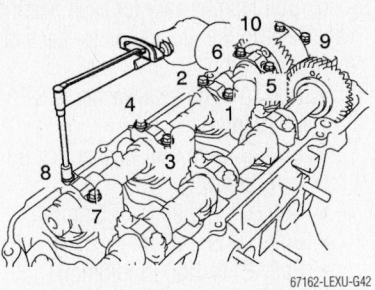

67162-LEXU-G42

Bolt torque procedure LH camshaft No. 4—3.3L (3MZ-FE) Engine

✱✱ WARNING

Do Not pry out camshaft

✱✱ WARNING

Do Not damage contact surface of the cylinder head that receives the shaft thrust.

  b. Remove the oil seal from camshaft
To install:
  6. Install RH No 2 Camshaft then RH Camshaft No. 1 using same procedure

✱✱ WARNING

Since the clearance of the camshaft is small, the camshaft must be kept level while bring installed. If the camshaft is not kept level, the cylinder head or camshaft may be damaged.

  a. Apply engine oil to the thrust portion and journal of camshaft
  b. No. 2 camshaft at 90 degree angel to the timing mark (2 dot marks) on the head.
  c. Multi-purpose grease to new oil seal
  d. Oil seal to camshaft

➡ **Do NOT turn over the oil seal lip**

➡ **Insert oil seal until it stops**

  e. Seal packing to bearing cap No. 1
➡ **Install bearing cap No. 1 within 5 minutes after applying seal packing**

➡ **Do NOT expose seal packing to engine oil within 2 hours after installation**

  f. 5 bearing caps in their proper locations
  g. Apply light coat of oil to the threads of the bearing caps
  h. Tighten the 10 bearing cap bolts in required sequence. Torque to: 12 ft. lbs. (16 Nm)

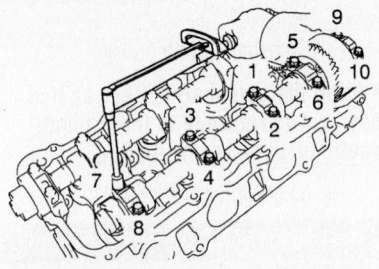

67162-LEXU-G40

Bolt torque procedure RH camshaft No. 1—3.3L (3MZ-FE) Engine

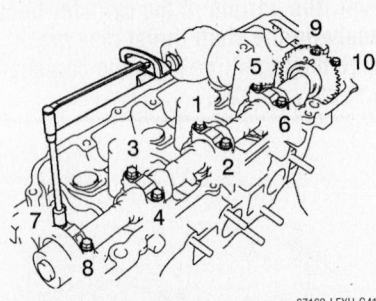

67162-LEXU-G41

Bolt torque procedure LH camshaft No. 3—3.3L (3MZ-FE) Engine

  7. Install LH camshaft No.3 and LH camshaft No. 4 using the above procedure
  8. Install or connect the following:
- Inner timing belt cover
- RH then LH camshaft timing pulleys. Torque to: 92 ft. lbs. (125 Nm)
- Timing belt idlers, tensioners, guide and belt
- RH engine mounting bracket
- Timing belt covers
- Crankshaft pulley
- Alternator bracket
- RH engine mounting stay
- Engine stabilizing rod
- Inspect or adjust valve lash
- A/C compressor and PS drive belts
- Cylinder head valve covers
- Ignition coil assembly
- Intake air surge tank

- Emission control valve set
- Air cleaner assembly
- Vacuum hoses
- V-bank cover sub-assembly
- Front suspension upper center brace
- RF wheel
- Engine coolant. Check for leaks
- System initialization

Valve Lash

ADJUSTMENT

3.0L (1MZ-FE) Engine

  1. Before servicing the vehicle, refer to the precautions in the beginning of this section.

➡ **Adjust the valve clearance when the engine is cold.**

  2. Remove or disconnect the following:
- Negative battery cable
- Accelerator/throttle cable from the throttle linkage
- Air intake chamber
- Cylinder head covers

  3. Turn the crankshaft pulley and align it's groove with the timing mark **0** of the No. 1 timing cover.
  4. Check that the valve lash adjusters on the No. 1 intake are loose and the

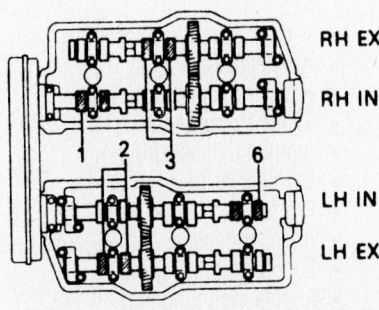

7923LG28

Adjust these valves FIRST—3.0L (1MZ-FE) engine

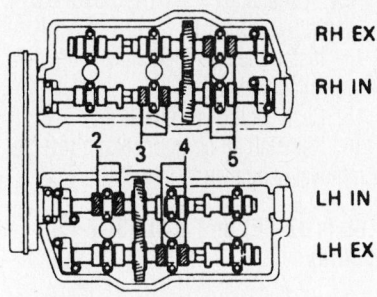

Adjust these valves SECOND—3.0L (1MZ-FE) engine

7923LG29

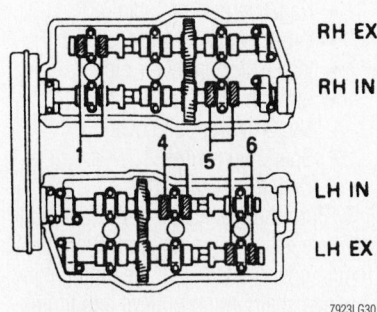

Adjust these valves THIRD—3.0L (1MZ-FE) engine

7923LG30

exhaust are tight. If not, turn the crankshaft on complete revolution (360 degrees).

5. Measure the clearance between the valve lash adjuster and the camshaft. Record the measurements on valves No. 1, 2, 3 and 6.

6. The intake valve clearance cold is 0.006–0.010 in. (0.15–0.25mm).

7. The exhaust valve clearance cold is 0.010–0.014 in. (0.25–0.35mm).

8. Turn the crankshaft ⅔ of a revolution (240 degrees) and check the clearance on valves No. 2, 3, 4 and 5 and record.

9. Turn the crankshaft another ⅔ of a revolution and check valves; No. 1, 4, 5 and 6 and record.

10. Remove or disconnect the following:
   • Adjusting shim and turn the crankshaft to position the cam lobe of the camshaft on the adjusting valve upward. Press down the valve lash adjuster with the proper tool and place the proper tool between the camshaft and the valve lash adjuster. Remove the tool.
   • Adjusting shim with the proper tool.

11. Determine the thickness of the replacement shim as follows:
   a. T: Thickness of the used shim
   b. A: Measured valve clearance
   c. N: Thickness of new shim

d. Intake: N = T + (A−0.006−0.010 in. (0.15–0.25mm))

e. Exhaust: N = T + (A−0.010−0.014 in. (0.25–0.35mm))

12. Install the specified valve shim on the valve lash adjuster

13. Recheck the valve clearance.

14. Install the cylinder head covers and intake chamber.

15. Connect the negative battery cable.

3.0L (2JZ-GE) Engine (Except 2004)

➡**Adjust the valve lash when the engine is cold.**

1. Before servicing the vehicle, refer to the precautions in the beginning of this section.

2. Remove or disconnect the following:
   • Negative battery cable
   • Engine cover
   • Intake and air cleaner assemblies
   • Throttle body and accelerator/throttle cable from the throttle linkage
   • Wiring and hoses required to access cylinder head covers (mark each one for reinstallation)
   • No. 3 timing belt cover
   • Ignition coils and wires
   • Spark plugs

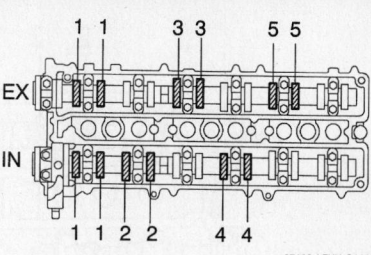

67162-LEXU-G111

Identifying the valves to be adjusted first—3.0L (2JZ-GE) Engine (2004 Model)

   • Cylinder head covers

3. Turn the crankshaft pulley and align its groove with the timing mark **0** of the No. 1 timing cover.

4. Check that the timing marks on the camshaft sprockets are in alignment with the marks on the No. 4 timing cover. If not, turn the crankshaft 1 complete revolution (360 degrees).

5. For this step, check only valves as indicated in appropriate illustration. Adjust these valves first.

6. Measure the clearance between the valve lash adjuster and the camshaft. Record the measurements.
   a. The intake valve lash cold is 0.006–0.010 in. (0.15–0.25mm).
   b. The exhaust valve lash cold is 0.010–0.014 in. (0.25–0.35mm).

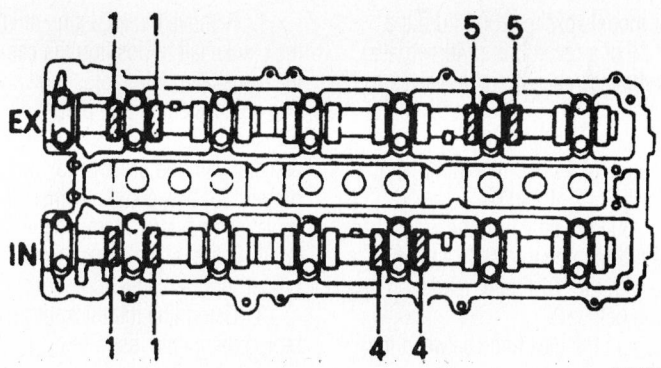

Adjust these valves FIRST—3.0L (2JZ-GE) Engine (except 2004)

7923LG32

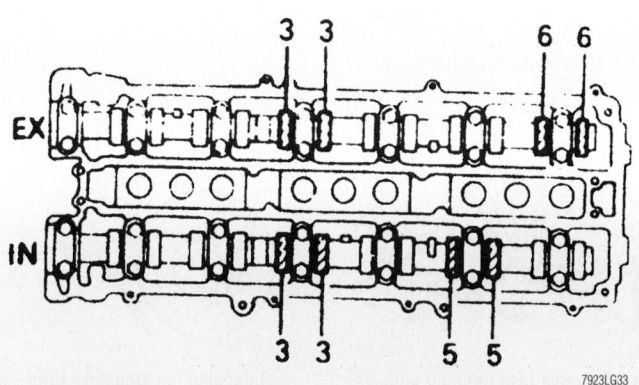

Adjust these valves SECOND—3.0L (2JZ-GE) engine (except 2004)

7923LG33

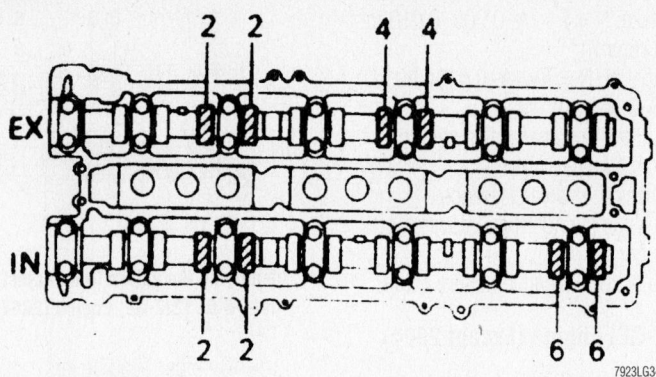

Adjust these valves THIRD—3.0L (2JZ-GE) engine (except 2004)

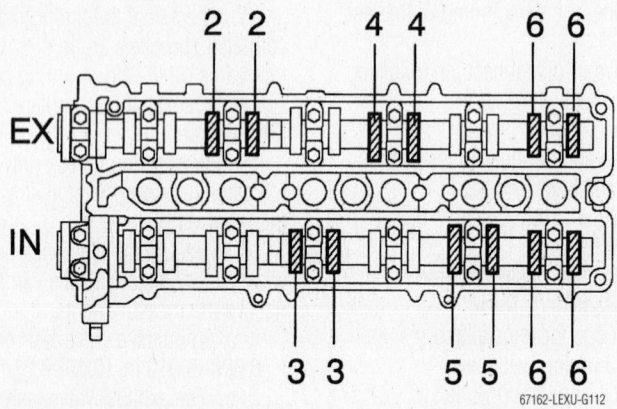

Identifying the valves to be adjusted second—3.0L (2JZ-GE) engine (2004)

7. For models except 2004, turn the crankshaft ⅔ of a revolution (240 degrees) and check the clearance on other valves as indicated.

8. On models except 2004, turn the crankshaft another ⅔ of a revolution and check valves: No. 2, 4 and 6.

9. For 2004 models, turn the crankshaft pulley one full revolution (360°) and align the groove with the "0" timing marks on the No. 1 timing belt cover.

10. Measure the clearance between the valve lash adjuster and the camshaft. Record the measurements.

   a. The intake valve lash cold is 0.006–0.010 in. (0.15–0.25mm).

   b. The exhaust valve lash cold is 0.010–0.014 in. (0.25–0.35mm).

11. Remove the adjusting shim and turn the crankshaft to position the cam lobe of the camshaft on the adjusting valve upward. The notches should be perpendicular to the camshaft. Press down the valve lash adjuster with the proper tool and place the proper tool between the camshaft and the valve lash adjuster. Remove the tool.

12. Remove the adjusting shim with the proper tool (a magnetic finger).

13. Determine the thickness of the replacement shim as follows:

   a. T = Thickness of the used shim

   b. A = Measured valve lash

   c. N = Thickness of new shim

   d. Intake: N = T + (A–0.006–0.010 in. (0.15–0.25mm))

   e. Exhaust: N = T + (A–0.010–0.014 in. (0.25–0.35mm))

14. Install the specified valve shim on the valve lash adjuster.

15. Recheck the valve lash.

16. Reinstall all components in reverse of removal sequence.

4.0L (1UZ-FE) and 4.3L (3UZ-FE) Engines

1. Before servicing the vehicle, refer to the precautions in the beginning of this section.

2. Remove or disconnect the following, as applicable:

   - Negative battery cable
   - V-bank cover
   - Intake air connector pipe
   - Ignition coils
   - No. 3 timing belt covers
   - Spark plug wires
   - Cylinder head covers

3. Turn the crankshaft pulley and align its groove with the timing mark **0** of the No. 1 timing cover. Check that the timing marks of the camshaft timing pulleys and timing belt rear plates are aligned. If not, turn the crankshaft 1 revolution (360 degrees) and align the mark.

4. Measure the clearance between the valve lash adjuster and the camshaft on the valves, as illustrated, in the first sequence. Record the measurements.

   a. The intake valve lash cold is 0.006–0.010 in. (0.15–0.25mm).

   b. The exhaust valve lash cold is 0.010–0.014 in. (0.25–0.35mm).

5. Turn the crankshaft 1 full revolution (360 degrees) and align the mark.

6. Measure the clearance between the

RH Bank

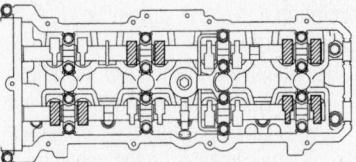

LH Bank

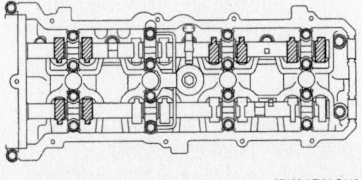

Adjust these valves FIRST—4.0L (1UZ-FE) and 4.3L (3UZ-FE) Engines

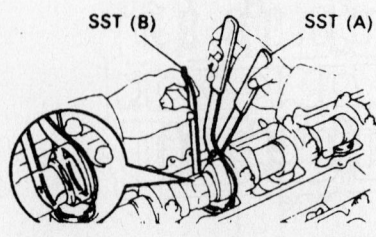

Press down the valve lash adjuster with a special tool—3.0L (2JZ-GE) engine

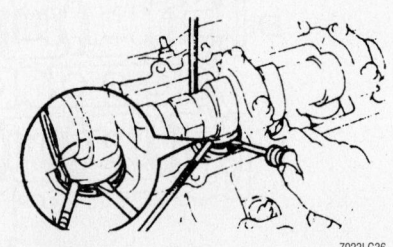

Removing the adjusting shim—3.0L (2JZ-GE) engine

RH Bank

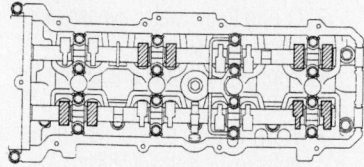

LH Bank

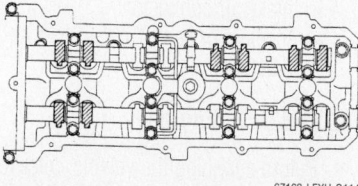

67162-LEXU-G114

Adjust these valves SECOND—4.0L (1UZ-FE) and 4.3L (3UZ-FE) Engines

valve lash adjuster and the camshaft on the valves, as illustrated, in the second sequence. Record these measurements.

7. If necessary, remove the camshafts.

8. Remove the adjusting shim and turn the crankshaft to position the cam lobe of the camshaft on the adjusting valve upward. Position the hole in the shim toward the outside of the cylinder head. Press down the valve lash adjuster with the proper tool and place the proper tool between the camshaft and the valve lash adjuster. Remove the tool.

9. Remove the adjusting shim with the proper tool.

10. Determine the thickness of the replacement shim as follows:

    a. T = Thickness of the used shim

    b. A = Measured valve lash

    c. N = Thickness of new shim

    d. Intake: N = T + (A−0.006−0.010 in. (0.15−0.25mm))

    e. Exhaust: N = T + (A−0.010−0.014 in. (0.25−0.35mm))

11. Recheck the valve lash. Install the cylinder head covers.

12. Connect the spark plug wires and install the No. 3 timing belt covers.

13. Install or reconnect all other components in reverse of removal procedure.

14. Connect the negative battery cable.

3.3 L (3MZ-FE) Engine

1. Before servicing the vehicle, refer to the precautions in the beginning of this section.

2. Remove or disconnect the following:

- Negative battery cable. Wait at least 90 seconds before performing any other work

- Front suspension upper brace center
- V-bank cover sub-assembly
- Air cleaner assembly
- Emission control valve set
- Intake air surge tank
- Ignition coil assembly
- Cylinder head valve covers

3. Turn the crankshaft pulley and align its groove with the timing mark **0** of the No. 1 timing cover. Check that the timing marks of the camshaft timing pulleys and timing belt rear plates are aligned. If not, turn the crankshaft 1 revolution (360 degrees) and align the mark.

4. Measure the clearance between the

RH Bank:

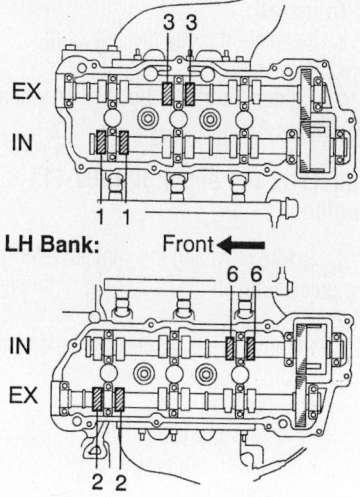

LH Bank: Front ◄───

67162-LEXU-G43

Measuring and adjust these valves first—3.3 L (3MZ-FE) Engine

RH Bank:

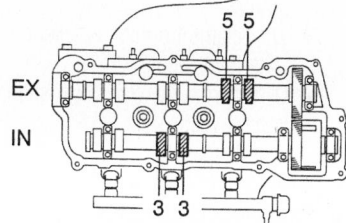

LH Bank: Front ◄───

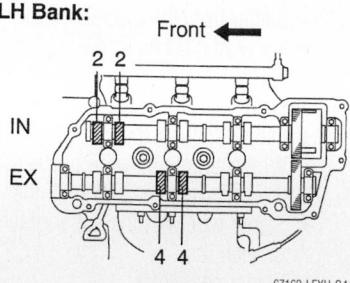

67162-LEXU-G44

Measuring and adjust these valves second—3.3 L (3MZ-FE) Engine

valve lash adjuster and the camshaft on the valves in the first sequence and record.

    a. The intake valve clearance cold is 0.006–0.010 in. (0.15–0.25mm).

    b. The exhaust valve clearance cold is 0.010–0.014 in. (0.25–0.35mm).

5. Turn the crankshaft ⅔ revolution and (240 degrees).

6. Measure the clearance between the valve lash adjuster and the camshaft on the valves in the second sequence and record.

    a. The intake valve clearance cold is 0.006–0.010 in. (0.15–0.25mm).

    b. The exhaust valve clearance cold is 0.010–0.014 in. (0.25–0.35mm).

7. Turn the crankshaft ⅔ revolution and (240 degrees).

8. Measure the clearance between the valve lash adjuster and the camshaft on the valves in the third sequence and record.

    a. The intake valve clearance cold is 0.006–0.010 in. (0.15–0.25mm).

    b. The exhaust valve clearance cold is 0.010–0.014 in. (0.25–0.35mm).

9. Adjust the valve lash using the following procedure

10. Remove the adjusting shim and turn the crankshaft to position the cam lobe of the camshaft on the adjusting valve upward. Position the hole in the shim toward the outside of the cylinder head. Press down the valve lash adjuster with the proper tool and place the proper tool between the camshaft and the valve lash adjuster. Remove the tool.

11. Remove the adjusting shim with the proper tool.

12. Determine the thickness of the replacement shim as follows:

RH Bank:

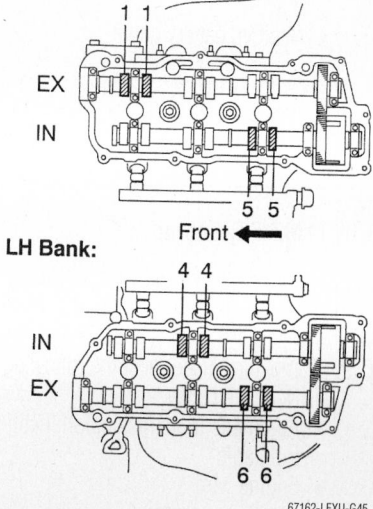

LH Bank: Front ◄───

67162-LEXU-G45

Measuring and adjust these valves third—3.3 L (3MZ-FE) Engine

Front of No. 1 and No. 2 Cylinders:

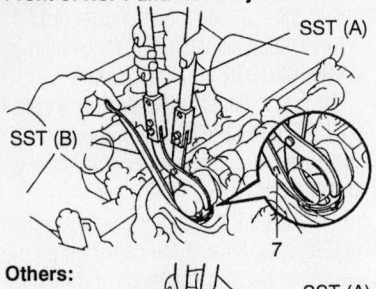

Others:

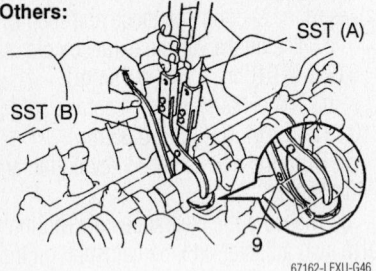

67162-LEXU-G46

Using special tool to remove the adjusting shim—3.3 L (3MZ-FE) Engine

a. T = Thickness of the used shim
b. A = Measured valve clearance
c. N = Thickness of new shim
d. Intake: N = T + (A–0.006–0.010 in. (0.15–0.25mm))
e. Exhaust: N = T + (A–0.010–0.014 in. (0.25–0.35mm))

➡**Place the adjusting shim on the valve lifter with the imprinted number facing down.**

13. Reinstall the following
  - Cylinder head valve covers
  - Ignition coil assembly
  - Intake air surge tank
  - Emission control valve set
  - Air cleaner assembly
  - Vacuum hoses
  - Front suspension upper center brace
  - Negitive battery cable

Starter Motor

REMOVAL & INSTALLATION

3.0L (1MZ-FE) Engine

1. Before servicing the vehicle, refer to the precautions in the beginning of this section.
2. Remove or disconnect the following:
  - Negative cable
  - Automatic transmission shift control cable
  - Engine wire
  - Starter connector
  - Nut, and disconnect the starter wire
  - 2 bolts, automatic transmission shift control cable clamp and starter

To install:
3. Installation is the reversal of the removal process.
4. Torque the starter bolts to 27 ft. lbs. (37 Nm).

3.0L (2JZ-GE) Engine

1. Before servicing the vehicle, refer to the precautions in the beginning of this section.
2. Remove or disconnect the following:
  - Negative cable
  - Starter connector
  - Nut, and disconnect the starter wire
  - 2 bolts and starter

To install:
3. Installation is the reversal of the removal procedure.
4. Tighten the starter bolts to 27 ft. lbs. (37 Nm).

4.0L (1UZ-FE) and 4.3L (3UZ-FE) Engines

1. Before servicing the vehicle, refer to the precautions in the beginning of this section.
2. Remove or disconnect the following:
  - Drain engine coolant
  - V-bank cover
  - Accelerator cable
  - Intake air connector
  - Throttle Body
  - Upper and lower intake manifold assembly
  - Rear water bypass joint
  - Water bypass pipe
  - Water bypass pipe from the water pump
  - Wire clamp from the bracket on the water bypass pipe
  - O-ring from the water bypass pipe
  - Water bypass pipe bracket from the water bypass pipe
  - 2 bolts holding the starter to the cylinder block
  - Starter connector
  - Starter from the cylinder block
  - Nut, and disconnect the starter wire
  - Starter

To install:
3. Install or connect the following:
  - Wire clamp to the wire bracket with the bolt. Tighten to 87 inch lbs.
  - Starter wire with the nut. Tighten to 87 inch lbs.
  - Starter connector
  - Starter with the 2 bolts. Torque the bolts to 29 ft. lbs. (39 Nm).
  - Water bypass pipe bracket to the water bypass pipe
  - O-ring to the water bypass pipe
  - Water bypass pipe
  - Wire clamp to the bracket on the water bypass pipe
  - Water bypass pipe bolts. Torque the bolts to 13 ft. lbs. (18 Nm).
  - Rear water bypass joint
  - Intake manifold assembly
  - Throttle body
  - Intake air connector
  - Accelerator cable
  - V-bank cover

3.3L (3MZ-FE) Engine

1. Before servicing the vehicle, refer to the precautions in the beginning of this section.
2. Remove or disconnect the following:
  - RH radiator side deflector
  - Air cleaner inlet assembly, air cleaner assembly and bracket
  - Battery and battery tray
  - Remove starter connector
  - Starter wire
  - Two bolts securing starter to engine
  - Starter assembly

To install:
3. Install or connect the following:
  - Starter assembly with 2 bolts; torque to 26 ft. lbs. (37 Nm)
  - Starter wires
  - Battery and battery tray
  - Air cleaner assembly, bracket and inlet
  - Radiator side deflector
4. System initialization

Oil Pan

REMOVAL & INSTALLATION

3.0L (1MZ-FE) Engine

1. Before servicing the vehicle, refer to the precautions in the beginning of this section.
2. Drain the engine oil.
3. Remove or disconnect the following:
  - Negative battery cable from the battery.
  - Right front wheel
  - Fender apron seal
  - Engine undercover
  - Front exhaust pipe
  - Front exhaust pipe bracket from the No. 1 oil pan
  - Flywheel housing undercover
  - 10 bolts and 2 nuts to the No. 2 oil pan

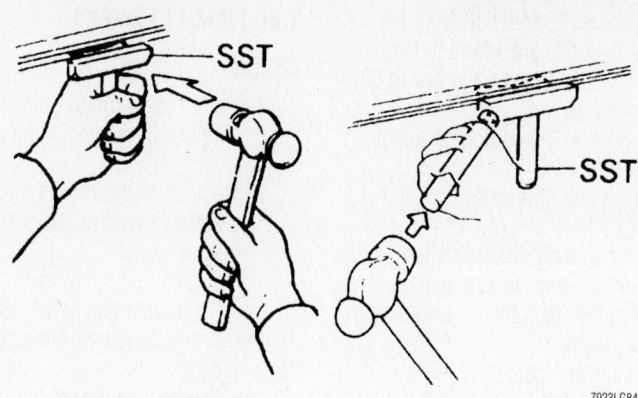

Use the special tool to break the seal and remove the oil pan—3.0L (1MZ-FE) engine

4. Insert a blade between the No. 1 and No. 2 oil pans. Tap the tool sideways to break the seal and remove the pan. Clean the surfaces of the oil pans.

- Oil strainer and gasket from the engine by removing the 3 nuts.
- No. 1 oil pan as follows. Make a note of the position of the each bolt. When replacing the bolts into the oil pan, place each bolt in the position from which it was removed.
- Baffle plate from the No. 1 oil pan

To install:

5. Clean all mating surfaces of the oil pans. Using a non-residue solvent, clean both sealing surfaces to the oil pan.

6. Install or connect the following:

- Baffle plate to the No. 1 oil pan and tighten to 69 inch lbs. (8 Nm)
- No. 1 oil pan. Apply RTV sealant to the oil pan and engine block.

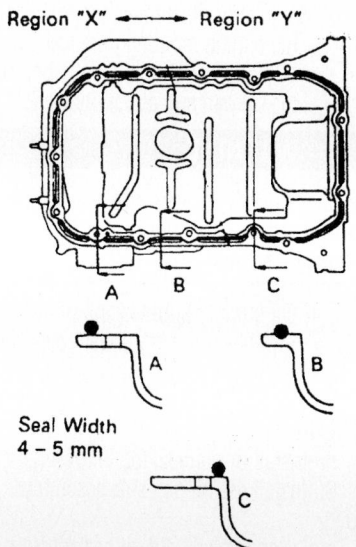

Apply sealant as shown to the No. 1 (upper) oil pan—3.0L (1MZ-FE) engine

Uniformly tighten the bolts and nuts in several passes to: 10mm: 69 inch lbs. (8 Nm); 12mm: 14 ft. lbs. (20 Nm); 14mm: 27 ft. lbs. (37 Nm)

- Flywheel housing undercover with the 2 bolts. Tighten the bolts to 69 inch lbs. (8 Nm).
- Oil strainer with the 3 nuts. Tighten the nuts to 69 inch lbs. (8 Nm).
- No. 2 oil pan. Apply RTV sealant to the oil pan and engine block. Uniformly tighten the bolts and nuts in several passes, to 69 inch lbs. (8 Nm).
- Flywheel housing undercover
- Front exhaust pipe bracket to the No. 1 oil pan. Tighten the bolts to 15 ft. lbs. (21 Nm).
- Front exhaust pipe. 4 pipe-to-pipe nuts: 46 ft. lbs. (62 Nm); front exhaust pipe-to-the center exhaust pipe bolts and nuts: 41 ft. lbs. (56 Nm); bracket bolts: 14 ft. lbs. (19 Nm); support stay bolts: 22 ft. lbs. (30 Nm).
- Engine undercover
- Right fender apron seal
- The right front wheel and lower the vehicle
- Engine with oil

3.0L (2JZ-GE) Engine

➡The No. 1 oil pan cannot be removed with the engine in the vehicle. The engine/transmission assembly must be removed. See ENGINE ASSEMBLY section. The manufacturer does not provide any on vehicle information for the No. 2 oil pan removal and installation. If only the No. 2 oil pan is being serviced, the engine/transmission assembly can remain in the vehicle.

1. Before servicing the vehicle, refer to the precautions in the beginning of this section.

2. Remove or disconnect the following:

- Engine/transmission assembly
- Timing belt
- Idler pulley
- Crankshaft timing pulley
- Oil dipstick and guide
- Oil sensor lead
- 4 attaching bolts and lift off the oil level sensor. Be careful not to drop this sensor.
- 14 bolts (16 bolts for GS 300) and 2 nuts and pry off the lower (No. 2) oil pan. Be careful not to damage the No. 1 pan while performing this procedure.
- Bolt and 2 nuts and drop down the oil strainer and gasket
- 5 bolts and 2 nuts and drop down the baffle plate
- 22 bolts and the carefully pry off the upper (No. 1) oil pan
- O-ring from the cylinder block

To install:

3. Install or connect the following:

- New O-ring in the block and scrape off any old sealant
- A ⅛ inch (3–4mm) bead of RTV sealant to the pan mating surface
- Upper pan. 12mm bolts: 15 ft. lbs. (21 Nm); 14mm bolts to 29 ft. lbs. (39 Nm)
- Baffle plate and oil strainer.

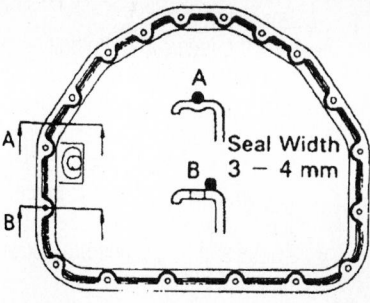

Lower oil pan sealant application—3.0L (2JZ-GE) engine

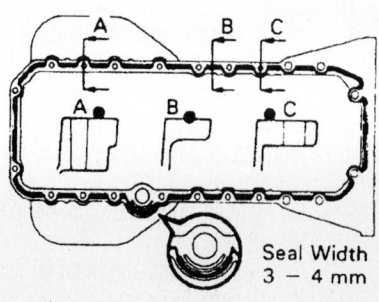

Upper oil pan sealant application—3.0L (2JZ-GE) engine

Tighten them both to 78 inch lbs. (9 Nm).

- Lower pan in the same manner as the upper pan and tighten the bolts to 78 inch lbs. (9 Nm)
- Oil level sensor and tighten it to 48 inch lbs. (5 Nm)
- Oil dipstick and guide
- Timing pulleys and belt
- Transmission to the engine
- Engine and transmission
- All fluids

4.0L (1UZ-FE) and 4.3L (3UZ-FE) Engines

LS 430

1. Before servicing the vehicle, refer to the precautions in the beginning of this section.
2. Remove or disconnect the following:
   - Engine/transmission assembly
   - Remove the timing belt
   - Idler pulleys
   - Crankshaft timing pulley
   - Oil dipstick and guide
   - Oil level sensor lead
   - 4 bolts and lift off the oil level sensor. Be careful not to drop this sensor.
   - Oil filter and the bracket assembly by removing the stud bolt and 2 nuts
   - Engine Crankshaft Position (CKP) sensor connector
   - Sensor by removing the bolt
   - 12 bolts and 2 nuts to the No. 2 oil pan. Use a gasket cutting tool to separate the No. 2 (lower) oil pan. Be careful not to damage the No. 1 pan while performing this procedure.
   - 2 bolts and 3 nuts and drop down the baffle plate
   - Oil strainer by removing the bolts and nuts
   - Bolts, then carefully pry off the No. 1 oil pan. There are slots for inserting the prybar.

To install:
3. Install or connect the following:
   - No. 1 pan. Apply a ⅛ inch (3–4mm) bead sealant to the pan mating surface. Bolts: 10mm: 66 inch lbs. (8 Nm); 12mm: 21 ft. lbs. (28 Nm)
   - Oil strainer. Bolts and nuts: 66 inch lbs. (8 Nm)
   - Baffle plate. Bolts and nuts: 66 inch lbs. (8 Nm)
   - No. 2 pan in the same manner as the No. 1 oil pan and tighten the

bolts to 66 inch lbs. (8 Nm). Be sure the bolts are 14mm in length.
- CKP sensor. Tighten the bolt to 56 inch lbs. (6 Nm).
- New O-ring in position on the oil filter bracket
- Bracket and tighten the bolt and nuts to 13 ft. lbs. (18 Nm)
- Wiring to the pressure switch
- Oil level sensor and tighten the 4 bolts to 48 inch lbs. (5 Nm). Use a new gasket.
- Dipstick and guide
- Timing belt pulleys and the timing belt components
- Transaxle to the engine
- Engine and transaxle
- All fluids

GS 430

➡The No. 1 oil pan cannot be removed with the engine in the vehicle. The engine and transmission must be removed as a unit, then separated. It may be possible to remove the No. 2 oil pan from the vehicle while the engine is still in the vehicle.

1. Before servicing the vehicle, refer to the precautions in the beginning of this section.
2. Remove or disconnect the following:
   - Engine/transmission assembly
   - Oil dipstick and guide
   - 12 bolts and 2 nuts. Use a gasket-cutting tool to separate the No. 2 (lower) oil pan. Be careful not to damage the No. 1 pan while performing this procedure.
   - 6 bolts and 2 nuts; remove the baffle plate
   - 16 bolts, then carefully pry off the No. 1 oil pan

➡There are slots for inserting the pry-bar.

To install:
3. Install or connect the following:
   - No. 1 pan. Apply a ⅛ inch (3–4mm) bead on RTV sealant to the pan mating surface. Bolts: 12mm: 69 inch lbs. (8mm); 14mm: 20 ft. lbs. (28 Nm)
   - Baffle plate. Torque bolts and nuts to 69 inch lbs. (8 Nm).
   - RTV sealant to the pan mating surface
   - No. 2 oil pan. Torque bolts to 69 inch lbs. (8 Nm)
   - Dipstick and guide
   - Engine/transaxle assembly
   - All fluids

3.3L (3MZ-FE) Engine

ES 330

1. Before servicing the vehicle, refer to the precautions in the beginning of this section.
2. Remove or disconnect the following:
   - Battery negative cable
   - RF wheel
   - Engine under covers
   - RH front fender apron seal
   - A/C compressor, alternator and PS drive belts
   - Engine stabilizer rod
   - Engine mounting stay No2.
   - Alternator bracket
   - Crankshaft pulley
   - Timing belt
   - Crankshaft timing pulley
   - Exhaust pipe and support brackets
   - Oil gauge guide
   - Alternator belt adjusting bar
   - A/C compressor/clutch assembly
   - A/C compressor mounting bracket
3. Separate FR engine mounting insulator

➡Do NOT remove the FR engine mounting at this time

   a. Remove bolt and disconnect the power steering return hose
   b. Remove the 4 nuts
   c. Place a wooden block underneath the engine
   d. Jack up the engine and remove the engine mounting insulator

✳✳ WARNING

Be careful not to damage the oil pan

4. Remove or disconnect the following:
   - RH engine mounting bracket
   - 10 bolts and 2 nuts, gently pry off the oil pan sub assembly No.2.

✳✳ WARNING

Be careful not to damage the oil pan flange area or the contact surface of the engine block.

- Oil strainer sub-assembly
- Flywheel housing under cover.
- Oil pan sub-assembly No. 1
- Engine oil level sensor

To install:
5. Install or connect the following:
6. Install the oil pan sub-assembly No. 1
   a. Remove any old oil sealant from contact surface. Clean the surface thoroughly.

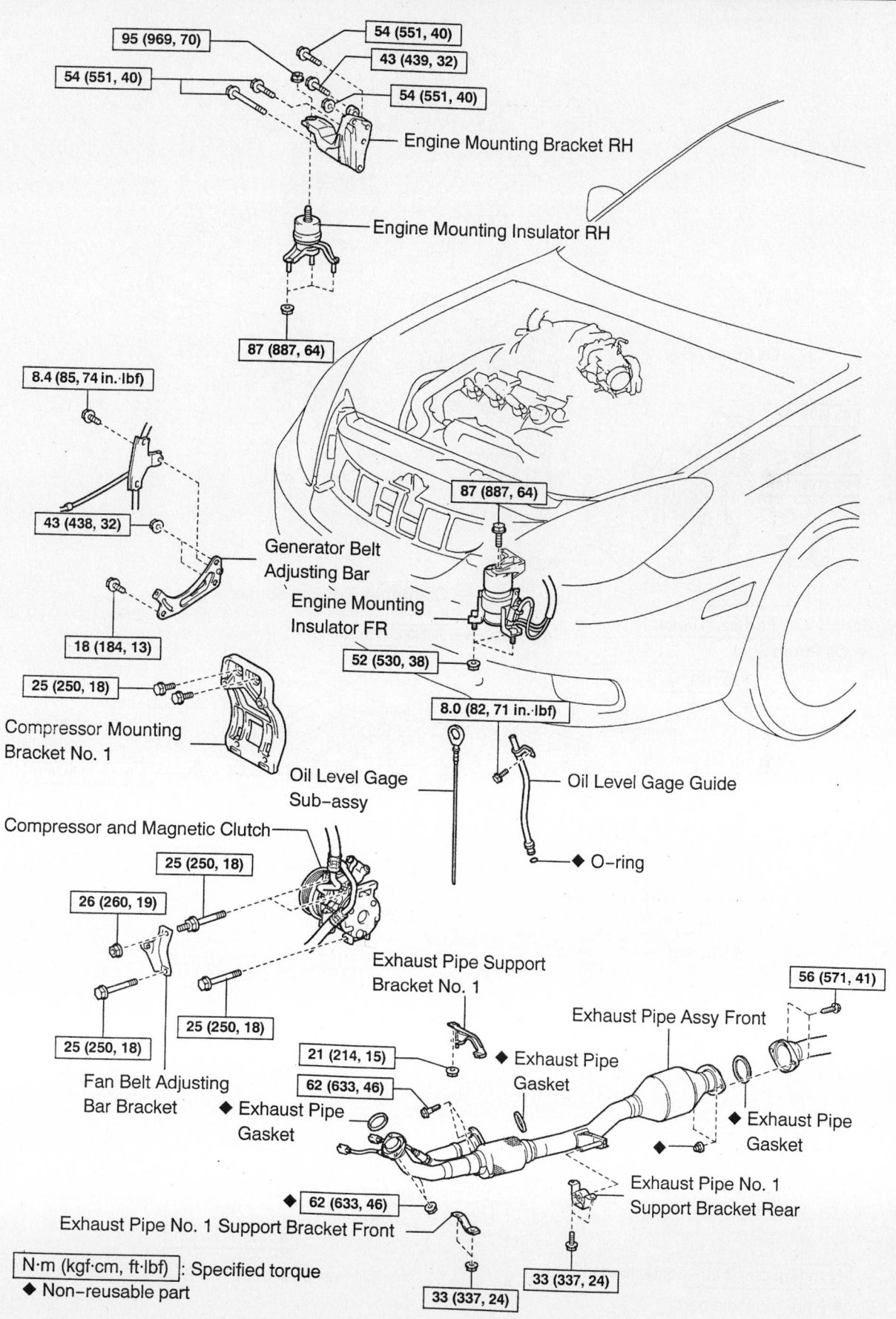

95 (969, 70)

54 (551, 40)

54 (551, 40)

43 (439, 32)

54 (551, 40)

Engine Mounting Bracket RH

Engine Mounting Insulator RH

87 (887, 64)

8.4 (85, 74 in.·lbf)

87 (887, 64)

43 (438, 32)

Generator Belt Adjusting Bar

Engine Mounting Insulator FR

18 (184, 13)

52 (530, 38)

25 (250, 18)

8.0 (82, 71 in.·lbf)

Compressor Mounting Bracket No. 1

Oil Level Gage Sub–assy

Oil Level Gage Guide

◆ O–ring

Compressor and Magnetic Clutch

25 (250, 18)

26 (260, 19)

Exhaust Pipe Support Bracket No. 1

56 (571, 41)

Exhaust Pipe Assy Front

21 (214, 15)

◆ Exhaust Pipe Gasket

25 (250, 18)

62 (633, 46)

◆ Exhaust Pipe Gasket

25 (250, 18)

Fan Belt Adjusting Bar Bracket

◆ Exhaust Pipe Gasket

◆ Exhaust Pipe Gasket

62 (633, 46)

Exhaust Pipe No. 1 Support Bracket Rear

Exhaust Pipe No. 1 Support Bracket Front

33 (337, 24)

| N·m (kgf·cm, ft·lbf) |: Specified torque

◆ Non–reusable part

33 (337, 24)

67162-LEXU-G47

Exploded view, component removal for oil pan—3.3L (3MZ-FE) Engine

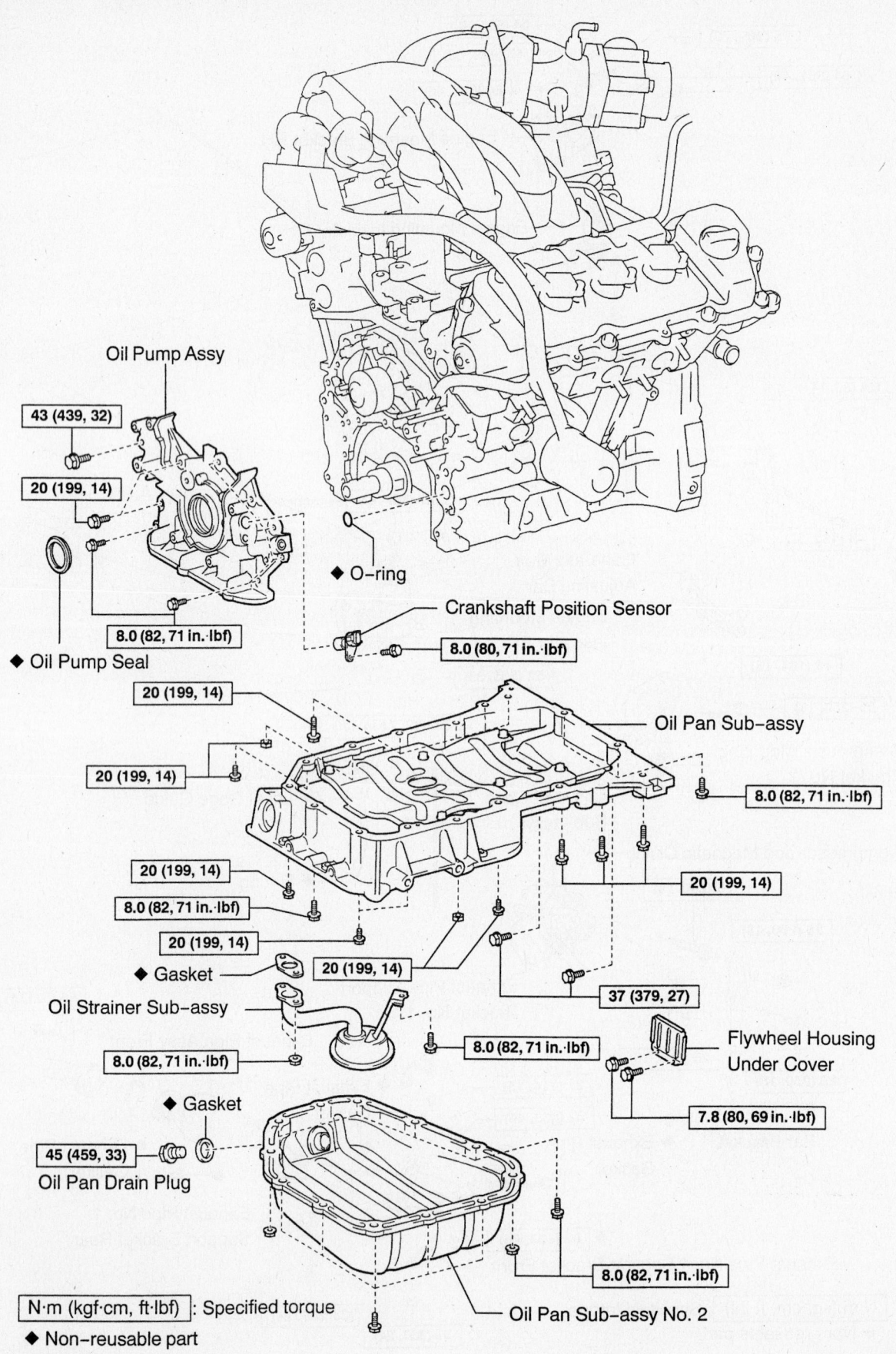

Oil Pump Assy

43 (439, 32)

20 (199, 14)

8.0 (82, 71 in.·lbf)

◆ Oil Pump Seal

◆ O-ring

Crankshaft Position Sensor

8.0 (80, 71 in.·lbf)

20 (199, 14)

20 (199, 14)

Oil Pan Sub-assy

8.0 (82, 71 in.·lbf)

20 (199, 14)

8.0 (82, 71 in.·lbf)

20 (199, 14)

20 (199, 14)

◆ Gasket

Oil Strainer Sub-assy

8.0 (82, 71 in.·lbf)

8.0 (82, 71 in.·lbf)

37 (379, 27)

Flywheel Housing Under Cover

7.8 (80, 69 in.·lbf)

◆ Gasket

45 (459, 33)

Oil Pan Drain Plug

Oil Pan Sub-assy No. 2

8.0 (82, 71 in.·lbf)

N·m (kgf·cm, ft·lbf) : Specified torque

◆ Non-reusable part

67162-LEXU-G48

Exploded view, removal of oil pan from engine—3.3L (3MZ-FE) Engine

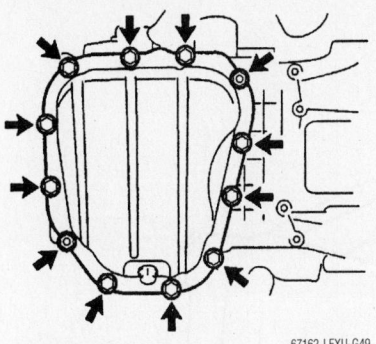

Oil pan sub-assembly No. 2—3.3L (3MZ-FE) Engine

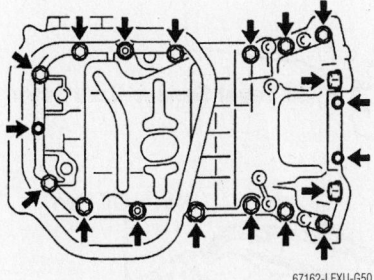

Oil pan sub-assembly No 1.—3.3L (3MZ-FE) Engine

  b. Apply a continuous bead of sealant 0.12 to 0.16 in (3 to 4 mm) around the block surface, making certain to surround the bolt holes.

  c. Install the oil pan within 3 minutes after applying the sealant

➡ **Do NOT expose sealant to engine oil within 2 hours after installing**

  d. Install the oil pan using the 17 bolts and 2 nuts. Tighten uniformly in several steps.
Torque to:
  • 10 mm head 71 in. lbs (8.0 Nm)
  • 12 mm head 14 ft. lbs.(20 Nm)
  • 14 mm head 27 ft. lbs (37 Nm)
  a. Engine oil level sensor
 7. Oil strainer assembly
 8. Oil pan sub assembly No. 2
  a. Remove any old oil sealant from contact surface.
   Clean the surface thoroughly.
  b. Apply a continuous bead of sealant 0.16 to 0.20 in. (3 to 4 mm) around the block surface, making certain to surround the bolt holes.
  c. Install the oil pan within 3 minutes after applying the sealant

➡ **Do NOT expose sealant to engine oil within 2 hours after installing**

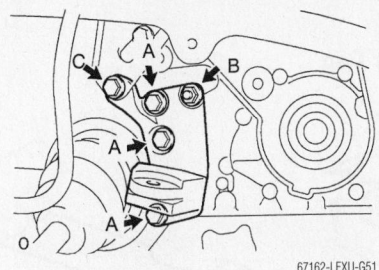

Engine mounting bracket tightening procedure—3.3L (3MZ-FE) Engine

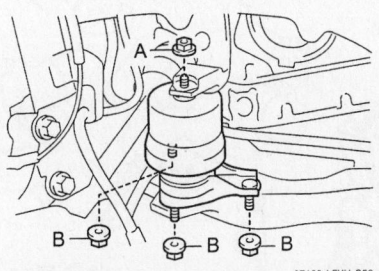

Engine mounting insulator tightening procedure—3.3L (3MZ-FE) Engine

  d. Install the oil pan using the 10 bolts and 2 nuts. Torque to 71 inch lbs. (8.0 Nm)
 9. Install or connect the following
  • RH engine mounting bracket; torque bolts "A" & "B" to 40 ft. lbs (54 Nm) and bolt "C" to 32 ft. lbs (43 Nm)
  • RH engine mounting insulator; torque nut "A" to 70 ft. lbs. (95 Nm) and nut "B" to 64 ft. lbs (87 Nm)
  • FR engine mounting insulator; torque bolt to 64 ft. lbs. (87 Nm) and nut to 38 ft. lbs (52 Nm)
  • A/C compressor mounting bracket; torque to 18 ft. lbs (25 Nm)
  • A/C compressor/clutch assembly; torque to 18 ft. lbs. (25 Nm)
  • Alternator belt adjusting bar; torque bolt to 18 ft. lbs (25 Nm) and nut to 19 ft. lbs. (26 Nm)
  • Oil level gage guide
  • Exhaust pipes and support brackets
  • Crankshaft timing pulley
  • Timing belt assembly
  • Crankshaft pulley
  • Alternator bracket
  • Engine mounting stay
  • Engine stabilizer rod
  • Alternator, PS pump and A/C drive belts
  • RH fender apron seal
  • RF wheel

  • Engine under covers
  • Negative battery cable
 10. System initialization

Oil Pump

REMOVAL & INSTALLATION

3.0L (1MZ-FE) Engine

 1. Before servicing the vehicle, refer to the precautions in the beginning of this section.
 2. Remove or disconnect the following:

  • Negative battery cable from the battery
  • Right front wheel
  • Fender apron seal
  • Engine undercover
  • Engine oil
  • Front exhaust pipe
  • Front exhaust pipe bracket from the No. 1 oil pan
  • Alternator drive belt from the engine
  • Air conditioning compressor from the engine, without disconnecting the compressor lines
  • Power steering pump drive belt and adjusting strut
  • Timing belt from the engine
  • Timing belt pulleys
  • Rear timing belt cover from the engine by removing the wire clamps and 6 bolts
  • Air conditioning compressor hous-

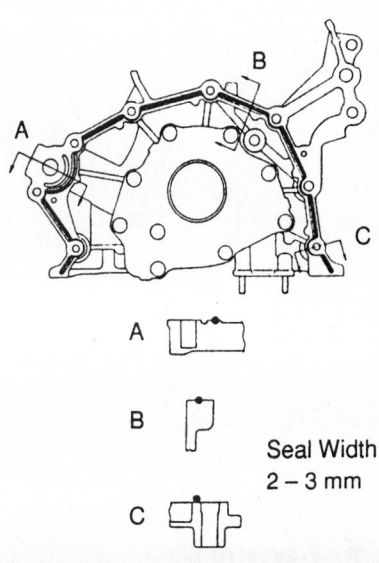

Apply sealant to the mounting surface of the oil pump in the areas shown—3.0L (1MZ-FE) engine

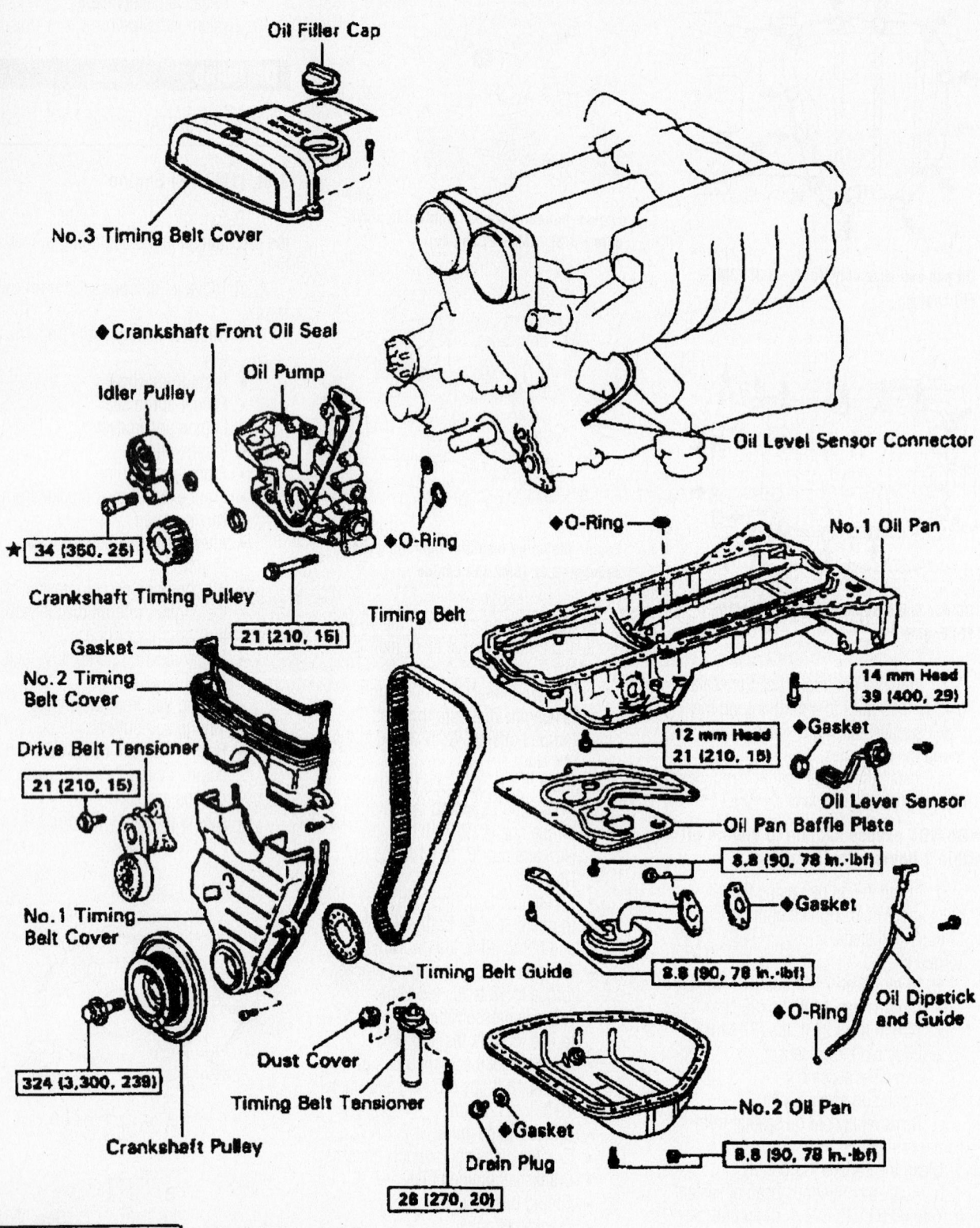

Oil Filler Cap

No.3 Timing Belt Cover

◆Crankshaft Front Oil Seal

Oil Pump

Idler Pulley

Oil Level Sensor Connector

★ 34 (350, 25)

Crankshaft Timing Pulley

◆O-Ring

◆O-Ring

No.1 Oil Pan

Gasket

21 (210, 15)

Timing Belt

14 mm Head 39 (400, 29)

No.2 Timing Belt Cover

12 mm Head 21 (210, 15)

◆Gasket

Drive Belt Tensioner

21 (210, 15)

Oil Lever Sensor

Oil Pan Baffle Plate

8.8 (90, 78 in.·lbf)

◆Gasket

No.1 Timing Belt Cover

Timing Belt Guide

8.8 (90, 78 in.·lbf)

◆O-Ring

Oil Dipstick and Guide

324 (3,300, 239)

Dust Cover

Timing Belt Tensioner

No.2 Oil Pan

Crankshaft Pulley

◆Gasket

Drain Plug

8.8 (90, 78 in.·lbf)

28 (270, 20)

N·m (kgf·cm, ft·lbf) : Specified torque

◆ Non-reusable part

★ Precoated part

Exploded view of the oil pump and related component mountings—3.0L (2JZ-GE) engine

7923LG90

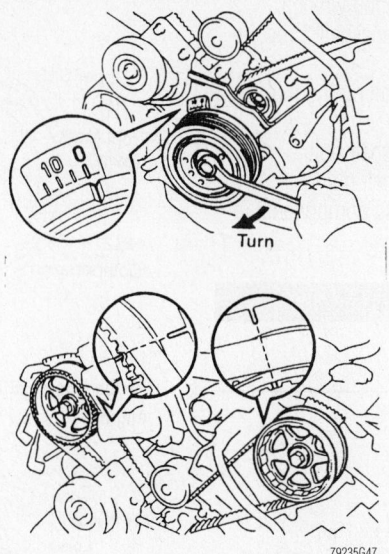

Timing belt sprocket mark alignment for belt installation—Lexus 4.0L (1UZ-FE) engine

a. Using the proper tool, slightly turn the right side timing pulley clockwise. Align the installation mark of the timing belt with the timing mark of the camshaft timing pulley and hang the timing belt on the right side camshaft pulley.

b. Using the proper tool, align the timing marks of the right side camshaft pulley and the timing belt rear plate.

c. Check that the timing belt has tension between the crankshaft timing pulley and the right side camshaft pulley.

14. The timing belt tensioner must be set prior to installation. The tensioner can be set as follows:

a. Place a plate washer between the tensioner and a block. Using a suitable press, press in the pushrod using 220–2205 lbs. (100–1000kg) of pressure.

b. Align the holes of the pushrod and housing, pass the proper tool (0.05 in. Allen wrench) through the holes to keep the setting position of the pushrod.

c. Release the press and install the dust boot on the tensioner.

15. Install the tensioner and tighten the bolts to 20 ft. lbs. (26 Nm). Remove the tool from the tensioner.

16. Turn the crankshaft pulley two complete revolutions from TDC-to-TDC. Always turn the crankshaft clockwise. Check that each pulley aligns with the timing marks.

17. Install all remaining components in the reverse order of removal.

3.3L (3MZ-FE) Engine

1. Remove or disconnect the following:
   • RH front fender apron seal
   • A/C drive belt
   • Engine stabilizing rod
   • Engine mounting stay
   • Alternator bracket
   • Crankshaft pulley
   • Upper and lower timing belt covers
   • Engine mounting bracket (If necessary)

2. Remove the timing belt
   • Set the No 1 cylinder to TDC/compression
   • Temporarily install the crankshaft pulley bolt
   • Turn the crankshaft clockwise, the align the timing mark on the crankshaft timing pulley with the mark on the oil pump body
   • Verify that the timing marks on the camshaft timing pulleys align with the timing marks on the inside timing belt cover. If not, turn the crankshaft on revolution (360 degrees).
   • If reusing the timing belt make certain there are 3 location marks on the timing belt corresponding with the marks on the timing gears.
   • Set the No 1 cylinder to approximately 60 degrees BTDC/compression. Turn the crankshaft counterclockwise by approximately 60 degrees.

✳✳ WARNING

If the timing belt is disengaged, having the crankshaft pulley set at the wrong angle can cause contact of the piston with the valves, causing damage to the valves. Always set the crankshaft pulley at the correct angle.

   • Remove timing belt tensioner

➡**Do NOT reinstall the timing belt tensioner with the plunger extended.**

3. Remove the timing belt in the following order.
   • No 1 idler pulley
   • RH camshaft timing pulley
   • No. 2 idler pulley
   • LH camshaft timing pulley
   • Water pump pulley
   • Crankshaft timing pulley

4. Inspect the timing belt

➡**do not reuse the timing belt if there is evidence of fraying, oil contamination, cracking, tooth damage or wear, or belt distortion. If there is any doubt in the belt condition, replace the belt.**

To install:

5. After inspecting the pulleys for wear and checking for oil leaks install the timing belt using the following procedure.

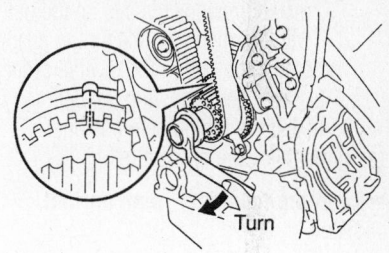

Aligning crankshaft pulley—3.3L (3MZ-FE) engine

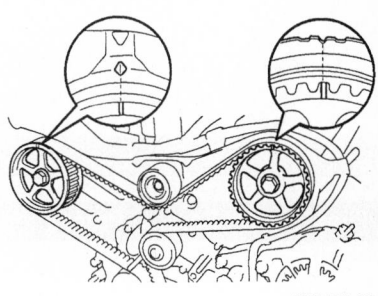

Aligning camshaft pulleys—3.3L (3MZ-FE) engine

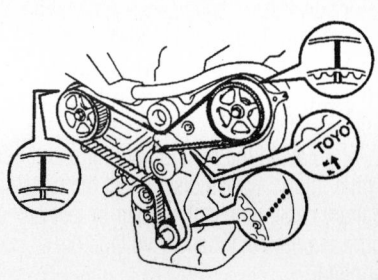

Marking the timing belt for reuse—3.3L (3MZ-FE) engine

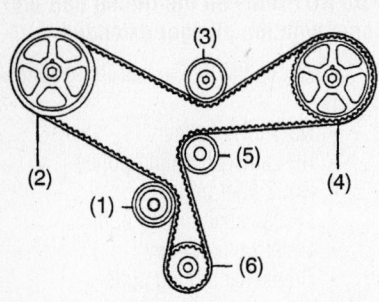

67162-LEXUS-G58

Removal of timing belt from pulleys—3.3L (3MZ-FE) engine

- Temporarily install the crankshaft pulley bolt
- Make sure the crankshaft pulley is set at 60° BTDC
- Align the timing marks on the camshaft pulley with the respective marks on the inside timing cover.
- If reusing the old belt, align the marks previously made on the belt with the marks on the timing gears.
- Install the belt in the reverse order used when removing the belt

6. Install the timing belt tensioner using the following procedure

➡ **Keep the tensioner in an upright position**

- Slowly depress the push rod and align the hole with the hole in the housing
- Insert a 1.5 mm hexagon wrench through the holes to maintain the setting position of the push rod.
- Install the tensioner with the 2 bolts and torque to 20 ft. lbs (27 Nm).

- Remove the hexagon wrench
7. Slowly turn the crankshaft 2 revolutions clockwise.
8. Check to see that all timing marks are in alignment.
9. Remove the crankshaft bolt
10. Install the timing belt guide
11. Install all remaining components in the reverse order of removal.

Piston and Ring

POSITIONING

RH Piston

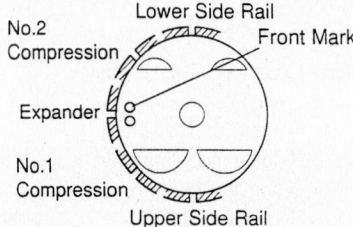

LH Piston

9307LG06

Piston ring positioning—3.0L (1MZE-FE) engine

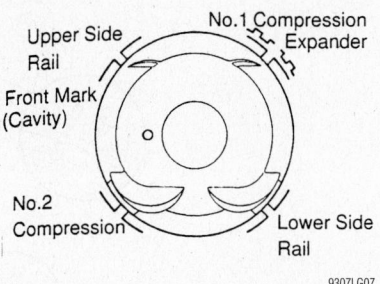

9307LG07

Piston ring positioning—3.0L (2JZ-GE) engine

RH Piston

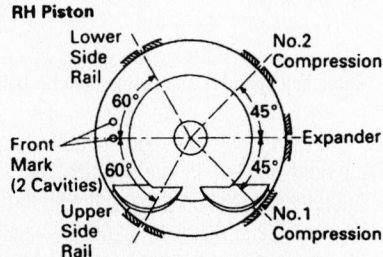

LH Piston

7923AG45

Piston ring positioning—4.0L (1UZ-FE) and 4.3L (3UZ-FE) engines

FUEL SYSTEM

Fuel System Service Precautions

Safety is the most important factor when performing not only fuel system maintenance but any type of maintenance. Failure to conduct maintenance and repairs in a safe manner may result in serious personal injury or death. Maintenance and testing of the vehicle's fuel system components can be accomplished safely and effectively by adhering to the following rules and guidelines.

- To avoid the possibility of fire and personal injury, always disconnect the negative battery cable unless the repair or test procedure requires that battery voltage be applied.
- Always relieve the fuel system pressure prior to disconnecting any fuel system

component (injector, fuel rail, pressure regulator, etc.), fitting or fuel line connection. Exercise extreme caution whenever relieving fuel system pressure, to avoid exposing skin, face and eyes to fuel spray. Please be advised that fuel under pressure may penetrate the skin or any part of the body that it contacts.

- Always place a shop towel or cloth around the fitting or connection prior to loosening to absorb any excess fuel due to spillage. Ensure that all fuel spillage (should it occur) is quickly removed from engine surfaces. Ensure that all fuel soaked cloths or towels are deposited into a suitable waste container.
- Always keep a dry chemical (Class B) fire extinguisher near the work area.
- Do NOT allow fuel spray or fuel vapors

to come into contact with a spark or open flame.

- Always use a back-up wrench when loosening and tightening fuel line connection fittings. This will prevent unnecessary stress and torsion to fuel line piping.
- Always replace worn fuel fitting O-rings with new. Do NOT substitute fuel hose or equivalent, where a fuel pipe is installed.

Fuel System Pressure

RELIEVING

1. Before servicing the vehicle, refer to the precautions in the beginning of this section.
2. Remove the fuse for the electronic fuel pump.

ing bracket by removing the 3 bolts.
- 10 bolts and 2 nuts to the No. 2 oil pan
- No. 2 oil pan from the engine
- Oil strainer and gasket from the engine by removing the 3 nuts
- No. 1 oil pan
- Baffle plate from the No. 1 oil pan
- Crankshaft Position (CKP) sensor by removing the connector and bolt
- Oil pump. Make a note of the position of the each bolt. When replacing the bolts into the oil pump body, place each bolt in the position from which it was removed.
- O-ring from the cylinder block
- Plug, gasket, spring and relief valve from the oil pump body

To install:
3. Install or connect the following:
- Driven rotors, drive, pump body cover, then install the 9 screws
- Relief valve, spring, gasket and the plug to the oil pump body
- New O-ring on the cylinder block
- RTV sealant to the oil pump as shown
- Pump on the engine block. Be sure to engage the spline teeth of the oil pump drive gear with the large teeth of the crankshaft.
- The 9 bolts to the oil pump and uniformly tighten the bolts in several passes. Tighten the bolts to: 10mm: 69 inch lbs. (8 Nm); 12mm: 14 ft. lbs. (20 Nm).
- CKP sensor and bolt. Tighten the bolt to 69 inch lbs. (8 Nm).
- Baffle plate to the No. 1 oil pan and tighten to 69 inch lbs. (8 Nm).
- No. 1 oil pan Uniformly tighten the bolts and nuts in several passes: 10mm–69 inch lbs. (8 Nm); 12mm–14 ft. lbs. (20 Nm); 14mm–27 ft. lbs. (37 Nm)
- Flywheel housing undercover with the 2 bolts. Tighten the bolts to 69 inch lbs. (8 Nm).
- Oil strainer with the 3 nuts. Tighten the nuts to 69 inch lbs. (8 Nm).
- No. 2 oil pan
- RTV sealant to the oil pan and engine block
- No. 2 oil pan. Uniformly tighten the bolts and nuts in several passes to 69 inch lbs. (8 Nm).
- Remaining components
- Right front wheel and lower the vehicle
- Engine with oil
- Negative battery cable to the battery

3.0L (2JZ-GE) Engine

1. Before servicing the vehicle, refer to the precautions in the beginning of this section.
2. Remove or disconnect the following:
- Engine and transmission
- Timing belt
- Idler pulley
- Crankshaft timing pulley
- Oil dipstick and tube
- Oil level sensor
- No. 2 (lower) oil pan
- Oil strainer by removing the bolt and 2 nuts
- Oil baffle plate by removing the 6 bolts
- No. 1 (upper) oil pan by removing the 22 bolts. Take note of bolt size and placement for correct re-installation.
- 9 mounting bolts to the oil pump body. Carefully drive the pump off the cylinder block using a brass drift.
- 2 O-rings

To install:
3. Install or connect the following:
- 2 new O-rings in the cylinder block
- A 1/8 inch (3–4mm) bead of RTV sealant around the pump mating surface, taking great care around the oil passages
- Pump and tighten the bolts to 15 ft. lbs. (21 Nm)
- A new O-ring on the block
- RTV sealant around the No. 1 oil pan
- No. 1 oil pan. Bolts: 12mm–15 ft. lbs. (21 Nm); 14mm–29 ft. lbs. (39 Nm)
- Oil baffle plate and tighten the nuts and bolts to 78 inch lbs. (9 Nm)
- Oil strainer and tighten the nuts and bolts to 78 inch lbs. (9 Nm)
- RTV sealant around the No. 2 oil pan

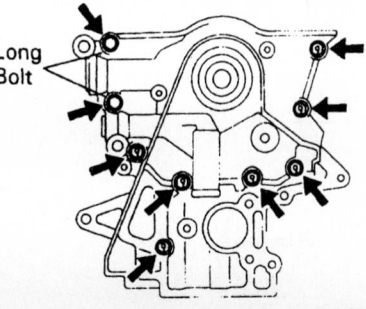

Long Bolt

7923LG91

Oil pump mounting bolt installation locations—3.0L (2JZ-GE) engine

- No. 2 oil pan and tighten the bolts to 78 inch lbs. (9 Nm)
- Oil lever sensor with a new gasket and tighten the bolts to 48 inch lbs. (6 Nm)
- Oil dipstick with a new O-ring
- Remaining components
- All fluids
- Negative battery cable

4.0L (1UZ-FE) and 4.3L (3UZ-FE) Engines

➡The oil pump cannot be removed with the engine in the vehicle. The engine and transmission must be removed as a unit, then separated.

1. Before servicing the vehicle, refer to the precautions in the beginning of this section.
2. Remove or disconnect the following:
- Engine/transmission assembly
- Timing belt
- Idler pulleys
- Crankshaft timing pulley
- Oil dipstick and guide
- Oil level sensor lead
- 4 bolts and lift off the oil level sensor. Be careful not to drop this sensor.
- Main Oxygen (O$_2$) sensor bracket, if necessary
- Oil filter and filter bracket assembly by removing the stud bolt and 2 nuts
- Engine Crankshaft Position (CKP) sensor. Remove the sensor by removing the bolt.
- 12 bolts and 2 nuts from the No. 2 oil pan
- No. 2 (lower) oil pan. Use a gasket-cutting tool
- 2 bolts and 3 nuts and drop down the baffle plate
- Oil strainer
- No. 1 oil pan. There are slots for inserting the prybar.
- 8 bolts holding the oil pump to the engine

➡Make certain to observe bolt position during removal. The bolts are different lengths and sizes. Record their position for proper reassembly.

- Oil pump from the engine block
- O-ring from the block

To install:

➡Prior to installing the oil pump, lubricate the gears with clean engine oil.

3. Install or connect the following:

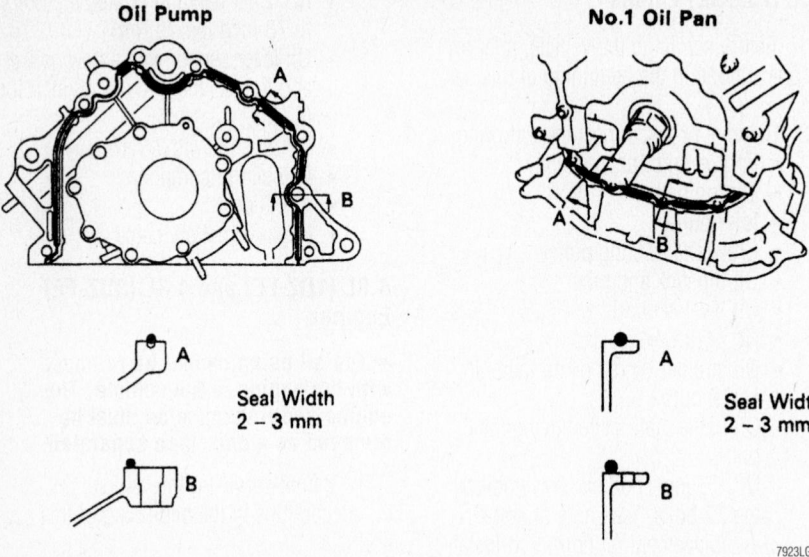

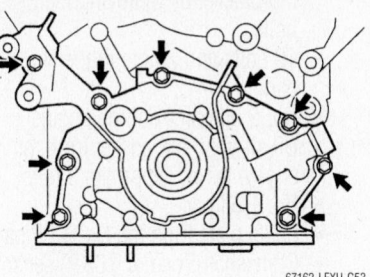

Oil pump removal bolt procedure—3.3L (3MZ-FE) Engine

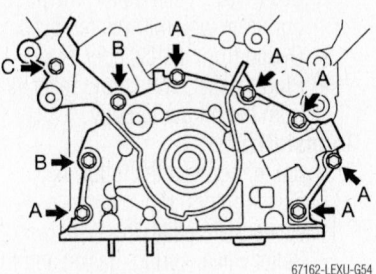

Oil pump bolt installation procedure—3.3L (3MZ-FE) Engine

Apply sealant to the oil pump and the No. 1 oil pan, as shown, before installing the oil pump—4.0L (1UZ-FE) and 4.3L (3UZ-FE) Engines

- A 2–3mm wide (0.08–0.12 in.) bead of RTV sealant to the oil pump
- New O-ring in position on the block
- Oil pump on the engine
- The 8 bolts in their correct locations. Tighten the bolts with 12mm heads to 12 ft. lbs. (16 Nm) and the bolts with 14mm heads to 22 ft. lbs. (30 Nm).
- A ⅛ inch (3–4mm) bead of RTV sealant to the pan mating surface.
- No. 1 pan. Bolts–10mm: 66 inch lbs. (8 Nm); 12mm: 21 ft. lbs. (28 Nm)
- Oil strainer and tighten the bolts to 66 inch lbs. (8 Nm)
- Baffle plate and tighten the bolts and nuts to 66 inch lbs. (8 Nm)
- Remaining components
- Engine/transaxle

3.3L (3MZ-FE) Engine

1. Follow engine oil pan removal procedure then perform the following steps.
2. Remove or disconnect the following:
   - Crankshaft position sensor
   - Oil pump assembly
   - 9 bolts
3. Remove oil pump by prying between the oil pump and bearing cap
4. Remove the oil ring

To install:

5. Install the oil pump seal using proper driver.
   a. Tap in the seal until it is flush with the oil pump body
   b. Apply multi-purpose grease to the seal lip.
6. Install oil pump assembly
   a. Remove any old sealant from the mating surfaces
   b. Apply a light coat of clean engine oil to the O-ring, then place it on the engine block.
   c. Thoroughly clean the mating surface of any oil or old sealant
   d. Apply a continuous bead of sealant on the oil pump body, making certain to surround the bolt holes.
   e. Install the oil pump within 3 minutes after applying the sealant.

➡ **Do NOT expose the sealant to engine oil within 2 hours after installing**

   f. Align the key of the oil pump drive gear with the keyway located on the crankshaft, then slide the oil pump into place.
   g. Install the oil pump with the 9 bolts. Tighten the bolts uniformly in several steps. Torque to: Bolt A 71 in. lbs (8.0 Nm), Bolt B14 ft lbs. (20 Nm), Bolt C 32 ft. lbs. (43 Nm)
7. Install crankshaft position sensor
8. Install oil pans using oil pan installation procedure

Timing Belt

REMOVAL & INSTALLATION

3.0L (1MZ-FE) Engine

1. Before servicing the vehicle, refer to the precautions in the beginning of this section.

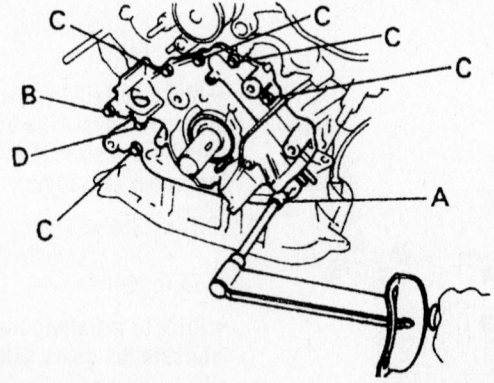

4.0L engine oil pump mounting bolt locations, according to bolt lengths—(A) 1.97 in. (50mm), (B) 4.17 in. (106mm), (C) 1.18 in. (30mm) and (D) 1.57 in. (40mm)

2. Remove all necessary components for access to the upper timing belt cover. Remove the 8 bolts and lift off the upper (No. 2) cover.

3. Paint matchmarks on the timing belt at all points where it meshes with the pulleys and the lower timing cover.

4. Set the No. 1 cylinder to Top Dead Center (TDC) of the compression stroke and check that the timing marks on the camshaft timing pulleys are aligned with those on the No. 3 timing cover. If not, turn the engine 1 complete revolution (360 degrees) and check again.

5. Remove or disconnect the following:
  • Timing belt tensioner and the dust boot

6. Turn the right camshaft pulley clockwise slightly to release tension, then remove the timing belt from the pulleys.
  • Upper (No. 3) and lower (No. 1) timing belt covers
  • Timing belt guide
  • Timing belt from the engine

➡ **If the timing belt is to be reused, draw a directional arrow on the timing belt in the direction of engine rotation (clockwise) and place matchmarks on the timing belt and crankshaft gear to match the drilled mark on the pulley.**

To install:

➡ **If the old timing belt is being reinstalled, be sure the directional arrow is facing in the original direction and that the belt and crankshaft gear matchmarks are properly aligned.**

7. Install the lower (No. 1) timing cover and tighten the bolts.

8. Set the No. 1 cylinder to TDC again. Turn the right camshaft until the knock pin hole is aligned with the timing mark on the No. 3 belt cover. Turn the left pulley until the marks on the pulley are aligned with the mark on the No. 3 timing cover.

9. Check that the mark on the belt matches with the edge of the lower cover. If not, shift it on the crank pulley until it does. Turn the left pulley clockwise a bit and align the mark on the timing belt with the timing mark on the pulley. Slide the belt over the left pulley. Now move the pulley until the marks on it align with the one on the No. 3 cover. There should be tension on the belt between the crankshaft pulley and the left camshaft pulley.

10. Align the installation mark on the timing belt with the mark on the right side camshaft pulley. Hang the belt over the pulley with the flange facing inward. Align the timing marks on the right pulley with the

one on the No. 3 cover and slide the pulley onto the end of the camshaft. Move the pulley until the camshaft knock pin hole is aligned with the groove in the pulley, then install the knock pin. Tighten the bolt to 55 ft. lbs. (75 Nm).

11. Position a plate washer between the timing belt tensioner and the block, then press in the pushrod until the holes are aligned between it and the housing. Slide a 0.05 in. Allen wrench through the hole to keep the pushrod set. Install the dust boot, and then install the tensioner. Tighten the bolts to 20 ft. lbs. (26 Nm). Don't forget to pull out the Allen wrench.

12. Turn the crankshaft clockwise 2 complete revolutions and check that all marks are still in alignment. If they aren't, remove the timing belt and start over again.

13. Install the remaining components.

3.0L (2JZ-GE) Engine

1. Before servicing the vehicle, refer to the precautions in the beginning of this section.

2. Remove all necessary components for access to the upper timing belt covers. Using a 5mm Allen wrench, remove the 9 bolts and lift off the two upper (No. 2 and No. 3) timing belt covers.

3. Rotate the crankshaft pulley clockwise so its groove is aligned with the **0** mark in the No. 1 (lower) timing cover. Check that the timing marks on the camshaft timing sprockets are aligned with the marks on the No. 4 (inner) cover. If not, rotate the crankshaft 1 complete revolution (360 degrees).

4. Alternately loosen the 2 tensioner mounting bolts and remove them, the tensioner and the dust boot. Slide the timing belt off of the 2 camshaft sprockets. Its a good idea to matchmark the belt to the pulleys.

5. Ensuring the timing belt is securely supported, hold the crankshaft pulley with a spanner wrench and loosen the mounting bolt. Remove the bolt and the pulley.

6. Remove or disconnect the following:
  • 5 bolts, then lift off the lower No. 1 timing belt cover
  • Timing belt guide
  • Timing belt

➡ **If the timing belt is to be reused, draw a directional arrow on the timing belt in the direction of engine rotation (clockwise) and place matchmarks on the timing belt and crankshaft gear to match the drilled mark on the pulley.**

To install:

7. Install the timing belt on the crankshaft timing pulley and the idler pulleys.

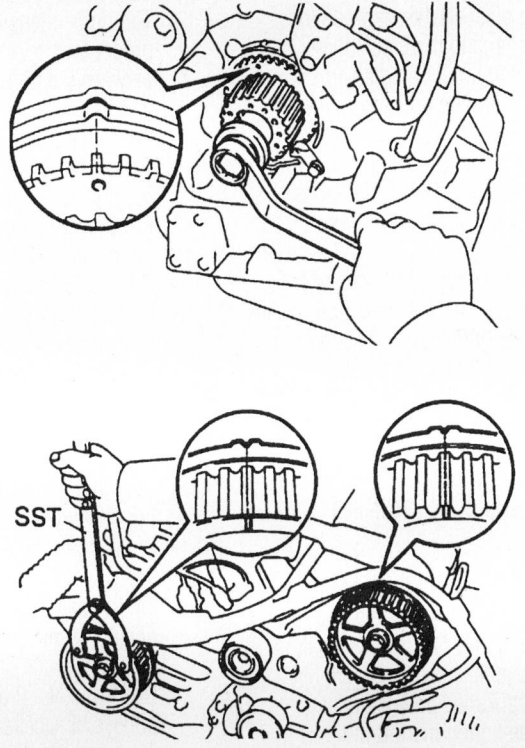

Camshaft and crankshaft pulley positioning for timing belt installation—3.0L (1MZ-FE) engine

79235G45

79235G46

Set the engine to TDC by aligning the marks before removing the lower timing cover—Lexus 3.0L (2JZ-GE) engine

➡If the old timing belt is being reinstalled, be sure the directional arrow is facing in the original direction and that the belt and crankshaft gear matchmarks are properly aligned.

8. Install the timing belt guide. Install the lower (No. 1) timing cover and tighten the bolts.

9. Align the crankshaft pulley set key with the key groove on the pulley and slide the pulley on. Tighten the bolt to 239 ft. lbs. (324 Nm).

10. Set the No. 1 cylinder to TDC again. Turn the camshaft until the sprocket timing marks are aligned with the timing marks on the No. 4 belt cover.

11. Check that the marks on the belt matches with those on the sprockets, then slide it over the sprockets. If not, shift it on the crank pulley until it does.

12. Position a plate washer between the timing belt tensioner and the block, then press in the pushrod until the holes are aligned between it and the housing. Slide a 1.5mm Allen wrench through the hole to keep the pushrod set. Install the dust boot, then install the tensioner. Tighten the bolts to 20 ft. lbs. (26 Nm). Don't forget to pull out the Allen wrench.

13. Turn the crankshaft clockwise two complete revolutions and check that all marks are still in alignment. If they aren't, remove the timing belt and start over again.

14. Position new gaskets, then install the upper (No. 2 and No. 3) timing covers.

4.0L (1UZ-FE) and 4.3L (3UZ-FE) Engines

1. Remove all necessary components for access to the right-hand side No. 3 and No. 2, and left-hand side No. 2 timing belt covers, then remove the covers.

2. Turn the crankshaft pulley and align it's groove with the timing mark **0** of the No. 1 timing cover. Check that the timing marks of the camshaft timing pulleys and timing

---

RH Bank

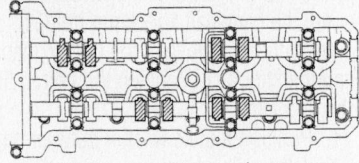

LH Bank

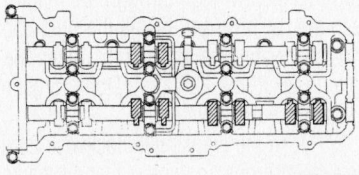

67162-LEXU-G115

Exploded view of the timing belt and cover assembly and related components—4.3L (3UZ-FE) engine shown

belt rear plates are aligned. If not, turn the crankshaft 1 full revolution (360 degrees).

3. Loosen crankshaft pulley bolt, then set the No. 1 cylinder to about 50°ATDC on the compression stroke. With the crankshaft pulley notch aligned with the **0** mark on the timing belt cover, turn the crankshaft pulley about 50°clockwise and put timing mark of the crank pulley in line with the centers of the pulley bolt and the No. 2 timing belt idler pulley bolt.

4. Remove or disconnect the following, as applicable for the vehicle:
- Timing belt tensioner. Using the proper tool, loosen the tension between the left side and right side timing pulleys by slightly turning the left side camshaft clockwise.
- Timing belt from the camshaft timing pulleys
- Power steering pump pulley
- Alternator
- Drive belt tensioner
- Bolt and timing pulleys, using the proper tool
- Bolt and the crankshaft pulley with the proper tool.
- Fan bracket
- Hydraulic pump, on SC 400
- Mounting bolts and the No. 1 timing belt cover
- 2 upper and lower timing belt covers
- Timing belt guide (No. 1 crank position sensor plate)
- Timing belt
- No. 1 and No. 2 timing belt idler pulleys, if necessary.

➡If the timing belt is to be reused, draw a directional arrow on the timing

---

belt in the direction of engine rotation (clockwise) and place matchmarks on the timing belt and crankshaft gear to match the drilled mark on the pulley.

To install:

5. Align the installation mark on the timing belt with the drilled mark of the crankshaft timing pulley. Install the timing belt on the crankshaft timing pulley, No. 1 idler pulley and the No. 2 idler pulley.

➡If the old timing belt is being reinstalled, be sure the directional arrow is facing in the original direction and that the belt and crankshaft gear matchmarks are properly aligned.

6. Install the timing belt guide (No. 1 crank angle sensor plate) with the cup side facing forward. Replace the timing belt cover spacer.

7. Install the No. 1 timing belt cover and tighten the mounting bolts. On the SC400, install the hydraulic pump. Install the fan bracket.

8. Align the pulley set key on the crankshaft with the key groove of the pulley. Install the pulley, using the proper tool to tap in the pulley. Tighten the pulley bolt to 181 ft. lbs. (245 Nm).

9. Align the knock pin on the right side camshaft with the knock pin of the timing pulley. Slide on the timing pulley with the right side mark facing forward. Tighten the bolt to 80 ft. lbs. (108 Nm).

10. Align the knock pin on the left side camshaft with the knock pin of the timing pulley. Slide on the timing pulley with the left side mark facing forward. Tighten the bolt to 80 ft. lbs. (108 Nm).

11. Turn the crankshaft pulley and align its groove with the **0** timing mark on the No. 1 timing belt cover. Using the proper tool, turn the crankshaft timing pulley and align the timing marks of the camshaft timing pulley and the timing belt rear plate.

12. Install the timing belt to the left side camshaft timing pulley by:

   a. Using the proper tool, slightly turn the left side timing pulley clockwise. Align the installation mark of the timing belt with the timing mark of the camshaft timing pulley and hang the timing belt on the left side camshaft pulley.

   b. Using the proper tool, align the timing marks of the left side camshaft pulley and the timing belt rear plate.

   c. Check that the timing belt has tension between crankshaft timing pulley and the left side camshaft pulley.

13. Install the timing belt to the right side camshaft timing pulley by:

3. Start the engine until the engine stalls.

4. Disconnect the negative battery terminal.

5. Place a catch-pan under the joint to be disconnected. A large quantity of fuel may be released when the joint is opened.

6. Wear eye or full-face protection.

7. Place a shop towel over the area and slowly release the joint using a wrench of the correct size.

8. Allow the any fuel left in the line to bleed off slowly before fully disconnecting the joint.

9. Plug the opened lines immediately to prevent fuel spillage or the entry of dirt.

10. Dispose of the released fuel properly.

11. After connecting fuel lines, install the fuse for the fuel pump and start the engine.

12. Check for leaks and repair as needed.

Fuel Filter

REMOVAL & INSTALLATION

The fuel filter on the ES 300 is located under the hood, on the driver's side, by the fenderwell.

The fuel filter LS 430 is located under the vehicle on the left side before the rear axle. The fuel filter on the GS 300 and GS 430is located under the vehicle, next to the left rear exhaust resonator.

1. Before servicing the vehicle, refer to the precautions in the beginning of this section.

2. Disconnect the negative battery cable. Wait at least 90 seconds before performing any other work.

3. On the GS 300, remove the rear body protector.

4. Slowly loosen the lower flare nut fitting until all the pressure is relieved and all the fuel is collected.

5. Loosen the union bolt on the upper portion of the filter and remove the banjo fitting and 2 metal gaskets. Discard the gaskets.

6. Loosen the fuel filter bracket bolt, remove the fuel line with the flared nut from the filter and pull the filter from the mounting bracket.

To install:

7. Install or connect the following:
- A new fuel filter to the vehicle and tighten the bracket bolt
- Banjo fitting with a new metal gasket on each side
- Union bolt. Tighten the union bolt to 22 ft. lbs. (30 Nm).

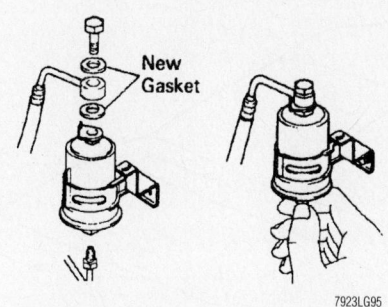

Exploded view of a typical fuel line connection at the filter

- Flare nut to the lower connection. Tighten the flare nut to 22 ft. lbs. (30 Nm).

8. On the GS 300, install the body protector.

9. Lower the vehicle if raised.

10. Remove the drain pan and/or rags and connect the negative battery cable.

11. Start the engine and visually inspect the upper and lower connections for leaks.

Fuel Pump

REMOVAL & INSTALLATION

ES 300 and IS 300

1. Before servicing the vehicle, refer to the precautions in the beginning of this section.

2. Relieve the fuel system pressure.

3. With the ignition switch in the **LOCK** position, disconnect the negative battery terminal.

4. Remove or disconnect the following:
- Rear seat cushion
- Fuel pump connector
- Floor service hole cover

➡**Do NOT lift the fuel pump assembly up using the wiring harness.**

- Fuel filler cap
- Fuel outlet pipe and the return hose from the pump bracket
- 8 screws and lift out the pump/bracket assembly with gasket
- Fuel pump lead wire
- Lower end of the pump off the bracket
- Fuel hose from the pump and remove the pump
- Rubber cushion from the pump

To install:

5. Install or connect the following:
- Filter and rubber cushion on the new pump
- Pump on the bracket
- Fuel hose and the wire connector on the pump
- Pump, using a new gasket and tighten the 8 screws to 35 inch lbs. (4 Nm)
- Fuel pipe and return hose to the pump and tighten the bolts to 22 ft. lbs. (29 Nm)

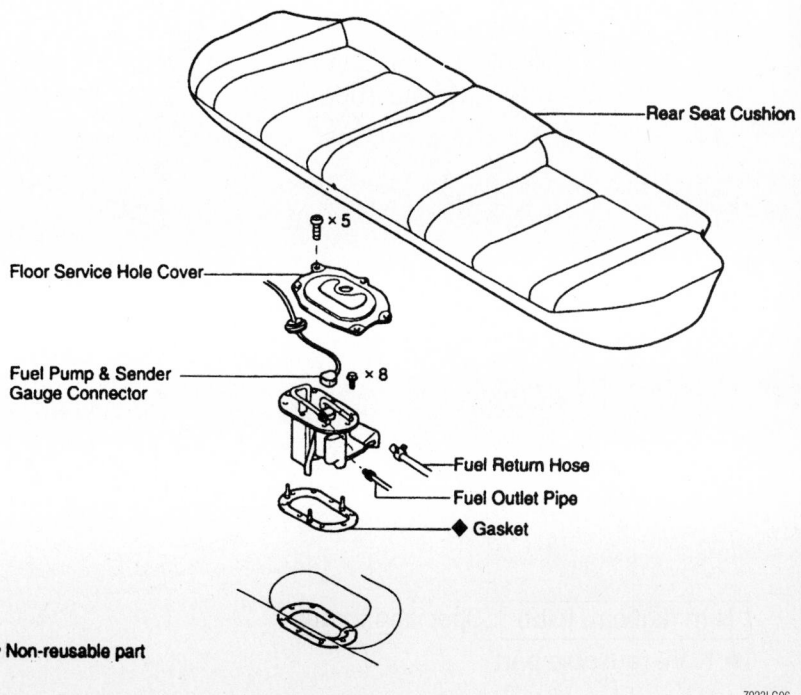

Non-reusable part

Exploded view of the fuel pump assembly—ES 300

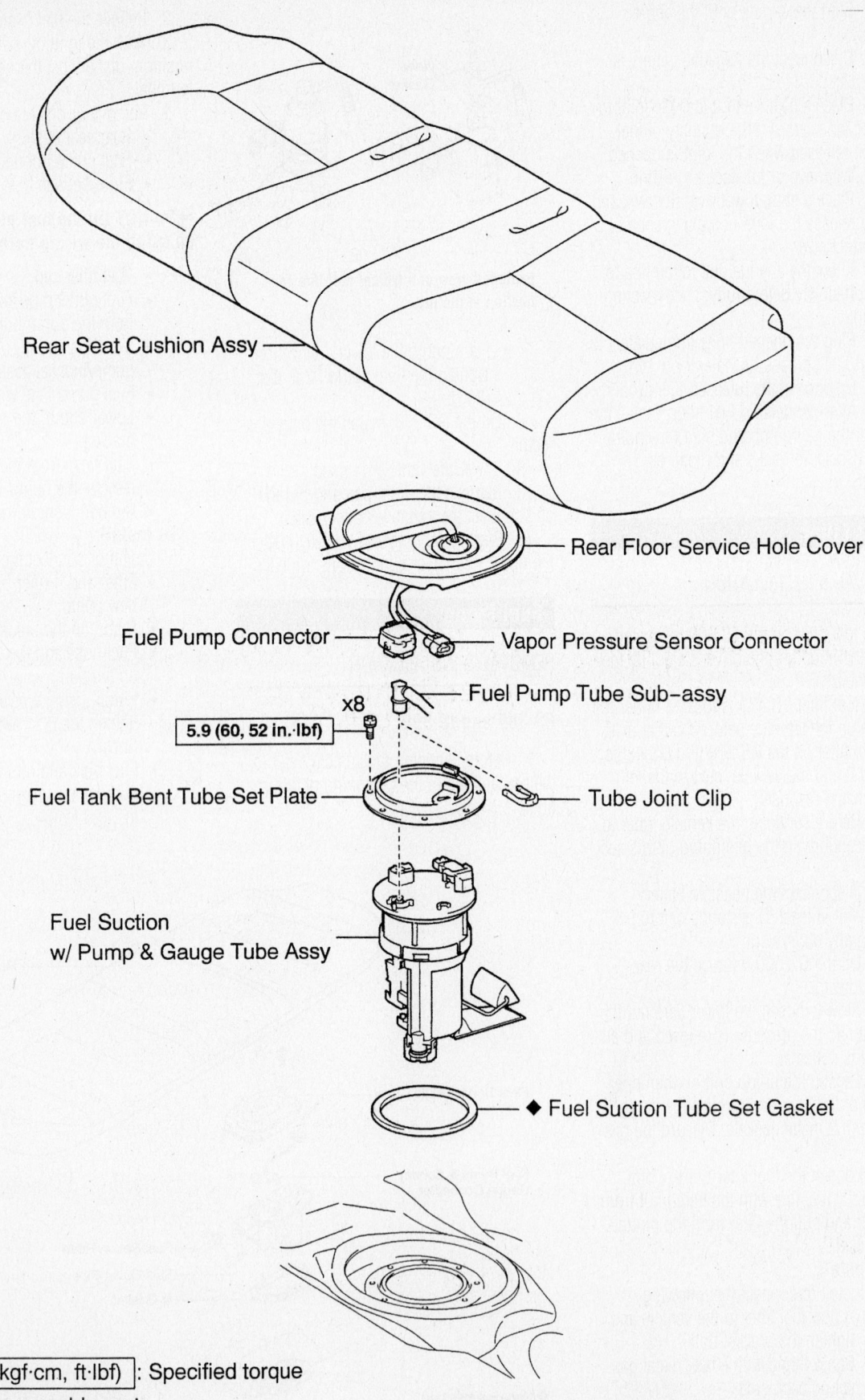

Rear Seat Cushion Assy

Rear Floor Service Hole Cover

Fuel Pump Connector — Vapor Pressure Sensor Connector

Fuel Pump Tube Sub–assy

x8

5.9 (60, 52 in.·lbf)

Fuel Tank Bent Tube Set Plate — Tube Joint Clip

Fuel Suction
w/ Pump & Gauge Tube Assy

◆ Fuel Suction Tube Set Gasket

N·m (kgf·cm, ft·lbf) : Specified torque

◆ Non–reusable part

67162-LEXU-G60

Exploded view for fuel pump removal—ES 330

- Wire
- Service cover, and replace the rear seat
- Negative battery cable
6. Start the engine and check for leaks.

Except ES 300 and IS 300

1. Before servicing the vehicle, refer to the precautions in the beginning of this section.
2. Remove or disconnect the following:
- Negative battery cable. Wait at least 90 seconds before performing any other work.
- Trunk floor mat
- Trunk trim cover
- Fuel pump electrical connector
- Rear seat bottom and seat back
- Partition cover
- Mounting bolts and remove the fuel pump set plate
- 3 nuts and disconnect the fuel pump bracket from the tank
- Fuel hose from the bracket
- Pump, bracket and set plate as an assembly

To install:
3. Install or connect the following:
- A new gasket on the set plate
- Fuel hose to the pump and bracket
- Pump and bracket assembly with the 3 nuts; tighten the nuts to 48 inch lbs. (5 Nm). Install the set plate and tighten the bolts to 26 inch lbs. (3 Nm).
- Panel partition
- Rear seat cushion and back
- Fuel pump electrical connector
- Trim panel
- Spare tire and the trunk floor mat
- Negative battery cable
4. Start the engine; check the fuel system for leaks

SC 430 AND LS 430

1. Before servicing the vehicle, refer to the precautions in the beginning of this section.
2. Remove or disconnect the following:
- Negative battery cable. Wait at least 90 seconds before performing any other work.
- Rear seat bottom and seat back
- Fuel pump tube
- 8 bolts and fuel tank vent tube set plate
- Fuel pump and sensor gauge assembly
- Fuel suction hose and support
- Fuel pump cushion rubber

- Fuel pressure w/jet pump regulator assembly
- Fuel suction plate w/sender gauge
- Fuel pump and filter

To install:
3. To install, reverse the removal procedure. Install a new O-ring on the fuel jet pump regulator assembly and a new gasket on the fuel pump/sender gauge assembly. Torque the 8 bolts securing the fuel tank set plate to 52 in. lbs (6.0 Nm).
4. Inspect fuel pump operation and check for fuel leaks.

ES 330

1. Before servicing the vehicle, refer to the precautions in the beginning of this section.
2. Remove or disconnect the following:
- Negative battery cable. Wait at least 90 seconds before performing any other work.
- Rear seat bottom and seat back
- Rear floor service hole cover
- Fuel suction w/pump and gauge tube assembly
- 8 bolts and fuel tank vent tube set plate
- Vapor pressure sensor assembly
- Fuel suction hose and support
- Fuel pump cushion rubber
- Fuel sender gauge assembly
- Fuel suction plate w/sender gauge
- Fuel pump harness
- Fuel pump
- Fuel pump filter. Pry out clip and remove from pump
- Fuel pressure regulator assembly

To install:
3. To install, reverse the removal procedure. Install a new O-ring on the fuel suction w/pump and gauge assembly. Torque the 8 bolts securing the fuel tank set plate to 52 inch lbs. (6.0 Nm).
4. Inspect fuel pump operation and check for fuel leaks.

Fuel Injector

REMOVAL & INSTALLATION

3.0L (1MZ-FE) Engine

1. Before servicing the vehicle, refer to the precautions in the beginning of this section.
2. Remove or disconnect the following:
- 3 cap nuts, using a 5mm hexagon wrench. Loosen the V-bank cover fastener counterclockwise.
- V-Bank cover

- Air cleaner hose with resonator
- air intake chamber assembly
- Injector connectors
- Air assist hoses and pipe
- No. 1 fuel pipe and remove the fuel hose clamp
- No. 1 fuel pipe (fuel tube connector) from the fuel filter outlet
- 5 bolts and delivery pipes together with the 6 injectors and No. 1 fuel pipe.
- 4 spacers from the intake manifold
- 6 injectors form he delivery pipes
- 2 O-rings and 2 grommets from each injector

To install:
3. Install or connect the following:
- New insulator and grommet to each injector
- New O-rings, coated with gasoline, to each injector
- A light coat of gasoline on the place where a delivery pipe touches an O-ring of the injector
- Injector, while turning it clockwise, into the delivery pipe

➡ **Position the injector connector outward.**

- The 4 spacers in position on the intake manifold
- A light coat of gasoline on the place where the intake manifold touches an O-ring
- The delivery pipe and fuel pipe together with the 6 injectors in position on the intake manifold. Position the injector connector outward.
- 4 bolts holding the delivery pipes to the intake manifold, temporarily
- Bolt holding the No. 1 fuel pipe to the intake manifold, temporarily

➡ **Check that the injectors rotate smoothly. If the injectors do not rotate smoothly, the probable cause is incorrect installation of the O-rings. Replace the O-rings.**

- 4 bolts holding the delivery pipes to the intake manifold. Torque to 14 ft. lbs. (19.5 Nm)
- No. 1 fuel pipe (fuel tube connector) to the fuel filter
- Fuel hose clamp to the fuel filter with a "click" sound. After installing the clamp, check that the clamp is fixed by pulling up the clamp.
- Air assist hoses
- Injector connectors
- Air intake chamber assembly

- Air cleaner hose with resonator
- V-bank cover, using a 5mm hexagon wrench with the 3 cap nuts
- Press down the V-bank cover.

3.0L (2JZ-GE) Engine

1. Before servicing the vehicle, refer to the precautions in the beginning of this section.
2. Remove or disconnect the following:
- Air intake chamber
- Fuel pressure pulsation damper
- Engine wire from intake manifold
- Bolt holding the engine wire protector to the body
- 6 injector connectors
- Camshaft Position (CMP) sensor connector
- Throttle Position (TP) sensor connector
- Vacuum Switching Valve (VSV) connector for Evaporative Emission (EVAP) control
- VSV connector for Acoustic Control Induction System (ACIS)
- 3 nuts holding the engine wire protector to the intake manifold

➡**Be careful not to drop the injectors when removing the delivery pipe.**

- 3 bolts and delivery pipe together with the 6 injectors
- The injectors from the delivery pipe
- O-rings, insulator and grommet from each injector
- 3 spacers from the intake manifold

To install:

3. Install or connect the following:
- A new insulator and grommet to each injector
- A light coat of gasoline on the place where a delivery pipe touches an O-ring of the injector
- Injector, while turning clockwise and counterclockwise, into the delivery pipe

➡**Position the injector connector outward.**

- The 3 spacers in position on the intake manifold
- A light coat of gasoline on the place where an intake manifold touches an O-ring
- Injectors together with the delivery pipe and 3 bolts in position on the intake manifold. Check that the injectors rotate smoothly. Position the injector connector upward.

➡**If the injectors do not rotate smoothly, the probable cause is incor-**

rect installation of the O-rings. Replace the O-rings.

4. Tighten the 3 bolts holding the delivery pipe to the intake manifold. Tighten the bolts to 15 ft. lbs. (21 Nm).
5. Install or connect the following:
- Engine wire protector with the 3 nuts
- 6 injector connectors

➡**The No. 1, No. 3 and No. 5 injector connectors are dark gray, and the No. 2, No. 4, and the No. 6 injectors connectors are brown.**

6. Install or connect the following:
- Camshaft position sensor connector
- Throttle position sensor connector
- VSV connector for EVAP
- Bolt holding the engine wire protector to the body

4.0L (1UZ-FE) and 4.3L (3UZ-FE) Engines

1. Before servicing the vehicle, refer to the precautions in the beginning of this section.
2. Remove or disconnect the following:
- V-bank cover
- Intake air connector
- Accelerator cable
- Fuel pressure pulsation dampers.
- VVT sensor connectors
- Vacuum Switching Valve (VSV) for Evaporative Emissions (EVAP)
- 2 nuts and accelerator cable bracket
- 2 nuts and accelerator cable bracket
- 3 V-bank cover brackets
- VSV connector for Acoustic Control Induction System (ACIS) from the No. 1 V-bank cover bracket
- 4 bolts and 3 V-bank cover brackets
- Engine wire from the delivery pipe
- 2 wire clamps from the wire clamp bracket on the right-hand deliver pipe
- 8 injector connectors
- 4 nuts holding the delivery pipe to the intake manifold
- 2 delivery pipes and 8 injectors assembly and 4 spacers
- 2 O-rings, grommet and insulator from each injector

To install:

3. Install or connect the following:
- A new insulator and grommet to each injector
- A light coat of gasoline to new O-

rings and install them to each injector
- A light coat of gasoline on the place where a delivery pipe touches an O-ring of the injector
- Injector, while turning the clockwise and counterclockwise, into the delivery pipe

➡**Position the injector connector outward.**

- The 4 spacers in position on the intake manifold
- A light coat of gasoline on the place where an intake manifold touches an O-ring
- The delivery pipes in position on the intake manifold
- Temporarily, the 3 bolts holding the delivery pipe to the intake manifold

➡**Check that the injectors rotate smoothly. If the injectors do not rotate smoothly, the probable cause is incorrect installation of the O-rings. Replace the O-rings.**

4. Tighten the 3 bolts holding the delivery pipe to the intake manifold. Tighten the bolts to 15 ft. lbs. (21 Nm).
5. Install or connect the following:
- Engine wire protector with the 3 nuts
- Injector connectors
- Remaining components

3.3L (3MZ-FE) Engine

1. Before servicing the vehicle, refer to the precautions in the beginning of this section.
2. Remove or disconnect the following:
- Negative battery cable
- Suspension upper center brace
- V-bank cover sub assembly
- Air cleaner assembly
- Emission control valve set
3. Remove intake air surge tank
4. Remove or disconnect the following:
- Throttle motor connector
- Water by-pass hoses
- Union check valve hose
- Ventilation hose
- Pressure feed hose
- Engine hangers
- Surge tank stays
- Bond cable connector
- Emission control valve bracket
- Intake air surge tank
- Gasket from intake air surge tank
5. Remove fuel pipe assembly
6. Remove or disconnect the following:
- Fuel pulsation damper and gasket

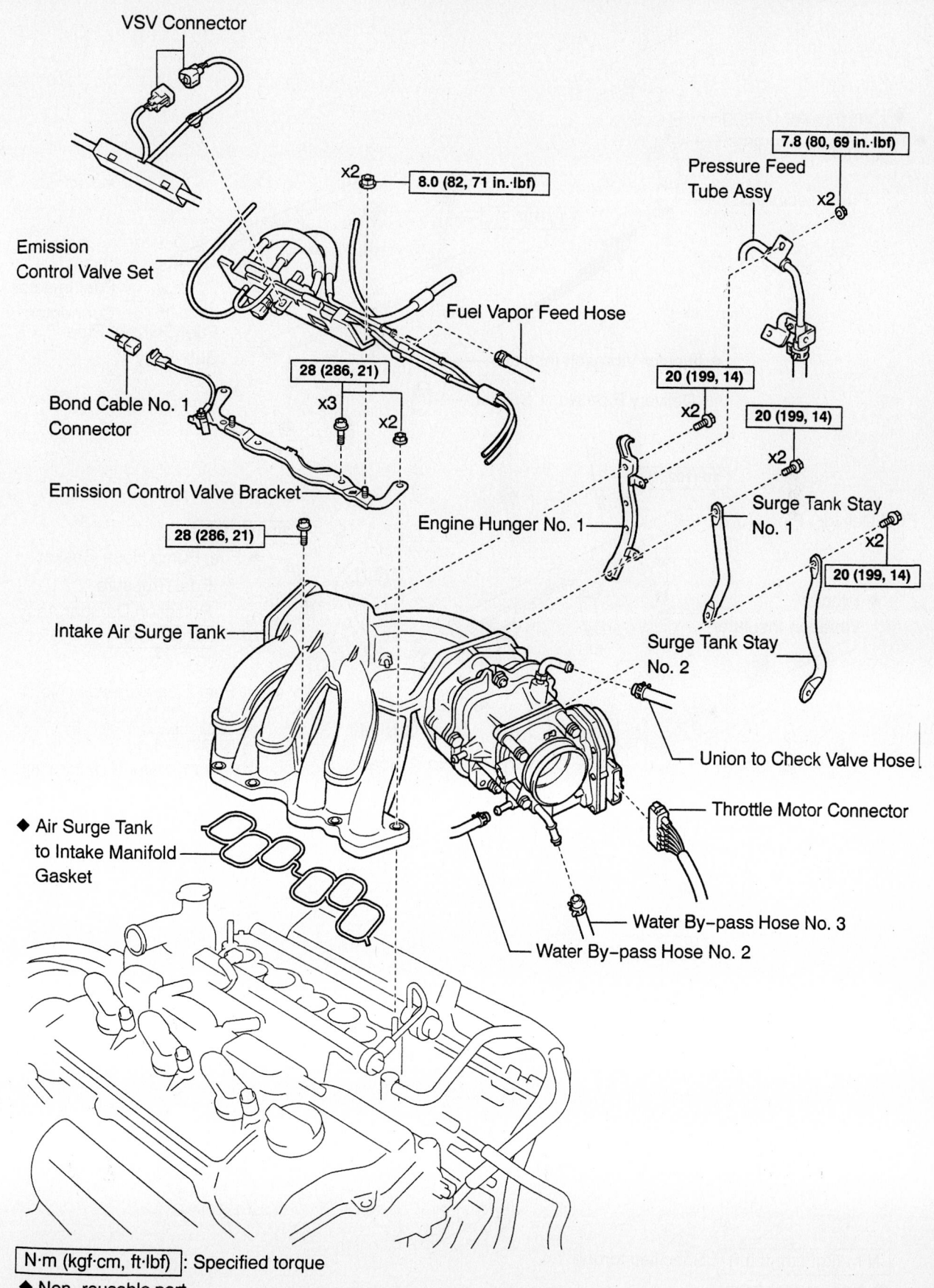

VSV Connector

7.8 (80, 69 in.·lbf)

Pressure Feed Tube Assy

x2

x2 8.0 (82, 71 in.·lbf)

Emission Control Valve Set

Fuel Vapor Feed Hose

Bond Cable No. 1 Connector

28 (286, 21)

x3

x2

20 (199, 14)

x2

20 (199, 14)

x2

Emission Control Valve Bracket

Engine Hunger No. 1

Surge Tank Stay No. 1

28 (286, 21)

Surge Tank Stay No. 2

x2

20 (199, 14)

Intake Air Surge Tank

Union to Check Valve Hose

Throttle Motor Connector

◆ Air Surge Tank to Intake Manifold Gasket

Water By-pass Hose No. 3

Water By-pass Hose No. 2

N·m (kgf·cm, ft·lbf) : Specified torque

◆ Non-reusable part

67162-LEXU-G61

Exploded view of Intake Air Surge tank—3.3L (3MZ-FE) Engine

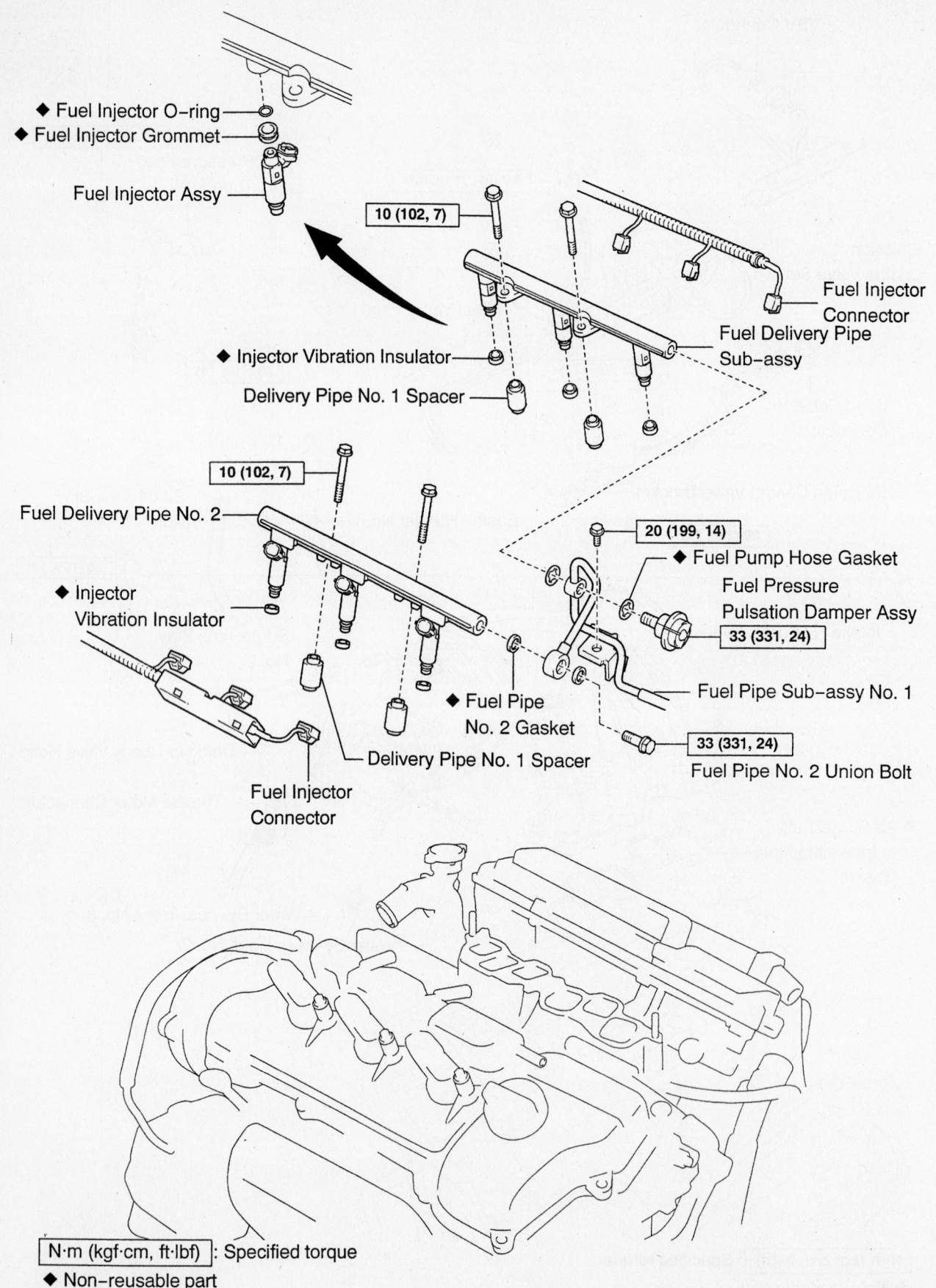

◆ Fuel Injector O-ring

◆ Fuel Injector Grommet

Fuel Injector Assy

10 (102, 7)

Fuel Injector Connector

Fuel Delivery Pipe Sub-assy

◆ Injector Vibration Insulator

Delivery Pipe No. 1 Spacer

10 (102, 7)

Fuel Delivery Pipe No. 2

◆ Injector Vibration Insulator

20 (199, 14)

◆ Fuel Pump Hose Gasket

Fuel Pressure Pulsation Damper Assy

33 (331, 24)

◆ Fuel Pipe No. 2 Gasket

Delivery Pipe No. 1 Spacer

Fuel Injector Connector

Fuel Pipe Sub-assy No. 1

33 (331, 24)

Fuel Pipe No. 2 Union Bolt

N·m (kgf·cm, ft·lbf) : Specified torque

◆ Non-reusable part

Exploded view of fuel injector delivery pipe and injectors—3.3L (3MZ-FE) Engine

67162-LEXU-G62

- Fuel pipe union bolt and 2 gaskets
- Bolt and separate the fuel pipe
7. Remove the fuel injector assembly
8. Remove or disconnect the following:
  - 6 fuel injector connectors
  - 4 bolts, then remove the fuel injector delivery pipes

✳✳ WARNING

Be careful not to drop the fuel injectors when removing the fuel delivery pipes.

- 4 delivery pipe spacers from intake manifold
- 6 insulators from the intake manifold
- fuel injector from the fuel delivery pipes.

To install:
9. Install new fuel injector assembly
10. Install or connect the following
  - A new insulator and grommet to each injector
  - A light coat of gasoline to new O-

rings and install them to each injector
- A light coat of gasoline on the place where a delivery pipe touches an O-ring of the injector
- Injector, while turning the clockwise and counterclockwise, into the delivery pipe
- The 6 insulators and 4 spacers in position on the intake manifold
- A light coat of gasoline on the place where an intake manifold touches an O-ring
- The delivery pipes in position on the intake manifold
- Temporarily, the 4 bolts holding the delivery pipe to the intake manifold

➥**Check that the injectors rotate smoothly. If the injectors do not rotate smoothly, the probable cause is incorrect installation of the O-rings. Replace the O-rings.**

- 4 bolts. Tighten bolts uniformly. Torque to: 7 ft. lbs. (10 Nm)

- 6 fuel injector connectors
11. Install fuel pipe assembly
12. Install or connect the following
  - 2 gaskets and fuel pipe union bolt
  - 2 gaskets and fuel pressure pulsation damper
  - Fuel pipe with bolt
13. Install intake air surge tank
14. Install or connect the following
  - New gaskets to intake air surge tank
  - Intake air surge tank and emission control valve bracket. Torque: 21 ft. lbs. (28 Nm)
  - 4 bolts. Torque: 21 ft. lbs. (28 Nm)
  - 2 bolts on surge tank stay. Torque: 21 ft. lbs. (28 Nm)
  - Engine hangers
  - Pressure feed tube
15. Remaining components in reverse order of removal procedure
16. Install engine coolant and check for leaks
17. Perform system initialization

DRIVE TRAIN

Transmission Assembly

REMOVAL & INSTALLATION

Automatic

GS 300 (EXC. 2004)

1. Before servicing the vehicle, refer to the precautions in the beginning of this section.
2. Turn the ignition switch to the **LOCK** position and disconnect the negative battery cable. Wait at least 90 seconds or longer before doing any work on the vehicle.
3. Remove or disconnect the following:
  - Transmission level gauge
  - Transmission dipstick and tube
  - Throttle cable from the throttle body
  - Oxygen (O2) sensor from the exhaust system.
  - Left and right tail pipes
  - Front and center exhaust pipe
  - Exhaust heat insulator
  - Rear center floor crossmember brace
  - Shift control rod from the shift lever
  - Driveshaft
  - Overdrive and direct clutch speed sensor
  - No. 1 Vehicle Speed Sensor (VSS)
  - No. 2 VSS
  - Solenoid wire

- Park/Neutral Position (PNP) switch
- Wiring from the starter
- 2 oil cooler union nuts
- Oil cooler hoses from the oil cooler pipes
- Front oil cooler pipe bracket
- Center and rear oil cooler pipe brackets
- 2 oil cooler pipes
- Torque converter inspection plate
- Torque converter bolts
4. Support the transmission with a suitable jack.
5. Support the engine with a jack and a block of wood.
6. Remove or disconnect the following:
  - Rear transmission mount
  - Wiring harness clamps
  - Starter
  - 9 transmission mounting bolts and transmission

To install:
7. Install or connect the following:
  - Transmission and tighten the bolts to 53 ft. lbs. (52 Nm)
  - Starter and tighten the bolts to 27 ft. lbs. (37 Nm)
  - Rear transmission mount and tighten the bolts to 19 ft. lbs. (25 Nm)
  - And tighten the torque converter bolts to 30 ft. lbs. (41 Nm) while rotating the crankshaft
  - Converter inspection plate

- 2 oil cooler pipes
- Center and rear oil cooler pipe brackets
- Front oil cooler pipe bracket and tighten to 49 inch lbs. (5.5 Nm)
- Oil cooler hoses to the oil cooler pipes
- 2 oil cooler union nuts and tighten to 32 ft. lbs. (44 Nm)
- Wiring to the starter
- Transmission electrical connectors
- Driveshaft
- Remaining components
- Negative battery cable
- Transmission level gauge
8. Fill the transmission to the proper level with recommended type fluid.

GS 300 (2004)

1. Before servicing the vehicle, refer to the precautions in the beginning of this section.
2. Turn the ignition switch to the **LOCK** position and disconnect the negative battery cable. Wait at least 90 seconds or longer before doing any work on the vehicle.
3. From the engine compartment, remove or disconnect the following:
  - Transmission level gauge
  - Transmission dipstick and tube
  - Air cleaner, MAF meter, and intake air connector pipe assembly
  - Intake manifold with catalytic converters

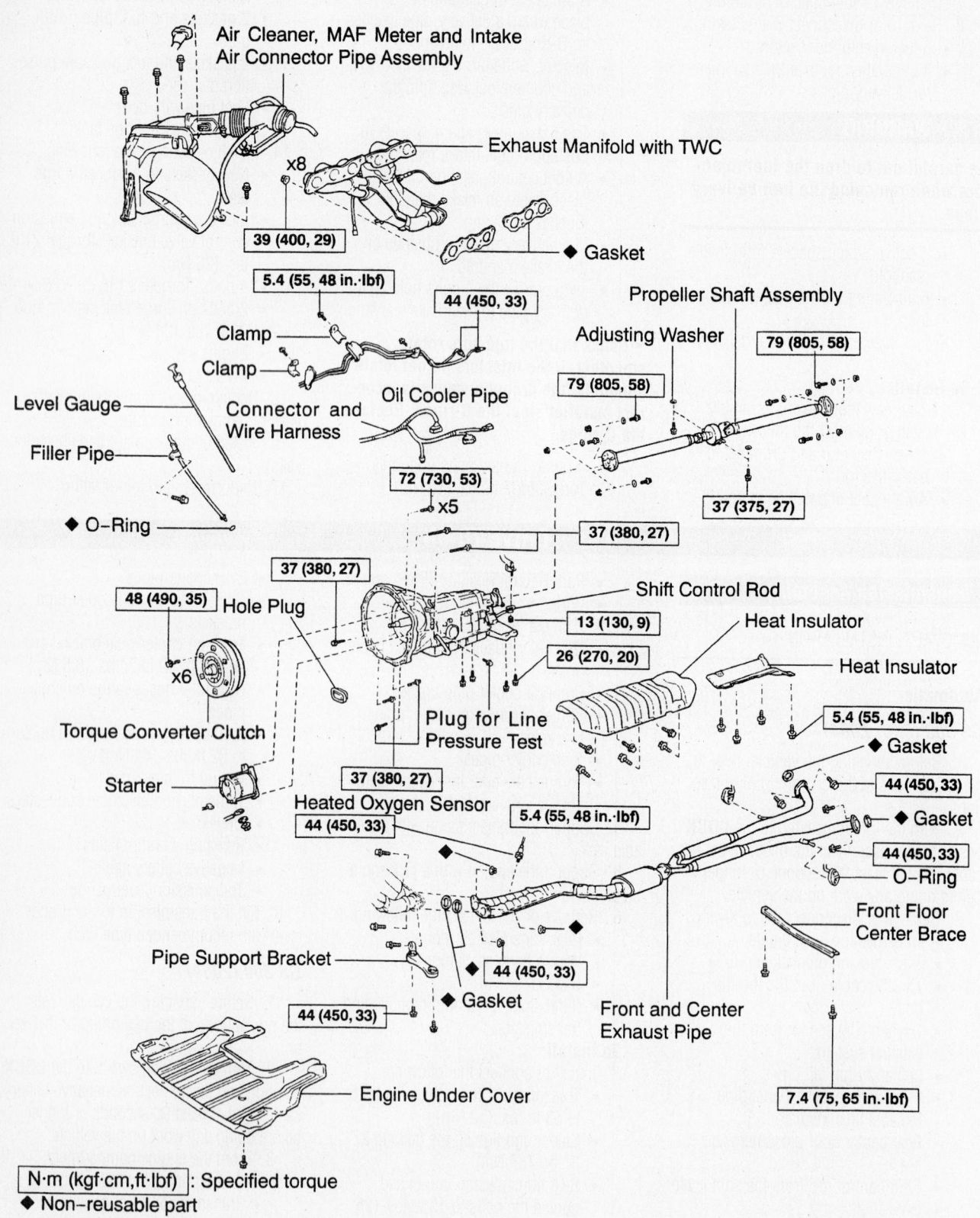

Air Cleaner, MAF Meter and Intake Air Connector Pipe Assembly

Exhaust Manifold with TWC

x8

39 (400, 29)

◆ Gasket

5.4 (55, 48 in.·lbf)

44 (450, 33)

Propeller Shaft Assembly

Adjusting Washer

79 (805, 58)

79 (805, 58)

Clamp

Clamp

Oil Cooler Pipe

Level Gauge

Connector and Wire Harness

Filler Pipe

72 (730, 53)

x5

37 (375, 27)

◆ O-Ring

37 (380, 27)

37 (380, 27)

Shift Control Rod

Hole Plug

48 (490, 35)

13 (130, 9)

Heat Insulator

Heat Insulator

26 (270, 20)

x6

5.4 (55, 48 in.·lbf)

Torque Converter Clutch

Plug for Line Pressure Test

◆ Gasket

44 (450, 33)

Starter

37 (380, 27)

Heated Oxygen Sensor

5.4 (55, 48 in.·lbf)

◆ Gasket

44 (450, 33)

44 (450, 33)

O-Ring

Front Floor Center Brace

Pipe Support Bracket

44 (450, 33)

◆ Gasket

Front and Center Exhaust Pipe

44 (450, 33)

Engine Under Cover

7.4 (75, 65 in.·lbf)

N·m (kgf·cm,ft·lbf) : Specified torque
◆ Non-reusable part

Exploded view of automatic transmission components—GS300 2004

4. Raise vehicle and remove or disconnect the following:
- Engine under cover
- Front and center exhaust pipe
- If necessary, oxygen sensor from the exhaust system.
- 2 exhaust heat insulators
- Driveshaft
- Shift control rod from end of shift lever
- Oil cooler pipes
- Torque converter inspection plate
- Torque converter bolts (through plate access hole)

5. Support the transmission with a suitable jack.

6. Support the front of the engine with a jack and a block of wood.

7. Remove or disconnect the following:
- Rear transmission mount (tilt end of transmission downward)
- Wiring harness connectors and wiring harness
- Starter
- 9 transmission mounting bolts and transmission (note position of longer bolts for reinstallation)

To install:

8. Inspect torque converter position in the transmission housing, using a straight-edge placed across the transmission case. If torque converter-to-transmission clearance is less than 0.004 in. (0.1 mm), check for proper installation of the torque converter.

9. Position the transmission into the vehicle.

10. Install or connect the following:
- Transmission and tighten the 17 mm head bolts to 53 ft. lbs. (72 Nm) and 14 mm head bolts to 27 ft. lbs. (37 Nm)
- Starter and tighten the bolts to 27 ft. lbs. (37 Nm)
- Wiring harness and connectors
- Rear transmission mount and tighten the bolts to 20 ft. lbs. (26 Nm)
- And tighten the torque converter bolts to 35 ft. lbs. (48 Nm) while rotating the crankshaft
- Converter inspection plate
- 2 oil cooler pipes
- Center and rear oil cooler pipe brackets
- Shift control rod to shift lever; torque nut to 9 ft. lbs. (13 Nm)
- Driveshaft
- 2 heat insulators
- Front and center exhaust pipes, with new gaskets; torque pipe support bracket bolts to 33 ft. lbs. (44 Nm), exhaust pipe connector nuts

to 33 ft. lbs. (44 Nm), and center bracket bolts to 65 inch lbs. (7.4 Nm).

✷✷ CAUTION

Twist oxygen sensor wiring 3-1/2 turns counterclockwise before installing oxygen sensor; this will prevent wires from being twisted after installation.

- If removed, install oxygen sensor to 33 ft. lbs. (44 Nm).
- Engine under cover
- Exhaust manifold, with new gaskets; torque nuts to 29 ft. lbs. (39 Nm)
- Intake air connector pipe, MAF meter, and air cleaner
- Filler pipe with new O-ring
- Transmission level gauge
- Negative battery cable

11. Fill the transmission to the proper level with Type T-IV fluid.

IS 300

1. Before servicing the vehicle, refer to the precautions in the beginning of this section.

2. Drain the cooling system.
- Negative battery cable
- Transmission oil dipstick and tube
- Air cleaner
- Mass Air Flow (MAF) sensor
- Exhaust manifold
- Engine under covers
- Left front floor center cover
- No. 1 rear floor board
- Upper radiator hose
- Exhaust front pipe
- Exhaust center pipe
- Shift control rod
- Drive shaft
- Oil cooler lines
- Torque converter mounting bolts (through access hole)
- Rear transmission mount cross-member. Support the transmission with a jack.
- Transmission wiring harness connectors
- Starter motor
- Transmission flange bolts
- Transmission

To install:

3. Installation is the reverse of the removal procedure, while using the following torque values:
- 17mm transmission flange bolts: 53 ft. lbs. (72 Nm)
- 14mm transmission flange bolts:

27 ft. lbs. (37 Nm)
- Starter motor bolts: 27 ft. lbs. (37 Nm)
- Rear transmission mount cross-member bolts: 19 ft. lbs. (25 Nm)
- Torque converter bolts: 35 ft. lbs. (48 Nm)
- Oil cooler lines: 33 ft. lbs. (44 Nm)
- Drive shaft center support bolts: 36 ft. lbs. (49 Nm)
- Drive shaft U-joint flange bolts: 54 ft. lbs. (74 Nm)
- Shift control rod nuts: 108 inch lbs. (13 Nm)
- Exhaust manifold nuts: 29 ft. lbs. (39 Nm)

LS 430

1. Before servicing the vehicle, refer to the precautions in the beginning of this section.

2. Remove or disconnect the following:
- Negative battery cable. Wait at least 90 seconds before performing any other work.
- Transmission dipstick and tube
- Throttle cable
- Driveshaft
- Engine undercover
- Shift control rod
- Exhaust pipe support bracket by removing the 2 bolts
- Catalytic converters by removing the 6 nuts
- Both side heat insulators
- Oil cooler tube clamps and disconnect the tubes
- Torque converter inspection plate by removing the 2 bolts
- Torque converter bolts

3. Support the transmission with a suitable jack.

4. Remove or disconnect the following:
- Rear transmission mount
- Overdrive direct clutch speed sensor connector
- Vehicle Speed Sensor (VSS) connector
- Park/Neutral Position (PNP) switch connector
- Solenoid connector
- 3 wiring harness clamps from the bracket on the transmission.
- 10 transmission mounting bolts and the transmission.

To install:

5. Install or connect the following:
- Transmission and tighten the bolts as follows: 14mm–27 ft. lbs. (37 Nm); 17mm–53 ft. lbs. (72 Nm)
- 3 wiring harness clamp to the bracket on the transmission

- Solenoid connector
- PNP switch connector
- VSS connector
- Overdrive direct clutch speed sensor connector
- Rear transmission mount and tighten the bolts to 19 ft. lbs. (20 Nm) and the nuts to 10 ft. lbs. (13 Nm)
- Torque converter bolts to 30 ft. lbs. (41 Nm)
- Converter inspection plate
- Support from the transmission
- Oil cooler pipes and tighten the union nuts to 32 ft. lbs. (44 Nm)
- Side heat insulators
- Catalytic converters with new gaskets and new nuts. Tighten the nuts to 46 ft. lbs. (62 Nm).
- Exhaust pipe support bracket with the 2 bolts and tighten the bolts to 32 ft. lbs. (44 Nm)
- Shift control rod
- Engine undercover
- Driveshaft
- Throttle control cable
- Transmission tube and dipstick
- Negative battery cable

6. Fill the transmission to the proper level with recommended type fluid.

GS 430 (EXC. 2004)

1. Before servicing the vehicle, refer to the precautions in the beginning of this section.

2. Remove or disconnect the following:
- Negative battery cable.
- V-bank cover, if equipped
- Automatic transmission oil level gauge if equipped
- Transmission dipstick and tube
- Throttle cable and clamps
- Exhaust pipe and converters
- Exhaust heat insulator
- Rear center floor crossmember brace
- Shift control rod
- Driveshaft

➡ **The bolts inserted from the driveshaft side should not be removed.**

- The electrical harness from the transmission.
- Oil cooler tube clamp and disconnect the tubes
- Lower engine cover
- Torque converter inspection plate
- Torque converter bolts

3. Support the transmission with a suitable jack.

4. Remove or disconnect the following:
- Starter, if necessary

- Rear transmission mount
- Transmission mounting bolts and the transmission.

To install:

5. Before installing the transmission, use calipers and a straightedge to check the distance between the installed surface of the torque converter and the front edge of the transmission case. Correct distance is 0.673 in. (17.1mm). If this distance is not correct, check the torque converter installation.

6. Install or connect the following:
- Transmission and tighten the bolts to 14mm: 29 ft. lbs. (39 Nm); 17mm: 42 ft. lbs. (57 Nm)
- If removed, the starter. Tighten the bolts to 27 ft. lbs. (37 Nm).
- Rear transmission mount and tighten the bolts to 19 ft. lbs. (20 Nm)
- And tighten the torque converter bolts to 25 ft. lbs. (33 Nm) while rotating the crankshaft
- Converter inspection plate
- Lower engine cover
- Oil cooler lines and tighten the lines to 25 ft. lbs. (34 Nm)
- Oil cooler pipe bracket and tighten the bolt
- Transmission electrical connectors
- Shift control rod and adjust the shift linkage. Tighten the nut to 12 ft. lbs. (16 Nm).
- Rear center floor crossmember brace. Tighten the bolts to 108 inch lbs. (13 Nm).
- Heat insulator
- Transmission filler tube and dipstick
- Front exhaust pipe and converters with new gaskets
- Throttle control cable
- Driveshaft. Flange bolts: 58 ft. lbs. (79 Nm). Center bearing support bolts: 36 ft. lbs. (49 Nm). Adjusting nut: 35 ft. lbs. (48 Nm).
- Crossmember brace and tighten to 96 inch lbs. (13 Nm)
- Automatic transmission oil level gauge
- V-bank cover
- Negative battery cable

7. Adjust the PNP switch
8. Fill the transmission with Dexron®II.

GS 430 (2004)

1. Before servicing the vehicle, refer to the precautions in the beginning of this section.

2. Remove or disconnect the following:
- Negative battery cable.

- Transmission oil level dipstick
- Front and center exhaust pipes (disconnect oxygen sensors)
- Both catalytic converters
- Exhaust heat insulator
- Transmission fluid filler pipe
- Driveshaft
- Shift control rod from end of shift lever
- Oil cooler pipes
- Torque converter inspection plate
- Torque converter bolts

3. Support the transmission with a suitable jack.

4. Remove or disconnect the following:
- Engine rear mount (4 bolts)
- Wiring and connectors (tilt transmission rear downward)
- Transmission mounting bolts and the transmission (place jack and wood block under front end of engine)

To install:

5. Before installing the transmission, use calipers and a straightedge to check the distance between the installed surface of the torque converter and the front edge of the transmission case. Correct distance is 0.673 in. (17.1mm). If this distance is not correct, check the torque converter installation.

6. Position transmission to the vehicle and install or connect the following:
- Transmission and tighten the bolts to 14mm: 27 ft. lbs. (37 Nm); 17mm: 53 ft. lbs. (72 Nm)
- Wiring and connectors
- Rear engine mount and tighten the bolts to 20 ft. lbs. (26 Nm)
- Torque converter bolts to 35 ft. lbs. (48 Nm) while rotating the crankshaft
- Converter inspection plate; torque bolts to 13 ft. lbs. (18 Nm)
- Oil cooler lines and tighten the clamp bolts to 48 inch lbs. (5.4 Nm)
- Oil cooler pipe union nuts to 33 ft. lbs. (44 Nm)
- Shift control rod; torque nut to 9 ft. lbs. (13 Nm).
- Driveshaft
- Transmission fluid filler pipe, with new O-ring
- Heat insulator
- Both catalytic converters, with new gaskets; torque nuts to 46 ft. lbs. (62 Nm)
- Tail pipe to center exhaust pipe, with new gaskets; torque bolts to 33 ft. lbs. (44 Nm)
- Exhaust pipe support bracket bolts to 33 ft. lbs. (44 Nm)

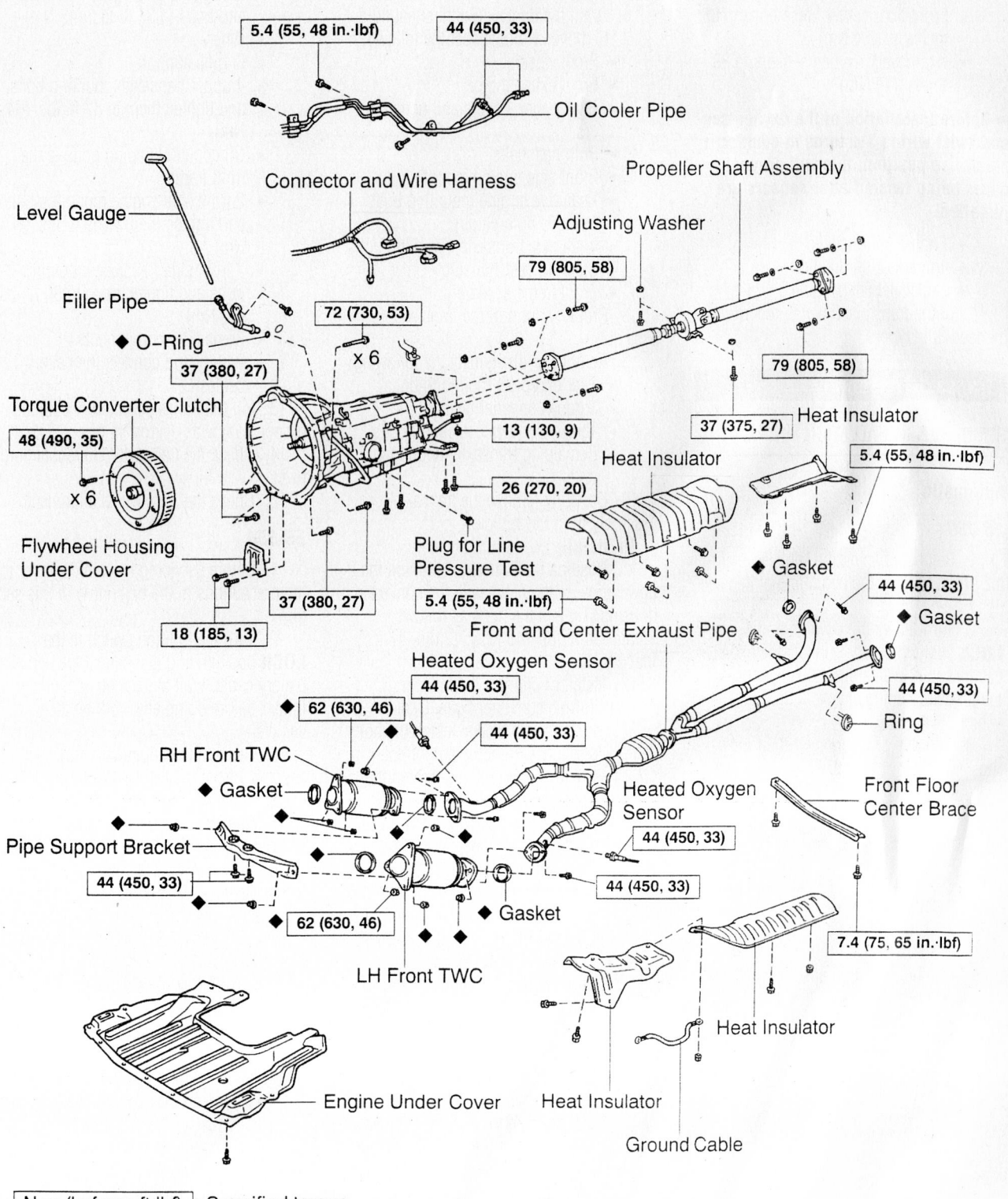

5.4 (55, 48 in.·lbf)

44 (450, 33)

Oil Cooler Pipe

Connector and Wire Harness

Propeller Shaft Assembly

Adjusting Washer

Level Gauge

79 (805, 58)

Filler Pipe

72 (730, 53)
×6

◆ O-Ring

79 (805, 58)

37 (380, 27)

Heat Insulator

Torque Converter Clutch

13 (130, 9)

37 (375, 27)

48 (490, 35)

Heat Insulator

5.4 (55, 48 in.·lbf)

×6

26 (270, 20)

Flywheel Housing
Under Cover

Plug for Line
Pressure Test

37 (380, 27)

◆ Gasket

5.4 (55, 48 in.·lbf)

44 (450, 33)

18 (185, 13)

Front and Center Exhaust Pipe

◆ Gasket

Heated Oxygen Sensor

44 (450, 33)

44 (450, 33)

62 (630, 46)

Ring

RH Front TWC

44 (450, 33)

◆ Gasket

Heated Oxygen
Sensor

Front Floor
Center Brace

Pipe Support Bracket

44 (450, 33)

44 (450, 33)

44 (450, 33)

62 (630, 46)

◆ Gasket

LH Front TWC

7.4 (75, 65 in.·lbf)

Heat Insulator

Engine Under Cover

Heat Insulator

Ground Cable

N·m (kgf·cm, ft·lbf) : Specified torque

◆ Non-reusable part

67162-LEXU-G66

Exploded view of the transmission components for removal—GS 430 (2004 application shown)

- Exhaust pipe-to-catalytic converters, with new gaskets, to 33 ft. lbs. (44 Nm)
- Front floor center brace bolts to 65 inch lbs. (7.4 Nm)
- If removed, oxygen sensors to 33 ft. lbs. (44 Nm)

➡ **Before installation of the oxygen sensor, twist wiring 3½ turns in counterclockwise position; this will prevent wires being twisted after sensors are installed.**

- Engine under cover
- Transmission dipstick
- Negative battery cable

7. Fill the transmission with recommended fluid type.

Transaxle Assembly

REMOVAL & INSTALLATION

Automatic

ES 300

1. Before servicing the vehicle, refer to the precautions in the beginning of this section.
2. Turn the ignition switch to the LOCK position and disconnect the negative battery cable. Wait at least 90 seconds or longer before doing any work on the vehicle.
3. Remove or disconnect the following:
- Battery
- Air cleaner assembly
- Throttle cable from the throttle body
- Cruise control actuator cover and detach the connector
- Ground wire
- Starter
- Vehicle Speed Sensor (VSS) connectors
- Direct clutch speed sensor
- Park/Neutral Position (PNP) switch connector on the transaxle
- Solenoid connector on the transaxle
- Shift control cable
- Oil cooler hoses
- 2 front side transaxle mounting bolts
- 2 front engine mounting bolts
- Oil cooler line mounting bolts from the front frame
- 3 upper transaxle to engine mounting bolts
- Install an engine support fixture. Tie

steering gear housing to engine support fixture.
5. Raise and safely support the vehicle.
6. Drain the transaxle/differential fluid.
7. Remove or disconnect the following:
- Front wheels
- Front exhaust pipe
- Engine side covers and undercovers
- Both halfshafts
- Front side engine mounting nut
- Rear side engine mounting bolts (remove hole plugs)
- 4 left side transaxle mounting bolts
- Steering gear housing
- Front frame assembly
8. Properly support the transaxle assembly.
9. Remove or disconnect the following:
- Rear end plate mounting bolts
- Torque converter cover
- Torque converter retaining bolts
- Remaining transaxle mounting bolts
10. Carefully remove the transaxle assembly from the vehicle.

To install:
11. Position the transaxle, aligning the 2 dowel pins on the block with the converter housing. Tighten the bolts as follows: 10mm–34 ft. lbs. (46 Nm); 12mm–47 ft. lbs. (64 Nm).
12. Install or connect the following:
- Torque converter bolts. Coat the threads with sealer. Install the bolts starting with the green bolt followed by the rest and tighten the bolts evenly to 20 ft. lbs. (27 Nm).
- End plate and tighten the bolts to 27 ft. lbs. (37 Nm).
- Front frame assembly and tighten the fasteners as follows: 12mm–24 ft. lbs. (32 Nm); 19mm–134 ft. lbs. (181 Nm); nut–27 ft. lbs. (36 Nm).
- 2 fender liner set screws
- Steering gear to the frame and tighten the bolts and nuts to 134 ft. lbs. (181 Nm)
- Sway bar brackets and toque the bolts to 14 ft. lbs. (19 Nm)
- Left transaxle mounting bolts and tighten them to 38 ft. lbs. (52 Nm)
- Rear side mounting bolts and nuts and tighten them to 48 ft. lbs. (66 Nm). Install the plugs.
- Front engine mounting nut and tighten it to 59 ft. lbs. (80 Nm)
- Halfshafts
- Right and left engine side covers
- Lower engine cover
- Exhaust pipe to the engine with new gaskets and tighten the nuts to

46 ft. lbs. (62 Nm). Connect the exhaust pipe to the converter with a new gasket and tighten the nuts and bolts to 32 ft. lbs. (43 Nm).
- Wheel
- Engine support
- 4 upper transaxle mounting bolts and tighten them to 47 ft. lbs. (64 Nm)
- Oil cooler clamping bolts to the front frame
- 2 front side engine mounting bolts and tighten them to 59 ft. lbs. (80 Nm)
- 2 front side transaxle mounting bolts and tighten them to 59 ft. lbs. (80 Nm)
- Remaining components
- Battery and connect the battery cables
13. Fill the transaxle/differential to the proper level with Dexron®II, or equivalent.
14. Check the transaxle/differential fluid level.
15. Check the front wheel alignment.

ES 330

1. Before servicing the vehicle, refer to the precautions in the beginning of this section.
2. Turn the ignition switch to the LOCK position and disconnect the negative battery cable. Wait at least 90 seconds or longer before doing any work on the vehicle.
3. Remove the engine and transaxle assembly from the vehicle, following procedures for engine removal.
4. Remove both front driveshaft assemblies.
5. With engine/transaxle assembly on the workbench, or in a suitable holding fixture, remove or disconnect the following:
- Transmission control cable bracket No. 2
- Wiring harness and clamp
- Starter
- Connectors for transmission wire, park/neutral position switch, and 2 transmission revolution sensors
- Transmission control cable bracket No. 1
- Oil cooler tube clamp (near control cable bracket)
- Oil filler tube
- Oil cooler inlet and outlet tubes
- Front right engine mount bracket
- 2 bolts and flywheel housing under cover
- Transaxle mounting bolts (through access hole)
- Transaxle from engine

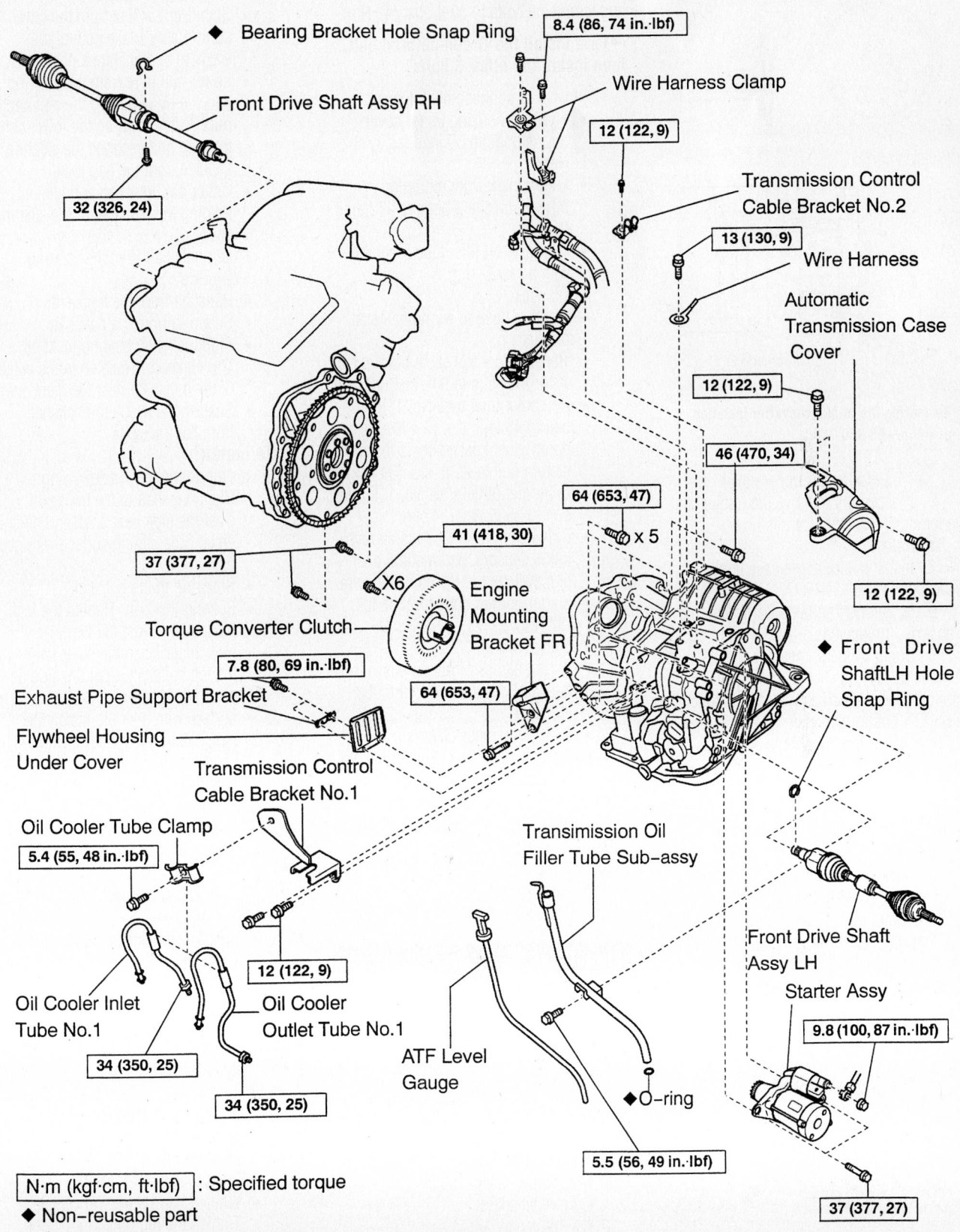

Bearing Bracket Hole Snap Ring

8.4 (86, 74 in.·lbf)

Wire Harness Clamp

Front Drive Shaft Assy RH

12 (122, 9)

Transmission Control Cable Bracket No.2

32 (326, 24)

13 (130, 9)

Wire Harness

Automatic Transmission Case Cover

12 (122, 9)

46 (470, 34)

37 (377, 27)

41 (418, 30)

64 (653, 47)

12 (122, 9)

X6

Torque Converter Clutch

Engine Mounting Bracket FR

Front Drive ShaftLH Hole Snap Ring

7.8 (80, 69 in.·lbf)

Exhaust Pipe Support Bracket

64 (653, 47)

Flywheel Housing Under Cover

Transmission Control Cable Bracket No.1

Oil Cooler Tube Clamp

Transmission Oil Filler Tube Sub-assy

5.4 (55, 48 in.·lbf)

Front Drive Shaft Assy LH

Starter Assy

12 (122, 9)

Oil Cooler Inlet Tube No.1

Oil Cooler Outlet Tube No.1

9.8 (100, 87 in.·lbf)

34 (350, 25)

34 (350, 25)

ATF Level Gauge

◆ O-ring

5.5 (56, 49 in.·lbf)

N·m (kgf·cm, ft·lbf) : Specified torque

◆ Non-reusable part

37 (377, 27)

67162-LEXU-G67

Exploded view of the transaxle assembly components for removal—ES 330

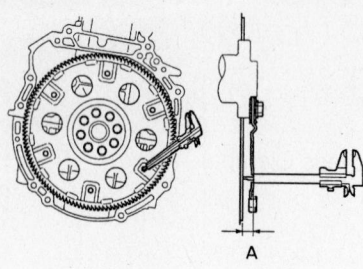

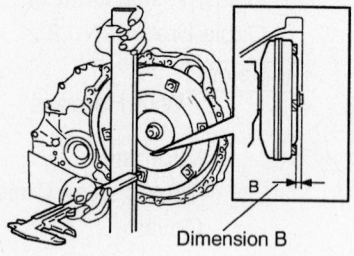

Dimension B

67162-LEXU-G68

Measuring the torque converter installed position—ES 330

- Transaxle case upper cover

6. If necessary, remove the torque converter clutch assembly at this time.

To install:

7. Install the transaxle case upper cover.

8. Install the torque converter into the transaxle, if removed.

9. Using calipers, measure the installed dimension "A", as shown. Then, using a straightedge and calipers, measure dimension "B". Ensure dimension "B" is at least 0.25" (1mm) more than dimension "A". If not, recheck the torque converter installation.

10. Install the automatic transaxle assembly to the engine and torque the bolts as follows:

- Bolt A: 47 ft. lbs. (64 Nm)
- Bolt B: 34 ft. lbs. (46 Nm)
- Bolt C: 27 ft. lbs. (37 Nm)

11. Apply a few drops of thread lock compound to 2 threads of each of the 6

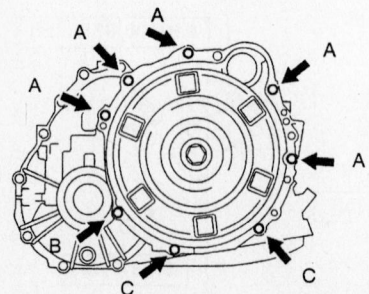

67162-LEXU-G69

Identifying the automatic transaxle mounting bolts for specific tightening requirements—ES 330

torque converter mounting bolts. Install the bolts through the cover plate access hole and torque the bolts to 30 ft. lbs. (41 Nm).

➡ **First install the Green-colored bolt, then install the other 5 bolts.**

12. Install or connect the following:
- Flywheel housing under cover; torque bolts to 69 inch lbs. (7.8 Nm)
- Engine right front mounting bracket; torque 3 bolts to 47 ft. lbs. (64 Nm)
- Transaxle oil filler tube assembly, with a new O-ring
- Dipstick
- Breather hose to wiring harness bracket
- Transaxle control cable bracket No. 1; torque bolts to 9 ft. lbs. (12 Nm)
- Oil cooler inlet tube No. 1; torque bolt to 48 inch lbs. (5.4 Nm) temporarily until assembly is installed, then torque to 25 ft. lbs. (34 Nm)
- Oil cooler outlet tube No. 1; torque bolt to 25 ft. lbs. (34 Nm)
- Connectors for transmission revolution sensors, park/neutral position switch, and transmission wire.
- Starter; torque bolts to 27 ft. lbs. (37 Nm)
- Starter wiring
- Wiring harness (ground strap) to transaxle; torque bolt to 9 ft. lbs. (12 Nm)
- 2 wiring harness clamps
- Transaxle control cable bracket No. 2; torque bolt to 9 ft. lbs. (12 Nm)

13. Install both front driveshafts.

14. Install the engine assembly, with transaxle, into the vehicle.

15. Refill the transaxle with recommended fluid type.

16. Check the front wheel alignment.

Halfshaft

REMOVAL & INSTALLATION

ES 300

1. Before servicing the vehicle, refer to the precautions in the beginning of this section.

2. Remove or disconnect the following:
- Negative battery cable
- Front wheel(s)
- Front fender apron seal
- Transaxle fluid
- Tie rod end from the steering knuckle by removing the cotter pin

and nut. Separate the tie rod from the steering knuckle.
- Stabilizer bar link from the lower control arm. Make note of the washers and cushions positions.
- Lower ball joint from the steering knuckle by removing the bolt and 2 nuts. Push down on the lower control arm and separate the steering knuckle from the ball joint.
- Cotter pin, lock cap and locknut holding the halfshaft to the steering knuckle
- Left halfshaft from the steering knuckle
- Halfshaft from the transaxle
- Snapring from the halfshaft
- Right halfshaft bearing lockbolt. The lockbolt is located in the center of the halfshaft, near the dampener.
- Snapring and pull the halfshaft from the transaxle

To install:

3. Install or connect the following:
- Right halfshaft to the transaxle. Coat the side gear shaft and differential case sliding surface with gear oil.
- Snapring to the halfshaft
- Bearing lockbolt. Tighten the lockbolt to 24 ft. lbs. (32 Nm).
- New snapring to the inner spline of the left halfshaft. Coat the side gear shaft and differential case sliding surface with gear oil. Install the halfshaft to the transaxle with the snapring opening facing down. The halfshaft should click into place when installing.
- Halfshaft to the steering knuckle, then install the locknut. Tighten the locknut to 217 ft. lbs. (294 Nm).
- Lock cap and a new cotter pin to the halfshaft
- Steering knuckle to the lower ball joint. Install the 2 nuts and bolt. Tighten the nuts and bolt to 94 ft. lbs. (127 Nm).
- Stabilizer bar link to the lower control arm. Tighten the nut to 29 ft. lbs. (39 Nm).
- Tie rod to the steering knuckle and tighten the nut to 36 ft. lbs. (49 Nm). Install a new cotter pin to the tie rod end.
- Front fender apron seal
- Wheel(s) and lower the vehicle. Tighten the lug nuts to 76 ft. lbs. (103 Nm).
- Transaxle fluid
- Negative battery cable

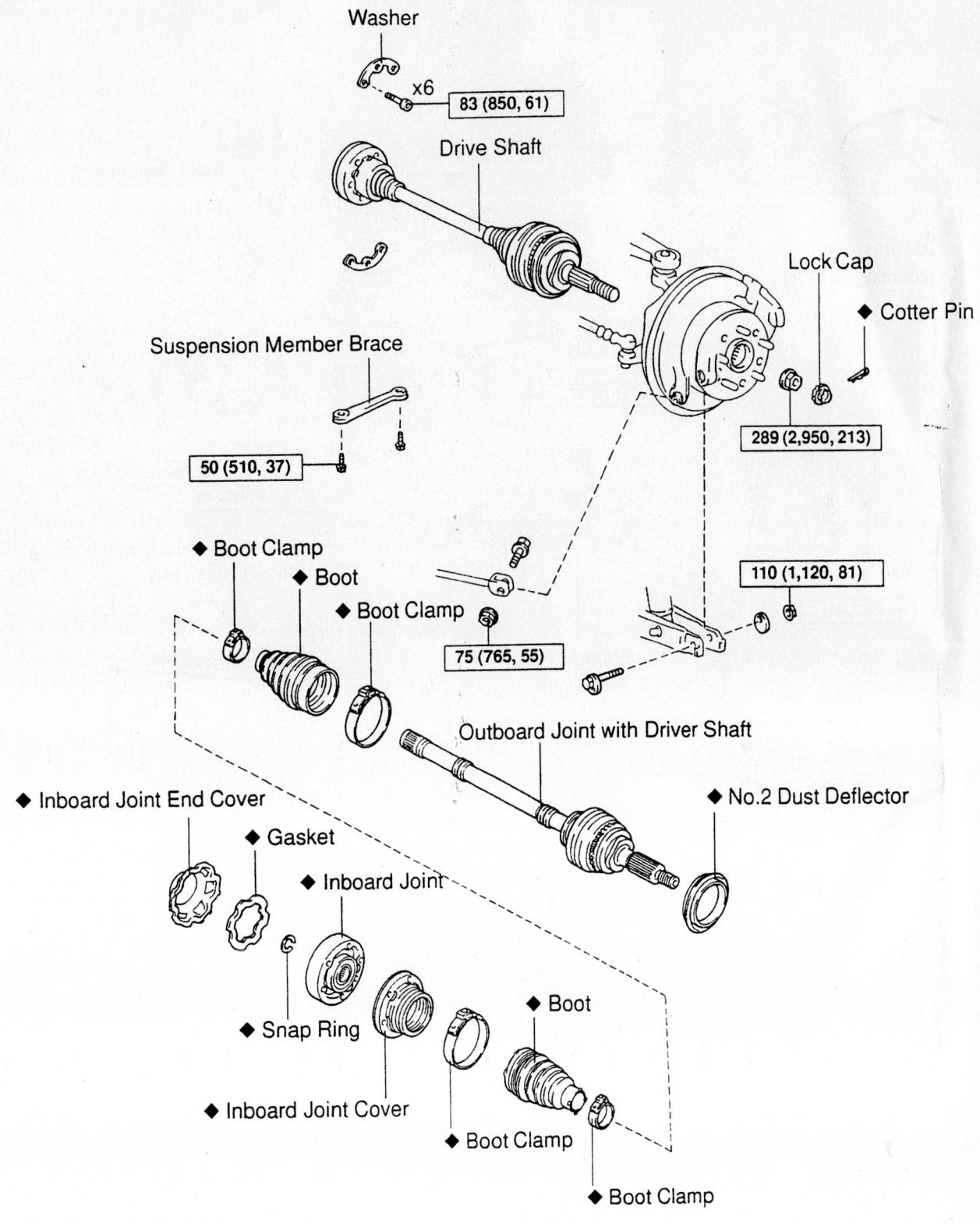

Washer

x6

83 (850, 61)

Drive Shaft

Lock Cap

◆ Cotter Pin

Suspension Member Brace

289 (2,950, 213)

50 (510, 37)

◆ Boot Clamp

◆ Boot

◆ Boot Clamp

110 (1,120, 81)

75 (765, 55)

Outboard Joint with Driver Shaft

◆ Inboard Joint End Cover

◆ No.2 Dust Deflector

◆ Gasket

◆ Inboard Joint

◆ Snap Ring

◆ Boot

◆ Inboard Joint Cover

◆ Boot Clamp

◆ Boot Clamp

N·m (kgf·cm, ft·lbf) : Specified torque
◆ Non–reusable part

9301LG12

Exploded view of the rear halfshaft and related components—except ES 300

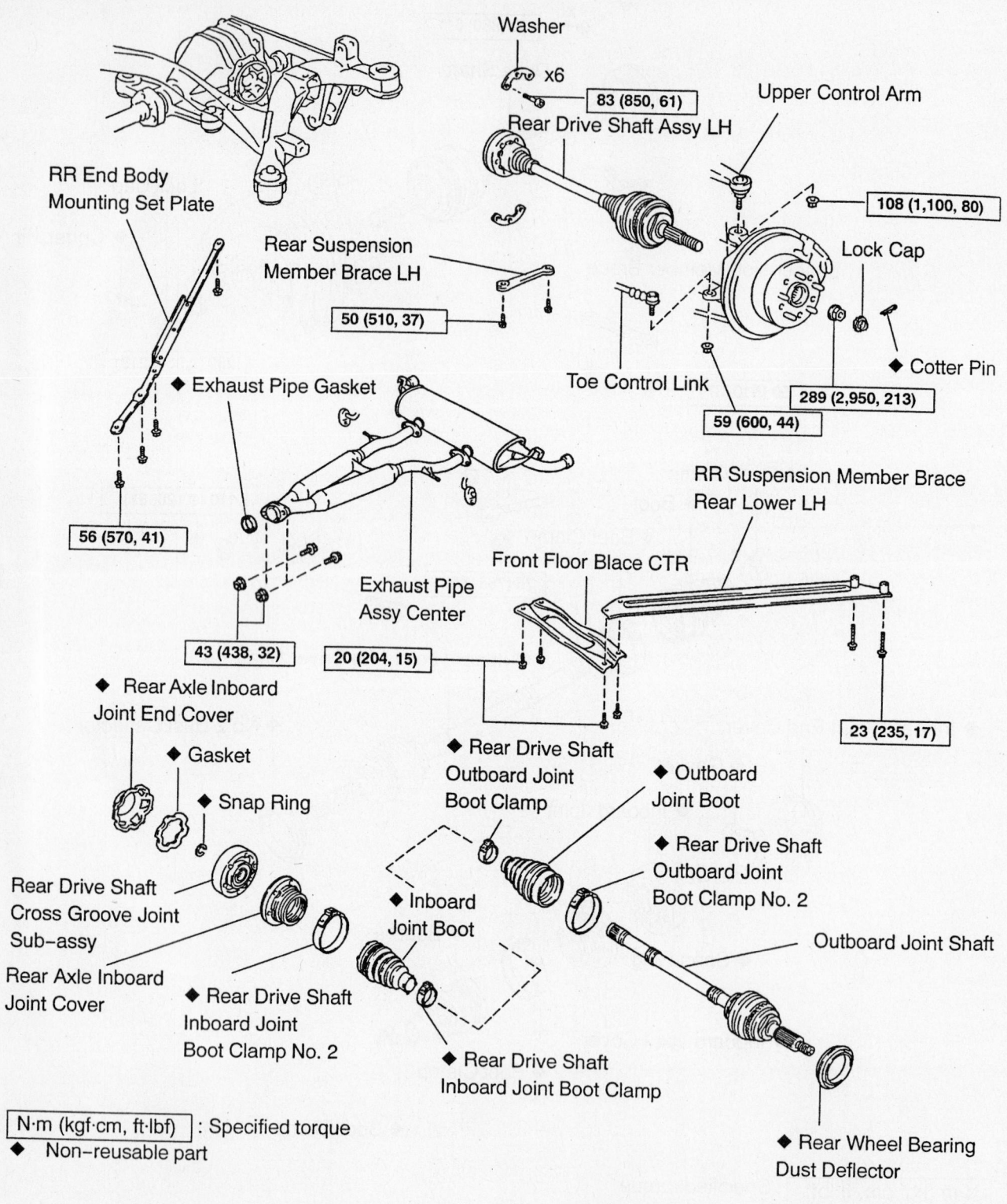

Washer

x6

83 (850, 61)

Rear Drive Shaft Assy LH

Upper Control Arm

108 (1,100, 80)

RR End Body Mounting Set Plate

Lock Cap

Rear Suspension Member Brace LH

50 (510, 37)

◆ Cotter Pin

Toe Control Link

289 (2,950, 213)

59 (600, 44)

◆ Exhaust Pipe Gasket

RR Suspension Member Brace Rear Lower LH

56 (570, 41)

Front Floor Blace CTR

Exhaust Pipe Assy Center

20 (204, 15)

23 (235, 17)

43 (438, 32)

◆ Rear Axle Inboard Joint End Cover

◆ Rear Drive Shaft Outboard Joint Boot Clamp

◆ Outboard Joint Boot

◆ Gasket

◆ Rear Drive Shaft Outboard Joint Boot Clamp No. 2

◆ Snap Ring

Rear Drive Shaft Cross Groove Joint Sub–assy

◆ Inboard Joint Boot

Outboard Joint Shaft

Rear Axle Inboard Joint Cover

◆ Rear Drive Shaft Inboard Joint Boot Clamp No. 2

◆ Rear Drive Shaft Inboard Joint Boot Clamp

N·m (kgf·cm, ft·lbf) : Specified torque
◆ Non–reusable part

◆ Rear Wheel Bearing Dust Deflector

67162-LEXU-G70

Exploded view rear drive shaft assembly—SC 430

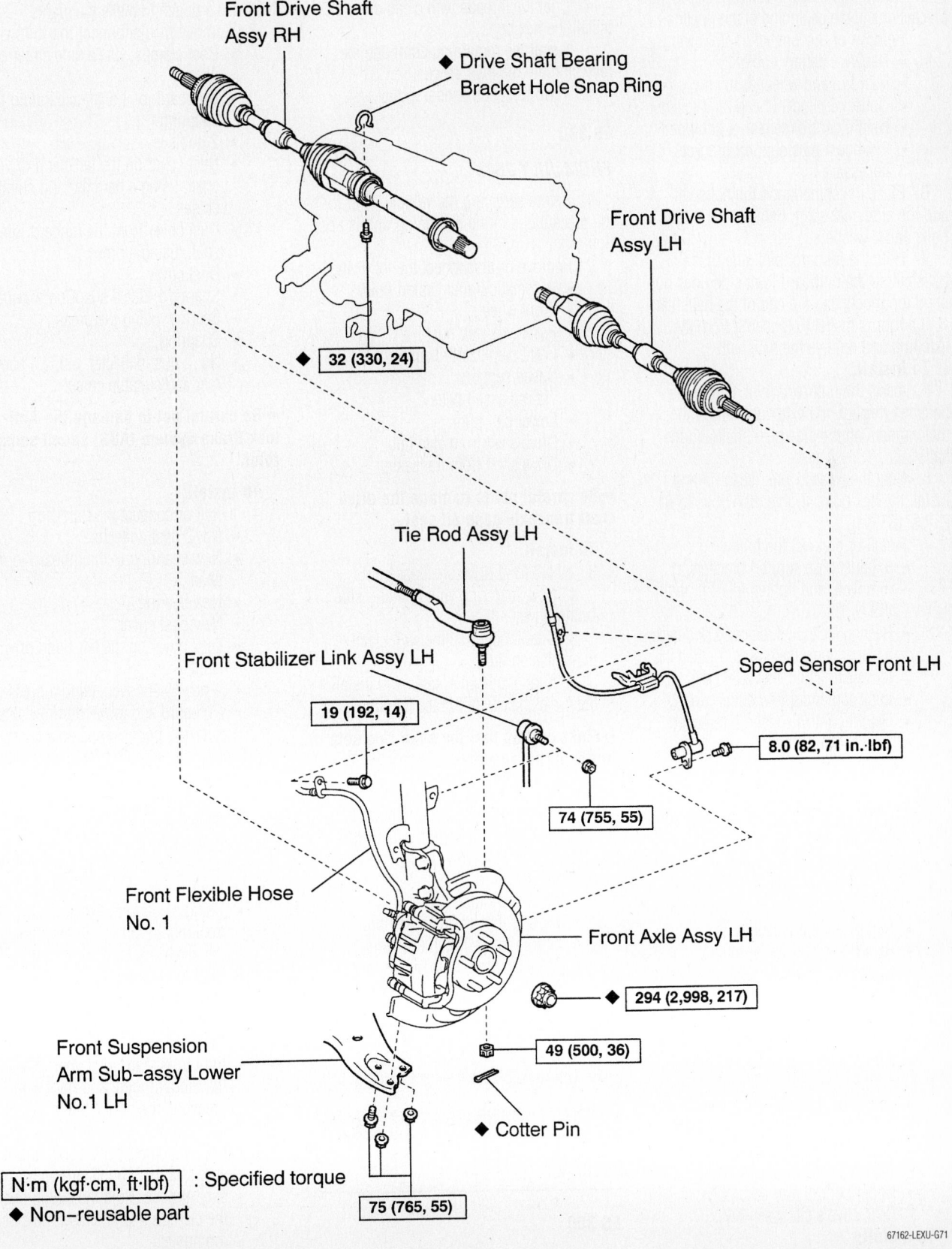

Front Drive Shaft Assy RH

◆ Drive Shaft Bearing Bracket Hole Snap Ring

Front Drive Shaft Assy LH

◆ 32 (330, 24)

Tie Rod Assy LH

Front Stabilizer Link Assy LH

Speed Sensor Front LH

19 (192, 14)

8.0 (82, 71 in.·lbf)

74 (755, 55)

Front Flexible Hose No. 1

Front Axle Assy LH

◆ 294 (2,998, 217)

49 (500, 36)

Front Suspension Arm Sub-assy Lower No.1 LH

◆ Cotter Pin

N·m (kgf·cm, ft·lbf) : Specified torque

◆ Non-reusable part

75 (765, 55)

67162-LEXU-G71

Exploded view drive shaft assembly—ES 330

Except ES 300, SC 430 and ES 330

1. Before servicing the vehicle, refer to the precautions in the beginning of this section.

2. Remove or disconnect the following:
- Negative battery cable
- Rear tire and wheel assembly
- Cotter pin, locknut cap, and locknut
- Height control sensor, if equipped
- 2 exhaust pipe support brackets, if necessary

3. Place matchmarks on the halfshaft and the side gear shaft. Remove the 6 hex bolts and 2 washers.

4. Hold the inboard joint side of the halfshaft so the outboard joint side does not bend too much. Tap the end of the halfshaft with a rubber mallet to loosen it from the axle hub and remove the halfshaft.

To install:

5. Insert the outboard joint side of the halfshaft through the axle hub. Align the matchmarks on the side gear shaft and the halfshaft.

6. Coat the threads with clean oil and install the hex bolts. Tighten the bolts to 61 ft. lbs. (83 Nm).

7. Install or connect the following:
- Exhaust pipe support brackets, if removed, and tighten to 14 ft. lbs. (19 Nm)
- Bearing locknut, if removed, and have a helper apply the brakes. Tighten the locknut to 213 ft. lbs. (289 Nm).
- Lock cap and a new cotter pin
- Height control sensor, if removed
- Rear tire and wheel assembly
- Negative battery cable

SC 430

1. Before servicing the vehicle, refer to the precautions in the beginning of this section.

2. Remove or disconnect the following:
- Negative battery cable
- Rear tire and wheel assembly
- Cotter pin, locknut cap, and locknut
- Front center floor brace
- 2 exhaust pipe support brackets, if necessary
- Rear lower suspension member brace
- End body mounting set plate
- Rear suspension member brace
- Rear wheel speed sensor
- Toe control link assembly
- Upper control arm assembly
- Rear drive shaft assembly

To install:

3. Insert the outboard joint side of the halfshaft through the axle hub. Align the

matchmarks on the side gear shaft and the halfshaft.

4. Coat the threads with clean oil and install the hex bolts.

5. Install the remaining components reversing the removal procedure.

6. Perform speed sensor signal test.

ES 330

FROM JULY 2003

1. Before servicing the vehicle, refer to the precautions in the beginning of this section.

2. Remove or disconnect the following:
- Automatic transmission fluid
- Front wheels
- Front axle hub nut
- Front stabilizer link assembly
- Speed sensors
- Tie rod assemblies
- Lower ball joint
- Drive shaft from axle hub
- Drive shaft from transaxle

➡**Be careful not to damage the drive shaft transaxle case oil seal.**

To install:

3. Install front drive shaft assembly

a. New snap ring with opening side facing down

b. Coat inboard spline with clean transmission fluid

c. Align the shaft splines and install the drive shaft with a brass hammer

➡**Make certain that the snap ring sets firmly in the transaxle**

4. Install or connect the following:
- Lower ball joint to spindle assembly. Torque to 55 ft. lbs. (75 Nm)
- Tie rod assembly. Torque to 36 ft. lbs. (49 Nm)
- Speed sensors
- Front stabilizer link assembly Torque to 55 ft. lbs. (74 Nm)
- Front axle hub nut. Torque to 217 ft. lbs. (294 Nm)
- Front wheels
- Automatic transmission fluid

5. Align the front wheels

6. Check ABS speed sensor signal

CV-Joints

OVERHAUL

ES 300

1. Before servicing the vehicle, refer to the precautions in the beginning of this section.

2. Once the driveshaft is removed from

the vehicle, place matchmarks on the outboard and inboard joints and the shaft. Do NOT use a punch to make the marks.

3. Remove or disconnect the following:
- Boot clamps. Use a side cutter or pliers
- Outboard joint shaft expanding the snapring
- 2 boots
- Dust cover on the left-hand driveshaft, using a hammer and suitable chisel
- Dust cover from the inboard joint shaft, using a press
- Dust cover
- Snapring. Use a snapring expander.
- Bearing, using the press
- Snapring
- No. 2 dust deflector, using a hammer and suitable chisel

➡**Be careful not to damage the Anti-lock Brake System (ABS) speed sensor rotor.**

To install:

4. Install or connect the following:
- No. 2 dust deflector
- New snapring to the inboard joint shaft
- New bearing
- New dust cover
- Dust cover on the left-hand driveshaft
- A new dust cover, using a press
- Outboard and inboard joint boots and new boot clamps as a temporary measure. Before installing the boot, place 3 new clamps to the small boot ends and large end (wheel side) and install it to the driveshaft.
- Inboard joint shaft to outboard joint shaft
- Using a snapring expander, put in the inboard joint shaft expanding the snapring
- Boot to outboard joint, before assembling the boot, pack the outboard joint and boot with grease in the boot kit
- Boot to inboard joint shaft. Pack the inboard joint, and boot with grease in the boot kit, and install the boot to the inboard joint shaft.
- Boot clamps to both boots, make sure that the 2 boots are on the shaft groove. Hold the clamp near the clamp's free end over the closing hooks.

5. Secure the clamp by drawing the closing hooks together. Secure the clamp onto the boot.

GS 300, GS 430 and IS 300

1. Before servicing the vehicle, refer to the precautions in the beginning of this section.

2. Remove or disconnect the following:
- End cover. Use nuts and bolts to keep the inboard joint together. Hand tighten only.
- 4 Boot clamps, using a side cutter or pliers
- Inboard joint, place matchmarks on the inboard joint and driveshaft; do not use punch marks.
- Snapring, using a snapring expander
- Using a press, the inboard joint from the driveshaft
- Inboard and outboard boot
- Inboard joint cover from the inboard joint

To install:

3. Install or connect the following:
- Inboard and outboard joint boots
- New No. 2 Dust deflector, using a press

➡ **Be careful not to damage the Anti-lock Brake System (ABS) speed sensor rotor.**

- Inboard joint
- Inner race to the cage so that the indented beveled part of the inner race is on the opposite side to the beveled top of the cage
- Outer race so that the indented side of the outer race is facing the same side as the beveled surface of the cage

4. Match the narrow projections of the inner race with the wide projections of the outer race.

5. Tilt the cage and inner race to the side and insert the balls one by one.

6. Install or connect the following:
- New boots and new boot clamps, temporarily
- 4 new boot clamps to the boots
- 2 boots to the driveshaft
- Inboard joint cover and apply RTV to the inboard joint cover

7. Remove grease from the surface of the inboard joint facing the cover.

8. Align the bolt holes of the cover with those of the inboard joint, then insert the hexagon bolts.

9. Use a plastic hammer to tap the rim of the inboard joint cover into place

10. To install the inboard joint, align the matchmarks placed before removal.

11. Using a brass bar and hammer, tap the inboard joint onto the driveshaft.

12. Install or connect the following:
- New snapring
- Boots to joints, pack with the proper grease. 3.5–3.7 oz. (100–105g)
- New boot clamps to both boots
- 6 hexagon bolts and washers from the end cover side, install the 6 nuts to the boot side.

LS 430 and SC 430

1. Before servicing the vehicle, refer to the precautions in the beginning of this section.

2. Using a suitable prytool, remove the end cover.

3. Use nuts and bolts to keep the inboard joint together. Hand tighten only.

4. Remove or disconnect the following:
- 4 Boot clamps, using a side cutter or pliers
- Inboard joint, place matchmarks on the inboard joint and driveshaft; do not use punch marks.
- Snapring, using a snapring expander
- Using a press, the inboard joint from the driveshaft
- Inboard joint cover from the inboard joint
- Inboard and outboard boot
- No. 2 dust deflector, using a suitable chisel and hammer
- New No. 3 dust deflector, using a press

To install:

5. If the joint has come apart, reassemble it in the following order.
   a. Align the matchmarks placed before removal
   b. Inner race to the cage so that the indented beveled part of the inner race is on the opposite side to the beveled top of the cage.
   c. Outer race so that the indented side of the outer race is facing the same side as the beveled surface of the cage.

6. Match the narrow projections of the inner race with the wide projections of the outer race.

7. Tilt the cage and inner race to the side and insert the balls one by one.

8. Install or connect the following:
- New boots and new boot clamps, temporarily
- 4 new boot clamps to the boots
- 2 boots to the driveshaft
- Inboard joint cover and apply Formed In Place Gasket (FIPG) to the inboard joint cover. Avoid applying an excessive amount to the surface

9. Remove grease from the surface of the inboard joint facing the cover

10. Align the bolt holes of the cover with those of the inboard joint, then insert the hexagon bolts.

11. Use a plastic hammer to tap the rim of the inboard joint cover into place

12. To install the inboard joint, align the matchmarks placed before removal.

13. Using a brass bar and hammer, tap the inboard joint onto the driveshaft.

14. Install or connect the following:
- New snapring
- Boots to joints, pack with 3.5–3.7 oz. (100–105g) of grease
- A new gasket, with the side with adhesive on it facing toward the outer race side of the inboard joint
- 6 hexagon bolts an washer from the end cover side
- 6 nuts to the boot side

15. Check that the claw of the end cover touches the inboard joint

ES 330

FROM JULY 2003

1. Remove drive shaft

2. Remove inboard drive axle clamps and slide boot off the inboard joint.
   a. Mark the joint and housing for proper alignment during installation.
   b. Remove the snap clip with snap ring expander
   c. Using and brass bar and hammer remove the tripod joint from the shaft
   d. Remove the boot from the shaft

3. Remove outboard drive axle clamps and slide boot off the inboard joint.
   a. Mark the joint and housing for proper alignment during installation.
   b. Remove the snap clip with snap ring expander
   c. Using and brass bar and hammer remove the tripod joint from the shaft
   d. Remove the boot from the shaft

➡ **Clean all the old grease from the joints.**

To install:

4. Install new outboard joint
   a. Slip new boot with clamps over the shaft
   b. Pack the boot with grease
   c. Install the new joint over the shaft
   d. Install new snap rings on the shaft

➡ **Do NOT reuse the old snap rings**

   e. Install joint housing over the joint making certain to align the marks
   f. Slide the boot over the joint and position on joint housing

g. Crimp the clamps using special tool

5. Install new inboard joint

  a. Slip new boot with clamps over the shaft

  b. Pack the boot with grease

  c. Install the new joint over the shaft

  d. Install new snap rings on the shaft

➡ **Do NOT reuse the old snap rings**

  e. Install joint housing over the joint making certain to align the marks

  f. Slide the boot over the joint and position on joint housing

  g. Crimp the clamps using special tool

6. Install new snap ring on inboard shaft before installing in transaxle

7. Install drive shaft assembly

Axle Shaft, Bearing and Seal

REMOVAL & INSTALLATION

GS 300, GS 430 and IS 300

1. Drain the gear oil.
2. Remove or disconnect the following:
   - Rear driveshaft

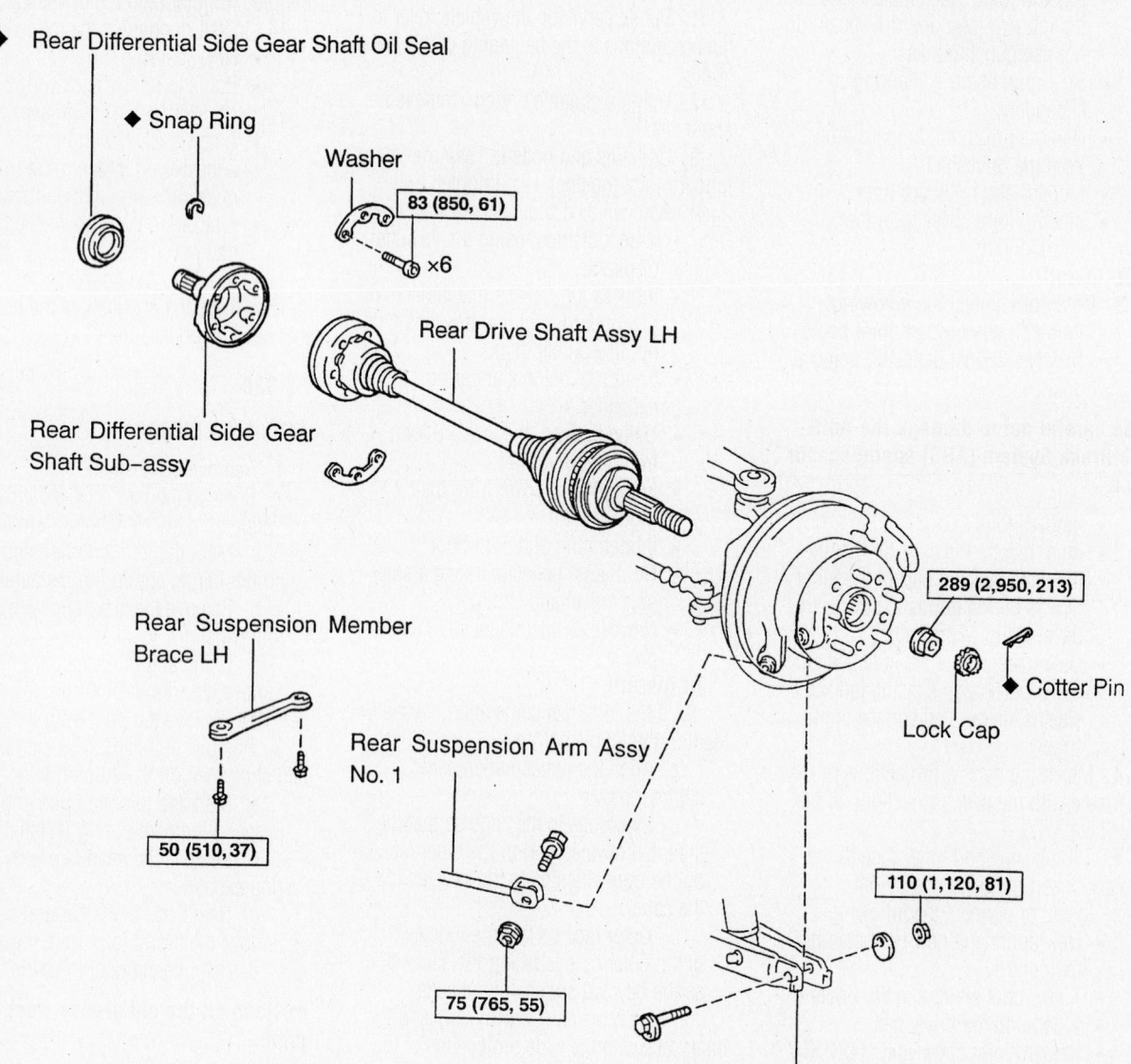

◆ Rear Differential Side Gear Shaft Oil Seal

◆ Snap Ring

Washer

83 (850, 61) ×6

Rear Drive Shaft Assy LH

Rear Differential Side Gear Shaft Sub–assy

Rear Suspension Member Brace LH

50 (510, 37)

Rear Suspension Arm Assy No. 1

75 (765, 55)

289 (2,950, 213)

Lock Cap

◆ Cotter Pin

110 (1,120, 81)

Rear Suspension Arm Assy No. 2

N·m (kgf·cm, ft·lbf) : Specified torque

◆ Non–reusable part

Expanded view, side gear shaft oil seal—SC 430

67162-LEXU-G72

- Side gear shaft
- Snapring from the side gear, using a suitable tool
- Side gear shaft oil seal

To install:

- Side gear shaft oil seal
- New oil seal

3. Check installation of side gear shaft. There should be 0.08–0.12 inch (2–3mm) of play in the axial direction. Check that the side gear shaft will not come out by pulling on it.

LS 430

1. Remove or disconnect the following:
   - Gear oil
   - Rear driveshaft
   - Side gear shaft
   - Snapring from the side gear
   - Side gear shaft oil seal

To install:

2. Install or connect the following:
   - Side gear shaft oil seal
   - New oil seal
   - Multi-Purpose (MP) grease to the oil seal lip
   - Side gear shaft
   - Snapring from the side gear, using a suitable tool
   - Side gear shaft oil seal
   - Gear oil

SC 430

1. Remove or disconnect the following:
   - Rear wheel
   - Differential oil
   - Rear suspension member brace
   - Rear suspension arm
   - Speed sensor
   - Matchmark and remove drive shaft

- Snap ring
- Differential side gear shaft
- Side gear seal using proper puller

To install:

2. Install or connect the following:
   - New snap ring on shaft
   - New seal using proper seal driver
   - Rear differential side gear shaft

➡**Take care not to damage the seal**

- Rear drive shaft assembly making certain to align the matchmarks
- Speed sensor
- Temporarily tighten rear suspension arm No. 2

➡**Do NOT torque nut**

- Temporarily tighten rear suspension arm No. 1

➡**Do NOT torque nut**

- Rear wheels
- Differential oil. Synthetic gear oil GL-5 75W-90 or equivalent

➡**Stabilize the rear suspension**

- Cross member brace

3. Fully tighten suspension arm No. 2. Torque: 81 ft. lbs. (110 Nm)

4. Fully tighten suspension arm No. 1. Torque: 55 ft. lbs (75 Nm)

5. Inspect and adjust rear wheel alignment

6. Speed sensor check.

ES 330

1. Remove or disconnect the following:
   - Front wheels
   - Engine under cover
   - Transaxle fluid
   - Drive shaft assembly

- Transaxle housing oil seal using proper puller
- Differential side bearing retainer oil seal using proper puller

To install:

2. Install or connect the following:
   - Differential side bearing oil seal using proper driver

➡**Coat seal lip with MP grease**

- Transaxle housing oil seal using proper driver

➡**Coat seal lip with MP grease**

- Install drive shaft assembly
- Front wheels
- Automatic Transaxle fluid

3. Check ABS speed sensor signal

Pinion Seal

REMOVAL & INSTALLATION

GS 300 and GS 430

1. Drain the gear oil.
2. Remove the driveshaft and the companion flange
3. Remove the oil seal
4. Check oil slinger

To install:

5. Installation is the reversal of the removal procedure.

LS 430, and SC 430

1. Remove or disconnect:
   - Gear oil
   - Driveshaft
   - Companion flange
   - Oil seal and slinger

To install:

2. Install or connect:
   - Oil slinger
   - New oil seal

➡**Oil seal drive-in depth: 0.079 inch (2.0mm).**

- Multi-Purpose (MP) grease to the oil seal lip
- Companion flange on the shaft
- Gear oil on the threads of a new nut. Torque to 80 ft. lbs. (108 Nm).

3. Adjust the drive pinion preload as necessary, stake drive the pinion nut, and install the driveshaft.

4. Fill the differential with hypoid gear oil

IS 300

1. Before servicing the vehicle, refer to the precautions in the beginning of this section.

2. Remove or disconnect the following:

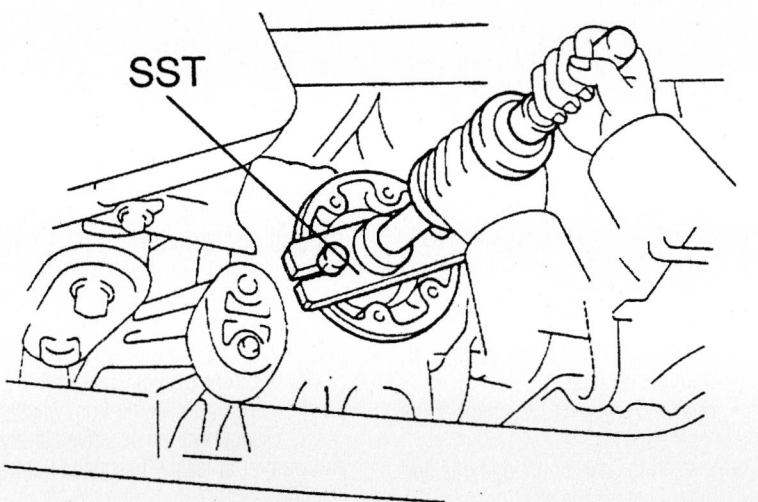

SST

Removing differential side shaft—SC 430

- Driveshaft
- Rear wheels
- Rear brake calipers
- Pinion flange
- Pinion seal

➡ **The rear brake calipers must be removed so that there is no additional drag when measuring pinion bearing preload.**

To install:

3. Install or connect the following:
   - Pinion seal and flange

- New pinion flange nut

4. Rotate the pinion flange occasionally while tightening the flange nut to make sure the pinion bearings seat correctly. Do NOT exceed 249 ft. lbs. (338 Nm).

5. Take frequent bearing preload torque readings.

6. The pinion bearing preload specifications are as follows:
   a. Used bearings: 4.3–6.9 inch lbs. (0.49–0.78 Nm).
   b. New bearings: 8.7–13.9 inch lbs. (0.98–1.57 Nm).

✳✳ CAUTION

Never loosen the pinion nut to reduce bearing preload. If it is necessary to reduce bearing preload, install a new collapsible spacer and pinion nut.

7. Install or connect the following:
   - Driveshaft
   - Brake calipers
   - Rear wheels

8. Fill the differential with gear lubricant and check for leaks.

STEERING AND SUSPENSION

Air Bag

✳✳ CAUTION

These vehicles are equipped with an air bag system. The system must be disabled before performing service on or around system components, steering column, instrument panel components, wiring and sensors. Failure to follow safety and disabling procedures could result in accidental air bag deployment, possible personal injury and unnecessary system repairs.

PRECAUTIONS

Several precautions must be observed when handling the inflator module to avoid accidental deployment and possible personal injury.

- Never carry the inflator module by the wires or connector on the underside of the module.

- When carrying a live inflator module, hold securely with both hands and ensure that the bag and trim cover are pointed away.

- Place the inflator module on a bench or other surface with the bag and trim cover facing up.

- With the inflator module on the bench, never place anything on or close to the module which may be thrown in the event of an accidental deployment.

DISARMING

To avoid personal injury when working on vehicles equipped with an air bag, the negative battery cable must be disconnected and at least 90 seconds must elapse before working on the system. Failure to do so may result in deployment of the air bag.

ARMING

To rearm the air bag system, simply reconnect the battery cable(s).

Power Rack and Pinion Steering Gear

REMOVAL & INSTALLATION

ES 300

1. Before servicing the vehicle, refer to the precautions in the beginning of this section.

2. Remove or disconnect the following:
   - Negative battery cable and wait at least 90 seconds before working on the vehicle to disarm the air bag.
   - Front wheels
   - Left and right front fender apron seals
   - Cotter pin and nut holding the steering knuckle to the tie rod end. Using a tie rod puller, disconnect the tie rod end from the steering knuckle.

3. Place matchmarks on the intermediate shaft and the control valve shaft.

4. Loosen the upper bolt and remove the lower bolt holding the control valve shaft to the intermediate shaft. Disconnect the intermediate shaft from steering rack housing.

5. Remove or disconnect the following:
   - Tube clamp
   - Return line and the pressure line from the control valve housing
   - 4 stabilizer bar bolts and 2 nuts. Position the stabilizer bar out of the way. Do NOT remove the sway bar from the vehicle.
   - Heated Oxygen (HO_2) sensor (bank 1 sensor 1)
   - 2 steering gear mounting bolts and nuts. Remove the steering gear through the left side of the vehicle.

To install:

6. Install or connect the following:
   - Steering gear on the vehicle and install the 2 mounting bolts and nuts. Tighten the nuts and bolts to 134 ft. lbs. (181 Nm).
   - HO_2 sensor. Tighten the sensor to 33 ft. lbs. (44 Nm).
   - Stabilizer bar bolts and nuts and tighten as follows: Bolts: 14 ft. lbs. (19 Nm); Nuts: 29 ft. lbs. (39 Nm).
   - Pressure and return lines and tighten the connectors to 18 ft. lbs. (25 Nm)
   - Tube clamp and tighten the nut to 84 inch lbs. (10 Nm)
   - Intermediate shaft to the steering rack and tighten the retaining bolts to 26 ft. lbs. (35 Nm)
   - Tie rods to the steering knuckles with the castellated nuts. Tighten the nut to 36 ft. lbs. and install a new cotter pin. The prongs of the cotter pin should be firmly wrapped around the flats of the nut.
   - Front fender apron seals by installing the 2 bolts
   - Front wheels and lower the vehicle
   - Power steering fluid
   - Negative battery cable

7. Release the steering wheel
8. Bleed the system
9. Check for leaks, adjust the toe-in and check the steering wheel center point

GS 300 and GS 430

1. Before servicing the vehicle, refer to the precautions in the beginning of this section.

2. Set front wheels in straight-ahead position, then remove the front wheels.

3. Disconnect the negative battery cable and wait for at least 90 seconds before proceeding.

4. Remove the steering wheel pad:

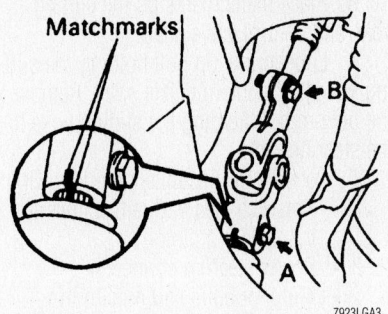

Matchmarks

7923LGA3

Matchmarking the intermediate shaft to the control valve shaft—GS 300 and GS 430

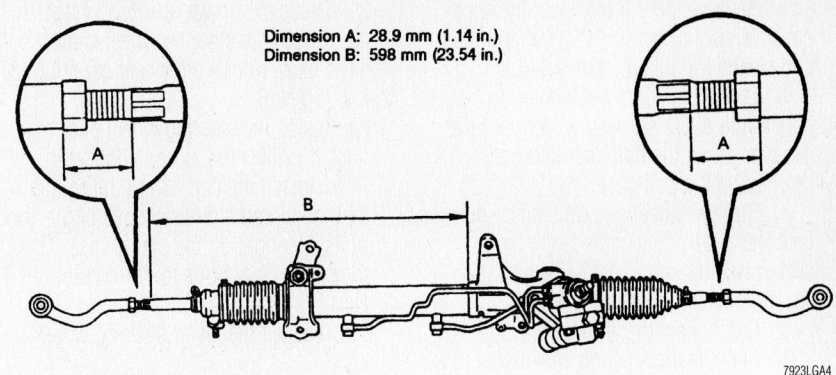

Dimension A: 28.9 mm (1.14 in.)
Dimension B: 598 mm (23.54 in.)

7923LGA4

Centering the rack and pinion—GS 300 and GS 430

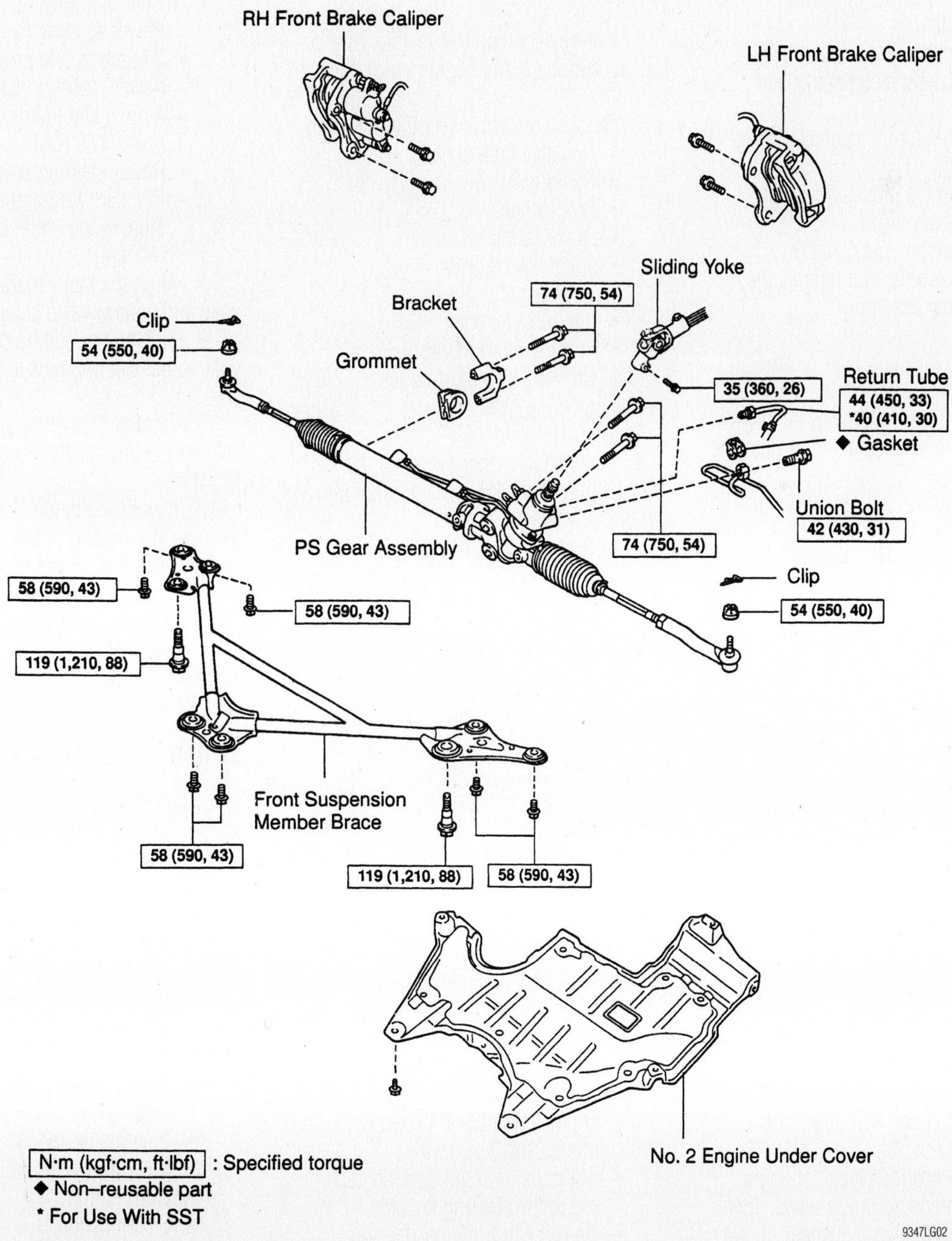

RH Front Brake Caliper

LH Front Brake Caliper

Sliding Yoke

74 (750, 54)

Clip

Bracket

54 (550, 40)

Grommet

35 (360, 26)

Return Tube
44 (450, 33)
*40 (410, 30)

◆ Gasket

Union Bolt
42 (430, 31)

74 (750, 54)

58 (590, 43)

58 (590, 43)

PS Gear Assembly

Clip

54 (550, 40)

119 (1,210, 88)

Front Suspension
Member Brace

58 (590, 43)

119 (1,210, 88)   58 (590, 43)

No. 2 Engine Under Cover

N·m (kgf·cm, ft·lbf) : Specified torque
◆ Non—reusable part
* For Use With SST

9347LG02

Exploded view of the steering gear mounting—IS 300

a. Remove the 2 lower steering wheel covers (3.0L) or the No. 2 cover and the pad switch modulator cover (4.3L).

b. Loose the Torx® head bolts on either side of the steering wheel until the grooves along the bolt circumference catches on the bolt case.

c. Pull the wheel pad out and disconnect the airbag connector. Store the airbag face up and in a safe place.

5. Remove the steering wheel:

a. Disconnect the wiring connector.

b. Remove the steering wheel set nut.

c. Make matchmarks across steering wheel and main shaft assembly.

6. Disconnect the left and right tie rod ends, placing matchmarks on screw portion for proper reinstallation.

7. Disconnect the intermediate shaft assembly after making matchmarks as shown.

8. Remove the front suspension member brace.

9. Disconnect the pressure feed and return power steering pipes.

10. Disconnect the tube clamp(s).

11. Remove the bolts and remove the power steering gear assembly.

To install:

12. Center the rack and pinion gear assembly as shown.

13. Install the steering gear assembly to its mounting position. Install and torque the bolts as follows:

- Except 2004: 72 ft. lbs. (98 Nm)
- 2004: 48 ft. lbs. (65 Nm)

14. Reconnect the tube clamp(s); torque tube clamps and tighten the bolt to 12 ft. lbs. (17 Nm).

15. Connect the steering gear to the intermediate shaft assembly and torque the bolts above and below the U-joint to 26 ft. lbs. (35 Nm).

16. Reconnect the pressure feed tube and the return tube, with a new gasket on the union bolts. Torque the bolts to 36 ft. lbs. (49 Nm).

17. Reconnect the tie rod ends, installing them to their original positions as marked.

18. Install the front wheels and position them in a straight-ahead position.

19. Install and center the spiral cable:

a. Turn the spiral cable counterclockwise by hand until it becomes hard to turn.

b. Then, rotate the cable clockwise about 3 turns to align the marks at the bottom of the assembly.

20. Install the steering wheel, aligning the matchmarks made during removal. Temporarily install and tighten the wheel set nut. Connect the connector.

21. Bleed the power steering system.

22. Check the steering wheel center point; if okay, torque the wheel set nut to 37 ft. lbs. (50 Nm).

23. Install the steering wheel pad:

a. Connect the airbag connector.

b. Install the pad, after confirming that the circumference groove of the Torx bolts is caught on the screw case. Then, torque the Torx bolts to 78 inch lbs. (8.8 Nm).

c. Install the lower steering wheel covers.

24. Reconnect the negative battery cable.

25. Check the front wheel alignment.

IS 300

1. Before servicing the vehicle, refer to the precautions in the beginning of this section.

2. Remove or disconnect the following:

- Negative battery cable
- Steering wheel
- Front wheels
- Brake calipers
- Outer tie rod ends
- Engine under cover
- Intermediate shaft
- Front subframe brace
- Pressure and return lines
- Steering gear

To install:

3. Installation is the reverse of the removal procedure, while using the following torque values:

- Steering gear mounting bracket bolts: 54 ft. lbs. (74 Nm)
- Fluid return line: 30 ft. lbs. (40 Nm)
- Fluid pressure line: 31 ft. lbs. (42 Nm)
- Front subframe brace large bolts: 88 ft. lbs. (119 Nm)
- Front subframe brace small bolts: 43 ft. lbs. (58 Nm)
- Tie rod end nuts: 40 ft. lbs. (54 Nm)
- Steering wheel nut: 26 ft. lbs. (35 Nm)

LS 430 (except 2004)

1. Before servicing the vehicle, refer to the precautions in the beginning of this section.

2. Remove or disconnect the following:

- Wheel(s)
- Engine undercover by removing the 8 bolts and 5 screws
- Cotter pin and nut holding each tie rod to the steering knuckle
- Tie rod end from the steering knuckle with a tie rod end puller

3. Place matchmarks on the sliding yoke and control valve shaft.

4. Loosen the top bolt holding the sliding yoke to the intermediate shaft. Remove the bottom bolt holding the sliding yoke to the steering rack.

5. Remove or disconnect the following:

- Pressure feed and return lines to the rack and pinion
- Power steering connector
- 4 mount bolts and nuts to the power steering rack
- 2 brackets and grommets
- Power steering rack from the vehicle

To install:

6. Install or connect the following:

- Power steering rack to the vehicle.
- 2 brackets and grommets to the power steering rack.
- 4 bolts and tighten the bolts to 56 ft. lbs. (76 Nm).
- Power steering solenoid connector
- Pressure feed and return tubes. Tighten the union bolt to 36 ft. lbs. (49 Nm).

7. Align the matchmarks on the sliding yoke and control valve shaft.

8. Tighten the bolt holding the sliding yoke to the steering rack to 26 ft. lbs. (35 Nm).

9. Tighten the bolt holding the sliding yoke to the intermediate shaft to 26 ft. lbs. (35 Nm).

10. Install or connect the following:

- Tie rod end to the steering knuckle. Tighten the nut to 48 ft. lbs. (65 Nm). Install a new cotter pin.
- Engine undercover
- Wheel(s)

11. Bleed the power steering system and check the front end alignment.

SC 430 AND 2004 LS 430

1. Before servicing the vehicle, refer to the precautions in the beginning of this section.

2. Place the front wheels facing straight ahead.

3. Remove or disconnect the following:

- Negative battery cable. Wait at least 90 seconds before performing any work.
- Front wheels
- Steering wheel switch & volume case
- Steering wheel pad
- Steering wheel column lower cover
- Horn button
- Steering wheel
- Brake caliper

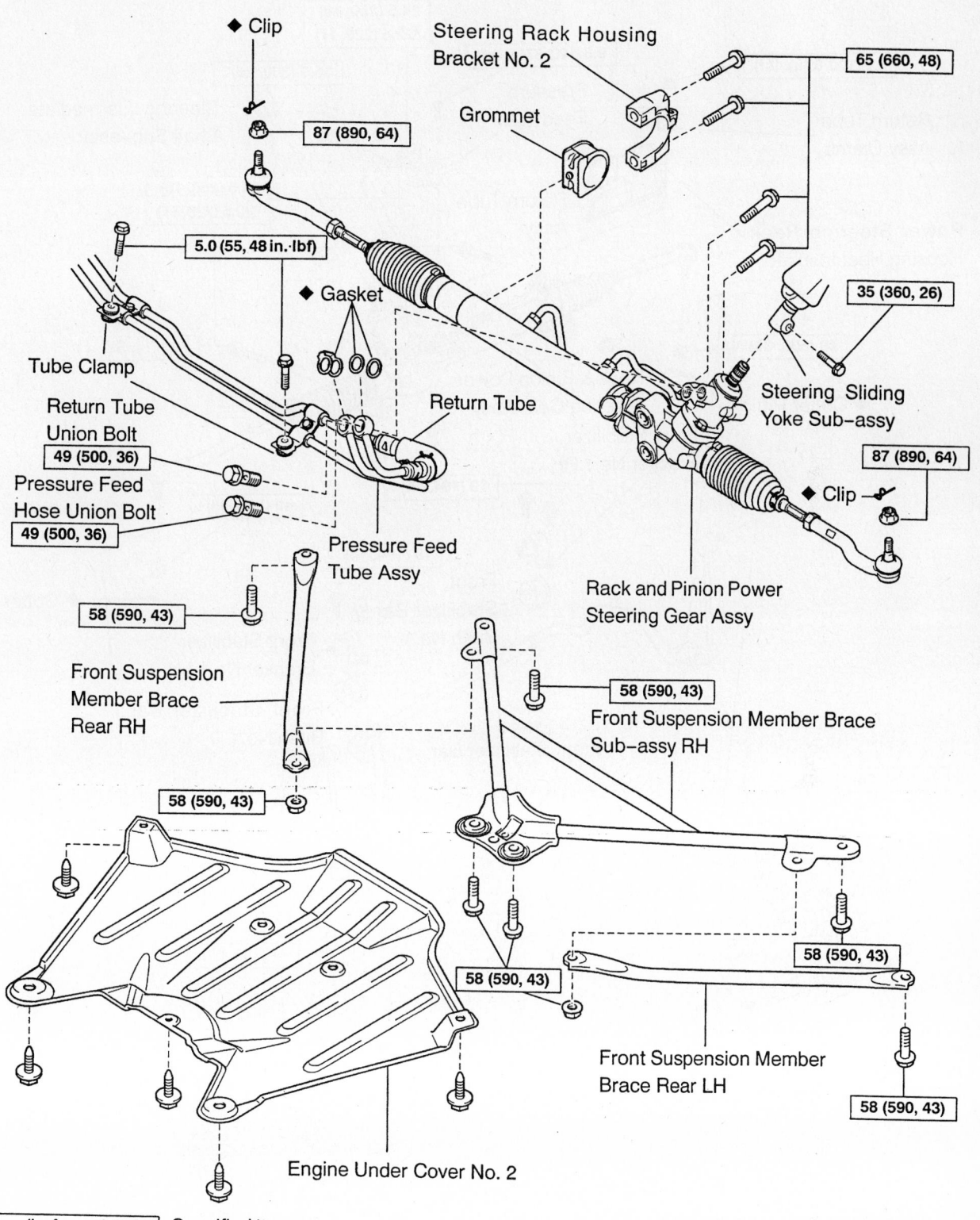

◆ Clip

Steering Rack Housing
Bracket No. 2

65 (660, 48)

Grommet

87 (890, 64)

5.0 (55, 48 in.·lbf)

◆ Gasket

Return Tube

35 (360, 26)

Tube Clamp

Steering Sliding
Yoke Sub-assy

Return Tube
Union Bolt

49 (500, 36)

Pressure Feed
Hose Union Bolt

49 (500, 36)

87 (890, 64)

◆ Clip

Pressure Feed
Tube Assy

Rack and Pinion Power
Steering Gear Assy

58 (590, 43)

Front Suspension
Member Brace
Rear RH

58 (590, 43)

Front Suspension Member Brace
Sub-assy RH

58 (590, 43)

58 (590, 43)

58 (590, 43)

Front Suspension Member
Brace Rear LH

58 (590, 43)

Engine Under Cover No. 2

N·m (kgf·cm, ft·lbf) : Specified torque

◆ Non-reusable part

67162-LEXU-G74

Exploded view power steering gear assembly—SC 430

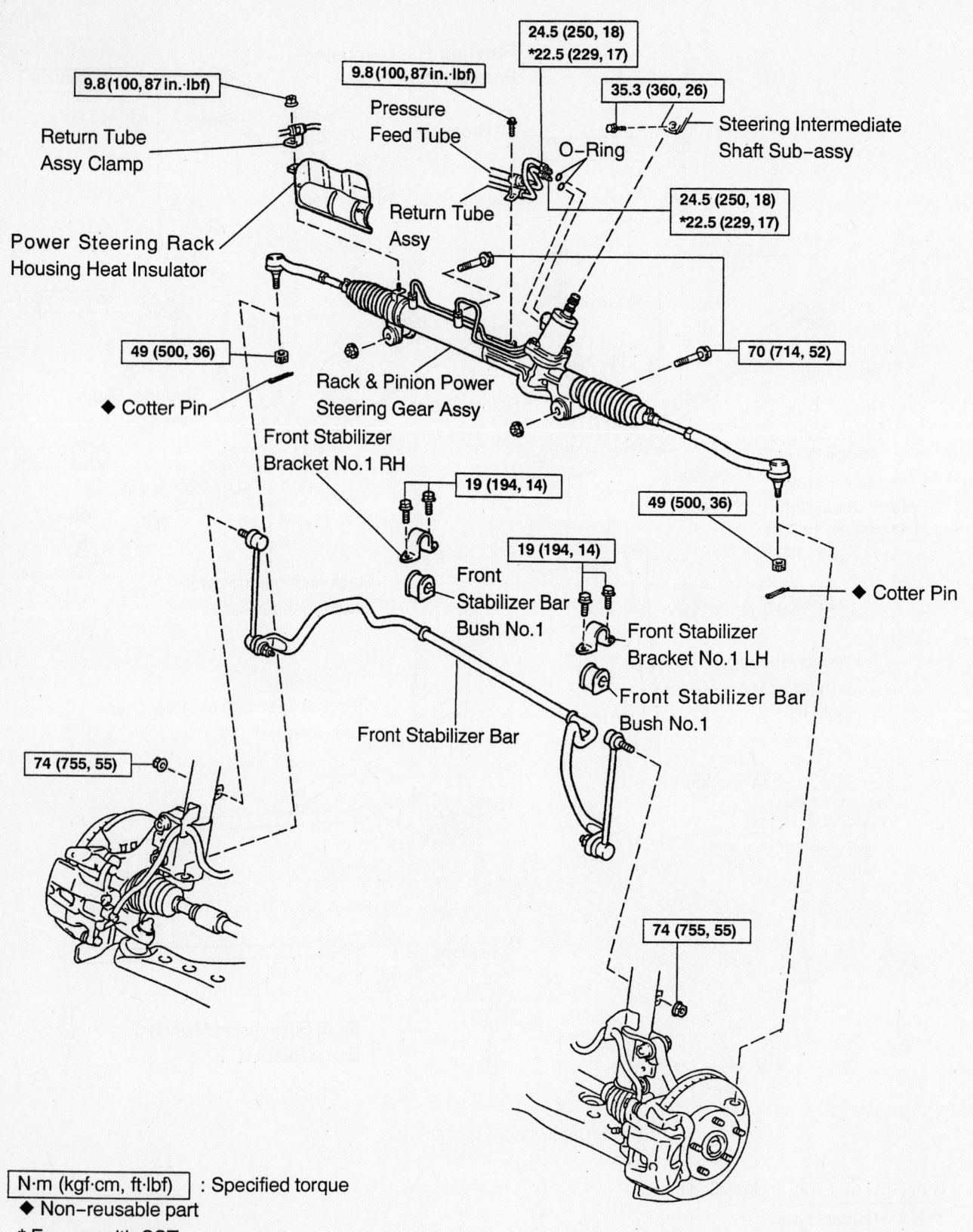

9.8 (100, 87 in.·lbf)

Return Tube Assy Clamp

9.8 (100, 87 in.·lbf)

Pressure Feed Tube

24.5 (250, 18)
*22.5 (229, 17)

35.3 (360, 26)

Steering Intermediate Shaft Sub-assy

O-Ring

Return Tube Assy

24.5 (250, 18)
*22.5 (229, 17)

Power Steering Rack Housing Heat Insulator

49 (500, 36)

70 (714, 52)

◆ Cotter Pin

Rack & Pinion Power Steering Gear Assy

Front Stabilizer Bracket No.1 RH

19 (194, 14)

49 (500, 36)

Front Stabilizer Bar Bush No.1

19 (194, 14)

◆ Cotter Pin

Front Stabilizer Bracket No.1 LH

Front Stabilizer Bar Bush No.1

Front Stabilizer Bar

74 (755, 55)

74 (755, 55)

N·m (kgf·cm, ft·lbf) : Specified torque

◆ Non-reusable part

* For use with SST

67162-LEXU-G75

Exploded view power steering gear assembly—ES 330

- Tie rods from lower ball joint
- Engine under cover
- Front suspension member braces
- Steering slide yoke. Match-mark with control valve shaft before removal
- Oil feed tubes

4. Remove Rack & Pinion Assembly
5. Remove or disconnect the following:
   - Tube clamps on rack assembly
   - Connector assembly
   - 4 gear assembly set bolts
   - Steering gear
   - Steering rack housing bracket
6. Match-mark and remove tie rod assemblies

To install:

7. Install or connect the following:
   - Tie rod assemblies. Align match marks
   - Oil feed tubes
   - Rack & Pinion gear assembly with 4 set bolts; torque to 48 ft. lbs (65 Nm)
   - Steering sliding yoke assembly
   - Pressure and return tubes
   - Tie rod assemblies to lower ball joint; torque to 64 ft. lbs. (87 Nm)
   - Brake calipers
   - Front suspension member braces
   - Engine under cover
   - Center spiral cable
   - Temporarily tighten steering wheel assembly
   - Center steering wheel and fully tighten set nut
   - Horn button
   - Steering wheel covers
   - Steering pad modulator switch
   - Negative battery cable
   - Bleed power steering system
   - Front wheels
8. Inspect toe in and adjust as necessary
9. Test drive

ES 330

1. Before servicing the vehicle, refer to the precautions in the beginning of this section.
2. Place the front wheels facing straight ahead.
3. Remove or disconnect the following:
   - Negative battery cable. Wait at least 90 seconds before performing any work.
   - Front wheels
   - Steering wheel switch & volume case
   - Steering wheel pad
   - Steering wheel column lower cover
   - Horn button

- Steering wheel
- Tie rods from steering knuckle
- Steering intermediate shaft Match-mark with control valve shaft before removal
- Front stabilizer link assembly
- Front stabilizer brackets
- Oil feed tubes

4. Remove Rack & Pinion Assembly
5. Remove or disconnect the following:
   - Tube clamps on rack assembly
   - Connector assembly
   - 2 bolts and 2 nuts
   - Steering gear
6. Power steering rack housing heat insulator
7. Right & Left turn pressure tubes from the rack assembly
8. Match-mark and remove tie rod assemblies

To install:

9. Install or connect the following:
   - Tie rod assemblies. Align match marks
   - Oil feed tubes
   - Steering rack heat insulator
   - Rack and pinion gear assembly with 2 bolts and 2 nuts; torque to 52 ft. lbs (70 Nm)
   - Pressure and return tubes
   - Front stabilizer brackets
   - Front stabilizer link assembly
   - Steering intermediate shaft
   - Tie rod assemblies to 36 ft. lbs. (49 Nm)
   - Center spiral cable
   - Temporarily tighten steering wheel assembly
   - Center steering wheel and fully tighten set nut
   - Horn button
   - Steering wheel covers
   - Steering pad modulator switch
   - Negative battery cable
   - Bleed power steering system
   - Front wheels
10. Inspect toe in and adjust as necessary
11. Test drive

Strut and Coil Spring

REMOVAL & INSTALLATION

Front

ES 300

1. Before servicing the vehicle, refer to the precautions in the beginning of this section.
2. Remove or disconnect the following:

- Negative battery cable
- Tire and wheel assembly
- If equipped with an Anti-lock Brake System (ABS), the ABS speed sensor connector
- Brake line from the strut housing
- Strut assembly from the steering knuckle
- 3 upper mounting nuts from the strut tower
- Strut assembly

✴✴ CAUTION

Do NOT remove the center nut to the strut at this time. The spring on the strut is under high pressure and can cause serious injury.

3. Temporarily install the bolt and nuts to the lower bracket of the strut to support it and secure the strut in a vise.
4. Compress the coil spring
5. Remove or disconnect the following:
   - Spring seat
   - Upper strut retaining nut
   - Suspension support
   - Upper insulator
   - Spring
   - Bumper
   - Insulator

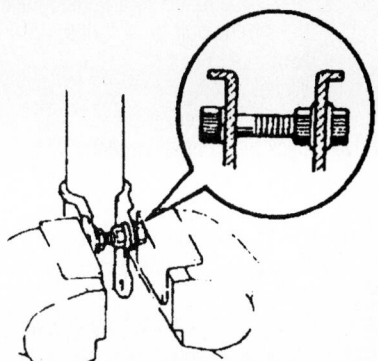

7923LGA5

Temporarily install the support nuts and bolt to the strut—ES 300

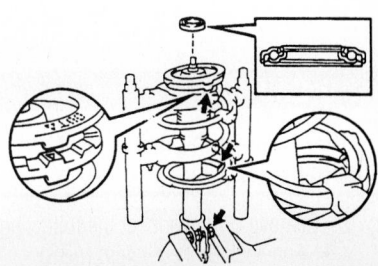

7923LGA6

Align the out mark of the upper spring seat with the mark on the upper insulator—ES 300

To install:

6. Install or connect the following:
- Lower insulator
- Bumper to the piston rod
- Coil spring end into the gap of the lower seat
- Upper insulator
- Upper support to the piston rod, aligning it with the groove in the strut rod

7. Install or connect the following:
- Spring seat. Tighten the new upper strut retaining nut to 36 ft. lbs. (49 Nm).

8. Remove the strut from the vise and disassemble the securing nuts and bolt.

9. Rotate the upper support so the lowest bolt on the support aligns with the projection part of the lower spring.

10. Install or connect the following:
- Strut and tighten the strut to body bolts to 59 ft. lbs. (80 Nm)
- Strut to the steering knuckle and tighten the bolts to 156 ft. lbs. (211 Nm)

11. Run the brake hose through the brake hose bracket and install the clip.

12. Install or connect the following:
- ABS speed sensor and tighten the mounting bolt to 48 inch lbs. (5 Nm)
- Brake line to the strut housing and tighten the bolt to 22 ft. lbs. (29 Nm)
- Wheel
- Negative battery cable

13. Check the front alignment.

GS 300 AND GS 430

1. Before servicing the vehicle, refer to the precautions in the beginning of this section.

2. Remove or disconnect the following:
- Negative battery cable.
- Front wheel
- Brake caliper, leaving the line attached

3. Loosen the 3 upper strut mounting nuts.

4. Loosen, but do not remove, the upper strut rod nut.

❋❋ CAUTION

Do NOT remove the upper strut nut at this time.

5. Remove or disconnect the following:
- Anti-lock Brake System (ABS) speed sensor and harness
- Upper suspension arm from the steering knuckle

- Stabilizer bar from the link and remove the bracket
- Strut from the lower suspension arm.
- 3 upper strut mounting nuts and remove the strut

6. Compress the coil spring.

7. Remove or disconnect the following:
- Piston rod locknut
- Suspension support, coil spring and bumper

8. If disposing the strut, perform the following procedure:
   a. Fully extend the strut rod.
   b. Drill a hole near the bottom of the shock to remove the gas inside.

❋❋ CAUTION

The gas is harmless, but be careful of chips that may fly up when the gas is released.

To install:

9. Install or connect the following:
- Spring bumper
- Coil spring
- Suspension support to the rod and temporarily install a new nut

10. Turn the suspension support so one of the bolts on the support faces the same direction as shown in the illustration.

➡**Align the bolt so a line drawn between the rod and bolt would be at 90° to the direction of the lower bushing.**

11. Install or connect the following:
- Spring compressor
- Strut and tighten the upper retaining nuts to 41 ft. lbs. (56 Nm)
- New upper strut rod nut to 20 ft. lbs. (27 Nm)
- Strut to the lower arm and temporarily tighten the nut and bolt
- Stabilizer bar bracket and tighten the bolts to 21 ft. lbs. (28 Nm)
- The stabilizer bar to the link and tighten the bolts to 29 ft. lbs. (39 Nm)

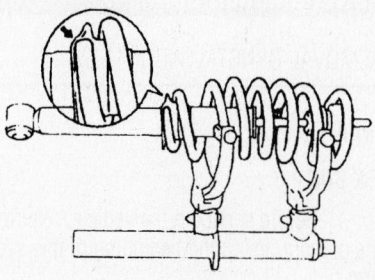

7923LGA7

Matching the spring to the seat

- Upper suspension arm to the steering knuckle. Tighten the nut to 64 ft. lbs. (87 Nm) and install a new cotter pin.
- ABS speed sensor and tighten the bolt to 69 inch lbs. (8 Nm)
- Caliper
- Wheel

12. Bounce the vehicle several times to stabilize the suspension.

13. Tighten the lower strut bolt and nut to 116 ft. lbs. (157 Nm).

14. Check the front wheel alignment.

IS 300

1. Before servicing the vehicle, refer to the precautions in the beginning of this section.

2. Remove or disconnect the following:
- Front wheel
- Wheel speed sensor and harness clamp
- Upper ball joint
- Level control sensor link
- Stabilizer bar link
- Lower strut bolt
- Upper strut mount cap
- Upper strut mount nuts
- Strut assembly

3. Install a suitable spring compressor and remove the center nut.

4. Remove the upper strut mount and the coil spring.

To install:

5. Installation is the reverse of the removal procedure, while using the following torque values:
- Upper strut mount center nut: 25 ft. lbs. (34 Nm)
- Upper strut mounting nuts: 26 ft. lbs. (35 Nm)
- Lower strut mount bolt: 47 ft. lbs. (64 Nm)
- Upper ball joint nut: 50 ft. lbs. (65 Nm)
- Stabilizer bar link nut: 36 ft. lbs. (49 Nm)

LS 430—WITHOUT AIR SUSPENSION

1. Before servicing the vehicle, refer to the precautions in the beginning of this section.

2. Remove or disconnect the following:
- Tire and wheel assembly
- Steering knuckle from the upper ball joint
- Strut assembly from the lower strut bracket
- Strut cover from the upper strut mount
- 3 mounting nuts and remove the strut assembly with the coil spring from the vehicle.

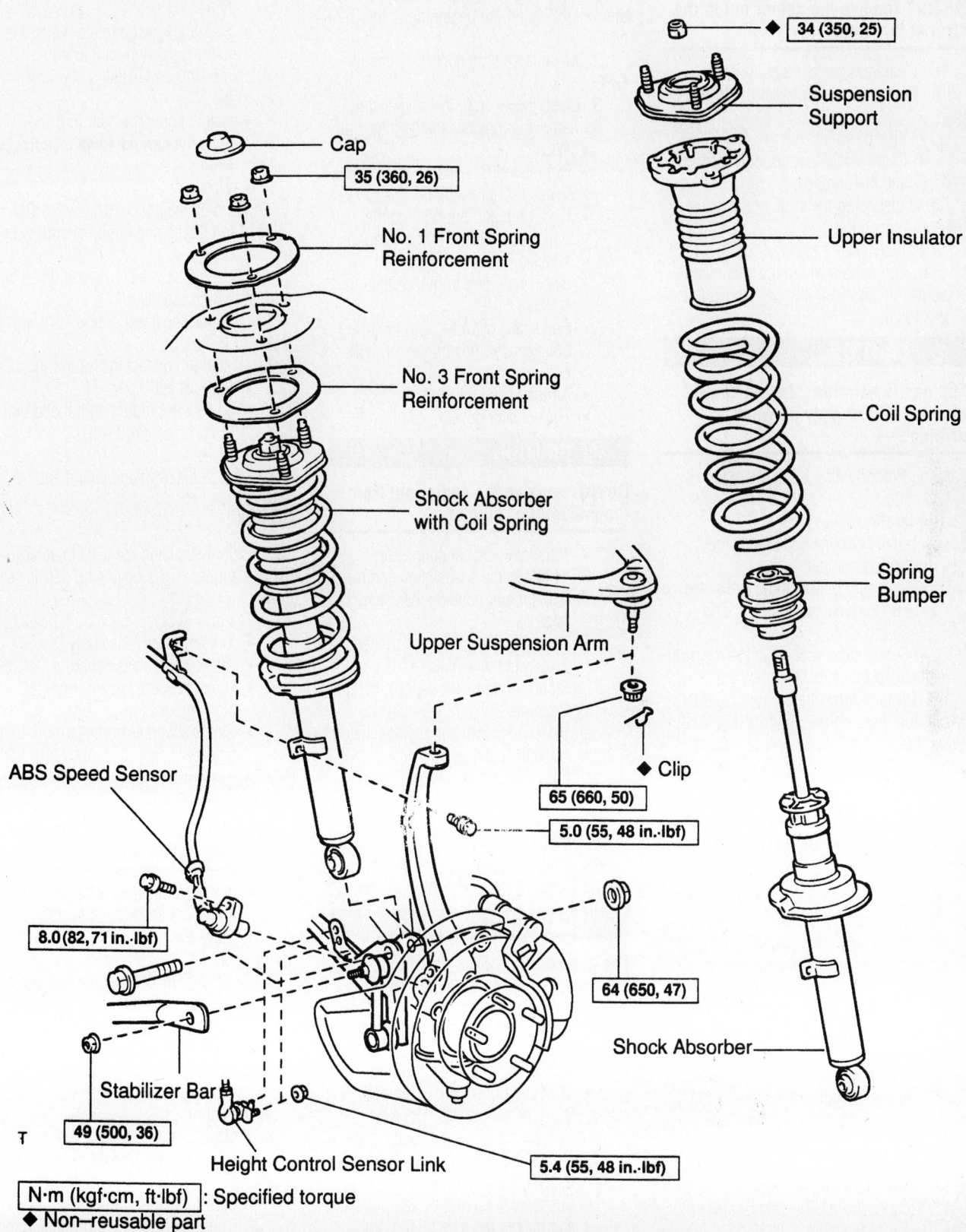

34 (350, 25)

Suspension Support

Cap

35 (360, 26)

No. 1 Front Spring Reinforcement

Upper Insulator

No. 3 Front Spring Reinforcement

Coil Spring

Shock Absorber with Coil Spring

Upper Suspension Arm

Spring Bumper

Clip

65 (660, 50)

5.0 (55, 48 in.·lbf)

ABS Speed Sensor

8.0 (82, 71 in.·lbf)

64 (650, 47)

Stabilizer Bar

Shock Absorber

49 (500, 36)

Height Control Sensor Link

5.4 (55, 48 in.·lbf)

N·m (kgf·cm, ft·lbf) : Specified torque
◆ Non–reusable part

Exploded view of the front strut assembly mounting—IS 300

❋❋ CAUTION

Do NOT remove the center nut to the strut at this time.

3. Compress the coil spring.
4. Remove or disconnect the following:

- Piston rod locknut
- Suspension support, coil spring and the bumper

5. If disposing the strut, perform the following procedure:

a. Fully extend the strut rod.

b. Drill a hole within the shaded area shown in the illustration to remove the gas inside.

❋❋ CAUTION

The gas is harmless, but be careful of chips that may fly up when drilling.

c. Properly dispose of the strut assembly.

To install:

6. Install or connect the following:

- Spring bumper
- Coil spring. Match the end of the coil into the recess of the strut spring seat.
- Suspension support to the rod and temporarily install a new nut

7. Turn the suspension support so one of the bolts on the support faces the same direction as shown in the illustration.

➡**Align the bolt so a line drawn between the rod and bolt would be at 90° to the direction of the lower bushing.**

8. Tighten the strut rod nut to 20 ft. lbs. (27 Nm) and install the cap.
9. Remove the spring compressor
10. Install or connect the following:

- Strut and tighten the upper retaining nuts to 43 ft. lbs. (58 Nm)
- Strut to the lower bracket and temporarily install the nut and bolt
- Upper control arm to the steering knuckle. Tighten the nut to 48 ft. lbs. (65 Nm) and install a new cotter pin.
- Wheel

11. Lower the vehicle.
12. Bounce the vehicle several times to stabilize the suspension.
13. Tighten the lower strut bolt and nut to 116 ft. lbs. (157 Nm).
14. Check the front wheel alignment.

LS 430—WITH AIR SUSPENSION

1. Before servicing the vehicle, refer to the precautions in the beginning of this section.
2. Move the height control switch to **OFF**.
3. Bleed the air from the suspension.
4. Remove or disconnect the following:

- Wheel
- Height control sensor link from the lower strut bracket
- Cotter pin and nut holding the upper control arm to the steering knuckle
- Upper ball joint from the steering knuckle
- Pneumatic cylinder from the lower bracket by removing the through-bolt
- Air tube from the strut
- The actuator cover

❋❋ CAUTION

Do NOT remove the center nut from the pneumatic cylinder.

- Actuator electrical connector
- 2 bolts to the suspension control actuator and position the actuator aside
- 3 upper mounting nuts and the strut from the vehicle

5. If disposing the strut perform the following procedure:

a. Using a screwdriver, remove the air from inside the cylinder.

b. Fully extend the cylinder.

c. Drill a hole in the cylinder at a point above 1.57 in. (40mm) from the bottom of the strut assembly. This will release the gas charge in the strut. Do NOT puncture the pneumatic cylinder.

❋❋ CAUTION

The gas coming out is harmless, but be careful of chips that may fly up while drilling.

To install:

6. Install or connect the following:

- Strut and tighten the upper mounting nuts to 43 ft. lbs. (58 Nm)
- Suspension control actuator and tighten the bolts. Tighten the 2 nuts to 13 ft. lbs. (17 Nm).
- Suspension control actuator cover and tighten the nuts to 43 ft. lbs. (58 Nm)
- 2 new O-rings to the air tube. Install the tube and tighten it to 13 ft. lbs. (17 Nm). Install the grommet.

- The strut to the lower strut bracket and temporarily install the nut and bolt
- Steering knuckle to the upper ball joint. Tighten the nut to 48 ft. lbs. (65 Nm) and install a new cotter pin.
- Height control sensor link and tighten a new nut to 48 inch lbs. (5 Nm)
- Wheel

7. Turn the height control switch **ON**.
8. Start the engine to fill the strut with air.
9. Bounce the vehicle several times to normalize the suspension.
10. Support the lower control arm with a jack.
11. Install or connect the following:

- Front wheel
- Lower strut mounting nut and bolt to 76 ft. lbs. (106 Nm)
- Wheel

12. Check the front end alignment.

SC 430

1. Before servicing the vehicle, refer to the precautions in the beginning of this section.
2. Remove or disconnect the following:

- Tire and wheel assembly
- Speed sensor assembly
- Upper suspension arm from the steering knuckle
- Shock absorber from the mounting bracket

❋❋ CAUTION

Loosen the piston rod lock nut. Do Not remove at this time

- 3 nuts front spring support reinforcement and strut assembly

3. Remove the shock absorber from the spring assembly
4. Remove or disconnect the following:

a. Compress the coil spring with proper spring compressor

b. Remove the piston lock nut

c. Remove the front suspension support assembly, front coil spring insulator and coil spring.

To install:

5. Install the shock into the spring assembly

a. Install the spring insulator

b. Compress the coil spring using the proper spring compressor

c. Install spring making sure that the spring seats properly

d. Temporarily install a new lock nut

SHOCK ABSORBER ASSY FRONT:

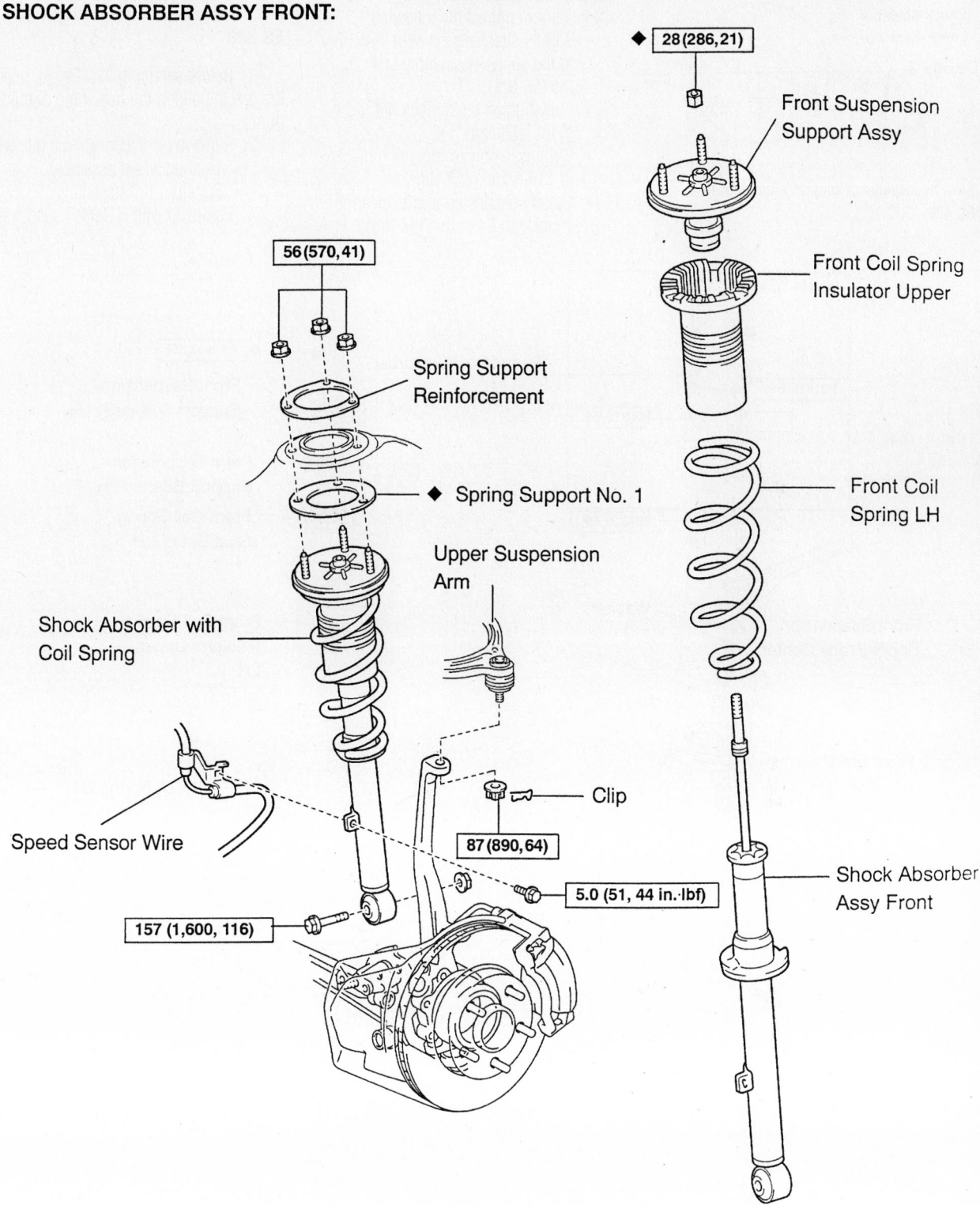

◆ 28 (286, 21)

Front Suspension Support Assy

Front Coil Spring Insulator Upper

Front Coil Spring LH

56 (570, 41)

Spring Support Reinforcement

◆ Spring Support No. 1

Upper Suspension Arm

Shock Absorber with Coil Spring

Clip

Speed Sensor Wire

87 (890, 64)

5.0 (51, 44 in.·lbf)

Shock Absorber Assy Front

157 (1,600, 116)

| N·m (kgf·cm, ft·lbf) | : Specified torque
◆ Non-reusable part

67162-LEXU-G76

Exploded view of front strut assembly—SC 430

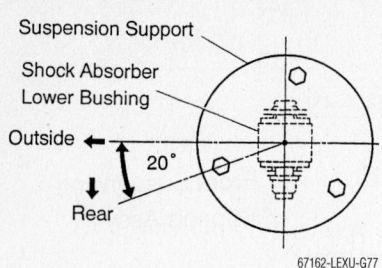

Suspension Support
Shock Absorber Lower Bushing
Outside
20°
Rear

67162-LEXU-G77

Strut to suspension support alignment—SC 430

e. Align the suspension support with the shock absorber lower bolt

6. Install or connect the following:
- 3 bolts attaching the strut assembly to the support assembly to 41 ft. lbs (56 Nm)
- Fully tighten piston lock nut to 21 ft. lbs. (28 Nm)
- Strut assembly to mounting bracket to 116 ft lbs. (157 Nm)
- Upper suspension arm to steering knuckle to 64 ft. lbs. (87 Nm)

- Speed sensor
- Tire & wheel

ES 330

1. Before servicing the vehicle, refer to the precautions in the beginning of this section.

2. Remove or disconnect the following:
- Tire and wheel assembly
- Stabilizer link assembly

3. Equipped with H-Tems® suspension

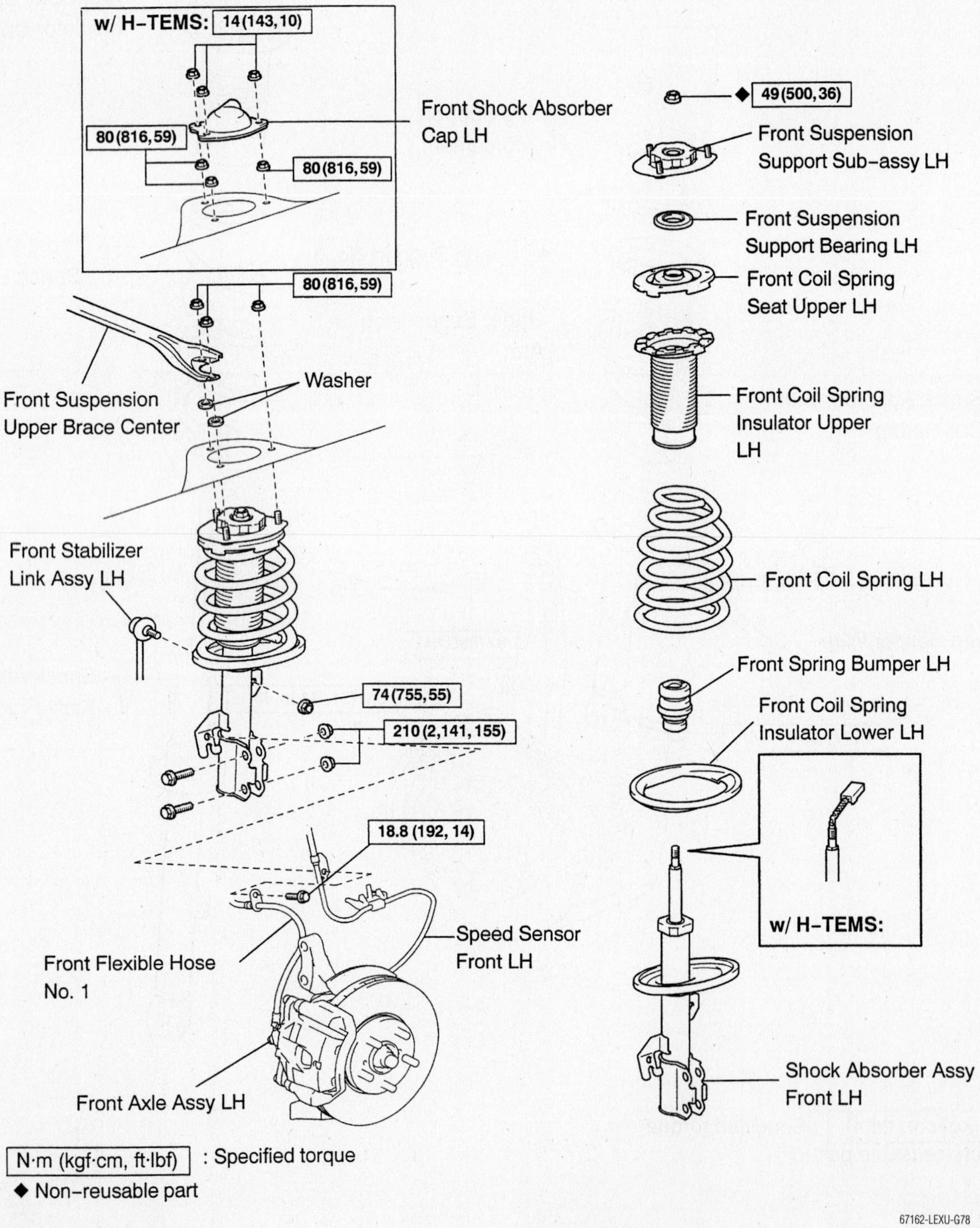

w/ H-TEMS: 14 (143, 10)

80 (816, 59)

80 (816, 59)

80 (816, 59)

Front Shock Absorber Cap LH

Front Suspension Upper Brace Center

Washer

Front Stabilizer Link Assy LH

74 (755, 55)

210 (2,141, 155)

18.8 (192, 14)

Front Flexible Hose No. 1

Speed Sensor Front LH

Front Axle Assy LH

◆ 49 (500, 36)

Front Suspension Support Sub-assy LH

Front Suspension Support Bearing LH

Front Coil Spring Seat Upper LH

Front Coil Spring Insulator Upper LH

Front Coil Spring LH

Front Spring Bumper LH

Front Coil Spring Insulator Lower LH

w/ H-TEMS:

Shock Absorber Assy Front LH

N·m (kgf·cm, ft·lbf) : Specified torque
◆ Non-reusable part

Exploded view front suspension—ES 330

67162-LEXU-G78

a. Disconnect the wiring connector

b. Disconnect the 3 nuts, harness clamp and shock absorber cap

c. With special tool loosen the piston rod nut. Do NOT remove nut at this time.

4. W/out H-Tems®

5. Remove or disconnect the following:
- Loosen lock nut on piston rod. Do NOT remove at this time
- Brake hose from strut

- Speed sensor
- 2 bolts and nuts on strut mount on steering knuckle
- Strut assembly

6. Secure strut in a vise and compress the spring with a spring compressor

7. Remove the piston rod nut and remove the support assembly

8. Remove the shock absorber from the spring assembly

To install:

9. Install shock absorber into the spring coil assembly.

a. Install coil spring insulator on the piston rod

b. Install coil spring

c. Align the coil spring to properly fit in lower seat

d. Install upper coil spring insulator

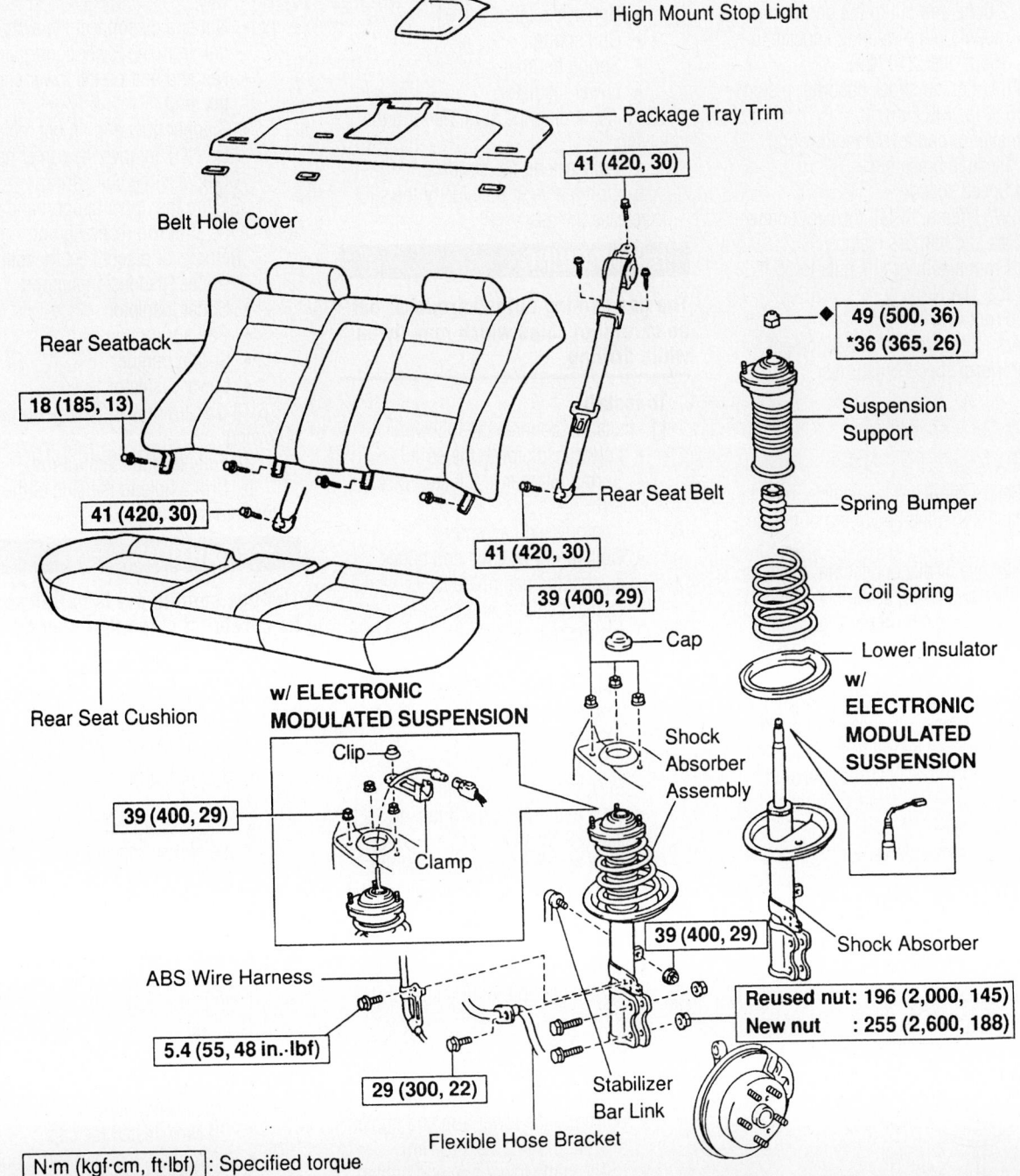

High Mount Stop Light

Package Tray Trim

41 (420, 30)

Belt Hole Cover

Rear Seatback

18 (185, 13)

41 (420, 30)

Rear Seat Cushion

Rear Seat Belt

41 (420, 30)

39 (400, 29)

Cap

49 (500, 36)
*36 (365, 26)

Suspension Support

Spring Bumper

Coil Spring

Lower Insulator

w/ ELECTRONIC MODULATED SUSPENSION

Clip

39 (400, 29)

Clamp

Shock Absorber Assembly

w/ ELECTRONIC MODULATED SUSPENSION

Shock Absorber

39 (400, 29)

ABS Wire Harness

5.4 (55, 48 in.·lbf)

29 (300, 22)

Flexible Hose Bracket

Stabilizer Bar Link

Reused nut: 196 (2,000, 145)
New nut   : 255 (2,600, 188)

N·m (kgf·cm, ft·lbf) : Specified torque

◆ Non–reusable part

* For use with SST

7923LGA0

Exploded view of the rear strut and coil spring mounting—ES 300

e. Install upper coil spring seat with the mark facing to the outside

f. Install new support bearing

g. Install the front suspension support with the mark facing to the outside of the vehicle.

h. Temporarily install the piston rod nut

10. Install or connect the following:
- 3 nuts to the upper side of the strut assembly. Torque: 59 ft lbs. (80 Nm)
- 2 bolts and nut to the on the strut mount on the steering knuckle to 155 ft. lbs (210 Nm)

11. Fully tighten shock absorber piston nut to 36 ft. lbs (49 Nm)

12. Install or connect the following:
- Flexible brake hose
- Speed sensor
- W/H-Tems, Install the cap, connector and harness clamp
- Front stabilizer link nuts to 55 ft. lbs. (74 Nm)
- Front wheel

13. Adjust the front wheel alignment

14. Perform speed sensor test

Rear

ES 300

1. Before servicing the vehicle, refer to the precautions in the beginning of this section.

2. Remove or disconnect the following:
- Tire and wheel assembly
- Load sensing proportioning valve spring assembly from the lower arm
- Anti-lock Brake System (ABS) speed sensor harness and brake line from the strut assembly
- Stabilizer bar link from the strut

3. Loosen the 2 nuts attaching the strut to the axle carrier.

4. Support the axle carrier.

5. Remove or disconnect the following:
- Rear seat back and package tray trim

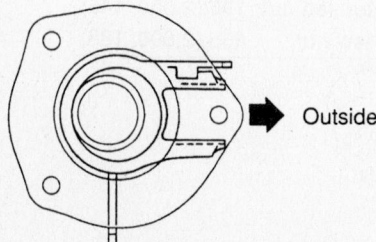

7923LGB1

Position the upper suspension support as shown when assembling the strut—ES 300

- Upper mounting nuts
- 2 lower mounting bolts and remove the strut assembly

6. Compress the coil spring.

7. Temporarily install a bolt and 2 nuts on the bracket at the lower end of the strut and secure it in a vise.

8. Secure the upper support and remove the strut rod retaining nut.

9. Remove or disconnect the following:
- Upper suspension support
- Upper insulator
- Coil spring
- Spring bumper
- Lower insulator

10. If discarding the strut, perform the following:
a. Fully extend the strut rod.
b. Drill a hole in the side of the strut to release the gas.

✲✲ WARNING

The gas coming out is harmless, but be careful of chips which may fly up while drilling.

To install:

11. Install or connect the following:
- Lower insulator to the strut
- Spring bumper to the strut piston rod
- Compressed coil spring
- Coil spring with the end butted against the gap in the lower seat
- Upper insulator and support matching the bolt of the support with the cut-off part of the insulator
- Upper suspension support
- New strut piston rod nut to 36 ft. lbs. (49 Nm)
- Spring compressor
- Strut rod piston nut cap
- Strut and tighten the 3 nuts to 29 ft. lbs. (39 Nm)
- Strut to the axle carrier. Coat the nuts with engine oil and tighten the nuts and bolts to 188 ft. lbs. (255 Nm)
- ABS harness to the strut and tighten the bolt to 48 inch lbs. (6 Nm)
- Brake line to the strut and tighten the retaining nut to 22 ft. lbs. (29 Nm)
- Spring to the lower arm and tighten the nut to 10 ft. lbs. (13 Nm)
- LSPV to the lower arm and tighten the nut to 108 inch lbs. (12 Nm)
- Rear wheel
- Rear seat and package tray

GS 300 AND GS 430 (EXCEPT 2004)

1. Before servicing the vehicle, refer to the precautions in the beginning of this section.

2. Remove or disconnect the following:
- Front trunk compartment trim cover
- Wheel(s)
- Brake caliper support bracket from the rear axle carrier by removing the 2 bolts. Leave the brake line connected and position it out of the way.
- Nut and disconnect the sway bar link from the lower control arm
- Nut and bolt on the lower end of the strut
- 3 upper nuts and lift out the strut. Do NOT remove the center nut.

3. Compress the coil spring.

4. Secure the upper support and remove the strut rod retaining nut.

5. Remove or disconnect the following:
- Upper suspension support
- Upper insulator
- Coil spring
- Spring bumper
- Lower insulator

6. If discarding the strut, perform the following:
a. Fully extend the strut rod.
b. Drill a hole in the side of the strut to drain the gas inside

✲✲ CAUTION

The gas coming out is harmless, but be careful of chips that may fly up while drilling.

To install:

7. Install or connect the following:
- Lower insulator to the strut
- Spring bumper to the strut piston rod
- Compressed coil spring. Position the coil spring with the end butted against the gap in the lower seat.
- Upper insulator and suspension support
- Upper suspension support and tighten a new strut piston rod nut to 20 ft. lbs. (27 Nm)

8. Remove the spring compressor.

9. Install or connect the following:
- Strut to the vehicle and tighten the 3 upper mounting nuts to 14 ft. lbs. (20 Nm). Install the cap.
- And tighten the lower strut bolt and nut to 101 ft. lbs. (137 Nm)
- Sway bar link to the lower control arm. Tighten the nut to 33 ft. lbs. (44 Nm).

- Brake caliper to the rear axle carrier by installing the 2 bolts. Tighten the bolts to 77 ft. lbs. (104 Nm).
- Wheel(s)
- Trunk compartment cover trim

10. Check and adjust the vehicle alignment as necessary.

GS 300 AND GS 430 (2004)

1. Before servicing the vehicle, refer to the precautions in the beginning of this section.

2. Remove the corresponding rear wheel(s).

3. Remove the luggage compartment trim front cover.

4. Remove the rear fender apron seal.

5. Remove the rear lower suspension arm:

    a. Remove 2 bolts, nuts and the No. 1 lower suspension arm (trailing arm).

    b. Remove the bolt and nuts and disconnect the stabilizer bar link (and height control link, if equipped) from the No. 2 lower suspension arm.

    c. Remove the bolt and nut and disconnect the strut from the No. 2 lower suspension arm.

    d. Place matchmarks on the adjusting cam and on the No. 2 lower suspension arm.

    e. Remove the nut and adjusting cams.

    f. Remove the bolt, nut and washer and the No. 2 lower suspension arm.

6. Remove the 3 upper retaining nuts. Loosen, but Do NOT remove the center stud nut.

7. Remove the lower strut mounting bolt.

8. Remove the strut assembly, with the coil spring.

9. Compress the coil spring in a suitable spring compressor.

➥**Do NOT remove these items with an impact wrench; it will damage the spring compressor.**

10. Remove or disconnect the following:
- Suspension support nut
- Washer
- 2 cushions
- Collar
- Suspension support
- Upper insulator
- Lower cup
- Spring bumper

11. Carefully release the spring compressor and remove the coil spring.

12. If the shock absorber is being replaced, fully extend the shock absorber rod and drill a hole to discharge the gas

from the cylinder. Drill this hole about 5-8 in. (130-185mm) from lower mounting bolt hole.

To install:

13. With a new shock absorber in place, install the suspension support and compress the coil spring.

14. With a non-impact wrench, install the coil spring to the shock absorber, fitting the lower end of the spring into the recent of the spring seat on the shock absorber.

15. Install the spring bumper, lower cup, cushion, collar, upper insulator, suspension support, cushion and washer onto the shock absorber. Temporarily tighten a new nut.

16. Rotate the suspension support so the rod and one of the bolts on the suspension support are aligned with the lower shock mounting hole such that while the lower mounting bolt hole is in proper position, the 3 studs on the top of the strut assembly will align with the holes in the body.

17. Carefully remove the spring compressor.

18. Install the strut assembly into the vehicle. Torque the 2 bolts on top of the coil spring to 13 ft. lbs. (18 Nm), the 3 nuts on the top mounting studs to 47 ft. lbs. (64 Nm), and the nut in the center of the upper strut mounting to 20 ft. lbs. (27 Nm).

19. Install the bolt, nut and washer and the No. 2 lower suspension arm. Torque bolt to 81 ft. lbs. (110 Nm).

20. Position both adjusting cams, referencing the matchmarks made during removal. Install and torque the retaining nuts to 81 ft. lbs. (110 Nm).

21. Install the strut assembly lower mounting bolt to the No. 2 lower suspension arm. Torque the nut to 81 ft. lbs. (110 Nm).

22. Install the stabilizer bar link (and height control link, if equipped) to the No. 2 lower suspension arm. Torque the bolt and nut to 22 ft. lbs. (30 Nm).

23. Install the No. 1 lower suspension arm (trailing arm) and torque the bolts and nuts to 55 ft. lbs. (75 Nm).

24. Install the rear fender apron seal.

25. Install the luggage compartment trim front cover.

26. Install the rear wheel(s). Torque the wheel nuts to 76 ft. lbs. (103 Nm).

LS 430—WITHOUT AIR SUSPENSION

1. Before servicing the vehicle, refer to the precautions in the beginning of this section.

2. Remove or disconnect the following:
- Rear seat cushion and seat back.
- Tray trim
- Tire and wheel assembly

- Rear halfshaft
- Stabilizer bar link from the stabilizer bar
- Anti-lock Brake System (ABS) speed sensor and wiring harness
- Brake caliper bracket from the axle carrier, leaving the brake line connected. Suspend the brake caliper aside with a piece of wire.
- Nut on the lower side of the strut. Do NOT remove the bolt.
- Rear axle assembly with a lifting device
- Strut cap by removing the 3 nuts
- 3 mounting nuts holding the strut assembly to the strut tower. Do NOT remove the center bolt.

✳✳ CAUTION

Do NOT remove the center nut to the strut at this time.

- Bolt on the lower side of the strut assembly
- Strut assembly with the coil spring

3. Compress the coil spring.

4. Secure the strut housing in a vise.

5. Remove or disconnect the following:
- Strut rod retaining nut
- Upper suspension support
- Upper insulator
- Coil spring
- Spring bumper
- Lower insulator

6. If discarding the strut, perform the following:

    a. Fully extend the strut rod.

    b. Drill a hole in the strut (about 1 in. above the strut lower mount) and drain the gas inside

✳✳ CAUTION

The gas coming out is harmless, but be careful of chips which may fly up while drilling.

To install:

7. Install or connect the following:
- Lower insulator to the strut.
- Spring bumper to the strut piston rod
- Coil spring
- Upper insulator and support
- Upper suspension support

8. Temporarily install the upper strut rod retaining nut.

9. Rotate the suspension support so that the rod and one of the bolts on the suspension support are aligned with the lower bushing.

10. Remove the spring compressor.

11. Install or connect the following:
- Strut assembly to the vehicle and tighten the 3 nuts to 47 ft. lbs. (64 Nm). Tighten the strut rod retaining nut to 20 ft. lbs. (27 Nm).
- Strut assembly cap and install the 3 nuts
- Strut to the rear axle carrier. Install the bolt from the rear of the vehicle and temporarily tighten the nut.
- Brake caliper and tighten the mounting bolts to 77 ft. lbs. (104 Nm)
- ABS speed sensor and wiring harness
- Stabilizer link to the stabilizer bar and tighten the nut to 48 ft. lbs. (65 Nm)
- Rear halfshaft
- Tire and wheel assembly

12. Bounce the vehicle up and down to stabilize the suspension.

13. Support the rear axle assembly with a lifting device. Tighten the lower strut bolt to 101 ft. lbs. (137 Nm).

14. Install or connect the following:
- Rear seat cushion and rear seat back.
- Package tray trim

15. Check the wheel alignment.

LS 430—WITH AIR SUSPENSION (EXCEPT 2004)

1. Before servicing the vehicle, refer to the precautions in the beginning of this section.

2. Bleed the air system from the suspension.

3. Remove or disconnect the following:
- Rear seat cushion and seat back
- Package tray trim
- Trunk trim panel. Move the height control switch, located in the trunk area, to the **OFF** position.
- Tire and wheel assembly
- Rear halfshaft
- Stabilizer links from the stabilizer bar
- Anti-lock Brake System (ABS) speed sensor and wiring harness
- Brake caliper bracket from the rear axle carrier. Do NOT disconnect the brake line.

4. Place matchmarks on the height control sensor link and bracket. Disconnect the height control sensor link from the No. 1 lower control arm.

5. Support the rear axle assembly with a lifting device.

6. Remove or disconnect the following:

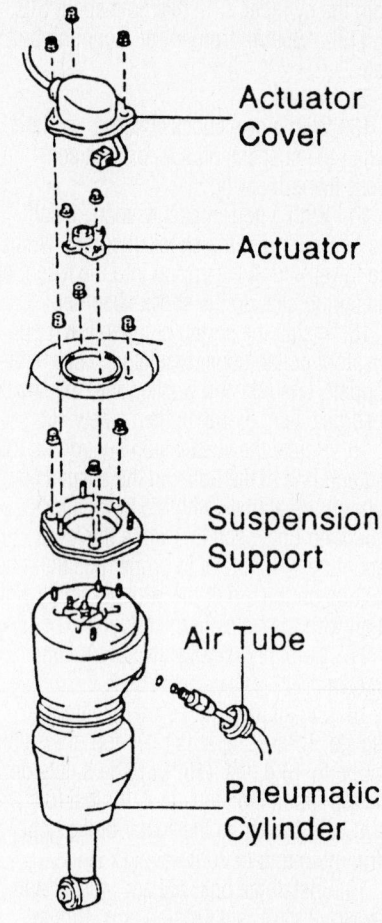

Pneumatic cylinder (strut) component overview (air suspension)

7923LGB2

- Nut on the lower side of the shock absorber. Do NOT remove the bolt.
- Grommet and disconnect the air tube from the shock absorber
- Actuator cover from the strut tower by removing the 3 nuts
- Actuator electrical connector from the top of the strut
- Actuator by removing the 2 nuts
- 3 upper mounting nuts holding the strut to the strut tower

7. Lower the rear axle assembly

8. Remove or disconnect the following:
- Bolt on the lower side of the shock absorber
- Pneumatic cylinder strut assembly from the vehicle
- Suspension support from the strut assembly by removing the 3 nuts

9. If discarding the pneumatic cylinder, perform the following:
   a. Using a screwdriver, depressurize the air from inside the cylinder.
   b. Drill a hole in the shaded area shown in the illustration and remove the gas inside.

To install:

10. Install or connect the following:
- Suspension support to the pneumatic cylinder (strut) and tighten the nuts to 27 ft. lbs. (36 Nm)
- Strut assembly to the vehicle and tighten the upper mounting nuts to 47 ft. lbs. (64 Nm)

11. Match the holes in the pneumatic cylinder with the holes in the suspension control actuator.

12. Install or connect the following:
- Actuator and tighten the mounting nuts to 69 inch lbs. (8 Nm)
- Actuator cover and tighten the 3 nuts to 18 ft. lbs. (25 Nm)
- New O-rings and connect the air line to the shock absorber. Tighten the fitting to 13 ft. lbs. (18 Nm).
- Strut to the rear axle carrier. Insert the bolt from the vehicle's rear and temporarily tighten the nut.
- Height control sensor link to the No. 1 lower control arm. Mounting nut: 48 inch lbs. (5 Nm).
- Rear brake caliper to the rear axle carrier and tighten the mounting bolts to 77 ft. lbs. (104 Nm)
- ABS speed sensor and wiring harness
- Stabilizer bar link and tighten the nut to 48 ft. lbs. (65 Nm)
- Halfshaft
- Actuator electrical connector to the top of the strut
- Tire

13. Move the height control switch to the **ON** position. Start the engine and fill the pneumatic cylinder with air.

14. Bounce the vehicle up and down several times to stabilize the suspension.

15. Turn the suspension height control to the **OFF** position.

16. Remove the tire and wheel assembly.

17. Support the rear axle carrier with a lifting device. Tighten the lower strut bolt to 101 ft. lbs. (137 Nm).

18. Install or connect the following:
- Package tray trim
- Rear seat cushion and seat back

19. Turn the suspension control switch to the **ON** position.

20. Check the wheel alignment.

Coil Spring:

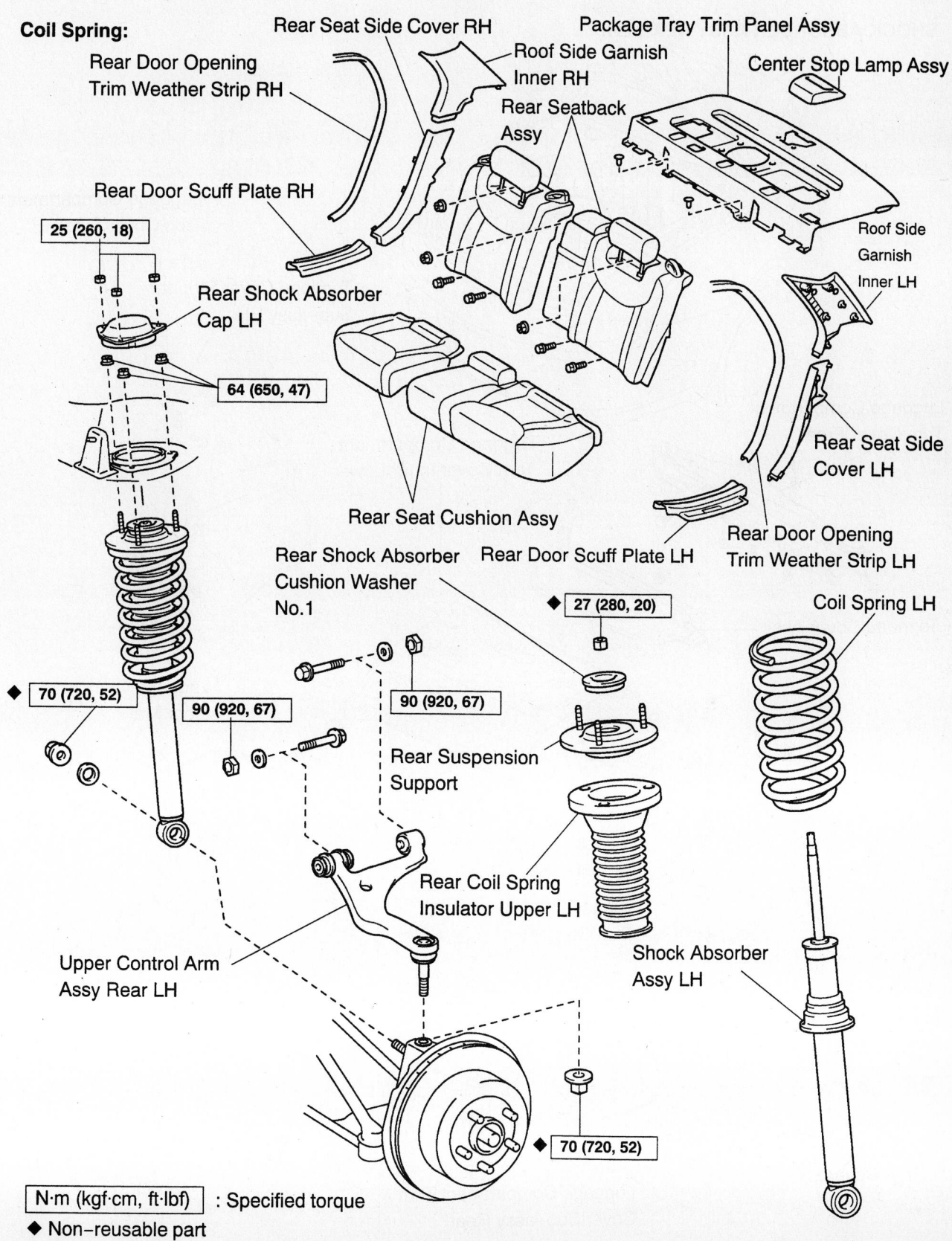

Rear Seat Side Cover RH

Package Tray Trim Panel Assy

Rear Door Opening Trim Weather Strip RH

Roof Side Garnish Inner RH

Center Stop Lamp Assy

Rear Seatback Assy

Rear Door Scuff Plate RH

25 (260, 18)

Rear Shock Absorber Cap LH

Roof Side Garnish Inner LH

64 (650, 47)

Rear Seat Side Cover LH

Rear Shock Absorber Cushion Washer No.1

Rear Seat Cushion Assy

Rear Door Scuff Plate LH

Rear Door Opening Trim Weather Strip LH

◆ 27 (280, 20)

Coil Spring LH

◆ 70 (720, 52)

90 (920, 67)

90 (920, 67)

Rear Suspension Support

Upper Control Arm Assy Rear LH

Rear Coil Spring Insulator Upper LH

Shock Absorber Assy LH

◆ 70 (720, 52)

N·m (kgf·cm, ft·lbf) : Specified torque

◆ Non-reusable part

67162-LEXU-G90

Expanded view rear suspension—LS 430

SHOCK ABSORBER ASSY REAR LH:

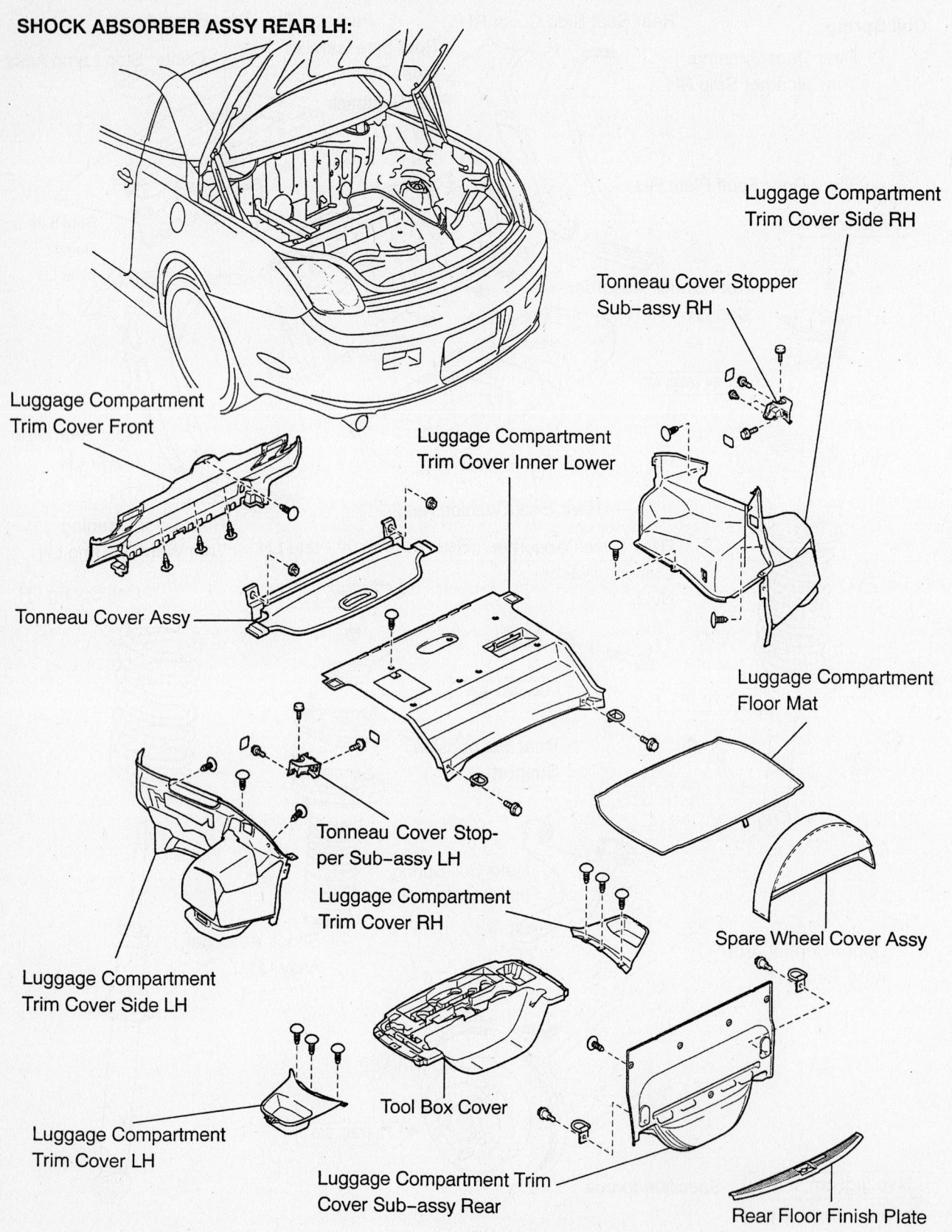

Luggage Compartment
Trim Cover Side RH

Tonneau Cover Stopper
Sub-assy RH

Luggage Compartment
Trim Cover Front

Luggage Compartment
Trim Cover Inner Lower

Tonneau Cover Assy

Luggage Compartment
Floor Mat

Tonneau Cover Stop-
per Sub-assy LH

Luggage Compartment
Trim Cover RH

Spare Wheel Cover Assy

Luggage Compartment
Trim Cover Side LH

Tool Box Cover

Luggage Compartment
Trim Cover LH

Luggage Compartment Trim
Cover Sub-assy Rear

Rear Floor Finish Plate

67162-LEXU-G79

Exploded view of trunk access for rear struts—SC 430

SHOCK ABSORBER ASSY REAR LH:

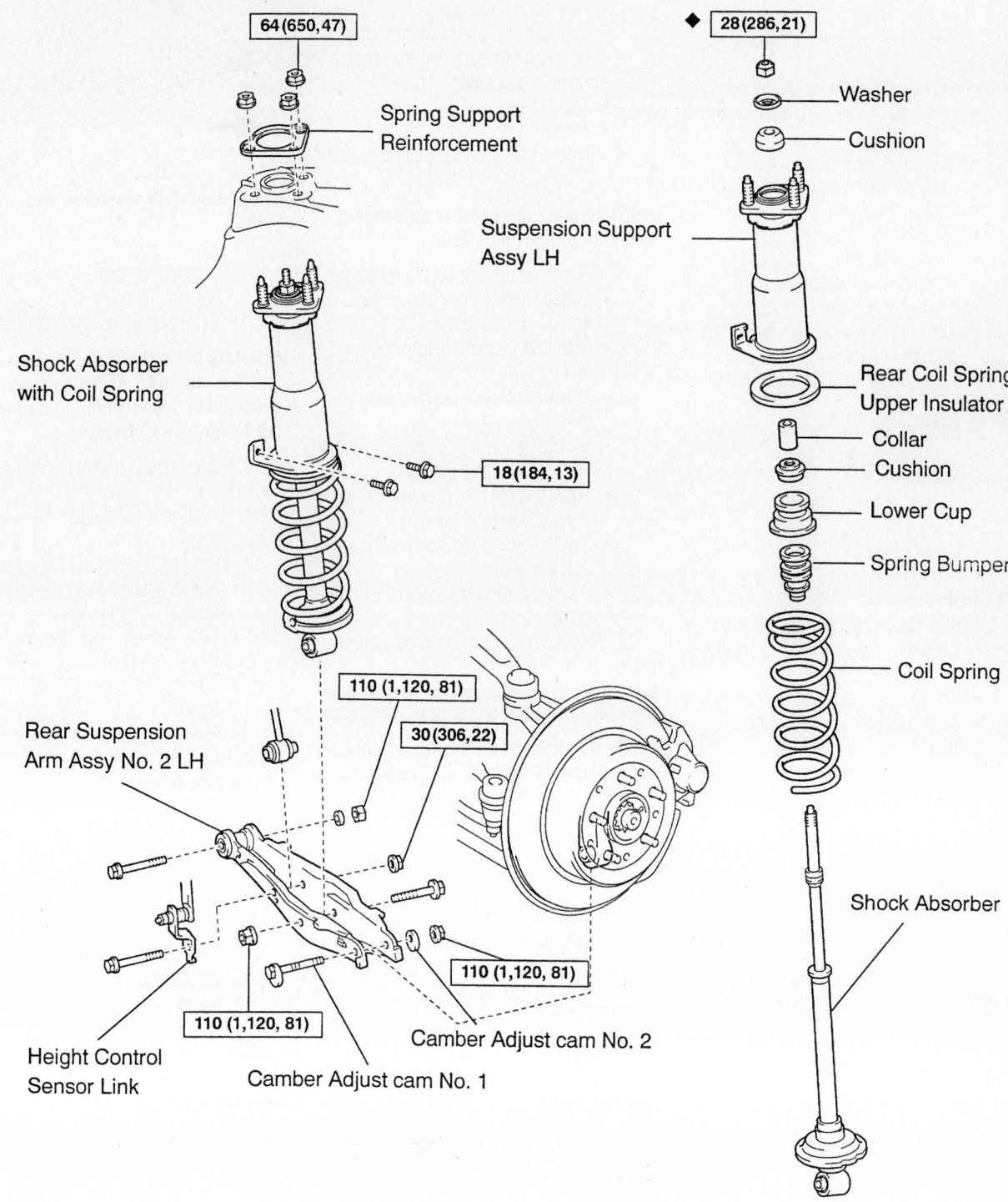

64 (650, 47)

Spring Support
Reinforcement

◆ 28 (286, 21)

Washer

Cushion

Suspension Support
Assy LH

Shock Absorber
with Coil Spring

18 (184, 13)

Rear Coil Spring
Upper Insulator

Collar

Cushion

Lower Cup

Spring Bumper

Coil Spring

Rear Suspension
Arm Assy No. 2 LH

110 (1,120, 81)

30 (306, 22)

Shock Absorber

110 (1,120, 81)

Height Control
Sensor Link

110 (1,120, 81)

Camber Adjust cam No. 1

Camber Adjust cam No. 2

N·m (kgf·cm, ft·lbf) : Specified torque
◆ Non-reusable part

67162-LEXU-G80

Exploded view of rear strut assembly—SC 430

2004 LS 430—WITH AIR SUSPENSION

1. Before servicing the vehicle, refer to the precautions in the beginning of this section.

2. Bleed the air system from the suspension.

3. Remove or disconnect the following:
- Rear seat cushion and seat back
- Scuff plates
- Door opening weather strip
- Rear seat side covers
- Roof side garnish
- Center stop light assembly
- Package tray trim
- Tire and wheel assembly
- 3 nuts and remove strut cap
- Turn control actuator clockwise 40 degrees and remove

4. Remove shock absorber cylinder assembly
- Support rear axle carrier with jack
- Rear tire house cover
- Air tube
- 3 upper nuts on assembly
- Lower side nut
- Lower assembly
- O-rings

To install:

5. Install or connect the following:
- 2 new O-rings and install connector No 2
- Temporarily install lower side nut
- Raise axle carrier and install air tube
- Control actuator until response is felt, turn approx. 40 degrees
- Control arm connector
- Shock absorber cap with 3 nuts. Tighten to 18 ft. lbs (25 Nm)

➡ **Start engine and allow air struts to fill. Checks for leaks**

- Rear wheels

➡ **Lower vehicle until all wheels are on the ground. Jounce several times to stabilize.**

6. Remove the rear wheel and support
7. Jack up the rear axle carrier
8. Install or connect the following:
- Tighten the lower nut. Tighten to 52 ft. lbs. (70 Nm)
- Rear wheel
9. Adjust vehicle height
10. Adjust rear wheel alignment.

SC 430

1. Before servicing the vehicle, refer to the precautions in the beginning of this section.

2. Raise the rear of the vehicle and support it with safety stands.

3. Remove or disconnect the following:
- Tonneau stoppers
- Rear floor finish plate
- Luggage compartment floor mat
- Luggage compartment trim covers
- Spare wheel cover
- Tool box cover
- Rear wheels

4. Remove rear suspension arm assembly

➡ **Match mark the camber adjustment cams for proper installation**

5. Remove or disconnect the following:
- Loosen nuts on rear suspension arm. Do NOT remove
- Stabilizer link and height control link
- shock absorber from suspension arm
- suspension arm from axle carrier and suspension member
- Loosen center nut of suspension support. Do NOT remove.
- 3 nuts holding the suspension support assembly
- 2 bolts holding the assembly to the frame

6. Remove the shock from the assembly

✱✱ WARNING

Using a suitable coil spring compressor, compress the coil spring before removing the following components

7. Remove or disconnect the following:
- Nut on suspension support previously loosened
- Washer, cushion, collar, suspension support assembly, upper insulator, lower cup, cushion, spring bumper and coil spring.

To install:

8. Install the shock absorber in assembly.

✱✱ WARNING

Using a suitable coil spring compressor, compress the coil spring before installing the following components

9. Install or connect the following:
- Spring into the spring seat

➡ **Make certain that spring seats properly in the seat**

- Spring bumper, lower cup, cushion, collar, upper insulator, suspension

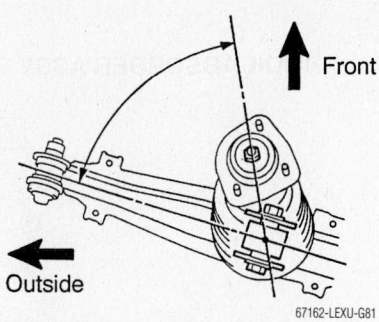

Alignment of the suspension support—SC 430

support, cushion and washer to shock
- Temporarily install center nut

➡ **Rotate the suspension support so that the rod and 1 of the bolts on the suspension support are aligned with the lower shock absorber.**

10. Release and remove the coil spring compressor
11. Install rear shock absorber with spring (Strut)
12. Install or connect the following:
- 3 nuts attaching the spring support to 47 ft. lbs (64 Nm)
- Fully tighten center lock nut to 13 ft. lbs. (18 Nm)
- 2 bolts attaching assembly to the frame
- Rear suspension arm bolts to 81 ft. lbs. (110 Nm)
- Stabilizer link and height control link
- Shock absorber to suspension arm bolt to 81 ft lbs. (110 Nm)
- Suspension arm to axle carrier with camber adjusting cams to 81 ft lbs. (Nm)
- Rear wheel
- All internal trunk covers, trims and plates

13. Inspect and adjust rear wheel alignment

ES 330

1. Before servicing the vehicle, refer to the precautions in the beginning of this section.

2. Raise the rear of the vehicle and support it with safety stands.

3. Remove or disconnect the following:
- Rear wheels
- Rear seat cushion and seat back
- Rood side garnish
- Door opening weather strip trim

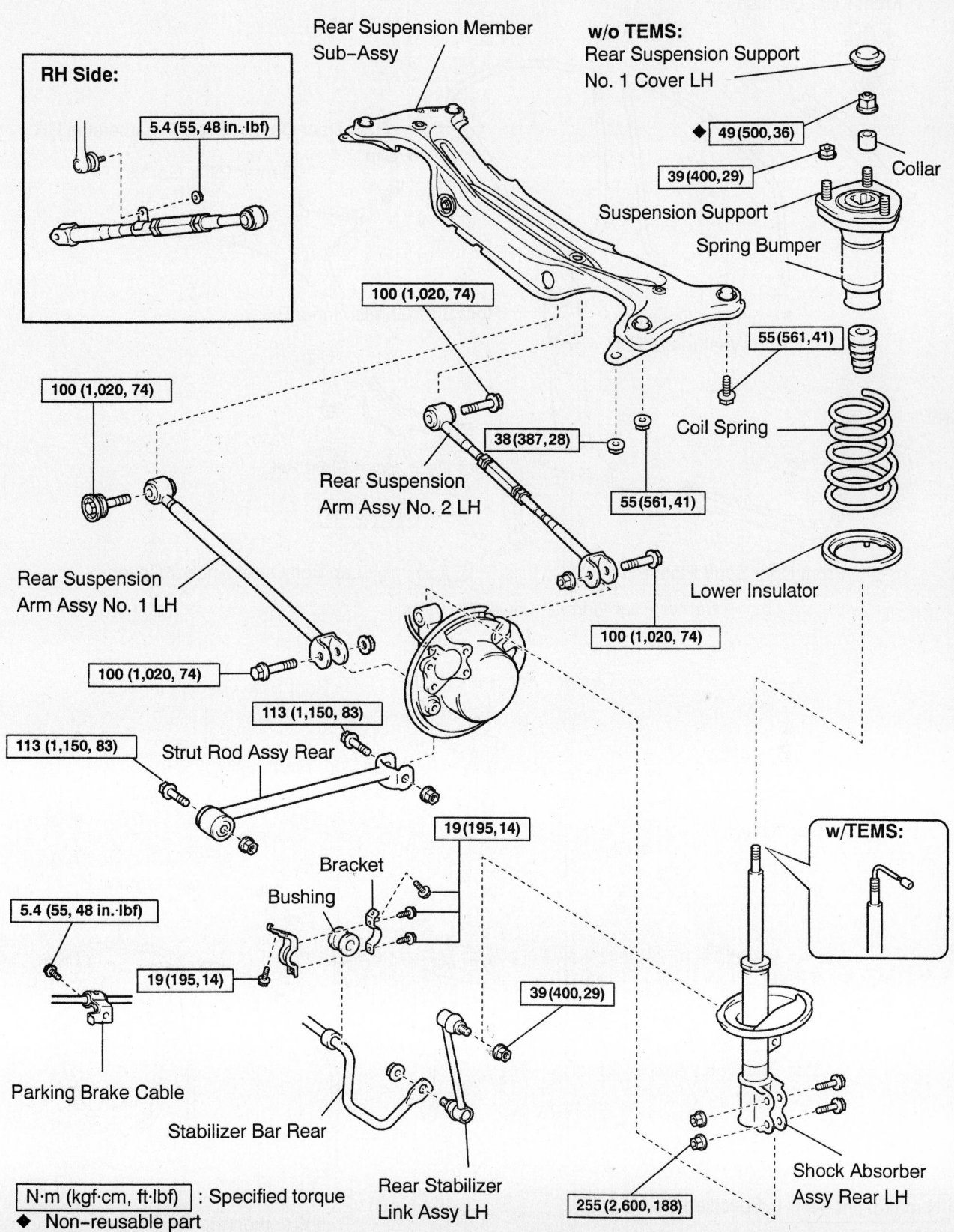

RH Side:

5.4 (55, 48 in.·lbf)

Rear Suspension Member Sub-Assy

w/o TEMS:
Rear Suspension Support No. 1 Cover LH

◆ 49 (500, 36)

39 (400, 29)

Collar

Suspension Support

Spring Bumper

100 (1,020, 74)

38 (387, 28)

55 (561, 41)

Coil Spring

55 (561, 41)

100 (1,020, 74)

Rear Suspension Arm Assy No. 2 LH

100 (1,020, 74)

Rear Suspension Arm Assy No. 1 LH

Lower Insulator

100 (1,020, 74)

113 (1,150, 83)

113 (1,150, 83)

Strut Rod Assy Rear

19 (195, 14)

Bracket

Bushing

5.4 (55, 48 in.·lbf)

19 (195, 14)

Parking Brake Cable

39 (400, 29)

Stabilizer Bar Rear

w/TEMS:

Shock Absorber Assy Rear LH

Rear Stabilizer Link Assy LH

255 (2,600, 188)

N·m (kgf·cm, ft·lbf) : Specified torque
◆ Non-reusable part

67162-LEXU-G82

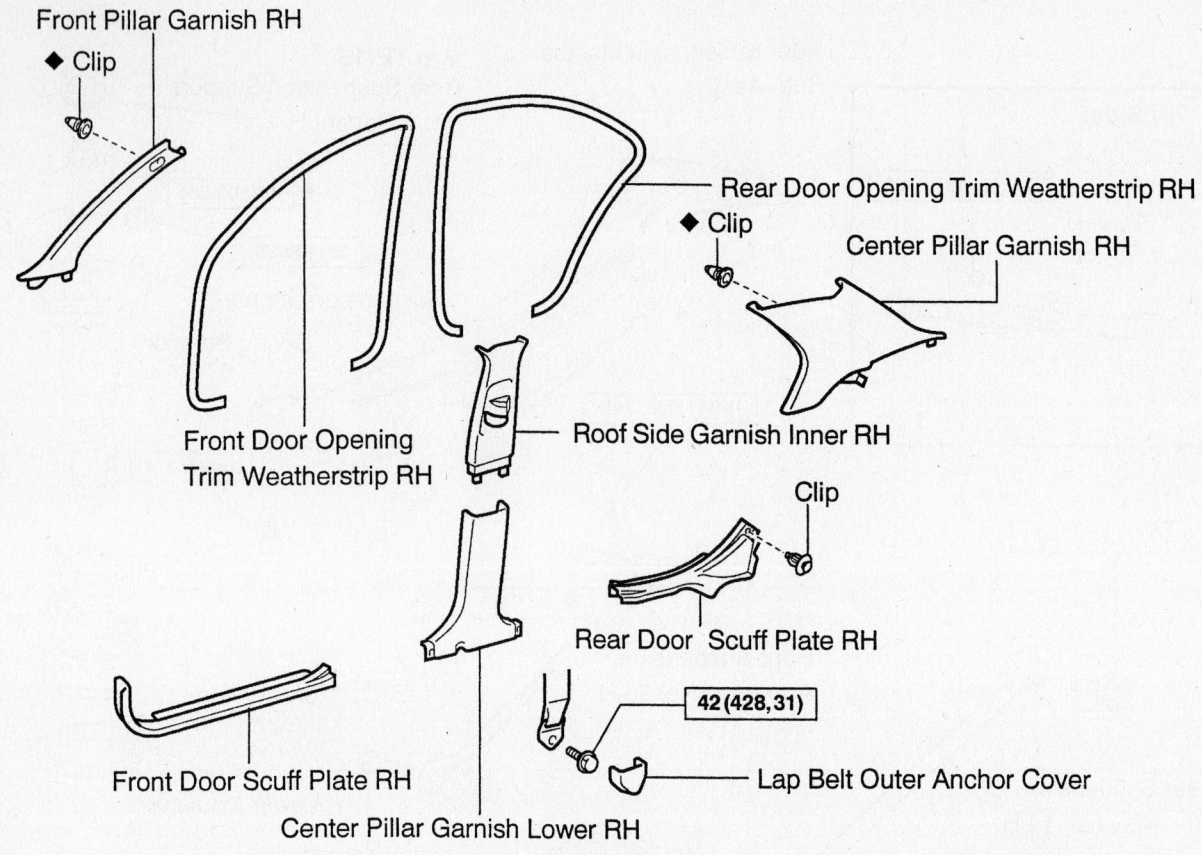

Front Pillar Garnish RH

◆ Clip

Rear Door Opening Trim Weatherstrip RH

◆ Clip

Center Pillar Garnish RH

Front Door Opening Trim Weatherstrip RH

Roof Side Garnish Inner RH

Clip

Rear Door Scuff Plate RH

42 (428, 31)

Lap Belt Outer Anchor Cover

Front Door Scuff Plate RH

Center Pillar Garnish Lower RH

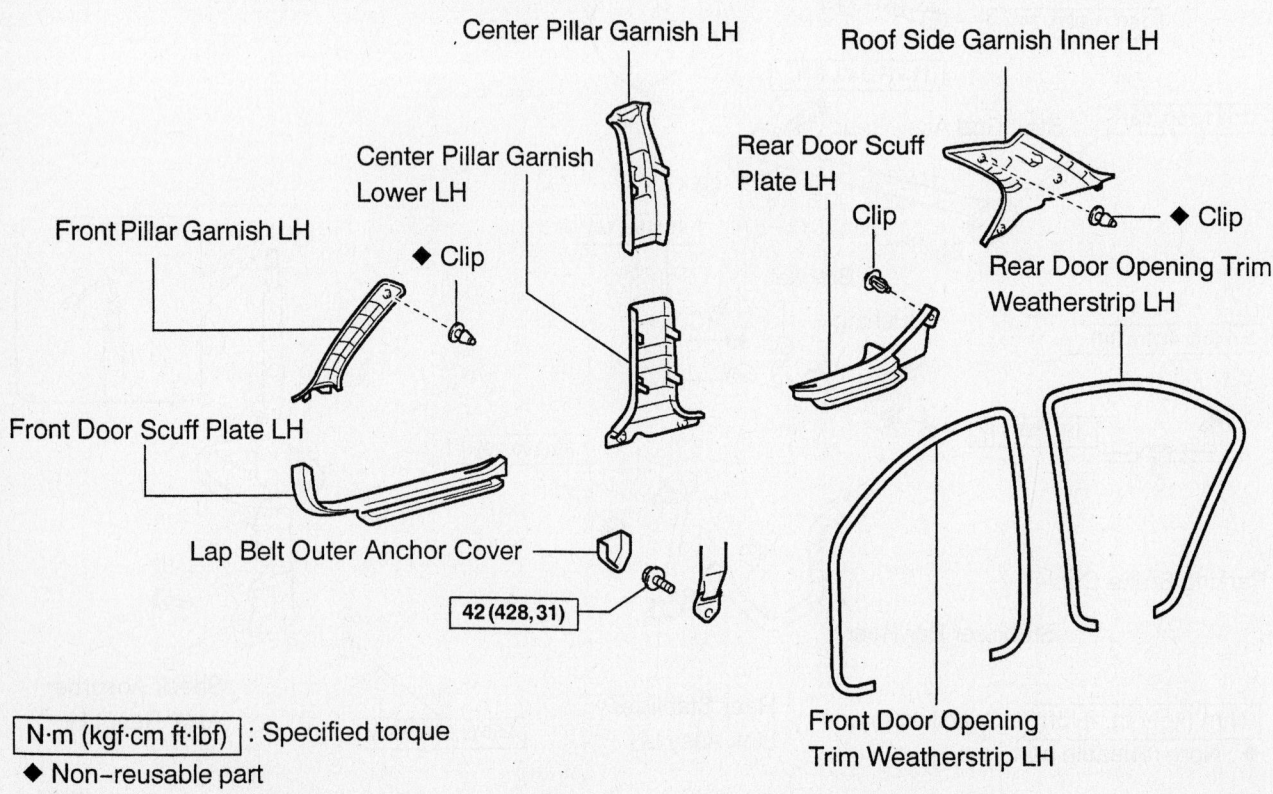

Center Pillar Garnish LH

Roof Side Garnish Inner LH

Center Pillar Garnish Lower LH

Rear Door Scuff Plate LH

Clip

◆ Clip

Front Pillar Garnish LH

◆ Clip

Rear Door Opening Trim Weatherstrip LH

Front Door Scuff Plate LH

Lap Belt Outer Anchor Cover

42 (428, 31)

Front Door Opening Trim Weatherstrip LH

N·m (kgf·cm ft·lbf) : Specified torque

◆ Non-reusable part

67162-LEXU-G83

Exploded view, Roof headlining assembly—ES 330

- Center stop light assembly (w/o sun shade)
- Rear shoulder belt cover
- Package tray trim panel
- Rear seat 3 point seat belt outer assembly
- Rear stabilizer bar link
- Rear suspension support No. 1 cover
- 2 bolts for brake flex hose and ABS wire harness
- Loosen 2 bolts on lower shock absorber
- Loosen center support nut. Do NOT remove
- 3 nuts from center suspension support
- Lower rear axle carrier and remove the 2 bolts loosened earlier
- Rear shock absorber and coil assembly

4. Remove the shock from the assembly

✳✳ WARNING

Using a suitable coil spring compressor, compress the coil spring before removing the following components

5. Remove or disconnect the following:
- Nut on suspension support previously loosened
- Collar, suspension support assembly, spring bumper and coil spring.
- Lower insulator

To install:

6. Install the shock absorber in assembly

✳✳ WARNING

Using a suitable coil spring compressor, compress the coil spring before installing the following components

7. Install or connect the following:
- Spring into the spring seat

➡**Make certain that spring seats properly in the seat**

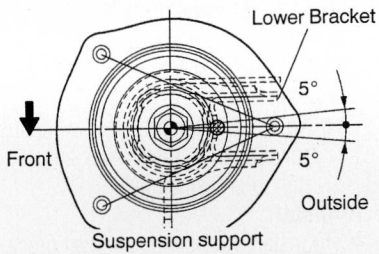

Aligning support bracket—ES 330

- Lower insulator, coil spring, suspension support assembly, collar
- Temporarily install center nut

➡**Align the suspension support with the shock absorber lower bracket**

8. Release and remove the coil spring compressor
9. Install rear shock absorber with spring (Strut)
10. Install or connect the following:
- 3 nuts attaching the spring support to 29 ft. lbs. (39 Nm)
- 2 bolts attaching sock to the rear axle carrier assembly to 188 ft. lbs. (255 Nm)
- Flex hose and ABS speed sensor wire harness
- Center nut on suspension support to 36 ft. lbs. (49 Nm)
- Rear suspension support cover
- Rear stabilizer link to 29 ft. lbs. (39 Nm)
- Rear wheels
- Reinstall interior components in reverse order of removal

11. Inspect and align the rear wheels

Upper Ball Joint

REMOVAL & INSTALLATION

The upper ball joint is an integral part of the upper arm and is not replaced separately. The upper ball joint replacement is accomplished by replacing the upper arm.

Upper Control Arm

REMOVAL & INSTALLATION

GS 300 and GS 430 (Except 2004)

1. Before servicing the vehicle, refer to the precautions in the beginning of this section.
2. Remove or disconnect the following:
- Negative battery cable
- Wheel

3. Loosen the 3 upper strut mounting nuts.
4. Loosen, but do not remove, the upper strut rod nut.

✳✳ CAUTION

Do NOT completely remove the upper strut nut at this time.

5. Remove or disconnect the following:
- Brake caliper, leaving the line

attached and secure it out of the way
- Anti-lock Brake System (ABS) speed sensor and harness
- Cotter pin and nut from the upper control arm
- Upper control arm from the steering knuckle
- Stabilizer bar from the link and remove the bracket
- Cotter pin and nut from the lower control arm
- Strut from the lower suspension arm
- 3 upper strut mounting nuts and remove the strut
- Mounting bolts holding the upper control arm to the frame
- Upper control arm from the vehicle

To install:

6. Install or connect the following:
- Upper suspension arm and tighten the mounting bolts to 39 ft. lbs. (53 Nm)
- Strut and tighten the upper retaining nuts to 41 ft. lbs. (56 Nm). Tighten the new upper strut rod nut to 20 ft. lbs. (27 Nm).
- Strut to the lower arm and temporarily tighten the nut and bolt
- Stabilizer bar bracket and tighten the bolts to 21 ft. lbs. (28 Nm)
- Stabilizer bar to the link and tighten the bolts to 29 ft. lbs. (39 Nm)
- Upper suspension arm to the steering knuckle. Tighten the nut to 64 ft. lbs. (87 Nm) and install a new cotter pin.
- ABS speed sensor and tighten the bolt to 69 inch lbs. (8 Nm)
- Caliper
- Front wheel

7. Lower the vehicle.
8. Bounce the vehicle several times to stabilize the suspension.
9. Tighten the lower strut bolt and nut to 116 ft. lbs. (157 Nm).
10. Check the front wheel alignment.

GS 300 and GS 430 (2004)

1. Raise and support the front of the vehicle or raise vehicle on a hoist.
2. Remove the front wheel(s).
3. Remove the front strut and coil spring assembly.
4. Remove the 2 bolts and remove the upper control arm.

To install:

5. Install the upper control arm into position and install the 2 retaining bolts. Torque the bolts to 39 ft. lbs. (53 Nm).

6. Install the front strut and coil spring assembly

7. Install the front wheel(s). Torque the wheel nuts to 76 ft. lbs. (103 Nm).

IS 300

1. Before servicing the vehicle, refer to the precautions in the beginning of this section.

2. Remove or disconnect the following:

- Front wheel
- Strut and spring assembly
- Inner bolts and the control arm

To install:

3. Install or connect the following:

- Control arm and tighten the inner bolts to 44 ft. lbs. (59 Nm)
- Strut and spring assembly
- Front wheel

LS 430 and SC 430

1. Before servicing the vehicle, refer to the precautions in the beginning of this section.

2. Raise and safely support the vehicle.

3. Remove or disconnect the following:

- Wheel
- Strut or if equipped with air suspension, remove the pneumatic cylinder
- Anti-lock Brake System (ABS) speed sensor wire harness from the upper control arm by removing the bolt.
- Mounting bolts holding the upper control arm to the vehicle
- Upper control arm

To install:

4. Install or connect the following:

- Upper control arm and tighten the 2 mounting bolts to (except SC 430) 83 ft. lbs. (113 Nm), or (SC 430) 39 ft. lbs. (53 Nm)
- ABS speed sensor wire harness to the upper control arm with the attaching bolt
- Strut, or if equipped with air suspension, install the pneumatic cylinder
- Wheel

5. Lower the vehicle.

6. Check and adjust the wheel alignment as necessary.

CONTROL ARM BUSHING REPLACEMENT

The control arm bushings are serviced with the control arm as an assembly.

Lower Control Arm

REMOVAL & INSTALLATION

ES 300

1. Before servicing the vehicle, refer to the precautions in the beginning of this section.

2. Remove or disconnect the following:

- Negative battery cable
- Front wheel(s)
- Side fender apron seal
- Steering knuckle with the axle hub, from the vehicle
- Dust deflector from the knuckle
- Cotter pin and the nut from the ball joint stud

3. Remove the lower ball joint from the steering knuckle.

To install:

4. Install the lower ball joint onto the steering knuckle and tighten nut to 90 ft. lbs. (123 Nm). Install new cotter pin.

5. Align the hole in the dust deflector with the ABS speed sensor. Using the appropriate driver, install a new dust deflector.

6. Install or connect the following:

- Steering knuckle and hub onto the vehicle
- Fender apron seal
- Front wheel(s)
- Negative battery cable

GS 300 and GS 430 (Except 2004)

1. Before servicing the vehicle, refer to the precautions in the beginning of this section.

2. Remove or disconnect the following:

- Negative battery cable
- Wheel(s)
- Caliper, leaving the brake line connected and suspend it out of the way

✱✱ WARNING

Never allow the brake caliper to hang freely from the brake hose.

- Rotor
- Anti-lock Brake System (ABS) speed sensor and harness
- Tie rod end from the arm on the lower ball joint
- Cotter pin and nut. Disconnect the upper control arm from the steering knuckle.
- Cotter pin and nut. Disconnect the

steering knuckle from the lower control arm.

- Steering knuckle and ball joint assembly from the vehicle
- 2 ball joint mounting bolts, then remove the ball joint from the steering knuckle

To install:

3. Install or connect the following:

- Ball joint and tighten the bolts to 83 ft. lbs. (113 Nm)
- Steering knuckle to the lower and upper suspension arms. Tighten the lower control arm nut to 95 ft. lbs. (127 Nm) and install a new cotter pin. Tighten the upper control arm to 64 ft. lbs. (87 Nm) and install a new cotter pin.
- Tie rod end to the ball joint arm. Tighten the nut to 64 ft. lbs. (87 Nm) and install a new cotter pin.
- Rotor
- Caliper
- ABS speed sensor and harness. Tighten the sensor retaining bolt to 69 inch lbs. (8 Nm).
- Wheel(s)
- Negative battery cable

4. Check the front wheel alignment.

GS 300 and GS 430 (2004)

1. Before servicing the vehicle, refer to the precautions in the beginning of this section.

2. Raise the vehicle on a hoist, so that front suspension components are hanging and accessible.

3. Remove or disconnect the following:

- Front wheels.
- Engine under covers.
- Brake caliper(s); Do NOT disconnect the brake hose; hang the caliper without stress on the hose
- Tie rod end from steering knuckle
- Stabilizer bar link from stabilizer bar
- Height control sensor link, if equipped, from shock absorber bracket
- Shock absorber lower mount
- Lower control arm set bolts (loosen only)
- Lower ball joint from No. 2 lower control arm (lower suspension arm)
- Steering gear assembly
- Strut bar bracket
- No. 1 lower control arm (lower suspension arm); matchmark adjusting cam to crossmember

To install:

4. To install, reverse the removal procedure, noting the following torque settings:

- No. 1 lower control arm (lower suspension arm) bolt to shock absorber bracket: 44 ft. lbs. (59 Nm)
- No. 1 lower control arm (lower suspension arm) adjusting cam bolt and nut to crossmember: 127 ft. lbs. (172 Nm)
- Strut bar bracket bolts: 43 ft. lbs. (58 Nm)
- Strut bar bracket nut: 112 ft. lbs. (152 Nm)

- No. 2 lower control arm (lower suspension arm) nuts: 122 ft. lbs. (164 Nm)
- Lower shock absorber mounting bolt and nut: 116 ft. lbs. (157 Nm)
- Stabilizer bar link nut: 83 ft. lbs. (113 Nm)
- Stabilizer bar link-to-stabilizer bar bolt and nut: 43 ft. lbs. (55 Nm)
- Tie rod end nut: 64 ft. lbs. (87 Nm)
- Brake caliper bolts: 87 ft. lbs. (118 Nm)

- Front wheel nuts: 76 ft. lbs. (103 Nm)

IS 300

1. Before servicing the vehicle, refer to the precautions in the beginning of this section.

2. Remove or disconnect the following:
- Front wheel
- Engine under covers
- Level control sensor link

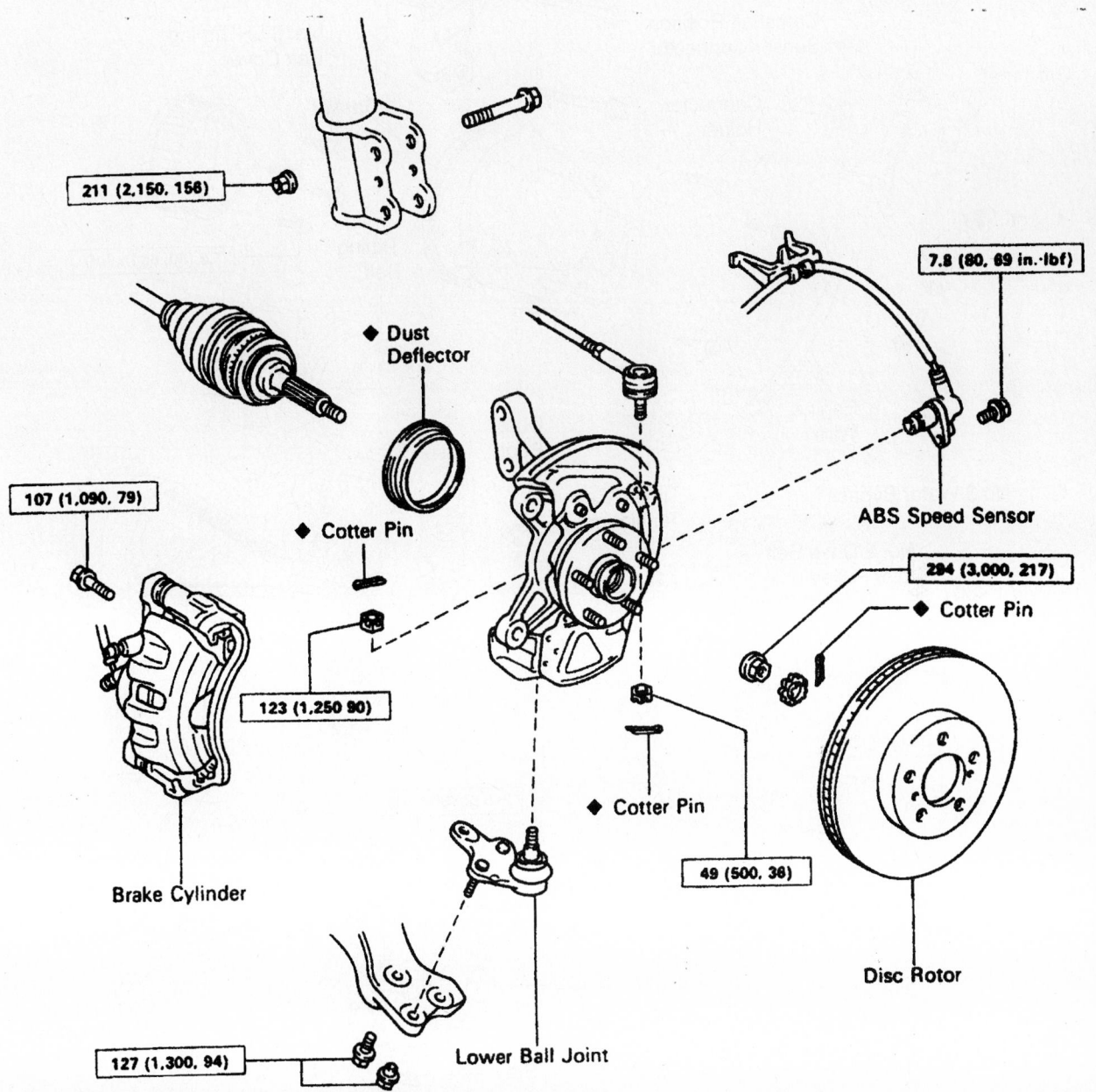

211 (2,150, 156)

7.8 (80, 69 in.·lbf)

◆ Dust Deflector

ABS Speed Sensor

107 (1,090, 79)

◆ Cotter Pin

294 (3,000, 217)

◆ Cotter Pin

123 (1,250 90)

◆ Cotter Pin

49 (500, 36)

Brake Cylinder

Disc Rotor

127 (1,300, 94)

Lower Ball Joint

N·m (kgf·cm, ft·lbf) : Specified torque

◆ Non-reusable part

7923LGB3

Exploded view of the lower suspension—ES 300

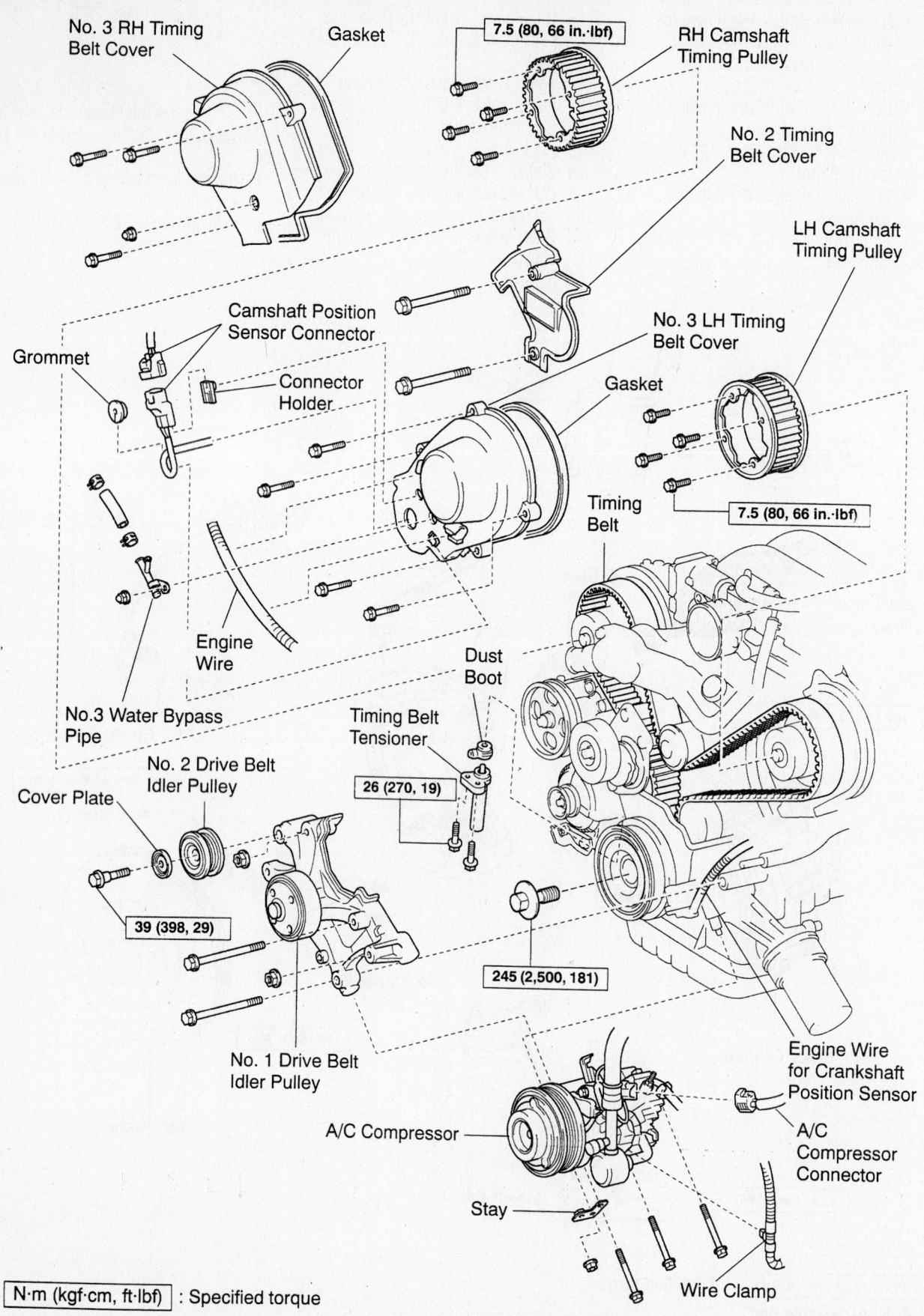

No. 3 RH Timing Belt Cover

Gasket

7.5 (80, 66 in.·lbf)

RH Camshaft Timing Pulley

No. 2 Timing Belt Cover

LH Camshaft Timing Pulley

Grommet

Camshaft Position Sensor Connector

Connector Holder

No. 3 LH Timing Belt Cover

Gasket

Timing Belt

7.5 (80, 66 in.·lbf)

Engine Wire

No.3 Water Bypass Pipe

Dust Boot

Timing Belt Tensioner

26 (270, 19)

Cover Plate

No. 2 Drive Belt Idler Pulley

39 (398, 29)

245 (2,500, 181)

No. 1 Drive Belt Idler Pulley

A/C Compressor

Engine Wire for Crankshaft Position Sensor

A/C Compressor Connector

Stay

Wire Clamp

N·m (kgf·cm, ft·lbf) : Specified torque

Exploded view of the front lower control arm and related components—GS 300 and GS 430 (2004 models)

67162-LEXU-G116

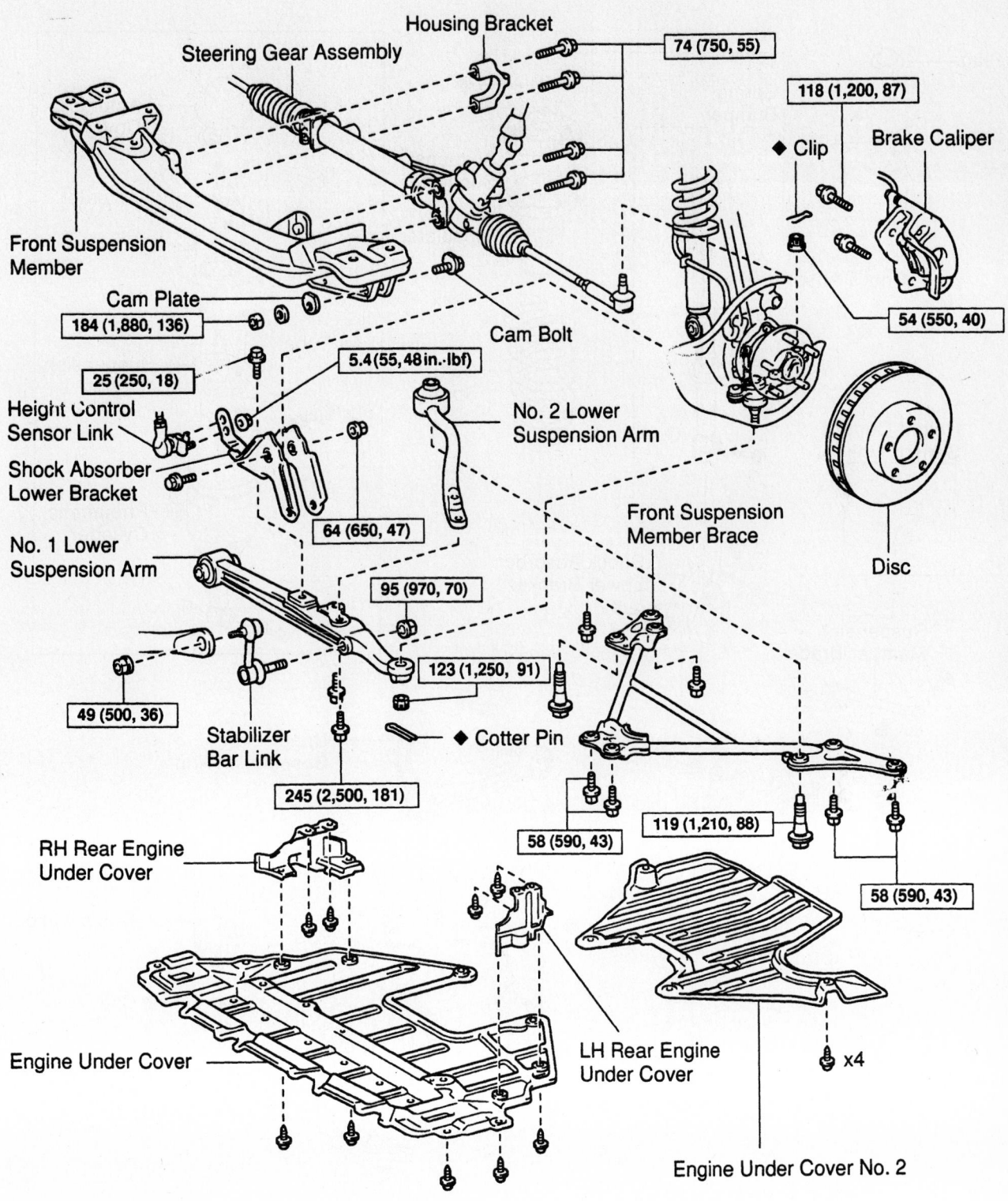

Steering Gear Assembly

Housing Bracket

74 (750, 55)

118 (1,200, 87)

◆ Clip

Brake Caliper

Front Suspension Member

Cam Plate

184 (1,880, 136)

Cam Bolt

54 (550, 40)

25 (250, 18)

5.4 (55, 48 in.·lbf)

Height Control Sensor Link

No. 2 Lower Suspension Arm

Shock Absorber Lower Bracket

64 (650, 47)

Front Suspension Member Brace

Disc

No. 1 Lower Suspension Arm

95 (970, 70)

49 (500, 36)

123 (1,250, 91)

◆ Cotter Pin

Stabilizer Bar Link

245 (2,500, 181)

119 (1,210, 88)

RH Rear Engine Under Cover

58 (590, 43)

58 (590, 43)

Engine Under Cover

LH Rear Engine Under Cover

x4

Engine Under Cover No. 2

N·m (kgf·cm, ft·lbf) : Specified torque

◆ Non–reusable part

9347LG06

Exploded view of the front suspension—IS 300

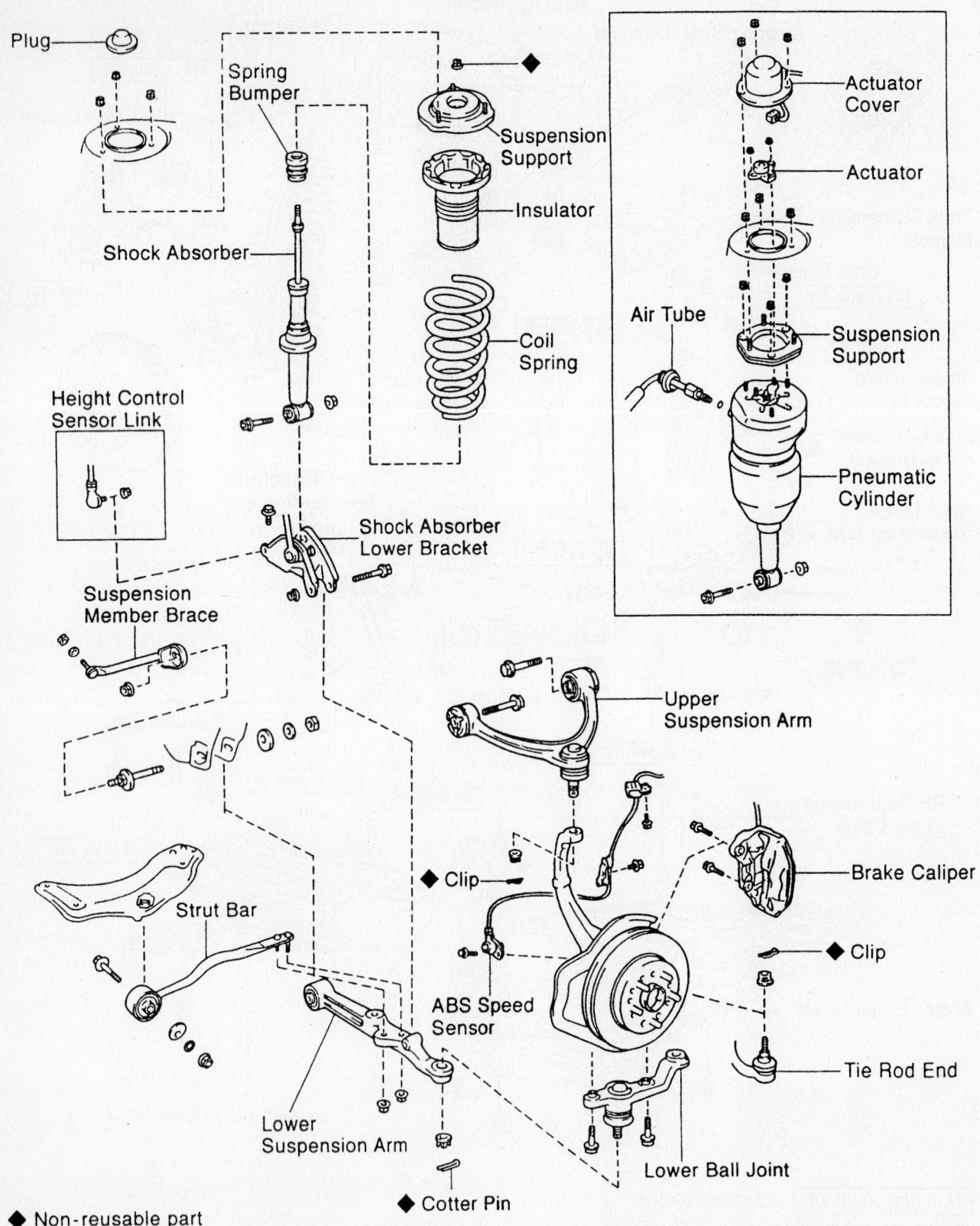

Plug

Spring Bumper

Suspension Support

Insulator

Shock Absorber

Coil Spring

Height Control Sensor Link

Actuator Cover

Actuator

Air Tube

Suspension Support

Pneumatic Cylinder

Shock Absorber Lower Bracket

Suspension Member Brace

Upper Suspension Arm

Brake Caliper

◆ Clip

Strut Bar

◆ Clip

ABS Speed Sensor

Tie Rod End

Lower Suspension Arm

◆ Cotter Pin

Lower Ball Joint

◆ Non-reusable part

Exploded view of the lower ball joint mounting—LS 430

7923LGB4

- Front subframe brace
- No. 2 lower control arm
- Brake caliper and rotor
- Outer tie rod end
- Stabilizer bar link
- Lower strut bolt
- Lower ball joint
- Steering gear
- No. 1 lower control arm

To install:

3. Installation is the reverse of the removal procedure, while using the following torque values:

- No. 1 lower control arm bolt: 136 ft. lbs. (184 Nm)
- Lower ball joint nut: 91 ft. lbs. (123 Nm)
- Outer tie rod end nut: 40 ft. lbs. (54 Nm)
- Brake caliper bolts: 87 ft. lbs. (118 Nm)
- No. 2 control arm-to-No. 1 control arm bolts: 181 ft. lbs. (245 Nm)
- Front subframe brace small bolts: 43 ft. lbs. (58 Nm)

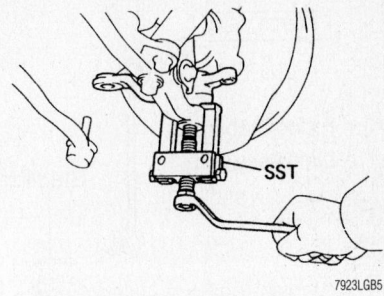

Disconnecting the ball joint from the lower suspension arm—LS 430

FRONT SUSPENSION LOWER ARM ASSY:

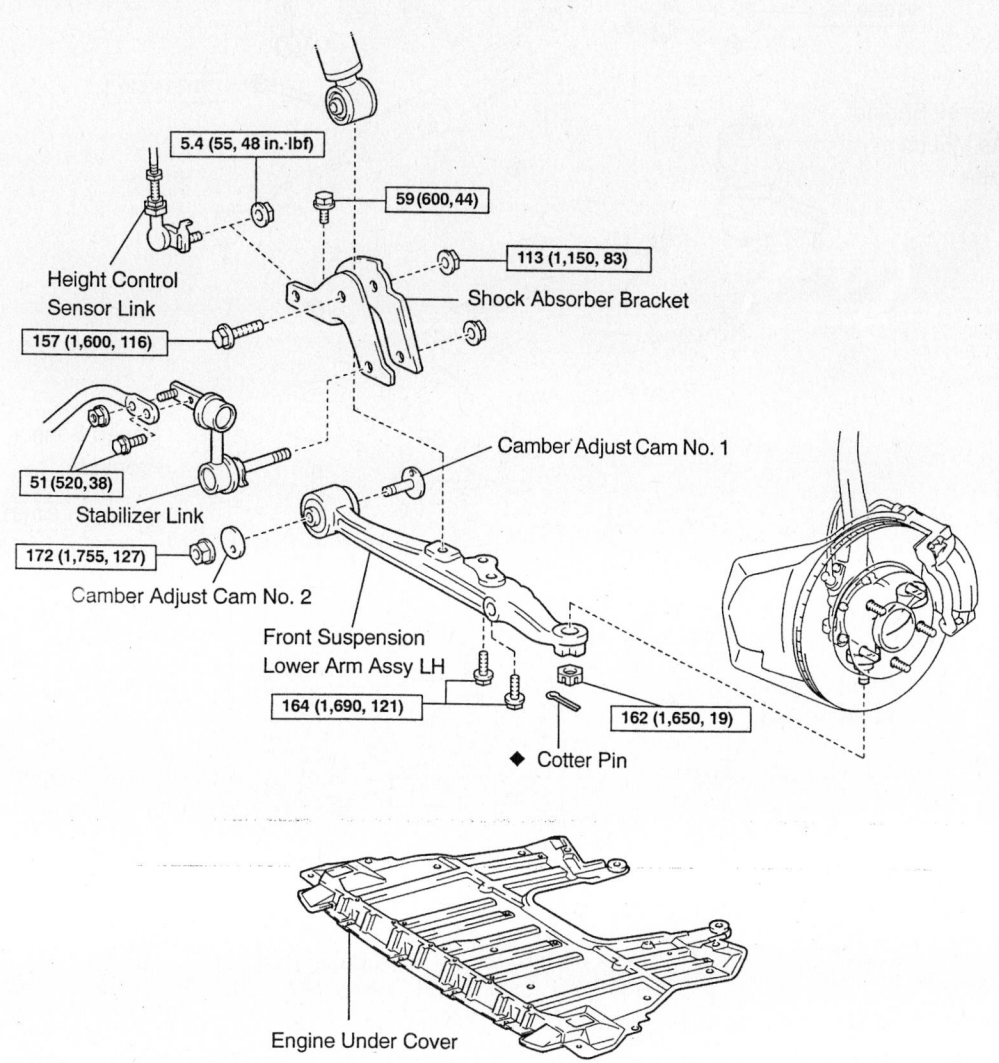

5.4 (55, 48 in.·lbf)

59 (600, 44)

113 (1,150, 83)

Height Control Sensor Link

157 (1,600, 116)

Shock Absorber Bracket

Camber Adjust Cam No. 1

51 (520, 38)

Stabilizer Link

172 (1,755, 127)

Camber Adjust Cam No. 2

Front Suspension Lower Arm Assy LH

164 (1,690, 121)

162 (1,650, 19)

◆ Cotter Pin

Engine Under Cover

N·m (kgf·cm, ft·lbf) : Specified torque
◆ Non-reusable part

67162-LEXU-G85

Expanded view, lower front suspension arm—SC 430

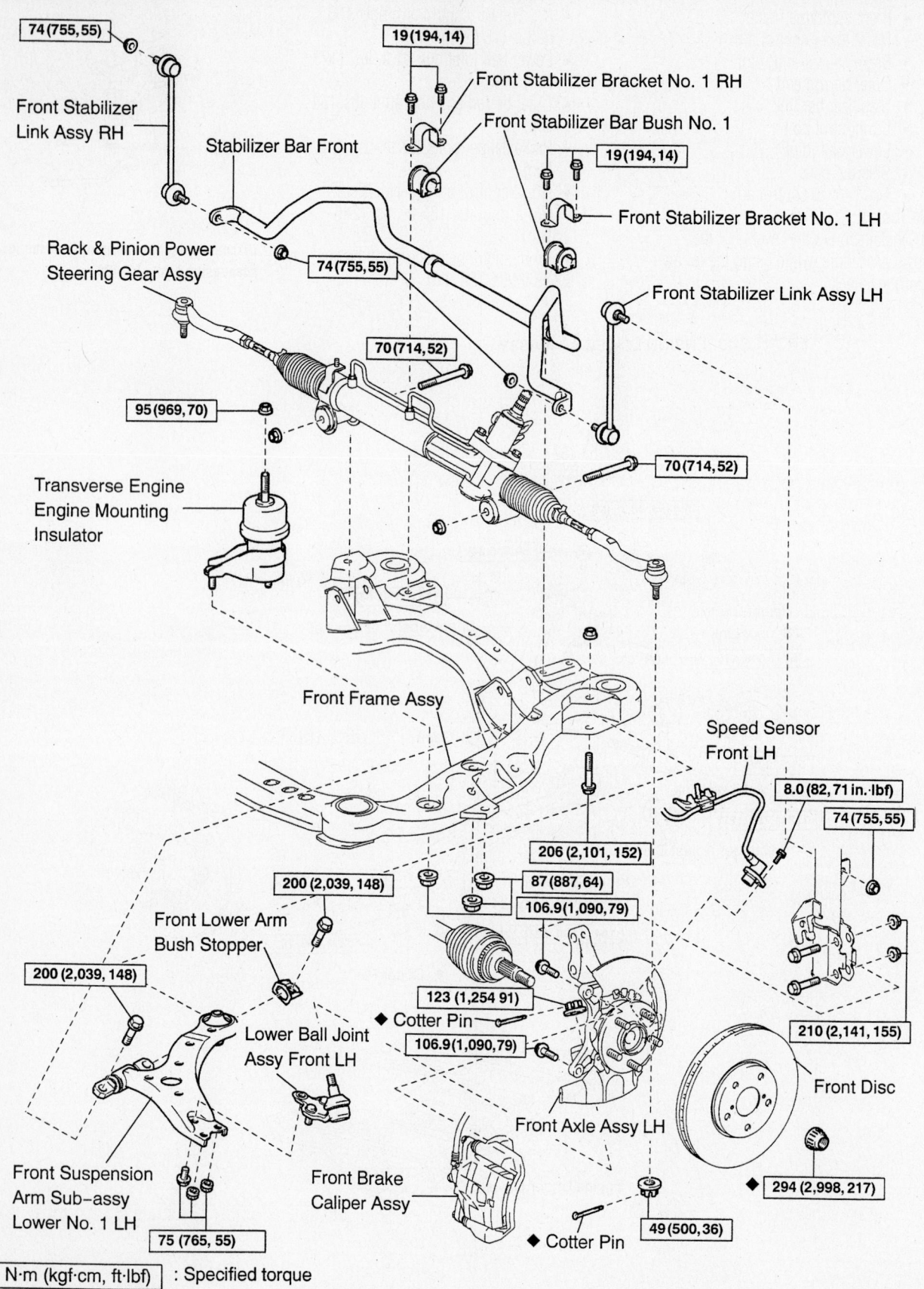

74 (755, 55)

Front Stabilizer
Link Assy RH

19 (194, 14)

Front Stabilizer Bracket No. 1 RH

Stabilizer Bar Front

Front Stabilizer Bar Bush No. 1

19 (194, 14)

Rack & Pinion Power
Steering Gear Assy

Front Stabilizer Bracket No. 1 LH

74 (755, 55)

Front Stabilizer Link Assy LH

70 (714, 52)

95 (969, 70)

Transverse Engine
Engine Mounting
Insulator

70 (714, 52)

Front Frame Assy

Speed Sensor
Front LH

8.0 (82, 71 in.·lbf)

74 (755, 55)

206 (2,101, 152)

200 (2,039, 148)

Front Lower Arm
Bush Stopper

87 (887, 64)

106.9 (1,090, 79)

200 (2,039, 148)

123 (1,254 91)

Cotter Pin

106.9 (1,090, 79)

210 (2,141, 155)

Lower Ball Joint
Assy Front LH

Front Disc

Front Axle Assy LH

Front Suspension
Arm Sub–assy
Lower No. 1 LH

Front Brake
Caliper Assy

294 (2,998, 217)

75 (765, 55)

Cotter Pin

49 (500, 36)

N·m (kgf·cm, ft·lbf) : Specified torque

◆ Non–reusable part

67162-LEXU-G86

Expanded view, front suspension—ES 330

- Front subframe large bolts: 88 ft. lbs. (119 Nm)

LS 430

1. Before servicing the vehicle, refer to the precautions in the beginning of this section.

2. If equipped with air suspension, move the height control switch (located in the trunk) to the **OFF** position.

3. Remove or disconnect the following:
- Tire and wheel assembly
- Anti-lock Brake System (ABS) speed sensor and wiring harness from the steering knuckle
- Brake caliper support bracket by removing the 2 bolts. Leave the brake line connected. Support the caliper aside by using a piece of wire.

4. Loosen the 2 lower ball joint mounting bolts.

➡ **Do NOT remove the bolts.**

5. Remove or disconnect the following:
- Clip and nut from the tie rod end
- Tie rod end from the steering arm with the proper tool
- Lower ball joint mounting bolts from the steering knuckle
- Cotter pin and nut from the lower ball joint
- Lower ball joint from the lower control arm

To install:

6. Install or connect the following:
- Ball joint to the lower control arm. Tighten the nut to 112 ft. lbs. (152 Nm) and install a new cotter pin.
- Mounting bolts, temporarily, holding the ball joint to the steering knuckle
- Tie rod end to the steering knuckle. Tighten the nut to 48 ft. lbs. (65 Nm) and install a new cotter pin.
- Lower ball joint bolts to 83 ft. lbs. (113 Nm)
- Brake caliper support bracket and tighten the 2 bolts to 87 ft. lbs. (118 Nm)
- ABS speed sensor and wiring harness to the steering knuckle
- Wheel

7. Lower the vehicle.

8. Turn the height control switch **ON**.

SC 430

1. Before servicing the vehicle, refer to the precautions in the beginning of this section.

2. Remove or disconnect the following:
- Front wheel
- Engine under covers
- Loosen 2 bolts on the suspension lower arm
- Shock absorber from mounting bracket
- Height control sensor
- Front stabilizer link
- Separate rack and pinion gear assembly

3. Remove front suspension lower arm

4. Remove or disconnect the following:
- Ball joint cotter pin and bolt
- Lower ball joint from the lower arm assembly
- Shock absorber bracket
- Lower arm assembly

5. Matchmark the front and rear adjustment cams to the body and then remove the nuts and adjusting cams.

6. Lift out the lower control arm.

To install:

7. Install or connect the following:
- Shock absorber bracket. Torque: 44 ft. lbs. (59 Nm)
- Lower control arm to the body and temporarily install the adjusting cams and nuts. Do NOT tighten the nuts at this time.
- Lower control arm to the knuckle and tighten the ball joint nut to 119 ft. lbs. (162 Nm). Install a new cotter pin.
- 2 bolts on lower suspension arm to 121 ft. lbs. (164 Nm)
- Rack and pinion steering gear assembly nuts to 48 ft. lbs. (65 Nm)
- Front stabilizer link bolts and nuts (to stabilizer bar) to 38 ft. lbs. (51 Nm) and (to stabilizer link) to 116 ft. lbs. (157 Nm)
- Shock absorber assembly bolt to 116 ft. lbs. (157 Nm)
- Height control sensor
- Front wheel

8. Inspect and adjust front wheel alignment

9. Adjust height control sensor

ES 330

➡ **Removal of the lower control arm requires the removal of the engine and transaxle.**

1. Before servicing the vehicle, refer to the precautions in the beginning of this section.

2. Remove or disconnect the following:
- Transverse engine mounting insulator

- Engine and transaxle assembly
- 2 bolts on front side of suspension arm
- Bolt and nut on rear side of suspension arm and lower arm
- Lower bush stopper

To install:

3. Install or connect the following:
- Lower bush arm stopper
- 2 bolt on the front side to 148 ft. lbs. (200 Nm)
- Rear side bolt and nut to 152 ft. lbs. (206 Nm)
- Transverse engine mounting insulator to 64 ft. lbs. (87 Nm)
- Engine and transaxle assembly

CONTROL ARM BUSHING REPLACEMENT

The control arm bushings are serviced with the control arm as an assembly.

Wheel Bearings

ADJUSTMENT

Check the backlash in bearing shaft direction and the axle hub deviation. Maximum for backlash should be 0.0020 in. (0.05mm) and for axle hub deviation 0.020 in. (0.05mm).

➡ **The front and rear wheel bearings are non-adjustable. If the wheel bearing is out of specifications, replace the wheel bearing.**

REMOVAL & INSTALLATION

Front

ES 300

1. Before servicing the vehicle, refer to the precautions in the beginning of this section.

2. Remove or disconnect the following:
- Negative battery cable
- Front wheels
- Fender apron seal
- Cotter pin and lock cap from the end of the halfshaft
- Halfshaft locknut.
- Brake caliper and use a wire to support it out of the way

✳✳ WARNING

Never allow the caliper to hang freely from the brake hose.

- Rotor
- Anti-lock Brake System (ABS)

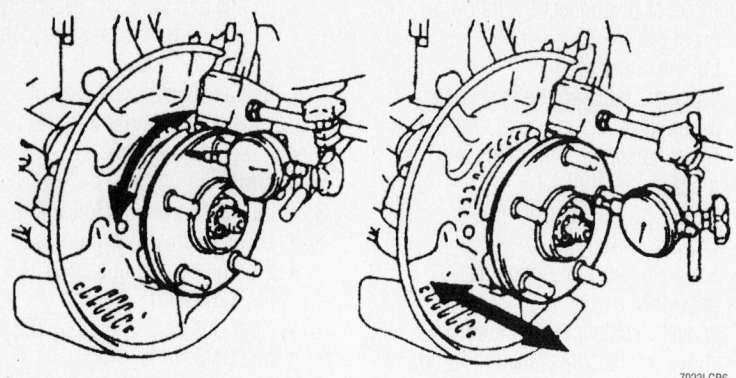

Checking wheel bearings for excessive play

speed sensor from the steering knuckle
- Nuts on the lower end of the strut
- Tie rod end from the steering knuckle
- Lower control arm from the ball joint
- Driveshaft from the axle hub
- 2 nuts on the lower end of the strut
- Steering knuckle

3. Clamp the steering knuckle in a vise with soft jaws to protect the knuckle.
- Dust deflector from the hub
- Ball joint from the steering knuckle
- Hub from the knuckle
- Inner race from the hub
- Dust cover
- Snapring
- Bearing from the steering knuckle

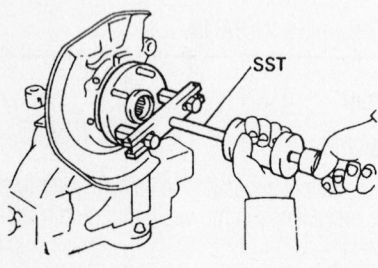

Removing the axle hub from the steering knuckle—ES 300

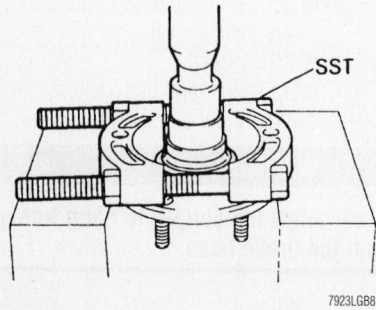

Remove the inner race from the hub—ES 300

To install:
4. Install or connect the following:
- Bearing into the knuckle
- Snapring
- Dust cover. Tighten the 4 bolts to 74 inch lbs. (8.3 Nm).
- Hub into the steering knuckle
- Lower ball joint to the steering knuckle. Tighten the nut to 90 ft. lbs. (123 Nm) and install a new cotter pin.
- Dust deflector
- Knuckle on the lower strut
- Lower ball joint to the lower arm. Tighten the bolts to 94 ft. lbs. (127 Nm).
- Tie rod end to the steering knuckle. Tighten the nut to 36 ft. lbs. (49 Nm).
- Nuts on the lower strut to 156 ft. lbs. (211 Nm)
- ABS speed sensor. Tighten the mounting bolt to 69 inch lbs. (8 Nm).
- Rotor
- Caliper. Tighten the mounting bolts to 79 ft. lbs. (107 Nm).
- Axle locknut. Tighten the nut to 217 ft. lbs. (294 Nm). Install the lock cap and a new cotter pin.
- Front fender apron seal
- Wheel

5. Turn the wheel by hand, verify that the wheel turns without noise and without binding.
6. Lower the vehicle.

GS 300 AND GS 430

1. Before servicing the vehicle, refer to the precautions in the beginning of this section.
2. Remove or disconnect the following:
- Negative battery cable
- Front wheel
- Caliper, leaving the brake line connected and suspend it out of the way

> **✳✳ WARNING**
>
> **Never allow the brake caliper to hang freely from the brake hose.**

- Rotor
- Anti-lock Brake System (ABS) speed sensor and harness
- Tie rod from the arm on the lower ball joint
- Upper suspension arm from the steering knuckle
- Steering knuckle from the lower control arm
- Ball joint from the steering knuckle
- Front hub grease cap

3. Clamp the hub in a soft jaw vise.
4. Using a hammer and chisel, loosen the staked part of the locknut.
5. Remove or disconnect the following:
- Locknut
- ABS speed sensor rotor

➡**Do NOT scratch the serrations of the sensor rotor.**

- Brake dust cover bolts and shift the cover toward the outside.
- Hub from the steering knuckle
- Inner bearing race from the hub shaft
- Oil seal from the knuckle
- Bearing snapring from the steering knuckle
- Bearing from the steering knuckle

To install:
6. Install or connect the following:
- New bearing into the steering knuckle

➡**If the inner race and balls come loose from the bearing outer race, be sure to install them on the same side as before.**

- Snapring
- New outside inner race and tap in the new seal. Tap the seal until it is flush with the end surface of the steering knuckle.
- Brake dust cover to the knuckle and tighten the bolts to 74 inch lbs. (8 Nm)
- Hub into the steering knuckle
- ABS speed sensor rotor
- Axle hub locknut. Tighten the nut to 147 ft. lbs. (199 Nm) and stake it.
- Grease cap to the steering knuckle by tapping lightly around the circumference of the cap with a hammer
- Ball joint to the steering knuckle. Tighten the 2 bolts to 83 ft. lbs. (113 Nm).

- Steering knuckle to the upper and lower suspension arms. Tighten the upper nut to 64 ft. lbs. (87 Nm) and the lower nut to 95 ft. lbs. (127 Nm). Install a new cotter pin on the lower nut. Install the clip on the upper suspension arm nut.
- Tie rod end to the steering knuckle. Tighten the nut to 64 ft. lbs. (87 Nm) and install a new cotter pin.
- Rotor, disc brake pads and the brake caliper
- ABS speed sensor and harness. Tighten the sensor retaining bolt to 69 inch lbs. (8 Nm).
- Wheel

7. Lower the vehicle and connect the negative battery cable.

8. Check the front wheel alignment.

LS 430 AND SC 430 (EXCEPT 2004 LS 430)

1. Before servicing the vehicle, refer to the precautions in the beginning of this section.

2. If equipped with air suspension, move the height control switch in the trunk area to the **OFF** position.

3. Remove or disconnect the following:

- Front tire and wheel assembly
- Brake caliper bracket from the steering knuckle, leaving the brake line connected. Support the caliper with a piece of wire.
- Brake rotor
- Anti-lock Brake System (ABS) speed sensor from the steering knuckle
- Steering knuckle from the lower ball joint by removing the 2 bolts
- Steering knuckle from the upper ball joint
- Steering knuckle with the axle hub from the vehicle
- Grease cap from the hub
- Nut and the speed sensor rotor
- 4 bolts and shift the brake dust cover towards the hub side
- Axle hub from the steering knuckle
- Outside inner race from the axle
- Oil seal from the steering knuckle
- Snapring and bearing from the steering knuckle

To install:

4. Install or connect the following:

- Bearing in the steering knuckle
- Snapring
- Inner race (outside)
- New oil seal until it is flush with the end surface of the steering knuckle

- Brake dust cover to the steering knuckle and tighten the bolts to 74 inch lbs. (8.4 Nm)
- Axle hub to the steering knuckle
- ABS speed sensor
- New nut on the axle shaft. Tighten the nut to 147 ft. lbs. (199 Nm). Stake the nut and install the grease cap.
- Steering knuckle to the lower ball joint and tighten the bolts to 83 ft. lbs. (113 Nm)
- Steering knuckle to the upper ball joint and tighten the nut to 48 ft. lbs. (65 Nm)
- Brake rotor
- Brake caliper and tighten the 2 bolts to 87 ft. lbs. (118 Nm)
- Speed sensor to the steering knuckle
- Front tire and wheel assembly

5. If equipped with air suspension, turn the height control switch to the **ON** position.

LS 430 (2004)

1. Before servicing the vehicle, refer to the precautions in the beginning of this section.

2. Remove or disconnect the following:

- Front wheels
- Skid control wire
- Disc brake caliper assembly
- 4 bolts and front axle hub assembly
- Skid control sensor

To install:

3. Install or connect the following:

- Skid control sensor
- 4 bolts and front axle hub assembly. Tighten to 51 ft. lbs (69 Nm)
- 2 bolts and front disc brake caliper assembly. Tighten to 58 ft. lbs. (78 Nm)
- Skid control sensor wire
- Front wheel

4. Inspect and adjust front wheel alignment

5. Check ABS sensor signal

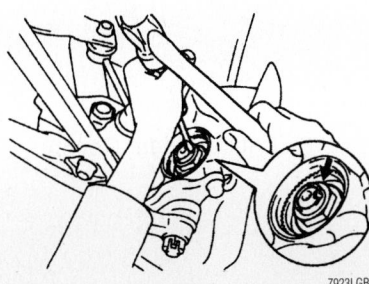

7923LGB9

Axle hub nut is located on the inboard side of the knuckle assembly—LS 430

ES 330

1. Before servicing the vehicle, refer to the precautions in the beginning of this section.

2. Remove or disconnect the following:

- Front tire and wheel assembly
- Brake caliper support bracket, leaving the brake line connected
- Rotor by removing the 2 screws
- Anti-lock Brake System (ABS) speed sensor
- Cotter pin and nut and disconnect the tie rod from the steering knuckle
- Cotter pin and nut
- Steering knuckle from the upper control arm
- Clip and nut and press the knuckle off the lower control arm
- Steering knuckle from the vehicle
- Hub bearing cap from the steering knuckle
- Hub nut
- ABS sensor rotor
- 4 bolts and shift the brake dust shield toward the hub (outside)
- Axle hub from the knuckle
- Inner bearing race from the axle hub
- Oil seal
- Snapring and bearing

To install:

3. Press the bearing into the knuckle. If the inner race and balls come loose from the outer race, be sure to install them on the same side as before.

4. Install or connect the following:

- Snapring and inner race, then tap in a new oil seal until it is flush with the end surface of the knuckle
- Brake dust cover and tighten the bolts to 74 inch lbs. (8.3 Nm)
- Hub into the knuckle
- Speed sensor
- New locknut and tighten it to 147 ft. lbs. (199 Nm), 217 ft lbs (294 Nm) on ES 330. Stake the nut with a chisel. Tap the bearing cap into place.
- Knuckle to the upper control arm and tighten the nut to 76 ft. lbs. (103 Nm). Install a new cotter pin
- Knuckle to the lower control arm and tighten the nut to 92 ft. lbs. (125 Nm). Install a new clip.
- Tie rod end to the steering knuckle with the nut. Tighten the nut to 36 ft. lbs. (49 Nm). Install a new cotter pin.
- Rotor by installing the 2 screws

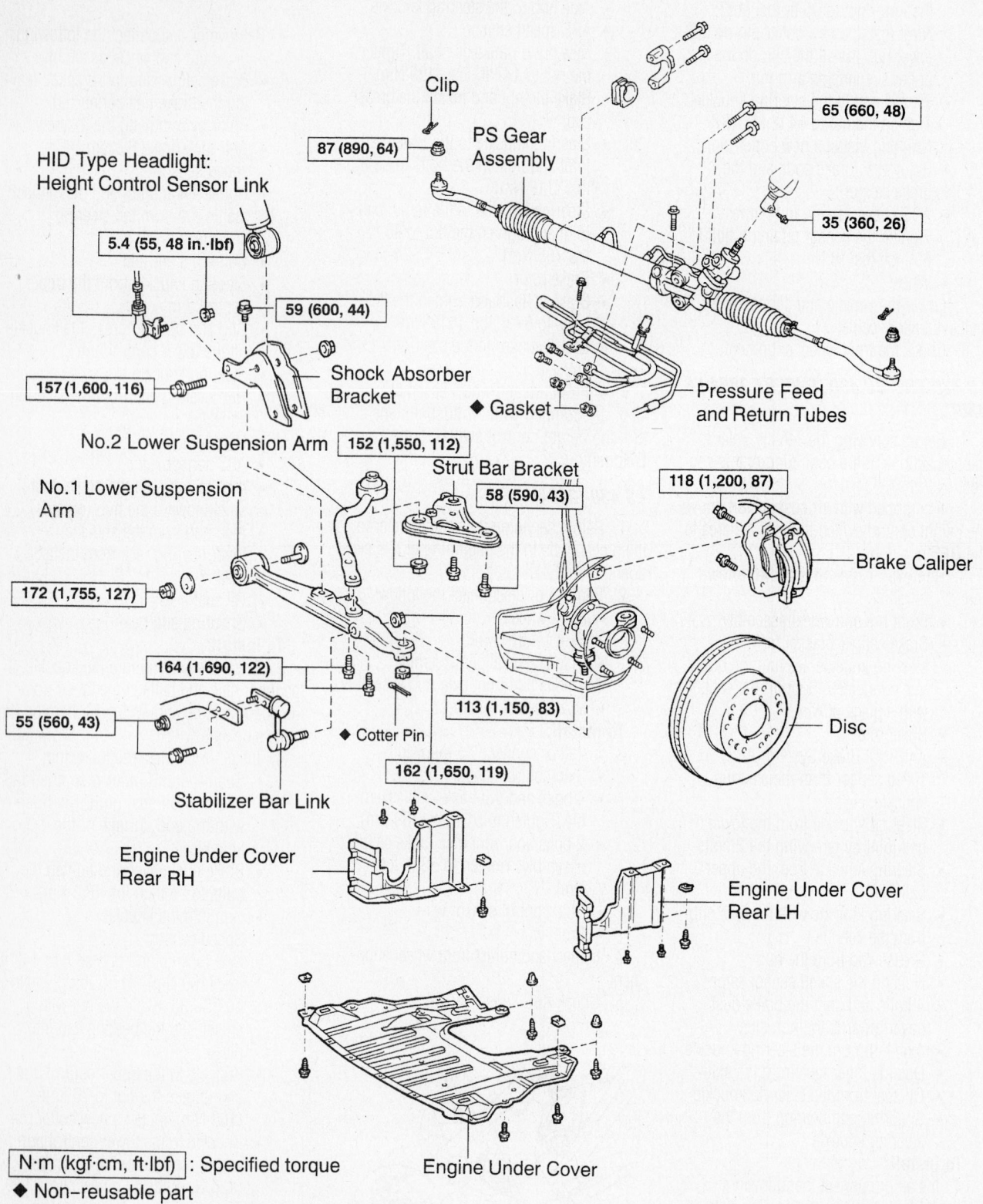

Clip

PS Gear Assembly

87 (890, 64)

65 (660, 48)

35 (360, 26)

HID Type Headlight:
Height Control Sensor Link

5.4 (55, 48 in.·lbf)

59 (600, 44)

157 (1,600, 116)

Shock Absorber Bracket

◆ Gasket

Pressure Feed and Return Tubes

No.2 Lower Suspension Arm

152 (1,550, 112)

Strut Bar Bracket

58 (590, 43)

118 (1,200, 87)

No.1 Lower Suspension Arm

Brake Caliper

172 (1,755, 127)

164 (1,690, 122)

55 (560, 43)

◆ Cotter Pin

113 (1,150, 83)

162 (1,650, 119)

Disc

Stabilizer Bar Link

Engine Under Cover Rear RH

Engine Under Cover Rear LH

N·m (kgf·cm, ft·lbf) : Specified torque

◆ Non–reusable part

Engine Under Cover

67162-LEXU-G117

Exploded view front hub and bearing assembly—LS 430 (2004)

- Caliper support bracket and tighten the bolt to 87 ft. lbs. (118 Nm)
- Speed sensor to the knuckle and tighten the bolt to 69 inch lbs. (8 Nm)
- Front wheel and tighten the lug nuts to 76 ft. lbs. (103 Nm)

5. Lower the vehicle.
6. Check the front end alignment and ABS speed sensor signal.

IS 300

1. Before servicing the vehicle, refer to the precautions in the beginning of this section.
2. Remove or disconnect the following:
   - Front wheel
   - Brake caliper and rotor
   - Wheel speed sensor
   - Upper and lower ball joints
   - Steering knuckle from the vehicle
   - Grease cap
   - Hub locknut
   - Brake dust cover
   - Wheel speed sensor pulse ring
3. Press the hub out of the wheel bearing.
4. Remove the grease seal and the snapring, then press the wheel bearing out of the steering knuckle.

To install:

➡**Use a new hub locknut for assembly.**

5. Installation is the reverse of the removal procedure, while using the following torque values:
   - Hub locknut: 108 ft. lbs. (147 Nm)
   - Brake dust cover bolts: 74 inch lbs. (8.3 Nm)
   - Lower ball joint bolts: 83 ft. lbs. (113 Nm)
   - Upper ball joint nut: 50 ft. lbs. (65 Nm)
   - Brake caliper support bolts: 87 ft. lbs. (118 Nm)
   - Wheel lug nuts: 76 ft. lbs. (103 Nm)

Rear

ES 300

1. Before servicing the vehicle, refer to the precautions in the beginning of this section.
2. Raise and safely support the vehicle.
3. Remove or disconnect the following:
   - Rear tire and wheel assembly
   - If equipped with rear disc brakes, the caliper mounting bolts. Leave the brake line connected and suspend the assembly out of the way.
   - Brake rotor or drum

- 4 bolts and pull off the rear axle hub
- O-ring

➡**If it is necessary to replace the hub or bearing, replace the components as an assembly.**

To install:

4. Install or connect the following:
   - Hub on the carrier and tighten the bolts to 59 ft. lbs. (80 Nm)
   - Rotor or drum
   - Caliper, f equipped with rear disc brakes and tighten the bolts to 34 ft. lbs. (64 Nm)
   - Wheel

GS 300 AND GS 430

1. Before servicing the vehicle, refer to the precautions in the beginning of this section.
2. Remove or disconnect the following:
   - Negative battery cable.
   - Rear tire and wheel assembly
   - Brake caliper support from the rear axle carrier and support it with a piece of wire
3. Place matchmarks on the disc brake rotor and the axle hub.
4. Remove or disconnect the following:
   - Brake rotor
   - Speed sensor
   - Rear halfshaft
   - Parking brake shoes
   - Parking brake cable

- Strut rod

5. Place matchmarks on the adjusting cam and rear control crossmember.
   - Nut, adjusting cam and the washer to the No. 1 control arm
   - No. 1 lower control arm from the crossmember
   - Loosen the nut holding the lower control arm to the axle carrier
   - No. 2 lower control arm from the axle carrier
   - Nut, then remove the No. 2 lower control arm from the axle carrier
   - Nut holding the upper control arm to the axle carrier
   - Axle carrier
   - Nut holding the No. 1 control arm to the axle carrier
   - No. 1 lower control arm from the axle carrier
   - Dust deflector
   - Axle hub from the carrier
   - Backing plate
   - Inner race (outside)
   - Oil seal
   - Snapring
   - Bearing

To install:

6. Install or connect the following:
   - Bearing to the axle carrier

➡**If the inner races come loose from the bearing outer race, be sure to install them on the same side as before.**

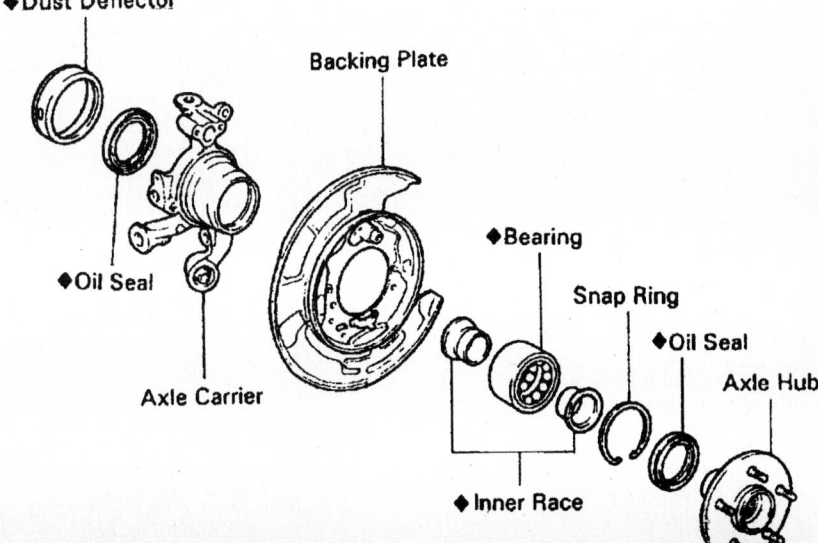

Exploded view of the axle carrier—GS 300

◆Dust Deflector

◆Oil Seal

Axle Carrier

Backing Plate

◆Bearing

Snap Ring

◆Oil Seal

Axle Hub

◆Inner Race

◆ Non-reusable part

7923LGC1

- Snapring. Install the inner race (outside) and a new oil seal.
- Backing plate. Install the inner race (inside) and press in the axle hub with the proper tools.
- Inner oil seal. Align the holes for the speed sensor in the dust deflector and axle carrier. Install the dust deflector.
- No. 1 lower arm to the axle carrier and install a new nut. Tighten the nut to 43 ft. lbs. (59 Nm).
- Upper control arm to the axle carrier. Tighten the new nut and bolt to 80 ft. lbs. (109 Nm).
- No. 2 lower control arm to the axle carrier and tighten a new nut to 110 ft. lbs. (150 Nm)
- No. 1 lower control arm to the rear crossmember. Tighten the nut to 136 ft. lbs. (184 Nm).
- Strut rod to the axle carrier. Tighten the nuts and bolts to 134 ft. lbs. (184 Nm).
- Parking brake cable and slide the backing plate to the inside. Install the hex bolt and tighten it to 132 ft. lbs. (180 Nm).
- Shoe guide plate set bolt. Tighten the bolt to 13 ft. lbs. (18 Nm).
- 4 hub bolts and tighten them to 19 ft. lbs. (26 Nm)
- Bolts at the speed sensor and tighten them to 69 inch lbs. (8 Nm)
- Parking brake shoes
- Halfshafts. Apply the brakes and tighten the locknut to 213 ft. lbs. (289 Nm).
- Brake rotor
- Brake caliper support to the rear axle carrier. Tighten the bolts to 77 ft. lbs. (104 Nm).
- Rear tire and wheel assembly
- Negative battery cable

7. Lower the vehicle and bounce it a few times to stabilize the suspension.

LS 430 (EXCEPT 2004 LS 430)

1. Before servicing the vehicle, refer to the precautions in the beginning of this section.

2. If equipped with air suspension, move the height control switch in the trunk area to the **OFF** position.

3. Remove or disconnect the following:
- Negative battery cable
- Rear wheel(s)
- Height control sensor link from the lower control arm
- Anti-lock Brake System (ABS) speed sensor and wiring harness
- Brake caliper bracket from the rear

axle carrier by removing the 2 bolts. Support the caliper with a piece of wire.
- Brake rotor
- Parking brake shoes and cable
- Cotter pin, lock cap and the nut holding the halfshaft to the rear axle
- Suspension member brace by removing the 2 bolts
- Halfshaft bolts and washers
- Halfshaft from the vehicle
- Strut rod

4. Place matchmarks on the adjusting cam and body for the No. 1 control arm.

5. Remove or disconnect the following:
- Nut and adjusting cam
- Nut on the axle carrier side of the No. 1 lower control arm
- Separate the control arm from the axle carrier
- No. 1 lower control arm
- Stabilizer bar link from the No. 2 lower control arm.

6. Place matchmarks on the adjusting cam and body.

7. Remove or disconnect the following:
- Nut and adjusting cam from the No. 2 lower control arm
- Nut and bolt holding the No. 2 lower control arm to the axle carrier
- No. 2 control arm from the vehicle
- Nut and bolt on the lower side of the strut assembly
- 2 upper control arm set nuts and bolts

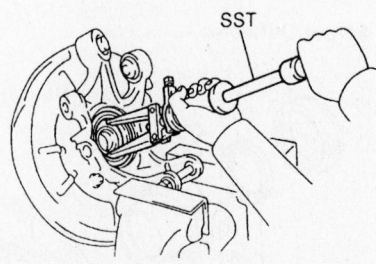

Removing the oil seal (inner)—LS 430

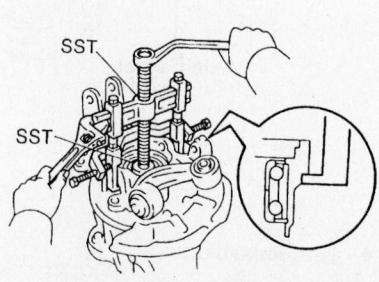

Removing the axle hub from the axle carrier—LS 430

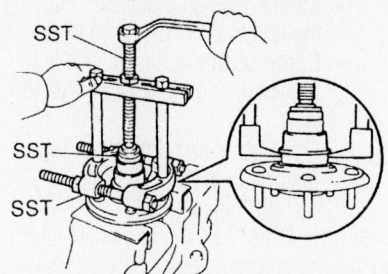

7923LGC4

Removing the inner race (outside) from the axle hub—LS 430

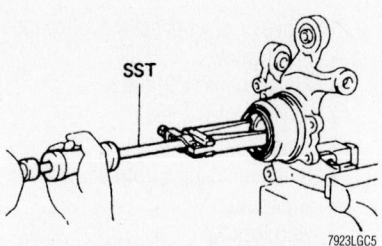

7923LGC5

Removing the oil seal (outer)—LS 430

- Axle carrier with the upper control arm

8. Secure the axle carrier in a vise.

9. Remove or disconnect the following:
- Nut holding the upper control arm to the axle carrier and remove the control arm
- Dust deflector. Use a suitable pry-tool.
- Oil seal

10. Remove the 2 bolts and nuts and shift the backing the plate towards the hub side (outside).

11. Remove or disconnect the following:
- Axle hub
- Backing plate.
- Inner race (outside) from the axle hub
- Oil seal (outer) from the axle
- Snapring from inside the axle housing
- Bearing from the axle housing

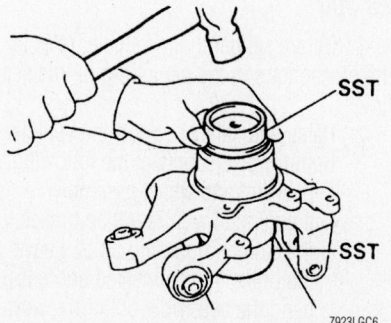

7923LGC6

Installing the oil seal (outer)—LS 430

To install:

12. Install or connect the following:
   - New bearing to the axle housing
   - Snapring to the axle carrier, using snapring pliers
   - New outer oil seal. Coat the oil seal lip with multipurpose grease.
   - Backing plate to the axle housing. Do NOT install the bolts or nuts at this time.
   - Inner race (inside) to the axle housing

   - Axle hub to the axle housing
   - Backing plate in position. Tighten the bolts and nuts to 43 ft. lbs. (59 Nm).
   - New oil seal (inner) to the axle housing. Coat the oil seal lip with multipurpose grease.
   - New dust deflector. Be sure to align the hose for the ABS speed sensor in the dust deflector and axle carrier.
   - Upper control arm to the axle car-

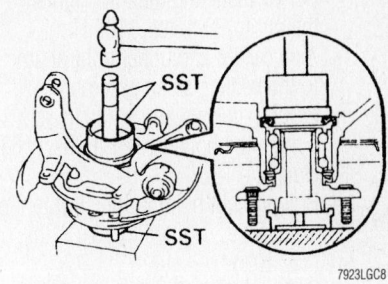

Installing the oil seal (inner)—LS 430

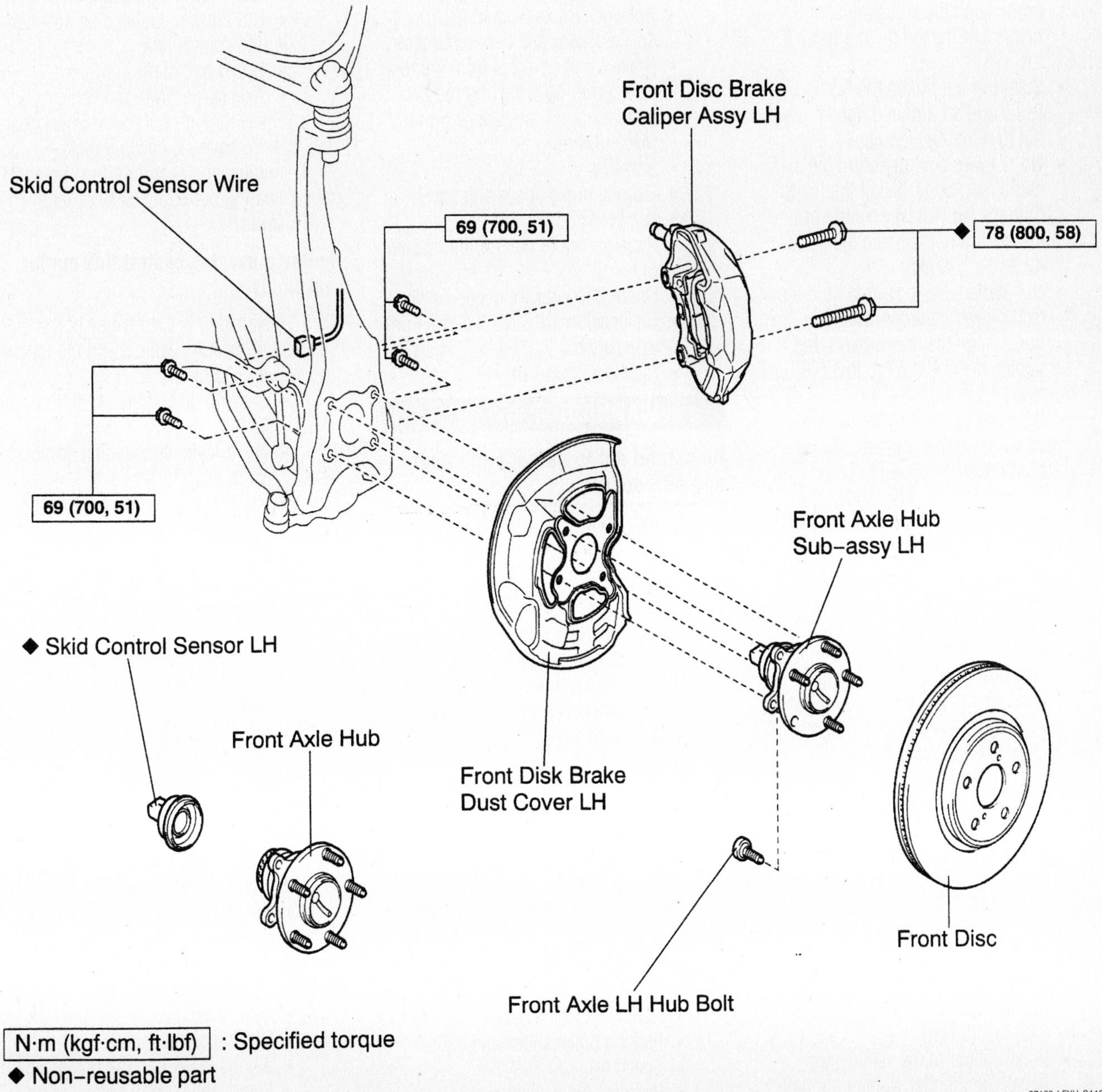

7923LGC8

Skid Control Sensor Wire

Front Disc Brake Caliper Assy LH

69 (700, 51)

◆ 78 (800, 58)

69 (700, 51)

◆ Skid Control Sensor LH

Front Axle Hub

Front Disk Brake Dust Cover LH

Front Axle Hub Sub–assy LH

Front Disc

Front Axle LH Hub Bolt

N·m (kgf·cm, ft·lbf) : Specified torque

◆ Non–reusable part

67162-LEXU-G118

Exploded view rear carrier assembly, hub and bearing—

rier by installing the nut. Tighten the nut to 80 ft. lbs. (108 Nm).
- Axle carrier and upper control arm to the vehicle as an assembly
- 2 upper control arm set bolts and tighten the bolts to 121 ft. lbs. (164 Nm)
- Bolt and nut holding the strut to the axle carrier. Tighten to 101 ft. lbs. (137 Nm).
- Bolt and nut connecting the No. 2 lower control arm to the axle carrier. Tighten the bolt to 60 ft. lbs. (81 Nm).
- Nut and adjusting cam to hold the No. 2 lower control arm to the body. Align the adjusting cam marks and tighten the nut to 57 ft. lbs. (78 Nm).
- Stabilizer bar link to the No. 2 lower control arm and tighten the nut to 48 ft. lbs. (65 Nm)
- No. 1 lower control arm to the axle carrier and body. Install the nut to hold the No. 1 lower control arm to the axle carrier. Tighten the nut to 43 ft. lbs. (59 Nm).
- Nut and adjusting cam to hold the No. 1 lower control arm to the body. Align the matchmarks and tighten the nut to 57 ft. lbs. (78 Nm).
- Strut rod to the axle carrier and body. Install the bolt and nut to hold the strut rod to the body. Tighten to 57 ft. lbs. (78 Nm).
- Bolt and nut to hold the strut rod to the axle carrier. Tighten to 136 ft. lbs. (184 Nm)
- Parking brake shoes and cable
- Outboard joint side of the halfshaft and align the matchmarks on the side gear shaft and the halfshaft. Coat the threads with clean oil and install the hexagon bolts. Tighten bolts to 61 ft. lbs. (83 Nm).
- Suspension member brace with the 2 bolts. Tighten the 2 bolts to 37 ft. lbs. (50 Nm).
- Nut to hold the halfshaft to the rear axle. Tighten the nut to 213 ft. lbs. (289 Nm).
- Lock cap and cotter pin
- Brake disc to the axle hub with the matchmarks aligned. Install the 2 screws and tighten the screws to 48 inch lbs. (5 Nm).
- Brake caliper to the vehicle and install the 2 bolts. Tighten the bolts to 77 ft. lbs. (104 Nm).
- ABS speed sensor and wiring harness

- Height control sensor link with the matchmarks aligned. Tighten the nut to 48 inch lbs. (5 Nm).
- Rear wheel(s)
- Negative battery cable

13. Lower the vehicle and turn **ON** the air suspension switch.

LS 430 (2004)

1. Before servicing the vehicle, refer to the precautions in the beginning of this section.

2. Remove or disconnect the following:
- Rear wheel(s)
- Height control sensor link from the lower control arm
- Anti-lock Brake System (ABS) speed sensor and wiring harness
- Brake caliper bracket from the rear axle carrier by removing the 2 bolts. Support the caliper with a piece of wire.
- Brake rotor
- Parking brake shoes and cable
- Toe control link
- Separate upper control arm assembly
- Shock absorber from axle carrier
- Suspension arms No 1 and 2 from axle carrier
- Rear axle from carrier

✷✷ WARNING

Be careful not to damage the boot and ABS sensor

- Wheel bearing dust deflector
- 4 bolts and axle & bearing hub assembly

To install:

3. Install or connect the following:
- 4 bolts and axle & bearing assembly. Tighten to 48 ft. lbs (65 Nm)
- Rear wheel bearing dust deflector
- Rear drive axle
- Upper control arm. Temporarily tighten
- No 2 and No 1 suspension arms. Tighten to 52 ft. lbs. (70 Nm)
- Shock absorber. Temporarily tighten
- Toe control link. Tighten to 37 ft. lbs. (50 Nm)
- Fully tighten upper control arm. Tighten to 52 ft. lbs. (70 Nm)
- Height control sensor
- Parking brake shoes and cable
- Speed sensor
- Rear disc brake caliper
- Fully tighten shock with new nut to 52 ft. lbs. (70 Nm)

- Rear tire
4. Inspect and adjust rear wheel alignment
5. Check ABS speed sensor signal

IS 300

1. Before servicing the vehicle, refer to the precautions in the beginning of this section.

2. Remove or disconnect the following:
- Rear wheel
- Wheel speed sensor
- Axle halfshaft
- Brake caliper and rotor
- Parking brake shoes
- Parking brake cable
- No. 1 lower suspension arm bolt
- No. 2 lower suspension arm bolt
- Toe control link
- Upper ball joint
- Axle carrier from the vehicle

3. Press the hub out of the wheel bearing, then remove the backing plate.

4. Remove the snapring, then press the wheel bearing out of the axle carrier.

To install:

➡**Use a new toe control link nut for assembly.**

5. Installation is the reverse of the removal procedure, while using the following torque values:
- Backing plate bolts: 43 ft. lbs. (59 Nm)
- No. 1 lower suspension arm bolt: 55 ft. lbs. (75 Nm)
- No. 2 lower suspension arm bolt: 81 ft. lbs. (110 Nm)
- Toe control link nut: 36 ft. lbs. (49 Nm)
- Upper ball joint nut: 80 ft. lbs. (108 Nm)
- Brake caliper support bolts: 77 ft. lbs. (104 Nm)
- Rear wheel lug nuts: 76 ft. lbs. (103 Nm)

SC 430

1. Before servicing the vehicle, refer to the precautions in the beginning of this section.

2. Remove or disconnect the following:
- Rear tire and wheel assembly
- Brake caliper support bracket
- Brake rotor
- Speed sensor
- Rear halfshaft
- Parking brake shoes
- 2 bolts at the parking brake cable. Remove the 2 hub bolts and the hex bolt. Slide the backing plate to the outside and disconnect the parking brake cable.

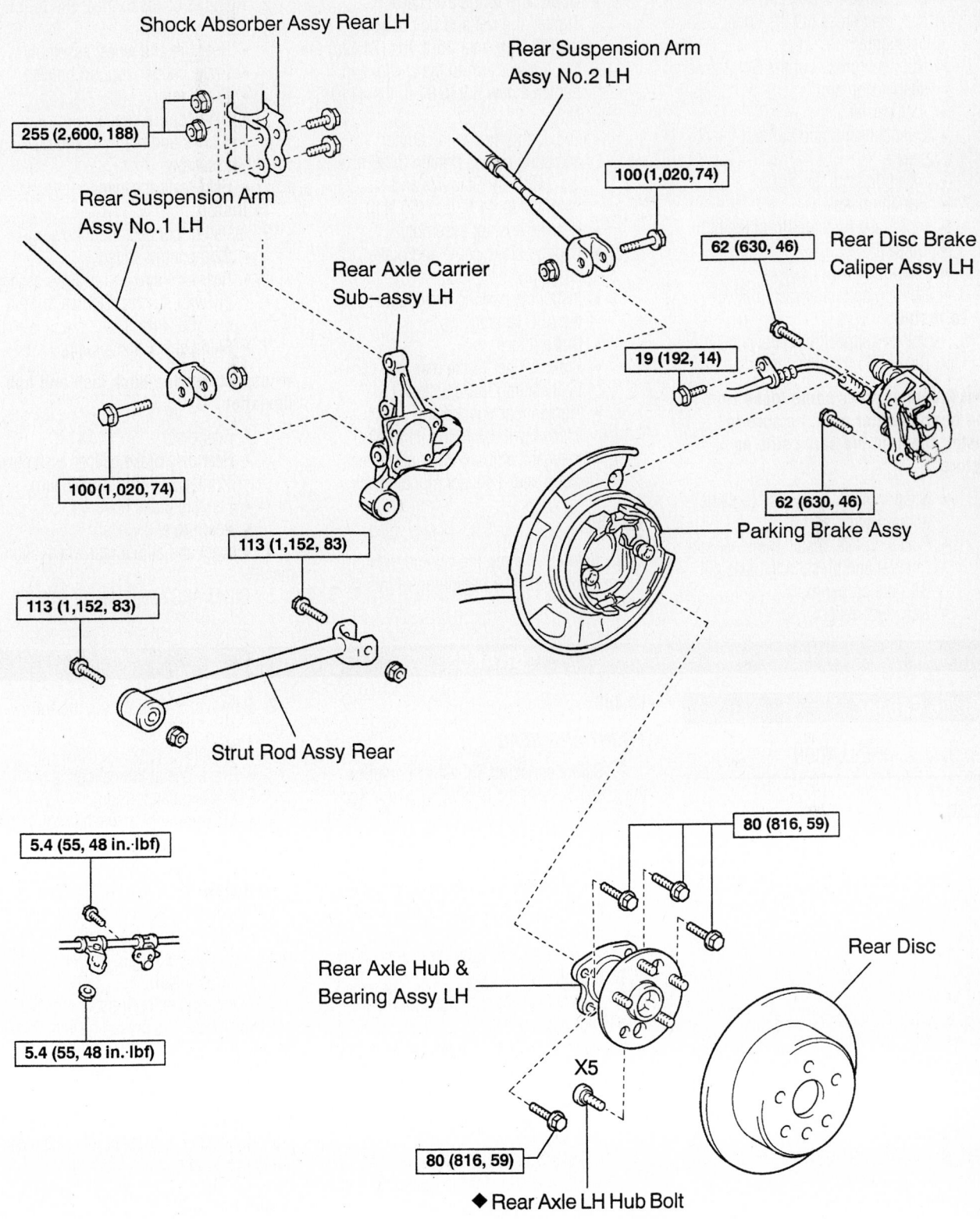

Shock Absorber Assy Rear LH

Rear Suspension Arm Assy No.2 LH

255 (2,600, 188)

Rear Suspension Arm Assy No.1 LH

Rear Axle Carrier Sub-assy LH

100 (1,020, 74)

62 (630, 46)

Rear Disc Brake Caliper Assy LH

19 (192, 14)

100 (1,020, 74)

113 (1,152, 83)

62 (630, 46)

Parking Brake Assy

113 (1,152, 83)

Strut Rod Assy Rear

80 (816, 59)

5.4 (55, 48 in.·lbf)

Rear Disc

Rear Axle Hub & Bearing Assy LH

5.4 (55, 48 in.·lbf)

X5

80 (816, 59)

◆ Rear Axle LH Hub Bolt

N·m (kgf·cm, ft·lbf) : Specified torque

◆ Non-reusable part

67162-LEXU-G87

Exploded view rear Axle carrier and hub assembly—ES 330

- Strut rod at the axle carrier
- Nut, then press out the upper suspension arm.
- Nut, then press out the No. 2 lower suspension arm
- Axle carrier
- Dust deflector and pull out the oil seal
- Axle hub from the carrier
- Backing plate
- Inner race (outside) from the hub
- Oil seal
- Snapring
- Bearing and inner race (inside)

To install:

3. Install or connect the following:
   - Bearing to the axle carrier

➡ **If the inner races come loose from the bearing outer race, be sure to install them on the same side as before.**

- Snapring, the inner race (outside) and a new oil seal
- Backing plate. Install the inner race (inside) and press in the axle hub with the proper tools.
- New dust deflector

- Upper arm to the axle carrier. Tighten the nut and bolt to 65 ft. lbs (88 Nm), rear 55 ft. lbs (74 Nm)
- No. 2 lower arm to the carrier and tighten a new nut to 81 ft. lbs. (110 Nm).
- Toe control link with camber adjusting cams. Tighten to 36 ft. lbs. (49 Nm). Stabilize and retighten to 44 ft. lbs. (59 Nm).
- Rear drive shaft assembly
- Parking brake cable and brake assembly
- Install the parking brake shoes and the ABS sensor.
- Brake rotor
- Brake caliper to the rear axle carrier by installing the 2 bolts.
- Tighten rear suspension arm assembly to 81 ft. lbs. (110 Nm)

4. Inspect and adjust to rear alignment
5. Perform speed sensor signal check.

ES 330

1. Before servicing the vehicle, refer to the precautions in the beginning of this section.

2. Remove or disconnect the following:
   - Rear tire and wheel assembly
   - Brake caliper support bracket
   - Brake rotor
   - Skid control sensor wire
   - 4 bolts and rear hub and bearing assembly
   - Skid control sensor

To install:

3. Install or connect the following:
   - Skid control sensor
   - Rear axle hub and bearing assembly with 4 bolts. Tighten to 59 ft. lbs. (80 Nm)
   - Skid control sensor wire

➡ **Inspect bearing back lash and hub deviation**

- Rear disc
- Rear disc brake caliper assembly. Tighten to 46 ft. lbs (62 Nm)
- Flexible brake hose
- Rear tire and wheel

4. inspect and adjust rear wheel alignment
5. Check ABS speed sensor signal

BRAKES

Brake Caliper

REMOVAL & INSTALLATION

ES 300

FRONT AND REAR

1. Before servicing the vehicle, refer to the precautions in the beginning of this section.

2. Remove or disconnect the following:
   - Wheels
   - Brake hose from the caliper
   - Bolts that attach the caliper to the torque plate
   - Caliper assembly by lifting the bottom

To install:

3. Grease the caliper slides and bolts with lithium grease.

4. Install or connect the following:
   - Caliper. Torque the bolts to 25 ft. lbs. (34 Nm).
   - Brake hose to the caliper using 2 new washers. Torque the union bolt to 21 ft. lbs. (29 Nm).

5. Fill the master cylinder to the proper level and bleed the brake system.

IS 300

FRONT AND REAR

1. Before servicing the vehicle, refer to the precautions in the beginning of this section.

2. Remove or disconnect the following:
   - Wheels
   - Brake hose from the caliper
   - Bolts that attach the caliper to the torque plate
   - Caliper assembly by lifting the bottom

To install:

3. Grease the caliper slides and bolts with lithium grease.

4. Install or connect the following:
   - Front caliper. Torque the bolts to 25 ft. lbs. (34 Nm).
   - Rear caliper. Torque the bolts to 77 ft. lbs. (34 Nm).
   - Brake hose to the caliper using 2 new washers. Torque the union bolt to 22 ft. lbs. (30 Nm).

5. Fill the master cylinder to the proper level and bleed the brake system.

GS 300 and GS 430

FRONT AND REAR

1. Before servicing the vehicle, refer to the precautions in the beginning of this section.

2. Remove or disconnect the following:
   - Wheels
   - Brake line at the caliper
   - Anti-squeal springs
   - Mounting bolts, while holding the sliding pin with a wrench
   - Caliper assembly

To install:

3. Install or connect the following:
   - Caliper. Hold the sliding pin and tighten the mounting bolts to 25 ft. lbs. (34 Nm).
   - Anti-squeal springs
   - Connect the brake line with 2 new gaskets and tighten the union bolt to 22 ft. lbs. (30 Nm)

4. Bleed the brake system.
   - Wheels

5. Check and if necessary fill the master cylinder reservoir.

LS 430 and SC 430

FRONT

1. Before servicing the vehicle, refer to the precautions in the beginning of this section.

2. Remove or disconnect the following:
   - Front wheels

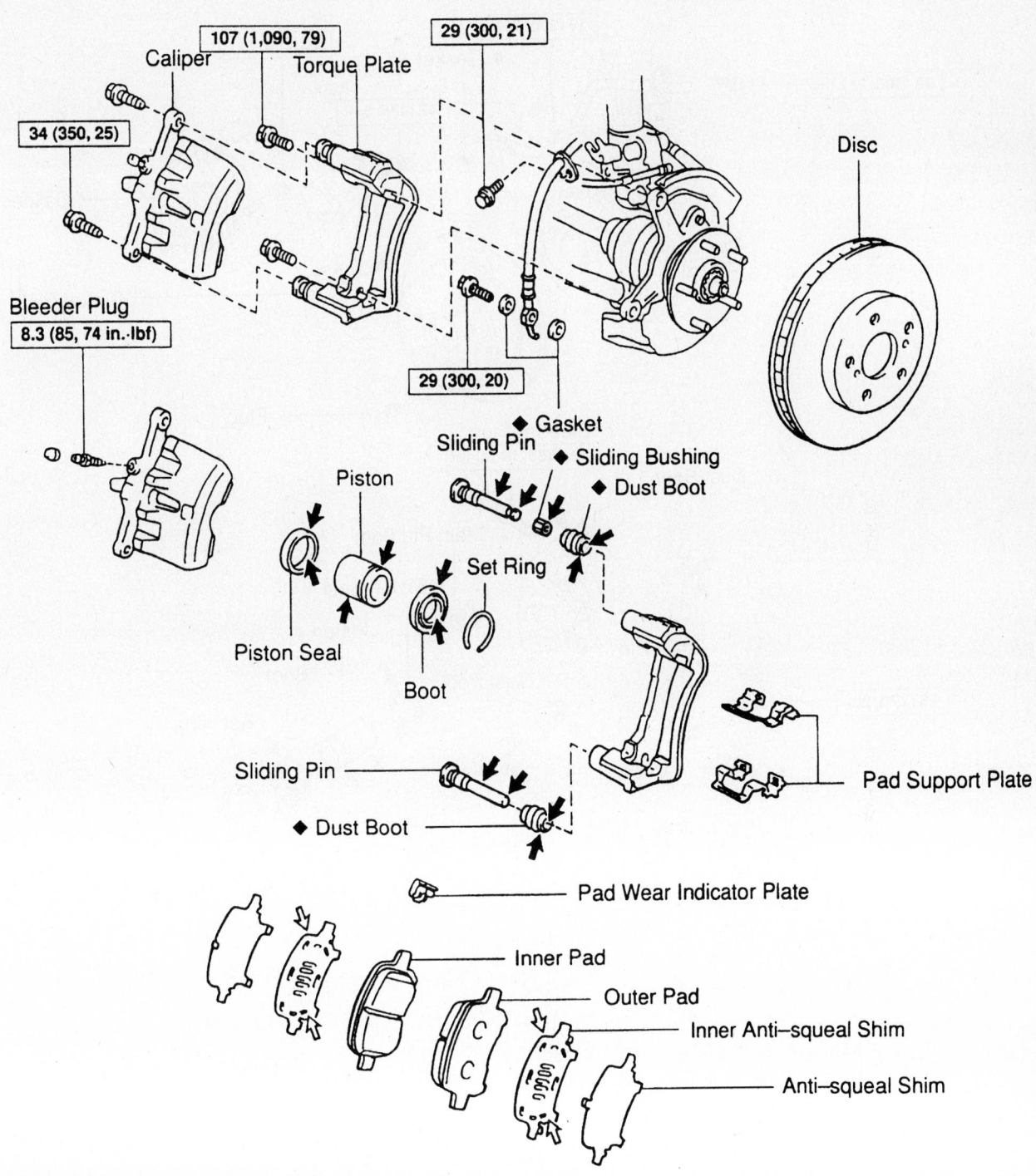

Caliper
Torque Plate
107 (1,090, 79)
29 (300, 21)
34 (350, 25)
Disc

Bleeder Plug
8.3 (85, 74 in.·lbf)

29 (300, 20)

◆ Gasket
Sliding Pin
Piston
◆ Sliding Bushing
◆ Dust Boot
Set Ring
Piston Seal
Boot

Sliding Pin
◆ Dust Boot
Pad Support Plate

Pad Wear Indicator Plate

Inner Pad
Outer Pad
Inner Anti–squeal Shim
Anti–squeal Shim

N·m (kgf·cm, ft·lbf) : Specified torque
◆ Non–reusable part
◀ Lithium soap base glycol grease
◁ Disc brake grease

93016G18

Front disc brakes—ES 300

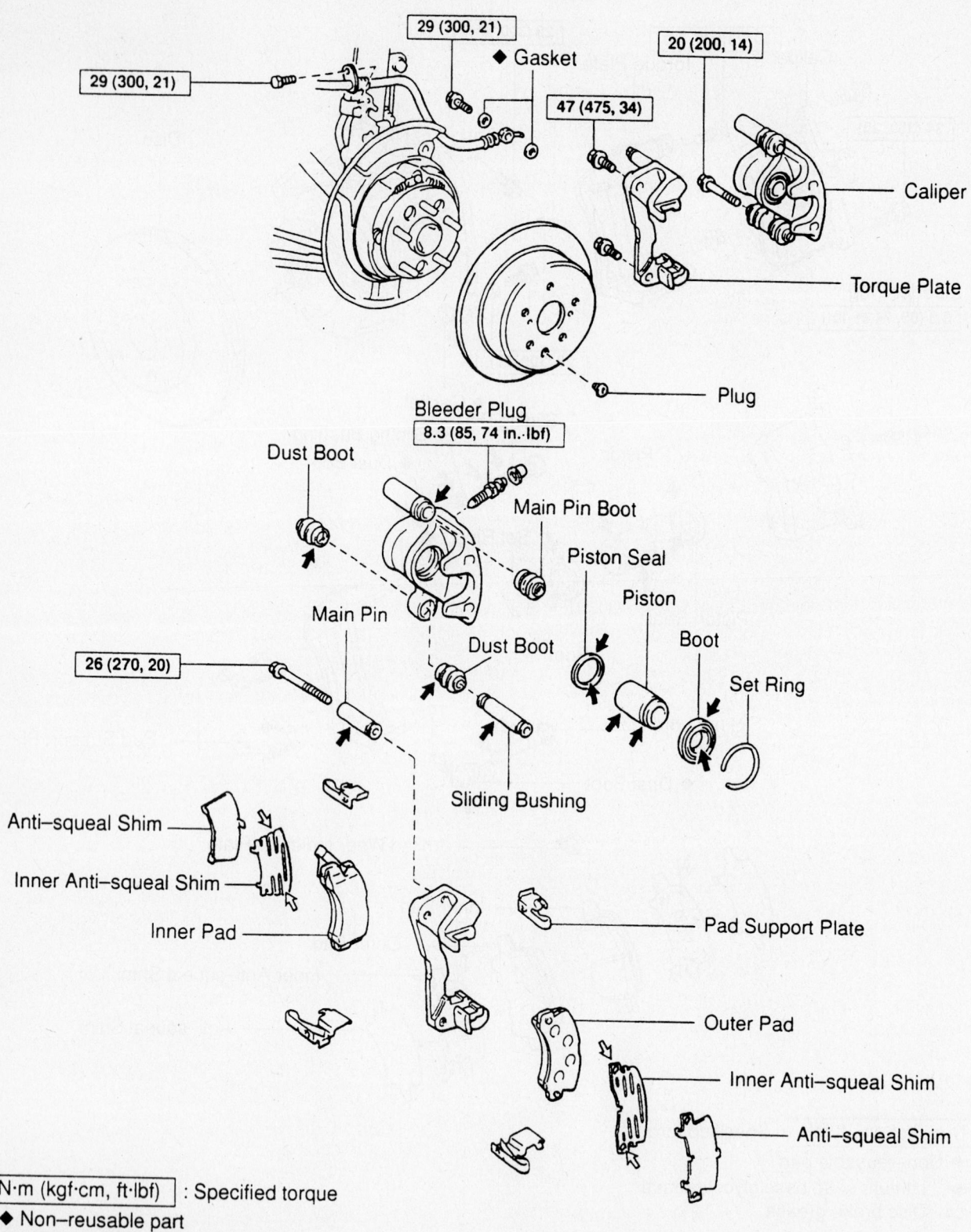

29 (300, 21)

29 (300, 21)

◆ Gasket

20 (200, 14)

47 (475, 34)

Caliper

Torque Plate

Plug

Bleeder Plug
8.3 (85, 74 in.·lbf)

Dust Boot

Main Pin Boot

Piston Seal

Piston

Boot

Set Ring

Main Pin

Dust Boot

26 (270, 20)

Sliding Bushing

Anti–squeal Shim

Inner Anti–squeal Shim

Inner Pad

Pad Support Plate

Outer Pad

Inner Anti–squeal Shim

Anti–squeal Shim

N·m (kgf·cm, ft·lbf) : Specified torque

◆ Non–reusable part

◀ Lithium soap base glycol grease

◁ Disc brake grease

93016G19

Rear disc brakes—ES 300

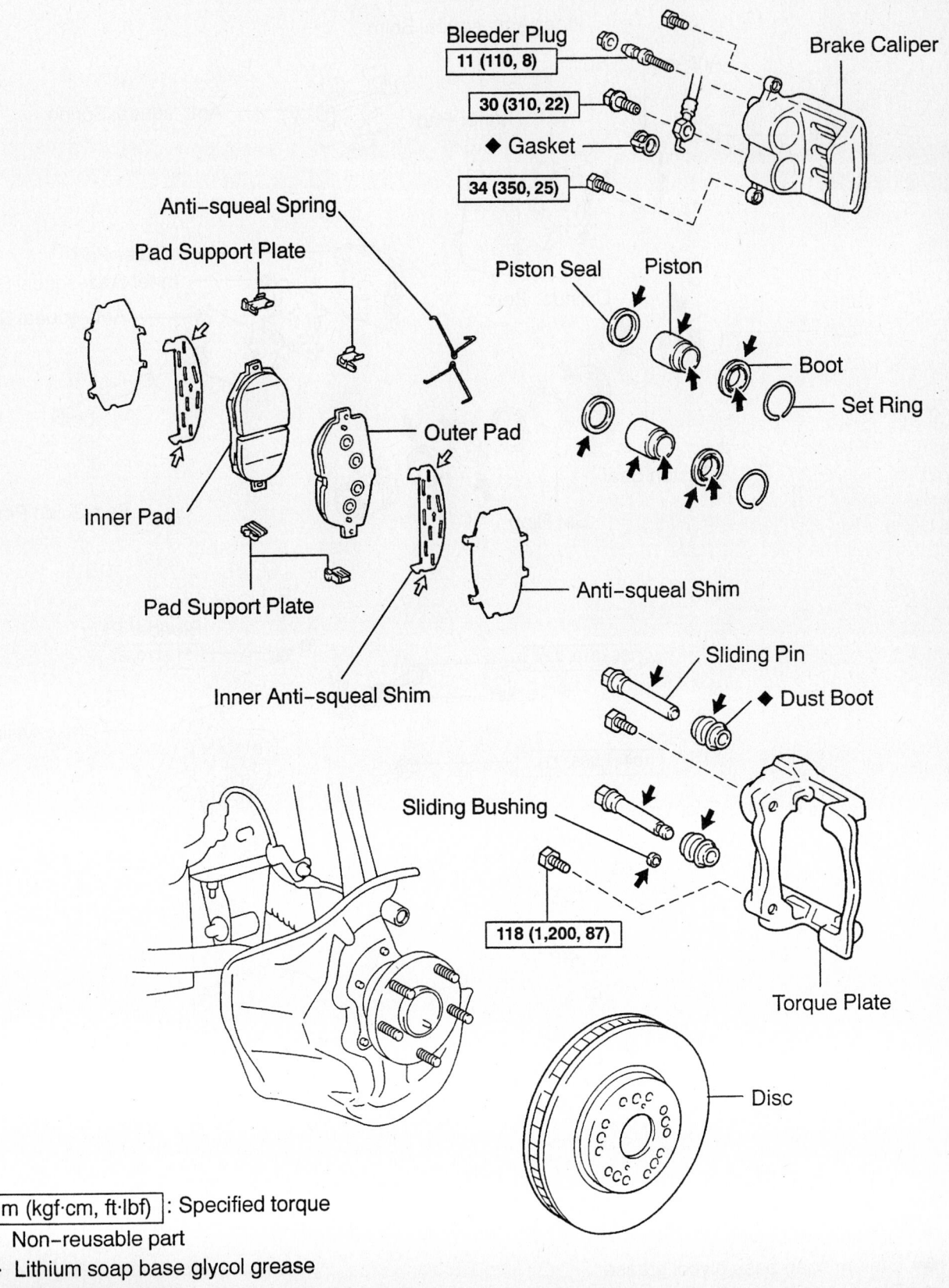

Bleeder Plug
11 (110, 8)

30 (310, 22)

◆ Gasket

34 (350, 25)

Brake Caliper

Anti-squeal Spring

Pad Support Plate

Piston Seal   Piston

Boot

Set Ring

Outer Pad

Inner Pad

Anti-squeal Shim

Pad Support Plate

Sliding Pin

◆ Dust Boot

Inner Anti-squeal Shim

Sliding Bushing

118 (1,200, 87)

Torque Plate

Disc

N·m (kgf·cm, ft·lbf) : Specified torque

◆   Non-reusable part

►   Lithium soap base glycol grease

⇨   Disc brake grease

67162-LEXU-G88

Front disc brakes—IS 300

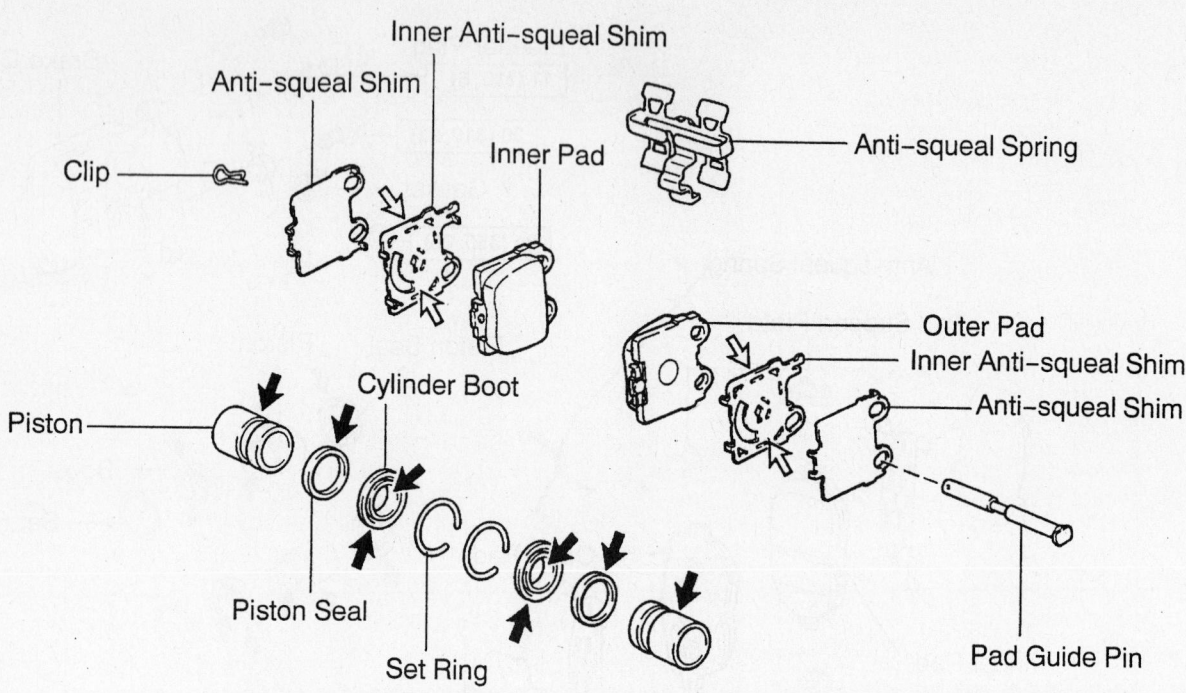

Clip

Anti-squeal Shim

Inner Anti-squeal Shim

Inner Pad

Anti-squeal Spring

Outer Pad

Inner Anti-squeal Shim

Anti-squeal Shim

Piston

Cylinder Boot

Piston Seal

Set Ring

Pad Guide Pin

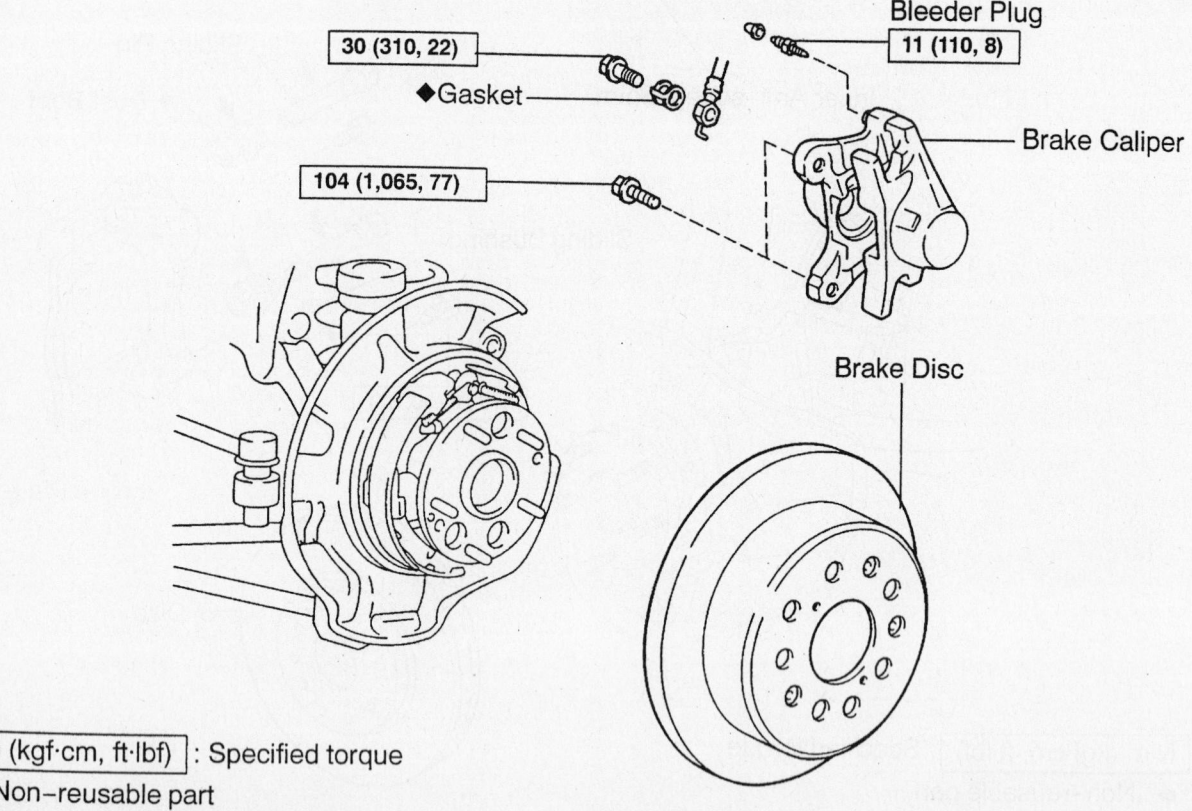

Bleeder Plug

30 (310, 22)

11 (110, 8)

◆Gasket

Brake Caliper

104 (1,065, 77)

Brake Disc

N·m (kgf·cm, ft·lbf) : Specified torque

◆ Non-reusable part

◀ Lithium soap base glycol grease

◁ Disc brake grease

67162-LEXU-G89

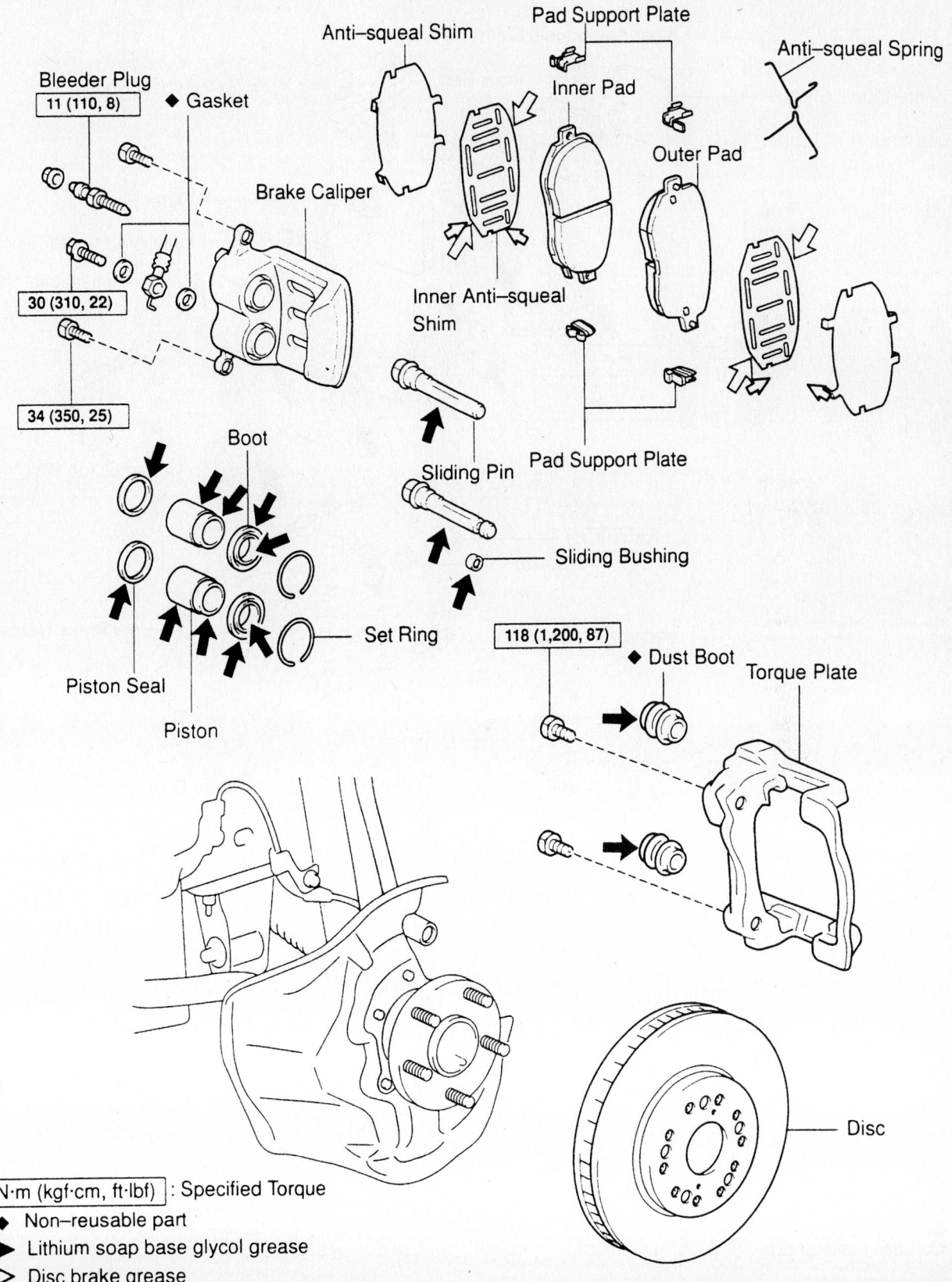

Anti–squeal Shim

Pad Support Plate

Anti–squeal Spring

Bleeder Plug

11 (110, 8)

◆ Gasket

Inner Pad

Brake Caliper

Outer Pad

30 (310, 22)

Inner Anti–squeal Shim

34 (350, 25)

Boot

Sliding Pin

Pad Support Plate

Piston Seal

Set Ring

Sliding Bushing

Piston

118 (1,200, 87)

◆ Dust Boot

Torque Plate

Disc

N·m (kgf·cm, ft·lbf) : Specified Torque

◆ Non–reusable part

➤ Lithium soap base glycol grease

⇨ Disc brake grease

93016G20

Front disc brakes—GS 300 and GS 430

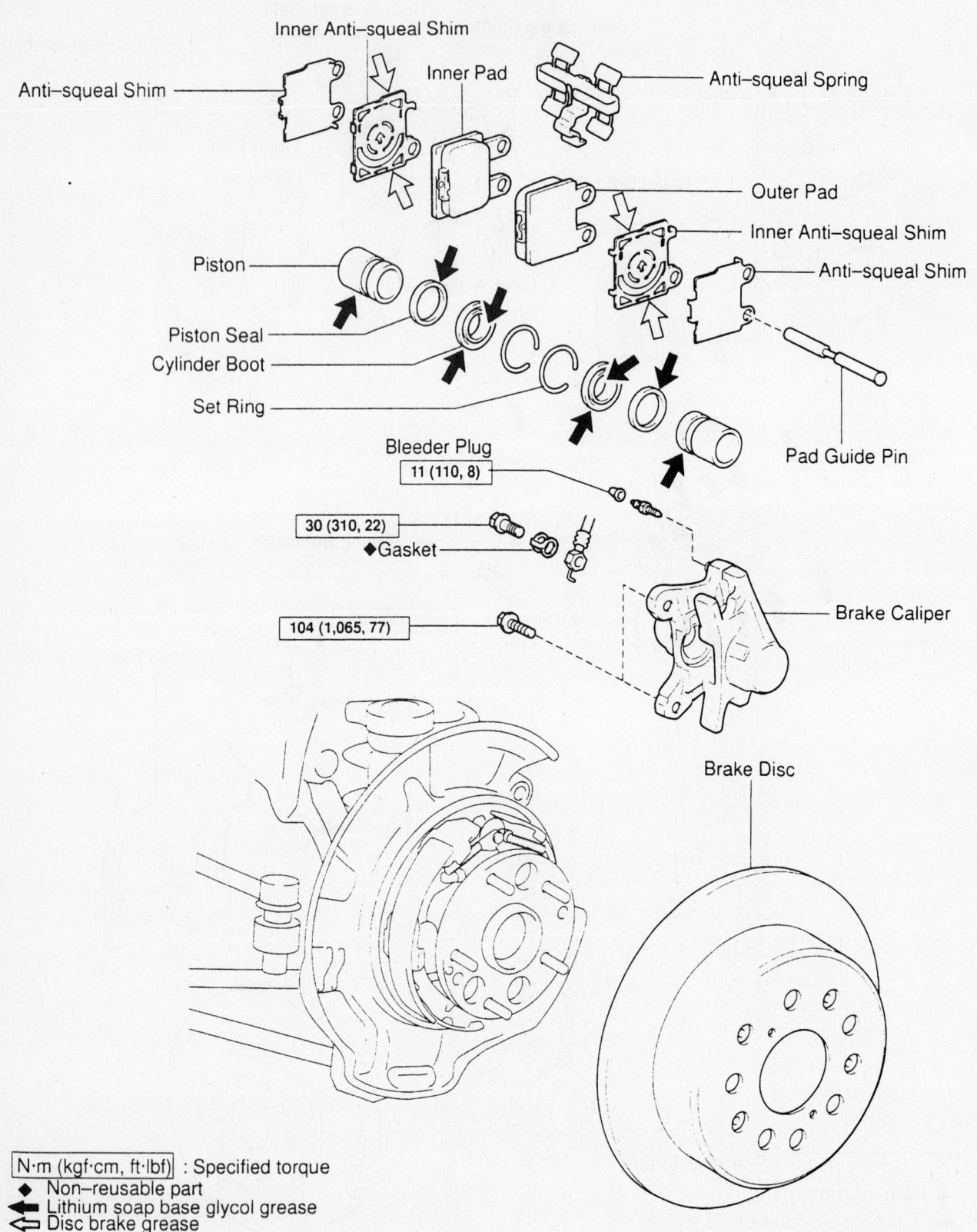

Anti–squeal Shim

Inner Anti–squeal Shim

Inner Pad

Anti–squeal Spring

Outer Pad

Inner Anti–squeal Shim

Anti–squeal Shim

Piston

Piston Seal

Cylinder Boot

Set Ring

Pad Guide Pin

Bleeder Plug
11 (110, 8)

30 (310, 22)
◆Gasket

104 (1,065, 77)

Brake Caliper

Brake Disc

N·m (kgf·cm, ft·lbf) : Specified torque
◆ Non–reusable part
← Lithium soap base glycol grease
⇐ Disc brake grease

93016G21

Rear disc brakes—GS 300 and GS 430

- Brake line at the caliper
- 2 bolts to the holding the caliper to the steering knuckle
- Caliper assembly

To install:

3. Install or connect the following:
- Caliper. Tighten the 2 bolts to 87 ft. lbs. (118 Nm).
- Brake line with 2 new gaskets and tighten the union to 29 ft. lbs. (39 Nm)

4. Refill the reservoir as necessary and bleed the brake system.

REAR

1. Before servicing the vehicle, refer to the precautions in the beginning of this section.

2. Remove or disconnect the following:
- Rear wheels
- Brake line at the caliper, then plug it
- Mounting bolts and the caliper assembly

To install:

3. Temporarily install the caliper on the torque plate with the 2 installation bolts.

4. Hold the sliding pin and tighten the mounting bolts to (except SC 430) 25 ft. lbs. (34 Nm),
Sc 430 tighten to 77 ft. lbs. (104 Nm).

5. Connect the brake line with 2 new gaskets and tighten the union to 29 ft. lbs. (39 Nm).

6. Refill the reservoir as necessary and bleed the brake system.

ES 330

FRONT AND REAR

1. Before servicing the vehicle, refer to the precautions in the beginning of this section.

2. Install or connect the following:
- Wheels
- Brake line at the caliper by removing the union bolt and 2 gaskets
- Mounting bolts, while holding the sliding pin with a wrench
- Caliper from the caliper support

To install:

3. Install or connect the following:
- Caliper to the caliper support
- Caliper bolts. Hold the sliding pin and tighten the mounting bolts to 25 ft. lbs. (34 Nm).
- Brake line with (2) new gaskets and tighten the union to 22 ft. lbs. (30 Nm)

4. Bleed the brake system.
- Wheels

5. Check and fill the master cylinder reservoir, if needed.

Disc Brake Pads

REMOVAL & INSTALLATION

ES 300 and IS 300

FRONT AND REAR

1. Before servicing the vehicle, refer to the precautions in the beginning of this section.

2. Remove or disconnect the following:
- Wheels
- Lower installation bolt
- Caliper and suspend it securely. Do NOT disconnect the fluid line.
- Brake pads and retainers

To install:

3. Install or connect the following:
- 2 pads so that the wear indicator plate is facing upward. Do NOT allow oil or grease to get in the rubbing face.

4. Draw out a small amount of brake fluid from the brake reservoir. Press in the caliper piston with a suitable tool.
- Caliper. Torque the sliding main pin to 25 ft. lbs. (34 Nm).
- Wheels

5. Check the fluid level in the master cylinder and add as necessary.

GS 300

FRONT AND REAR

1. Before servicing the vehicle, refer to the precautions in the beginning of this section.

2. Remove or disconnect the following:
- Wheels

3. Hold the sliding pin on the lower mounting bolt and remove the bolt. Swivel the caliper upward and out of the way.
- Anti-squeal springs
- Brake pads, retainers and anti-squeal shims

To install:

4. Install or connect the following:
- Pad support plates and the pad wear indicator plate on the inside pad
- Both pads (and anti-squeal shims) with the wear indicator plates facing downward

5. Compress the caliper pistons and install the caliper.
- Anti-squeal springs
- Hold the sliding pin and tighten the

mounting bolts to 25 ft. lbs. (34 Nm)
- Wheels

6. Check the brake fluid level in the reservoir.

SC 430

FRONT

1. Before servicing the vehicle, refer to the precautions in the beginning of this section.

2. Remove or disconnect the following:
- Front wheel
- 2 bolts and remove the disc brake caliper assembly

➡ **Support the caliper. Do NOT allow to hang by the brake hose**

- 2 anti-squeal springs
- Pads with anti-squeal shims
- Disc brake support plates
- Slide pins

To install:

3. Install or connect the following:
- Slide pins
- Disc brake pad support plate
- Anti-squeal shims to each pad

4. Using a suitable tool, compress the piston carefully in the cylinder bores
- Inner pad with the wear indicator plate facing upward
- Install outer pad
- Caliper assembly and 2 bolts. Tighten to 25 ft. lbs (34 Nm)
- Front wheel

REAR

1. Before servicing the vehicle, refer to the precautions in the beginning of this section.

2. Remove or disconnect the following:
- Rear wheel
- Anti-squeal springs
- Clip and guide pin
- Disc pads
- 4 anti-squeal shims from each pad

To install:

3. Install or connect the following:
- Apply disc brake grease to both sides of the inner anti-squeal shims
- Install 2 shims on each pad

➡ **Make sure that the arrows on the anti-squeal shims face the direction of wheel rotation.**

4. Using a suitable tool, compress the piston carefully in the cylinder bores
- Install pads
- Anti- squeal springs
- Rear wheel

ES 330

FRONT

1. Before servicing the vehicle, refer to the precautions in the beginning of this section.

2. Remove or disconnect the following:
   - Front wheel
   - 2 bolts and remove the disc brake caliper assembly

➡**Support the caliper. Do NOT allow to hang by the brake hose**

   - Pads with anti-squeal shims
   - Disc brake support plates
   - Slide pins

To install:

3. Install or connect the following:
   - Slide pins
   - Disc brake pad support plate
   - Anti-squeal shims to each pad

4. Using a suitable tool, compress the piston carefully in the cylinder bores
   - Inner pad with the wear indicator plate facing upward
   - Install outer pad
   - Caliper assembly and 2 bolts. Tighten to 25 ft. lbs (34 Nm)
   - Front wheel

REAR

1. Before servicing the vehicle, refer to the precautions in the beginning of this section.

2. Remove or disconnect the following:
   - Rear wheel
   - Anti-squeal springs
   - Clip and guide pin
   - Disc pads
   - 4 anti-squeal shims from each pad

To install:

3. Install or connect the following:
   - Apply disc brake grease to both sides of the inner anti-squeal shims
   - Install 2 shims on each pad

➡**Make sure that the arrows on the anti-squeal shims face the direction of wheel rotation.**

4. Using a suitable tool, compress the piston carefully in the cylinder bores
   - Install pads
   - Anti- squeal springs
   - Rear wheel

SPECIFICATIONS AND MAINTENANCE CHARTS

ENGINE AND VEHICLE IDENTIFICATION

Engine							Model Year	
Code ①	Liters (cc)	Cu. In.	Cyl.	Fuel Sys.	Engine Type	Eng. Mfg.	Code ②	Year
2UZ-FE	4.7 (4664)	285	8	SFI	DOHC	Toyota	1	2001
							2	2002
							3	2003
							4	2004
							5	2005

SFI: Sequential Fuel Injection

DOHC: Double Overhead Camshaft

① Stamped on the left side of the engine block

② 10th digit of the Vehicle Identification Number (VIN)

67162-X470-C01

GENERAL ENGINE SPECIFICATIONS

Year	Model	Engine Displacement Liters	Engine Series ID	Net Horsepower @ rpm	Net Torque @ rpm (ft. lbs.)	Bore x Stroke (in.)	Compression Ratio	Oil Pressure @ rpm
2001	LX470	4.7	2UZ-FE	245@4800	315@3400	3.70x3.31	9.6:1	45-65@3000
2002	LX470	4.7	2UZ-FE	245@4800	315@3400	3.70x3.31	9.6:1	45-65@3000
2003	LX470	4.7	2UZ-FE	245@4800	315@3400	3.70x3.31	9.6:1	45-65@3000
	GX470	4.7	2UZ-FE	235@4800	320@3400	3.70x3.31	9.6:1	45-65@3000
2004	LX470	4.7	2UZ-FE	245@4800	315@3400	3.70x3.31	9.6:1	45-65@3000
	GX470	4.7	2UZ-FE	245@4800	315@3400	3.70x3.31	9.6:1	45-65@3000

SFI: Sequential Fuel Injection

67162-X470-C02

ENGINE TUNE-UP SPECIFICATIONS

Year	Engine Displacement Liters	Engine ID	Spark Plug Gap (in.)	Ignition Timing (deg.)*	Fuel Pump (psi)	Idle Speed (rpm) MT	Idle Speed (rpm) AT	Valve Clearance Intake	Valve Clearance Exhaust
2001	4.7	2UZ-FE	0.031	5-15B	38-44	—	650-750	0.006-0.010	0.010-0.014
2002	4.7	2UZ-FE	0.031	5-15B	38-44	—	650-750	0.006-0.010	0.010-0.014
2003	4.7	2UZ-FE	0.031	5-15B	38-44	—	650-750	0.006-0.010	0.010-0.014
2004	4.7	2UZ-FE	0.043	5-15B	38-44	—	650-750	0.006-0.010	0.010-0.014

NOTE: The Vehicle Emission Control Information label often reflects specification changes made during production.

The label figures must be used if they differ from those in this chart.

B: Before top dead center

* With terminals TC and E1 connected to DLC1

67162-X470-C03

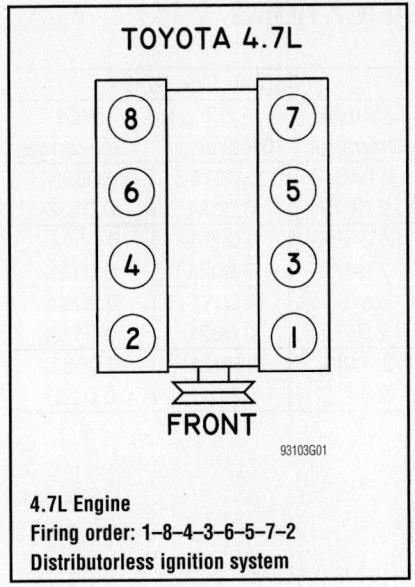

TOYOTA 4.7L

FRONT

93103G01

4.7L Engine
Firing order: 1–8–4–3–6–5–7–2
Distributorless ignition system

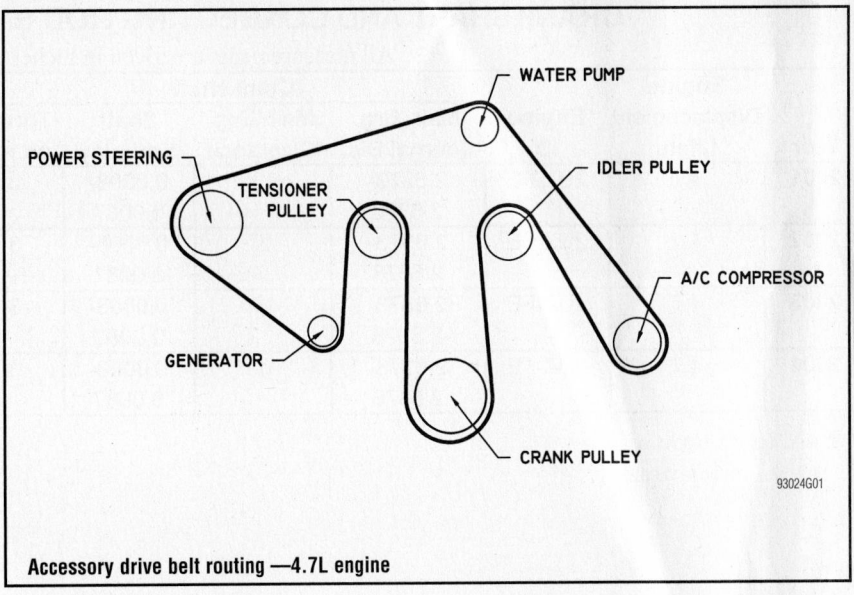

Accessory drive belt routing —4.7L engine

CAPACITIES

Year	Model	Engine Displacement Liters	Engine ID	Engine Oil with Filter (qts.)	Transmission (pts.) 5-Spd	Transmission (pts.) Auto.*	Transfer Case (pts.)	Drive Axle Front (pts.)	Drive Axle Rear (pts.)	Fuel Tank (gal.)	Cooling System (qts.)
2001	LX470	4.7	2UZ-FE	7.2	—	3.6	2.8	3.6	6.8	25.4	①
2002	LX470	4.7	2UZ-FE	6.6	—	4.2	2.6	2.4	7.7	25.4	12.3
2003	LX470	4.7	2UZ-FE	6.6	—	4.2	2.6	2.4	7.7	25.4	12.3
	GX470	4.7	2UZ-FE	6.5	—	6.4	3.0	3.9	6.4	23.0	13.6
2004	LX470	4.7	2UZ-FE	7.2	—	6.4	2.8	3.4	7.0	25.4	②
	GX470	4.7	2UZ-FE	6.5	—	6.4	3.0	3.9	6.4	23.0	13.6

*After draining, add the following amounts, then fill to the cold full line

① With rear heater: 9.5
   Without rear heater: 8.5

② Without rear heater: 15.6
   With rear heater: 16.2

67162-X470-C04

VALVE SPECIFICATIONS

Year	Engine Displacement Liters	Engine ID	Seat Angle (deg.)	Face Angle (deg.)	Spring Test Pressure (lbs. @ in.)	Spring Installed Height (in.)	Stem-to-Guide Clearance (in.) Intake	Stem-to-Guide Clearance (in.) Exhaust	Stem Diameter (in.) Intake	Stem Diameter (in.) Exhaust
2001	4.7	2UZ-FE	45	44.5	45.9-50.7@ 1.378	1.380	0.0010- 0.0024	0.0012- 0.0026	0.2154- 0.2159	0.2152- 0.2157
2002	4.7	2UZ-FE	45	44.5	45.9-50.7@ 1.378	1.380	0.0010- 0.0024	0.0012- 0.0026	0.2154- 0.2159	0.2152- 0.2157
2003	4.7	2UZ-FE	45	44.5	45.9-50.7@ 1.378	1.380	0.0010- 0.0024	0.0012- 0.0026	0.2154- 0.2159	0.2152- 0.2157
2004	4.7	2UZ-FE	45	44.5	45.9-50.7@ 1.378	1.380	0.0010- 0.0024	0.0012- 0.0026	0.2154- 0.2159	0.2152- 0.2157

67162-X470-C05

CRANKSHAFT AND CONNECTING ROD SPECIFICATIONS

All measurements are given in inches.

Year	Engine Displacement Liters	Engine ID	Crankshaft Main Brg. Journal Dia.	Crankshaft Main Brg. Clearance	Crankshaft Shaft End-play	Crankshaft Thrust on No.	Connecting Rod Journal Diameter	Connecting Rod Oil Clearance	Connecting Rod Side Clearance
2001	4.7	2UZ-FE	2.6373-2.6378	①	0.0008-0.0087	3	2.0465-2.0472	0.0011-0.0021	0.0063-0.0138
2002	4.7	2UZ-FE	2.6373-2.6378	①	0.0008-0.0087	3	2.0465-2.0472	0.0011-0.0021	0.0063-0.0138
2003	4.7	2UZ-FE	2.6373-2.6378	①	0.0008-0.0087	3	2.0465-2.0472	0.0011-0.0021	0.0063-0.0138
2004	4.7	2UZ-FE	2.6373-2.6378	①	0.0008-0.0087	3	2.0465-2.0472	0.0011-0.0021	0.0063-0.0138

① Nos. 1 and 2: 0.0011-0.0018
All others: 0.0016-0.0023

67162-X470-C06

PISTON AND RING SPECIFICATIONS

All measurements are given in inches.

Year	Engine Displacement Liters	Engine ID	Piston Clearance	Ring Gap Top Comp.	Ring Gap Bottom Comp.	Ring Gap Oil Control	Ring Side Clearance Top Comp.	Ring Side Clearance Bottom Comp.	Ring Side Clearance Oil Control
2001	4.7	2UZ-FE	0.0035-0.0044	0.0118-0.0157	0.0157-0.0217	0.0051-0.0150	0.0012-0.0031	0.0012-0.0028	SNUG
2002	4.7	2UZ-FE	0.0035-0.0044	0.0118-0.0157	0.0157-0.0217	0.0051-0.0150	0.0012-0.0031	0.0012-0.0028	SNUG
2003	4.7	2UZ-FE	0.0035-0.0044	0.0118-0.0157	0.0157-0.0217	0.0051-0.0150	0.0012-0.0031	0.0012-0.0028	SNUG
2004	4.7	2UZ-FE	0.0035-0.0044	0.0118-0.0157	0.0157-0.0217	0.0051-0.0150	0.0012-0.0031	0.0012-0.0028	SNUG

67162-X470-C07

TORQUE SPECIFICATIONS

All readings in ft. lbs.

Year	Engine Displacement Liters	Engine ID	Cylinder Head Bolts	Main Bearing Bolts	Rod Bearing Bolts	Crankshaft Damper Bolts	Flywheel Bolts	Manifold Intake	Manifold Exhaust	Spark Plugs	Oil Pan Drain Plug
2001	4.7	2UZ-FE	①	②	③	181	④	13	33	13	29
2002	4.7	2UZ-FE	①	②	③	181	④	13	33	13	29
2003	4.7	2UZ-FE	①	②	③	181	④	13	33	13	29
2004	4.7	2UZ-FE	①	②	③	181	④	13	33	13	29

① Step 1: 24
  Step 2: Plus 90 degrees
  Step 3: Plus 90 degrees

② Step 1: 20 ft. lbs.
  Step 2: Plus 90 degrees

③ Step 1: 18 ft. lbs.
  Step 2: Plus 90 degrees

④ Step 1: 35 ft. lbs.
  Step 2: Plus 90 degrees

67162-X470-C08

WHEEL ALIGNMENT

Year	Model	Caster Range (+/-Deg.)	Caster Preferred Setting (Deg.)	Camber Range (+/-Deg.)	Camber Preferred Setting (Deg.)	Toe-in (in.)	Steering Axis Inclination (Deg.)
2001	LX470	0.75	+3.08	0.75	0	0+/-0.08	12.25+/-0.75
2002	LX470	0.75	①	0.75	+0.13	0.05+/-0.08	10.635+/-0.75
2003	LX470	0.75	①	0.75	+0.13	0.05+/-0.08	10.635+/-0.75
	GX470	0.75	+3.28	0.75	-0.02	0.08+/-0.16	12.48+/-0.75
2004	LX470	0.75	+3.08	0.75	0	0+/-0.08	12.25+/-0.75
	GX470	0.75	+3.28	0.75	-0.02	0.08+/-0.16	12.48+/-0.75

Note: All alignment specifications are based on nominal ride height and standard tires

① P245/70R16: +2.95
   P265/70R16: +3.00

67162-X470-C09

TIRE, WHEEL AND BALL JOINT SPECIFICATIONS

Year	Model	OEM Tires Standard	OEM Tires Optional	Tire Pressures (psi) Front	Tire Pressures (psi) Rear	Wheel Size	Ball Joint Inspection	Lugnut Torque (ft. lbs.)
2001	LX470	P275/70HR16	None	32	32	8-JJ	①	97
2002	LX470	P275/70HR16	None	32	32	8-JJ	①	97
2003	LX470	P275/60HR18	None	NA	NA	8-JJ	②	97
	GX470	P265/65SR17	None	32	32	7.5	③	NA
2004	LX470	P275/60HR18	None	NA	NA	8-JJ	④	97
	GX470	P265/65SR17	None	32	32	7.5	③	NA

OEM: Original Equipment Manufacturer

PSI: Pounds Per Square Inch

STD: Standard

OPT: Optional

NA: Not Available

① Replace if any measurable movement is found.

② Upper ball joint turning torque: 6-39 inch lbs.
   Lower ball joint turning torque: 1-22 inch lbs.
   Lower ball joint excessive play: 0.020 in.

③ Upper arm ball joint turning torque: 40 inch lbs. or less
   Lower arm ball joint turning torque: 27 inch lbs or less

④ Upper arm ball joint turning torque: 9-39 inch lbs.
   Lower arm ball joint turning torque: 2.6-26 inch lbs.

67162-X470-C10

BRAKE SPECIFICATIONS
All measurements in inches unless noted

Year	Model		Brake Disc Original Thickness	Brake Disc Minimum Thickness	Brake Disc Maximum Runout	Minimum Lining Thickness	Brake Caliper Bracket Bolts (ft. lbs.)	Brake Caliper Mounting Bolts (ft. lbs.)
2001	LX470	F	1.260	1.181	0.0028	0.039	76	90
		R	0.709	0.611	0.0040	0.039	—	65
2002	LX470	F	1.260	1.181	0.0028	0.039	—	90
		R	0.709	0.611	0.0040	0.039	—	20
2003	LX470	F	1.260	1.181	0.0028	0.039	—	90
		R	0.709	0.611	0.0040	0.039	—	20
	GX 470	F	1.102	1.024	0.0020	0.039	—	91
		R	0.709	0.630	0.0079	0.039	—	77
2004	LX470	F	1.260	1.181	0.0028	0.039	—	90
		R	0.709	0.611	0.0040	0.039	—	20
	GX 470	F	1.102	1.024	0.0020	0.039	—	91
		R	0.709	0.630	0.0079	0.039	—	77

F: Front
R: Rear

67162-X470-C11

SCHEDULED MAINTENANCE INTERVALS
Lexus GX470 and LX470

TO BE SERVICED	TYPE OF SERVICE	VEHICLE MILEAGE INTERVAL (x1000)												
		7.5	15	22.5	30	37.5	45	52.5	60	67.5	75	82.5	90	97.5
Engine oil & filter	R	✓	✓	✓	✓	✓	✓	✓	✓	✓	✓	✓	✓	✓
Automatic transmission fluid & filter	S/I		✓		✓		✓		✓		✓		✓	
Ball joints & dust covers	S/I		✓		✓		✓		✓		✓		✓	
Bolts & nuts on chassis & body	S/I		✓		✓		✓		✓		✓		✓	
Brake line pipes & hoses	S/I		✓		✓		✓		✓		✓		✓	
Brake pads & discs	S/I		✓		✓		✓		✓		✓		✓	
Propeller shaft grease	S/I		✓		✓		✓		✓		✓		✓	
Steering knuckle & chassis grease	S/I		✓		✓		✓		✓		✓		✓	
Steering linkage	S/I		✓		✓		✓		✓		✓		✓	
Transfer and differential oil	S/I		✓		✓		✓		✓		✓		✓	
Air cleaner filter	R				✓				✓				✓	
Spark plugs ①	R				✓				✓				✓	
Drive belts	S/I				✓				✓				✓	
Exhaust pipes & mountings	S/I				✓				✓				✓	
Fuel lines & connections	S/I				✓				✓				✓	
Engine coolant	R						✓				✓			
Charcoal canister	R								✓					
Fuel tank cap gasket	R								✓					
Heated oxygen sensors (exc. Cal.) ②	R													

R: Replace   S/I: Service or Inspect
① Platinum plugs, replace every 100,000 miles.
② Heated oxygen sensors (except Calif.): replace every 80,000 miles.

FREQUENT OPERATION MAINTENANCE (SEVERE SERVICE)

If a vehicle is operated under any of the following conditions it is considered severe service:

- Extremely dusty areas.
- 50% or more of the vehicle operation is in 32°C (90°F) or higher temperatures, or constant operation in temperatures below 0°C (32°F).
- Prolonged idling (vehicle operation in stop and go traffic).
- Frequent short running periods (engine does not warm to normal operating temperatures).
- Police, taxi, delivery usage or trailer towing usage.

Air cleaner filter: service or inspect every 3750 miles.

Engine oil & filter: replace every 3750 miles.

Ball joints & dust covers: service or inspect every 7500 miles.

Bolts & nuts on chassis & body: service or inspect every 7500 miles.

Brake pads & discs (front & rear): service or inspect every 7500 miles.

Steering knuckle & chassis grease: service or inspect every 7500 miles.

Steering linkage: service or inspect every 7500 miles.

Propeller shaft grease: service or inspect every 7500 miles.

Exhaust pipes & mountings: service or inspect every 15,000 miles.

67162-X470-C12

PRECAUTIONS

Before servicing any vehicle, please be sure to read all of the following precautions, which deal with personal safety, prevention of component damage, and important points to take into consideration when servicing a motor vehicle:

• Never open, service or drain the radiator or cooling system when the engine is hot; serious burns can occur from the steam and hot coolant.

• Observe all applicable safety precautions when working around fuel. Whenever servicing the fuel system, always work in a well-ventilated area. Do not allow fuel spray or vapors to come in contact with a spark, open flame, or excessive heat (a hot drop light, for example). Keep a dry chemical fire extinguisher near the work area. Always keep fuel in a container specifically designed for fuel storage; also, always properly seal fuel containers to avoid the possibility of fire or explosion. Refer to the additional fuel system precautions later in this section.

• Fuel injection systems often remain pressurized, even after the engine has been turned **OFF**. The fuel system pressure must be relieved before disconnecting any fuel lines. Failure to do so may result in fire and/or personal injury.

• Brake fluid often contains polyglycol ethers and polyglycols. Avoid contact with the eyes and wash your hands thoroughly after handling brake fluid. If you do get brake fluid in your eyes, flush your eyes with clean, running water for 15 minutes. If eye irritation persists, or if you have taken brake fluid internally, IMMEDIATELY seek medical assistance.

• The EPA warns that prolonged contact with used engine oil may cause a number of skin disorders, including cancer. You should make every effort to minimize your exposure to used engine oil. Protective gloves should be worn when changing oil. Wash your hands and any other exposed skin areas as soon as possible after exposure to used engine oil. Soap and water, or waterless hand cleaner should be used.

• All new vehicles are now equipped with an air bag system. The system must be disabled before performing service on or around system components, steering column, instrument panel components, wiring and sensors. Failure to follow safety and disabling procedures could result in accidental air bag deployment, possible personal injury and unnecessary system repairs.

• Always wear safety goggles when working with, or around, the air bag system. When carrying a non-deployed air bag, be sure the bag and trim cover are pointed away from your body. When placing a non-deployed air bag on a work surface, always face the bag and trim cover upward, away from the surface. This will reduce the motion of the module if it is accidentally deployed. Refer to the additional air bag system precautions later in this section.

• NEVER disconnect the negative battery cable with the ignition **ON** or the engine running. Removing power from the computer control module with the ignition **ON** may destroy the module.

• Clean, high quality brake fluid from a sealed container is essential to the safe and

proper operation of the brake system. You should always buy the correct type of brake fluid for your vehicle. If the brake fluid becomes contaminated, completely flush the system with new fluid. Never reuse any brake fluid. Any brake fluid that is removed from the system should be discarded. Also, do not allow any brake fluid to come in contact with a painted surface; it will damage the paint.

• Never operate the engine without the proper amount and type of engine oil; doing so WILL result in severe engine damage.

• Timing belt maintenance is extremely important. Many models utilize an interference-type, non-freewheeling engine. If the timing belt breaks, the valves in the cylinder head may strike the pistons, causing potentially serious (also time-consuming and expensive) engine damage. Refer to the maintenance interval charts in the front of this manual for the recommended replacement interval for the timing belt, and to the timing belt section for belt replacement and inspection.

• Disconnecting the negative battery cable on some vehicles may interfere with the functions of the on-board computer system(s) and may require the computer to undergo a relearning process once the negative battery cable is reconnected.

• When servicing drum brakes, only disassemble and assemble one side at a time, leaving the remaining side intact for reference.

• Only an MVAC-trained, EPA-certified automotive technician should service the air conditioning system or its components.

ENGINE REPAIR

➡**Disconnecting the negative battery cable on some vehicles may interfere with the functions of the on board computer system. The computer may undergo a relearning process once the negative battery cable is reconnected.**

Alternator

REMOVAL

1. Before servicing the vehicle, refer to the precautions in the beginning of this section.
2. Drain the cooling system.
3. Remove or disconnect the following:
   • Negative battery cable
   • Accessory drive belt
   • Engine under cover

   • Radiator
   • Power steering pump pulley
   • Alternator harness connectors
   • Alternator

INSTALLATION

1. Install or connect the following:
   • Alternator. Tighten the fasteners to 29 ft. lbs. (39 Nm).
   • Alternator harness connectors
   • Power steering pump pulley
   • Radiator
   • Engine under cover
   • Accessory drive belt
   • Negative battery cable
2. Fill the cooling system.
3. Start the engine and check for leaks.

Ignition Timing

ADJUSTMENT

All engines are equipped with a Distributorless Ignition System (DIS). No timing adjustment is possible.

Engine Assembly

REMOVAL & INSTALLATION

GX470

1. Remove the transmission.
2. Remove the hood.
3. Remove the V-bank cover.

4. Remove the air cleaner assembly.

5. Remove the under-covers.

6. Remove the radiator.

7. Remove the fan shroud.

8. Tag and disconnect all hoses, pipes and wires necessary for engine removal.

9. Remove the fan.

10. Remove the power steering pump and secure it out of the way.

11. Remove the alternator and secure it out of the way.

12. Remove the compressor and secure it out of the way.

13. Remove the transmission filler tube.

14. Remove the oil level sending unit.

15. Remove the exhaust manifolds.

16. Attach a crane and equalizer to the engine.

17. Support the weight of the engine with the crane and remove the mount bolts.

18. Remove the engine.

19. Installation is the reverse of removal. Observe the following torques:
- Engine mount bolts: 28 ft. lbs. (38 Nm)
- Exhaust manifold nuts: 33 ft. lbs. (44 Nm)
- Oil level sending unit: 11 ft. lbs. 15 Nm)
- Fan bolts: 21 ft. lbs. (29 Nm)
- Compressor: bolt, 34 ft. lbs. (47 Nm); nut, 18 ft. lbs. (25 Nm)
- Power steering pump: 32 ft. lbs. (43 Nm)
- Hood: 10 ft. lbs. (13 Nm)

LX470

1. Before servicing the vehicle, refer to the precautions in the beginning of this section.

2. Relieve the fuel system pressure.

3. Drain the cooling system.

4. Drain the engine oil.

5. Remove or disconnect the following:
- Battery and tray
- Hood
- Engine appearance cover
- Air intake pipe
- Engine under covers
- Coolant recovery tank
- Radiator hoses
- Radiator and fan shroud
- Accessory drive belt
- Cooling fan and pulley
- Powertrain Control Module (PCM) harness connectors and pass the wiring harness through the firewall
- Accelerator cable
- Power steering vacuum hoses
- Alternator harness connectors
- Heater hoses

- Engine control wiring harness and grommet at the firewall
- Ground cable connector
- Fuel lines
- Evaporative Emissions (EVAP) canister hoses
- Wire clamp at right inner fender
- Negative battery cable at the relay box and right inner fender
- Positive battery cable
- Center console
- Transmission shift lever assembly
- Transfer case shift lever and rod
- Exhaust front pipes
- Stabilizer bar
- Front and rear driveshafts
- A/C compressor
- Power steering pump

6. Attach a hoist to the engine lifting eyes.

7. Remove or disconnect the following:
- Transfer case skid plate
- Left and right motor mounts
- Transmission mount crossmember

8. Attach a hoist to the engine lifting eyes and raise the powertrain out of the vehicle.

To install:

9. Lower the powertrain into the vehicle.

10. Install or connect the following:
- Transmission mount crossmember. Tighten the bolts to 37 ft. lbs. (50 Nm) and the nuts to 55 ft. lbs. (74 Nm).
- Transfer case skid plate
- Left and right motor mounts. Tighten the fasteners to 22 ft. lbs. (30 Nm).
- Power steering pump. Tighten the bolts to 13 ft. lbs. (17 Nm).
- A/C compressor. Tighten the bolts to 36 ft. lbs. (49 Nm).
- Front driveshaft. Tighten the fasteners to 59 ft. lbs. (80 Nm).
- Rear driveshaft. Tighten the fasteners to 78 ft. lbs. (106 Nm).
- Stabilizer bar. Tighten the bracket bolts to 13 ft. lbs. (18 Nm) and the link nuts to 18 ft. lbs. (25 Nm).
- Exhaust front pipes
- Transfer case shift lever and rod
- Transmission shift lever assembly
- Center console
- Positive battery cable
- Negative battery cable at the relay box and right inner fender
- Wire clamp at right inner fender
- EVAP canister hoses
- Fuel lines
- Ground cable connector

- Engine control wiring harness and grommet at the firewall
- Heater hoses
- Alternator harness connectors
- Power steering vacuum hoses
- Accelerator cable
- PCM harness connectors
- Cooling fan and pulley
- Accessory drive belt
- Radiator and fan shroud
- Radiator hoses
- Coolant recovery tank
- Engine under covers
- Air intake pipe
- Engine appearance cover
- Hood
- Battery and tray

11. Fill the crankcase to the correct level.

12. Fill the cooling system.

13. Start the engine and check for leaks.

Heater Core

REMOVAL & INSTALLATION

LX 470

FRONT HEATER

1. Disconnect the negative battery cable.

2. Drain the cooling system into a clean container for reuse.

3. Disconnect the heater hoses from the heater core.

4. Remove the steering wheel by performing the following procedure:

a. Position the front wheels facing straight-ahead.

b. Remove the steering wheel side covers.

c. Using a Torx® wrench, loosen the 2 screws located at each side of the steering wheel until the screw's circumference groove catches on the screw case.

d. Pull the air bag module from the steering wheel and disconnect the electrical connector.

✳✳ CAUTION

Place the air bag module in a safe place with the front side facing upward.

e. Remove the steering wheel nut.

f. Place alignment marks on the steering wheel and the main shaft.

g. Using a steering wheel puller, press the steering wheel from the steering column.

5. Remove the instrument panel and

reinforcement by performing the following procedure:

a. Remove the front door scuff plates, the cowl side trim and the front door opening trim.

b. At the driver's side, remove the 2 assist grip plugs, the 2 screws and assist grip and the front pillar garnish.

c. At the passenger's side, remove the 4 assist grip plugs, the 4 screws, the 2 assist grips and the front pillar garnish.

d. Remove the instrument cluster finish panel.

e. Remove the 2 screws and the hood lock control cable.

f. Remove the 2 screws and the fuel lid control cable lever.

g. Remove the lower No. 1 panel screw and the panel.

h. Remove the lower left side panel.

i. Remove the 3 steering column cover screws and the covers.

j. At the steering column, disconnect the electrical connectors; then, remove the clamp, the 3 screws and the combination switch.

k. Remove the No. 2 heater-to-register duct screw and the duct.

l. Remove the steering column-to-instrument panel bolts and the steering column.

m. At the combination meter, disconnect the electrical connectors; then, remove the 4 screws and the combination meter.

n. Remove the glove compartment door stoppers, the 2 screws and the glove box door.

o. At the passenger's side air bag module, remove the No. 1 undercover, pull the air bag connector up from the undercover and disconnect it; then, remove the air bag.

❊❊ CAUTION

Place the air bag module in a safe place with the front side facing upward.

p. Remove the 3 lower No. 2 panel screws and the panel.

q. Remove the center cluster; then, pry the center cluster from the dash by prying the 8 clips in the following order:
• Left side
• Right side
• Top left side
• Top right side

r. Remove the 4 radio screws, pull the radio outward, disconnect the electrical connectors and remove the radio.

s. At the rear console panel, remove the transfer shift lever knob; then, pry the panel upward disengaging the 4 clips (2 on each side) and remove the panel.

t. At the rear of the console, remove the 2 rear end panel-to-console screws; then, pry the end panel rearward disengaging the 2 clips and remove the panel.

u. If not equipped with a rear air conditioning system, disconnect the connector and control cable; then, remove the 3 rear heater control panel screws and the panel.

v. Remove the 4 rear console box-to-chassis screws/bolts and the console box.

w. Remove the center lower cluster finish panel by prying panel rearward disengaging the 5 clips; then, disconnect the electrical connector.

x. Remove the 2 front console-to-chassis bolts/screws, disengage the 2 clips and remove the console.

y. At the instrument panel, disconnect the junction connectors (the connectors can be disconnected by loosening the bolts), the instrument panel-to-chassis 8 bolts and 2 nuts. Using an assistant, remove the instrument panel.

z. Disconnect the electrical connector and remove the ECM.

aa. Remove the No. 3 and No. 4 heater-to-register ducts.

bb. Remove the floor brace, the No. 1 brace and the reinforcement.

6. Remove the evaporator housing by performing the following procedure:

a. Discharge and recover the air conditioning system refrigerant.

b. Remove the air conditioning liquid line clamp.

c. Remove the air conditioning suction line clamp.

d. Disconnect both air conditioning lines and plug the openings to prevent contamination. Discard the 4 O-rings.

e. Remove the antenna relay electrical connector, the 2 screws and the relay.

f. Remove the evaporator housing-to-chassis 4 screws/2 nuts and the housing.

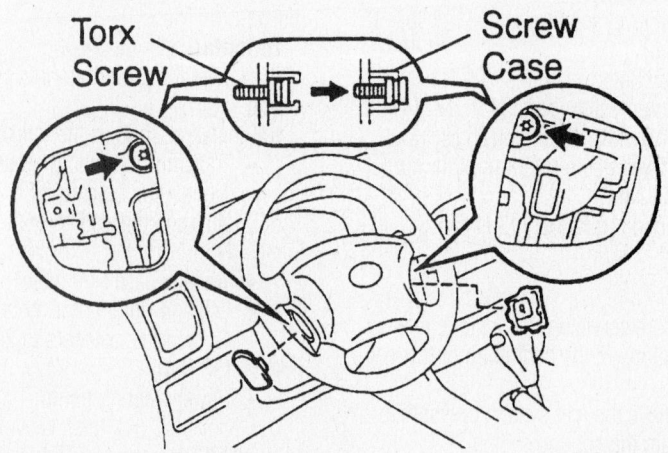

View the steering wheel's Torx® bolts—Lexus LX 470

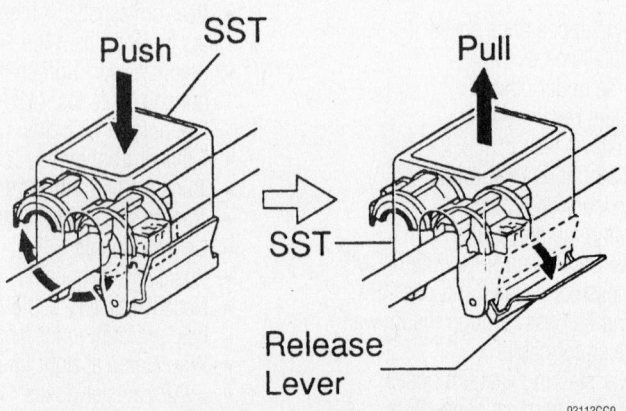

View the air conditioning line clamp removal tool—Lexus LX 470

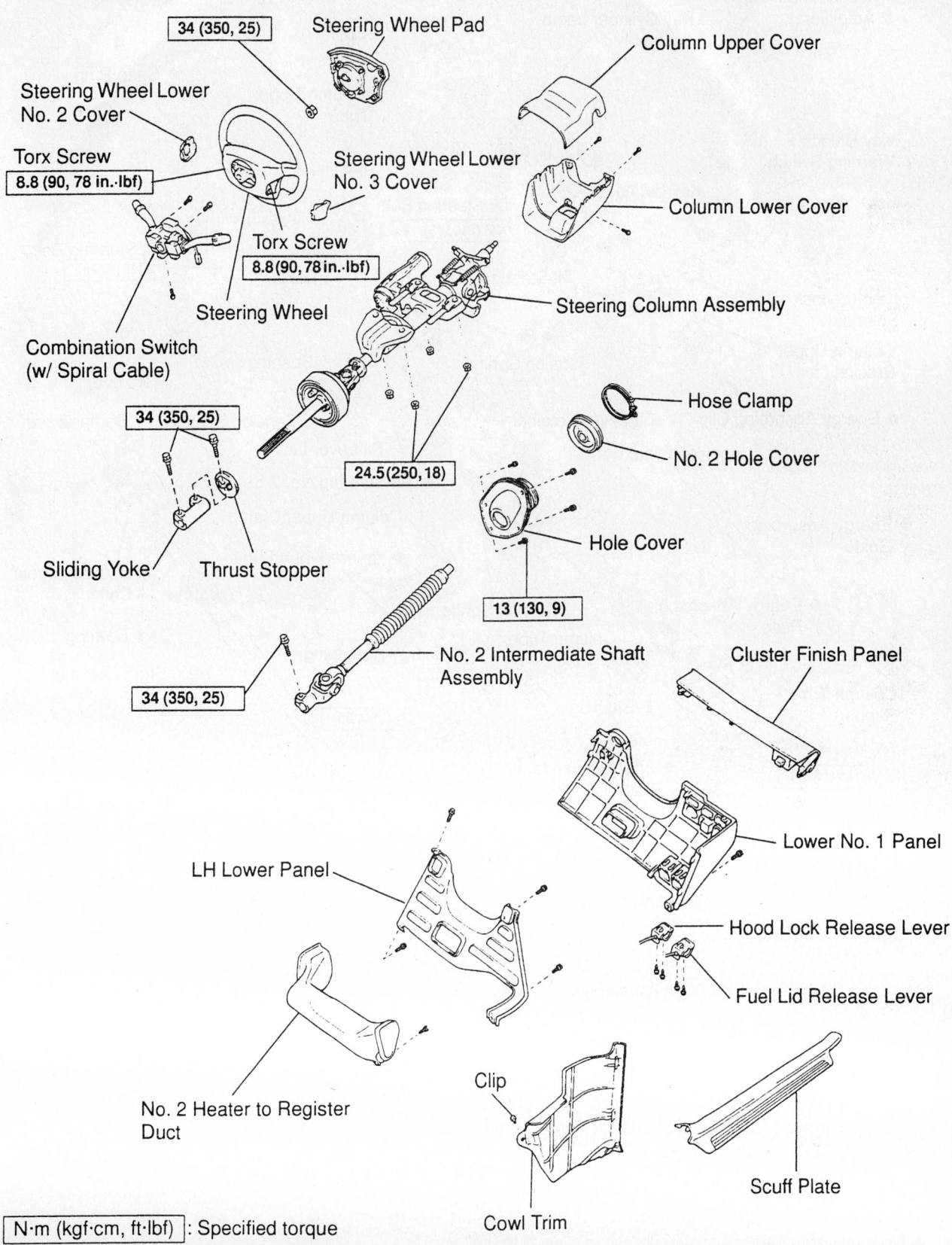

34 (350, 25)

Steering Wheel Pad

Column Upper Cover

Steering Wheel Lower
No. 2 Cover

Torx Screw
8.8 (90, 78 in.·lbf)

Steering Wheel Lower
No. 3 Cover

Column Lower Cover

Torx Screw
8.8 (90, 78 in.·lbf)

Steering Wheel

Combination Switch
(w/ Spiral Cable)

Steering Column Assembly

Hose Clamp

34 (350, 25)

No. 2 Hole Cover

24.5 (250, 18)

Hole Cover

Sliding Yoke

Thrust Stopper

13 (130, 9)

Cluster Finish Panel

34 (350, 25)

No. 2 Intermediate Shaft
Assembly

Lower No. 1 Panel

LH Lower Panel

Hood Lock Release Lever

Fuel Lid Release Lever

Clip

No. 2 Heater to Register
Duct

Scuff Plate

Cowl Trim

N·m (kgf·cm, ft·lbf) : Specified torque

93113GG5

Exploded view the steering column—Lexus LX 470 (Part 1 of 2)

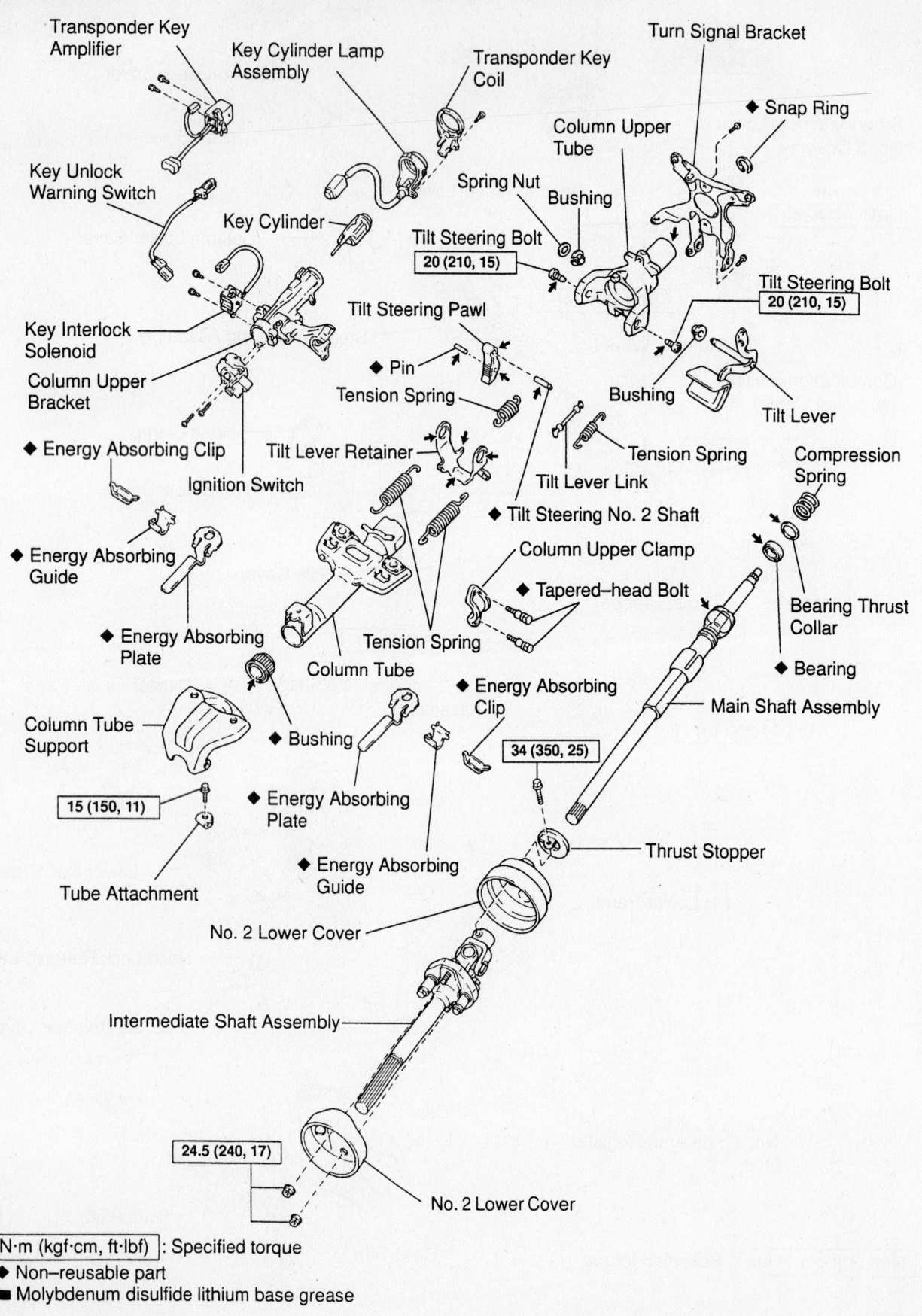

Transponder Key Amplifier

Key Cylinder Lamp Assembly

Transponder Key Coil

Turn Signal Bracket

Column Upper Tube

◆ Snap Ring

Key Unlock Warning Switch

Key Cylinder

Spring Nut

Bushing

Tilt Steering Bolt
20 (210, 15)

Tilt Steering Bolt
20 (210, 15)

Key Interlock Solenoid

Tilt Steering Pawl

◆ Pin

Tension Spring

Bushing

Tilt Lever

Column Upper Bracket

◆ Energy Absorbing Clip

Ignition Switch

Tilt Lever Retainer

Tension Spring

Tilt Lever Link

Compression Spring

◆ Energy Absorbing Guide

◆ Tilt Steering No. 2 Shaft

Column Upper Clamp

Bearing Thrust Collar

◆ Energy Absorbing Plate

◆ Tapered–head Bolt

◆ Bearing

◆ Energy Absorbing Clip

Main Shaft Assembly

Column Tube Support

Tension Spring

Column Tube

◆ Bushing

34 (350, 25)

15 (150, 11)

◆ Energy Absorbing Plate

Thrust Stopper

Tube Attachment

◆ Energy Absorbing Guide

No. 2 Lower Cover

Intermediate Shaft Assembly

24.5 (240, 17)

No. 2 Lower Cover

N·m (kgf·cm, ft·lbf) : Specified torque
◆ Non–reusable part
Molybdenum disulfide lithium base grease

93113GG6

Exploded view the steering column—Lexus LX 470 (Part 2 of 2)

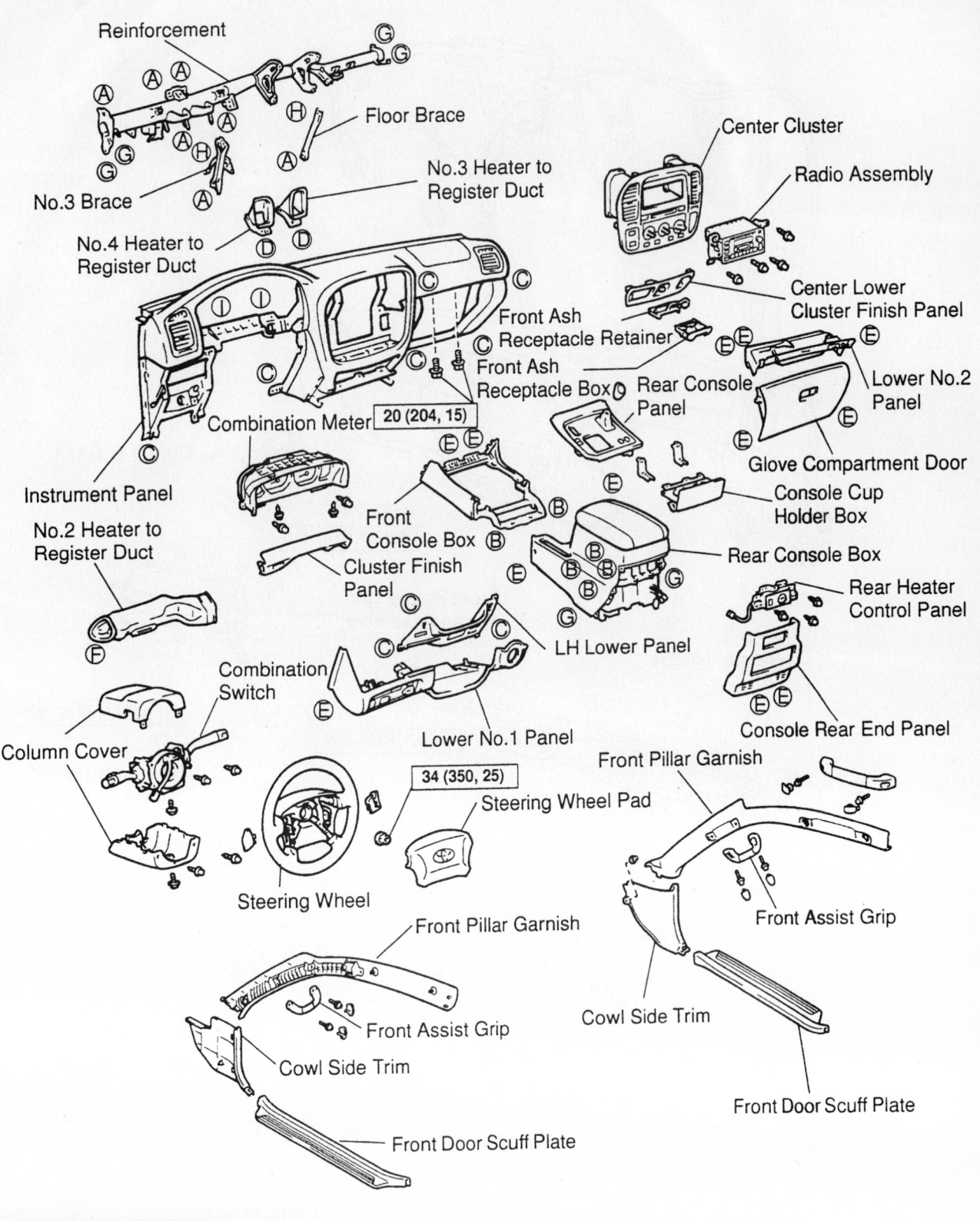

Reinforcement

No.3 Brace

No.4 Heater to
Register Duct

Floor Brace

No.3 Heater to
Register Duct

Center Cluster

Radio Assembly

Center Lower
Cluster Finish Panel

Front Ash
Receptacle Retainer

Front Ash
Receptacle Box

Rear Console
Panel

Lower No.2
Panel

Glove Compartment Door

Console Cup
Holder Box

Rear Console Box

Rear Heater
Control Panel

Console Rear End Panel

Combination Meter   20 (204, 15)

Instrument Panel

No.2 Heater to
Register Duct

Front
Console Box

Cluster Finish
Panel

LH Lower Panel

Lower No.1 Panel

Combination
Switch

Column Cover

Steering Wheel

34 (350, 25)
Steering Wheel Pad

Front Pillar Garnish

Front Assist Grip

Cowl Side Trim

Front Door Scuff Plate

Front Pillar Garnish

Front Assist Grip

Cowl Side Trim

Front Door Scuff Plate

N·m (kgf·cm, ft·lbf) : Specified torque

93113GG7

Exploded view the instrument panel and related components—LX 470

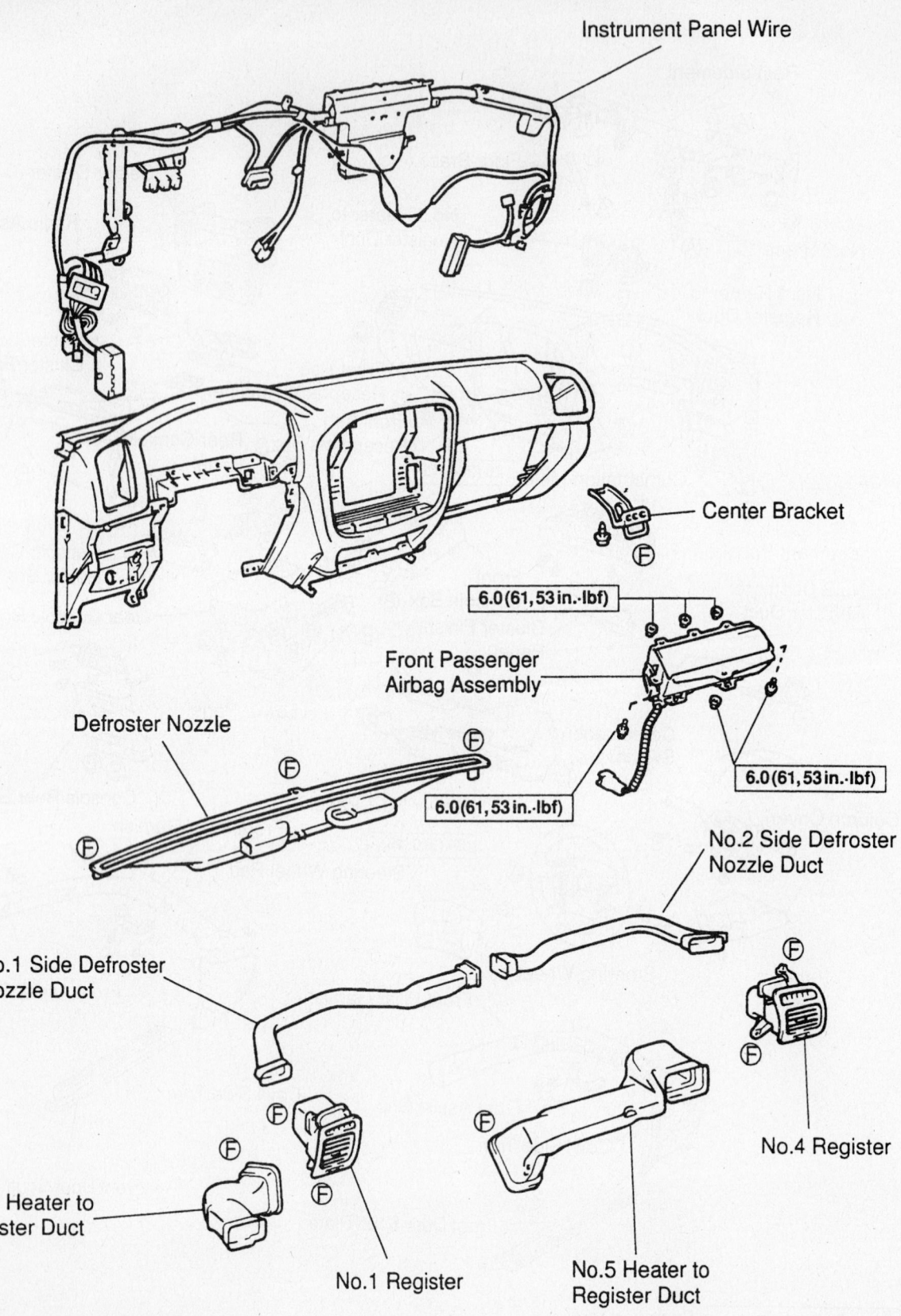

Instrument Panel Wire

Center Bracket

6.0 (61, 53 in.·lbf)

Front Passenger
Airbag Assembly

6.0 (61, 53 in.·lbf)

6.0 (61, 53 in.·lbf)

Defroster Nozzle

No.2 Side Defroster
Nozzle Duct

No.1 Side Defroster
Nozzle Duct

No.4 Register

No.1 Heater to
Register Duct

No.1 Register

No.5 Heater to
Register Duct

N·m (kgf·cm, ft·lbf) : Specified torque

93113GG8

Exploded view the front ventilation ducts and related components—Lexus LX 470

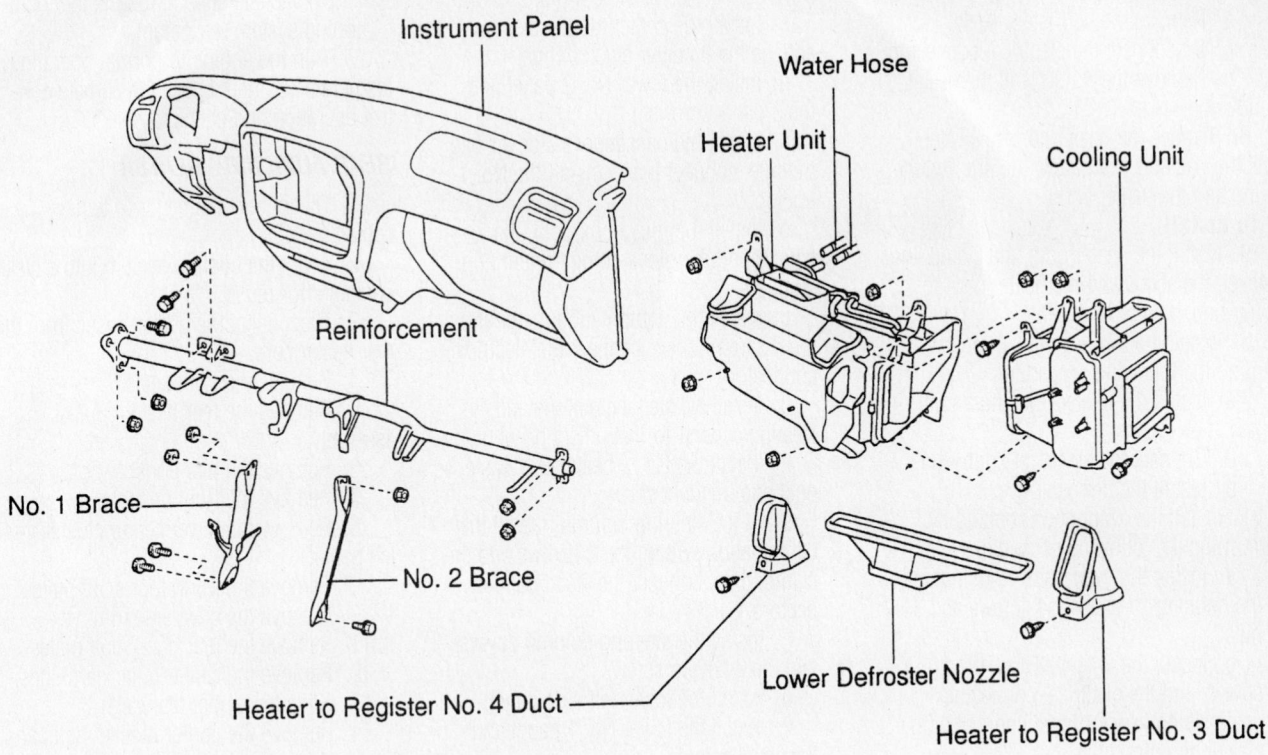

Instrument Panel

Water Hose

Heater Unit

Cooling Unit

Reinforcement

No. 1 Brace

No. 2 Brace

Heater to Register No. 4 Duct

Lower Defroster Nozzle

Heater to Register No. 3 Duct

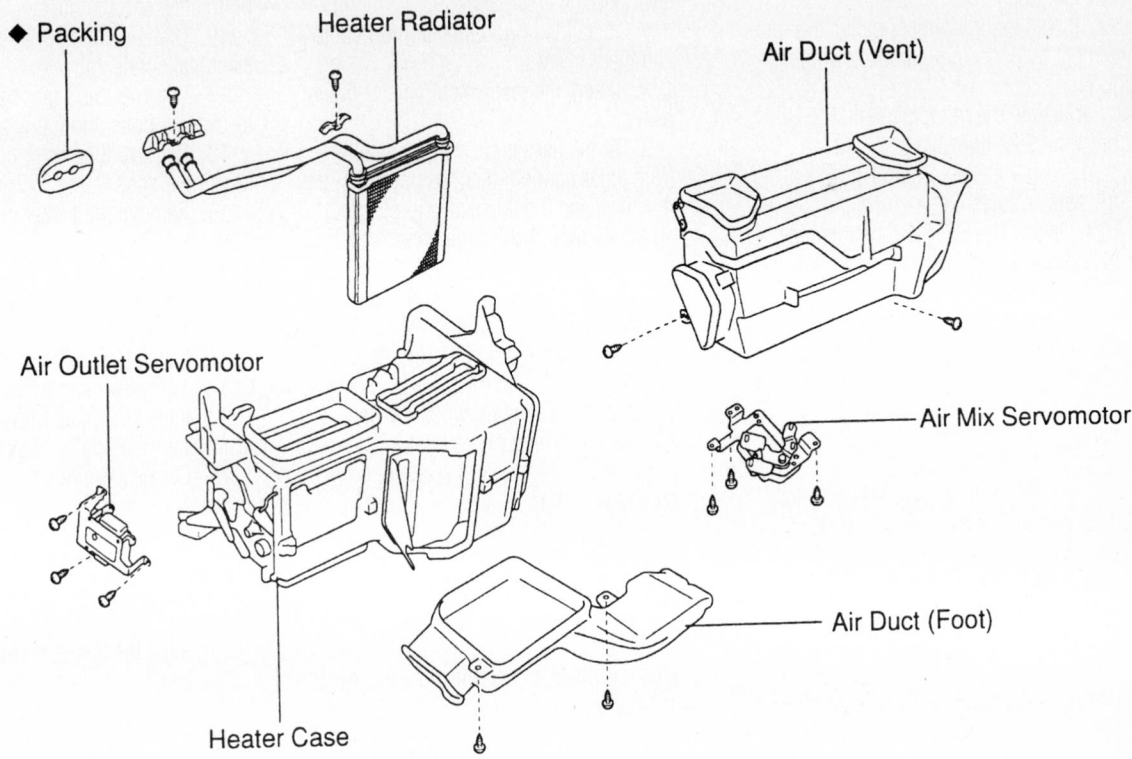

◆ Packing

Heater Radiator

Air Duct (Vent)

Air Outlet Servomotor

Air Mix Servomotor

Air Duct (Foot)

Heater Case

◆ Non–reusable part

93113GG0

Exploded view the front heater core, heater housing, evaporator housing and related components—Lexus LX 470

7. Remove the heater housing by performing the following procedure:

    a. Remove the defroster nozzle.

    b. Disconnect the electrical connector.

    c. Remove the 4 nuts and the heater housing.

8. Remove the heater core-to-heater housing packing, the screw, the bracket, the clamp and the heater core.

To install:

9. Install the heater core, the clamp, the bracket, the screw and the heater core-to-heater housing packing.

10. Install the heater housing by performing the following procedure:

    a. Install the heater housing and the 4 nuts.

    b. Connect the electrical connector.

    c. Install the defroster nozzle.

11. Install the evaporator housing by performing the following procedure:

    a. Install the evaporator housing and the housing-to-chassis 4 screws and 2 nuts.

    b. Install the antenna relay, the 2 screws and the electrical connector.

    c. Using new O-rings, connect both air conditioning lines.

    d. Install the air conditioning liquid line and suction line clamp.

12. Install the instrument panel and reinforcement by performing the following procedure:

    a. Install the reinforcement, the No. 1 brace and the floor brace.

    b. Install the No. 3 and No. 4 heater-to-register ducts.

    c. Install the ECM and connect the electrical connector.

    d. Using an assistant, install the instrument panel, connect the junction connectors, the instrument panel-to-chassis 8 bolts and 2 nuts.

    e. Install the front the console, engage the 2 clips and install the 2 console-to-chassis bolts/screws.

    f. Connect the electrical connector; then, install the center lower cluster finish panel by engaging the 5 clips.

    g. Install the console box and the 4 rear console box-to-chassis screws/bolts.

    h. If not equipped with a rear air conditioning system, install rear heater control panel, the 3 panel screws; then, connect the connector and control cable.

    i. Install the rear of the console and engage the 2 clips; then, install the 2 rear end panel-to-console screws.

    j. Install the rear console panel and engage the 4 clips (2 on each side); then, install the transfer shift lever knob.

    k. Install the radio, connect the electrical connectors and the 4 radio screws.

    l. Install the center cluster and engage the 8 center cluster clips.

    m. Install the lower No. 2 panel and the 3 panel screws.

    n. Install the passenger's side air bag module, connect it and install the No. 1 undercover.

    o. Install the glove box door, the 2 screws and the glove compartment door stoppers.

    p. Install the combination meter and the 4 screws; then, connect the electrical connectors.

    q. Install the steering column and the steering column-to-instrument panel bolts.

    r. Install the No. 2 heater-to-register duct and the duct screw.

    s. At the steering column, install the combination switch, the 3 screws and the clamp; then, connect the electrical connectors.

    t. Install the steering column covers and the 3 covers screws.

    u. Install the lower left side panel.

    v. Install the lower No. 1 panel and the panel screw.

    w. Install the fuel lid control cable lever and the 2 screws.

    x. Install the hood lock control cable and the 2 screws.

    y. Install the instrument cluster finish panel.

    z. At the passenger's side, install the front pillar garnish, the 2 assist grips, the 4 screws and the 4 assist grip plugs.

    aa. At the driver's side, install the front pillar garnish, assist grip, the 2 screws and the 2 assist grip plugs.

    bb. Install the front door scuff plates, the cowl side trim and the front door opening trim.

13. Install the steering wheel by performing the following procedure:

    a. Install the steering wheel to the steering column.

    b. Align the steering wheel-to-main shaft marks.

    c. Install the steering wheel nut and torque to 25 ft. lbs. (34 Nm).

    d. Install the air bag module to the steering wheel and connect the electrical connector.

    e. Using a Torx® wrench, tighten the 2 screws located at each side of the steering wheel to 78 inch lbs. (8.8 Nm).

    f. Install the steering wheel side covers.

14. Connect the heater hoses to the heater core.

15. Refill the cooling system.

16. Connect the negative battery cable.

    a. Evacuate and charge the air conditioning system refrigerant.

17. Run the engine to normal operating temperatures; then, check the climate control operation and check for leaks.

REAR AUXILIARY HEATER

1. Disconnect the negative battery cable.

2. Drain the cooling system into a clean container for reuse.

3. Disconnect the heater hoses from the rear heater core.

4. Remove the front seats.

5. Remove the rear heater control assembly.

6. Remove the rear console box.

7. Remove the front console box cover.

8. Remove the lower center cluster finish panel.

9. Remove the front door scuff plates.

10. Remove the cowl side trim.

11. Remove the rear door scuff plates.

12. Remove the center pillar garnishes.

13. Slide the carpet rearward.

14. Remove the cooler bracket bolts and the bracket.

15. Remove the rear heater duct bolt/screw and the duct.

16. Disconnect the rear heater housing electrical connector.

17. Remove the 3 rear heater housing-to-chassis bolts and the heater housing.

18. Remove the heater core-to-heater housing 3 screws and 2 clamps.

19. Remove the heater core from the heater housing.

To install:

20. Install the heater core to the heater housing.

21. Install the heater core-to-heater housing 3 screws and 2 clamps.

22. Install the heater housing and the 3 rear heater housing-to-chassis bolts.

23. Connect the rear heater housing electrical connector.

24. Install the rear heater duct and the duct bolt/screw.

25. Install the cooler bracket and the bracket bolts.

26. Slide the carpet rearward.

27. Install the center pillar garnishes.

28. Install the rear door scuff plates.

29. Install the cowl side trim.

30. Install the front door scuff plates.

31. Install the lower center cluster finish panel.

32. Install the front console box cover.

33. Install the rear console box.

34. Install the rear heater control assembly.

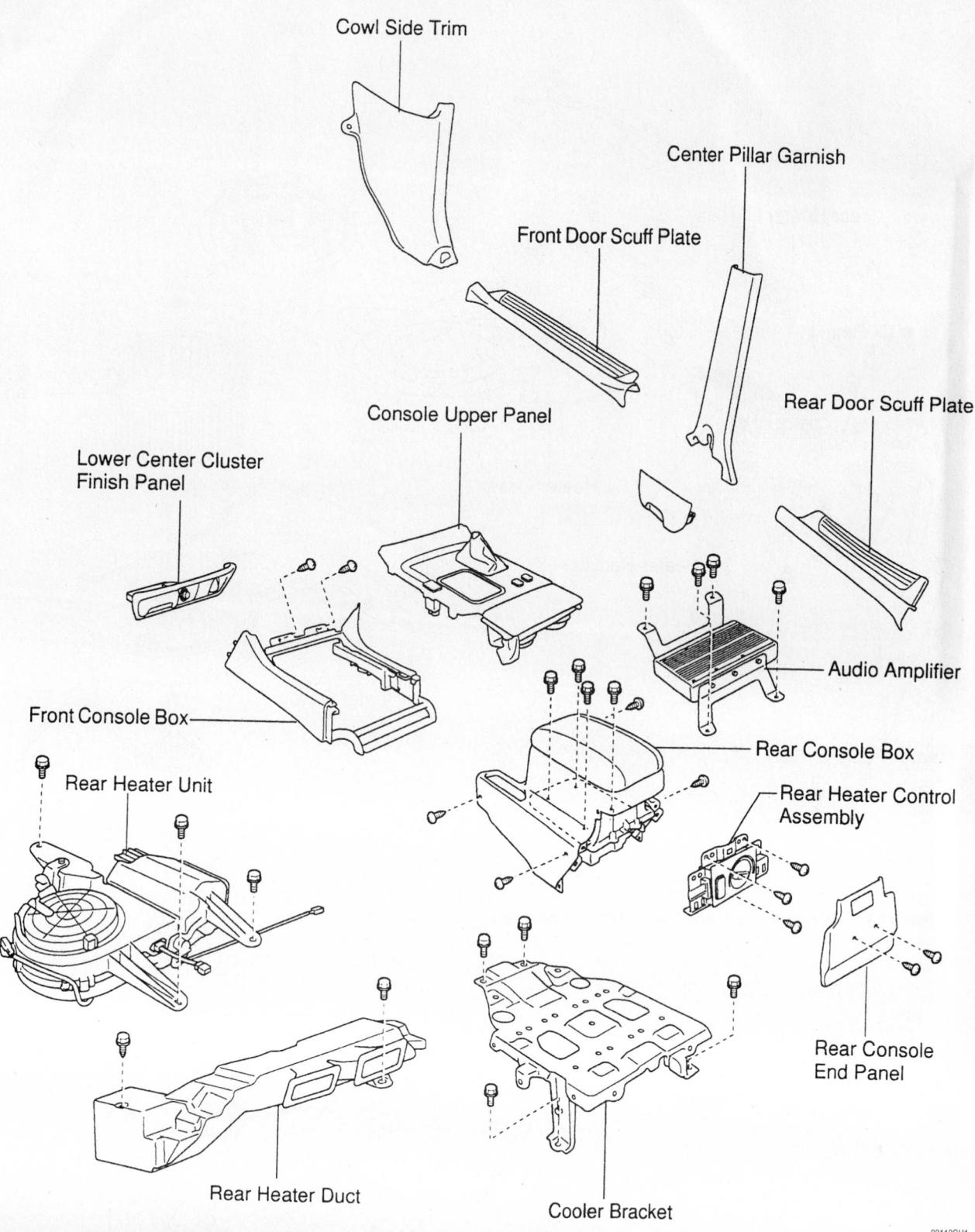

Cowl Side Trim

Center Pillar Garnish

Front Door Scuff Plate

Rear Door Scuff Plate

Console Upper Panel

Lower Center Cluster
Finish Panel

Audio Amplifier

Front Console Box

Rear Console Box

Rear Heater Unit

Rear Heater Control
Assembly

Rear Console
End Panel

Rear Heater Duct

Cooler Bracket

93113GH1

Exploded view of the rear heater housing and related components—Lexus LX 470

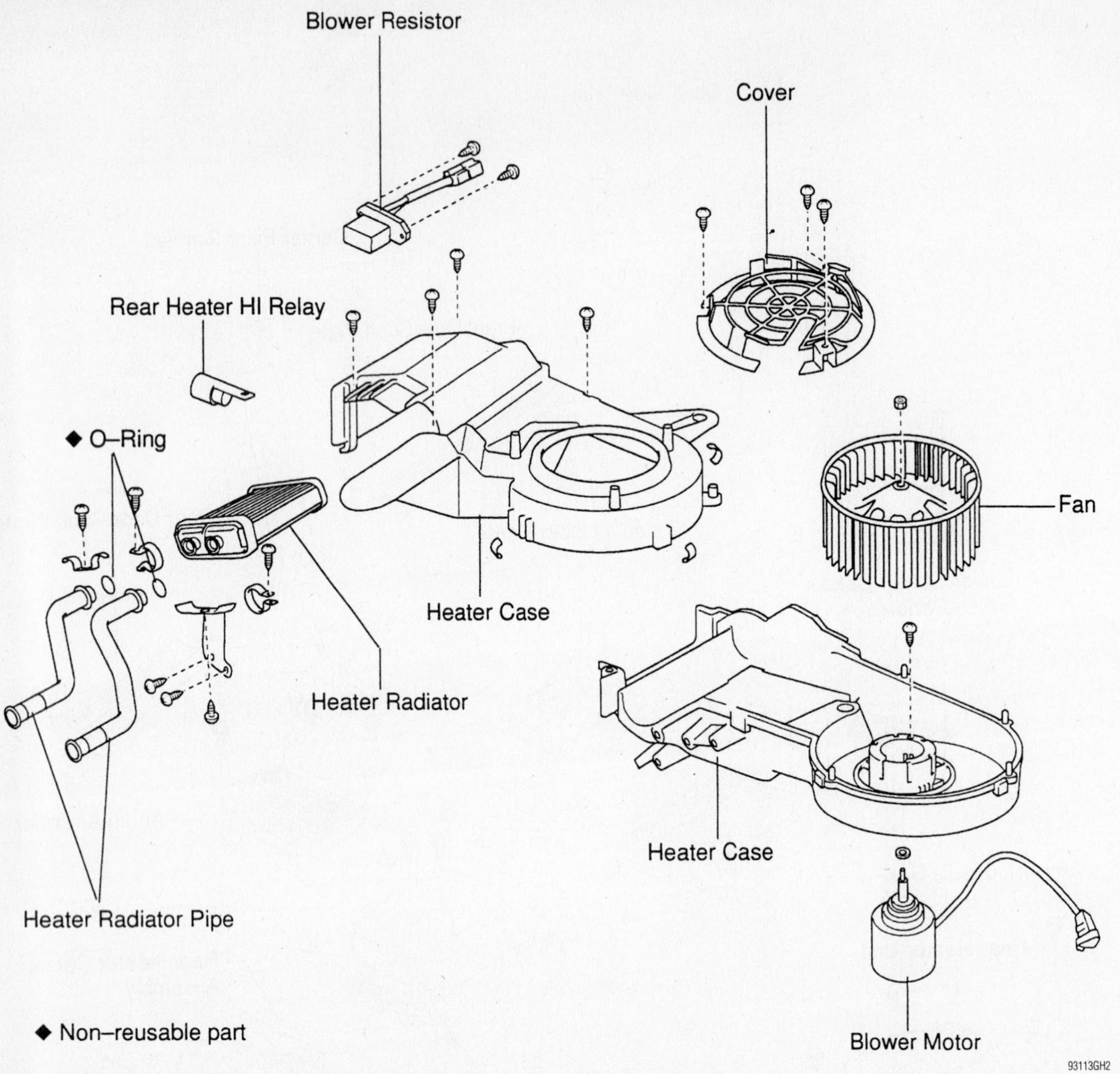

Blower Resistor

Cover

Rear Heater HI Relay

◆ O–Ring

Fan

Heater Case

Heater Radiator

Heater Case

Heater Radiator Pipe

Blower Motor

◆ Non–reusable part

93113GH2

Exploded view of the rear heater core, heater housing and related components—Lexus LX 470

35. Install the front seats.
36. Connect the heater hoses to the rear heater core.
37. Refill the cooling system.
38. Connect the negative battery cable.

Water Pump

REMOVAL & INSTALLATION

1. Before servicing the vehicle, refer to the precautions in the beginning of this section.
2. Drain the cooling system.
3. Remove or disconnect the following:
   • Negative battery cable

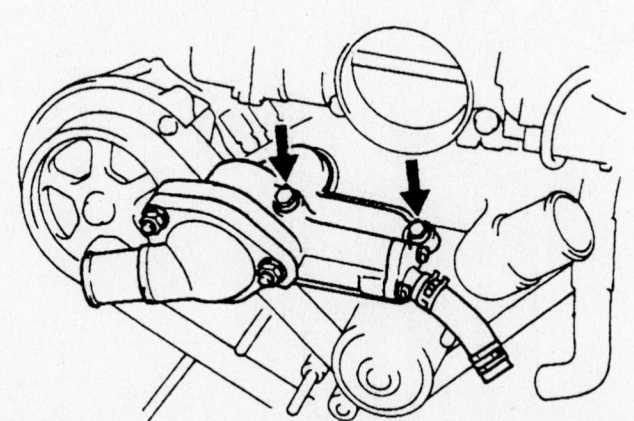

7924SG40

Water inlet housing attaching bolts

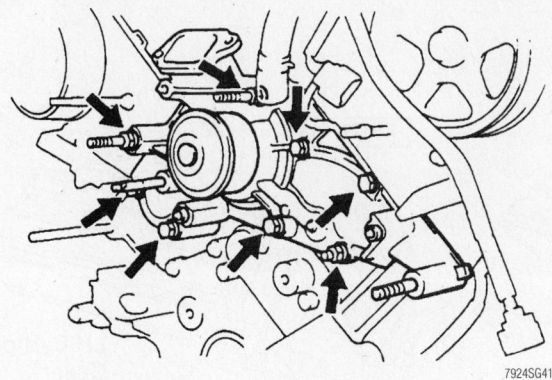

Water pump mounting bolts, stud bolts and nut locations

7924SG41

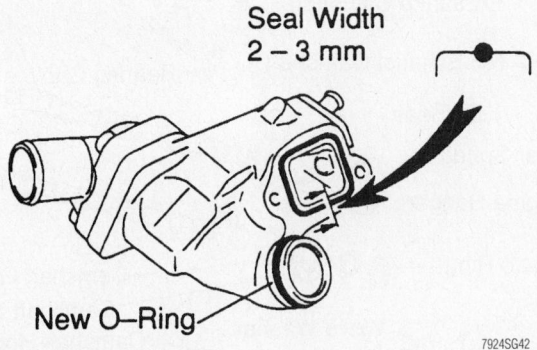

Seal Width
2 – 3 mm

New O–Ring

7924SG42

Water inlet housing sealant application

- Timing belt.
- No. 2 idler pulley
- Radiator hose
- Bypass hose
- Water inlet housing assembly
- Water pump

To install:

4. Install or connect the following:
- Water pump. Use a new gasket and tighten the bolts to 15 ft. lbs. (21 Nm). Tighten the stud bolt and nut to 13 ft. lbs. (18 Nm).
- Water inlet housing assembly. Use a new O-ring and apply sealant as shown. Tighten the bolts to 13 ft. lbs. (18 Nm).
- Bypass hose
- Radiator hose
- No. 2 idler pulley
- Timing belt
- Negative battery cable

5. Fill the cooling system.
6. Start the engine and check for leaks.

Cylinder Head

REMOVAL & INSTALLATION

1. Before servicing the vehicle, refer to the precautions in the beginning of this section.

2. Drain the cooling system.
3. Relieve the fuel system pressure.
4. Remove or disconnect the following:
- Battery and tray
- Engine appearance cover
- Engine under covers
- Air intake assembly
- Accessory drive belt
- A/C compressor and bracket
- Cooling fan and bracket
- Radiator
- Idler pulley
- Front covers
- Timing belt.
- Camshaft sprockets
- Camshaft Position (CMP) sensor
- Power steering pump
- Exhaust front pipes
- Transmission dipstick tube
- Ignition coils
- Rear timing belt covers
- Fuel lines
- Intake manifold
- Water inlet housing assembly
- Front and rear water bypass joints
- Engine lifting eyes
- Oil dipstick tube
- Valve covers
- Camshafts
- Cylinder heads with the exhaust

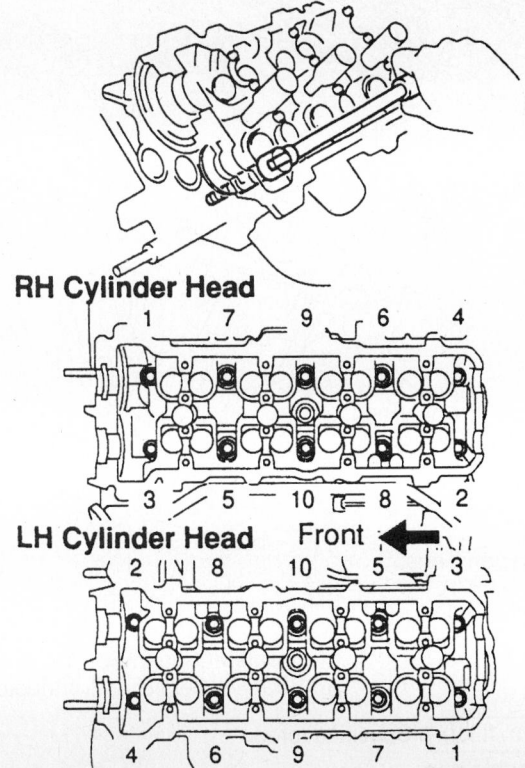

RH Cylinder Head

LH Cylinder Head   Front

7924SG43

Cylinder head loosening sequence

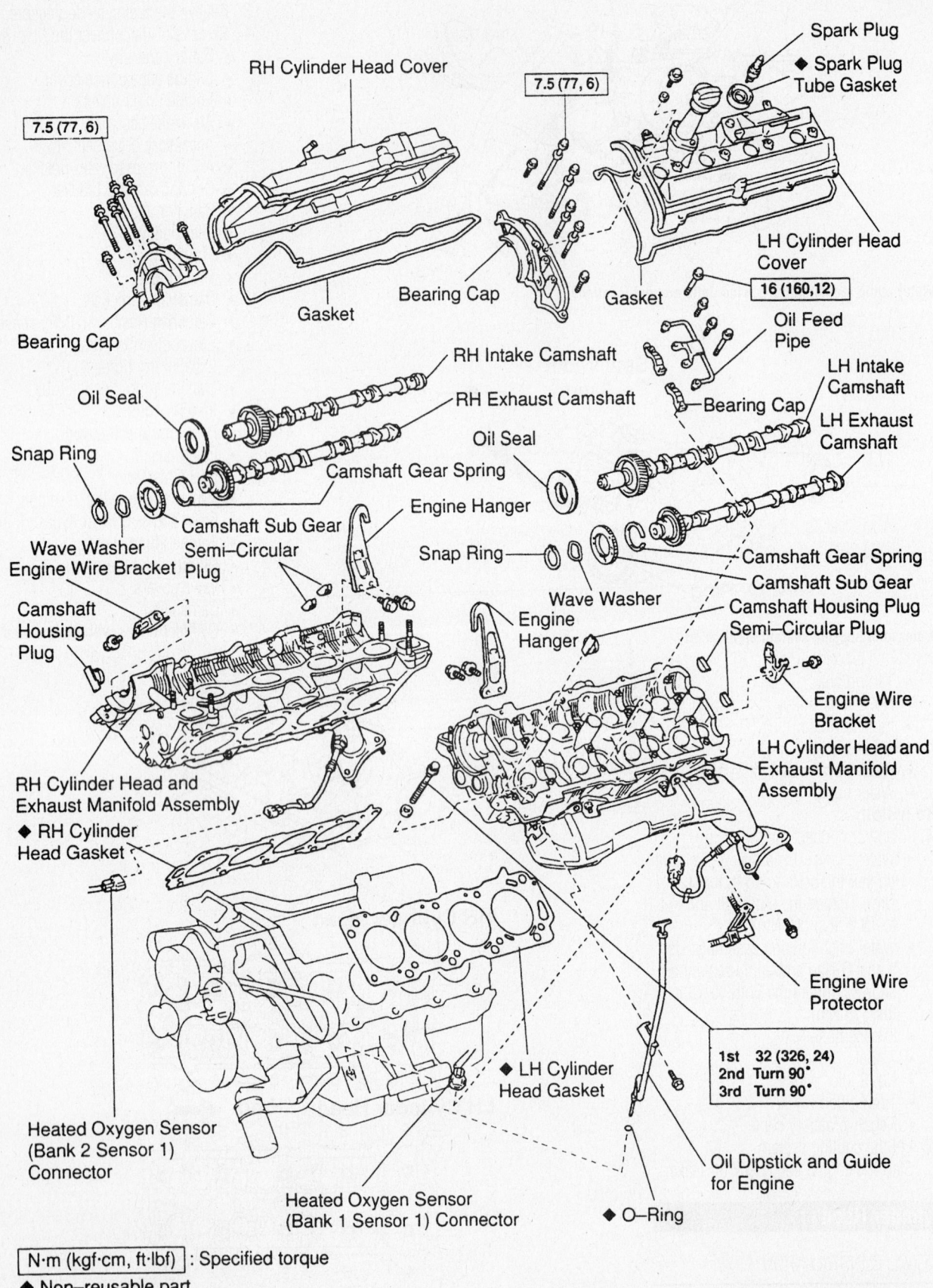

7.5 (77, 6)

RH Cylinder Head Cover

7.5 (77, 6)

Spark Plug

◆ Spark Plug Tube Gasket

7.5 (77, 6)

Bearing Cap

Gasket

LH Cylinder Head Cover

16 (160,12)

Bearing Cap

Gasket

Oil Feed Pipe

Bearing Cap

Oil Seal

Snap Ring

RH Intake Camshaft

RH Exhaust Camshaft

Oil Seal

Camshaft Gear Spring

Engine Hanger

LH Intake Camshaft

LH Exhaust Camshaft

Wave Washer
Engine Wire Bracket

Camshaft Sub Gear
Semi–Circular Plug

Snap Ring

Wave Washer

Engine Hanger

Camshaft Gear Spring

Camshaft Sub Gear

Camshaft Housing Plug
Semi–Circular Plug

Camshaft Housing Plug

Engine Wire Bracket

RH Cylinder Head and
Exhaust Manifold Assembly

LH Cylinder Head and
Exhaust Manifold Assembly

◆ RH Cylinder Head Gasket

◆ LH Cylinder Head Gasket

Engine Wire Protector

1st   32 (326, 24)
2nd Turn 90°
3rd   Turn 90°

Heated Oxygen Sensor
(Bank 2 Sensor 1)
Connector

Heated Oxygen Sensor
(Bank 1 Sensor 1) Connector

◆ O–Ring

Oil Dipstick and Guide
for Engine

N·m (kgf·cm, ft·lbf) : Specified torque

◆ Non–reusable part

7924SG49

Exploded view of the cylinder head mounting—4.7L LX470

RH Cylinder Head

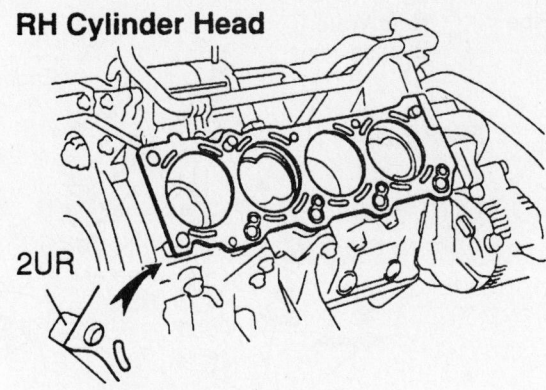

2UR

LH Cylinder Head

2UL

7924SG47

Cylinder head gasket identification

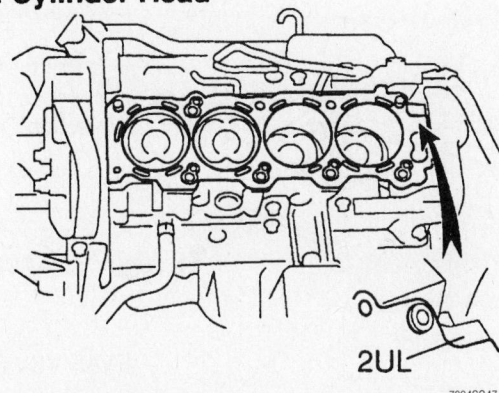

RH Cylinder Head

LH Cylinder Head   Front

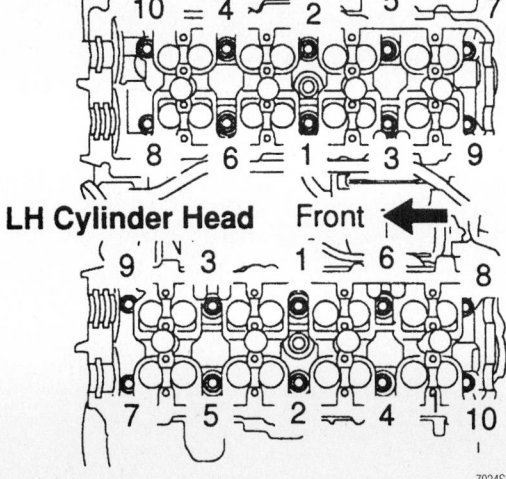

7924SG46

Cylinder head torque sequence

manifolds attached. Loosen the bolts in the sequence shown.

To install:

5. Install the cylinder heads with new gaskets. Tighten the bolts in sequence as follows:
    a. Step 1: 24 ft. lbs. (32 Nm)
    b. Step 2: Plus 180 degrees

6. Install or connect the following:
    - Camshafts
    - Valve covers
    - Oil dipstick tube
    - Engine lifting eyes
    - Front and rear water bypass joints
    - Water inlet housing assembly
    - Intake manifold
    - Fuel lines
    - Rear timing belt covers
    - Ignition coils
    - Transmission dipstick tube
    - Exhaust front pipes
    - Power steering pump
    - CMP sensor
    - Camshaft sprockets
    - Timing belt
    - Front covers
    - Idler pulley
    - Radiator
    - Cooling fan and bracket
    - A/C compressor and bracket
    - Accessory drive belt
    - Air intake assembly
    - Engine under covers
    - Engine appearance cover
    - Battery and tray

7. Fill the cooling system.

8. Start the engine and check for leaks.

Intake Manifold

REMOVAL & INSTALLATION

1. Before servicing the vehicle, refer to the precautions in the beginning of this section.

2. Drain the cooling system.

3. Relieve the fuel system pressure.

4. Remove or disconnect the following:
    - Negative battery cable
    - Engine appearance cover
    - Accelerator cable
    - Throttle Position (TP) sensor connector
    - Accelerator pedal position sensor
    - Throttle motor connector
    - Evaporative Emissions (EVAP) vacuum switching valve connector
    - Fuel injector connectors
    - Engine Coolant Temperature (ECT) sensor connector
    - ETC gauge sender connector

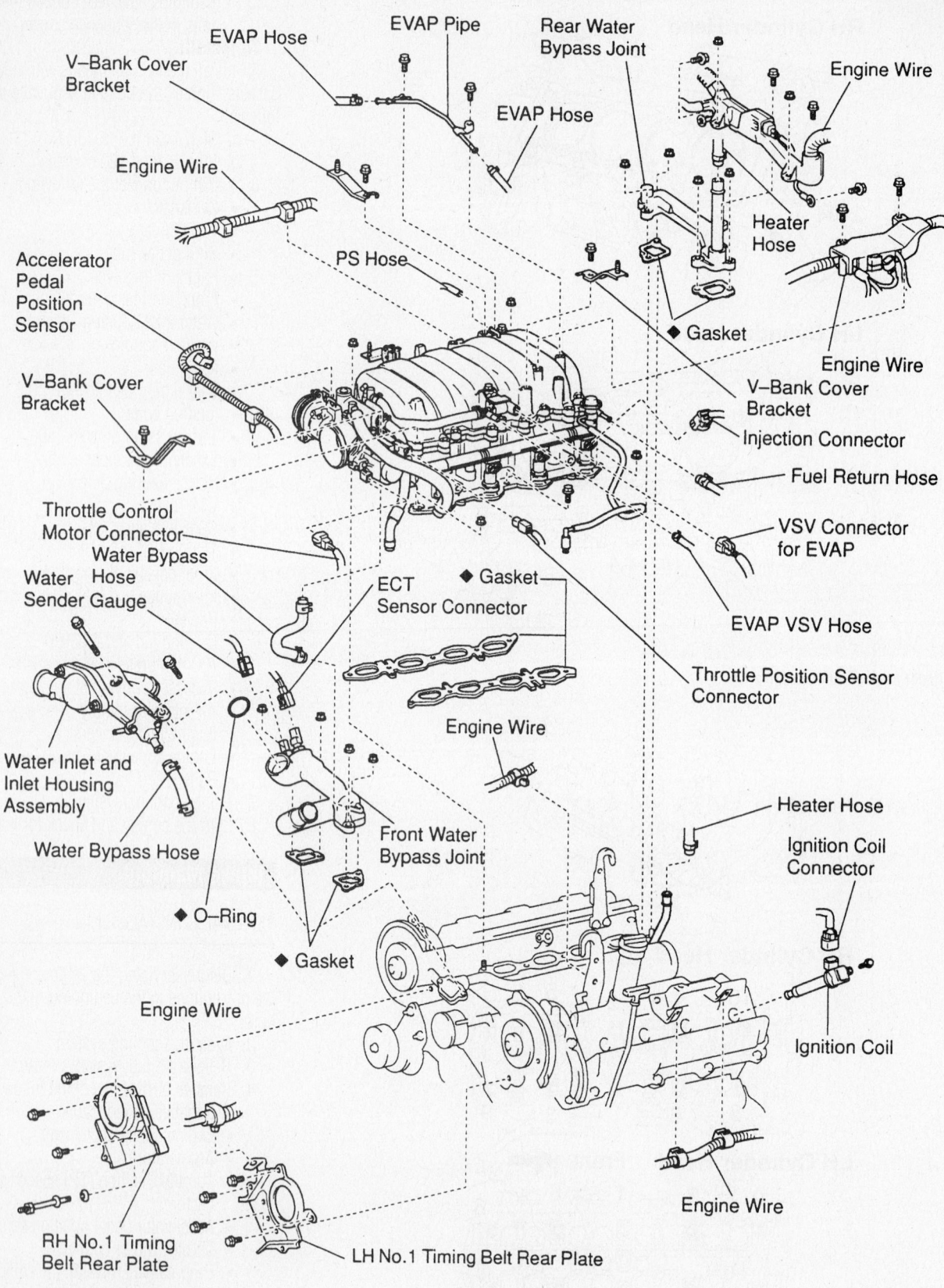

EVAP Hose

EVAP Pipe

Rear Water Bypass Joint

Engine Wire

V–Bank Cover Bracket

EVAP Hose

Engine Wire

Heater Hose

Accelerator Pedal Position Sensor

PS Hose

◆ Gasket

Engine Wire

V–Bank Cover Bracket

Injection Connector

V–Bank Cover Bracket

Fuel Return Hose

Throttle Control Motor Connector

Water Bypass Hose

VSV Connector for EVAP

Water Sender Gauge

ECT Sensor Connector

◆ Gasket

EVAP VSV Hose

Water Inlet and Inlet Housing Assembly

Throttle Position Sensor Connector

Water Bypass Hose

Engine Wire

◆ O–Ring

Front Water Bypass Joint

Heater Hose

Ignition Coil Connector

◆ Gasket

Engine Wire

Ignition Coil

Engine Wire

RH No.1 Timing Belt Rear Plate

LH No.1 Timing Belt Rear Plate

◆ Non–reusable part

Exploded of the intake manifold mounting

7924SG50

- Heated Oxygen (HO2S) sensor connectors
- Fuel pressure regulator vacuum hose
- Positive Crankcase Ventilation (PCV) valve and hose
- EVAP hoses
- Power steering vacuum hoses
- Water bypass hose
- Engine control wiring harness clamps
- Cylinder head ground cables
- Intake manifold wire harness protector
- EVAP pipe
- Engine appearance cover brackets
- Intake manifold

To install:
5. Install or connect the following:
- Intake manifold. Tighten the fasteners to 13 ft. lbs. (18 Nm).
- Engine appearance cover brackets
- EVAP pipe
- Intake manifold wire harness protector
- Cylinder head ground cables
- Engine control wiring harness clamps
- Water bypass hose
- Power steering vacuum hoses
- EVAP hoses
- PCV valve and hose
- Fuel pressure regulator vacuum hose
- HO2S sensor connectors
- ETC gauge sender connector
- ECT sensor connector
- Fuel injector connectors
- EVAP vacuum switching valve connector
- Throttle motor connector
- Accelerator pedal position sensor
- TP sensor connector
- Accelerator cable
- Engine appearance cover
- Negative battery cable
6. Fill the cooling system.
7. Start the engine and check for leaks.

Exhaust Manifold

REMOVAL & INSTALLATION

1. Before servicing the vehicle, refer to the precautions in the beginning of this section.
2. Attach a hoist to the engine lifting eyes.
3. Remove or disconnect the following:
- Negative battery cable
- Heated Oxygen (HO2S) sensor connectors

- Exhaust manifold heat shield
- Exhaust front pipe
- Motor mount
- Motor mount bracket
- Exhaust manifold

To install:

➡**Use new exhaust manifold nuts for assembly.**

4. Install or connect the following:
- Exhaust manifold. Tighten the nuts to 32 ft. lbs. (44 Nm).
- Motor mount bracket. Tighten the bolts to 27 ft. lbs. (36 Nm).
- Motor mount. Tighten the fasteners to 22 ft. lbs. (30 Nm).
- Exhaust front pipe. Tighten the nuts to 46 ft. lbs. (62 Nm).
- Exhaust manifold heat shield
- HO2S sensor connectors
- Negative battery cable
5. Start the engine and check for leaks.

Front Crankshaft Seal

REMOVAL & INSTALLATION

1. Before servicing the vehicle, refer to the precautions in the beginning of this section.
2. Drain the cooling system.
3. Remove or disconnect the following:
- Negative battery cable
- Engine under cover
- Engine appearance cover
- Air intake assembly
- Accessory drive belt
- Cooling fan and pulley
- Radiator
- Drive belt idler pulley
- Camshaft Position (CMP) sensor connector
- Upper timing covers
- Oil cooler pipe
- Center timing cover
- A/C compressor
- Cooling fan bracket
- Crankshaft pulley
- Lower timing cover
- Timing belt.
- Crankshaft timing sprocket
- Front crankshaft seal

To install:
4. Install the oil seal so that it is flush with the oil pump housing.
5. Install or connect the following:
- Crankshaft timing sprocket
- Timing belt
- Lower timing cover
- Crankshaft pulley. Tighten the bolt to 181 ft. lbs. (245 Nm).

- Cooling fan bracket. Tighten the 12mm bolts to 12 ft. lbs. (16 Nm) and the 14mm bolts to 24 ft. lbs. (32 Nm).
- A/C compressor
- Center timing cover
- Oil cooler pipe
- Upper timing covers
- CMP sensor connector
- Drive belt idler pulley. Tighten the bolt to 27 ft. lbs. (37 Nm).
- Radiator
- Cooling fan and pulley. Tighten the nuts to 16 ft. lbs. (21 Nm).
- Accessory drive belt
- Air intake assembly
- Engine appearance cover
- Engine under cover
- Negative battery cable
6. Fill the cooling system.
7. Start the engine and check for leaks.

Camshaft and Valve Lifters

REMOVAL & INSTALLATION

1. Before servicing the vehicle, refer to the precautions in the beginning of this section.
2. Drain the cooling system.
3. Relieve the fuel system pressure.
4. Remove or disconnect the following:
- Negative battery cable
- Engine under covers
- Engine appearance cover
- Air intake hose
- Accessory drive belt
- Cooling fan
- Radiator
- Idler pulley
- Upper and middle timing belt covers
- A/C compressor
- Cooling fan bracket
- Alternator
- Accessory drive belt tensioner
5. Set the engine to Top Dead Center (TDC) with the camshaft sprocket timing marks aligned with the rear cover timing marks.
6. Rotate the crankshaft to 50 degrees After TDC as shown. The crankshaft pulley timing mark should align with the center of the No. 2 idler pulley bolt.
7. Remove or disconnect the following:
- Crankshaft pulley
- Lower timing cover
- Timing belt.
- Camshaft timing sprockets
- Camshaft Position (CMP) sensor

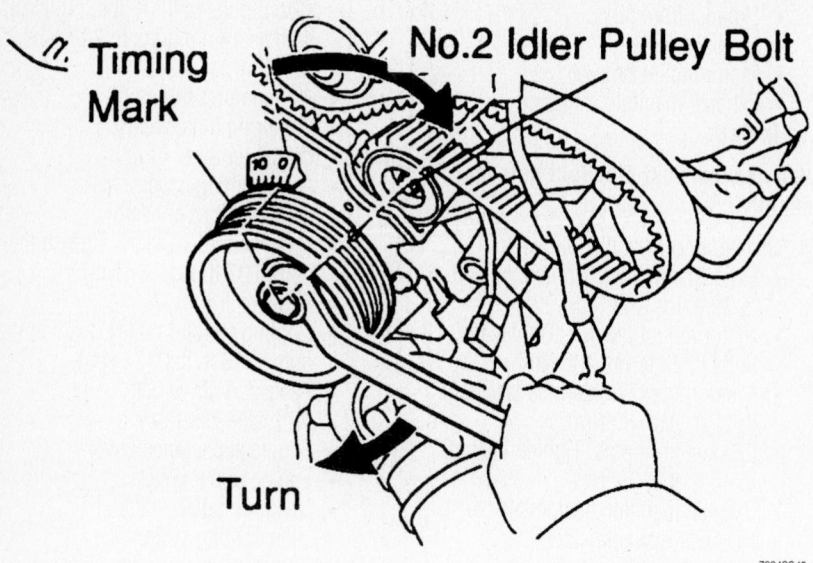

Setting the crankshaft to 50 degrees ATDC

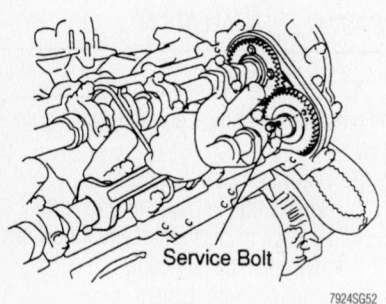

Camshaft service bolt installation

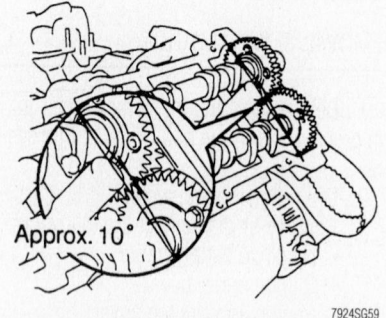

Right bank camshaft timing mark (1 dot marks) alignment

- Ignition coils
- Valve cover
- Timing belt rear covers

8. Rotate the right bank camshafts as necessary to access the exhaust camshaft sub-gear service bolt hole and install a 6mm x 1.0mm bolt.

➡**Keep all valvetrain components in order for assembly.**

9. Align the right bank camshaft 1 dot timing marks to a **10** degree angle as shown.

10. Loosen the bearing cap bolts in sequence and in several passes.

11. Remove the right bank camshafts.

12. Rotate the left bank camshafts as necessary to access the exhaust camshaft sub-gear service bolt hole and install a 6mm x 1.0mm bolt.

13. Align the left bank camshaft 2 dot timing marks as shown.

14. Loosen the bearing cap bolts in sequence and in several passes.

15. Remove the left bank camshafts.

16. Remove the valve lifters and shims.

To install:

17. Ensure that the crankshaft is at 50 degrees After TDC.

18. Install or connect the following:
- Valve lifters and shims in their original positions
- Right bank camshafts with the 1 dot timing marks at 10 degrees
- Left bank camshafts with the 2 dot timing marks aligned
- Left and right bank camshaft bearing caps in their original positions. Apply sealant to the front bearing caps as shown.
- Camshaft oil seals

19. The bearing cap bolts vary in length and are identified as follows:
- A: 3.70 inches (94mm)
- B: 2.83 inches (72mm)
- C: 0.98 inches (25mm)

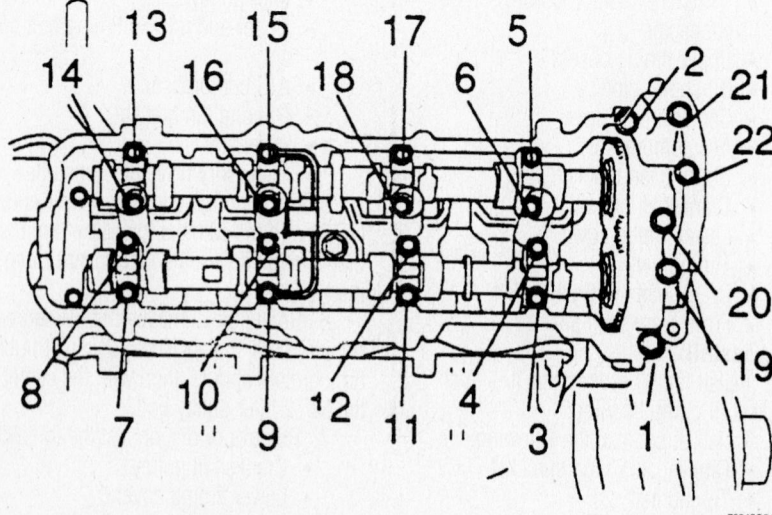

Right bank camshaft bearing cap loosening sequence

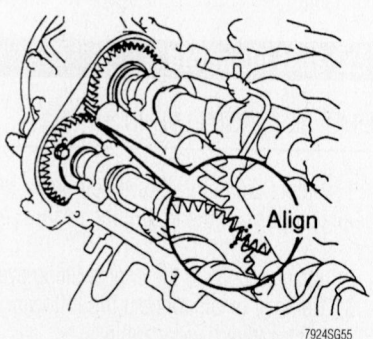

Left bank camshaft timing mark (2 dot marks) alignment

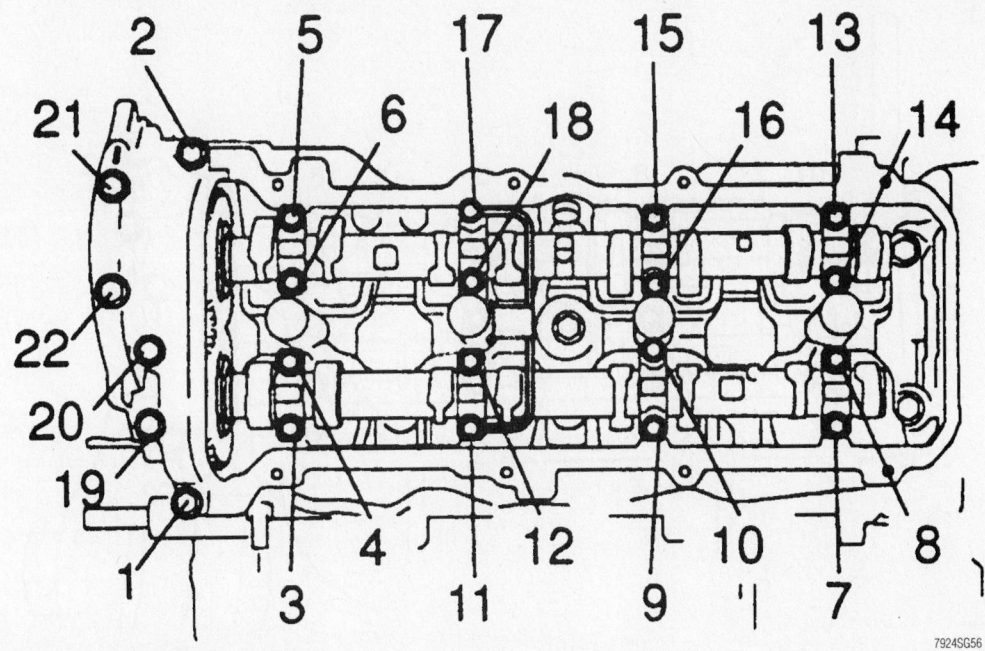

7924SG56

Left bank camshaft bearing cap loosening sequence

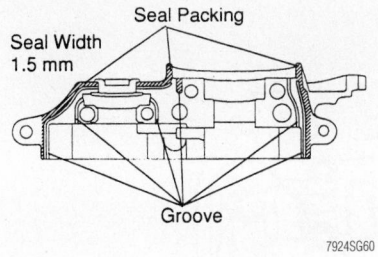

7924SG60

Apply a 1.5mm bead of sealant to the front bearing caps

- D: 2.05 inches (52mm)
- E: 1.50 inches (38mm)

20. Bolts in positions **A**, **B** and **C** are installed dry.

21. Lubricate the threads and under the contact flange for bolts in positions **D** and **E**.

22. Install oil feed pipes and the bearing cap bolts according to position in the illustrations.

23. Tighten the camshaft bearing bolts in sequence and in several passes to the following specifications:
- Bolt C: 66 inch lbs. (7.5 Nm)
- All others: 12 ft. lbs. (16 Nm)

24. Remove the service bolts from the exhaust camshaft gears.

25. Install or connect the following:
- Timing belt rear covers
- Valve cover
- Ignition coils
- CMP sensor

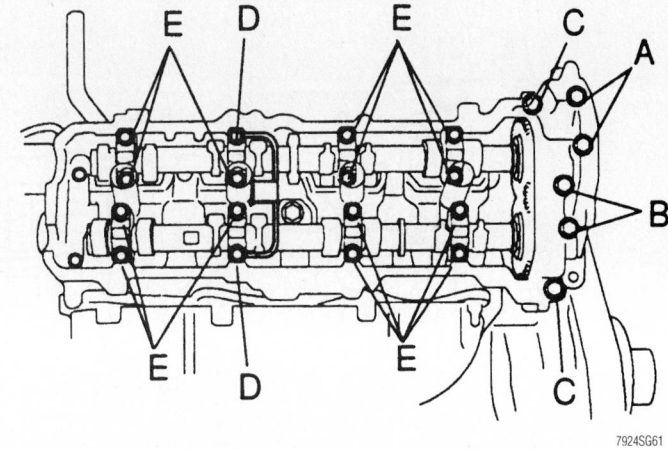

7924SG61

Right bank bearing cap bolt location

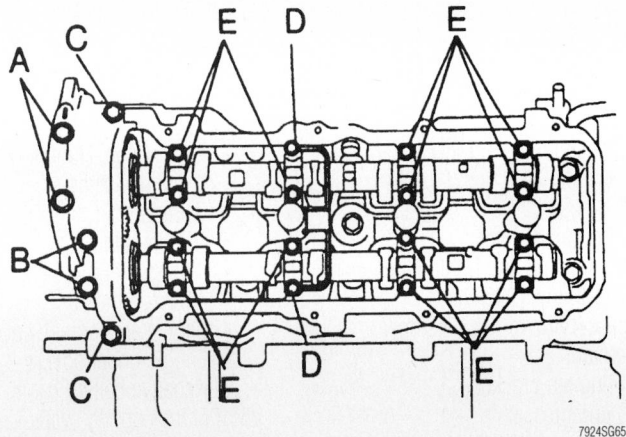

7924SG65

Left camshaft bearing cap bolt locations

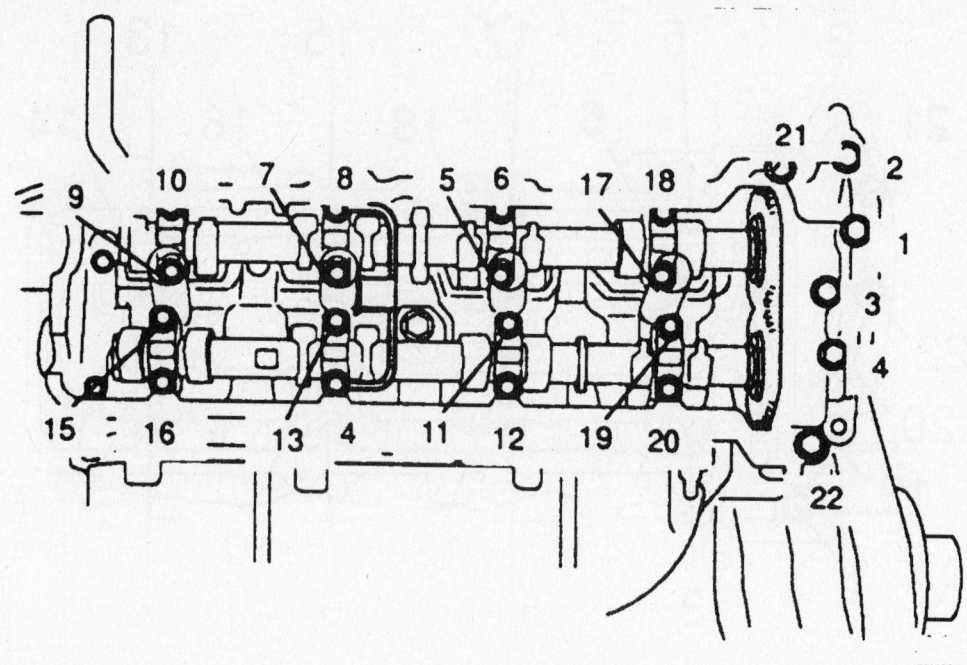

Right bank camshaft bearing cap bolt torque sequence

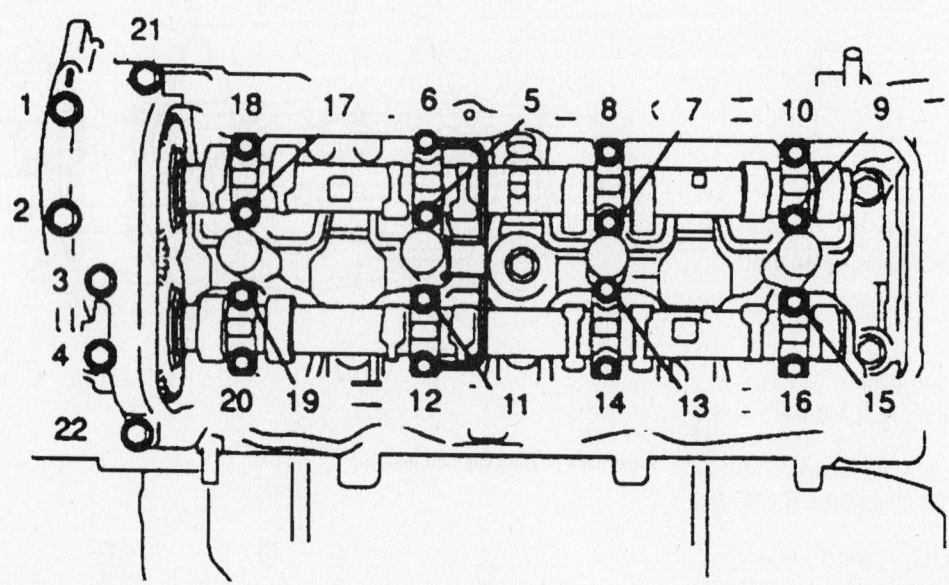

Left bank camshaft bearing cap bolt torque sequence

- Camshaft timing sprockets. Tighten the bolts to 80 ft. lbs. (108 Nm).
- Timing belt
- Lower timing cover
- Crankshaft pulley. Tighten the bolt to 181 ft. lbs. (245 Nm).
- Accessory drive belt tensioner
- Alternator
- Cooling fan bracket
- A/C compressor
- Upper and middle timing belt covers

- Idler pulley. Tighten the bolt to 27 ft. lbs. (37 Nm).
- Radiator
- Cooling fan
- Accessory drive belt
- Air intake hose
- Engine appearance cover
- Engine under covers
- Negative battery cable
26. Fill the cooling system.
27. Start the engine and check for leaks.

Valve Lash

ADJUSTMENT

➡**Measure valve clearance with the engine cold.**

1. Before servicing the vehicle, refer to the precautions in the beginning of this section.

2. Drain the cooling system.

Intake valve clearance shim selection chart

Intake valve clearance (Cold):
0.15 – 0.25 mm (0.006 – 0.010 in.)

EXAMPLE:
The 2.300 mm (0.0906 in.) shim is installed, and the measured clearance is 0.440 mm (0.0173 in.). Replace the 2.300 mm (0.0906 in.) shim with a No. 54 shim.

New shim thickness
mm (in.)

Shim No.	Thickness	Shim No.	Thickness	Shim No.	Thickness
00	2.000 (0.0787)	28	2.280 (0.0898)	56	2.560 (0.1008)
02	2.020 (0.0795)	30	2.300 (0.0906)	58	2.580 (0.1016)
04	2.040 (0.0803)	32	2.320 (0.0913)	60	2.600 (0.1024)
06	2.060 (0.0811)	34	2.340 (0.0921)	62	2.620 (0.1031)
08	2.080 (0.0819)	36	2.360 (0.0929)	64	2.640 (0.1039)
10	2.100 (0.0827)	38	2.380 (0.0937)	66	2.660 (0.1047)
12	2.120 (0.0835)	40	2.400 (0.0945)	68	2.680 (0.1055)
14	2.140 (0.0843)	42	2.420 (0.0953)	70	2.700 (0.1063)
16	2.160 (0.0850)	44	2.440 (0.0961)	72	2.720 (0.1071)
18	2.180 (0.0858)	46	2.460 (0.0969)	74	2.740 (0.1079)
20	2.200 (0.0866)	48	2.480 (0.0976)	76	2.760 (0.1087)
22	2.220 (0.0874)	50	2.500 (0.0984)	78	2.780 (0.1094)
24	2.240 (0.0882)	52	2.520 (0.0992)	80	2.800 (0.1102)
26	2.260 (0.0890)	54	2.540 (0.1000)		

Shim selection matrix (installed shim thickness across top, measured clearance down the left side). The body of the chart lists the new shim number to install for each intersection.

Installed shim thickness mm (in.): 2.000 (0.0787) through 2.800 (0.1102)

Measured clearance mm (in.):

Measured clearance mm (in.)
0.000 – 0.030 (0.0000 – 0.0012)
0.031 – 0.050 (0.0012 – 0.0020)
0.051 – 0.070 (0.0020 – 0.0028)
0.071 – 0.090 (0.0028 – 0.0035)
0.091 – 0.110 (0.0036 – 0.0043)
0.111 – 0.130 (0.0044 – 0.0051)
0.131 – 0.149 (0.0052 – 0.0059)
0.150 – 0.250 (0.0059 – 0.0098)
0.251 – 0.270 (0.0099 – 0.0106)
0.271 – 0.290 (0.0107 – 0.0114)
0.291 – 0.310 (0.0115 – 0.0122)
0.311 – 0.330 (0.0122 – 0.0130)
0.331 – 0.350 (0.0130 – 0.0138)
0.351 – 0.370 (0.0138 – 0.0146)
0.371 – 0.390 (0.0146 – 0.0154)
0.391 – 0.410 (0.0154 – 0.0161)
0.411 – 0.430 (0.0162 – 0.0169)
0.431 – 0.450 (0.0170 – 0.0177)
0.451 – 0.470 (0.0178 – 0.0185)
0.471 – 0.490 (0.0185 – 0.0193)
0.491 – 0.510 (0.0193 – 0.0201)
0.511 – 0.530 (0.0201 – 0.0209)
0.531 – 0.550 (0.0209 – 0.0217)
0.551 – 0.570 (0.0217 – 0.0224)
0.571 – 0.590 (0.0225 – 0.0232)
0.591 – 0.610 (0.0233 – 0.0240)
0.611 – 0.630 (0.0241 – 0.0248)
0.631 – 0.650 (0.0248 – 0.0256)
0.651 – 0.670 (0.0256 – 0.0264)
0.671 – 0.690 (0.0264 – 0.0272)
0.691 – 0.710 (0.0272 – 0.0280)
0.711 – 0.730 (0.0280 – 0.0287)
0.731 – 0.750 (0.0288 – 0.0295)
0.751 – 0.770 (0.0296 – 0.0303)
0.771 – 0.790 (0.0304 – 0.0311)
0.791 – 0.810 (0.0311 – 0.0319)
0.811 – 0.830 (0.0319 – 0.0327)
0.831 – 0.850 (0.0327 – 0.0335)
0.851 – 0.870 (0.0335 – 0.0343)
0.871 – 0.890 (0.0343 – 0.0350)
0.891 – 0.910 (0.0351 – 0.0358)
0.911 – 0.930 (0.0359 – 0.0366)
0.931 – 0.950 (0.0367 – 0.0374)
0.951 – 0.970 (0.0374 – 0.0382)
0.971 – 0.990 (0.0382 – 0.0390)
0.991 – 1.010 (0.0390 – 0.0398)
1.011 – 1.030 (0.0398 – 0.0406)
1.031 – 1.050 (0.0406 – 0.0413)

7924SG71

Exhaust valve clearance shim selection chart

New shim thickness — mm (in.)

Shim No.	Thickness	Shim No.	Thickness	Shim No.	Thickness
00	2.000 (0.0787)	28	2.280 (0.0898)	56	2.560 (0.1008)
02	2.020 (0.0795)	30	2.300 (0.0906)	58	2.580 (0.1016)
04	2.040 (0.0803)	32	2.320 (0.0913)	60	2.600 (0.1024)
06	2.060 (0.0811)	34	2.340 (0.0921)	62	2.620 (0.1031)
08	2.080 (0.0819)	36	2.360 (0.0929)	64	2.640 (0.1039)
10	2.100 (0.0827)	38	2.380 (0.0937)	66	2.660 (0.1047)
12	2.120 (0.0835)	40	2.400 (0.0945)	68	2.680 (0.1055)
14	2.140 (0.0843)	42	2.420 (0.0953)	70	2.700 (0.1063)
16	2.160 (0.0850)	44	2.440 (0.0961)	72	2.720 (0.1071)
18	2.180 (0.0858)	46	2.460 (0.0969)	74	2.740 (0.1079)
20	2.200 (0.0866)	48	2.480 (0.0976)	76	2.760 (0.1087)
22	2.220 (0.0874)	50	2.500 (0.0984)	78	2.780 (0.1094)
24	2.240 (0.0882)	52	2.520 (0.0992)	80	2.800 (0.1102)
28	2.260 (0.0890)	54	2.540 (0.1000)		

Exhaust valve clearance (Cold):
0.25 – 0.35 mm (0.010 – 0.014 in.)

EXAMPLE:

The 2.300 mm (0.0906 in.) shim is installed, and the measured clearance is 0.440 mm (0.0173 in.). Replace the 2.300 mm (0.0906 in.) shim with a No. 44 shim.

Installed shim thickness mm (in.): columns from 2.000 (0.0787) through 2.800 (0.1102) in 0.020 mm (approx.) increments.

Measured clearance mm (in.):

Measured clearance mm (in.)
0.000–0.030 (0.0000–0.0012)
0.031–0.050 (0.0012–0.0020)
0.051–0.070 (0.0020–0.0028)
0.071–0.090 (0.0028–0.0035)
0.091–0.110 (0.0036–0.0043)
0.111–0.130 (0.0044–0.0051)
0.131–0.150 (0.0052–0.0059)
0.151–0.170 (0.0059–0.0067)
0.171–0.190 (0.0067–0.0075)
0.191–0.210 (0.0075–0.0083)
0.211–0.230 (0.0083–0.0091)
0.231–0.249 (0.0091–0.0098)
0.250–0.350 (0.0098–0.0138)
0.351–0.370 (0.0138–0.0146)
0.371–0.390 (0.0146–0.0154)
0.391–0.410 (0.0154–0.0161)
0.411–0.430 (0.0162–0.0169)
0.431–0.450 (0.0170–0.0177)
0.451–0.470 (0.0178–0.0185)
0.471–0.490 (0.0185–0.0193)
0.491–0.510 (0.0193–0.0201)
0.511–0.530 (0.0201–0.0209)
0.531–0.550 (0.0209–0.0217)
0.551–0.570 (0.0217–0.0224)
0.571–0.590 (0.0225–0.0232)
0.591–0.610 (0.0233–0.0240)
0.611–0.630 (0.0241–0.0248)
0.631–0.650 (0.0248–0.0256)
0.651–0.670 (0.0256–0.0264)
0.671–0.690 (0.0264–0.0272)
0.691–0.710 (0.0272–0.0280)
0.711–0.730 (0.0280–0.0287)
0.731–0.750 (0.0288–0.0295)
0.751–0.770 (0.0296–0.0303)
0.771–0.790 (0.0304–0.0311)
0.791–0.810 (0.0311–0.0319)
0.811–0.830 (0.0319–0.0327)
0.831–0.850 (0.0327–0.0335)
0.851–0.870 (0.0335–0.0343)
0.871–0.890 (0.0343–0.0350)
0.891–0.910 (0.0351–0.0358)
0.911–0.930 (0.0359–0.0366)
0.931–0.950 (0.0367–0.0374)
0.951–0.970 (0.0374–0.0382)
0.971–0.990 (0.0382–0.0390)
0.991–1.010 (0.0390–0.0398)
1.011–1.030 (0.0398–0.0406)
1.031–1.050 (0.0406–0.0413)
1.051–1.070 (0.0414–0.0421)
1.071–1.090 (0.0422–0.0429)
1.091–1.110 (0.0430–0.0437)
1.111–1.130 (0.0437–0.0445)
1.131–1.150 (0.0445–0.0453)

79245G72

3. Remove or disconnect the following:
   - Negative battery cable
   - Ignition coils
   - Valve covers
4. Set the engine to the top of the compression stroke with the valves closed for the cylinder to be measured.
5. Check the valve clearance. The valve clearance specifications are as follows:
   - Intake: 0.006–0.010 in. (0.15–0.25mm)
   - Exhaust: 0.010–0.014 in. (0.25–0.35mm)
6. Record the measurements for each valve.
7. When all valve clearances have been measured, remove the camshafts.
8. Remove the valve shims and measure them. Note this measurement along with the clearance measurement recorded earlier.
9. Using the valve clearance and shim thickness measurements, find replacement shims in the Adjusting Shim Selection charts.
10. Install or connect the following:
    - Replacement valve shims
    - Camshafts
    - Valve covers
    - Ignition coils
    - Negative battery cable
11. Fill the cooling system.
12. Start the engine and check for leaks.

Starter Motor

REMOVAL & INSTALLATION

1. Before servicing the vehicle, refer to the precautions in the beginning of this section.
2. Drain the cooling system.
3. Relieve the fuel system pressure.
4. Remove or disconnect the following:
   - Negative battery cable
   - Engine appearance cover
   - Air intake tube
   - Intake manifold
   - Starter motor mounting bolts
   - Starter wiring connectors
   - Starter motor

To install:

5. Install or connect the following:
   - Starter motor
   - Starter wiring connectors. Tighten the cable nut to 86 inch lbs. (10 Nm).
   - Starter motor mounting bolts. Tighten the bolts to 29 ft. lbs. (39 Nm).
   - Intake manifold
   - Air intake tube
   - Engine appearance cover
   - Negative battery cable
6. Fill the cooling system.
7. Start the engine and check for leaks.

Timing Belt

REMOVAL & INSTALLATION

1. Disconnect the negative battery cable.
2. Raise and safely support the vehicle.
3. Remove the oil pan protector and the engine under cover.
4. Drain the cooling system and store the coolant for refilling purposes.
5. Lower the vehicle and remove the battery clamp cover.
6. From the top of the engine, remove the fuel return hose, the engine cover nuts/bolts and the cover.
7. Remove the air cleaner and the intake air connector assembly.
8. Remove the cooling fan pulley by performing the following procedures:
   a. Loosen the 4 fan clutch-to-fan pulley nuts.
   b. Using a box-end wrench on the serpentine drive belt tensioner bolt, rotate the tensioner counterclockwise and remove the drive belt.

➡The serpentine drive belt tensioner bolt is a left-hand thread.

   c. Remove the fan clutch-to-fan pulley nuts, the fan, the clutch assembly and the fan pulley.
9. Remove the radiator by performing the following procedures:
   a. Disconnect the upper, lower and reservoir hoses from the radiator.
   b. Disconnect and plug the automatic transmission oil cooler at the radiator. Disconnect the automatic transmission oil cooler hoses from the fan shroud clamp.
   c. Remove the radiator reservoir tank.
   d. Remove the fan shroud-to-radiator bolts and the shroud.
   e. Remove the 2 upper radiator-to-chassis nuts.
   f. Remove the middle radiator-to-chassis nut/bolts and brackets.
   g. Carefully, lift the radiator from the vehicle.
10. Remove the serpentine drive belt idler pulley bolt, cover plate and pulley.
11. Remove the right side (No. 3) timing belt cover.
12. Remove the left side (No. 3) timing belt cover by performing the following procedures:
   a. Disconnect the engine wire from both wire clamps.
   b. Disconnect the camshaft position sensor wire from the wire clamp on the left-side (No.3) timing belt cover.
   c. Disconnect the sensor connector from the connector bracket.
   d. Disconnect the sensor connector.
   e. Remove the wire grommet from the left-side (No. 3) timing belt cover.
   f. Remove the oil cooler tube bolts and tube.
13. Remove the middle (No. 2) timing belt cover bolts and cover.
14. Remove the cooling fan bracket nuts/bolts and bracket.

➡If reusing the timing belt, make sure that there are 3 installation marks on the belt; if there are none, install them.

15. Using the Crankshaft Pulley Holding tool 09213-70010, Bolt tool 90105-08076 and Companion Flange Holding tool 09330-00021, or equivalent, loosen the crankshaft pulley bolt.
16. Position the No. 1 cylinder to approximately 50 degrees After Top Dead Center (ATDC) of the compression stroke by performing the following procedures:
   a. Rotate the crankshaft pulley (CLOCKWISE) to align its groove with the timing mark "0" on the lower (No. 1) timing belt cover.
   b. Check that the camshaft sprocket timing marks are aligned with the rear timing belt plate marks; if not, rotate the crankshaft 1 revolution (360 degrees).
   c. Rotate the crankshaft pulley approximately 50 degrees (CLOCKWISE) and align the crankshaft pulley timing mark between the centers of the crankshaft pulley bolt and the idler pulley bolt.

⁂ WARNING

If the timing belt is disengaged, having the crankshaft pulley in the wrong angle can cause the valve to come into contact with the piston when removing the camshaft pulley.

17. Remove the crankshaft pulley bolt.

➡If reusing the timing belt and the installation marks have disappeared, place new installation marks on the timing belt to match the camshaft timing sprocket marks.

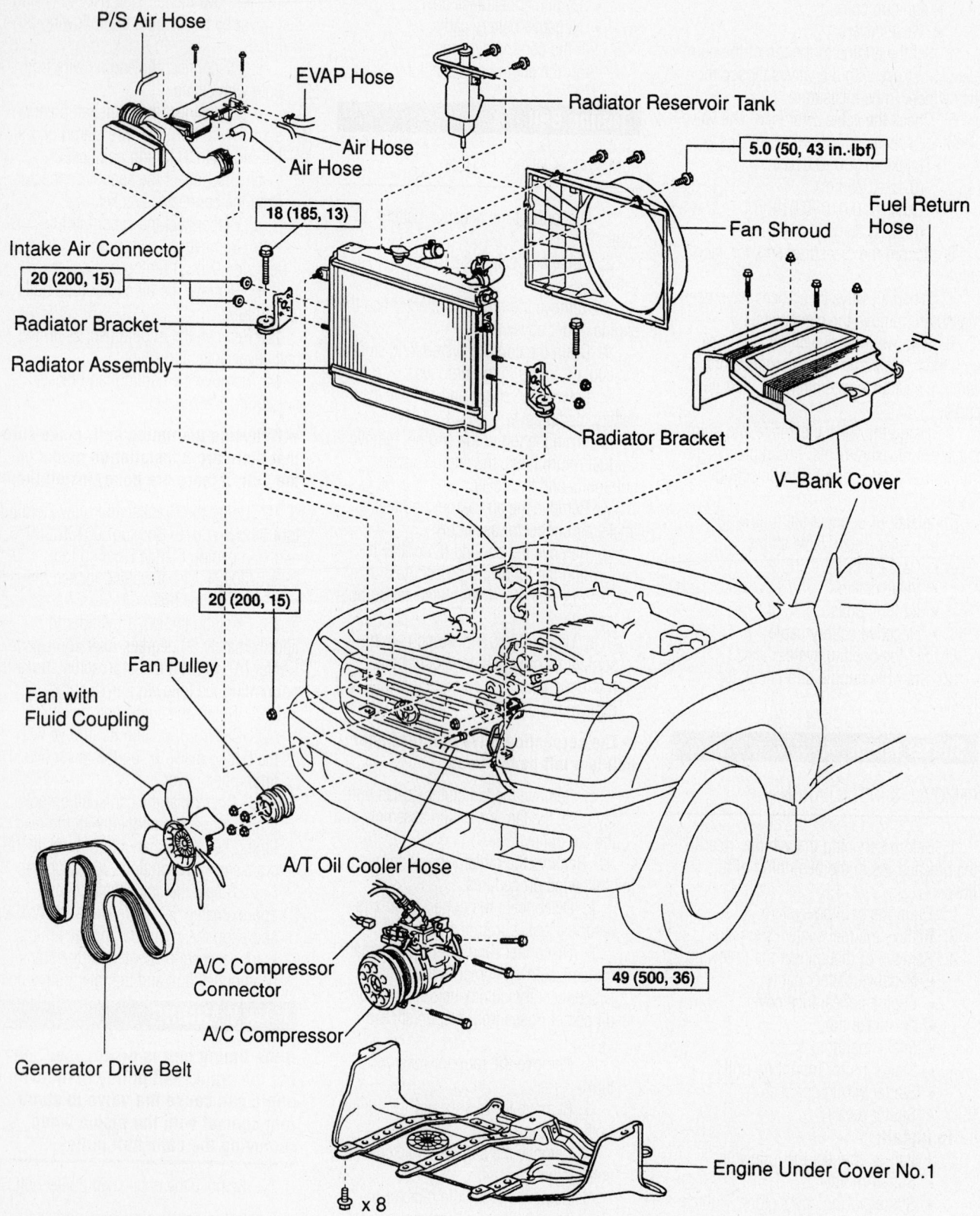

P/S Air Hose

EVAP Hose

Radiator Reservoir Tank

Air Hose

Air Hose

5.0 (50, 43 in.·lbf)

Fuel Return Hose

Intake Air Connector

Fan Shroud

18 (185, 13)

20 (200, 15)

Radiator Bracket

Radiator Assembly

Radiator Bracket

V-Bank Cover

20 (200, 15)

Fan Pulley

Fan with Fluid Coupling

A/T Oil Cooler Hose

A/C Compressor Connector

49 (500, 36)

A/C Compressor

Generator Drive Belt

Engine Under Cover No.1

x 8

Exploded view of vehicle components for timing belt replacement

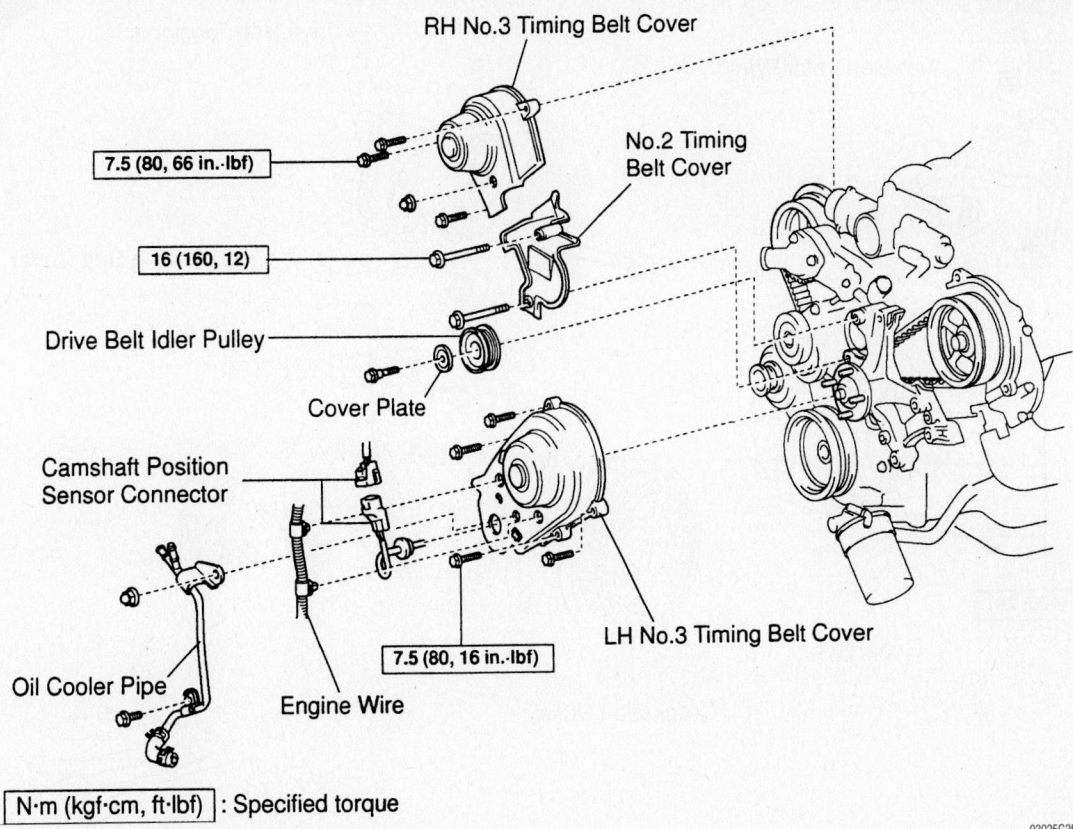

RH No.3 Timing Belt Cover

No.2 Timing Belt Cover

7.5 (80, 66 in.·lbf)

16 (160, 12)

Drive Belt Idler Pulley

Cover Plate

Camshaft Position Sensor Connector

LH No.3 Timing Belt Cover

7.5 (80, 16 in.·lbf)

Oil Cooler Pipe

Engine Wire

N·m (kgf·cm, ft·lbf) : Specified torque

93025G25

Exploded view of upper timing belt covers

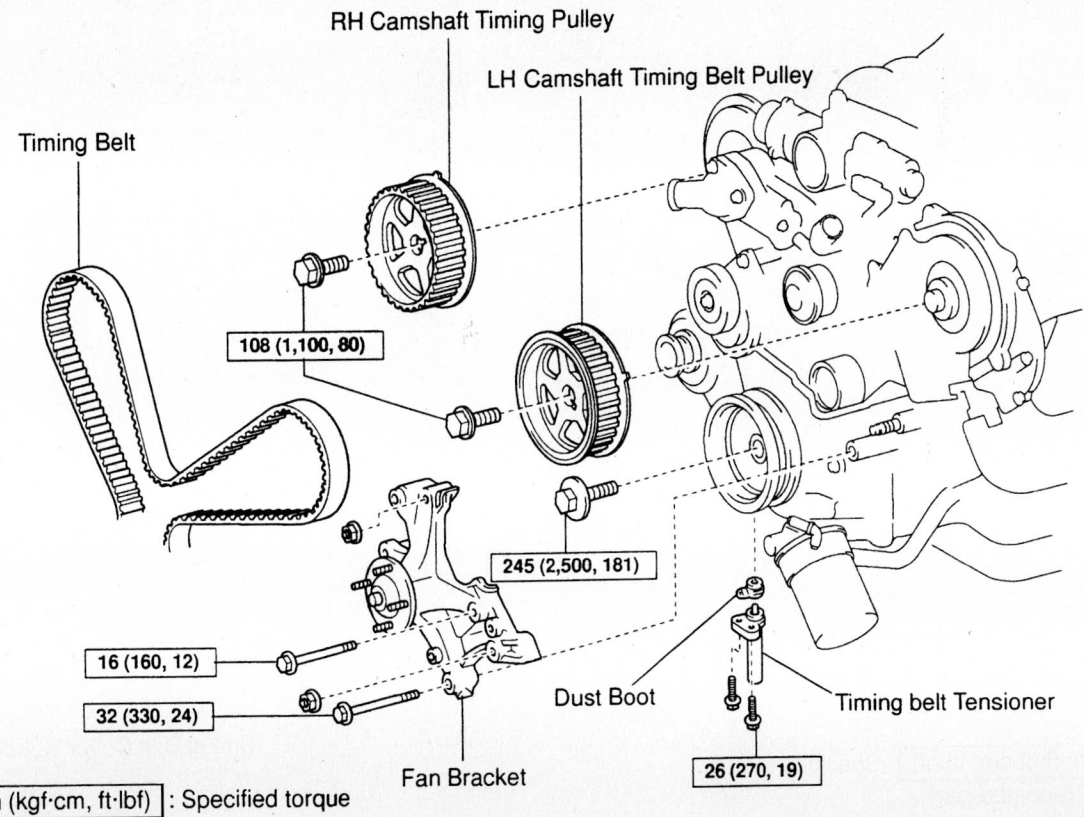

RH Camshaft Timing Pulley

LH Camshaft Timing Belt Pulley

Timing Belt

108 (1,100, 80)

245 (2,500, 181)

16 (160, 12)

32 (330, 24)

Dust Boot

Timing belt Tensioner

Fan Bracket

26 (270, 19)

N·m (kgf·cm, ft·lbf) : Specified torque

93025G26

Exploded view of upper timing sprockets and components

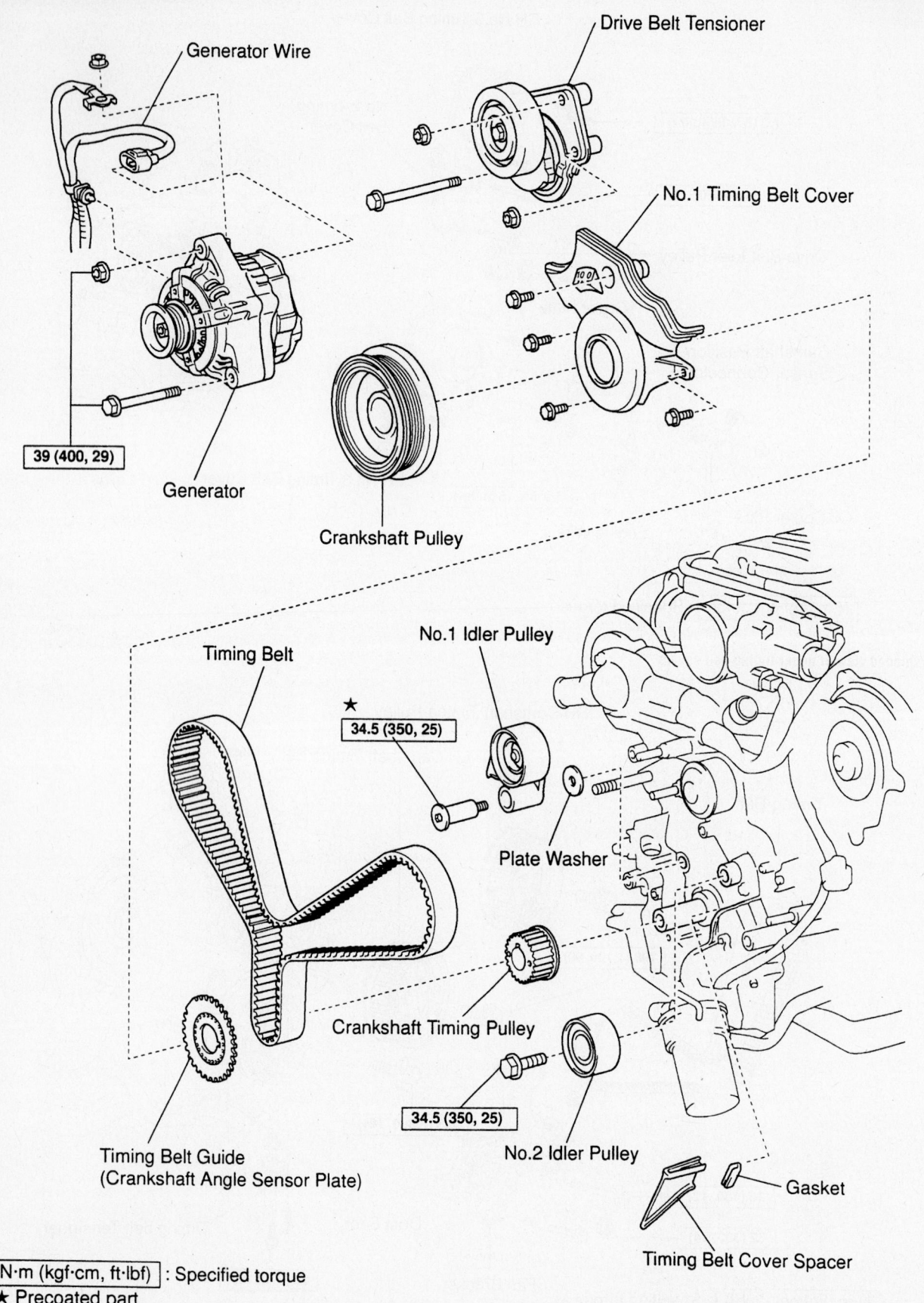

Generator Wire

Drive Belt Tensioner

No.1 Timing Belt Cover

39 (400, 29)

Generator

Crankshaft Pulley

Timing Belt

No.1 Idler Pulley

★ 34.5 (350, 25)

Plate Washer

Crankshaft Timing Pulley

Timing Belt Guide
(Crankshaft Angle Sensor Plate)

34.5 (350, 25)

No.2 Idler Pulley

Gasket

Timing Belt Cover Spacer

N·m (kgf·cm, ft·lbf) : Specified torque
★ Precoated part

Exploded view of lower timing belt cover, sprockets and components

93025G27

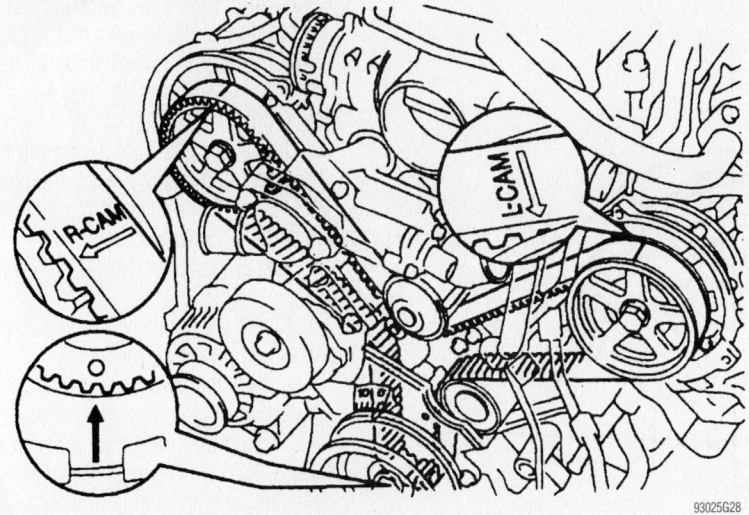

Alignment of timing belt with the timing sprockets

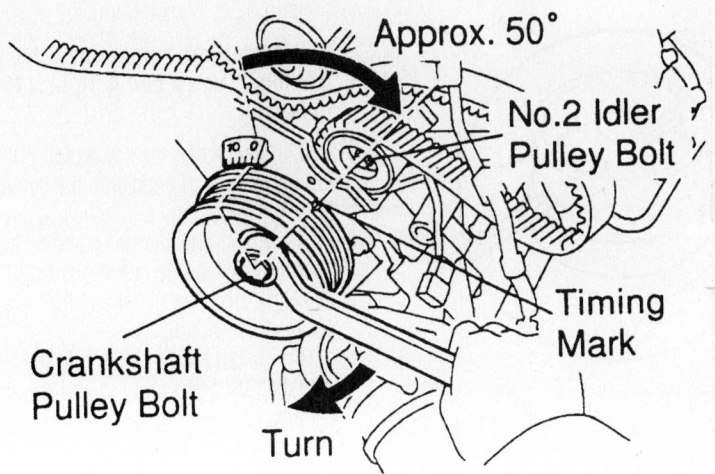

Aligning of crankshaft pulley timing mark with the center line of the crankshaft pulley bolt and the idler pulley bolt

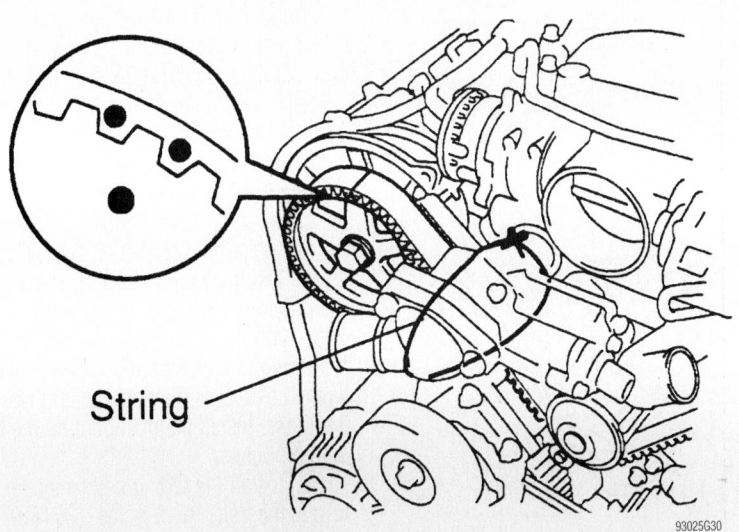

Securing the timing belt with string and matchmarking the camshaft with the timing belt

➡ **To avoid meshing the timing sprocket and the timing belt, secure one with a string; then, place matchmarks on the timing belt and the right-side camshaft timing sprocket.**

18. Remove the timing belt tensioner bolts and the tensioner.

19. Using the Camshaft Holding tool 09960-10010, or equivalent, slightly turn the left-side camshaft sprocket clockwise to loosen the tension spring. Then, disconnect the timing belt from the camshaft sprockets.

20. Remove the alternator by performing the following procedures:

a. Disconnect the electrical connector from the alternator.

b. Remove the rubber cap/nut and disconnect the battery wire from the alternator.

c. Disconnect the wire clamp from the alternator cord clip.

d. Remove the alternator-to-engine nuts/bolts and the alternator.

21. Remove the serpentine drive belt tensioner nuts/bolts and the tensioner.

22. Using the Crankshaft Puller Assembly tool 09950-50012, or equivalent, press the crankshaft pulley from the crankshaft.

✳✳ WARNING

DO NOT rotate the crankshaft pulley.

23. Remove the lower (No. 1) timing belt cover bolts and the cover.

24. Remove the timing belt guide, spacer and the timing belt.

To install:

➡ **With the timing belt removed, this is a perfect opportunity to inspect and/or replace the water pump.**

25. Inspect the timing belt tensioner by performing the following procedures:

a. Inspect the seal for leakage; if leakage is suspected, replace the tensioner.

b. Using both hands to hold the tensioner facing upward, strongly press the pushrod against a solid surface. If the pushrod moves, replace the tensioner.

✳✳ WARNING

Never hold the tensioner with the pushrod facing downward.

c. Measure the pushrod protrusion from the housing end, it should be 0.413–0.453 in. (10.5–11.5mm). If the protrusion is not as specified, replace the tensioner.

26. Temporarily install the timing belt by performing the following procedures:

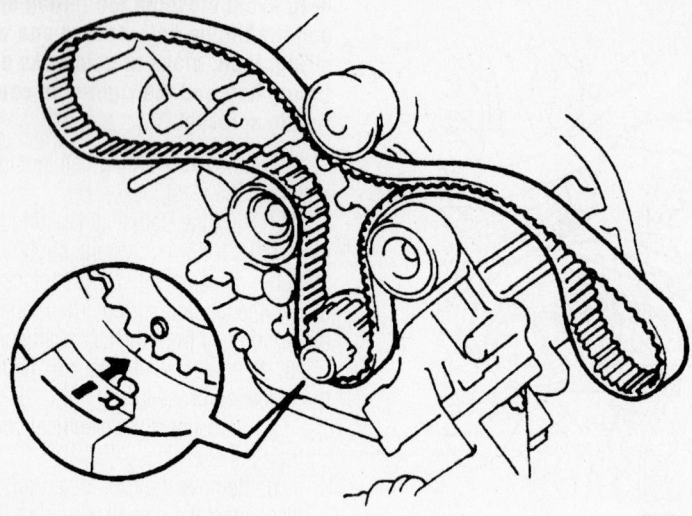

Installing the timing belt on the crankshaft sprocket

93025G31

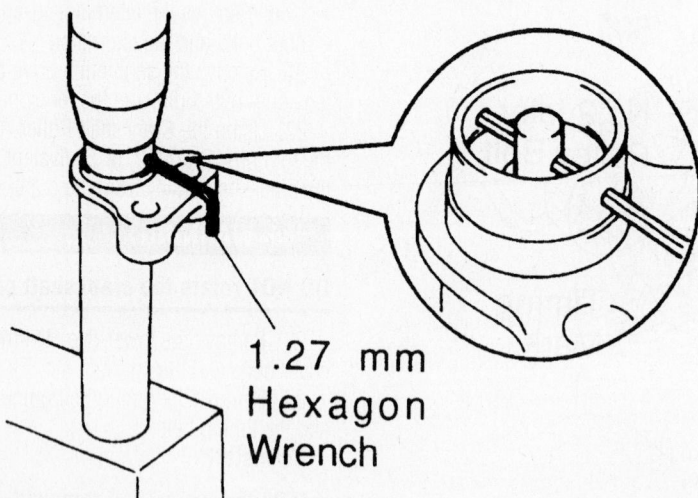

1.27 mm Hexagon Wrench

Securing the timing belt tensioner pushrod

93025G32

93025G33

Checking the TDC alignment marks after rotating the crankshaft 2 revolutions

a. Align the timing belt's installation mark with the crankshaft timing sprocket.

b. Install the timing belt on the crankshaft timing sprocket, the No. 1 idler pulley and the No. 2 idler pulley.

27. Install the gasket to the timing belt cover spacer and install the cover spacer.

28. Install the timing belt guide with the cup side facing outward.

29. Install the lower (No. 1) timing belt cover.

30. Install the crankshaft pulley by performing the following procedures:

a. Align the crankshaft pulley with the crankshaft key.

b. Using the Crankshaft Installer tool 09223-46011, or equivalent, and a hammer, tap the crankshaft pulley into position.

31. Install the serpentine drive belt tensioner and torque the tensioner-to-engine bolts to 12 ft. lbs. (16 Nm).

➡**To install the serpentine drive belt tensioner, use a bolt 4.18 in. (106mm) in length.**

32. Check that the crankshaft pulley's timing mark is aligned with the centers of the idler pulley and crankshaft pulley bolts.

33. Install the alternator and torque the alternator-to-engine nuts/bolts to 29 ft. lbs. (39 Nm). Connect the alternator's electrical connectors and clip.

34. Install the timing belt to the left-side camshaft by performing the following procedures:

a. Rotate the left-side camshaft pulley to align the timing belt installation mark with the camshaft sprocket's timing mark and slide the belt onto the camshaft timing sprocket.

b. Using the Camshaft Holding tool 09960-10010, or equivalent, slightly turn the left-side camshaft sprocket counterclockwise to place tension on the timing belt between the crankshaft sprocket and the camshaft sprocket.

35. Rotate the right-side camshaft pulley to align the timing belt installation mark with the camshaft sprocket's timing mark and slide the belt onto the camshaft timing sprocket.

36. Using a vertical press, slowly press the pushrod into the housing using 200–2205 lbs. (981–9807 N) until the holes align, then, install a 1.27mm Allen® wrench to secure the pushrod and release the press. Install the dust boot on the tensioner housing.

37. Install the timing belt tensioner and torque the bolts to 19 ft. lbs. (26 Nm).

38. Using a pair of pliers, remove the Allen® wrench from the tensioner housing.

39. Check the valve timing by performing the following procedure:

a. Temporarily install the crankshaft pulley bolt.

b. Slowly, rotate the crankshaft pulley 2 revolutions (CLOCKWISE) and realign the TDC marks.

➡️ **If the pulley/sprocket timing marks do not realign, remove the timing belt and reinstall it.**

40. Using the Crankshaft Pulley Holding tool 09213-70010, Bolt tool 90105-08076 and Companion Flange Holding tool 09330-00021, or equivalent, torque the crankshaft pulley bolt to 181 ft. lbs. (245 Nm).

41. Install the cooling fan bracket and torque the 12mm (head size) bolt to 12 ft. lbs. (16 Nm) and the 14mm (head size) bolt to 24 ft. lbs. (32 Nm).

42. Install the air conditioning compressor.

43. Install the middle (No. 2) timing belt cover and torque the bolts to 12 ft. lbs. (16 Nm).

44. Install the upper right-side (No. 3) timing belt cover and torque the bolts to 66 inch lbs. (7.5 Nm).

45. Install the upper left-side (No. 3) timing belt cover by performing the following procedures:

a. Install the oil cooler tube and bolt.

b. Feed the Camshaft Position Sensor (CPS) through the left-side (No. 3) timing belt cover hole.

c. Install the left-side (No. 3) timing belt cover and torque the bolts to 66 inch lbs. (7.5 Nm).

d. Install the wire grommet to the left-side (No. 3) timing belt cover.

e. Install the sensor connector to the connector bracket and connect the sensor connector.

f. Install the sensor wire and the engine wire to the clamps on the left-side (No. 3) timing belt cover.

46. Install the drive belt idler pulley and cover plate; then, torque the pulley bolt to 27 ft. lbs. (37 Nm).

47. To complete the installation, reverse the removal procedures.

48. Refill the cooling system and connect the negative battery cable.

Oil Pan

REMOVAL & INSTALLATION

1. Before servicing the vehicle, refer to the precautions in the beginning of this section.

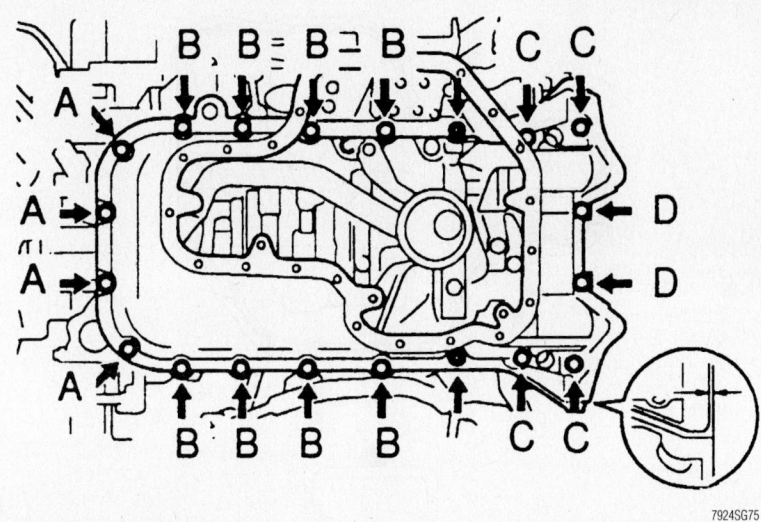

Upper oil pan bolt location

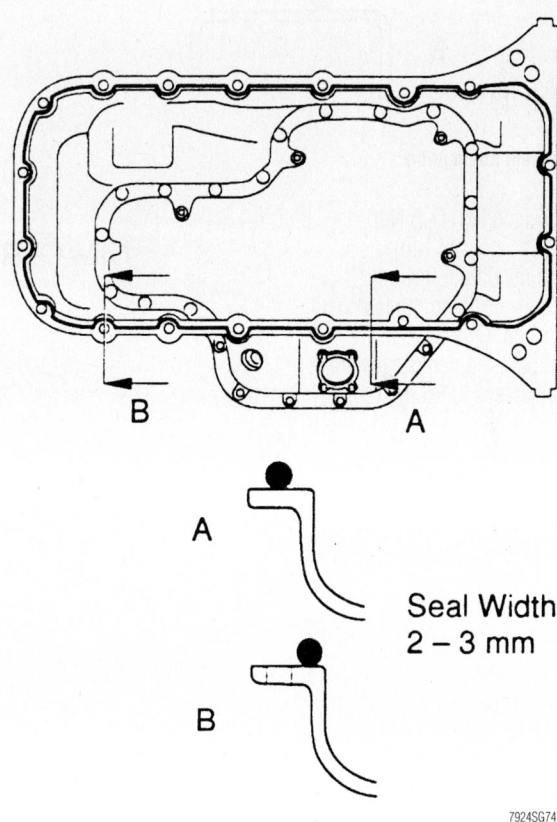

Seal Width 2 – 3 mm

Upper oil pan sealant application

2. Remove the engine from the vehicle and mount it on a stand.

3. Remove or disconnect the following:
- Oil dipstick tube
- Lower oil pan
- Oil pan baffle
- Upper oil pan

To install:

4. The upper oil pan bolts are different lengths and are identified as follows:

- A: 0.79 inch (20mm) w/10mm head
- B: 0.98 inch (25mm) w/12mm head
- C: 2.36 inch (60mm) w/12mm head
- D: 1.38 inch (35mm) w/10mm head

5. Apply silicone sealant to the upper oil pan as shown.

6. Install the upper oil pan and tighten the fasteners in several passes to the following specifications:

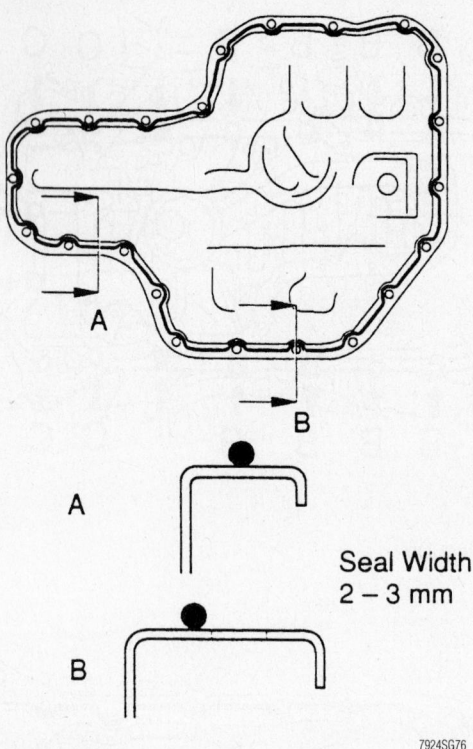

Lower oil pan sealant application

- 10mm: 66 inch lbs. (7.5 Nm)
- 12mm: 21 ft. lbs. (28 Nm)
7. Install or connect the following:
  - Oil pan baffle. Tighten the fasteners to 66 inch lbs. (7.5 Nm).
  - Lower oil pan. Tighten the fasteners in several passes to 66 inch lbs. (7.5 Nm).
  - Oil dipstick tube
8. Install the engine.

Oil Pump

REMOVAL & INSTALLATION

1. Before servicing the vehicle, refer to the precautions in the beginning of this section.
2. Remove the engine from the vehicle and mount it on a stand.
3. Remove or disconnect the following:
   - Front cover
   - Timing belt.
   - Timing belt idler pulleys
   - Crankshaft timing sprocket
   - Oil dipstick tube
   - Oil filter and bracket
   - Crankshaft Position (CKP) sensor
   - Oil pan and baffle
   - Oil pump pickup tube
   - Oil pump
To install:
4. The upper oil pan bolts are different lengths and are identified as follows:

- A: 1.38 inch (35mm) w/12mm head
- B: 1.97 inch (50mm) w/12mm head
- C: 4.17 inch (106mm) w/12mm head
- D: 1.57 inch (40mm) w/14mm head
- E: 1.18 inch (30mm) w/6mm hex head
5. Install a new O-ring on the engine block.
6. Apply silicone sealant to the oil pump housing as shown.
7. Install the oil pump. Tighten the bolts in several passes to the following specifications:

- 12mm: 11 ft. lbs. (15.5 Nm)
- 14mm: 22 ft. lbs. (30.5 Nm)
- 6mm Hex: 11 ft. lbs. (15.5 Nm)
8. Install or connect the following:
  - Oil pump pickup tube. Tighten the bolts to 66 inch lbs. (7.5 Nm).
  - Oil pan and baffle
  - CKP sensor
  - Oil filter and bracket. Tighten the bolts to 13 ft. lbs. (18 Nm).
  - Oil dipstick tube
  - Crankshaft timing sprocket

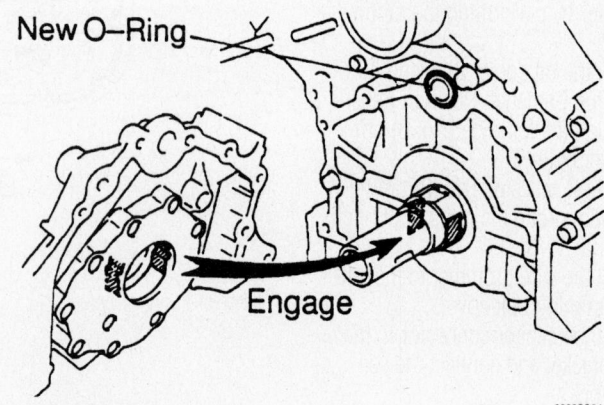

Location of the O-ring seal

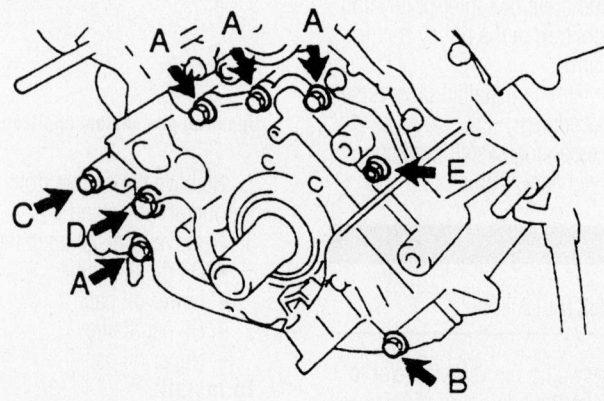

Oil pump bolt location

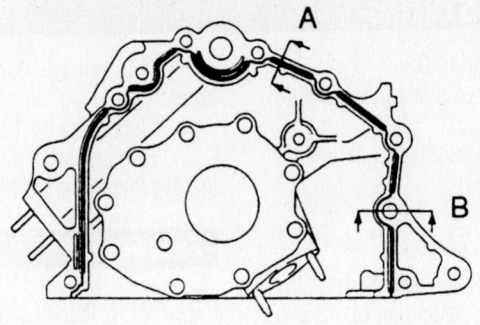

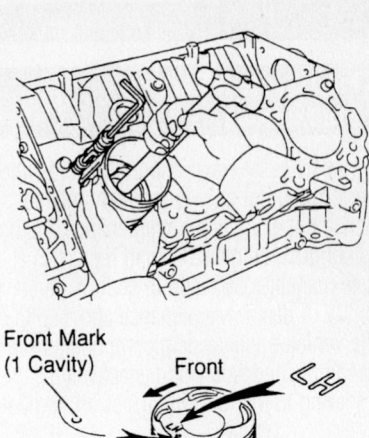

Seal Width
2 – 3 mm

7924SG78

Oil pump housing sealant application

- Timing belt idler pulleys
- Timing belt
- Front cover
9. Install the engine.

Rear Main Seal

REMOVAL & INSTALLATION

1. Before servicing the vehicle, refer to the precautions in the beginning of this section.

2. Remove the transmission and flywheel from the vehicle.

3. Cut off the rubber lip portion of the seal with a sharp knife.

4. Pry out the oil seal.
To install:

5. Install the rear main seal so that it is flush with the seal retainer housing.

6. Install or connect the following:
- Flywheel/driveplate. Tighten the bolts to 35 ft. lbs. (48 Nm) plus a 90 degree turn.
- Transmission

Piston and Ring

POSITIONING

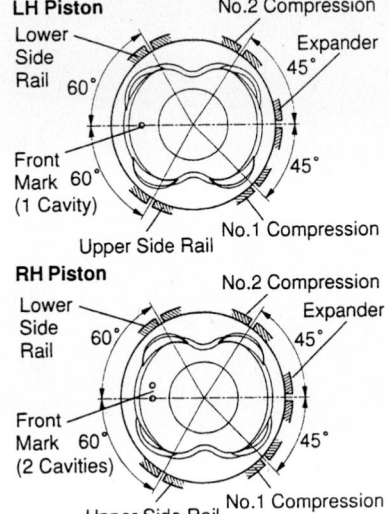

9302AG07

Piston ring positioning

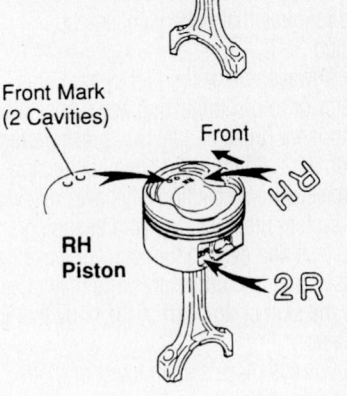

9302AG08

Piston positioning

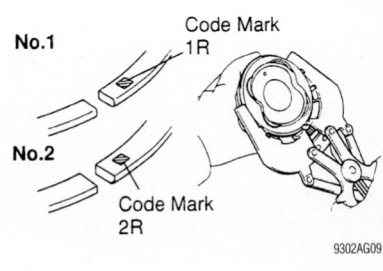

9302AG09

Piston ring identification

FUEL SYSTEM

Fuel System Service Precautions

Safety is the most important factor when performing not only fuel system maintenance but any type of maintenance. Failure to conduct maintenance and repairs in a safe manner may result in serious personal injury or death. Maintenance and testing of the vehicle's fuel system components can be accomplished safely and effectively by adhering to the following rules and guidelines.

• To avoid the possibility of fire and personal injury, always disconnect the negative battery cable unless the repair or test procedure requires that battery voltage be applied.

• Always relieve the fuel system pressure prior to disconnecting any fuel system component (injector, fuel rail, pressure regulator, etc.), fitting or fuel line connection. Exercise extreme caution whenever relieving fuel system pressure, to avoid exposing skin, face and eyes to fuel spray. Please be advised that fuel under pressure may penetrate the skin or any part of the body that it contacts.

• Always place a shop towel or cloth around the fitting or connection prior to loosening to absorb any excess fuel due to spillage. Ensure that all fuel spillage (should it occur) is quickly removed from engine surfaces. Ensure that all fuel soaked cloths or towels are deposited into a suitable waste container.

• Always keep a dry chemical (Class B) fire extinguisher near the work area.

• Do not allow fuel spray or fuel vapors to come into contact with a spark or open flame.

• Always use a back-up wrench when loosening and tightening fuel line connection fittings. This will prevent unnecessary stress and torsion to fuel line piping.

• Always replace worn fuel fitting O-rings with new. Do not substitute fuel hose or equivalent, where fuel pipe is installed.

Fuel System Pressure

RELIEVING

GX470

1. Remove the fuel pump relay from the engine compartment relay block.
2. Start the engine and let it run until it shuts off.

3. Turn the ignition to OFF.
4. Try to start the engine and make sure it won't start.
5. Disconnect the negative battery cable.
6. Install the relay.

LX470

1. Before servicing the vehicle, refer to the precautions in the beginning of this section.
2. Disconnect the fuel pump connector near the fuel tank.
3. Start the engine and allow it to run until it stalls. Crank the engine for a few seconds to relieve additional fuel pressure.
4. Disconnect the negative battery cable.
5. When repairs are complete, connect the negative battery cable.

Fuel Filter

REMOVAL & INSTALLATION

GX470

The fuel filter is part of the fuel pump module unit and is not a normally replaced item.

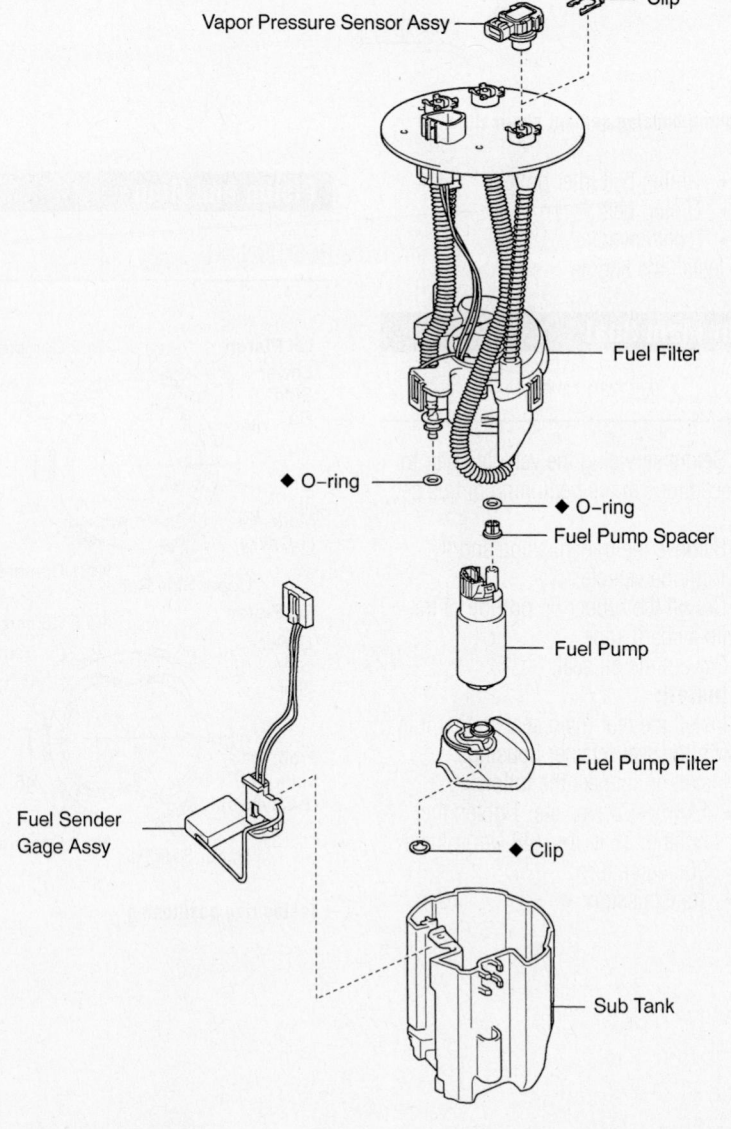

◆ Non-reusable part

67162-X470-G15

Fuel pump components—GX470

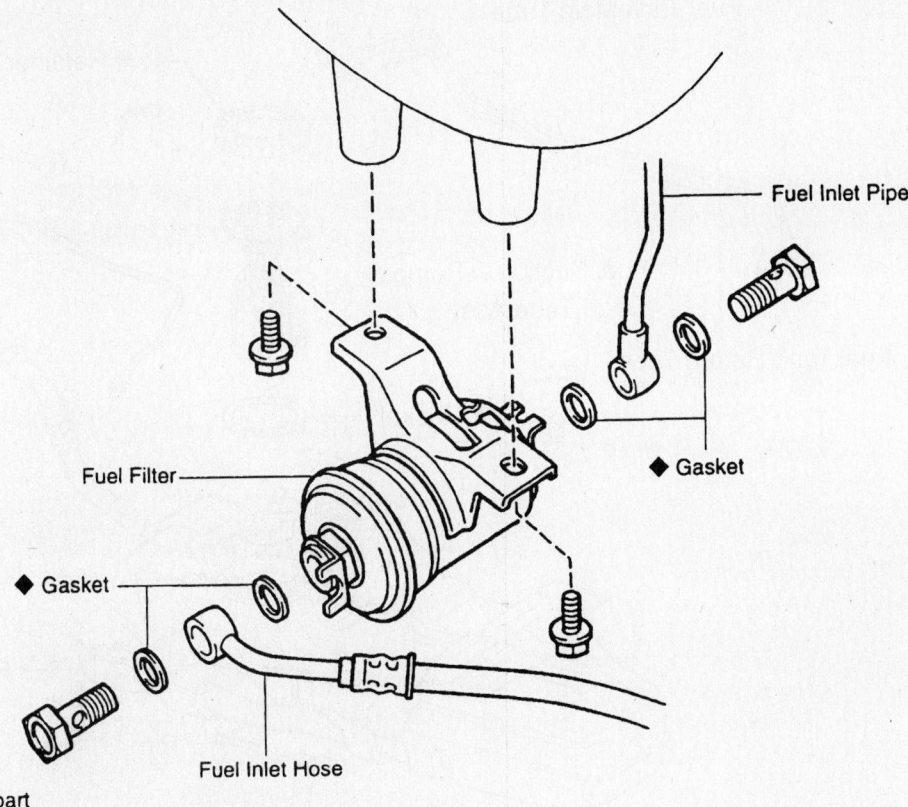

Fuel Inlet Pipe

◆ Gasket

Fuel Filter

◆ Gasket

Fuel Inlet Hose

◆ Non-reusable part

7924SG28

Always use new gaskets when replacing the fuel filter—LX470

LX470

1. Before servicing the vehicle, refer to the precautions in the beginning of this section.
2. Relieve the fuel system pressure.
3. Remove or disconnect the following:
   - Negative battery cable
   - Fuel lines
   - Fuel filter

To install:

4. Install the fuel filter.
5. Use new washers and tighten the fuel line bolts to the following specifications:
   - Banjo bolt fittings: 21 ft. lbs. (29 Nm) for 2001–03; 25 ft. lbs. (34 Nm) for 2004
   - Flare nut fitting: 28 ft. lbs. (38 Nm)
6. Connect the negative battery cable.
7. Start the engine and check for leaks.

Fuel Pump

REMOVAL & INSTALLATION

GX470

1. Disconnect the battery ground.
2. Remove the rear seat.
3. Remove the rear floor step cover.

4. Remove the left quarter scuff plate.
5. Remove the service hole cover.
6. Disconnect the wiring.
7. Remove the filler hose.
8. Drain the fuel.
9. Remove the skid plate.
10. Disconnect the fuel lines.
11. Disconnect the breather hose.
12. Disconnect the vent hose.
13. Remove the fuel tank.
14. Disconnect the supply and return lines.
15. Unscrew the lockring, using a tool such as SST09808-14020.

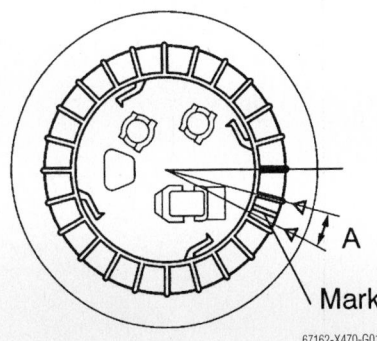

Mark

A

67162-X470-G01

Position the lockring marks as shown— GX470

16. Remove the pump module from the tank.
17. Installation is the reverse of removal. Make sure that all fuel fittings are secure. Use a new gasket and new lockring. Start the lockring and make a reference mark before turning it. Turn the locking 1 full turn by hand. Using the SST, turn the lockring one more full turn. And position the mark on the locking as shown.

LX470

1. Before servicing the vehicle, refer to the precautions in the beginning of this section.
2. Relieve the fuel system pressure.
3. Remove or disconnect the following:
   - Negative battery cable
   - Rear seats
   - Door sill trim plates
   - Carpeting and floor mats
   - Access panel
   - Fuel pump harness connector
   - Fuel lines
   - Fuel pump module

To install:

4. Install or connect the following:
   - Fuel pump module. Tighten the bolts to 35 inch lbs. (4 Nm).
   - Fuel lines

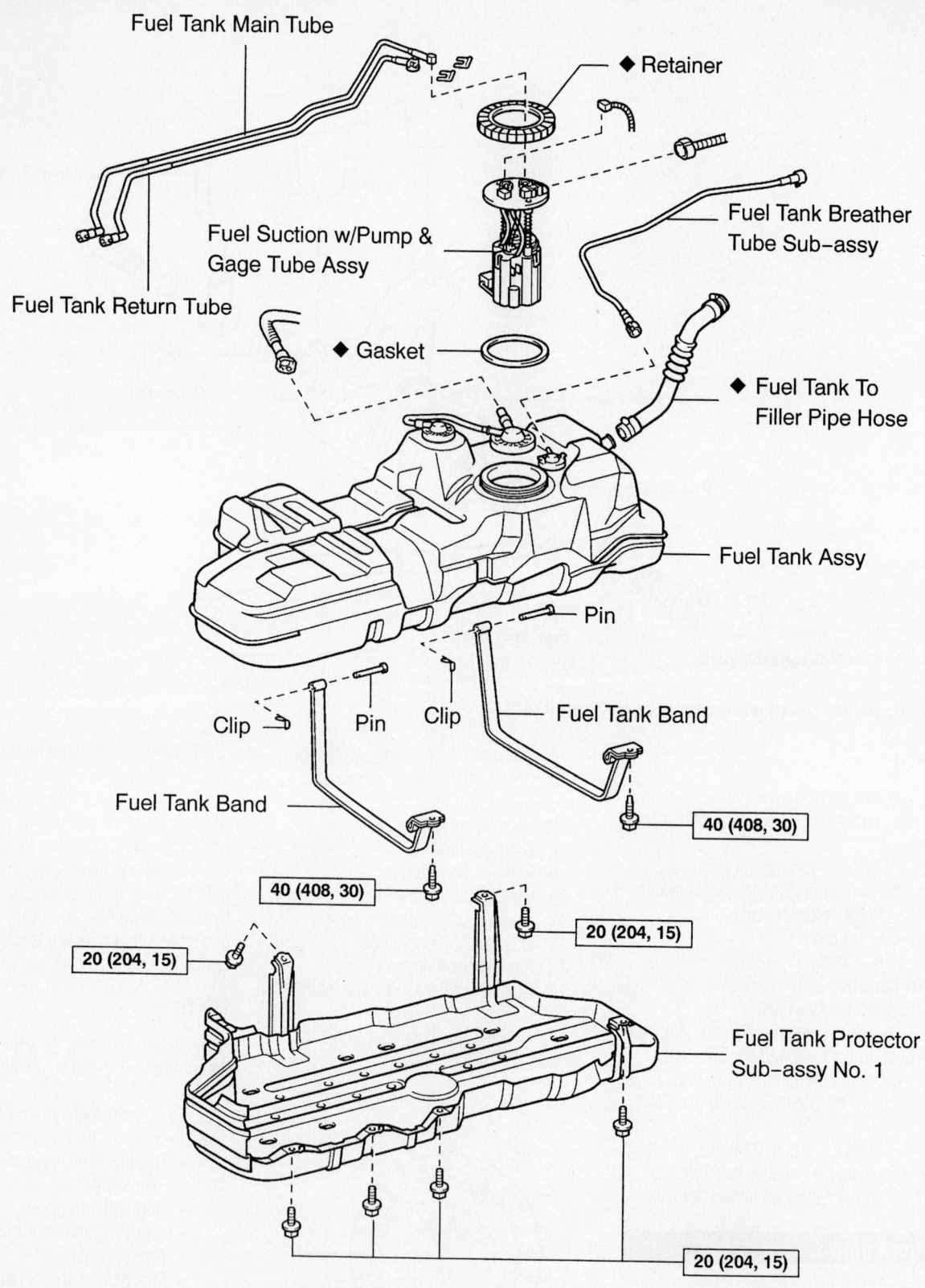

Fuel Tank Main Tube

◆ Retainer

Fuel Suction w/Pump & Gage Tube Assy

Fuel Tank Breather Tube Sub-assy

Fuel Tank Return Tube

◆ Gasket

◆ Fuel Tank To Filler Pipe Hose

Fuel Tank Assy

Pin

Clip

Pin

Clip

Fuel Tank Band

Fuel Tank Band

40 (408, 30)

40 (408, 30)

20 (204, 15)

20 (204, 15)

Fuel Tank Protector Sub-assy No. 1

20 (204, 15)

N·m (kgf·cm, ft·lbf) : Specified torque

◆ Non-reusable part

Fuel tank and related parts—GX470

67162-X470-G02

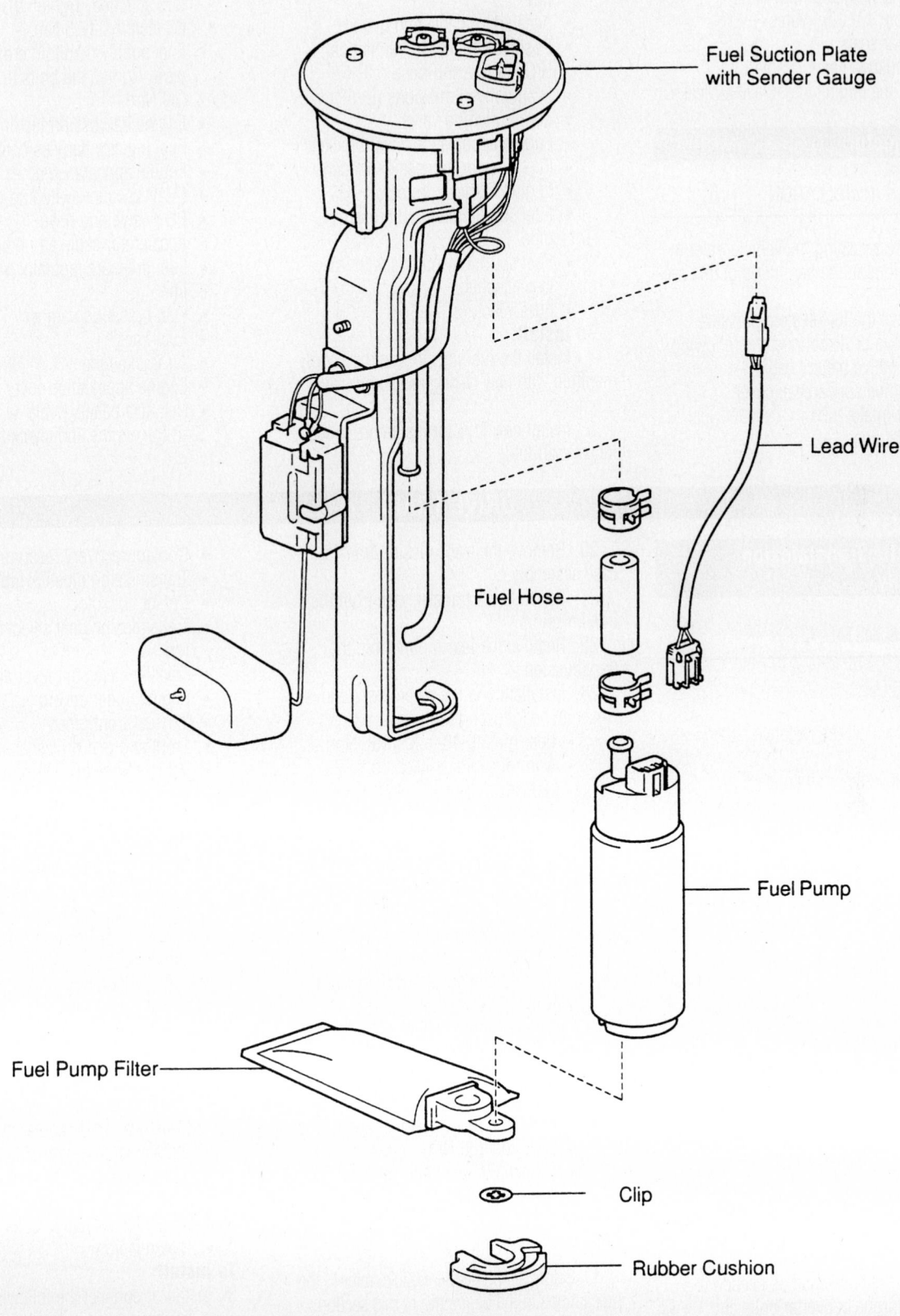

Fuel Suction Plate
with Sender Gauge

Lead Wire

Fuel Hose

Fuel Pump

Fuel Pump Filter

Clip

Rubber Cushion

◆ Non-reusable part

7924SG81

Exploded view of the fuel pump and related components—LX470

- Fuel pump harness connector
- Access panel
- Carpeting and floor mats
- Door sill trim plates
- Rear seats
- Negative battery cable

5. Start the engine and check for leaks.

Fuel Injector

REMOVAL & INSTALLATION

1. Before servicing the vehicle, refer to the precautions in the beginning of this section.
2. Relieve the fuel system pressure.
3. Remove or disconnect the following:
   - Negative battery cable
   - Engine appearance cover
   - Air intake tube
   - Fuel lines

- Fuel pulsation damper
- Fuel pressure regulator vacuum line
- Accelerator cable and bracket
- Positive Crankcase Ventilation (PCV) valve and hose
- Evaporative Emissions (EVAP) vacuum switching valve
- Engine appearance cover brackets
- Fuel injector harness connectors
- Engine harness protector
- Fuel supply manifold crossover pipe
- Fuel supply manifolds with injectors attached
- Fuel injectors

To install:

4. Install the fuel injectors to the supply manifold with new O-ring seals and new grommets.
5. Install new injector insulators to the intake manifold.

6. Install or connect the following:
   - Fuel supply manifolds with injectors attached. Tighten the bolts to 66 inch lbs. (7.5 Nm).
   - Fuel supply manifold crossover pipe. Tighten the bolts to 29 ft. lbs. (39 Nm).
   - Engine harness protector
   - Fuel injector harness connectors
   - Engine appearance cover brackets
   - EVAP vacuum switching valve
   - PCV valve and hose
   - Accelerator cable and bracket
   - Fuel pressure regulator vacuum line
   - Fuel pulsation damper
   - Fuel lines
   - Air intake tube
   - Engine appearance cover
   - Negative battery cable
7. Start the engine and check for leaks.

DRIVE TRAIN

Automatic Transmission Assembly

REMOVAL & INSTALLATION

GX470

1. Disconnect the negative battery cable.
2. Remove the shift knob.
3. Remove the upper trim panels.
4. Remove the console.
5. Remove the snapring and remove the transfer case lever.
6. Remove the engine under-covers.
7. Remove the front suspension member brackets.
8. Disconnect the oxygen sensor.
9. Remove the exhaust pipe.
10. Remove the driveshafts.
11. Remove the drain plug.
12. Remove the transmission control cable.
13. Support the transmission with a transmission jack.
14. Remove the crossmember.
15. Disconnect all wires and lines as necessary.
16. Disconnect the breather hose.
17. Remove the bellhousing cover.
18. Turn the crankshaft as needed to access the torque converter bolts and remove them.
19. Remove the transmission-to-engine bolts.

20. Remove the transmission/transfer case assembly.
21. Separate the transfer case from the transmission.
22. Remove the rear mount from the transmission.
23. Installation is the reverse of removal. Observe the following torques:
   - Rear mount: 48 ft. lbs. (65 Nm)
   - Control cable bracket: 19 ft. lbs. (25 Nm)
   - Transfer case-to-transmission: 17 ft. lbs. (24 Nm)
   - Transmission-to-engine: 17mm bolts, 53 ft. lbs. (71 Nm); 14mm bolts, 27 ft. lbs. (37 Nm)
   - Torque converter bolts: 35 ft. lbs. (48 Nm)
   - Bellhousing cover: 13 ft. lbs. (18 Nm)
   - Crossmember-to-frame: 53 ft. lbs. (72 Nm)
   - Transmission-to-crossmember: 13 ft. lbs. (18 Nm)
   - Front and rear driveshaft flanges: 65 ft. lbs. (88 Nm)
   - Suspension member brackets: 24 ft. lbs. (33 Nm)

LX470

1. Before servicing the vehicle, refer to the precautions in the beginning of this section.
2. Remove or disconnect the following:
   - Battery and tray
   - Air intake assembly
   - Cooling fan and shroud

- Coolant recovery reservoir
- Transmission dipstick tube
- Center console
- Transmission gear select lever and rod
- Transfer case shift lever and rod
- Engine under covers
- Exhaust front pipes
- Front and rear driveshafts
- Vehicle Speed (VSS) sensor connectors
- Overdrive clutch speed sensor connector
- Solenoid harness connector
- Transmission fluid temperature sensor connector
- Park/Neutral Position (PNP) switch connector
- Center differential lock indicator switch connector
- L4 solenoid valve position switch connector
- Motor actuator connector
- Torque converter
- Transmission oil cooler lines
- Transmission mount crossmember. Support the transmission with a jack.
- Transmission flange bolts
- Transmission

To install:
3. Install or connect the following:
   - Transmission. Tighten the flange bolts to 53 ft. lbs. (72 Nm).
   - Transmission mount crossmember. Tighten the bolts to 37 ft. lbs. (50

Nm) and the nuts to 54 ft. lbs. (74 Nm).

- Transmission oil cooler lines
- Torque converter. Tighten the bolts to 35 ft. lbs. (48 Nm).
- Motor actuator connector
- L4 solenoid valve position switch connector
- Center differential lock indicator switch connector
- PNP switch connector
- Transmission fluid temperature sensor connector
- Solenoid harness connector
- Overdrive clutch speed sensor connector
- VSS sensor connectors
- Front driveshaft. Tighten the fasteners to 59 ft. lbs. (80 Nm).
- Rear driveshaft. Tighten the fasteners to 78 ft. lbs. (106 Nm).
- Exhaust front pipes
- Engine under covers
- Transfer case shift lever and rod
- Transmission gear select lever and rod
- Center console
- Transmission dipstick tube
- Coolant recovery reservoir
- Cooling fan and shroud
- Air intake assembly
- Battery and tray

4. Check the transmission and transfer case fluid levels and adjust as necessary.

Transfer Case Assembly

REMOVAL & INSTALLATION

GX470

1. Drain the fluid.
2. Remove the skid plate.
3. Remove the transmission.
4. Separate the transfer case from the transmission.
5. Installation is the reverse of removal. Torque the bolts to 17 ft. lbs. (24 Nm).

LX470

1. Before servicing the vehicle, refer to the precautions in the beginning of this section.
2. Drain the transfer case oil.
3. Remove or disconnect the following:
- Transfer case protector
- Front and rear driveshafts
- Transfer case shift lever rod
- Ground cable
- Transmission mount crossmember. Support the transmission with a jack.

- Transfer case vent hose
- Vehicle Speed (VSS) sensor connector
- Center differential lock indicator switch connector
- Motor actuator connectors
- Transfer case adapter bolts
- Transfer case

To install:

4. Install or connect the following:
- Transfer case. Tighten the adapter bolts to 51 ft. lbs. (69 Nm).
- Motor actuator connectors
- Center differential lock indicator switch connector
- VSS sensor connector
- Transfer case vent hose
- Transmission mount crossmember. Tighten the bolts to 37 ft. lbs. (50 Nm) and the nuts to 54 ft. lbs. (74 Nm).
- Ground cable
- Transfer case shift lever rod
- Front driveshaft. Tighten the fasteners to 59 ft. lbs. (80 Nm).
- Rear driveshaft. Tighten the fasteners to 78 ft. lbs. (106 Nm).
- Transfer case protector

5. Fill the transfer case to the correct level.

Halfshaft

REMOVAL & INSTALLATION

GX470

1. Before servicing the vehicle, refer to the precautions in the beginning of this section.
2. Remove the wheel.
3. Drain the differential oil.
4. Remove the cotter pin and cap, then remove the hub nut.
5. Remove the speed sensor wiring harness. Remove the sensor.
6. Remove the tie rod end from the knuckle.

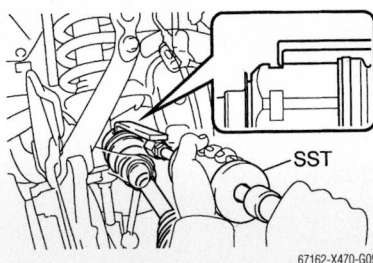

67162-X470-G09

Remove the halfshaft using a slidehammer and adapter

7. Remove the 2 bolts and separate the lower arm from the ball joint.
8. Remove the halfshaft using a slidehammer and adapter. Keep the halfshaft level when carrying it.

To install:

9. Coat the inboard end splines of the halfshaft with clean ATF.
10. Align the splines and drive the halfshaft into place with a brass drift.
11. Install a new snapring with the opening facing down.
12. Install the sensor. Torque to 10 ft. lbs. (13 Nm). Connect the wire harness.
13. Connect the arm to the ball joint. Torque to 166 ft. lbs. (225 Nm).
14. Connect the tie rod end. Torque to 67 ft. lbs. (91 Nm). The nut can be advanced up to 60 degrees to align the cotter pin hole.
15. Install the hub nut. Torque to 173 ft. lbs. (235 Nm). Install the cap and a new cotter pin.
16. Fill the differential.
17. Install the wheel. Torque to 83 ft. lbs. (112 Nm).

LX470

1. Before servicing the vehicle, refer to the precautions in the beginning of this section.
2. Remove or disconnect the following:
- Front wheel
- Brake caliper
- Grease cap
- Snapring
- Wheel speed sensor and wire harness
- Steering knuckle arm
- Lower ball joint
- Upper ball joint
- Steering knuckle
- Axle halfshaft

To install:

➡ **Use new split pins, snaprings and circlips for assembly.**

3. Install or connect the following:
- Axle halfshaft
- Steering knuckle
- Upper ball joint. Tighten the nut to 81 ft. lbs. (110 Nm).
- Lower ball joint. Tighten the nut to 117 ft. lbs. (159 Nm).
- Steering knuckle arm. Tighten the bolts to 108 ft. lbs. (147 Nm).
- Wheel speed sensor and wire harness
- Snapring
- Grease cap
- Brake caliper
- Front wheel

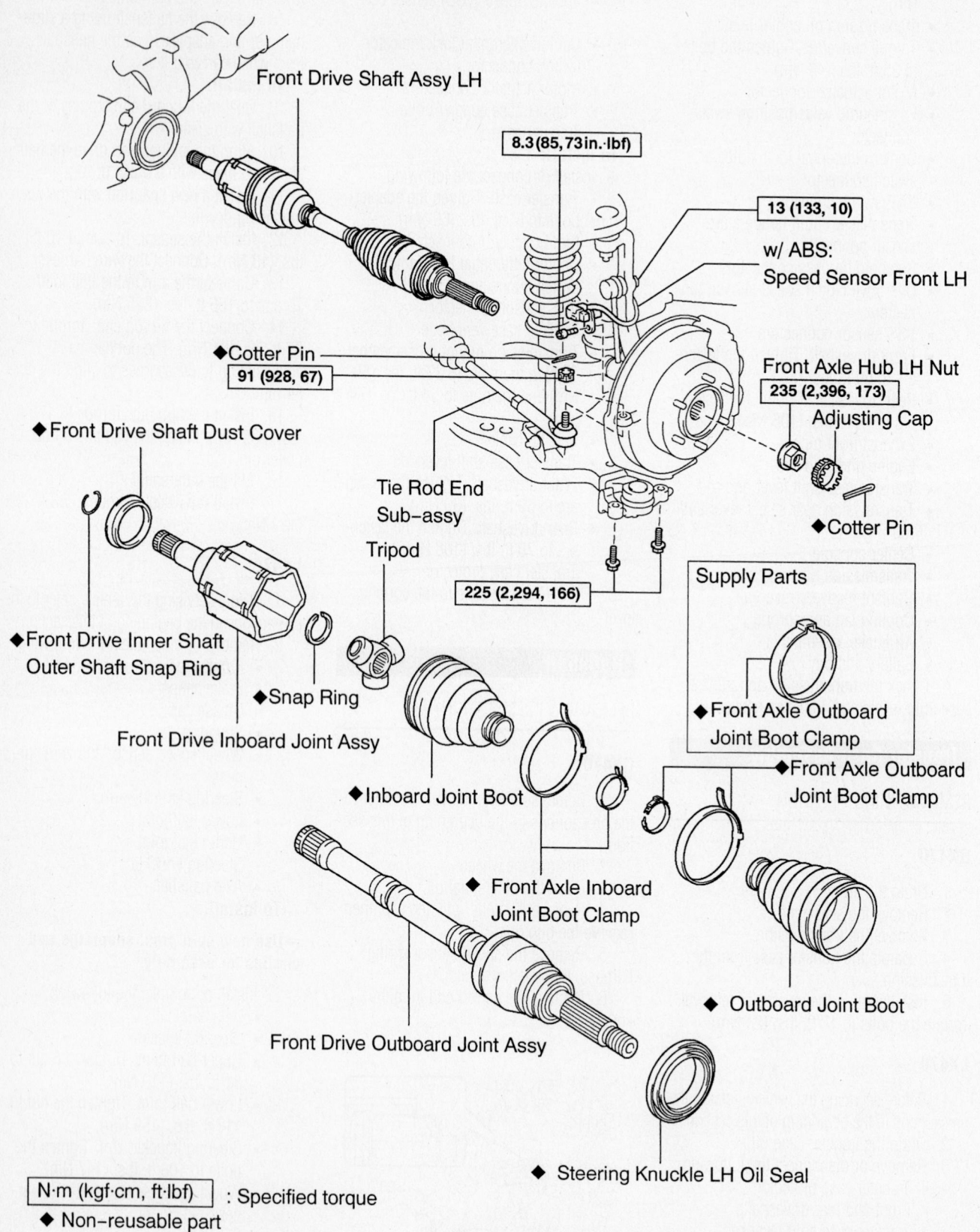

Front Drive Shaft Assy LH

8.3 (85, 73 in.·lbf)

13 (133, 10)

w/ ABS:
Speed Sensor Front LH

◆Cotter Pin
91 (928, 67)

◆Front Drive Shaft Dust Cover

Front Axle Hub LH Nut
235 (2,396, 173)
Adjusting Cap

Tie Rod End
Sub-assy

◆Cotter Pin

Tripod

◆Front Drive Inner Shaft
Outer Shaft Snap Ring

◆Snap Ring

Front Drive Inboard Joint Assy

Supply Parts

◆Front Axle Outboard
Joint Boot Clamp

◆Inboard Joint Boot

◆Front Axle Outboard
Joint Boot Clamp

225 (2,294, 166)

◆ Front Axle Inboard
Joint Boot Clamp

◆ Outboard Joint Boot

Front Drive Outboard Joint Assy

◆ Steering Knuckle LH Oil Seal

N·m (kgf·cm, ft·lbf) : Specified torque
◆ Non-reusable part

67162-X470-G08

Front halfshaft, left side shown—GX470

LH side:

RH side:

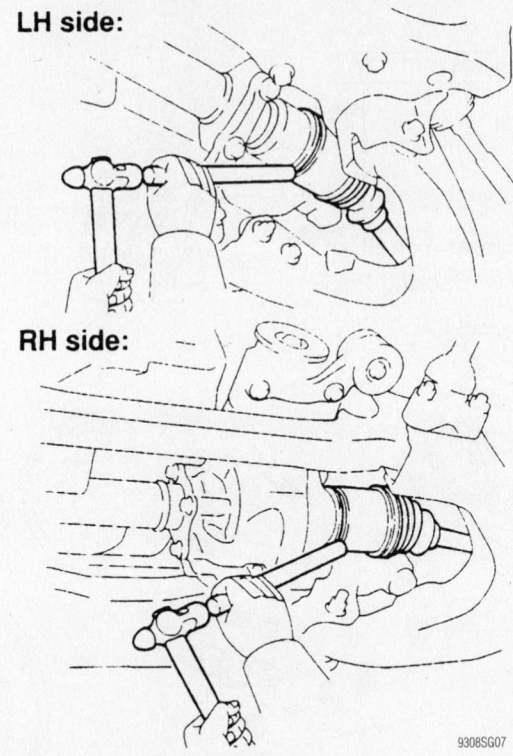

Axle halfshaft removal—LX470

CV-Joints

OVERHAUL

Outer CV-Joint

The outer CV-joint is serviced with the axle shaft as an assembly. The outer CV-joint boot can be serviced by removing the inner CV-joint.

Inner CV-Joint

1. Before servicing the vehicle, refer to the precautions in the beginning of this section.
2. Remove or disconnect the following:
   - Halfshaft from the vehicle
   - Grease boot clamps
   - Outer race snapring
   - Outer race
   - Shaft snapring
   - Inner race, cage and balls

To install:

3. Install or connect the following:
   - Inner race, cage and balls
   - Shaft snapring
   - Outer race
   - Outer race snapring
4. Fill the outer race and the grease boot with CV-joint grease and tighten the boot clamps.
5. Install the axle halfshaft.

Spindle Bearings

REMOVAL, PACKING AND INSTALLATION

LX470

1. Before servicing the vehicle, refer to the precautions in the beginning of this section.
2. Remove or disconnect the following:
   - Front wheel
   - Brake caliper
   - Grease cap
   - Snapring
   - Hub drive flange
   - Locknut
   - Lockwasher
   - Adjusting nut
   - Outer bearing

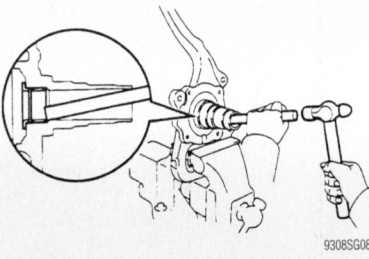

Removing the oil seal, bushing and spindle bearing—LX470

- Wheel hub
- Disc brake dust shield
- Wheel speed sensor and harness
- Outer tie rod end
- Upper ball joint
- Lower ball joint
- Steering knuckle
- Oil seal, bushing and spindle bearing

To install:

3. Coat the spindle bearing and bushing with lithium grease.
4. Fill the spindle cavity with lithium grease.
5. Press the spindle bearing and bushing into the spindle.
6. Install or connect the following:
   - Oil seal
   - Steering knuckle
   - Upper ball joint. Tighten the nut to 81 ft. lbs. (110 Nm).
   - Lower ball joint. Tighten the nut to 117 ft. lbs. (159 Nm).
   - Outer tie rod end. Tighten the nut to 91 ft. lbs. (122 Nm).
   - Wheel speed sensor and harness
   - Disc brake dust shield. Tighten the bolts to 13 ft. lbs. (18 Nm).
   - Wheel hub
   - Outer bearing
   - Adjusting nut. Adjust the wheel bearings.
   - Lockwasher
   - Locknut. Tighten the nut to 47 ft. lbs. (64 Nm).
   - Hub drive flange. Tighten the nuts to 26 ft. lbs. (35 Nm).
   - Snapring
   - Grease cap
   - Brake caliper
   - Front wheel

Axle Shaft, Bearing and Seal

REMOVAL & INSTALLATION

Rear

GX470

1. Remove the wheel.
2. Remove the speed sensor.
3. Remove the caliper.
4. Remove the rotor.
5. Remove the parking brake assembly.
6. Remove the 4 nuts and pull out the axle shaft with backing plate.
7. Remove the oil seal with a slidehammer.
8. Installation is the reverse of removal. Torque the nuts to 89 ft. lbs. (120 Nm).

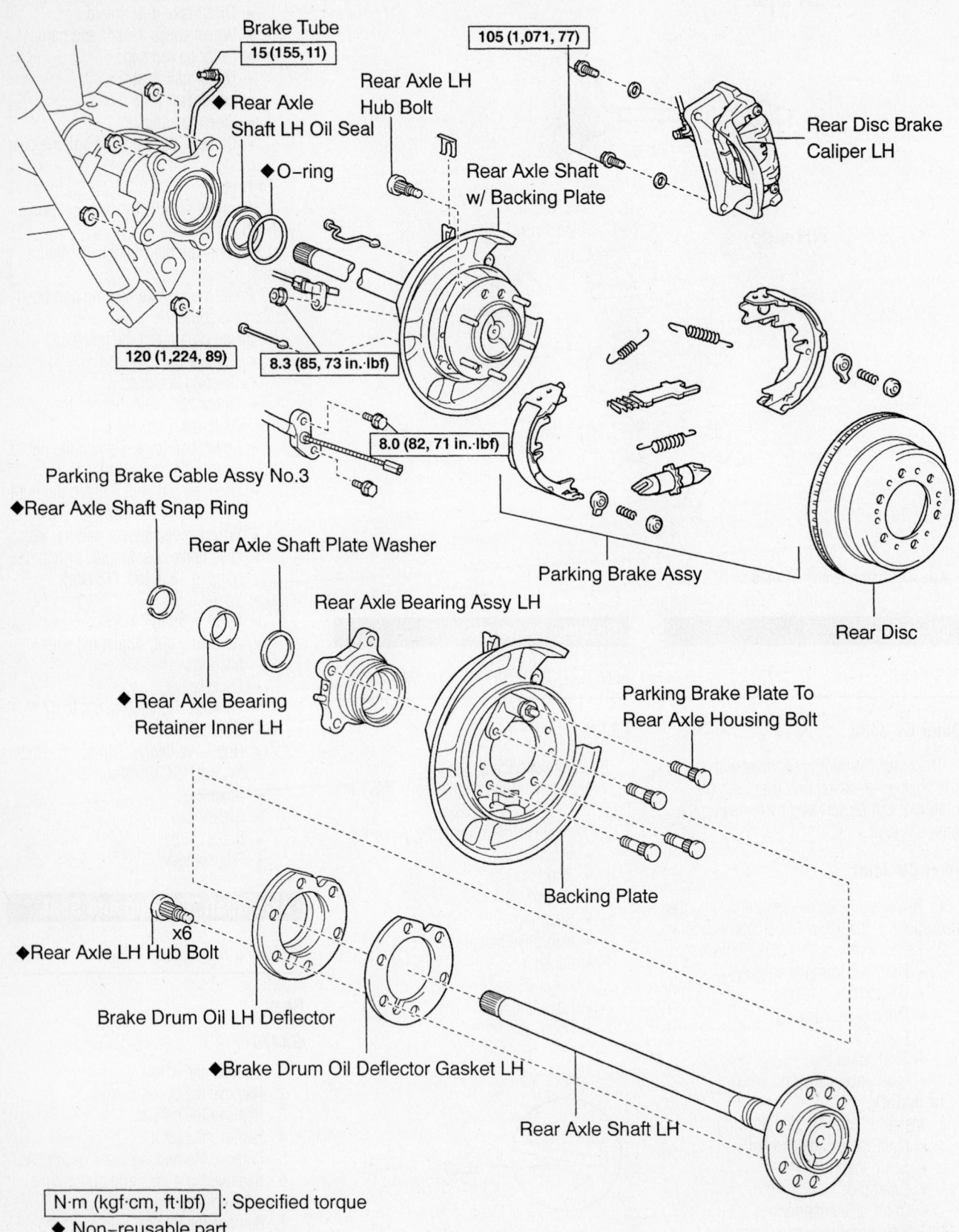

Brake Tube
15 (155, 11)

◆ Rear Axle
Shaft LH Oil Seal

◆ O-ring

Rear Axle LH
Hub Bolt

105 (1,071, 77)

Rear Disc Brake
Caliper LH

Rear Axle Shaft
w/ Backing Plate

120 (1,224, 89)

8.3 (85, 73 in.·lbf)

8.0 (82, 71 in.·lbf)

Parking Brake Cable Assy No.3

◆Rear Axle Shaft Snap Ring

Rear Axle Shaft Plate Washer

Rear Axle Bearing Assy LH

◆Rear Axle Bearing
Retainer Inner LH

Parking Brake Assy

Rear Disc

Parking Brake Plate To
Rear Axle Housing Bolt

Backing Plate

x6

◆Rear Axle LH Hub Bolt

Brake Drum Oil LH Deflector

◆Brake Drum Oil Deflector Gasket LH

Rear Axle Shaft LH

N·m (kgf·cm, ft·lbf) : Specified torque

◆ Non-reusable part

67162-X470-G11

Rear axle shaft and related parts—GX470

LX470

1. Before servicing the vehicle, refer to the precautions in the beginning of this section.

2. Remove or disconnect the following:
- Rear wheel
- Brake caliper and rotor
- Parking brake shoes and hardware
- Bearing case nuts
- Axle shaft and bearing assembly

3. Separate the backing plate from the bearing case by removing the serrated bolts.

4. Grind a flat spot on the wheel speed sensor rotor and retainer, then split them with a hammer and chisel.

5. Remove the axle snapring.

6. Press the axle bearing case, bearing and retainer off of the axle.

7. Press the axle bearing from the bearing case.

8. Remove or disconnect the following:
- Backing plate
- Axle housing oil seal
- Bearing case oil seal

To install:

9. Press the wheel bearing into the bearing case.

10. Install the bearing case to the backing plate with the serrated bolts.

11. Install or connect the following:
- Bearing case oil seal
- Axle housing oil seal
- Axle shaft to backing plate and bearing assembly
- Bearing retainer
- Axle snapring
- Wheel speed sensor rotor and retainer
- Axle shaft and bearing assembly to the axle housing. Tighten the nuts to 91 ft. lbs. (123 Nm).
- Parking brake shoes and hardware
- Brake caliper and rotor
- Rear wheel

Pinion Seal

REMOVAL & INSTALLATION

Front

GX470

1. Before servicing the vehicle, refer to the precautions in the beginning of this section.

2. Remove the wheels.

3. Remove the engine under-covers.

4. Remove the front driveshaft.

5. Remove the pinion nut.

6. Remove the companion flange with a puller.

7. Remove the oil seal with a seal puller.

8. Remove the oil slinger.

9. Remove the bearing with a puller.

10. Remove the oil storage ring.

11. Remove the spacer and discard it.

To install:

12. Install a new spacer.

13. Install the oil storage ring using a brass drift.

14. Install the bearing.

15. Install the slinger.

16. Using a seal driver, install the new oil seal. Drive the seal into a depth of 4.35mm +/- 0.45mm.

17. Install the companion flange. Coat the threads of a new flange nut with gear oil. Hold the flange and torque the nut to 273 ft. lbs. (370 Nm).

18. Using an inch-pound torque wrench, check the preload. Preload for a new bearing should be 9-14 inch lbs.; for a used bearing, 4.3-7 inch lbs. If not, a new spacer must be installed.

19. When preload is correct, stake the nut.

20. Install the driveshaft. Torque the bolts to 65 ft. lbs. (88 Nm).

21. Fill the differential.

22. Install the under-covers.

LX470

1. Before servicing the vehicle, refer to the precautions in the beginning of this section.

2. Remove or disconnect the following:
- Driveshaft
- Front wheels
- Front brake calipers

➡**The front brake calipers must be removed so that there is no additional drag when measuring pinion bearing preload.**

3. Use an inch lb. torque wrench and measure the amount of torque required to maintain pinion rotation through several revolutions.

4. Remove or disconnect the following:
- Pinion flange
- Oil seal
- Oil slinger
- Pinion bearing and race
- Oil storage ring
- Collapsible spacer

To install:

➡**Use a new collapsible spacer and flange nut for assembly.**

5. Install or connect the following:
- Collapsible spacer
- Oil storage ring
- Pinion bearing and race
- Pinion seal
- Pinion flange. Tighten the nut to 80 ft. lbs. (108 Nm).

6. Rotate the pinion flange occasionally while tightening the flange nut to make sure the pinion bearings seat correctly.

7. Take frequent bearing preload torque readings. Tighten the flange nut to achieve the preload torque readings originally recorded. Do not exceed 249 ft. lbs. (338 Nm) torque when tightening the pinion flange nut.

✳✳ CAUTION

Never loosen the pinion nut to reduce bearing preload. If it is necessary to reduce bearing preload, install a new collapsible spacer and pinion nut.

8. Install or connect the following:
- Front brake calipers
- Front wheels
- Driveshaft. Tighten the fasteners to 59 ft. lbs. (80 Nm).

9. Fill the differential with gear lubricant and check for leaks.

Rear

GX470

1. Before servicing the vehicle, refer to the precautions in the beginning of this section.

2. Remove the wheels.

3. Remove the engine under-covers.

4. Remove the front driveshaft.

5. Remove the pinion nut.

6. Remove the companion flange with a puller.

7. Remove the oil seal with a seal puller.

8. Remove the oil slinger.

9. Remove the bearing with a puller.

10. Remove the oil storage ring.

11. Remove the spacer and discard it.

To install:

12. Install a new spacer.

13. Install the oil storage ring using a brass drift.

14. Install the bearing.

15. Install the slinger.

16. Using a seal driver, install the new oil seal. Drive the seal into a depth of 1.00mm +/- 0.45mm.

17. Install the companion flange. Coat the threads of a new flange nut with gear oil. Hold the flange and torque the nut to 273 ft. lbs. (370 Nm).

18. Using an inch-pound torque wrench, check the preload. Preload for a new bearing should be 9-15 inch lbs.; for a used

bearing, 5-7.5 inch lbs. If not, a new spacer must be installed.

19. When preload is correct, stake the nut.

20. Install the driveshaft. Torque the bolts to 65 ft. lbs. (88 Nm).

21. Fill the differential.

22. Install the under-covers.

LX470

1. Before servicing the vehicle, refer to the precautions in the beginning of this section.

2. Remove or disconnect the following:
   • Driveshaft
   • Rear wheels
   • Rear brake calipers

➡ **The rear brake calipers must be removed so that there is no additional drag when measuring pinion bearing preload.**

3. Use an inch lb. torque wrench and measure the amount of torque required to maintain pinion rotation through several revolutions.

4. Remove or disconnect the following:
   • Pinion flange
   • Oil seal
   • Oil slinger
   • Pinion bearing and race
   • Collapsible spacer

To install:

➡ **Use a new collapsible spacer and flange nut for assembly.**

5. Install or connect the following:
   • Collapsible spacer
   • Pinion bearing and race
   • Pinion seal
   • Pinion flange. Tighten the nut to 181 ft. lbs. (245 Nm).

6. Rotate the pinion flange occasionally while tightening the flange nut to make sure the pinion bearings seat correctly.

7. Take frequent bearing preload torque readings. Tighten the flange nut to achieve the preload torque readings originally recorded. Do not exceed 326 ft. lbs. (441 Nm) torque when tightening the pinion flange nut.

✳✳ CAUTION

Never loosen the pinion nut to reduce bearing preload. If it is necessary to reduce bearing preload, install a new collapsible spacer and pinion nut.

8. Install or connect the following:
   • Rear brake calipers
   • Rear wheels
   • Driveshaft. Tighten the fasteners to 78 ft. lbs. (106 Nm).

9. Fill the differential with gear lubricant and check for leaks.

STEERING AND SUSPENSION

Air Bag

✳✳ CAUTION

Some vehicles are equipped with an air bag system. The system must be disarmed before performing service on, or around, system components, the steering column, instrument panel components, wiring and sensors. Failure to follow the safety precautions and the disarming procedure could result in accidental air bag deployment, possible injury and unnecessary system repairs.

PRECAUTIONS

Several precautions must be observed when handling the inflator module to avoid accidental deployment and possible personal injury.

• Never carry the inflator module by the wires or connector on the underside of the module.

• When carrying a live inflator module, hold securely with both hands and ensure that the bag and trim cover are pointed away.

• Place the inflator module on a bench or other surface with the bag and trim cover facing up.

• With the inflator module on the bench, never place anything on or close to the module which may be thrown in the event of an accidental deployment.

DISARMING

To avoid personal injury when working on vehicles equipped with an air bag, the negative battery cable must be disconnected and at least 90 seconds must elapse before working on the system. Failure to do so may result in deployment of the air bag.

Power Rack and Pinion Steering Gear

REMOVAL & INSTALLATION

GX470

1. Before servicing the vehicle, refer to the precautions in the beginning of this section.

2. Disconnect the battery ground cable.

3. Place the front wheels in the straight ahead position.

4. Remove the horn pad.

5. Remove the steering wheel.

6. Remove the lower steering column cover.

7. Remove the turn signal switch.

8. Remove the spiral cable assembly.

9. Remove the front wheels.

10. Remove the engine under-covers.

11. Remove the stabilizer bar.

12. Remove the tie rod ends from the knuckle.

13. Remove the steering intermediate shaft.

14. Disconnect the pressure and return lines.

15. Remove the 2 bolts and remove the steering gear assembly.

To install:

16. Position the gear and install the 2 bolts. Torque to 74 ft. lbs. (100 Nm).

➡ **The nuts have detents. Never turn the nuts, just the bolts.**

17. Install the stabilizer bar. Torque the end links to 52 ft. lbs. (70 Nm); the clamp bolts to 30 ft. lbs. (40 Nm).

18. Connect the return line. Use a torque wrench with SST 09023-12700, or equivalent. The torque wrench should have a fulcrum length of 300mm. Torque to 31 ft. lbs. (42 Nm).

19. Connect the pressure line at the subframe. Torque to 21 ft. lbs. (28 Nm).

20. Connect the pressure line to the gear. Use a torque wrench with SST 09023-12700, or equivalent. The torque wrench should have a fulcrum length of 300mm. Torque to 31 ft. lbs. (42 Nm).

21. Connect the intermediate shaft. Torque to 26 ft. lbs. (36 Nm).

22. Connect the tie rod ends. Torque to 67 ft. lbs. (91 Nm).

23. Install the under-covers.

24. The remainder of installation is the reverse of removal.

LX470

1. Before servicing the vehicle, refer to the precautions in the beginning of this section.

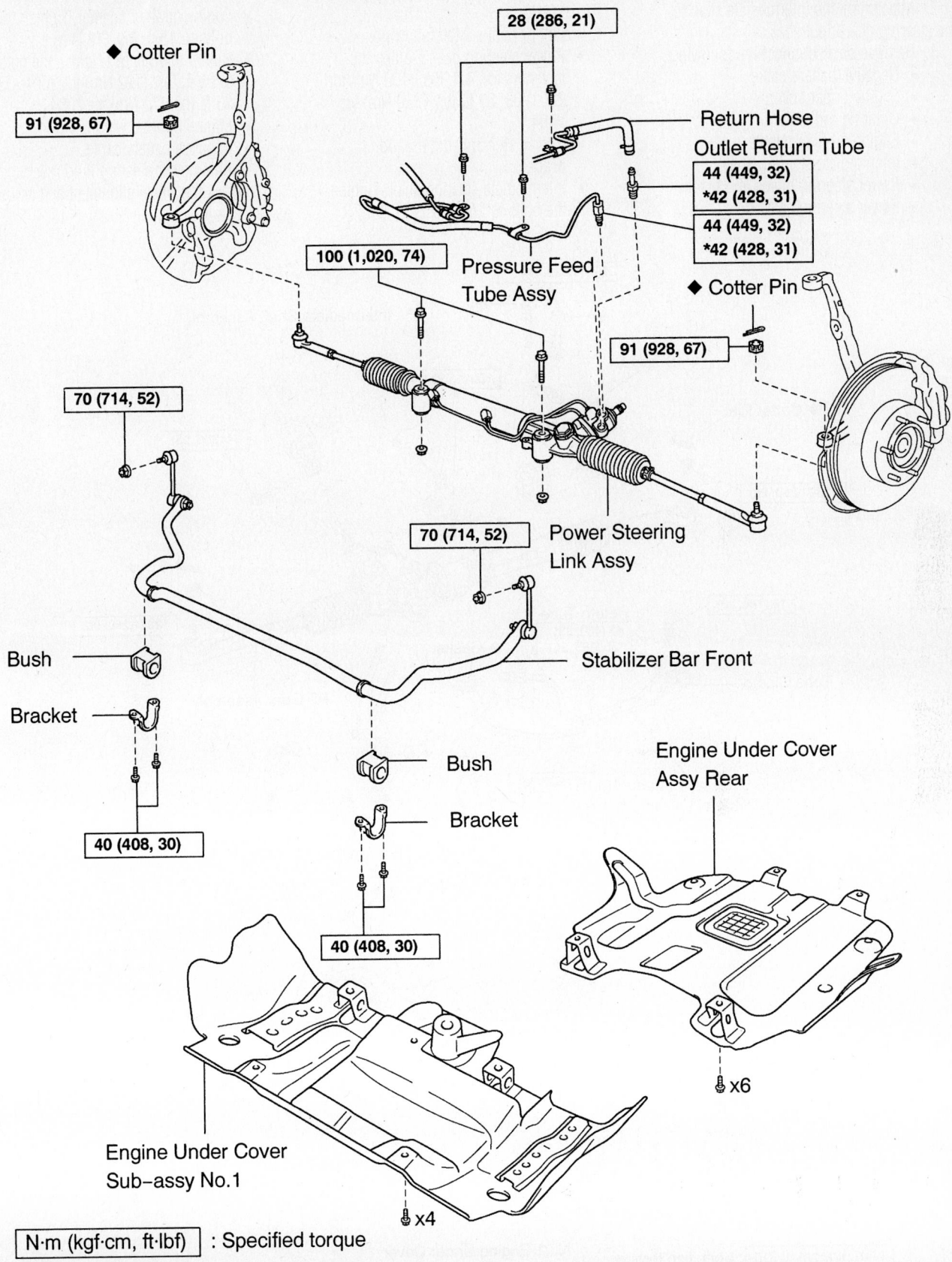

28 (286, 21)

◆ Cotter Pin

91 (928, 67)

Return Hose
Outlet Return Tube

44 (449, 32)
*42 (428, 31)

44 (449, 32)
*42 (428, 31)

100 (1,020, 74)

Pressure Feed
Tube Assy

◆ Cotter Pin

91 (928, 67)

70 (714, 52)

70 (714, 52)

Power Steering
Link Assy

Bush

Bracket

Stabilizer Bar Front

40 (408, 30)

Bush

Bracket

Engine Under Cover
Assy Rear

40 (408, 30)

x6

Engine Under Cover
Sub-assy No.1

x4

N·m (kgf·cm, ft·lbf) : Specified torque

◆ Non-reusable part

* For use with SST

67162-X470-G14

Steering gear and related parts—GX470

2. Matchmark the intermediate shaft to the steering gear input shaft.

3. Remove or disconnect the following:
- Negative battery cable
- Engine under covers
- Outer tie rod ends
- Engine oil filter adapter
- Intermediate steering shaft
- Power steering hoses and bracket
- Power steering gear

To install:

4. Install or connect the following:
- Power steering gear. Tighten the fasteners to 74 ft. lbs. (100 Nm) for 2001–03; 89 ft. lbs. (120 Nm) for 2004.
- Power steering hoses and bracket
- Intermediate steering shaft. Tighten the bolts to 25 ft. lbs. (34 Nm).

- Engine oil filter adapter. Tighten the bolts to 13 ft. lbs. (18 Nm).
- Outer tie rod ends. Tighten the nuts to 90 ft. lbs. (122 Nm) for 2001–03; 53 ft. lbs. (72 Nm) for 2004.
- Engine under covers
- Negative battery cable

5. Fill the power steering fluid reservoir.

6. Check the wheel alignment and adjust as necessary.

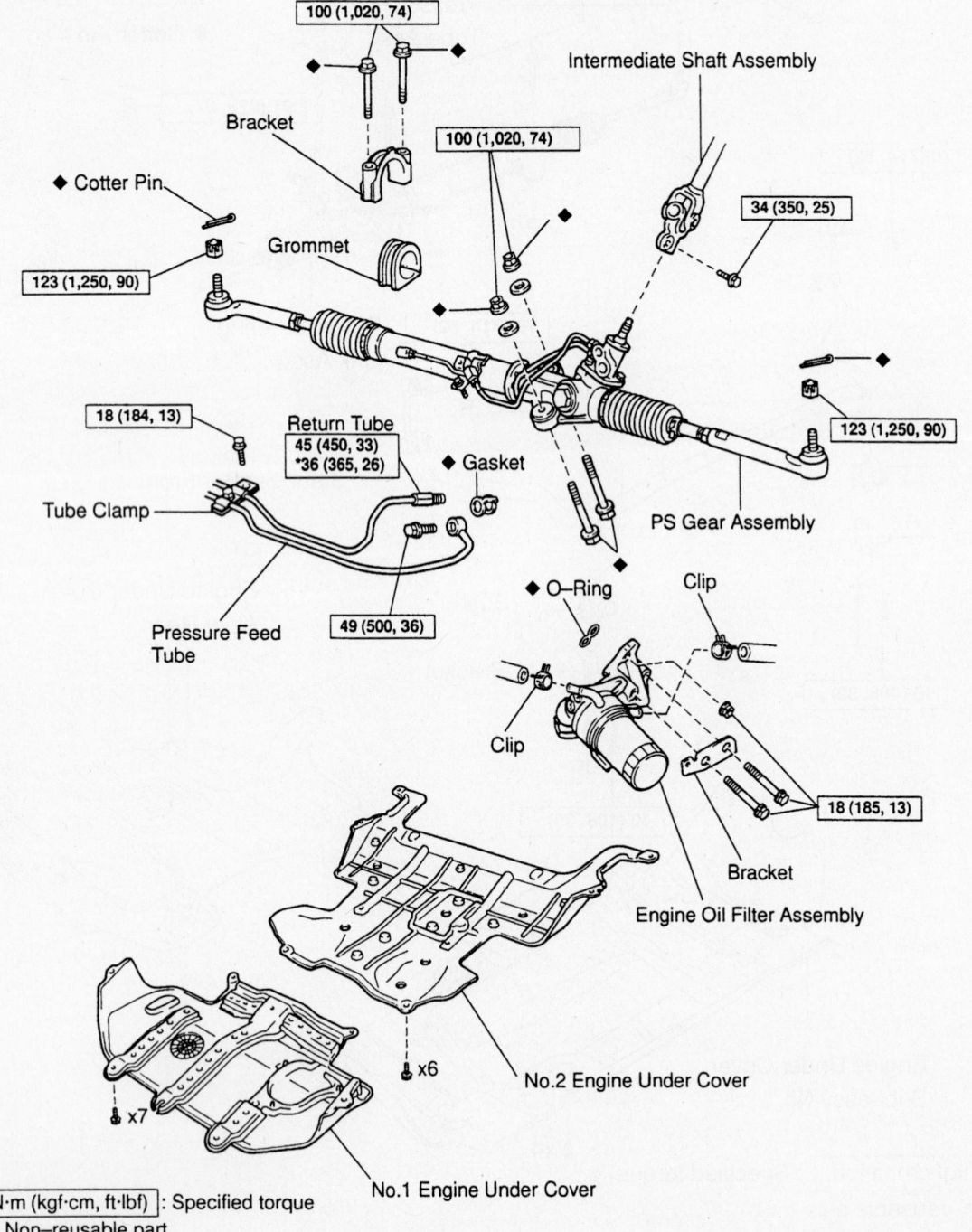

Exploded view of the rack and pinion steering gear mounting—LX470

7924SG90

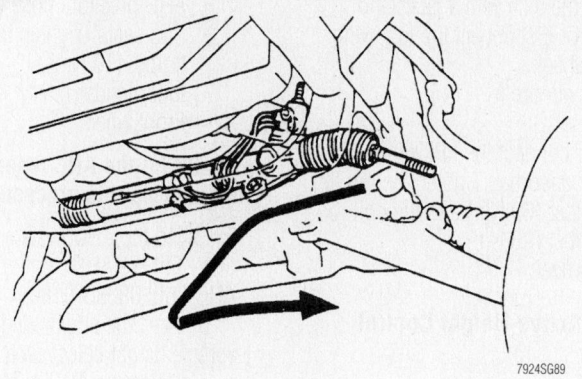

Power rack and pinion steering gear removal—LX470

7924SG89

REMOVAL & INSTALLATION

GX470

1. Before servicing the vehicle, refer to the precautions in the beginning of this section.
2. Remove the wheel.
3. Remove the stabilizer bar.
4. Remove the clamps and connector.
5. Remove the wire bracket.
6. Remove the lower strut bolt.

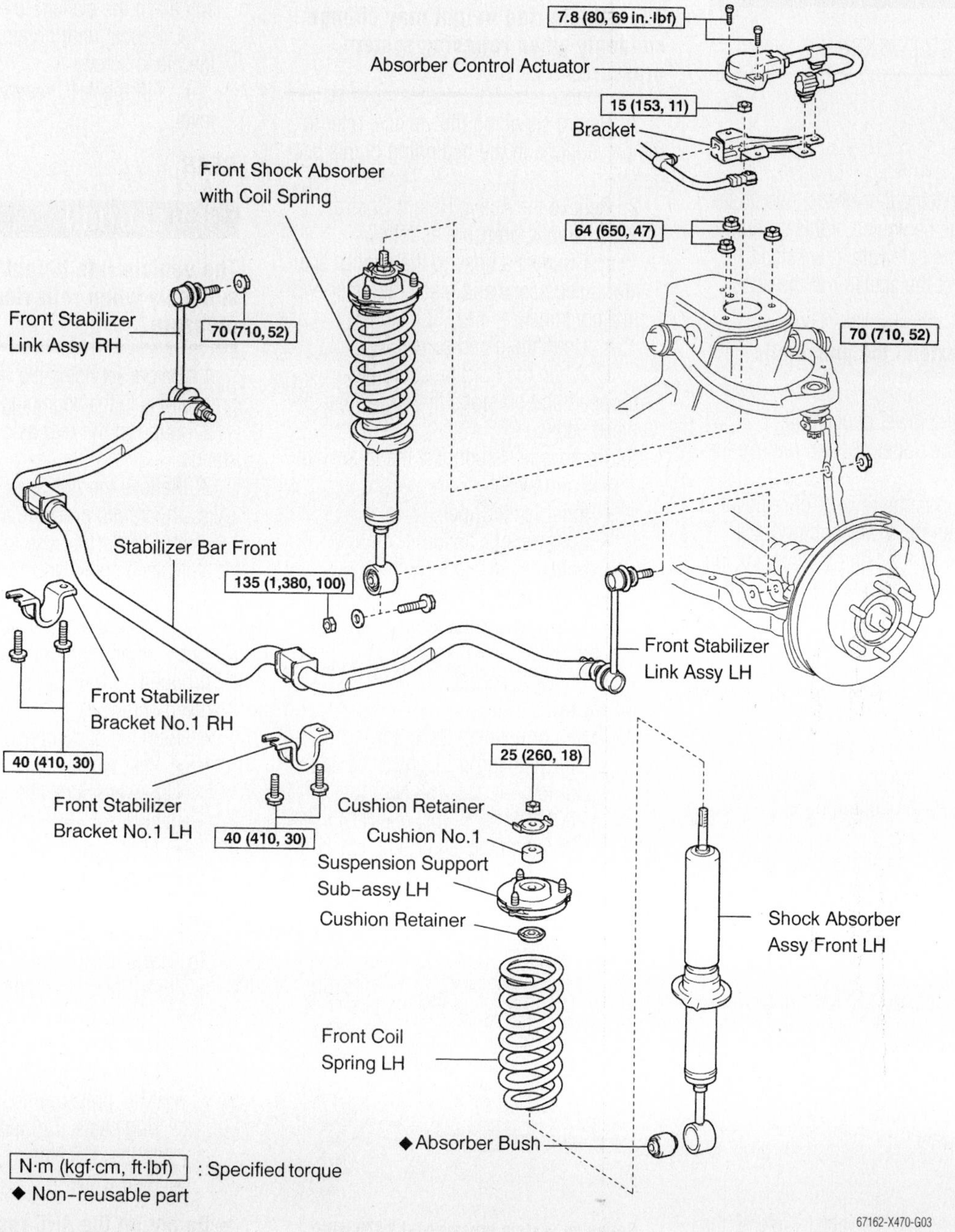

7.8 (80, 69 in.·lbf)

Absorber Control Actuator

15 (153, 11)

Bracket

64 (650, 47)

Front Shock Absorber with Coil Spring

Front Stabilizer Link Assy RH

70 (710, 52)

70 (710, 52)

Stabilizer Bar Front

135 (1,380, 100)

Front Stabilizer Link Assy LH

Front Stabilizer Bracket No.1 RH

40 (410, 30)

Front Stabilizer Bracket No.1 LH

40 (410, 30)

25 (260, 18)

Cushion Retainer

Cushion No.1

Suspension Support Sub–assy LH

Cushion Retainer

Shock Absorber Assy Front LH

Front Coil Spring LH

◆Absorber Bush

N·m (kgf·cm, ft·lbf) : Specified torque
◆ Non–reusable part

67162-X470-G03

Front strut and related components—GX470

7. Remove the 3 upper strut nuts.

8. Remove the strut.

9. Installation is the reverse of removal. Do not fully tighten the lower strut bolt until the vehicle is resting on the ground and the suspension has been jounced a few times. Observe the following torques:

- Upper nuts: 47 ft. lbs. (64 Nm)
- Bracket nut: 11 ft. lbs. (15 Nm)
- Stabilizer bar links: 52 ft. lbs. (70 Nm)
- Wheel: 83 ft. lbs. (112 Nm)
- Lower strut bolt: 100 ft. lbs. (135 Nm)

Shock Absorbers

REMOVAL & INSTALLATION

GX470

REAR

1. Before servicing the vehicle, refer to the precautions in the beginning of this section.

2. Support the axle with a jackstand.

3. Disconnect the actuator at the shock absorber.

➡ **Don't over-extend the pneumatic shock.**

4. Remove the lower shock bolt.

5. Remove the upper nut and remove the shock.

6. Installation is the reverse of removal. Don't fully tighten the lower bolt until the vehicle is on the ground and the suspension jounced a few times. Torque the upper nut to 18 ft. lbs. (25 Nm); the lower bolt to 72 ft. lbs. (98 Nm).

LX470 Without Active Height Control

FRONT

1. Before servicing the vehicle, refer to the precautions in the beginning of this section.

2. Support the axle with a jackstand.

3. Remove or disconnect the following:

- Front wheel
- Shock absorber

To install:

4. Install or connect the following:

- Shock absorber. Tighten the nut to 51 ft. lbs. (69 Nm) for 2001–03; 57 ft. lbs. (78 Nm) for 2004; and the bolt to 100 ft. lbs. (135 Nm).
- Front wheel

REAR

1. Before servicing the vehicle, refer to the precautions in the beginning of this section.

2. Support the axle with a jackstand.

3. Remove or disconnect the following:

- Rear wheel
- Shock absorber

To install:

4. Install or connect the following:

- Shock absorber. Tighten the nut to 51 ft. lbs. (69 Nm) and the bolt to 72 ft. lbs. (98 Nm).
- Rear wheel

LX470 With Active Height Control

FRONT

✱✱ CAUTION

The vehicle ride height may change suddenly when relieving system pressure.

1. Before servicing the vehicle, refer to the precautions in the beginning of this section.

2. Relieve the Active Height Control (AHC) hydraulic pressure as follows:

  a. Connect a hose to the control actuator bleed screw and place the other end in a container.

  b. Open the bleed screw.

  c. When the fluid pressure has dropped and oil stops flowing, close the bleed screw.

3. Remove or disconnect the following:

- Front wheel
- Inner fender liner
- Lower shock absorber mounting bolt
- AHC pressure hose
- Upper shock absorber mounting nut
- Shock absorber

To install:

4. Install or connect the following:

- Shock absorber. Tighten the upper nut to 51 ft. lbs. (68 Nm) for 2001–03; 57 ft. lbs. (78 Nm) for 2004; and the lower bolt to 101 ft. lbs. (135 Nm).

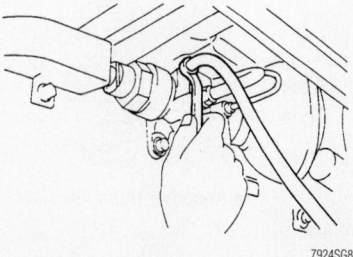

Relieving system pressure—LX470 with Active Height Control (AHC)

- AHC pressure hose with new O-ring seals. Tighten the bolts to 13 ft. lbs. (18 Nm).
- Inner fender liner
- Front wheel

➡ **Do not let the AHC reservoir run empty during this procedure.**

5. Bleed the AHC system as follows:

  a. Fill the AHC system reservoir with AHC fluid 08886-01805.

  b. Start the engine and push **N** on the vehicle height select switch.

  c. When the AHC pump stops, turn the engine **OFF**.

  d. Open the bleed screw and allow any air in the system to escape.

  e. Repeat until no air is expelled from the bleed screw.

  f. Fill the AHC reservoir to the correct level.

REAR

✱✱ CAUTION

The vehicle ride height may change suddenly when relieving system pressure.

1. Before servicing the vehicle, refer to the precautions in the beginning of this section.

2. Support the rear axle with a jack or stands.

3. Relieve the Active Height Control (AHC) hydraulic pressure as follows:

  a. Connect a hose to the control actuator bleed screw and place the other end in a container.

  b. Open the bleed screw.

  c. When the fluid pressure has dropped and oil stops flowing, close the bleed screw.

4. Remove or disconnect the following:

- Rear wheel
- Lower shock absorber mounting bolt
- AHC pressure hose
- Upper shock absorber mounting nut
- Shock absorber

To install:

5. Install or connect the following:

- Shock absorber. Tighten the upper nut to 51 ft. lbs. (68 Nm) and the lower bolt to 72 ft. lbs. (98 Nm).
- AHC pressure hose with new O-ring seals. Tighten the bolts to 13 ft. lbs. (18 Nm).
- Rear wheel

➡ **Do not let the AHC reservoir run empty during this procedure.**

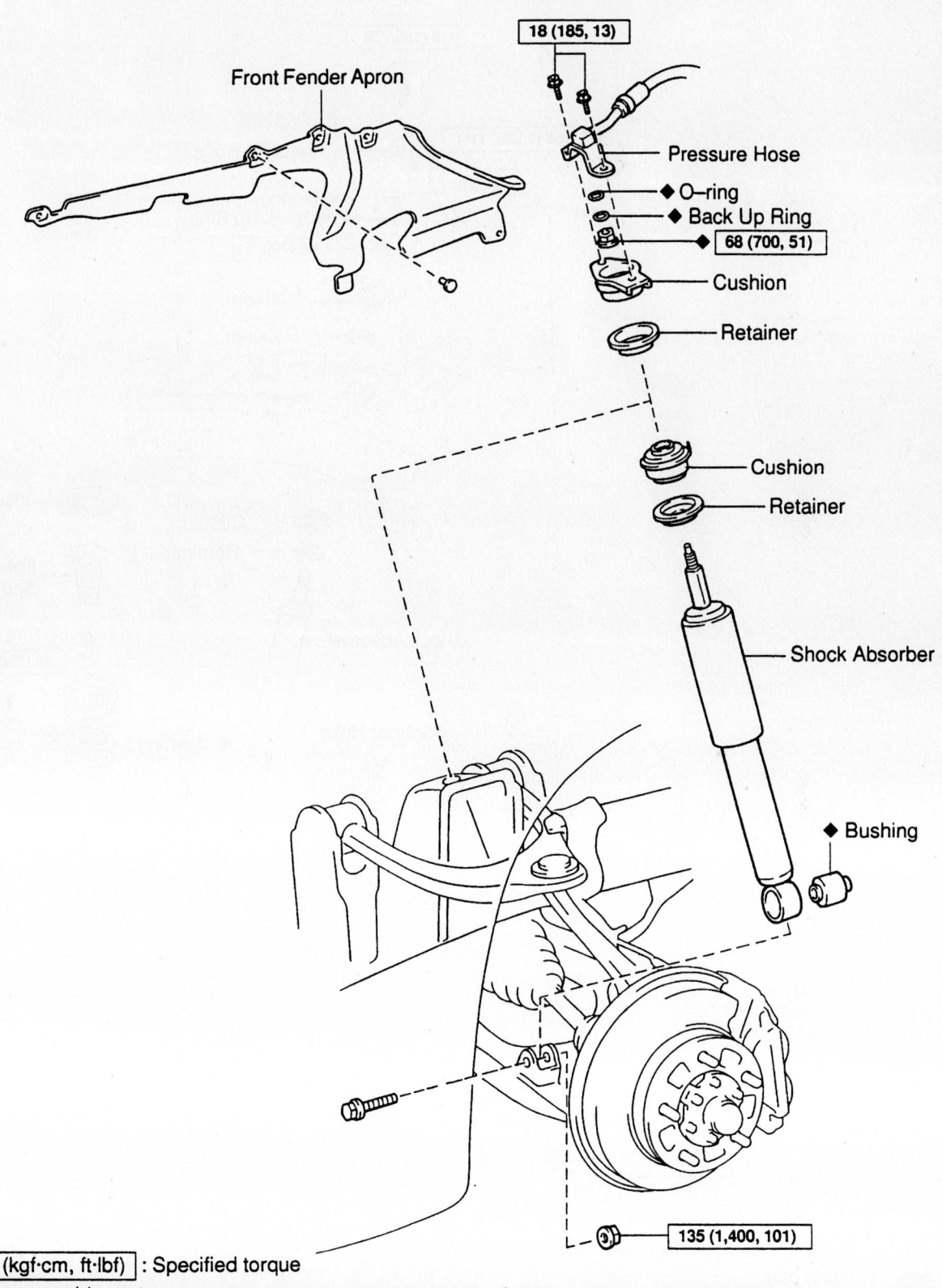

18 (185, 13)

Front Fender Apron

Pressure Hose

◆ O–ring
◆ Back Up Ring
◆ 68 (700, 51)

Cushion

Retainer

Cushion

Retainer

Shock Absorber

◆ Bushing

135 (1,400, 101)

N·m (kgf·cm, ft·lbf) : Specified torque
◆ Non–reusable part

7924SG86

Exploded view of the front shock absorber mounting—LX470 models with Active Height Control (AHC)

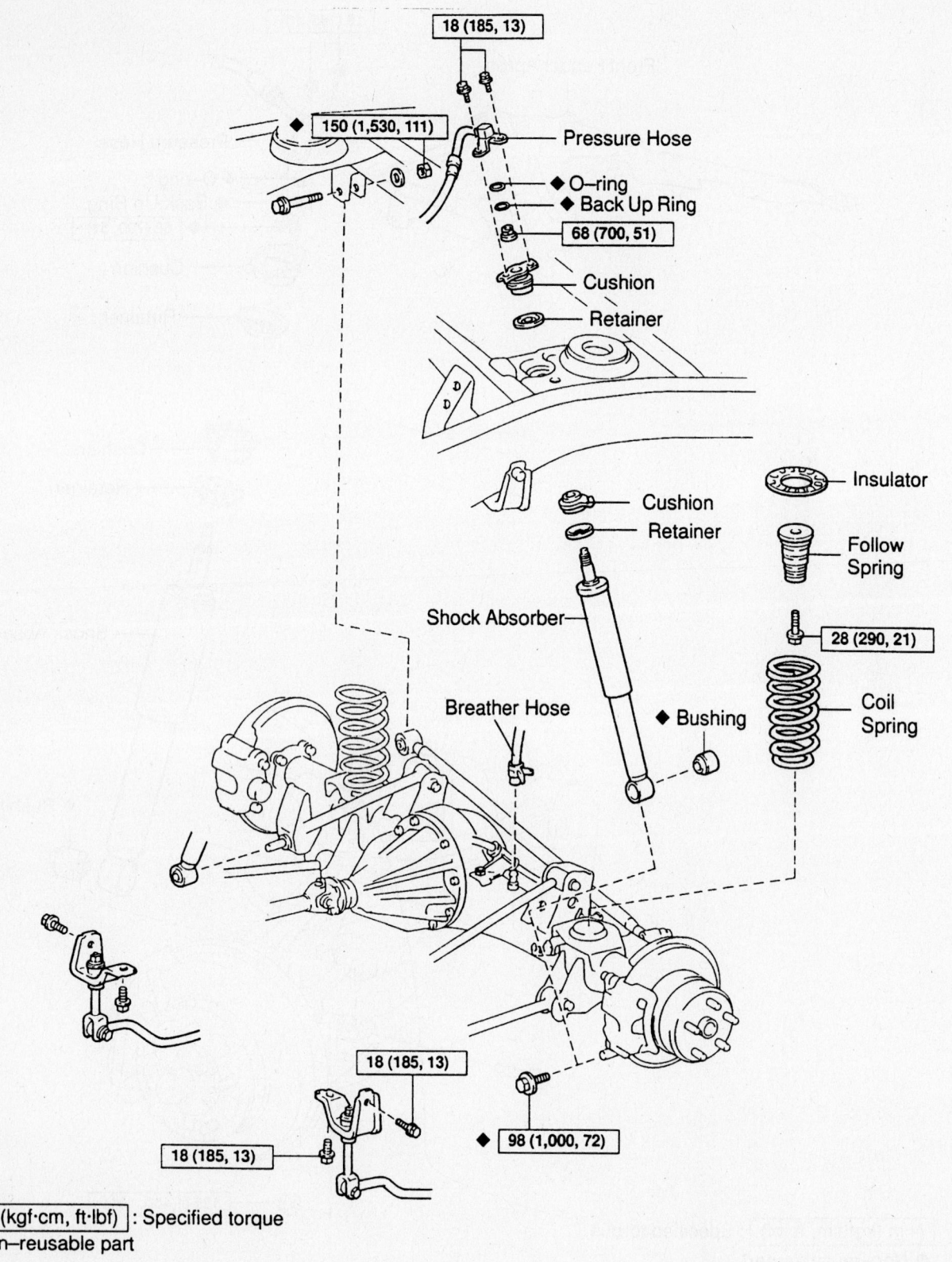

18 (185, 13)

150 (1,530, 111)

Pressure Hose

◆ O-ring

◆ Back Up Ring

68 (700, 51)

Cushion

Retainer

Insulator

Cushion

Retainer

Follow Spring

Shock Absorber

Breather Hose

28 (290, 21)

◆ Bushing

Coil Spring

18 (185, 13)

18 (185, 13)

◆ 98 (1,000, 72)

18 (185, 13)

N·m (kgf·cm, ft·lbf) : Specified torque
◆ Non-reusable part

7924SG87

Exploded view of the rear shock absorber mounting—LX470 with Active Height Control (AHC)

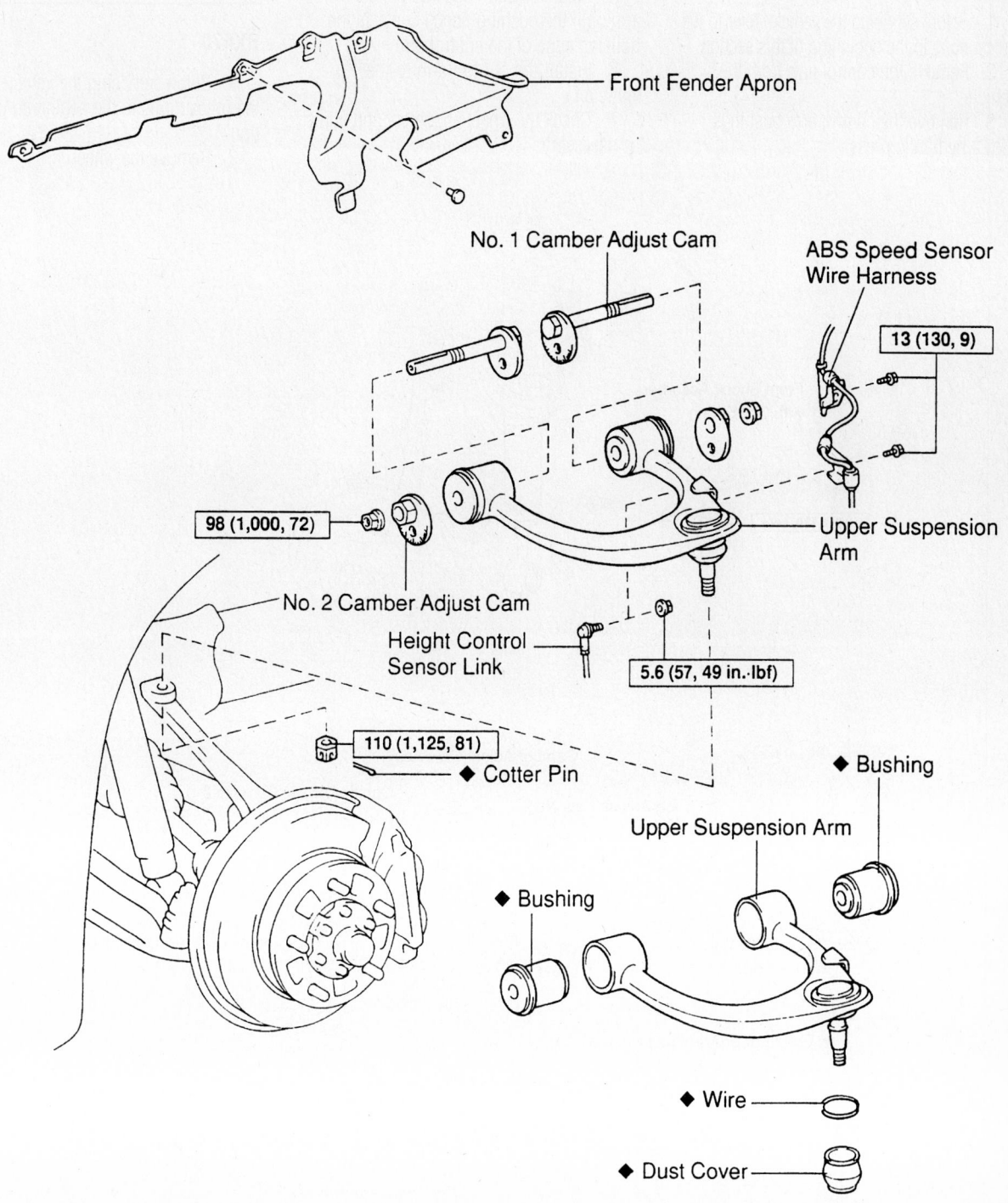

Front Fender Apron

No. 1 Camber Adjust Cam

ABS Speed Sensor
Wire Harness

13 (130, 9)

98 (1,000, 72)

Upper Suspension
Arm

No. 2 Camber Adjust Cam

Height Control
Sensor Link

5.6 (57, 49 in.·lbf)

110 (1,125, 81)

◆ Cotter Pin

◆ Bushing

Upper Suspension Arm

◆ Bushing

◆ Wire

◆ Dust Cover

N·m (kgf·cm, ft·lbf) : Specified torque

◆ Non–reusable part

9302SG01

Exploded view of the upper control arm and related components—LX470

CONTROL ARM BUSHING REPLACEMENT

LX470

1. Before servicing the vehicle, refer to the precautions in the beginning of this section.
2. Remove the control arm from the vehicle.
3. Remove the control arm bushings with a hydraulic press.

To install:

4. Lubricate the control arm bushings with liquid soap.
5. Press the bushings into the control arm until the bushing flange contacts the housing edge of the control arm.
6. Install the control arm to the vehicle.
7. Check the wheel alignment and adjust as necessary.

Lower Control Arm

REMOVAL & INSTALLATION

GX470

1. Before servicing the vehicle, refer to the precautions in the beginning of this section.
2. Remove the wheel.

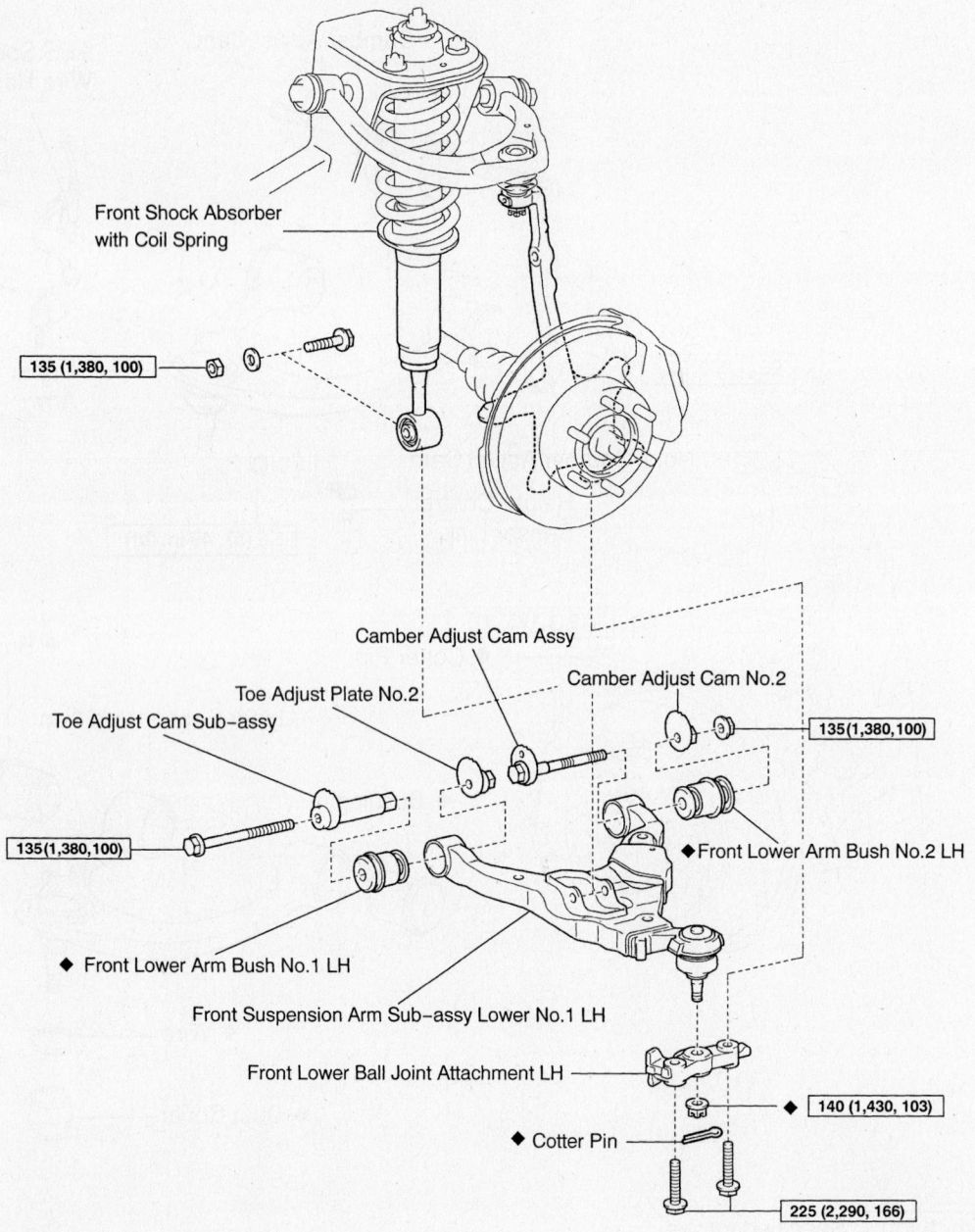

Front Shock Absorber with Coil Spring

135 (1,380, 100)

Camber Adjust Cam Assy

Toe Adjust Plate No.2

Camber Adjust Cam No.2

Toe Adjust Cam Sub–assy

135(1,380,100)

135(1,380,100)

◆ Front Lower Arm Bush No.2 LH

◆ Front Lower Arm Bush No.1 LH

Front Suspension Arm Sub–assy Lower No.1 LH

Front Lower Ball Joint Attachment LH

◆ Cotter Pin

140 (1,430, 103)

225 (2,290, 166)

N·m (kgf·cm, ft·lbf) : Specified torque

◆ Non–reusable part

Lower control arm and related parts—GX470

67162-X470-G05

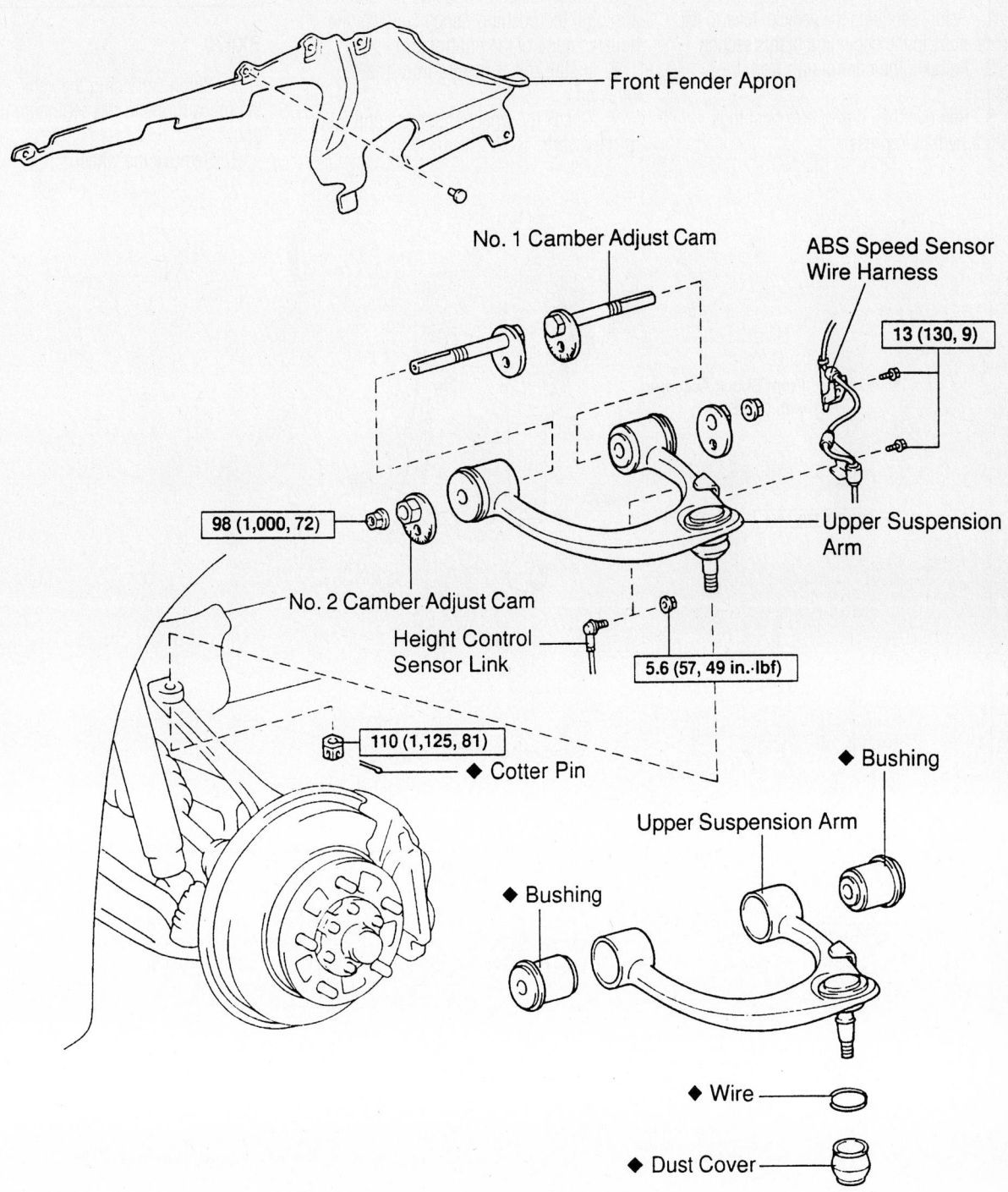

Front Fender Apron

No. 1 Camber Adjust Cam

ABS Speed Sensor
Wire Harness

13 (130, 9)

98 (1,000, 72)

Upper Suspension
Arm

No. 2 Camber Adjust Cam

Height Control
Sensor Link

5.6 (57, 49 in.·lbf)

110 (1,125, 81)

◆ Cotter Pin

◆ Bushing

Upper Suspension Arm

◆ Bushing

◆ Wire

◆ Dust Cover

N·m (kgf·cm, ft·lbf) : Specified torque

◆ Non–reusable part

9302SG01

Exploded view of the upper control arm and related components—LX470

CONTROL ARM BUSHING REPLACEMENT

LX470

1. Before servicing the vehicle, refer to the precautions in the beginning of this section.
2. Remove the control arm from the vehicle.
3. Remove the control arm bushings with a hydraulic press.

To install:

4. Lubricate the control arm bushings with liquid soap.
5. Press the bushings into the control arm until the bushing flange contacts the housing edge of the control arm.
6. Install the control arm to the vehicle.
7. Check the wheel alignment and adjust as necessary.

Lower Control Arm

REMOVAL & INSTALLATION

GX470

1. Before servicing the vehicle, refer to the precautions in the beginning of this section.
2. Remove the wheel.

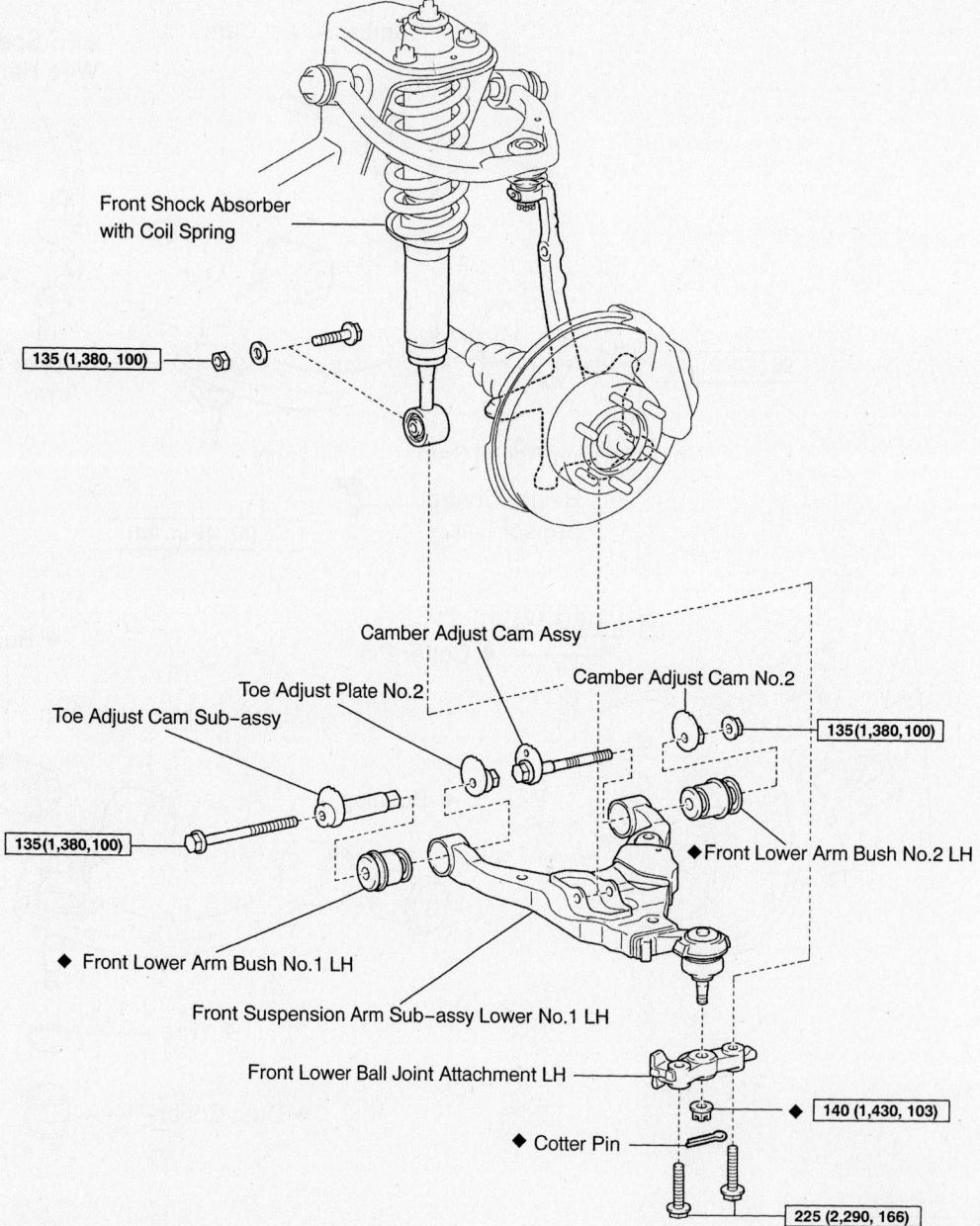

Front Shock Absorber with Coil Spring

135 (1,380, 100)

Camber Adjust Cam Assy

Toe Adjust Plate No.2

Camber Adjust Cam No.2

Toe Adjust Cam Sub-assy

135 (1,380, 100)

135 (1,380, 100)

◆ Front Lower Arm Bush No.2 LH

◆ Front Lower Arm Bush No.1 LH

Front Suspension Arm Sub-assy Lower No.1 LH

Front Lower Ball Joint Attachment LH

◆ 140 (1,430, 103)

◆ Cotter Pin

225 (2,290, 166)

N·m (kgf·cm, ft·lbf) : Specified torque

◆ Non-reusable part

67162-X470-G05

Lower control arm and related parts—GX470

6. Bleed the AHC system as follows:

a. Fill the AHC system reservoir with AHC fluid 08886-01805.

b. Start the engine and push **N** on the vehicle height select switch.

c. When the AHC pump stops, turn the engine **OFF**.

d. Open the bleed screw and allow any air in the system to escape.

e. Repeat until no air is expelled from the bleed screw.

f. Fill the AHC reservoir to the correct level.

Front Coil Spring

REMOVAL & INSTALLATION

GX470

1. Remove the strut.

2. Place the strut in a compressor, such as SST 09727-30021, and compress the spring.

3. Hold the rod and remove the nut.

➡**Don't use an impact wrench.**

4. Remove the bushing retainer.
5. Remove the upper bushing.
6. Remove the support.
7. Remove the lower bushing retainer.
8. Remove the spring.
9. Remove the lower bushing.

To install:

10. Install the new lower bushing.
11. Compress the spring and install it.
12. Install the bushing retainer.
13. Install the suspension support.
14. Install the upper bushing.
15. Install the retainer.
16. Align the support, rod and bushing as shown. Install the locknut and torque to 18 ft. lbs. (25 Nm).
17. Release the spring from the compressor and check the alignment of the parts.
18. Install the strut.

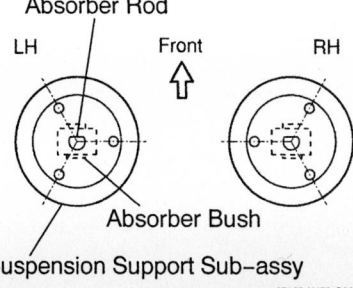

Absorber Rod

LH   Front   RH

Absorber Bush

Suspension Support Sub-assy

67162-X470-G06

Align the support, rod and bushing as shown

Rear Coil Spring

REMOVAL & INSTALLATION

LX470

1. Before servicing the vehicle, refer to the precautions in the beginning of this section.
2. Support the vehicle at the frame.
3. Support the axle with a floor jack.
4. Remove or disconnect the following:
   • Rear wheel
   • Shock absorber
   • Stabilizer bar brackets
   • Lateral control rod
   • Coil spring

To install:

5. Install or connect the following:
   • Coil spring
   • Lateral control rod. Tighten the axle housing bolt to 181 ft. lbs. (245 Nm) for 2001–03; 110 ft. lbs. (149 Nm) for 2004.
   • Stabilizer bar brackets. Tighten the bolts to 13 ft. lbs. (18 Nm)
   • Shock absorber
   • Rear wheel

Rear Air Spring

REMOVAL & INSTALLATION

GX470

1. Before servicing the vehicle, refer to the precautions in the beginning of this section.

2. Remove the wheel.

3. Support the frame with jackstands and allow the axle to hang.

4. Disconnect the height control tube.

5. Disconnect the clip on the underside of the air spring. If the clip is difficult to remove, thread a wire through the hole and pull it. Discharge the air from the air spring to retract it.

6. Turn the unit 90 degrees and remove it from the axle.

➡**Don't manually extend the unit.**

7. Installation is the reverse of removal. Use new O-rings on the height control tube.

Torsion Bars

REMOVAL & INSTALLATION

LX470

1. Before servicing the vehicle, refer to the precautions in the beginning of this section.

2. Remove or disconnect the following:
   • Front wheel
   • Engine under cover

3. Measure dimension **A** as shown between the adjustment bolt head and the frame.

4. Loosen the adjusting bolt until all spring tension is relieved.

5. Measure dimension **B** as shown between the adjustment bolt head and the frame.

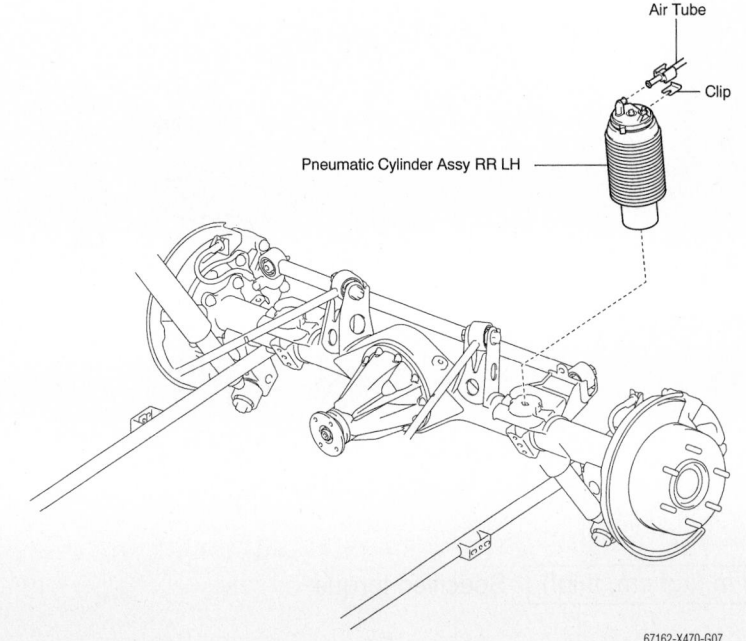

Air Tube

Clip

Pneumatic Cylinder Assy RR LH

67162-X470-G07

Rear air spring—GX470

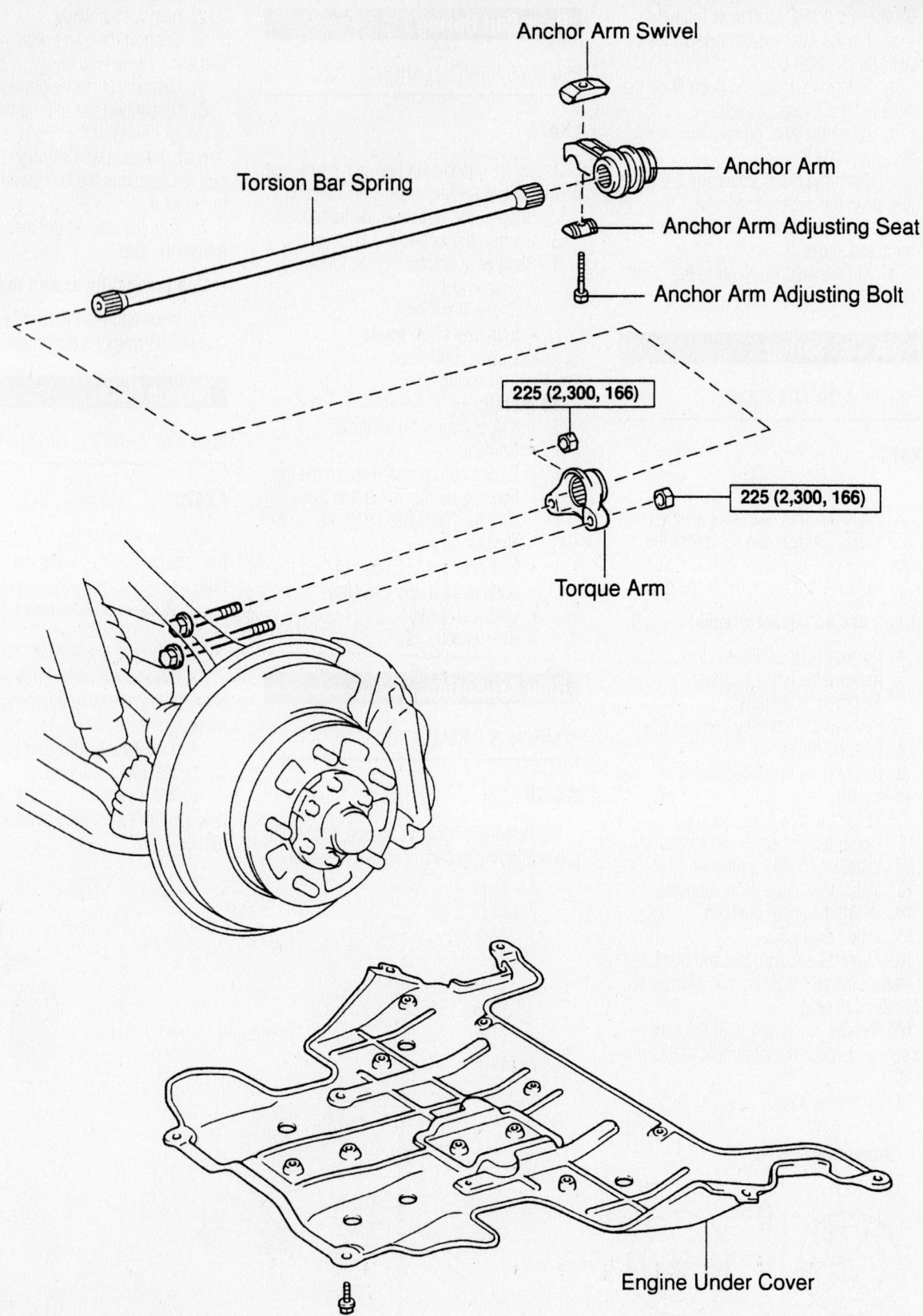

Anchor Arm Swivel

Anchor Arm

Anchor Arm Adjusting Seat

Anchor Arm Adjusting Bolt

Torsion Bar Spring

225 (2,300, 166)

225 (2,300, 166)

Torque Arm

Engine Under Cover

N·m (kgf·cm, ft·lbf) : Specified torque

9308SG10

Torsion bar mounting exploded view—LX470

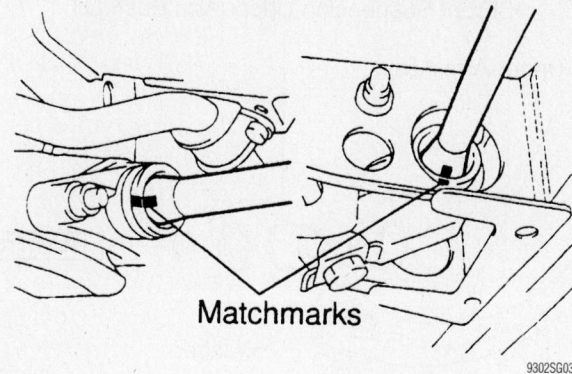

Matchmark the torsion bar to the anchor arm and torque arm—LX470

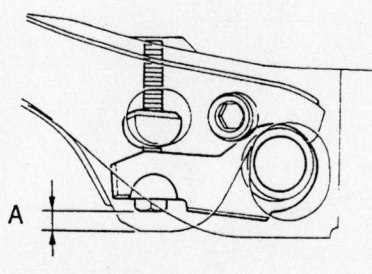

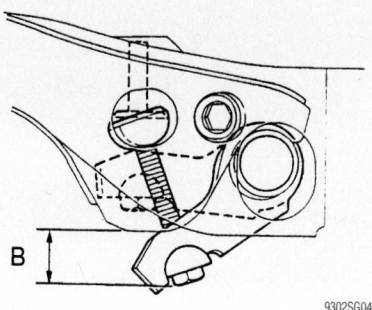

Reference measurements A and B—LX470

Front:

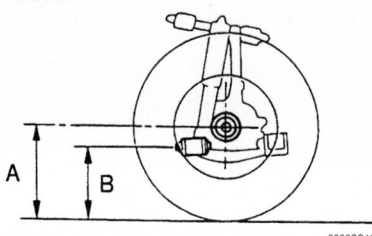

Ride height measurements A and B—LX470

6. Remove or disconnect the following:
    • Adjustment bolt, swivel and seat
    • Torsion bar and anchor arm. Separate the anchor arm from the torsion bar.
    • Torque arm

To install:

7. Install or connect the following:
    • Torque arm. Tighten the fasteners to 166 ft. lbs. (225 Nm).
    • Torsion bar and anchor arm. Align the matchmarks.
    • Adjustment bolt, swivel and seat
8. Check that dimension **B** is close to the measurement made at disassembly.
9. If installing a new torsion bar, tighten the adjustment bolt until dimension **A** is as follows:
    • Left torsion bar: 0.315–0.984 inches (8–25mm)
    • Right torsion bar: 0.079–0.709 inches (2–18mm)
10. If installing the original torsion bar, tighten the adjustment bolt until dimension **A** is close to the measurement made at disassembly.
11. Install or connect the following:
    • Engine under cover
    • Front wheel
12. Place the vehicle on a flat, level surface and check the vehicle curb height as follows:
    a. Step 1: Measure dimension **A** between the spindle center and the ground.
    b. Step 2: Measure dimension **B** between the lower control arm front bolt center and the ground.
    c. Step 3: Turn the adjusting bolt so that **A** minus **B** is equal to 2.795 inches (71mm).

Upper Ball Joint

REMOVAL & INSTALLATION

The upper ball joint is serviced with the upper control arm as an assembly.

Lower Ball Joint

REMOVAL & INSTALLATION

The lower ball joint is serviced with the lower control arm as an assembly.

Upper Control Arm

REMOVAL & INSTALLATION

GX470

1. Before servicing the vehicle, refer to the precautions in the beginning of this section.
2. Remove the wheel.
3. Disconnect the skid control wire.
4. Support the lower arm with a jack.
5. Remove the cable bracket.
6. Disconnect the ball joint from the knuckle.
7. Remove the through-bolt, washers and nut.
8. Remove the arm.
9. Installation is the reverse of removal. Don't fully tighten the through-bolt until the vehicle is on the ground and the suspension is jounced a few times.
    • Ball joint nut: 81 ft. lbs. (110 Nm)
    • Through-bolt: 85 ft. lbs. (115 Nm)

LX470

1. Before servicing the vehicle, refer to the precautions in the beginning of this section.
2. Remove or disconnect the following:
    • Front wheel
    • Inner fender liner
    • Wheel speed sensor harness
    • Upper ball joint
    • Adjustment cam bolts
    • Upper control arm

To install:

3. Install or connect the following:
    • Upper control arm. Tighten the adjustment cam bolts to 72 ft. lbs. (98 Nm).
    • Upper ball joint. Tighten the nut to 81 ft. lbs. (110 Nm).
    • Wheel speed sensor harness. Tighten the bolts to 10 ft. lbs. (13 Nm).
    • Inner fender liner
    • Front wheel
4. Check the wheel alignment and adjust as necessary.

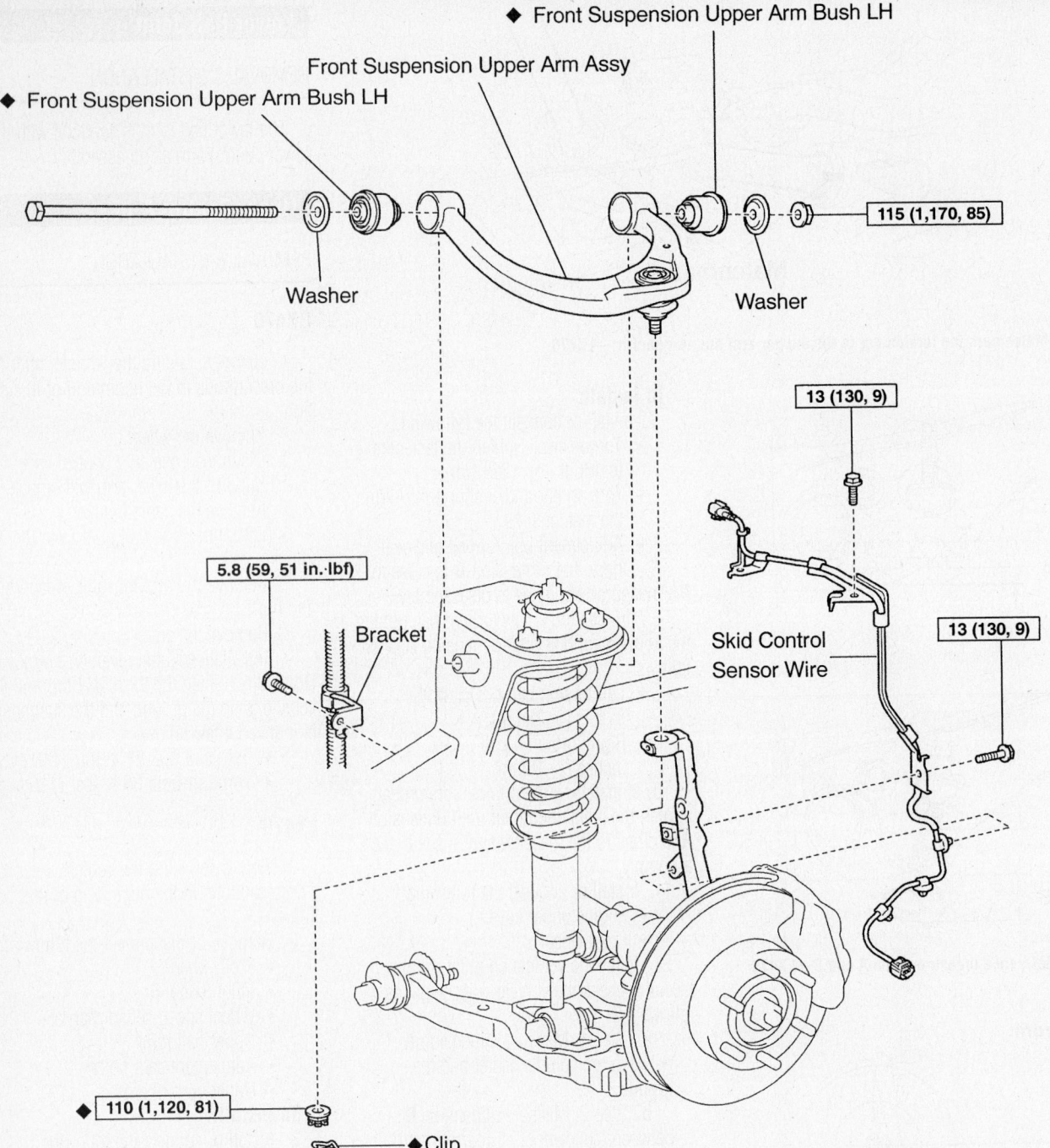

◆ Front Suspension Upper Arm Bush LH

Front Suspension Upper Arm Assy

◆ Front Suspension Upper Arm Bush LH

Washer

Washer

115 (1,170, 85)

13 (130, 9)

5.8 (59, 51 in.·lbf)

Bracket

Skid Control
Sensor Wire

13 (130, 9)

110 (1,120, 81)

◆Clip

N·m (kgf·cm, ft·lbf) : Specified torque

◆ Non-reusable part

67162-X470-G04

Upper control arm and related parts—GX470

3. Support the lower arm with a jack.

4. Remove the lower strut bolt.

5. Remove the 2 bolts and separate the lower ball joint attachment from the knuckle.

6. Place matchmarks on the camber adjusting cam and toe adjusting cam.

7. Remove the 2 nuts and remove the arm along with the cams.

8. Installation is the reverse of removal. Align all matchmarks. Use new nuts and

cotter pins. Don't fully tighten the control arm bolts until the vehicle is on the ground and the suspension jounced a few times. Observe the following torques:

- Lower ball joint stud: 103 ft. lbs. (140 Nm)
- Lower ball joint attachment bolts: 166 ft. lbs. (225 Nm)
- Lower arm bolts: 100 ft. lbs. (135 Nm)

LX470

1. Before servicing the vehicle, refer to the precautions in the beginning of this section.

2. Remove or disconnect the following:
- Front wheel
- Engine under cover
- Torsion bar
- Stabilizer bar link
- Shock absorber

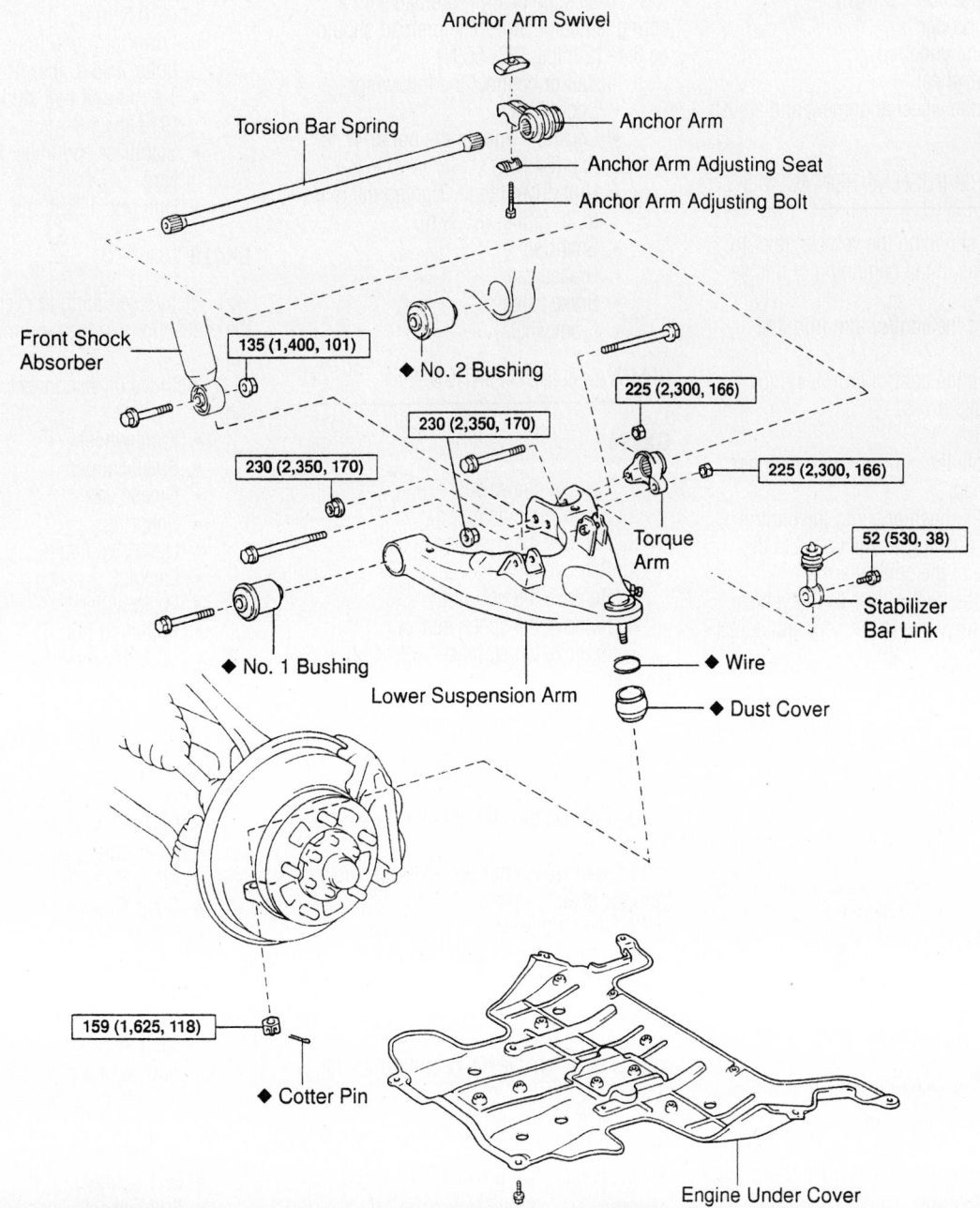

Anchor Arm Swivel

Torsion Bar Spring

Anchor Arm

Anchor Arm Adjusting Seat

Anchor Arm Adjusting Bolt

Front Shock Absorber

135 (1,400, 101)

◆ No. 2 Bushing

225 (2,300, 166)

230 (2,350, 170)

230 (2,350, 170)

225 (2,300, 166)

Torque Arm

52 (530, 38)

Stabilizer Bar Link

◆ No. 1 Bushing

◆ Wire

Lower Suspension Arm

◆ Dust Cover

159 (1,625, 118)

◆ Cotter Pin

Engine Under Cover

N·m (kgf·cm, ft·lbf) : Specified torque

◆ Non−reusable part

9302SG02

Exploded view of the lower control arm and related components—2001 LX470

- Lower ball joint
- Lower control arm

To install:

3. Install or connect the following:
- Lower control arm. Tighten the bolts to 170 ft. lbs. (230 Nm) for 2001–03; 123 ft. lbs. (167 Nm) for 2004.
- Lower ball joint. Tighten the nut to 117 ft. lbs. (159 Nm).
- Shock absorber
- Stabilizer bar link. Tighten the bolt to 38 ft. lbs. (52 Nm).
- Torsion bar
- Engine under cover
- Front wheel

4. Check the wheel alignment and adjust as necessary.

CONTROL ARM BUSHING REPLACEMENT

1. Before servicing the vehicle, refer to the precautions in the beginning of this section.

2. Remove the control arm from the vehicle.

3. Remove the control arm bushings with a hydraulic press.

To install:

4. Lubricate the control arm bushings with liquid soap.

5. Press the bushings into the control arm until the bushing flange contacts the housing edge of the control arm.

6. Install the control arm to the vehicle.

7. Check the wheel alignment and adjust as necessary.

Front Wheel Bearing

ADJUSTMENT

GX470

1. Before servicing the vehicle, refer to the precautions in the beginning of this section.

No adjustment is possible. Check for axle hub backlash and axle hub deviation. If either exceeds 0.0020 in., replace the bearing.

LX470

1. Before servicing the vehicle, refer to the precautions in the beginning of this section.

2. Remove or disconnect the following:
- Front wheel
- Brake caliper
- Grease cap
- Snapring
- Hub drive flange
- Locknut
- Lockwasher

3. Tighten the adjusting nut to 43 ft. lbs. (59 Nm) while rotating the hub to seat the bearings.

4. Loosen the adjusting nut.

5. Tighten the adjusting nut to 48 inch lbs. (5.4 Nm) and check that the bearing has no play.

6. Check the bearing preload with a spring tension gauge. The preload should be 6.4–12.6 lbs. (28–56 N).

7. Install or connect the following:
- Lockwasher
- Locknut. Tighten the nut to 47 ft. lbs. (64 Nm).
- Hub drive flange. Tighten the nuts to 26 ft. lbs. (35 Nm).
- Snapring
- Grease cap
- Brake caliper
- Front wheel

REMOVAL & INSTALLATION

GX470

1. Remove the wheel.
2. Remove the caliper.
3. Remove the hub grease cap.
4. Remove the cotter pin.
5. Remove the hub nut.
6. Remove the speed sensor.
7. Remove the stabilizer links from the knuckles.
8. Remove the tie rod end from the knuckle.
9. Remove the lower arm from the knuckle.
10. Remove the upper arm from the knuckle.
11. Remove the hub/knuckle assembly from the shaft.
12. Mount the assembly in a vise.
13. Remove the knuckle oil seal.
14. Remove the 4 bolts and remove the hub assembly from the knuckle.
15. Using SST 09710-30021 and its components, remove the bearing from the hub.
16. Remove the oil seal.

To install:

17. Using a seal driver, install a new seal.

➡**Take care to avoid damage to the spacer.**

18. Press a new bearing into the hub.

19. Coat a new O-ring with MP grease and install it in the hub.

20. Attach the hub to the knuckle. Torque to 59 ft. lbs. (80 Nm).

21. Install a new knuckle oil seal.

22. The remainder of installation is the reverse of removal. Observe the following torques:
- Upper arm ball stud nut: 81 ft. lbs. (110 Nm)
- Lower arm ball joint attachment bolts: 166 ft. lbs. (225 Nm)
- Tie rod end ball stud nut: 67 ft. lbs. (91 Nm)
- Stabilizer end links: 52 ft. lbs. (70 Nm)
- Hub nut: 173 ft. lbs. (235 Nm)

LX470

1. Before servicing the vehicle, refer to the precautions in the beginning of this section.

2. Remove or disconnect the following:
- Front wheel
- Brake caliper
- Grease cap
- Snapring
- Hub drive flange
- Locknut
- Lockwasher
- Adjusting nut
- Outer bearing
- Wheel hub
- Inner grease seal
- Inner bearing

To install:

3. Install or connect the following:
- Inner bearing
- Inner grease seal
- Wheel hub
- Outer bearing
- Adjusting nut. Adjust the wheel bearings.
- Lockwasher
- Locknut. Tighten the nut to 47 ft. lbs. (64 Nm).
- Hub drive flange. Tighten the nuts to 26 ft. lbs. (35 Nm).
- Snapring
- Grease cap
- Brake caliper
- Front wheel

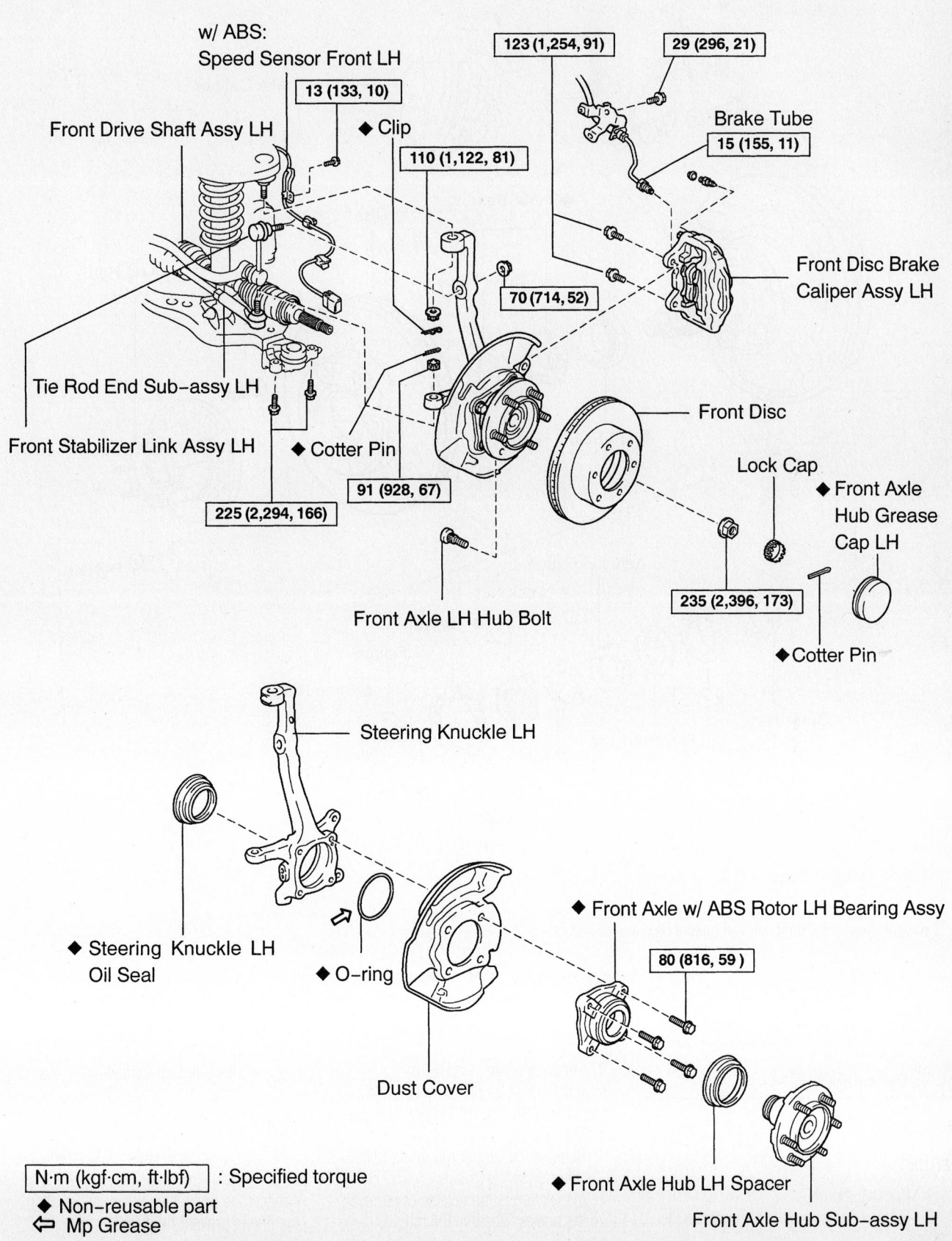

w/ ABS:
Speed Sensor Front LH

Front Drive Shaft Assy LH

13 (133, 10)

◆ Clip

110 (1,122, 81)

123 (1,254, 91)

29 (296, 21)

Brake Tube

15 (155, 11)

Front Disc Brake
Caliper Assy LH

70 (714, 52)

Tie Rod End Sub-assy LH

Front Stabilizer Link Assy LH

◆ Cotter Pin

225 (2,294, 166)

91 (928, 67)

Front Disc

Lock Cap

◆ Front Axle
Hub Grease
Cap LH

235 (2,396, 173)

Front Axle LH Hub Bolt

◆ Cotter Pin

Steering Knuckle LH

◆ Steering Knuckle LH
Oil Seal

◆ O-ring

◆ Front Axle w/ ABS Rotor LH Bearing Assy

80 (816, 59)

Dust Cover

N·m (kgf·cm, ft·lbf) : Specified torque
◆ Non-reusable part
⇐ Mp Grease

◆ Front Axle Hub LH Spacer

Front Axle Hub Sub-assy LH

67162-X470-G10

Front hub and related parts—GX470

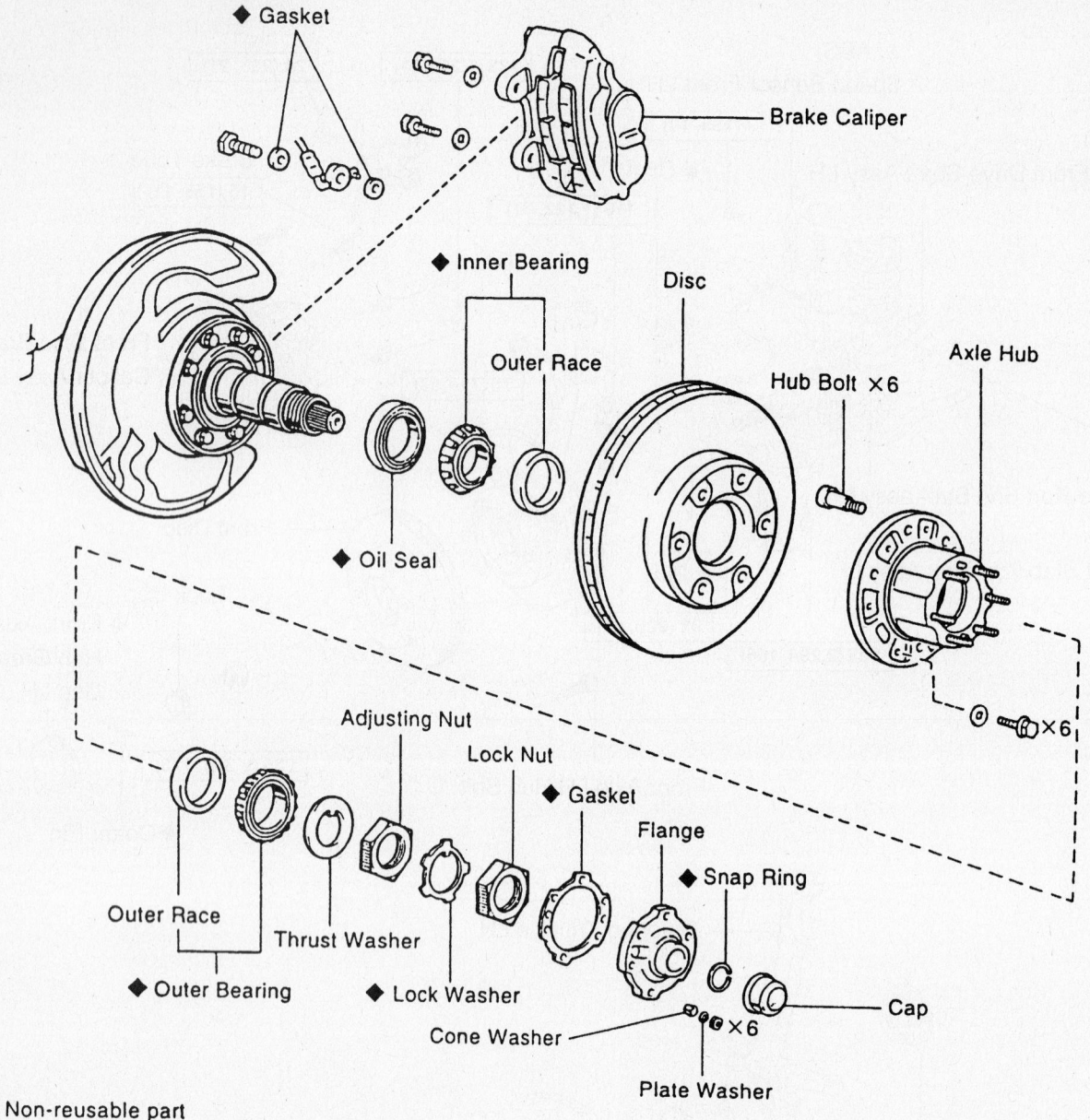

Gasket

Brake Caliper

Inner Bearing

Disc

Outer Race

Hub Bolt ×6

Axle Hub

Oil Seal

Adjusting Nut

Lock Nut

Gasket

Flange

Snap Ring

Outer Race

Thrust Washer

Outer Bearing

Lock Washer

Cap

Cone Washer

Plate Washer

◆ **Non-reusable part**

7924SG31

Exploded view of the front hub and related components—LX470

BRAKES

Brake Caliper

REMOVAL & INSTALLATION

GX470

FRONT

1. Remove the wheel.
2. Remove the anti-rattle spring from the caliper.

3. Remove the clips and anti-rattle pins.
4. Lift out the pads and shims.
5. If the caliper is being replaced, disconnect the brake line. Plug the line to prevent fluid loss.
6. Remove the caliper mounting bolts. Lift off the caliper.
7. Installation is the reverse of removal. Bleed the brakes. Observe the following torques:

- Caliper mounting bolts: 91 ft. lbs. (123 Nm)
- Brake line-to-caliper: 11 ft. lbs. (15 Nm)

REAR

1. Remove the wheel.
2. If the caliper is being replaced, disconnect the brake line at the caliper. Plug the line to prevent fluid loss.

3. Remove the 2 caliper attaching pins and lift off the caliper.

4. Installation is the reverse of removal. Torque the caliper pins to 65 ft. lbs. (88 Nm). Bleed the brakes if the line was disconnected.

LX470

FRONT

1. Disconnect the negative battery cable from the battery.

2. Raise and support the vehicle safely.

3. Remove the wheels.

4. Disconnect the brake hose from the caliper by removing the union bolt and 2 gaskets. Plug the end of the hose to prevent loss of fluid.

5. Remove the bolts that attach the caliper to the torque plate.

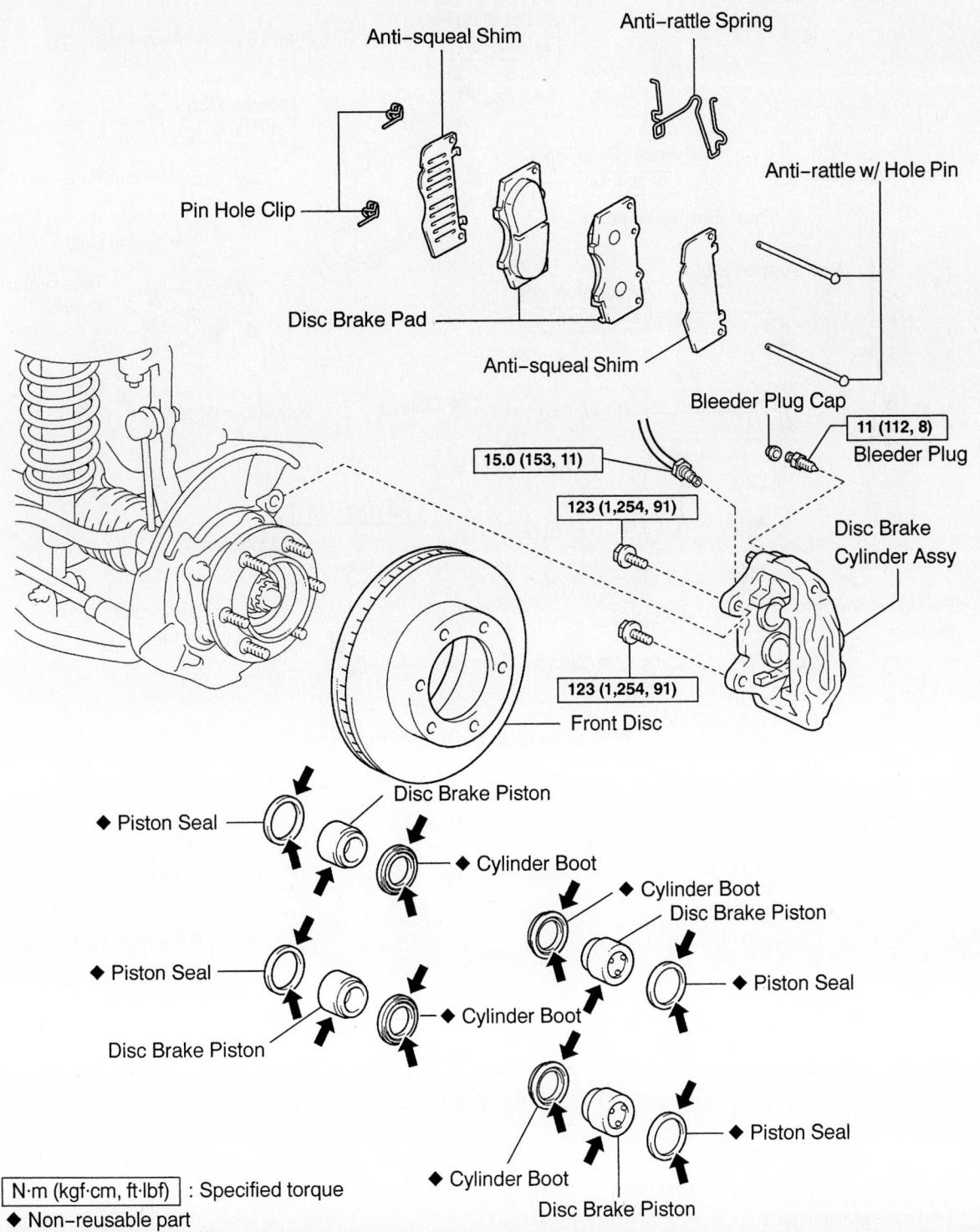

Anti-squeal Shim

Anti-rattle Spring

Pin Hole Clip

Anti-rattle w/ Hole Pin

Disc Brake Pad

Anti-squeal Shim

Bleeder Plug Cap

11 (112, 8)

Bleeder Plug

15.0 (153, 11)

123 (1,254, 91)

Disc Brake Cylinder Assy

123 (1,254, 91)

Front Disc

Disc Brake Piston

◆ Piston Seal

◆ Cylinder Boot

◆ Cylinder Boot

Disc Brake Piston

◆ Piston Seal

◆ Piston Seal

Disc Brake Piston

◆ Cylinder Boot

◆ Cylinder Boot

◆ Piston Seal

◆ Cylinder Boot

Disc Brake Piston

N·m (kgf·cm, ft·lbf) : Specified torque

◆ Non-reusable part

← Lithium soap base glycol grease

67162-X470-G12

Front brake components—GX470

6. Lift the bottom of the caliper up and remove the caliper assembly.

To install:

7. Grease the caliper slides and bolts with lithium grease or equivalent. Install the caliper and secure with the bolts. Torque the bolts to 90 ft. lbs. (123 Nm).

8. Connect the brake hose to the caliper, using 2 new washers. Make sure the flexible hose lock is securely in the lock hole of the caliper. Torque the union bolt to 22 ft. lbs. (30 Nm).

9. Fill the brake system to the proper level and bleed the brake system.

10. Install the tire and wheel assembly.

11. Top off the brake fluid level in the master cylinder. Check for leaks and proper brake operation.

12. Connect the negative battery cable to the battery.

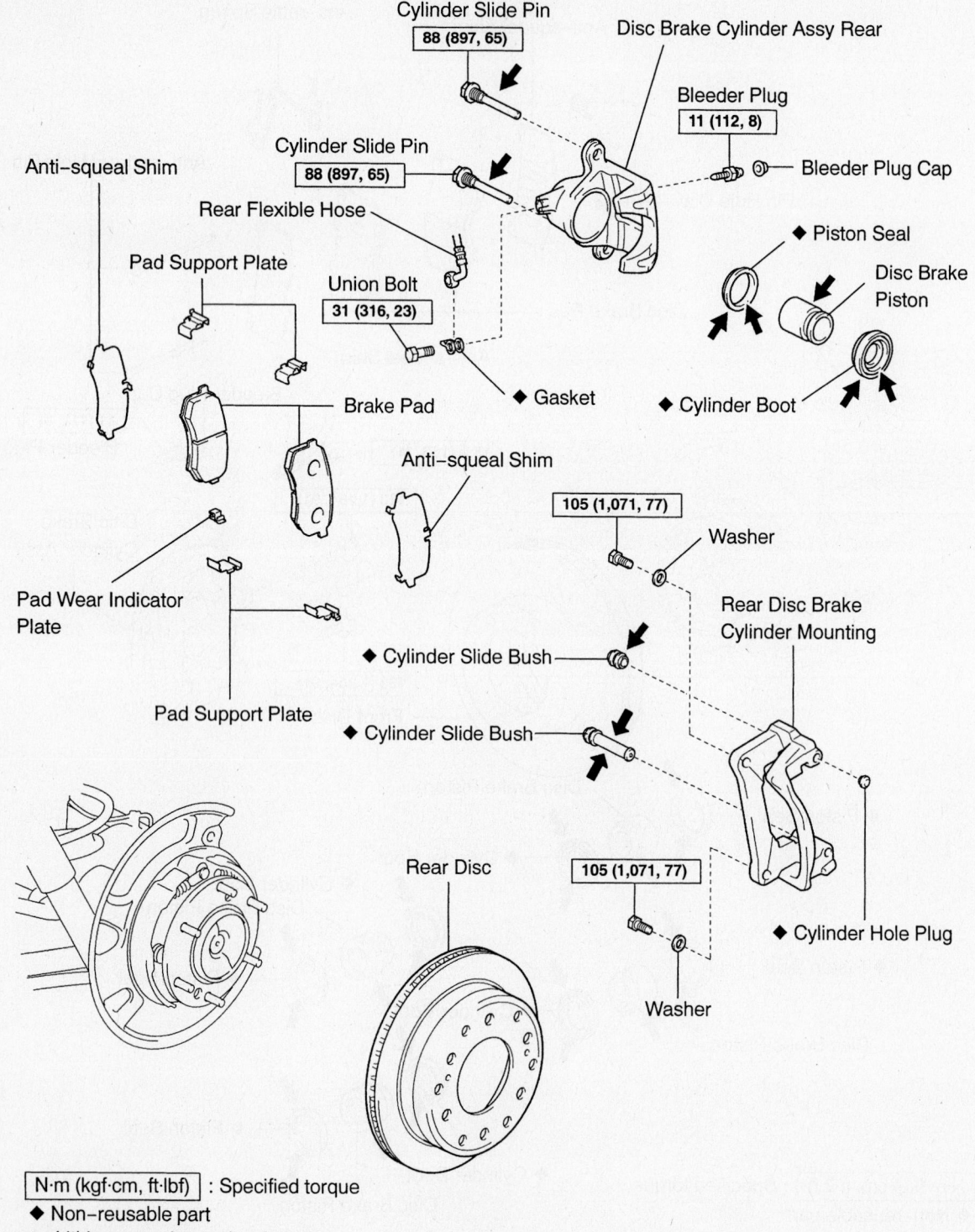

Cylinder Slide Pin 88 (897, 65)

Disc Brake Cylinder Assy Rear

Bleeder Plug 11 (112, 8)

Bleeder Plug Cap

Anti-squeal Shim

Cylinder Slide Pin 88 (897, 65)

Rear Flexible Hose

Pad Support Plate

Union Bolt 31 (316, 23)

Piston Seal

Disc Brake Piston

Brake Pad

Gasket

Cylinder Boot

Anti-squeal Shim

105 (1,071, 77)

Washer

Pad Wear Indicator Plate

Rear Disc Brake Cylinder Mounting

Cylinder Slide Bush

Pad Support Plate

Cylinder Slide Bush

Rear Disc

105 (1,071, 77)

Cylinder Hole Plug

Washer

N·m (kgf·cm, ft·lbf) : Specified torque

◆ Non-reusable part

◀ Lithium soap base glycol grease

67162-X470-G13

Rear disc brake components—GX470

REAR

1. Remove the brake line from the caliper.
2. Hold the siding pin and remove the 2 bolts.
3. Remove the caliper from the torque plate.
4. Remove the pads and shims.
5. Remove the pad support plates.
6. Installation is the reverse of removal. Torque the caliper bolts to 20 ft. lbs. (26 Nm). Torque the brake line union bolt to 22 ft. lbs. (30 Nm).

Disc Brake Pads

REMOVAL & INSTALLATION

GX470

FRONT

1. Remove the wheel.
2. Remove the anti-rattle spring from the caliper.
3. Remove the clips and anti-rattle pins.
4. Lift out the pads and shims.
5. Installation is the reverse of removal.

REAR

1. Remove the wheel.
2. If the caliper is being replaced, dis-connect the brake line at the caliper. Plug the line to prevent fluid loss.
3. Remove the 2 caliper attaching pins and lift off the caliper.
4. Remove the pads, shims and wear indicator plate and support plates.
5. Installation is the reverse of removal. The wear indicator is installed downward. Torque the caliper pins to 65 ft. lbs. (88 Nm). Bleed the brakes if the line was disconnected.

LX470

FRONT

1. Raise the vehicle and support it safely.
2. Remove the wheels.
3. Remove the clip, pins and anti-rattle spring.
4. Withdraw the pads and remove the anti-squeal shims.

To install:

5. Before installing the new pads, check the disc thickness and disc runout.
6. Siphon out a small amount of brake fluid from the reservoir.
7. Press in the pistons with a hammer handle or equivalent.
8. Apply disc brake grease to both sides of the inner anti-squeal shim. Install the anti-squeal shims to the new pads.

9. Install the pads.
10. Install the anti-rattle springs and pins. Install the clip.
11. Install the wheels.
12. Check and adjust the fluid level. Apply the brake pedal several times.
13. Road-test the vehicle for proper operation.

REAR

1. Raise the vehicle and support it safely.
2. Remove the wheels.
3. Remove the brake caliper and sus-pend it so the hose is not stretched.
4. Remove the brake pads, anti-squeal shim, pad support plates and wear indica-tors.

To install:

5. Before installing the new pads, check the disc thickness and disc runout.
6. Install the pad support plates.
7. Install the pad wear indicator plate to each pads.
8. Install the anti-squeal shim to the outer pad. Install the pads so the wear indi-cator plate is facing upward.
9. Install the brake caliper.
10. Install the wheels.
11. Apply the brake pedal several times.
12. Road-test the vehicle for proper operation.

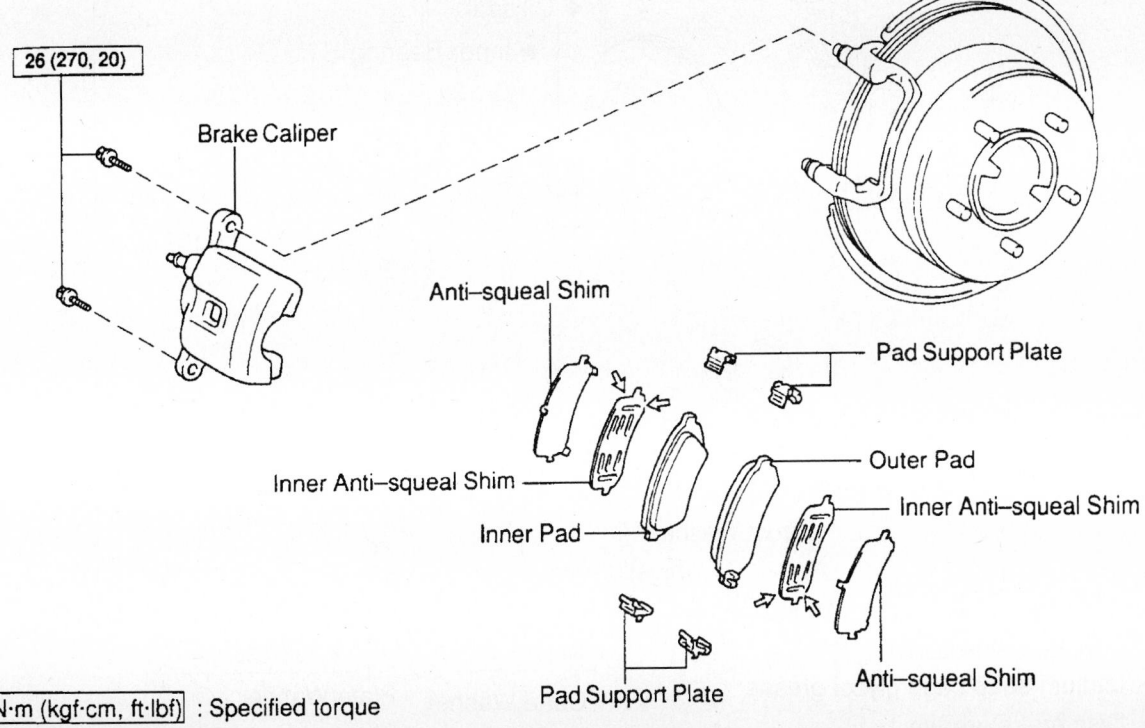

26 (270, 20)

Brake Caliper

Anti-squeal Shim

Pad Support Plate

Inner Anti-squeal Shim

Inner Pad

Outer Pad

Inner Anti-squeal Shim

Pad Support Plate

Anti-squeal Shim

N·m (kgf·cm, ft·lbf) : Specified torque
⇨ Disc brake grease

93026G83

Exploded view of the rear disc brake components—LX470

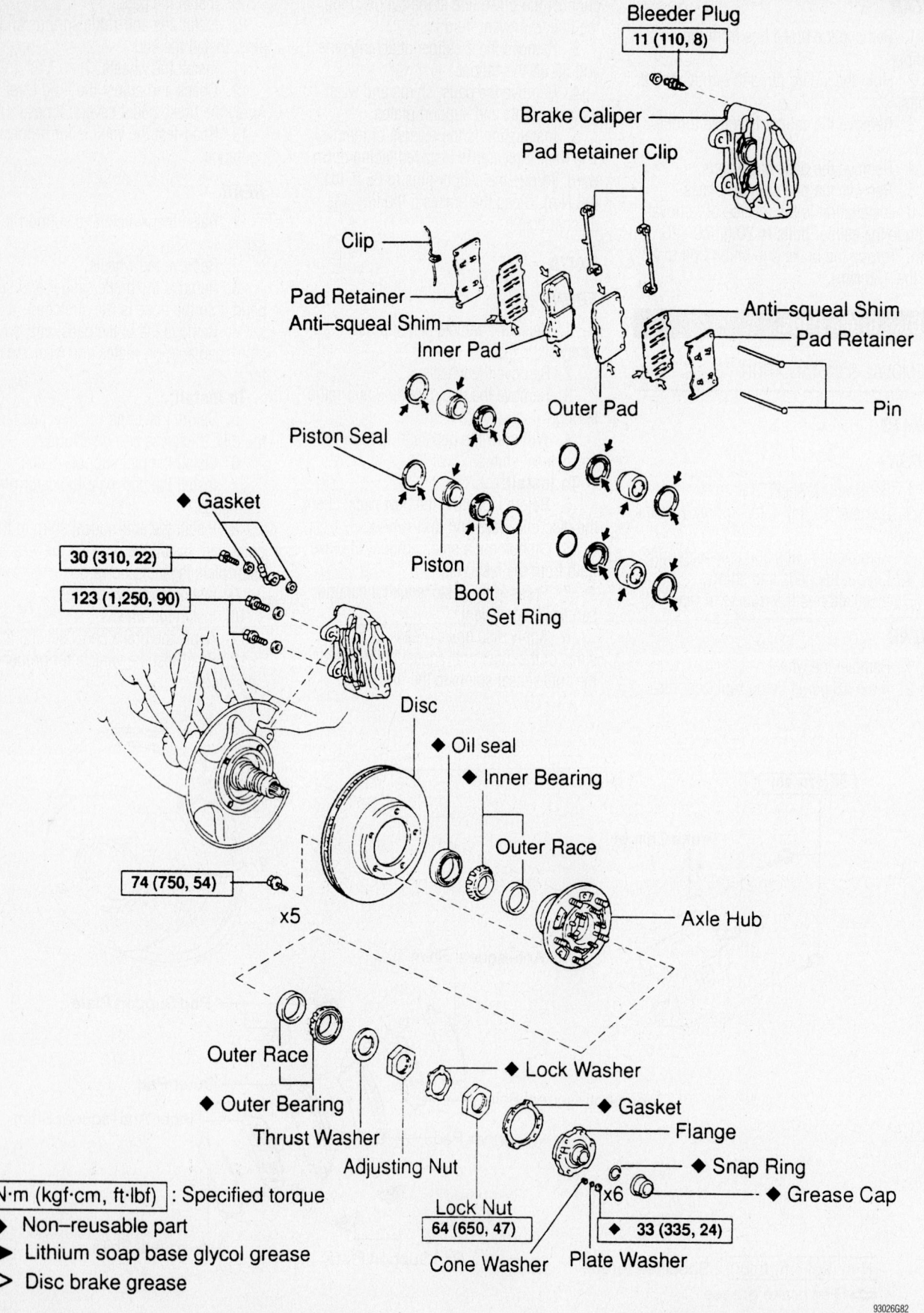

Bleeder Plug
11 (110, 8)

Brake Caliper

Pad Retainer Clip

Clip

Pad Retainer

Anti–squeal Shim

Inner Pad

Outer Pad

Anti–squeal Shim

Pad Retainer

Pin

Piston Seal

Gasket

30 (310, 22)

123 (1,250, 90)

Piston

Boot

Set Ring

Disc

Oil seal

Inner Bearing

Outer Race

74 (750, 54)

x5

Axle Hub

Outer Race

Lock Washer

Outer Bearing

Gasket

Thrust Washer

Flange

Adjusting Nut

Snap Ring

Lock Nut
64 (650, 47)

x6

Grease Cap

33 (335, 24)

Cone Washer

Plate Washer

N·m (kgf·cm, ft·lbf) : Specified torque

◆ Non–reusable part

► Lithium soap base glycol grease

▷ Disc brake grease

Exploded view of the front disc brake components—LX470

93026G82

SPECIFICATIONS AND MAINTENANCE CHARTS

ENGINE AND VEHICLE IDENTIFICATION

Code ①	Liters (cc)	Cu. In.	Cyl.	Fuel Sys.	Engine Type	Eng. Mfg.
1MZ-FE	3.0 (2995)	183	6	SFI	DOHC	Toyota
3MZ-FE	3.3 (NA)	NA	6	SFI	DOHC	Toyota

NA: Information not available

SFI: Sequential Fuel Injection

DOHC: Double Overhead Camshaft

① Stamped on the left side of the engine block

② 10th digit of the Vehicle Identification Number (VIN)

Code ②	Year
1	2001
2	2002
3	2003
4	2004
5	2005

67162-X300-C01

GENERAL ENGINE SPECIFICATIONS

Year	Model	Engine Displacement Liters	Engine Series ID	Net Horsepower @ rpm	Net Torque @ rpm (ft. lbs.)	Bore x Stroke (in.)	Compression Ratio	Oil Pressure @ rpm
2001	RX300	3.0	1MZ-FE	220@5800	222@4400	3.44x3.27	10.5:1	43-78@3000
2002	RX300	3.0	1MZ-FE	220@5800	222@4400	3.44x3.27	10.5:1	43-78@3000
2003	RX300	3.0	1MZ-FE	220@5800	222@4400	3.44x3.27	10.5:1	43-78@3000
2004	RX330	3.3	3MZ-FE	230@5600	242@3600	NA	NA	36-78@3000

NA: Information not Available

67162-X300-C02

ENGINE TUNE-UP SPECIFICATIONS

Year	Engine Displacement Liters	Engine ID	Spark Plug Gap (in.)	Ignition Timing (deg.)*	Fuel Pump (psi)	Idle Speed (rpm)	Valve Clearance Intake	Valve Clearance Exhaust
2001	3.0	1MZ-FE	0.043	8-12B	44-50	650-750	0.006-0.010	0.010-0.014
2002	3.0	1MZ-FE	0.039-0.043	8-12B	44-50	650-750	0.006-0.010	0.010-0.014
2003	3.0	1MZ-FE	0.039-0.043	8-12B	44-50	650-750	0.006-0.010	0.010-0.014
2004	3.3	3MZ-FE	0.039-0.043	8-12B	44-50	650-750	0.006-0.010	0.010-0.014

NOTE: The Vehicle Emission Control Information label often reflects specification changes made during production.

The label figures must be used if they differ from those in this chart.

B: Before top dead center

* With terminals TC and CG of DLC3 connected

67162-X300-C03

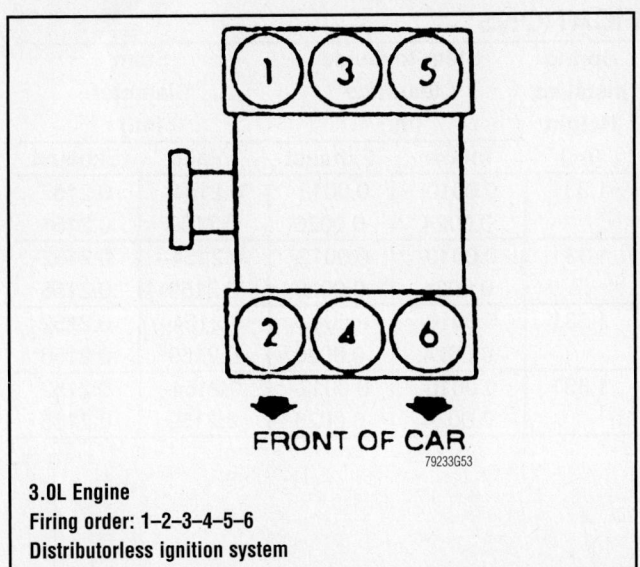

3.0L Engine
Firing order: 1–2–3–4–5–6
Distributorless ignition system

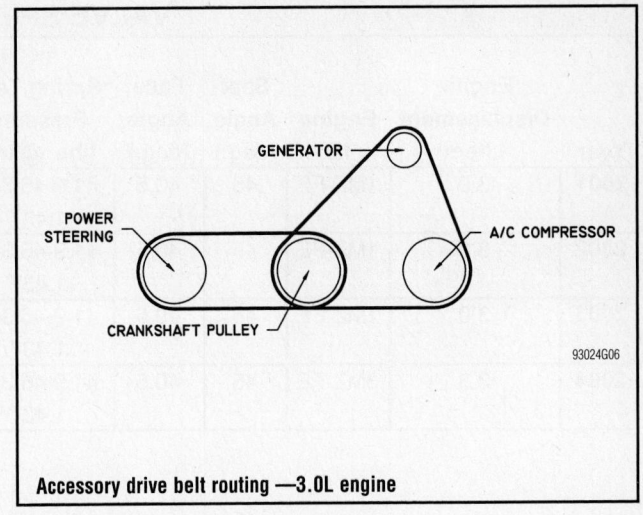

Accessory drive belt routing —3.0L engine

CAPACITIES

Year	Model	Engine Displacement Liters	Engine ID	Engine Oil with Filter (qts.)	Transaxle (pts)*	Transfer Case (pts.)	Rear Drive Axle (pts.)	Fuel Tank (gal.)	Cooling System (qts.)
2001	RX300	3.0	1MZ-FE	5.0	①	2.0	1.9	17.2	9.5
2002	RX300	3.0	1MZ-FE	5.0	①	2.0	1.9	17.2	9.5
2003	RX300	3.0	1MZ-FE	5.0	①	2.0	1.9	19.8	9.5
2004	RX330	3.3	3MZ-FE	5.0	②	2.0	1.9	19.2	10.3

*After draining, add the following amounts, then, fill to the cold full line.

① U140E Transaxle:

Dry Fill: 17.44 pts.

Drain and Refill: 7.4 pts.

U140F Transaxle:

Dry Fill: 19.34 pts.

Drain and Refill: 8.6 pts.

② U151E Transaxle

Dry Fill: 18.6 pts.

Drain and Refill: 7.4 pts.

U151F Transaxle

Dry Fill: 19.0 pts.

Drain and Refill: 7.6 pts.

67162-X300-C04

VALVE SPECIFICATIONS

Year	Engine Displacement Liters	Engine ID	Seat Angle (deg.)	Face Angle (deg.)	Spring Test Pressure (lbs. @ in.)	Spring Installed Height (in.)	Stem-to-Guide Clearance (in.)		Stem Diameter (in.)	
							Intake	Exhaust	Intake	Exhaust
2001	3.0	1MZ-FE	45	40.5	41.9-46.3@ 1.437	1.331	0.0010- 0.0024	0.0012- 0.0026	0.2154- 0.2159	0.2152 0.2156
2002	3.0	1MZ-FE	45	40.5	41.9-46.3@ 1.437	1.331	0.0010- 0.0024	0.0012- 0.0026	0.2154- 0.2159	0.2152 0.2156
2003	3.0	1MZ-FE	45	40.5	41.9-46.3@ 1.437	1.331	0.0010- 0.0024	0.0012- 0.0026	0.2154- 0.2159	0.2152 0.2156
2004	3.3	3MZ-FE	45	40.5	41.9-46.3@ 1.437	1.331	0.0010- 0.0024	0.0012- 0.0026	0.2154- 0.2159	0.2152 0.2156

67162-X300-C05

CRANKSHAFT AND CONNECTING ROD SPECIFICATIONS

All measurements are given in inches.

Year	Engine Displacement Liters	Engine ID	Crankshaft				Connecting Rod		
			Main Brg. Journal Dia.	Main Brg. Oil Clearance	Shaft End-play	Thrust on No.	Journal Diameter	Oil Clearance	Side Clearance
2001	3.0	1MZ-FE	2.4011- 2.4016	①	0.0016- 0.0095	2	2.0863- 2.0866	0.0015- 0.0025	0.0059- 0.0188
2002	3.0	1MZ-FE	2.4011- 2.4016	①	0.0016- 0.0095	2	2.0863- 2.0866	0.0015- 0.0025	0.0059- 0.0188
2003	3.0	1MZ-FE	2.4011- 2.4016	①	0.0016- 0.0095	2	2.0863- 2.0866	0.0015- 0.0025	0.0059- 0.0188
2004	3.3	3MZ-FE	2.4011- 2.4016	①	0.0016- 0.0095	2	2.0863- 2.0866	0.0015- 0.0026	0.0059- 0.0118

① Journals 1 and 4: 0.0006 - 0.0013 in.
  Journals 2 and 3: 0.0010 - 0.0018 in.

67162-X300-C06

PISTON AND RING SPECIFICATIONS

All measurements are given in inches.

Year	Engine Displ. Liters	Engine ID	Piston Clearance	Ring Gap			Ring Side Clearance		
				Top Comp.	Bottom Comp.	Oil Control	Top Comp.	Bottom Comp.	Oil Control
2001	3.0	1MZ-FE	0.0033- 0.0042	0.0098- 0.0138	0.0138- 0.0177	0.0059- 0.0157	0.0008- 0.0028	0.0008- 0.0024	SNUG
2002	3.0	1MZ-FE	0.0033- 0.0042	0.0098- 0.0138	0.0138- 0.0177	0.0059- 0.0157	0.0008- 0.0028	0.0008- 0.0024	SNUG
2003	3.0	1MZ-FE	0.0033- 0.0042	0.0098- 0.0138	0.0138- 0.0177	0.0059- 0.0157	0.0008- 0.0028	0.0008- 0.0024	SNUG
2004	3.3	3MZ-FE	0.0013- 0.0023	0.0118- 0.0138	0.0197- 0.0236	0.0059- 0.0157	0.0012- 0.0031	0.0008- 0.0024	0.0012- 0.0043

67162-X300-C07

TORQUE SPECIFICATIONS

All readings in ft. lbs.

Year	Engine Displacement Liters	Engine ID	Cylinder Head Bolts	Main Bearing Bolts	Rod Bearing Bolts	Crankshaft Damper Bolts	Flywheel Bolts	Manifold Intake	Manifold Exhaust	Spark Plugs	Oil Pan Drain Plug
2001	3.0	1MZ-FE	①	②	③	159	61	11	36	13	33
2002	3.0	1MZ-FE	①	②	③	159	61	11	36	13	33
2003	3.0	1MZ-FE	①	②	③	159	61	11	36	13	33
2004	3.3	3MZ-FE	①	②	③	162	61	11	36	18	33

① Step 1: 12 point bolts to 40 ft. lbs.

   Step 2: 12 point bolts plus 90 degrees

   Step 3: Hex head recessed bolt to 13 ft. lbs.

② Step 1: 12 point cap bolts to 16 ft. lbs.

   Step 2: 12 point cap bolts plus 90 degrees

   Step 3: Hex head side bolts to 20 ft. lbs.

③ Step 1: 18 ft. lbs.

   Step 2: Plus 90 degrees

67162-X300-C08

WHEEL ALIGNMENT

Year	Model		Caster Range (+/-Deg.)	Caster Preferred Setting (Deg.)	Camber Range (+/-Deg.)	Camber Preferred Setting (Deg.)	Toe-in (in.)
2001	RX300	F	0.75	+2.08	0.75	-0.33	0.04+/-0.08
		2WD R	—	—	0.75	-0.33	0.08+/-0.08
		4WD R	—	—	0.75	-0.35	0.12+/-0.08
2002	RX300	F	0.75	+2.08	0.75	-0.33	0.04+/-0.08
		2WD R	—	—	0.75	-0.33	0.08+/-0.08
		4WD R	—	—	0.75	-0.35	0.12+/-0.08
2003	RX300	F	0.75	+2.08	0.75	-0.33	0.04+/-0.08
		2WD R	—	—	0.75	-0.33	0.08+/-0.08
		4WD R	—	—	0.75	-0.35	0.12+/-0.08
2004	RX330 without air suspension	2WD F	0.75	+2.85	0.75	-0.67	0+/-0.08
		4WD F	0.75	+2.83	0.75	-0.58	0+/-0.08
		2WD R	—	—	0.75	-1.33	0.12+/-0.08
		4WD R	—	—	0.75	-0.83	0.12+/-0.08
	with air suspension	2WD F	0.75	+2.85	0.75	-0.67	0+/-0.08
		4WD F	0.75	+2.67	0.75	-0.62	0+/-0.08
		2WD R	—	—	0.75	-1.35	0.12+/-0.08
		4WD R	—	—	0.75	-0.92	0.12+/-0.08

67162-X300-C09

TIRE, WHEEL AND BALL JOINT SPECIFICATIONS

Year	Model	OEM Tires Standard	OEM Tires Optional	Tire Pressures (psi) Front	Tire Pressures (psi) Rear	Wheel Size	Ball Joint Inspection	Lugnut Torque (ft. lbs.)
2001	RX300	P255/70HR16	None	30	30	6.5-JJ	①	70
2002	RX300	P255/70HR16	None	30	30	6.5-JJ	①	70
2003	RX300	P255/70HR16	None	30	30	6.5-JJ	①	70
2004	RX330	P255/65R17	P235/55R18	30	30	6.5-JJ	①	76

OEM: Original Equipment Manufacturer

PSI: Pounds Per Square Inch

STD: Standard

OPT: Optional

① Replace if any measurable movement is found.

67162-X300-C10

BRAKE SPECIFICATIONS
All measurements in inches unless noted

Year	Model		Brake Disc Original Thickness	Brake Disc Minimum Thickness	Brake Disc Maximum Runout	Minimum Lining Thickness	Brake Caliper Bracket Bolts (ft. lbs.)	Brake Caliper Mounting Bolts (ft. lbs.)
2001	RX300	F	1.020	1.024	0.0020	0.039	79	25
		R	0.394	0.354	0.0059	0.039	34	14
2002	RX300	F	1.020	1.024	0.0020	0.039	79	25
		R	0.394	0.354	0.0059	0.039	34	14
2003	RX300	F	1.020	1.024	0.0020	0.039	79	25
		R	0.394	0.354	0.0059	0.039	34	14
2004	RX330	F	1.020	1.024	0.0020	0.039	77	25
		R	0.394	0.335	0.0059	0.039	58	32

F: Front

R: Rear

67162-X300-C11

SCHEDULED MAINTENANCE INTERVALS

LEXUS—RX300

TO BE SERVICED	TYPE OF SERVICE	VEHICLE MILEAGE INTERVAL (x1000)												
		7.5	15	22.5	30	37.5	45	52.5	60	67.5	75	82.5	90	97.5
Engine oil & filter	R	✓	✓	✓	✓	✓	✓	✓	✓	✓	✓	✓	✓	✓
Automatic transmission fluid	S/I		✓		✓		✓		✓		✓		✓	
Ball joints & dust covers	S/I		✓		✓		✓		✓		✓		✓	
Bolts & nuts on chassis & body	S/I		✓		✓		✓		✓		✓		✓	
Brake linings & drums	S/I		✓		✓		✓		✓		✓		✓	
Brake line pipes & hoses	S/I		✓		✓		✓		✓		✓		✓	
Brake pads & discs (front & rear)	S/I		✓		✓		✓		✓		✓		✓	
Propeller shaft grease	S/I		✓		✓		✓		✓		✓		✓	
Steering knuckle & chassis grease	S/I		✓		✓		✓		✓		✓		✓	
Steering linkage	S/I		✓		✓		✓		✓		✓		✓	
Air cleaner filter	R				✓				✓				✓	
Spark plugs	R				✓				✓				✓	
Drive belts	S/I				✓				✓				✓	
Exhaust pipes & mountings	S/I				✓				✓				✓	
Fuel lines & connections	S/I				✓				✓				✓	
Engine coolant	R						✓				✓			
Charcoal canister	R								✓					
Fuel tank cap gasket	R								✓					
Oxygen sensors (exc. Calif.)①	R													

R: Replace   S/I: Service or Inspect

① Heated oxygen sensors (except Calif.): replace every 80,000 miles.

FREQUENT OPERATION MAINTENANCE (SEVERE SERVICE)

If a vehicle is operated under any of the following conditions it is considered severe service:

- Extremely dusty areas.

- 50% or more of the constant operation is in 32°C (90°F) or higher temperatures, or in temperatures below 0°C (32°F).

- Prolonged idling (vehicle operation in stop and go traffic).

- Frequent short running periods (engine does not warm to normal operating temperatures).

- Police, taxi, delivery usage or trailer towing usage.

Air cleaner filter: service or inspect every 3750 miles

Engine oil & filter: replace every 3750 miles.

Ball joints & dust covers: service or inspect every 7500 miles.

Bolts & nuts on chassis & body: service or inspect every 7500 miles.

Brake pads & discs (front & rear): service or inspect every 7500 miles.

Steering knuckle & chassis grease: service or inspect every 7500 miles.

Steering linkage: service or inspect every 7500 miles.

Exhaust pipes & mountings: service or inspect every 15,000 miles.

67162-X300-C12

PRECAUTIONS

Before servicing any vehicle, please be sure to read all of the following precautions, which deal with personal safety, prevention of component damage, and important points to take into consideration when servicing a motor vehicle:

• Never open, service or drain the radiator or cooling system when the engine is hot; serious burns can occur from the steam and hot coolant.

• Observe all applicable safety precautions when working around fuel. Whenever servicing the fuel system, always work in a well-ventilated area. Do not allow fuel spray or vapors to come in contact with a spark, open flame, or excessive heat (a hot drop light, for example). Keep a dry chemical fire extinguisher near the work area. Always keep fuel in a container specifically designed for fuel storage; also, always properly seal fuel containers to avoid the possibility of fire or explosion. Refer to the additional fuel system precautions later in this section.

• Fuel injection systems often remain pressurized, even after the engine has been turned **OFF**. The fuel system pressure must be relieved before disconnecting any fuel lines. Failure to do so may result in fire and/or personal injury.

• Brake fluid often contains polyglycol ethers and polyglycols. Avoid contact with the eyes and wash your hands thoroughly after handling brake fluid. If you do get brake fluid in your eyes, flush your eyes with clean, running water for 15 minutes. If eye irritation persists, or if you have taken brake fluid internally, IMMEDIATELY seek medical assistance.

• The EPA warns that prolonged contact with used engine oil may cause a number of skin disorders, including cancer. You should make every effort to minimize your exposure to used engine oil. Protective gloves should be worn when changing oil. Wash your hands and any other exposed skin areas as soon as possible after exposure to used engine oil. Soap and water, or waterless hand cleaner should be used.

• All new vehicles are now equipped with an air bag system. The system must be disabled before performing service on or around system components, steering column, instrument panel components, wiring and sensors. Failure to follow safety and disabling procedures could result in accidental air bag deployment, possible personal injury and unnecessary system repairs.

• Always wear safety goggles when working with, or around, the air bag system. When carrying a non-deployed air bag, be sure the bag and trim cover are pointed away from your body. When placing a non-deployed air bag on a work surface, always face the bag and trim cover upward, away from the surface. This will reduce the motion of the module if it is accidentally deployed. Refer to the additional air bag system precautions later in this section.

• NEVER disconnect the negative battery cable with the ignition **ON** or the engine running. Removing power from the computer control module with the ignition **ON** may destroy the module.

• Clean, high quality brake fluid from a sealed container is essential to the safe and proper operation of the brake system. You should always buy the correct type of brake fluid for your vehicle. If the brake fluid becomes contaminated, completely flush the system with new fluid. Never reuse any brake fluid. Any brake fluid that is removed from the system should be discarded. Also, do not allow any brake fluid to come in contact with a painted surface; it will damage the paint.

• Never operate the engine without the proper amount and type of engine oil; doing so WILL result in severe engine damage.

• Timing belt maintenance is extremely important. Many models utilize an interference-type, non-freewheeling engine. If the timing belt breaks, the valves in the cylinder head may strike the pistons, causing potentially serious (also time-consuming and expensive) engine damage. Refer to the maintenance interval charts in the front of this manual for the recommended replacement interval for the timing belt, and to the timing belt section for belt replacement and inspection.

• Disconnecting the negative battery cable on some vehicles may interfere with the functions of the on-board computer system(s) and may require the computer to undergo a relearning process once the negative battery cable is reconnected.

• When servicing drum brakes, only disassemble and assemble one side at a time, leaving the remaining side intact for reference.

• Only an MVAC-trained, EPA-certified automotive technician should service the air conditioning system or its components.

ENGINE REPAIR

➡**Disconnecting the negative battery cable on some vehicles may interfere with the functions of the on board computer system. The computer may undergo a relearning process once the negative battery cable is reconnected.**

Alternator

REMOVAL

On some models, the alternator is mounted very low on the engine. On these models, it may be necessary to remove the gravel shield and work from beneath the vehicle in order to gain access to the alternator. Replacing the alternator while the

engine is cold is recommended. A hot engine can result in personal injury.

1. Before servicing the vehicle, refer to the precautions in the beginning of this section.
2. Remove or disconnect the following:
   • Alternator electrical connectors
   • Wiring harness from the clip
   • Pivot bolt
   • Plate washer
   • Adjusting lockbolt
   • Drive belt
   • Alternator

To install:

3. Install or connect the following:
   • Alternator
   • Drive belt. Tension the belt to

170–180 lbs. for a new belt or 95–135 lbs. for a used belt.
   • Adjusting lockbolt. Tighten the bolt to 13 ft. lbs. (18 Nm).
   • Plate washer
   • Pivot bolt. Tighten the bolt to 41 ft. lbs. (56 Nm) for the 3.0L and 43 ft. lbs. (58 Nm) for the 3.3L.
   • Wiring harness from the clip
   • Alternator electrical connectors

Ignition Timing

ADJUSTMENT

All engines are equipped with a Distributorless Ignition System (DIS). No timing adjustment is possible.

Engine Assembly

REMOVAL & INSTALLATION

3.0L Engine

1. Before servicing the vehicle, refer to the precautions in the beginning of this section.

2. Matchmark the hood position.
3. Remove or disconnect the following:
   - Hood
   - Wiper and blade assembly
   - Top cowl seal and panel
   - Window washer hoses from the ventilator louvers
   - Left and right ventilator louvers

- Heater air duct
4. Properly relieve the fuel system pressure.
5. Remove or disconnect the following:
   - Both battery cables
   - Battery and tray
6. Drain the engine coolant.
7. Drain the engine oil.

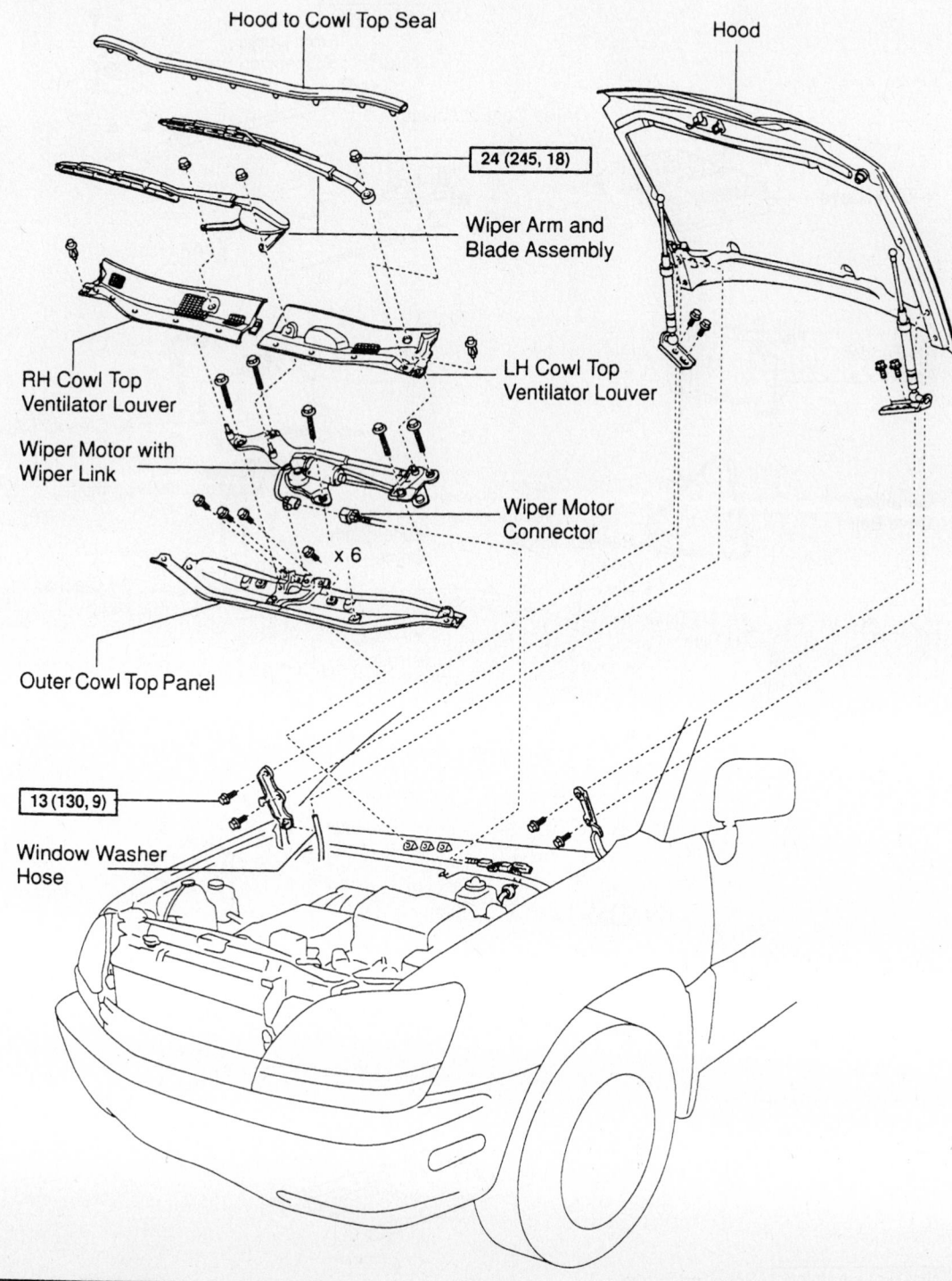

Hood to Cowl Top Seal

Hood

24 (245, 18)

Wiper Arm and Blade Assembly

RH Cowl Top Ventilator Louver

LH Cowl Top Ventilator Louver

Wiper Motor with Wiper Link

Wiper Motor Connector

x 6

Outer Cowl Top Panel

13 (130, 9)

Window Washer Hose

N·m (kgf·cm, ft·lbf) : Specified torque

7924ZG83

Exploded view of the top cowl and related components—RX300

8. Remove or disconnect the following:
- Intake air cleaner and case assembly
- Cruise control actuator, if equipped
- Upper suspension brace
- Upper and lower radiator hoses
- Radiator

- Automatic transmission oil cooler lines
- Any connectors, hoses and sensors that would interfere with engine removal
- Engine Control Module (ECM) engine wiring harness from inside

the glove box; then, pull the harness into the engine compartment
- Compressor

➡**It may be necessary to remove the air conditioning compressor lines in order to remove the engine.**

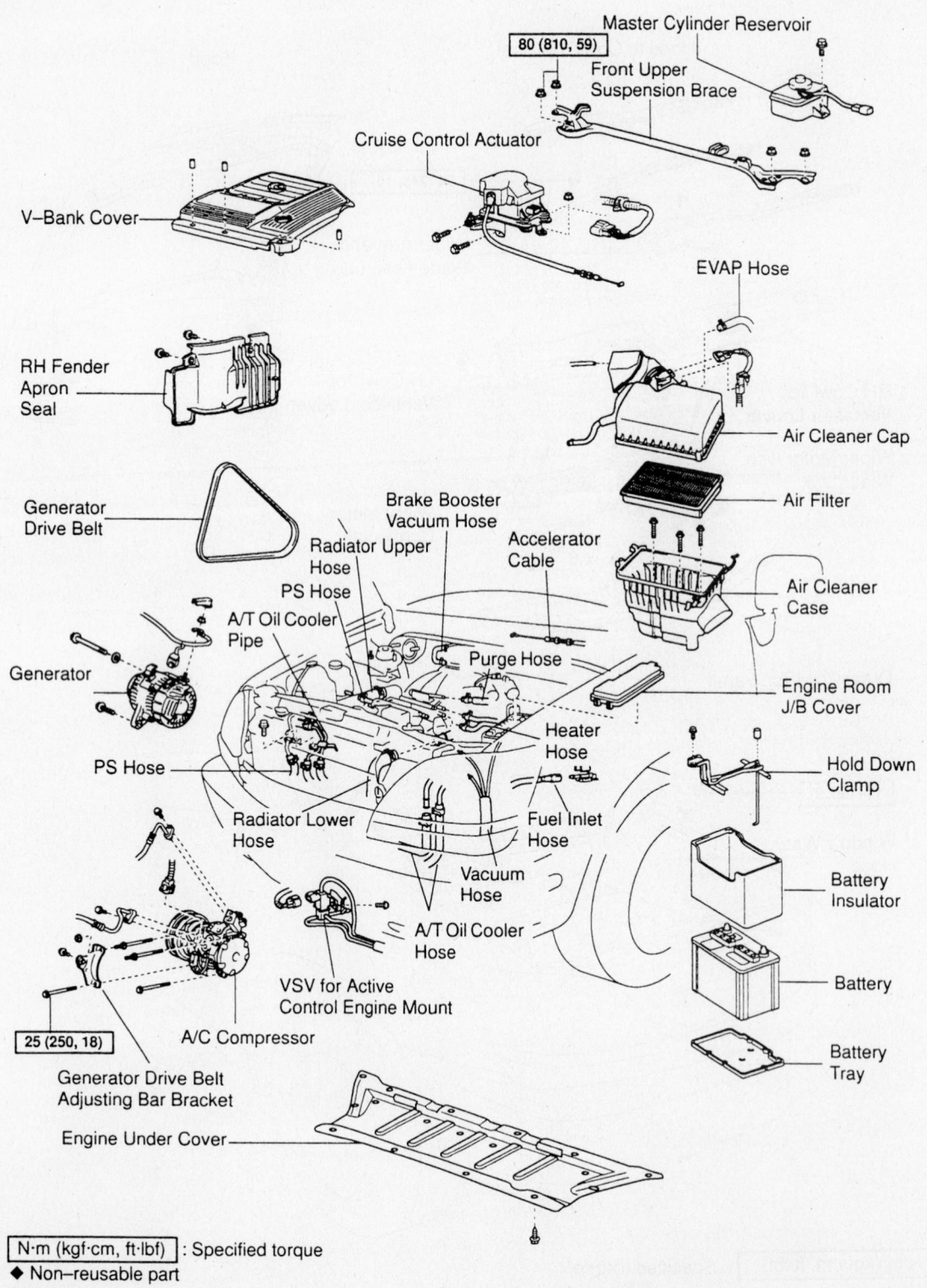

N·m (kgf·cm, ft·lbf) : Specified torque
◆ Non–reusable part

Exploded view of engine pre-removal components—3.0L

7924ZG84

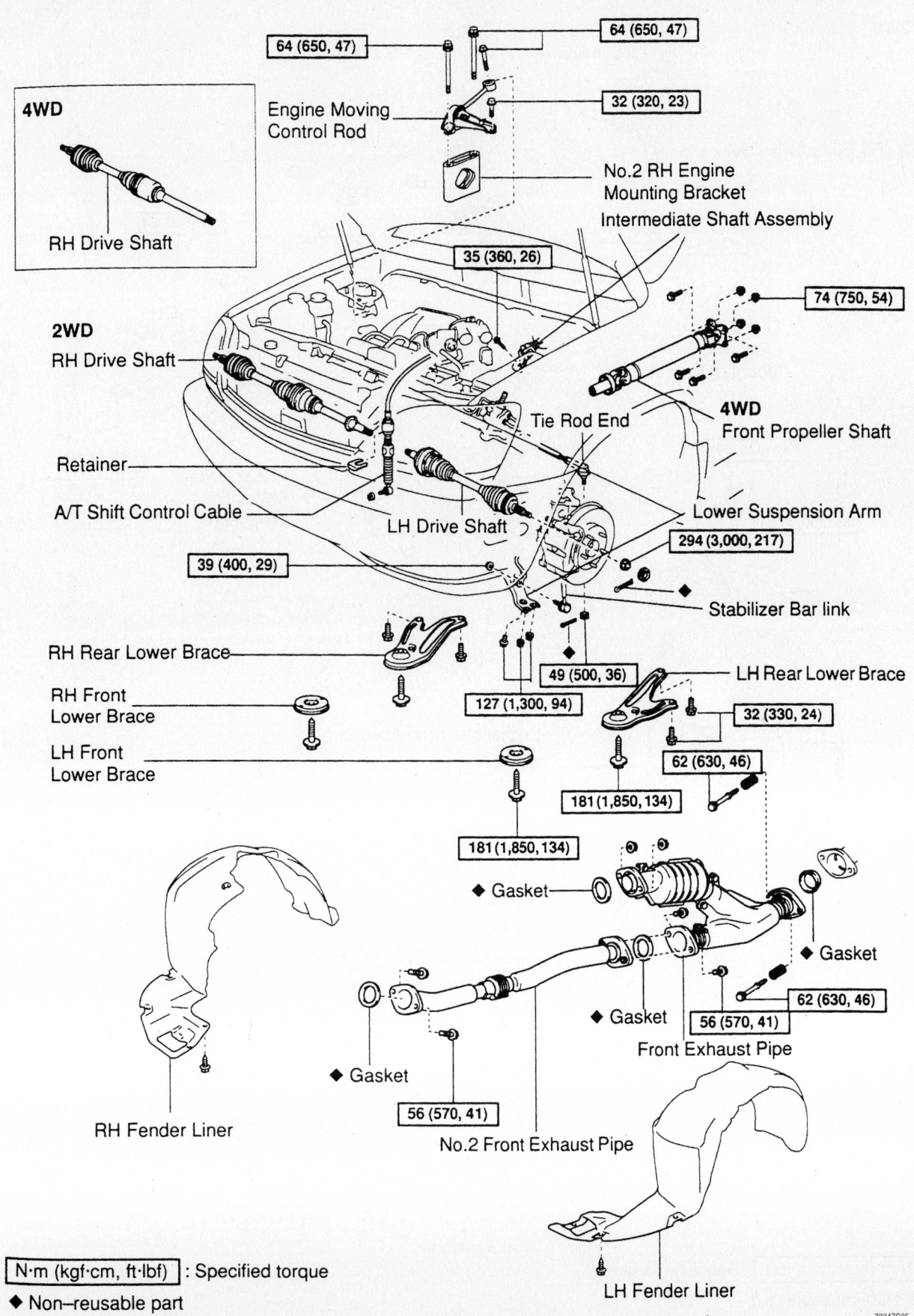

4WD

RH Drive Shaft

64 (650, 47)

64 (650, 47)

32 (320, 23)

Engine Moving
Control Rod

No.2 RH Engine
Mounting Bracket

Intermediate Shaft Assembly

35 (360, 26)

74 (750, 54)

2WD

RH Drive Shaft

4WD
Front Propeller Shaft

Tie Rod End

Retainer

A/T Shift Control Cable

LH Drive Shaft

Lower Suspension Arm

294 (3,000, 217)

Stabilizer Bar link

39 (400, 29)

RH Rear Lower Brace

LH Rear Lower Brace

RH Front
Lower Brace

49 (500, 36)

127 (1,300, 94)

32 (330, 24)

LH Front
Lower Brace

62 (630, 46)

181 (1,850, 134)

181 (1,850, 134)

◆ Gasket

62 (630, 46)

◆ Gasket

◆ Gasket

56 (570, 41)

Front Exhaust Pipe

RH Fender Liner

◆ Gasket

56 (570, 41)

No.2 Front Exhaust Pipe

LH Fender Liner

N·m (kgf·cm, ft·lbf) : Specified torque

◆ Non–reusable part

7924ZG85

Exploded view of engine removal and installation tightening specifications of the related components—3.0L

2WD

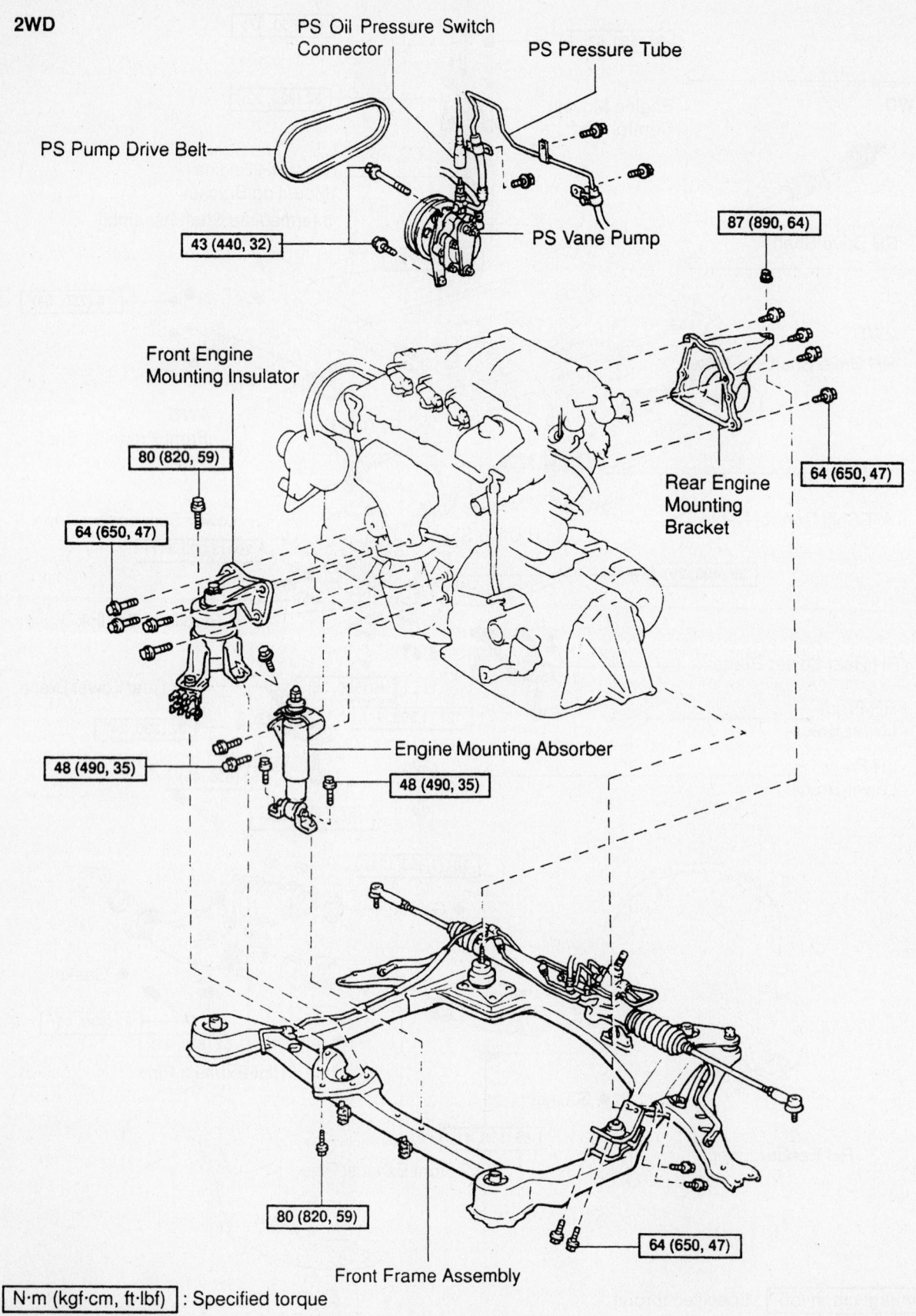

PS Oil Pressure Switch Connector

PS Pressure Tube

PS Pump Drive Belt

PS Vane Pump

43 (440, 32)

87 (890, 64)

Front Engine Mounting Insulator

80 (820, 59)

64 (650, 47)

Rear Engine Mounting Bracket

64 (650, 47)

48 (490, 35)

Engine Mounting Absorber

48 (490, 35)

80 (820, 59)

Front Frame Assembly

64 (650, 47)

N·m (kgf·cm, ft·lbf) : Specified torque

◆ Non–reusable part

7924ZG86

Exploded view of the suspension component removal and installation for engine removal—2WD RX300

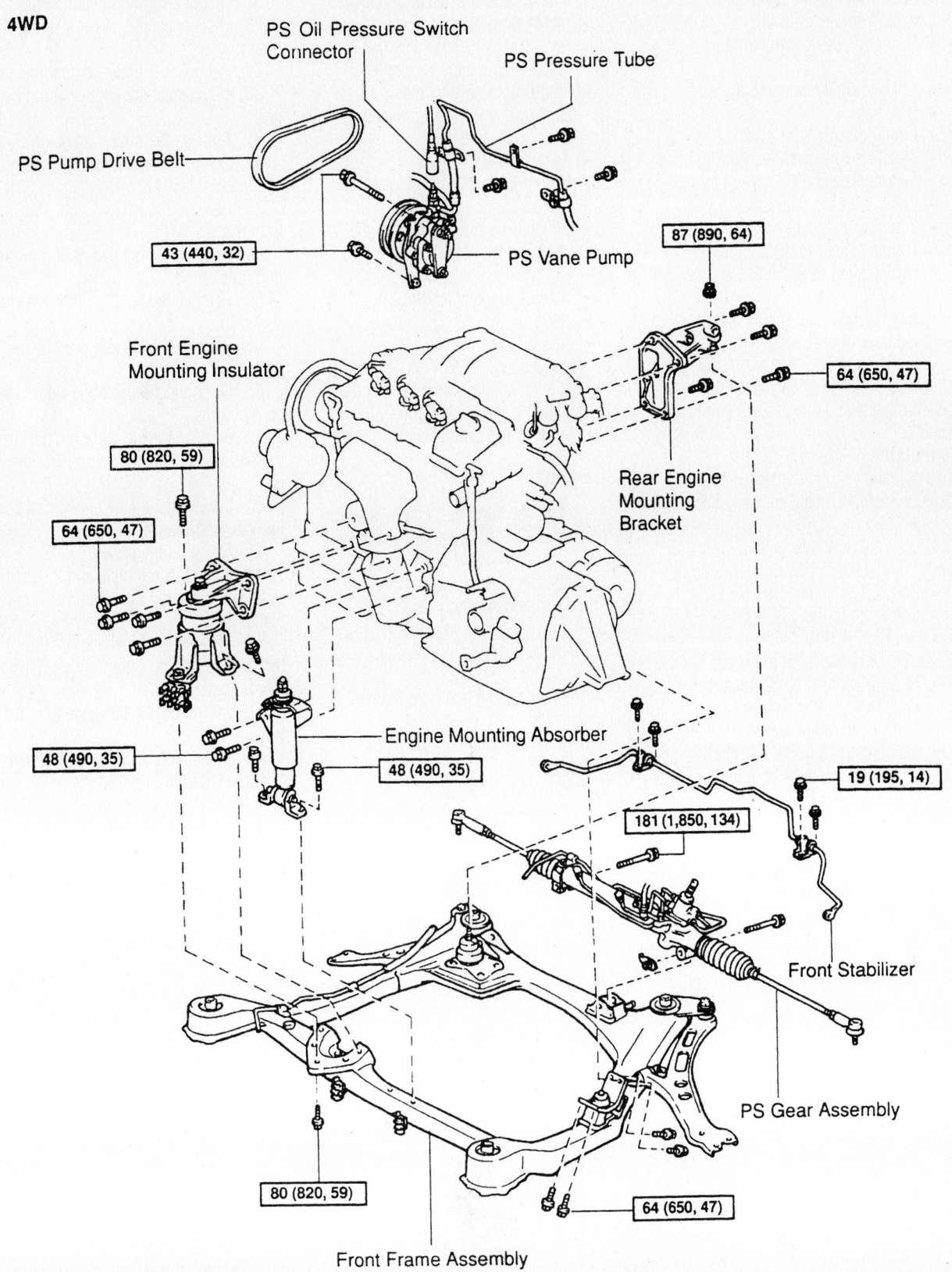

4WD

PS Oil Pressure Switch Connector

PS Pressure Tube

PS Pump Drive Belt

PS Vane Pump

43 (440, 32)

87 (890, 64)

64 (650, 47)

Front Engine Mounting Insulator

Rear Engine Mounting Bracket

80 (820, 59)

64 (650, 47)

48 (490, 35)

Engine Mounting Absorber

48 (490, 35)

19 (195, 14)

181 (1,850, 134)

Front Stabilizer

PS Gear Assembly

80 (820, 59)

64 (650, 47)

Front Frame Assembly

N·m (kgf·cm, ft·lbf) : Specified torque

◆ Non–reusable part

7924ZG87

Exploded view of the suspension component removal and installation for engine removal—4WD RX300

- Automatic transmission shifter cable from the transaxle
- Header pipes from the exhaust manifolds
- Left and right fender apron seals
- Halfshafts
- Front driveshaft, for 4WD
- Stabilizer links and the steering intermediate shaft
- Power steering pump
- Engine undercover
- Engine hanger to the engine
- Engine sling device to the engine hangers
- Right-hand motor mount and moving control rod
- Front suspension lower braces

9. Lower the engine, transaxle and front suspension member as an assembly from the vehicle.

To install:

10. Raise the engine, transaxle and front suspension member as an assembly into the vehicle.

11. Install the front suspension lower braces, and tighten the fasteners, as follows:
- Bolt A: 134 ft. lbs. (181 Nm)
- Bolt B: 24 ft. lbs. (32 Nm)
- Nut C: 27 ft. lbs. (36 Nm)

12. Install or connect the following:
- Moving control rod. Tighten the bolts to 47 ft. lbs. (64 Nm).
- Right-hand motor mount. Tighten the bolts to 23 ft. lbs. (32 Nm).
- Engine sling device from the engine hangers
- Engine undercover
- Power steering pump hoses
- Stabilizer links and the steering intermediate shaft
- Front driveshaft, for 4WD
- Halfshafts
- Left and right fender apron seals
- Header pipes to the exhaust manifolds
- Automatic transmission shifter cable to the transaxle
- Air conditioning compressor to the engine

13. Push the wiring harness into the glove box.

14. Install or connect the following:
- ECM
- Any connectors, hoses and sensors that were removed
- Automatic transmission oil cooler lines
- Upper and lower radiator hoses and fit the radiator
- Front upper suspension brace. Tighten the nuts to 59 ft. lbs. (80 Nm).

- Cruise control actuator, if removed
- Intake air cleaner and case assembly

15. Fill the engine oil to proper level.
16. Fill the engine with coolant.
17. Install or connect the following:
- Battery tray and battery
- Battery cables
- Heater air duct
- Left and right ventilator louvers
- Window washer hoses from the ventilator louvers
- Top cowl seal and panel
- Wiper and blade assembly
- Hood
- New oil filter

18. Refill the engine with oil.
19. Refill the engine with engine coolant.
20. Install the engine undercovers.
21. Start the engine and check for leaks.

3.3L Engine

1. Before servicing the vehicle, refer to the precautions in the beginning of this section.

2. Drain the coolant, engine oil, transfer case fluid and transmission fluid.

3. Remove the front wheels.

4. Remove the engine undercover assembly.

5. Remove the left and right fender splash shields.

6. Remove the left and right fender apron seals.

7. Remove the wiper arms.

8. Remove the cowl top ventilator louver.

9. Remove the wiper linkage.

10. Remove the cowl top panel outer sub-assembly.

11. Remove the V-bank cover.

12. Remove the battery.

13. Remove the air cleaner assembly.

14. Remove the A/C compressor drive belt.

15. Remove the alternator.

16. Remove the engine roll brace.

17. Remove the front engine mount bracket.

18. Remove the alternator bracket.

19. Remove the alternator belt adjusting bar.

20. Remove the magnetic clutch from the A/C compressor.

21. Remove the transmission control cable.

22. Disconnect the check valve hose.

23. Disconnect the fuel vapor feed hose.

24. Disconnect the fuel pipes.

25. Disconnect the heater hoses.

26. Disconnect the radiator hoses.

27. Disconnect the oil cooler hoses.

28. Disconnect the power steering hoses.

29. Remove the glove compartment door.

30. Disconnect the engine wiring harness.

31. Remove the driveshaft (4wd).

32. Remove the exhaust pipes and brackets.

33. Remove the left and right stabilizer bar links.

34. Remove the left and right front axle hub nuts.

35. Remove the left and right speed sensors.

36. Remove the left and right tie rod ends.

37. Disconnect the left and right lower control arms.

38. Remove the left and right halfshafts.

39. Disconnect the steering intermediate shaft.

40. Disconnect the height control sensor link (air suspension).

41. Attach a lifting crane.

42. Remove the 6 bolts and 2 nuts, then, remove the left and right frame side rail plates.

43. Remove the 6 bolts and 2 nuts, then, remove the left and right front suspension rear braces.

44. Lift the engine/transaxle from the vehicle.

45. Installation is the reverse of removal. Observe the following torques:
- Engine hanger: 14 ft. lbs. (20 Nm)
- Alternator bracket: 43 ft. lbs. (58 Nm)
- Right engine mount bracket: 40 ft. lbs. (54 Nm)
- Manifold stay: 36 ft. lbs. (49 Nm)
- Right rear engine mount bracket: 47 ft. lbs. (64 Nm)
- Front frame nuts: 70 ft. lbs. (95 Nm)
- Front right engine mount insulator nut: 64 ft. lbs. (87 Nm)
- Right rear engine mount insulator bolts: 55 ft. lbs. (75 Nm)
- Steering link: 52 ft. lbs. (70 Nm)
- Stabilizer bar: 21 ft. lbs. (29 Nm)
- Power steering pump adjusting bar: 32 ft. lbs. (43 Nm)
- Power steering pressure tube nuts: 69 inch lbs. (8 Nm)
- Left and right frame side rail plates: single end bolts 63 ft. lbs. (85 Nm); double end bolts and nuts: 24 ft. lbs. (32 Nm)
- Left and right front suspension member rear braces: single end bolts 63 ft. lbs. (85 Nm); double

end bolts and nuts: 24 ft. lbs. (32 Nm)
- Height control sensor link: 48 inch lbs. (5 Nm)

Heater Core

REMOVAL & INSTALLATION

RX300

FRONT HEATER

1. Disconnect the negative battery cable.
2. Drain the cooling system into a clean container for reuse.
3. Disconnect the heater hoses from the heater core.
4. Remove the steering wheel by performing the following procedure:
   a. Position the front wheels facing straight-ahead.
   b. Remove the steering wheel side covers.
   c. Using a Torx® wrench, loosen the 2 screws located at each side of the steering wheel until the screw's circumference groove catches on the screw case.
   d. Pull the air bag module from the steering wheel and disconnect the electrical connector.

✳✳ CAUTION

Place the air bag module in a safe place with the front side facing upward.

   e. Remove the steering wheel nut.
   f. Place alignment marks on the steering wheel and the main shaft.
   g. Using a steering wheel puller, press the steering wheel from the steering column.
5. Remove the instrument panel and reinforcement by performing the following procedure:
   a. Remove the front door scuff plates.
   b. Remove the cowl side boards.
   c. Remove the front door trim covers.
   d. Remove the front pillar garnish by disengaging the 5 clips. If equipped with a tweeter speaker, disconnect the electrical connector.
   e. Remove the steering column covers-to-steering column screws and the covers.
   f. Remove the combination switch-to-steering column screws, disconnect the electrical connector(s) and remove the combination switch.
   g. Remove the 2 hood open lever screws and the hood open lever.

   h. Remove the 2 lower finish panel bolts and disengage the panel from the 3 clips.
   i. Remove the 2 No. 1 safety pad insert bolts and the insert.
   j. Remove the 2 No. 2 finish panel bolts and disengage the panel from the 4 clips.
   k. In the left side of the glove compartment, pry out the glove box door finish plate and disconnect the air bag module connector.
   l. Remove the glove box 3 nuts and 2 screws and the glove box.
   m. Remove the center cluster finish panel by disengaging the claw (bottom center) and 4 clips (1 at each corner).
   n. Remove the ashtray, the 2 ashtray receptacle box screws.
   o. Remove the 4 lower center cluster finish panel screws and disconnect the connector.
   p. Remove the clock, the No. 1 and No. 2 registers from the panel.
   q. Remove the 3 cluster finish panel screws, disengage the 8 clips and remove the panel.
   r. Remove the combination meter.
   s. Remove the radio assembly.
   t. Remove the heater control assembly.
   u. Remove 2 passenger's side air bag module bolts; then, disconnect and remove the air bag module.

✳✳ CAUTION

Place the air bag module in a safe place with the front side facing upward.

   v. Remove the instrument panel-to-chassis 5 bolts and nut.
   w. Remove the audio amplifier.
   x. Remove the No. 1 and No. 2 braces.
   y. Remove the No. 2 cowl brace.
   z. Remove the instrument panel reinforcement.
6. Remove the evaporator housing by performing the following procedure:
   a. Discharge and recover the air conditioning system refrigerant.
   b. In the engine compartment, remove the refrigerant lines-to-cowl connector bolts; then, disconnect the lines and discard the O-rings.
   c. Disconnect the electrical connector at the evaporator housing.
   d. Disconnect the wiring harness clamp.
   e. Remove the evaporator housing-to-chassis 2 rivets, 3 bolts and nut.
   f. Remove the evaporator housing.

7. Remove the 4 defroster nozzle nuts and the nozzle.
8. Disconnect and remove the theft deterrent and the wireless door lock ECUs.
9. Release the 2 air duct claws and the air duct.
10. Remove the 2 heater housing-to-chassis rivets and the heater housing.

➡**When installing the heater housing, use new screws in place of the rivets.**

11. Remove the heater core-to-heater housing cover.
12. Remove both heater core screws and clamps; then, remove the heater core.
   To install:
13. Install the heater core and both heater core screws and clamps.
14. Install the heater core-to-heater housing cover.

➡**When installing the heater housing, use new screws in place of the rivets.**

15. Install the heater housing-to-chassis and the 2 heater housing screws.
16. Release the air duct and the air duct claws.
17. Connect and install the theft deterrent and the wireless door lock ECUs.
18. Install the defroster nozzle and the 4 nozzle nuts.
19. Install the evaporator housing by performing the following procedure:
   a. Install the evaporator housing.
   b. Install the evaporator housing-to-chassis 2 rivets, 3 bolts and nut.
   c. Connect the wiring harness clamp.
   d. Connect the electrical connector at the evaporator housing.
   e. In the engine compartment, use new O-rings and install the refrigerant lines-to-cowl connector and install the bolts.
20. Install the instrument panel and reinforcement by performing the following procedure:
   a. Install the instrument panel reinforcement.
   b. Install the No. 2 cowl brace.
   c. Install the No. 1 and No. 2 braces.
   d. Install the audio amplifier.
   e. Install the instrument panel-to-chassis 5 bolts and nut.
   f. Connect and install the air bag module and the 2 passenger's side air bag module bolts.
   g. Install the heater control assembly.
   h. Install the radio assembly.
   i. Install the combination meter.
   j. Install the cluster finish panel, engage the 8 clips and install the panel screws.

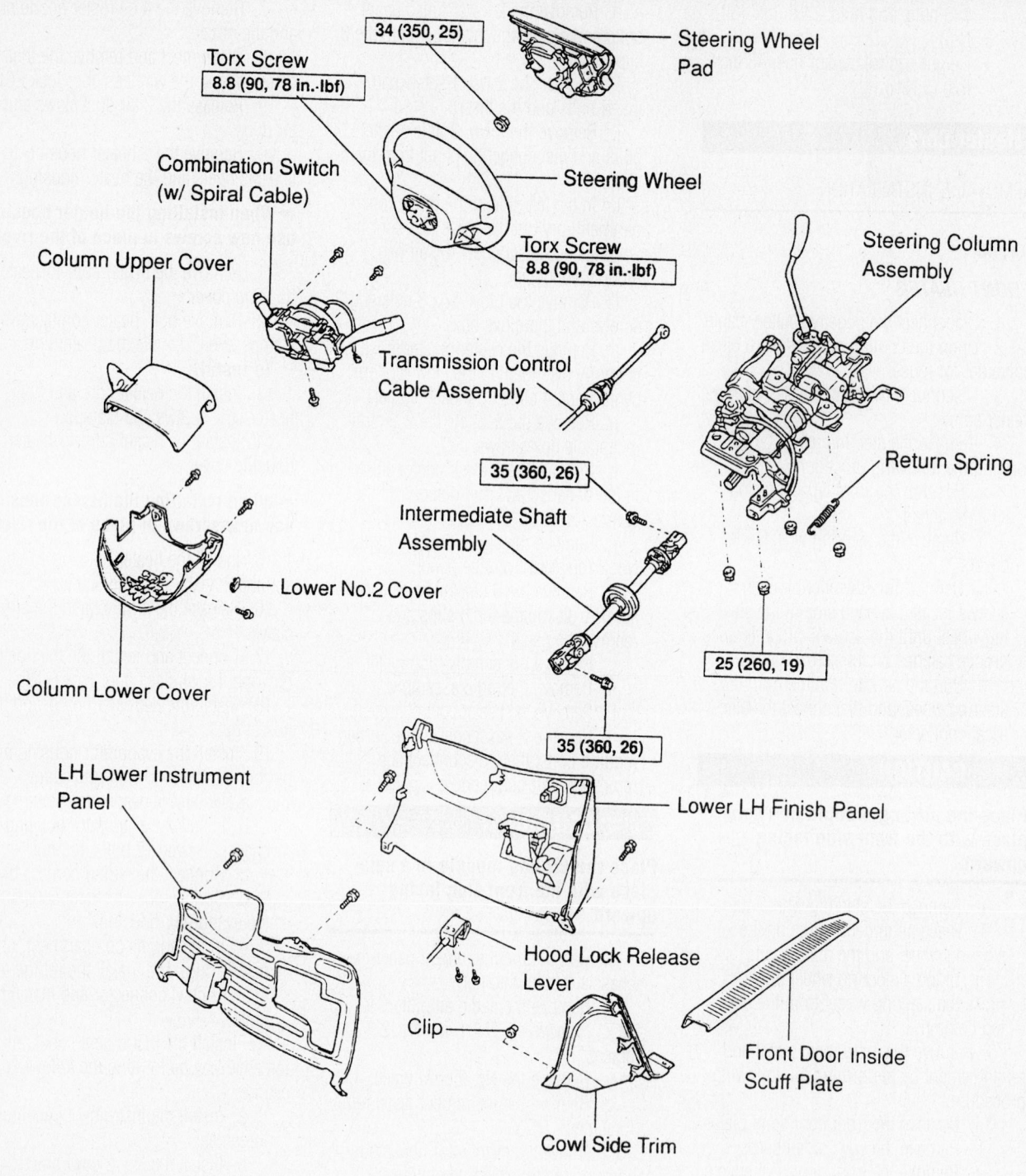

Torx Screw
8.8 (90, 78 in.-lbf)

34 (350, 25)

Steering Wheel Pad

Combination Switch (w/ Spiral Cable)

Steering Wheel

Column Upper Cover

Torx Screw
8.8 (90, 78 in.-lbf)

Steering Column Assembly

Transmission Control Cable Assembly

Return Spring

35 (360, 26)

Intermediate Shaft Assembly

25 (260, 19)

Lower No.2 Cover

Column Lower Cover

35 (360, 26)

LH Lower Instrument Panel

Lower LH Finish Panel

Hood Lock Release Lever

Clip

Front Door Inside Scuff Plate

Cowl Side Trim

N·m (kgf·cm, ft·lbf) : Specified torque

93113GH3

Exploded view of the steering wheel, steering column and related components—Lexus RX300

k. Install the No. 1 and No. 2 registers and the clock to the panel.

l. Connect the lower center cluster finish panel connector and install the 4 lower center cluster finish panel screws.

m. Install the 2 ashtray receptacle box screws and the ashtray.

n. Install the center cluster finish panel by engaging the 4 clips (1 at each corner) and the claw (bottom center).

o. Install the glove box and the glove box 3 nuts and 2 screws.

p. In the left side of the glove compartment, connect the air bag module connector and install the glove box door finish plate.

q. Install the No. 2 finish panel, engage the 4 panel clips and install the 3 panel bolts.

r. Install the No. 1 safety pad insert and the 2 insert bolts.

s. Install the finish panel, engage the 3 finish panel clips and install 2 lower finish panel bolts.

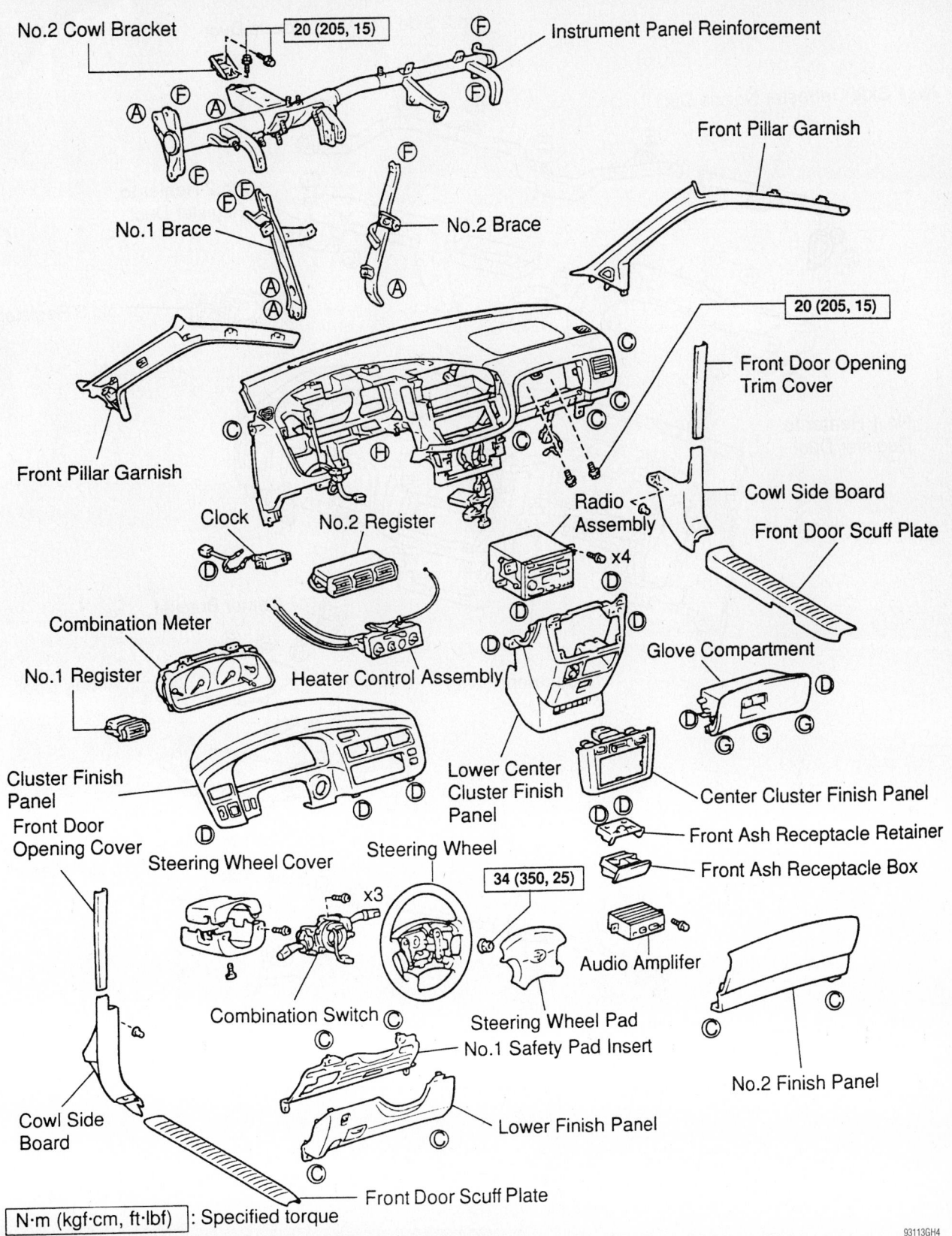

No.2 Cowl Bracket

20 (205, 15)

Instrument Panel Reinforcement

Front Pillar Garnish

No.1 Brace

No.2 Brace

20 (205, 15)

Front Door Opening Trim Cover

Front Pillar Garnish

Cowl Side Board

Front Door Scuff Plate

Clock

No.2 Register

Radio Assembly

x4

Combination Meter

Glove Compartment

No.1 Register

Heater Control Assembly

Center Cluster Finish Panel

Cluster Finish Panel

Lower Center Cluster Finish Panel

Front Ash Receptacle Retainer

Front Door Opening Cover

Steering Wheel Cover

Steering Wheel

Front Ash Receptacle Box

34 (350, 25)

Audio Amplifer

Combination Switch

Steering Wheel Pad

No.1 Safety Pad Insert

No.2 Finish Panel

Cowl Side Board

Lower Finish Panel

Front Door Scuff Plate

N·m (kgf·cm, ft·lbf) : Specified torque

93113GH4

Exploded view of the instrument panel and related components—Lexus RX300

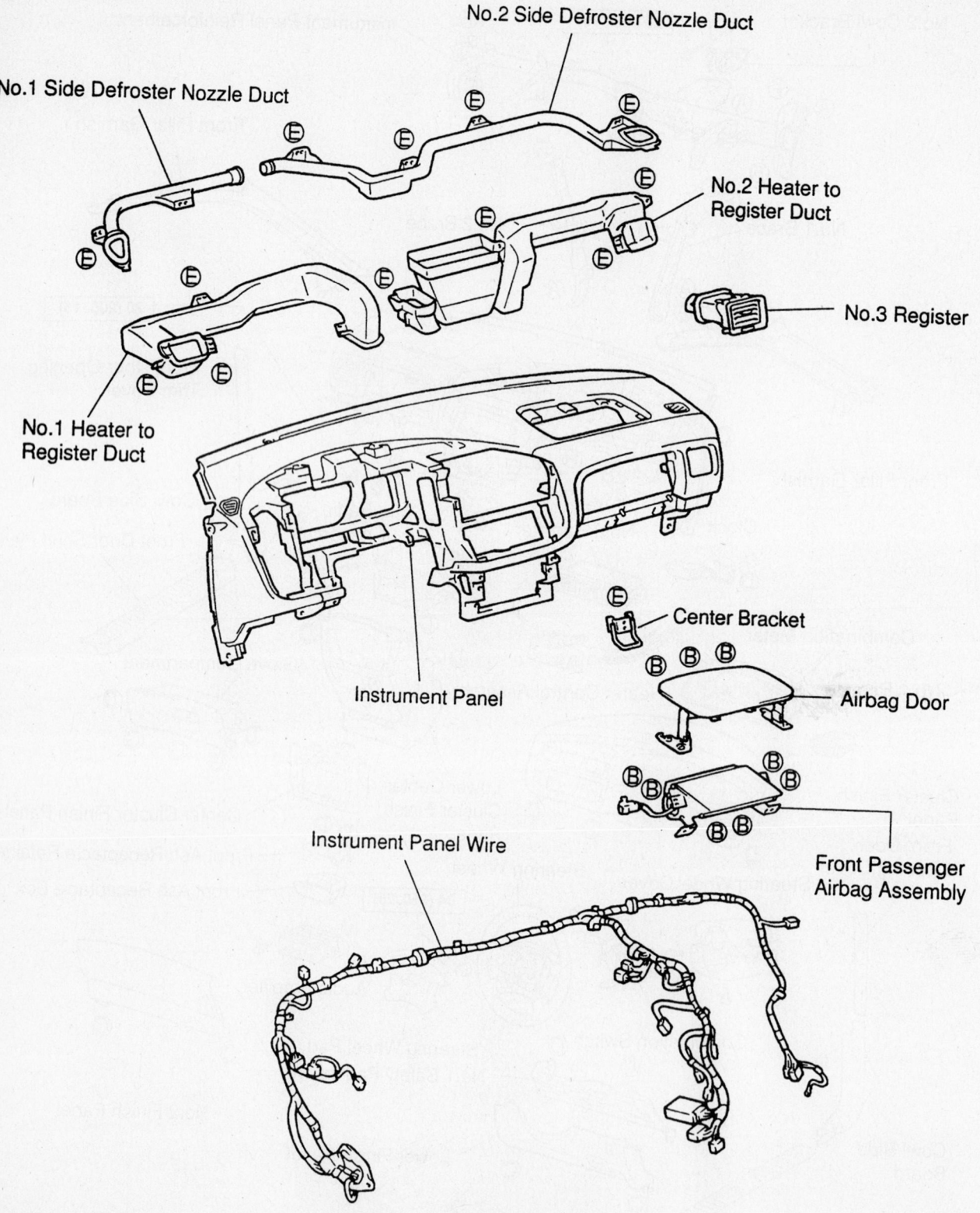

No.2 Side Defroster Nozzle Duct

No.1 Side Defroster Nozzle Duct

No.2 Heater to Register Duct

No.1 Heater to Register Duct

No.3 Register

Instrument Panel

Center Bracket

Airbag Door

Front Passenger Airbag Assembly

Instrument Panel Wire

93113GH5

Exploded view of the ventilation system and related components—Lexus RX300

t. Install the hood open lever and the 2 hood open lever screws.

u. Install the combination switch, connect the electrical connector(s) and install the combination switch-to-steering column screws.

v. Install the steering column covers and the covers-to-steering column screws.

w. Install the front pillar garnish by engaging the 5 clips. If equipped with a tweeter speaker, connect the electrical connector.

x. Install the front door trim covers.

y. Install the cowl side boards.

z. Install the front door scuff plates.

21. Install the steering wheel by performing the following procedure:

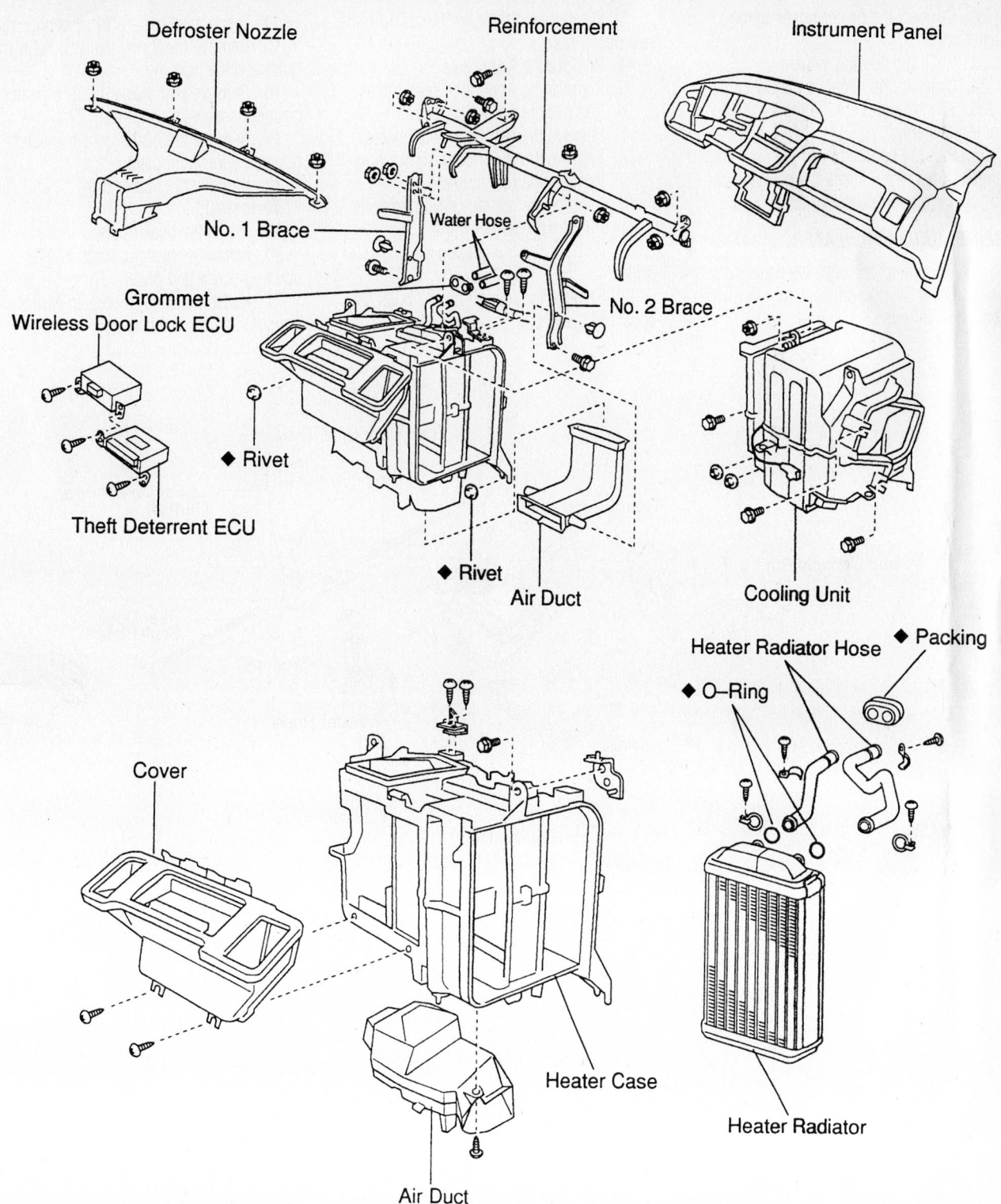

Defroster Nozzle

Reinforcement

Instrument Panel

No. 1 Brace

Water Hose

No. 2 Brace

Grommet

Wireless Door Lock ECU

◆ Rivet

Theft Deterrent ECU

◆ Rivet

Air Duct

Cooling Unit

Heater Radiator Hose

◆ Packing

◆ O-Ring

Cover

Heater Case

Heater Radiator

Air Duct

◆ Non-reusable part

93113GH6

Exploded view of the heater core, heater housing, evaporator housing and related components—Lexus RX300

a. Install the steering wheel to the steering column.

b. Align the steering wheel-to-main shaft marks.

c. Install the steering wheel nut and torque the nut to 25 ft. lbs. (34 Nm).

d. Install the air bag module to the steering wheel and connect the electrical connector.

e. Using a Torx® wrench, tighten the steering wheel screws to 78 inch lbs. (8.8 Nm).

f. Install the steering wheel side covers.

22. Connect the heater hoses to the heater core.

23. Refill the cooling system.

24. Connect the negative battery cable.

25. Evacuate and charge the air conditioning system.

26. Run the engine to normal operating temperatures; then, check the climate control operation and check for leaks.

REAR AUXILIARY HEATER

1. Disconnect the negative battery cable.

2. Drain the cooling system into a clean container for reuse.

3. Disconnect the heater hoses from the rear heater core.

4. Remove the front seats.

5. Remove the front door scuff plates.

6. Remove the cowl side trim.

7. Remove the rear door scuff plates.

8. Remove the lower door scuff plates.

9. Remove the rear console box.

10. Remove the left side air outlet grille.

11. Pull the carpet rearward.

12. Remove the 3 clips and the air outlet grille.

13. Remove the rear air duct 2 bolts, 2 clips and the duct.

14. Disconnect the electrical connectors.

15. Remove the 3 rear heater housing bolts and the housing.

16. Remove both heater core-to-heater housing screws and clamps.

17. Remove the heater core-to-heater housing screw and plate.

18. Remove the heater core.

To install:

19. Install the heater core.

20. Install the heater core-to-heater housing screw and plate.

21. Install both heater core-to-heater housing screws and clamps.

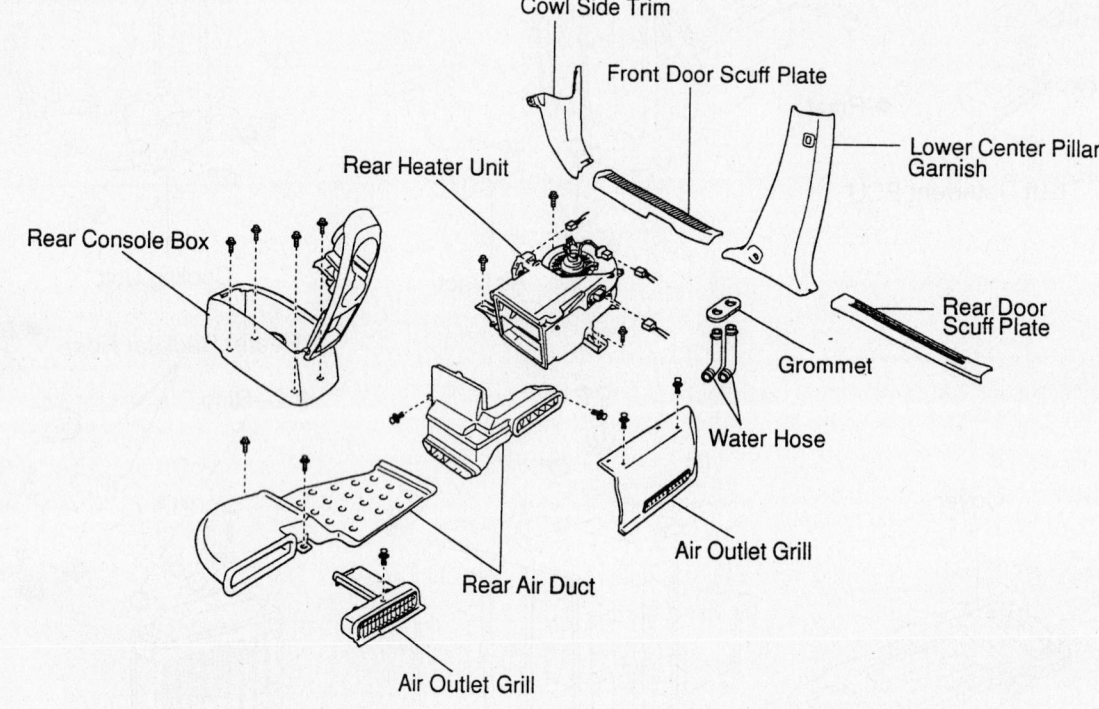

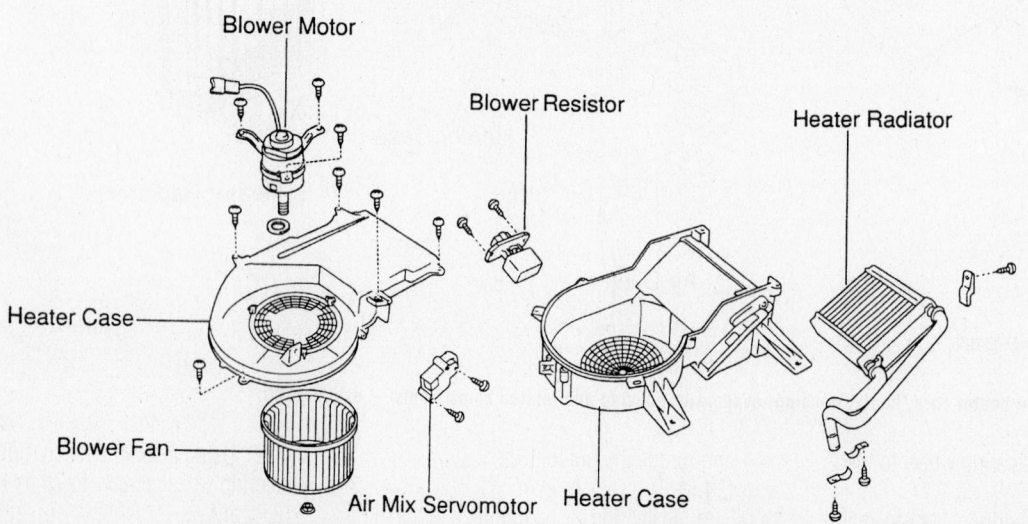

Exploded view of the rear heater core, the rear heater housing and related components—Lexus RX300

22. Install the rear heater housing and the 3 housing bolts.
23. Connect the electrical connectors.
24. Install the rear air duct and the duct 2 bolts and 2 clips.
25. Install the 3 clips and the air outlet grille.
26. Move the carpet forward.
27. Install the left side air outlet grille.
28. Install the rear console box.

29. Install the lower door scuff plates.
30. Install the rear door scuff plates.
31. Install the cowl side trim.
32. Install the front door scuff plates.
33. Install the front seats.
34. Connect the heater hoses to the rear heater core.
35. Refill the cooling system.
36. Connect the negative battery cable.

Water Pump

REMOVAL & INSTALLATION

3.0L and 3.3L

1. Before servicing the vehicle, refer to the precautions in the beginning of this section.

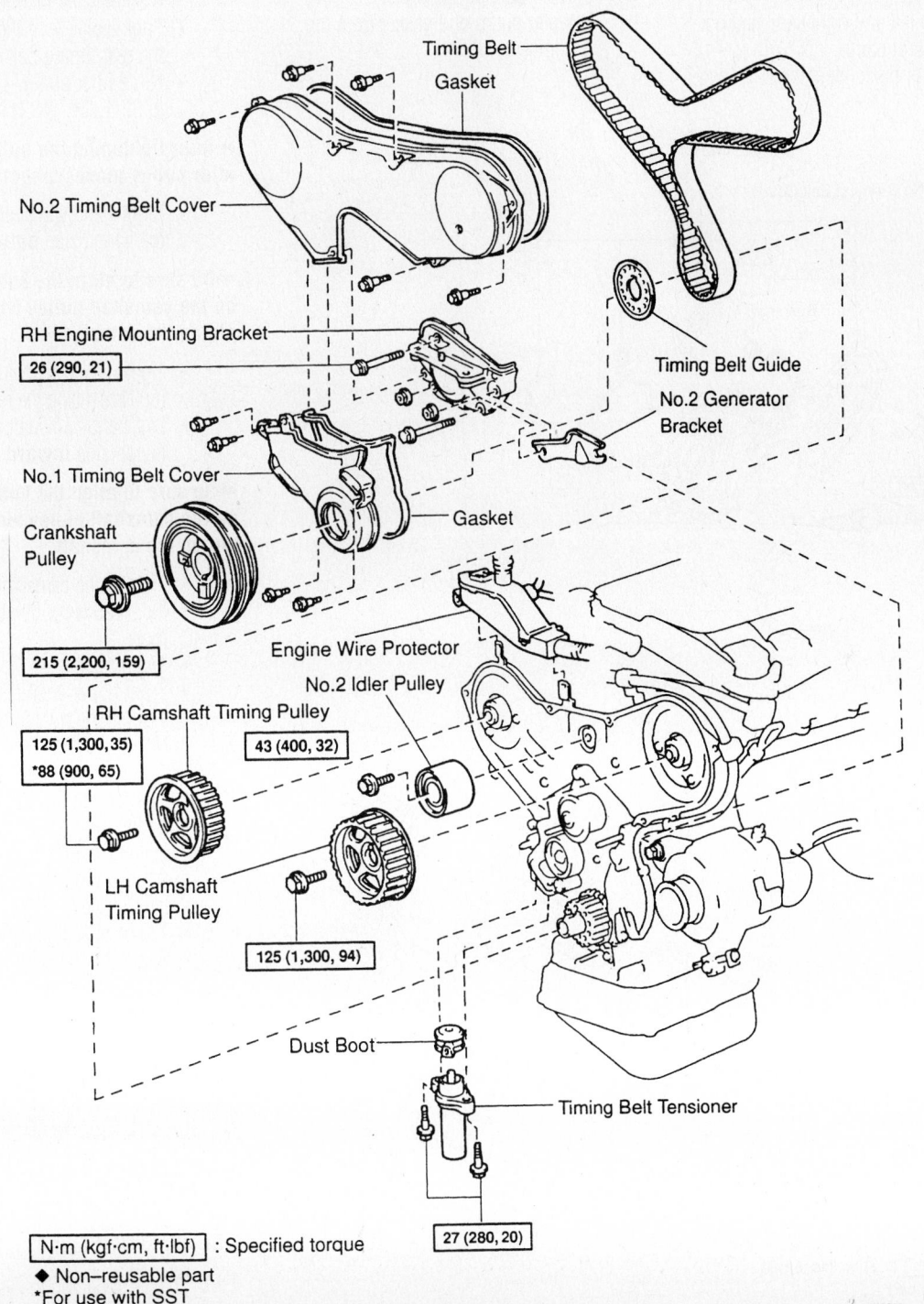

Timing Belt
Gasket
No.2 Timing Belt Cover
Timing Belt Guide
No.2 Generator Bracket
RH Engine Mounting Bracket
26 (290, 21)
No.1 Timing Belt Cover
Crankshaft Pulley
Gasket
215 (2,200, 159)
Engine Wire Protector
No.2 Idler Pulley
RH Camshaft Timing Pulley
43 (400, 32)
125 (1,300, 35)
*88 (900, 65)
LH Camshaft Timing Pulley
125 (1,300, 94)
Dust Boot
Timing Belt Tensioner
27 (280, 20)

N·m (kgf·cm, ft·lbf) : Specified torque
◆ Non–reusable part
*For use with SST

7924ZG15

Exploded view of the components to gain access to the water pump—3.0L

2. Disconnect the negative battery cable.

3. Drain the engine coolant.

4. Remove or disconnect the following:
- Wiper and blade assembly
- Top cowl seal and panel
- Window washer hoses, from the ventilator louvers
- Left and right ventilator louvers
- Heater air duct
- Front upper suspension brace
- Timing belt

5. Mark the left and right camshaft pulleys with a touch of paint.

6. Remove or disconnect the following:

- Right and left camshaft pulleys bolts
- Pulleys from the engine

➡ **Be sure not to mix up the pulleys.**

- No. 2 idler pulley by removing the bolt
- 3 clamps and engine wire from the rear timing belt cover
- 6 No. 3 timing belt cover-to-engine bolts
- Water pump nuts/bolts
- Water pump and gasket from the engine

To install:

7. Check that the water pump turns smoothly. Also check the air hole for coolant leakage.

8. Apply liquid sealer to the gasket, water pump and engine block.

9. Install or connect the following:
- Water pump, using a new gasket. Tighten the nuts/bolts to 53 inch lbs. (6 Nm).
- Rear timing belt cover. Tighten the 6 bolts to 74 inch lbs. (9 Nm).
- Engine wire with the 3 clamps to the rear timing belt cover
- No. 2 idler pulley. Tighten the bolt to 32 ft. lbs. (43 Nm).

➡ **After tightening the bolt, be sure the idler pulley moves smoothly.**

- Right-hand camshaft pulley, with the flange side **outward**.

➡ **Be sure to align the knock pin hole on the camshaft pulley with the knock pin on the camshaft.**

- Tighten the camshaft bolt to 65 ft. lbs. (88 Nm), using the removal tools
- Left-hand camshaft pulley, with the flange side **inward**.

➡ **Be sure to align the knock pin hole on the camshaft pulley with the knock pin on the camshaft.**

- Tighten the camshaft bolt to 94 ft. lbs. (125 Nm), using the removal tools
- Timing belt
- Front upper suspension brace. Tighten the nuts to 59 ft. lbs. (80 Nm).

10. Fill the engine coolant.

11. Install or connect the following:
- Heater air duct
- Left and right ventilator louvers
- Window washer hoses to the ventilator louvers
- Top cowl seal and panel
- Wiper and blade assembly
- Negative battery cable

12. Start the engine.

13. Top off the engine coolant and check for leaks.

Cylinder Head

REMOVAL & INSTALLATION

3.0L and 3.3L Engines

1. Before servicing the vehicle, refer to the precautions in the beginning of this section.

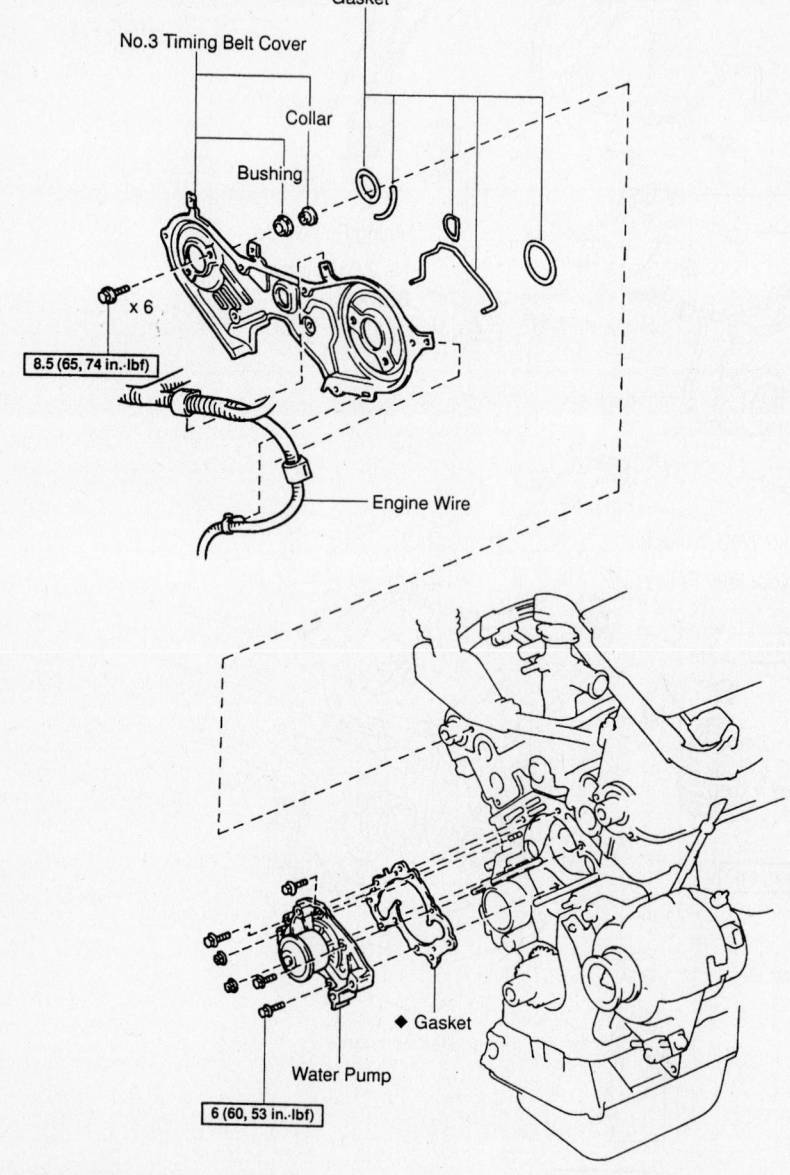

No.3 Timing Belt Cover
Gasket
Collar
Bushing
x 6
8.5 (65, 74 in.·lbf)
Engine Wire
Gasket
Water Pump
6 (60, 53 in.·lbf)

N·m (kgf·cm, ft·lbf) : Specified torque
◆ Non–reusable part

Exploded view of the water pump and related components—3.0L

7924ZG16

2. Remove or disconnect the following:
- Wiper and blade assembly
- Top cowl seal and panel
- Window washer hoses from the ventilator louvers
- Left and right ventilator louvers
- Heater air duct

3. Relieve the fuel pressure.

4. Remove or disconnect the following:
- Turn the ignition key to the **OFF** position
- Negative battery cable

➡**Wait at least 90 seconds from the time the negative battery was disconnected to start work.**

5. Drain the cooling system.

6. Remove or disconnect the following:
- Accelerator and throttle cables, if equipped with an automatic transaxle
- Air cleaner cover, air flow meter and the air duct
- Front upper suspension brace
- Cruise control actuator and bracket, if equipped
- 2 engine ground straps
- Right engine mounting support
- Radiator hoses
- 2 heater hoses
- Fuel feed and return lines from the fuel rail assembly
- Pressure hose from the hydraulic motor
- V-bank cover

7. Disconnect the following vacuum hoses:
- Fuel pressure control Vacuum Switching Valve (VSV)
- Fuel pressure regulator
- Cylinder head rear plate
- Intake air control valve VSV
- Exhaust Gas Recirculation (EGR) vacuum modulator
- EGR valve

8. Disconnect the following wiring and hoses:
- Intake air control valve
- Fuel pressure regulator
- EGR VSV

9. Remove the 2 nuts and the emission control valve set.

10. Disconnect the following hoses;
- Brake booster vacuum hose
- PCV hose
- Intake air control valve vacuum hose

11. Remove or disconnect the following:
- Data Link Connector (DLC) from the mounting bracket
- 2 ground straps from the intake chamber

- Hydraulic motor pressure hose from the intake chamber
- Right Oxygen (O_2) sensor connector from the power steering pressure tube
- 2 nuts and the power steering pressure tube from the intake chamber
- Both power steering air hoses
- Engine hanger and the intake chamber support
- EGR pipe and gaskets

12. Disconnect the following wiring:
- Throttle Position (TP) sensor connector
- Idle Air Control (IAC) valve connector
- EGR gas temperature connector
- Air conditioning idle up connector

13. Disconnect the following vacuum hoses:
- 2 vacuum hoses from the Thermal Vacuum Valve (TVV)
- Vacuum hose from the cylinder head rear plate
- Vacuum hose from the charcoal canister

14. Remove or disconnect the following:
- Air assist hose and the 2 water bypass hoses
- Air intake chamber
- Left engine wiring harness and move it aside
- Wiring harness from the rear of the engine
- Right engine wiring harness and move it aside
- Ignition coils and move them aside
- Timing belt
- Camshaft pulleys and the timing belt rear cover
- Cylinder head rear plate
- Water inlet pipe
- Air assist hose and vacuum hose

- Intake manifold and fuel rail assembly
- Water outlet
- EGR pipe from the right exhaust manifold
- Front exhaust pipe and exhaust manifolds
- Dipstick assembly and the power steering pump bracket
- Valve covers and the Camshaft Position (CMP) sensor
- Camshafts

15. Be sure the engine is at/or near ambient temperature and remove the 2 (1 on each head) 8mm recessed hex bolts. Loosen and remove the 8 head bolts evenly, in 3 passes, in the reverse order of the installation sequence. Carefully lift the head from the engine; if necessary to pry the head loose, take great care not to damage the mating surfaces. Place the head on wood blocks in a clean work area.

➡**If the cylinder head bolts are loosened out of sequence, warpage or cracking could result.**

16. Remove the cylinder head gasket. With a gasket scraper, carefully remove all the old gasket material from the cylinder head and engine block surfaces.

To install:

17. Place the new cylinder head gasket onto the cylinder block.

18. Install the cylinder head, in sequence, using several steps, as follows:
- Cylinder head onto the gasket
- Cylinder head bolts lubricated with clean engine oil
- Tighten the bolts in sequence in 3 steps to 40 ft. lbs. (54 Nm).

➡**If any bolt does not meet the torque, replace it.**

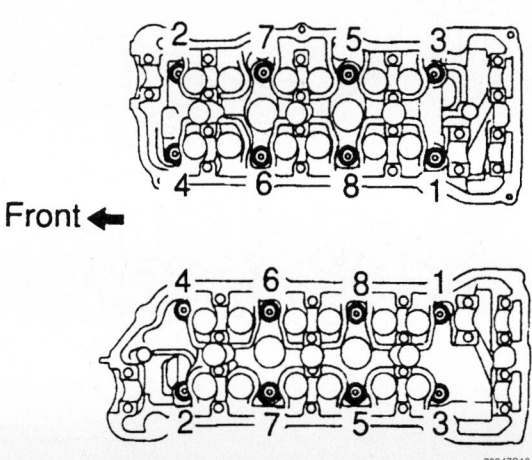

Front ⬅

7924ZG19

Cylinder head bolt loosening sequence—3.0L and 3.3L engines

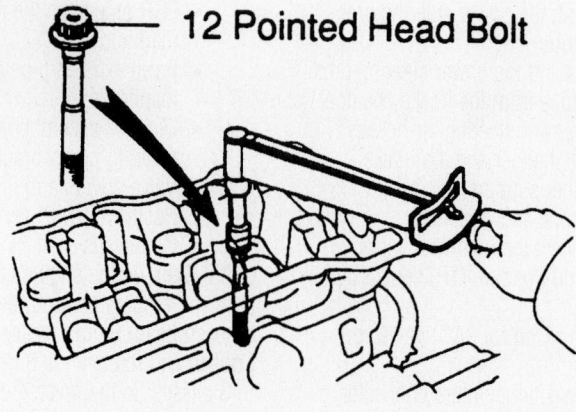

12 Pointed Head Bolt

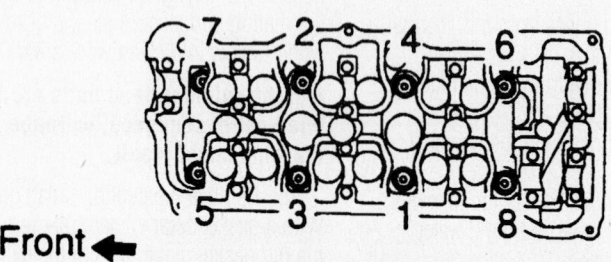

Front ←

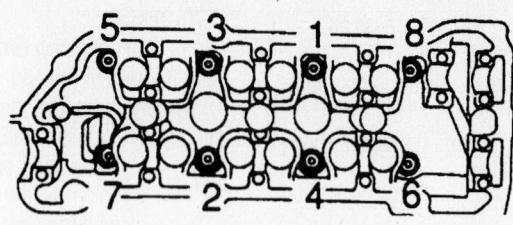

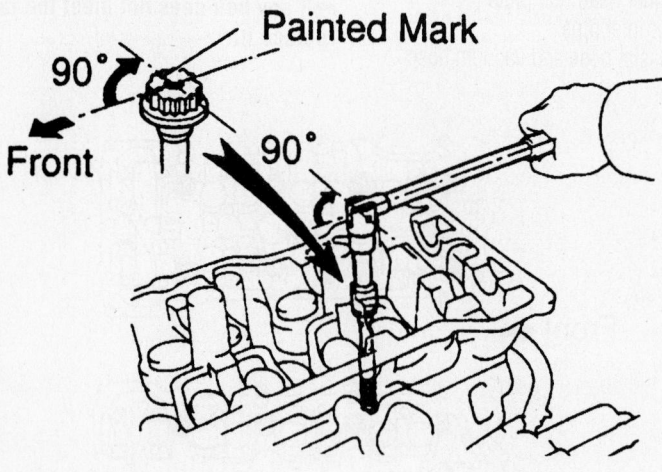

90°
Painted Mark
Front
90°

• Mark the forward edge of each bolt with paint, then tighten each bolt, in proper sequence, an additional 90 degrees.
• Check that each painted mark is now at a 90 degrees angle to the front

➡ **The paint mark applied to the bolt in the 9 o'clock position and should now be in the 12 o'clock position.**

• Remaining 8mm bolts, lubricated with engine oil. Tighten both bolts to 13 ft. lbs. (18 Nm).
19. Install the camshafts.
20. Check and adjust the valves.
21. Apply sealant to the cylinder heads where the camshaft supports meet the cylinder heads.
22. Install or connect the following:
• Cylinder head covers, using new gaskets
• Dipstick and power steering pump bracket
• Exhaust manifolds. Tighten the nuts to 36 ft. lbs. (49 Nm).
• EGR pipe to the right exhaust manifold
• Water outlet
• Intake manifold and the fuel rail assembly. Tighten the intake manifold nuts/bolts to 11 ft. lbs. (15 Nm).
• Air assist hose and the 2 water bypass hoses
• Water inlet pipe and cylinder head rear plate
• Timing belt rear cover and camshaft pulleys
• Timing belt
• Spark plugs and ignition coils
• Right engine wiring harness
• Wiring harness to the rear of the engine
• Left engine wiring harness
• Air intake chamber
• EGR pipe, using new gaskets
23. Connect the following vacuum hoses:
• The 2 TVV vacuum hoses
• The vacuum hose to the rear cylinder head plate
• Charcoal canister vacuum hose
24. Connect the following electrical wiring:
• TP sensor connector
• IAC valve connector
• EGR gas temperature connector
• Air conditioning idle up connector
25. Install or connect the following:
• Engine hanger and the intake chamber support
• Both power steering air hoses

7924ZG20

Cylinder head bolt tightening sequence—3.0L and 3.3L engines engine

- Power steering pressure tube to the intake chamber
- O$_2$sensor connector to the pressure tube.
- Both ground straps, to the intake chamber
- DLC to the bracket
26. Connect the following hoses:
- Power brake booster vacuum hose
- PCV hose
- IAC valve vacuum hose
27. Install or connect the following:
- Emission control valve set and related vacuum hoses and connectors
- V-bank cover
- Pressure hose to the hydraulic motor
- Fuel lines to the fuel rail assembly
- Heater and radiator hoses
- Right engine mounting support
- both engine ground straps
- Upper front suspension brace, if removed. Tighten the nuts to 59 ft. lbs. (80 Nm).
- Cruise control actuator and bracket
- Air cleaner, air flow meter and air duct assembly
- Accelerator and throttle cables
28. Fill the cooling system.
29. Install or connect the following:
- Negative battery cable
- Heater air duct
- Left and right ventilator louvers
- Window washer hoses from the ventilator louvers
- Top cowl seal and panel
- Wiper and blade assembly
30. Start the engine and check for leaks.
31. Bleed the air from the cooling system.
32. Road test the vehicle and check for unusual noise, shock, slippage, correct shift points and smooth operation.
33. Recheck the coolant and engine oil levels.

Intake Manifold

REMOVAL & INSTALLATION

3.0L and 3.3L Engines

1. Before servicing the vehicle, refer to the precautions in the beginning of this section.
2. Remove or disconnect the following:
- Wiper and blade assembly
- Top cowl seal and panel
- Window washer hoses from the ventilator louvers
- Left and right ventilator louvers

- Heater air duct
- Front upper suspension brace
3. Properly relieve the fuel system pressure.
4. Remove the battery and battery tray.
5. Drain and recycle the engine coolant.
6. Remove or disconnect the following:
- Accelerator cable
- Throttle cable
- Air cleaner cap assembly
- Any wiring or hoses interfering with removal
- Right side engine mount stay
- Radiator and heater hoses in the way of the intake manifold removal
- V-bank cover
- All the vacuum hose and wiring for the emission control valve set
- Air intake chamber and discard the gasket
- Exhaust Gas Recirculation (EGR) pipe and discard the gaskets
- Hydraulic motor pressure hose from the air intake chamber
- Engine wiring harnesses from the left side, right side, rear and No. 3 timing belt cover
- Front exhaust pipe, if necessary
- Timing belt, camshaft timing pulleys, No. 2 idler pulley and No. 3 timing belt cover
- Cylinder head rear plate
- 2 bolts, nuts and plate washers with the intake manifold.

➡The delivery pipes with injectors will be attached to the manifold.

- Other fuel related components such as the No. 2 fuel pipe and pulsation damper, if needed
- Delivery pipes from the intake manifold
7. Clean and inspect the intake manifold mating surfaces. Scrape all old gasket material off.

To install:
8. Install or connect the following:
- Delivery pipes with injectors to the intake manifold.

➡Be sure to place 4 spacers in position on the manifold. Temporarily install 4 bolts to retain the delivery pipes to the manifold. Inspect the injectors for smooth rotation.

- Tighten the delivery pipes bolts to 84 inch lbs. (10 Nm), once the injectors are properly seated
- No. 2 fuel pipe with union bolts and gaskets. Tighten the bolts to 24 ft. lbs. (32 Nm).
- No. 1 fuel pipe with pulsation

damper, using 4 new gaskets. Tighten the damper to 35 ft. lbs. (32 Nm) and the bolt to 11 ft. lbs. (15 Nm).
- Fuel pressure regulator, if removed
- Intake manifold. Tighten the 9 bolts and 2 nuts in a crisscross pattern to 11 ft. lbs. (15 Nm).

➡Be sure the gasket is in place properly prior to tightening.

9. Retighten the water outlet mounting nuts/bolts to 11 ft. lbs. (15 Nm), if loosened.
10. Install or connect the following:
- Air assist hose and water inlet pipe, using a new O-ring, by applying a small amount of soapy water. Tighten the fastener(s) to 14 ft. lbs. (20 Nm).
- Ground strap
- Vacuum hoses removed to the air intake chamber and vacuum tank
- Any remaining components, using new gaskets. Tighten the air intake chamber nuts/bolts to 32 ft. lbs. (43 Nm), the EGR pipe nuts to 108 inch lbs. (12 Nm) and the emission control valve set to 69 inch lbs. (8 Nm).
- Air cleaner assembly
- Heater hoses
- Battery and tray
- Throttle cable with bracket onto the throttle body
- Accelerator cable, by adjusting it, if equipped with an automatic transaxle
- Front upper suspension brace. Tighten the nuts to 59 ft. lbs. (80 Nm).
11. Refill the cooling system
12. Install or connect the following:
- Negative battery cable
- Heater air duct
- Left and right ventilator louvers
- Window washer hoses from the ventilator louvers
- Top cowl seal and panel
- Wiper and blade assembly
13. Start the engine and inspect for leaks.

Exhaust Manifold

REMOVAL & INSTALLATION

3.0L and 3.3L Engines

FRONT MANIFOLD

➡Removing the oil filter helps gain access to a lower bolt in the front exhaust manifold.

1. Before servicing the vehicle, refer to the precautions in the beginning of this section.

2. Remove or disconnect the following:
- Negative battery cable
- Engine undercovers
- Front exhaust pipe from the exhaust manifolds, by removing the nuts

➡ **Check for access to some of the manifold lower bolts, if so remove any possible.**

- Heated Oxygen (HO₂) sensor
- Exhaust manifold stay, by removing the bolt and nut
- Remaining exhaust manifold nuts; then, separate the exhaust manifold from the engine

To install:

3. Install or connect the following:
- Exhaust manifold, using a new gasket. Uniformly, tighten the bolts to 36 ft. lbs. (49 Nm).
- Exhaust manifold stay. Tighten the nut/bolt to 15 ft. lbs. (20 Nm).
- Heated Oxygen (HO₂) sensor to the exhaust manifold
- Front exhaust pipe to the exhaust manifold, using a new gasket. Tighten both nuts to 46 ft. lbs. (62 Nm).
- Engine undercovers
- Negative battery cable

REAR MANIFOLD

1. Before servicing the vehicle, refer to the precautions in the beginning of this section.

2. Remove or disconnect the following:
- Negative battery cable
- Engine undercovers
- Front exhaust pipe from both exhaust manifolds, from below the engine
- Exhaust Gas Recirculation (EGR) pipe from the rear exhaust manifold, by removing the 4 nuts
- Heated Oxygen (HO₂) sensor wiring, from the right exhaust manifold
- Exhaust manifold stay
- 6 exhaust manifold nuts and the exhaust manifold

To install:

3. Install or connect the following:
- Exhaust manifold to the engine, using a new gasket. Tighten the 6 nuts to 36 ft. lbs. (49 Nm).
- Exhaust manifold stay. Tighten the nut/bolt to 15 ft. lbs. (20 Nm).

- HO₂sensor wiring to the exhaust manifold
- EGR pipe to the exhaust manifold and the engine, using new gaskets. Tighten the 4 nuts to 108 inch lbs. (12 Nm).
- Front exhaust pipe to the exhaust manifold, use a new gasket. Tighten both nuts to 46 ft. lbs. (62 Nm).
- Engine undercovers
- Negative battery cable

Front Crankshaft Seal

REMOVAL & INSTALLATION

3.0L and 3.3L Engines

1. Before servicing the vehicle, refer to the precautions in the beginning of this section.

2. Remove or disconnect the following:
- Engine coolant reservoir tank and the alternator belt
- Right front wheel and the splash shield
- Power steering pump drive belt, by loosening both bolts
- Both ground wire connectors
- Right engine mounting stay
- Engine moving control rod and the No. 2 right engine mount bracket

➡ **To extract the engine bracket and control rod, raise the engine slightly.**

- No. 2 alternator bracket
- Crankshaft pulley bolt, using a pry-bar and wrench or Crankshaft Pulley Holding tool 09213-54015 and Flange Holding tool 09330-00021
- Crankshaft pulley, using a puller
- No. 1 timing belt cover

3. Remove the No. 2 timing belt cover, as follows:
- Engine wire protector from the No. 3 (rear) timing belt cover
- Engine wire protector clamp from the No. 3 timing belt cover
- 5 bolts from the No. 2 timing belt cover
- No. 2 cover

To install:

4. Install or connect the following:
- No. 2 timing belt cover, using a new gasket

➡ **Install it evenly to the part of the belt cover shaded black. After installation, press down on it so that the adhesive sticks to the belt cover firmly.**

- No. 2 timing belt cover. Tighten the 5 bolts to 74 inch lbs. (8 Nm).

- Engine wire protector clamp to the No. 3 timing belt cover
- Engine wire protector to the No. 3 timing belt cover with the bolt
- No. 3 timing belt cover, using a new gasket
- Tighten the 4 No. 1 timing belt cover bolts to 74 inch lbs. (8 Nm).
- Crankshaft pulley. Tighten the bolt to 159 ft. lbs. (215 Nm).
- No. 2 alternator bracket. Tighten the nut to 21 ft. lbs. (28 Nm). Do not tighten the pivot bolt at this time.
- No. 2 right engine mounting bracket and the moving control rod
- Right engine mount stay
- Both ground wire connectors
- Drive belts by adjusting them
- Coolant reservoir
- Right front splash shield and wheel
- Negative battery cable

5. Start the vehicle and check for any leaks.

6. Recheck the ignition timing.

Camshaft and Valve Lifters

REMOVAL & INSTALLATION

3.0L and 3.3L Engines

1. Before servicing the vehicle, refer to the precautions in the beginning of this section.

2. Remove or disconnect the following:
- Timing belt and idler pulley
- Camshaft timing pulleys
- Cylinder head covers

➡ **The thrust clearance on both the intake and exhaust camshafts is very small; the camshafts must be kept level during removal. If the camshafts are removed without being kept level, the camshaft may be caught in the cylinder head, causing the head to break or the camshaft to seize.**

3. Remove the exhaust and intake camshafts from the right side cylinder head, as follows:

a. Turn the camshaft with a wrench until the 2 pointed marks drive and driven gears are aligned. (The right camshaft gears have 2 marks apiece; the left side camshaft gears have 1 mark each.)

b. Secure the exhaust camshaft sub-gear to the main gear using a service bolt. A bolt 0.63–0.79 in. (16–20mm) long with a 6mm thread diameter and a 1mm pitch is recommended. When

Intake

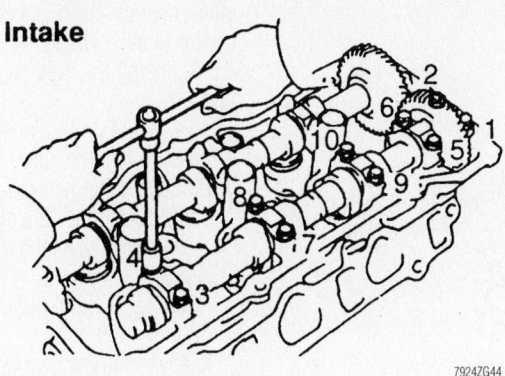

7924ZG44

Right intake camshaft bearing cap bolt loosening sequence

Exhaust

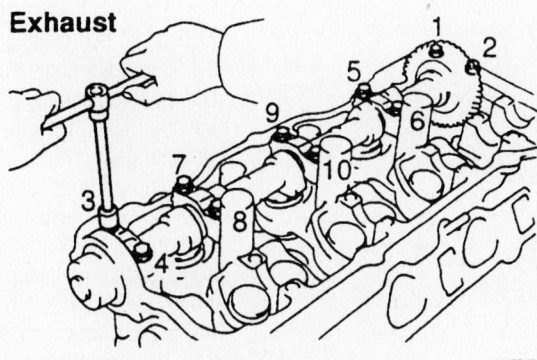

7924ZG45

Right side exhaust camshaft bearing cap bolt loosening sequence

Intake

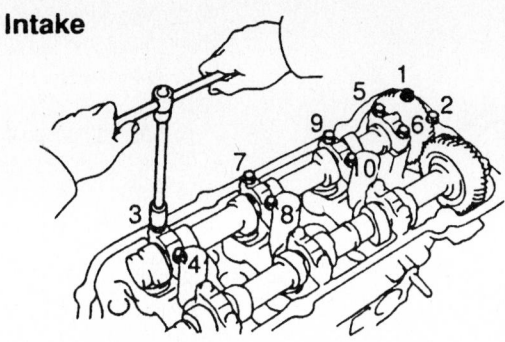

7924ZG46

Left intake camshaft bearing cap bolt loosening sequence

Exhaust

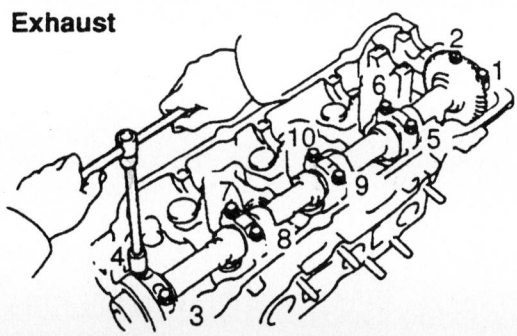

7924ZG47

Left side exhaust camshaft bearing cap bolt loosening sequence

removing the exhaust camshaft be sure the sub-gear is not loaded; all the force must be eliminated.

 c. Uniformly loosen and remove the exhaust camshaft bearing cap bolts in several passes and in the proper sequence. Remove the 8 bearing cap bolts and remove the caps, keeping them in the correct order.

 d. Remove the exhaust camshaft from the engine.

 e. Uniformly loosen and remove the 10 bearing cap bolts in several passes, in the proper sequence. Remove the bearing caps, keeping them in order, remove the oil seal, then lift out the intake camshaft.

4. Remove the exhaust and intake camshafts from the left side cylinder head, as follows:

 a. Turn the camshaft with a wrench until the pointed marks on the drive and driven gears are aligned. (The right camshaft gears have 2 marks apiece; the left side camshaft gears have 1 mark each.)

 b. Secure the exhaust camshaft sub-gear to the main gear using a service bolt. A bolt 16–20mm long with a 6mm thread diameter and a 1mm pitch is recommended. When removing the exhaust camshaft be sure the sub-gear is not loaded; all the force must be eliminated.

 c. Uniformly loosen and remove the exhaust camshaft bearing cap bolts in several passes and in the proper sequence. Remove the 8 bearing cap bolts and remove the caps. Keep the caps in the correct order.

 d. Remove the exhaust camshaft from the engine.

 e. Uniformly loosen and remove the 10 bearing cap bolts in several passes, in the reverse order of the installation sequence. Remove the bearing caps, keeping them in order, remove the oil seal, then lift out the intake camshaft.

5. Remove the valve lifter shims and hydraulic lifters. Identify each lifter and shim as it is removed so it can be reinstalled in the same position. If the lifters are to be reused, store them upside down in a sealed container.

To install:

6. Install the valve lifters into their original positions and install the shims. Check valve clearance and replace the shims as necessary.

7. When reinstalling, remember that the camshafts must be handled carefully and kept straight and level to avoid damage.

8. Before installing the camshafts in

Exhaust

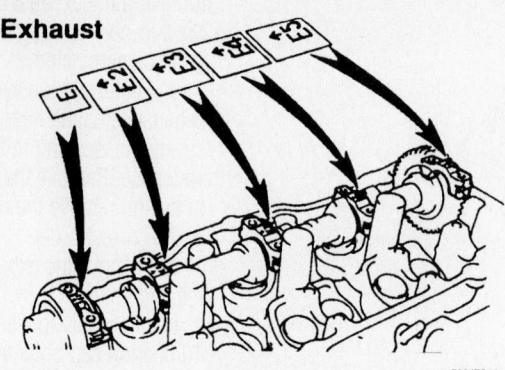

7924ZG48

Right exhaust bearing caps must be placed in their proper locations

Exhaust

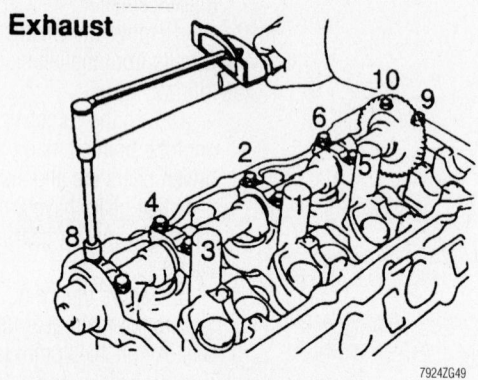

7924ZG49

Right exhaust camshaft bearing cap bolt tightening sequence

Intake

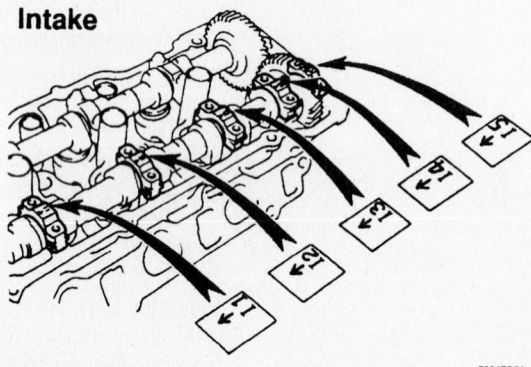

7924ZG50

Right intake bearing caps must be placed in their proper locations

Intake

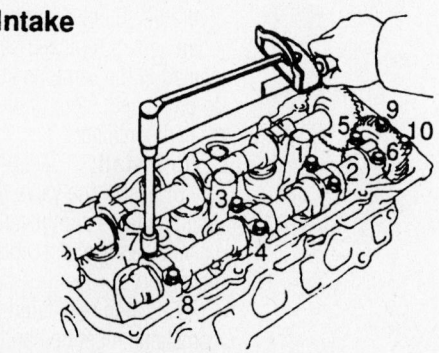

7924ZG51

Right intake camshaft bearing cap bolt tightening sequence

either cylinder head, apply multi-purpose grease to each camshaft.

9. Install the right camshafts, as follows:

a. Position the intake camshaft on the head so that the alignment marks are at a 90 degrees angle from vertical. The mark should be at the "3 o'clock" position.

b. Apply sealant to the No. 1 bearing cap.

c. Apply a light coat of clean engine oil to the bolt threads and under the bolt head. Install the bearing caps to their proper position. Tighten the bolts evenly and in several passes to 12 ft. lbs. (16 Nm) in the proper sequence.

d. Position the exhaust camshaft on the head so that the alignment marks are at a 90 degrees angle from vertical. The mark should be at the "9 o'clock" position and must align with the marks on the other gear.

e. Apply a light coat of clean engine oil to the bolt threads and under the bolt head. Install the bearing caps to their proper position. Tighten the bolts evenly and in several passes to 12 ft. lbs. (16 Nm) in the proper sequence.

f. Remove the service bolt.

10. Install the left camshafts, as follows:

a. Position the intake camshaft on the head so that the alignment mark is at a 90 degrees angle from vertical. The mark should be at the "9 o'clock" position.

b. Apply sealant to the No. 1 bearing cap.

c. Apply a light coat of clean engine oil to the bolt threads and under the bolt head. Install the bearing caps to their proper position. Tighten the bolts evenly and in several passes to 12 ft. lbs. (16 Nm) in the proper sequence.

d. Position the exhaust camshaft on the head so that the alignment marks are at a 90 degrees angle from vertical. The mark should be at the "3 o'clock" position and must align with the marks on the other gear.

e. Apply a light coat of clean engine oil to the bolt threads and under the bolt head. Install the bearing caps to their proper position. Tighten the bolts evenly and in several passes to 12 ft. lbs. (16 Nm) in the proper sequence.

f. Remove the service bolt.

11. Install or connect the following:

• New camshaft oil seals, lubricated with multi-purpose grease
• No. 3 (rear) timing belt cover
• Camshaft timing gears
• Idler pulley, timing belt and covers

Exhaust

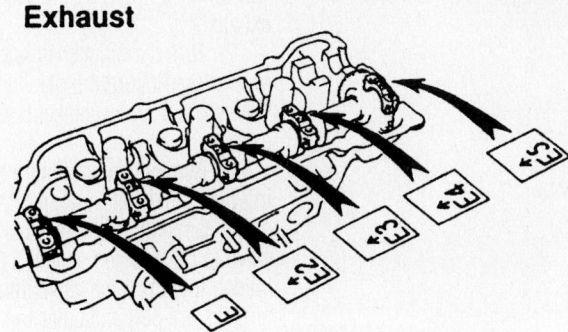

Exhaust

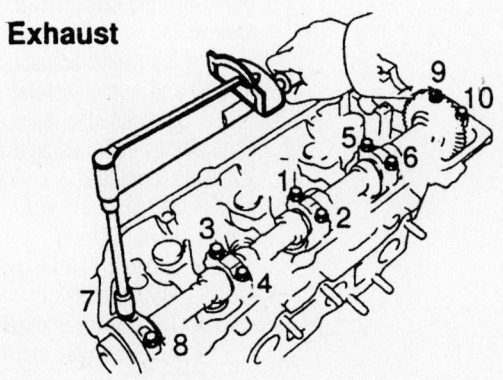

Left exhaust bearing caps locations and bolt tightening sequence

7924ZG52

Intake

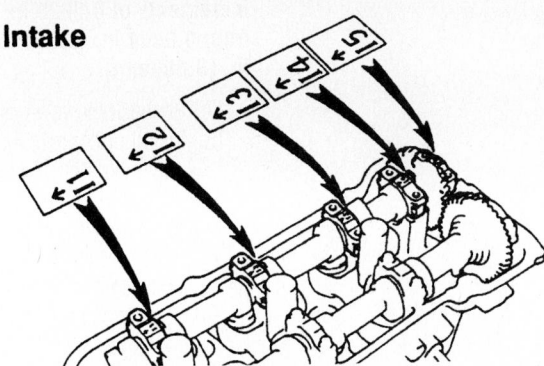

Intake

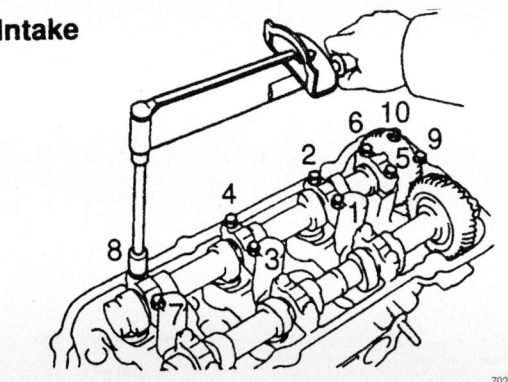

7924ZG53

Left intake camshaft bearing cap locations and bolt tightening sequence

12. Check and adjust the valve clearance.

13. Install the cylinder head (valve) covers.

14. Start the engine. Check the ignition timing.

15. Test drive the vehicle.

16. Check all fluid levels.

Valve Lash

ADJUSTMENT

3.0L and 3.3L Engines

➡**Adjust the valve clearance when the engine is cold.**

1. Before servicing the vehicle, refer to the precautions in the beginning of this section.

2. Remove or disconnect the following:
- Negative battery cable. If equipped with an air bag, wait at least 90 seconds before proceeding.
- Accelerator/throttle cable from the throttle linkage
- Air cleaner cover, air flow meter and air duct assembly
- V-bank cover
- Emission control valve set
- Air intake chamber
- Engine harness from the injectors and the ignition coils
- Ignition coils and keep them in order for reassembly
- Spark plugs
- Cylinder head covers

3. Turn the crankshaft pulley and align its groove with the timing mark **0** of the No. 1 timing cover.

4. Check that the valve lifters on the No. 1 intake are loose and the No. 1 exhaust are tight. If not, turn the crankshaft 1 complete revolution (360 degrees).

➡**All measurements should be written down. These recorded measurements will need to be used in conjunction with a mathematical formula to determine the thickness of the replacement shims.**

5. Measure the clearance between the valve lifters and the camshaft. Record the measurements on valves No. 1 and 6 intake; No. 2 and 3 exhaust.

　a. The intake valve clearance cold is 0.006–0.010 in. (0.15–0.25mm).

　b. The exhaust valve clearance cold is 0.010–0.014 in. (0.25–0.35mm).

6. Turn the crankshaft ⅔ of a revolution (240 degrees). Record the measurements on

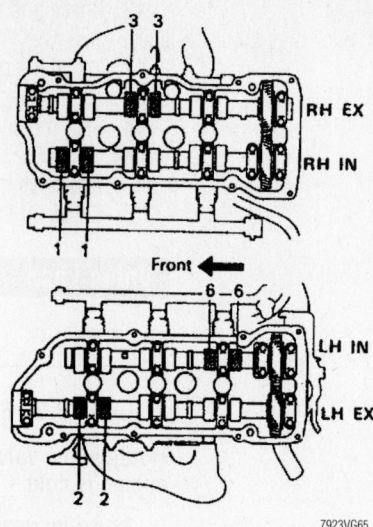

Adjust these valves during the 1st step

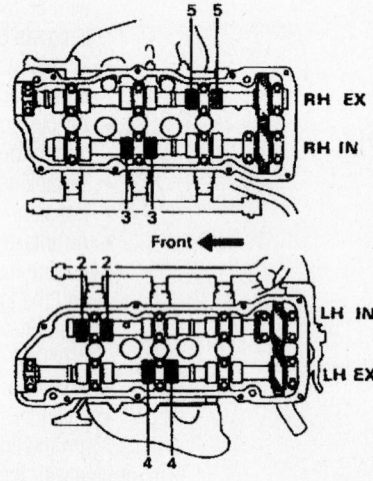

Adjust these valves during the 2nd step

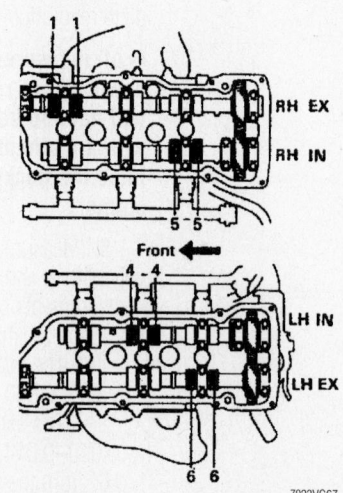

Adjust these valves during the 3rd step

valves No. 2 and 3 intake; No. 4 and 5 exhaust.

7. Turn the crankshaft another ⅔ of a revolution. Record the measurements on valves No. 4 and 5 intake; No. 1 and 6 exhaust.

8. Remove the adjusting shim by turning the crankshaft to position the cam lobe of the camshaft in the up position on the valve to be adjusted. Using a small thin flat bladed tool, turn the valve lifter so that the notches are perpendicular to the camshaft. Press down the valve lifter with tool 09248-55010 part A. Place too 09248-55010 part B between the camshaft and the valve lifter; remove part A.

9. Remove the adjusting shim with a magnet and a small screwdriver.

10. Determine the replacement adjusting shim size by either using the charts or the following formulas:
- Intake: $N = T + (A—0.008$ in./0.020mm)
- Exhaust: $N = T + (A—0.012$ in./0.30mm)
- T = Thickness of removed shim
- A = Measured valve clearance
- N = Thickness of new shim

11. Select a new shim with a thickness as close as possible to the calculated value. Install the new replacement shim.

➡ **Shims are available in 17 sizes in increments of 0.0020 in. (0.050mm), from 0.0984 in. (2.500mm) to 0.1299 in. (3.300mm).**

12. Recheck the valve clearance.
13. Install or connect the following:
- Cylinder head covers
- Spark plugs and the ignition coils
- Engine wiring harness to the injectors and the coils
- Intake chamber
- Emission control valve set
- V-bank cover
- Air flow meter, air duct and air cleaner cover
- Negative battery cable

Starter Motor

REMOVAL & INSTALLATION

3.0L

1. Before servicing the vehicle, refer to the precautions in the beginning of this section.

2. Remove or disconnect the following:
- Battery
- Battery tray

3. Remove or disconnect the cruise control actuator, if equipped, as follows:
- Actuator connector and clamp
- 3 bolts and the actuator with the bracket

4. Remove or disconnect the following:
- Automatic transaxle shift control cable
- Engine wiring
- Starter electrical connectors
- Both bolts, shift control cable clamp and the starter

To install:

5. Install or connect the following:
- Starter and the shift control cable clamp. Tighten the bolts to 27 ft. lbs. (37 Nm).
- Starter electrical connectors
- Engine wiring
- Automatic transaxle shift control cable

6. Install or connect the following, if equipped with cruise control:
- 3 bolts and the actuator with the bracket
- Actuator connector and clamp

7. Install or connect the following:
- Battery tray
- Battery

3.3L Engine

1. Before servicing the vehicle, refer to the precautions in the beginning of this section.
2. Remove the air cleaner assembly and inlet tubes.
3. Remove the air cleaner bracket.
4. Remove the wiring from the starter.
5. Remove the 2 bolts and lower the starter.
6. Installation is the reverse of removal. Torque the starter bolts to 26 ft. lbs. (37 Nm).

Oil Pan

REMOVAL & INSTALLATION

3.0L and 3.3L Engines

1. Before servicing the vehicle, refer to the precautions in the beginning of this section.
2. Remove or disconnect the following:
- Right front wheel
- Fender apron seal
- Engine undercover

3. Drain the engine oil from the engine.
4. Remove or disconnect the following:
- Front exhaust pipe
- Front exhaust pipe bracket from the No. 1 oil pan

- Flywheel housing undercover
- 10 bolts and 2 nuts to the No. 2 oil pan

5. Insert the blade of the Oil Pan Seal Cutting tool 09032-00100 between the No. 1 and No. 2 oil pans. Clean the surfaces of the oil pans.
6. Remove or disconnect the following:
- 3 oil strainer nuts and gasket

7. Remove the No. 1 oil pan, as follows:
- 2 bolts and the flywheel housing undercover
- 17 bolts and 2 nuts to the No. 1 oil pan

➡**Make a note of the position of the each bolt. When replacing the bolts into the oil pan, place each bolt in the position from which it was removed.**

- Oil pan, by prying the portions between the cylinder block and the oil pan

➡**Be careful not to damage the contact surfaces.**

- Baffle plate from the No. 1 oil pan

To install:

8. Clean all mating surfaces of the oil pans.
9. Install the baffle plate to the No. 1 oil pan and tighten to 69 inch lbs. (8 Nm).
10. Install the No. 1 oil pan, as follows:
   a. Using a non residue solvent, clean both sealing surfaces to the oil pan.
   b. Apply liquid sealant to the oil pan and engine block.
   c. Install the oil pan with the 17 bolts and 2 nuts. Uniformly tighten the bolts and nuts in several passes.
   d. Tighten the No. 1 oil pan bolts, as follows:
   - 10mm head bolt: 69 inch lbs. (8 Nm)
   - 12mm head bolt: 14 ft. lbs. (20 Nm)
   - 14mm head bolt: 27 ft. lbs. (37 Nm)
   e. Install the flywheel housing undercover with the 2 bolts. Tighten the bolts to 69 inch lbs. (8 Nm).

11. Install the oil strainer with the 3 nuts. Tighten the nuts to 69 inch lbs. (8 Nm).
12. Install the No. 2 oil pan, as follows:
   a. Using a non residue solvent, clean both sealing surfaces to the oil pan.
   b. Apply liquid sealant to the oil pan and engine block.
   c. Install the No. 2 oil pan with the 10 bolts and 2 nuts. Uniformly tighten the bolts and nuts in several passes. Tighten the bolts to 69 inch lbs. (8 Nm).

13. Install or connect the following:
- Flywheel housing undercover
- Front exhaust pipe bracket to the No. 1 oil pan. Tighten the bolts to 15 ft. lbs. (21 Nm).

14. Install the front exhaust pipe, as follows:
- Temporarily install the 3 new gaskets and the front exhaust pipe with the 2 bolts and 6 nuts
- Tighten the 4 exhaust manifolds-to-front exhaust pipe nuts to 46 ft. lbs. (62 Nm).
- Tighten the both front exhaust pipe-to-center exhaust pipe nuts/bolts to 41 ft. lbs. (56 Nm).
- Bracket. Tighten both bolts to 14 ft. lbs. (19 Nm).
- Support stay. Tighten both bolts to 22 ft. lbs. (29 Nm).

15. Install or connect the following:
- Engine undercover
- Right fender apron seal
- Right front wheel

16. Fill the engine with oil.
17. Start the engine and check for leaks.

Oil Pump

REMOVAL & INSTALLATION

3.0L and 3.3L Engines

1. Before servicing the vehicle, refer to the precautions in the beginning of this section.
2. Remove or disconnect the following:
- Oil pan
- Crankshaft Position (CKP) sensor
- 9 oil pump bolts

➡**Make a note of the position of the each bolt. When replacing the bolts into the oil pump body, place each bolt in the position from which it was removed.**

- Oil pump body, by prying between the oil pump and main bearing cap
- O-ring from the cylinder block
- Plug, gasket, spring and relief valve from the oil pump body
- 9 screws, pump body cover, drive and driven rotors

To install:

3. Install or connect the following:
- Driven rotors, drive, pump body cover, using the 9 screws
- Oil pump relief valve, spring, gasket and the plug to the oil pump body
- New O-ring on the cylinder block

4. Using a non residue solvent, clean both sealing surfaces to the oil pump.

5. Apply liquid sealant to the oil pump and engine block.

6. Install or connect the following:
- Oil pump

➡ **Be sure to engage the splined teeth of the oil pump drive gear with the large teeth of the crankshaft.**

- 9 oil pump bolts. Tighten the bolts in several passes to 69 inch lbs. (8 Nm), for 10mm; 14 ft. lbs. (20 Nm), for 12mm; 32 ft. lbs. (43 Nm) for 14mm
- CKP sensor. Tighten the bolt to 69 inch lbs. (8 Nm).
- Baffle plate to the No. oil pan. Tighten to 69 inch lbs. (8 Nm).
- No. 1 oil pan, oil strainer and No. 2 oil pan

7. Refill the engine with oil.

8. Start the engine and inspect for leaks.

9. Recheck the engine oil level.

Rear Main Seal

REMOVAL & INSTALLATION

3.0L and 3.3L Engines

If the rear oil seal retainer is not installed to the block, use a tapered ended screwdriver and hammer to remove the oil seal. Apply multi-purpose grease to the new oil seal lip. Using a seal driver, tap the seal into place. Be careful not to install it slantwise.

1. Before servicing the vehicle, refer to the precautions in the beginning of this section.

If the rear oil seal retainer is installed on the cylinder block, using a knife, cut off the lip of the seal. Using a taped ended prytool, pry the old seal out of the retainer. Inspect the oil seal lip contacting surface of the crankshaft for cracks or damage. Apply mul-

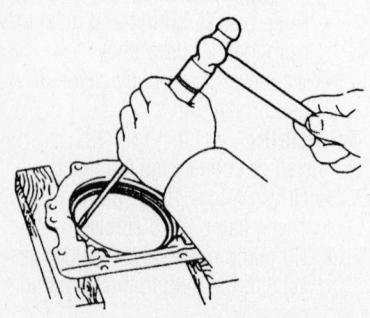

Carefully tap the old seal from the retainer

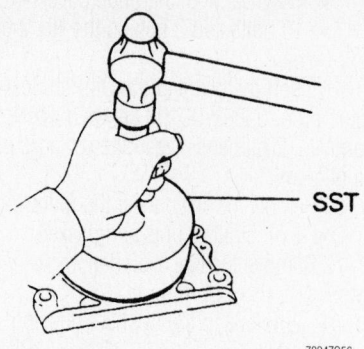

Use the proper sized driver to seat the seal

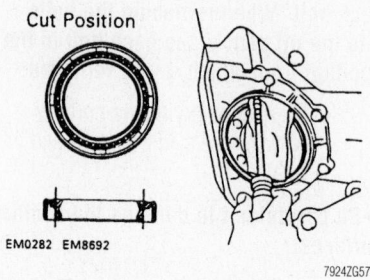

Cut off the oil seal lip, then pry the seal out of the retaining plate

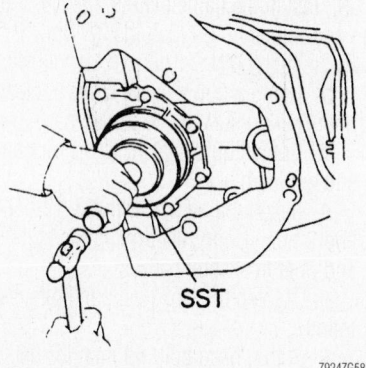

Tap a new seal into place

tipurpose grease to the new oil seal, then tap the seal in place with a seal installer. Be careful not to install the seal slantwise.

Timing Belt

REMOVAL & INSTALLATION

1. Before servicing the vehicle, refer to the precautions in the beginning of this section.

2. Remove the right front wheel.

3. Remove the wiper arms.

4. Remove the wiper linkage.

5. Remove the top cowl panel.

6. Remove the engine undercovers.

7. Remove the right front fender apron seal.

8. Remove the A/C compressor drive belt.

9. Remove the power steering pump belt.

10. Remove the engine roll control rod.

11. Remove the right side engine mount stay.

12. Remove the alternator bracket.

13. Remove the crankshaft pulley.

14. Remove the upper belt cover.

15. Remove the right engine mount bracket.

16. Remove the no. 2 timing belt guide.

17. Set the no.1 cylinder to TDC compression.

18. Temporarily install the crank pulley bolt. Turn the crankshaft clockwise to align the timing mark on the crankshaft timing pulley with the notch in the oil pump body. Check that the timing marks on the camshaft pulleys are aligned with the notches on the inner belt cover. If not, rotate the crankshaft 360 degrees clockwise.

➡ **If the timing belt is re-used, check that the 3 original installation marks are visible on the belt as shown. If not, paint three new marks on the belt.**

19. Turn the crankshaft counterclockwise by 60 degrees. Make sure that the belt is still engaged.

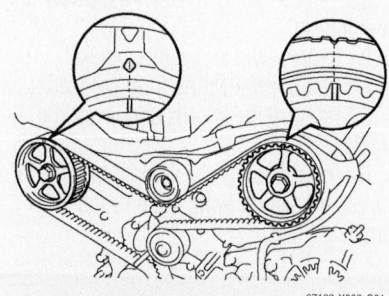

Check that the timing marks on the camshaft pulleys are aligned with the notches on the inner belt cover

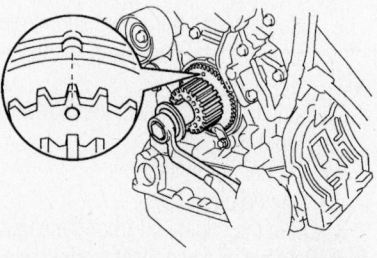

Turn the crankshaft clockwise to align the timing mark on the crankshaft timing pulley with the notch in the oil pump body

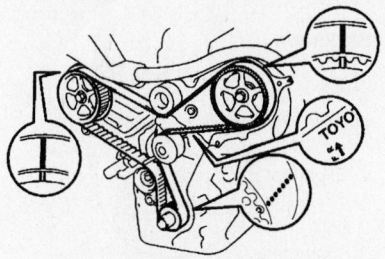

67162-X300-G02

If the timing belt is re-used, check that the 3 original installation marks are visible on the belt as shown

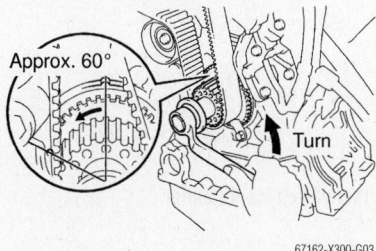

67162-X300-G03

Turn the crankshaft counterclockwise by 60 degrees

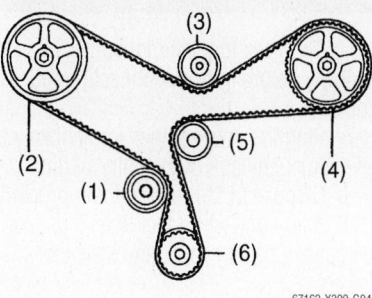

67162-X300-G04

Remove the belt from the pulleys in this order

20. Remove the tensioner.

21. Remove the belt from the pulleys in this order:

- Lower idler pulley
- Right camshaft pulley
- Upper idler pulley
- Left camshaft pulley
- Water pump pulley
- Crankshaft timing pulley

22. If the belt is being re-used, check it for wear or damage; don't twist it or turn it inside-out. If there is any doubt as to it's condition, replace it.

To install:

23. Clean all the pulleys.

24. Turn the crankshaft another 60 degrees counterclockwise.

25. Turn the camshaft pulleys back into alignment so the marks align with the notches on the inner cover.

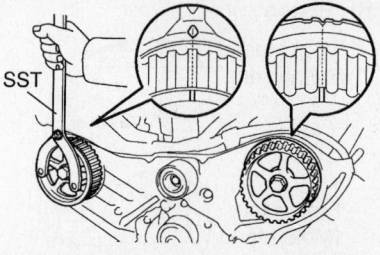

67162-X300-G05

Turn the camshaft pulleys back into alignment so the marks align with the notches on the inner cover

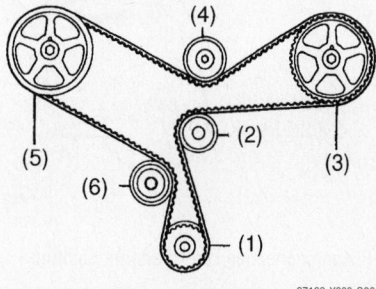

67162-X300-G06

Install the belt in this order

26. Turn the crankshaft back so that the timing mark aligns with the notch on the oil pump.

27. Align the installation marks on the belt with the timing marks on the pulleys.

28. Install the belt in this order:

- Crankshaft
- Water pump
- Left camshaft
- Upper idler
- Right camshaft
- Lower idler

29. Set the tensioner in a press and collapse the plunger. Do not apply more that 2,205 lbs (9.8 kN) of force. Insert a suitable metal rod through the holes to hold the plunger in position.

30. Install the tensioner and torque the 2 bolts alternately to 20 ft. lbs. (27 Nm).

✳✳ WARNING

Be sure to tighten to bolts alternately and evenly so the tensioner seats flat.

31. Remove the metal rod from the tensioner.

32. Turn the crankshaft 2 full revolutions clockwise (720 degrees), and align the timing mark on the crank pulley with the notch on the oil pump.

33. Check the timing marks on the camshaft pulleys for alignment with the notches on the inner cover. If they do not

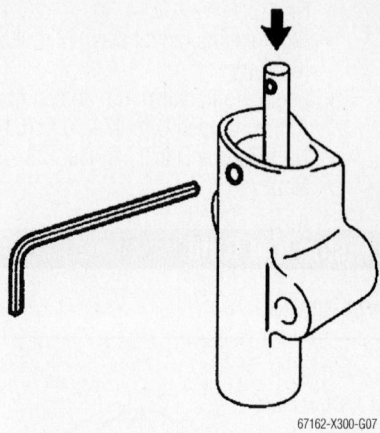

67162-X300-G07

Set the tensioner in a press and collapse the plunger. Do not apply more that 2,205 lbs (9.8 kN) of force. Insert a suitable metal rod through the holes to hold the plunger in position

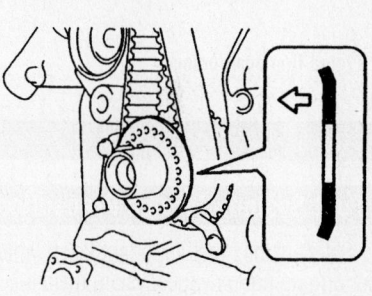

67162-X300-G08

Install the timing belt guide with the cupped side facing front

align, remove the belt and align the mismatched mark(s).

34. The remainder of installation is the reverse of removal. Observe the following torques:

- Right engine mount bracket: 21 ft. lbs. (28 Nm)
- Right engine mount insulator: 70 ft. lbs. (95 Nm)
- Timing belt covers: 75 inch lbs. (8.5 Nm)
- Crankshaft pulley: 162 ft. lbs. (220 Nm)

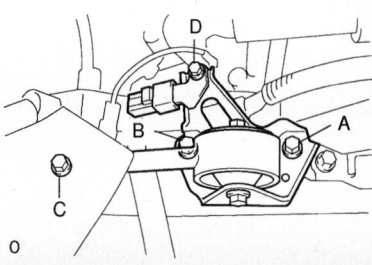

67162-X300-G10

Tighten the engine roll control rod bolts in this order

- Alternator bracket: 21 ft. lbs. (28 Nm)
- Right engine mount stay: 47 ft. lbs. (64 Nm)
- Engine roll control rod: tighten first A, then B, then C to 47 ft. lbs. (64 Nm). Torque D to 17 ft. lbs. (23 Nm)

Piston and Ring

POSITIONING

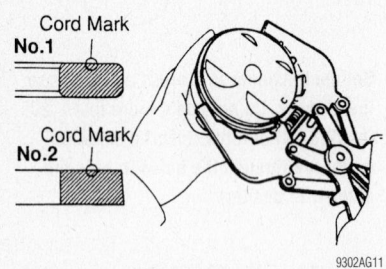

Piston ring positioning

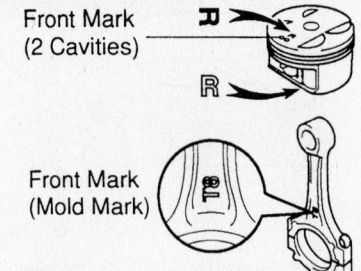

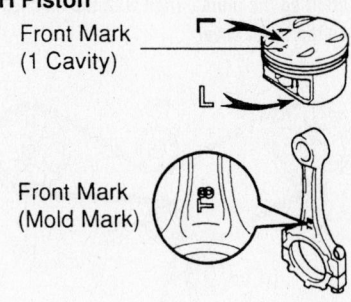

Piston/connecting rod-to-engine positioning

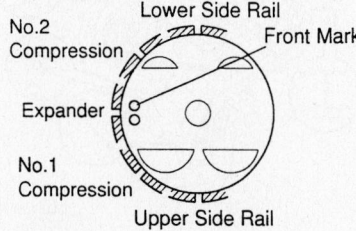

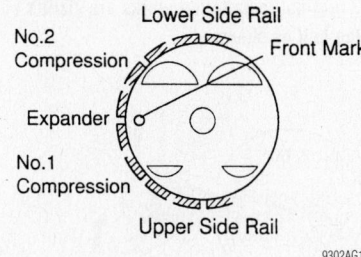

Piston ring identification

FUEL SYSTEM

Fuel System Service Precautions

Safety is the most important factor when performing not only fuel system maintenance but any type of maintenance. Failure to conduct maintenance and repairs in a safe manner may result in serious personal injury or death. Maintenance and testing of the vehicle's fuel system components can be accomplished safely and effectively by adhering to the following rules and guidelines.

- To avoid the possibility of fire and personal injury, always disconnect the negative battery cable unless the repair or test procedure requires that battery voltage be applied.
- Always relieve the fuel system pressure prior to disconnecting any fuel system component (injector, fuel rail, pressure regulator, etc.), fitting or fuel line connection. Exercise extreme caution whenever relieving fuel system pressure, to avoid exposing skin, face and eyes to fuel spray. Please be advised that fuel under pressure may penetrate the skin or any part of the body that it contacts.
- Always place a shop towel or cloth around the fitting or connection prior to loosening to absorb any excess fuel due to spillage. Ensure that all fuel spillage (should it occur) is quickly removed from engine surfaces. Ensure that all fuel soaked cloths or towels are deposited into a suitable waste container.
- Always keep a dry chemical (Class B) fire extinguisher near the work area.
- Do not allow fuel spray or fuel vapors to come into contact with a spark or open flame.
- Always use a back-up wrench when loosening and tightening fuel line connection fittings. This will prevent unnecessary stress and torsion to fuel line piping.
- Always replace worn fuel fitting O-rings with new. Do not substitute fuel hose or equivalent, where fuel pipe is installed.

Fuel System Pressure

RELIEVING

RX300

1. Before servicing the vehicle, refer to the precautions in the beginning of this section.
2. Disconnect the negative battery terminal.
3. Place a catch-pan under the joint to be disconnected. A large quantity of fuel may be released when the joint is opened.
4. Wear eye or full face protection.
5. Place a shop towel over the area and slowly loosen the joint using a wrench of the correct size. Use a back-up wrench if needed.

6. Allow the fuel left in the line to bleed off slowly before fully disconnecting the joint.
7. Plug the opened lines immediately to prevent fuel spillage or the entry of dirt.
8. Dispose of the released fuel properly.
9. After connecting fuel lines, connect the negative battery cable and start the engine.
10. Check for leaks and repair as needed.

RX330

1. Before servicing the vehicle, refer to the precautions in the beginning of this section.
2. Disconnect the fuel pump wire at the pump.
3. Start the engine. After the engine stops, turn the ignition switch to OFF.
4. Disconnect the negative battery terminal.
5. Connect the fuel pump.

Fuel Filter

REMOVAL & INSTALLATION

RX300

1. Before servicing the vehicle, refer to the precautions in the beginning of this section.

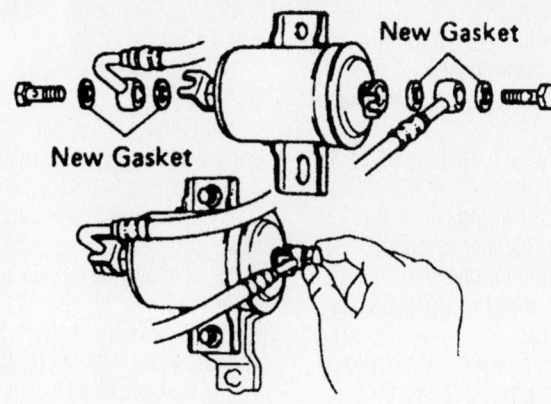

7924ZG59

Exploded view of the fuel filter—RX300

2. Disconnect the negative battery cable.
3. Relieve the fuel system pressure.

➡**The fuel filter is located in the engine compartment, at the inlet line to the fuel rail.**

4. Remove or disconnect the following:
   • Inlet and outlet lines from the filter
   • Fuel filter

To install:

5. Install or connect the following:
   • Fuel filter, using new O-rings. Tighten the lines to 22 ft. lbs. (29 Nm).
   • Negative battery cable
6. Start the engine and check for leaks.

RX330

The fuel filter is part of the fuel suction tube/fuel pump assembly and is located in the fuel tank. It is not a normally serviced item.

Fuel Pump

REMOVAL & INSTALLATION

RX300

1. Before servicing the vehicle, refer to the precautions in the beginning of this section.
2. Relieve the fuel system pressure.
3. Remove or disconnect the following:
   • Negative battery cable
   • Left-hand rear seat assembly
   • Floor service hole by pulling back the carpet; then, remove the 4 screws
   • Fuel pump and sender gauge connector

➡**Loosen the fuel cap to relieve any fuel pressure within the tank.**

   • Fuel pipe union bolt and both gaskets
   • Fuel pump outlet pipe
   • Return vent hose from the fuel pump
   • 8 fuel pump bolts and the pump assembly from the tank

To install:

4. Install or connect the following:
   • Fuel pump to the fuel tank. Tighten the 8 bolts to 31 inch lbs. (3.5 Nm).
   • Return vent hose to the fuel pump
   • Outlet pipe to the fuel pump, using new gaskets. Tighten the union bolts to 22 ft. lbs. (29 Nm).
   • Fuel pump and sender gauge connector
   • Floor hole cover with the 4 screws
   • Carpet
   • Left rear seat assembly
   • Negative battery cable
   • Fuel cap
5. Start the vehicle and check for leaks.

RX330

1. Before servicing the vehicle, refer to the precautions in the beginning of this section.
2. Relieve the fuel system pressure.
3. Remove the deck board assembly.
4. Remove the rear seats.
5. Remove the left side door scuff plate.
6. Remove the left side trim cover.
7. Remove the left rear seat side cover.
8. Remove the carpet.
9. Remove the rear floor service cover.
10. Disconnect the fuel pump wiring.
11. Disconnect the fuel hose.
12. Remove the fuel pump locking ring. Special tool 09808-14020 is available for this job.
13. Remove the pump assembly from the tank.

14. Installation is the reverse of removal. When installing the locking ring, tighten it first by hand, then, using the special tool turn the ring 1½ turn. The triangle mark on the lockring must be positioned between the **A** and **MAX** marks on the fuel tank.

✳✳ WARNING

No other type of tool should be used for this operation.

Fuel Injector

REMOVAL & INSTALLATION

RX300

1. Before servicing the vehicle, refer to the precautions in the beginning of this section.
2. Remove or disconnect the following:
   • Outer front cowl top panel assembly
   • Air cleaner cap with hose
   • Negative battery cable. Work must be started approximately 90 seconds or longer after the negative battery cable has been disconnected, if equipped with an air bag.
   • Coolant
   • Accelerator and throttle cables
   • V-bank cover
   • Emission valve control set
   • No. 2 EGR pipe
   • Hydraulic motor pressure pipe from the water inlet and air inlet chamber
   • Air intake chamber assembly
   • Injector wiring
   • Air assist pipe from the bracket on the No. 1 fuel pipe
   • Air assist hoses from the intake manifold
   • Fuel return hose from the No. 1 fuel pipe
   • Fuel inlet hose for the fuel filter
   • 2 union bolts holding the No. 2 fuel pipe to the delivery pipes
   • Fuel return hose from the fuel pressure regulator
   • Union bolt for the right hand delivery pipe, 2 gaskets, 2 bolts, left hand delivery pipe together with the 3 injectors and the No. 2 fuel pipe
   • Union bolt for the delivery pipe and 2 gaskets from the No. 2 fuel pipe
   • The 3 bolts, right hand delivery pipe together with the 3 injectors and the No. 1 fuel pipe
   • The 4 spacers from the intake manifold
   • The 6 injectors from the delivery pipes

- The two O-rings and two grommets from each injector

To install:

3. Install or connect the following:

- 2 new grommets to each injector
- New O-rings, with a light coat of fuel, to each injector
- Injectors
- The 4 spacers on the intake manifold
- Right hand delivery pipe and the No. 1 fuel pipe together with the 3 injectors in position on the intake manifold
- Bolt holding the right side delivery pipe, temporarily, to the intake manifold
- Left hand delivery pipe and the No. 2 fuel pipe together with the 3 injectors in position on the intake manifold
- Fuel return hose to the fuel pressure regulator

4. Temporarily install the 2 bolts holding the left hand delivery pipe to the intake manifold.

5. Temporarily install the No. 2 fuel pipe to the left side delivery pipe with the union bolt and 2 new gaskets.

6. Check that the injectors rotate smoothly. If they do not, replace the O-rings.

7. Position the injector connector outward. Tighten the 4 bolts holding the delivery pipes to the intake manifold and tighten to 7 ft. lbs. (10 Nm). Tighten the bolt holding the No. 1 fuel pipe to the intake manifold to 14 ft. lbs. (20 Nm). Tighten the 2 union bolts holding the no. 2 fuel pipe to the delivery pipes to 24 ft. lbs. (32 Nm).

8. Install or connect the following:

- Fuel inlet and return hoses. Union bolt: 22 ft. lbs. (30 Nm).

- Fuel return hose to the No. 1 fuel pipe. Pass the fuel return hose under the heater hoses.
- Air assist hoses to the intake manifold
- Air assist pipe to the bracket on the No. 1 fuel pipe
- Fuel injector wiring connectors
- Air intake chamber assembly
- Hydraulic motor pressure pipe to the intake chamber. Bolts: 69 inch lbs. (8 Nm)
- No. 2 EGR pipe with new gaskets, tighten to 9 ft. lbs. (12 Nm)
- Emission control valve set
- V-bank cover
- Air cleaner hose
- Throttle and accelerator cables
- Coolant
- Air cleaner cap with hose
- Outer front cowl top panel assembly
- Negative battery cable

RX330

1. Before servicing the vehicle, refer to the precautions in the beginning of this section.

2. Relieve the fuel system pressure.

3. Drain the coolant.

4. Remove the wiper arms.

5. Remove the wiper linkage.

6. Remove the fender-to-cowl side seals.

7. Remove the rain sensor.

8. Remove the front shock absorber caps (air suspension).

9. Remove the 4 set nuts from the strut (w/o air suspension).

10. Remove the cowl top outer panel.

11. Remove the 6 set nuts from the shock absorber.

12. Remove the V-bank cover.

13. Remove the air cleaner assembly and inlet tubes.

14. Remove the emission control valve set.

15. Remove the upper intake manifold (intake air surge tank). Discard the gasket.

16. Remove the fuel pipe sub-assembly.

17. Disconnect the wiring at the injectors.

18. Remove the 4 bolts and 2 delivery pipe along with the injectors.

19. Remove the delivery pipe spacers and insulators from the manifold.

20. Pull each injector from the pipe.

To install:

21. Install new O-rings on each injector. Apply a light coating of gasoline to the O-rings and mating points on the pipes.

22. Using a twisting motion, install the injectors on the pipes.

➡ **Be careful to avoid twisting the O-rings. After installation, check that the injectors turn smoothly. If not, use new O-rings.**

23. Install the pipes and injectors.

24. Loosely install the bolts and make sure that the injectors still turn freely. If not, replace the O-rings.

25. Torque the bolts to 84 inch lbs. (10 Nm).

26. The remainder of installation is the reverse of removal. Observe the following torques:

- Fuel line union bolt: 24 ft. lbs. (33 Nm)
- Pulsation damper: 24 ft. lbs. (33 Nm)
- Fuel feed pipe: 14 ft. lbs. (20 Nm)
- Upper intake manifold (air surge tank): 21 ft. lbs. (28 Nm)
- Upper intake manifold stays: 14 ft. lbs. (20 Nm)

DRIVE TRAIN

Automatic Transmission/ Transaxle Assembly

REMOVAL & INSTALLATION

RX300

1. Before servicing the vehicle, refer to the precautions in the beginning of this section.

2. Remove or disconnect the following:

- Hood
- Wiper and blade assembly
- Top cowl seal and panel

- Window washer hoses, from the ventilator louvers
- Left and right ventilator louvers
- Heater air duct
- Battery and tray
- Throttle cable
- Front upper suspension brace
- Cruise control actuator with its bracket, if equipped
- Starter
- Shift control cable
- Driveshaft, for 4WD
- Body-to-engine ground strap
- Park/Neutral Position (PNP) switch,

solenoid and ATF temperature connectors
- 5 upper transaxle-to-engine mounting bolts
- Front wheel
- Engine undercover
- Halfshafts
- Front exhaust pipe
- Stabilizer bar
- Both steering gear mounting bolts and support it in the vehicle
- Shift control cable from its bracket
- Power steering pipe and the oil cooler clamps from the frame

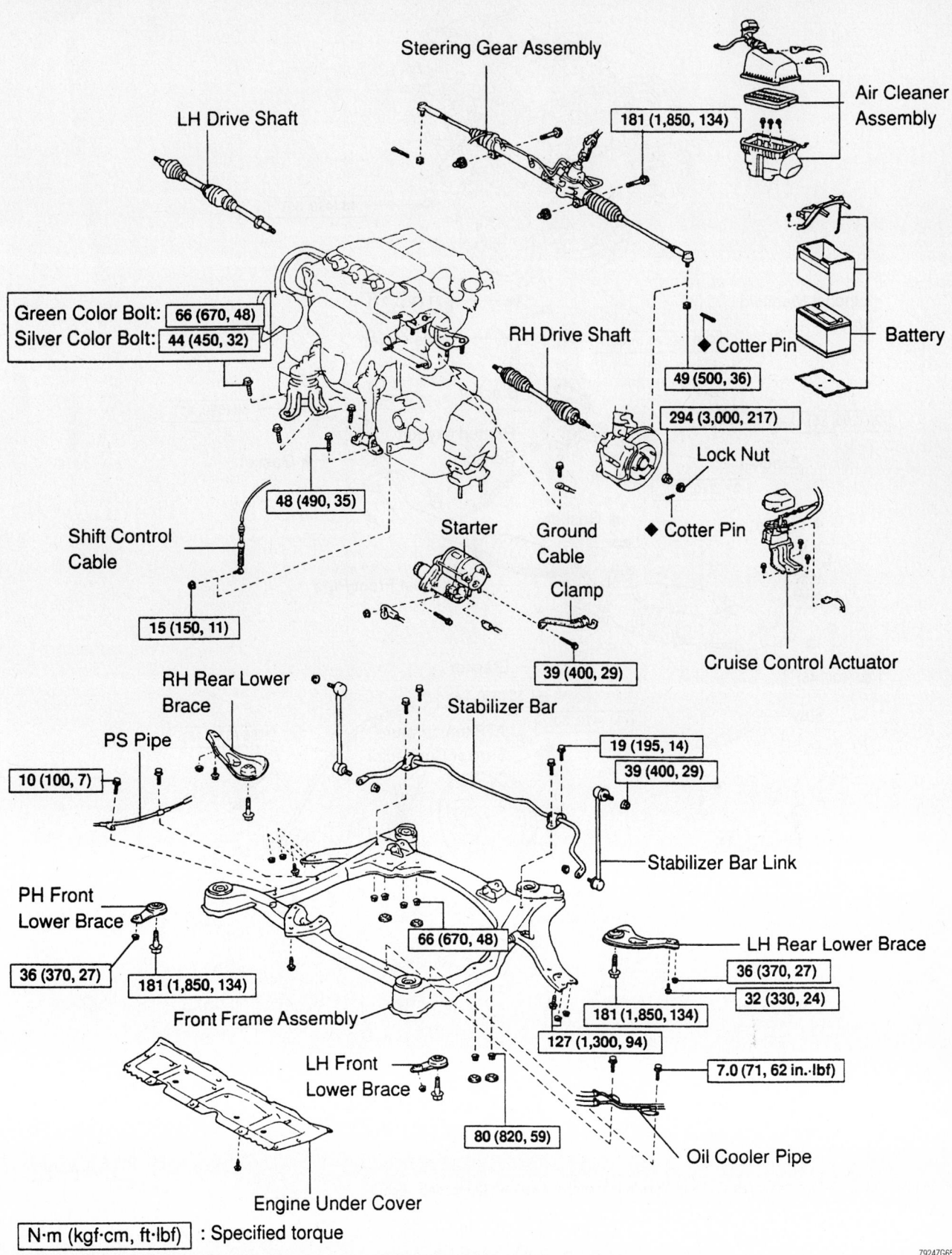

Steering Gear Assembly

LH Drive Shaft

181 (1,850, 134)

Air Cleaner Assembly

Green Color Bolt: 66 (670, 48)
Silver Color Bolt: 44 (450, 32)

RH Drive Shaft

◆ Cotter Pin

Battery

49 (500, 36)

294 (3,000, 217)

Lock Nut

48 (490, 35)

Shift Control Cable

Starter

Ground Cable

◆ Cotter Pin

Clamp

15 (150, 11)

39 (400, 29)

Cruise Control Actuator

RH Rear Lower Brace

Stabilizer Bar

19 (195, 14)

39 (400, 29)

PS Pipe

10 (100, 7)

Stabilizer Bar Link

PH Front Lower Brace

66 (670, 48)

LH Rear Lower Brace

36 (370, 27)

181 (1,850, 134)

36 (370, 27)

32 (330, 24)

Front Frame Assembly

181 (1,850, 134)

127 (1,300, 94)

7.0 (71, 62 in.·lbf)

LH Front Lower Brace

80 (820, 59)

Oil Cooler Pipe

Engine Under Cover

N·m (kgf·cm, ft·lbf) : Specified torque

7924ZG65

Exploded view of the transaxle removal and installation components—RX300 models

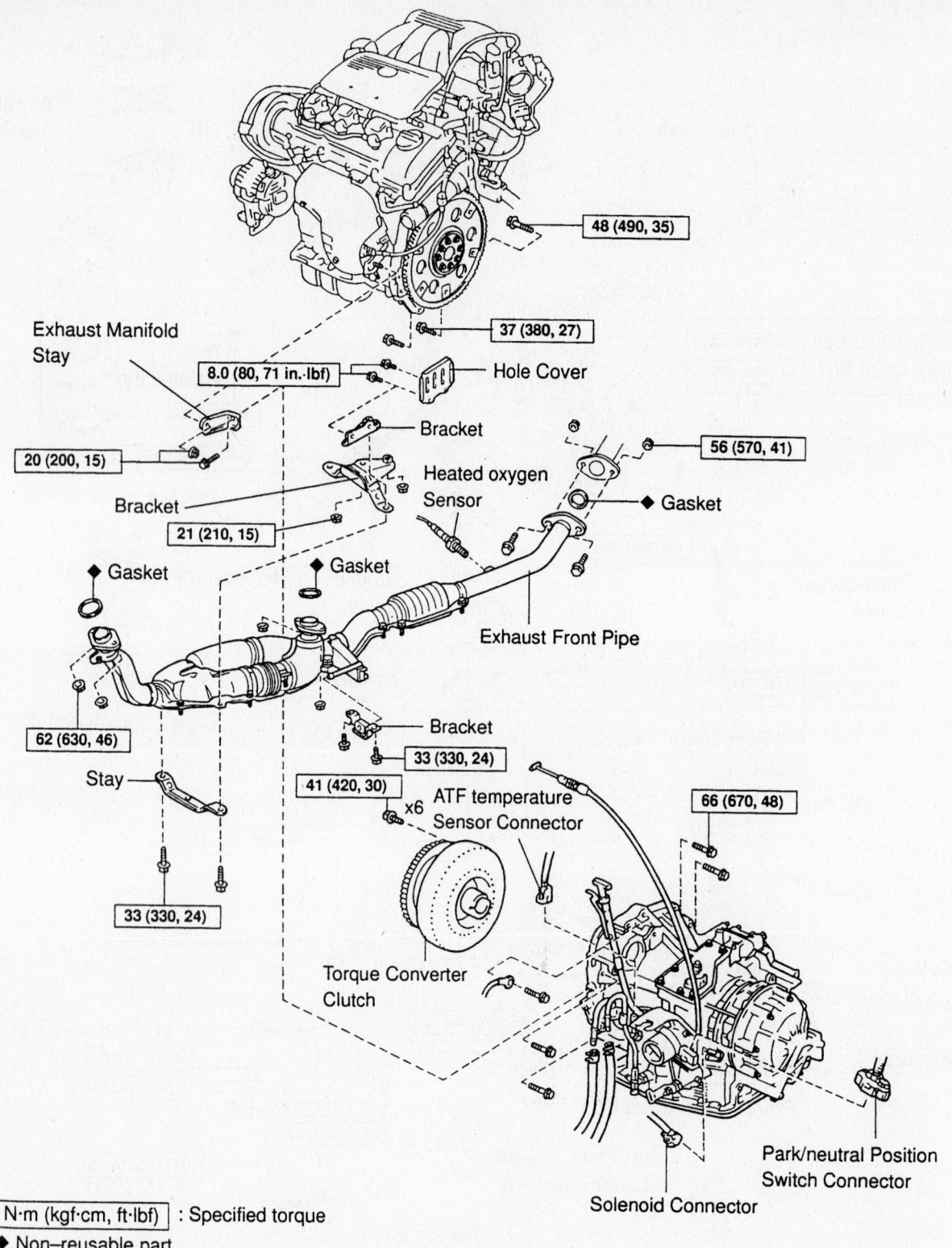

48 (490, 35)

37 (380, 27)

Exhaust Manifold Stay

8.0 (80, 71 in.·lbf)

Hole Cover

Bracket

20 (200, 15)

56 (570, 41)

Bracket

Heated oxygen Sensor

◆ Gasket

21 (210, 15)

◆ Gasket

◆ Gasket

Exhaust Front Pipe

Bracket

62 (630, 46)

33 (330, 24)

Stay

41 (420, 30)

66 (670, 48)

x6

ATF temperature Sensor Connector

33 (330, 24)

Torque Converter Clutch

Park/neutral Position Switch Connector

Solenoid Connector

N·m (kgf·cm, ft·lbf) : Specified torque

◆ Non–reusable part

79242G66

Exploded view of the transaxle removal and installation components—RX300 models, Cont.

- Both left-side transaxle mounting nuts
- Rear-side engine mounting nuts
- Engine shock absorber mounting bolts
- 3 front-side engine mounting bolts

3. Attach an engine sling to the engine hangers in order to support the engine weight.

4. Remove or disconnect the following:
- Front frame mounting bolts and the frame

- Transaxle oil cooler lines

5. Support the transaxle with a transmission jack.
- Torque converter access cover
- 6 torque converter mounting bolts

- 3 lower transaxle-to-engine mounting bolts
- Engine from the transaxle

To install:

6. Install or connect the following:
   - Transaxle
   - 3 lower transaxle-to-engine mounting bolts and tighten to the illustrated value.
   - Torque converter-to-flexplate bolts, starting with the black bolt, then the other 5.

7. The rest of installation is the reverse of the removal referring to the illustrations for the tightening specifications.

RX330

1. Remove the engine/transaxle assembly. See Engine Removal and Installation, earlier in this chapter.
2. Remove the halfshafts.
3. Disconnect the wiring.
4. Remove the starter.
5. Remove the cables and hoses.
6. Remove the filler tube.
7. Remove the front engine mount bracket.
8. Remove the flywheel housing undercover.
9. Turn the crankshaft to gain access to the torque converter bolts. There is one green bolt.
10. Separate the transaxle from the engine.
11. Remove the right stiffener plate.
12. Separate the transfer case from the transaxle.
13. Installation is the reverse of removal. Observe the following torques:
    - Transaxle-to-engine: bolts A 47 ft. lbs (64 Nm); bolts B 34 ft. lbs. (46 Nm); bolts C 27 ft. lbs. (37 Nm)
    - Torque converter bolts (use a thread locking compound such as Three Bond 1324): 30 ft. lbs. (41 Nm)

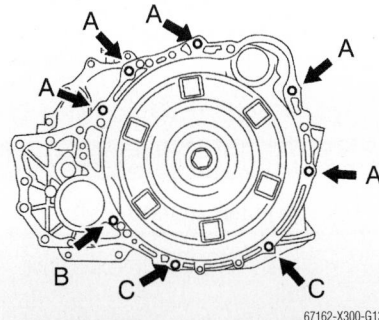

Transaxle-to-engine bolts—RX330

67162-X300-G13

➡**Install the green bolt first.**

- Undercover: 69 inch lbs. (8 Nm)
- Stiffener plate: 25 ft. lbs. (34 Nm)
- Engine mount bracket: 47 ft. lbs. (64 Nm)
- Transfer case-to-transaxle: 51 ft. lbs. (69 Nm)

Transfer Case Assembly

The transfer case for the RX300 is part of the transmission/transaxle assembly and is serviced with those units.

Front Halfshaft

REMOVAL & INSTALLATION

1. Before servicing the vehicle, refer to the precautions in the beginning of this section.
2. Remove or disconnect the following:
   - Front wheels
   - Cotter pin and locknut cap

➡**Have an assistant depress the brake pedal and loosen the bearing locknut.**

- Engine undercover
- Fender apron seal
- Tie rod end, from the steering knuckle
- Steering knuckle, from the lower control arm
- Halfshaft from the axle hub, using a plastic hammer
- Cover the outer boot with a rag
- Halfshaft from the transaxle, using the proper tools

To install:

3. Reverse the removal procedures to complete installation, tightening fasteners to specifications.
4. Fill the transaxle with gear oil, install the fender apron, check front end alignment and test drive.

➡**If the cotter pin holes do not align, always correct by tightening the nut until the next hole aligns.**

5. Install a new cotter pin.

Rear Halfshaft

REMOVAL & INSTALLATION

RX300

1. Before servicing the vehicle, refer to the precautions in the beginning of this section.

2. Remove or disconnect the following:
   - Negative battery cable
   - Rear wheels
   - Anti-lock Brake System (ABS) speed sensor from the axle assembly by removing the bolt, if equipped
   - Cotter pin, lock cap and the nut holding the halfshaft to the axle carrier

3. Place matchmarks on the halfshaft and differential side gear shaft.
4. Remove or disconnect the following:
   - 4 nuts, washers and the halfshaft from the differential
   - Halfshaft from the axle carrier

To install:

5. Install or connect the following:
   - Halfshaft into the axle carrier. Tighten the 4 nuts to 51 ft. lbs. (69 Nm).
   - Halfshaft. Tighten the locknut to 159 ft. lbs. (216 Nm).
   - ABS sensor
   - Rear wheels
   - Negative battery cable

RX330

1. Before servicing the vehicle, refer to the precautions in the beginning of this section.
2. Remove the rear wheel.
3. Disconnect and remove the speed sensor.
4. Remove the axle shaft nut.
5. Disconnect the height control sensor.
6. Disconnect the rear control arms.
7. Disconnect the strut rod.
8. Push the rear axle carrier sub-assembly towards the outside and separate the shaft from the carrier.
9. Remove the shaft, keeping it level.

To install

10. Align the shaft splines and install the shaft with a brass bar and hammer.

➡**Set the snapring with the opening side facing downward. Keep the shaft level.**

11. Push the carrier towards the inside and insert the shaft.
12. Connect the control arms and strut rod with the fasteners hand-tight. Tighten all fasteners with the suspension loaded.
13. The remainder of installation is the reverse of removal. Observe the following torques:
    - Axle shaft nut: 217 ft. lbs. (294 Nm)
    - Wheel: 76 ft. lbs. (103 Nm)
    - Control arms: 83 ft. lbs. (112 Nm)
    - Strut: 133 ft. lbs. (180 Nm)

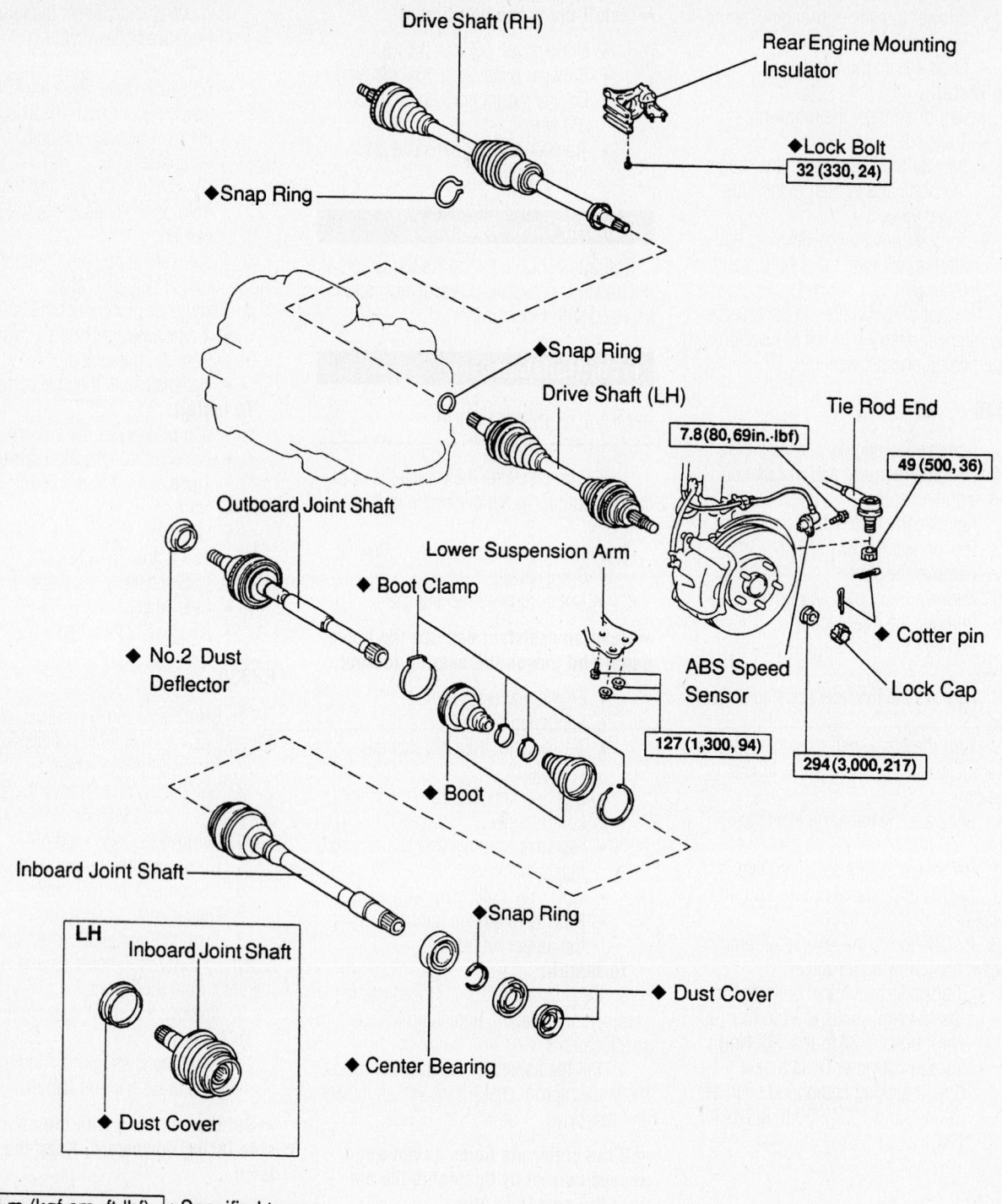

Drive Shaft (RH)

Rear Engine Mounting Insulator

◆Snap Ring

◆Lock Bolt
32 (330, 24)

◆Snap Ring

Drive Shaft (LH)

Tie Rod End

7.8 (80, 69 in.·lbf)

49 (500, 36)

Outboard Joint Shaft

Lower Suspension Arm

◆ Boot Clamp

◆ No.2 Dust Deflector

◆ Cotter pin

Lock Cap

ABS Speed Sensor

127 (1,300, 94)

294 (3,000, 217)

◆ Boot

Inboard Joint Shaft

◆Snap Ring

◆ Dust Cover

LH

Inboard Joint Shaft

◆ Center Bearing

◆ Dust Cover

N·m (kgf·cm, ft·lbf) : Specified torque
◆ Non–reusable part

7924ZG73

Exploded view of front halfshaft

CV-Joints

OVERHAUL

1. Before servicing the vehicle, refer to the precautions in the beginning of this section.

2. Remove the inboard and outboard joint boot clamps.
3. Disassemble the inboard joint tulip, as follows:
• Matchmark the tri-pot, inboard joint tulip or center driveshaft to the driveshaft

※※ WARNING

Do not use punch marks.

• Inboard joint tulip from the drive-shaft
4. Remove the inboard and outboard joint clamps.

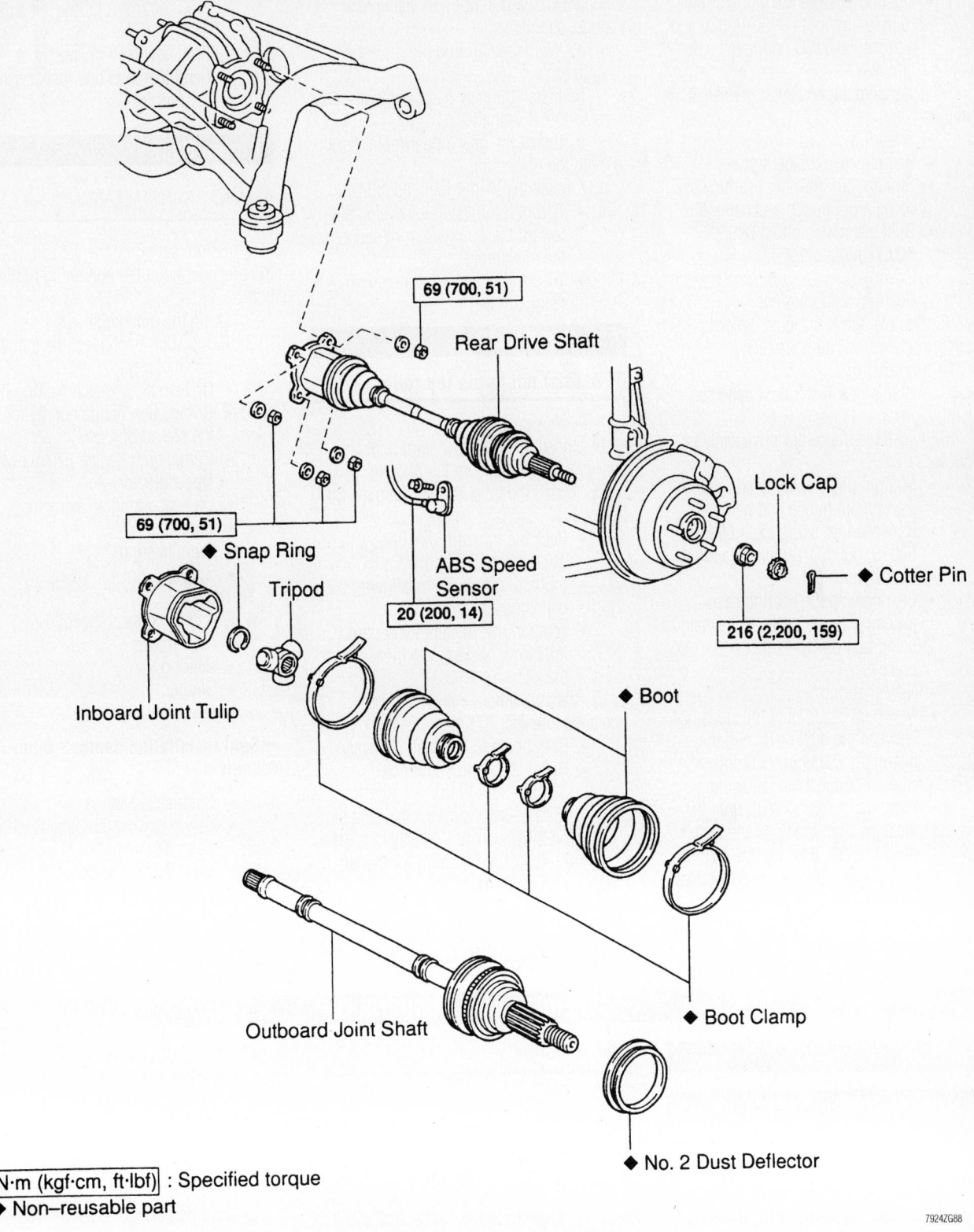

69 (700, 51)

Rear Drive Shaft

69 (700, 51)

◆ Snap Ring

Tripod

Inboard Joint Tulip

ABS Speed
Sensor

20 (200, 14)

Lock Cap

◆ Cotter Pin

216 (2,200, 159)

◆ Boot

◆ Boot Clamp

Outboard Joint Shaft

◆ No. 2 Dust Deflector

N·m (kgf·cm, ft·lbf) : Specified torque
◆ Non–reusable part

7924ZG88

Exploded view of the rear halfshaft—RX300 model with 4WD

5. Remove the tri-pot joint, as follows:
- Snapring
- Matchmark the tri-pot joint to the driveshaft
- Tri-pot joint, using a brass bar and hammer

✳✳ WARNING

Do not tap the roller.

6. Remove or disconnect the following:
- Inboard and outboard joint boots

➡**Do not disassemble the outboard joint.**

- Dust cover from the center drive-shaft, using a press, for 2WD on the right side

- Dust cover from the inboard joint tulip, using tool 09950-00020 and a press, for 2WD on the left side and 4WD
7. Disassemble the center driveshaft, as follows:
  - Snapring
  - Bearing case, using a press
  - Straight pin from the bearing case, using a pin punch and hammer
  - Dust cover, using tool 09950-00020 and a press
  - Snapring
  - Bearing, using a press
8. Remove the No. 2 dust deflector, using a screwdriver and hammer.

To assemble:
9. Install a new No. 2 dust deflector, using a press.
10. Assemble the center driveshaft, as follows:
  - Straight pin into the bearing case, using a pin punch and hammer
  - New bearing, using tools 09959-60010, 09950-70010 and a press
  - New snapring
  - Bearing with the bearing case assembly to the center driveshaft, using tool 09710-30021 and a press
  - New snapring
  - New dust cover, until the clearance between the dust cover and the bearing is 0.039 in. (1.0mm)
11. Install or connect the following:
  - Right dust cover (2WD), until the distance from the tip of the center drive is 3.39–3.34 in. (86–87mm) to the inner edge of the dust cover
  - Left side dust cover (2WD and 4WD), using a press
12. Temporarily install new

outboard/inboard joint boots using new clamps, as follows:
  a. Warp tape around the driveshaft splines.
  b. Install the new outboard joint boot onto the driveshaft.
  c. Install the new inboard joint boot onto the driveshaft.
13. Install the tri-pot joint, as follows:
  - Tri-pot joint, face the beveled side toward the outboard joint and align the matchmarks
  - Tri-pot joint onto the driveshaft, using a press

✳✳ WARNING

Be careful not to tap the roller.

  - New snapring
14. Install the outboard joint boot packed with grease from the boot kit.
15. Install the inboard joint tulip, as follows:
  - Pack the inboard joint boot with grease from the boot kit
  - Inboard joint tulip, by aligning the matchmarks
  - Temporarily, install the inboard joint boot packed with grease from the kit
16. Install the boot clamps to both boots, as follows:
  - Both boots to the shaft grooves
  - Halfshaft length should be 33.055–33.449 in. (839.6–849.6mm) for the right side on 2WD with A/T, 21.397–21.791 in. (543.5–553.5mm) for the left side on 2WD with A/T, 19.929–20.323 in. (506.2–516.2mm) for the right side on 4WD or 19.803–20.197 in.

(503–511mm) for the left side on 4WD
  - Both new boot clamps boot
  - Bend the band and lock it using a screwdriver

Pinion Seal

REMOVAL & INSTALLATION

1. Before servicing the vehicle, refer to the precautions in the beginning of this section.
2. Drain the differential oil.
3. Remove or disconnect the following:
  - Exhaust pipe
  - Driveshaft by matchmarking it
  - Companion flange nut, by loosen the staked portion
  - Companion flange, using a screw-type extractor
  - Oil seal, using an extractor
  - Slinger
  - Front bearing
  - Spacer

To install:
4. Install or connect the following
  - New spacer
  - Bearing
  - Slinger
  - New seal

➡ **Seal installation depth: 2.0mm +/- 0.3mm**

  - Companion flange
  - New nut. Coat the threads with clean differential oil. Torque the nut to 80 ft. lbs. (108Nm).
5. The remainder of installation is the reverse of removal.

STEERING AND SUSPENSION

Air Bag

✳✳ CAUTION

Some vehicles are equipped with an air bag system. The system must be disarmed before performing service on, or around, system components, the steering column, instrument panel components, wiring and sensors. Failure to follow the safety precautions and the disarming procedure could result in accidental air bag deployment, possible injury and unnecessary system repairs.

PRECAUTIONS

Several precautions must be observed when handling the inflator module to avoid accidental deployment and possible personal injury.
  - Never carry the inflator module by the wires or connector on the underside of the module.
  - When carrying a live inflator module, hold securely with both hands and ensure that the bag and trim cover are pointed away.
  - Place the inflator module on a bench or other surface with the bag and trim cover facing up.

  - With the inflator module on the bench, never place anything on or close to the module which may be thrown in the event of an accidental deployment.

DISARMING

To avoid personal injury when working on vehicles equipped with an air bag, the negative battery cable must be disconnected and at least 90 seconds must elapse before working on the system. Failure to do so may result in deployment of the air bag.

Power Rack And Pinion Steering Gear

REMOVAL & INSTALLATION

RX300

1. Before servicing the vehicle, refer to the precautions in the beginning of this section.

2. Remove or disconnect the following:
   - Negative battery cable

➡**Wait at least 90 seconds before working on the vehicle to allow the Supplemental Restraint System (SRS) system to disarm.**

   - Right and left side fender apron seals
   - Right and left tie rod ends

3. Place matchmarks on the intermediate shaft.

4. Remove or disconnect the following:
   - Pinch bolt and the intermediate shaft out from under the vehicle
   - Power steering line clamp
   - Pressure and feed lines
   - Stabilizer bar, unbolt it but do not remove it
   - Heated Oxygen (HO$_2$) sensor

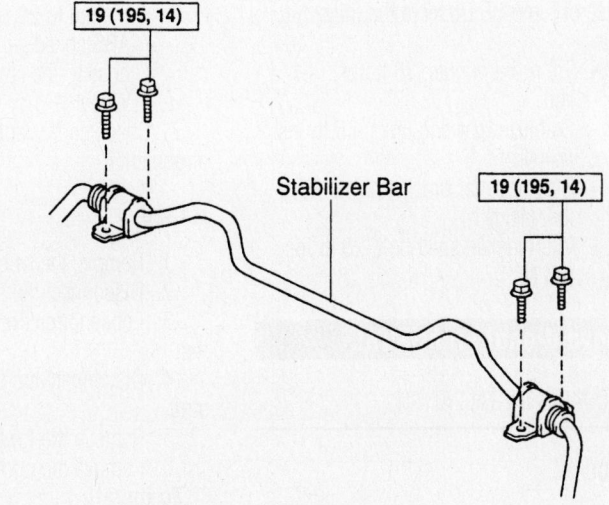

19 (195, 14)

Stabilizer Bar

19 (195, 14)

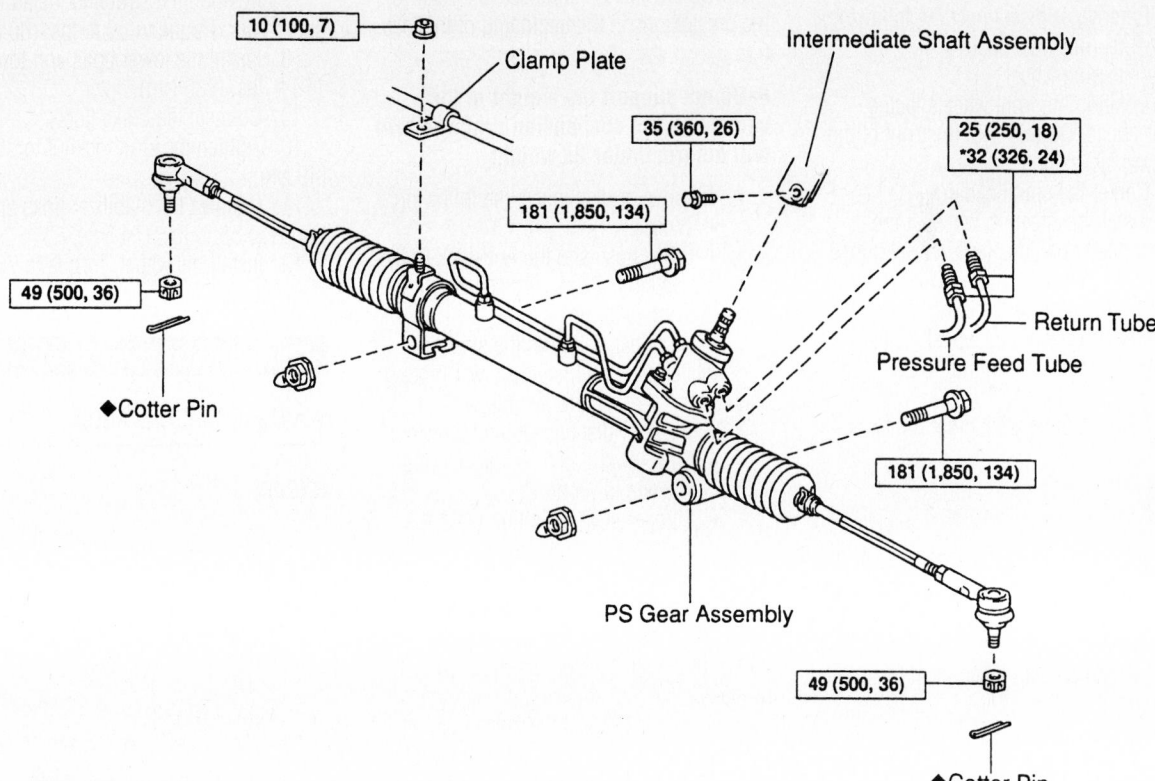

10 (100, 7)

Clamp Plate

Intermediate Shaft Assembly

35 (360, 26)

25 (250, 18)
*32 (326, 24)

181 (1,850, 134)

49 (500, 36)

◆Cotter Pin

Return Tube

Pressure Feed Tube

181 (1,850, 134)

PS Gear Assembly

49 (500, 36)

◆Cotter Pin

N·m (kgf·cm, ft·lbf) : Specified torque
◆ Non–reusable part
* For use with SST

Exploded view of the power steering gear and related components—RX300 models

7924ZG76

- Both gear assembly set bolts and nuts, by lifting the stabilizer bar
- Gear assembly from the left side of the vehicle

To install:

5. Install or connect the following:
- Gear assembly to the left side of the vehicle

✷✷ WARNING

Be careful not to damage the power steering lines.

- Tighten the gear assembly set bolts and nuts to 134 ft. lbs. (181 Nm), by lifting the stabilizer bar
- HO$_2$sensor
- Stabilizer bar. Tighten the bolt to 14 ft. lbs. (19 Nm) and the nut to 29 ft. lbs. (39 Nm).
- Pressure and feed return lines. Tighten them to 18 ft. lbs. (25 Nm).
- Line clamps. Tighten the nut to 84 inch lbs. (10 Nm).
- Intermediate shaft, by aligning the joint and main shaft matchmarks. Tighten to 26 ft. lbs. (35 Nm).
- Tie rod ends
- Fender apron seals. Securely tighten the bolts.

6. Remove or disconnect the following:
- Steering wheel pad
- Steering wheel

7. Position the front wheels facing straight-ahead. Do this with the front of the vehicle on jackstands.

8. Center the spiral cable.

9. Install the steering wheel at the straight-ahead position. Temporarily tighten the wheel set nut. Attach the wiring.

10. Bleed the power steering system.

11. Check the steering wheel center point. Tighten the steering nut to 26 ft. lbs. (35 Nm).

12. Check and/or adjust the front wheel alignment.

RX330

1. Before servicing the vehicle, refer to the precautions in the beginning of this section.

2. Center the steering wheel.

3. Matchmark and disconnect the intermediate shaft.

4. Remove the wheels.

5. Separate the tie rods.

6. Disconnect the stabilizer bar end links.

7. Remove the front exhaust pipe.

8. Remove the stabilizer bar brackets.

9. Remove the height control sensor.

10. Disconnect the pressure and return lines.

11. Remove the bolts and nuts and remove the steering rack assembly.

To install:

12. Position the assembly and install the bolts and nuts. Torque the nuts to 52 ft. lbs. (70 Nm).

13. Connect the pressure and lines. Torque to 16 ft. lbs. (22 Nm). Torque the clamp bolt to 87 inch lbs. (10 Nm).

14. The remainder of installation is the reverse of removal. Observe the following torques:
- Tie rod end nuts: 36 ft. lbs. (49 Nm)
- Stabilizer bar end links: 55 ft. lbs. (74 Nm)
- Stabilizer bar bracket bolts: 12 ft. lbs. (16 Nm)
- Intermediate shaft bolt: 26 ft. lbs. (35 Nm)

Conventional Front Strut

REMOVAL & INSTALLATION

RX300

1. Before servicing the vehicle, refer to the precautions in the beginning of this section.

➡**Do not support the weight of the vehicle on the suspension arm; the arm will deform under its weight.**

2. Remove or disconnect the following:
- Wheel
- Brake hose and the Anti-lock Brake System (ABS) speed sensor wire from the strut
- Sway bar link from the strut

3. Matchmark on the strut lower bracket and camber adjust cam, if equipped.

4. Remove or disconnect the following:
- Strut lower end from the steering knuckle lower arm
- 3 upper strut mounting plate-to-upper wheel arch nuts
- Strut

To install:

5. Align the upper suspension support hole with the strut piston or end, so they fit properly.

6. Install or connect the following:
- Strut piston rod end to the upper suspension support. Tighten the new nut to 29–40 ft. lbs. (39–54 Nm).

➡**Do not use an impact wrench to tighten the nut.**

- Lubricate the suspension support bearing with multi-purpose grease.
- Pack the upper support space with multi-purpose grease, also, after installation.
- Tighten the 3 suspension support-to-wheel arch nuts to 47 ft. lbs. (64 Nm).
- Tighten the strut-to-steering knuckle arm bolts to 156 ft. lbs. (211 Nm).
- Sway bar link to the strut. Tighten the nut to 29 ft. lbs. (39 Nm).
- ABS speed sensor and the brake hose to the strut, if equipped
- Wheel

7. Check and/or adjust the front wheel alignment.

RX330

1. Remove the wheel.

2. Disconnect the stabilizer bar link.

3. Loosen, don't remove, the strut locknut.

4. Disconnect the brake hose from the strut.

5. Remove the lower mounting bolts.

6. Remove the upper retaining nuts.

To install:

7. Position the strut and install the upper nuts. Torque to 59 ft. lbs. (80 Nm).

8. Install the lower bolts and torque to 170 ft. lbs. (230 Nm).

9. Connect the brake line.

10. Tighten the strut locknut to 36 ft. lbs. (49 Nm).

11. Connect the stabilizer links and torque to 55 ft. lbs. (74 Nm).

12. Install the wheel. Torque to 76 ft. lbs. (103 Nm).

Conventional Rear Strut

REMOVAL & INSTALLATION

RX300

1. Before servicing the vehicle, refer to the precautions in the beginning of this section.

2. Remove or disconnect the following:
- Negative battery cable
- Deck side cover
- Rear wheels
- Anti-lock Brake System (ABS) sensor from the strut bracket
- Flexible brake hose from the strut
- Sway bar link from the strut
- Loosen the 2 lower strut mounting bolts

3. Support the rear axle carrier with a jack.

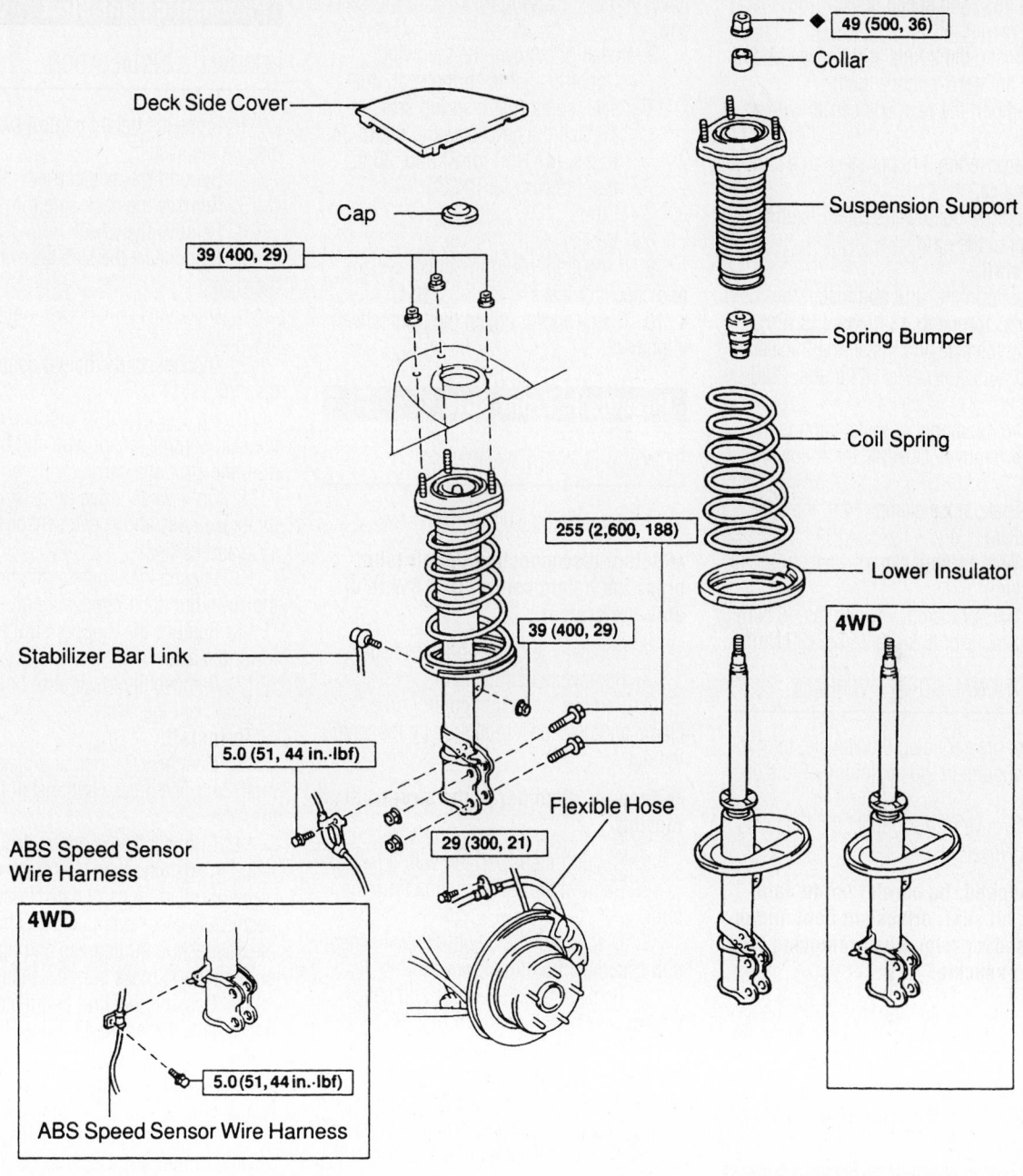

49 (500, 36)
Collar
Deck Side Cover
Suspension Support
Cap
Spring Bumper
39 (400, 29)
Coil Spring
255 (2,600, 188)
Lower Insulator
Stabilizer Bar Link
39 (400, 29)
4WD
5.0 (51, 44 in.·lbf)
ABS Speed Sensor
Wire Harness
Flexible Hose
29 (300, 21)
4WD
5.0 (51, 44 in.·lbf)
ABS Speed Sensor Wire Harness

N·m (kgf·cm, ft·lbf) : Specified torque

◆ Non–reusable part

7924ZG89

Exploded view of the rear strut assembly—RX300

4. Remove or disconnect the following:
   • 3 upper strut mounting nuts
   • Strut, by lower the rear axle
To install:
5. Install or connect the following:
   • Strut
   • Both lower strut mounting bolts, but do not tighten
   • Axle carrier by aligning the 3 upper mounting studs. Tighten the nuts to 29 ft. lbs. (39 Nm).

6. Lower the axle carrier.
7. Install or connect the following:
   • Tighten both lower mounting bolts to 188 ft. lbs. (255 Nm).
   • Sway bar link. Tighten the nut to 29 ft. lbs. (39 Nm).
   • Flexible brake hose and the ABS sensor to the strut
   • Rear wheels and the deck side cover
   • Negative battery cable

RX330

1. Remove the tonneau cover.
2. Remove the left side deck trim cover.
3. Remove the rear wheels.
4. Remove the stabilizer link from the strut.
5. On 2-wheel drive models, disconnect the skid control sensor wire and brake hose from the strut and carrier.
6. On 4-wheel drive models, disconnect

the brake hose and speed sensor from the strut and carrier.

7. Loosen the 2 nuts at the lower end of the strut, but don't remove them.

8. Support the rear axle carrier with a jack.

9. Remove the 3 upper strut nuts and lower the axle.

10. Remove the lower strut bolts and nuts. Lift out the strut.

To install:

11. Position the strut and install the 3 upper nuts. Torque to 43 ft. lbs. (58 Nm).

12. Lift the axle and install the 2 lower bolts and nuts. Torque to 133 ft. lbs. (180 Nm).

13. The remainder of installation is the reverse of removal. Observe the following torques:

- Brake hose clamp: 14 ft. lbs. (19 Nm)
- Wire-to-strut clamp: 44 inch lbs. (5 Nm)
- Sensor clamp: 71 inch lbs. (8 Nm)
- Stabilizer link: 29 ft. lbs. (39 Nm)

CONVENTIONAL STRUT OVERHAUL

1. Before servicing the vehicle, refer to the precautions in the beginning of this section.

2. Remove or disconnect the following:
- Wheel

➡**If equipped, be careful not to damage the oil seal, driveshaft boot and/or speed sensor rotor when removing the steering knuckle.**

- Shock absorber (strut assembly)

3. Install a nut/bolt to the bracket at the lower portion of the strut assembly and secure it in a vise.

4. Compress the coil spring with a spring compressor.

✱✱ CAUTION

The proper tools must be used for this procedure. The spring on the strut is under high pressure and can cause serious injury if not properly removed and installed.

5. Remove or disconnect the following:
- Center retaining nut, by holding the spring seat
- Support, dust seal, spring seat, insulator and spring from the strut assembly

To install:

6. Install the spring bumper and lower insulator to the strut assembly.

7. Compress the coil spring and fit the

lower end of the spring into the spring seat gap.

8. Install or connect the following:
- Upper insulator, spring seat, dust seal, support and spring seat. Tighten the new retaining nut to 34 ft. lbs. (47 Nm) for RX300; 36 ft. lbs. (49 Nm) for RX330.
- Strut
- Wheel

9. If required, bleed the brake system and check for leaks.

10. Check and/or adjust the front wheel alignment.

Pneumatic Front Strut

REMOVAL & INSTALLATION

1. Remove the wheels.

➡**Before disconnecting the air tube, press the height control OFF SW to disable the system.**

2. Remove the cowl top silencer pad.

3. Remove the strut cap.

4. Remove the height control tube clamp and disconnect the tube by loosening the nut.

➡**Keep the chamber of the strut from moving.**

5. Support the lower arm with a jack.

6. Remove the stabilizer link from the strut.

7. Disconnect the height control sensor sub-assembly from the lower arm.

8. Remove the brake hose and speed sensor wire from the strut.

9. Remove the nuts from the 2 lower strut bolts, but leave the bolts in place.

10. Remove the 3 upper strut nuts.

11. Lower the jack slowly until the strut is free, then remove the bolts and lift out the strut.

To install:

12. Position the strut and install the bolts from the front side. Torque the nuts to 170 ft. lbs. (230 Nm).

13. Raise the arm and position the upper end. Install the nuts and torque to 59 ft. lbs. (80 Nm).

14. Install the brake hose and sensor wire and torque the bolt to 14 ft. lbs. (19 Nm).

15. Connect the stabilizer link. Torque the nut to 55 ft. lbs. (74 Nm).

16. The remainder of installation is the reverse of removal.

Pneumatic Rear Strut

REMOVAL & INSTALLATION

1. Press the height control switch to disable the system.

2. Support the axle carrier with a jack.

3. Remove the deck side trim cover.

4. Remove the wheel.

5. Disconnect the stabilizer link from the strut.

6. Disconnect the height control sensor at the strut.

7. Disconnect the height control tube at the strut.

8. On 2-wheel drive models, disconnect the skid control sensor wire and brake hose from the strut and carrier.

9. On 4-wheel drive models, disconnect the brake hose and speed sensor from the strut and carrier.

10. Loosen the 2 nuts at the lower end of the strut, but don't remove them.

11. Remove the 3 upper strut nuts and lower the axle.

12. Remove the lower strut bolts and nuts. Lift out the strut.

To install:

13. Coat new O-rings and plate with multi-purpose grease and install them on the tube.

14. Connect the height control tube. It helps to push the tube into place with a piece of rolled up cardboard. Push the connector into place until a click is heard. Turn the connection 90 degrees and lightly pull on the tube to make sure it's secure.

15. Position the strut and install the 3 upper nuts. Torque to 43 ft. lbs. (58 Nm).

16. Raise the axle and install the lower bolts and nuts. Torque to 133 ft. lbs. (180 Nm).

17. The remainder of installation is the reverse of removal. Observe the following torques:

- Brake hose clamp: 14 ft. lbs. (19 Nm)
- Wire-to-strut clamp: 44 inch lbs. (5 Nm)
- Sensor clamp: 71 inch lbs. (8 Nm)
- Stabilizer link: 29 ft. lbs. (39 Nm)

Lower Ball Joint

REMOVAL & INSTALLATION

RX300

1. Before servicing the vehicle, refer to the precautions in the beginning of this section.

2. Remove or disconnect the following:
- Wheel
- Steering knuckle with the axle hub
- Dust deflector, by prying it from the knuckle
- Cotter pin and nut from the ball joint
- Ball joint from the steering knuckle, by removing the 2 bolts
- Lower ball joint, using a Ball Joint Separator tool 09628-62011

To install:

3. Install or connect the following:
- Lower ball joint. Tighten the nut to 76 ft. lbs. (103 Nm) and both bolts to 94 ft. lbs. (127 Nm).
- New cotter pin
- Wheel

RX330

1. Remove the wheel.
2. Remove the axle hub nut.
3. Disconnect the speed sensor.
4. Remove the caliper and suspend it out of the way.
5. Remove the rotor.
6. Disconnect the tie rod end.
7. Remove the lower arm.
8. Pull the knuckle from the halfshaft.
9. Remove the bolts securing the ball joint to the arm.
10. Installation is the reverse of removal. Observe the following torques:
- Ball joint-to-arm: 94 ft. lbs. (127 Nm)
- Ball joint-to-knuckle: 91 ft. lbs. (123 Nm)
- Arm-to-frame: 148 ft. lbs. (200 Nm)
- Stabilizer bar link: 55 ft. lbs. (74 Nm)
- Caliper support bolts: 77 ft. lbs. (104 Nm)
- Caliper pins: 25 ft. lbs. (34 Nm)
- Hub nut: 217 ft. lbs. (294 Nm)

Lower Control Arm

REMOVAL & INSTALLATION

RX300

1. Before servicing the vehicle, refer to the precautions in the beginning of this section.
2. Remove or disconnect the following:
- Engine/transaxle assembly
- Transverse engine mount insulator
- 2 front and 1 rear lower arm mount bolts
- Lower arm
3. Installation is the reverse of removal. Observe the following torques:

- Front side bolts: 148 ft. lbs. (200Nm)
- Rear arm bolt/nut: 152 ft. (206Nm)
- Insulator: 64 ft. lbs. (87Nm)

RX330

1. Remove the engine/transaxle assembly.
2. Remove the transverse engine mounting insulator.
3. Remove the 3 bolts securing the arm to the engine support member.
4. Remove the front lower arm bush stopper.
5. Remove the ball joint-to-arm bolts.
6. Installation is the reverse of removal. Observe the following torques:
- 2 short arm-to-support bolts: 148 ft. lbs. (200 Nm)
- 1 long arm-to-support bolt: 152 ft. lbs. (206 Nm)
- Ball joint-to-arm: 94 ft. lbs. (127 Nm)
- Transverse engine mounting insulator: 64 ft. lbs. (87 Nm)

Front Wheel Bearing

REMOVAL & INSTALLATION

RX300

The front wheel bearings and 4wd rear wheel bearings are serviced as a unit with the hub and are not replaceable. See the Halfshaft Removal and Installation procedure.

RX330
REMOVAL

1. Remove the wheel.
2. Remove the hub nut.
3. Remove the caliper and suspend it out of the way.
4. Remove the brake rotor.
5. Disconnect the tie rod end from the knuckle.
6. Disconnect the control arm from the ball joint.
7. Remove the halfshaft from the knuckle.
8. Remove the lower strut-to-knuckle bolts and remove the hub/knuckle assembly.
9. Remove the ball joint.

DISASSEMBLY

1. Remove the inner seal from the hub.
2. Remove the snap ring from the hub.
3. Mount the knuckle in a vise and, using a slidehammer, remove the hub from the knuckle.
4. Remove the outer seal.

5. Using a press, the bearings and races can now be replaced.
6. Replace the snap ring and seals.
7. Press the knuckle onto the hub assembly.

INSTALLATION

1. Install the ball joint. Torque to 91 ft. lbs. (123 Nm). Advance the nut as much as 60 degrees to align the cotter pin hole. Use a new cotter pin.
2. Attach the knuckle assembly to the strut. Torque to 170 ft. lbs. (230 Nm).
3. The remainder of installation is the reverse of removal. Observe the following torques:
- Caliper support: 77 ft. lbs. (104 Nm)
- Hub nut: 217 ft. lbs. (294 Nm)
- Caliper pins: 25 ft. lbs. (34 Nm)

Rear Wheel Bearing

REMOVAL & INSTALLATION

RX300

The 4wd rear wheel bearings are serviced as a unit with the hub and are not replaceable. See the Halfshaft Removal and Installation procedure.

WITH 2WD

1. Before servicing the vehicle, refer to the precautions in the beginning of this section.
2. Remove or disconnect the following:
- Rear wheel
- Flexible brake hose from the rear strut assembly
- Brake caliper and support it on using wire
- Brake rotor
3. Remove or disconnect the following:
- Anti-lock Brake System (ABS) speed sensor connector
- 4 rear axle hub assembly nuts
- Hub assembly

To install:

4. Install or connect the following:
- New hub assembly. Tighten the nuts to 59 ft. lbs. (80 Nm).
- ABS speed sensor
- Brake rotor
- Brake caliper. Tighten the mounting bolts to 34 ft. lbs. (47 Nm).
- Flexible brake hose to the rear strut assembly. Tighten the mounting bolt to 21 ft. lbs. (29 Nm).
- Wheels
5. Test drive the vehicle.

RX330

1. Remove the wheel.
2. Remove the speed sensor from the axle carrier.
3. Remove the axle shaft nut.
4. Remove the caliper and support assembly and suspend it out of the way.

5. Check the bearing backlash. It should not exceed 0.0020 in. (0.05mm). If it does, it must be areplaced.
6. Remove the 4 bolts and remove the hub/bearing assembly.
7. Installation is the reverse of removal. Observe the following torques:

- Hub/bearing assembly bolts: 55 ft. lbs. (75 Nm)
- Caliper support: 58 ft. lbs. (78 Nm)
- Brake hose clamp: 14 ft. lbs. (19 Nm)
- Hub nut: 217 ft. lbs. (294 Nm)

BRAKES

Brake Caliper

REMOVAL & INSTALLATION

RX300

FRONT

1. Disconnect the negative battery cable from the battery.
2. Raise and support the vehicle safely.
3. Remove the wheels.
4. Disconnect the brake hose from the caliper by removing the union bolt and 2 gaskets. Plug the end of the hose to prevent loss of fluid.
5. Remove the caliper mounting bolts.
6. Remove the caliper, pads, shims and support plates.
   To install:
7. Grease the caliper slides and bolts with lithium grease or equivalent. Install the support plates, shims, pads and caliper and secure with the bolts. Torque the bolts to 25 ft. lbs. (34 Nm).
8. Connect the brake hose to the caliper, using 2 new washers. Make sure the flexible hose lock is securely in the lock hole of the caliper. Torque the union bolt to 21 ft. lbs. (29 Nm).
9. Fill the brake system to the proper level and bleed the brake system.
10. Install the tire and wheel assembly.
11. Top off the brake fluid level in the master cylinder. Check for leaks and proper brake operation.
12. Connect the negative battery cable to the battery.

REAR

1. Disconnect the brake line from the caliper.
2. Remove the caliper mounting bolt.
3. Remove the caliper, pads, shims and support plates.
4. Remove the main pin.
5. Installation is the reverse of removal. Torque the main pin to 20 ft. lbs. (26 Nm), the caliper bolt to 14 ft. lbs. (20 Nm), and the union bolt to 21 ft. lbs. (29 Nm).

RX330

FRONT

1. Disconnect the brake line from the caliper and plug it.
2. Hold the caliper slide pins and remove the mounting bolts.
3. Lift off the caliper.
4. Remove the pads and anti-squeal shims.
5. Remove the wear indicator from the inner pad.
6. Installation is the reverse of removal. Grease the caliper slides and bolts with lithium grease or equivalent. Apply disc brake grease to the anti-squeal shims. Torque the caliper bolts to 25 ft. lbs. (34 Nm); the brake line union bolt to 21 ft. lbs. (29 Nm).

REAR

1. Disconnect the brake line from the caliper and plug it.
2. Remove the caliper mounting bolts.
3. Lift off the caliper.
4. Remove the pads and anti-squeal shims.
5. Remove the wear indicators from each pad.
6. Installation is the reverse of removal. Grease the caliper slides and bolts with lithium grease or equivalent. Apply disc brake grease to the anti-squeal shims. Torque the caliper bolts to 32 ft. lbs. (43 Nm); the brake line union bolt to 21 ft. lbs. (29 Nm).

Disc Brake Pads

REMOVAL & INSTALLATION

RX300

FRONT

1. Hold the sliding pin and remove the lower bolt.
2. Lift the caliper up and secure it.
3. Remove the pads, 4 shims and wear indicator plate. Remove the 2 pad support plates.

➡The support plates can be reused, provided they have sufficient rebound, are not deformed or cracked, show no

signs of wear and are cleaned of all rust and debris.

To install:

➡Always use new shims and wear indicators, even when re-installing the original pads.

4. Install a wear indicator plate on the inner pad.
5. Apply disc brake grease to both sides of the inner anti-squeal shims and install the shims.
6. Install the inner pad with the wear indicator plate facing upwards.
7. Install the outer pad.
8. Install the caliper. Torque the bolt to 25 ft. lbs.

REAR

1. On 2WD, unbolt the brake hose from the shock absorber.
2. Remove the caliper installation bolt from the torque plate.
3. Lift the caliper up and secure it.
4. Remove the pads, 4 shims and 4 support plates.

➡The support plates can be reused, provided they have sufficient rebound, are not deformed or cracked, show no signs of wear and are cleaned of all rust and debris.

To install:

➡Always use new shims and wear indicators, even when re-installing the original pads.

5. Install the support plates.
6. Apply disc brake grease to both sides of the anti-squeal shims.
7. Install the anti-squeal shims.
8. Install the inner pad with the wear indicator plate facing upwards.
9. Install the outer pad.
10. Install the caliper. Torque the bolt to 14 ft. lbs. (20 Nm).
11. On 2WD, connect the line to the shock absorber. Torque the bolt to 21 ft. lbs. (29 Nm).

RX330

FRONT

1. Disconnect the brake line from the caliper and plug it.

2. Hold the caliper slide pins and remove the mounting bolts.
3. Lift off the caliper.
4. Remove the pads and anti-squeal shims.

5. Remove the wear indicator from the inner pad.
6. Installation is the reverse of removal. Grease the caliper slides and bolts with lithium grease or equivalent. Apply disc

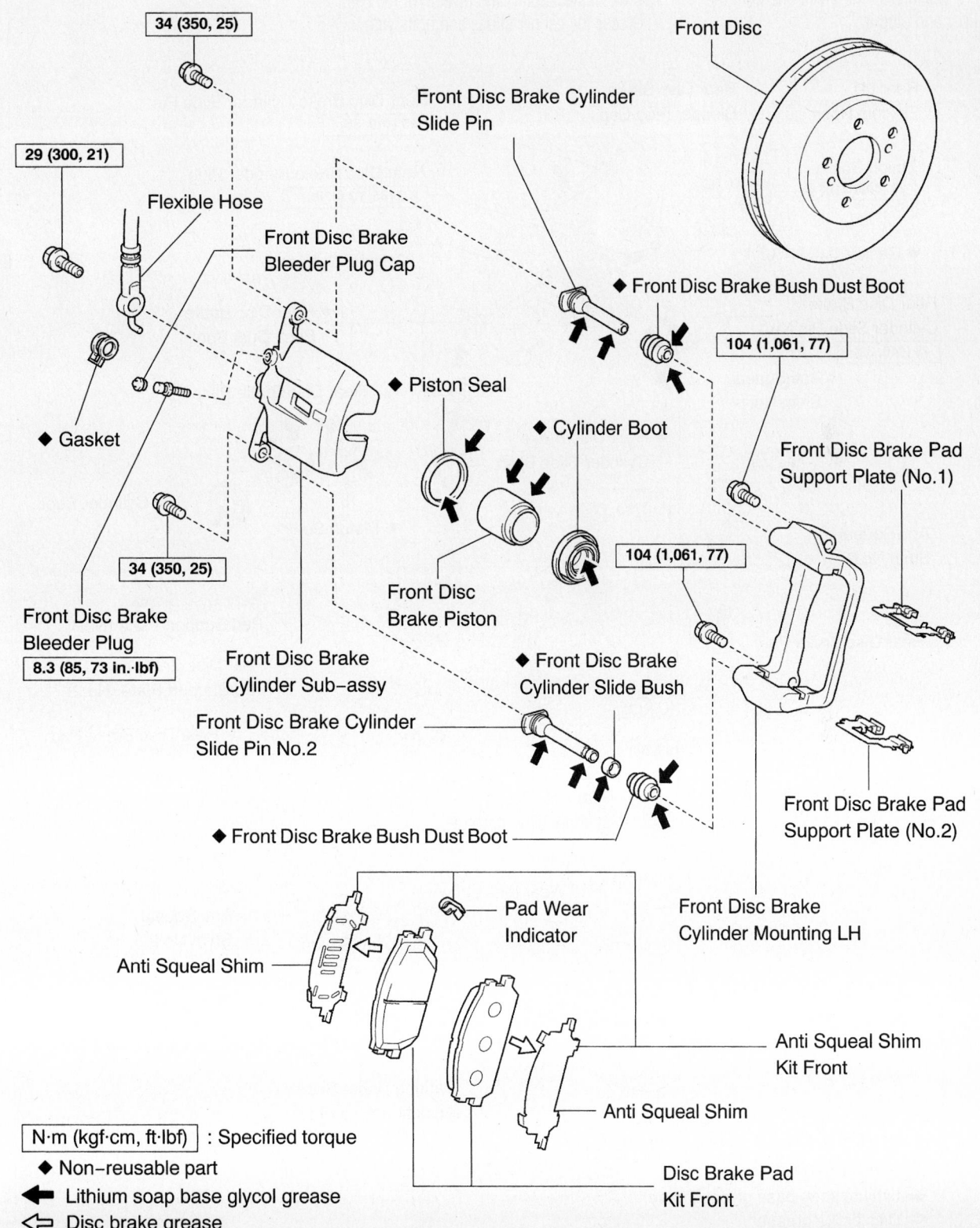

34 (350, 25)

29 (300, 21)

Flexible Hose

Front Disc Brake
Bleeder Plug Cap

Front Disc Brake Cylinder
Slide Pin

Front Disc

◆ Front Disc Brake Bush Dust Boot

104 (1,061, 77)

◆ Piston Seal

◆ Cylinder Boot

◆ Gasket

Front Disc Brake Pad
Support Plate (No.1)

34 (350, 25)

104 (1,061, 77)

Front Disc
Brake Piston

Front Disc Brake
Bleeder Plug

8.3 (85, 73 in.·lbf)

Front Disc Brake
Cylinder Sub-assy

◆ Front Disc Brake
Cylinder Slide Bush

Front Disc Brake Cylinder
Slide Pin No.2

Front Disc Brake Pad
Support Plate (No.2)

◆ Front Disc Brake Bush Dust Boot

Front Disc Brake
Cylinder Mounting LH

Pad Wear
Indicator

Anti Squeal Shim

Anti Squeal Shim
Kit Front

Anti Squeal Shim

N·m (kgf·cm, ft·lbf) : Specified torque

◆ Non-reusable part

◀ Lithium soap base glycol grease

◁ Disc brake grease

Disc Brake Pad
Kit Front

67162-X300-G11

Front disc brake components—RX330

brake grease to the anti-squeal shims. Torque the caliper bolts to 25 ft. lbs. (34 Nm); the brake line union bolt to 21 ft. lbs. (29 Nm).

REAR

1. Disconnect the brake line from the caliper and plug it.

2. Remove the caliper mounting bolts.
3. Lift off the caliper.
4. Remove the pads and anti-squeal shims.
5. Remove the wear indicators from each pad.
6. Installation is the reverse of removal. Grease the caliper slides and bolts with lithium grease or equivalent. Apply disc brake grease to the anti-squeal shims. Torque the caliper bolts to 32 ft. lbs. (43 Nm); the brake line union bolt to 21 ft. lbs. (29 Nm).

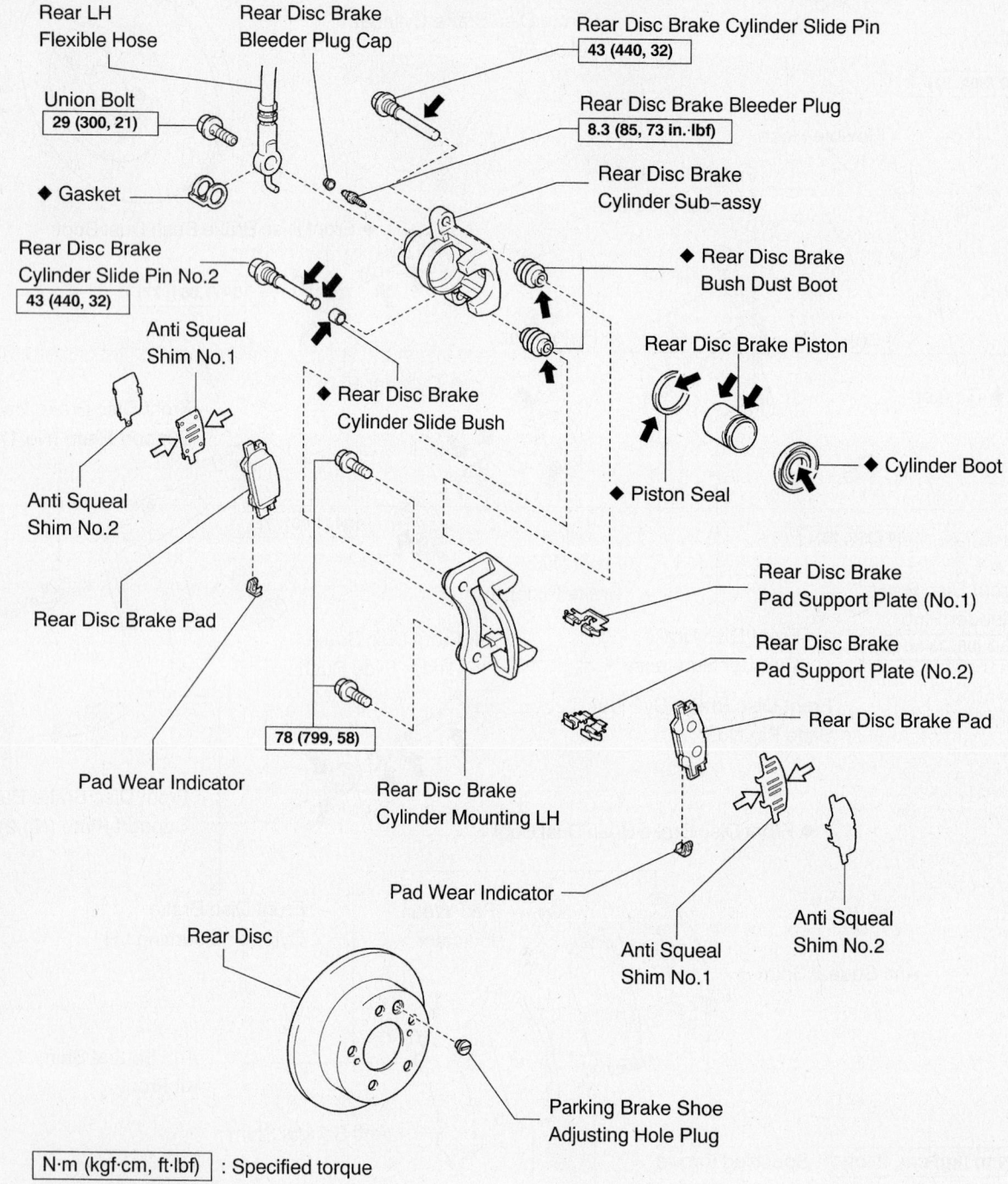

N·m (kgf·cm, ft·lbf) : Specified torque

◆ Non-reusable part

◄ Lithium soap base glycol grease

⇐ Disc brake grease

Rear disc brake components—RX330

67162-X300-G12

MAZDA

25

626 • Mazda 3 • Mazda 6 • Miata • Millenia • Protégé

SPECIFICATIONS AND MAINTENANCE CHARTS

ENGINE AND VEHICLE IDENTIFICATION

Code ①	Liters (cc)	Cu. In.	Cyl.	Fuel Sys.	Engine Type	Eng. Mfg.	Code ②	Year
ZM	1.6 (1597)	97.4	4	MPFI	DOHC	Mazda	1	2001
BP	1.8 (1839)	112.2	4	MPFI	DOHC	Mazda	2	2002
FP	1.8 (1839)	112.2	4	MPFI	DOHC	Mazda	3	2003
FS	2.0 (1991)	121.5	4	MPFI	DOHC	Mazda	4	2004
LF	2.0 (1999)	121.9	4	MPFI	DOHC	Mazda	5	2005
KJ	2.3 (2254)	137.2	6	MPFI	DOHC	Mazda		
L3	2.3 (2261)	137.9	4	MPFI	DOHC	Mazda		
KL	2.5 (2496)	152.3	6	MPFI	DOHC	Mazda		
AJ	3.0 (2999)	181.1	6	MPFI	DOHC	Mazda		

MPFI: Multi-Point Fuel Injection

DOHC: Double Over Head Cam

① Located above the starter

② 10th digit of the Vehicle Identification Number (VIN)

67162-MAZC-C01

GENERAL ENGINE SPECIFICATIONS

Year	Model	Engine Displacement Liters (VIN)	Net Horsepower @ rpm	Net Torque @ rpm (ft. lbs.)	Bore x Stroke (in.)	Com- pression Ratio	Oil Pressure @ rpm
2001	626 ES	2.0 (FS)	125@5500	127@3300	3.27x3.62	9.0:1	57-71@3000
	626 ES-V6	2.5 (KL)	165@6000	161@5000	3.33x2.92	9.5:1	49-71@3000
	626 LX	2.0 (FS)	125@5500	127@3300	3.27x3.62	9.0:1	57-71@3000
	626 LX-V6	2.5 (KL)	165@6000	161@5000	3.33x2.92	9.5:1	49-71@3000
	Miata	1.8 (BP)	①	②	3.27x3.35	9.5:1	43-57@3000
	Millenia	2.5 (KL)	170@5800	160@4800	3.33x2.92	9.2:1	49-71@3000
	Millenia S	2.3 (KJ)	210@5300	210@3500	3.16x2.92	10.0:1	44-66@3000
	Protege DX	1.6 (ZM)	③	④	3.07x3.29	9.0:1	43-57@3000
	Protege ES	1.8 (FP)	⑤	⑥	3.27x3.35	9.1:1	43-57@3000
	Protege LX	1.6 (ZM)	③	④	3.07x3.29	9.0:1	43-57@3000
2002	626 ES	2.0 (FS)	125@5500	127@3300	3.27x3.62	9.0:1	57-71@3000
	626 ES-V6	2.5 (KL)	165@6000	161@5000	3.33x2.92	9.5:1	49-71@3000
	626 LX	2.0 (FS)	125@5500	127@3300	3.27x3.62	9.0:1	57-71@3000
	626 LX-V6	2.5 (KL)	165@6000	161@5000	3.33x2.92	9.5:1	49-71@3000
	Miata	1.8 (BP)	142@7000	125@5500	3.3x3.4	10.0:1	43-56@3000
	Millenia	2.5 (KL)	170@5800	160@4800	3.33x2.92	9.2:1	49-71@3000
	Millenia S	2.3 (KJ)	210@5300	210@3500	3.16x2.92	10.0:1	44-66@3000
	Protege DX	1.6 (ZM)	③	④	3.07x3.29	9.0:1	43-57@3000
	Protege LX	1.6 (ZM)	③	④	3.07x3.29	9.0:1	43-57@3000
	Protege5	2.0 (FS)	125@5500	127@3300	3.27x3.62	9.1:1	57-71@3000
2003-04	626 ES	2.0 (FS)	125@5500	127@3300	3.27x3.62	9.0:1	57-71@3000
	626 ES-V6	2.5 (KL)	165@6000	161@5000	3.33x2.92	9.5:1	49-71@3000
	626 LX	2.0 (FS)	125@5500	127@3300	3.27x3.62	9.0:1	57-71@3000
	626 LX-V6	2.5 (KL)	165@6000	161@5000	3.33x2.92	9.5:1	49-71@3000
	Miata	1.8 (BP)	142@7000	125@5500	3.3x3.4	10.0:1	43-56@3000
	Millenia	2.5 (KL)	170@5800	160@4800	3.33x2.92	9.2:1	49-71@3000
	Millenia S	2.3 (KJ)	210@5300	210@3500	3.16x2.92	10.0:1	44-66@3000
	Protege DX	1.6 (ZM)	③	④	3.07x3.29	9.0:1	43-57@3000
	Protege LX	1.6 (ZM)	③	④	3.07x3.29	9.0:1	43-57@3000
	Protege5	2.0 (FS)	125@5500	127@3300	3.27x3.62	9.1:1	57-71@3000
	Mazda 3 (I)	2.0 (LF)	⑦	⑧	344x3.27	10.0:1	33.9-75.5@3000
	Mazda 3 (S)	2.3 (L3)	160@6500	150@4500	344x3.70	9.7:1	57-94@3000
	Mazda 6 (I)	2.3 (L3)	160@6000	155@4000	344x3.70	9.7:1	57-94@3000
	Mazda 6 (S)	3.0 (AJ)	220@6300	192@5000	3.50x3.13	10.0:1	20@45@1500

EFI: Electronic Fuel Injection

① LEV states: 138@6500
  Except LEV states: 140@6500

② LEV states: 117@5000
  Except LEV states: 119@5500

③ LEV states: 103@5500
  Except LEV states: 105@5500

④ LEV states: 106@4000
  Except LEV states: 107@4000

⑤ LEV states: 120@6000
  Except LEV states: 122@6000

⑥ LEV states: 119@4000
  Except LEV states: 120@4000

⑦ Partial Zero Emission Vehicle (PZEV): 144@6500
  Except PZEV states: 148@6500

⑧ Partial Zero Emission Vehicle (PZEV): 132@4500
  Except PZEV states: 135@4500

ENGINE TUNE-UP SPECIFICATIONS

Year	Engine Displacement Liters (VIN)	Spark Plug Gap (in.)	Ignition Timing (deg.) MT	Ignition Timing (deg.) AT	Fuel Pump (psi)	Idle Speed (rpm) MT	Idle Speed (rpm) AT	Valve Clearance In.	Valve Clearance Ex.
2001	1.6 (ZM)	0.040-0.043	6-18B	6-18B	39-45	650-750	650-750	0.010-0.011	0.010-0.011
	1.8 (BP)	0.040-0.043	6-18B	6-18B	53-61	750-850	750-850	0.008-0.009	0.012-0.013
	1.8 (FP)	0.040-0.043	6-18B	6-18B	39-45	650-750	650-750	0.009-0.012	0.009-0.012
	2.0 (FS)	0.040-0.043	6-18B	6-18B	37-45	550-850	500-800	0.008-0.0011	0.008-0.0011
	2.3 (KJ)	28-31	6-8B	6-8B	41-48	600-700	600-700	0.011-0.012	0.011-0.012
	2.5 (KL)	①	②	②	39-45	600-700	600-700	③	④
2002	1.6 (ZM)	0.040-0.043	6-18B	6-18B	39-45	650-750	650-750	0.010-0.011	0.010-0.011
	1.8 (BP)	0.040-0.043	6-18B	6-18B	53-61	750-850	750-850	0.008-0.0011	0.012-0.013
	2.0 (FS)	0.040-0.043	6-18B	6-18B	37-45	550-850	500-800	0.008-0.0011	0.008-0.0011
	2.3 (KJ)	28-31	6-8B	6-8B	41-48	600-700	600-700	0.011-0.012	0.011-0.012
	2.5 (KL)	①	②	②	39-45	600-700	600-700	③	④
2003-04	1.6 (ZM)	0.040-0.043	6-18B	6-18B	39-45	650-750	650-750	0.010-0.011	0.010-0.011
	1.8 (BP)	0.040-0.043	6-18B	6-18B	53-61	750-850	750-850	0.008-0.0011	0.012-0.013
	2.0 (FS)	0.040-0.043	6-18B	6-18B	37-45	550-850	500-800	0.008-0.0011	0.008-0.0011
	2.0 (LF)	0.049-0.053	8B	8B	36	600-700	650-750	0.008-0.0011	0.008-0.011
	2.3 (L3)	0.049-0.053	8B	8B	29	600-700	650-750	0.008-0.0011	0.008-0.011
	2.3 (KJ)	28-31	8B	8B	41-48	600-700	600-700	0.011-0.012	0.011-0.012
	2.5 (KL)	①	②	②	39-45	600-700	600-700	③	④
	3.0 (AJ)	0.049-0.053	10B	10B	36	700-800	500-800	—	—

NOTE: The Vehicle Emission Control Information label often reflects specification changes made during production. The label figures must be used if they differ from those in this chart.

B: Before top dead center

HYD: Hydraulic

① 626 models: 0.028-0.031
Millenia models: 0.039-0.043

② 626 models: 4-16
Millenia models: 9-11

③ 626 models: 0.0097-0.012
Millenia models: maintenace free

④ 626 models: 0.010-0.013
Millenia models: maintenace free

67162-MAZC-C03

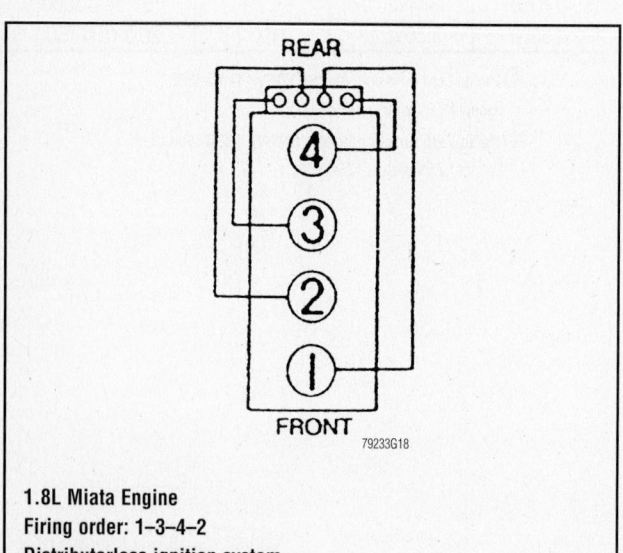

1.8L Miata Engine
Firing order: 1–3–4–2
Distributorless ignition system

79233G18

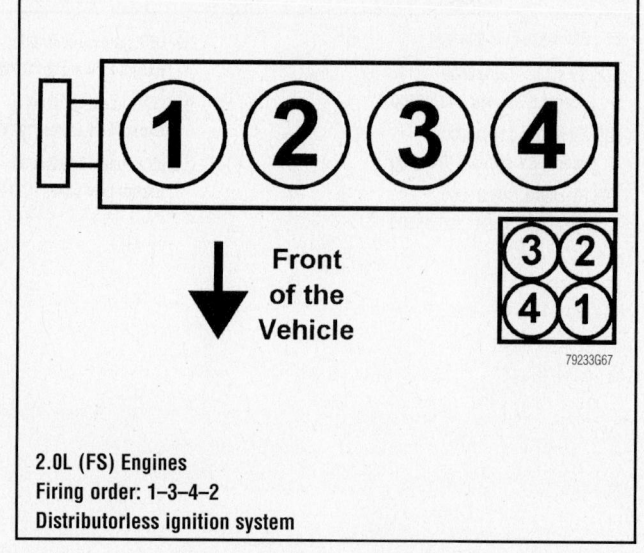

2.0L (FS) Engines
Firing order: 1–3–4–2
Distributorless ignition system

79233G67

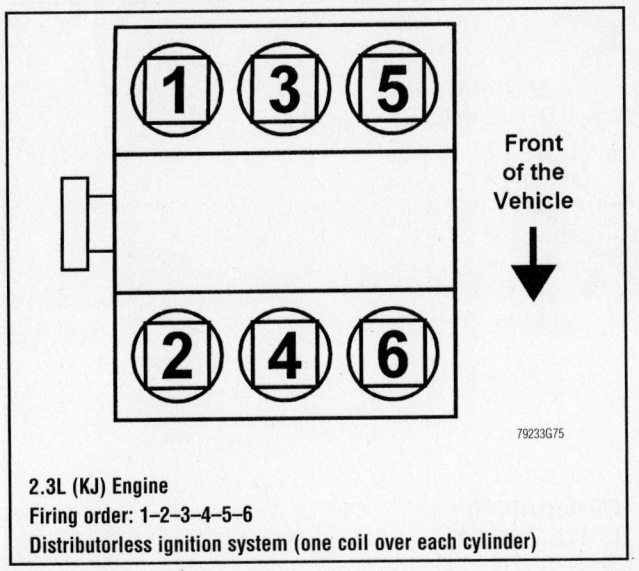

2.3L (KJ) Engine
Firing order: 1–2–3–4–5–6
Distributorless ignition system (one coil over each cylinder)

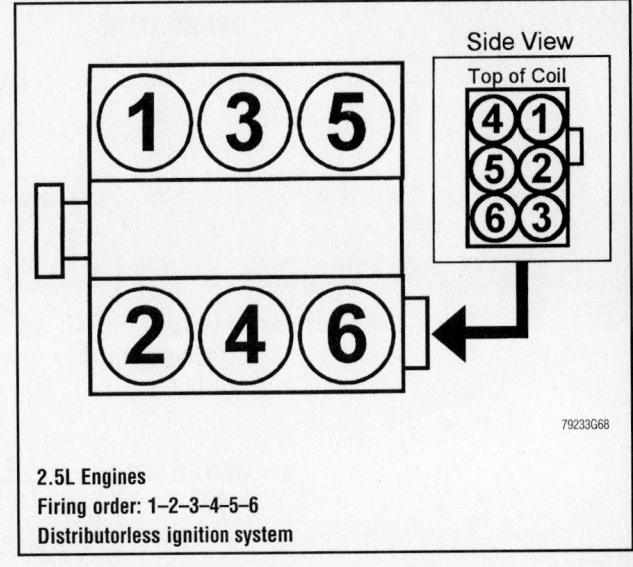

2.5L Engines
Firing order: 1–2–3–4–5–6
Distributorless ignition system

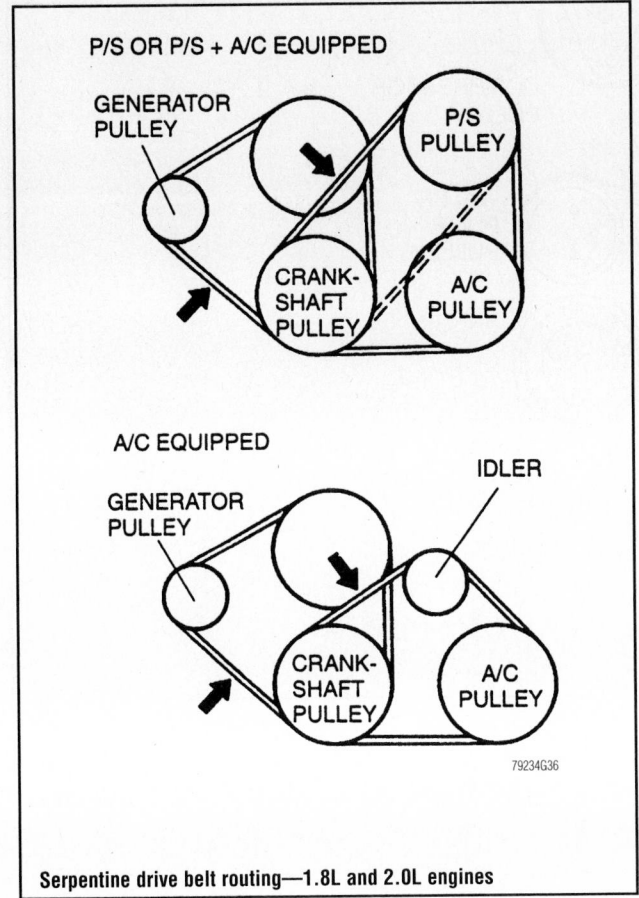

Serpentine drive belt routing—1.8L and 2.0L engines

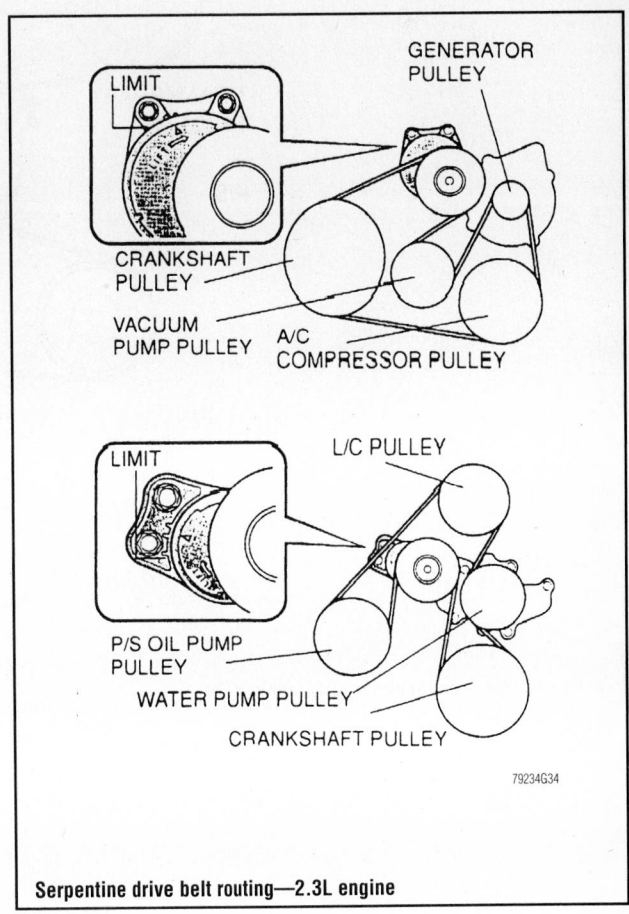

Serpentine drive belt routing—2.3L engine

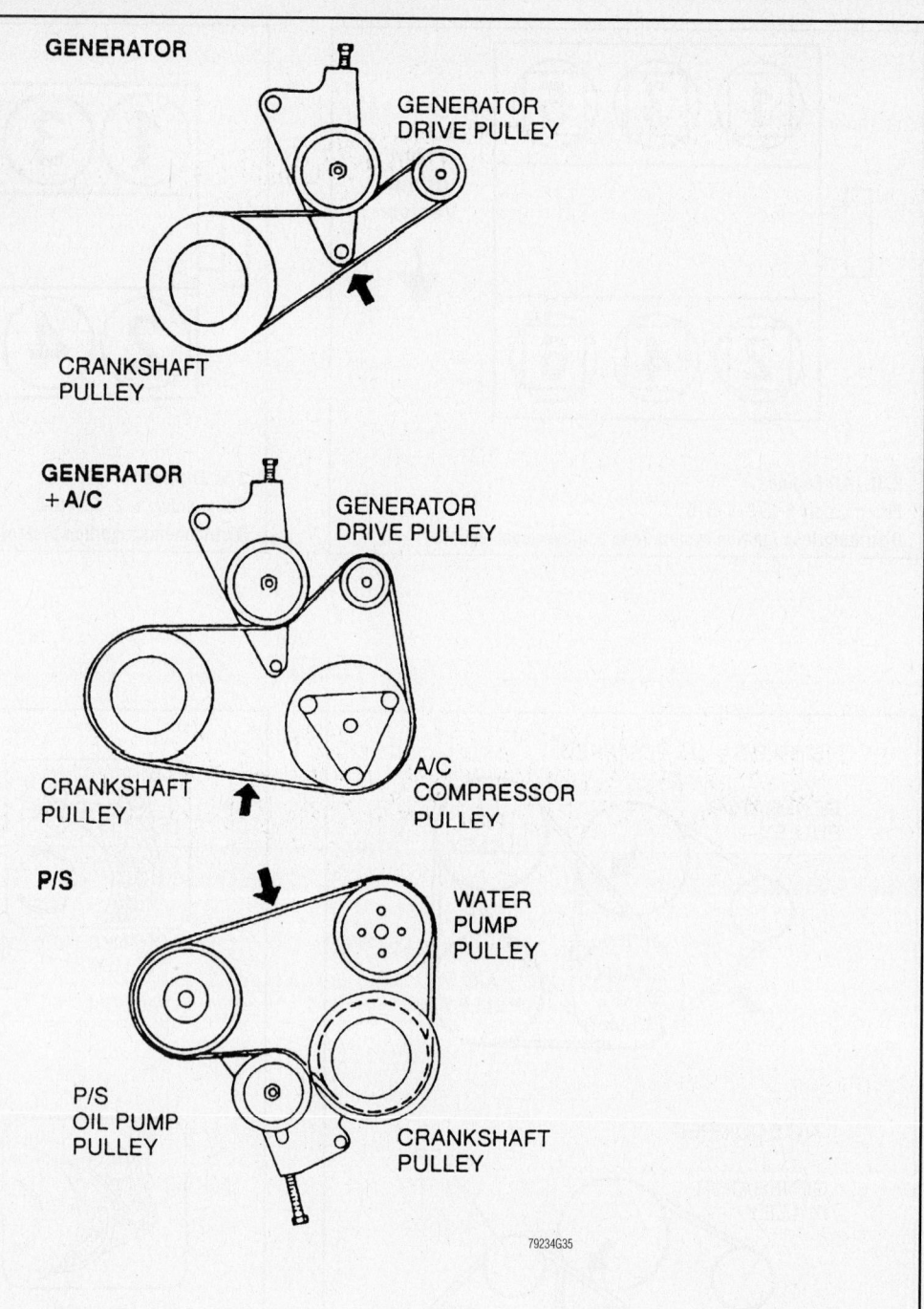

GENERATOR

GENERATOR DRIVE PULLEY

CRANKSHAFT PULLEY

GENERATOR + A/C

GENERATOR DRIVE PULLEY

CRANKSHAFT PULLEY

A/C COMPRESSOR PULLEY

P/S

WATER PUMP PULLEY

P/S OIL PUMP PULLEY

CRANKSHAFT PULLEY

79234G35

Serpentine drive belt routing—2.5L engines

CAPACITIES

Year	Model	Engine Displacement Liters (VIN)	Engine Oil with Filter (qts.)	Transmission (pts.)		Drive Axle		Fuel Tank (gal.)	Cooling System (qts.)
				5-Spd	Auto.	Front (pts.)	Rear (pts.)		
2001	626 ES	2.0 (FS)	3.7	5.8	18.6	①	—	16.9	7.9
	626 ES-V6	2.5 (KL)	4.2	6.0	17.0	①	—	16.9	7.9
	626 LX	2.0 (FS)	3.7	5.8	18.6	①	—	16.9	7.9
	626 LX-V6	2.5 (KL)	4.2	6.0	17.0	①	—	16.9	7.9
	Miata	1.8 (BP)	4.0	4.2	13.5	—	2.1	12.7	6.3
	Millenia	2.5 (KL)	4.2	—	16.9	①	—	18.0	7.9
	Millenia S	2.3 (KJ)	4.3	—	16.9	①	—	18.0	7.9
	Protege DX	1.6 (ZM)	3.4	5.8	15.2	①	—	13.2	7.9
	Protege ES	1.8 (FP)	5.0	5.6	15.2	①	—	13.2	9.9
	Protege LX	1.6 (ZM)	3.4	5.8	15.2	①	—	13.2	7.9
2002	626 ES	2.0 (FS)	3.7	5.8	18.6	①	—	16.9	7.9
	626 ES-V6	2.5 (KL)	4.2	6.0	17.0	①	—	16.9	7.9
	626 LX	2.0 (FS)	3.7	5.8	18.6	①	—	16.9	7.9
	626 LX-V6	2.5 (KL)	4.2	6.0	17.0	①	—	16.9	7.9
	Miata	1.8 (BP)	4.0	②	14.2	—	2.1	12.7	6.3
	Millenia	2.5 (KL)	4.2	—	16.9	①	—	18.0	7.9
	Millenia S	2.3 (KJ)	4.3	—	16.9	①	—	18.0	7.9
	Protege DX	1.6 (ZM)	3.4	5.8	15.2	①	—	13.2	7.9
	Protege LX	2.0 (FS)	3.7	5.8	15.2	①	—	14.5	7.9
	Protege5	2.0 (FS)	3.7	6.0	17.6	①	—	14.5	7.9
2003-04	626 ES	2.0 (FS)	3.7	5.8	18.6	①	—	16.9	7.9
	626 ES-V6	2.5 (KL)	4.2	6.0	17.0	①	—	16.9	7.9
	626 LX	2.0 (FS)	3.7	5.8	18.6	①	—	16.9	7.9
	626 LX-V6	2.5 (KL)	4.2	6.0	17.0	①	—	16.9	7.9
	Miata	1.8 (BP)	4.0	②	14.2	—	2.1	12.7	6.3
	Millenia	2.5 (KL)	4.2	—	16.9	①	—	18.0	7.9
	Millenia S	2.3 (KJ)	4.3	—	16.9	①	—	18.0	7.9
	Protege DX	1.6 (ZM)	3.4	5.8	15.2	①	—	13.2	7.9
	Protege LX	2.0 (FS)	3.7	5.8	15.2	①	—	14.5	7.9
	Protege5	2.0 (FS)	3.7	6.0	17.6	①	—	14.5	7.9
	Mazda 3 (I)	2.0 (LF)	4.5	6.06	15.2	①	—	14.5	7.9
	Mazda 3 (S)	2.3 (L3)	4.5	6.06	15.2	①	—	14.5	7.9
	Mazda 6 (I)	2.3 (L3)	4.5	6.06	15.2	①	—	14.5	7.9
	Mazda 6 (S)	3.0 (AJ)	6.0	4.8	19.4	①	—	14.5	7.9

NOTE: All capacities are approximate. Add fluid gradually and check to be sure a proper fluid level is obtained.

① Included in transaxle

② 5-speed: 4.2 qts. Included in transaxle

6-speed: 3.8 qts

67162-MAZC-C04

VALVE SPECIFICATIONS

Year	Engine Displacement Liters (VIN)	Seat Angle (deg.)	Face Angle (deg.)	Maximum out of Square (in.)	Spring Free Length (in.)	Stem-to-Guide Clearance (in.)		Stem Diameter (in.)	
						Intake	Exhaust	Intake	Exhaust
2001	1.6 (ZM)	NA	NA	NA	NA	NA	NA	NA	NA
	1.8 (BP)	45	45	0.062	①	0.0010-0.0023	0.0012-0.0025	0.2351-0.2356	0.2349-0.2354
	1.8 (FP)	NA	NA	NA	NA	NA	NA	NA	NA
	2.0 (FS)	45	45	0.061	1.732	0.0010-0.0023	0.0012-0.0025	0.2351-0.2356	0.2349-0.2354
	2.3 (KJ)	NA	45	0.062	1.413	0.0010-0.0023	0.0012-0.0025	0.2351-0.2356	0.2349-0.2354
	2.5 (KL)	45	45	0.064	1.847	0.0010-0.0023	0.0012-0.0025	0.2351-0.2356	0.2349-0.2354
2002	1.6 (ZM)	NA	NA	NA	NA	NA	NA	NA	NA
	1.8 (BP)	45	45	0.062	①	0.0010-0.0023	0.0012-0.0025	0.2351-0.2356	0.2349-0.2354
	2.0 (FS)	45	45	0.061	1.732	0.0010-0.0023	0.0012-0.0025	0.2351-0.2356	0.2349-0.2354
	2.3 (KJ)	NA	45	0.062	1.413	0.0010-0.0023	0.0012-0.0025	0.2351-0.2356	0.2349-0.2354
	2.5 (KL)	45	45	0.064	1.847	0.0010-0.0023	0.0012-0.0025	0.2351-0.2356	0.2349-0.2354
2003-04	1.6 (ZM)	NA	NA	NA	NA	NA	NA	NA	NA
	1.8 (BP)	45	45	0.062	①	0.0010-0.0023	0.0012-0.0025	0.2351-0.2356	0.2349-0.2354
	2.0 (FS)	45	45	0.061	1.732	0.0010-0.0023	0.0012-0.0025	0.2351-0.2356	0.2349-0.2354
	2.0 (LF)	NA	NA	NA	NA	NA	NA	NA	NA
	2.3 (KJ)	NA	45	0.062	1.413	0.0010-0.0023	0.0012-0.0025	0.2351-0.2356	0.2349-0.2354
	2.3 (L3)	NA	NA	NA	NA	NA	NA	NA	NA
	2.5 (KL)	45	45	0.064	1.847	0.0010-0.0023	0.0012-0.0025	0.2351-0.2356	0.2349-0.2354
	3.0 (AJ)	NA	NA	NA	NA	NA	NA	NA	NA

NA: Not Available

① Intake: 1.80 in.
  Exhaust: 1.903 in.

67162-MAZC-C05

CRANKSHAFT AND CONNECTING ROD SPECIFICATIONS

All measurements are given in inches.

Year	Engine Displacement Liters (VIN)	Crankshaft				Connecting Rod		
		Main Brg. Journal Dia.	Main Brg. Oil Clearance	Shaft End-play	Thrust on No.	Journal Diameter	Oil Clearance	Side Clearance
2001	1.6 (ZM)	NA	NA	NA	NA	NA	NA	NA
	1.8 (BP)	1.9661-1.9667	0.0008-0.0014	0.0032-0.0111	4	1.7693-1.7699	0.0008-0.0017	0.0044-0.0103
	1.8 (FP)	NA	NA	NA	NA	NA	NA	NA
	2.0 (FS)	2.2022-2.2029	①	0.0031-0.0111	4	1.8874-1.8880	0.0005-0.0015	0.0043-0.0103
	2.3 (KJ)	2.4385-2.4391	0.0015-0.0022	0.0032-0.0111	4	2.0843-2.0848	0.0010-0.0016	0.0071-0.0157
	2.5 (KL)	2.4385-2.4391	0.0015-0.0022	0.0032-0.0111	4	2.0843-2.0848	0.0010-0.0016	0.0071-0.0157
2002	1.6 (ZM)	NA	NA	NA	NA	NA	NA	NA
	1.8 (BP)	1.9661-1.9667	0.0008-0.0014	0.0032-0.0111	4	1.7693-1.7699	0.0008-0.0017	0.0044-0.0103
	2.0 (FS)	2.2022-2.2029	①	0.0031-0.0111	4	1.8874-1.8880	0.0005-0.0015	0.0043-0.0103
	2.3 (KJ)	2.4385-2.4391	0.0015-0.0022	0.0032-0.0111	4	2.0843-2.0848	0.0010-0.0016	0.0071-0.0157
	2.5 (KL)	2.4385-2.4391	0.0015-0.0022	0.0032-0.0111	4	2.0843-2.0848	0.0010-0.0016	0.0071-0.0157
2003-04	1.6 (ZM)	NA	NA	NA	NA	NA	NA	NA
	1.8 (BP)	1.9661-1.9667	0.0008-0.0014	0.0032-0.0111	4	1.7693-1.7699	0.0008-0.0017	0.0044-0.0103
	2.0 (LF)	NA	NA	NA	NA	NA	NA	NA
	2.0 (FS)	2.2022-2.2029	①	0.0031-0.0111	4	1.8874-1.8880	0.0005-0.0015	0.0043-0.0103
	2.3 (L3)	NA	NA	NA	NA	NA	NA	NA
	2.3 (KJ)	2.4385-2.4391	0.0015-0.0022	0.0032-0.0111	4	2.0843-2.0848	0.0010-0.0016	0.0071-0.0157
	2.5 (KL)	2.4385-2.4391	0.0015-0.0022	0.0032-0.0111	4	2.0843-2.0848	0.0010-0.0016	0.0071-0.0157
	3.0 (AJ)	NA	NA	NA	NA	NA	NA	NA

NA: Not Avilable

① No. 1, 2, 4 & 5: 0.0009-0.0020 in.
No. 3: 0.0012-0.0022 in.

67162-MAZC-C06

PISTON AND RING SPECIFICATIONS
All measurements are given in inches.

Year	Engine Displacement Liters (VIN)	Piston Clearance	Ring Gap			Ring Side Clearance		
			Top Compression	Bottom Compression	Oil Control	Top Compression	Bottom Compression	Oil Control
2001	1.6 (ZM)	NA	NA	NA	NA	NA	NA	NA
	1.8 (BP)	0.0010-0.0014	0.006-0.011	0.006-0.011	0.008-0.027	0.0012-0.0026	0.0012-0.0027	0.0030-0.0060
	1.8 (FP)	NA	NA	NA	NA	NA	NA	NA
	2.0 (FS)	0.0015-0.0020	0.006-0.012	0.006-0.012	0.008-0.028	0.0014-0.0026	0.0014-0.0026	SNUG
	2.3 (KJ)	0.0004-0.0014	0.006-0.010	0.010-0.014	0.008-0.030	0.0014-0.0025	0.0012-0.0025	0.0028-0.0062
	2.5 (KL)	0.0012-0.0022	0.006-0.012	0.010-0.016	0.008-0.028	0.0008-0.0025	0.0012-0.0025	0.0008-0.0020
2002	1.6 (ZM)	NA	NA	NA	NA	NA	NA	NA
	1.8 (BP)	0.0010-0.0014	0.006-0.011	0.006-0.011	0.008-0.027	0.0012-0.0026	0.0012-0.0027	0.0030-0.0060
	2.0 (FS)	0.0015-0.0020	0.006-0.012	0.006-0.012	0.008-0.028	0.0014-0.0026	0.0014-0.0026	SNUG
	2.3 (KJ)	0.0004-0.0014	0.006-0.010	0.010-0.014	0.008-0.030	0.0014-0.0025	0.0012-0.0025	0.0028-0.0062
	2.5 (KL)	0.0012-0.0022	0.006-0.012	0.010-0.016	0.008-0.028	0.0008-0.0025	0.0012-0.0025	0.0008-0.0020
2003-04	1.6 (ZM)	NA	NA	NA	NA	NA	NA	NA
	1.8 (BP)	0.0010-0.0014	0.006-0.011	0.006-0.011	0.008-0.027	0.0012-0.0026	0.0012-0.0027	0.0030-0.0060
	2.0 (LF)	NA	NA	NA	NA	NA	NA	NA
	2.0 (FS)	0.0015-0.0020	0.006-0.012	0.006-0.012	0.008-0.028	0.0014-0.0026	0.0014-0.0026	SNUG
	2.3 (L3)	NA	NA	NA	NA	NA	NA	NA
	2.3 (KJ)	0.0004-0.0014	0.006-0.010	0.010-0.014	0.008-0.030	0.0014-0.0025	0.0012-0.0025	0.0028-0.0062
	2.5 (KL)	0.0012-0.0022	0.006-0.012	0.010-0.016	0.008-0.028	0.0008-0.0025	0.0012-0.0025	0.0008-0.0020
	3.0 (AJ)	NA	NA	NA	NA	NA	NA	NA

NA: Not Available

TORQUE SPECIFICATIONS
All readings in ft. lbs.

Year	Engine Displacement Liters (VIN)	Cylinder Head Bolts	Main Bearing Bolts	Rod Bearing Bolts	Crankshaft Damper Bolts	Flywheel Bolts	Manifold Intake	Manifold Exhaust	Spark Plugs	Oil Pan Drain Plug
2001	1.6 (ZM)	①	NA	NA	116-122	71-76	14-18	15-20	11-16	65-87
	1.8 (BP)	56-60	40-43	35-36	116-130	71-76	17-21	29-33	11-16	65-87
	1.8 (FP)	①	NA	NA	116-122	71-76	14-18	15-20	11-16	65-87
	2.0 (FS)	①	②	③	116-122	71-76	14-18	④	11-16	65-87
	2.3 (KJ)	⑤	⑥	③	116-122	45-49	14-18	14-18	11-16	65-87
	2.5 (KL)	⑤	⑦	③	116-122	45-49	14-18	12-22	11-16	65-87
2002	1.6 (ZM)	①	NA	NA	116-122	71-76	14-18	15-20	11-16	65-87
	1.8 (BP)	56-60	40-43	35-36	116-130	71-76	17-21	29-33	11-16	65-87
	2.0 (FS)	①	②	③	116-122	71-76	14-18	④	11-16	65-87
	2.3 (KJ)	⑤	⑥	③	116-122	45-49	14-18	14-18	11-16	65-87
	2.5 (KL)	⑤	⑦	③	116-122	45-49	14-18	12-22	11-16	65-87
2003-04	1.6 (ZM)	①	NA	NA	116-122	71-76	14-18	15-20	11-16	65-87
	1.8 (BP)	56-60	40-43	35-36	116-130	71-76	17-21	29-33	11-16	65-87
	2.0 (FS)	①	②	③	116-122	71-76	14-18	④	11-16	65-87
	2.0 (LF)	⑧	NA	NA	⑨	⑩	⑪	32-47	8-10	80-88
	2.3 (L3)	⑧	NA	NA	⑨	⑩	⑪	32-47	8-10	80-88
	2.3 (KJ)	⑤	⑥	③	116-122	45-49	14-18	14-18	11-16	65-87
	2.5 (KL)	⑤	⑦	③	116-122	45-49	14-18	12-22	11-16	65-87
	3.0 (AJ)	⑤	NA	NA	⑫	⑬	⑭	14-18	⑮	80-88

NA: Not Available

① Step 1: 13-16 ft. lbs.
  Step 2: Tighten 85-95 degees
  Step 3: Repeat step 2

② Step 1: 16 ft. lbs.
  Step 2: Tighten each bolt 90 degrees

③ Nuts: 15-20 ft. lbs
  Bolts: 12-16 ft. lbs.

④ Nuts: 20 ft. lbs
  Bolts: 22 ft. lbs.

⑤ Step 1: 17-19 ft. lbs.
  Step 2: Tighten 85-95 degees
  Step 3: Repeat step 2

⑥ Step 1: Inner bolts: 17-19 ft. lbs.
  Step 2: Outer bolts: 13.5-15.5 ft. lbs.
  Step 3: Inner bolt Nos. 1-3: Tighten each bolt 70 degrees
  Step 4: Inner bolt No. 4: Tighten each bolt 80 degrees
  Step 5: Outer bolts: Tighten each bolt 60 degrees
  Step 6: Repeat Step 3-5

⑦ Step 1: Inner bolts: 17-18 ft. lbs.; Outer bolts: 13-15 ft. lbs.
  Step 2: Inner bolt Nos. 1-3: Tighten each bolt 70 degrees
  Step 3: Inner bolt No. 4: Tighten each bolt 80 degrees
  Step 4: Outer bolts: Tighten each bolt 60 degrees
  Step 5: Repeat Step 2

⑧ Step 1: 97 inch lbs.
  Step 2: Tighten 12.5 ft. lbs
  Step 3: Tighten 34.6 ft. lbs.
  Step 4: Tighten 88 degrees
  Step 5: Tighten 88 degrees

⑨ 71-77 ft. lbs. plus 87-93 degrees

⑩ Manual transmission: 80-85 ft. lbs.
  Automatic transmission: 71-76 ft. lbs.

⑪ Mazda 3 with the 2.0L (LF) and 2.3L (L3) engines: 142-177 ft. lbs.
  Mazda 6 with the 2.3L (L3) engine: 71-101 inch lbs.

⑫ Step 1: 88 ft. lbs.
  Step 2: loosen one full turn
  Step 3: Tighten 35-39 ft. lbs.
  Step 4: Tighten 85-95 degrees

⑬ G35M-R manual or FN4A-EL automatic transmission: 80-85 ft. lbs.
  A65M-R manual or JA5A-EL automatic transmission: 54-64 ft. lbs.

⑭ Lower intake: 72-106 inch lbs.
  Upper intake plenum: 72-106 inch lbs.

⑮ 79-177 inch lbs.

67162-MAZC-C08

WHEEL ALIGNMENT

Year	Model		Caster Range (+/-Deg.)	Caster Preferred Setting (Deg.)	Camber Range (+/-Deg.)	Camber Preferred Setting (Deg.)	Toe-in (in.)
2001	626 ①	F	1.00	+2.13	1.00	-0.70	0.12 +/- 0.16
		R	—	—	1.00	-0.10	0.12 +/- 0.16
	626 ②	F	1.00	+2.15	1.00	-0.70	0.12 +/- 0.16
		R	—	—	1.00	-0.10	0.12 +/- 0.16
	626 ③	F	1.00	+2.05	1.00	-0.70	0.12 +/- 0.16
		R	—	—	1.00	-0.10	0.12 +/- 0.16
	Miata	F	1.00	+5.75	1.00	+0.05	0.12 +/- 0.16
		R	—	—	1.00	-0.75	0.12 +/- 0.16
	Millenia	F	1.00	+2.23	0.75	-0.19	0.12 +/- 0.16
		R	—	—	1.00	-0.31	0.12 +/- 0.16
	Protégé	F	1.00	+1.88	1.00	-0.75	0.08 +/- 0.16
		R	—	—	1.00	-0.52	0.08 +/- 0.16
2002	626 ①	F	1.00	+2.13	1.00	-0.70	0.12 +/- 0.16
		R	—	—	1.00	-0.10	0.12 +/- 0.16
	626 ②	F	1.00	+2.15	1.00	-0.70	0.12 +/- 0.16
		R	—	—	1.00	-0.10	0.12 +/- 0.16
	626 ③	F	1.00	+2.05	1.00	-0.70	0.12 +/- 0.16
		R	—	—	1.00	-0.10	0.12 +/- 0.16
	Miata	F	1.00	+5.75	1.00	+0.05	0.12 +/- 0.16
		R	—	—	1.00	-0.75	0.12 +/- 0.16
	Millenia	F	1.00	+2.23	0.75	-0.19	0.12 +/- 0.16
		R	—	—	1.00	-0.31	0.12 +/- 0.16
	Protégé	F	1.00	+1.88	1.00	-0.75	0.08 +/- 0.16
		R	—	—	1.00	-0.52	0.08 +/- 0.16
2003-04	626 ①	F	1.00	+2.13	1.00	-0.70	0.12 +/- 0.16
		R	—	—	1.00	-0.10	0.12 +/- 0.16
	626 ②	F	1.00	+2.15	1.00	-0.70	0.12 +/- 0.16
		R	—	—	1.00	-0.10	0.12 +/- 0.16
	626 ③	F	1.00	+2.05	1.00	-0.70	0.12 +/- 0.16
		R	—	—	1.00	-0.10	0.12 +/- 0.16
	Miata	F	1.00	+5.75	1.00	+0.05	0.12 +/- 0.16
		R	—	—	1.00	-0.75	0.12 +/- 0.16
	Millenia	F	1.00	+2.23	0.75	-0.19	0.12 +/- 0.16
		R	—	—	1.00	-0.31	0.12 +/- 0.16
	Mazda 3	F	1.00	+2.58	1.00	-0.40	0.04 +/- 0.12
		R	—	—	1.00	0	0.08 +/- 0.16
	Mazda 6	F	1.00	④	1.00	⑤	0.12 +/- 0.16
		R	—	—	1.00	⑥	0.08 +/- 0.16
	Protégé	F	1.00	+1.88	1.00	-0.75	0.08 +/- 0.16
		R	—	—	1.00	-0.52	0.08 +/- 0.16

① With 14 in. wheels

② With 15 in. wheels

③ With 16 in. wheels

③ With 16 in. wheels

④ Except wagon: 0 degrees 11', plus or minus 0 degrees 22'

   Wagon: 3 degrees 33' plus or minus 1 degree

⑤ Except wagon:-0 degrees 16', plus or minus 1 degree

   Wagon: -0 degrees 16', plus or minus 1 degree

⑥ Except wagon:-1 degree 09', plus or minus 1 degree

   Wagon: -1 degree 02', plus or minus 1 degree

TIRE, WHEEL AND BALL JOINT SPECIFICATIONS

| Year | Model | OEM Tires | | Tire Pressures (psi) | | Wheel Size | Ball Joint Inspection | Lug Nut |
		Standard	Optional	Front	Rear			
2001	Millenia	P215/55R16	P215/50R17	32	29	Std: 6 1/2-JJ Opt:7-JJ	④	65-87
	Miata	P195/50R14	P205/45VR16	26	26	6-JJ	① ②	65-87
	Protégé 1.6L	P185/65R14	None	32	29	6-JJ	8-43 in. ①	65-87
	Protégé 1.8L	P185/65R14	P195/50VR15	32	32	Std: 5.5-JJ Opt: 6-JJ	8-43 in. ①	65-87
	626 2.0L	P205/60R15	P205/55R16	32	36	6-JJ	③	65-87
	626 2.5L	P205/60R15	P205/55R16	32	36	6-JJ	③	65-87
2002	Millenia	P215/55R16	P215/50R17	32	29	Std: 6 1/2-JJ Opt:7-JJ	④	65-87
	Miata	P195/50R14	P205/45VR16	26	26	6-JJ	① ②	65-87
	Protégé 1.6L	P185/65R14	None	32	32	6-JJ	8-43 in. ①	65-87
	Protégé 2.0L	P185/65R14	P195/50VR15	32	32	Std: 5.5-JJ	8-43 in. ①	65-87
	626 2.0L	P205/60R15	P205/55R16	32	36	6-JJ	③	65-87
	626 2.5L	P205/60R15	P205/55R16	32	36	6-JJ	③	65-87
2003-04	Millenia	P215/55R16	P215/50R17	32	29	Std: 6 1/2-JJ Opt:7-JJ	④	65-87
	Miata	P195/50R14	P205/45VR16	26	26	6-JJ	① ②	65-87
	Protégé 1.6L	P185/65R14	None	32	32	6-JJ	8-43 in. ①	65-87
	Protégé 2.0L	P185/65R14	P195/50VR15	32	32	Std: 5.5-JJ	8-43 in. ①	65-87
	Mazda 3	⑤	⑤	32	32	⑥	⑦	65-87
	Mazda 6	P205/60R16	P215/50R17	32	32	⑧	⑨	65-87

OEM: Original Equipment Manufacturer

PSI: Pounds Per Square Inch

STD: Standard

OPT: Optional

① Torque required in ft. lbs. to rotate ball joint using a pull scale

② Lower arm ball rotation torque: 0.78-4.29
   Upper arm ball rotation torque: 0.7-5.0

③ Lower arm ball rotation torque: 0.3-11

④ Lower arm ball rotation torque: 0.7-7.7
   Upper arm ball rotation torque: 0.7-8.8
   Upper leading link rotation torque: 0.7-4.4

⑤ Standard Steel rim: P195/65R15
   Alluminum alloy: P205/55R16

⑥ Standard Steel rim: 15 x 6 J
   Alluminum alloy: 16 x 6 1/2 J or 17 x 6 1/2 J

⑦ Lower arm ball rotation torque, front: 3-10 ft. lbs.

⑧ Standard Steel rim: 16 x 6 1/2JJ
   Alluminum alloy: 16 x 7 JJ or 17 x 7 JJ

⑨ Lower arm ball rotation torque, front: 10.5-19.7 inch lbs.
   Lower arm ball rotation torque, rear: 8.86-19.6 inch lbs.
   Upper arm ball rotation torque: 13.2 inch lbs. (max)

67162-MAZC-C10

BRAKE SPECIFICATIONS
All measurements in inches unless noted

Year	Model		Brake Disc Original Thickness	Brake Disc Minimum Thickness	Brake Disc Maximum Runout	Brake Drum Original Inside Diameter	Brake Drum Max. Wear Limit	Brake Drum Maximum Machine Diameter	Minimum Lining Thickness	Brake Caliper Bracket Bolts (ft. lbs.)	Brake Caliper Mounting Bolts (ft. lbs.)
2001	626	F	0.940	0.870	0.002	—	—	—	0.080	58-75	22-28
		R	0.390	0.310	0.002	9.00	NA	9.059	0.040	34-49	26-28
	Miata	F	0.790	①	0.002	—	—	—	0.080	37-50	58-65
		R	0.350	0.310	0.002	—	—	—	0.040	37-50	26-28
	Millenia	F	1.100	1.000	0.002	—	—	—	0.080	47-62	47-62
		R	0.370	0.300	0.002	—	—	—	0.080	37-50	28-36
	Protege	F	NA	②	0.002	—	—	—	③	58-75	④
		R	—	—	—	7.87	7.993	NA	0.040	26-28	26-28
2002	626	F	0.940	0.870	0.002	—	—	—	0.080	58-75	22-28
		R	0.390	0.310	0.002	9.00	NA	9.059	0.040	34-49	26-28
	Miata	F	0.790	①	0.002	—	—	—	0.080	37-50	58-65
		R	0.350	0.310	0.002	—	—	—	0.040	37-50	26-28
	Millenia	F	1.100	1.000	0.002	—	—	—	0.080	47-62	47-62
		R	0.370	0.300	0.002	—	—	—	0.080	37-50	28-36
	Protege	F	NA	②	0.002	—	—	—	③	58-75	④
		R	NA	0.310	0.002	7.87	7.993	NA	0.040	26-28	26-28
2003-04	626	F	0.940	0.870	0.002	—	—	—	0.080	58-75	22-28
		R	0.390	0.310	0.002	9.00	NA	9.059	0.040	34-49	26-28
	Miata	F	0.790	①	0.002	—	—	—	0.080	37-50	58-65
		R	0.350	0.310	0.002	—	—	—	0.040	37-50	26-28
	Millenia	F	1.100	1.000	0.002	—	—	—	0.080	47-62	47-62
		R	0.370	0.300	0.002	—	—	—	0.080	37-50	28-36
	Mazda 3	F	NA	0.910	0.002	—	—	—	0.079	75-87	19-22
		R	NA	0.350	0.002	—	—	—	0.079	43-56	19-22
	Mazda 6	F	NA	0.910	0.002	—	—	—	0.079	58-75	36-39
		R	NA	0.310	0.002	—	—	—	0.079	37-49	27-36
	Protege	F	NA	②	0.002	—	—	—	③	58-75	④
		R	NA	0.310	0.002	7.87	7.993	NA	0.040	26-28	26-28

NA: Not Avilable

F: Front

R: Rear

① With 15 inch wheel: 0.071 in.
   With 16 inch wheel: 0.079 in.

② With 1.6L engine: 0.780 in.
   With 1.8L engine: 0.870 in.
   With 2.0L engine: 0.870 in.

③ With 1.6L engine: 0.059 in.
   With 1.8L engine: 0.080 in.
   With 2.0L engine: 0.079 in.

④ Type A: 37-39
   Type B: 16-23

67162-MAZC-C11

SCHEDULED MAINTENANCE INTERVALS
Mazda Car—Except Mazda 3 and Mazda 6

TO BE SERVICED	TYPE OF SERVICE	VEHICLE MILEAGE INTERVAL (x1000)												
		7.5	15	22.5	30	37.5	45	52.5	60	67.5	75	82.5	90	97.5
Engine oil & filter	R	✓	✓	✓	✓	✓	✓	✓	✓	✓	✓	✓	✓	✓
Air cleaner element	R				✓				✓				✓	
Engine coolant ①	R				✓				✓				✓	
Spark plugs	R				✓				✓				✓	
Automatic transaxle fluid	S/I				✓				✓				✓	
Bolts & nuts on chassis & body	S/I				✓				✓				✓	
Brake lines, hoses & connections	S/I				✓				✓				✓	
Cooling system	S/I				✓				✓				✓	
Disc brakes	S/I				✓				✓				✓	
Drive belts (Millenia ②)	S/I				✓				✓				✓	
Drive shaft dust boots	S/I				✓				✓				✓	
Exhaust system heat shield	S/I				✓				✓				✓	
Front & rear suspension ball joints	S/I				✓				✓				✓	
Fuel lines & hoses	S/I				✓				✓				✓	
Idle speed	S/I					✓			✓				✓	
Steering operation & linkages	S/I					✓			✓				✓	
Engine timing belt ③ ④	R								✓					
Fuel filter	R								✓					
Valve Clearance	I								✓					
Manual transmission	R								✓					
Hose & tube for emission	S/I								✓					

R: Replace   S/I: Service or Inspect

① Replace initially at 45,000 miles & every 30,000 miles thereafter except on 626. On 626, replace at 105,000 miles & every 30,000 miles therafter

② (Millenia KJ engine): replace every 105,000 miles

③ Except Miata and Millenia KJ engine; inspect every 60,000 miles & replace at 105,000 miles (if not replaced previously).

③ Miata and Millenia KJ engine replace at 60,000 miles (if not replaced previously).

FREQUENT OPERATION MAINTENANCE (SEVERE SERVICE)

If a vehicle is operated under any of the following conditions it is considered severe service

- Extremely dusty areas.

- 50% or more of the vehicle operation is in 32°C (90°F) or higher temperatures, or constant operation in temperatures below 0°C (32°F).

- Prolonged idling (vehicle operation in stop and go traffic).

- Frequent short running periods (engine does not warm to normal operating temperatures).

- Police, taxi, delivery usage or trailer towing usage.

Oil & oil filter: change every 5000 miles.

Oil & oil filter (Puerto Rico): change every 3000 miles.

Air cleaner element: service or inspect every 15,000 miles

Automatic transaxle fluid: service or inspect every 15,000 miles.

Bolts & nuts on chassis & body: tighten every 15,000 miles.

Disc brakes: service or inspect every 15,000 miles.

SCHEDULED MAINTENANCE INTERVALS
Mazda Car—Mazda 3 and Mazda 6

TO BE SERVICED	TYPE OF SERVICE	VEHICLE MILEAGE INTERVAL (x1000)												
		7.5	15	22.5	30	37.5	45	52.5	60	67.5	75	82.5	90	97.5
Engine oil & filter	R	✓	✓	✓	✓	✓	✓	✓	✓	✓	✓	✓	✓	✓
Air cleaner element	R					✓					✓			
Engine coolant ① ②	R													
Spark plugs	R										✓			
Bolts & nuts on chassis & body	S/I				✓				✓				✓	
Brake lines, hoses & connections	S/I				✓				✓				✓	
Cooling system	S/I				✓				✓				✓	
Disc brakes	S/I				✓				✓				✓	
Drive belts ③	S/I				✓				✓				✓	
Drive shaft dust boots	S/I				✓				✓				✓	
Exhaust system heat shield	S/I				✓				✓				✓	
Front & rear suspension ball joints	S/I				✓				✓				✓	
Fuel lines & hoses	S/I				✓				✓				✓	
Steering operation & linkages	S/I				✓				✓				✓	
Fuel filter	R								✓					
Valve Clearance ④	I								✓					
Hose & tube for emission	S/I								✓					

R: Replace   S/I: Service or Inspect

① Mazda 6: replace initially at 105,000 miles or 6 months and every 30, 000 miles or 24 months thereafter

① Mazda 3: replace initially at 60,000 miles or 4 years and every 24 months thereafter

③ 2.0L LF engine and 2.3L L3 engine: inspect every 37, 500 miles. 3.0L AJ engine: inspect every 30,000 miles

④ 2.0L LF and 2.3L L3 engine: Audible inspect every 75, 000 miles and if noisy adjust.

FREQUENT OPERATION MAINTENANCE (SEVERE SERVICE)

If a vehicle is operated under any of the following conditions it is considered severe service

- Extremely dusty areas.

- 50% or more of the vehicle operation is in 32°C (90°F) or higher temperatures, or constant operation in temperatures below 0°C (32°F).

- Prolonged idling (vehicle operation in stop and go traffic).

- Frequent short running periods (engine does not warm to normal operating temperatures).

- Police, taxi, delivery usage or trailer towing usage.

Oil & oil filter: change every 5000 miles.

Oil & oil filter (Puerto Rico): change every 3000 miles.

Air cleaner element: service or inspect every 15,000 miles

Automatic transaxle fluid: service or inspect every 15,000 miles.

Bolts & nuts on chassis & body: tighten every 15,000 miles.

Disc brakes: service or inspect every 15,000 miles.

67162-MAZC-C13

PRECAUTIONS

Before servicing any vehicle, please be sure to read all of the following precautions, which deal with personal safety, prevention of component damage, and important points to take into consideration when servicing a motor vehicle:

• Never open, service or drain the radiator or cooling system when the engine is hot; serious burns can occur from the steam and hot coolant.

• Observe all applicable safety precautions when working around fuel. Whenever servicing the fuel system, always work in a well-ventilated area. Do not allow fuel spray or vapors to come in contact with a spark, open flame or excessive heat (a hot drop light, for example). Keep a dry chemical fire extinguisher near the work area. Always keep fuel in a container specifically designed for fuel storage; also, always properly seal fuel containers to avoid the possibility of fire or explosion. Refer to the additional fuel system precautions later in this section.

• Fuel injection systems often remain pressurized, even after the engine has been turned **OFF**. The fuel system pressure must be relieved before disconnecting any fuel lines. Failure to do so may result in fire and/or personal injury.

• Brake fluid often contains polyglycol ethers and polyglycols. Avoid contact with the eyes and wash your hands thoroughly after handling brake fluid. If you do get brake fluid in your eyes, flush your eyes with clean, running water for 15 minutes. If eye irritation persists, or if you have taken brake fluid internally, IMMEDIATELY seek medical assistance.

• The EPA warns that prolonged contact with used engine oil may cause a number of skin disorders, including cancer! You should make every effort to minimize your exposure to used engine oil. Protective gloves should be worn when changing oil. Wash your hands and any other exposed skin areas as soon as possible after exposure to used engine oil. Soap and water, or waterless hand cleaner should be used.

• All new vehicles are now equipped with an air bag system. The system must be disabled before performing service on or around system components, steering column, instrument panel components, wiring and sensors. Failure to follow safety and disabling procedures could result in accidental air bag deployment, possible personal injury and unnecessary system repairs.

• Always wear safety goggles when working with, or around, the air bag system. When carrying a non-deployed air bag, be sure the bag and trim cover are pointed away from your body. When placing a non-deployed air bag on a work surface, always face the bag and trim cover upward, away from the surface. This will reduce the motion of the module if it is accidentally deployed. Refer to the additional air bag system precautions later in this section.

• Clean, high quality brake fluid from a sealed container is essential to the safe and proper operation of the brake system. You should always buy the correct type of brake fluid for your vehicle. If the brake fluid becomes contaminated, completely flush the system with new fluid. Never reuse any brake fluid. Any brake fluid that is removed from the system should be discarded. Also, do not allow any brake fluid to come in contact with a painted surface; it will damage the paint.

• Never operate the engine without the proper amount and type of engine oil; doing so WILL result in severe engine damage.

• Timing belt maintenance is extremely important! Many models utilize an interference-type, non-freewheeling engine. If the timing belt breaks, the valves in the cylinder head may strike the pistons, causing potentially serious (also time-consuming and expensive) engine damage.

• Disconnecting the negative battery cable on some vehicles may interfere with the functions of the on-board computer system(s) and may require the computer to undergo a relearning process once the negative battery cable is reconnected.

• When servicing drum brakes, only disassemble and assemble one side at a time, leaving the remaining side intact for reference.

• Only an MVAC-trained, EPA-certified automotive technician should service the air conditioning system or its components.

ENGINE REPAIR

Distributor

REMOVAL

1. Before servicing the vehicle, refer to the precautions in the beginning of this section.

2. Remove or disconnect the following:
   • Negative battery cable
   • Distributor cap and position it aside, leaving the ignition wires connected
   • Distributor electrical connector(s) from the side of the distributor

3. Using a wrench on the crankshaft pulley, rotate the crankshaft to position the No. 1 piston on Top Dead Center (TDC) of the compression stroke; the crankshaft pulley mark should align with the timing indicator and the distributor rotor should point towards the No. 1 spark plug wire tower position of the cap.

4. Using chalk or paint, mark the position of the distributor housing on the cylinder head. Also mark the position of the distributor rotor in relation to the distributor housing.

5. Remove distributor hold-down bolt(s).

6. On distributors attached to the end of the cylinder head (or inline with the camshaft), remove it by pulling it straight outward.

7. On distributors attached to the side of the cylinder head (or perpendicular with the

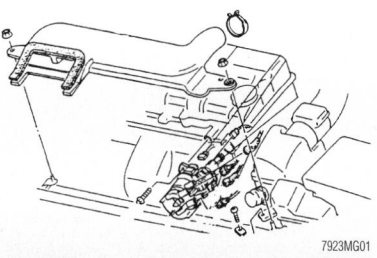

Exploded view of a typical side mounted distributor

7923MG01

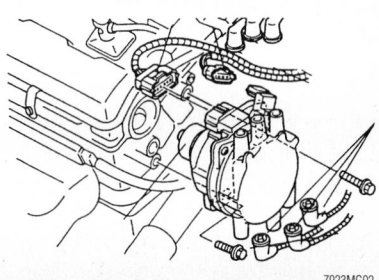

Exploded view of a typical end or inline mounted distributor

7923MG02

camshaft), slowly pull it outward while watching the rotor. These distributors are gear driven and as you remove it, the gears will disengage inside the engine, causing the rotor to rotate. when the rotor stops moving, stop pulling outward. Re-align the distributor body-to-cylinder head match-mark (do not push it back in to do this, simply rotate the body to align the marks). Place a third mark indicating the new rotor position-to-distributor body relation. When installing the distributor, align this mark and the body-to-head mark to properly position the distributor.

8. Inspect the O-ring on the distributor housing and replace it, if it is damaged or worn.

INSTALLATION

Engine Not Disturbed

1. Using engine oil, lubricate the O-ring.
2. Install or connect the following:
   • Distributor

➡**Be sure to engage the distributor drive gear or tangs with the camshaft gear or slot. Align the mark that was made on the distributor housing with the mark that was made on the cylinder head.**

   • Distributor hold-down bolt(s)
   • Electrical connector(s)
   • Distributor cap
   • Negative battery cable
3. Start the engine and check or adjust the ignition timing.

Engine Disturbed

1. Remove or disconnect the following:
   • Spark plug wire from the No. 1 cylinder spark plug
   • Spark plug from the No. 1 cylinder
2. Press a thumb over the spark plug hole.
3. Using a wrench on the crankshaft pulley, rotate the crankshaft until pressure is felt at the spark plug hole, indicating the piston is approaching TDC on the compression stroke. Continue rotating the crankshaft until the crankshaft pulley mark aligns with the timing cover indicator.
4. Place the distributor rotor in position so that it aligns with the No. 1 spark plug wire tower on the distributor cap.
5. Using engine oil, lubricate the O-ring.
6. Install or connect the following:
   • Distributor

➡**Be sure to engage the distributor drive gear or tangs with the camshaft**

gear or slot. Align the mark that was made on the distributor housing with the mark that was made on the cylinder head.

   • Distributor hold-down bolt(s)
   • Electrical connector(s)
   • Distributor cap
   • Spark plug in the No. 1 cylinder and connect the spark plug wire
   • Negative battery cable
7. Start the engine and check or adjust the ignition timing.

Alternator

REMOVAL

Miata

1. Before servicing the vehicle, refer to the precautions in the beginning of this section.
2. Remove or disconnect the following:
   • Negative battery cable
   • Intake manifold bracket
   • Electrical connectors from the alternator
   • Alternator bolts
   • Alternator

Mazda 3

1. Before servicing the vehicle, refer to the precautions in the beginning of this section.
2. Remove the battery cover and disconnect the negative battery cable.
3. Remove the under cover and splash shield as an assembly.
4. Remove the plug hole plate.
5. Remove the drive belt.
6. Disconnect the alternator electrical connections.

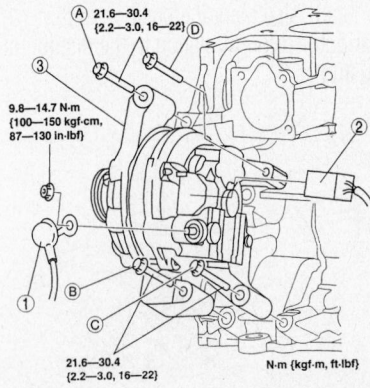

21.6—30.4
{2.2—3.0, 16—22}

9.8—14.7 N·m
{100—150 kgf·cm, 87—130 in·lbf}

21.6—30.4
{2.2—3.0, 16—22}

N·m {kgf·m, ft·lbf}

67162-MAZC-G01

Location of alternator mounting bolts A, B, C, and D—Mazda 3 models

7. Position the coolant overflow tank to one side to facilitate alternator removal.
8. Remove the bolts A, B, C, and D in order and the alternator. Refer to the illustration for bolt location.

Mazda 6

2.3L (L3) ENGINE

1. Before servicing the vehicle, refer to the precautions in the beginning of this section.
2. Before servicing the vehicle, refer to the precautions in the beginning of this section.
3. Disconnect the negative battery cable.
4. Remove the under cover.
5. Remove the drive belt.
6. Remove the alternator duct and heat insulator.
7. Disconnect the alternator electrical connections.
8. Remove the bolts A, B and C. Refer to the illustration for bolt location.

3.0L (AJ) ENGINE

1. Before servicing the vehicle, refer to the precautions in the beginning of this section.
2. Before servicing the vehicle, refer to the precautions in the beginning of this section.
3. Disconnect the negative battery cable.
4. Remove the under cover.
5. Remove the drive belt.
6. Remove the right hand three way catalytic converter and front pipe.
7. Disconnect the alternator electrical connections.
8. Remove the bolts and the alternator.

Protégé

1. Before servicing the vehicle, refer to the precautions in the beginning of this section.
2. Remove or disconnect the following:
   • Negative battery cable
   • Electrical connectors from the alternator
   • Alternator drive belt
   • Alternator pivot and adjusting bar bolts
   • Alternator

626

2.0L (FS) ENGINES

1. Before servicing the vehicle, refer to the precautions in the beginning of this section.
2. Remove or disconnect the following:

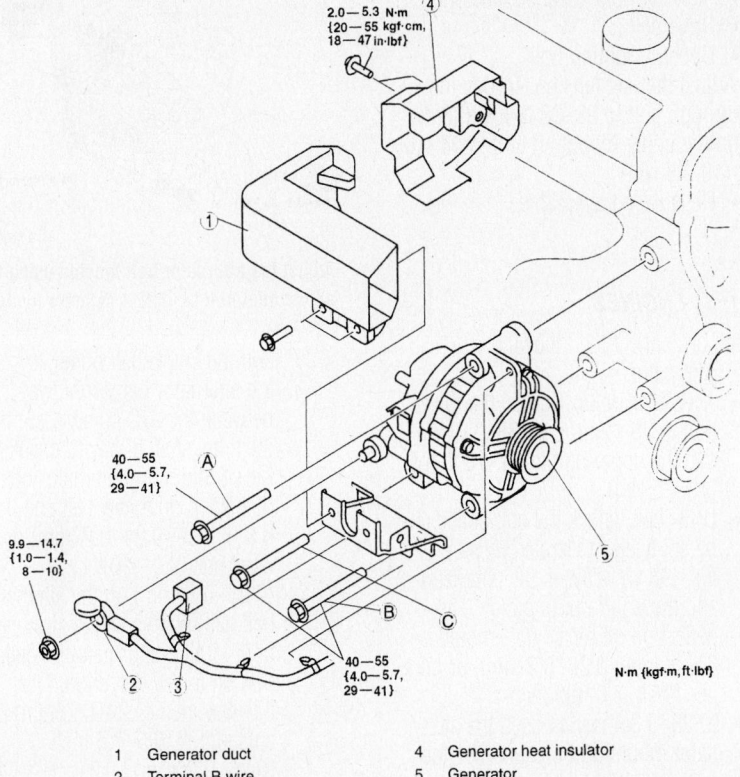

1 Generator duct
2 Terminal B wire
3 Generator connector
4 Generator heat insulator
5 Generator

67162-MAZC-G110

Location of alternator mounting bolts A, B and C—Mazda 6 models with 2.3L (L3) engine

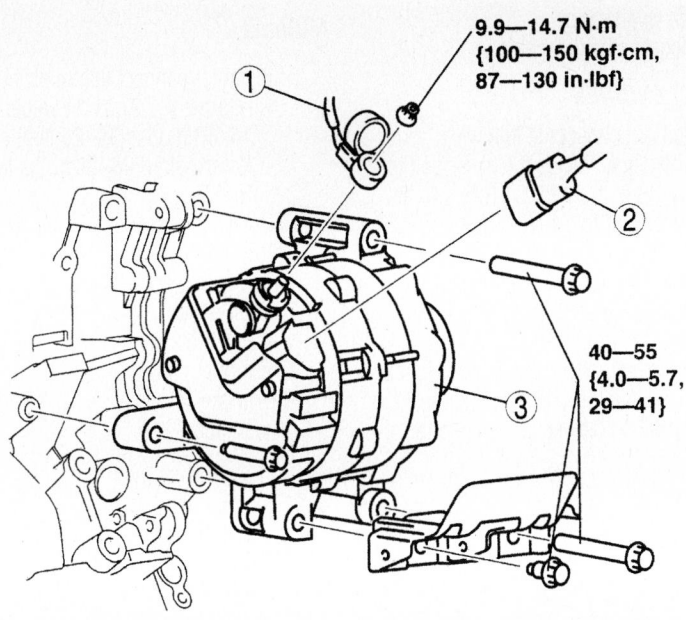

N·m {kgf·m, ft·lbf}

1 Terminal B wire
2 Connector
3 Generator

67162-MAZC-G111

Alternator mounting—Mazda 6 models with 3.0L (AJ) engine

- Negative battery cable
- Alternator upper mounting bolt
- Alternator adjusting bolt
- Drive belt from the alternator pulley
- Transverse member
- Electrical connectors from the alternator
- Front exhaust pipe at the catalytic converter and suspend it on a piece of wire
- Oxygen Sensor (O$_2$S)
- 3 exhaust manifold flange nuts and the hold-down bracket clamp
- Exhaust pipe
- Alternator lower through-bolt
- Alternator

2.5L (KL) ENGINES

1. Before servicing the vehicle, refer to the precautions in the beginning of this section.

2. Remove or disconnect the following:
- Negative battery cable
- Fresh air duct
- Radiator upper bracket
- Condenser fan, if equipped
- Electrical connectors from the alternator
- Loosen the belt tensioner locknut and tension adjusting bolt
- Alternator upper mounting bolt
- Right splash shield
- Drive belt from the alternator pulley
- A/C compressor and support it aside, leaving the refrigerant lines connected, if necessary
- Alternator through-bolt
- Alternator

Millenia

1. Before servicing the vehicle, refer to the precautions in the beginning of this section.

2. Disconnect the negative battery cable.

3. If equipped with the 2.3L (KJ) engine, remove the front charge air cooler, radiator upper seal board and the condenser fan assembly.

4. Remove or disconnect the following:
- Intake air system on 2.4 (KL) engine
- Electrical connectors from the alternator
- Right splash shield
- Drive belt from the alternator pulley
- A/C compressor and support it aside, leaving the refrigerant lines connected
- Upper and lower alternator mounting bolts
- Alternator

INSTALLATION

Miata

1. Install or connect the following:
   - Alternator
   - Alternator bolts. Tighten the lower bolts to 38 ft. lbs. (51 Nm) and the upper bolt 18 ft. lbs. (25 Nm).
   - Electrical connectors to the alternator
   - Intake manifold bracket
   - Negative battery cable

Mazda 3

1. Place the alternator in position matching the fixing and engine side holes and hand tighten the bolts A, B, C and D in alphabetical order. Refer to the illustration for bolt locations.

2. Attach the alternator electrical connections. Tighten bolts A, B, C and D in alphabetical order to 16–22 ft. lbs. (21–30 Nm). Refer to the illustration for bolt locations.

3. Install the drive belt.

4. Install the under cover and splash shield.

5. Install the plug hole plate.

6. Connect the negative battery cable and install the battery cover.

Mazda 6

2.3L (L3) ENGINE

1. Install the alternator, tighten bolt A temporarily, then tighten the bolts in the following order; B, A and C to 29–41 ft. lbs. (40–55 Nm).

2. Connect the alternator electrical connections.

3. Install the heat insulator and tighten the bolt to 18–47 inch lbs. (2–5 Nm).

4. Install the alternator duct

5. Install the under cover.

6. Install the drive belt.

7. Connect the negative battery cable.

3.0L (AJ) ENGINE

1. Install the alternator and tighten the bolts to 29–41 ft. lbs. (40–55 Nm).

2. Connect the alternator electrical connections.

3. Install the right hand three way catalytic converter and front pipe.

4. Install the drive belt.

5. Install the under cover.

6. Connect the negative battery cable.

Protégé

1. Install or connect the following:
   - Alternator with the pivot bolt

- Alternator electrical connectors
- Drive belt
- Upper mounting bolt

2. Adjust the belt tension. Torque the lower through bolt to 28–38 ft. lbs. (38–51 Nm) and the upper mounting bolt to 14–18 ft. lbs. (19–26 Nm).
   - Negative battery cable

626

2.0L (FS) ENGINES

Install or connect the following:
- Alternator
- Alternator bolts. Tighten the lower bolt to 28–38 ft. lbs. (38–51 Nm) and the upper bolt to 14–18 ft. lbs. (19–25 Nm).
- Drive belt, check the belt deflection by applying moderate pressure 22 lbs. (98 N) between the alternator and the water pump pulley. The deflection on a new belt should be 6.5–7 inch (0.26–0.27mm) or on a used belt, 7.0–9.0 inch (0.28–0.35mm). Loosen the alternator mounting bolts and use the adjusting bolt to adjust the belt tension to the correct specification.
- Electrical connector and terminal wire
- Front exhaust pipe
- Transverse pipe
- Negative battery cable

2.5L (KL) ENGINES

1. Install or connect the following:
   - Alternator. Torque the lower bolt to 28–38 ft. lbs. (38–51 Nm) and the upper bolt to 14–18 ft. lbs. (19–25 Nm).
   - A/C compressor, if removed. Torque the bolts to 11–15 ft. lbs. (15–21 Nm).
   - Drive belt, check the belt deflection by applying moderate pressure 22 lb. (98 N) between the alternator

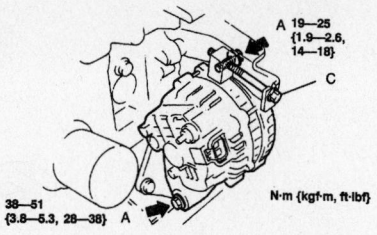

N·m {kgf·m, ft-lbf}

42356-MAZC-G01

Adjust the alternator belt tension using the adjustment bolt C—626 4 cylinder models

and the crankshaft pulley to check the alternator belt or midway between the A/C compressor and the crankshaft pulley to check the Air conditioning/alternator belt. The deflection on a new belt should be 6.0–7 inch (0.24–0.27mm) or on a used belt, 7.0–8.0 inch (0.28–0.31mm) on the alternator belt. On models with a air conditioning/alternator belt; the deflection on a new belt should be 5.5–6.5 inch (0.22–0.25mm) or on a used belt, 6.5–7.5 inch (0.26–0.29mm) Loosen the alternator mounting bolts and use the adjusting bolt to adjust the belt tension to the correct specification.
- Negative battery cable

Millenia

1. Install or connect the following:
   - Alternator. Torque the upper bolt to 14–18 ft. lbs. (19–25 Nm) and the lower bolt to 28–38 ft. lbs. (38–51 Nm).
   - A/C compressor. Torque the bolts to 12–16 ft. lbs. (16–22 Nm).
   - Right splash shield
   - Drive belt
   - Electrical connectors to the alternator
   - Intake air system on 2.4 (KL) engine

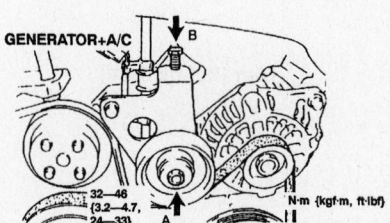

GENERATOR+A/C

32–46 {3.2–4.7, 24–33} N·m {kgf·m, ft-lbf}

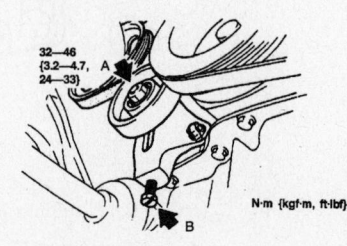

P/S

32–46 {3.2–4.7, 24–33} A

N·m {kgf·m, ft-lbf}

42356-MAZC-G02

Adjust the alternator belt tension using the adjustment bolt B—626 6 cylinder models

2. If equipped with the 2.3L engine, install the condenser fan assembly, radiator upper seal board and the front charge air cooler using new O-rings. Torque the mounting bolts to 12–16 ft. lbs. (16–22 Nm).

3. Connect the negative battery cable.

Ignition Timing

ADJUSTMENT

Except Millenia 2.5L (KL) Engines

1. Before servicing the vehicle, refer to the precautions in the beginning of this section.

➡**If the information given in the following procedures differs from that on the emission information label located in the engine compartment, follow the directions given on the label. The label often reflects production changes made during the model year.**

The timing is controlled by the computer. Ignition timing adjustment is not possible or necessary.

2. If the timing is still not within specification. the following components may be defective:

- Camshaft position (CMP) sensor
- Crankshaft Position (CKP) sensor
- Throttle Position (TP) sensor
- Engine Coolant Temperature (ECT) sensor
- Neutral switch if equipped with a manual transaxle
- Clutch switch if equipped with a manual transaxle
- Transaxle range switch if equipped with an automatic transaxle

3. If the above components are normal, replace the Powertrain Control Module (PCM).

Millenia 2.5L (KL) Engine

1. Before servicing the vehicle, refer to the precautions in the beginning of this section.

2. Let the engine warm to normal operating temperature.

3. Apply the parking brake. If equipped with a manual transaxle, place the shifter in the neutral position. If equipped with an automatic transaxle, place the shift lever in **P**.

4. Start the engine and allow it to come to normal operating temperature. Be sure all accessories are **OFF**.

5. Wait until the cooling fan stops,

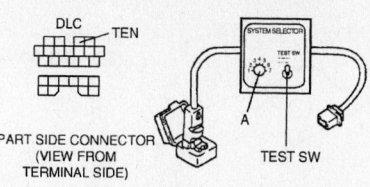

Jumper the connections shown on the data link connector and system selector tool

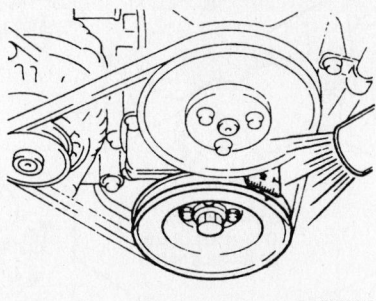

Connect an inductive timing light and aim it at the crankshaft pulley. Read the pulley mark against the scale

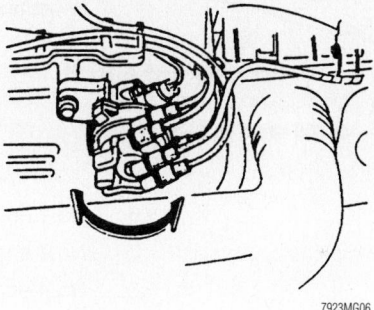

If adjustment is necessary, loosen the distributor lockbolts and rotate it until the mark is aligned

then, connect the scan tool to the Data Link Connector 2 (DLC2) and access the RPM PID.

6. Locate the timing marks on the crankshaft pulley and timing belt lower cover. The engine may have to be cranked slightly to see the mark on the crankshaft pulley.

7. Check the idle speed and adjust, if necessary.

8. Connect a jumper wire between the TEN terminal and the GND terminal at the underhood diagnosis connector.

9. Connect the system selector tool to the DLC and set switch **A** to position **1**.

10. Set the test switch to **SELF TEST**.

11. Connect an inductive timing light according to the manufacturer's instructions.

12. Start the engine and allow the idle to stabilize. Aim the timing light at the timing marks. The timing should be 9–11 degrees Before Top Dead Center (TDC).

13. Loosen the distributor lockbolts just enough to turn the distributor. While aiming the timing light at the timing marks, turn the distributor until the marks are aligned. Tighten the distributor lockbolts to 14–18 ft. lbs. (19–25 Nm) and recheck the timing.

14. The ignition timing is now set. Disconnect the jumper wire from the DLC

15. Remove all test equipment.

Engine Assembly

REMOVAL & INSTALLATION

626

2.0L (FS) ENGINE

➡**The procedure for pulling the engine requires removing the transaxle along with it. As a result, when the halfshafts are pulled from the transaxle, a special plug/side gear holding tool is recommended.**

1. Before servicing the vehicle, refer to the precautions in the beginning of this section.

2. Properly relieve the fuel system pressure.

3. Drain the engine oil.

4. Drain the transaxle fluid.

5. Drain the cooling system.

6. Remove or disconnect the following:
- Negative battery cable
- Radiator
- Air cleaner assembly
- Accelerator cable
- Fuel hoses
- Front exhaust pipe
- Any rods, pipes or cables related to the transaxle that would hinder removal
- Battery
- Fuse box
- Power steering pump with the lines still attached and set aside
- A/C compressor with the lines still attached and position aside
- Cruise actuator connector, actuator retainers and the actuator
- Halfshafts
- Number 5 engine mount retainers. Refer to the accompanying illustration for location.
- Engine mount member (2) retainers. Refer to the accompanying illustration for location.
- Number 1 engine mount stay

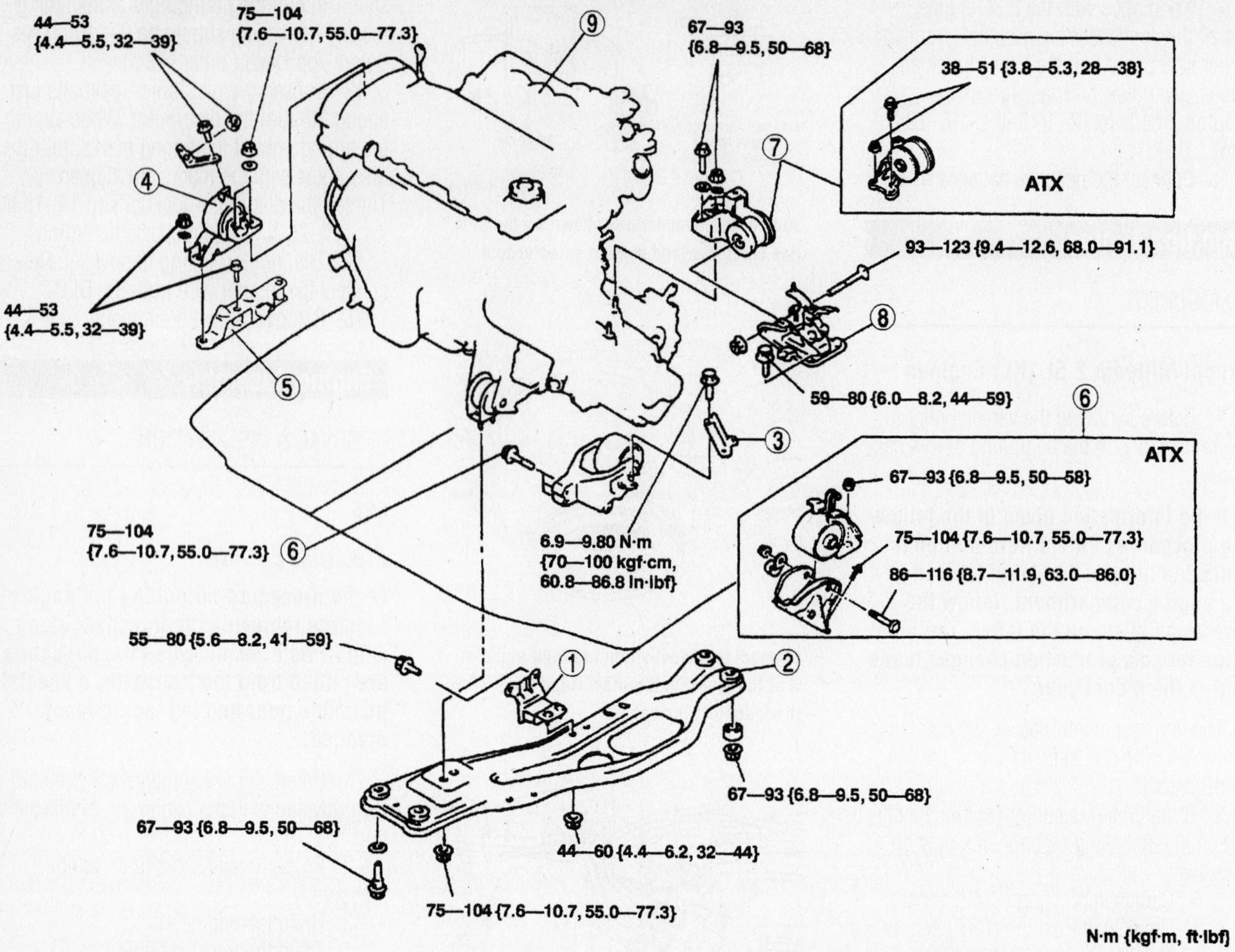

44—53 {4.4—5.5, 32—39}

75—104 {7.6—10.7, 55.0—77.3}

67—93 {6.8—9.5, 50—68}

38—51 {3.8—5.3, 28—38}

ATX

93—123 {9.4—12.6, 68.0—91.1}

44—53 {4.4—5.5, 32—39}

59—80 {6.0—8.2, 44—59}

67—93 {6.8—9.5, 50—58}

ATX

75—104 {7.6—10.7, 55.0—77.3}

86—116 {8.7—11.9, 63.0—86.0}

75—104 {7.6—10.7, 55.0—77.3}

6.9—9.80 N·m {70—100 kgf·cm, 60.8—86.8 In·lbf}

55—80 {5.6—8.2, 41—59}

67—93 {6.8—9.5, 50—68}

67—93 {6.8—9.5, 50—68}

44—60 {4.4—6.2, 32—44}

75—104 {7.6—10.7, 55.0—77.3}

N·m {kgf·m, ft·lbf}

1	No.5 engine mount rubber	6	No.1 engine mount bolt and nut
2	Engine mount member	7	No.4 engine mount rubber
3	No.1 engine mount stay bracket	8	No.4 engine mount bracket
4	No.3 engine mount rubber	9	Engine, transaxle
5	No.3 engine bracket		

42356-MAZC-G03

Location of the engine mounting components and their torque specifications—626 models equipped with the 2.0L (FS) engine

bracket retainers. Refer to the accompanying illustration for location.

• Number 3 engine mount rubber retainers. Refer to the accompanying illustration for location.

• Number 3 engine bracket retainers. Refer to the accompanying illustration for location.

• Number 1 engine mount nut and bolt. Refer to the accompanying illustration for location.

• Number 4 engine mount rubber retainers. Refer to the accompanying illustration for location.

• Number 4 engine bracket retainers. Refer to the accompanying illustration for location.

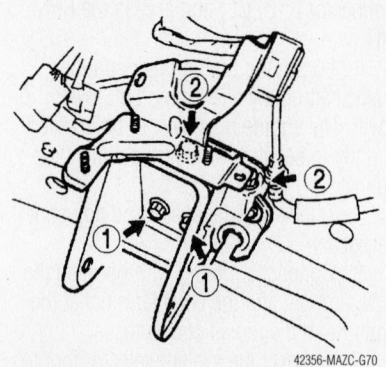

42356-MAZC-G70

Tighten the number 4 engine bracket retainers in the sequence shown—626 models equipped with the 2.0L (FS) engine

• Engine and transaxle assembly from the vehicle.

To install:

7. Installation is the reverse of removal. Tighten the fasteners to the specifications shown in the accompanying illustration.

8. When possible, leave the engine mounting nuts/bolts loose (hand tight) until all mounts are aligned and bolted. This may help in aligning the engine and transmission assembly in the vehicle.

9. Install or connect the following:

• Number 4 engine bracket retainers. Tighten the fasteners to the specifications shown in the accompanying illustration.

• Number 4 engine mount rubber retainers. Tighten the fasteners to

Free play
1.5—4.0 mm {0.06—0.15 in}

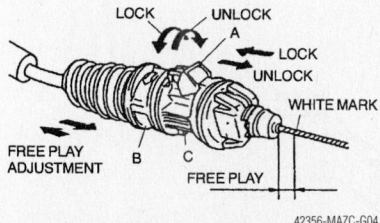

Accelerator cable adjustment components—626 models equipped with the 2.0L (FS) engine

the specifications shown in the accompanying illustration.

- Number 1 engine mount nut and bolt. Tighten the fasteners to the specifications shown in the accompanying illustration.
- Number 3 engine bracket retainers. Tighten the fasteners to the specifications shown in the accompanying illustration.
- Number 3 engine mount rubber retainers. Tighten the fasteners to the specifications shown in the accompanying illustration.
- Number 1 engine mount stay bracket retainers. Tighten the fasteners to the specifications shown in the accompanying illustration.
- Engine mount member (2) retainers. Tighten the fasteners to the specifications shown in the accompanying illustration.
- Number 5 engine mount retainers. Tighten the fasteners to the specifications shown in the accompanying illustration.

10. When connecting the accelerator cable, perform the following adjustment:

a. Move the white locking tab to the unlock position.

b. Turn stopper B to the to the unlock position.

➡**If stopper B will not unlock, it may be necessary to carefully bend back tab C out a little using a suitable pry tool.**

c. Push or pull the cable housing directly behind the spring.

d. Turn the stopper B to the lock position.

e. Measure the free play which should be 0.06–0.15 inch (1.5–4mm) and make sure the cable free play is within specification.

f. Move the white locking tab to the

lock position and check for proper accelerator operation.

11. Fill the engine and the transaxle with the proper type and amount of fluids. Fill the cooling system.

12. Connect the negative battery cable, start the engine and check for leaks.

13. Check the ignition timing and the idle speed.

14. Check all fluid levels.

2.5L (KL) ENGINE

➡**The procedure for pulling the engine requires removing the transaxle along with it. As a result, when the halfshafts are pulled from the transaxle, a special plug/side gear holding tool is recommended.**

1. Before servicing the vehicle, refer to the precautions in the beginning of this section.

2. Properly relieve the fuel system pressure.

3. Drain the engine oil.

4. Drain the transaxle fluid.

5. Drain the cooling system.

6. Remove or disconnect the following:

- Negative battery cable
- Radiator
- Air cleaner assembly
- Accelerator cable
- Fuel hoses
- Front exhaust pipe
- Any rods, pipes or cables related to the transaxle that would hinder removal
- Battery
- Power steering pump with the lines still attached and set aside
- A/C compressor with the lines still attached and position aside
- Cruise actuator connector, actuator retainers and the actuator
- Halfshafts
- Number 5 engine mount retainers. Refer to the accompanying illustration for location.
- Engine mount member (2) retainers. Refer to the accompanying illustration for location.
- Number 1 engine mount stay bracket retainers. Refer to the accompanying illustration for location.
- Number 3 engine mount rubber retainers. Refer to the accompanying illustration for location.
- Number 1 engine mount nut and bolt. Refer to the accompanying illustration for location.

- Number 4 engine mount rubber retainers. Refer to the accompanying illustration for location.
- Number 4 engine bracket retainers. Refer to the accompanying illustration for location.
- Engine and transaxle assembly from the vehicle.

To install:

7. Installation is the reverse of removal. Tighten the fasteners to the specifications shown in the accompanying illustration.

8. When possible, leave the engine mounting nuts/bolts loose (hand tight) until all mounts are aligned and bolted. This may help in aligning the engine and transmission assembly in the vehicle.

9. Install or connect the following:

- Number 4 engine bracket retainers. Tighten the fasteners to the specifications shown in the accompanying illustration.
- Number 4 engine mount rubber retainers. Tighten the fasteners to the specifications shown in the accompanying illustration.
- Number 1 engine mount nut and bolt. Tighten the fasteners to the specifications shown in the accompanying illustration.
- Number 3 engine mount rubber retainers. Tighten the fasteners to the specifications shown in the accompanying illustration.
- Number 1 engine mount stay bracket retainers. Tighten the fasteners to the specifications shown in the accompanying illustration.
- Engine mount member (2) retainers. Tighten the fasteners to the specifications shown in the accompanying illustration.
- Number 5 engine mount retainers. Tighten the fasteners to the specifications shown in the accompanying illustration.

10. When connecting the accelerator cable, perform the following adjustment:

a. Move the white locking tab to the unlock position.

b. Turn stopper B to the to the unlock position.

➡**If stopper B will not unlock, it may be necessary to carefully bend back tab C out a little using a suitable pry tool.**

c. Push or pull the cable housing directly behind the spring.

d. Turn the stopper B to the lock position.

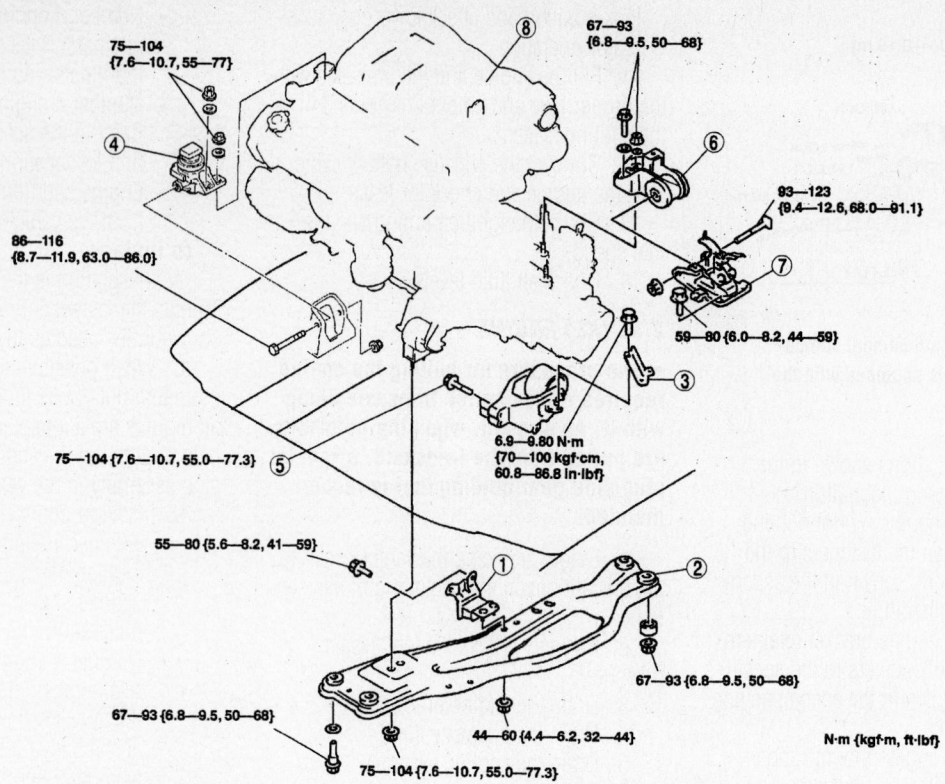

75—104
{7.6—10.7, 55—77}

67—93
{6.8—9.5, 50—68}

86—116
{8.7—11.9, 63.0—86.0}

93—123
{9.4—12.6, 68.0—81.1}

59—80 {6.0—8.2, 44—59}

75—104 {7.6—10.7, 55.0—77.3}

6.9—9.80 N·m
{70—100 kgf·cm,
60.8—86.8 in·lbf}

55—80 {5.6—8.2, 41—59}

67—93 {6.8—9.5, 50—68}

67—93 {6.8—9.5, 50—68}

44—60 {4.4—6.2, 32—44}

75—104 {7.6—10.7, 55.0—77.3}

N·m {kgf·m, ft·lbf}

1	No.5 engine mount rubber	
2	Engine mount member	
3	No.1 engine mount stay bracket	
4	No.3 engine mount rubber	
5	No.1 engine mount bolt and nut	
6	No.4 engine mount rubber	
7	No.4 engine mount bracket	
8	Engine, transaxle	

No.4 Engine Mount Bracket Installation Note
● Tighten the bolt in the order shown.

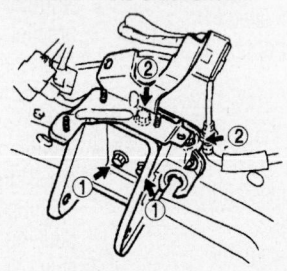

42356-MAZC-G05

Location of the engine mounting components and their torque specifications—626 models equipped with the 2.5L (KL) engine

Free play
1.5—4.0 mm {0.06—0.15 in}

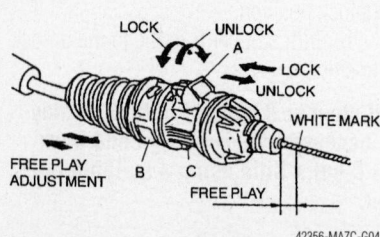

LOCK UNLOCK
A
LOCK
UNLOCK
WHITE MARK
FREE PLAY
ADJUSTMENT
B C
FREE PLAY

42356-MAZC-G04

Accelerator cable adjustment components—626 models equipped with the 2.5L (KL) engine

e. Measure the free play which should be 0.06–0.15 inch (1.5–4mm) and make sure the cable free play is within specification.

f. Move the white locking tab to the lock position and check for proper accelerator operation.

11. Fill the engine and the transaxle with the proper type and amount of fluids. Fill the cooling system.

12. Connect the negative battery cable, start the engine and check for leaks.

13. Check the ignition timing and the idle speed.

14. Check all fluid levels.

Millenia

2.5L (KL) ENGINE

➡**The procedure for pulling the engine requires removing the transaxle along with it. As a result, when the halfshafts are pulled from the transaxle, a special plug/side gear holding tool is recommended.**

1. Before servicing the vehicle, refer to the precautions in the beginning of this section.

2. Properly relieve the fuel system pressure.

3. Drain the engine oil.
4. Drain the transaxle fluid.
5. Drain the cooling system.
6. Remove or disconnect the following:
   - Hood
   - Front wheels
   - Splash shield
   - Battery clamp, cover, battery, battery carrier and battery air duct
   - Air cleaner assembly
   - Upper seal board
   - Radiator reservoir hose and the reservoir

- Condenser fan motor connector and the fan
- Cooling fan connector and fan
- Oil cooler hose
- Radiator hose, bracket and the radiator
- Accelerator cable
- Drive belt
- A/C compressor with the lines still attached and position aside
- Power steering pump with the lines still attached and position aside
- Selector cable

- Front exhaust pipe
- Tie rod end ball joint
- Upper and lower ball joints
- Halfshafts
- Joint shaft

7. Support the engine using a suitable support device.
   - Number 1 engine mount stay. Refer to the accompanying illustration for location.
   - Number 1 engine mount bracket. Refer to the accompanying illustration for location.

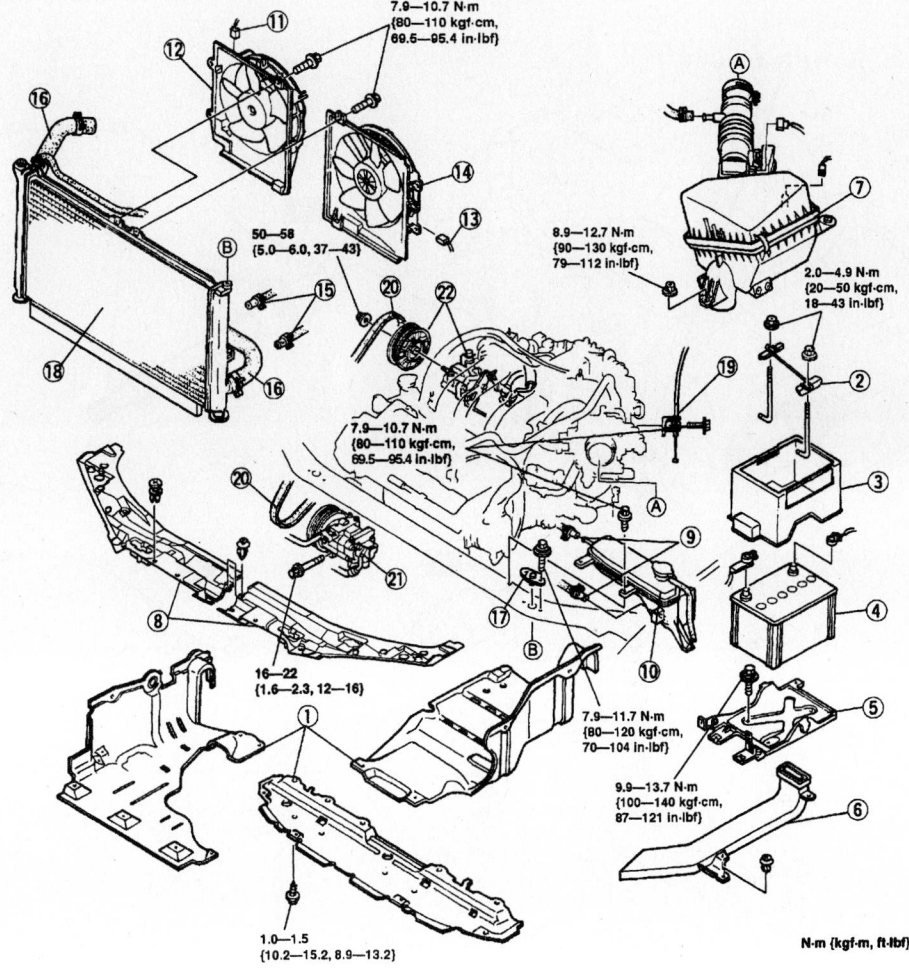

1	Splash shield	
2	Battery clamp	
3	Battery cover	
4	Battery	
5	Battery carrier	
6	Battery air duct	
7	Air cleaner component	
8	Upper seal board	
9	Radiator reservoir hose	
10	Radiator reservoir	
11	Condenser fan motor connector	
12	Condenser fan component	
13	Cooling fan motor connector	
14	Cooling fan component	
15	Oil cooler hose	
16	Radiator hose	
17	Radiator bracket	
18	Radiator	
19	Accelerator cable	
20	Drive belt	
21	A/C compressor	
22	P/S oil pump	

42356-MAZC-GDA

Location of the engine mounting components and their torque specifications (part 1 of 3)—Millenia models equipped with the 2.5L (KL) engine

• Engine mount member retainers. Remove the bolts **A** first and then the bolts **B** and remove the member. Refer to the accompanying illustration for location.

✳✳ CAUTION

Engine load can damage the number 4 mount bolts holes when removing the bolts so make sure all weight is off the mount.

8. Attach a hoist to the engine, take the weight of the engine with the hoist and remove the engine support device, Lift up the engine/transaxle assembly slightly until the number 3 and 4 mounts are free from engine weight.

• Number 4 engine mount bracket. Refer to the accompanying illustration for location.
• Number 3 engine mount sub bracket. Refer to the accompanying illustration for location.
• Engine and transaxle assembly from the vehicle. Be careful the powertrain assembly does not swing and strike the vehicle causing damage

To install:

9. Installation is the reverse of removal. Tighten the fasteners to the specifications shown in the accompanying illustration.

10. When possible, leave the engine mounting nuts/bolts loose (hand tight) until all mounts are aligned and bolted. This may help in aligning the engine and transmission assembly in the vehicle.

11. Install or connect the following:
• Number 4 engine bracket retainers and hand tighten the retainers

12. Using the hoist lower the powertrain assembly and hand tighten the number 3 mount sub bracket retainers

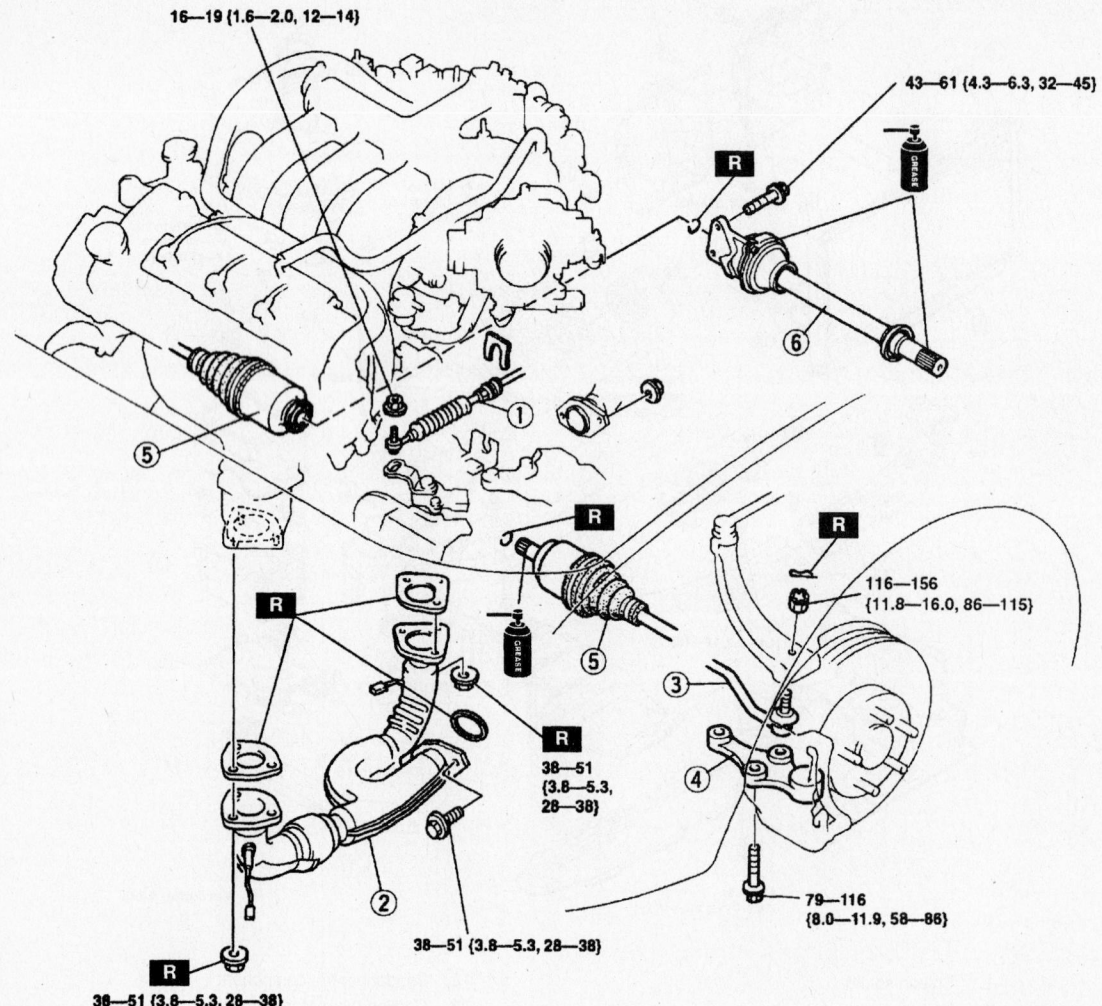

N·m {kgf·m, ft·lbf}

1	Selector cable	5	Drive shaft
2	Front pipe	6	Joint shaft
3	Upper lateral link ball joint		
4	Lower ball joint		

42356-MAZC-GDB

Location of the engine mounting components and their torque specifications (part 2 of 3)—Millenia models equipped with the 2.5L (KL) engine

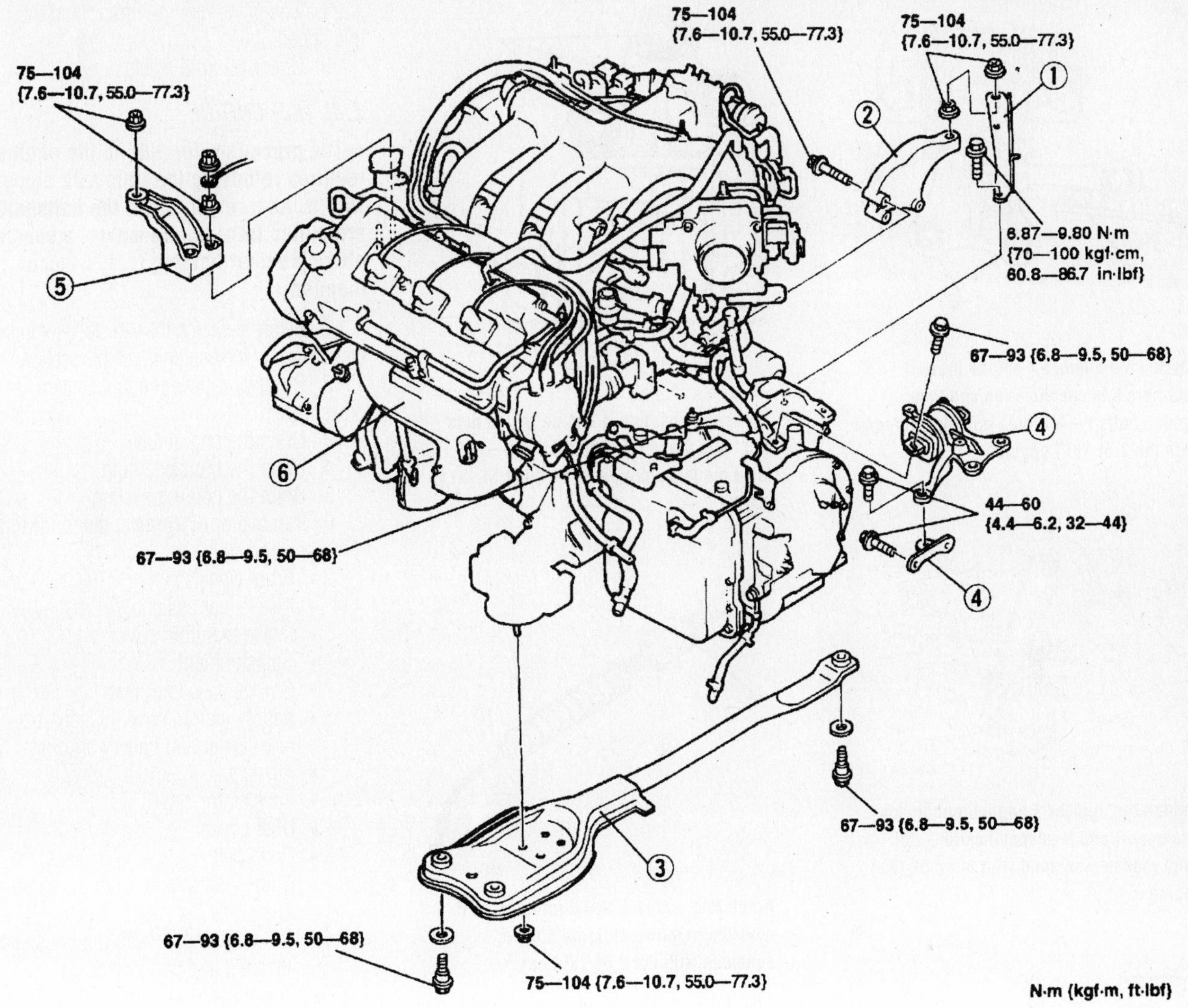

75—104
{7.6—10.7, 55.0—77.3}

75—104
{7.6—10.7, 55.0—77.3}

75—104
{7.6—10.7, 55.0—77.3}

①

②

6.87—9.80 N·m
{70—100 kgf·cm,
60.8—86.7 in·lbf}

67—93 {6.8—9.5, 50—68}

④

75—104
{7.6—10.7, 55.0—77.3}

⑤

44—60
{4.4—6.2, 32—44}

④

⑥

67—93 {6.8—9.5, 50—68}

③

67—93 {6.8—9.5, 50—68}

67—93 {6.8—9.5, 50—68}

75—104 {7.6—10.7, 55.0—77.3}

N·m {kgf·m, ft·lbf}

1	No.1 engine mount stay	4	No.4 engine mount bracket
2	No.1 engine mount bracket	5	No.3 engine mount sub bracket
3	Engine mount member	6	Engine and transaxle

42356-MAZC-GDC

Location of the engine mounting components and their torque specifications (part 3 of 3)—Millenia models equipped with the 2.5L (KL) engine

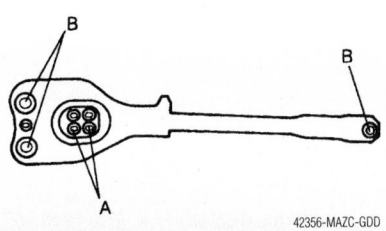

42356-MAZC-GDD

Remove the bolts A first and then the bolts B and remove the engine mount member—Millenia models equipped with the 2.5L (KL) engine

• Number 4 engine bracket retainers and tighten the bolts **E** to 44 ft. lbs. (60 Nm) and hand tighten bolts **F**.

13. Remove the engine hoist and install the engine support device

14. Install the engine mount member and tighten bolts **B** to 68 ft. lbs. (93 Nm) and hand tighten bolts **A**.

15. Install the number 1 bracket and tighten the retainers to 77 ft. lbs. (104 Nm) and remove the engine support.

16. Install the number 1 engine mount stay and tighten retainers **G** to 86 inch lbs. (10 Nm) and **H** to 77 ft. lbs. (104 Nm).

17. Tighten the number 2 engine mount bolts **A** to 77 ft. lbs. (104 Nm).

18. Tighten the number 3 engine mount sub bracket nuts **D** to 77 ft. lbs. (104 Nm).

19. Tighten the number 4 engine mount nuts and bolts **C** and **F** to 68 ft. lbs. (93 Nm)

• Joint shaft. Install a new circlip with the opening facing up. Tighten the bolts to 32–45 ft. lbs. (43–61 Nm).

• Halfshafts. Install new circlips on the inner CV-joint stub shafts, if equipped, and the intermediate shaft. Grease the shaft splines

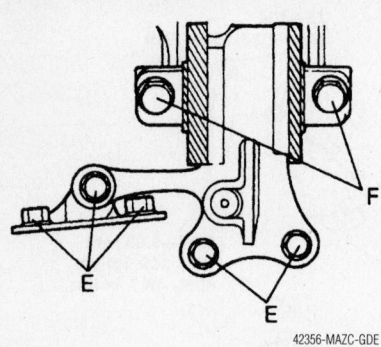

42356-MAZC-GDE

Tighten the number 4 engine bracket retainers E to specification and hand tighten bolts F—Millenia models equipped with the 2.5L (KL) engine

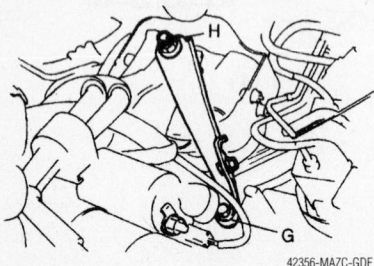

42356-MAZC-GDF

Tighten the number 1 engine mount stay retainers G and H to specification—Millenia models equipped with the 2.5L (KL) engine

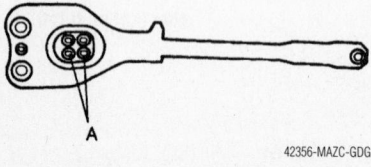

42356-MAZC-GDG

Tighten the number 2 engine mount bolts A to specification—Millenia models equipped with the 2.5L (KL) engine

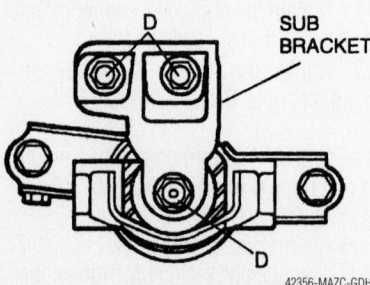

42356-MAZC-GDH

Tighten the number 3 engine mount sub bracket nuts D to specification—Millenia models equipped with the 2.5L (KL) engine

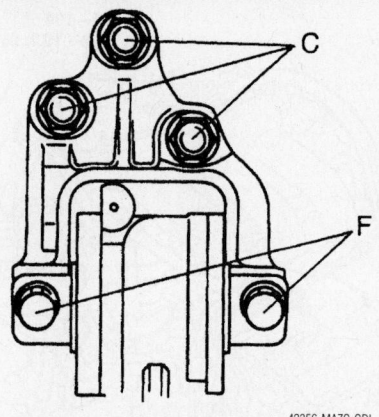

42356-MAZC-GDI

Tighten the number 4 engine mount nuts and bolts C and F to specification—Millenia models equipped with the 2.5L (KL) engine

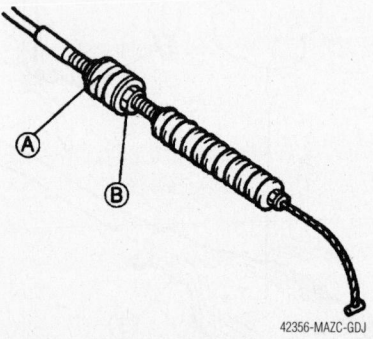

42356-MAZC-GDJ

Accelerator cable A and throttle cable B adjustment nuts—Millenia models equipped with the 2.5L (KL) engine

before installing the halfshaft/intermediate shaft into the trans-axle.
- Ball joints
- Front pipe
- Selector cable
- Power steering pump
- A/C Compressor
- Drive belt

20. When connecting the accelerator cable, perform the following adjustment:

   a. Measure the free play of the cable, it should be 0.04–0.11 inch (1–3mm), if not turn adjustment nut **A** until the specification is reached.

   b. Fully depress the accelerator pedal and check the throttle is wide open, if not adjust using nut **B**.

21. Install the remaining components in the reverse of removal.

22. Fill the engine and the transaxle with the proper type and amount of fluids. Fill the cooling system.

23. Connect the negative battery cable, start the engine and check for leaks.

24. Check the ignition timing and the idle speed.

25. Check all fluid levels.

2.3L (KJ) ENGINE

➡The procedure for pulling the engine requires removing the transaxle along with it. As a result, when the halfshafts are pulled from the transaxle, a special plug/side gear holding tool is recommended.

1. Before servicing the vehicle, refer to the precautions in the beginning of this section.

2. Properly relieve the fuel system pressure.

3. Drain the engine oil.

4. Drain the transaxle fluid.

5. Drain the cooling system.

6. Remove or disconnect the following:
- Hood
- Front wheels
- Air cleaner assembly
- Intake manifold cover
- Splash shield
- Charge air cooler duct
- Battery clamp, cover, battery, battery carrier and battery air duct
- Air duct
- Accelerator cable
- Dust cover
- Drive belt
- Crankshaft and vacuum pump pulleys
- A/C compressor with the lines still attached and position aside
- Power steering pump with the lines still attached and position aside
- Radiator grille
- Upper seal board
- Radiator reservoir
- Cooling fan connector and fan
- Condenser fan motor connector and the fan
- Oil cooler hose
- Radiator hose, bracket and the radiator
- Selector cable
- Front exhaust pipe
- Upper and lower ball joints
- Halfshafts
- Joint shaft

7. Support the engine using a suitable support device.
- Number 1 engine mount bracket. Refer to the accompanying illustration for location.
- Engine mount member retainers. Remove the bolts **A** first and then the bolts **B** and remove the member. Refer to the accompanying illustration for location.

> ❋❋ **CAUTION**

Engine load can damage the number 4 mount bolts holes when removing the bolts so make sure all weight is off the mount.

8. Attach a hoist to the engine, take the weight of the engine with the hoist and remove the engine support device, Lift up the engine/transaxle assembly slightly until the number 3 and 4 mounts are free from engine weight.
- Number 4 engine mount bracket.

Refer to the accompanying illustration for location.
- Number 3 engine mount sub bracket. Refer to the accompanying illustration for location.
- Engine and transaxle assembly from the vehicle. Be careful the powertrain assembly does not swing and strike the vehicle causing damage

To install:

9. Installation is the reverse of removal. Tighten the fasteners to the specifications shown in the accompanying illustration.

10. When possible, leave the engine mounting nuts/bolts loose (hand tight) until all mounts are aligned and bolted. This may help in aligning the engine and transmission assembly in the vehicle.

11. Install or connect the following:
- Number 4 engine bracket retainers and hand tighten the retainers

12. Using the hoist lower the powertrain assembly and hand tighten the number 3 mount sub bracket retainers
- Number 4 engine bracket retainers and tighten the bolts **E** to 44 ft. lbs. (60 Nm) and hand tighten bolts **F**.

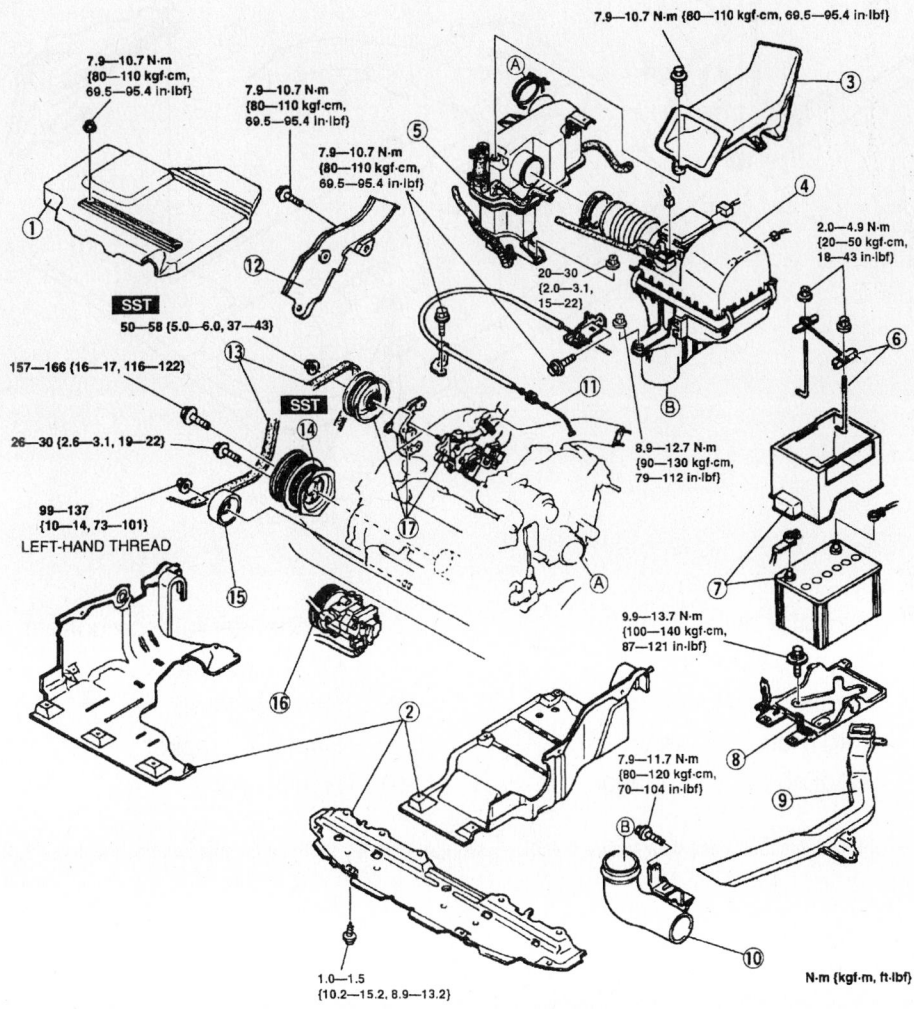

1	Dynamic chamber cover	
2	Splash shield	
3	Charge air cooler air duct	
4	Air cleaner component	
5	Resonator	
6	Battery clamp	
7	Battery and battery cover	
8	Battery carrier	
9	Battery air duct	
10	Air duct	
11	Accelerator cable	
12	Dust cover	
13	Drive belt	
14	Crankshaft pulley	
15	Vacuum pump pulley	
16	A/C compressor	
17	P/S oil pump	

42356-MAZC-GDK

Location of the engine mounting components and their torque specifications (part 1 of 4)—Millenia models equipped with the 2.3L (KJ) engine

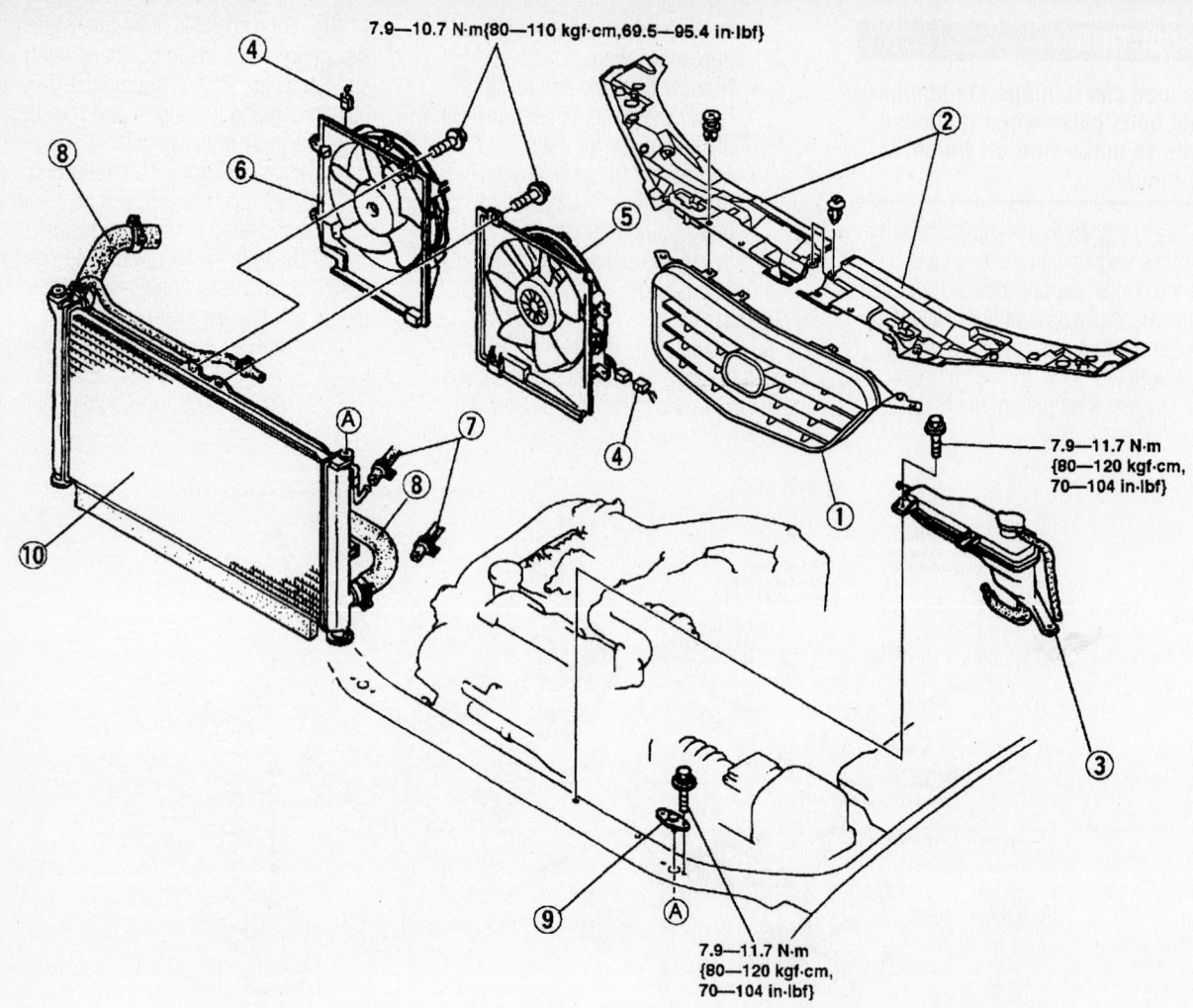

7.9—10.7 N·m{80—110 kgf·cm,69.5—95.4 in·lbf}

7.9—11.7 N·m
{80—120 kgf·cm,
70—104 in·lbf}

7.9—11.7 N·m
{80—120 kgf·cm,
70—104 in·lbf}

1	Radiator grille	6	Condenser fan component
2	Upper seal board	7	Oil cooler hose
3	Radiator reservoir	8	Radiator hose
4	Fan motor connector	9	Radiator bracket
5	Cooling fan component	10	Radiator

42356-MAZC-GDL

Location of the engine mounting components and their torque specifications (part 2 of 4)—Millenia models equipped with the 2.3L (KJ) engine

13. Remove the engine hoist and install the engine support device

14. Install the engine mount member and tighten bolts **B** to 68 ft. lbs. (93 Nm) and hand tighten bolts **A**.

15. Install the number 1 bracket and tighten the retainers to 77 ft. lbs. (104 Nm) and remove the engine support.

16. Install the number 1 engine mount stay and tighten retainers **G** to 68 inch lbs. (10 Nm) and **H** to 77 ft. lbs. (104 Nm).

17. Tighten the number 2 engine mount bolts **A** to 77 ft. lbs. (104 Nm).

18. Tighten the number 3 engine mount sub bracket nuts **D** to 77 ft. lbs. (104 Nm).

19. Tighten the number 4 engine mount nuts and bolts **C** and **F** to 68 ft. lbs. (93 Nm)

20. Install or connect the following:
- Joint shaft. Install a new circlip with the opening facing up. Tighten the bolts to 32–45 ft. lbs. (43–61 Nm).
- Halfshafts. Install new circlips on the inner CV-joint stub shafts, if equipped, and the intermediate shaft. Grease the shaft splines before installing the halfshaft/intermediate shaft into the transaxle.
- Ball joints

- Front pipe
- Selector cable
21. Power steering pump
- A/C Compressor
- Drive belt
22. When connecting the accelerator cable, perform the following adjustment:

   a. Measure the free play of the cable, it should be 0.04–0.11 inch (1–3mm), if not turn adjustment nut **A** until the specification is reached.

   b. Fully depress the accelerator pedal and check the throttle is wide open, if not adjust using nut **B**.

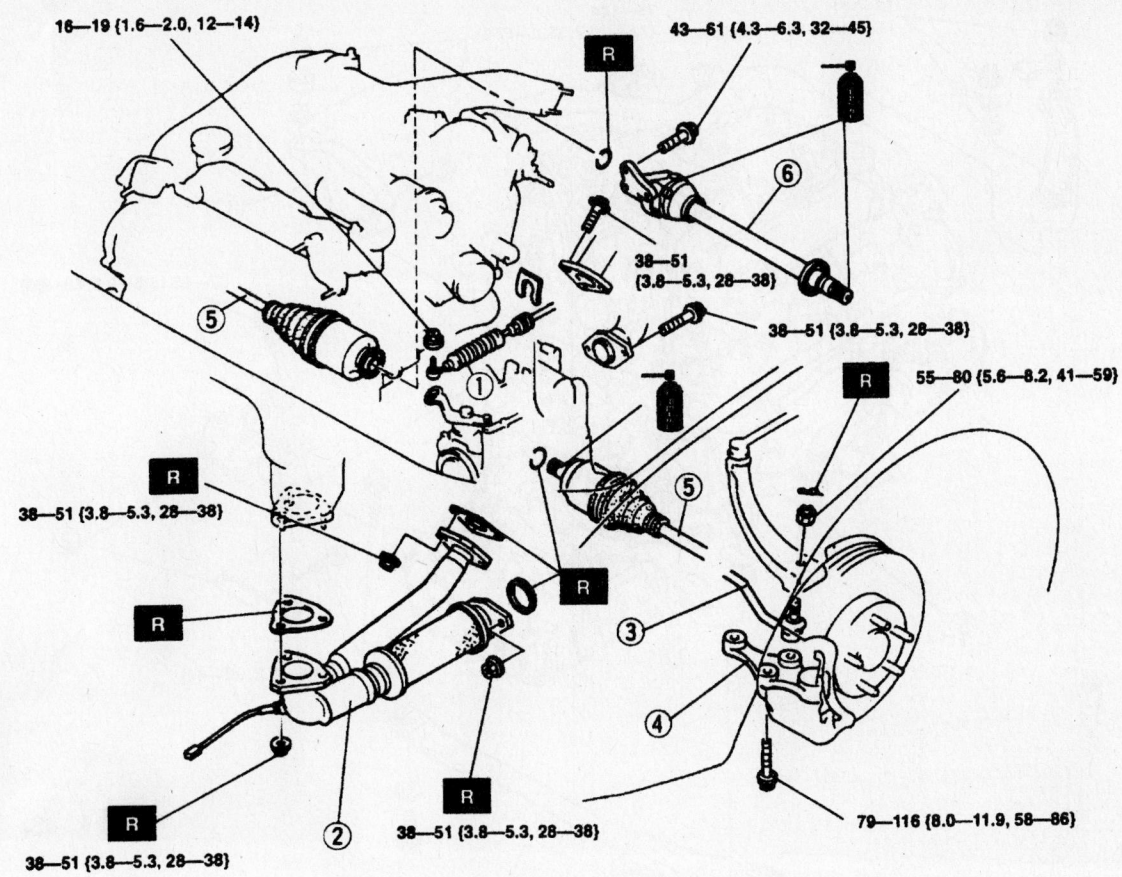

16—19 {1.6—2.0, 12—14}

43—61 {4.3—6.3, 32—45}

38—51 {3.8—5.3, 28—38}

38—51 {3.8—5.3, 28—38}

55—80 {5.6—8.2, 41—59}

38—51 {3.8—5.3, 28—38}

79—116 {8.0—11.9, 58—86}

38—51 {3.8—5.3, 28—38}

38—51 {3.8—5.3, 28—38}

38—51 {3.8—5.3, 28—38}

N·m {kgf·m, ft·lbf}

1	Selector cable	4	Lower ball joint
2	Front pipe	5	Drive shaft
3	Upper lateral link ball joint	6	Joint shaft

42356-MAZC-GDM

Location of the engine mounting components and their torque specifications (part 3 of 4)—Millenia models equipped with the 2.3L (KJ) engine

23. Install the remaining components in the reverse of removal.

24. Fill the engine and the transaxle with the proper type and amount of fluids. Fill the cooling system.

25. Connect the negative battery cable, start the engine and check for leaks.

26. Check the ignition timing and the idle speed.

27. Check all fluid levels.

Miata

1. Before servicing the vehicle, refer to the precautions in the beginning of this section.

2. Properly relieve the fuel system pressure.

3. Drain the engine oil.

4. Drain the cooling system.

5. Remove or disconnect the following:
- Negative battery cable
- Air cleaner assembly
- Radiator
- Accelerator cable and bracket from the throttle body
- Fuel hose
- Vacuum hoses and engine harness connectors
- Heater hose
- Accessory drive belts

- Power steering pump and move it aside without disconnecting the hydraulic hoses
- Air conditioner compressor and move it aside without disconnecting the refrigerant lines
- Transmission

6. Disconnect the following electrical connectors, if equipped:
- Steering pressure sensor electrical connector
- Throttle Position (TP) sensor electrical connector
- Idle Air Control (IAC) valve electrical connector

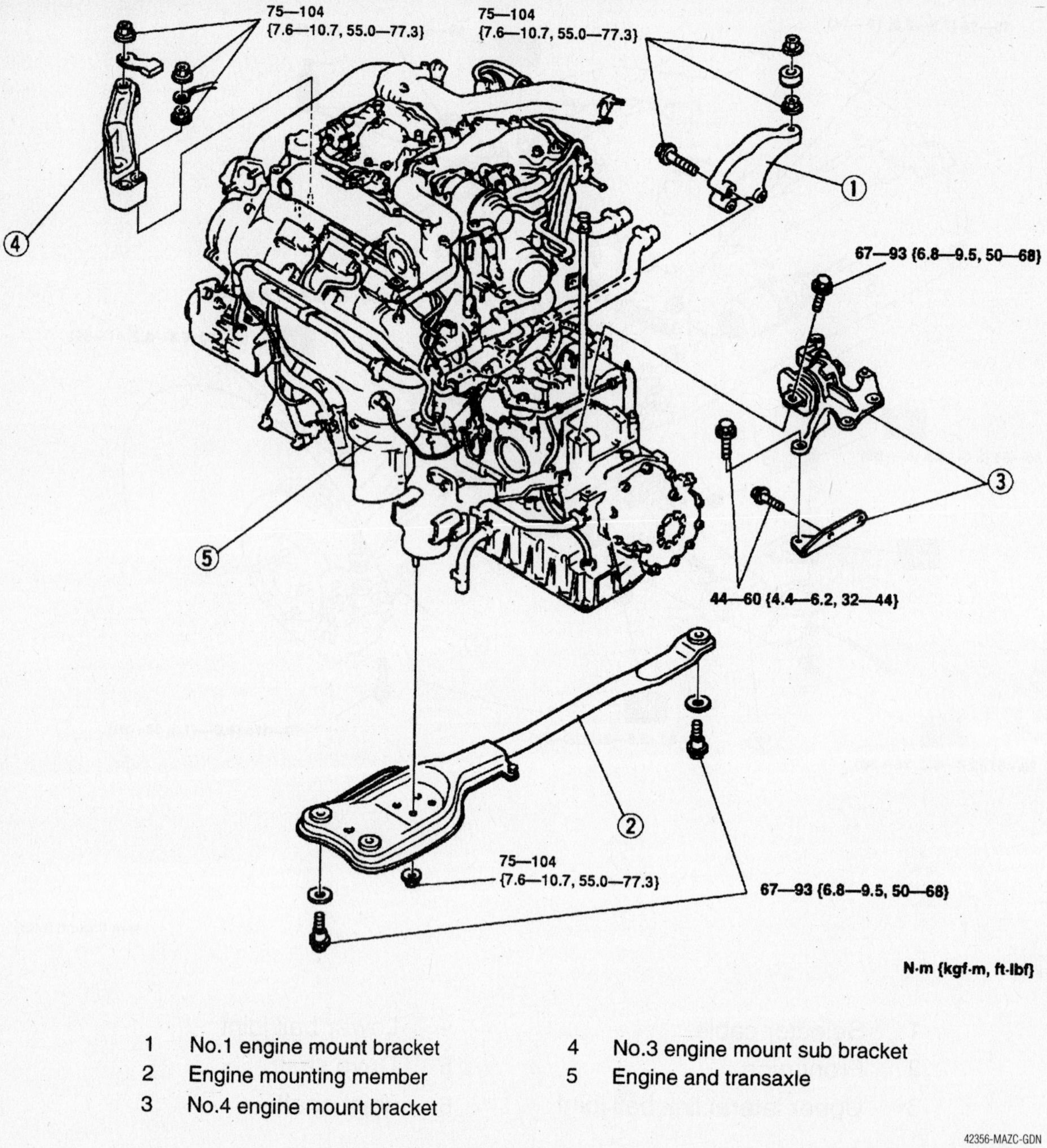

75—104
{7.6—10.7, 55.0—77.3}

75—104
{7.6—10.7, 55.0—77.3}

④

①

67—93 {6.8—9.5, 50—68}

③

44—60 {4.4—6.2, 32—44}

⑤

②

75—104
{7.6—10.7, 55.0—77.3}

67—93 {6.8—9.5, 50—68}

N·m {kgf·m, ft·lbf}

1 No.1 engine mount bracket	4 No.3 engine mount sub bracket
2 Engine mounting member	5 Engine and transaxle
3 No.4 engine mount bracket	

42356-MAZC-GDN

Location of the engine mounting components and their torque specifications (part 4 of 4)—Millenia models equipped with the 2.3L (KJ) engine

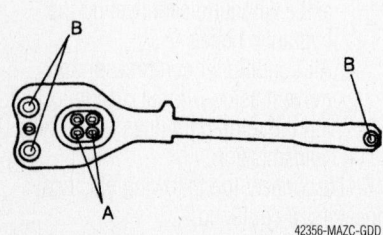

42356-MAZC-GDD

Remove the bolts A first and then the bolts B and remove the engine mount member—Millenia models equipped with the 2.3L (KJ) engine

- Heated Oxygen (HO2S) sensor electrical connector
- Ignition coil electrical connectors
- Crankshaft Position (CKP) sensor electrical connector
- Ground electrical connectors
- Fuel injector electrical connectors
- Alternator electrical connectors
- Oil pressure sensor electrical connector
- Starter electrical connectors

7. Remove or disconnect the following:
- Exhaust pipe from the exhaust

manifold and install suitable lifting equipment onto the engine.
- Engine from the vehicle

To install:

8. Install or connect the following:
- Engine assembly by tilting the engine downward and aligning the engine mounts with the crossmember holes. Torque the nuts to 42–57 ft. lbs. (57–78 Nm).
- Exhaust pipe to the manifold using a new gasket. Torque the nuts to 34 ft. lbs. (46 Nm).
- All vacuum hoses

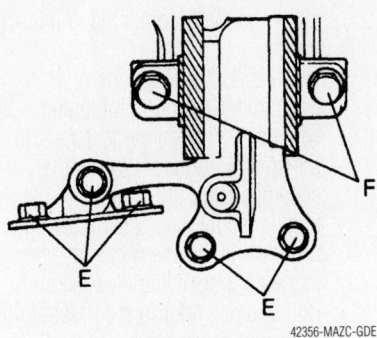

42356-MAZC-GDE

Tighten the number 4 engine bracket retainers E to specification and hand tighten bolts F—Millenia models equipped with the 2.3L (KJ) engine

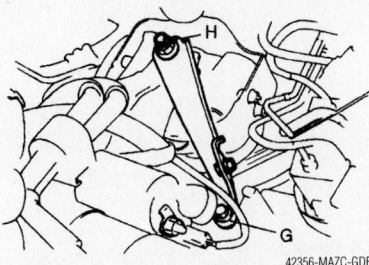

42356-MAZC-GDF

Tighten the number 1 engine mount stay retainers G and H to specification—Millenia models equipped with the 2.3L (KJ) engine

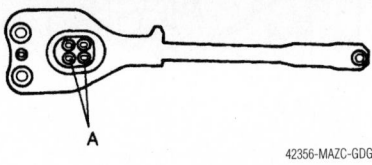

42356-MAZC-GDG

Tighten the number 2 engine mount bolts A to specification—Millenia models equipped with the 2.3L (KJ) engine

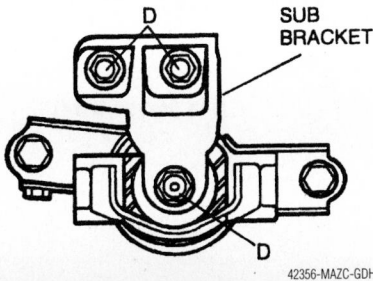

42356-MAZC-GDH

Tighten the number 3 engine mount sub bracket nuts D to specification—Millenia models equipped with the 2.3L (KJ) engine

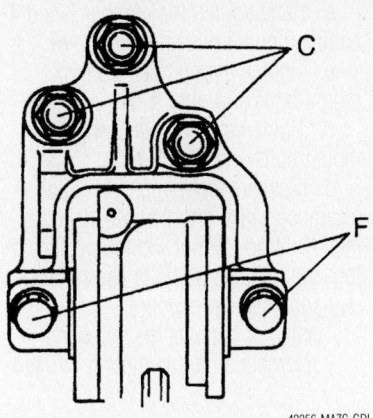

42356-MAZC-GDI

Tighten the number 4 engine mount nuts and bolts C and F to specification—Millenia models equipped with the 2.3L (KJ) engine

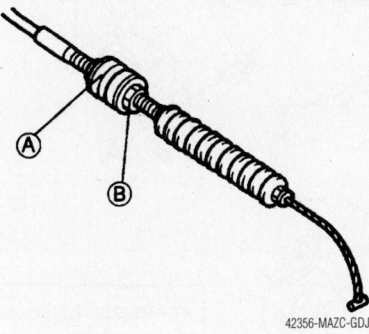

42356-MAZC-GDJ

Accelerator cable A and throttle cable B adjustment nuts—Millenia models equipped with the 2.3L (KJ) engine

9. Connect the following electrical connectors, if equipped:
- Steering pressure sensor electrical connector
- TP sensor electrical connector
- IAC valve electrical connector
- HO$_2$S electrical connector
- Ignition coil electrical connectors
- CKP sensor electrical connector
- Ground electrical connectors
- Fuel injector electrical connectors
- Alternator electrical connectors
- Oil pressure sensor electrical connector
- Starter electrical connectors

10. Install or connect the following:
- Transmission
- Air conditioner compressor and power steering pump
- Drive belt(s)
- Any remaining vacuum hoses
- Radiator and fans and all cooling system hoses
- Accelerator cable and bracket
- Air cleaner assembly

- Negative battery cable
11. Fill and bleed the cooling system.
12. Fill the engine and transmission.
13. Start the engine and check for leaks.
14. Check the ignition timing and idle speed.

Protégé

2.0L (FS) ENGINE

➡The procedure for pulling the engine requires removing the transaxle along with it. As a result, when the halfshafts are pulled from the transaxle, a special plug/side gear holding tool is recommended.

1. Before servicing the vehicle, refer to the precautions in the beginning of this section.
2. Properly relieve the fuel system pressure.
3. Drain the engine oil.
4. Drain the transaxle fluid.
5. Drain the cooling system.
6. Remove or disconnect the following:
- Negative battery cable
- Radiator
- Air cleaner assembly
- Accelerator cable
- Fuel hoses
- Front exhaust pipe
- Any rods, pipes or cables related to the transaxle that would hinder removal
- Battery
- Fuse box
- Power steering pump with the lines still attached and set aside
- A/C compressor with the lines still attached and position aside
- Halfshafts
- Roll dampener
7. Support the engine with a suitable engine support device.
- Engine mount member retainers. Refer to the accompanying illustration for location.
- Number 3 engine mount. Refer to the accompanying illustration for location.
- Number 1 engine mount. Refer to the accompanying illustration for location.
- Number 2 engine mount. Refer to the accompanying illustration for location.
- Number 4 engine mount. Refer to the accompanying illustration for location.
- Engine and transaxle assembly from the vehicle.

To install:

8. Installation is the reverse of removal. Tighten the fasteners to the specifications shown in the accompanying illustration.

9. When possible, leave the engine mounting nuts/bolts loose (hand tight) until all mounts are aligned and bolted. This may help in aligning the engine and transmission assembly in the vehicle.

10. Install the number 4 engine mount. Tighten the fasteners to the specifications shown in the accompanying illustration as follows:

a. Hand tighten the number 3 and 4 engine mount bolts and nuts **A–M**.

b. Tighten the number 4 engine mount bolts and nuts **A–H**.

c. Tighten the number 3 engine mount bolts and nuts **I–N**.

d. Measure the number 4 mount clearance which should be 0.12–0.15 inch (3–4mm). If not within specification, loosen the nuts and bolts and retorque using the same procedure.

11. Install or connect the following:
- Number 1 engine mount. Tighten

the fasteners to the specifications shown in the accompanying illustration.

- Number 3 engine mount, refer to the number 4 mount tightening sequence. Tighten the fasteners to the specifications shown in the accompanying illustration.
- Engine mount member. Tighten the fasteners to the specifications shown in the accompanying illustration.
- Roll damper, refer to the illustration or proper assembly.

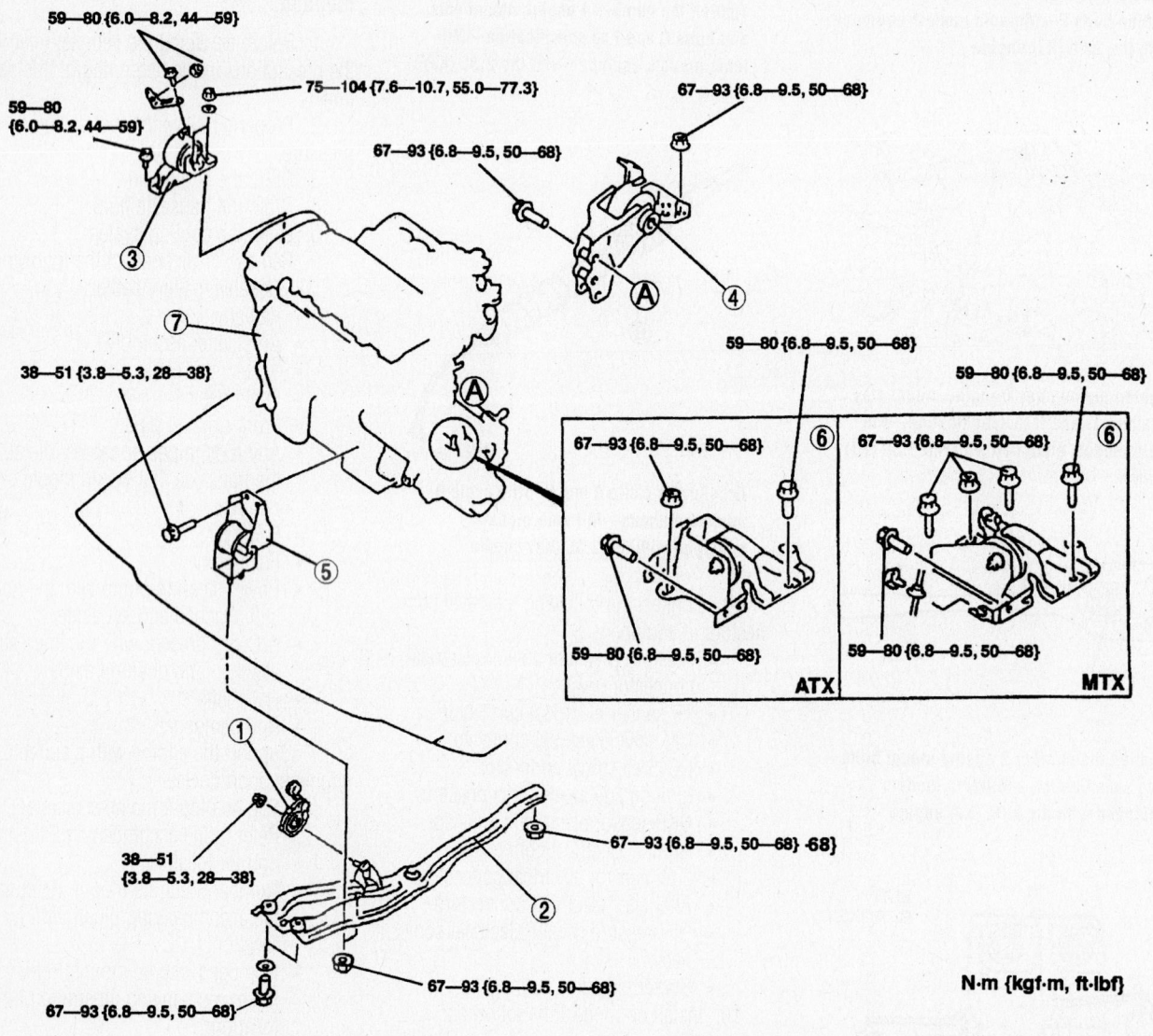

1	Roll damper
2	Engine mount member
3	No.3 Engine mount
4	No.1 Engine mount
5	No.2 Engine mount
6	No.4 Engine mount
7	Engine, transaxle

N·m {kgf·m, ft·lbf}

42356-MAZC-GGA

Location of the engine mounting components and their torque specifications—Protégé models equipped with the 2.0L (FS) engine

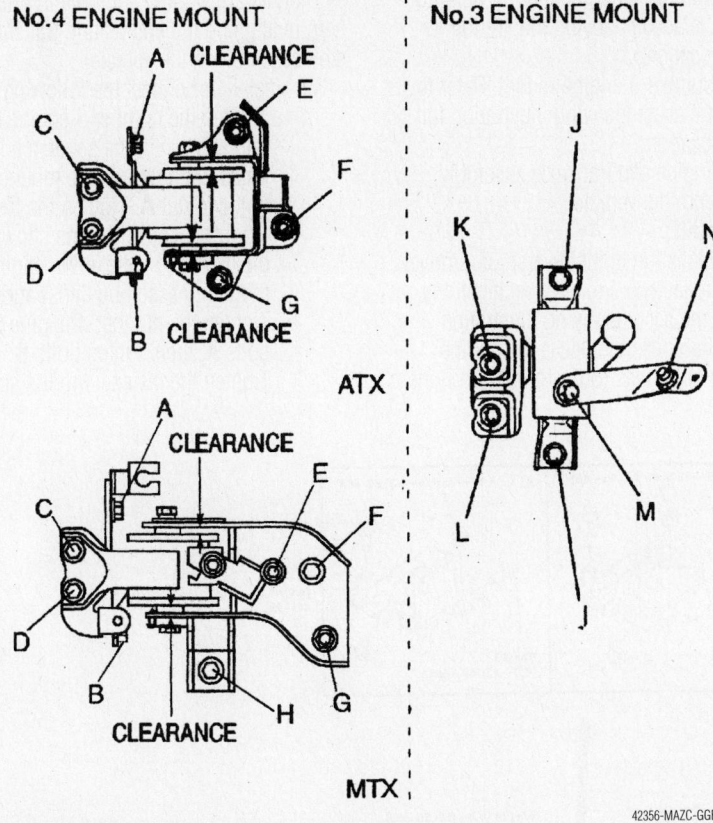

Tighten the number 4 engine bracket retainers in the sequence shown—Protégé models equipped with the 2.0L (FS) engine

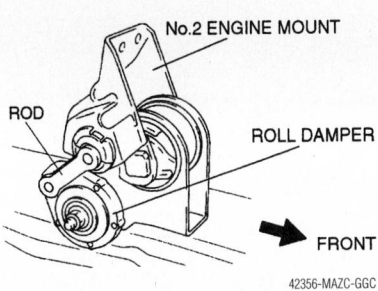

42356-MAZC-GGC

Assemble the roll damper as illustrated—Protégé models equipped with the 2.0L (FS) engine

- Halfshafts. Install new circlips on the inner CV-joint stub shafts, if equipped, and the intermediate shaft. Grease the shaft splines before installing the halfshaft/intermediate shaft into the transaxle.
- A/C compressor
- Power steering pump
- Fuse box
- Battery
- Any rods, pipes or cables related to the transaxle
- Front exhaust pipe. Always install

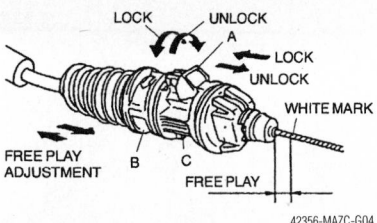

42356-MAZC-G04

Accelerator cable adjustment components—Protégé models equipped with the 2.0L (FS) engine

new gaskets and/or O-rings. Use new self-locking nuts, especially on the exhaust.
- Fuel hoses
12. When connecting the accelerator cable, perform the following adjustment:
    a. Move the white locking tab to the unlock position.
    b. Turn stopper B to the to the unlock position.

➡️ If stopper B will not unlock, it may be necessary to carefully bend back tab C out a little using a suitable pry tool.

c. Push or pull the cable housing directly behind the spring.
    d. Turn the stopper B to the lock position.
    e. Measure the free play which should be 0.04–0.11 inch (1–3mm) and make sure the cable free play is within specification.
    f. Move the white locking tab to the lock position and check for proper accelerator operation.
13. Install or connect the following:
    - Air cleaner assembly
    - Radiator
    - Negative battery cable
14. Fill the engine and the transaxle with the proper type and amount of fluids. Fill the cooling system.
15. Connect the negative battery cable, start the engine and check for leaks.
16. Check the ignition timing and the idle speed.
17. Check all fluid levels.

1.6L (ZM) ENGINE

➡️ The procedure for pulling the engine requires removing the transaxle along with it. As a result, when the halfshafts are pulled from the transaxle, a special plug/side gear holding tool is recommended.

1. Before servicing the vehicle, refer to the precautions in the beginning of this section.
2. Properly relieve the fuel system pressure.
3. Drain the engine oil.
4. Drain the transaxle fluid.
5. Drain the cooling system.
6. Remove or disconnect the following:
    - Battery
    - Air cleaner, air hose and resonance chamber
    - Front exhaust pipe
    - Accelerator cable and bracket
    - Heater and vacuum hoses
    - Radiator
    - Drive belt
    - Fuel hoses
    - Any rods, pipes or cables related to the transaxle that would hinder removal
    - Halfshafts
    - Power steering pump with the lines still attached and set aside
    - A/C compressor with the lines still attached and position aside
    - Air cleaner bracket
    - Battery carrier bracket
7. Support the engine with a suitable engine support device.

- Number 1 engine mount. Refer to the accompanying illustration for location.
- Number 2 engine mount nut **A**. Refer to the accompanying illustration for location.
- Engine mount member retainers. Refer to the accompanying illustration for location.
- Number 2 engine mount. Refer to the accompanying illustration for location.

8. Remove the engine support device and attach a hoist and chain to the engine.

- Number 4 engine mount. Refer to the accompanying illustration for location.
- Number 3 engine mount. Refer to the accompanying illustration for location.
- Engine and transaxle assembly from the vehicle.

To install:

9. Installation is the reverse of removal. Tighten the fasteners to the specifications shown in the accompanying illustration.

10. When possible, leave the engine mounting nuts/bolts loose (hand tight) until

all mounts are aligned and bolted. This may help in aligning the engine and transmission assembly in the vehicle.

11. Install or connect the following:

- Tighten the number 3 engine mount bolt and nut in the sequence illustrated, then tighten the mount stay bolt and nut **A**. Tighten the fasteners to the specifications shown in the exploded view of the engine mounting assembly illustration.
- Tighten the number 4 engine mount bolts **A**, then tighten bolts **B**. Tighten the fasteners to the specifi-

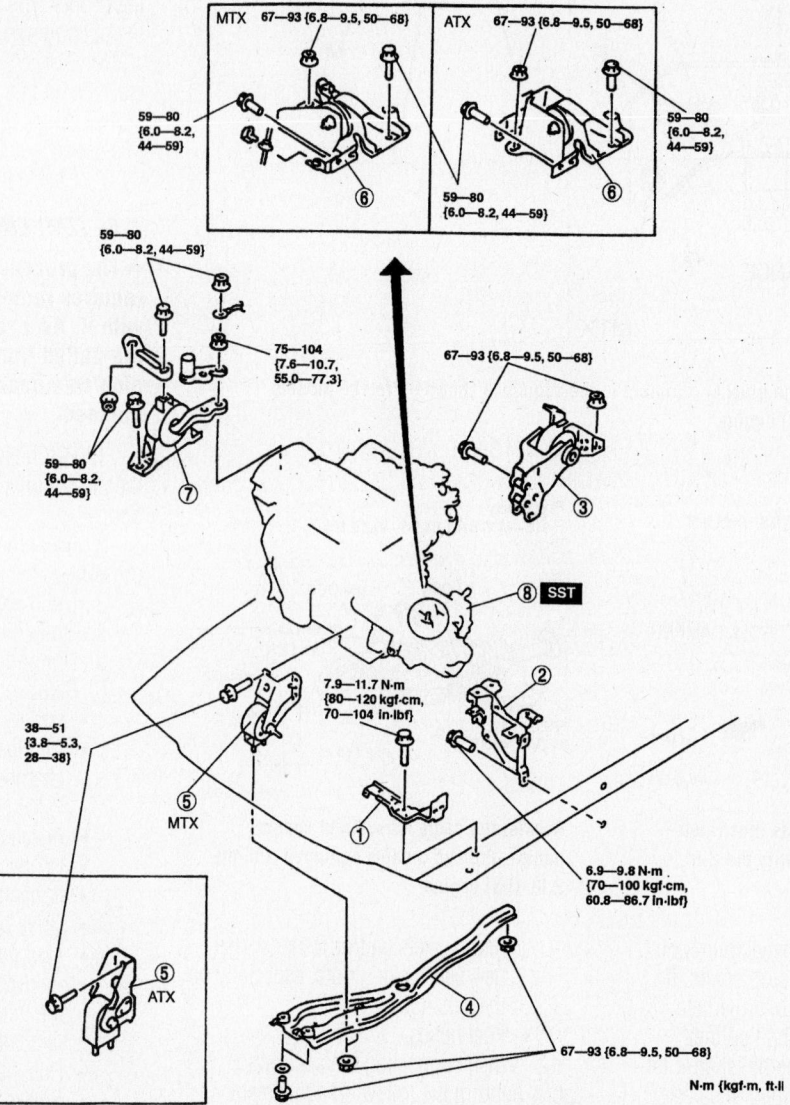

1	Air cleaner bracket	5	No.2 engine mount
2	Battery carrier bracket	6	No.4 engine mount
3	No.1 engine mount	7	No.3 engine mount
4	Engine mount member	8	Engine, transaxle

42356-MAZC-GHD

Location of the engine mounting components and their torque specifications—Protégé models equipped with the 1.6L (ZM) engine

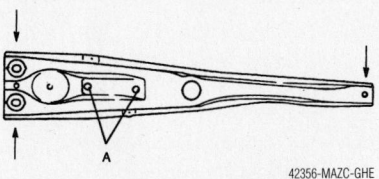

Remove the number 2 engine mount nut retainers A, then the engine mount retainers—Protégé models equipped with the 1.6L (ZM) engine

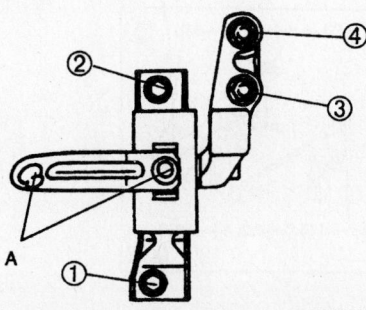

Tighten the number 3 engine mount bolt and nut in this sequence and the mount stay bolt A—Protégé models equipped with the 1.6L (ZM) engine

MTX

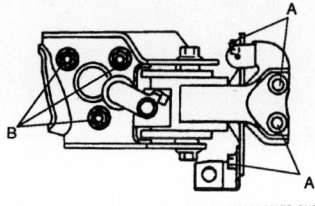

Number 4 engine mount retainer torque sequence on models with a manual transaxle—Protégé models equipped with the 1.6L (ZM) engine

ATX

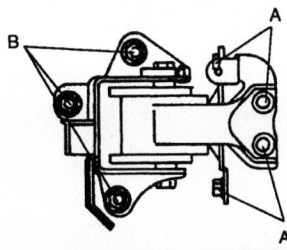

Number 4 engine mount retainer torque sequence on models with a automatic transaxle—Protégé models equipped with the 1.6L (ZM) engine

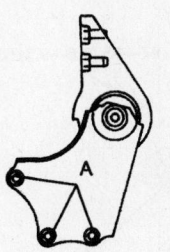
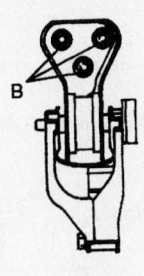

Number 1 engine mount retainer torque sequence—Protégé models equipped with the 1.6L (ZM) engine

cations shown in the exploded view of the engine mounting assembly illustration.

12. Remove the hoist and chain and install the engine support device.

- Number 2 engine mount nut **A**, then tighten the mount member bolt and nut. Tighten the fasteners to the specifications shown in the exploded view of the engine mounting assembly illustration.
- Number 1 engine mount bolts **A** and then the bolt **B**. Tighten the fasteners to the specifications shown in the accompanying illustration.
- Battery carrier bracket
- Air cleaner bracket
- A/C compressor
- Power steering pump
- Halfshafts
- Any rods, pipes or cables related to the transaxle that were removal
- Fuel hoses
- Drive belt
- Radiator
- Heater and vacuum hoses
- Accelerator cable and bracket
- Front exhaust pipe
- Air cleaner, air hose and resonance chamber
- Battery

13. Fill the engine and the transaxle with the proper type and amount of fluids. Fill the cooling system.

14. Connect the negative battery cable, start the engine and check for leaks.

15. Check the ignition timing and the idle speed.

16. Check all fluid levels.

1.8L (FP) ENGINE

➡The procedure for pulling the engine requires removing the transaxle along with it. As a result, when the halfshafts are pulled from the transaxle, a special plug/side gear holding tool is recommended.

1. Before servicing the vehicle, refer to the precautions in the beginning of this section.

2. Properly relieve the fuel system pressure.

3. Drain the engine oil.

4. Drain the transaxle fluid.

5. Drain the cooling system.

6. Remove or disconnect the following:

- Negative battery cable
- Radiator
- Air cleaner
- Accelerator cable
- Fuel hoses
- Front exhaust pipe
- Any rods, pipes or cables related to the transaxle that would hinder removal
- Battery
- Fuse box
- Drive belt
- Power steering pump with the lines still attached and set aside
- A/C compressor with the lines still attached and position aside
- Halfshafts

7. Support the engine with a suitable engine support device.

- Engine mount member retainers
- Number 3 engine mount. Refer to the accompanying illustration for location.
- Number 4 engine mount. Refer to the accompanying illustration for location.
- Number 1 engine mount. Refer to the accompanying illustration for location.
- Number 2 engine mount. Refer to the accompanying illustration for location.

8. Remove the engine support device and attach a hoist and chain to the engine.

- Engine and transaxle assembly from the vehicle.

To install:

9. Installation is the reverse of removal. Tighten the fasteners to the specifications shown in the accompanying illustration.

10. When possible, leave the engine mounting nuts/bolts loose (hand tight) until all mounts are aligned and bolted. This may help in aligning the engine and transmission assembly in the vehicle.

11. Install or connect the following:

- Engine and transaxle assembly
- Number 2 engine mount . Tighten the fasteners to the specifications shown in the accompanying illustration.

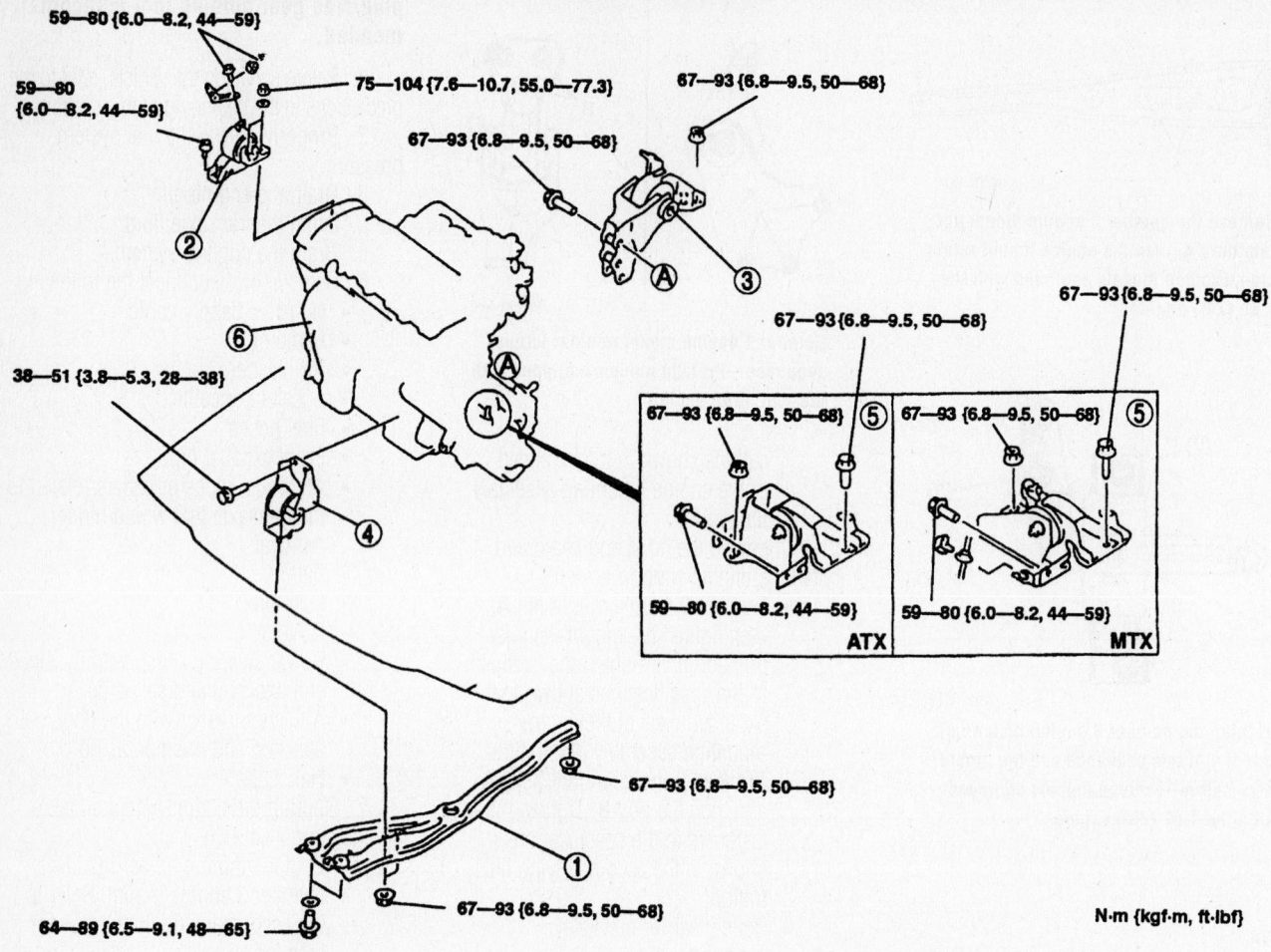

59—80 {6.0—8.2, 44—59}

59—80 {6.0—8.2, 44—59}

75—104 {7.6—10.7, 55.0—77.3}

67—93 {6.8—9.5, 50—68}

67—93 {6.8—9.5, 50—68}

38—51 {3.8—5.3, 28—38}

67—93 {6.8—9.5, 50—68}

67—93 {6.8—9.5, 50—68}

67—93 {6.8—9.5, 50—68}

67—93 {6.8—9.5, 50—68}

67—93 {6.8—9.5, 50—68}

59—80 {6.0—8.2, 44—59}

59—80 {6.0—8.2, 44—59}

ATX

MTX

67—93 {6.8—9.5, 50—68}

67—93 {6.8—9.5, 50—68}

64—89 {6.5—9.1, 48—65}

N·m {kgf·m, ft·lbf}

1	Engine mount member	4	No.2 Engine mount
2	No.3, No.4 Engine mount	5	Engine and transaxle
3	No.1 Engine mount		

42356-MAZC-GJA

Location of the engine mounting components and their torque specifications—Protégé models equipped with the 1.8L (FP) engine

- Number 1 engine mount. Tighten the fasteners to the specifications shown in the accompanying illustration.
- Hand tighten the number 3 and 4 engine mount bolts **A–M**, tighten number 4 bolts and nuts **A–G**, then tighten the number 3 mount bolts and nuts **H–M**. Tighten the fasteners to the specifications shown in the exploded view of the engine mounting assembly illustration. Measure the number 4 mount clearance, if not 0.12–0.15 inch (3–4mm), repeat the tightening sequence.
- Engine mount member. Tighten the fasteners to the specifications shown in the accompanying illus-

tration and remove the support device.
- Halfshafts
- Power steering pump
- A/C compressor
- Fuse box
- Battery
- Any rods, pipes or cables related to the transaxle that were removed
- Front exhaust pipe
- Fuel hoses
- Accelerator cable
- Air cleaner
- Radiator
- Drive belt
- Negative battery cable

12. Fill the engine and the transaxle with the proper type and amount of fluids. Fill the cooling system.

13. Connect the negative battery cable, start the engine and check for leaks.

14. Check the ignition timing and the idle speed.

15. Check all fluid levels.

Mazda 3

1. Before servicing the vehicle, refer to the precautions in the beginning of this section.

2. Remove the plug hole plate by lifting off and removing the plug hole plate from the areas shown in the accompanying illustration.

3. Remove the air hose and air cleaner assembly.

4. Remove the battery cover, duct, clamp, battery and battery tray.

5. Relieve the fuel system pressure and disconnect the fuel hoses.

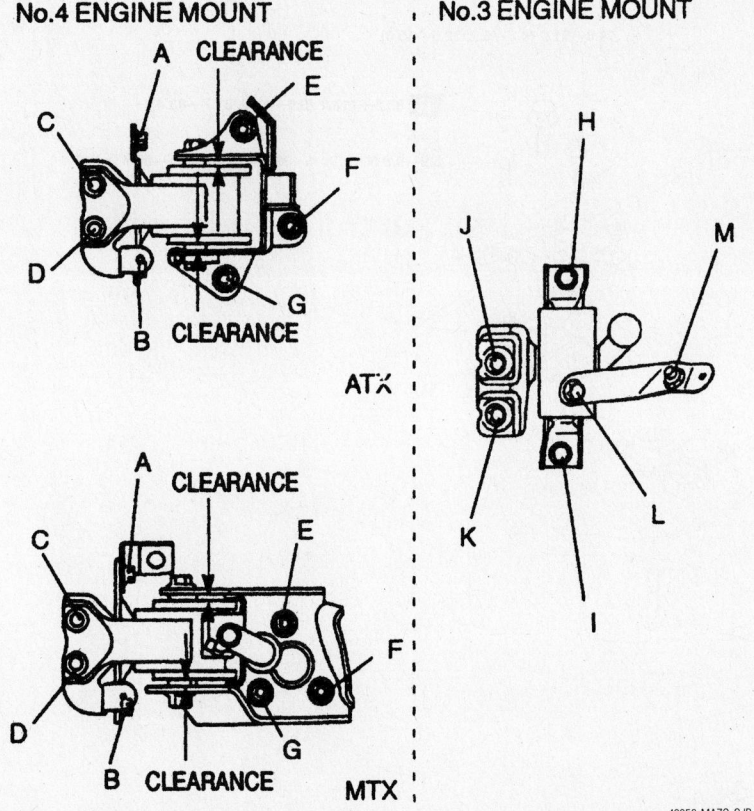

Number 3 and 4 engine mount retainer torque sequence—Protégé models equipped with the 1.8L (FP) engine

shown in the accompanying illustration and remove the No. 3 engine mount.

25. Remove the battery bracket.

26. Remove the No. 4 engine mount.

27. Remove the engine and transaxle assembly from the vehicle as an assembly.

To install:

28. Installation is the reverse of removal, please keep in mind the following:

29. When installing the No. 4 engine mount, secure the engine and transaxle using an engine jack and attachment as shown in the accompanying illustration. Install the No. 1 and No. 4 engine mounts, do not tighten the retainers at this time.

30. Use a new No. 4 mount bolt and tighten it to 61–87 ft. lbs. (83–113 Nm).

31. Tighten the No. 4 engine mount and battery bracket nuts and bolts in the sequence shown in the accompanying illustration. Tighten the retainers (1) to 32–45 ft. lbs. (44–61 Nm) and the retainers (2) to 61–86 inch lbs. (7–10 Nm).

32. Tighten the No. 3 mount bracket stud bolts to 62–115 inch lbs. (7–13 Nm).

33. Tighten the No. 3 joint bracket nuts and bolts to 55–77 ft. lbs. (74–105 Nm).

34. Remove the jack and the attachment and tighten the No. 1 mount bolts in the sequence illustrated to 69–86 ft. lbs. (93–116 Nm).

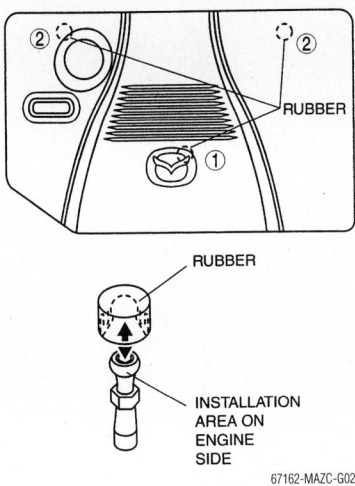

Plug hole plate locations—Mazda 3 models

6. Remove the accelerator cable and bracket.

7. Remove the front wheels, under cover and splash shields.

8. Remove the A/C drive belt.

9. Remove the A/C compressor with the lines still attached and wire the compressor out of the way.

10. Drain the transaxle fluid.

11. Drain the coolant.

12. Disconnect the brake booster vacuum hose.

13. Remove the exhaust system member.

14. Remove the front crossmember, front stabilizer bar, lower control arm, steering gear and the No. 1 engine mount.

15. Remove the drive shafts.

16. Remove the coolant over flow tank with the hose attached and wire it to one side.

17. Remove the cooling fan assembly.

18. If equipped with an automatic transaxle, disconnect the transaxle fluid lines, selector cable and wiring harness.

19. If equipped with a manual transaxle, remove the shift cable and the clutch release cylinder with the line still attached.

20. Disconnect the heater hoses.

21. Remove the radiator hoses.

22. Remove the exhaust system main silencer.

23. Remove the main fuse block connector by releasing the tab in the order shown in the accompanying illustration and pulling the lock lever up and remove the connector.

24. Secure the engine and transaxle using an engine jack and attachment as

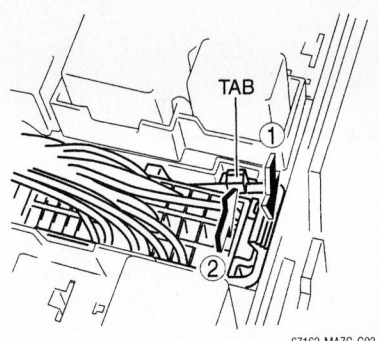

Removing the main fuse block connector—Mazda 3 models

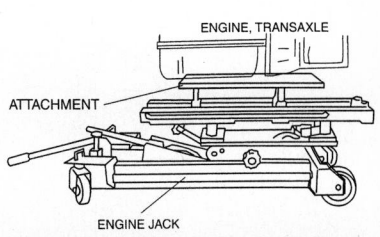

Secure the engine and transaxle using an engine jack and attachment as shown before removing the No. 3 and No. 4 engine mounts—Mazda 3 models

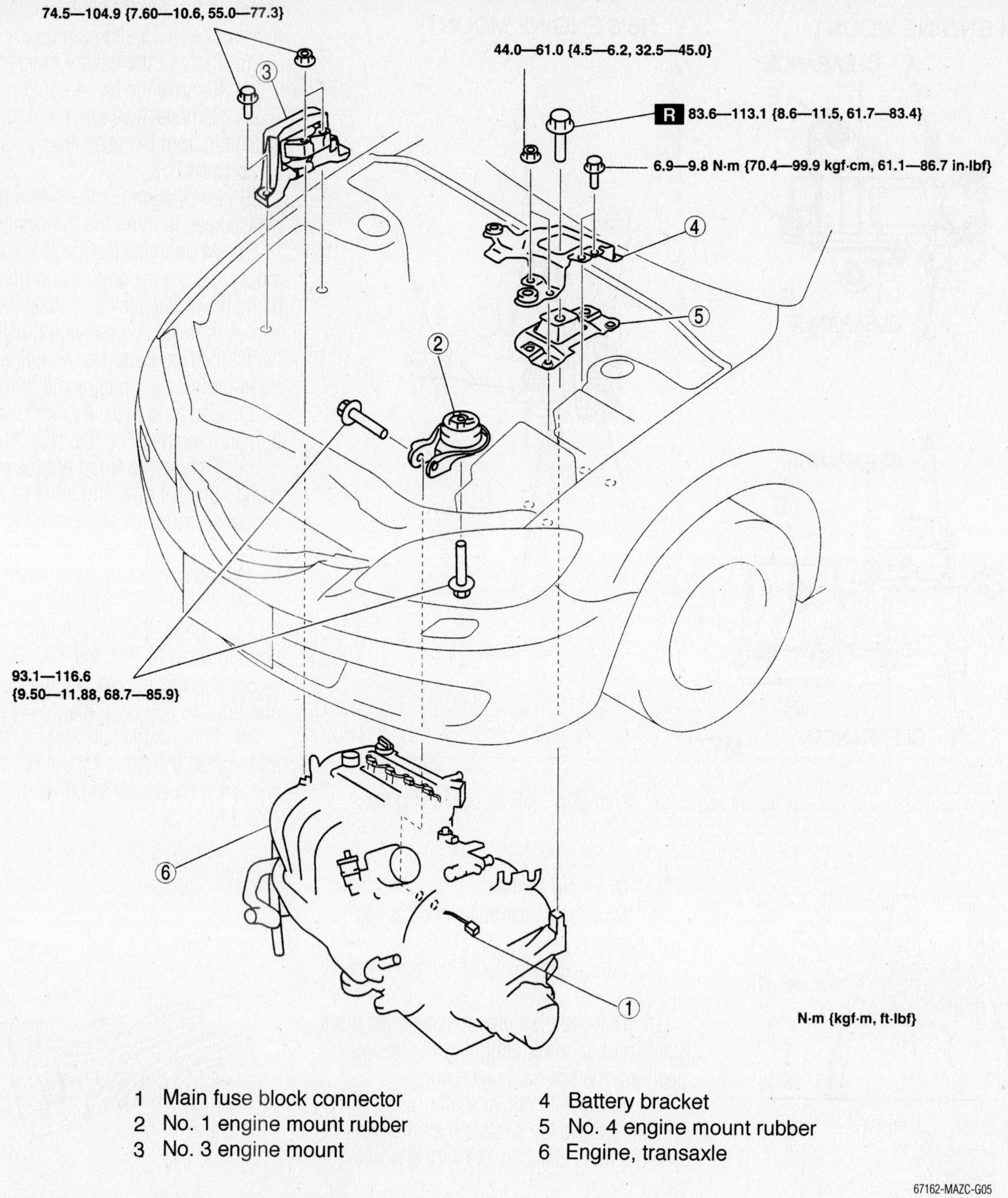

74.5—104.9 {7.60—10.6, 55.0—77.3}

44.0—61.0 {4.5—6.2, 32.5—45.0}

R 83.6—113.1 {8.6—11.5, 61.7—83.4}

6.9—9.8 N·m {70.4—99.9 kgf·cm, 61.1—86.7 in·lbf}

③

④

⑤

②

93.1—116.6
{9.50—11.88, 68.7—85.9}

⑥

①

N·m {kgf·m, ft·lbf}

1	Main fuse block connector	4	Battery bracket
2	No. 1 engine mount rubber	5	No. 4 engine mount rubber
3	No. 3 engine mount	6	Engine, transaxle

Exploded view of the engine mounting—Mazda 3 models

67162-MAZC-G05

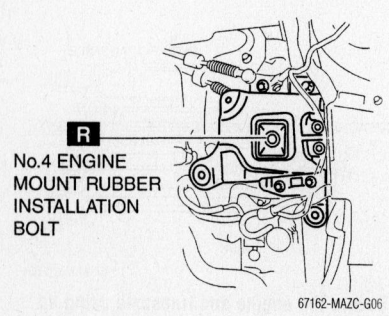

R No.4 ENGINE
MOUNT RUBBER
INSTALLATION
BOLT

67162-MAZC-G06

Location of the No. 4 engine mount rubber bolt—Mazda 3 models

35. Fill the engine and the transaxle with the proper type and amount of fluids. Fill the cooling system.

36. Connect the negative battery cable, start the engine and check for leaks.

37. Check the ignition timing and the idle speed.

38. Check all fluid levels.

39. Check the vehicle alignment.

Mazda 6

2.3L (L3) ENGINE

1. Before servicing the vehicle, refer to the precautions in the beginning of this section.

2. Remove the battery and battery tray.

3. Remove the shroud panel.

4. Remove the radiator.

5. Drain the transaxle fluid.

6. Remove the plug hole plate.

7. Remove the power steering pump with the lines attached and set aside.

8. Remove the A/C compressor with the lines still attached and wire the compressor out of the way.

9. Remove the joint shaft as follows:

   a. Remove the ABS sensor.

   b. Separate the tie rod ends from the knuckle.

   c. Remove the damper fork bolt.

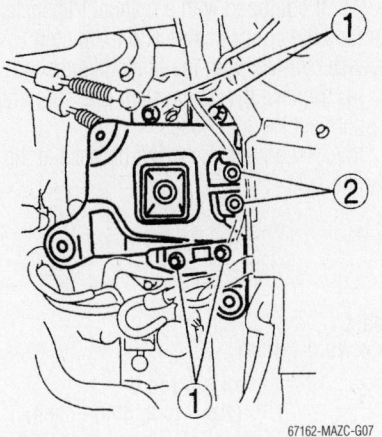

67162-MAZC-G07

Tighten the No. 4 engine mount and battery bracket nuts and bolts in this sequence—Mazda 3 models

d. Separate the lower control arm front and rear ball joints.

e. Remove the stabilizer bar link nut and the link from the damper fork.

f. Remove the joint shaft bracket bolt.

g. Remove the halfshafts.

h. Disconnect the right halfshaft from

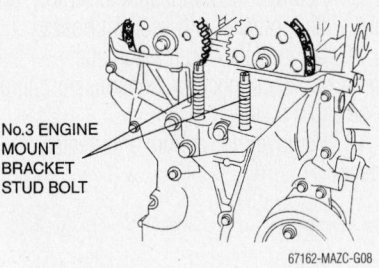

No.3 ENGINE MOUNT BRACKET STUD BOLT

67162-MAZC-G08

Location of the No. 3 engine mount bracket stud bolts—Mazda 3 models

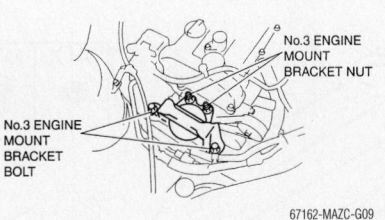

No.3 ENGINE MOUNT BRACKET NUT

No.3 ENGINE MOUNT BRACKET BOLT

67162-MAZC-G09

Location of the No. 3 engine joint bracket nuts and bolts—Mazda 3 models

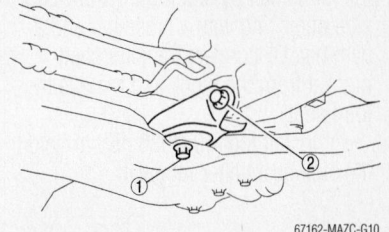

67162-MAZC-G10

Tighten the No. 1 mount bolts in the sequence illustrated—Mazda 3 models

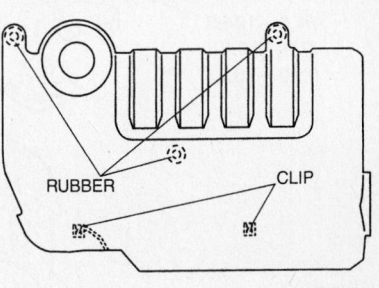

RUBBER   CLIP

67162-MAZC-G112

Plug hole plate assembly—Mazda 6 with 2.3L (L3) engine

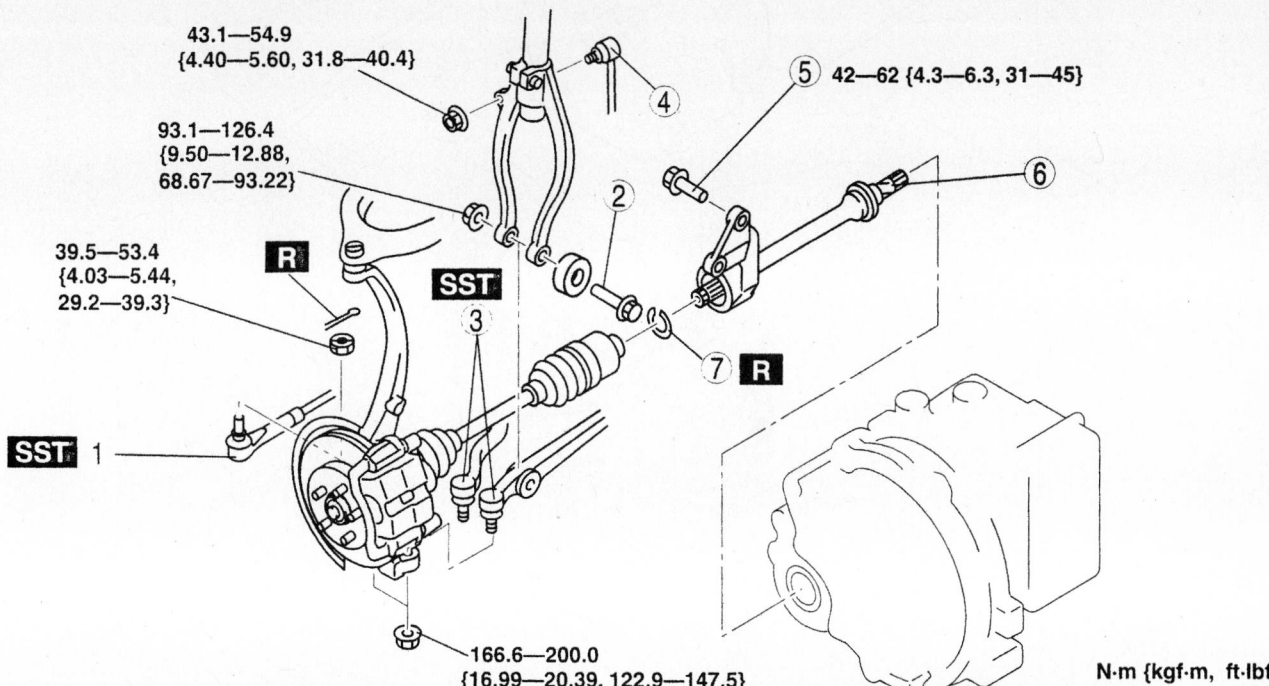

43.1—54.9
{4.40—5.60, 31.8—40.4}

93.1—126.4
{9.50—12.88, 68.67—93.22}

39.5—53.4
{4.03—5.44, 29.2—39.3}

42—62 {4.3—6.3, 31—45}

166.6—200.0
{16.99—20.39, 122.9—147.5}

N·m {kgf·m, ft·lbf}

1	Tie-rod end ball joint	5	Joint shaft bracket bolt
2	Bolt	6	Joint shaft
3	Lower arm (front, rear) ball joint	7	Clip
4	Stabilizer control link		

67162-MAZC-G125

Exploded view of the joint shaft assembly—Mazda 6 models with the 2.3L (L3) engine

the joint shaft by tapping the transaxle side outer ring with a brass bar and a hammer. Disconnect the joint shaft bracket from the block and remove the joint shaft.

i. Install tool 49 G030 455 to hold the side gears after removal.

10. Remove the air cleaner assembly, air intake duct, bracket and vacuum hoses.

11. On models with a automatic transaxle, disconnect the transmission fluid hose and the selector cable.

12. Remove the vacuum and heater hoses

13. If equipped with a manual transaxle, remove the control cable and the clutch release cylinder with the line still attached.

14. Relieve the fuel system pressure and disconnect the fuel hoses.

15. Disconnect the wring harness at the engine side.

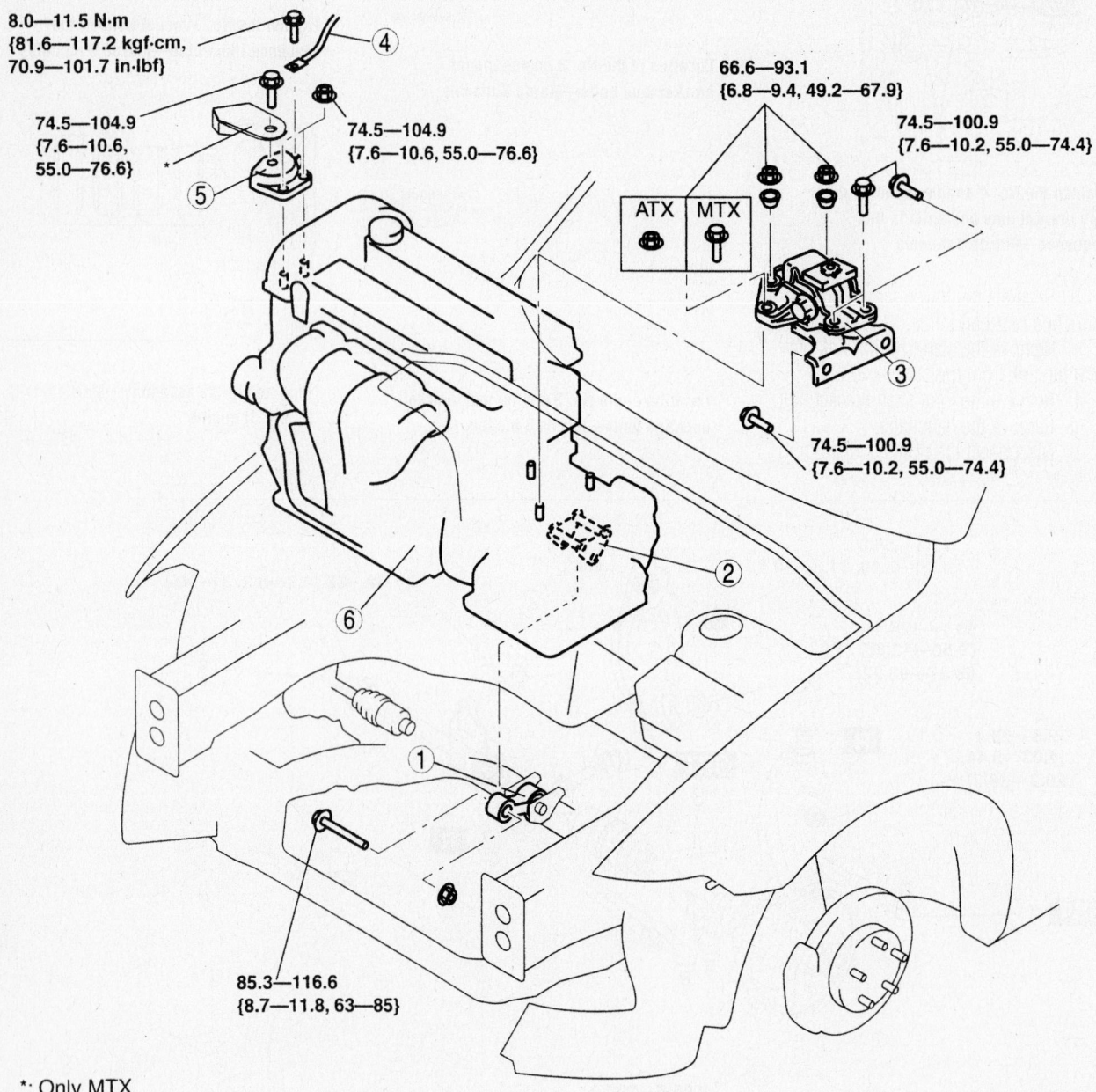

8.0—11.5 N·m
{81.6—117.2 kgf·cm,
70.9—101.7 in·lbf}

74.5—104.9
{7.6—10.6,
55.0—76.6} *

74.5—104.9
{7.6—10.6, 55.0—76.6}

66.6—93.1
{6.8—9.4, 49.2—67.9}

74.5—100.9
{7.6—10.2, 55.0—74.4}

ATX   MTX

74.5—100.9
{7.6—10.2, 55.0—74.4}

85.3—116.6
{8.7—11.8, 63—85}

*: Only MTX

1   No.1 Engine mount rubber

2   No.1 Engine mount bracket

3   No.4 Engine mount bracket and No.4 Engine mount rubber

4   Engine ground

5   No.3 Engine joint bracket

6   Engine, transaxle

Exploded view of the engine mounting—Mazda 6 models with the 2.3L (L3) engine

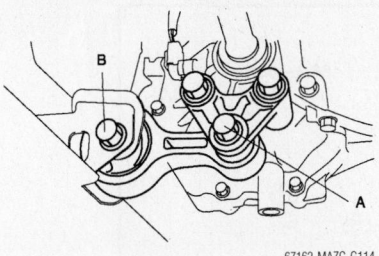

Locations of bolts A and B on the No. 1 mount rubber—Mazda 6 models with the 2.3L (L3) engine

16. Remove the front exhaust pipe.

17. Remove the No. 1 mount rubber bolt A on the engine mount bracket side, loosen bolt B on the crossmember side until about 3 pitches are showing. DO NOT remove the N0, 1 mount rubber from the vehicle.

18. Remove the No. 1 engine mount bracket.

19. Secure the engine and transaxle using a hoist.

20. Remove the No. 4 mount bracket and rubber as a unit.

21. Remove the engine ground.

22. Remove the No. 3 engine joint bracket.

23. Remove the engine and transaxle from the vehicle as an assembly.

24. Remove the engine–to–transaxle bolts and separate the engine from the transaxle.

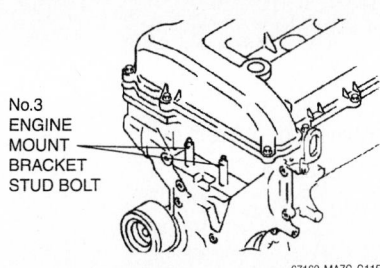

No. 3 engine mount bracket stud bolt locations—Mazda 6 models with the 2.3L (L3) engine

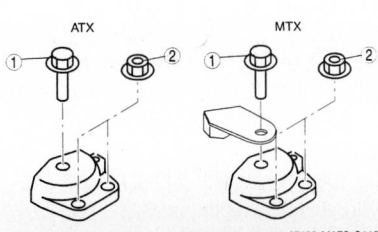

Tighten the No. 3 joint bracket bolt and nut in the order shown—Mazda 6 models with the 2.3L (L3) engine

To install:

25. Installation is the reverse of removal, please keep in mind the following:

26. On models with an automatic transaxle, tighten the engine–to–transaxle bolts to 28–38 ft. lbs. (37–52 Nm)

27. On models with a manual transaxle, tighten the engine–to–transaxle bolts to 28–36 ft. lbs. (37–50 Nm)

28. When installing the No. 3 engine joint bracket, tighten the mount bracket stud bolt to 62–115 inch lbs. (7–13 Nm) and the joint bracket bolt and nut in the order illustrated to 55–76 ft. lbs. (74–105 Nm).

29. Install the No. 4 engine mount bracket and rubber. Tighten the No.4 mount bracket and rubber bolts and nuts in the order illustrated as follows:

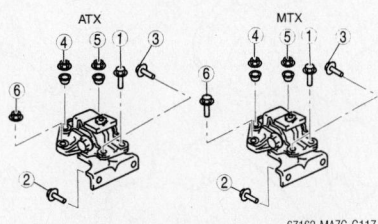

Tighten the No. 4 mount bracket and rubber bolts and nuts in the order shown—Mazda 6 models with the 2.3L (L3) engine

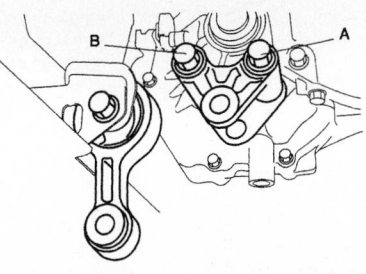

Tighten the No. 1 mount bracket bolts in the order shown—Mazda 6 models with the 2.3L (L3) engine

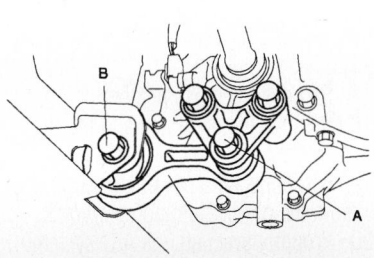

Tighten the No. 1 mount rubber bolts in the order shown—Mazda 6 models with the 2.3L (L3) engine

a. Bolts numbered 1, 2 and 3: 43–58 ft. lbs. (58–80 Nm).

b. Bolts numbered 4, 5 and 6: 49–68 ft. lbs. (66–93 Nm).

30. Install the No. 1 engine mount bracket as follows:

a. Install and tighten bolt A to 68–86 ft. lbs. (93–116 Nm). refer to the illustration for bolt location.

b. Install and tighten bolt B to 68–86 ft. lbs. (93–116 Nm). refer to the illustration for bolt location.

31. Install the No. 1 engine mount rubber as follows:

a. Install and tighten bolt A to 63–85 ft. lbs. (85–116 Nm). refer to the illustration for bolt location.

b. Install and tighten bolt B to 68–86 ft. lbs. (93–116 Nm). refer to the illustration for bolt location.

32. Fill the engine and the transaxle with the proper type and amount of fluids. Fill the cooling system.

33. Connect the negative battery cable, start the engine and check for leaks.

34. Check the ignition timing and the idle speed.

35. Check all fluid levels.

36. Check the vehicle alignment.

3.0L (AJ) ENGINE

1. Before servicing the vehicle, refer to the precautions in the beginning of this section.

2. Remove the under cover.

3. Drain and recycle the engine coolant.

4. Drain the transaxle fluid.

5. Remove the battery and battery tray.

6. Remove the air cleaner assembly, air intake duct, bracket and vacuum hoses.

7. On models with a automatic transaxle, disconnect the transmission fluid hose and the selector cable.

8. If equipped with a manual transaxle, remove the shift and selector cables and the clutch release cylinder with the line still attached.

9. Relieve the fuel system pressure and disconnect the fuel hoses.

10. Disconnect the engine wiring and powertrain Control Module (PCM) harness connections.

11. Remove the joint shaft as follows:

a. Remove the ABS sensor.

b. Remove the halfshaft lock nut.

c. Separate the tie rod ends from the knuckle.

d. Remove the damper fork.

e. Separate the lower control arm front and rear ball joints.

f. Remove the stabilizer bar link nut and the link from the damper fork.

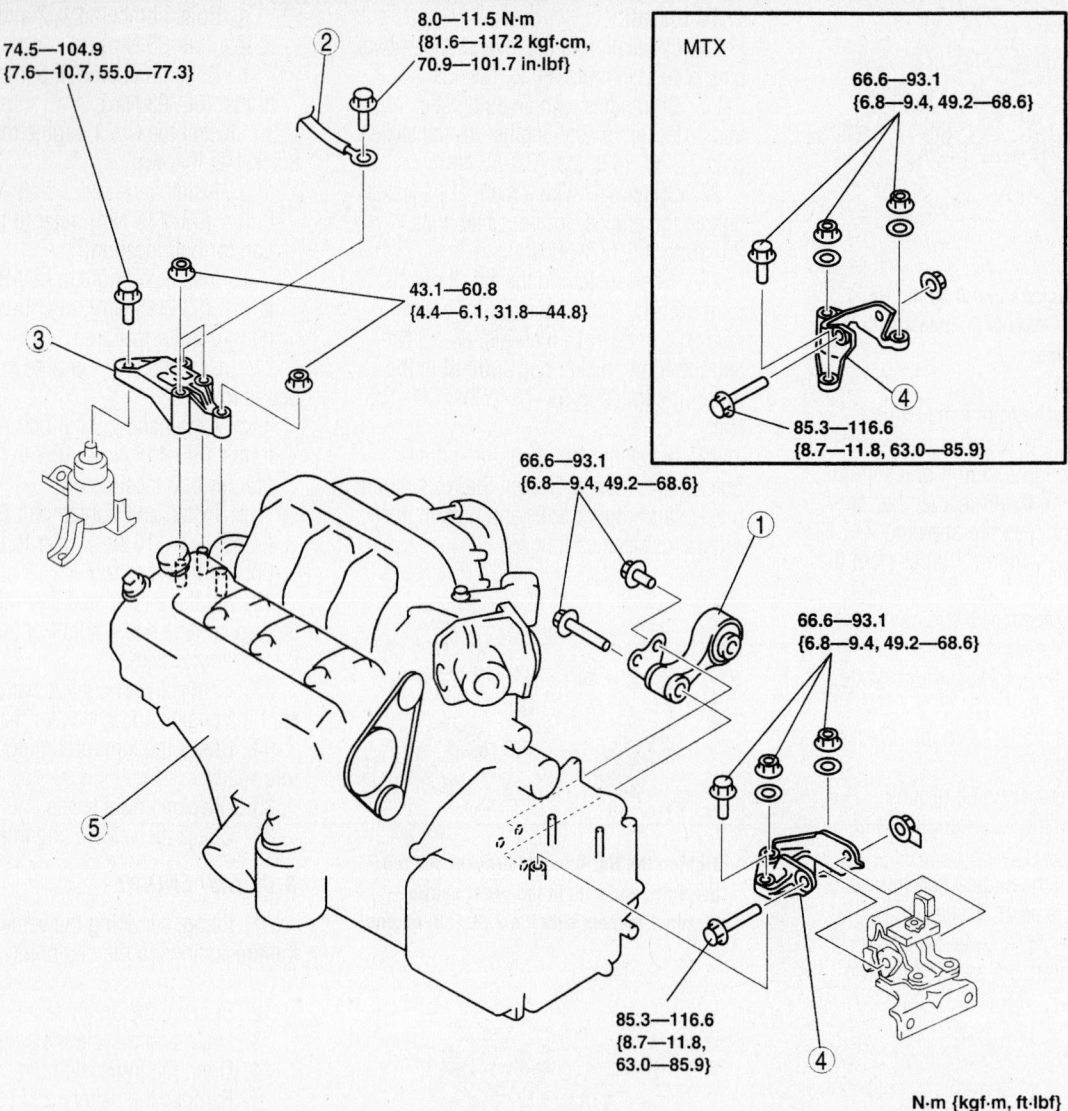

74.5—104.9
{7.6—10.7, 55.0—77.3}

8.0—11.5 N·m
{81.6—117.2 kgf·cm,
70.9—101.7 in·lbf}

MTX

66.6—93.1
{6.8—9.4, 49.2—68.6}

85.3—116.6
{8.7—11.8, 63.0—85.9}

43.1—60.8
{4.4—6.1, 31.8—44.8}

66.6—93.1
{6.8—9.4, 49.2—68.6}

66.6—93.1
{6.8—9.4, 49.2—68.6}

85.3—116.6
{8.7—11.8,
63.0—85.9}

N·m {kgf·m, ft·lbf}

1 No.1 engine mount rubber
2 Engine ground
3 No.3 engine joint bracket

4 No.4 engine mount bracket
5 Engine, transaxle

67162-MAZC-G120

Exploded view of the engine mounting—Mazda 6 models with the 3.0L (AJ) engine

g. Remove the joint shaft bracket bolt.

h. Remove the halfshafts.

i. Disconnect the left halfshaft from the joint shaft by inserting a prybar between the transaxle and the halfshaft outer ring. Disconnect the joint shaft bracket from the block and remove the joint shaft.

j. Install tool 49 G030 455 to hold the side gears after removal

12. Remove the transverse member.

13. Remove the steering gear and linkage assembly from the front crossmember and use mechanics wire to support the gear and linkage away from the crossmember.

14. Lower front shock absorber bolt.

15. No. 1 engine mount center bolt.

16. Support the crossmember with a jack and remove the nuts and the crossmember bracket.

17. Remove the crossmember.

18. Remove the drive belt and the dipstick tube.

19. Disconnect the power steering hoses and drain the power steering reservoir.

20. Properly evacuate the A/C system using approved equipment.

21. Remove the A/c compressor.

22. Remove the left hand three way catalytic converter.

23. Remove the front exhaust pipe.

24. Remove the radiator and heater hoses.

25. Remove the No. 1 engine mount rubber.

26. Disconnect the engine ground.

27. Support the engine using an engine jack and attachment as shown in the illustration.

28. Remove the No. 3 joint bracket and the No. 4 engine mount bracket.

29. Remove the No. 4 engine mount bracket.

30. Remove the engine and transaxle from the vehicle as an assembly.

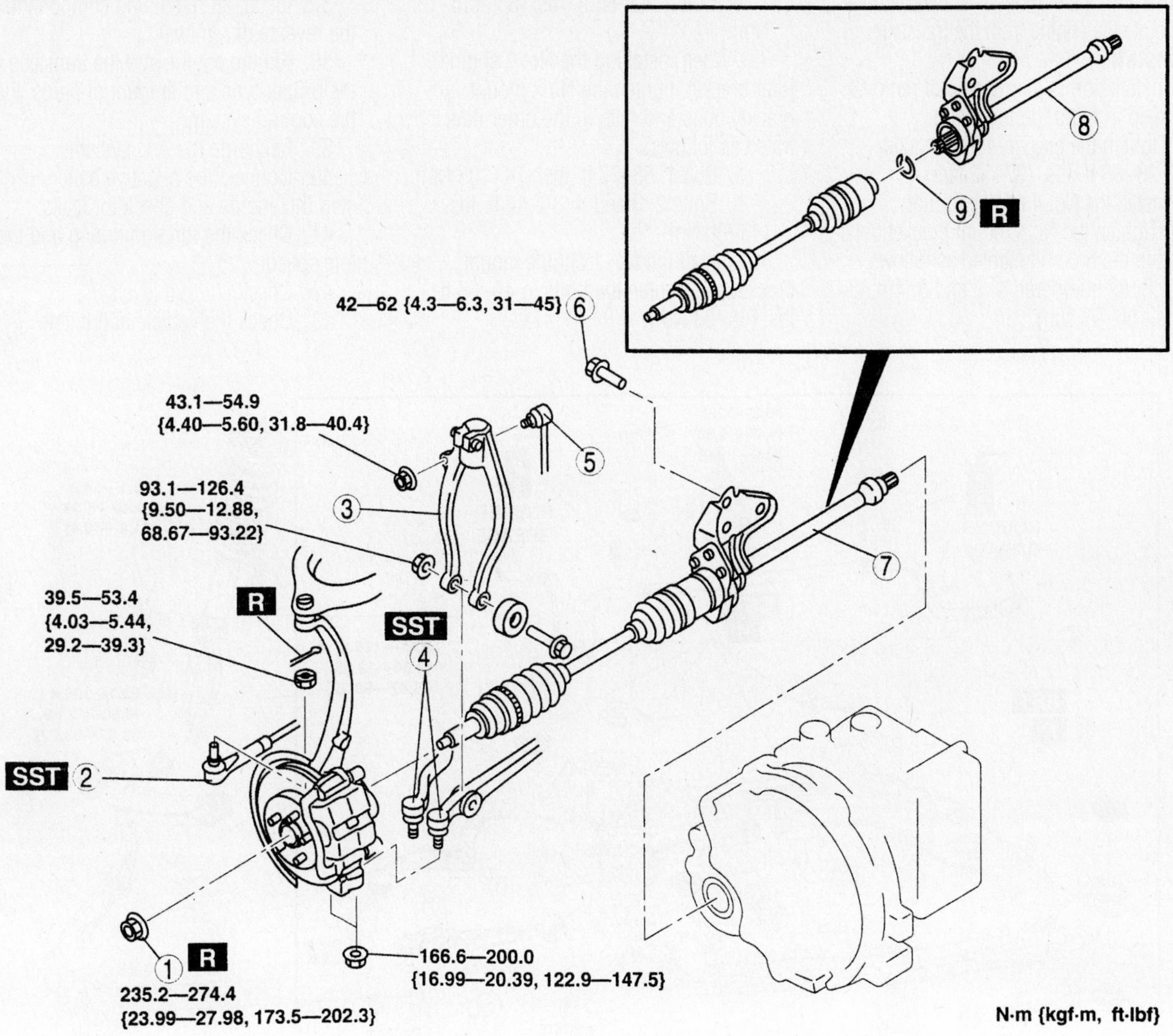

42—62 {4.3—6.3, 31—45}
6
43.1—54.9
{4.40—5.60, 31.8—40.4}
5
93.1—126.4
{9.50—12.88,
68.67—93.22}
3
39.5—53.4
{4.03—5.44,
29.2—39.3}
R
SST
4
SST 2
7
8
9 R
1 R
235.2—274.4
{23.99—27.98, 173.5—202.3}
166.6—200.0
{16.99—20.39, 122.9—147.5}
N·m {kgf·m, ft·lbf}

1	Locknut	5	Stabilizer control link
2	Tie-rod end ball joint	6	Joint shaft bracket bolt
3	Damper fork	7	Drive shaft and joint shaft
4	Lower arm (front, rear) ball joint	8	Joint shaft
		9	Clip

67162-MAZC-G121

Exploded view of the joint shaft assembly—Mazda 6 models with the 3.0L (AJ) engine

ENGINE, TRANSAXLE
ATTACHMENT
ENGINE JACK

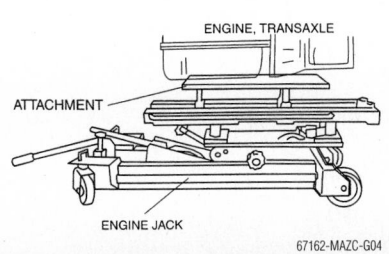

67162-MAZC-G04

Support the engine using an engine jack and attachment as shown—Mazda 6 models with the 3.0L (AJ) engine

ATX
MTX
1
2
3
4

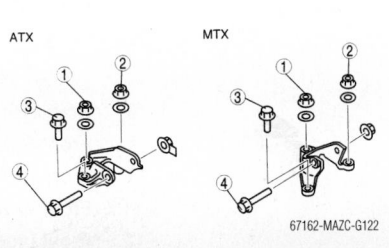

67162-MAZC-G122

Tighten the No. 4 engine joint bracket as shown—Mazda 6 models with the 3.0L (AJ) engine

No.3 ENGINE
MOUNT RUBBER
1
2
3
4
No.3 ENGINE
JOINT BRACKET

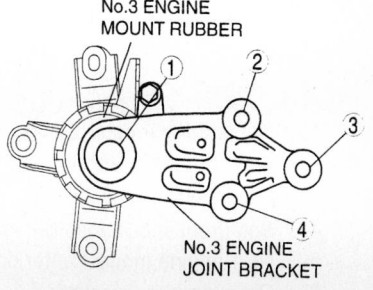

67162-MAZC-G123

Tighten the No. 3 engine joint bracket as shown—Mazda 6 models with the 3.0L (AJ) engine

31. Remove the engine–to–transaxle bolts and separate the engine from the transaxle.

To install:

32. Installation is the reverse of removal, please keep in mind the following:

33. Tighten the engine–to–transaxle bolts to 28–38 ft. lbs. (37–52 Nm)

34. Install the No. 4 engine mount bracket. Tighten the No.4 mount bracket bolts and nuts in the order illustrated as follows:

    a. Bolts numbered 1, 2 and 3: 49–68 ft. lbs. (66–93 Nm).

    b. Bolt 4: 63–86 ft. lbs. (85–116 Nm).

35. When installing the No. 3 engine joint bracket, tighten the No.3 mount bracket bolts and nuts in the order illustrated as follows:

    a. Bolt 1: 55–77 ft. lbs. (74–105 Nm).

    b. Bolts 2, 3 and 4: 31–44 ft. lbs. (43–60 Nm).

36. Install the No. 1 engine mount bracket and tighten the bolts to 49–68 ft. lbs. (66–93 Nm).

37. Install all remaining components in the reverse of removal.

38. Fill the engine and the transaxle with the proper type and amount of fluids. Fill the cooling system.

39. Recharge the A/C system.

40. Connect the negative battery cable, start the engine and check for leaks.

41. Check the ignition timing and the idle speed.

42. Check all fluid levels.

43. Check the vehicle alignment.

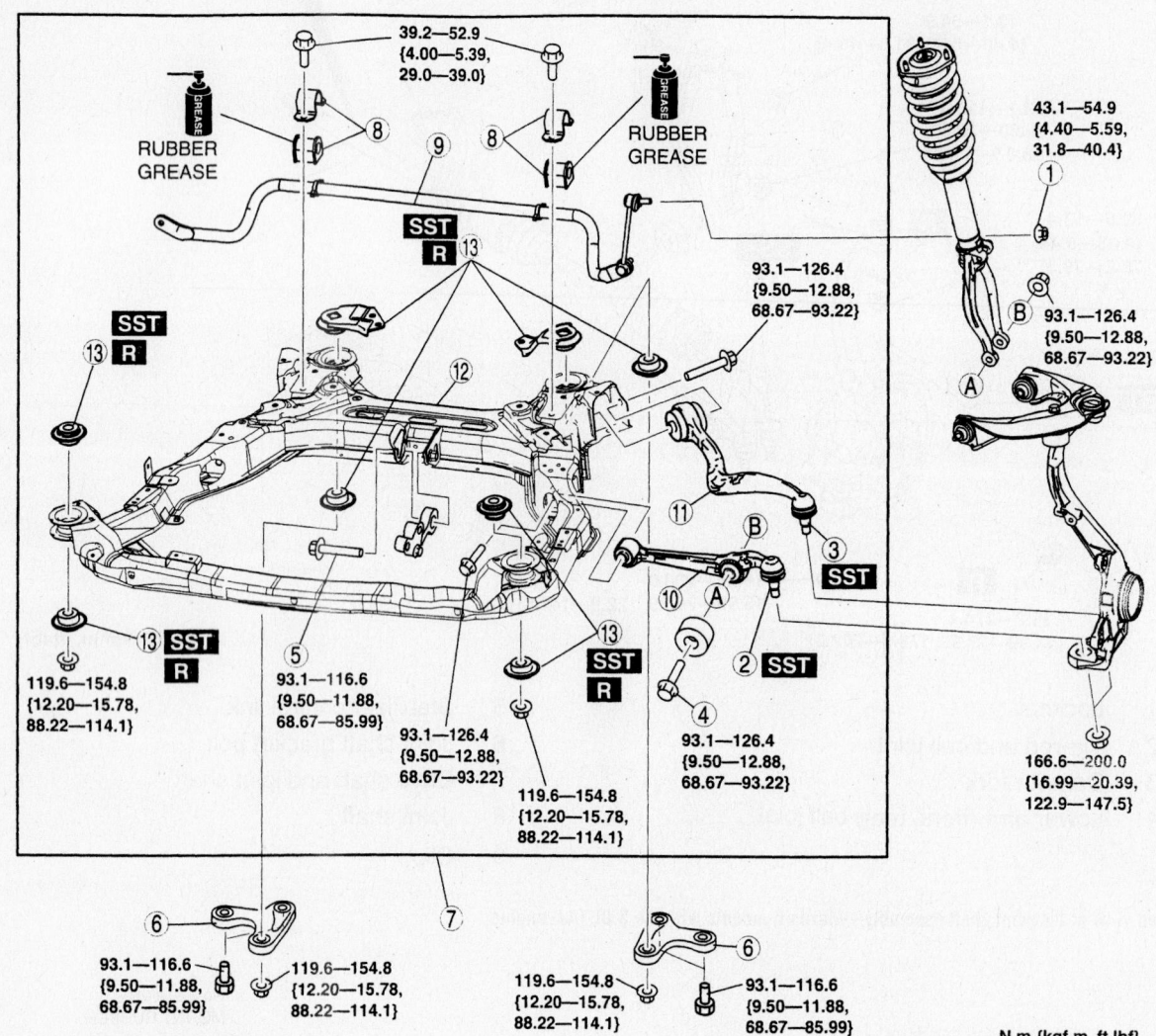

1	Nut (stabilizer control link)
2	Front lower arm (front) ball joint
3	Front lower arm (rear) ball joint
4	Bolt (front shock absorber lower side)
5	No.1 engine mount center bolt
6	Crossmember bracket
7	Crossmember component
8	Stabilizer bracket and bushing
9	Front Stabilizer
10	Front lower arm (front)
11	Front lower arm (rear)
12	Front crossmember
13	Front crossmember bushing

67162-MAZC-G124

Exploded view of the front crossmember, related components and their torque specifications—Mazda 6 models with the 3.0L (AJ) engine

Water Pump

REMOVAL & INSTALLATION

1.6L (ZM) Engines

1. Before servicing the vehicle, refer to the precautions in the beginning of this section.
2. Drain the cooling system.
3. Remove or disconnect the following:
   - Negative battery cable

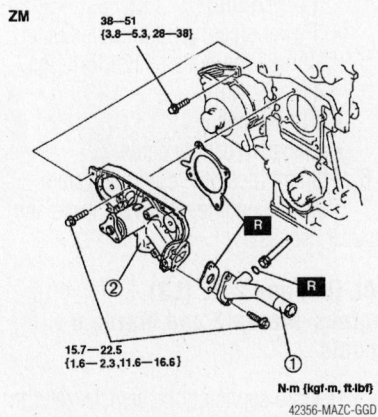

ZM

38—51
{3.8—5.3, 28—38}

15.7—22.5
{1.6— 2.3,11.6—16.6}

N·m {kgf·m, ft·lbf}

42356-MAZC-GGD

Exploded view of the water pump assembly—1.6L (ZM) engines

- Fresh air duct
- Exhaust manifold insulator
- Timing belt
- Power steering oil pump with the lines attached and set aside
- A/C compressor with the lines attached and set aside
- Water inlet pipe
- Water pump mounting bolts
- Water pump

To install:

4. Clean all gasket mating surfaces.
5. Install or connect the following:
   - Water pump using a new gasket. Torque the main bolts to 12–17 ft. lbs. (16–23 Nm) and the upper left hand bolt to 28–38 ft. lbs. (38–51 Nm).
   - Water inlet pipe with new gasket and O–ring and tighten the bolts to 12–17 ft. lbs. (16–23 Nm)
   - A/C compressor
   - Power steering oil pump
   - Timing belt
   - Exhaust manifold insulator
   - Fresh air duct
   - Negative battery cable
6. Fill and bleed the cooling system.
7. Start the engine, check for leaks and repair if necessary.

1.8L (FP) Engines

1. Before servicing the vehicle, refer to the precautions in the beginning of this section.
2. Drain the cooling system.
3. Remove or disconnect the following:
   - Negative battery cable
   - Timing belt
   - Power steering oil pump adjuster
   - Water pump mounting bolts
   - Water pump

To install:

4. Clean all gasket mating surfaces.
5. Install or connect the following:
   - Water pump using a new gasket. Make sure the gasket is installed with the sealing ring facing the pump. Torque the bolts to 14–18 ft. lbs. (19–25 Nm).
   - Power steering oil pump adjuster
   - Timing belt
   - Negative battery cable
6. Fill and bleed the cooling system.
7. Start the engine, check for leaks and repair if necessary.

1.8L (BP) Engines

1. Before servicing the vehicle, refer to the precautions in the beginning of this section.
2. Disconnect the negative battery cable.
3. Drain the engine coolant.
4. Remove or disconnect the following:
   - Timing belt covers and timing belt
   - Power steering pump, leaving the lines attached and set aside
   - Idler, on models not equipped with power steering
   - Coolant inlet pipe and gasket
   - Water pump

To install:

5. Clean all gasket mating surfaces.
6. Using a new gasket, install the water pump on the engine. Torque the mounting bolts to 14–18 ft. lbs. (19–25 Nm). Torque

19—25
{1.9—2.6, 14—18}

19—25
{1.9—2.6, 14—18}

N·m {kgf·m, ft·lbf}

1   P/S oil pump adjuster

2   Water pump

42356-MAZC-GJC

Exploded view of the water pump assembly—1.8L (FP) engines

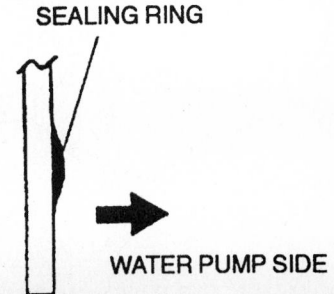

SEALING RING

WATER PUMP SIDE

42356-MAZC-GJD

Make sure the gasket is installed with the sealing ring facing the pump—1.8L (FP) engines

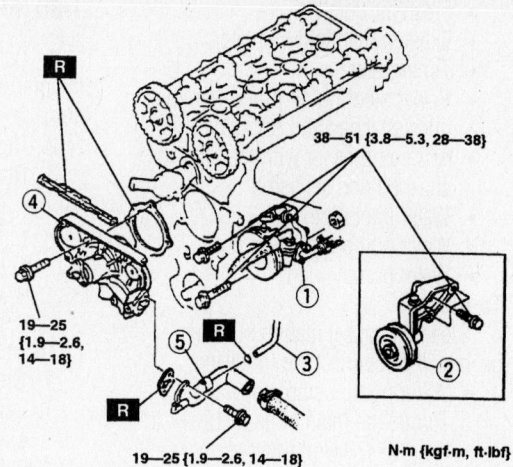

1. P/S oil pump
2. Idler (without P/S oil pump)
3. Water hose
4. Water pump
5. Water inlet pipe

38—51 {3.8—5.3, 28—38}

19—25 {1.9—2.6, 14—18}

19—25 {1.9—2.6, 14—18}

N·m {kgf·m, ft·lbf}

42356-MAZC-G94

Exploded view of the water pump assembly—1.8L (BP) engines

the bolt from the water pump to the alternator bracket to 28–38 ft. lbs. (38–51 Nm).

7. Install or connect the following:
- Coolant inlet pipe with a new gasket. Torque the coolant inlet pipe bolts to 14–18 ft. lbs. (19–25 Nm).
- Idler, on models not equipped with power steering
- Power steering pump
- Timing belt and the timing belt covers

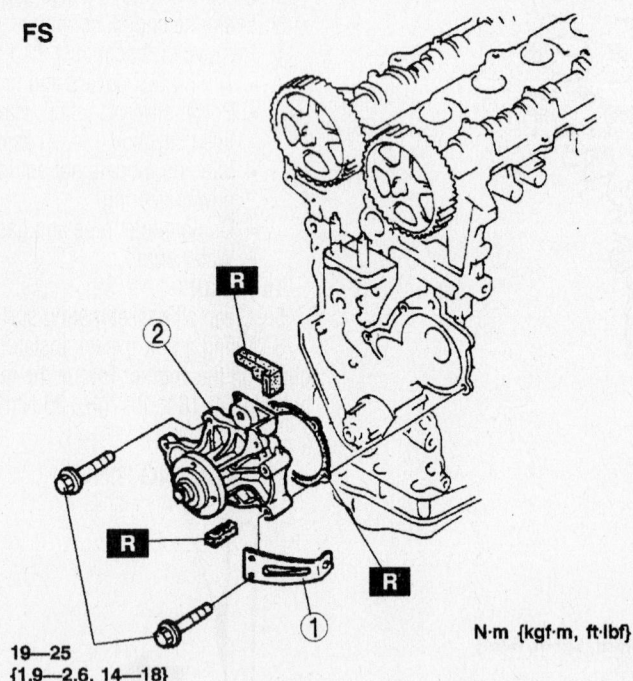

FS

19—25 {1.9—2.6, 14—18}

N·m {kgf·m, ft·lbf}

1   P/S oil pump adjuster
2   Water pump

42356-MAZC-G05A

Exploded view of the water pump assembly—2.0L (FS) engines

- Negative battery cable
8. Fill and bleed the cooling system.
9. Start the engine and bring to normal operating temperature. Check for leaks.

2.0L (FS) Engines

1. Before servicing the vehicle, refer to the precautions in the beginning of this section.

2. Drain the cooling system.
3. Remove or disconnect the following:
- Negative battery cable
- Timing belt
- Power steering oil pump adjuster
- Water pump mounting bolts
- Water pump

To install:
4. Clean all gasket mating surfaces.
5. Install or connect the following:
- Water pump using a new gasket. Torque the bolts to 14–18 ft. lbs. (19–25 Nm).
- Power steering oil pump adjuster. Torque the bolts to 12–16 ft. lbs. (16–22 Nm).
- Timing belt
- Negative battery cable
6. Fill and bleed the cooling system.
7. Start the engine, check for leaks and repair if necessary.

2.0L (LF) and 2.3L (L3) Engines–Mazda 3 and Mazda 6 Models

1. Before servicing the vehicle, refer to the precautions in the beginning of this section.

2. Remove the battery cover and disconnect the negative battery cable.
3. Remove the under cover and splash shield as an assembly.
4. Drain the cooling system.
5. Position the coolant reservoir tank aside with the hose still attached.
6. Remove the plug hole plate.
7. Loosen the water pump pulley bolt and position the drive belt aside.
8. Remove the water pump pulley.
9. Remove the water pump bolts, the pump and the O–ring.

To install:
10. Clean the water pump mating surfaces.
11. Install a new O–ring and the water pump. Tighten the bolts to 71–101 inch lbs. (8–11 Nm).
12. Install the water pump pulley and tighten the tighten the bolts to 12–17 ft. lbs. (17–23 Nm).
13. Install the drive belt.
14. Install the plug hole plate.
15. Install the coolant reservoir tank.
16. Install the splash shield and under cover.
17. Connect the battery cable and install the cover.
18. Fill the cooling system.
19. Start the engine, check for leaks and repair if necessary.

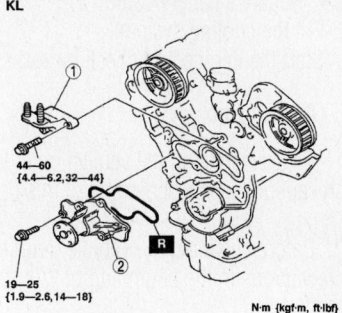

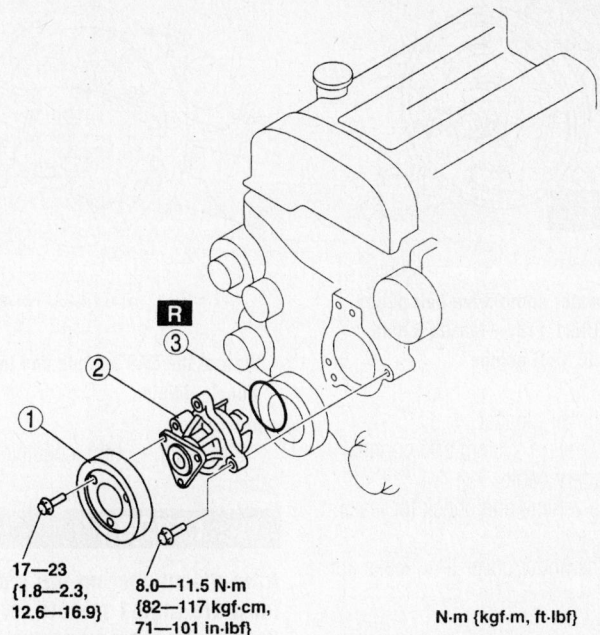

1   Water pump pulley
2   Water pump
3   O-ring

67162-MAZC-G11

Water pump mounting and related components—Mazda 3 and Mazda 6 models equipped with the 2.0L (LF) and 2.3L (L3) engines

1   No.3 engine mount bracket
2   Water pump

42356-MAZC-G06

Exploded view of the water pump assembly—2.5L (KL) engines

- Water pump bolts
- Water pump

To install:

4. Clean the mating surfaces.
5. Install or connect the following:
   - Water pump using a new gasket. Tighten the bolts to 14–18 ft. lbs. (19–25 Nm).
   - Number 3 engine mount bracket and tighten the bolts to 32–44 ft. lbs. (44–60 Nm)
   - Timing belt

2.3L (KJ) Engine

1. Before servicing the vehicle, refer to the precautions in the beginning of this section.
2. Drain the cooling system.
3. Remove or disconnect the following:
   - Negative battery cable
   - Timing belt and water pump together

To install:

4. Clean the mating surfaces.
5. Install or connect the following:
   - Water pump using a new gasket along with the timing belt. Torque the bolts to 14–18 ft. lbs. (19–25 Nm).
   - Negative battery cable
6. Fill the cooling system.
7. Start the engine, check for leaks and repair if necessary.

2.5L (KL) Engine

1. Before servicing the vehicle, refer to the precautions in the beginning of this section.
2. Disconnect the negative battery cable.
3. Drain the cooling system.
   - Timing belt
   - Number 3 engine mount bracket

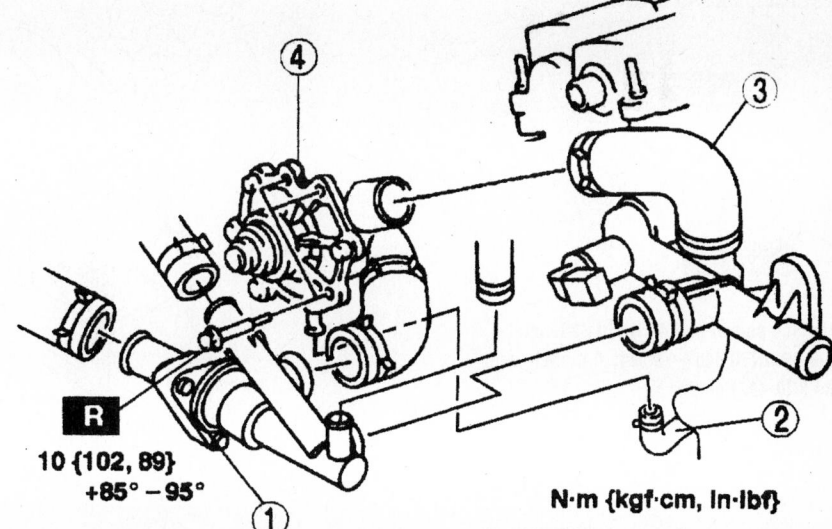

1   Thermostat case
2   Heater hose
3   Water outlet pipe
4   Water pump

67162-MAZC-G126

Exploded view of the water pump and related components—Mazda 6 models with the 3.0L (AJ) engine

- Negative battery cable
6. Fill the cooling system.
7. Start the engine and check for leaks.

3.0L (AJ) Engine

1. Before servicing the vehicle, refer to the precautions in the beginning of this section.

2. Before servicing the vehicle, refer to the precautions in the beginning of this section.

3. Disconnect the negative battery cable.

4. Remove the under cover.

5. Drain the cooling system.

6. Remove the air cleaner.

7. Remove the water pump drive belt pulley as follows:

 a. Replace part of tool 49 UN30 3009 with tool 49 UN30 3457.

 b. Install the tools as illustrated and remove the water pump drive pulley.

8. Remove the thermostat case, heater hose and the water outlet pipe.

9. Remove the water pump bolts and the pump.

To install:

10. Install the water pump and tighten the bolts 89 inch lbs. (10 Nm), plus an additional 85–95 degrees.

11. Install the water outlet pipe, heater hose and thermostat case.

12. Install a new water pump drive belt pulley using tool 49 UN21 1185.

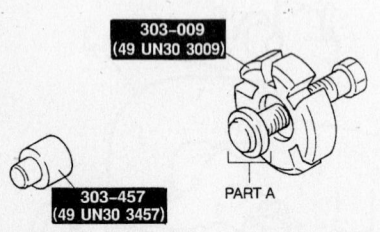

Replace part of tool 49 UN30 3009 with tool 49 UN30 3457—Mazda 6 models with the 3.0L (AJ) engine

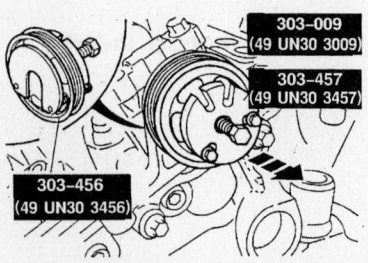

Install the tools as illustrated and remove the water pump drive pulley—Mazda 6 models with the 3.0L (AJ) engine

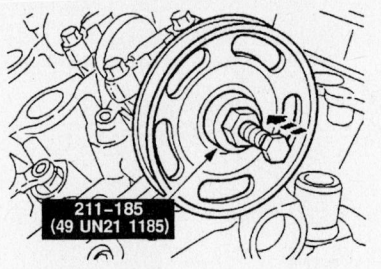

Install a new water pump drive belt pulley using tool 49 UN21 1185—Mazda 6 models with the 3.0L (AJ) engine

13. Install the air cleaner.

14. Fill the cooling system and connect the negative battery cable.

15. Start the vehicle and check for leaks, repair if necessary.

16. Install the under cover if no leaks are found.

Heater Core

REMOVAL & INSTALLATION

Miata

1. Before servicing the vehicle, refer to the precautions in the beginning of this section.

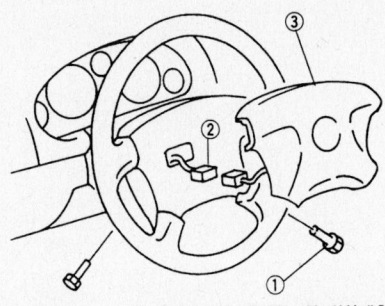

7.9—11.7 N·m {80—120 kgf·cm, 70—104 in·lbf}

View of the SAS module and the steering wheel—Miata

2. Disconnect the negative battery cable.

✳✳ CAUTION

After disconnecting the battery, wait for more than 1 minute for the SAS to deplete its stored energy.

3. Drain the cooling system into a clean container for reuse.

4. Disconnect the heater hoses from the heater core.

5. Discharge and recover the air conditioning system refrigerant.

6. At the driver's side, remove the SAS

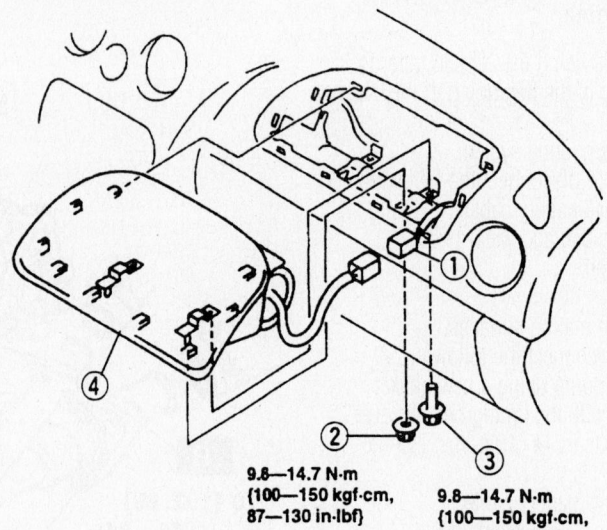

9.8—14.7 N·m {100—150 kgf·cm, 87—130 in·lbf}   9.8—14.7 N·m {100—150 kgf·cm, 87—130 in·lbf}

1	Connector
2	Nut
3	Bolt
4	Passenger-side air bag module

View of the passenger's side SAS module—Miata

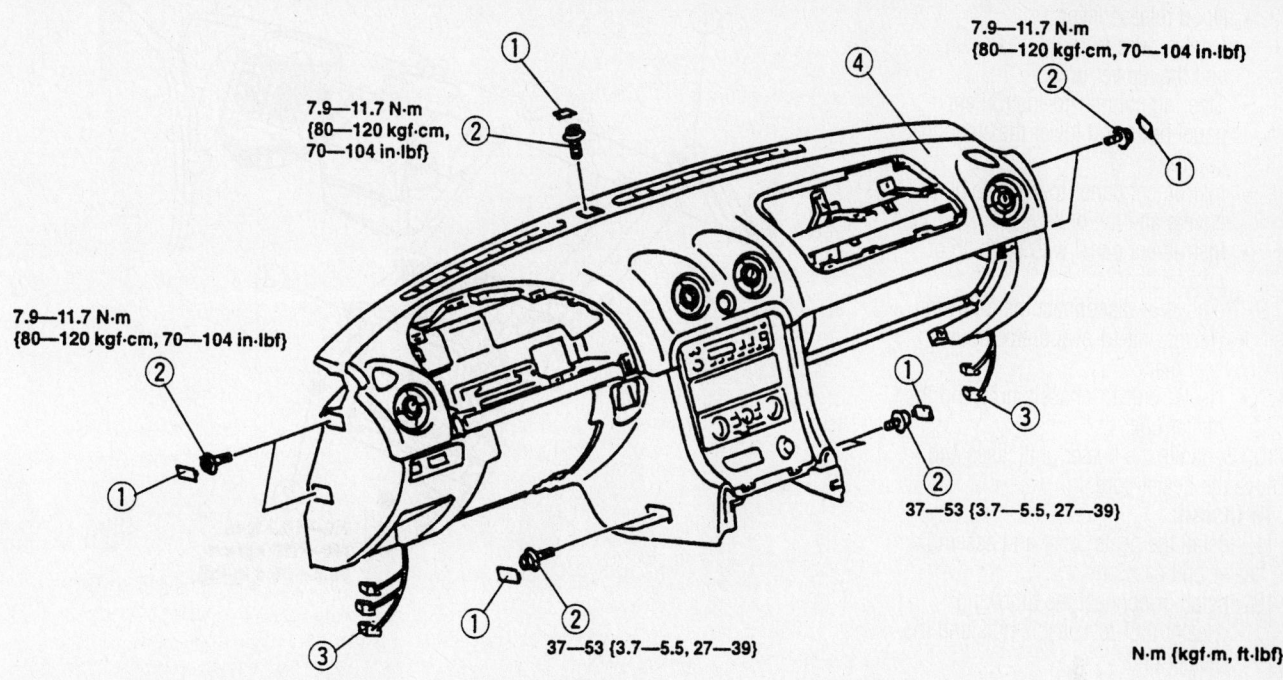

7.9—11.7 N·m
{80—120 kgf·cm, 70—104 in·lbf}

7.9—11.7 N·m
{80—120 kgf·cm, 70—104 in·lbf}

7.9—11.7 N·m
{80—120 kgf·cm, 70—104 in·lbf}

37—53 {3.7—5.5, 27—39}

37—53 {3.7—5.5, 27—39}

N·m {kgf·m, ft·lbf}

1	Cover	3	Connector
2	Bolt	4	Dashboard

93112GH2

View of the instrument panel—Miata

module and the steering wheel by removing or disconnecting the following:
- Place the wheel in the straight-ahead position and turn the ignition switch to LOCK
- Cover clips at both sides of the steering wheel
- Steering wheel-to-SAS module bolts
- SAS module from the steering wheel and disconnect the electrical connector
- Steering wheel-to-column nut
- Steering wheel from the steering column using a suitable puller

7. At the passenger's side, remove the SAS module by removing or disconnecting the following:
- Glove compartment and the glove compartment cover
- SAS module-to-dash bolts
- SAS module and disconnect the electrical connector
- Console

8. Remove the instrument cluster by removing or disconnecting the following:
- A-pillar trim at both sides
- Lower panel
- Instrument cluster hood
- Instrument cluster-to-dash screws and the instrument cluster

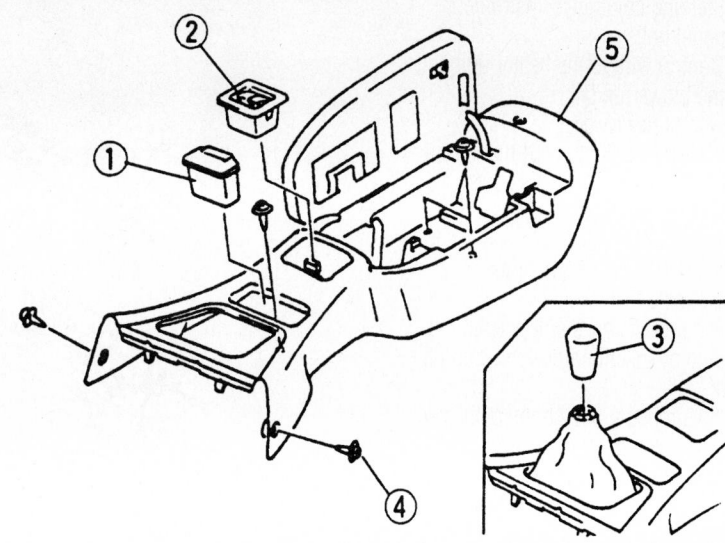

1	Ashtray
2	Power window switch
3	Shift lever knob (MT)
4	Screw
5	Console

93112GH3

View of the console—Miata

- Hood release lever
- Control wire from the heater unit and the blower unit
- Steering column-to-instrument panel bolts and lower the steering column
- Instrument panel-to-chassis bolt covers and the bolts
- Instrument panel with the help of an assistant

9. Remove or disconnect the following:
- Heater unit-to-evaporator housing seal plate
- Heater unit-to-chassis nuts and the heater unit

10. Separate the heater unit cases and remove the heater core.

To install:

11. Install the heater core and assemble the heater unit cases.

12. Install or connect the following:
- Heater unit-to-chassis nuts and the heater unit
- Heater unit-to-evaporator housing seal plate

13. Install the instrument cluster by installing or connecting the following:
- Instrument panel with the help of an assistant
- Instrument panel-to-chassis bolt covers and the bolts
- Steering column-to-instrument panel bolts
- Control wire to the heater unit and the blower unit
- Hood release lever
- Instrument cluster and the instrument cluster-to-dash screws
- Instrument cluster hood
- Lower panel
- A-pillar trim at both sides
- Console

14. At the passenger's side, install the SAS module by installing or connecting the following:
- SAS module and connect the electrical connector
- SAS module-to-dash bolts
- Glove compartment cover and the glove compartment

15. At the driver's side, install the SAS module and the steering wheel by installing or connecting the following:
- Steering wheel-to-column nut. Torque the steering wheel nut to 29–36 ft. lbs. (40–49 Nm).
- SAS module to the steering wheel and connect the electrical connector
- Steering wheel-to-SAS module bolts. Torque the steering column-to-SAS module bolts to 70–104 inch lbs. (8–12 Nm).

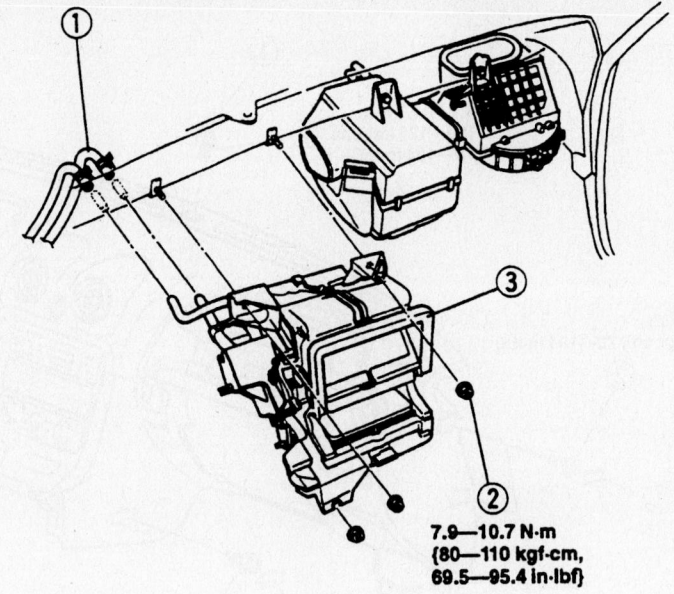

7.9—10.7 N·m
{80—110 kgf-cm,
69.5—95.4 in-lbf}

1	Heater hose
2	Nut
3	Heater unit

93112GH4

View of the heater unit and the heater unit-to-evaporator unit seal plate—Miata

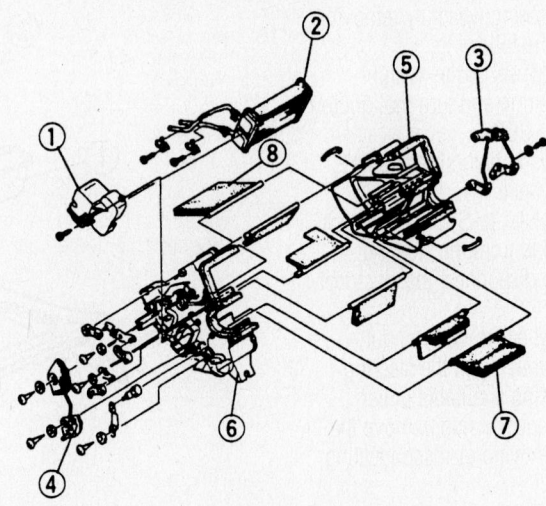

1	Cover
2	Heater core
3	Air mix link
4	Airflow mode link
5	Case (RH)
6	Case (LH)
7	Air mix door
8	Airflow mode door

93112GH5

Exploded view of the heater core and heater unit assembly—Miata

- Cover clips at both sides of the steering wheel

16. Connect the heater hoses to the heater core.

17. Refill the cooling system.

18. Connect the negative battery cable.

19. Evacuate, charge and leak test the air conditioning system refrigerant.

20. Operate the engine to normal operating temperatures; then, check the climate control operation and check for leaks.

Millenia

1. Before servicing the vehicle, refer to the precautions in the beginning of this section.

2. Disconnect the negative battery cable.

✳✳ CAUTION

After disconnecting the battery, wait for more than 1 minute for the SAS to deplete its stored energy.

3. Drain the cooling system into a clean container for reuse.

4. Discharge and recover the air conditioning system refrigerant.

5. At the driver's side, remove the SAS module and the steering wheel by removing or disconnecting the following:

- Place the wheel in the straight-ahead position and turn the ignition switch to LOCK
- Cover clips at both sides of the steering wheel
- Steering wheel-to-SAS module clips
- SAS module from the steering wheel and disconnect the electrical connector
- Steering wheel-to-column nut
- Steering wheel from the steering column using a suitable puller

6. At the passenger's side, remove the SAS module by removing or disconnecting the following:

- Glove compartment and the glove compartment cover
- SAS module-to-dash bolts
- SAS module and disconnect the electrical connector

7. Remove the rear console by removing or disconnecting the following:

- Rear console's box
- Bake boot
- Center panel
- Rear console

8. Remove or disconnect the following:

- A-pillar trim at both sides
- Undercover at the passenger's side

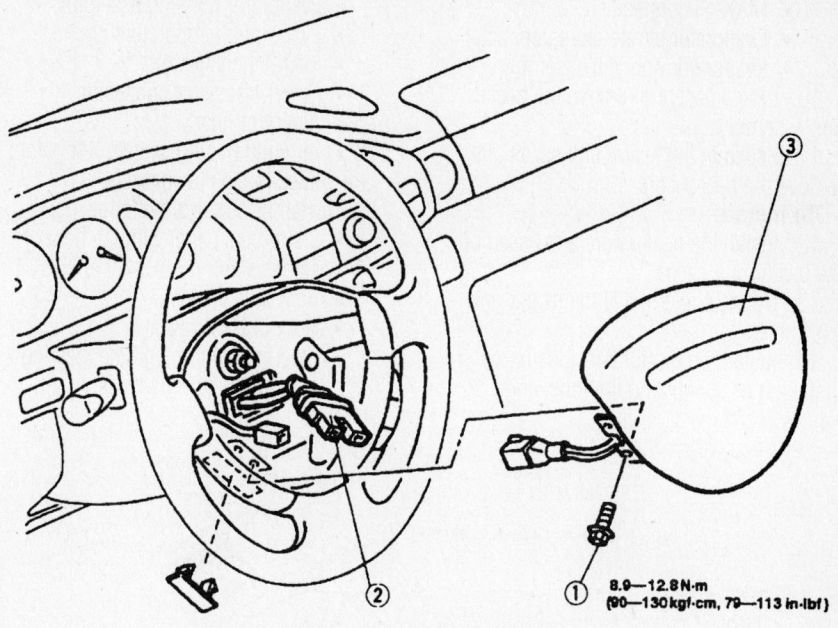

1	Bolt
2	Connector
3	Driver-side air bag module

93112GE5

Exploded view of the SAS module and steering wheel assembly—Millenia

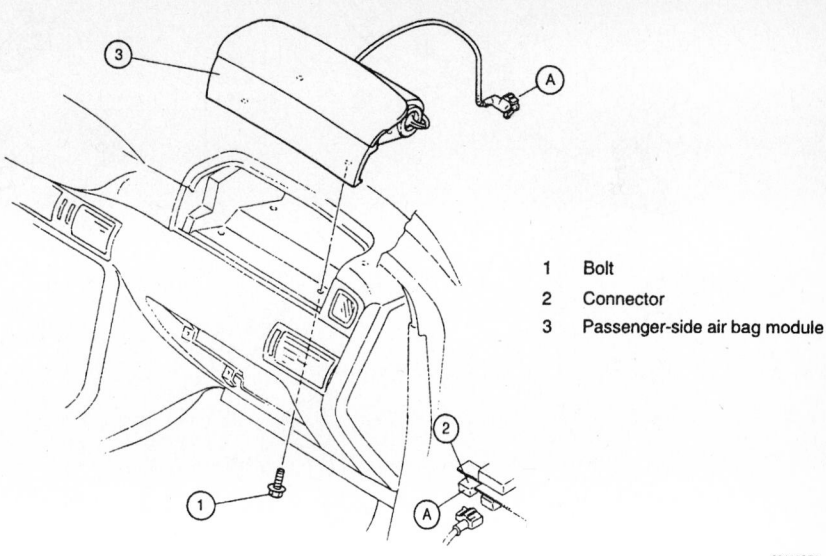

1	Bolt
2	Connector
3	Passenger-side air bag module

93112GE6

Exploded view of the passenger's side SAS module—Millenia

- Upper and lower steering column covers
- Electrical connectors and remove the combination switch from the steering column
- Meter hood
- Electrical connectors and remove the instrument cluster
- Steering column-to-chassis bolts and the steering column

- Hood release lever
- Both side panels
- Instrument panel with the help of an assistant

9. Remove the evaporator housing by removing or disconnecting the following:

- Air conditioning system refrigerant lines and discard the gaskets
- Aspirator hose
- Power transistor connector

- MAX-HI connector
- Evaporator temperature connector
- Evaporator housing assembly

10. Disconnect and remove the heater unit assembly.

11. Separate the heater unit cases and remove the heater core.

To install:

12. Install the heater core and assemble the heater unit cases.

13. Connect and install the heater unit assembly.

14. Install the evaporator housing by installing or connecting the following:

- Evaporator housing assembly
- Evaporator temperature connector
- MAX-HI connector
- Power transistor connector
- Aspirator hose
- Air conditioning system refrigerant lines using new gaskets

15. Install or connect the following:

- Instrument panel with the help of an assistant
- Both side panels
- Hood release lever
- Steering column and the steering column-to-chassis bolts

- Instrument cluster and connect the electrical connectors
- Meter hood
- Combination switch to the steering column and connect the electrical connectors
- Upper and lower steering column covers
- Undercover at the passenger's side
- A-pillar trim at both sides

16. Install the rear console by installing or connecting the following:

- Rear console
- Center panel

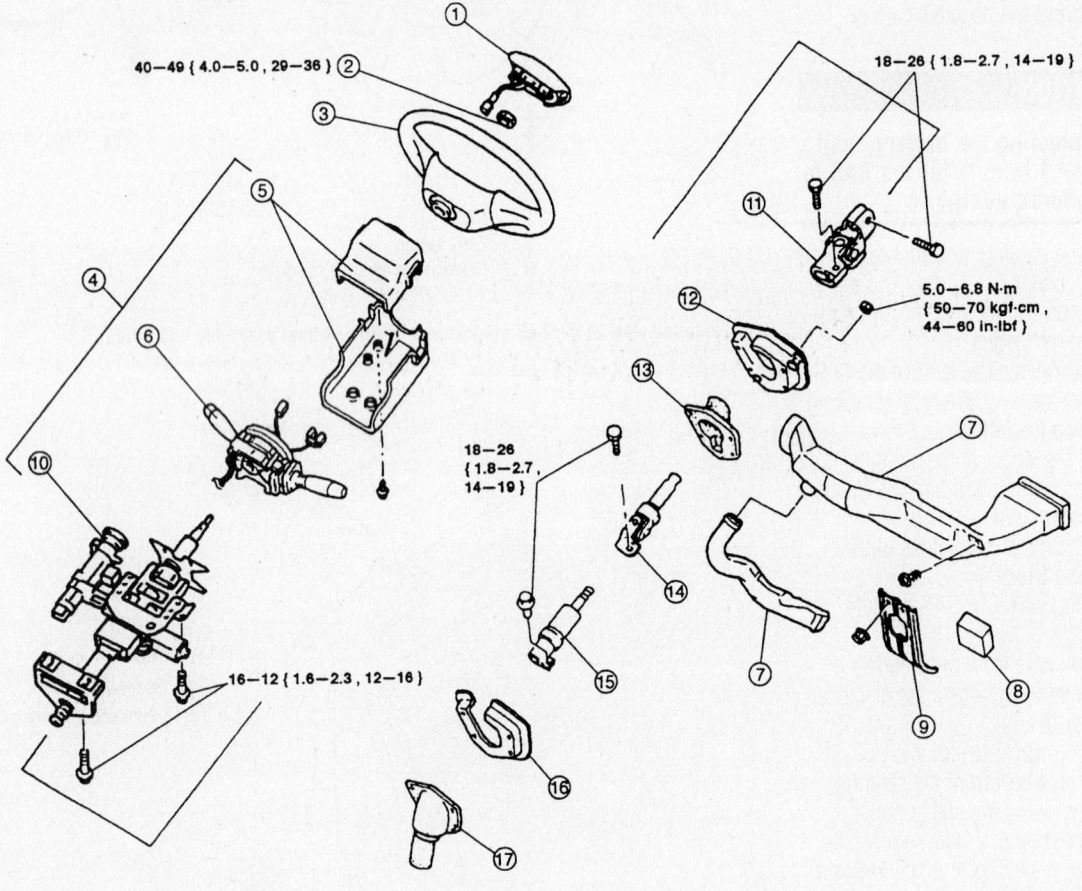

1	Air bag module
2	Locknut
3	Steering wheel
4	Dashboard, console, and steering shaft component
5	Column cover
6	Combination switch
7	Air duct
8	Flasher unit
9	Bracket
10	Steering shaft component
11	Universal joint (intermediate shaft)
12	Cover
13	Shaft seal
14	Intermediate shaft
15	Collapsible shaft
16	Set plate
17	Dust cover

N·m { kgf·m , ft·lbf }

93112GE9

Exploded view of the steering wheel and steering column assembly—Millenia

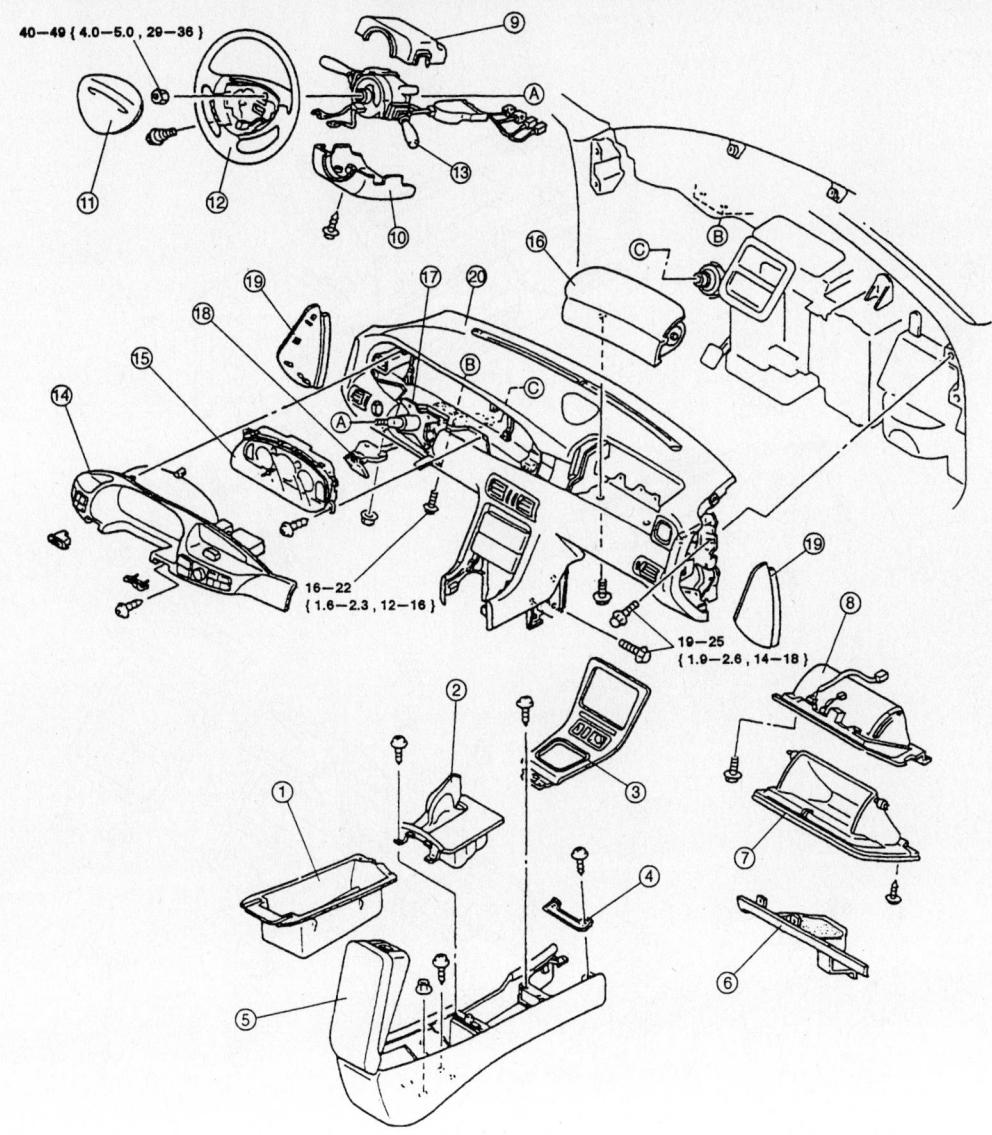

40–49 { 4.0–5.0 , 29–36 }

16—22
{ 1.6—2.3 , 12—16 }

19—25
{ 1.9—2.6 , 14—18 }

1	Rear console box	8	Glove compartment cover	15	Instrument cluster
2	Brake boot	9	Upper column cover	16	Passenger-side air bag module
3	Center panel	10	Lower column cover	17	Steering shaft
4	Bracket	11	Driver-side air bag module	18	Hood release lever
5	Rear console	12	Steering wheel	19	Side panel
6	Under cover	13	Combination switch	20	Dashboard
7	Glove compartment	14	Meter hood		

93112GE8

Exploded view of the instrument panel and rear console assemblies—Millenia

- Brake boot
- Rear console's box

17. At the passenger's side, install the SAS module by installing or connecting the following:

- SAS module and connect the electrical connector
- SAS module-to-dash bolts
- Glove compartment cover and the glove compartment

18. At the driver's side, install the SAS

module and the steering wheel by installing or connecting the following:

- Steering wheel to the steering column
- Steering wheel-to-column nut. Torque the nut to 29–36 ft. lbs. (40–49 Nm).
- SAS module to the steering wheel and connect the electrical connector. Torque the bolts to 79–113 inch lbs. (9–13 Nm).

- Steering wheel-to-SAS module clips
- Cover clips at both sides of the steering wheel

19. Refill the cooling system.
20. Connect the negative battery cable.
21. Evacuate, charge and leak test the air conditioning system refrigerant.
22. Operate the engine to normal operating temperatures; then, check the climate control operation and check for leaks.

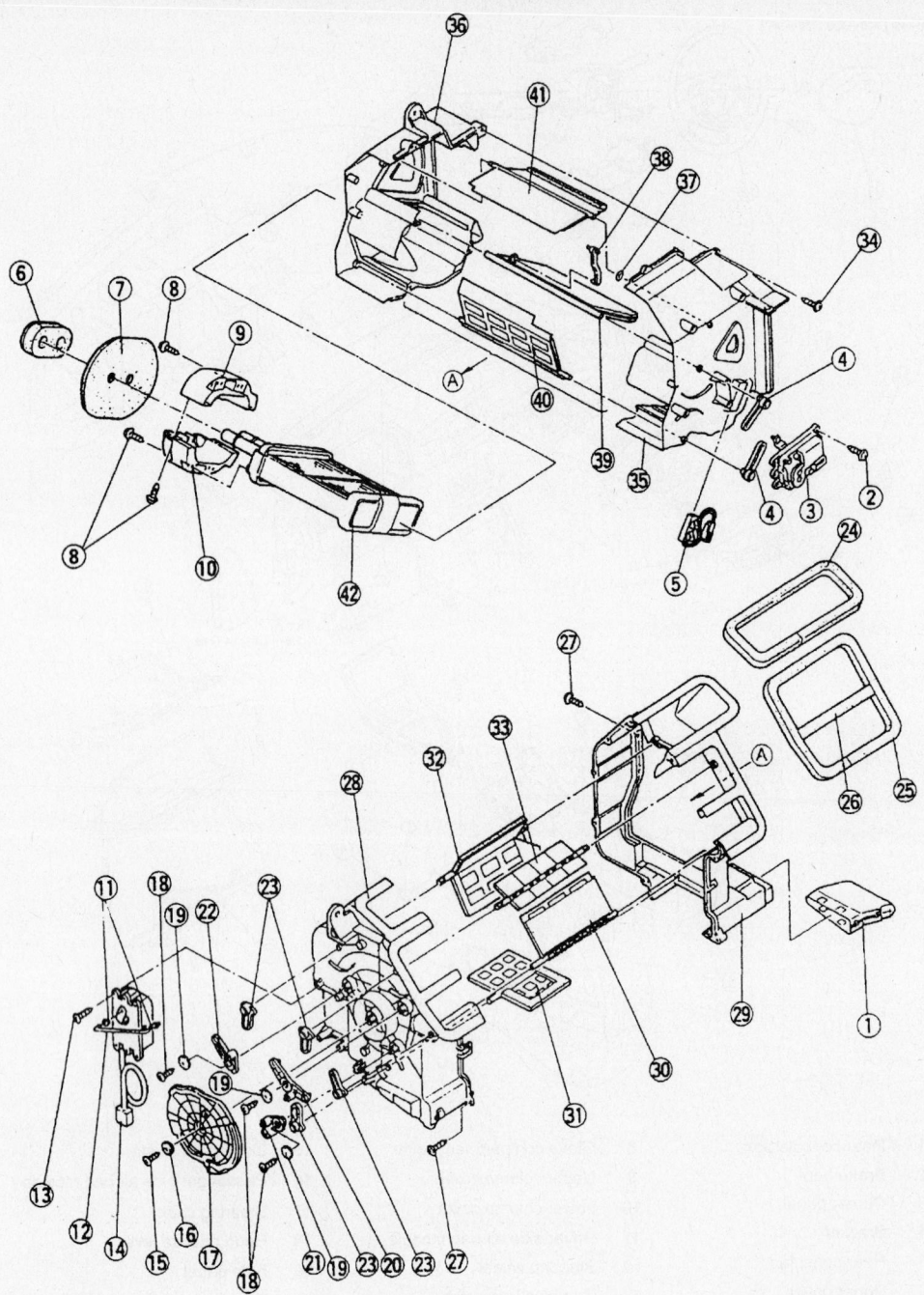

1	Heat duct	16	Link collar	30	Vent door
2	Tapping screw	17	Airflow mode main link	31	Heat door
3	Air mix actuator	18	Tapping screw	32	Defroster door
4	Air mix crank	19	Link collar	33	Side vent door
5	Water temperature sensor	20	Airflow mode sub link (VENT)	34	Tapping screw
6	Polyurethane foam (thick)	21	Airflow mode sub link (HEAT)	35	Heater case (1)
7	Polyurethane foam (thin)	22	Airflow mode sub link (DEFROSTER)	36	Heater case (2)
8	Tapping screw	23	Airflow mode crank	37	Collar
9	Heater core bracket (1)	24	Polyurethane protector (DEFROSTER)	38	Air mix rod
10	Heater core bracket (2)	25	Polyurethane protector (VENT)	39	Air mix main door
11	Rod stopper	26	Polyurethane protector (SIDE VENT)	40	Air mix sub door
12	Airflow mode rod	27	Tapping screw	41	Air mix guide door
13	Tapping screw	28	Heater case (4)	42	Heater core
14	Airflow mode actuator	29	Heater case (3)		
15	Tapping screw				

93112GE7

Exploded view of the heater core and heater case assembly—Millenia

Protégé

1. Before servicing the vehicle, refer to the precautions in the beginning of this section.

2. Disconnect the negative battery cable.

✳✳ CAUTION

After disconnecting the battery, wait for more than 1 minute for the SAS to deplete its stored energy.

3. Drain the cooling system into a clean container for reuse.

4. Disconnect the heater hoses from the heater core.

5. Discharge and recover the air conditioning system refrigerant.

6. Place the wheel in the straight-ahead position and turn the ignition switch to LOCK.

7. At the driver's side, remove the SAS module and the steering wheel by removing or disconnecting the following:

- Cover clips at both sides of the steering wheel
- Steering wheel-to-SAS module bolts
- SAS module from the steering wheel and disconnect the electrical connector
- Steering wheel-to-column nut
- Steering wheel from the steering column using a suitable puller

8. At the passenger's side, remove the SAS module by removing or disconnecting the following:

- Glove compartment and the glove compartment cover
- SAS module-to-dash bolts
- SAS module and disconnect the electrical connector

9. Remove the console by removing or disconnecting the following:

- Shift lever knob, if equipped with a manual transmission
- Console's cover
- Console-to-chassis screws and console

10. Remove the combination switch by removing or disconnecting the following:

- Steering column cover
- Electrical connectors and remove the combination switch-to-steering column bolts and the combination switch

11. Remove the instrument cluster by:

- Meter hood
- Instrument cluster-to-dash panel screws
- Electrical connectors and remove the instrument cluster

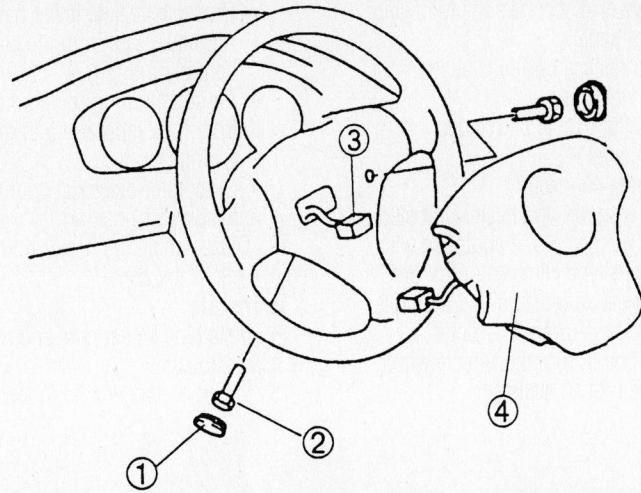

7.9—11.7 N·m {80—120 kgf·cm, 70—104 in·lbf}

1　Cap
2　Bolt
3　Connector
4　Driver-side air bag module

93112GG4

Exploded view of the SAS module and the steering wheel assembly—Protégé

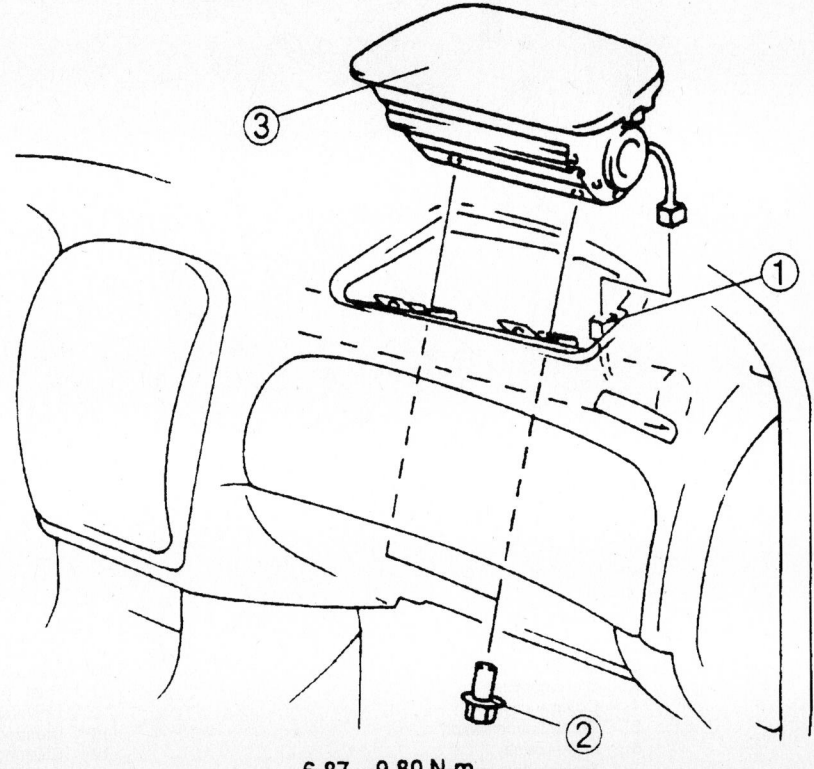

6.87—9.80 N·m
{70—100 kgf·cm, 60.8—86.7 in·lbf}

93112GG5

Exploded view of the passenger's side SAS module—Protégé

12. Remove or disconnect the following:
- Lower panel
- Hood release cable installation nut
- Side wall trim
- "A" pillar trim at both sides
- Side panel
- Antenna connector
- Blower motor and heater unit electrical connectors, if equipped with the wire-type climate control unit
- Electrical connectors and the bolts
- Dashboard-to-chassis bolts
- Dashboard from the vehicle with the help of an assistant

- Passenger's side lower panel
- Air intake wire from the climate control unit
- Air conditioning refrigerant lines from the evaporator and discard the O-rings
- Evaporator electrical connector(s)
- Evaporator housing

13. Disassemble the heater housing and remove the heater core.

To install:

14. Install the heater core and assemble the heater housing.

15. Install or connect the following:

- Evaporator housing
- Evaporator electrical connector(s)
- Air conditioning refrigerant lines to the evaporator using new O-rings
- Air intake wire to the climate control unit
- Passenger's side lower panel
- Dashboard to the vehicle with the help of an assistant
- Dashboard-to-chassis bolts
- Electrical connectors and the bolts
- Blower motor and heater unit electrical connectors, if equipped with the wire-type climate control unit

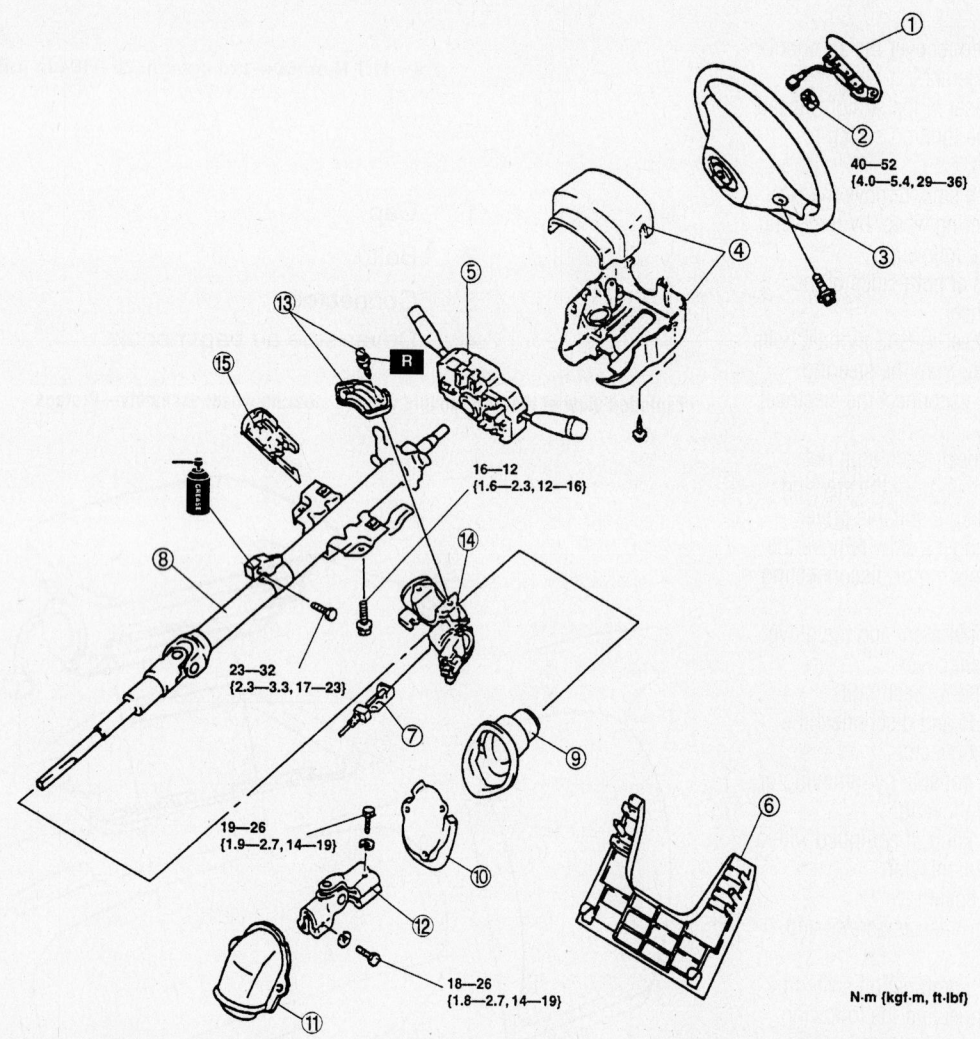

1   Air bag module
2   Locknut
3   Steering wheel
4   Column cover
5   Combination switch
6   Lower panel
7   Key interlock cable
8   Steering shaft

9   Shaft seal
10   Set plate
11   Dust cover
12   Universal joint
13   Steering lock mounting bolts and bracket
14   Steering lock component
15   Cylinder outer component

N·m {kgf·m, ft·lbf}

Exploded view of the steering column assembly—Protégé

93112GG6

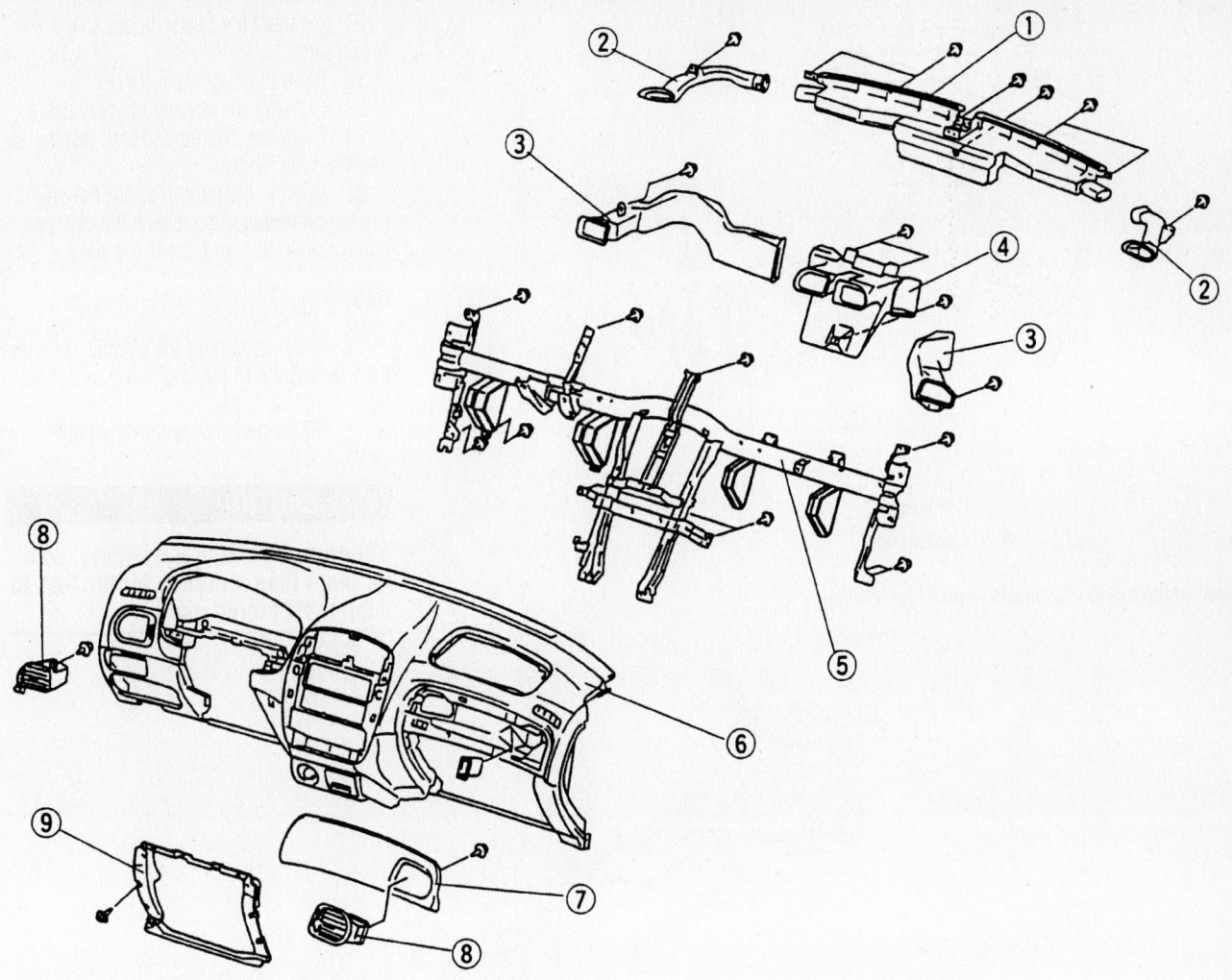

1	Defroster nozzle	6	Crush pad
2	Side demister duct	7	Pad
3	Duct	8	Ventilator grille
4	Center duct	9	Passenger-side lower panel
5	Dashboard member		

93112GG7

Exploded view of the dashboard assembly—Protégé

- Antenna connector
- Side panel
- A-pillar trim at both sides
- Side wall trim
- Hood release cable installation nut
- Lower panel

16. Install the instrument cluster by installing or connecting the following:

- Instrument cluster and connect the electrical connectors
- Instrument cluster-to-dash panel screws
- Meter hood

17. Install the combination switch by installing or connecting the following:

- Electrical connectors and install the combination switch-to-steering column bolts and the combination switch
- Steering column cover

18. Install the console by installing or connecting the following:

- Console and the console-to-chassis screws
- Console's cover
- Shift lever knob, if equipped with a manual transmission

19. At the passenger's side, install the SAS module by installing or connecting the following:

- SAS module and connect the electrical connector
- SAS module-to-dash bolts
- Glove compartment and the glove compartment cover

20. At the driver's side, install the SAS module and the steering wheel by installing or connecting the following:

- Steering wheel-to-column nut
- SAS module from the steering wheel and connect the electrical connector
- Steering wheel-to-SAS module bolts
- Cover clips at both sides of the steering wheel

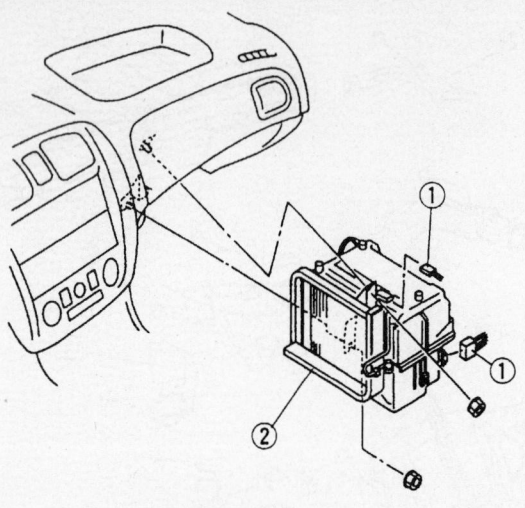

1   Connector
2   Cooling unit

93112GG8

View of the evaporator housing assembly—Protégé

21. Connect the heater hoses to the heater core.

22. Refill the cooling system.

23. Connect the negative battery cable.

24. Evacuate, charge and leak test the air conditioning system.

25. Operate the engine to normal operating temperatures; then, check the climate control operation and check for leaks.

626

1. Before servicing the vehicle, refer to the precautions in the beginning of this section.

2. Disconnect the negative battery cable.

❋❋ CAUTION

After disconnecting the battery, wait for more than 1 minute for the SAS to deplete its stored energy.

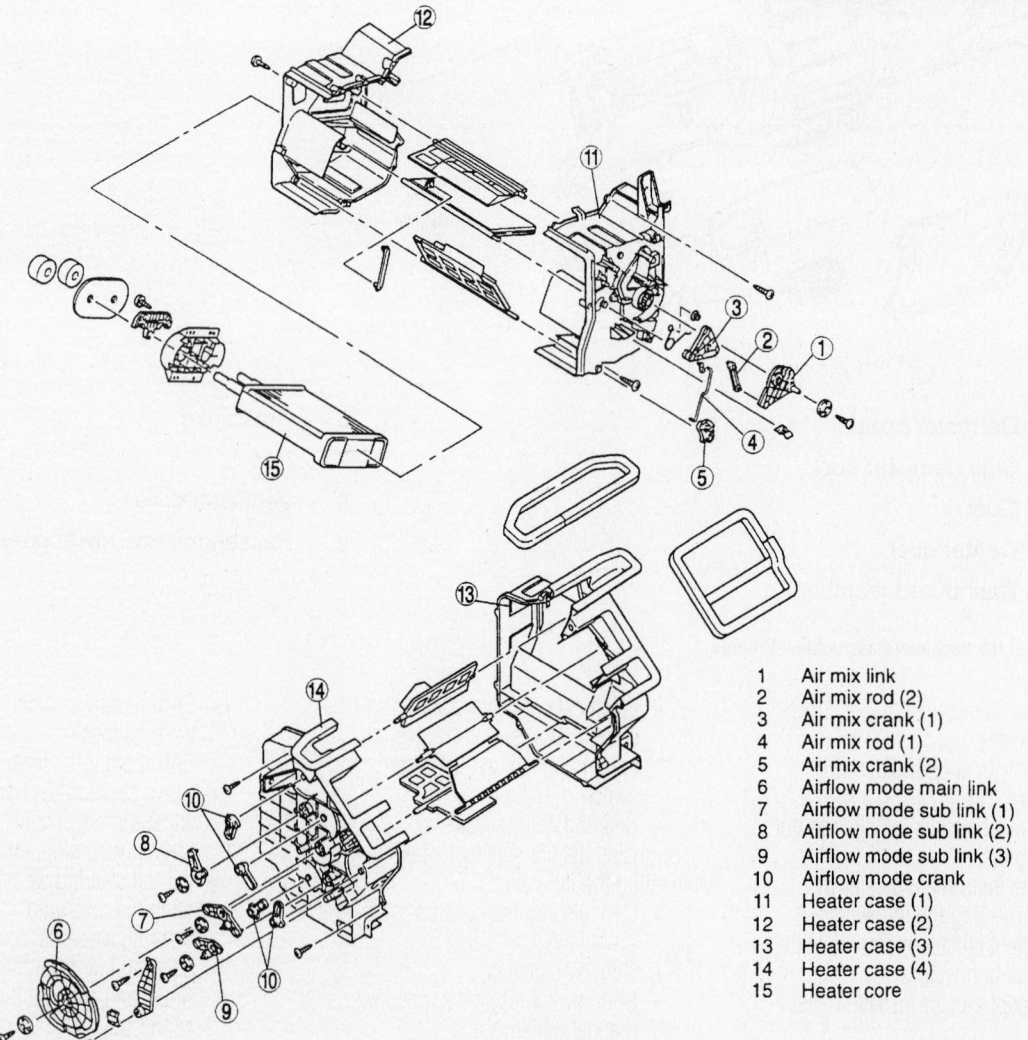

1	Air mix link
2	Air mix rod (2)
3	Air mix crank (1)
4	Air mix rod (1)
5	Air mix crank (2)
6	Airflow mode main link
7	Airflow mode sub link (1)
8	Airflow mode sub link (2)
9	Airflow mode sub link (3)
10	Airflow mode crank
11	Heater case (1)
12	Heater case (2)
13	Heater case (3)
14	Heater case (4)
15	Heater core

93112GG9

Exploded view of the heater housing assembly—Protégé

3. Drain the cooling system into a clean container for reuse.

4. Discharge and recover the air conditioning system refrigerant.

5. Place the wheel in the straight-ahead position and turn the ignition switch to LOCK.

6. At the driver's side, remove the SAS module and the steering wheel removing or disconnecting the following:
- Cover clips at both sides of the steering wheel
- Steering wheel-to-SAS module bolts
- SAS module from the steering wheel and disconnect the electrical connector
- Steering wheel-to-column nut
- Steering wheel from the steering column using a suitable puller

7. At the passenger's side, remove the SAS module by removing or disconnecting the following:
- Glove compartment and the glove compartment cover
- SAS module-to-dash bolts
- SAS module and disconnect the electrical connector

8. Remove the instrument cluster by removing or disconnecting the following:
- Instrument cluster meter hood
- Instrument cluster-to-dash screws, the instrument cluster and disconnect the electrical connectors

9. Remove the climate control assembly by removing or disconnecting the following:
- Climate control meter hood
- Climate control assembly screws
- Climate control assembly and disconnect the electrical connector and the assembly

10. Remove the audio unit by removing or disconnecting the following:
- Hole covers by inserting a small tape-wrapped flathead screwdriver into the slot and carefully pry off the hole covers
- Using 2 removal tools 49 UN01 050 or equivalent, insert them into sides of the audio unit.
- Slide audio unit outward and forward
- Electrical connectors and the antenna jack

11. Remove the instrument panel by removing or disconnecting the following:
- Upper center panel cover
- Dash panel-to-chassis bolts
- Dash panel with the help of an assistant

12. Remove the evaporator housing by removing or disconnecting the following:

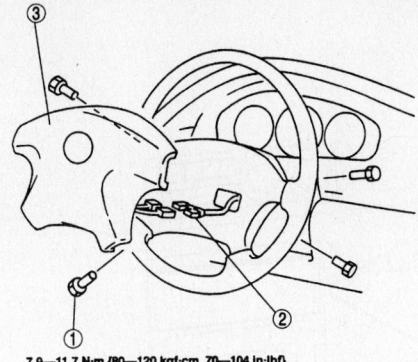

7.9—11.7 N·m {80—120 kgf·cm, 70—104 in·lbf}

1	Bolt
2	Connector
3	Driver-side air bag module

93112GH6

Exploded view of the steering wheel and SAS module—626

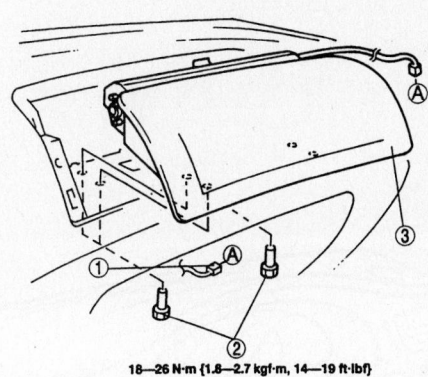

18—26 N·m {1.8—2.7 kgf·m, 14—19 ft·lbf}

1	Connector
2	Bolt
3	Passenger-side air bag module

93112GH7

Exploded view of the passenger's side SAS module—626

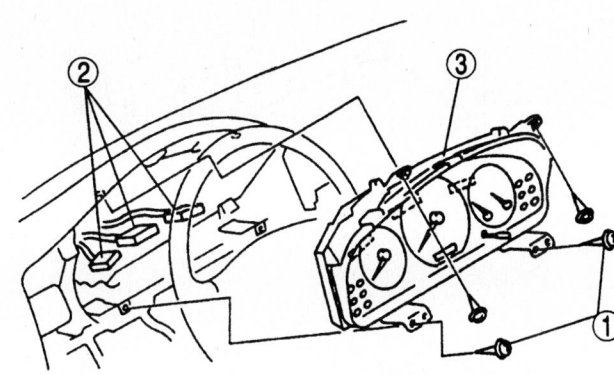

1	**Screw**
2	**Connector**
3	**Instrument cluster**

93112GH8

View of the instrument cluster assembly—626

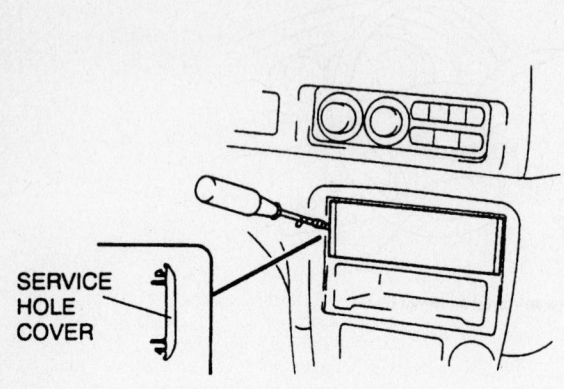

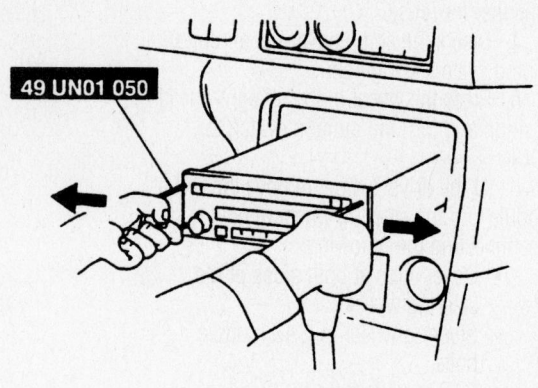

Removing the audio unit—626

93112GH9

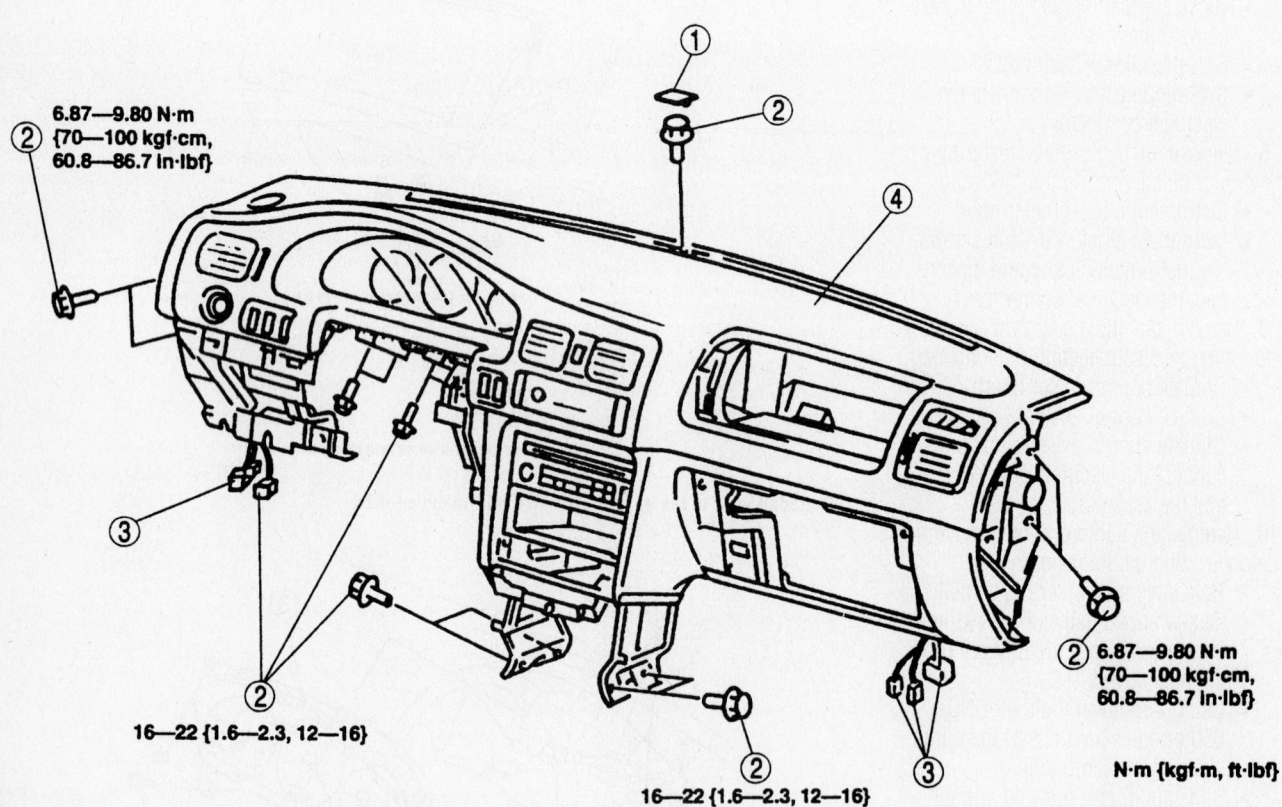

6.87—9.80 N·m
{70—100 kgf·cm,
60.8—86.7 in·lbf}

6.87—9.80 N·m
{70—100 kgf·cm,
60.8—86.7 in·lbf}

16—22 {1.6—2.3, 12—16}

16—22 {1.6—2.3, 12—16}

N·m {kgf·m, ft·lbf}

| 1 | Cover | 3 | Connector |
| 2 | Bolt | 4 | Dashboard |

93112GH0

View of the instrument panel assembly—626

- Center lower panel
- Refrigerant lines from the air conditioning evaporator and discard the gaskets
- Blower motor assembly, if necessary
- Evaporator assembly fasteners and remove the evaporator assembly

13. Remove or disconnect the following:
- Rear heater duct
- Heater housing fasteners and the heater housing
- Airflow mode actuator

14. Separate the heater housing and remove the heater core.

To install:

15. Install the heater core and assemble the heater housing.

16. Install or connect the following:
- Airflow mode actuator
- Heater housing and the heater housing fasteners
- Rear heater duct

17. Install the evaporator housing by installing or connecting the following:
- Center lower panel
- Refrigerant lines to the air conditioning evaporator using new gaskets

- Blower motor assembly, if necessary
- Evaporator assembly and the evaporator assembly fasteners

18. Install the instrument panel by installing or connecting the following:
- Upper center panel cover
- Dash panel-to-chassis bolts
- Dash panel with the help of an assistant

19. Install the audio unit by installing or connecting the following:
- Hole covers
- Audio unit
- Electrical connectors and the antenna jack

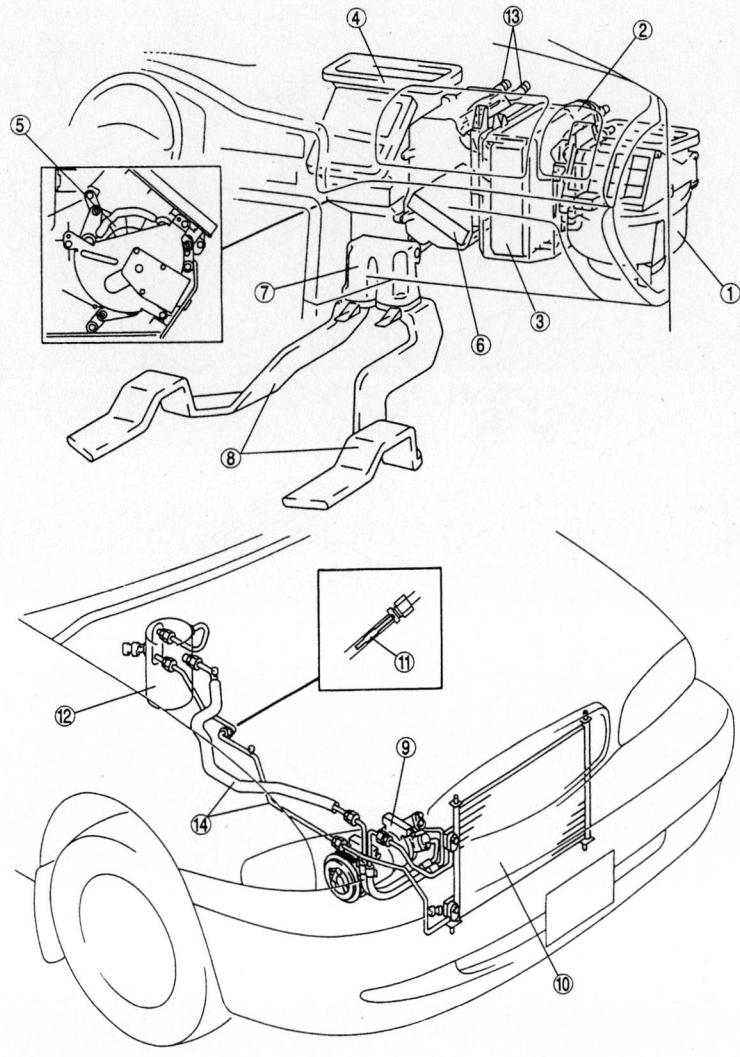

1	Blower unit	8	Rear heat duct (CANADA only)
2	Cooling unit	9	A/C compressor
3	Evaporator	10	Condenser
4	Heater unit	11	Orifice tube
5	Airflow mode main link	12	Accumulator tank
6	Heater core	13	Heater hose
7	Rear duct (CANADA only)	14	Refrigerant lines

93112GI1

View of the heater and air conditioning housing assemblies—626

20. Install the climate control assembly by installing or connecting the following:

- Climate control meter hood
- Climate control assembly screws
- Climate control assembly and connect the electrical connector

21. Install the instrument cluster by installing or connecting the following:

- Instrument cluster meter hood
- Instrument cluster-to-dash screws, the instrument cluster and connect the electrical connectors

22. At the passenger's side, install the SAS module by installing or connecting the following:

- Glove compartment and the glove compartment cover
- SAS module-to-dash bolts
- SAS module and connect the electrical connector

23. At the driver's side, install the SAS module and the steering wheel by installing or connecting the following:

- Steering wheel and the steering wheel-to-column nut. Torque the steering wheel nut to 29–36 ft. lbs. (40–49 Nm).
- SAS module to the steering wheel

and connect the electrical connector. Torque the steering column-to-SAS module bolts to 70–104 inch lbs. (8–12 Nm).

- Steering wheel-to-SAS module clips
- Cover clips, at both sides of the steering wheel

24. Refill the cooling system.

25. Connect the negative battery cable.

26. Evacuate, charge and leak test the air conditioning system refrigerant.

27. Operate the engine to normal operating temperatures; then, check the climate control system and check for leaks.

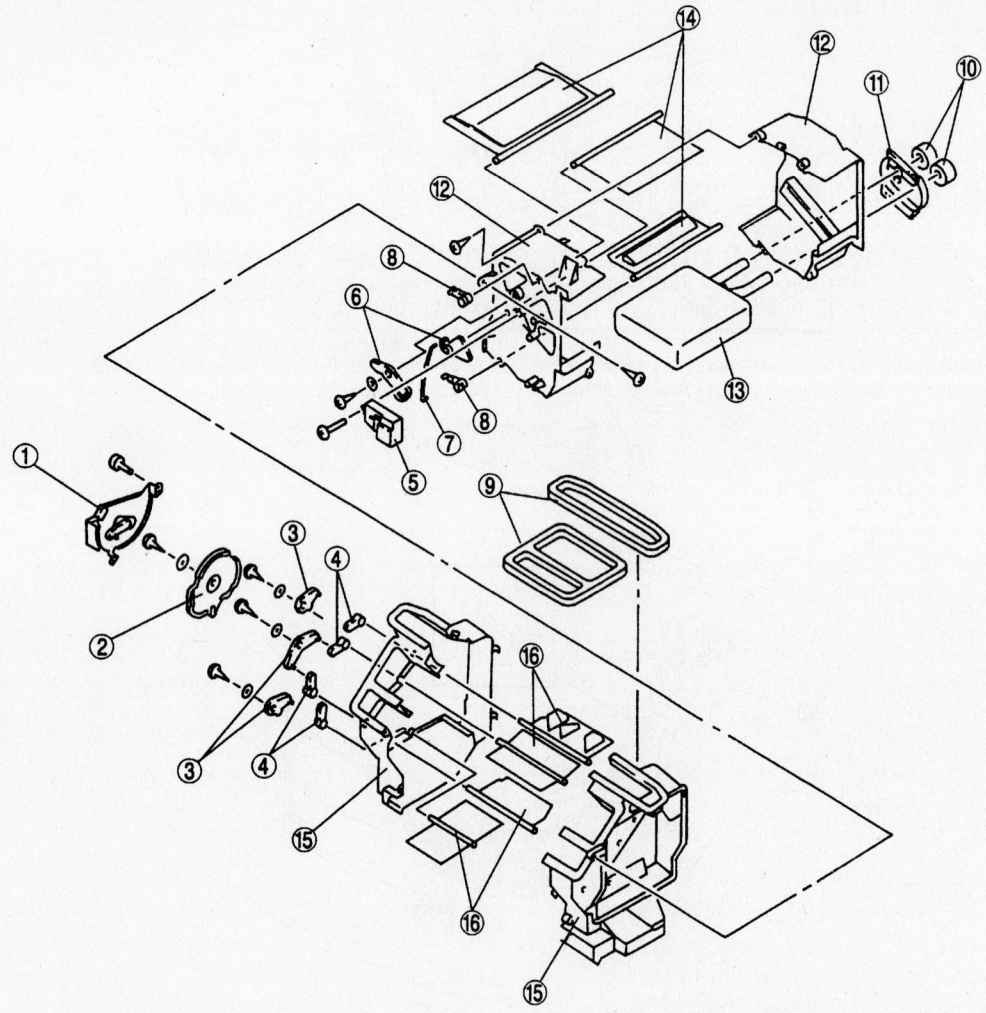

1	Airflow mode actuator	9	Polyurethane protector
2	Airflow mode main link	10	Seal
3	Airflow mode sub link	11	Cover
4	Airflow mode crank	12	Case
5	Air mix actuator	13	Heater core
6	Air mix link	14	Air mix door
7	Air mix rod	15	Case
8	Air mix crank	16	Airflow mode door

93112GI2

Exploded view of the heater core and heater housing assembly—626

Rocker Arm (Valve) Cover

REMOVAL & INSTALLATION

626 and Protégé

2.0L (FS) ENGINE

1. Before servicing the vehicle, refer to the precautions in the beginning of this section.
2. Remove or disconnect the following:
   - Negative battery cable
   - Any components that would interfere with cover removal
   - Rocker arm cover bolts in the sequence illustrated
   - Rocker arm cover and discard the gasket
3. Clean all mating surfaces of any residual gasket material.

To install:

4. Apply silicone sealant to cylinder head at the areas illustrated.

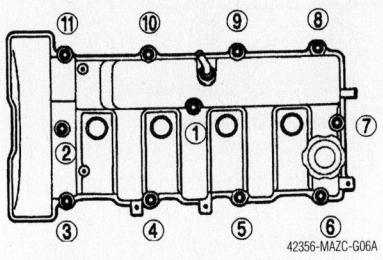

Remove the rocker arm cover bolts in the sequence shown—2.0L (FS) engine

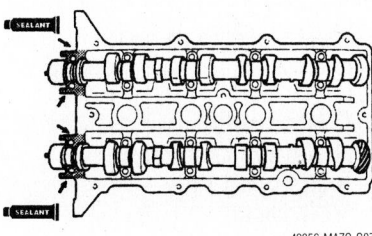

Apply silicone sealant to cylinder head at the areas illustrated—2.0L (FS) engine

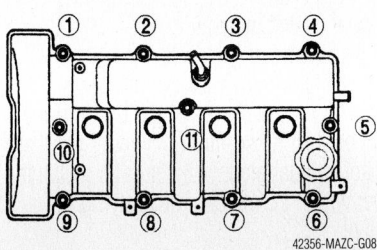

Tighten the rocker arm cover bolts in the sequence shown—2.0L (FS) engine

5. Install or connect the following:
   - Rocker arm cover with a new gasket. Torque the bolts in the sequence illustrated to 95 inch lbs. (11 Nm).
   - Remaining components removed to facilitate the rocker arm cover removal
   - Negative battery cable

1.6L (ZM) Engines

1. Before servicing the vehicle, refer to the precautions in the beginning of this section.
2. Remove or disconnect the following:
   - Negative battery cable
   - Spark plug wires
   - Vent hose
   - Positive Crankcase Ventilation (PCV) hose
   - Rocker arm cover and discard the gasket
3. Clean all mating surfaces of any residual gasket material.

To install:

4. Install or connect the following:
5. Apply a 0.12–0.15 inch (3–4mm) bead of silicone sealant to cylinder head at the areas illustrated.
6. Install or connect the following:
   - Rocker arm cover with a new gasket. Torque the bolts in the sequence illustrated to 95 inch lbs. (11 Nm).
   - Remaining components removed to facilitate the rocker arm cover removal
   - Negative battery cable

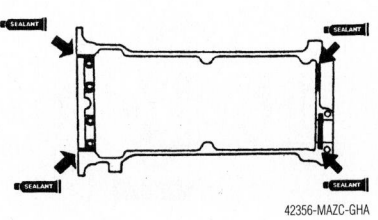

Apply silicone sealant to cylinder head at the areas illustrated—1.6L (ZM) engine

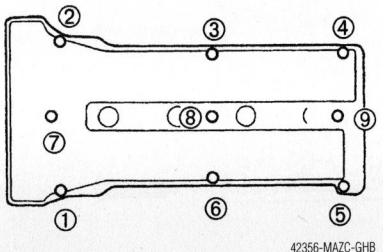

Tighten the rocker arm cover bolts in the sequence shown—1.6L (ZM) engine

1.8L (FP) Engines

1. Before servicing the vehicle, refer to the precautions in the beginning of this section.
2. Remove or disconnect the following:
   - Negative battery cable
   - Spark plug wires
   - Vent hose
   - Positive Crankcase Ventilation (PCV) hose
   - Rocker arm cover bolts in the sequence illustrated
   - Rocker arm cover and discard the gasket
3. Clean all mating surfaces of any residual gasket material.

To install:

4. Install or connect the following:
5. Apply a bead of silicone sealant to cylinder head at the areas illustrated.
6. Install or connect the following:
   - Rocker arm cover with a new gasket. Torque the bolts in the

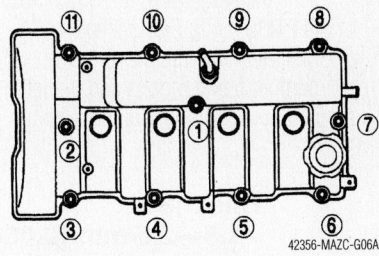

Remove the valve cover bolts in sequence—1.8L (FP) engines

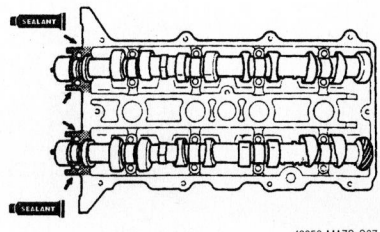

Apply silicone sealant to cylinder head at the areas illustrated—1.8L (FP) engine

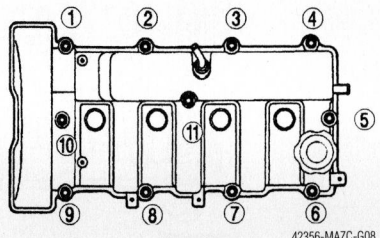

Tighten the rocker arm cover bolts in the sequence shown—1.8L (FP) engine

sequence illustrated to 86 inch lbs. (10 Nm).
- Remaining components removed to facilitate the rocker arm cover removal
- Negative battery cable

2.5L (KL) ENGINE

1. Before servicing the vehicle, refer to the precautions in the beginning of this section.
2. Remove or disconnect the following:
   - Negative battery cable
   - Any components that would interfere with cover removal
   - Rocker arm cover bolts
   - Rocker arm cover and discard the gasket
3. Clean all mating surfaces of any residual gasket material.

To install:

4. Apply silicone sealant to cylinder head at the areas illustrated.
5. Install or connect the following:
   - Rocker arm cover with a new gasket. Torque the bolts in the sequence illustrated to 95 inch lbs. (11 Nm) using 2 or 3 steps. Retighten the right hand cover number 5 and 6 bolts and the left hand cover number 6 and 7 bolts.

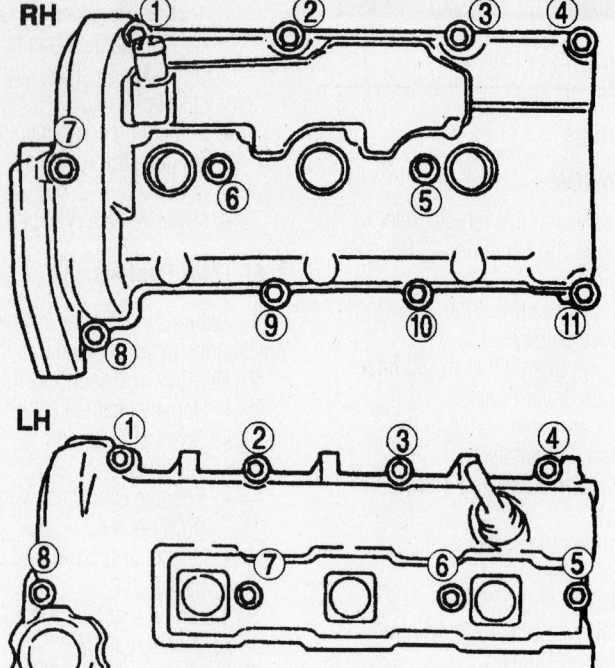

42356-MAZC-G10

Tighten the rocker arm cover bolts in the sequence shown—2.5L (KL) engine

Thickness
1.5—2.5 mm {0.060—0.098 in}

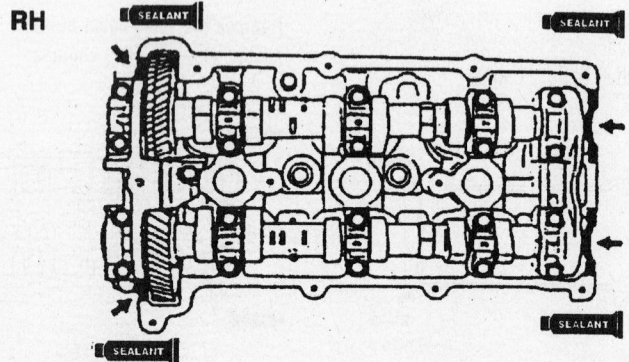

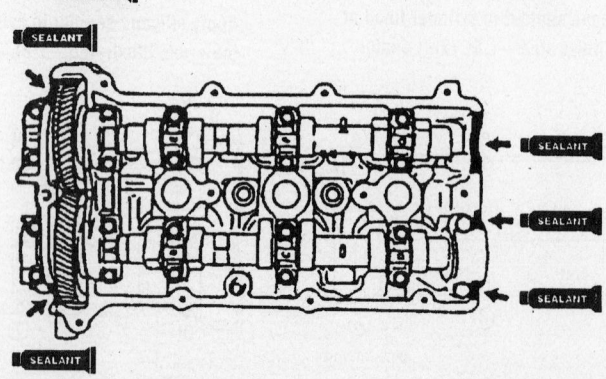

42356-MAZC-G09

Apply silicone sealant to cylinder head at the areas illustrated—2.5L (KL) engine

- Remaining components removed to facilitate the rocker arm cover removal
- Negative battery cable

Mazda 3 and Mazda 6 Models

2.0L (LF) AND 2.3L (L3) ENGINES

1. Before servicing the vehicle, refer to the precautions in the beginning of this section.
2. Remove or disconnect the following:
   - Negative battery cable
   - Spark plug wires
   - Vent hose
   - Positive Crankcase Ventilation (PCV) hose
   - Rocker arm cover and discard the gasket
3. Clean all mating surfaces of any residual gasket material.

To install:

4. Install or connect the following:
5. Apply a 0.16–0.24 inch (4–7mm) bead of silicone sealant to cylinder head at the areas illustrated. Make sure to install the cover within 10 minutes of applying the sealant.
6. Install or connect the following:
   - Rocker arm cover with a new gasket. Torque the bolts in the

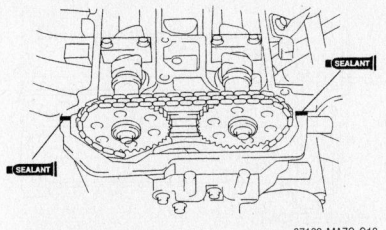

67162-MAZC-G13

Apply silicone sealant to cylinder head at the areas illustrated–All Mazda 3 models and Mazda 6 models with the 2.3L (L3) engine models

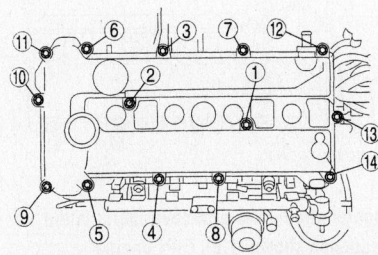

67162-MAZC-G14

Tighten the rocker arm cover bolts in the sequence shown–All Mazda 3 models and Mazda 6 models with the 2.3L (L3) engine models

sequence illustrated to 71–92 inch lbs. (8–10 Nm) on all Mazda 3 engines or 71–101inch lbs. (8–8–11.5 Nm) on Mazda 6 models equipped with the 2.3L (L3) engine.
- Remaining components removed to facilitate the rocker arm cover removal
- Negative battery cable

3.0L (AJ) ENGINE

1. Before servicing the vehicle, refer to the precautions in the beginning of this section.
2. Remove or disconnect the following:
- Negative battery cable
- Any components that would interfere with cover removal
- Rocker arm cover bolts in the sequence illustrated
- Rocker arm cover and discard the gasket
3. Clean all mating surfaces of any residual gasket material.
To install:
4. Apply silicone sealant to cylinder head at the areas illustrated.
5. Install or connect the following:
- Rocker arm cover with a new gasket.
- Oil control valve with the cylinder

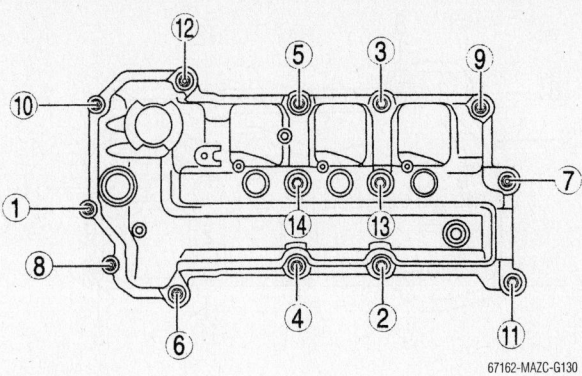

67162-MAZC-G130

Remove the left hand rocker arm cover bolts in the sequence shown—Mazda 6 models with the 3.0L (AJ) engine

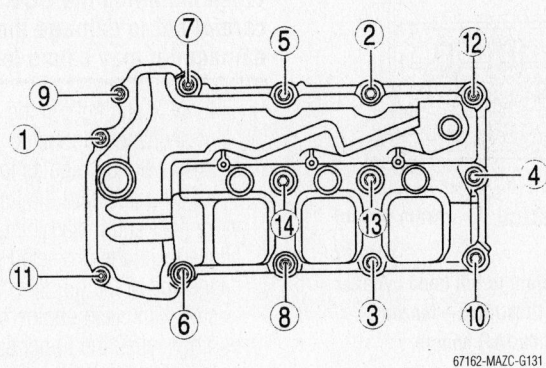

67162-MAZC-G131

Remove the right hand rocker arm cover bolts in the sequence shown—Mazda 6 models with the 3.0L (AJ) engine

head cover raised as shown in the accompanying illustration, being careful not to let the valve retaining bolt slip into the timing chain cover when installing and tighten the valve bolt to 71–106 inch lbs. (8–12 Nm). Torque the bolts in the sequence illustrated to 71–106 inch lbs. (8–12 Nm).

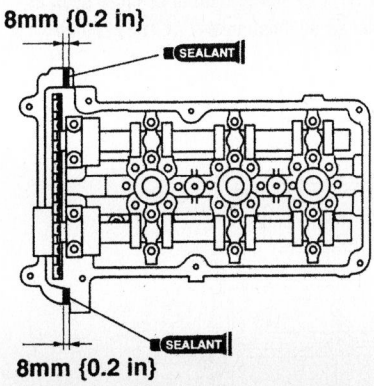

8mm {0.2 in}

8mm {0.2 in}

67162-MAZC-G132

Apply silicone sealant to right hand cylinder head at the areas illustrated—Mazda 6 models with the 3.0L (AJ) engine

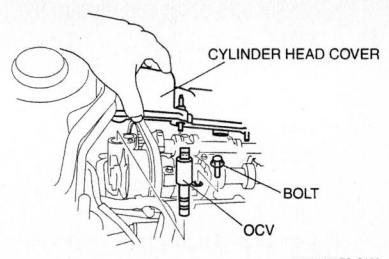

CYLINDER HEAD COVER

BOLT

OCV

67162-MAZC-G133

Install the oil control valve on the right hand side—Mazda 6 models with the 3.0L (AJ) engine

- Remaining components removed to facilitate the rocker arm cover removal
- Negative battery cable

Miata

1.8L (BP) ENGINES

1. Before servicing the vehicle, refer to the precautions in the beginning of this section.
2. Remove or disconnect the following:
- Negative battery cable
- Upper radiator hose

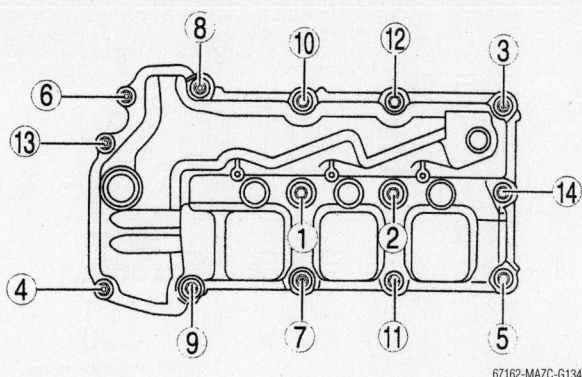

67162-MAZC-G134

Right hand cylinder head cover torque sequence—Mazda 6 models with the 3.0L (AJ) engine

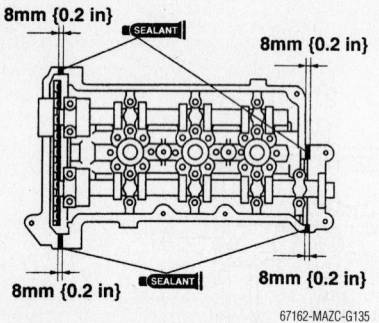

67162-MAZC-G135

Apply silicone sealant to left hand cylinder head at the areas illustrated—Mazda 6 models with the 3.0L (AJ) engine

- Water hose
- Oil pipe
- Oil Control Valve (OCV)
- Rocker arm cover and discard the gasket

3. Clean all mating surfaces of any residual gasket material.

To install:

4. Install or connect the following:
- Rocker arm cover with a new gasket.
- Temporarily tighten the cover bolts **A**, refer to the illustration for location. Torque the bolts in sequence to 80 inch lbs. (9 Nm).

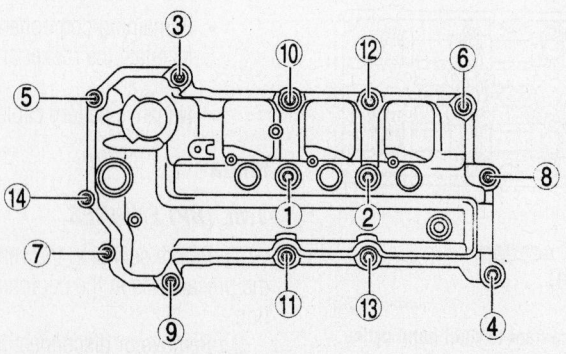

67162-MAZC-G136

Left hand cylinder head cover torque sequence—Mazda 6 models with the 3.0L (AJ) engine

✳✳ CAUTION

When installing the OCV valve, be careful not to damage the O-ring, if damaged it may cause leaking.

- OCV and tighten the bolts in the sequence illustrated
5. Install the oil pipe as follows:
   a. Oil pipe. Hold the frame of the OCV valve filter and install the OCV so it is aligned with the projected part on the flange end of the oil pipe. Coat the new washer with clean engine oil and temporarily install the upper and side oil pipe and position the pipe.
   b. **A** using several passes to 95 inch lbs. (11 Nm).

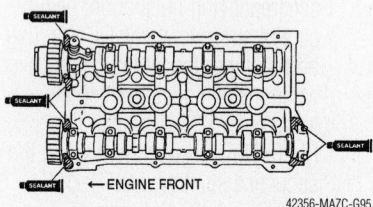

42356-MAZC-G95

Apply silicone sealant to cylinder head at the areas illustrated–1.8L (BP) engine

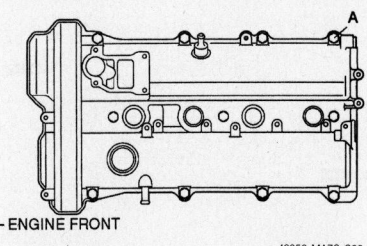

42356-MAZC-G96

Temporarily tighten the cover bolts A–1.8L (BP) engine

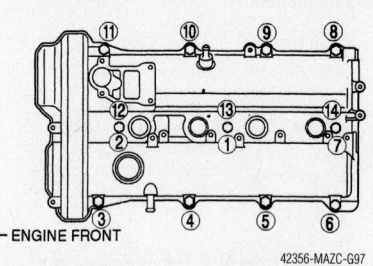

42356-MAZC-G97

Tighten the rocker arm cover bolts in the sequence shown–1.8L (BP) engine

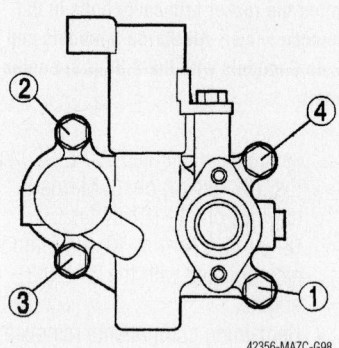

42356-MAZC-G98

Tighten the OCV bolts in the sequence shown–1.8L (BP) engine

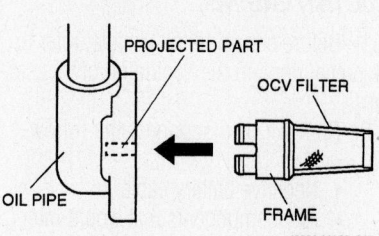

42356-MAZC-G99

Hold the frame of the OCV valve filter and install the OCV so it is aligned with the projected part on the flange end of the oil pipe–1.8L (BP) engine

   c. Tighten oil pipe bolts **B** using several passes to 61 inch lbs. (7 Nm).
   d. Tighten oil pipe bolts **C** using several passes to 13 ft. lbs. (17 Nm).
   e. Tighten oil pipe bolts 1, 2, 3, 4 and 7 using several passes to 95 inch lbs. (11 Nm).

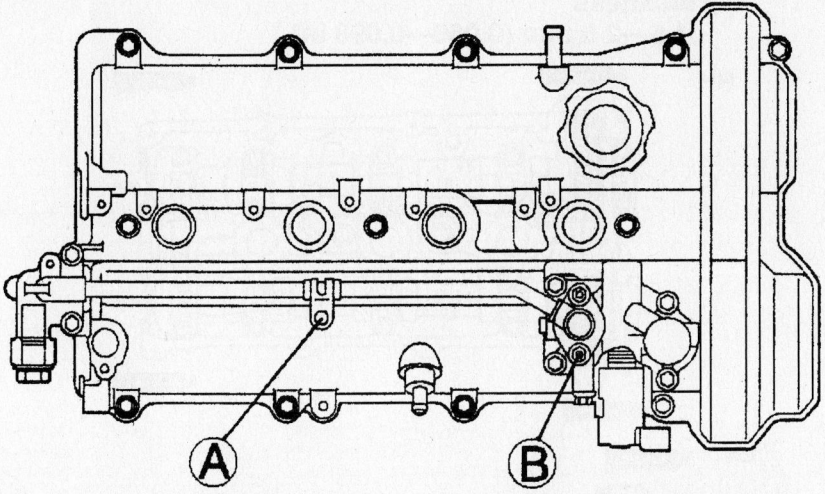

ENGINE FRONT →

42356-MAZC-GAA

Tighten oil pipe bolts A and B using several passes to the specification in the text—1.8L (BP) engine

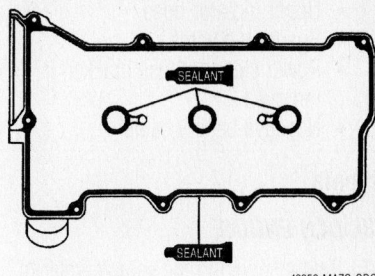

42356-MAZC-GDO

Apply a 0.004–0.07 inch (1–2mm) bead of sealant to rocker arm cover—Millenia models equipped with the 2.3L (KJ) engine

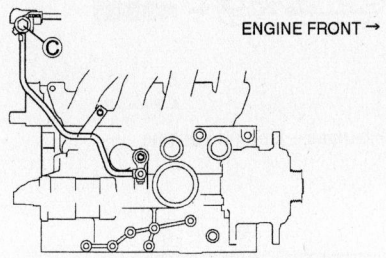

ENGINE FRONT →

42356-MAZC-GAB

Tighten oil pipe bolts C using several passes to the specification in the text—1.8L (BP) engine

f. Tighten oil pipe bolt 5 using several passes to 17 ft. lbs. (23 Nm).

g. Tighten oil pipe bolt 6 using several passes to 34 ft. lbs. (47 Nm).

6. Install or connect the following:
- Water hose

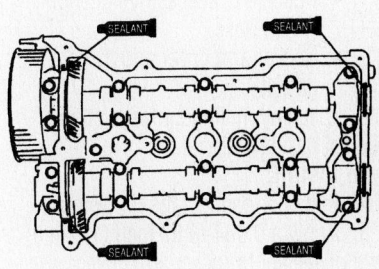

42356-MAZC-GDP

Apply a 0.059–0.098 inch (1.5–2.5mm) to the areas shown—Millenia models equipped with the 2.3L (KJ) engine

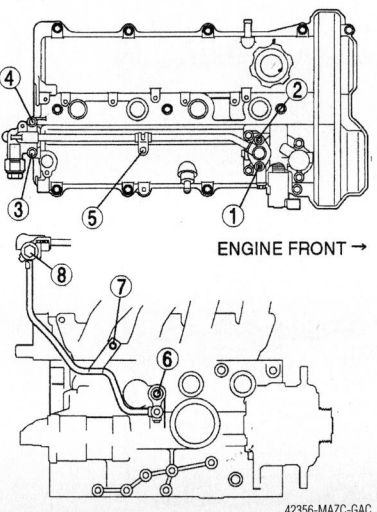

ENGINE FRONT →

42356-MAZC-GAC

Tighten oil pipe bolts using several passes to the specification in the text—1.8L (BP) engine

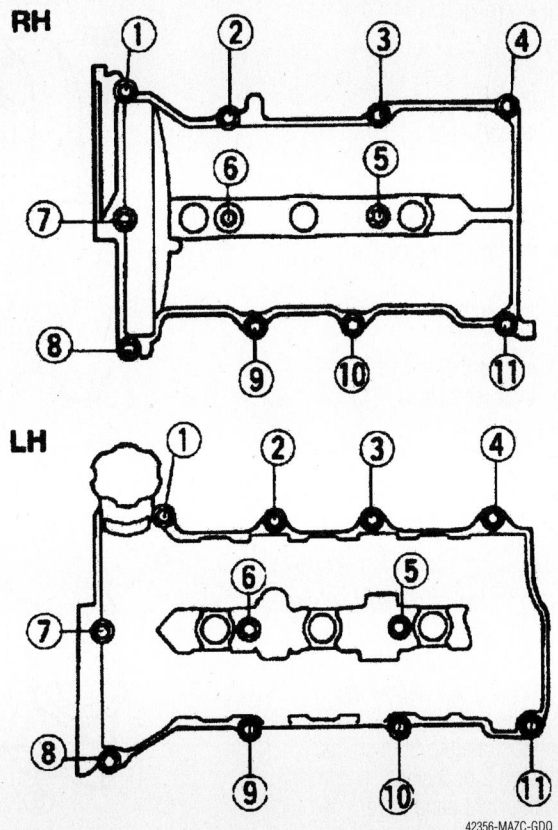

42356-MAZC-GDQ

Rocker arm cover bolt sequence—Millenia models equipped with the 2.3L (KJ) engine

- Upper radiator hose
- Spark plug wires
- Power steering hose bracket, if removed
- Negative battery cable

Millenia

2.3L (KJ) ENGINE

1. Before servicing the vehicle, refer to the precautions in the beginning of this section.

2. Remove or disconnect the following:
- Negative battery cable
- Ignition coils
- Vent hose
- Positive Crankcase Ventilation (PCV) hose
- Rocker arm cover and discard the gasket

3. Clean all mating surfaces of any residual gasket material.

To install:

4. Install or connect the following:

5. Apply a 0.004–0.07 inch (1–2mm) bead of sealant to rocker arm cover.
- New rocker arm gasket

6. Apply a 0.059–0.098 inch (1.5–2.5mm) to the areas illustrated
- Rocker arm cover. Torque the bolts in sequence to 86 inch lbs. (10 Nm), then retighten in sequence to 86 inch lbs. (10 Nm).
- PCV hose
- Vent hose
- Ignition coils
- Negative battery cable

2.5L (KL) ENGINE

1. Before servicing the vehicle, refer to the precautions in the beginning of this section.

2. Remove or disconnect the following:
- Negative battery cable
- Any components that would interfere with cover removal.
- Rocker arm cover bolts
- Rocker arm cover and discard the gasket

3. Clean all mating surfaces of any residual gasket material.

To install:

4. Apply silicone sealant to cylinder head at the areas illustrated.

5. Install or connect the following:
- Rocker arm cover with a new gasket. Torque the bolts in the sequence illustrated to 95 inch lbs. (11 Nm) using 2 or 3 steps. Retighten the right hand cover number 5 and 6 bolts and the left hand cover number 6 and 7 bolts.

Thickness
1.5—2.5 mm {0.060—0.098 in}

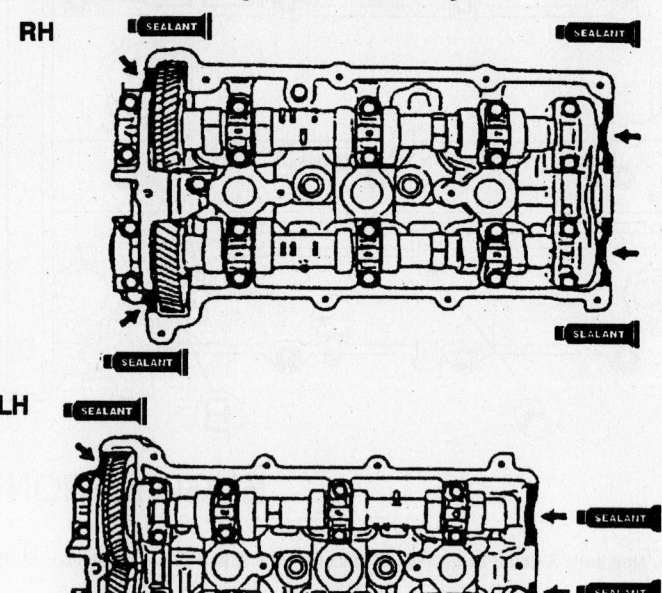

42356-MAZC-G09

Apply silicone sealant to cylinder head at the areas illustrated—2.5L (KL) engine

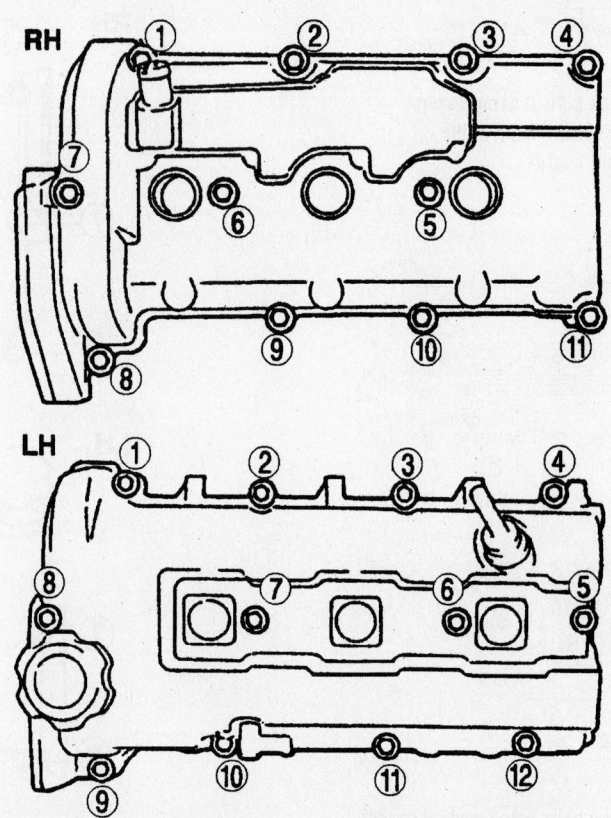

42356-MAZC-G10

Tighten the rocker arm cover bolts in the sequence shown—2.5L (KL) engine

- Remaining components removed to facilitate the rocker arm cover removal
- Negative battery cable

Cylinder Head

REMOVAL & INSTALLATION

1.8L (BP) Engines

1. Before servicing the vehicle, refer to the precautions in the beginning of this section.
2. Relieve the fuel system pressure.
3. Drain the cooling system.
4. Remove or disconnect the following:
   - Negative battery cable
   - Timing belt
   - Air cleaner assembly and front pipe
   - Exhaust manifold
   - Vacuum hoses
   - All engine harness connectors necessary to access the cylinder head
   - Fuel hose
   - Intake manifold bracket
   - Accelerator cable bracket
   - Cylinder head bolts, in 2–3 steps, in sequence
   - Cylinder head

To install:

5. Thoroughly, clean the cylinder head and the block contact surfaces. Examine the head gasket and check the cylinder head for cracks. Check the cylinder head for warpage using a feeler gauge and straightedge. The maximum allowable distortion is 0.004 inch (0.10mm).

6. Clean the cylinder head bolts and the threads in the block. Be sure the bolts turn freely in the block.

7. Install new head gasket on the engine block.

8. Install the cylinder head.

9. Lubricate the bolt threads and seat surfaces with clean engine oil and install them as follows:

   a. Torque the bolts in 2–3 steps to 56–60 ft. lbs. (75–81 Nm) in the proper sequence.

10. Install or connect the following:
    - Accelerator cable bracket
    - Intake manifold bracket
    - Fuel hose
    - All engine harness connectors removed to access the cylinder head
    - Vacuum hoses
    - Exhaust manifold using a new gasket
    - Front pipe and air cleaner assembly

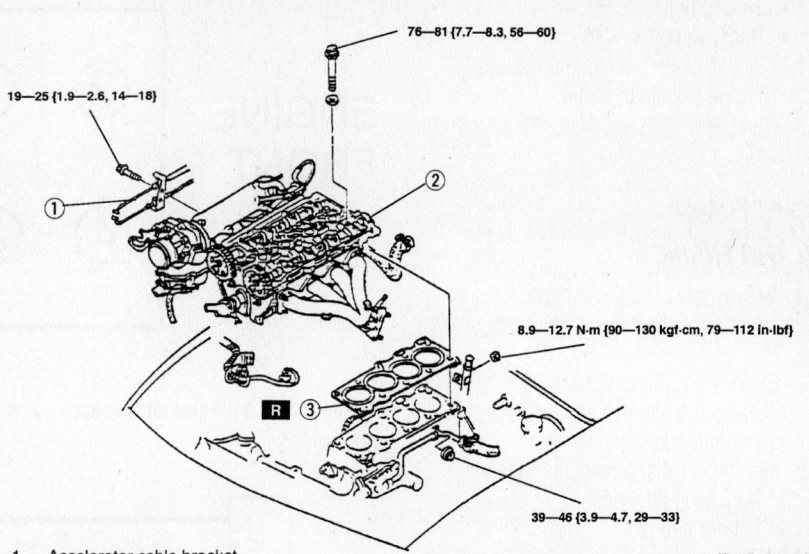

1   Accelerator cable bracket
2   Cylinder head
3   Cylinder head gasket

N·m {kgf·m, ft·lbf}

42356-MAZC-GAE

Exploded view of the cylinder head assembly—1.8L (BP) engines

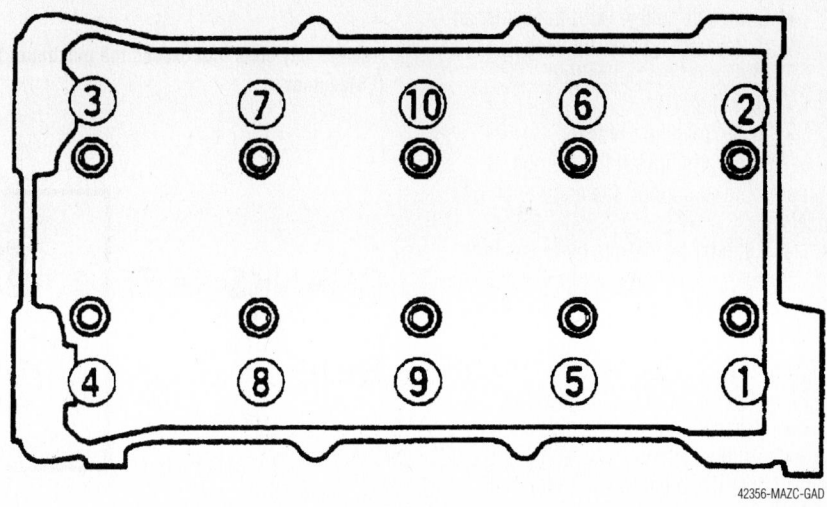

42356-MAZC-GAD

Cylinder head loosening sequence—1.8L (BP) engines

7923MG16

Cylinder head bolt tightening sequence—1.8L (BP) engines

- Timing belt
- Negative battery cable

11. Fill and bleed the cooling system.

12. Change the oil and filter.

13. Run the engine and check for proper operation.

626 and Protégé

2.0L (FS) ENGINE

1. Before servicing the vehicle, refer to the precautions in the beginning of this section.

2. Drain the cooling system.

3. Relieve the fuel system pressure.

4. Drain the cooling system.

5. Remove or disconnect the following:
- Negative battery cable
- Timing belt
- Front exhaust pipe
- Air cleaner assembly
- Power steering pump with the lines attached and set aside, if necessary on Protégé models
- Accelerator cable
- Fuel hose
- Ignition coil
- Camshaft pulley. Hold the camshaft pulley with a back–up wrench while loosening the pulley bolt.
- Camshaft
- Intake manifold bracket

6. Temporarily install the Number 3 engine mount to support the engine, on 626 models.
- Cylinder head bolts by loosening them, in 2–3 steps, in the sequence illustrated
- Cylinder head

To install:

7. Install or connect the following:
- New cylinder head gasket
- Cylinder head

8. Apply clean engine oil to the bolt threads and seating faces.

9. Install new cylinder head bolts and torque in 2–3 steps, in sequence, to 13–16 ft. lbs. (17–22 Nm). If reusing old bolts, which is not recommended, make sure the maximum length of the bolt is 4.154 inch (105.5mm). The standard length of the bolt should be 4.103–4.125 inch (104.2–104.8mm).

10. Paint a mark on the edge of each cylinder head bolt to use as a reference. Turn each bolt, in sequence, 85–95 degrees. Again, turn each bolt, in sequence, an additional 85–95 degrees.

11. Support the engine assembly with a suitable lifting device and remove the number 3 mount.

12. Install or connect the following:

ENGINE FRONT

Cylinder head bolt removal sequence—2.0L (FS) engines

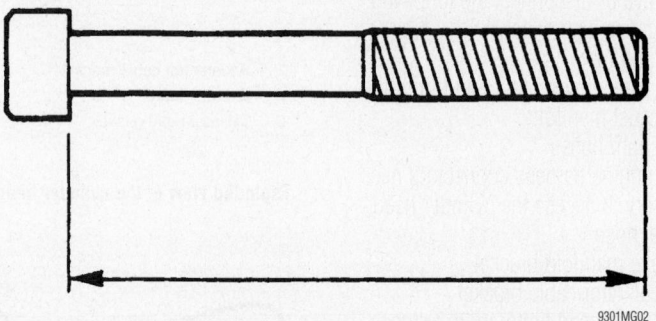

Replace any bolts that exceed the maximum length of 4.154 inch (105.5mm)—2.0L (FS)engines

CRANKSHAFT PULLEY SIDE

Cylinder head bolt tightening sequence—2.0L (FS) engines

- Intake manifold bracket
- Camshafts
- Camshaft pulley, make sure the camshaft sprocket pulleys are positioned with the pins facing up. Hold the camshaft pulley with a back–up wrench while tightening the pulley bolt to 37–44 ft. lbs. (50–60 Nm).
- Ignition coil
- Fuel hose
- Accelerator cable
- Power steering pump, if removed
- Air cleaner assembly
- Front exhaust pipe
- Timing belt

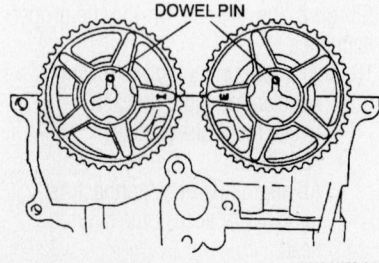

Make sure the camshaft sprocket pulleys are positioned with the pins facing up—2.0L (FS) engine

- Negative battery cable
13. Fill and bleed the cooling system.
14. Change the oil and filter.
15. Run the engine and check for proper operation.

1.6L (ZM) ENGINES

1. Before servicing the vehicle, refer to the precautions in the beginning of this section.

2. Properly relieve the fuel system pressure.

3. Drain the engine coolant.

4. Remove or disconnect the following:
- Negative battery cable
- Timing belt
- Front exhaust pipe
- Exhaust manifold insulator and the Exhaust Gas Recirculation (EGR) pipe
- Fresh air duct and air cleaner assembly
- Accelerator cable and bracket
- All vacuum hoses
- Engine wiring harness connectors
- Fuel supply and return hoses
- Intake manifold support bracket
- Heater hoses
- Camshaft pulleys by holding the them with a wrench
- Camshafts
- Cylinder head bolts in sequence
- Cylinder head

To install:

5. Thoroughly, clean the cylinder head and the block contact surfaces.

6. Clean the cylinder head bolts and the threads in the block. Be sure the bolts turn freely in the block.

7. Measure the length of the cylinder head bolts, as shown, maximum bolt length is 3.956 inch (100.5mm).

8. Install or connect the following:
- New head gasket
- Cylinder head

9. Torque the cylinder head bolts, in sequence, as follows.
   a. Step 1: 13–16 ft. lbs. (17–22 Nm).
   b. Step 2: Turn 85–95 degrees.
   c. Step 3: Turn an additional 85–95 degrees.

10. Install or connect the following:
- Camshafts
- Camshaft pulleys, install the pulleys so that the **I** mark on the intake side or **E** mark on the exhaust side are facing up. Torque the bolts to 37–44 ft. lbs. (50–60 Nm).
- Heater hoses
- Intake manifold support bracket
- Fuel hoses
- Engine wiring harness connector

Cylinder head bolt removal sequence—1.6L (ZM) and 1.8L (FP) engines

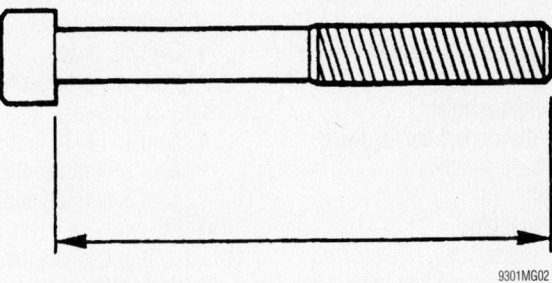

Replace any bolts that exceed the maximum length—1.6L (ZM) and 1.8L (FP) engines

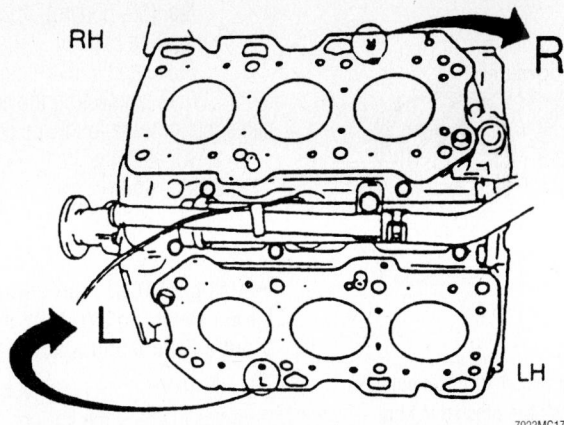

Cylinder head gasket positioning—6-cylinder engines

Cylinder head bolt tightening sequence—1.6L (ZM) and 1.8L (FP) engines

- Any vacuum hoses that were removed
- Accelerator cable and bracket
- Air cleaner assembly and fresh air duct
- EGR system
- Exhaust manifold insulator and front exhaust pipe
- Timing belt
- Negative battery cable

11. Fill the cooling system.

12. Start the vehicle, check for leaks and repair if necessary.

1.8L (FP) ENGINES

1. Before servicing the vehicle, refer to the precautions in the beginning of this section.

2. Properly relieve the fuel system pressure.

3. Drain the engine coolant.

4. Remove or disconnect the following:
- Negative battery cable
- Timing belt
- Front exhaust pipe
- Air cleaner assembly
- Power steering pump and bracket, if necessary, and move it aside leaving the hoses attached
- Accelerator cable

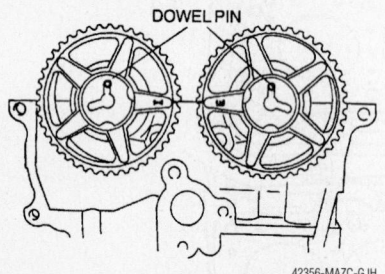

DOWEL PIN

42356-MAZC-GJH

Make sure the camshaft sprocket pulleys are positioned with the pins facing up— 1.8L (FP) engine

**Free play
1.5—4.0 mm {0.06—0.15 in}**

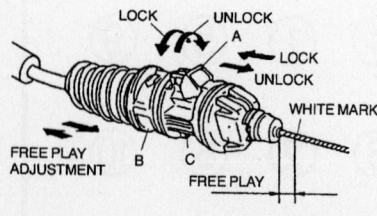

LOCK UNLOCK
A
LOCK
UNLOCK
WHITE MARK
FREE PLAY
ADJUSTMENT
B C
FREE PLAY

42356-MAZC-G04

Accelerator cable adjustment components—Protégé models equipped with the 1.8L (FP) engine

- Fuel supply and return hoses
- Ignition coil
- Camshaft pulleys by holding the them with a wrench
- Camshafts
- Cylinder head bolts in sequence
- Cylinder head

To install:

5. Thoroughly, clean the cylinder head and the block contact surfaces.

6. Clean the cylinder head bolts and the threads in the block. Be sure the bolts turn freely in the block.

7. Measure the length of the cylinder head bolts, as shown, maximum bolt length is 4.153 inch (105.5mm).

8. Install or connect the following:
- New head gasket
- Cylinder head

9. Torque the cylinder head bolts, in sequence, as follows.
  a. Step 1: 13–16 ft. lbs. (17–22 Nm).
  b. Step 2: Turn 85–95 degrees.
  c. Step 3: Turn an additional 85–95 degrees.

10. Install or connect the following:
- Camshafts
- Camshaft pulleys, install the pulleys so that the dowels are facing up. Torque the bolts to 37–44 ft. lbs. (50–60 Nm).
- Ignition coil
- Fuel supply and return hoses

11. When connecting the accelerator cable, perform the following adjustment:
  a. Move the white locking tab A to the unlock position.
  b. Turn stopper B to the to the unlock position.

➡**If stopper B will not unlock, it may be necessary to carefully bend back tab C out a little using a suitable pry tool.**

  c. Push or pull the cable housing directly behind the spring.
  d. Turn the stopper B to the lock position.
  e. Measure the free play which should be 0.04–0.11 inch (1–3mm) and make sure the cable free play is within specification.
  f. Move the white locking tab to the lock position and check for proper accelerator operation.
- Power steering pump and bracket
- Air cleaner assembly
- Front exhaust pipe
- Timing belt
- Negative battery cable

12. Fill the cooling system.

13. Start the vehicle, check for leaks and repair if necessary.

2.5L (KL) ENGINE

1. Before servicing the vehicle, refer to the precautions in the beginning of this section.

2. Drain the cooling system.

3. Relieve the fuel system pressure.

4. Drain the cooling system.

5. Remove or disconnect the following:

- Negative battery cable
- Timing belt
- Front exhaust pipe
- Air cleaner assembly
- Intake manifold
- Fuel hose
- Cylinder head cover
- Camshaft pulley. Hold the camshaft pulley with a back–up wrench while loosening the pulley bolt.
- Upper radiator hose
- Number 3 engine mount bracket
- Seal plate
- Water outlet
- Camshaft
- Alternator stay

6. Temporarily install the Number 3 engine mount to support the engine.
- Cylinder head bolts by loosening them, in 2–3 steps, in the sequence illustrated
- Cylinder head

To install:

7. Install or connect the following:
- New cylinder head gasket
- Cylinder head

8. Apply clean engine oil to the bolt threads and seating faces.

9. Install new cylinder head bolts and torque in 2–3 steps, in sequence, to 17–19 ft. lbs. (23–26 Nm). If reusing old bolts, which is not recommended; make sure the maximum length of the bolt is 5.217 inch (132.5mm). The standard length of the bolt should be 5.166–5.188 inch (131.2–131.8mm).

10. Paint a mark on the edge of each cylinder head bolt to use as a reference. Turn each bolt, in sequence, 85–95 degrees. Again, turn each bolt, in sequence, an additional 85–95 degrees.

11. Support the engine assembly with a suitable lifting device and remove the number 3 mount.

12. Install or connect the following:
- Alternator stay
- Camshaft
- Water outlet
- Seal plate
- Number 3 engine mount bracket
- Upper radiator hose
- Camshaft pulley. Hold the camshaft

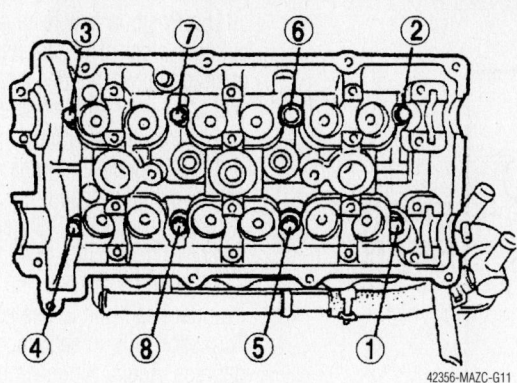

Cylinder head bolt removal sequence—2.5L (KL) engines

42356-MAZC-G11

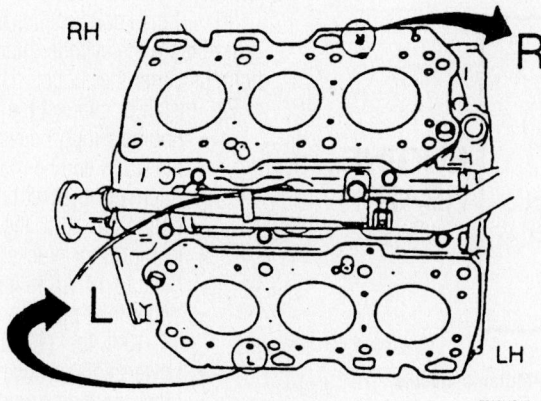

Cylinder head gasket positioning—2.5L (KL) engines

7923MG17

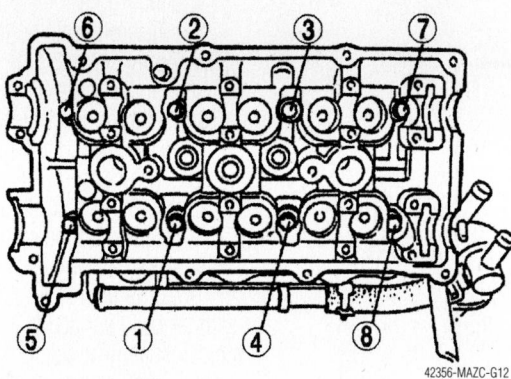

Cylinder head bolt tightening sequence—2.5L (KL) engines

42356-MAZC-G12

pulley with a back–up wrench while tightening the pulley bolt to 90–97 ft. lbs. (123–132 Nm).

- Cylinder head cover
- Fuel hose
- Intake manifold
- Air cleaner assembly
- Front exhaust pipe
- Timing belt
- Negative battery cable

13. Fill and bleed the cooling system.
14. Change the oil and filter.
15. Run the engine and check for proper operation.

Millenia

2.3L (KJ) ENGINES

1. Before servicing the vehicle, refer to the precautions in the beginning of this section.
2. Relieve the fuel system pressure.
3. Drain the engine coolant.
4. Remove or disconnect the following:
   - Negative battery cable
   - Oxygen (O_2S) sensor connectors
   - Exhaust pipe-to-manifold nuts and lower the exhaust pipes
   - Right-hand 3-way catalytic converter

- Compressor (supercharger)
- Intake manifold
- Timing belt covers and timing belt
- Spacer and O-ring from the front of the camshaft
- Ignition coils
- Cylinder head cover mounting bolts, in 5–6 steps, using the reverse of the tightening sequence
- Cylinder head cover
- Camshaft sprockets

5. Turn the camshafts so the knock pins are aligned with the marks on the camshaft caps. This will reduce the pressure on the adjustment shims.

6. Note the markings on the camshaft caps prior to removal, so they can be reinstalled in the same positions. The right-hand (rear) caps are marked with numbers and the left-hand (front) caps are marked with letters.

7. Loosen the front camshaft cap bolts in sequence, in 5–6 steps. Remove the front camshaft caps. remove the remaining camshaft cap bolts in the proper sequence. Remove the caps, being sure to remove the thrust caps last. Do not damage the cylinder head thrust bearing support.

8. Remove or disconnect the following:
   - Camshafts and oil seals
   - Lifters and adjustment shims

9. Identify and mark each lifter as it is removed so it can be reinstalled in the same position.

10. Remove or disconnect the following:
   - Lower radiator hose
   - Water inlet pipe
   - Compressor bracket
   - Alternator bracket bolt to gain additional clearance
   - Rubber insulator from the left-hand cylinder head

11. Temporarily install the No. 3 engine mount, which was removed with the timing belt, to support the engine.

12. Remove or disconnect the following:
   - Engine support device
   - Cylinder head bolts, in 2–3 steps, in the sequence illustrated
   - Cylinder heads
   - Oil control plug O-rings

13. Clean all gasket mating surfaces. Inspect the cylinder head for damage, cracks, and water and oil leakage. Check the head gasket surface for distortion using a straightedge and feeler gauge. Maximum allowable distortion is 0.004 inch (0.10mm).

To install:

14. Apply clean engine oil to the O-rings, and install them onto the oil control plugs.

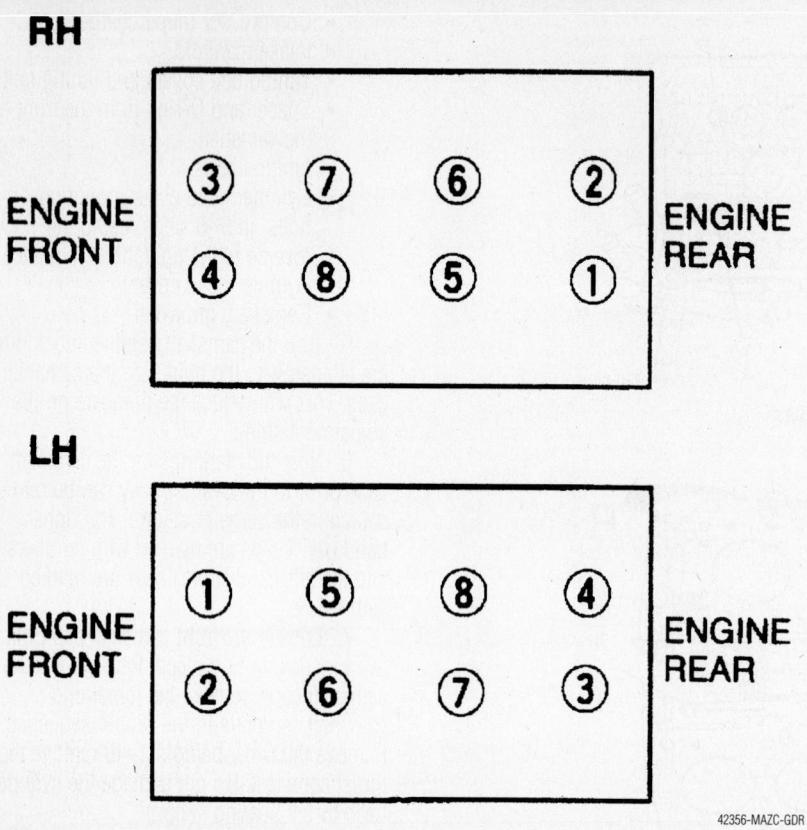

Loosen the cylinder head bolts in this sequence using several passes—Millenia models equipped with the 2.3L (KJ) engine

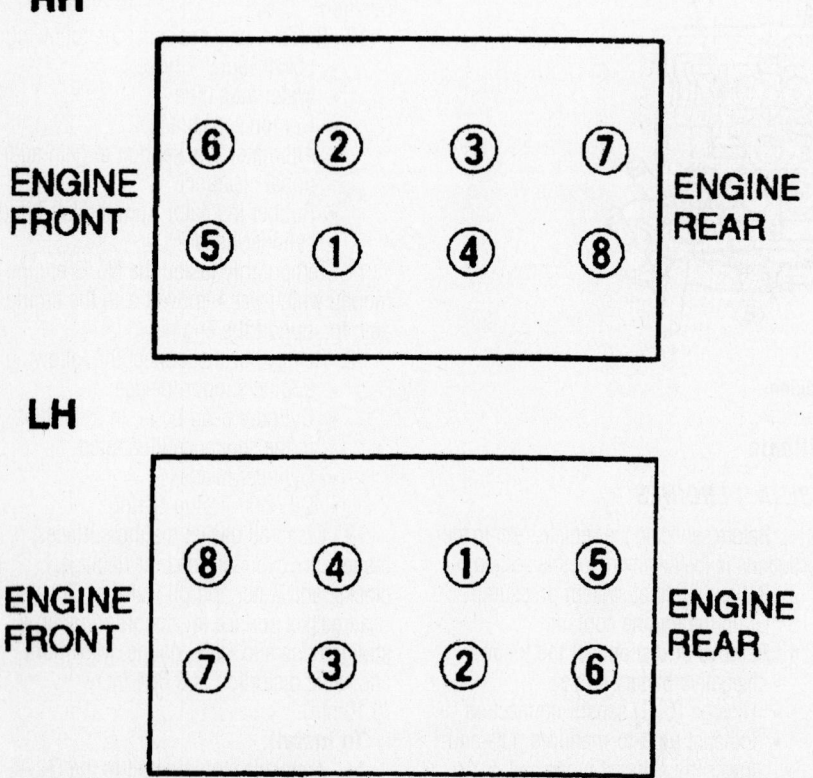

Cylinder head bolt torque sequence—Millenia models equipped with the 2.3L (KJ) engine

15. Position new head gaskets on the cylinder block. The gaskets cannot be interchanged between sides and are marked **R** and **L** for right and left side.

16. Install the cylinder heads.

17. Apply clean engine oil to the threads of new cylinder head bolts and install. Torque the bolts in 2–3 steps, in sequence, to 17–19 ft. lbs. (23–26 Nm).

18. Paint a mark on the edge of each cylinder head bolt to use as a reference. Turn each bolt, in sequence, 85–95 degrees. Again, turn each bolt, in sequence, an additional 85–95 degrees.

19. Install the rubber insulator onto the left-hand cylinder head.

20. Fit the knock sensor harness into the drill hole on the cylinder block. Pass the harness under the rubber insulator.

21. Install or connect the following:
- Engine support device and remove the No. 3 engine mount
- Alternator bracket bolt. Torque the bolt to 12–16 ft. lbs. (16–22 Nm).
- Compressor bracket. Torque the bolts to 14–18 ft. lbs. (19–25 Nm).
- Water inlet pipe. Torque the bolts to 14–18 ft. lbs. (19–25 Nm).
- Lower radiator hose
- Lifters in their original positions by lubricating them with engine oil. Verify that they move smoothly in their bore
- New oil seals on the camshafts

22. Apply clean engine oil to the camshaft lobes, journals and supports.

23. Install or connect the following:
- Camshafts so the gear marks align
- Thrust caps. Torque the bolts, in 5–6 steps, until they are fully seated on the cylinder head

24. Apply silicone sealant, at a thickness of 0.06–0.09 inch (1.5–2.5mm), to the cylinder head surface in the area forward of the camshaft gear cavity.

25. Install or connect the following:
- Remaining camshaft caps in their original positions. Torque the caps, in sequence, in 5 equal steps, with the final step being 100–125 inch lbs. (11–14 Nm).
- New camshaft oil seal lubricated with engine oil. Tap the seal in evenly with a Seal Installer tool 49 F401 337A with a final protrusion of 0–0.02 inch (0–0.5mm). Tap in a new blind cap
- Camshaft sprockets. Torque the bolts to 91–103 ft. lbs. (123–140 Nm).

26. Measure and adjust the valve clearances.

27. Remove any sealant and gasket material from the cylinder head cover contact surfaces.

28. Apply silicone sealant to the cylinder head in the area adjacent to the front and rear camshaft caps.

29. Install or connect the following:
- Cylinder head cover using a new gasket. Torque the bolts in 5–6 steps, in sequence, to 44–78 inch lbs. (5–9 Nm).
- Distributor using a new O-ring
- Ignition coils
- Spacer using a new O-ring. Torque

the bolt to 14–18 ft. lbs. (19–25 Nm).
- Timing belt and timing belt cover
- Intake manifold
- Compressor (supercharger)
- Right-hand 3-way catalytic converter
- Exhaust pipes to the manifolds. Torque the nuts to 28–38 ft. lbs. (38–51 Nm).
- O$_2$S connectors
- Negative battery cable

30. Fill and bleed the coolant system.
31. Run the engine and check for leaks.

2.5L (KL) ENGINE

1. Before servicing the vehicle, refer to the precautions in the beginning of this section.
2. Drain the cooling system.
3. Relieve the fuel system pressure.
4. Drain the cooling system.
5. Remove or disconnect the following:

- Negative battery cable
- Timing belt
- Air cleaner assembly
- Spark plug wires
- Distributor

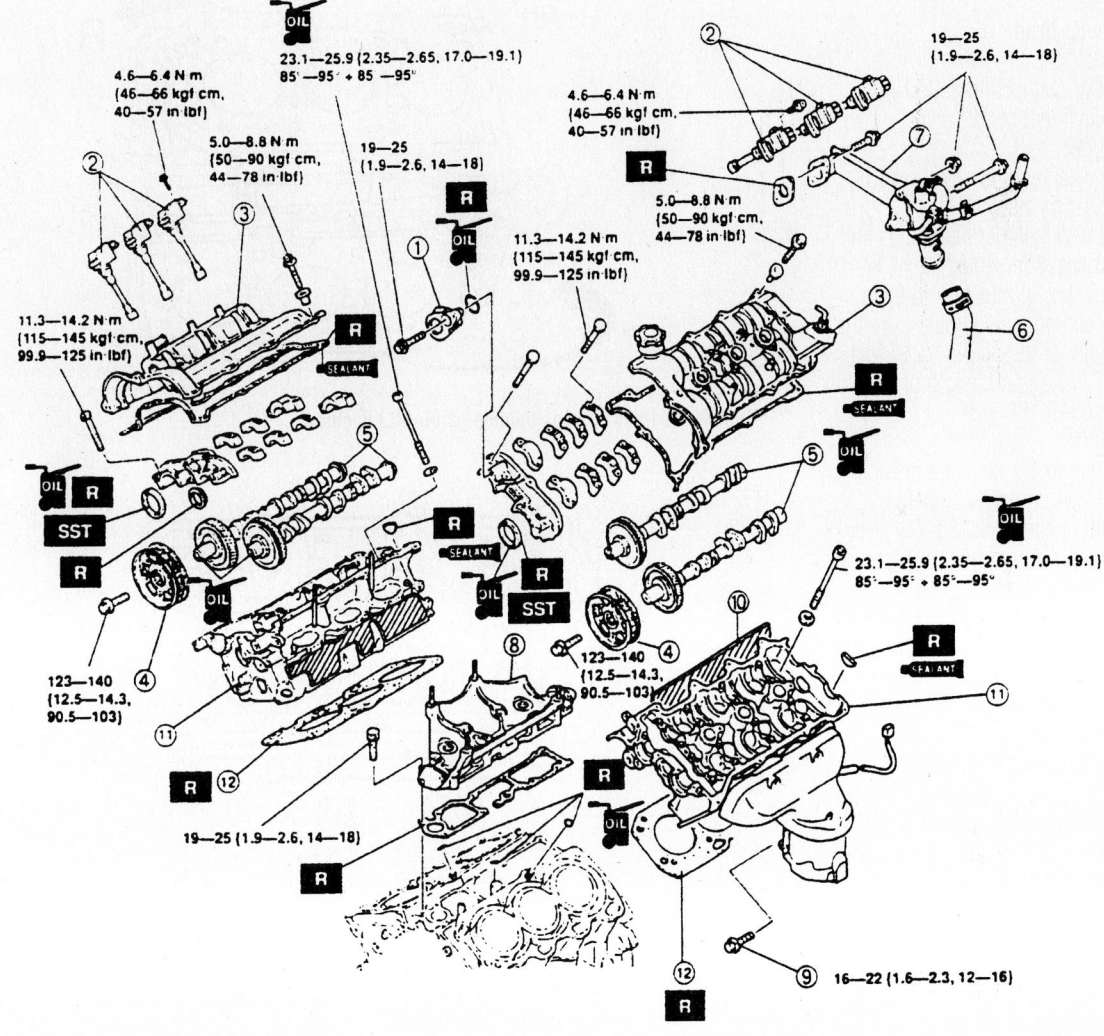

N·m (kgf·m, ft·lbf)

1. Spacer
2. Ignition coil
3. Cylinder head cover
4. Camshaft pulley
5. Camshaft
6. Lower radiator hose
7. Water inlet pipe
8. Lysholm compressor bracket
9. Generator bolt
10. Rubber insulator (LH)
11. Cylinder head
12. Cylinder head gasket

Exploded view of the cylinder head and related components—2.3L engine

7923MG24

- Intake manifold
- Upper radiator hose
- Water outlet
- Number 3 engine mount bracket
- Seal plate
- Front exhaust pipe
- Alternator stay
- Cylinder head cover
- Camshaft pulley. Hold the camshaft pulley with a back–up wrench while loosening the pulley bolt.
- Camshaft

6. Temporarily install the Number 3 engine mount to support the engine.

- Cylinder head bolts by loosening them, in 2–3 steps, in the sequence illustrated
- Cylinder head

To install:

7. Install or connect the following:
- New cylinder head gasket
- Cylinder head

8. Apply clean engine oil to the bolt threads and seating faces.

9. Install new cylinder head bolts and torque in 2–3 steps, in sequence, to 17–19 ft. lbs. (23–26 Nm). If reusing old bolts, which is not recommended; make sure the maximum length of the bolt is 5.217 inch (132.5mm). The standard length of the bolt should be 5.166–5.188 inch (131.2–131.8mm).

10. Paint a mark on the edge of each cylinder head bolt to use as a reference. Turn each bolt, in sequence, 85–95 degrees. Again, turn each bolt, in sequence, an additional 85–95 degrees.

11. Support the engine assembly with a suitable lifting device and remove the number 3 mount.

12. Install or connect the following:
- Camshaft
- Camshaft pulley. Hold the camshaft pulley with a back–up wrench while tightening the pulley bolt to 90–97 ft. lbs. (123–132 Nm).
- Cylinder head cover
- Alternator stay
- Front exhaust pipe
- Seal plate
- Number 3 engine mount bracket
- Water outlet
- Upper radiator hose
- Intake manifold
- Distributor and spark plug wires
- Air cleaner assembly
- Timing belt
- Negative battery cable

13. Fill and bleed the cooling system.
14. Change the oil and filter.
15. Run the engine and check for proper operation.

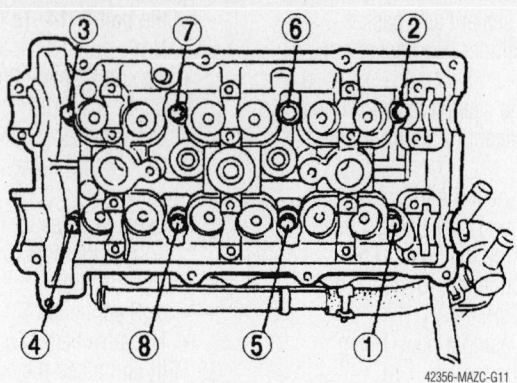

Cylinder head bolt removal sequence—2.5L (KL) engines

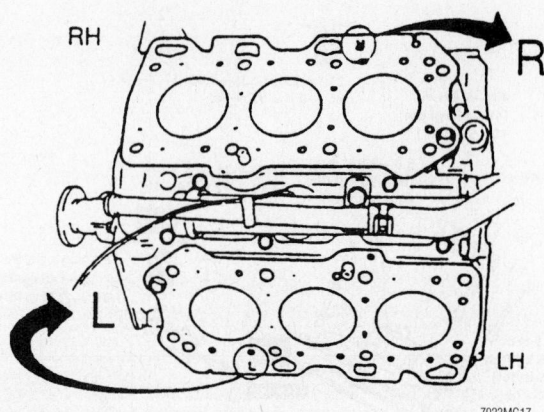

Cylinder head gasket positioning—2.5L (KL) engines

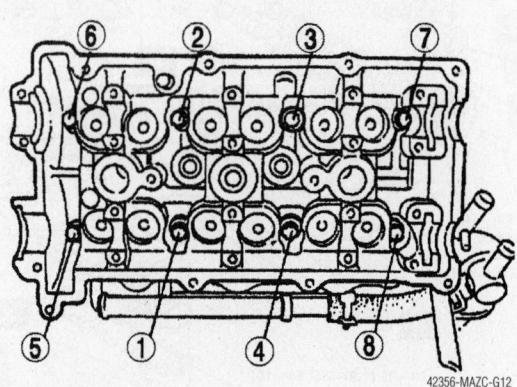

Cylinder head bolt tightening sequence—2.5L (KL) engines

Mazda 3

1. Before servicing the vehicle, refer to the precautions in the beginning of this section.

2. Disconnect the negative battery cable.

3. Remove the timing chain.

4. Remove the intake manifold.

5. Disconnect the following components:
- Warm Up–Three Way Converter (WU–TWC)
- Upper radiator hose
- Water and heater hose
- Wiring harness

6. Support the engine using an engine jack and attachment as shown in the illustration.

7. Remove the camshafts.

➡The cylinder head and camshaft caps are numbered and must be reassembled in their original locations. When removing a when removed, keep the caps with the cylinder head they were removed from, do not switch the caps.

8. Loosen the cylinder head bolts using 2–3 passes in the sequence illustrated.

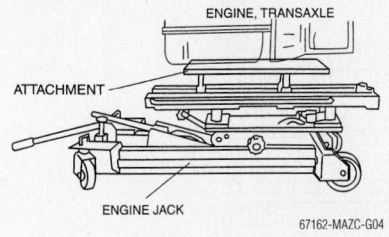

Support the engine using an engine jack and attachment as shown—Mazda 3 models

9. Once all the bolts have been removed, remove the cylinder head and gasket.

To install:

10. If reusing old bolts, which is not recommended; measure the length of each head bolt at the locations illustrated. If any bolt exceeds 5.717–5.740 inch (145.2–145.8mm) replace the bolt.

11. Install a new cylinder head gasket.

12. Install the cylinder head.

13. Apply clean engine oil to the bolt threads and seating faces.

14. Install new cylinder head bolts and torque in 2–3 steps, in sequence as follows:

 a. Step 1: Tighten the bolts in sequence to 27–97 inch lbs. (3–11 Nm).

 b. Step 2: Tighten the bolts in sequence to 9.5–12.5 ft. lbs. (13–17 Nm).

 c. Step 3: Tighten the bolts in sequence to 32–34.5 ft. lbs. (43–47 Nm).

 d. Step 4: Paint a mark on the edge of each cylinder head bolt to use as a reference. Turn each bolt, in sequence, 88–92 degrees.

 e. Step 5: Turn each bolt, in sequence, 88–92 degrees.

15. Install the camshafts and caps in their original positions.

16. Install the remaining components in the reverse order of removal.

17. Fill and bleed the cooling system.

18. Change the oil and filter.

19. Run the engine and check for proper operation.

Mazda 6

2.3L (L3) ENGINE

1. Before servicing the vehicle, refer to the precautions in the beginning of this section.

2. Disconnect the negative battery cable.

3. Remove the timing chain.

4. Remove the ignition coil.

5. Unbolt the alternator and move it aside with the connectors still attached.

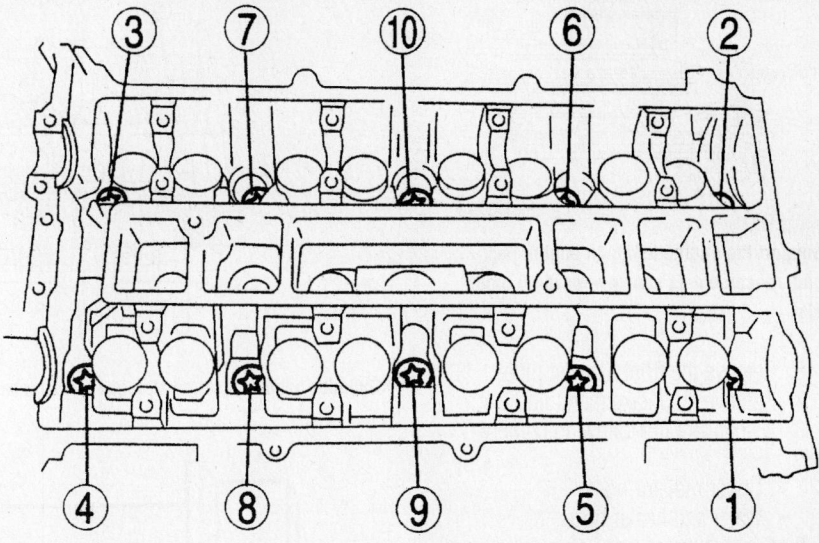

Cylinder head bolt removal sequence—Mazda 3 models

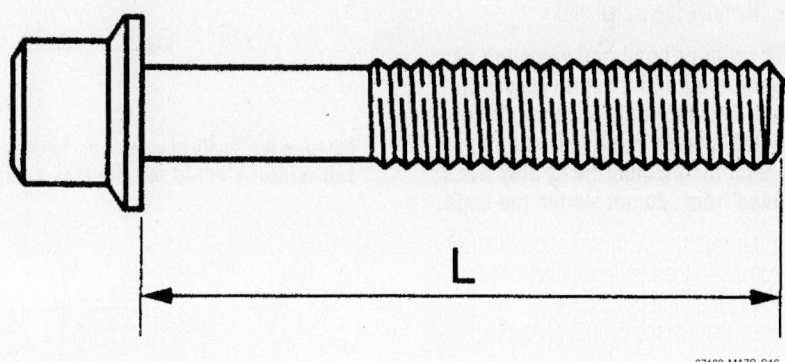

Measure the length of each head bolt at the locations illustrated, if reusing the old bolts. If any bolt exceeds 5.717–5.740 inch (145.2–145.8mm) replace the bolt—Mazda 3 models

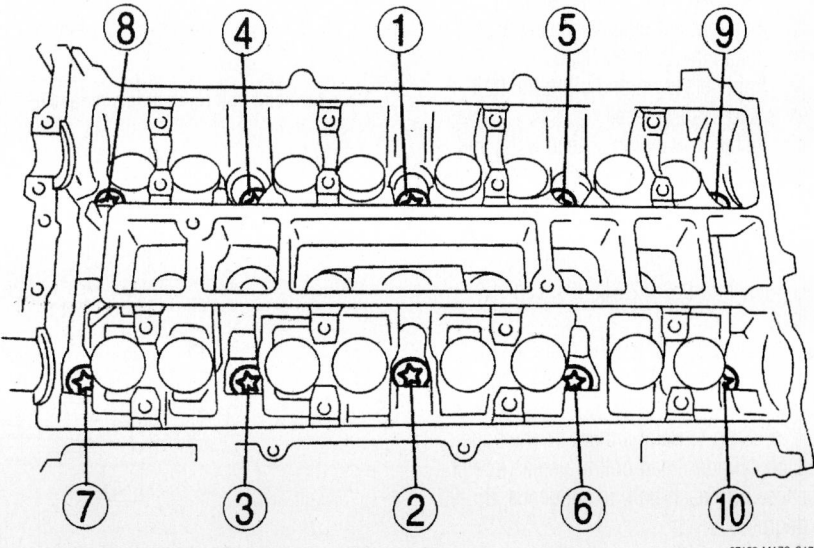

Cylinder head bolt tightening sequence—Mazda 3 models

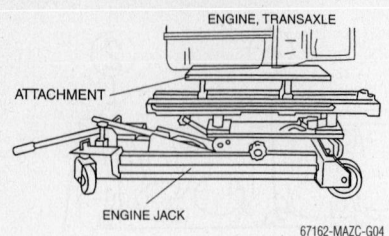

67162-MAZC-G04

Support the engine using an engine jack and attachment as shown—Mazda 6 models

6. Remove the front exhaust pipe.

7. Remove the intake manifold.

8. Disconnect the following components:
- Upper radiator hose
- Water and heater hose

9. Support the engine using an engine jack and attachment as shown in the illustration.

10. Remove the oil control valve.

11. Remove the camshafts.

➡**The cylinder head and camshaft caps are numbered and must be reassembled in their original locations. When removing a when removed, keep the caps with the cylinder head they were removed from, do not switch the caps.**

12. Loosen the cylinder head bolts using 2–3 passes in the sequence illustrated.

13. Once all the bolts have been removed, remove the cylinder head and gasket.

To install:

14. If reusing old bolts, which is not recommended; measure the length of each head bolt at the locations illustrated. If any bolt exceeds 5.717–5.740 inch (145.2–145.8mm) replace the bolt.

15. Install a new cylinder head gasket.

16. Install the cylinder head.

17. Apply clean engine oil to the bolt threads and seating faces.

18. Install new cylinder head bolts and torque in 2–3 steps, in sequence as follows:

a. Step 1: Tighten the bolts in sequence to 27–97 inch lbs. (3–11 Nm).

b. Step 2: Tighten the bolts in sequence to 9.5–12.5 ft. lbs. (13–17 Nm).

c. Step 3: Tighten the bolts in sequence to 32–34.5 ft. lbs. (43–47 Nm).

d. Step 4: Paint a mark on the edge of each cylinder head bolt to use as a reference. Turn each bolt, in sequence, 85–92 degrees.

e. Step 5: Turn each bolt, in sequence, 85–92 degrees.

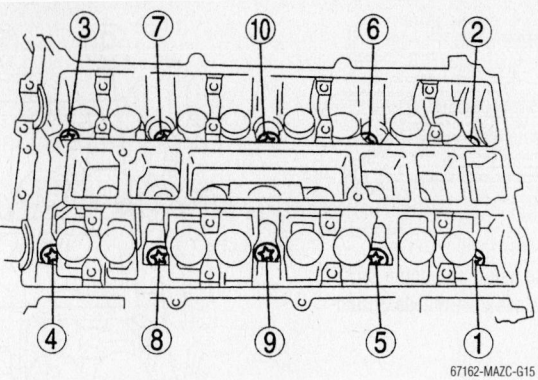

67162-MAZC-G15

Cylinder head bolt removal sequence—Mazda 6 models

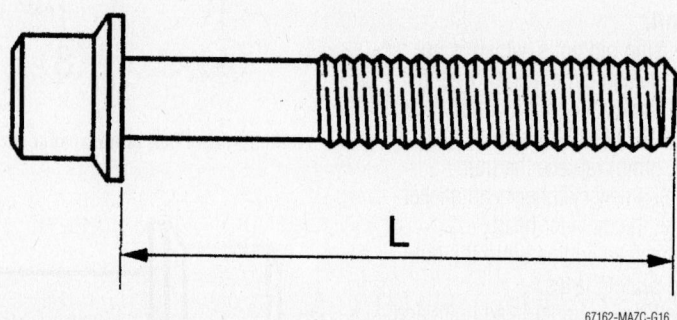

67162-MAZC-G16

Measure the length of each head bolt at the locations illustrated, if reusing the old bolts. If any bolt exceeds 5.717–5.740 inch (145.2–145.8mm) replace the bolt—Mazda 6 models

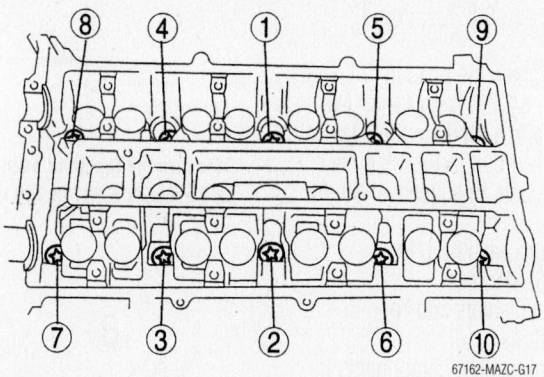

67162-MAZC-G17

Cylinder head bolt tightening sequence—Mazda 6 with the 2.3L (L3) engine

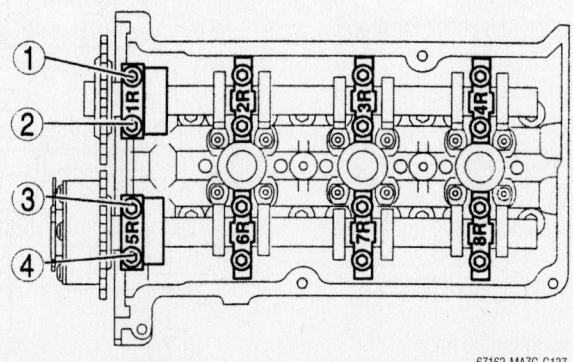

67162-MAZC-G137

If removing the right hand camshaft caps, remove thrust caps 1R and 5R first—Mazda 6 with the 3.0L (AJ) engine

19. Install the camshafts and caps in their original positions.

20. Install the remaining components in the reverse order of removal.

21. Fill and bleed the cooling system.

22. Change the oil and filter.

23. Run the engine and check for proper operation.

3.0L (AJ) ENGINE

1. Before servicing the vehicle, refer to the precautions in the beginning of this section.

2. Properly relieve the fuel system pressure.

3. Drain the cooling system.

4. Disconnect the negative battery cable.

5. Remove the water pump.

6. Remove the timing chain.

7. Support the engine using an engine jack and attachment.

8. Remove the water bypass tube stud bolt and bolt and disconnect the tube from the cylinder head.

9. Remove the camshaft(s).

10. Remove the rocker arm(s).

11. Remove the camshafts.

➡ If removing the right hand cylinder head and camshaft caps, remove thrust caps 1R and 5R first. Do not loosen any of the other cap bolts until the thrust caps are removed or you could damage the thrust caps. If removing the left hand cylinder head and camshaft caps, remove thrust caps 1L and 6L first. Do not loosen any of the other cap bolts until the thrust caps are removed or you could damage the thrust caps.

12. Loosen the camshaft caps using 7–8 passes in sequence after first removing the thrust caps to allow the camshafts to be slowly raised. Keep the caps in the order they were removed so they are re-installed in their original positions.

13. Loosen the cylinder head bolts using 2–3 passes in the sequence illustrated.

14. Once all the bolts have been removed, remove the cylinder head and gasket.

To install:

15. Install a new head gasket and the cylinder head.

16. Lubricate the cylinder head bolt threads.

17. Torque the cylinder head bolts in the proper sequence as follows:

   a. Step 1: Tighten the bolts in sequence 24–28 ft. lbs. (32–38 Nm).

   b. Step 2: Tighten the bolts in sequence an additional 85–95 degrees.

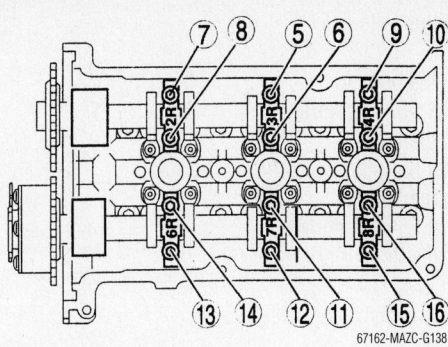

Loosen the right hand camshaft caps using 7–8 passes in this sequence after first removing the thrust caps—Mazda 6 with the 3.0L (AJ) engine

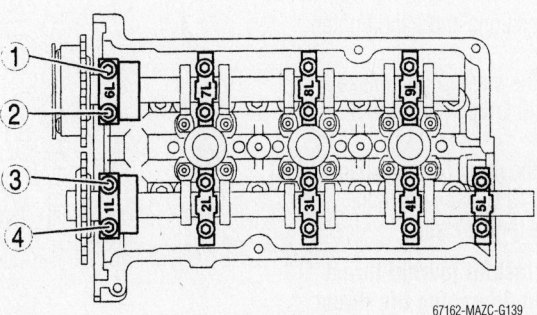

If removing the left hand camshaft caps, remove thrust caps 1L and 6L first—Mazda 6 with the 3.0L (AJ) engine

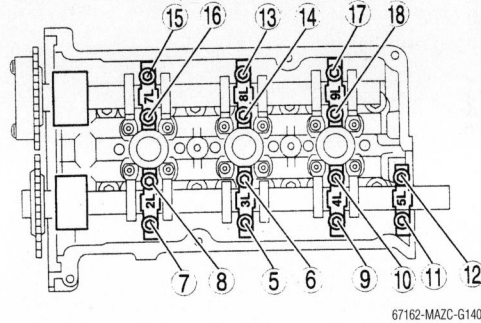

Loosen the left hand camshaft caps using 7–8 passes in this sequence after first removing the thrust caps—Mazda 6 with the 3.0L (AJ) engine

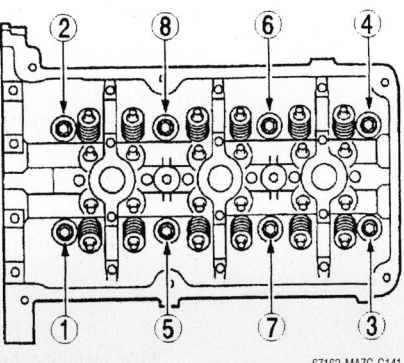

Cylinder head bolt loosening sequence—Mazda 6 with the 3.0L (AJ) engine

c. Step 3: Loosen the bolts in sequence one full turn.

d. Step 4: Tighten the bolts in sequence 24–28 ft. lbs. (32–38 Nm).

e. Step 5: Tighten the bolts in sequence an additional 85–95 degrees.

f. Step 6: Tighten the bolts in sequence an additional 85–95 degrees.

18. Install the camshafts in their original positions.

19. Install the left hand camshaft as follows:

a. Place the crankshaft keyway at the 11 O'clock position by rotating the crankshaft clockwise.

b. Install the camshaft and the caps in their original positions and hand tighten the bolts.

c. Position the mark on the intake camshaft to the 9 O'clock position as illustrated.

d. Position the mark on the exhaust camshaft to the 12 O'clock position.

e. Install the timing chain.

➡ **Tighten the camshaft journal thrust caps last to avoid damaging the thrust caps.**

f. Tighten the camshaft caps using several passes to 71–106 inch lbs. (8–12 Nm). After adjust the camshaft endplay using the thrust caps 1L and 6L, tighten the remaining caps.

20. Install the right hand camshaft as follows:

a. Install the camshaft and the caps in their original positions and hand tighten the bolts.

b. Place the crankshaft keyway at the 3 O'clock position.

c. Position the mark on the exhaust camshaft to the 12 O'clock position by rotating the crankshaft clockwise.

d. Position the mark on the intake camshaft to the 3 O'clock position as illustrated.

e. Install the timing chain.

➡ **Tighten the camshaft journal thrust caps last to avoid damaging the thrust caps.**

f. Tighten the camshaft caps using several passes to 71–106 inch lbs. (8–12 Nm). After adjust the camshaft endplay using the thrust caps 1R and 5R, tighten the remaining caps.

21. Install the water bypass tube.

22. Remove the engine jack and attachment.

23. Install the water pump.

24. Fill the cooling system.

25. Fill and bleed the cooling system.

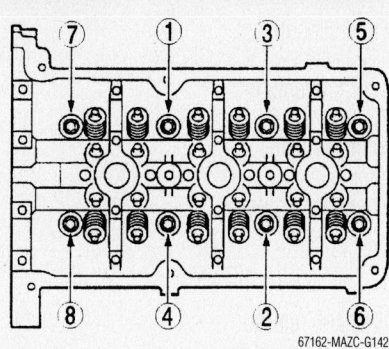

Cylinder head bolt torque sequence—Mazda 6 with the 3.0L (AJ) engine

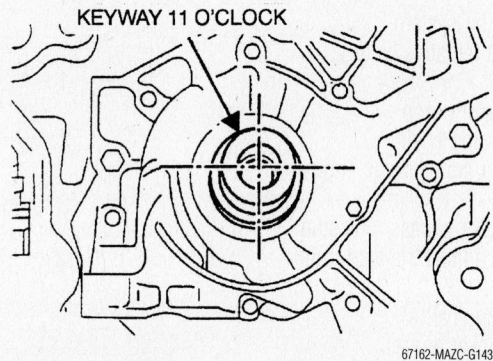

When installing the left hand camshaft place the crankshaft keyway at the 11 O'clock position—Mazda 6 with the 3.0L (AJ) engine

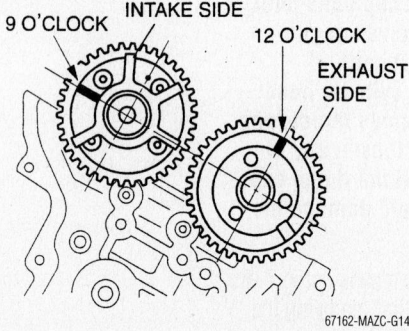

Install the left hand camshaft and position the mark on the intake camshaft to the 9 O'clock and the exhaust camshaft to the 12 O'clock position—Mazda 6 with the 3.0L (AJ) engine

When installing the right hand camshaft place the crankshaft keyway at the 3 O'clock position—Mazda 6 with the 3.0L (AJ) engine

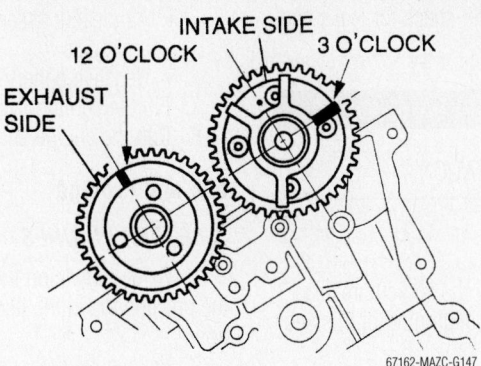

Install the right hand camshaft and position the mark on the intake camshaft to the 3 O'clock and the exhaust camshaft to the 12 O'clock position—Mazda 6 with the 3.0L (AJ) engine

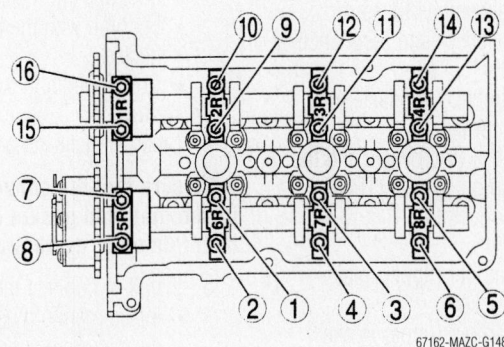

Right side camshaft cap torque sequence—Mazda 6 with the 3.0L (AJ) engine

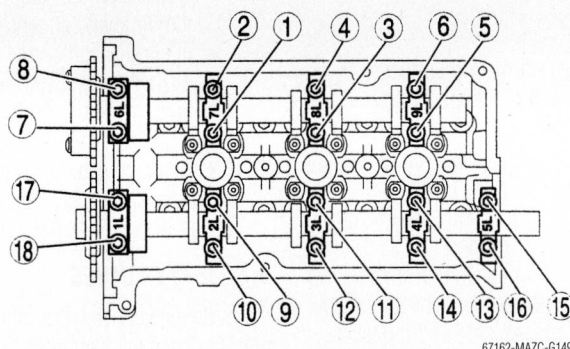

Left side camshaft cap torque sequence—Mazda 6 with the 3.0L (AJ) engine

26. Change the oil and filter.
27. Run the engine and check for proper operation.

Rocker Arms/Shafts

REMOVAL & INSTALLATION

All Mazda engines covered in this manual are not equipped with rocker arms/shafts, the camshafts directly actuate the valves through a bucket type cam follower.

Supercharger

REMOVAL & INSTALLATION

2.3L Engine

1. Before servicing the vehicle, refer to the precautions in the beginning of this section.
2. Relieve the fuel system pressure.
3. Drain the cooling system.
4. Remove or disconnect the following:
   • Negative battery cable

- Dynamic chamber cover
- Charge air cooler air duct
- Vacuum hoses and electrical connectors from the air cleaner housing
- Air cleaner assembly
- Fresh air ducts
- Mass Air Flow (MAF) sensor and the air intake hose from the throttle body
- Resonator
- Right-hand charge air cooler
- Left-hand charge air cooler
- Accelerator cable
- Vacuum hoses from the rear of the intake manifold and Exhaust Gas Recirculation (EGR) valve
- EGR valve
- Air intake pipe assembly
- Charge air cooler pipe
- Fuel supply line at the fuel rails and discard the copper washers
- Fuel and vacuum lines from the fuel pressure regulator
- Coolant hoses
- Wiring harness from the intake manifold
- Intake manifold mounting nuts and bolts in 2–3 steps
- Intake manifold
- Fuel hoses and electrical connectors from the throttle body
- Throttle body
- Drive belt from the compressor (supercharger)
- Mounting bolts from the compressor
- Compressor

To install:

5. Clean all gasket mating surfaces.
6. Position the rubber shield for the compressor onto the compressor using double sided adhesive tape.
7. Install or connect the following:
   • Compressor. Torque the nuts to 14–18 ft. lbs. (19–25 Nm).

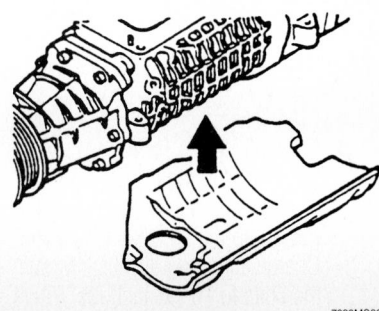

When installing the compressor, ensure that the rubber insulating pad is temporarily affixed to the compressor

- Compressor drive belt
- Throttle body. Torque the bolts to 14–18 ft. lbs. (19–25 Nm).
- Fuel hoses and electrical connectors
- Intake manifold using a new gasket. Torque the bolts in 2–3 steps, from the center to the ends, to 14–18 ft. lbs. (19–25 Nm).
- Wiring harness onto the intake manifold
- Coolant hoses
- Fuel and vacuum lines to the fuel pressure regulator
- Fuel supply line to the fuel rail using new copper crush washers
- Charge air cooler pipe
- Position the air intake pipe assembly using new gaskets

8. Hand-tighten the nuts/bolts in the order shown until the air intake pipe contacts the intake manifold. Verify that the rubber gaskets are not twisted or distorted. Torque the bolts marked **A** to 70–95 inch lbs. (8–11 Nm) and all others, in sequence, to 14–18 ft. lbs. (19–25 Nm).

9. Install or connect the following:
- EGR valve using a new gasket
- Vacuum hoses to the intake manifold and EGR valve
- Accelerator cable and adjust it
- Left and right-hand charge air coolers using new gaskets. Hand-tighten the nuts/bolts in the order shown until the air intake pipes and charge air coolers contact the intake manifold. Verify that the rubber

gaskets are not twisted or distorted.

10. Torque the charge air cooler bolts to:
   a. Marked **A**: 44–78 inch lbs. (5–9 Nm).
   b. Marked **B**: to 70–95 inch lbs. (8–11 Nm).
   c. All others, in sequence: 14–18 ft. lbs. (19–25 Nm).

11. Install or connect the following:
- Resonator. Torque the bolts to 12–16 ft. lbs. (16–22 Nm).
- Air intake hose onto the throttle body
- MAF sensor
- Fresh air ducts
- Air cleaner assembly
- Vacuum hoses and electrical connectors to the air cleaner housing
- Charge air cooler air duct. Torque the bolts to 70–95 inch lbs. (8–11 Nm).
- Dynamic chamber cover
- Negative battery cable

12. Fill the cooling system.

13. Start the vehicle, check for leaks and repair if necessary.

Intake Manifold

REMOVAL & INSTALLATION

1.8L (BP) Engines

1. Before servicing the vehicle, refer to the precautions in the beginning of this section.
2. Relieve the fuel system pressure.
3. Drain the cooling system.
4. Remove or disconnect the following:
- Negative battery cable
- Intake Air Temperature (IAT) sensor
- Air cleaner assembly, Mass Air Flow (MAF) sensor and resonance chamber
- Throttle (automatic transaxle only) and accelerator cables
- Idle Air Control (IAC) valve and the Throttle Position Sensor (TPS) electrical connectors
- Throttle body
- Dynamic chamber (upper intake manifold) bracket
- Dynamic chamber (upper intake manifold) and gasket
- Exhaust Gas Recirculation (EGR) pipe
- Intake manifold and discard the gasket

To install:

5. Clean all gasket mating surfaces.
6. Install or connect the following:
- Intake manifold using a new gasket, make sure the convex side of the gasket is facing up. Torque the bolts to 17–21 ft. lbs. (19–25 Nm).
- EGR pipe to the manifold
- Dynamic chamber (upper intake manifold) using new gaskets, make sure the convex side of the gasket is facing up. Tighten the bolts to 14–18 ft. lbs. (19–25 Nm).
- Dynamic chamber (upper intake manifold) bracket and tighten the bolts 95 inch lbs. (11 Nm). Tighten bolts firmly, then tighten the chamber side bolt before tightening the fuel rail side bolt.
- Throttle body using a new gasket. Tighten the bolts to 14–18 ft. lbs. (19–25 Nm).
- Electrical connectors for the IAC valve and the TPS
- Connect and adjust the throttle and accelerator cables
- IAT sensor

- Air cleaner assembly, MAF sensor and ducts
- Negative battery cable

7. Fill the cooling system.
8. Run the engine and check for leaks.

626 and Protégé

1.6L (ZM) ENGINES

1. Before servicing the vehicle, refer to the precautions in the beginning of this section.
2. Relieve the fuel system pressure. Drain the cooling system.
3. Remove or disconnect the following:
- Negative battery cable
- Fuel hoses from the fuel rail
- Fuel injector electrical connectors
- Fuel rail with the injectors connected
- Components in the order illustrated

To install:

4. Clean all gasket mating surfaces.

➡ **Be sure that the convex side of the intake manifold gasket is facing the manifold side, as shown.**

5. Install or connect the following:
- Intake manifold using a new gasket, make sure the convex side of the gasket faces the manifold. Torque the bolts to 14–18 ft. lbs. (19–25 Nm).
- Components in the reverse order of the removal sequence
- Fuel rail. Tighten the bolts to 14–18 ft. lbs. (19–25 Nm).
- Fuel lines and electrical connectors to the fuel rail
- Negative battery cable

6. Fill the cooling system.
7. Run the engine and check for leaks.

1.8L (FP) ENGINES

1. Before servicing the vehicle, refer to the precautions in the beginning of this section.
2. Relieve the fuel system pressure. Drain the cooling system.
3. Remove or disconnect the following:
- Negative battery cable
- Fresh air duct and resonance chamber
- Intake Air Temperature (IAT) sensor
- Air cleaner and filter
- Mass Air Flow (MAF) sensor
- Air hose
- Accelerator cable
- Throttle body
- VICS solenoid valve bracket and valve
- Dynamic chamber
- Intake manifold and gasket

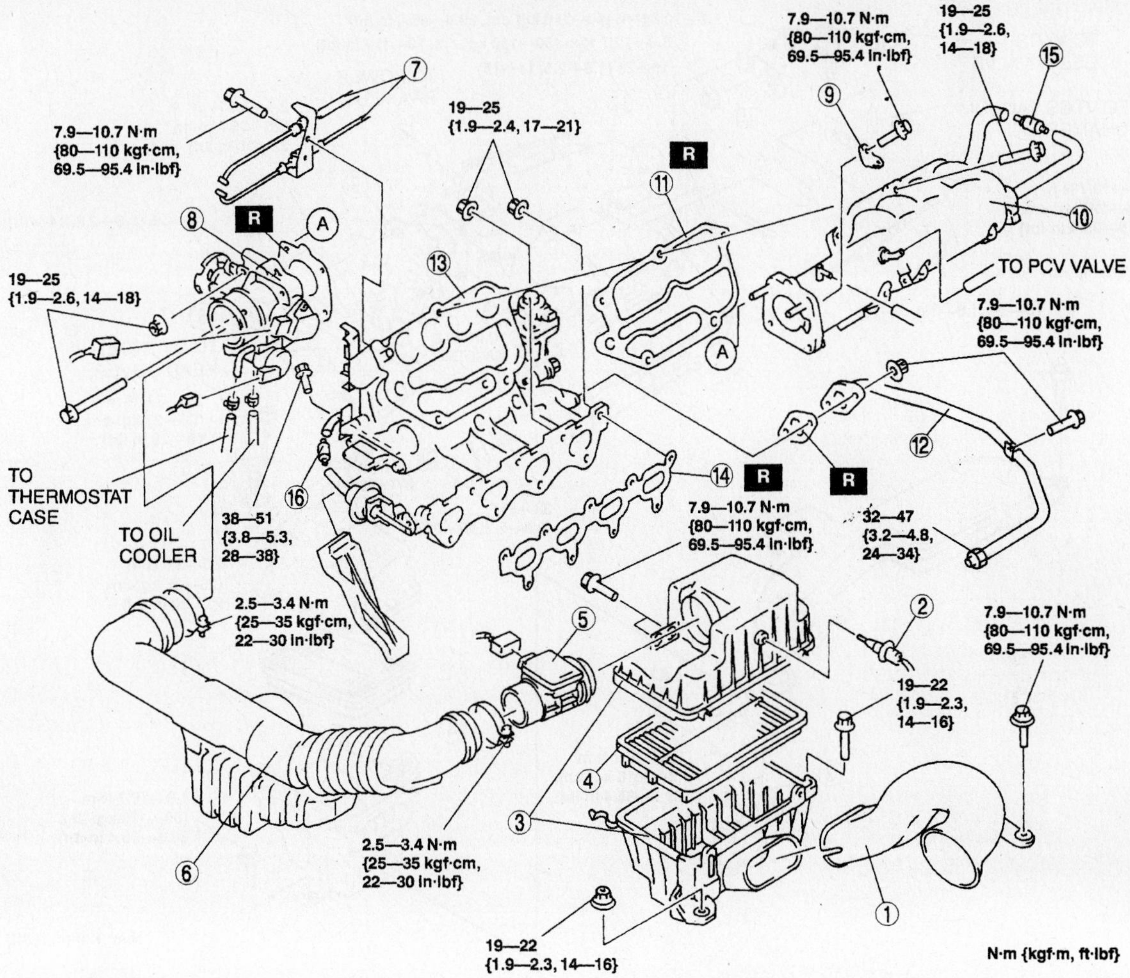

7.9—10.7 N·m
{80—110 kgf·cm,
69.5—95.4 In·lbf}

19—25
{1.9—2.6,
14—18}

7.9—10.7 N·m
{80—110 kgf·cm,
69.5—95.4 In·lbf}

19—25
{1.9—2.4, 17—21}

TO PCV VALVE

7.9—10.7 N·m
{80—110 kgf·cm,
69.5—95.4 In·lbf}

19—25
{1.9—2.6, 14—18}

TO
THERMOSTAT
CASE

TO OIL
COOLER

38—51
{3.8—5.3,
28—38}

7.9—10.7 N·m
{80—110 kgf·cm,
69.5—95.4 In·lbf}

32—47
{3.2—4.8,
24—34}

2.5—3.4 N·m
{25—35 kgf·cm,
22—30 In·lbf}

7.9—10.7 N·m
{80—110 kgf·cm,
69.5—95.4 In·lbf}

19—22
{1.9—2.3,
14—16}

2.5—3.4 N·m
{25—35 kgf·cm,
22—30 In·lbf}

19—22
{1.9—2.3, 14—16}

N·m {kgf·m, ft·lbf}

1	Fresh-air duct	9	Dynamic chamber bracket
2	IAT sensor	10	Dynamic chamber
3	Air cleaner (ACL)	11	Dynamic chamber gasket
4	Air cleaner (ACL) element	12	EGR pipe
5	MAF sensor	13	Intake manifold
6	Air hose	14	Intake manifold gasket
7	Accelerator cable (and throttle cable (AT only))	15	VTCS check valve (one-way)
8	Throttle body (TB)	16	Delay valve

42356-MAZC-GAF

Exploded view of the intake manifold and related components—1.8L (BP) engine

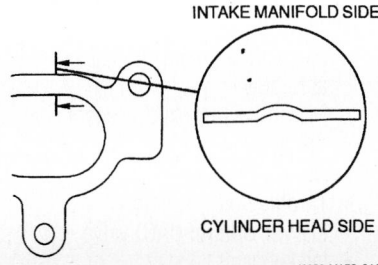

INTAKE MANIFOLD SIDE

CYLINDER HEAD SIDE

42356-MAZC-GAG

Make sure the convex side of the gasket is facing up when installing the intake manifold gasket—1.8L (BP) engine

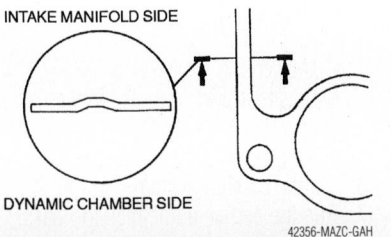

INTAKE MANIFOLD SIDE

DYNAMIC CHAMBER SIDE

42356-MAZC-GAH

Make sure the convex side of the gasket is facing up when installing the upper intake manifold gasket—1.8L (BP) engine

To install:

4. Clean all gasket mating surfaces.

➡**Be sure that the convex side of the intake manifold and dynamic gaskets are positioned as shown.**

5. Install or connect the following:
   - Intake manifold using a new gasket, make sure the convex side of the gasket faces the manifold. Torque the bolts to 14–18 ft. lbs. (19–25 Nm).
   - Dynamic chamber
   - VICS solenoid valve and bracket
   - Throttle body

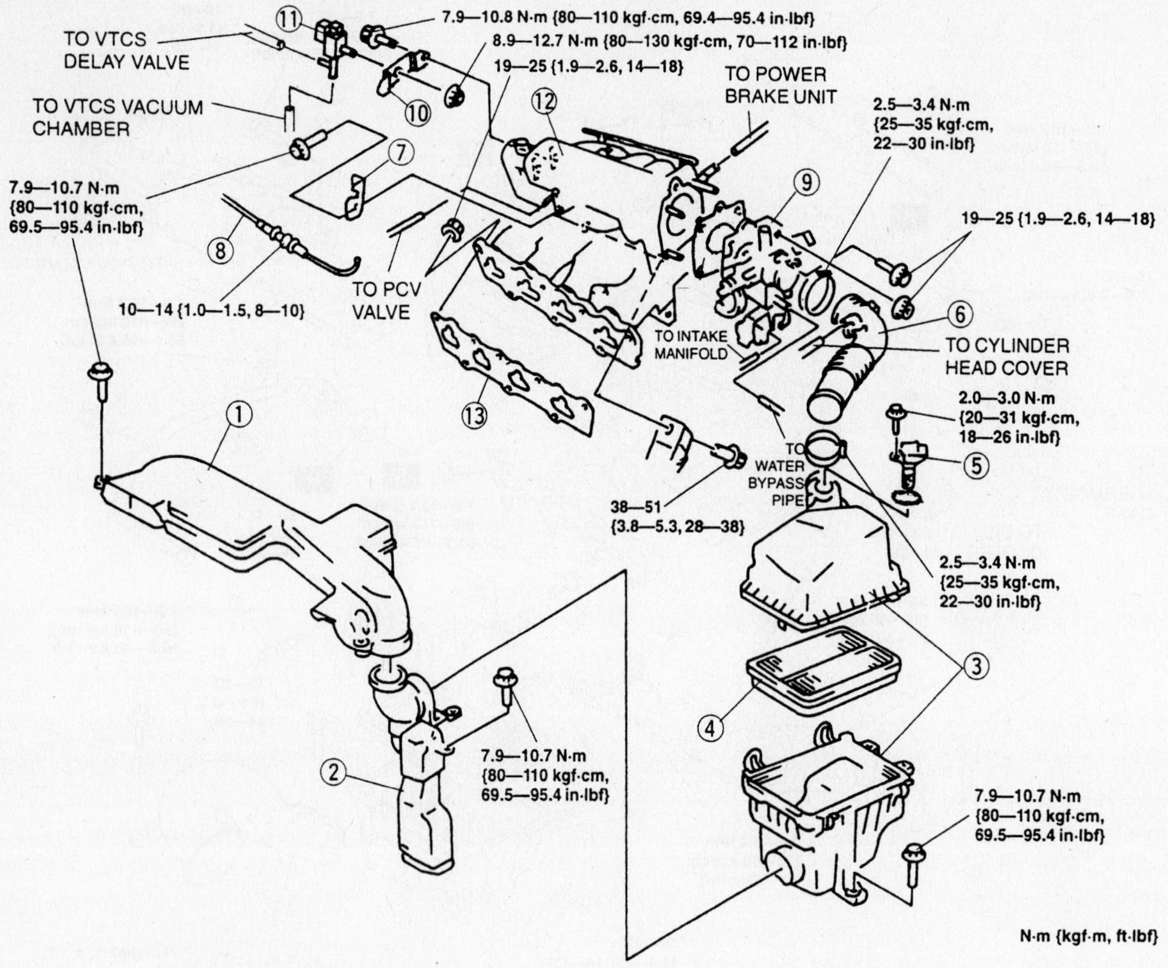

TO VTCS
DELAY VALVE

TO VTCS VACUUM
CHAMBER

7.9—10.8 N·m {80—110 kgf·cm, 69.4—95.4 in·lbf}

8.9—12.7 N·m {80—130 kgf·cm, 70—112 in·lbf}

19—25 {1.9—2.6, 14—18}

TO POWER
BRAKE UNIT

2.5—3.4 N·m
{25—35 kgf·cm,
22—30 in·lbf}

7.9—10.7 N·m
{80—110 kgf·cm,
69.5—95.4 in·lbf}

19—25 {1.9—2.6, 14—18}

10—14 {1.0—1.5, 8—10}

TO PCV
VALVE

TO INTAKE
MANIFOLD

TO CYLINDER
HEAD COVER

TO WATER
BYPASS
PIPE

2.0—3.0 N·m
{20—31 kgf·cm,
18—26 in·lbf}

38—51
{3.8—5.3, 28—38}

2.5—3.4 N·m
{25—35 kgf·cm,
22—30 in·lbf}

7.9—10.7 N·m
{80—110 kgf·cm,
69.5—95.4 in·lbf}

7.9—10.7 N·m
{80—110 kgf·cm,
69.5—95.4 in·lbf}

N·m {kgf·m, ft·lbf}

1	Fresh-air duct	8	Accelerator cable
2	Resonance chamber	9	Throttle body
3	Air cleaner	10	VTCS solenoid valve bracket
4	Air cleaner element	11	VTCS solenoid valve
5	MAF sensor (Integrated with IAT sensor)	12	Intake manifold
6	Air hose	13	Intake manifold gasket
7	Accelerator cable bracket		

9301MG03

Exploded view of the intake manifold, illustrating the removal and installation components with tightening values—1.6L engine

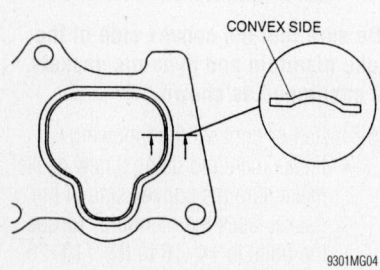

CONVEX SIDE

9301MG04

Cross-sectional view of the intake manifold gasket—1.6L (ZM) engines

- Accelerator cable
- Air hose
- MAF sensor
- Air cleaner and filter
- IAT sensor
- Fresh air duct and resonance chamber
- Negative battery cable
6. Fill the cooling system.
7. Run the engine and check for leaks.

626 WITH 2.0L (FS) ENGINE

1. Before servicing the vehicle, refer to the precautions in the beginning of this section.
2. Relieve the fuel system pressure.

3. Drain the cooling system.
4. Remove or disconnect the following:
- Negative battery cable
- Fresh air duct and air cleaner assembly
- Mass Air Flow (MAF) sensor
- Air hose
- Accelerator cable
- Throttle body assembly
- Idle Air Control (IAC) valve
- Intake manifold stay
- Intake manifold and discard the gasket
5. Clean the mating surfaces of any gasket material

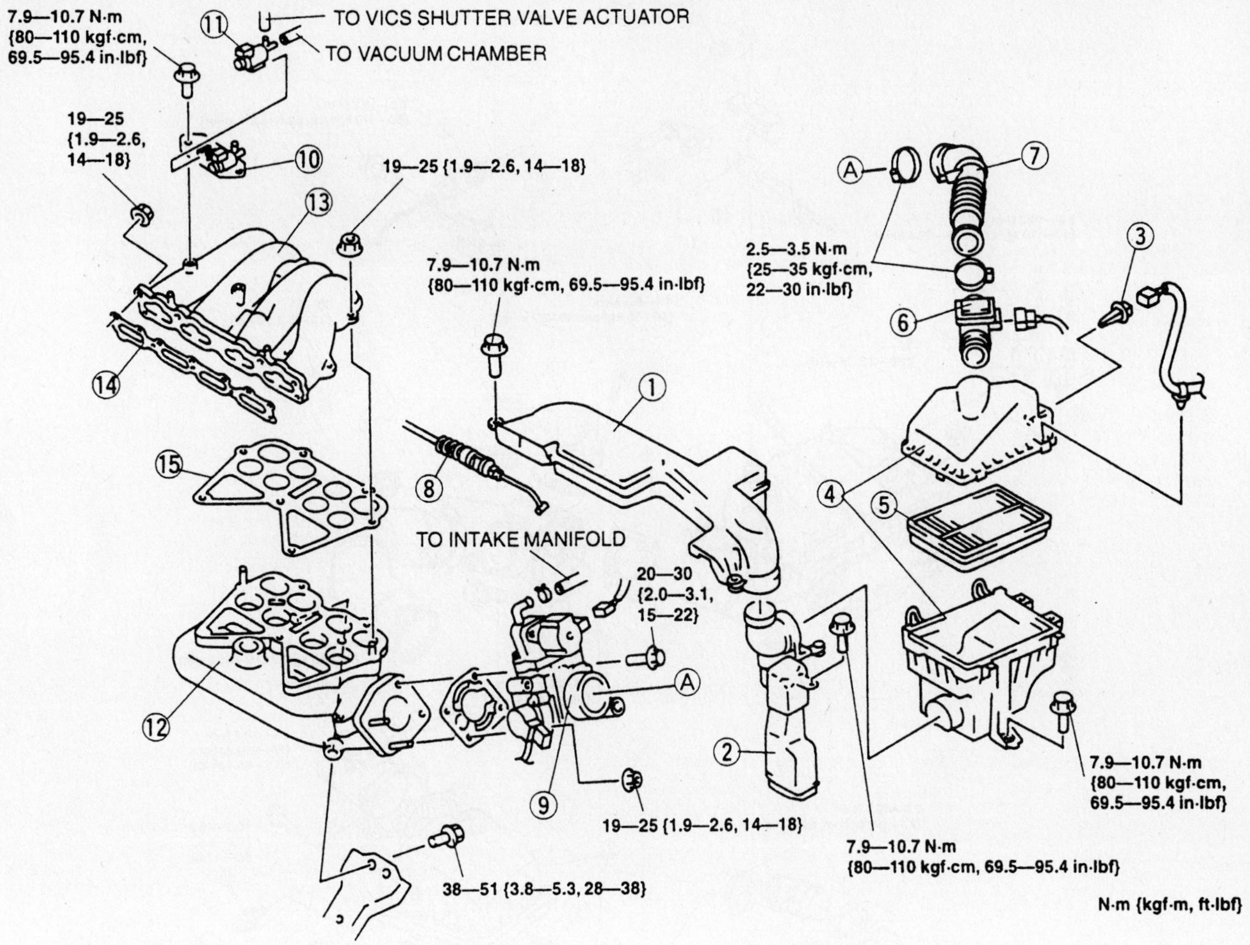

7.9—10.7 N·m
{80—110 kgf·cm,
69.5—95.4 in·lbf}

TO VICS SHUTTER VALVE ACTUATOR
TO VACUUM CHAMBER

19—25
{1.9—2.6,
14—18}

19—25 {1.9—2.6, 14—18}

7.9—10.7 N·m
{80—110 kgf·cm, 69.5—95.4 in·lbf}

2.5—3.5 N·m
{25—35 kgf·cm,
22—30 in·lbf}

TO INTAKE MANIFOLD

20—30
{2.0—3.1,
15—22}

19—25 {1.9—2.6, 14—18}

7.9—10.7 N·m
{80—110 kgf·cm,
69.5—95.4 in·lbf}

7.9—10.7 N·m
{80—110 kgf·cm, 69.5—95.4 in·lbf}

38—51 {3.8—5.3, 28—38}

N·m {kgf·m, ft·lbf}

1	Fresh-air duct	9	Throttle body
2	Resonance chamber	10	VICS solenoid valve bracket
3	IAT sensor	11	VICS solenoid valve
4	Air cleaner	12	Dynamic chamber
5	Air cleaner element	13	Intake manifold
6	MAF sensor	14	Intake manifold gasket
7	Air hose	15	Dynamic chamber gasket
8	Accelerator cable		

9301MG05

Exploded view of the intake manifold assembly components with tightening values—1.8L (FP) engine

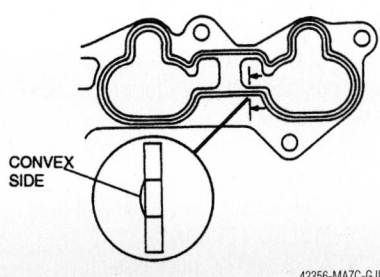

CONVEX SIDE

42356-MAZC-GJI

Be sure that the convex side of the intake manifold gasket is facing the manifold side—1.8L (FP) engine

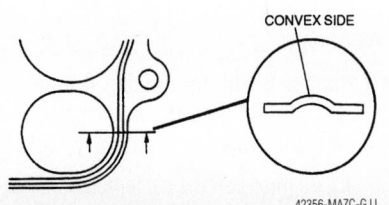

CONVEX SIDE

42356-MAZC-GJJ

Cross-sectional view of the dynamic chamber gasket—1.8L (FP) engines

To install:

6. Install or connect the following:
    *   Intake manifold with a new gasket. Make certain that the convex side of the new gasket faces the intake manifold. Torque the bolts to 18 ft. lbs. (25 Nm).
    *   Intake manifold stay
    *   IAC valve. Torque the bolt to 39–57 inch lbs. (4–6 Nm).
    *   Throttle body. Torque the nuts to 18 ft. lbs. (25 Nm).

7. When connecting the accelerator cable, perform the following adjust-ment:

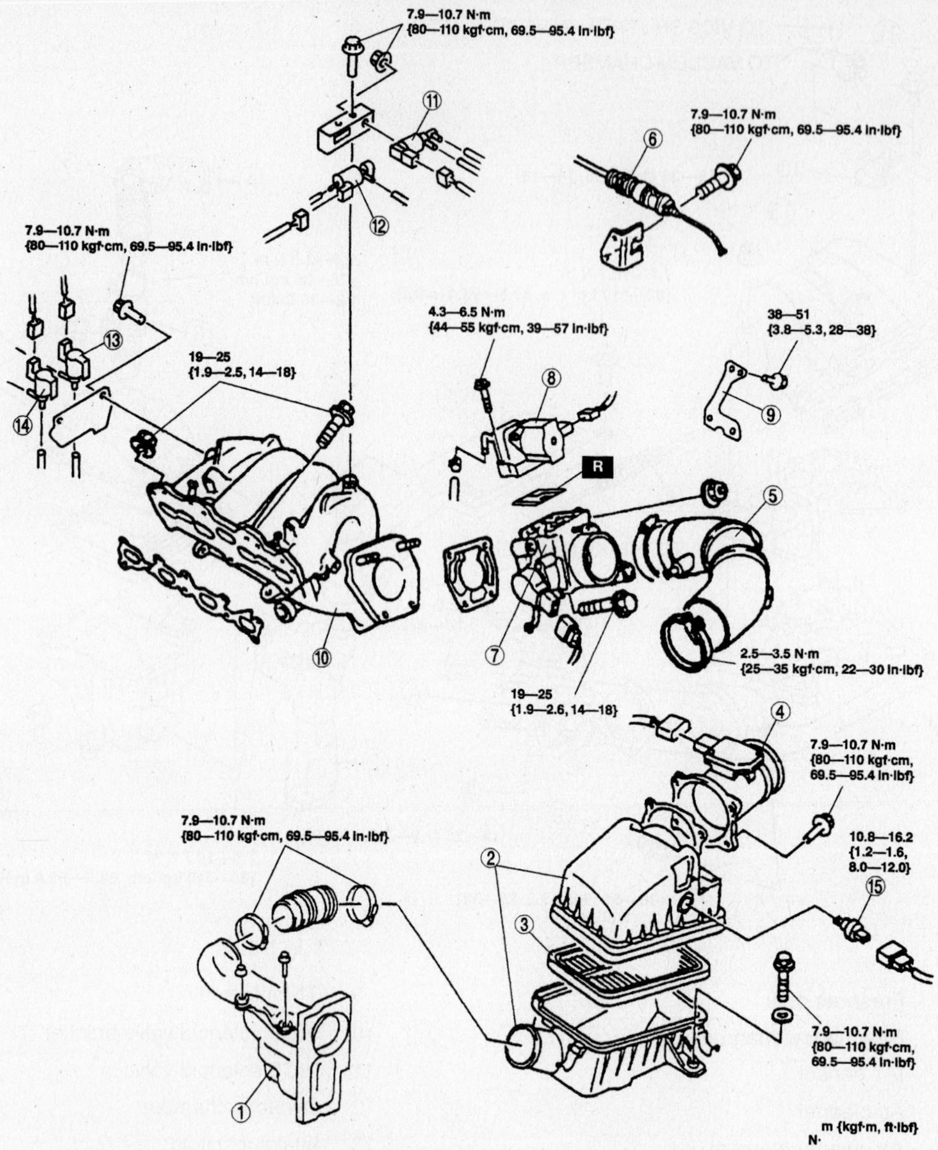

7.9—10.7 N·m
{80—110 kgf·cm, 69.5—95.4 in·lbf}
7.9—10.7 N·m
{80—110 kgf·cm, 69.5—95.4 in·lbf}
7.9—10.7 N·m
{80—110 kgf·cm, 69.5—95.4 in·lbf}
4.3—6.5 N·m
{44—55 kgf·cm, 39—57 in·lbf}
38—51
{3.8—5.3, 28—38}
19—25
{1.9—2.5, 14—18}
R
2.5—3.5 N·m
{25—35 kgf·cm, 22—30 in·lbf}
19—25
{1.9—2.6, 14—18}
7.9—10.7 N·m
{80—110 kgf·cm,
69.5—95.4 in·lbf}
7.9—10.7 N·m
{80—110 kgf·cm, 69.5—95.4 in·lbf}
10.8—16.2
{1.2—1.6,
8.0—12.0}
7.9—10.7 N·m
{80—110 kgf·cm,
69.5—95.4 in·lbf}
N·m {kgf·m, ft·lbf}

1	Fresh-air duct	6	Accelerator cable	11	Purge solenoid valve
2	Air cleaner	7	Throttle body	12	PRC solenoid valve
3	Air cleaner element	8	IAC valve	13	EGR boost solenoid valve
4	Mass air flow sensor	9	Intake manifold stay	14	VTCS solenoid valve
5	Air hose	10	Intake manifold	15	Intake air temperature sensor

42356-MAZC-G13

Exploded view of the intake manifold assembly—626 with 2.0L (FS) engine

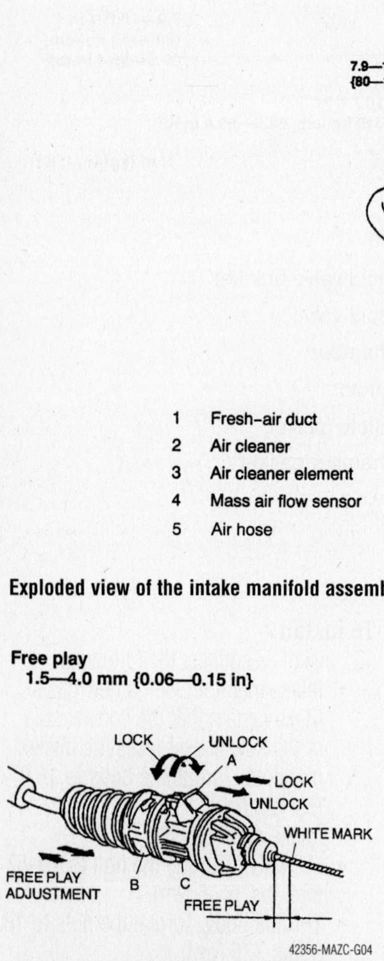

Free play
1.5—4.0 mm {0.06—0.15 in}
LOCK
UNLOCK
A
LOCK
UNLOCK
WHITE MARK
FREE PLAY
ADJUSTMENT
B
C
FREE PLAY
42356-MAZC-G04

Accelerator cable adjustment components—626 and Protégé models equipped with the 2.0L (FS) engine

a. Move the white locking tab to the unlock position.

b. Turn stopper B to the to the unlock position.

➡If stopper B will not unlock, it may be necessary to carefully bend back tab C out a little using a suitable pry tool.

c. Push or pull the cable housing directly behind the spring.

d. Turn the stopper B to the lock position.

e. Measure the free play which should be 0.06–0.15 inch (1.5–4mm) and make sure the cable free play is within specification.

f. Move the white locking tab to the lock position and check for proper accelerator operation.

8. Install or connect the following:
   - Air hose
   - MAF sensor. Torque the bolt to 95 inch lbs. (11 Nm).
   - Air cleaner/air duct assembly
   - Negative battery cable
9. Fill the cooling system.
10. Start the vehicle, check for leaks and repair if necessary.

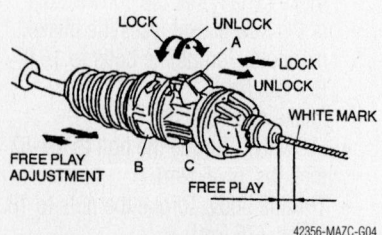

PROTÉGÉ WITH 2.0L (FS) ENGINE

1. Before servicing the vehicle, refer to the precautions in the beginning of this section.
2. Relieve the fuel system pressure.
3. Drain the cooling system.
4. Remove or disconnect the following:
   - Negative battery cable
   - Fresh air duct and resonance assembly
   - Intake Air Temperature (IAT) sensor
   - Air cleaner and filter
   - Mass Air Flow (MAF) sensor
   - Air hose
   - Accelerator cable
   - Throttle body assembly
   - Solenoid valve bracket
   - PRC solenoid valve
   - Dynamic chamber
   - Intake manifold and discard the gasket

5. Clean the mating surfaces of any gasket material

To install:

6. Install or connect the following:
   - Intake manifold with a new gasket. Make certain that the convex side of the new gasket faces the intake manifold. Torque the bolts to 18 ft. lbs. (25 Nm).
   - Dynamic chamber
   - PRC solenoid valve
   - Solenoid valve bracket
   - Throttle body assembly. Torque the bolts and nuts to 18 ft. lbs. (25 Nm).

7. When connecting the accelerator cable, perform the following adjustment:
   a. Move the white locking tab to the unlock position.
   b. Turn stopper B to the to the unlock position.

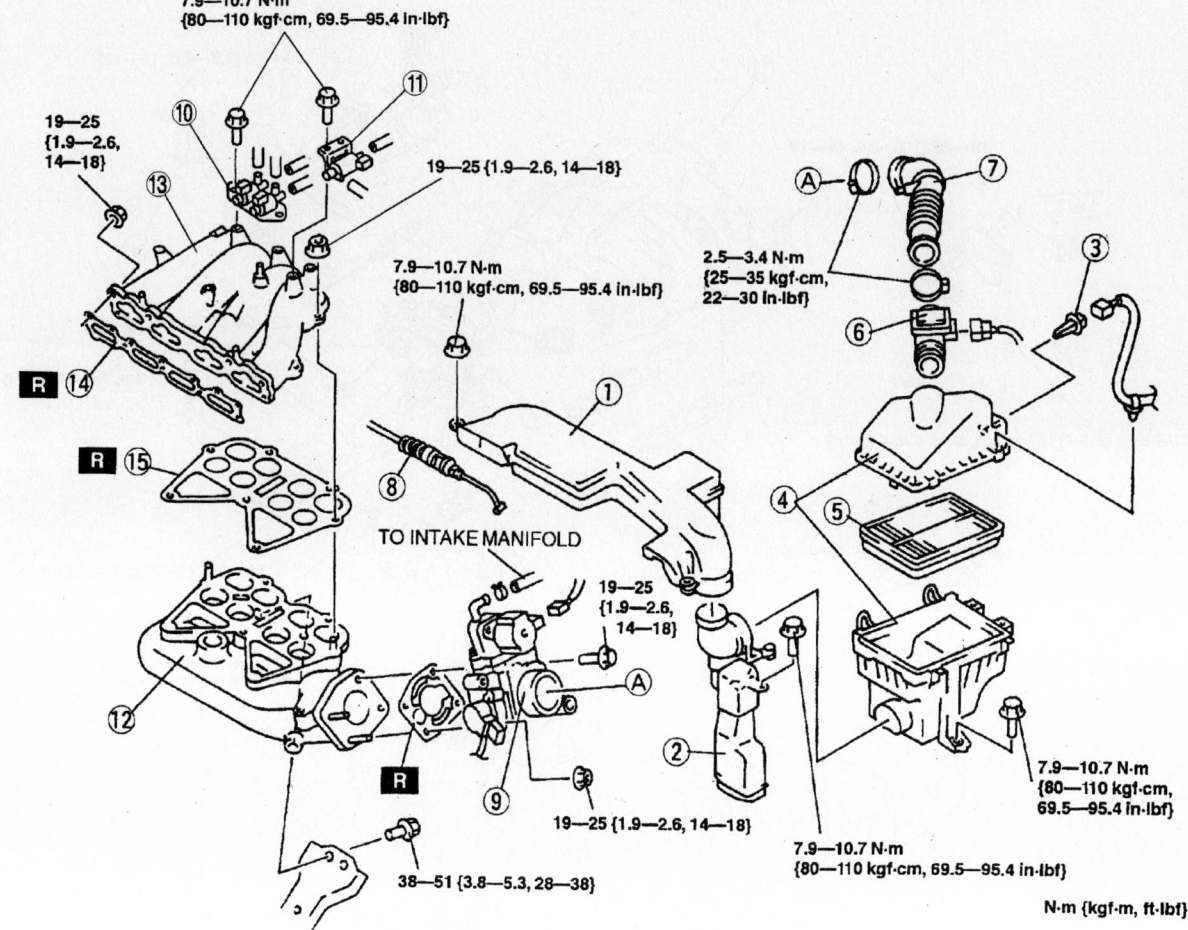

1	Fresh-air duct	
2	Resonance chamber	
3	IAT sensor	
4	Air cleaner	
5	Air cleaner element	
6	MAF sensor	
7	Air hose	
8	Accelerator cable	
9	Throttle body	
10	Solenoid valve bracket	
11	PRC solenoid valve	
12	Dynamic chamber	
13	Intake manifold	
14	Intake manifold gasket	
15	Dynamic chamber gasket	

Exploded view of the intake manifold assembly—Protégé with 2.0L (FS) engine

42356-MAZC-GHC

➡ **If stopper B will not unlock, it may be necessary to carefully bend back tab C out a little using a suitable pry tool.**

  c. Push or pull the cable housing directly behind the spring.

  d. Turn the stopper B to the lock position.

  e. Measure the free play which should be 0.06–0.15 inch (1.5–4mm) and make sure the cable free play is within specification.

  f. Move the white locking tab to the lock position and check for proper accelerator operation.

8. Install or connect the following:

- Air hose
- AF sensor
- Air filter and cleaner
- IAT sensor
- Fresh air duct and resonance assembly
- Negative battery cable

9. Fill the cooling system.

10. Start the vehicle, check for leaks and repair if necessary.

2.5L (KL) ENGINE

1. Before servicing the vehicle, refer to the precautions in the beginning of this section.

2. Relieve the fuel system pressure.

3. Drain the cooling system.

4. Remove or disconnect the following:
- Negative battery cable
- Fresh air duct and air cleaner assembly
- Mass Air Flow (MAF) sensor
- Air hose
- Accelerator cable
- Throttle body assembly
- Idle Air Control (IAC) valve
- Intake manifold stay
- Intake manifold and discard the gasket

5. Clean the mating surfaces of any gasket material

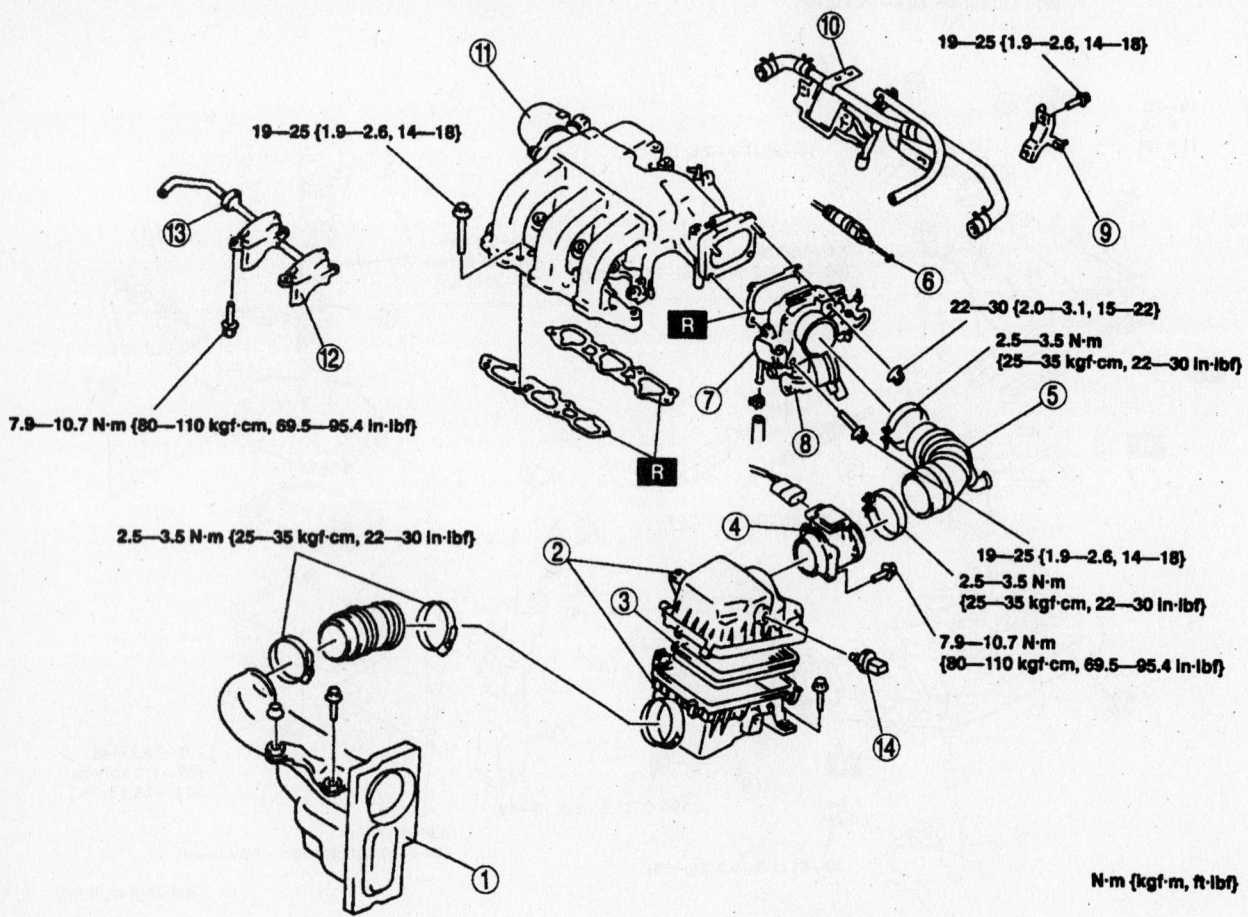

1	Fresh–air duct	8	IAC valve
2	Air cleaner	9	Intake manifold stay
3	Air cleaner element	10	Ventilation pipe
4	Mass air flow sensor	11	Intake manifold component
5	Air hose	12	Vacuum chamber
6	Accelerator cable	13	VRIS check valve (one–way)
7	Throttle body	14	Intake air temperature sensor

42356-MAZC-G14

Exploded view of the intake manifold assembly—626 with 2.5L (KL) engine

To install:

6. Install or connect the following:
   - Intake manifold with a new gasket. Torque the bolts to 18 ft. lbs. (25 Nm).
   - Intake manifold stay
   - IAC valve.
   - Throttle body. Torque the nuts to 18 ft. lbs. (25 Nm).
   - Accelerator cable
   - Air hose
   - MAF sensor. Torque the bolt to 95 inch lbs. (11 Nm).
   - Air cleaner/air duct assembly
   - Negative battery cable

7. Fill the cooling system.

8. Start the vehicle, check for leaks and repair if necessary.

Millenia

2.3L ENGINE

1. Before servicing the vehicle, refer to the precautions in the beginning of this section.

2. Relieve the fuel system pressure.

3. Drain the cooling system.

4. Remove or disconnect the following:
   - Negative battery cable
   - Dynamic chamber cover
   - Charge air cooler air duct
   - Vacuum hoses and electrical connectors from the air cleaner housing
   - Air cleaner assembly
   - Fresh air ducts
   - Mass Air Flow (MAF) sensor and the air intake hose from the throttle body
   - Resonator
   - Right-hand charge air cooler
   - Left-hand charge air cooler
   - Accelerator cable
   - Vacuum hoses from the rear of the intake manifold and Exhaust Gas Recirculation (EGR) valve
   - EGR valve
   - Air intake pipe assembly
   - Charge air cooler pipe

   - Fuel supply line at the fuel rails and discard the copper washers
   - Fuel and vacuum lines from the fuel pressure regulator
   - Coolant hoses
   - Wiring harness from the intake manifold
   - Intake manifold mounting nuts and bolts in 2–3 steps
   - Intake manifold

5. If necessary, label and disconnect the fuel hoses and electrical connectors from the throttle body. Remove the throttle body.

To install:

6. Clean all gasket mating surfaces.

7. Install or connect the following:
   - Throttle body, if removed. Tighten the nuts/bolts to 14–18 ft. lbs. (19–25 Nm).
   - Fuel hoses and electrical connectors.
   - Intake manifold using new gaskets. Tighten the nuts/bolts in 2–3 steps, from the center to the ends, to 14–18 ft. lbs. (19–25 Nm).

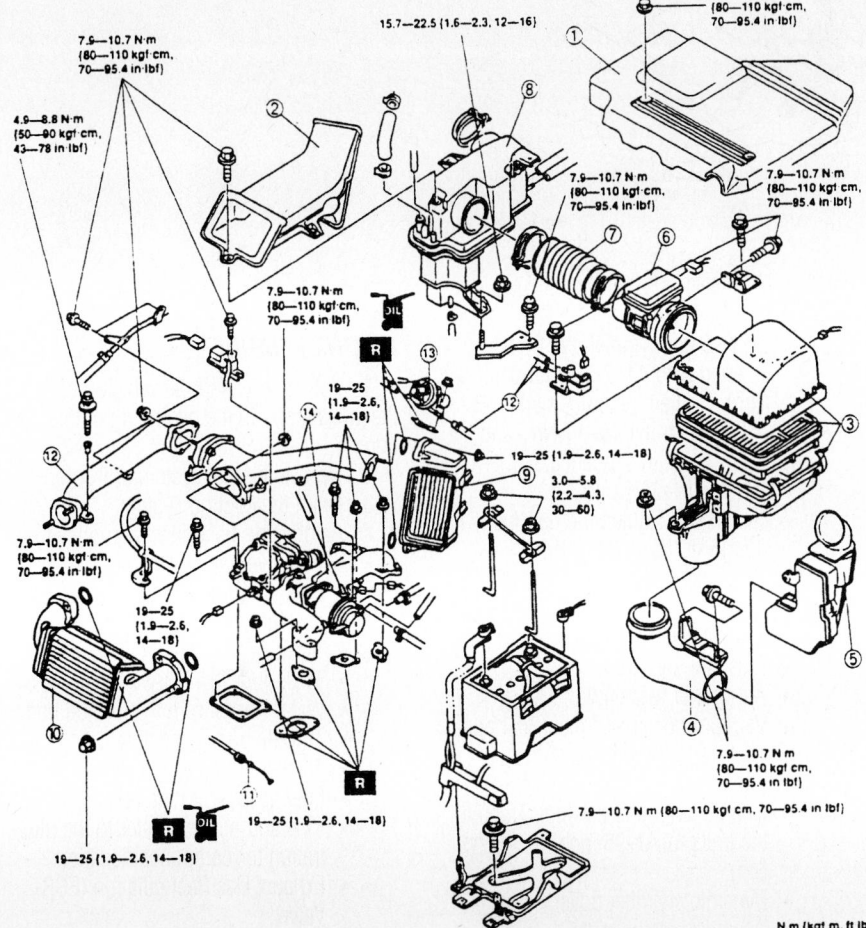

1. Dynamic chamber cover
2. Charge air cooler air duct
3. Air cleaner assembly
4. Air duct
5. Fresh air duct
6. Mass air flow sensor
7. Air intake hose
8. Resonator
9. Charge air cooler (RH)
10. Charge air cooler (LH)
11. Accelerator cable
12. Vacuum hose assembly
13. EGR control valve
14. Air intake pipe assembly

Exploded view of the intake manifold assembly (1 of 2)—2.3L engine

7923MG31

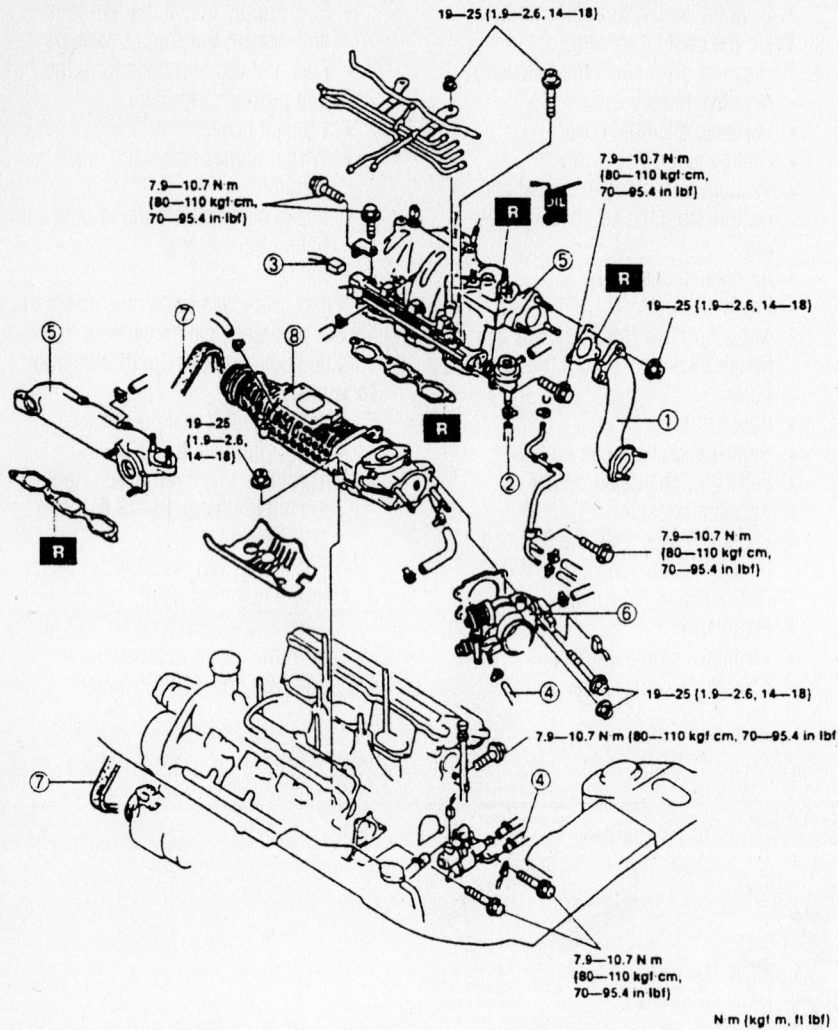

19—25 (1.9—2.6, 14—18)

7.9—10.7 N·m
(80—110 kgf·cm,
70—95.4 in·lbf)

7.9—10.7 N·m
(80—110 kgf·cm,
70—95.4 in·lbf)

19—25 (1.9—2.6, 14—18)

19—25
(1.9—2.6,
14—18)

7.9—10.7 N·m
(80—110 kgf·cm,
70—95.4 in·lbf)

19—25 (1.9—2.6, 14—18)

7.9—10.7 N·m (80—110 kgf·cm, 70—95.4 in·lbf)

7.9—10.7 N·m
(80—110 kgf·cm,
70—95.4 in·lbf)

N·m (kgf·m, ft·lbf)

7923MG32

1. Charge air cooler pipe
2. Fuel hose
3. Fuel distributor connector
4. Coolant hose
5. Intake manifold assembly
6. Throttle body assembly
7. Drive belt
8. Lysholm compressor

Exploded view of the intake manifold assembly (2 of 2)—2.3L engine

- Wiring harness onto the intake manifold
- Coolant hoses
- Fuel and vacuum lines to the fuel pressure regulator
- Fuel supply line to the fuel rail using new copper washers
- Charge air cooler pipe
- Air intake pipe assembly using new gaskets. Verify that the rubber gaskets are not twisted or distorted. Tighten the bolts marked **A** to 70–95 inch lbs. (8–11 Nm) and all other bolts, in sequence, to 14–18 ft. lbs. (19–25 Nm).
- EGR valve using a new gasket
- Vacuum hoses to the intake manifold and EGR valve
- Accelerator cable, adjust as necessary
- Left and right-hand charge air coolers using new gaskets. Verify that the rubber gaskets are not twisted

or distorted. Tighten the bolts marked **A** to 44–78 inch lbs. (5–9 Nm). Tighten the bolts marked **B** to 70–95 inch lbs. (8–11 Nm) and all other bolts, in sequence, to 14–18 ft. lbs. (19–25 Nm).
- Resonator. Tighten the nuts/bolts to 12–16 ft. lbs. (16–22 Nm).
- Air intake hose onto the throttle body
- MAF sensor
- Fresh air and air ducts
- Air cleaner assembly
- Vacuum hoses and electrical connectors to the air cleaner housing
- Charge air cooler air duct. Tighten the bolts to 70–95 inch lbs. (8–11 Nm).
- Dynamic chamber cover
- Negative battery cable
8. Fill the cooling system.
9. Run the engine and check for leaks.

2.5L (KL) ENGINE

1. Before servicing the vehicle, refer to the precautions in the beginning of this section.
2. Relieve the fuel system pressure.
3. Drain the cooling system.
4. Remove or disconnect the following:
   - Negative battery cable
   - Fresh air duct and air cleaner assembly
   - Air hose
   - Mass Air Flow (MAF) sensor
   - Water hoses from the throttle body
   - Throttle body assembly
   - Accelerator cable
   - Fuel hoses from the rail
   - Pipe and harness, refer to the illustration for component location
   - Exhaust Gas Recirculation (EGR) Valve
   - EGR pipe
   - Intake manifold and discard the gasket

5. Clean the mating surfaces of any gasket material

To install:

6. Install or connect the following:
- Intake manifold with a new gasket. Torque the bolts to 18 ft. lbs. (25 Nm).
- EGR pipe and valve
- Pipe and harness, refer to the illustration for component location
- Fuel hoses to the rail
- Accelerator cable

- Throttle body assembly
- Water hoses to the throttle body. Torque the nuts to 18 ft. lbs. (25 Nm).
- MAF sensor. Torque the bolt to 95 inch lbs. (11 Nm).
- Air hose
- Fresh air duct and air cleaner assembly
- Negative battery cable

7. Fill the cooling system.
8. Start the vehicle, check for leaks and repair if necessary.

Mazda 3

1. Before servicing the vehicle, refer to the precautions in the beginning of this section.
2. Drain the cooling system.
3. Relieve the fuel system pressure.
4. Remove the plug hole plate by lifting off and removing the plug hole plate from the areas shown in the accompanying illustration.
5. Remove the battery cover and battery duct.

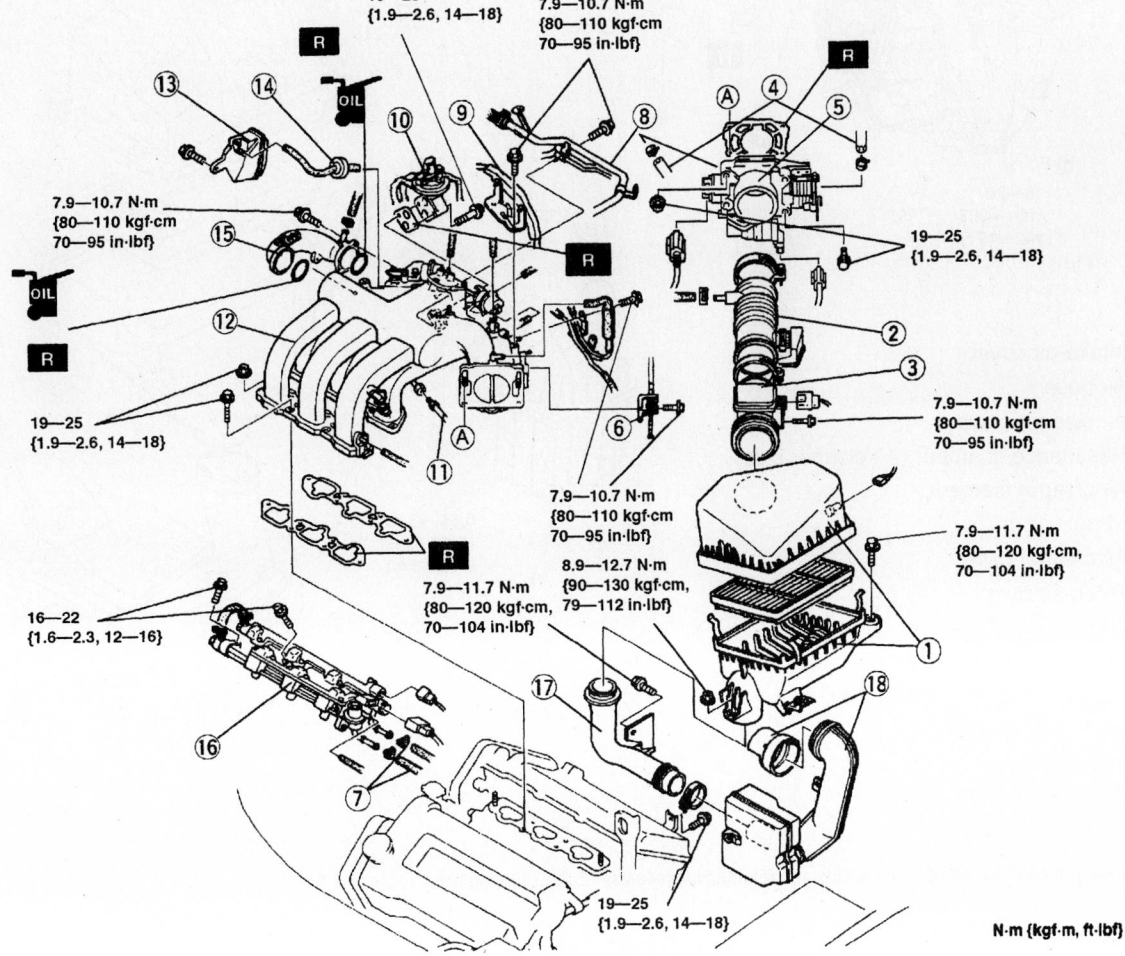

1	Air cleaner	10	EGR valve
2	Air intake hose	11	EGR pipe
3	MAF sensor	12	Intake manifold
4	Water hose	13	Vacuum chamber
5	Throttle body component	14	Check valve
6	Accelerator cable	15	Air intake pipe
7	Fuel hose	16	Fuel distributor component
8	Pipe	17	Air duct
9	Harness	18	Fresh-air duct

42356-MAZC-GDT

Exploded view of the intake manifold assembly—Millenia with 2.5L (KL) engine

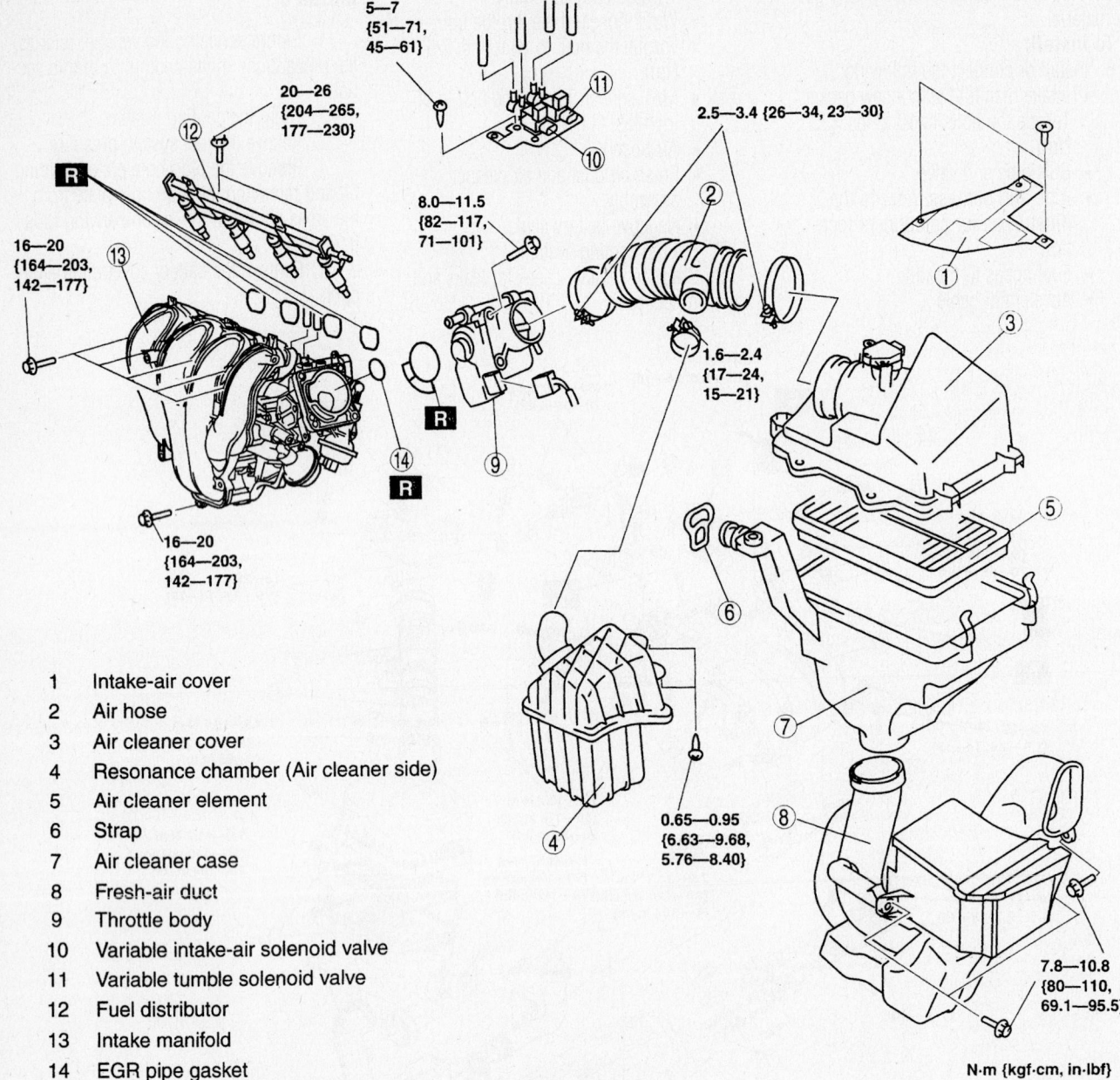

1. Intake-air cover
2. Air hose
3. Air cleaner cover
4. Resonance chamber (Air cleaner side)
5. Air cleaner element
6. Strap
7. Air cleaner case
8. Fresh-air duct
9. Throttle body
10. Variable intake-air solenoid valve
11. Variable tumble solenoid valve
12. Fuel distributor
13. Intake manifold
14. EGR pipe gasket

N·m {kgf·cm, in·lbf}

67162-MAZC-G18

Exploded view of the intake manifold assembly and related components—2.0L (LF) engines

6. Remove the under cover and disconnect the negative battery cable.

7. Remove the intake air cover, the air hose and the air cleaner element.

8. Remove the resonance chamber and air filter element.

9. Remove the strap and the air cleaner case.

10. Remove the fresh air duct.

11. Remove the throttle body.

12. Remove the variable intake air solenoid valve and the variable tumble solenoid valve.

13. Remove the fuel rail and injectors as an assembly.

14. Disconnect the vacuum hose from the intake manifold.

15. Remove the engine oil dipstick tube.

16. Remove the intake manifold bolts, the manifold and gasket.

17. Discard the gasket.

18. Clean the mating surfaces of any gasket material.

19. If necessary, remove the Exhaust Gas Recirculation (EGR) pipe gasket.

To install:

20. If removed, install the EGR pipe gasket.

21. Install a new gasket and the manifold. Tighten the manifold bolts to 142–177 inch lbs. (16–20 Nm).

22. Install the engine oil dipstick tube.

23. Connect the vacuum hose to the intake manifold.

24. Install the fuel rail and injectors.

25. Install the variable intake air solenoid valve and the variable tumble solenoid valve.

26. Install the throttle body and tighten the bolts to 71–101 inch lbs. (8–11.5 Nm).

27. Install the fresh air duct.

28. Verify the rubber mounts on the battery support bracket are still in place. Insert the air cleaner case into the rubber mounts,

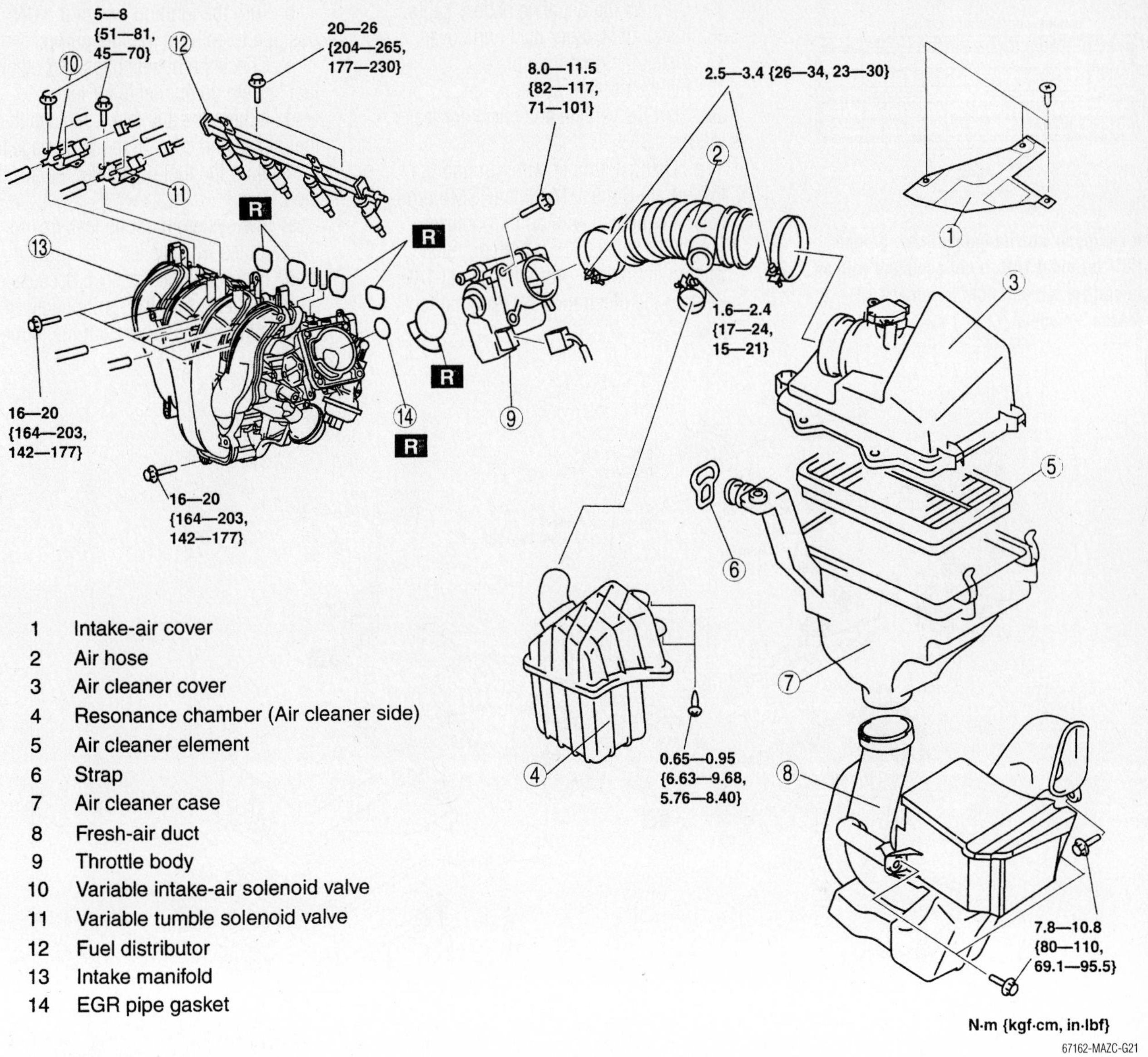

5—8
{51—81,
45—70}
20—26
{204—265,
177—230}
8.0—11.5
{82—117,
71—101}
2.5—3.4 {26—34, 23—30}
R
R
R
R
1.6—2.4
{17—24,
15—21}
16—20
{164—203,
142—177}
16—20
{164—203,
142—177}
0.65—0.95
{6.63—9.68,
5.76—8.40}
7.8—10.8
{80—110,
69.1—95.5}
N·m {kgf·cm, in·lbf}
67162-MAZC-G21

1 Intake-air cover
2 Air hose
3 Air cleaner cover
4 Resonance chamber (Air cleaner side)
5 Air cleaner element
6 Strap
7 Air cleaner case
8 Fresh-air duct
9 Throttle body
10 Variable intake-air solenoid valve
11 Variable tumble solenoid valve
12 Fuel distributor
13 Intake manifold
14 EGR pipe gasket

Exploded view of the intake manifold assembly and related components—Mazda 3 with the 2.3L (L3) engine

RUBBER
RUBBER
INSTALLATION
AREA ON
ENGINE
SIDE
67162-MAZC-G02

Plug hole plate locations—Mazda 3 models

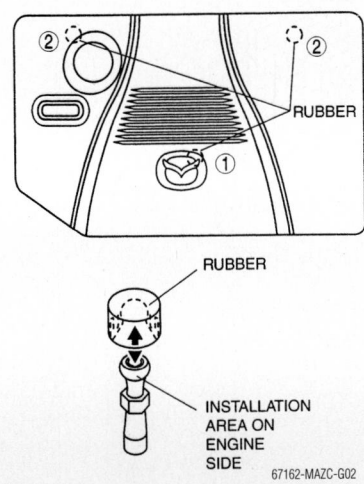

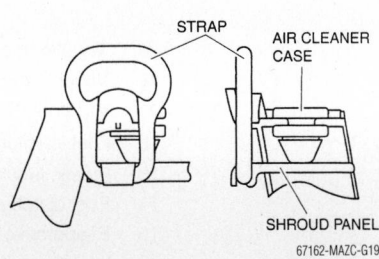

using soapy water if necessary to ease installation.

29. Use the strap to secure the shroud panel and the air cleaner case as shown in the accompanying illustration.

30. Install the air cleaner element and the resonance chamber.

31. Install the air cleaner cover.

32. Install the air hose, make sure to align the alignment marks on the throttle body and the air hose.

33. Install the intake air cover.

34. Install the under cover.

STRAP
AIR CLEANER
CASE
SHROUD PANEL
67162-MAZC-G19

Use the strap to secure the shroud panel and the air cleaner case—Mazda 3 models

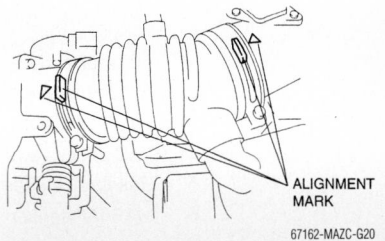

ALIGNMENT
MARK
67162-MAZC-G20

Make sure to align the alignment marks on the throttle body and the air hose—Mazda 3 models

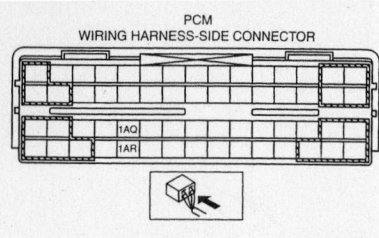

PCM
WIRING HARNESS-SIDE CONNECTOR

1AQ
1AR

67162-MAZC-G22

If equipped with an immobilizer, ground PCM terminal 1AR, if not equipped with an imobilizer, ground PCM terminal 1AQ—Mazda 3 models

35. Connect the negative battery cable.
36. Install the battery duct and cover.
37. Install the plug hole plate.
38. Fill the cooling system.
39. Start the vehicle and check for leaks as follows:

a. Using a jumper wire, ground the Powertrain Control Module (PCM) terminals. If equipped with an immobilizer, ground terminal 1AR, if not equipped with an imobilizer, ground terminal 1AQ. Refer to the illustration for terminal location.

b. Turn the ignition switch the ON position to activate the fuel pump.

c. Check the hoses, clips and other fuel system components for leaks.

d. If there are any leaks, replace the fuel hoses and clips. If the is damage to the seal on the fuel pipe side, replace the pipe.

e. The system must be leak free for five minutes with the terminal grounded. If any component is replaced because of a system, leak, turn the ignition key OFF; remove the jumper wire from the termi-

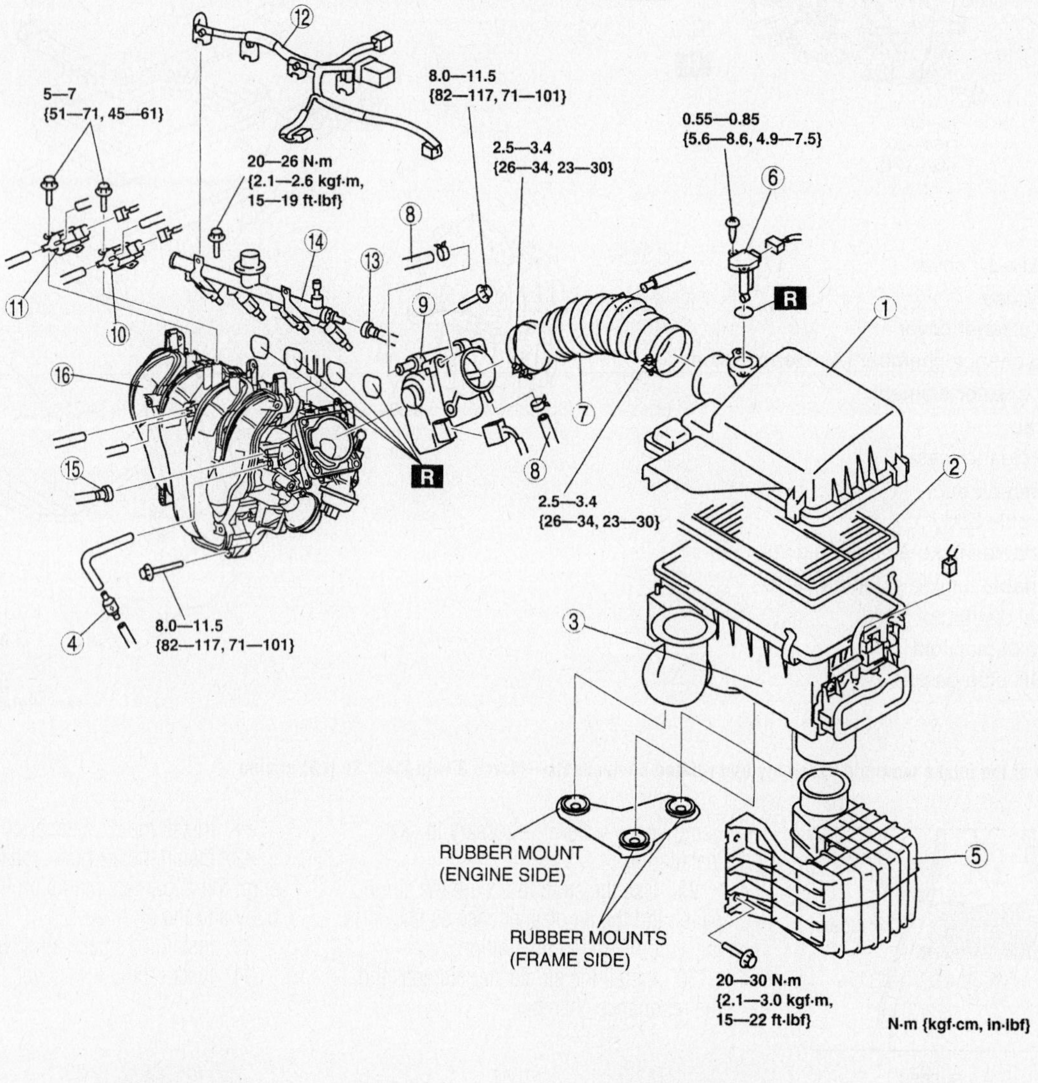

5—7
{51—71, 45—61}

8.0—11.5
{82—117, 71—101}

0.55—0.85
{5.6—8.6, 4.9—7.5}

20—26 N·m
{2.1—2.6 kgf·m,
15—19 ft·lbf}

2.5—3.4
{26—34, 23—30}

2.5—3.4
{26—34, 23—30}

8.0—11.5
{82—117, 71—101}

RUBBER MOUNT
(ENGINE SIDE)

RUBBER MOUNTS
(FRAME SIDE)

20—30 N·m
{2.1—3.0 kgf·m,
15—22 ft·lbf}

N·m {kgf·cm, in·lbf}

1	Air cleaner cover	9	Throttle body
2	Air cleaner element	10	Variable tumble control solenoid valve
3	Air cleaner case	11	VIS control solenoid valve
4	VAD check valve (one-way)	12	Fuel injector connector
5	Resonance chamber	13	Plastic fuel hose
6	MAF/IAT sensor	14	Fuel distributor
7	Air hose	15	Evaporative hose
8	Water hose	16	Intake manifold

67162-MAZC-G150

Exploded view of the intake manifold assembly and related components—Mazda 6 with the 2.3L (L3) engine

nal. reapply the jumper wire, turn the ignition On and check for leaks.

Mazda 6

2.3L (L3) ENGINE

1. Before servicing the vehicle, refer to the precautions in the beginning of this section.
2. Drain the cooling system.
3. Relieve the fuel system pressure.
4. Disconnect the negative battery cable.
5. Remove the intake air cover and element.

6. Remove the air cleaner case.
7. Remove the VAD check valve.
8. Remove the left front mud guard and resonance chamber.
9. Remove the Mass Airflow (MAF) and Intake Air Temperature (IAT) sensors.
10. Remove the air hose.
11. Remove the water hose.
12. Remove the throttle body.
13. Remove the variable intake air solenoid valve and the variable tumble solenoid valve.
14. Remove the fuel rail and injectors as an assembly.
15. Disconnect the evaporative hose and vacuum hose from the intake manifold.

16. Remove the intake manifold bolts, the manifold and gasket.
17. Discard the gasket.
18. Clean the mating surfaces of any gasket material.

To install:

19. Install a new gasket and the manifold. Tighten the manifold bolts to 171–101 inch lbs. (8–11.5 Nm).
20. Install the engine oil dipstick tube.
21. Connect the vacuum hose and evaporative hose to the intake manifold.
22. Install the fuel rail and injectors.
23. Install the variable intake air solenoid valve and the variable tumble solenoid valve.

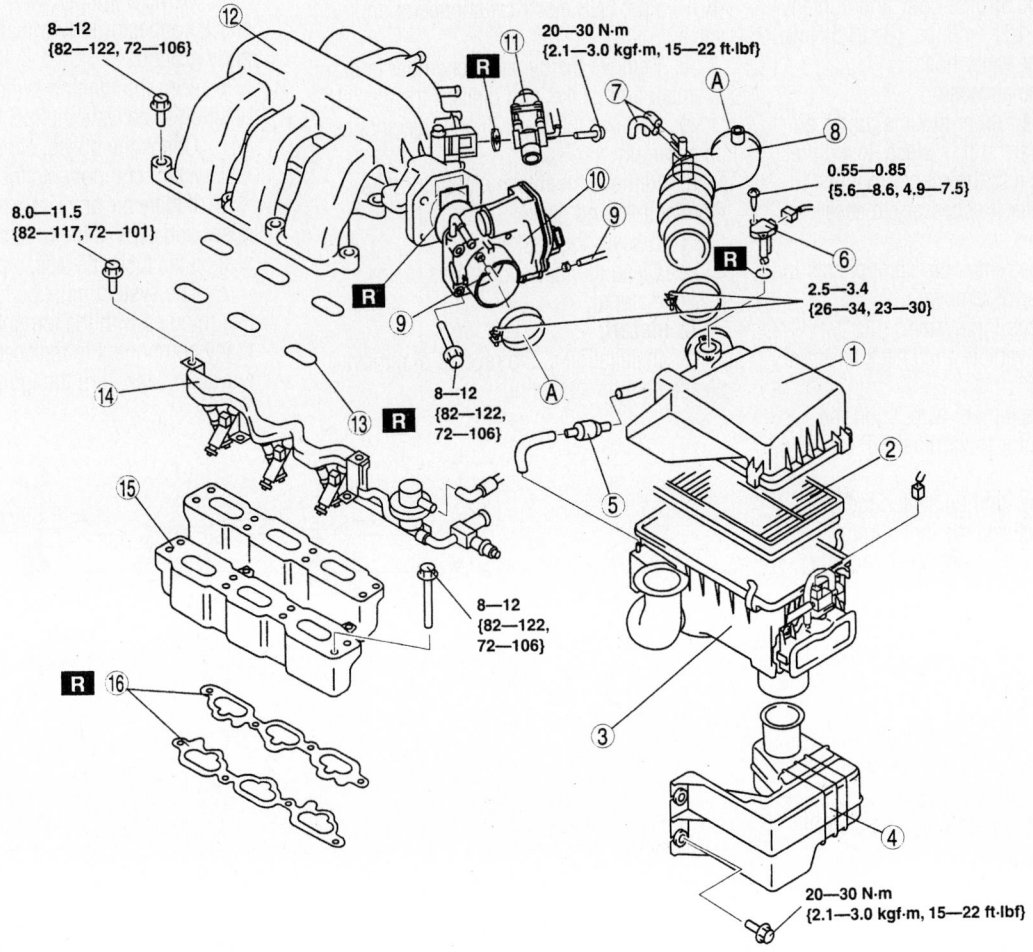

1	Air cleaner cover	9	Water hose
2	Air cleaner element	10	Throttle body
3	Air cleaner case	11	EGR valve
4	Resonance chamber	12	Dynamic chamber
5	VAD check valve (one-way)	13	Dynamic chamber gasket
6	MAF/IAT sensor	14	Fuel distributor
7	Vacuum hose (purge solenoid valve)	15	Intake manifold
8	Air hose	16	Intake manifold gasket

67162-MAZC-G152

Exploded view of the intake manifold assembly and related components—Mazda 6 with the 3.0L (AJ) engine

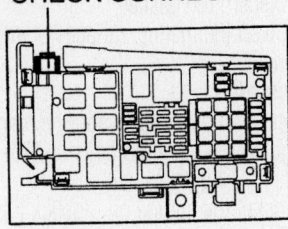

CHECK CONNECTOR

MAIN FUSE BLOCK

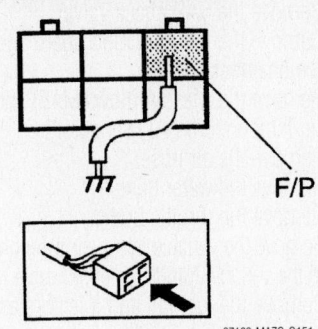

CHECK CONNECTOR

F/P

67162-MAZC-G151

Using a jumper wire, short the check connector to terminal F/P to a body ground—Mazda 6 models

24. Install the throttle body and tighten the bolts to 71–101 inch lbs. (8–11.5 Nm).

25. Install the water hose.

26. Install the air hose.

27. Verify the rubber mounts on the air cleaner bracket are still in place. Insert the air cleaner case into the rubber mounts, using soapy water if necessary to ease installation.

28. Install the remaining components in the reverse order of removal.

29. Fill the cooling system.

30. Start the vehicle and check for leaks as follows:

   a. Using a jumper wire, short the check connector to terminal F/P to a body ground.

   b. Turn the ignition switch the ON position to activate the fuel pump.

   c. Check the hoses, clips and other fuel system components for leaks.

   d. If there are any leaks, replace the fuel hoses and clips. If the is damage to the seal on the fuel pipe side, replace the pipe.

   e. The system must be leak free for five minutes with the terminal grounded. If any component is replaced because of a system, leak, turn the ignition key OFF; remove the jumper wire from the terminal. reapply the jumper wire, turn the ignition ON and check for leaks.

3.0L (AJ) ENGINE

1. Remove the left front mud guard and resonance chamber.

2. Remove the Mass Airflow (MAF) and Intake Air Temperature (IAT) sensors.

3. Remove the purge solenoid valve vacuum hose.

4. Remove the air hose.

5. Remove the water hose.

6. Remove the throttle body.

7. Remove the upper radiator hose, disconnect the Exhaust Gas Recirculation (EGR) pipe, EGR electrical connector and valve.

8. Remove the dynamic chamber bolts, chamber and gasket and discard the gasket.

9. Remove the fuel rail and injectors as an assembly.

10. Remove the intake manifold bolts, the manifold and gasket.

11. Discard the gasket.

12. Clean the mating surfaces of any gasket material.

To install:

13. Installation is the reverse of removal, please note the following:

   a. Install a new gasket and the manifold. Tighten the manifold bolts in the sequence illustrated to 72–106 inch lbs. (8–12 Nm).

   b. Install a new dynamic chamber gasket and the chamber. Tighten the bolts in the sequence illustrated to 72–106 inch lbs. (8–12 Nm).

   c. Install the EGR valve and tighten the retainers to the specifications shown in the accompanying illustration.

   d. Connect the EGR connector, and pipe.

14. Fill the cooling system.

15. Start the vehicle and check for leaks as follows:

   a. Using a jumper wire, short the check connector to terminal F/P to a body ground.

   b. Turn the ignition switch the ON position to activate the fuel pump.

   c. Check the hoses, clips and other fuel system components for leaks.

   d. If there are any leaks, replace the fuel hoses and clips. If the is damage to the seal on the fuel pipe side, replace the pipe.

   e. The system must be leak free for five minutes with the terminal grounded. If any component is replaced because of a system, leak, turn the ignition key OFF;

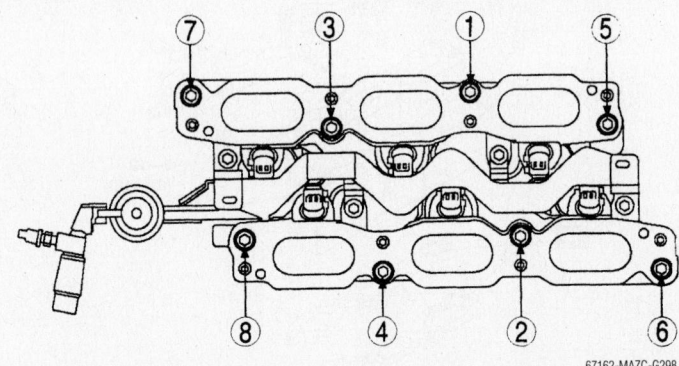

67162-MAZC-G298

Tighten the lower intake manifold bolts in this sequence—Mazda 6 with the 3.0L (AJ) engine

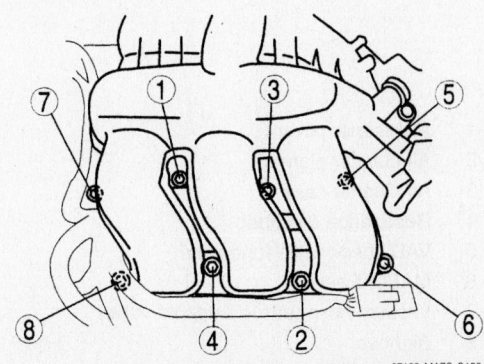

67162-MAZC-G153

Tighten the dynamic chamber bolts in this sequence—Mazda 6 with the 3.0L (AJ) engine

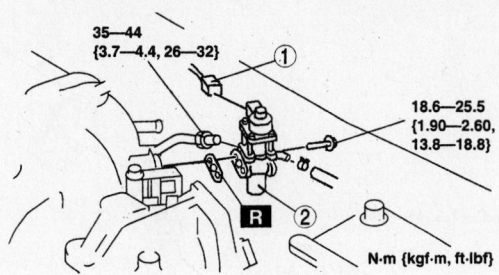

35—44
{3.7—4.4, 26—32}

① ②

18.6—25.5
{1.90—2.60,
13.8—18.8}

R

N·m {kgf·m, ft·lbf}

1 EGR valve connector
2 EGR valve

67162-MAZC-G154

Exploded view of the EGR assembly mounting and components—Mazda 6 with the 3.0L (AJ) engine

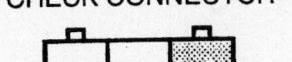

CHECK CONNECTOR

CHECK CONNECTOR

MAIN FUSE BLOCK

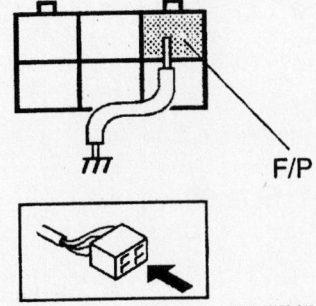

F/P

67162-MAZC-G151

Using a jumper wire, short the check connector to terminal F/P to a body ground—Mazda 6 models

remove the jumper wire from the terminal. reapply the jumper wire, turn the ignition ON and check for leaks.

Exhaust Manifold

REMOVAL & INSTALLATION

Miata

1.8L (BP) ENGINES

1. Before servicing the vehicle, refer to the precautions in the beginning of this section.
2. Remove or disconnect the following:
   - Negative battery cable
   - Air cleaner and air hose
   - Exhaust manifold heat shield bolts and the heat shield
   - Oxygen (O2S) sensor electrical connector
   - Exhaust pipe-to-exhaust manifold nuts and discard them
   - Exhaust Gas Recirculation (EGR) pipe from the exhaust manifold
   - Exhaust manifold nuts and bolts and discard the nuts
   - Exhaust manifold

To install:
3. Clean all gasket mating surfaces.
4. Install or connect the following:
   - Exhaust manifold. Torque the bolts to 29–33 ft. lbs. (39–46 Nm).
   - Exhaust pipe. Torque the new nuts to 38 ft. lbs. (52 Nm).
   - O2S connector
   - EGR pipe. Torque the pipe to 34 ft. lbs. (47 Nm).
   - Heat shield. Torque the bolts to 95 inch lbs. (11 Nm).
   - Air hose and air cleaner
   - Negative battery cable

626 and Protégé

1.6L (ZM) AND 1.8L (FP) ENGINES

1. Before servicing the vehicle, refer to the precautions in the beginning of this section.
2. Remove or disconnect the following:
   - Negative battery cable
   - Air cleaner and hose assembly
   - Water bypass pipe-to-engine block bolt, if equipped
   - Exhaust Gas Recirculation (EGR) pipe

- Oxygen Sensor (O2S) from the exhaust system
- Front exhaust pipe from the Warm-Up Three Way Catalytic (WU-TWC) converter
- Exhaust manifold insulator
- WU-TWC converter from the manifold
- Exhaust manifold

To install:
3. Be sure all gasket mating surfaces are clean prior to assembly.
4. Tighten the components following the illustration.
5. Install or connect the following:
   - Exhaust manifold
   - WU-TWC converter to the manifold
   - Exhaust manifold insulator
   - Front exhaust pipe from the WU-TWC converter
   - O2S to the exhaust system
   - EGR pipe
   - Water bypass pipe-to-engine block bolt
   - Air cleaner and hose assembly
   - Negative battery cable

2.0L (FS) ENGINE

1. Before servicing the vehicle, refer to the precautions in the beginning of this section.
2. Remove or disconnect the following:
   - Negative battery cable
   - Exhaust manifold insulator
   - Oxygen Sensor (O2S) electrical connector
   - O2S, if necessary
   - Exhaust Gas Recirculation (EGR) pipe, if equipped
   - Exhaust pipe flange nuts
   - Exhaust pipe from the manifold
   - Exhaust manifold

To install:
3. Be sure all gasket mating surfaces are clean prior to assembly.
4. Install or connect the following:
   - Exhaust manifold using a new gasket. Torque the bolts to 22 ft. lbs. (30 Nm) and the nuts to 20 ft. lbs. (28 Nm).
   - Exhaust pipe flange using a new gasket. Torque the nuts to 38 ft. lbs. (51 Nm).
   - EGR pipe, if equipped
   - O2S, if necessary
   - O2S electrical connector
   - Exhaust manifold insulator
   - Negative battery cable

2.5L ENGINE

1. Before servicing the vehicle, refer to the precautions in the beginning of this section.

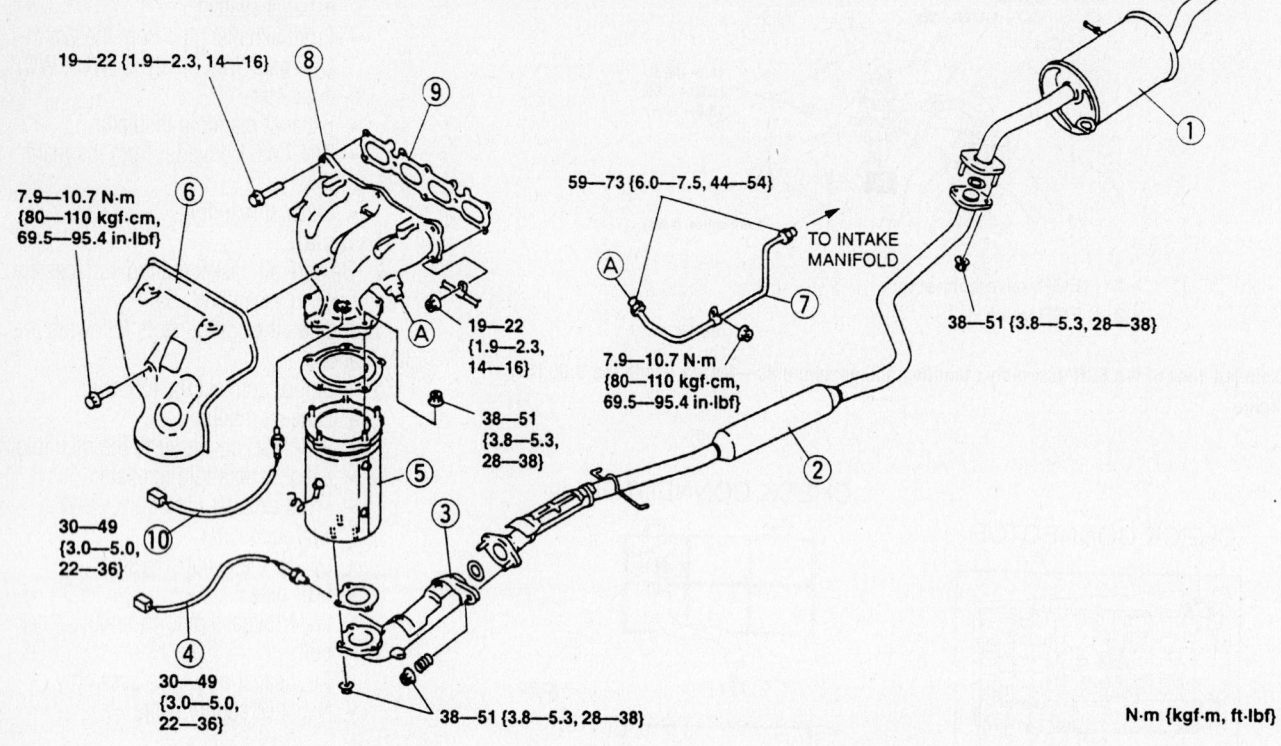

19—22 {1.9—2.3, 14—16}

7.9—10.7 N·m
{80—110 kgf·cm,
69.5—95.4 in·lbf}

59—73 {6.0—7.5, 44—54}

TO INTAKE
MANIFOLD

38—51 {3.8—5.3, 28—38}

19—22
{1.9—2.3,
14—16}

7.9—10.7 N·m
{80—110 kgf·cm,
69.5—95.4 in·lbf}

38—51
{3.8—5.3,
28—38}

30—49
{3.0—5.0,
22—36}

30—49
{3.0—5.0,
22—36}

38—51 {3.8—5.3, 28—38}

N·m {kgf·m, ft·lbf}

9301MG06

1	Main silencer	6	Exhaust manifold insulator
2	Presilencer	7	EGR Pipe
3	Front pipe	8	Exhaust manifold
4	HO2S (Rear)	9	Exhaust manifold gasket
5	WU-TWC	10	HO2S (Front)

Exploded view of the exhaust system—1.6L (ZM) engines

2. Remove or disconnect the following:
- Negative battery cable
- Oxygen (O2S) sensor connectors
- Front and rear exhaust pipe nuts and lower the exhaust system

➡**Both pipes must be disconnected, even if only one manifold is to be removed.**

- Exhaust Gas Recirculation (EGR) pipe, if equipped; if removing the rear (right side) manifold
- 3Heat shield
- 2 nuts and 5 bolts and the exhaust manifold

To install:
3. Clean all gasket mating surfaces.
4. Install or connect the following:
- Exhaust manifold using a new gasket. Torque the nuts and bolts to 12–22 ft. lbs. (16–22 Nm).
- Heat shield. Torque the bolts to 95 inch lbs. (11 Nm).

- EGR pipe, if equipped; if installing the rear (right side) manifold
- Exhaust pipes using new gaskets. Torque the nuts to 38 ft. lbs. (51 Nm).
- O2S connectors
- Negative battery cable

Millenia

2.3L (KJ) ENGINE

1. Before servicing the vehicle, refer to the precautions in the beginning of this section.
2. Remove or disconnect the following:
- Negative battery cable
- Front and rear exhaust pipe nuts and lower the exhaust system

➡**Both pipes must be disconnected, even if only one manifold is to be removed.**

- Exhaust Gas Recirculation (EGR) pipe, if removing the rear (right side) manifold

- Charge air cooler and coolant/condenser fans, If removing the front (left side) manifold
- Front and rear Oxygen Sensor (O2S) connectors
- 3 heat shield bolts and the heat shield
- Exhaust manifold

To install:
3. Clean all gasket mating surfaces.
4. Install or connect the following:
- Exhaust manifold using a new gasket. Torque the bolts to 14–18 ft. lbs. (18–24 Nm).
- Heat shield. Torque the bolts to 70–95 inch lbs. (8–11 Nm).
- O2S connectors
- Coolant/condenser fans and the charge air cooler, if installing the front (left side) manifold. Torque the bolts to 14–18 ft. lbs. (19–25 Nm).
- EGR pipe, if installing the rear (right side) manifold

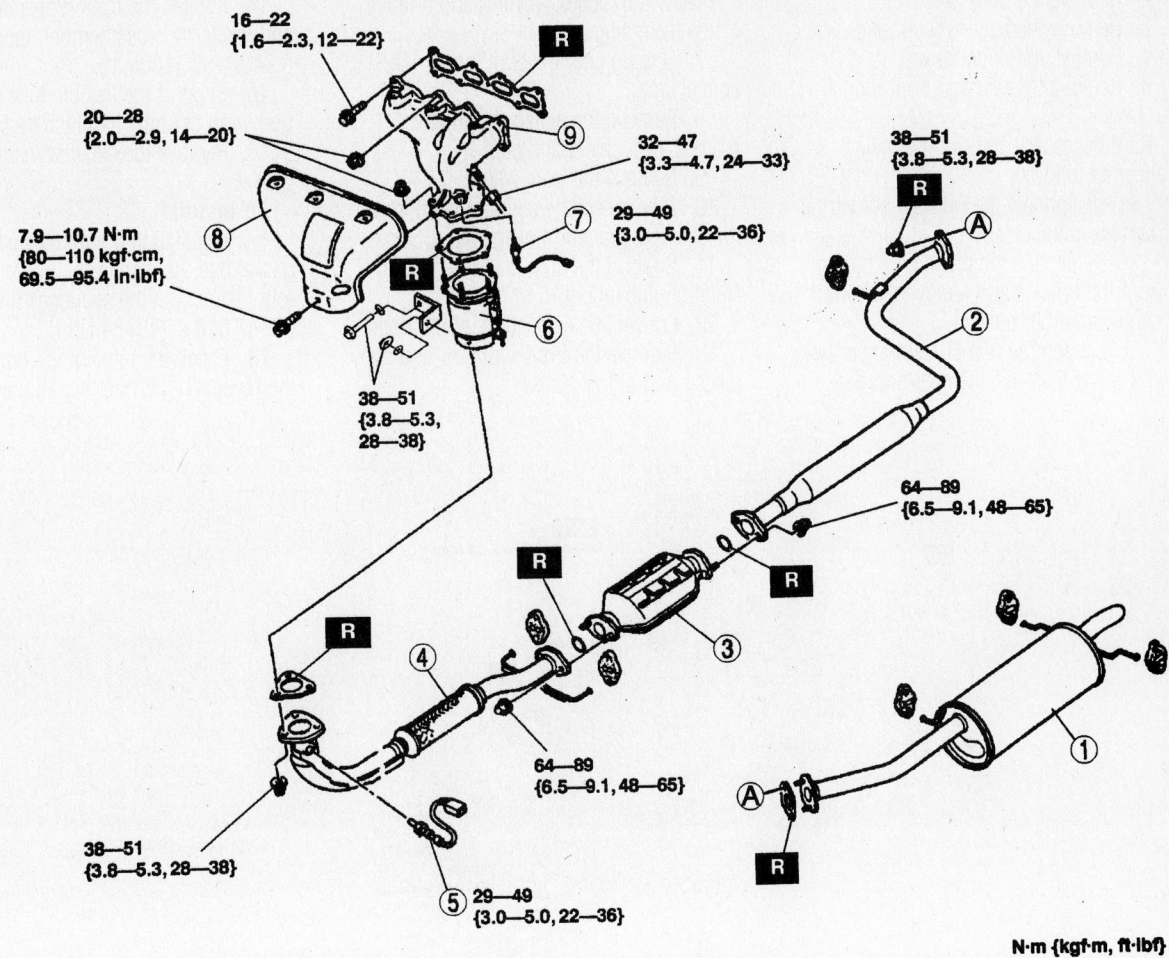

16—22
{1.6—2.3, 12—22}

20—28
{2.0—2.9, 14—20}

7.9—10.7 N·m
{80—110 kgf·cm,
69.5—95.4 in·lbf}

32—47
{3.3—4.7, 24—33}

29—49
{3.0—5.0, 22—36}

38—51
{3.8—5.3, 28—38}

38—51
{3.8—5.3,
28—38}

64—89
{6.5—9.1, 48—65}

38—51
{3.8—5.3, 28—38}

64—89
{6.5—9.1, 48—65}

29—49
{3.0—5.0, 22—36}

N·m {kgf·m, ft·lbf}

1	Main silencer	6	Warm up three way catalytic converter
2	Presilencer	7	Heated oxygen sensor (Front)
3	Three way catalytic converter	8	Exhaust manifold insulator
4	Front pipe	9	Exhaust manifold
5	Heated oxygen sensor (Rear)		

42356-MAZC-G15

Exploded view of the exhaust system—626 shown, Protégé similar

- Exhaust pipes using new gaskets. Torque the nuts to 28–38 ft. lbs. (38–51 Nm).
- Negative battery cable

2.5L (KL) ENGINE

1. Before servicing the vehicle, refer to the precautions in the beginning of this section.
2. Remove or disconnect the following:
   - Negative battery cable
   - Oxygen Sensor (O2S) connectors
   - Front and rear exhaust pipe nuts and lower the exhaust system

➡**Both pipes must be disconnected, even if only one manifold is to be removed.**

- Exhaust Gas Recirculation (EGR) pipe, if equipped; if removing the rear (right side) manifold
- Heat shield
- Nuts and bolts and the exhaust manifold

To install:

3. Clean all gasket mating surfaces.
4. Install or connect the following:
   - Exhaust manifold using a new gasket. Torque the nuts and bolts to 12–16 ft. lbs. (16–22 Nm).

- Heat shield. Torque the bolts to 95 inch lbs. (11 Nm).
- EGR pipe, if equipped; if installing the rear (right side) manifold
- Exhaust pipes using new gaskets. Torque the nuts to 38 ft. lbs. (51 Nm).
- O2S connectors
- Negative battery cable

Mazda 3

2.0L (LF) ENGINE

1. Before servicing the vehicle, refer to the precautions in the beginning of this section.

2. Remove the plug hole plate.

3. Remove the battery cover and duct.

4. Remove the under cover.

5. Remove the rear and front tunnel members.

6. If necessary, remove the main silencer as follows:

   a. Disconnect the ABS sensor wiring harness connector.

   b. Disengage the brake pipe mount from the bracket and remove the lower shock absorber bolts.

   c. Loosen the rear crossmember bolts and lower the crossmember approxi-

mately 2.8 inches (70mm) and remove the main silencer.

7. Unplug both oxygen Sensor (O2S) connectors.

8. Remove the exhaust manifold bracket, heat shield and clip.

9. Remove the front wheels.

10. Disconnect the steering shaft from the steering gear and linkage side.

11. Support the engine and remove the No. 1 engine mount.

12. Loosen the exhaust manifold bolts.

13. Remove the front stabilizer and stabilizer control link bolts.

14. Loosen the front crossmember bolts and lower the crossmember approximately 3.94 inches (100mm).

15. Support the flexible pipe with a support wrap or splint as illustrated.

16. Remove the exhaust manifold and gasket.

To install:

17. Installation a new exhaust manifold gasket and the manifold . Tighten the manifold retainers in the sequence illustrated to 32–47 ft. lbs. (43–64 Nm).

18. Install the remaining components in the reverse order of removal, using the

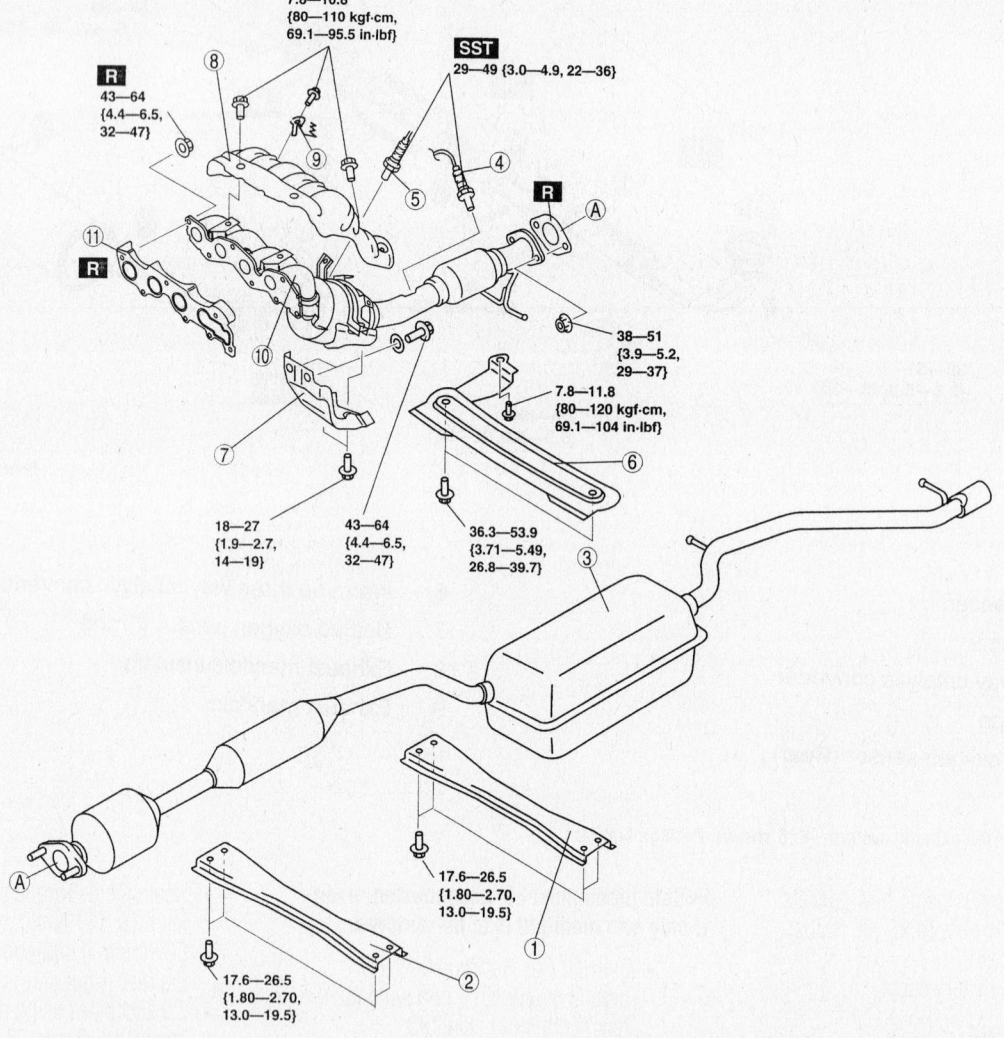

1   Rear tunnel member
2   Front tunnel member
3   Main silencer
4   Rear heated oxygen sensor
5   Front heated oxygen sensor
6   Member
7   Exhaust manifold bracket
8   Exhaust manifold insulator
9   Clip
10 Exhaust manifold
11 Exhaust manifold gasket

67162-MAZC-G23

Exploded view of the exhaust manifold assembly and related components–Non California emissions models—Mazda 3 models

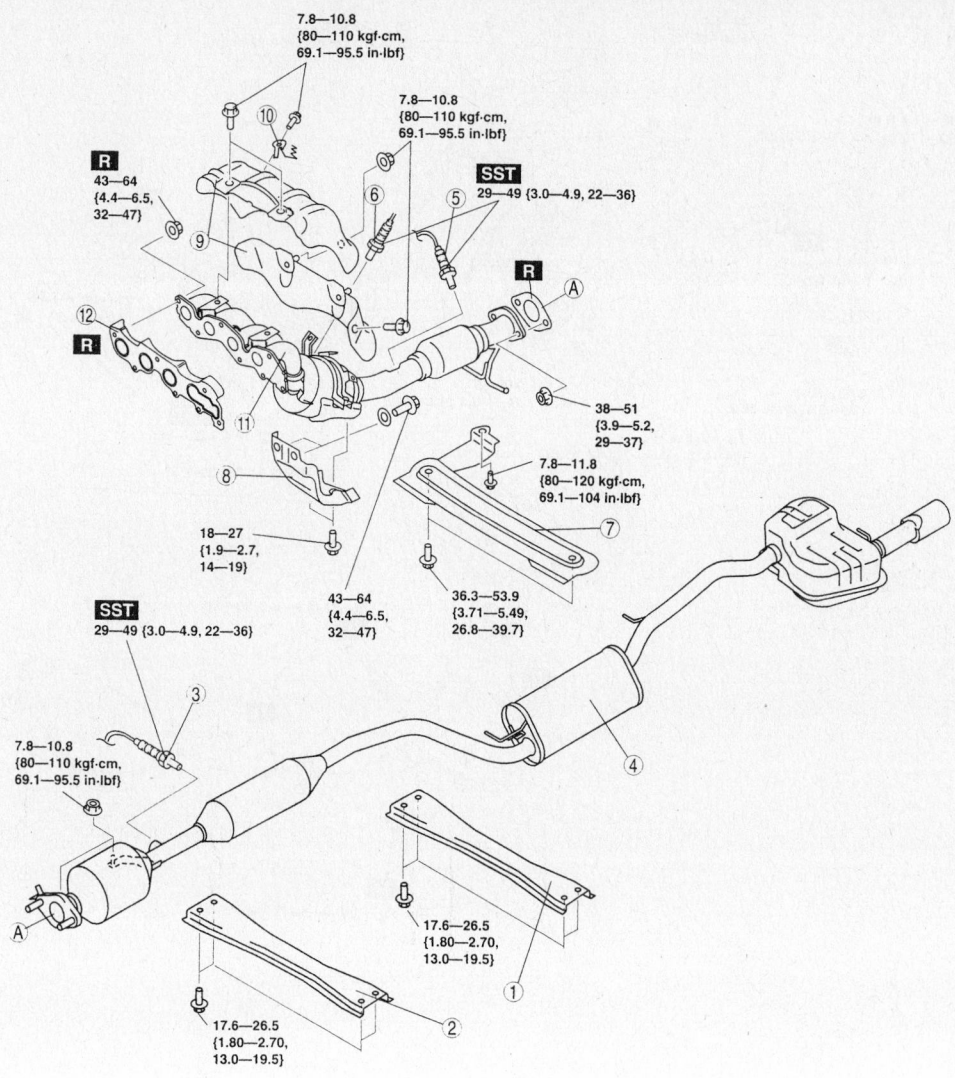

7.8—10.8
{80—110 kgf·cm,
69.1—95.5 in·lbf}

7.8—10.8
{80—110 kgf·cm,
69.1—95.5 in·lbf}

R
43—64
{4.4—6.5,
32—47}

SST 29—49 {3.0—4.9, 22—36}

R
A

R

38—51
{3.9—5.2,
29—37}

7.8—11.8
{80—120 kgf·cm,
69.1—104 in·lbf}

18—27
{1.9—2.7,
14—19}

43—64
{4.4—6.5,
32—47}

36.3—53.9
{3.71—5.49,
26.8—39.7}

SST
29—49 {3.0—4.9, 22—36}

7.8—10.8
{80—110 kgf·cm,
69.1—95.5 in·lbf}

17.6—26.5
{1.80—2.70,
13.0—19.5}

17.6—26.5
{1.80—2.70,
13.0—19.5}

N·m {kgf·m, ft·lbf}

1	Rear tunnel member	7	Member
2	Front tunnel member	8	Exhaust manifold bracket
3	Rear heated oxygen sensor	9	Exhaust manifold insulator
4	Main silencer	10	Clip
5	Middle heated oxygen sensor	11	Exhaust manifold
6	Front heated oxygen sensor	12	Exhaust manifold gasket

67162-MAZC-G24

Exploded view of the exhaust manifold assembly and related components–California emissions models–Mazda 3 models

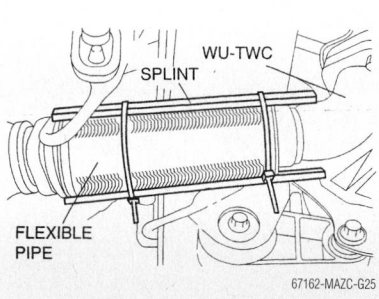

WU-TWC

SPLINT

FLEXIBLE
PIPE

67162-MAZC-G25

Support the flexible pipe with a support wrap or splint—Mazda 3 models

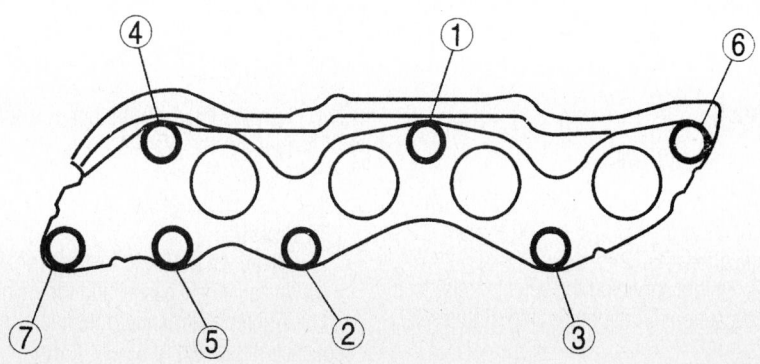

67162-MAZC-G26

Tighten the manifold retainers in this sequence—Mazda 3 models

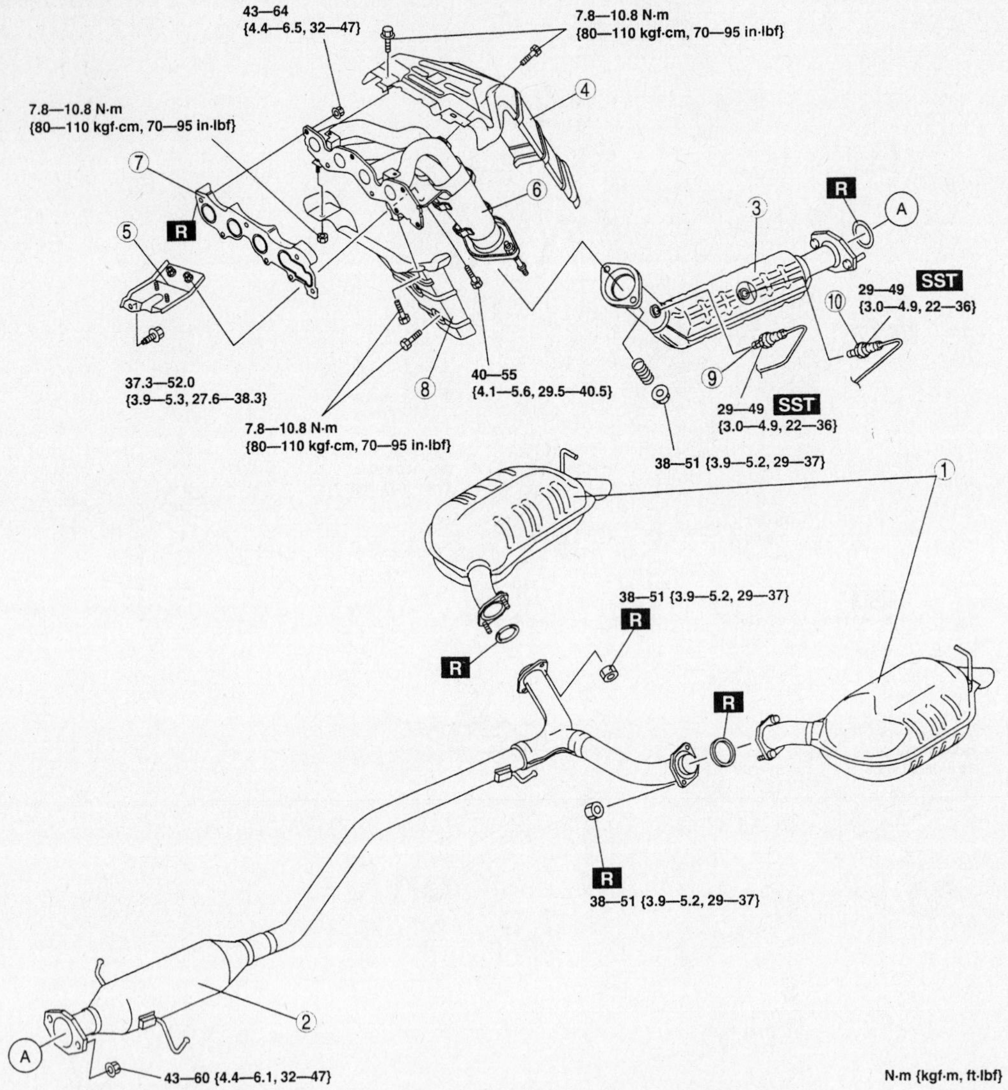

43—64
{4.4—6.5, 32—47}

7.8—10.8 N·m
{80—110 kgf·cm, 70—95 in·lbf}

7.8—10.8 N·m
{80—110 kgf·cm, 70—95 in·lbf}

R

29—49 **SST**
{3.0—4.9, 22—36}

R Ⓐ

10

29—49 **SST**
{3.0—4.9, 22—36}

37.3—52.0
{3.9—5.3, 27.6—38.3}

40—55
{4.1—5.6, 29.5—40.5}

7.8—10.8 N·m
{80—110 kgf·cm, 70—95 in·lbf}

38—51 {3.9—5.2, 29—37}

① ② ③ ④ ⑤ ⑥ ⑦ ⑧ ⑨

38—51 {3.9—5.2, 29—37}

R

R

R

38—51 {3.9—5.2, 29—37}

N·m {kgf·m, ft·lbf}

Ⓐ

43—60 {4.4—6.1, 32—47}

1	Main silencer	6	Exhaust manifold
2	Presilencer	7	Exhaust manifold gasket
3	TWC	8	Exhaust manifold insulator (lower)
4	Exhaust manifold insulator (upper)	9	HO2S (front)
5	Bracket	10	HO2S (rear)

67162-MAZC-G155

Exploded view of the exhaust manifold assembly mounting and related components—Mazda 6 with the 2.3L (L3) engine

exploded view of the exhaust system for component location and torque specifications.

Mazda 6

2.3L (L3) ENGINE

1. Refer to the exploded view illustration for component location and torque specifications.

2. Before servicing the vehicle, refer to the precautions in the beginning of this section.

3. Disconnect the negative battery cable.

4. Remove the main silencer and presilencer.

5. Remove the Three Way Converter (TWC).

6. Remove the manifold upper insulator.

7. Remove the bracket.

8. Remove the exhaust manifold and gasket. Discard the gasket.

To install:

9. Clean the manifold mating surfaces.

10. Install a new gasket and the manifold.

11. Tighten the manifold retainers in the sequence illustrated to 32–47 ft. lbs. (43–64 Nm).

12. Install the bracket as follows:

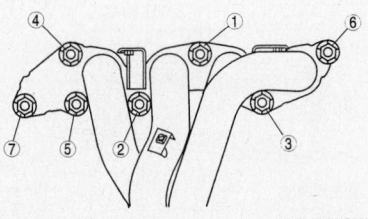

67162-MAZC-G156

Exhaust manifold torque sequence— Mazda 6 with the 2.3L (L3) engine

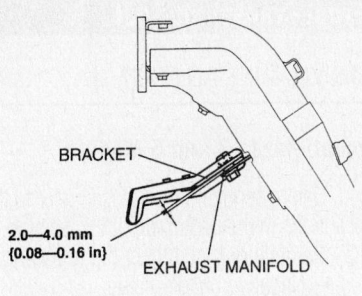

BRACKET

2.0—4.0 mm
{0.08—0.16 in}

EXHAUST MANIFOLD

67162-MAZC-G157

Measure the gap between the manifold and bracket—Mazda 6 with the 2.3L (L3) engine

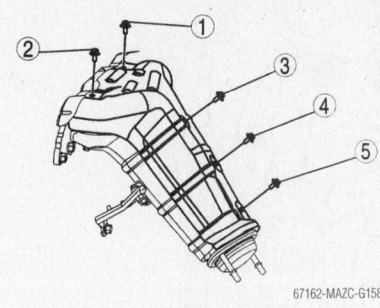

67162-MAZC-G158

Tighten the manifold upper insulator bolts in this sequence—Mazda 6 with the 2.3L (L3) engine

a. Install the bracket and hand tighten the manifold side bolts.

b. Measure the gap between the manifold and bracket it should be 0.08–0.18 inch (2–4mm).

c. Tighten the cylinder block side bolt to 27–38 ft. lbs. (37–52 Nm).

d. Tighten the manifold side bolts to 29–40 ft. lbs. (40–55 Nm).

13. Tighten the manifold upper insulator bolts in the sequence illustrated to 70–95 inch lbs. (7–10 Nm).

14. Install the remaining components on the reverse order of removal, referring to the illustration for torque specifications.

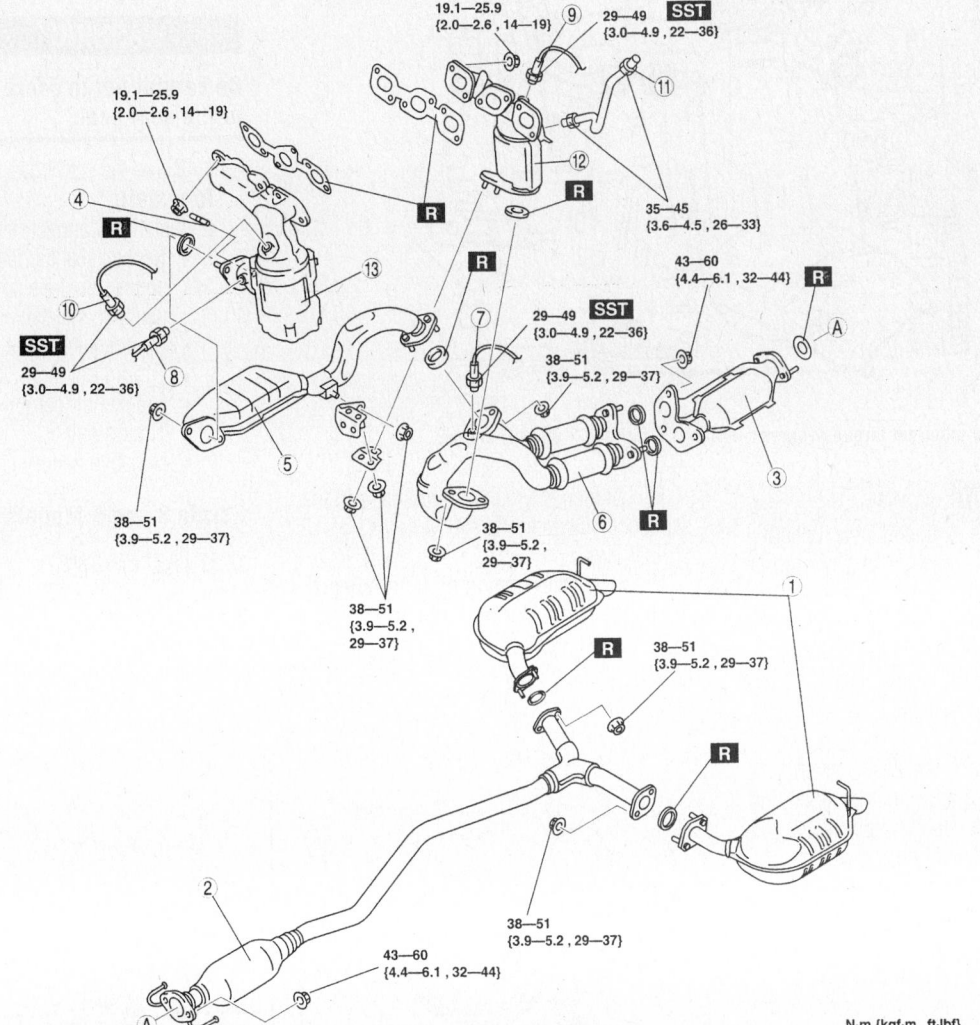

N·m {kgf·m, ft·lbf}

1	Main silencer	7	HO2S (RR)
2	Presilencer	8	HO2S (LR)
3	TWC (RH)	9	HO2S (RF)
4	Stud bolt	10	HO2S (LF)
5	TWC (LH)	11	EGR pipe
6	Front pipe	12	Exhaust manifold (RH)
		13	Exhaust manifold (LH)

67162-MAZC-G159

Exploded view of the exhaust manifold assembly mounting and related components—Mazda 6 with the 3.0L (AJ) engine

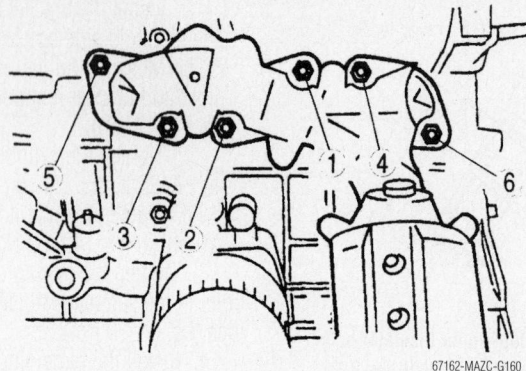

Left side exhaust manifold torque sequence—Mazda 6 with the 3.0L (AJ) engine

67162-MAZC-G160

Right side exhaust manifold torque sequence—Mazda 6 with the 3.0L (AJ) engine

67162-MAZC-G161

3.0L (AJ) ENGINE

1. Refer to the exploded view illustration for component location and torque specifications.

2. Before servicing the vehicle, refer to the precautions in the beginning of this section.

3. Disconnect the negative battery cable.

4. Remove the main silencer and presilencer.

5. Remove the Three Way Converter (TWC) from the left or right side depending on which manifold is being replaced.

6. Remove the stud bolt.

7. Remove the front pipe.

8. Unplug both Heated oxygen Sensor (HO2S) connectors.

9. Remove the Exhaust Gas Recirculation (EGR) pipe.

10. Remove the exhaust manifold and gasket, from the left or right side depending on which manifold is being replaced. If removing the right hand side manifold, remove the alternator bracket first.

11. Discard the manifold gasket.

To install:

12. Clean the manifold mating surfaces.

13. Install a new gasket and the manifold.

14. Tighten the manifold retainers in the sequence illustrated to 14–19 ft. lbs. (19–26 Nm).

15. Install the remaining components on the reverse order of removal, referring to the illustration for torque specifications.

Front Crankshaft Seal

REMOVAL & INSTALLATION

Except Mazda 3 and 6 Models

1. Before servicing the vehicle, refer to the precautions in the beginning of this section.

2. Remove or disconnect the following:

- Negative battery cable
- Timing belt
- Crankshaft damper bolt and damper
- Timing belt sprocket
- Sprocket key from the crankshaft
- Oil seal from the engine block using a prybar

✳✳ WARNING

Be careful not to score the crankshaft or the seal seat.

3. Clean the seal bore.

To install:

4. Install or connect the following:

- New oil seal lubricated with clean engine oil, drive it into the engine using an installation tool until it seats
- Sprocket key onto the crankshaft
- Timing belt sprocket
- Crankshaft damper
- Timing belt
- Negative battery cable

Mazda 3 and 6 Models

2.3L (L3) ENGINE

1. Before servicing the vehicle, refer to the precautions in the beginning of this section.

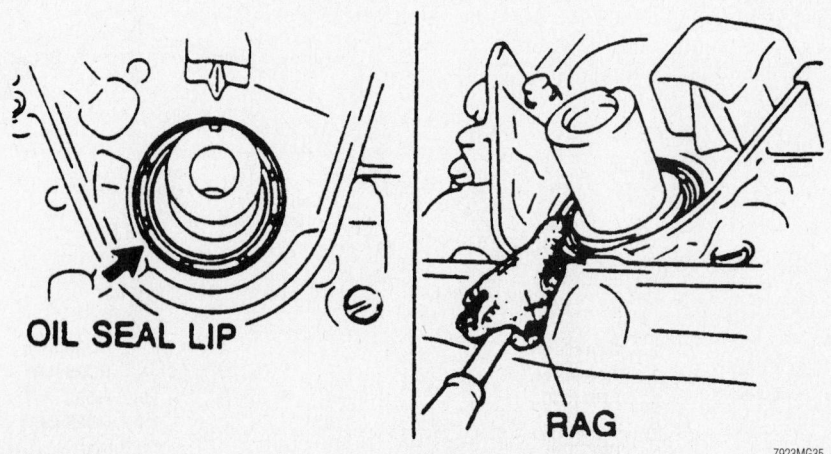

OIL SEAL LIP

RAG

7923MG35

Remove the front engine seal by cutting the seal lip, then, so as not to damage the crankshaft, carefully pry the seal out with a prybar—except Mazda 3 and 6 models

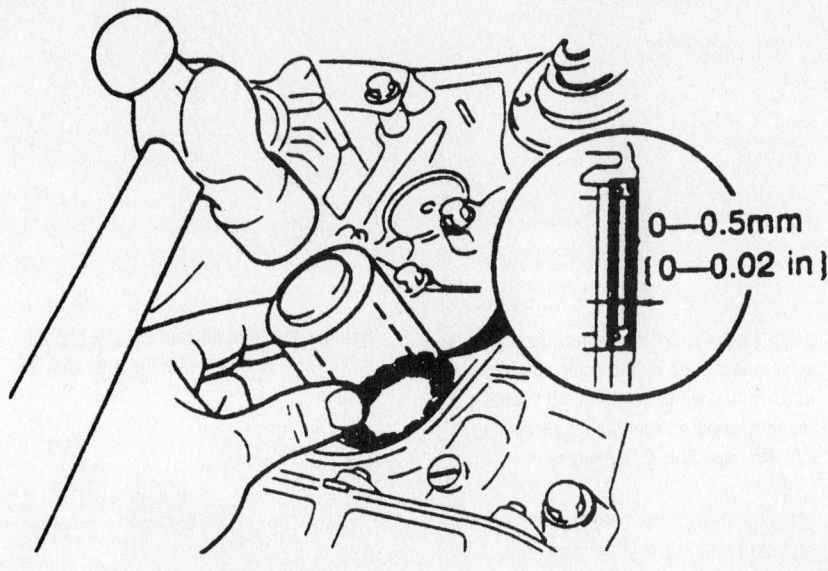

7923MG36

Install the seal using an appropriate driver, which fits over the crankshaft snout and presses on the outside edge of the seal—except Mazda 3 and 6 models

2. Remove the plug hole plate and bracket.

3. Remove the battery cover.

4. Disconnect the negative battery cable.

5. Disconnect the wiring harness.

6. Remove the ignition coils, spark plugs, valve cover and accessory drive belt.

7. Remove the front wheels.

8. Remove the under cover and splash shield, if equipped.

9. On Mazda 6 models, remove the right hand halfshaft from the joint shaft.

10. Remove the Crankshaft Position (CKP) sensor.

11. Remove the crankshaft pulley bolt and pulley.

12. Remove the cylinder block lower blind plug.

13. Install tool 49 JE01 061.

14. Turn the crankshaft clockwise until the crankshaft is in the number 1 cylinder Top Dead Center position (the balance weight should be attached to tool 49 JE01 061).

15. Hold the crankshaft pulley using the tools illustrated and remove the bolt.

16. Remove the crankshaft pulley.

17. Cut the seal lip using a suitable knife and using a suitable prytool wrapped in a rag, remove the seal.

To install:

18. Coat the new seal with clean engine oil.

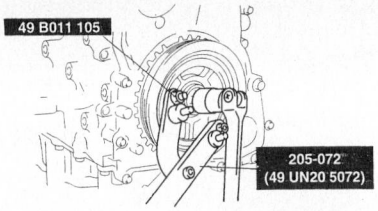

67162-MAZC-G28

Install tool 49 JE01 061 Hold the crankshaft pulley using the tools illustrated and remove the bolt—2.0L (LF) and 2.3L (L3) engines

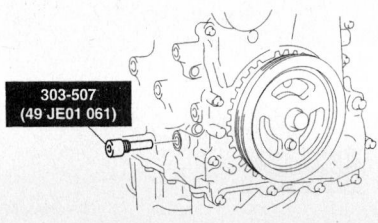

67162-MAZC-G27

Install tool 49 JE01 061—2.0L (LF) and 2.3L (L3) engines

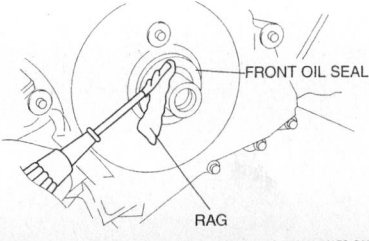

67162-MAZC-G29

Cut the seal lip using a suitable knife and using a suitable prytool wrapped in a rag, remove the seal—2.0L (LF) and 2.3L (L3) engines

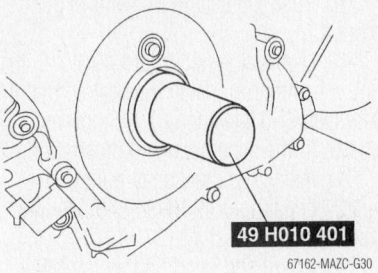

67162-MAZC-G30

Use seal installer tool 49 H010 401 and a hammer to install the seal—2.0L (LF) and 2.3L (L3) engines

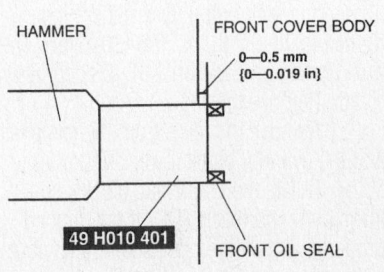

67162-MAZC-G31

The seal should be recessed 0–0.019 inch (0–0.5mm) as shown when properly installed—2.0L (LF) and 2.3L (L3) engines

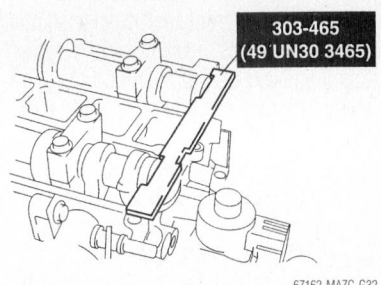

67162-MAZC-G32

Install tool 49 UN30 3465 on the camshaft—2.0L (LF) and 2.3L (L3) engines

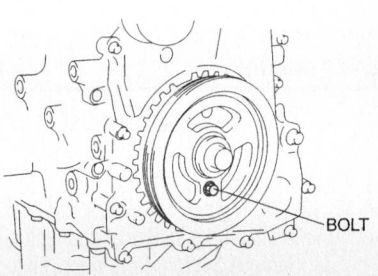

67162-MAZC-G33

Install and hand tighten the M6 x 1.0 bolt—2.0L (LF) and 2.3L (L3) engines

19. Push the seal in by hand to get it started.

20. Use seal installer tool 49 H010 401 and a hammer to install the seal so that it is recessed 0–0.019 inch (0–0.5mm) as shown in the accompanying illustration.

21. Install the crankshaft pulley.

22. Install tool 49 UN30 3465 as illustrated on the camshaft.

23. Install and hand tighten the M6 x 1.0 bolt as illustrated.

24. Turn the crankshaft clockwise until the crankshaft is in number 1 TDC (the balance weight should be attached to the tool).

25. Hold the crankshaft pulley and tighten the lock bolt in two steps. First tighten to 71–77 ft. lbs. (96–104 Nm). Then final tighten an additional 87–93 degrees.

26. Remove the M6 x 1.0 bolt.

27. Remove the tools from the camshaft and the cylinder block lower blind plug.

28. Rotate the crankshaft clockwise 2 turns until you reach TDC. If not aligned properly, loosen the crankshaft pulley lock bolt and reinstall the lock bolt again using the above procedure and special tools.

29. Install the cylinder block blind plug and tighten to 13–16 ft. lbs. (18–22 Nm).

30. When installing the CKP sensor perform the following:

a. Remove the cylinder block lower blind plug.

b. Install tool 49 JE01 061.

c. Turn the crankshaft clockwise until the crankshaft is in the number 1 cylinder Top Dead Center position (the balance weight should be attached to tool 49 JE01 061).

d. Using a ruler, mark the center line on the pulse wheel teeth on the crank pulley which is located at the 9th tooth counting counterclockwise from the empty space. Refer to the illustration for more detail.

✳✳ CAUTION

If you do not mark the center line correctly this will cause improper engine control for the ignition and fuel system, so be sure to mark the line carefully.

e. Install the CKP sensor making sure the mark on the sensor is aligned with the mark on the pulse wheel. Tighten the sensor bolt to 49–66 inch lbs. (5.5–7.5 Nm)and attach the electrical connector.

f. Remove the tool from the cylinder block lower blind plug.

g. Rotate the crankshaft clockwise 2 turns until you reach TDC. If not aligned

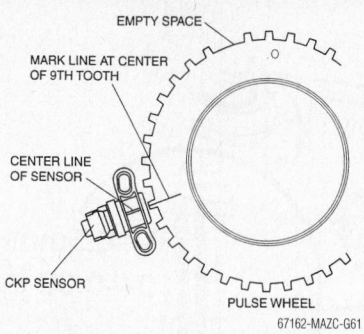

Using a ruler, mark the center line on the pulse wheel teeth on the crank pulley which is located at the 9th tooth counting counterclockwise from the empty space— 2.0L (LF) and 2.3L (L3) engines

properly, loosen the crankshaft pulley lock bolt and reinstall the lock bolt.

h. Install the cylinder block blind plug and tighten to 13–16 ft. lbs. (18–22 Nm).

31. Install the remaining components in the reverse order of removal.

32. Start the vehicle and check for leaks, repair if necessary.

3.0L (AJ) ENGINE

1. Before servicing the vehicle, refer to the precautions in the beginning of this section.

2. Disconnect the negative battery cable.

3. Remove the accessory drive belt.

4. Hold the crankshaft pulley using the tools illustrated and remove the pulley bolt.

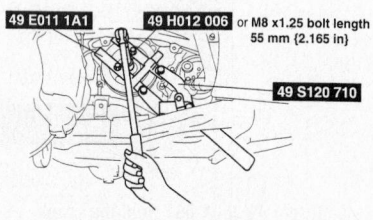

Hold the crankshaft pulley using the tools illustrated and remove the pulley bolt— Mazda 6 with the 3.0L (AJ) engine

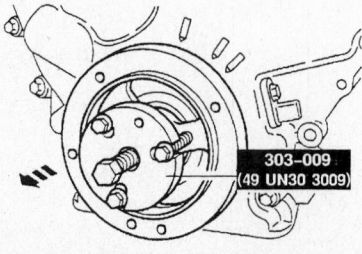

Remove the crankshaft pulley using a suitable puller such as the one shown— Mazda 6 with the 3.0L (AJ) engine

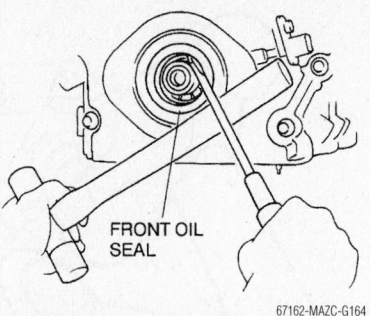

Remove the front oil seal using a prytool as shown—Mazda 6 with the 3.0L (AJ) engine

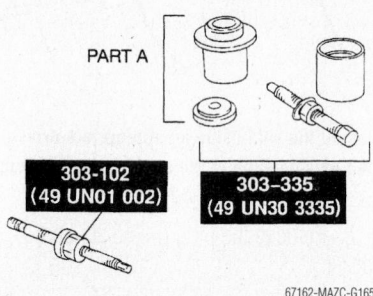

Assemble the seal using part (A) of tool 49 UN01 002 and tool 49 UN01 002 as shown—Mazda 6 with the 3.0L (AJ) engine

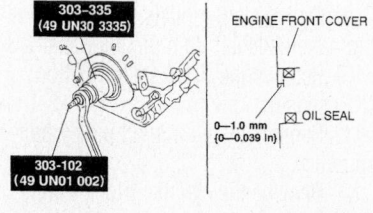

Install the seal using the installation tools until the seal is recessed 0–0.039 inch (0–1mm)—Mazda 6 with the 3.0L (AJ) engine

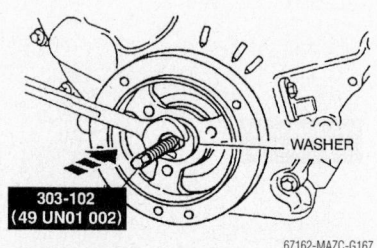

Install the pulley using the tools shown— Mazda 6 with the 3.0L (AJ) engine

5. Remove the A/C compressor and set aside with the lines still attached.

6. Remove the crankshaft pulley using a suitable puller such as the one illustrated.

7. Remove the front oil seal using a prytool as illustrated.

To install:

8. Assemble the seal using part (A) of tool 49 UN01 002 and tool 49 UN01 002 as illustrated.

9. Apply clean oil to the seal and push the seal in by hand.

10. Install the seal using the installation tools until the seal is recessed 0–0.039 inch (0–1mm).

11. Seal the crankshaft pulley using a silicone sealant.

12. Install the pulley using the tools illustrated.

13. While holding the pulley, tighten the crankshaft pulley bolt as follows:

   a. Tighten to 88 ft. lbs. (120 Nm).

   b. Loosen one full turn.

   c. Tighten to 35–39 ft. lbs. (47–53 Nm).

   d. Tighten 85–95 degrees.

14. Install the A/C compressor.

15. Install the accessory drive belt.

16. Connect the negative battery cable.

17. Start the vehicle and check for leaks, repair if necessary.

Camshaft

REMOVAL & INSTALLATION

1.6L (ZM) and 1.8L (FP) Engines

1. Before servicing the vehicle, refer to the precautions in the beginning of this section.

2. Remove or disconnect the following:
   - Negative battery cable
   - Spark plug wires
   - Spark plugs
   - Cylinder head cover hoses, if equipped
   - Cylinder head cover bolts
   - Cylinder head cover
   - Timing belt
   - Camshaft by holding it with a wrench on the hexagon cast into the camshaft
   - Sprocket bolts
   - Sprockets

➡**Label the caps so they can be reinstalled in their original positions.**

   - Camshaft cap bolts by loosening in 2–3 steps in the sequence shown
   - Camshaft caps

Camshaft cap bolt loosening sequence—1.6L (ZM) engines

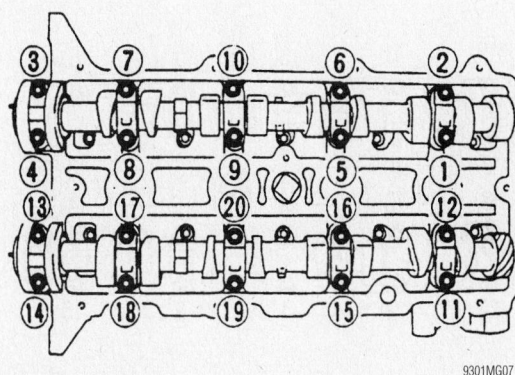

Camshaft cap bolt loosening sequence—1.8L (FP) engines

To install:

✳✳ CAUTION

Because there is little thrust clearance, the camshaft must be held in the horizontal position during installation. If not excessive force will be applied to the thrust area causing a burr on the receiving area of the cylinder head journal. Make sure to use the following procedure to avoid damage.

3. Install or connect the following:
   - Lubricate the camshaft journals and lobes with clean engine oil
   - Camshafts onto the head facing the cam noses at the no. 1 and 3 cylinders as illustrated. Make sure the camshaft sliding surface is free of sealant.

4. Apply a 0.04 inch (1mm) bead of silicone sealant to the areas illustrated in the illustration. Do not allow any sealant on the camshaft journals.

5. Install or connect the following:
   - Camshaft caps in their original positions
   - Cap bolts and hand tighten bolts 5, 7, 2 and 4

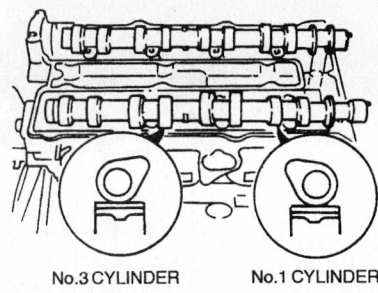

No.3 CYLINDER   No.1 CYLINDER

Install the camshafts onto the head facing the cam noses at the no. 1 and 3 cylinders as shown—1.6L (ZM) and 1.8L (FP) engines

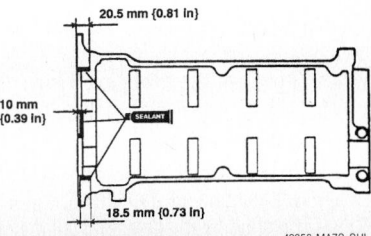

Apply silicone sealant to the cylinder head in the positions shown—1.6L (ZM) engines

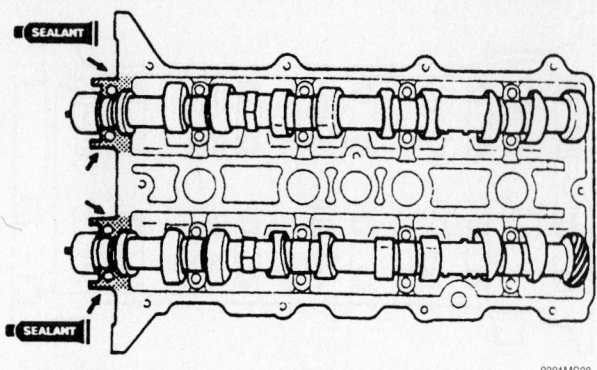

Apply silicone sealant to the cylinder head in the positions shown—1.8L (FP) engines

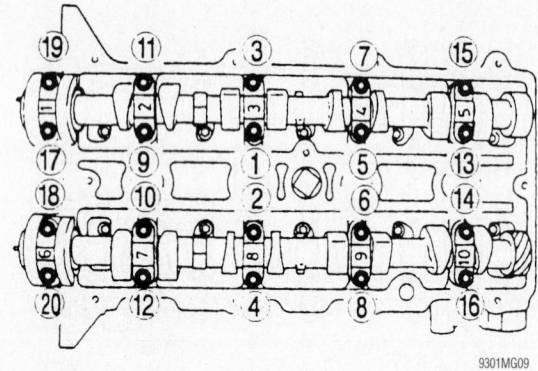

Camshaft cap bolt tightening sequence—1.6L (ZM) and 1.8L (FP) engines

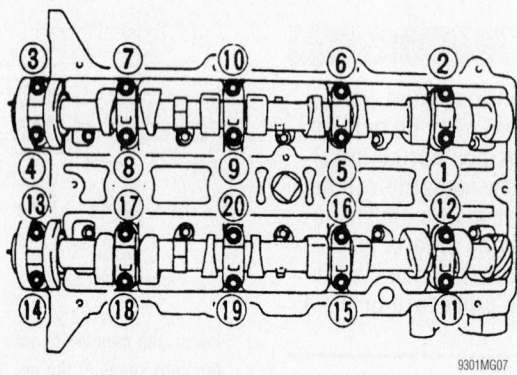

Camshaft cap bolt loosening sequence—2.0L (FS) engines

- Cap bolts a few turns in the sequence illustrated

6. Make sure the camshaft settles horizontally when the bearing cap bolts at the number 3 journals are tightened. Torque the bolts in 2–3 steps to 125 inch lbs. (14 Nm) in the proper sequence.

- New camshaft seal by lubricating it with engine oil. Tap the seal into position, using a seal installer, until it is recessed into the cylinder head 0.01 inch (0.4mm) on the 1.6L (ZM) engines and 0.012–0.027 inch (0.3–0.7mm) on 1.8L (FP) engines.

7. Turn the camshafts until the dowel pins face straight up.

8. Install or connect the following:
- Camshaft sprockets. Torque the bolts to 44 ft. lbs. (60 Nm) by holding the camshaft with the wrench on the cast hexagon.
- Remaining components

2.0L (FS) Engine

1. Before servicing the vehicle, refer to the precautions in the beginning of this section.

2. Drain the cooling system.

3. Relieve the fuel system pressure.

4. Drain the cooling system.

5. Remove or disconnect the following:

- Negative battery cable
- Timing belt
- Front exhaust pipe
- Air cleaner assembly
- Accelerator cable
- Fuel hose
- Ignition coil
- Camshaft pulley. Hold the camshaft pulley with a back–up wrench while loosening the pulley bolt.

➡**Label the caps so they can be reinstalled in their original positions.**

- Camshaft cap bolts by loosening in 2–3 steps in the sequence shown
- Camshaft caps

To install:

6. Install or connect the following:
- Camshafts making sure to lubricate the journals and lobes with clean engine oil. Place the camshafts onto the cylinder head facing the cam noses at the No. 1 and No. 3 cylinders as illustrated.

7. Apply silicone sealant to the cylinder head on the front camshaft cap mating surfaces. Do not allow any sealant on the camshaft journals.

8. Install or connect the following:
- Camshaft caps in their original positions
- Hand tighten cap bolts marked 5, 5, 2 and 4, refer to the torque sequence illustration for location
- Cap bolts. Torque the bolts in 2–3 steps to 125 inch lbs. (14 Nm) in the proper sequence.

9. Make sure the camshaft is settled horizontally when the 2 bearing cap bolts at the number 3 journal are tightened.

- New camshaft seal by lubricating it with engine oil. Tap the seal into position, using a seal installer, until

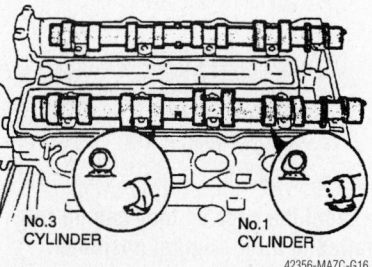

Install the camshafts so the cam noses are positioned at the No. 1 and No. 3 cylinders—2.0L (FS) engine

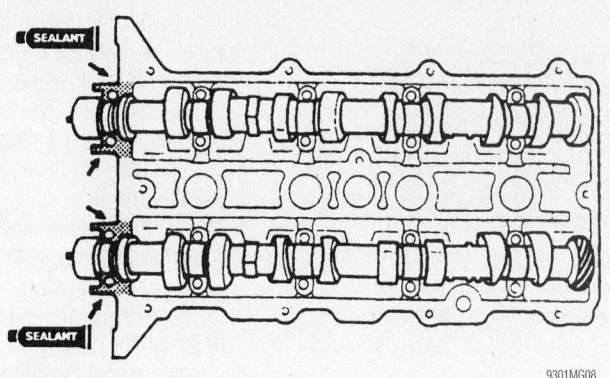

Apply silicone sealant to the cylinder head in the positions shown—2.0L (FS) engines

9301MG08

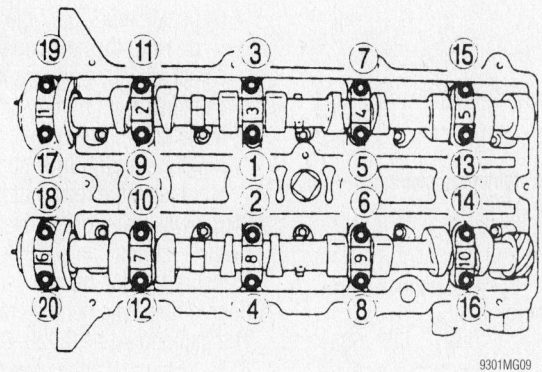

Camshaft cap bolt tightening sequence—2.0L (FS) engines

9301MG09

it is recessed into the cylinder head 0.012–0.027 inch (0.3–0.7mm.

10. Turn the camshafts until the dowel pins face straight up.

11. Install or connect the following:
- Camshaft sprockets. Torque the bolts to 44 ft. lbs. (60 Nm) by holding the camshaft with the wrench on the cast hexagon.
- Ignition coil
- Fuel hose
- Accelerator cable
- Air cleaner assembly
- Front exhaust pipe
- Timing belt
- Negative battery cable

12. Fill and bleed the cooling system.

13. Change the oil and filter.

14. Run the engine and check for proper operation.

1.8L (BP) Engines

1. Before servicing the vehicle, refer to the precautions in the beginning of this section.

2. Remove or disconnect the following:
- Negative battery cable
- Timing belt
- Variable valve timing actuator and

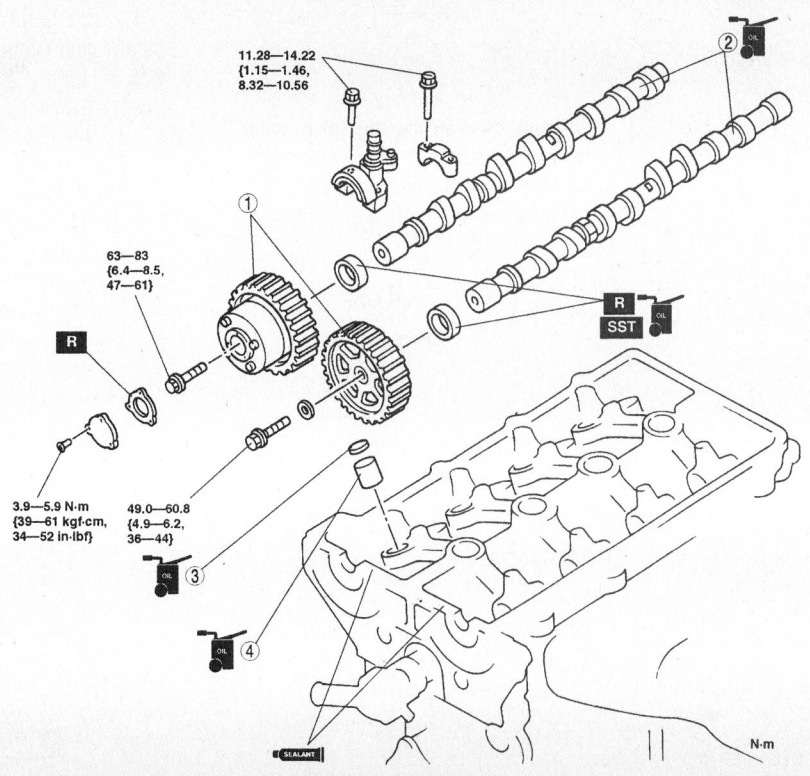

1	Variable valve timing actuator and Camshaft Pulley	3	Adjustment shim
2	Camshaft	4	Tappet

42356-MAZC-GAI

Exploded view of the camshaft assembly and related components—1.8L (BP) engine

camshaft pulley. Hold the hexagonal part of the camshaft with a wrench to prevent the camshaft from turning and remove the actuator and pulley bolt.

➡**Label the caps so they can be reinstalled in their original positions.**

- Camshaft cap bolts by loosening in 2–3 steps in the sequence shown
- Camshaft caps
- Lifters and adjustment shims, if necessary

➡**Identify and mark each lifter as it is removed so it can be reinstalled in the same position.**

To install:
3. Install or connect the following:
- Lifters in their original positions by lubricating them with engine oil

➡**Verify that they move smoothly in their bore.**

4. Lubricate the camshaft journals and lobes with clean engine oil.

5. Install the camshafts so the cam projections of cylinders 1 and 3 face in the direction as illustrated.

6. Apply silicone sealant to the cylinder head on the front camshaft cap mating surfaces. Do not allow any sealant on the camshaft journals.

7. Install or connect the following:

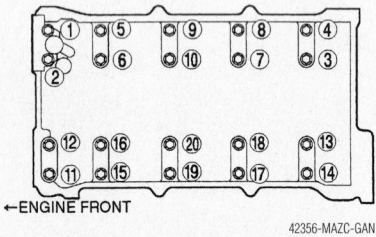

42356-MAZC-GAN

Camshaft cap bolt loosening sequence——1.8L (BP) engine

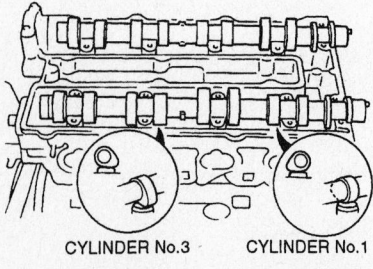

42356-MAZC-GAJ

Install the camshafts so the cam projections of cylinders 1 and 3 face in the direction shown—1.8L (BP) engine

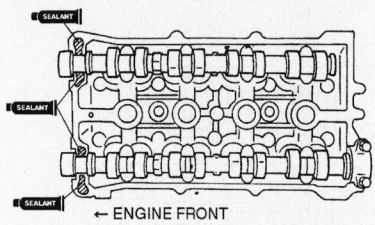

42356-MAZC-GAK

Apply silicone sealant to the cylinder head in the positions shown—1.8L (BP) engine

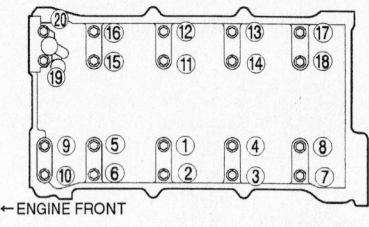

42356-MAZC-GAL

Camshaft cap bolt tightening sequence—1.8L (BP) engine

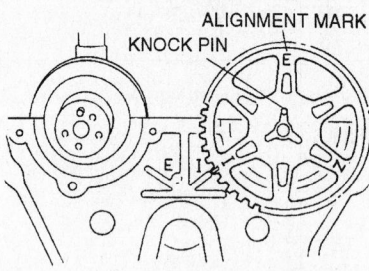

42356-MAZC-GAM

When installing the variable valve timing actuator and camshaft pulley, make sure the knock pin and alignment marks are aligned as shown—1.8L (BP) engine

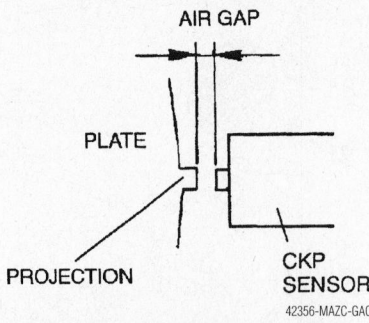

42356-MAZC-GAO

Check the Crankshaft position (CKP) sensor gap—1.8L (BP) engine

- Camshaft caps in their original positions
- Cap bolts. Torque the bolts in 2–3 steps to 125 inch lbs. (14 Nm) in the proper sequence.

8. Make sure the camshaft is settled

horizontally when the 2 bearing cap bolts at the number 3 journal are tightened.

- New camshaft seal by lubricating it with engine oil. Tap the seal into position, using a seal installer, until it is flush with the edge of the camshaft.

9. Install the variable valve timing actuator and camshaft pulley as follows:

a. Rotate the camshaft and face the knock pin to the position illustrated.

b. Install the camshaft so that the alignment mark of the pulley faces as illustrated.

c. Install the camshaft so the knock pin of the camshaft is connected to the camshaft knock pin hole of the actuator.

d. Hold the hexagonal part of the camshaft with a wrench to prevent the camshaft from turning and tighten the actuator and pulley bolt. Refer to the illustration for torque specifications.

10. Install the remaining components in the reverse order of removal and inspect the Crankshaft position (CKP) sensor gap. Measure the gap between each 4 projections of the plate behind crankshaft pulley. The gap should be 0.020–0.059 inch (0.5–1.5mm). If not as specified, adjust.

2.3L (KJ) Engine

1. Before servicing the vehicle, refer to the precautions in the beginning of this section.

2. Relieve the fuel system pressure.
3. Drain the engine coolant.
4. Remove or disconnect the following:
- Negative battery cable
- Oxygen (O_2S) sensor connectors
- Exhaust pipe-to-manifold nuts and lower the exhaust pipes
- Right-hand 3-way catalytic converter
- Compressor (supercharger)
- Intake manifold
- Timing belt covers and timing belt
- Spacer and O-ring from the front of the camshaft

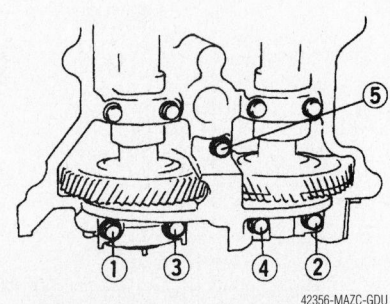

42356-MAZC-GDU

Loosen the front camshaft cap bolts using this sequence—2.3L (KJ) engine

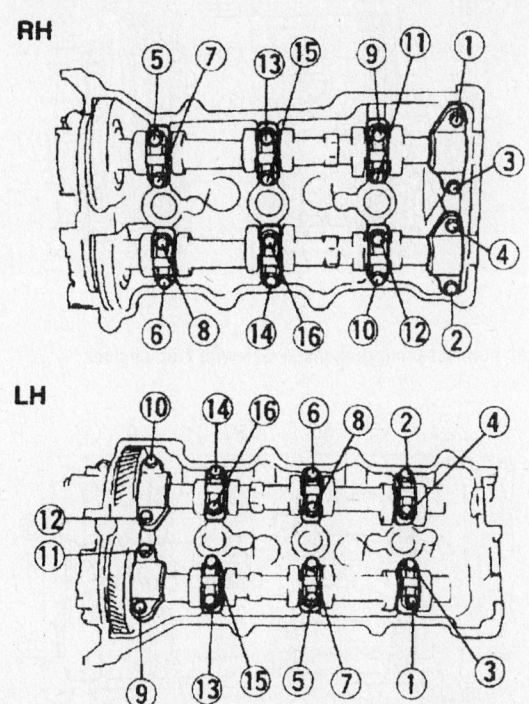

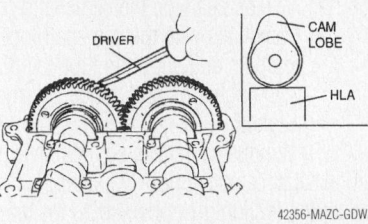

The intake cam lobes of the number 1 cylinder on the right hand side and the number 2 cylinder on the left hand side face straight up—2.3L (KJ) engine

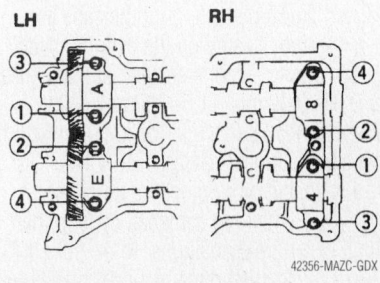

Tighten the front camshaft cap bolts in this sequence—2.3L (KJ) engine

Loosen the remaining camshaft cap bolts using this sequence—2.3L (KJ) engine

- Ignition coils
- Cylinder head cover mounting bolts, in 5–6 steps, using the reverse of the tightening sequence
- Cylinder head cover
- Camshaft sprockets

5. Turn the camshafts so the knock pins are aligned with the marks on the camshaft caps. This will reduce the pressure on the adjustment shims.

6. Note the markings on the camshaft caps prior to removal, so they can be reinstalled in the same positions. The right-hand (rear) caps are marked with numbers and the left-hand (front) caps are marked with letters.

7. Loosen the front camshaft cap bolts in sequence, in 5–6 steps. Remove the front camshaft caps. remove the remaining camshaft cap bolts in the proper sequence. Remove the caps, being sure to remove the thrust caps last. Do not damage the cylinder head thrust bearing support.

8. Remove or disconnect the following:
- Camshafts and oil seals
- Lifters and adjustment shims

9. Identify and mark each lifter as it is removed so it can be reinstalled in the same position.

To install:

10. Install or connect the following:
- Lifters in their original positions, lubricated with engine oil. Verify that they move smoothly in their bore.

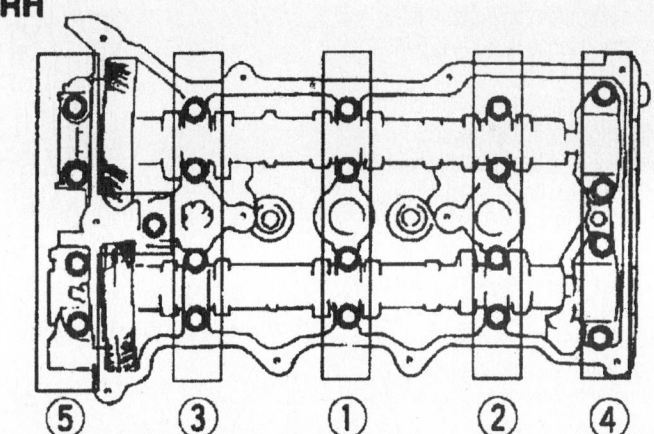

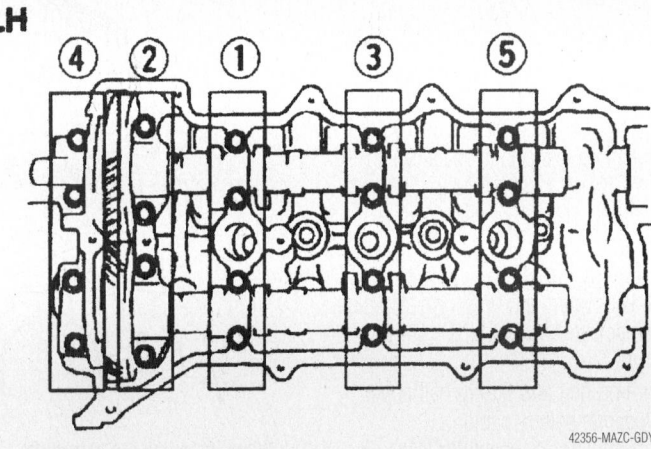

Tighten the camshaft cap bolts using this sequence—2.3L (KJ) engine

- New oil seals on the camshafts
- Camshafts with the camshaft lobes, journals and supports lubricated with engine oil and the gear marks aligned

11. Install the camshafts so the intake and exhaust camshaft gear marks align. Adjust the friction gear position so the tappets are not lifted using the cam lobes. The intake cam lobes of the number 1 cylinder on the right hand side and the number 2 cylinder on the left hand side face straight up.

- Front trust caps. Torque the bolts, in 5–6 steps, until the caps are fully seated on the cylinder head.

12. Apply silicone sealant, at a thickness of 0.06–0.09 inch (1.5–2.5mm), to the cylinder head surface in the area forward of the camshaft gear cavity.

13. Install or connect the following:

- Remaining camshaft caps in their original positions. Make sure the camshaft remains horizontal as the camshaft bolts marked **1** in the illustration are tightened. Torque the bolts, in sequence, in 5 equal steps, with the final step being 100–125 inch lbs. (11–14 Nm). Then retighten the bolts in the same sequence to 100–125 inch lbs. (11–14 Nm).
- New camshaft oil seal lubricated with engine oil. Tap the seal in evenly with a Seal Installer tool 49 F401 337A with a final protrusion of 0–0.02 inch (0–0.5mm).
- New blind cap by tapping it in
- Camshaft sprockets. Torque the bolts to 91–103 ft. lbs. (123–140 Nm).

14. Measure and adjust valve clearances.

15. Remove any sealant and gasket material from the cylinder head cover contact surfaces.

16. Apply silicone sealant to the cylinder head in the area adjacent to the front and rear camshaft caps.

17. Install or connect the following:

- New gasket on the cylinder head
- Cylinder head cover. Torque the cover bolts in 5–6 steps, in sequence, to 44–78 inch lbs. (5–9 Nm).
- Distributor using a new O-ring
- Ignition coils
- Intake manifold
- Spacer using a new O-ring. Torque the bolt to 14–18 ft. lbs. (19–25 Nm).
- Timing belt and timing belt cover
- Negative battery cable

18. Start the engine, check for leaks and repair if necessary.

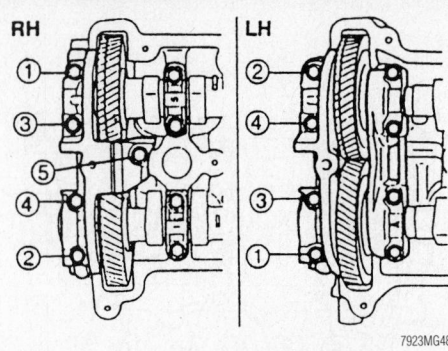

Front camshaft cap bolt loosening sequence—626 with 2.5L engines

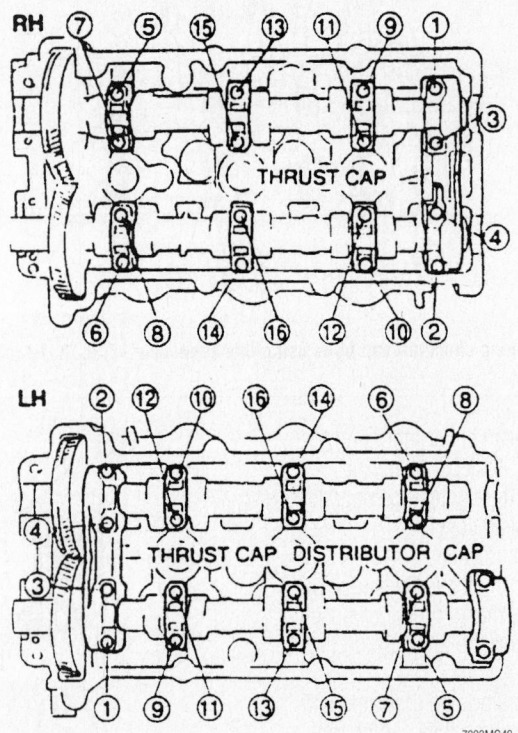

Camshaft cap bolt loosening sequence—626 with 2.5L (KL) engines

When installing the camshafts, ensure that the marks on the cam gears are aligned—2.3L and 2.5L engines

2.5L (KL) Engine

626

1. Before servicing the vehicle, refer to the precautions in the beginning of this section.
2. Drain the cooling system.
3. Relieve the fuel system pressure.
4. Drain the cooling system.
5. Remove or disconnect the following:
   - Negative battery cable
   - Timing belt
   - Front exhaust pipe
   - Air cleaner assembly
   - Intake manifold
   - Fuel hose
   - Cylinder head cover
   - Camshaft sprocket bolt by holding the camshaft with a wrench on the hexagon cast into the camshaft
   - Upper radiator hose
   - Number 3 engine mount bracket
   - Seal plate
   - Water outlet
6. Turn the camshaft, using a wrench on the cast hexagon, until the camshaft knock pin is aligned with the cylinder head marks.

➡**Do not remove the camshaft caps when the camshaft lobe is pressing on a lifter, as the thrust journal support may become damaged.**

7. Loosen the front camshaft cap bolts in 5–6 steps, in the proper sequence. Bolt **A** is only on the right cylinder head. Remove the front camshaft cap.
8. Mark the position of the camshaft caps so they can be reinstalled in their original locations. Loosen the remaining camshaft cap bolts in 5–6 steps, in the proper sequence, then remove the caps.
9. Remove or disconnect the following:
   - Camshafts
   - Lifters and adjustment shims, if necessary

➡**Identify and mark each lifter as it is removed so it can be reinstalled in the same position.**

To install:

10. Install or connect the following:
    - Lifters in their original positions lubricated with engine oil. Verify that they move smoothly in their bore.
    - Camshafts by lubricating the camshaft journals, lobes and gears with clean engine oil and aligning the timing marks

➡**The thrust plate positions for the right and left cylinder head camshafts are different.**

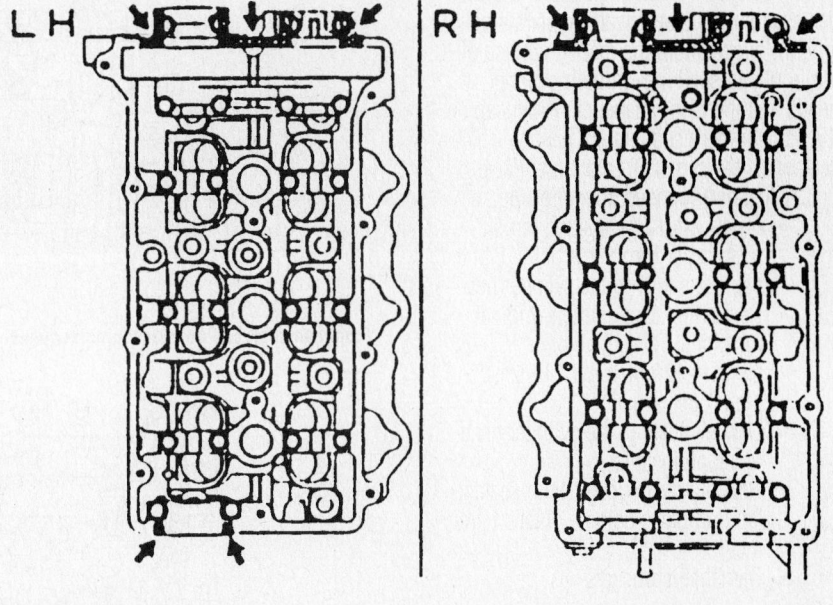

Put silicone sealant on the cylinder head at the positions shown—626 with 2.5L (KL) engines

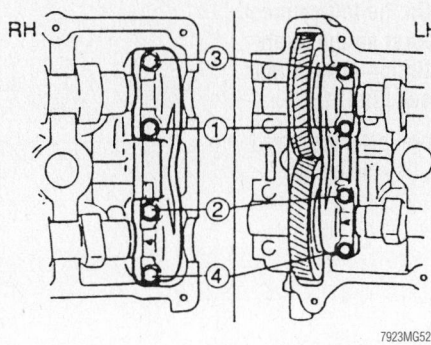

Front camshaft cap bolt tightening sequence—626 with 2.5L (KL) engines

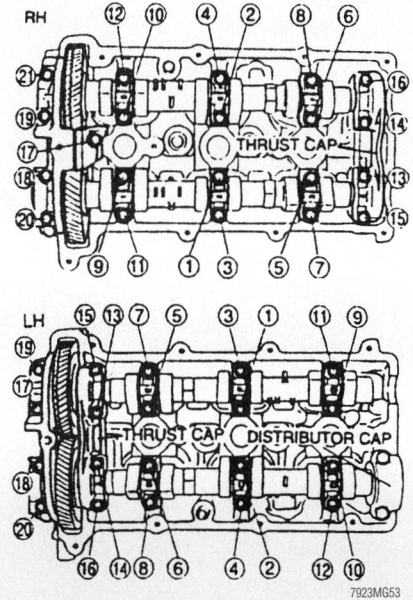

Camshaft cap bolt tightening sequence—626 with 2.5L (KL) engines

11. Be sure the camshaft cap and cylinder head surfaces are clean. Apply a small amount of sealant to the mating surface of the front camshaft cap on both cylinder heads and the rear exhaust camshaft cap on the left cylinder head. Do not get any sealant on the camshaft rotating surfaces.

12. Install or connect the following:
- Front camshaft caps and thrust plate caps. Torque the bolts until the cap seats fully to the cylinder head.
- Remaining caps in their original locations. Torque the bolts in 5–6 steps to 126 inch lbs. (14 Nm), in the proper sequence.
- New oil seals in the cylinder head using an installer
- New blind cap coated with sealant, tap it in place using a plastic hammer
- Camshaft sprockets

➡ **On the right cylinder head, install the sprocket so the R mark can be seen and the timing mark aligns with the camshaft knock pin. On the left cylinder head, install the sprocket so the L mark can be seen and the timing mark aligns with the camshaft knock pin.**

- Camshaft sprocket bolts lubricated with engine oil, by holding the camshaft with a wrench on the cast hexagon. Torque the bolt to 97 ft. lbs. (132 Nm).
- Water outlet
- Seal plate
- Number 3 engine mount bracket
- Upper radiator hose
- Camshaft pulley. Hold the camshaft pulley with a back–up wrench while tightening the pulley bolt to 90–97 ft. lbs. (123–132 Nm).
- Cylinder head cover
- Fuel hose
- Intake manifold
- Air cleaner assembly
- Front exhaust pipe
- Timing belt
- Negative battery cable

13. Fill and bleed the cooling system.
14. Change the oil and filter.
15. Run the engine and check for proper operation.

MILLENIA

1. Before servicing the vehicle, refer to the precautions in the beginning of this section.
2. Drain the cooling system.
3. Relieve the fuel system pressure.
4. Drain the cooling system.
5. Remove or disconnect the following:

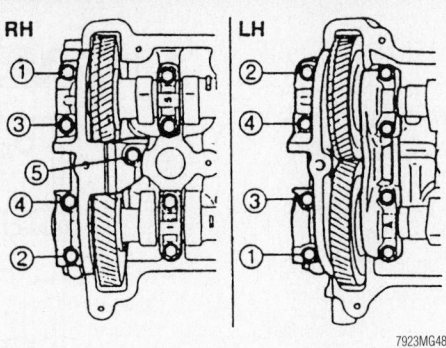

Front camshaft cap bolt loosening sequence—2.5L engines

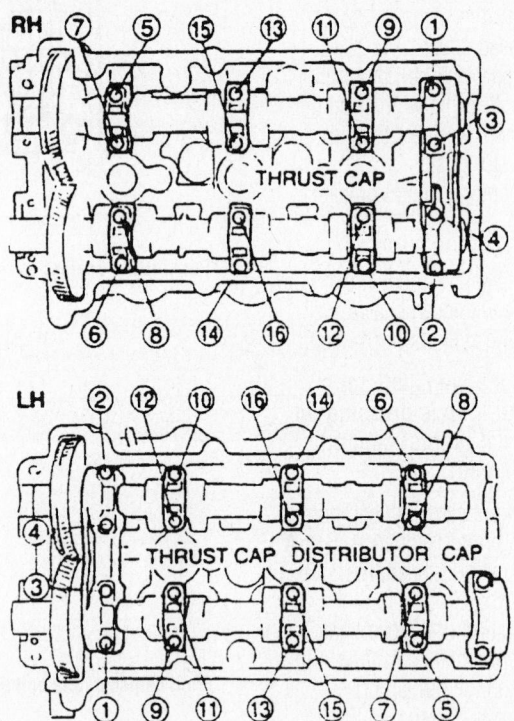

Camshaft cap bolt loosening sequence—2.5L (KL) engines

When installing the camshafts, ensure that the marks on the cam gears are aligned—2.5L (KL) engines

- Negative battery cable
- Timing belt
- Air cleaner assembly
- Spark plug wires
- Distributor
- Intake manifold
- Upper radiator hose
- Water outlet
- Number 3 engine mount bracket
- Seal plate
- Front exhaust pipe
- Alternator stay
- Cylinder head cover
- Camshaft pulley. Hold the camshaft pulley with a back-up wrench while loosening the pulley bolt.

6. Turn the camshaft, using a wrench on the cast hexagon, until the camshaft knock pin is aligned with the cylinder head marks.

➡️**Do not remove the camshaft caps when the camshaft lobe is pressing on a lifter, as the thrust journal support may become damaged.**

7. Loosen the front camshaft cap bolts in 5–6 steps, in the proper sequence. Bolt **A** is only on the right cylinder head. Remove the front camshaft cap.

8. Mark the position of the camshaft caps so they can be reinstalled in their original locations. Loosen the remaining camshaft cap bolts in 5–6 steps, in the proper sequence, then remove the caps.

9. Remove or disconnect the following:
- Camshafts
- Lifters and adjustment shims, if necessary

➡️**Identify and mark each lifter as it is removed so it can be reinstalled in the same position.**

To install:

10. Install or connect the following:
- Lifters in their original positions lubricated with engine oil. Verify that they move smoothly in their bore.
- Camshafts by lubricating the camshaft journals, lobes and gears with clean engine oil and aligning the timing marks

➡️**The thrust plate positions for the right and left cylinder head camshafts are different.**

11. Be sure the camshaft cap and cylinder head surfaces are clean. Apply a small amount of sealant to the mating surface of the front camshaft cap on both cylinder heads and the rear exhaust camshaft cap on the left cylinder head. Do not get any sealant on the camshaft rotating surfaces.

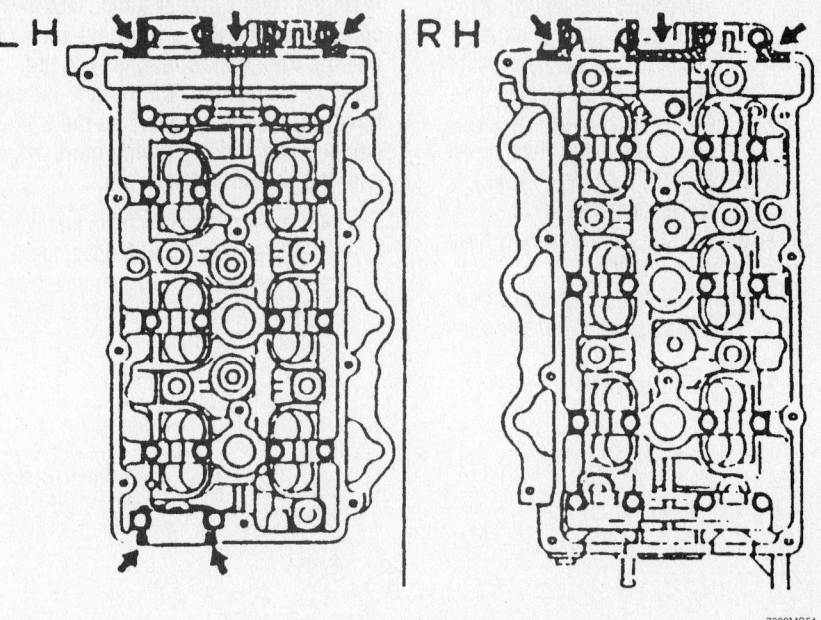

Put silicone sealant on the cylinder head at the positions shown—2.5L (KL) engines

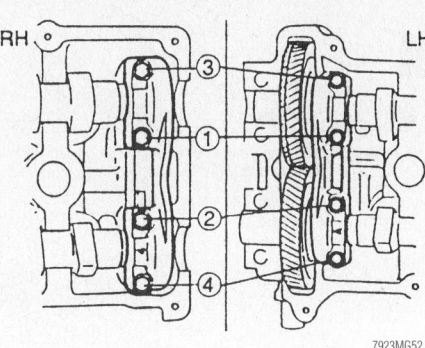

Front camshaft cap bolt tightening sequence—2.5L (KL) engines

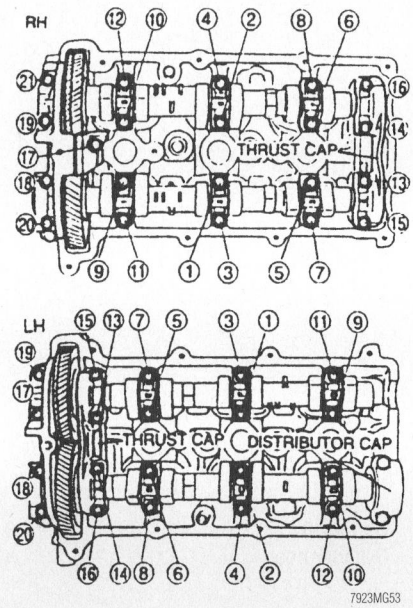

Camshaft cap bolt tightening sequence—2.5L (KL) engines

12. Install or connect the following:
- Front camshaft caps and thrust plate caps. Torque the bolts until the cap seats fully to the cylinder head.
- Remaining caps in their original locations. Torque the bolts in 5–6 steps to 126 inch lbs. (14 Nm), in the proper sequence.
- New oil seals in the cylinder head using an installer
- New blind cap coated with sealant, tap it in place using a plastic hammer
- Camshaft sprockets

➡ On the right cylinder head, install the sprocket so the R mark can be seen and the timing mark aligns with the camshaft knock pin. On the left cylinder head, install the sprocket so the L mark can be seen and the timing mark aligns with the camshaft knock pin.

- Camshaft sprocket bolts lubricated with engine oil, by holding the camshaft with a wrench on the cast hexagon. Torque the bolt to 97 ft. lbs. (132 Nm).
- Cylinder head cover
- Alternator stay

- Front exhaust pipe
- Seal plate
- Number 3 engine mount bracket
- Water outlet
- Upper radiator hose
- Intake manifold
- Distributor and spark plug wires
- Air cleaner assembly
- Timing belt
- Negative battery cable

13. Fill and bleed the cooling system.
14. Change the oil and filter.
15. Run the engine and check for proper operation.

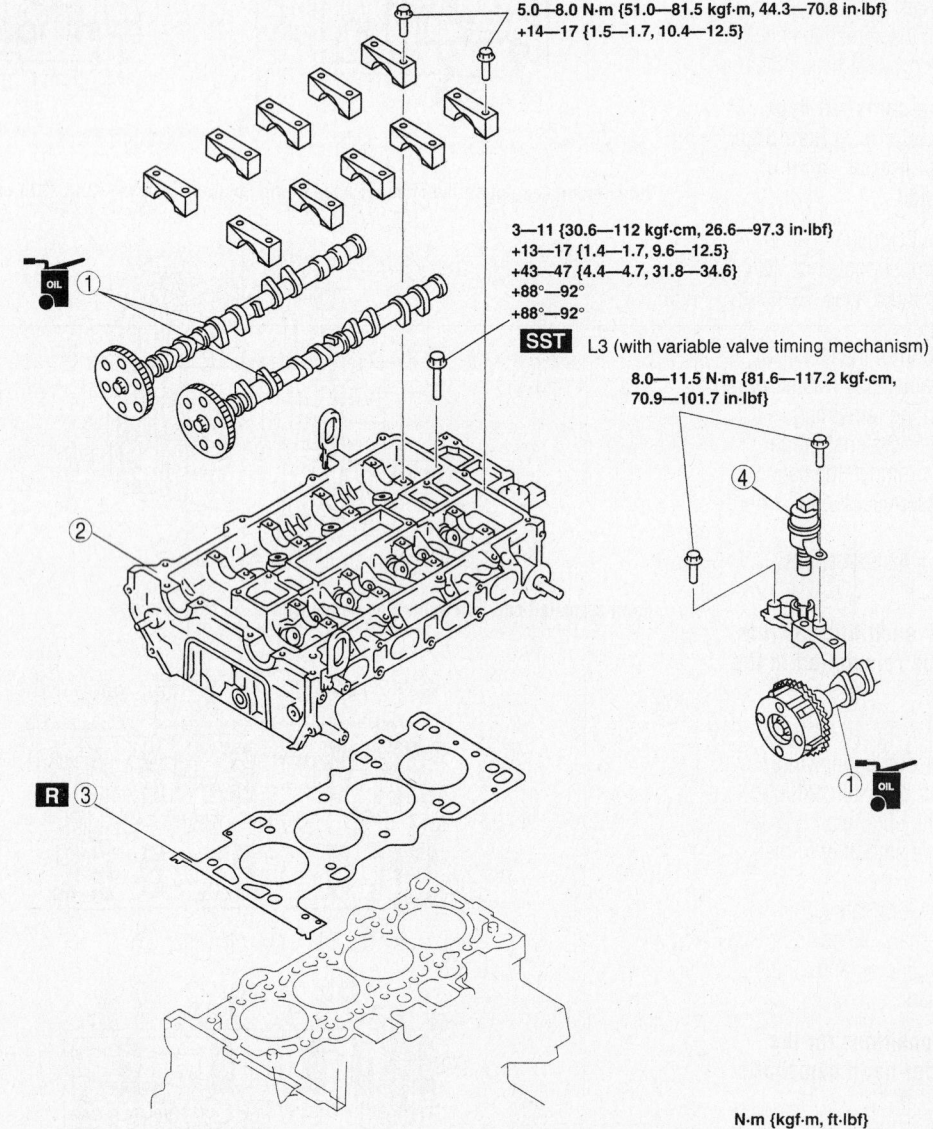

5.0—8.0 N·m {51.0—81.5 kgf·m, 44.3—70.8 in·lbf}
+14—17 {1.5—1.7, 10.4—12.5}

3—11 {30.6—112 kgf·cm, 26.6—97.3 in·lbf}
+13—17 {1.4—1.7, 9.6—12.5}
+43—47 {4.4—4.7, 31.8—34.6}
+88°—92°
+88°—92°

SST L3 (with variable valve timing mechanism)

8.0—11.5 N·m {81.6—117.2 kgf·cm, 70.9—101.7 in·lbf}

N·m {kgf·m, ft·lbf}

1 Camshaft
2 Cylinder head
3 Cylinder head gasket
4 Oil control valve (OCV) (L3 (with variable valve timing mechanism))

67162-MAZC-G34

Exploded view of the camshaft and cylinder head components—Mazda 3 models

MAZDA 3

1. Before servicing the vehicle, refer to the precautions in the beginning of this section.

2. Disconnect the negative battery cable.

3. Remove the timing chain.

4. Remove the intake manifold.

5. Disconnect the following components:
   - Warm Up–Three Way Converter (WU–TWC)
   - Upper radiator hose
   - Water and heater hose
   - Wiring harness

6. Support the engine using an engine jack and attachment as shown in the illustration.

7. Remove the camshafts.

➡**The cylinder head and camshaft caps are numbered and must be reassembled in their original locations. When removing a when removed, keep the caps with the cylinder head they were removed from, do not switch the caps.**

8. Loosen the camshaft cap bolts using 2–3 passes in the sequence illustrated.

9. Once all the bolts have been removed, remove the caps, keeping them in the original positions and remove the camshafts.

To install:

10. Set the cam position of the number 1 cylinder at Top Dead center (TDC) and install the camshaft.

11. Temporarily install the camshaft caps in their original positions.

12. Tighten the camshaft cap bolt in two steps:
   a. Step 1: 44–71 inch lbs. (5–8 Nm).
   b. Step 2: 10.4–12.5 ft. lbs. (14–17 Nm).

13. Remove the engine jack and attachment.

14. Connect the following components:
   - Wiring harness
   - Water and heater hose

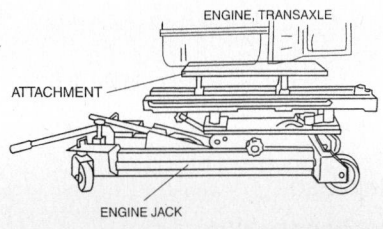

Support the engine using an engine jack and attachment as shown—Mazda 3 models

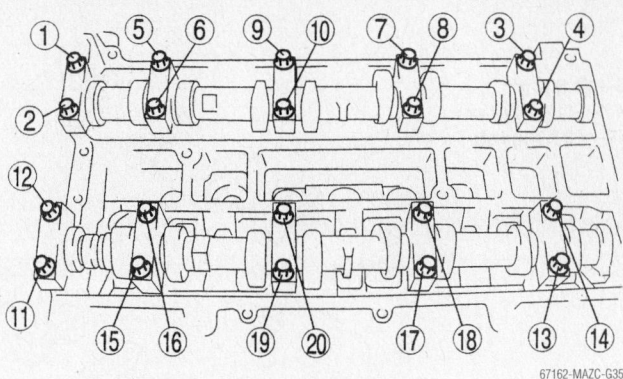

Camshaft cap bolt removal sequence—Mazda 3 models

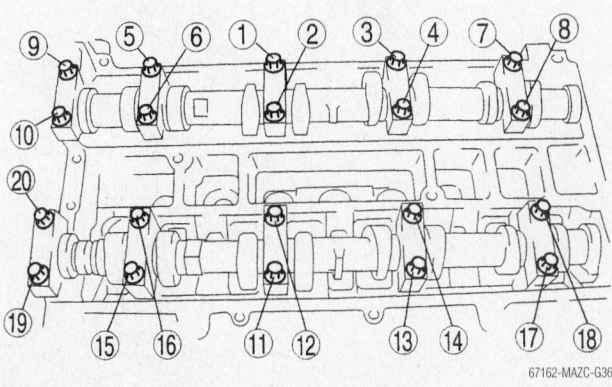

Camshaft cap bolt tightening sequence—Mazda 3 models

   - Upper radiator hose
   - WU–TWC converter

15. Install the intake manifold.

16. Install the timing chain.

17. Connect the negative battery cable.

18. Fill and bleed the cooling system.

19. Change the oil and filter.

20. Run the engine and check for proper operation.

Mazda 6

2.3L (L3) ENGINE

1. Before servicing the vehicle, refer to the precautions in the beginning of this section.

2. Disconnect the negative battery cable.

3. Remove the timing chain.

4. Remove the ignition coil and spark plug wires.

5. Disconnect the alternator and move it aside.

6. Remove the front exhaust pipe.

7. Remove the intake manifold.

8. Remove the upper radiator hose, bypass and heater hoses.

9. Support the engine using an engine jack and attachment as shown in the illustration.

10. Remove the oil control valve.

11. Remove the camshafts.

➡**The cylinder head and camshaft caps are numbered and must be reassembled in their original locations. When removing a when removed, keep the caps with the cylinder head they were removed from, do not switch the caps.**

12. Loosen the camshaft cap bolts using 2–3 passes in the sequence illustrated.

13. Once all the bolts have been removed, remove the caps, keeping them in the original positions and remove the camshafts.

To install:

14. Set the cam position of the number 1 cylinder at Top Dead center (TDC) and install the camshaft.

15. Temporarily install the camshaft caps in their original positions.

16. Tighten the camshaft cap bolt in two steps:
   a. Step 1: 44–71 inch lbs. (5–8 Nm).
   b. Step 2: 10.4–12.5 ft. lbs. (14–17 Nm).

17. Remove the engine jack and attachment.

18. Install the remaining components in the reverse order of removal.

19. Connect the negative battery cable.

20. Fill and bleed the cooling system.

21. Change the oil and filter.

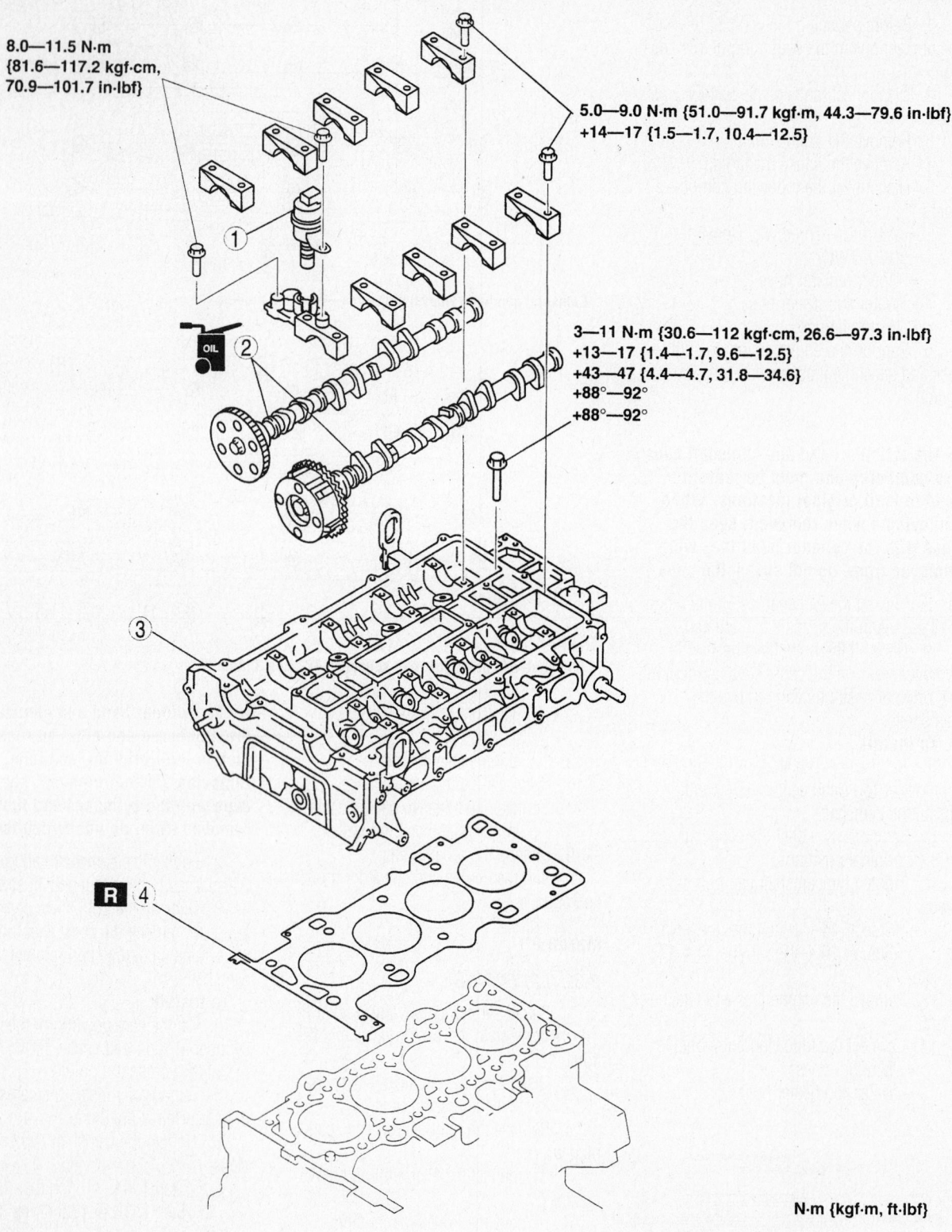

8.0—11.5 N·m
{81.6—117.2 kgf·cm,
70.9—101.7 in·lbf}

5.0—9.0 N·m {51.0—91.7 kgf·m, 44.3—79.6 in·lbf}
+14—17 {1.5—1.7, 10.4—12.5}

3—11 N·m {30.6—112 kgf·cm, 26.6—97.3 in·lbf}
+13—17 {1.4—1.7, 9.6—12.5}
+43—47 {4.4—4.7, 31.8—34.6}
+88°—92°
+88°—92°

N·m {kgf·m, ft·lbf}

| 1 | Oil control valve (OCV) | 3 | Cylinder head |
| 2 | Camshaft | 4 | Cylinder head gasket |

67162-MAZC-G168

Exploded view of the camshaft and cylinder head components—Mazda 6 models with the 2.3L (L3) engine

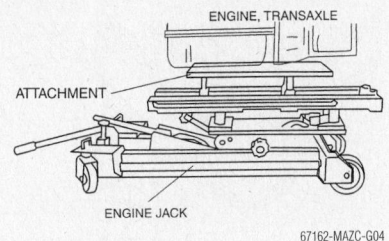

Support the engine using an engine jack and attachment as shown

22. Run the engine and check for proper operation.

3.0L (AJ) ENGINE

1. Before servicing the vehicle, refer to the precautions in the beginning of this section.

2. Properly relieve the fuel system pressure.

3. Drain the cooling system.

4. Disconnect the negative battery cable.

5. Remove the water pump.

6. Remove the timing chain.

7. Support the engine using an engine jack and attachment.

8. Remove the water bypass tube stud bolt and bolt and disconnect the tube from the cylinder head.

9. Remove the camshaft(s).

10. Remove the rocker arm(s).

11. Remove the camshafts.

➡If removing the right hand camshaft caps, remove thrust caps 1R and 5R first. Do not loosen any of the other cap bolts until the thrust caps are removed or you could damage the thrust caps. If removing the left hand camshaft caps, remove thrust caps 1L and 6L first. Do not loosen any of the other cap bolts until the thrust caps are removed or you could damage the thrust caps.

12. Loosen the camshaft caps using 7–8 passes in sequence after first removing the thrust caps to allow the camshafts to be slowly raised. Keep the caps in the order they were removed so they are re-installed in their original positions.

To install:

13. Install the camshafts in their original positions.

14. Install the left hand camshaft as follows:

　a. Place the crankshaft keyway at the 11 O'clock position by rotating the crankshaft clockwise.

　b. Install the camshaft and the caps in their original positions and hand tighten the bolts.

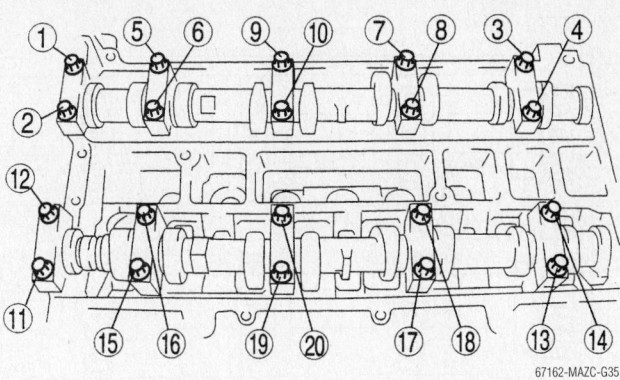

Camshaft cap bolt removal sequence—Mazda 6 models with the 2.3L (L3) engine

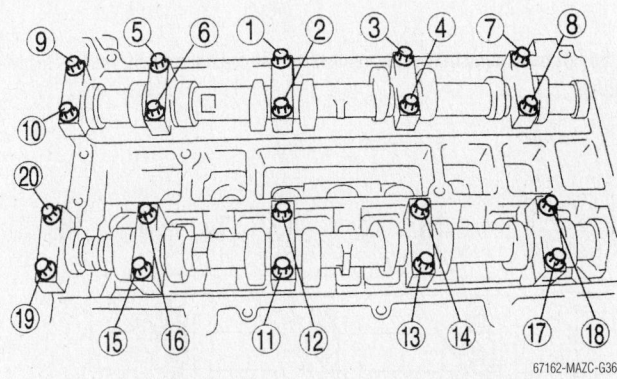

Camshaft cap bolt tightening sequence—Mazda 6 models with the 2.3L (L3) engine

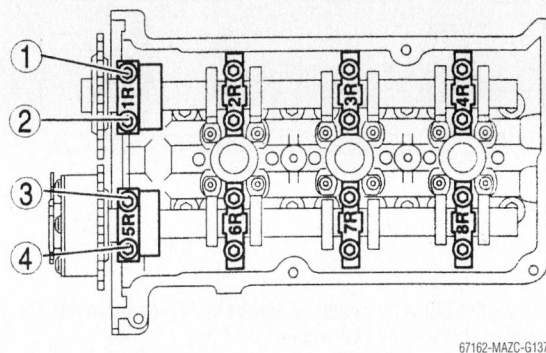

If removing the right hand camshaft caps, remove thrust caps 1R and 5R first—Mazda 6 with the 3.0L (AJ) engine

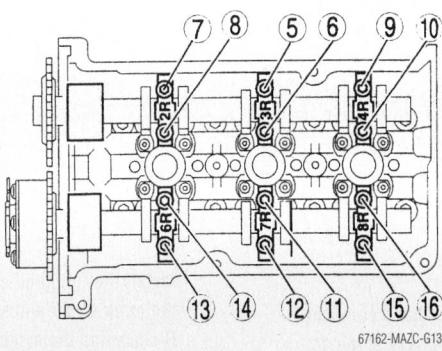

Loosen the right hand camshaft caps using 7–8 passes in this sequence after first removing the thrust caps—Mazda 6 with the 3.0L (AJ) engine

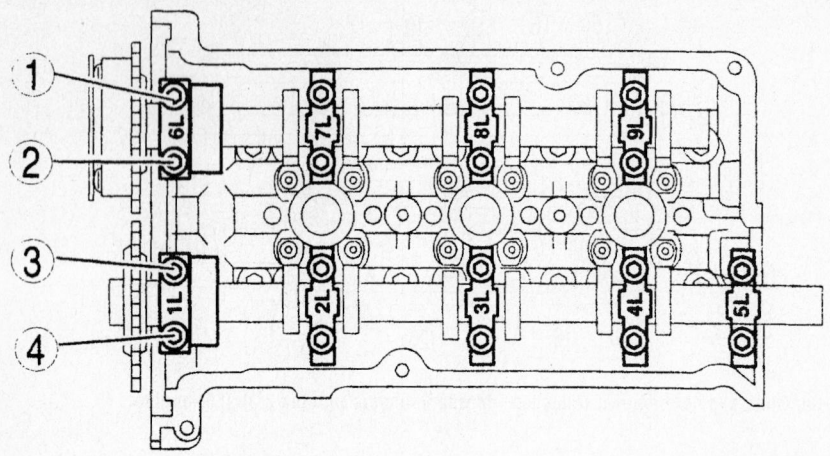

67162-MAZC-G139

If removing the left hand camshaft caps, remove thrust caps 1L and 6L first—Mazda 6 with the 3.0L (AJ) engine

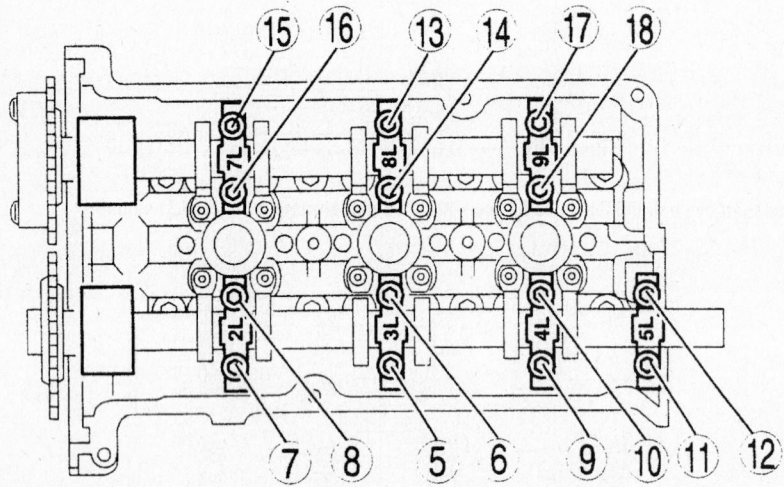

67162-MAZC-G140

Loosen the left hand camshaft caps using 7–8 passes in this sequence after first removing the thrust caps—Mazda 6 with the 3.0L (AJ) engine

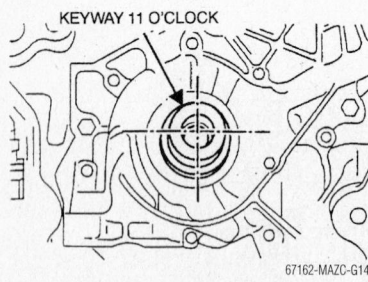

67162-MAZC-G143

When installing the left hand camshaft place the crankshaft keyway at the 11 O'clock position—Mazda 6 with the 3.0L (AJ) engine

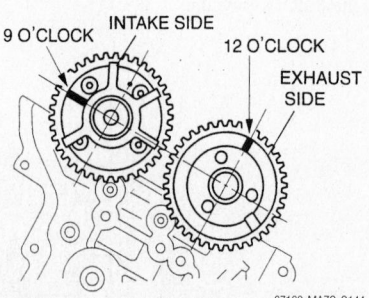

67162-MAZC-G144

Install the left hand camshaft and position the mark on the intake camshaft to the 9 O'clock and the exhaust camshaft to the 12 O'clock position—Mazda 6 with the 3.0L (AJ) engine

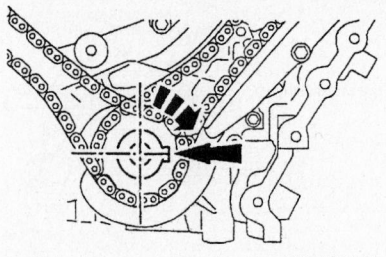

67162-MAZC-G145

When installing the right hand camshaft place the crankshaft keyway at the 3 O'clock position—Mazda 6 with the 3.0L (AJ) engine

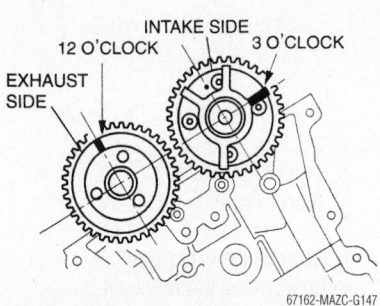

67162-MAZC-G147

Install the right hand camshaft and position the mark on the intake camshaft to the 3 O'clock and the exhaust camshaft to the 12 O'clock position—Mazda 6 with the 3.0L (AJ) engine

    c. Position the mark on the intake camshaft to the 9 O'clock position as illustrated.

    d. Position the mark on the exhaust camshaft to the 12 O'clock position.

    e. Install the timing chain.

➡**Tighten the camshaft journal thrust caps last to avoid damaging the thrust caps.**

    f. Tighten the camshaft caps using several passes to 71–106 inch lbs. (8–12 Nm). After adjust the camshaft endplay using the thrust caps 1L and 6L, tighten the remaining caps.

15. Install the right hand camshaft as follows:

    a. Install the camshaft and the caps in their original positions and hand tighten the bolts.

    b. Place the crankshaft keyway at the 3 O'clock position.

    c. Position the mark on the exhaust camshaft to the 12 O'clock position by rotating the crankshaft clockwise.

    d. Position the mark on the intake camshaft to the 3 O'clock position as illustrated.

    e. Install the timing chain.

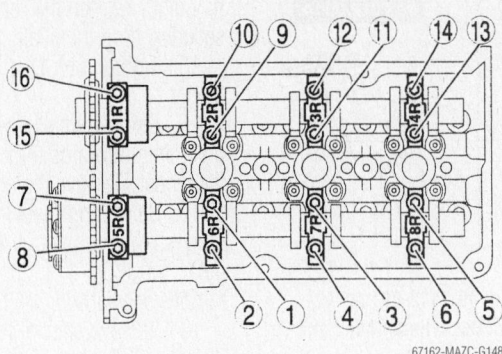

Right side camshaft cap torque sequence—Mazda 6 with the 3.0L (AJ) engine

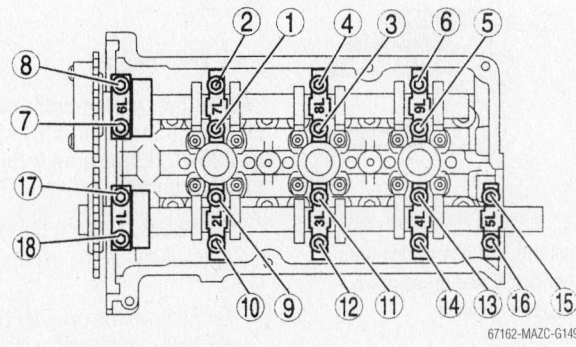

Left side camshaft cap torque sequence—Mazda 6 with the 3.0L (AJ) engine

➡**Tighten the camshaft journal thrust caps last to avoid damaging the thrust caps.**

    f. Tighten the camshaft caps using several passes to 71–106 inch lbs. (8–12 Nm). After adjust the camshaft endplay using the thrust caps 1R and 5R, tighten the remaining caps.

16. Install the water bypass tube.
17. Remove the engine jack and attachment.
18. Install the water pump.
19. Fill the cooling system.
20. Fill and bleed the cooling system.
21. Change the oil and filter.
22. Run the engine and check for proper operation.

Valve Lash

ADJUSTMENT

    These engines use solid cam followers with a removable adjustment shim. The valve lash clearance is measured with the original shim installed and checked against the specification. If adjustment is necessary, the original shim is removed, and a thicker or thinner shim is installed to obtain the proper clearance. Special tools are required

in order to adjust the shim without removing the camshaft.

1.8L (BP) Engine

➡**With the engine cold, standard valve clearance is 0.08–0.009 inch (0.18–0.24mm) on intake side and 0.012–0.013 inch (0.28–0.34mm) on the exhaust side.**

1. Before servicing the vehicle, refer to the precautions in the beginning of this section.
2. Remove the cylinder head cover.
3. Measure the valve clearance by turning the crankshaft clockwise until the No. 1 piston is at Top Dead Center (TDC).
4. Measure the valve clearance at **A**. If the clearance exceeds specifications, replace the adjustment shim.
5. Turn the crankshaft clockwise until the cam on the camshaft requiring the adjustment is positioned straight up.
6. Turn the crankshaft clockwise 360 degrees until the No. 4 piston is at TDC. Measure the valve clearance at **B**. If the clearance exceeds specifications, replace the adjustment shim.
7. Repeat this procedure for all the camshafts.
8. Turn the crankshaft clockwise until

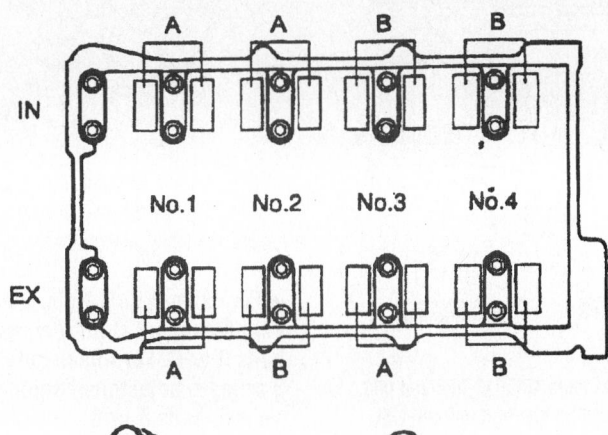

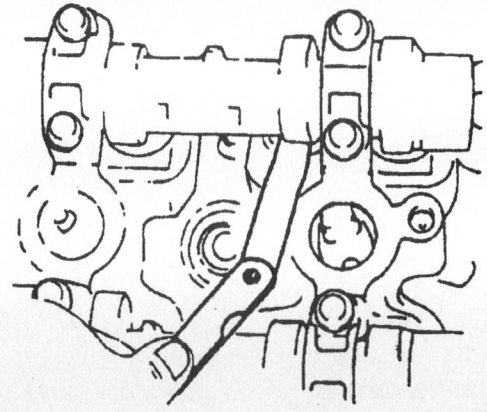

Valve clearance checking positions—1.8L (BP) engines

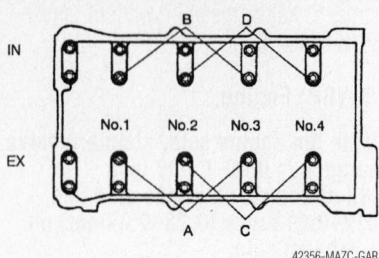

Cam bearing cap bolt removal positions—1.8L (BP) engines

the cam on the camshaft requiring the adjustment is positioned straight up.

9. Remove the camshaft cap bolts one pair at a time as follows:

   a. For exhaust side No. 1, 2 and 3 cylinder adjustment shim removal use **A**.

   b. For intake side No. 1, 2 and 3 cylinder adjustment shim removal use **B**.

   c. For exhaust side No. 2, 3 and 4 cylinder adjustment shim removal use **C**.

   d. For intake side No. 2, 3 and 4 cylinder adjustment shim removal use **D**.

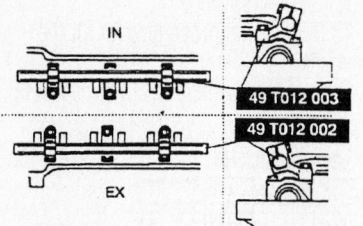

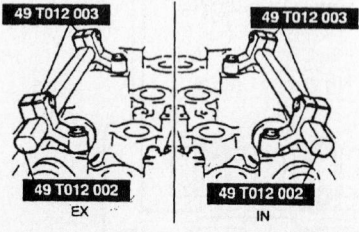

Install special tools 49-T012-002 and 003, using the camshaft cap bolt holes—1.8L (BP) engines

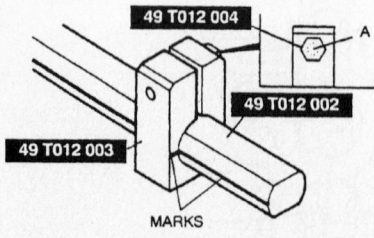

Align the mark on the 49-T012-002 (shaft) with the mark on the 49-T012-003 (clamp—1.8L (BP) engines

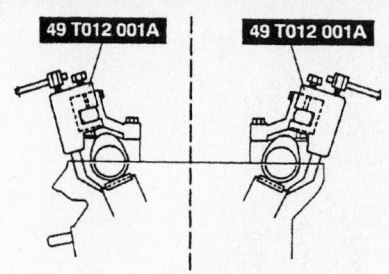

Position special tool 49-T012-001 A toward the center of the cylinder head and mount it on the shaft where the adjustment shim needs replacement—1.8L (BP) engines

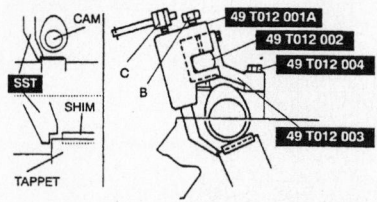

Tighten the mounting bolt B securing it on the shaft. Tighten bolt C and press down the tappet—1.8L (BP) engines

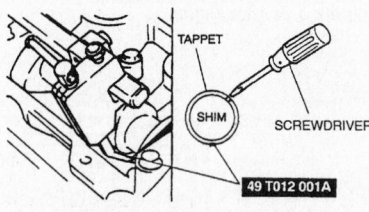

Using a small prytool, pry the adjustment shim upwards through the notch on the tappet—1.8L (BP) engines

➤ **For exhaust side No's. 2 and 3, cylinder adjustment shim removal; remove bolts A or C. For intake side No's 2 and 3 cylinder adjustment shim removal, remove bolts B or B.**

10. Install special tools 49-T012-002 and 003, using the camshaft cap bolt holes. Tighten the bolts to 100–125 inch lbs. (11–14 Nm).

11. Align the mark on the 49-T012-002 (shaft) with the mark on the 49-T012-003 (clamp). Tighten special tool 49-T012-004 (bolt) to secure the shaft.

12. Position special tool 49-T012-001A toward the center of the cylinder head and mount it on the shaft where the adjustment shim needs replacement.

13. Position the notch of the tappet to allow a small prytool to be inserted.

14. Set the special tool on the tappet by

its notch. Tighten the mounting bolt **B** securing it on the shaft.

15. Tighten bolt **C**, and press down the tappet.

16. Using a small prytool, pry the adjustment shim upwards through the notch on the tappet. Remove the shim with a magnet.

17. Select and install the proper adjustment shim. Loosen bolt **C** to allow the tappet to move up, and loosen bolt **B** to remove special tool 49-T012-001A.

18. Remove special tools 49-T012-002, 003 and 004, and torque the camshaft cap bolts to 100–125 inch lbs. (11–14 Nm).

19. Repeat the procedure for all necessary adjustment shims. Check the valve clearance.

2.0L (FS) Engine

➤ **With the engine cold, standard valve clearance is 0.089–0.0116 inch (0.225–0.295mm) on intake and exhaust sides.**

1. Before servicing the vehicle, refer to the precautions in the beginning of this section.

2. Remove the cylinder head cover.

3. Measure the valve clearance by turning the crankshaft clockwise until the No. 1 piston is at Top Dead Center (TDC).

4. Measure the valve clearance at **A**. If the clearance exceeds specifications, replace the adjustment shim.

5. Turn the crankshaft clockwise 360 degrees until the No. 4 piston is at TDC. Measure the valve clearance at **B**. If the clearance exceeds specifications, replace the adjustment shim.

6. Repeat this procedure for all the camshafts.

7. Turn the crankshaft clockwise until the cam on the camshaft requiring the adjustment is positioned straight up.

8. Remove the camshaft cap bolts as follows:

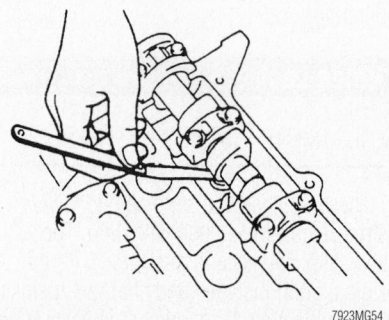

Ensure that the cam lobe faces away from the follower when checking the valve clearance

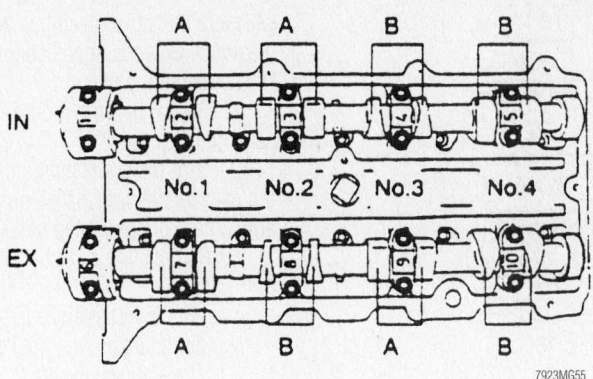

Valve clearance checking positions—2.0L (FS) engines

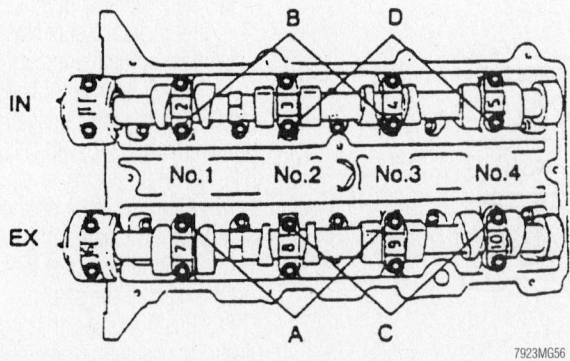

Cam bearing cap bolt removal positions—2.0L (FS) engines

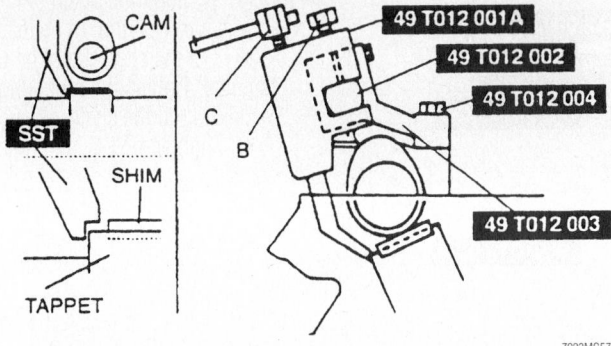

Mount the tappet depressor tool onto the shaft above the tappet that needs adjustment—2.0L (FS) engines

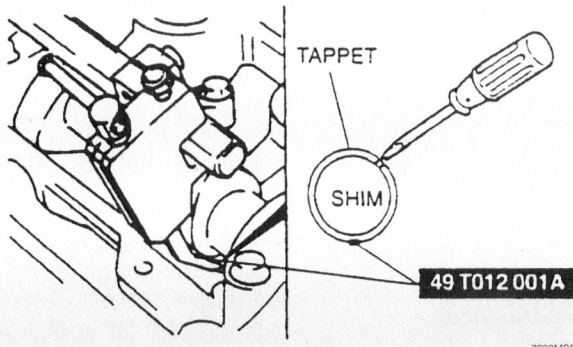

With the tappet depressed, use a small prytool to remove the adjustment shim—2.0L (FS) engines

a. For exhaust side No. 1, 2 and 3 cylinder adjustment shim removal use **A**.

b. For intake side No. 1, 2 and 3 cylinder adjustment shim removal use **B**.

c. For exhaust side No. 2, 3 and 4 cylinder adjustment shim removal use **C**.

d. For intake side No. 2, 3 and 4 cylinder adjustment shim removal use **D**.

9. Install special tools 49-T012-002 and 003, using the camshaft cap bolt holes. Tighten the bolts to 100–125 inch lbs. (11–14 Nm).

10. Align the mark on the 49-T012-002 (shaft) with the mark on the 49-T012-003 (clamp). Tighten special tool 49-T012-004 (bolt) to secure the shaft.

11. Position special tool 49-T012-001A toward the center of the cylinder head and mount it on the shaft where the adjustment shim needs replacement.

12. Position the notch of the tappet to allow a small prytool to be inserted.

13. Set the special tool on the tappet by its notch. Tighten the mounting bolt **B** securing it on the shaft.

14. Tighten bolt **C**, and press down the tappet.

15. Using a small prytool, pry the adjustment shim upwards through the notch on the tappet. Remove the shim with a magnet.

16. Select and install the proper adjustment shim. Loosen bolt **C** to allow the tappet to move up, and loosen bolt **B** to remove special tool 49-T012-001A.

17. Remove special tools 49-T012-002, 003 and 004, and torque the camshaft cap bolts to 100–125 inch lbs. (11–14 Nm).

18. Repeat the procedure for all necessary adjustment shims. Check the valve clearance.

1.6L (ZM) and 1.8 (FP) Engines

➡With the engine cold, standard valve clearance is 0.010–0.012 inch (0.25–0.31mm) on 1.6L (ZM) engines or 0.010–0.001 inch (0.225–0.295mm) on 1.8L (FP) engines on both the intake and exhaust sides.

1. Before servicing the vehicle, refer to the precautions in the beginning of this section.

2. Remove the cylinder head cover.

3. Measure the valve clearance by turning the crankshaft clockwise until the No. 1 piston is at Top Dead Center (TDC).

4. Measure the valve clearance at **A**. If the clearance exceeds specifications, replace the adjustment shim.

5. Turn the crankshaft clockwise 360 degrees until the No. 4 piston is at TDC.

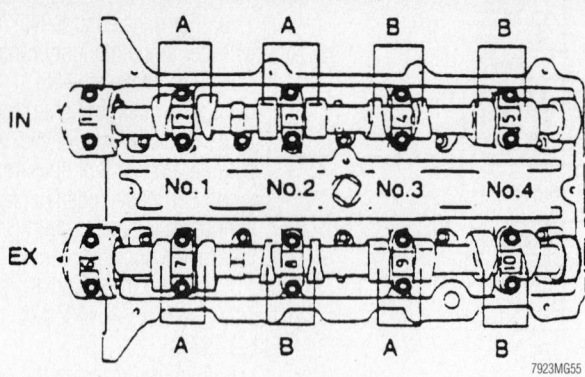

Valve clearance checking positions—4-cylinder engines

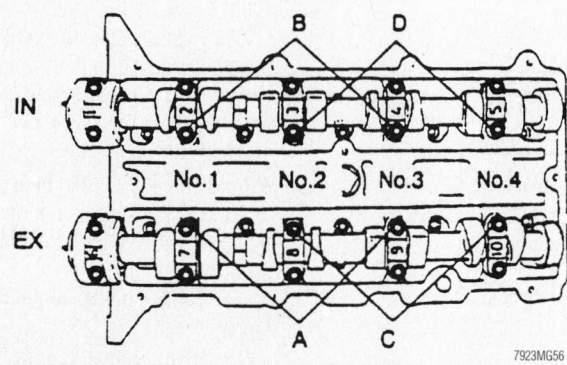

Cam bearing cap bolt removal positions—4-cylinder engine

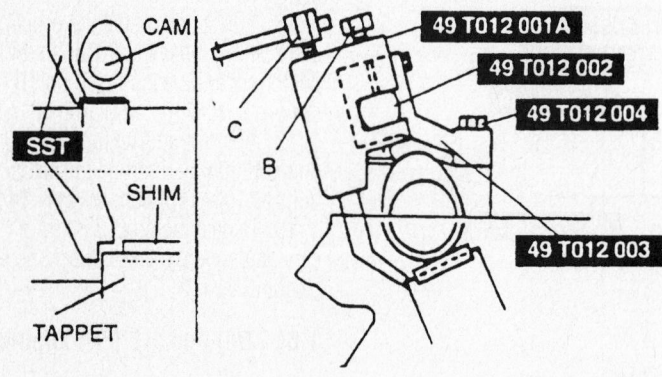

Mount the tappet depressor tool onto the shaft above the tappet that needs adjustment

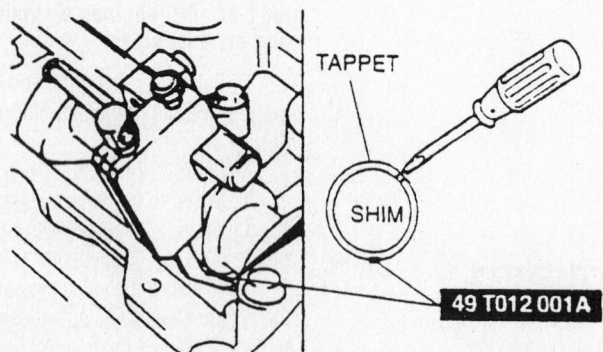

With the tappet depressed, use a small prytool to remove the adjustment shim

Measure the valve clearance at **B**. If the clearance exceeds specifications, replace the adjustment shim.

6. Repeat this procedure for all the camshafts.

7. Turn the crankshaft clockwise until the cam on the camshaft requiring the adjustment is positioned straight up.

8. Remove the camshaft cap bolts as follows:

   a. For exhaust side No. 1, 2 and 3 cylinder adjustment shim removal use **A**.

   b. For intake side No. 1, 2 and 3 cylinder adjustment shim removal use **B**.

   c. For exhaust side No. 2, 3 and 4 cylinder adjustment shim removal use **C**.

   d. For intake side No. 2, 3 and 4 cylinder adjustment shim removal use **D**.

9. Install special tools 49-T012-002 and 003, using the camshaft cap bolt holes. Tighten the bolts to 100–125 inch lbs. (11–14 Nm).

10. Align the mark on the 49-T012-002 (shaft) with the mark on the 49-T012-003 (clamp). Tighten special tool 49-T012-004 (bolt) to secure the shaft.

11. Position special tool 49-T012-001A toward the center of the cylinder head and mount it on the shaft where the adjustment shim needs replacement.

12. Position the notch of the tappet to allow a small prytool to be inserted.

13. Set the special tool on the tappet by its notch. Tighten the mounting bolt **B** securing it on the shaft.

14. Tighten bolt **C**, and press down the tappet.

15. Using a small prytool, pry the adjustment shim upwards through the notch on the tappet. Remove the shim with a magnet.

16. Select and install the proper adjustment shim. Loosen bolt **C** to allow the tappet to move up, and loosen bolt **B** to remove special tool 49-T012-001A.

17. Remove special tools 49-T012-002, 003 and 004, and torque the camshaft cap bolts to 100–125 inch lbs. (11–14 Nm).

18. Repeat the procedure for all necessary adjustment shims. Check the valve clearance.

2.3L (KJ) and 2.5L (KL) Engines

This procedure for the 2.5L (KL) engine valve adjustment is for 626 models only. The Millenia model equipped with the 2.5L engine is equipped with hydraulic lash adjusters and no adjustment is possible or necessary.

1. Before servicing the vehicle, refer to the precautions in the beginning of this section.

➡With the engine cold, standard valve clearance on the 2.5L (KL) engine is 0.0097–0.0124 inch (0.245–0.311mm) on intake side and 0.0105–0.0131 inch (0.265–0.335mm) on the exhaust side. On the 2.3L (KJ) the measurement is 0.011–0.012 inch (0.27–0.33mm) on both the intake and exhaust sides.

2. Measure the valve clearance by turning the crankshaft clockwise until the No. 1 piston is at Top Dead Center (TDC).

3. Measure the valve clearance at **A**. Turn the crankshaft clockwise 240 degrees until the No. 3 piston is at TDC. Measure the valve clearance at **B**. Turn the crankshaft clockwise 240 degrees until the No. 5 piston is at TDC. Measure the valve clearance at **C**.

➡If the valve clearance exceeds the standard, replace the adjustment shim.

4. Turn the crankshaft clockwise until the cam, on the camshaft requiring the adjustment shim replacement, is positioned straight up.

5. Camshaft cap bolts as follows:
   a. For right-hand exhaust side shim removal use **A**.
   b. For right-hand intake side shim removal use **B**.
   c. For left-hand intake side shim removal use **C**.
   d. For left-hand exhaust side shim removal use **D**.

6. Install special tools 49-T012-002 and 003, using the camshaft cap bolt holes.

7. Align the mark on the 49-T012-002 (shaft) with the mark on the 49-T012-003 (clamp).

8. Position special tool 49-T012-001 toward the center of the cylinder head and mount it on the shaft where the adjustment shim needs replacement.

9. Position the notch of the tappet to allow a small prytool to be inserted.

10. Set the special tool on the tappet by its notch. Tighten the mounting bolt **B** securing it on the shaft.

11. Tighten bolt **C** and press down the tappet.

12. Using a small prytool, pry the adjustment shim upwards through the notch on the tappet. Remove the shim with a magnet.

13. Select and install the proper adjustment shim. Loosen bolt **C** to allow the tappet to move up and loosen bolt B to remove special tool 49-T012-001.

14. Remove special tools 49-T012-002, 003 and 004 and tighten the camshaft cap bolts to 100–125 inch lbs. (11–14 Nm).

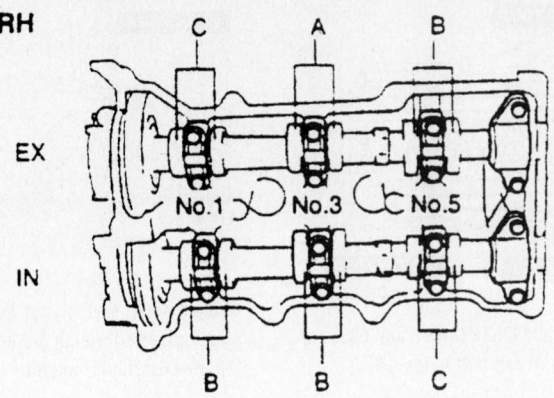

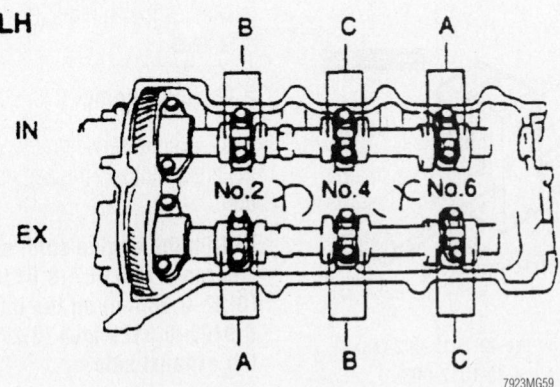

Valve clearance checking positions—6-cylinder engine

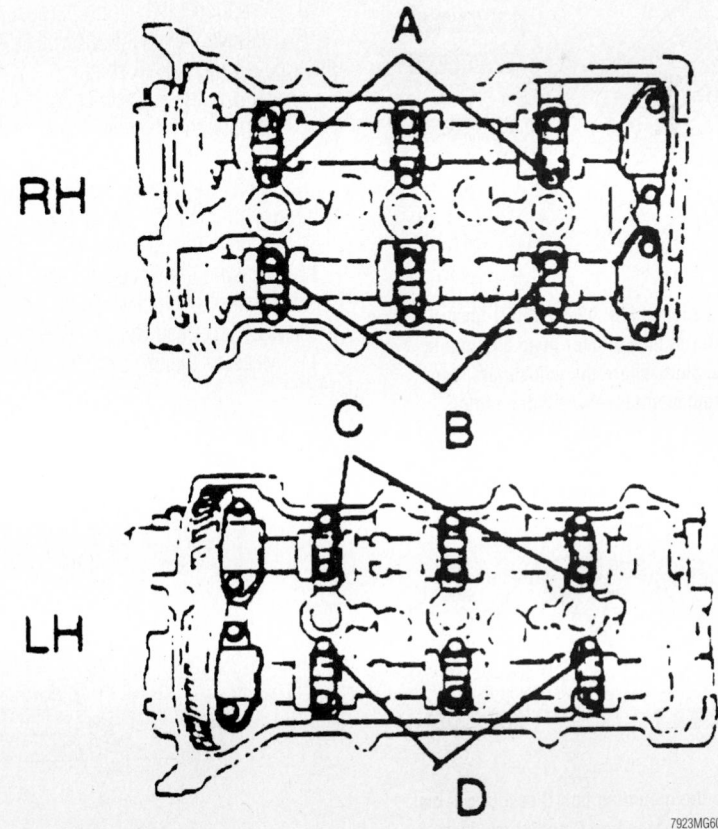

Camshaft cap bolt removal positions—6-cylinder engine—refer to text

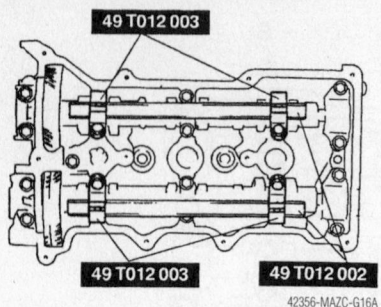

42356-MAZC-G16A

Install special tools 49-T012-002 and 003, using the camshaft cap bolt holes—6-cylinder engine

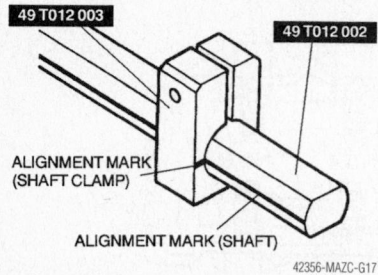

42356-MAZC-G17

Align the mark on the 49-T012-002 (shaft) with the mark on the 49-T012-003 (clamp)—6-cylinder engine

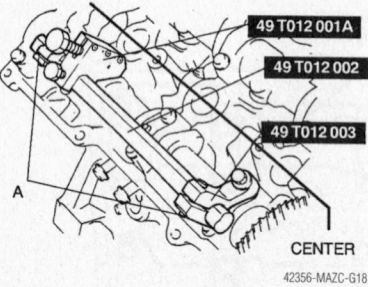

42356-MAZC-G18

Position special tool 49-T012-001 toward the center of the cylinder head and mount it on the shaft where the adjustment shim needs replacement—6-cylinder engine

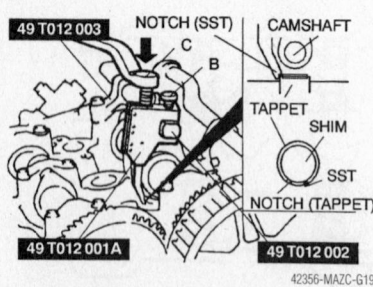

42356-MAZC-G19

Tighten the mounting bolt B securing it on the shaft. Tighten bolt C and press down the tappet—6-cylinder engine

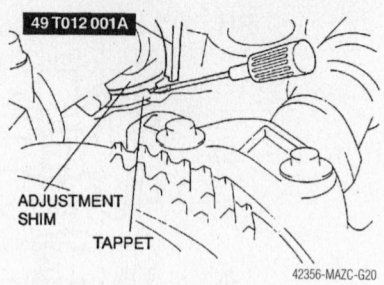

42356-MAZC-G20

Using a small prytool, pry the adjustment shim upwards through the notch on the tappet—6-cylinder engine

15. Repeat the procedure for all necessary adjustment shims. Check the valve clearance.

2.0L (LF) Engine

1. Before servicing the vehicle, refer to the precautions in the beginning of this section.

➡ With the engine cold, standard valve clearance is 0.087–0.0110 inch (0.22–0.28mm) on the intake side and 0.0107–0.0129 inch (0.27–0.33mm) on the exhaust side.

2. Remove the plug hole plate.
3. Remove the battery cover and disconnect the negative battery cable.
4. Disconnect the wiring harness.
5. Remove the right front wheel, engine under cover and splash shield.
6. Remove the ignition coils.
7. Remove the Positive Crankcase Ventilation (PCV) hose, if equipped.
8. Disconnect the oil control valve connector, if equipped.
9. Remove the cylinder head cover.
10. Measure the valve clearance by turning the crankshaft clockwise until the No. 1 piston is at Top Dead Center (TDC).
11. Measure the valve clearance at **A**. If

the clearance exceeds specifications, replace the adjustment shim.

12. Turn the crankshaft clockwise 360 degrees until the No. 4 piston is at TDC. Measure the valve clearance at **B**. If the clearance exceeds specifications, replace the adjustment shim.

13. Repeat this procedure for all the camshafts.

14. Remove the right side halfshaft from the joint shaft on Mazda 6 models.

15. Remove the engine front cover lower and upper blind plugs and the cylinder block lower blind plug.

16. Install tool 49 JE01 061.

17. Turn the crankshaft clockwise until the crankshaft is in the number 1 cylinder Top Dead Center position (the balance weight should be attached to tool 49 JE01 061).

18. Loosen the timing chain as follows:
   a. Unlock the chain tensioner ratchet using a screwdriver.
   b. Turn the exhaust camshaft clockwise using a wrench on the cast hexagon portion of the camshaft and loosen the timing chain.
   c. Place a M6 x 1.0 length bolt: 0.99–1.37 inch (25–35mm) at the engine front cover upper blind plug and secure the timing chain guide at the location where the chain tension is released.

19. Hold the exhaust camshaft using a wrench on the cast hexagon portion of the camshaft and remove the exhaust camshaft sprocket.

➡ The cylinder head and camshaft caps are numbered and must be reassembled in their original locations. When removing a when removed, keep the caps with the cylinder head they were removed from, do not switch the caps.

20. Loosen the camshaft cap bolts using 2–3 passes in the sequence illustrated.

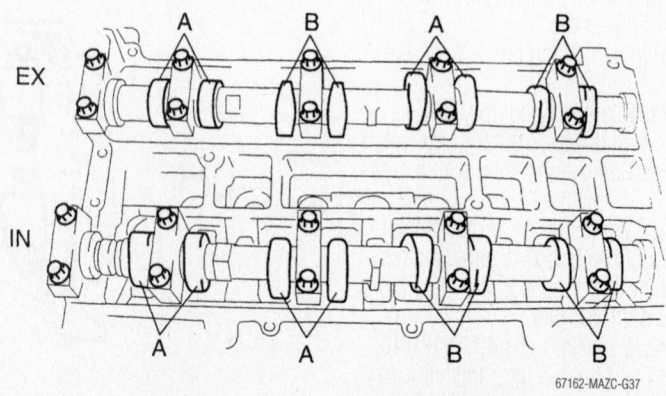

67162-MAZC-G37

Valve clearance checking positions—2.0L (LF) and 2.3L (L3) engines

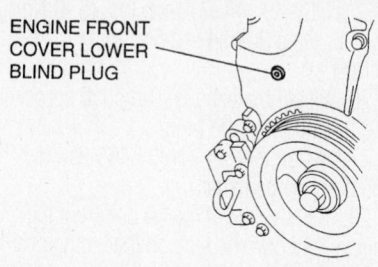

67162-MAZC-G38

Remove the engine front cover lower blind plug—2.0L (LF) and 2.3L (L3) engines

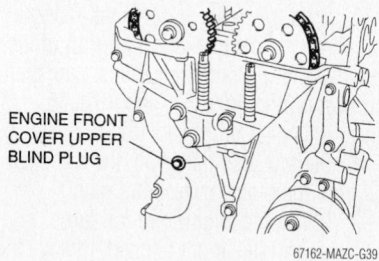

67162-MAZC-G39

Remove the engine front cover upper blind plug—2.0L (LF) and 2.3L (L3) engines

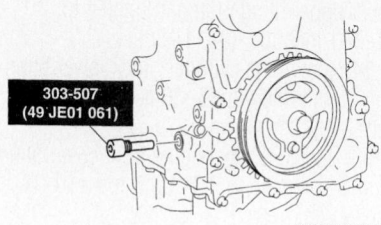

67162-MAZC-G27

Install tool 49 JE01 061—2.0L (LF) and 2.3L (L3) engines

21. Once all the bolts have been removed, remove the caps, keeping them in the original positions and remove the camshafts.

22. Remove the tappets and select the proper adjustment shim. The formula for selecting a shim is as follows:

   a. Removed shim thickness + measured valve clearance–standard valve clearance 0.0098 inch (0.25mm) for the intake and 0.0118 inch (0.30mm) for the exhaust.

23. Install the adjustment shim.

24. Set the cam position of the number 1 cylinder at Top Dead center (TDC) and install the camshaft.

25. Temporarily install the camshaft caps in their original positions.

26. Tighten the camshaft cap bolt in two steps:

   a. Step 1: 44–71 inch lbs. (5–8 Nm).

   b. Step 2: 10.4–12.5 ft. lbs. (14–17 Nm).

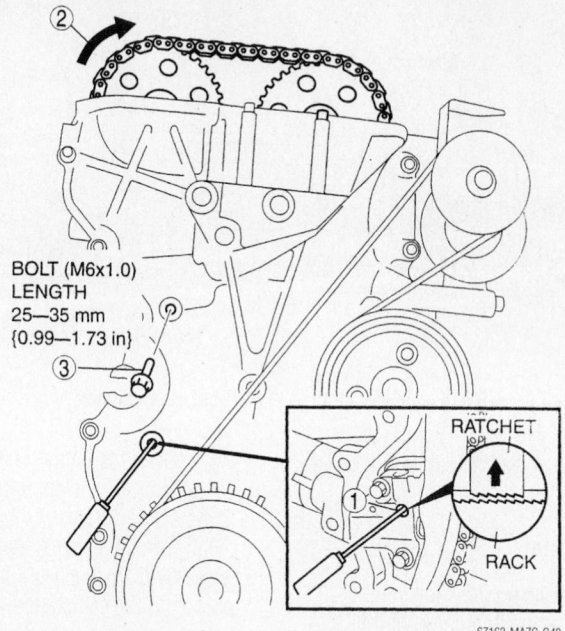

67162-MAZC-G40

Loosen the timing chain tension, refer to the text for procedure—2.0L (LF) and 2.3L (L3) engines

27. Install the exhaust camshaft sprocket and hand tighten the bolt.

28. Install tool 49 UN30 3465 as illustrated on the camshaft.

29. Remove the M6 x 1.0 bolt from the engine front cover to tension the timing chain.

30. Turn the crankshaft clockwise until the crankshaft is in the number 1 cylinder Top Dead Center position (the balance weight should be attached to tool 49 JE01 061).

31. Hold the exhaust camshaft using a wrench on the cast hexagon portion of the camshaft and tighten the exhaust camshaft sprocket bolt to 51–55 ft. lbs. (69–75 Nm).

32. Remove the tools from the camshaft and from the block lower blind plug.

33. Rotate the engine two full turns clockwise to TDC. If not aligned, remove the crankshaft pulley lockbolt, remove the tappet and repeat the adjustment shim selection and replacement procedure.

34. Apply sealant to the engine front cover upper blind plug and tighten to 71–101 inch lbs. (8–11 Nm).

35. Install the cylinder block lower blind plug and tighten to 13–16 ft. lbs. (18–22 Nm).

36. Install a new engine front cover blind plug and tighten to 7.4–10 ft. lbs. (10–14 Nm).

37. Install the right side halfshaft from the joint shaft on Mazda 6 models.

38. Install the cylinder head cover.

39. Install the PCV hose, if equipped.

40. Connect the oil control valve connector, if equipped.

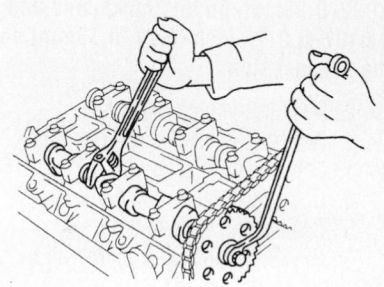

67162-MAZC-G41

Hold the exhaust camshaft using a wrench on the cast hexagon portion of the camshaft and remove the exhaust camshaft sprocket—2.0L (LF) and 2.3L (L3) engines

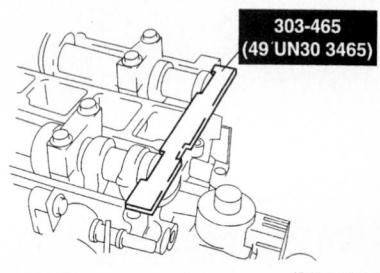

67162-MAZC-G32

Install tool 49 UN30 3465 on the camshaft—2.0L (LF) and 2.3L (L3) engines

41. Install the ignition coils.

42. Connect the wiring harness.

43. Connect the negative battery cable and install the battery cover.

44. Install the plug hole plate.

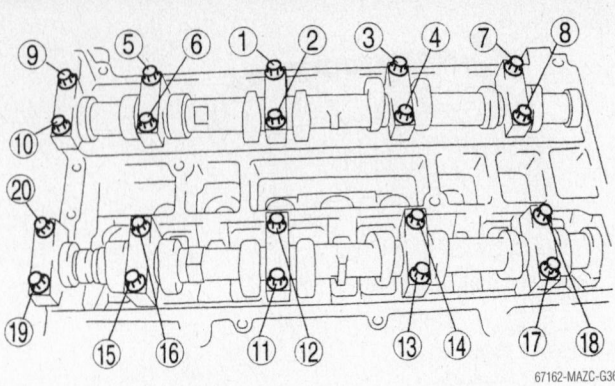

Camshaft cap bolt tightening sequence—2.0L (LF) and 2.3L (L3) engines

67162-MAZC-G36

45. Install the splash shield, under cover and wheel.

2.3L (L3) Engine

1. Before servicing the vehicle, refer to the precautions in the beginning of this section.

➡ **With the engine cold, standard valve clearance is 0.087–0.0110 inch (0.22–0.28mm) on the intake side and 0.0107–0.0129 inch (0.27–0.33mm) on the exhaust side.**

2. Remove the plug hole plate.

3. Remove the battery cover and disconnect the negative battery cable.

4. Disconnect the wiring harness.

5. Remove the ignition coils.

6. Disconnect the Oil Control Valve (OCV) connector.

7. Remove the right front wheel, under cover and splash shield.

8. Remove the Positive Crankcase Ventilation (PCV) hose.

9. Remove the cylinder head cover.

10. Measure the valve clearance by turning the crankshaft clockwise until the No. 1 piston is at Top Dead Center (TDC).

11. Measure the valve clearance at **A**. If the clearance exceeds specifications, replace the adjustment shim.

12. Turn the crankshaft clockwise 360 degrees until the No. 4 piston is at TDC. Measure the valve clearance at **B**. If the clearance exceeds specifications, replace the adjustment shim.

13. Repeat this procedure for all the camshafts.

14. Remove the engine front cover lower and upper blind plugs and the cylinder block lower blind plug.

15. Install tool 49 JE01 061.

16. Turn the crankshaft clockwise until the crankshaft is in the number 1 cylinder Top Dead Center position (the balance weight should be attached to tool 49 JE01 061).

17. Loosen the timing chain as follows:

a. Unlock the chain tensioner ratchet using a screwdriver.

b. Turn the exhaust camshaft clockwise using a wrench on the cast hexagon portion of the camshaft and loosen the timing chain.

c. Place a M6 x 1.0 length bolt: 0.99–1.37 inch (25–35mm) at the engine front cover upper blind plug and secure the timing chain guide at the location where the chain tension is released.

18. Hold the exhaust camshaft using a wrench on the cast hexagon portion of the camshaft and remove the exhaust camshaft sprocket.

➡ **The cylinder head and camshaft caps are numbered and must be reassembled in their original locations. When removing a when removed, keep the caps with the cylinder head they were removed from, do not switch the caps.**

19. Loosen the camshaft cap bolts using 2–3 passes in the sequence illustrated.

20. Once all the bolts have been removed, remove the caps, keeping them in the original positions and remove the camshafts.

21. Remove the tappets and select the proper adjustment shim. The formula for selecting a shim is as follows:

a. Removed shim thickness + measured valve clearance—standard valve clearance 0.0098 inch (0.25mm) for the intake and 0.0118 inch (0.30mm) for the exhaust.

22. Install the adjustment shim.

23. Set the cam position of the number 1 cylinder at Top Dead center (TDC) and install the camshaft.

24. Temporarily install the camshaft caps in their original positions.

25. Tighten the camshaft cap bolt in two steps:

a. Step 1: 44–71 inch lbs. (5–8 Nm).

b. Step 2: 10.4–12.5 ft. lbs. (14–17 Nm).

26. Install the exhaust camshaft sprocket and hand tighten the bolt.

27. Install tool 49 UN30 3465 as illustrated on the camshaft.

28. Remove the M6 x 1.0 bolt from the engine front cover to tension the timing chain.

29. Turn the crankshaft clockwise until the crankshaft is in the number 1 cylinder Top Dead Center position (the balance weight should be attached to tool 49 JE01 061).

30. Hold the exhaust camshaft using a wrench on the cast hexagon portion of the camshaft and tighten the exhaust camshaft sprocket bolt to 51–55 ft. lbs. (69–75 Nm).

31. Remove the tools from the camshaft and from the block lower blind plug.

32. Rotate the engine two full turns clockwise to TDC. If not aligned, remove the crankshaft pulley lockbolt, remove the tappet and repeat the adjustment shim selection and replacement procedure.

33. Apply sealant to the engine front cover upper blind plug and tighten to 71–101 inch lbs. (8–11 Nm).

34. Install the cylinder block lower blind plug and tighten to 13–16 ft. lbs. (18–22 Nm).

35. Install a new engine front cover blind plug and tighten to 7.4–10 ft. lbs. (10–14 Nm).

36. Install the cylinder head cover.

37. Install the PCV hose.

38. Connect the OCV connector.

39. Install the ignition coils.

40. Connect the wiring harness.

41. Connect the negative battery cable and install the battery cover.

42. Install the plug hole plate.

43. Install the under cover, splash shield and wheel.

3.0L (AJ) Engine

1. Before servicing the vehicle, refer to the precautions in the beginning of this section.

2. Remove or disconnect the following:
- Negative battery cable
- Camshaft followers
- Hydraulic lash adjusters

➡ **Mark the position of the hydraulic lash adjusters to assure they are assembled in their original position**

3. Inspect the adjusters for scoring or uneven wear in the bore and replace them as required.

To install:

4. Install or connect the following:
   - Hydraulic lash adjusters after lubricating them with clean engine oil
   - Camshaft followers
   - Negative battery cable

Starter Motor

REMOVAL & INSTALLATION

Protégé

1. Remove or disconnect the following:
   - Negative battery cable
   - Air cleaner
   - Intake manifold support bracket bolts
   - Starter electrical connectors
   - Starter

To install:

2. Install or connect the following:
   - Starter and loosely tighten the lower starter mounting bolt
   - Starter electrical connectors
   - Intake manifold support bracket. Torque the bolts to 28–38 ft. lbs. (38–51 Nm).
   - Starter bolts. Torque the bolts 28–38 ft. lbs. (38–51 Nm). The upper mounting bolts must be tightened first.
   - Air cleaner
   - Negative battery cable

626

2.0L (FS) ENGINE

1. Remove or disconnect the following:
   - Battery
   - Air cleaner assembly
   - Transverse member
   - Intake manifold bracket
   - Wiring at the starter
   - Starter

To install:

2. Install or connect the following:
   - Starter. Torque the bolts to 38 ft. lbs. (51 Nm).
   - Wiring at the starter
   - Intake manifold bracket
   - Transverse member
   - Air cleaner assembly
   - Battery

2.5L (KL) ENGINE

1. Remove or disconnect the following:
   - Battery
   - Air cleaner assembly
   - Transverse member
   - Fuel filter with the hose still attached and set aside
   - Oil filler pipe
   - Transaxle selector cable from the automatic transaxle and remove the cable bracket
   - Wiring at the starter
   - Starter

To install:

2. Install or connect the following:
   - Starter. Torque the bolts to 38 ft. lbs. (51 Nm).
   - Wiring at the starter
   - Cable bracket and connect the transaxle selector cable to the transaxle
   - Oil filler pipe
   - Fuel filter
   - Transverse member
   - Air cleaner assembly
   - Battery

Miata

1. Remove or disconnect the following:
   - Negative battery cable
   - Air cleaner assembly
   - Dipstick tube
   - Intake manifold bracket
   - Wiring at the starter
   - Starter

To install:

2. Install or connect the following:
   - Starter. Torque the bolts to 33 ft. lbs. (46 Nm).
   - Wiring at the starter
   - Intake manifold bracket. Torque the bolts to 38 ft. lbs. (51 Nm).
   - Dipstick tube
   - Air cleaner assembly
   - Negative battery cable

Millenia

2.3L (KJ) ENGINE

1. Remove or disconnect the following:
   - Negative battery cable
   - Charge air cooler duct
   - Battery clamp, box and battery
   - Battery tray
   - Rear charge air cooler
   - Pipe bracket
   - Starter electrical connectors
   - Starter

To install:

2. Install or connect the following:
   - Starter. Torque the bolts to 28–38 ft. lbs. (38–51 Nm).
   - B-terminal wire
   - S-terminal wire
   - Pipe bracket. Torque the bolt to 14–18 ft. lbs. (19–25 Nm).
   - Rear charge air cooler using new O-rings. Torque the nuts to 14–18 ft. lbs. (19–25 Nm).
   - Battery tray
   - Battery, box and clamp
   - Charge air cooler duct
   - Negative battery cable

2.5L (KL) ENGINE

1. Remove or disconnect the following:
   - Negative battery cable
   - Battery clamp, box and battery
   - Battery tray
   - Shift cable from the selector lever
   - Cable from the bracket by squeeze the lock tabs
   - Electrical connectors from the starter solenoid
   - 2 selector cable bracket bolts and the bracket
   - 2 nuts and the bolt from the starter bracket and the bracket
   - Starter electrical connectors
   - Starter

To install:

2. Install or connect the following:
   - Starter. Torque the bolts to 28–38 ft. lbs. (38–51 Nm).
   - B-terminal wire.
   - S-terminal wire to the solenoid
   - Starter bracket
   - Selector cable bracket. Torque the bolts to 5–7 ft. lbs. (7–9 Nm).
   - Starter solenoid electrical connectors
   - Shift cable into the cable bracket and into the selector lever
   - Battery tray
   - Battery, box and clamp
   - Negative battery cable

Mazda 3

1. Before servicing the vehicle, refer to the precautions in the beginning of this section.

2. Remove the battery cover and disconnect the negative battery cable.

3. Remove the under cover.

4. If equipped with a manual transaxle, remove the clutch release cylinder with the line still attached.

5. Remove the starter electrical connections and wiring harness bracket.

6. Remove the starter bolts and the starter.

To install:

7. Installation is the reverse of removal. Tighten the wiring harness bracket retainers to 70–104 inch lbs. (8–11 Nm) and the starter bolts to 28–38 ft. lbs. (38–51 Nm).

Mazda 6

2.3L (L3) ENGINE

1. Before servicing the vehicle, refer to the precautions in the beginning of this section.

2. Remove the battery cover and disconnect the negative battery cable.

3. Remove the under cover.

4. If equipped with a manual transaxle, remove the clutch release cylinder with the line still attached.

5. Remove the starter electrical connection.

6. Remove the starter bolts and the starter.

To install:

7. Installation is the reverse of removal. Tighten the starter bolts to 29–37 ft. lbs. (38–51 Nm).

3.0L (AJ) ENGINE

1. Before servicing the vehicle, refer to the precautions in the beginning of this section.

2. Remove the battery and tray.

3. Disconnect the selector cable and bracket.

4. Remove the starter electrical connection.

5. Remove the starter bolts and the starter.

To install:

6. Installation is the reverse of removal. Tighten the starter bolts to 28–38 ft. lbs. (38–51 Nm).

Oil Pan

REMOVAL & INSTALLATION

1.8L (BP) Engines

1. Before servicing the vehicle, refer to the precautions in the beginning of this section.

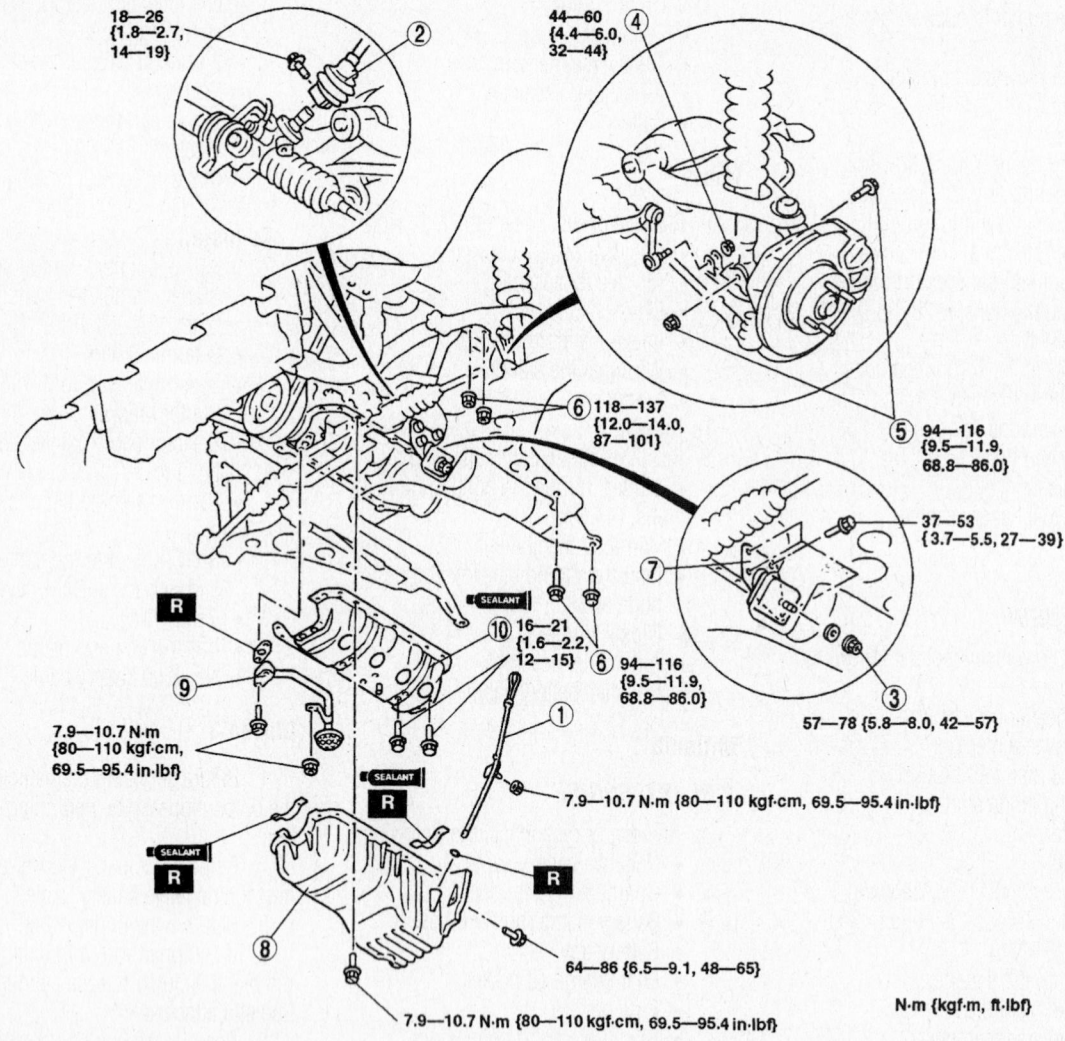

1	Dipstick and pipe	6	Crossmember bolt and nut
2	Intermediate shaft	7	Engine mount
3	Engine mount nut	8	Oil pan
4	Stabilizer control link nut	9	Oil strainer
5	Shock absorber bolt and nut	10	MBSP

Exploded view of the oil pan and related components—1.8L (BP) engine

42356-MAZC-GAY

2. Drain the engine oil.

3. Remove or disconnect the following:
- Negative battery cable
- Air cleaner assembly
- Wheel speed sensor
- Dipstick tube
- Intermediate shaft

4. Attach an engine to the hoist, loosen the oil pan bolts, remove the engine mounting bolts and raise the engine slightly using a hoist.
- Stabilizer link nut
- Shock absorber-to-knuckle nut and bolt

❉❉ CAUTION

When removing the crossmember, be careful do to damage the brake hoses, A/C pipes and power steering pipes when lowering the crossmember.

5. Support the crossmember with a transmission jack, remove the crossmember retainers. Separate the steering intermediate shaft from the pinion shaft and lower the crossmember until the clearance between the oil pan and the steering gear exceeds 5.12 inch (130mm).

6. Remove or disconnect the following:
- Engine mount
- Oil pan bolts and the oil pan using a seal cutter, then insert a flat pry tool into the locations illustrated.

To install:

7. Clean the oil pan. Clean all dirt, oil, gasket and old sealant from the oil pan and cylinder block contact surfaces.

8. Apply a continuous bead of silicone sealant to the contact surfaces of the new oil pan gaskets as illustrated.

9. Install new gaskets into the oil pump body and rear cover facing the notches as illustrated.

10. Apply a 0.079 inch (2mm) continuous bead of silicone sealant on the oil pan as illustrated.

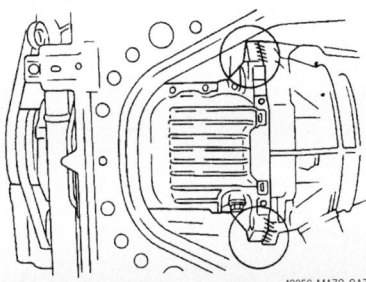

Remove the oil pan by inserting a flat pry tool into the locations shown—1.8L (BP) engine

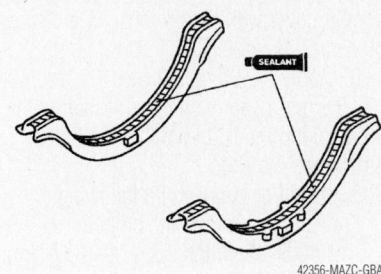

Apply a continuous bead of silicone sealant to the contact surfaces of the new oil pan gaskets as shown—1.8L (BP) engine

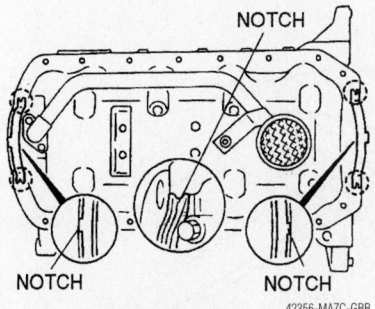

Install new gaskets into the oil pump body and rear cover facing the notches as shown—1.8L (BP) engine

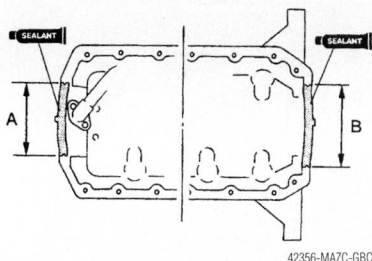

Apply a 0.079 inch (2mm) continuous bead of silicone sealant on the oil pan as shown—1.8L (BP) engine

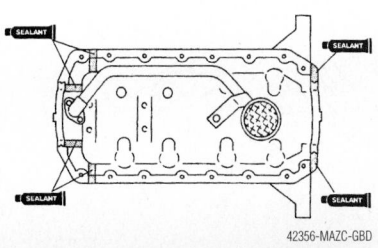

Apply a continuous bead of silicone sealant on the oil pan-to-block areas as shown—1.8L (BP) engine

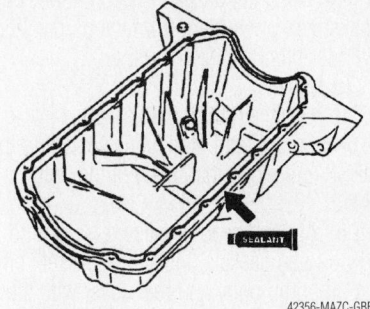

Apply a 0.099–0.137 inch (2.5–3.5mm) continuous bead of silicone sealant on the oil pan along the inside of the bolt holes and overlap the ends—1.8L (BP) engine

11. Apply a continuous bead of silicone sealant on the oil pan-to-block areas as illustrated.

12. Apply a 0.099–0.137 inch (2.5–3.5mm) continuous bead of silicone sealant on the oil pan along the inside of the bolt holes and overlap the ends.

13. Install or connect the following:
- New oil pan gaskets
- Oil pan. Torque the vertical bolts to 70–95 inch lbs. (8–11 Nm) and the horizontal bolts to 48–65 ft. lbs. (64–86 Nm).
- Engine mount

14. Raise the crossmember into position and tighten the bolts to 15 ft. lbs. (21 Nm) and the nuts to 101 ft lbs. (137 Nm).
- Shock absorber lower nut and bolt and tighten to 86 ft. lbs. (116 Nm)
- Stabilizer link nut and tighten to 44 ft. lbs. (60 Nm)
- Engine mount nut and tighten to 57 ft. lbs. (78 Nm)
- Intermediate shaft and tighten the bolt to 19 ft. lbs. (26 Nm)
- Dipstick tube
- Wheel speed sensor
- Air cleaner assembly
- Negative battery cable

15. Fill the engine with clean oil.

16. Start the vehicle, check for leaks and repair if necessary.

1.6L (ZM) Engines

1. Before servicing the vehicle, refer to the precautions in the beginning of this section.

2. Drain the engine oil.

3. Remove or disconnect the following:
- Negative battery cable
- Oxygen (O_2S) sensors
- Front exhaust pipe
- Integrated stiffener (1)
- Oil pan (2) and oil strainer (3) and the Main Bearing Support Plate (MBSP) (4)

4. Clean the oil pan. Clean all dirt, oil, gasket and old sealant from the oil pan and cylinder block contact surfaces.

To install:

5. Apply a 0.099–0.137 inch (2.5–3.5mm) bead of silicone sealant to the MBSP and along the inside of the bolt holes.

6. Apply a bead of silicone sealant to new oil pan gaskets. Install the gaskets onto the oil pump body and rear cover with the projections in the notches as illustrated.

7. Apply a 0.079 inch (2mm) bead of silicone sealant to area of the oil pan gaskets marked by **A and B** in the illustration.

Install the gaskets onto the oil pump body and rear cover with the projections in the notches as illustrated.

8. Apply a 0.099–0.137 inch (2.5–3.5mm) bead of silicone sealant to the oil pan along the inside of the bolt holes and overlap the ends.

9. Install or connect the following:
- Oil pan. Torque the bolts to 69.5–95.4 inch lbs. (7.9–10.7 Nm).

10. Install the integrated stiffener and hand tighten bolt **A**, then bolt **B**.

11. Tighten bolt **C** to 38 ft. lbs. (52 Nm).

12. Tighten bolt **D** to 38 ft. lbs. (52 Nm).

13. Tighten bolt **A** to 38 ft. lbs. (52 Nm).

14. Tighten bolt **B** to 38 ft. lbs. (52 Nm).
- Front exhaust pipe
- O$_2$S
- Negative battery cable

15. Fill the engine with clean oil.

16. Start the vehicle, check for leaks and repair if necessary.

1.8L (FP) Engines

1. Before servicing the vehicle, refer to the precautions in the beginning of this section.

2. Drain the engine oil.

3. Remove or disconnect the following:

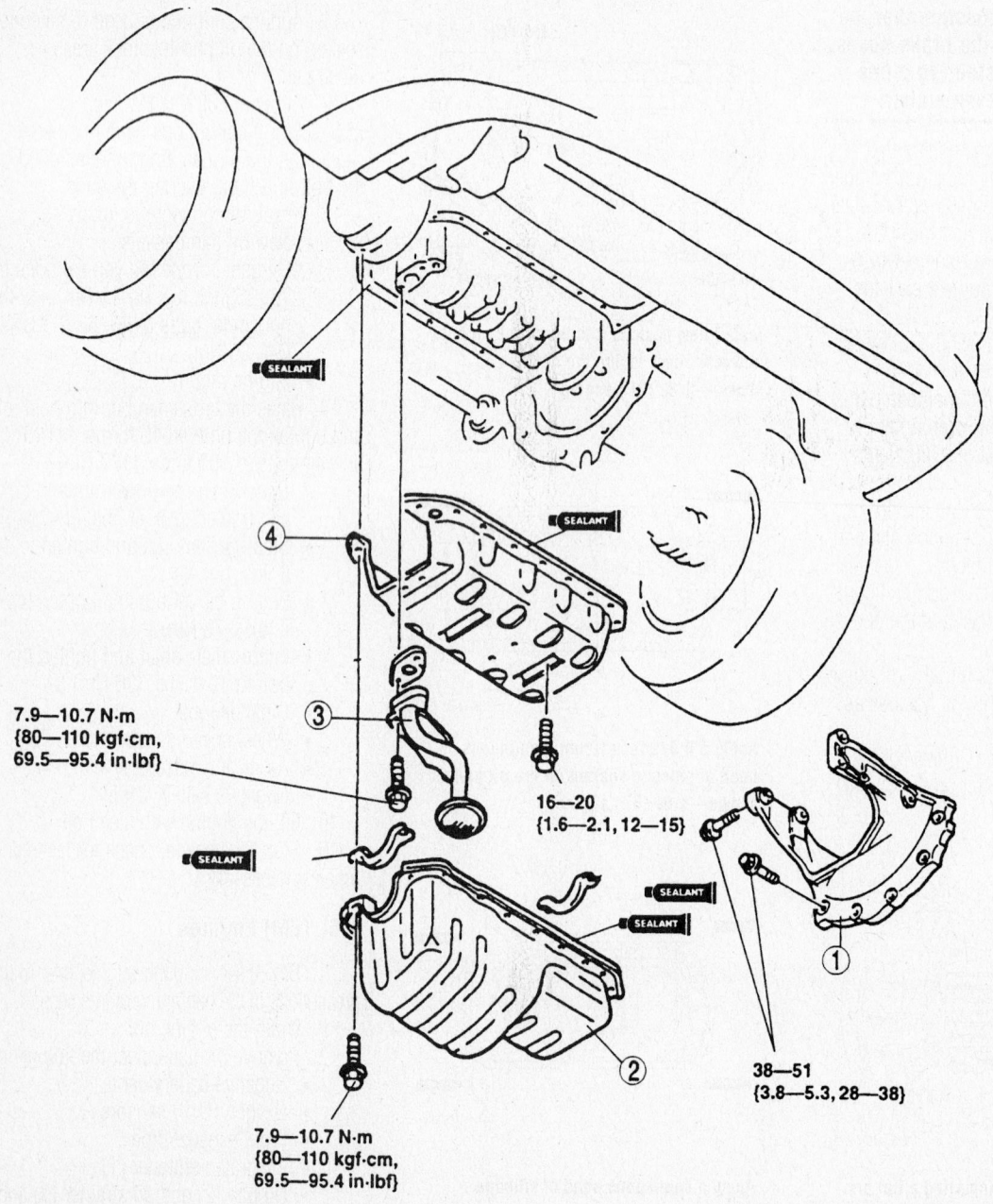

7.9—10.7 N·m
{80—110 kgf·cm,
69.5—95.4 in·lbf}

16—20
{1.6—2.1, 12—15}

38—51
{3.8—5.3, 28—38}

7.9—10.7 N·m
{80—110 kgf·cm,
69.5—95.4 in·lbf}

N·m {kgf·m, ft·lbf}

9301MG10

Exploded view of the oil pan and related components—1.6L engine

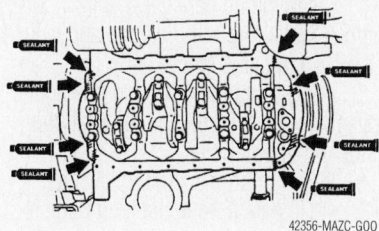

Apply a bead of silicone sealant to area as illustrated—1.6L (ZM) engines

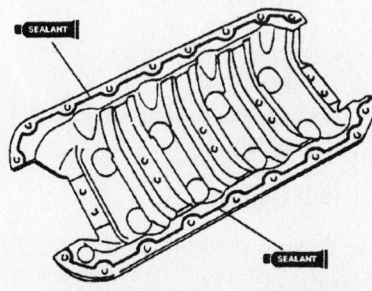

Apply a 0.099–0.137 inch (2.5–3.5mm) bead of silicone sealant to the MBSP and along the inside of the bolt holes—1.6L (ZM) engines

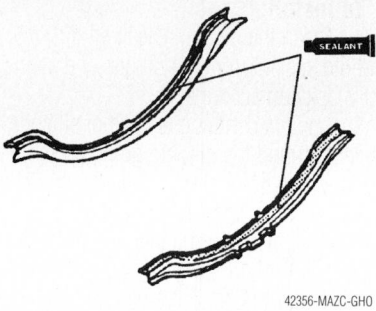

Apply a bead of silicone sealant to new oil pan gaskets . . .

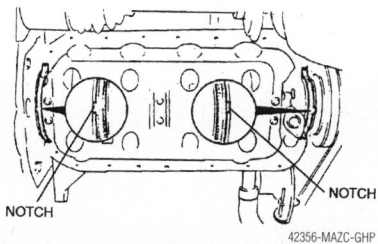

. . . and install the gaskets onto the oil pump body and rear cover with the projections in the notches as illustrated—1.6L (ZM) engines

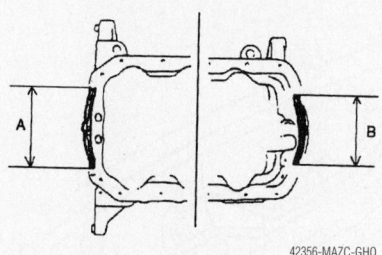

Apply a 0.079 inch (2mm) bead of silicone sealant to area of the oil pan gaskets marked by A and B—1.6L (ZM) engines

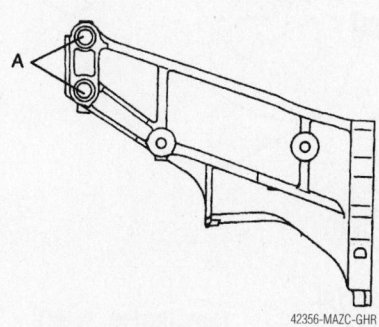

Location of the integrated stiffener bolt A—1.6L (ZM) engines

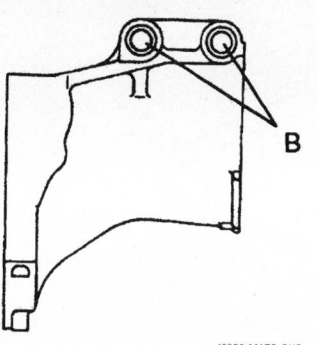

Location of the integrated stiffener bolt B—1.6L (ZM) engines

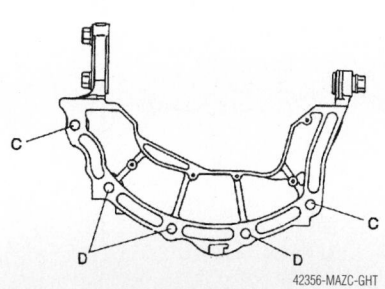

Location of the integrated stiffener bolts C and D—1.6L (ZM) engines

- Negative battery cable
- Oxygen (O_2S) sensors
- Front exhaust pipe
- Oil pan

4. Clean the oil pan. Clean all dirt, oil, gasket and old sealant from the oil pan and cylinder block contact surfaces.

To install:

5. Apply a continuous bead of silicone sealant on the gaskets and around the oil pan, going on the inside of the bolt holes.

6. Install or connect the following:
- New oil pan gasket
- Oil pan. Torque the bolts to 14–18 ft. lbs. (19–25 Nm).
- Front exhaust pipe
- O_2S
- Negative battery cable

7. Fill the engine with clean oil.

8. Start the vehicle, check for leaks and repair if necessary.

2.0L (FS) Engine

1. Before servicing the vehicle, refer to the precautions in the beginning of this section.

2. Drain the engine oil.

3. Remove or disconnect the following:
- Negative battery cable
- Oxygen (O_2S) sensor
- Front exhaust pipe
- Oil pan bolts and the oil pan

To install:

4. Clean the oil pan. Clean all dirt, oil and old sealant from the oil pan and cylinder block contact surfaces.

5. Apply a continuous bead of silicone sealant around the oil pan, going on the inside of the bolt holes.

6. Install or connect the following:
- Oil pan. Torque the bolts to 14–18 ft. lbs. (19–25 Nm).
- Front pipe. Torque the nuts to 28–38 ft. lbs. (38–51 Nm).
- O_2S connector
- Negative battery cable

7. Fill the engine with clean oil.

8. Start the engine, check for leaks and repair if necessary.

2.5L (KL) Engine

1. Before servicing the vehicle, refer to the precautions in the beginning of this section.

2. Drain the engine oil.

3. Remove or disconnect the following:
- Negative battery cable
- Right hand splash shield on Millenia models
- Oxygen (O_2S) sensor
- Front exhaust pipe
- Oil pan bolts and the oil pan

FS

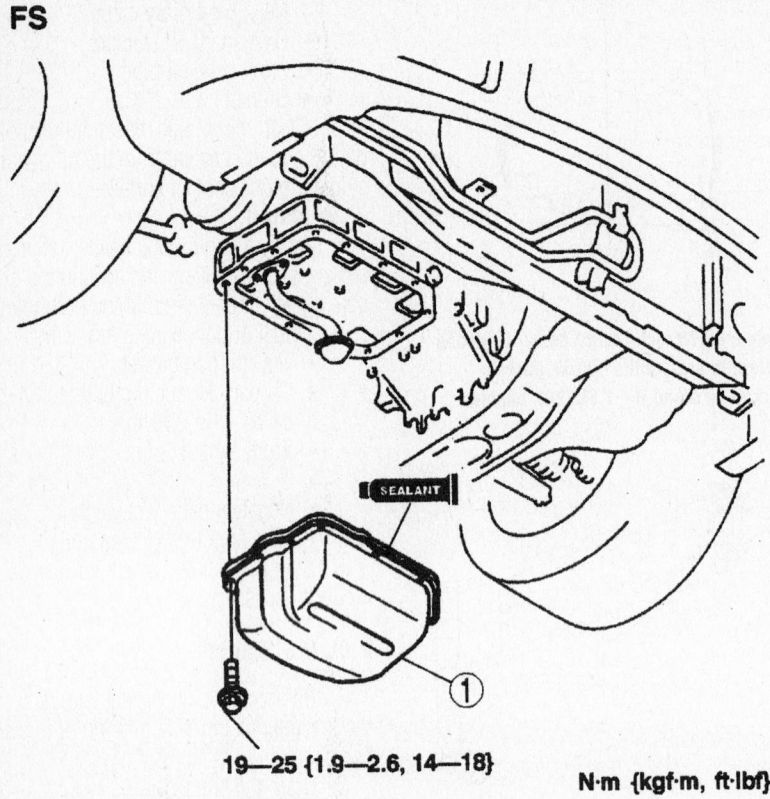

19—25 {1.9—2.6, 14—18}

N·m {kgf·m, ft·lbf}

42356-MAZC-G21

Exploded view of the oil pan and related components for the 2.0L (FS) engines

KL

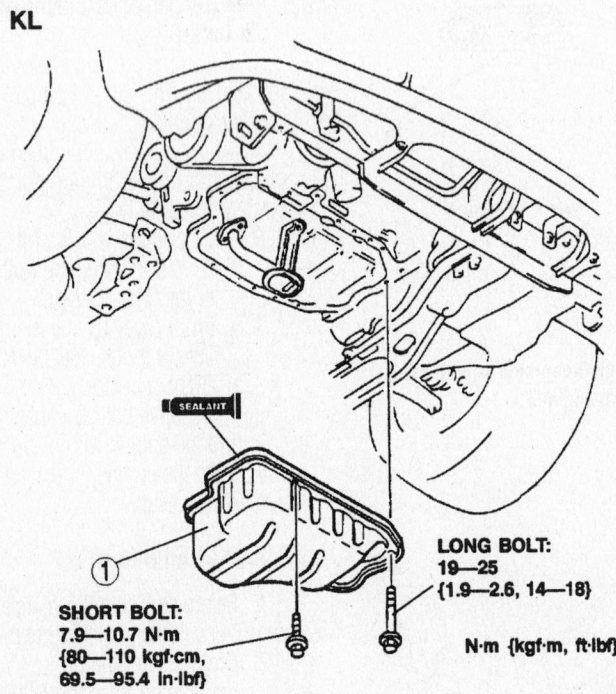

SHORT BOLT:
7.9—10.7 N·m
{80—110 kgf·cm,
69.5—95.4 in·lbf}

LONG BOLT:
19—25
{1.9—2.6, 14—18}

N·m {kgf·m, ft·lbf}

1   Oil pan

42356-MAZC-G72

Exploded view of the oil pan and related components—2.5L (KL) engines, 626 model shown
Millenia similar

To install:

4. Clean the oil pan. Clean all dirt, oil and old sealant from the oil pan and cylinder block contact surfaces.

5. Apply a continuous bead of silicone sealant around the oil pan, going on the inside of the bolt holes.

6. Install or connect the following:
- Oil pan. Torque the long bolts to 14–18 ft. lbs. (19–25 Nm) and the short bolts to 70–95 inch lbs. (8–11 Nm).
- Front pipe
- O₂S connector
- Right hand splash shield on Millenia models
- Negative battery cable

7. Fill the engine with clean oil.

8. Start the engine, check for leaks and repair if necessary.

2.3L (KJ) Engines

1. Before servicing the vehicle, refer to the precautions in the beginning of this section.

2. Drain the engine oil.

3. Remove or disconnect the following:
- Negative battery cable
- Passenger side splash shield
- Oxygen (O₂S) sensor
- Front exhaust pipe
- Oil pan bolts and the oil pan

To install:

4. Clean the oil pan. Clean all dirt, oil and old sealant from the oil pan and cylinder block contact surfaces.

5. Apply a continuous bead of silicone sealant around the oil pan, going on the inside of the bolt holes.

6. Install or connect the following:
- Oil pan. Torque the long bolts to 14–18 ft. lbs. (19–25 Nm) and the short bolts to 70–95 inch lbs. (8–11 Nm).
- Front pipe. Torque the nuts to 28–38 ft. lbs. (38–51 Nm).
- O₂S connector
- Splash shield
- Negative battery cable

7. Fill the engine with clean oil.

8. Start the engine, check for leaks and repair if necessary.

Mazda 3

1. Before servicing the vehicle, refer to the precautions in the beginning of this section.

2. Remove the battery cover and disconnect the negative battery cable.

3. Remove the engine under cover and splash shield as an assembly.

4. Remove the right front wheel.

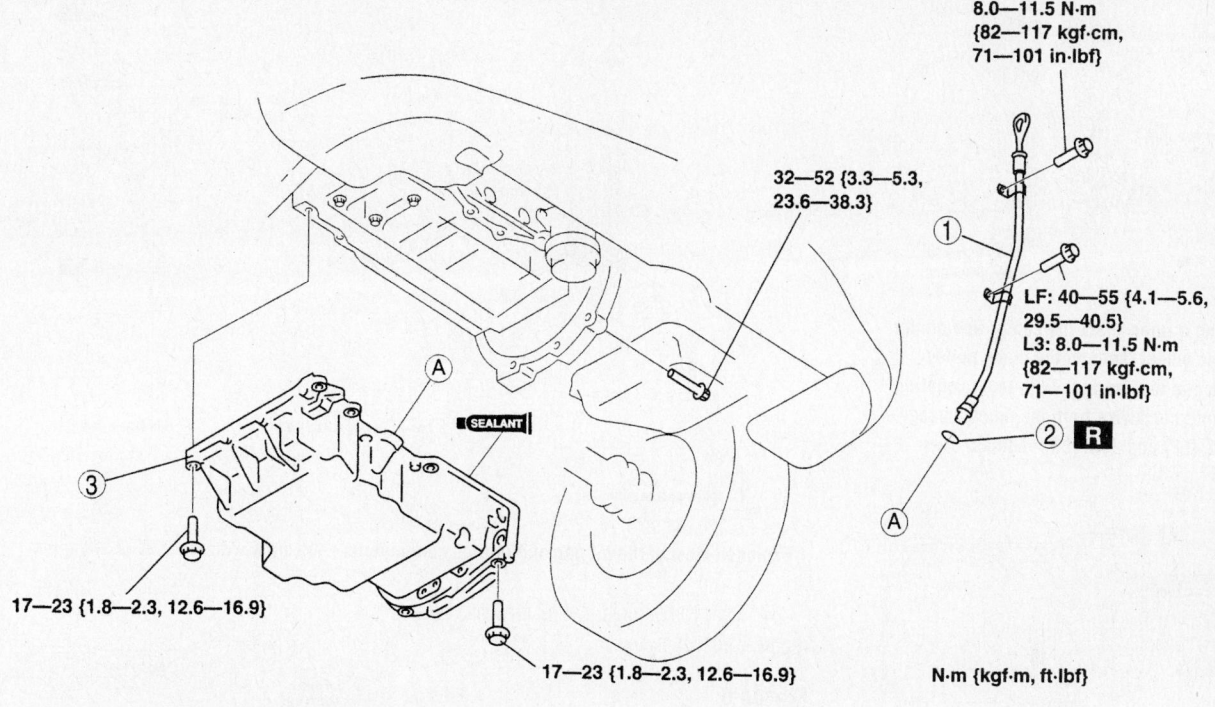

8.0—11.5 N·m
{82—117 kgf·cm,
71—101 in·lbf}

32—52 {3.3—5.3,
23.6—38.3}

LF: 40—55 {4.1—5.6,
29.5—40.5}
L3: 8.0—11.5 N·m
{82—117 kgf·cm,
71—101 in·lbf}

SEALANT

17—23 {1.8—2.3, 12.6—16.9}

17—23 {1.8—2.3, 12.6—16.9}

N·m {kgf·m, ft·lbf}

1	Oil level gauge pipe	3	Oil pan
2	O-ring		

67162-MAZC-G42

Exploded view of the oil pan and related components—Mazda 3 with the 2.0L (LF) and 2.3L (L3) engines

5. Remove the plug hole plate.
6. Drain the engine oil.
7. Remove the drive belt.
8. Remove the coolant reservoir and position aside with the lines attached.
9. Remove the A/C compressor and position aside with the lines attached.
10. Remove the ignition coil and position the accelerator cable bracket aside.
11. Remove the Crankshaft Position (CKP) sensor.
12. Remove the engine front cover. Refer to the timing chain removal procedure in this section.

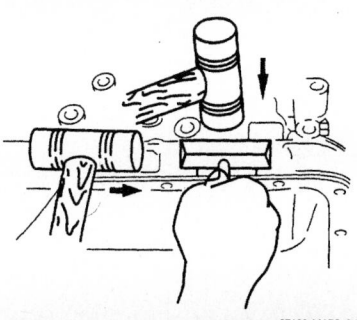

67162-MAZC-G43

Use a separator tool such as the one illustrated to separate the oil pan from the block—2.0L (LF) and 2.3L (L3) engines

13. Remove the dipstick tube pipe and O-ring.
14. Remove the oil pan bolts and use a separator tool such as the one illustrated to separate the oil pan from the block.
15. Remove the oil pan.

To install:

16. Clean the oil pan mating surfaces.
17. Apply a 0.087–0.126 inch (2.2–3.2mm) bead of sealant around the perimeter of the oil pan as illustrated.
18. Use a square ruler to align the oil pan and the block junction side on the engine front cover side as illustrated.
19. Install the oil pan within 5 minutes of applying the sealant.
20. Install the lower oil pan bolts and tighten in sequence to 12.6–16.9 ft. lbs. (17–23 Nm) and the oil pan–to–transaxle bolts to 23–38 ft. lbs. (32–52 Nm).
21. When installing the CKP sensor perform the following:
   a. Remove the cylinder block lower blind plug.
   b. Install tool 49 JE01 061.
   c. Turn the crankshaft clockwise until the crankshaft is in the number 1 cylinder Top Dead Center position (the balance weight should be attached to tool 49 JE01 061).

   d. Using a ruler, mark the center line on the pulse wheel teeth on the crank pulley which is located at the 9th tooth counting counterclockwise from the empty space. Refer to the illustration for more detail.

⁂ CAUTION

If you do not mark the center line correctly this will cause improper engine control for the ignition and fuel system, so be sure to mark the line carefully.

   e. Install the CKP sensor making sure the mark on the sensor is aligned with the mark on the pulse wheel. Tighten the sensor bolt to 49–66 inch lbs. (5.5–7.5 Nm) and attach the electrical connector.
   f. Remove the tool from the cylinder block lower blind plug.
   g. Rotate the crankshaft clockwise 2 turns until you reach TDC. If not aligned properly, loosen the crankshaft pulley lock bolt and reinstall the lock bolt.
   h. Install the cylinder block blind plug and tighten to 13–16 ft. lbs. (18–22 Nm).
22. Install the remaining components in the reverse order of removal.
23. Fill the engine with clean oil.

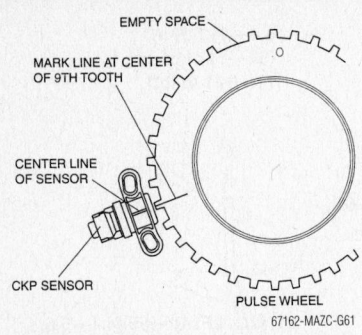

Using a ruler, mark the center line on the pulse wheel teeth on the crank pulley which is located at the 9th tooth counting counterclockwise from the empty space—2.0L (LF) and 2.3L (L3) engines

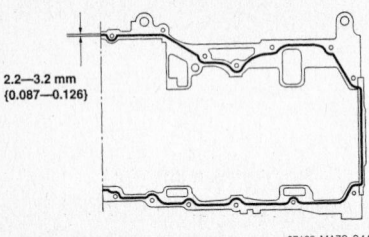

Apply a 0.087–0.126 inch (2.2–3.2mm) bead of sealant around the perimeter of the oil pan—Mazda 3 with the 2.0L (LF) and 2.3L (L3) engines

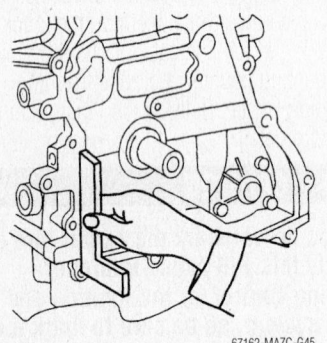

Use a square ruler to align the oil pan and the block junction side on the engine front cover side—2.0L (LF) and 2.3L (L3) engines

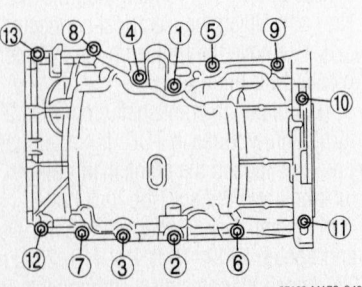

Oil pan bolt torque sequence—2.0L (LF) and 2.3L (L3) engines

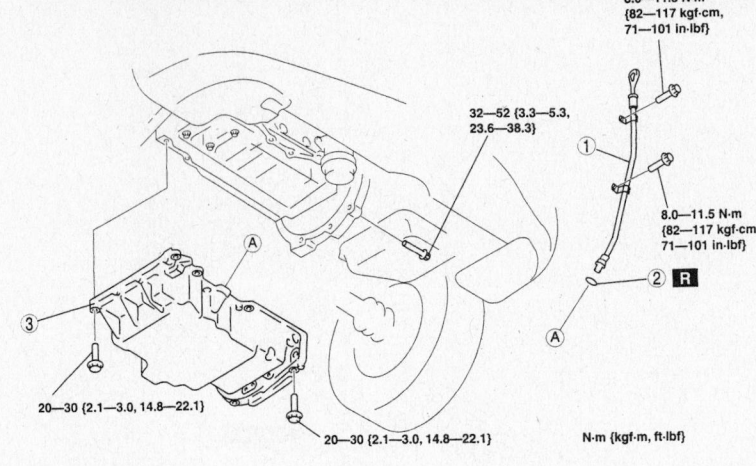

| 1 | Oil level gauge pipe | 2 | Oil pan |

Exploded view of the oil pan and related components—Mazda 6 with the 2.3L (L3) engine

24. Start the vehicle, check for leaks and repair if necessary.

Mazda 6

2.3L (L3) ENGINE

1. Before servicing the vehicle, refer to the precautions in the beginning of this section.
2. Disconnect the negative battery cable.
3. Remove the engine under cover.
4. Remove the right front wheel.
5. Drain the engine oil.

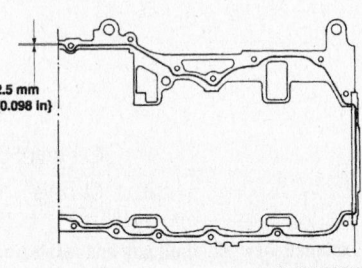

Apply a 0.098 inch (2.5mm) bead of sealant around the perimeter of the oil pan—Mazda 6 with the 2.3L (L3) engine

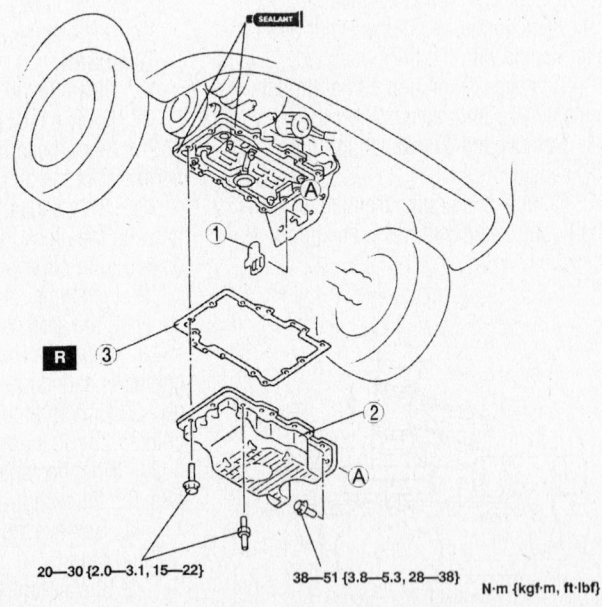

| 1 | End plate cover | 3 | Oil pan gasket |
| 2 | Oil Pan | | |

Exploded view of the oil pan and related components—Mazda 6 with the 3.0L (AJ) engine

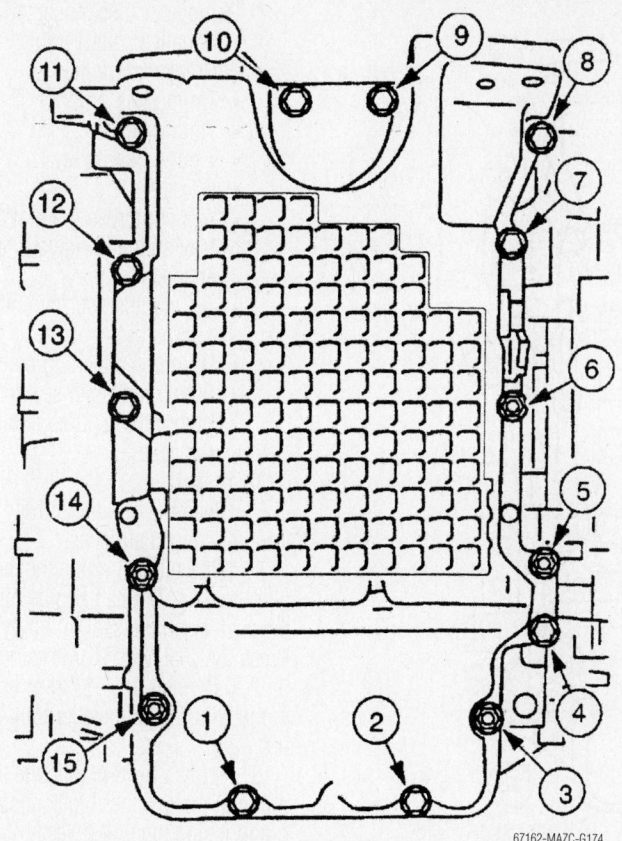

Loosen the oil pan retainers using 2–3 passes in the order shown—Mazda 6 with the 3.0L (AJ) engine

67162-MAZC-G174

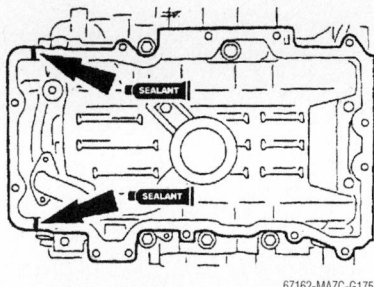

67162-MAZC-G175

Apply a 0.39 inch (10mm) bead of silicone sealer to the oil pan at the locations shown—Mazda 6 with the 3.0L (AJ) engine

6. Remove the engine front cover. Refer to the timing chain removal procedure in this section.

7. Remove the dipstick tube pipe and O-ring.

8. Remove the oil pan bolts and use a separator tool to separate the oil pan from the block.

9. Remove the oil pan.

To install:

10. Clean the oil pan mating surfaces.

11. Apply a 0.098 inch (2.5mm) bead of

sealant around the perimeter of the oil pan as illustrated.

12. Use a square ruler to align the oil pan and the block junction side on the engine front cover side.

13. Install the oil pan within 5 minutes of applying the sealant.

14. Install the lower oil pan bolts and tighten in sequence to 15–22 ft. lbs. (20–30 Nm) and the oil pan–to–transaxle bolts to 23–38 ft. lbs. (32–52 Nm).

15. Install the remaining components in the reverse order of removal.

16. Fill the engine with clean oil.

17. Start the vehicle, check for leaks and repair if necessary.

3.0L (AJ) Engine

1. Before servicing the vehicle, refer to the precautions in the beginning of this section.

2. Drain the engine oil.

3. Disconnect the negative battery cable.

4. Remove the right hand three way converter.

5. Remove the end plate cover.

6. Remove the oil pan–to–transaxle bolts.

7. Loosen the oil pan retainers using 2–3 passes in the order illustrated.

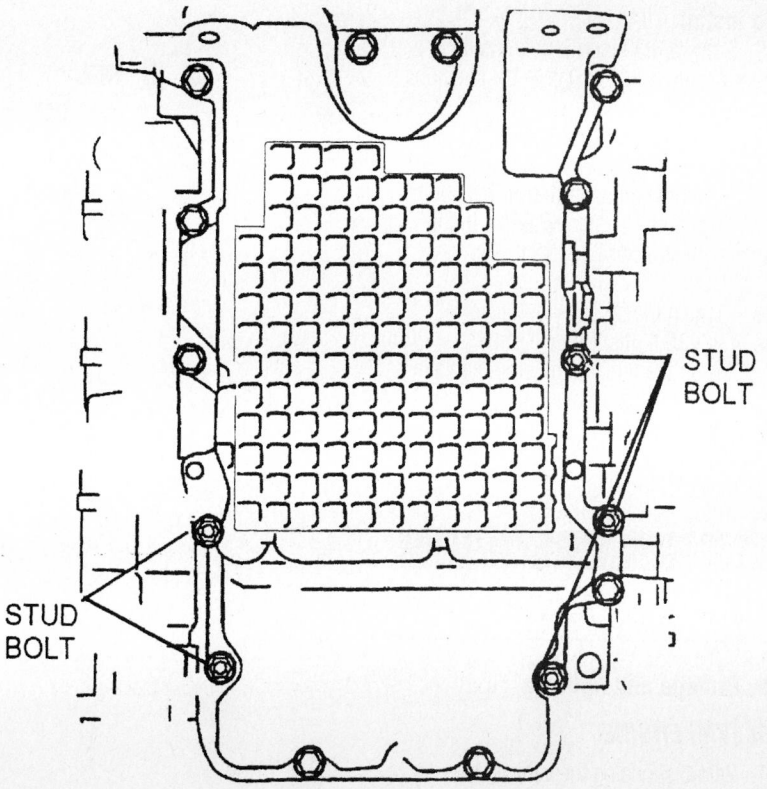

STUD BOLT

STUD BOLT

67162-MAZC-G176

Location of the oil pan stud bolts—Mazda 6 with the 3.0L (AJ) engine

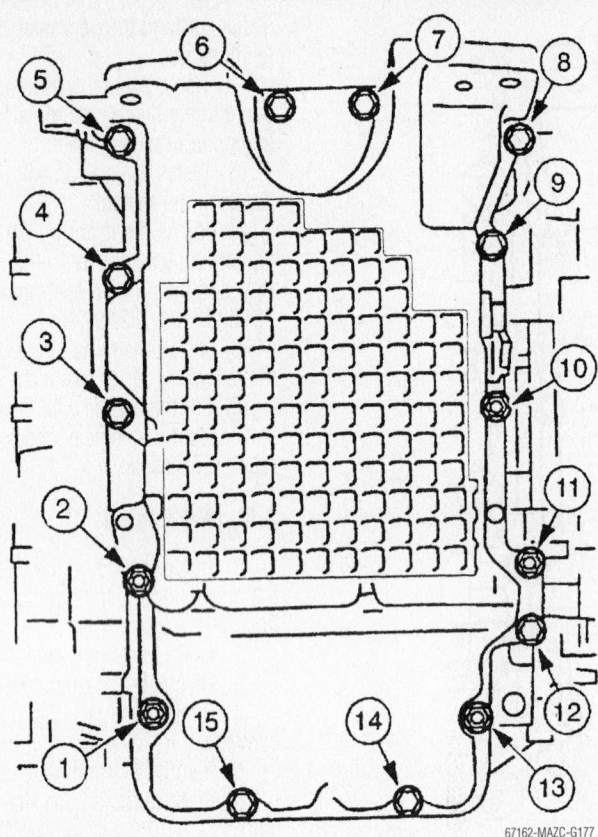

Oil pan torque sequence—Mazda 6 with the 3.0L (AJ) engine

2. Remove or disconnect the following:
- Negative battery cable
- Crankshaft pulley
- Timing belt cover
- Timing belt
- Timing belt pulley
- Oil pan
- A/C compressor and move it aside, leaving the refrigerant lines attached
- A/C compressor mounting bracket
- Alternator
- Oil pump attaching bolts
- Front crankshaft seal from the oil pump, if the pump is being replaced

To install:

3. Clean the oil, dirt and old sealant from all contact surfaces.

4. If the oil seal was removed from the oil pump, apply clean engine oil to the lip of the seal. Push the seal in lightly be hand. Press the seal, with a protrusion of 0.02–0.04 inch (0.5–1.0mm) into the oil pump with a Seal Installer tool 49 B014 001.

5. Apply a bead of silicone to the oil pump at the cylinder block contact surface, going inside the bolt holes.

6. Install or connect the following:

8. Thoroughly clean the gasket mating surfaces.

To install:

9. Apply a 0.39 inch (10mm) bead of silicone sealer to the oil pan at the locations illustrated.

10. Install the oil pan retainers and hand tighten.

11. Install the oil pan–to–transaxle bolts.

12. Tighten the oil pan retainers in the sequence illustrated to 15–22 ft. lbs. (20–30 Nm).

13. Tighten the oil pan–to–transaxle bolts to 28–38 ft. lbs. (38–51 Nm)

14. Install the remaining components in the reverse order of removal.

15. Fill the engine with clean oil.

16. Start the vehicle, check for leaks and repair if necessary.

Oil Pump

REMOVAL & INSTALLATION

626, Protégé and Millenia

1.6L (ZM) ENGINE

1. Before servicing the vehicle, refer to the precautions in the beginning of this section.

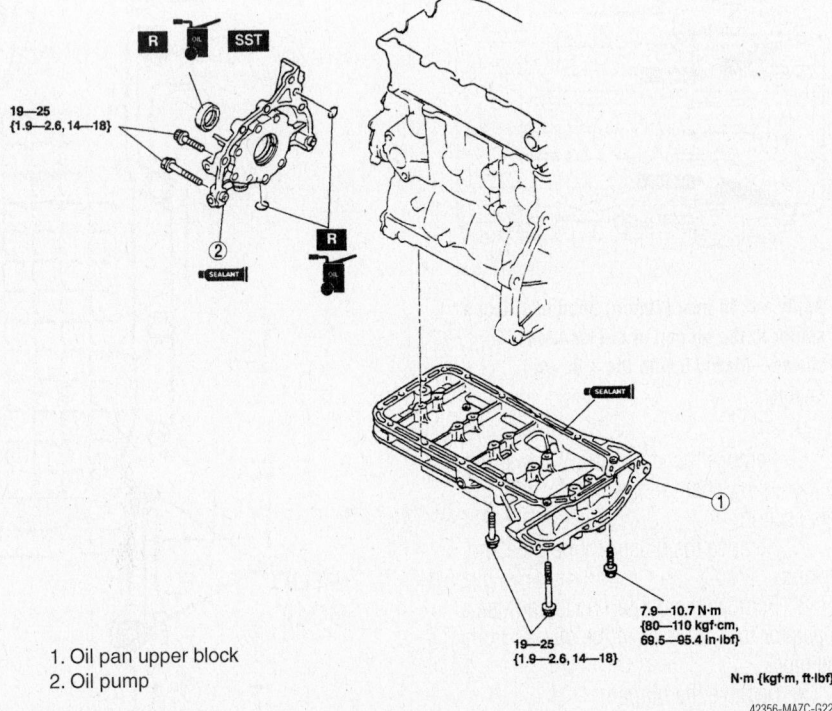

1. Oil pan upper block
2. Oil pump

Exploded view of the oil pump and related components for the 1.8L (FP) and 2.0L (FS) engines

- New O-rings on the oil pump
- Oil pump. Torque the bolts to 14–18 ft. lbs. (19–25 Nm).
- Alternator
- Air conditioning compressor bracket
- Air conditioning compressor
- Oil pan
- Timing belt pulley
- Timing belt
- Timing belt cover
- Crankshaft pulley
- Negative battery cable

7. Fill the engine with clean oil.

8. Start the vehicle, check for leaks and repair if necessary.

1.8L (FP) AND 2.0L (FS) ENGINE

1. Before servicing the vehicle, refer to the precautions in the beginning of this section.

2. Remove or disconnect the following:
- Negative battery cable
- Crankshaft pulley
- Timing belt cover
- Timing belt
- Timing belt pulley
- Oil pan
- A/C compressor and move it aside, leaving the refrigerant lines attached
- A/C compressor mounting bracket
- Alternator
- Transaxle
- Two bolts at the rear of the oil pan upper block, on 1.8L (FP) engines or if equipped with a manual transaxle on 2.0L (FS) engines
- Rubber caps at the bottom surface, then remove the two bolts from the upper block through the holes uncovered by removing the caps, if equipped an automatic transaxle, on 2.0L (FS) engines
- Oil pan upper block bolts using 2–3 steps in the sequence illustrated

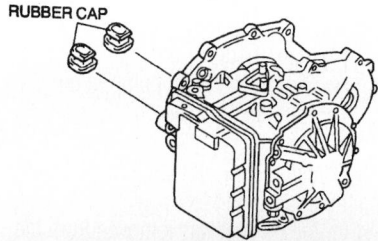

RUBBER CAP

42356-MAZC-G23

Remove the rubber caps at the bottom surface of the upper block, if equipped an automatic transaxle—2.0L (FS) engines

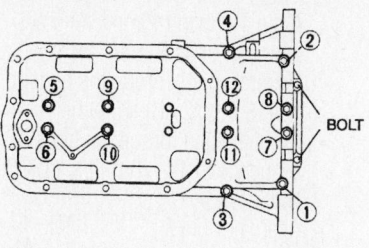

BOLT

42356-MAZC-G24

Remove the oil pan upper block bolts using 2–3 steps in the sequence illustrated—1.8L (FP) and 2.0L (FS) engines

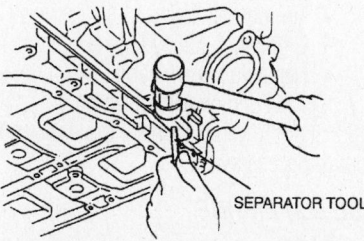

SEPARATOR TOOL

42356-MAZC-G25

Remove the oil pan upper block using a rubber mallet and a suitable separator tool—1.8L (FP) and 2.0L (FS) engines

- Oil pan upper block using a rubber mallet and a suitable separator tool
- Oil pump body bolts and the oil pump body
- Seal from the oil pump, if the pump is being replaced

To install:

3. Clean the oil, dirt and old sealant from all contact surfaces.

4. If the oil seal was removed from the oil pump, apply clean engine oil to the lip of the seal. Push the seal in lightly be hand. Press the seal, with a protrusion of 0–0.019 inch (0–0.5mm) into the oil pump.

5. Apply 0.04–0.07 inch (1–2mm) bead of sealant to the oil pump at the locations illustrated

6. Install or connect the following:
- Oil pump. Torque the bolts to 14–18 ft. lbs. (19–25 Nm).
- Oil pan

7. Apply a bead of sealant 0.08–0.11 inch (2–3mm) the mating surface of the upper block as illustrated, place the block assembly into position and tighten the bolts in sequence illustrated to 18 ft. lbs. (25 Nm) using 2–3 steps.
- Two bolts at the rear of the oil pan upper block, on 1.8L (FP) engines or if equipped with a manual transaxle on 2.0L (FS) engines and torque to 95 inch lbs. (11 Nm) and install the caps

- Two bolts to the upper block through the holes uncovered by removing the caps, if equipped an automatic transaxle, tighten the bolts to 95 inch lbs. (11 Nm)
- Transaxle, if equipped with a manual transaxle
- Air conditioning compressor bracket
- Air conditioning compressor
- Oil pan
- Timing belt pulley
- Timing belt
- Timing belt cover
- Crankshaft pulley
- Negative battery cable

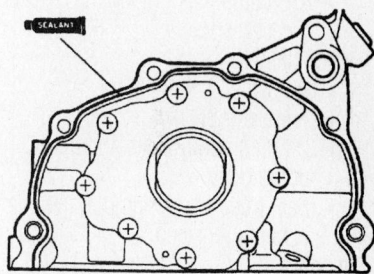

SEALANT

42356-MAZC-G26

Apply 0.04–0.07 inch (1–2mm) bead of sealant to the oil pump—1.8L (FP) and 2.0L (FS) engines

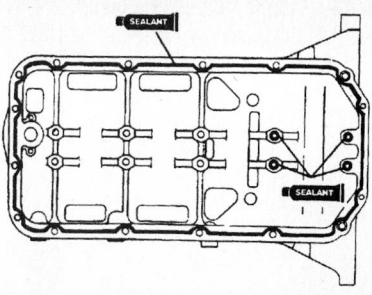

SEALANT **SEALANT**

42356-MAZC-G27

Apply a bead of sealant 0.08–0.11 inch (2–3mm) the mating surface of the upper block—1.8L (FP) and 2.0L (FS) engines

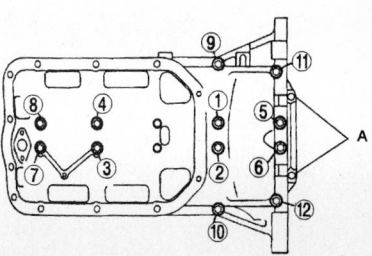

42356-MAZC-G28

Tighten the upper block bolts in this sequence—1.8L (FP) and 2.0L (FS)

8. Fill the engine with clean oil.

9. Start the vehicle, check for leaks and repair if necessary.

2.5L (KL) ENGINE

1. Before servicing the vehicle, refer to the precautions in the beginning of this section.

2. Remove or disconnect the following:
- Negative battery cable
- Crankshaft pulley
- Timing belt cover
- Timing belt
- Timing belt pulley
- Oil pan
- A/C compressor and move it aside, leaving the refrigerant lines attached
- A/C compressor mounting bracket, if necessary
- Alternator
- Oil baffle
- Alternator bracket
- Oil pump body bolts and the oil pump body
- Seal from the oil pump, if the pump is being replaced

To install:

3. Clean the oil, dirt and old sealant from all contact surfaces.

4. If the oil seal was removed from the oil pump, apply clean engine oil to the lip of the seal. Push the seal in lightly be hand. Press the seal, with a protrusion of 0–0.019 inch (0–0.5mm) into the oil pump.

5. Apply clean engine oil to the NEW O–ring and install it in the pump

6. Install or connect the following:
- Oil pump bolts in the locations illustrated noting the location and

length of the bolts. The long bolts **A** are 1.6 inch (40mm) long and the short bolts **B** are 1 inch (25mm) long. Torque the bolts to 14–18 ft. lbs. (19–25 Nm).
- Oil baffle and torque the bolts in sequence to 14–18 ft. lbs. (19–25 Nm)
- Alternator bracket
- Air conditioning compressor bracket
- Air conditioning compressor
- Oil pan
- Timing belt pulley
- Timing belt
- Timing belt cover
- Crankshaft pulley
- Negative battery cable

7. Fill the engine with clean oil.

8. Start the vehicle, check for leaks and repair if necessary.

2.3L (KJ) ENGINE

1. Before servicing the vehicle, refer to the precautions in the beginning of this section.

2. Remove or disconnect the following:
- Negative battery cable
- Crankshaft pulley
- Timing belt cover
- Timing belt
- Timing belt pulley
- Oil pan
- Alternator
- Power steering pump and bracket, move the pump aside leaving the refrigerant lines attached
- Oil baffle
- A/C compressor and move it aside, leaving the refrigerant lines attached

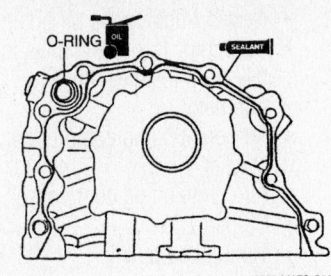

Apply clean engine oil to the NEW O–ring and install it in the pump—2.5L (KL) engines

Bolt length
A: 40 mm {1.6 in}
B: 25 mm {1.0 in}

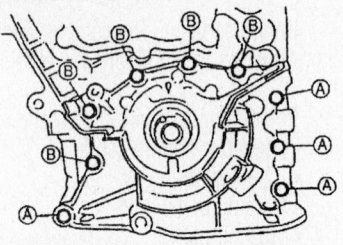

The oil pump bolts are different lengths, make sure to install them where indicated. The long bolts A are 1.6 inch (40mm) long and the short bolts B are 1 inch (25mm) long—2.5L (KL) engines

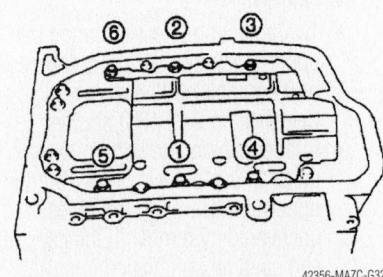

Tighten the oil baffle and torque the bolts in sequence—2.5L (KL) engines

- A/C compressor mounting bracket, if necessary
- Vacuum pump
- Oil pump body bolts and the oil pump body
- Seal from the oil pump, if the pump is being replaced

To install:

3. Clean the oil, dirt and old sealant from all contact surfaces.

4. If the oil seal was removed from the oil pump, apply clean engine oil to the lip of the seal. Push the seal in lightly be hand. Press the seal, with a protrusion of 0–0.02 inch (0–0.7mm) into the oil pump.

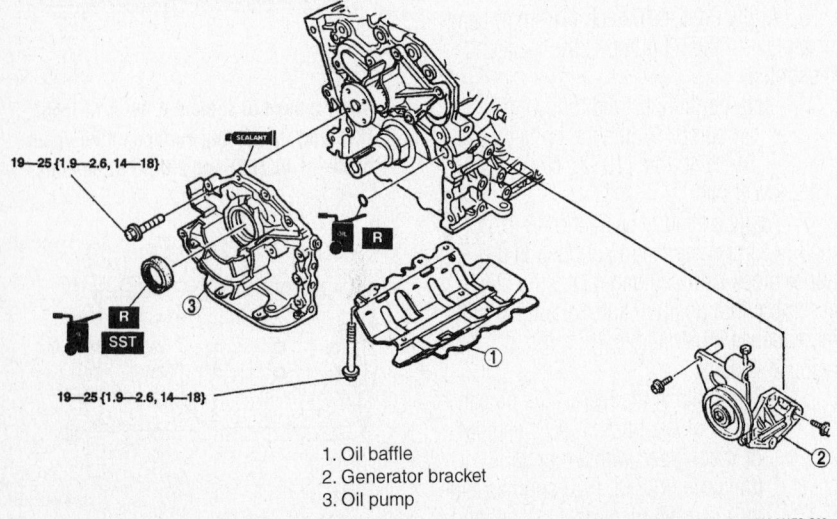

19—25 {1.9—2.6, 14—18}

19—25 {1.9—2.6, 14—18}

1. Oil baffle
2. Generator bracket
3. Oil pump

Exploded view of the oil pump and related components for the 2.5L (KL) engines

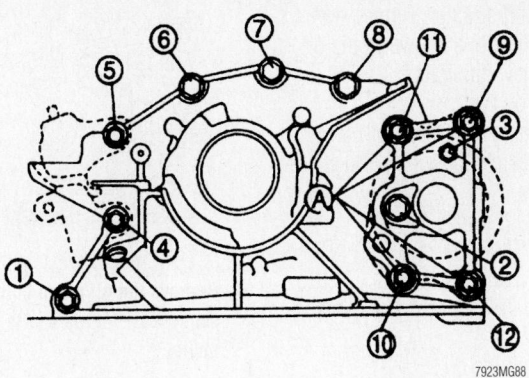

Tightening the oil pump mounting bolts in sequence—2.3L (KJ) engines

5. Apply a 0.04–0.07 inch (1–2mm) bead of silicone sealant to the contact surface of the pump.

6. Apply clean engine oil to the NEW O–ring and install it in the pump

7. Install or connect the following:
- Oil pump. Torque the bolts, in sequence, to 15–22 ft. lbs. (20–30 Nm) for the **A** bolts and to 14–18 ft. lbs. (19–25 Nm) for all other bolts.
- Vacuum pump
- Air conditioning compressor bracket
- Oil baffle and torque the bolts in sequence to 14–18 ft. lbs. (19–25 Nm)
- Power steering pump and bracket
- Alternator
- Air conditioning compressor
- Oil pan
- Timing belt pulley
- Timing belt
- Timing belt cover
- Crankshaft pulley
- Negative battery cable

8. Fill the engine with clean oil.

9. Start the vehicle, check for leaks and repair if necessary.

Miata

1.8L (BP) ENGINES

1. Before servicing the vehicle, refer to the precautions in the beginning of this section.

2. Remove or disconnect the following:
- Negative battery cable
- Crankshaft pulley
- Timing belt cover
- Timing belt
- Crankshaft sprocket
- A/C compressor and move it aside, leaving the refrigerant lines attached
- A/C compressor mounting bracket
- Alternator

- Oil pump bolts and the oil pump
- Front crankshaft seal from the oil pump, if the pump is being replaced

To install:

3. Clean the oil, dirt and old sealant from all contact surfaces.

4. If the oil seal was removed from the oil pump, apply clean engine oil to the lip of the seal. Push the seal in lightly be hand. Press the seal, with a protrusion of 0–0.02 inch (0–0.5mm) into the oil pump.

5. Apply a bead of silicone to the oil pump body-to-cylinder block contact surface, going inside the bolt holes.

6. Install or connect the following:
- Oil pump. Torque the bolts to 14–18 ft. lbs. (19–25 Nm).
- Air conditioning compressor bracket
- Air conditioning compressor

- Crankshaft sprocket
- Timing belt
- Timing belt cover
- Crankshaft pulley
- Negative battery cable

7. Fill the engine with clean oil.

8. Start the vehicle, check for leaks and repair if necessary.

Mazda 3 and 6 Models

2.0L (LF) AND 2.3L (L3) ENGINES

1. Before servicing the vehicle, refer to the precautions in the beginning of this section.

2. Remove the oil pan.

3. Remove the oil strainer.

4. Remove the oil pump chain guide, tensioner, spring and chain.

5. Remove the oil pump sprocket using tool 49 G032 354 to stop the pump rotating.

6. Remove the oil pump bolts and the pump.

To install:

7. Install the oil pump and tighten the bolts in two steps in the sequence illustrated. Tighten the bolts in sequence to 71–88 inch lbs. (8–10 Nm and final tighten to 15.2–18.4 Nm).

8. Install the oil pump sprocket using tool 49 G032 354 to stop the pump rotating. Tighten the bolts to the specifications shown in the oil pump exploded view illustration.

9. Install the oil pump chain, spring, tensioner and guide. Tighten the bolts to the specifications shown in the oil pump exploded view illustration.

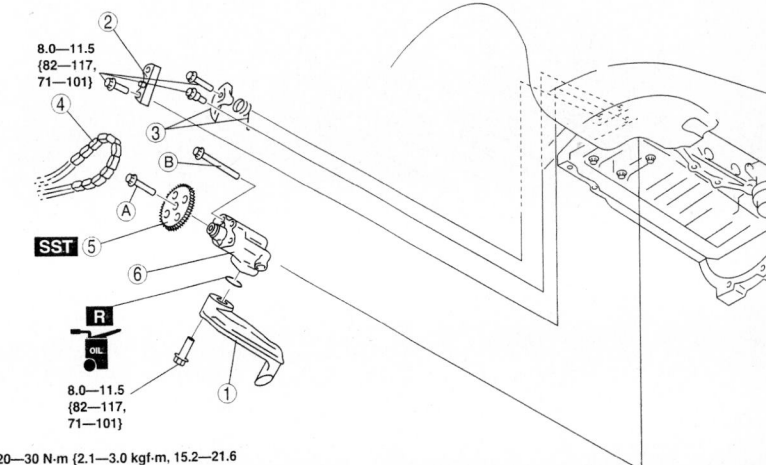

```
8.0—11.5
{82—117,
71—101}
```

```
8.0—11.5
{82—117,
71—101}
```

Ⓐ : 20—30 N·m {2.1—3.0 kgf·m, 15.2—21.6
Ⓑ : 8—10 {82—101, 71—88} + 20—25 {204—254,177—221}

N·m {kgf·cm, in·lbf}

1	Oil strainer	4	Oil pump chain
2	Oil pump chain guide	5	Oil pump sprocket
3	Oil pump chain tensioner and spring component	6	Oil pump

67162-MAZC-G47

Exploded view of the oil pump and related components—2.0L (LF) and 2.3L (L3) engines

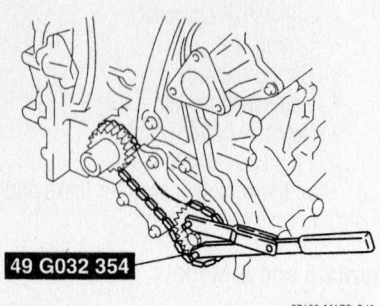

49 G032 354

67162-MAZC-G48

Remove and install the oil pump sprocket using tool 49 G032 354 to stop the pump rotating—2.0L (LF) and 2.3L (L3) engines

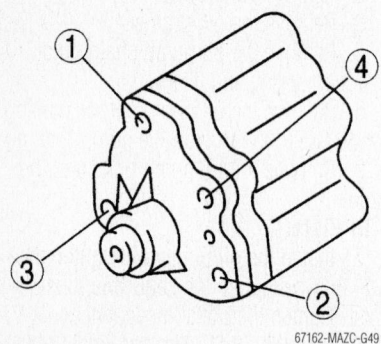

67162-MAZC-G49

Oil pump torque sequence—2.0L (LF) and 2.3L (L3) engines

10. Install the oil strainer. Tighten the bolts to the specifications shown in the oil pump exploded view illustration.

11. Install the oil pan.

12. Fill the engine with clean oil.

13. Start the vehicle, check for leaks and repair if necessary.

3.0L (AJ) ENGINE

1. Before servicing the vehicle, refer to the precautions in the beginning of this section.

2. Drain the engine oil.

3. Disconnect the negative battery cable.

4. Remove the oil pan.

5. Remove the timing chain.

6. Remove the oil strainer.

7. Remove the oil pump bolts in the sequence illustrated.

8. Thoroughly clean the gasket mating surfaces.

To install:

9. Install the oil pump and tighten the bolts in the proper sequence. Torque the bolts to 71–106 inch lbs. (8–12 Nm).

10. Install the oil strainer. Tighten the strainer bolts to 71–106 inch lbs. (8–12 Nm) and the oil strainer stay nut to 133 inch lbs. (15 Nm) plus an additional 45 degree turn.

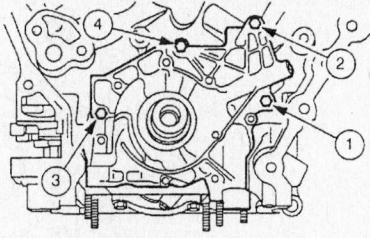

67162-MAZC-G179

Remove the oil pump bolts in the proper sequence—Mazda 6 with the 3.0L (AJ) engine

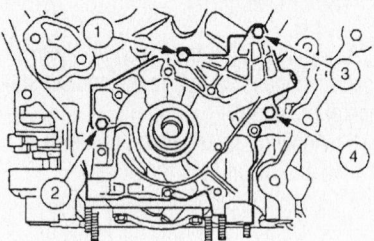

67162-MAZC-G180

Tighten the oil pump bolts in the proper sequence—Mazda 6 with the 3.0L (AJ) engine

11. Install the remaining components in the reverse order of removal.

12. Refill the engine with clean oil.

13. Start the engine and check for leaks; repair if necessary.

Rear Main Seal

REMOVAL & INSTALLATION

Mazda 3 and 6 Models

1. Before servicing the vehicle, refer to the precautions in the beginning of this section.

2. Remove or disconnect the following:
   - Negative battery cable
   - Transaxle/transmission assembly
   - Clutch/flywheel assembly, if equipped with a manual transaxle/transmission
   - Flexplate/shim plates, if equipped with an automatic transaxle/transmission

3. Cut the oil seal lip with a knife. Install a rag to the housing and using a prytool, carefully pry the oil seal from the oil seal housing.

4. Clean the gasket mounting surfaces.

To install:

5. Clean the oil seal housing. Coat the oil seal and the housing with clean engine oil.

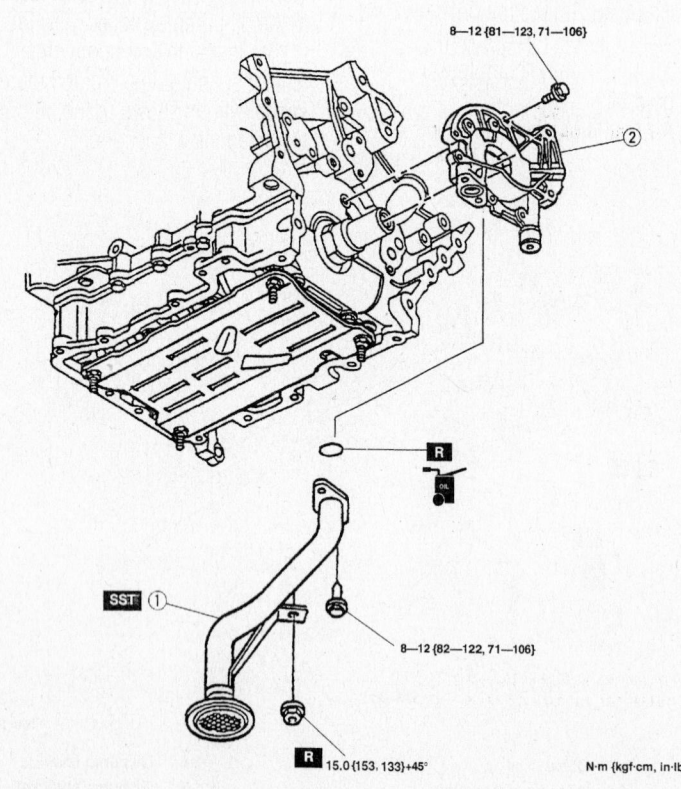

8—12 {81—123, 71—106}

8—12 {82—122, 71—106}

15.0 {153, 133}+45°

N·m {kgf·cm, in-lbf}

| 1 | Oil strainer | 2 | Oil pump |

67162-MAZC-G178

Exploded view of the oil pump and related components—Mazda 6 with the 3.0L (AJ) engine

6. Install or connect the following:
- New oil seal into the housing by tapping it evenly into place with a hammer and a seal installer until it is flush with the edge of the rear cover
- Clutch/flywheel assembly or the flexplate, as applicable
- Transaxle/transmission
- Negative battery cable

Mazda 3 and 6 Models

2.0L (LF) AND 2.3L (L3) ENGINES

1. Before servicing the vehicle, refer to the precautions in the beginning of this section.
2. Remove the transmission.
3. Remove the flywheel.
4. Remove the rear seal housing bolts.
5. Remove the seal.

To install:

6. Apply a 0.16–0.23 inch (4–6mm) bead of silicone sealant to the seal mating surfaces as illustrated. Install the seal within 10 minutes of applying the sealant.
7. Apply a coat of clean engine oil to the new seal lip.
8. Using a suitable installer tool, install the seal as illustrated.
9. Tighten the seal housing bolts in sequence to 71–101 inch lbs. (8–11.5 Nm).

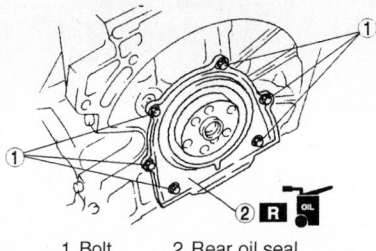

1 Bolt 2 Rear oil seal

67162-MAZC-G181

Rear oil seal housing bolt removal sequence—2.0L (LF) and 2.3L (L3) engines

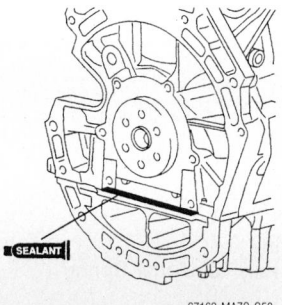

67162-MAZC-G50

Apply a 0.16–0.23 inch (4–6mm) bead of silicone sealant to the seal mating surfaces—2.0L (LF) and 2.3L (L3) engines

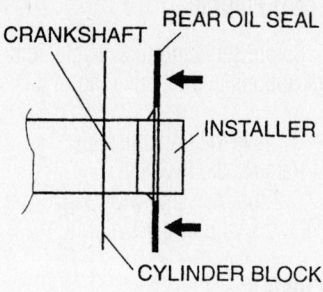

67162-MAZC-G51

Using a suitable installer tool, install the seal as shown—2.0L (LF) and 2.3L (L3) engines

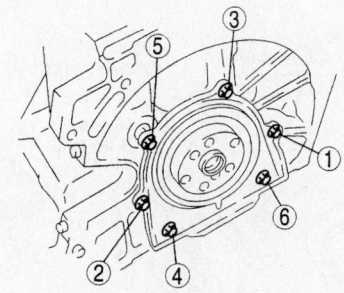

67162-MAZC-G52

Rear oil seal housing bolt torque sequence—2.0L (LF) and 2.3L (L3) engines

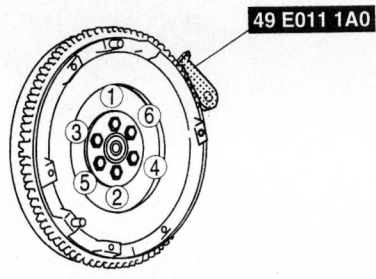

67162-MAZC-G53

Flywheel bolt torque sequence—2.0L (LF) and 2.3L (L3) engines with a manual transaxle

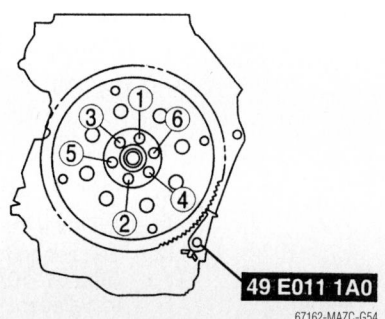

67162-MAZC-G54

Flywheel bolt torque sequence—2.0L (LF) and 2.3L (L3) engines with an automatic transaxle

10. Install the flywheel, use tool 49 E011 1A0 to prevent the unit from turning. If reusing the old bolts coat them with a thread locking compound. If install new bolts no locking compound is needed.
11. On Mazda 3 models, tighten bolts in sequence 80–85 ft. lbs. (108–116 Nm) on models equipped with a manual transaxle. On models equipped with an automatic transaxle, tighten the bolts to 80–86 ft. lbs. (96–103 Nm).
12. On Mazda 6 models, if equipped with a G35M-R manual transaxle, tighten the bolts to 79–84 ft. lbs. (108–115 Nm). If equipped with a A65M-R manual transaxle, tighten the bolts to 54–64 ft. lbs. (73–87 Nm). On models equipped with an FN4A-EL automatic transaxle, tighten the bolts to 80–86 ft. lbs. (96–103 Nm). On models equipped with an JA5A-EL automatic transaxle, tighten the bolts to 54–64 ft. lbs. (73–87 Nm).
13. Install the transaxle.

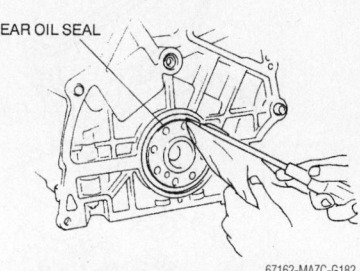

67162-MAZC-G182

Remove the seal using a suitable pry-tool—Mazda 6 with the 3.0L (AJ) engine

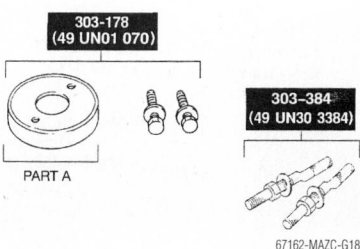

67162-MAZC-G183

Assemble the new oil seal with part (A) of tool 49 UN01 070 and tool 49 UN30 3384—Mazda 6 with the 3.0L (AJ) engine

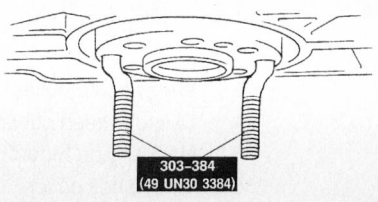

67162-MAZC-G184

Install the studs of tool 49 UN30 3384—Mazda 6 with the 3.0L (AJ) engine

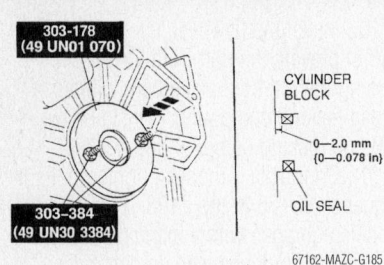

303-178
(49 UN01 070)

303-384
(49 UN30 3384)

CYLINDER
BLOCK

0—2.0 mm
(0—0.078 in)

OIL SEAL

67162-MAZC-G185

compress the seal into the bore by tightening the nuts on tool 49 UN30 3384 until the seal is recessed 0–0.078 inch (0–2mm)—Mazda 6 with the 3.0L (AJ) engine

3.0L (AJ) Engine

1. Before servicing the vehicle, refer to the precautions in the beginning of this section.
2. Remove the transmission.
3. Remove the flywheel.
4. Cut the rear seal with a razor and remove the seal using a suitable pry-tool.

To install:

5. Assemble the new oil seal with part (A) of tool 49 UN01 070 and tool 49 UN30 3384.

6. Install the studs of tool 49 UN30 3384 as illustrated.
7. Coat the oil seal with clean engine oil.
8. Push the seal in by hand and install part (A) of tool 49 UN01 070 and compress the seal into the bore by tightening the nuts on tool 49 UN30 3384 until the seal is recessed 0–0.078 inch (0–2mm).
9. Remove the tools.
10. Install the flywheel and tighten bolts in sequence 80–85 ft. lbs. (108–116 Nm).
11. Install the transaxle.

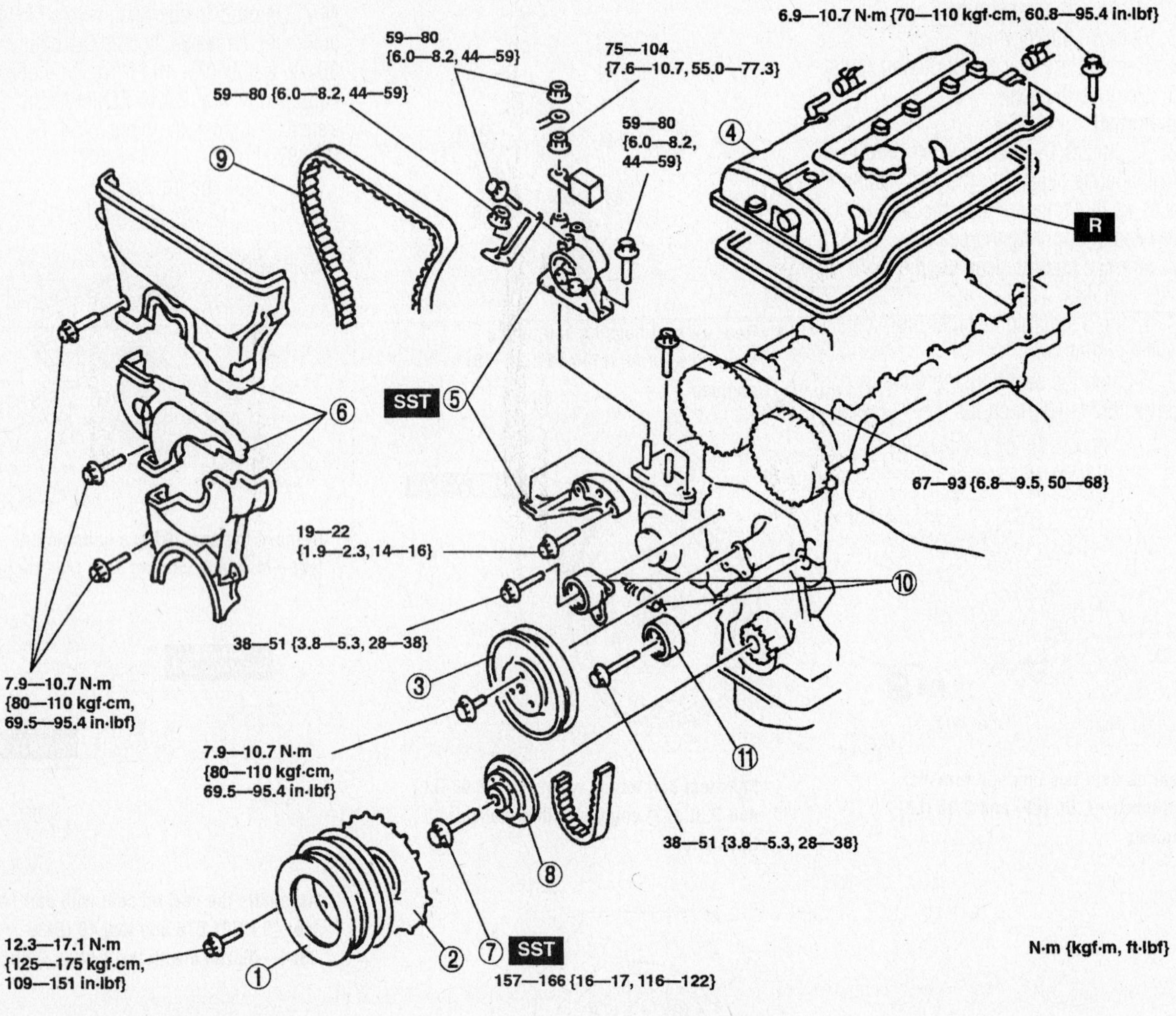

59—80
{6.0—8.2, 44—59}

59—80 {6.0—8.2, 44—59}

75—104
{7.6—10.7, 55.0—77.3}

59—80
{6.0—8.2, 44—59}

6.9—10.7 N·m {70—110 kgf·cm, 60.8—95.4 in·lbf}

⑨

④

R

SST ⑤

⑥

67—93 {6.8—9.5, 50—68}

19—22
{1.9—2.3, 14—16}

⑩

38—51 {3.8—5.3, 28—38}

③

⑪

7.9—10.7 N·m
{80—110 kgf·cm,
69.5—95.4 in·lbf}

7.9—10.7 N·m
{80—110 kgf·cm,
69.5—95.4 in·lbf}

⑧

38—51 {3.8—5.3, 28—38}

12.3—17.1 N·m
{125—175 kgf·cm,
109—151 in·lbf}

①

②

⑦ SST

157—166 {16—17, 116—122}

N·m {kgf·m, ft·lbf}

1	Crankshaft pulley	7	Pulley lock bolt
2	Plate	8	Pulley boss
3	Water pump pulley	9	Timing belt
4	Cylinder head cover	10	Tensioner, tensioner spring (See 01–10A–11 Tensioner, Tensioner Spring Installation Note)
5	No.3 engine mount		
6	Timing belt cover	11	Idler

42356-MAZC-GIA

Exploded view of the timing belt assembly—1.6L (ZM) engines

Timing Belt

REMOVAL & INSTALLATION

626 and Protégé

1.6L (ZM) ENGINE

1. Before servicing the vehicle, refer to the precautions in the beginning of this section.
2. Drain the cooling system.
3. Remove or disconnect the following:
   - Negative battery cable
   - Camshaft Position (CMP) sensor
   - Ignition coils
   - Drive belt
   - Crankshaft pulley and plate
   - Water pump pulley
   - Rocker arm cover and discard the gasket
4. Support the engine with a suitable support device and remove the number 3 engine mount.
   - Timing belt cover
   - Pulley using tool 49 D011 102 to prevent crankshaft rotation and remove the pulley boss
5. Install the pulley boss on the crankshaft and tighten the bolt.
6. Turn the crankshaft until the timing marks on the crankshaft and camshaft sprockets are aligned. Face the camshaft pulley marks **I** and **E** straight up, then align the timing marks with the horizontal surface on the cylinder head. The pin on the pulley boss must face upward. Hold the crankshaft pulley boss with a suitable tool and remove the pulley lockbolt, being careful not to rotate the crankshaft. Remove the crankshaft pulley boss.

➡**Protect the tensioner with a shop towel before prying on it. Do not rotate the crankshaft after the timing belt has been removed.**

7. Mark the direction of rotation on the timing belt. Loosen the tensioner lockbolt

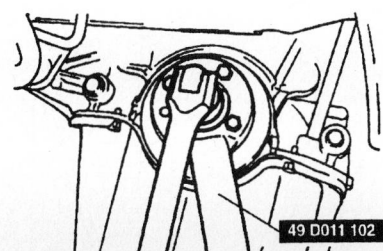

Remove pulley bolt and boss by using tool 49 D011 102 to prevent crankshaft rotation—1.6L (ZM) engines

and pry the tensioner outward. Tighten the lockbolt with the tensioner spring fully extended. Remove the timing belt.

8. Remove the tensioner and spring. If necessary, remove the idler pulley.
9. Inspect the belt for wear, peeling, cracking, hardening or signs of oil contamination. Inspect the tensioner for free and smooth rotation. Check the tensioner spring free length; it should not exceed 2.43 inch (68mm). Inspect the sprocket teeth for wear or damage. Replace parts, as necessary.

To install:

10. Install the crankshaft sprocket bolt and temporarily tighten.
11. Install the tensioner and tensioner spring. Install the spring with the damper rubber closing face on the right side. Temporarily tighten the tensioner lockbolt with the tensioner spring fully extended.
12. Face the **I** and **E** marks of the camshaft pulley marks straight up, then align the timing marks with the horizontal surface on the cylinder head. Refer to the illustration for timing mark alignment.
13. Install the timing belt so there is no looseness, refer to the illustration for location as follows:
   - Crankshaft pulley
   - Idler pulley
   - Exhaust camshaft pulley
   - Intake camshaft pulley
   - Tensioner
14. Make sure no pressure other than the tensioner spring is applied to the belt. If reusing the old belt, be sure it is installed in the same direction of rotation.
15. Temporarily install the pulley boss and lockbolt.
16. Turn the crankshaft 1⅚ turns clockwise and align the crankshaft sprocket tim-

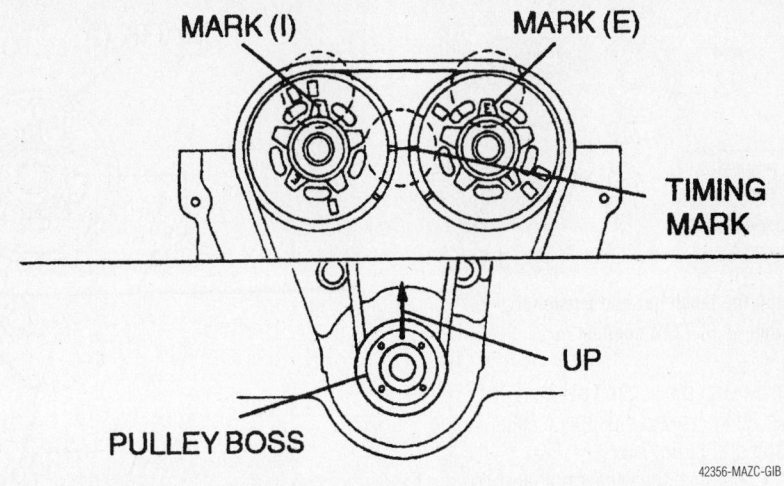

Turn the crankshaft until the timing marks on the crankshaft and camshaft sprockets are aligned. The pin on the pulley boss must face upward—1.6L (ZM) engines

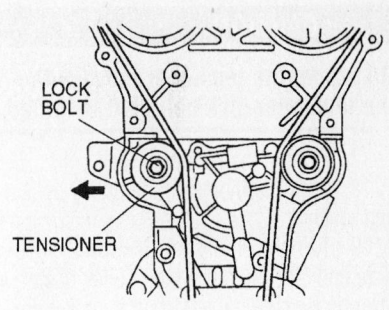

Loosen the tensioner lockbolt and pry the tensioner outward—1.6L (ZM) engines

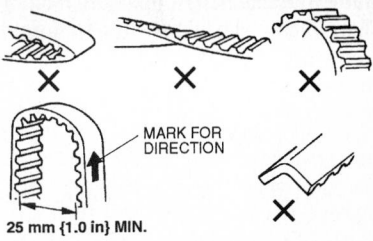

Do not to bend, twist or get oil or other contaminates on the belt as this will damage the belt, if reusing the belt; mark the direction of rotation prior to removal—1.6L (ZM) engines

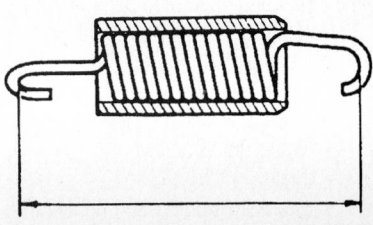

Check the tensioner spring free length; it should not exceed 2.43 inch (68mm)—1.6L (ZM) engines

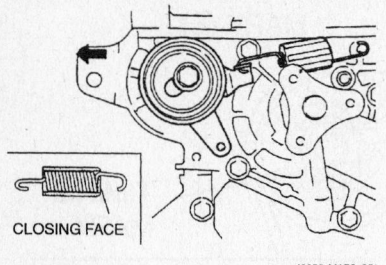

Install the tensioner and tensioner spring—1.6L (ZM) engines

ing mark with the tension set mark for proper belt tension adjustment. Remove the lockbolt and pulley boss.

- Remove the pulley bolt and boss, and verify the timing belt pulley mark is still aligned with the tensioner set mark.

✳✳ CAUTION

Do not let the tensioner move with the tensioner lock bolt as it is turned.

17. Tighten the tensioner lock bolt.
18. Temporarily install the pulley boss and lockbolt.
19. Turn the crankshaft 2⅙ turns clockwise and face the pin on the pulley boss upright. Be sure the camshaft sprocket timing marks are aligned. If they are not, repeat the alignment steps.

➡ **The timing marks are aligned normally if the camshaft pulley marksIandEare facing straight up, the timing marks are aligned to the horizontal surface on the cylinder head.**

20. Apply approximately 22 lbs. (10kg) pressure to the timing belt at a point midway between the camshaft sprockets. The belt should deflect 0.24–0.29 inch (6–7.5mm). If the deflection is not correct, repeat the alignment and tensioning procedure.
21. Install the pulley boss and lockbolt. Tighten the bolt to 116–122 ft. lbs. (157–166 Nm).
22. Install the timing belt covers and tighten the bolts to 95 inch lbs. (11 Nm).
23. Install the number 3 mount and remove the engine support device.
24. Install or connect the following:

- Rocker arm cover with a new gasket
- Water pump pulley
- Plate
- Crankshaft pulley and tighten the bolts to 109–151 inch lbs. (12–17 Nm)
- Drive belt
- Ignition coils

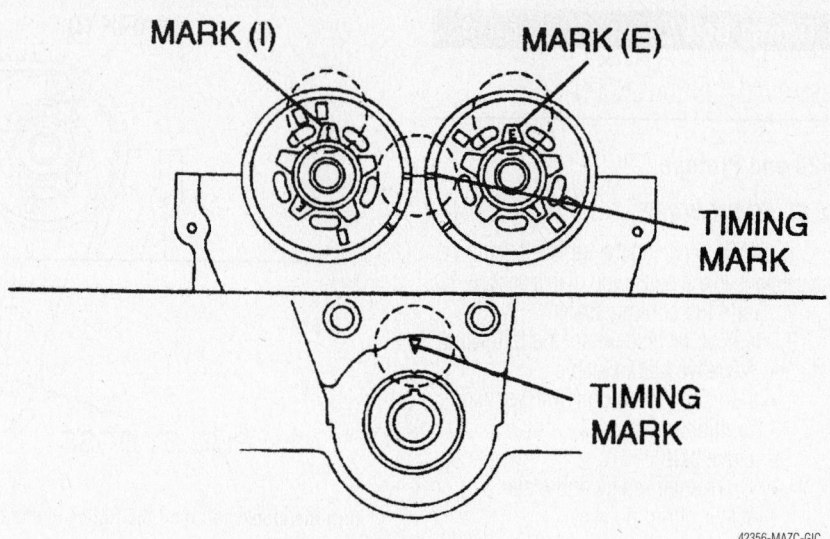

Verify the timing marks are aligned as shown before installing the belt—1.6L (ZM) engines

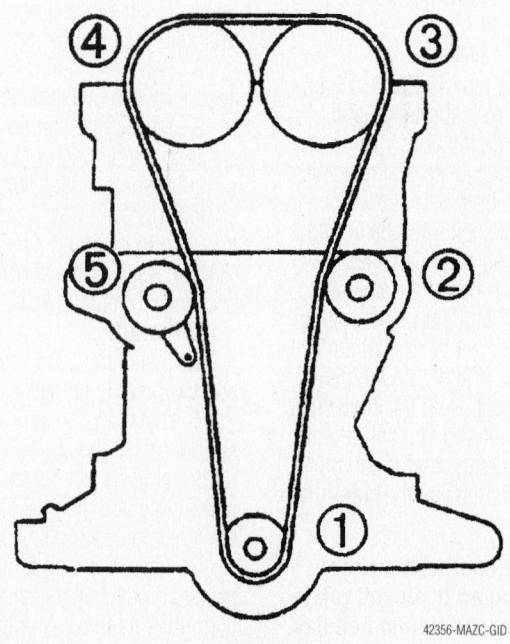

Install the timing belt in this order—1.6L (ZM) engines

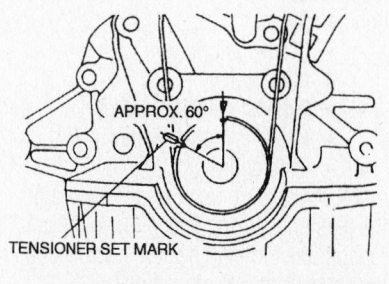

Turn the crankshaft 1⅚ turns clockwise and align the crankshaft sprocket timing mark with the tension set mark for proper belt tension adjustment—1.6L (ZM)

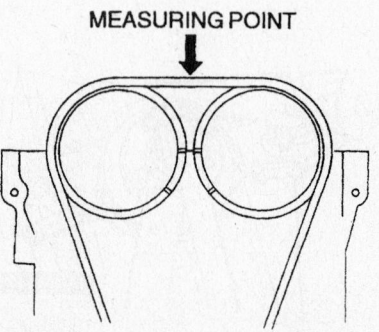

The belt should deflect 0.24–0.29 inch (6–7.5mm)—1.6L (ZM) engines

- CMP sensor and check the air gap between the sensor and the plate teeth using a brass feeler gauge. If the gap is not 0.020–0.059 inch (0.5–1.5mm) adjust the sensor until the proper gap is reached.
- Negative battery cable

1.8L (FP) AND 2.0L (FS) ENGINES

1. Before servicing the vehicle, refer to the precautions in the beginning of this section.
2. Refer to the illustration of the exploded view of the timing belt assembly for component location and torque specifications.
3. Remove or disconnect the following:
   - Negative battery cable
   - Crankshaft Position (CKP) sensor
   - Camshaft Position (CMP) sensor on Protégé models
   - Spark plugs
   - Power steering pump and bracket, leave the lines attached and position the pump aside on 2.0L (FS) engines
   - Drive belt
   - Water pump pulley

✳✳ CAUTION

The CKP sensor rotor is on the rear of the crankshaft pulley and can be damaged if the pulley is not removed carefully.

- Crankshaft pulley using special tools 49 G011 103, 49 E011 1A1 and 49 S120 710. Refer to the illustration for tool positioning.

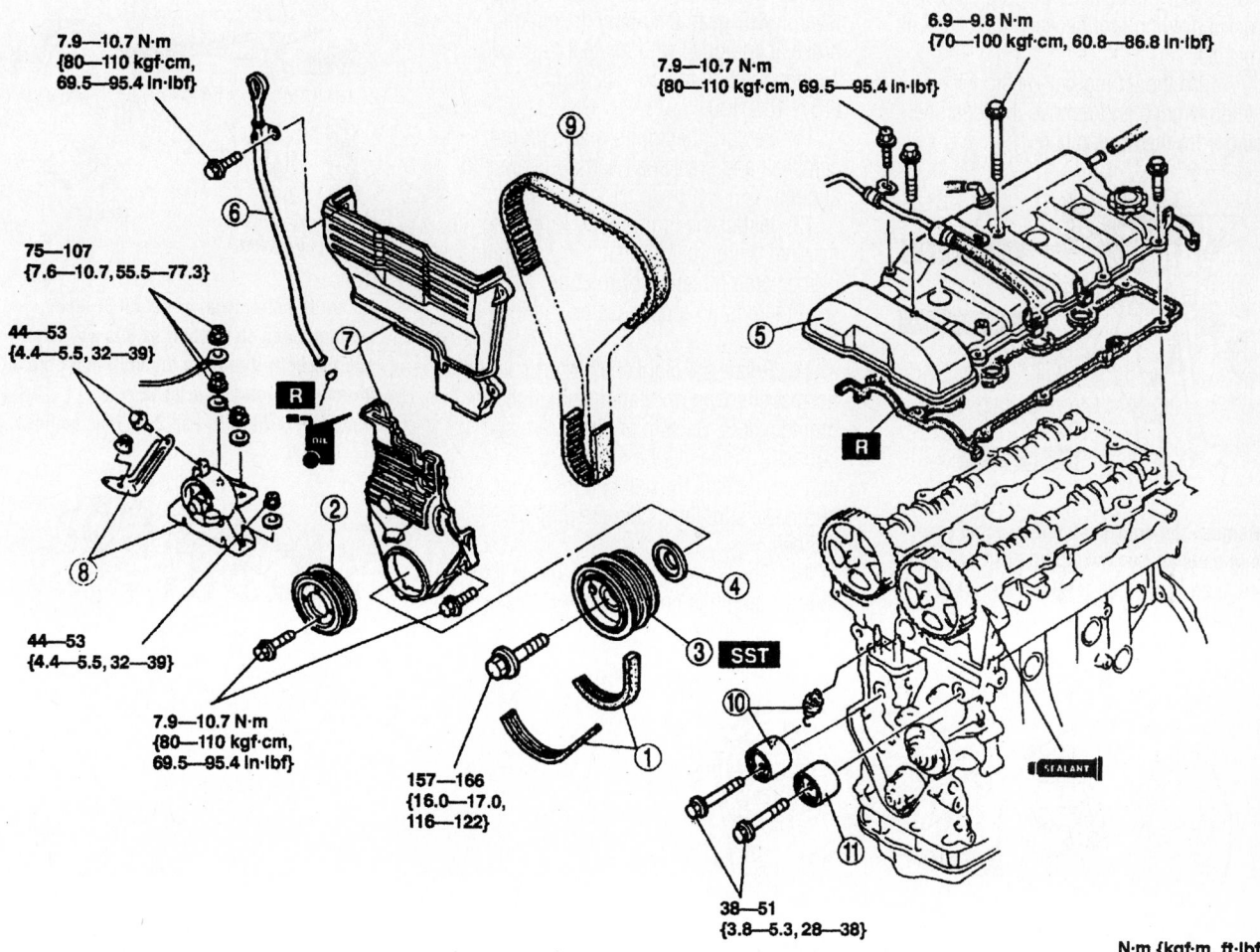

1	Drive belt
2	Water pump pulley
3	Crankshaft pulley
4	Guide plate
5	Cylinder head cover
6	Dipstick and pipe
7	Timing belt cover
8	No.3 engine mount rubber
9	Timing belt
10	Tensioner, tensioner spring
11	Idler

Exploded view of the timing belt assembly–2.0L (FS) engines shown, 1.8L (FP) engines are similar

- Guide plate from behind the crankshaft pulley
- Rocker arm cover
- Oil dipstick and tube
- Timing belt cover

4. Support the engine assembly with a hoist and remove the number 3 engine mount rubber.

5. Turn the crankshaft until the timing mark on the crankshaft sprocket aligns with the timing mark on the oil pump and the camshaft sprocket timing marks **E** and **I** align on the camshaft sprockets. Refer to the illustration for proper timing mark alignment.

6. Lower the vehicle. Insert a camshaft sprocket holding tool between the camshaft sprockets.

7. Turn the timing belt tensioner with an Allen wrench and remove the tensioner spring from the hook pinch

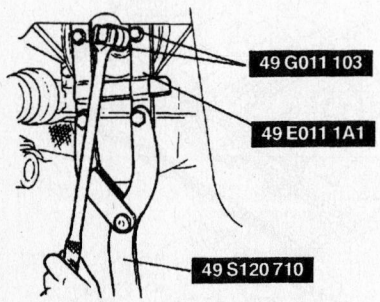

Remove the crankshaft pulley using special tools 49 G011 v103, 49 E011 1A1 and 49 S120 710—1.8L (FP) and 2.0L (FS) engines

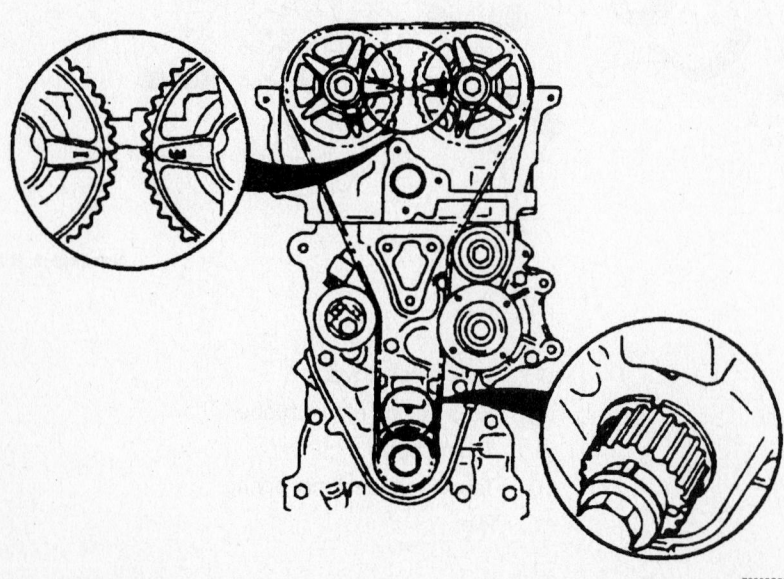

When properly aligned for belt removal, the cam gear marks should face each other—1.8L (FP) and 2.0L (FS) engines

✳✳ CAUTION

Be careful not to bend, twist or get oil or other contaminates on the belt as this will damage the belt.

8. If the timing belt is to be reused, mark the direction of rotation on the timing belt. Remove the timing belt.

9. Rotate the tensioner, if the tension rotates with no resistance or does not rotate; replace the tensioner

To install:

10. Install the crankshaft sprocket bolt. Install the flywheel locking tool, if equipped with automatic transaxle, or place the shift lever in **4th** gear and apply the parking brake, if equipped with manual transaxle. Tighten the bolt to 116–122 ft. lbs. (157–166 Nm).

11. Be sure the timing marks on the camshaft and crankshaft sprockets are still aligned.

12. Install the timing belt. If reusing the original timing belt, be sure it is installed in the same direction of rotation. Make sure there is no looseness on the idler side.

13. Rotate the crankshaft 2 turns in the normal direction of rotation and align the timing marks. Be sure all marks are still correctly aligned. If the marks are not aligned, remove the belt and then reinstall it and make sure the marks are properly aligned.

14. Check the tensioner spring length, if the free length is not 1.44 inch (36.5mm), replace the spring.

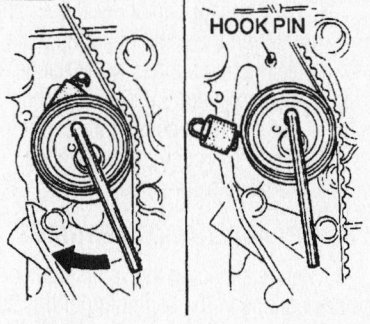

Remove the tensioner spring from the hook pin—1.8L (FP) and 2.0L (FS)

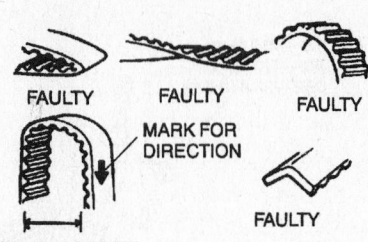

Do not to bend, twist or get oil or other contaminates on the belt as this will damage the belt, if reusing the belt; mark the direction of rotation prior to removal—1.8L (FP) and 2.0L (FS) engines

Rotate the tensioner, if the tension rotates with no resistance or does not rotate; replace the tensioner—1.8L (FP) and 2.0L (FS) engines

✳✳ CAUTION

Do not use tension other than that supplied by the tension spring or damage could occur

15. Turn the tensioner clockwise with an Allen wrench and install the tensioner spring. Remove the holding tool from between the camshaft sprockets.

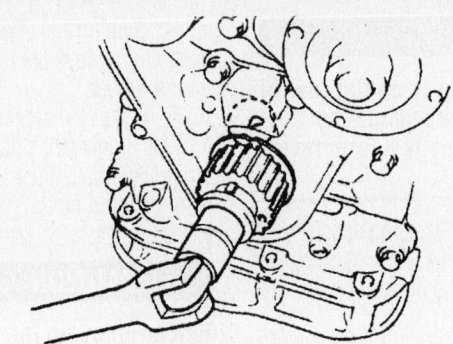

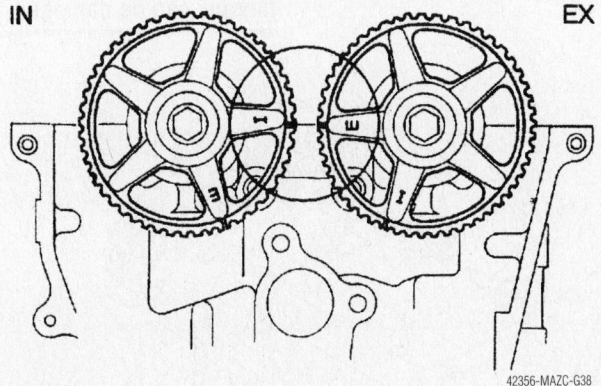

Be sure the timing marks on the camshaft and crankshaft sprockets are still aligned—1.8L (FP) and 2.0L (FS) engines

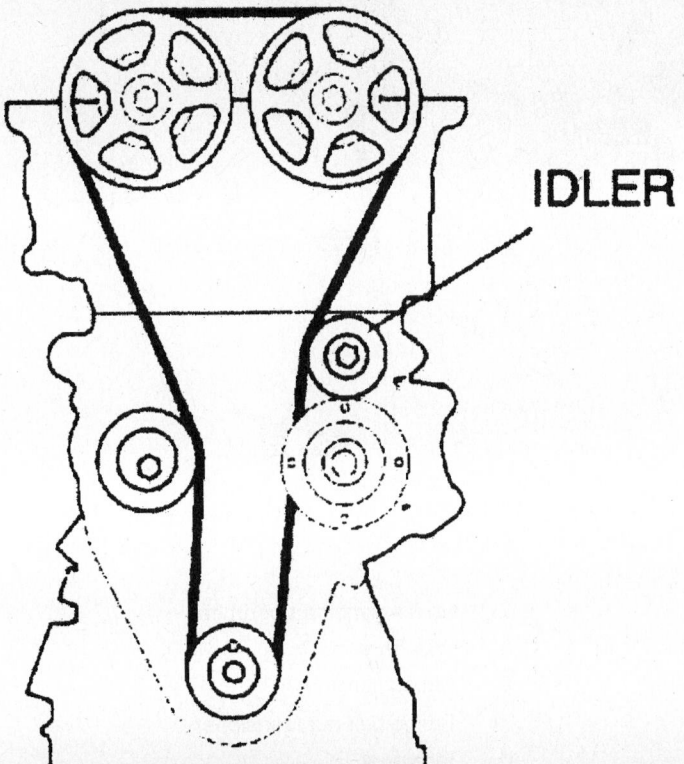

Make sure there is no looseness on the idler side when the belt is installed—1.8L (FP) and 2.0L (FS) engines

Check the tensioner spring length, if the free length is not 1.44 inch (36.5mm); replace the spring—1.8L (FP) and 2.0L (FS) engines

16. Rotate the crankshaft 2 turns in the normal direction of rotation and align the timing marks. Be sure all marks are still correctly aligned.

17. Install the number 3 engine mount rubber and tighten the fasteners to the specifications shown in the illustration. Remove the engine hoist.

18. Install or connect the following:
- Timing belt cover. Refer to the exploded view of the timing belt components for torque specifications.
- Oil dipstick and tube. Refer to the exploded view of the timing belt components for torque specifications.
- Rocker arm cover
- Guide plate

❄❄ CAUTION

The CKP sensor rotor is on the rear of the crankshaft pulley and can be damaged if the pulley is not installed carefully.

- Crankshaft pulley using special tools 49 G011 v103, 49 E011 1A1 and 49 S120 710 and tighten the bolt to 116–122 ft. lbs. (157–166 Nm).
- Water pump pulley. Refer to the exploded view of the timing belt components for torque specifications.
- Drive belt
- Power steering bracket and pump, if removed
- Spark plugs
- CKP sensor
- Negative battery cable

2.5L (KL) ENGINE

1. Before servicing the vehicle, refer to the precautions in the beginning of this section.

2. Refer to the illustration of the exploded view of the timing belt assembly for component location and torque specifications.

3. Remove or disconnect the following:
- Negative battery cable
- Crankshaft Position (CKP) sensor
- Spark plugs
- Power steering pump, leave the lines attached and position the pump aside
- Drive belt
- Water pump pulley
- Idler pulley bracket

✳✳ CAUTION

The CKP sensor rotor is on the rear of the crankshaft pulley and can be damaged if the pulley is not removed carefully.

- Crankshaft pulley using special tools 49 E011 1A1 and 49 S120 710. Refer to the illustration for tool positioning.
- Rocker arm cover
- Oil dipstick and tube
- Timing belt cover

4. Support the engine assembly with a hoist and remove the number 3 engine mount rubber.

5. Install the crankshaft pulley bolt and turn the crankshaft clockwise and align the timing marks as illustrated. The number 1 piston should be at Top Dead Center (TDC) of the compression stroke.

✳✳ CAUTION

When removing the number 1 idler pulley bolt, hold the pulley so that the threads are aligned or the threads can be damaged.

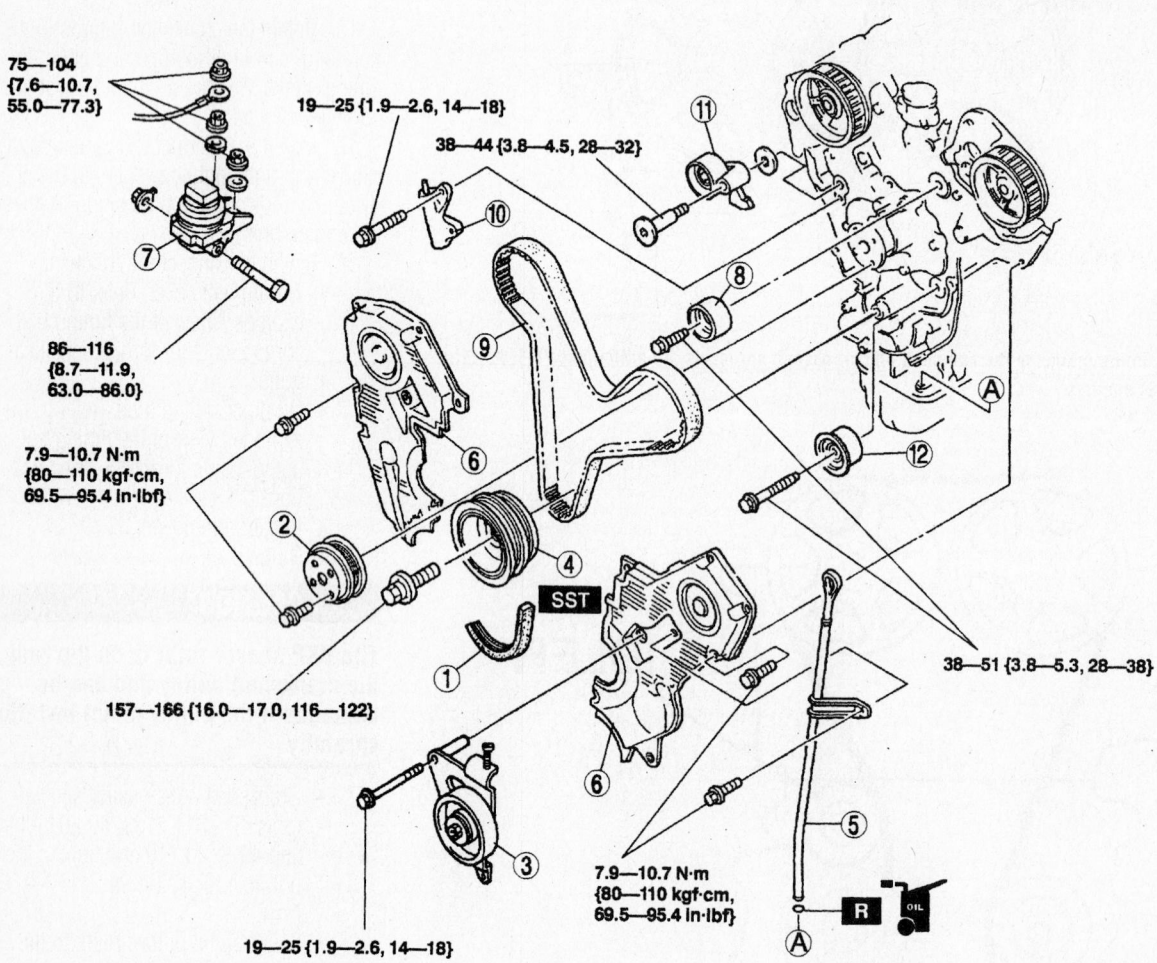

N·m {kgf·m, ft·lbf}

1	Drive belt	7	No.3 engine mount rubber
2	Water pump pulley	8	No.1 idler
3	Idler pulley bracket	9	Timing belt
4	Crankshaft pulley	10	Timing belt auto tensioner
5	Dipstick and pipe	11	Tensioner
6	Timing belt cover	12	No.2 idler

42356-MAZC-G41

Exploded view of the timing belt assembly—2.5L (KL) engines—626 models

6. Remove the number 1 idler pulley

✷✷ CAUTION

When removing the auto tensioner bolts, hold the tensioner so that the threads are aligned or the threads can be damaged.

7. Loosen the auto tensioner bolts and remove the lower bolt.

✷✷ CAUTION

Be careful not to bend, twist or get oil or other contaminates on the belt as this will damage the belt.

8. If the timing belt is to be reused, mark the direction of rotation on the timing belt. Remove the timing belt.

9. Remove the belt.

10. If necessary, remove the auto tensioner.

To install:

✷✷ CAUTION

There are two type of auto tensioners and they are interchangeable.

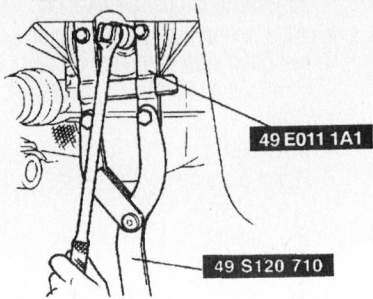

Remove the crankshaft pulley using special tools 49 E011 1A1 and 49 S120 710—2.5L (KL) engines

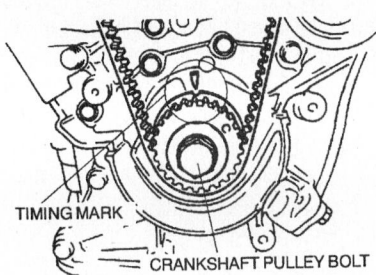

Turn the crankshaft clockwise and align the timing marks as illustrated. The number 1 piston should be at Top Dead Center (TDC) of the compression stroke.—2.5L (KL) engines

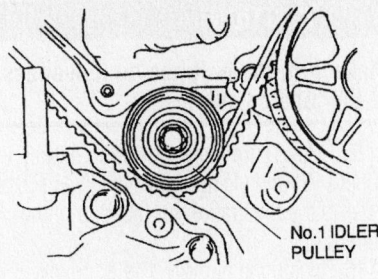

Remove the number 1 idler pulley.—2.5L (KL) engines

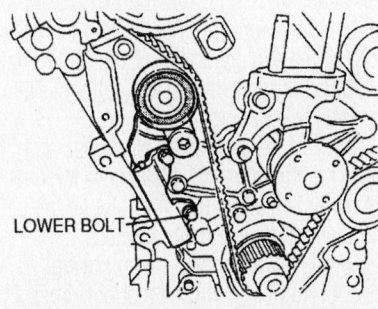

Loosen the auto tensioner bolts and remove the lower bolt—2.5L (KL) engines

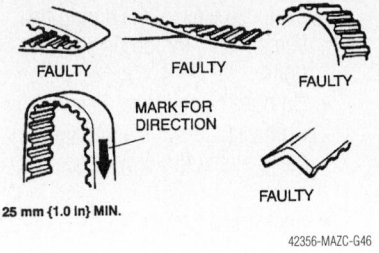

Do not to bend, twist or get oil or other contaminates on the belt as this will damage the belt, if reusing the belt; mark the direction of rotation prior to removal—2.5L (KL) engines

11. If removed install the auto tensioner as follows:

    a. Measure the tensioner rod projection length. If the length exceeds 0.563–0.594 inch (14.3–15.1mm) on type **A** tensioners or 0.473–0.511 inch (12–13mm) on type **B** tensioners replace the tensioner. Refer to the illustration to distinguish tensioner type.

    b. Inspect the tensioner for leakage and replace if defective.

✷✷ CAUTION

Do not apply pressure of more than 2,200 lb. (9.8 Kn) to the tensioner or damage will occur.

    c. Using a press, slowly press in the tensioner rod.

    d. Insert a 0.055 inch (1.4mm) diameter pin to hold the rod in position.

    e. Install the tensioner and hand tighten the bolt.

12. Turn the camshafts clockwise and align the timing marks as illustrated.

13. Install the crankshaft pulley bolt and turn the crankshaft clockwise and align the marks as illustrated.

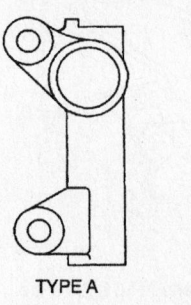 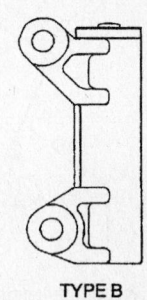

There are two type of auto tensioners and they are interchangeable—2.5L (KL) engines

Projection (Free length)
  Type A: 14.3–15.1 mm {0.563—0.594 in}
  Type B: 12.0—13.0 mm {0.473—0.511 in}

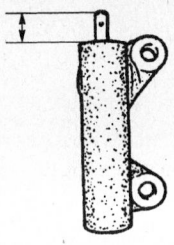

Measure the tensioner rod projection length and replace if it exceeds specification—2.5L (KL) engines

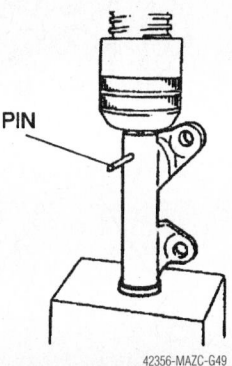

Using a press, slowly press in the tensioner rod and insert a 0.055 inch (1.4mm) diameter pin to hold the rod in position—2.5L (KL) engines

14. With the number 1 idler pulley removed, install the belt on the pulleys in this order:

- Timing belt pulley
- Number 2 idler pulley
- Left hand camshaft pulley
- Tensioner pulley
- Right hand camshaft pulley

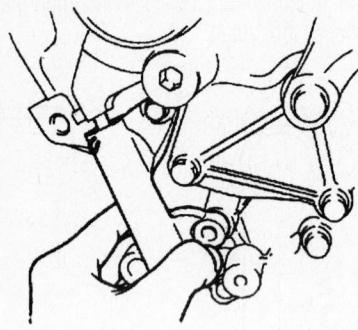

Install the tensioner and hand tighten the bolt—2.5L (KL) engines

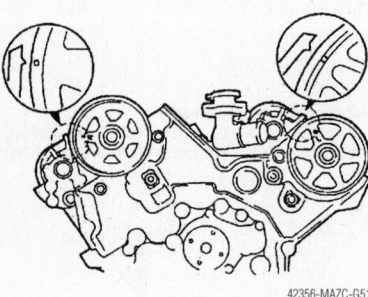

Turn the camshafts clockwise and align the timing marks—2.5L (KL) engines

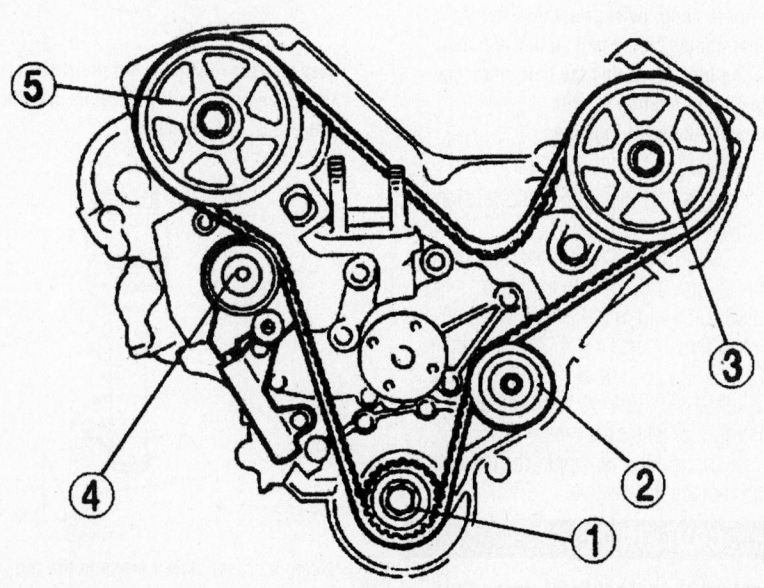

Make sure there is tension between pulleys (3) and (1) and pulleys (1) and (5)—2.5L (KL) engines

---

❋❋ CAUTION

Make sure the belt has no looseness at the tension side.

15. Make sure there is tension between pulleys (3) and (1) and pulleys (1) and (5). Refer to the illustration for location of the pulleys.

16. Install the number 1 idler pulley while applying the pressure on the timing belt. Tighten the bolt to 28–38 ft. lbs. (38–51 Nm).

17. Push the auto tensioner in the direction of arrow (refer to illustration) and hand tighten the lower bolt, then tighten the bolts. Refer to the exploded view of the timing belt components for torque specifications. Then remove the retaining pin.

18. Turn the crankshaft clockwise and align the timing marks. Make sure the timing marks are aligned, if not remove the belt and repeat the installation process.

19. Install or connect the following:

- Number 3 engine mount rubber. Refer to the exploded view of the timing belt components for torque specifications. Remove the engine hoist.
- Timing belt cover. Refer to the exploded view of the timing belt components for torque specifications.
- Oil dipstick and tube. Refer to the exploded view of the timing belt components for torque specifications.
- Rocker arm cover

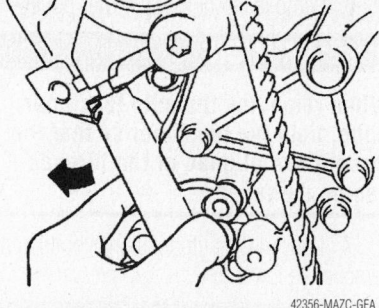

Push the auto tensioner in the direction of arrow and hand tighten the lower bolt—2.5L (KL) engine

❋❋ CAUTION

The CKP sensor rotor is on the rear of the crankshaft pulley and can be damaged if the pulley is not removed carefully.

- Crankshaft pulley using special tools 49 E011 1A1 and 49 S120 710. Refer to the illustration for tool positioning. Tighten the bolt to 16–122 ft. lbs. (157–166 Nm).
- Idler pulley bracket. Refer to the exploded view of the timing belt components for torque specifications.
- Water pump pulley. Refer to the exploded view of the timing belt components for torque specifications.
- Drive belt
- Power steering pump
- Spark plugs
- CKP sensor
- Negative battery cable

Millenia

2.3L (KJ) ENGINE

1. Before servicing the vehicle, refer to the precautions in the beginning of this section.

2. Refer to the illustration of the exploded view of the timing belt assembly for component location and torque specifications.

3. Remove or disconnect the following:

- Negative battery cable
- Right front wheel
- Splash shield from the right hand side
- Dust cover
- Drive belt

❋❋ CAUTION

The Crankshaft Position (CKP) sensor rotor is on the rear of the crankshaft pulley and can be damaged if the pulley is not removed carefully.

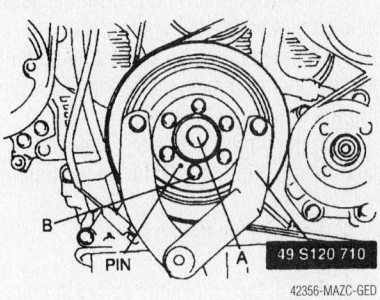

42356-MAZC-GED

Remove the crankshaft pulley using special tools 49 E120 710—2.3L (KJ) engines

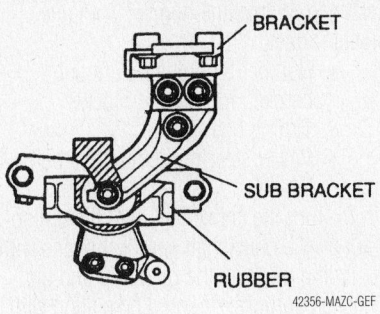

42356-MAZC-GEF

Remove the number 3 engine mount sub bracket, rubber and bracket—2.3L (KJ) engines

4. Turn the crankshaft pulley clockwise until the pin on the pulley is facing down.
   - Crankshaft pulley using special tools 49 E120 710. Refer to the illustration for tool positioning.
   - Power steering pump, leave the lines attached and position the pump aside
   - Water pump pulley
   - Alternator drive belt tensioner
   - Camshaft Position (CMP) sensor
   - Oil dipstick and tube
   - Vacuum pipe
   - Upper radiator hose
5. Support the engine assembly with a

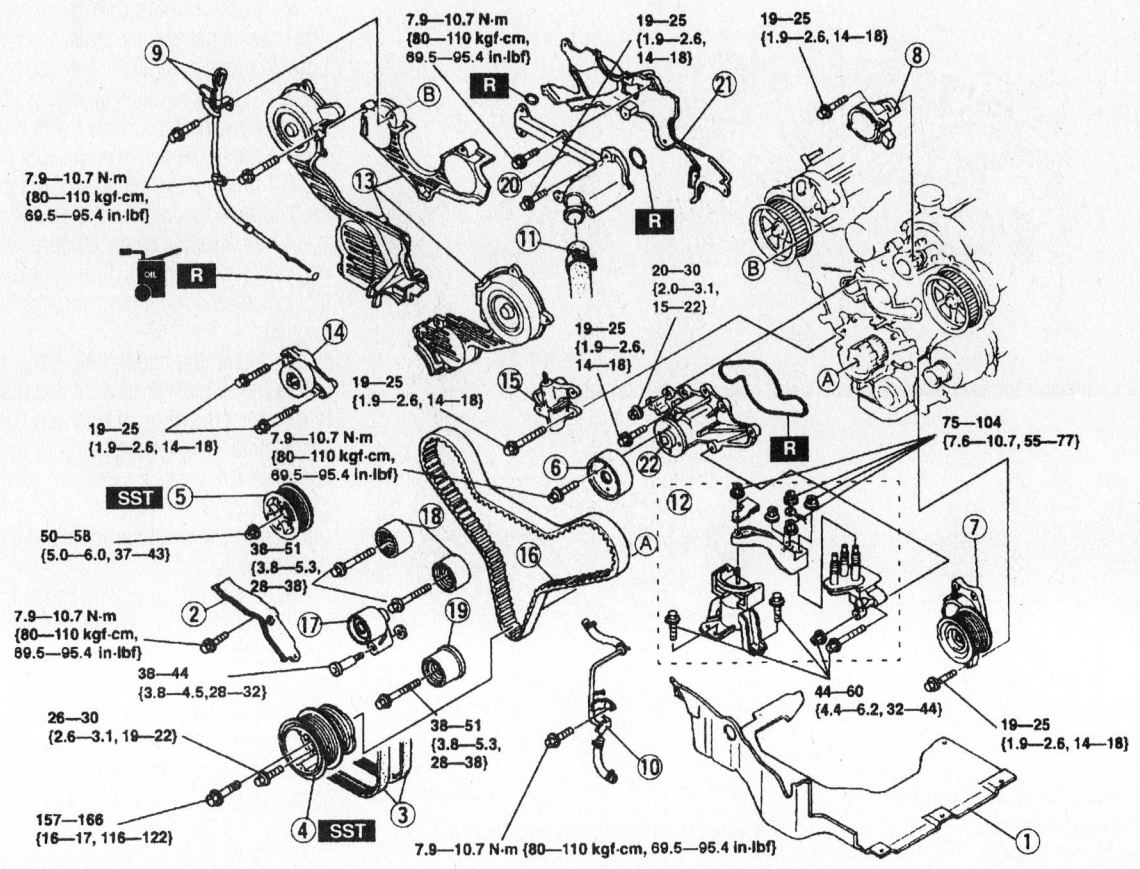

N·m {kgf·m, ft·lbf}

1	Splash shield (RH)	12	No.3 engine mount
2	Dust cover	13	Timing belt cover
3	Drive belt	14	Drive belt auto tensioner (P/S)
4	Crankshaft pulley	15	Timing belt auto tensioner
5	P/S oil pump pulley	16	Timing belt
6	Water pump pulley	17	Tensioner pulley
7	Drive belt auto tensioner (Generator)	18	No.1 idler pulley
8	CMP sensor	19	No.2 idler pulley
9	Dipstick and pipe	20	Water outlet pipe
10	Vacuum pipe	21	Seal plate
11	Upper radiator hose	22	Water pump

42356-MAZC-GEC

Exploded view of the timing belt assembly—2.3L (KJ) engines

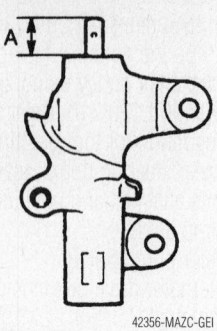

42356-MAZC-GEI

Check the tension rod projection, it should be 0.563–0.594 inch (14.3–15.1mm)–2.3L (KJ) engines

hoist and remove the number 3 engine mount rubber.

- Number 3 engine mount sub bracket, rubber and bracket
- Timing belt cover
- Power steering pump drive belt tensioner

6. Turn the crankshaft until the timing mark on the crankshaft sprocket aligns with the timing mark on the oil pump and the camshaft sprocket timing marks align with the marks on the cylinder head. The No. 1 piston should be at Top Dead Center (TDC) of the compression stroke.

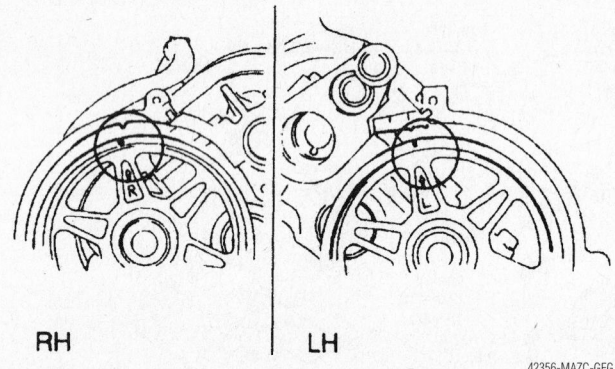

RH   LH

42356-MAZC-GEG

Be sure the camshaft sprocket timing marks are still aligned–2.3L (KJ) engines

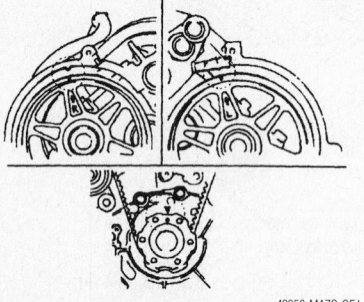

42356-MAZC-GEJ

Check the timing belt mark alignment once all the components are installed–2.3L (KJ) engines

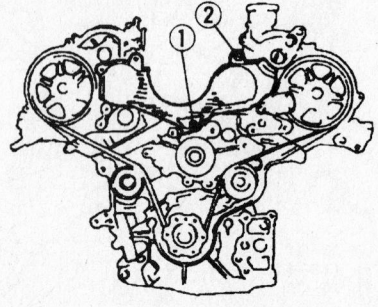

42356-MAZC-GEL

Tighten the upper timing belt cover in this sequence–2.3L (KJ) engine

7. Remove the two bolts from the automatic tensioner, removing the lower one first. Keep the bolt holes aligned by holding the tensioner to reduce the chance of stripping the threads on the bolts.

8. If the timing belt is to be reused, mark the direction of rotation on the timing belt.

9. Remove the timing belt.

To install:

10. Install the crankshaft sprocket bolt. Install the flywheel locking tool. Tighten the bolt to 116–122 ft. lbs. (157–166 Nm).

11. Check the tension rod projection, it should be 0.563–0.594 inch (14.3–15.1mm), if not, replace the tensioner

12. Position the automatic tensioner in a press. Set a flat washer under the tensioner body to prevent damage to the body plug.

13. Compress the tensioner until the hole in the piston is aligned with the 2nd hole in the tensioner case. Insert a 0.055 inch (1.4mm) diameter wire or pin through the 2nd hole to keep the piston compressed.

14. Be sure the camshaft sprocket timing marks are still aligned. Turn the crankshaft clockwise until the timing sprocket is aligned.

15. Install the timing belt. If the original belt is being reused, be sure it is installed in the same direction of rotation. The order of installation is: timing belt (crankshaft) sprocket, No. 2 idler pulley, left-hand camshaft sprocket, both No. 1 idler pulleys, right-hand camshaft sprocket and the tensioner pulley.

16. Install the automatic belt tensioner and tighten the bolts to 14–18 ft. lbs. (19–25 Nm). Remove the wire or pin from the tensioner.

17. Turn the crankshaft clockwise, until the crankshaft sprocket timing mark is again at TDC. This should place all of the belt slack in the automatic tensioner portion of the belt.

18. Rotate the crankshaft two turns in the normal direction of rotation and align the

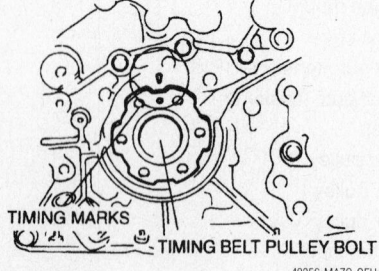

TIMING MARKS   TIMING BELT PULLEY BOLT

42356-MAZC-GEH

Turn the crankshaft clockwise, until the crankshaft sprocket timing mark is again at TDC–2.3L (KJ) engines

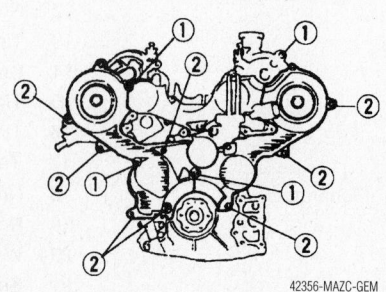

42356-MAZC-GEM

Tighten the right and left hand timing belt cover in this sequence–2.3L (KJ) engine

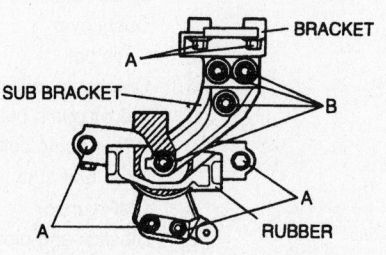

BRACKET
SUB BRACKET
A
B
A
A
RUBBER

42356-MAZC-GEK

Install the number 3 engine mount rubber, bracket and sub bracket and tighten to specification–2.3L (KJ) engine

timing marks. Be sure all marks are still correctly aligned.

19. Inspect timing belt deflection, 0.24–0.31 inch (6–8mm), between the crankshaft sprocket and the tensioner pulley. If it is out of specification, replace the auto-tensioner.

20. Install or connect the following:
- Power steering pump drive belt tensioner
- Upper timing belt cover and tighten the bolts to 95 inch lbs. (11 Nm) in the sequence illustrated
- Right and left hand timing belt cover

and tighten the bolts to 95 inch lbs. (11 Nm) in the sequence illustrated
- Number 3 engine mount rubber, bracket and sub bracket. Tighten bolts **A** to 44 ft. lbs. (60 Nm) and bolts **B** to 77 ft. lbs. (104 Nm).
- Upper radiator hose
- Vacuum pipe
- Oil dipstick and tube
- CMP sensor
- Alternator drive belt tensioner
- Water pump pulley
- Power steering pump

21. Remove the timing belt pulley **A**,

install the crankshaft pulley. Hand tighten bolts **A** and **B**. Tighten the bolt **A** to 116–122 ft. lbs. (157–166 Nm) and bolt **B** to 22 ft. lbs. (30 Nm).
- Drive belt
- Dust cover
- Splash shield from the right hand side
- Right front wheel
- Negative battery cable

2.5L (KL) ENGINE

1. Before servicing the vehicle, refer to the precautions in the beginning of this section.

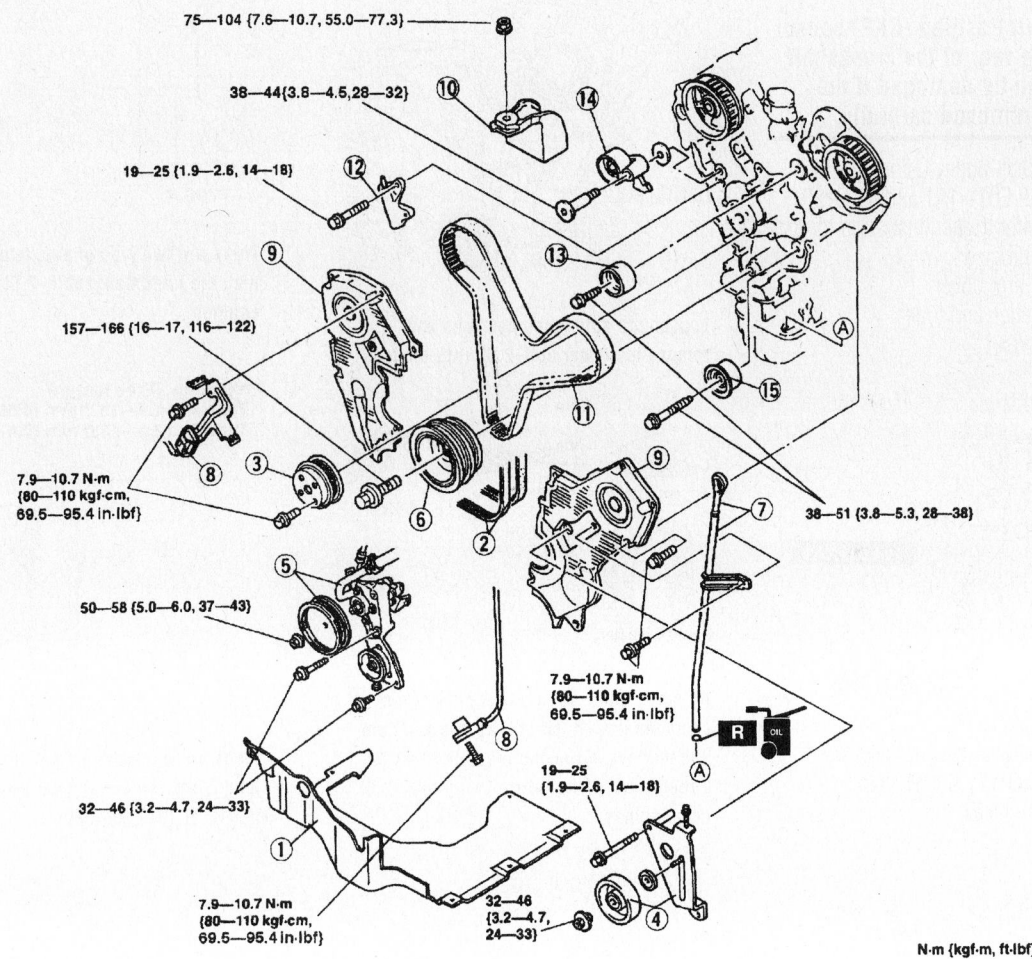

1 Splash shield (RH)
2 Drive belt
3 Water pump pulley
4 Idler pulley and bracket
5 P/S oil pump
6 Crankshaft pulley
7 Dipstick and pipe
8 CKP sensor
9 Timing belt cover
10 No.3 engine mount sub bracket
11 Timing belt
12 Timing belt auto tensioner
13 No.1 idler pulley
14 Tensioner pulley
15 No.2 idler pulley

42356-MAZC-GEB

Exploded view of the timing belt assembly–2.5L (KL) engines—Millenia models

2. Refer to the illustration of the exploded view of the timing belt assembly for component location and torque specifications.

3. Remove or disconnect the following:
- Negative battery cable
- Splash shield from the right hand side
- Drive belt
- Water pump pulley
- Idler pulley and bracket
- Power steering pump, leave the lines attached and position the pump aside

✳✳ CAUTION

The Crankshaft Position (CKP) sensor rotor is on the rear of the crankshaft pulley and can be damaged if the pulley is not removed carefully.

- Crankshaft pulley using special tools 49 E011 1A1 and 49 S120 710. Refer to the illustration for tool positioning.
- Rocker arm cover
- Oil dipstick and tube
- CKP sensor

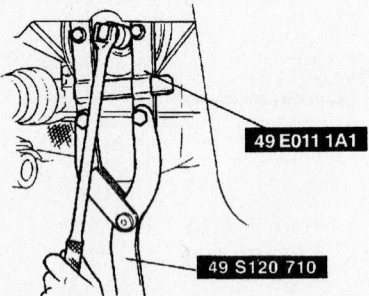

Remove the crankshaft pulley using special tools 49 E011 1A1 and 49 S120 710–2.5L (KL) engines

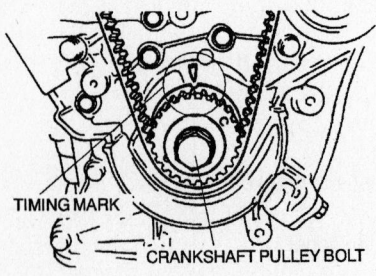

Turn the crankshaft clockwise and align the timing marks as illustrated. The number 1 piston should be at Top Dead Center (TDC) of the compression stroke.–2.5L (KL) engines

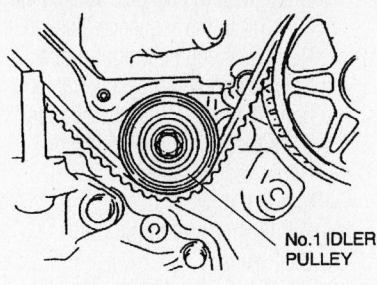

Remove the number 1 idler pulley.–2.5L (KL) engines

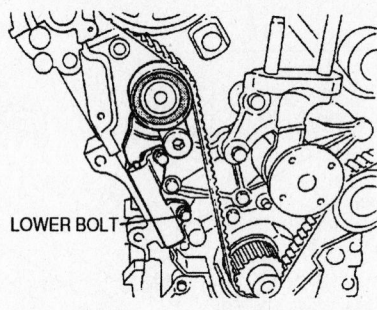

Loosen the auto tensioner bolts and remove the lower bolt–2.5L (KL) engines

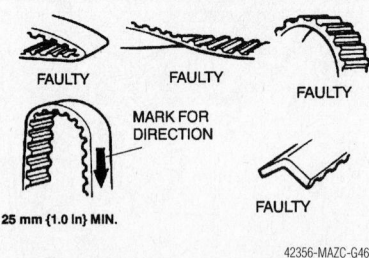

Do not to bend, twist or get oil or other contaminates on the belt as this will damage the belt, if reusing the belt; mark the direction of rotation prior to removal–2.5L (KL) engines

- Timing belt cover

4. Support the engine assembly with a hoist and remove the number 3 engine mount rubber.

5. Install the crankshaft pulley bolt and turn the crankshaft clockwise and align the timing marks as illustrated. The number 1 piston should be at Top Dead Center (TDC) of the compression stroke.

✳✳ CAUTION

When removing the auto tensioner bolts, hold the tensioner so that the threads are aligned or the threads can be damaged.

6. Loosen the auto tensioner bolts and remove the lower bolt.

✳✳ CAUTION

When removing the number 1 idler pulley bolt, hold the pulley so that the threads are aligned or the threads can be damaged.

7. Remove the number 1 idler pulley

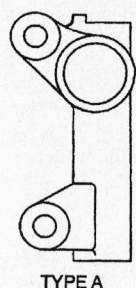

There are two type of auto tensioners and they are interchangeable–2.5L (KL) engines

Projection (Free length)
Type A: 14.3—15.1 mm {0.563—0.594 in}
Type B: 12.0—13.0 mm {0.473—0.511 in}

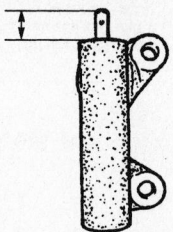

Measure the tensioner rod projection length and replace if it exceeds specification–2.5L (KL) engines

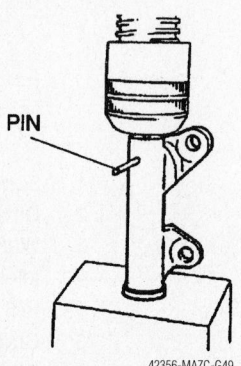

Using a press, slowly press in the tensioner rod and insert a 0.055 inch (1.4mm) diameter pin to hold the rod in position–2.5L (KL) engines

✳✳ CAUTION

Be careful not to bend, twist or get oil or other contaminates on the belt as this will damage the belt.

8. If the timing belt is to be reused, mark the direction of rotation on the timing belt. Remove the timing belt.

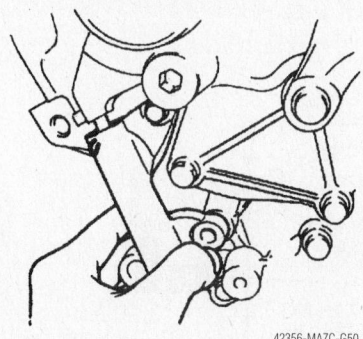

Install the tensioner and hand tighten the bolt—2.5L (KL) engines

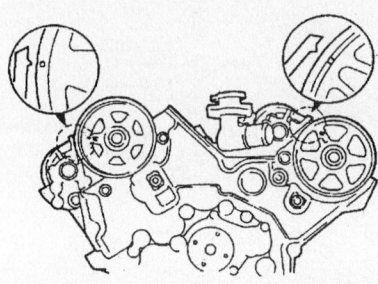

Turn the camshafts clockwise and align the timing marks—2.5L (KL) engines

9. Remove the belt.
10. If necessary, remove the auto tensioner.

To install:

✳✳ CAUTION

There are two type of auto tensioners and they are interchangeable.

11. If removed install the auto tensioner as follows:

  a. Measure the tensioner rod projection length. If the length exceeds 0.0.512–0.551 inch (13–14mm) on type **A** tensioners or 0.563–0.594 inch (14.3–15.1mm) on type **B** tensioners replace the tensioner. Refer to the illustration to distinguish tensioner type.

  b. Inspect the tensioner for leakage and replace if defective.

✳✳ CAUTION

Do not apply pressure of more than 2,200 lb. (9.8 Kn) to the tensioner or damage will occur.

  c. Using a press, slowly press in the tensioner rod.

  d. Insert a pin whose diameter is 0.055 inch (1.4mm) on type **A** tensioners or 0.079 inch (2mm) on type **B** tensioners to hold the rod in position.

  e. Install the tensioner and hand tighten the bolt.

12. Turn the camshafts clockwise and align the timing marks as illustrated.

13. Install the crankshaft pulley bolt and turn the crankshaft clockwise and align the marks as illustrated.

14. With the number 1 idler pulley removed, install the belt on the pulleys in this order:

- Timing belt pulley
- Number 2 idler pulley
- Left hand camshaft pulley
- Tensioner pulley
- Right hand camshaft pulley

✳✳ CAUTION

Make sure the belt has no looseness at the tension side.

15. Make sure there is tension between pulleys (3) and (1) and pulleys (1) and (5). Refer to the illustration for location of the pulleys.

16. Install the number 1 idler pulley while applying the pressure on the timing belt. Tighten the bolt to 28–38 ft. lbs. (38–51 Nm).

17. Push the auto tensioner in the direction of arrow (refer to illustration) and hand tighten the lower bolt, then tighten the bolts. Refer to the exploded view of the timing belt components for torque specifications. Then remove the retaining pin.

18. Turn the crankshaft clockwise and align the timing marks. Make sure the timing marks are aligned, if not remove the belt and repeat the installation process.

19. Install or connect the following:

- Number 3 engine mount rubber. Refer to the exploded view of the timing belt components for torque specifications. Remove the engine hoist.
- Timing belt cover. Refer to the exploded view of the timing belt components for torque specifications.
- Oil dipstick and tube. Refer to the exploded view of the timing belt components for torque specifications.
- CKP sensor
- Rocker arm cover

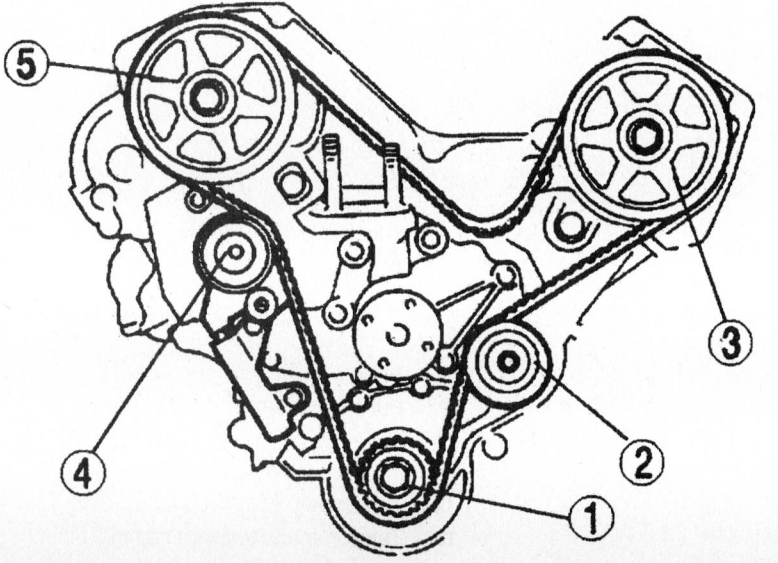

Make sure there is tension between pulleys (3) and (1) and pulleys (1) and (5)—2.5L (KL) engines

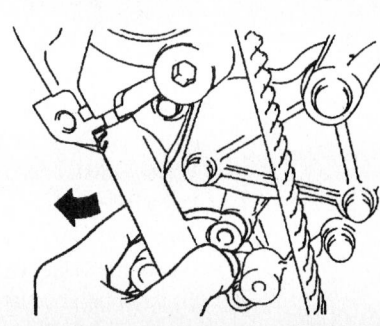

Push the auto tensioner in the direction of arrow and hand tighten the lower bolt—2.5L (KL) engine

✳✳ CAUTION

The CKP sensor rotor is on the rear of the crankshaft pulley and can be damaged if the pulley is not removed carefully.

- Crankshaft pulley using special tools 49 E011 1A1 and 49 S120 710. Refer to the illustration for tool positioning. Tighten the bolt to 16–122 ft. lbs. (157–166 Nm).

- Power steering pump
- Idler pulley bracket. Refer to the exploded view of the timing belt components for torque specifications.
- Water pump pulley. Refer to the exploded view of the timing belt components for torque specifications.
- Drive belt
- Splash shield
- Negative battery cable

Miata

1.8L (BP) ENGINE

1. Before servicing the vehicle, refer to the precautions in the beginning of this section.
2. Drain the cooling system.
3. Remove or disconnect the following:
   - Negative battery cable
   - Front suspension lower bar
   - Air pipe
   - Drive belt

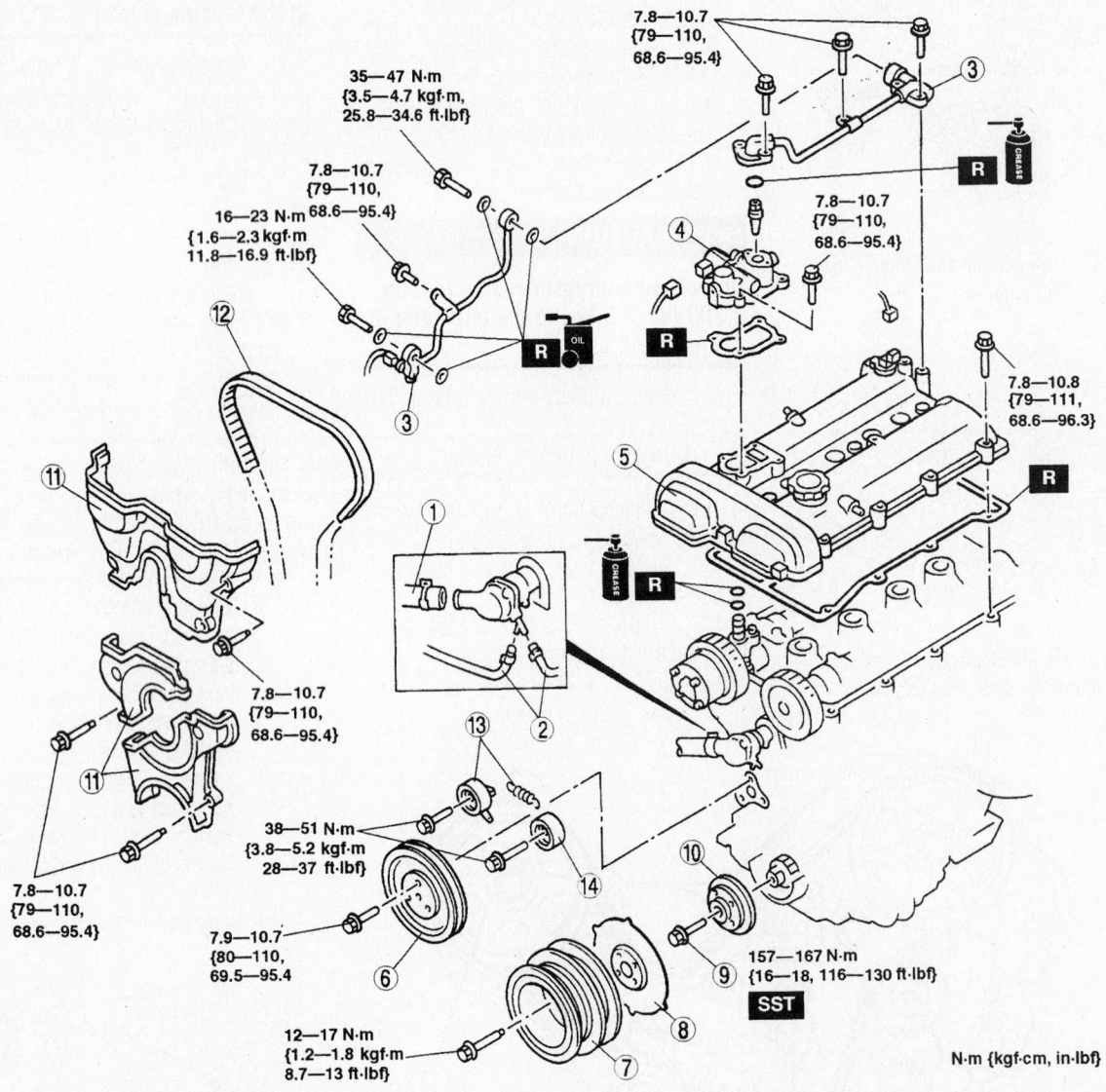

1	Upper radiator hose	8	Plate
2	Water hose	9	Pulley lock bolt
3	Oil pipe	10	Pulley boss
4	Oil control valve (OCV) case	11	Timing belt cover
5	Cylinder head cover	12	Timing belt
6	Water pump pulley	13	Tensioner and tensioner spring
7	Crankshaft pulley	14	Idler

Exploded view of the timing belt assembly—1.8L (BP) engines

42356-MAZC-GBF

- Crankshaft Position (CKP) sensor
- High tension lead and the ignition coil
- Spark plugs
- Upper radiator hose
- Water hose
- Oil pipe
- Oil Control Valve (OCV)
- Rocker arm cover and discard the gasket
- Water pump pulley
- Crankshaft pulley and plate
- Pulley bolt and boss by using tool 49 D011 102 to prevent crankshaft rotation
- Timing belt cover

4. Install the pulley boss on the crankshaft and tighten the bolt.

5. Turn the crankshaft until the timing marks on the crankshaft and camshaft sprockets are aligned. The pin on the pulley boss must face upward. Hold the crankshaft pulley boss with a suitable tool and remove the pulley lockbolt, being careful not to rotate the crankshaft. Remove the crankshaft pulley boss.

➡ **Protect the tensioner with a shop towel before prying on it. Do not rotate the crankshaft after the timing belt has been removed.**

6. Mark the direction of rotation on the timing belt. Loosen the tensioner lockbolt and pry the tensioner outward. Tighten the lockbolt with the tensioner spring fully extended. Remove the timing belt.

7. Remove the tensioner and spring. If necessary, remove the idler pulley.

8. Inspect the belt for wear, peeling, cracking, hardening or signs of oil contamination. Inspect the tensioner for free and smooth rotation. Check the tensioner spring free length; it should not exceed 2.31 inch (59.2mm). Inspect the sprocket teeth for wear or damage. Replace parts, as necessary.

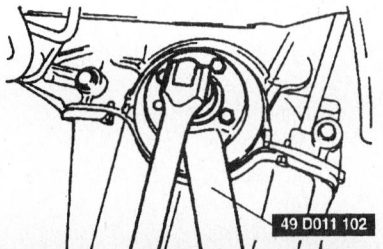

42356-MAZC-GBG

Remove pulley bolt and boss by using tool 49 D011 102 to prevent crankshaft rotation—1.8L (BP) engines

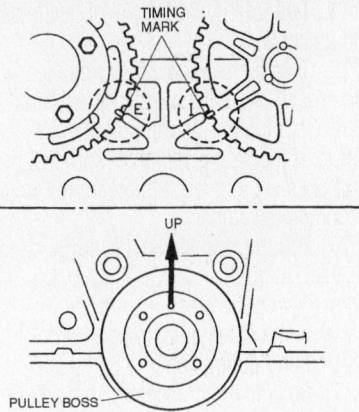

42356-MAZC-GBH

Turn the crankshaft until the timing marks on the crankshaft and camshaft sprockets are aligned. The pin on the pulley boss must face upward—1.8L (BP) engines

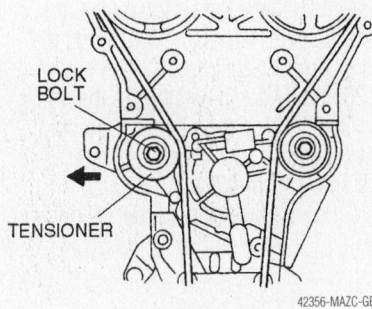

42356-MAZC-GBI

Loosen the tensioner lockbolt and pry the tensioner outward—1.8L (BP) engines

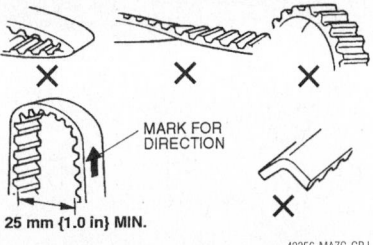

42356-MAZC-GBJ

Do not to bend, twist or get oil or other contaminates on the belt as this will damage the belt, if reusing the belt; mark the direction of rotation prior to removal—1.8L (BP) engines

To install:

9. Install the crankshaft sprocket bolt and temporarily tighten.

10. If removed, install the idler pulley and tighten the bolt to 37 ft. lbs. (51 Nm).

11. Install the tensioner and tensioner spring. Pry the tensioner outward and temporarily tighten the tensioner lock-

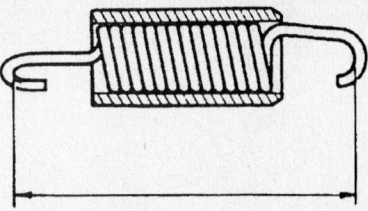

42356-MAZC-GBK

Check the tensioner spring free length; it should not exceed 2.31 inch (59.2mm)—1.8L (BP) engines

bolt with the tensioner spring fully extended.

12. Be sure the crankshaft sprocket timing mark is aligned with the mark on the oil pump housing. Be sure the camshaft sprocket timing marks are aligned with the marks on the seal plate. Refer to the illustration for timing mark alignment.

13. Install the timing belt so there is no looseness at the idler pulley side or between the camshaft sprockets. If reusing the old belt, be sure it is installed in the same direction of rotation.

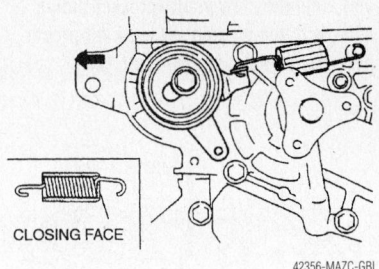

42356-MAZC-GBL

Install the tensioner and tensioner spring—1.8L (BP) engines

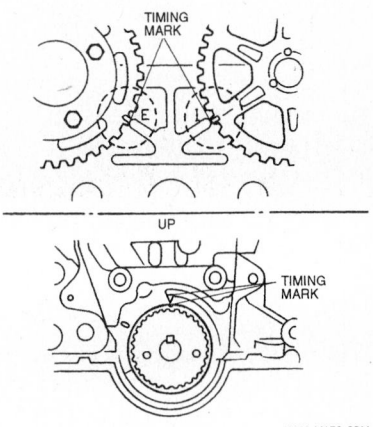

42356-MAZC-GBM

Be sure the crankshaft sprocket timing mark is aligned with the mark on the oil pump housing. Be sure the camshaft sprocket timing marks are aligned with the marks on the seal plate—1.8L (BP) engines

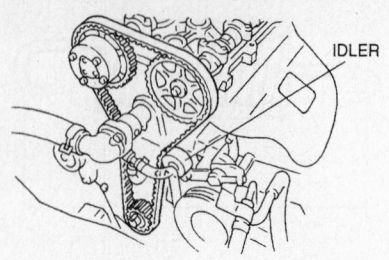

Install the timing belt so there is no looseness at the idler pulley side or between the camshaft sprockets—1.8L (BP) engines

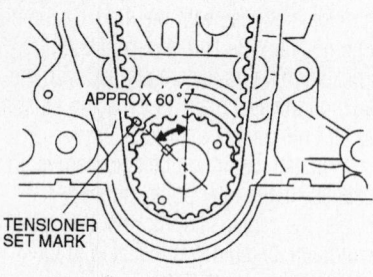

Turn the crankshaft 1⅚ turns clockwise and align the crankshaft sprocket timing mark with the tension set mark for proper belt tension adjustment—1.8L (BP)

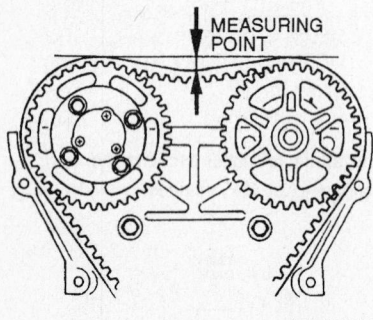

The belt should deflect 0.35–0.45 inch (9.0–11.5mm)—1.8L (BP) engines

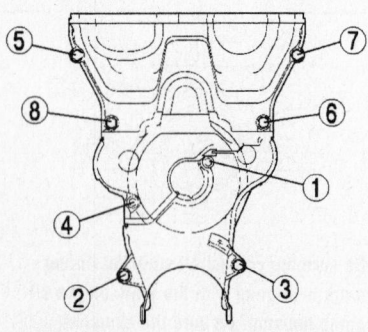

Tighten the timing belt cover using this sequence—1.8L (BP) engines

14. Temporarily install the pulley boss and lockbolt.

15. Turn the crankshaft 2 turns clockwise and align the crankshaft sprocket timing mark. Face the pin on the pulley boss upright. Be sure the camshaft sprocket timing marks are aligned. If they are not, repeat the alignment steps.

16. Turn the crankshaft 1⅚ turns clockwise and align the crankshaft sprocket timing mark with the tension set mark for proper belt tension adjustment. Remove the lockbolt and pulley boss.

17. Be sure the crankshaft sprocket timing mark is aligned with the tension set mark. Loosen the tensioner lockbolt, and allow the spring to apply tension to the belt. Tighten the tensioner lockbolt to 37 ft. lbs. (51 Nm).

18. Install the pulley boss and lockbolt.

19. Turn the crankshaft 2⅙ turns clockwise and be sure the timing marks are correctly aligned. Make sure the pin on the pulley boss is straight up.

20. Apply approximately 22 lbs. (10kg) pressure to the timing belt at a point midway between the camshaft sprockets. The belt should deflect 0.35–0.45 inch (9.0–11.5mm). If the deflection is not cor-

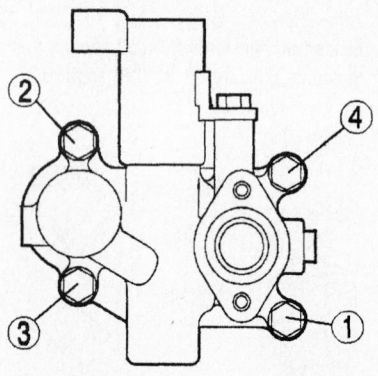

Tighten the OCV bolts in the sequence shown—1.8L (BP) engine

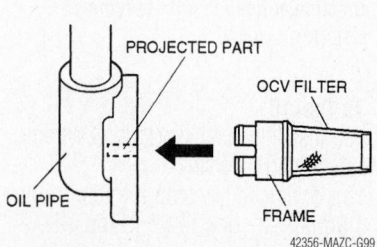

Hold the frame of the OCV valve filter and install the OCV so it is aligned with the projected part on the flange end of the oil pipe—1.8L (BP) engine

rect, repeat the alignment and tensioning procedure.

21. Install the timing belt covers and tighten the bolts in sequence to 95 inch lbs. (11 Nm).

22. Install the valve cover and spark plugs, along with all other applicable components.

23. Hold the pulley boss with a suitable tool, and tighten the lockbolt to 116–130 ft. lbs. (157–167 Nm).

24. Install or connect the following:
- Crankshaft plate and pulley
- Water pump pulley
- Rocker arm cover with a new gasket
- Rocker arm cover with a new gasket
- Temporarily tighten the cover bolts **A**, refer to the illustration for location. Torque the bolts in sequence to 80 inch lbs. (9 Nm).

✳✳ CAUTION

When installing the OCV valve, be careful not to damage the O–ring, if damaged it may cause leaking.

- OCV and tighten the bolts in the sequence illustrated

25. Install the oil pipe as follows:

a. Oil pipe. Hold the frame of the OCV valve filter and install the OCV so it is aligned with the projected part on the flange end of the oil pipe. Coat the new washer with clean engine oil and temporarily install the upper and side oil pipe and position the pipe.

b. **A** using several passes to 95 inch lbs. (11 Nm).

c. Tighten oil pipe bolts **B** using several passes to 61 inch lbs. (7 Nm).

d. Tighten oil pipe bolts **C** using several passes to 13 ft. lbs. (17 Nm).

e. Tighten oil pipe bolts 1, 2, 3, 4 and 7 using several passes to 95 inch lbs. (11 Nm).

f. Tighten oil pipe bolt 5 using several passes to 17 ft. lbs. (23 Nm).

g. Tighten oil pipe bolt 6 using several passes to 34 ft. lbs. (47 Nm).

26. Install or connect the following:
- Water hose
- Upper radiator hose
- Spark plugs
- High tension lead and the ignition coil
- CKP sensor
- Drive belt
- Air pipe
- Front suspension lower bar
- Negative battery cable

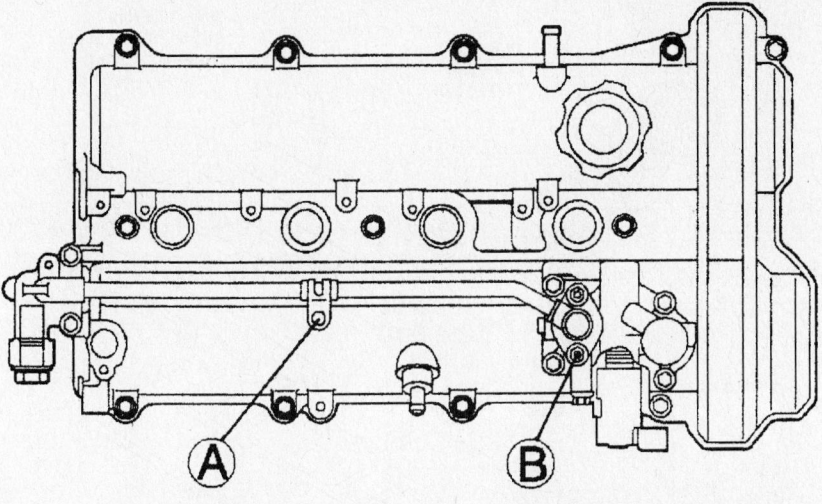

ENGINE FRONT →

42356-MAZC-GAA

Tighten oil pipe bolts A and B using several passes to the specification in the text—1.8L (BP) engine

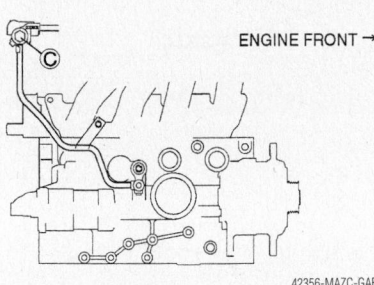

ENGINE FRONT →

42356-MAZC-GAB

Tighten oil pipe bolts C using several passes to the specification in the text—1.8L (BP) engine

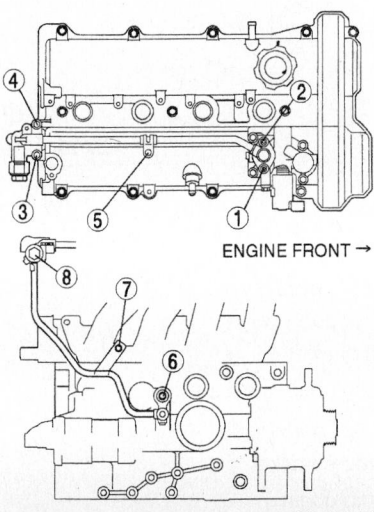

ENGINE FRONT →

42356-MAZC-GAC

Tighten oil pipe bolts using several passes to the specification in the text—1.8L (BP) engine

Timing Chain

REMOVAL & INSTALLATION

Mazda 3

2.0L (LF) AND 2.3L (L3) ENGINES

1. Before servicing the vehicle, refer to the precautions in the beginning of this section.
2. Remove the plug hole plate and bracket.
3. Remove the accelerator cable and bracket.
4. Remove the battery cover and disconnect the negative battery cover.
5. Remove the ignition coils.
6. Remove the right hand front wheel.
7. Remove the engine under cover and splash shields.
8. Remove the Crankshaft Position (CKP) sensor.
9. Remove the accessory drive belt.
10. Remove the A/C compressor and set aside with the lines attached.
11. Remove the coolant reservoir tank and set aside with the lines still attached.
12. Remove the cylinder head cover.
13. Remove the cylinder block lower blind plug.
14. Install tool 49 JE01 061.
15. Turn the crankshaft clockwise until the crankshaft is in the number 1 cylinder Top Dead Center position (the balance weight should be attached to tool 49 JE01 061).
16. Hold the crankshaft pulley using the tools illustrated and remove the bolt.

17. Remove the crankshaft pulley.
18. Remove the water pump pulley.
19. Remove the drive belt tensioner.
20. Remove the No. 3 engine mount and joint bracket as follows:
   a. Install two suitable pieces of wood between the front fender panel and upper apron reinforcement as illustrated. The wood size should be approximately 1.38 inch (35mm) on 4 door models or 2.36 inch (60mm) on 5 door models.
21. Install an engine support device such as 49 E017 5A0.
22. Remove the engine front cover and oil seal.
23. Unlock the chain tensioner using a suitable tool to slowly compress the tensioner piston. Insert a 0.059 inch (1.5mm) wire or a paper clip to hold the piston in its compressed position.
24. Remove the tensioner arm, chain guide and timing chain.

To install:

25. Install tool 49 UN30 3465 as illustrated on the camshaft.
26. Install the timing chain and remove the paper clip or wire retaining the tensioner piston to apply tension to the chain.
27. Install the timing chain guide and tighten the bolts to 71–101 inch lbs. (8–11.5 Nm).
28. Install the tensioner arm and tighten the bolts to 71–101 inch lbs. (8–11.5 Nm).
29. Install a new oil seal in the front cover as follows:
   a. Coat the new seal with clean engine oil.
   b. Push the seal in by hand to get it started.
   c. Use seal installer tool 49 H010 401 and a hammer to install the seal so that it is recessed 0–0.019 inch (0–0.5mm) as shown in the accompanying illustration.
30. Apply sealant to the engine front cover. At point A the bead should be 0.087–0.125 inch (2.2–3.2mm) thick and at point B the bead should be 0.059–0.098 inch (1.5–2.5mm) thick. Refer to the accompanying illustration for the locations of points A and B. No sealant is needed at the points marked C on 2.3L (L3) engines with variable valve timing.
31. Install the cover within 10 minutes of apply the sealant. Tighten the cover bolts as follows:
   a. Bolts 1 through 18: In sequence to 71–101 inch lbs. (8–11.5 Nm).
   b. Bolts 19 through 22: In sequence to 29.7–40.5 ft. lbs. (40–55 Nm).
   c. Bolt 23: 14.8–22 ft. lbs. (20–30 Nm).

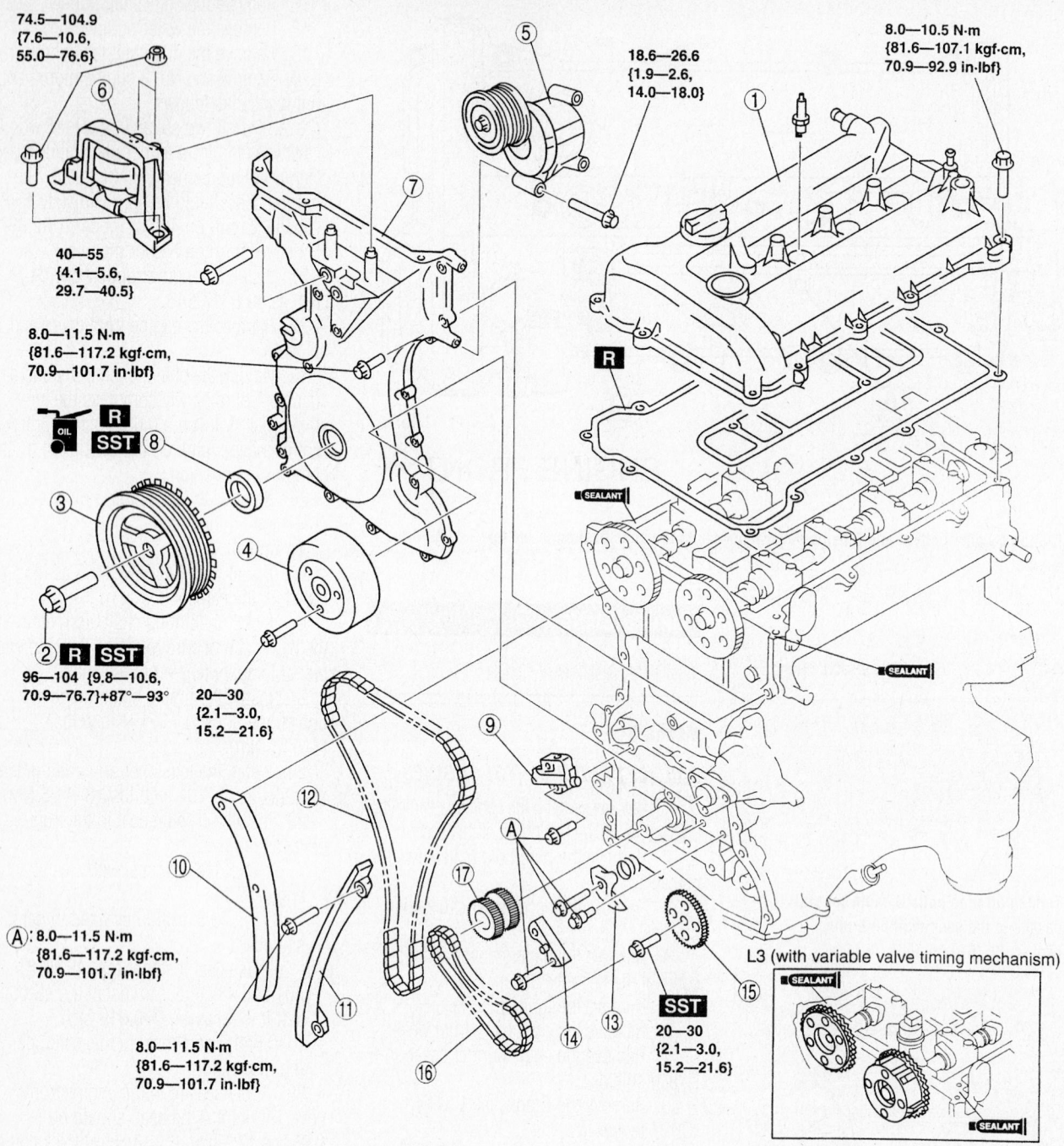

74.5—104.9
{7.6—10.6,
55.0—76.6}

18.6—26.6
{1.9—2.6,
14.0—18.0}

8.0—10.5 N·m
{81.6—107.1 kgf·cm,
70.9—92.9 in·lbf}

40—55
{4.1—5.6,
29.7—40.5}

8.0—11.5 N·m
{81.6—117.2 kgf·cm,
70.9—101.7 in·lbf}

96—104 {9.8—10.6,
70.9—76.7}+87°—93°

20—30
{2.1—3.0,
15.2—21.6}

Ⓐ 8.0—11.5 N·m
{81.6—117.2 kgf·cm,
70.9—101.7 in·lbf}

8.0—11.5 N·m
{81.6—117.2 kgf·cm,
70.9—101.7 in·lbf}

SST
20—30
{2.1—3.0,
15.2—21.6}

L3 (with variable valve timing mechanism)

N·m {kgf·m, ft·lbf}

1	Cylinder head cover	9	Chain tensioner
2	Crankshaft pulley lock bolt	10	Tensioner arm
3	Crankshaft pulley	11	Chain guide
4	Water pump pulley	12	Timing chain
5	Drive belt auto tensioner	13	Oil pump chain tensioner
6	No.3 engine mount rubber and No.3 engine joint bracket	14	Oil pump chain guide
7	Engine front cover	15	Oil pump sprocket
8	Front oil seal	16	Oil pump chain
		17	Crankshaft sprocket

67162-MAZC-G55

Exploded view of the timing chain assembly and related components—Mazda 3 with the 2.0L (LF) and 2.3L (L3) engines

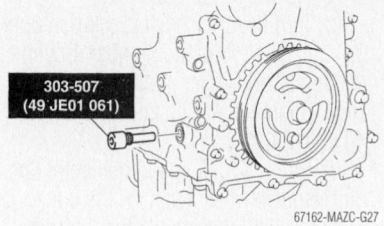

67162-MAZC-G27

Install tool 49 JE01 061—Mazda 3 with the 2.0L (LF) and 2.3L (L3) engines

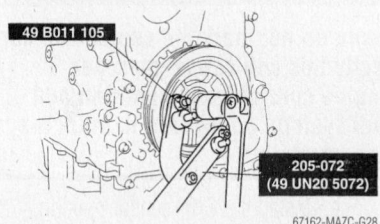

67162-MAZC-G28

Install tool 49 JE01 061 to hold the crankshaft pulley using the tools illustrated and remove the bolt—Mazda 3 with the 2.0L (LF) and 2.3L (L3) engines

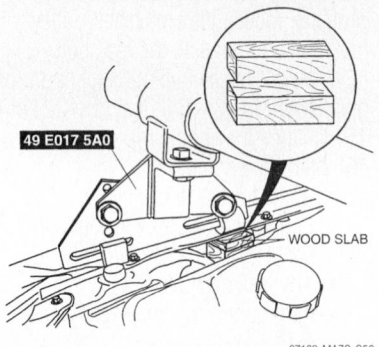

67162-MAZC-G56

Install two suitable pieces of wood between the front fender panel and upper apron reinforcement—Mazda 3 with the 2.0L (LF) and 2.3L (L3) engines

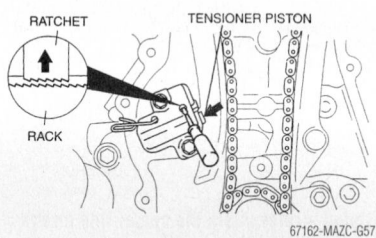

67162-MAZC-G57

Compressing and retaining the chain tensioner piston—Mazda 3 with the 2.0L (LF) and 2.3L (L3) engines

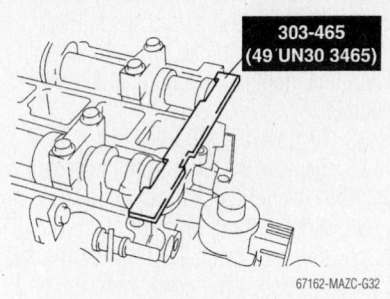

67162-MAZC-G32

Install tool 49 UN30 3465 on the camshaft—Mazda 3 with the 2.0L (LF) and 2.3L (L3) engines

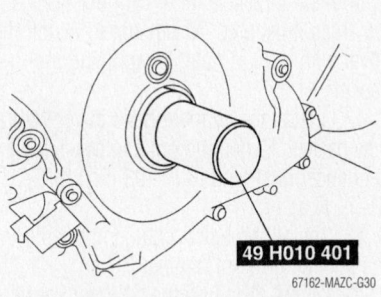

67162-MAZC-G30

Use seal installer tool 49 H010 401 and a hammer to install the seal—Mazda 3 with the 2.0L (LF) and 2.3L (L3) engines

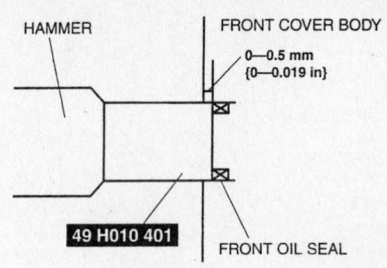

67162-MAZC-G31

The seal should be recessed 0–0.019 inch (0–0.5mm) as shown when properly installed—Mazda 3 with the 2.0L (LF) and 2.3L (L3) engines

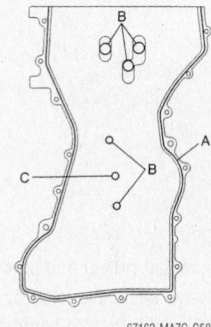

67162-MAZC-G58

Apply a bead of sealant to the engine front cover at the locations shown. Refer to the text for the bead thickness—Mazda 3 with the 2.0L (LF) and 2.3L (L3) engines

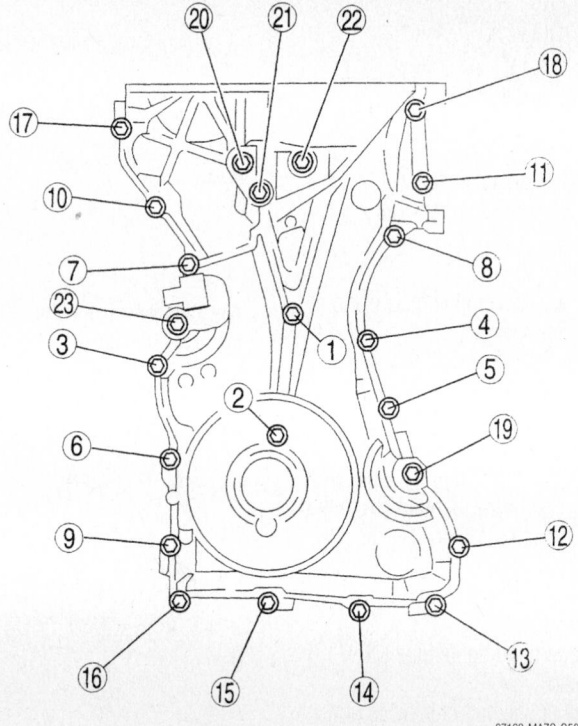

67162-MAZC-G59

Engine front cover bolts tightening sequence—Mazda 3 with the 2.0L (LF) and 2.3L (L3) engines

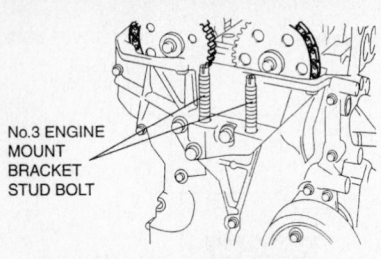

No.3 ENGINE MOUNT BRACKET STUD BOLT

67162-MAZC-G08

Location of the No. 3 engine mount bracket stud bolts—Mazda 3 with the 2.0L (LF) and 2.3L (L3) engines

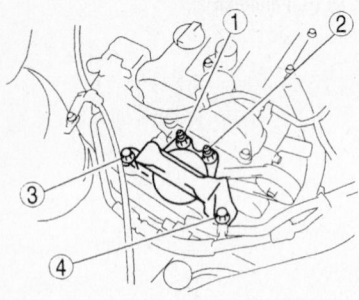

67162-MAZC-G60

Tighten the mount rubber and bracket nuts and bolts in the sequence—Mazda 3 with the 2.0L (LF) and 2.3L (L3) engines

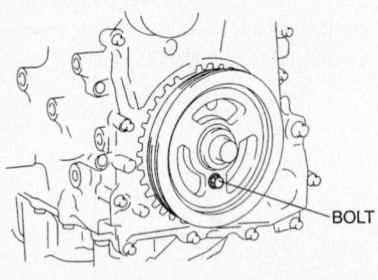

BOLT

67162-MAZC-G33

Install and hand tighten the M6 x 1.0 bolt—Mazda 3 with the 2.0L (LF) and 2.3L (L3) engines

32. Install the No. 3 engine mount and joint bracket as follows:

   a. Tighten the No. 3 mount bracket stud bolts to 62–115 inch lbs. (7–13 Nm). Tighten the stud bolt with the mount nut loosened.

   b. Hand tighten the mount rubber and bracket nuts and bolts, then tighten in the sequence illustrated to 55–77 ft. lbs. (74–105 Nm).

33. Install tool 49 UN30 3465 as illustrated on the camshaft.

34. Install and hand tighten the M6 x 1.0 bolt as illustrated.

35. Turn the crankshaft clockwise until the crankshaft is in number 1 TDC (the balance weight should be attached to the tool).

36. Hold the crankshaft pulley and tighten the lock bolt in two steps. First tighten to 71–77 ft. lbs. (96–104 Nm). Then final tighten an additional 87–93 degrees.

37. Remove the M6 x 1.0 bolt.

38. Remove the tools from the camshaft and the cylinder block lower blind plug.

39. Rotate the crankshaft clockwise 2 turns until you reach TDC. If not aligned properly, loosen the crankshaft pulley lock bolt and reinstall the lock bolt again using the above procedure and special tools.

40. Install the cylinder block blind plug and tighten to 13–16 ft. lbs. (18–22 Nm).

41. Apply a 0.16–0.24 inch (4–7mm) bead of silicone sealant to cylinder head at the areas illustrated. Make sure to install the cover within 10 minutes of applying the sealant.

42. Install the cylinder head cover with a new gasket. Torque the bolts in the sequence illustrated to 71–93 inch lbs. (8–11 Nm).

43. Install the remaining components in the reverse order of removal.

44. When installing the CKP sensor perform the following:

   a. Remove the cylinder block lower blind plug.

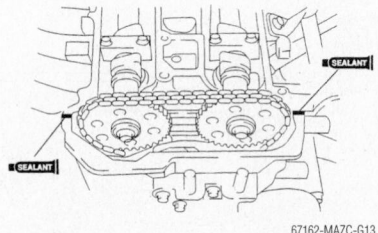

SEALANT

SEALANT

67162-MAZC-G13

Apply silicone sealant to cylinder head at the areas illustrated—Mazda 3 with the 2.0L (LF) and 2.3L (L3) engines

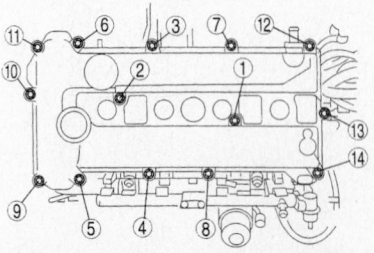

67162-MAZC-G14

Tighten the cylinder head cover bolts in the sequence shown—Mazda 3 with the 2.0L (LF) and 2.3L (L3) engines

   b. Install tool 49 JE01 061.

   c. Turn the crankshaft clockwise until the crankshaft is in the number 1 cylinder Top Dead Center position (the balance weight should be attached to tool 49 JE01 061).

   d. Using a ruler, mark the center line on the pulse wheel teeth on the crank pulley which is located at the 9th tooth counting counterclockwise from the empty space. Refer to the illustration for more detail.

❊❊ CAUTION

If you do not mark the center line correctly this will cause improper engine control for the ignition and fuel system, so be sure to mark the line carefully.

   e. Install the CKP sensor making sure the mark on the sensor is aligned with the mark on the pulse wheel. Tighten the sensor bolt to 49–66 inch lbs. (5.5–7.5 Nm)and attach the electrical connector.

   f. Remove the tool from the cylinder block lower blind plug.

   g. Rotate the crankshaft clockwise 2 turns until you reach TDC. If not aligned properly, loosen the crankshaft pulley lock bolt and reinstall the lock bolt.

   h. Install the cylinder block blind plug and tighten to 13–16 ft. lbs. (18–22 Nm).

45. Change the engine oil.

46. Start the engine.

47. Inspect for the following:

- Pulley and belt for run–out and contact
- Any leaking fluids.
- Ignition timing, idle speed and exhaust emissions
- All remaining components for proper operation.

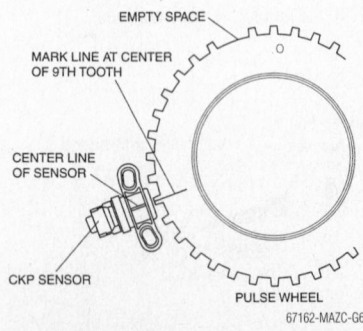

EMPTY SPACE

MARK LINE AT CENTER OF 9TH TOOTH

CENTER LINE OF SENSOR

CKP SENSOR

PULSE WHEEL

67162-MAZC-G61

Using a ruler, mark the center line on the pulse wheel teeth on the crank pulley which is located at the 9th tooth counting counterclockwise from the empty space— Mazda 3 with the 2.0L (LF) and 2.3L (L3) engines

Mazda 6

2.3L (L3) ENGINE

1. Before servicing the vehicle, refer to the precautions in the beginning of this section.
2. Disconnect the negative battery cable.
3. Remove the cylinder head cover.
4. Remove the cylinder block lower blind plug.
5. Install tool 49 JE01 061.
6. Turn the crankshaft clockwise until the crankshaft is in the number 1 cylinder Top Dead Center position (the balance weight should be attached to tool 49 JE01 061).
7. Hold the crankshaft pulley using the tools illustrated and remove the bolt.
8. Remove the crankshaft pulley.

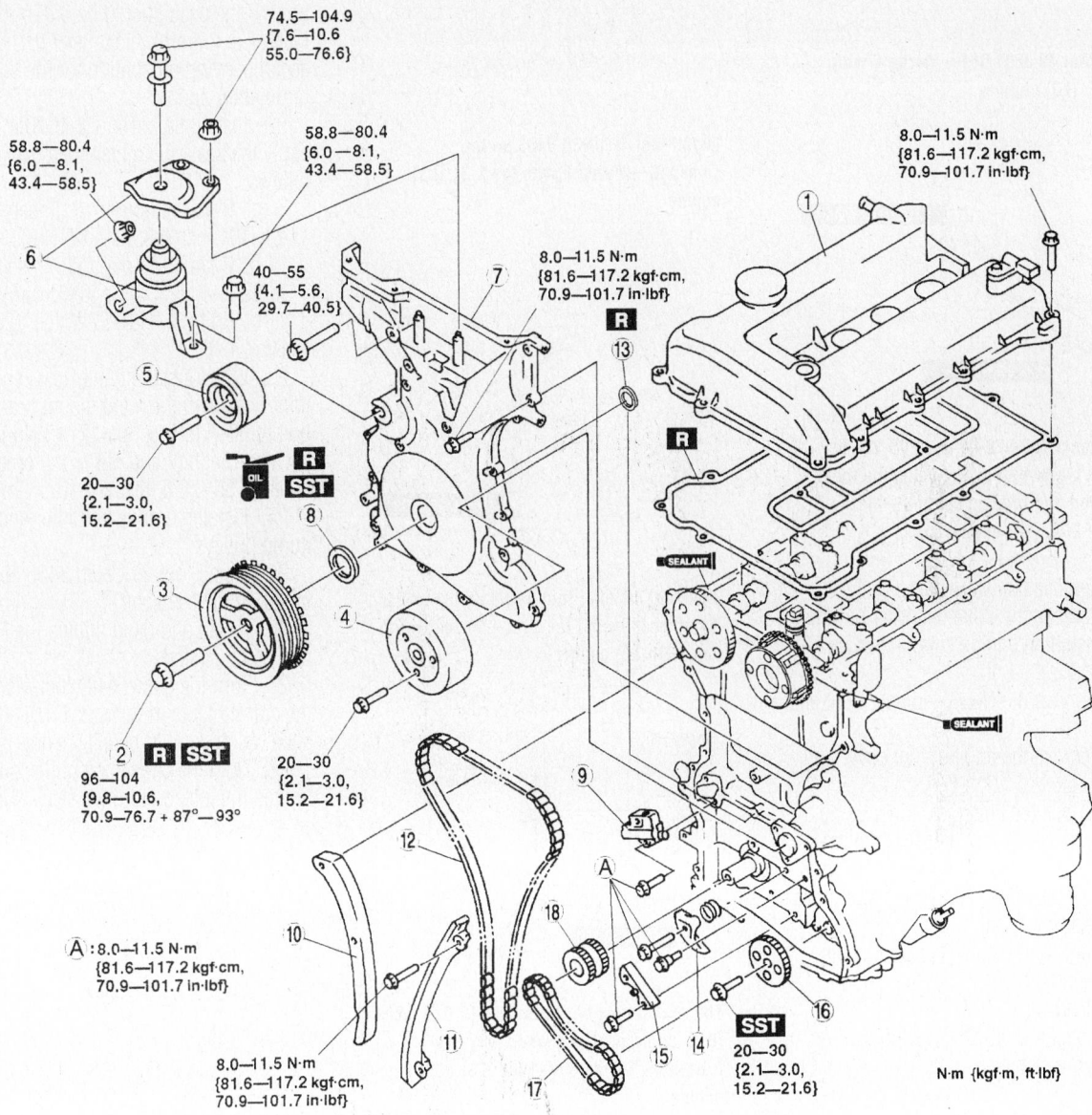

1	Cylinder head cover	10	Tensioner arm
2	Crankshaft pulley lock bolt	11	Chain guide
3	Crankshaft pulley	12	Timing chain
4	Water pump pulley	13	Seal
5	Drive belt idler pulley	14	Oil pump chain tensioner
6	No.3 engine mount rubber and No.3 engine joint bracket	15	Oil pump chain guide
7	Engine front cover	16	Oil pump sprocket
8	Front oil seal	17	Oil pump chain
9	Chain tensioner	18	Crankshaft sprocket

Exploded view of the timing chain assembly and related components—Mazda 6 with the 2.3L (L3) engine

67162-MAZC-G186

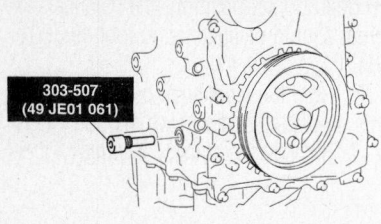

Install tool 49 JE01 061—Mazda 6 with the 2.3L (L3) engine

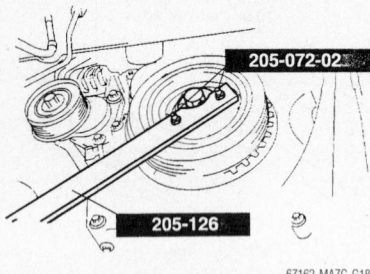

Install tools 205-072-02 and 205-126 to hold the crankshaft pulley using the tools illustrated and remove the bolt—Mazda 6 with the 2.3L (L3) engine

9. Remove the water pump pulley.

10. Remove the drive belt idler pulley.

11. Install an engine support device such as 49 E017 5A0.

12. Remove the No. 3 engine mount and joint.

13. Remove the engine front cover and oil seal.

14. Unlock the chain tensioner using a suitable tool to slowly compress the tensioner piston. Insert a 0.059 inch (1.5mm) wire or a paper clip to hold the piston in its compressed position.

15. Remove the tensioner arm, chain guide and timing chain.

To install:

16. Install tool 49 UN30 3465 as illustrated on the camshaft.

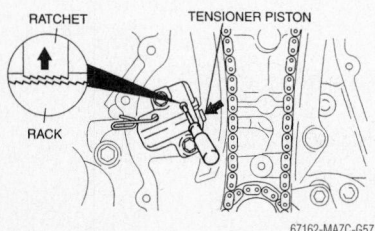

Compressing and retaining the chain tensioner piston—Mazda 6 with the 2.3L (L3) engine

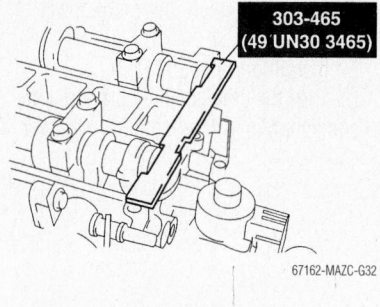

Install tool 49 UN30 3465 on the camshaft—Mazda 6 with the 2.3L (L3) engine

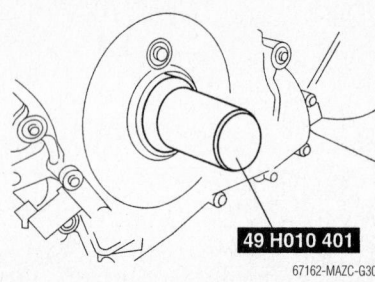

Use seal installer tool 49 H010 401 and a hammer to install the seal—Mazda 6 with the 2.3L (L3) engine

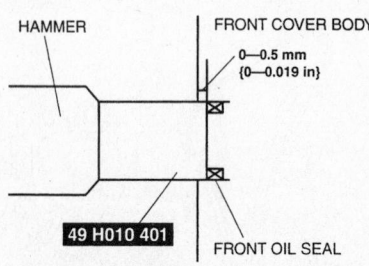

The seal should be recessed 0–0.019 inch (0–0.5mm) as shown when properly installed—Mazda 6 with the 2.3L (L3) engine

17. Install the timing chain and remove the paper clip or wire retaining the tensioner piston to apply tension to the chain.

18. Install the timing chain guide and tighten the bolts to 71–101 inch lbs. (8–11.5 Nm).

19. Install the tensioner arm and tighten the bolts to 71–101 inch lbs. (8–11.5 Nm).

20. Install a new oil seal in the front cover as follows:

a. Coat the new seal with clean engine oil.

b. Push the seal in by hand to get it started.

c. Use seal installer tool 49 H010 401 and a hammer to install the seal so that it is recessed 0–0.019 inch (0–0.5mm) as shown in the accompanying illustration.

21. Apply sealant to the engine front cover. At point A the bead should be 0.087–0.125 inch (2.2–3.2mm) thick and at point B the bead should be 0.059–0.098 inch (1.5–2.5mm) thick. Refer to the accompanying illustration for the locations of points A and B.

22. Install the cover within 10 minutes of apply the sealant. Tighten the cover bolts as follows:

a. Bolts 1 through 18: In sequence to 71–101 inch lbs. (8–11.5 Nm).

b. Bolts 19 through 22: In sequence to 29.7–40.5 ft. lbs. (40–55 Nm).

c. Bolt 23: 14.8–22 ft. lbs. (20–30 Nm).

23. When installing the No. 3 engine joint bracket, tighten the mount bracket stud bolt to 62–115 inch lbs. (7–13 Nm) and the joint bracket bolt and nut in the order illustrated to 55–76 ft. lbs. (74–105 Nm).

24. Install the drive belt idler and water pump pulleys.

25. Install tool 49 UN30 3465 as illustrated on the camshaft.

26. Install and hand tighten the M6 x 1.0 bolt as illustrated.

27. Turn the crankshaft clockwise until the crankshaft is in number 1 TDC (the balance weight should be attached to the tool).

28. Hold the crankshaft pulley and tighten the lock bolt in two steps. First

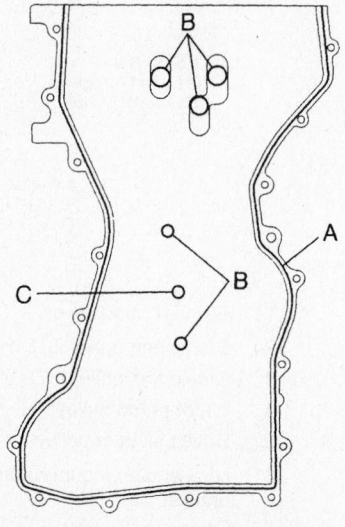

Apply a bead of sealant to the engine front cover at the locations shown. Refer to the text for the bead thickness—Mazda 6 with the 2.3L (L3) engine

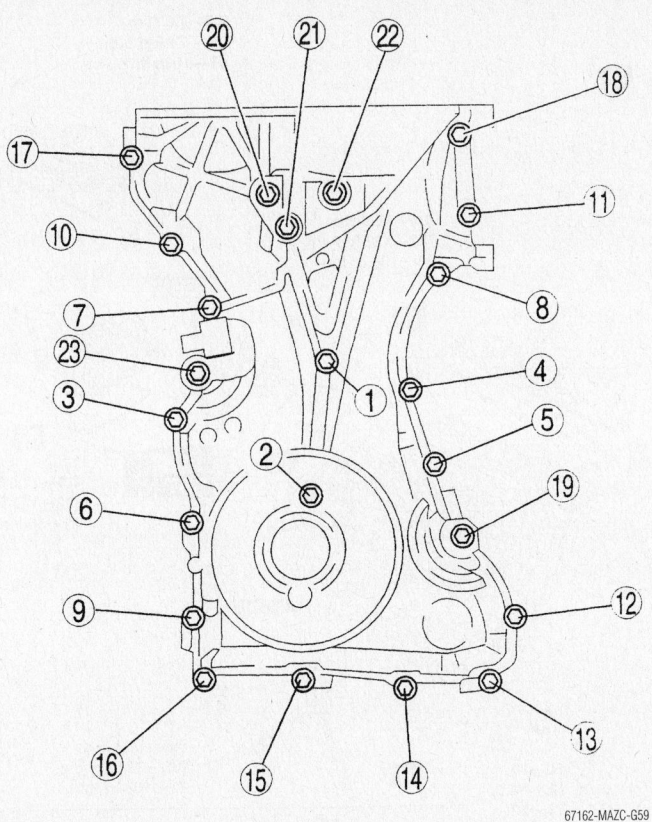

Engine front cover bolts tightening sequence—Mazda 6 with the 2.3L (L3) engine

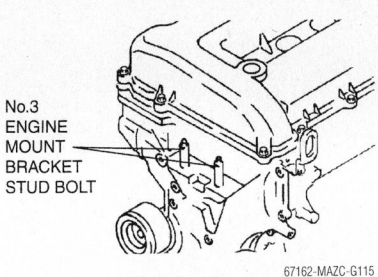

No. 3 engine mount bracket stud bolt locations—Mazda 6 models with the 2.3L (L3) engine

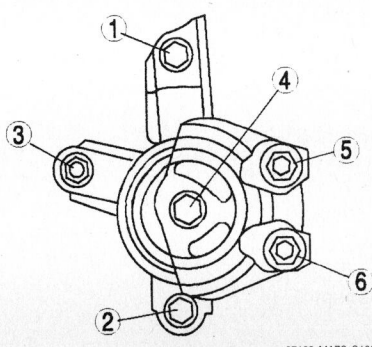

Tighten the No. 3 joint bracket bolts and nuts in the order shown—Mazda 6 models with the 2.3L (L3) engine

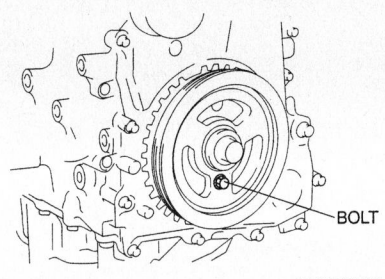

Install and hand tighten the M6 x 1.0 bolt—Mazda 6 with the 2.3L (L3) engine

tighten to 71–77 ft. lbs. (96–104 Nm). Then final tighten an additional 87–93 degrees.

29. Remove the M6 x 1.0 bolt.

30. Remove the tools from the camshaft and the cylinder block lower blind plug.

31. Rotate the crankshaft clockwise 2 turns until you reach TDC. If not aligned properly, loosen the crankshaft pulley lock bolt and reinstall the lock bolt again using the above procedure and special tools.

32. Install the cylinder block blind plug and tighten to 13–16 ft. lbs. (18–22 Nm).

33. Install the cylinder head cover with a new gasket. Torque the bolts in the sequence illustrated to 71–93 inch lbs. (8–11 Nm).

34. Install the remaining components in the reverse order of removal.

35. Change the engine oil.

36. Start the engine.

37. Inspect for the following:
   • Pulley and belt for run–out and contact
   • Any leaking fluids.
   • Ignition timing, idle speed and exhaust emissions
   • All remaining components for proper operation.

3.0L (AJ) ENGINE

1. Before servicing the vehicle, refer to the precautions in the beginning of this section.

2. Disconnect the negative battery cable.

3. Remove the upper intake manifold (dynamic chamber).

4. Remove the ignition coils.

5. Remove the accessory drive belt.

6. Remove the power steering pump and bracket with the lines still attached and set aside.

7. Remove the A/C compressor with the lines still attached and set aside.

8. Unbolt the alternator with the connector still attached and wire it aside.

9. Remove the right hand halfshaft.

10. Remove the front crossmember as follows:
   a. Remove the stabilizer bar nut.
   b. Remove the front lower arm front and rear ball joints from the knuckle.
   c. Remove the front shock absorber lower bolt.
   d. Remove the No. 1 engine mount center bolt.

✷✷ WARNING

Support the crossmember with a suitable jack making sure it is securely damaged or it could fall causing injury or damage.

   e. Support the crossmember with a jack and remove the crossmember bracket.
   f. Remove the crossmember assembly retainers and lower the jack to remove the crossmember.

11. Remove the left hand three way converter.

12. Remove the oil pan.

13. Remove the plug hole plate.

14. Remove the water pump drive belt by rotating the belt tensioner counterclockwise.

15. Remove the water pump drive belt pulley as follows:

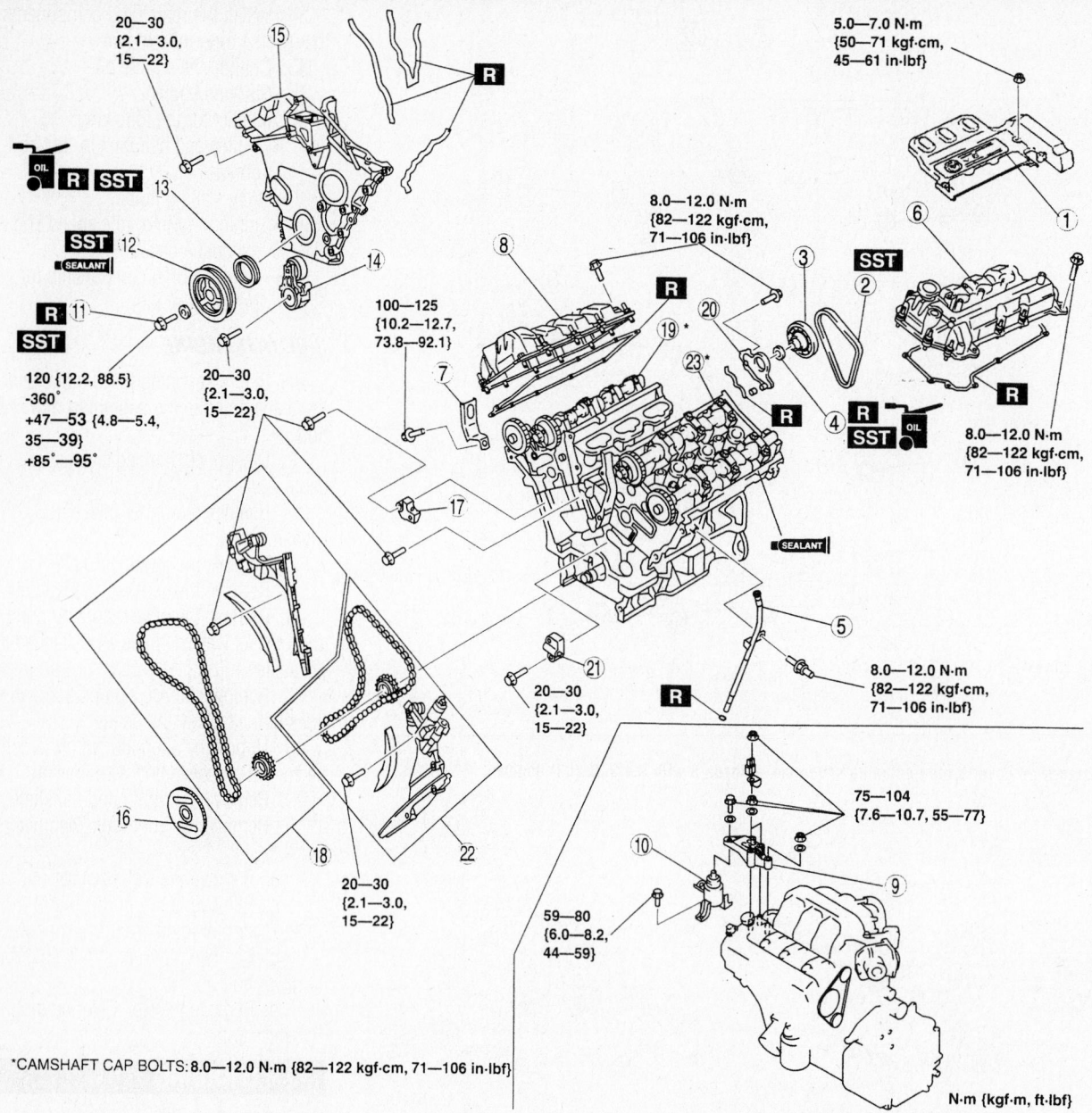

20—30
{2.1—3.0,
15—22}

5.0—7.0 N·m
{50—71 kgf·cm,
45—61 in·lbf}

8.0—12.0 N·m
{82—122 kgf·cm,
71—106 in·lbf}

100—125
{10.2—12.7,
73.8—92.1}

20—30
{2.1—3.0,
15—22}

120 {12.2, 88.5}
-360°
+47—53 {4.8—5.4,
35—39}
+85°—95°

8.0—12.0 N·m
{82—122 kgf·cm,
71—106 in·lbf}

8.0—12.0 N·m
{82—122 kgf·cm,
71—106 in·lbf}

20—30
{2.1—3.0,
15—22}

75—104
{7.6—10.7, 55—77}

20—30
{2.1—3.0,
15—22}

59—80
{6.0—8.2,
44—59}

N·m {kgf·m, ft·lbf}

*CAMSHAFT CAP BOLTS: 8.0—12.0 N·m {82—122 kgf·cm, 71—106 in·lbf}

1	Plug hole plate	13	Front oil seal
2	Water pump drive belt	14	Front drive belt auto tensioner
3	Water pump drive belt pulley	15	Engine front cover
4	Camshaft oil seal	16	CKP sensor pulse wheel
5	Oil level gauge pipe	17	Chain tensioner (RH)
6	Cylinder head cover (LH)	18	Timing chain component (RH)
7	Engine hanger (RH)	19	Camshaft cap (RH)
8	Cylinder head cover (RH)	20	Camshaft oil seal housing
9	No.3 engine joint bracket	21	Chain tensioner (LH)
10	No.3 engine mount rubber	22	Timing chain component (LH)
11	Crankshaft pulley lock bolt	23	Camshaft cap (LH)
12	Crankshaft pulley		

67162-MAZC-G189

Exploded view of the timing chain assembly and related components—Mazda 6 with the 3.0L (AJ) engine

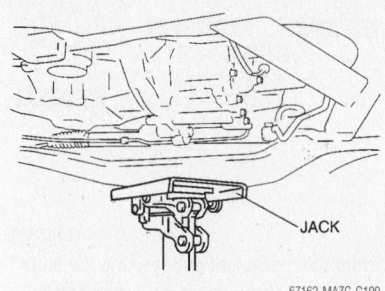

JACK

67162-MAZC-G190

Support the crossmember with a jack and remove the crossmember bracket—Mazda 6 with the 3.0L (AJ) engine

a. Replace part of tool 49 UN30 3009 with tool 49 UN30 3457.

b. Install the tools as illustrated and remove the water pump drive pulley.

16. Cut the camshaft oil seal using a razor knife and remove the seal using the tools illustrated.

17. Remove the oil dipstick tube.

18. Remove the left rocker arm cover.

19. Raise the vehicle and remove the engine hanger bolts and the hanger from below.

20. Remove the right rocker arm cover.

21. Remove the No. 3 joint bracket as follows:

a. Install the right hand engine hanger.

b. Use a M10 x 1.25 0.984 (25mm) length bolt to attach tool 49 UN30 3050 as illustrated. Tighten the bolt to 73–92 ft. lbs. (100–125 Nm).

c. Suspend the engine using the engine support device and remove the joint bracket.

22. Remove the No. 3 mount rubber.

23. Remove the A/C compressor and set aside with the lines still attached.

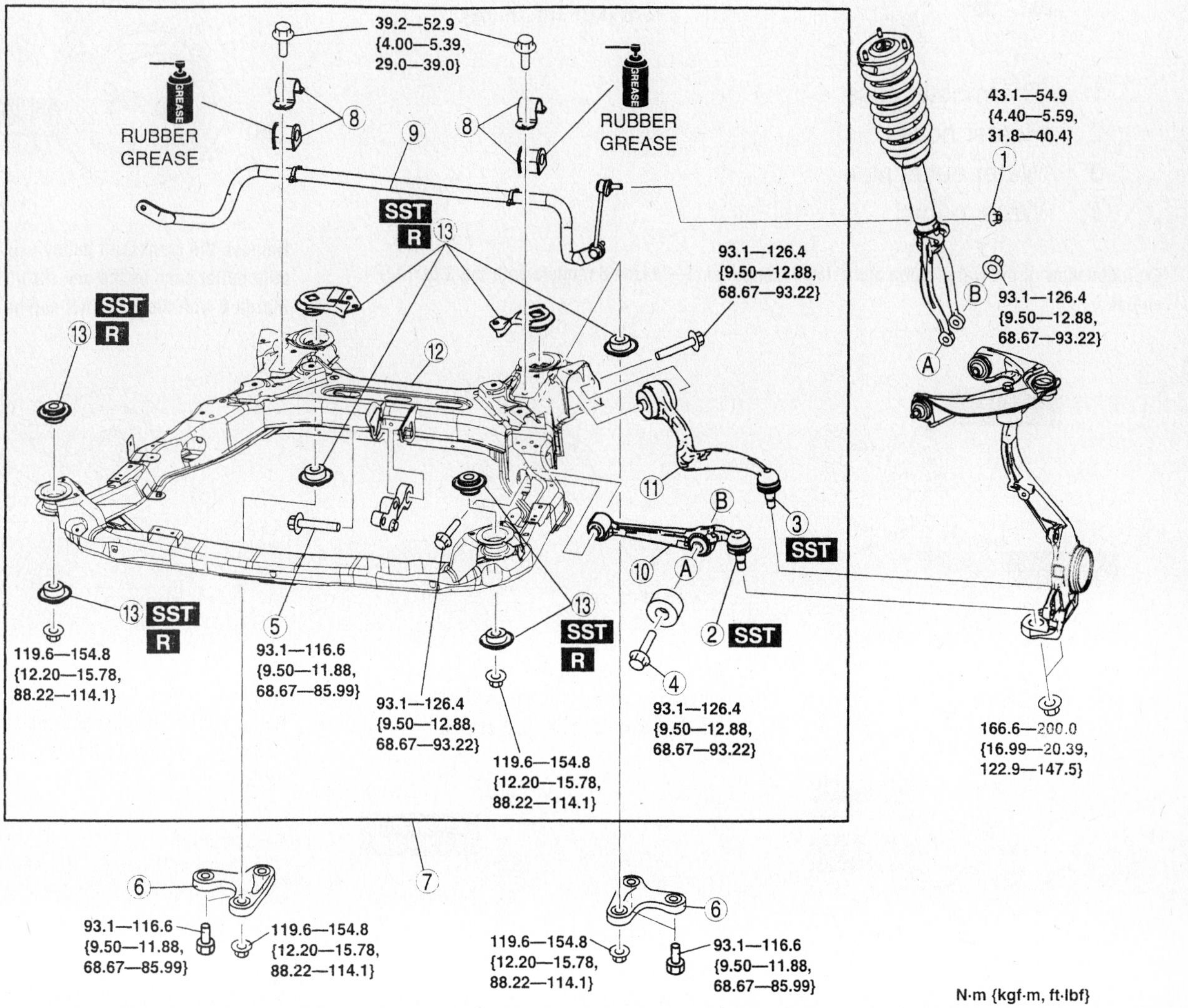

39.2—52.9
{4.00—5.39,
29.0—39.0}

RUBBER GREASE

RUBBER GREASE

43.1—54.9
{4.40—5.59,
31.8—40.4}

SST R

SST R

93.1—126.4
{9.50—12.88,
68.67—93.22}

93.1—126.4
{9.50—12.88,
68.67—93.22}

119.6—154.8
{12.20—15.78,
88.22—114.1}

93.1—116.6
{9.50—11.88,
68.67—85.99}

SST R

93.1—126.4
{9.50—12.88,
68.67—93.22}

SST R

119.6—154.8
{12.20—15.78,
88.22—114.1}

SST

166.6—200.0
{16.99—20.39,
122.9—147.5}

93.1—126.4
{9.50—12.88,
68.67—93.22}

SST

93.1—116.6
{9.50—11.88,
68.67—85.99}

119.6—154.8
{12.20—15.78,
88.22—114.1}

119.6—154.8
{12.20—15.78,
88.22—114.1}

93.1—116.6
{9.50—11.88,
68.67—85.99}

N·m {kgf·m, ft·lbf}

1	Nut (stabilizer control link)	8	Stabilizer bracket and bushing
2	Front lower arm (front) ball joint	9	Front Stabilizer
3	Front lower arm (rear) ball joint	10	Front lower arm (front)
4	Bolt (front shock absorber lower side)	11	Front lower arm (rear)
5	No.1 engine mount center bolt	12	Front crossmember
6	Crossmember bracket	13	Front crossmember bushing
7	Crossmember component		

67162-MAZC-G124

Exploded view of the crossmember assembly and related components—Mazda 6 with the 3.0L (AJ) engine

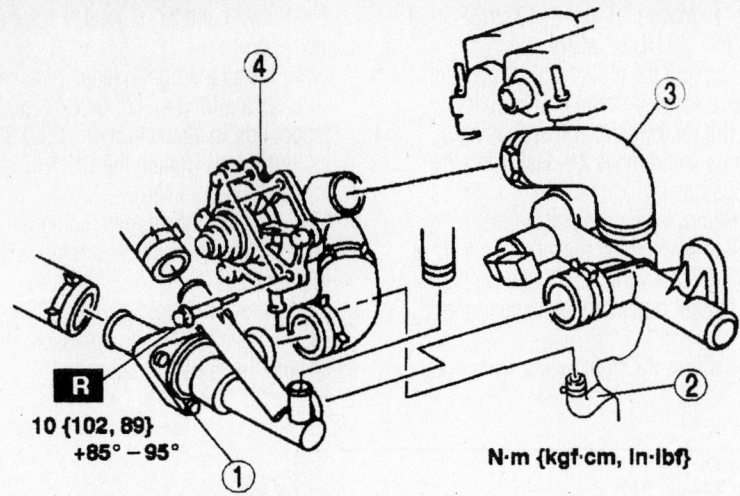

10 {102, 89}
+85° – 95°

N·m {kgf·cm, in·lbf}

1   Thermostat case
2   Heater hose
3   Water outlet pipe
4   Water pump

67162-MAZC-G126

Exploded view of the water pump and related components—Mazda 6 models with the 3.0L (AJ) engine

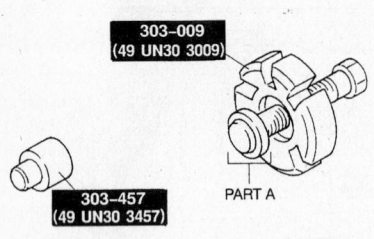

67162-MAZC-G127

Replace part of tool 49 UN30 3009 with tool 49 UN30 3457—Mazda 6 models with the 3.0L (AJ) engine

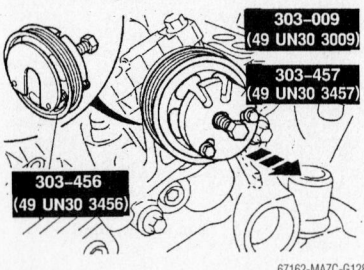

67162-MAZC-G128

Install the tools as illustrated and remove the water pump drive pulley—Mazda 6 models with the 3.0L (AJ) engine

24. Remove the crankshaft pulley using a suitable puller such as the one illustrated.
25. Remove the front oil seal using a prytool as illustrated.
26. Remove the drive belt tensioner.

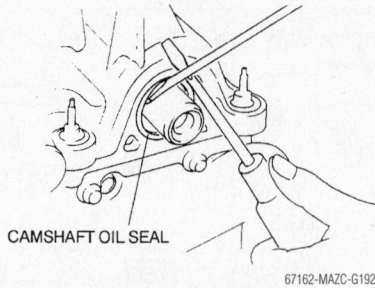

CAMSHAFT OIL SEAL

67162-MAZC-G192

Cut the camshaft oil seal using a razor knife and remove the seal as shown—Mazda 6 with the 3.0L (AJ) engine

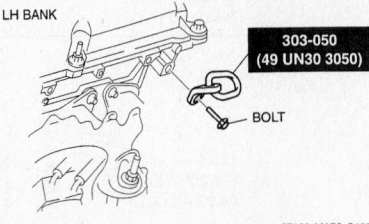

LH BANK

303–050
(49 UN30 3050)

BOLT

67162-MAZC-G193

Use a M10 x 1.25 0.984 (25mm) length bolt to attach tool 49 UN30 3050 as shown—Mazda 6 with the 3.0L (AJ) engine

27. Remove the engine front cover bolts in the sequence illustrated.
28. Remove the Crankshaft position (CKP) sensor pulse wheel.
29. Remove the right side chain tensioner and chain assembly as follows:

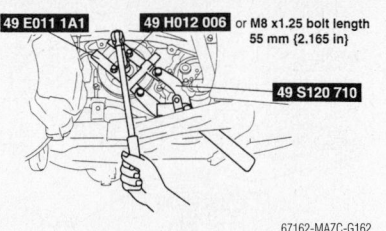

67162-MAZC-G162

Hold the crankshaft pulley using the tools illustrated and remove the pulley bolt—Mazda 6 with the 3.0L (AJ) engine

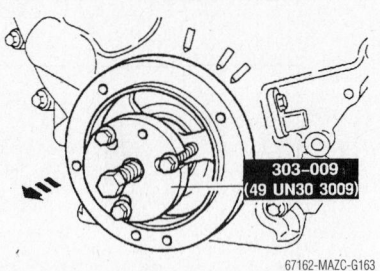

67162-MAZC-G163

Remove the crankshaft pulley using a suitable puller such as the one shown—Mazda 6 with the 3.0L (AJ) engine

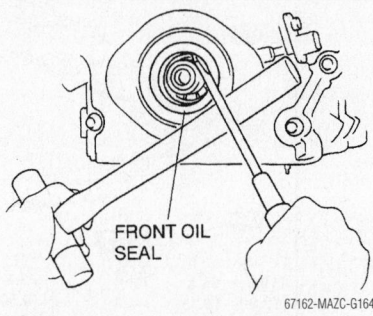

FRONT OIL SEAL

67162-MAZC-G164

Remove the front oil seal using a prytool as shown—Mazda 6 with the 3.0L (AJ) engine

   a. Rotate the crankshaft clockwise so the crankshaft keyway is at the 3 O'clock position. The camshafts should be in the neutral position.

➡ Do not rotate the crankshaft counterclockwise or you may bind the chains causing engine damage.

   b. Remove the right side chain in this order:
  • Chain tensioner
  • Tensioner arm
  • Timing chain
  • Chain guide
  • Timing chain crankshaft sprocket

30. Remove the right side camshaft caps. Refer to the camshaft removal procedure earlier in this section

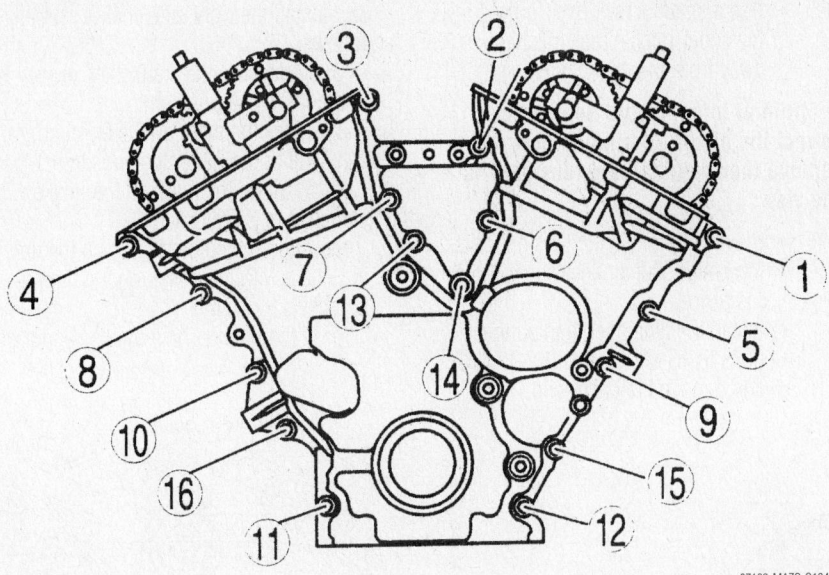

67162-MAZC-G194

Remove the engine front cover bolts in the sequence shown—Mazda 6 with the 3.0L (AJ) engine

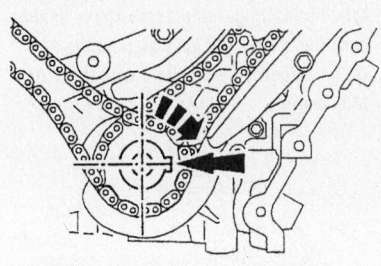

67162-MAZC-G195

Rotate the crankshaft clockwise so the crankshaft keyway is at the 3 O'clock position when removing the right side chain—Mazda 6 with the 3.0L (AJ) engine

31. Remove the left side chain tensioner and chain assembly as follows:

a. Rotate the crankshaft clockwise 1⅔ turns until the keyway is at the 11 O'clock position.

b. Remove the chain in this order:
• Chain tensioner
• Tensioner arm
• Timing chain
• Chain guide
• Timing chain crankshaft sprocket

32. Remove the left side camshaft caps. Refer to the camshaft removal procedure earlier in this section

To install:

33. Install the left side chain assembly as follows:

a. Make sure the crankshaft keyway is at the 11 O'clock position.

b. Position the mark on the intake camshaft to 9 O'clock.

c. Place the mark on the exhaust camshaft at the 12 O'clock position.

d. Align the colored links on the timing chain with the marks on the timing sprockets.

e. If the timing chain marks cannot be seen, use a marker or paint pen to mark the crankshaft and camshaft marks on the chain as follows:
• Mark any link as the crankshaft link
• Count 29 links from the crankshaft mark and place a mark to be used as the exhaust camshaft sprocket mark
• Continue counting to the 42 link mark and mark this link as the intake camshaft sprocket mark

34. Install the left hand chain in this order:
• Timing chain crankshaft sprocket
• Chain guide
• Timing chain
• Tensioner arm
• Chain tensioner

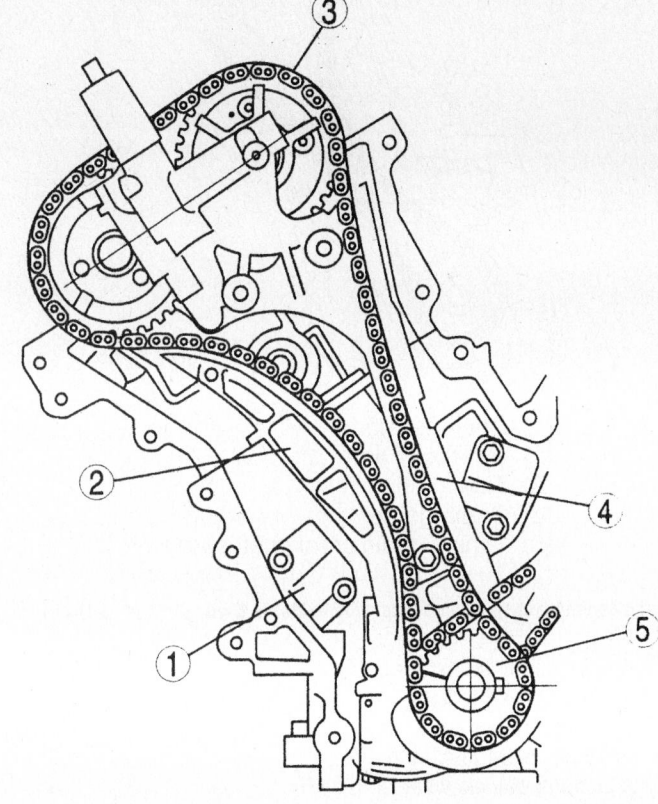

(1) Chain tensioner
(2) Tensioner arm
(3) Timing chain
(4) Chain guide
(5) Timing chain crankshaft sprocket

67162-MAZC-G196

Remove right side timing chain in numerical order—Mazda 6 with the 3.0L (AJ) engine

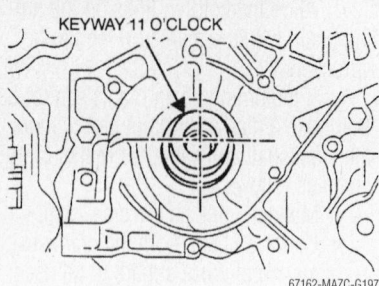

67162-MAZC-G197

Rotate the crankshaft clockwise 1⅔ turns until the keyway is at the 11 O'clock when removing the left side chain—Mazda 6 with the 3.0L (AJ) engine

- Use a small screwdriver to hold the tensioner ratchet lock mechanism away from the ratchet stem

➡ **Minimal force should be used to retract the piston, if binding occurs, remove then reinstall the tensioner is the vise.**

- Slowly compress the tensioner piston and install a paper clip to hold the piston
- Install the tensioner, tighten the bolts to 15–22 ft. lbs. (20–30 Nm) and remove the paper clip

35. Install the right side chain assembly as follows:
    a. Make sure the crankshaft keyway is at the 3 O'clock position.
    b. Place the mark on the exhaust camshaft at the 12 O'clock position.
    c. Position the mark on the intake camshaft to 3 O'clock.
    d. Align the colored links on the timing chain with the marks on the timing sprockets.
    e. If the timing chain marks cannot be

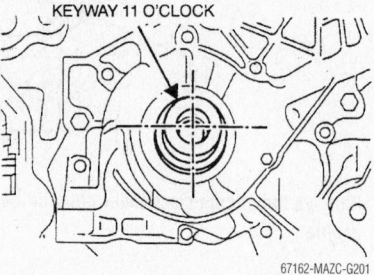

67162-MAZC-G201

When installing the left side chain, make sure the crankshaft keyway is at the 11 O'clock position—Mazda 6 with the 3.0L (AJ) engine

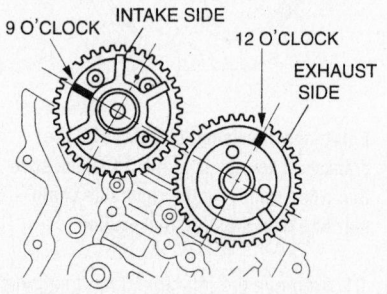

67162-MAZC-G202

When installing the left side chain, Position the mark on the intake camshaft to 9 O'clock and the mark on the exhaust camshaft at the 12 O'clock position— Mazda 6 with the 3.0L (AJ) engine

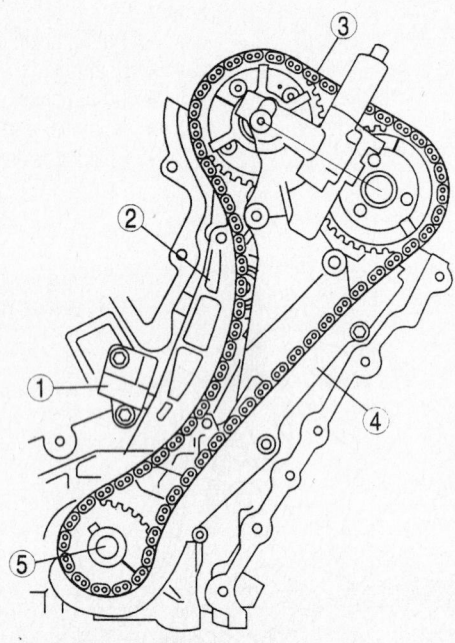

(1) Chain tensioner
(2) Tensioner arm
(3) Timing chain
(4) Chain guide
(5) Timing chain crankshaft sprocket

67162-MAZC-G198

Remove left side timing chain in numerical order—Mazda 6 with the 3.0L (AJ) engine

➡ **The chain guide should be installed to the actuator and allowed to hang freely when the bolts are installed. Do not hold the guide up when installing the bolts. The actuator causes wear to an O–ring and holding the guide will increase wear.**

    a. Install the left side camshaft caps. Refer to the camshaft removal procedure earlier in this section
    b. Install the left side chain tensioner as follows:
- Place the tensioner in a vise with jaw protectors

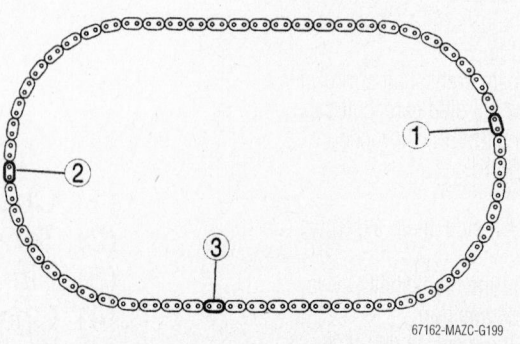

67162-MAZC-G199

If the timing chain marks cannot be seen, use a marker or paint pen to mark the crankshaft and camshaft marks on the chain. Refer to the text for marking steps.—Mazda 6 with the 3.0L (AJ) engine

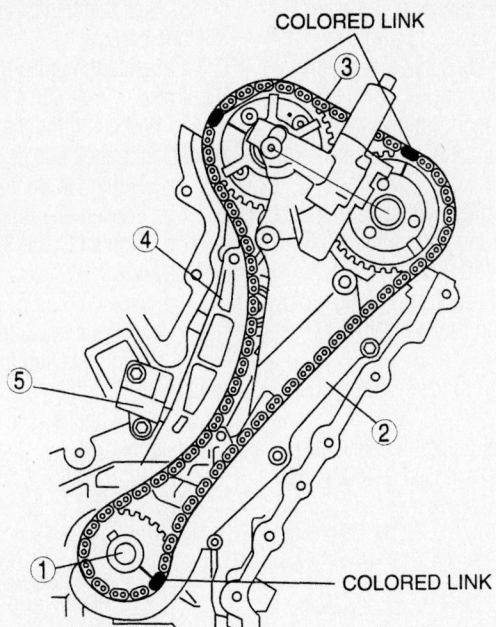

COLORED LINK

COLORED LINK

(1) Timing chain crankshaft sprocket
(2) Chain guide
(3) Timing chain
(4) Tensioner arm
(5) Chain tensioner

67162-MAZC-G200

Install the left side timing chain in numerical order and align the colored links with the timing marks on the sprockets—Mazda 6 with the 3.0L (AJ) engine

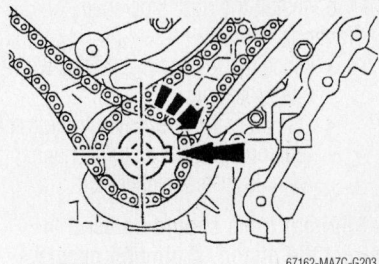

67162-MAZC-G203

When installing the right side chain, make sure the crankshaft keyway is at the 3 O'clock position—Mazda 6 with the 3.0L (AJ) engine

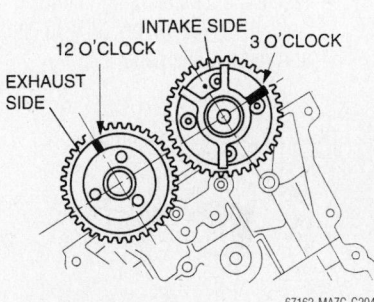

67162-MAZC-G204

When installing the right side chain, Position the mark on the exhaust camshaft to 12 O'clock and the mark on the intake camshaft at the 3 O'clock position—Mazda 6 with the 3.0L (AJ) engine

seen, use a marker or paint pen to mark the crankshaft and camshaft marks on the chain as follows:

- Mark any link as the crankshaft link
- Count 29 links from the crankshaft mark and place a mark to be used as the exhaust camshaft sprocket mark
- Continue counting to the 42 link mark and mark this link as the intake camshaft sprocket mark

36. Install the left hand chain in this order:

- Timing chain crankshaft sprocket
- Chain guide
- Timing chain
- Tensioner arm
- Chain tensioner

➡The chain guide should be installed to the actuator and allowed to hang freely when the bolts are installed. Do not hold the guide up when installing the bolts. The actuator causes wear to an O–ring and holding the guide will increase wear.

 a. Install the right side camshaft caps. Refer to the camshaft removal procedure earlier in this section

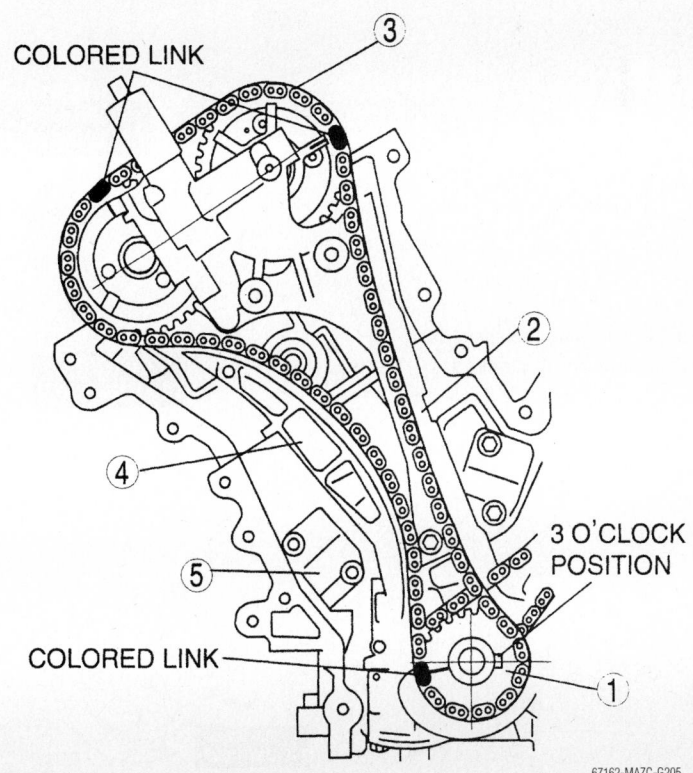

COLORED LINK

3 O'CLOCK POSITION

COLORED LINK

67162-MAZC-G205

Install the right side timing chain in numerical order and align the colored links with the timing marks on the sprockets—Mazda 6 with the 3.0L (AJ) engine

b. Install the right side chain tensioner as follows:

- Place the tensioner in a vise with jaw protectors
- Use a small screwdriver to hold the tensioner ratchet lock mechanism away from the ratchet stem

➡ **Minimal force should be used to retract the piston, if binding occurs, remove then reinstall the tensioner in the vise.**

- Slowly compress the tensioner piston and install a paper clip to hold the piston
- Install the tensioner, tighten the bolts to 15–22 ft. lbs. (20–30 Nm) and remove the paper clip.

37. Place the CKP sensor pulse wheel with the **Front** mark facing towards you, using the keyway on the same side as the empty space as illustrated.

38. Apply a 0.24 inch (6mm) bead of silicone sealant to the locations illustrated and install the front cover.

39. Tighten the front cover bolts in the sequence illustrated and tighten the retainers to 15–20 ft. lbs. (20–30 Nm).

40. Assemble the seal using part (A) of tool 49 UN01 002 and tool 49 UN01 002 as illustrated.

41. Apply clean oil to the seal and push the seal in by hand.

42. Install the seal using the installation tools until the seal is recessed 0–0.039 inch (0–1mm).

43. Seal the crankshaft pulley using a silicone sealant.

44. Install the pulley using the tools illustrated.

45. While holding the pulley, tighten the crankshaft pulley bolt as follows:

a. Tighten to 88 ft. lbs. (120 Nm).
b. Loosen one full turn.
c. Tighten to 35–39 ft. lbs. (47–53 Nm).
d. Tighten 85–95 degrees.

46. Install the no. 3 engine mount rubber.

47. When installing the No. 3 engine joint bracket, tighten the No.3 mount bracket bolts and nuts in the order illustrated as follows:

a. Bolt 1: 55–77 ft. lbs. (74–105 Nm).

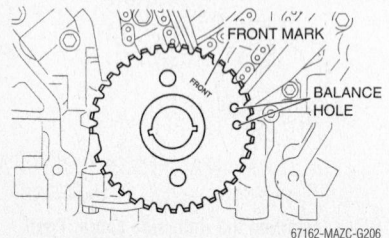

Place the CKP sensor pulse wheel with the Front mark facing towards you, using the keyway on the same side as the empty space as shown—Mazda 6 with the 3.0L (AJ) engine

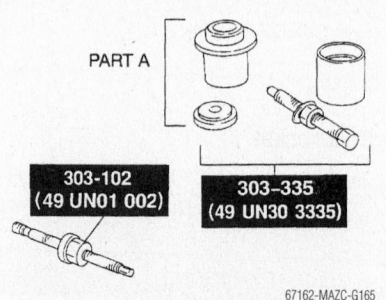

Assemble the seal using part (A) of tool 49 UN01 002 and tool 49 UN01 002 as shown—Mazda 6 with the 3.0L (AJ) engine

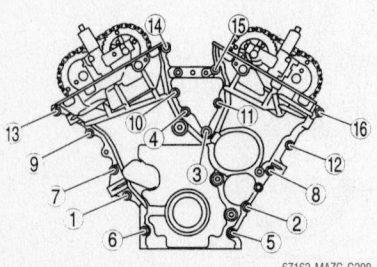

Tighten the No. 3 engine joint bracket as shown—Mazda 6 models with the 3.0L (AJ) engine

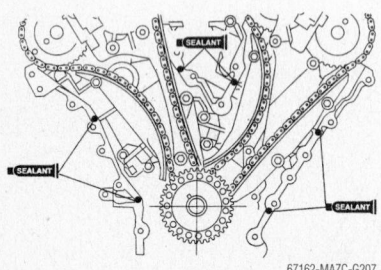

Apply a 0.24 inch (6mm) bead of silicone sealant to the locations illustrated and install the front cover—Mazda 6 with the 3.0L (AJ) engine

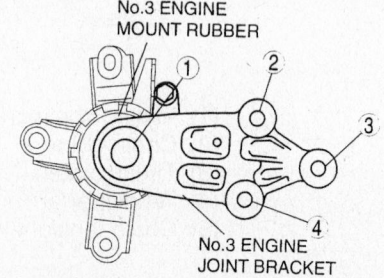

Install the seal using the installation tools until the seal is recessed 0–0.039 inch (0–1mm)—Mazda 6 with the 3.0L (AJ) engine

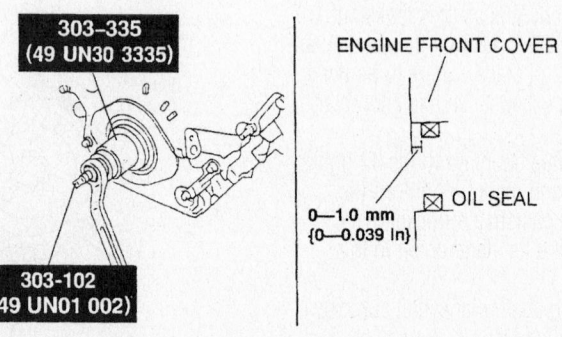

Tighten the front cover bolts in this sequence—Mazda 6 with the 3.0L (AJ) engine

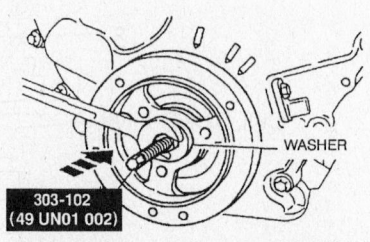

Install the pulley using the tools shown—Mazda 6 with the 3.0L (AJ) engine

b. Bolts 2, 3 and 4: 31–44 ft. lbs. (43–60 Nm).

48. Install the right hand valve cover.

49. Install the right hand engine hanger.

50. Install the left hand valve cover.

51. Install the dipstick tube.

52. Apply clean engine oil to the camshaft seal and install using the tools illustrated. The seal should be recessed 0.10–0.11 inch (2.5–3mm).

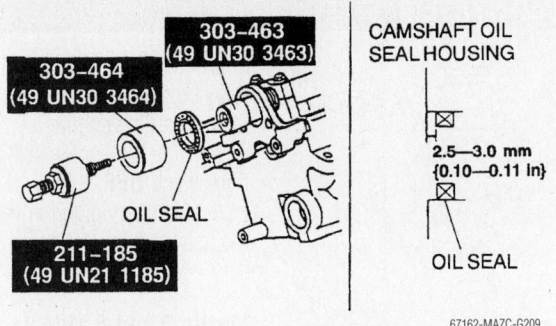

Install the camshaft oil seal—Mazda 6 with the 3.0L (AJ) engine

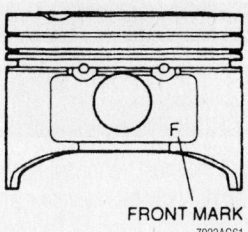

FRONT MARK
7923AG61

Piston-to-engine block mark location on the piston—Mazda 1.8L (BP) engines

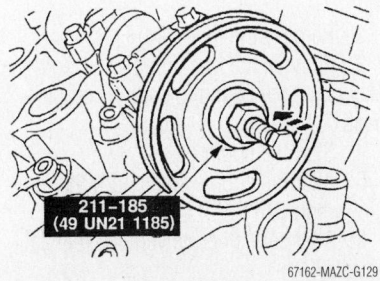

Install a new water pump drive belt pulley using tool 49 UN21 1185—Mazda 6 models with the 3.0L (AJ) engine

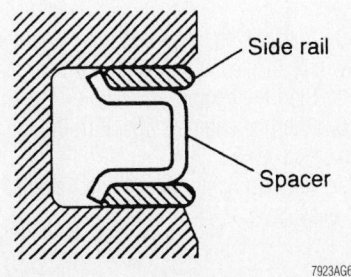

Compression ring identification and positioning—Mazda engines

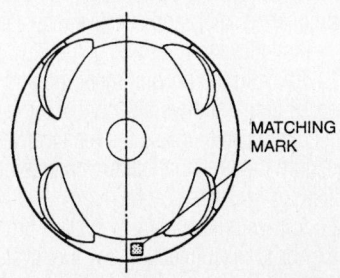

MATCHING MARK
7923AG56

Piston-to-engine block mark location on the piston face—Mazda 2.0L (FS) and 2.5L (KL) engines

53. Install a new water pump drive belt pulley using tool 49 UN21 1185.
54. Install the water pump drive belt.
55. Install the plug hole plate.
56. Install the remaining components in the reverse order of removal.
57. Chain the engine oil and filter.
58. Check all fluid levels and replenish as necessary.
59. Road test the vehicle.

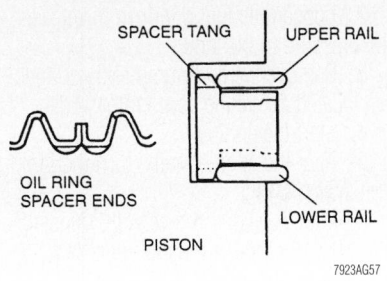

Upper, spacer and lower oil ring identification and positioning—Mazda engines

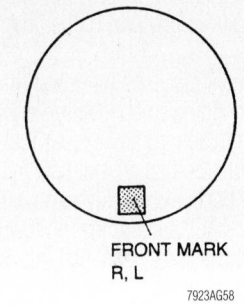

FRONT MARK
R, L
7923AG58

Piston-to-engine positioning mark location—Mazda 2.3L (KJ) engine

Piston and Ring

POSITIONING

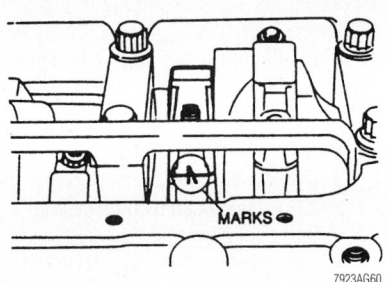

Before removing the caps from the connecting rods, be sure to matchmark them—Mazda engines

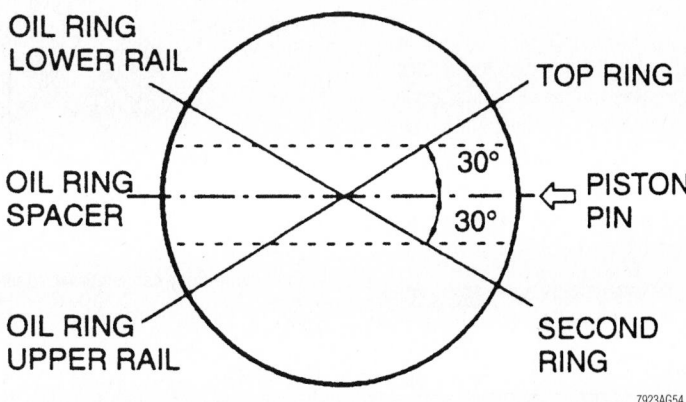

Piston ring end-gap spacing—Mazda engines

FUEL SYSTEM

Fuel System Service Precautions

Safety is the most important factor when performing not only fuel system maintenance but any type of maintenance. Failure to conduct maintenance and repairs in a safe manner may result in serious personal injury or death. Maintenance and testing of the vehicle's fuel system components can be accomplished safely and effectively by adhering to the following rules and guidelines.

1. To avoid the possibility of fire and personal injury, always disconnect the negative battery cable unless the repair or test procedure requires that battery voltage be applied.

2. Always relieve the fuel system pressure prior to disconnecting any fuel system component (injector, fuel rail, pressure regulator, etc.), fitting or fuel line connection. Exercise extreme caution whenever relieving fuel system pressure, to avoid exposing skin, face and eyes to fuel spray. Please be advised that fuel under pressure may penetrate the skin or any part of the body that it contacts.

3. Always place a shop towel or cloth around the fitting or connection prior to loosening to absorb any excess fuel due to spillage. Ensure that all fuel spillage (should it occur) is quickly removed from engine surfaces. Ensure that all fuel soaked cloths or towels are deposited into a suitable waste container.

4. Always keep a dry chemical (Class B) fire extinguisher near the work area.

5. Do not allow fuel spray or fuel vapors to come into contact with a spark or open flame.

6. Always use a back-up wrench when loosening and tightening fuel line connection fittings. This will prevent unnecessary stress and torsion to fuel line piping. Always follow the proper torque specifications.

7. Always replace worn fuel fitting O-rings with new. Do not substitute fuel hose where fuel pipe is installed.

Fuel System Pressure

RELIEVING

626 and Protégé

1. Before servicing the vehicle, refer to the precautions in the beginning of this section.

2. Remove the filler cap.

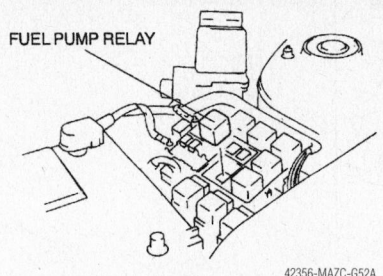

Fuel pump relay location—626 and with 2.0L and 2.5L engines

3. Remove the fuel pump relay from the relay box, located in the engine compartment.

4. Start the engine.

5. After the engine stalls, turn the ignition switch **OFF**.

6. After servicing the vehicle, reinstall the relay.

Miata

1. Before servicing the vehicle, refer to the precautions in the beginning of this section.

2. Loosen the fuel filler cap to release the pressure in the tank.

3. Remove the fuel pump relay connector, located above the accelerator pedal.

4. Start the engine.

5. After the engine stalls, turn the ignition switch **OFF**.

6. After servicing the vehicle, reinstall the relay and tighten the fuel filler cap.

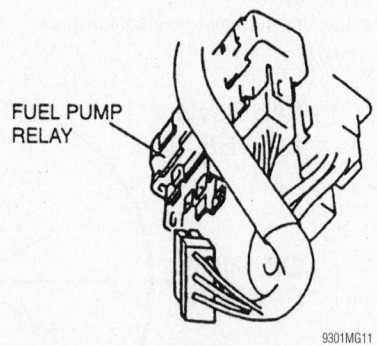

Fuel pump relay connector location—Miata

Millenia

1. Before servicing the vehicle, refer to the precautions in the beginning of this section.

2. If necessary for clearance, remove the cruise control actuator and position aside on 2.3L (KJ) engines.

3. Remove the fuel pump relay from the relay box.

4. Start the engine.

5. After the engine stalls, turn the ignition switch **OFF**.

6. After servicing the vehicle, reinstall the relay and the cruise actuator, if necessary.

Mazda 3 and 6 Models

1. Before servicing the vehicle, refer to the precautions in the beginning of this section.

2. Remove the filler cap.

3. Remove the fuel pump relay from the relay box, located in the engine compartment.

4. Start the engine.

5. After the engine stalls, turn the ignition switch **OFF**.

6. After servicing the vehicle, reinstall the relay.

Fuel Filter

The fuel filter on most Mazda cars can be located on a bracket in the left rear of the engine compartment, next to or beneath the brake master cylinder fluid reservoir or as part of the fuel pump assembly.

On the Millenia, the fuel filter is located beneath an access cover in the trunk. Access to the cover is achieved by removing the trunk mat to expose the cover.

The filter on Mazda 3 models is part of the fuel pump assembly.

REMOVAL & INSTALLATION

626 and Protégé

1. Before servicing the vehicle, refer to the precautions in the beginning of this section.

2. Properly relieve the fuel system pressure.

3. Remove or disconnect the following:
- Negative battery cable
- Cruise control actuator with the harness and cable connected on 2.0L (FS) engines
- Air cleaner assembly, 2.5L (KL) engines
- Harness from the filter
- Fuel hoses from the filter
- Fuel filter

To install:

4. Installation is the reverse of removal.

5. Pressurize the fuel system and check all connections for leaks.

Miata

1. Before servicing the vehicle, refer to the precautions in the beginning of this section.
2. Properly relieve the fuel system pressure.
3. Remove or disconnect the following:
   - Negative battery cable
4. Raise and safely support the rear of the vehicle.
   - Fuel filter protector
   - Fuel lines by squeezing the tabs. Cover the lines to prevent leakage and contamination.
   - Filter bracket and the filter. Mark the filter and bracket to aid for correct installation

To install:

5. Install or connect the following:
   - Filter and bracket aligning the marks made during removal. Tighten the bracket bolts to 52 inch lbs. (6 Nm).
   - Fuel lines making sure the tabs on the fittings are firmly engaged by pulling lightly on the lines
   - Fuel filter protector
   - Negative battery cable
6. Start the vehicle and check for leaks, then lower the vehicle.

Miata

1. Before servicing the vehicle, refer to the precautions in the beginning of this section.
2. Properly relieve the fuel system pressure.
3. Remove or disconnect the following:
   - Negative battery cable
   - Fuel lines by squeezing the tabs. Cover the lines to prevent leakage and contamination.
   - Filter bracket and the filter. Mark the filter and bracket to aid for correct installation

To install:

4. Install or connect the following:
   - Filter and bracket aligning the marks made during removal
   - Fuel lines making sure the tabs on the fittings are firmly engaged by pulling lightly on the lines
   - Negative battery cable
5. Start the vehicle and check for leaks, then lower the vehicle.

Millenia

1. Before servicing the vehicle, refer to the precautions in the beginning of this section.
2. Insure the ignition switch is **OFF**.

3. Relieve the fuel system pressure.
4. Remove or disconnect the following:
   - Negative battery cable
   - Trunk mat
   - Service hole cover
   - Fuel lines from both ends of the fuel filter
   - Fuel filter and bracket
   - Fuel filter from the mounting bracket

To install:

5. Install or connect the following:
   - Filter in the bracket. Torque the nut to 70–95 inch lbs. (8–11 Nm).
   - Fuel lines to the filter
   - Service hole cover
   - Trunk mat
   - Negative battery cable
6. Run the engine and check for any fuel leaks.

Fuel Pump

REMOVAL & INSTALLATION

Protégé

1. Before servicing the vehicle, refer to the precautions in the beginning of this section.
2. Relieve the fuel pressure.
3. Remove or disconnect the following:
   - Negative battery cable
   - Rear seat cushion
   - Fuel pump/sending unit electrical connector
   - Fuel pump/sending unit access cover
   - Fuel supply and return hoses from the fuel pump/sending unit
   - Fuel pump/sending unit from the fuel tank
   - Sending unit electrical connector
   - Sending unit from the fuel pump assembly

To install:

4. Install or connect the following:
   - Sending unit to the fuel pump assembly
   - Sending unit electrical connector
   - Fuel pump/sending unit into the fuel tank with a new gasket
   - Fuel supply and return lines
   - Access cover
   - Sending unit electrical connector
   - Rear seat cushion
   - Negative battery cable
5. Start the engine and check fuel leaks.

Miata

1. Before servicing the vehicle, refer to the precautions in the beginning of this section.

2. Properly relieve the fuel pressure.
3. Remove or disconnect the following:
   - Negative battery cable
   - Rear package trim
   - Service hole cover
   - Fuel pump cover
   - Fuel pump connector
   - Fuel hoses
   - Fuel pump and gauge sender unit as an assembly
   - Fuel pump from the sender bracket

To install:

4. Install or connect the following:
   - New O-ring set
   - Fuel pump to the sender bracket

➡ **Pull the fuel pump down so that it is tight against the bracket.**

   - Fuel pump and gauge sender unit as an assembly
   - Fuel hoses
   - Fuel pump connector
   - Fuel pump cover
   - Service hole cover
   - Rear package trim
   - Negative battery cable

626

1. Before servicing the vehicle, refer to the precautions in the beginning of this section.
2. Relieve the fuel system pressure.
3. Drain the fuel from the tank.
4. Remove or disconnect the following:
   - Negative battery cable
   - Fuel pump electrical connector
   - Hoses from the fuel tank
   - Pressure control valve
   - Fuel tank pressure sensor
   - Fuel pipe
   - Presilencer insulator
   - Fuel tank strap while supporting the fuel tank
   - Fuel tank
   - All fuel hoses from the fuel pump unit
   - Fuel pump ring retainers and the ring
   - Fuel pump and gaskets from the fuel tank

To install:

5. Install or connect the following:
   - Fuel pump using a new gasket
   - Fuel pump ring and tighten the fasteners to 19 inch lbs. (2 Nm)
   - Fuel hoses to the fuel pump
   - Fuel tank
   - Fuel tank strap while supporting the fuel tank. Refer to the illustration for torque specifications.
   - Presilencer insulator

SST
1.1—2.1 N·m
{11—22 kgf·cm, 10—19 in·lbf}
R

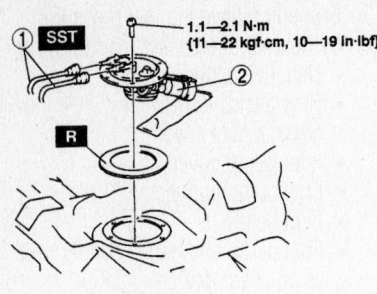

1 Fuel pipe
2 Fuel pump

42356-MAZC-G53

Exploded view of the fuel pump assembly—626

- Fuel pipe
- Fuel tank pressure sensor
- Pressure control valve
- Hoses to the fuel tank
- Fuel pump electrical connector
- Negative battery cable

6. Add a minimum of 10 gallons of fuel to the tank and check for leaks.

Millenia

1. Before servicing the vehicle, refer to the precautions in the beginning of this section.
2. Relieve the fuel system pressure.
3. Remove or disconnect the following:
   - Negative battery cable
   - Rear seat cushion
4. Drain the fuel from the tank.
   - Service hole cover
   - Fuel pump electrical connector
   - Hoses from the fuel tank
   - Fuel tank strap while supporting the fuel tank
   - Evaporative hose
   - Fuel tank
   - All fuel hoses from the fuel pump unit
   - Fuel pump ring using tool 49 T042 011

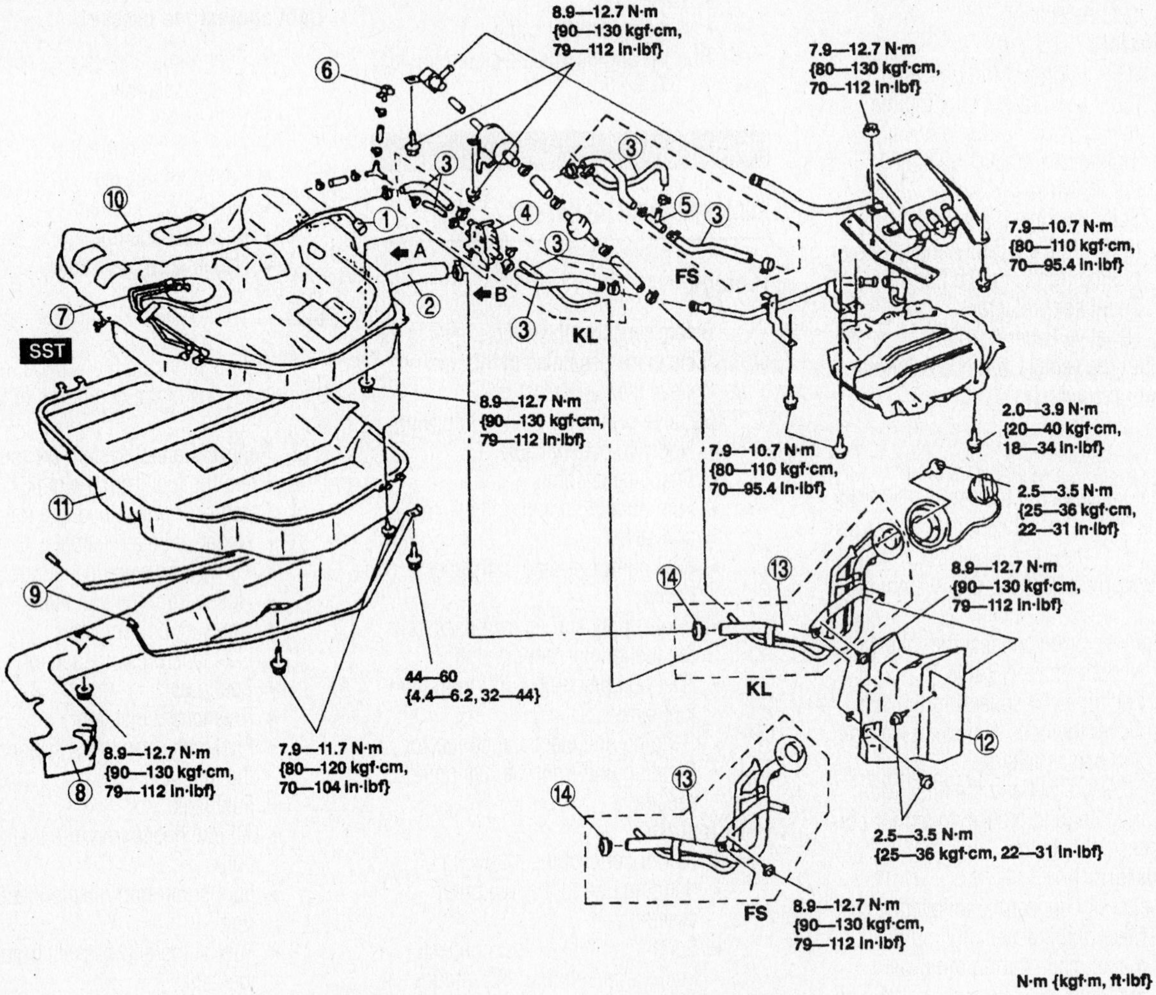

8.9—12.7 N·m
{90—130 kgf·cm,
79—112 in·lbf}
7.9—12.7 N·m
{80—130 kgf·cm,
70—112 in·lbf}
7.9—10.7 N·m
{80—110 kgf·cm,
70—95.4 in·lbf}
A
B
FS
KL
SST
8.9—12.7 N·m
{90—130 kgf·cm,
79—112 in·lbf}
2.0—3.9 N·m
{20—40 kgf·cm,
18—34 in·lbf}
7.9—10.7 N·m
{80—110 kgf·cm,
70—95.4 in·lbf}
2.5—3.5 N·m
{25—36 kgf·cm,
22—31 in·lbf}
8.9—12.7 N·m
{90—130 kgf·cm,
79—112 in·lbf}
44—60
{4.4—6.2, 32—44}
KL
8.9—12.7 N·m
{90—130 kgf·cm,
79—112 in·lbf}
7.9—11.7 N·m
{80—120 kgf·cm,
70—104 in·lbf}
2.5—3.5 N·m
{25—36 kgf·cm, 22—31 in·lbf}
FS
8.9—12.7 N·m
{90—130 kgf·cm,
79—112 in·lbf}
N·m {kgf·m, ft·lbf}

1	Fuel pump connector	8	Presilencer insulator
2	Joint hose	9	Fuel tank strap
3	Evaporative hose	10	Fuel tank
4	Pressure control valve	11	Fuel tank protector
5	Hose joint	12	Fuel-filler pipe insulator
6	Fuel tank pressure sensor	13	Fuel-filler pipe
7	Fuel pipe	14	Nonreturn valve

42356-MAZC-G54

Exploded view of the fuel tank assembly—626

49 T042 001
FUEL TANK
UNION NUT
42356-MAZC-GEO

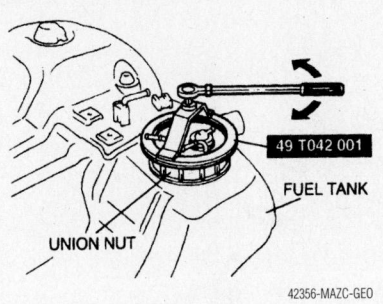

Remove the fuel pump ring—Millenia

- Fuel pump and gaskets from the fuel tank

To install:

5. Install or connect the following:
   - Fuel pump using a new gasket
   - Fuel pump ring and tighten the fasteners to 75 ft. lbs. (102 Nm)
   - Fuel hoses to the fuel pump
   - Fuel tank
   - Evaporative hose
   - Fuel tank strap while supporting the fuel tank. Tighten the strap bolts to 44 ft. lbs. (60 Nm).
   - Hoses to the fuel tank
   - Fuel pump electrical connector
   - Service hole cover
   - Rear seat cushion
   - Negative battery cable
6. Add a minimum of 10 gallons of fuel to the tank and check for leaks.

Mazda 3

1. Before servicing the vehicle, refer to the precautions in the beginning of this section.

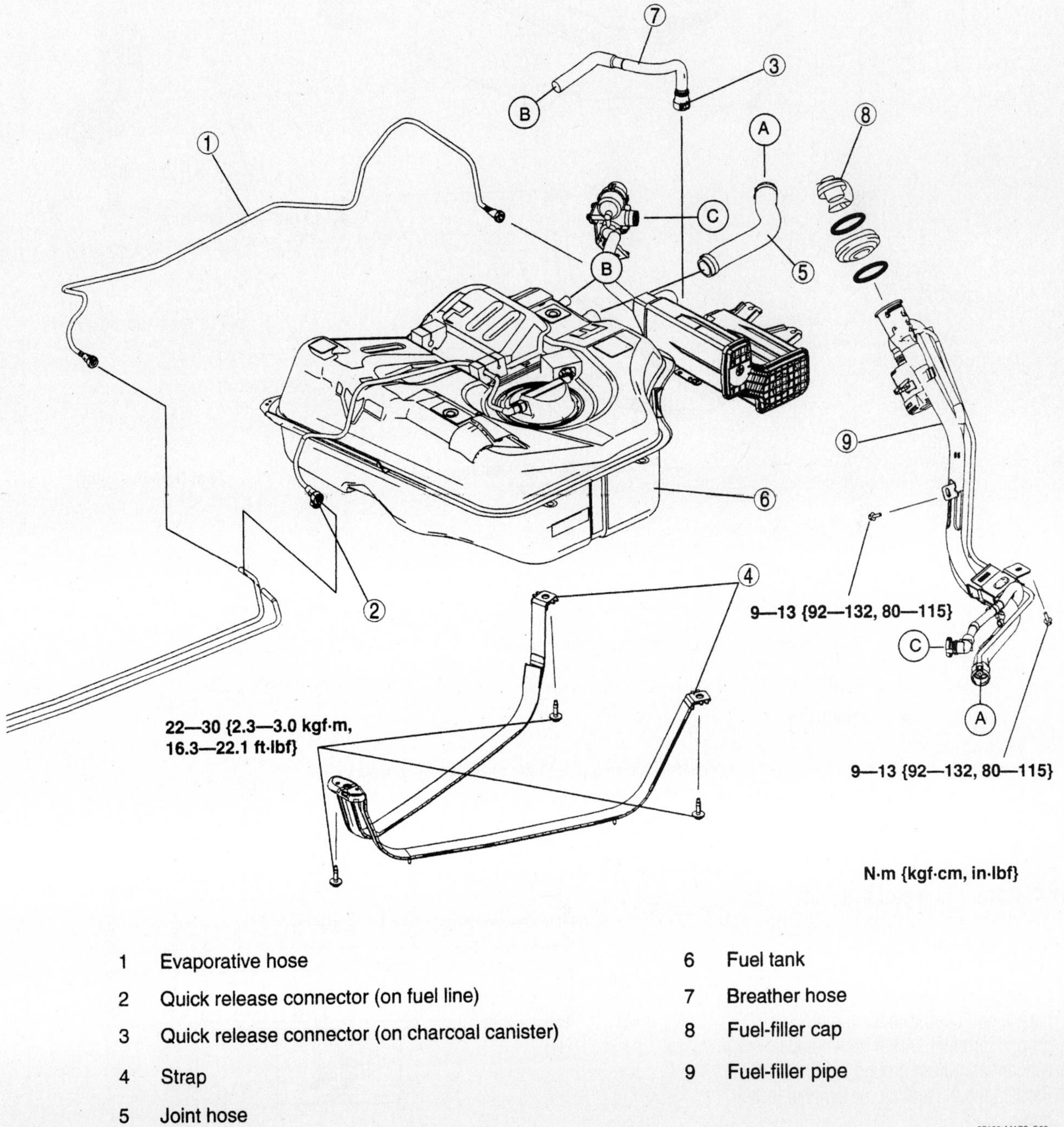

1
2
3
4
5
6
7
8
9
A
B
C
9—13 {92—132, 80—115}
22—30 {2.3—3.0 kgf·m, 16.3—22.1 ft·lbf}
9—13 {92—132, 80—115}
N·m {kgf·cm, in·lbf}
1 Evaporative hose
2 Quick release connector (on fuel line)
3 Quick release connector (on charcoal canister)
4 Strap
5 Joint hose
6 Fuel tank
7 Breather hose
8 Fuel-filler cap
9 Fuel-filler pipe
67162-MAZC-G62

Exploded view of the fuel tank and related components—Mazda 3 California emission models

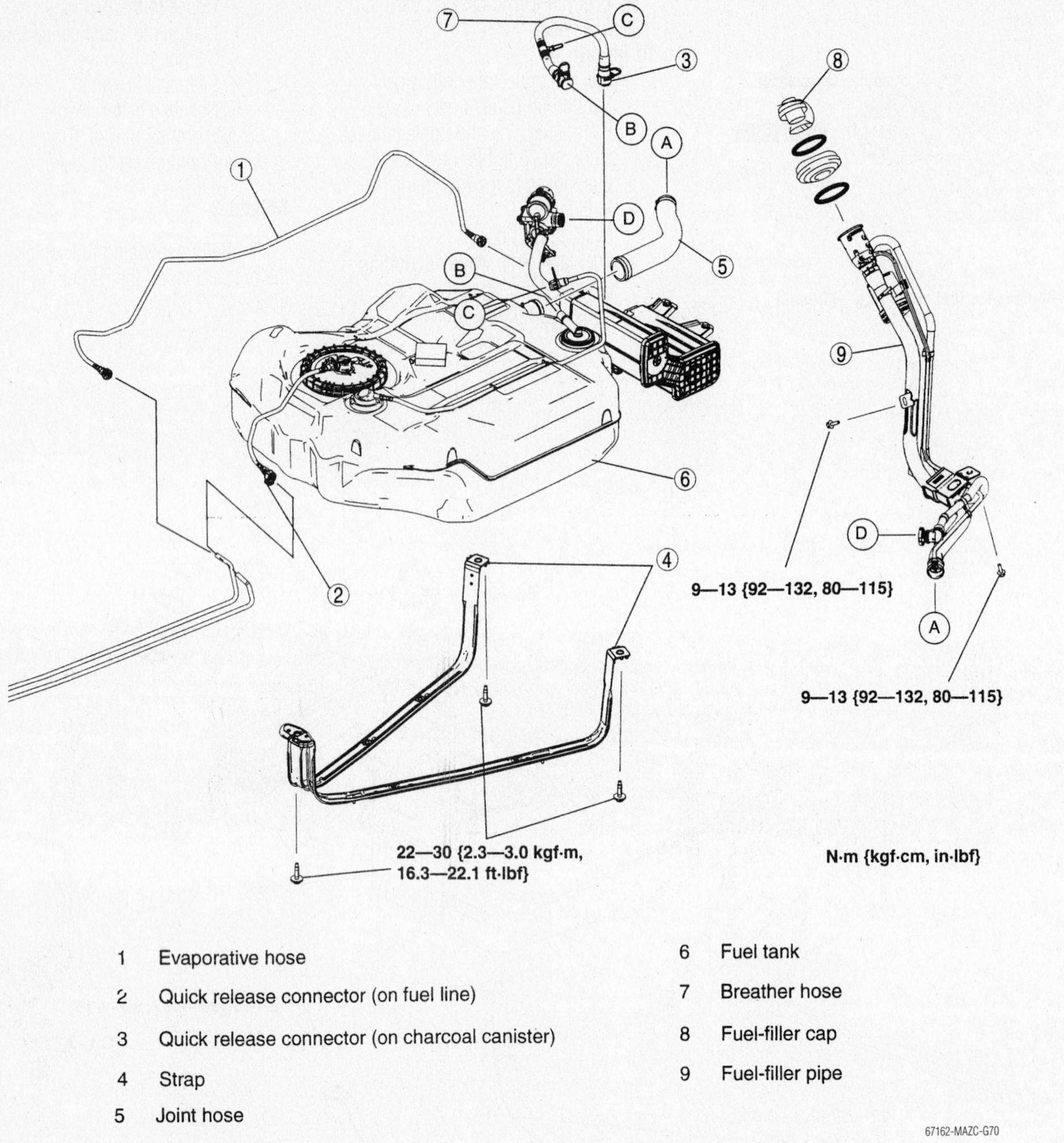

9—13 {92—132, 80—115}

9—13 {92—132, 80—115}

22—30 {2.3—3.0 kgf·m,
16.3—22.1 ft·lbf}

N·m {kgf·cm, in·lbf}

1	Evaporative hose	6	Fuel tank
2	Quick release connector (on fuel line)	7	Breather hose
3	Quick release connector (on charcoal canister)	8	Fuel-filler cap
4	Strap	9	Fuel-filler pipe
5	Joint hose		

67162-MAZC-G70

Exploded view of the fuel tank and related components—Mazda 3 non–California emission models

2. Relieve the fuel system pressure.

3. Disconnect the quick release connector in the engine compartment.

4. Attach a long hose to the disconnect fuel pipe and drain the fuel into a suitable container as follows:

a. Using a jumper wire, ground the Powertrain Control Module (PCM) terminals. If equipped with an immobilizer, ground terminal 1AR, if not equipped with an imobilizer, ground terminal 1AQ. Refer to the illustration for terminal location.

b. Turn the ignition switch the ON position to activate the fuel pump.

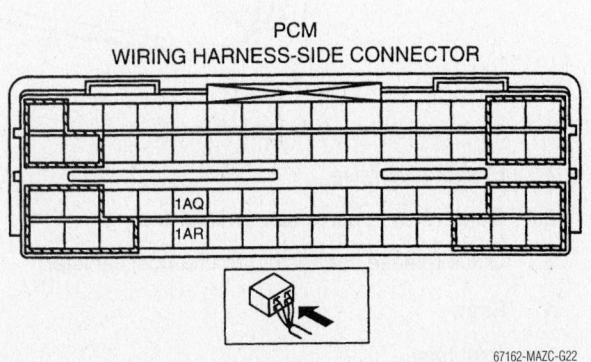

PCM
WIRING HARNESS-SIDE CONNECTOR

1AQ
1AR

67162-MAZC-G22

If equipped with an immobilizer, ground PCM terminal 1AR, if not equipped with an imobilizer, ground PCM terminal 1AQ—Mazda 3 models

✳✳ CAUTION

The fuel pump can be damaged if all fuel is removed from the tank, monitor the hose and stop when no fuel is being discharged from the tank.

c. Once no more fuel is being discharged, turn the ignition OFF.

5. Disconnect the negative battery cable.

6. Remove the rear seat cushion and remove the pump service hole cover.

7. Disconnect the fuel pump connector and remove the charcoal canister.

8. Lower the exhaust main silencer so the insulator can be removed and remove the insulator.

9. Remove the rear left hand under cover.

10. Disconnect the evaporator hose, quick release fuel connector on the fuel line and charcoal canister.

11. Support the fuel tank and remove the strap.

12. Remove the filler pipe bolt, loosen the tie band and pull down on the filler pipe to disconnect the joint hose.

13. Lower and remove the fuel tank.

14. Disconnect the fuel release connector.

15. Use tool 49 F042 001 to remove the pump cap on non–California emission models.

16. Use a brass drift and a hammer to remove the fuel pump retaining ring on California emission models.

17. Remove the fuel pump assembly.

To install:

18. Install the fuel pump assembly.

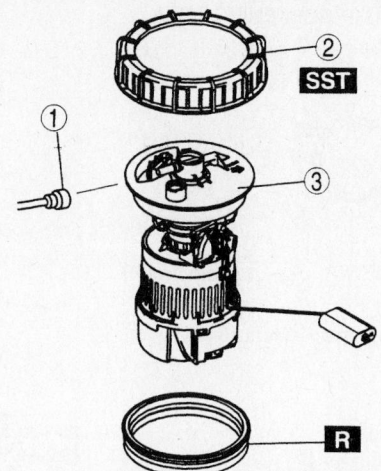

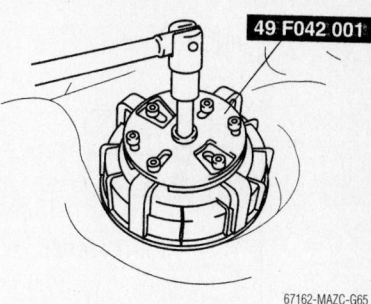

1 Quick release connector

2 Fuel pump cap

3 Fuel pump unit

67162-MAZC-G64

Exploded view of the pump assembly and related components—Mazda 3 California emission models

19. Clean all gasoline from the pump gasket to prevent it from turning during installation.

20. On non–California emission models, align the pump and tank assembly as illustrated.

21. Use tool 49 F042 001 to tighten the pump cap on non–California emission models. Tighten the cap to 59–66 ft. lbs. (80–90 Nm). If the torque cannot be reached, replace the pump cap and gasket.

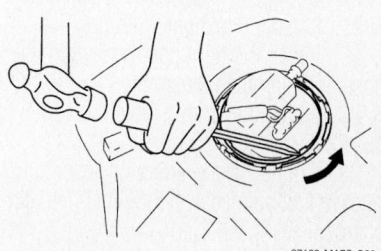

67162-MAZC-G65

Use tool 49 F042 001 to remove the pump cap—Mazda 3 non–California emission models

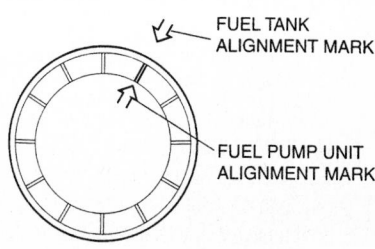

67162-MAZC-G66

Use a brass drift and a hammer to remove the fuel pump retaining ring—Mazda 3 non–California emission models

If the torque cannot still be reached, replace the fuel tank.

22. On California emission models, use a brass drift and a hammer to install the fuel pump retaining ring

23. Before installing the tank apply 1.7

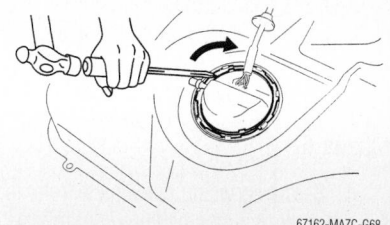

FUEL TANK ALIGNMENT MARK

FUEL PUMP UNIT ALIGNMENT MARK

67162-MAZC-G67

Align the pump and tank assembly as shown—Mazda 3 non–non–California emission models

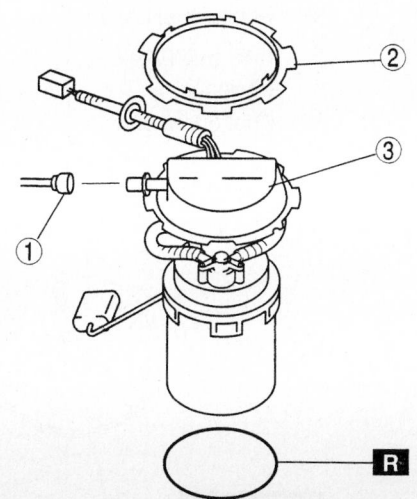

1 Quick release connector

2 Retainer ring

3 Fuel pump unit

67162-MAZC-G63

Exploded view of the pump assembly and related components—Mazda 3 non–California emission models

67162-MAZC-G68

Use a brass drift and a hammer install the fuel pump retaining ring—Mazda 3 non–California emission models

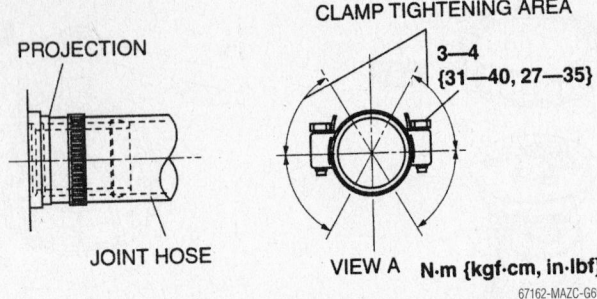

Install the joint hose and align the hose and clamp as shown—Mazda 3

inch Hg (5.9 kPa) of pressure to the tank to check for leakage around the pump.

24. Connect the fuel release connector.

25. Raise the fuel tank into position.

26. Install the joint hose and align the hose and clamp as illustrated.

27. Install the strap. Tighten the bolts to 16–22 ft. lbs. (22–30 Nm).

28. Connect the evaporator hose, quick release fuel connector on the fuel line and charcoal canister.

29. Install the rear left hand under cover.

30. Install insulator and the exhaust main silencer.

31. Install the charcoal canister and connect the fuel pump connector.

32. Install the pump service hole cover and rear seat.

33. Disconnect the negative battery cable.

34. Start the vehicle and check for leaks as follows:

   a. Using a jumper wire, ground the Powertrain Control Module (PCM) terminals. If equipped with an immobilizer, ground terminal 1AR, if not equipped with an imobilizer, ground terminal 1AQ. Refer to the illustration for terminal location.

   b. Turn the ignition switch the ON position to activate the fuel pump.

   c. Check the hoses, clips and other fuel system components for leaks.

   d. If there are any leaks, replace the fuel hoses and clips. If the is damage to the seal on the fuel pipe side, replace the pipe.

   e. The system must be leak free for five minutes with the terminal grounded. If any component is replaced because of a system, leak, turn the ignition key OFF; remove the jumper wire from the terminal. reapply the jumper wire, turn the ignition On and check for leaks.

Mazda 6

1. Before servicing the vehicle, refer to the precautions in the beginning of this section.

2. Relieve the fuel system pressure.

3. Disconnect the negative battery cable.

4. Remove the rear seat cushion.

5. Remove the service hole cover.

6. Disconnect the fuel pump electrical connector.

7. Disconnect the hoses from the fuel tank.

➡ **The fuel pump cap can be damaged if there is a gap between the removal/installation tool 49 F042 001. Make sure the tool is attached securely to the cap and that no gap exists between them.**

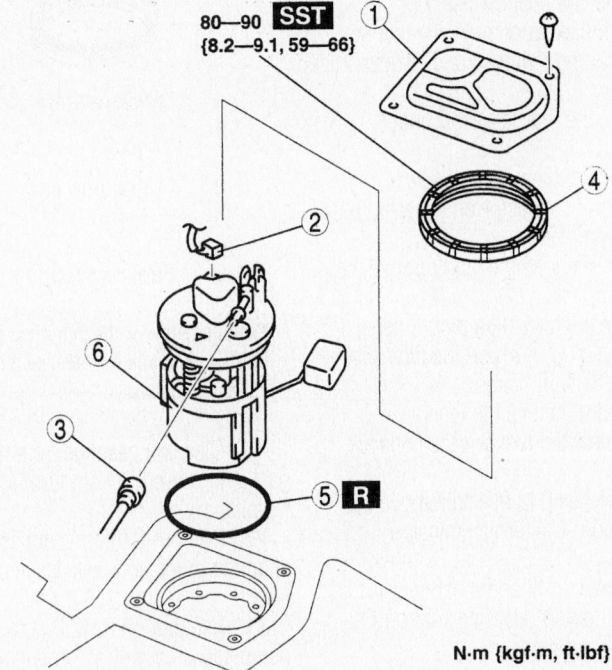

1	Service hole cover
2	Connector
3	Plastic fuel hose
4	Fuel pump cap
5	Packing
6	Fuel pump unit

Exploded view of the fuel pump assembly—Mazda 6 models

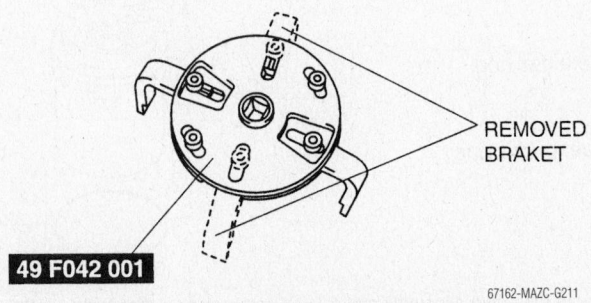

Remove the fuel pump ring using tool 49 F042 001—Mazda 6 models

8. Remove the fuel pump ring using tool 49 F042 001.

9. Remove the fuel pump and gaskets from the fuel tank.

To install:

10. Install the fuel pump using a new gasket.

➡The fuel pump cap can be damaged if there is a gap between the removal/installation tool 49 F042 001. Make sure the tool is attached securely to the cap and that no gap exists between them.

11. Align the fuel pump assembly alignment marks and the floating lines as illustrated.

12. Align the tightening start positions of the cap and the retainer notch as illustrated and tighten them one full turn by hand. If the cap cannot be tightened by hand, remove the cap and make sure the cap is not damaged or misaligned on the retainer or cap and retighten.

13. Make sure the alignment mark and the floating lines are still aligned, tighten the cap to 59–99 ft. lbs. (80–135 Nm) or a rotation angle of 50–140 degrees with the total angle of step 2 and 3 being 410–500 degrees.

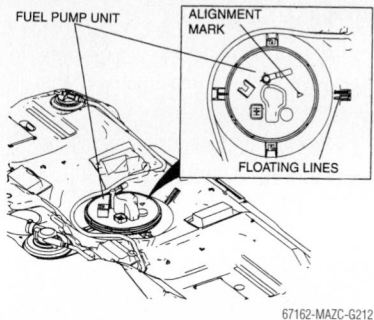

67162-MAZC-G212

Align the fuel pump assembly alignment marks and the floating lines—Mazda 6 models

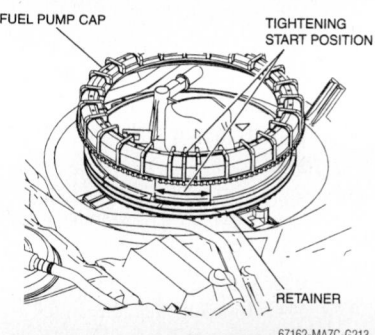

67162-MAZC-G213

Align the tightening start positions of the cap and the retainer notch—Mazda 6 models

Fuel Injector

REMOVAL & INSTALLATION

✳✳ CAUTION

Fuel injection systems remain under pressure after the engine has been turned OFF. Properly relieve fuel pressure before disconnecting any fuel lines. Failure to do so may result in fire or personal injury. Do not allow fuel spray or fuel vapors to come in contact with a spark or open flame. Keep a dry chemical fire extinguisher nearby. Never store fuel in an open container due to risk of fire or explosion.

626

2.0L (FS) ENGINE

1. Before servicing the vehicle, refer to the precautions in the beginning of this section.

2. Refer to the illustration for component location and torque specifications.

3. Relieve the fuel system pressure.

4. Remove or disconnect the following:
- Negative battery cable
- Fuel injector wiring harness
- Fuel lines at the fuel rail
- Hose from the pressure regulator
- Fuel rail with the injectors attached

- Fuel injectors, grommets and O-rings from the fuel rail
- O-rings from the fuel injectors

To install:

5. Install or connect the following:
- New O-rings and grommets lubricated with engine oil on the fuel injectors.
- Insulators and injectors on the intake manifold
- Grommets and the fuel rail onto the injectors. Torque the bolts to 14–18 ft. lbs. (19–25 Nm).
- Fuel lines to the fuel rail
- Fuel injector wiring harness
- Negative battery cable

6. Turn the ignition switch **ON** to pressurize the fuel system.

7. Check for leaks and correct as necessary, before starting the engine.

2.5L (KL) ENGINE

1. Before servicing the vehicle, refer to the precautions in the beginning of this section.

2. Relieve the fuel system pressure.

3. Remove or disconnect the following:
- Negative battery cable
- Fuel injector wiring harness
- Fuel lines at the fuel rail
- Fuel rail
- Accumulated connector
- Spacer
- Fuel injectors, grommets and O-rings from the fuel rail
- O-rings from the fuel injectors

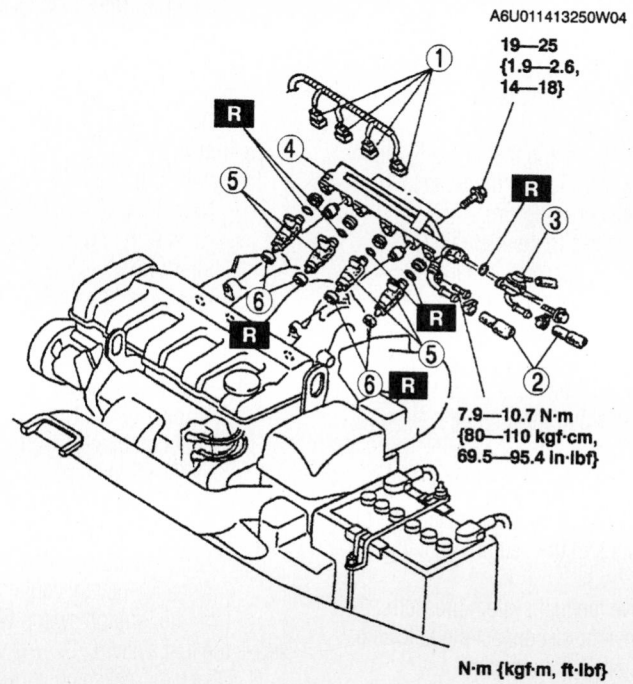

A6U011413250W04

19—25
{1.9—2.6,
14—18}

7.9—10.7 N·m
{80—110 kgf·cm,
69.5—95.4 In·lbf}

N·m {kgf·m, ft·lbf}

42356-MAZC-G56

Exploded view of the fuel rail and injector assembly—626 2.0L (FS)

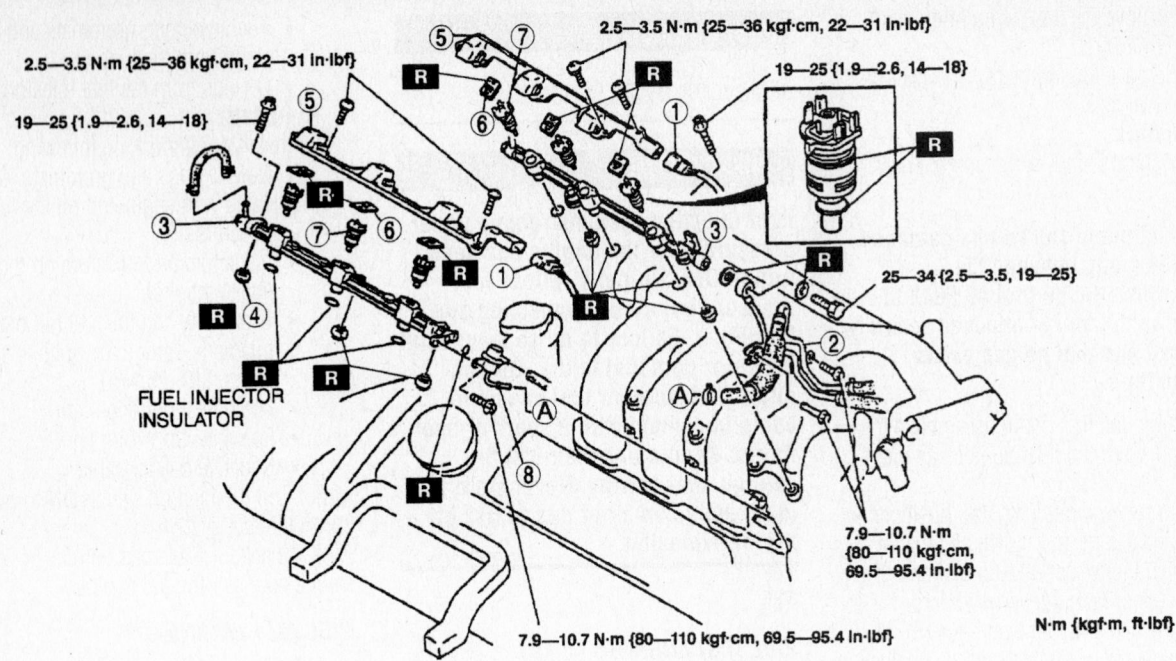

1	Connector	5	Accumulated connector
2	Fuel hose and fuel pipe	6	Spacer
3	Fuel distributor	7	Fuel injector
4	Insulator	8	Pressure regulator

Exploded view of the fuel rail and injector assembly—626 2.5L (KL) engine

To install:

4. Install or connect the following:

- New O-rings and grommets lubricated with engine oil on the fuel injectors. Fit the injector squarely into the rail while using a twisting motion. Fit the injector tab into the notch on the rail.
- Insulators and injectors on the intake manifold
- Grommets and the fuel rail onto the injectors. Torque the bolts to 14–18 ft. lbs. (19–25 Nm).
- Fuel lines to the fuel rail
- Fuel injector wiring harness
- Negative battery cable

5. Turn the ignition switch **ON** to pressurize the fuel system.

6. Check for leaks and correct as necessary, before starting the engine.

Miata

1. Before servicing the vehicle, refer to the precautions in the beginning of this section.

2. Relieve the fuel system pressure.

3. Remove or disconnect the following:

- Negative battery cable
- Upper intake manifold
- Fuel lines at the fuel rail

- Vacuum hose from the fuel pressure regulator
- Fuel rail mounting bracket
- Fuel rail mounting bolts, spacers and insulators
- Fuel rail, with the injectors attached
- Fuel injectors, grommets and O-rings from the fuel rail
- O-rings from the fuel injectors

To install:

4. Install or connect the following:

- New O-rings and grommets lubricated with engine oil on the fuel injectors
- Insulators and injectors on the intake manifold
- Grommets and the fuel rail onto the injectors. Torque the bolts to 14–18 ft. lbs. (19–25 Nm).
- Fuel line bracket. Torque the bolts to 70–95 inch lbs. (8–11 Nm).
- Fuel lines to the fuel rail
- Fuel injector wiring harness
- Upper intake manifold
- Negative battery cable

5. Turn the ignition switch **ON** to pressurize the fuel system.

6. Check for leaks and correct as necessary, before starting the engine.

Protégé

1. Before servicing the vehicle, refer to the precautions in the beginning of this section.

2. Relieve the fuel system pressure.

3. Remove or disconnect the following:

- Negative battery cable
- Accelerator cables and the cable bracket
- Fuel injector connectors and wiring harness
- Fuel lines at the fuel rail
- Vacuum hose from the fuel pressure regulator
- Fuel rail mounting bolts, spacers and insulators
- Fuel rail, with the injectors attached
- Fuel injectors, grommets and O-rings from the fuel rail
- O-rings from the fuel injectors

To install:

4. Install or connect the following:

- New O-rings and grommets lubricated with engine oil on the fuel injectors
- Insulators and injectors on the intake manifold
- Grommets and the fuel rail onto the injectors. Torque the bolts to 14–18 ft. lbs. (19–25 Nm).

13. Temporarily attach a new clip to the injector groove.

➡**When the clip is correctly installed, the central area of the injector and clip finger positions are aligned.**

14. While firmly holding the injector, push the clip into the injector until the clip stops. Make sure the injector connector is correctly positioned.

15. Press the injector into the cup until the cup contacts the lower surface of the cup. Make sure the injector and clip are installed properly and the clip is hooked into the injector cup notch.

16. Install the injectors and fuel rail as an assembly.

17. Install the fuel rail bolts and tighten to 15–19 ft. lbs. (20–26 Nm).

18. Attach the fuel lines to the rail.

19. Start the vehicle and check for leaks as follows:

a. Using a jumper wire, short the check connector to terminal F/P to a body ground.

b. Turn the ignition switch the ON position to activate the fuel pump.

c. Check the hoses, clips and other fuel system components for leaks.

d. If there are any leaks, replace the fuel hoses and clips. If the is damage to the seal on the fuel pipe side, replace the pipe.

e. The system must be leak free for five minutes with the terminal grounded. If any component is replaced because of a system, leak, turn the ignition key OFF; remove the jumper wire from the terminal. reapply the jumper wire, turn the ignition On and check for leaks.

20. Install the remaining components.

3.0L (AJ) ENGINE

1. Before servicing the vehicle, refer to the precautions in the beginning of this section.

2. Release the fuel system pressure.

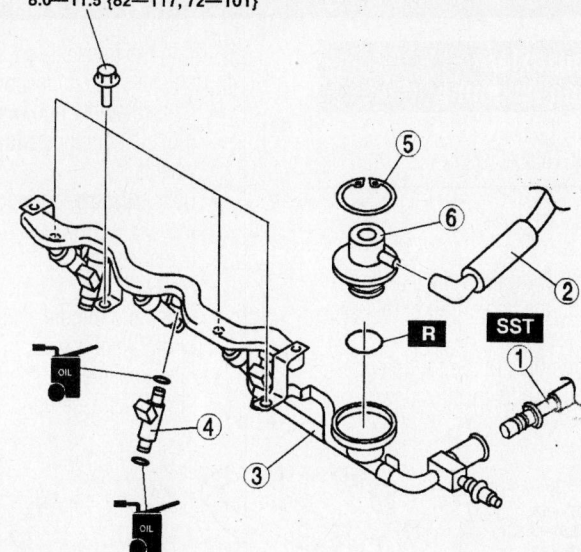

1 Plastic fuel hose
2 Hose
3 Fuel distributor
4 Fuel injector
5 Snap ring
6 Pulsation damper

Exploded view of the fuel rail and injector assembly—Mazda 6 models with the 3.0L (AJ) engine

3. Disconnect the negative battery cable.
4. Remove the upper intake manifold (dynamic chamber).
5. Disconnect the fuel injector connectors.
6. Disconnect the fuel hoses from the rail.
7. Remove the rail and injectors as an assembly.
8. Gently twist the fuel injector out of the manifold
9. Check the O-rings and replace if damaged.

To install:
10. Install the fuel injector(s) using new O-rings lubricated with clean engine oil.

11. Attach the fuel injector into the supply manifold.
12. Install the fuel rail and injectors as an assembly. Tighten the rail bolts to 72–101 inch . lbs. (8–11 Nm).
13. Connect the fuel hoses to the rail.
14. Install the injector connectors.
15. Install the upper intake manifold.
16. Connect the negative battery cable.
17. Start the vehicle and check for leaks as follows:

a. Using a jumper wire, short the check connector to terminal F/P to a body ground.

b. Turn the ignition switch the ON position to activate the fuel pump.

c. Check the hoses, clips and other fuel system components for leaks.

d. If there are any leaks, replace the fuel hoses and clips. If the is damage to the seal on the fuel pipe side, replace the pipe.

e. The system must be leak free for five minutes with the terminal grounded. If any component is replaced because of a system, leak, turn the ignition key OFF; remove the jumper wire from the terminal. reapply the jumper wire, turn the ignition On and check for leaks.

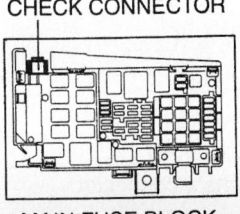

Using a jumper wire, short the check connector to terminal F/P to a body ground—Mazda 6 models

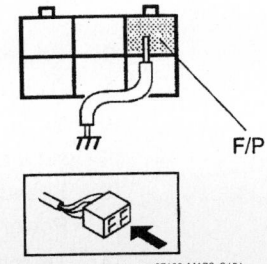

DRIVE TRAIN

Manual Transaxle/Transmission Assembly

REMOVAL & INSTALLATION

Miata

1. Before servicing the vehicle, refer to the precautions in the beginning of this section.

2. Refer to the illustration for component location and torque specifications.

3. Drain the transaxle oil.
4. Remove or disconnect the following:
   • Crossmember and bracket, on models equipped with 16 inch wheels
   • Rear crossbar, on models equipped with 16 inch wheels
   • Under cover
   • Starter
   • Front and middle exhaust pipes
   • Shifter knob
   • Rear console and insulator
   • Shift lever assembly
   • Dust boot
   • Front crossbar
   • Drive shaft
   • Clutch release cylinder
   • Back up light switch connector
   • Neutral safety switch connector
   • Speedometer sensor connector
   • Wiring harness from the Power Plant Frame (PPF)
5. Support the transmission with a jack.
   • PPF bracket
   • Differential side bolts and pry out the spacer

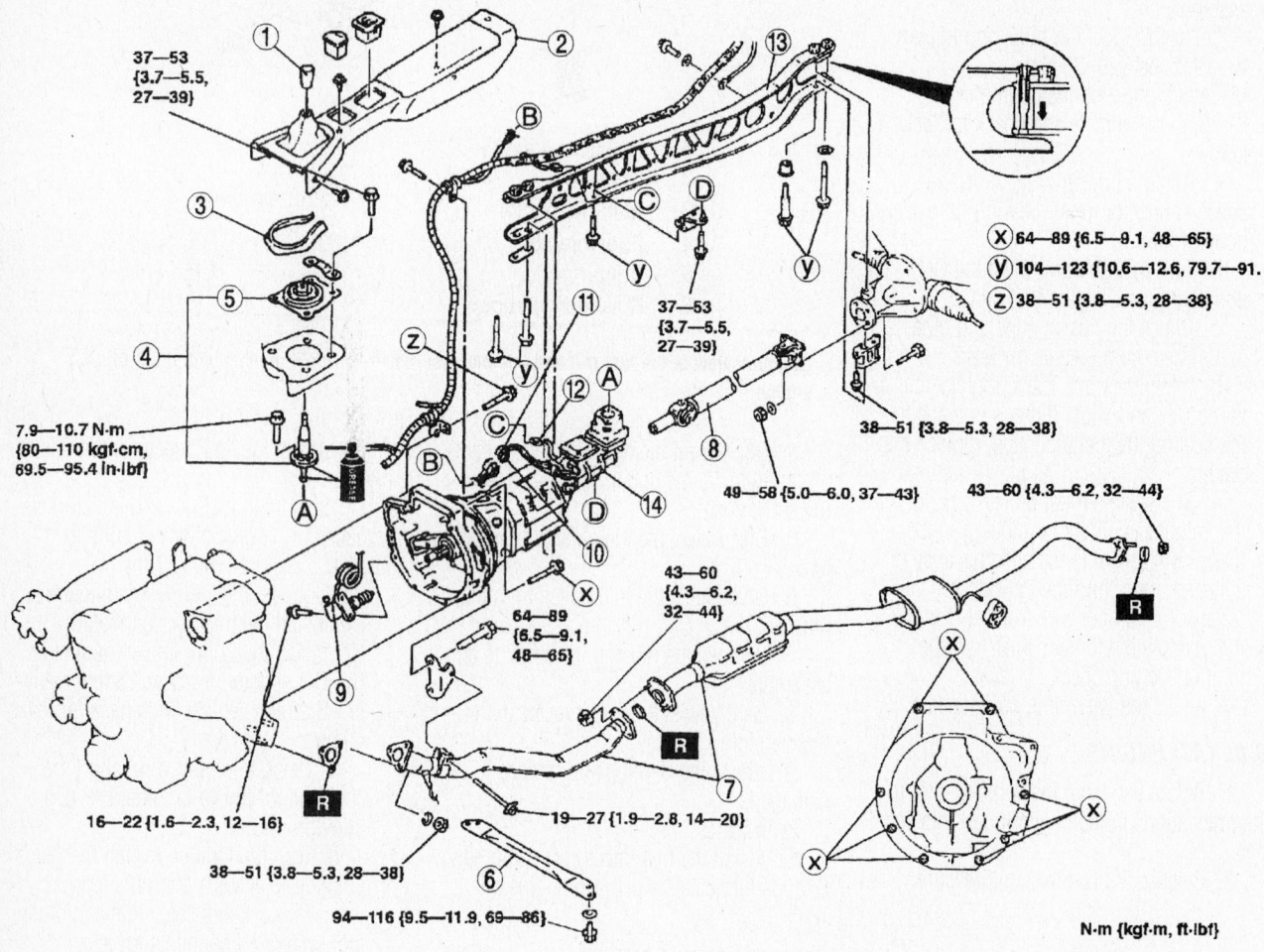

1	Shift lever knob	8	Propeller shaft
2	Rear console	9	Clutch release cylinder
3	Insulation	10	Back-up light switch connector
4	Shift lever component	11	Neutral switch connector
5	Dust boot	12	Speedometer sensor connector
6	Front crossbar	13	Power plant frame (PPF)
7	Front pipe and middle pipe	14	Transmission

Exploded view of the M15M–D manual transmission assembly—Miata

42356-MAZC-GBS

✲✲ CAUTION

Removing the PPF spacers will reduce the performance of the PPF. If the spacers are removed, replace the PPF as an assembly.

- Differential mounting spacer
- Transmission side bolts and the PPF
- Transmission bolts and the transmission

To install:

6. Tilt the engine by pushing up on the

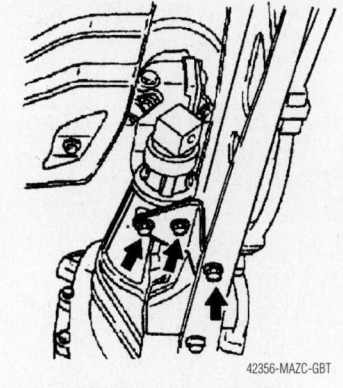

42356-MAZC-GBT

Remove the PPF bracket—Miata

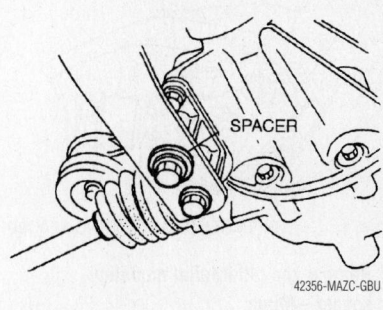

42356-MAZC-GBU

Remove the differential side bolts and pry out the spacer—Miata

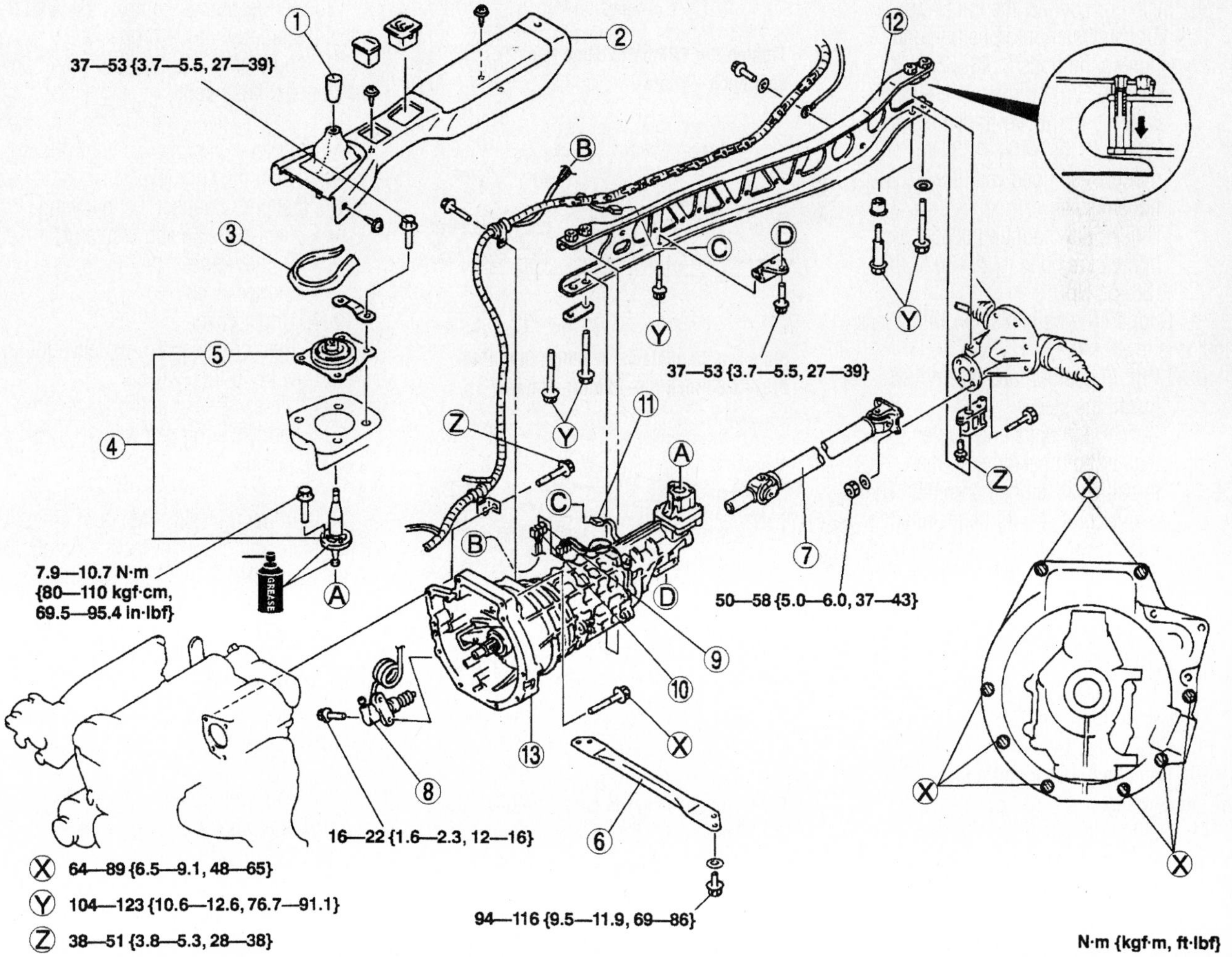

37—53 {3.7—5.5, 27—39}

37—53 {3.7—5.5, 27—39}

7.9—10.7 N·m
{80—110 kgf·cm,
69.5—95.4 in·lbf}

50—58 {5.0—6.0, 37—43}

16—22 {1.6—2.3, 12—16}

Ⓧ 64—89 {6.5—9.1, 48—65}

Ⓨ 104—123 {10.6—12.6, 76.7—91.1}

Ⓩ 38—51 {3.8—5.3, 28—38}

94—116 {9.5—11.9, 69—86}

N·m {kgf·m, ft·lbf}

1	Shift lever knob	8	Clutch release cylinder
2	Rear console	9	Back-up light switch connector
3	Insulator	10	Neutral switch connector
4	Shift lever component	11	Speedometer sensor connector
5	Dust boot	12	Power plant frame (PPF)
6	Front crossbar	13	Transmission
7	Propeller shaft		

Exploded view of the Y16M–D manual transmission assembly—Miata

42356-MAZC-GCD

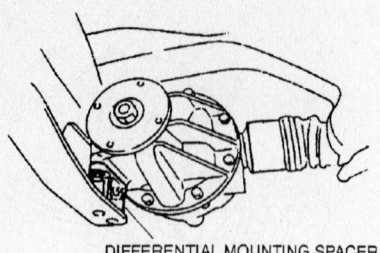

DIFFERENTIAL MOUNTING SPACER
42356-MAZC-GBV

Remove the differential mounting spacer—Miata

front oil pan with a wooden block placed on a transmission jack.

7. Support the transmission with a jack.

8. Install or connect the following:
- Transmission into position and tighten the bolts to 48–65 ft. lbs. (64–89 Nm) on models equipped with the M15M–D transmission or 76–91 ft. lbs. (104–123 Nm) on models equipped with the Y16M–D transmission.
- Differential mounting spacer and tighten the bolts to 28–38 ft. lbs. (38–51 Nm).

9. Support the transmission with a jack until it is level.
- PPF in position and if removed install the sleeve
- Spacer and bolts and tighten the reamer bolt making sure the threading is aligned correctly. The reamer bolt should be installed in the forward hole. Tighten the outer bolts making sure the threads are properly aligned. Tighten all bolts in the sequence illustrated to 91 ft. lbs. (123 Nm).
- PPF bracket and tighten bolts **A** to 91 ft. lbs. (123 Nm) and bolts **B** to 39 ft. lbs. (53 Nm).

10. Remove the jack and connect the wiring harness.

11. Using a straightedge and Vernier caliper measure the distance **A** which should be 2.37–2.83 inch (60–72mm). If

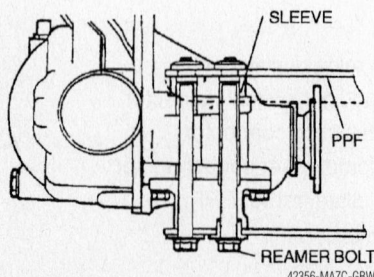

SLEEVE

PPF

REAMER BOLT
42356-MAZC-GBW

The reamer bolt should be installed in the forward hole—Miata

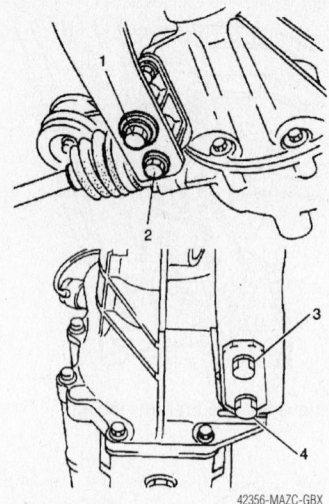

42356-MAZC-GBX

Tighten the PPF bolts using this sequence—Miata

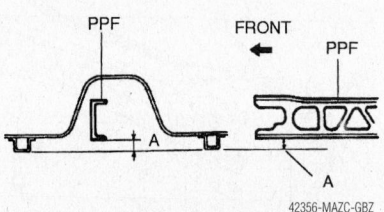

PPF   FRONT   PPF

A   A

42356-MAZC-GBZ

Measure the distance A which should be 2.37–2.83 inch (60–72mm)—Miata

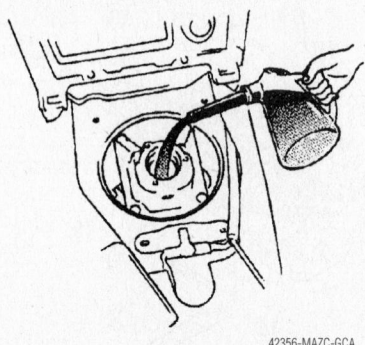

42356-MAZC-GCA

Fill the change control case with the specified amount of the proper oil—Miata

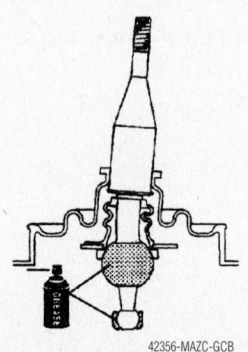

42356-MAZC-GCB

Apply grease to the shift lever components—Miata

the distance is not as specified, reposition the PPF to the transmission.

12. Install or connect the following:
- Speedometer sensor connector
- Neutral safety switch connector
- Back up light switch connector
- Clutch release cylinder
- Drive shaft
- Front crossbar
- Dust boot

➡**The change control assembly must be filled with 4.9–5.8 cubic inch (80–95cc) of GL–4 or 5 oil if the extension housing was removed or the transmission overhauled.**

13. Fill the change control case with the specified amount of the proper oil.

14. Apply grease to the shift lever components as illustrated.

15. Apply sealant to the contact surfaces of the shift lever and change control case.

16. Install the shift lever.

17. Install or connect the following:
- Rear console and insulator
- Shifter knob
- Starter
- Under cover
- Rear crossbar, on models equipped with 16 inch wheels
- Crossmember and bracket, on models equipped with 16 inch wheels
- Front and middle exhaust pipes

18. Fill the transaxle with fluid. Road test the vehicle and check for leaks. Top off all fluids as needed.

626

2.0L (FS) ENGINE

1. Before servicing the vehicle, refer to the precautions in the beginning of this section.

2. Refer to the illustration for component location and torque specifications.

3. Drain the transaxle oil.

4. Remove or disconnect the following:
- Battery and battery box
- Air cleaner assembly and fresh air duct
- Wheels
- Splash shields
- Starter
- Neutral safety switch connector
- Back up light switch connector
- Vehicle Speed Sensor (VSS) connector
- Clutch release cylinder
- Transverse member
- Extension bar
- Change control rod

FS ENGINE

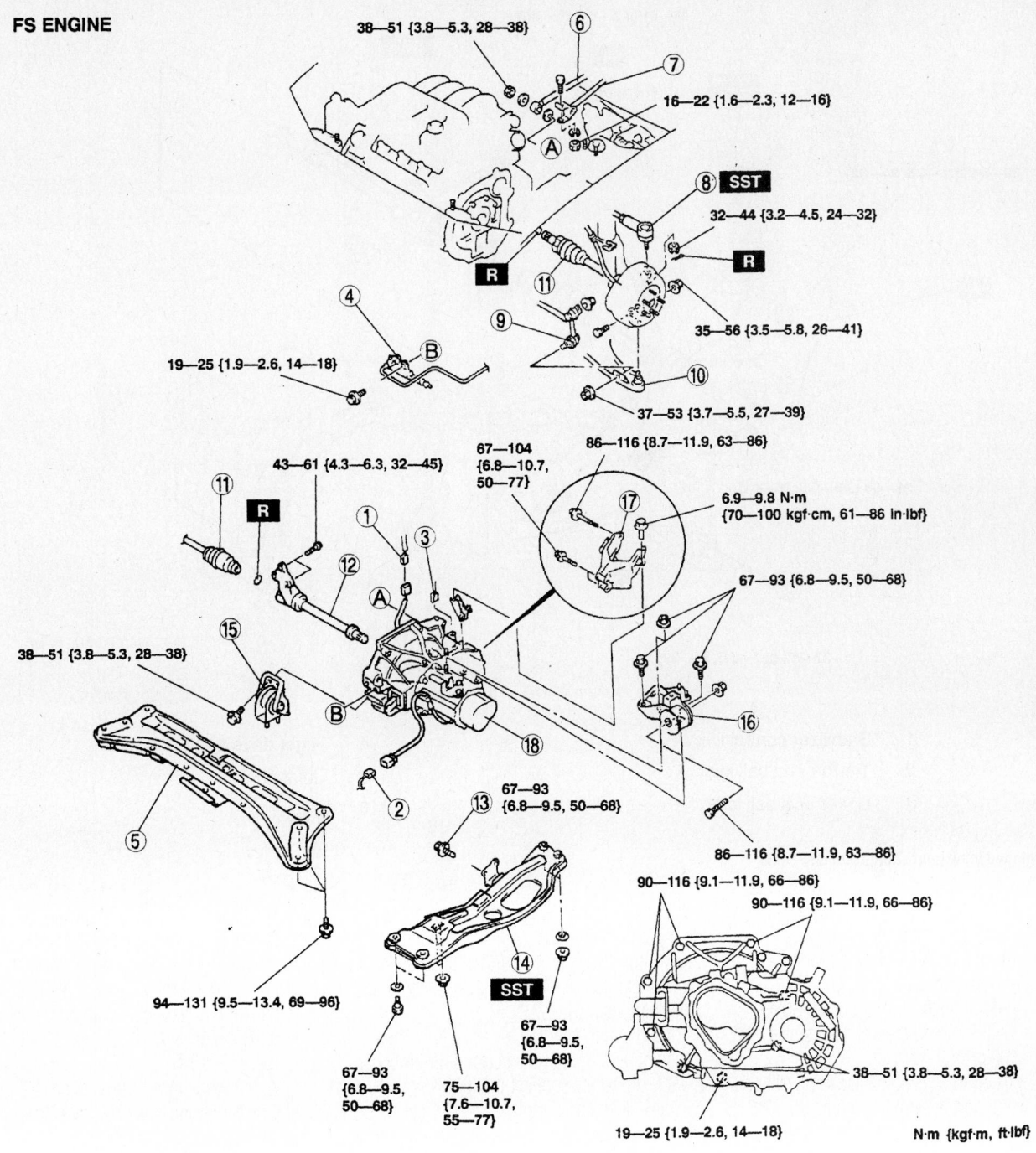

38—51 {3.8—5.3, 28—38}
16—22 {1.6—2.3, 12—16}
32—44 {3.2—4.5, 24—32}
35—56 {3.5—5.8, 26—41}
19—25 {1.9—2.6, 14—18}
37—53 {3.7—5.5, 27—39}
86—116 {8.7—11.9, 63—86}
67—104 {6.8—10.7, 50—77}
43—61 {4.3—6.3, 32—45}
6.9—9.8 N·m {70—100 kgf·cm, 61—86 in·lbf}
67—93 {6.8—9.5, 50—68}
38—51 {3.8—5.3, 28—38}
86—116 {8.7—11.9, 63—86}
90—116 {9.1—11.9, 66—86}
90—116 {9.1—11.9, 66—86}
67—93 {6.8—9.5, 50—68}
94—131 {9.5—13.4, 69—96}
67—93 {6.8—9.5, 50—68}
75—104 {7.6—10.7, 55—77}
67—93 {6.8—9.5, 50—68}
38—51 {3.8—5.3, 28—38}
19—25 {1.9—2.6, 14—18}
N·m {kgf·m, ft·lbf}

SST
R

1	Neutral switch connector	10	Lower arm ball joint
2	Back-up light switch connector	11	Drive shaft
3	Vehicle speedometer sensor connector	12	Joint shaft
4	Clutch release cylinder	13	No.5 engine mount bolt
5	Transverse member	14	Engine mount member
6	Extension bar	15	No.2 engine mount
7	Change control rod	16	No.4 engine mount rubber
8	Tie-rod end ball joint	17	No.1 engine mount bracket
9	Stabilizer control link	18	Transaxle

42356-MAZC-G57

Exploded view of the manual transaxle assembly mounting—626 with 2.0L (FS) engine

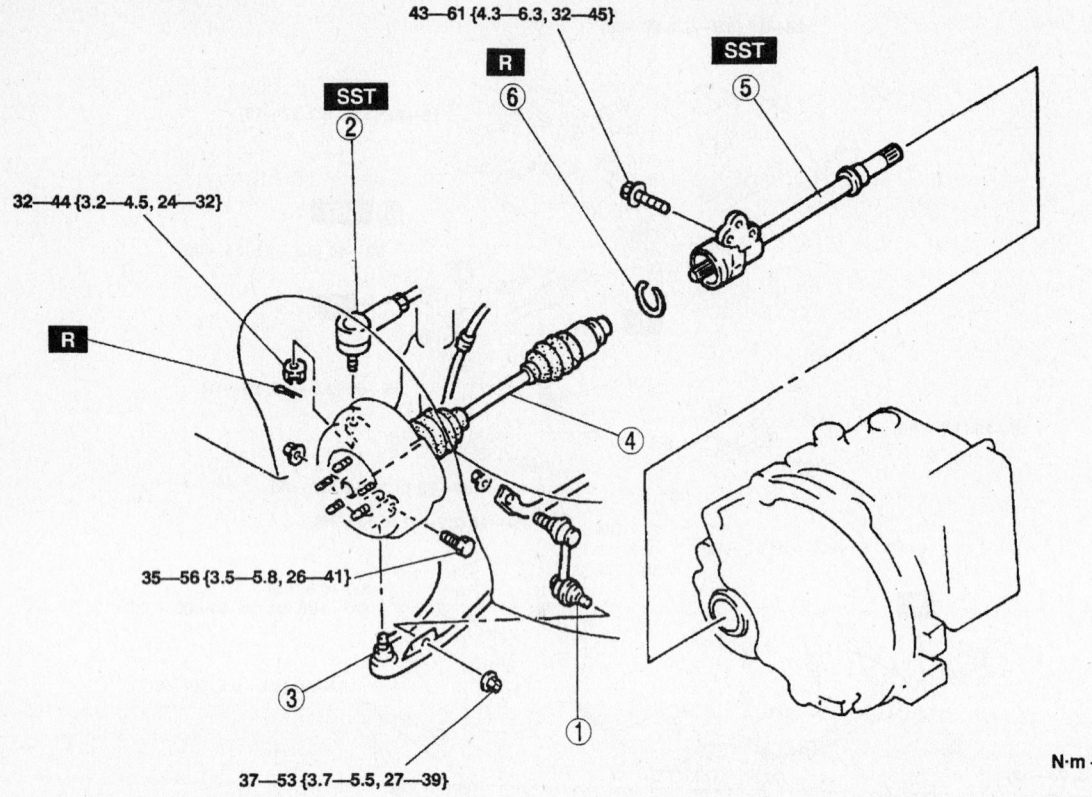

43—61 {4.3—6.3, 32—45}

32—44 {3.2—4.5, 24—32}

35—56 {3.5—5.8, 26—41}

37—53 {3.7—5.5, 27—39}

N·m {kgf·m, ft·lbf}

1	Stabilizer control link	4	Right drive shaft and axle
2	Tie-rod end ball joint	5	Joint shaft
3	Lower arm ball joint	6	Clip

42356-MAZC-G58

Exploded view joint shaft assembly—626

- Tie rod ends from the knuckle
- Stabilizer bar link
- Lower control arm ball joint
- Halfshafts
- Joint shaft, install tool 49 G030 455 to hold the side gears after removal

5. Support the engine assembly with a engine support assembly such as 49 E017 5A0.

- Number 5 engine mount bolt
- Engine mount member
- Number 2 engine mount
- Number 4 engine mount rubber
- Number 1 engine mount bracket

6. Loosen the engine support assembly and lean the engine towards the transaxle.

7. Support the transaxle with a jack, remove the transaxle bolts and the transaxle.

To install:

8. Place the transaxle onto a jack and raise into position.

9. Install the transaxle bolts and tighten the upper bolts to 66–86 ft. lbs. (90–116 Nm), the lower bolts (except very bottom bolt) to 28–38 ft. lbs. (38–51 Nm) and the very bottom bolt to 14–18 ft. lbs. (19–25 Nm). Refer to the exploded view illustration for bolt locations.

10. Lean the engine towards its normal position and tighten the support assembly.

11. Install or connect the following:
- Number 1 engine mount bracket. Refer to the illustration for component location and torque specifications.
- Number 4 engine mount rubber. Refer to the illustration for component location and torque specifications.
- Number 2 engine mount. Refer to the illustration for component location and torque specifications.

12. Install the engine mount member as follows while referring to the illustration for bolt locations:

a. Position the direction indicator on the mount member bushings facing towards the front side and install the bushings onto the mount.

b. Put the number 2 engine mount stud bolts through the mount member installation holes. Install mount member bolts **A** and nuts **A** and tighten to 50–68 ft. lbs. (67–93 Nm).

c. Loosely tighten the number 2 engine mount nuts and remove the engine support assembly.

d. Tighten the number 2 engine mount nuts **B** to 55–77 ft. lbs. (75–104 Nm).

13. Install or connect the following:
- Number 5 engine mount bolt. Refer to the illustration for component location and torque specifications.
- Joint shaft. Install a new circlip with the opening facing up. Hand tighten the bolts **A**, then tighten the bolts to 32–45 ft. lbs. (43–61 Nm). Refer to the illustration for bolt location.
- Halfshafts

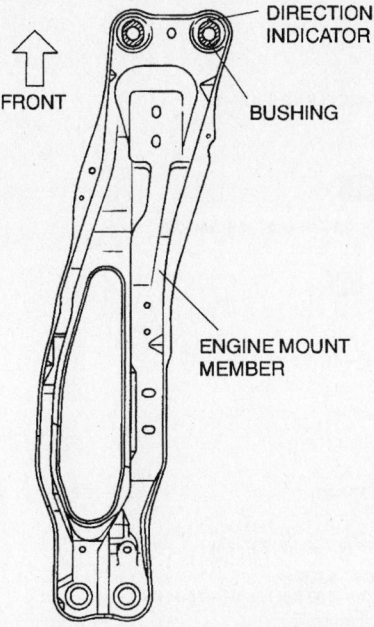

Position the direction indicator on the mount member bushings facing towards the front side and install the bushings onto the mount—626

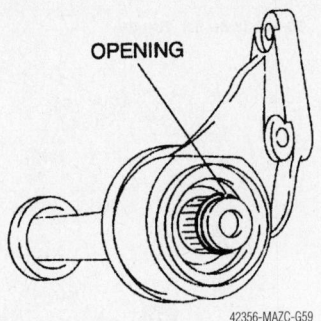

When installing the joint shaft make sure to install a new clip with the opening facing up—626

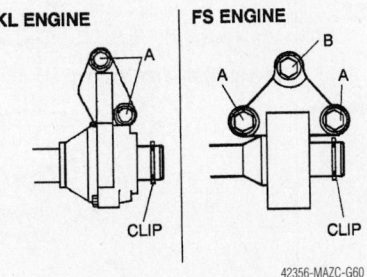

Tighten the joint shaft bolts as shown—626

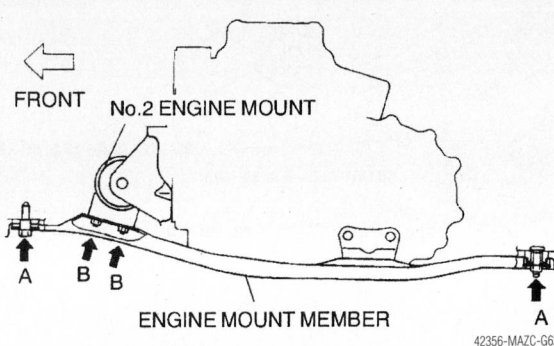

Location of the front mount member bolt locations—626

- Lower control arm ball joint
- Stabilizer bar link
- Tie rod ends to the knuckle and tighten the nut to 24–32 ft. lbs. (32–44 Nm) and install a new cotter pinch
- Change control rod
- Extension bar
- Transverse member
- Clutch release cylinder
- VSS connector
- Back up light switch connector
- Neutral safety switch connector
- Starter
- Splash shields
- Wheels
- Air cleaner assembly and fresh air duct
- Battery and battery box

14. Fill the transaxle with fluid. Road test the vehicle and check for leaks. Top off all fluids as needed.

2.5L (KL) ENGINE

1. Before servicing the vehicle, refer to the precautions in the beginning of this section.

2. Refer to the illustration for component location and torque specifications.

3. Drain the transaxle oil.

4. Remove or disconnect the following:
- Battery and battery box
- Air cleaner assembly and fresh air duct
- Wheels
- Splash shields
- Neutral safety switch connector
- Back up light switch connector

- Vehicle Speed Sensor (VSS) connector
- Starter
- Clutch release cylinder
- Transverse member
- Extension bar
- Change control rod
- Tie rod ends from the knuckle
- Stabilizer bar link
- Lower control arm ball joint
- Halfshafts
- Joint shaft, install tool 49 G030 455 to hold the side gears after removal

5. Support the engine assembly with a engine support assembly such as 49 E017 5A0.

- Engine mount member
- Number 2 engine mount
- Number 4 engine mount rubber
- Under cover
- Number 1 engine mount bracket

6. Loosen the engine support assembly and lean the engine towards the transaxle.

7. Support the transaxle with a jack, remove the transaxle bolts and the transaxle.

8. Remove the number 5 mount

To install:

9. Install the number 5 mount. Refer to the illustration for component location and torque specifications.

10. Place the transaxle onto a jack and raise into position.

11. Install the transaxle bolts and tighten the upper 4 bolts and the 3 side bolts to 50–73 ft. lbs. (68–99 Nm) and the lower 4 bolts to 28–38 ft. lbs. (38–51 Nm). Refer to the exploded view illustration for bolt locations.

12. Lean the engine towards its normal position and tighten the support assembly.

13. Install or connect the following:
- Number 1 engine mount bracket. Refer to the illustration for component location and torque specifications.
- Under cover
- Number 4 engine mount rubber. Refer to the illustration for component location and torque specifications.
- Number 2 engine mount. Refer to the illustration for component location and torque specifications.

14. Install the engine mount member as follows while referring to the illustration for bolt locations:

a. Position the direction indicator on the mount member bushings facing towards the front side and install the bushings onto the mount.

KL ENGINE

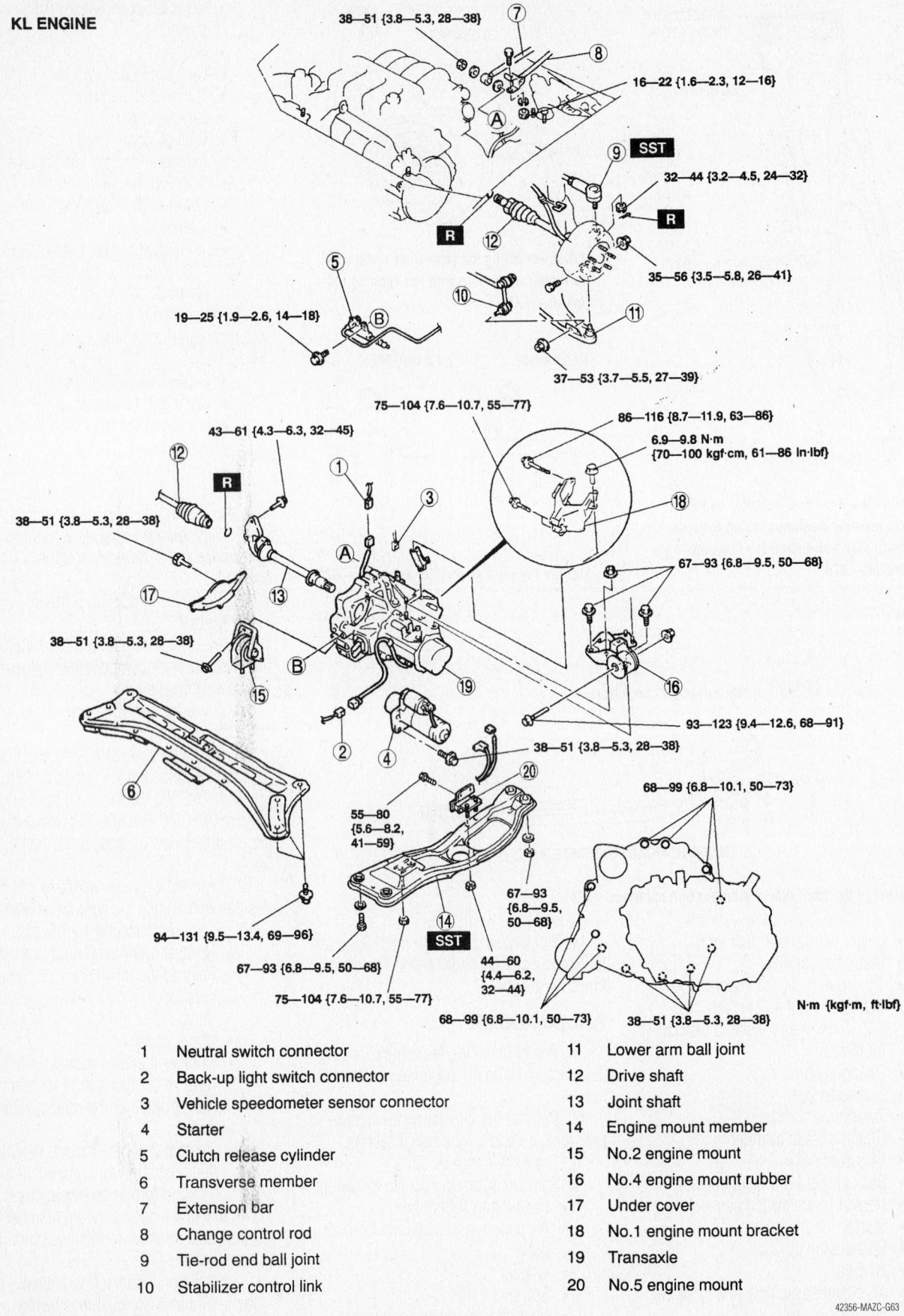

1	Neutral switch connector
2	Back-up light switch connector
3	Vehicle speedometer sensor connector
4	Starter
5	Clutch release cylinder
6	Transverse member
7	Extension bar
8	Change control rod
9	Tie-rod end ball joint
10	Stabilizer control link
11	Lower arm ball joint
12	Drive shaft
13	Joint shaft
14	Engine mount member
15	No.2 engine mount
16	No.4 engine mount rubber
17	Under cover
18	No.1 engine mount bracket
19	Transaxle
20	No.5 engine mount

42356-MAZC-G63

Exploded view of the manual transaxle assembly mounting—626 with 2.5L (KL) engine

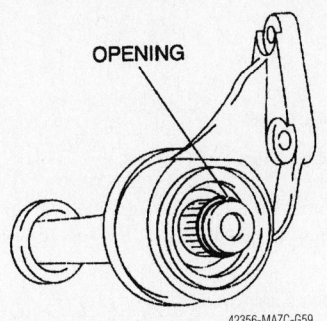

OPENING

42356-MAZC-G59

When installing the joint shaft make sure to install a new clip with the opening facing up—626

b. Put the number 2 engine mount stud bolts through the mount member installation holes. Install mount member bolts **A** and nuts **A** and tighten to 50–68 ft. lbs. (67–93 Nm).

c. Loosely tighten the number 2 engine mount nutsand remove the engine support assembly.

d. Tighten the number 2 engine mount nuts **B** to 55–77 ft. lbs. (75–104 Nm).

15. Install or connect the following:

• Number 5 engine mount bolt. Refer to the illustration for component location and torque specifications.
• Joint shaft. Install a new circlip with the opening facing up. Hand tighten the bolts **A**, then tighten the bolts to 32–45 ft. lbs. (43–61 Nm). Refer to the illustration for bolt location.
• Halfshafts
• Lower control arm ball joint
• Stabilizer bar link
• Tie rod ends to the knuckle and tighten the nut to 24–32 ft. lbs. (32–44 Nm) and install a new cotter pinch
• Change control rod
• Extension bar
• Transverse member
• Clutch release cylinder
• Starter
• VSS connector
• Back up light switch connector
• Neutral safety switch connector
• Starter
• Splash shields
• Wheels
• Air cleaner assembly and fresh air duct
• Battery and battery box

16. Fill the transaxle with fluid. Road test the vehicle and check for leaks. Top off all fluids as needed.

Protégé

G15M–R TRANSAXLE

1. Before servicing the vehicle, refer to the precautions in the beginning of this section.

2. Refer to the illustration for component location and torque specifications.

3. Drain the transaxle oil.

4. Remove or disconnect the following:

• Battery and battery box
• Air cleaner assembly and fresh air duct
• Wheels
• Splash shields
• Exhaust Gas Recirculation (EGR) pipe
• Front exhaust pipe and three way catalytic converter
• Starter
• Speedometer sensor connector
• Neutral safety switch connector
• Back up light switch connector
• Clutch release cylinder
• Extension bar
• Change control rod
• Tie rod ends from the knuckle
• Lower control arm ball joint
• Halfshafts
• Joint shaft, install tool 49 B027 006 to hold the side gears after removal

5. Support the engine assembly with a engine support assembly such as 49 E017 5A0.

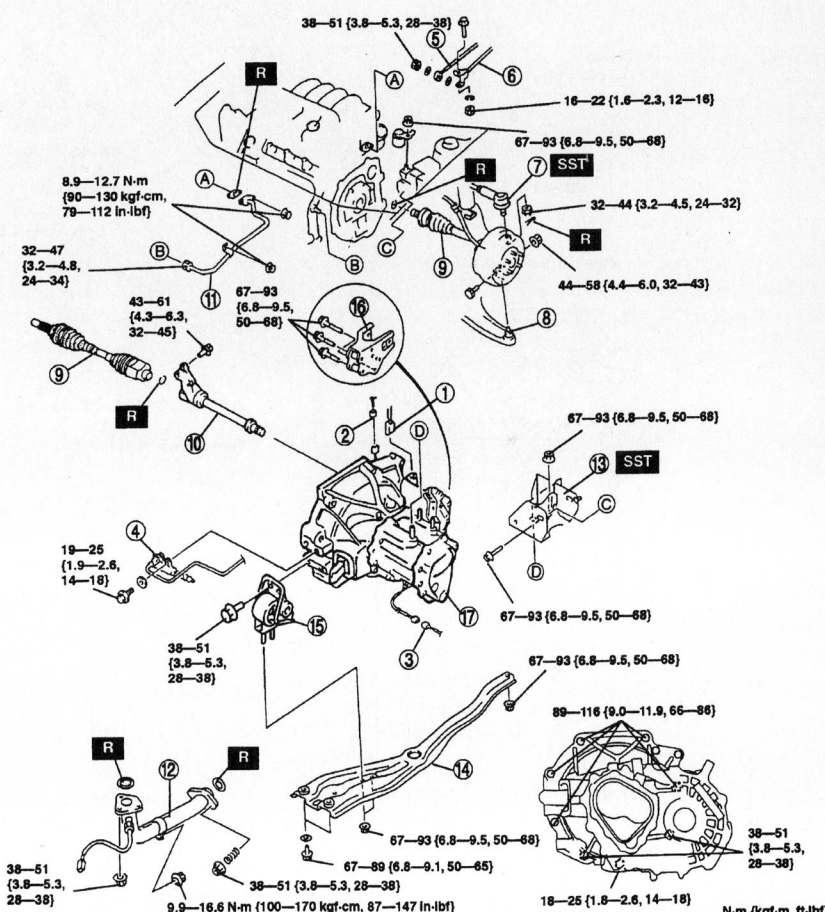

1	Speedometer sensor connector	10	Joint shaft
2	Neutral switch connector	11	EGR pipe
3	Back-up light switch connector	12	Front pipe
4	Clutch release cylinder	13	No.4 engine mount bracket
5	Extension bar	14	Engine mount member
6	Change control rod	15	No.2 engine mount
7	Tie-rod end ball joint	16	No.1
8	Lower arm ball joint	17	MTX
9	Drive shaft		

42356-MAZC-GQQ

Exploded view of the manual transaxle assembly mounting—2000–01 Protégé with G15M–R transaxle

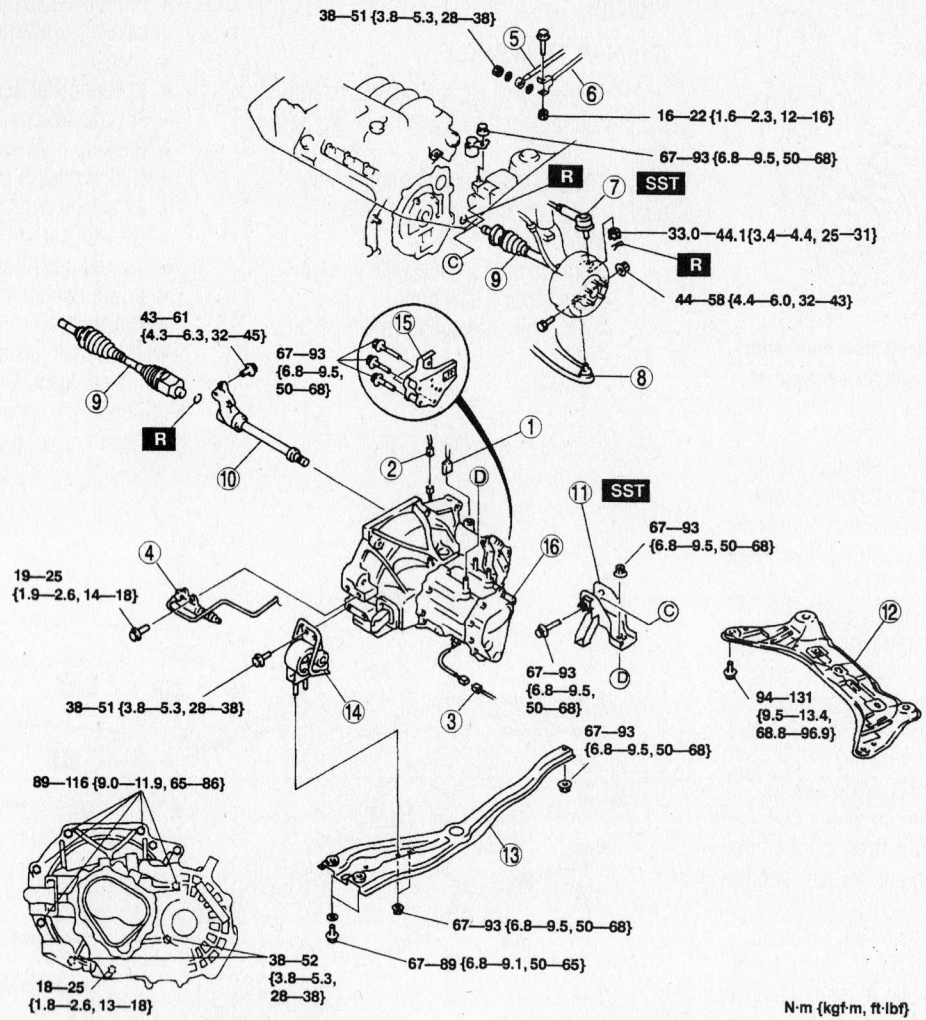

38—51 {3.8—5.3, 28—38}

16—22 {1.6—2.3, 12—16}

67—93 {6.8—9.5, 50—68}

R **SST**

33.0—44.1{3.4—4.4, 25—31}

R

44—58 {4.4—6.0, 32—43}

43—61 {4.3—6.3, 32—45}

67—93 {6.8—9.5, 50—68}

R

19—25 {1.9—2.6, 14—18}

67—93 {6.8—9.5, 50—68}

SST

67—93 {6.8—9.5, 50—68}

67—93 {6.8—9.5, 50—68}

94—131 {9.5—13.4, 68.8—96.9}

38—51 {3.8—5.3, 28—38}

67—93 {6.8—9.5, 50—68}

89—116 {9.0—11.9, 65—86}

67—93 {6.8—9.5, 50—68}

67—89 {6.8—9.1, 50—65}

38—52 {3.8—5.3, 28—38}

18—25 {1.8—2.6, 13—18}

N·m {kgf·m, ft·lbf}

1	Speedometer sensor connector	9	Drive shaft
2	Neutral switch connector	10	Joint shaft
3	Back-up light switch connector	11	No.4 engine mount bracket
4	Clutch release cylinder	12	Transverse member
5	Extension bar	13	Engine mount member
6	Change control rod	14	No.2 engine mount
7	Tie-rod end ball joint	15	No.1 engine mount bracket
8	Lower arm ball joint	16	MTX

42356-MAZC-GIM

Exploded view of the manual transaxle assembly mounting—2002 Protégé with G15M–R transaxle

- Number 4 engine mount
- Engine mount member
- Number 2 engine mount
- Number 1 engine mount bracket

6. Loosen the engine support assembly and lean the engine towards the transaxle.

7. Support the transaxle with a jack, remove the transaxle bolts and the transaxle.

To install:

8. Place the transaxle onto a jack and raise into position.

9. Install the transaxle bolts and tighten the upper bolts to 65–86 ft. lbs. (89–116 Nm), the lower bolts except the lowest bolt to 28–38 ft. lbs. (38–51 Nm). Tighten the lowest bolt to 13–18 ft. lbs. (18–25 Nm). Refer to the exploded view illustration for bolt locations.

10. Install or connect the following:
- Number 1 engine mount bracket. Refer to the illustration for component location and torque specifications.
- Number 2 engine mount. Refer to the illustration for component location and torque specifications.

11. Install the engine mount member as follows while referring to the illustration for component locations:

a. Position the direction indicator on the mount member bushings as illustrated.

b. Put the number 2 engine mount stud bolts through the mount member installation holes. Install mount member bolts **A** and nuts **B** and tighten to 50–68 ft. lbs. (67–93 Nm).

c. Tighten bolts **C** to 50–68 ft. lbs. (67–93 Nm).

d. By aligning the holes on the stud bolts, install the number 4 mount bracket. Align hole of the bracket with the rubber and hand tighten bolt **D**, then nuts **E** and bolt **D** to 50–68 ft. lbs. (67–93 Nm).

12. Install or connect the following:
- Joint shaft. Install a new circlip with the opening facing up. Hand tighten the bolts **A**, then tighten the bolts to 32–45 ft. lbs. (43–61 Nm).

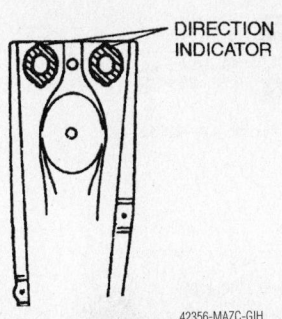

Position the direction indicator on the mount member and bushings—Protégé with both manual and automatic transaxles

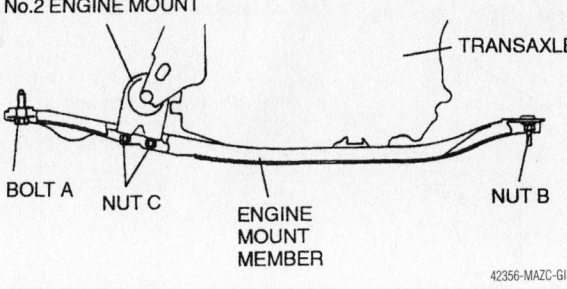

Location of the front mount member bolt locations—Protégé with both manual and automatic transaxles

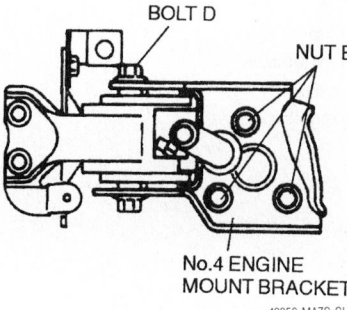

Number 4 mount fastener locations— 2000—01 Protégé with G15M–R transaxle

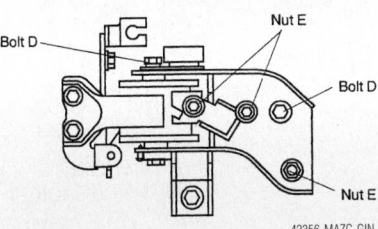

Number 4 mount fastener locations—2002 Protégé with G15M–R transaxle

Refer to the illustration for bolt location.
- Halfshafts
- Lower control arm ball joint
- Tie rod ends to the knuckle

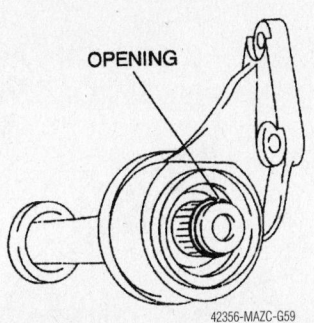

When installing the joint shaft make sure to install a new clip with the opening facing up—Protégé with both manual and automatic transaxles

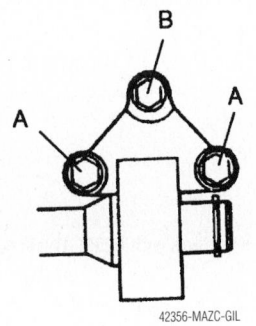

Tighten the joint shaft bolts as shown— Protégé with both manual and automatic transaxles

- Change control rod
- Extension bar
- Clutch release cylinder
- Neutral safety switch connector
- Back up light switch connector
- Speedometer connector
- Starter
- EGR pipe
- Splash shields

- Wheels
- Air cleaner assembly and fresh air duct
- Battery and battery box

13. Fill the transaxle with fluid. Road test the vehicle and check for leaks. Top off all fluids as needed.

F25M–R TRANSAXLE

1. Before servicing the vehicle, refer to the precautions in the beginning of this section.
2. Refer to the illustration for component location and torque specifications.
3. Drain the transaxle oil.
4. Remove or disconnect the following:
- Battery and battery box
- Air cleaner assembly and fresh air duct
- Wheels
- Splash shields
- Exhaust Gas Recirculation (EGR) pipe
- Front exhaust pipe and three way catalytic converter
- Starter
- Speedometer sensor connector
- Neutral safety switch connector
- Back up light switch connector
- Clutch release cylinder
- Extension bar
- Change control rod
- Tie rod ends from the knuckle
- Lower control arm ball joint
- Halfshafts
- Joint shaft, install tool 49 B027 006 to hold the side gears after removal

5. Support the engine assembly with a engine support assembly such as 49 E017 5A0.
- Number 4 engine mount
- Engine mount member
- Number 2 engine mount
- Number 1 engine mount bracket

6. Loosen the engine support assembly and lean the engine towards the transaxle.
7. Support the transaxle with a jack, remove the transaxle bolts and the transaxle.

To install:
8. Place the transaxle onto a jack and raise into position.
9. Install the transaxle bolts and tighten the upper bolts to 48–65 ft. lbs. (64–89 Nm), the lower bolts to 28–38 ft. lbs. (38–51 Nm). Refer to the exploded view illustration for bolt locations.
10. Install or connect the following:
- Number 1 engine mount bracket.

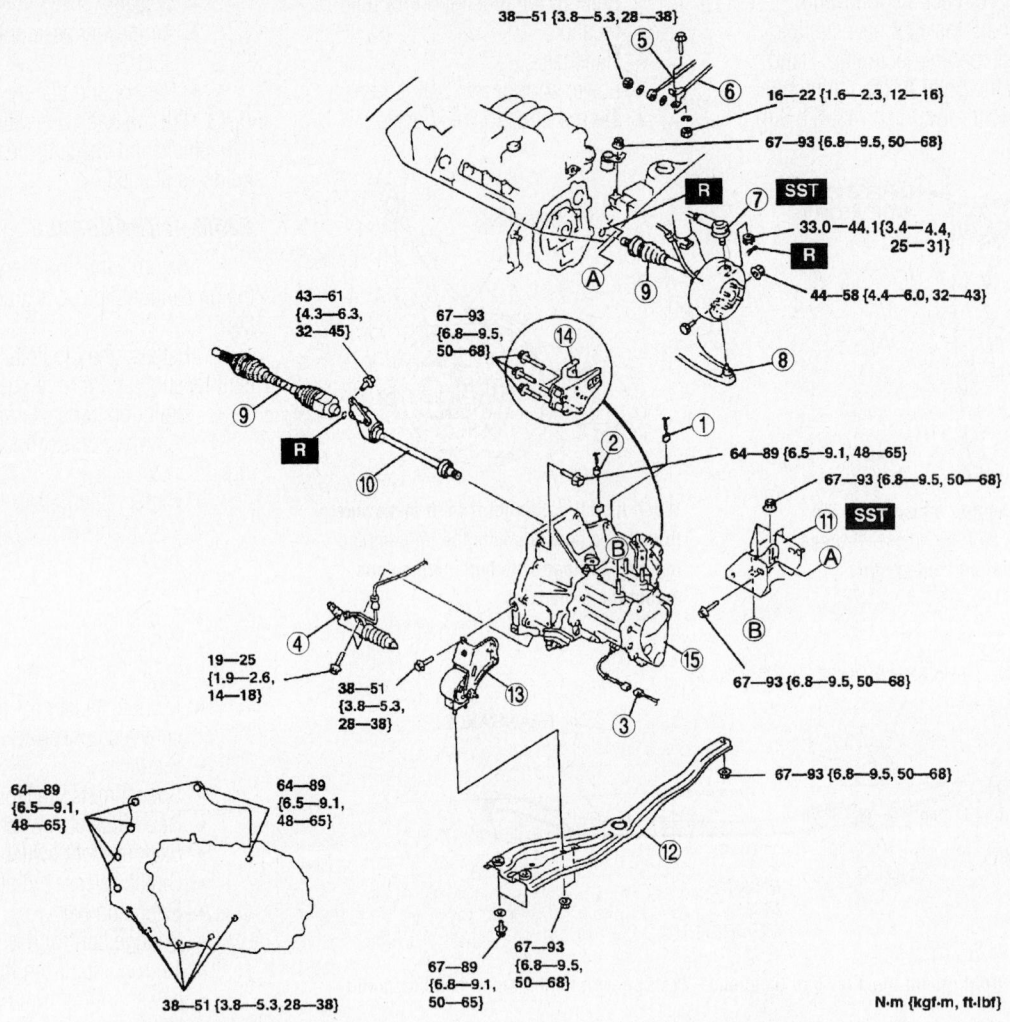

Exploded view of the manual transaxle assembly mounting—Protégé with F25M–R transaxle

1	Speedometer sensor connector
2	Neutral switch connector
3	Back-up light switch connector
4	Clutch release cylinder
5	Extension bar
6	Change control rod
7	Tie-rod end ball joint
8	Lower arm ball joint

9	Drive shaft
10	Joint shaft
11	No.4 engine mount bracket
12	Engine mount member
13	No.2 engine mount
14	No.1 engine mount bracket
15	MTX

42356-MAZC-GIG

Refer to the illustration for component location and torque specifications.

• Number 2 engine mount. Refer to the illustration for component location and torque specifications.

11. Install the engine mount member as follows while referring to the illustration for bolt locations:

a. Position the direction indicator on the mount member bushings as illustrated.

b. Put the number 2 engine mount stud bolts through the mount member installation holes. Install mount member bolts **A** and nuts **B** and tighten to 50–68 ft. lbs. (67–93 Nm).

c. Tighten bolts **C** to 50–68 ft. lbs. (67–93 Nm).

d. By aligning the holes on the stud bolts, install the number 4 mount bracket. Align hole of the bracket with the rubber and hand tighten bolt **D**, then nuts **E** and bolt **D** to 50–68 ft. lbs. (67–93 Nm).

12. Install or connect the following:

• Joint shaft. Install a new circlip with the opening facing up. Hand tighten the bolts **A**, then tighten the bolts to 32–45 ft. lbs. (43–61 Nm).

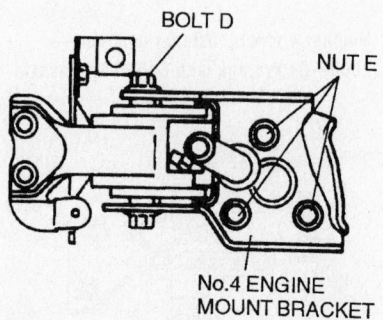

Number 4 mount fastener locations—Protégé with F25M–R transaxle

42356-MAZC-GIJ

Refer to the illustration for bolt location.
- Halfshafts
- Lower control arm ball joint
- Tie rod ends to the knuckle
- Change control rod
- Extension bar
- Clutch release cylinder
- Neutral safety switch connector

- Back up light switch connector
- Speedometer connector
- Starter
- EGR pipe
- Splash shields
- Wheels
- Air cleaner assembly and fresh air duct
- Battery and battery box

13. Fill the transaxle with fluid. Road test the vehicle and check for leaks. Top off all fluids as needed.

Mazda 3

1. Before servicing the vehicle, refer to the precautions in the beginning of this section.

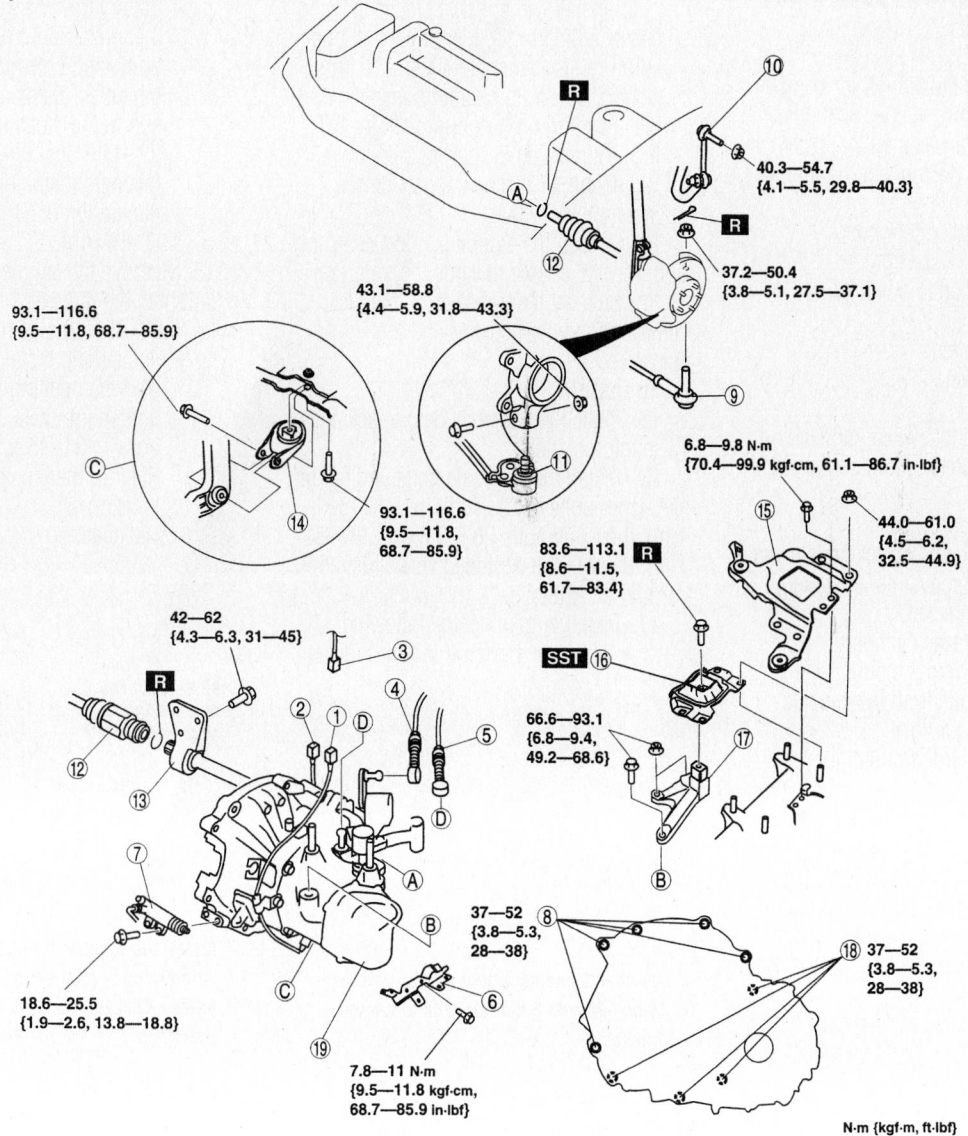

#	Description	#	Description
1	Back-up light switch connector	10	Stabilizer control link
2	Neutral switch connector	11	Lower arm ball joint
3	Vehicle speed sensor connector (without ABS)	12	Drive shaft
4	Select cable	13	Joint shaft
5	Shift cable	14	No.1 engine mount rubber
6	Harness bracket	15	Battery tray bracket
7	Clutch release cylinder	16	No.4 engine mount rubber
8	Transaxle mounting bolt (upper side)	17	No.4 engine mount bracket
9	Tie-rod end ball joint	18	Transaxle mounting bolt (lower side)
		19	Manual transaxle

67162-MAZC-G75

Exploded view of the manual transaxle assembly mounting—Mazda 3 models

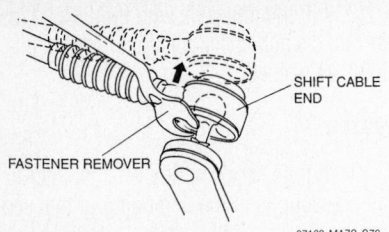

67162-MAZC-G76

Remove the shift cable and select cable using a fastener remover—Mazda 3 models

2. Refer to the illustration for component location and torque specifications.

3. Drain the transaxle oil.

4. Remove or disconnect the following:
- Battery and battery box
- Air cleaner assembly
- Exhaust manifold insulator
- Wheels
- Splash shields
- Under cover
- Starter
- Back up light switch connector
- Neutral safety switch connector, if equipped
- Vehicle speed sensor
- Shift cable and select cable using a fastener remover as illustrated.
- Harness bracket
- Clutch release cylinder
- Transaxle upper mount bolts
- Tie rod ends from the knuckle
- Stabilizer bar link
- Lower control arm ball joint
- Halfshafts

5. Remove the joint shaft as follows:

a. Disconnect the right halfshaft from the joint shaft by tapping the transaxle side outer ring with a brass bar and a hammer. Disconnect the joint shaft bracket from the block and remove the joint shaft.

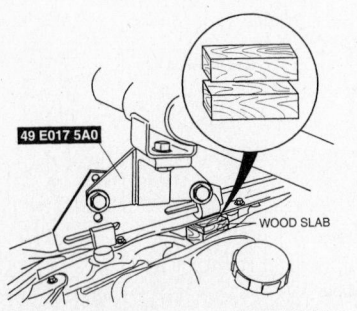

67162-MAZC-G56

Install two suitable pieces of wood between the front fender panel and upper apron reinforcement—Mazda 3 with a manual transaxle

b. Install tool 49 G030 455 to hold the side gears after removal

6. Support the engine assembly with a engine support assembly such as 49 E017 5A0.

7. Remove or disconnect the following:
- Number 1 engine mount rubber
- Battery tray bracket

8. Remove the number 4 engine mount and joint bracket as follows:

a. Install two suitable pieces of wood between the front fender panel and upper apron reinforcement as illustrated. The wood size should be approximately 1.38 inch (35mm) on 4 door models or 2.36 inch (60mm) on 5 door models.

9. Install an engine support device such as 49 E017 5A0.

10. Loosen the engine support assembly and lean the engine towards the transaxle.

11. Support the transaxle with a jack, remove the transaxle bolts and the transaxle.

To install:

12. Place the transaxle onto a jack and raise into position.

13. Install the transaxle bolts and tighten the upper bolts to 28–38 ft. lbs. (37–52 Nm), the lower bolts 28–38 ft. lbs. (37–52 Nm). Refer to the exploded view illustration for bolt locations.

14. Install or connect the following:
- Number 4 engine mount bracket

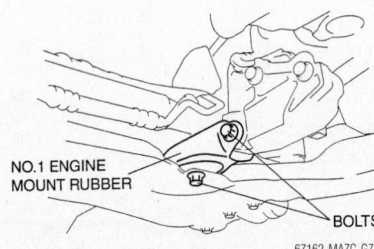

67162-MAZC-G77

Number 1 engine mount rubber bolt locations—Mazda 3 models with a manual transaxle

to the transaxle case and tighten the bolts to 49–68 ft. lbs. (66–93 Nm)
- Number 1 engine mount rubber to the crossmember and hand tighten the bolts
- Number 4 engine mount rubber with the body stud passing through the holes and tighten the bolts to 49–62 ft. lbs. (66–93 Nm)
- Battery tray bracket on the number 4 engine mount rubber with the body stud bolts passing through the holes. Tighten retainers identified in the illustration as 1 to 32–45 ft. lbs. (44–61 Nm) and the retainers identified as 2 in the same illustration to 61–86 inch lbs. (7–10 Nm).

15. Remove the engine support device and tighten the number 1 mount rubber bolts to 68–85 ft. lbs. (116 Nm).
- Joint shaft. Install a new circlip with the opening facing up. Hand tighten the bolts, then tighten the bolts to 31–45 ft. lbs. (42–62 Nm). Refer to the illustration for bolt location.
- Halfshafts

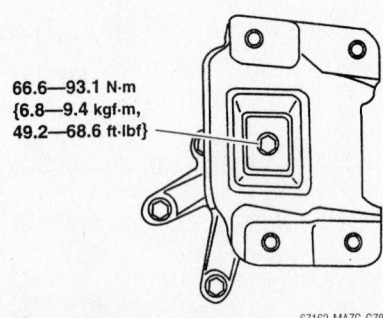

66.6—93.1 N·m
{6.8—9.4 kgf·m,
49.2—68.6 ft·lbf}

67162-MAZC-G78

Install the number 4 engine mount rubber with the body stud passing through the holes—Mazda 3 models with a manual transaxle

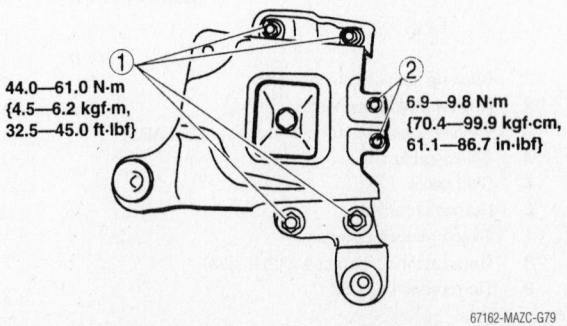

44.0—61.0 N·m
{4.5—6.2 kgf·m,
32.5—45.0 ft·lbf}

6.9—9.8 N·m
{70.4—99.9 kgf·cm,
61.1—86.7 in·lbf}

67162-MAZC-G79

Install the battery tray bracket on the number 4 engine mount rubber with the body stud bolts passing through the holes—Mazda 3 models with a manual transaxle

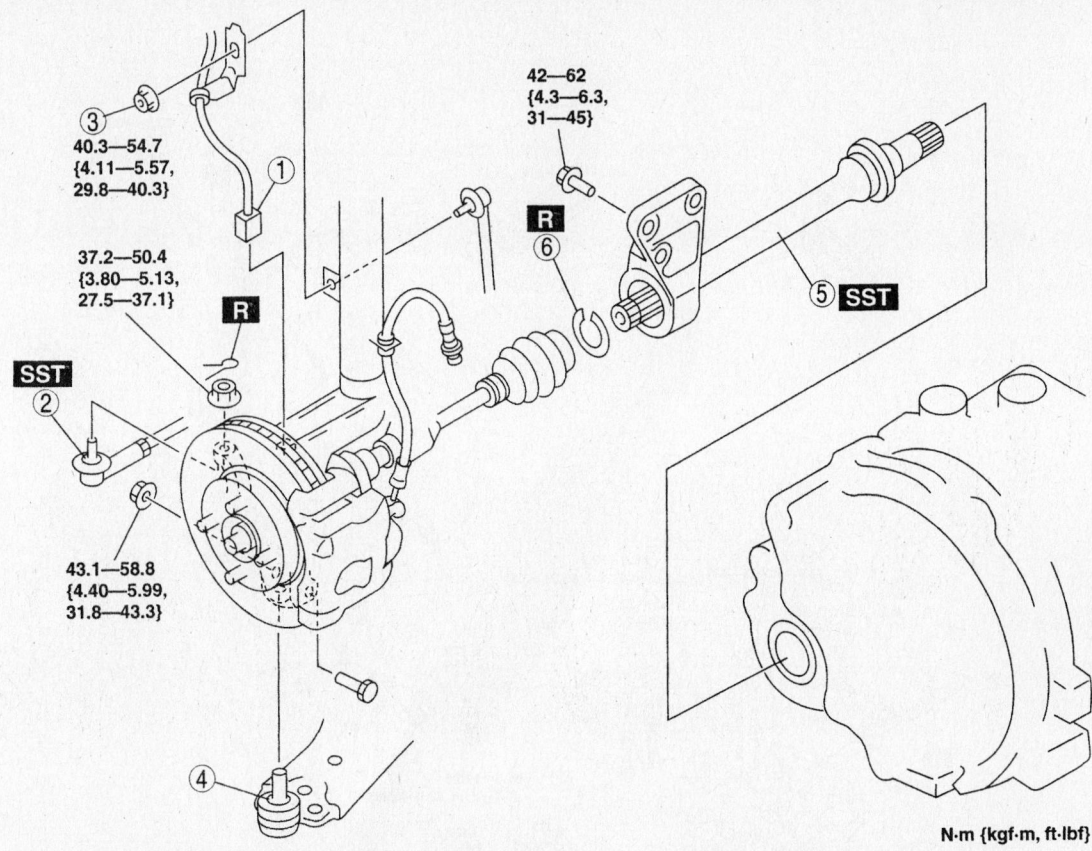

40.3—54.7
{4.11—5.57,
29.8—40.3}

37.2—50.4
{3.80—5.13,
27.5—37.1}

42—62
{4.3—6.3,
31—45}

43.1—58.8
{4.40—5.99,
31.8—43.3}

N·m {kgf·m, ft·lbf}

1	ABS wheel-speed sensor connector	4	Front lower arm ball joint
2	Tie-rod end ball joint	5	Joint shaft
3	Stabilizer control link upper nut	6	Clip

67162-MAZC-G80

Exploded view of the joint shaft assembly—Mazda 3

- Lower control arm ball joint. Tighten the retainers to 32–43 ft. lbs. (43–59 Nm).
- Stabilizer bar link. Tighten the retainers to 30–40 ft. lbs. (40–54 Nm).
- Tie rod ends to the knuckle. Tighten the retainers to 27–37 ft. lbs. (37–50 Nm).
- Clutch release cylinder. Tighten the retainers to 14–19 ft. lbs. (18–25 Nm).
- Harness bracket
- Shift and select cables
- Vehicle speed sensor
- Neutral safety switch connector
- Back up light switch connector
- Starter
- Under cover
- Splash shields
- Wheels
- Exhaust manifold insulator
- Air cleaner assembly
- Battery tray and battery

16. Fill the transaxle with fluid. Road test the vehicle and check for leaks. Top off all fluids as needed.

Mazda 6

G35M-R TRANSAXLE

1. Before servicing the vehicle, refer to the precautions in the beginning of this section.

2. Refer to the illustration for component location and torque specifications.

3. Drain the transaxle oil.

4. Remove or disconnect the following:
   - Battery and battery tray
   - Air cleaner assembly
   - Wheels
   - Splash shields
   - Under cover
   - Steering gear, linkage and pipe assembly bolts from the cross-member and using mechanics wire, position the steering gear and linkage assembly aside.
   - Heated Oxygen (HO$_2$S) connector
   - Back up light switch connector
   - Neutral safety switch connector
   - Harness bracket
   - Vehicle speed sensor
   - Shift cable and select cable
   - Selector cable bracket
   - Clutch release cylinder
   - Starter
   - Endplate cover
   - Transaxle upper mount bolts
   - Lower control arm ball joints
   - Damper fork
   - Tie rod ends from the knuckle
   - Stabilizer bar link
   - Halfshafts.

5. Disconnect the right halfshaft from the joint shaft by tapping the transaxle side outer ring with a brass bar and a hammer. Disconnect the joint shaft bracket from the block and remove the joint shaft.

6. Install tool 49 G030 455 to hold the side gears after removal.

7. Support the engine assembly with a

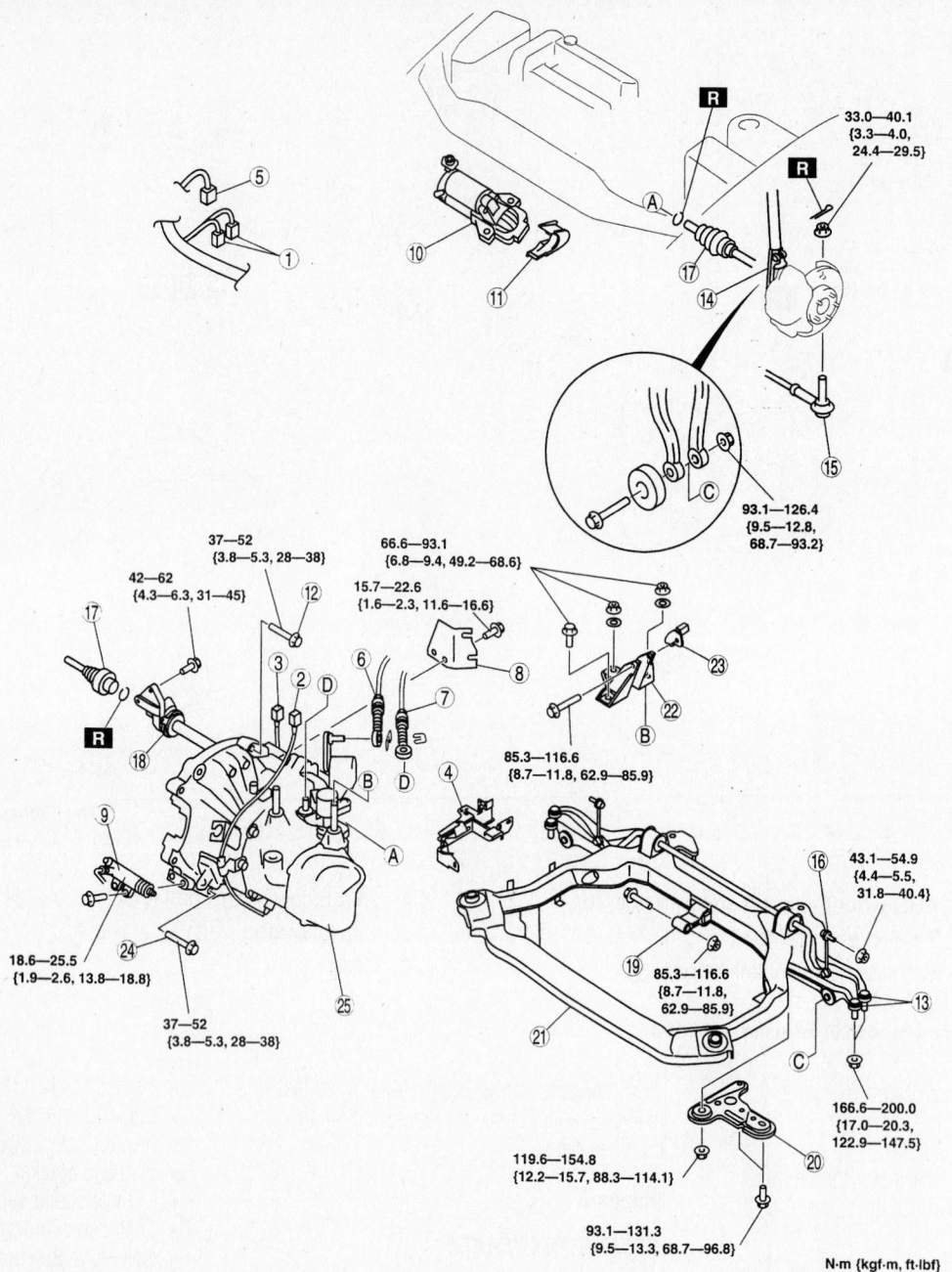

N·m {kgf·m, ft·lbf}

1	HO2S connector	14	Damper fork
2	Back-up light switch connector	15	Tie-rod end ball joint
3	Neutral switch connector	16	Stabilizer control link
4	Harness bracket	17	Drive shaft
5	Vehicle speedometer sensor connector (Without ABS)	18	Joint shaft
6	Selector cable	19	No.1 engine mount
7	Shift cable	20	Crossmember bracket
8	Selector cable bracket	21	Crossmember component
9	Clutch release cylinder	22	No.4 engine mount
10	Starter	23	Dynamic damper
11	Endplate cover	24	Transaxle mounting bolt (lower side)
12	Transaxle mounting bolt (upper side)	25	Manual transaxle
13	Lower arm (front, rear) ball joint		

67162-MAZC-G216

Exploded view of the G35M-R manual transaxle assembly mounting—Mazda 6 models

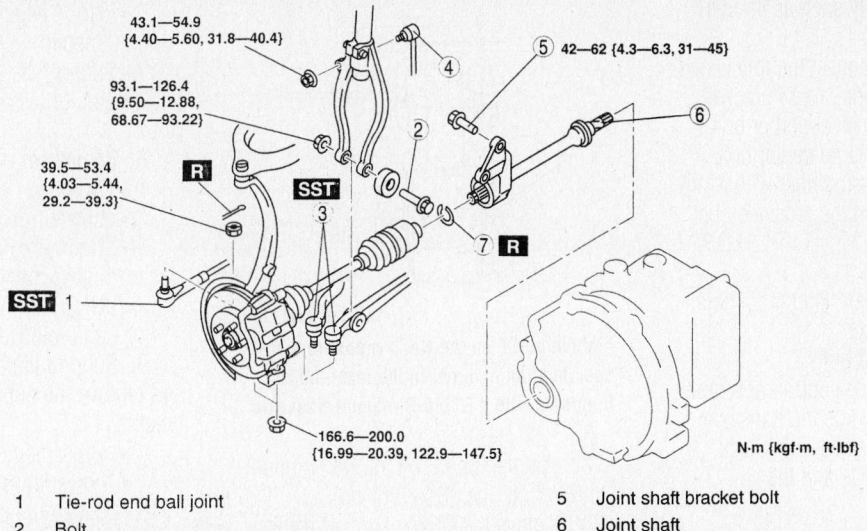

43.1—54.9
{4.40—5.60, 31.8—40.4}

93.1—126.4
{9.50—12.88,
68.67—93.22}

39.5—53.4
{4.03—5.44,
29.2—39.3}

42—62 {4.3—6.3, 31—45}

166.6—200.0
{16.99—20.39, 122.9—147.5}

N·m {kgf·m, ft·lbf}

1	Tie-rod end ball joint	5	Joint shaft bracket bolt
2	Bolt	6	Joint shaft
3	Lower arm (front, rear) ball joint	7	Clip
4	Stabilizer control link		

67162-MAZC-G125

Exploded view of the joint shaft assembly—Mazda 6 models with the 2.3L (L3) engine

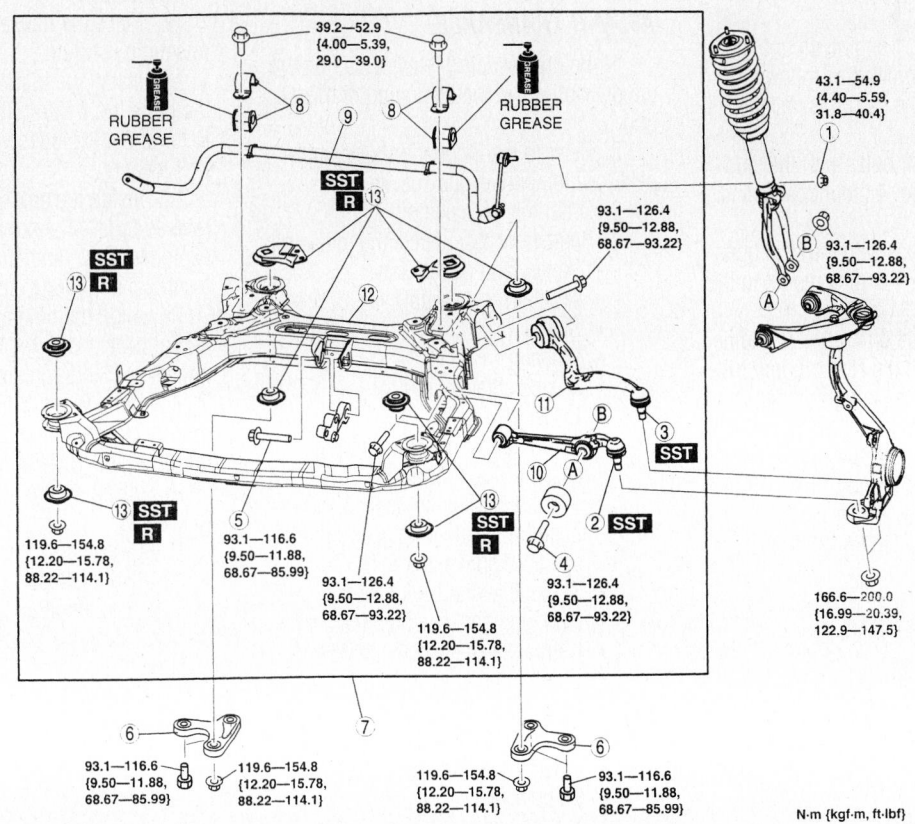

39.2—52.9
{4.00—5.39,
29.0—39.0}

RUBBER
GREASE

RUBBER
GREASE

43.1—54.9
{4.40—5.59,
31.8—40.4}

93.1—126.4
{9.50—12.88,
68.67—93.22}

93.1—126.4
{9.50—12.88,
68.67—93.22}

166.6—200.0
{16.99—20.39,
122.9—147.5}

119.6—154.8
{12.20—15.78,
88.22—114.1}

93.1—116.6
{9.50—11.88,
68.67—85.99}

93.1—126.4
{9.50—12.88,
68.67—93.22}

93.1—126.4
{9.50—12.88,
68.67—93.22}

119.6—154.8
{12.20—15.78,
88.22—114.1}

119.6—154.8
{12.20—15.78,
88.22—114.1}

93.1—116.6
{9.50—11.88,
68.67—85.99}

119.6—154.8
{12.20—15.78,
88.22—114.1}

93.1—116.6
{9.50—11.88,
68.67—85.99}

N·m {kgf·m, ft·lbf}

1	Nut (stabilizer control link)	8	Stabilizer bracket and bushing
2	Front lower arm (front) ball joint	9	Front Stabilizer
3	Front lower arm (rear) ball joint	10	Front lower arm (front)
4	Bolt (front shock absorber lower side)	11	Front lower arm (rear)
5	No.1 engine mount center bolt	12	Front crossmember
6	Crossmember bracket	13	Front crossmember bushing
7	Crossmember component		

67162-MAZC-G124

Exploded view of the front crossmember, related components and their torque specifications—Mazda 3 and 6 models

engine support assembly such as 49 E017 5A0.

8. Remove or disconnect the following:
- Number 1 engine mount bracket
- Lower front shock absorber bolt
- No. 1 engine mount center bolt

9. Support the crossmember with a jack and remove the nuts and the crossmember bracket.
- Crossmember

10. Remove the number 4 engine mount
- Dynamic damper
- Lower transaxle bolts

11. Loosen the engine support assembly and lean the engine towards the transaxle.

12. Support the transaxle with a jack, remove the transaxle bolts and the transaxle.

To install:

13. Place the transaxle onto a jack and raise into position.

14. Install the transaxle bolts and tighten the bolts to 28–38 ft. lbs. (37–52 Nm). Refer to the exploded view illustration for bolt locations.

15. Install the No. 1 engine mount and No. 4 engine mount bracket as follows:

    a. Make sure the mount is installed as illustrated.

    b. Align the bolt holes with the stud bolts, install the No. 4 mount bracket to the transaxle.

    c. Align the holes with the stud bolts and install the No. 1 engine mount to the transaxle.

    d. Align the hole on the No. 4 engine mount bracket with the No. 4 mount rubber on the car and hand tighten bolt D. Refer to the illustration for bolt locations.

16. Handtighten bolts B, C and A.

    a. Tighten bolts B and C then bolt A to 49–68 ft. lbs. (66–93 Nm).

    b. Tighten bolt D to 63–86 ft. lbs. (85–116 Nm).

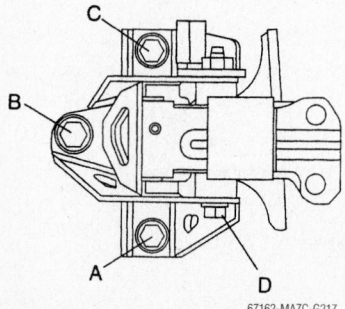

67162-MAZC-G217

Install the No. 1 mount and No. 4 mount bracket and tighten the bolts in the sequence and specifications outlined in the text—Mazda 6 models with a G35M-R manual transaxle

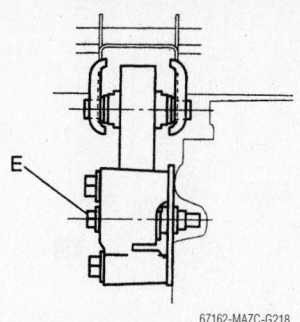

67162-MAZC-G218

Tighten bolt E on the No. 1 mount to the specification outlined in the text—Mazda 6 models with a G35M-R manual transaxle

    c. Tighten bolt E on the No. 1 mount to 63–86 ft. lbs. (85–116 Nm).

17. Remove the engine support device.

18. Install the remaining components in the reverse order of removal.

19. Fill the transaxle with fluid. Road test the vehicle and check for leaks. Top off all fluids as needed.

A65M-R TRANSAXLE

1. Before servicing the vehicle, refer to the precautions in the beginning of this section.

2. Refer to the illustration for component location and torque specifications.

3. Drain the transaxle oil.

4. Remove or disconnect the following:
- Battery and battery tray
- Air cleaner assembly
- Starter
- Wheels
- Splash shields
- Under cover
- Steering gear, linkage and pipe assembly bolts from the crossmember and using mechanics wire, position the steering gear and linkage assembly aside.
- Shift cable and select cable
- Back up light switch connector
- Neutral safety switch connector
- Clutch release cylinder
- Ground harness
- Transaxle upper mount bolts
- Lower control arm ball joints
- Damper fork
- Tie rod ends from the knuckle
- Stabilizer bar link
- Halfshafts

5. Disconnect the left halfshaft from the joint shaft by inserting a prybar between the transaxle and the halfshaft outer ring. Disconnect the joint shaft bracket from the block and remove the joint shaft.

6. Install tool 49 G030 455 to hold the side gears after removal

- No. 4 engine mount bracket and rubber

7. Support the engine assembly with a engine support assembly such as 49 E017 5A0.

8. Remove or disconnect the following:
- Number 1 engine mount bracket
- Lower front shock absorber bolt
- No. 1 engine mount. Remove the intake manifold and attach tool 49 UN30 3050 to the head as shown and remove the mount.

9. Support the crossmember with a jack and remove the nuts and the crossmember bracket.
- Crossmember
- Lower transaxle bolts

10. Loosen the engine support assembly and lean the engine towards the transaxle.

11. Support the transaxle with a jack, remove the transaxle bolts and the transaxle.

To install:

12. Place the transaxle onto a jack and raise into position.

13. Install the transaxle bolts and tighten the bolts to 28–38 ft. lbs. (37–52 Nm). Refer to the exploded view illustration for bolt locations.

14. Install the crossmember in the reverse order of removal and refer to the exploded view illustration for component locations and torque specifications.

15. Align the hole of the No. 1 engine mount rubber with the bolt hole of the transaxle. Hand tighten bolt A and then tighten bolts B and C to 49–68 ft. lbs. (66–93 Nm) and then tighten bolt A to 49–68 ft. lbs. (66–93 Nm).

16. Install the No. 4 engine mount as follows:

    a. Make sure the No. 4 mount is installed as illustrated.

    b. Hand tighten bolts A and B.

    c. Align the holes on the contact area of the front frame with bolt C hole.

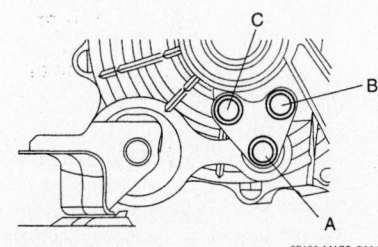

67162-MAZC-G221

Tighten the No. 1 mount in the sequence and specifications outlined in the text—Mazda 6 models with a A65M-R manual transaxle

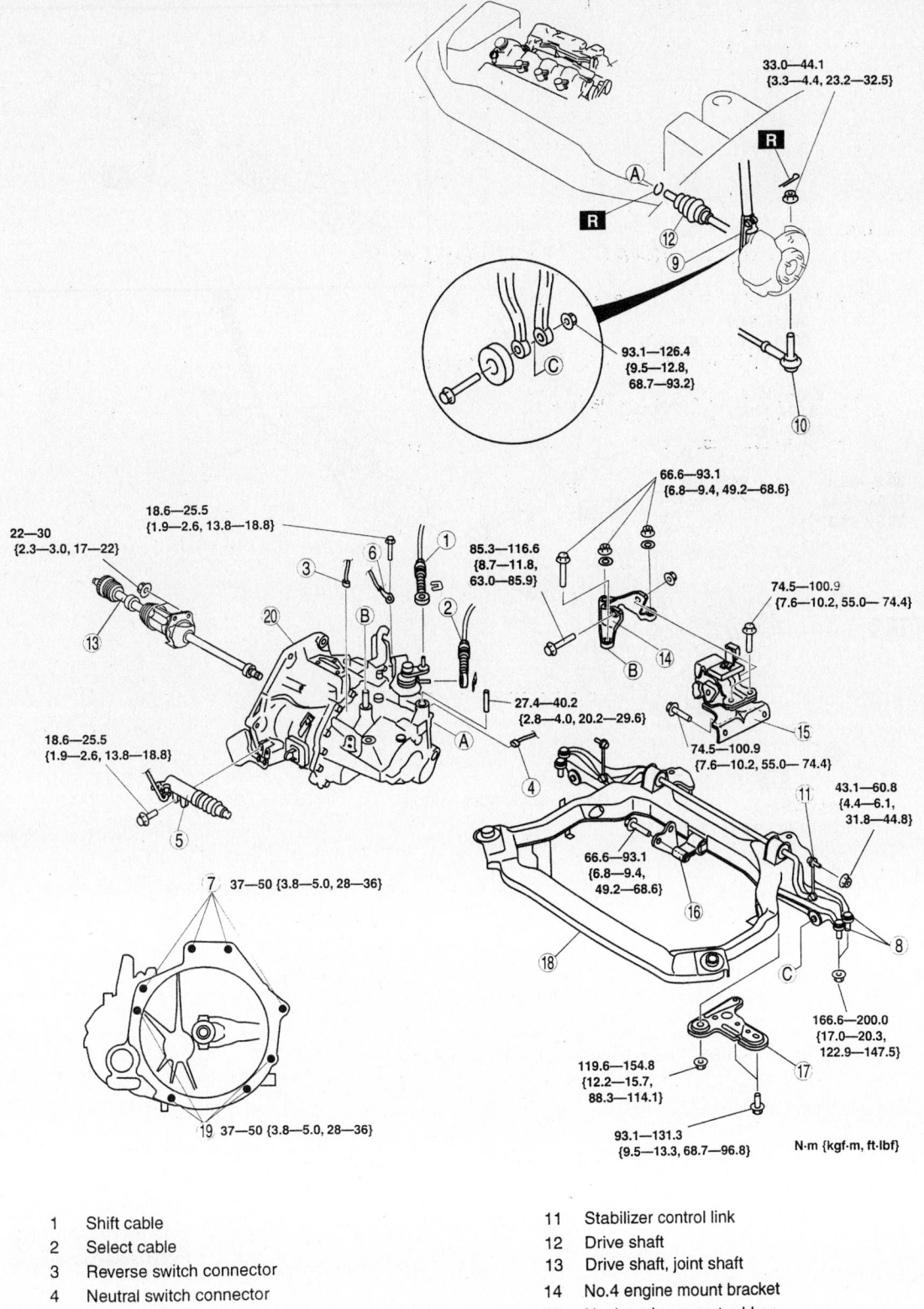

1	Shift cable
2	Select cable
3	Reverse switch connector
4	Neutral switch connector
5	Clutch release cylinder
6	GND harness
7	Transaxle mounting bolt (upper side)
8	Lower arm (front, rear) ball joint
9	Damper fork
10	Tie-rod end ball joint
11	Stabilizer control link
12	Drive shaft
13	Drive shaft, joint shaft
14	No.4 engine mount bracket
15	No.4 engine mount rubber
16	No.1 engine mount
17	Crossmember bracket
18	Crossmember
19	Transaxle mounting bolt (lower side)
20	Manual transaxle

67162-MAZC-G219

Exploded view of the A65M-R manual transaxle assembly mounting—Mazda 6 models

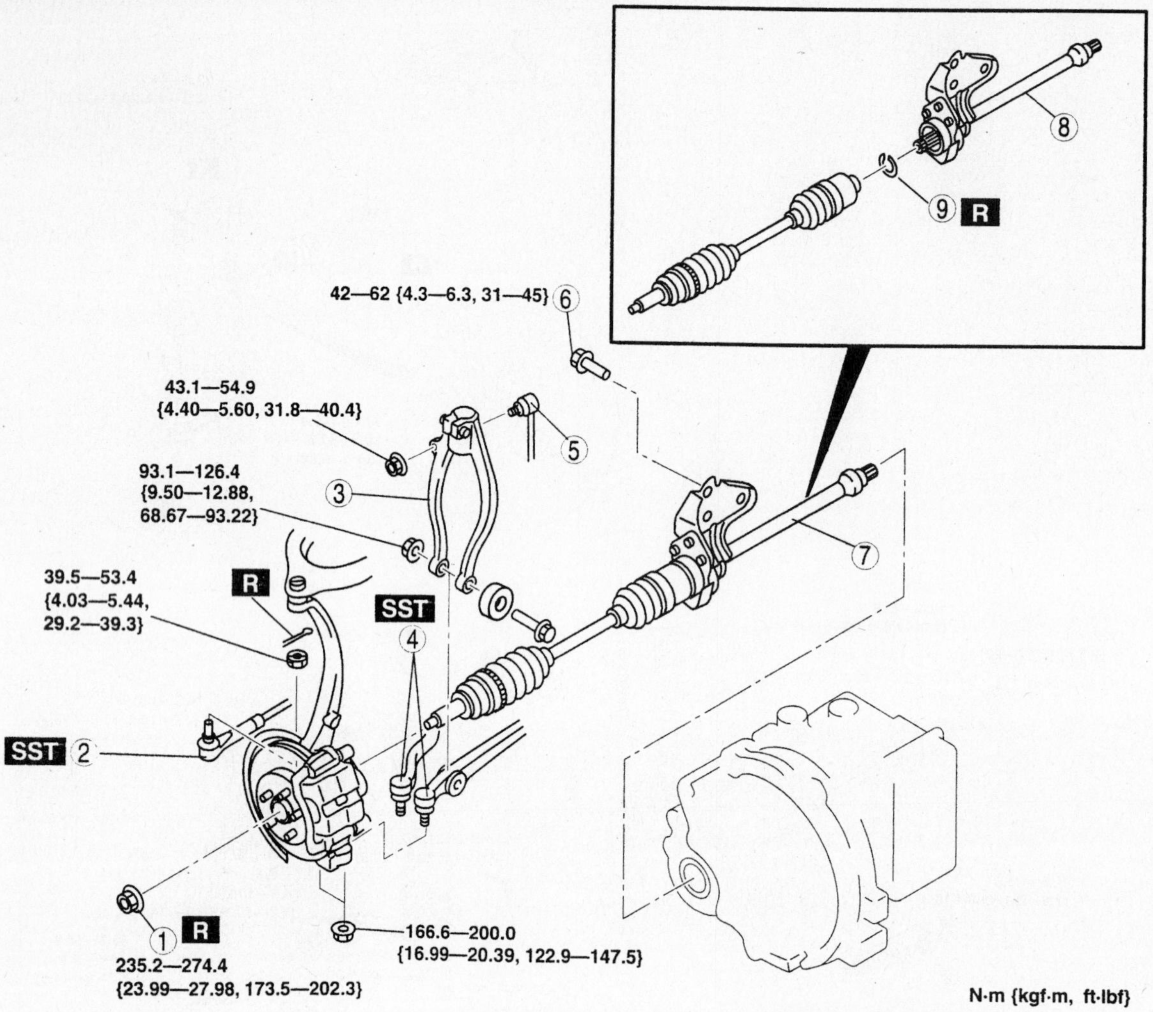

42—62 {4.3—6.3, 31—45} 6

43.1—54.9
{4.40—5.60, 31.8—40.4}

93.1—126.4
{9.50—12.88,
68.67—93.22}

39.5—53.4
{4.03—5.44,
29.2—39.3}

SST 2

1

235.2—274.4
{23.99—27.98, 173.5—202.3}

166.6—200.0
{16.99—20.39, 122.9—147.5}

N·m {kgf·m, ft·lbf}

1	Locknut	5	Stabilizer control link
2	Tie-rod end ball joint	6	Joint shaft bracket bolt
3	Damper fork	7	Drive shaft and joint shaft
4	Lower arm (front, rear) ball joint	8	Joint shaft
		9	Clip

67162-MAZC-G121

Exploded view of the joint shaft assembly—Mazda 6 models with the 3.0L (AJ) engine

d. Tighten bolt A and then B to 55–74 ft. lbs. (74–100 Nm).

e. Tighten bolt C and then D to 55–74 ft. lbs. (74–100 Nm).

f. Make sure the No. 4 mount is installed as illustrated.

g. Handtighten bolt F.

h. Raise the transaxle with a floor jack and align the hole on the No. 4 engine mount bracket with the stud bolts on the transaxle.

i. Install bolts F, G and H and tighten in that sequence to 49–68 ft. lbs. (66–93 Nm).

j. Tighten bolt E to 63–86 ft. lbs. (85–116 Nm).

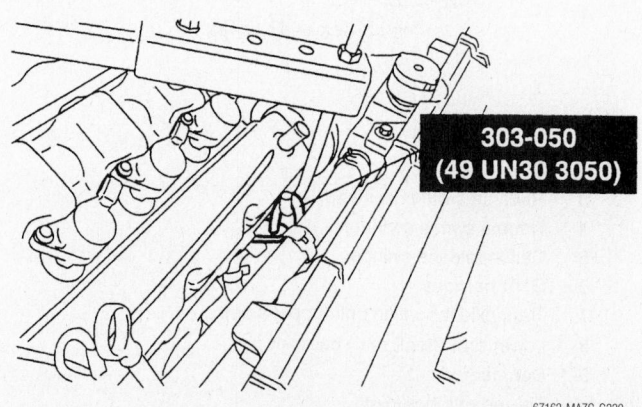

303-050
(49 UN30 3050)

67162-MAZC-G220

When removing the No. 1 engine mount, remove the intake manifold and attach tool 49 UN30 3050 to the head as shown—Mazda 6 models with the A65M-R manual transaxle

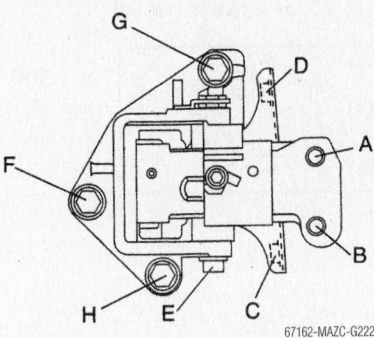

Install the No. 4 mount and tighten bolts A thru D in the sequence and specifications outlined in the text—Mazda 6 models with a A65M-R manual transaxle

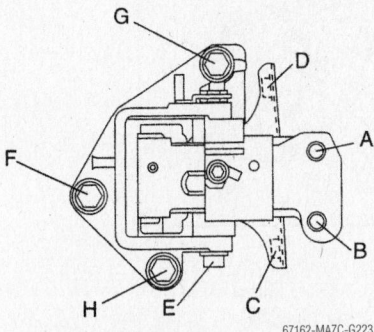

When installing the No. 4 mount, tighten bolts F, G and H, then bolt E in the sequence and specifications outlined in the text—Mazda 6 models with a A65M-R manual transaxle

17. Remove the engine support device.
18. Install the remaining components in the reverse order of removal.
19. Fill the transaxle with fluid. Road test the vehicle and check for leaks. Top off all fluids as needed.

Automatic Transaxle/ Transaxle Assembly

REMOVAL & INSTALLATION

Protégé

1. Before servicing the vehicle, refer to the precautions in the beginning of this section.
2. Refer to the illustration for component location and torque specifications.
3. Drain the transaxle oil.
4. Remove or disconnect the following:
- Wheels
- Splash shields
- Battery and battery box

- Air cleaner assembly and fresh air duct
- Exhaust Gas Recirculation (EGR) pipe
- Front exhaust pipe and three way catalytic converter
- Speedometer sensor connector
- Transmission range switch connector
- Input/turbine speed sensor connector
- Transaxle connector
- Harness bracket
- Battery tray bracket
- Oil dipstick tube
- Oil hose
- Brake hose clip and ABS sensor bracket
- Tie rod ends from the knuckle
- Lower arm bolt
- Stabilizer link nut
- Lower arm
- Transverse member
- Halfshafts
- Joint shaft, install tool 49 G030 455 to hold the side gears after removal
- Selector cable
- Intake manifold stay
- Starter
- Torque converter nuts

5. Support the engine assembly with a engine support assembly such as 49 E017 5A0.

- Number 4 engine mount
- Number 1 engine mount bracket
- Roll damper on models with the 2.0L (FS) engine
- Engine mount member
- Number 2 engine mount

6. Loosen the engine support assembly and lean the engine towards the transaxle.
7. Support the transaxle with a jack, remove the transaxle bolts and the transaxle.

To install:
8. Place the transaxle onto a jack and raise into position.
9. Install the transaxle bolts and tighten the bolts as follows:
   a. Bolts **A** to 65 ft. lbs. (89 Nm).
   b. Bolts **B** to 86 ft. lbs. (116 Nm).
   c. Bolts **C** to 38 ft. lbs. (51 Nm).
   d. Bolts **D** to 18 ft. lbs. (25 Nm).
10. Install or connect the following:
- Number 2 engine mount. Refer to the illustration for component location and torque specifications.
11. Install the engine mount member as follows while referring to the illustration for bolt locations:
   a. Position the direction indicator on

the mount member bushings as illustrated.
   b. Put the number 2 engine mount stud bolts through the mount member installation holes. Install mount member bolts **A** and tighten to 50–68 ft. lbs. (67–93 Nm).
   c. Tighten the nuts **B** to 50–68 ft. lbs. (67–93 Nm) on 1.6L (ZM) and 1.8L (FP) engines or 63–86 ft. lbs. (86–116 Nm) on 2.0L (FS) engines.
   d. By aligning the holes on the stud bolts, install the number 4 mount bracket. Align hole of the bracket with the rubber and hand tighten bolt **A**. Tighten nut **B** and tighten bolt **A** to 50–68 ft. lbs. (67–93 Nm).
12. Install or connect the following:
- Roll damper, on 2.0L (FS) engines
- Number 1 engine mount
- Torque converter nuts
- Starter
- Intake manifold stay
- Selector cable
- Joint shaft. Install a new circlip with the opening facing up. Hand tighten the bolts **A**, then tighten the bolts to 32–45 ft. lbs. (43–61 Nm). Refer to the illustration for bolt location.
- Halfshafts
- Transverse member

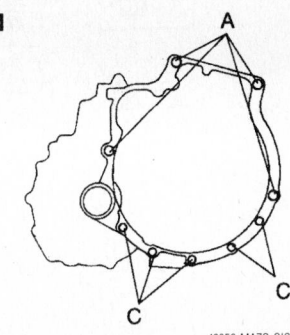

Automatic transaxle assembly bolt torque sequence—1.6L (ZM) engine

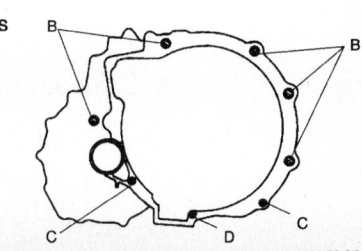

Automatic transaxle assembly bolt torque sequence—1.8L (FP) and 2.0L (FS) engines

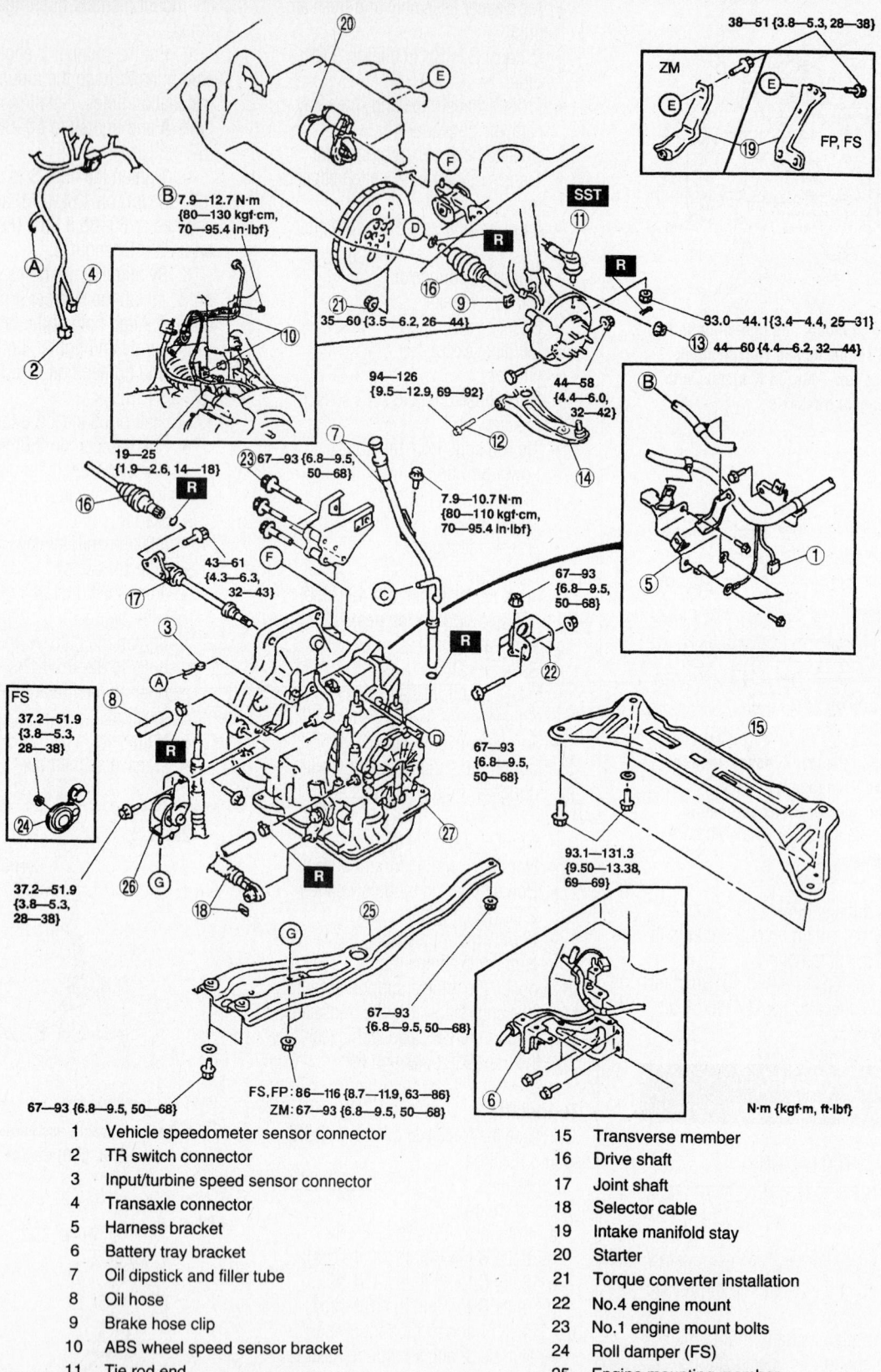

Exploded view of the automatic transaxle assembly mounting—Protégé

N·m {kgf·m, ft·lbf}

1   Vehicle speedometer sensor connector
2   TR switch connector
3   Input/turbine speed sensor connector
4   Transaxle connector
5   Harness bracket
6   Battery tray bracket
7   Oil dipstick and filler tube
8   Oil hose
9   Brake hose clip
10 ABS wheel speed sensor bracket
11 Tie rod end
12 Bolt
13 Stabilizer control link nut
14 Lower arm

15 Transverse member
16 Drive shaft
17 Joint shaft
18 Selector cable
19 Intake manifold stay
20 Starter
21 Torque converter installation
22 No.4 engine mount
23 No.1 engine mount bolts
24 Roll damper (FS)
25 Engine mounting member
26 No.2 engine mount
27 Transaxle

42356-MAZC-GIP

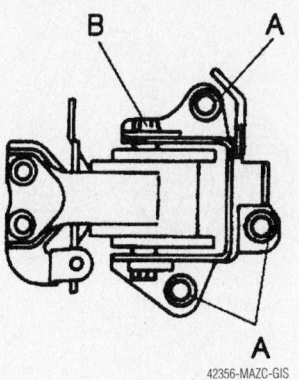

42356-MAZC-GIS

Number 4 mount fastener locations—Protégé with automatic transaxle

- Lower arm
- Stabilizer link nut
- Lower arm bolt
- Tie rod ends to the knuckle
- ABS sensor bracket and brake hose clip
- Oil hose
- Oil dipstick tube
- Battery tray bracket
- Harness bracket
- Transaxle connector
- Input/turbine speed sensor connector
- Transmission range switch connector
- Speedometer sensor connector

- Front exhaust pipe and three way catalytic converter
- EGR pipe
- Air cleaner assembly and fresh air duct
- Battery and battery box
- Splash shields
- Wheels

13. Fill the transaxle with fluid. Road test the vehicle and check for leaks. Top off all fluids as needed.

Miata

1. Refer to the illustration for component location and torque specifications.

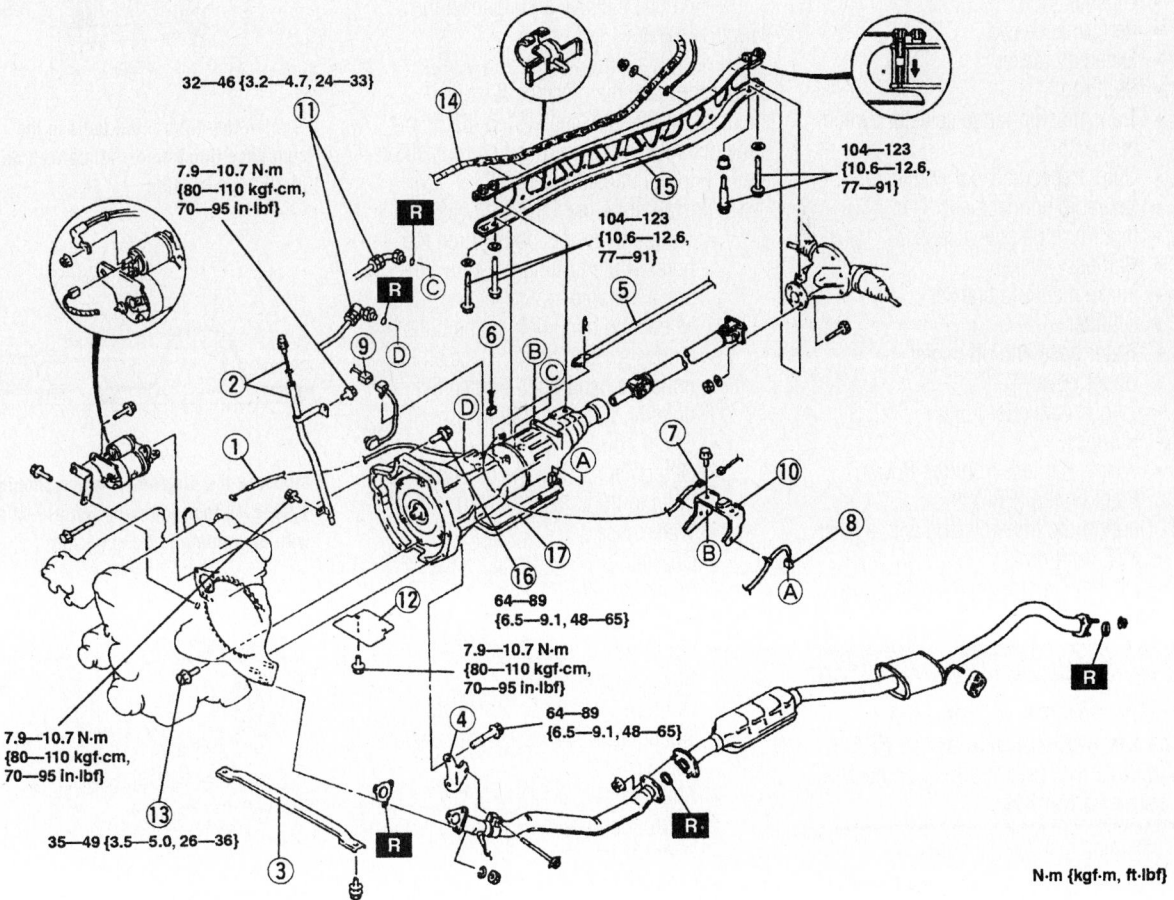

N·m {kgf·m, ft·lbf}

1	Throttle cable	10	Harness bracket
2	Filler tube, dipstick	11	Oil pipe
3	Performance rod	12	Undercover
4	Exhaust bracket	13	Torque converter bolts
5	Shift rod	14	Harness
6	TR switch connector	15	Power plant frame
7	Output speed sensor connector	16	Transmission mount bolts
8	Solenoid connector	17	Transmission
9	Input/turbine speed sensor		

Exploded view of the automatic transmission assembly—Miata

42356-MAZC-GXX

2. Before servicing the vehicle, refer to the precautions in the beginning of this section.

3. Refer to the illustration for component location and torque specifications.

4. Drain the transaxle oil, coolant and engine oil.

5. Remove or disconnect the following:
- Negative battery cable
- Crossmember and bracket, on models equipped with 16 inch wheels
- Rear crossbar, on models equipped with 16 inch wheels
- Exhaust system
- Drive shaft
- Throttle cable
- Dipstick tube
- Performance rod
- Exhaust bracket
- Shift rod
- Transmission range selector switch connector
- Output speed sensor connector
- Solenoid connector
- Input/turbine speed sensor
- Harness bracket
- Intake manifold bracket
- Oil filter
- Water hose and oil cooler
- Under cover
- Torque converter bolts
- Harness
- Wiring harness from the Power Plant Frame (PPF)

6. Support the transmission with a jack.
- PPF front bolts
- Differential side bolts and pry out the spacer

✳✳ CAUTION

Removing the PPF spacers will reduce the performance of the PPF. If the spacers are removed, replace the PPF as an assembly.

- Differential mounting spacer
- Transmission side bolts and the PPF
- Transmission bolts and the transmission; keep the torque converter end slightly elevated during removal

To install:

7. Support the transmission with a jack.

8. Install or connect the following:
- Transmission into position, make sure the torque converter side is slightly elevated, once aligned, install and tighten the bolts to 48–65 ft. lbs.
- Differential mounting spacer and

tighten the bolts to 28–38 ft. lbs. (38–51 Nm).

9. Support the transmission with a jack until it is level.
- PPF in position and if removed install the sleeve
- Spacer and bolts and tighten the reamer bolt making sure the threading is aligned correctly. The reamer bolt should be installed in the forward hole. Tighten the outer bolts making sure the threads are properly aligned. Tighten all bolts in the sequence illustrated to 91 ft. lbs. (123 (Nm).
- PPF front bolts and tighten to 91 ft. lbs. (123 Nm)

10. Remove the jack and connect the wiring harness.

11. Using a straightedge and Vernier caliper measure the distance **A** which should be 1.99–2.46 inch (50.5–62.5mm). If the distance is not as specified, reposition the PPF to the transmission.

12. Install or connect the following:
- Torque converter bolts. Align the holes while turning the converter, use a suitable tool to lock the flywheel and hand tighten the bolts in a criss–cross pattern. Once all the bolts are hand tight, tighten the bolts to 36 ft. lbs. (49 Nm).
- Under cover
- Oil cooler, tighten the oil cooler bolts to 28 ft. lbs. (39 Nm)
- Water hose and
- Oil filter
- Intake manifold bracket
- Harness bracket
- Input/turbine speed sensor
- Solenoid connector
- Output speed sensor connector
- Transmission range selector switch connector
- Shift rod
- Exhaust bracket
- Performance rod
- Dipstick tube
- Throttle cable. Measure the free-play on the cable, if the free-play is not 0.04–0.11 inch (1–3mm), then adjust by turning the lock nuts until the desired specification is reached and tighten the lock nuts to 10 ft. lbs. (14 Nm).
- Drive shaft
- Exhaust system
- Rear crossbar, on models equipped with 16 inch wheels
- Crossmember and bracket, on models equipped with 16 inch wheels

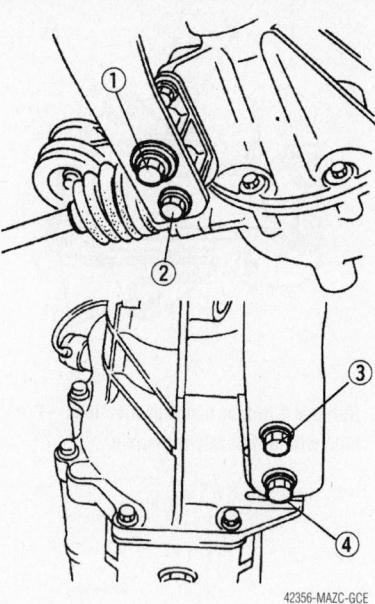

Tighten the differential bolts in the sequence illustrated—Miata with an automatic transmission

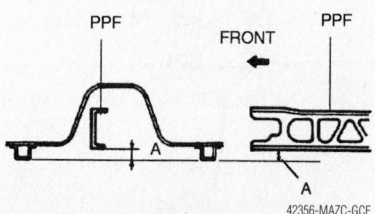

Measure the distance A which should be 1.99–2.46 inch (50.5–62.5mm)—Miata with an automatic transmission

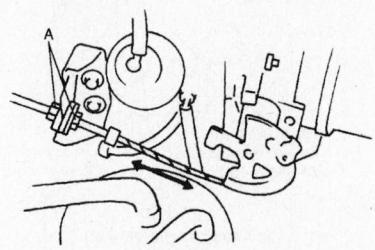

Adjust the free-play on the throttle and adjust using the lock nuts A—Miata with an automatic transmission

- Negative battery cable

13. Fill the transmission with fluid. Road test the vehicle and check for leaks. Top off all fluids as needed.

626

LA4A–EL TRANSXLE

1. Refer to the illustration for component location and torque specifications.

2. Before servicing the vehicle, refer to the precautions in the beginning of this section.

3. Refer to the illustration for component location and torque specifications.

4. Drain the transaxle oil.

5. Remove or disconnect the following:
- Negative battery cable
- Air cleaner assembly and fresh air duct
- Wheels
- Splash shields
- Ground cable from the transaxle
- Bracket
- Solenoid body connector
- Transaxle range switch connector
- Turbine speed shaft connector
- Oil filler tube
- Fuel filter nut
- Selector cable
- Number 1 engine mount nut, rubber and bracket

6. Support the engine assembly with a engine support assembly such as 49 E017 5A0.
- Number 4 mount
- Transverse member
- Vehicle Speed Sensor (VSS) connector
- Engine mount member
- Tie rod ends from the knuckle
- Stabilizer bar link
- Lower control arm
- Halfshafts
- Joint shaft, install tool 49 G030 455 to hold the side gears after removal
- Intake manifold bracket
- Starter

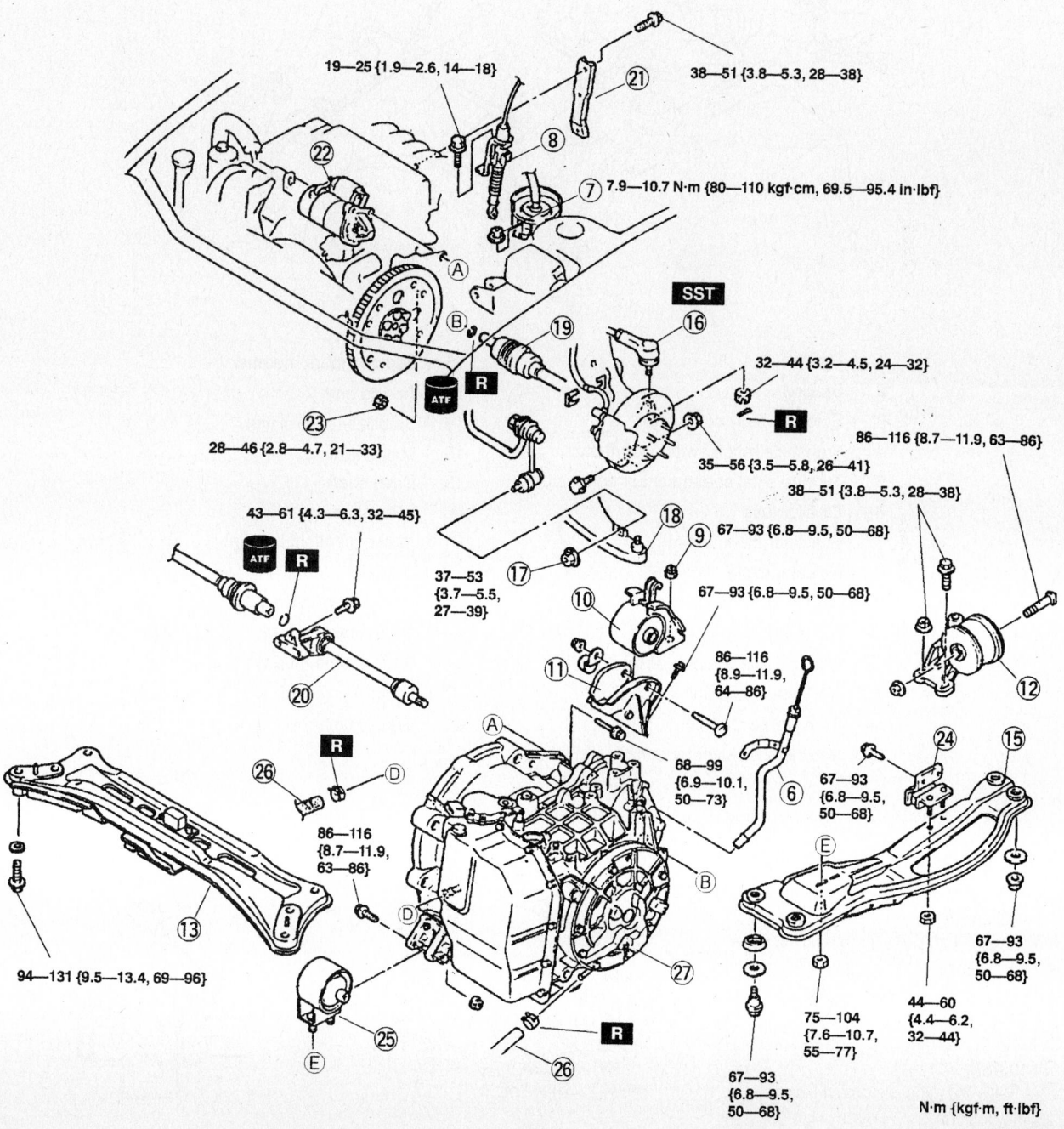

Exploded view of the LA4A–EL automatic transaxle assembly (1 OF 2)—626

42356-MAZC-G73

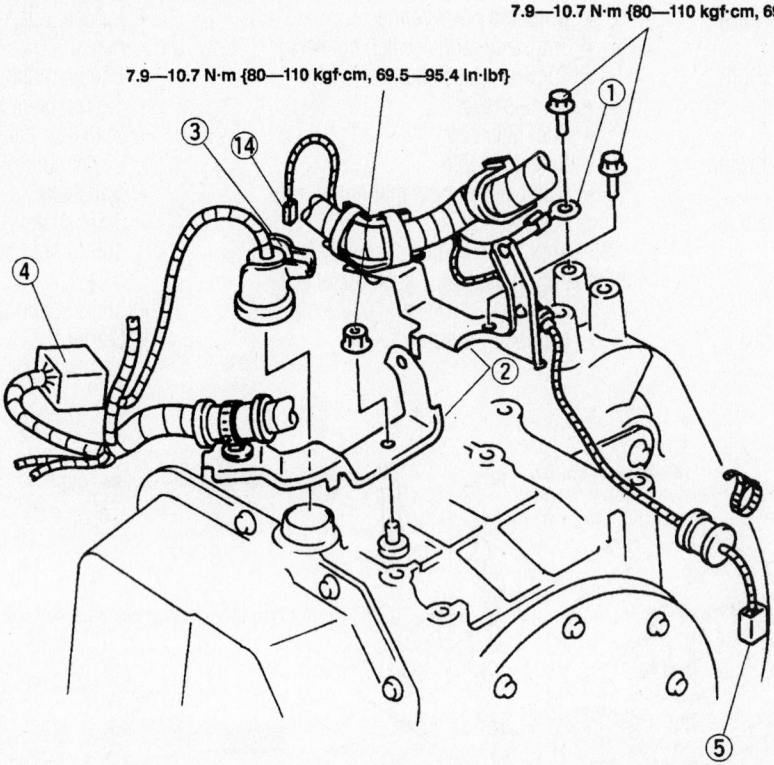

7.9—10.7 N·m {80—110 kgf·cm, 69.5—95.4 In·lbf}

7.9—10.7 N·m {80—110 kgf·cm, 69.5—95.4 In·lbf}

1	Ground	15	Engine mount member
2	Bracket	16	Tie-rod end
3	Solenoid body connector	17	Stabilizer control link
4	Transaxle range switch connector	18	Lower arm
5	Turbine shaft speed sensor connector	19	Drive shaft
6	Oil filler tube	20	Joint shaft
7	Fuel filter mounting nut	21	Intake manifold bracket
8	Selector cable	22	Starter
9	No.1 engine mount nut	23	Torque converter nut
10	No.1 engine mount rubber	24	No.5 engine mount
11	No.1 engine mount bracket	25	No.2 engine mount
12	No.4 engine mount	26	Oil hose
13	Transverse member	27	Transaxle
14	Vehicle speedometer sensor connector		

42356-MAZC-G74

Exploded view of the LA4A–EL automatic transaxle assembly (2 OF 2)—626

- Torque converter nuts
- Number 5 and 2 engine mounts
- Oil hose

7. Loosen the engine support assembly and lean the engine towards the transaxle.

8. Support the transaxle with a jack, remove the transaxle bolts and the transaxle.

To install:

9. Place the transaxle onto a jack and raise into position.

10. Install the transaxle bolts and tighten the bolts **A** to 66–86 ft. lbs. (90–116 Nm),

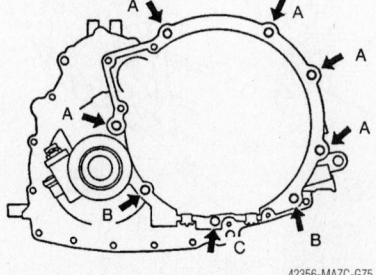

42356-MAZC-G75

LA4A–EL automatic transaxle assembly bolt locations—626

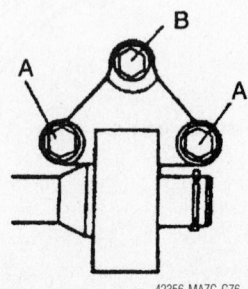

42356-MAZC-G76

Bolt locations on the joint shaft—EL automatic transaxle assembly

bolts **B** to 28–38 ft. lbs. (38–51 Nm) and bolts **C** to 14–18 ft. lbs. (19–25 Nm). Refer to the illustration for bolt locations.

- Hand tighten the number 4 mount retainers

11. Lean the engine towards its normal position and tighten the support assembly.

12. Install or connect the following:

- Oil hose
- Number 5 and 2 engine mounts. Refer to the illustration for component location and torque specifications.
- Torque converter nuts and tighten to 21–33 ft. lbs. (28–46 Nm)
- Starter
- Intake manifold bracket. Refer to the illustration for component location and torque specifications.
- Joint shaft. Install a new circlip with the opening facing up. Hand tighten the bolts **A**, then tighten the bolts to 32–45 ft. lbs. (43–61 Nm). Refer to the illustration for bolt location.
- Halfshafts
- Lower control arm
- Stabilizer bar link. Refer to the illustration for component location and torque specifications.
- Tie rod ends to the knuckle. Refer to the illustration for component location and torque specifications.

13. Install the engine mount member as follows while referring to the illustration for bolt locations:

a. Position the direction indicator on the mount member bushings facing towards the front side and install the bushings onto the mount.

b. Put the number 2 engine mount stud bolts through the mount member installation holes. Install mount member bolts **A** and nuts **A** and tighten to 50–68 ft. lbs. (67–93 Nm).

c. Loosely tighten the number 2

engine mount nuts and remove the engine support assembly.

d. Tighten the number 2 engine mount nuts **B** to 55–77 ft. lbs. (75–104 Nm).

14. Install or connect the following:

- VSS connector
- Transverse member. Refer to the illustration for component location and torque specifications.
- Number 4 mount. Refer to the illustration for component location and torque specifications.
- Number 1 engine mount nut, rubber and bracket. Refer to the illustration for component location and torque specifications.
- Selector cable

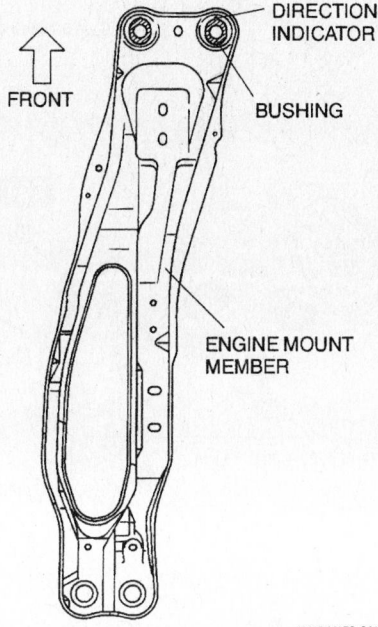

42356-MAZC-G61

Position the direction indicator on the mount member bushings facing towards the front side and install the bushings onto the mount—626

- Fuel filter nut
- Oil filler tube
- Turbine speed shaft connector
- Transaxle range switch connector
- Solenoid body connector
- Bracket
- Ground cable from the transaxle
- Splash shields
- Wheels
- Air cleaner assembly and fresh air duct
- Negative battery cable

15. Fill the transmission with fluid.

16. Start the engine and check for leaks and proper operation.

GFA4A–EL TRANSAXLE

1. Before servicing the vehicle, refer to the precautions in the beginning of this section.

2. Refer to the illustration for component location and torque specifications.

3. Drain the transaxle oil.

4. Remove or disconnect the following:

- Battery
- Air cleaner assembly and fresh air duct
- Wheels
- Splash shields
- Selector cable clip
- Selector cable from the manual shaft lever, pull out the cable from the bracket and remove the cable
- Transaxle range switch connector
- Solenoid valve connector
- Oxygen (O$_2$S) connector
- Vehicle Speed Sensor (VSS) connector
- Input/Turbine speed shaft connector
- Starter harness
- Engine mount stay
- Fuel filter bolts
- Oil hose and breather hose

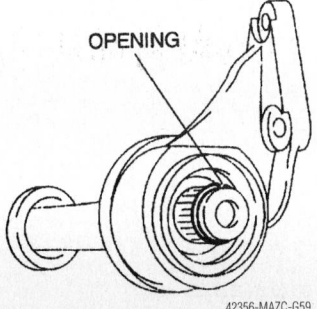

42356-MAZC-G59

When installing the joint shaft make sure to install a new clip with the opening facing up—626

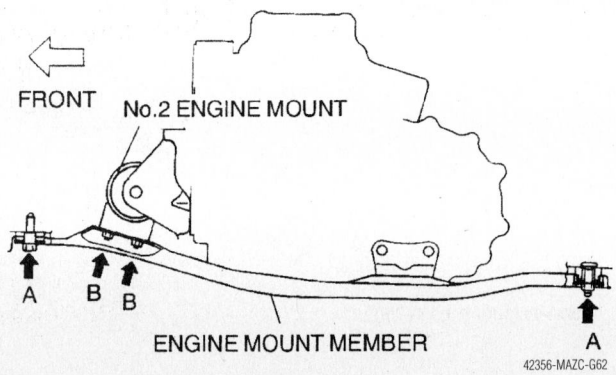

42356-MAZC-G62

Location of the front mount member bolt locations—626

- Lower control arm
- Tie rod ends from the knuckle
- Stabilizer bar link
- Transverse member
- Halfshafts
- Joint shaft, install tool 49 G030 455 to hold the side gears after removal
- Number 5 mount rubber
- Number 1 engine mount nut, rubber and bracket

5. Support the engine assembly with a engine support assembly such as 49 E017 5A0.

- Engine mount member
- Number 2 engine mount
- Under cover
- Torque converter nuts
- Number 4 engine mount

6. Loosen the engine support assembly and lean the engine towards the transaxle.

7. Support the transaxle with a jack, remove the transaxle bolts and the transaxle.

To install:

8. Place the transaxle onto a jack and raise into position.

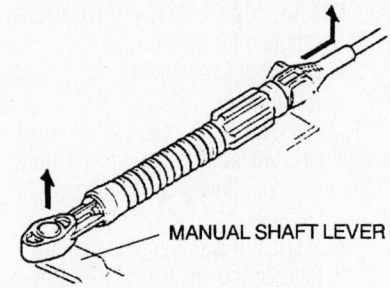

MANUAL SHAFT LEVER
42356-MAZC-G80

Remove the selector cable GF4A-EL automatic transaxle assembly—626

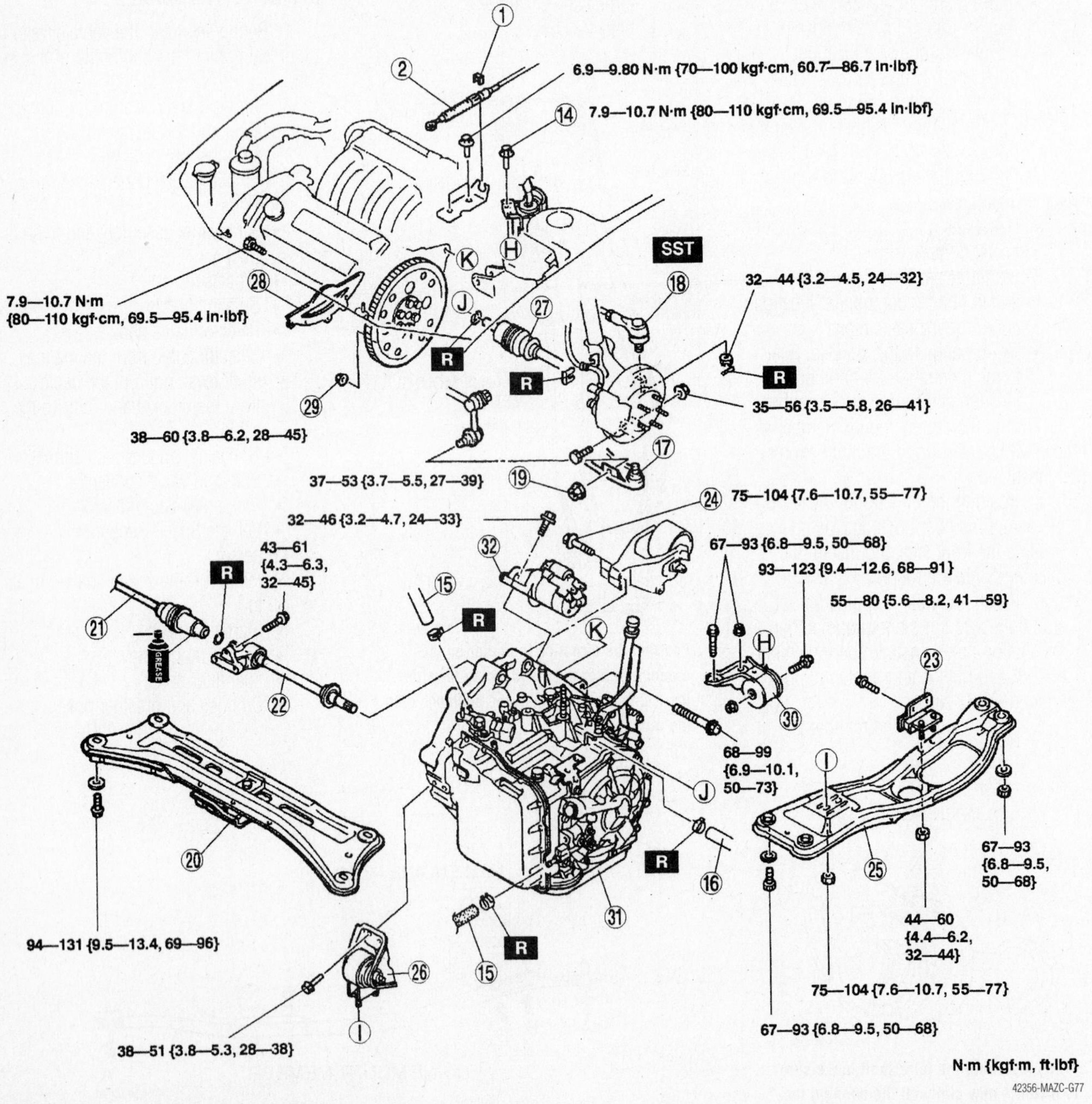

1
2
6.9—9.80 N·m {70—100 kgf·cm, 60.7—86.7 In·lbf}
14
7.9—10.7 N·m {80—110 kgf·cm, 69.5—95.4 In·lbf}
K
H
SST
28
7.9—10.7 N·m
{80—110 kgf·cm, 69.5—95.4 In·lbf}
18
32—44 {3.2—4.5, 24—32}
J
27
R
R
R
29
35—56 {3.5—5.8, 26—41}
38—60 {3.8—6.2, 28—45}
17
37—53 {3.7—5.5, 27—39}
19
24
75—104 {7.6—10.7, 55—77}
32—46 {3.2—4.7, 24—33}
43—61
{4.3—6.3,
32—45}
R
32
67—93 {6.8—9.5, 50—68}
93—123 {9.4—12.6, 68—91}
55—80 {5.6—8.2, 41—59}
15
R
21
GREASE
K
H
23
22
30
68—99
{6.9—10.1,
50—73}
I
J
20
R
16
25
67—93
{6.8—9.5,
50—68}
31
94—131 {9.5—13.4, 69—96}
R
44—60
{4.4—6.2,
32—44}
26
15
75—104 {7.6—10.7, 55—77}
I
67—93 {6.8—9.5, 50—68}
38—51 {3.8—5.3, 28—38}
N·m {kgf·m, ft·lbf}
42356-MAZC-G77

Exploded view of the GF4A–EL automatic transaxle assembly (1 OF 2)—626

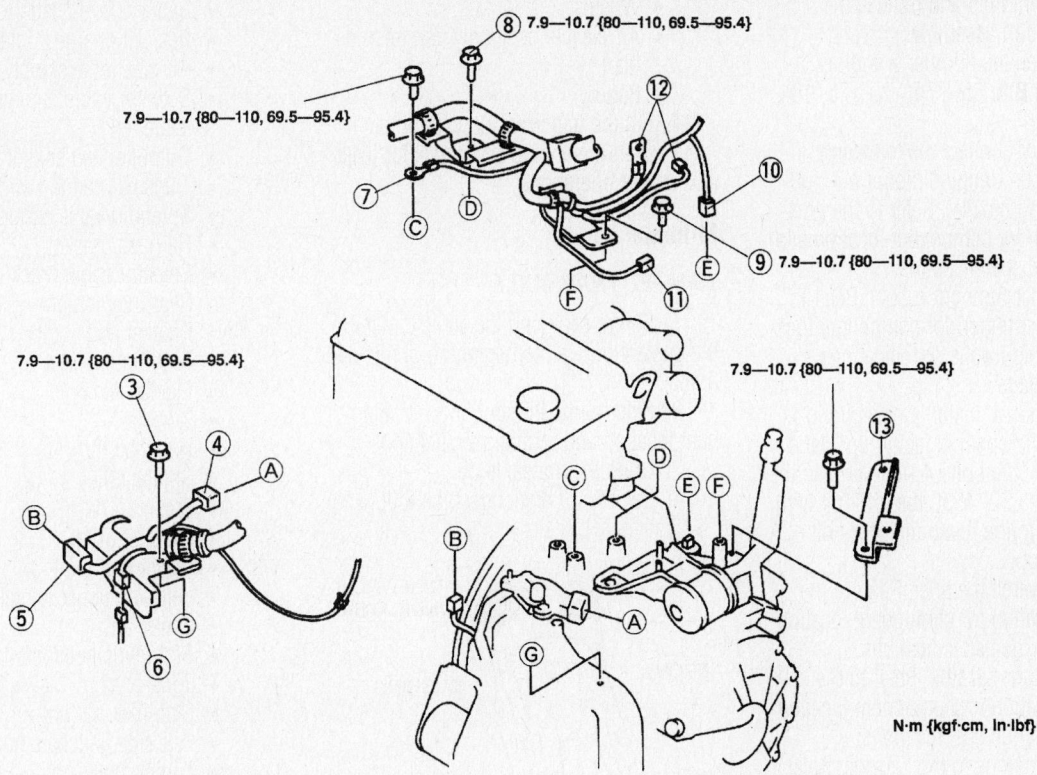

7.9—10.7 {80—110, 69.5—95.4}

N·m {kgf·cm, in·lbf}

1	Clip	17	Lower arm
2	Selector cable	18	Tie-rod end
3	Bolt	19	Stabilizer control link nut
4	Transaxle range switch connector	20	Transverse member
5	Solenoid valve connector	21	Drive shaft
6	Oxygen sensor connector	22	Joint shaft
7	Ground	23	No.5 engine mount rubber
8	Bolt	24	No.1 engine mount bolts
9	Bolt	25	Engine mount member
10	Vehicle speedometer sensor connector	26	No.2 engine mount
11	Input/turbine speed sensor connector	27	Drive shaft
12	Starter harness	28	Undercover
13	Engine mount stay	29	Torque converter nuts
14	Fuel filter mounting bolts	30	No.4 engine mount
15	Oil hose	31	Transaxle
16	Breather hose	32	Starter

42356-MAZC-G78

Exploded view of the GF4A–EL automatic transaxle assembly (2 OF 2)—626

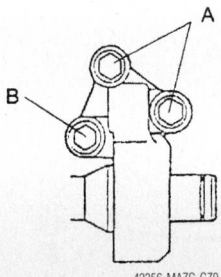

42356-MAZC-G79

Bolt locations on the joint shaft—GF4A–EL automatic transaxle assembly

9. Install the transaxle bolts and tighten the bolts to 50–73 ft. lbs. (68–99 Nm).
  • Hand tighten the number 4 mount retainers, then tighten to 55–77 ft. lbs. (75–104 Nm)
10. Lean the engine towards its normal position and tighten the support assembly.
11. Install or connect the following:
  • Torque converter nuts
  • Under cover
  • Number 2 engine mount

12. Install the engine mount member as follows while referring to the illustration for bolt locations:
  a. Position the direction indicator on the mount member bushings facing towards the front side and install the bushings onto the mount.
  b. Put the number 2 engine mount stud bolts through the mount member installation holes. Install mount member bolts **A** and nuts **A** and tighten to 50–68 ft. lbs. (67–93 Nm).
  c. Loosely tighten the number 2

engine mount nuts and remove the engine support assembly.

d. Tighten the number 2 engine mount nuts **B** to 55–77 ft. lbs. (75–104 Nm).

13. Install or connect the following:

- Number 1 engine mount nut, rubber and bracket. Refer to the illustration for component location and torque specifications.
- Number 5 mount rubber. Refer to the illustration for component location and torque specifications.
- Halfshafts
- Joint shaft. Install a new circlip with the opening facing up. Hand tighten the bolts **A**, then tighten the bolts to 32–45 ft. lbs. (43–61 Nm). Refer to the illustration for bolt location.
- Transverse member. Refer to the illustration for component location and torque specifications.
- Stabilizer bar link. Refer to the illustration for component location and torque specifications.
- Tie rod ends to the knuckle. Refer to the illustration for component location and torque specifications.
- Lower control arm
- Oil hose and breather hose
- Fuel filter bolts
- Engine mount stay. Refer to the illustration for component location and torque specifications.
- Starter harness
- Input/Turbine speed shaft connector
- VSS connector
- O2S connector
- Solenoid valve connector
- Transaxle range switch connector
- Selector cable to the bracket, install the clip and attach the cable to the manual shaft lever
- Splash shields

- Wheels
- Air cleaner assembly and fresh air duct
- Battery

14. Fill the transmission with fluid.

15. Start the engine and check for leaks and proper operation.

Millenia

LJ4A–EL TRANSAXLE

1. Before servicing the vehicle, refer to the precautions in the beginning of this section.

2. Refer to the illustration for component location and torque specifications.

3. Drain the transaxle oil.

4. Remove or disconnect the following:

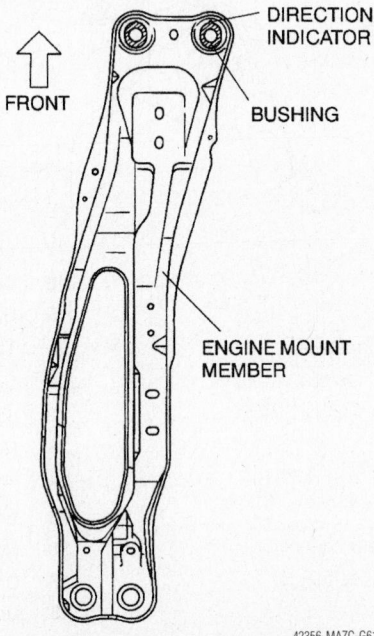

42356-MAZC-G61

Position the direction indicator on the mount member bushings facing towards the front side and install the bushings onto the mount—626

- Battery cover, battery and tray
- Rear air charge cooler duct
- Air cleaner assembly
- Selector cable clip and nut and the cable
- Output speed sensor connector
- Transaxle range switch connector
- Solenoid valve connector
- Harness bracket
- Selector cable bracket
- Rear intercooler
- Bracket and starter
- Front intercooler
- Electric fan
- Wheels
- Halfshaft locknut
- Splash shields
- Fresh air duct
- Exhaust pipe
- Lower ball joint
- Left side upper lateral link
- Lower arm
- Right side halfshaft
- Joint shaft
- Stabilizer bar link
- Left side halfshaft
- Timing belt
- Drive belts
- Power steering pump leaving the lines attached and set aside
- A/C compressor leaving the lines attached and set aside
- Selector rod
- Transmission dipstick tube

5. Support the engine assembly with a engine support assembly such as 49 E017 5A0.

- Loosen the number 3 engine mount bolt (refer to illustration) and remove the number 4 mount
- Number 1 engine mount nut, rubber and bracket
- Number 2 engine mount nut
- Under cover
- Torque converter bolts
- Number 1 engine mount bolt

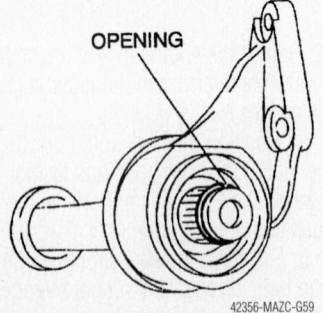

42356-MAZC-G59

When installing the joint shaft make sure to install a new clip with the opening facing up—626

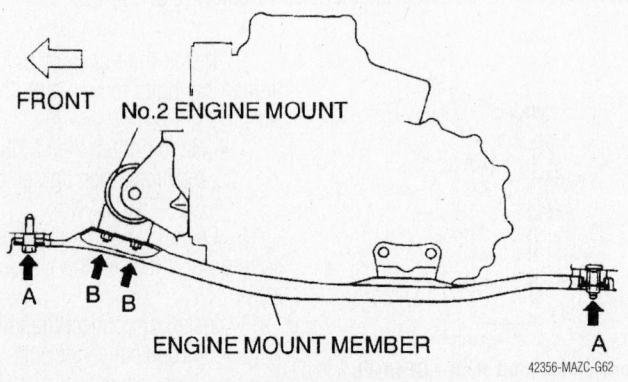

42356-MAZC-G62

Location of the front mount member bolt locations—626

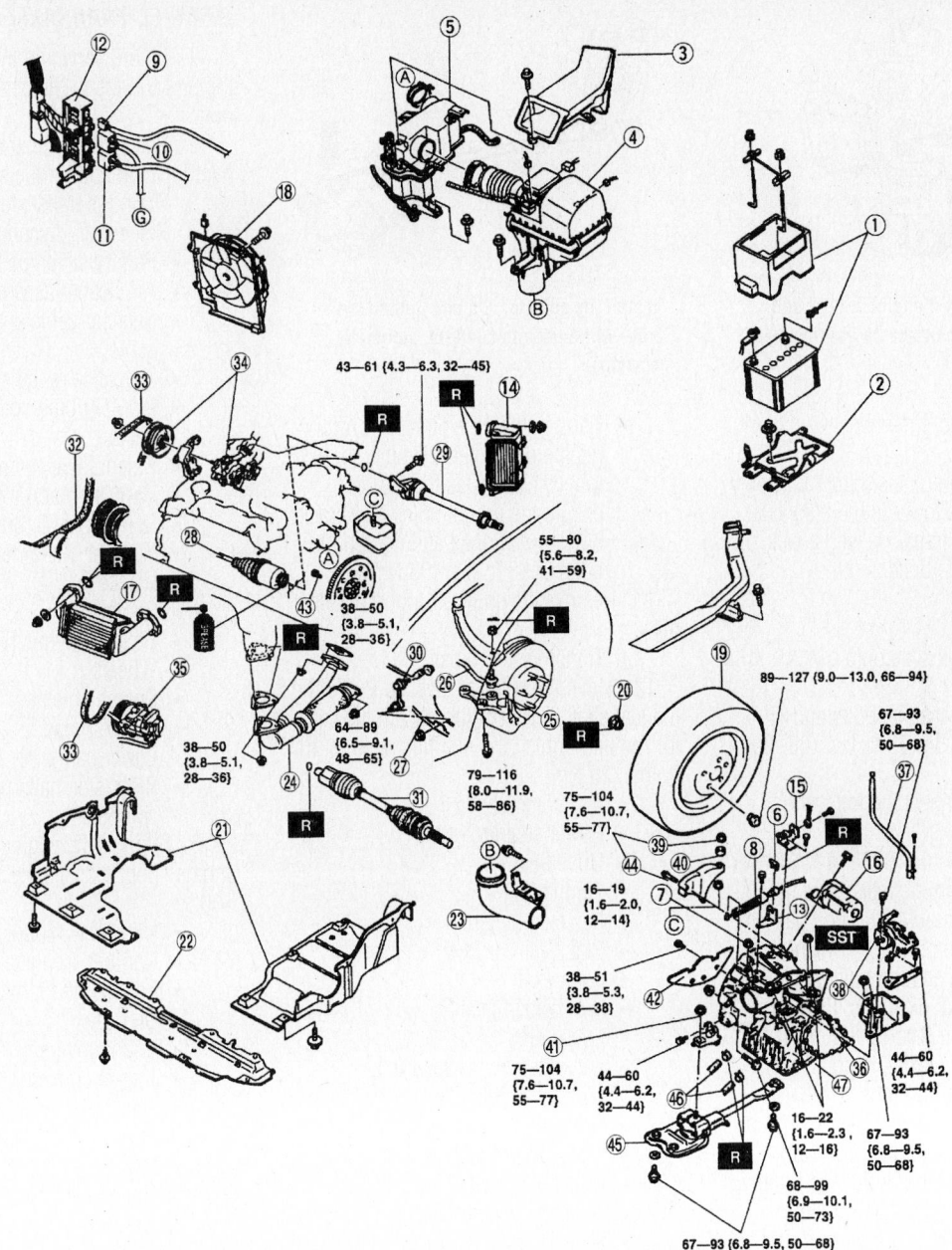

N·m (kgf·m, ft·lbf)

1	Battery and battery cover	17	Front intercooler
2	Battery carrier	18	Electric coolant fan component
3	Rear change air cooler air duct	19	Wheel and tires
4	Air cleaner component	20	Locknut
5	Resonance chamber	21	Splash shield
6	Clip	22	Splash shield
7	Nut	23	Fresh air duct
8	Selector cable	24	Exhaust pipe
9	Output speed sensor connector	25	Lower ball joint
10	Transaxle range switch connector	26	Upper lateral link (left side)
11	Solenoid valve connector	27	Lower arm
12	Harness bracket	28	Drive shaft (right side)
13	Selector cable bracket	29	Joint shaft
14	Rear intercooler	30	Stabilizer control link
15	Bracket	31	Drive shaft (left side)
16	Starter	32	Timing belt

33	P/S, A/C Drive belt
34	P/S oil pump
35	A/C compressor
36	Selector rod
37	ATF filler tube
38	No.4 engine mount
39	No.1 engine mount nut
40	No.1 engine mount damper
41	No.2 engine mount nut
42	Under cover
43	Torque converter bolts
44	No.1 engine mount bolt
45	Engine mounting member
46	Oil hose
47	Transaxle

42356-MAZC-GEPQ

Exploded view of the LJ4A–EL automatic transaxle assembly—Millenia

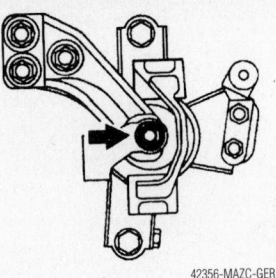

42356-MAZC-GER

Loosen the number 3 engine mount bolt LJ4A–EL automatic transaxle assembly—Millenia

- Engine mount member
- Oil hose

6. Loosen the engine support assembly and lean the engine towards the transaxle.

7. Support the transaxle with a jack, remove the transaxle bolts and the transaxle.

To install:

8. Place the transaxle onto a jack and raise into position.

9. Install the transaxle bolts and tighten the bolts to 50–73 ft. lbs. (68–99 Nm).

10. Install or connect the following:
- Oil hose
- Engine mount member and number 2 engine mount, make sure the number 2 mount stud bolt passes through the number 2 mount bracket installation hole and tighten the bolt **A** to 68 ft. lbs. (93 Nm). Tighten bolt **B** to 77 ft. lbs. (104 Nm). Refer to the illustration for component location.
- Number 3 engine mount bolt and tighten to 77 ft. lbs. (104 Nm)
- Torque converter bolts and tighten to 28–38 ft. lbs. (38–51 Nm)
- Under cover
- Number 2 engine mount nut
- Number 1 engine mount nut, rubber and bracket. Tighten the bolt to 77 ft. lbs. (104 Nm).

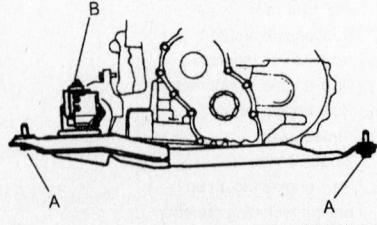

42356-MAZC-GES

Install the engine mount member bolts and tighten to specification—LJ4A–EL automatic transaxle assembly—Millenia with LJ4A–EL automatic transaxle

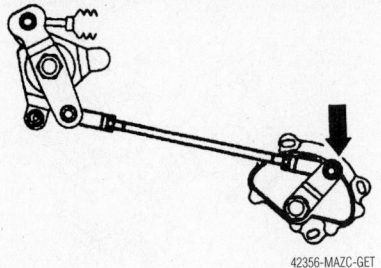

42356-MAZC-GET

Install the selector rod and tighten the nut—Millenia with LJ4A–EL automatic transaxle

- Hand tighten the number 4 mount retainers. Tighten the bolts to 44 ft. lbs. (60 Nm), use the jack to ensure the mount holes and bracket holes are aligned and tighten the bolts to 68 ft. lbs. (93 Nm).

11. Remove the engine support assembly.
- Transmission dipstick tube

12. Move the selector lever to **P** and the manual shaft to **P** position. Install the selector rod and tighten the nut illustrated to 16 ft. lbs. (22 Nm).
- A/C compressor
- Power steering pump
- Drive belts
- Timing belt
- Left side halfshaft
- Stabilizer bar link
- Joint shaft
- Right side halfshaft
- Lower arm
- Left side upper lateral link
- Lower ball joint
- Exhaust pipe
- Fresh air duct
- Splash shields
- Halfshaft locknut
- Wheels
- Electric fan
- Front intercooler
- Bracket and starter
- Rear intercooler
- Selector cable bracket
- Harness bracket
- Solenoid valve connector
- Transaxle range switch connector
- Output speed sensor connector
- Selector cable clip and nut and the cable
- Air cleaner assembly
- Rear air charge cooler duct
- Battery tray, battery and cover

13. Fill the transmission with fluid.

14. Start the engine and check for leaks and proper operation.

GF4A–EL TRANSAXLE

1. Before servicing the vehicle, refer to the precautions in the beginning of this section.

2. Refer to the illustration for component location and torque specifications.

3. Drain the transaxle oil.

4. Remove or disconnect the following:
- Battery cover, battery and tray
- Air cleaner assembly
- Selector cable clip and nut and the cable
- Transaxle range switch connector
- Input/Turbine speed sensor connector
- Solenoid valve connector
- Harness bracket
- Selector cable bracket
- Number 1 engine mount stay bracket
- Starter
- Heated Oxygen (HO$_2$S) connector
- Wheels
- Splash shields
- Exhaust pipe
- Lower ball joint
- Right side halfshaft
- Joint shaft

5. Support the engine assembly with a engine support assembly such as 49 E017 5A0.
- Number 4 mount
- Number 1 engine mount nut
- Engine mount member
- Oil hose
- Under covers
- Torque converter bolts
- Left side halfshaft
- Number 1 engine mount bolt

6. Loosen the engine support assembly and lean the engine towards the transaxle.

7. Support the transaxle with a jack, remove the transaxle bolts and the transaxle.

To install:

8. Place the transaxle onto a jack and raise into position.

9. Install the transaxle bolts and tighten the bolts to 50–73 ft. lbs. (68–99 Nm).

10. Install or connect the following:
- Number 1 engine mount bolt. Make sure the holes align before tightening. Tighten the bolt to 77 ft. lbs. (104 Nm).
- Left side halfshaft
- Torque converter bolts and tighten to 32–44 ft. lbs. (44–60 Nm)
- Under covers
- Oil hose
- Engine mount member and number 2 engine mount, make sure

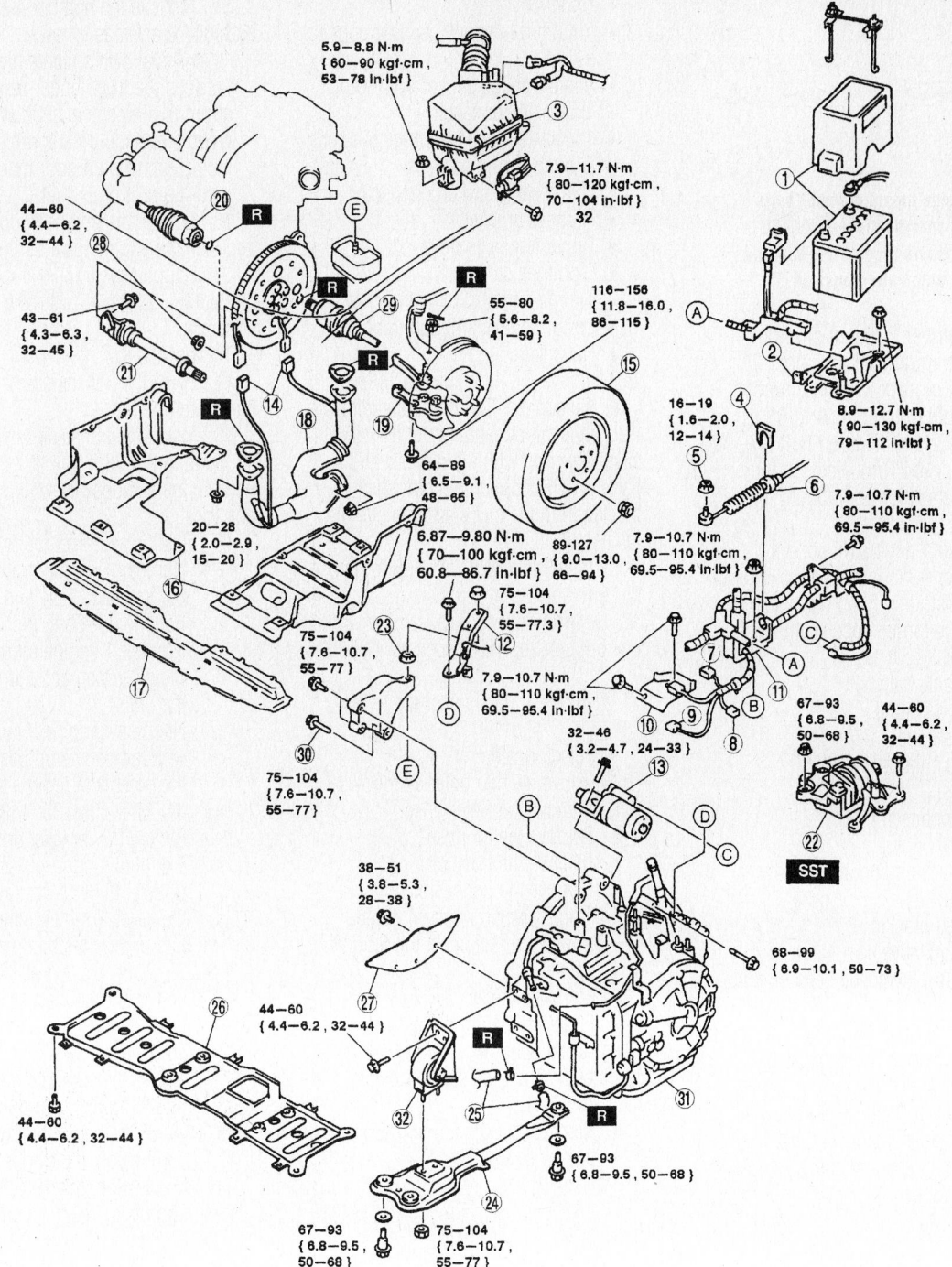

5.9—8.8 N·m
{ 60—90 kgf·cm ,
53—78 in·lbf }

7.9—11.7 N·m
{ 80—120 kgf·cm ,
70—104 in·lbf }
32

44—60
{ 4.4—6.2 ,
32—44 }

43—61
{ 4.3—6.3 ,
32—45 }

55—80
{ 5.6—8.2 ,
41—59 }

116—156
{ 11.8—16.0 ,
86—115 }

16—19
{ 1.6—2.0 ,
12—14 }

8.9—12.7 N·m
{ 90—130 kgf·cm ,
79—112 in·lbf }

7.9—10.7 N·m
{ 80—110 kgf·cm ,
69.5—95.4 in·lbf }

20—28
{ 2.0—2.9 ,
15—20 }

64—89
{ 6.5—9.1 ,
48—65 }

6.87—9.80 N·m
{ 70—100 kgf·cm ,
60.8—86.7 in·lbf }

89—127
{ 9.0—13.0 ,
66—94 }

7.9—10.7 N·m
{ 80—110 kgf·cm ,
69.5—95.4 in·lbf }

75—104
{ 7.6—10.7 ,
55—77.3 }

75—104
{ 7.6—10.7 ,
55—77 }

7.9—10.7 N·m
{ 80—110 kgf·cm ,
69.5—95.4 in·lbf }

32—46
{ 3.2—4.7 , 24—33 }

67—93
{ 6.8—9.5 ,
50—68 }

44—60
{ 4.4—6.2 ,
32—44 }

75—104
{ 7.6—10.7 ,
55—77 }

SST

68—99
{ 6.9—10.1 , 50—73 }

38—51
{ 3.8—5.3 ,
28—38 }

44—60
{ 4.4—6.2 , 32—44 }

44—60
{ 4.4—6.2 , 32—44 }

67—93
{ 6.8—9.5 ,
50—68 }

67—93
{ 6.8—9.5 ,
50—68 }

75—104
{ 7.6—10.7 ,
55—77 }

1 Battery and battery cover	12 No.1 engine mount bracket stay	23 No.1 engine mount nut
2 Battery carrier	13 Starter	24 Engine mounting member
3 Air cleaner component	14 Heated oxygen sensor connector	25 Oil hose
4 Clip	15 Wheel and tire	26 Undercover
5 Nut	16 Splash shield	27 Undercover
6 Selector cable	17 Splash shield	28 Torque converter nut
7 Transaxle range switch connector	18 Front exhaust pipe	29 Drive shaft (left)
8 Input/turbine speed sensor connector	19 Lower ball joint	30 No.1 engine mount bolt
9 Solenoid valve connector	20 Drive shaft (right)	31 Transaxle
10 Harness bracket	21 Joint shaft	32 No.2 engine mount
11 Selector cable bracket	22 No.4 engine mount	

Exploded view of the GF4A–EL automatic transaxle assembly—Millenia

42356-MAZC-GEUV

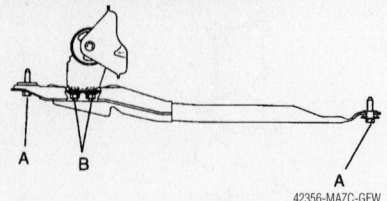

Install the engine mount member bolts and tighten to specification—LJ4A–EL automatic transaxle assembly—Millenia with GF4A–EL automatic transaxle

the number 2 mount stud bolt passes through the number 2 mount bracket installation hole and tighten the bolt **A** to 68 ft. lbs. (93 Nm). Tighten bolt **B** to 77 ft. lbs. (104 Nm). Refer to the illustration for component location.

- Number 1 engine mount nut
- Number 4 mount retainers. Refer to the illustration for component location and torque specifications.
- Joint shaft
- Right side halfshaft
- Lower ball joint
- Exhaust pipe
- Splash shields
- Wheels
- HO2S connector
- Starter
- Number 1 engine mount stay bracket
- Harness bracket
- Solenoid valve connector
- Input/Turbine speed sensor connector
- Transaxle range switch connector
- Selector cable nut and clip
- Air cleaner assembly
- Battery tray, battery and cover

11. Fill the transmission with fluid.

12. Start the engine and check for leaks and proper operation.

Mazda 3

1. Before servicing the vehicle, refer to the precautions in the beginning of this section.

2. Refer to the illustration for component location and torque specifications.

3. Drain the transaxle oil.

4. Remove or disconnect the following:
- Battery duct and cover
- Negative battery cable
- Battery, tray and box
- Air cleaner assembly
- Exhaust manifold insulator
- Wheels
- Splash shields

- Under cover
- Input/turbine speed sensor connector
- Vehicle speed sensor connector
- Transaxle connector
- Transaxle range switch connector
- Ground wiring harness
- Oil pressure switch connector
- Harness bracket
- Upper transaxle bolts
- Stabilizer bar link
- Tie rod ends from the knuckle
- Lower control arm ball joint
- Halfshafts

5. Remove the joint shaft as follows:

a. Disconnect the right halfshaft from the joint shaft by tapping the transaxle side outer ring with a brass bar and a hammer. Disconnect the joint shaft bracket from the block and remove the joint shaft.

b. Install tool 49 G030 455 to hold the side gears after removal

6. Support the engine assembly with a engine support assembly such as 49 E017 5A0.

7. Remove or disconnect the following:

- Selector cable

8. Remove the oil cooler as follows:
a. Remove the water hose.
b. Remove the oil hose.
c. Remove the hose clamp.
d. Remove the connector bolt, packing, oil pipe, packing and oil cooler.

9. Remove or disconnect the following:
- Starter
- End plate cover
- Torque converter bolts, through the starter opening

10. Install an engine support device such as 49 E017 5A0.
- Number 1 engine mount rubber
- Battery tray bracket

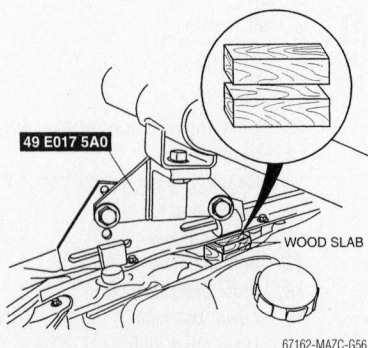

Install two suitable pieces of wood between the front fender panel and upper apron reinforcement—Mazda 3 with an automatic transaxle

11. Remove the number 4 engine mount and joint bracket as follows:

a. Install two suitable pieces of wood between the front fender panel and upper apron reinforcement as illustrated. The wood size should be approximately 1.38 inch (35mm) on 4 door models or 2.36 inch (60mm) on 5 door models.

12. Loosen the engine support assembly and lean the engine towards the transaxle.

13. Support the transaxle with a jack, remove the transaxle bolts and the transaxle.

To install:

14. Place the transaxle onto a jack and raise into position.

15. Install the transaxle bolts and tighten the bolts to 28–38 ft. lbs. (37–52 Nm). Refer to the exploded view illustration for bolt locations.

16. Install or connect the following:
- Number 4 engine mount bracket to the transaxle case and tighten the bolts to 49–68 ft. lbs. (66–93 Nm)
- Number 1 engine mount rubber to the crossmember and hand tighten the bolts
- Number 4 engine mount rubber with the body stud passing through the holes and tighten the bolts to 61–87 ft. lbs. (83–113 Nm)
- Battery tray bracket on the number 4 engine mount rubber with the body stud bolts passing through the holes. Tighten retainers identified in the illustration as 2 to 32–45 ft. lbs. (44–61 Nm) and the retainers identified as 1 in the same illustration to 61–86 inch lbs. (7–10 Nm).

17. Remove the engine support device and tighten the number 1 mount rubber bolts to 68–85 ft. lbs. (116 Nm).
- Torque converter bolts and tighten to 25–44 ft. lbs. (34–60 Nm)
- End plate cover
- Starter

18. Install the oil cooler in the reverse order of removal keeping in mind the fol-

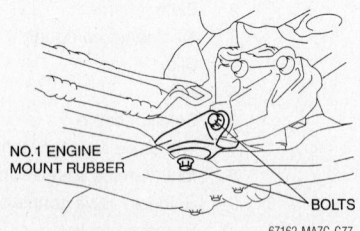

Number 1 engine mount rubber bolt locations—Mazda 3 models with an automatic transaxle

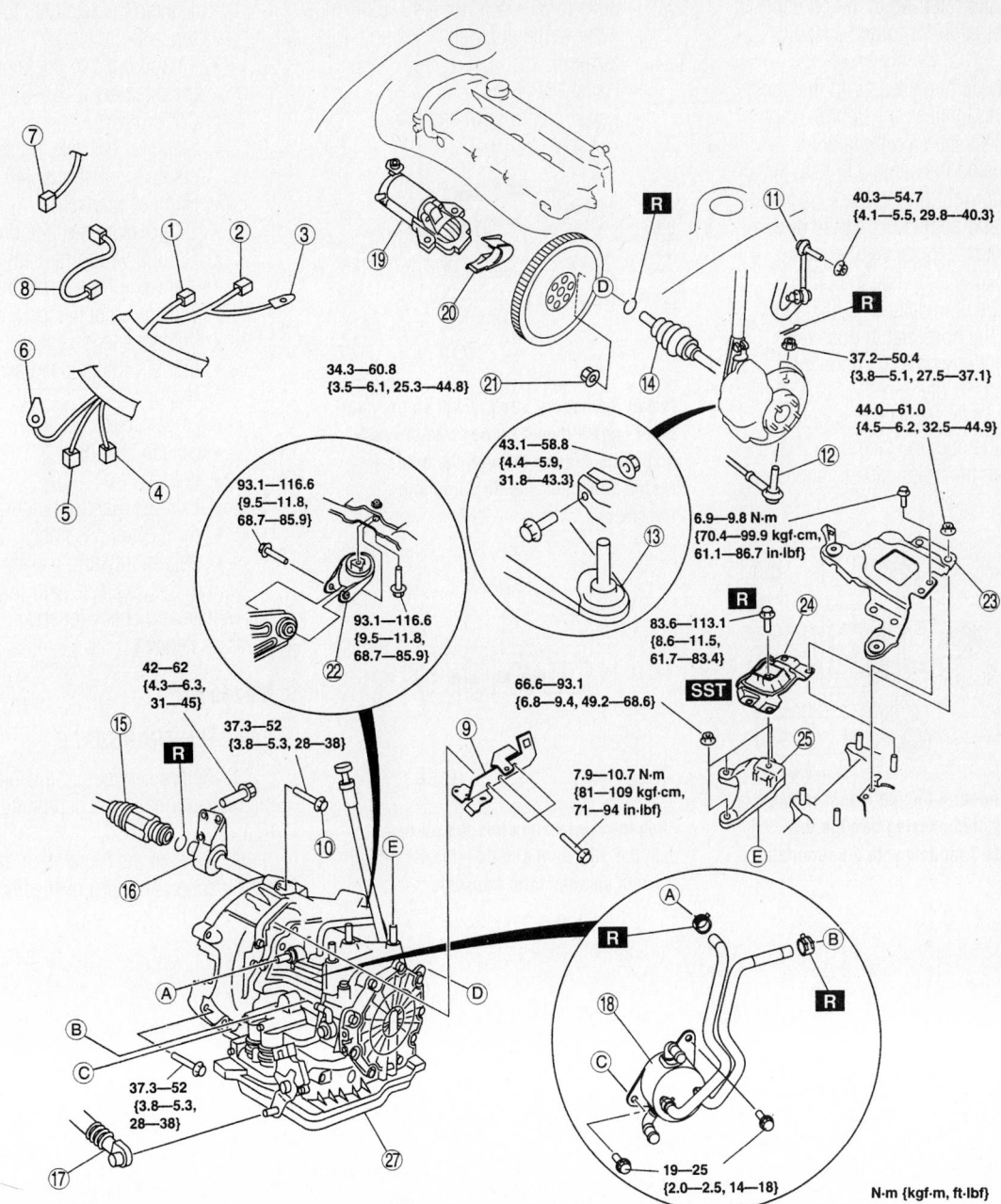

40.3—54.7
{4.1—5.5, 29.8—40.3}

37.2—50.4
{3.8—5.1, 27.5—37.1}

44.0—61.0
{4.5—6.2, 32.5—44.9}

34.3—60.8
{3.5—6.1, 25.3—44.8}

93.1—116.6
{9.5—11.8,
68.7—85.9}

93.1—116.6
{9.5—11.8,
68.7—85.9}

43.1—58.8
{4.4—5.9,
31.8—43.3}

6.9—9.8 N·m
{70.4—99.9 kgf·cm,
61.1—86.7 in·lbf}

83.6—113.1
{8.6—11.5,
61.7—83.4}

42—62
{4.3—6.3,
31—45}

37.3—52
{3.8—5.3, 28—38}

66.6—93.1
{6.8—9.4, 49.2—68.6}

7.9—10.7 N·m
{81—109 kgf·cm,
71—94 in·lbf}

37.3—52
{3.8—5.3,
28—38}

19—25
{2.0—2.5, 14—18}

N·m {kgf·m, ft·lbf}

1	Input/turbine speed sensor connector	15	Drive shaft
2	VSS connector	16	Joint shaft
3	GND wiring harness	17	Selector cable
4	Transaxle connector	18	Oil cooler
5	TR switch connector	19	Starter
6	GND wiring harness	20	End plate cover
7	Oil pressure switch connector (for oil filter)	21	Torque converter installation nuts
8	Oil pressure switch connector (for L3 ATX)	22	No.1 engine mount rubber
9	Harness bracket	23	Battery tray bracket
10	Transaxle mounting bolt (upper side)	24	No.4 engine mount rubber
11	Stabilizer control link	25	No.4 engine mount bracket
12	Tie-rod end ball joint	26	Transaxle mounting bolt (lower side)
13	Lower arm ball joint	27	Transaxle
14	Drive shaft		

67162-MAZC-G81

Exploded view of the automatic transaxle assembly mounting—Mazda 3 models

lowing steps and referring to the oil cooler exploded illustration for torque specifications:

a. Apply compressed air to the cooler side opening to clear any debris. Apply the air for less than a minute only.

b. If reusing the same oil hose, install a new clamp right on the mark left by the old clamp and apply force in the direction of the arrow to secure the clamp. Align the marks and slide the oil hose onto the pipe until seated as illustrated. Make sure the hose clamp does not interfere with any other component.

- Selector cable
- Joint shaft. Install a new circlip with the opening facing up. Hand tighten the bolts, then tighten the

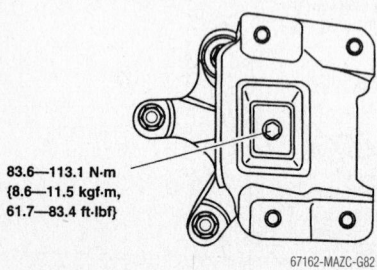

83.6—113.1 N·m
{8.6—11.5 kgf·m,
61.7—83.4 ft·lbf}

67162-MAZC-G82

Install the number 4 engine mount rubber with the body stud passing through the holes—Mazda 3 models with an automatic transaxle

bolts to 31–45 ft. lbs. (42–62 Nm). Refer to the illustration for bolt location.
- Halfshafts
- Lower control arm ball joint.

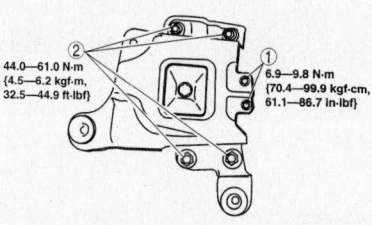

44.0—61.0 N·m
{4.5—6.2 kgf·m,
32.5—44.9 ft·lbf}

6.9—9.8 N·m
{70.4—99.9 kgf·cm,
61.1—86.7 in·lbf}

67162-MAZC-G83

Install the battery tray bracket on the number 4 engine mount rubber with the body stud bolts passing through the holes—Mazda 3 models with an automatic transaxle

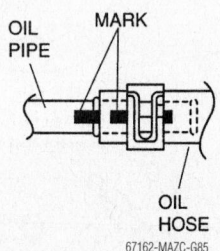

OIL PIPE   MARK

OIL HOSE

67162-MAZC-G85

Align the marks and slide the oil hose onto the pipe until seated—Mazda 3 models with an automatic transaxle

Tighten the retainers to 31–43 ft. lbs. (43–59 Nm).
- Tie rod ends to the knuckle. Tighten the retainers to 27–37 ft. lbs. (37–50 Nm).
- Stabilizer bar link. Tighten the retainers to 30–40 ft. lbs. (40–54 Nm).
- Harness bracket
- Oil pressure switch connector
- Ground wiring harness
- Transaxle range switch connector
- Transaxle connector
- Vehicle speed sensor
- Input/Turbine speed sensor connector
- Under cover
- Splash shields
- Wheels
- Exhaust manifold insulator
- Air cleaner assembly
- Battery tray and battery

19. Fill the transaxle with fluid. Road test the vehicle and check for leaks. Top off all fluids as needed.

Mazda 6

FN4A–EL TRANSAXLE

1. Before servicing the vehicle, refer to the precautions in the beginning of this section.

2. Refer to the illustration for component location and torque specifications.

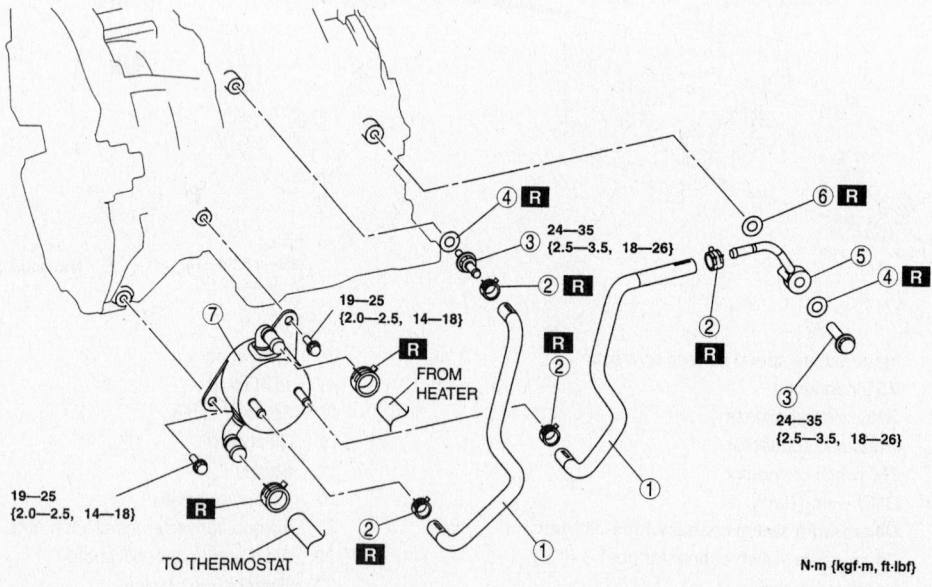

19—25
{2.0—2.5, 14—18}

24—35
{2.5—3.5, 18—26}

FROM HEATER

24—35
{2.5—3.5, 18—26}

19—25
{2.0—2.5, 14—18}

TO THERMOSTAT

N·m {kgf·m, ft·lbf}

1	Oil hose	5	Oil pipe
2	Hose clamp	6	Packing
3	Connector bolt	7	Oil cooler
4	Packing		

67162-MAZC-G84

Exploded view of the oil cooler assembly and related components—Mazda 3 models with an automatic transaxle

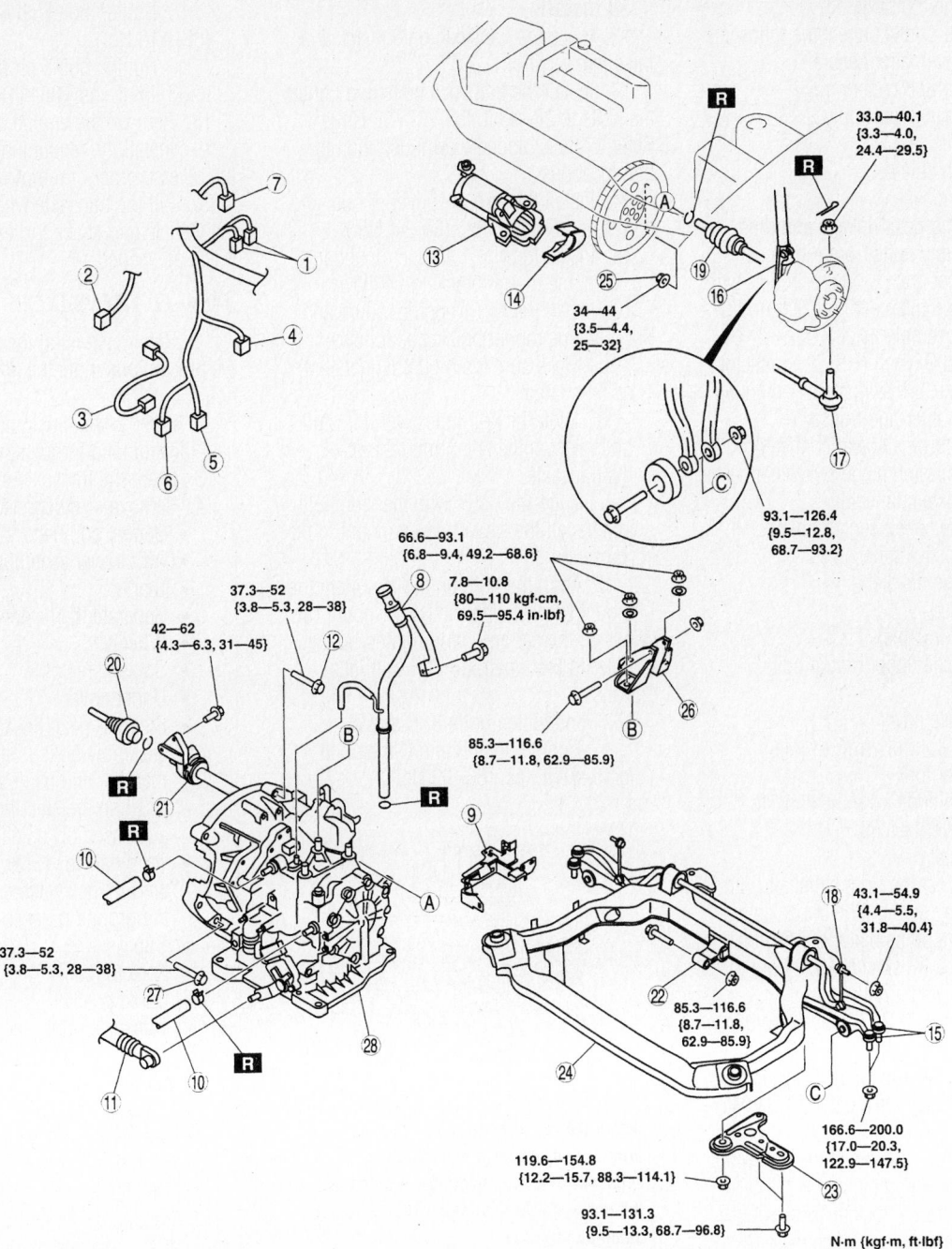

1	HO2S connector
2	Oil pressure switch connector (for oil filter)
3	Oil pressure switch connector (for ATX)
4	Input/turbine speed sensor connector
5	TR switch connector
6	Transaxle connector
7	VSS connector (Without ABS)
8	Oil dipstick and filler tube
9	Harness bracket
10	Oil hose
11	Selector cable
12	Transaxle mounting bolt (Upper side)
13	Starter
14	Endplate cover
15	Lower arm (front, rear) ball joint
16	Damper fork
17	Tie-rod end ball joint
18	Stabilizer control link
19	Drive shaft
20	Drive shaft
21	Joint shaft
22	No.1 engine mount
23	Crossmember bracket
24	Crossmember
25	Torque converter installation nuts
26	No.4 engine mount
27	Transaxle mounting bolt (lower side)
28	Transaxle

67162-MAZC-G224

Exploded view of the FN4A–EL automatic transaxle assembly mounting—Mazda 6 models

3. Drain the transaxle oil.

4. Remove or disconnect the following:

- Negative battery cable
- Battery and battery tray
- Air cleaner assembly
- Wheels
- Splash shields
- Under cover
- Steering gear, linkage and pipe assembly bolts from the cross-member and using mechanics wire, position the steering gear and linkage assembly aside.
- Heated Oxygen (HO2S) connector
- Oil pressure switch connectors for the oil filter and transaxle
- Input/turbine speed connector
- Transmission range switch connector
- Transaxle connector
- Vehicle speed sensor connector
- Oil dipstick tube
- Harness bracket
- Oil hose
- Selector cable
- Transaxle upper mount bolts
- Starter
- Endplate cover
- Lower control arm ball joints
- Damper fork
- Tie rod ends from the knuckle
- Stabilizer bar link
- Halfshafts

5. Disconnect the right halfshaft from the joint shaft by tapping the transaxle side outer ring with a brass bar and a hammer. Disconnect the joint shaft bracket from the block and remove the joint shaft.

6. Install tool 49 G030 455 to hold the side gears after removal.

7. Support the engine assembly with a engine support assembly such as 49 E017 5A0.

8. Remove or disconnect the following:

- Number 1 engine mount
- Lower front shock absorber bolt
- No. 1 engine mount center bolt

9. Support the crossmember with a jack and remove the nuts and the crossmember bracket.

- Crossmember
- Torque converter nuts, hold the crankshaft pulley to prevent the flywheel from turning while loosening the nuts and remove the nuts through the starter motor opening
- Number 4 engine mount
- Lower transaxle bolts

10. Loosen the engine support assembly and lean the engine towards the transaxle.

11. Support the transaxle with a jack, remove the transaxle bolts and the transaxle.

To install:

12. Place the transaxle onto a jack and raise into position.

13. Install the transaxle bolts and tighten the bolts to 28–38 ft. lbs. (37–52 Nm). Refer to the exploded view illustration for bolt locations.

14. Install the torque converter nuts and tighten to 25–32 ft. lbs. (34–44 Nm).

15. Install the No. 1 engine mount and No. 4 engine mount bracket as follows:

16. Install the No. 1 engine mount and No. 4 engine mount bracket as follows:

   a. Make sure the mount is installed as illustrated.

   b. Align the bolt holes with the stud bolts, install the No. 4 mount bracket to the transaxle.

   c. Align the holes with the stud bolts and install the No. 1 engine mount to the transaxle.

   d. Align the hole on the No. 4 engine mount bracket with the No. 4 mount rubber on the car and hand tighten bolt D. Refer to the illustration for bolt locations.

17. Handtighten bolts B, C and A.

   a. Tighten bolts B and C then bolt A to 49–68 ft. lbs. (66–93 Nm).

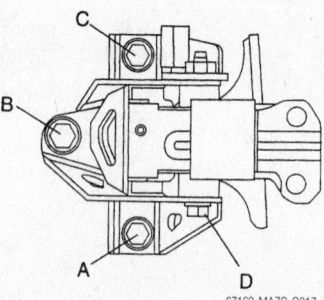

67162-MAZC-G217

Install the No. 1 mount and No. 4 mount bracket and tighten the bolts in the sequence and specifications outlined in the text—Mazda 6 models with a FN4A-EL automatic transaxle

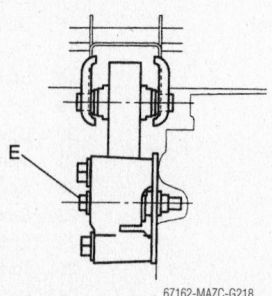

67162-MAZC-G218

Tighten bolt E on the No. 1 mount to the specification outlined in the text—Mazda 6 models with a FN4A-EL automatic transaxle

   b. Tighten bolt D to 63–86 ft. lbs. (85–116 Nm).

   c. Tighten bolt E on the No. 1 mount to 63–86 ft. lbs. (85–116 Nm).

18. Remove the engine support device.

19. Install the remaining components in the reverse order of removal.

20. Fill the transaxle with fluid. Road test the vehicle and check for leaks. Top off all fluids as needed.

JA5A–EL TRANSAXLE

1. Before servicing the vehicle, refer to the precautions in the beginning of this section.

2. Refer to the illustration for component location and torque specifications.

3. Drain the transaxle oil.

4. Remove or disconnect the following:

- Battery and battery tray
- Air cleaner assembly
- Starter
- Separate the heater pipe
- Wheels
- Splash shields
- Under cover
- Steering gear, linkage and pipe assembly bolts from the cross-member and using mechanics wire, position the steering gear and linkage assembly aside.
- Terminal No. 1, No. 2, Transmission range switch connectors and the ground harness
- Clip
- Selector cable
- Cable bracket
- Transaxle upper mount bolts
- Tranxle oil hoses (cooler lines)
- Lower control arm ball joints
- Damper fork
- Tie rod ends from the knuckle
- Stabilizer bar link
- Halfshafts

5. Disconnect the left halfshaft from the joint shaft by inserting a prybar between the transaxle and the halfshaft outer ring. Disconnect the joint shaft bracket from the block and remove the joint shaft.

6. Install tool 49 G030 455 to hold the side gears after removal

- No. 4 engine mount

7. Support the engine assembly with a engine support assembly such as 49 E017 5A0.

8. Remove or disconnect the following:

- Number 1 engine mount bracket
- Lower front shock absorber bolt
- No. 1 engine mount. Remove the intake manifold and attach tool 49 UN30 3050 to the head as shown and remove the mount.

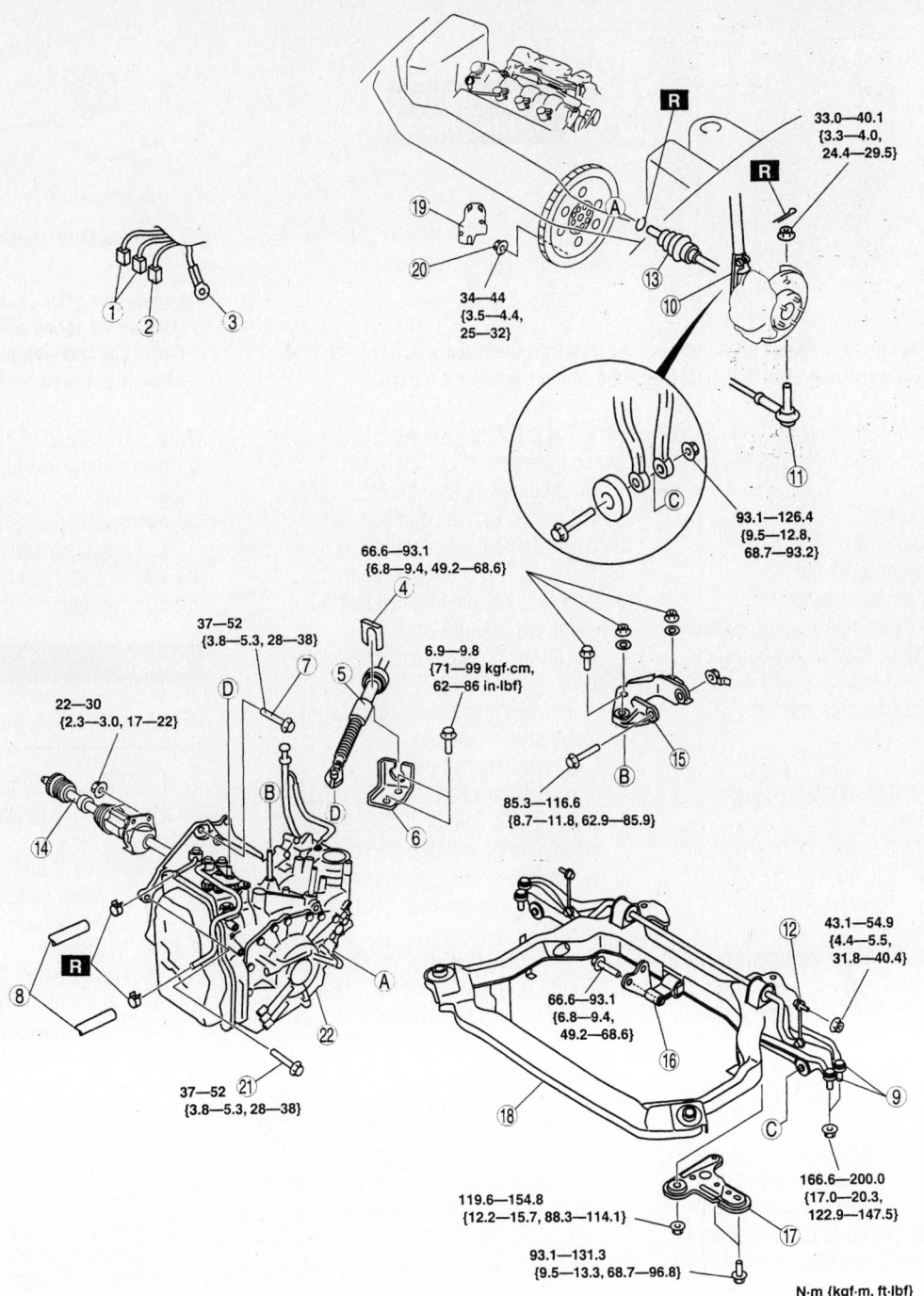

33.0—40.1
{3.3—4.0,
24.4—29.5}

34—44
{3.5—4.4,
25—32}

93.1—126.4
{9.5—12.8,
68.7—93.2}

66.6—93.1
{6.8—9.4, 49.2—68.6}

37—52
{3.8—5.3, 28—38}

6.9—9.8
{71—99 kgf·cm,
62—86 in·lbf}

22—30
{2.3—3.0, 17—22}

85.3—116.6
{8.7—11.8, 62.9—85.9}

43.1—54.9
{4.4—5.5,
31.8—40.4}

66.6—93.1
{6.8—9.4,
49.2—68.6}

37—52
{3.8—5.3, 28—38}

166.6—200.0
{17.0—20.3,
122.9—147.5}

119.6—154.8
{12.2—15.7, 88.3—114.1}

93.1—131.3
{9.5—13.3, 68.7—96.8}

N·m {kgf·m, ft·lbf}

1	Terminal component No.1, No.2 connector	12	Stabilizer control link
2	TR switch connector	13	Drive shaft
3	GND harness	14	Drive shaft, joint shaft
4	Clip	15	No.4 engine mount
5	Selector cable	16	No.1 engine mount
6	Cable bracket	17	Crossmember bracket
7	Transaxle mounting bolt (Upper side)	18	Crossmember
8	Oil hose	19	Endplate cover
9	Lower arm (front, rear) ball joint	20	Torque converter installation nuts
10	Damper fork	21	Transaxle mounting bolt (lower side)
11	Tie-rod end ball joint	22	Transaxle

67162-MAZC-G225

Exploded view of the JA5A—EL automatic transaxle assembly mounting—Mazda 6 models

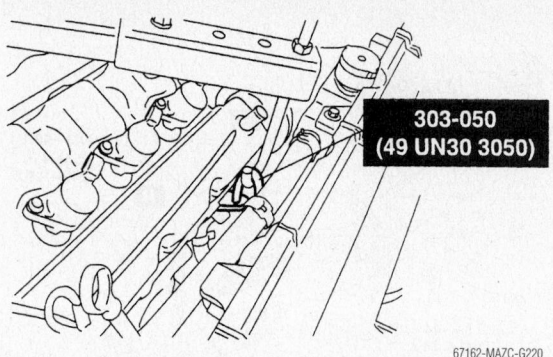

303-050
(49 UN30 3050)

67162-MAZC-G220

When removing the No. 1 engine mount, remove the intake manifold and attach tool 49 UN30 3050 to the head as shown—Mazda 6 models with the JA5A-EL automatic transaxle

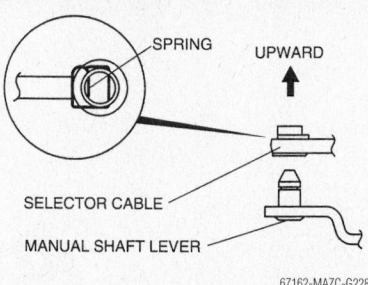

67162-MAZC-G228

Connect the selector lever to manual shaft lever so the spring side of the cable end is facing upwards—Mazda 6 models with a JA5A-EL automatic transaxle

9. Support the crossmember with a jack and remove the nuts and the crossmember bracket.
- Crossmember
- Endplate cover
- Torque converter nuts
- Lower transaxle bolts

10. Loosen the engine support assembly and lean the engine towards the transaxle.

11. Support the transaxle with a jack, remove the transaxle bolts and the transaxle.

To install:

12. Place the transaxle onto a jack and raise into position.

13. Install the transaxle bolts and tighten the bolts to 28–38 ft. lbs. (37–52 Nm). Refer to the exploded view illustration for bolt locations.

14. Align the holes for the torque converter nuts by rotating the converter and then lock the drive plate by inserting a suitable tool through the starter motor opening. Hand tighten the nuts, then final tighten them to 25–32 ft. lbs. (34–44 Nm) using 2–3 passes.

15. Install the crossmember in the reverse order of removal and refer to the exploded view illustration for component locations and torque specifications. Refer to the exploded view of the crossmember assembly in the engine removal procedure

for the Mazda 6 models equipped the 3.0L (AJ) engine earlier in this section.

16. Align the hole of the No. 1 engine mount rubber with the bolt hole of the transaxle. Hand tighten bolt A and then tighten bolts B and C to 49–68 ft. lbs. (66–93 Nm) and then tighten bolt A to 49–68 ft. lbs. (66–93 Nm).

17. Install the No. 4 engine mount as follows:
   a. Make sure the No. 4 mount is installed as illustrated.
   b. Hand tighten bolt D.
   c. Raise the transaxle with a floor jack and align the hole on the No. 4 engine mount bracket with the stud bolts on the transaxle.
   d. Hand tighten bolt A and nuts B and C.
   e. Tighten bolts B, C to 49–68 ft. lbs. (66–93 Nm) and then tighten bolt A to 49–68 ft. lbs. (66–93 Nm).
   f. Tighten bolt D to 62–86 ft. lbs. (85–116 Nm).

18. Remove the engine support device.

19. Connect the selector lever to manual shaft lever so the selector cable does not have a load on it and the spring side of the cable end is facing upwards as illustrated.

Make sure the end of the manual shaft lever sticks out of the end of the selector cable.

20. Install the remaining components in the reverse order of removal.

21. Fill the transaxle with fluid. Road test the vehicle and check for leaks. Top off all fluids as needed.

Clutch

REMOVAL & INSTALLATION

1. Before servicing the vehicle, refer to the precautions in the beginning of this section.

2. Remove or disconnect the following:
- Negative battery cable
- Clutch release cylinder
- Transaxle
- Rubber boot
- Clutch release collar
- Clutch release fork
- Pressure plate loosening the bolts one turn each in a criss–cross pattern
- Clutch disc

3. Inspect the pilot bearing. If it is worn or damaged and does not turn easily by hand, remove it using a puller/slide hammer.

4. Check the flywheel surface for scoring, cracks or burning and machine or replace, as necessary.

5. Install Holder tool 49 E011 1A0 to keep the flywheel from turning. Loosen the flywheel bolts evenly and gradually in a crisscross pattern. Remove the flywheel.

6. Inspect the clutch release bearing for wear. Replace it if it sticks or does not turn easily.

7. Inspect the release fork for wear or damage and replace as necessary.

To install:

8. Lubricate the release fork fingers and pivot with molybdenum grease and install in the release fork boot.

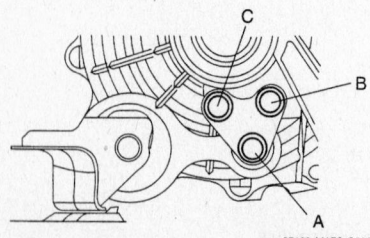

67162-MAZC-G226

Tighten the No. 1 mount in the sequence and specifications outlined in the text—Mazda 6 models with a JA5A-EL automatic transaxle

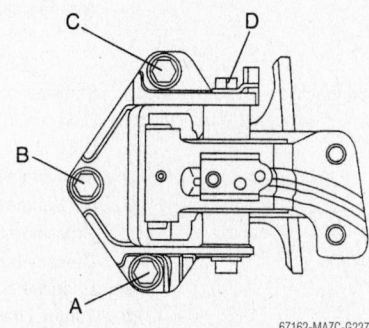

67162-MAZC-G227

Location of the No. 4 mount bolts, tighten in the sequence and specifications outlined in the text—Mazda 6 models with a JA5A-EL automatic transaxle

9. Install or connect the following:
- Clutch release bearing on the release fork
- New pilot bearing in the flywheel, if removed, using an installation tool

10. Be sure the flywheel mounting surface and the crankshaft or eccentric shaft mounting surfaces are clean. Remove any old sealant from the flywheel bolt hole threads and the flywheel bolts.
- Flywheel
- Sealant to the flywheel bolt threads and install them hand tight
- Flywheel holding tool. Tighten the bolts, in a crisscross pattern as follows:

a. Except Mazda 3, 6 and 2.5L (KL) engines to 71–75 ft. lbs. (97–102 Nm).

b. On the 2.5L (KL) engines, tighten to 45–49 ft. lbs. (61–66 Nm).

c. Mazda 3 models, tighten to 79–85 ft. lbs. (108–116 Nm).

d. Mazda 6 models equipped with the G35M-R transaxle, tighten to 79–84 ft. lbs. (108–115 Nm).

e. Mazda 6 models equipped with the A65M-R transaxle, tighten to 54–64 ft. lbs. (73–87 Nm).

11. Install or connect the following:
- Small amount of molybdenum grease to the clutch disc splines
- Clutch disc on the flywheel with the spring side toward the transaxle
- An alignment tool in the pilot bearing to position the clutch disc
- Clutch pressure plate by aligning the dowel holes with the flywheel dowels
- Pressure plate. Gradually, torque the bolts, in a crisscross pattern, to 19 ft. lbs. (26 Nm).

12. Remove the alignment tool.
- Clutch release fork
- Clutch release collar
- Rubber boot
- Transaxle
- Clutch release cylinder
- Negative battery cable

Hydraulic Clutch System

BLEEDING

1. Before servicing the vehicle, refer to the precautions in the beginning of this section.

2. Remove the rubber cap from the bleeder screw on the release cylinder.

3. Place a bleeder tube over the end of the bleeder screw.

4. Submerge the other end of the tube in a jar half filled with hydraulic brake fluid.

5. Slowly pump the clutch pedal fully and allow it to return slowly, several times.

6. While pressing the clutch pedal to the floor, loosen the bleeder screw until the fluid starts to run out. Then, close the bleeder screw. Keep repeating this Step, while watching the hydraulic fluid in the jar. As soon as the air bubbles disappear, close the bleeder screw.

7. During the bleeding procedure the reservoir must be kept at least ¾ full.

Halfshafts

REMOVAL & INSTALLATION

Except Mazda 3 and Mazda 6 Models

1. Before servicing the vehicle, refer to the precautions in the beginning of this section.

2. Drain the transaxle oil.

3. Remove or disconnect the following:
- Wheels
- ABS sensor, if equipped
- Splash shield, if equipped

4. Raise the staked portion of the hub locknut with a hammer and chisel.

5. Lock the hub by applying the brakes and remove the nut.

6. Remove or disconnect the following:
- Stabilizer bar from the lower control arm
- Cotter pin and nut from the tie rod end ball stud
- Tie rod end from the knuckle
- Transverse member, on the 626
- Lower ball joint pinch bolt and nut
- Lower ball joint from the knuckle

7. If removing the left side shaft on 626 with automatic transaxle, proceed as follows:

a. Suspend the engine using engine support tool 49 G017 5A0.

b. Remove the engine mount member.

8. Position a prybar between the inner CV-joint and transaxle case. Carefully pry the halfshaft from the transaxle being careful not to damage the oil seal. If equipped with a right side intermediate shaft, insert the prybar between the halfshaft and intermediate shaft and tap on the bar to uncouple them.

9. Pull outward on the hub/knuckle assembly, push the outer CV-joint stub shaft through the hub and remove the halfshaft. If the halfshaft is stuck in the hub, install the old hub nut to protect the stub shaft threads. Tap on the nut, using only a soft mallet, to remove the halfshaft.

➡ **Install plug tool 49 G030 455 into the transaxle after removing the halfshaft, to keep the differential side gear in position. If the gear becomes positioned incorrectly, the differential may have to be removed to realign the gear.**

10. Remove the intermediate shaft, if necessary, by removing the support bearing bolts and pulling the shaft from the transaxle.

To install:

11. Install or connect the following:
- New circlip on the end of the intermediate shaft, if removed, with the end gap facing upward.
- Intermediate shaft in the transaxle, being careful not to damage the oil seals
- Intermediate shaft support bearing bolts. Torque in sequence, to 45 ft. lbs. (61 Nm).
- New circlip on the end of the halfshaft, with the end gap facing upward
- Halfshaft into the transaxle, being careful not to damage the oil seal

➡ **If equipped, push the halfshaft into the intermediate shaft.**

- Other end of the halfshaft through the hub. Loosely install a new locknut

12. If installing the left side shaft on 626 with automatic transaxle, proceed as follows:

a. Install the engine mount member. Torque the mount member-to-body bolts to 66 ft. lbs. (89 Nm).

b. Torque the front mount-to-mount member nuts to 77 ft. lbs. (104 Nm) and the side mount bolts to 44 ft. lbs. (60 Nm).

c. Remove the engine support tool.

13. Install or connect the following:
- Lower ball joint into the knuckle. Torque the pinch bolt to 40 ft. lbs. (54 Nm).
- Transverse member, on 626. Torque the bolts to 96 ft. lbs. (132 Nm).
- Tie rod end to the steering knuckle. Torque the nut to 42 ft. lbs. (57 Nm) on all except 626 and Millenia or to 32 ft. lbs. (44 Nm) for 626 and Millenia.
- New cotter pin. Tighten the nut, if necessary, to align the ball stud hole with the nut castellation.
- Stabilizer bar to the lower control arm
- Splash shield
- Wheels

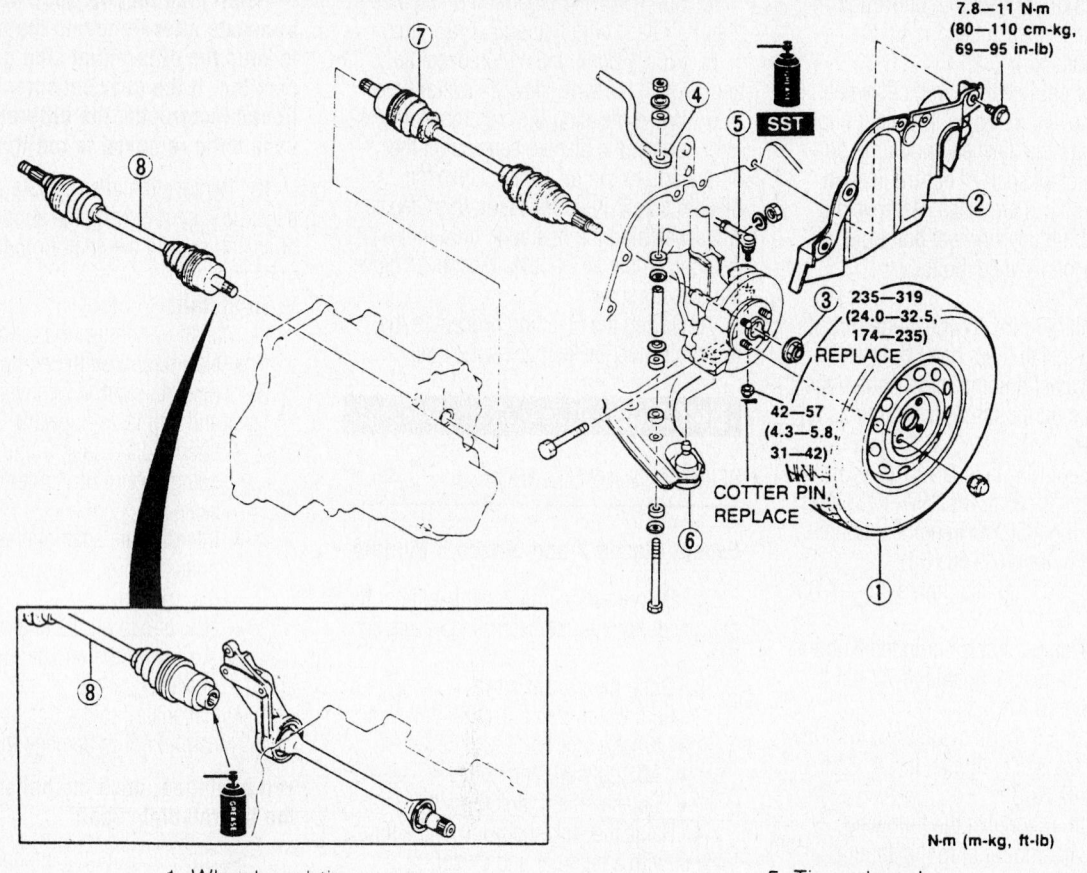

1. Wheel and tire
2. Splash shield
3. Locknut
4. Stabilizer
5. Tie-rod end
6. Lower ball joint
7. Left driveshaft
8. Right driveshaft

7923MG76

Exploded view of a typical halfshaft mounting—Except Mazda 3

- New hub nut. Torque it to 174–235 ft. lbs. (235–318 Nm). After tightening, stake the locknut using a hammer and dull bladed chisel.
14. Fill the transaxle.

Mazda 3 Models

1. Before servicing the vehicle, refer to the precautions in the beginning of this section.
2. Drain the transaxle oil.
3. Remove or disconnect the following:
   - Wheels
   - ABS sensor, if equipped
   - Splash shield, if equipped
   - Halfshaft lockbolt
4. Install a spare bolt onto the halfshaft, tap the bolt with a copper hammer and separate the halfshaft from the knuckle.
5. Remove or disconnect the following:
   - Stabilizer bar from the lower control arm
   - Cotter pin and nut from the tie rod end ball stud

   - Tie rod end from the knuckle
   - Lower ball joint pinch bolt and nut
   - Lower ball joint from the knuckle
6. Position a prybar between the inner CV-joint and transaxle case. Carefully pry the halfshaft from the transaxle being careful not to damage the oil seal.
7. Pull outward on the hub/knuckle assembly, push the outer CV-joint stub shaft through the hub and remove the halfshaft.

➡ **Install plug tool 49 G030 455 into the transaxle after removing the halfshaft, to keep the differential side gear in position. If the gear becomes positioned incorrectly, the differential may have to be removed to realign the gear.**

To install:

8. Install or connect the following:
   - New circlip on the end of the halfshaft, with the end gap facing upward
   - Halfshaft into the transaxle, being careful not to damage the oil seal

➡ **If equipped, push the halfshaft into the intermediate shaft.**

   - Other end of the halfshaft through the hub. Loosely install a new lockbolt
   - Lower ball joint into the knuckle. Torque the pinch bolt to 31–43 ft. lbs. (43–58 Nm).
   - Stabilizer bar to the lower control arm. Tighten to 29–40 ft. lbs. (40–54 Nm).
   - Tie rod end to the steering knuckle. Torque the nut to 27–37 ft. lbs. (37–50 Nm.)
   - Tighten the nut, if necessary, to align the ball stud hole with the nut castellation.
   - Halfhaft lockbolt. Torque it to 23–28 ft. lbs. (31–38 Nm), then tighten an additional 85–95 degrees.
   - ABS sensor
   - Splash shield
   - Wheels
9. Fill the transaxle.

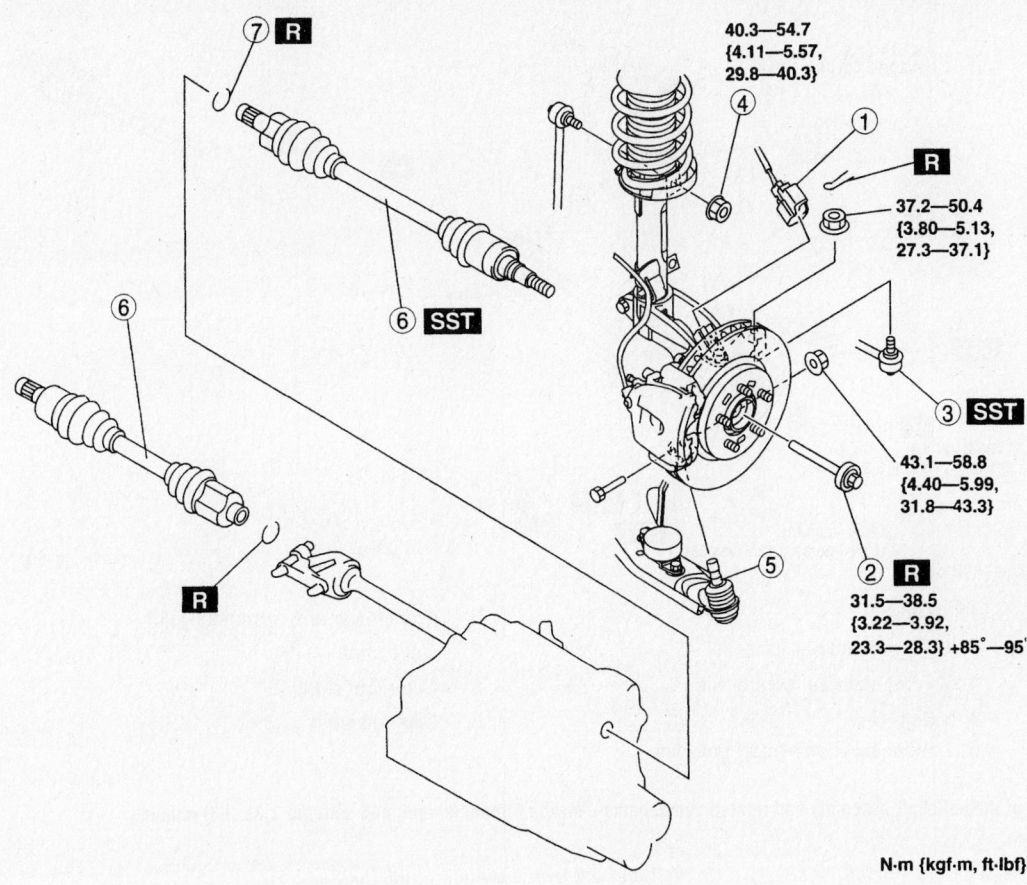

40.3—54.7
{4.11—5.57,
29.8—40.3}

37.2—50.4
{3.80—5.13,
27.3—37.1}

43.1—58.8
{4.40—5.99,
31.8—43.3}

31.5—38.5
{3.22—3.92,
23.3—28.3} +85°—95°

N·m {kgf·m, ft·lbf}

1	ABS wheel-speed sensor connector	4	Stabilizer control link upper nut
2	Lockbolt	5	Front lower arm ball joint
3	Tie-rod end ball joint	6	Drive shaft
		7	Clip

67162-MAZC-G86

Exploded view of a typical halfshaft mounting—Mazda 3

Mazda 6 Models

2.3L (L3) ENGINES

1. Before servicing the vehicle, refer to the precautions in the beginning of this section.

2. Drain the transaxle oil.

3. Raise the staked portion of the hub locknut with a hammer and chisel.

4. Lock the hub by applying the brakes and remove the nut.

5. Remove or disconnect the following:
   • Wheels
   • Tie rod end from the knuckle
   • Stabilizer bar link from the damper fork
   • Bolt attaching the fork to the knuckle
   • Front and rear lower arm ball joints

6. Install a spare nut onto the drive shaft so the nut is flush with the end of the drive shaft and tap the nut with a copper hammer to separate the halfshaft from the hub assembly.

➡The halfshaft edges can be quite sharp and can damage the oil seal. be careful when removing the halfshaft not to cause any damage.

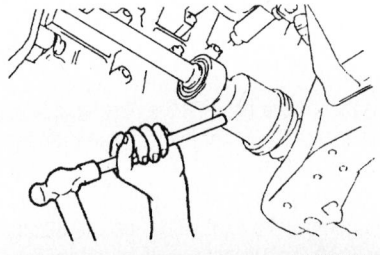

67162-MAZC-G230

Remove the right side halfshaft by tapping on a bar positioned between the halfshaft and joint shaft—Mazda 6 models equipped with the 2.3L (L3) engine

7. If removing the left side halfshaft, insert a prybar between the outer ring and the transaxle and pry the halfshaft from the transaxle.

8. If removing the right side halfshaft, separate it from the joint shaft by tapping on a bar positioned between the halfshaft and joint shaft.

➡Install plug tool 49 G030 455 into the transaxle after removing the halfshaft, to keep the differential side gear in position. If the gear becomes positioned incorrectly, the differential may have to be removed to realign the gear.

To install:

9. Install a new circlip on the end of the shaft, if removed, with the end gap facing upward. Measure the outer diameter of the clip, if it exceeds 1.19–1.23 inch (30–31mm) replace the clip.

10. Install the left side shaft as follows:

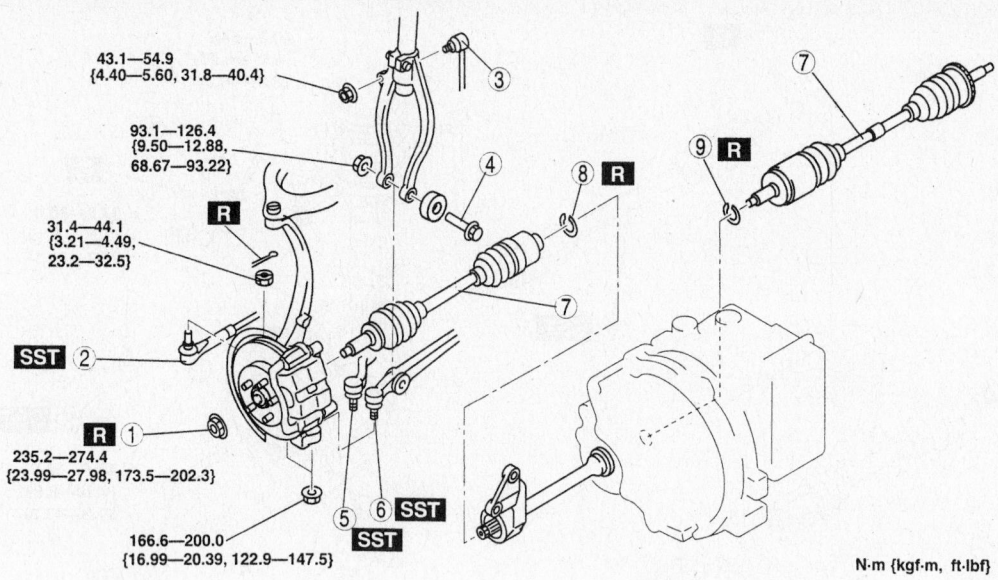

43.1—54.9
{4.40—5.60, 31.8—40.4}

93.1—126.4
{9.50—12.88,
68.67—93.22}

31.4—44.1
{3.21—4.49,
23.2—32.5}

235.2—274.4
{23.99—27.98, 173.5—202.3}

166.6—200.0
{16.99—20.39, 122.9—147.5}

N·m {kgf·m, ft·lbf}

1 Locknut	6 Front lower arm (front) ball joint
2 Tie-rod end ball joint	7 Drive shaft
3 Front stabilizer control link	8 Clip (right side)
4 Bolt	9 Clip (left side)
5 Front lower arm (rear) ball joint	

67162-MAZC-G229

Exploded view of the halfshaft assembly and related components—Mazda 6 models equipped with the 2.3L (L3) engine

   a. Insert the shaft into the hub.
   b. Apply clean engine oil to the oil seal lip.
   c. Push the shaft into the transaxle and once the shaft clicks into place, pull on the outer ring to make sure the clip is engaged.
11. Install the right side shaft as follows:
   a. Insert the shaft into the hub.
   b. Insert the halfshaft into the joint shaft and once the shaft clicks into place, pull on the outer ring to make sure the clip is engaged.
12. Install the remaining components in the reverse order of removal. Refer to the exploded view of the halfshaft assembly and related components for torque specifications.
13. Install a new hub nut. Torque it to 173–202 ft. lbs. (235–274 Nm). After tightening, stake the locknut using a hammer and dull bladed chisel with an indent at least 0.02 inch (0.5mm).
14. Fill the transaxle.

3.0L (AJ) ENGINES
1. Before servicing the vehicle, refer to the precautions in the beginning of this section.
2. Drain the transaxle oil.
3. Remove the knuckle and hub assembly.

4. Raise the staked portion of the hub locknut with a hammer and chisel.
5. Lock the hub by applying the brakes and remove the nut.
6. Remove or disconnect the following:
 • Wheels
 • Tie rod end from the knuckle
 • Damper fork
 • Front and rear lower arm ball joints
 • Stabilizer bar link from the damper fork
7. Install a spare nut onto the drive shaft so the nut is flush with the end of the drive shaft and tap the nut with a copper hammer to separate the halfshaft from the hub assembly.

➡The halfshaft edges can be quite sharp and can damage the oil seal. be careful when removing the halfshaft not to cause any damage.

8. If removing the left side halfshaft, insert a prybar between the outer ring and the transaxle and pry the halfshaft from the transaxle.
9. Remove the joint shaft bracket bolt and remove the right side halfshaft. Secure the shaft in a vise and insert a prybar between the halfshaft and joint shaft and separate them.

➡Install plug tool 49 G030 455 into the transaxle after removing the halfshaft,

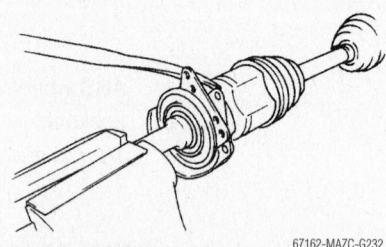

67162-MAZC-G232

Remove the right side halfshaft by securing the shaft in a vise, inserting a prybar between the halfshaft and joint shaft and separating them—Mazda 6 models equipped with the 3.0L (AJ) engine

to keep the differential side gear in position. If the gear becomes positioned incorrectly, the differential may have to be removed to realign the gear.

To install:
10. Install a new circlip on the end of the shaft, if removed, with the end gap facing upward. Measure the outer diameter of the clip, if it exceeds 1.26–1.30 inch (32–33mm) replace the clip.
11. Install the left side shaft as follows:
   a. Insert the shaft into the hub.
   b. Apply clean engine oil to the oil seal lip.
   c. Push the shaft into the transaxle and once the shaft clicks into place, pull

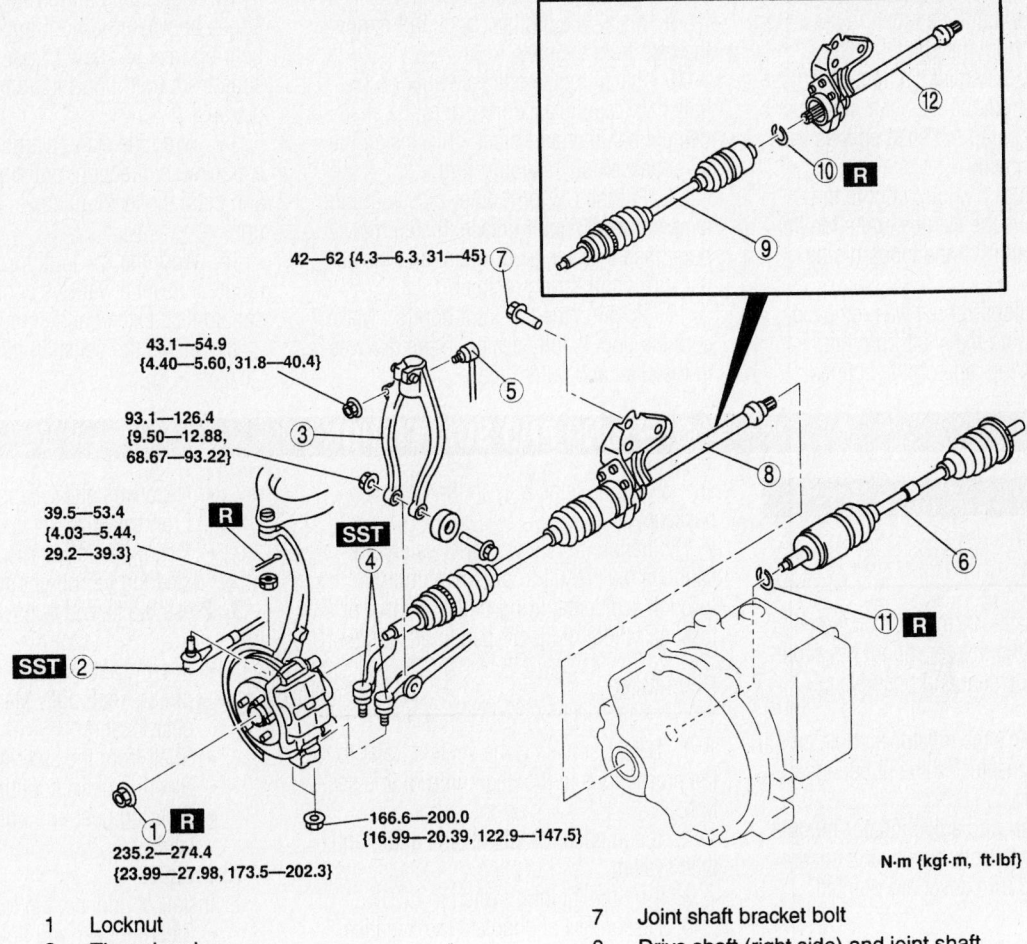

42—62 {4.3—6.3, 31—45}

43.1—54.9
{4.40—5.60, 31.8—40.4}

93.1—126.4
{9.50—12.88,
68.67—93.22}

39.5—53.4
{4.03—5.44,
29.2—39.3}

166.6—200.0
{16.99—20.39, 122.9—147.5}

235.2—274.4
{23.99—27.98, 173.5—202.3}

N·m {kgf·m, ft·lbf}

1	Locknut	7	Joint shaft bracket bolt
2	Tie-rod end	8	Drive shaft (right side) and joint shaft
3	Damper fork	9	Drive shaft (right side)
4	Lower arm (front, rear) ball joint	10	Clip (right side)
5	Stabilizer control link	11	Clip (left side)
6	Drive shaft (left side)	12	Joint shaft

67162-MAZC-G231

Exploded view of the halfshaft assembly and related components—Mazda 6 models equipped with the 3.0L (AJ) engine

on the outer ring to make sure the clip is engaged.

12. Install the right side shaft as follows:

a. Place the halfshaft and joint shaft assembly in a vise.

b. Insert the halfshaft into the joint shaft and use a plastic hammer to attach the joint shaft as illustrated.

c. Tighten the joint shaft bracket bolt to 31–45 ft. lbs. (42–62 Nm).

13. Install the remaining components in the reverse order of removal. Refer to the exploded view of the halfshaft assembly and related components for torque specifications.

- New hub nut. Torque it to 173–202 ft. lbs. (235–274 Nm). After tightening, stake the locknut using a hammer and dull bladed chisel with an indent at least 0.02 inch (0.5mm).

14. Fill the transaxle.

CV-Joints

OVERHAUL

1. Before servicing the vehicle, refer to the precautions in the beginning of this section.

Two types of CV-joints are used. The inboard CV-joints are the tri-Pot type. All outboard CV-joints are Birfield type. The Birfield CV-joint cannot be disassembled; if an outboard CV-joint boot needs replacement, the inboard CV-joint must be removed. If the outboard CV-joint needs to be replaced, replace the entire halfshaft as an assembly.

2. Remove the halfshaft from the vehicle and clamp it in a vise equipped with jaw caps, to prevent damage to the machined surfaces. Do not allow the vise to contact the boot or its clamps.

3. Remove the large boot clamp from the inboard CV-joint, using side cutters. After removing the clamp, roll the boot back over the shaft.

➡**Check the grease for contamination by rubbing it between 2 fingers. Any gritty feeling indicates a contaminated CV-joint, in which case the entire CV-joint must be disassembled, cleaned and inspected. If the grease is not contaminated and the CV-joint has been operating satisfactorily, continue with the boot replacement procedure and add the required lubricant.**

4. Paint alignment marks on the outer race and shaft for assembly reference. Remove the wire ring bearing retainer and remove the outer race.

5. Paint alignment marks on the tri-pot bearing and shaft for assembly reference.

Remove the tri-pot bearing snapring and, using a brass drift and hammer, remove the tri-pot bearing from the shaft.

6. Remove the small clamp and remove the inner boot from the halfshaft. If the boot is to be reused, wrap the shaft splines with tape before removing.

7. If the outer CV-joint boot is to be replaced, remove the clamps and slide the boot off the shaft from the inboard side.

To install:

8. If the outboard boot was removed, slide the boot onto the shaft from the inboard side. Wrap tape on the splines

before installing to protect the boot.

9. Install the inboard boot and remove the tape from the shaft.

10. Install the tri-pot assembly on the halfshaft. Tap the assembly onto the shaft using a hammer and brass drift. Install the tri-pot assembly retaining ring.

11. Fill the CV-joint outer race with high temperature CV-joint grease. Install the outer race over the tri-pot joint and install the wire ring bearing retainer.

12. Position the CV-joint boot(s). Make sure the boot is fully seated in the grooves in the shaft and outer race.

13. Insert a small prybar with rounded edges between the boot and the outer bearing race to allow trapped air to escape from the boot. Install new boot clamps.

14. Wrap the clamps around the boots in a clockwise direction, pull tight with pliers and bend the locking tabs to secure in position.

15. Work the CV-joint through its full range of travel at various angles. The joint should flex, extend and compress smoothly.

16. Install the halfshaft into the vehicle.

STEERING AND SUSPENSION

Air Bag

PRECAUTIONS

Several precautions must be observed when handling the inflator module to avoid accidental deployment and possible personal injury.

1. Never carry the inflator module by the wires or connector on the underside of the module.

2. When carrying a live inflator module, hold securely with both hands, and ensure that the bag and trim cover are pointed away.

3. Place the inflator module on a bench or other surface with the bag and trim cover facing up.

4. With the inflator module on the bench, never place anything on or close to the module which may be thrown in the event of an accidental deployment.

5. An air bag is an explosive device. Handle with extreme caution.

6. Always disconnect the battery and the air bag connector before removing the steering wheel or beginning work on the air bag system.

7. Air bag components must not be repaired or opened. Always use new parts, including the wiring harness.

8. Always place a removed air bag unit with the horn pad facing up. Put it in a safe place where it will not be disturbed.

9. The air bag unit must not be exposed to grease, fluids, or cleaning agents.

10. The air bag unit must not be exposed to temperatures above 194°F (90°C) at any time. Even the heat of a soldering iron can damage or ignite the charge.

11. Storage and transport of air bags is subject to rules governing explosive devices

and should be done only in the original package.

12. Failure to follow proper safety precautions may result in personal injury through accidental firing of the air bag, or through failure of the air bag in an accident.

DISARMING

1. Before servicing the vehicle, refer to the precautions in the beginning of this section.

2. If equipped, deactivate the audio anti-theft system.

3. Turn the ignition switch to LOCK.

4. Disconnect and isolate the negative battery cable and wait for more than 1 minute to allow the backup power supply to deplete its stored power.

ARMING

1. Before servicing the vehicle, refer to the precautions in the beginning of this section.

2. Connect the negative battery cable, turn the ignition switch **ON** and verify the air bag warning light cones on for 6 seconds. If the light does not illuminate there are problems with the system.

3. If equipped, activate the audio anti-theft system.

Rack and Pinion Steering Gear

REMOVAL & INSTALLATION

Manual

MIATA

1. Before servicing the vehicle, refer to the precautions in the beginning of this section.

2. Remove or disconnect the following:

- Negative battery cable
- Front wheels
- Cotter pins and nuts from both steering tie rod ends

3. Press the tie rod out from the knuckle arm.

- Intermediate shaft to steering gear pinion shaft bolt. Mark the shaft-to-gear location.
- Shaft from the steering gear
- Steering gear mounting nuts
- Steering gear and linkage from the vehicle

To install:

4. Install or connect the following:

- Steering gear and linkage to the vehicle. Torque the bolts in sequence to 55–77 ft. lbs. (75–104 Nm).
- Steering shaft to the steering gear pinion shaft, align the marks made during removal and tighten the bolt to 19 ft. lbs. (26 Nm)
- Tie rod ends to the knuckle arm. Torque the nuts to 32–41 ft. lbs. (43–56 Nm).
- New cotter pins
- Wheels
- Negative battery cable

5. Check and/or adjust the front end alignment.

Power

MIATA

1. Before servicing the vehicle, refer to the precautions in the beginning of this section.

2. Remove or disconnect the following:

- Negative battery cable
- Front wheels
- Cotter pins and nuts from both steering tie rod ends

3. Press the tie rod out from the knuckle arm.

- Intermediate shaft to steering gear pinion shaft bolt. Mark the shaft-to-gear location.
- Shaft from the steering gear
- Pressure and return lines
- Steering gear mounting nuts
- Steering gear and linkage from the vehicle

To install:

4. Install or connect the following:
- Steering gear and linkage to the vehicle. Torque the bolts in sequence to 55–77 ft. lbs. (75–104 Nm).
- Steering shaft to the steering gear

pinion shaft, align the marks made during removal and tighten the bolt to 19 ft. lbs. (26 Nm).
- Pressure and return lines. Tighten the pressure line fitting to 34 ft. lbs. (47 Nm).
- Tie rod ends to the knuckle arm. Torque the nuts to 32–41 ft. lbs. (43–56 Nm).
- New cotter pins
- Wheels
- Negative battery cable

5. Check and/or adjust the front end alignment.

626

1. Before servicing the vehicle, refer to the precautions in the beginning of this section.

2. Remove or disconnect the following:
- Negative battery cable
- Front wheels
- Cotter pins and nuts from both steering tie rod ends

3. Press the tie rod out from the knuckle arm.

4. Remove or disconnect the following:
- Transverse member

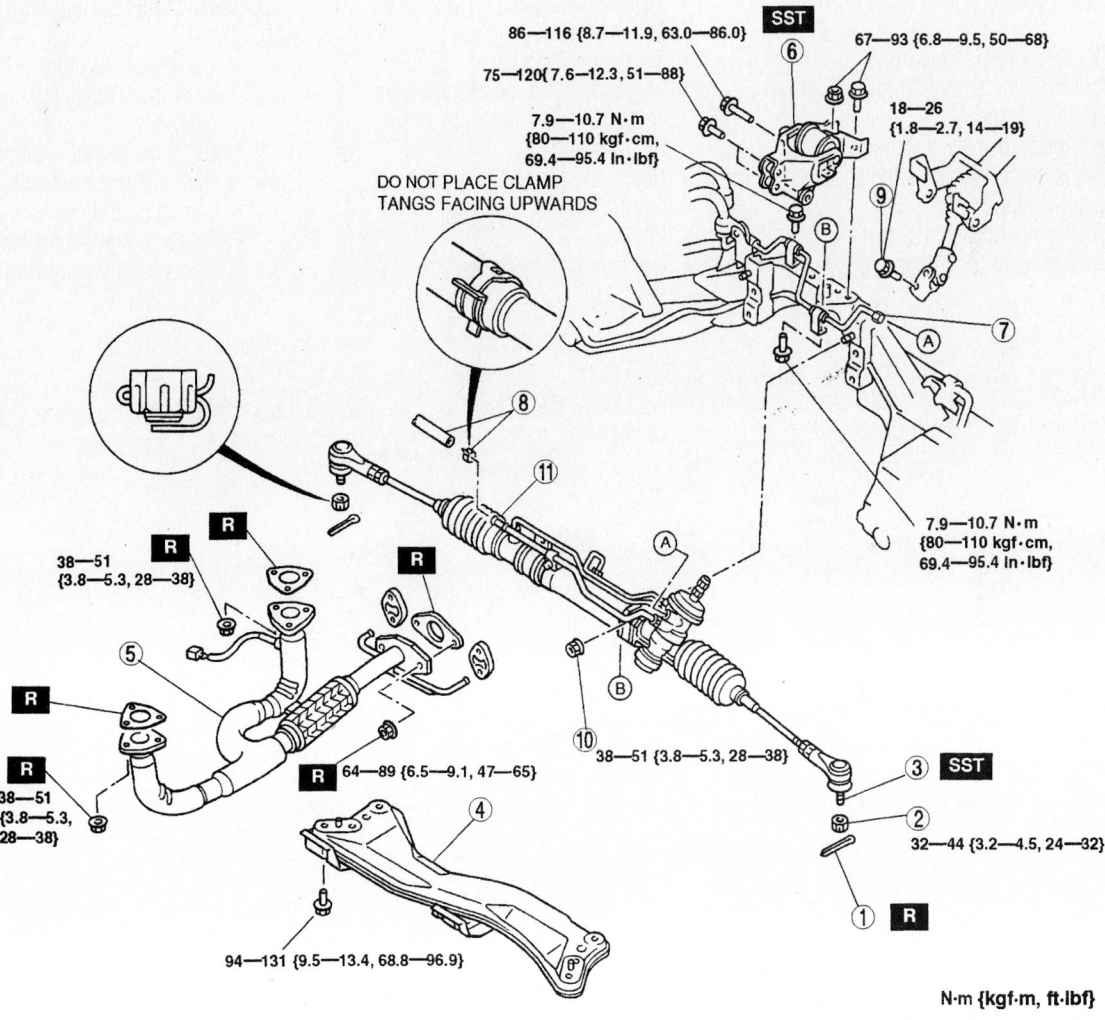

N·m {kgf·m, ft·lbf}

1	Cotter pin	7	Pressure pipe
2	Nut	8	Return hose and clamp
3	Tie-rod end ball joint	9	Bolt (intermediate shaft)
4	Transverse member	10	Mounting bracket nut and bolt
5	Front exhaust pipe (KL engine)	11	Steering gear and linkage
6	No.1 engine mount component		

Exploded view of the power steering gear assembly—626

42356-MAZC-G81

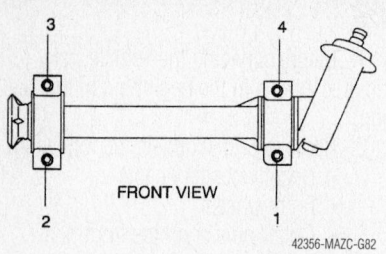

Tighten the power steering gear assembly bracket retainers in this sequence—626

- Front exhaust pipe on 2.5L (KL) engines

5. Support the engine assembly with a engine support assembly such as 49 E017 5A0.

- Number 1 engine mount
- Pressure line and return pipe from the steering gear
- Intermediate shaft to steering gear pinion shaft bolt. Mark the shaft-to-gear location.
- Shaft from the steering gear
- Steering gear mounting nuts

- Steering gear from the right of the vehicle

To install:

6. Install or connect the following:
- Steering gear to the vehicle. Torque the nuts/bolts in sequence to 28–38 ft. lbs. (37–52 Nm).
- Steering shaft to the steering gear pinion shaft, align the marks made during removal and tighten the bolt to 19 ft. lbs. (26 Nm)
- Pressure line and return hose to the steering gear
- Number 1 engine mount and remove the engine support
- Front exhaust pipe on 2.5L (KL) engines
- Transverse member
- Tie rod ends to the knuckle arm. Torque the nuts to 24–32 ft. lbs. (32–44 Nm).
- New cotter pins and check the power steering fluid level
- Wheels
- Negative battery cable

7. Check and/or adjust the front end alignment.

PROTÉGÉ

1. Before servicing the vehicle, refer to the precautions in the beginning of this section.

2. Remove or disconnect the following:
- Negative battery cable
- Front wheels
- Intermediate shaft to steering gear pinion shaft bolt. Mark the shaft-to-gear location.
- Cotter pins and nuts from both steering tie rod ends

3. Press the tie rod out from the knuckle arm.

- Transverse member, if necessary on some models to facilitate removal

4. Support the engine assembly with a suitable engine support assembly.

- Engine mount member
- Pressure line, return pipe and clamp from the steering gear

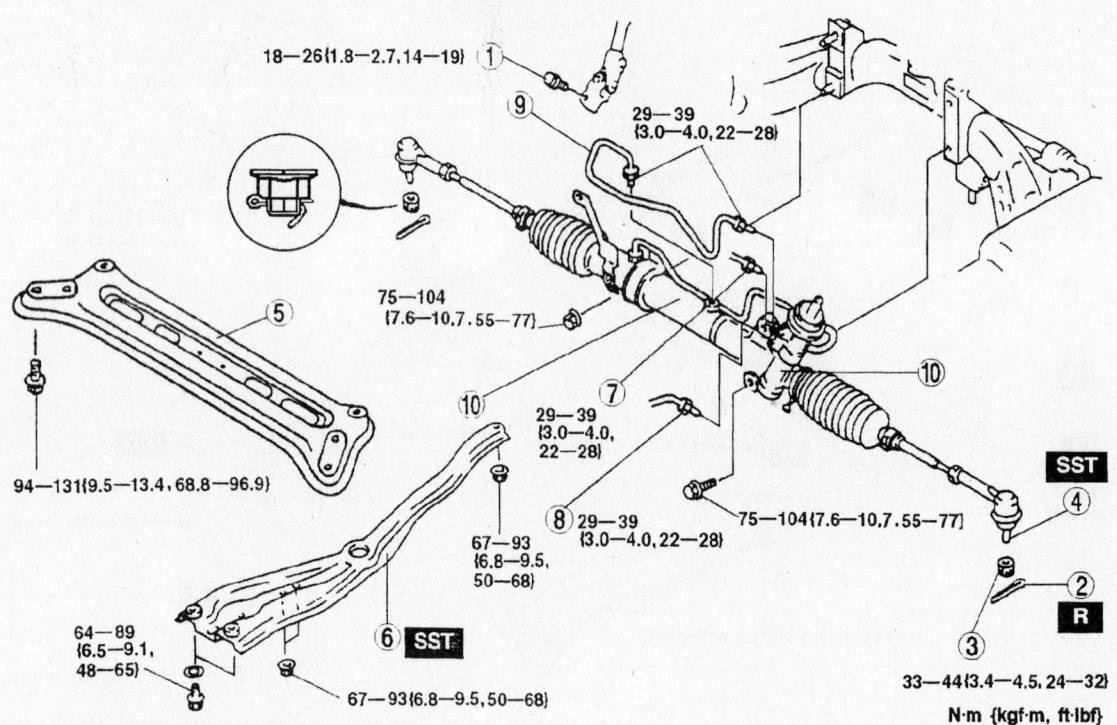

1	Bolt (intermediate shaft)
2	Cotter pin
3	Nut
4	Tie-rod end ball joint
5	Transverse member (ZM (ATX), FS)
6	Engine mount member
7	Pressure pipe
8	Return pipe and clamp
9	Oil pipe
10	Steering gear and linkage

Exploded view of the power steering gear assembly—Protégé

- Oil pipe
- Shaft from the steering gear
- Steering gear mounting nuts
- Steering gear from the right of the vehicle

To install:

5. Install or connect the following:
   - Steering gear to the vehicle. Torque the nuts/bolts 55–77 ft. lbs. (75–104 Nm).
   - Steering shaft to the steering gear pinion shaft, align the marks made during removal
   - Oil pipe

- Pressure line, return pipe and clamp to the steering gear
- Engine mount member and remove the engine support
- Transverse member, if removed and tighten the bolts to 68–97 ft. lbs. (94–131 Nm).
- Tie rod ends to the knuckle arm. Torque the nuts to 24–32 ft. lbs. (32–44 Nm).
- New cotter pins
- Tighten the intermediate shaft bolt to 19 ft. lbs. (26 Nm). Check the power steering fluid level.

- Wheels
- Negative battery cable

6. Check and/or adjust the front end alignment.

MILLENIA

1. Before servicing the vehicle, refer to the precautions in the beginning of this section.

2. Remove or disconnect the following:
   - Negative battery cable
   - Front wheels
   - Cotter pins and nuts from both steering tie rod ends

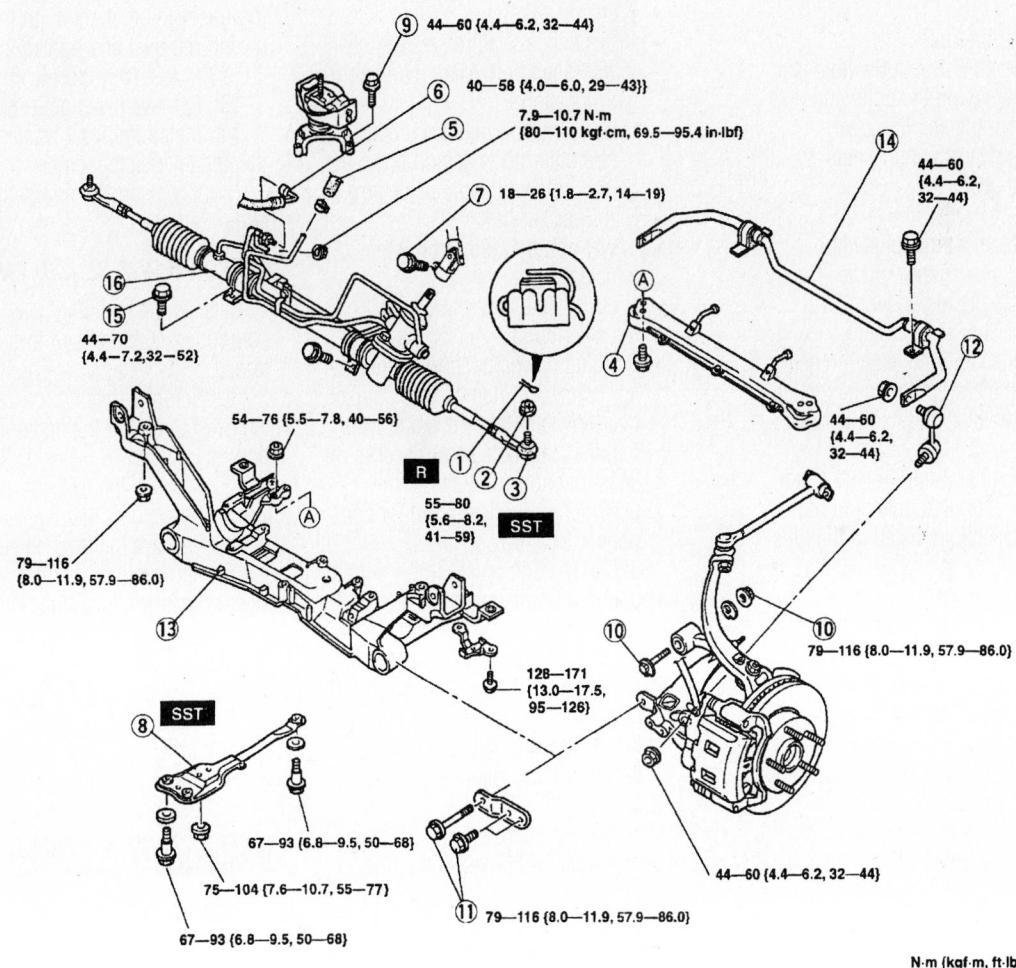

1	Cotter pin	9	Bolts (engine mount No.1)
2	Nut	10	Bolt and nut (upper lateral link)
3	Tie-rod end ball joint	11	Lower arm bolt (crossmember side)
4	Transverse member	12	Stabilizer control link
5	Return hose	13	Crossmember
6	Pressure pipe	14	Stabilizer
7	Bolt (intermediate shaft)	15	Mounting bracket bolts
8	Engine mount member	16	Steering gear and linkage

Exploded view of the power steering gear assembly—Millenia

3. Press the tie rod out from the knuckle arm.

4. Remove or disconnect the following:
- Transverse member
- Pressure line and return pipe from the steering gear
- Intermediate shaft to steering gear pinion shaft bolt. Mark the shaft-to-gear location.

5. Support the engine assembly with a engine support assembly such as 49 E017 5A0.
- Engine mount member
- Number 1 engine mount bolts
- Upper lateral link bolt and nut
- Lower arm bolt on the crossmember side
- Stabilizer bar link
- Crossmember. Support with a jack before removing the bolts and nuts.
- Shaft from the steering gear
- Steering gear mounting nuts
- Steering gear

To install:

6. Install or connect the following:
- Steering gear to the vehicle. Torque the nuts/bolts in sequence to 23–52 ft. lbs. (44–70 Nm).
- Steering shaft to the steering gear pinion shaft, align the marks made during removal and hand tighten the bolt
- Crossmember. Support with a jack before and tighten bolts and nuts to the specifications listed in the illustration.
- Stabilizer bar link
- Lower arm bolt on the crossmember side
- Upper lateral link bolt and nut
- Number 1 engine mount bolts
- Engine mount member

7. Remove the engine support assembly.
- Align the marks made during removal of the steering shaft to the steering gear pinion shaft bolt and tighten the bolt to 19 ft. lbs. (26 Nm)
- Pressure line and return pipe to the steering gear

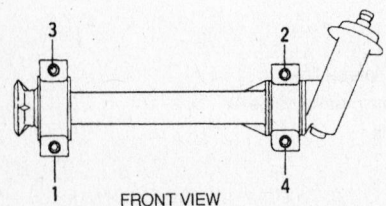

Tighten the power steering gear assembly bracket retainers in this sequence—Millenia

- Transverse member
- Tie rods to the knuckle arm. Install the nuts and tighten to 59 ft. lbs. (80 Nm) and install new cotter pins.
- Front wheels
- Negative battery cable

8. Check and/or adjust the front end alignment.

MAZDA 3

1. Before servicing the vehicle, refer to the precautions in the beginning of this section.

2. Remove or disconnect the following:
- Negative battery cable
- Front wheels
- Intermediate shaft to steering gear pinion shaft bolt. Mark the shaft-to-gear location.
- Cotter pins and nuts from both steering tie rod ends and press the tie rod out from the knuckle arm
- Stabilizer bar link
- Lower arm ball joint from the knuckle
- Pressure line and return pipe from the steering gear
- Bolt and power steering angle sensor connector. Remove the coolant over flow tank and set aside with the lines still attached to access the sensor connector. Refer to the exploded view illustration for component location.

3. Support the engine assembly with a engine support assembly such as 49 E017 5A0.
- Crossmember bolts. Support with a jack before removing the bolts and nuts.
- Front crossmember, stabilizer, lower arm and steering gear as an assembly

4. Remove the stabilizer bar, the steering gear, insulator and linkage retaining bolts and remove from the crossmember.

To install:

5. Attach the steering gear and insulator to the crossmember and tighten the insulator bolts to 69–95 inch lbs. (8–10 Nm) and the steering gear bolts to 55–79 ft. lbs. (74–105 Nm).

6. Install the stabilizer bar to the crossmember and tighten the bolts to 30–40 ft. lbs. (40–55 Nm).

7. Install the lower arm to the crossmember, tighten the side mounting bolt to 96–110 ft. lbs. (130–150 Nm) and the inner arm bolt to 72–97 ft. lbs. (97–132 Nm).

8. Install the front crossmember, stabilizer, lower arm and steering gear as an

assembly, using a transmission jack to raise the assembly into position. Refer to the exploded view illustration for the crossmember bolt locations and torque values.

9. Install the power steering angle sensor connector, the bolt and the coolant reservoir.

10. Install the power steering return hose and pressure pipe. Tighten the pipe retainer to 21–32 ft. lbs. (29–44 Nm).

11. Install the stabilizer shaft link nut upper link, and the tie rod end ball joint. Tighten the stabilizer bar link upper nut to 30–40 ft. lbs. (40–54 Nm) and the ball joint nut to 27–37 ft. lbs. (37–50 Nm).

12. Install the steering shaft to the steering gear pinion shaft, align the marks made during removal and tighten the bolt to 14–18 ft. lbs. (19–25.5 Nm).

13. Remove the engine support assembly.

14. Install the front wheels and connect the negative battery cable.

15. Check and/or adjust the front end alignment.

MAZDA 6 W/2.3L (L3) ENGINE

Refer to the exploded view illustration for component location and torque specifications.

1. Before servicing the vehicle, refer to the precautions in the beginning of this section.

2. Remove the ABS speed sensor.

3. Disconnect the negative battery cable.

4. Remove the front wheels.

5. Remove the intermediate shaft to steering gear pinion shaft bolt. Mark the shaft-to-gear location.

6. Remove the cotter pins and nuts from both steering tie rod ends

7. Press the tie rod out from the knuckle arm.

8. Disconnect the pressure line and return hose from the steering gear.

✳✳ CAUTION

Do not remove the crossmember nuts completely as this may cause the crossmember to fall. Leave the nuts threaded on the studs while loosening.

9. Support the crossmember with a jack.

10. Loosen the crossmember nuts but do not remove, lower the jack and crossmember assembly enough to access to steering gear retainers.

11. Remove the steering gear and linkage from the left side.

To install:

12. Install the steering gear and linkage.

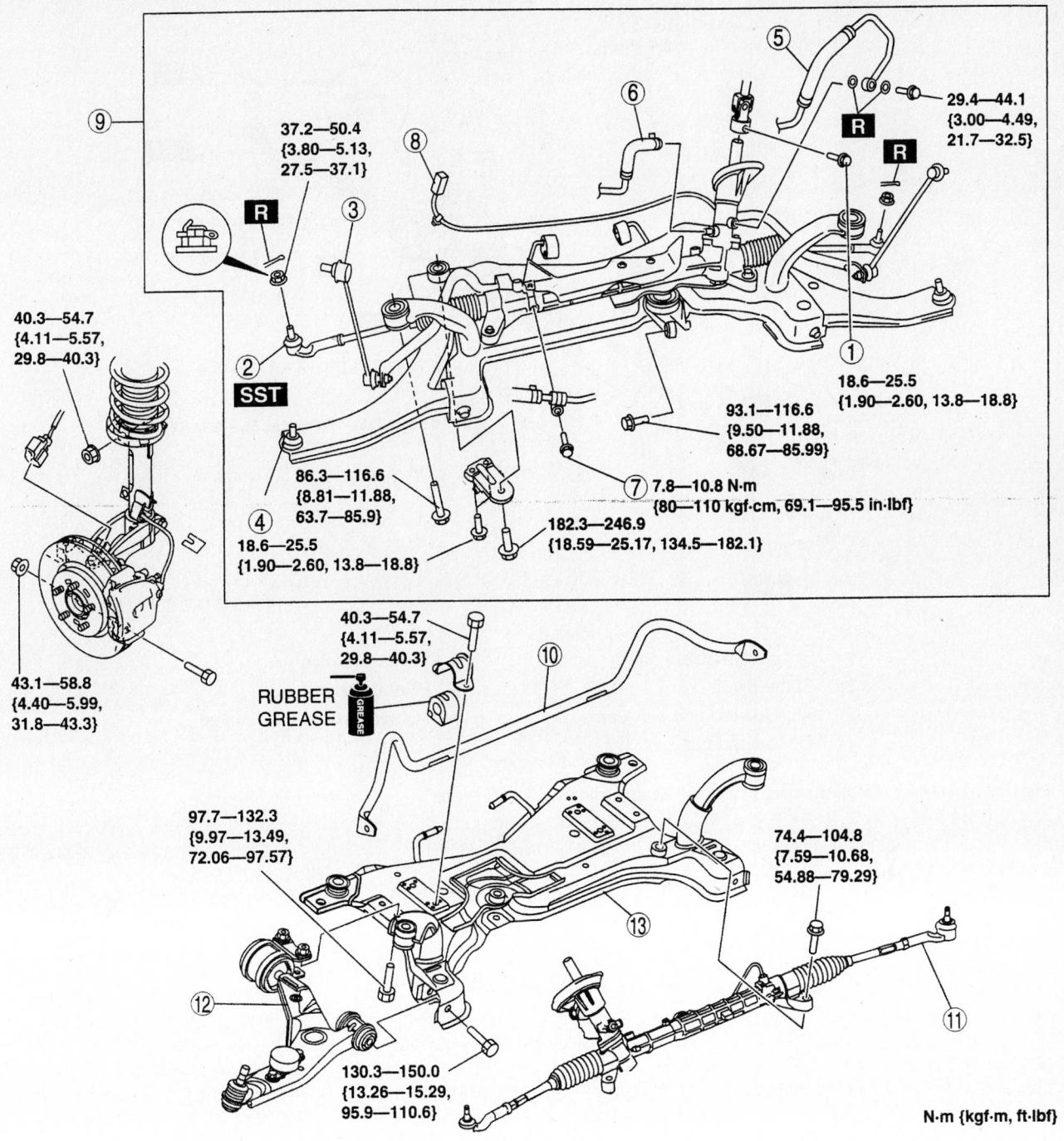

N·m {kgf·m, ft·lbf}

1   Bolt (intermediate shaft)
2   Tie-rod end ball joint
3   Stabilizer control link upper nut
4   Lower arm ball joint
5   Pressure pipe (gear side)
6   Return hose (gear side)
7   Bolt
8   P/S angle sensor connector

9   Front crossmember component, steering gear and linkage component
10 Front stabilizer
11 Steering gear and linkage
12 Front lower arm
13 Front crossmember

67162-MAZC-G87

Exploded view of the steering gear/front crossmember assembly and related components—Mazda 3

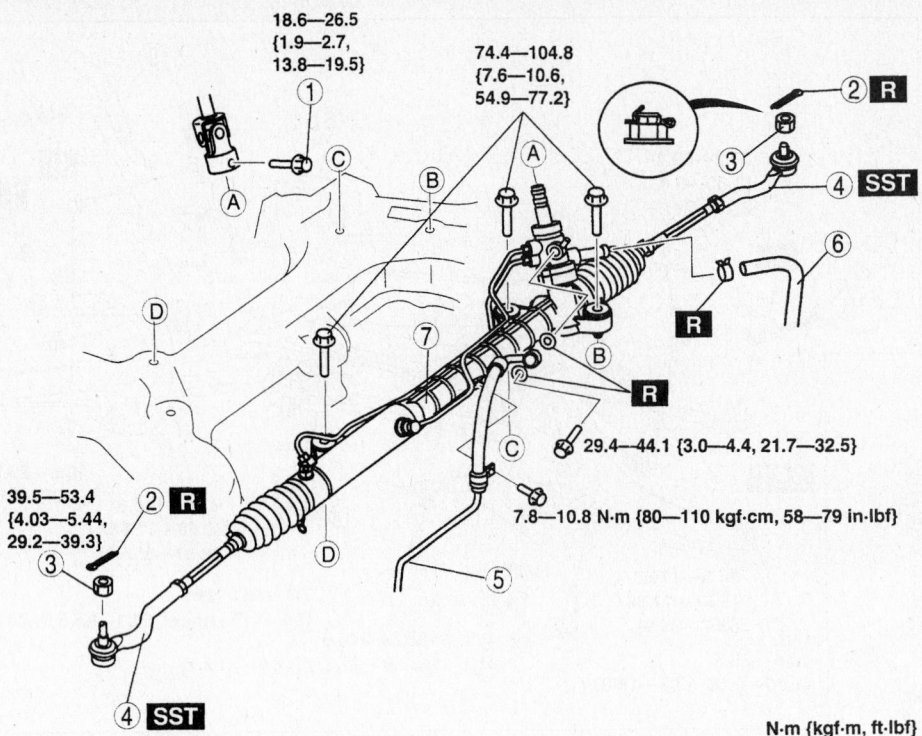

1 Bolt (intermediate shaft)
2 Cotter pin
3 Nuts (tie-rod end ball joint)
4 Tie-rod end ball joint
5 Pressure pipe
6 Return hose
7 Steering gear and linkage

67162-MAZC-G233

Exploded view of the power steering gear assembly—Mazda 6 with the 2.3L (L3) engine

13. Tighten the gear mounting bolts in the sequence illustrated to 55–77 ft. lbs. (74–104 Nm).

14. Raise the crossmember assembly into position and tighten the nuts to 89–114 ft. lbs. (119–154 Nm) and the bolts to 68–86 ft. lbs. (93–116 Nm).

15. Install the return hose and pressure pipe. Tighten the pipe banjo bolt to 21–32 ft. lbs. (29–44 Nm).

16. Install the tie rod ends to the knuckle arm. Torque the nuts to 29–39 ft. lbs. (39–53 Nm). Install new cotter pins .

17. Aling the intermediate shaft–to–gear marks made during removal and tighten the bolt to 13–19 ft. lbs. (18–26 Nm). Check the power steering fluid level.

18. Install the abs sensor and the wheels.

19. Connect the negative battery cable.

20. Check and/or adjust the front end alignment.

MAZDA 6 W/3.0L (AJ) ENGINE

Refer to the exploded view illustration for component location and torque specifications.

1. Before servicing the vehicle, refer to the precautions in the beginning of this section.

2. Disconnect the negative battery cable.

3. Remove the ABS speed sensor.

4. Remove the front wheels, under cover and splash shield.

5. Separate the stabilizer bar link at the shock absorber side.

6. Separate the front lower arm front and rear ball joints from the knuckle.

7. Remove the lower side shock absorber bolt.

8. Remove the intermediate shaft to steering gear pinion shaft bolt. Mark the shaft-to-gear location.

9. Remove the cotter pins and nuts from both steering tie rod ends

10. Press the tie rod out from the knuckle arm.

11. Remove the heat shield bolts.

12. Disconnect the pressure line and return hose from the steering gear.

13. Remove the No. 1 engine mount center bolt.

✳✳ CAUTION

Support the crossmember with a jack and make sure the jack is attached securely to the crossmember before removing the bracket.

14. Support the crossmember with a jack.

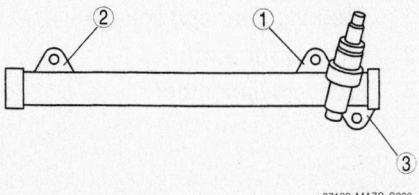

67162-MAZC-G299

Exploded view of the power steering gear assembly—Mazda 6 models

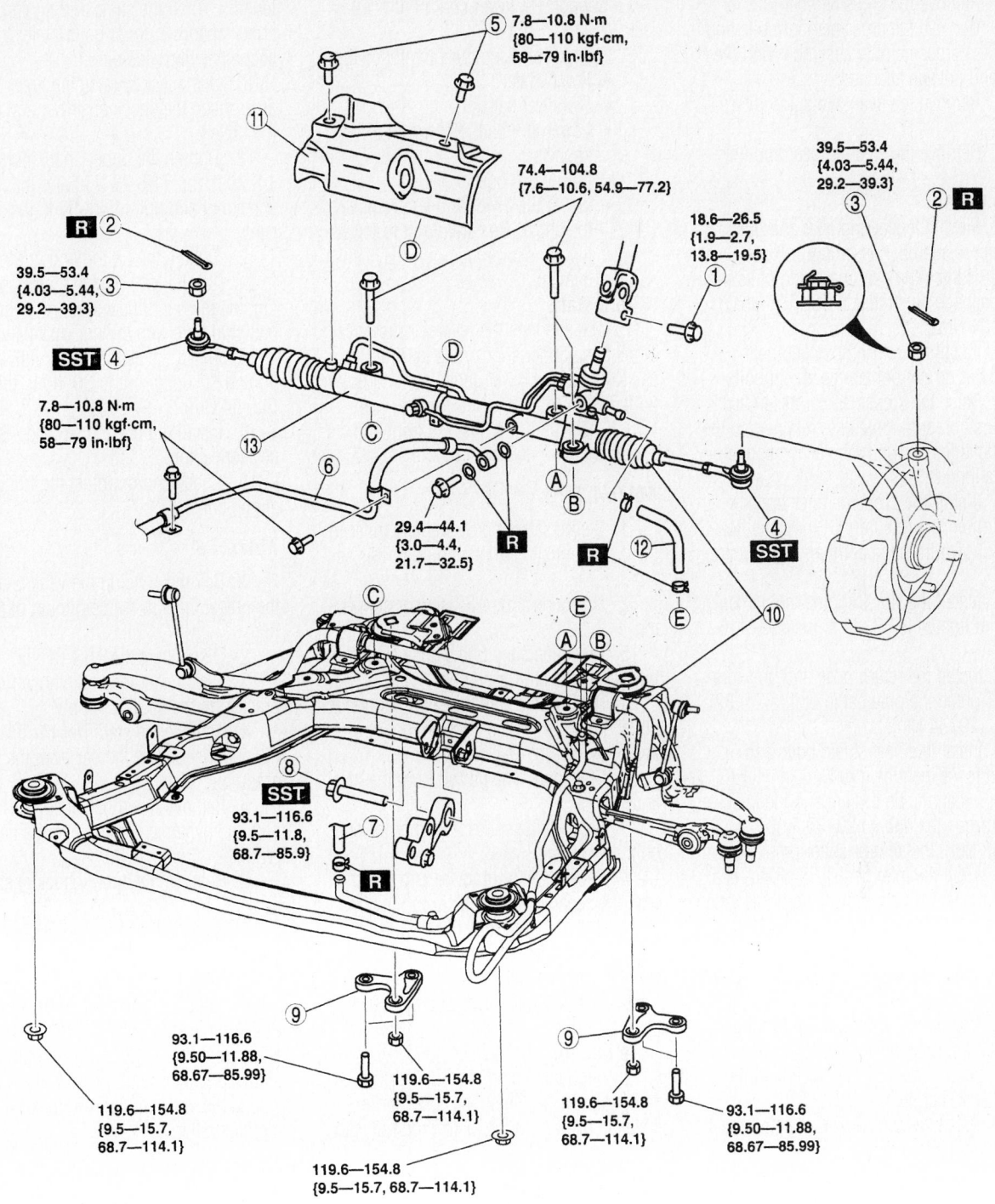

5 — 7.8—10.8 N·m
{80—110 kgf·cm,
58—79 in·lbf}

74.4—104.8
{7.6—10.6, 54.9—77.2}

18.6—26.5
{1.9—2.7,
13.8—19.5}

39.5—53.4
{4.03—5.44,
29.2—39.3}

39.5—53.4
{4.03—5.44,
29.2—39.3}

7.8—10.8 N·m
{80—110 kgf·cm,
58—79 in·lbf}

29.4—44.1
{3.0—4.4,
21.7—32.5}

93.1—116.6
{9.5—11.8,
68.7—85.9}

93.1—116.6
{9.50—11.88,
68.67—85.99}

119.6—154.8
{9.5—15.7,
68.7—114.1}

119.6—154.8
{9.5—15.7,
68.7—114.1}

93.1—116.6
{9.50—11.88,
68.67—85.99}

119.6—154.8
{9.5—15.7,
68.7—114.1}

119.6—154.8
{9.5—15.7, 68.7—114.1}

N·m {kgf·m, ft·lbf}

1	Bolt (intermediate shaft)	8	No.1 engine mount center bolt
2	Cotter pin	9	Crossmember bracket
3	Nuts (tie-rod end ball joint)	10	Crossmember component, steering gear and linkage
4	Tie-rod end ball joint	11	Insulator
5	Insulator bolts	12	Return hose
6	Pressure pipe	13	Steering gear and linkage
7	Return hose		

67162-MAZC-G234

Exploded view of the power steering gear assembly—Mazda 6 with the 3.0L (AJ) engine

15. Remove the crossmember bracket.

16. Remove the crossmember bolts and lower the crossmember assembly with the gear and linkage attached.

17. Remove the heat shield and return hose.

18. Remove the steering gear and linkage.

To install:

19. Install the steering gear and linkage to the crossmember assembly.

20. Tighten the gear mounting bolts in the sequence illustrated to 55–77 ft. lbs. (74–104 Nm).

21. Install the return hose and heat shield but do not tighten the shield bolts.

22. Raise the crossmember assembly into position and tighten the retainers to the specifications shown in the accompanying illustration.

23. Install the crossmember bracket and tighten the retainers to the specifications shown in the accompanying illustration.

24. Install the No. 1 engine mount center bolt and tighten to 68–86 ft. lbs. (93–116 Nm).

25. Install the return hose and pressure pipe. Tighten the pipe banjo bolt to 21–32 ft. lbs. (29–44 Nm).

26. Install the heat shield bolts and tighten to 79 inch lbs. (10 Nm).

27. Install the tie rod ends to the knuckle arm. Torque the nuts to 29–39 ft. lbs. (39–53 Nm). Install new cotter pins .

28. Align the intermediate shaft–to–gear marks made during removal and tighten the bolt to 13–19 ft. lbs. (18–26 Nm). Check the power steering fluid level.

29. Install the lower side shock absorber bolt.

30. Attach the front lower arm front and rear ball joints to the knuckle.

31. Attach the stabilizer bar link to the shock absorber side.

32. Install the abs sensor, wheels, splash shield and under cover.

33. Connect the negative battery cable.

34. Check and/or adjust the front end alignment.

Strut

REMOVAL & INSTALLATION

Front

MIATA

1. Before servicing the vehicle, refer to the precautions in the beginning of this section.

2. Support the lower control arm with a jack.

3. Remove or disconnect the following:
- Front wheel
- Stabilizer bar nut
- Lower arm ball joint from the knuckle
- Lower arm bolts
- Lower the lower arm and remove the shock. Be careful not to lower the arms too much or it may cause damage.

To install:

4. Installation is the reverse of removal. Tighten the upper shock nuts to 26 ft. lbs. (36 Nm) and the lower shock nut and bolt to 86 ft. lbs. (116 Nm).

5. Check and/or adjust the front end alignment.

MAZDA 3

1. Before servicing the vehicle, refer to the precautions in the beginning of this section.

2. Disconnect the ABS sensor connector.

3. Disconnect the brake hose from the bracket on the strut assembly.

4. Remove the stabilizer bar link upper nut and disconnect the bar from the strut assembly.

5. Disconnect the tie rod from the knuckle.

6. Remove the shock absorber lower bolt.

7. Loosen the front lower arm inner bolt, the separate the shock from the hub assembly by tapping the knuckle with a hammer or mallet being careful not to damage any components.

8. Remove the shock absorber assembly.

To install:

9. Align the piston rod nut with the center part where the shock absorber is installed by placing the piston rod nut with

lengths (A) all the same and tighten the shock absorber upper bolts. Refer to the accompanying illustration.

10. Use a jack to raise the lower control arm, attach the shock absorber and tighten the bolts

11. Tighten the upper shock nuts to 14–20 ft. lbs. (19–28 Nm) and the lower shock nut and bolt to 40–54 ft. lbs. (54–74 Nm).

12. Tighten the tie rod end nut to 27–37 ft. lbs. (37–50 Nm), install a new cotter pin. Tighten the nut, if necessary, to align the ball stud hole with the nut castellation.

13. Install the stabilizer bar link and tighten the upper nut to 30–40 ft. lbs. (40–54 Nm).

14. Install the brake hose and ABS sensor connector.

15. Check and/or adjust the front end alignment.

MAZDA 6

1. Before servicing the vehicle, refer to the precautions in the beginning of this section.

2. Disconnect the ABS sensor.

3. Disconnect the brake hose from the bracket on the strut assembly.

4. Remove the stabilizer bar link upper nut and disconnect the bar from the strut assembly.

5. Remove the shock absorber lower bolt, the dynamic damper and the damper fork.

6. Remove the upper shock absorber nuts.

7. Remove the shock absorber assembly.

To install:

8. Install the shock absorber assembly. Position the stud bolts at a 27–33 degree angle from where the stabilizer bar is installed (center line), towards the inner side of the vehicle. Refer to the accompanying illustration.

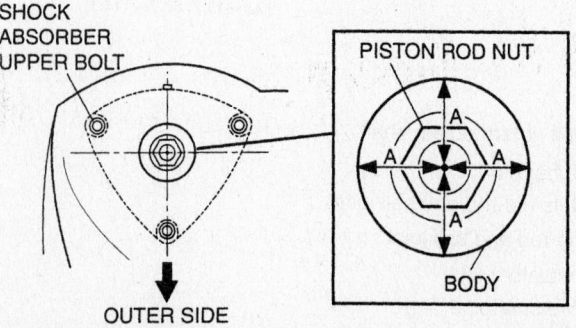

67162-MAZC-G89

Align the piston rod nut with the center part where the shock absorber is installed by placing the piston rod nut with lengths (A) all the same—Mazda 3

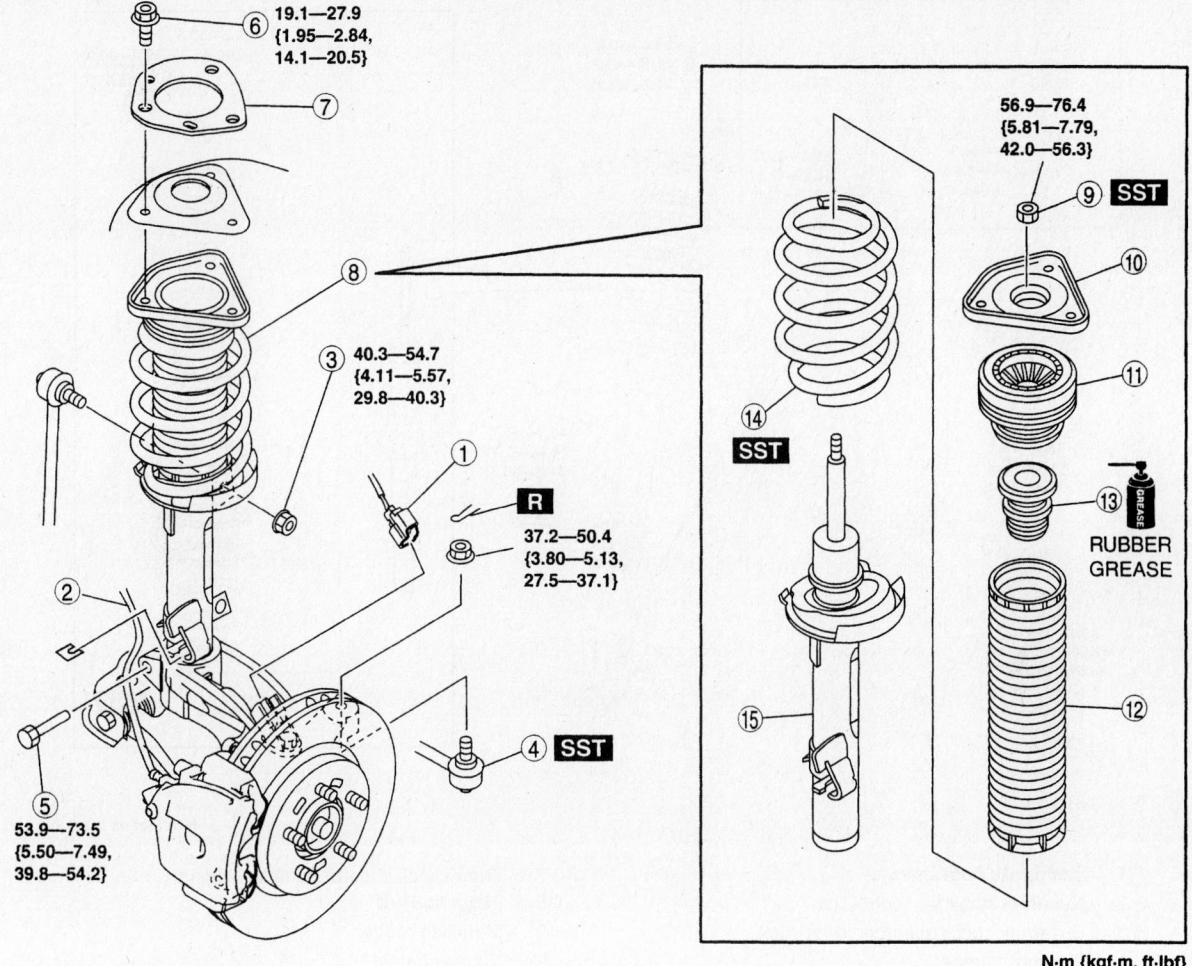

1   ABS wheel-speed sensor wiring harness connector
2   Brake hose
3   Stabilizer control link upper nut
4   Tie-rod end ball joint
5   Shock absorber lower bolt
6   Shock absorber upper bolt
7   Stiffener
8   Shock absorber and coil spring

9   Piston rod nut
10 Mounting rubber
11 Bearing
12 Dust boot
13 Bound stopper
14 Coil spring
15 Front shock absorber

67162-MAZC-G88

Exploded view of the front strut and spring assembly components—Mazda 3

9. Tighten the upper nuts to 21–29 ft. lbs. (29–39 Nm).

10. Install the damper fork and align the gap of the fork with the projection of the damper as illustrated and install and tighten the bolt to 31–38 ft. lbs. (43–52 Nm)

11. Install the dynamic damper and the lower shock bolt and tighten to 68–93 ft. lbs. (93–126 Nm).

12. Install the stabilizer bar link and tighten the nut to 31–40 ft. lbs. (43–54 Nm).

13. Install the brake hose, tighten the bolt to 13–18 ft. lbs. (18–25 Nm).

14. Install the sensor.

15. Check and/or adjust the front end alignment.

626

1. Before servicing the vehicle, refer to the precautions in the beginning of this section.

2. Support the lower control arm with a jack.

3. Remove or disconnect the following:
   • Front wheel
   • Brake hose and/or ABS sensor harness to the strut bolts or clips

   • Actuator from the top of the strut, if equipped with Automatic Adjusting Suspension (AAS)

4. Paint alignment marks on the upper strut mounting block and strut tower, and on the lower strut mount-to-steering knuckle so the strut can be reinstalled in the same position.

5. Remove or disconnect the following:
   • Upper strut mounting nuts
   • Strut-to-knuckle bolts
   • Strut assembly

To install:

6. Install or connect the following:

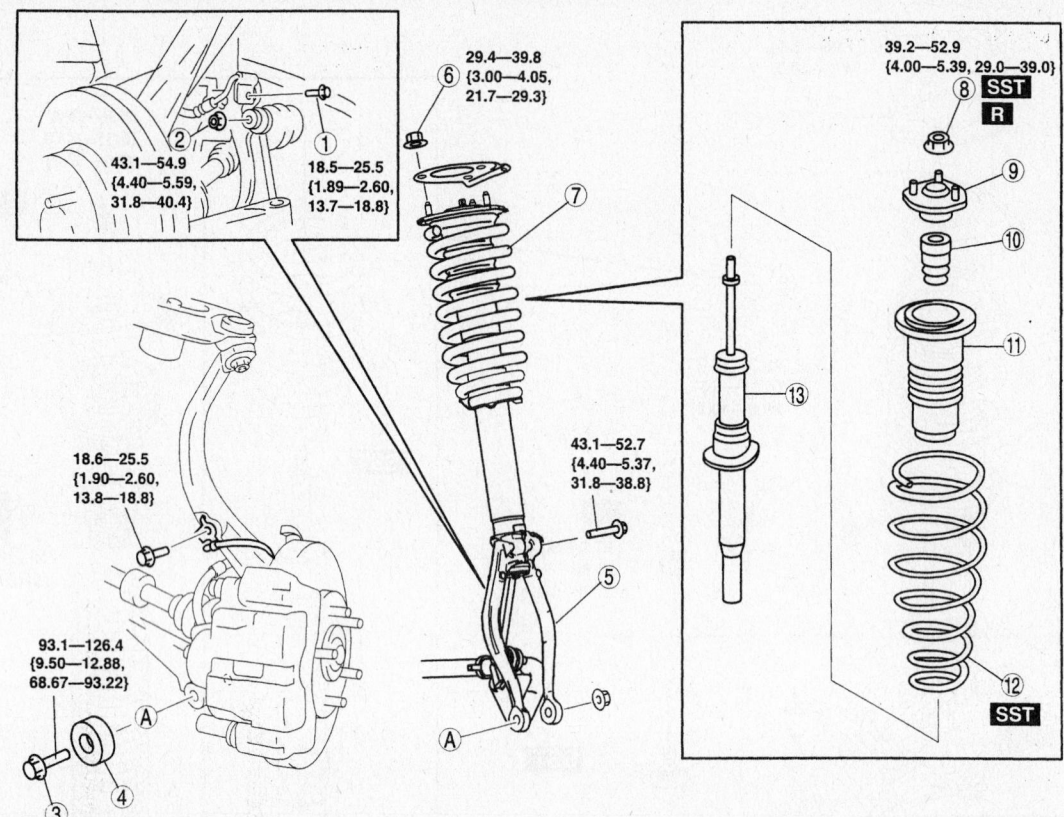

29.4—39.8
{3.00—4.05,
21.7—29.3}

43.1—54.9
{4.40—5.59,
31.8—40.4}

18.5—25.5
{1.89—2.60,
13.7—18.8}

39.2—52.9
{4.00—5.39, 29.0—39.0}

18.6—25.5
{1.90—2.60,
13.8—18.8}

43.1—52.7
{4.40—5.37,
31.8—38.8}

93.1—126.4
{9.50—12.88,
68.67—93.22}

N·m {kgf·m, ft·lbf}

1	Bolt (brake hose bracket)	7	Front shock absorber and coil spring
2	Nut (front stabilizer control link)	8	Piston rod nut
3	Bolt (front shock absorber lower side)	9	Mounting rubber
4	Dynamic damper	10	Bound stopper
5	Damper fork	11	Dust boot
6	Nut (front shock absorber upper side)	12	Coil spring
		13	Front shock absorber

67162-MAZC-G235

Exploded view of the front strut and spring assembly components—Mazda 6

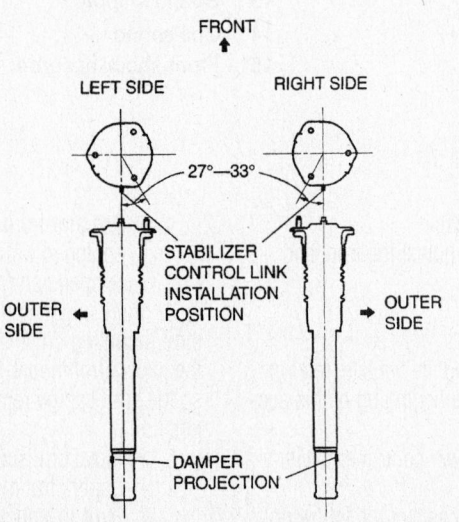

FRONT

LEFT SIDE   RIGHT SIDE

27°—33°

STABILIZER
CONTROL LINK
INSTALLATION
POSITION

OUTER
SIDE

OUTER
SIDE

DAMPER
PROJECTION

67162-MAZC-G236

Position the stud bolts at a 27-33 degree angle from where the stabilizer bar is installed (center line), towards the inner side of the vehicle when installing the front shock assembly—Mazda 6

- Strut into the strut tower, aligning the paint marks made during removal
- Upper mounting nuts and tighten to 34–46 ft. lbs. (47–62 Nm)
- Strut-to-knuckle bolts and tighten to 55–61 ft. lbs. (74–83 Nm)
- Actuator and engage the electrical connector, if equipped with AAS
- Brake hose and/or ABS sensor harness clips or bolts
- Wheel

7. Check and/or adjust the front end alignment.

PROTÉGÉ

1. Before servicing the vehicle, refer to the precautions in the beginning of this section.

2. Support the lower control arm with a jack.

3. Remove or disconnect the following:
- Front wheel

- Brake hose and/or ABS sensor harness to the strut bolts or clips
- Stabilizer link nut
- Strut-to-knuckle bolt
- Upper strut mounting nuts
- Stiffener, if equipped
- Sheet
- Strut assembly

To install:

4. Install or connect the following:

- Face the mounting block direction indicator towards the rear outboard position and install the shock assembly.
- Sheet
- Stiffener, if equipped
- Upper mounting nuts and tighten to 34–46 ft. lbs. (47–62 Nm)
- Strut-to-knuckle bolts and tighten to 68–93 ft. lbs. (94–126 Nm)
- Stabilizer link nut and tighten to 32–44 ft. lbs. (47–60 Nm)
- Brake hose and/or ABS sensor harness clips or bolts
- Wheel

5. Check and/or adjust the front end alignment.

MILLENIA

1. Before servicing the vehicle, refer to the precautions in the beginning of this section.

2. Support the lower control arm with a jack.

3. Remove or disconnect the following:

- Front wheel
- Brake hose and/or ABS sensor harness to the strut bolts or clips

4. Paint alignment marks on the upper strut mounting block and strut tower, and on the lower strut mount-to-steering knuckle so the strut can be reinstalled in the same position.

5. Remove or disconnect the following:

- Upper strut mounting nuts
- Stiffener
- Strut-to-knuckle bolts
- Strut assembly

To install:

6. Install or connect the following:

- Stiffener. Make sure the word **LH or RH** faces up and the parting area faces the inside of the vehicle.
- Strut into the strut tower, aligning the paint marks made during removal
- Upper mounting nuts and tighten to 38–49 ft. lbs. (51–67 Nm)
- Strut-to-knuckle bolts and tighten to 73–101 ft. lbs. (99–137 Nm)
- Brake hose and/or ABS sensor harness clips or bolts

- Wheel

7. Check and/or adjust the front end alignment.

Rear

MIATA

1. Before servicing the vehicle, refer to the precautions in the beginning of this section.

2. Support the lower control arm with a jack.

3. Remove or disconnect the following:

- Rear wheel
- Stabilizer bar nut
- Lower arm ball joint from the knuckle
- Lower arm bolts
- Lower the lower and upper arms and remove the shock. Be careful not to lower the arms too much or it may cause damage.

To install:

4. Installation is the reverse of removal. Tighten the upper shock nuts to 26 ft. lbs. (36 Nm) and the lower shock nut and bolt to 70 ft. lbs. (95 Nm).

5. Check and/or adjust the front end alignment.

MAZDA 3

1. Before servicing the vehicle, refer to the precautions in the beginning of this section.

2. Support the rear axle assembly with a jack.

3. Remove or disconnect the following:

- Rear wheel(s)
- Top strut nuts

➡ **The suspension will drop when the weight lifts off the wheels.**

- Bottom strut mount bolt(s)
- Strut assembly

To install:

4. Install or connect the following:

- Strut assembly
- Bottom strut mount bolt(s) and tighten to 56–74 ft. lbs. (76–101 Nm)
- Upper mounting nuts and tighten to 15–21 ft. lbs. (21–28 Nm)
- Rear wheel(s)

MAZDA 6

1. Before servicing the vehicle, refer to the precautions in the beginning of this section.

2. Support the rear axle assembly with a jack.

3. Remove or disconnect the following:

- Rear wheel(s)

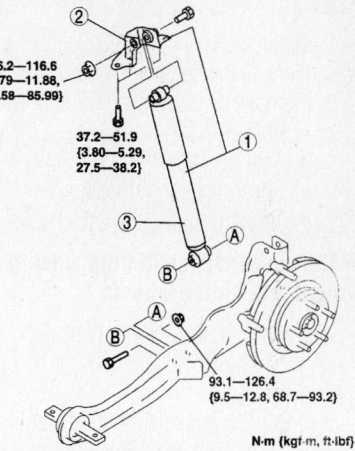

86.2—116.6
{8.79—11.88,
63.58—85.99}

37.2—51.9
{3.80—5.29,
27.5—38.2}

93.1—126.4
{9.5—12.8, 68.7—93.2}

N-m {kgf·m, ft·lbf}

1 Rear shock absorber and bracket
2 Bracket
3 Rear shock absorber

67162-MAZC-G237

Exploded view of the rear shock absorber assembly—Mazda 6

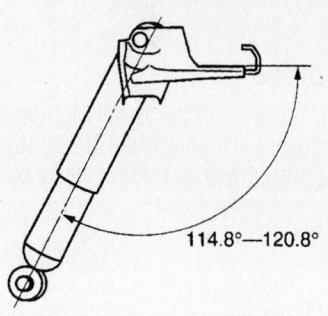

114.8°—120.8°

67162-MAZC-G238

Position the bracket and rear shock absorber assembly as shown when installing—Mazda 6

- ABS sensor
- Top strut and bracket retainers
- Bracket

➡ **The suspension will drop when the weight lifts off the wheels.**

- Bottom strut mount bolt(s)
- Strut assembly

To install:

4. Install or connect the following:

- Strut assembly
- Bracket and position the strut and bracket as illustrated. Tighten the bracket bolts to 27–38 ft. lbs. (37–51 Nm).
- Bottom strut mount bolt(s) and tighten to 68–93 ft. lbs. (93–126 Nm)
- Upper mounting retainers and tighten to 63–85 ft. lbs. (86–116 Nm)
- ABS sensor
- Rear wheel(s)

626

1. Before servicing the vehicle, refer to the precautions in the beginning of this section.
2. Remove or disconnect the following:
   - Rear wheel(s)
   - Rear package trim
   - Speed sensor harness
   - Top strut nuts

➡ **The suspension will drop when the weight lifts off the wheels.**

   - Bottom strut mount bolt(s)
   - Strut assembly

To install:

3. Install or connect the following:
   - Strut assembly
   - Bottom strut mount bolt(s) and tighten to 55–61 ft. lbs. (74–83 Nm)
   - Upper mounting nuts and tighten to 33–46 ft. lbs. (47–62 Nm)
   - Speed sensor harness
   - Rear package trim
   - Rear wheel(s)

PROTÉGÉ

1. Before servicing the vehicle, refer to the precautions in the beginning of this section.
2. Remove or disconnect the following:
   - Rear wheel(s)
   - Speed sensor harness
   - Rear seat belt on 4SD models or trunk side trim on the 5HB.
   - Clip and brake hose
   - Stabilizer link nut
   - Bottom strut mount bolt
   - Cap
   - Top strut nuts

➡ **The suspension will drop when the weight lifts off the wheels.**

   - Strut assembly

To install:

3. Install or connect the following:
   - Strut assembly
   - Upper mounting nuts and tighten to 34–46 ft. lbs. (47–62 Nm)
   - Cap
   - Bottom strut mount bolt and tighten to 68–93 ft. lbs. (94–126 Nm)
   - Stabilizer link nut and tighten to 32–44 ft. lbs. (44–60 Nm)
   - Brake hose and clip
   - Rear seat belt on 4SD models or trunk side trim on the 5HB.
   - Speed sensor harness
   - Rear wheel(s)

MILLENIA

1. Before servicing the vehicle, refer to the precautions in the beginning of this section.
2. Remove or disconnect the following:
   - Rear wheel(s)
   - Rear package front trim
   - Speed sensor harness
   - Lower lateral link ball joint
   - Top strut nuts
   - Stiffener

➡ **The suspension will drop when the weight lifts off the wheels.**

   - Bottom strut mount bolt(s)
   - Strut assembly

To install:

3. Install or connect the following:
   - Stiffener. Make sure the word **OUT** faces up and the parting area faces the inside of the vehicle.
   - Strut assembly
   - Bottom strut mount bolt(s) and tighten to 76–101 ft. lbs. (102–137 Nm)
   - Upper mounting nuts and tighten to 34–46 ft. lbs. (47–62 Nm)
   - Lower lateral link ball joint, tighten the nut to 115 ft. lbs. (156 Nm) and install a new cotter pin
   - Speed sensor harness
   - Rear package front trim
   - Rear wheel(s)

Coil Spring

REMOVAL & INSTALLATION

626 and Millenia

1. Before servicing the vehicle, refer to the precautions in the beginning of this section.
2. Remove or disconnect the following:
   - Strut from the vehicle
   - Cap from the top of the strut, if not equipped with Automatic Adjusting Suspension (AAS)
   - Piston rod upper nut 1 turn but do not remove it
3. Place the lower end of the strut in the vise.
4. Install a coil spring compressor and compress the coil spring.

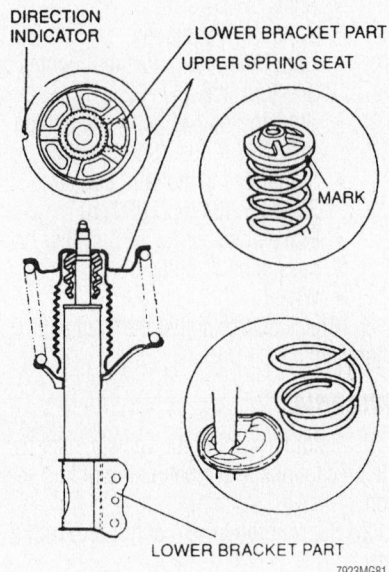

Be sure the end of the coil spring is in the step of the lower seat—626

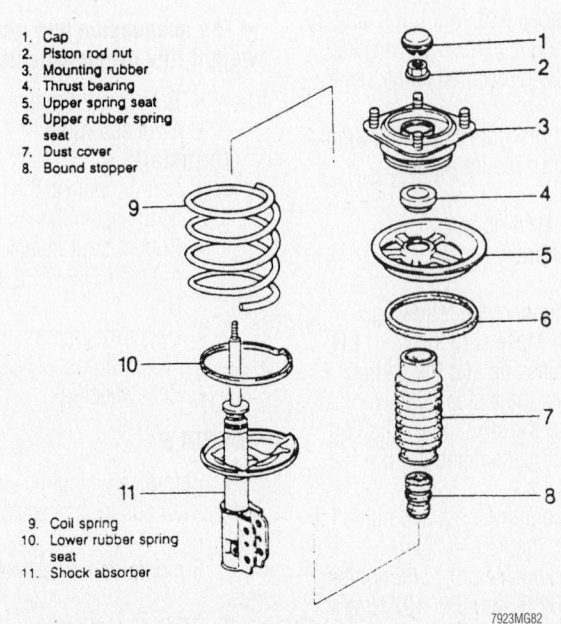

1. Cap
2. Piston rod nut
3. Mounting rubber
4. Thrust bearing
5. Upper spring seat
6. Upper rubber spring seat
7. Dust cover
8. Bound stopper
9. Coil spring
10. Lower rubber spring seat
11. Shock absorber

Exploded view of the front strut assembly—rear strut is similar

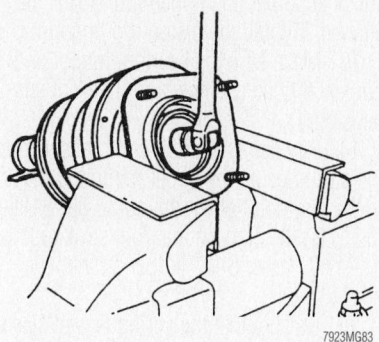

Secure the upper strut mount in a vise and loosen the piston rod nut one turn but do not remove it

✳✳ CAUTION

Failure to fully compress the spring and hold it securely can be extremely dangerous.

5. Remove or disconnect the following:
- Upper strut nut
- Slowly release the coil spring tension
- Suspension support, dust seal, spring seat, spring insulators, coil spring and bumper

6. While pushing on the piston rod, be sure that the pull stroke is even and that there is no unusual noise or resistance. Also inspect for any oil leakage around the piston rod.

7. Push the piston rod in, then release it. Be sure that the return rate is constant.

8. If the shock absorber does not operate as described, replace it.

To assemble:

9. Install or connect the following:
- Strut assembly into a vise
- Bump stopper and dust boot onto the piston rod
- Temporarily install the upper spring seat, seat rubber and spring. Mark the seat, shock and spring assem-

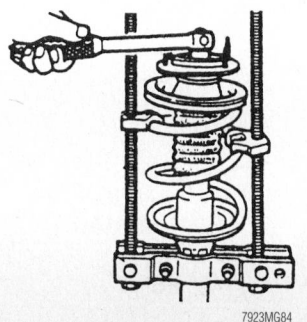

Use a coil spring compressor and relieve the spring tension from the upper mount, then remove the piston rod nut

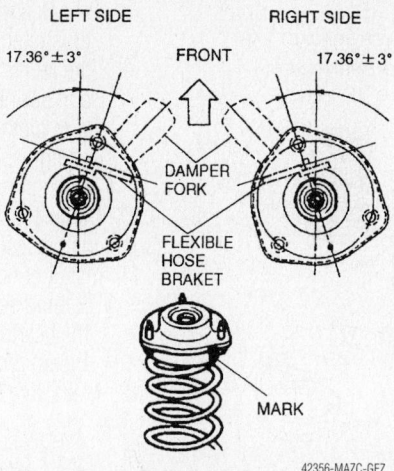

Mark the front seat, shock and spring assembly as illustrated for reassembly—Millenia

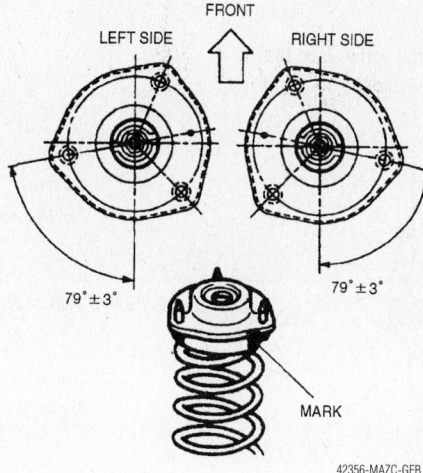

Mark the rear seat, shock and spring assembly as illustrated for reassembly—Millenia

bly as illustrated for reassembly. Align the marks of the upper seat and coil spring. Protect the assembly with cloth and install the spring compressor.
- Coil spring

10. Compress the coil spring with the spring compressor

11. Install or connect the following:
- Rubber seat, the spring upper seat, the bearing and the mounting block
- Piston rod upper nut

12. Be sure that the spring upper seat notched portion is facing inward and tighten the piston rod upper nut.

13. Remove the spring compressor from the strut. Secure the upper mounting block in the vise and tighten the nut to 40–59 ft. lbs. (55–80 Nm) on 626 models and 24–33 ft. lbs. (32–46 ft. lbs. on Millenia models.

14. Be sure that the spring is well seated in the upper seats.

15. Install the strut to the vehicle.

Protégé

1. Before servicing the vehicle, refer to the precautions in the beginning of this section.

2. Remove or disconnect the following:
- Strut from the vehicle
- Piston rod upper nut 1 turn but do not remove it

3. Place the lower end of the strut in the vise.

4. Install a coil spring compressor and compress the coil spring.

✳✳ CAUTION

Failure to fully compress the spring and hold it securely can be extremely dangerous.

5. Remove or disconnect the following:
- Upper strut nut
- Slowly release the coil spring tension
- Suspension support, dust seal,

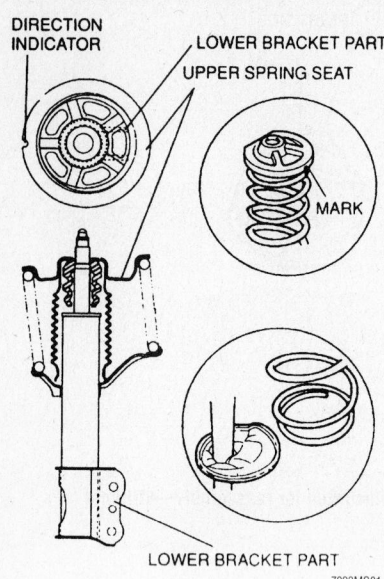

Be sure the end of the coil spring is in the step of the lower seat—Protégé

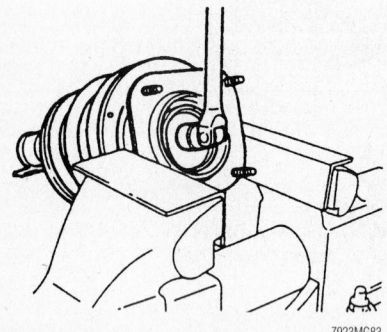

Secure the upper strut mount in a vise and loosen the piston rod nut one turn but do not remove it—Protégé

spring seat, spring insulators, coil spring and bumper

6. While pushing on the piston rod, be sure that the pull stroke is even and that there is no unusual noise or resistance. Also inspect for any oil leakage around the piston rod.

7. Push the piston rod in, then release it. Be sure that the return rate is constant.

8. If the shock absorber does not operate as described, replace it.

To assemble:

9. Install or connect the following:
   • Strut assembly into a vise
   • Bump stopper and dust boot onto the piston rod
   • Temporarily install the upper spring seat, seat rubber and spring so that the lower end of the spring is seated on the step of the lower spring seat. Mark the seat,

shock and spring assembly as illustrated for reassembly. Align the marks of the upper seat and coil spring. Protect the assembly with cloth and install the spring compressor.
   • Coil spring

10. Compress the coil spring with the spring compressor

11. Install or connect the following:
   • Rubber seat, the spring upper seat, the bearing and the mounting block
   • Piston rod upper nut

12. Make sure that the marks on the shock and spring assembly are aligned. Tighten the piston rod upper nut.

13. Make sure the bearing, rubber and nut are aligned as shown on the front suspension.

14. Remove the spring compressor from the strut. Secure the upper mounting block in the vise and tighten the nut to 58–81 ft. lbs. (79–109 Nm) on the front strut and 41–49 ft. lbs. (55–67 ft. lbs. on the rear strut.

15. Be sure that the spring is well seated in the upper seats.

16. Install the strut to the vehicle.

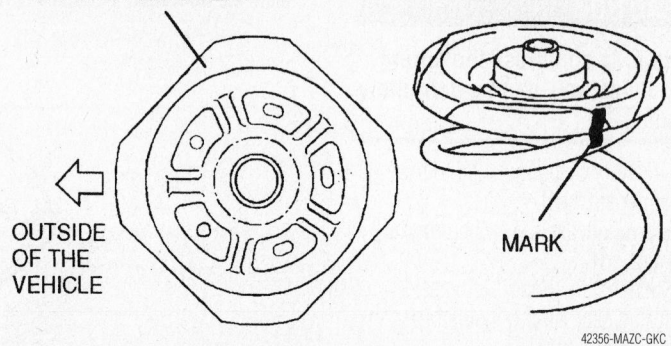

Mark the front seat, shock and spring assembly as illustrated for reassembly—Protégé

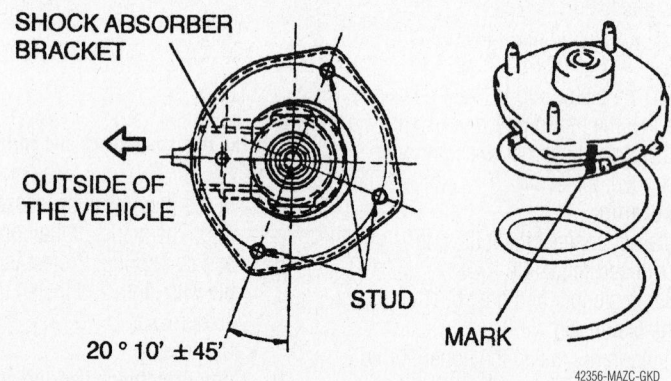

Mark the rear seat, shock and spring assembly as illustrated for reassembly—Protégé

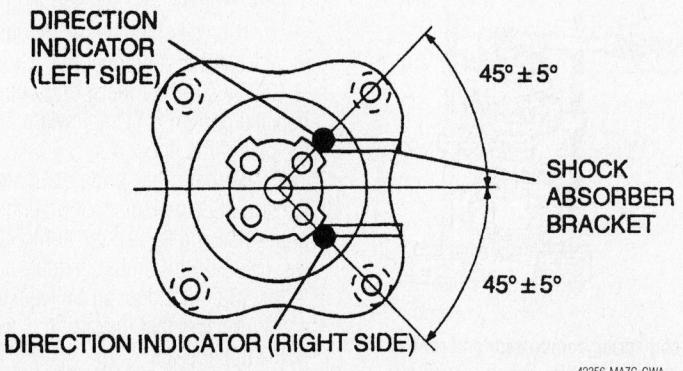

Make sure the bearing, rubber and nut are aligned as shown on the front suspension—Protégé

1. Stabilizer nut
2. Retainer, bushing and spacer
3. Stabilizer bolt
4. Bolt, washer
5. Bolt
6. Bolt, nut
7. Nut
8. Washer
9. Lower control arm bushing (rear)
10. Nut
11. Bolt
12. Lower arm ball joint
13. Ball joint dust boot
14. Lower arm bushing (front)
15. Lower arm

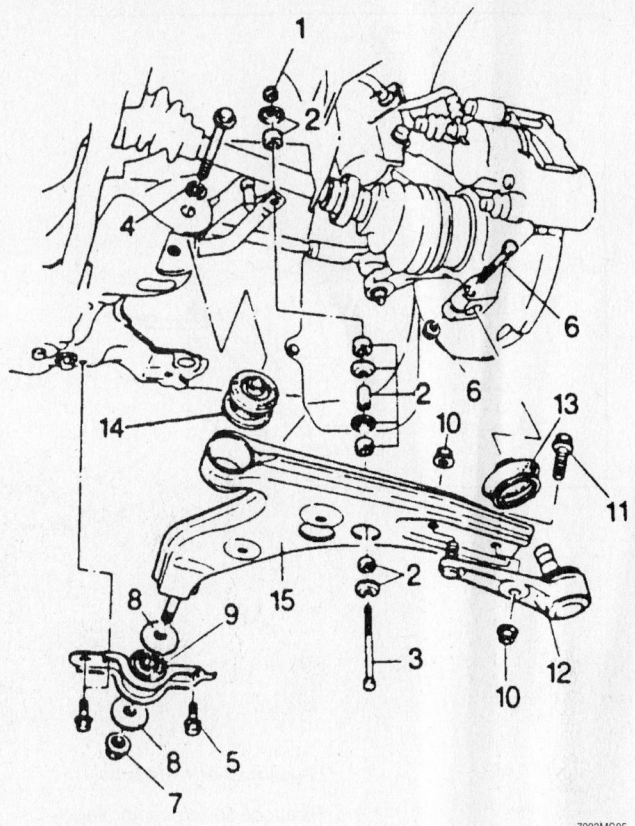

Exploded view of a common lower control arm with replaceable ball joint

- Ball joint to lower control arm bolt and nut

To install:

3. Install or connect the following:
- Ball joint to lower control arm. Torque the bolt to 86 ft. lbs. (116 Nm).
- Ball joint to the knuckle
- Ball joint to the steering knuckle. Torque the bolt to 57 ft. lbs. (77 Nm).
- Wheel

4. Check and/or adjust the front wheel alignment.

Upper Control Arm

REMOVAL & INSTALLATION

Miata

FRONT

1. Before servicing the vehicle, refer to the precautions in the beginning of this section.
2. Remove or disconnect the following:
- Wheel
3. Support the lower control arm with a jack.
4. Remove or disconnect the following:

- Cotter pin
- Upper ball joint nut by loosening it
- Ball joint by pressing it from the knuckle
- Upper ball joint nut
- Lower strut mounting bolt
- Upper control arm bolt and nut
- Upper control arm

To install:

5. Install the upper control arm.
6. Loosely tighten the bolt and nut.
7. Loosely install the lower strut mounting bolt.
- Ball joint to the knuckle. Torque nut to 47–60 ft. lbs. (63–82 Nm).
- New cotter pin to ball joint nut
- Wheel
8. Torque upper control arm bolt to 87–101 ft. lbs. (118–137 Nm) and the lower strut mounting bolt to 69–86 ft. lbs. (94–116 Nm).
9. Check and/or adjust the front wheel alignment.

REAR

1. Before servicing the vehicle, refer to the precautions in the beginning of this section.
2. Remove or disconnect the following:
- Wheel

3. Support the lower control arm with a jack.
4. Remove or disconnect the following:
- Upper control arm bolts and nuts
- Upper control arm

To install:

5. Install the upper control arm.
6. Loosely tighten the bolt and nut.
- Wheel
7. Torque upper control arm nuts and bolts to 40–56 ft. lbs. (54–76 Nm).
8. Check and/or adjust the front wheel alignment.

Mazda 3

1. Before servicing the vehicle, refer to the precautions in the beginning of this section.
2. Remove or disconnect the following:
- Wheel
3. Support the lower control arm with a jack.
4. Remove or disconnect the following:
- Coil spring
- Upper control arm bolts and nuts
- Upper control arm

To install:

5. Install the upper control arm.
6. Loosely tighten the bolt and nut.
- Wheel

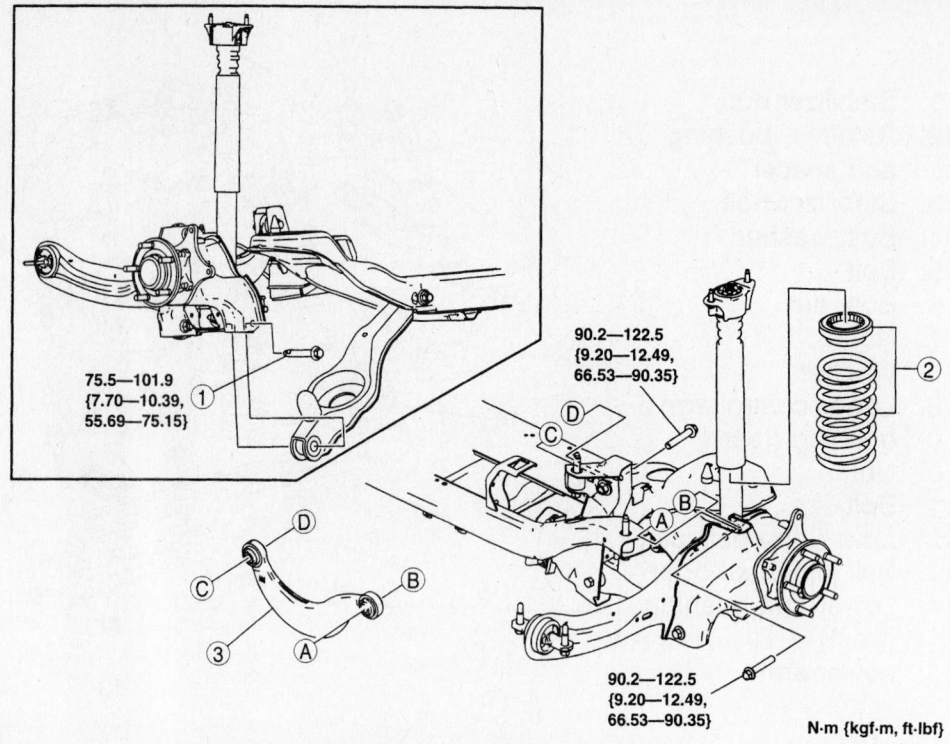

75.5—101.9
{7.70—10.39,
55.69—75.15}

90.2—122.5
{9.20—12.49,
66.53—90.35}

90.2—122.5
{9.20—12.49,
66.53—90.35}

N·m {kgf·m, ft·lbf}

1   Rear lower arm outer bolt

2   Rear coil spring component

3   Rear upper arm

67162-MAZC-G95

Exploded view of the rear upper control arm assembly and related components—Mazda 3

7. Torque upper control arm nuts and bolts to 55–75 ft. lbs. (75–101 Nm).

8. Check and/or adjust the front wheel alignment.

Mazda 6

FRONT

1. Before servicing the vehicle, refer to the precautions in the beginning of this section.

2. Remove the wheel.

3. Remove the ABS sensor.

4. Remove the brake hose bracket bolt.

5. Remove the stabilizer bar link nut.

6. Support the lower control arm with a jack.

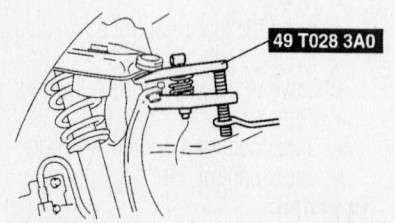

49 T028 3A0

67162-MAZC-G241

Separate the upper arm ball joint from the knuckle using a separator tool—Mazda 6

7. Separate the upper arm ball joint from the knuckle using a separator tool such as 49 T028 3A0.

8. Remove the upper arm rear bolts and the arm.

To install:

9. Install the upper arm and the bolts.

10. Attach the ball joint to the knuckle. Tighten the ball joint bolt to 29–39 ft. lbs. (39–53 Nm) and the rear upper arm bolts to 36–49 ft. lbs. (49–66 Nm).

11. Install the stabilizer bar link nut and tighten to 31–40 ft. lbs. (43–54 Nm).

12. Install the brake hose bracket bolt and tighten to 13–18 ft. lbs. (18–25 Nm).

13. Install the ABS sensor and in the wheel.

REAR

1. Before servicing the vehicle, refer to the precautions in the beginning of this section.

2. Remove the wheel.

3. Remove the ABS sensor.

4. Disconnect the parking brake cable.

5. Support the rear trailing link with a jack.

6. Remove the rear shock absorber lower bolt.

7. Remove the front trailing link bolt.

8. Remove the rear upper arm bolt and the arm.

To install:

9. Install the arm and the rear bolt.

10. Attach the bolt to the trailing link front side.

11. Tighten the rear upper bolt to 68–93 ft. lbs. (93–126 Nm) and the trailing link bolt to 63–86 ft. lbs. (86–116 Nm).

12. Install the lower shock absorber bolt and tighten to 68–93 ft. lbs. (93–126 Nm).

13. Attach the parking brake cable.

14. Install the ABS sensor.

15. Install the wheel.

CONTROL ARM BUSHING REPLACEMENT

All Mazda's use a pressed in control arm bushing, and the pressing can be done using two appropriately sized sockets (a press socket and a catch socket) and a large bench vise.

1. Position the control arm and the 2 sockets into a vise.

2. Position the press socket onto the control arm bushing.

3. Position the catch socket onto the control arm, opposite of the press socket.

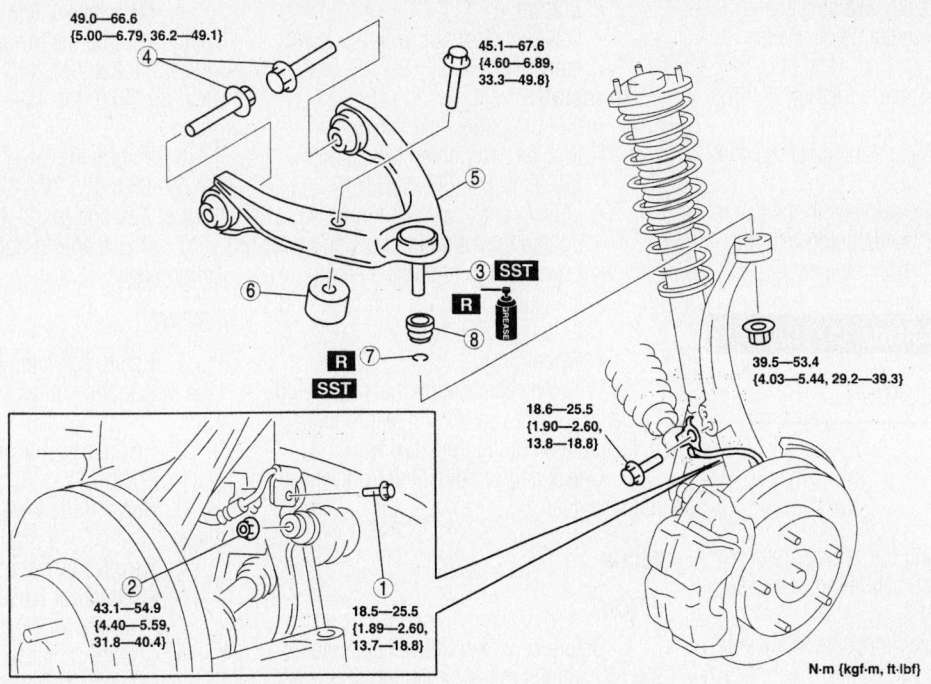

49.0—66.6
{5.00—6.79, 36.2—49.1}

45.1—67.6
{4.60—6.89,
33.3—49.8}

39.5—53.4
{4.03—5.44, 29.2—39.3}

18.6—25.5
{1.90—2.60,
13.8—18.8}

43.1—54.9
{4.40—5.59,
31.8—40.4}

18.5—25.5
{1.89—2.60,
13.7—18.8}

N·m {kgf·m, ft·lbf}

1 Bolt (brake hose bracket
2 Nut (stabilizer control link)
3 Front upper arm ball joint
4 Bolt (front upper arm)
5 Front upper arm
6 Dynamic Vamper
7 Clip
8 Dust boot

67162-MAZC-G240

Exploded view of the front upper control arm assembly—Mazda 6

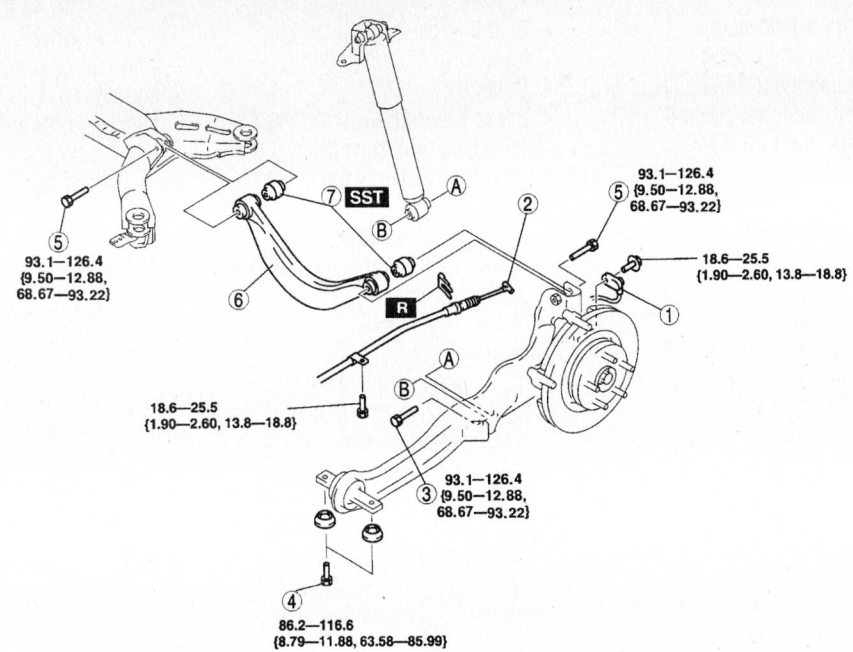

93.1—126.4
{9.50—12.88,
68.67—93.22}

93.1—126.4
{9.50—12.88,
68.67—93.22}

18.6—25.5
{1.90—2.60, 13.8—18.8}

18.6—25.5
{1.90—2.60, 13.8—18.8}

93.1—126.4
{9.50—12.88,
68.67—93.22}

86.2—116.6
{8.79—11.88, 63.58—85.99}

N·m {kgf·m, ft·lbf}

1 ABS wheel-speed sensor
2 Parking brake cable
3 Bolt (rear shock absorber lower side)
4 Bolt (trailing link front side)
5 Bolt (rear upper arm)
6 Rear upper arm
7 Rear upper arm bushing

67162-MAZC-G242

Exploded view of the rear upper control arm assembly—Mazda 6

4. Tighten the bench vise slowly and press the bushing into the catch socket.

To install:

5. Apply soapy water to the new control arm bushing.

6. Position the bushing against the control arm.

7. Using the same sockets, in the same positions, press the new bushing into the control arm.

Lower Control Arm

REMOVAL & INSTALLATION

626

FRONT

1. Before servicing the vehicle, refer to the precautions in the beginning of this section.

2. Remove or disconnect the following:
   • Wheel
   • Lower ball joint pinch bolt from the steering knuckle
   • Stabilizer bar link from the lower control arm
   • Lower control arm bolts and nuts
   • Lower control arm with the lower ball joint

To install:

3. Install or connect the following:
   • Lower control arm and loosely tighten the mounting nuts and bolts
   • Stabilizer link to the lower control arm. Torque the nut to 27–39 ft. lbs. (37–53 Nm).
   • Lower ball joint to the steering knuckle. Torque the bolt to 26–41 ft. lbs. (35–56 Nm).
   • Wheel

4. With the vehicle at normal rode height, tighten the lower control arm mounting bolts. Torque the front bushing through-bolt to 40–59 ft. lbs. (55–80 Nm) and the rear bushing strap bolts to 69–96 ft. lbs. (94–131 Nm).

5. Check and/or adjust the front wheel alignment.

Protégé

FRONT

1. Before servicing the vehicle, refer to the precautions in the beginning of this section.

2. Remove or disconnect the following:
   • Wheel
   • Rear lower arm bolt
   • Lower ball joint pinch bolt from the steering knuckle

   • Bracket
   • Lower control arm with the lower ball joint

To install:

3. Install or connect the following:
   • Bracket and tighten the bolts to 68–97 ft. lbs. (94–126 Nm).
   • Lower control arm and loosely tighten the mounting nuts and bolts
   • Lower ball joint to the steering knuckle. Torque the bolt to 32–43 ft. lbs. (44–58 Nm).
   • Wheel

4. With the vehicle at normal rode height, tighten the lower control arm rear mounting bolts to 69–94 ft. lbs. (94–126 Nm).

5. Check and/or adjust the front wheel alignment.

Miata

FRONT

1. Before servicing the vehicle, refer to the precautions in the beginning of this section.

2. Remove or disconnect the following:
   • Wheel
   • Undercover
   • Cotter pin and nut from the tie-rod end
   • Tie-rod end from the steering knuckle
   • Stabilizer bar link bolt and the lower strut mounting bolt
   • Cotter pin
   • Lower ball joint by loosening it

3. With the nut protecting the ball joint stud, separate the stud from the knuckle. Remove the nut.
   • Dust boot, if necessary
   • Lower control arm bolts, nuts and adjusting cams
   • Lower control arm

To install:

4. Install or connect the following:
   • Dust boot, if removed using a press. Always fill the inside of the new boot with grease prior to installation.
   • Lower control arm by loosely tightening the bolts and nuts
   • Lower ball joint to the knuckle. Torque the nut to 42–57 ft. lbs. (57–77 Nm).
   • New cotter pin
   • Lower strut and stabilizer link bolts by loosely tightening them
   • Tie-rod end to the steering knuckle. Torque the nut to 22–32 ft. lbs. (30–44 Nm).
   • New cotter pin
   • Wheel and undercover

5. With the vehicle at normal rode height, torque the lower control arm bolts to 69–83 ft. lbs. (94–113 Nm) for the inner and 69–86 ft. lbs. (94–112 Nm) for the outer.

6. Torque the lower strut mounting bolt to 69–86 ft. lbs. (94–116 Nm) and the stabilizer link bolt to 32–44 lbs. (44–60 Nm).

7. Check and/or adjust the front wheel alignment.

REAR

1. Before servicing the vehicle, refer to the precautions in the beginning of this section.

2. Remove or disconnect the following:
   • Wheel

3. Support the lower control arm with a jack.

4. Remove or disconnect the following:
   • Stabilizer bar link nut
   • Lower shock absorber bolt
   • Lower control arm bolts and nuts
   • Lower control arm

To install:

5. Install or connect the following:
   • Lower control arm and loosely tighten the bolts and nuts
   • Lower shock absorber bolt and tighten to 70 ft. lbs. (85 Nm)
   • Stabilizer bar link nut and tighten to 32–44 ft. lbs. (44–60 Nm)
   • Wheel

6. Torque lower control arm inner nuts and bolts to 54–70 ft. lbs. (73–95 Nm) and the outer nuts and bolts to 47–54 ft. lbs. (63–74 Nm).

7. Check and/or adjust the front wheel alignment.

Millenia

1. Before servicing the vehicle, refer to the precautions in the beginning of this section.

2. Remove or disconnect the following:
   • Transverse member
   • Power steering return hose and pressure pipe
   • Intermediate steering shaft bolt

3. Support the engine from the top.
   • Engine mount member
   • Bolts for engine mount No. 1
   • Lower strut mounting bolt
   • Tie-rod end from the steering knuckle
   • Upper lateral link ball joint
   • Lower ball joint
   • Stabilizer control link nut
   • Gusset

4. Support the crossmember using a jack.

- Crossmember mounting nuts and lower the crossmember to gain clearance
- Lower arm assembly

To install:

5. Install or connect the following:
- Lower arm assembly to the vehicle
- Crossmember mounting bolts
- Gusset. Torque the bolts to 58–86 ft. lbs. (79–116 Nm).
- Stabilizer control link. Torque the nut to 32–44 ft. lbs. (44–60 Nm).
- Lower ball joint. Torque the bolts to 58–86 ft. lbs. (79–116 Nm) and the nut to 86–115 ft. lbs. (116–156 Nm).
- Upper lateral link nut and bolt. Torque to 58–86 ft. lbs. (79–116 Nm).
- Tie-rod end. Torque the nut to 41–59 ft. lbs. (55–80 Nm).
- Strut lower mounting bolt. Torque to 73–101 ft. lbs. (98–137 Nm).

- No. 1 engine mount. Torque the bolts to 32–44 ft. lbs. (44–60 Nm).
- Engine mount member. Torque the bolts to 50–68 ft. lbs. (67–93 Nm) and the nuts to 55–77.3 ft. lbs. (75–104 Nm).
6. Remove the engine support tool.
- Intermediate steering shaft. Torque the bolt to 14–19 ft. lbs. (18–26 Nm).
- Power steering pressure pipe and return hose
- Transverse member
7. Check the power steering fluid and fill to proper level, bleed if necessary.
8. Check and/or adjust the front end alignment.

Mazda 3

FRONT

1. Before servicing the vehicle, refer to the precautions in the beginning of this section.

2. Remove or disconnect the following:
- Wheel
- Cotter pin
- Lower ball joint by loosening it
3. With the nut protecting the ball joint stud, separate the stud from the knuckle. Remove the nut.
- Cotter pin and nut from the tie-rod end
- Tie-rod end from the steering knuckle

→**If removing the right side lower arm, move the engine and transaxle slightly towards the front side of the vehicle so the engine does not interfere with the lower arm rear side bolt removal.**

- No. 1 engine mount center bolt.
- Engine and transaxle assembly slightly towards the front of the vehicle if necessary to remove the right lower arm
- Lower arm rear side bolt and the lower arm.
- Dust boot, if necessary

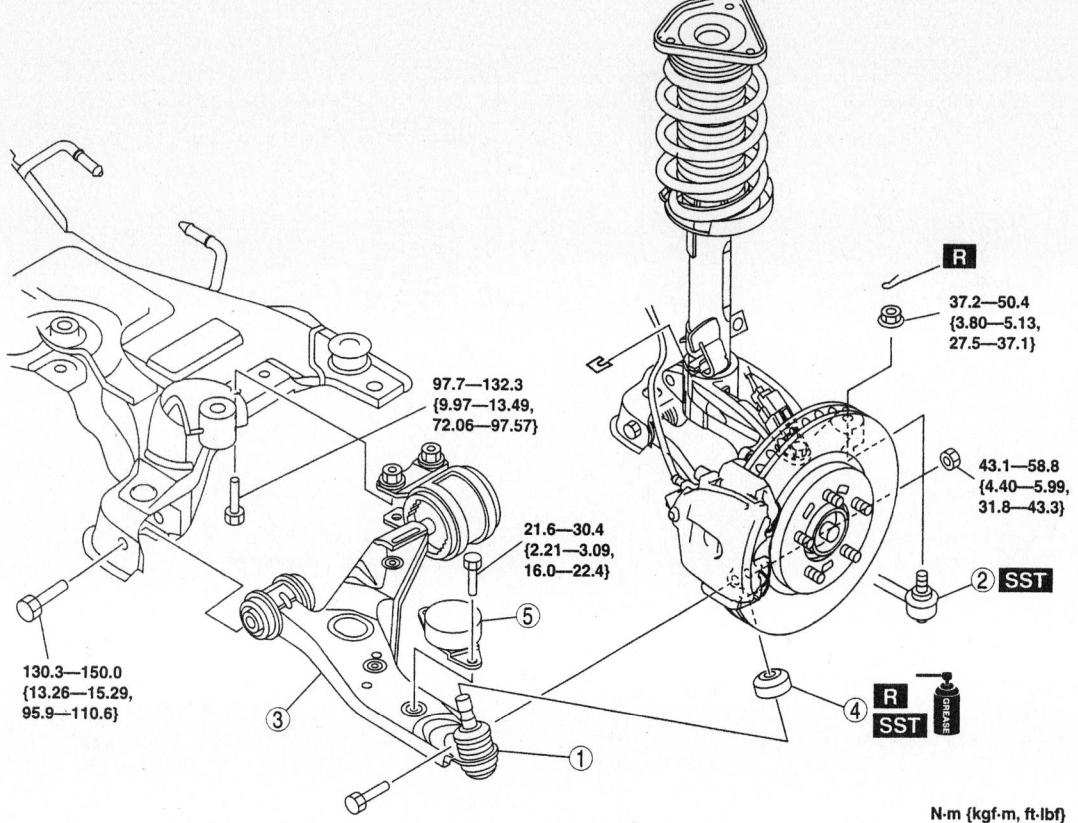

N·m {kgf·m, ft·lbf}

1	Front lower arm ball joint	4	Dust boot
2	Tie-rod end ball joint	5	Dynamic damper
3	Front lower arm		

67162-MAZC-G96

Exploded view of the front lower control arm assembly and related components—Mazda 3

To install:

4. Install or connect the following:
- Dust boot, if removed using a press. Always fill the inside of the new boot with grease prior to installation.
- Lower control arm by loosely tightening the bolts and nuts
- No. 1 engine mount bolt and tighten to 68–86 ft. lbs. (93–116 Nm).

5. Tighten the lower control arm bolts to the specifications shown in the accompanying illustration.

6. Install the remaining components in the reverse order of removal, refer to the exploded view of the lower control arm assembly and related components illustration for component location and torque specifications.

7. Check and/or adjust the front wheel alignment.

REAR

1. Before servicing the vehicle, refer to the precautions in the beginning of this section.

2. Remove the rear wheels.

3. Remove the rear crossmember bracket.

4. Remove the rear stabilizer.

5. Support the rear axle with a jack, loosen the rear lower control arm inner bolt and remove the outer bolt.

6. Remove the rear coil spring and upper spring seat rubber.

7. Remove the rear lower arm inner bolt and the arm.

To install:

8. Install the rear arm and the inner bolt and hand tighten.

9. Install the upper seat rubber and the coil spring.

10. Raise the axle into position with the jack.

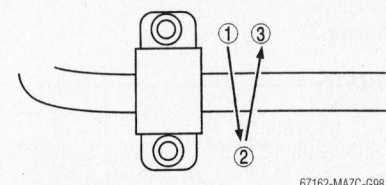

Tighten the rear stabilizer bracket bolts in this sequence—Mazda 3

67162-MAZC-G98

11. Install the rear lower arm outer bolt and tighten the outer bolt to 55–75 ft. lbs. (75–102 Nm) and the inner bolt to 59–74 ft. lbs. (80–100 Nm).

12. Install the rear stabilizer, tighten the bracket bolts in the sequence illustrated to 30–40 ft. lbs. (40–54 Nm). Tighten the link bolts to 30–40 ft. lbs. (40–54 Nm).

13. Insatl the crossmember bracket and wheels.

14. Check and/or adjust the wheel alignment.

Mazda 6

The Mazda 6 models have two lower control arms in the front suspension assembly, they are identified as the front lower control arm (Front side) and front lower control arm (rear side). They also have one rear suspension lower control arm, which will be identified as rear lower control arm.

FRONT ARM (FRONT SIDE)

Refer to the accompanying illustration for component locations and torque specifications.

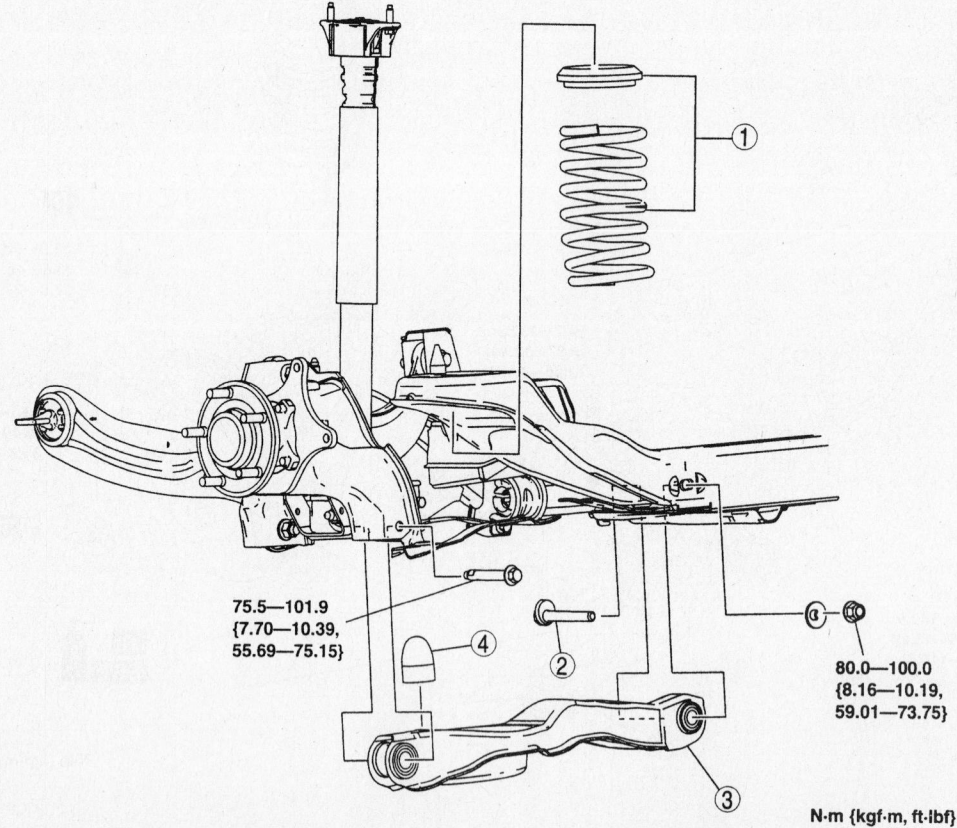

75.5—101.9
{7.70—10.39,
55.69—75.15}

80.0—100.0
{8.16—10.19,
59.01—73.75}

N·m {kgf·m, ft·lbf}

1	Rear coil spring component	3	Rear lower arm
2	Rear lower arm inner bolt	4	Bound stopper

67162-MAZC-G97

Exploded view of the rear lower control arm assembly and related components—Mazda 3

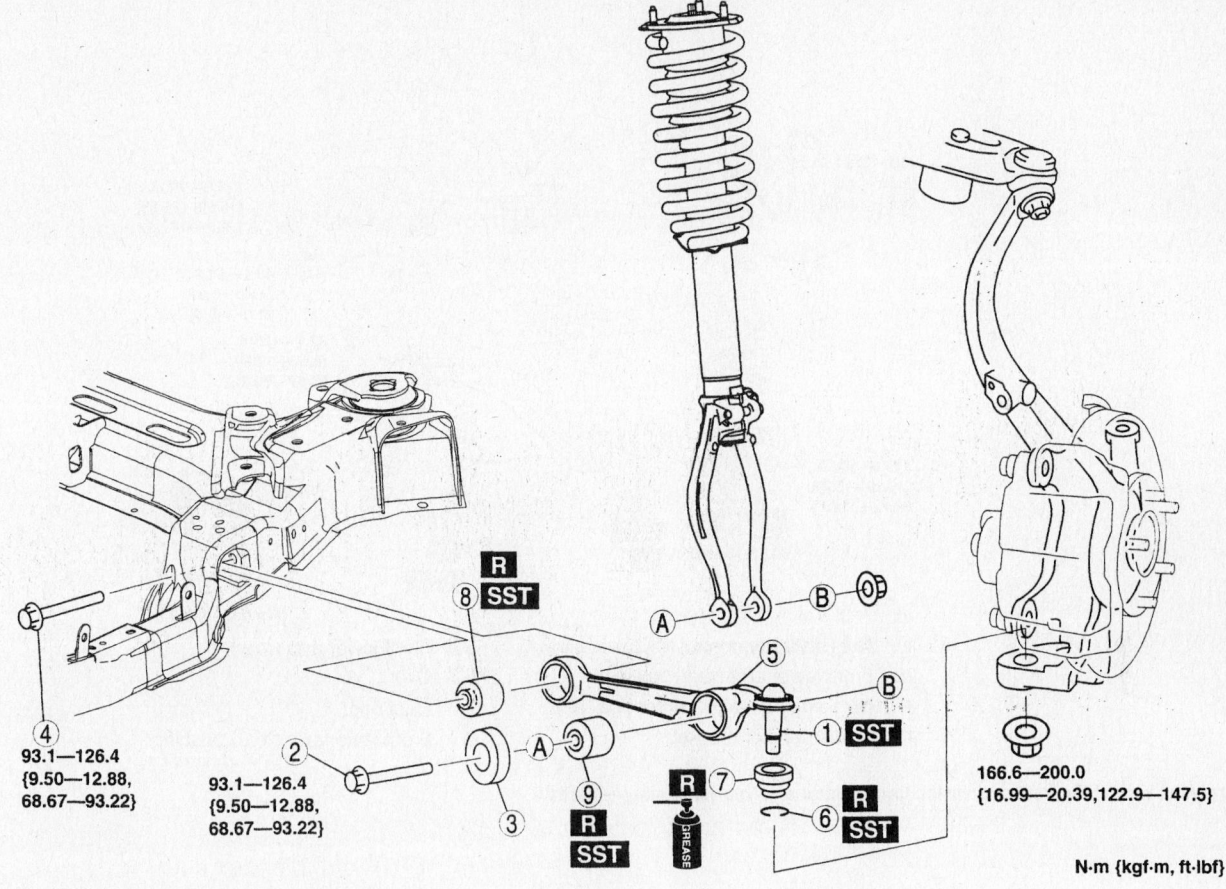

93.1—126.4
{9.50—12.88,
68.67—93.22}

93.1—126.4
{9.50—12.88,
68.67—93.22}

166.6—200.0
{16.99—20.39,122.9—147.5}

N·m {kgf·m, ft·lbf}

1 Front lower arm (front) ball joint
2 Bolt (front shock absorber lower side)
3 Dynamic damper
4 Bolt (front lower arm inner side)
5 Front lower arm (front) component
6 Clip
7 Dust boot
8 Front lower arm (front) bushing (inner side)
9 Front lower arm (front) bushing (outer side)

67162-MAZC-G250

Exploded view of the front suspension lower control arm (front) assembly—Mazda 6

1. Before servicing the vehicle, refer to the precautions in the beginning of this section.

2. Remove the three way converter on models with the 3.0L (AJ) engine.

3. Separate the lower control arm (front) ball joint from the knuckle.

4. Remove the front shock lower bolt and the dynamic damper.

5. Remove the lower inner side arm bolt.

6. Remove the front lower arm (front).

To install:

7. Installation is the reverse of removal. refer to the illustration for bolt torque specifications.

FRONT ARM (REAR SIDE)

Refer to the accompanying illustration for component locations and torque specifications.

1. Before servicing the vehicle, refer to the precautions in the beginning of this section.

2. Remove the wheels, engine under cover and splash shield.

3. Remove the steering gear and linkage bolts, pipe assembly from the crossmember and wire the assembly aside.

4. Remove the stabilizer bar link nut.

5. Separate the lower control arm (rear) ball joint from the knuckle.

6. Remove the No. 1 center bolt.

7. Support the crossmember with a jack and remove the crossmember bracket.

8. Remove the front lower arm (rear).

To install:

9. Installation is the reverse of removal. refer to the illustration for bolt torque specifications.

REAR LOWER CONTROL ARM

1. Before servicing the vehicle, refer to the precautions in the beginning of this section.

2. Support the lower control arm with a floor jack.

3. Remove the rear stabilizer bar bolt from the lower link side.

4. Loosen the inner bolt of the rear lower control arm.

5. Remove the outer bolts of the rear lower control arm.

6. Lower the control arm and remove the coil spring.

7. Remove the rear lower arm inner side bolt and the arm.

To install:

8. Install the rear lower arm and the inner side bolt. Tighten the bolt to 63–85 ft. lbs. (86–116 Nm).

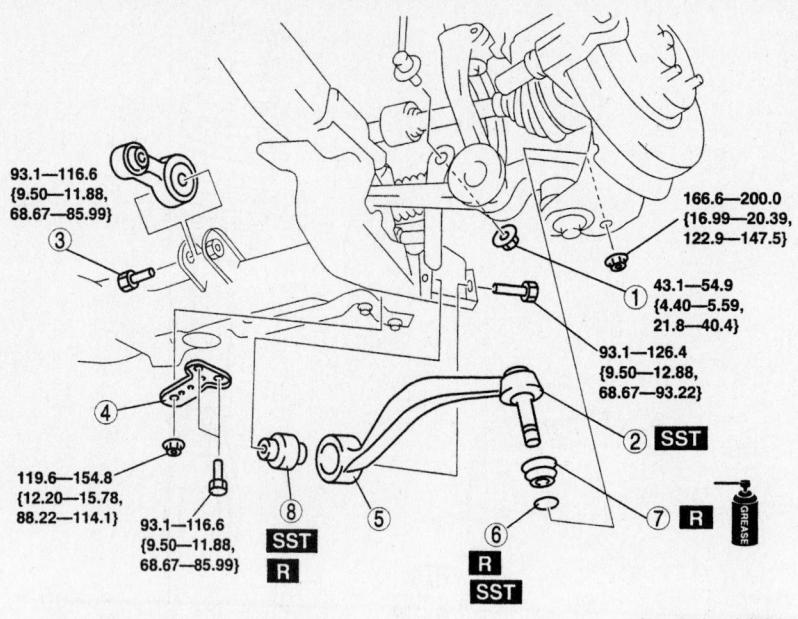

93.1—116.6
{9.50—11.88,
68.67—85.99}

166.6—200.0
{16.99—20.39,
122.9—147.5}

43.1—54.9
{4.40—5.59,
21.8—40.4}

93.1—126.4
{9.50—12.88,
68.67—93.22}

119.6—154.8
{12.20—15.78,
88.22—114.1}

93.1—116.6
{9.50—11.88,
68.67—85.99}

N·m {kgf·m, ft·lbf}

1	Nut (stabilizer control link lower side)	5	Front lower arm (rear)
2	Front lower arm (rear) ball joint	6	Clip
3	No.1 engine mount center bolt	7	Dust boot
4	Crossmember bracket	8	Front lower arm (rear) bushing

67162-MAZC-G251

Exploded view of the front suspension lower control arm (rear) assembly—Mazda 6

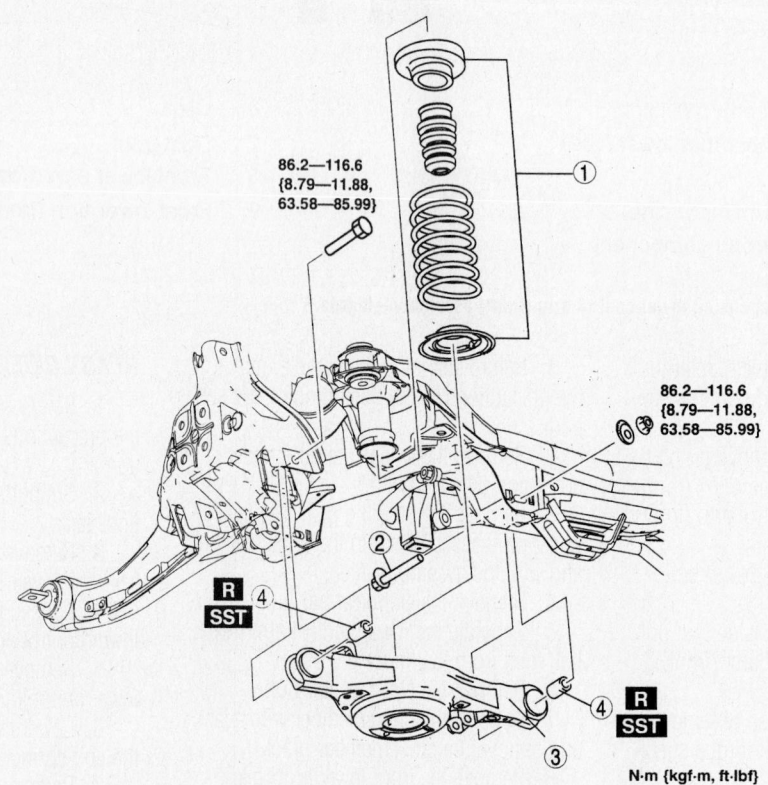

86.2—116.6
{8.79—11.88,
63.58—85.99}

86.2—116.6
{8.79—11.88,
63.58—85.99}

N·m {kgf·m, ft·lbf}

1	Rear coil spring component	3	Rear lower arm
2	Bolt (rear lower arm inner side)	4	Rear lower arm bushing

67162-MAZC-G249

Exploded view of the rear suspension lower control arm assembly—Mazda 6

9. Install the coil spring with the small outer diameter side faces downwards and raise the lower control arm using the jack.

10. Install the lower arm outer and inner bolts and tighten the outer bolt first then the inner bolt to 63–86 ft. lbs. (86–116 Nm).

11. Install the rear stabilizer link bolt and tighten to 31–44 ft. lbs. (43–60 Nm).

CONTROL ARM BUSHING REPLACEMENT

Except Protégé and Mazda 6 Models

1. Before servicing the vehicle, refer to the precautions in the beginning of this section.

All Mazda's use a pressed in control arm bushing, and the pressing can usually be done using 2 appropriately sized sockets (a press socket and a catch socket) and a large vise.

2. Position the control arm and the 2 sockets into a vise.

3. Position the press socket onto the control arm bushing.

4. Position the catch socket onto the control arm, opposite of the press socket.

5. Tighten the vise slowly and press the bushing into the catch socket.

To install:

6. Apply soapy water to the new control arm bushing.

7. Position the bushing against the control arm.

8. Using the same sockets, in the same positions, press the new bushing into the control arm.

Protégé

1. Before servicing the vehicle, refer to the precautions in the beginning of this section.

2. Position the control arm into a vise. Cut away the projecting rubber from the front bushing. Use tool 49 B034 2A2 to remove the bushing.

3. Remove the rear bushing using a press and tools 49 B034 212 and 49 G034 202.

To install:

4. When installing the rear bushing on the lower arm, align the mark of the lower arm and the small projection of the lower arm bushing (rear) as illustrated, set the arm onto the press and press the bushings into position.

5. Apply soapy water to the new control arm bushing.

6. Install the rear bushing using a press and tools 49 B034 212 and 49 G034 211.

7. Position the control arm into a vise.

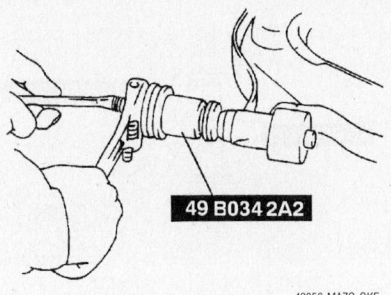

Use tool 49 B034 2A2 to remove and install the front bushing—Protégé

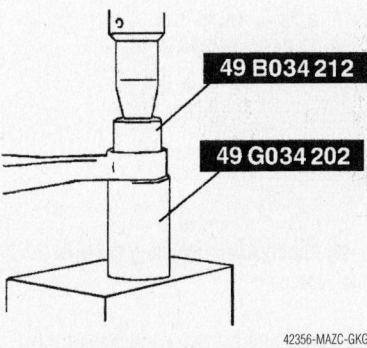

Use tools 49 B034 212 and 49 G034 202 to remove the rear bushing—Protégé

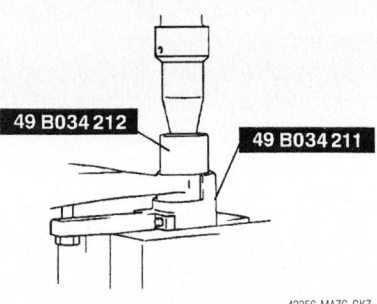

Use tools 49 B034 212 and 49 G034 211 to install the rear bushing—Protégé

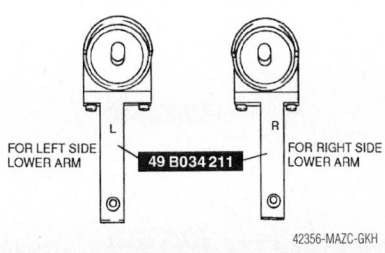

Align the mark of the lower arm and the small projection of the lower arm bushing (rear) when installing the rear bushing—Protégé

Use tool 49 B034 2A2 to pull the bushing into the arm bore.

Mazda 6

The Mazda 6 models have two lower control arms in the front suspension assembly, they are identified as the front lower control arm (Front side) and front lower control arm (rear side). They also have one rear suspension lower control arm, which will be identified as Rear lower control arm.

FRONT ARM (FRONT SIDE)

1. Before servicing the vehicle, refer to the precautions in the beginning of this section.

2. Remove the front lower control arm (front).

3. Press the bushing from the front lower control arm (front) inner side using the tools illustrated.

4. Press the bushing from the front lower control arm (front) outer side using the tools illustrated.

To install:

5. Mark the front bushing for the outer side on the front lower arm (front) as illustrated.

6. Press the bushing into the arm up to the marking made on the bushing using the

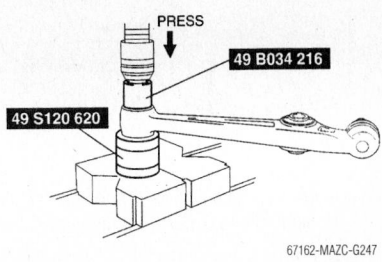

Press the bushing from/into the front lower control arm (front) inner side using the removal/installer tools shown—Mazda 6

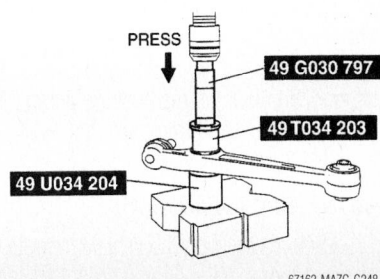

Press the bushing from/into the front lower control arm (front) outer side using the removal/installer tools shown—Mazda 6

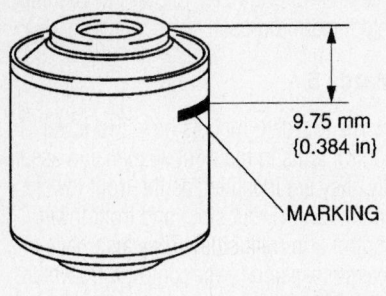

9.75 mm
{0.384 in}

MARKING

67162-MAZC-G245

Mark the front bushing for the outer side on the front lower arm (front) as shown—Mazda 6

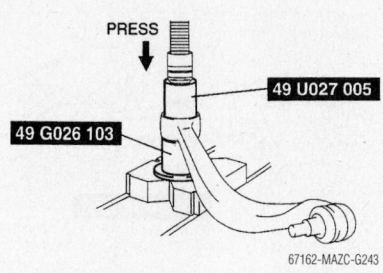

PRESS

49 U027 005

49 G026 103

67162-MAZC-G243

Press the bushing from/into the front lower control arm (rear) using the removal/installer tools shown—Mazda 6

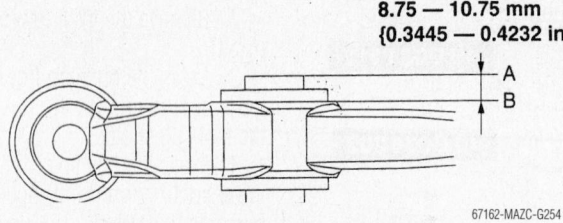

8.75 — 10.75 mm
{0.3445 — 0.4232 in}

A
B

67162-MAZC-G254

Make sure the clearance is 0.3445–0.4234 inch (8.75–10.75mm) when installing the front bushing for the outer side on the front lower arm (front)—Mazda 6

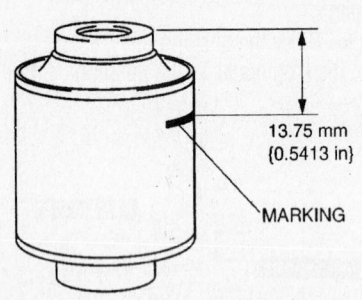

13.75 mm
{0.5413 in}

MARKING

67162-MAZC-G246

Mark the front bushing for the inner side on the front lower arm (front) as shown—Mazda 6

tools and press used to remove the bushing. Make sure the clearance is 0.3445–0.4234 inch (8.75–10.75mm) as illustrated.

7. Mark the front bushing for the inner side on the front lower arm (front) as illustrated.

8. Press the bushing into the arm up to the marking made on the bushing using the tools and press used to remove the outer side bushing.

FRONT ARM (REAR SIDE)

1. Before servicing the vehicle, refer to the precautions in the beginning of this section.

2. Remove the front lower control arm (rear).

3. Press the bushing from the front

lower control arm (rear) using the tools illustrated. remove the arm from the press and use a hammer to remove the bushing completely.

To install:

4. Mark the rear bushing on the front lower arm (rear) as illustrated.

5. Press the bushing into the arm up to the marking made on the bushing using the

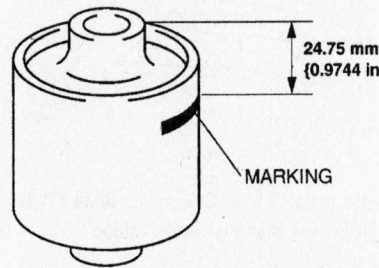

24.75 mm
{0.9744 in}

MARKING

67162-MAZC-G255

Mark the rear bushing on the front lower arm (rear) as shown—Mazda 6

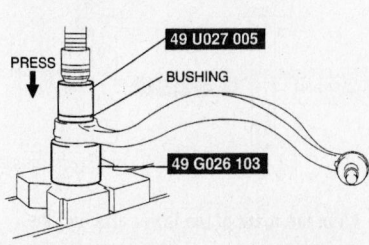

49 U027 005

BUSHING

PRESS

49 G026 103

67162-MAZC-G256

Install the rear bushing on the front lower arm (rear) as shown—Mazda 6

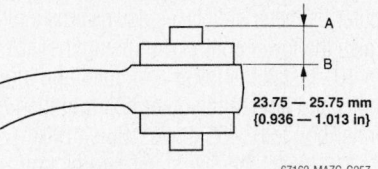

A
B

23.75 — 25.75 mm
{0.936 — 1.013 in}

67162-MAZC-G257

The rear bushing on the front lower arm (rear) should be 0.936–1.013 inch (23–25mm) once the bushing is seated properly—Mazda 6

tools and press used to remove the bushing.

6. The clearance should be 0.936–1.013 inch (23–25mm) as illustrated once the bushing is seated properly.

REAR LOWER CONTROL ARM

1. Before servicing the vehicle, refer to the precautions in the beginning of this section.

2. Remove the lower control arm.

3. Press the bushing from the arm using the tools illustrated.

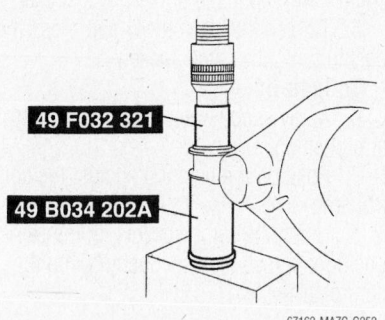

49 F032 321

49 B034 202A

67162-MAZC-G252

Press the bushing from the rear suspension lower control arm using the tools shown—Mazda 6

To install:

4. Press the bushing into the arm using the tools illustrated.

5. Install the lower control arm.

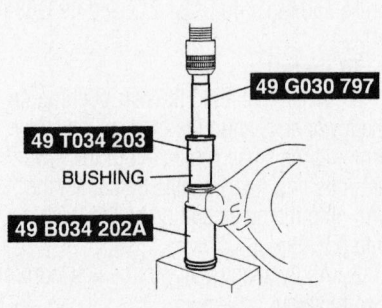

49 G030 797

49 T034 203

BUSHING

49 B034 202A

67162-MAZC-G253

Press the bushing into the rear suspension lower control arm using the tools shown—Mazda 6

Wheel Bearings

ADJUSTMENT

The front and rear wheel bearings are not adjustable. If the bearings become loose or make noise, they must be replaced.

REMOVAL & INSTALLATION

626

FRONT

1. Before servicing the vehicle, refer to the precautions in the beginning of this section.

2. Refer to the illustration for component location and torque specifications.
3. Remove or disconnect the following:
   - Wheels
   - Halfshaft axle nut, unstake the nut prior to removal
   - Brake caliper and rotor
   - Tie rod end from the knuckle

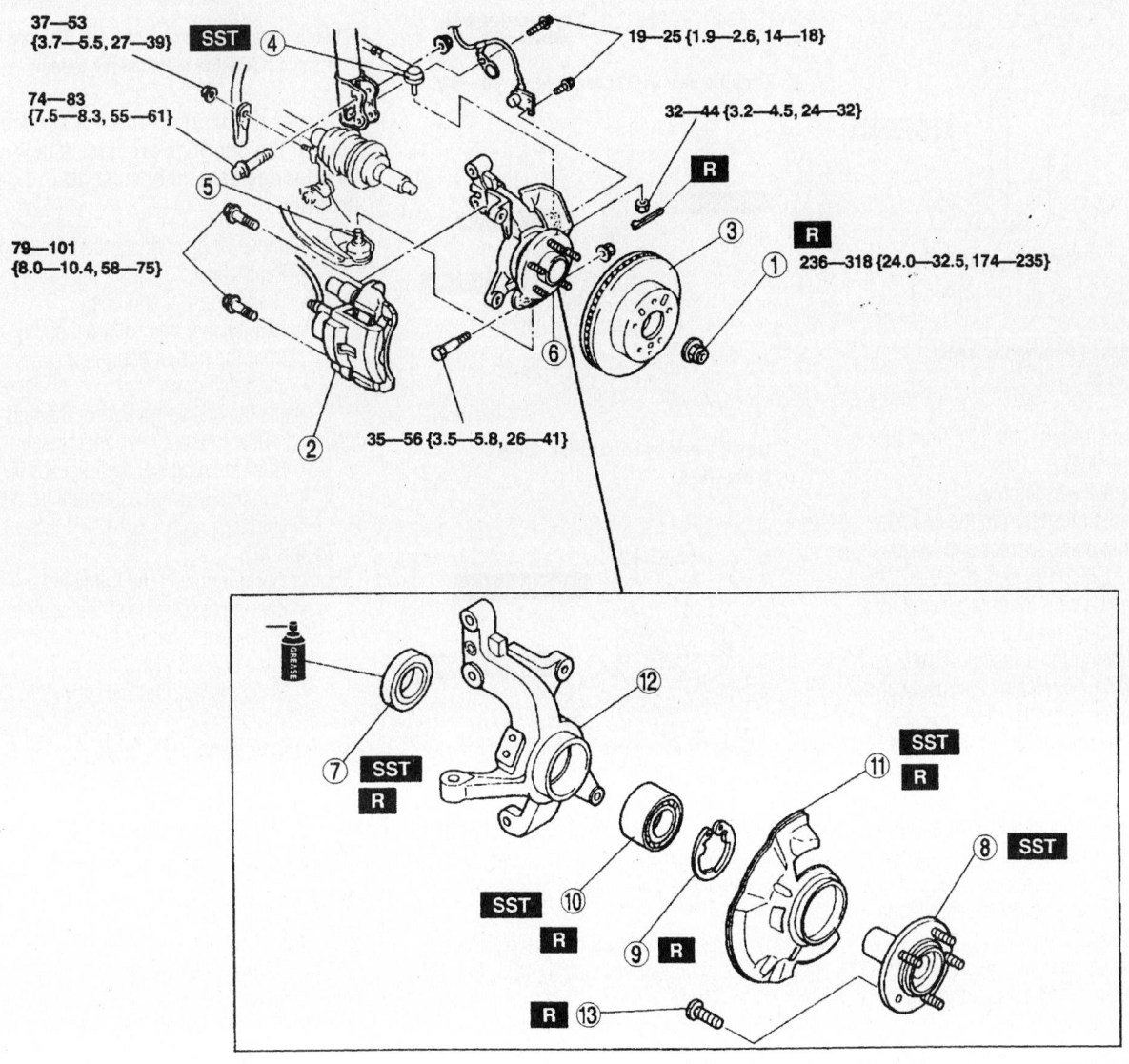

N·m {kgf·m, ft·lbf}

1	Locknut	8	Wheel hub component
2	Brake caliper component	9	Retaining ring
3	Disc plate	10	Wheel bearing
4	Tie-rod end ball joint	11	Dust cover
5	Lower arm ball joint	12	Steering knuckle
6	Wheel hub, steering knuckle, dust cover	13	Hub bolt
7	Oil seal		

Exploded view of the front wheel bearing and knuckle assembly—626

42356-MAZC-G83

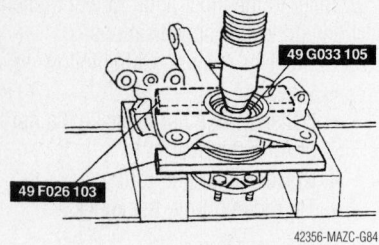

Use a press to remove the hub—626

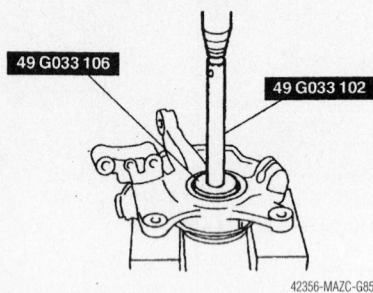

Use a press to remove the wheel bearing—626

- Control arm ball joint from the knuckle
- Knuckle assembly
- Inner oil seal from the knuckle
- Hub using a press and Mazda tools 49 G033 103 and 49 G033 105. If the bearing inner race remains in the hub, grind a section of the bearing inner race until 0.02 inch (0.5mm) remains and use a chisel to remove it.
- Bearing from the hub using a press and Mazda tools 49 G033 102 and 49 G033 106
- Brake dust shield, if it is being replaced. Mark the cover and knuckle for replacement purposes and use a chisel to remove the shield.

4. Clean and inspect all parts but do not wash or clean the wheel bearing. The bearing must be replaced.

To install:

5. Using Mazda Press tools 49 G033 107a and 49 F027 009, install a new dust shield cover assembly to the knuckle, if removed.

6. Using press tools 49 G033 797, 49 F027 004 and 49 F027 003; press a new wheel bearing into the knuckle assembly.

7. Install or connect the following:
- Wheel bearing retaining ring

8. Using press tools 49 G033 105, 49 F027 009; press in the hub assembly.
- New oil seal using installation tool 49 V001 795
- Knuckle assembly, tighten the

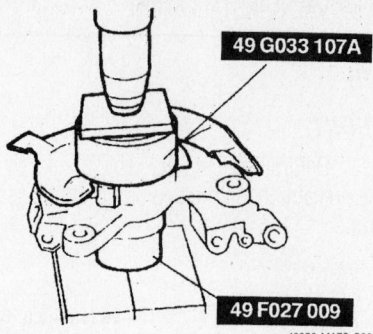

Install a new dust cover, if removed—626

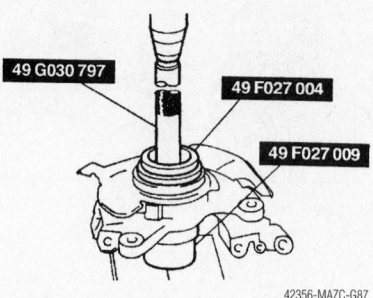

Install a new wheel bearing using a press—626

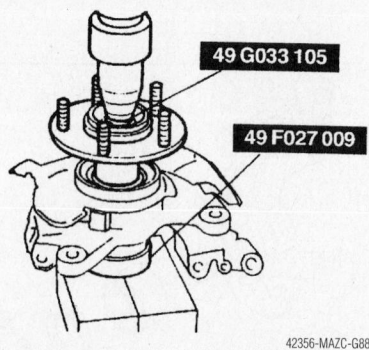

Install a wheel hub using a press—626

upper bolt to 61 ft. lbs. (83 Nm) and the lower bolt to 41 ft. lbs. (56 Nm)
- Control arm ball joint to the knuckle
- Tie rod end to the knuckle
- Brake rotor and caliper
- Halfshaft axle nut, tighten the nut to 174–235 ft. lbs. (236–318 Nm) and stake the nut
- Wheels

REAR

1. Before servicing the vehicle, refer to the precautions in the beginning of this section.

2. Refer to the illustration for component location and torque specifications.

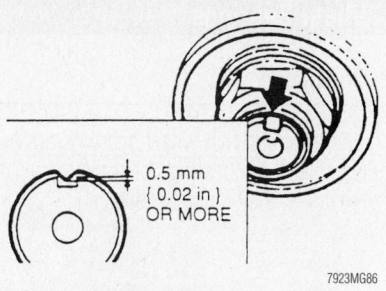

When installing the new hub nut, be sure to stake it into the notch on the spindle

➡The wheel bearings are not service-able. If the bearings are bad, a new hub/bearing assembly must be installed.

3. Remove or disconnect the following:
- Rear wheels
- Brake drum, if equipped
- Rear caliper and rotor assembly from the hub, if equipped
- Hub dust cover

4. Raise the staked portion of the hub retaining nut with a hammer and chisel.
- Hub retaining nut and discard it
- Hub and bearing assembly from the spindle

To install:

5. Install or connect the following:
- Bearing assembly on the spindle. Torque the new nut to 131–173 ft. lbs. (177–235 Nm).
- Stake the nut into the groove in the spindle
- Dust cover
- Assemble the brakes
- Rear wheel

Protégé

FRONT

1. Before servicing the vehicle, refer to the precautions in the beginning of this section.

2. Refer to the illustration for component location and torque specifications.

3. Remove or disconnect the following:
- Wheels
- Halfshaft axle nut, Unstake the nut prior to removal
- Brake caliper and rotor
- Tie rod end from the knuckle
- Control arm ball joint from the knuckle
- Knuckle assembly
- Inner oil seal from the knuckle
- Hub using a press and Mazda tools 49 G030 727, 49 G033 102 and 49 F026 103. If the bearing inner race remains in the hub, grind a section

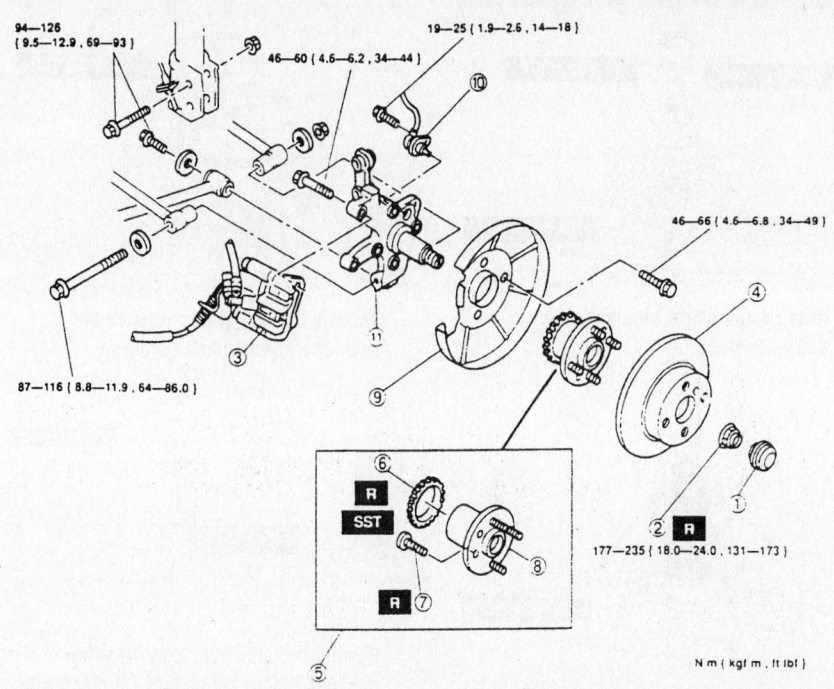

1. Hub cap
2. Locknut
3. Brake caliper assembly
4. Disc plate
5. Wheel hub assembly
   Inspect for damage
   Inspect bearing for damage and rough rotation
6. ABS sensor rotor
7. Hub bolt
8. Wheel hub
9. Dust cover
   Inspect for damage and cracks
10. ABS wheel-speed sensor
11. Hub spindle
    Inspect for damage and cracks

N·m (kgf·m , ft·lbf)

7923MG87

Exploded view of the rear wheel hub and bearing assembly (disc brake model shown, drum is similar)—626

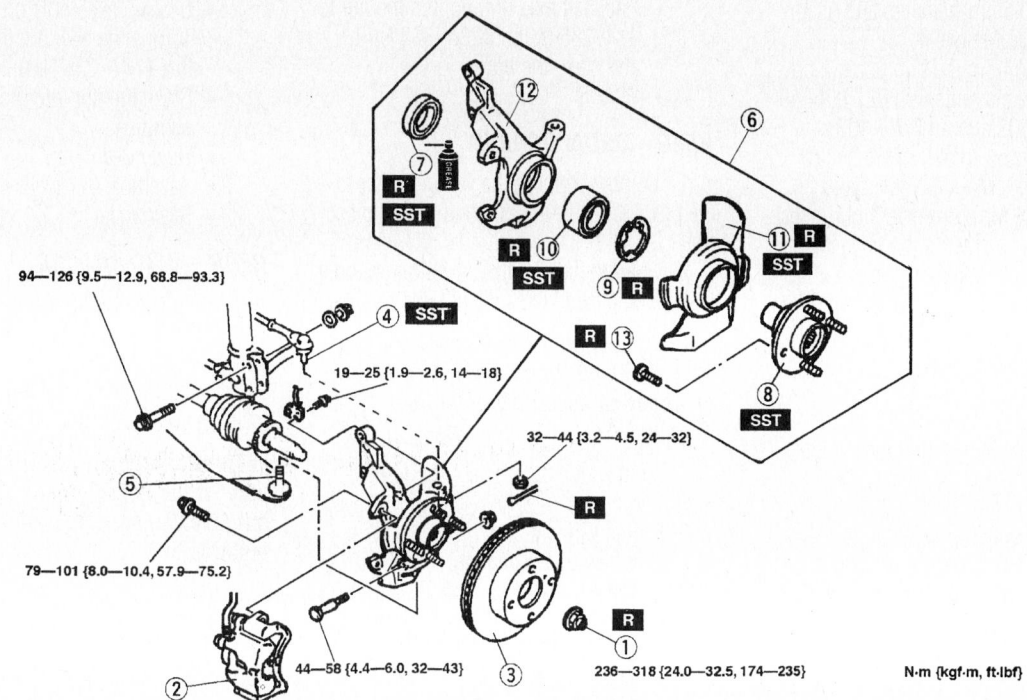

N·m {kgf·m, ft·lbf}

42356-MAZC-GKI

1	Locknut	8	Wheel hub component
2	Brake caliper component	9	Retaining ring
3	Disc plate	10	Wheel bearing
4	Tie-rod end	11	Dust cover
5	Lower arm ball joint	12	Steering knuckle
6	Wheel hub, steering knuckle, dust cover	13	Hub bolt
7	Oil seal		

Exploded view of the front wheel bearing and knuckle assembly—Protégé

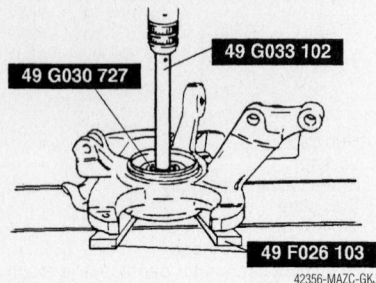

Use a press to remove the hub—Protégé

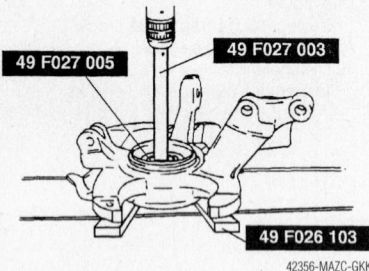

Use a press to remove the wheel bearing—Protégé

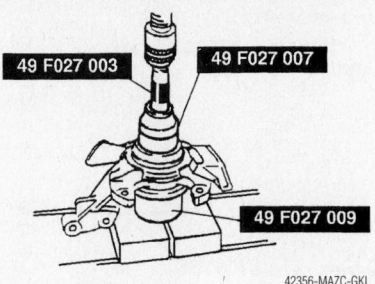

Install a new wheel bearing using a press—Protégé

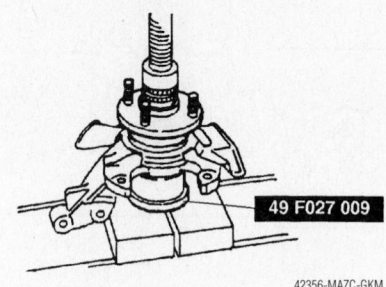

Install a wheel hub using a press—Pro-

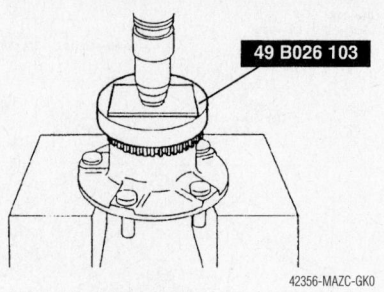

Position the ABS sensor rotor on the hub—Protégé with drum brakes

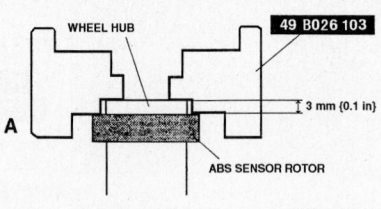

Position tool 49 B026 103 so that the marking A faces the bottom—Protégé with drum brakes

of the bearing inner race until 0.02 inch (0.5mm) remains and use a chisel to remove it.
- Bearing from the hub using a press and Mazda tools 49 F027 005, 49 F027 003 and 49 F026 103
- Brake dust shield, if it is being replaced. Mark the cover and knuckle for replacement purposes and use a chisel to remove the shield.

4. Clean and inspect all parts but do not wash or clean the wheel bearing. The bearing must be replaced.

To install:

5. Using Mazda Press tools 49 E033 101 and 49 F027 009, install a new dust shield cover assembly to the knuckle, if removed.

6. Using press tools 49 F027 003, 49 F027 007 and 49 F027 009; press a new wheel bearing into the knuckle assembly.

7. Install or connect the following:
- Wheel bearing retaining ring

8. Using press tool 49 F027 009 press in the hub assembly.
- New oil seal using installation tool 49 V001 795
- Knuckle assembly, tighten the upper bolt to 93 ft. lbs. (126 Nm) and the lower bolt to 43 ft. lbs. (58 Nm)
- Control arm ball joint to the knuckle
- Tie rod end to the knuckle

- Brake rotor and caliper
- Halfshaft axle nut, tighten the nut to 174–235 ft. lbs. (236–318 Nm) and stake the nut
- Wheels

REAR—DRUM BRAKES

1. Before servicing the vehicle, refer to the precautions in the beginning of this section.

2. Refer to the illustration for component location and torque specifications.

➥**The wheel bearings are not serviceable. If the bearings are bad, a new hub/bearing assembly must be installed.**

3. Remove or disconnect the following:
- Rear wheels
- Wheel speed sensor

4. Raise the staked portion of the hub retaining nut with a hammer and chisel.
- Hub retaining nut and discard it
- Brake drum
- Hub and bearing assembly from the spindle with the ABS sensor rotor, if equipped with ABS
- ABS sensor rotor using a chisel

To install:

5. Position the ABS sensor rotor on the hub as illustrated.

6. Position tool 49 B026 103 so that the marking **A** faces the bottom.

7. Press the rotor onto the hub.

8. Install or connect the following:
- Bearing assembly on the spindle. Torque the new nut to 131–173 ft. lbs. (177–235 Nm).
- Stake the nut into the groove in the spindle
- Dust cover
- Assemble the brakes
- Rear wheel

REAR—DISC BRAKES

1. Before servicing the vehicle, refer to the precautions in the beginning of this section.

2. Refer to the illustration for component location and torque specifications.

➥**The wheel bearings are not serviceable. If the bearings are bad, a new hub/bearing assembly must be installed.**

3. Remove or disconnect the following:
- Rear wheels
- Wheel speed sensor

4. Raise the staked portion of the hub retaining nut with a hammer and chisel.
- Hub retaining nut and discard it
- Brake caliper and rotor
- Hub and bearing assembly from the spindle with the ABS sensor rotor, if equipped with ABS
- ABS sensor rotor using a chisel

To install:

5. Position the ABS sensor rotor on the hub as illustrated.

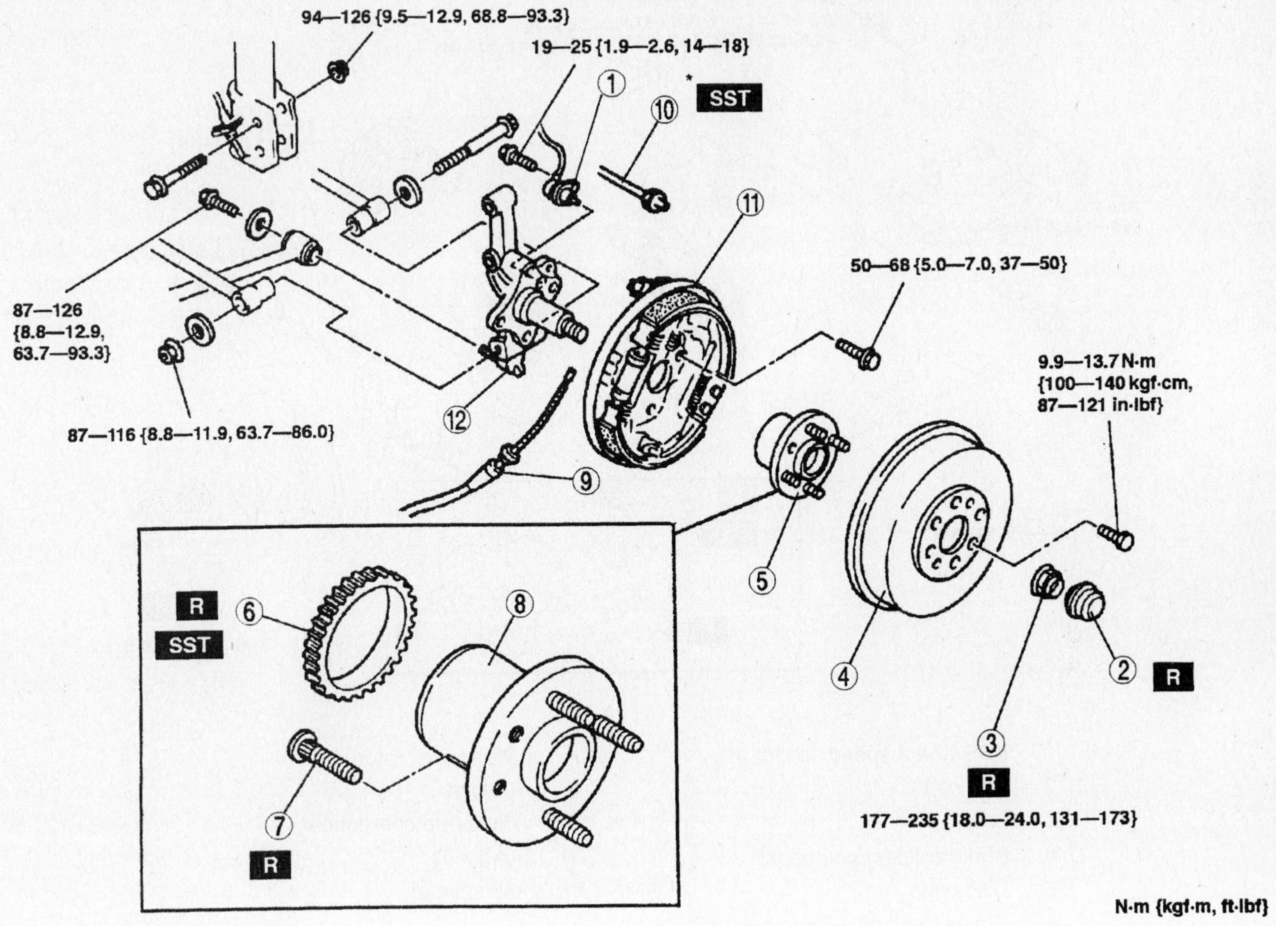

94—126 {9.5—12.9, 68.8—93.3}

19—25 {1.9—2.6, 14—18}

SST

50—68 {5.0—7.0, 37—50}

87—126
{8.8—12.9,
63.7—93.3}

9.9—13.7 N·m
{100—140 kgf·cm,
87—121 in·lbf}

87—116 {8.8—11.9, 63.7—86.0}

R
SST

R

R

177—235 {18.0—24.0, 131—173}

N·m {kgf·m, ft·lbf}

* 49 0259 770B (Flare nut wrench)

1	ABS-wheel speed sensor	7	Hub bolt
2	Hub cap	8	Wheel hub
3	Locknut	9	Parking brake cable
4	Brake drum	10	Brake pipe
5	Wheel hub and ABS sensor rotor (with ABS)	11	Rear brake component
6	ABS sensor rotor (with ABS)	12	Hub spindle

42356-MAZC-GKO

Exploded view of the rear wheel hub and bearing assembly—Protégé with drum brakes

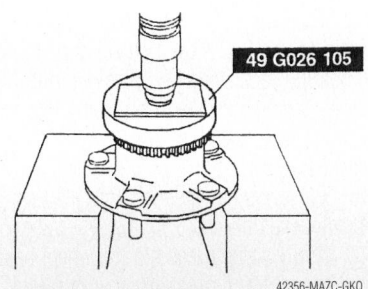

49 G026 105

42356-MAZC-GKQ

Position the ABS sensor rotor on the hub—Protégé with disc brakes

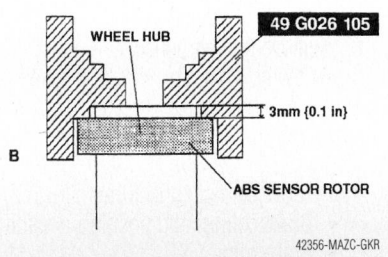

WHEEL HUB
49 G026 105
3mm {0.1 in}
B
ABS SENSOR ROTOR

42356-MAZC-GKR

Position tool 49 B026 105 so that the marking B faces the bottom—Protégé with disc brakes

6. Position tool 49 B026 105 so that the marking **B** faces the bottom.
7. Press the rotor onto the hub.
8. Install or connect the following:
   - Bearing assembly on the spindle. Torque the new nut to 131–173 ft. lbs. (177–235 Nm).
   - Stake the nut into the groove in the spindle
   - Dust cover
   - Assemble the brakes
   - Rear wheel

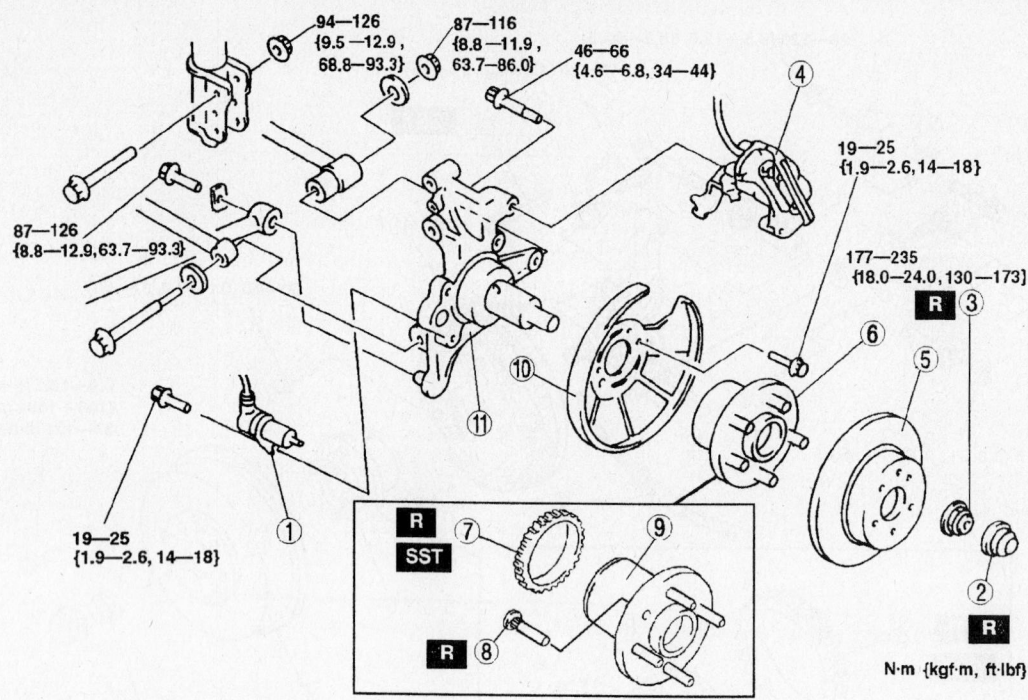

94—126
{9.5—12.9,
68.8—93.3}

87—116
{8.8—11.9,
63.7—86.0}

46—66
{4.6—6.8, 34—44}

19—25
{1.9—2.6, 14—18}

87—126
{8.8—12.9, 63.7—93.3}

177—235
{18.0—24.0, 130—173]

19—25
{1.9—2.6, 14—18}

N·m {kgf·m, ft·lbf}

1	ABS wheel-speed sensor	7	ABS sensor rotor
2	Hub cap	8	Hub bolt
3	Locknut	9	Wheel hub component
4	Brake caliper component	10	Dust cover
5	Disc plate	11	Hub spindle
6	Wheel hub and ABS sensor rotor		

42356-MAZC-GKN

Exploded view of the rear wheel hub and bearing assembly—Protégé with disc brakes

Miata

FRONT

1. Before servicing the vehicle, refer to the precautions in the beginning of this section.

➡ **The wheel bearings are not serviceable. If the bearings are bad, a new hub/bearing assembly must be installed.**

2. Remove or disconnect the following:
   • Front wheels
   • Hub center dust cap
3. Raise the staked portion of the hub locknut with a hammer and chisel. Lock the hub by applying the brakes and loosen the nut.
   • Brake caliper
   • Brake rotor
   • Hub locknut and discard it
   • Hub from the spindle

To install:
4. Install or connect the following:

   • Hub to the spindle
   • Hub locknut. Torque the nut to 123–159 ft. lbs. (167–215 Nm).
   • Stake the new nut into the spindle groove using a dull chisel
   • Brake rotor
   • Brake caliper
   • Hub center dust cap
   • Front wheels

REAR

1. Before servicing the vehicle, refer to the precautions in the beginning of this section.
2. Remove the rear wheels.
3. Raise the staked portion of the hub locknut with a hammer and chisel. Lock the hub by applying the brakes and loosen the nut.
4. Remove or disconnect the following:
   • Brake caliper and position it aside
   • Brake rotor
   • Hub locknut and discard it
   • Anti-lock Brake System (ABS) wheel speed sensor, if equipped

   • Speed sensor bracket from the rear knuckle
   • Upper and lower knuckle mounting through bolts
   • Knuckle assembly from vehicle
   • Rear bearing oil seal by prying it from the knuckle
   • Wheel hub by pressing it from the knuckle
   • Retaining snapring from within the knuckle
   • Bearing assembly from the knuckle, once the wheel hub is removed

5. The inner race of the bearing may remain on the hub. Use a chisel and move the bearing race away from the rear hub flange. Once there is enough clearance, press the race from the hub.

To install:
6. Press the new bearing into the rear knuckle assembly.
7. Apply some grease to the wheel bearing inner race and press the rear hub into the bearing. Make sure to position a suitable support on the backside of the bearing.

8. Install or connect the following:
- Retaining snapring to the knuckle
- New rear oil seal by lubricating it with grease and pressing it into the rear knuckle
- Knuckle assembly to the vehicle
- Upper and lower knuckle mounting through bolts. Torque the upper bolt to 34–49 ft. lbs. (47–66 Nm) and the lower bolt to 47–54 ft. lbs. (63–74 Nm).
- Speed sensor bracket to the rear knuckle
- ABS wheel speed sensor, if equipped
- New hub locknut. Torque the nut to 174–235 ft. lbs. (236–318 Nm).
- Stake the new nut into the spindle groove using a dull chisel
- Brake rotor

- Brake caliper
- Rear wheels

Millenia

FRONT

1. Before servicing the vehicle, refer to the precautions in the beginning of this section.
2. Refer to the illustration for component location and torque specifications.
3. Remove or disconnect the following:
- Wheels
- Halfshaft axle nut, Unstake the nut prior to removal.
- Brake caliper and rotor
- Tie rod end from the knuckle
- Upper leading link ball joint
- Upper lateral link ball joint

- Lower ball joint
- Knuckle assembly
- Snap ring
- Hub using a press and Mazda tools 49 F026 102 and 49 W017 101. If the bearing inner race remains in the hub, grind a section of the bearing inner race until 0.02 inch (0.5mm) remains and use a chisel to remove it.
- Bearing from the hub using a press and Mazda tools 49 G0797, 49 G026 102, 49 E033 101 and 49 W017 101
- Brake dust shield, if it is being replaced. Mark the cover and knuckle for replacement purposes and use a chisel to remove the shield.

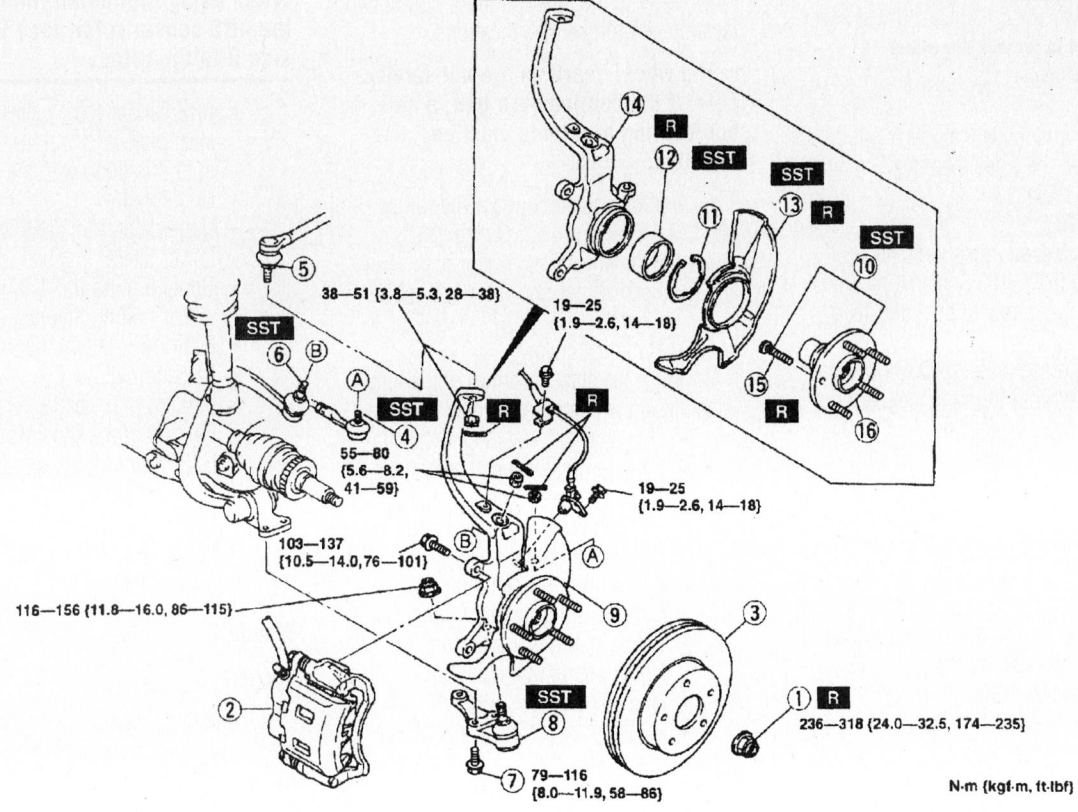

1 Locknut	9 Front wheel hub, steering knuckle
2 Brake caliper component	10 Wheel hub component
3 Disc plate	11 Snap ring
4 Tie-rod end ball joint	12 Wheel bearing
5 Upper leading link ball joint	13 Dust cover
6 Upper lateral link ball joint	14 Steering knuckle
7 Lower ball joint bolt	15 Hub bolt
8 Lower ball joint	16 Wheel hub

42356-MAZC-GFC

Exploded view of the front wheel bearing and knuckle assembly—Millenia

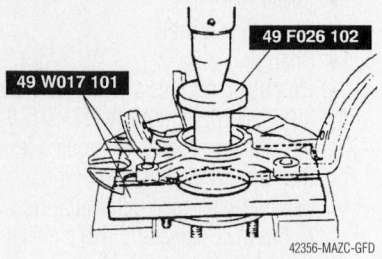

Use a press to remove the hub—Millenia

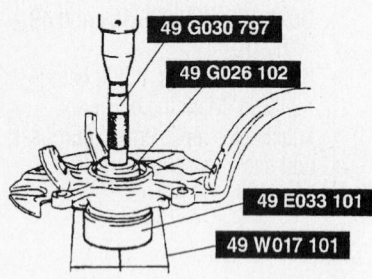

Use a press to remove the wheel bearing—Millenia

4. Clean and inspect all parts but do not wash or clean the wheel bearing. The bearing must be replaced.

To install:

5. Using Mazda Press tools 49 G033 107a and 49 G026 103, install a new dust shield cover assembly to the knuckle, if removed.

6. Using press tools 49 G026 103 and 49 F026 102, press a new wheel bearing into the knuckle assembly.

7. Install or connect the following:
- Wheel bearing retaining ring

8. Using press tools 49 G033 105, 49 F027 009; press in the hub assembly.
- Snap ring
- Knuckle assembly, tighten the upper bolt to the specifications shown in the illustration
- Lower ball joint
- Upper lateral link ball joint
- Upper leading link ball joint
- Tie rod end to the knuckle
- Brake caliper and rotor

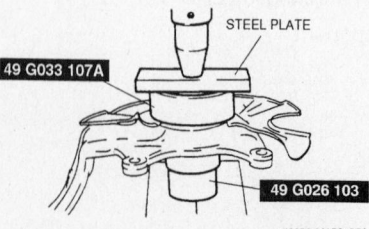

Install a new wheel bearing using a press -Millenia

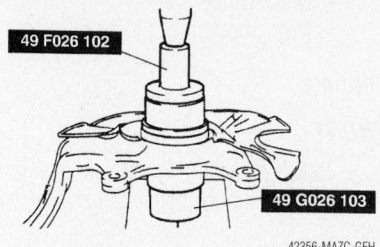

Install a wheel hub using a press—Millenia

- Halfshaft axle nut, tighten the nut to 174–235 ft. lbs. (236–318 Nm) and stake the nut
- Wheels

REAR

1. Before servicing the vehicle, refer to the precautions in the beginning of this section.

2. Refer to the illustration for component location and torque specifications.

➡ The wheel bearings are not serviceable. If the bearings are bad, a new hub/bearing assembly must be installed.

3. Remove or disconnect the following:
- Rear wheels
- Rear caliper and rotor assembly from the hub
- Hub dust cover

4. Raise the staked portion of the hub retaining nut with a hammer and chisel.
- Hub retaining nut and discard it
- Parking brake shoe assembly
- Parking brake cable
- Backing plate
- Rear lower lateral link ball joint
- Lower trailing link ball joint
- Upper trailing link ball joint
- Lower lateral link ball joint
- Upper lateral link ball joint
- Hub spindle
- ABS sensor rotor using a chisel
- Hub bolts
- Hub and bearing assembly from the spindle

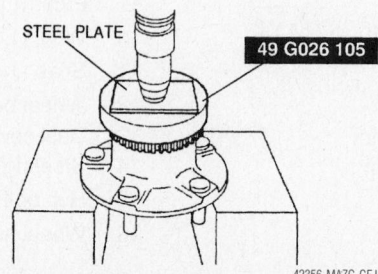

Install the ABS rotor—Millenia

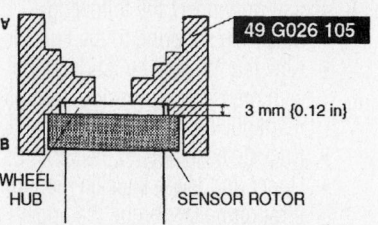

3 mm {0.12 in}

WHEEL HUB   SENSOR ROTOR

When using the installation tool for the ABS sensor rotor, face the carved side B to the rotor—Millenia

To install:
- Hub and bearing assembly on the spindle
- Hub bolts and tighten to the specifications shown in the illustration

✳✳ CAUTION

When using the installation tool for the ABS sensor rotor, face the carved side B to the rotor.

- ABS sensor rotor using a press, steel plate and tool 49 G02610 until it is flush with the hub.
- Hub spindle
- Upper lateral link ball joint
- Lower lateral link ball joint
- Upper trailing link ball joint
- Lower trailing link ball joint
- Rear lower lateral link ball joint
- Backing plate
- Parking brake cable
- Parking brake shoe assembly
- Hub retaining nut and discard it
- Hub dust cover
- Rear caliper and rotor assembly from the hub, if equipped
- Rear wheels

Mazda 3

FRONT

1. Before servicing the vehicle, refer to the precautions in the beginning of this section.

2. Refer to the illustration for component location and torque specifications.

3. Remove or disconnect the following:
- Wheels
- ABS sensor connector and the sensor
- Halfshaft lockbolt
- Brake caliper and rotor
- Tie rod end from the knuckle
- Lower control arm ball joint from the knuckle
- Stabilizer bar link nut
- Knuckle assembly
- Hub bearing using a press and

55—80{5.6—8.2,41—59}

44—60{4.4—6.2,32—44}

116—156
{11.8—16.0,86—115}

50—68
{5.0—7.0,37—50}

75—104
{7.6—10.7,55—77}

38—51{3.8—5.3,28—38}

19—25{1.9—2.6,14—18}

55—80{5.6—8.2,41—59}

28—40
{2.8—4.1,
21—29}

177—235
{18.0—24.0,
131—173}

N·m {kgf·m, ft·lbf}

1	Brake caliper component	10	Lower trailing link ball joint
2	Hub cap	11	Upper trailing link ball joint
3	Locknut	12	Lower lateral link ball joint
4	Disc plate	13	Upper lateral link ball joint
5	Wheel hub component	14	Hub spindle
6	Parking brake shoe component	15	ABS sensor rotor
7	Parking brake cable	16	Hub bolt
8	Back plate	17	Wheel hub
9	Rear lower lateral link outer ball joint		

42356-MAZC-GFI

Exploded view of the rear hub and spindle assembly—Millenia

Mazda tools 49 G030 795 and 49 B033 1A0

4. Clean and inspect all parts but do not wash or clean the wheel bearing. The bearing must be replaced.

To install:

5. Using Mazda Press tools 49 H034 201 and 49 B033 1A0, press a new wheel bearing into the knuckle assembly. Make sure the installation tool engages to the bearing outer race properly to avoid damage.

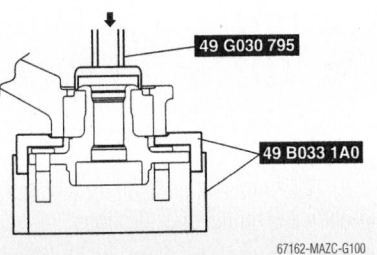

67162-MAZC-G100

Use a press and the tools illustrated to disassemble the hub/bearing assembly— Mazda 3

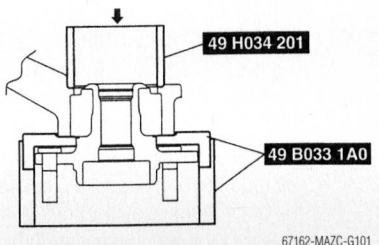

67162-MAZC-G101

Use a press and the tools illustrated to assemble the hub/bearing assembly— Mazda 3

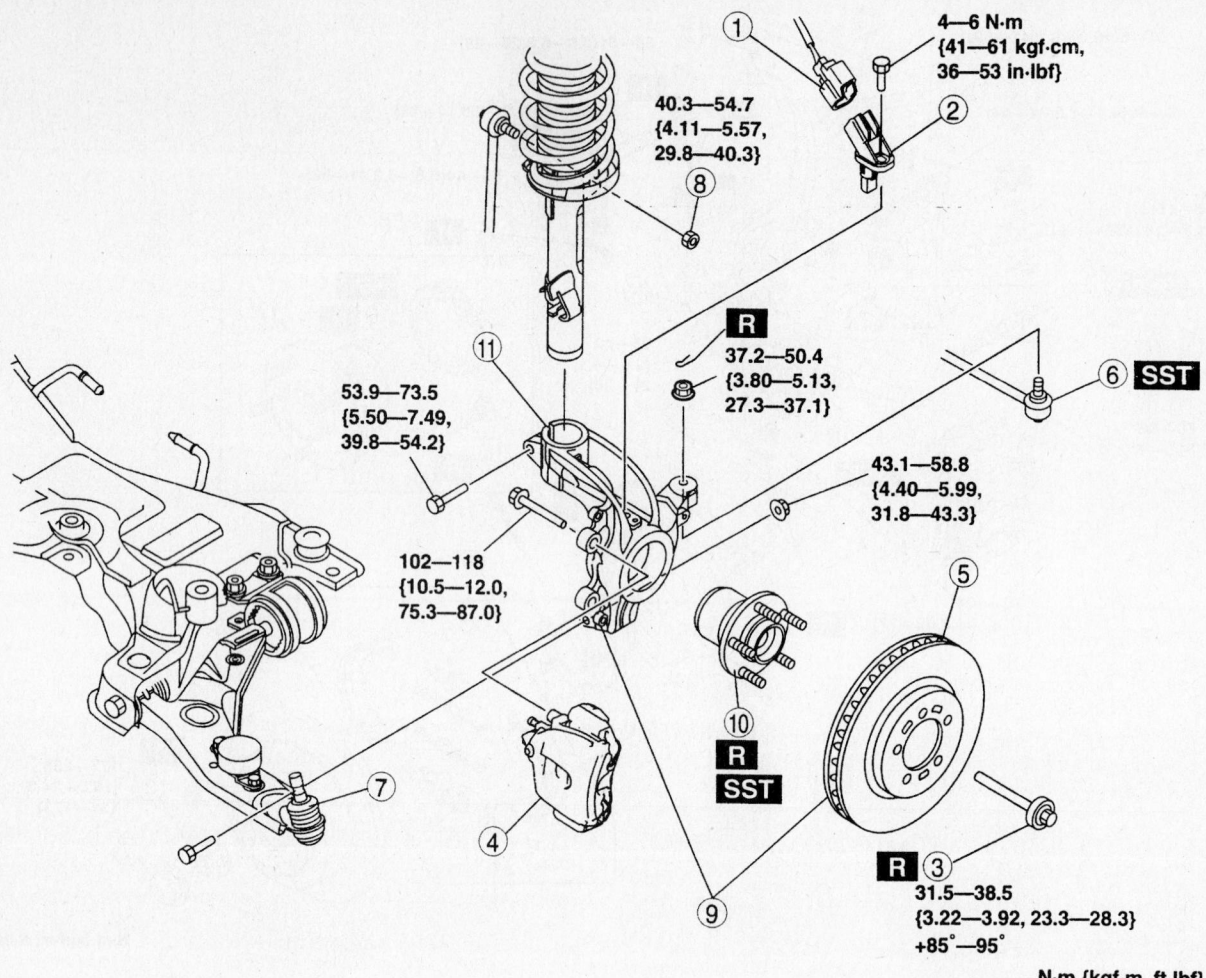

4—6 N·m
{41—61 kgf·cm,
36—53 in·lbf}

40.3—54.7
{4.11—5.57,
29.8—40.3}

37.2—50.4
{3.80—5.13,
27.3—37.1}

53.9—73.5
{5.50—7.49,
39.8—54.2}

43.1—58.8
{4.40—5.99,
31.8—43.3}

102—118
{10.5—12.0,
75.3—87.0}

31.5—38.5
{3.22—3.92, 23.3—28.3}
+85°—95°

N·m {kgf·m, ft·lbf}

1	ABS wheel-speed sensor connector	7	Front lower arm ball joint
2	ABS wheel-speed sensor	8	Stabilizer control link upper nut
3	Lockbolt	9	Wheel hub, steering knuckle component
4	Brake caliper component	10	Wheel hub component
5	Disc plate	11	Steering knuckle
6	Tie-rod end ball joint		

67162-MAZC-G99

Exploded view of the front wheel bearing and knuckle assembly—Mazda 3

6. Install or connect the following:
- Press the hub onto the knuckle
- Knuckle assembly, tighten the bolts to the specifications shown in the accompanying illustration
- Control arm ball joint to the knuckle tighten, the bolts to the specifications shown in the accompanying illustration
- Tie rod end to the knuckle, tighten the bolts to the specifications shown in the accompanying illustration
- Brake rotor and caliper
- New halfshaft lockbolt and tighten

to 23–28 ft. lbs. (31–38 Nm), plus an additional 85–95 degrees
- ABS sensor and connector
- Wheels

REAR

1. Before servicing the vehicle, refer to the precautions in the beginning of this section.

2. Refer to the illustration for component location and torque specifications.

➡The wheel bearings are not service-able. If the bearings are bad, a new hub/bearing assembly must be installed.

3. Remove or disconnect the following:
- Rear wheels
- Wheel speed sensor connector and sensor
- Rear parking brake cable
- Brake hose grommet and move the hose aside
- Brake caliper and rotor

To install:

4. Install or connect the following:
- Hub/bearing assembly. Torque the new nut to 36–48 ft. lbs. (49–65 Nm).
- Brake assembly
- Rear parking brake cable. Pass the

FRONT

Mazda 6

1. Before servicing the vehicle, refer to the precautions in the beginning of this section.
2. Refer to the illustration for component location and torque specifications.
3. Remove or disconnect the following:
   • Wheels
   • ABS sensor
   • Halfshaft axle nut, Unstake the nut prior to removal.
   • Brake caliper and rotor
   • Tie rod end from the knuckle
   • Damper fork—to—control arm bolt
   • Front lower control arm ball joints
   • Front upper arm ball joint
   • Wheel hub dust cover
   • Hub bolts and the hub
   • Hub using a press and Mazda tools 49 F026 10, 49 G033 102 and 49 G033 105. If the bearing inner race remains in the hub, grind a section of the bearing inner race until 0.02 inch (0.5mm) remains and use a chisel to remove it.
   • Snap ring

Use a press to remove the hub—Mazda 6

67162-MAZC-G259

(tools: 49 F026 103, 49 G033 105, 49 G033 102)

Install the dust shield—Mazda 6

67162-MAZC-G261

(tools: 49 F027 009, 49 S033 101)

Use a press to remove the wheel bearing—Mazda 6

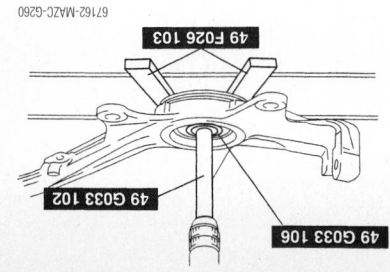

67162-MAZC-G260

(tools: 49 F026 103, 49 G033 102, 49 G033 106)

• Bearing from the hub using a press and the tools illustrated
• Brake dust shield, if it is being replaced. Mark the cover and knuckle for replacement purposes and use a chisel to remove the shield.

4. Clean and inspect all parts but do not wash or clean the wheel bearing. The bearing must be replaced.

To install:

5. Using the tools illustrated, install a new dust shield cover assembly to the knuckle, if removed.
6. Using the tools illustrated, press a new wheel bearing into the knuckle assembly and install the snap ring.
7. Install or connect the following:
   • Wheel bearing retaining ring
8. Using the tools illustrated, press in the hub assembly.
   • Wheel hub dust cover
   • Front upper arm ball joint and tighten the nut to 29-39 ft. lbs. (39-53 Nm)
   • Front lower arm ball joints and tighten the nuts to 122-147 ft. lbs. (166-200 Nm)
   • Damper fork bolt and tighten to 68-93 ft. lbs. (93-126 Nm)
   • Tie rod end to the knuckle and tighten the nut to 29-39 ft. lbs. (39-53 Nm) and install a new cotter pin
   • Brake caliper and rotor
   • Halfshaft axle nut, tighten the nut to 173-202 ft. lbs. (235-274 Nm) and stake the nut
   • Wheels

---

installing—Mazda 3

   • Pass the cable inside the rear wheel speed sensor wiring harness as illustrated.
   • Wheel speed sensor and conector
   • Rear wheel

Pass the cable inside the rear wheel speed sensor wiring harness when installing—Mazda 3

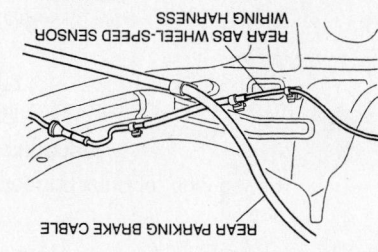

REAR ABS WHEEL-SPEED SENSOR WIRING HARNESS

REAR PARKING BRAKE CABLE

67162-MAZC-G103

Exploded view of the rear wheel bearing assembly and related components—Mazda 3

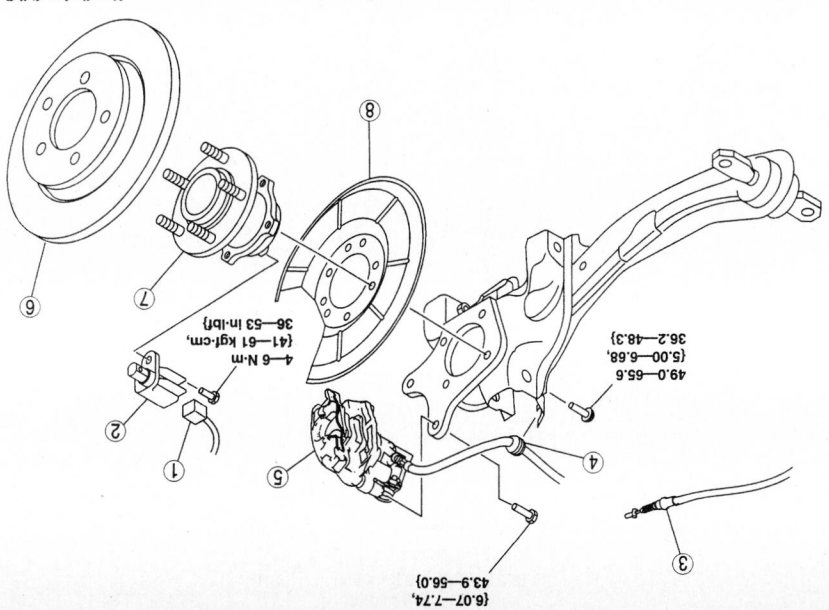

67162-MAZC-G102

1. ABS wheel-speed sensor connector
2. ABS wheel-speed sensor
3. Rear parking brake cable
4. Brake hose
5. Brake caliper component
6. Disc plate
7. Wheel hub component
8. Dust cover

N·m {kgf·m, ft·lbf}

4—6 N·m {41—61 kgf·cm, 36—53 in·lbf}

49.0—65.6 {5.00—6.68, 36.2—48.3}

59.5—76.0 {6.07—7.74, 43.9—56.0}

18.6—25.5
{1.90—2.60, 13.8—18.8}
4 SST
8 SST
39.5—53.4
{4.03—5.44,
29.2—39.3}
93.1—126.4
{9.50—12.88,
68.67—93.22}
9
5
18.6—25.5
{1.90—2.60,
13.8—18.8}
14
R
SST
12
R
11
R
39.5—53.4
{4.03—5.44,
29.2—39.3}
13
R
SST
10
6
SST
7
SST
78.4—101.9
{8.00—10.39,
57.83—75.15}
9
166.6—200.0
{16.99—20.39, 122.9—147.5}
1 R
235.2—274.4
{23.99—27.98,
173.5—202.3}
15 SST
R
2
3
N·m {kgf·m, ft·lbf}

1	Locknut	9	Wheel hub, steering knuckle, dust cover
2	Brake caliper component	10	Wheel hub component
3	Disc plate	11	Retaining ring
4	Tie-rod end ball joint	12	Wheel bearing
5	Bolt	13	Dust cover
6	Front lower arm (front) ball joint	14	Steering knuckle
7	Front lower arm (rear) ball joint	15	Hub bolt
8	Front upper arm ball joint		

67162-MAZC-G258

Exploded view of the front wheel bearing and knuckle assembly—Mazda 6

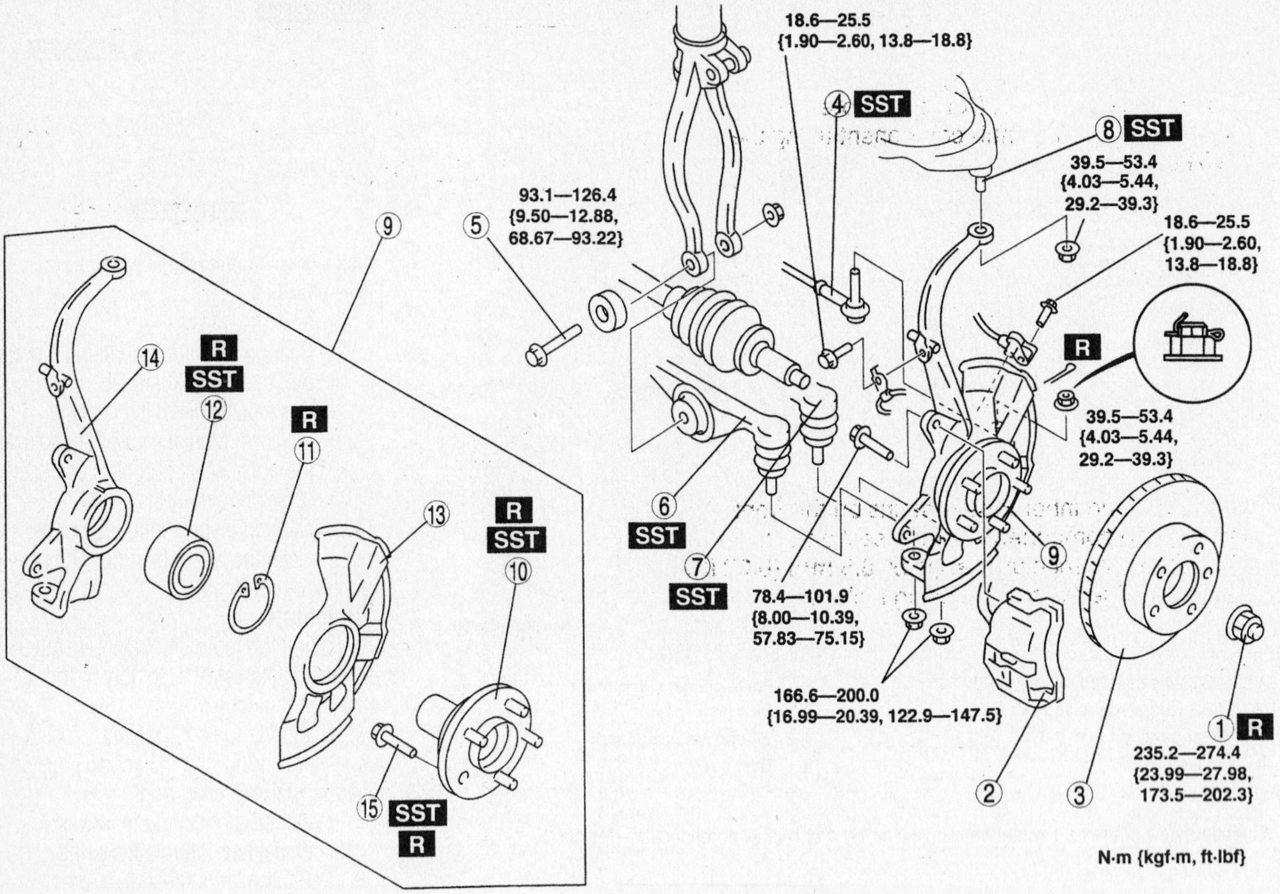

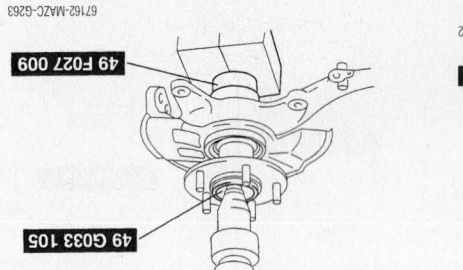

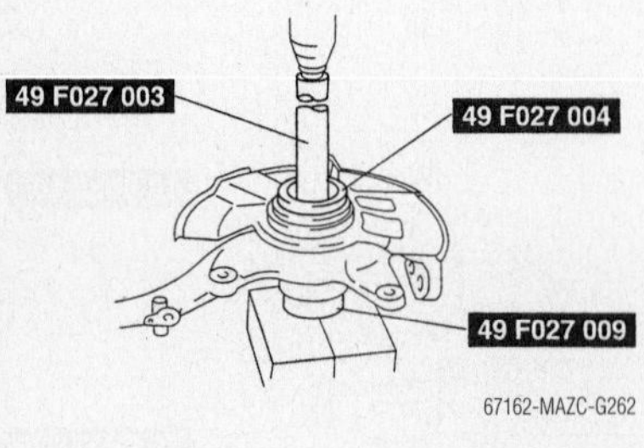

49 F027 003
49 F027 004
49 F027 009

67162-MAZC-G262

Install a new wheel bearing using a press—Mazda 6

49 G033 105
49 F027 009

67162-MAZC-G263

Install a wheel hub using a press—Mazda 6

REAR

1. Before servicing the vehicle, refer to the precautions in the beginning of this section.
2. Refer to the illustration for component location and torque specifications.

➡The wheel bearings are not serviceable. If the bearings are bad, a new hub/bearing assembly must be installed.

3. Remove or disconnect the following:
   - Hub cap
4. Raise the staked portion of the hub retaining nut with a hammer and chisel.

When using the installation tool for the ABS sensor rotor, face the carved side B to the bottom—Mazda 6

67162-MAZC-G266

- Hub bolts and tighten to the specifications shown in the illustration

To install:
- Hub and bearing assembly on the spindle

- Rear wheels
- ABS sensor
- Parking brake cable
- Rear caliper and rotor assembly
- Hub retaining nut and discard it
- ABS sensor rotor using a chisel
- Hub bolts
- Hub and bearing assembly from the spindle

Remove the ABS sensor rotor using a chisel—Mazda 6

67162-MAZC-G265

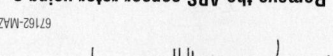
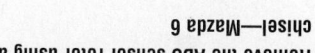

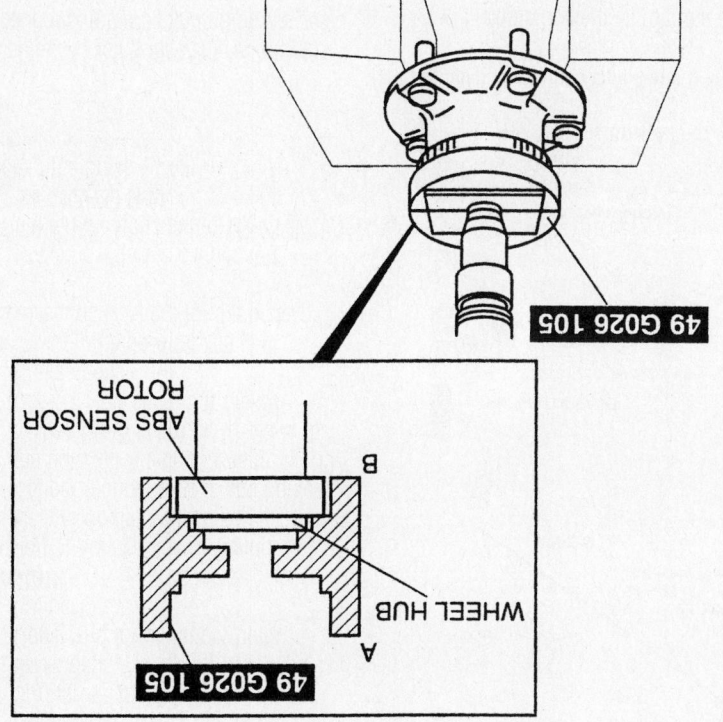

49 G026 105

ABS SENSOR ROTOR

WHEEL HUB

A

B

49 G026 105

Exploded view of the rear hub and spindle assembly—Mazda 6

67162-MAZC-G264

1	Hub cap
2	Locknut
3	Parking brake cable
4	Brake caliper component
5	Disc plate
6	Wheel hub component
7	Wheel hub
8	Hub bolt
9	ABS sensor rotor (with ABS)
10	Dust cover
11	Hub spindle

N·m {kgf·m, ft·lbf}

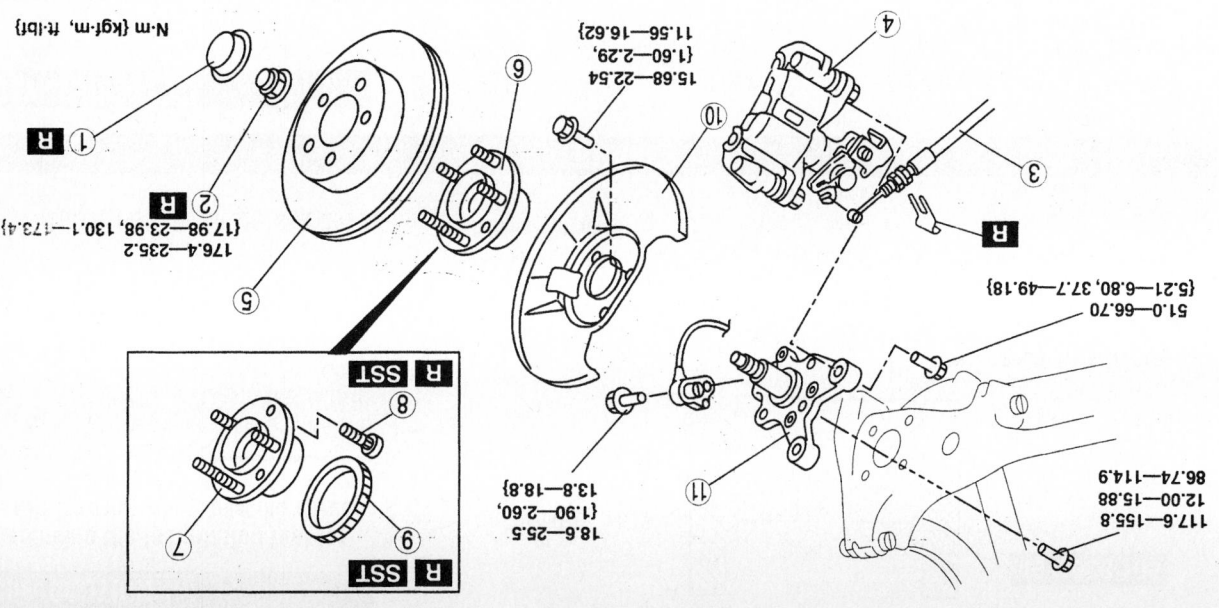

117.6–155.8
{12.00–15.88,
86.74–114.9}

51.0–66.70
{5.21–6.80, 37.7–49.18}

15.68–22.54
{1.60–2.29,
11.56–16.62}

18.6–25.5
{1.90–2.60,
13.8–18.8}

176.4–235.2
{17.98–23.98, 130.1–173.4}

BRAKES

Brake-Caliper

REMOVAL & INSTALLATION

Protégé

FRONT

1. Before servicing the vehicle, refer to the precautions in the beginning of this section.
2. Remove or disconnect the following:
   • Wheels
   • Flexible brake hose from the caliper
   • Cap on type **A** brakes or bolt on typebrakes
   • Spring on type **A** brakes
   • Brake pads
   • Upper and lower caliper bolts
   • Caliper

To install:

3. Install or connect the following:
   • Caliper on the brake disc
   • Caliper mounting bolts and tighten the bolts on **A** type brakes to 37–39 ft. lbs. (50–53 N·m) or 16–23 ft. lbs. (22–31 N·m) on type **B** brakes
   • Brake pads
   • Spring on type **A** brakes
4. Replace the washers for the brake line.
   • Brake hose to the caliper and tighten the hose nut to 16–21 ft. lbs. (22–29 N·m)
5. Bleed the brake system.
6. Install the wheels.

REAR

1. Before servicing the vehicle, refer to the precautions in the beginning of this section.
2. Remove or disconnect the following:
   • Wheels
   • Parking brake cable from the cable bracket and the operating lever

3. Turn the manual adjustment gear counterclockwise with an Allen wrench to pull the caliper piston inward (turn until it stops).
4. Install or connect the following:

To install:
   • Caliper
   • Caliper mounting bolts
   • Flexible brake line from the caliper
   • Caliper
   • Torque the caliper mount bolts to 26–28 ft. lbs. (35–39 N·m).

Exploded view of the type A and B brake systems—Protégé with disc brakes

1 Flexible hose
2 Cap (type A only)
3 Guide pin
4 Caliper
5 M-spring (type A only)
6 Disc pad
7 Guide plate
8 Mounting support
9 Disc plate

42356-MAZC-GKT

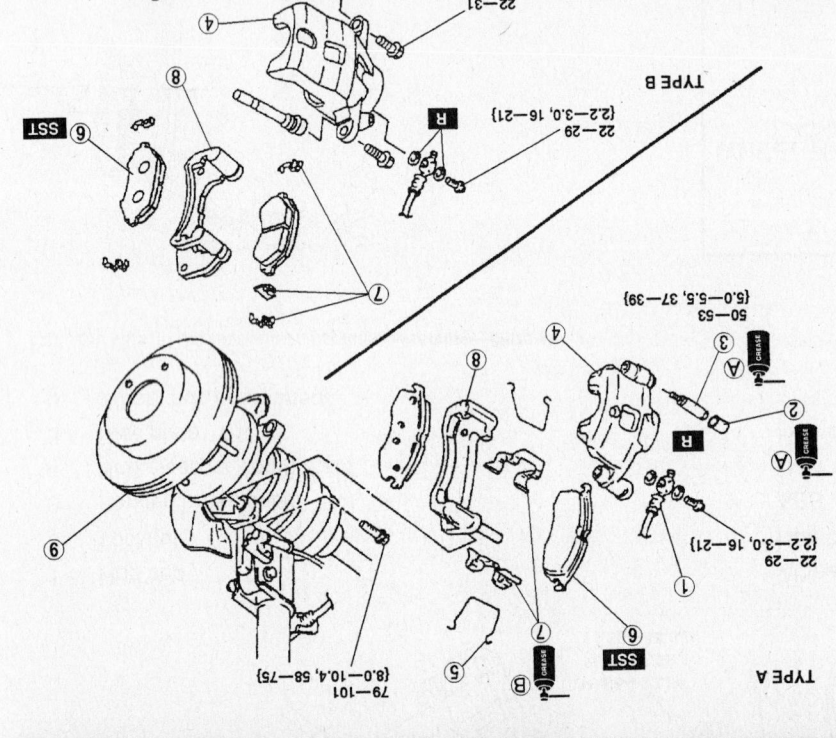

RUBBER GREASE

BRAKE GREASE

N·m {kgf·m, ft·lbf}

TYPE B
22–29 {2.2–3.0, 16–21}
22–31 {2.2–3.2, 16–23}

TYPE A
22–29 {2.2–3.0, 16–21}
50–53 {5.0–5.5, 37–39}
79–101 {8.0–10.4, 58–75}

✱✱ CAUTION

When using the installation tool for the ABS sensor rotor, face the carved side B to the rotor.

• ABS sensor rotor using a press and tools illustrated. Make sure there is a 0.12 inch (3mm) gap as illustrated.
• Remaining components in the reverse order of removal, referring to the illustration for component location and torque specifications.

When installing the ABS sensor rotor, make sure there is a 0.12 inch (3mm) gap as shown B to the bottom—Mazda 6

ABS SENSOR ROTOR
WHEEL HUB
3 mm {0.12 in}
49 G026 105
B
A

67162-MAZC-G267

626

FRONT

1. Before servicing the vehicle, refer to the precautions in the beginning of this section.
2. Remove or disconnect the following:
   - Wheels
   - Flexible brake hose from the caliper
   - Lower caliper bolt and pivot the caliper upward. Slide the top of the caliper off of the top pin and remove it from the vehicle.

To install:
3. Lubricate the caliper pin and slide the caliper onto the guide pinch Pivot the caliper over the brake pads.
4. Install or connect the following:
   - Brake hose to the caliper and tighten the hose nut to 16–21 ft. lbs. (22–29 Nm)
   - Caliper mounting bolt and tighten the bolt to 22–28 ft. lbs. (30–41 Nm)
   - Brake hose. Torque the line bolt to 16–22 ft. lbs. (22–30 Nm).
5. Install the wheels and bleed the brake system.

Exploded view of rear brake systems—Protégé with rear disc brakes

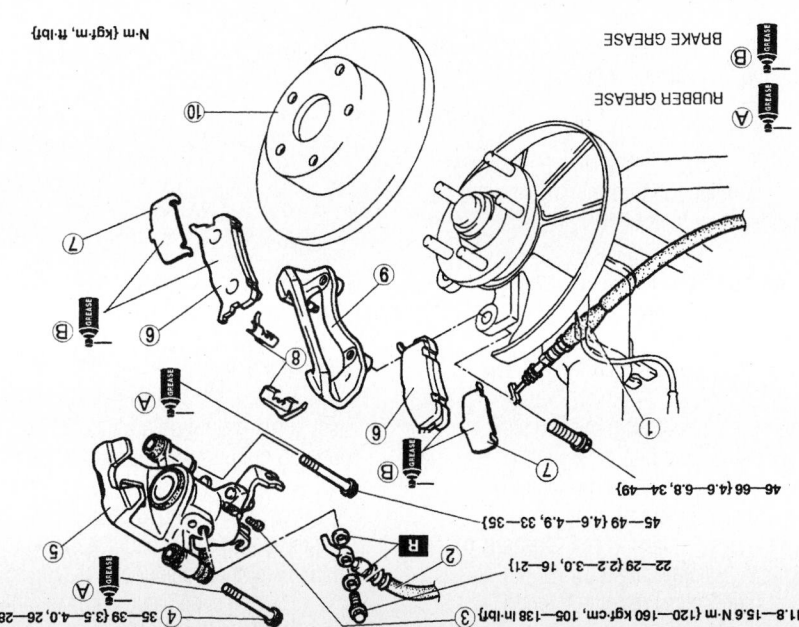

1 Parking brake cable, clip		6 Disc pad	
2 Flexible hose		7 Shim	
3 Screw plug		8 Guide plate	
4 Lock bolt		9 Mounting support	
5 Caliper		10 Disc plate	

A RUBBER GREASE
B BRAKE GREASE

11.8–15.6 N·m {120–160 kgf·cm, 105–138 in-lbf}
35–39 {3.5–4.0, 26–28}
22–29 {2.2–3.0, 16–21}
45–49 {4.6–4.9, 33–35}
46–66 {4.6–6.8, 34–49}

N·m {kgf·m, ft-lbf}

42356-MAZC-GWW

Front disc brakes—626

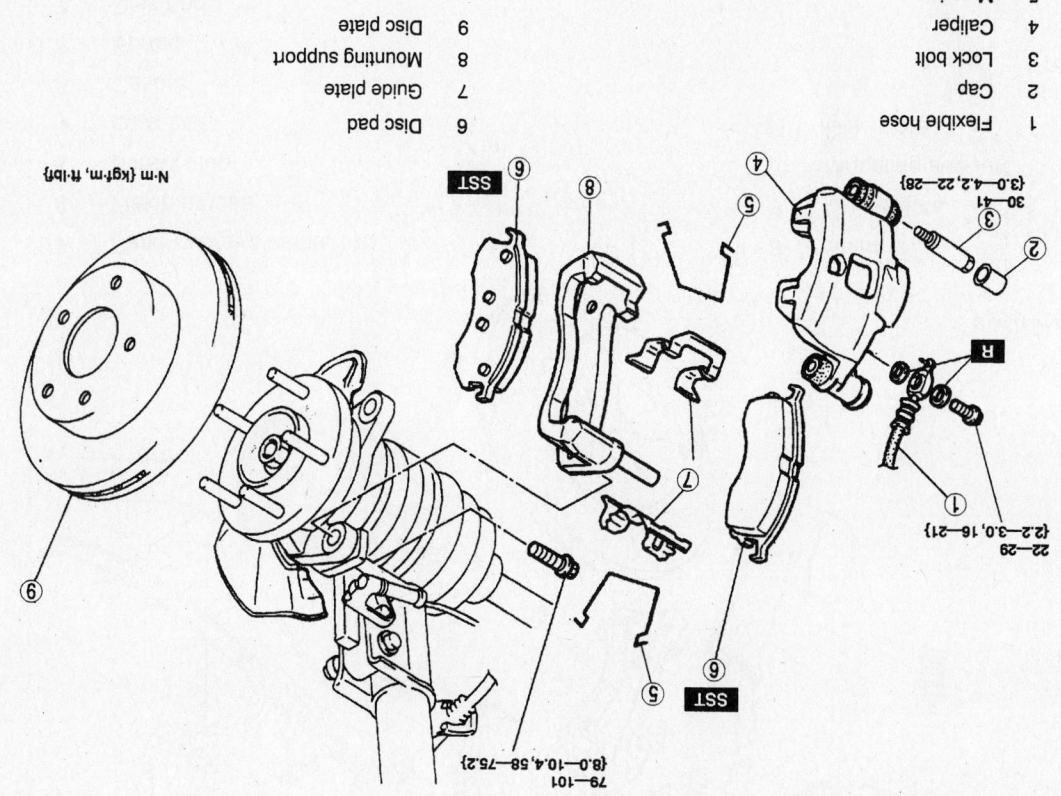

1 Flexible hose		6 Disc pad	
2 Cap		7 Guide plate	
3 Lock bolt		8 Mounting support	
4 Caliper		9 Disc plate	
5 M-spring			

30–41 {3.0–4.2, 22–28}
22–29 {2.2–3.0, 16–21}
79–101 {8.0–10.4, 58–75.2}

N·m {kgf·m, ft-lbf}

42356-MAZC-G89

42356-MAZC-G91

7 Disc pad
6 Spring
5 Caliper
4 Lock bolt
3 Screw plug
2 Flexible hose
1 Parking brake cable, clip

11 Disc plate
10 Mounting support
9 Guide plate
8 Shim

N·m {kgf·m, ft·lbf}

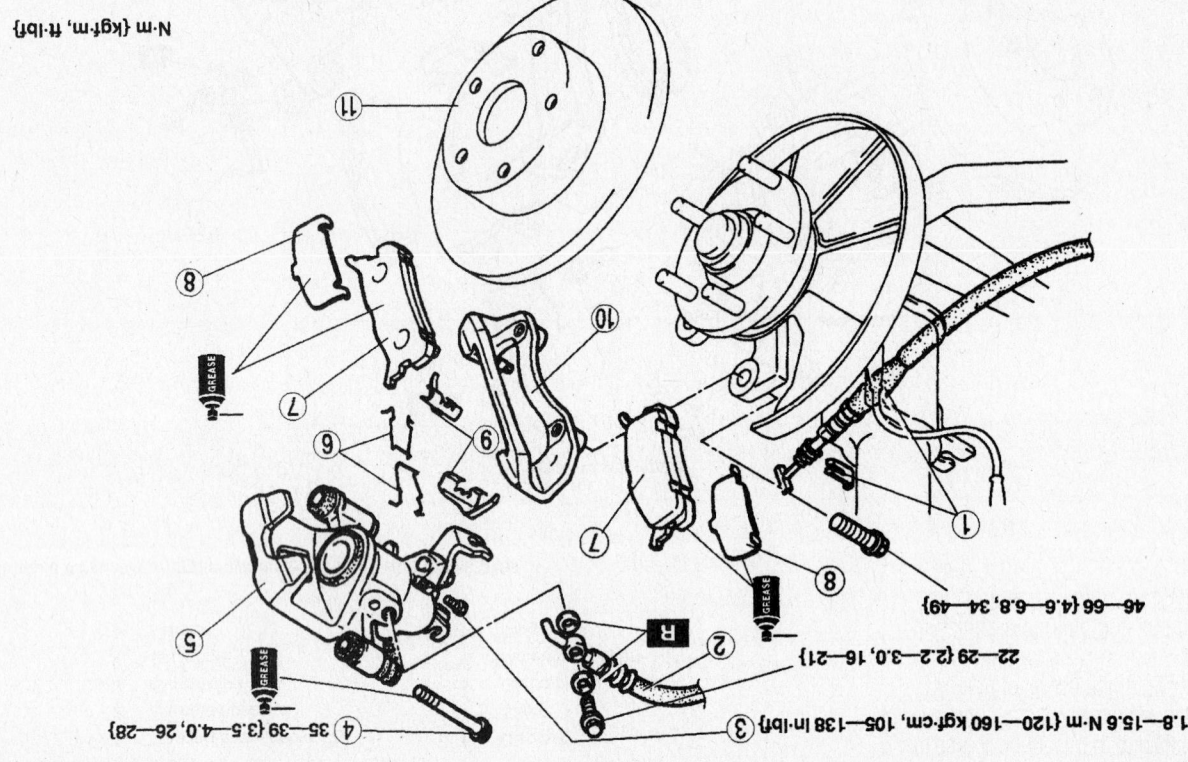

46-66 {4.6-6.8, 34-49}

22-29 {2.2-3.0, 16-21}

11.8-15.6 N·m {120-160 kgf·cm, 105-138 in·lbf}

35-39 {3.5-4.0, 26-28}

5. Bleed the brake system and inspect the brake system for proper operation.
6. Install the wheels.

REAR
1. Before servicing the vehicle, refer to the precautions in the beginning of this section.
2. Remove the wheels. Loosen the parking brake cable adjustment from inside the vehicle.
3. Remove or disconnect the following:
• Parking brake cable from the cable bracket and the operating lever
• Flexible brake line from the caliper assembly
• Caliper upper mounting bolt and pivot the caliper downward. Slide the caliper off of the guide pinch
• Caliper

To install:
4. Lubricate the caliper pin and slide the caliper onto the guide pinch Pivot the caliper over the brake pads.

Miata

FRONT
1. Before servicing the vehicle, refer to the precautions in the beginning of this section.
2. Remove or disconnect the following:
• Wheels

5. Install or connect the following:
• Brake hose to the caliper and tighten the hose nut to 16-21 ft. lbs. (22-29 Nm)
• Upper caliper mounting bolt and tighten the bolt to 26-28 ft. lbs. (35-39 Nm)
• Parking brake cable to the cable bracket and the operating lever
6. Bleed the brake system.
7. Install the wheels.

REAR
1. Before servicing the vehicle, refer to the precautions in the beginning of this section.

Flexible brake hose from the caliper
• Upper and lower caliper bolts
• Caliper from the vehicle

To install:
3. Install or connect the following:
• Caliper on the brake disc
• Caliper mounting bolts and tighten the bolts to 58-65 ft. lbs. (79-88 Nm)
4. Replace the washers for the brake line.
• Brake hose to the caliper and tighten the hose nut to 16-21 ft. lbs. (22-29 Nm)
5. Bleed the brake system and inspect the brake system for proper operation.
6. Install the wheels.

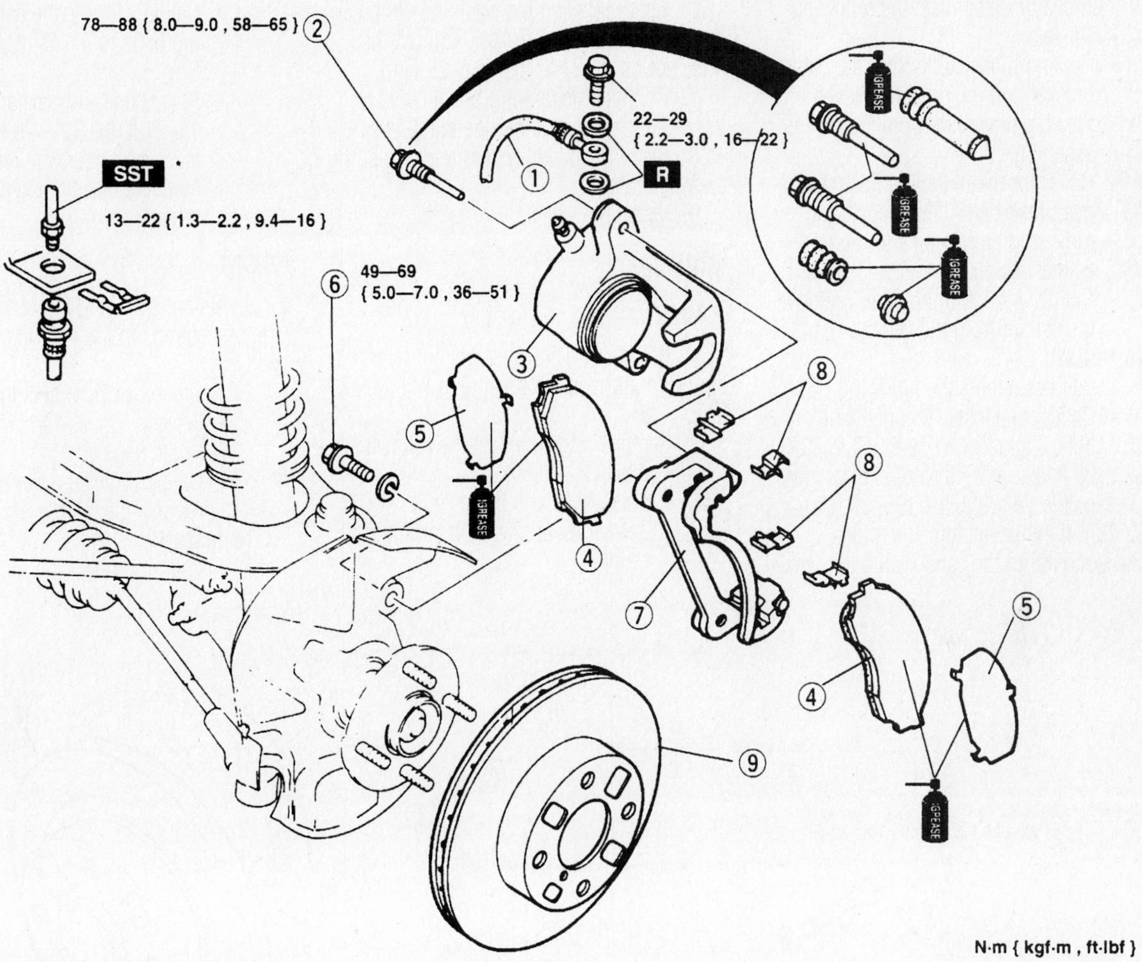

78—88 { 8.0—9.0 , 58—65 } ②

SST

13—22 { 1.3—2.2 , 9.4—16 }

22—29
{ 2.2—3.0 , 16—22 }

① **R**

49—69
{ 5.0—7.0 , 36—51 }
⑥

③

⑤

⑧

⑧

④

⑦

⑤

④

⑨

N·m { kgf·m , ft·lbf }

1	Brake hose
2	Connecting bolt
3	Caliper
4	Disc pad
5	Shim

6	Bolt
7	Mounting support
8	Guide plate
9	Disc plate

Disc Plate Removal Note
- Mark the wheel hub bolt and disc plate before removal for reference during installation.

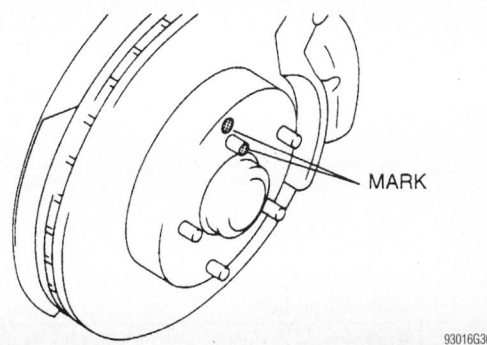

MARK

93016G30

Front disc brakes—Miata

2. Loosen the parking brake cable adjustment from inside the vehicle.

3. Remove or disconnect the following:
- Wheels
- Parking brake cable from the cable bracket and the operating lever
- Flexible brake line from the caliper assembly
- Cover for the manual adjustment gear. Insert an Allen wrench and turn counterclockwise to retract the caliper piston.
- Caliper mounting bolts and remove the caliper from the vehicle

To install:

4. Install or connect the following:
- Caliper and install the mounting bolts. Tighten the upper bolt to 33–36 ft. lbs. (45–49 Nm), and the lower bolt to 26–28 ft. lbs. (35–39 Nm).

5. Turn the manual adjustment gear clockwise to return the caliper until the brake pads just touch the disc, then turn counterclockwise ⅓ of a turn. Replace the cover.

6. After replacing the washers, connect the brake hose to the caliper. Tighten the hose nut to 16–21 ft. lbs. (22–29 Nm).

7. Connect the parking brake cable to the cable bracket and the operating lever.

8. Bleed the brake system and adjust the parking brake.

9. Install the wheels.

Millenia

FRONT

1. Before servicing the vehicle, refer to the precautions in the beginning of this section.

2. Remove or disconnect the following:
- Wheels
- Brake hose and brake pipe.
- Caliper mounting bolts and the caliper

To install:

3. Install or connect the following:
- Caliper. Torque the caliper mounting bolts to 47–62 ft. lbs. (63–84 Nm).
- Brake hose and pipe and tighten to 15–21 ft. lbs. (22–29 Nm). Fill the master cylinder with clean brake fluid and bleed the hydraulic system.
- Wheels

REAR

1. Before servicing the vehicle, refer to the precautions in the beginning of this section.

2. Remove or disconnect the following:
- Wheels
- Brake hose
- Caliper bracket mounting bolts and remove the caliper

To install:

3. Install or connect the following:

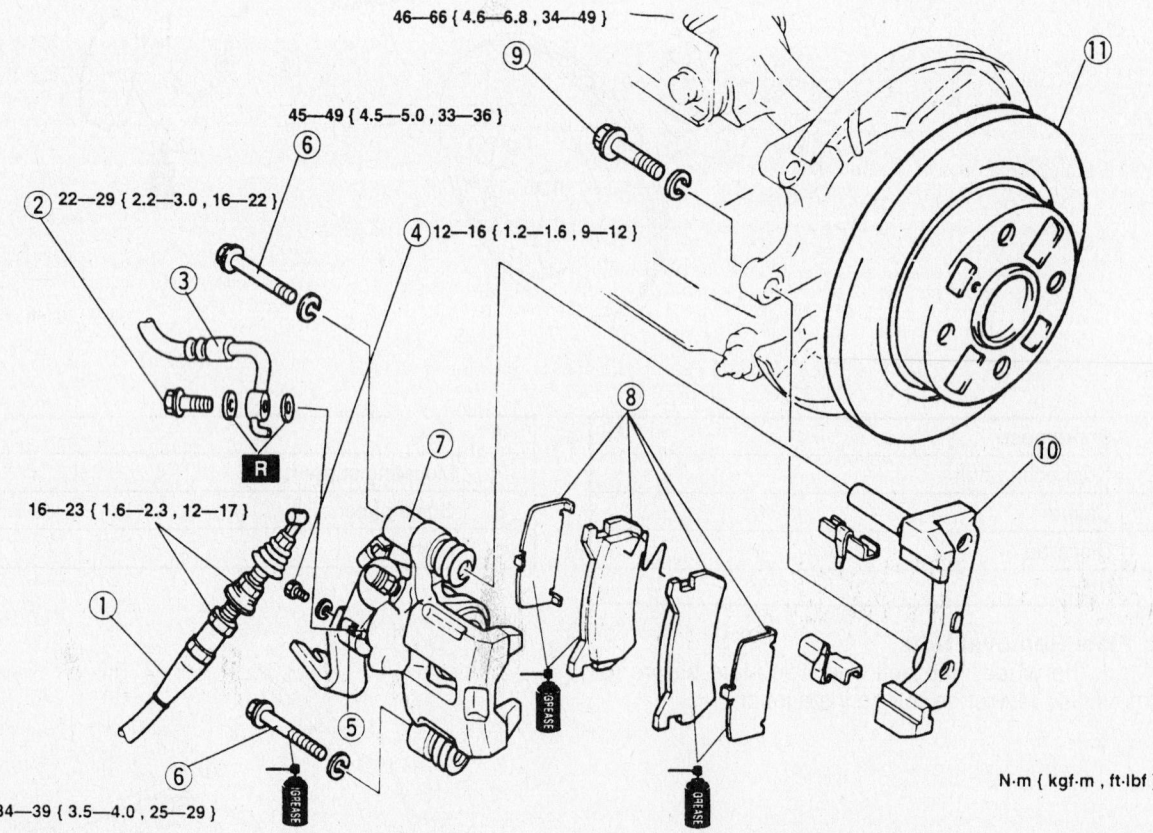

46—66 { 4.6—6.8 , 34—49 }

45—49 { 4.5—5.0 , 33—36 }

22—29 { 2.2—3.0 , 16—22 }

12—16 { 1.2—1.6 , 9—12 }

16—23 { 1.6—2.3 , 12—17 }

34—39 { 3.5—4.0 , 25—29 }

N·m { kgf·m , ft·lbf }

1	Parking brake cable	5	Manual adjustment gear ☞ Removal Note ☞ Installation Note
2	Connecting bolt	6	Lock bolt
3	Brake hose		
4	Plug		

Rear disc brakes—Miata

93016G31

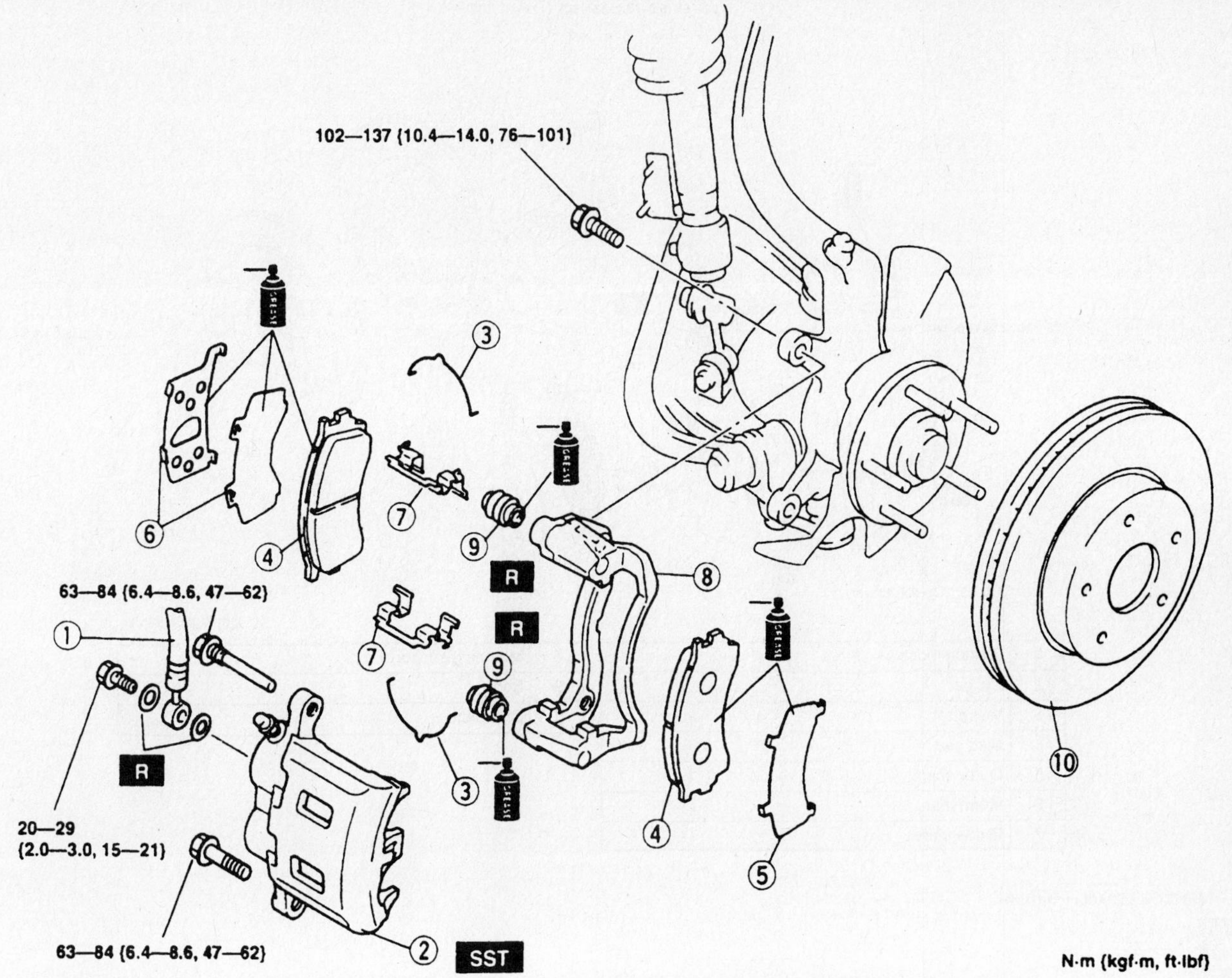

102—137 (10.4—14.0, 76—101)

63—84 (6.4—8.6, 47—62)

20—29
(2.0—3.0, 15—21)

63—84 (6.4—8.6, 47—62)

N·m (kgf·m, ft·lbf)

1	Flexible hose	7	Guide plate
2	Caliper	8	Mounting support
3	V-spring	9	Boot
4	Disc pad	10	Disc plate
5	Outer shim		
6	Inner shim		

93016G32

Front disc brakes—Millenia

- Caliper. Torque the caliper mounting bolts to 12–17 ft. lbs. (16–23 Nm).
- Brake hose and pipe and tighten to 16–21 ft. lbs. (22–29 Nm). Fill the master cylinder with clean brake fluid and bleed the hydraulic system.
- Wheels

Mazda 3

FRONT

1. Before servicing the vehicle, refer to the precautions in the beginning of this section.

2. Remove or disconnect the following:
   - Wheels
   - Brake hose
   - Retaining clip
   - Cap from the caliper bolts
   - Caliper mounting bolts and the caliper

To install:

3. Install or connect the following:
   - Caliper. Torque the caliper mounting bolts to 19–22 ft. lbs. (25–30 Nm). Install the bolt caps.
   - Brake hose to the caliper. Tighten the flare nut while holding the hose at location A shown in the accompanying illustration with an open

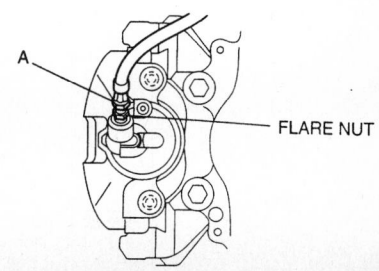

A — FLARE NUT

67162-MAZC-G105

Tighten the brake hose flare nut while holding the hose at point A—Mazda 3

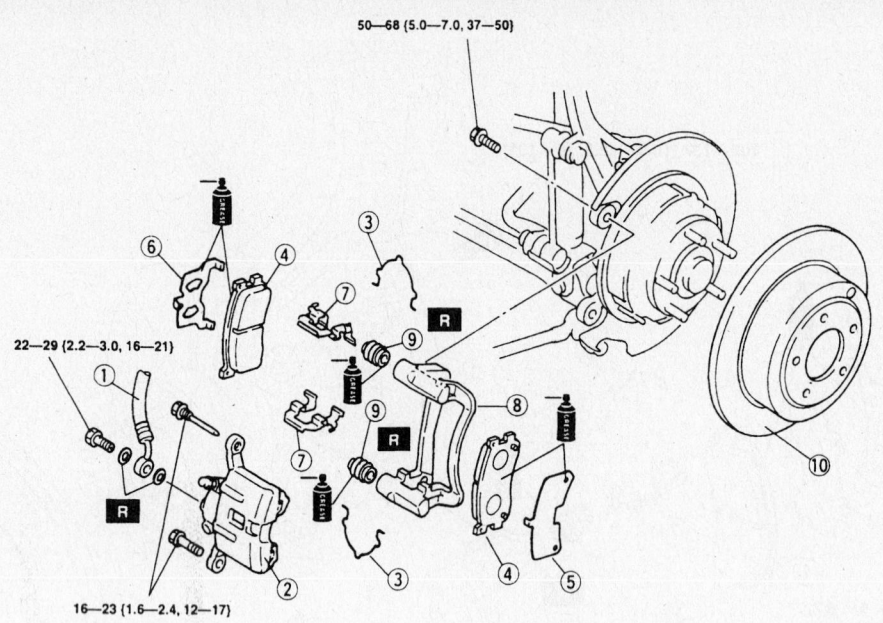

50—68 (5.0—7.0, 37—50)

22—29 {2.2—3.0, 16—21}

16—23 {1.6—2.4, 12—17}

N·m {kgf·m, ft·lbf}

1	Flexible hose	8	Mounting support
2	Caliper	9	Boot
3	V-spring	10	Disc plate
4	Disc pad		
5	Outer shim		
6	Inner shim		
7	Guide plate		

93016G33

Rear disc brakes—Millenia

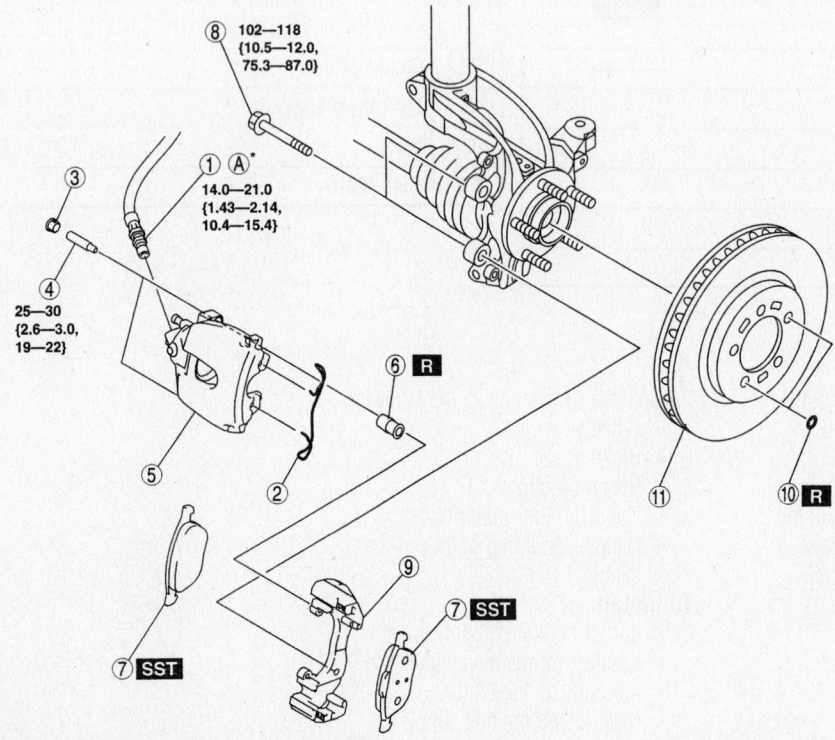

8 102—118
{10.5—12.0,
75.3—87.0}

1 Ⓐ*
14.0—21.0
{1.43—2.14,
10.4—15.4}

25—30
{2.6—3.0,
19—22}

Ⓐ* COMMERCIALLY AVAILABLE FLARE NUT WRENCH
(FLARE NUT ACROSS FLAT 13 mm {0.51 in})

N·m {kgf·m, ft·lbf}

1	Brake hose
2	Retaining clip
3	Cap
4	Bolt
5	Caliper
6	Boot
7	Disc pad
8	Bolt
9	Mounting support
10	Washer
11	Disc plate

67162-MAZC-G104

Exploded view of the front disc brakes—Mazda 3

end wrench. Tighten the nut to 10–15 ft. lbs. (14–21 Nm) and make sure the brake hose is not twisted, if it is unfasten the flare nut and retighten making sure the brake line remains straight. Fill the master cylinder with clean brake fluid and bleed the hydraulic system.

- Retaining clip
- Wheels

4. Pump the brake pedal several times to seat the pads.

REAR

1. Before servicing the vehicle, refer to the precautions in the beginning of this section.

2. Remove or disconnect the following:
- Wheels
- Parking brake cable
- Brake pipe from the hose and the clip
- Pad retaining clip
- Caps from the caliper bolts
- Caliper brake hose
- Caliper mounting bolts and the caliper

To install:

3. Install or connect the following:
- Caliper. Torque the caliper mounting bolts to 19–22 ft. lbs. (25–30 Nm). Install the bolt caps.
- Brake hose and pipe and tighten to 14–16 ft. lbs. (19–23 Nm). Reinstall a new clip and fill the master cylinder with clean brake fluid and bleed the hydraulic system.

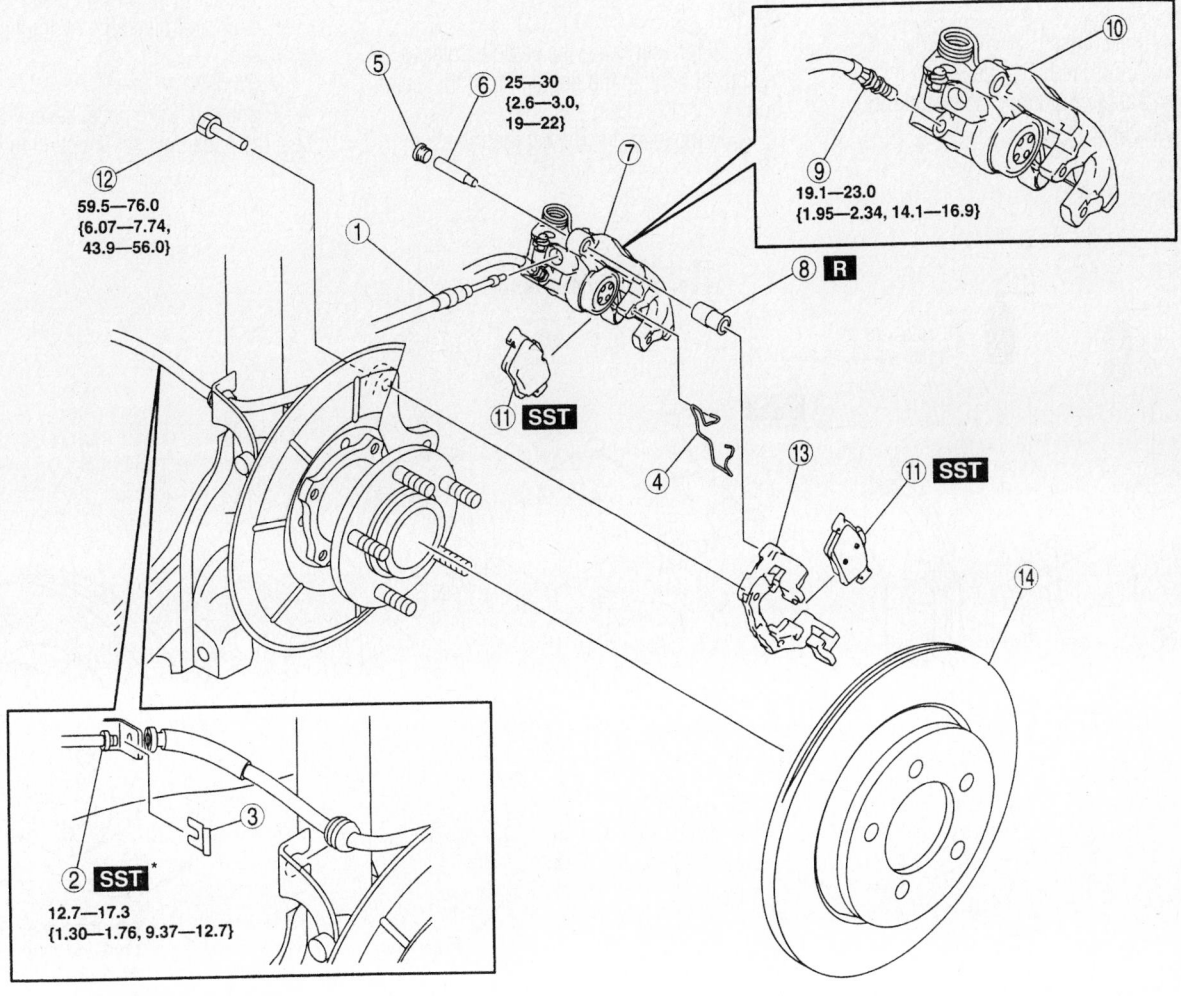

25—30
{2.6—3.0,
19—22}

59.5—76.0
{6.07—7.74,
43.9—56.0}

19.1—23.0
{1.95—2.34, 14.1—16.9}

12.7—17.3
{1.30—1.76, 9.37—12.7}

SST * SST (49 0259 770B)
(FLARE NUT ACROSS FLAT 10 mm {0.39 in})

N·m {kgf·m, ft·lbf}

1	Parking brake cable		8	Boot
2	Brake pipe		9	Brake hose
3	Clip		10	Caliper
4	Retaining clip		11	Disc pad
5	Cap		12	Bolt
6	Bolt		13	Mounting support
7	Caliper, brake hose		14	Disc plate

67162-MAZC-G106

Exploded view of the rear disc brakes—Mazda 3

- Pad retaining clip and parking brake cable
- Wheels

4. Pump the brake pedal several times to seat the pads. Inspect the parking brake lever stroke and brake drag and adjust as necessary.

Mazda 6

FRONT

1. Before servicing the vehicle, refer to the precautions in the beginning of this section.

2. Remove or disconnect the following:
- Wheels
- Brake hose
- Cap from the caliper bolts
- Caliper mounting bolts and the caliper

To install:

3. Install or connect the following:
- Caliper. Torque the caliper mounting bolts to 36—39 ft. lbs. (49—54 Nm). Install the bolt caps.
- Brake hose to the caliper. Tighten the bolt to 16—21 ft. lbs. (21—29 Nm) and make sure the brake hose is not twisted. Fill the master cylinder with clean brake fluid and bleed the hydraulic system.
- Wheels

4. Pump the brake pedal several times to seat the pads.

REAR

1. Before servicing the vehicle, refer to the precautions in the beginning of this section.

2. Remove or disconnect the following:

- Wheels
- Parking brake cable clip
- Caliper brake hose bolt
- Caliper brake hose
- Caliper mounting bolts and the caliper

To install:

3. Install or connect the following:
- Caliper. Torque the caliper mounting bolts to 27—36 ft. lbs. (37—49 Nm).
- Brake hose and pipe and tighten to 16—21 ft. lbs. (21—29 Nm).
- Parking brake cable clip and fill the master cylinder with clean brake fluid and bleed the hydraulic system.
- Wheels

4. Pump the brake pedal several times to seat the pads. Inspect the parking brake

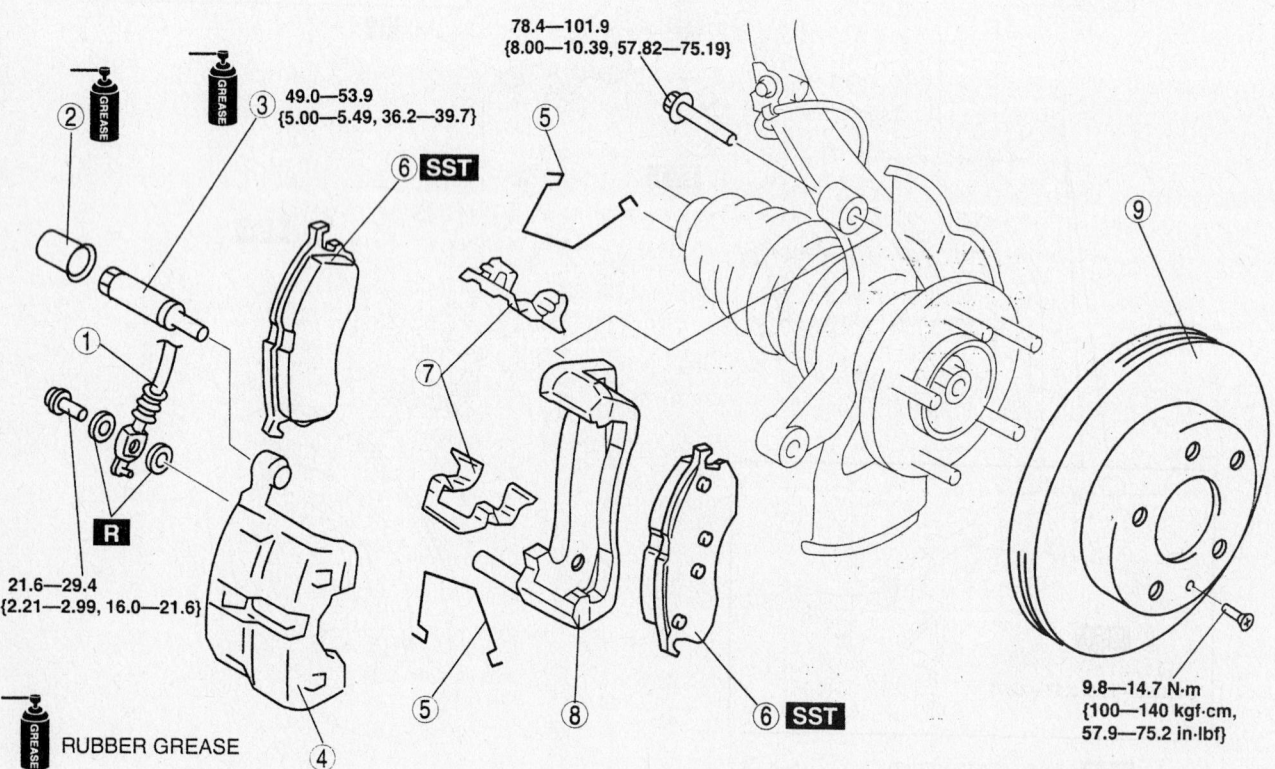

78.4—101.9
{8.00—10.39, 57.82—75.19}

49.0—53.9
{5.00—5.49, 36.2—39.7}

21.6—29.4
{2.21—2.99, 16.0—21.6}

RUBBER GREASE

9.8—14.7 N·m
{100—140 kgf·cm,
57.9—75.2 in·lbf}

N·m {kgf·m, ft·lbf}

1	Flexible hose	6	Disc pad
2	Cap	7	Guide plate
3	Guide pin	8	Mounting support
4	Caliper	9	Disc plate
5	M-spring		

67162-MAZC-G268

Exploded view of the front disc brakes—Mazda 6

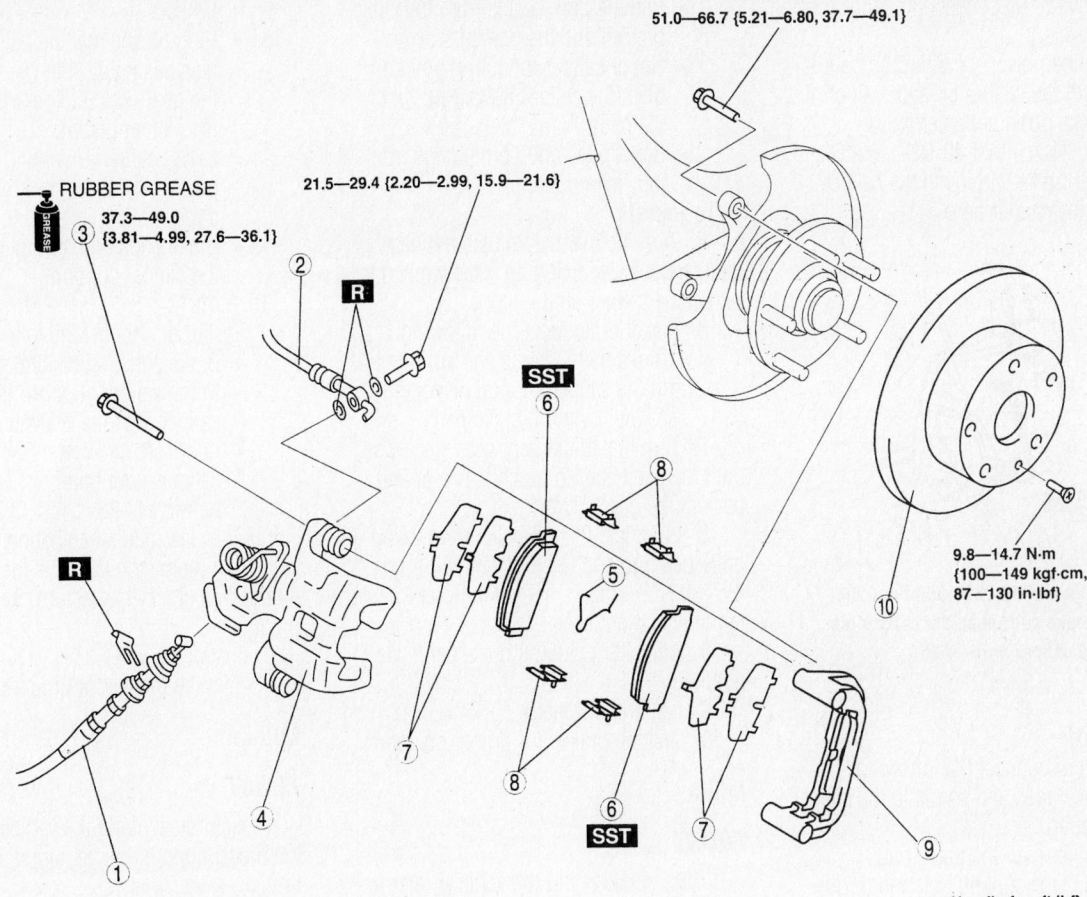

51.0—66.7 {5.21—6.80, 37.7—49.1}

RUBBER GREASE
37.3—49.0
{3.81—4.99, 27.6—36.1}

21.5—29.4 {2.20—2.99, 15.9—21.6}

9.8—14.7 N·m
{100—149 kgf·cm,
87—130 in·lbf}

N·m {kgf·m, ft·lbf}

1	Parking brake cable, clip	6	Disc pad
2	Flexible hose	7	Shim
3	Bolt	8	Guide plate
4	Caliper	9	Mounting support
5	Spring	10	Disc plate

67162-MAZC-G269

Exploded view of the rear disc brakes—Mazda 6

lever stroke and brake drag and adjust as necessary.

Disc Brake Pads

REMOVAL & INSTALLATION

Protégé

FRONT

1. Before servicing the vehicle, refer to the precautions in the beginning of this section.
2. Remove or disconnect the following:
   • Wheels
   • Cap on type **A** brakes or bolt on typebrakes
   • Caliper
   • Spring on type **A** brakes
   • Brake pads

To install:

3. Install or connect the following:
   • Caliper on the brake disc
   • Caliper mounting bolts and tighten the bolts on **A** type brakes to 37–39 ft. lbs. (50–53 Nm) or 16–23 ft. lbs. (22–31 Nm) on type **B** brakes
   • Brake pads
   • Spring on type **A** brakes
4. Install the wheels.

REAR

1. Before servicing the vehicle, refer to the precautions in the beginning of this section.
2. Remove or disconnect the following:
   • Wheels
   • Parking brake cable from the cable bracket and the operating lever
3. Turn the manual adjustment gear counterclockwise with an Allen wrench to pull the caliper piston inward (turn until it stops).
   • Caliper
   • Pads

To install:

4. Install or connect the following:
   • Pads
   • Caliper. Torque the caliper mount bolts to 26–28 ft. lbs. (35–39 Nm).
   • Parking brake cable
5. Install the wheels and bleed the brake system.

626

FRONT

1. Before servicing the vehicle, refer to the precautions in the beginning of this section.
2. Remove some of the brake fluid from the master cylinder reservoir.

3. Remove or disconnect the following:
- Wheels
- Caliper lower mounting bolt and pivot the caliper up and support it
- Brake pads, shims and pin

4. Using Mazda tool 49-0221-600C and the old inner brake pad, push the caliper piston into the caliper bore.

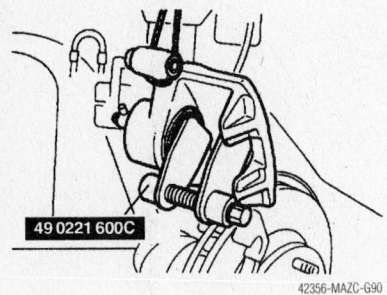

49 0221 600C

42356-MAZC-G90

Using Mazda tool 49-0221-600C and the old inner brake pad, push the caliper piston into the caliper bore—626

To install:

5. Install or connect the following:
- Brake pads and shims to the caliper support
- Caliper over the brake pads
- Caliper mounting bolt and torque to 22–28 ft. lbs. (30–41 Nm)
- Wheels

6. Test the brakes for proper operation.

REAR

1. Before servicing the vehicle, refer to the precautions in the beginning of this section.

2. Remove some of the brake fluid from the master cylinder reservoir.

3. Loosen the parking brake cable adjustment from inside the vehicle.

4. Remove or disconnect the following:
- Wheels.

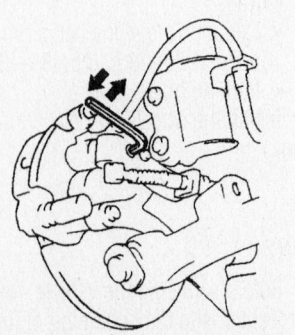

42356-MAZC-G92

Location of the manual adjustment gear on the rear caliper—626

- Parking brake cable from the cable bracket and the operating lever
- Upper caliper mounting bolt and pivot the caliper downward off of the pads
- Brake pads and spring clips from the caliper support

To install:

5. Turn the manual adjustment gear counterclockwise using an Allen wrench to retract the caliper until it stops.

6. Install or connect the following:
- Brake pads, shims and spring clips to the caliper support. Pivot the caliper over the brake pads.

7. Turn the adjustment gear clockwise until the pads start to touch the rotor and back of the gear ⅓ turn.

8. Lubricate the top caliper mounting bolt, then tighten the bolt to 26–28 ft. lbs. (35–39 Nm). Attach the parking brake cable to the operating lever.

9. Adjust the parking brake cable, as required.

10. Install the wheels.

11. Test the brakes for proper operation.

Miata

FRONT

1. Before servicing the vehicle, refer to the precautions in the beginning of this section.

2. Remove some of the brake fluid from the master cylinder reservoir.

3. Remove or disconnect the following:
- Wheels
- Lower lockbolt, and pivot the brake caliper upwards
- Spring, if equipped with 15 inch wheels
- Brake pads and shim

4. Using Mazda piston compressor tool 49-0221-600C and the old inner brake pad, push the caliper piston into the caliper bore.

To install:
- Brake pads and shims onto the caliper
- Caliper
- Spring, if equipped with 15 inch wheels
- Lockbolt and tighten to 58–65 ft. lbs. (78–88 Nm)
- Wheels

5. Test the brakes for proper operation.

REAR

1. Before servicing the vehicle, refer to the precautions in the beginning of this section.

2. Remove some of the brake fluid from the master cylinder reservoir.

3. Remove or disconnect the following:

- Wheels
- Plug for the manual adjustment gear. Using an Allen wrench turn the gear counterclockwise to retract the caliper piston.
- Lower caliper mounting bolt and pivot the caliper upward off of the pads
- Brake pads and spring clips from the caliper support

To install:

4. Install or connect the following:
- Brake pads, shims and spring clips to the caliper support. Pivot the caliper over the brake pads.
- Lubricate and install the lower caliper mounting bolt. Tighten the bolt to 26–28 ft. lbs. (35–39 Nm).

5. Turn the manual adjusting gear clockwise until the piston contacts the brake disc, then turn it clockwise ⅓ turn. Replace the plug.

6. Install the wheels.

7. Test the brakes for proper operation.

Millenia

FRONT

1. Before servicing the vehicle, refer to the precautions in the beginning of this section.

2. Remove or disconnect the following:
- Wheels
- Bottom caliper lock pin and swing the caliper upwards
- V-springs the pads and shims

To install:

3. Install or connect the following:
- Brake pads, shims and V-springs

4. Press the caliper pistons back into their cylinders.
- Calipers. Torque the lock pin to 47–62 ft. lbs. (63–84 Nm).
- Wheels

REAR

1. Before servicing the vehicle, refer to the precautions in the beginning of this section.

2. Remove or disconnect the following:
- Wheels
- Lock pin and rotate the caliper upwards
- V-springs, pads and the shims from the pads

To install:

3. Press the caliper piston back into the cylinder.

4. Install or connect the following:
- Pads, shims, and V-springs
- Caliper and torque the lock pin to 37–50 ft. lbs. (50–68 Nm)
- Wheels

Mazda 3

FRONT

1. Before servicing the vehicle, refer to the precautions in the beginning of this section.
2. Remove or disconnect the following:
   - Wheels
   - Pad retaining clip
   - Cap from the bottom caliper bolt
   - Lower caliper mounting bolt and swing the caliper up
   - Brake pads

To install:

3. Press the caliper piston back into the cylinder using tool 49 0221 600C.
4. Install or connect the following:
   - Outer pad to the mounting support and inner pad to the caliper
   - Caliper. Torque the caliper lower mounting bolt to 19–22 ft. lbs. (25–30 Nm). Install the bolt cap.
   - Retaining clip
   - Wheels
5. Pump the brake pedal several times to seat the pads.

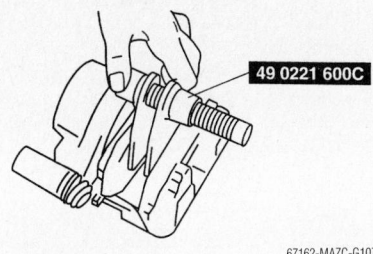

Press the front caliper piston back into the cylinder—Mazda 3

REAR

1. Before servicing the vehicle, refer to the precautions in the beginning of this section.
2. Remove or disconnect the following:
   - Wheels
   - Parking brake cable
   - Pad retaining clip
   - Caps from the caliper bolts
   - Caliper mounting bolts and the caliper
   - Outer brake pad from the mount support and pull the inner pad from the caliper

To install:

3. Install the out pad to the mounting support and clean the piston area.
4. Using tool 49 F043 002, turn the piston clockwise slowly until the piston is fully seated in its bore.
5. Align the inner side pad spring with the piston groove and insert the pad. Refer to the illustration for spring location and inner pad installation arrow.

6. Install or connect the following:
   - Caliper. Torque the caliper mounting bolts to 19–22 ft. lbs. (25–30 Nm). Install the bolt caps.
   - Pad retaining clip and parking brake cable
   - Wheels
7. Pump the brake pedal several times to seat the pads. Inspect the parking brake lever stroke and brake drag and adjust as necessary.

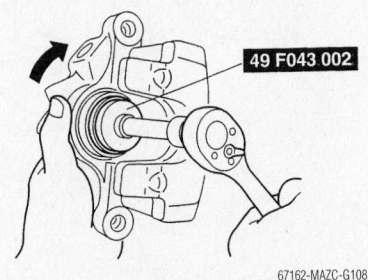

Turn the rear caliper piston clockwise slowly until the piston is fully seated in its bore—Mazda 3

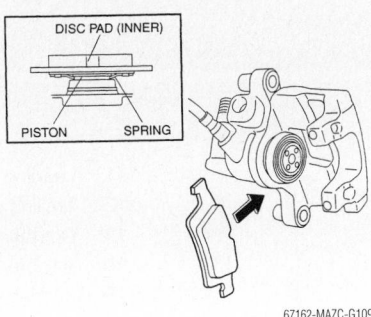

Align the inner side pad spring with the piston groove and insert the pad—Mazda 3

Mazda 6

FRONT

1. Before servicing the vehicle, refer to the precautions in the beginning of this section.
2. Remove or disconnect the following:
   - Wheels
   - Brake pipe bolts
   - Top caliper bolt cap and bolt and swing the caliper downwards
   - M-springs the pads and shims

To install:

3. Install or connect the following:
4. Press the caliper pistons back into their cylinders using tool 49 0221 600C
   - Brake pads, shims and M-springs
   - Caliper. Torque the bolt to 36–39 ft. lbs. (49–52 Nm) and install the cap.
   - Wheels

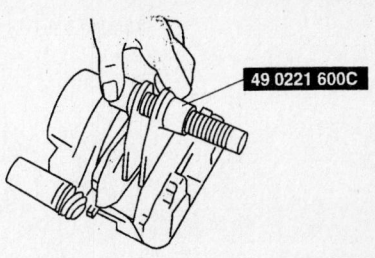

Press the front caliper pistons back into their bores using tool 49 0221 600C—Mazda 6

REAR

1. Before servicing the vehicle, refer to the precautions in the beginning of this section.
2. Remove or disconnect the following:
   - Wheels
   - Parking brake cable clip
   - Upper caliper bolt and rotate the caliper downwards
   - V-springs, pads and the shims from the pads

To install:

3. Press the rear caliper pistons back into their bores using tool 49 FA18 602.
4. Install or connect the following:
   - Pads, shims, and springs
   - Caliper and torque the bolt to 27–36 ft. lbs. (37–49 Nm)
   - Wheels

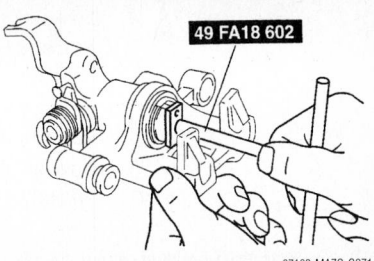

Press the rear caliper pistons back into their bores using tool 49 FA18 602—Mazda 6

Brake Drums

REMOVAL & INSTALLATION

All Models

1. Before servicing the vehicle, refer to the precautions in the beginning of this section.
2. Remove or disconnect the following:
   - Rear wheel
   - Screw securing the rear brake

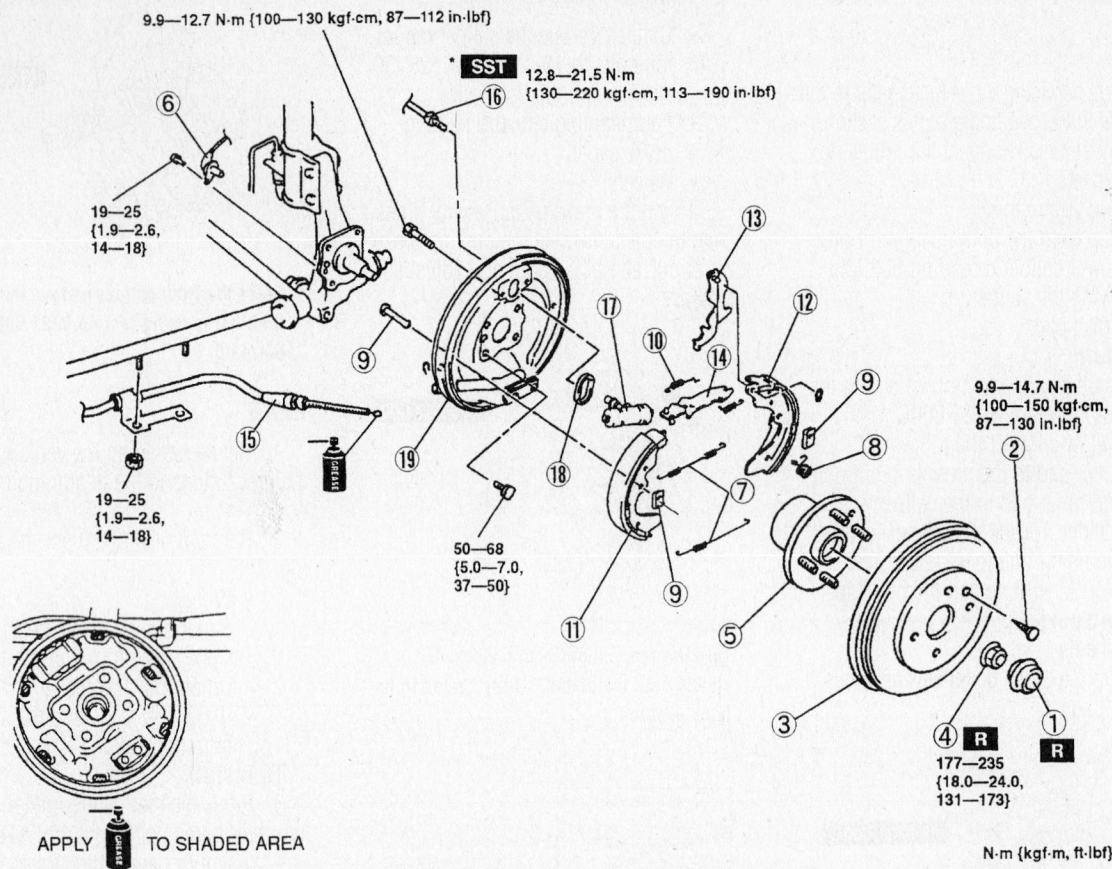

9.9—12.7 N·m {100—130 kgf·cm, 87—112 in·lbf}

* SST 12.8—21.5 N·m {130—220 kgf·cm, 113—190 in·lbf}

19—25 {1.9—2.6, 14—18}

19—25 {1.9—2.6, 14—18}

50—68 {5.0—7.0, 37—50}

9.9—14.7 N·m {100—150 kgf·cm, 87—130 in·lbf}

177—235 {18.0—24.0, 131—173}

N·m {kgf·m, ft·lbf}

APPLY GREASE TO SHADED AREA
*49 0259 770B

1	Hub cap	11	Leading shoe
2	Screw	12	Trailing shoe
3	Brake drum	13	Operating lever
4	Locknut	14	Adjuster
5	Wheel hub	15	Parking brake cable
6	ABS wheel-speed sensor (if equipped)	16	Brake pipe
7	Return spring	17	Wheel cylinder
8	Lever spring	18	O-ring
9	Hold pin and hold spring	19	Backing plate
10	Anti-rattle spring		

42356-MAZC-GKU

Typical rear drum brake assembly, Protégé shown others similar

drum, and pull the brake drum outward to remove

To install:

3. Install or connect the following:
   - Drum and retaining screw
   - Wheels and adjust the rear brakes as necessary

Brake Shoes

REMOVAL & INSTALLATION

1. Before servicing the vehicle, refer to the precautions in the beginning of this section.

2. Remove or disconnect the following:
   - Wheel and the brake drum
   - ABS speed sensor, if equipped
   - Parking brake cable from the backside of the brake backing plate
   - Upper return spring
   - Hold pin and the spring from the lower shoe (leading side)
   - Lower return spring and the anti-rattle spring and shoe from the lower shoe (leading side)
   - Hold pin and spring and remove the upper brake shoe (trailing side)

To install:

3. Install or connect the following:
   - Upper (trailing side) brake shoe to

the operating lever and then to the wheel cylinder and backing plate
   - Brake shoe hold spring and hold pin
   - Anti-rattle spring
   - Lower return spring to both brake shoes
   - Leading side brake shoe to the operating lever and then to the wheel cylinder and anchor plate
   - Hold spring and hold pin to the leading side brake shoe
   - Upper return spring
   - Brake drum
   - ABS speed sensor, if equipped
   - Wheels and adjust the brakes

MAZDA

B-Series

26

SPECIFICATIONS AND MAINTENANCE CHARTS

ENGINE AND VEHICLE IDENTIFICATION

Code ①	Engine Liters (cc)	Cu. In.	Cyl.	Fuel Sys.	Type	Eng. Mfg.
D	2.3 (2261)	138	4	MFI	DOHC	Ford
C	2.5 (2500)	152	4	MFI	SOHC	Ford
U	3.0 (2999)	183	6	MFI	OHV	Ford
E	4.0 (4000)	244	6	MFI	SOHC	Ford

Model Year Code ②	Year
1	2001
2	2002
3	2003
4	2004
5	2005

MFI: Multi-port Fuel Injection

OHV: Overhead Valve

DOHC: Dual Overhead Camshafts

SOHC: Single Overhead Camshaft

① 8th digit of the Vehicle Identification Number (VIN)

② 10th digit of the Vehicle Identification Number (VIN)

67162-BSER-C01

GENERAL ENGINE SPECIFICATIONS

Year	Model	Engine Displacement Liters	Engine (VIN)	Net Horsepower @ rpm	Net Torque @ rpm (ft. lbs.)	Bore x Stroke (in.)	Compression Ratio	Oil Pressure @ rpm
2001	B2300	2.3	D	143@5200	154@3750	3.44x3.70	NA	29-39@3000
	B2500	2.5	C	147@5000	147@5000	3.50x3.14	9.3:1	40-60@2500
	B3000	3.0	U	147@5000	162@3250	3.50x3.14	9.3:1	40-60@2500
	B4000	4.0	E	160@4000	225@2500	3.81x3.39	9.0:1	40-60@2000
2002	B2300	2.3	D	143@5200	154@3750	3.44x3.70	NA	29-39@3000
	B3000	3.0	U	147@5000	147@5000	3.50x3.14	9.3:1	40-60@2500
	B4000	4.0	E	160@4000	225@2500	3.81x3.39	9.0:1	40-60@2000
2003	B2300	2.3	D	143@5200	154@3750	3.44x3.70	NA	29-39@3000
	B3000	3.0	U	147@5000	147@5000	3.50x3.14	9.3:1	40-60@2500
	B4000	4.0	E	160@4000	225@2500	3.81x3.39	9.0:1	40-60@2000
2004	B2300	2.3	D	143@5200	154@3750	3.44x3.70	NA	29-39@3000
	B3000	3.0	U	147@5000	147@5000	3.50x3.14	9.3:1	40-60@2500
	B4000	4.0	E	160@4000	225@2500	3.81x3.39	9.0:1	40-60@2000

MFI: Multi-port Fuel Injection

67162-BSER-C02

GASOLINE ENGINE TUNE-UP SPECIFICATIONS

Year	Engine Displacement Liters	Engine VIN	Spark Plug Gap (in.)	Ignition Timing (deg.) ①		Fuel Pump (psi)	Idle Speed (rpm)		Valve Clearance	
				MT	AT		MT	AT	In.	Ex.
2001	2.3	D	0.041-0.045	10B	10B	64-72	①	①	HYD	HYD
	2.5	C	0.044	10B	10B	64-72	①	①	HYD	HYD
	3.0	U	0.044	10B	10B	64-72	①	①	HYD	HYD
	4.0	E	0.052-0.056	10B	10B	64-72	①	①	HYD	HYD
2002	2.3	D	0.041-0.045	10B	10B	64-72	①	①	HYD	HYD
	3.0	U	0.042-0.046	10B	10B	64-72	①	①	HYD	HYD
	4.0	E	0.062-0.068	10B	10B	64-72	①	①	HYD	HYD
2003	2.3	D	0.041-0.045	10B	10B	64-72	①	①	HYD	HYD
	3.0	U	0.042-0.046	10B	10B	64-72	①	①	HYD	HYD
	4.0	E	0.062-0.068	10B	10B	64-72	①	①	HYD	HYD
2004	2.3	D	0.041-0.045	10B	10B	64-72	①	①	HYD	HYD
	3.0	U	0.042-0.046	10B	10B	64-72	①	①	HYD	HYD
	4.0	E	0.062-0.068	10B	10B	64-72	①	①	HYD	HYD

NOTE: The Vehicle Emission Control Information label often reflects specification changes changes made during production. The label figures must be used if they differ from those in this chart.

B: Before top dead center

HYD: Hydraulic

NA: Not Available

① Electronically controlled and cannot be adjusted

67162-BSER-C03

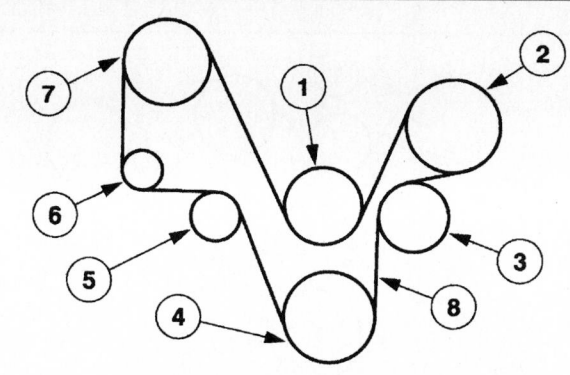

1 Fan pulley
2 Power steering pump pulley
3 Water pump pulley
4 Crankshaft pulley
5 Belt tensioner pulley
6 Generator pulley
7 A/C clutch pulley
8 Drive belt

67197-RANG-G97

Accessory drive belt routing—2.3L with A/C

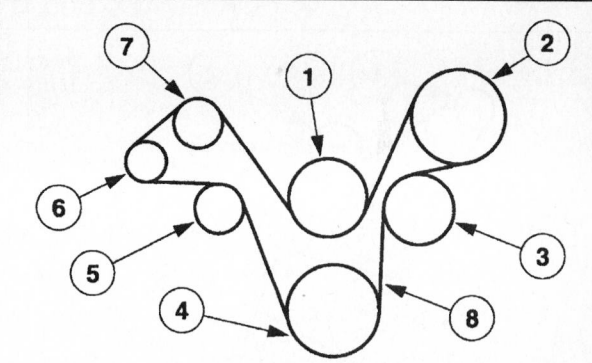

1 Fan pulley
2 Power steering pump pulley
3 Water pump pulley
4 Crankshaft pulley
5 Belt tensioner pulley
6 Generator pulley
7 Belt idler pulley
8 Drive belt

67197-RANG-G98

Accessory drive belt routing—2.3L without A/C

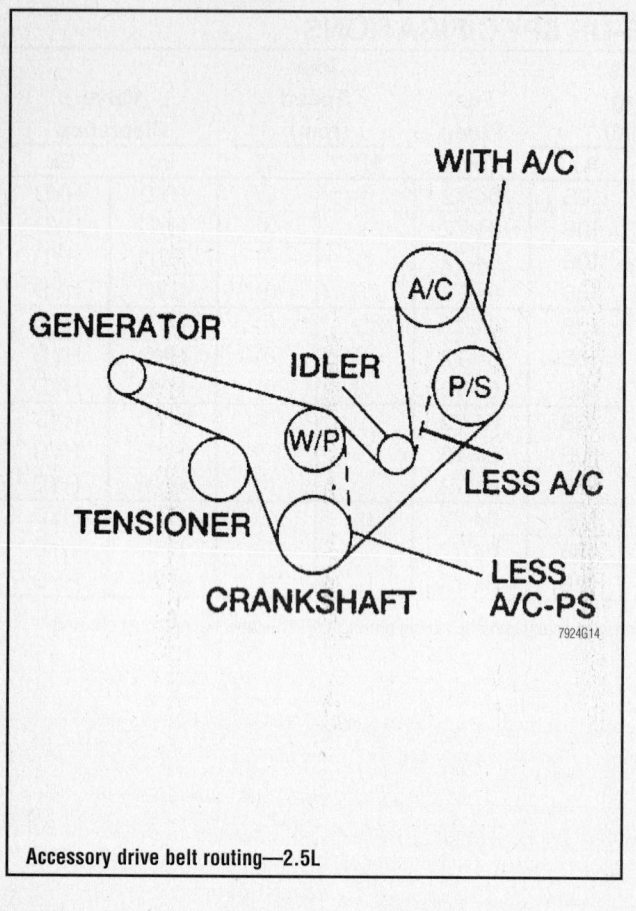

Accessory drive belt routing—2.5L

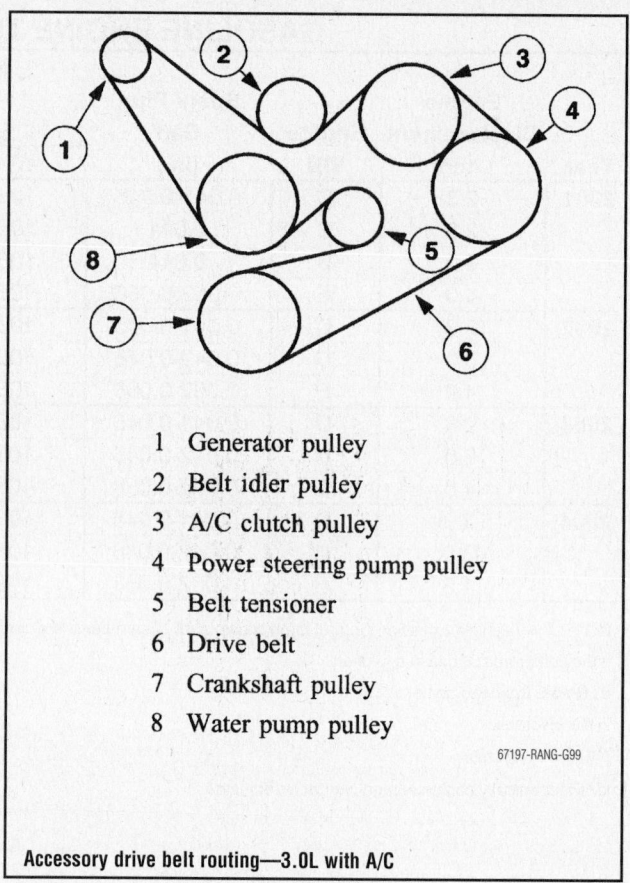

1 Generator pulley
2 Belt idler pulley
3 A/C clutch pulley
4 Power steering pump pulley
5 Belt tensioner
6 Drive belt
7 Crankshaft pulley
8 Water pump pulley

67197-RANG-G99

Accessory drive belt routing—3.0L with A/C

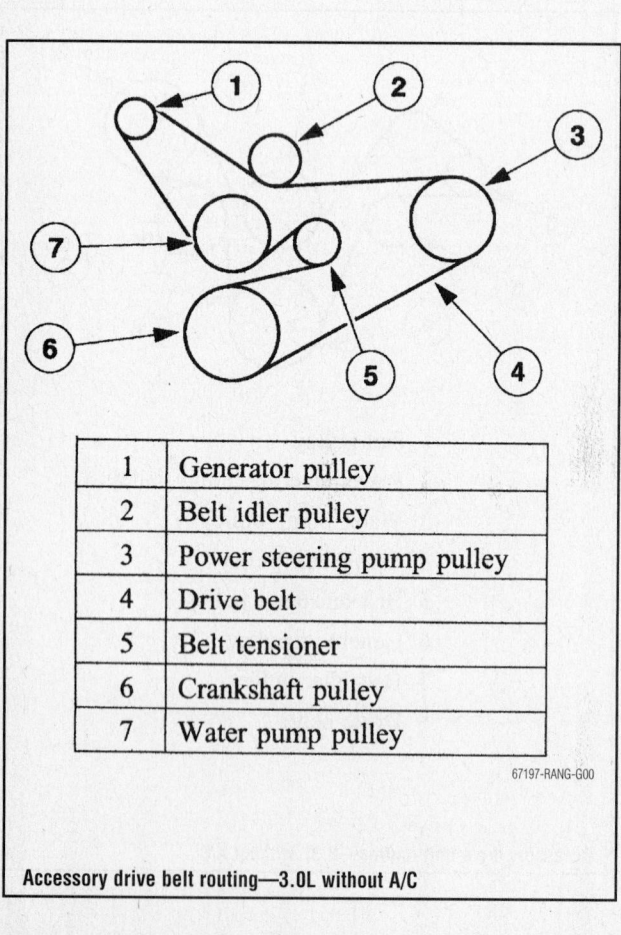

1	Generator pulley
2	Belt idler pulley
3	Power steering pump pulley
4	Drive belt
5	Belt tensioner
6	Crankshaft pulley
7	Water pump pulley

67197-RANG-G00

Accessory drive belt routing—3.0L without A/C

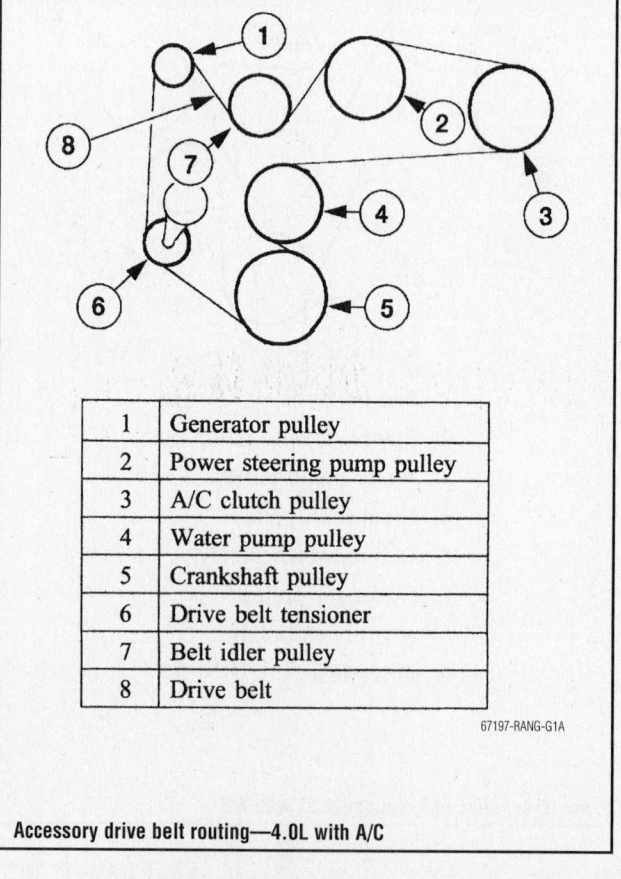

1	Generator pulley
2	Power steering pump pulley
3	A/C clutch pulley
4	Water pump pulley
5	Crankshaft pulley
6	Drive belt tensioner
7	Belt idler pulley
8	Drive belt

67197-RANG-G1A

Accessory drive belt routing—4.0L with A/C

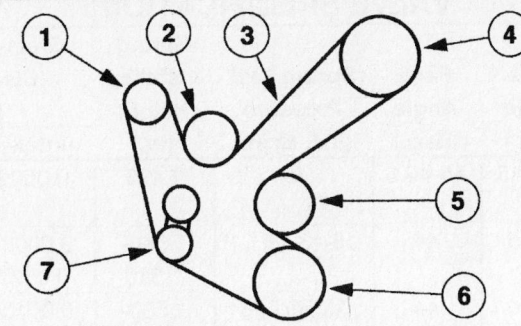

1	Generator pulley
2	Belt idler pulley
3	Drive belt
4	Power steering pump pulley
5	Water pump pulley
6	Crankshaft pulley
7	Belt tensioner

67197-RANG-G2A

Accessory drive belt routing—4.0L without A/C

CAPACITIES

Year	Model	Engine Displacement Liters	Engine VIN	Engine Oil with Filter (qts.)	Transmission (pts.) 5-Spd	Transmission (pts.) Auto.	Transfer Case (pts.)	Drive Axle Front (pts.)	Drive Axle Rear (pts.)*	Fuel Tank (gal.)	Cooling System (qts.)
2001	B2300	2.3	D	4.0	3.0	19.8	—	—	5.0	①	②
	B2500	2.5	C	5.0	3.0	19.8	—	—	5.0	①	②
	B3000	3.0	U	4.5	3.0	③	2.5	3.3	5.0	①	④
	B4000	4.0	E	5.0	3.0	③	2.5	3.25	5.0	①	⑤
2002	B2300	2.3	D	4.0	3.0	19.8	—	—	5.0	①	②
	B3000	3.0	U	4.5	3.0	③	2.5	2.7	5.0	①	④
	B4000	4.0	E	5.0	3.0	③	2.5	2.7	5.0	①	⑤
2003	B2300	2.3	D	4.0	3.0	19.8	—	—	5.0	①	②
	B3000	3.0	U	4.5	3.0	③	2.5	2.7	5.0	①	④
	B4000	4.0	E	5.0	3.0	③	2.5	2.7	5.0	①	⑤
2004	B2300	2.3	D	4.0	3.0	19.8	—	—	5.0	①	②
	B3000	3.0	U	4.5	3.0	③	2.5	3.25	5.0	①	④
	B4000	4.0	E	5.0	3.0	③	2.5	3.25	5.0	①	⑤

NOTE: All capacities are approximate. Add fluid gradually and check to be sure a proper fluid level is obtained.

* For limited slip axles, add 4 oz. of friction modifier, exc. for 1-ton models

① Std: 16.5
   Long Wheelbase: 20.0
   Super Cab: 19.5

② w/MT: 11.2
   w/AT: 10.9

③ 2wd: 20.0
   4wd: 20.6

④ w/MT: 15.2
   w/AT: 14.8

⑤ w/MT: 13.5
   w/AT: 13.2

67162-BSER-C04

VALVE SPECIFICATIONS

Year	Engine Displacement Liters	Engine VIN	Seat Angle (deg.)	Face Angle (deg.)	Spring Test Pressure (lbs. @ in.)	Spring Installed Height (in.)	Stem-to-Guide Clearance (in.)		Stem Diameter (in.)	
							Intake	Exhaust	Intake	Exhaust
2001	2.3	D	44.5-45	45-45.5	①	1.492	0.0009	0.0011	0.2153-0.2159	0.2151-0.2157
	2.5	C	44.75	44	118-132@1.16	1.540-1.580	0.0008-0.0027	0.0018-0.0037	0.2746-0.2754	0.2736-0.2744
	3.0	U	45	44	185@1.16	1.580-1.610	0.0010-0.0027	0.0015-0.0032	0.3126-0.3134	0.3121-0.3129
	4	E	45	45	202-225@1.413-1.445	1.569-1.601	0.0010-0.0020	0.0010-0.0020	0.2740-0.2748	0.2730-0.2740
2002	2.3	D	45	45	①	1.492	0.0009	0.0011	0.2153-0.2159	0.2151-0.2157
	3.0	U	45	44	185@1.16	1.580-1.610	0.0010-0.0027	0.0015-0.0032	0.3126-0.3134	0.3121-0.3129
	4.0	E	45	45	202-225@1.413-1.445	1.569-1.601	0.0010-0.0020	0.0010-0.0030	0.2742-0.2748	0.2736-0.2742
2003	2.3	D	45	45	①	1.492	0.0009	0.0011	0.2153-0.2159	0.2151-0.2157
	3.0	U	45	44	185@1.16	1.580-1.610	0.0010-0.0027	0.0015-0.0032	0.3126-0.3134	0.3121-0.3129
	4.0	E	45	45	202-225@1.413-1.445	1.569-1.601	0.0010-0.0020	0.0010-0.0030	0.2742-0.2748	0.2736-0.2742
2004	2.3	D	45	45	①	1.492	0.0009	0.0011	0.2153-0.2159	0.2151-0.2157
	3.0	U	45	44	185@1.16	1.580-1.610	0.0010-0.0027	0.0015-0.0032	0.3126-0.3134	0.3121-0.3129
	4.0	E	45	45	202-225@1.413-1.445	1.569-1.601	0.0010-0.0020	0.0010-0.0030	0.2742-0.2748	0.2736-0.2742

① Intake: 97.0@1.201
Exhaust: 93.3@1.201

67162-BSER-C05

CRANKSHAFT AND CONNECTING ROD SPECIFICATIONS

All measurements are given in inches.

Year	Engine Displacement Liters	Engine VIN	Crankshaft				Connecting Rod		
			Main Brg. Journal Dia.	Main Brg. Oil Clearance	Shaft End-play	Thrust on No.	Journal Diameter	Oil Clearance	Side Clearance
2001	2.3	D	2.0465-2.2059	0.0007-0.0013	0.0080-0.0160	3	1.9606-1.9685	0.0010-0.0020	0.0767-0.1200
	2.5	C	2.2051-2.2059	0.0008-0.0015	0.0040-0.0080	3	2.0464-2.0472	0.0008-0.0015	0.0035-0.0115
	3.0	U	2.5190-2.5198	0.0010-0.0014	0.0040-0.0080	3	2.1253-2.1261	0.0010-0.0014	0.0060-0.0140
	4.0	E	2.2430-2.2440	0.0008-0.0015	0.0020-0.0125	3	2.7252-2.7260	0.0003-0.0024	0.0036-0.0106
2002	2.3	D	2.0465-2.2059	0.0007-0.0013	0.0080-0.0160	3	1.9606-1.9685	0.0010-0.0020	0.0767-0.1200
	3.0	U	2.5190-2.2059	0.0010-0.0015	0.0040-0.0080	3	2.1253-2.0472	0.0010-0.0015	0.0060-0.0115
	4.0	E	2.2430-2.2440	0.0008-0.0015	0.0020-0.0125	3	2.7252-2.7260	0.0003-0.0024	0.0036-0.0106
2003	2.3	D	2.0465-2.2059	0.0007-0.0013	0.0080-0.0160	3	1.9606-1.9685	0.0010-0.0020	0.0767-0.1200
	3.0	U	2.5190-2.2059	0.0010-0.0015	0.0040-0.0080	3	2.1253-2.0472	0.0010-0.0015	0.0060-0.0115
	4.0	E	2.2430-2.2440	0.0008-0.0015	0.0020-0.0125	3	2.7252-2.7260	0.0003-0.0024	0.0036-0.0106
2004	2.3	D	2.0465-2.2059	0.0007-0.0013	0.0080-0.0160	3	1.9606-1.9685	0.0010-0.0020	0.0767-0.1200
	3.0	U	2.5190-2.2059	0.0010-0.0015	0.0040-0.0080	3	2.1253-2.0472	0.0010-0.0015	0.0060-0.0115
	4.0	E	2.2430-2.2440	0.0008-0.0015	0.0020-0.0125	3	2.7252-2.7260	0.0003-0.0024	0.0036-0.0106

67162-BSER-C06

PISTON AND RING SPECIFICATIONS
All measurements are given in inches.

Year	Engine Displacement Liters	Engine VIN	Piston Clearance	Ring Gap			Ring Side Clearance		
				Top Compression	Bottom Compression	Oil Control	Top Compression	Bottom Compression	Oil Control
2001	2.3	D	0.0009-0.0017	0.006-0.0012	0.012-0.0180	0.007-0.0270	0.0008-0.0013	0.0004-0.0011	0.0025-0.0054
	2.5	C	0.0010-0.0020	0.008-0.014	0.013-0.019	0.010-0.030	0.0014-0.0030	0.0014-0.0030	SNUG
	3.0	U	0.0012-0.0023	0.010-0.020	0.010-0.020	0.010-0.049	0.0602-0.0612	0.0602-0.0612	SNUG
	4.0	E	0.0008-0.0019	0.015-0.023	0.015-0.023	0.015-0.055	0.0010-0.0030	0.0010-0.0030	SNUG
2002	2.3	D	0.0009-0.0017	0.006-0.0012	0.012-0.0180	0.007-0.0270	0.0008-0.0013	0.0004-0.0011	0.0025-0.0054
	3.0	U	0.0012-0.0023	0.010-0.020	0.010-0.020	0.010-0.049	0.0602-0.0612	0.0602-0.0612	SNUG
	4.0	E	0.0008-0.0019	0.015-0.023	0.015-0.023	0.015-0.055	0.0010-0.0030	0.0010-0.0030	SNUG
2003	2.3	D	0.0009-0.0017	0.006-0.0012	0.012-0.0180	0.007-0.0270	0.0008-0.0013	0.0004-0.0011	0.0025-0.0054
	3.0	U	0.0012-0.0023	0.010-0.020	0.010-0.020	0.010-0.049	0.0602-0.0612	0.0602-0.0612	SNUG
	4.0	E	0.0008-0.0019	0.015-0.023	0.015-0.023	0.015-0.055	0.0010-0.0030	0.0010-0.0030	SNUG
2004	2.3	D	0.0009-0.0017	0.006-0.0012	0.012-0.0180	0.007-0.0270	0.0008-0.0013	0.0004-0.0011	0.0025-0.0054
	3.0	U	0.0012-0.0023	0.010-0.020	0.010-0.020	0.010-0.049	0.0602-0.0612	0.0602-0.0612	SNUG
	4.0	E	0.0008-0.0019	0.015-0.023	0.015-0.023	0.015-0.055	0.0010-0.0030	0.0010-0.0030	SNUG

67162-BSER-C07

TORQUE SPECIFICATIONS
All readings in ft. lbs.

	Engine Displacement Liters	Engine VIN	Cylinder Head Bolts	Main Bearing Bolts	Rod Bearing Bolts	Crankshaft Damper Bolts	Flywheel Bolts	Manifold		Spark Plugs	Oil Pan Drain Plug
								Intake *	Exhaust		
2001	2.3	D	①	NA	NA	②	③	13	40	9	21
	2.5	C	④	⑤	⑥	103-133	54-64	19-28	14-21	5-10	NA
	3.0	U	⑦	60	26	107	54-64	24	⑧	8-10	NA
	4.0	E	⑨	72	⑩	⑪	75-85	7	16	15	18
2002	2.3	D	①	NA	NA	⑫	⑬	13	40	9	21
	3.0	U	⑦	60	26	107	54-64	24	⑧	8-10	NA
	4.0	E	⑭	72	⑩	⑪	75-85	7	16	13	18
2003	2.3	D	①	NA	NA	⑫	⑬	13	40	9	21
	3.0	U	⑦	60	26	107	54-64	24	⑧	8-10	NA
	4.0	E	⑭	72	⑩	⑪	75-85	7	16	13	18
2004	2.3	D	①	NA	NA	⑫	⑬	13	40	9	21
	3.0	U	⑦	60	26	107	59	21	⑧	11	NA
	4.0	E	⑭	72	⑩	⑪	⑮	7	16	13	18

NA: Information not available

* NOTE: Applies to Lower Manifold only.

① Step 1: 44 inch lbs.
  Step 2: 11 ft. lbs.
  Step 3: 33 ft. lbs.
  Step 4: +90 degrees
  Step 5: +90 degrees

② Step 1: 30 ft. lbs.
  Step 2: +90 degrees

③ Step 1: 51-59 ft. lbs.
  Step 2: 50 ft. lbs.
  Step 3: 83 ft. lbs.

④ Step 1: 52 ft. lbs.
  Step 2: recheck at 52 ft. lbs.
  Step 3: +90 degrees

⑤ Step 1: 51-59 ft. lbs.
  Step 2: 76-84 ft. lbs.

⑥ Step 1: 26-30 ft. lbs.
  Step 2: 31-36 ft. lbs.

⑦ Step 1: 59 ft. lbs.
  Step 2: Back off 1 full turn
  Step 3: 34-40 ft. lbs.
  Step 4: 63-73 ft. lbs.

⑧ Step 1: 89 inch lbs.
  Step 2: 17 ft. lbs.

⑨ Step 1: 24 ft. lbs.
  Step 2: plus 90 degrees

⑩ Step 1: 15 ft. lbs.
  Step 2: +90 degrees

⑪ Step 1: 37 ft. lbs.
  Step 2: plus 90 degrees

⑫ Step 1: 74 ft. lbs.
  Step 2: +90 degrees

⑬ Step 1: 37 ft. lbs.
  Step 2: 50 ft. lbs.
  Step 3: 83 ft. lbs.

⑭ 8mm bolts: 24 ft. lbs.
  12mm bolts: 24 ft. lbs. +80 degrees, +80 degrees more

⑮ Step 1: 10 ft.lbs.
  Step 2: 52 ft. lbs.

67162-BSER-C08

WHEEL ALIGNMENT

| Year | Model | | Caster | | Camber | | Toe-in (in.) |
			Range (+/-Deg.)	Preferred Setting (Deg.)	Range (+/-Deg.)	Preferred Setting (Deg.)	
2000	B Series	2wd	1.0	①	0.70	-0.50	0.06+/-0.25
		4wd	1.0	②	0.70	-0.50	0.12+/-0.25
2001	B Series	2wd	1.0	④	0.70	-0.50	0.06+/-0.25
		4wd	1.0	②	0.70	-0.50	0.12+/-0.25
2002	B Series	2wd	1.0	④	0.70	-0.50	0.06+/-0.25
		4wd	1.0	②	0.70	-0.50	0.12+/-0.25
		Rear	—	—	0.75	0	0+/-0.30
2003	B Series	2wd	1.0	④	0.70	-0.50	0.06+/-0.25
		4wd	1.0	②	0.70	-0.50	0.12+/-0.25
		Rear	—	—	0.75	0	0+/-0.30
2004	Ranger	2wd	1.0	⑤	0.70	⑥	0.06+/-0.25
		4wd	1.0	⑦	0.70	-0.50	0.12+/-0.25
		Rear	—	—	0.75	0	0+/-0.30

① Left: +4.1
Right: +4.5

② Left: +3.9
Right: +4.4

③ Left: +4.6
Right: +5.0

④ Left: +4.0
Right: +4.4

⑤ Left: +3.0
Right: +3.4

⑥ Left: -0.5
Right: -0.7

⑦ Left: +2.9
Right: +3.6

TIRE, WHEEL AND BALL JOINT SPECIFICATIONS

| Year | Model | OEM Tires | | Tire Pressures (psi) | | Wheel Size | Ball Joint Inspection | Lugnut Torque (ft. lbs.) |
		Standard	Optional	Front	Rear			
2000	B-Series 2wd	P205/75R14SL	P225/70R14SL	35	35	6-JJ	0.030 in. ①	100
	B-Series 4wd	P215/75R15SL	P235/75R15SL	35	35	6-JJ	0.030 in. ①	100
2001	B-Series 2wd	P205/75R14SL	P225/70R14SL	35	35	6-JJ	0.030 in. ①	100
	B-Series 4wd	P215/75R15SL	P235/75R15SL	35	35	6-JJ	0.030 in. ①	100
2002	B-Series 2wd	P235/75R15	none	②	②	7-JJ	0.030 in. ①	100
	B-Series 4wd	P245/75R16	none	②	②	7-JJ	0.030 in. ①	100
2003	B-Series 2wd	P235/75R15	none	②	②	7-JJ	0.030 in. ①	100
	B-Series 4wd	P245/75R16	none	②	②	7-JJ	0.030 in. ①	100
2004	B2300	P225/70R15	none	②	②	7-JJ	0.030 in. ①	100
	B3000	P225/70R15	P235/75R15SL	②	②	7-JJ	0.030 in. ①	100
	B4000	P235/75R15	P245/75R16	②	②	7-JJ	0.030 in. ①	100

OEM: Original Equipment Manufacturer

PSI: Pounds Per Square Inch

STD: Standard

OPT: Optional

① Both upper and lower

② See placard on door post

67162-BSER-C10

BRAKE SPECIFICATIONS

All measurements in inches unless noted

| Year | Model | Brake Disc | | | Brake Drum Diameter | | | Minimum Lining Thickness | Brake Caliper | |
		Original Thickness	Minimum Thickness	Maximum Runout	Original Inside Diameter	Max. Wear Limit	Maximum Machine Diameter		Bracket Bolts (ft. lbs.)	Mounting Bolts (ft. lbs.)
2001	B-Series ①	0.850	0.810	0.0030	9.00	9.09	9.06	0.030	72-97	21-26
	②	0.850	0.810	0.0030	10.00	10.09	10.06	0.030	72-97	21-26
	③	0.850	0.810	0.0030	10.00	10.09	10.06	0.030	72-97	21-26
2002	B-Series ①	0.850	0.810	0.0030	9.00	9.09	9.06	0.030	85	24
	②	0.850	0.810	0.0030	10.00	10.09	10.06	0.030	85	24
	③	0.850	0.810	0.0030	10.00	10.09	10.06	0.030	85	24
2003	B-Series ①	0.850	0.810	0.0030	9.00	9.09	9.06	0.030	85	24
	②	0.850	0.810	0.0030	10.00	10.09	10.06	0.030	85	24
	③	0.850	0.810	0.0030	10.00	10.09	10.06	0.030	85	24
2004	B-Series	NA	④	NA	NA	NA	⑤	⑥	85	24

NOTE: Due to changes made during production, refer to manufacturer's specifications if they differ from those in this chart

NA: Not Available

F: Front

R: Rear

① With 9 inch brakes

② 4x2 with 10 inch brakes

③ 4x4 with 10 inch brakes

④ Molded into the disc

⑤ Molded into the drum

⑥ Front: 0.04 in.

   Rear: 0.03 in.

67162-BSER-C11

SCHEDULED MAINTENANCE INTERVALS
2001-02 B-SERIES

TO BE SERVICED	TYPE OF SERVICE	5	10	15	20	25	30	35	40	45	50	55	60	65
Engine oil & filter	R	✓	✓	✓	✓	✓	✓	✓	✓	✓	✓	✓	✓	✓
Tires	Rotate	✓	✓	✓	✓	✓	✓	✓	✓	✓	✓	✓	✓	✓
Auto trans. fluid	I			✓			✓			✓			✓	
Brake pads/shoes	I			✓			✓			✓			✓	
Coolant hoses	S/I			✓			✓			✓			✓	
Steering linkage	I			✓			✓			✓			✓	
Cabin air filter	R			✓			✓			✓			✓	
Ball joints (2wd)	L			✓			✓			✓			✓	
Exhaust system	I						✓						✓	
Engine air filter	R						✓						✓	
Fuel filter ①	R						✓						✓	
Auto trans fluid (4-speed)	R						✓						✓	
Green coolant ②	R									✓				
Wheel bearings (2wd)	L												✓	
Manual trans. fluid	R												✓	
Spark plugs	R	every 100,000 miles												
PCV valve	R	every 100,000 miles												
Orange coolant	R	every 150,000 miles												
Auto trans fluid (5-speed)	R	every 150,000 miles												
Differential fluid	R	every 150,000 miles												
Accessory drive belts	R	every 150,000 miles												
Transfer case fluid	R	every 150,000 miles												

R: Replace   S: Service   I: Inspect   L: Lubricate

① Recommended, but not required in Calif.

② Change every 30,000 miles or 36 months thereafter

FREQUENT OPERATION MAINTENANCE (SEVERE SERVICE)

If a vehicle is operated under any of the following conditions it is considered severe service:

- Towing a trailer or using a camper or car-top carrier.

- Repeated short trips of less than 5 miles in temperatures below freezing, or trips of less than 10 miles in any temperature.

- Extensive idling or low-speed driving for long distance as in heavy commercial use, such as delivery, taxi or police cars.

- Operating on rough, muddy or salt-covered roads.

- Operating on unpaved or dusty roads.

- Driving in extremely hot (over 90°) conditions.

Engine oil & filter: replace every 3000 miles.
Air cleaner filter: service or inspect every 6000 miles.
Exhaust system: check every 6000 miles.
Automatic transmission fluid & filter: change every 30,000 miles.
Transfer case fluid: change every 60,000 miles
Fule filter: change every 15,000 miles
Spark plugs: change every 60,000 miles
2wd front wheel bearings: lubricate every 30,000 miles

67162-BSER-C12

SCHEDULED MAINTENANCE INTERVALS
2003 B-SERIES

TO BE SERVICED	TYPE OF SERVICE	VEHICLE MILEAGE INTERVAL (x1000)												
		5	10	15	20	25	30	35	40	45	50	55	60	65
Engine oil & filter	R	✓	✓	✓	✓	✓	✓	✓	✓	✓	✓	✓	✓	✓
Tires	Rotate	✓	✓	✓	✓	✓	✓	✓	✓	✓	✓	✓	✓	✓
Auto trans. fluid	I			✓			✓			✓			✓	
Brake pads/shoes	I			✓			✓			✓			✓	
Coolant hoses	S/I			✓			✓			✓			✓	
Steering linkage	I			✓			✓			✓			✓	
Suspension and driveshaft	I			✓			✓			✓			✓	
Cabin air filter	R			✓			✓			✓			✓	
Ball joints (2wd)	I			✓			✓			✓			✓	
Exhaust system	I						✓						✓	
Engine air filter	R						✓						✓	
Fuel filter ①	R						✓						✓	
Auto trans fluid ②	R						✓						✓	
Green coolant	R									✓				
Accessory drive belts	I	every 100,000 miles												
Spark plugs	R	every 100,000 miles												
PCV valve	R	every 100,000 miles												
Yellow coolant ③	R	first 100,000 miles or 5 years												
Auto trans fluid and filter ④	R	every 150,000 miles												
Rear axle fluid	R	every 150,000 miles												
Accessory drive belts ⑤	R	every 150,000 miles												

R: Replace   S: Service   I: Inspect   L: Lubricate

① Recommended, but not required in Calif.

② If equipped with AX4S, 4F50N, 4R100, 4F27E

③ Every 3 years or 50,000 miles thereafter

④ All except AX4S, 4F50N, 4R100, 4F27E

⑤ If not previously replaced

Special Operating Condition Requirements

When towing a trailer or using a camper or car-top carrier:
Change engine oil and install a new oil filter every 4,800 km (3,000 miles) or 3 months.
Change transfer case fluid every 96,000 km (60,000 miles).
Change manual transmission fluid as required.
Inspect and lubricate U-joints as required.

During extensive idling and/or low speed driving for long distances, as in heavy commercial use such as delivery, taxi, patrol car or livery:
Change engine oil and install a new oil filter, lube front lower control arm and steering linkage ball joints with
zerk fittings (if equipped) every 4,800 km (3,000 miles) or 3 months.
Inspect brake system and check battery electrolyte level (Patrol cars) every 8,000 km (5,000 miles).
Install a new fuel filter every 24,000 km (15,000 miles).
Change automatic transmission fluid, lubricate 4x2 wheel bearings,
 install new grease seals and adjust bearings every 48,000 km (30,000 miles).
Install new spark plugs and change transfer case fluid every 96,000 km (60,000 miles).
Install a new cabin air filter as required.

When operating in dusty conditions such as unpaved or dusty roads:
Change engine oil and install a new oil filter every 4,800 km (3,000 miles) or 3 months.
Install a new fuel filter every 24,000 km (15,000 miles).
Change automatic transmission fluid every 48,000 km (30,000 miles).
Change transfer case fluid every 96,000 km (60,000 miles).
Install a new engine air filter as required.
Install a new cabin air filter as required.

When operating in off-road conditions:
Change automatic transmission fluid every 48,000 km (30,000 miles).
Change transfer case fluid every 96,000 km (60,000 miles).
Install a new cabin air filter as required.
Inspect and lubricate U-joints.
Inspect and lubricate steering linkage ball joints with zerk fittings.

SCHEDULED MAINTENANCE INTERVALS
2004 B-SERIES

TO BE SERVICED	TYPE OF SERVICE	VEHICLE MILEAGE INTERVAL (x1000)												
		5	10	15	20	25	30	35	40	45	50	55	60	65
Engine oil & filter	R	✓	✓	✓	✓	✓	✓	✓	✓	✓	✓	✓	✓	✓
Tires	Rotate	✓	✓	✓	✓	✓	✓	✓	✓	✓	✓	✓	✓	✓
Auto trans. fluid	I			✓			✓			✓			✓	
Brake pads/shoes	I			✓			✓			✓			✓	
Coolant hoses	S/I			✓			✓			✓			✓	
Steering linkage	I			✓			✓			✓			✓	
Suspension and driveshaft	I			✓			✓			✓			✓	
Cabin air filter	R			✓			✓			✓			✓	
Ball joints (2wd)	I/L			✓			✓			✓			✓	
Exhaust system	I						✓						✓	
Engine air filter	R						✓						✓	
Fuel filter	R						✓						✓	
Auto trans fluid ①	R						✓						✓	
Wheel bearings (2wd)	R	at 150,000 miles, if not previously replaced, including seals												
Coolant, exc. Premium Gold	R	every 105,000 miles												
Premium Gold coolant ③	R	at 5 years or 100,000 miles												
Spark plugs	R	every 100,000 miles												
PCV valve	R	every 100,000 miles												
Auto trans fluid ④	R	every 150,000 miles												
Fuel injectors	Clean	every 100,000 miles												
Rear axle fluid	R	every 150,000 miles												
Accessory drive belts	I	every 100,000 miles												
Accessory drive belts	R	every 150,000 miles, if not previously replaced												

R: Replace   S: Service   I: Inspect   L: Lubricate

① Vehicles equipped with AX4S, 4F50N, 4R100, 4F27E

② Change every 30,000 miles or 36 months thereafter

③ After the initial change, every 3 years or 50,000 miles

④ Except vehicles equipped with AX4S, 4F50N, 4R100, 4F27E

Special Operating Condition Requirements

When towing a trailer or using a camper or car-top carrier:

Change engine oil and install a new oil filter every 4,800 km (3,000 miles), 3 months or 200 hours of engine operation (whichever occurs first).

Change transfer case fluid every 96,000 km (60,000 miles).

Change manual transmission fluid as required.

Inspect and lubricate U-joints as required.

During extensive idling and/or low speed driving for long distances, as in heavy commercial use such as delivery, taxi, patrol car or livery:

Change engine oil and install a new oil filter every 4,800 km (3,000 miles), 3 months or 200 hours of engine operation (whichever occurs first).

Lube front lower control arm and steering linkage ball joints with zerk fittings (if equipped) every 4,800 km (3,000 miles) or 3 months.

Inspect brake system and check battery electrolyte level (Patrol cars) every 8,000 km (5,000 miles).

Install a new fuel filter every 24,000 km (15,000 miles).

Change automatic transmission fluid, lubricate 4x2 wheel bearings, install new grease seals and adjust bearings every 48,000 km (30,000 miles). If equipped, change the in-line service installed transmission fluid filter.

Install new spark plugs and change transfer case fluid every 96,000 km (60,000 miles).

Install a new cabin air filter as required.

67162-BSER-C14

SCHEDULED MAINTENANCE INTERVALS
2004 Mazda B-Series
Footnotes Continued

When operating in dusty conditions such as unpaved or dusty roads:

Change engine oil and install a new oil filter every 4,800 km (3,000 miles) or 3 months.

Install a new fuel filter every 24,000 km (15,000 miles).

Change automatic transmission fluid every 48,000 km (30,000 miles). If equipped, change the in-line service installed transmission fluid filter.

Change transfer case fluid every 96,000 km (60,000 miles).

Install a new engine air filter as required.

Install a new cabin air filter as required.

When operating in off-road conditions:

Change automatic transmission fluid every 48,000 km (30,000 miles). If equipped, change the in-line service installed transmission fluid filter.

Change transfer case fluid every 96,000 km (60,000 miles).

Install a new cabin air filter as required.

Inspect and lubricate U-joints.

Inspect and lubricate steering linkage ball joints with zerk fittings.

67162-BSER-C15

PRECAUTIONS

Before servicing any vehicle, please be sure to read all of the following precautions, which deal with personal safety, prevention of component damage, and important points to take into consideration when servicing a motor vehicle:

• Never open, service or drain the radiator or cooling system when the engine is hot; serious burns can occur from the steam and hot coolant.

• Observe all applicable safety precautions when working around fuel. Whenever servicing the fuel system, always work in a well-ventilated area. Do not allow fuel spray or vapors to come in contact with a spark, open flame, or excessive heat (a hot drop light, for example). Keep a dry chemical fire extinguisher near the work area. Always keep fuel in a container specifically designed for fuel storage; also, always properly seal fuel containers to avoid the possibility of fire or explosion. Refer to the additional fuel system precautions later in this section.

• Fuel injection systems often remain pressurized, even after the engine has been turned **OFF**. The fuel system pressure must be relieved before disconnecting any fuel lines. Failure to do so may result in fire and/or personal injury.

• Brake fluid often contains polyglycol ethers and polyglycols. Avoid contact with the eyes and wash your hands thoroughly after handling brake fluid. If you do get brake fluid in your eyes, flush your eyes with clean, running water for 15 minutes. If eye irritation persists, or if you have taken brake fluid internally, IMMEDIATELY seek medical assistance.

• The EPA warns that prolonged contact with used engine oil may cause a number of skin disorders, including cancer! You should make every effort to minimize your exposure to used engine oil. Protective gloves should be worn when changing oil. Wash your hands and any other exposed skin areas as soon as possible after exposure to used engine oil. Soap and water, or waterless hand cleaner should be used.

• All new vehicles are now equipped with an air bag system, often referred to as a Supplemental Restraint System (SRS) or Supplemental Inflatable Restraint (SIR) system. The system must be disabled before performing service on or around system components, steering column, instrument panel components, wiring and sensors. Failure to follow safety and disabling procedures could result in accidental air bag deployment, possible personal injury and unnecessary system repairs.

• Always wear safety goggles when working with, or around, the air bag system. When carrying a non-deployed air bag, be sure the bag and trim cover are pointed away from your body. When placing a non-deployed air bag on a work surface, always face the bag and trim cover upward, away from the surface. This will reduce the motion of the module if it is accidentally deployed. Refer to the additional air bag system precautions later in this section.

• Clean, high quality brake fluid from a sealed container is essential to the safe and proper operation of the brake system. You should always buy the correct type of brake fluid for your vehicle. If the brake fluid becomes contaminated, completely flush the system with new fluid. Never reuse any brake fluid. Any brake fluid that is removed from the system should be discarded. Also, do not allow any brake fluid to come in contact with a painted surface; it will damage the paint.

• Never operate the engine without the proper amount and type of engine oil; doing so WILL result in severe engine damage.

• Timing belt maintenance is extremely important! Many models utilize an interference-type, non-freewheeling engine. If the timing belt breaks, the valves in the cylinder head may strike the pistons, causing potentially serious (also time-consuming and expensive) engine damage. Refer to the maintenance interval charts in the front of this manual for the recommended replacement interval for the timing belt, and to the timing belt section for belt replacement and inspection.

• Disconnecting the negative battery cable on some vehicles may interfere with the functions of the on-board computer system(s) and may require the computer to undergo a relearning process once the negative battery cable is reconnected.

• When servicing drum brakes, only disassemble and assemble one side at a time, leaving the remaining side intact for reference.

ENGINE REPAIR

Alternator

REMOVAL

2.3L Engines

1. Before servicing the vehicle, refer to the precautions in the beginning of this section.

2. Remove or disconnect the following:

• Battery ground cable
• Air cleaner outlet tube
• Accessory drive belt
• Mounting bolts and alternator
• Nut and the electrical connectors.

3. To install, reverse the removal procedure. Torque all mounting bolts to 18 ft. lbs. (25Nm).

2.5L Engines

1. Before servicing the vehicle, refer to the precautions in the beginning of this section.

2. Remove or disconnect the following:

• Negative battery cable
• Drive belt

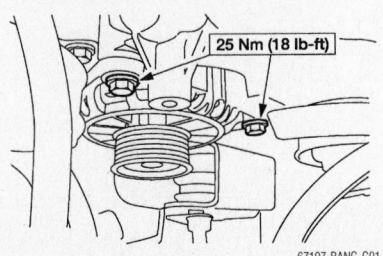

25 Nm (18 lb-ft)

67197-RANG-G01

Alternator mounting bolts—2.3L engine

• Electrical connections to the alternator
• Alternator

To install:

3. Install or connect the following:

• Alternator. Torque the bolts to 40 ft. lbs. (55 Nm).

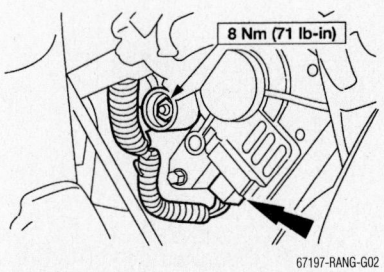

8 Nm (71 lb-in)

67197-RANG-G02

Alternator wiring connections—2.3L engine

- Electrical connectors to the alternator
- Drive belt
- Negative battery cable

3.0L Engines

1. Before servicing the vehicle, refer to the precautions in the beginning of this section.

2. Remove or disconnect the following:
- Negative battery cable
- Air cleaner outlet tube
- Drive belt
- Electrical connectors from the alternator
- Wiring harness to alternator push pin
- Alternator

To install:

3. Install or connect the following:
- Alternator. Torque the bolts to 40 ft.

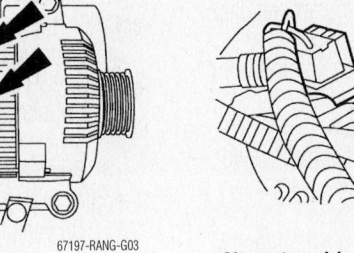

Alternator wiring connections—3.0L engine

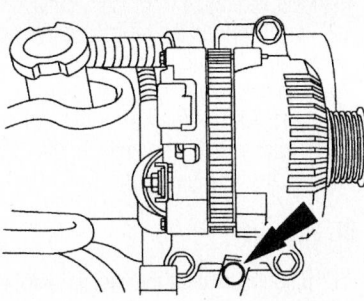

Wiring harness pin-type retainer—3.0L engine

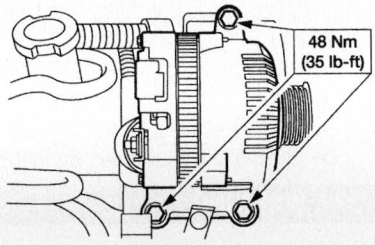

Alternator mounting bolts—3.0L engine

lbs. (55 Nm) for 2001–03; 35 ft. lbs for (48 Nm) fir 2004.
- Pushpin for the alternator wiring harness
- Electrical connectors to the alternator
- Drive belt
- Air cleaner outlet tube
- Negative battery cable

4.0L OHC Engine

1. Before servicing the vehicle, refer to the precautions in the beginning of this section.

2. Remove or disconnect the following:
- Negative battery cable
- Air cleaner outlet tube
- Accessory drive belt
- Electrical connectors
- Wiring harness-to-generator pin-type retainer

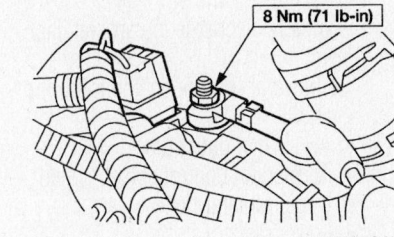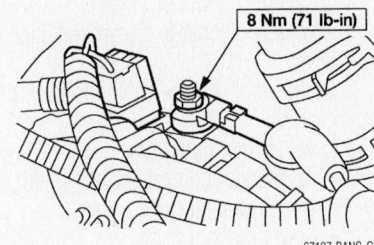

Alternator wiring connections—4.0L SOHC engine

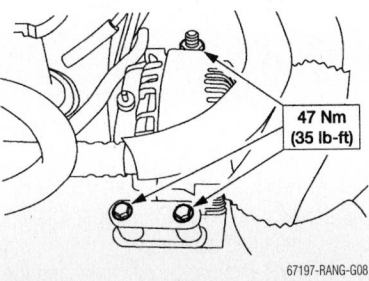

Alternator wiring pin-type retainer—4.0L SOHC engine

Alternator mounting bolts—4.0L SOHC engine

- Stud bolt, the bolts and the alternator

3. To install, reverse the removal procedure. Torque all mounting bolts to 35 ft. lbs. (47Nm).

Ignition Timing

ADJUSTMENT

The ignition timing is preset and is not adjustable.

Engine Assembly

REMOVAL & INSTALLATION

2.3L Engines

1. Before servicing the vehicle, refer to the precautions in the beginning of this section.

2. Relieve the fuel system pressure.

3. Drain the cooling system.

4. Drain the engine oil.

5. Properly discharge the air conditioning system.

6. Remove or disconnect the following:
- Hood
- Accelerator control snow shield
- Air cleaner tube
- Upper radiator hose
- Lower radiator hose
- Fan and shroud
- PCM electrical connector. Remove the retaining nut on the harness clamp. Position the harness on the engine.
- Ground stud for the PCM
- Heater hoses
- All vacuum hoses
- Coolant reservoir hoses
- Air conditioning compressor clutch
- MAF electrical connector
- Air conditioning compressor manifold, plug the lines and the compressor ports
- Accelerator and speed control cables
- Power steering return hose
- PSP switch electrical connector
- High pressure power steering hose
- Fuel supply hose
- 42-pin electrical connector
- VMV vacuum regulator solenoid supply hose
- Evaporative purge hose
- Brake booster vacuum hose and the engine ground strap
- Positive battery cable
- Solenoid control wire at the starter

- Starter wiring harness clamp bolt and position it out of the way.
- RH splash shield
- Alternator electrical connections
- Block heater electrical connector
- Front heated oxygen sensor electrical connector at the bell housing
- Oil pressure sensor electrical connector
- Engine wiring pushpins and position the engine wiring harnesses out of the way.
- Oil filter
- With automatic transmission, the bolt retaining the transmission cooling tubes to the engine. Remove the bracket.
- Transmission dust shield
- Starter motor
- Heated oxygen sensor electrical connector at the rear of the transmission
- Transmission wiring harness
- Vehicle speed sensor, transmission range sensor, backup light switch and the transmission electrical connectors. Disconnect the pushpins and position the harness forward to the engine.
- Oil filter adapter

➡ **Leave two side bolts in until the engine is ready to be removed.**

- Nine of the transmission-to-engine bolts
- With automatic transmission, the transmission fluid indicator and tube assembly
- Starter dust shield

➡ **Mark one stud and the flexplate for assembly reference.**

- With automatic transmission, the four torque converter nuts

7. Support the transmission with a floor jack.
8. Support the engine with a floor crane using a spreader bar.

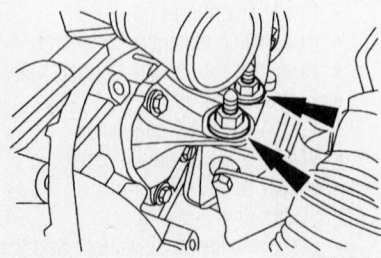

67197-RANG-G09

Engine support insulator nuts, left side shown, right side similar—2.3L engine

9. Remove the two side transmission-to-engine bolts.
10. Remove the four engine support insulator.
11. Remove the engine from the vehicle.
12. Installation is the reverse of removal. Observe the following torques:

- Torque converter bolts: 26 ft. lbs. (35Nm)
- Nine transmission-to-engine bolts 35 ft. lbs. (48Nm)
- Oil filter adapter: 18 ft. lbs. (25Nm)
- Starter: 30 ft. lbs. (40Nm)
- Engine support nuts: 75 ft. lbs. (102Nm)

2.5L Engines

1. Before servicing the vehicle, refer to the precautions in the beginning of this section.
2. Relieve the fuel system pressure.
3. Drain the cooling system.
4. Drain the engine oil.
5. Properly discharge the air conditioning system.
6. Remove or disconnect the following:
- Hood
- Air cleaner outlet tube
- Accelerator control splash shield
- Upper radiator hose
- Lower radiator hose
- The two bolts and position the fan shroud on the fan
- Water pump pulley, fan and shroud
- Radiator overflow tube
- Transmission cooler lines
- Mass air flow sensor
- Heater hoses
- Power connection from the alternator
- Vacuum connection at the vacuum reservoir
- Throttle body heater hose
- Air conditioning cycling switch
- Connector from the PCM
- Ground wire stud from the powertrain control module
- Power steering cut-out switch
- Peanut fitting from the air conditioning condenser core
- Air conditioning high pressure cut-out switch
- Air conditioning manifold hose
- Accelerator cable and speed control cable
- Brake booster vacuum hose and vacuum tube at the upper intake manifold assembly
- EVR vacuum supply hose and the vacuum reservoir vacuum line
- Fuel line

- Air conditioning compressor
- Power steering pressure and return hoses
- Nut retaining the wiring harness
- Block heater
- Engine ground cable
- With automatic transmission, the two transmission harness connectors
- With manual transmission, the transmission harness and the heated oxygen sensor
- Starter motor
- Nut retaining the starter harness and transmission cooler brackets
- Three way catalytic converter

➡ **The torque converter nuts are accessed through the starter motor hole.**

- With automatic transmission, remove the four torque converter nuts and six bolts

7. Support the transmission with a floor jack.
8. Remove the differential pressure feedback EGR sensor.
9. Support the engine with a floor crane.
10. Remove the two upper transmission-to-engine bolts.
11. On vehicles equipped with automatic transmission, disconnect then separate the heated exhaust gas oxygen sensor connector from the bracket located on the bell housing. Remove the four nuts.
Remove the engine from the vehicle.
12. Installation is the reverse of removal. Observe the following torques:

- Engine mount nuts: 85 ft. lbs. (115Nm)
- Transmission-to-engine bolts: 39 ft. lbs. (51 Nm)

3.0L Engines

1. Before servicing the vehicle, refer to the precautions in the beginning of this section.
2. Relieve the fuel system pressure.
3. Drain the cooling system.
4. Drain the engine oil.
5. Properly discharge the air conditioning system.
6. Remove or disconnect the following:
- Hood
- Air cleaner outlet tube
- Upper and the lower radiator hoses

✻✻ WARNING

The fan clutch has left-hand threads.

- The fan clutch and blade as an assembly
- Drive belt
- Fan shroud
- Radiator
- Air conditioning manifold and tube. Remove the nut and position the line aside.
- Air conditioning compressor wiring
- Air conditioning compressor and the air conditioning compressor mounting bracket
- Heater hoses
- Ground cable
- Fuel lines
- Snow shield
- Accelerator cable and the speed control actuator cable
- All vacuum lines
- 42-pin connector
- Powertrain control module connector
- Nut from the powertrain control module harness
- Stud bolt and the powertrain control module ground strap
- Alternator wiring and position aside
- Both heated oxygen sensors
- Transmission harness connectors
- MAF sensor
- LH heated oxygen sensor
- Dual converter Y pipe
- Starter motor and the starter grounding stud bolt
- Torque converter nuts
- 8 transmission-to-engine bolts

7. Install the lifting eyes.
8. Remove the four nuts.
9. Support the transmission.
10. Remove the engine from the vehicle.
11. Installation is the reverse of removal. Observe the following torques:

- Engine mount nuts: 80 ft. lbs. (109Nm)
- Transmission-to-engine bolts: 33 ft. lbs. (45Nm)
- Torque converter nuts: 2001–03 26

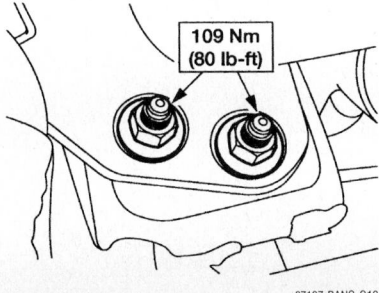

67197-RANG-G10

Engine support nuts, one side shown— 3.0L engine

ft. lbs. (35Nm); 2004 35 ft. lbs. (47 Nm).

4.0L OHC Engines

1. Before servicing the vehicle, refer to the precautions in the beginning of this section.

✶✶ CAUTION

If the fuel supply manifold is used as a leverage device, damage may occur to the supply manifold. Care must be taken when working around the fuel supply manifold.

2. Remove or disconnect the following:

- Accelerator cable from engine
- Speed control cable from engine
- Radiator, the fan blade, and the fan shroud
- Accessory bracket bolts and position bracket aside
- Alternator wiring
- Wiring harness retainer and position generator wiring away from engine
- Engine electrical connector
- PCM connector
- PCM ground wire
- Engine ground wire
- Brake booster vacuum hose
- Air conditioning high pressure switch electrical connector
- Bolt and position the air conditioning lines aside

➡**Heater hose will be removed with engine.**

- Heater hoses
- Fuel line
- Starter motor
- Engine oil
- Oil drain plug
- Transmission portion of wiring harness
- RH and LH heated oxygen sensor connectors
- Transmission control connector
- Output shaft speed sensor connector
- Digital transmission range sensor connector
- Catalyst monitor sensor electrical connector
- Transmission/transfer case portion of the wiring harness from any routing clips or pushpins. Route transmission/transfer case portion of the wiring harness to top of engine.
- Bolt, and position the transmission cooling line bracket aside

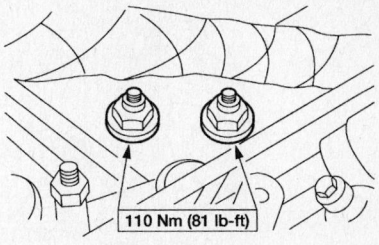

67197-RANG-G11

Left side engine insulator nuts—4.0L SOHC engine

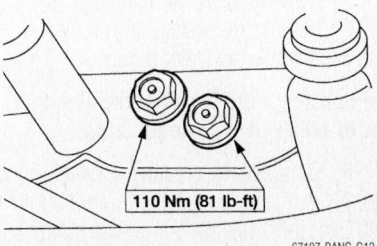

67197-RANG-G12

Right side engine insulator nuts—4.0L SOHC engine

- Air conditioning line bracket nut and position it aside
- Power steering return hose
- Power steering pressure hose
- Vapor management valve hose connector
- Eight bolts and the LH and the RH engine support insulator nuts

➡**The lifting eyes should be installed on the exhaust manifold studs for number three and number four cylinders.**

3. Install the lifting eyes.
4. Install the spreader bar to the lifting eyes.
5. Attach a floor crane to the spreader bar and remove the engine.
6. Installation is the reverse of removal. Observe the following torques:

- Left and right engine insulator nuts: 81 ft. lbs. (110 Nm)
- Engine mount nuts: 59 ft. lbs. (80Nm)
- Transmission-to-engine bolts: 35 ft. lbs. (47Nm)
- Torque converter nuts: 35 ft. lbs. (47Nm)

Water Pump

REMOVAL & INSTALLATION

2.3L Engine

1. Before servicing the vehicle, refer to the precautions in the beginning of this section.

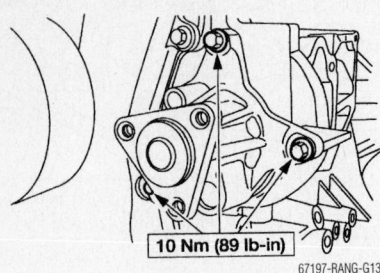

10 Nm (89 lb-in)

67197-RANG-G13

Water pump mounting bolts—2.3L engine

2. Drain the cooling system.
3. Remove the drive belt.
4. Remove the water pump pulley.
5. Remove the water pump.

➡ **Lubricate the water pump O-ring, with MERPOL®, or equivalent..**

6. To install, reverse the removal procedure. Torque the water pump mount bolts to 89 inch lbs. (10Nm). Torque the pulley bolts to 18 ft. lbs. (25Nm).

2.5L Engines

1. Before servicing the vehicle, refer to the precautions in the beginning of this section.
2. Drain the cooling system.
3. Remove or disconnect the following:
   - Negative battery cable
   - Drive belt
   - Fan clutch and shroud
   - Water pump pulley
   - Heater hose from the water pump inlet tube
   - Lower radiator hose
   - Water pump inlet tube
   - Water pump and discard the gasket

To install:

4. Clean the mating surface with the water pump connects to the engine.
5. Install or connect the following:
   - Water pump with a new O-ring. Torque the bolts to 15 ft. lbs. (20 Nm).
   - Inlet tube. Torque the bolts to 89 inch lbs. (10 Nm).
   - Lower radiator hose
   - Water pump pulley
   - Heater hose from the water pump inlet tube
   - Fan clutch and shroud
   - Drive belt
   - Negative battery cable
6. Fill the cooling system.
7. Start the vehicle and check for leaks, repair if necessary.

3.0L Engines

1. Before servicing the vehicle, refer to the precautions in the beginning of this section.
2. Drain the cooling system.
3. Remove or disconnect the following:

   - Negative battery cable
   - Air cleaner outlet tube
   - Fan and radiator shroud
   - Water bypass tube
   - Drive belt
   - Heater hose
   - Water pump pulley
   - Lower radiator hose
   - Air conditioning compressor and bracket assembly and move them aside
   - Water pump

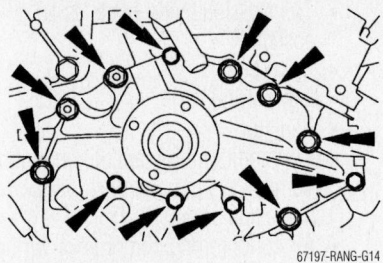

67197-RANG-G14

Water pump mounting bolts—3.0L engine

To install:

4. Clean the mating surfaces where the water pump attaches to the engine.
5. Install or connect the following:
   - Water pump. Torque the bolts to 89 inch lbs. (10 Nm).
   - Air conditioning compressor mounting bracket. Torque the bolts to 44 ft. lbs. (61 Nm).
   - Water pump pulley. Torque the bolts to 20 ft. lbs. (28 Nm).
   - Drive belt
   - Heater hose
   - Lower radiator hose
   - Fan and shroud
   - Air cleaner outlet tube
   - Negative battery cable
6. Fill the cooling system.
7. Start the vehicle and check for leaks, repair if necessary.

4.0L OHC Engine

1. Before servicing the vehicle, refer to the precautions in the beginning of this section.
2. Drain the cooling system.
3. Remove or disconnect the following:

   - Fan shroud
   - Accessory drive belt
   - Idler pulley
   - Water bypass hose
   - Heater hose
   - Lower radiator hose
   - Water pump pulley
   - Water pump

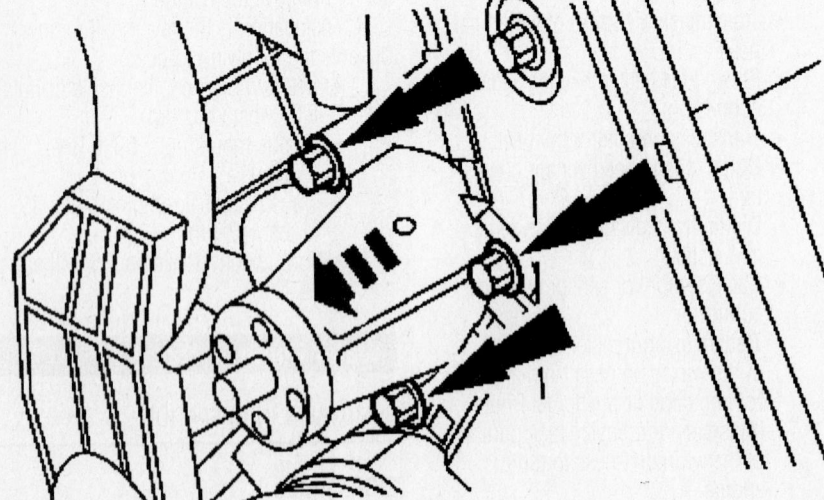

9308EG05

Exploded view of the water pump 2.5L engines

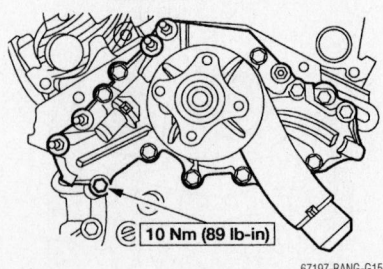

10 Nm (89 lb-in)

67197-RANG-G15

Water pump mounting bolts—4.0L SOHC engine

4. Clean all the sealing surfaces.

5. To install, reverse the removal procedure. Torque the water pump bolts to 89 inch lbs. (10Nm). Torque the pulley bolts to 18 ft. lbs. (25Nm).

Heater Core

REMOVAL & INSTALLATION

2001–03

1. Before servicing the vehicle, refer to the precautions in the beginning of this section.

2. Disconnect the negative battery cable.

3. Drain the cooling system into a clean container for reuse.

4. Remove the steering column by performing the following procedure:

　a. Position the front wheels in the straight-ahead direction.

　b. At the both sides of the steering wheel, remove the cover plugs, the steering wheel-to-air bag module screws, disconnect the air bag electrical connector and carefully remove the air bag module.

　c. Remove the steering wheel-to-steering column nut.

　d. Using a steering wheel puller, press the steering wheel from the steering column.

　e. Remove the parking brake release handle screws and move the release handle aside.

　f. Remove the hood release screws and move the hood release aside.

　g. Remove the 2 instrument panel-to-steering column cover screws and the cover.

　h. Remove the instrument panel steer-

ing column opening reinforcement bolts and the reinforcement.

　i. Remove the ignition switch bolt and disconnect the ignition switch electrical connector.

　j. At the base of the steering column, disconnect the electrical connectors.

　k. If equipped with an automatic transmission, remove the transmission range indicator bolt and the cable.

　l. If equipped with an automatic transmission, disconnect the shift cable from the steering column shift tube lever and the steering column bracket.

　m. Disconnect the brake shift interlock solenoid electrical connector.

　n. Remove the air bag sliding contact.

　o. Remove the upper intermediate steering shaft-to-column shaft bolt and discard the bolt.

　p. Remove the lower steering column-to-instrument panel nuts and the steering column.

5. Remove the instrument panel by performing the following procedure:

　a. Remove the parking brake release handle screws and move the release handle aside.

　b. Disconnect the Brake Pedal Position (BPP) switch electrical connector.

　c. Remove both front door scuff plates.

　d. Remove the push pins and remove both cowl side trim panels.

　e. At the right side cowl panel, disconnect the electrical connectors and ground wires.

　f. Remove both sides windshield garnish moldings.

　g. Remove the instrument panel fuse door.

　h. Disconnect the power distribution box from its bracket and move it aside.

　i. In the engine compartment, loosen the bulkhead wiring harness bolts and disconnect the electrical connectors.

　j. Pull the bulkhead electrical connector handle and disconnect the wiring harness.

　k. Remove the passenger's side air bag module-to-instrument panel screws, disconnect the electrical connector and remove the air bag module.

　l. Disconnect the blend door actuator's electrical connector.

　m. Disconnect the climate control vacuum harness connector.

　n. Disconnect the radio's antenna connector.

　o. Remove the glove compartment.

　p. Remove the instrument panel defroster grille.

　q. Remove the upper instrument panel bolts.

　r. Under the steering column, remove the instrument panel brace bolt.

　s. Remove both the right and left instrument panel-to-cowl bolts.

　t. Pull the instrument panel away from the dash.

　u. Loosen the instrument panel-to-body harness bolt and disconnect the harness.

　v. Using an assistant, remove the instrument panel.

6. Remove the evaporator core by performing the following procedure:

　a. Discharge and recover the air conditioning system refrigerant.

　b. Remove the refrigerant lines from the evaporator core. Discard the O-rings.

　c. If equipped, remove the air conditioning vacuum reservoir tank/bracket screws and reposition the tank.

　d. If equipped, disconnect the speed control servo connector; then, remove the bolt and reposition the speed control servo.

　e. If equipped with a 3.0L or 4.0L engine, remove the support bracket.

　f. Disengage the windshield washer hose retainer and move it aside.

　g. Disconnect the vacuum hose and the retainer; then, move the hose aside.

　h. Remove the passenger's compartment nut.

　i. At the back of the engine, remove the hose support bolts.

　j. Remove the evaporator housing-to-chassis nuts.

　k. Remove the air conditioning accumulator bracket screws.

　l. Remove the evaporator housing cover screws, clips and the cover.

　m. Remove the evaporator core from the housing.

7. Disconnect the heater hoses from the heater core.

8. Remove the heater housing plenum chamber nuts and the plenum chamber.

9. Remove the heater core-to-heater housing screws and the cover.

10. Remove the heater core.

To install:

11. Install the heater core.

12. Install the heater core-to-heater housing cover and the cover screws.

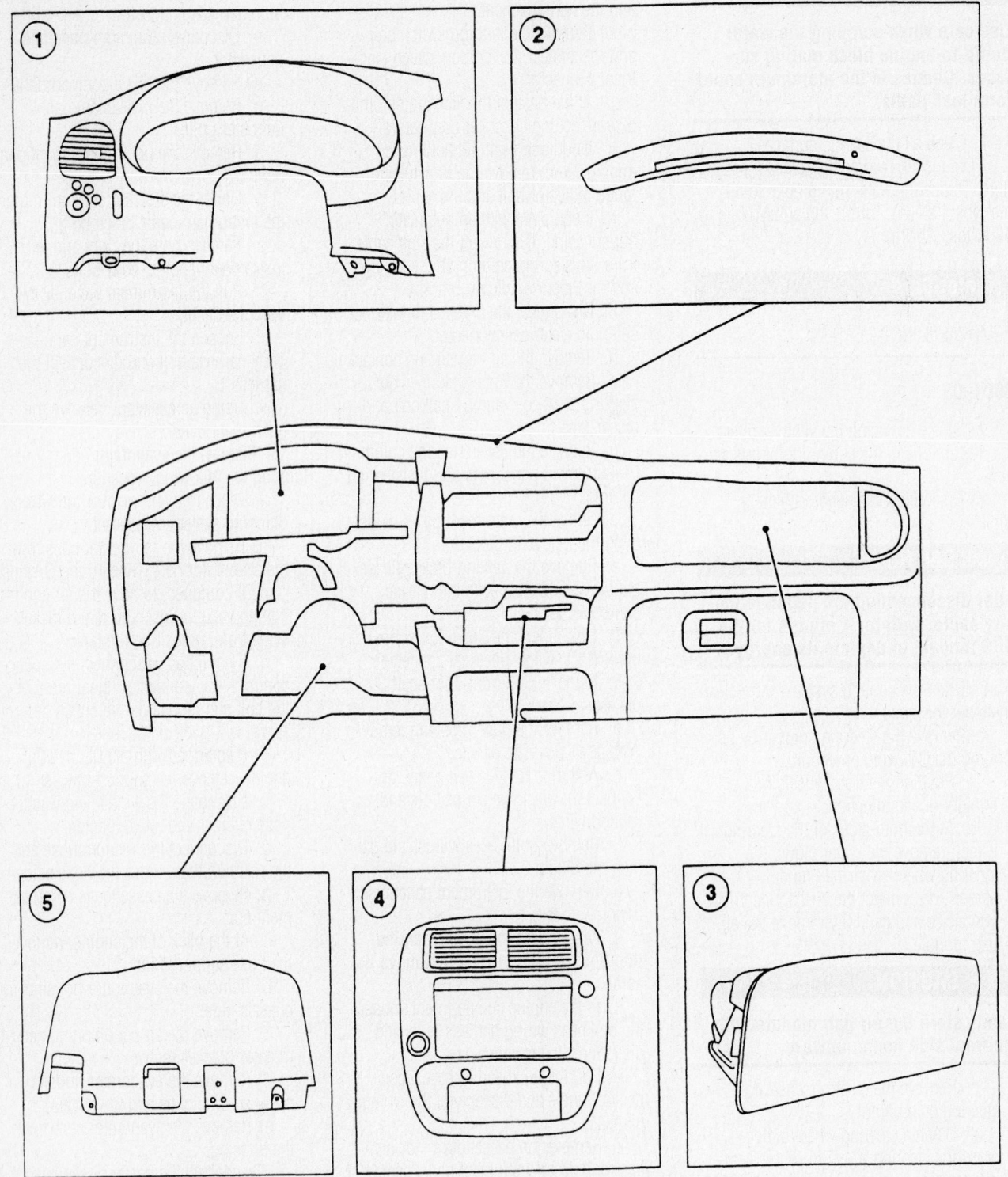

1 Instrument Panel Finish Panel
2 Instrument Panel Defroster
   Opening Grille Assembly
3 Passenger Side Air Bag Module
4 Instrument Panel Center Finish Panel
5 Instrument Panel Steering
   Column Cover

93113GL3

View of the instrument panel components—2001–03

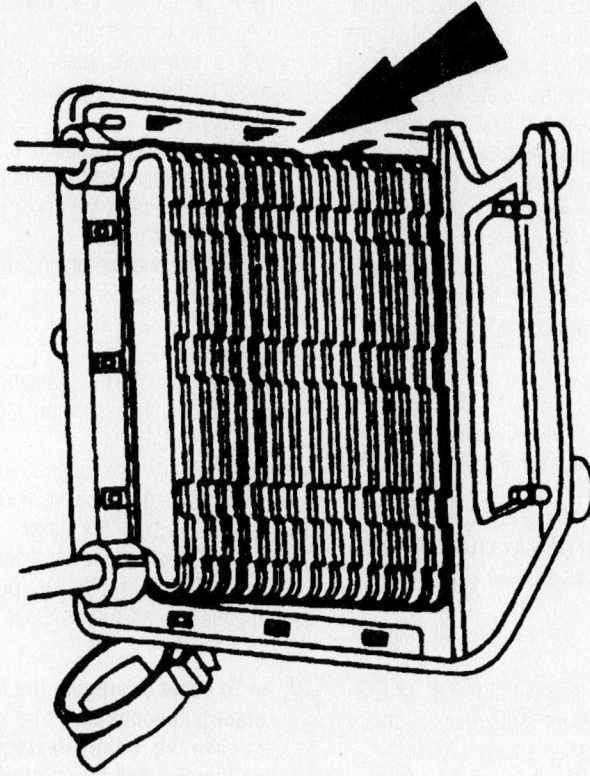

View of the evaporator core—2001–03

93113GL4

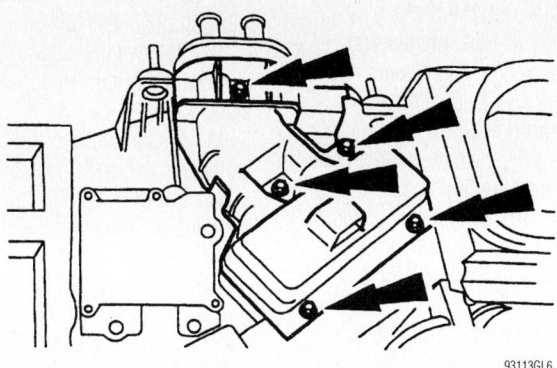

View of the heater core cover—2001–03

93113GL6

View of the heater core—2001–03

93113GL5

13. Install the heater housing plenum chamber and the plenum chamber nuts.

14. Connect the heater hoses to the heater core.

15. Install the evaporator core by performing the following procedure:

   a. Install the evaporator core to the housing.

   b. Install the evaporator housing cover, clips and the cover screws.

   c. Install the air conditioning accumulator bracket screws.

   d. Install the evaporator housing-to-chassis nuts.

   e. At the back of the engine, install the hose support bolts.

   f. Install the passenger's compartment nut.

   g. Connect the vacuum hose and the retainer.

   h. Engage the windshield washer hose retainer.

   i. If equipped with a 3.0L or 4.0L engine, install the support bracket.

   j. If equipped, install the speed control servo bolt and connect the connector.

   k. If equipped, install the air conditioning vacuum reservoir tank and the bracket screws.

   l. Using new O-rings, install the refrigerant lines to the evaporator core.

16. Install the instrument panel by performing the following procedure:

   a. Using an assistant, install the instrument panel.

   b. Connect the harness and tighten the instrument panel-to-body harness bolt.

   c. Push the instrument panel toward the dash.

   d. Install both the right and left instrument panel-to-cowl bolts.

   e. Under the steering column, install the instrument panel brace bolt.

   f. Install the upper instrument panel bolts.

   g. Install the instrument panel defroster grille.

   h. Install the glove compartment.

   i. Connect the radio's antenna connector.

   j. Connect the climate control vacuum harness connector.

   k. Connect the blend door actuator's electrical connector.

   l. Install the passenger's side air bag module, connect the electrical connector and torque the air bag module-to-instrument panel screws to 67–92 inch lbs. (7.6–10.4 Nm).

   m. Connect the bulkhead electrical connector handle wiring harness.

n. In the engine compartment, connect the electrical connectors and tighten the bulkhead wiring harness bolts.

o. Connect the power distribution box to its bracket.

p. Install the instrument panel fuse door.

q. Install both sides windshield garnish moldings.

r. At the right side cowl panel, connect the electrical connectors and ground wires.

s. Install both cowl side trim panels and the push pins.

t. Install both front door scuff plates.

u. Connect the brake pedal position (BPP) switch electrical connector.

v. Install the parking brake release handle and the release handle aside screws.

17. Install the steering column by performing the following procedure:

18. Install the steering column by performing the following procedure:

a. Install the lower steering column and the steering column-to-instrument panel nuts; then, torque the nuts to 10–13 ft. lbs. (13–17 Nm).

b. Using a new bolt, install the upper intermediate steering shaft-to-column shaft bolt and torque to 19–25 ft. lbs. (26–34 Nm).

c. Install the air bag sliding contact.

d. Connect the brake shift interlock solenoid electrical connector.

e. If equipped with an automatic transmission, connect the shift cable from the steering column shift tube lever and the steering column bracket.

f. If equipped with an automatic transmission, install the transmission range indicator cable and bolt.

g. At the base of the steering column, connect the electrical connectors.

h. Connect the ignition switch electrical connector and install the ignition switch bolt.

i. Install the instrument panel steering column opening reinforcement and the reinforcement bolts.

j. Install the instrument panel-to-steering column cover and the 2 cover screws.

k. Install the hood release and the hood release screws.

l. Install the parking brake release handle and the release handle screws.

m. Install the steering wheel to the steering column.

n. Install the steering wheel-to-steering column nut and torque the nut to 25–34 ft. lbs. (34–46 Nm).

o. At the both sides of the steering wheel, install the air bag module, connect the air bag electrical connector, install the steering wheel-to-air bag module screws and the cover plugs.

19. Refill the cooling system.

20. Connect the negative battery cable.

21. Evacuate and charge the air conditioning system.

22. Run the engine to normal operating temperatures; then, check the climate control operation and check for leaks.

2004

1. Before servicing the vehicle, refer to the precautions in the beginning of this section.

2. Recover the refrigerant.

3. Remove the suction accumulator.

4. On trucks with the 2.3L engine:

a. Remove the A/C compressor.

b. Remove the engine oil indicator and tube.

5. On vehicles with the 3.0L or 4.0L engine Position the coolant reservoir and windshield washer reservoir aside.

6. Detach the speed control servo.

7. Disconnect the blower motor and blower motor resistor electrical connectors.

8. Disconnect the heater hoses. Using suitable tools, clamp-off the heater

9. Detach the pin-type retainer and position aside the windshield washer hose.

10. Disconnect and detach the heater control valve vacuum hose.

11. Disconnect the vacuum supply hose near the evaporator core housing.

12. Disconnect the condenser to evaporator line spring lock coupling from the passenger compartment evaporator core inlet.

13. Disconnect the vacuum hose connector and remove the nut.

14. Remove the nuts and the evaporator core housing.

15. Remove the driver side and passenger side air bag modules.

16. Remove the front seats.

17. Remove the screws and position the parking brake release handle aside.

18. Remove the left and right lower cowl kick panels.

19. Position the parking brake assembly aside.

20. Remove the screws and position the hood release handle aside.

21. Remove the instrument panel steering column cover.

22. Remove the instrument panel steering column opening cover reinforcement.

23. Disconnect the brake pedal position

(BPP) switch electrical connector from the steering column shaft.

24. If equipped, disconnect the clutch pedal position (CPP) switch electrical connector.

25. If equipped, disconnect the shift cable from the steering column.

26. Disconnect the upper intermediate steering shaft.

27. Remove the left and right side garnish moldings.

a. Remove the bolt covers and remove the bolts.

b. Remove the assist handle.

c. Remove the windshield side garnish molding.

28. Remove the door moldings.

29. On regular cab vehicles

a. Remove the screws.

b. Remove the scuff plate.

30. Disconnect the electrical connectors and the ground wire from the RH side lower kick panel.

➡ **To avoid damaging the bulkhead electrical connectors, be sure the release tab is fully depressed before pulling release lever into the disconnect position.**

31. Disconnect the LH side bulkhead electrical connector. Press the release tab and pull the release lever.

32. Remove the audio unit. Insert the removal tool. Remove and support the audio unit.

33. Disconnect the audio unit electrical connector and antenna cable.

34. Lower the glove compartment. Press the release tabs inward while lowering the compartment.

35. Through the glove compartment opening, disconnect the blend door actuator electrical connector.

36. Through the glove compartment opening, disconnect the climate control vacuum harness connector.

37. Raise and secure the glove compartment. Press the release tabs inward while raising the glove compartment.

38. Remove the instrument panel defroster opening grille.

39. Remove the instrument panel cowl top bolts.

40. If equipped, remove the floor console.

41. If not equipped with the high-series floor console, remove the cup holders.

42. Release the clips and remove the restraints control module (RCM) cover.

43. If not equipped with high-series floor console, remove the consolette mat.

44. Remove the screws and remove the restraints control module (RCM) cover.

45. Remove the consolette base.
46. Remove the gearshift lever.
47. Remove the screws and remove the manual transmission consolette.
48. Disconnect the RCM electrical connector.
49. Pull the floor carpeting back.
50. On 4x2 vehicles disconnect the instrument panel main harness.
51. On 4x4 vehicles disconnect the instrument panel main harness. From underneath the vehicle, release the instrument panel main harness at the transfer case.
52. Remove the instrument panel side finish panel.
53. Disconnect the door harness electrical connector.
54. Remove the RH side instrument panel bolt. If necessary, transfer the components to the new instrument panel.
55. Remove the LH instrument panel cowl side bolts.
56. Remove the instrument panel.
57. Remove the powertrain control module (PCM).

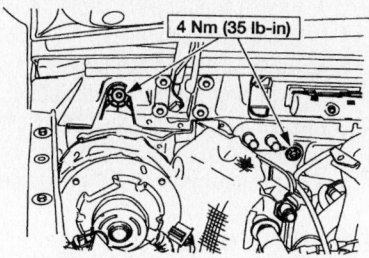

2004 evaporator core housing bolts

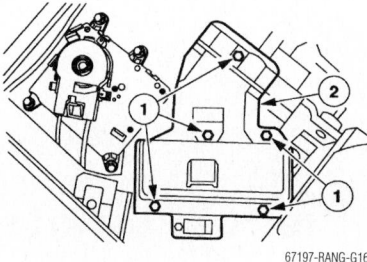

2004 heater core cover bolts

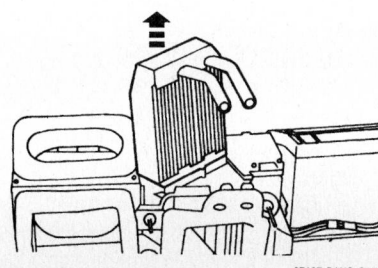

2004 heater core removal

58. Remove the PCM heat sink.
59. Remove the four nuts from the engine side of the dash panel. Position the plenum chamber on the vehicle floor.
60. Remove the heater core cover.
61. Remove the heater core.
62. To install, reverse the removal procedure. During installation, be sure to install a new oval foam seal around the heater core inlet and outlet tubes.
63. Lubricate the refrigerant system with the correct amount of clean PAG oil or equivalent. Install new O-ring seals lubricated in clean mineral oil. Lubricate the coolant hoses with plain water only, if needed.
64. Evacuate, leak test, and charge the refrigerant system.

Cylinder Head

REMOVAL & INSTALLATION

2.3L Engines

1. Before servicing the vehicle, refer to the precautions in the beginning of this section.
2. Relieve the fuel system pressure.
3. Drain the cooling system.
4. Properly discharge the air conditioning system.
5. Remove or disconnect the following:
   - Negative battery cable
   - Drive belt.
   - Engine oil level indicator assembly.
   - Engine oil level indicator.
   - Engine oil level indicator tube.
   - Water outlet tube.
   - Water outlet tube.
   - Air conditioning compressor.

➡**The generator will be removed with the accessory bracket.**

   - Accessory bracket.
   - Right motor mount.
   - Coolant hose from the thermostat.
   - Coolant hose from the EGR valve.
   - Coolant tube assembly.
   - Exhaust manifold and gasket.
   - Block heater (if so equipped).
   - Water outlet.
   - EGR valve.
   - Power steering pump and reservoir as an assembly.
   - Idle air control (IAC) valve.
   - Throttle position (TP) sensor.
   - Manifold absolute pressure (MAP) sensor.
   - Swirl control valve monitor electrical connector.

   - CKP sensor and the wiring harness pin-type retainers.
   - Knock sensor (KS).
   - Electric thermostat.
   - Swirl control valve.
   - CMP sensor electrical connector and disconnect the PCV hose from the intake manifold.
   - Engine wiring harness pin-type retainers from the intake manifold.
   - Engine wiring harness connector bracket. Position the engine wiring harness aside.
   - EGR tube.
   - Fuel supply line clip from the front of the intake manifold. Disconnect the vacuum hose from the intake manifold.
   - Intake manifold assembly.
   - Fuel injector electrical connectors. Detach the wiring harness pin-type retainers.
   - Ignition coil and the cylinder head temperature (CHT) sensor electrical connectors.
   - Engine wiring harness anchors from the valve cover studs. Remove the engine wiring harness.
   - Ignition coil.
   - Bypass hose.
   - Thermostat housing.
   - Knock sensor and the engine vent cover.
   - Left motor mount.
   - Fuel injector supply manifold with the injectors and the ground strap.
   - Water pump pulley.
   - Water pump.
   - CMP sensor.
   - CHT sensor.
   - Spark plugs.
   - Valve cover.
   - CKP sensor.
   - Crankshaft vibration damper

➡**There is one front cover bolt behind the cooling fan drive pulley. To remove this bolt, align one of the cooling fan drive pulley access holes with the bolt head to access the bolt.**

   - Front cover.
   - Timing chain tensioner.
   - Timing chain guides.
   - Timing chain assembly.

➡**Use a wrench on the flats between cylinders No. 1 and No. 2 to hold the camshaft in place.**

   - Camshaft drive sprockets.
   - Oil pump chain tensioner and guide.

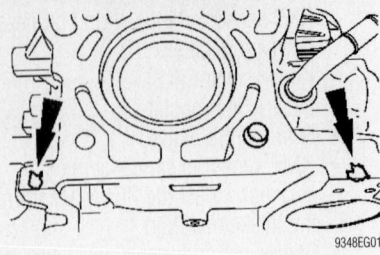
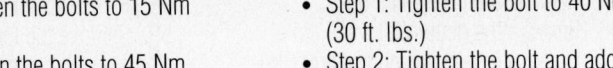

RTV sealer application—2.3L cylinder head

➡The oil pump chain sprocket must be held in place.

• Oil pump chain and sprockets.

➡Note the position of the lobes on the No. 1 cylinder before removing the camshafts for assembly reference.

6. Loosen the camshaft bearing cap bolts in sequence, one turn at a time. Repeat the first step until all tension is released from the camshaft bearing caps. Remove the camshaft bearing caps.

7. Remove or disconnect the following:
• Camshafts.
• Cylinder head bolts and the cylinder head.
• Cylinder head gasket.

8. Installation is the reverse of removal. Apply RTV sealer to the places shown. The head must be installed within 4 minutes of application. Observe the following torques:

a. Cylinder head:
• Step 1: Tighten the bolts to 5 Nm (44 inch lbs.)

• Step 2: Tighten the bolts to 15 Nm (11 ft. lbs.)
• Step 3: tighten the bolts to 45 Nm (33 ft. lbs.)
• Step 4: Tighten the bolts an additional 90 degrees (¼ turn)
• Step 5: Tighten the bolts an additional 90 degrees (¼ turn)

b. Camshafts:

➡Install the camshafts with the alignment notches in the camshaft lined up so the camshaft alignment plate can be installed without rotating the camshafts. Make sure the lobes on the No. 1 cylinder are in the same position as noted in the disassembly procedure. Rotating the camshafts, or installing the camshafts 180 degrees out of position can cause severe damage to the valves and pistons. Lubricate the camshaft journals and bearing caps with clean engine oil. Install the camshafts and bearing caps. Tighten the bolts in the sequence shown in three stages.

• Step 1: Tighten the camshaft bearing caps one turn at a time until tight.
• Step 2: Tighten the bolts to 7 Nm (62 inch lbs.)
• Step 3: Tighten the bolts to 16 Nm (12 ft. lbs.)

c. Crankshaft vibration damper:

➡Do not reuse the crankshaft pulley bolt. Tighten the bolt in two stages.

• Step 1: Tighten the bolt to 40 Nm (30 ft. lbs.)
• Step 2: Tighten the bolt and additional 90 degrees (¼ turn).

2.5L Engines

1. Before servicing the vehicle, refer to the precautions in the beginning of this section.
2. Relieve the fuel system pressure.
3. Drain the cooling system.
4. Properly discharge the air conditioning system.
5. Remove or disconnect the following:
• Negative battery cable
• Loosen the water pump pulley bolts
• Drive belt
• Water pump pulley
• Fan and clutch assembly
• Intake manifolds
• Ignition wires from the spark plugs
• Spark plugs
• Oil level indicator tube
• Exhaust Gas Recirculation (EGR) valve to the exhaust manifold tube
• Valve cover
• Engine control wiring from the air conditioning compressor
• Air conditioning compressor mounting bracket with the power steering pump attached and move them aside
• Engine control sensor wiring from the alternator
• Lower radiator hose
• Water pump inlet tube
• Upper radiator hose
• Alternator
• Alternator mounting bracket
• Ignition wire and bracket
• Outer timing belt cover
• Timing belt
• Exhaust manifold
• Cylinder head and discard the bolts and the gasket

To install:

6. Clean the mating surface where the cylinder head attaches to the engine.
7. Install a new gasket and the cylinder head.
8. Torque the new cylinder head bolts in stages as follows:
a. Step 1: 51 ft. lbs. (70 Nm).
b. Step 2: An additional 51 ft. lbs. (70 Nm).
c. Step 3: Plus and additional 90 degrees.
9. Install or connect the following:
• Exhaust manifold. Torque the bolts to 15 ft. lbs. (20 Nm) plus an additional 30 ft. lbs. (40 Nm).

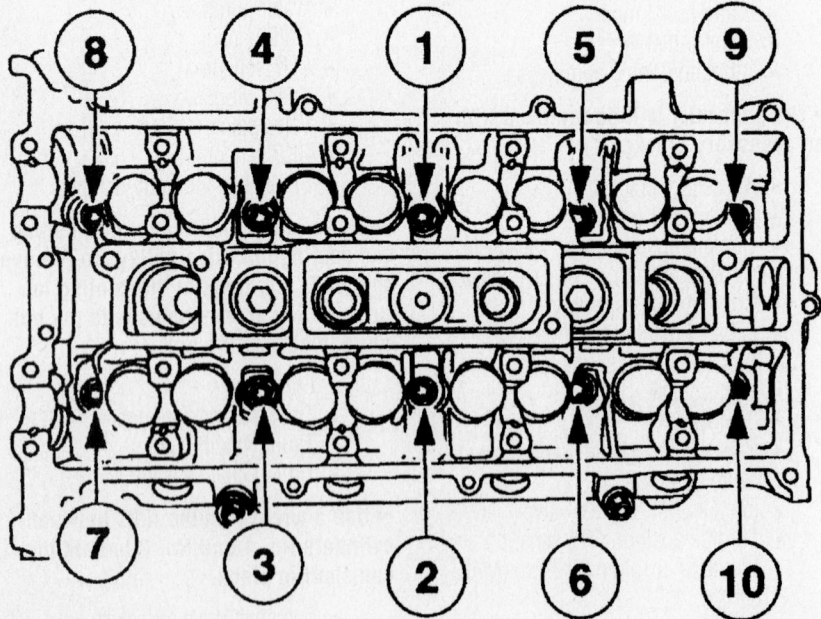

Head bolt torque sequence—2.3L

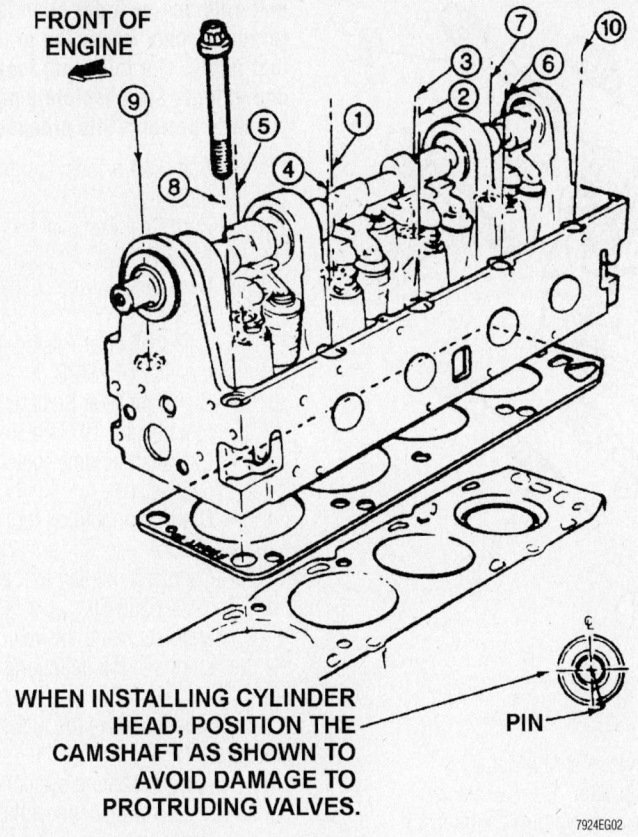

FRONT OF ENGINE

WHEN INSTALLING CYLINDER HEAD, POSITION THE CAMSHAFT AS SHOWN TO AVOID DAMAGE TO PROTRUDING VALVES.

PIN

7924EG02

Cylinder head bolt torque sequence—2.5L engines

- Timing belt tensioner and timing belt
- Timing belt cover
- Ignition wires and coil

10. Install the alternator bracket. Torque the bolts in 4 stages as follows:

    a. Step 1: Hand tighten bolt No. 1.

    b. Step 2: Torque bolt No. 2 to 40 ft. lbs. (55 Nm).

    c. Step 3: Torque bolt No. 3 to 40 ft. lbs. (55 Nm).

    d. Step 4: Torque bolt No. 1 to 40 ft. lbs. (55 Nm).

11. Install or connect the following:

- Water pump inlet tube with a new O-ring to the water pump. Torque the bolts to 89 inch lbs. (10 Nm).
- Lower radiator hose
- Alternator
- Upper radiator hose and heater hose
- Air conditioning compressor mounting bracket with the power steering pump attached. Torque the bolts to 40 ft. lbs. (55 Nm).
- Air conditioning compressor. Torque the bolts to 20 ft. lbs. (28 Nm).
- Water pump pulley and fan clutch. Hand tighten the bolts

- Drive belt. When the belt is positioned properly, torque the fan clutch bolts to 16 ft. lbs. (23 Nm).
- Fan shroud
- Sparks plugs
- Oil level indicator tube
- Engine control sensor wiring
- Upper intake manifold
- EGR valve to the exhaust manifold tube. Torque the bolts to 34 ft. lbs. (47 Nm).
- EGR transducer to the rear of the engine
- Negative battery cable

12. Recharge the air conditioning system.

13. Filling the cooling system.

14. Start the vehicle and check for leaks, repair if necessary.

3.0L Engine

➡ It may be easier to remove the engine from the vehicle. If removing the engine, refer to the engine removal procedure in this section.

1. Before servicing the vehicle, refer to the precautions in the beginning of this section.

2. Evacuate the air conditioning system.

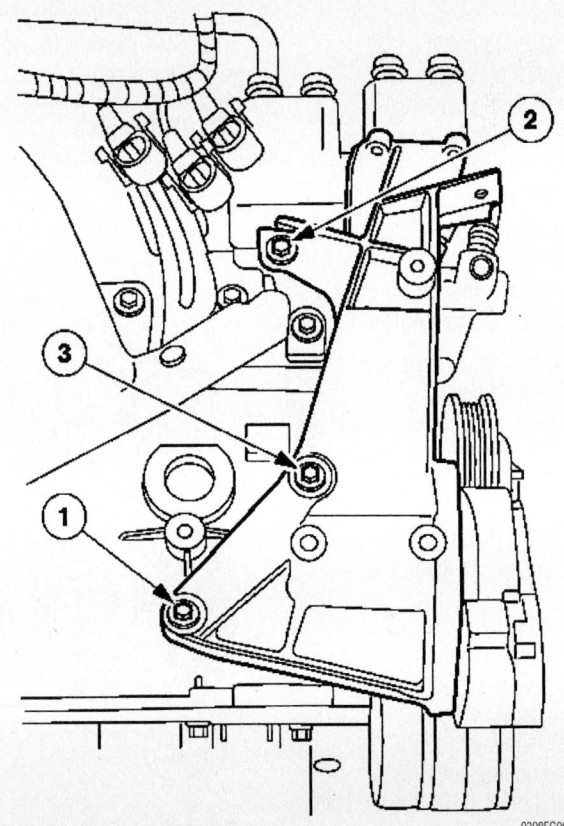

9308EG06

Alternator bracket bolt tightening sequence 2.5L engines

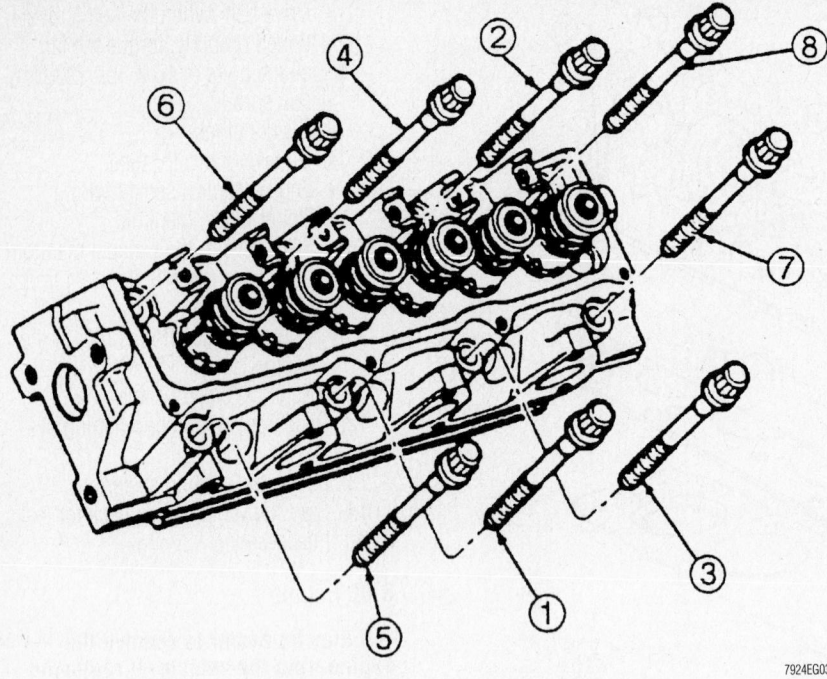

Cylinder head bolt torque sequence 3.0L engine

3. Drain the cooling system.
4. Drain the engine oil.
5. Remove or disconnect the following:
   - Negative battery cable
   - Lower intake manifold
   - Air conditioning compressor
   - Alternator
   - Power steering pump
   - Alternator mounting bracket
   - Air conditioning compressor mounting bracket
   - Exhaust manifolds
   - Cylinder head and discard the bolts and gasket

To install:

➡ The "V" in the cylinder head gasket must face the front of the engine.

6. Clean the mating surfaces where the head attaches to the engine.
7. Install a new cylinder head gasket and the cylinder head to the engine.
8. Torque the new cylinder head bolts in stages as follows:
   a. Step 1: 59 ft. lbs. (80 Nm).
   b. Step 2: Loosen the bolts one full turn.
   c. Step 3: 40 ft. lbs. (55 Nm).
   d. Step 4: 63 ft. lbs. (85 Nm).
9. Install or connect the following:
   - Lower intake manifold
   - Exhaust manifold
   - Air conditioning compressor mounting bracket. Torque the bolts to 44 ft. lbs. (66 Nm).
   - Alternator mounting bracket

   - Power steering pump
   - Alternator
   - Air conditioning compressor
   - Negative battery cable
10. Fill the engine with clean oil.
11. Fill the cooling system.
12. recharge the air conditioning system
13. Start the vehicle and check for leaks, repair if necessary.

4.0L OHC Engine

1. Before servicing the vehicle, refer to the precautions in the beginning of this section.

➡ If only one cylinder head is to be removed, only follow the procedures that apply. The following tools, or their equivalents are absolutely necessary to properly perform this procedure:

- Cam Chain Tensioner tool T97T-6K254-A
- Cam Gear Removal tool T97T-6256-F
- Cam Gear Torque adapter T97T-6256-G
- Camshaft Gear Positioning/Holding tool T97T-6256-B
- Camshaft Gear Positioning/Holding tool adapter T97T-6256-A
- Camshaft holding tool T97T-6256-C
- Crankshaft holding tool T97T-6303-A
- Camshaft holding tool adapter T97T-6256-D

2. Before servicing the vehicle, refer to the precautions in the beginning of this section.
3. Properly relieve the fuel system pressure.
4. Drain the cooling system.
5. Remove or disconnect the following:
   - Negative battery cable
   - Lower intake manifold
   - Fan blade and shroud
   - Valve cover
   - Roller followers, if equipped
   - Drive belt
   - Upper radiator hose and tube
   - Alternator electrical connectors
   - Alternator mounting bracket
   - Engine accessory bracket and move it aside

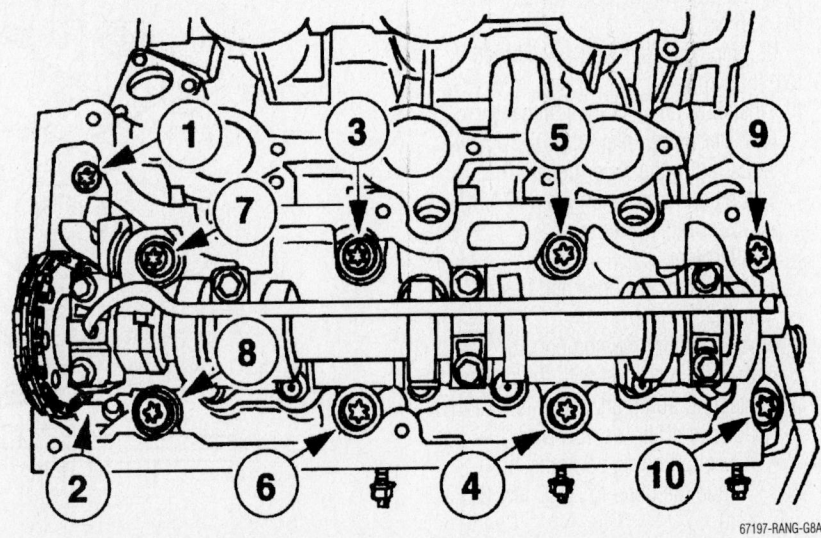

The correct cylinder head bolt loosening sequence must be used to prevent warpage–4.0L SOHC engine

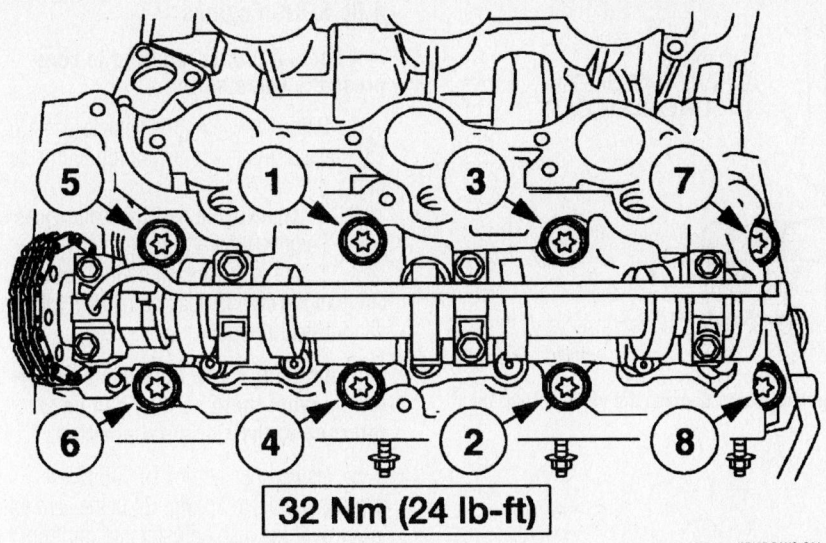

32 Nm (24 lb-ft)

67197-RANG-G9A

Cylinder head bolt torque sequence 4.0L SOHC engine

- Camshaft Position (CMP) electrical connector
- Crankshaft Position (CKP) sensor electrical connector
- Engine Coolant Temperature (ECT) sensor electrical connector
- Coil pack electrical connector
- Exhaust Gas Recirculation (EGR) valve electrical connector
- EGR valve bracket and move it aside
- Heater hoses
- Fuel injector electrical connectors
- Water bypass hose
- Thermostat housing
- Spark plug wires
- Fuel injection supply manifold
- Fuel injectors
- Crankcase vent separator spring
- Oil dipstick housing
- Exhaust manifold
- Hydraulic chain tensioner
- Cassette retaining bolt
- Camshaft sprocket
- Cylinder head and discard the gasket

To install:

6. Thoroughly clean all gasket mating surfaces. Remove all traces of old gasket material, oil, grease or dirt.

7. Insure that the rubber band is holding the right-hand chain to the cassette.

8. Install a new head gasket and the cylinder head.

9. Torque the new cylinder head bolts in sequence as follows:

   a. Step 1: 26 ft. lbs. (34 Nm).
   b. Step 2: Plus 90 degrees.
   c. Step 3: Plus an additional 90 degrees.

10. Install or connect the following:

- Camshaft sprocket in the cassette and make certain that the camshaft sprocket turns freely on the camshaft
- Cassette retaining bolt. Torque the bolt to 89 inch lbs. (10 Nm).
- Exhaust manifold
- Oil level indicator tube. Torque the bolt to 18 ft. lbs. (25 Nm).
- Crankcase vent separator and spring
- Thermostat housing. Torque the bolts to 8 ft. lbs. (11 Nm).
- Water bypass hose
- Heater hoses
- EGR bracket. Torque the bolt to 89 inch lbs. (10 Nm).
- EGR tube. Torque the nut to 30 ft. lbs. (40 Nm).
- ECT sensor electrical connector
- Electrical harness retainer. Torque the bolt to 89 inch lbs. (10 Nm).
- CKP and CMP electrical connectors
- Accessory bracket. Torque the bolts to 31 ft. lbs. (42 Nm).
- Alternator mounting bracket. Torque the bolts to 31 ft. lbs. (42 Nm).
- Alternator and electrical connectors
- Drive belt
- Fan shroud
- Roller followers
- Valve cover
- Lower intake manifold
- Negative battery cable

11. Change the engine oil and filter.
12. Refill the cooling system.
13. Start the engine and check for leaks, repair if necessary.

REMOVAL & INSTALLATION

2.3L Engines

This DOHC engine does not employ rocker arms. The camshafts bear directly on the lifters. For lifter removal, remove the camshafts.

2.5L Engines

➡**A special tool is required to compress the valve spring.**

1. Before servicing the vehicle, refer to the precautions in the beginning of this section.

2. Disconnect the negative battery cable.

3. Remove the valve cover.

4. Rotate the camshaft so that the base circle of the cam is against the cam follower you intend to remove.

➡**If removing more than one cam follower, label them so they can be returned to their original position.**

5. Using special tool T88T-6565-BH depress the valve spring. Slide the cam follower over the lash adjuster and out from under the camshaft.

To install:

6. Compress the valve spring and slide the roller follower into position.

7. Release the tension from the spring.

8. Install the valve cover and connect the negative battery cable.

3.0L Engines

1. Before servicing the vehicle, refer to the precautions in the beginning of this section.

2. Remove or disconnect the following:

- Negative battery cable
- Rocker arm covers
- Retaining bolt at each rocker arm

3. The rocker arm and pushrod may then be removed from the engine. Keep all rocker arms and pushrods in order so they may be installed in their original locations.

To install:

4. Lubricate the rocker arm assemblies with SAE 50W engine oil.

5. Ensure that the fulcrums are properly seated into the cylinder head. Torque the rocker arm fulcrum bolts to 19 ft. lbs. (26 Nm).

6. Install the rocker arm covers and connect the negative battery cable.

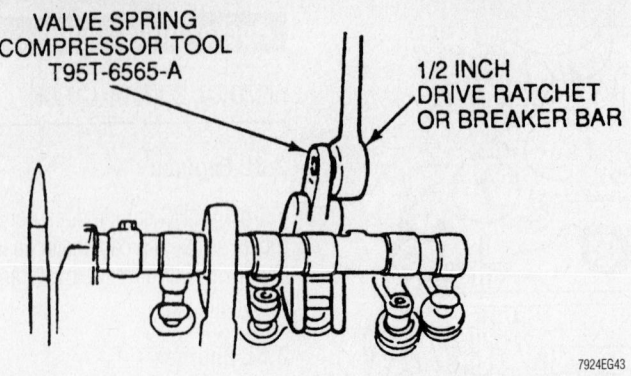

To remove the cam follower (rocker arm), use the special tool to depress the valve spring, then remove the cam follower—2.5L engines

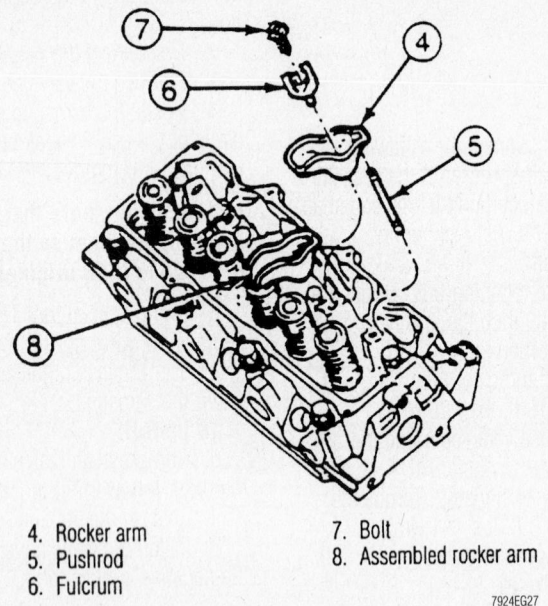

4. Rocker arm
5. Pushrod
6. Fulcrum

7. Bolt
8. Assembled rocker arm

Exploded view of the rocker arm assembly—3.0L engine

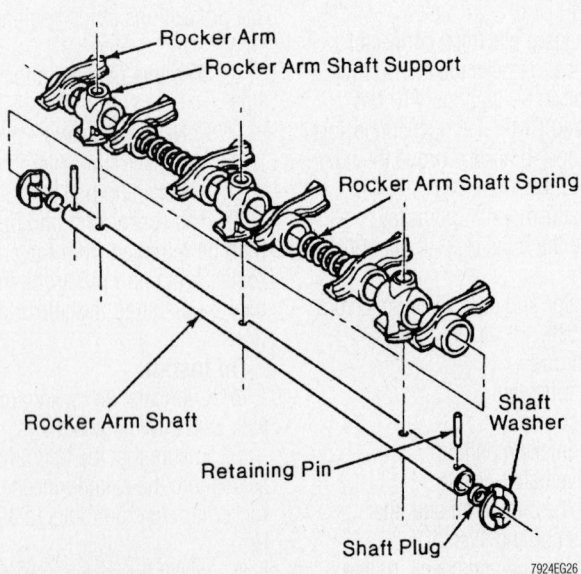

Rocker arm and shaft assembly—4.0L SOHC engine

4.0L SOHC Engines

➡**A special tool is required to compress the valve spring.**

1. Before servicing the vehicle, refer to the precautions in the beginning of this section.
2. Disconnect the negative battery cable.
3. Remove the valve cover.
4. Rotate the camshaft so that the base circle of the cam is against the cam follower you intend to remove.

➡**If removing more than one cam follower, label them so they can be returned to their original position.**

5. Using special tool T97T-6565-A depress the valve spring. Slide the cam follower over the lash adjuster and out from under the camshaft.

To install:

6. Compress the valve spring and slide the roller follower into position.
7. Release the tension from the spring.
8. Install the valve cover and connect the negative battery cable.

Intake Manifold

REMOVAL & INSTALLATION

2.3L Engine

1. Before servicing the vehicle, refer to the precautions in the beginning of this section.
2. Relieve the fuel system pressure.
3. Drain the cooling system.
4. Properly discharge the air conditioning system.
5. Remove or disconnect the following:
   • Negative battery cable
   • Water outlet tube.
   • Water outlet tube.

➡**The alternator will be removed with the accessory bracket.**

   • Accessory bracket.
   • Coolant hose from the thermostat.
   • Coolant hose from the EGR valve.
   • Coolant tube assembly.
   • Block heater (if so equipped).
   • Water outlet.
   • EGR valve.
   • Idle air control (IAC) valve.
   • Throttle position (TP) sensor.
   • Manifold absolute pressure (MAP) sensor.
   • Swirl control valve monitor electrical connector.
   • Electric thermostat.

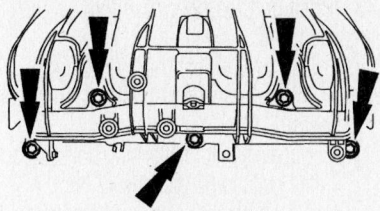

67197-RANG-G19

2.3L engine intake manifold bolts

- Swirl control valve.
- CMP sensor electrical connector and disconnect the PCV hose from the intake manifold.
- Engine wiring harness pin-type retainers from the intake manifold.
- Engine wiring harness connector bracket. Position the engine wiring harness aside.
- EGR tube.
- Fuel supply line clip from the front of the intake manifold. Disconnect the vacuum hose from the intake manifold.
- Intake manifold assembly.

6. Installation is the reverse of removal. Torque the bolts to 13 ft. lbs. (18Nm). There is no special torque sequence.

2.5L Engine

1. Before servicing the vehicle, refer to the precautions in the beginning of this section.

2. Remove or disconnect the following:
- Negative battery cable
- Intake air temperature (IAT) sensor
- Air cleaner outlet tube
- Accelerator control splash shield
- Engine control sensor wiring from the throttle position (TP) sensor and the idle air control (IAC) valve
- Accelerator cable and speed control cable, if equipped
- Accelerator cable bracket
- Crankcase vent hose from the valve cover
- Vacuum hoses from the intake manifold vacuum tee
- Heater hose from the intake manifold
- Exhaust Gas Recirculation (EGR) tube
- EGR valve
- Upper intake manifold and discard the gasket

To install:

3. Install a new upper intake manifold gasket.

4. Install the upper intake manifold. Torque the bolts in sequence as follows:
   a. Step 1: 89 inch lbs. (10 Nm).

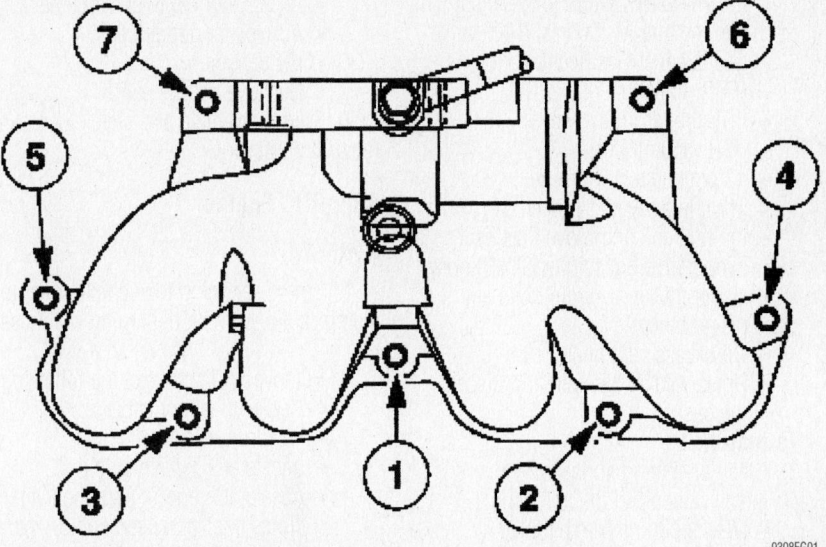

9308EG01

Tighten the lower manifold bolts in the sequence shown—2.5L engine

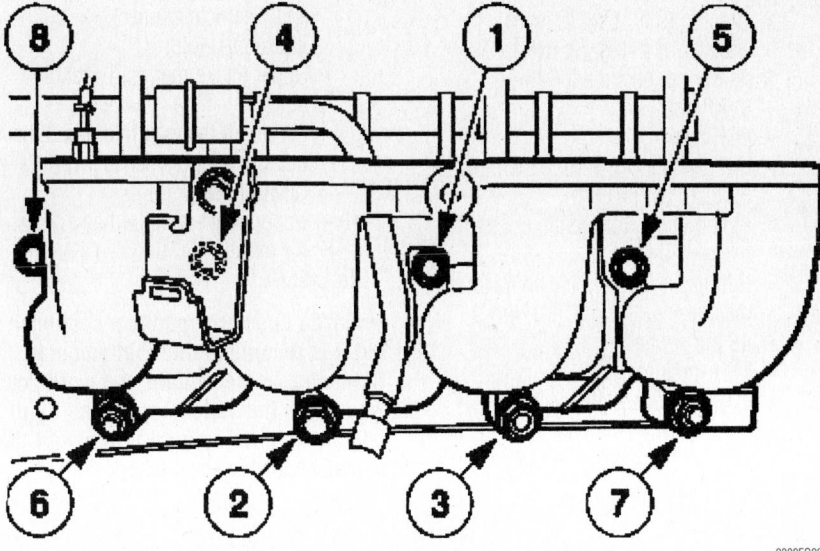

9308EG02

Tighten the upper manifold bolts in the sequence shown—2.5L engine

   b. Step 2: 28 ft. lbs. (38 Nm).

5. Install or connect the following:
- EGR valve. Torque the bolts to 22 ft. lbs. (30 Nm).
- EGR valve tube. Torque the bolts to 34 ft. lbs. (47 Nm).
- Heater hoses to the intake manifold
- Vacuum hoses to the tee
- Crankcase vent hose to the valve cover
- Accelerator cable bracket. Torque the bolts to 20 ft. lbs. (28 Nm).
- Accelerator cable and speed control cable, if equipped
- IAC valve and TP sensor electrical connectors
- Air cleaner outlet tube
- IAT sensor
- Negative battery cable

6. Start the engine and check for leaks, repair if necessary.

3.0L Engine

1. Before servicing the vehicle, refer to the precautions in the beginning of this section.

2. Remove or disconnect the following:
- Negative battery cable
- Intake Air Temperature (IAT) sensor
- Air cleaner outlet tube
- Accelerator control splash shield
- Accelerator cable and speed control cable, if equipped
- Engine control sensor wiring from

the Throttle Position (TP) sensor and the Idle Air Control (IAC) valve and Exhaust Gas Recirculation (EGR) transducer
- EGR tube from the valve
- EGR vacuum lines
- 42 pin connector bracket
- Throttle body and gasket
- Ignition coil and move it aside
- Evaporative Emissions (EVAP) hose
- Upper intake manifold bolts and discard them
- Crankcase vent hose
- Upper intake manifold and discard the gasket

To install:
3. Clean all mating surfaces.
4. Install or connect the following:
- New upper intake manifold
- Intake manifold
- Crankcase vent hose

5. Torque the upper intake manifold bolts in sequence as follows:
  a. Step 1: 15 ft. lbs. (20 Nm).
  b. Step 2: 18 ft. lbs. (25 Nm).
6. Install or connect the following:
- EVAP hose
- EGR tube
- EGR transducer. Torque the bolts to 89 inch lbs. (10 Nm).
- Ignition coil. Torque the bolts to 15 ft. lbs. (20 Nm).
- New gasket and throttle body. Torque the bolts to 22 ft. lbs. (30 Nm).
- 42 pin connector
- EGR vacuum lines
- EGR transducer, IAC valve and TP sensor electrical connectors
- Accelerator cable and speed control, if equipped

- Accelerator cable splash shield
- Air cleaner outlet tube
- IAT sensor
- Negative battery cable

7. Start the vehicle and check for leaks, repair if necessary.

4.0L OHC Engine

2001–02

1. Before servicing the vehicle, refer to the precautions in the beginning of this section.
2. Remove or disconnect the following:
- Negative battery cable
- Air cleaner-to-intake tube
- Accelerator splash shield
- Accelerator and, if equipped with cruise control, speed control cables from the throttle control cam
- Accelerator cable retaining bracket from the upper intake manifold
- Label and disengage all vacuum and electrical connections on the intake manifold.
- Upper intake manifold attaching bolts
- Lift up on the manifold and remove both fuel Vapor Management Valve (VMV) hoses
- Upper intake manifold and discard the gasket

To install:

➡ Mazda does not specify a sequence, but it is recommended that you start tightening in the middle and work your way out to the ends. Repeat the tightening sequence several times until the specified torque is reached.

3. Position the upper manifold on the lower manifold.
4. Install or connect the following:
- Attach both VMV hoses to the manifold
- Upper manifold attaching bolts. Torque the bolts to 62 inch lbs. (7 Nm).
- Attach any vacuum and electrical connections that were removed
- Accelerator cable bracket to the intake and the cable (or cables if equipped with cruise control) to the throttle cam
- Accelerator splash shield
- Air cleaner-to-intake supply tube
- Negative battery cable

5. Start the vehicle and check for leaks, repair if necessary.

2003–04

1. Disconnect the battery ground cable.
2. Remove the bolts and the shield.
3. Remove the air cleaner outlet pipe.
4. Disconnect the idle air control (IAC) valve, throttle position (TP) sensor electrical connectors and the TP sensor wiring pin-type retainer.
5. Disconnect the MAF sensor wiring pin-type retainer.
6. Detach the accelerator and speed control cables from the throttle body.
7. Detach the accelerator and speed control cables from the bracket, and position the cables aside.
8. Disconnect the exhaust gas recirculation (EGR) valve vacuum hose and tube fitting.
9. Disconnect the EGR vacuum regulator solenoid valve electrical connector and vacuum hose.
10. Disconnect the hose.
11. Loosen the clamp and disconnect the brake booster vacuum hose.

> ✳✳ **CAUTION**

It is important to twist the spark plug wire boots while pulling upward to avoid possible damage to the spark plug wire.

➡ Mark the spark plug wire locations before removing them.

12. Disconnect the RH spark plug wires from the coil. Remove the spark plug wire routing clip pin-type retainer and position the wires aside.
13. Remove the wiring harness bracket retainer, then position the wiring harness aside.
14. Remove the accelerator cable routing clip pin-type retainer and the wiring harness pin-type retainer.

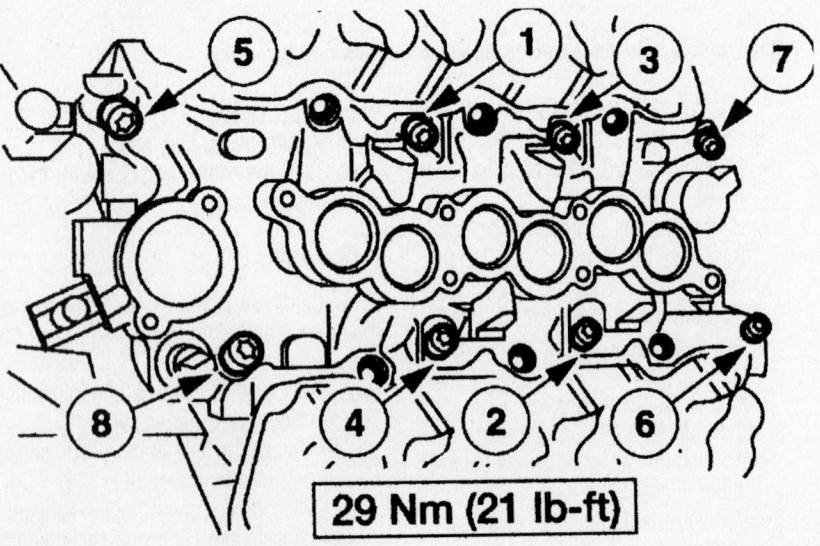

29 Nm (21 lb-ft)

67197-RANG-G4A

Tighten the lower manifold bolts in the sequence shown—3.0L engine

10 Nm (89 lb-in)

67197-RANG-G20

Intake manifold installation–2003–04 4.0L SOHC engine

15. Remove the bolts.
16. Remove the bolts and position the coil and bracket aside.
17. Disconnect the vacuum hoses.
18. Remove the nut.
19. Disconnect the powertrain control module (PCM) electrical connector.
20. Remove the retainer and position the ground wires aside.
21. Detach the electrical connector retainer.
22. Remove the intake manifold bolts and lift up the intake manifold.
23. Remove the heated positive crankcase ventilation (PCV) hose retainers and remove the heated PCV fitting.
24. Remove the intake manifold.
25. To install, reverse the removal procedure. Torque the fasteners to 89 inch lbs. (10 Nm).

Exhaust Manifold

REMOVAL & INSTALLATION

2.3L Engine

1. Before servicing the vehicle, refer to the precautions in the beginning of this section.
2. Remove or disconnect the following:
   - Negative battery cable
   - Exhaust flange nuts
   - Drive belt
   - Coolant
   - Upper radiator hose and the engine reservoir hose
   - Air conditioning compressor
   - Heater hose
   - Oil indicator and the upper bolt for the tube assembly
   - Lower bolt and remove the oil indicator tube assembly
   - Front radiator tube
3. Remove the pushpins and position the right inner fender splash shield out of the way.
4. Remove or disconnect the following:

   - Alternator electrical connectors
   - Lower front end accessory drive (FEAD) mounting bolts
   - Upper mounting bolt and the FEAD assembly
   - Two nuts and position the coolant tube out of the way
   - Exhaust manifold
   - Exhaust manifold gasket

To install:

5. Install or connect the following:
   - Exhaust manifold gasket

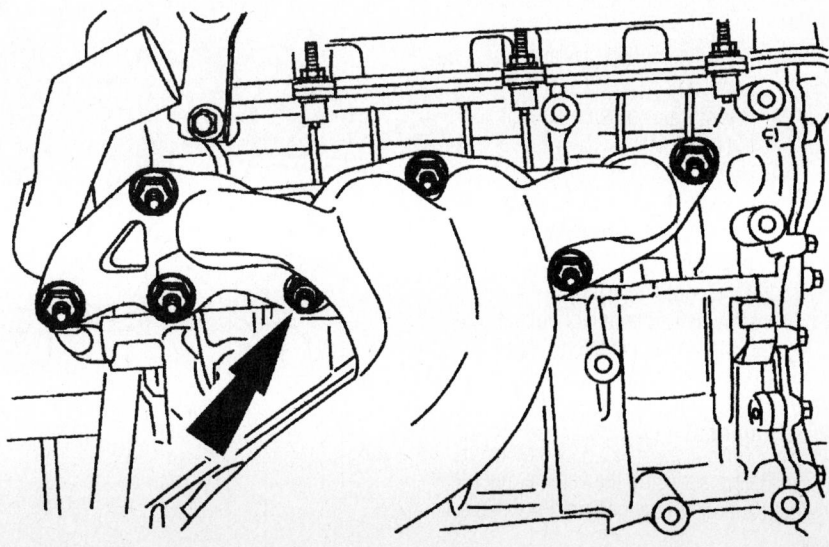

67197-RANG-G21

Exhaust manifold retaining nuts—2.3L engine

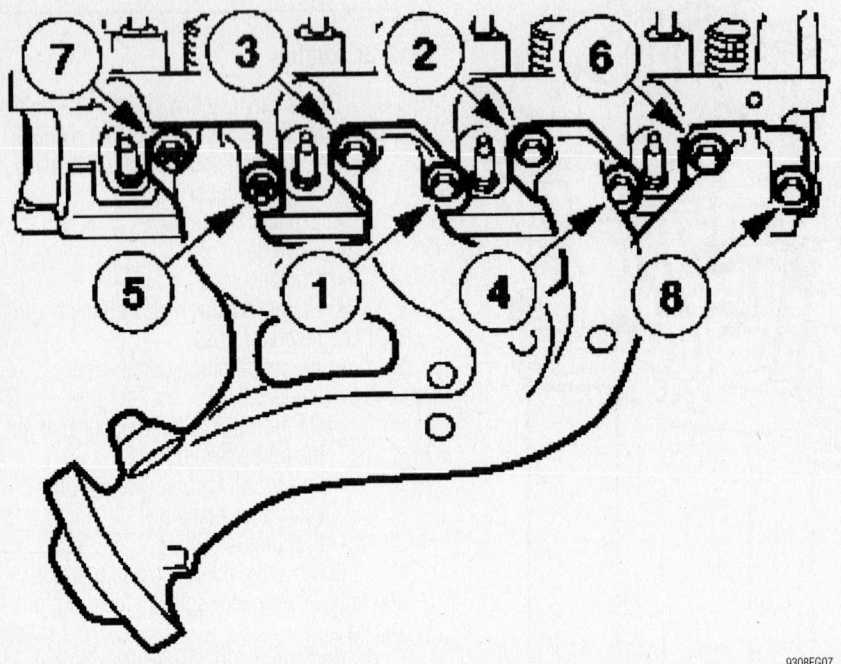

Tighten the exhaust manifold bolts in 2 stages—2.5L engine

- Negative battery cable
- Intake Air Temperature (IAT) sensor
- Air cleaner outlet tube
- Differential Pressure Feedback (DPFE) sensor and move it aside
- Exhaust Gas Recirculation (EGR) transducer lines at the tube
- Loosen and remove the EGR valve-to-exhaust manifold tube
- Catalytic converter from the exhaust manifold
- Rear engine lifting eye nuts
- Exhaust manifold and discard the gasket

To install:

3. Clean the mating surfaces on the exhaust manifold and the cylinder head.

4. Install a new gasket and the exhaust manifold. Torque the bolts in sequence as follows:
   a. 16 ft. lbs. (23 Nm).
   b. 59 ft. lbs. (80 Nm).

5. Install or connect the following:
   - Rear engine lifting eye. Torque the bolts to 15 ft. lbs. (20 Nm).
   - Catalytic converter to the exhaust manifold
   - EGR valve to the exhaust manifold tube

- Exhaust manifold and the nuts
- Coolant tube and the nuts
- FEAD assembly and the upper mounting bolts
- Lower FEAD mounting bolts
- Alternator electrical connectors
- Right inner splash shield and pushpins
- Upper radiator tube and install the bolts
- Oil indicator tube assembly and the lower bolt
- Oil indicator tube upper bolt and the oil indicator
- Heater water hose
- Air conditioning compressor
- Upper radiator hose and the engine reservoir hose

6. Fill the cooling system.
7. Install the serpentine drive belt.
8. Install the exhaust flange nuts.
9. Connect the battery ground cable.

2.5L Engines

1. Before servicing the vehicle, refer to the precautions in the beginning of this section.

2. Remove or disconnect the following:

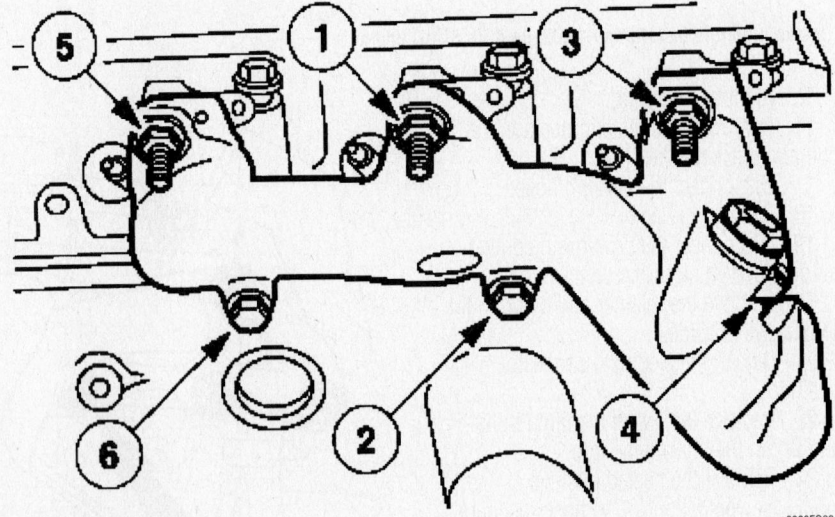

Tighten the exhaust manifold bolts in sequence–3.0L left side

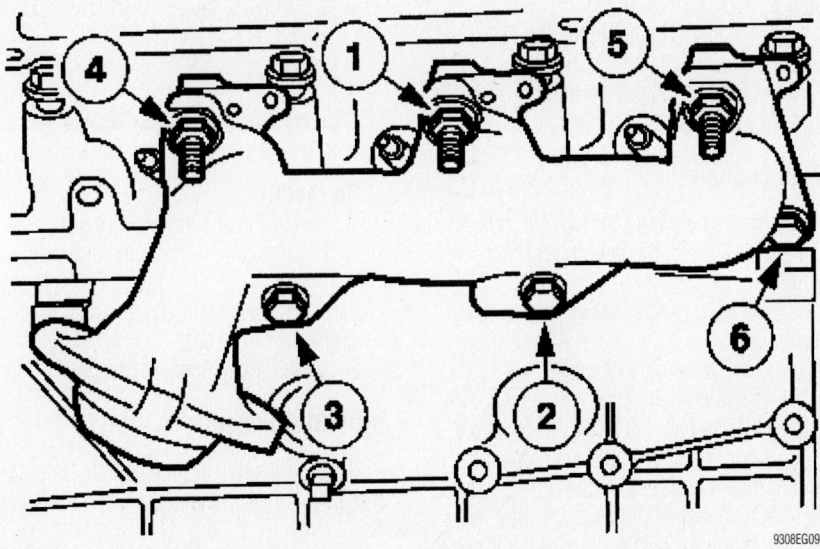

9308EG09

Tighten the right side exhaust manifold bolts in the proper sequence–3.0L

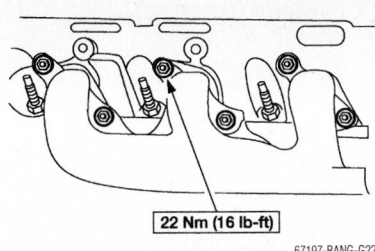

22 Nm (16 lb-ft)

67197-RANG-G22

Left side exhaust manifold retaining nuts—4.0L SOHC engine

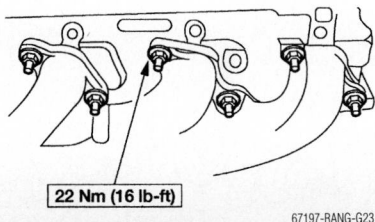

22 Nm (16 lb-ft)

67197-RANG-G23

Right side exhaust manifold retaining nuts—4.0L SOHC engine

- EGR transducer lines
- DPFE sensor
- Air cleaner outlet tube
- IAT sensor
- Negative battery cable

6. Start the vehicle and check for leaks, repair if necessary.

3.0L Engine

LEFT SIDE

1. Before servicing the vehicle, refer to the precautions in the beginning of this section.

2. Install or connect the following:
- Negative battery cable
- Exhaust flange nuts
- Exhaust Gas Recirculation (EGR) valve from the exhaust manifold tube
- Oil lever indicator and bracket
- Exhaust manifold and discard the gasket

To install:

3. Clean the mating surfaces for the exhaust manifold and cylinder head.

4. Install a new gasket and the exhaust manifold. Torque the bolts in sequence to:
   a. 89 inch lbs. (10 Nm).
   b. 15 ft. lbs. (20 Nm).

5. Install or connect the following:
- Oil lever indicator tube and bracket. Torque the bolt to 12 ft. lbs. (16 Nm).
- EGR valve to the exhaust manifold tube. Torque the fastener to 26 ft. lbs. (35 Nm).
- Exhaust flange. Torque the nuts to 25 ft. lbs. (34 Nm).
- Negative battery cable

6. Start the vehicle and check for leaks, repair if necessary.

RIGHT SIDE

1. Before servicing the vehicle, refer to the precautions in the beginning of this section.

2. Remove or disconnect the following:
- Negative battery cable
- Exhaust manifold flange
- Ignition coil support bracket
- Exhaust manifold and discard the gasket

To install:

3. Clean the mating surfaces for the exhaust manifold and cylinder head

4. Install a new gasket and the exhaust manifold. Torque the bolts is sequence to:
   a. 89 inch lbs. (10 Nm).
   b. 18 ft. lbs. (25 Nm).

5. Install or connect the following:
- Ignition coil support bracket. Torque the bolts to 15 ft. lbs. (20 Nm).
- Exhaust flange nuts. Torque the nuts to 33 ft. lbs. (46 Nm).
- Negative battery cable

6. Start the vehicle and check for leaks, repair if necessary.

4.0L OHC Engine

1. Before servicing the vehicle, refer to the precautions in the beginning of this section.

2. Remove or disconnect the following:
- Negative battery cable
- Exhaust inlet pipe-to-manifold attaching bolts
- Differential Pressure Feedback EGR (DPFE) transducer hoses, left side manifold only
- Exhaust Gas Recirculation (EGR) tube from the manifold and valve, left side manifold only
- Exhaust manifold and discard the gasket

To install:

3. Clean the gasket mating surfaces.

4. Install or connect the following:
- New gasket and the exhaust mani-

fold. Torque the bolts to 16 ft. lbs. (22 Nm).

- EGR tube to the manifold. Torque the fastener to 30 ft. lbs. (40 Nm) left side manifold only
- DPFE transducer hoses, left side manifold only
- Exhaust inlet pipe-to-manifold attaching bolts. Torque the bolts to 30 ft. lbs. (40 Nm).
- Negative battery cable

5. Start the vehicle and check for leaks, repair if necessary.

Front Crankshaft Seal

REMOVAL & INSTALLATION

2.3L Engines

1. Before servicing the vehicle, refer to the precautions in the beginning of this section.
2. Remove or disconnect the following:

- Negative battery cable
- Crankshaft pulley

✳✳ WARNING

Use care not to damage the engine front cover or the crankshaft when removing the seal.

- Crankshaft front oil seal by prying the seal out of the front cover

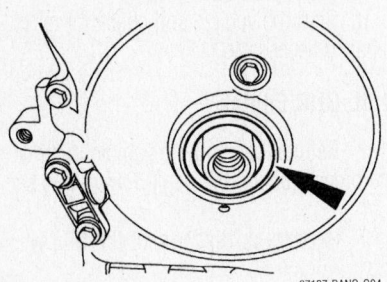

Front crankshaft seal—2.3L engine

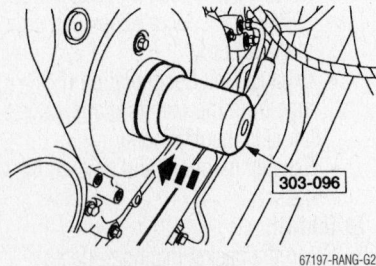

Front crankshaft seal installation—2.3L engine

To install:

3. Using the special tool, install the crankshaft front oil seal.
4. Install the crankshaft pulley.
5. Tighten the crankshaft damper in two stages:

- Step 1: Tighten to 40 Nm (30 ft. lbs.)
- Step 2: Tighten an additional 90 degrees

2.5L Engines

1. Before servicing the vehicle, refer to the precautions in the beginning of this section.
2. Remove or disconnect the following:

- Negative battery cable
- Timing belt cover
- Drive belt

3. Align the crankshaft and camshaft timing marks and remove the timing belt.

- Crankshaft pulley center bolt and slide the pulley off of the crankshaft
- Crankshaft key

✳✳ WARNING

Do not damage the crankshaft sealing surface while removing the oil seal.

- Crankshaft Seal Remover tool T74P-6700-B on the crankshaft and into the oil seal.
- Oil seal and clean the seal journal

To install:

4. Apply clean engine oil to the rubber lip of the new seal to aid installation.
5. Using Cam Bearing Adapter Tube T72C-6250, or equivalent, and crankshaft center bolt, carefully install the new oil seal until flush with the engine.
6. Install or connect the following:

- Key and crankshaft pulley, washer and bolt. Torque the bolt to 92–121 ft. lbs. (125–165 Nm).
- Timing belt
- Timing belt and cover
- Drive belt
- Negative battery cable

7. Start the vehicle and check for leaks, repair if necessary.

3.0L Engine

1. Before servicing the vehicle, refer to the precautions in the beginning of this section.
2. Remove the crankshaft vibration damper.
3. Remove the crankshaft front seal.

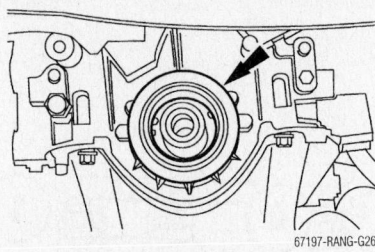

Front crankshaft seal—3.0L engine

To install:

4. Install the crankshaft front seal.

a. Lubricate the seal lip with clean engine oil.

b. Using a seal installer, install the crankshaft front seal.

5. Install the crankshaft damper.

4.0L OHC Engine

1. Before servicing the vehicle, refer to the precautions in the beginning of this section.
2. Remove or disconnect the following:

- Negative battery cable
- Crankshaft pulley

3. Using a seal remover, remove the crankshaft front oil seal.

To install:

4. Lubricate the seal lip with clean engine oil.
5. Using a seal driver, install the crankshaft front oil seal.
6. Install the crankshaft pulley.

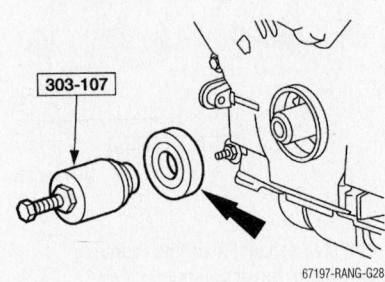

Front crankshaft seal removal—4.0L SOHC engine

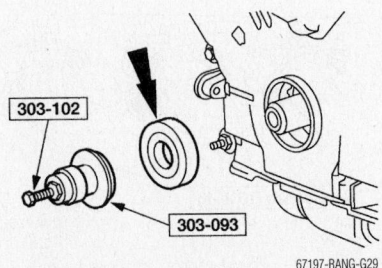

Front crankshaft seal installation—4.0L SOHC engine

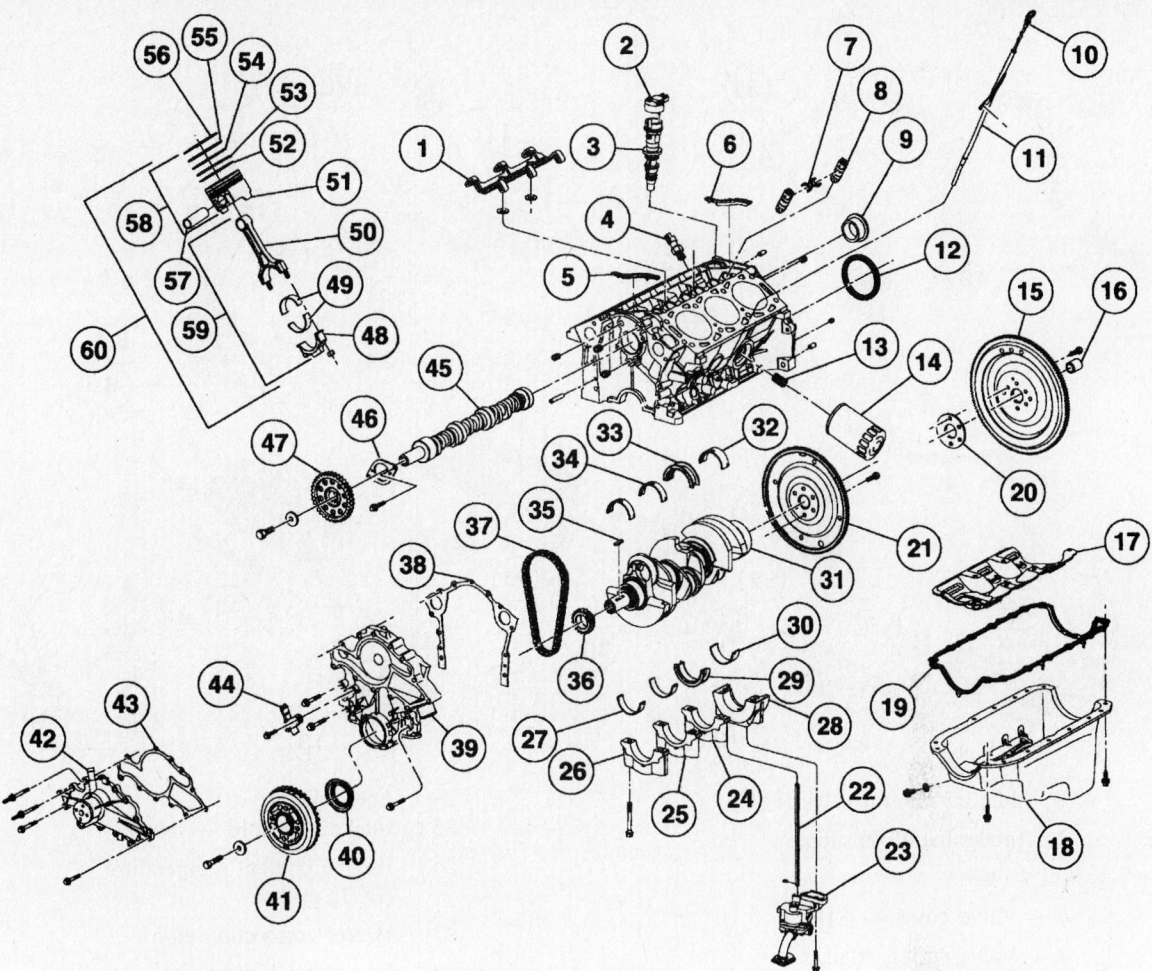

1	Valve tappet guide plate	21	Flywheel (automatic transmission)	42	Coolant pump
2	Camshaft position sensor	22	Oil pump drive	43	Coolant pump gasket
3	Camshaft synchronizer	23	Oil pump assembly	44	Crankshaft position (CKP) sensor
4	Oil pressure sender	24	Crankshaft thrust bearing cap	45	Camshaft
5	Lower intake manifold gasket	25	Crankshaft main bearing cap	46	Camshaft thrust plate
6	Lower intake manifold gasket	26	Crankshaft front main bearing caps	47	Camshaft sprocket
7	Rocker arm center cover	27	Crankshaft main bearing	48	Connecting rod bearing cap
8	Valve tappet	28	Crankshaft rear main bearing cap	49	Connecting rod bearings
9	Camshaft rear bearing cover	29	Crankshaft thrust bearing (lower)	50	Piston connecting rod
10	Oil level indicator	30	Crankshaft rear main bearing (lower)	51	Piston
11	Oil level indicator tube	31	Crankshaft	52	Oil ring (lower)
12	Rear main seal	32	Crankshaft rear main bearing (upper)	53	Oil ring baffle
13	Oil filter mounting insert	33	Crankshaft thrust bearing (upper)	54	Oil ring (upper)
14	Oil filter	34	Crankshaft main bearings (upper)	55	Piston compression ring (lower)
15	Flexplate (manual transmission)	35	Key	56	Piston ring (upper)
16	Pilot shaft bearing (manual transmission)	36	Crankshaft timing sprocket	57	Piston pin
17	Windage tray	37	Timing chain	58	Piston assembly
18	Oil pan	38	Engine front cover gasket	59	Connecting rod assembly
19	Oil pan gasket	39	Engine front cover	60	Piston and connecting rod assembly
20	Flywheel reinforcement plate	40	Crankshaft front seal		
		41	Crankshaft pulley (harmonic balancer)		

Lower engine components—3.0L engine

67197-RANG-G30

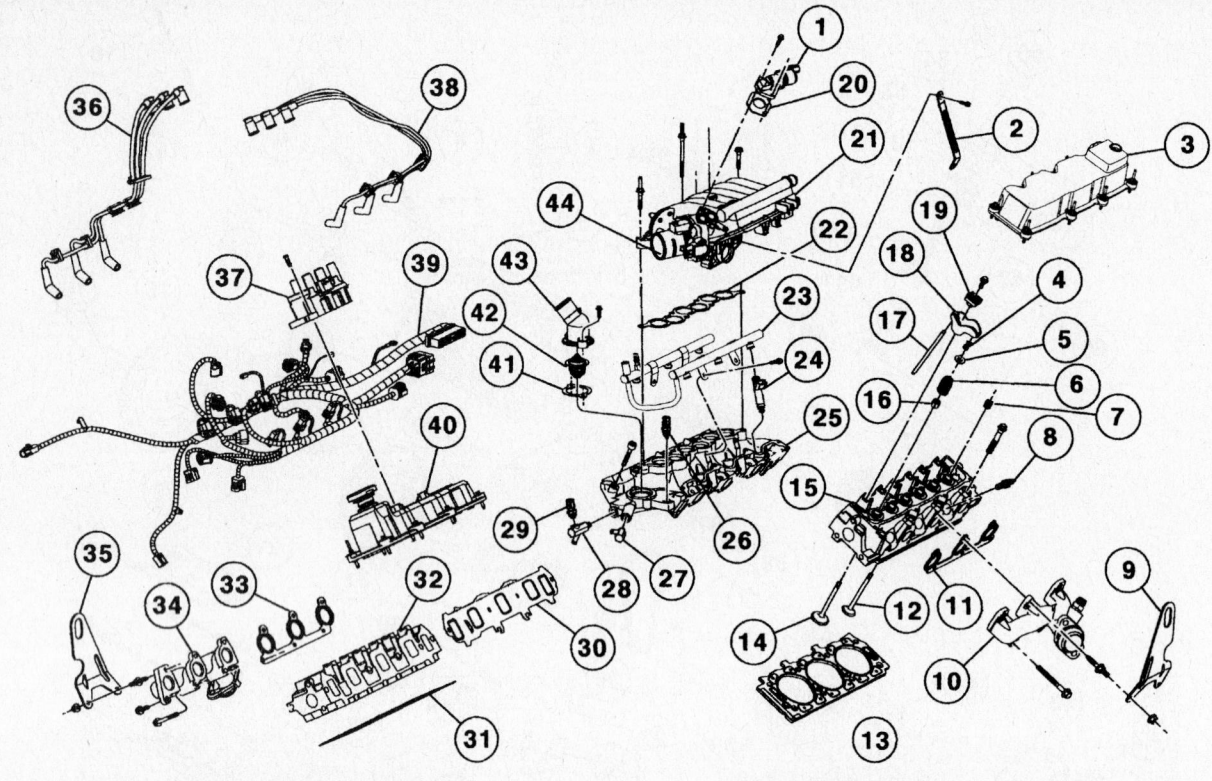

Major upper engine components—3.0L engine

1	Idle air control valve (IAC)	24	Fuel injector
2	Intake manifold support bracket	25	Intake manifold — lower
3	Valve cover — LH	26	Engine coolant temperature (ECT) sensor
4	Valve spring retainer key	27	Heater connection elbow
5	Valve spring retainer	28	Engine coolant hose connection
6	Valve spring	29	Engine coolant temperature (ECT) sensor
7	Valve stem seal	30	Lower intake manifold gasket
8	Spark plug	31	Cylinder head gasket — RH
9	Engine lifting bracket	32	Cylinder head — RH
10	Exhaust manifold — LH	33	Exhaust manifold gasket — RH
11	Exhaust manifold gasket	34	Exhaust manifold — RH
12	Exhaust valve	35	Engine lifting bracket
13	Cylinder head gasket	36	Ignition wire set — RH
14	Intake valve	37	Ignition coil
15	Cylinder head	38	Ignition wire set — LH
16	Valve stem seal	39	Engine wiring harness assembly
17	Pushrod	40	Valve cover — RH
18	Rocker arm	41	Thermostat housing gasket
19	Rocker arm pivot	42	Thermostat
20	Idle air control valve gasket	43	Thermostat housing
21	Intake manifold — upper	44	Throttle body
22	Upper intake manifold gasket		
23	Fuel injection supply manifold		

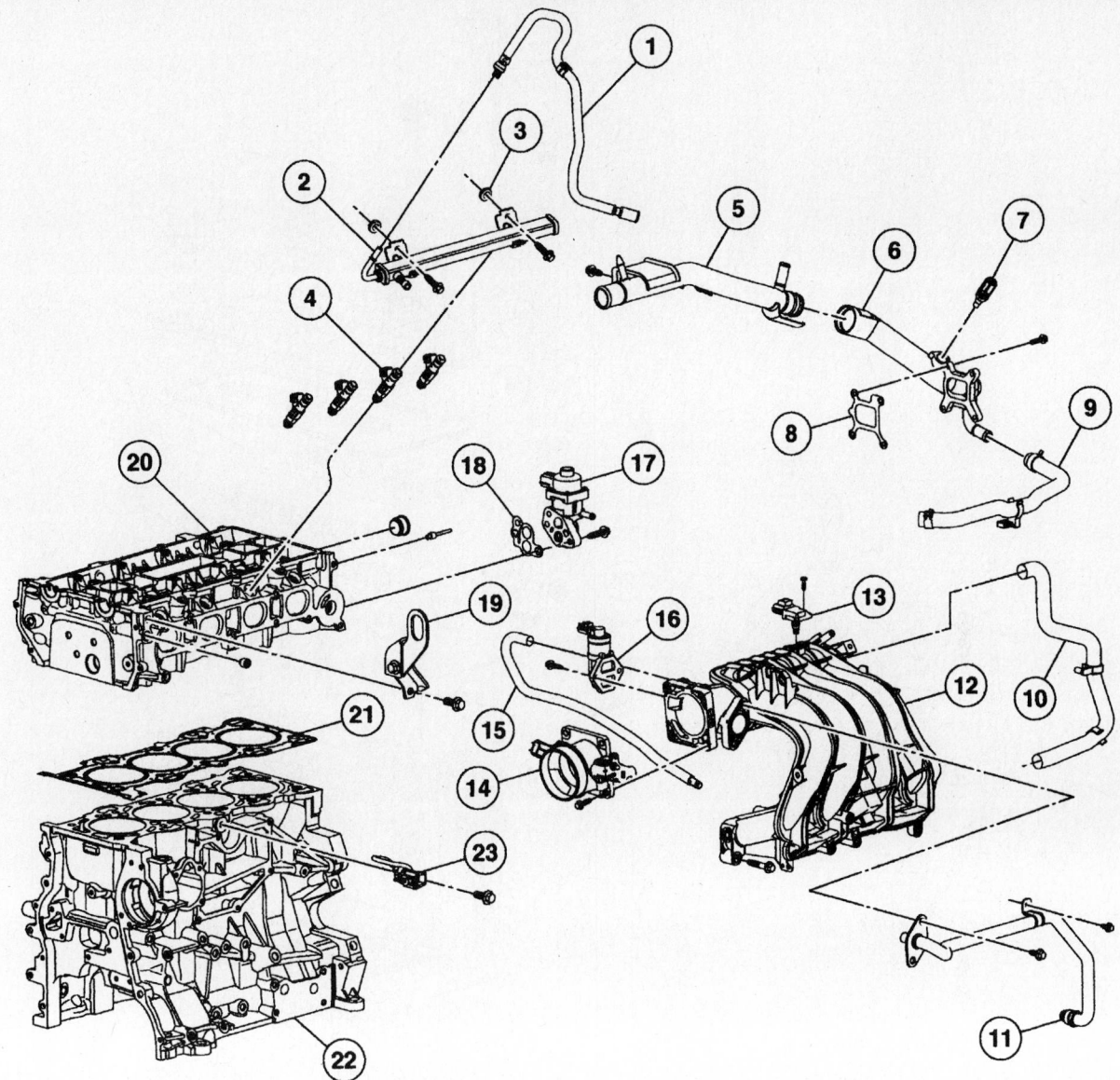

1	High pressure fuel line (supply)	11	EGR valve-to-exhaust manifold tube
2	Fuel injection supply manifold	12	Intake manifold
3	Spacer (2 req'd)	13	Absolute pressure sensor
4	Fuel injector (4 req'd)	14	Throttle body assembly
5	Water outlet adapter assembly (front)	15	Evaporative emission hose
6	Water outlet adapter assembly (rear)	16	Idle air control valve assembly
7	Water temperature indicator sender unit	17	EGR valve
8	Water outlet connector gasket	18	EGR valve mounting gasket
9	Water bypass hose and clamp assembly	19	Engine lifting eye
		20	Cylinder head
10	Crankcase ventilation tube	21	Cylinder head gasket
		22	Cylinder block
		23	Knock sensor

67197-RANG-G32

Intake manifold components—2.3L engine

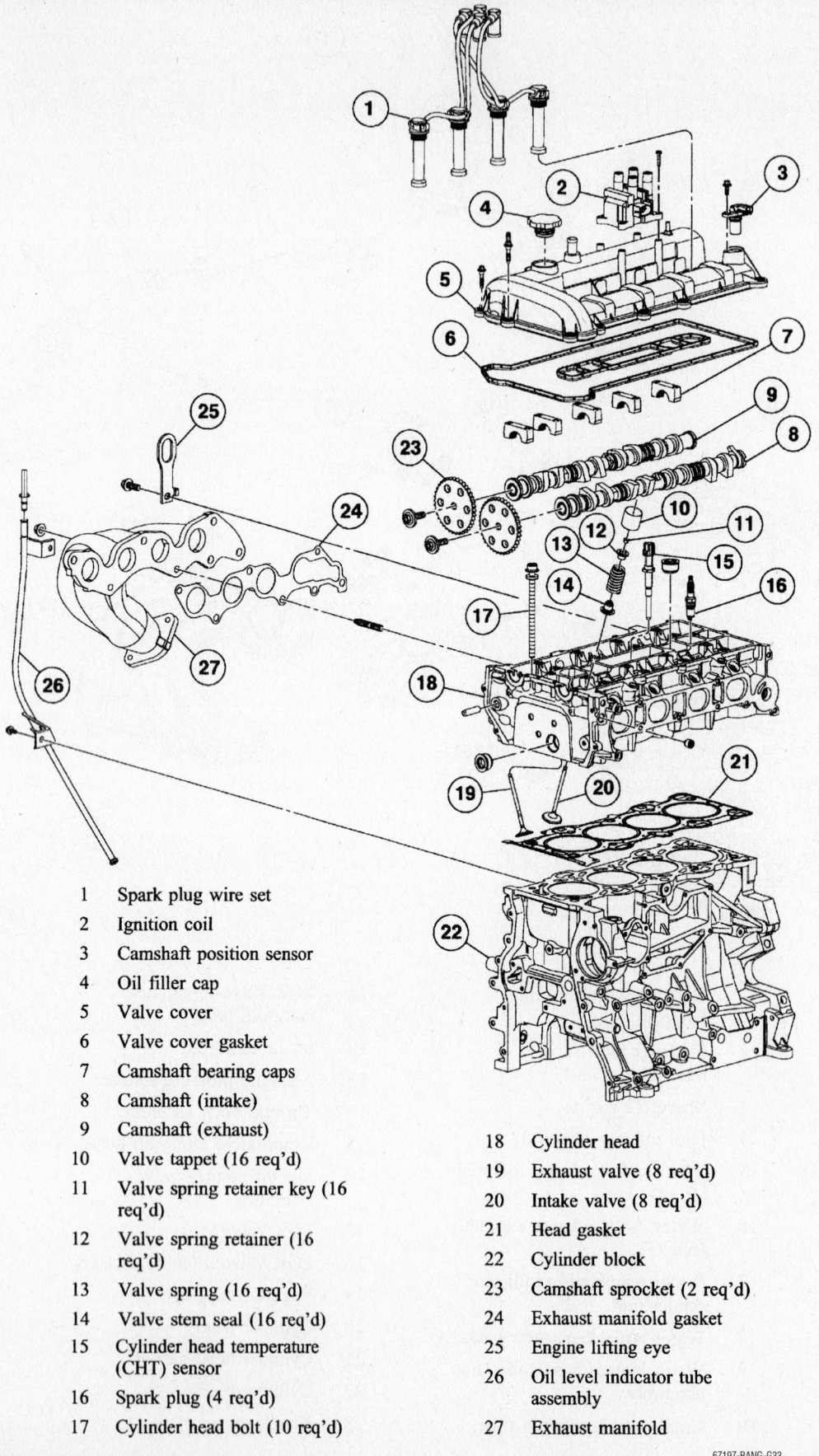

1. Spark plug wire set
2. Ignition coil
3. Camshaft position sensor
4. Oil filler cap
5. Valve cover
6. Valve cover gasket
7. Camshaft bearing caps
8. Camshaft (intake)
9. Camshaft (exhaust)
10. Valve tappet (16 req'd)
11. Valve spring retainer key (16 req'd)
12. Valve spring retainer (16 req'd)
13. Valve spring (16 req'd)
14. Valve stem seal (16 req'd)
15. Cylinder head temperature (CHT) sensor
16. Spark plug (4 req'd)
17. Cylinder head bolt (10 req'd)

18. Cylinder head
19. Exhaust valve (8 req'd)
20. Intake valve (8 req'd)
21. Head gasket
22. Cylinder block
23. Camshaft sprocket (2 req'd)
24. Exhaust manifold gasket
25. Engine lifting eye
26. Oil level indicator tube assembly
27. Exhaust manifold

67197-RANG-G33

Cylinder head and related components—2.3L engine

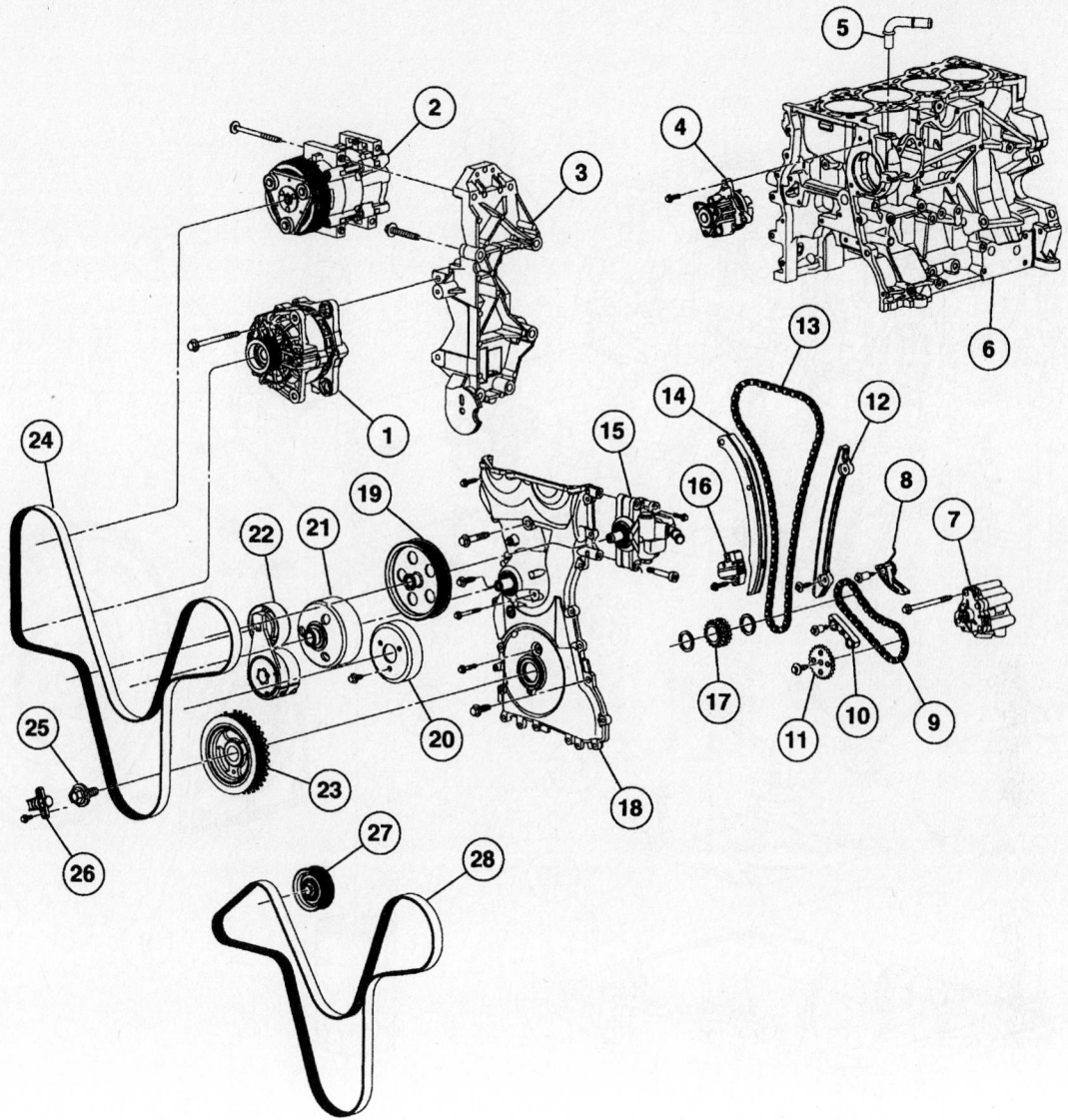

1	Generator	16	Timing chain tensioner
2	A/C compressor	17	Crankshaft sprocket
3	Front end accessory drive (FEAD) mounting bracket	18	Engine front cover
4	Water pump	19	Power steering pump pulley
5	Water bypass tube	20	Water pump pulley
6	Cylinder block	21	Fan drive pulley
7	Oil pump	22	Drive belt tensioner
8	Oil pump chain tensioner	23	Crankshaft damper
9	Oil pump chain	24	Accessory drive belt (with A/C)
10	Oil pump chain guide	25	Crankshaft pulley bolt
11	Oil pump drive gear	26	Crankshaft position sensor
12	Timing chain guide	27	Drive belt pulley idler (without A/C)
13	Timing chain	28	Accessory drive belt (without A/C)
14	Timing chain tensioner arm		
15	Power steering pump		

Engine block front components—2.3L engine

67197-RANG-G34

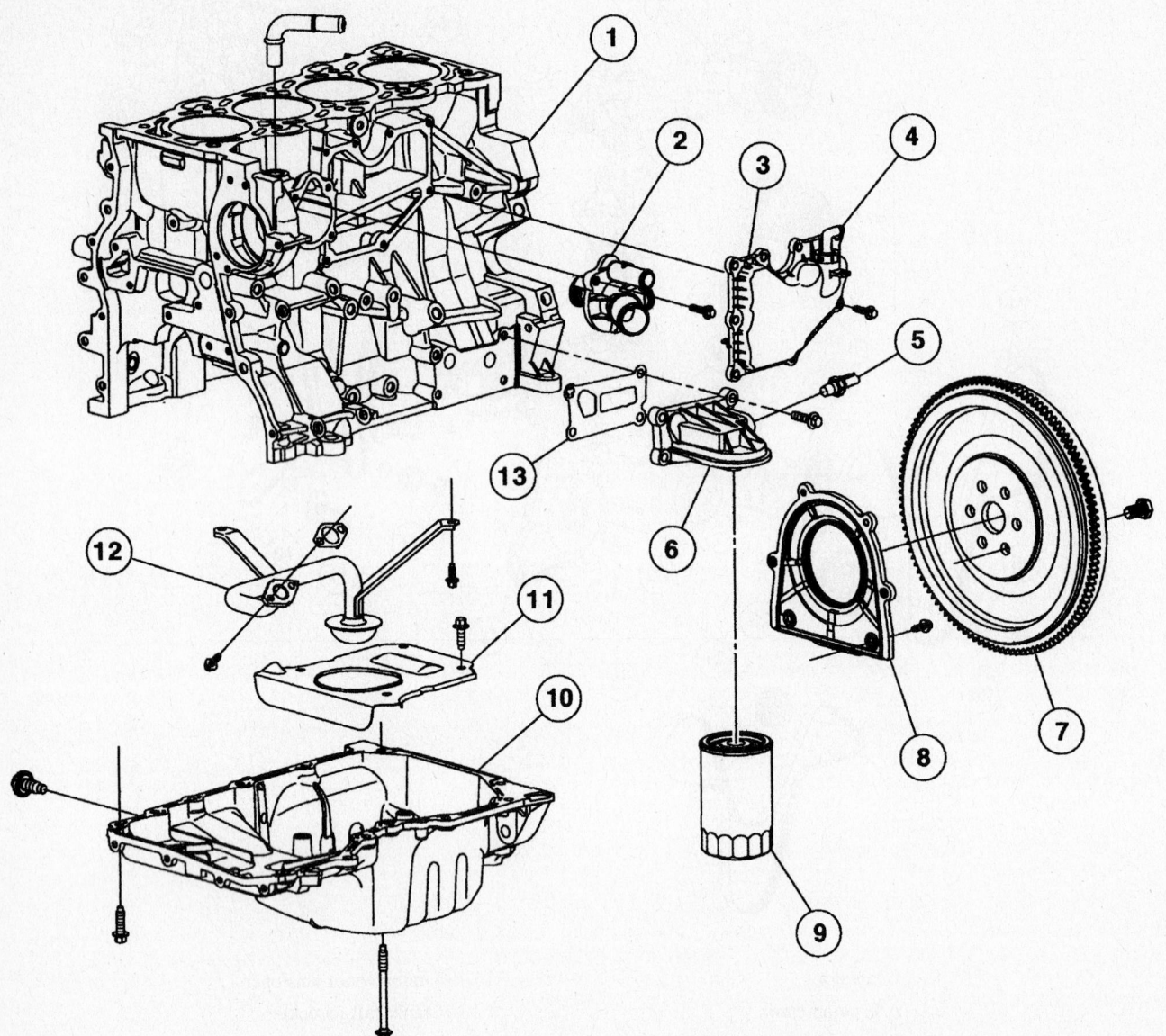

1 Cylinder block
2 Water thermostat assembly
3 Crankcase vent oil separator
4 PCV valve
5 Oil pressure sensor
6 Oil filter adapter
7 Flywheel
8 Crankshaft rear oil seal and retainer
9 Oil filter
10 Oil pan
11 Oil pan baffle
12 Oil pump screen and pickup tube
13 Oil filter adapter gasket

67197-RANG-G35

Lower engine block components—2.3L engine

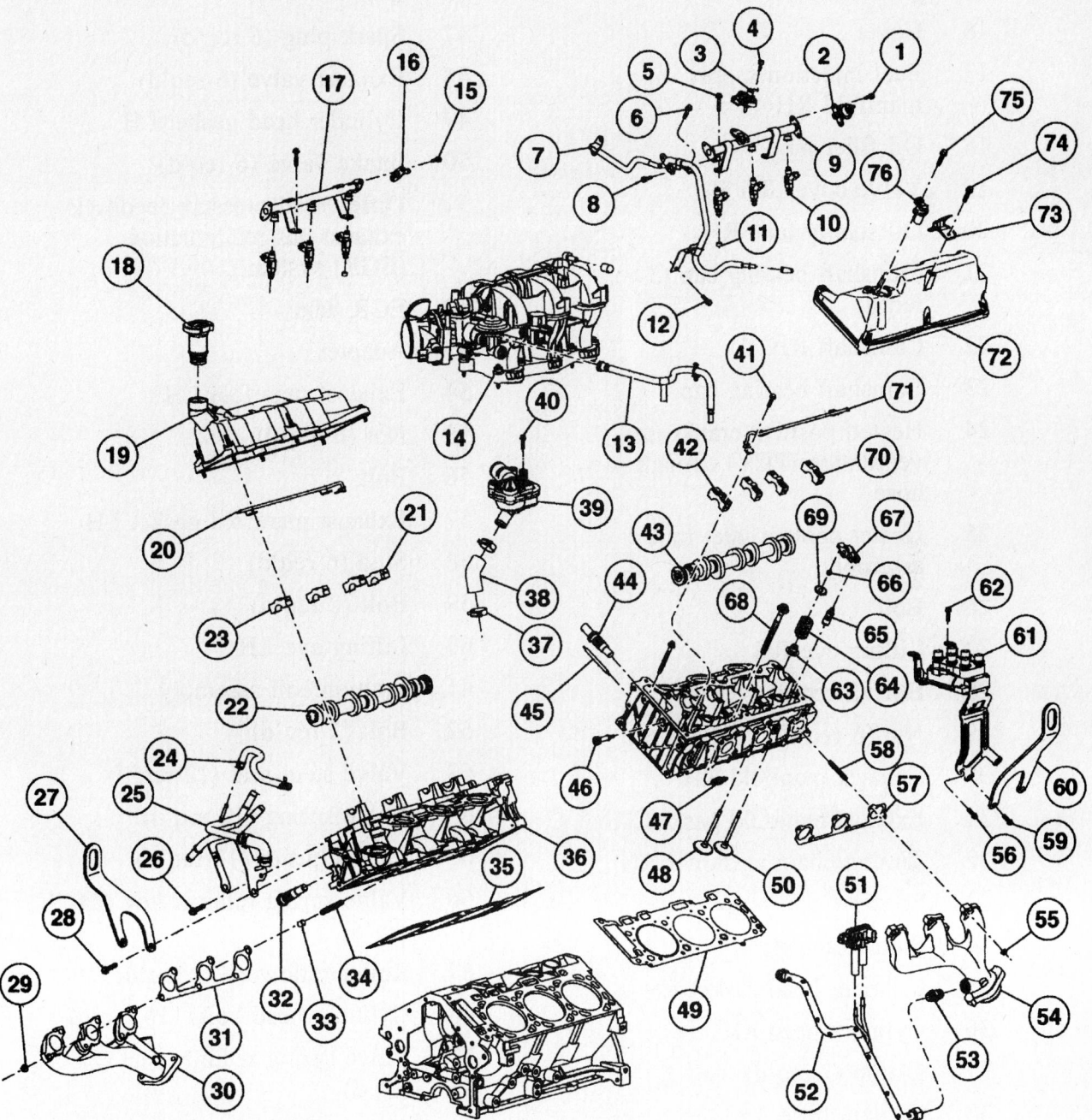

1   Bolt (2 req'd)

2   Fuel injection pulse damper

3   Fuel pressure/temperature
    sensor

4   Bolt (2 req'd)

5   Bolt (4 req'd)

6   Bolt

7   Fuel supply tube

8   Bolt (4 req'd)

9   Fuel injection supply
    manifold LH

10 Fuel injector (6 req'd)

11 Fuel injector adapter (6 req'd)

12 Bolt

13 Fuel vapor tube

14 Intake manifold

67197-RANG-G36A

Upper engine components—4.0L SOHC engine

15	Bolt (2 req'd)
16	Cover
17	Fuel injection supply manifold RH
18	Oil filler pipe
19	Valve cover RH
20	Oil supply tube RH
21	Camshaft bearing cap (3 req'd)
22	Camshaft RH
23	Camshaft bearing cap
24	Heated positive crankcase ventilation (PCV) coolant hose
25	Heater coolant inlet tube assembly
26	Bolt
27	Lifting eye RH
28	Bolt (2 req'd)
29	Nut (6 req'd)
30	Exhaust manifold RH
31	Exhaust manifold gasket RH
32	Timing chain tensioner RH
33	Spacer
34	Stud (6 req'd)
35	Cylinder head gasket RH
36	Cylinder head RH
37	Clamp (2 req'd)
38	Coolant hose
39	Thermostat housing
40	Bolt (3 req'd)
41	Bolt (16 req'd)
42	Camshaft bearing cap
43	Camshaft LH
44	Timing chain tensioner LH
45	Oil volume reduction plug
46	Bolt
47	Spark plug (6 req'd)
48	Exhaust valve (6 req'd)
49	Cylinder head gasket LH
50	Intake valve (6 req'd)
51	Differential pressure feedback exhaust gas recirculation (EGR) system
52	EGR tube
53	Adapter
54	Exhaust manifold LH
55	Nut (6 req'd)
56	Bolt
57	Exhaust manifold gasket LH
58	Stud (6 req'd)
59	Bolt (2 req'd)
60	Lifting eye LH
61	Ignition coil assembly
62	Bolt (2 req'd)
63	Valve stem seal (12 req'd)
64	Valve spring (12 req'd)
65	Lash adjuster (12 req'd)
66	Valve spring retainer key (24 req'd)
67	Roller follower (12 req'd)
68	Cylinder head bolts (16 req'd)
69	Valve spring retainer seat (12 req'd)
70	Camshaft bearing cap (3 req'd)
71	Oil supply tube LH
72	Valve cover LH
73	Bracket
74	Bolt
75	Bolt
76	Camshaft position (CMP) sensor

67197-RANG-G36B

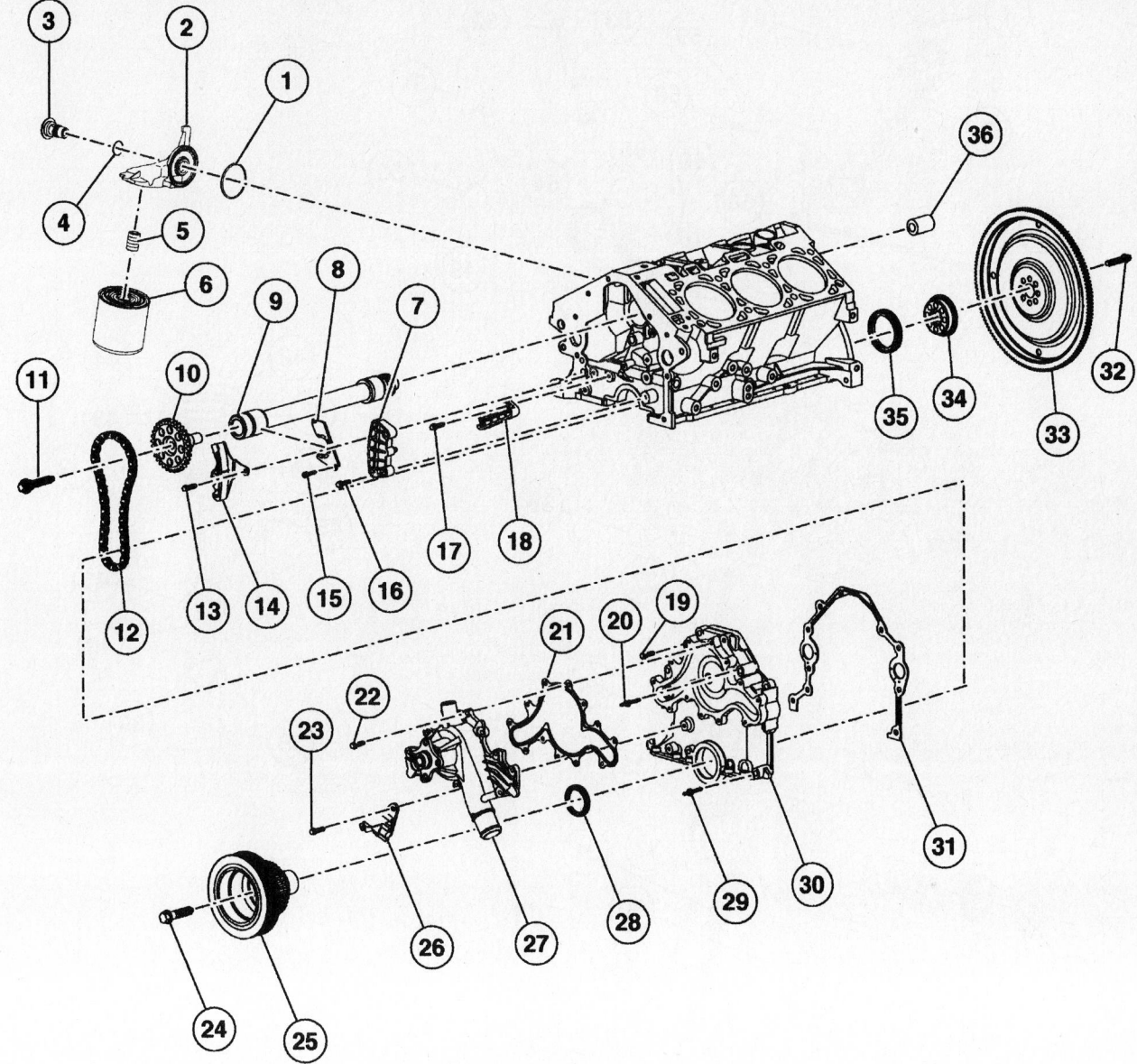

1	Gasket	14	Jackshaft chain tensioner	26	Crankshaft position (CKP) sensor
2	Oil filter adapter	15	Bolt (2 req'd)	27	Coolant pump
3	Bolt	16	Bolt (2 req'd)	28	Crankshaft front seal
4	O-ring	17	Bolt (2 req'd)	29	Stud (4 req'd)
5	Adapter	18	Balance shaft chain guide (4x4 only)	30	Front cover
6	Oil filter	19	Bolt (5 req'd)	31	Gasket
7	Jackshaft chain guide	20	Stud	32	Bolt (8 req'd)
8	Jackshaft thrust plate	21	Gasket	33	Flexplate
9	Jackshaft	22	Bolt (12 req'd)	34	Spacer
10	Jackshaft sprocket	23	Bolt (2 req'd)	35	Crankshaft rear seal
11	Bolt	24	Bolt	36	Spacer
12	Jackshaft chain	25	Crankshaft pulley		
13	Bolt (2 req'd)				

Front and rear engine components—4.0L SOHC engine

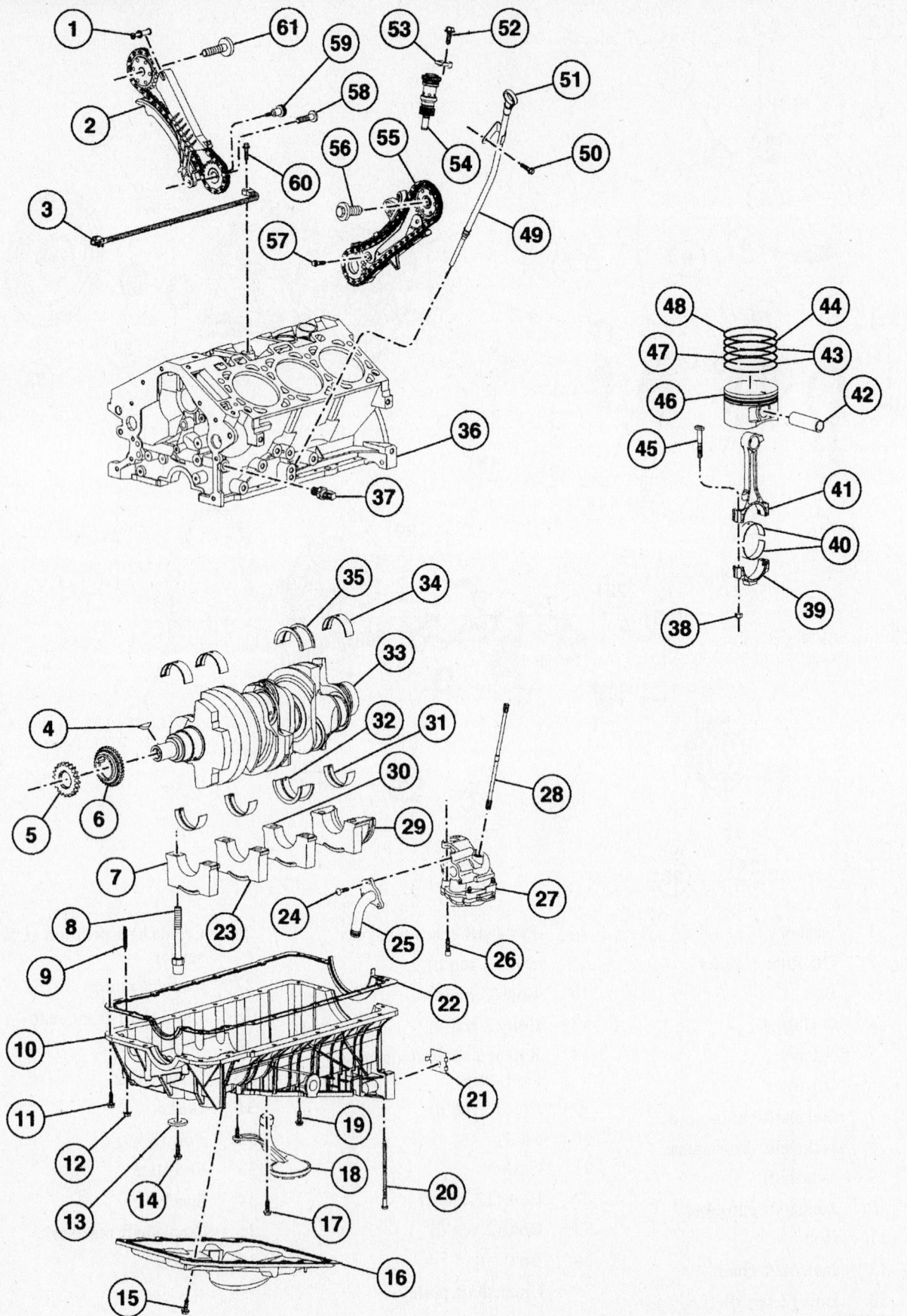

Lower engine components—4.0L SOHC engine

67197-RANG-G38A

1	Bolt
2	Camshaft drive cassette RH
3	Knock sensor (KS)
4	Crankshaft key
5	Sprocket
6	Sprocket
7	Crankshaft front main bearing cap
8	Bolt (8 req'd)
9	Stud (2 req'd)
10	Cylinder block cradle
11	Bolt (20 req'd)
12	Nut
13	Gasket
14	Bolt
15	Bolt
16	Oil pan
17	Bolt (6 req'd)
18	Oil pump screen and pickup tube
19	Bolt
20	Bolt
21	Spacer
22	Gasket
23	Crankshaft front intermediate main bearing cap
24	Screw (3 req'd)
25	Oil pickup adapter
26	Bolt (2 req'd)
27	Oil pump
28	Oil pump intermediate shaft
29	Crankshaft rear main bearing cap
30	Crankshaft center main bearing cap
31	Crankshaft lower main bearing (3 req'd)
32	Crankshaft lower center main bearing
33	Crankshaft
34	Crankshaft upper main bearing (3 req'd)
35	Crankshaft upper center main bearing
36	Block
37	Oil pressure sensor
38	Nut (12 req'd)
39	Connecting rod bearing cap (6 req'd)
40	Connecting rod bearings
41	Connecting rod
42	Piston pin
43	Piston ring, oil control (2 req'd)
44	Piston ring, lower compression
45	Bolt (2 req'd)
46	Piston
47	Piston ring, oil control spacer
48	Piston ring, upper compression
49	Oil level indicator tube
50	Bolt
51	Oil level indicator
52	Bolt
53	Clamp
54	Oil pump drive assembly
55	Camshaft drive cassette LH
56	Bolt
57	Bolt
58	Bolt
59	Bolt
60	Bolt
61	Bolt

Camshaft and Valve Lifters

➡ **Although Mazda suggests that this component is removable while the engine is installed in the vehicle, depending on the particular options with which your truck is equipped, working clearance may be extremely tight and this procedure may be much easier to perform with the engine removed. Before commencing, read through this procedure and make certain enough clearance, or working room, exists with the engine in the vehicle; if there is not enough space, the engine should be removed.**

REMOVAL & INSTALLATION

2.3L Engine

1. Before servicing the vehicle, refer to the precautions in the beginning of this section.
2. Relieve the fuel system pressure.
3. Drain the cooling system.
4. Properly discharge the air conditioning system.
5. Remove or disconnect the following:
   - Negative battery cable
   - Drive belt.
   - Engine oil level indicator assembly.
   - Engine oil level indicator.
   - Engine oil level indicator tube.
   - Water outlet tube.
   - Water outlet tube.
   - Air conditioning compressor.

➡ **The generator will be removed with the accessory bracket.**

   - Accessory bracket.
   - Right motor mount.
   - Coolant hose from the thermostat.
   - Coolant hose from the EGR valve.
   - Coolant tube assembly.
   - Exhaust manifold and gasket.
   - Block heater (if so equipped).
   - Water outlet.
   - EGR valve.
   - Power steering pump and reservoir as an assembly.
   - Idle air control (IAC) valve.
   - Throttle position (TP) sensor.
   - Manifold absolute pressure (MAP) sensor.
   - Swirl control valve monitor electrical connector.
   - CKP sensor and the wiring harness pin-type retainers.
   - Knock sensor (KS).
   - Electric thermostat.
   - Swirl control valve.

   - CMP sensor electrical connector and disconnect the PCV hose from the intake manifold.
   - Engine wiring harness pin-type retainers from the intake manifold.
   - Engine wiring harness connector bracket. Position the engine wiring harness aside.
   - EGR tube.
   - Fuel supply line clip from the front of the intake manifold. Disconnect the vacuum hose from the intake manifold.
   - Intake manifold assembly.
   - Fuel injector electrical connectors. Detach the wiring harness pin-type retainers.
   - Ignition coil and the cylinder head temperature (CHT) sensor electrical connectors.
   - Engine wiring harness anchors from the valve cover studs. Remove the engine wiring harness.
   - Ignition coil.
   - Bypass hose.
   - Thermostat housing.
   - Knock sensor and the engine vent cover.
   - Left motor mount.
   - Fuel injector supply manifold with the injectors and the ground strap.
   - Water pump pulley.
   - Water pump.
   - CMP sensor.
   - CHT sensor.

   - Spark plugs.
   - Valve cover.
   - CKP sensor.
   - Crankshaft vibration damper

➡ **There is one front cover bolt behind the cooling fan drive pulley. To remove this bolt, align one of the cooling fan drive pulley access holes with the bolt head to access the bolt.**

   - Front cover.
   - Timing chain tensioner.
   - Timing chain guides.
   - Timing chain assembly.

➡ **Use a wrench on the flats between cylinders No. 1 and No. 2 to hold the camshaft in place.**

   - Camshaft drive sprockets.
   - Oil pump chain tensioner and guide.

➡ **The oil pump chain sprocket must be held in place.**

   - Oil pump chain and sprockets.

➡ **Note the position of the lobes on the No. 1 cylinder before removing the camshafts for assembly reference.**

6. Loosen the camshaft bearing cap bolts in sequence, one turn at a time. Repeat the first step until all tension is released from the camshaft bearing caps. Remove the camshaft bearing caps.
7. Remove the camshafts.
8. Installation is the reverse of removal.

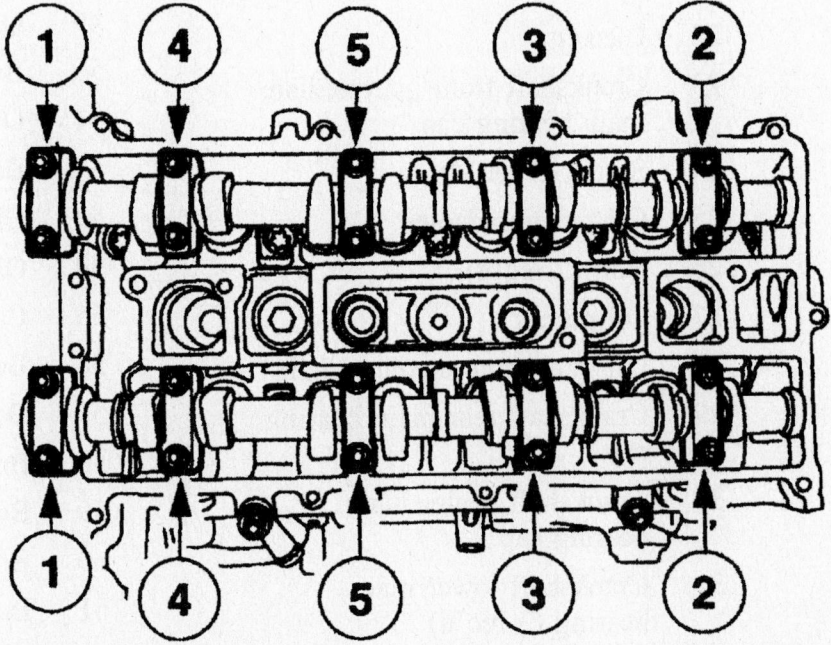

Camshaft cap loosening sequence—2.3L

67197-RANG-G5A

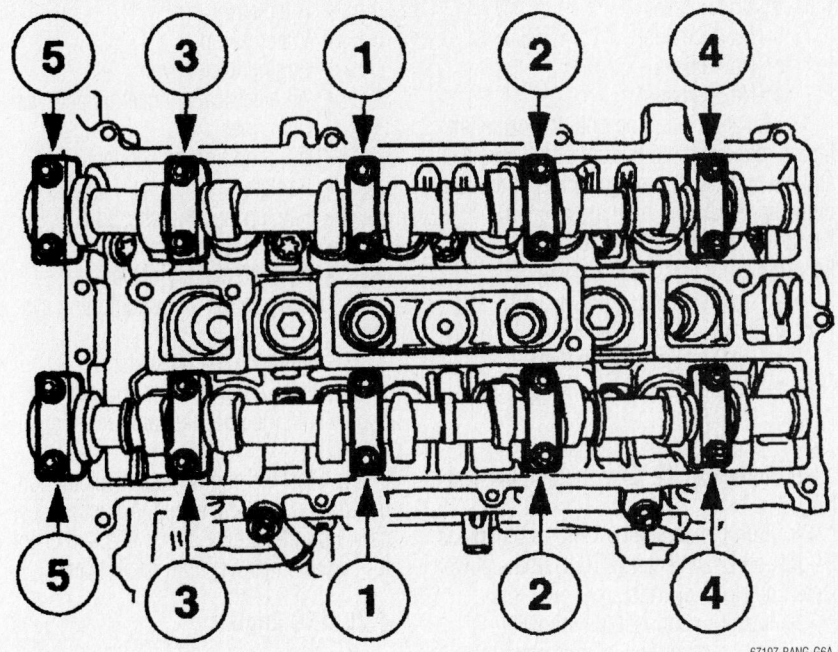

67197-RANG-G6A

Camshaft cap torque sequence—2.3L

➡Install the camshafts with the alignment notches in the camshaft lined up so the camshaft alignment plate can be installed without rotating the camshafts. Make sure the lobes on the No. 1 cylinder are in the same position as noted in the disassembly procedure. Rotating the camshafts, or installing the camshafts 180 degrees out of position can cause severe damage to the valves and pistons. Lubricate the camshaft journals and bearing caps with clean engine oil. Install the camshafts and bearing caps. Tighten the bolts in the sequence shown in three stages.

- Step 1: Tighten the camshaft bearing caps one turn at a time until tight.
- Step 2: Tighten the bolts to 7 Nm (62 inch lbs.)
- Step 3: Tighten the bolts to 16 Nm (12 ft. lbs.)
  a. Crankshaft vibration damper:

➡Do not reuse the crankshaft pulley bolt. Tighten the bolt in two stages.

- Step 1: Tighten the bolt to 40 Nm (30 ft. lbs.)
- Step 2: Tighten the bolt and additional 90 degrees (1/4 turn).

2.5L Engines

1. Drain the cooling system.
2. Before servicing the vehicle, refer to the precautions in the beginning of this section.

3. Remove or disconnect the following:
- Negative battery cable
- Air cleaner
- Spark plug wires and retainers
- Vacuum lines
- Drive belts
- Alternator and bracket
- Upper radiator hose
- Radiator shroud
- Fan blades
- Water pump pulley
- Fan shroud

4. Align the engine timing marks at Top Dead Center (TDC) for No. 1 cylinder. Remove the timing belt.
- Valve covers
- Rocker arms (camshaft followers)
- Camshaft drive gear and belt guide using a suitable puller. Remove the front oil seal with Front Seal Replacer T74P-6150-A
- Camshaft retainer located on the rear mounting stand
- Front motor mount bolts
- Lower radiator hose from the radiator
- Automatic transmission cooler lines, if equipped

5. Position a piece of wood on a floor jack and raise the engine carefully as far as it will go. Place blocks of wood between the engine mounts and crossmember pedestals.

6. Remove camshaft by carefully withdrawing it toward the front of the engine. Caution should be used to prevent damage to cam bearings, lobes and journals.

7. Check the camshaft journals and lobes for wear. Inspect the cam bearings, if worn (unless the proper bearing installing tool is on hand), the cylinder head must be removed for new bearings to be installed by a machine shop.

To install:

8. Install or connect the following:
- Camshaft and lower the engine to its original position
- Transmission cooler lines, if equipped
- Lower radiator hose
- Front motor mount. Torque the bolts to 65 ft. lbs. (88 Nm).
- Camshaft retainer on the rear mounting stand
- Camshaft drive gear and belt guide
- Valve covers
- Timing belt. Make certain that the timing marks are properly aligned
- Fan shroud
- Water pump pulley
- Fan blades
- Upper radiator hose and radiator shroud
- Alternator and bracket
- Drive belts
- Vacuum lines
- Spark plugs wires and retainers
- Air cleaner
- Negative battery cable

9. Fill the cooling system.
10. Start the engine and check for leaks, repair if necessary.

3.0L Engine

1. Before servicing the vehicle, refer to the precautions in the beginning of this section.
2. Properly relieve the fuel system pressure.
3. Drain the cooling system.
4. Drain the engine oil.
5. Evacuate the air conditioning system.
6. Remove or disconnect the following:
- Negative battery cable
- Air cleaner hoses
- Fan, spacer and shroud
- Radiator

7. Rotate the crankshaft so that No. 1 piston is at Top Dead Center (TDC) on the compression stroke.
- Air conditioning condenser
- Fuel lines from the fuel supply manifold
- Vacuum hoses
- Electrical wiring
- Engine front cover
- Water pump
- Alternator

- Power steering pump. Do not disconnect the hoses
- Air conditioning compressor. Do not disconnect the hoses
- Throttle body
- Fuel injection wire harness

8. Turn the engine by hand to TDC of the power stroke on No. 1 cylinder.

- Spark plug wires from the plugs
- Distributor cap with the spark plug wires as an assembly, if equipped

9. Matchmark the rotor, distributor body and engine. Disconnect the distributor wiring harness and remove the distributor, if equipped.

- Rocker arm covers
- Intake manifold
- Loosen the rocker arm bolts enough to pivot the rocker arms out of the way and remove the pushrods. Identify them for installation
- Lifters and identify them for installation
- Crankshaft pulley/damper
- Starter
- Oil pan
- Camshaft gear attaching bolt and washer, then slide the gear off the camshaft
- Camshaft thrust plate

10. Carefully slide the camshaft out of the engine block, using caution to avoid any damage to the camshaft bearings.

To install:

11. Oil the camshaft journals and cam lobes with heavy SJ engine oil (50W). Install the spacer ring with the chamfered side toward the camshaft, then insert the camshaft key.

12. Install or connect the following:

- Camshaft using caution to avoid any damage to the camshaft bearings
- Thrust plate. Torque the screws to 84 inch lbs. (10 Nm).

13. Rotate the camshaft and crankshaft as necessary to align the timing marks. Install the camshaft gear and chain. Torque the bolt to 46 ft. lbs. (62 Nm).

14. Coat the tappets with 50W engine oil and place them in their original locations.

15. Apply 50W engine oil to both ends of the pushrods. Install the pushrods in their original locations.

16. Pivot the rocker arms into position. Torque the fulcrum bolts to 96 inch lbs. (11 Nm).

17. Rotate the engine until both timing marks are at the top of their sprockets and aligned. Torque the following fulcrum bolts to 18 ft. lbs. (24 Nm):

a. No.1 intake.
b. No.2 exhaust.
c. No.4 intake.
d. No.5 exhaust.

18. Rotate the engine until the camshaft timing mark is at the bottom of the sprocket and the crankshaft timing mark is at the top of the sprocket, and both are aligned. Torque the following fulcrum bolts to 18 ft. lbs. (24 Nm):

a. No.1 exhaust.
b. No.2 intake.
c. No.3 intake and exhaust.
d. No.4 exhaust.
e. No.5 intake.
f. No.6 intake and exhaust.

19. Torque all the bolts to 24 ft. lbs. (33 Nm).

20. Turn the engine by hand to 0 degrees Before Top Dead center (BTDC) of the power stroke on No. 1 cylinder.

21. Install or connect the following:

- Engine front cover and water pump assembly
- Oil pan
- Crankshaft damper/pulley and tighten the retaining bolt to 107 ft. lbs. (145 Nm).
- Intake manifold
- Starter
- Crankshaft pulley and damper
- Rocker arm covers
- Rotor and distributor cap, if equipped
- Spark plug wires
- Fuel lines to the fuel supply manifold
- Fuel injection wire harness
- Throttle body
- Air conditioning compressor
- Power steering pump

- Alternator
- Water pump
- Engine front cover
- All electrical connectors and vacuum lines
- Air conditioning condenser
- Radiator
- Fan, spacer and shroud
- Air cleaner hoses
- Negative battery cable

22. Recharge the air conditioning system.

23. Refill the cooling system.

24. Replace the oil filter and refill the engine with the specified amount of engine oil.

25. Start the engine and check the ignition timing and idle speed. Adjust if necessary. Run the engine at fast idle and check for coolant, fuel, vacuum or oil leaks.

4.0L OHC Engine

1. Before servicing the vehicle, refer to the precautions in the beginning of this section.

2. Remove or disconnect the following:

- Negative battery cable for safety
- Valve cover
- Hydraulic camshaft tensioner

➡ **The right-hand camshaft sprocket bolt uses left-hand threads.**

3. For the right-hand camshaft use the Cam Gear Torque Adapter tool T97T-6256-F, to remove the camshaft sprocket bolt.

4. For the left-hand camshaft, remove the sprocket bolt.

➡ **When removing the followers, label them so that they may be returned to their original positions.**

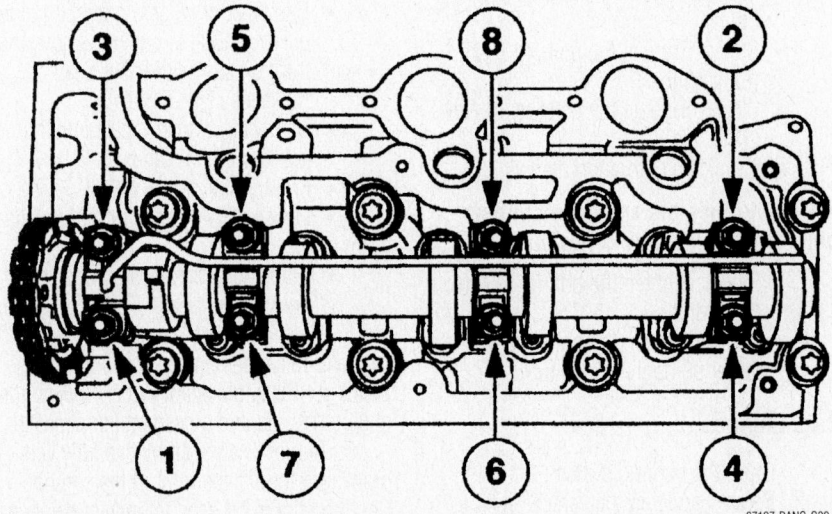

67197-RANG-G39

Camshaft bolt removal sequence—4.0L SOHC engine

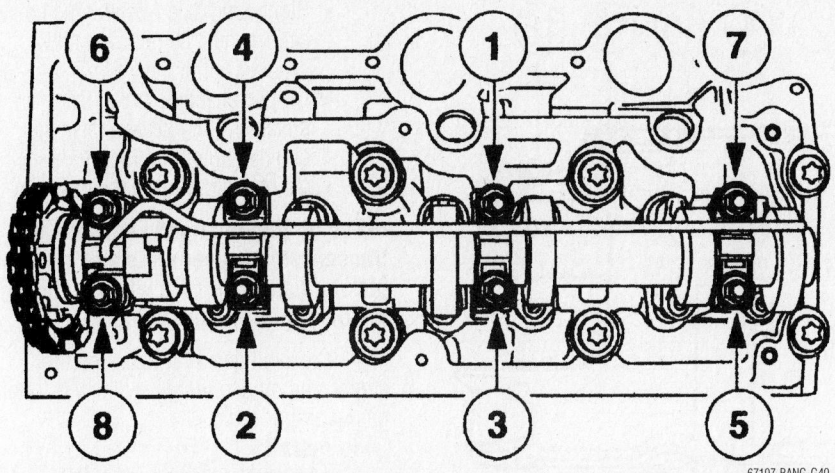

Camshaft bolt tightening sequence—4.0L SOHC engine

67197-RANG-G40

5. Using the Valve Spring Compressor tool ST1330-A, remove the camshaft roller followers.

6. Remove or disconnect the following:
- Camshaft bearing cap bolts and the oil rail
- Camshaft

To install:

7. Lubricate all of the moving parts with SAE 50W engine oil.

8. Install camshaft onto the cylinder head.

9. Position the oil rail and install the bearing caps and bolts. Torque the bolts in 2 steps:

a. Step 1—53.5 inch lbs. (6 Nm).

b. Step 2—11–12.5 ft. lbs. (15–17 Nm).

10. Install or connect the following:
- Camshaft followers
- Camshaft sprocket bolt and hand tighten the bolt
- Camshaft Chain Tensioner T97T-6K254-A in the hole that the hydraulic chain tensioner was in

11. Turn the crankshaft one revolution clockwise until No. 1 piston is Top Dead Center (TDC).

12. Install or connect the following:
- Crankshaft Holding tool T97T-6303-A on the crankshaft to keep it from turning
- Position the timing slot on the rear of the camshaft to fit Camshaft Holding tool T97T-6256-C and install the holding tool on the rear of the head
- Camshaft Gear Holding tool T97T-6256-B and Camshaft Gear Holding tool T97T-6256-A on the front of the cylinder head to securely hold the camshaft gear
- Tighten the camshaft sprocket bolt to 63 ft. lbs. (85 Nm).

13. Remove the Camshaft Chain Tensioner tool and install the hydraulic chain tensioner, tighten the tensioner to 35–39 ft. lbs. (47–53 Nm).

14. Remove the special tools from the engine.

15. Install or connect the following:
- Valve cover
- Negative battery cable

16. Start the engine check for leaks and repair if necessary.

Oil Pan

REMOVAL & INSTALLATION

2.3L Engine

1. Before servicing the vehicle, refer to the precautions in the beginning of this section.

2. Drain the engine oil.

3. Remove or disconnect the following:
- Engine from the vehicle
- Engine oil level indicator assembly
- Engine oil pan bolts and oil pan

To install:

4. Clean and inspect all mating surfaces.

➡ **The oil pan must be installed and the bolts tightened with four minutes of applying the silicone gasket and sealant.**

5. Apply a 2.5 mm bead of silicone gasket and sealant to the oil pan. Install the oil pan. Tighten the oil pan in the sequence shown.

6. Lubricate the O-ring with clean engine oil and install the engine oil level indicator assembly.

7. Install the engine into the vehicle.

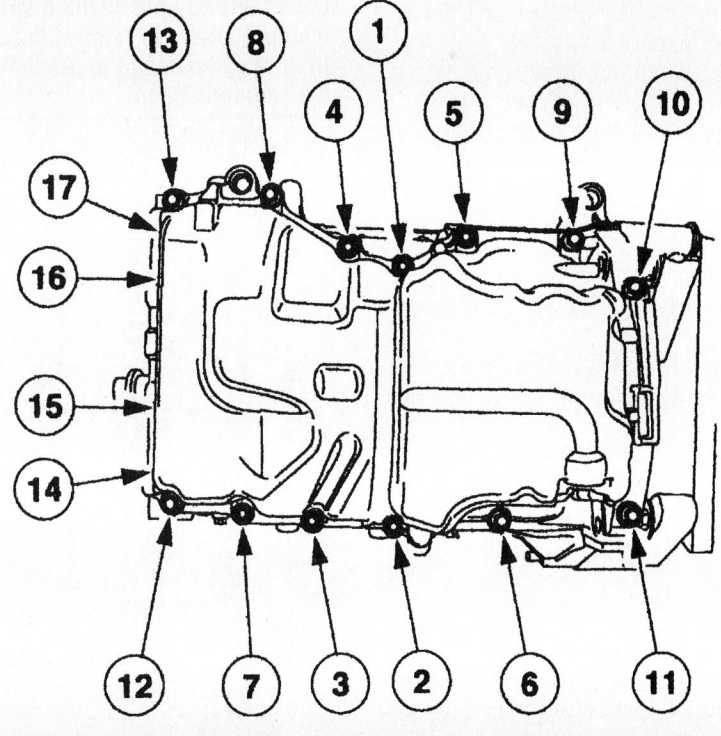

25 Nm (18 lb-ft)

Oil pan torque sequence—2.3L

67197-RANG-G7A

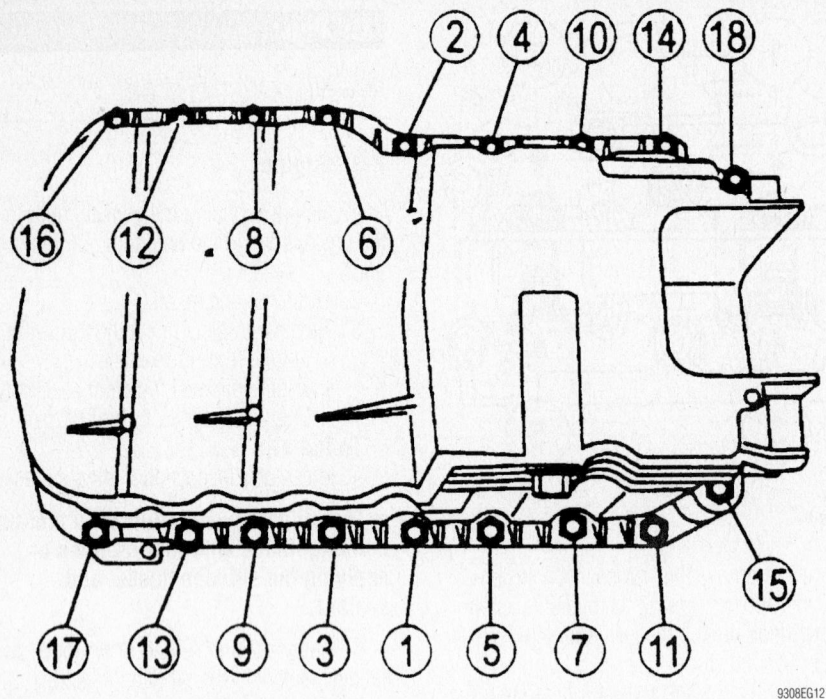

Tighten the oil pan bolts in sequence–2.5L engines

9308EG12

2.5L Engines

1. Before servicing the vehicle, refer to the precautions in the beginning of this section.

2. Drain the engine oil.

3. Remove or disconnect the following:
- Negative battery cable
- Engine from the vehicle and place it on a suitable engine stand
- Oil pan and discard the gasket

To install:

4. Clean the mating surface on the oil pan.

5. Install or connect the following:
- Oil pan gasket
- Apply a bead of silicone sealant to the oil pan
- Oil pan. Torque the bolts in sequence to 141 inch lbs. (16 Nm).
- Engine
- Negative battery cable

6. Fill the engine with clean oil.

7. Start the vehicle and check for leaks, repair if necessary.

3.0L Engine

2WD

1. Before servicing the vehicle, refer to the precautions in the beginning of this section.

2. Drain the engine oil.

3. Remove or disconnect the following:
- Negative battery cable
- Oil level dipstick tube

- Fan shroud. Leave the fan shroud over the fan assembly
- Motor mount nuts from the frame

※※ WARNING

On models equipped with distributor ignition, failure to remove the distributor will damage or break it when the engine is lifted.

- Starter
- Transmission inspection cover

- Right hand axle I-beam. The brake caliper must be removed and secured out of the way.
- Oil pan attaching bolts, using a suitable lifting device, raise the engine about 2 in. (5cm)
- Oil pan and discard the gasket

➡ **The oil pan fits tightly between the transmission spacer plate and oil pump pick-up tube. Use care when removing the oil pan from the engine.**

4. Clean all gasket surfaces on the engine and oil pan. Remove all traces of old gasket and/or sealer.

To install:

5. Apply a ⅛ (4mm) bead of RTV sealer to the junctions of the rear main bearing cap and block, and the front cover and block. The sealer sets in 15 minutes, so work quickly!

6. Apply adhesive to the gasket surfaces and install the oil pan gasket.

7. Install or connect the following:
- Oil pan on the engine block. Torque the bolts EVENLY to 9 ft. lbs. (12 Nm) working from the center to the end position on the oil pan.
- Right hand axle I-beam
- Brake caliper
- Transmission inspection cover
- Starter
- Fan shroud
- Motor mount retaining nuts
- Oil level dipstick tube
- Negative battery cable

8. Fill the engine with clean oil.

9. Start the vehicle and check for leaks, repair if necessary.

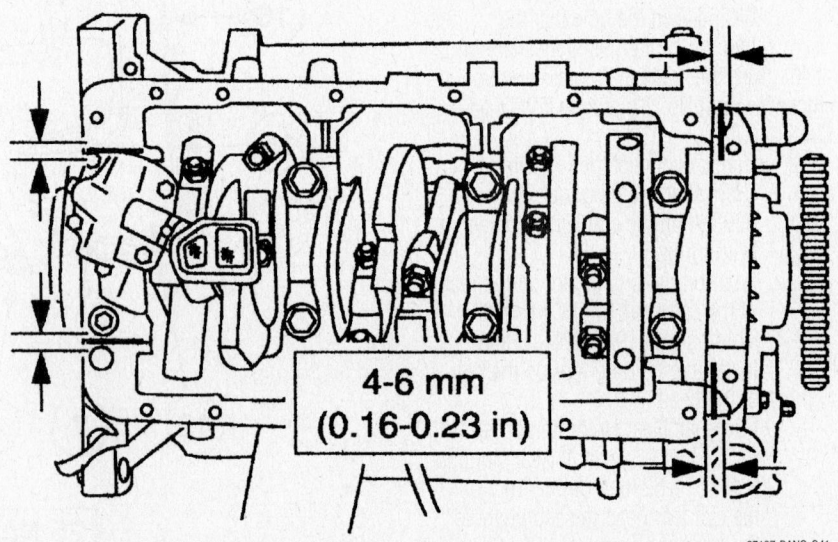

4-6 mm
(0.16-0.23 in)

67197-RANG-G41

Oil pan sealer application—3.0L engine

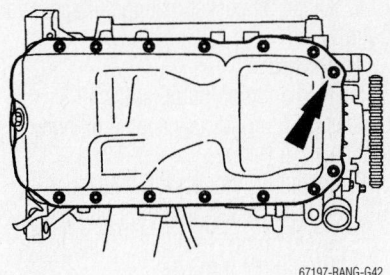

Oil pan bolts—3.0L engine with 2-wheel drive

67197-RANG-G42

4WD

1. Before servicing the vehicle, refer to the precautions in the beginning of this section.
2. Drain the engine oil.
3. Remove or disconnect the following:
   - Negative battery cable
   - Engine from the vehicle and place it on a suitable engine stand
   - Oil pan and discard the gasket

To install:

4. Install or connect the following:

- New oil pan gasket and secure the gasket with trim adhesive
- Oil pan. Torque the bolts to 9 ft. lbs. (12 Nm).
- Engine
- Negative battery cable

5. Fill the engine with clean oil.
6. Start the vehicle and check for leaks, repair if necessary.

4.0L OHC Engine

➡The 4.0L SOHC engine does not use an oil pan in the conventional sense. There is a separate access panel that unbolts from what would be considered the oil pan (which is now known as the ladder frame).

1. Before servicing the vehicle, refer to the precautions in the beginning of this section.
2. Drain the engine oil.
3. Remove or disconnect the following:
   - Negative battery cable
   - Oil pan and discard the gasket

To install:

4. Install or connect the following:
   - New gasket and oil pan. Torque the bolts to 80 inch lbs. (9 Nm).
   - Negative battery cable

5. Fill the engine with clean oil.
6. Start the vehicle and check for leaks, repair if necessary.

Oil Pump

REMOVAL & INSTALLATION

2.3L Engine

➡The oil pump is located on the front of the engine and is turned by the timing belt.

1. Before servicing the vehicle, refer to the precautions in the beginning of this section.
2. Remove or disconnect the following:
   - Negative battery cable
   - Timing chain
   - Oil pump chain and sprockets
   - Oil pan
   - Oil pump pickup tube and gasket
   - Oil pump assembly and gasket

To install:

3. Turn the crankshaft clockwise to position the No. 1 piston.
4. Remove the plug bolt.
5. Install the Engine Timing Peg 303-507.

➡Clean the gasket surface with metal surface cleaner.

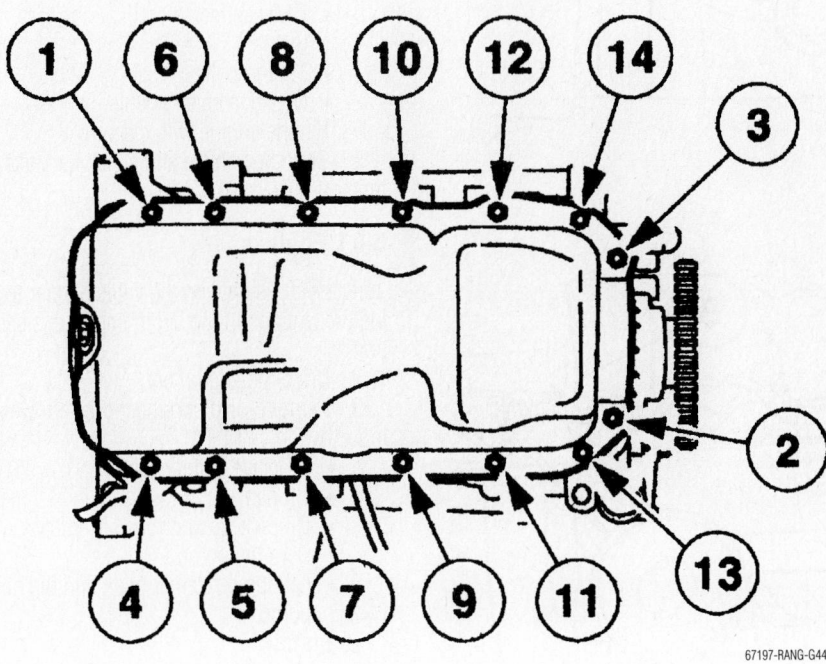

Oil pan bolts—3.0L engine with 4-wheel drive

67197-RANG-G44

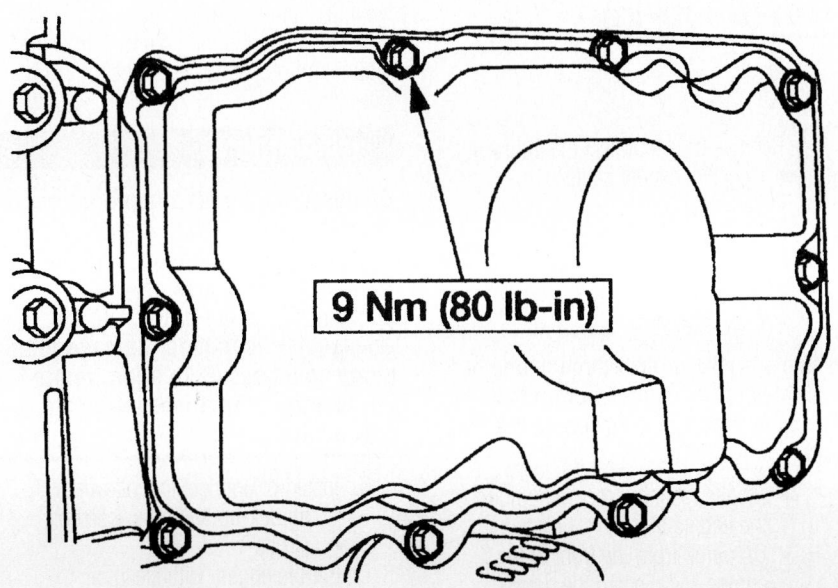

9 Nm (80 lb-in)

67197-RANG-G45

Oil pan installation—4.0L SOHC engine

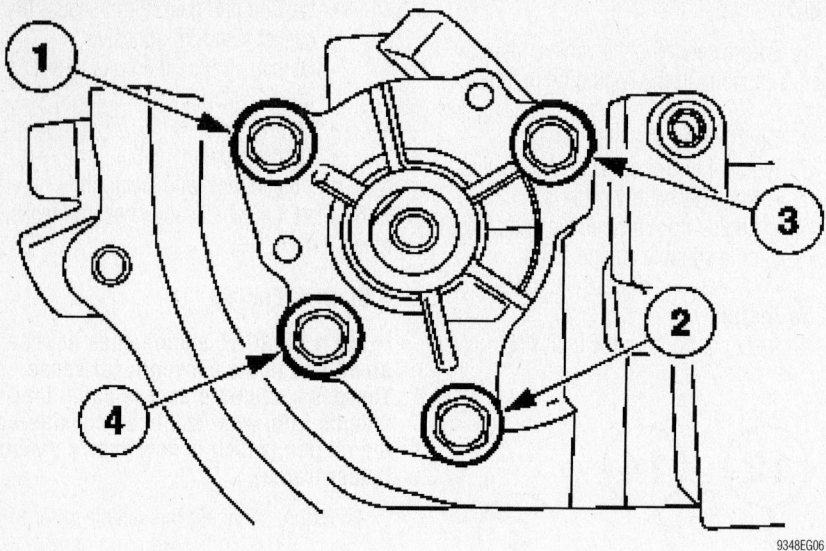

Oil pump torque sequence—2.3L

9348EG06

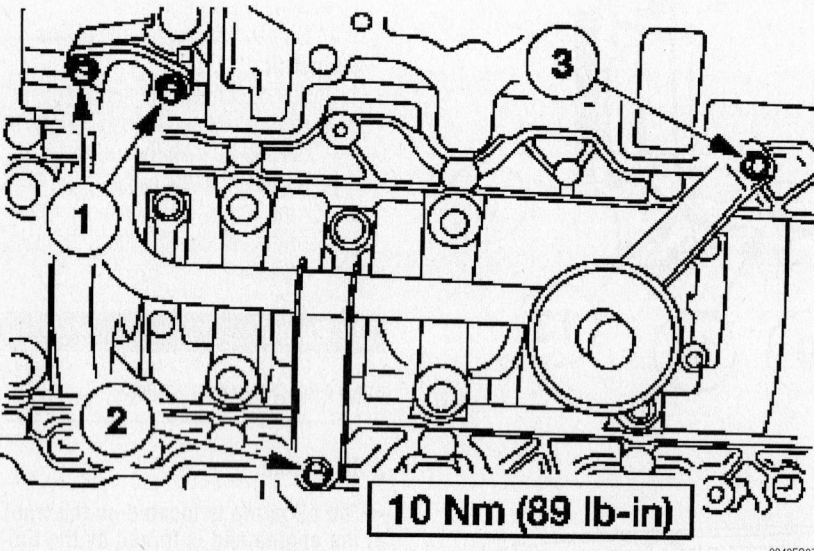

10 Nm (89 lb-in)

9348EG07

Oil pump pickup tube torque sequence—2.3L

6. Install a new gasket and the oil pump assembly. Tighten the bolts in the sequence shown in two stages.
- Step 1: Tighten the bolts to 10 Nm (80 inch lbs.)
- Step 2: Tight the bolts to 23 Nm (17 ft. lbs.)

7. Install a new oil pump pickup tube gasket and the pickup tube. Tighten the bolts in the sequence shown

2.5L Engines

➡The oil pump is located on the front of the engine and is turned by the timing belt.

1. Before servicing the vehicle, refer to the precautions in the beginning of this section.

2. Remove or disconnect the following:
- Negative battery cable
- Timing belt
- Camshaft Position (CMP) sensor electrical connector
- Oil pump sprocket
- CMP sensor

➡Use a prybar or drift through one of the holes in the pump sprocket to keep it from turning while loosening the bolt.

- 4 bolts retaining the oil pump to the engine block
- Oil pump from the front of the engine and discard the gasket

3. Inspect the oil pump and O-rings and replace as necessary.

4. Clean all gasket mating surfaces thoroughly.

To install:

5. Prime the oil pump and with 8 ounces (236 ml) of new engine oil and lubricate the O-rings.

6. Install or connect the following:
- New gasket on the oil pump
- Oil pump. Torque the bolts to 89 inch lbs. (10 Nm).
- CMP sensor. Torque the bolts to 61 inch lbs. (7 Nm).
- Oil pump sprocket bolt. Torque the bolt to 40 ft. lbs. (55 Nm).
- CMP sensor electrical connector
- Timing belt
- Negative battery cable

7. Fill the engine with clean oil.

8. Start the vehicle and check for leaks, repair if necessary.

3.0 L Engines

1. Before servicing the vehicle, refer to the precautions in the beginning of this section.

2. Drain the engine oil.

3. Remove or disconnect the following:
- Negative battery cable
- Oil pan
- Oil pick-up and tube assembly from the pump
- Oil pump retainer bolts and the oil pump

To install:

4. Prime the oil pump with clean engine oil by filling either the inlet or outlet port. Rotate the pump shaft to distribute the oil within the pump body.

5. Install the oil pump and tighten the mounting bolts to 30–40 ft. lbs. (41–54 Nm).

❈❈ WARNING

Do not force the oil pump if it does not seat readily. The oil pump driveshaft may be misaligned with the distributor or shaft assembly. If the pump is tightened down with the driveshaft misaligned, damage to the pump could occur. To align, rotate the intermediate driveshaft into a new position.

6. Install or connect the following:
- Oil pick-up and tube assembly
- Oil pan

7. Fill the engine with clean oil.

8. Start the vehicle and check for leaks, repair if necessary.

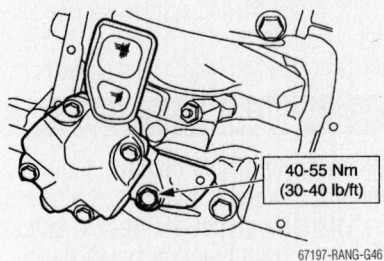

Oil pump installed—3.0L engine

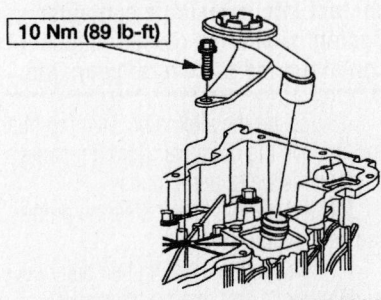

10 Nm (89 lb-ft)

Oil pump installation—4.0L SOHC engine

4.0L SOHC Engines

➡**The oil pump cannot be removed with the engine in the vehicle.**

1. Before servicing the vehicle, refer to the precautions in the beginning of this section.
2. Drain the engine oil.
3. Remove or disconnect the following:
   - Engine from the vehicle
   - Oil pan
   - Unbolt the oil pick-up tube
   - The 8 ladder frame bolts that were under the oil pan
   - The 2 rear outer ladder frame bolts
   - The 7 left-hand and the 8 right-hand ladder frame bolts
   - The ladder frame from the engine
   - The 2 oil pump attaching bolts and the pump.

To install:

4. Submerge the pump in clean engine oil to prime it.
5. Install or connect the following:
   - The ladder frame on the engine
   - The 8 right-hand and 7 left-hand ladder frame bolts
   - The 2 rear outer and the 8 frame bolts under the pan
   - The oil pump. Torque the bolts to 13–15 ft. lbs. (17–21 Nm).
   - Oil pick-up tube
   - Oil pan
   - Engine to the vehicle
   - Negative battery cable
6. Fill the engine with clean oil.

7. Start the vehicle and check for leaks, repair if necessary.

Rear Main Seal

REMOVAL & INSTALLATION

2.3L Engine

1. Before servicing the vehicle, refer to the precautions in the beginning of this section.
2. Remove or disconnect the following:
   - Flywheel or flexplate
   - Bolts and the crankshaft rear oil seal

To install

3. Install or connect the following:
   - Rear oil seal on the Crankshaft Rear Main Oil Seal Installer
   - Crankshaft Rear Main Oil Seal Installer and the crankshaft rear oil seal on the crankshaft
4. Tighten the bolts in the sequence shown to 10 Nm (89 inch lbs.)
5. Remove the Crankshaft Rear Main Oil Seal Installer.
6. Install the flywheel or flexplate.

2.5L Engine

1. Before servicing the vehicle, refer to the precautions in the beginning of this section.
2. Remove the flywheel or flexplate

➡**Clean the crankshaft rear oil seal and cylinder block prior to removing the rear oil seal.**

3. Screw in the Jet Plug Remover.
4. Remove the seal.

To install

➡**Apply 5W-30 motor oil to seal and seal edge.**

5. Install crankshaft rear oil seal on Rear Main Seal Replacer.
6. Install the Rear Main Seal Replacer and the crankshaft rear oil seal on the crankshaft.
7. Alternate bolt tightening to crankshaft rear oil seal.
8. Install the flywheel or flexplate.

3.0L Engines

1. Before servicing the vehicle, refer to the precautions in the beginning of this section.
2. Remove the flexplate or flywheel.

✶✶ WARNING

Use care to avoid scratching or damaging the oil seal surface or leakage may occur.

3. Using a sharp awl, punch one hole into the crankshaft rear oil seal metal surface between the seal lip and the cylinder block.
4. Screw the threaded end of the special tool into the oil seal. Use the special tool to remove the crankshaft rear oil seal.

To install:

5. Lubricate the outer lips and the inner seal on the crankshaft rear oil seal with clean engine oil.
6. Using the special tool, install the crankshaft rear oil seal. Alternate bolt tightening to correctly seat the crankshaft rear oil seal.
7. Install the flexplate or flywheel.

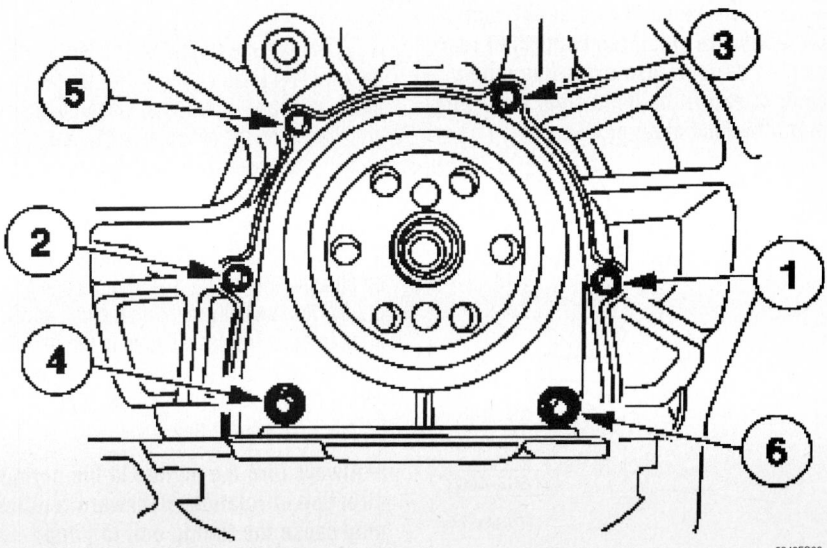

Rear main seal torque sequence—2.3L

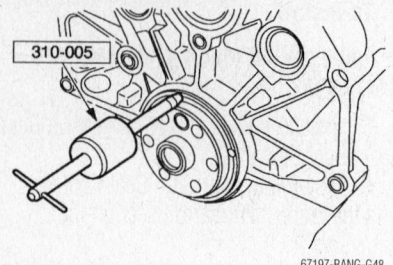

Rear main seal removal—3.0L engine

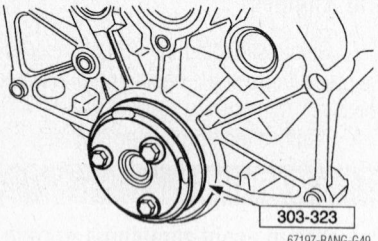

Rear main seal installation—3.0L engine

4.0L SOHC Engine

1. Before servicing the vehicle, refer to the precautions in the beginning of this section.

2. Remove the flexplate or flywheel.

❋❋ WARNING

Avoid scratching or damaging the oil crankshaft seal running surface during removal of the crankshaft rear oil seal.

3. Using the special tool, remove the crankshaft rear oil seal.

To install:

➡Be sure the crankshaft rear sealing surface is clean and free of any rust or corrosion. To clean the crankshaft rear sealing surface, use extra-fine emery cloth or extra-fine 0000 steel wool with metal surface cleaner.

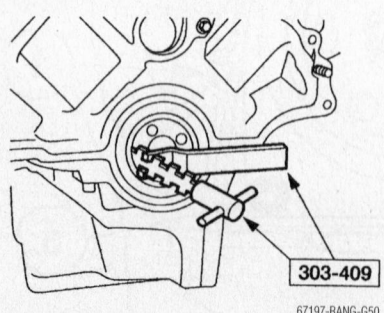

Rear main seal removal—4.0L SOHC engine

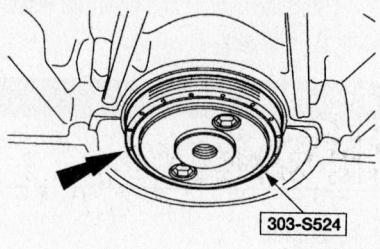

Front part of installation tool—4.0L SOHC engine

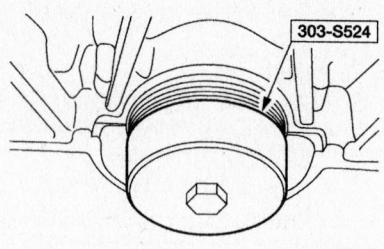

Rear main seal installation—4.0L SOHC engine

4. Lubricate the crankshaft rear oil seal with clean engine oil and install on the special tool.

5. Using the special tool, install the crankshaft rear oil seal.

6. Install the flexplate or flywheel.

Timing Belt and Cover

REMOVAL & INSTALLATION

2.5L Engine

1. Before servicing the vehicle, refer to the precautions in the beginning of this section.

2. Rotate the engine so that No. 1 cylinder is at Top Dead Center (TDC) on the compression stroke. Check that the timing marks are aligned on the camshaft and crankshaft pulleys. An access plug is provided in the cam belt cover so that the camshaft timing can be checked without removal of the cover or any other parts. Set the crankshaft to TDC by aligning the timing mark on the crank pulley with the TDC mark on the belt cover. Look through the access hole in the belt cover to be sure that the timing mark on the cam drive sprocket is aligned with the pointer on the inner belt cover.

➡Always turn the engine in the normal direction of rotation. Backward rotation may cause the timing belt to jump time, due to the arrangement of the belt tensioner.

3. Drain cooling system. Remove the upper radiator hose as necessary. Remove the fan blade and water pump pulley bolts.

❋❋ CAUTION

When draining the coolant, keep in mind that cats and dogs are attracted by ethylene glycol antifreeze, and are likely to drink any that is left in an uncovered container or in puddles on the ground. This will prove fatal in sufficient quantity. Always drain the coolant into a sealable container. Coolant should be reused unless it is contaminated or several years old.

4. Loosen the alternator retaining bolts and remove the drive belt from the pulleys. Remove the water pump pulley.

5. Remove the power steering pump and set it aside.

6. Remove the 4 timing belt outer cover retaining bolts and remove the cover. Remove the crankshaft pulley and belt guide.

7. Loosen the belt tensioner pulley assembly, then position a camshaft belt adjuster tool T74P-6254-A, or equivalent, on the tension spring roll pin and retract the belt tensioner away from the timing belt. Tighten the adjustment bolt to lock the tensioner in the retracted position.

8. If the belt is to be reused, mark the direction of rotation on the belt for installation reference.

9. Remove the timing belt.

To install:

10. Install the new belt over the crankshaft sprocket and then counterclockwise over the auxiliary and camshaft sprockets, making sure the lugs on the belt properly engage the sprocket teeth on the pulleys. Be careful not to rotate the pulleys when installing the belt.

11. Release the timing belt tensioner pulley, allowing the tensioner to take up the belt slack. If the spring does not have enough tension to move the roller against the belt (belt hangs loose), it might be necessary to manually push the roller against the belt and tighten the bolt.

➡The spring cannot be used to set belt tension; a wrench must be used on the tensioner assembly.

❋❋ WARNING

If any binding is felt when adjusting the timing belt tension by turning the crankshaft, STOP turning the engine, because the pistons may be hitting the valves.

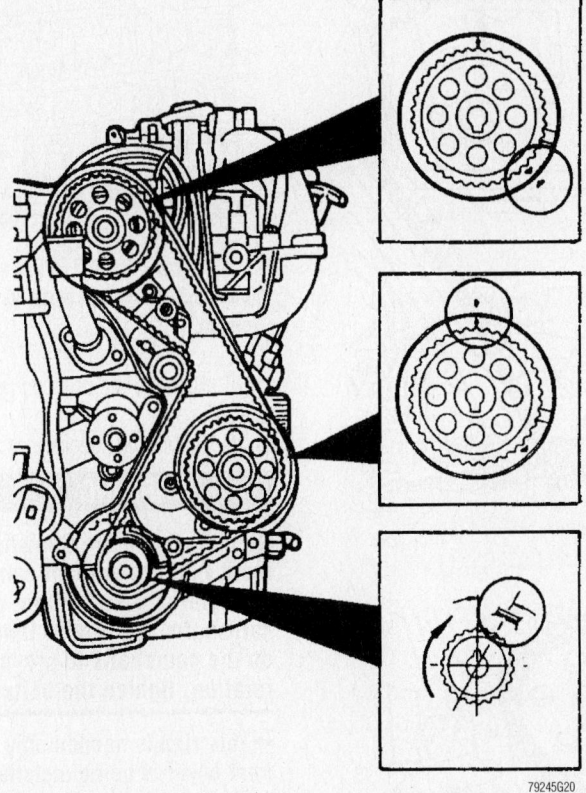

Camshaft, auxiliary shaft and crankshaft timing belt sprocket alignment mark locations—2.5L engines

12. Rotate the crankshaft 2 complete turns by hand (in the normal direction of rotation) to remove slack from the belt. Tighten the tensioner adjustment to 26–33 ft. lbs. (35–45 Nm) and pivot bolts to 30–40 ft. lbs. (40–55 Nm). Be sure the belt is seated properly on the pulleys and that the timing marks are still in alignment when No. 1 cylinder is again at TDC/compression.

13. Install the crankshaft pulley and belt guide.

14. Install the timing belt cover.

15. Install the water pump pulley and fan blades. Install the upper radiator hose if necessary. Refill the cooling system.

16. Install the accessory drive belts.

17. Start the engine and check the ignition timing. Adjust the timing, if necessary.

Timing Chain, Sprockets, Front Cover and Seal

REMOVAL & INSTALLATION

2.3L Engine

1. Before servicing the vehicle, refer to the precautions in the beginning of this section.

2. Remove or disconnect the following:
- Negative battery cable
- Fan and shroud
- Drive belt
- Valve cover

3. Set No. 1 piston to TDC and install the Camshaft Alignment Plate 303-376, or equivalent.

4. Remove the plug for the crankshaft timing peg.

5. Install the Crankshaft Timing Peg 303-507, or equivalent.

6. Install an M6 bolt into the crankshaft pulley to verify the engine timing.

7. Remove or disconnect the following:
- Camshaft pulley
- Crankshaft position sensor
- Crankshaft position sensor
- Belt tensioner
- Water pump pulley
- Power steering high pressure hose. Remove the nylon O-ring.
- Power steering return hose
- Power steering pump

➡This step is needed only if a new front cover is being installed.

8. Using a three-jaw puller, remove the fan drive pulley.

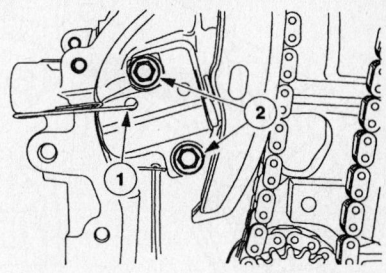

67197-RANG-G53

Timing chain tensioner removal; 1-paper clip, 2-bolts—2.3L engine

➡There is one bolt behind the cooling fan drive pulley. This bolt can be accessed by lining up one of the holes in the pulley with the bolt.

9. Remove the bolts and the engine front cover.

10. Compress the timing chain tensioner and remove the tensioner.

11. Remove the right-hand timing chain guide.

12. Remove the timing chain.

13. Remove the bolts and the left-hand timing chain guide.

✳✳ WARNING

Do not rely on the Camshaft Alignment Plate to prevent camshaft rotation. Damage to the tool or the camshaft can occur.

14. If necessary, remove the bolts and the camshaft sprockets. Use the flats on the camshaft to prevent camshaft rotation.

To install:

15. Remove the special tool.

✳✳ WARNING

Do not rotate the camshafts. Damage to the valves and pistons can occur.

If the camshaft sprockets were not removed, use the flats on the camshafts to prevent camshaft rotation and loosen the sprocket bolts.

16. If removed, install the camshaft sprockets and the bolts. Do not tighten the bolts at this time.

17. Install or connect the following:
- Left-hand timing chain guide and bolts
- Timing chain
- Right-hand timing chain guide
- Timing chain tensioner and release the piston
- Timing chain tensioner and the bolts

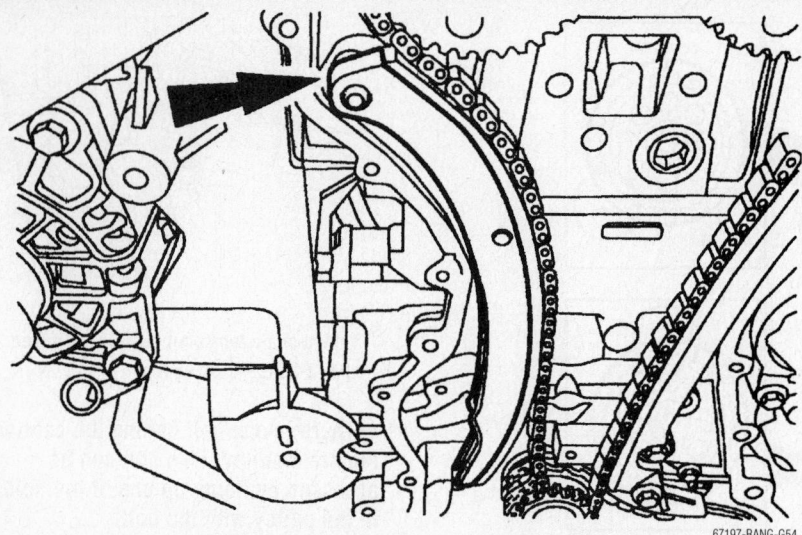

Right side timing chain guide—2.3L engine

67197-RANG-G54

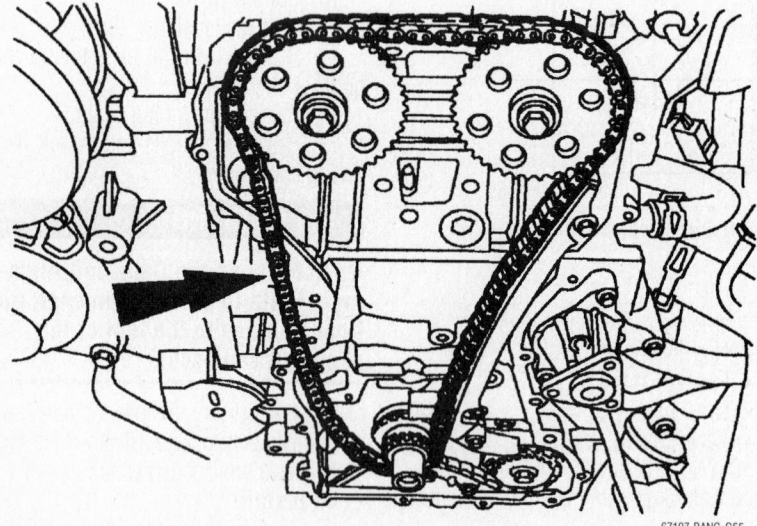

Timing chain removal—2.3L engine

67197-RANG-G55

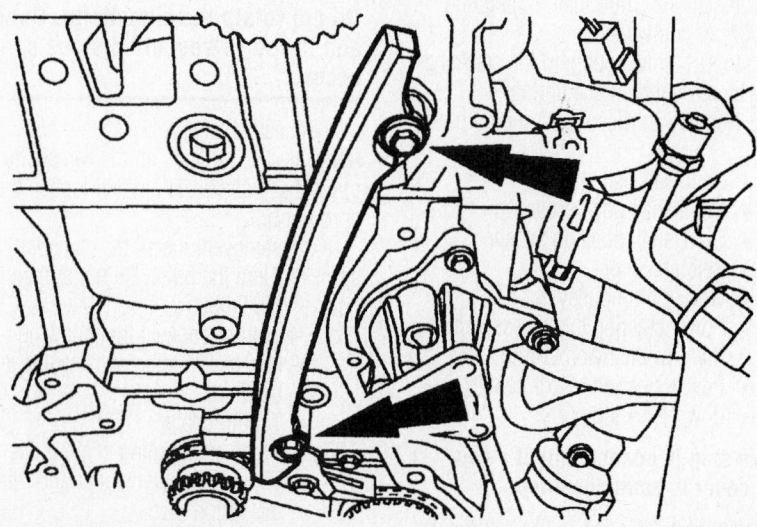

Left timing chain guide—2.3L engine

67197-RANG-G56

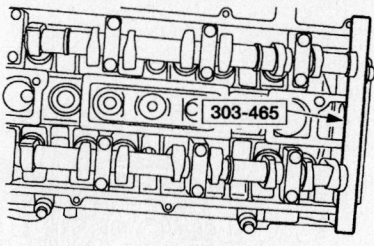

67197-RANG-G57

Camshaft alignment plate installed—2.3L engine

18. Remove the drill rod to release the piston.

19. Install the special tool.

☀ WARNING

Do not rely on the Camshaft Alignment Plate to prevent camshaft rotation. Damage to the tool or the camshafts can result. Using the flats on the camshafts to prevent camshaft rotation, tighten the bolts.

➡ This step is needed only if a new front cover is being installed.

20. Install the fan drive pulley using a nut and bolt with flat washers.

21. Clean and inspect the mounting surfaces of the engine and the front cover.

➡ The engine front cover must be installed and the bolts tightened within four minutes of applying the silicone gasket and sealant.

22. Apply a 2.5 mm bead of silicone gasket and sealant to the cylinder head and oil pan joint areas. Apply a 2.5 mm bead of silicone gasket and sealant to the front cover.

23. Install the front cover. Tighten the bolts in the sequence shown, to the following specifications:

- Step 1: 8 mm bolts to 10 Nm (89 inch lbs.)
- Step 2: 10 mm bolts to 25 Nm (18 ft. lbs.)
- Step 3: 13 mm bolts to 48 Nm (35 ft. lbs.)

24. Install or connect the following:

- Power steering pump and lower retaining bolt
- Power steering return hose
- New nylon O-ring and install the high pressure line.
- Water pump pulley
- Belt tensioner

➡ Do not reuse the crankshaft damper bolt.

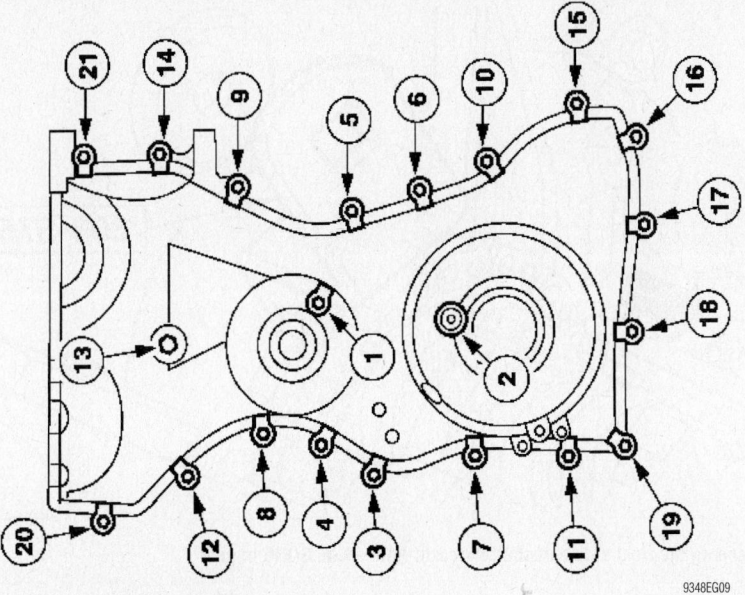

Front cover torque sequence—2.3L

- Crankshaft pulley and hand-tighten the bolt

25. Install an M6 bolt in the crankshaft pulley. Tighten the crankshaft retaining bolt in two stages.
- Step 1: 40 Nm (30 ft. lbs.)
- Step 2: Rotate the bolt an additional 90 degrees.

26. Install the crankshaft position sensor, do not tighten the bolts at this time.

27. Adjust the crankshaft position sensor with the Alignment Tool, and tighten the mounting bolts.

28. Connect the crankshaft position sensor electrical connector.

29. Remove the M6 bolt from the crankshaft pulley.

30. Remove the Crankshaft Timing Peg 303-507.

31. Install the plug.

32. Remove the Camshaft Alignment Plate 303-376.

33. Install the valve cover.

34. Install the drive belt.

35. Install the fan and shroud.

36. Connect the battery ground cable.

3.0L Engines

1. Before servicing the vehicle, refer to the precautions in the beginning of this section.

2. Remove or disconnect the following:
- Negative battery cable
- Engine front cover
- Rotate the crankshaft and align the timing marks
- Timing chain tensioner, 4.0L engine only
- Sprocket bolt

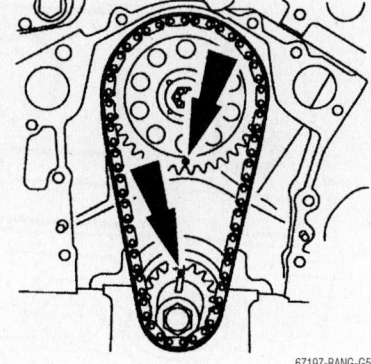

Timing mark alignment—3.0L engine

- Timing chain, camshaft sprocket and crankshaft sprocket as an assembly

To install:

3. Install or connect the following:
- Timing chain, camshaft and crankshaft sprockets as an assembly

4. Align the timing marks.

5. Install or connect the following:
- Timing chain tensioner
- Sprocket bolt. Torque the bolt to 51 ft. lbs. (70 Nm).
- Engine front cover
- Negative battery cable

4.0L OHC Engine

TIMING DRIVE COMPONENTS

1. Before servicing the vehicle, refer to the precautions in the beginning of this section.

2. With the vehicle in neutral, position it on a hoist.

3. Remove the intake manifold.

4. Remove the fuel supply manifold.

5. Remove the accessory drive belt.

6. Remove the thermostat housing.

7. Remove the roller followers.

➡**You must retime the LH and RH camshafts when either camshaft is disturbed.**

Turn the crankshaft clockwise to position the number one cylinder at top dead center (TDC).

➡**The special tool must be installed on the damper and should contact the engine block to position the engine at TDC.**

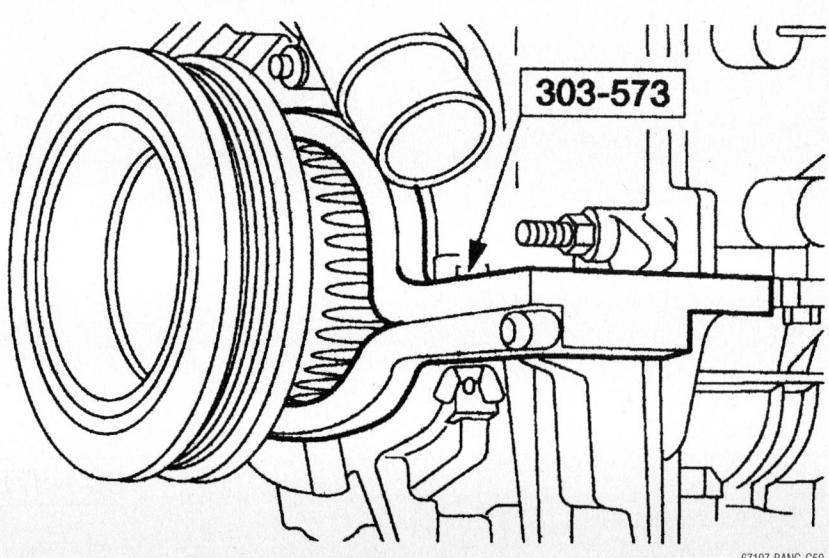

TDC positioning tool installed—4.0L SOHC engine

8. Install the special tool.

➡ **The right-hand camshaft sprocket bolt is a left-hand threaded bolt.**

➡ **If necessary, use camshaft gear torque adapter to loosen the camshaft sprocket bolt.**

9. Using the special tool, loosen the RH camshaft sprocket bolt.

➡ **The camshaft timing slots are off-center.**

10. Position the camshaft timing slots below the centerline of the camshaft to correctly fit the special tools. Install the special tools on the front of the RH cylinder head.

11. Remove the RH lower splash shield.

12. Remove the RH camshaft tensioner.

➡ **Leave the top two special tool clamp bolts loose.**

13. Install the special tools on the rear of the RH cylinder head.

14. Install the special tool.

➡ **The right-hand camshaft sprocket bolt is a left-hand threaded bolt.**

➡ **If necessary, use camshaft gear torque adapter to tighten the camshaft sprocket bolt.**

15. Tighten the bolts.

16. Tighten the special tool top two clamp bolts to 10 Nm (89 inch lbs.).

17. Tighten the camshaft bolt.

18. Install the RH camshaft tensioner.

19. Install the RH lower splash shield.

20. Remove the LH camshaft tensioner.

21. Install the special tools on the front of the LH cylinder head and tighten the top two clamp bolts to 10 Nm (89 inch lbs.).

22. Loosen the LH camshaft sprocket bolt.

23. Loosen the top two clamp bolts on the special tool to allow the camshaft sprocket to rotate freely.

➡ **The camshaft timing slots are off-center.**

24. Position the camshaft timing slots below the centerline of the camshaft to correctly fit the special tools. Install the special tools on the rear of the LH cylinder head.

25. Install the special tool.

26. Tighten the bolts.

27. Tighten the special tool top two clamp bolts to 10 Nm (89 inch lbs.).

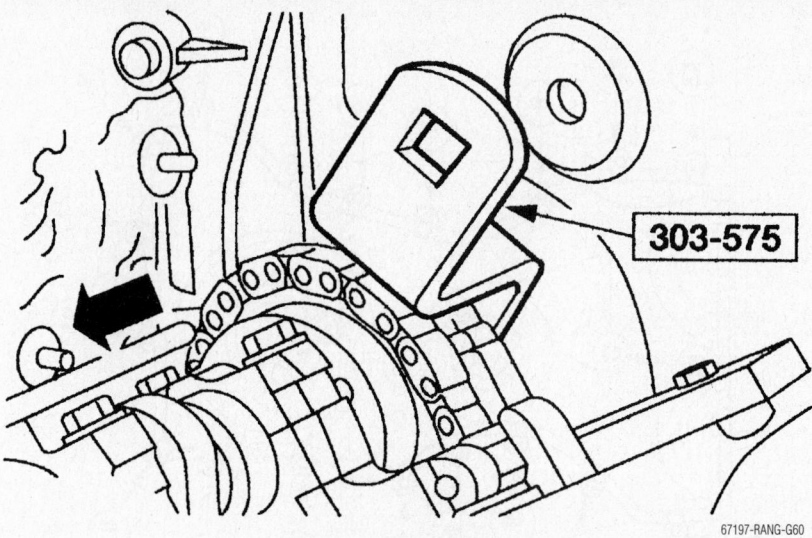

Loosening the right side camshaft sprocket bolt—4.0L SOHC engine

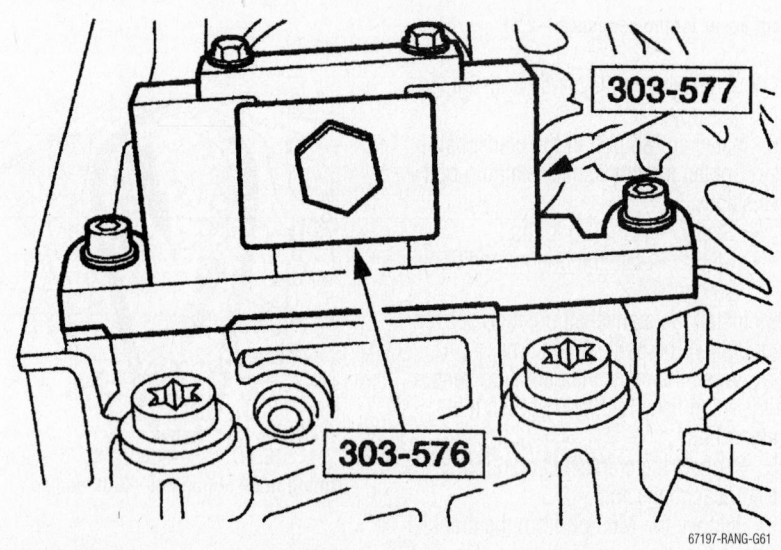

Camshaft holding tool installed—4.0L SOHC engine

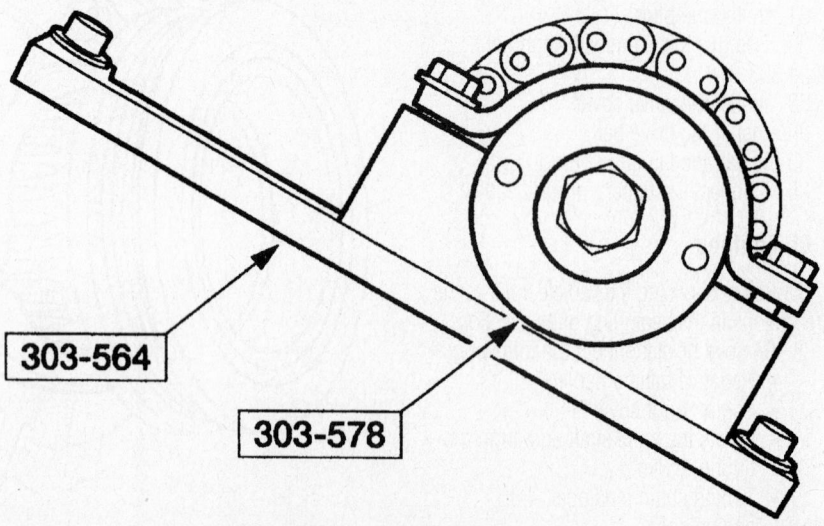

Camshaft gear holding tool and adapter—4.0L SOHC engine

28. Tighten the camshaft bolt.
29. Install the LH camshaft tensioner.
30. Install the roller followers.
31. Install the thermostat housing.
32. Install the accessory drive belt.
33. Install the fuel supply manifold.
34. Install the intake manifold.
35. Install the RH valve cover.
36. Install the LH valve cover.

Piston and Ring

POSITIONING

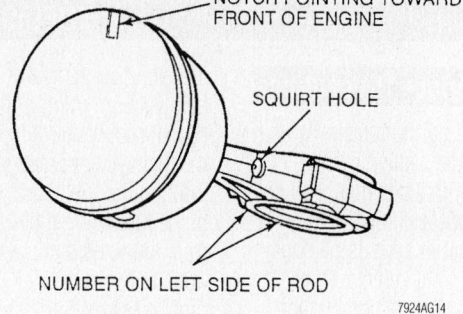

Piston and connecting rod positioning on 2.5L engines

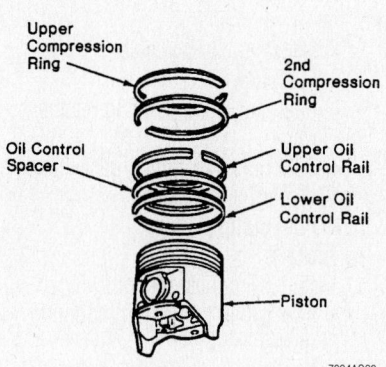

Piston ring positioning

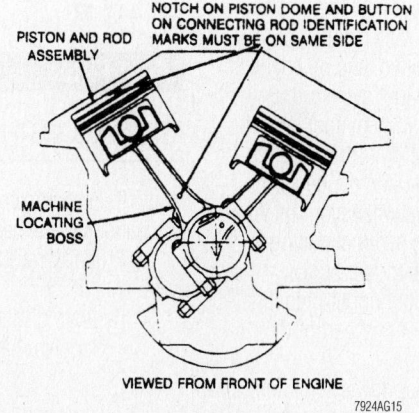

Piston and connecting rod positioning on 3.0L engines

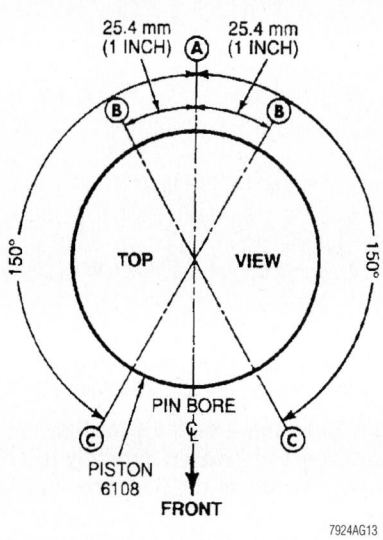

Piston ring end gap spacing

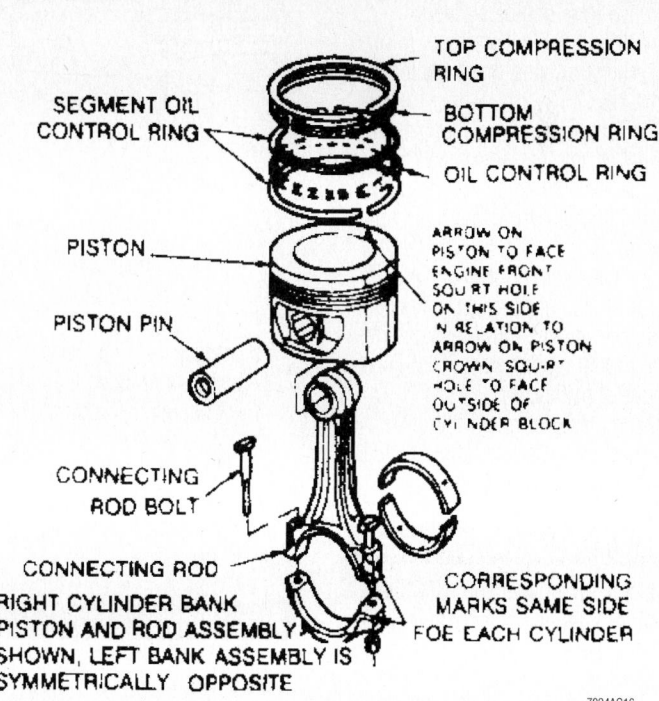

Piston and connecting rod positioning on 4.0L engines

FUEL SYSTEM

Fuel System Service Precautions

Safety is the most important factor when performing not only fuel system maintenance, but any type of maintenance. Failure to conduct maintenance and repairs in a safe manner may result in serious personal injury or death. Work on a vehicle's fuel system components can be accomplished safely and effectively by adhering to the following rules and guidelines.

• To avoid the possibility of fire and personal injury, always disconnect the negative battery cable unless the repair or test procedure requires that battery voltage be applied.

• Always relieve the fuel system pressure prior to disconnecting any fuel system component (injector, fuel rail, pressure regulator, etc.) fitting or fuel line connection. Exercise extreme caution whenever relieving fuel system pressure, to avoid exposing your skin, face and eyes to fuel spray. Please be advised that fuel under pressure may penetrate the skin or any part of the body that it contacts.

• Always place a shop towel or cloth around the fitting or connection prior to loosening to absorb any excess fuel due to spillage. Ensure that all fuel spillage is quickly remove from engine surfaces. Ensure that all fuel-soaked cloths or towels are deposited into a flame-proof waste container with a lid.

• Always keep a dry chemical (Class B) fire extinguisher near the work area.

• Do not allow fuel spray or fuel vapors to come into contact with a light bulb, spark or open flame.

• Always use a second wrench when loosening or tightening fuel line connection fittings. This will prevent unnecessary stress and torsion to fuel piping. Always follow the proper torque specifications.

• Always replace worn fuel fitting O-rings with new ones. Do not substitute fuel hose where rigid pipe is installed.

Relieving Fuel System Pressure

All Sequential Fuel Injection (SFI) injected engines are equipped with a pressure relief valve located on the fuel supply manifold. Remove the fuel tank cap and attach fuel pressure gauge T80L-9974-B, or equivalent, to the valve to release the fuel pressure. Be sure to drain the fuel into a suitable container and to avoid gasoline spillage. If a pressure gauge is not available, disconnect the vacuum hose from the

fuel pressure regulator and attach a hand-held vacuum pump. Apply about 25 in. Hg (84 kPa) of vacuum to the regulator to vent the fuel system pressure into the fuel tank through the fuel return hose. Note that this procedure will remove the fuel pressure from the lines, but not the fuel. Take precautions to avoid the risk of fire and use clean rags to soak up any spilled fuel when the lines are disconnected.

An alternate method of relieving the fuel system pressure involves disconnecting the inertia switch.

Fuel Filter

REMOVAL & INSTALLATION

1. Before servicing the vehicle, refer to the precautions in the beginning of this section.
2. Properly relieve the fuel system pressure.
3. Remove or disconnect the following:
   • Negative battery cable

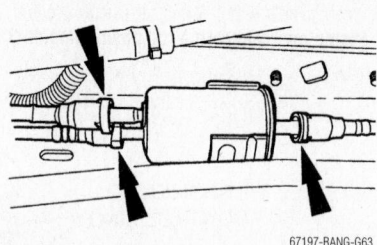

Fuel filter connections

67197-RANG-G63

   • Push connect and R-clip fittings from the fuel filter
   • Fuel filter

To install:

4. Install or connect the following:
   • Fuel filter. Torque the nut to 17 ft. lbs. (23 Nm).
   • R-clip and push connect fittings
   • Negative battery cable

5. Start the vehicle, check for leaks and repair if necessary.

Fuel Pump

REMOVAL & INSTALLATION

2001–03

1. Before servicing the vehicle, refer to the precautions in the beginning of this section.

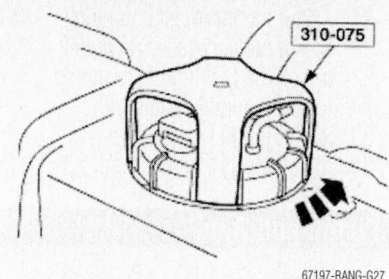

310-075

Lock ring wrench—2001–03 models

67197-RANG-G27

2. Properly relieve the fuel system pressure.
3. Remove or disconnect the following:
   • Negative battery cable
   • Fuel tank
   • Fuel tank pump locking retainer ring
   • Fuel pump mounting gasket and discard the gasket
   • Fuel pump

To install:

4. Install or connect the following:
   • Fuel pump and a new mounting gasket
   • Fuel tank pump locking retainer ring. Torque the ring to 66 ft. lbs. (90 Nm).
   • Fuel tank
   • Negative battery cable

5. Start the vehicle, check for leaks and repair if necessary.

2004

1. Before servicing the vehicle, refer to the precautions in the beginning of this section.
2. Remove the fuel tank.
3. Clean the area around the fuel pump mounting flange.
4. Using the special tool, remove the fuel tank pump assembly locking retainer ring.

✷✷ WARNING

The fuel pump assembly must be removed and handled carefully to avoid damage to the float arm and filter.

5. Remove the fuel pump assembly.
6. Remove and discard the fuel pump mounting gasket.

To install:

7. Clean the fuel pump mounting flange and the fuel tank mounting surface.

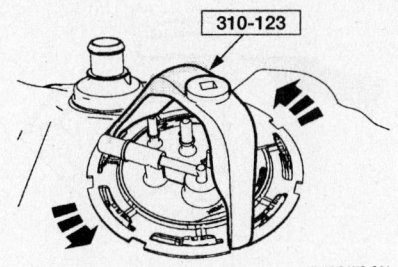

Lock ring tool—2004 models

8. Install a new fuel pump mounting gasket.

9. Install the fuel pump and sender assembly with the float toward the rear of the tank. Align the arrows molded into the tank and flange.

10. Install the locking ring while compressing the pump assembly into the tank.

11. Using the special tool, tighten the fuel pump assembly locking ring retainer ring until it locks in place.

12. Install the fuel tank.

Fuel Injectors

REMOVAL & INSTALLATION

2.3L Engine

1. Before servicing the vehicle, refer to the precautions in the beginning of this section.

2. Properly relieve the fuel system pressure.

3. Remove or disconnect the following:

- Negative battery cable
- Upper intake manifold
- Fuel injector connectors
- Fuel injector harness from the fuel injector supply manifold
- Fuel line spring lock
- Fuel line
- Fuel injection supply manifold
- Fuel injector retaining clip
- Fuel injector

✳✳ WARNING

Use O-ring seals that are made of special fuel-resistant material. Use of ordinary O-ring seals can cause the fuel system to leak. Do not reuse the O-ring seals.

4. Installation is the reverse of removal. Install new O-rings. Lubricate the O-rings with clean engine oil. Torque the supply manifold bolts to 18 ft. lbs. (25 Nm).

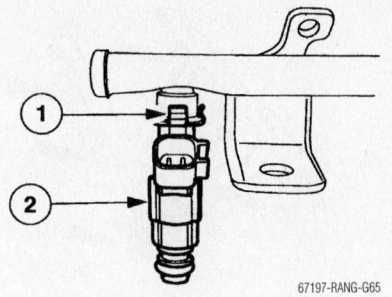

Fuel injector-to-fuel rail installation. 1-retaining clip; 2-injector—2.3L engine

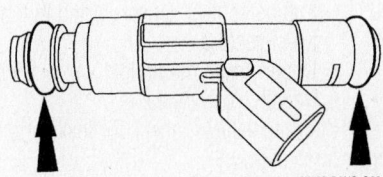

Fuel injector O-rings—2.3L engine

2.5L Engines

1. Before servicing the vehicle, refer to the precautions in the beginning of this section.

2. Properly relieve the fuel system pressure.

3. Remove or disconnect the following:

- Negative battery cable
- Fuel injection supply manifold
- Fuel injectors by gently twisting them
- Inspect the O-rings and replace as needed

Fuel injector wiring connectors—3.0L engine

To install:

4. Install or connect the following:

- Fuel injectors
- Fuel injector supply manifold
- Negative battery cable

5. Start the vehicle, check for leaks and repair if necessary.

3.0L and 4.0L Engines

1. Before servicing the vehicle, refer to the precautions in the beginning of this section.

2. Properly relieve the fuel system pressure.

3. Remove or disconnect the following:

- Negative battery cable
- Upper intake manifold
- Engine control sensor wiring from the fuel injectors
- Fuel lines
- Fuel injection supply manifold and injectors as an assembly
- Vacuum line

Fuel rail bolts—3.0L engine

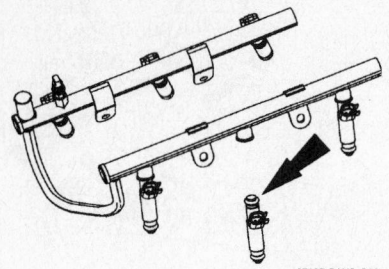

Fuel rail and injectors—3.0L engine

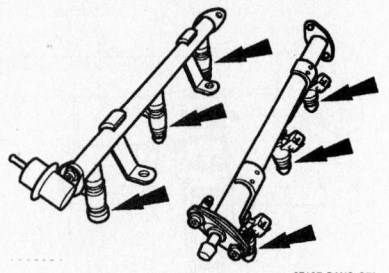

Fuel rail and injectors—4.0L engine

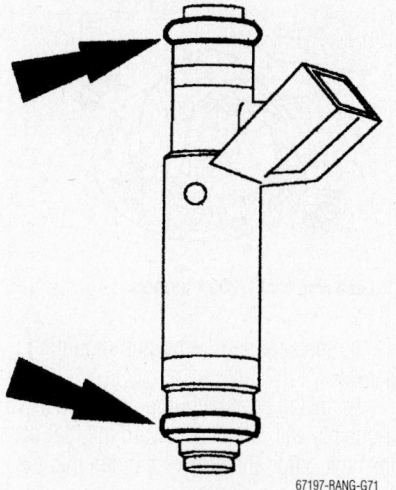

Fuel injector O-rings—4.0L shown; 3.0L similar

- Fuel injectors from the supply manifold
- Inspect the O-rings and replace them as needed

To install:

4. Install or connect the following:
  - Fuel injectors
  - Vacuum line
  - Fuel injection supply manifold.

Torque the bolts to 89 inch lbs. (10 Nm).
  - Fuel line
  - Engine control sensor wiring to the fuel injectors
  - Upper intake manifold
  - Negative battery cable

5. Start the vehicle, check for leaks and repair if necessary.

DRIVE TRAIN

Transmission Assembly

REMOVAL & INSTALLATION

Manual Transmission

2001–03

1. Before servicing the vehicle, refer to the precautions in the beginning of this section.

2. Remove or disconnect the following:
  - Negative battery cable
  - Upper gearshift lever and the outer gearshift lever boot and console assembly as an assembly

3. If transmission disassembly is necessary, remove the drain plug, and drain the transmission fluid. Install the drain plug after draining all of the fluid.

4. Remove or disconnect the following:
  - Electrical connector from the reverse lamp switch
  - Electrical connector from the vehicle speed sensor (VSS)
  - Heated oxygen sensor (HO2S) electrical connector from the bracket
  - Wiring harness from the bracket
  - Electrical connectors from the heated oxygen sensors (HO2S)
  - Starter motor

➡The driveshaft centering socket yoke fits tightly on the rear axle pinion flange pilot. Never hammer on the driveshaft or any of its components to

disconnect the yoke from the flange. Pry only in the area shown, with a suitable tool, to disconnect the yoke from the flange.

➡If equipped, always disconnect the front driveshaft from the transfer case first. Otherwise, the weight of the driveshaft can cause the boot to tear.

- Rear driveshaft, and the front driveshaft, if so equipped
- Bolts retaining the exhaust inlet crossover pipe to the exhaust manifold
- Bolts retaining the catalytic converter to the muffler. Discard the exhaust converter outlet gasket.
- Exhaust hanger from the insulator. Position the exhaust assembly aside.
- On 4-wheel drive vehicles, the transfer case
- Clutch hydraulic line from the clutch slave cylinder

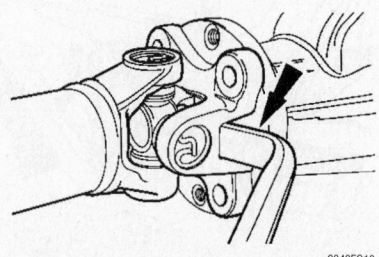

Pry here for driveshaft removal

✳✳ WARNING

Secure the transmission to the jack with a suitable safety strap. Failure to follow these instructions may result in personal injury.

5. Using a suitable transmission jack, support the transmission. Secure the transmission to the jack with a suitable safety strap.

6. Loosen, but do not remove the nuts retaining the transmission insulator to the crossmember.

7. Remove the six bolts retaining the crossmember to the frame.

8. Remove the nuts and the crossmember.

➡Lower the transmission enough to gain access to the upper bolts retaining the transmission to the engine.

9. Remove the nine bolts retaining the transmission to the engine.

10. Remove the transmission from the vehicle.

11. Installation is the reverse of removal. Observe the following torques:
  - Transmission-to-engine bolts: 44 ft. lbs. (60Nm)
  - Crossmember-to-frame: 46 ft. lbs. (63Nm)
  - Transmission insulator-to-crossmember: 72 ft. lbs. (98Nm)

2004

2-WHEEL DRIVE

1. Before servicing the vehicle, refer to the precautions in the beginning of this section.

2. Remove the upper gearshift lever, the outer gearshift lever boot and the console as an assembly.

3. With the vehicle in NEUTRAL, position it on a hoist.

4. If transmission disassembly is required, remove the drain plug and drain the transmission fluid. Install the drain plug after draining all the fluid.

5. To maintain initial driveshaft balance, index-mark the driveshaft yoke to the axle flange, so they can be installed in their original positions. Remove the rear driveshaft.

6. Disconnect the wire harness from the crossmember.

7. Place a suitable jack under the transmission. Secure the transmission to the jack with a safety strap.

8. Remove the six crossmember bolts.

9. Remove the transmission mount nuts and the crossmember.

10. Remove the heated oxygen senor (HO2S) bracket nut and the bracket from the extension housing.

11. Remove the transmission mount bolts and the transmission mount.

12. For 3.0L and 4.0L engines, remove the catalytic converter Y-pipe.

13. Disconnect the vehicle speed sensor (VSS) electrical connector and the reverse lamp switch electrical connector. Then unclip the wiring harness from the transmission.

14. Remove the starter motor. Using mechanics wire, position the starter aside.

15. Using the special tool, disconnect the clutch hydraulic line.

16. Using a suitable jack, support the engine.

17. Lower the transmission enough to gain access to the upper transmission-to-engine bolts. Remove the nine transmission-to-engine bolts.

18. Pull the transmission rearward until the input shaft is clear of the pressure plate, then lower the transmission from the vehicle.

To install:

19. Before securing the engine to the transmission, connect the hydraulic line to the clutch slave cylinder.

20. Install the transfer case with a new gasket.

21. Tighten the bolts that retain the transfer case to the extension housing in a clockwise direction beginning with the upper LH bolt.

22. Install the front driveshaft with new bolts and washers and the rear driveshaft with new bolts.

23. Align the index marks when installing the front and rear driveshafts.

24. Check and, if necessary, fill the transmission with the specified type and quantity of fluid.

25. To install, reverse the removal procedure. Observe the following torques:

- Drain plug: 36 ft. lbs. (48 Nm)
- Crossmember bolts: 72 ft. lbs. (98Nm)
- Transmission mount nuts: 72 ft. lbs. (98 Nm)
- Oxygen sensor bracket nut: 29 ft. lbs. (39 Nm)
- transmission mount bolts: 73 ft. lbs. (99 Nm)
- Transmission-to-engine bolts: 44 ft. lbs. (60 Nm)

4-WHEEL DRIVE

1. Before servicing the vehicle, refer to the precautions in the beginning of this section.

2. Remove the upper gearshift lever, the outer gearshift lever boot and the console as an assembly.

3. With the vehicle in NEUTRAL, position it on a hoist.

4. Remove the skid plate, if equipped.

5. If transmission disassembly is required, remove the drain plug and drain the transmission fluid. Install the drain plug after draining all the fluid.

6. To maintain initial driveshaft balance, index-mark the front output shaft and the front driveshaft constant velocity (CV) joint. Index-mark the rear driveshaft yokes to the axle flange and on the transfer case flange.

7. Remove the transfer case.

8. Disconnect the wire harness from the crossmember.

9. Position a suitable jack under the transmission. Secure the transmission to the jack with a safety strap.

10. Remove the six crossmember bolts.

11. Remove the transmission mount nuts and the crossmember

12. Remove the heated oxygen senor (HO2S) bracket nut and the bracket from the extension housing.

13. Remove the catalytic converter Y-pipe.

14. Disconnect the vehicle speed sensor (VSS) electrical connector and the reverse lamp switch electrical connector. Then unclip the wiring harness from the transmission.

15. Remove the starter motor. Using mechanics wire, position the starter aside.

16. Using the special tool, disconnect the clutch hydraulic line.

17. Using a suitable jack, support the engine.

18. Lower the transmission enough to gain access to the upper transmission-to-engine bolts. Remove the nine transmission-to-engine bolts.

19. Pull the transmission rearward until the input shaft is clear of the pressure plate, then lower the transmission from the vehicle.

To install:

20. Before securing the engine to the transmission, connect the hydraulic line to the clutch slave cylinder.

21. Install the transfer case with a new gasket.

22. Tighten the bolts that retain the transfer case to the extension housing in a clockwise direction beginning with the upper LH bolt.

23. Install the front driveshaft with new bolts and washers and the rear driveshaft with new bolts.

24. Align the index marks when installing the front and rear driveshafts.

25. Check and, if necessary, fill the transmission with the specified type and quantity of fluid.

26. To install, reverse the removal procedure. Observe the following torques:

- Skid plate: 18 ft. lbs. (24 Nm)
- Drain plug: 36 ft. lbs. (48 Nm)
- Crossmember bolts: 46 ft. lbs. (63 Nm)
- Transmission mount nuts: 72 ft. lbs. (98 Nm)
- Oxygen sensor bracket: 29 ft. lbs. (39 Nm)
- Transmission-to-engine bolts: 44 ft. lbs. (60Nm)

Automatic Transmission

2001–03

1. Before servicing the vehicle, refer to the precautions in the beginning of this section.

2. Place the selector lever in NEUTRAL position.

3. Remove or disconnect the following:

- Negative battery cable
- Fluid level indicator
- The two bolts retaining the fan shroud to the radiator.

➡**If transmission disassembly is required, drain the transmission fluid.**

- With 4wd, the transfer case

➡️**Mark the driveshaft yoke and axle flange, so they may be installed in their original alignment.**

- Rear driveshaft
- Starter motor
- Torque converter access cover

➡️**Mark the torque converter and the flexplate for correct alignment at reinstallation.**

- The four converter nuts
- Shift cable
- Transmission wiring harness
- Three way catalytic converter
- Left HO$_2$S sensor
- Front exhaust crossover pipe
- Transmission cooler lines

4. Position a High-Lift Jack under the transmission. Raise and support the transmission.
5. Remove or disconnect the following:
- Crossmember.
- Transmission mount
- Transmission upper fill tube

➡️**Lower the High-Lift Transmission Jack to gain access to screws.**

- On 4x4 models, the vent tube assembly

✳✳ WARNING

Install the Torque Converter Holding Tool before lowering the transmission from the vehicle. Secure the transmission to the transmission jack with a safety chain. Failure to follow these instructions can result in personal injury.

6. Lower the transmission.
7. Installation is the reverse of removal. Observe the following torques:
- Transmission-to-engine bolts: 41 ft. lbs. (55Nm)
- Exhaust bracket bolts: 81 ft. lbs. (110Nm)
- Crossmember-to-frame: 87 ft. lbs. (118Nm)
- Transmission mount-to-crossmember: 81 ft. lbs. (110Nm)
- Converter-to-flexplate: 30 ft. lbs. (40Nm)
- Rear driveshaft-to-flange bolts: 95 ft. lbs. (129Nm)

2004

1. Before servicing the vehicle, refer to the precautions in the beginning of this section.

➡️**If the transmission is to be removed for a period of time, support the engine with a safety stand and a wood block.**

2. With the vehicle in NEUTRAL, position it on a hoist.

➡️**When the battery has been disconnected and reconnected, some abnormal drive symptoms can occur while the vehicle relearns its adaptive strategy.**

3. Disconnect the battery ground cable
4. On 4.0L SOHC vehicles, remove the fluid level indicator tube bolt and remove the tube and indicator.
5. If transmission disassembly is required, drain the transmission fluid.
6. On 4x4 vehicles, remove the transfer case.
7. To maintain initial driveshaft balance, mark the driveshaft yoke and axle flange so they can be installed in their original alignment.
8. Remove the rear driveshaft.
   a. Remove the four bolts.
   b. Remove the driveshaft.
9. Remove the starter motor.
10. With a 2.3L engine:
   a. When removing the torque converter nuts, the crankshaft must be rotated only in the clockwise direction, otherwise engine damage can occur. The crankshaft, crankshaft sprocket and the pulley are fitted together by friction between the flange faces on each part. For that reason, the crankshaft sprocket can also be moved when the crankshaft pulley is turned in the counterclockwise direction.
   b. It may be necessary to gain access to the flexplate nuts through the wheel well.
   c. Mark the torque converter and the flexplate for correct alignment at reinstallation.
   d. Remove and discard the four torque converter nuts. Rotate the flexplate to access to all the nuts.
11. With 3.0L and 4.0L SOHC engines:
   a. Mark the torque converter and the flexplate for correct alignment at reinstallation. Remove the four nuts. Rotate the flexplate to access to all the nuts.
   b. Disconnect the shift cable.
   c. Disconnect the transmission wiring harness from the case.
12. Disconnect the transmission wiring harness. Remove the three way catalytic converter.
13. With a 2.3L engine, remove the rear engine cover plate.

➡️**Care should be taken not to bend or damage the cooler lines.**

14. Hold the case fitting and remove the transmission cooler lines.

15. Remove the nuts.
16. Position a transmission jack under the transmission. Raise and support the transmission.
17. Remove the crossmember.
18. Remove the transmission mount.
19. With a 2.3L engine, remove the rear vibration damper.
20. 2.3L and 3.0L engines, remove the transmission upper fill tube.
21. Lower the jack to gain access to screws. Remove the transmission-to-engine bolts.
22. With a 2.3L engine, remove the lower screws.
23. Remove the HO$_2$S connector bracket from the transmission.
24. On 4x4 vehicles, remove the vent tube assembly.

✳✳ CAUTION

The torque converter is heavy and may result in injury if it falls out of the transmission. Secure the torque converter in the transmission. Failure to follow these instructions may result in personal injury. Install a converter locking tool before lowering the transmission from the vehicle.

✳✳ CAUTION

Secure the transmission to the transmission jack with a safety chain. Failure to follow these instructions may result in personal injury.

25. Lower the transmission.
26. If the transmission is being overhauled or if installing a new or remanufactured transmission, carry out transmission fluid cooler backflushing and cleaning.

To install:

27. On 4x4 vehicles, install the vent tube assembly.

✳✳ CAUTION

Secure the transmission to the transmission jack with a safety chain. Failure to follow these instructions may result in personal injury.

28. Raise and position the transmission.
29. Remove the holding tool.
30. With the 4.0L SOHC engine:
   a. Align the flexplate to the converter marks made at removal.
   b. Install the transmission-to-engine screws. Torque to 35 ft. lbs. (48 Nm).
31. With the 2.3L and 3.0L engines

➡**Align the flexplate to the converter marks made at removal.**

a. Install the transmission-to-engine screws. Torque to 35 ft. lbs. (48 Nm).

b. Install the upper fluid filler tube and bracket screw.

32. With the 2.3L engine:

a. Install the lower transmission-to-engine screws. Torque to 35 ft. lbs. (48 Nm).

b. Install the HO2S connector bracket.

c. Install the rear engine cover plate.

33. On 4x4 vehicles, install the transfer case.

34. Install the exhaust bracket. Torque the bolts to 73 ft. lbs. (99 Nm).

35. Install the crossmember. Tighten the bolts to 74 ft. lbs. (101 Nm).

36. Install the transmission mount into the crossmember and torque the nuts to 73 ft. lbs. (99 Nm).

37. With 3.0L engines, install the rear vibration damper. Torque the nuts to 22 ft. lbs. (30 Nm).

38. With 3.0L and 4.0L SOHC engines:

➡**Prior to installing the cooler lines to the case, inspect the O-rings. If damaged new O-rings will need to be installed.**

a. Hold the case fitting and install the transmission cooler lines. Torque to 19 ft. lbs. (28 Nm).

b. Install four new torque converter nuts. Rotate the crankshaft as needed to gain access to all the nuts. Torque to 26 ft. lbs. (35 Nm).

39. With the 2.3L vehicles: Install four new torque converter nuts. Rotate the crankshaft as needed to gain access to all the nuts. Torque to 26 ft. lbs. (35 Nm).

➡**When installing the torque converter nuts, the crankshaft must be rotated only in the clockwise direction, otherwise engine damage can occur. The crankshaft, the crankshaft sprocket and the pulley are fitted together by friction between the flange faces on each part. For that reason, the crankshaft sprocket can also be moved when the crankshaft pulley is turned in the counterclockwise direction.**

40. Install the starter motor.

41. Install the catalytic converter assembly.

42. Position the transmission wiring harness in place.

43. Connect the transmission wiring harness.

44. Install the shift cable.

45. Align the driveshaft yoke and the axle shaft marks made at removal to maintain driveline balance. Install the rear driveshaft. Install the driveshaft bolts. Torque to 83 ft. lbs. (112 Nm).

46. Use the following guidelines for installing the in-line transmission fluid filter:

a. If the transmission was overhauled and the vehicle was equipped with an in-line fluid filter, install a new in-line fluid filter.

b. If the transmission was overhauled and the vehicle was not equipped with an in-line fluid filter, install a new in-line fluid filter kit.

c. If the transmission is being installed for a non-internal repair, do not install an in-line filter or filter kit.

d. If installing a new or re-manufactured transmission, install the in-line transmission fluid filter that is supplied.

e. Prior to lowering the vehicle, install a new in-line transmission filter or a filter kit.

47. With the 4.0L SOHC engine, install the transmission fill tube and indicator as an assembly.

➡**When the battery has been disconnected and reconnected, some abnormal drive symptoms can occur while the vehicle relearns its adaptive strategy.**

48. Connect the battery ground cable.

49. Fill the transmission with clean automatic transmission fluid to the specified level.

50. Check the transmission for correct operation.

51. Verify that the shift cable is correctly adjusted.

Clutch

REMOVAL & INSTALLATION

1. Before servicing the vehicle, refer to the precautions in the beginning of this section.

2. Remove or disconnect the following:
   - Negative battery cable
   - Transmission

➡**If the clutch disc and pressure plate are to be reinstalled, bolts must be removed evenly or permanent damage to the diaphragm spring will occur resulting in complete clutch release.**

   - Bolts, clutch pressure plate and the clutch disc.

➡**If the parts are to be reused, index-mark the clutch pressure plate to the flywheel.**

To installation:

3. Lubricate the transmission input shaft pilot bearing with front axle grease.

4. Using a suitable press, press downward on the pressure plate fingers until the adjusting ring moves freely.

5. Rotate the adjusting ring counterclockwise to compress the tension springs. Hold the adjusting ring in this position.

6. Release the pressure on the fingers.

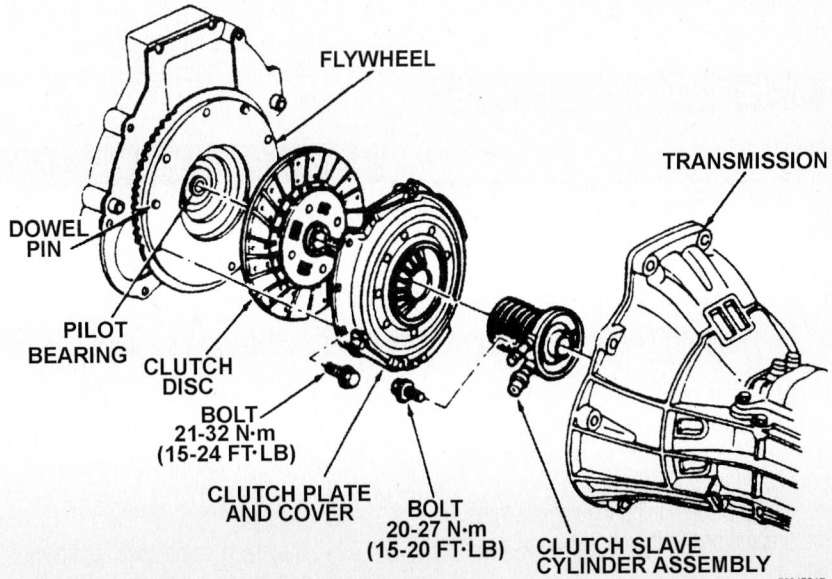

Clutch disc, pressure plate and bearing assembly

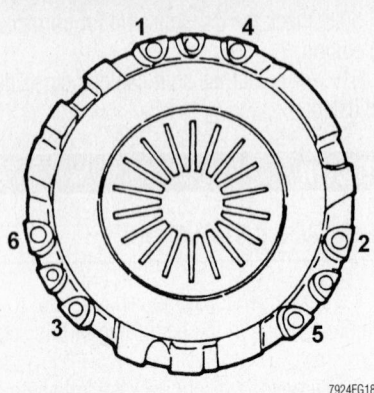

7924EG18

Tighten the bolts gradually in the correct sequence to avoid warping the pressure plate

The adjusting ring will stay in the reset position.

7. Position the clutch disc on the flywheel.

➡ **If reusing the clutch pressure plate and flywheel, align the marks made during removal.**

8. Align the clutch disc and the clutch pressure plate. Install the bolts and tighten in a star pattern sequence to 24 ft. lbs. (35Nm) for 2001–02; 20 ft. lbs. (27 Nm) for 2003–04 models.
- Install the transmission.

ADJUSTMENT

Because the clutch is hydraulically driven, there is no adjustment required.

In the event the clutch pedal develops a squeak or uneven feel when depressing, spray the pedal bushing assembly with penetrating oil and work the pedal back-and-forth.

Hydraulic Clutch System

BLEEDING

The following procedure is recommended for bleeding the clutch hydraulic system installed on the vehicle. It is recommended that the original clutch tube, with quick-connect fitting be replaced when servicing the hydraulic system, because air can be trapped in the quick-connect fitting and prevent complete bleeding of the system. The replacement tube does not include a quick-connect fitting.

1. Before servicing the vehicle, refer to the precautions in the beginning of this section.
2. Clean the dirt and grease from the dust cap.

3. Remove the cap and diaphragm and fill the reservoir to the top with approved brake fluid C6AZ-19542-AA or BA, (ESA-M6C25-A).

➡ **To keep brake fluid from entering the clutch housing, route a suitable rubber tube of appropriate inside diameter from the bleed screw to a container.**

4. Loosen the bleed screw, located in the slave cylinder body, next to the inlet connection. Fluid will now begin to move from the master cylinder down the tube to the slave cylinder.

➡ **The reservoir must be kept full at all times during the bleeding operation, to ensure no additional air enters the system.**

5. Observe the bleed screw outlet. When the slave cylinder is full, a steady stream of fluid will flow from the outlet port. Tighten the bleed screw.
6. Depress the clutch pedal to the floor and hold for 1–2 seconds. Release the pedal as rapidly as possible. The pedal must be released completely. Pause for 1–2 seconds. Repeat 10 times.
7. Check the fluid level in the reservoir. The fluid should be level with the step when the diaphragm is removed.
8. Hold the pedal to the floor, slightly open the bleed screw to allow any additional air to escape. Close the bleed screw, then release the pedal.
9. Check the fluid in the reservoir. The hydraulic system should now be fully bled, and should actuate the clutch.
10. Check the vehicle by starting, pushing the clutch pedal to the floor and selecting reverse gear. There should be no grating of gears. If there is, and the hydraulic system still contains air; repeat the bleeding procedure.

Transfer Case Assembly

REMOVAL & INSTALLATION

2001–03

1. Before servicing the vehicle, refer to the precautions in the beginning of this section.
2. Place the transmission in neutral.
3. Remove or disconnect the following:
- Skid plate
- Damper
- Transfer case harness connector and position it aside
4. If transfer case disassembly is neces-

sary, remove the drain plug and drain the fluid.

➡ **Index-mark the front output shaft assembly and the front driveshaft constant velocity (CV) joint. Always disconnect the front driveshaft from the transfer case first. Otherwise, the weight of the driveshaft can pinch the boot between the shaft and the boot can and cause the boot to tear.**

- Front driveshaft from the transfer case and position the driveshaft aside. Remove and discard the bolts and washers.

➡ **Index-mark the front flange on the rear driveshaft and the flange on the transfer case.**

- Rear driveshaft

➡ **Secure the transfer case to the jack with safety straps.**

5. Position a high lift jack under the transfer case.
6. Remove or disconnect the following:
- Five bolts retaining the transfer case to the extension housing
- Transfer case rearward and off of the transmission output shaft
7. Remove and discard the front extension housing gasket and clean the mating surfaces.

To install:

8. Installation is the reverse of removal. Take note of the following:
- Install the transfer case with a new gasket.
- Tighten the bolts that retain the transfer case to the extension housing in a clockwise direction beginning with the upper LH bolt.
- Install the front and the rear driveshafts with new bolts. If new bolts are not available, coat the threads of the original bolts with Thread-lock and Sealer E0AZ-19554-AA or equivalent meeting Mazda specification WSK-M2G351-A5.
- When installing the front driveshaft, always connect it to the axle first and then connect it to the transfer case.
- Align the index marks when installing the front and rear driveshafts.
9. Observe the following torques:
- Nut retaining the flange to the rear output shaft: 262 ft. lbs. (355Nm)
- Bolt retaining the rear driveshaft to the flange: 82 ft. lbs. (111Nm)
- Bolt retaining the motor assembly

and connector to the transfer case cover: 89 inch lbs. (10Nm)
- Bolt retaining the skid plate to the frame: 18 ft. lbs. (24Nm)
- Bolt retaining the damper to the transfer case: 30 ft. lbs. (40Nm)
- Bolt retaining the driveshaft CV joint to the front output shaft assembly: 22 ft. lbs. (30Nm)
- Bolt retaining the transfer case to the extension housing: 40 ft. lbs. (54Nm)
- Bolt retaining the front adapter to the transfer case: 30 ft. lbs. (40Nm)
- Bolt retaining the transfer case to the transfer case cover: 27 ft. lbs. (36Nm)
- Drain plug: 18 ft. lbs. (24Nm)
- Fill plug: 18 ft. lbs. (24 Nm)

2004

1. Before servicing the vehicle, refer to the precautions in the beginning of this section.

2. With the vehicle in NEUTRAL, raise and support the vehicle.

3. Remove the skid plate.

4. Remove the damper, if so equipped.

5. Disconnect the transfer case harness connector and position it aside.

6. If transfer case disassembly is necessary, remove the drain plug and drain the fluid. Install the drain plug when all of the fluid has drained.

➡**Index-mark the front output shaft assembly and the front driveshaft constant velocity (CV) joint.**

✳✳ WARNING

Always disconnect the front driveshaft from the transfer case first. Otherwise, the weight of the driveshaft can pinch the boot between the shaft and the boot can and cause the boot to tear.

7. Index-mark the front output shaft assembly and the front driveshaft constant velocity (CV) joint.

8. Remove and discard the bolts and washers.

9. Disconnect the front driveshaft from the transfer case and position the driveshaft aside.

➡**Index-mark the front flange on the rear driveshaft and the flange on the transfer case.**

10. Remove the rear driveshaft.

✳✳ CAUTION

Secure the transfer case to the jack with safety straps.

11. Position a high lift jack under the transfer case.

12. Remove the five bolts retaining the transfer case to the extension housing.

13. Slide the transfer case rearward and off of the transmission output shaft.

14. Remove and discard the front extension housing gasket, and clean the mating surfaces.

To install:

15. Install the transfer case with a new gasket.

16. Tighten the bolts that retain the transfer case to the extension housing in a clockwise direction beginning with the upper LH bolt.

17. Install the front driveshaft with new bolts and washers and the rear driveshaft with new bolts. If new bolts are not available, coat the threads of the original bolts with Threadlock and Sealer E0AZ-19554-AA, or equivalent.

➡**When installing the front driveshaft, always connect it to the axle first and then connect it to the transfer case.**

➡**Align the index marks when installing the front and rear driveshafts.**

18. The remainder of installation is the reverse of the removal procedure.

19. Check and, if necessary, fill the

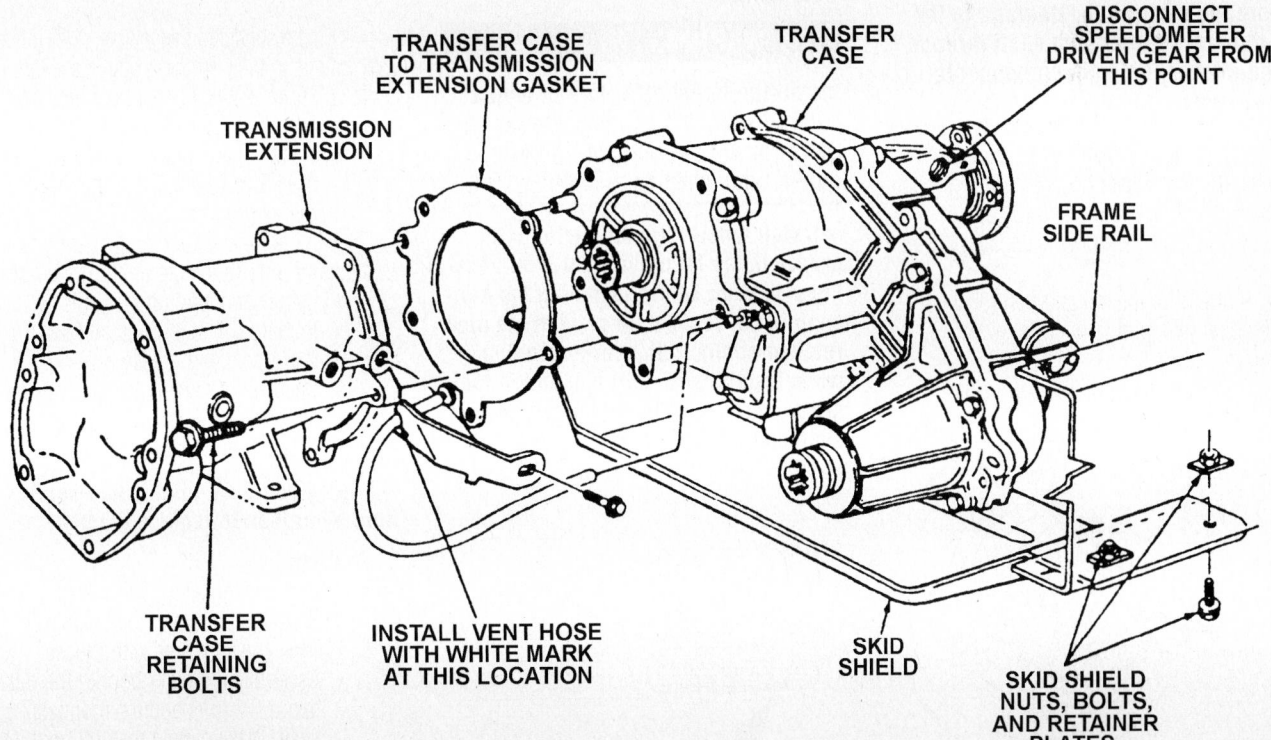

Exploded view of the 13-54 electronic shift transfer case-to-transmission mounting

7924EG20

transfer case with the specified type and quantity of fluid.

Halfshaft

REMOVAL & INSTALLATION

1. Before servicing the vehicle, refer to the precautions in the beginning of this section.

2. With the vehicle in NEUTRAL, raise and support the vehicle.

3. Remove the front wheel and tire assembly.

➡ Do not reuse the torque prevailing design hub nut and washer assembly.

4. Remove and discard the hub nut and washer assembly.

❉ WARNING

Do not allow the disc brake caliper to hang suspended from the brake hose. Provide a suitable support.

5. Remove the front disc brake caliper, anchor plate, and pads as an assembly, and position the assembly aside.

6. Remove the brake disc.

❉ WARNING

Do not use a hammer to separate the outboard front wheel halfshaft joint from the wheel hub. Damage to the outboard CV joint stub shaft threads and internal CV joint components may result.

7. Using the special tool, separate the outboard front wheel halfshaft joint from the wheel hub. Remove the special tool.

8. Support the front suspension lower arm.

9. Remove the nut and bolt retaining the upper ball joint to the front wheel knuckle.

10. Rotate the front wheel knuckle.

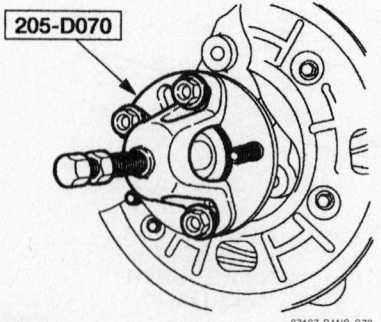

205-D070

Halfshaft removal tool

67197-RANG-G72

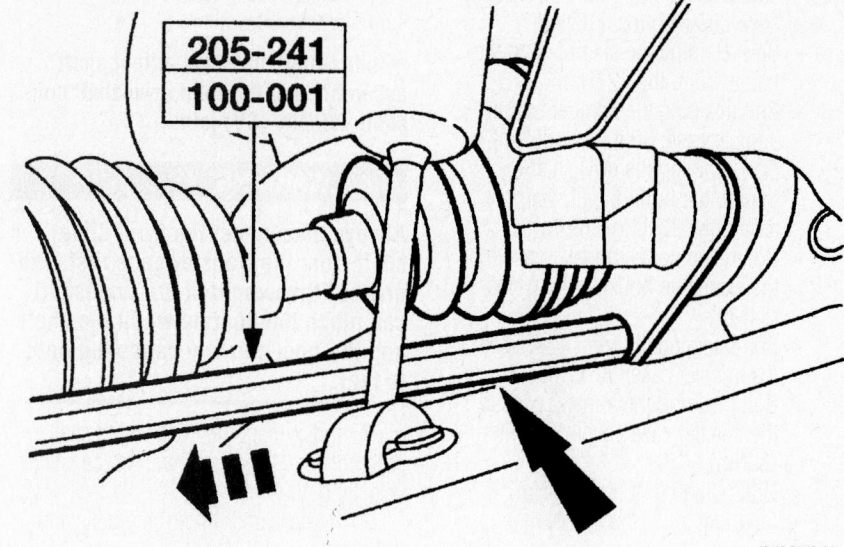

205-241
100-001

Separating the halfshaft from the axle housing

67197-RANG-G73

11. Compress the outboard front wheel halfshaft joint.

12. Remove the outboard front wheel halfshaft joint from the wheel hub.

13. Using the special tools, 205-241 and 100-001, or equivalent, separate the inboard front wheel halfshaft joint from the front axle housing.

14. Remove the halfshaft assembly from the vehicle with both hands. Do not damage the axle seal.

To install:

❉ WARNING

Install the halfshaft with a new hub nut and washer assembly. Do not use power or impact tools to tighten the hub nut and washer assembly.

➡ Install a new retainer circlip in the groove in the LH inboard CV joint housing stub shaft before installing the halfshaft in the vehicle. To prevent the new retainer circlip from over-expanding when installing it, start one end in the groove and work the circlip over the shaft and into the groove.

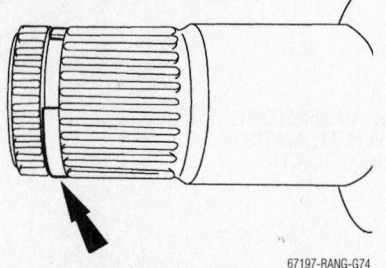

Circlip installed

67197-RANG-G74

15. To install, reverse the removal procedure. Observe the following torques:
- Hub nut: 162 ft. lbs. (220 Nm)
- Upper ball joint-to-knuckle nut: 41 ft. lbs. (55 Nm)

CV-Joints

OVERHAUL

2001–03

1. Before servicing the vehicle, refer to the precautions in the beginning of this section.

2. Remove or disconnect the following:
- Negative battery cable
- Halfshaft and place it in a vice with the inboard joint lower than the outboard joint

3. Cut the inner boot clamps with side cutters and remove the clamp from the boot.
- Larger boot end off the joint
- Inboard CV-joint bolts and separate the spacer and grease cap
- Snap-ring retaining the interconnecting shaft end to the CV-joint cage
- CV-joint and discard the washer

➡ The outboard CV-joint is non-serviceable other than to replace the boot.

To install:

4. Install or connect the following:
- Slide the boot over the shaft

5. Fill the CV-joint area with grease.
- Assemble the outer boot to the outboard CV-joint and interconnecting shaft. Make certain that the boot is seated in the grooves on the outer race and on the shaft

- New clamps to the boot
- New inner boot to the shaft
- New washer to the end of the shaft
- Assemble the inboard CV-joint to the interconnecting shaft spline until it rests on the washer
- Snap-ring

6. Fill the CV-joint area with grease.

- Boot into position and make certain that it is seated in the grooves on the boot adapter and the shaft
- New clamps and tighten the clamps with crimping pliers
- Spacer to the CV-joint end pilot. Torque the bolts to 25 ft. lbs. (34 Nm).
- Halfshaft
- Negative battery cable

2004

1. Before servicing the vehicle, refer to the precautions in the beginning of this section.

2. Remove the front wheel halfshaft. Do not damage the halfshaft boot.

3. Remove the two inboard boot clamps.

4. Slide the inboard halfshaft boot off the inboard CV joint housing.

5. Separate the CV joint from the CV joint housing.

6. Index-mark the shaft and the inboard CV joint for correct alignment during assembly.

7. Remove the snap ring.

8. Remove the CV joint.

9. Remove the inboard halfshaft boot from the shaft assembly.

10. Remove the front wheel excluder seal, if necessary. Discard the seal. Tap uniformly around the seal to separate it from the joint.

11. Remove the two outboard boot clamps.

12. Remove the outboard halfshaft boot.

13. If the grease is contaminated, clean and inspect the joint for wear. Install a new outboard CV joint and shaft assembly if worn/damaged.

14. Inspect the assembly for contaminated grease.

To assemble

15. Pack the outboard CV joint with grease. Use Mazda High Temp Constant Velocity Joint Grease E43Z-19590-A or equivalent meeting Mazda specification ESP-M1C207-A. Spread any remaining grease from the service kit evenly inside the outboard halfshaft boot.

16. Clean the halfshaft boot mounting surfaces of excess grease before positioning the halfshaft boot into place.

17. Position the outboard halfshaft boot.

18. Position the boot clamps on the outboard halfshaft boot.

19. Tighten the through-bolt until the installer is in the closed position.

20. Install the outboard CV joint boot clamps. There are special tools made for this procedure.

21. Position the boot clamp on the halfshaft.

22. Position the inboard halfshaft boot.

23. Align the index marks on the halfshaft and the CV joint.

24. Install the CV joint on the halfshaft.

25. Install the snap ring.

26. Lubricate the three CV joint needle bearings. Use Mazda High Temp Constant Velocity Joint Grease E43Z-19590-A or equivalent meeting Mazda specification ESP-M1C207-A.

27. Fill the inboard CV joint housing with 235 grams (8.3 oz.) of grease. Use

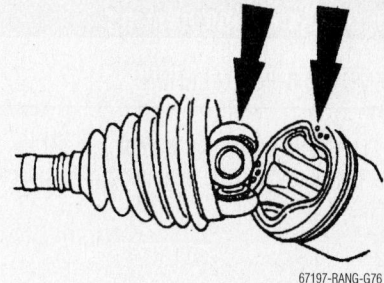

67197-RANG-G76

Separating the joint from the housing

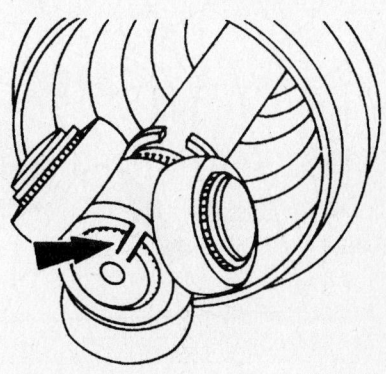

67197-RANG-G77

Make an alignment mark for reassembly

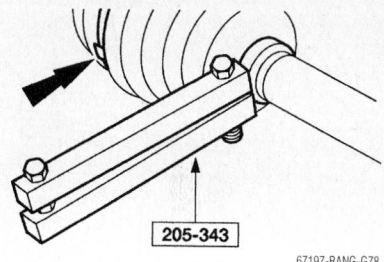

205-343

67197-RANG-G78

Boot clamp crimping tool

Mazda High Temp Constant Velocity Joint Grease E43Z-19590-A or equivalent meeting Mazda specification ESP-M1C207-A.

28. Position the CV joint housing onto the CV joint.

29. Remove any excess grease from the inboard halfshaft boot mating surface before positioning it into place.

30. Position the inboard halfshaft boot into place.

31. Position the boot clamp.

32. Insert a dulled screwdriver blade to relieve built-up air pressure in the halfshaft boot.

33. Using the special tool, install the inboard boot clamps.

34. Using the special tool, install the new front wheel excluder seal, if removed. Seat the metal ring at the seal's inner diameter flat against the CV joint housing.

35. Install the front wheel halfshaft.

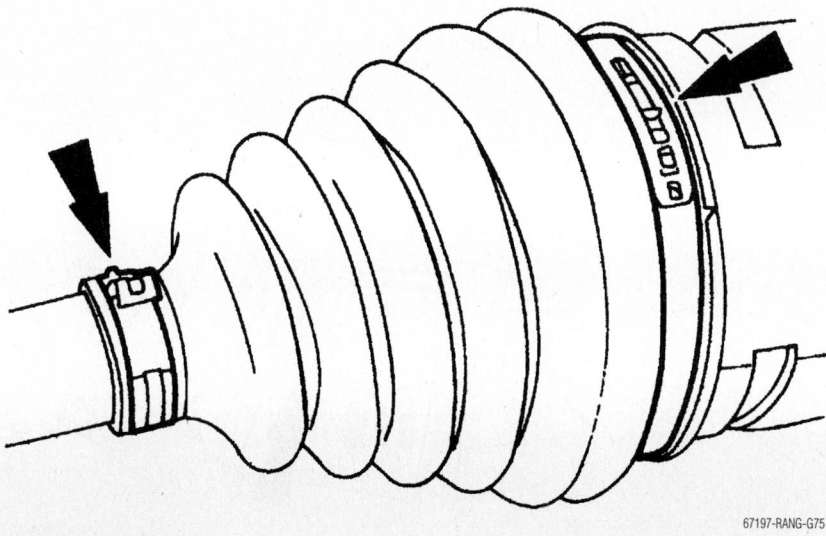

67197-RANG-G75

Inboard boot clamps

Front Axle Tube Bearing

REMOVAL & INSTALLATION

1. Before servicing the vehicle, refer to the precautions in the beginning of this section.

2. Remove or disconnect the following:
- Right-hand halfshaft
- Right-hand axle shaft
- Axle seal, with a slide hammer
- Axle tube bearing, with a slide hammer

3. Clean the bearing and seal surfaces of any foreign debris.

To install:

4. Use an axle bearing replacer and the handle to replace the RH axle tube bearing.

5. Check the bearing depth as shown.

6. Use an axle seal replacer and the handle to replace the axle tube seal.

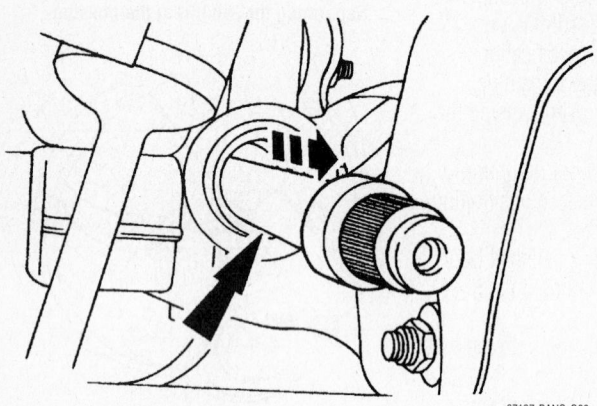

Right side axle shaft removal

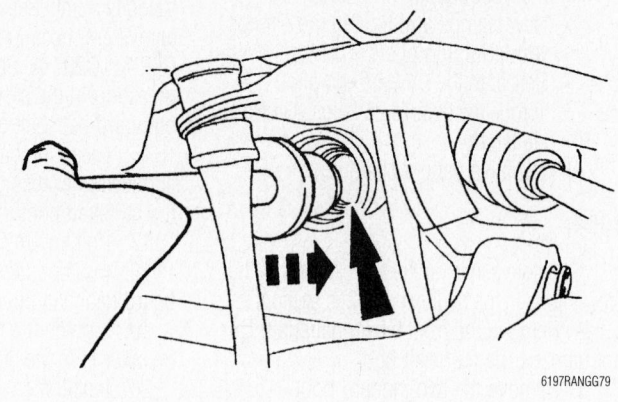

Axle tube bearing installation

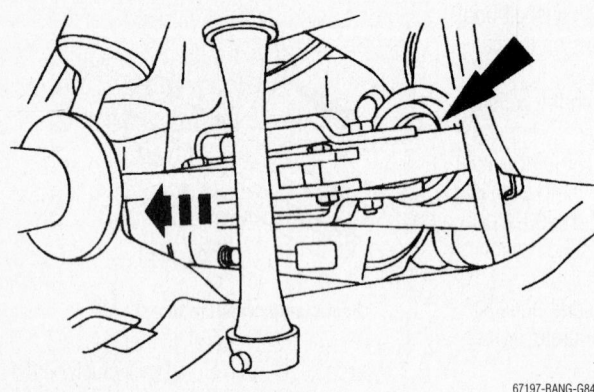

Axle seal removal

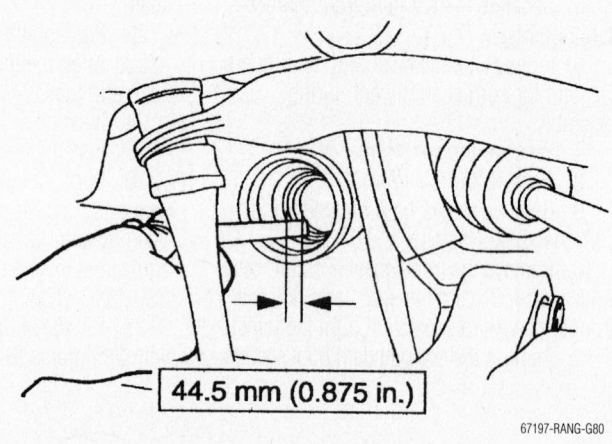

44.5 mm (0.875 in.)

Axle tube bearing depth

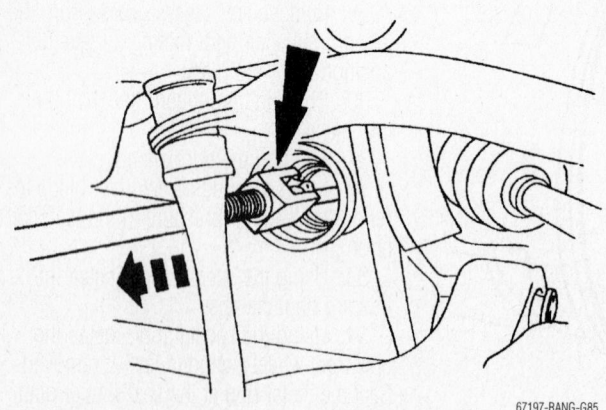

Axle tube bearing removal

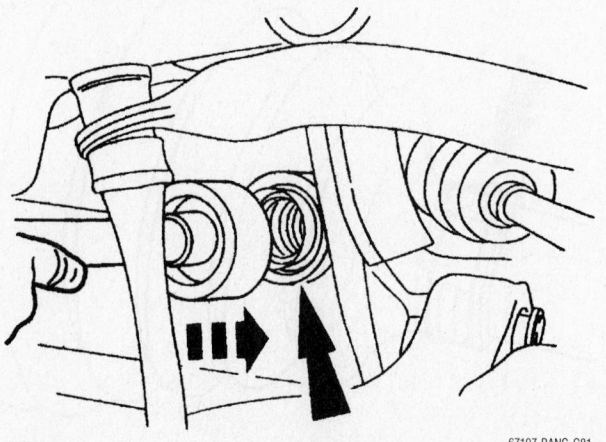

Axle tube seal installation

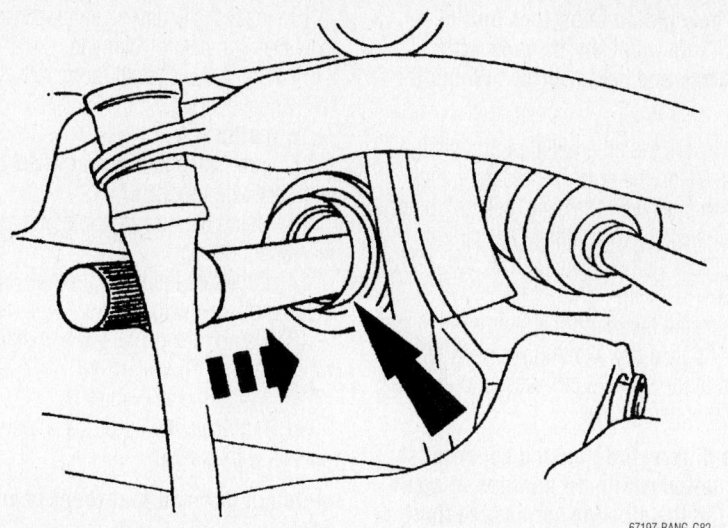

Axle shaft installation

67197-RANG-G82

➡ **Care should be taken not to damage the axle seal surface.**

7. Install the axle shaft.

8. Refill the front drive axle to proper level using SAE 80W90.

9. Install the RH halfshaft.

Rear Axle Shaft, Bearing and Seal

REMOVAL & INSTALLATION

2001–03

1. Before servicing the vehicle, refer to the precautions in the beginning of this section.

2. Drain the axle housing fluid.

3. Remove or disconnect the following:
   - Negative battery cable
   - Rear wheel
   - Brake drum
   - Wheel speed sensor, if equipped
   - Axle housing cover
   - Bearing retainer nuts
   - Axle shaft and bearing
   - Axle shaft inner oil seal

4. If equipped with ABS, grind a flat spot on the wheel speed sensor tone ring, then split the ring with a chisel.

5. Press the wheel bearing off the axle shaft.

6. Remove the bearing retainer and the outer oil seal.

To install:

7. Install or connect the following:
   - Outer oil seal to the bearing retainer
   - Bearing retainer to the axle shaft
   - Bearing and retainer ring pressed onto the axle shaft

   - Wheel speed sensor tone ring pressed onto the axle shaft, if equipped
   - Axle shaft inner oil seal
   - Axle shaft and bearing
   - Bearing retainer nuts. Tighten them to 17 ft. lbs. (23 Nm).
   - Wheel speed sensor, if equipped
   - Brake drum
   - Rear wheel
   - Negative battery cable

8. Fill the rear differential to the correct level.

2004

7½ INCH RING GEAR

1. Before servicing the vehicle, refer to the precautions in the beginning of this section.

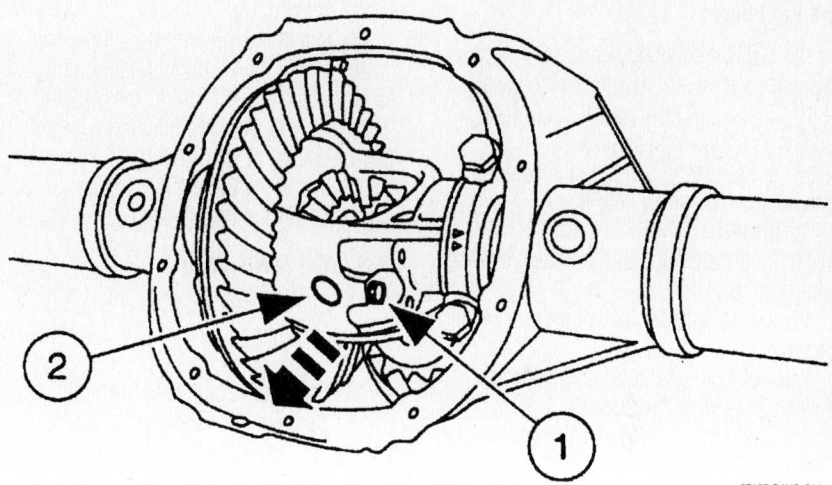

67197-RANG-G90

Differential pinion shaft removal. 1-lock bolt; 2-pinion shaft

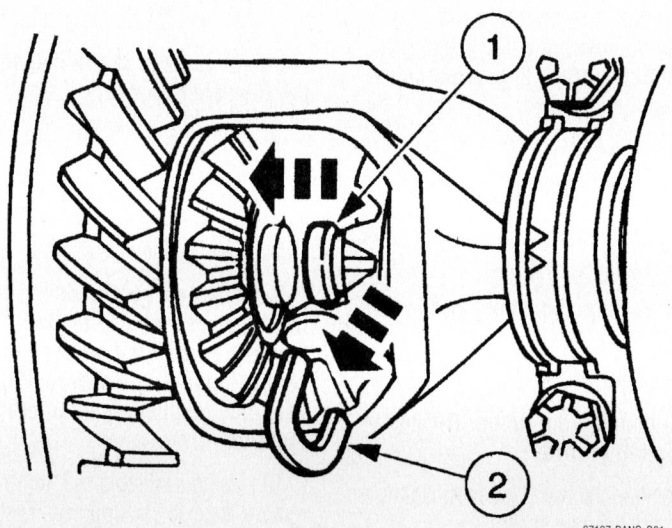

67197-RANG-G91

Removing the U-washers. 1-axle shaft; 2-U-washer

2. Raise and support the vehicle.

3. Remove the wheel and tire assembly.

4. Remove the 10 differential housing cover bolts and drain the lubricant from the rear axle housing.

5. Remove the differential housing cover.

6. Remove the rear brake drums.

7. Remove and discard the differential pinion shaft lock bolt.

8. Remove the differential pinion shaft.

➡ **Do not damage the rubber O-rings in the axle shaft grooves.**

9. Push the axle shafts inboard.

10. Remove the U-washers.

➡ **Do not damage the wheel bearing oil seal.**

11. Remove the two axle shafts.

➡ **If only a new seal needs to be installed, use care to avoid damaging the seal bore.**

12. Using a suitable seal remover, remove the axle shaft oil seal. Discard the oil seal.

13. Inspect the rear wheel bearing and axle shaft for wear or damage.

14. If necessary, using a slidehammer, remove the rear wheel bearing.

To install:

15. Lubricate the new rear wheel bearing with lubricant.

16. Using a driver, install the rear wheel bearing.

17. Lubricate the lip of the new wheel bearing oil seal with grease.

18. Using a driver, install the wheel bearing oil seal.

➡ **Make sure the machined surfaces on both the rear axle housing and the differential housing cover are clean and free of oil before installing the new silicone sealant. The inside of the rear axle must be covered when cleaning the machined surface to prevent contamination.**

19. Clean the gasket mating surface of the rear axle and the differential housing cover.

20. Lubricate the lip of the wheel bearing oil seal with grease.

➡ **Do not damage the wheel bearing oil seal.**

21. Install the axle shafts.

➡ **Do not damage the rubber O-rings in the axle shaft grooves.**

22. Position the two U-washers on the button end of the axle shafts.

23. Pull the axle shafts outward.

➡ **If a new pinion shaft lock bolt is unavailable, coat the threads with threadlock and sealer prior to installation.**

24. Install the differential pinion shaft.

a. Align the hole in the differential pinion shaft with the case lock bolt hole.

b. Install a new differential pinion shaft lock bolt. Torque to 20 ft. lbs. (33 Nm).

25. Install the rear brake drums.

26. Apply a new continuous bead of sealant of the specified thickness to the differential housing cover.

➡ **The differential housing cover must be installed within 15 minutes of application of the silicone, or new sealant must be applied. If possible, allow one hour before filling with lubricant to make sure the silicone sealant has correctly cured.**

27. Install the different housing cover.

28. Install the 10 differential housing cover bolts. Torque to 33 ft. lbs. (45Nm).

29. Fill the rear axle housing with 2.4 liters (5 pints) of lubricant.

30. Install the wheels and tires.

31. Lower the vehicle.

8.8 INCH RING GEAR

1. Before servicing the vehicle, refer to the precautions in the beginning of this section.

2. Raise and support the vehicle.

3. Remove the rear wheel and tire assembly.

4. Remove the brake drum.

5. Remove the differential housing cover and drain the lubricant.

6. Remove and discard the pinion shaft bolt.

7. Remove the differential pinion shaft.

➡ **Do not damage the rubber O-ring in the U-washer groove.**

8. Push the axle shaft inboard.

9. Remove the U-washer.

10. Do not damage the wheel bearing oil seal.

11. Remove the axle shaft.

➡ **If only a new seal needs to be installed, use care to avoid damaging the seal bore. If the wheel bearing oil seal is leaking, the differential housing vent may be plugged with foreign material.**

12. Using a suitable seal remover, remove the axle shaft oil seal. Discard the oil seal.

13. Inspect the rear wheel bearing and axle shaft for wear or damage.

14. Using the special tools, remove the rear wheel bearing.

To install:

15. Lubricate the new rear wheel bearing with rear axle lubricant.

16. Using the special tools, install the rear wheel bearing.

17. Lubricate the lip of the new wheel bearing oil seal with grease.

18. Using the special tools, install the wheel bearing oil seal.

19. Install the axle shaft.

20. Lubricate the lip of the wheel bearing oil seal with grease.

➡ **Do not damage the wheel bearing oil seal.**

21. Install the axle shaft.

22. Do not damage the rubber O-ring in the U-washer groove.

23. Position the U-washer on the button end of the axle shaft.

24. Pull the axle shaft outward.

➡ **If a new bolt is unavailable, coat the bolt threads with threadlock prior to installation.**

25. Align the bolt hole in the differential pinion shaft with the bolt hole in the case.

26. Install the new bolt. Torque to 22 ft. lbs. (33 Nm).

27. Install the brake drum.

28. Install the differential housing cover and fill the differential housing with the specified lubricant.

29. Install the rear wheel and tire assembly.

30. Lower the vehicle.

Front Pinion Seal

REMOVAL & INSTALLATION

➡ **This operation disturbs the differential pinion bearing preload. Carefully reset the preload during assembly.**

✳✳ CAUTION

The electrical power to the air suspension system must be shut off prior to hoisting, jacking or towing an air suspension vehicle. This can be accomplished by turning off the air suspension switch located in the rear jack storage area. Failure to do so can result in unexpected inflation or deflation of the air springs, which can result in shifting of the vehicle during these operations.

1. Before servicing the vehicle, refer to the precautions in the beginning of this section.

2. Index-mark the front driveshaft and pinion flange.

3. Remove or disconnect the following:
   - Front driveshaft from the pinion flange, and position it aside

➡**Do not allow the driveshaft to hang unsupported.**

4. Using a Nm (inch-pound) torque wrench, measure the torque required to maintain pinion rotation. Record the measurement.

5. Index-mark the pinion flange and the pinion stem.

6. Hold the pinion flange while removing the nut.

7. Place a drain pan under the differential housing.

8. Using a puller, remove the pinion flange.

9. Inspect the pinion flange for burrs and damage. Inspect the end of the pinion flange that contacts the bearing cone, the nut counterbore, and the seal surface for nicks. Discard the pinion flange as necessary.

10. Using a seal remover and impact slide hammer, remove the pinion seal.

11. Remove the front axle drive pinion shaft oil slinger and the differential pinion bearing.

12. Remove and discard the collapsible spacer.

To install:

13. Verify that the splines on the pinion stem are free of burrs. If burrs are evident, remove them with a fine crocus cloth. Work in a rotating motion to wipe the pinion clean.

14. Clean the pinion seal bore.

15. Install a new collapsible spacer.

16. Install the original differential pinion bearing and the front axle drive pinion shaft oil slinger.

17. Lubricate the pinion seal. Use Motorcraft SAE 80W90 Thermally Stable 4x4 Axle Lubricant meeting Mazda specification WSP-M2C197-A.

18. Install the pinion seal.

19. Lubricate the pinion flange splines. Use Motorcraft SAE 80W90 Thermally Stable 4x4 Axle Lubricant meeting Mazda specification WSP-M2C197-A.

➡**Never use a metal hammer on the pinion flange or install the flange with power tools. If necessary, use a plastic hammer to tap on a tight fitting flange.**

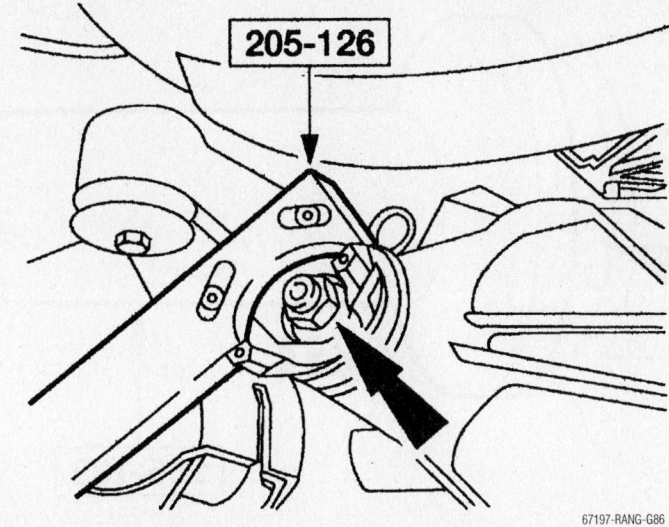

67197-RANG-G86

Holding the front axle pinion flange

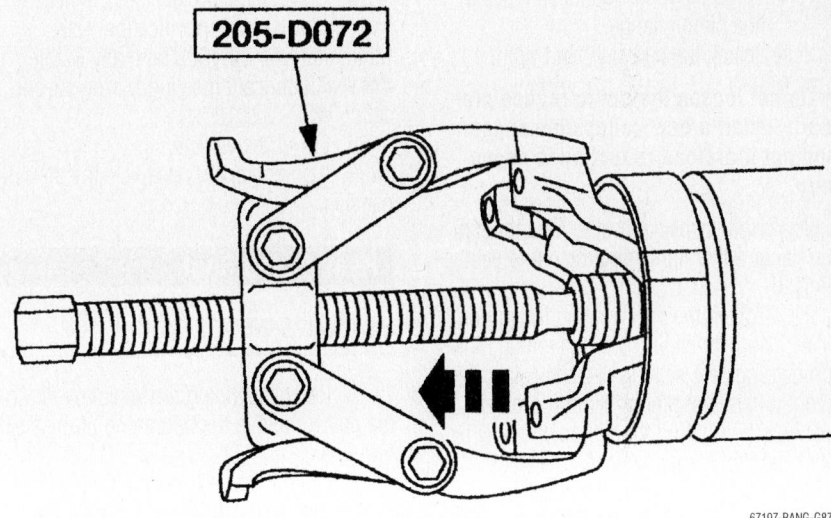

67197-RANG-G87

Removing the front axle pinion flange

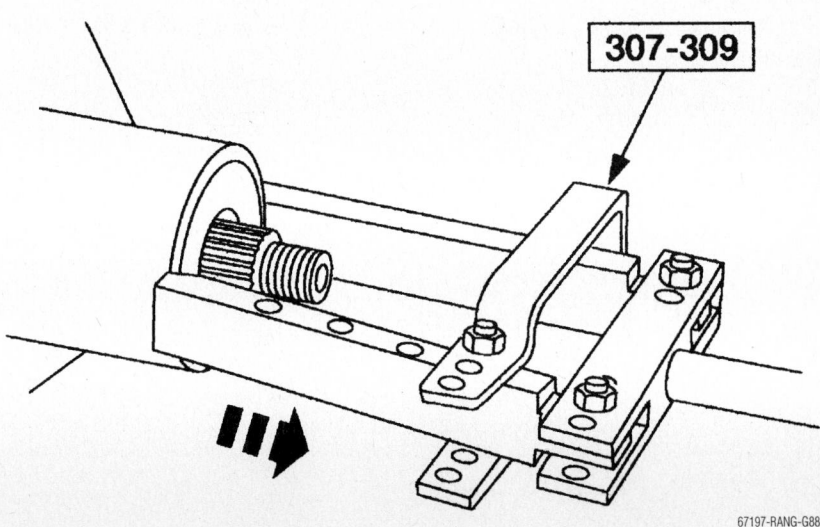

67197-RANG-G88

Removing the front axle pinion seal

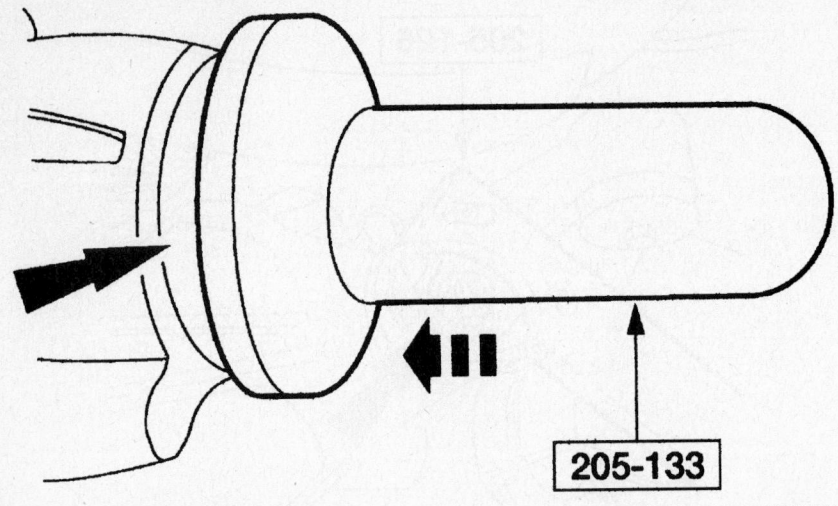

205-133

67197-RANG-G89

Installing the front axle pinion seal

- Align the index marks and install the pinion flange.
- Install the new nut hand-tight.

➡**Do not loosen the nut to reduce preload. Install a new collapsible spacer and nut if preload reduction is necessary.**

20. Use the special tool to hold the pinion flange while tightening the nut to set the preload.

21. Tighten the nut, rotating the pinion occasionally to ensure the differential pinion bearings are seating correctly. Take frequent differential pinion bearing preload readings by rotating the pinion with a Nm (inch-pound) torque wrench. The final reading must be 0.56 Nm (5 inch lbs.) more than the initial reading taken during removal.

22. Align the index marks and position the front driveshaft.

23. Install the universal joint spider retainers and bolts.

24. Check the fluid level and, if necessary, fill the axle to specification. Use Motorcraft SAE 80W90 Thermally Stable 4x4 Axle Lubricant meeting Mazda specification WSP-M2C197-A.

25. Lower the vehicle.

26. If so equipped, reactivate the air suspension.

Rear Pinion Seal

REMOVAL & INSTALLATION

1. Before servicing the vehicle, refer to the precautions in the beginning of this section.

2. Drain the axle housing fluid.

3. Remove or disconnect the following:
- Negative battery cable
- Rear wheels
- Driveshaft
- Brake calipers and pads or brake drum

➡**The brake calipers and pads or brake drum must be removed so that there is no additional drag when measuring pinion bearing preload.**

4. Use an inch lb. torque wrench and measure and record the amount of torque required to maintain pinion rotation through several revolutions.

5. Remove or disconnect the following:
- Pinion flange
- Pinion seal
- Pinion bearing
- Collapsible spacer

To install:

➡**Use a new collapsible spacer and flange nut for assembly.**

6. Install or connect the following:
- Collapsible spacer
- Pinion bearing
- Pinion seal
- Pinion flange

7. Rotate the pinion flange occasionally while tightening the flange nut to make sure the pinion bearings seat correctly.

8. Take frequent bearing preload torque readings. Tighten the flange nut to achieve the preload torque readings originally recorded.

✳✳ CAUTION

Never loosen the pinion nut to reduce bearing preload. If it is necessary to reduce bearing preload, install a new collapsible spacer and pinion nut.

9. Install or connect the following:
- Driveshaft
- Brake calipers and pads or brake drum
- Wheels
- Negative battery cable

10. Fill the differential with gear lubricant and check for leaks.

STEERING AND SUSPENSION

Air Bag

PRECAUTIONS

- Always wear safety glasses when servicing an air bag vehicle, and when handling an air bag.
- Never attempt to service the steering wheel or steering column on an air bag-equipped vehicle without first properly disarming the air bag system. The air bag system should be properly disarmed whenever ANY service procedure in this manual indicates that you should do so.
- When carrying a live air bag module, always make sure the bag and trim cover are pointed away from your body. In the unlikely event of an accidental deployment, the bag will then deploy with minimal chance of injury.
- When placing a live air bag on a bench or other surface, always face the bag and trim cover up, away from the surface. This will reduce the motion of the air bag if it is accidentally deployed.
- If you should come in contact with a deployed air bag, be advised that the air bag surface may contain deposits of sodium hydroxide, which is a product of the gas combustion and is irritating to the skin. Always wear gloves and safety glasses when handling a deployed air bag, and wash your hands with mild soap and water afterwards.

DISARMING THE SYSTEM

1. Before servicing the vehicle, refer to the precautions in the beginning of this section.
2. Disconnect the negative battery cable from the battery.
3. Disconnect the positive battery cable from the battery.
4. Wait 1 minute. This time is required for the back-up power supply in the air bag diagnostic monitor to completely drain. The system is now disarmed.

ARMING THE SYSTEM

1. Before servicing the vehicle, refer to the precautions in the beginning of this section.

2. Connect the positive battery cable.
3. Connect the negative battery cable.
4. Stand outside the vehicle and carefully turn the ignition to the **RUN** position. Be sure that no part of your body is in front of the air bag module on the steering wheel, to prevent injury in case of an accidental air bag deployment.
5. Ensure the air bag indicator light turns off after approximately 6 seconds. If the light does not illuminate at all, does not turn off, or starts to flash, test the system.

Power Rack and Pinion Steering Gear

REMOVAL & INSTALLATION

2001–02

> **※ WARNING**
>
> **If equipped, always turn off the Automatic Ride Control (ARC) service switch before lifting the vehicle off of the ground. Failure to do so could damage the ARC system components.**

1. Before servicing the vehicle, refer to the precautions in the beginning of this section.
2. Raise and safely support the front of the vehicle, block the rear wheels and apply the parking brake.
3. Start the engine then rotate the steering wheel from lock-to-lock and record the number of rotations.
4. Divide the number of rotations by 2. This gives the number of rotations to achieve true center of the steering. Turn the wheel in one direction to the full lock.
5. Turn the wheel in the opposite direction the number of turns equal to true steering (lock-to-lock number divided by 2).

> **※ WARNING**
>
> **Do not rotate the steering wheel when the shaft is disconnected from the steering gear as damage to the clock spring could occur.**

6. Drain the power steering fluid reservoir.
7. Remove or disconnect the following:
   - Negative battery cable
   - Bolt retaining the lower steering column shaft to the steering gear input shaft

- Stabilizer bar
- Quick-connect fittings for the power steering pressure and return hoses at the steering gear housing
- Nuts securing the power steering cooler and remove the cooler
- Outer tie rod ends
- Nuts, bolts and washer assemblies retaining the steering gear housing to the front crossmember
- Steering gear from the vehicle

To install:

8. Install or connect the following:
   - Position the steering gear to the front crossmember and install the nuts, bolts and washer assemblies. Torque to 94–127 ft. lbs. (128–172 Nm).
   - Power steering cooler retaining bolts
   - Power steering lines to the steering gear housing and torque the fittings to 20–25 ft. lbs. (27–34 Nm).
   - Outer tie rod ends and ensure that the steering shaft or gear input shaft has not been rotated
   - Intermediate shaft-to-steering input shaft retaining (pinch) bolt and torque the bolt to 30–42 ft. lbs. (41–56 Nm).
   - Negative battery cable
9. Fill the power steering pump reservoir.
10. Bleed the air from the power steering system.
11. Ensure that there are no leaks and the fluid is maintained at the proper level.
12. Check the alignment.

2003–04

2-WHEEL DRIVE

1. Before servicing the vehicle, refer to the precautions in the beginning of this section.
2. Turn the wheel to the straight-ahead position and turn the ignition switch to the OFF position.
3. Remove the front wheel and tire assemblies.
4. Remove the fluid cooler.
5. Remove and discard the cotter pins and nuts.

➡ **Do not damage the tie-rod boot when installing the special tool.**

6. Using special tool, separate the tie-rod ends from the wheel knuckles.

➡**Do not allow the intermediate shaft to rotate while it is disconnected from the steering gear or damage to the clockspring can result. If there is evidence that the intermediate shaft has rotated, the clockspring must be removed and recentered.**

7. Remove the pinch bolt and detach the intermediate shaft from the gear. Discard the bolt.

8. Remove the nut and disconnect the lines.

9. Plug or cap the power steering return hose, power steering pressure hose, and the steering gear ports to prevent the entry of dirt.

➡**Hold the tops of the steering gear to crossmember stud bolts to avoid damaging the steering gear fluid transfer tubes.**

10. Remove the rack mounting nuts.

11. Remove the mounting stud, nut, washer and stop assemblies.

12. Remove the steering gear. Clean the mounting surfaces.

13. To install, reverse the removal procedure.

   a. Install new seals on the power steering return hose and power steering pressure hose.

   b. The dished side of the washer faces downward.

   c. Install a new intermediate shaft pinch bolt.

14. Observe the following torques:
   - Pinch bolt: 35 ft. lbs. (48 Nm)
   - Pressure line bracket nut: 18 ft. lbs. (25 Nm)

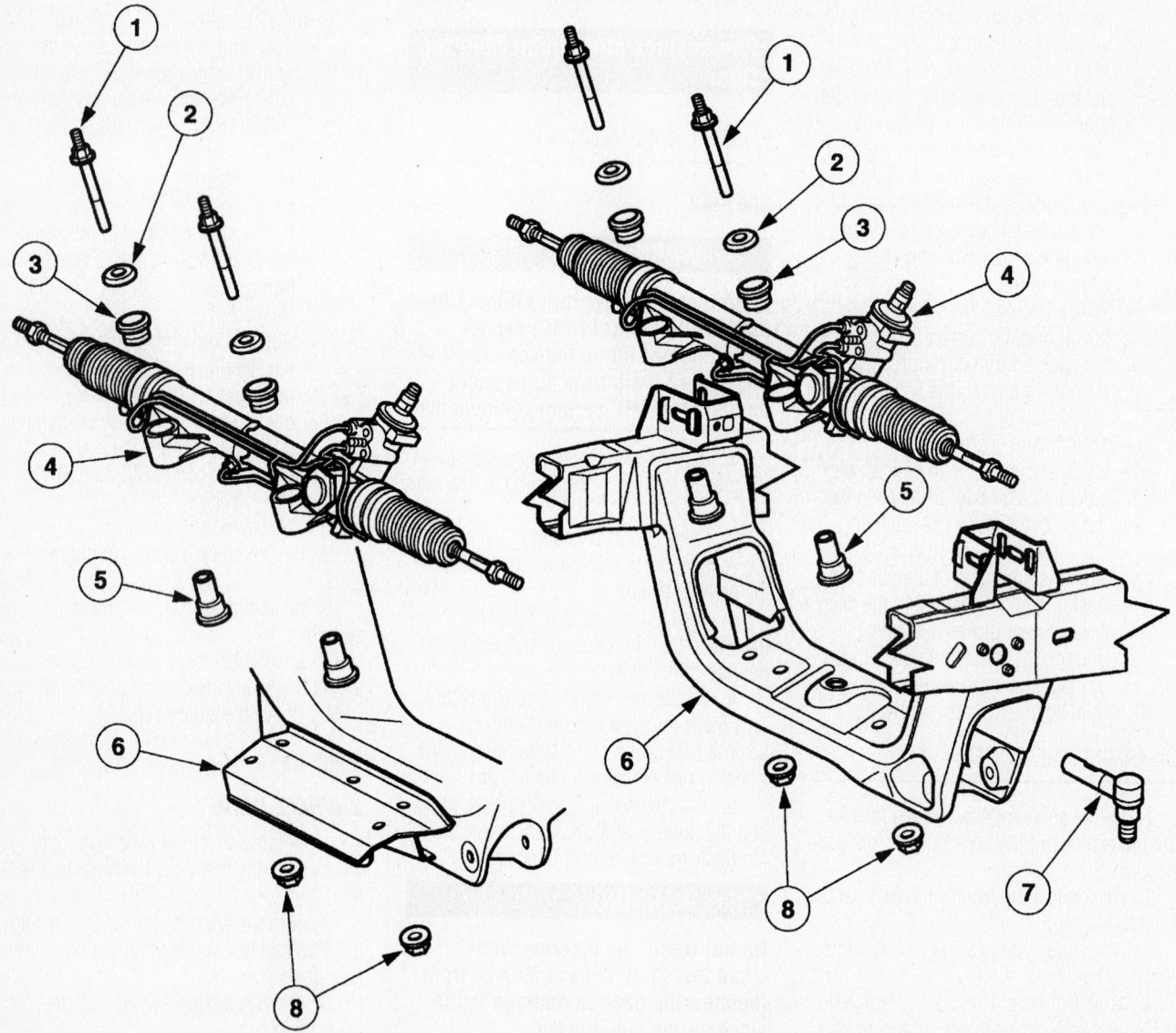

1	Stud	5	Insulator
2	Washer	6	Crossmember
3	Insulator	7	Tie-rod end — outer
4	Steering gear	8	Nut

Steering gear mounting—2-wheel drive and 4-wheel drive, 2003–04 models

67197-RANG-G91A

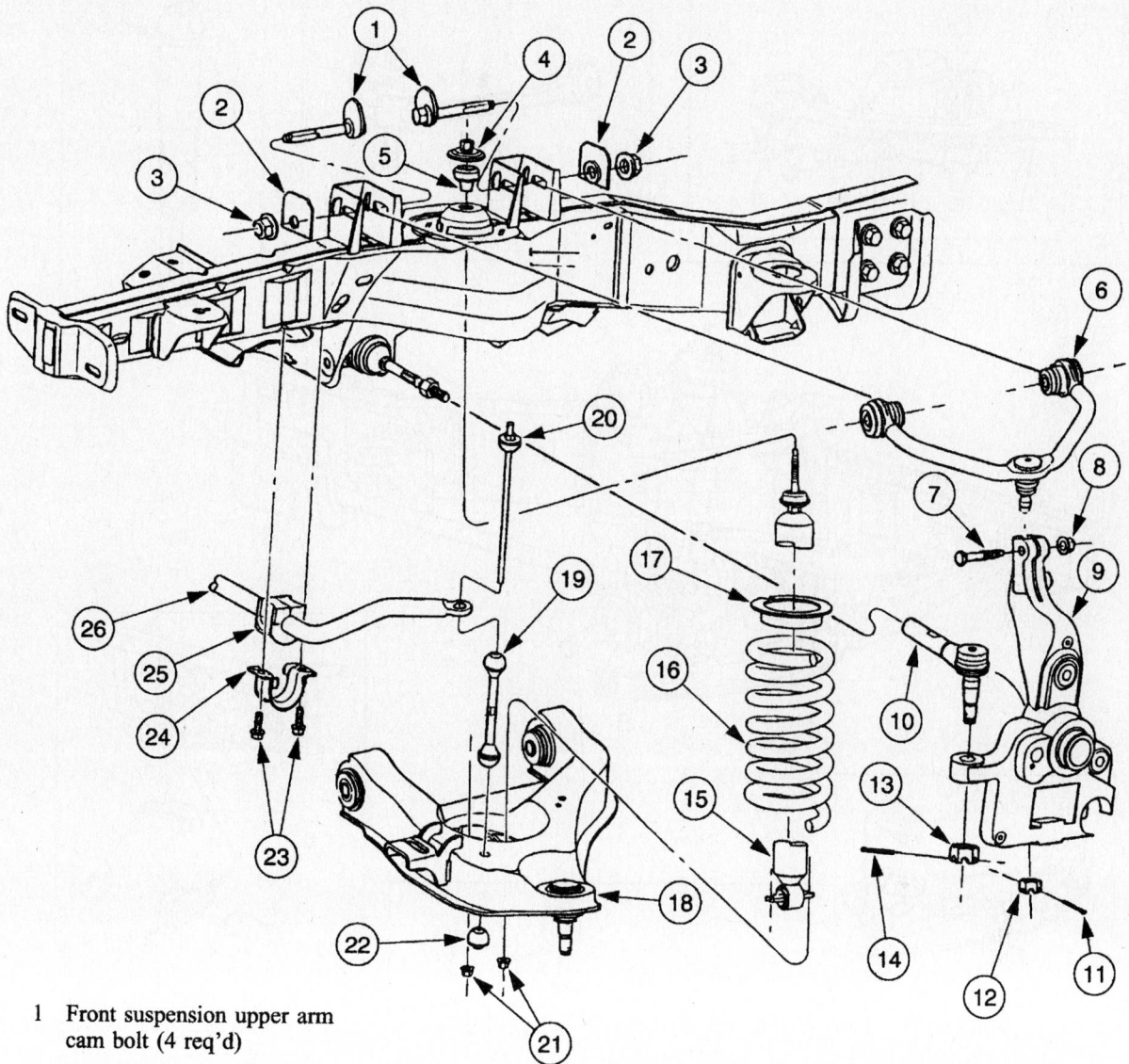

1 Front suspension upper arm
   cam bolt (4 req'd)

2 Front suspension upper arm
   cam assy (2 req'd)

3 Front suspension upper arm
   cam assy nut (2 req'd)

4 Front shock absorber upper
   nut/washer assy
   (2 req'd)

5 Front shock absorber upper
   bushing (2 req'd)

6 Front suspension upper arm

6 Front suspension upper arm

7 Front wheel spindle pinch
   bolt (2 req'd)

8 Front wheel spindle pinch
   bolt nut

9 Front wheel spindle

10 Tie-rod end

11 Cotter pin

12 Lower ball joint castellated
    nut (2 req'd)

13 Tie-rod end castellated nut
    (2 req'd)

14 Cotter pin

15 Front shock absorber

16 Front coil spring

17 Front spring insulator

18 Front suspension lower arm

18 Front suspension lower arm

19 Front stabilizer bar link

20 Front stabilizer bar stud and
    bushing assy (2 req'd)

21 Front shock absorber lower
    nut (4 req'd)

22 Front stabilizer bar nut and
    washer assy (2 req'd)

23 Front stabilizer bar mounting
    bolts (4 req'd)

24 Stabilizer bar bracket
    (2 req'd)

25 Front stabilizer bar bushing
    assy (2 req'd)

26 Front stabilizer bar

67197-RANG-G92

2-wheel drive front suspension—2004 models shown

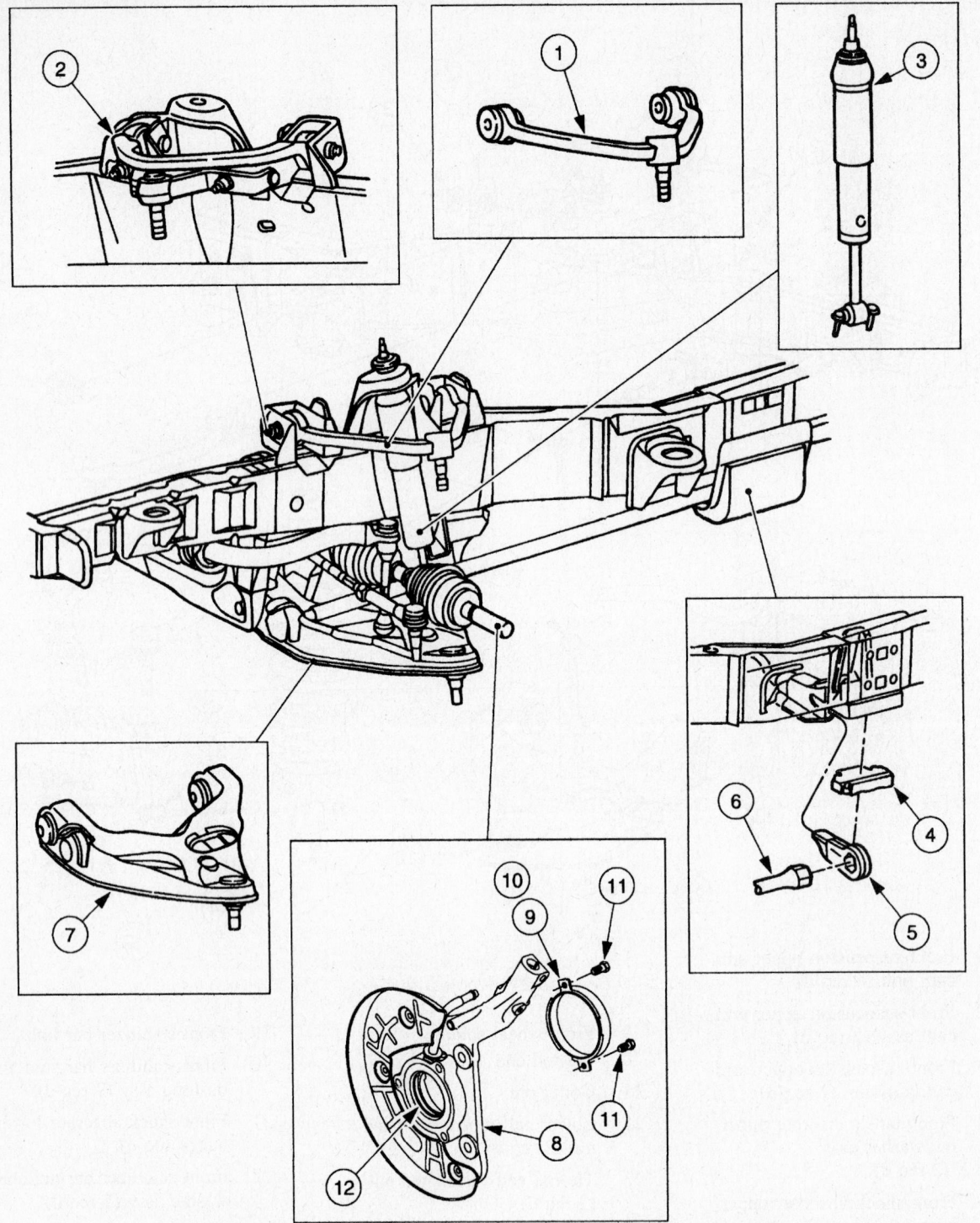

1 Upper arm, bushing and joint assembly (LH)

2 Upper arm, bushing and joint assembly (RH)

3 Shock assembly

4 Torsion bar adjuster plate

5 Torsion bar adjuster

6 Torsion bar (LH)

7 Lower arm, bushing and joint assembly (LH)

8 Knuckle assembly

9 Protection shield (RH)

10 Protection shield (LH)

11 Screw (self-tapping)

12 Oil seal

67197-RANG-G93

4-wheel drive front suspension—2004 models shown

- Rack retaining nuts: 111 ft. lbs. (150 Nm)

15. Fill and leak check the power steering system.

16. Check and, if necessary, adjust the wheel alignment.

4-WHEEL DRIVE

1. Before servicing the vehicle, refer to the precautions in the beginning of this section.

2. Turn the wheel to the straight-ahead position and turn the ignition switch to the OFF position.

3. Remove the wheel and tire assemblies.

4. Remove the fluid cooler.

5. Remove the four air deflector retaining screws.

6. Pull downward on the air deflector to disengage the retaining pins.

7. Loosen the LH tie-rod end jam nut.

8. Remove and discard the cotter pins and nuts.

➡**Do not damage the tie-rod boot when installing the special tool.**

9. Using a separator, separate the tie-rod ends from the wheel knuckles

10. Remove the LH tie-rod end. Count and record the number of turns required to remove the tie-rod end.

11. Remove the front stabilizer bar. Note or mark the driver side end of the sway bar for correct installation.

➡**Do not allow the intermediate shaft to rotate while it is disconnected from the steering gear or damage to the clockspring can result. If there is evidence that the intermediate shaft has rotated, the clockspring must be removed and recentered.**

12. Remove the pinch bolt and detach the intermediate shaft from the gear.

13. Remove the nut and disconnect the lines.

14. Plug the ends of all fluid lines removed and ports in the steering gear to prevent entry of dirt.

➡**Hold the tops of the steering gear to crossmember stud bolts to avoid damaging the steering gear fluid transfer tubes.**

15. Remove the steering rack nuts.

16. Remove the stud bolts and washers.

17. Remove the steering gear to crossmember insulator bushings.

18. Rotate the steering gear control valve housing toward the front of the vehicle.

19. Turn the steering gear input shaft to the right until the stop is reached.

20. Move the steering gear as far to the RH side of the vehicle as possible.

21. Move the LH front wheel spindle tie-rod forward to clear the frame crossmember.

22. Remove the steering gear from the vehicle.

To install:

23. Using special tool 211-027, install new seals on the power steering return hose and power steering pressure hose.

➡**Make sure the steering gear input shaft is turned to the left until the stop is reached.**

➡**Handle the steering gear with caution to avoid damage to fluid transfer tubes and to avoid dimples in tie-rod boots.**

24. Turn the steering gear input shaft to the right until the stop is reached. Note the number of turns required.

➡**Make sure the steering gear control valve housing is turned toward the front of the vehicle.**

25. Install the steering gear into the RH opening of the crossmember.

26. Move the steering gear as far to the RH side of the vehicle as possible.

27. Move the LH front wheel spindle tie-rod into the opening in the crossmember and move the steering gear into position.

28. To place the steering gear in the straight ahead position, turn the steering gear input shaft to the left by half the number of turns recorded previously.

29. Rotate the steering gear control valve housing toward the rear of the vehicle.

30. Install the steering gear to crossmember insulator bushings.

   a. The large end of the metal sleeve must be positioned downward.

   b. Check that the mounting surfaces on the crossmember are clean and free of foreign material.

31. Install the steering gear to crossmember washers and stud bolts. The dished side of the washer faces downward.

➡**Hold the tops of the steering gear to crossmember stud bolts to avoid damaging the steering gear fluid transfer tubes.**

32. Install the rack mounting nuts. Torque to 111 ft. lbs. (150 Nm).

33. Install the lines and tighten the pressure line bracket nut to 18 ft. lbs. (25 Nm).

➡**Do not allow the intermediate shaft to rotate while it is disconnected from the steering gear or damage to the clockspring can result. If there is evi-**

dence that the intermediate shaft has rotated, the clockspring must be removed and recentered.

34. Connect the intermediate shaft to the steering gear input shaft. Install a new lower steering column pinch bolt. Torque to 35 ft. lbs. (48 Nm).

35. Install the power steering fluid cooler.

36. Install the front stabilizer bar. Orient the front stabilizer bar as noted during removal.

37. Install the LH tie-rod end on the front wheel spindle tie-rod. Rotate the tie-rod end the number of turns recorded during removal.

38. Position the tie-rod ends on the steering knuckles. Install the castellated nuts and new cotter pins. Check that the brake dust shields are not bent and are not in contact with the outer tie-rod boot seals. Torque to 52 ft. lbs. (70 Nm).

39. Tighten the tie-rod end jam nut. Torque to 59 ft. lbs. (80 Nm).

40. Position the air deflector, and install the retaining screws.

41. Install the front wheel and tire assemblies.

42. Fill and leak check the system.

43. Check and, if necessary, adjust the wheel alignment.

Shock Absorber

REMOVAL & INSTALLATION

Front

➡**Low pressure gas shocks are charged with nitrogen gas. Do not attempt to open, puncture or apply heat to them. Prior to installing a new shock absorber, hold it upright and extend it fully. Invert it and fully compress and extend it at least 3 times. This will bleed trapped air.**

1. Before servicing the vehicle, refer to the precautions in the beginning of this section.

2. Remove or disconnect the following:
   - Negative battery cable
   - Upper shock-to-frame attaching nut, washer and insulator assembly
   - Lower shock-to-control arm attaching nuts
   - Slightly compress the shock absorber by hand and remove it from the vehicle

To install:

3. Install or connect the following:
   - Position the lower washer and insulator on the shock absorber rod

and position the shock absorber to the upper frame bracket mount
- Position the upper insulator and washer on the shock absorber rod and install the attaching nut loosely.
- Position the lower shock absorber mounting studs into the control arm and install the attaching nuts loosely.
- Torque the lower shock attaching nuts to 15–21 ft. lbs. (21–29 Nm), and the upper shock attaching bolts to 30–40 ft. lbs. (40–55 Nm).
- Negative battery cable

Rear

➡Low pressure gas shocks are charged with nitrogen gas. Do not attempt to open, puncture or apply heat to them. Prior to installing a new shock absorber, hold it upright and extend it fully. Invert it and fully compress and extend it at least 3 times. This will bleed trapped air.

1. Before servicing the vehicle, refer to the precautions in the beginning of this section.
2. Remove or disconnect the following:
   - Upper shock-to-frame attaching nut
   - Lower shock nut
   - Slightly compress the shock absorber by hand and remove it from the vehicle

To install:
3. Install or connect the following:
   - Shock absorber upper end and nut
   - Shock absorber lower end and nut
   - Torque the upper and lower shock attaching nuts to 53 ft. lbs. (72Nm)

Coil Spring

REMOVAL & INSTALLATION

1. Before servicing the vehicle, refer to the precautions in the beginning of this section.
2. Remove or disconnect the following:
   - Wheel and tire assembly
   - Shock absorber
   - Front stabilizer bar link nut
3. Use a coil spring compressor to compress the coil spring.
4. Remove the cotter pin and castellated nut.
5. Separate the lower ball joint from the front wheel spindle.
6. Position the front wheel spindle out of the way and remove the coil spring.

To install:

➡The end of the coil spring must cover the first hole and should not be visible in the second hole.

7. Install the coil spring in the lower arm.

❋❋ WARNING

Always install the cotter pin into the lower ball joint castellated nut from outboard to inboard. Failure to do so will result in damage to the wheel and tire assembly.

8. Install the lower ball joint.
9. Install the front stabilizer bar link nut.
10. Remove the Coil Spring Compressor.
11. Install the front shock absorber and the two lower nuts.
12. Install the upper shock absorber bushing and nut/washer assembly.
13. Install the wheel and tire assembly.

Leaf Springs

REMOVAL & INSTALLATION

1. Before servicing the vehicle, refer to the precautions in the beginning of this section.
2. Remove or disconnect the following:
   - Negative battery cable
   - Rear wheels
   - U-bolts from the rear spring plate
   - Hardware from the spring to bracket at the front of the rear spring
   - Upper and lower shackle bolts at the rear of the spring
   - Spring and shackle from the bracket

To install:
3. Install or connect the following:
   - Spring and shackle to the bracket
   - Upper and lower shackle bolts at the rear of the spring. Torque the nuts to 87 ft. lbs. (118 Nm).
   - U-bolts to the spring plate. Torque the nuts 83 ft. lbs. (113 Nm).
   - Rear wheels
   - Negative battery cable

Torsion Bar

REMOVAL & INSTALLATION

❋❋ CAUTION

The electrical power to the air suspension system must be shut off prior to hoisting, jacking or towing an air suspension vehicle. This can be accomplished by turning off the air

suspension switch located in the rear jack storage area. Failure to do so can result in unexpected inflation or deflation of the air springs or shocks, which can result in shifting of the vehicle during these operations.

1. Before servicing the vehicle, refer to the precautions in the beginning of this section.
2. Remove or disconnect the following:
3. Remove the torsion bar cover plate

➡Before relieving the torsion bar tension, measure and record the measurement of the torsion bar adjustment bolt. This measurement will be used as the preset depth for the new torsion bar adjustment bolt during installation.

4. Relieve the torsion bar tension.
   a. Position the Torsion Bar Tool and adapters.
   b. Tighten the Torsion Bar Tool until the torsion bar adjuster lifts off the adjustment bolt.

❋❋ CAUTION

The torsion bar adjustment bolt is coated with dry adhesive; and must be replaced if it is backed off or removed. Failure to do so can cause the adjustment bolt to loosen during operation and cause a loss of vehicle alignment.

   c. Remove the torsion bar adjustment bolt and nut.
   d. Loosen the Torsion Bar Tool until the tension is removed from the torsion bar.
5. Mark the torsion bar and the adjuster for proper installation.
6. Remove the torsion bar insulator.
7. Grasp the torsion bar, and pull it free from the front suspension lower arm.

To install:
8. Position the torsion bar and the torsion bar adjuster.
9. Align the marks on the torsion bar and the torsion bar adjuster, then install the torsion bar adjuster.
10. Position the torsion bar insulator.
11. Install the Torsion Bar Tool and the adapters.
12. Tighten the Torsion Bar Tool until the new adjustment bolt and nut can be installed.
13. Turn the adjustment bolt until the preliminary adjustment measurement (recorded length of the old adjustment bolt) is reached.

14. Install the torsion bar cover plate. Torque the bolts to 46 ft. lbs. (63Nm).

15. If equipped with air suspension, reactivate the system by turning on the air suspension switch.

16. Lower the vehicle.

17. Adjust the ride height.

18. Check the alignment.

Upper Ball Joint

REMOVAL & INSTALLATION

The ball joints are integral with the control arm. If the ball joint is defective, the entire control arm must be replaced.

Lower Ball Joint

REMOVAL & INSTALLATION

The ball joints are integral with the control arm. If the ball joint is defective, the entire control arm must be replaced.

Upper Control Arm

REMOVAL & INSTALLATION

Coil Spring Suspension

1. Before servicing the vehicle, refer to the precautions in the beginning of this section.

2. Remove or disconnect the following:
   - Wheel and tire assembly
   - Brake disc shield

3. Use a jack to support the front suspension lower arm.

4. Mark the position of the front suspension upper arm adjustment cams.

5. Remove the upper ball joint retaining nut and pinch bolt.

6. Separate the ball joint from the front wheel spindle.

7. Remove the front suspension upper arm.

8. Installation is the reverse of removal. Align the marks made during removal on the front suspension upper arm adjustment cam. The forward front suspension upper arm nut must be tightened first while the arm is held at the curb position ride height. Observe the following torques:
   - Control arm attaching nuts: 98 ft. lbs. (133Nm)
   - Pinch bolt: 46 ft. lbs. (63Nm)

Torsion Bar Suspension

2-WHEEL DRIVE

> ✳✳ **WARNING**
>
> **The electrical power to the air suspension system must be shut off prior to hoisting, jacking or towing an air suspension vehicle. This can be accomplished by turning off the air suspension switch located in the rear jack storage area. Failure to do so can result in unexpected inflation or deflation of the air springs or shocks, which can result in shifting of the vehicle during these operations.**

1. Before servicing the vehicle, refer to the precautions in the beginning of this section.

2. Raise the vehicle on a hoist.

3. Remove the wheel and tire assembly

4. Use a suitable jack stand to support the front suspension lower arm.

➡**To avoid possible damage to the front wheel spindle, secure the spindle to keep it from tilting before removing the pinch bolt and nut.**

5. Remove the pinch bolt and nut from the front wheel spindle.

6. Remove the front suspension upper arm:
   a. Remove the two nuts.
   b. Remove the two bolts.
   c. Remove the front suspension upper arm.

To install:

7. Position the front suspension upper arm.

8. Install the two bolts and two nuts. Torque to 83–112 ft. lbs. (113–153 Nm).

9. Position the front wheel spindle.

10. Install the pinch bolt and nut. Torque to 35–46 ft. lbs. (47–63 Nm).

11. Remove the jack stand from under the front suspension lower arm.

12. Install the tire and wheel assembly

➡**If equipped with air suspension, reactivate the system by turning on the air suspension switch.**

13. Lower the vehicle.

14. Inspect the front end ride height

15. Inspect and adjust the front end alignment

4-WHEEL DRIVE

1. Before servicing the vehicle, refer to the precautions in the beginning of this section.

2. Raise the vehicle on a hoist.

3. Remove the wheel and tire assembly.

4. Use a suitable jack stand to support the front suspension lower arm.

5. Remove the pinch bolt.

> ✳✳ **WARNING**
>
> **Before separating the front suspension upper arm from the front wheel knuckle, secure the front wheel knuckle to prevent it from tilting outward. Failure to do so can cause damage to the front axle shaft.**

6. Separate the front suspension upper arm from the front wheel knuckle.

7. Remove the front suspension upper arm:
   a. Remove the two nuts and alignment plates.
   b. Remove the two bolts and the cams.

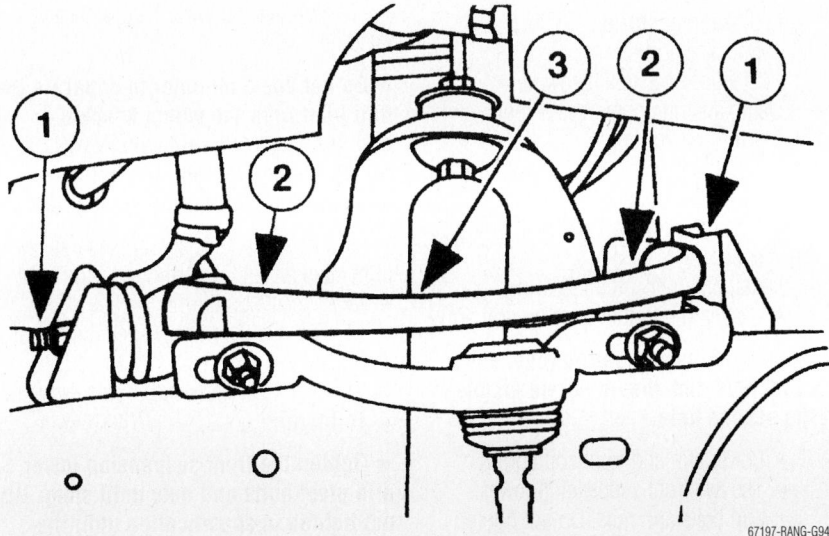

67197-RANG-G94

Upper control arm—2-wheel drive torsion bar suspension; 1-nuts; 2-bolts; 3-arm

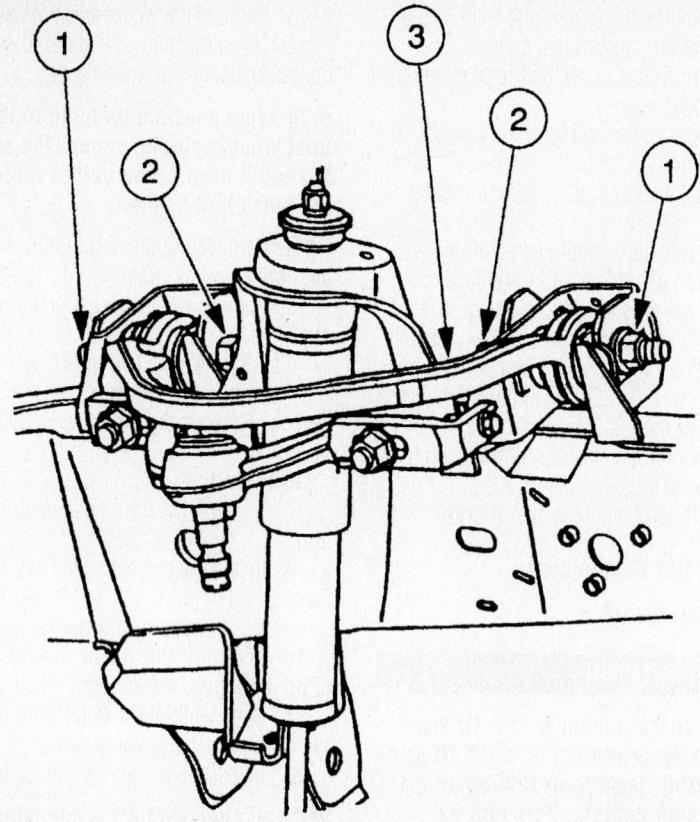

67197-RANG-G95

4-wheel drive upper control arm—2004 model shown

c. Remove the front suspension upper arm.

To Install:

➡**When installing the front suspension upper arm, replace the alignment plates with new alignment cams.**

8. Install the front suspension upper arm.

a. Position the front suspension arm bushing joint.

b. Install the two bolts, four cams and two nuts. Torque to 83–112 ft. lbs. (113–153 Nm).

9. Install the pinch bolt and nut.

a. Position the upper arm into the front wheel knuckle.

b. Install the pinch bolt and nut. Torque to 41 ft. lbs. (55 Nm).

10. Remove the jack stand from under the front suspension lower arm.

11. Install the wheel and tire assembly.

12. Lower the vehicle.

13. Check the wheel alignment.

UPPER CONTROL ARM BUSHING REPLACEMENT

The control arm bushings are not serviceable. If they require service, the upper or lower arm must be replaced.

Lower Control Arm

REMOVAL & INSTALLATION

Coil Spring Suspension

1. Before servicing the vehicle, refer to the precautions in the beginning of this section.

2. Remove or disconnect the following:
   - Negative battery cable
   - Front wheel
   - Brake rotor shield
   - Shock absorber
   - Stabilizer bar link hardware

3. Using a spring compressor tool, compress the coil spring.
   - Lower ball joint from the spindle
   - Lower control arm bolts
   - Lower control arm and coil spring

To install:

4. Install or connect the following:
   - Coil spring to the lower control arm

➡**The end of the coil spring must cover the first hole and should not be visible in the second hole.**

   - Lower arm and front coil spring
   - The two front suspension lower arm bolts and nuts. Do not tighten the nuts at this time.

➡**On the RH front suspension lower arm, install the rear bolt adjustment cam, and nut in the center of the frame slot.**

☀ CAUTION

Always install the cotter pin into the lower ball joint castellated nut from outboard to inboard, with the fingers bent together at a right angle. Failure to do so will result in damage to the wheel and tire assembly.

   - Lower ball joint. Torque the nut to 113 ft. lbs. (153Nm).

5. Remove the Coil Spring Compressor.

6. Install or connect the following:
   - Front stabilizer bar link nut. Torque the nut to 21 ft. lbs. (29Nm).
   - Shock absorber and the two lower nuts
   - Upper shock absorber bushing and nut/washer assembly

7. Support the lower control arm with a jackstand. Torque the bolts to 129 ft. lbs. (175 Nm).
   - Brake disc shield
   - Wheel and tire assembly

8. Inspect and adjust the front end alignment.

Torsion Bar Suspension

1. Before servicing the vehicle, refer to the precautions in the beginning of this section.

2. Raise the vehicle on a hoist.

3. Remove the wheel and tire assembly.

4. Remove the stabilizer link nut, washer and bushing.

5. Remove the front shock absorber-to-front suspension lower arm nuts.

6. Remove the torsion bar

7. Remove the lower ball joint castellated nut.

➡**Do not use a hammer to separate the ball joint from the wheel knuckle or damage to the wheel knuckle will result. Do not damage the ball joint boot while installing the special tool.**

8. Using the special tool, separate the front suspension lower arm from the front wheel knuckle/spindle.

9. Remove the front suspension lower arm bolts and nuts.

10. Remove the front suspension lower arm.

To install:

➡**Tighten the front suspension lower arm pivot bolts and nuts until snug. Do not tighten to specification until the installation procedure is complete.**

11. Position the front suspension lower arm to the front suspension crossmember.

12. Install the pivot bolts and nuts and tighten until snug.

✳✳ WARNING

Install the cotter pin into the lower ball joint from outboard to inboard with the fingers bent together at a right angle. Failure to do so will cause damage to the wheel and tire assembly.

13. Position the lower ball joint into the front wheel knuckle/spindle.

14. Install the new castellated nut. Torque to 83–112 ft. lbs. (113–153 Nm).

15. Install a new cotter pin.

16. Install the front shock absorber-to-front suspension lower arm nuts. Torque to 15–21 ft. lbs. (21–29 Nm).

17. Install the stabilizer link bushing, washer, and nut. Torque to 15–21 ft. lbs. (21–29 Nm).

➡**Whenever the torsion bar or torsion bar adjuster is removed, the vehicle ride height must be checked.**

18. Install the torsion bar.

19. Install the tire and wheel assembly.

20. Lower the vehicle.

21. Tighten the front suspension lower arm nuts. Torque to 111–148 ft. lbs. (150–200 Nm).

22. Inspect and adjust the front end alignment.

LOWER CONTROL ARM BUSHING REPLACEMENT

The control arm bushings are not serviceable. If they require service, the upper or lower arm must be replaced.

Wheel Bearings

ADJUSTMENT

2-Wheel Drive Vehicles

1. Before servicing the vehicle, refer to the precautions in the beginning of this section.

2. Remove the grease cap from the hub and wipe the excess grease from the end of the spindle. Remove the cotter pin and retainer. Discard the cotter pin.

3. Loosen the adjusting nut 3 turns.

✳✳ WARNING

Obtain running clearance between the disc brake rotor surface and shoe linings by rocking the entire wheel assembly in and out several times in order to push the caliper and brake pads away from the rotor. An alternate method to obtain proper running clearance is to tap lightly on the caliper housing. Be sure not to tap on any other area that may damage the disc brake rotor or the brake lining surfaces. Do not pry on the phenolic caliper piston. The running clearance must be maintained throughout the adjustment procedure. If proper clearance cannot be maintained, the caliper must be removed from its mounting.

4. While rotating the wheel assembly, tighten the adjusting nut to 17–25 ft. lbs. (23–34 Nm) in order to seat the bearings. Loosen the adjusting nut a half turn. Retighten the adjusting nut 18–20 inch lbs. (2.0–2.2 Nm).

5. Place the retainer on the adjusting nut. The castellations on the retainer must be in alignment with the cotter pin holes in the spindle. Once this is accomplished install a new cotter pin and bend the ends to insure its being locked in place.

6. Check for proper wheel rotation. If correct, install the grease cap.

7. Lower the vehicle and tighten the lug nuts to 100 ft. lbs., (136 Nm) if the wheel was removed. Before driving the vehicle, pump the brake pedal several times to restore normal brake pedal travel.

✳✳ CAUTION

If the wheel was removed, retighten the wheel lug nuts to specification after about 500 miles (804km) of driving. Failure to do this could result in the wheel coming off while the vehicle is in motion causing loss of vehicle control or collision.

4-Wheel Drive

1. Before servicing the vehicle, refer to the precautions in the beginning of this section.

2. Remove or disconnect the following:
   - Wheel assembly
   - Retainer washers from the lug nut studs and remove the automatic locking hub assembly from the spindle
   - Snapring and spacer from the end of the spindle shaft
   - Pull the locking cam assembly and the 2 plastic spacers off of the wheel bearing adjusting nut

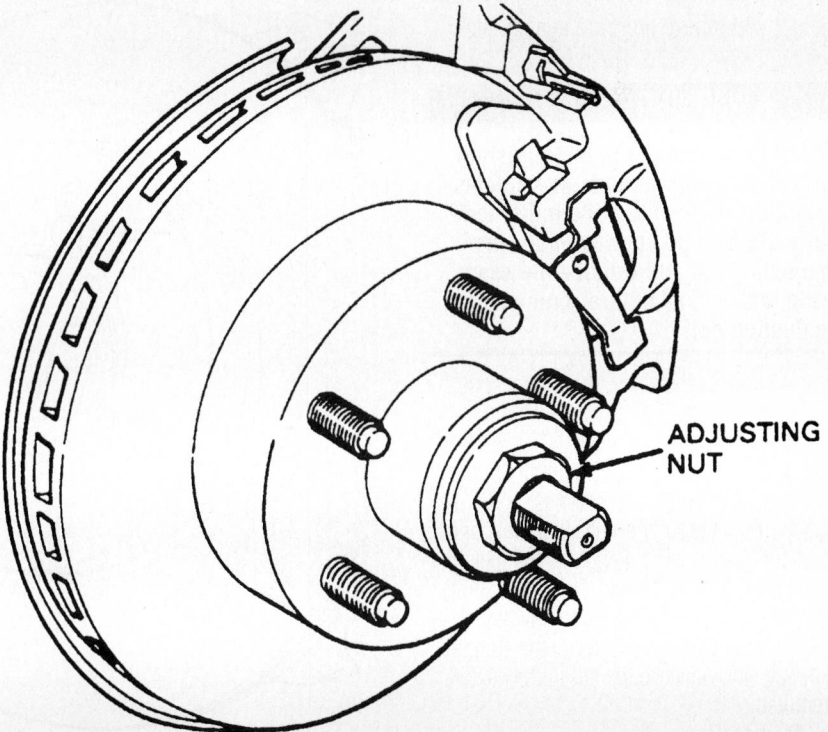

ADJUSTING NUT

7924EG34

Loosen the adjusting nut 3 turns, then rock the entire wheel assembly in-and-out to spread the brake pads before attempting to adjust the bearing—2wd vehicles

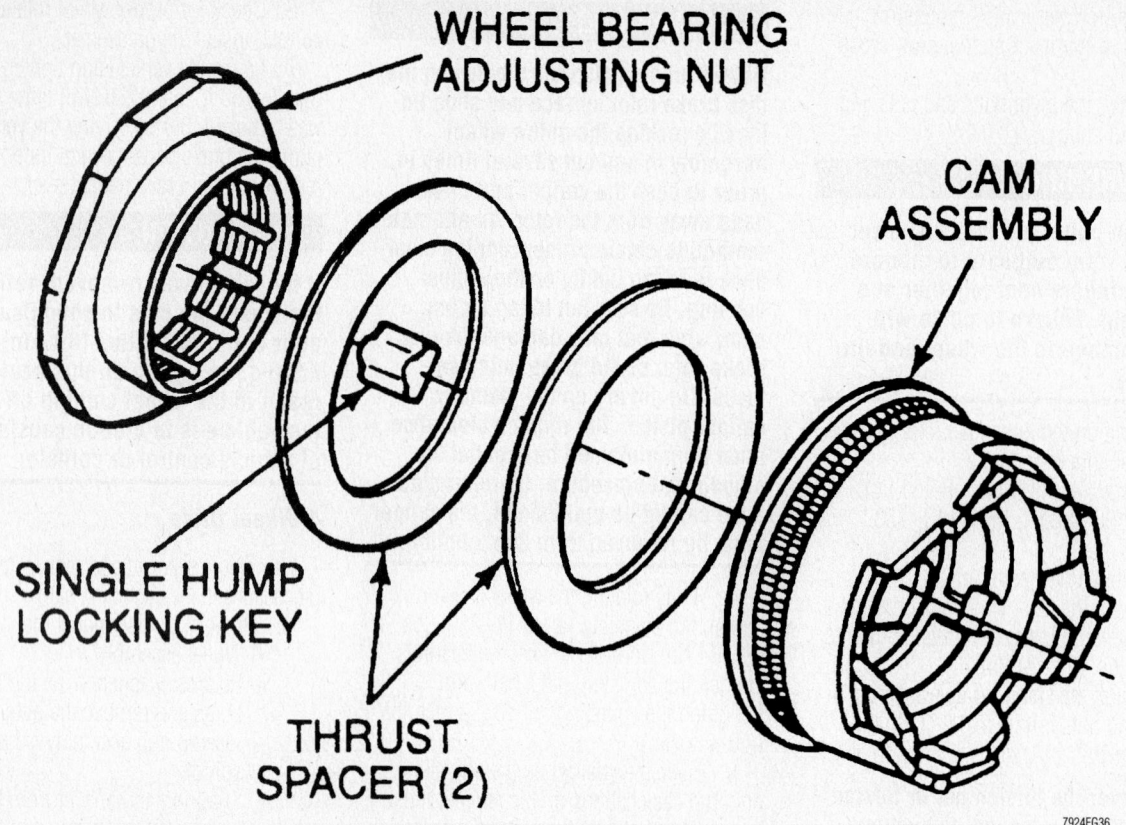

WHEEL BEARING ADJUSTING NUT

CAM ASSEMBLY

SINGLE HUMP LOCKING KEY

THRUST SPACER (2)

7924EG36

Exploded view of the wheel bearing adjusting nut and related components—automatic locking hub shown

3. Use a magnet and remove the locking key from under the adjusting nut. If required, rotate the adjusting nut slightly to relieve pressure against the locking key.

✳✳ WARNING

To prevent damage to the adjusting nut and spindle threads on vehicles equipped with automatic hubs, look into the spindle keyway under the adjusting nut and remove the separate locking key before removing the adjusting nut.

4. Loosen the wheel bearing locknut using a 2⅜ inch (60.3mm) hex socket, such as Hex Locknut Wrench T70T-4252-B.

5. Tighten the inner locknut to 35 ft. lbs. (47 Nm) to seat the bearings.

6. Spin the rotor and back off the inner locknut ¼ turn (90˚). Retighten the locknut to 16 inch lbs. (1.8 Nm).

7. Align the closest lug in the bearing adjusting nut with the center of the spindle keyway slot. Advance the nut to the next if required.

To install:

8. Separate locking key in the spindle keyway under the adjusting nut.

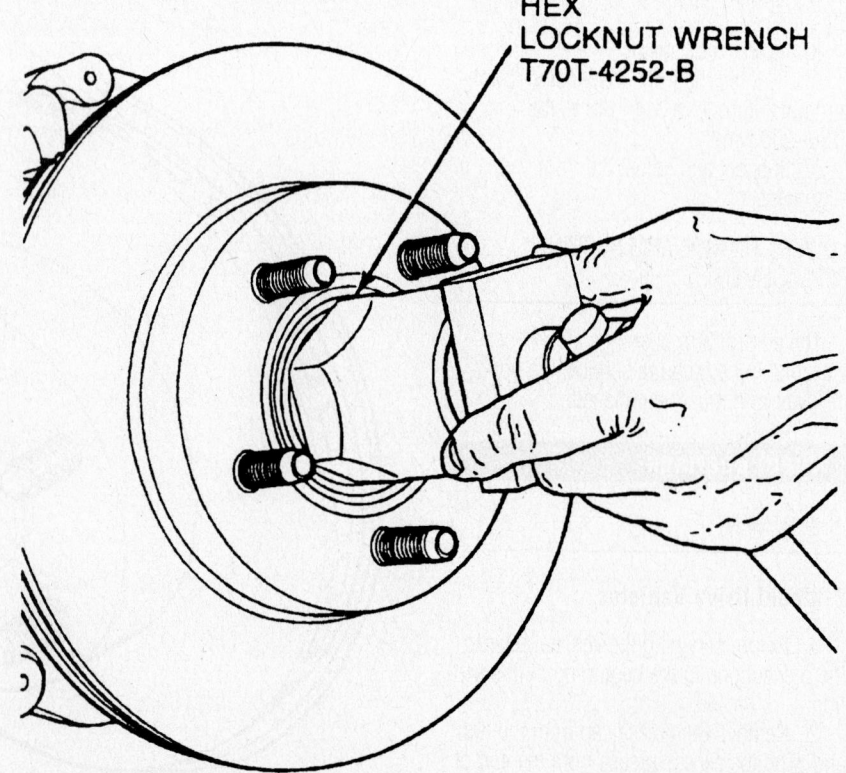

HEX LOCKNUT WRENCH T70T-4252-B

7924EG37

An oversize socket is needed to properly adjust the wheel bearing—automatic locking hub shown

Extreme care must be taken when aligning the adjusting nut with the center of the spindle keyway slot to prevent damage to the separate locking key. The wheel and tire assembly may come off while the vehicle is in motion if the key is damaged.

9. Install or connect the following:
   • 2 plastic thrust spacers and push or press the cam assembly onto the adjusting nut by lining up the keyway in the cam assembly with the separate locking key

✳✳ WARNING

Do not damage the locking key when installing the cam assembly.

   • Axle shaft spacer
   • Clip the snapring onto the end of the spindle
   • Manual hub assembly over the spindle. Install the retainer washers
   • Wheel assembly

10. Check the end-play of the wheel and tire assembly on the spindle. End-play should be 0.001–0.003 in. (0.025–0.076mm) and the maximum torque to rotate the hub should be 25 inch lbs. (2.8 Nm).

REMOVAL & INSTALLATION

2-Wheel Drive

1. Before servicing the vehicle, refer to the precautions in the beginning of this section.
2. Remove or disconnect the following:
   • Disc brake caliper anchor plate
   • Hub grease cap
   • Cotter pin
   • Nut retainer
   • Spindle nut
   • Wheel outer bearing retainer washer
   • Outer front wheel bearing
   • Brake disc and hub
   • Hub grease seal
   • Inner wheel bearing

To install:
3. Thoroughly clean and inspect the front wheel bearings and the brake disc and hub.
4. Lubricate the front wheel bearings.
5. Install the inner front wheel bearing.
6. Install a new wheel hub grease seal.
7. Position the brake disc and hub.
8. Assemble all parts and adjust the bearings.

4-Wheel Drive

1. Before servicing the vehicle, refer to the precautions in the beginning of this section.
2. Remove or disconnect the following:
   • Negative battery cable
   • Wheel assembly
   • Retainer washers from the lug nut studs and remove the automatic locking hub assembly from the spindle
   • Snapring and spacer from the end of the spindle shaft
   • Pull the locking cam assembly and the 2 plastic spacers off of the wheel bearing adjusting nut
3. Use a magnet and remove the locking key from under the adjusting nut. If required, rotate the adjusting nut slightly to relieve pressure against the locking key

✳✳ WARNING

To prevent damage to the adjusting nut and spindle threads on vehicles equipped with automatic hubs, look into the spindle keyway under the adjusting nut and remove the separate locking key before removing the adjusting nut.

   • Wheel bearing locknut using a 2⅜ inch (60.3mm) hex socket, such as Hex Locknut Wrench T70T-4252-B
   • Outer bearing cone and roller assembly from the hub
   • Hub and rotor from the spindle
   • Grease seal, using seal removal tool 1175-AC and discard
   • Inner bearing cone and roller assembly from the hub
4. Clean the inner and outer bearing assemblies in solvent. Inspect the bearings

and the cones for wear and damage. Replace defective parts, as required.
5. If the cups are worn or damaged, remove them with front hub remover tool T81P-1104-C and tool T77F-1102-A.
6. Wipe the old grease from the spindle. Check the spindle for excessive wear or damage. Replace defective parts, as required.

To install:
7. If the inner and outer cups were removed, use bearing driver handle tool T80-4000-W and replace the cups. Be sure to seat the cups properly in the hub.
8. Use a bearing packer tool and properly repack the wheel bearings with the proper grade and type of grease. If a bearing packer is not available, work as much of the grease as possible between the rollers and cages. Also, grease the cone surfaces.
9. Install or connect the following:
   • Inner bearing cone and roller assembly in the inner cup. A light film of grease should be included between the lips of the new grease seal.
   • Grease seal by driving in place with Hub Seal Replacer tool T83T-1175-B and Driver Handle T80T-4000-W
   • Hub and rotor assembly onto the spindle. Keep the hub centered on the spindle to prevent damage to the spindle and the retainer
   • Outer bearing cone and roller assembly
   • Rotor onto the spindle
   • Outer wheel bearing in the rotor
   • Adjusting nut. Torque the nut to 35 ft. lbs. (47 Nm) to seat the bearings. Adjust the bearing as needed.
   • Thrust spacers and press the cam assembly on the locknut by aligning the key in the fixed cam with the keyway of the front spindle
   • Axle shaft spacer
   • Snapring on the end of the shaft
   • Locking hub assembly over the front spindle
   • Align the 3 hub legs to the cam pockets and install the retainer washers
   • Wheel assembly
   • Negative battery cable

BRAKES

Brake Caliper

REMOVAL & INSTALLATION

1. Before servicing the vehicle, refer to the precautions in the beginning of this section.
2. Loosen the wheel lug nuts.
3. Raise and safely support the front of the vehicle. Remove the wheel.
4. Place an 8 in. (203mm) C-clamp on the caliper and tighten the clamp to bottom the caliper pistons in their bores. Remove the clamp.
5. Remove the two caliper slide pin bolts and lift the caliper from the anchor plate.

➡**Use care to retain as much of the original caliper slide pin grease as possible.**

6. Position the caliper on a frame member or suspend it with some wire. Do not allow the caliper to hang by the brake hose.
7. Disconnect and plug the brake hose at the caliper. Remove the caliper from the rotor.

To install:

8. Position the caliper over the brake pads and align the slide pin mounting holes.
9. Install the slide pin bolts and tighten them to 21–26 ft. lbs. (30–36 Nm).
10. Install the caliper brake hose using new washers. Tighten the bolt to 29 ft. lbs. (40
11. Install the wheel and snug the lug nuts.
12. Lower the vehicle and tighten the lug nuts to 100 ft. lbs. (135 Nm).

➡**The first couple of times you apply the brakes, the pedal may go to the floor. Continue to pump the brake pedal until it feels firm.**

13. Start the engine and apply the brakes several times to readjust the caliper pistons. Ensure that the pedal feels firm before operating the vehicle.

Disc Brake Pads

REMOVAL & INSTALLATION

1. Before servicing the vehicle, refer to the precautions in the beginning of this section.
2. Raise and safely support the front of the vehicle. Remove the wheel.

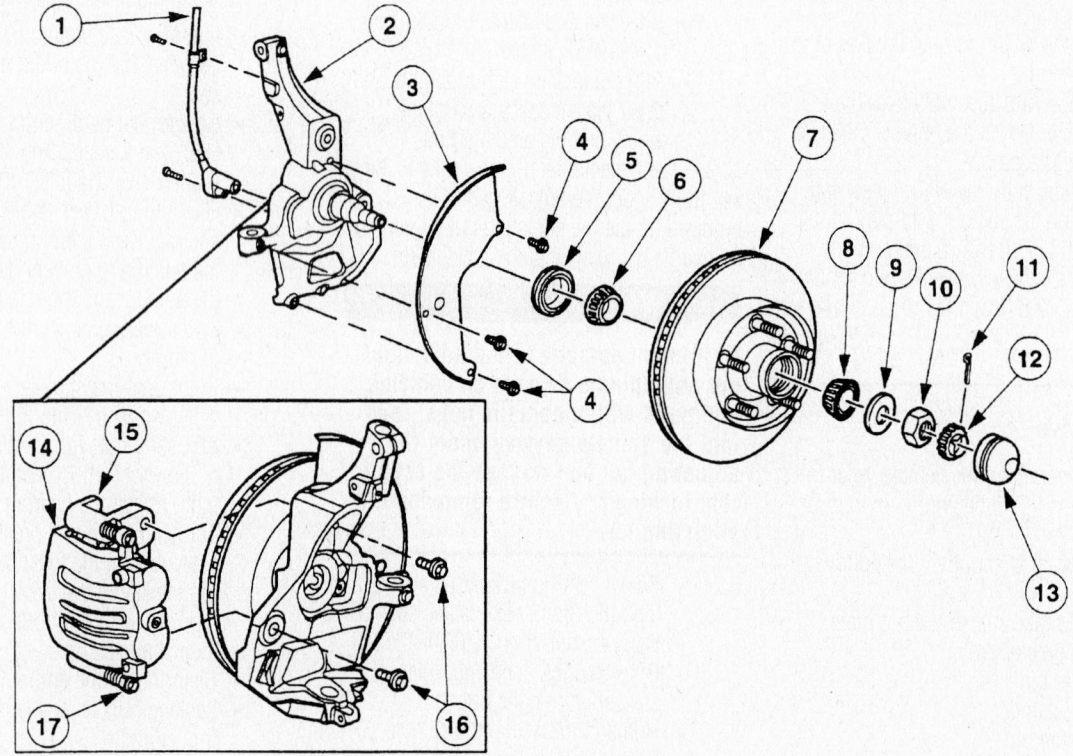

1 Front Brake Anti-Lock Sensor	7 Front Disc Brake Hub and Rotor	12 Nut Retainer
2 Front Wheel Spindle	8 Front Wheel Bearing	13 Hub Grease Cap
3 Front Disc Brake Rotor Shield	9 Front Wheel Outer Bearing Retainer Washer	14 Disc Brake Caliper
4 Rotor Shield Bolt	10 Hub Spindle Nut	15 Front Disc Brake Caliper Anchor Plate
5 Grease Seal	11 Cotter Pin	16 Caliper Anchor Plate Bolts
6 Front Wheel Bearing		17 Disc Brake Caliper Bolt

93026G22

Exploded view of the 2WD front disc brake assembly

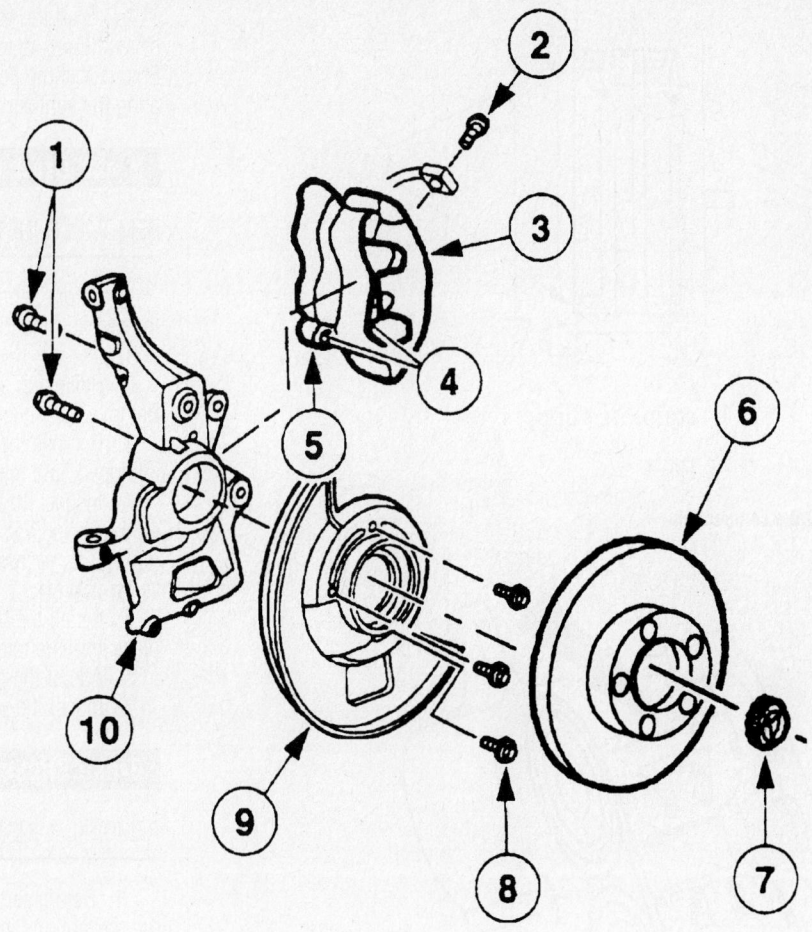

1 Front Disc Brake Caliper
   Anchor Plate Bolt (2 Req'd)

2 Front Brake Hose Bolt

3 Disc Brake Caliper

4 Pads

5 Front Disc Brake Caliper
   Anchor Plate

6 Front Disc Brake Rotor

7 Front Axle Wheel Hub
   Retainer

8 Front Disc Brake Rotor Shield
   Bolt (3 Req'd)

9 Front Disc Brake Rotor Shield

10 Front Wheel Knuckle

93026G23

Exploded view of the 4WD front disc brake assembly

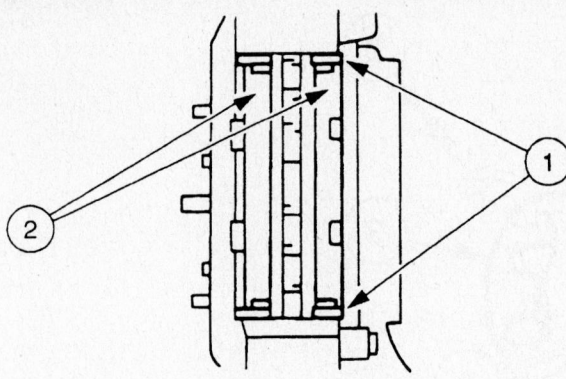

1 stainless slippers

2 pads

93026G24

Position of the front disc brake components

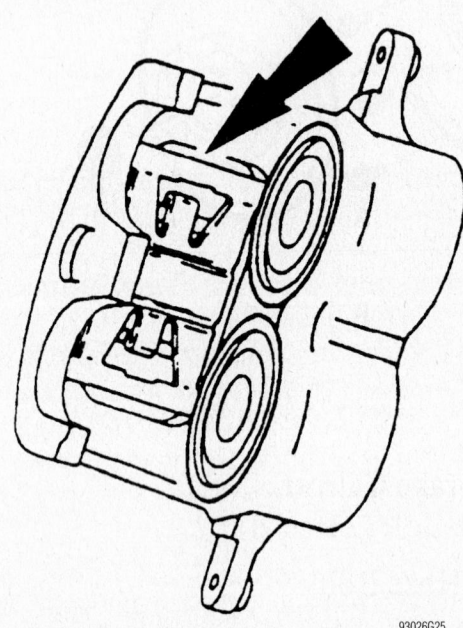

93026G25

View of the front disc brake anti-rattle spring

3. Place an 8 in. (203mm) C-clamp on the caliper and tighten the clamp to bottom the caliper pistons in their bores. Remove the clamp.

4. Remove the two caliper slide pin bolts and lift the caliper from the anchor plate.

➡**Use care to retain as much of the original caliper slide pin grease as possible.**

5. Position the caliper on a frame member or suspend it with some wire. Do not allow the caliper to hang by the brake hose.

6. Remove the brake pads and, if necessary, the anti-rattle clips from the anchor plate.

7. Remove the shims, if any, from the brake pads for re-use.

To install:

8. If removed, install the anti-rattle clips.

9. Install the brake pads to the anchor plate.

10. Position the caliper over the brake pads and align the slide pin mounting holes.

11. Install the slide pin bolts and tighten them to 21–26 ft. lbs. (30–36 Nm).

12. Install the wheel and snug the lug nuts.

13. Lower the vehicle and tighten the lug nuts to 100 ft. lbs. (135 Nm).

➡**The first couple of times you apply the brakes, the pedal may go to the floor. Continue to pump the brake pedal until it feels firm.**

14. Start the engine and apply the brakes several times to readjust the caliper pistons. Ensure that the pedal feels firm before operating the vehicle.

Brake Drums

REMOVAL & INSTALLATION

1. Before servicing the vehicle, refer to the precautions in the beginning of this section.

2. Raise and safely support the vehicle. Remove the wheel and tire assembly.

3. Remove the retaining nuts, if equipped, and remove the brake drum.

4. Inspect the brake drum surface for wear, scoring and runout. Machine or replace, as necessary.

To install:

5. Install the brake drum and secure in place with the retainer nuts, if equipped.

6. Adjust the rear brakes.

7. Install the wheel. Lower the vehicle.

Brake Shoes

REMOVAL & INSTALLATION

1. Before servicing the vehicle, refer to the precautions in the beginning of this section.

2. Raise and safely support the vehicle. Remove the wheel and tire assembly and the brake drum.

3. Pull backward on the adjusting lever cable to disengage the adjusting lever from the adjusting screw. Move the outboard side of the adjusting screw upward and back off the pivot nut as far as it will go.

4. Pull the adjusting lever, cable and automatic adjuster spring down and toward the rear to unhook the pivot hook from the large hole in the secondary shoe web. Do not pry the pivot hook from the hole.

5. Remove the automatic adjuster spring and adjusting lever.

6. Remove the secondary shoe-to-anchor spring using a suitable brake spring removal/installation tool. Using the tool, remove the primary shoe-to-anchor spring and unhook the cable anchor. Remove the anchor pin plate, if equipped.

7. Remove the cable guide from the secondary shoe.

8. Remove the shoe hold-down springs, shoes, adjusting screw, pivot nut and socket. Note the color and position of each hold-down spring so they can be reassembled in the same position.

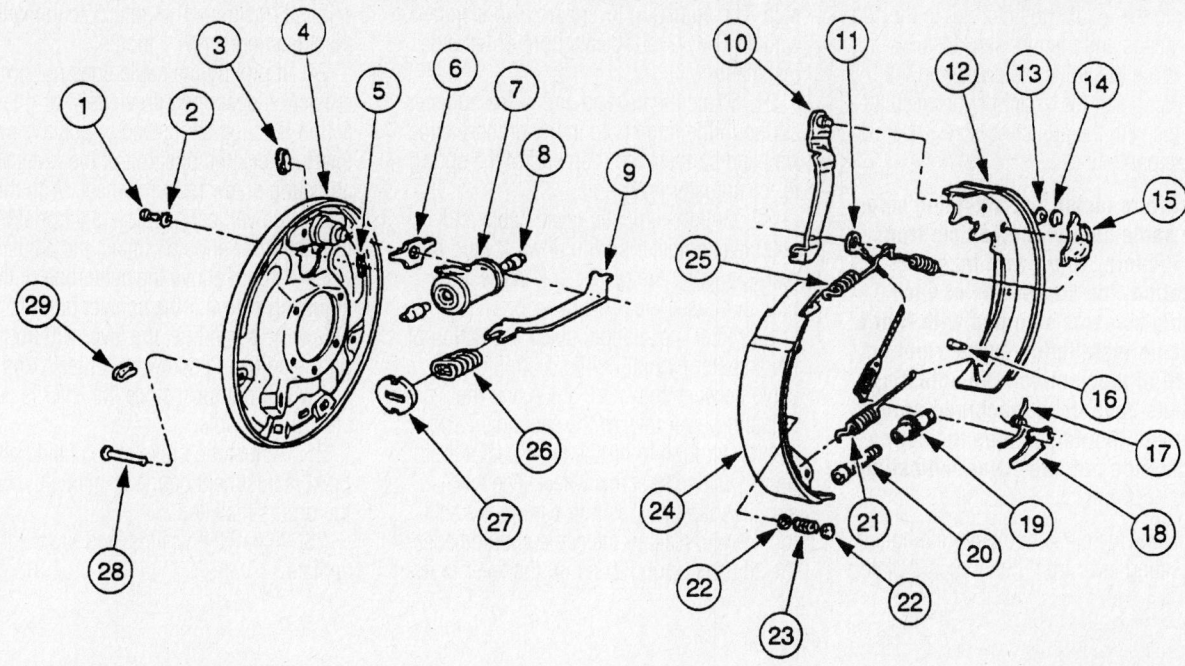

1 Wheel Cylinder-to-Backing Plate Bolt (2 Req'd)
2 Washer
3 Inspection Hole Cover
4 Brake Backing Plate
5 Lining Inspection Hole
6 Anchor Pin Guide Plate
7 Rear Wheel Cylinder
8 Wheel Cylinder Brake Shoe Link
9 Parking Brake Strut
10 Parking Brake Lever
11 Brake Shoe Adjusting Lever Cable

12 Rear Brake Shoe and Lining, Secondary
13 Washer
14 Parking Brake Lever Pin Retainer
15 Cable Guide
16 Adjusting Lever Pin
17 Adjusting Lever Return Spring
18 Brake Shoe Adjusting Lever
19 Brake Shoe Adjusting Screw Nut
20 Brake Adjuster Screw
21 Brake Shoe Adjusting Screw Spring

22 Brake Shoe Hold-Down Spring Cup
23 Brake Shoe Hold-Down Spring
24 Rear Brake Shoe and Lining, Primary
25 Brake Shoe Retracting Spring, Short
26 Parking Brake Link Spring
27 Parking Brake Spring Retainer
28 Brake Shoe Hold-Down Spring Pin
29 Brake Adjusting Hole Cover

93026G21

Exploded view of the rear brake shoes and components

9. Remove the parking brake link and spring. Disconnect the parking brake cable from the parking brake lever.

10. Remove the secondary brake shoe. On 9 in. (22.8cm) rear brakes, remove the parking brake lever from the shoe. On 10 in. (25.4cm) rear brakes, remove the retainer clip and spring washer and remove the parking brake lever.

To install:

11. Clean the backing plate ledge pads and sand lightly. Apply a light coating of high temperature lithium grease to the points where the brake shoes touch the backing plate. Lubricate the adjusting cable eye and the anchor pin area.

12. Install the parking brake lever on the secondary shoe. On 10 in. (25.4cm) brakes, secure with the spring washer and retaining clip.

13. Position the brake shoes on the backing plate and install the hold-down spring pins, springs and cups. Install the parking brake link, spring and washer. Connect the parking brake cable to the parking brake lever.

14. Install the anchor pin plate, if equipped, and place the cable anchor over the anchor pin with the crimped side toward the backing plate.

15. Install the primary shoe-to-anchor spring using the brake spring removal/installation tool.

16. Install the cable guide on the sec-ondary shoe with the flanged hole fitted into the hole in the secondary shoe. Thread the cable around the cable guide groove.

➡ **Make sure the cable is positioned in the groove and not between the guide and shoe web.**

17. Install the secondary shoe-to-anchor (long) spring.

➡ **Make sure the cable end is not cocked or binding on the anchor pin when installed. All parts should be flat on the anchor pin.**

18. Apply high temperature lithium grease to the threads and the socket end of the adjusting screw. Turn the adjusting

screw into the adjusting pivot nut to the end of the threads and then loosen, 1/2 turn.

19. Place the adjusting socket on the screw and install the assembly between the shoe ends with the adjusting screw nearest the secondary shoe.

➡**Be sure to install the adjusting screw on the same side of the vehicle from which it came. To prevent incorrect installation, the socket end of each adjusting screw is stamped with R or L, to indicate installation on the right or left side of the vehicle. The adjusting pivot nuts have lines machined around the body of the nut, 2 lines indicating the right side nut and 1 line indicating the left side nut.**

20. Hook the cable hook into the hole in the adjusting lever from the outboard plate side. The adjusting levers are also stamped with an **R** or **L** to indicate right or left side installation.

21. Place the hooked end of the adjuster spring in the large hole in the primary shoe web and connect the loop end of the spring to the adjuster lever hole.

22. Pull the adjuster lever, cable and automatic adjuster spring down toward the rear to engage the pivot hook in the large hole in the secondary shoe web.

23. After installation, check the action of the adjuster by pulling the section of the cable between the cable guide and the adjusting lever toward the secondary shoe web far enough to lift the lever past a tooth on the adjusting screw wheel. The lever should snap into position behind the next tooth and releasing the cable should cause the adjuster spring to return the lever to its original position. This return action will turn the adjusting screw 1 tooth.

24. If pulling the cable does not produce the action described previously, or if lever action is sluggish instead of positive and sharp, check the position of the lever on the adjusting screw toothed wheel. With the brake in a vertical position, anchor at the top, the lever should contact the adjusting wheel 1 tooth above the centerline of the adjusting screw. If the contact point is below the centerline, the lever will not lock on the adjusting screw wheel teeth and the screw will not turn, since the lever is actuated by the cable.

25. Adjust the brake shoes using either a brake adjustment gauge or manually with the drums installed.

26. Install the wheels, and lower the vehicle.

MAZDA

MPV

27

SPECIFICATIONS AND MAINTENANCE CHARTS

ENGINE AND VEHICLE IDENTIFICATION

		Engine							Model Year	
Code ①	Liters (cc)	Cu. In.	Cyl.	Fuel Sys.	Engine Type	Eng. Mfg.		Code ②	Year
GY	2.5 (2507)	153	6	SFI	DOHC	Mazda		1	2001
AJ	3.0 (NA)	NA	6	SFI	DOHC	Mazda		2	2002
								3	2003
								4	2004
								5	2005

MFI: Multi-port Fuel Injection

NA: Not available

SFI: Sequential Fuel Injection

SOHC: Single Overhead Camshaft

DOHC: Double Overhead Camshaft

OHV: Overhead Valve

① 8th digit of the Vehicle Identification Number (VIN)

② 10th digit of the Vehicle Identification Number (VIN)

67162-MMPV-C01

GENERAL ENGINE SPECIFICATIONS

Year	Model	Engine Displacement Liters	Engine ID	Net Horsepower @ rpm	Net Torque @ rpm (ft. lbs.)	Bore x Stroke (in.)	Com-pression Ratio	Oil Pressure @ rpm
2001	MPV	2.5	GY	170@6250 ①	165@4250	3.25x3.13	9.7:1	20-45@1500
2002	MPV	3.0	AJ	200@6200	200@3000	NA	NA	20-45@1500
2003	MPV	3.0	AJ	200@6200	200@3000	NA	NA	20-45@1500
2004	MPV	3.0	AJ	200@6200	200@3000	NA	NA	20-45@1500

NA: Not available

① California LEV: 160@6250

67162-MMPV-C02

ENGINE TUNE-UP SPECIFICATIONS

Year	Engine Displacement Liters	Engine ID	Spark Plug Gap (in.)	Ignition Timing (deg.) MT	Ignition Timing (deg.) AT	Fuel Pump (psi)	Idle Speed (rpm) MT	Idle Speed (rpm) AT	Valve Clearance Intake	Valve Clearance Exhaust
2001	2.5	GY	0.054	—	10B	37-41	—	①	HYD	HYD
2002	3.0	AJ	NA	—	10B	61-66	—	650-750	HYD	HYD
2003	3.0	AJ	NA	—	10B	61-66	—	650-750	HYD	HYD
2004	3.0	AJ	NA	—	10B	61-66	—	650-750	HYD	HYD

NOTE: The Vehicle Emission Control Information label often reflects specification changes made during production. The label figures must be used if they differ from those in this chart

B: Before top dead center

HYD: Hydraulic

NA: Information not available

① Refer to Vehicle's Emission Control Information label

67162-MMPV-C03

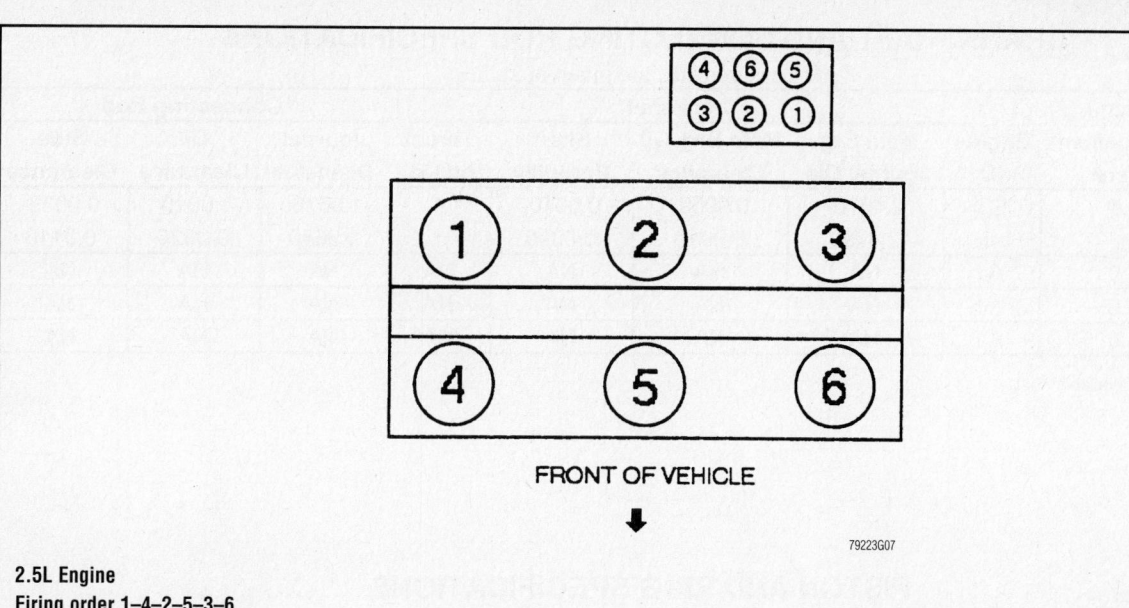

2.5L Engine
Firing order 1–4–2–5–3–6
Distributorless ignition system

FRONT OF VEHICLE

79223G07

CAPACITIES

Year	Model	Engine Displacement Liters	Engine ID	Engine Oil with Filter (qts.)	Transmission (pts.)		Transfer Case (pts.)	Drive Axle		Fuel Tank (gal.)	Cooling System (qts.)
					Manual	Auto.		Front (pts.)	Rear (pts.)		
2001	MPV	2.5	GY	6.0	—	20.6	—	—	—	18.5	①
2002	MPV	3.0	AJ	5.5	—	20.6	—	—	—	19.8	①
2003	MPV	3.0	AJ	5.5	—	20.6	—	—	—	19.8	①
2004	MPV	3.0	AJ	5.5	—	20.6	—	—	—	19.8	①

NOTE: All capacities are approximate. Add fluid gradually and ensure a proper fluid level is obtained.

① With rear heater: 12.7
Without rear heater: 10.8

67162-MMPV-C04

VALVE SPECIFICATIONS

Year	Engine Displacement Liters	Engine ID	Seat Angle (deg.)	Face Angle (deg.)	Spring Test Pressure (lbs. @ in.)	Spring Installed Height (in.)	Stem-to-Guide Clearance (in.)		Stem Diameter (in.)	
							Intake	Exhaust	Intake	Exhaust
2001	2.5	GY	44.75	45.5	153@1.18	1.570	0.0007-0.0027	0.0017-0.0037	0.2350-0.2358	0.2343-0.2350
2002	3.0	AJ	NA	NA	NA	NA	NA	NA	NA	NA
2003	3.0	AJ	NA	NA	NA	NA	NA	NA	NA	NA
2004	3.0	AJ	NA	NA	NA	NA	NA	NA	NA	NA

NA: Information not available

67162-MMPV-C05

CRANKSHAFT AND CONNECTING ROD SPECIFICATIONS

All measurements are given in inches.

Year	Engine Displacement Liters	Engine ID	Crankshaft				Connecting Rod		
			Main Brg. Journal Dia.	Main Brg. Oil Clearance	Shaft End-play	Thrust on No.	Journal Diameter	Oil Clearance	Side Clearance
2001	2.5	GY	2.4670-2.4790	0.0009-0.0019	0.0040-0.0090	4	1.9670-1.9680	0.0010-0.0025	0.0039-0.0118
2002	3.0	AJ	NA	NA	NA	NA	NA	NA	NA
2003	3.0	AJ	NA	NA	NA	NA	NA	NA	NA
2004	3.0	AJ	NA	NA	NA	NA	NA	NA	NA

NA: Information not available

67162-MMPV-C06

PISTON AND RING SPECIFICATIONS

All measurements are given in inches.

Year	Engine Displacement Liters	Engine ID	Piston Clearance	Ring Gap			Ring Side Clearance		
				Top Compression	Bottom Compression	Oil Control	Top Compression	Bottom Compression	Oil Control
2001	2.5	GY	0.0005-0.0009	0.004-0.010	0.011-0.017	0.006-0.026	0.0015-0.0029	0.0015-0.0033	SNUG
2002	3.0	AJ	NA	NA	NA	NA	NA	NA	NA
2003	3.0	AJ	NA	NA	NA	NA	NA	NA	NA
2004	3.0	AJ	NA	NA	NA	NA	NA	NA	NA

NA: Information not available

67162-MMPV-C07

TORQUE SPECIFICATIONS

All readings in ft. lbs.

Year	Engine Displacement Liters	Engine ID	Cylinder Head Bolts	Main Bearing Bolts	Rod Bearing Bolts	Crankshaft Damper Bolts	Flywheel Bolts	Manifold		Spark Plugs	Oil Pan Drain Plug
								Intake	Exhaust		
2001	2.5	GY	①	②	③	④	54-64	6-9	13-16	14	NA
2002	3.0	AJ	⑤	NA	NA	⑥	NA	7.5	18	8-14	16-22
2003	3.0	AJ	⑤	NA	NA	⑥	NA	7.5	18	8-14	16-22
2004	3.0	AJ	⑤	NA	NA	⑥	NA	7.5	18	8-14	16-22

NA: Information not available

① Step 1: 28-31 ft. lbs.
　Step 2: Plus 90 degrees
　Step 3: Loosen bolts 1 turn
　Step 4: 28-31 ft. lbs.
　Step 5: Plus 90 degrees
　Step 6: Plus 90 degrees
② Step 1: 12-43 inch lbs.
　Step 2: Outer cap bolts: 16-21 ft. lbs.
　Step 3: Inner cap bolts: 27-32 ft. lbs.
　Step 4: All cap bolts: 85-95 degrees
　Step 5: Remaining bolts: 15-22 ft. lbs.

③ 26-33 ft. lbs. plus 90-120 degrees
④ Step 1: 89 ft. lbs.
　Step 2: Loosen bolt
　Step 3: 35-39 ft. lbs.
　Step 4: Plus 85-95 degrees
⑤ Step 1: 28 ft. lbs.
　Step 2: plus 90 degrees
　Step 3: back off one full turn
　Step 4: 28 ft. lbs.
　Step 5: plus 90 degrees
　Step 6: plus an additional 90 degrees

⑥ Step 1: 88 ft. lbs.
　Step 2: Back off one full turn
　Step 3: 39 ft. lbs.
　Step 4: plus 90 degrees

67162-MMPV-C08

WHEEL ALIGNMENT

Year	Model		Caster ① Range (+/-Deg.)	Caster ① Preferred Setting (Deg.)	Camber ① Range (+/-Deg.)	Camber ① Preferred Setting (Deg.)	Toe-in (in.)	Kingpin Angle (Deg.)
2001	MPV	F	1.0	+1.70	1.0	-0.90	0.08+/-0.16	11.09
		R	—	—	1.0	-1.00	0.12+/-0.16	—
2002	MPV	F	1.0	+2.03	1.0	-0.23	0.08+/-0.16	11.18
		R	—	—	1.0	-1.00	0.12+/-0.16	—
2003	MPV	F	1.0	+1.98	1.0	-0.13	0.08+/-0.16	11.10
		R	—	—	1.0	-1.00	0.17+/-0.22	—
2004	MPV	F	1.0	+1.98	1.0	-0.13	0.08+/-0.16	11.10
		R	—	—	1.0	-1.00	0.17+/-0.22	—

① Empty vehicle

67162-MMPV-C09

TIRE, WHEEL AND BALL JOINT SPECIFICATIONS

Year	Model	OEM Tires Standard	OEM Tires Optional	Tire Pressures (psi) Front	Tire Pressures (psi) Rear	Wheel Size	Ball Joint Inspection	Lugnut Torque (ft. lbs.)
2001	MPV	P215/70R15	None	32	32	6-JJ	18-30 in. ①	66-86
2002	MPV	205/65R15	215/60R16 P215/60R17	②	②	std: 6-JJ opt: 6.5-JJ, 7-JJ	NA	108
2003	MPV	205/65R15	215/60R16 P215/60R17	②	②	std: 6-JJ opt: 6.5-JJ, 7-JJ	NA	108
2004	MPV	205/65R15	215/60R16 P215/60R17	②	②	std: 6-JJ opt: 6.5-JJ, 7-JJ	NA	108

NA: Information not available

OEM: Original Equipment Manufacturer

PSI: Pounds Per Square Inch

① Torque required in inch lbs. to rotate ball joint when removed from the knuckle

② See placard on vehicle

67162-MMPV-C10

BRAKE SPECIFICATIONS
All measurements in inches unless noted

Year	Model		Brake Disc Original Thickness	Brake Disc Minimum Thickness	Brake Disc Maximum Runout	Brake Drum Diameter Original Inside Diameter	Brake Drum Diameter Max. Wear Limit	Brake Drum Diameter Maximum Machine Diameter	Minimum Lining Thickness Front	Minimum Lining Thickness Rear	Brake Caliper Bracket Bolts (ft. lbs.)	Brake Caliper Mounting Bolts (ft. lbs.)
2001	MPV		1.100	1.030	0.002	10.000	①	10.050	0.080	0.04	66-79	62-68
2002	MPV		NA	1.030	0.002	①	①	10.060	0.080	NA	65-79	62-68
2003	MPV		NA	1.030	0.002	①	①	10.060	0.080	NA	65-79	62-68
2004	MPV	F	NA	1.030	0.002	—	—	—	0.080	—	65-79	62-68
		R	NA	0.630	0.002	①	①	10.060	—	②	36-51	28-36

NA: Information not available

① Stamped on drum

② Disc brakes: 0.08
   Drum brakes: 0.04

67162-MMPV-C11

SCHEDULED MAINTENANCE INTERVALS
MAZDA—2001 MPV

TO BE SERVICED	TYPE OF SERVICE	VEHICLE MILEAGE INTERVAL (x1000)												
		7.5	15	22.5	30	37.5	45	52.5	60	67.5	75	82.5	90	97.5
Engine oil & filter	R	✓	✓	✓	✓	✓	✓	✓	✓	✓	✓	✓	✓	✓
Air cleaner filter	R				✓				✓				✓	
Brake fluid	R				✓				✓				✓	
Spark plugs	R				✓				✓				✓	
Bolts & nuts on chassis & body	S/I				✓				✓				✓	
Cooling system	S/I				✓				✓				✓	
Disc brakes, brake lines, hoses & connections	S/I				✓				✓				✓	
Drive belt(s)	S/I				✓				✓				✓	
Driveshaft dust boots (4WD)	S/I				✓				✓				✓	
Exhaust system heat shields	S/I				✓				✓				✓	
Front suspension ball joints	S/I				✓				✓				✓	
Fuel lines & hoses	S/I				✓				✓				✓	
Idle speed	S/I		✓				✓				✓			
Steering operation & linkages	S/I				✓				✓				✓	
Engine coolant	R						✓				✓			
Timing belt (except Calif.)	R								✓					
Timing belt (Calif.)①	S/I								✓				✓	
Automatic transmission fluid & filter	R								✓					
Fuel filter & PCV valve	R								✓					
Emission hoses & tubes②	S/I								✓					
Ignition timing	S/I								✓					

R: Replace   S/I: Service or Inspect

① Timing belt (Calif.): replace at 105,000 miles, unless previously replaced.

② Emission hoses & tubes: replace at 80,000 miles.

FREQUENT OPERATION MAINTENANCE (SEVERE SERVICE)

If a vehicle is operated under any of the following conditions it is considered severe service:

- Extremely dusty areas.

- 50% or more of the vehicle operation is in 32°C (90°F) or higher temperatures, or constant operation in temperatures below 0°C (32°F).

- Prolonged idling (vehicle operation in stop and go traffic).

- Frequent short running periods (engine does not warm to normal operating temperatures).

- Police, taxi, delivery usage or trailer towing usage.

Air cleaner filter: service or inspect every 15,000 miles

Engine oil & filter: replace every 5000 miles.

Ball joints & dust covers: service or inspect every 7500 miles.

Bolts & nuts on chassis & body: tighten every 15,000 miles.

Spark plugs: replace every 15,000 miles.

Automatic transmission fluid & filter: replace every 30,000 miles.

Front & rear axle oil: replace every 30,000 miles.

Transfer case oil (4WD): replace every 30,000 miles.

67162-MMPV-C12

SCHEDULED MAINTENANCE INTERVALS
MAZDA—2002-03 MPV

TO BE SERVICED	TYPE OF SERVICE	VEHICLE MILEAGE INTERVAL (x1000)												
		7.5	15	22.5	30	37.5	45	52.5	60	67.5	75	82.5	90	97.5
Engine oil & filter	R	✓	✓	✓	✓	✓	✓	✓	✓	✓	✓	✓	✓	✓
Drive belt(s)	S/I				✓				✓				✓	
PCV valve	I								✓					
Spark plugs(platinum tip)	R	every 100,000 miles												
Air cleaner filter	R				✓				✓				✓	
Brake lines and hoses	I				✓				✓				✓	
Brake fluid	R				✓				✓				✓	
Brake pads/shoes	I				✓				✓				✓	
Bolts & nuts on chassis & body	S/I				✓				✓				✓	
Cooling system hoses	S/I				✓				✓				✓	
Driveshaft dust boots	S/I				✓				✓				✓	
Exhaust system heat shields	S/I				✓				✓				✓	
Front suspension ball joints	S/I				✓				✓				✓	
Fuel lines & hoses	S/I				✓				✓				✓	
Steering operation & linkages	S/I				✓				✓				✓	
Engine coolant	R						✓				✓			

R: Replace   S/I: Service or Inspect

FREQUENT OPERATION MAINTENANCE (SEVERE SERVICE)

If a vehicle is operated under any of the following conditions it is considered severe service:

- Extremely dusty areas.
- 50% or more of the vehicle operation is in 32°C (90°F) or higher temperatures, or constant operation in temperatures below 0°C (32°F).
- Prolonged idling (vehicle operation in stop and go traffic).
- Frequent short running periods (engine does not warm to normal operating temperatures).
- Police, taxi, delivery usage or trailer towing usage.

Air cleaner filter: service or inspect every 15,000 miles

Engine oil & filter: replace every 5000 miles.

Ball joints & dust covers: service or inspect every 7500 miles.

Bolts & nuts on chassis & body: tighten every 15,000 miles.

Automatic transmission fluid & filter: replace every 30,000 miles.

67162-MMPV-C13

SCHEDULED MAINTENANCE INTERVALS
MAZDA—2004 MPV

TO BE SERVICED	TYPE OF SERVICE	VEHICLE MILEAGE INTERVAL (x1000)												
		7.5	15	22.5	30	37.5	45	52.5	60	67.5	75	82.5	90	97.5
Engine oil & filter	R	✓	✓	✓	✓	✓	✓	✓	✓	✓	✓	✓	✓	✓
Drive belt(s)	S/I				✓				✓				✓	
PCV valve	I								✓					
Spark plugs(platinum tip)	R	every 100,000 miles												
Air cleaner filter	R					✓					✓			
Brake lines and hoses	I				✓				✓				✓	
Cabin air filter	R	Every 25,000 miles												
Disc brake pads	I		✓		✓		✓		✓		✓		✓	
Brum brake shoes	I				✓				✓				✓	
Tires	Rotate	✓	✓	✓	✓	✓	✓	✓	✓	✓	✓	✓	✓	✓
All locks and hinges	L	✓	✓	✓	✓	✓	✓	✓	✓	✓	✓	✓	✓	✓
Cooling system hoses	S/I				✓				✓				✓	
Driveshaft dust boots	S/I				✓				✓				✓	
Exhaust system heat shields	S/I				✓				✓				✓	
Front suspension ball joints	S/I				✓				✓				✓	
Fuel lines & hoses	S/I				✓				✓				✓	
Engine valve clearance	I ①													
Steering operation & linkages	S/I				✓				✓				✓	
Engine coolant	R ②								✓					

R: Replace   S/I: Service or Inspect   L: Lubricate

① Inspect audibly, and, if noisy, adjust

② Then, every 2 years afterward

FREQUENT OPERATION MAINTENANCE (SEVERE SERVICE)

If a vehicle is operated under any of the following conditions it is considered severe service:

- Extremely dusty areas.

- 50% or more of the vehicle operation is in 32°C (90°F) or higher temperatures, or constant operation in temperatures below 0°C (32°F).

- Prolonged idling (vehicle operation in stop and go traffic).

- Frequent short running periods (engine does not warm to normal operating temperatures).

- Police, taxi, delivery usage or trailer towing usage.

Air cleaner filter: service or inspect every 15,000 miles

Engine oil & filter: replace every 5000 miles.

Ball joints & dust covers: service or inspect every 7500 miles.

Bolts & nuts on chassis & body: tighten every 15,000 miles.

Automatic transmission fluid & filter: replace every 30,000 miles.

67162-MMPV-C14

PRECAUTIONS

Before servicing any vehicle, please be sure to read all of the following precautions, which deal with personal safety, prevention of component damage, and important points to take into consideration when servicing a motor vehicle:

• Never open, service or drain the radiator or cooling system when the engine is hot; serious burns can occur from the steam and hot coolant.

• Observe all applicable safety precautions when working around fuel. Whenever servicing the fuel system, always work in a well-ventilated area. Do not allow fuel spray or vapors to come in contact with a spark, open flame or excessive heat (a hot drop light, for example). Keep a dry chemical fire extinguisher near the work area. Always keep fuel in a container specifically designed for fuel storage; also, always properly seal fuel containers to avoid the possibility of fire or explosion. Refer to the additional fuel system precautions later in this section.

• Fuel injection systems often remain pressurized, even after the engine has been turned **OFF**. The fuel system pressure must be relieved before disconnecting any fuel lines. Failure to do so may result in fire and/or personal injury.

• Brake fluid often contains polyglycol ethers and polyglycols. Avoid contact with the eyes and wash your hands thoroughly after handling brake fluid. If you do get brake fluid in your eyes, flush your eyes with clean, running water for 15 minutes. If eye irritation persists, or if you have taken brake fluid internally, IMMEDIATELY seek medical assistance.

• The EPA warns that prolonged contact with used engine oil may cause a number of skin disorders, including cancer! You should make every effort to minimize your exposure to used engine oil. Protective gloves should be worn when changing oil. Wash your hands and any other exposed skin areas as soon as possible after exposure to used engine oil. Soap and water, or waterless hand cleaner should be used.

• All new vehicles are now equipped with an air bag system. The system must be disabled before performing service on or around system components, steering column, instrument panel components, wiring and sensors. Failure to follow safety and disabling procedures could result in accidental air bag deployment, possible personal injury and unnecessary system repairs.

• Always wear safety goggles when working with, or around, the air bag system. When carrying a non-deployed air bag, be sure the bag and trim cover are pointed away from your body. When placing a non-deployed air bag on a work surface, always face the bag and trim cover upward, away from the surface. This will reduce the motion of the module if it is accidentally deployed. Refer to the additional air bag system precautions later in this section.

• Clean, high quality brake fluid from a sealed container is essential to the safe and proper operation of the brake system. You should always buy the correct type of brake fluid for your vehicle. If the brake fluid becomes contaminated, completely flush the system with new fluid. Never reuse any brake fluid. Any brake fluid that is removed from the system should be discarded. Also, do not allow any brake fluid to come in contact with a painted surface; it will damage the paint.

• Never operate the engine without the proper amount and type of engine oil; doing so WILL result in severe engine damage.

• Timing belt maintenance is extremely important! Many models utilize an interference-type, non-freewheeling engine. If the timing belt breaks, the valves in the cylinder head may strike the pistons, causing potentially serious (also time-consuming and expensive) engine damage.

• Disconnecting the negative battery cable on some vehicles may interfere with the functions of the on-board computer system(s) and may require the computer to undergo a relearning process once the negative battery cable is reconnected.

• When servicing drum brakes, only disassemble and assemble one side at a time, leaving the remaining side intact for reference.

• Only an MVAC-trained, EPA-certified automotive technician should service the air conditioning system or its components.

ENGINE REPAIR

➡ **Disconnecting the negative battery cable on some vehicles may interfere with the functions of the on board computer system. The computer may undergo a relearning process once the negative battery cable is reconnected.**

Distributor

This engine is equipped with a Distributorless Ignition System (DIS).

Alternator

REMOVAL & INSTALLATION

2.5L Engine

1. Before servicing the vehicle, refer to the precautions in the beginning of this section.

2. Remove or disconnect the following:
• Negative battery cable
• Accessory drive belt
• Subframe transverse section
• Exhaust front pipe
• Right axle halfshaft and center shaft assembly
• Alternator harness connectors
• Center shaft support bracket
• Alternator

To install:

3. Install or connect the following:
• Alternator. Tighten the bolts to 29–41 ft. lbs. (40–50 Nm).
• Center shaft support bracket. Tighten the bolts to 32–45 ft. lbs. (43–61 Nm).
• Alternator harness connectors. Tighten the battery terminal nut to 87–130 inch lbs. (10–15 Nm).

• Right axle halfshaft and center shaft assembly
• Exhaust front pipe
• Subframe transverse section. Tighten the bolts to 69–96 ft. lbs. (94–131 Nm).
• Accessory drive belt
• Negative battery cable

3.0L Engine

1. Before servicing the vehicle, refer to the precautions in the beginning of this section.

2. Remove or disconnect the following:
• Negative battery cable
• Accessory drive belt
• Exhaust front pipe
• Right axle halfshaft and center shaft assembly
• Alternator harness connectors
• Center shaft support bracket
• Alternator

To install:

3. Install or connect the following:
- Alternator. Tighten the bolts to 29–41 ft. lbs. (40–50 Nm).
- Center shaft support bracket. Tighten the bolts to 32–45 ft. lbs. (43–61 Nm).
- Alternator harness connectors. Tighten the battery terminal nut to 87–130 inch lbs. (10–15 Nm).
- Right axle halfshaft and center shaft assembly
- Exhaust front pipe
- Accessory drive belt
- Negative battery cable

Engine

REMOVAL & INSTALLATION

2.5L Engine

1. Before servicing the vehicle, refer to the precautions in the beginning of this section.
2. Drain the cooling system.
3. Drain the engine oil.
4. Drain the transaxle.
5. Relieve the fuel system pressure.
6. Install a support fixture to the engine lifting eyes.
7. Remove or disconnect the following:
- Battery and tray
- Inner fender liners
- Axle halfshafts
- Air intake assembly
- Accelerator cable and bracket
- Gear select cable
- Transaxle dipstick tube
- Cruise control actuator
- Radiator
- Fuel line
- Brake booster vacuum line
- Powertrain Control Module (PCM) connector. Pull the harness through the firewall into the engine compartment.
- Exhaust front pipe
- Accessory drive belt
- Alternator and bracket
- Right engine mount bracket
- Power steering hoses
- Front engine mount
- Subframe center section
- Rear engine mount
- Left engine mount

8. Lower the powertrain from the vehicle.

To install:

9. Raise the powertrain into position.
10. Install or connect the following:
- Left engine mount
- Rear engine mount
- Subframe center section. Tighten the bolts to 48–65 ft. lbs. (64–89 Nm) and the nut to 50–67 ft. lbs. (67–93 Nm).
- Front engine mount
- Power steering hoses
- Right engine mount bracket

11. Check that the right engine mount stud is centered in the mount bracket with no tension applied to the stud by the bracket.

12. Tighten the engine mount fasteners as follows:
 a. Step 1: Left and rear engine mount through bolts to 63–86 ft. lbs. (85–116 Nm).
 b. Step 2: Front engine mount nuts to 50–67 ft. lbs. (67–93 Nm).
 c. Right engine mount bracket nuts to 56–76 ft. lbs. (75–104 Nm).

13. Install or connect the following:
- Alternator and bracket
- Accessory drive belt
- Exhaust front pipe
- PCM connector. Pull the harness through the firewall into the passenger compartment.
- Brake booster vacuum line
- Fuel line
- Radiator
- Cruise control actuator
- Transaxle dipstick tube
- Gear select cable
- Accelerator cable and bracket
- Air intake assembly
- Axle halfshafts
- Inner fender liners
- Battery and tray

14. Fill the crankcase and transaxle to the correct level.
15. Fill the cooling system.
16. Start the engine and check for leaks.

3.0L Engine

1. Before servicing the vehicle, refer to the precautions in the beginning of this section.
2. Drain the cooling system.
3. Drain the engine oil.
4. Drain the transaxle.
5. Relieve the fuel system pressure.
6. Disconnect the negative batter cable.
7. Install a support fixture to the engine lifting eyes.
8. Remove or disconnect the following:
- Timing chain plug hole plate
- Power steering hoses
- Splash shield
- Axle halfshafts
- Battery and tray
- Air intake assembly
- Accelerator cable and bracket
- Gear select cable
- Transaxle dipstick tube
- Cruise control actuator
- Radiator
- Fuel line
- Powertrain Control Module (PCM) connector. Pull the harness through the firewall into the engine compartment.
- Exhaust front pipe
- Accessory drive belt
- A/C compressor with lines still attached
- Alternator and bracket
- Engine mounts

To install:

9. Installation is the reverse of removal. Observe the following torques:
- Subframe center section. Tighten the bolts to 48–65 ft. lbs. (64–89 Nm) and the nut to 50–67 ft. lbs. (67–93 Nm).
- Left and rear engine mount through bolts to 63–86 ft. lbs. (85–116 Nm).
- Front engine mount nuts to 50–67 ft. lbs. (67–93 Nm).
- Right engine mount bracket nuts to 56–76 ft. lbs. (75–104 Nm).

Water Pump

REMOVAL & INSTALLATION

2.5L and 3.0L Engines

1. Before servicing the vehicle, refer to the precautions in the beginning of this section.
2. Drain the cooling system.
3. Remove or disconnect the following:
- Battery and tray
- Water pump drive belt
- Water pump drive pulley
- Thermostat housing
- Water pump belt tensioner
- Oil cooler hose
- Water outlet pipe
- Water pump

To install:

4. Install or connect the following:
- Water pump. Tighten the bolts to 89 inch lbs. (10 Nm) plus 90 degrees.
- Water outlet pipe
- Oil cooler hose
- Water pump belt tensioner
- Thermostat housing

- Water pump drive pulley
- Water pump drive belt
- Battery and tray

5. Fill the cooling system.
6. Start the engine and check for leaks.

Heater Core

REMOVAL & INSTALLATION

Front System

1. Disconnect the negative battery cable.

❋❋ CAUTION

After disconnecting the battery, wait for more than 1 minute for the air bag system to deplete its stored energy.

2. Drain the cooling system into a clean container for reuse.

3. Disconnect the heater hoses from the heater core.

4. Discharge and recover the air conditioning system refrigerant.

5. At the driver's side, remove the SAS module and the steering wheel by performing the following procedure:

a. Place the wheel in the straight-ahead position and turn the ignition switch to LOCK.

b. Remove the lower steering column cover.

c. Disconnect the clock spring connector.

d. Remove the steering wheel-to-SAS module bolts.

e. Carefully, lift the SAS module from the steering wheel.

❋❋ CAUTION

Place the SAS in a safe place with the module facing upward.

f. Remove the steering wheel-to-column nut.

g. Using a steering wheel puller, press the steering wheel from the steering column.

6. At the passenger's side, remove the SAS module by performing the following procedure:

a. Remove the glove compartment by sliding it to the left; then, pull the right side forward to remove the stopper and the pin, then, move it to the right to remove it.

b. Remove the SAS module-to-dash bolts.

c. Carefully, lift the SAS module and disconnect the electrical connector.

❋❋ CAUTION

Place the SAS in a safe place with the module facing upward.

7. Remove the instrument panel by performing the following procedure:

a. Remove the A-pillar trim from both sides.

b. Remove the front side trim and the side panel trim.

c. Remove the hood release handle and the lower panel.

d. Remove the meter hood and the instrument cluster.

e. Remove the steering column-to-instrument panel bolts and lower the steering column.

f. Disconnect the electrical connectors from the blower motor and the heater housing.

g. Remove the instrument panel hole cover and the instrument panel-to-chassis bolts.

h. Disconnect the electrical connectors.

i. Using an assistant, carefully remove the instrument panel.

8. Remove the heater housing-to-chassis nuts.

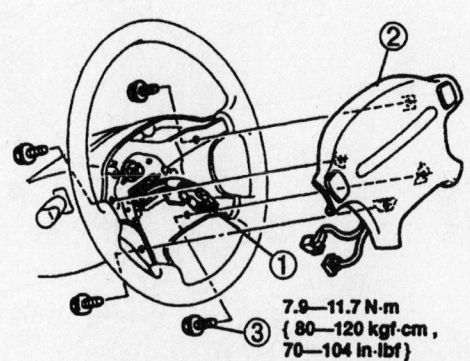

```
7.9—11.7 N·m
{ 80—120 kgf·cm ,
70—104 in·lbf}
```

1 **Connector**
2 **Driver-side air bag module**
3 **Bolt**

93113GF1

Exploded view of the steering wheel and SAS module

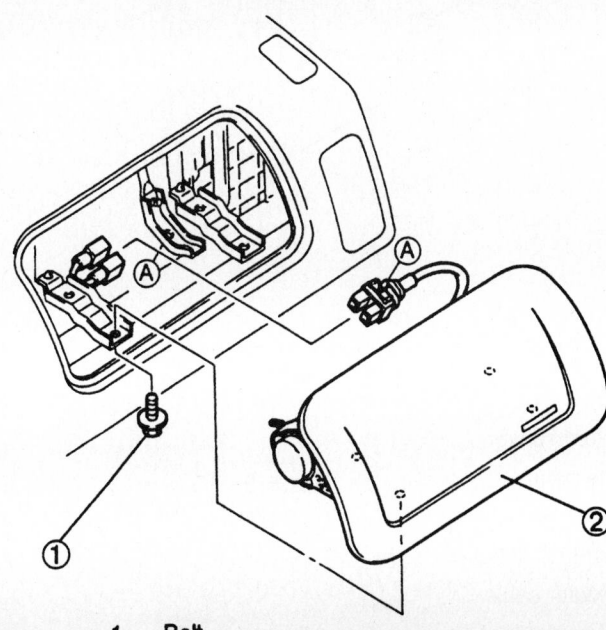

1 **Bolt**
2 **Passenger-side air bag module**

93113GF2

Exploded view of the passenger's side SAS module

9. Remove the heater housing-to-air conditioning housing fastener and seal plate.

10. Remove the heater housing.

11. Disassemble the heater housing and remove the heater core.

To install:

12. Install the heater core and assemble the heater housing.

13. Install the heater housing.

14. Install the heater housing-to-air conditioning housing seal plate and fastener.

15. Install the heater housing-to-chassis nuts.

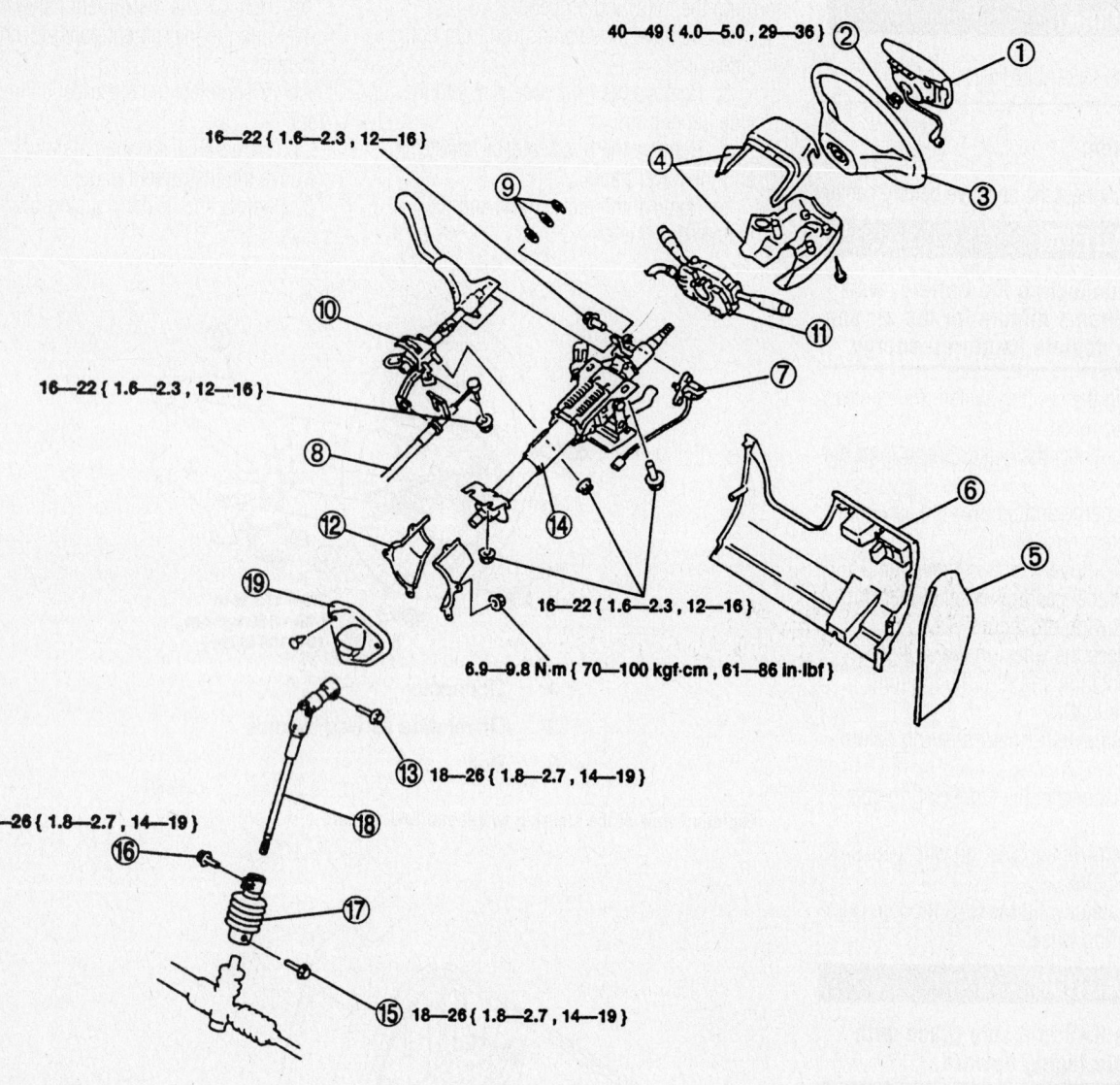

16—22 { 1.6—2.3 , 12—16 }

40—49 { 4.0—5.0 , 29—36 }

16—22 { 1.6—2.3 , 12—16 }

16—22 { 1.6—2.3 , 12—16 }

6.9—9.8 N·m { 70—100 kgf·cm , 61—86 in·lbf }

18—26 { 1.8—2.7 , 14—19 }

18—26 { 1.8—2.7 , 14—19 }

18—26 { 1.8—2.7 , 14—19 }

N·m { kgf·m , ft·lbf }

1	Air bag module
2	Locknut
3	Steering wheel
4	Column cover
5	Side panel
6	Lower panel
7	Shift-lock actuator
8	Selector cable
9	Retaining ring, wave washer, adjustment washer(s)
10	Selector lever component
11	Combination switch
12	Joint cover
13	Fixing bolt (steering shaft/intermediate shaft)
14	Steering shaft
15	Fixing bolt (universal joint/pinion shaft)
16	Fixing bolt (intermediate shaft/universal joint)
17	Universal joint
18	Intermediate shaft
19	Dust cover

Exploded view of the steering column and related components

93113GF3

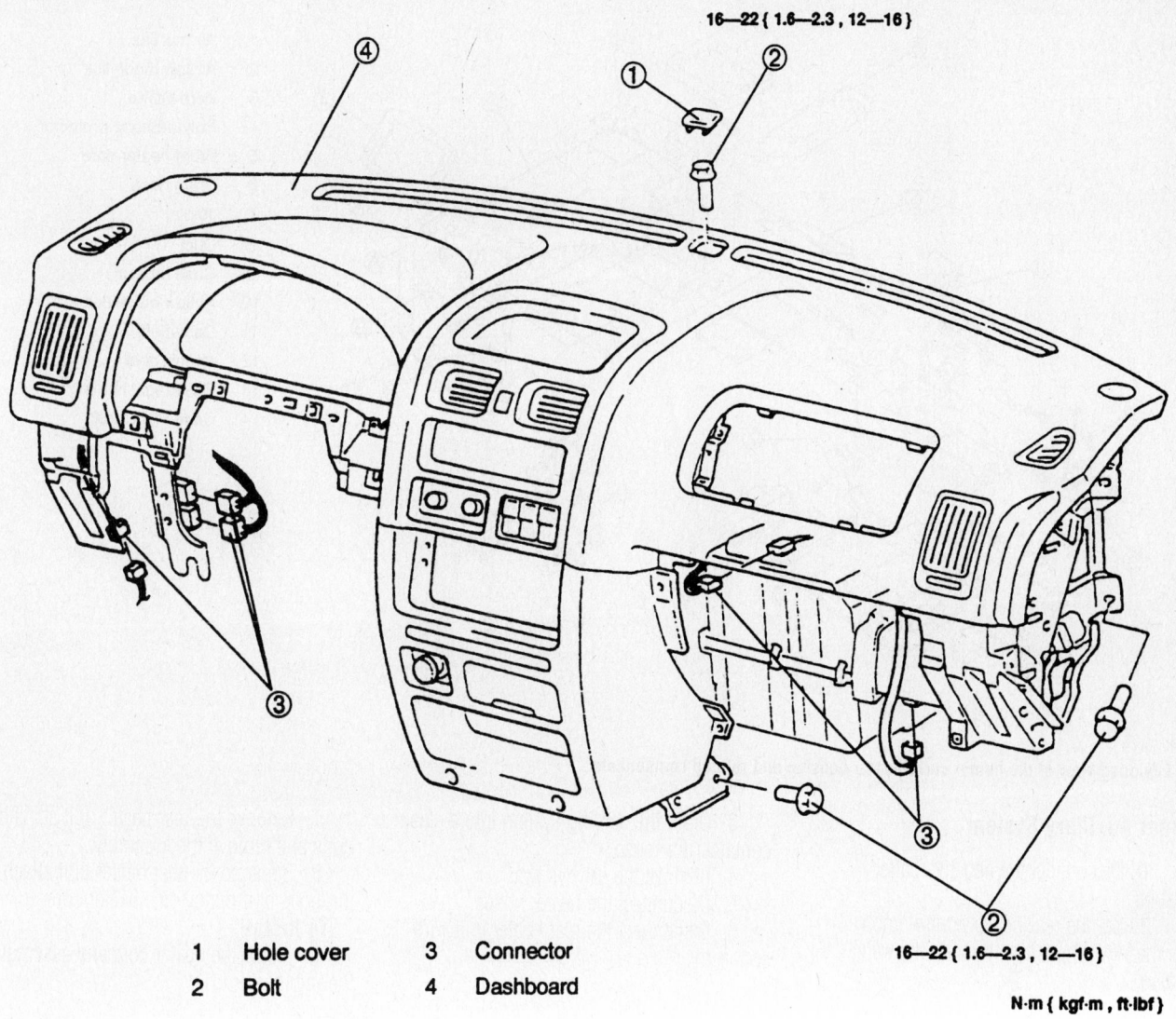

16—22 { 1.6—2.3 , 12—16 }

| 1 | Hole cover | 3 | Connector |
| 2 | Bolt | 4 | Dashboard |

16—22 { 1.6—2.3 , 12—16 }

N·m { kgf·m , ft·lbf }

93113GF4

View of the instrument panel and related components

16. Install the instrument panel by performing the following procedure:

a. Using an assistant, carefully, install the instrument panel.

b. Connect the electrical connectors.

c. Install the instrument panel-to-chassis bolts and the instrument panel hole cover.

d. Connect the electrical connectors to the blower motor and the heater housing.

e. Install the steering column and the steering column-to-instrument panel bolts. Torque the bolts to 12–16 ft. lbs. (16–22 Nm).

f. Install the instrument cluster and the meter hood.

g. Install the hood release handle and the lower panel.

h. Install the front side trim and the side panel trim.

i. Install the A-pillar trim to both sides.

17. At the passenger's side, install the SAS module by performing the following procedure:

a. Carefully, install the SAS module and connect the electrical connector.

b. Install the SAS module-to-dash bolts. Torque the bolts to 12–16 ft. lbs. (16–22 Nm).

c. Install the glove compartment by sliding it to the right; then, push the right side rearward to install the stopper and the pin, then, move it to the right.

18. At the driver's side, install the SAS module and the steering wheel by performing the following procedure:

a. Install the steering wheel and the steering wheel-to-column nut. Torque the steering wheel nut to 29–36 ft. lbs. (40–49 Nm).

b. Carefully, install the SAS module to the steering wheel.

c. Install the steering wheel-to-SAS module bolts. Torque the bolts to 70—104 inch lbs. (7.9–11.7 Nm).

d. Connect the clock spring connector.

e. Install the lower steering column cover.

19. Connect the heater hoses to the heater core.

20. Refill the cooling system.

21. Connect the negative battery cable.

22. Run the engine to normal operating temperatures; then, check the climate control operation and check for leaks.

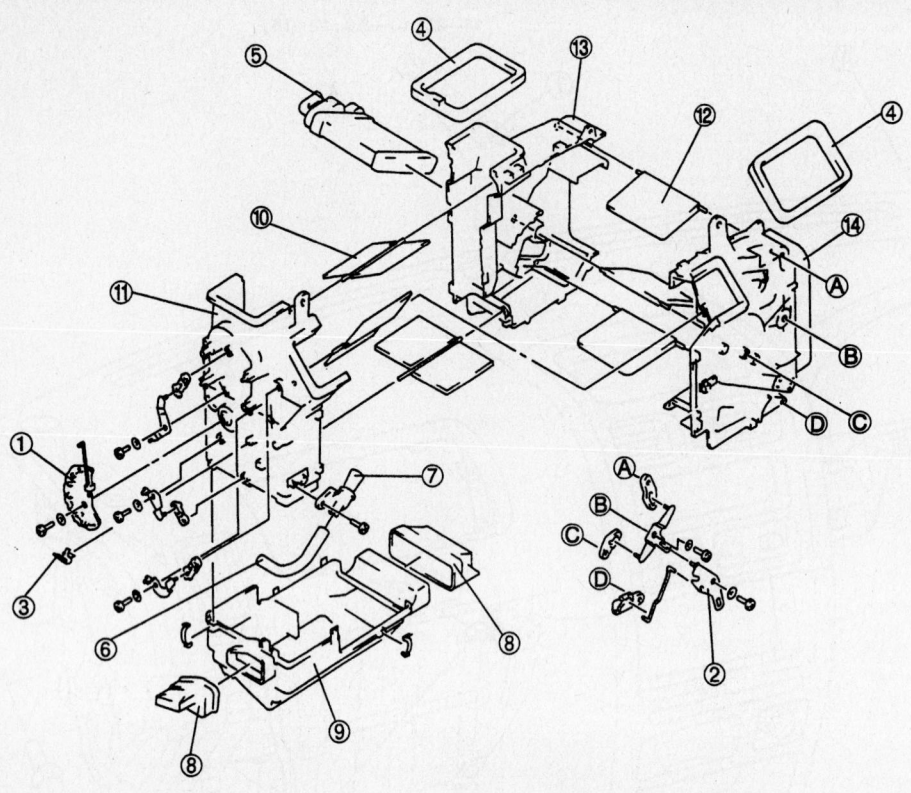

1. Air mix link
2. Airflow mode link
3. Wire clamp
4. Polyurethane protector
5. Front heater core
6. Drain hose
7. Joint
8. Duct
9. Case (bottom)
10. Airflow mode door
11. Case (left)
12. Air mix door
13. Case (right No.1)
14. Case (right No.2)

93113GF5

Exploded view of the heater core, heater housing and related components

Rear Auxiliary System

1. Disconnect the negative battery cable.

2. Set the rear heater control knob to the WARM position to open the water valve.

3. Drain the cooling system into a clean container for reuse.

4. Remove the driver's seat.

5. Disconnect the heater hoses.

6. Disconnect the rear heater unit wire connector.

7. Remove the rear heater unit attaching bolts and remove the assembly.

8. Separate the rear heater unit attaching bolts and remove the heater core.

To install:

9. Install the heater core and assemble the rear heater unit.

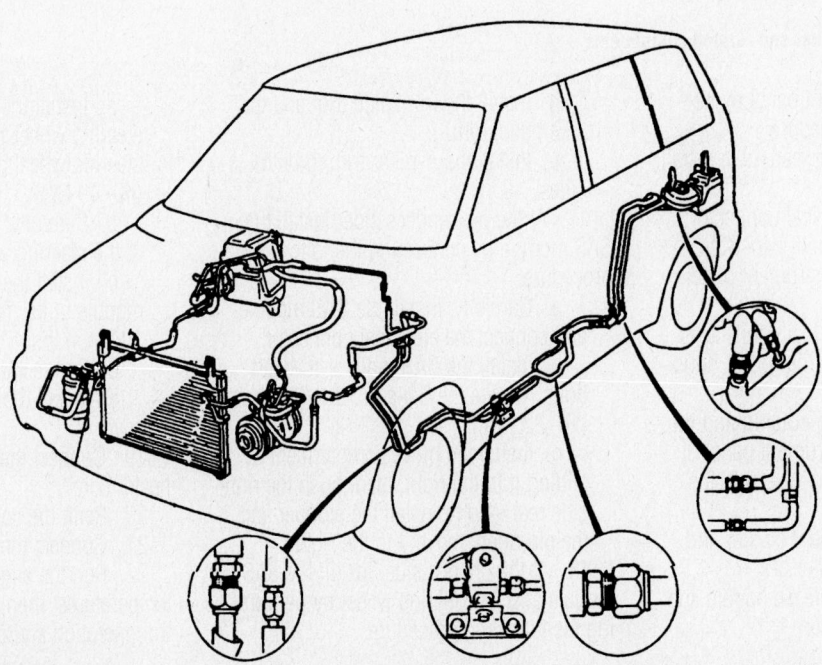

93113G42

View of the rear auxiliary heater assembly and related components—Mazda MPV

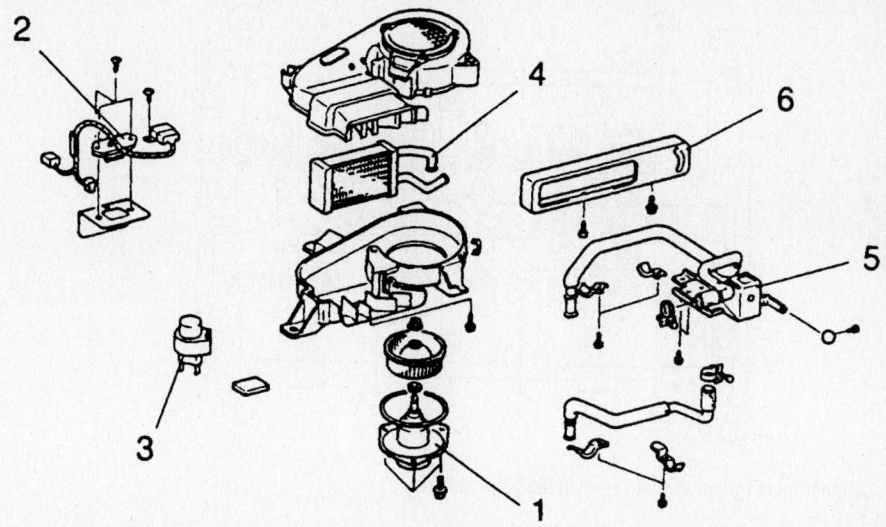

1. Rear heater blower motor
2. Resistor assembly
3. Rear heater relay
4. Heater core
5. Water valve
6. Switch panel

93113G43

Exploded view of the rear auxiliary heater unit—Mazda MPV

10. Install the assembly and the rear heater unit.

11. Connect the rear heater unit wire connector.

12. Connect the heater hoses to the heater core.

13. Install the driver's seat.

14. Refill the cooling system.

15. Connect the negative battery cable.

Cylinder Head

REMOVAL & INSTALLATION

2.5L Engine

1. Before servicing the vehicle, refer to the precautions in the beginning of this section.

2. Drain the cooling system.

3. Drain the engine oil.

4. Relieve the fuel system pressure.

5. Remove or disconnect the following:
- Battery and tray
- Accessory drive belt
- Water pump and drive pulley
- Intake manifold
- Power steering pump
- Intake Manifold Runner Control (IMRC) actuator
- Spark plug wires
- Ignition coil
- Heated Oxygen (HO2S) sensor connectors
- Exhaust front pipe
- Exhaust Gas Recirculation (EGR) pipe
- Exhaust manifolds
- Oil pan

- Alternator
- A/C compressor
- Valve covers
- Front cover
- Timing chains
- Camshafts
- Cylinder heads. Loosen the bolts in several passes and in the sequence shown.

To install:

➡ **The cylinder head bolts are a torque-to-yield design and must be replaced.**

6. Install the cylinder heads with new gaskets. Tighten the bolts in sequence as follows:

   a. Step 1: 28–31 ft. lbs. (37–43 Nm).
   b. Step 2: Plus 90 degrees.
   c. Step 3: Loosen one full turn.
   d. Step 4: 28–31 ft. lbs. (37–43 Nm).
   e. Step 5: Plus 90 degrees.

   f. Step 6: Plus 90 degrees.

7. Install or connect the following:
- Camshafts
- Timing chains
- Front cover
- Valve covers
- A/C compressor
- Alternator
- Oil pan
- Exhaust manifolds
- EGR pipe
- Exhaust front pipe
- HO2S sensor connectors
- Ignition coil
- Spark plug wires
- IMRC actuator
- Power steering pump
- Intake manifold
- Water pump and drive pulley
- Accessory drive belt
- Battery and tray

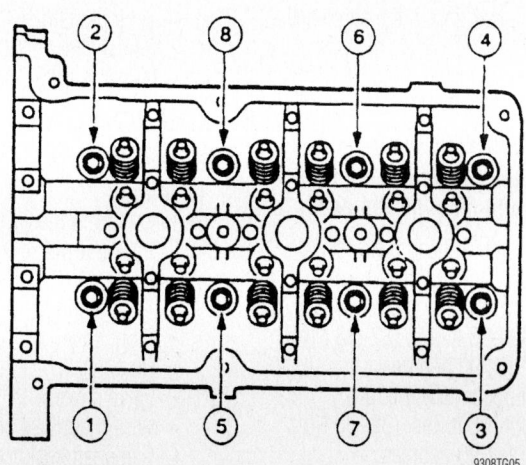

9308TG05

Cylinder head loosening sequence—2.5L and 3.0L engine

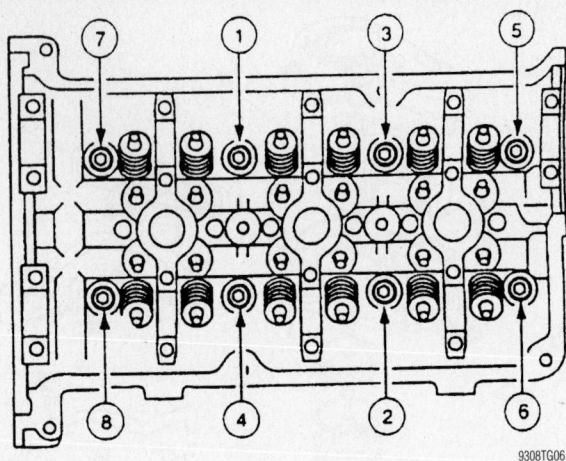

Cylinder head torque sequence—2.5L and 3.0L engines

9308TG06

8. Fill the crankcase to the correct level.
9. Fill the cooling system.
10. Start the engine and check for leaks.

3.0L Engine

1. Before servicing the vehicle, refer to the precautions in the beginning of this section.
2. Drain the cooling system.
3. Drain the engine oil.
4. Relieve the fuel system pressure.
5. Remove or disconnect the following:
   • Accessory drive belt
   • Water pump and drive pulley
   • Timing chains
   • No.3 engine mount rubber and joint bracket
   • Ventilation pipe
   • Water bypass tube
   • Camshafts

➡**Remove the Nos. 1 and 5 caps first. Don't loosen any other cap bolts until these caps are removed.**

   • Rocker arms
   • Cylinder heads. Loosen the bolts in several passes and in the sequence shown.

To install:
6. Installation is the reverse of removal. Observe the following torques:

➡**The cylinder head bolts are a torque-to-yield design and must be replaced.**

7. Install the cylinder heads with new gaskets. Tighten the bolts in sequence as follows:
   a. Step 1: 24–28 ft. lbs. (32–38 Nm).
   b. Step 2: Plus 90 degrees.
   c. Step 3: Loosen one full turn.
   d. Step 4: 24–28 ft. lbs. (32–38 Nm).
   e. Step 5: Plus 90 degrees.
   f. Step 6: Plus 90 degrees.

Rocker Arms/Shafts

REMOVAL & INSTALLATION

2.5L and 3.0L Engines

1. Before servicing the vehicle, refer to the precautions in the beginning of this section.
2. Relieve the fuel system pressure.
3. Drain the engine oil.
4. Remove or disconnect the following:
   • Negative battery cable
   • Intake manifold
   • Accessory drive belt
   • Power steering pump
   • Intake Manifold Runner Control (IMRC) actuator
   • Spark plug wires
   • Ignition coil
   • Water pump drive belt and pulley
   • Camshaft seal housing
   • Wiring harness connector bracket
   • Valve covers
   • Oil pan
   • Front cover
   • Timing chains
   • Camshafts
   • Rocker arms

➡**Keep all valvetrain components in order for assembly.**

To install:
5. Install or connect the following:
   • Rocker arms
   • Camshafts
   • Timing chains
   • Front cover
   • Oil pan
   • Valve covers
   • Wiring harness connector bracket
   • Camshaft seal housing
   • Water pump drive belt and pulley

   • Ignition coil
   • Spark plug wires
   • Intake Manifold Runner Control (IMRC) actuator
   • Power steering pump
   • Accessory drive belt
   • Intake manifold
   • Negative battery cable
6. Fill the crankcase to the correct level.
7. Start the engine and check for leaks.

Intake Manifold

REMOVAL & INSTALLATION

2.5L Engine

1. Before servicing the vehicle, refer to the precautions in the beginning of this section.
2. Relieve the fuel system pressure.
3. Remove or disconnect the following:
   • Negative battery cable
   • Air cleaner housing and fresh air duct
   • Mass Air Flow (MAF) sensor
   • Throttle body intake hose
   • Accelerator cable and bracket
   • Intake Manifold Runner Control (IMRC) cable at the IMRC housing
   • Throttle body
   • Exhaust Gas Recirculation (EGR) valve
   • Idle Air Control (IAC) valve
   • Pressure Regulator Control (PRC) solenoid
   • Intake manifold
   • Fuel lines
   • Fuel pressure regulator vacuum line
   • Fuel supply manifold
   • IMRC housing

To install:
4. Install or connect the following:
   • IMRC housing. Tighten the bolts in sequence to 72–105 inch lbs. (8–12 Nm).
   • Fuel supply manifold. Tighten the bolts to 72–105 inch lbs. (8–12 Nm).
   • Fuel pressure regulator vacuum line
   • Fuel lines
   • Intake manifold. Tighten the bolts in sequence to 72–105 inch lbs. (8–12 Nm).
   • PRC solenoid. Tighten the bolt to 45–61 inch lbs. (5–7 Nm).
   • IAC valve. Tighten the bolts to 72–105 inch lbs. (8–12 Nm).
   • EGR valve

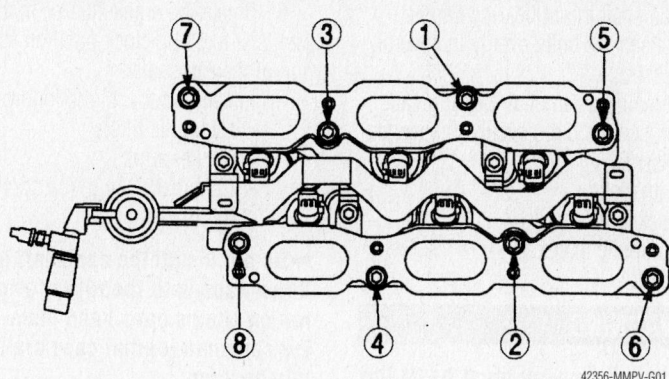

Intake manifold torque sequence—3.0L engine

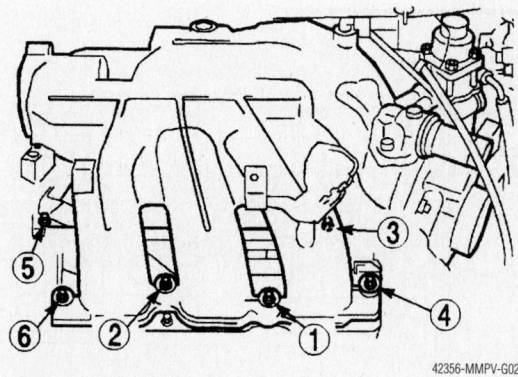

Dynamic chamber torque sequence—3.0L engine

- Throttle body. Tighten the bolts to 72–105 inch lbs. (8–12 Nm).
- IMRC cable
- Accelerator cable and bracket. Tighten the bolt to 71–94 inch lbs. (8–11 Nm).
- Throttle body intake hose
- MAF sensor
- Air cleaner housing and fresh air duct
- Negative battery cable

5. Start the engine and check for leaks.

3.0L Engine

1. Before servicing the vehicle, refer to the precautions in the beginning of this section.
2. Relieve the fuel system pressure.
3. Remove or disconnect the following:
   - Negative battery cable
   - Air cleaner housing and fresh air duct
   - Resonance chamber
   - Mass Air Flow (MAF) sensor
   - Throttle body intake hose
   - Accelerator cable and bracket
   - IMCC actuator
   - Throttle body
   - Exhaust Gas Recirculation (EGR) valve

- Idle Air Control (IAC) valve
- Dynamic chamber
- Intake manifold

To install:

4. Install or connect the following:
   - Intake manifold. Tighten the bolts in sequence to 72–105 inch lbs. (8–12 Nm).
   - Dynamic chamber. Tighten the bolts in sequence to 72–105 inch lbs. (8–12 Nm).

5. The remainder if installation is the reverse of removal.
6. Start the engine and check for leaks.

Exhaust Manifold

REMOVAL & INSTALLATION

2.5L and 3.0L Engines

1. Before servicing the vehicle, refer to the precautions in the beginning of this section.
2. Remove or disconnect the following:
   - Negative battery cable
   - Subframe transverse section
   - Heated Oxygen (HO$_2$S) sensor connectors
   - Exhaust front pipe

Right exhaust manifold torque sequence–2.5L and 3.0L engine

Left exhaust manifold torque sequence–2.5L and 3.0L engine

- Exhaust Gas Recirculation (EGR) pipe
- Exhaust manifolds

To install:

3. Install or connect the following:
   - Exhaust manifolds. Tighten the nuts in sequence to 14 ft. lbs. (20 Nm).
   - EGR pipe
   - Exhaust front pipe
   - HO2S sensor connectors
   - Subframe transverse section. Tighten the bolts to 69–96 ft. lbs. (94–131 Nm).
   - Negative battery cable
4. Start the engine and check for leaks.

Front Crankshaft Seal

REMOVAL & INSTALLATION

Refer to the Timing Chain, Sprockets, Front Cover and Seal procedure in this section.

Camshaft and Valve Lifters

REMOVAL & INSTALLATION

2.5L Engine

1. Before servicing the vehicle, refer to the precautions in the beginning of this section.
2. Relieve the fuel system pressure.
3. Drain the engine oil.
4. Remove or disconnect the following:
   - Negative battery cable
   - Intake manifold
   - Accessory drive belt
   - Power steering pump
   - Intake Manifold Runner Control (IMRC) actuator
   - Spark plug wires
   - Ignition coil
   - Exhaust front pipe
   - Oil pan
   - Alternator and bracket
   - A/C compressor
   - Water pump belt and drive pulley
   - Camshaft oil seal housing
   - Wiring harness connector bracket
   - Valve covers
   - Front cover
   - Timing chains

➡ **Keep all valvetrain components in order for assembly**

➡ **Remove the camshaft thrust bearing caps before loosening any of the other bearing cap bolts.**

- Camshaft thrust bearing caps. Loosen the bolts evenly in several passes.
- Remaining camshaft bearing caps. Loosen the bolts evenly in several passes.
- Camshafts
- Rocker arms
- Hydraulic lifters

To install:

※※ WARNING

The crankshaft keyway must be at the 11 o'clock position before reassembly. Failure to do so may lead to engine damage.

5. Rotate the crankshaft so that the keyway is at the 11 o'clock position for installation of the camshafts.
6. Install or connect the following:
   - Hydraulic lifters
   - Rocker arms
   - Camshafts. Align the sprocket timing marks.

➡ **Do not install the camshaft journal thrust caps until the rocker arms and timing chains have been installed and the camshaft journal caps are secured into position.**

- All camshaft journal caps except the thrust caps.

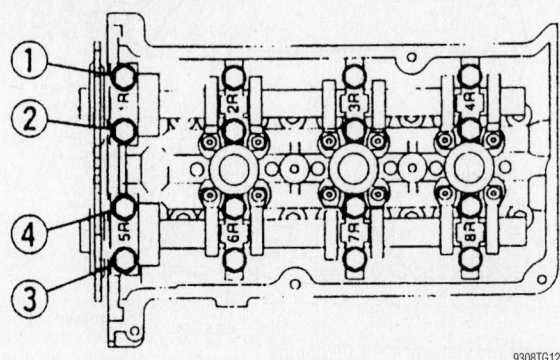

Right bank camshaft thrust cap loosening sequence—2.5L and 3.0L engine

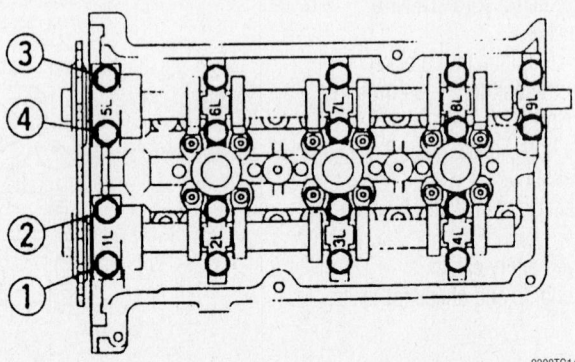

Left bank camshaft thrust cap loosening sequence—2.5L and 3.0L engine

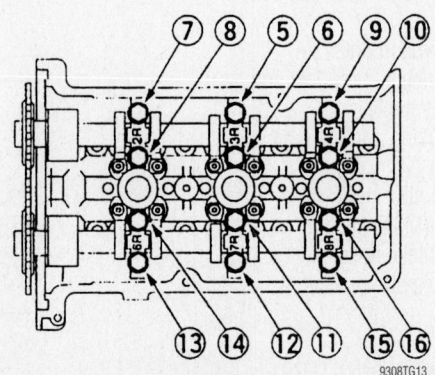

Right bank camshaft bearing cap loosening sequence—2.5L and 3.0L engine

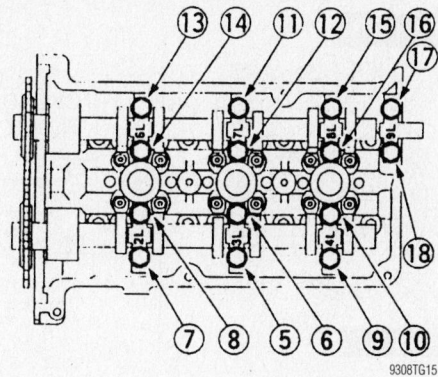

Left bank camshaft bearing cap loosening sequence—2.5L and 3.0L engine

- Timing chains. Tighten the camshaft journal cap bolts in reverse of the loosening order and in several steps to 71–106 inch lbs. (8–12 Nm).
- Thrust caps. Tighten the bolts to 71–106 inch lbs. (8–12 Nm).
- Front cover
- Valve covers
- Wiring harness connector bracket
- Camshaft oil seal housing
- Water pump belt and drive pulley
- A/C compressor
- Alternator and bracket
- Oil pan
- Exhaust front pipe
- Ignition coil
- Spark plug wires
- IMRC actuator
- Power steering pump
- Accessory drive belt
- Intake manifold
- Negative battery cable

7. Fill the crankcase to the correct level.

8. Fill the cooling system.

9. Start the engine and check for leaks.

3.0L Engine

1. Before servicing the vehicle, refer to the precautions in the beginning of this section.

2. Drain the cooling system.

3. Drain the engine oil.

4. Relieve the fuel system pressure.

5. Remove or disconnect the following:
- Accessory drive belt
- Water pump and drive pulley
- Timing chains
- No.3 engine mount rubber and joint bracket
- Ventilation pipe
- Water bypass tube
- Camshafts

➡ **Remove the Nos. 1 and 5 caps first. Don't loosen any other cap bolts until these caps are removed.**

- Rocker arms

To install:

6. Installation is the reverse of removal. Observe the following torques:

7. Install the camshaft caps. Tighten the bolts in sequence to 71–106 inch lbs.

Valve Lash

ADJUSTMENT

The engine covered in this section are equipped with hydraulic lash adjusters. Valve clearance adjustments are not possible.

Starter Motor

REMOVAL & INSTALLATION

2.5L and 3.0L Engines

1. Before servicing the vehicle, refer to the precautions in the beginning of this section.

2. Remove or disconnect the following:
- Battery and tray
- Air intake assembly
- Gear select cable
- Starter harness connectors
- Starter motor

To install:

3. Install or connect the following:
- Starter motor. Tighten the bolts to 28–38 ft. lbs. (38–51 Nm).
- Starter harness connectors. Tighten the battery cable nut to 87–104 inch lbs. (10–12 Nm).
- Gear select cable
- Air intake assembly
- Battery and tray

Oil Pan

REMOVAL & INSTALLATION

2.5L Engine

1. Before servicing the vehicle, refer to the precautions in the beginning of this section.

2. Drain the engine oil.

3. Remove or disconnect the following:
- Negative battery cable
- Subframe transverse section
- Exhaust front pipe
- Flywheel access panel
- Transaxle housing bolts
- Oil pan bolts. Loosen the bolts in reverse of the tightening sequence and in several steps.
- Oil pan

To install:

4. Apply a bead of silicone sealer to the gasket area where the pan meets the parting lines of the lower cylinder block and the front engine cover.

5. Install or connect the following:
- Oil pan. Use a new gasket, tighten the pan bolts in several passes to 15–22 ft. lbs. (20–30 Nm), then tighten the transaxle case bolts to 28–38 ft. lbs. (38–51 Nm).
- Flywheel access panel
- Exhaust front pipe
- Subframe transverse section. Tighten the bolts to 69–96 ft. lbs. (94–131 Nm).
- Negative battery cable

6. Fill the crankcase to the correct level.

7. Start the engine and check for leaks.

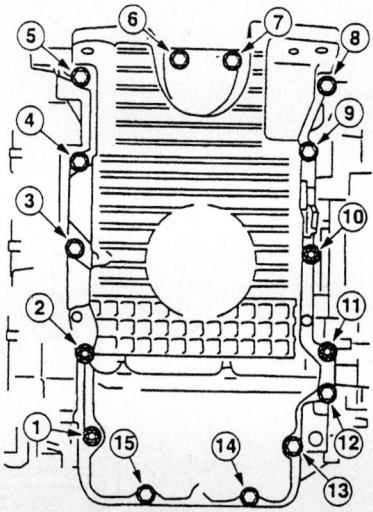

Oil pan torque sequence—2.5L and 3.0L engine

3.0L Engine

1. Before servicing the vehicle, refer to the precautions in the beginning of this section.
2. Drain the engine oil.
3. Remove or disconnect the following:
   - Negative battery cable
   - Exhaust front pipe
   - Flywheel access panel
   - Transaxle housing bolts
   - Oil pan bolts. Loosen the bolts in the sequence shown.
   - Oil pan

To install:

4. Apply a bead of silicone sealer to the gasket area where the pan meets the parting lines of the lower cylinder block and the front engine cover.
5. Install or connect the following:
   - Oil pan. Use a new gasket, tighten the pan bolts in several passes to 15–22 ft. lbs. (20–30 Nm), then tighten the transaxle case bolts to 28–38 ft. lbs. (38–51 Nm).
   - Flywheel access panel
   - Exhaust front pipe
   - Negative battery cable
6. Fill the crankcase to the correct level.
7. Start the engine and check for leaks.

Oil Pump

REMOVAL & INSTALLATION

2.5L and 3.0L Engine

1. Before servicing the vehicle, refer to the precautions in the beginning of this section.
2. Drain the engine oil.
3. Remove or disconnect the following:
   - Negative battery cable
   - Oil pan
   - Timing chains
   - Oil pump pick up tube

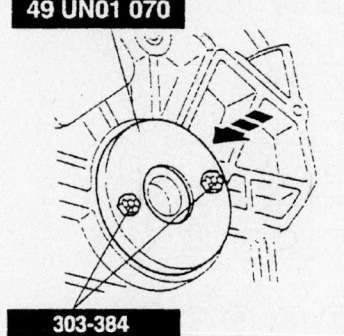
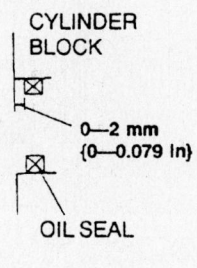
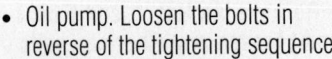

Rear main seal installation—2.5L and 3.0L engines

- Oil pump. Loosen the bolts in reverse of the tightening sequence.

To install:

4. Install or connect the following:
   - Oil pump. Tighten the bolts in sequence to 71–106 inch lbs. (8–12 Nm).
   - Oil pump pick up tube. Tighten the bolts to 71–106 inch lbs. (8–12 Nm) and the nut to 44 inch lbs. (5 Nm) plus 45 degrees.
   - Timing chains
   - Oil pan
   - Negative battery cable
5. Fill the crankcase to the correct level.
6. Start the engine and check for leaks.

Rear Main Seal

REMOVAL & INSTALLATION

2.5L and 3.0L Engines

1. Before servicing the vehicle, refer to the precautions in the beginning of this section.
2. Remove or disconnect the following:
   - Negative battery cable
   - Transaxle

- Flywheel
- Oil seal

To install:

3. Install or connect the following:
   - Oil seal. Press the seal in evenly with Special Service Tools 49 UN01 070 and 303-384 as shown.
   - Flywheel. Tighten the bolts to 54–64 ft. lbs. (73–87 Nm).
   - Transaxle
   - Negative battery cable
4. Start the engine and check for leaks.

Timing Chain, Sprockets, Front Cover and Seal

REMOVAL & INSTALLATION

2.5L and 3.0L Engines

1. Before servicing the vehicle, refer to the precautions in the beginning of this section.
2. Relieve the fuel system pressure.
3. Drain the engine oil.
4. Remove or disconnect the following:
   - Negative battery cable
   - Intake manifold
   - Accessory drive belt
   - Intake Manifold Runner Control (IMRC) actuator
   - Spark plug wires
   - Ignition coil
   - Exhaust front pipe
   - Oil pan
   - Alternator and bracket
   - A/C compressor
   - Wiring harness connector bracket
   - Water pump drive belt and pulley
   - Camshaft oil seal housing
   - Valve covers
   - Right motor mount and bracket
   - Crankshaft pulley
   - Front crankshaft seal
   - Front cover. Loosen the bolts in the sequence shown.

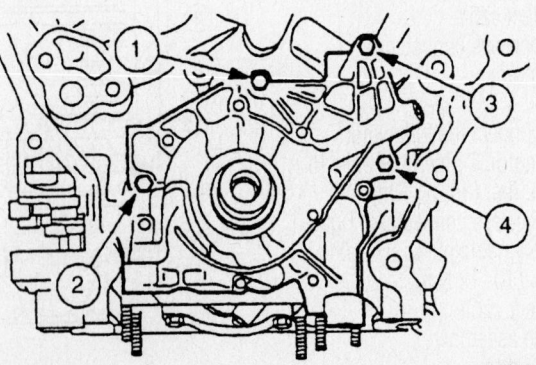

Oil pump torque sequence—2.5L and 3.0L engines

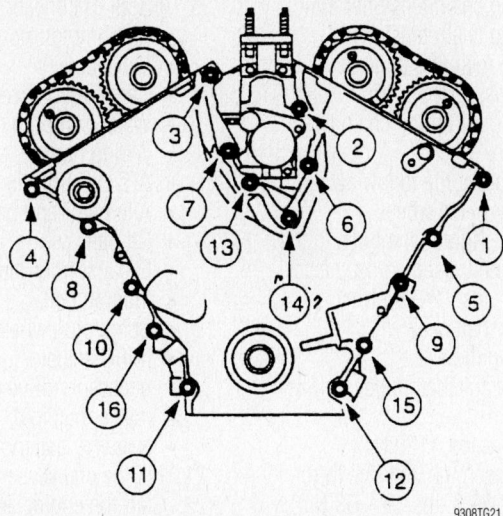

Front cover bolt removal sequence—2.5L and 3.0L engine

* Crankshaft Position (CKP) sensor pulse wheel

5. Rotate the crankshaft so that the keyway is at the 11 o'clock position to locate the crankshaft at TDC for No. 1 cylinder.

6. Verify that the alignment arrows on the camshafts are aligned. If not, rotate the crankshaft 1 complete revolution and recheck.

7. Rotate the crankshaft so that the keyway is at the 3 o'clock position. This positions the right cylinder head camshafts to the neutral position.

➡**Keep all valvetrain components in order for assembly.**

8. Remove or disconnect the following:
* Right timing chain tensioner
* Right timing chain tensioner arm
* Right timing chain and crankshaft timing sprocket
* Right bank camshafts

9. Rotate the crankshaft 1 and ⅔ turns and set the crankshaft keyway at the 11 o'clock position. This places the left bank camshafts in the neutral position.

10. Remove or disconnect the following:
* Left timing chain tensioner
* Left timing chain tensioner arm
* Left timing chain and crankshaft timing sprocket

To install:

11. Prepare the timing chain tensioners for installation as follows:
   a. Place the left chain tensioner in a vise.
   b. Using a small prytool, release and hold the timing chain tensioner ratchet/pawl mechanism through the access hole in the timing chain tensioner.
   c. Slowly compress the tensioner.
   d. Lock the piston with a 1.5mm wire or paperclip.
   e. Repeat for the right chain tensioner.

➡**Be sure that the crankshaft keyway is still at the 11 o'clock position.**

12. Install or connect the following:
* Left timing chain and crankshaft sprocket. Align the colored links

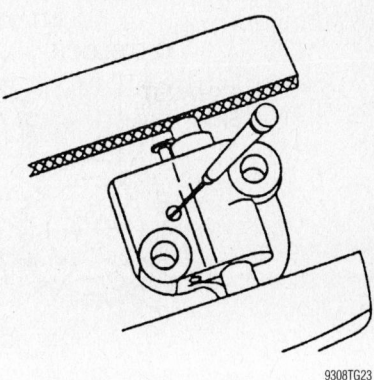

Using a thin prytool, release and hold the timing chain tensioner ratchet/pawl mechanism—2.5L and 3.0L engine

(1) Timing chain crankshaft sprocket
(2) Chain guide
(3) Timing chain
(4) Tensioner arm

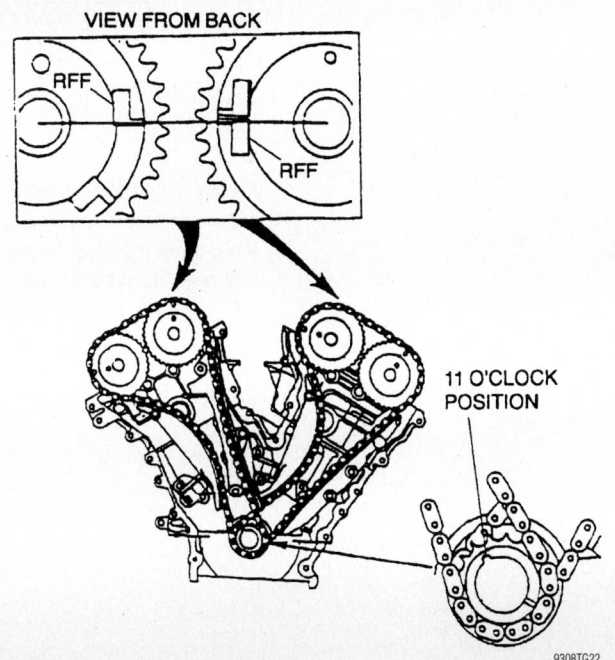

Camshaft alignment with the crankshaft in the 11 o'clock position—2.5L and 3.0L engine

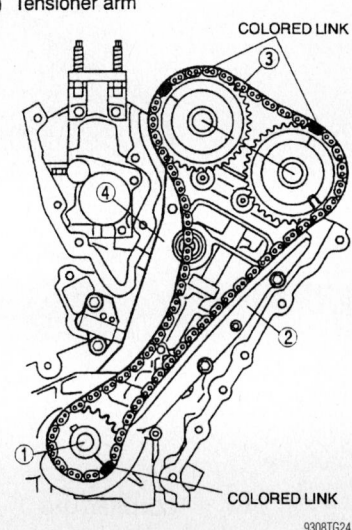

Left bank timing chain alignment—2.5L and 3.0L engine

with the index marks on the camshaft and crankshaft sprockets.
- Left timing chain tensioner arm
- Left timing chain tensioner. Tighten the retaining bolts to 15–22 ft. lbs. (20–30 Nm).

13. Remove the left timing chain tensioner retaining wire.
14. Rotate the crankshaft so that the keyway is at the 3 o'clock position.
15. Install the right bank camshafts with the exhaust camshaft index mark at 12 o'clock and the intake camshaft index mark at 3 o'clock as shown.
16. Install or connect the following:
- Right timing chain and crankshaft sprocket. Align the colored links with the index marks on the camshaft and crankshaft sprockets.

- Right timing chain tensioner arm
- Right timing chain tensioner. Tighten the retaining bolts to 15–22 ft. lbs. (20–30 Nm).

17. Remove the right timing chain tensioner retaining wire.
18. Install or connect the following:
- CKP sensor pulse wheel
- Front cover. Tighten the bolts in the reverse of the loosening sequence to 15–22 ft. lbs. (20–30 Nm).
- Front crankshaft seal
- Crankshaft pulley

19. Tighten the crankshaft pulley bolt as follows:
   a. Step 1: 88 ft. lbs. (120 Nm).
   b. Step 2: Loosen the bolt one turn.
   c. Step 3: 35–39 ft. lbs. (47–53 Nm).
   d. Step 4: Plus 85–95 degrees.

20. Install or connect the following:
- Right motor mount and bracket
- Valve covers
- Camshaft oil seal housing
- Water pump drive belt and pulley
- Wiring harness connector bracket
- A/C compressor
- Alternator and bracket
- Oil pan
- Exhaust front pipe
- Ignition coil
- Spark plug wires
- IMRC actuator
- Accessory drive belt
- Intake manifold
- Negative battery cable

21. Fill the crankcase to the correct level.
22. Start the engine and check for leaks.

Piston and Ring

POSITIONING

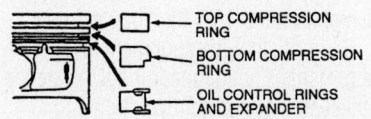

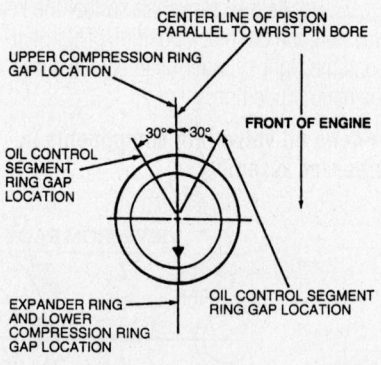

Piston ring positioning, end-gap spacing, and piston positioning. The small directional arrow must face the front of the engine—2.5L and 3.0L engine

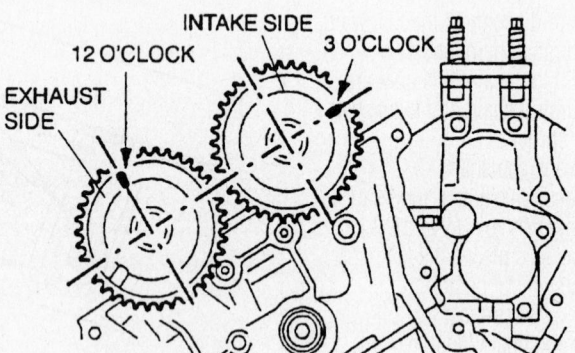

Right bank camshaft positioning—2.5L and 3.0L engine

(1) Chain guide
(2) Timing chain
(3) Tensioner arm

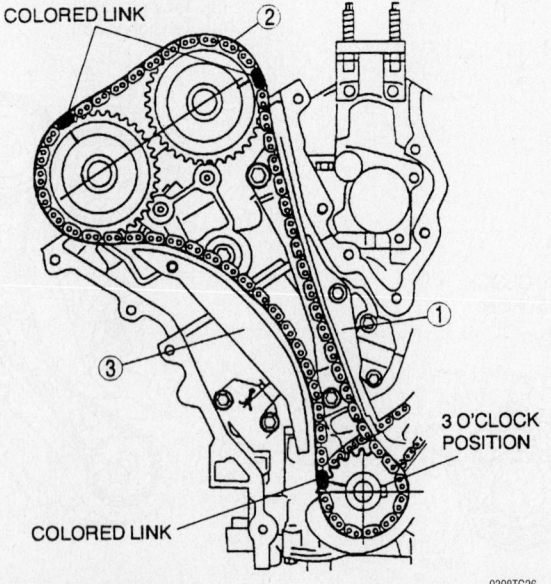

Right bank timing chain alignment—2.5L and 3.0L engine

FUEL SYSTEM

Fuel System Service Precautions

Safety is the most important factor when performing not only fuel system maintenance but any type of maintenance. Failure to conduct maintenance and repairs in a safe manner may result in serious personal injury or death. Maintenance and testing of the vehicle's fuel system components can be accomplished safely and effectively by adhering to the following rules and guidelines.

• To avoid the possibility of fire and personal injury, always disconnect the negative battery cable unless the repair or test procedure requires that battery voltage be applied.

• Always relieve the fuel system pressure prior to disconnecting any fuel system component (injector, fuel rail, pressure regulator, etc.), fitting or fuel line connection. Exercise extreme caution whenever relieving fuel system pressure, to avoid exposing skin, face and eyes to fuel spray. Please be advised that fuel under pressure may penetrate the skin or any part of the body that it contacts.

• Always place a shop towel or cloth around the fitting or connection prior to loosening to absorb any excess fuel due to spillage. Ensure that all fuel spillage (should it occur) is quickly removed from engine surfaces. Ensure that all fuel soaked cloths or towels are deposited into a suitable waste container.

• Always keep a dry chemical (Class B) fire extinguisher near the work area.

• Do not allow fuel spray or fuel vapors to come into contact with a spark or open flame.

• Always use a back-up wrench when loosening and tightening fuel line connection fittings. This will prevent unnecessary stress and torsion to fuel line piping. Always follow the proper tighten specifications.

• Always replace worn fuel fitting O-rings with new. Do not substitute fuel hose or equivalent, where fuel pipe is installed.

Fuel System Pressure

RELIEVING

1. Before servicing the vehicle, refer to the precautions in the beginning of this section.
2. Disconnect the fuel pump relay, located at the ECM.

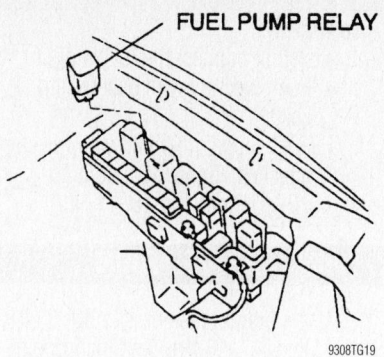

Fuel pump relay—2001

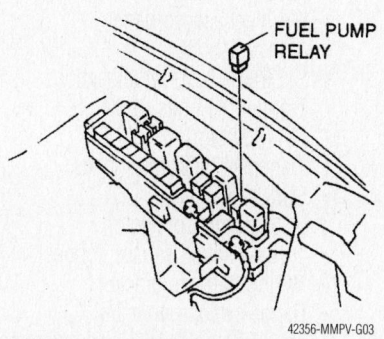

Fuel pump relay—2002

3. Start the engine.
4. After the engine stalls, crank the engine several times.
5. Turn the ignition switch **OFF**.
6. When repairs are complete, connect the fuel pump relay.

Fuel Filter

REMOVAL & INSTALLATION

The fuel filter is located in the fuel tank as part of the fuel pump module.

Fuel Pump

REMOVAL & INSTALLATION

1. Before servicing the vehicle, refer to the precautions in the beginning of this section.
2. Relieve the fuel system pressure.
3. Remove or disconnect the following:
   • Negative battery cable
   • Front seats
   • Center console
   • Door sill plates
   • Parking brake lever
   • Carpet
   • Access panel
   • Fuel lines
   • Fuel pump module harness connector
   • Fuel pump module

To install:

4. Install or connect the following:
   • Fuel pump module
   • Fuel pump module harness connector
   • Fuel lines
   • Access panel
   • Carpet
   • Parking brake lever
   • Door sill plates
   • Center console
   • Front seats
   • Negative battery cable
5. Start the engine and check for leaks.

Fuel Injector

REMOVAL & INSTALLATION

2.5L and 3.0L Engine

1. Before servicing the vehicle, refer to the precautions in the beginning of this section.

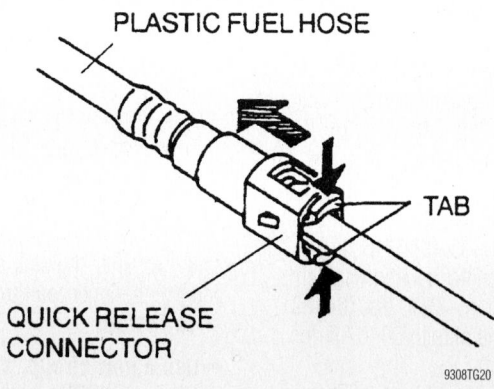

Fuel hose quick release connector

2. Relieve the fuel system pressure.
3. Remove or disconnect the following:
   - Negative battery cable
   - Air cleaner housing and fresh air duct
   - Mass Air Flow (MAF) sensor
   - Intake manifold
   - Fuel injector harness connectors
   - Fuel lines
   - Pressure regulator vacuum hose

- Fuel supply manifold with injectors attached
- Fuel injectors

To install:

4. Install or connect the following:
   - Fuel injectors with new O-ring seals
   - Fuel supply manifold with injectors attached. Tighten the bolts to 72–101 inch lbs. (8–11 Nm).

- Pressure regulator vacuum hose
- Fuel lines
- Fuel injector harness connectors
- Intake manifold
- MAF sensor
- Air cleaner housing and fresh air duct
- Negative battery cable
5. Start the engine and check for leaks.

DRIVE TRAIN

Automatic Transaxle Assembly

REMOVAL & INSTALLATION

1. Before servicing the vehicle, refer to the precautions in the beginning of this section.
2. Drain the transaxle fluid.
3. Attach a support fixture to the engine lifting eyes.
4. Remove or disconnect the following:
   - Battery and tray
   - Air cleaner assembly
   - Mass Air Flow (MAF) sensor
   - Front wheels
   - Inner fender liner
   - Starter motor
   - Transaxle solenoid valve connector
   - Range switch connector
   - Wiring harness bracket
   - Turbine speed sensor connector
   - Vehicle Speed (VSS) sensor connector
   - Shift cable
   - Transaxle oil cooler hoses
   - Subframe transverse section
   - Axle halfshafts
   - Flywheel access panel
   - Torque converter
   - Left engine mount bracket
   - Subframe center section
   - Rear engine mount
   - Transaxle flange bolts. Support the transaxle.
5. Lower the transaxle from the vehicle.

To install:

6. Install or connect the following:
   - Transaxle. Tighten the flange bolts to 28–38 ft. lbs. (38–51 Nm).
   - Rear engine mount. Tighten the bracket bolts to 50–68 ft. lbs. (67–93 Nm) and the through bolt to 63–86 ft. lbs. (86–116 Nm).
   - Subframe center section. Tighten the bolts to 48–65 ft. lbs. (64–89 Nm) and the nuts to 50–67 ft. lbs. (67–93 Nm).
   - Left engine mount bracket. Tighten

the bracket fasteners to 50–68 ft. lbs. (67–93 Nm) and the through bolt to 63–86 ft. lbs. (86–116 Nm).
   - Torque converter. Tighten the nuts to 26–36 ft. lbs. (35–49 Nm).
   - Flywheel access panel
   - Axle halfshafts
   - Subframe transverse section. Tighten the bolts to 69–97 ft. lbs. (94–131 Nm).
   - Transaxle oil cooler hoses
   - Shift cable
   - VSS sensor connector
   - Turbine speed sensor connector
   - Wiring harness bracket
   - Range switch connector
   - Transaxle solenoid valve connector
   - Starter motor
   - Inner fender liner
   - Front wheels
   - MAF sensor
   - Air cleaner assembly
   - Battery and tray
7. Fill the transaxle to the correct level.
8. Start the engine and check for leaks.

Halfshaft

REMOVAL & INSTALLATION

Left

1. Before servicing the vehicle, refer to the precautions in the beginning of this section.
2. Drain the transaxle fluid.
3. Remove or disconnect the following:
   - Front wheel
   - Wheel speed sensor
   - Hub locknut
   - Outer tie rod end
   - Lower ball joint
   - Stabilizer bar link
4. Separate the stub shaft from the hub and pry the inner joint from the transaxle.

To install:

➡ **Use a new circlip, split pin and locknut for assembly.**

5. Insert the stub shaft into the wheel hub.
6. Lubricate the oil seal with transaxle fluid, then push the axle halfshaft into the transaxle. Pull on the inner joint to confirm that the circlip is seated.
7. Install or connect the following:
   - Stabilizer bar link
   - Lower ball joint. Tighten the pinch bolt to 32–43 ft. lbs. (44–58 Nm).
   - Outer tie rod end. Tighten the nut to 24–32 ft. lbs. (32–44 Nm).
   - Hub locknut. Tighten the nut to 174–235 ft. lbs. (236–318 Nm) for 2001–02; 174–202 ft. lbs. (235–274 Nm) for 2003–04.
   - Wheel speed sensor
   - Front wheel

Right

✳✳ WARNING

Attempting to remove the right axle halfshaft while the center shaft support bracket is installed may result in damage to the center shaft support bracket.

1. Before servicing the vehicle, refer to the precautions in the beginning of this section.
2. Remove or disconnect the following:
   - Front wheel
   - Wheel speed sensor
   - Hub locknut
   - Brake caliper and rotor
   - Outer tie rod end
   - Lower ball joint
   - Strut bracket bolts
   - Steering knuckle. Separate the stub shaft from the wheel hub.
   - Center shaft support bracket
   - Axle halfshaft and center shaft assembly
3. Separate the axle halfshaft and the center shaft as follows:
   a. Step 1: Place the center shaft in a vise.

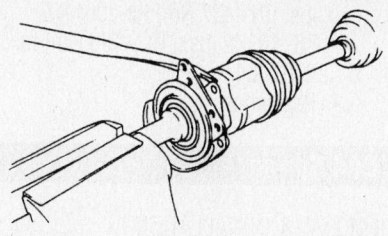

Separating the axle halfshaft from the center shaft

b. Step 2: Insert a pry tool between the center shaft and the axle halfshaft.

c. Step 3: Tap on the pry tool to separate the axle halfshaft from the center shaft.

To install:

➡**Use a new split pin, locknut, and new circlips for assembly.**

4. Place the axle halfshaft in a vise and install the center shaft by tapping it with a plastic hammer as shown.

5. Lubricate the oil seal with transaxle fluid, then push the center shaft into the transaxle. Pull on the inner joint to confirm that the circlip is seated.

6. Install or connect the following:
- Center shaft support bracket. Tighten the nuts to 16–22 ft. lbs. (22–30 Nm).

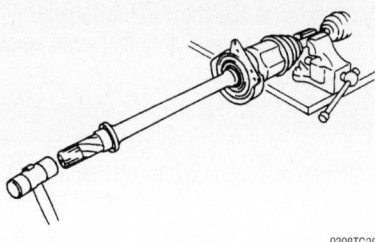

Installing the center shaft

- Steering knuckle. Guide the stub shaft into the wheel hub.
- Strut bracket bolts. Tighten the bolts to 76–90 ft. lbs. (103–122 Nm).
- Lower ball joint. Tighten the pinch bolt to 32–43 ft. lbs. (44–58 Nm).
- Outer tie rod end. Tighten the nut to 24–32 ft. lbs. (32–44 Nm).
- Hub locknut. Tighten the nut to 174–235 ft. lbs. (236–318 Nm) for 2001–02; 174–202 ft. lbs. (235–274 Nm) for 2003–04.
- Brake caliper and rotor. Tighten the caliper bracket bolts to 66–79 ft. lbs. (89–107 Nm).
- Wheel speed sensor
- Front wheel

7. Check the wheel alignment and adjust as necessary.

CV-Joint

OVERHAUL

Outer CV-Joint

The outer CV-joint is serviced with the axle halfshaft as an assembly. The outer CV-joint boot may be serviced by removing the inner joint.

Inner CV-Joint

1. Before servicing the vehicle, refer to the precautions in the beginning of this section.

2. Remove or disconnect the following:
- Axle halfshaft from the vehicle
- Inner CV-joint boot clamps
- Housing retainer clip
- CV-joint housing

- CV-joint balls and cage
- Snapring
- CV-joint inner race
- CV-joint boot

To install:

➡**Use new snaprings, clips, and boot clamps for assembly.**

3. Install or connect the following:
- CV-joint boot
- CV-joint inner race
- Snapring
- CV-joint balls and cage
- CV-joint housing
- Housing retainer clip

4. Fill the CV-joint housing and boot with CV-joint grease and tighten the boot clamps.

5. Install the axle halfshaft.

Inner Tri-Pot Joint

1. Before servicing the vehicle, refer to the precautions in the beginning of this section.

2. Remove or disconnect the following:
- Axle halfshaft from the vehicle
- Inner tri-pot joint boot clamps
- Tri-pot joint housing
- Snapring
- Tri-pot joint

To install:

➡**Use new snaprings, clips, and boot clamps for assembly.**

3. Install or connect the following:
- Tri-pot joint
- Snapring
- Tri-pot joint housing

4. Fill the tri-pot joint housing and boot with grease and tighten the boot clamps.

5. Install the axle halfshaft.

STEERING AND SUSPENSION

Air Bag

✳✳ CAUTION

Some vehicles are equipped with an air bag system. The system must be disarmed before performing service on, or around, system components, the steering column, instrument panel components, wiring and sensors. Failure to follow the safety precautions and the disarming procedure could result in accidental air bag deployment, possible injury and unnecessary system repairs.

PRECAUTIONS

Several precautions must be observed when handling the inflator module to avoid accidental deployment and possible personal injury.

- Never carry the inflator module by the wires or connector on the underside of the module.
- When carrying a live inflator module, hold securely with both hands, and ensure that the bag and trim cover are pointed away.
- Place the inflator module on a bench or other surface with the bag and trim cover facing up.

- With the inflator module on the bench, never place anything on or close to the module which may be thrown in the event of an accidental deployment.

DISARMING

1. Turn the ignition switch to the **LOCK** position.

2. Disconnect the negative battery cable and wait at least 1 minute to allow the back-up power supply to deplete its stored power.

3. When repairs are complete, connect the negative battery cable.

Power Rack and Pinion Steering Gear

REMOVAL & INSTALLATION

1. Before servicing the vehicle, refer to the precautions in the beginning of this section.

2. Attach a support fixture to the engine lifting eyes.

3. Remove or disconnect the following:
   - Front wheels
   - Wheel speed sensors
   - Steering shaft pinch bolt
   - Outer tie rod ends
   - Subframe transverse section
   - Subframe center section
   - Power steering pressure and return lines
   - Steering gear

To install:

4. Install or connect the following:
   - Steering gear. Tighten the fasteners in sequence to 55–77 ft. lbs. (75–104 Nm).
   - Power steering pressure and return lines
   - Subframe center section. Tighten the bolts to 48–65 ft. lbs. (64–89 Nm) and the nuts to 50–68 ft. lbs. (67–93 Nm).
   - Subframe transverse section. Tighten the bolts to 69–96 ft. lbs. (94–131 Nm).
   - Outer tie rod ends. Tighten the nuts to 24–32 ft. lbs. (32–44 Nm).
   - Steering shaft pinch bolt. Tighten the bolt to 14–19 ft. lbs. (19–26 Nm).
   - Wheel speed sensors
   - Front wheels

5. Fill the power steering reservoir.

6. Check the wheel alignment and adjust as necessary.

Strut

REMOVAL & INSTALLATION

1. Before servicing the vehicle, refer to the precautions in the beginning of this section.

2. Remove or disconnect the following:
   - Front wheel
   - Brake hose clip
   - Stabilizer bar link
   - Steering knuckle bolts
   - Upper strut mount nuts
   - Strut assembly

To install:

3. Install or connect the following:
   - Strut assembly. Tighten the upper strut mount nuts to 34–46 ft. lbs. (47–62 Nm) and the steering knuckle bolts to 76–90 ft. lbs. (103–122 Nm).
   - Stabilizer bar link. Tighten the nut to 32–44 ft. lbs. (44–60 Nm).
   - Brake hose clip
   - Front wheel

4. Check the wheel alignment and adjust as necessary.

Shock Absorber

REMOVAL & INSTALLATION

1. Before servicing the vehicle, refer to the precautions in the beginning of this section.

2. Support the rear axle with a jack or stands.

3. Remove or disconnect the following:
   - Rear wheel
   - Shock absorber

To install:

4. Install or connect the following:
   - Shock absorber. Tighten the upper nut to 56–76 ft. lbs. (76–102 Nm). Tighten the lower bolt to 71–94 ft.

lbs. (97–127 Nm) for 2001–02; 76–102 ft. lbs. (103–139 Nm) for 2003–04.
   - Rear wheel

Coil Spring

REMOVAL & INSTALLATION

Front

1. Before servicing the vehicle, refer to the precautions in the beginning of this section.

2. Remove the strut assembly from the vehicle.

3. Compress the coil spring and remove the piston rod nut.

4. Remove or disconnect the following:
   - Upper strut mount
   - Strut mount bearing
   - Spring upper seat
   - Coil spring

To install:

5. Install or connect the following:
   - Coil spring
   - Spring upper seat
   - Strut mount bearing
   - Upper strut mount. Tighten the piston rod nut to 66–94 ft. lbs. (90–127 Nm).

6. Remove the spring compressor and install the strut assembly to the vehicle.

7. Check the wheel alignment and adjust as necessary.

Rear

1. Before servicing the vehicle, refer to the precautions in the beginning of this section.

2. Support the vehicle at the frame and support the axle with a jack.

3. Remove or disconnect the following:
   - Rear wheels
   - Lateral rod
   - Shock absorber

4. Lower the rear axle and remove the coil springs.

To install:

5. Place the coil springs on the spring seats and raise the axle into position.

6. Install or connect the following:
   - Shock absorber. Tighten the upper nut to 56–76 ft. lbs. (76–102 Nm) and the lower bolt to 71–94 ft. lbs. (97–127 Nm).
   - Lateral rod. Tighten the fastener to 76–101 ft. lbs. (102–137 Nm).
   - Rear wheels

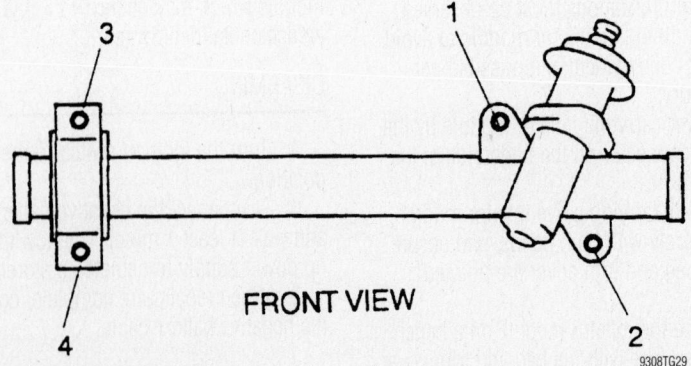

FRONT VIEW

9308TG29

Steering gear torque sequence

Lower Ball Joint

REMOVAL & INSTALLATION

The lower ball joint is serviced with the lower control arm as an assembly.

Lower Control Arm

REMOVAL & INSTALLATION

2001

1. Before servicing the vehicle, refer to the precautions in the beginning of this section.
2. Support the arm.
3. Remove or disconnect the following:
   • Front wheel
   • Lower ball joint
   • Front inner control arm bolt
   • Rear inner control arm bracket
   • Lower control arm

To install:

4. Install or connect the following:
   • Lower control arm. Tighten the front inner bolt to 69–93 ft. lbs. (94–126 Nm).
   • Rear inner control arm bracket. Tighten the nut to 69–97 ft. lbs. (94–131 Nm).
   • Lower ball joint. Tighten the pinch bolt to 32–43 ft. lbs. (44–58 Nm).
   • Front wheel
5. Check the wheel alignment and adjust as necessary.

2002–04

1. Before servicing the vehicle, refer to the precautions in the beginning of this section.
2. Support the arm.
3. Remove or disconnect the following:
   • Front wheel
   • Pivot bolt
   • Dynamic damper
   • Ball joint bolt
   • Ball joint bracket
   • Nut
   • Lower arm
4. Installation is the reverse of removal. Observe the following torques:
   • Nut: 76–97 ft. lbs. (103–131 Nm)
   • Bracket bolt: 32–43 ft. lbs. (43–59 Nm)
   • Pivot bolt: 74–101 ft. lbs. (101–137 Nm)

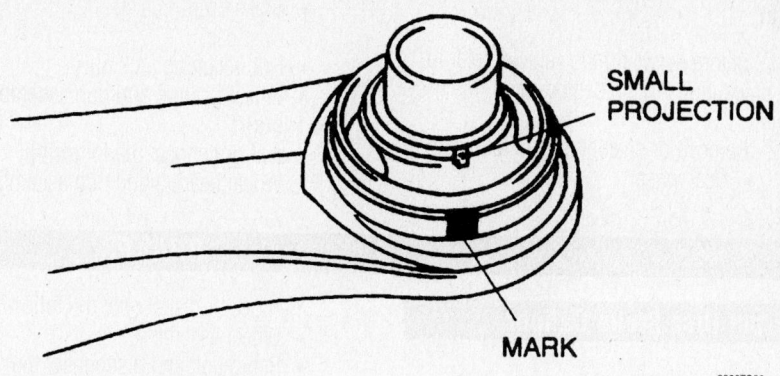

Rear bushing alignment marks

CONTROL ARM BUSHING REPLACEMENT

1. Before servicing the vehicle, refer to the precautions in the beginning of this section.
2. Remove the control arm from the vehicle.
3. Mark the control arm to indicate the alignment of the rear bushing as shown.
4. Remove the control arm bushings with a hydraulic press.

To install:

5. Lubricate the control arm bushings with liquid soap.
6. If replacing the rear bushing, align the direction marks as shown.
7. Press the bushings into the control arm until the bushing flange contacts the housing edge of the control arm.
8. Install the control arm to the vehicle.
9. Check the wheel alignment and adjust as necessary.

Wheel Bearing

ADJUSTMENT

1. Before servicing the vehicle, refer to the precautions in the beginning of this section.
2. Remove or disconnect the following:
   • Front wheel
   • Brake caliper and rotor
3. Position a dial indicator gauge against the wheel hub. Push and pull the wheel hub in and out and measure the end-play of the wheel bearing.
4. End-play should not exceed 0.002 in. (0.05mm).
5. If end-play is excessive, replace the hub retainer locknut and tighten it to specification. Recheck the end-play.
6. If end-play is not within specification, replace the wheel bearing assembly.

REMOVAL & INSTALLATION

Front

1. Before servicing the vehicle, refer to the precautions in the beginning of this section.
2. Remove or disconnect the following:
   • Front wheel
   • Brake caliper and rotor
   • Wheel speed sensor
   • Outer tie rod end
   • Lower ball joint
   • Hub retainer locknut
   • Strut bracket bolts
   • Steering knuckle
   • Inner oil seal
   • Hub
   • Snapring
   • Wheel bearing cartridge

To install:

➡**Use new locknuts, split pins and oil seals for assembly.**

3. Install or connect the following:
   • Wheel bearing cartridge
   • Snapring
   • Hub
   • Inner oil seal
   • Steering knuckle. Tighten the strut bracket bolts to 76–90 ft. lbs. (103–122 Nm).
   • Hub retainer locknut. Tighten the nut to 174–235 ft. lbs. (236–318 Nm) for 2001–02; 174–202 ft. lbs. (235–274 Nm) for 2003–04.
   • Lower ball joint. Tighten the pinch bolt to 32–43 ft. lbs. (44–58 Nm).
   • Outer tie rod end. Tighten the nut to 24–32 ft. lbs. (32–44 Nm).
   • Wheel speed sensor. Tighten the bolt to 14–18 ft. lbs. (19–25 Nm).
   • Brake caliper and rotor
   • Front wheel

Rear

1. Before servicing the vehicle, refer to the precautions in the beginning of this section.
2. Remove or disconnect the following:
   - Rear wheel

- Brake drum
- Dust cap
- Hub retaining lock nut
- Wheel bearing and hub assembly

To install:
3. Install or connect the following:
   - Wheel bearing and hub assembly.

Tighten the locknut to 131–173 ft. lbs. (177–235 Nm).
- Dust cap
- Brake drum
- Rear wheel

BRAKES

Front Disc Brake Caliper

REMOVAL & INSTALLATION

2001

1. Before servicing the vehicle, refer to the precautions in the beginning of this section.

2. Remove or disconnect the following:
   - Wheel assembly
   - Banjo bolt and disconnect the brake hose from the caliper. Plug the hose to prevent fluid leakage.
   - Caliper mounting bolt and pivot the caliper about the mounting pin and off the brake rotor
   - Caliper from the pin

3. Installation is the reverse of the removal procedure. Lubricate the caliper mounting bolts or bolt and pin prior to installation.
4. Tighten the caliper mounting bolt(s) to 61–69 ft. lbs. (83–93 Nm).
5. Bleed the brake system.

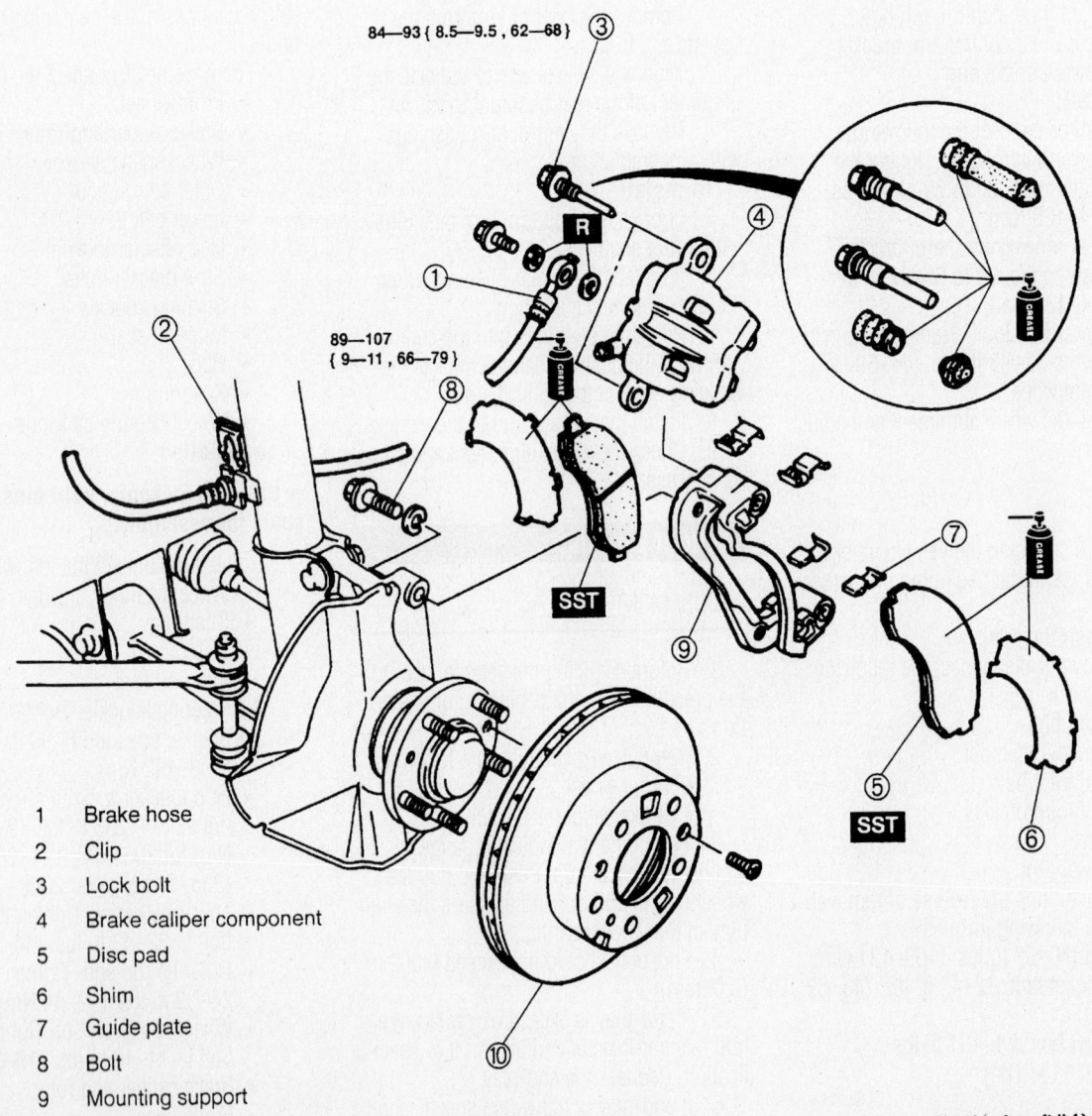

1	Brake hose
2	Clip
3	Lock bolt
4	Brake caliper component
5	Disc pad
6	Shim
7	Guide plate
8	Bolt
9	Mounting support
10	Disc plate

N·m { kgf·m , ft·lbf }

93026G35

Front disc brake assembly—2001

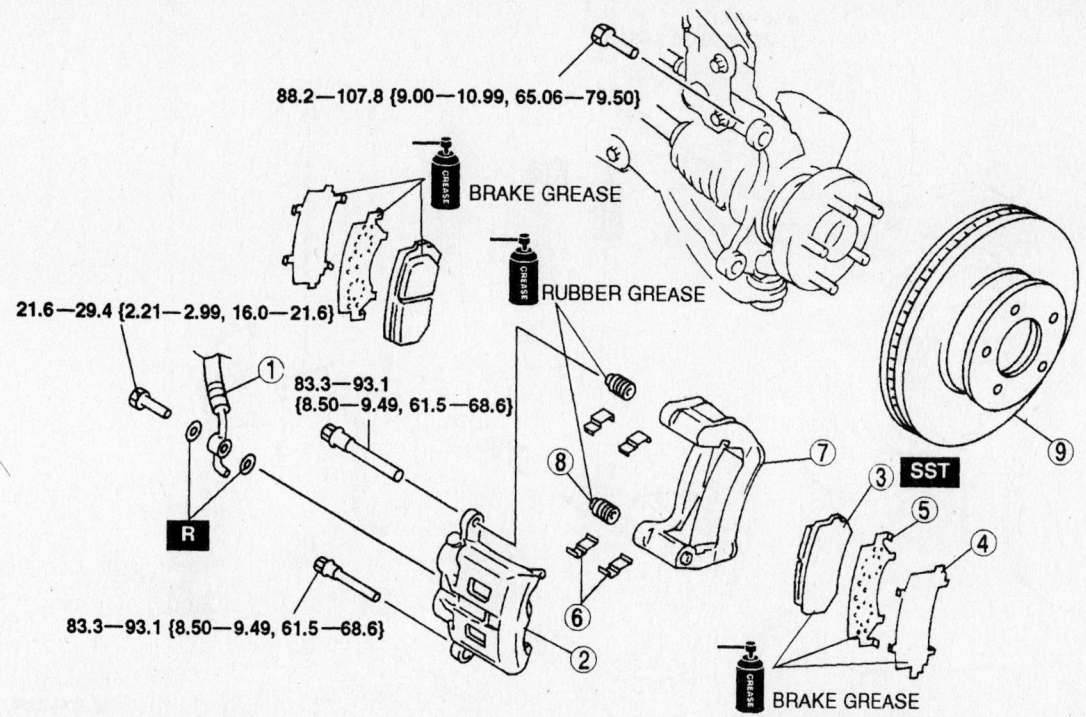

88.2—107.8 {9.00—10.99, 65.06—79.50}

BRAKE GREASE

RUBBER GREASE

21.6—29.4 {2.21—2.99, 16.0—21.6}

83.3—93.1 {8.50—9.49, 61.5—68.6}

R

83.3—93.1 {8.50—9.49, 61.5—68.6}

SST

BRAKE GREASE

N·m {kgf·m, ft·lbf}

1	Flexible hose
2	Caliper
3	Disc pad
4	Outer shim
5	Inner shim

6	Guide plate
7	Mounting support
8	Dust boot
9	Disc plate

42356-MMPV-G04

Front disc brake assembly—2002

2002–04

1. Before servicing the vehicle, refer to the precautions in the beginning of this section.
2. Remove or disconnect the following:
   - Wheel assembly
   - Banjo bolt and disconnect the brake hose from the caliper. Plug the hose to prevent fluid leakage.
   - Caliper mounting bolts
3. Installation is the reverse of the removal procedure. Lubricate the caliper mounting bolts or bolt and pin prior to installation.
4. Tighten the caliper mounting bolts to 61–69 ft. lbs. (83–93 Nm).
5. Bleed the brake system.

Front Disc Brake Pads

REMOVAL & INSTALLATION

1. Before servicing the vehicle, refer to the precautions in the beginning of this section.

2. Remove or disconnect the following:
   - Wheel assembly
   - Lower lock pin bolt from the caliper
3. Rotate the caliper upward and remove the brake pads, shims, guide plates and if equipped, the springs.
   To install:
4. Remove the master cylinder reservoir cap and remove about ½ of the fluid from the reservoir.
5. Using a large C-clamp and piece of wood, depress the caliper piston(s) until they bottom in their bores.
6. Install the shims, guide plates, new pads and if removed, the springs.
7. Reposition the caliper and install the lock pin bolt. Torque the lockbolt to 61–69 ft. lbs. 84–93 Nm).
8. Install the wheels, lower the vehicle, refill the master cylinder and depress the brake pedal a few times to restore pressure. Bleed the system if required.

Rear Disc Brake Caliper

REMOVAL & INSTALLATION

1. Before servicing the vehicle, refer to the precautions in the beginning of this section.
2. Raise and support the rear of the vehicle.
3. Remove the wheel(s).
4. If the caliper is being replaced or repaired, disconnect and plug the brake hose at the caliper.
5. Remove the caliper pins and lift off the caliper.
6. Installation is the reverse of removal. Torque the caliper pins to 28–36 ft. lbs. (37–49 Nm).

Rear Disc Brake Pads

REMOVAL & INSTALLATION

1. Before servicing the vehicle, refer to the precautions in the beginning of this section.

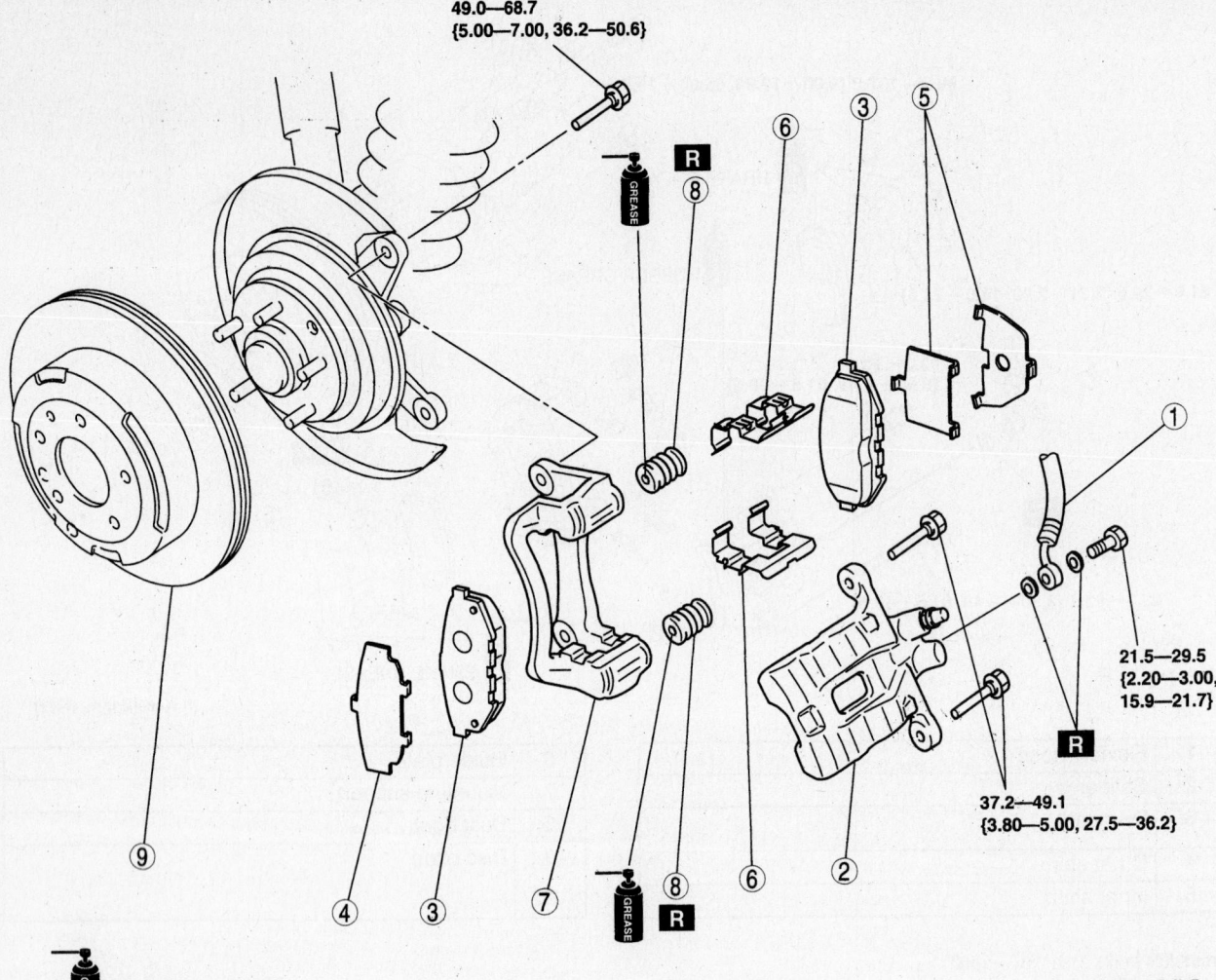

49.0—68.7
{5.00—7.00, 36.2—50.6}

21.5—29.5
{2.20—3.00,
15.9—21.7}

37.2—49.1
{3.80—5.00, 27.5—36.2}

RUBBER GREASE

N·m {kgf·m, ft·lbf}

1	Brake hose	5	Inner shim
2	Caliper	6	Guide plate
3	Disc pad	7	Mounting support
4	Outer shim	8	Dust boot
		9	Disc plate

67162-MMPV-G01

Rear disc brake components

2. Raise and support the rear of the vehicle.
3. Remove the wheel(s).
4. Remove the caliper.
5. Remove the brake pads, shims and guide plates.
6. Installation is the reverse of removal. Torque the caliper pins to 28–36 ft. lbs. (37–49 Nm).

Brake Drum

REMOVAL & INSTALLATION

1. Raise and safely support the vehicle. Remove the wheel and tire assembly.

2. Remove the screws, if equipped, and remove the brake drum.
3. Inspect the brake drum surface for wear, scoring and runout. Machine or replace, as necessary.
   To install:
4. Install the brake drum and secure in place with the screws, if equipped. Torque the screws to 10 ft. lbs. (14 Nm).
5. Adjust the rear brakes.
6. Install the wheel. Lower the vehicle.

Brake Shoes

REMOVAL & INSTALLATION

1. Raise and safely support the vehicle. Remove the wheel and tire assembly and the brake drum.
2. Pull backward on the adjusting lever cable to disengage the adjusting lever from the adjusting screw. Move the outboard side of the adjusting screw upward and back off the pivot nut as far as it will go.
3. Pull the adjusting lever, cable and automatic adjuster spring down and toward the rear to unhook the pivot hook from the

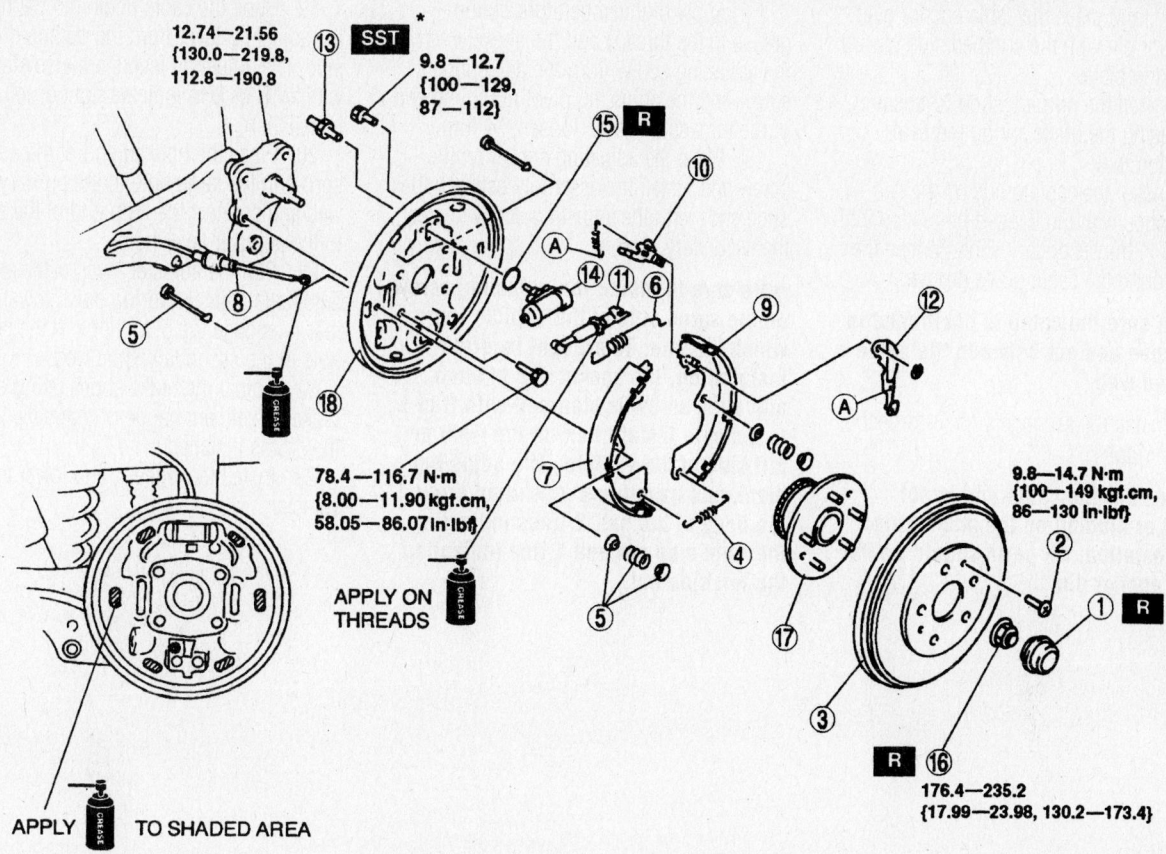

12.74—21.56
{130.0—219.8,
112.8—190.8

9.8—12.7
{100—129,
87—112}

78.4—116.7 N·m
{8.00—11.90 kgf·cm,
58.05—86.07 In·lbf}

APPLY ON
THREADS

APPLY GREASE TO SHADED AREA

9.8—14.7 N·m
{100—149 kgf·cm,
86—130 In·lbf}

176.4—235.2
{17.99—23.98, 130.2—173.4}

N·m {kgf·cm, In·lbf}

1	Hub cap		10	Ajuster lever
2	Screw		11	Adjuster component
3	Brake drum		12	Operating lever
4	Return spring		13	Brake pipe
5	Hold pin and hold spring		14	Wheel cylinder
6	Anti-rattle spring		15	O-ring
7	Leading shoe		16	Locknut
8	Parking brake cable		17	Wheel hub
9	Trailing shoe		18	Backing plate

42356-MMPV-G05

Exploded view of the rear brake shoes and components

large hole in the secondary shoe web. Do not pry the pivot hook from the hole.

4. Remove the automatic adjuster spring and adjusting lever.

5. Remove the secondary shoe-to-anchor spring using a suitable brake spring removal/installation tool. Using the tool, remove the primary shoe-to-anchor spring and unhook the cable anchor. Remove the anchor pin plate, if equipped.

6. Remove the cable guide from the secondary shoe.

7. Remove the shoe hold-down

springs, shoes, adjusting screw, pivot nut and socket. Note the color and position of each hold-down spring so they can be reassembled in the same position.

8. Remove the parking brake link and spring. Disconnect the parking brake cable from the parking brake lever.

9. Remove the secondary brake shoe. Remove the retainer clip and spring washer and remove the parking brake lever.

To install:

10. Clean the backing plate ledge pads and sand lightly. Apply a light coating of

high temperature lithium grease to the points where the brake shoes touch the backing plate. Lubricate the adjusting cable eye and the anchor pin area.

11. Install the parking brake lever on the secondary shoe.

12. Position the brake shoes on the backing plate and install the hold-down spring pins, springs and cups. Install the parking brake link, spring and washer. Connect the parking brake cable to the parking brake lever.

13. Install the anchor pin plate, if

equipped, and place the cable anchor over the anchor pin with the crimped side toward the backing plate.

14. Install the primary shoe-to-anchor spring using the brake spring removal/installation tool.

15. Install the cable guide on the secondary shoe with the flanged hole fitted into the hole in the secondary shoe. Thread the cable around the cable guide groove.

➡ **Make sure the cable is positioned in the groove and not between the guide and shoe web.**

16. Install the secondary shoe-to-anchor (long) spring.

➡ **Make sure the cable end is not cocked or binding on the anchor pin when installed. All parts should be flat on the anchor pin.**

17. Apply high temperature lithium grease to the threads and the socket end of the adjusting screw. Turn the adjusting screw into the adjusting pivot nut to the end of the threads and then loosen, ½ turn.

18. Place the adjusting socket on the screw and install the assembly between the shoe ends with the adjusting screw nearest the secondary shoe.

➡ **Be sure to install the adjusting screw on the same side of the vehicle from which it came. To prevent incorrect installation, the socket end of each adjusting screw is stamped with R or L, to indicate installation on the right or left side of the vehicle. The adjusting pivot nuts have lines machined around the body of the nut, 2 lines indicating the right side nut and 1 line indicating the left side nut.**

19. Hook the cable hook into the hole in the adjusting lever from the outboard plate side. The adjusting levers are also stamped with an **R** or **L** to indicate right or left side installation.

20. Place the hooked end of the adjuster spring in the large hole in the primary shoe web and connect the loop end of the spring to the adjuster lever hole.

21. Pull the adjuster lever, cable and automatic adjuster spring down toward the rear to engage the pivot hook in the large hole in the secondary shoe web.

22. Adjust the brake shoes using either a brake adjustment gauge or manually with the drums installed.

23. Install the wheels, and lower the vehicle.

SPECIFICATION AND MAINTENANCE CHARTS

VEHICLE AND ENGINE IDENTIFICATION CHART

		Engine					Model Year	
Code	Liters	Cu. In.	Cyl.	Fuel Sys.	Engine Type	Eng. Mfg.	Code	Year
①	1.3	79	②	MPI	Rotary	Mazda	4	2004

MPI: Multi point fuel injection

① 13B-MSP standard and high output versions.

② Twin rotary.

67162-MRX8-C01

GENERAL ENGINE SPECIFICATIONS

Year	Engine Displacement Liters	Engine VIN	Net Horsepower @ rpm	Net Torque @ rpm (ft. lbs.)	Bore x Stroke (in.)	Compression Ratio	Oil Pressure @ rpm
2004	1.3	①	197@7200	164@5000	③	10.0:1	51@3000
	1.3	②	238@8500	159@5500	③	10.0:1	51@3000

① 13B-MSP standard output.

② 13B-MSP high output.

③ Rotary chamber not measured.

67162-MRX8-C02

GASOLINE ENGINE TUNE-UP SPECIFICATIONS

Year	Engine Displacement Liters	Engine VIN	Spark Plugs Gap (in.)	Ignition Timing (deg.) MT	AT	Fuel Pump (psi)	Idle Speed (rpm) MT	AT	Valve Clearance In.	Ex.
2004	1.3	①	0.046-0.049	②	②	③	750-850	760-860	NA	NA

NOTE: The Vehicle Emission Control Information label often reflects specification changes changes made during production.

The label figures must be used if they differ from those in this chart.

NA: Not applicable.

① 13B-MSP standard and high output versions.

② Controlled by the Powertrain Control Module and cannot be adjusted.

③ Fuel line hold pressure 55-65 psi

67162-MRX8-C03

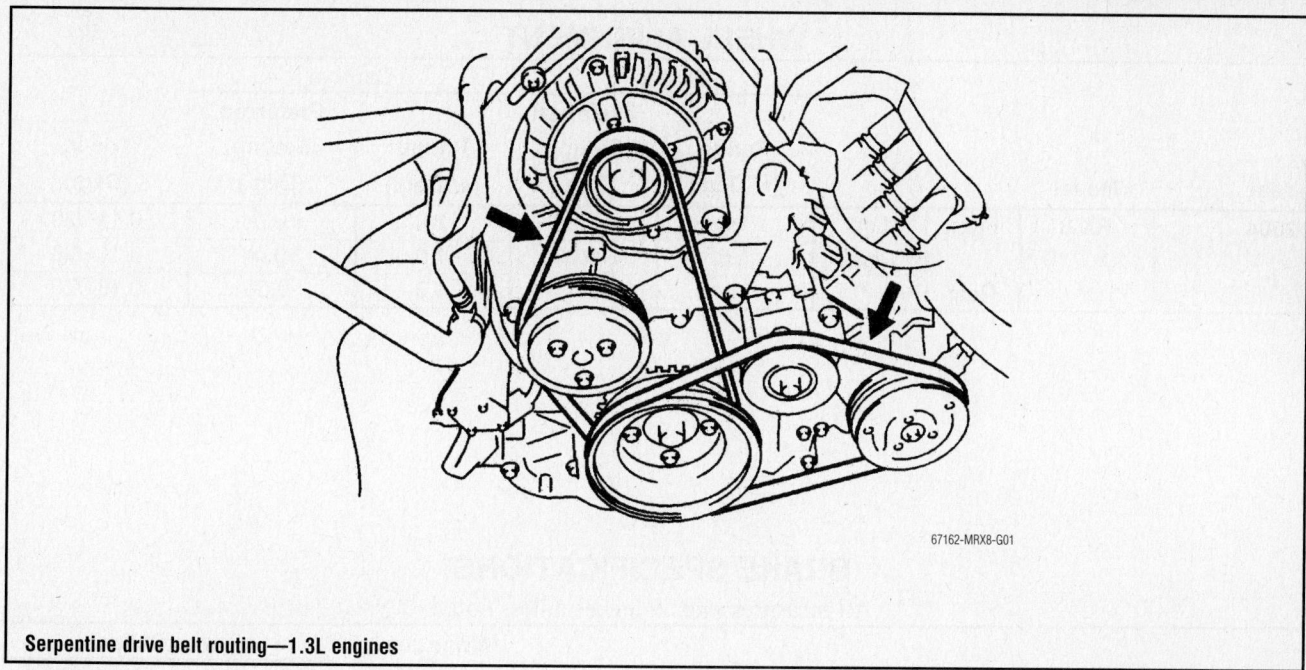

67162-MRX8-G01

Serpentine drive belt routing—1.3L engines

CAPACITIES

Year	Model	Engine Displacement Liters	Engine ID/VIN	Engine Oil with Filter (qts.)	Transmission (pts.)		Drive Axle		Fuel Tank (gal.)	Cooling System (qts.)
					6-Spd	Auto.	Front (pts.)	Rear (pts.)		
2004	RX-8	1.3	①	②	3.7	18.4	—	2.7	15.9	8.7

NOTE: All capacities are approximate. Add fluid gradually and check to be sure a proper fluid level is obtained.

① 13B-MSP standard and high output versions.

② Auto. Trans. 6.0 qts.
Man. Trans. 6.7 qts.

67162-MRX8-C04

TORQUE SPECIFICATIONS
All readings in ft. lbs.

Year	Engine VIN	Engine Displacement Liters	Cylinder Head Bolts	Main Bearing Bolts	Rod Bearing Bolts	Crankshaft Damper Bolts	Flywheel Locknut	Manifold		Spark Plugs	Oil Pan Drain Plug
								Intake	Exhaust		
2004	①	1.3	NA	NA	NA	NA	325	17	38	12	25

NA: Not applicable

① 13B-MSP standard and high output versions.

67162-MRX8-C05

WHEEL ALIGNMENT

Year	Model			Caster Range (+/-Deg.)	Caster Preferred Setting (Deg.)	Camber Range (+/-Deg.)	Camber Preferred Setting (Deg.)	Toe-in (Deg.)
2004	RX-8	Front	Left	1.0	+6.06	1.0	+0.04	0.11+/20
			Right	1.0	+6.06	1.0	+0.04	0.11+/20
		Rear		—	—	1.0	-0.56	0.16+/20

67162-MRX8-C06

BRAKE SPECIFICATIONS
All measurements in inches unless noted

Year	Model		Brake Disc Original Thickness	Brake Disc Minimum Thickness	Brake Disc Maximum Runout	Minimum Lining Thickness Front	Minimum Lining Thickness Rear	Brake Caliper Bracket Bolts (ft. lbs.)	Brake Caliper Mounting Bolts (ft. lbs.)
2004	RX-8	F	①	0.870	0.002	0.790	—	66	20
		R	①	0.630	0.002	—	0.079	43	184

① Not available

67162-MRX8-C07

TIRE AND WHEEL SPECIFICATIONS

Year	Model	OEM Tires Standard	OEM Tires Optional	Tire Pressures (psi) Front	Tire Pressures (psi) Rear	Wheel Size	Wheel Lug Nut Torque (Ft. Lbs.)
2004	RX-8	P225/55R16	P225/45R18	32	32	①	76

OEM: Original Equipment Manufacturer

PSI: Pounds Per Square Inch

① Not available

67162-MRX8-C08

SCHEDULED MAINTENANCE INTERVALS
MAZDA RX-8

TO BE SERVICED	TYPE OF SERVICE	VEHICLE MILEAGE INTERVAL (x1000)															
		7.5	15	22.5	30	37.5	45	52.5	60	67.5	75	82.5	90	97.5	105	112.5	120
Engine oil & filter	R	✓	✓	✓	✓	✓	✓	✓	✓	✓	✓	✓	✓	✓	✓	✓	✓
Cabin air filter	R				✓				✓				✓				✓
Engine coolant strength hoses & clamps	S/I				✓				✓				✓				✓
Air cleaner filter	R						✓						✓				✓
Brake fluid	R				✓				✓				✓				✓
Engine coolant ①	R								✓								✓
Spark plugs	R					✓						✓					✓
Drive belts	S/I				✓				✓				✓				✓
Exhaust system & heat shields	S/I				✓				✓				✓				✓
Manual transmission oil	R								✓								✓
Rear differential oil									✓								✓
Front & rear brakes	S/I			✓			✓			✓			✓				✓
Fuel filter	R														✓		✓

R: Replace   S/I: Service or Inspect

① Engine coolant: change initially at 60,000 miles and every 24 months thereafter.

FREQUENT OPERATION MAINTENANCE (SEVERE SERVICE)

 If a vehicle is operated under any of the following conditions it is considered severe service:

- Extremely dusty areas.

- 50% or more of the vehicle operation is in 32°C (90°F) or higher temperatures, or constant operation in temperatures below 0°C (32°F).

- Prolonged idling (vehicle operation in stop and go traffic.

- Frequent short running periods (engine does not warm to normal operating temperatures).

- Police, taxi, delivery usage or trailer towing usage.

Engine oil & filter: replace every 5000 miles.

Air cleaner filter: change every 35,000 miles.

Exhaust system: check every 30,000 miles.

PRECAUTIONS

Before servicing any vehicle, please be sure to read all of the following precautions, which deal with personal safety, prevention of component damage, and important points to take into consideration when servicing a motor vehicle:

• Never open, service or drain the radiator or cooling system when the engine is hot; serious burns can occur from the steam and hot coolant.

• Observe all applicable safety precautions when working around fuel. Whenever servicing the fuel system, always work in a well-ventilated area. Do not allow fuel spray or vapors to come in contact with a spark, open flame or excessive heat (a hot drop light, for example). Keep a dry chemical fire extinguisher near the work area. Always keep fuel in a container specifically designed for fuel storage; also, always properly seal fuel containers to avoid the possibility of fire or explosion. Refer to the additional fuel system precautions later in this section.

• Fuel injection systems often remain pressurized, even after the engine has been turned **OFF**. The fuel system pressure must be relieved before disconnecting any fuel lines. Failure to do so may result in fire and/or personal injury.

• Brake fluid often contains polyglycol ethers and polyglycols. Avoid contact with the eyes and wash your hands thoroughly after handling brake fluid. If you do get brake fluid in your eyes, flush your eyes with clean, running water for 15 minutes. If eye irritation persists, or if you have taken brake fluid internally, IMMEDIATELY seek medical assistance.

• The EPA warns that prolonged contact with used engine oil may cause a number of skin disorders, including cancer! You should make every effort to minimize your exposure to used engine oil. Protective gloves should be worn when changing oil. Wash your hands and any other exposed skin areas as soon as possible after exposure to used engine oil. Soap and water, or waterless hand cleaner should be used.

• All new vehicles are now equipped with an air bag system. The system must be disabled before performing service on or around system components, steering column, instrument panel components, wiring and sensors. Failure to follow safety and disabling procedures could result in accidental air bag deployment, possible personal injury and unnecessary system repairs.

• Always wear safety goggles when working with, or around, the air bag system. When carrying a non-deployed air bag, be sure the bag and trim cover are pointed away from your body. When placing a non-deployed air bag on a work surface, always face the bag and trim cover upward, away from the surface. This will reduce the motion of the module if it is accidentally deployed. Refer to the additional air bag system precautions later in this section.

• Clean, high quality brake fluid from a sealed container is essential to the safe and proper operation of the brake system. You should always buy the correct type of brake fluid for your vehicle. If the brake fluid becomes contaminated, completely flush the system with new fluid. Never reuse any brake fluid. Any brake fluid that is removed from the system should be discarded. Also, do not allow any brake fluid to come in contact with a painted surface; it will damage the paint.

• Never operate the engine without the proper amount and type of engine oil; doing so WILL result in severe engine damage.

• Disconnecting the negative battery cable on some vehicles may interfere with the functions of the on-board computer system(s) and may require the computer to undergo a relearning process once the negative battery cable is reconnected.

• When servicing drum brakes, only disassemble and assemble one side at a time, leaving the remaining side intact for reference.

• Only an MVAC-trained, EPA-certified automotive technician should service the air conditioning system or its components.

• On models equipped with the Dynamic Stability Control system, whenever the negative battery cable is disconnected, the steering angle sensor must be initialized to enable the DSC system. See the Steering Angle Initialization procedure in the Steering and Suspension section.

ENGINE REPAIR

Distributor

REMOVAL

1.3L Engine

The 1.3L engine is equipped with a distributorless ignition system.

Alternator

REMOVAL

1. Before servicing the vehicle, refer to the precautions in the beginning of this section.
2. Remove or disconnect the following:
• Negative battery cable
• Engine cover
• Rear engine cross brace
• Intake air duct

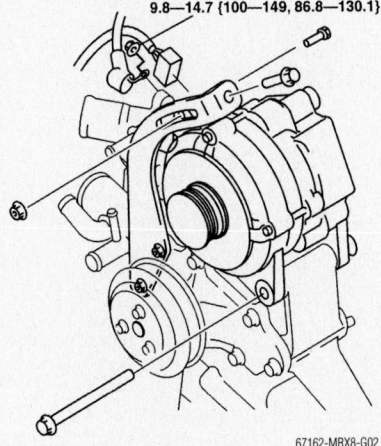

9.8—14.7 {100—149, 86.8—130.1}

67162-MRX8-G02

Alternator mounting—1.3L engine

• Accessory drive belt
• Electrical connectors from the alternator
• Alternator bolts
• Alternator

INSTALLATION

1. Installation is the reverse of the removal procedure, noting the following:
   a. Tighten the left side engine cross brace nut to 40 ft. lbs. (Nm) and the right side nut to 16 ft. lbs.

Ignition Timing

ADJUSTMENT

1. Before servicing the vehicle, refer to the precautions in the beginning of this section.

➡The ignition timing cannot be adjusted. To check the timing, connect a scan tool to the Data Link Connector (DLC). Connect a timing light to the front rotor housing on the leading side. Place the scan tool in the test mode. Start the engine and verify that the white alignment marks on the front cover and eccentric shaft plate are aligned.

REMOVAL & INSTALLATION

➡The procedure for pulling the engine requires removing the transaxle and front crossmember along with it. A suitable support fixture that will support the entire assembly must be used for removal.

1. Before servicing the vehicle, refer to the precautions in the beginning of this section.
2. Properly relieve the fuel system pressure.
3. Drain the engine oil.
4. Drain the transaxle fluid.
5. Drain the cooling system.
6. Place the front wheels in the straight ahead position.

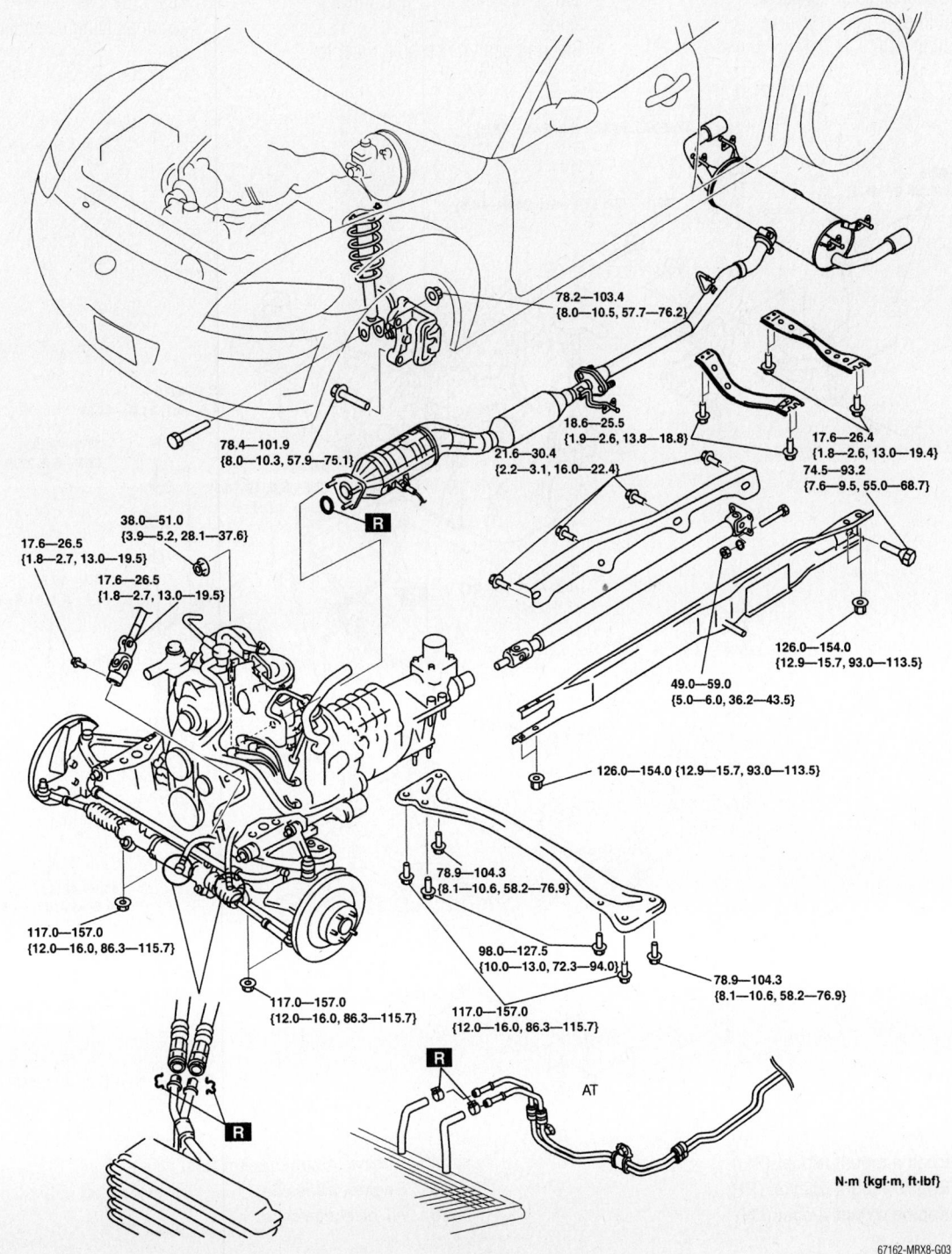

N·m {kgf·m, ft·lbf}

67162-MRX8-G03

Location of the engine mounting components and their torque specifications—RX-8 1.3L engine

7. Raise and support the vehicle.
8. Remove or disconnect the following:
- Front wheels
- Engine cover
- Rear engine cross brace
- Battery cover, battery, box and tray
- Air cleaner, intake duct and insulator
- Powertrain Control Module (PCM)
- Secondary Air Injection (AIR) pump
- Brake vacuum hose
- Charcoal canister connector
- Fuel lines
- Ignition coils

- Accessory drive belts
- A/C compressor and wire aside
- Engine wiring harness from main fuse block
- Engine splash shield
- ABS speed sensor connector
- Radiator, heater and coolant tank hoses
- Selector link on automatic transmission
- Manual transmission shift lever
- Clutch slave cylinder and wire aside
- Steering shaft pinch bolt. Refer to

the accompanying illustration for location.
- Oil lines. Refer to the accompanying illustration for location.
- Transmission cooler lines. Refer to the accompanying illustration for location.
- Brake caliper. Refer to the accompanying illustration for location.
- Lower strut bolt. Refer to the accompanying illustration for location.
- Rear crossmembers. Refer to the accompanying illustration for location.

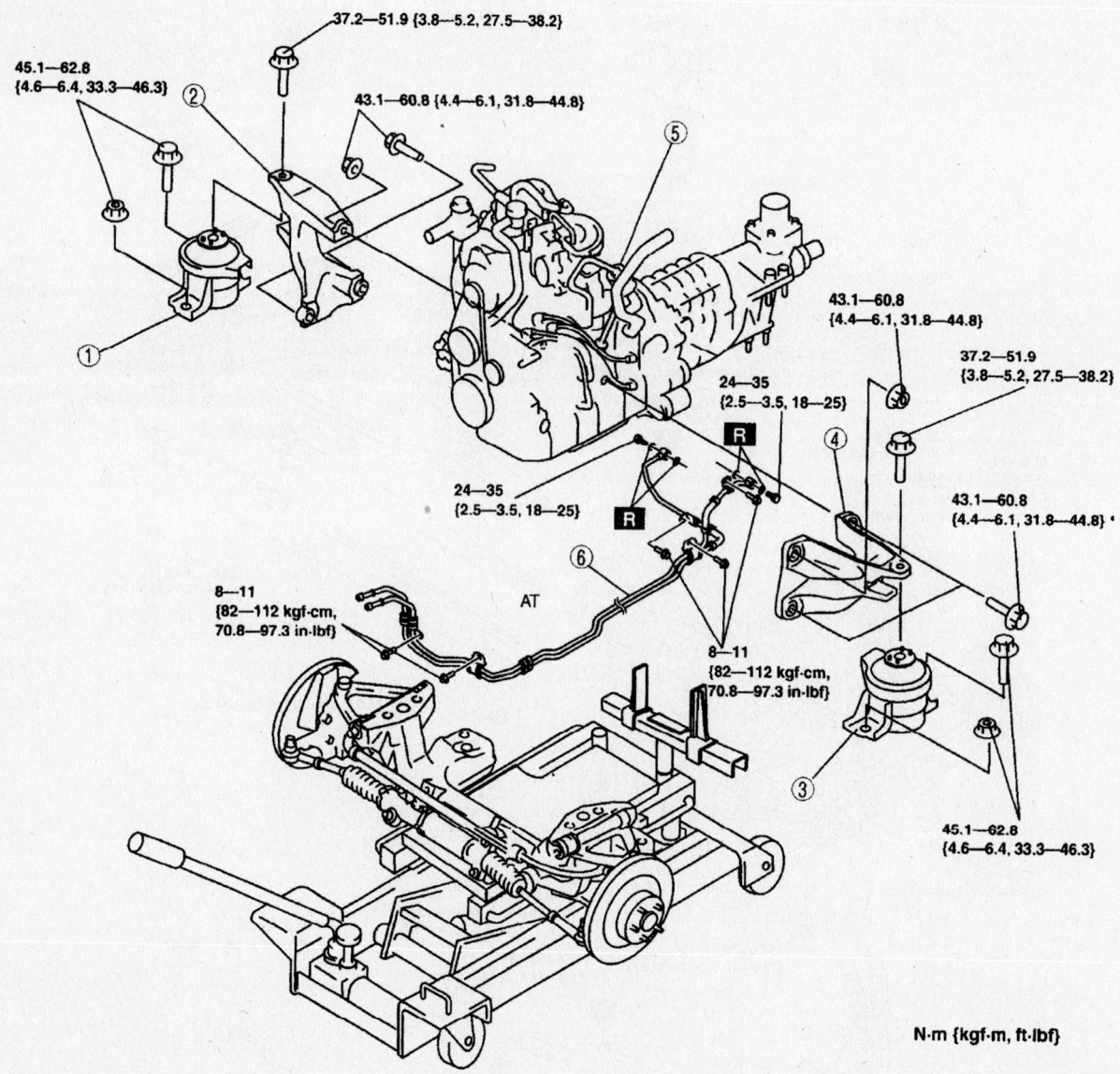

N·m {kgf·m, ft·lbf}

1	Engine mount rubber (RH)	4	Engine mount bracket (LH)
2	Engine mount bracket (RH)	5	Engine, transaxle
3	Engine mount rubber (LH)	6	AT oil cooler pipe

67162-MRX8-G04

Separating the engine/transmission assembly from the front crossmember— RX-8 1.3L engine

- Catalytic converter and exhaust system. Refer to the accompanying illustration for location.
- Heat insulator. Refer to the accompanying illustration for location.
- Driveshaft. Refer to the accompanying illustration for location.
- Front crossmember. Refer to the accompanying illustration for location.
- Power plant frame. Refer to the accompanying illustration for location.
- Engine, transmission and front suspension frame assembly

9. Use an engine hoist and separate the engine/transmission assembly from the front suspension crossmember using the steps shown in the accompanying illustration.

To install:

10. Installation is the reverse of removal. Tighten the fasteners to the specifications shown in the accompanying illustrations.

11. When possible, leave the engine mounting nuts/bolts loose (hand tight) until all mounts are aligned and bolted. This may help in aligning the engine and transmission assembly in the vehicle.

12. Fill the engine and the transaxle with the proper type and amount of fluids. Fill the cooling system.

13. Connect the battery cables.

14. On models with dynamic suspension, perform the steering angle sensor initialization procedure.

15. Check the ignition timing and the idle speed.

16. Check the front wheel alignment.

17. Check all fluid levels.

Water Pump

REMOVAL & INSTALLATION

1. Before servicing the vehicle, refer to the precautions in the beginning of this section.

2. Drain the cooling system.

3. Remove or disconnect the following:
- Battery cables, battery box and tray
- Loosen the water pump pulley bolt
- Drive belt
- Water pump pulley
- Front engine hangar
- Alternator strap
- Water pump and gasket

To install:

4. Installation is the reverse of removal. Tighten the fasteners to the specifications shown in the accompanying illustration.

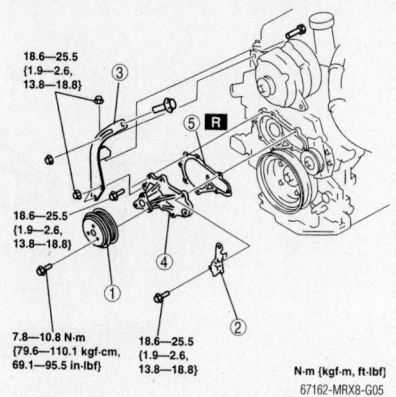

18.6—25.5 {1.9—2.6, 13.8—18.8}

18.6—25.5 {1.9—2.6, 13.8—18.8}

7.8—10.8 N·m {79.6—110.1 kgf-cm, 69.1—95.5 in-lbf}

18.6—25.5 {1.9—2.6, 13.8—18.8}

N·m {kgf-m, ft-lbf}

67162-MRX8-G05

Exploded view of the water pump assembly—1.3L engine

5. Fill and bleed the cooling system.

6. On models with dynamic suspension, perform the steering angle sensor initialization procedure.

7. Start the engine, check for leaks and repair if necessary.

Heater Core

REMOVAL & INSTALLATION

1. Before servicing the vehicle, refer to the precautions in the beginning of this section.

2. Place the ignition switch in the **LOCK** position.

3. Disconnect the negative battery cable.

✳✳ CAUTION

After disconnecting the battery, wait for more than 1 minute for the SAS to deplete its stored energy.

4. Drain the cooling system into a clean container for reuse.

5. Disconnect the heater hoses from the heater core.

6. Discharge and recover the air conditioning system refrigerant.

7. Remove or disconnect the following:
- Center console upper panel
- Ash tray panel
- Cigar lighter connector
- Ash tray light
- Storage compartment
- Front and rear center consoles
- Console under cover
- Glove box
- Lower scuff plate
- Lower door side trim plate
- Lower dashboard side panel
- Lower dashboard front panel

- Steering column upper cover
- Ignition key light
- Steering column lower cover

8. At the driver's side, remove the SAS module and the steering wheel by removing or disconnecting the following:
- Place the wheel in the straight-ahead position and turn the ignition switch to LOCK
- Cover clips at both sides of the steering wheel
- Steering wheel-to-SAS module bolts
- SAS module from the steering wheel and disconnect the electrical connector
- Steering wheel-to-column nut
- Steering wheel from the steering column using a suitable puller
- Steering column mounting bolts and lower the column
- Both A pillar trims
- Instrument panel-to-chassis fasteners in the order shown
- Instrument panel with the help of an assistant

9. Remove or disconnect the following:
- A/C unit

10. Separate the heater core from the A/C unit.

To install:

11. Install the heater core to the A/C unit.

12. Install or connect the following:
- A/C unit
- Instrument panel and tighten the fasteners as shown
- Both A pillar trims
- Raise the steering column and tighten the bolts to 14 ft. lbs. (19 Nm)
- Steering wheel to steering column
- Steering wheel-to-column nut and tighten to 33 ft. lbs. (45 Nm)
- SAS module to the steering wheel and connect the electrical connector
- Steering wheel-to-SAS module bolts and tighten to 70–103 inch lbs. (8–12 Nm)

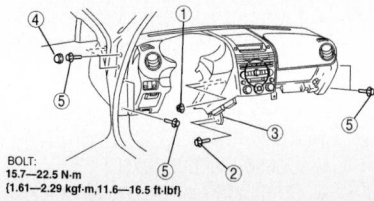

BOLT:
15.7—22.5 N·m
{1.61—2.29 kgf-m, 11.6—16.5 ft-lbf}

67162-MRX8-G06

Instrument panel fastener removal sequence—RX-8

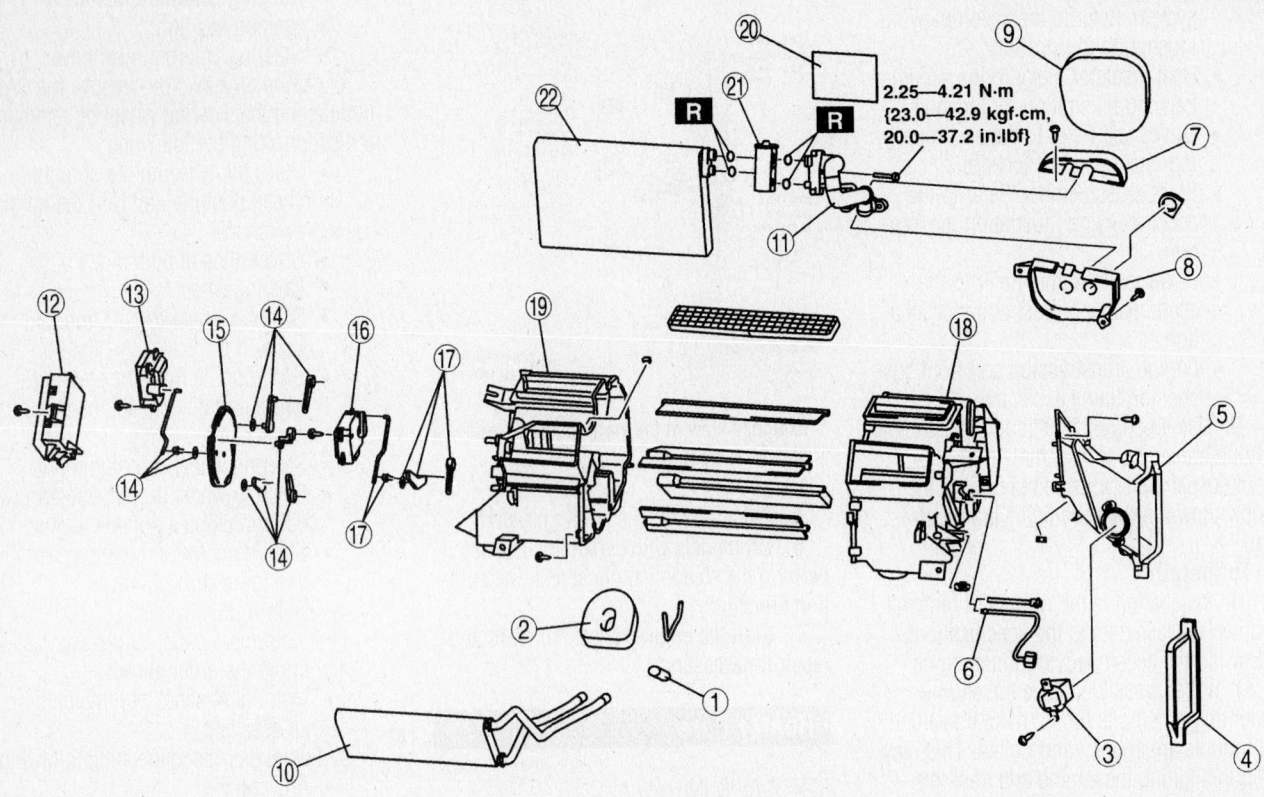

2.25—4.21 N·m
{23.0—42.9 kgf·cm,
20.0—37.2 in·lbf}

1	Drain hose	12	A/C amplifier
2	Polyurethane foam (1)	13	Airflow mode actuator
3	Resistor	14	Airflow mode link set
4	Adhesive polyurethane (1)	15	Airflow mode main link
5	Air duct	16	Air mix actuator
6	Evaporator temperature sensor	17	Air mix link set
7	Polyurethane foam (2)	18	A/C case (1)
8	Bracket (1)	19	A/C case (2)
9	Bracket (2)	20	Adhesive polyurethane (2)
10	Heater core	21	Expansion valve
11	Evaporator pipe	22	Evaporator

67162-MRX8-G07

Exploded view of the A/C unit with heater core—RX8

- Cover clips at both sides of the steering wheel
- Steering column lower cover
- Ignition key light
- Steering column upper cover
- Lower dashboard front panel
- Lower dashboard side panel
- Lower door side trim plate
- Lower scuff plate
- Glove box
- Console under cover
- Front and rear center consoles
- Storage compartment
- Ash tray light
- Cigar lighter connector

- Ash tray panel
- Center console upper panel

13. Connect the heater hoses to the heater core.

14. Refill the cooling system.

15. Connect the negative battery cable.

16. Evacuate, charge and leak test the air conditioning system refrigerant.

17. On models with dynamic suspension, perform the steering angle sensor initialization procedure.

18. Operate the engine to normal operating temperatures; then, check the climate control operation and check for leaks.

Intake Manifold

REMOVAL & INSTALLATION

1. Before servicing the vehicle, refer to the precautions in the beginning of this section.

2. Relieve the fuel system pressure.

3. Drain the cooling system.

4. Remove the engine and transmission assembly.

5. Remove or disconnect the following:
- Air hose
- Air cleaner cover

- Variable Fresh Air Duct (VFAD) solenoid valve on high output engines
- Vacuum chamber on high output engines
- Air cleaner case
- Throttle body
- Upper extension manifold
- Lower extension manifold on high output engines
- Oil filler pipe
- Air Injection Reactor (AIR) solenoid valve
- Secondary Shutter Valve (SSV) solenoid

- Variable Dynamic Effect Intake (VDI) solenoid valve.
- Air cleaner insulator
- Auxiliary Port Valve (APV) bracket and motor on high output engines
- Fresh air intake duct
- Fuel distributors
- Intake manifold and discard the gasket

To install:

6. Clean all gasket mating surfaces.
7. Apply clean oil to the APV valves as shown
8. Install or connect the following:

- Intake manifold using a new gasket in the sequence shown. Torque the bolts to 14–19 ft. lbs. (19–26 Nm). Retighten bolt no. 1.
- Fuel distributors and torque the bolts to 14–19 ft. lbs. (19–26 Nm)
- Fresh air intake duct and torque the bolts to 69–96 inch lbs. (8–11 Nm)
- Auxiliary Port Valve (APV) bracket and motor on high output engines
- Air cleaner insulator and torque the bolts to 69–96 inch lbs. (8–11 Nm)
- Variable Dynamic Effect Intake (VDI) solenoid valve.

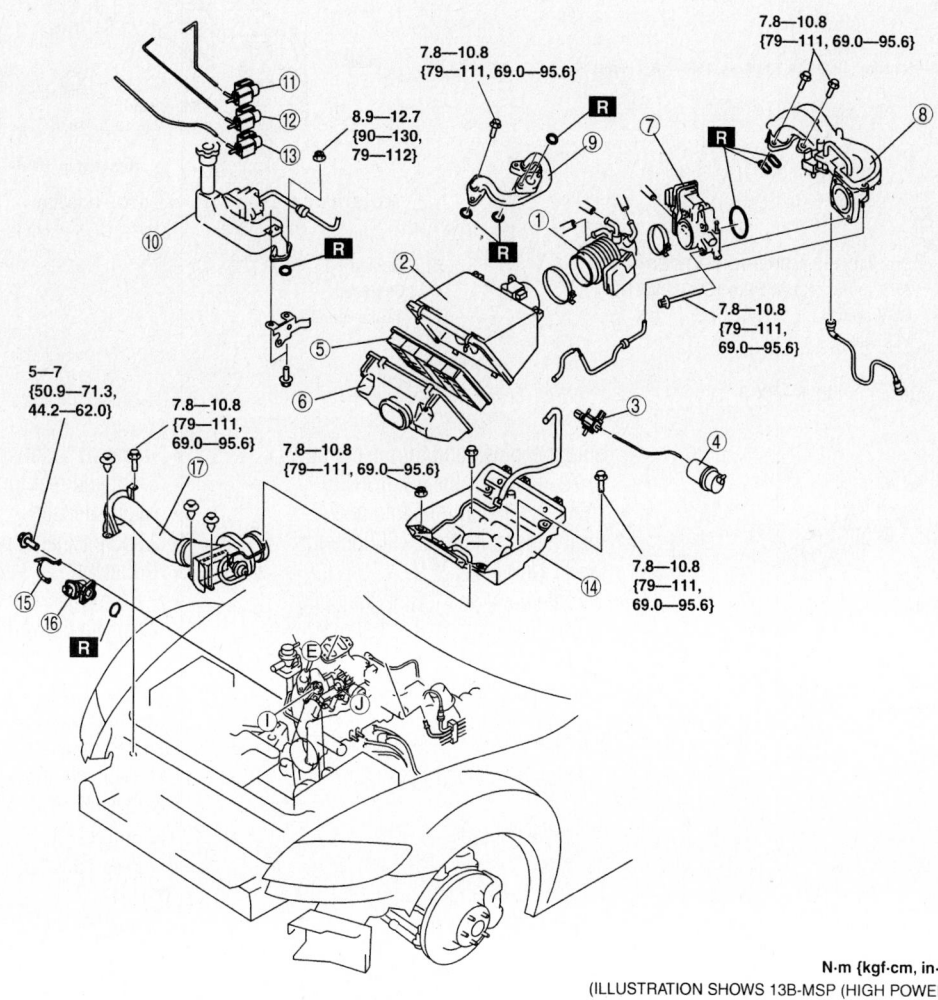

N·m {kgf·cm, in·lbf}
(ILLUSTRATION SHOWS 13B-MSP (HIGH POWER))

1	Air hose	10	Oil filler pipe
2	Air cleaner cover	11	AIR solenoid valve
3	VFAD solenoid valve (13B-MSP (High power))	12	SSV solenoid valve
4	Vacuum chamber (13B-MSP (High power))	13	VDI solenoid valve
5	Air cleaner element	14	Air cleaner insulator
6	Air cleaner case	15	Bracket (13B-MSP (High power))
7	Throttle body	16	APV motor (13B-MSP (High power))
8	Extension manifold (upper)	17	Fresh-air duct
9	Extension manifold (lower) (13B-MSP (High power))		

67162-MRX8-G08

Exploded view of the air intake system—RX-8

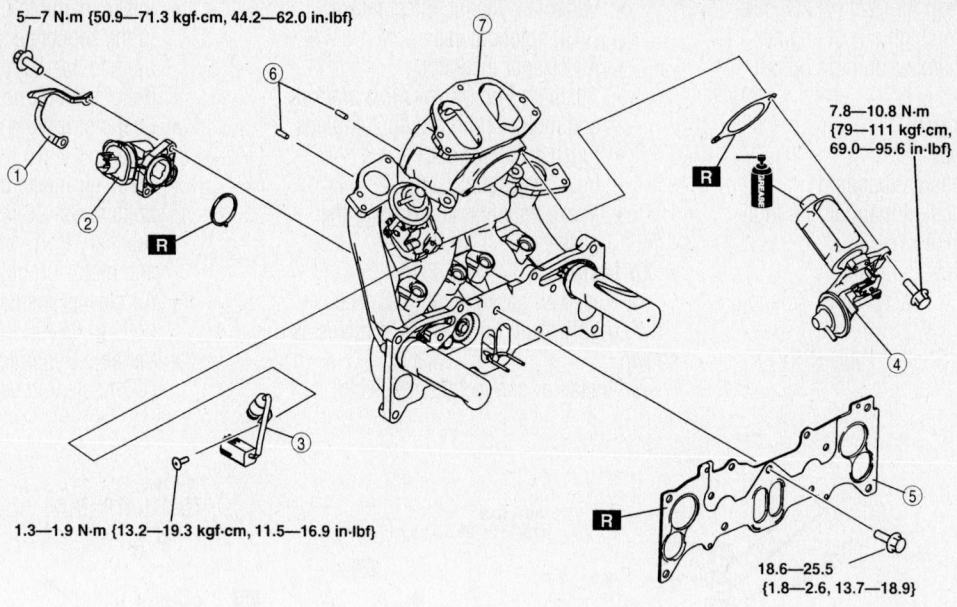

5—7 N·m {50.9—71.3 kgf·cm, 44.2—62.0 in·lbf}

7.8—10.8 N·m {79—111 kgf·cm, 69.0—95.6 in·lbf}

1.3—1.9 N·m {13.2—19.3 kgf·cm, 11.5—16.9 in·lbf}

18.6—25.5 {1.8—2.6, 13.7—18.9}

N·m {kgf·m, ft·lbf}

(ILLUSTRATION SHOWS 13B-MSP (HIGH POWER))

1	Bracket (13B-MSP (HIGH POWER))	5	Gasket
2	APV motor (13B-MSP (HIGH POWER))	6	Blind cap
3	SSV switch	7	Intake manifold
4	VDI valve		

67162-MRX8-G09

Exploded view of the intake manifold—RX-8

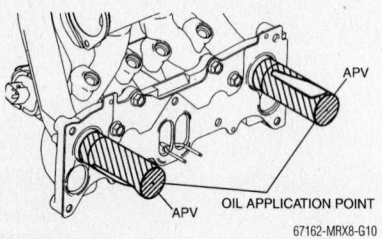

APV

OIL APPLICATION POINT

APV

67162-MRX8-G10

Applying oil to the APV valves—RX-8

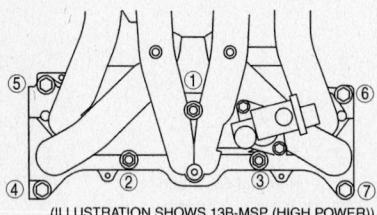

(ILLUSTRATION SHOWS 13B-MSP (HIGH POWER))

67162-MRX8-G11

Intake manifold tightening sequence—RX-8

- Secondary Shutter Valve (SSV) solenoid
- Air Injection Reactor (AIR) solenoid valve
- Oil filler pipe and torque the bolts to 79–112 inch lbs. (9–13 Nm)
- Lower extension manifold on high output engines and torque the bolts to 79–112 inch lbs. (9–13 Nm)
- Upper extension manifold and torque the bolts to 69–96 inch lbs. (8–11 Nm)
- Throttle body and torque the bolts to 69–96 inch lbs. (8–11 Nm)
- Air cleaner case
- Vacuum chamber on high output engines
- Variable Fresh Air Duct (VFAD) solenoid valve on high output engines
- Air cleaner cover
- Air hose
- Engine and transmission assembly
- Negative battery cable

9. Fill the cooling system.
10. On models with dynamic suspension, perform the steering angle sensor initialization procedure.
11. Run the engine and check for leaks.

Exhaust Manifold

REMOVAL & INSTALLATION

1. Before servicing the vehicle, refer to the precautions in the beginning of this section.
2. Remove or disconnect the following:

- Negative battery cable
- Front and rear tunnel crossmembers
- Main silencer
- Middle exhaust pipe
- Protector
- Rear oxygen sensor
- Catalytic converter
- Bracket
- Front oxygen sensor
- Air Injection reactor (AIR pipe)

3. Use a suitable overhead engine lift and support the engine.
4. Remove the right side engine mounting rubber and bracket.
5. Remove the exhaust manifold.

To install:
6. Clean all gasket mating surfaces.

➡**Use new self-locking nuts. The exhaust manifold gasket has crimps attached to it. Ensure that all the crimps are in place when installing the gasket, or the gasket will leak.**

7. Install or connect the following:
- Exhaust manifold. Torque the nuts to 31–44 ft. lbs. (43–61 Nm).
- Right side engine mounting bracket. Torque the bolts as shown.
- Air Injection reactor (AIR pipe)
- Bracket
- Catalytic converter

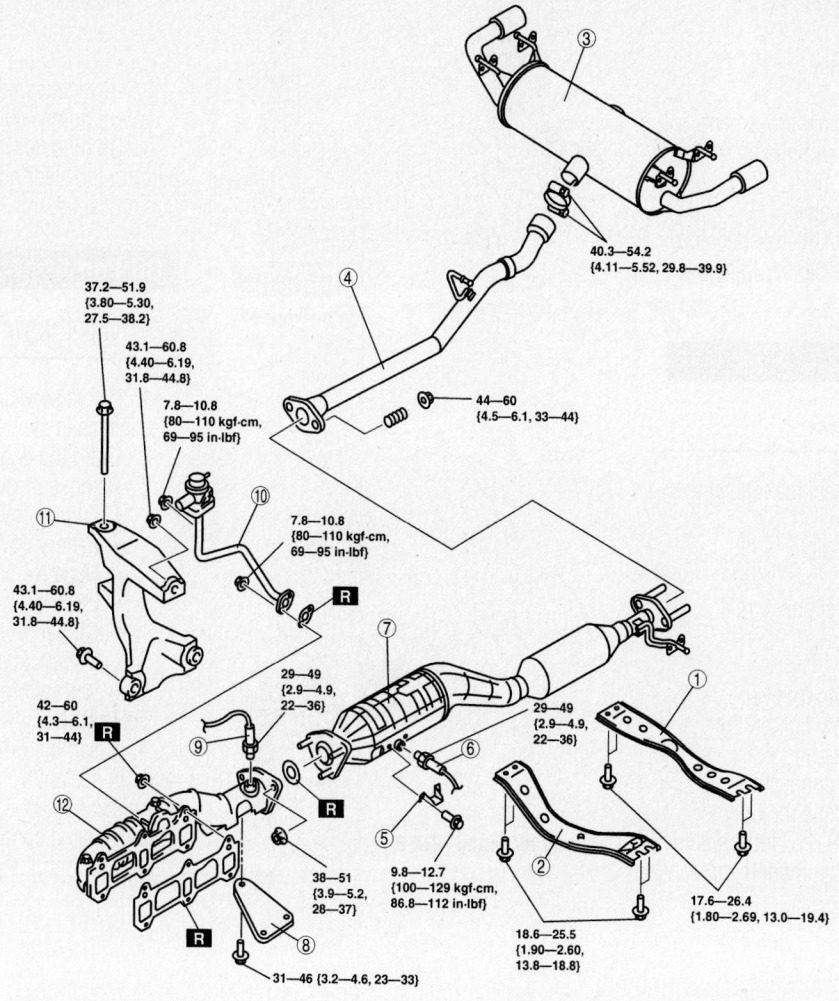

37.2—51.9
{3.80—5.30,
27.5—38.2}

43.1—60.8
{4.40—6.19,
31.8—44.8}

7.8—10.8
{80—110 kgf-cm,
69—95 in-lbf}

40.3—54.2
{4.11—5.52, 29.8—39.9}

44—60
{4.5—6.1, 33—44}

7.8—10.8
{80—110 kgf-cm,
69—95 in-lbf}

43.1—60.8
{4.40—6.19,
31.8—44.8}

42—60
{4.3—6.1,
31—44}

29—49
{2.9—4.9,
22—36}

29—49
{2.9—4.9,
22—36}

17.6—26.4
{1.80—2.69, 13.0—19.4}

38—51
{3.9—5.2,
28—37}

9.8—12.7
{100—129 kgf-cm,
86.8—112 in-lbf}

18.6—25.5
{1.90—2.60,
13.8—18.8}

31—46 {3.2—4.6, 23—33}

N·m {kgf·m, ft-lbf}

1	Rear tunnel member	7	Catalytic converter
2	Front tunnel member	8	Bracket
3	Main silencer	9	Front heated oxygen sensor
4	Middle pipe	10	AIR pipe
5	Protector	11	Engine mount bracket (RH)
6	Rear heated oxygen sensor	12	Exhaust manifold

67162-MRX8-G12

Exploded view of the exhaust system—RX-8

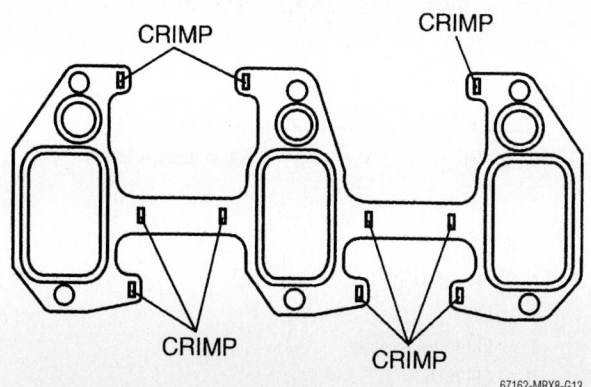

CRIMP   CRIMP

CRIMP   CRIMP

67162-MRX8-G13

Exhaust manifold gasket identification—RX-8

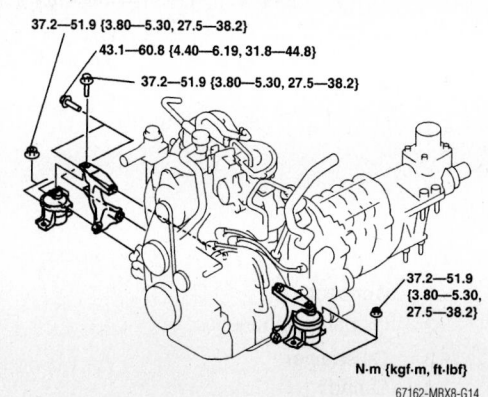

37.2—51.9 {3.80—5.30, 27.5—38.2}

43.1—60.8 {4.40—6.19, 31.8—44.8}

37.2—51.9 {3.80—5.30, 27.5—38.2}

37.2—51.9
{3.80—5.30,
27.5—38.2}

N·m {kgf·m, ft-lbf}

67162-MRX8-G14

Removing and installing the engine mount bracket—RX-8

- Rear oxygen sensor
- Protector
- Middle exhaust pipe
- Main silencer
- Front and rear tunnel crossmembers and torque the bolts to 14–19 ft. lbs. (19–26 Nm)
- Negative battery cable

8. On models with dynamic suspension, perform the steering angle sensor initialization procedure.

Starter Motor

REMOVAL & INSTALLATION

1. Remove or disconnect the following:
- Engine cover
- Negative battery cable
- Air cleaner
- Starter electrical connectors
- Starter

To install:

2. Install or connect the following:
- Starter and loosely tighten the lower starter mounting bolt
- Starter electrical connectors
- Starter bolts. Torque the bolts 14–18 ft. lbs. (19–25 Nm) on automatic transmission models, or

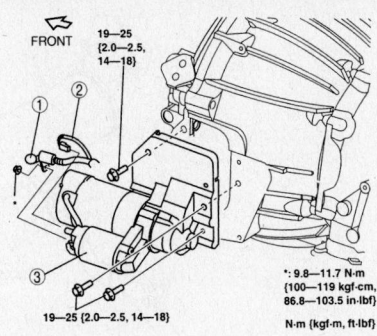

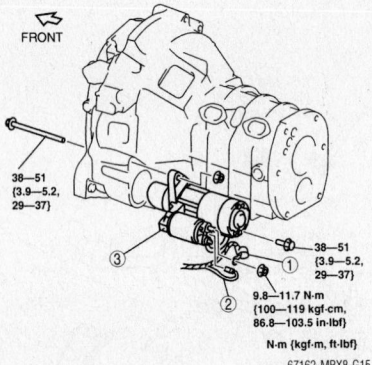

Starter mounting—RX-8

29–37 ft. lbs. (38–51 Nm) on manual transmission models.
- Air cleaner
- Negative battery cable
- Engine cover

3. On models with dynamic suspension, perform the steering angle sensor initialization procedure.

Oil Pan

REMOVAL & INSTALLATION

1. Before servicing the vehicle, refer to the precautions in the beginning of this section.
2. Drain the engine oil.
3. Remove or disconnect the following:
- Engine cover
- Battery cover
- Negative battery cable
- Electrical connector
- Oil pan bolts and the oil pan using a seal cutter, then insert a flat pry tool into the locations illustrated.

To install:

4. Clean the oil pan. Clean all dirt, oil and old sealant from the oil pan and cylinder block contact surfaces.

5. Apply a continuous bead of silicone sealant around the perimeter of the oil pan.

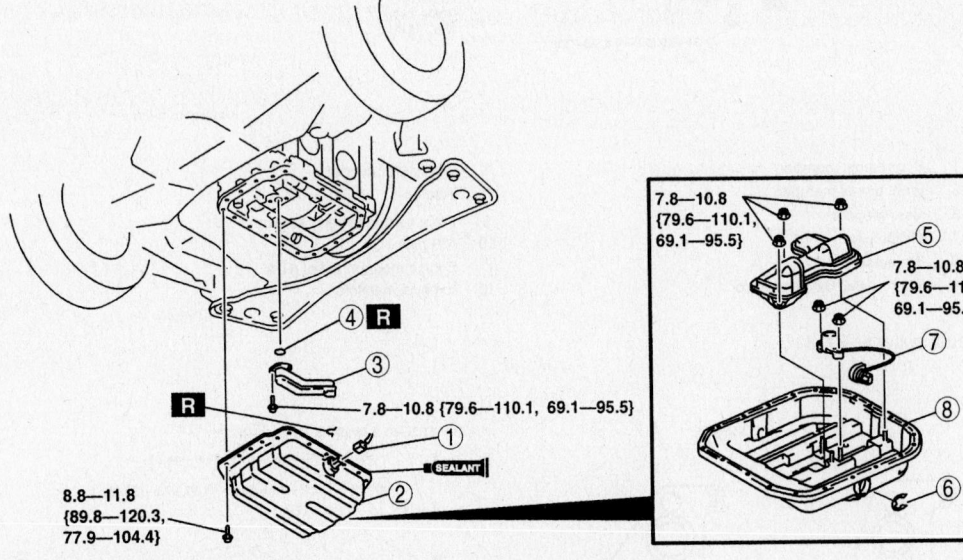

1	Connector	5 Oil baffle plate
2	Oil pan component	6 Clip
3	Oil strainer	7 Oil-level switch
4	O-ring	8 Oil pan

67162-MRX8-G16

Exploded view of the oil pan and related components—RX-8

6. Install the oil pan and tighten the bolts to 80–105 inch lbs. (9–12 Nm).
- Electrical connector
- Battery cover
- Engine cover

7. Fill the engine with clean oil.

8. Start the vehicle, check for leaks and repair if necessary.

9. On models with dynamic suspension, perform the steering angle sensor initialization procedure.

Oil Pump

REMOVAL & INSTALLATION

1. Before servicing the vehicle, refer to the precautions in the beginning of this section.

2. Drain the engine oil.

3. Remove or disconnect the following:
- Engine cover
- Battery cables
- Battery cover, battery box and tray
- Upper and lower extension manifolds
- Electrical connectors
- Oil pipe
- Gasket
- Oil pump and o-ring

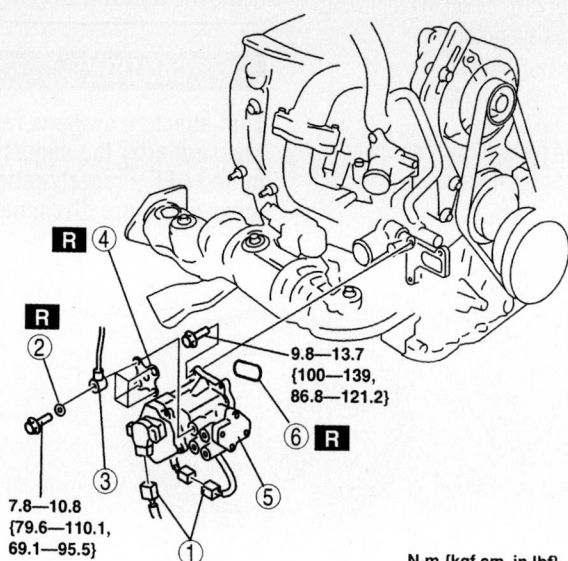

9.8—13.7
{100—139,
86.8—121.2}

7.8—10.8
{79.6—110.1,
69.1—95.5}

N·m {kgf·cm, in·lbf}

1   Connectors
2   Washer
3   Oil pipe (metering oil pump side)
4   Gasket
5   Metering oil pump
6   O-ring

67162-MRX8-G17

Exploded view of the oil pump mounting—RX-8

To install:

4. Clean the oil, dirt and old sealant from all contact surfaces.

5. Install or connect the following:
- New O-rings on the oil pump
- Oil pump. Torque the bolts to 87–122 inch lbs. (10–14 Nm).
- Gasket
- Oil pipe
- Electrical connectors
- Upper and lower extension manifolds
- Battery cover, battery box and tray
- Battery cables
- Engine cover

6. Fill the engine with clean oil.

7. Start the vehicle, check for leaks and repair if necessary.

8. On models with dynamic suspension, perform the steering angle sensor initialization procedure.

Rear Main Seal

REMOVAL & INSTALLATION

1. Before servicing the vehicle, refer to the precautions in the beginning of this section.

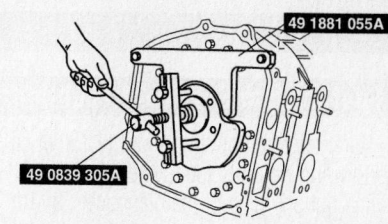

Removing the counterweight using special tools—RX-8

2. Remove or disconnect the following:
- Negative battery cable
- Transmission assembly
- Clutch/flywheel assembly, if equipped with a manual transaxle/transmission
- Flexplate/shim plates, if equipped with an automatic transaxle/transmission
- Using special tools 49 1881-055A and 49-0820-035, remove the counterweight locknut.
- Using special tools 49 1881-055A and 49-0839-305A, remove the counterweight
- Place a rag over the eccentric shaft and using a prytool, carefully pry the oil seal from the oil seal housing.

3. Clean the gasket mounting surfaces.

To install:

4. Clean the oil seal housing. Coat the lip of the oil seal and the housing with clean engine oil.

5. Install or connect the following:
- New oil seal into the housing by tapping it evenly into place with a hammer and a seal installer until it is flush with the edge of the rear cover
- Install the key into the eccentric shaft and install the counterweight
- Apply sealant to the seating face then install the locknut and loosely tighten.
- Lock the counterweight using the special tools, then tighten the locknut to 290–361 ft. lbs. (392–490 Nm).
- Clutch/flywheel assembly or the flexplate, as applicable
- Transaxle/transmission
- Negative battery cable

6. On models with dynamic suspension, perform the steering angle sensor initialization procedure.

FUEL SYSTEM

Fuel System Service Precautions

Safety is the most important factor when performing not only fuel system maintenance but any type of maintenance. Failure to conduct maintenance and repairs in a safe manner may result in serious personal injury or death. Maintenance and testing of the vehicle's fuel system components can be accomplished safely and effectively by adhering to the following rules and guidelines.

1. To avoid the possibility of fire and personal injury, always disconnect the negative battery cable unless the repair or test procedure requires that battery voltage be applied.

2. Always relieve the fuel system pressure prior to disconnecting any fuel system component (injector, fuel rail, pressure regulator, etc.), fitting or fuel line connection. Exercise extreme caution whenever relieving fuel system pressure, to avoid exposing skin, face and eyes to fuel spray. Please be advised that fuel under pressure may penetrate the skin or any part of the body that it contacts.

3. Always place a shop towel or cloth around the fitting or connection prior to loosening to absorb any excess fuel due to spillage. Ensure that all fuel spillage (should it occur) is quickly removed from engine surfaces. Ensure that all fuel soaked cloths or towels are deposited into a suitable waste container.

4. Always keep a dry chemical (Class B) fire extinguisher near the work area.

5. Do not allow fuel spray or fuel vapors to come into contact with a spark or open flame.

6. Always use a back-up wrench when loosening and tightening fuel line connection fittings. This will prevent unnecessary stress and torsion to fuel line piping. Always follow the proper torque specifications.

7. Always replace worn fuel fitting O-rings with new. Do not substitute fuel hose where fuel pipe is installed.

Fuel System Pressure

RELIEVING

1. Before servicing the vehicle, refer to the precautions in the beginning of this section.

2. Remove the filler cap.

3. Remove the fuel pump relay from the relay box, located in the main fuse block.

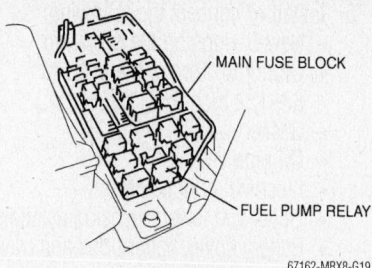

Fuel pump relay location—RX-8

4. Start the engine.

5. After the engine stalls, turn the ignition switch **OFF**.

6. After servicing the vehicle, reinstall the relay.

Fuel Pump

REMOVAL & INSTALLATION

1. Before servicing the vehicle, refer to the precautions in the beginning of this section.

2. Relieve the fuel system pressure.

3. Remove or disconnect the following:
   - Negative battery cable
   - Rear seat cushion

4. Drain the fuel from the tank.
   - Service hole cover
   - Fuel pump electrical connector
   - All fuel hoses from the fuel pump unit

   - Fuel pump ring using tool 49 T042 001
   - Fuel pump and gaskets from the fuel tank

To install:

5. Align the fuel pump alignment mark with the notch in the retainer and install the fuel pump using a new gasket.

6. Align the cap with the retainer and tighten one full turn by hand.

7. Tighten the cap with the special tool to 75 ft. lbs. (102 Nm).

8. Install or connect the following:
   - Fuel hoses to the fuel pump
   - Fuel pump electrical connector
   - Service hole cover
   - Rear seat cushion
   - Negative battery cable

9. Add a minimum of 10 gallons of fuel to the tank and check for leaks.

10. On models with dynamic suspension, perform the steering angle sensor initialization procedure.

Fuel Injector

REMOVAL & INSTALLATION

✳✳ CAUTION

Fuel injection systems remain under pressure after the engine has been turned OFF. Properly relieve fuel pressure before disconnecting any

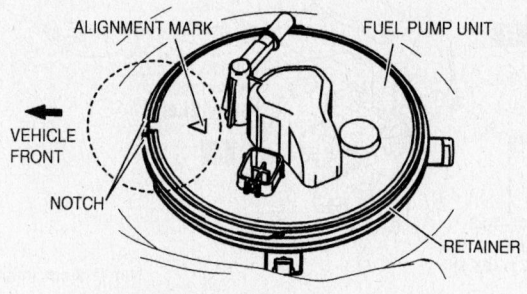

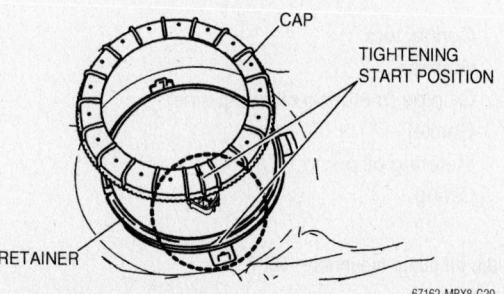

Installing the fuel the fuel pump ring–RX-8

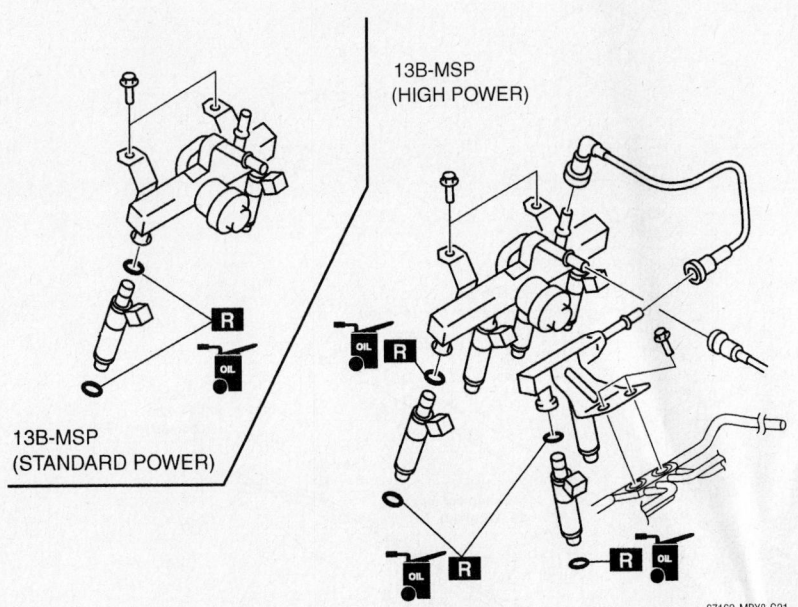

13B-MSP
(HIGH POWER)

13B-MSP
(STANDARD POWER)

67162-MRX8-G21

Exploded view of the fuel rail and injector assembly—RX-8

fuel lines. Failure to do so may result
in fire or personal injury. Do not
allow fuel spray or fuel vapors to
come in contact with a spark or open
flame. Keep a dry chemical fire
extinguisher nearby. Never store fuel
in an open container due to risk of
fire or explosion.

1. Before servicing the vehicle, refer to
the precautions in the beginning of this section.
2. Relieve the fuel system pressure.
3. Remove or disconnect the following:
- Negative battery cable
- Upper and lower extension manifolds

- Variable Dynamic Effect (VDI) actu-
ator and position out of the way
- Fuel injector wiring harness
- Fuel lines at the fuel rail
- Fuel distributor from intake mani-
fold and housing sides
- Fuel rail with the injectors
attached
- Fuel injectors, grommets and O-
rings from the fuel rail
- O-rings from the fuel injectors

To install:
4. Install or connect the following:
- New O-rings and grommets lubricated with engine oil on the fuel
injectors.
- Insulators and injectors on the
intake manifold
- Grommets and the fuel rail onto the
injectors. Torque the bolts to 14–18
ft. lbs. (19–25 Nm).
- Fuel lines to the fuel rail
- Fuel distributors
- Fuel injector wiring harness
- VDI actuator
- Negative battery cable
5. Turn the ignition switch **ON** to pressurize the fuel system.
6. Check for leaks and correct as necessary, before starting the engine.
7. On models with dynamic suspension,
perform the steering angle sensor initialization procedure.

DRIVE TRAIN

Manual Transmission Assembly

REMOVAL & INSTALLATION

1. Before servicing the vehicle, refer to
the precautions in the beginning of this sec-
tion.
2. Refer to the illustration for compo-
nent location and torque specifications.
3. Drain the transmission oil.
4. Remove or disconnect the following:
- Engine cover
- Negative battery cable
- Shifter knob
- Shifter panel
- Upper and lower shift insulators
- Shift lever
- Front and rear tunnel supports
- Oxygen sensor connector and
bracket
- Catalytic converter
- Middle exhaust pipe
- Rear silencer

- Exhaust manifold stay
- Heat insulator
- Starter
- Clutch slave cylinder
5. Support the transmission with a suitable jack
6. Support the rear differential with a
block of wood.
7. Place match marks on the driveshaft,
then remove the driveshaft.
8. Remove or disconnect the following:
- Power plant frame
- Back up light switch connector
- Neutral safety switch connector
- Transmission bolts and the transmission

To install:
9. Install the transmission and tighten
the mounting bolts to 28–38 ft. lbs. (37–52
Nm).
10. Support the transmission with a jack
and install the power plant frame. Temporarily tighten the bolts in the sequence shown.
11. Raise the front end of the frame and

adjust dimension A to 1.91–2.22 inches
(48.4–56.4mm).
12. Tighten the power plant frame bolts
in the sequence shown to 93–113 ft. lbs.
(126–154 Nm) for bolts 1 and 2 and 55–69
ft. lbs. (75–93 Nm) for bolt 3.
13. Install or connect the following:
- Neutral safety switch connector
- Back up light switch connector
- Clutch release cylinder
- Drive shaft
- Starter
- Heat insulator
- Exhaust manifold stay
- Rear silencer
- Middle exhaust pipe
- Catalytic converter
- Oxygen sensor connector and
bracket
- Front and rear tunnel supports
- Shift lever
14. Apply grease to the shift lever com-
ponents as illustrated.
15. Install or connect the following:

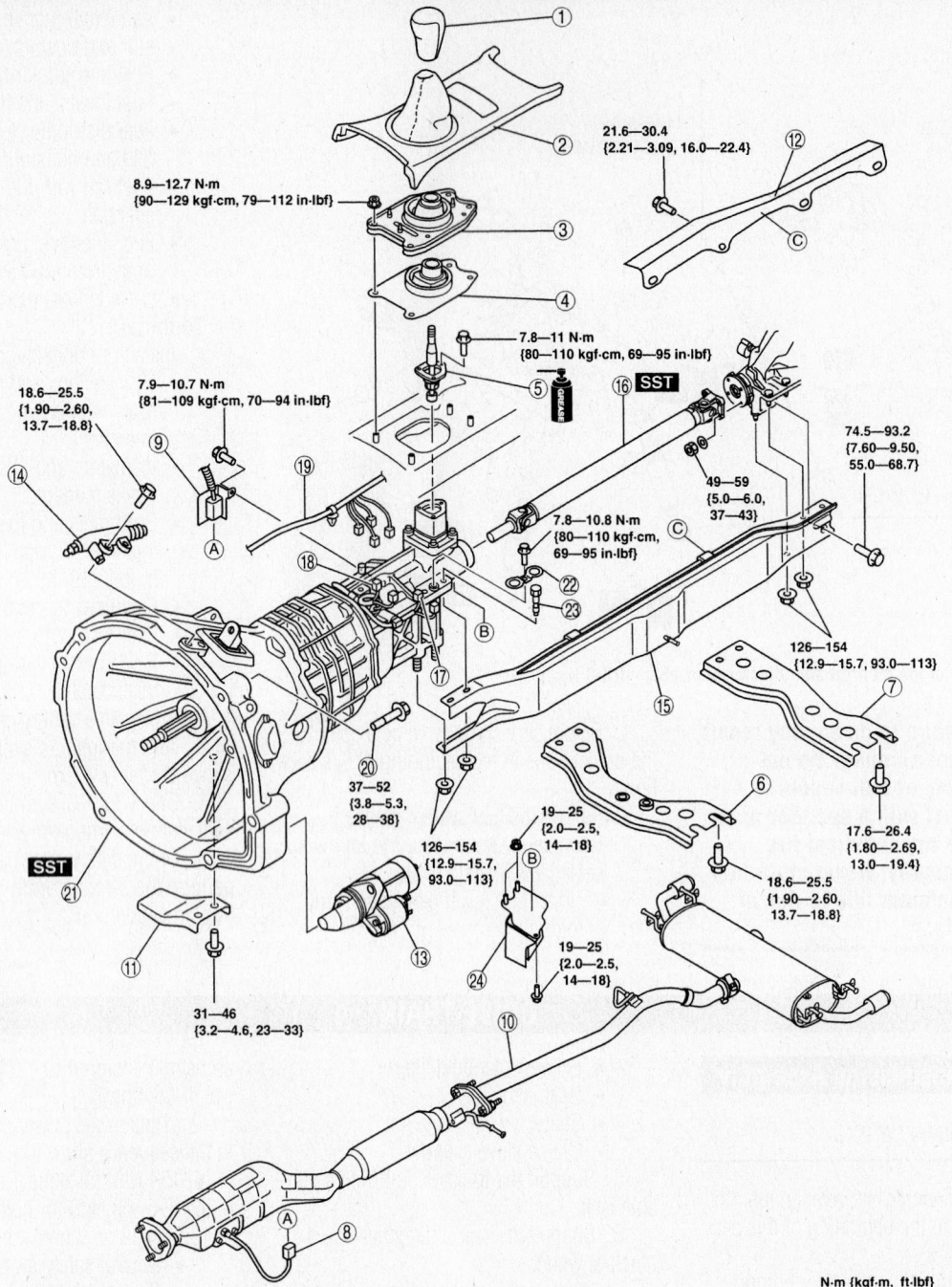

8.9—12.7 N·m
{90—129 kgf·cm, 79—112 in·lbf}

21.6—30.4
{2.21—3.09, 16.0—22.4}

7.8—11 N·m
{80—110 kgf·cm, 69—95 in·lbf}

7.9—10.7 N·m
{81—109 kgf·cm, 70—94 in·lbf}

18.6—25.5
{1.90—2.60,
13.7—18.8}

74.5—93.2
{7.60—9.50,
55.0—68.7}

49—59
{5.0—6.0,
37—43}

7.8—10.8 N·m
{80—110 kgf·cm,
69—95 in·lbf}

126—154
{12.9—15.7, 93.0—113}

37—52
{3.8—5.3,
28—38}

126—154
{12.9—15.7,
93.0—113}

19—25
{2.0—2.5,
14—18}

18.6—25.5
{1.90—2.60,
13.7—18.8}

17.6—26.4
{1.80—2.69,
13.0—19.4}

19—25
{2.0—2.5,
14—18}

31—46
{3.2—4.6, 23—33}

N·m {kgf·m, ft·lbf}

1	Shift lever knob	13	Starter
2	Upper panel	14	Clutch release cylinder
3	Shift insulator component (outer)	15	Power plant frame
4	Shift insulator component (inner)	16	Propeller shaft
5	Shift lever component	17	Back-up light switch connector
6	Front tunnel member	18	Neutral switch connector
7	Rear tunnel member	19	Wire
8	Heated oxygen sensor connector	20	Transmission installation bolt
9	Heated oxygen sensor connector bracket	21	Transmission
10	Catalytic converter, middle pipe, main silencer	22	Stopper
11	Exhaust manifold stay	23	Bolt
12	Heat insulator	24	Dynamic damper

67162-MRX8-G22

Exploded view of the manual transmission mounting—RX-8

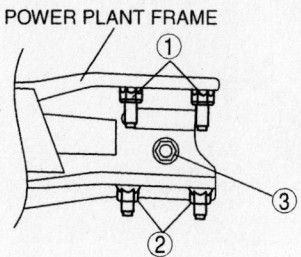

REAR DIFFERENTIAL SIDE

POWER PLANT FRAME

TRANSMISSION SIDE

POWER PLANT FRAME

67162-MRX8-G23

Tighten the power plant frame bolts using this sequence—RX-8

- Upper and lower shift insulators
- Shifter panel
- Shifter knob
- Negative battery cable
- Engine cover

16. Fill the transmission with fluid. Road test the vehicle and check for leaks. Top off all fluids as needed.

17. On models with dynamic suspension, perform the steering angle sensor initialization procedure.

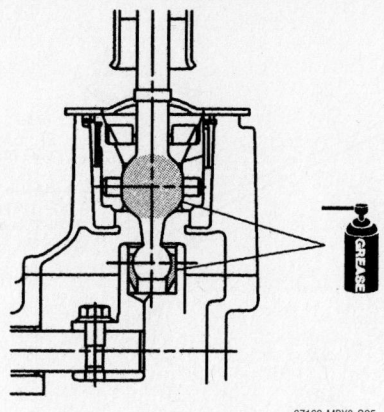

67162-MRX8-G25

Apply grease to the shift lever components—RX-8

Automatic Transmission Assembly

REMOVAL & INSTALLATION

1. Before servicing the vehicle, refer to the precautions in the beginning of this section.

2. Refer to the illustration for component location and torque specifications.

3. Drain the transaxle oil.

4. Remove or disconnect the following:
- Engine cover
- Front and rear tunnel supports
- Oxygen sensor connector
- Catalytic converter
- Middle exhaust pipe
- Rear silencer
- Exhaust manifold stay

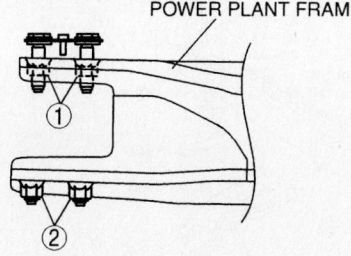

POWER PLANT FRAME

A

FRONT TUNNEL MEMBER

BOLT (M12 X 1.25)

67162-MRX8-G24

Measure the distance A which should be 1.91–2.22 inch (48.4–56.4mm)—RX-8

- Manual shaft lever after match marking its position
- Heat insulator
- Transverse member
- Starter
- Under cover
- Torque converter nuts
- Drive plate bolts
- Connector bolt
- Washer
- Oil pipe and hose
- Insulator
- Oil filler tube
- Transmission range switch connector
- Solenoid valve connector
- VSS connector
- Turbine sensor connector
- Connector wire loom
- Power plant frame

5. Support the transmission with a suitable jack

6. Support the rear differential with a block of wood.

7. Place match marks on the driveshaft, then remove the driveshaft.

8. Remove the transmission bolts and the transmission.

To install:

9. Install the transmission and tighten the mounting bolts to 28–38 ft. lbs. (37–52 Nm).

10. Support the transmission with a jack and install the power plant frame. Temporarily tighten the bolts in the sequence shown.

11. Raise the front end of the frame and adjust dimension A to 1.91–2.22 inches (48.4–56.4mm).

12. Tighten the power plant frame bolts in the sequence shown to 93–113 ft. lbs. (126–154 Nm) for bolts 1 and 2 and 55–69 ft. lbs. (75–93 Nm) for bolt 3.

13. Install the torque converter nuts and drive plate bolts and tighten them equally and evenly to 26–36 ft. lbs. (34–49 Nm).

14. Install or connect the following:
- Driveshaft
- Connector wire loom
- Turbine sensor connector
- VSS connector
- Solenoid valve connector
- Transmission range switch connector
- Oil filler tube
- Insulator
- Oil pipe and hose
- Washer
- Connector bolt
- Under cover
- Starter
- Transverse member
- Heat insulator
- Manual shaft lever
- Exhaust manifold stay
- Rear silencer

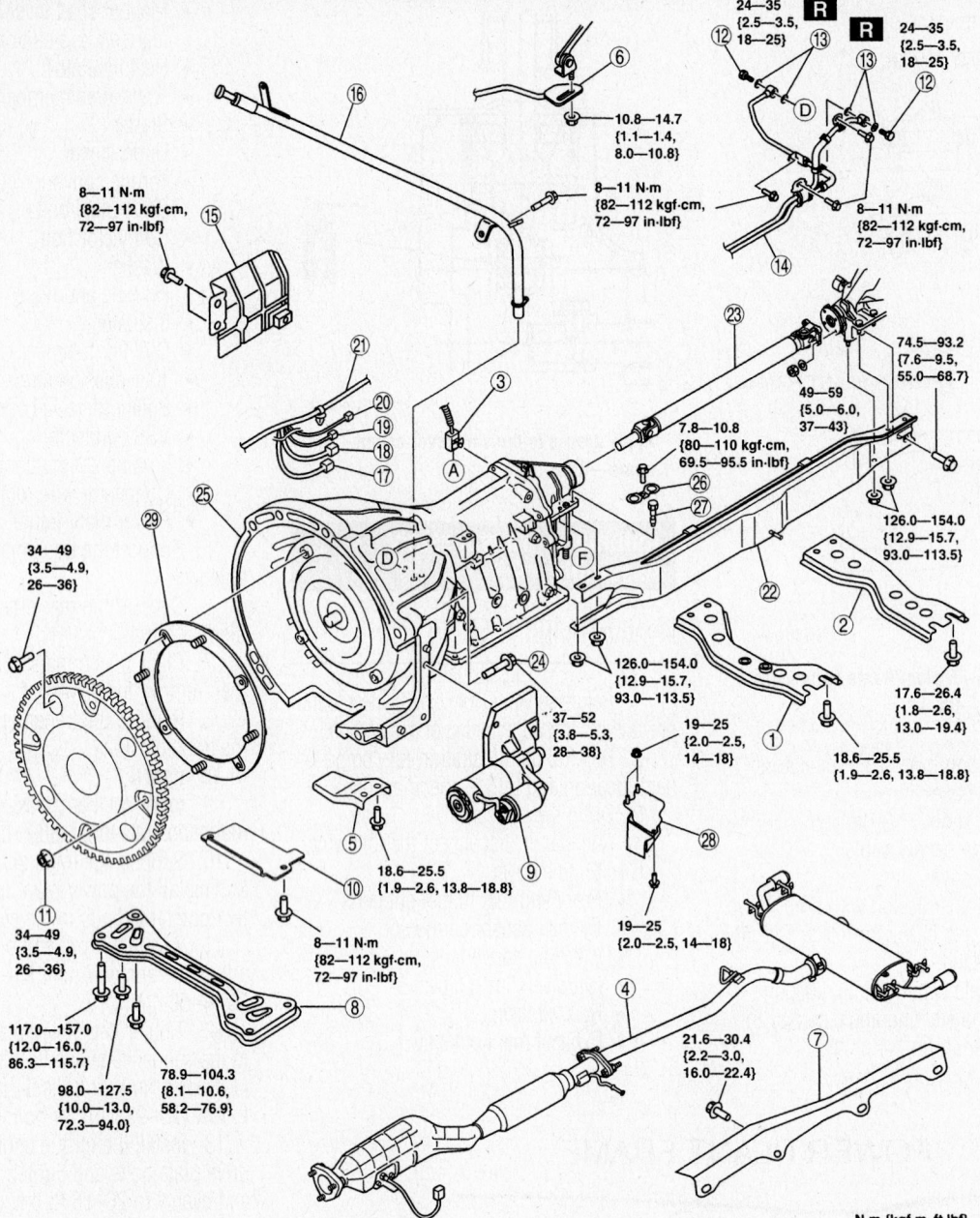

N·m {kgf·m, ft·lbf}

1	Front tunnel member	16	Oil filter tube, Dipstick
2	Rear tunnel member	17	TR switch connector
3	Heated oxygen sensor connector	18	Solenoid valve connector
4	Catalytic converter, middle pipe, main silencer	19	VSS connector
5	Exhaust manifold stay	20	Turbine sensor connector
6	Manual shaft lever component	21	Wire
7	Heat insulator	22	Power plant frame
8	Transverse member	23	Propeller shaft
9	Starter	24	Transmission installation bolt
10	Under cover	25	Transmission
11	Torque converter installation nuts	26	Stopper
12	Connector bolt	27	Bolt
13	Washer	28	Dynamic dumper
14	Oil pipe, oil hose	29	Driven plate
15	Insulator		

Exploded view of the automatic transaxle assembly mounting—RX-8

67162-MRX8-G26

REAR DIFFERENTIAL SIDE

POWER PLANT FRAME

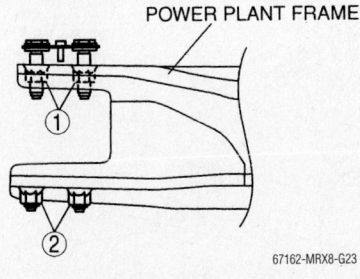

TRANSMISSION SIDE

POWER PLANT FRAME

67162-MRX8-G23

Tighten the power plant frame bolts using this sequence—RX-8

POWER PLANT FRAME

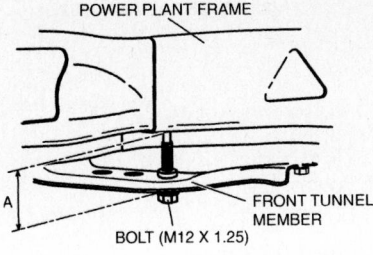

FRONT TUNNEL MEMBER

BOLT (M12 X 1.25)

67162-MRX8-G24

Measure the distance A which should be 1.91–2.22 inch (48.4–56.4mm)—RX-8

- Middle exhaust pipe
- Catalytic converter
- Oxygen sensor connector
- Front and rear tunnel supports
- Engine cover

15. Fill the transmission with fluid. Road test the vehicle and check for leaks. Top off all fluids as needed.

16. On models with dynamic suspension, perform the steering angle sensor initialization procedure.

Clutch

REMOVAL & INSTALLATION

1. Before servicing the vehicle, refer to the precautions in the beginning of this section.

2. Remove or disconnect the following:
- Negative battery cable
- Clutch release cylinder
- Transmission
- Rubber boot
- Clutch release collar
- Clutch release fork
- Pressure plate loosening the bolts one turn each in a criss–cross pattern
- Clutch disc

3. Inspect the pilot bearing. If it is worn or damaged and does not turn easily by hand, remove it using a puller/slide hammer.

4. Check the flywheel surface for scoring, cracks or burning and machine or replace, as necessary.

5. Install Holder tool 49 F011 101 to keep the flywheel from turning, remove the flywheel lock bolt and remove the flywheel.

6. Inspect the clutch release bearing for wear. Replace it if it sticks or does not turn easily.

7. Inspect the release fork for wear or damage and replace as necessary.

To install:

8. Lubricate the release fork fingers and pivot with molybdenum grease and install in the release fork boot.

9. Install or connect the following:
- Clutch release bearing on the release fork
- New pilot bearing in the flywheel, if removed, using a installation tool

10. Be sure the flywheel mounting surface and the crankshaft or eccentric shaft mounting surfaces are clean. Remove any old sealant from the flywheel lock bolt and threads.
- Flywheel
- Sealant to the flywheel lock bolt threads and install it tight
- Flywheel holding tool. Tighten the lock bolt to 289–361 ft. lbs. (392–490 Nm)..
- Small amount of molybdenum grease to the clutch disc splines
- Clutch disc on the flywheel with the spring side toward the transaxle
- An alignment tool in the pilot bearing to position the clutch disc
- Clutch pressure plate by aligning the dowel holes with the flywheel dowels
- Pressure plate. Gradually, torque the bolts, in a crisscross pattern, to 13–19 ft. lbs. (18–27 Nm).

11. Remove the alignment tool.
- Clutch release fork
- Clutch release collar
- Rubber boot

- Transaxle
- Clutch release cylinder
- Negative battery cable

12. On models with dynamic suspension, perform the steering angle sensor initialization procedure.

Hydraulic Clutch System

BLEEDING

1. Before servicing the vehicle, refer to the precautions in the beginning of this section.

2. Remove the rubber cap from the bleeder screw on the release cylinder.

3. Place a bleeder tube over the end of the bleeder screw.

4. Submerge the other end of the tube in a jar half filled with hydraulic brake fluid.

5. Slowly pump the clutch pedal fully and allow it to return slowly, several times.

6. While pressing the clutch pedal to the floor, loosen the bleeder screw until the fluid starts to run out. Then, close the bleeder screw. Keep repeating this Step, while watching the hydraulic fluid in the jar. As soon as the air bubbles disappear, close the bleeder screw.

7. During the bleeding procedure the reservoir must be kept at least ¾ full.

Halfshafts

REMOVAL & INSTALLATION

1. Before servicing the vehicle, refer to the precautions in the beginning of this section.

2. Drain the differential oil.

3. Remove or disconnect the following:
- Rear wheels

4. Raise the staked portion of the hub locknut with a hammer and chisel.

5. Lock the hub by applying the brakes and remove the nut.

6. Remove or disconnect the following:
- Parking brake cable
- Brake caliper and wire aside
- Later link upper ball joint
- Stabilizer link
- Lower ball joint
- Lower shock absorber bolt
- Trailing link upper ball joint
- Outer toe control link

7. Position a prybar between the inner CV-joint and differential case. Carefully pry the halfshaft from the transaxle being careful not to damage the oil seal.

8. Pull outward on the hub/knuckle

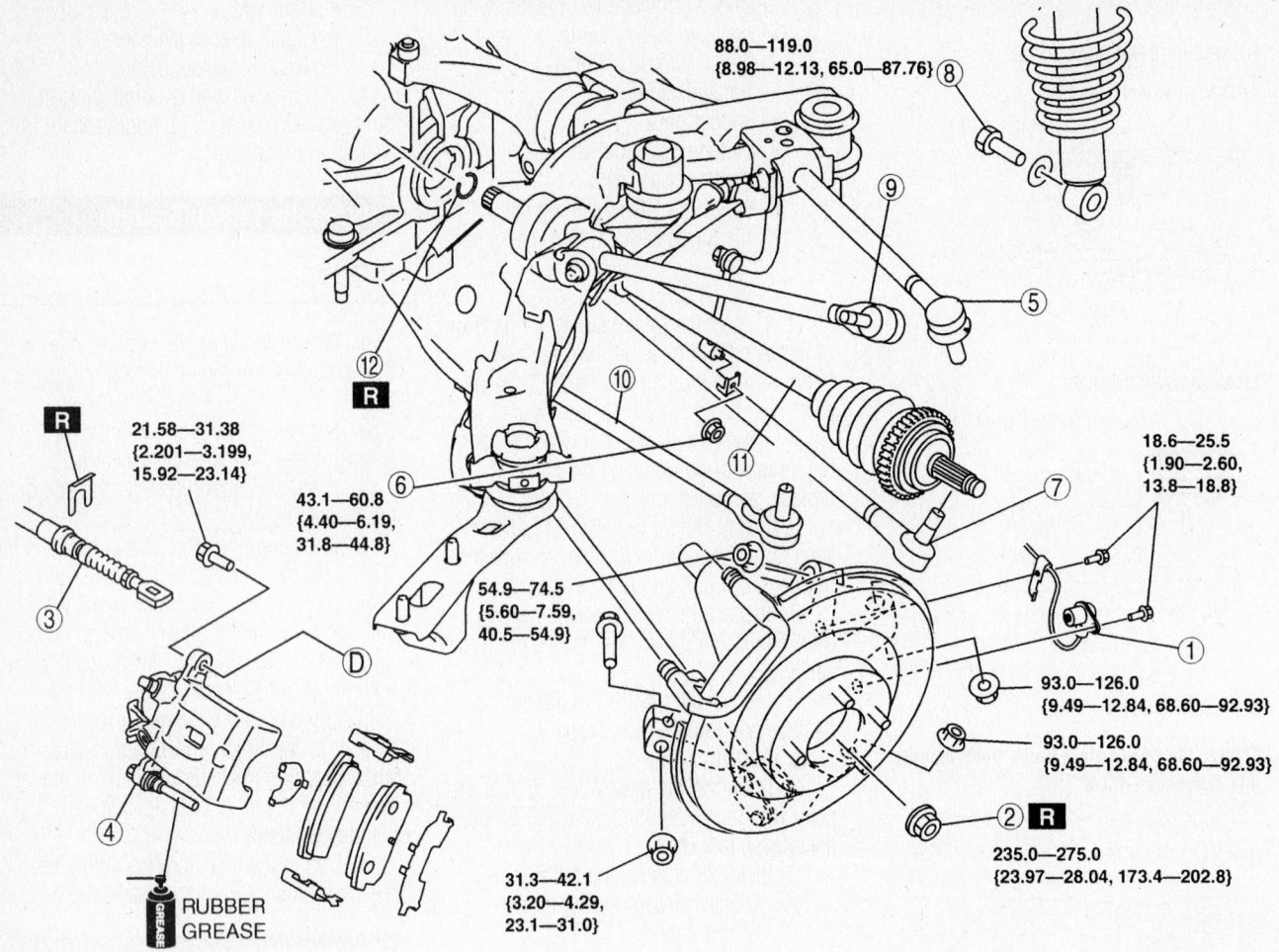

88.0—119.0
{8.98—12.13, 65.0—87.76}

21.58—31.38
{2.201—3.199,
15.92—23.14}

43.1—60.8
{4.40—6.19,
31.8—44.8}

54.9—74.5
{5.60—7.59,
40.5—54.9}

18.6—25.5
{1.90—2.60,
13.8—18.8}

93.0—126.0
{9.49—12.84, 68.60—92.93}

93.0—126.0
{9.49—12.84, 68.60—92.93}

31.3—42.1
{3.20—4.29,
23.1—31.0}

235.0—275.0
{23.97—28.04, 173.4—202.8}

RUBBER
GREASE

N·m {kgf·m, ft·lbf}

1 ABS wheel-speed sensor	7 Rear lateral link (lower) ball joint
2 Locknut	8 Shock absorber bolt (lower)
3 Parking brake cable	9 Rear trailing link (upper) ball joint
4 Brake caliper component	10 Toe control link (outer)
5 Rear lateral link (upper) ball joint	11 Rear drive shaft
6 Stabilizer control link (lower)	12 Clip

67162-MRX8-G27

Exploded view of rear halfshaft mounting—RX-8

assembly, push the outer CV-joint stub shaft through the hub and remove the half-shaft. If the halfshaft is stuck in the hub, install the old hub nut to protect the stub shaft threads. Tap on the nut, using only a soft mallet, to remove the halfshaft.

➡**Reinstall the rear lateral upper link to the knuckle to temporarily hold it in position.**

To install:

9. Install or connect the following:
  • New circlip on the end of the half-shaft, if removed, with the end gap facing upward.

  • Differential oil to the oil seal lip
  • Halfshaft into the diferential, being careful not to damage the oil seal
  • Other end of the halfshaft through the hub. Loosely install a new lock-nut
  • Outer toe control link
  • Trailing link upper ball joint
  • Lower shock absorber bolt
  • Lower ball joint
  • Stabilizer link
  • Later link upper ball joint
  • Brake caliper and wire aside
  • Parking brake cable
  • Wheels

  • New hub nut. Torque it to 174–203 ft. lbs. (235–275 Nm). After tightening, stake the locknut using a hammer and dull bladed chisel.
10. Fill the differential.

CV-Joints

OVERHAUL

1. Before servicing the vehicle, refer to the precautions in the beginning of this section.
2. Remove the halfshaft from the

vehicle and clamp it in a vise equipped with jaw caps, to prevent damage to the machined surfaces. Do not allow the vise to contact the boot or its clamps.

3. Remove the large boot clamp from the inboard CV-joint, using side cutters. After removing the clamp, roll the boot back over the shaft.

➡**Check the grease for contamination by rubbing it between 2 fingers. Any gritty feeling indicates a contaminated CV-joint, in which case the entire CV-joint must be disassembled, cleaned and inspected. If the grease is not contaminated and the CV-joint has been operating satisfactorily, continue with the boot replacement procedure and add the required lubricant.**

4. Paint alignment marks on the outer race and shaft for assembly reference. Remove the wire ring bearing retainer and remove the outer race.

5. Paint alignment marks on the tri-pot bearing and shaft for assembly reference.

Standard

		Drive shaft length (mm {in})
MT	Left side	792.6—802.6 {31.21—31.59}
	Right side	832.6—842.6 {32.78—33.17}
AT	Left side	791.1—801.1 {31.15—31.53}
	Right side	831.1—841.1 {32.71—33.11}

67162-MRX8-G28

Halfshaft length specifications—RX-8

Remove the tri-pot bearing snapring and, using a brass drift and hammer, remove the tri-pot bearing from the shaft.

6. Remove the small clamp and remove the inner boot from the halfshaft. If the boot is to be reused, wrap the shaft splines with tape before removing.

7. If the outer CV-joint boot is to be replaced, remove the clamps and slide the boot off the shaft from the inboard side.

To install:

8. If the outboard boot was removed, slide the boot onto the shaft from the inboard side. Wrap tape on the splines before installing to protect the boot.

9. Install the inboard boot and remove the tape from the shaft.

10. Install the tri-pot assembly on the halfshaft. Tap the assembly onto the shaft using a hammer and brass drift. Install the tri-pot assembly retaining ring.

11. Fill the CV-joint outer race with high temperature CV-joint grease. Install the outer race over the tri-pot joint and install the wire ring bearing retainer.

12. Position the CV-joint boot(s). Make sure the boot is fully seated in the grooves in the shaft and outer race.

13. Insert a small prybar with rounded edges between the boot and the outer bearing race to allow trapped air to escape from the boot. Install new boot clamps.

14. Wrap the clamps around the boots in a clockwise direction, pull tight with pliers and bend the locking tabs to secure in position.

15. Set the halfshaft length to the specification as shown.

16. Work the CV-joint through its full range of travel at various angles. The joint should flex, extend and compress smoothly.

17. Install the halfshaft into the vehicle.

STEERING AND SUSPENSION

Air Bag

PRECAUTIONS

Several precautions must be observed when handling the inflator module to avoid accidental deployment and possible personal injury.

1. Never carry the inflator module by the wires or connector on the underside of the module.

2. When carrying a live inflator module, hold securely with both hands, and ensure that the bag and trim cover are pointed away.

3. Place the inflator module on a bench or other surface with the bag and trim cover facing up.

4. With the inflator module on the bench, never place anything on or close to the module which may be thrown in the event of an accidental deployment.

5. An air bag is an explosive device. Handle with extreme caution.

6. Always disconnect the battery and the air bag connector before removing the steering wheel or beginning work on the air bag system.

7. Air bag components must not be repaired or opened. Always use new parts, including the wiring harness.

8. Always place a removed air bag unit with the horn pad facing up. Put it in a safe place where it will not be disturbed.

9. The air bag unit must not be exposed to grease, fluids, or cleaning agents.

10. The air bag unit must not be exposed to temperatures above 194°F (90°C) at any time. Even the heat of a soldering iron can damage or ignite the charge.

11. Storage and transport of air bags is subject to rules governing explosive devices and should be done only in the original package.

12. Failure to follow proper safety precautions may result in personal injury through accidental firing of the air bag, or through failure of the air bag in an accident.

DISARMING

1. Before servicing the vehicle, refer to the precautions in the beginning of this section.

2. If equipped, deactivate the audio anti-theft system.

3. Turn the ignition switch to LOCK.

4. Disconnect and isolate the negative battery cable and wait for more than 1 minute to allow the backup power supply to deplete its stored power.

ARMING

1. Before servicing the vehicle, refer to the precautions in the beginning of this section.

2. Connect the negative battery cable, turn the ignition switch **ON** and verify the air bag warning light cones on for 6 seconds. If the light does not illuminate there are problems with the system.

3. If equipped, activate the audio anti-theft system.

Steering Angle Sensor

INITIALIZATION

1. Connect the negative battery cable.

2. Turn the ignition switch to the **ON** position.

3. Confirm that the Dynamic Stability Control (DSC) indicator light comes on and the DSC OFF light flashes.

4. Turn the steering wheel to the right full lock position and then all the way to the left full lock position.

5. Confirm that the DSC OFF light goes out.

6. Turn the ignition off.

7. Turn the ignition on again and confirm that the DSC indicator light is out. If the light is not out, repeat the procedure.

8. Drive the vehicle for about 10 minutes and confirm that the ABS and DSC indicator lights are out.

Rack and Pinion Steering Gear

REMOVAL & INSTALLATION

1. Before servicing the vehicle, refer to the precautions in the beginning of this section.
2. Remove or disconnect the following:
   - Negative battery cable
   - Front wheels
   - Splash shield
   - Radiator bracket
   - Cotter pins and nuts from both steering tie rod ends

3. Press the tie rod out from the knuckle arm.
   - Intermediate shaft to steering gear pinion shaft bolt. Mark the shaft-to-gear location.
   - Shaft from the steering gear
   - Torque sensor connector
   - Electronic Power Steering (ESP) motor connector
   - Steering gear mounting nuts
   - Steering gear and linkage from the vehicle

To install:
4. Install or connect the following:
   - Steering gear and linkage to the

vehicle. Torque the bolts in sequence to 55–77 ft. lbs. (75–104 Nm).
   - Steering shaft to the steering gear pinion shaft, align the marks made during removal and tighten the bolt to 19 ft. lbs. (26 Nm)
   - Tie rod ends to the knuckle arm. Torque the nuts to 27–36 ft. lbs. (37–49 Nm).
   - Electronic Power Steering (ESP) motor connector
   - Torque sensor connector
   - New cotter pins
   - Radiator bracket

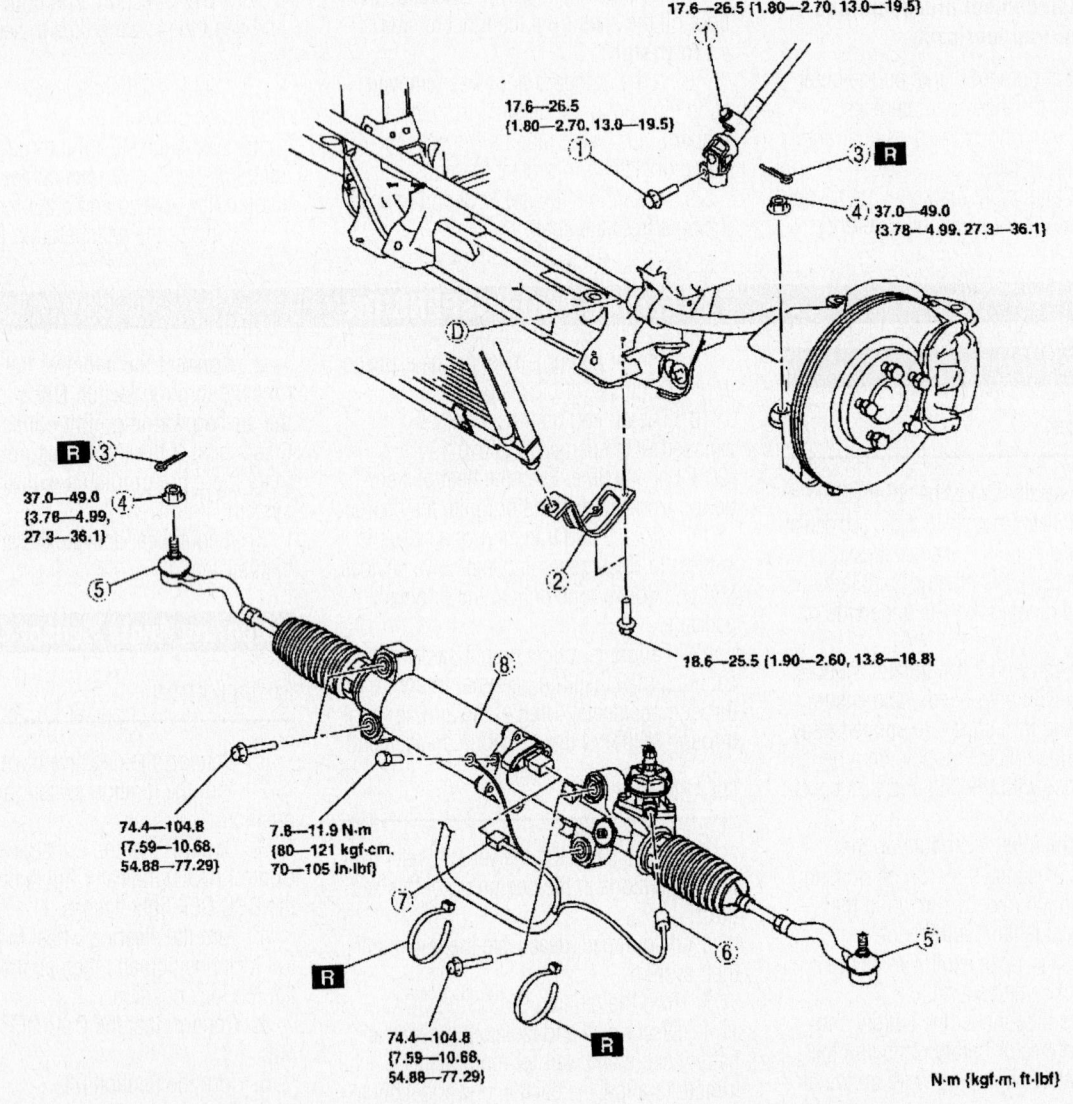

1	Bolt (intermediate shaft)	5	Tie-rod end
2	Radiator bracket	6	Torque sensor connector
3	Cotter pin	7	EPS motor connector
4	Locknut (tie-rod end)	8	Steering gear and linkage

67162-MRX8-G29

Exploded view of steering gear mounting—RX-8

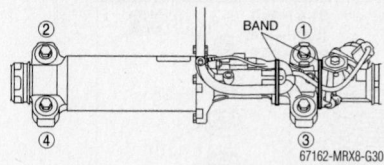

67162-MRX8-G30

Steering gear bolt tightening sequence—RX-8

- Splash shield
- Wheels
- Negative battery cable

5. Check and/or adjust the front end alignment.

6. On models with dynamic suspension, perform the steering angle sensor initialization procedure.

Shock Absorber

REMOVAL & INSTALLATION

Front

1. Before servicing the vehicle, refer to the precautions in the beginning of this section.

2. Remove the strut cross brace in the engine compartment and the upper shock mounting nuts.

3. Remove or disconnect the following:
- Front wheel
- Brake hose bracket
- Stabilizer bar nut
- Upper arm ball joint

4. Remove the upper and lower shock absorber mounting nuts and bolts and remove the shock absorber/coil spring assembly.

To install:

5. Installation is the reverse of removal. Tighten the upper shock nuts to 34–46 ft. lbs. (46–62 Nm) and the lower shock nut and bolt to 58–76 ft. lbs. (78–103 Nm)

6. Check and/or adjust the front end alignment.

Rear

1. Before servicing the vehicle, refer to the precautions in the beginning of this section.

2. Remove or disconnect the following:

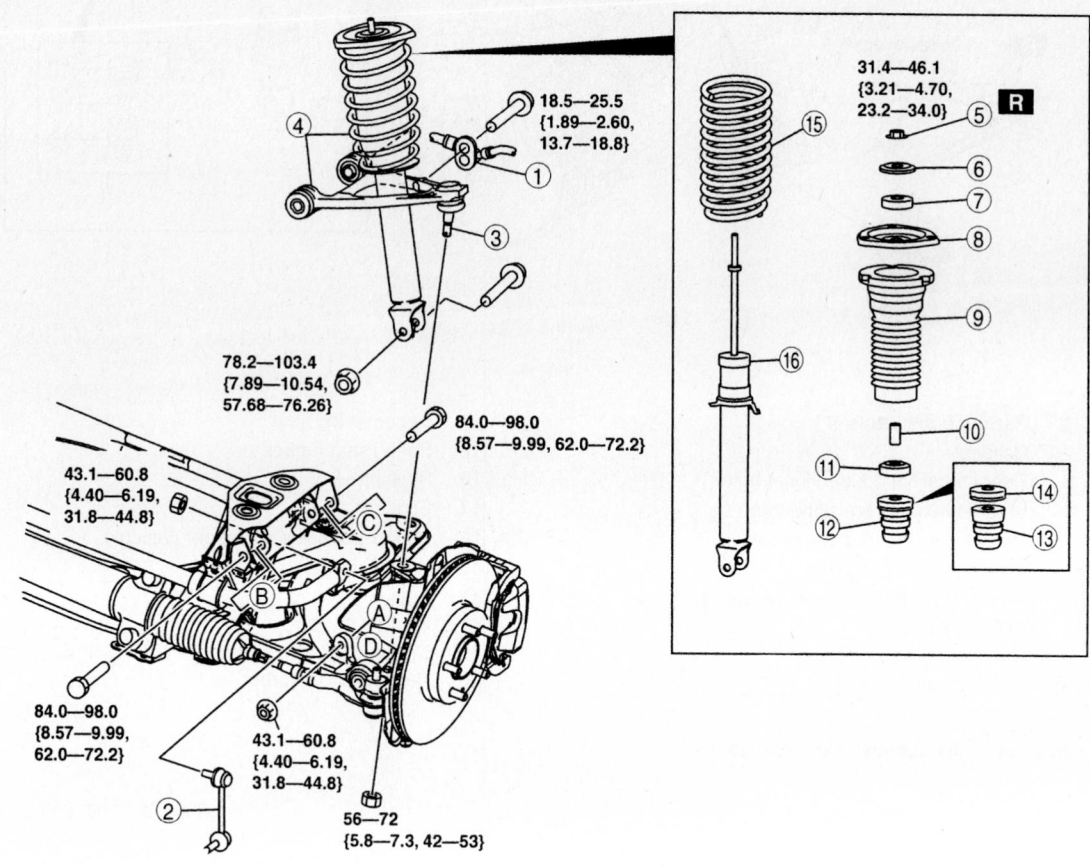

N·m {kgf·m, ft·lbf}

1	Brake hose bracket	9	Dust boot
2	Front stabilizer control link	10	Spacer
3	Front upper arm ball joint	11	Bushing
4	Front shock absorber and coil spring	12	Stopper casing and bound stopper
5	Piston rod nut	13	Bound stopper
6	Retainer	14	Stopper casing
7	Bushing	15	Coil spring
8	Upper spring seat	16	Front shock absorber

67162-MRX8-G31

Exploded view of front shock absorber mounting—RX-8

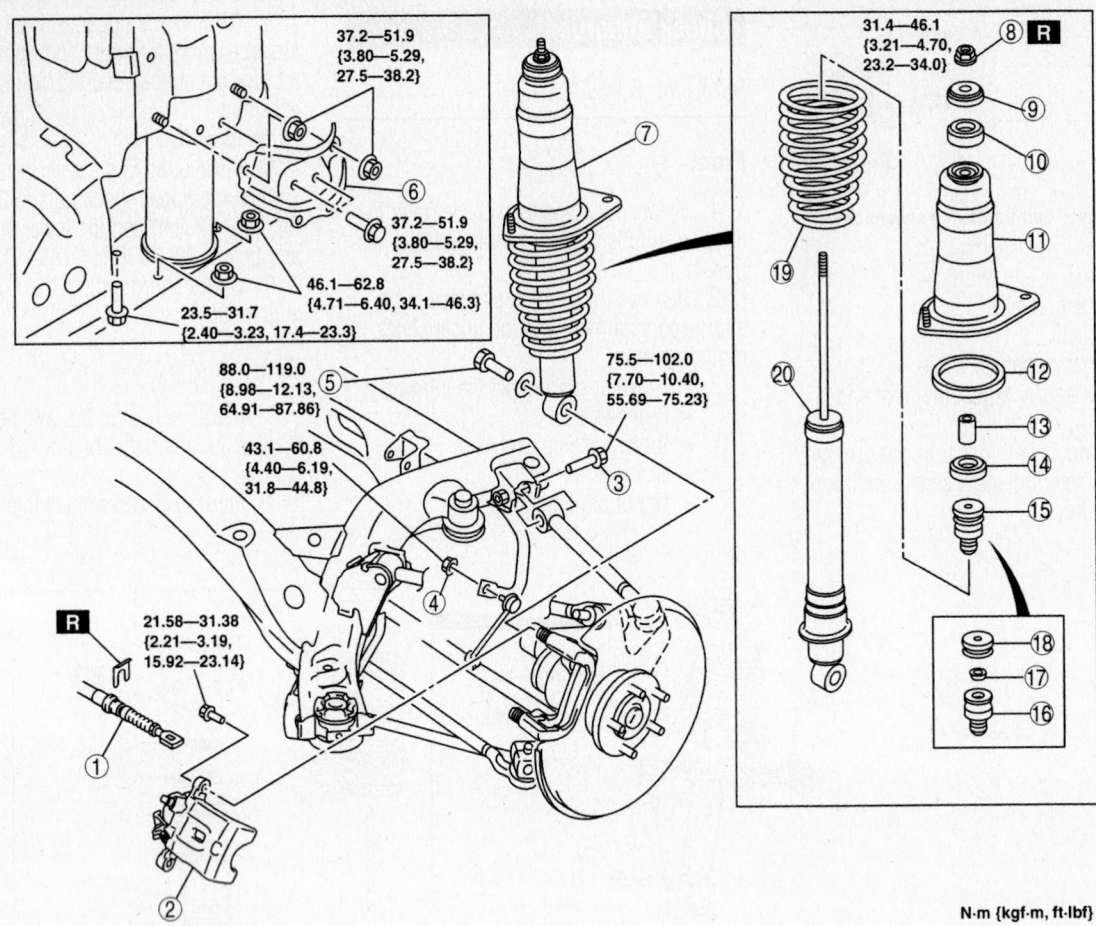

N·m {kgf·m, ft·lbf}

1	Parking brake cable	11	Upper spring seat
2	Caliper	12	Spring seat rubber
3	Rear lateral link (upper) inner bolt	13	Bushing
4	Stabilizer control link upper nut	14	Spacer
5	Rear shock absorber lower bolt	15	Bound stopper and stopper casing
6	Rear shock absorber bracket	16	Bound stopper
7	Rear shock absorber and coil spring	17	Collar
8	Piston rod nut	18	Stopper casing
9	Retainer	19	Coil spring
10	Bushing	20	Rear shock absorber

67162-MRX8-G32

Exploded view of rear shock absorber mounting—RX-8

- Rear wheel
- Brake caliper and wire aside
- Rear lateral link upper inner bolt
- Stabilizer bar upper nut
- Shock absorber lower bolt
- Inside trunk end and side trim panels
- Shock absorber upper bracket and mounting nuts
- Shock absorber/coil spring assembly

To install:

3. Installation is the reverse of removal. Tighten the upper shock nuts to 34–46 ft. lbs. (46–62 Nm) and the lower shock nut and bolt to 65–88 ft. lbs. (88–119 Nm). Tighten the shock bracket bolts and nuts to 28–38 ft. lbs. (37–52 Nm).

4. Check and/or adjust the front end alignment.

Coil Spring

REMOVAL & INSTALLATION

1. Before servicing the vehicle, refer to the precautions in the beginning of this section.

2. Remove or disconnect the following:

- Shock absorber/coil spring
- Place the shock/coil spring in a spring compressor and compress the spring
- Piston rod nut
- Upper retainer, bushing spring seat and rubber insulator
- Bushing, spacer, bound stopper and casing
- Coil spring

3. While pushing on the piston rod, be sure that the pull stroke is even and that there is no unusual noise or resistance. Also inspect for any oil leakage around the piston rod.

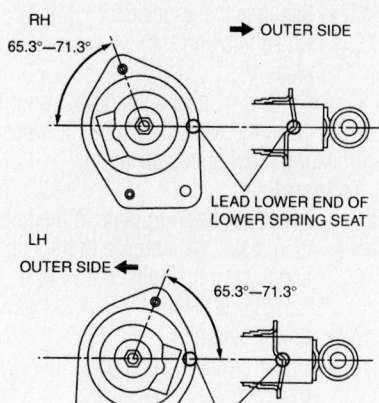

Mark the lower seat, shock and spring assembly as illustrated for reassembly–RX-8

4. Push the piston rod in, then release it. Be sure that the return rate is constant.

5. If the shock absorber does not operate as described, replace it.

To install:

6. Install or connect the following:
- Strut assembly into a vise
- Bound stopper and casing onto the piston rod
- Temporarily install the lower spring seat, seat rubber and spring. Mark the seat, shock and spring assembly as illustrated for reassembly. Align the marks of the upper seat and coil spring. Protect the assembly with cloth and install the spring compressor.
- Coil spring

7. Compress the coil spring with the spring compressor

8. Install or connect the following:
- Bushing, spacer, bound stopper and casing
- Upper retainer, bushing spring seat and rubber insulator
- Piston rod upper nut

9. Be sure that the spring upper seat notched portion is facing inward and tighten the piston rod upper nut to 23–34 ft. lbs. (32–46 Nm).

10. Be sure that the spring is well seated in the upper seats.

11. Install the shock to the vehicle.

Lower Ball Joint

REMOVAL & INSTALLATION

1. Before servicing the vehicle, refer to the precautions in the beginning of this section.

2. Remove or disconnect the following:
- Wheel
- Ball joint clip
- Ball joint using a ball joint remover

To install:

3. Install or connect the following:
- Ball joint to lower control arm using a ball joint installer
- Ball joint clip
- Wheel

4. Check and/or adjust the front wheel alignment.

Upper Ball Joint

REMOVAL & INSTALLATION

1. Before servicing the vehicle, refer to the precautions in the beginning of this section.

2. Remove or disconnect the following:
- Wheel
- Lower nut on the vehicle side
- Ball joint using a ball joint remover

To install:

3. Install or connect the following:
- Ball joint to lower control arm using a ball joint installer
- Tighten the nut to 42–53 ft. lbs. (56–72 Nm).
- Wheel

4. Check and/or adjust the front wheel alignment.

Upper Control Arm

REMOVAL & INSTALLATION

1. Before servicing the vehicle, refer to the precautions in the beginning of this section.

2. Remove or disconnect the following:
- Front wheel
- Brake caliper and wire aside
- Brake rotor
- Stabilizer bar nut
- Lower shock mounting bolt and nut
- Upper arm ball joint

3. Remove the front upper arm bolts, push down on the lower arm and remove the upper arm through the lower shock and lower arm gap.

To install:

4. Install the upper control arm.

5. Loosely tighten the bolt and nut.

6. Loosely install the lower strut mounting bolt.
- Upper arm ball joint. Torque the nut 42–53 ft. lbs. (56–72 Nm).
- Wheel

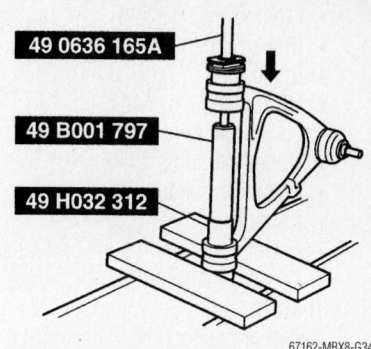

Removing upper control arm bushing – RX-8

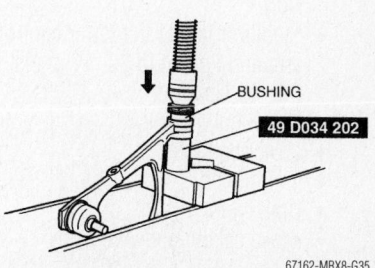

Installing upper control arm bushing – RX-8

7. Torque upper control arm bolt to 62–72 ft. lbs. (84–98 Nm) and the lower strut mounting bolt to 58–76 ft. lbs. (78–103 Nm).

8. Check and/or adjust the front wheel alignment.

CONTROL ARM BUSHING REPLACEMENT

1. Before servicing the vehicle, refer to the precautions in the beginning of this section.

2. Position the control arm into a vise. Cut away the projecting rubber from the bushing. Use tools 49 B0636 165A, 49 B0001 797 and 49 H032 312 to remove the bushing.

To install:

3. Apply soapy water to the new control arm bushing.

4. Install the rear bushing using and tool 49 D034 202.

Lower Control Arm

REMOVAL & INSTALLATION

1. Before servicing the vehicle, refer to the precautions in the beginning of this section.

2. Remove the strut cross brace in the engine compartment.

3. Remove or disconnect the following:

- Front wheel
- Brake caliper and wire aside
- Brake rotor
- Tie rod end
- Lower ball joint
- Upper ball joint
- Stabilizer link from the lower control arm
- Lower control arm bolts and nuts
- Lower control arm

To install:

4. Install or connect the following:
- Lower control arm
- Lower control arm bolt and nut and tighten to 62–72 ft. lbs. (84–98 Nm)
- Stabilizer link to the lower control arm and tighten to 32–45 ft. lbs. (43–61 Nm)
- Upper ball joint
- Lower ball joint
- Tie rod end
- Brake rotor
- Brake caliper and wire aside
- Front wheel

5. Check and/or adjust the front wheel alignment.

CONTROL ARM BUSHING REPLACEMENT

1. Before servicing the vehicle, refer to the precautions in the beginning of this section.

2. Position the control arm into a vise. Cut away the stopper rubber and the projecting rubber from the front bushing and press out the bushing

3. Remove the rear bushing using a press and tools 49 U034 204 and 49 S010 301.

4. Remove the shock absorber bushing using a press and tools 49 U034 204 and 49 S032 333.

To install:

5. When installing the rear bushing on the lower arm, align the mark of the lower arm and the holes of the bushing as illustrated, set the arm onto the press and press the bushings into position.

6. Apply soapy water to the new control arm bushing.

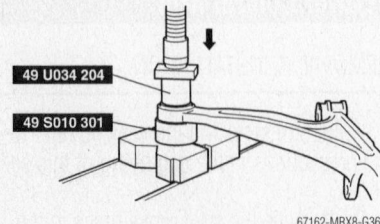

Removing and installing the lower control arm front bushing—RX-8

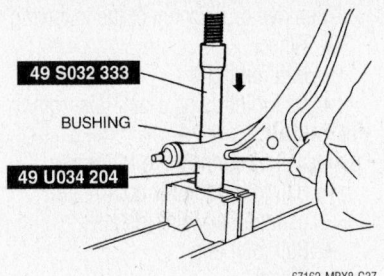

67162-MRX8-G37

Removing and installing the lower control arm shock absorber bushing—RX-8

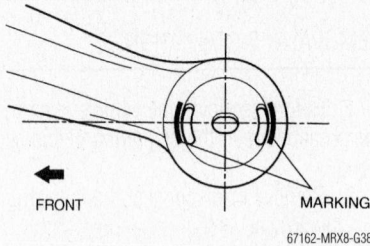

67162-MRX8-G38

Align the mark of the lower arm rear bushing when installing the bushing—RX-8

7. Install the rear bushing using a press and tools 49 U034 204 and 49 S010 301.

8. Install the front bushing using a vise.

9. Install the shock absorber bushing using a press and tools 49 U034 204 and 49 S032 333.

Wheel Bearings

ADJUSTMENT

The front and rear wheel bearings are not adjustable. If the bearings become loose or make noise, they must be replaced.

REMOVAL & INSTALLATION

Front

1. Before servicing the vehicle, refer to the precautions in the beginning of this section.

2. Refer to the illustration for component location and torque specifications.

3. Remove or disconnect the following:
- Wheels
- ABS wheel speed sensor
- Brake caliper and rotor
- Tie rod end from the knuckle
- Stabilizer bar link
- Upper control arm ball joint from the knuckle
- Upper arm bolt
- Lower ball joint
- Wheel hub and knuckle assembly

- Separate the knuckle from the wheel hub and remove the backing plate.

4. Clean and inspect all parts but do not wash or clean the wheel bearing. The wheel hub/bearing must be replaced.

To install:

- Install the backing plate to the hub, then install the knuckle to the hub and tighten the bolts to 40–44 ft. lbs. (54–60 Nm).
- Lower ball joint
- Upper arm bolt and tighten to 62–72 ft. lbs. (84–96 Nm).
- Upper control arm ball joint
- Stabilizer bar link and tighten to 32–45 ft. lbs. (43–61 Nm).
- Tie rod end
- ABS wheel speed sensor
- Brake caliper and rotor
- Wheels

Rear

1. Before servicing the vehicle, refer to the precautions in the beginning of this section.

2. Refer to the illustration for component location and torque specifications.

➡ **The wheel bearings are not serviceable. If the bearings are bad, a new hub/bearing assembly must be installed.**

3. Remove or disconnect the following:
- Rear wheels
- ABS wheel speed sensor

4. Raise the staked portion of the axle retaining nut with a hammer and chisel.

5. Remove or disconnect the following:
- Axle nut
- Parking brake cable
- Rear caliper and rotor assembly from the hub
- Rear lateral link upper ball joint
- Stabilizer bar lower link
- Rear lateral link lower ball joint
- Lower shock absorber bolt
- Rear trailing link lower outside bolt
- Rear trailing link upper ball joint
- Toe control link outside bolt
- Axle shaft
- Wheel hub/knuckle assembly

6. Press the wheel hub from the knuckle.

7. Press the bearing inner race from the wheel hub.

8. Press the wheel bearing from the rear knuckle.

To install:

9. Press the wheel bearing into the rear knuckle.

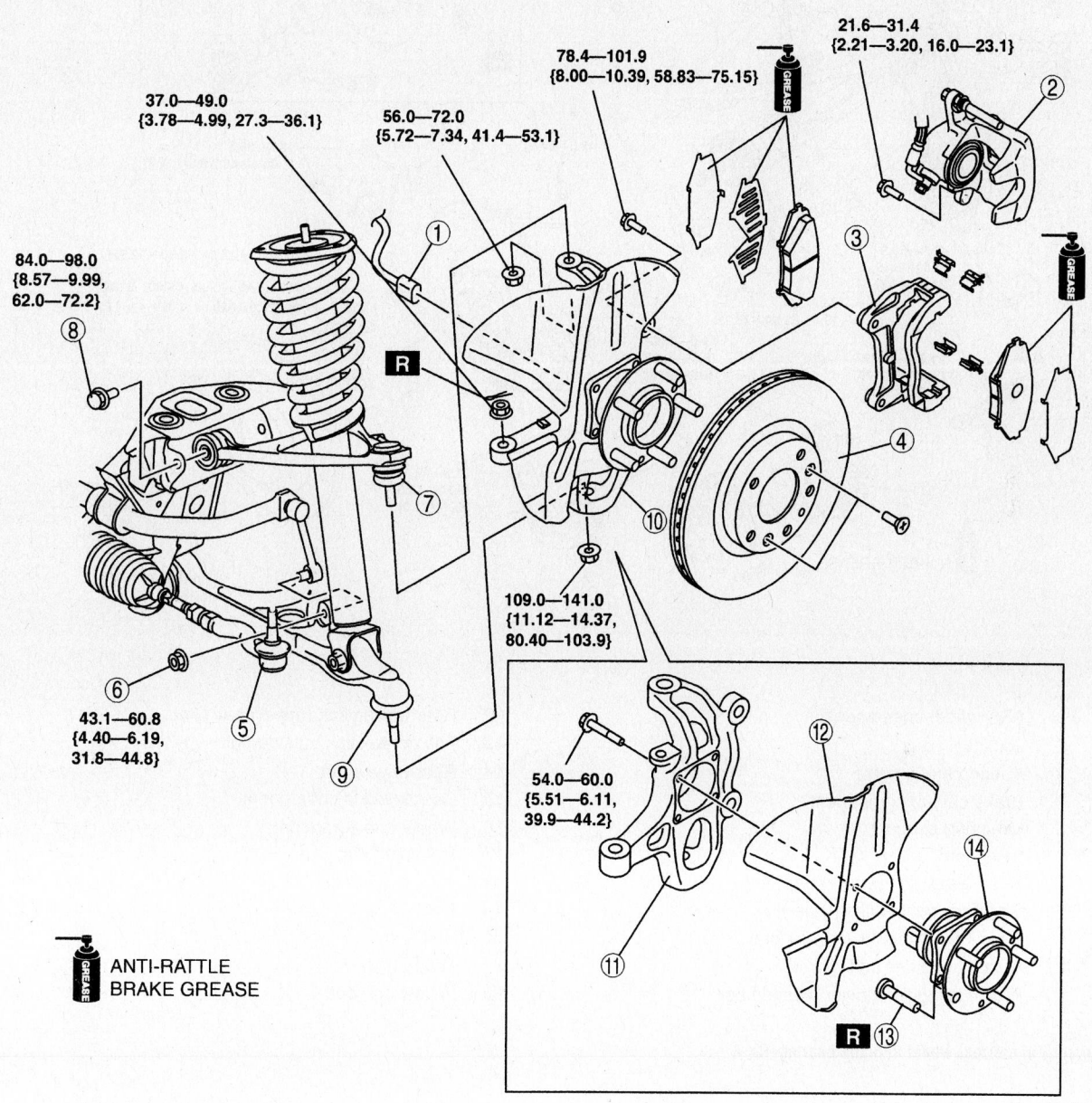

37.0—49.0
{3.78—4.99, 27.3—36.1}

78.4—101.9
{8.00—10.39, 58.83—75.15}

21.6—31.4
{2.21—3.20, 16.0—23.1}

56.0—72.0
{5.72—7.34, 41.4—53.1}

84.0—98.0
{8.57—9.99,
62.0—72.2}

43.1—60.8
{4.40—6.19,
31.8—44.8}

109.0—141.0
{11.12—14.37,
80.40—103.9}

54.0—60.0
{5.51—6.11,
39.9—44.2}

ANTI-RATTLE
BRAKE GREASE

N·m {kgf·m, ft·lbf}

1	ABS wheel-speed sensor connector	5	Tie-rod end
2	Brake caliper component	6	Stabilizer control link (lower)
3	Mounting support	7	Front upper arm ball joint
4	Disc plate	8	Front upper arm bolt

67162-MRX8-G39

Exploded view of the front wheel bearing and knuckle assembly–RX-8

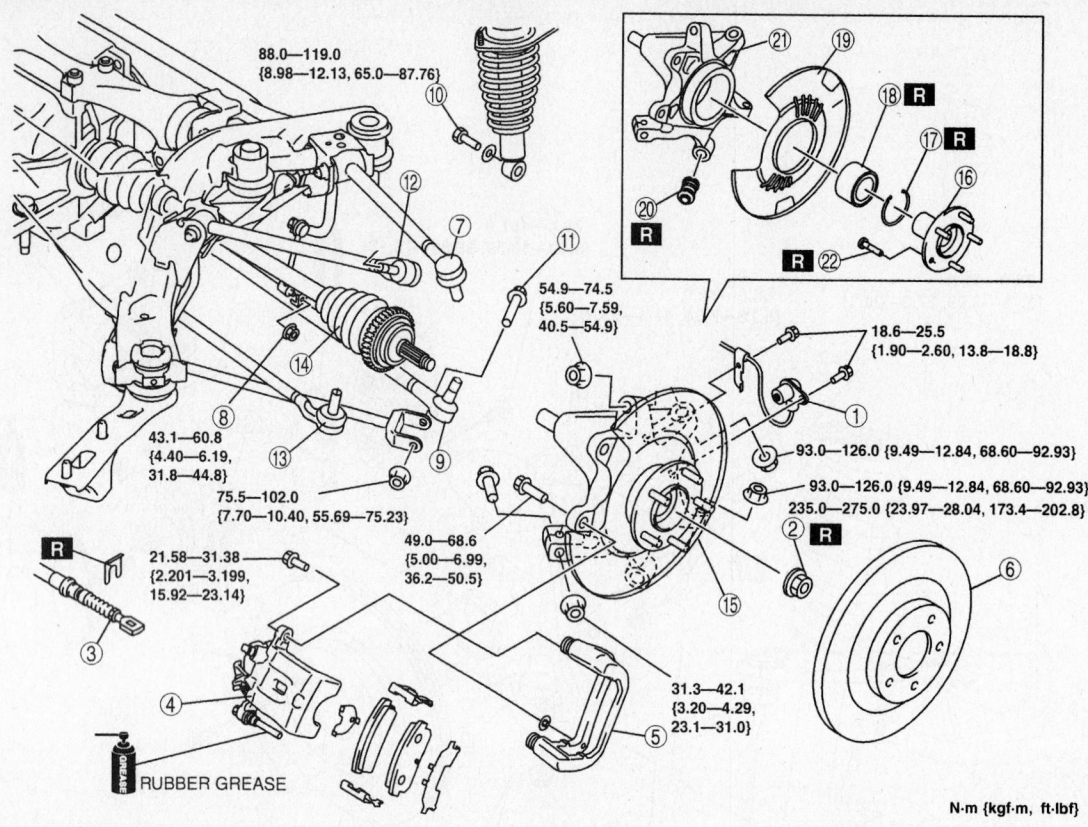

88.0—119.0
{8.98-12.13, 65.0—87.76}

54.9—74.5
{5.60-7.59,
40.5-54.9}

18.6—25.5
{1.90—2.60, 13.8—18.8}

43.1—60.8
{4.40-6.19,
31.8-44.8}

75.5—102.0
{7.70-10.40, 55.69-75.23}

93.0—126.0 {9.49—12.84, 68.60—92.93}

93.0—126.0 {9.49—12.84, 68.60—92.93}
235.0—275.0 {23.97—28.04, 173.4—202.8}

21.58—31.38
{2.201-3.199,
15.92-23.14}

49.0—68.6
{5.00-6.99,
36.2-50.5}

31.3—42.1
{3.20-4.29,
23.1-31.0}

RUBBER GREASE

N·m {kgf·m, ft·lbf}

1	ABS wheel-speed sensor	12	Rear trailing link (upper) ball joint
2	Locknut	13	Toe control link outside bolt
3	Parking brake cable	14	Rear drive shaft
4	Brake caliper component	15	Rear knuckle component
5	Mounting support	16	Wheel hub component
6	Disc plate	17	Retaining ring
7	Rear lateral link (upper) ball joint	18	Wheel bearing
8	Stabilizer control link (lower)	19	Dust cover
9	Rear lateral link (lower) ball joint	20	Bushing
10	Shock absorber bolt (lower)	21	Rear knuckle
11	Rear trailing link (lower) outside bolt	22	Wheel hub bolt

67162-MRX8-G40

Exploded view of the rear wheel hub and bearing–RX-8

10. Press the bearing inner race into the wheel hub.

11. Press the wheel hub into the knuckle.

12. Install or connect the following:
- Wheel hub/knuckle assembly. Tighten the bolts to 23–31 ft. lbs. (31–42 Nm).

- Axle shaft
- Toe control link outside bolt
- Rear trailing link upper ball joint
- Rear trailing link lower outside bolt and tighten to 56–75 ft. lbs. (76–102 Nm).
- Lower shock absorber bolt and tighten to 65–88 ft. lbs. (86–119 Nm).

- Rear lateral link lower ball joint
- Stabilizer bar lower link and tighten to 32–45 ft. lbs. (43–61 Nm).
- Rear lateral link upper ball joint
- Rear caliper and rotor assembly
- Parking brake cable
- Axle nut and tighten to 174–203 ft. lbs. (235–275 Nm).

BRAKES

Brake Caliper

REMOVAL & INSTALLATION

Front

1. Before servicing the vehicle, refer to the precautions in the beginning of this section.

2. Remove or disconnect the following:
- Wheels
- Flexible brake hose from the caliper
- Caliper bolt
- Caliper

To install:

3. Install or connect the following:
- Caliper on the brake disc
- Caliper mounting bolts and tighten the bolts 16–23 ft. lbs. (22–31 Nm)
- Brake hose to the caliper and tighten the hose nut to 16–21 ft. lbs. (22–29 Nm)

4. Bleed the brake system.
5. Install the wheels.

Rear

1. Before servicing the vehicle, refer to the precautions in the beginning of this section.

2. Remove or disconnect the following:
- Wheels
- Flexible brake line from the caliper assembly

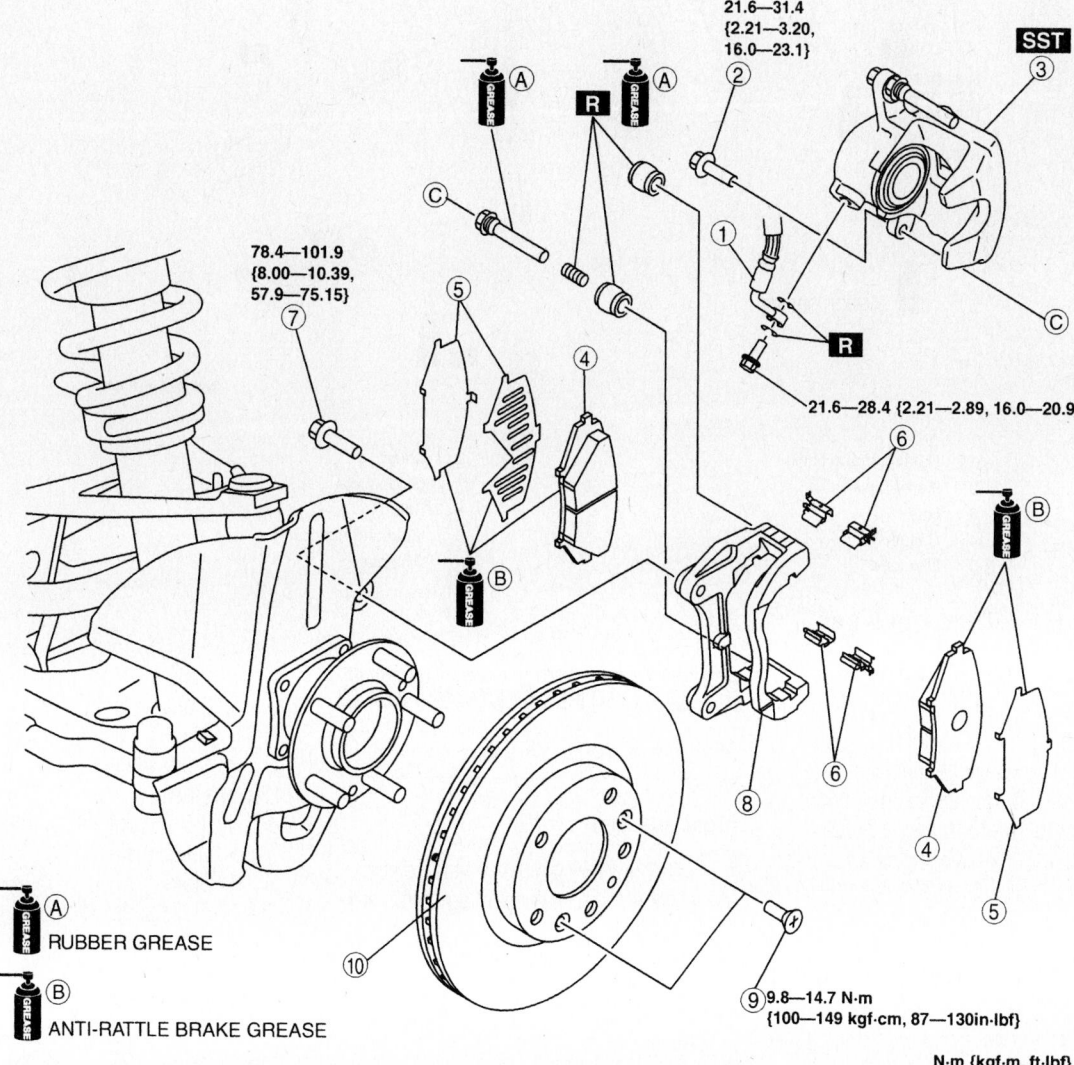

21.6—31.4
{2.21—3.20,
16.0—23.1}

78.4—101.9
{8.00—10.39,
57.9—75.15}

21.6—28.4 {2.21—2.89, 16.0—20.9}

9.8—14.7 N·m
{100—149 kgf·cm, 87—130in·lbf}

N·m {kgf·m, ft·lbf}

(A) RUBBER GREASE

(B) ANTI-RATTLE BRAKE GREASE

1	Brake hose	6	Guide plate
2	Bolt	7	Bolt
3	Caliper	8	Mounting support
4	Disc pad	9	Screw
5	Shim	10	Disc plate

Exploded view of the front brake caliper–RX-8

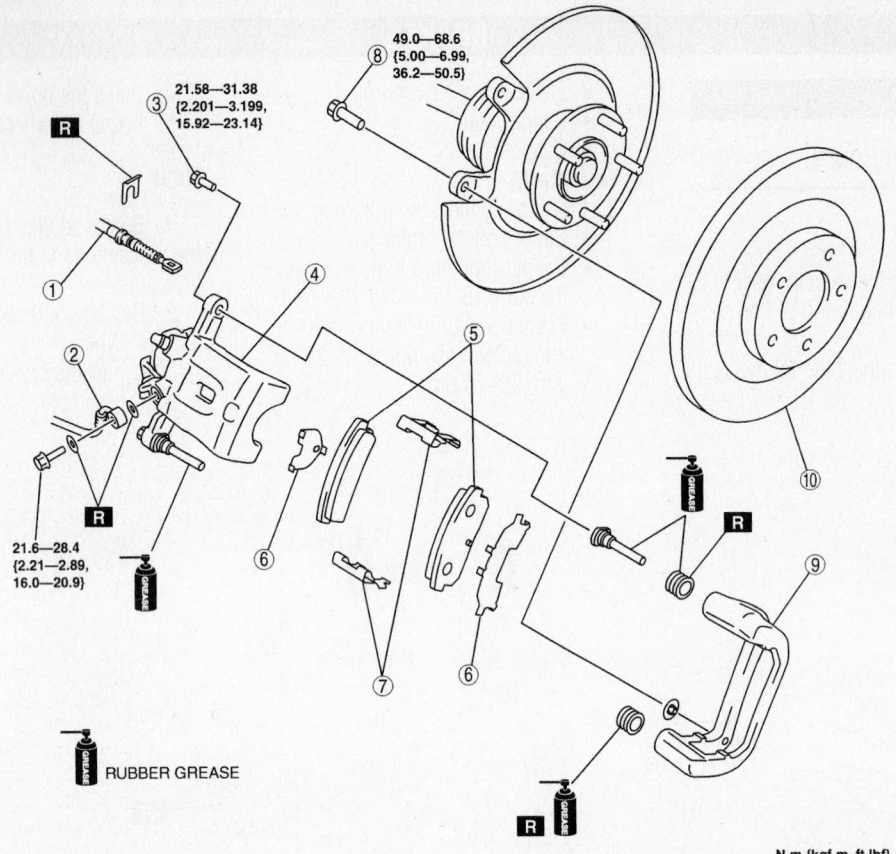

21.58—31.38
{2.201—3.199,
15.92—23.14}

49.0—68.6
{5.00—6.99,
36.2—50.5}

21.6—28.4
{2.21—2.89,
16.0—20.9}

RUBBER GREASE

N·m {kgf·m, ft·lbf}

1	Parking brake cable	6	Shim
2	Brake hose	7	Guide plate
3	Bolt	8	Bolt
4	Caliper	9	Mounting support
5	Disc pad	10	Disc plate

67162-MRX8-G42

Exploded view of rear brake system–RX-8

- Caliper mounting bolts
- Caliper

To install:

3. Install or connect the following:
- Caliper. Torque the caliper mount bolts to 16–23 ft. lbs. (22–32 Nm).
- Brake hose. Torque the line bolt to 16–22 ft. lbs. (22–30 Nm).
- Parking brake cable

4. Install the wheels and bleed the brake system.

Disc Brake Pads

REMOVAL & INSTALLATION

Front and Rear

1. Before servicing the vehicle, refer to the precautions in the beginning of this section.

2. Remove or disconnect the following:
- Wheels

- Caliper
- Brake pads
- Shim
- Guide plate

To install:

3. Install or connect the following:
- Guide plate
- Shim
- Brake pads
- Caliper and tighten the bolt to 16–23 ft. lbs. (22–32 Nm).

4. Install the wheels.

SPECIFICATIONS AND MAINTENANCE CHARTS

ENGINE AND VEHICLE IDENTIFICATION

Engine							Model Year	
Code ①	Liters	Cu. In.	Cyl.	Fuel Sys.	Engine Type	Eng. Mfg.	Code ②	Year
B	2.0	121	4	SFI	DOHC	Ford	1	2001
Z	2.3	137	4	SFI	DOHC	Ford	2	2002
1	3.0	182	6	SFI	DOHC	Ford	3	2003
							4	2004
							5	2005

SFI: Multi-port Fuel Injection

DOHC: Double Overhead Camshafts

① 8th digit of VIN

② 10th digit of VIN

67162-TRIB-C01

GENERAL ENGINE SPECIFICATIONS

Year	Model	Engine Displacement Liters	Engine VIN	Net Horsepower @ rpm	Net Torque @ rpm (ft. lbs.)	Bore x Stroke (in.)	Com-pression Ratio	Oil Pressure @ rpm
2001	Tribute	2.0	B	127@5500	135@4500	3.34x3.46	9.6:1	54-80 ①
	Tribute	3.0	1	200@5500	200@4500	3.50x3.13	10.0:1	45 ①
2002	Tribute	2.0	B	127@5500	135@4500	3.34x3.46	9.6:1	54-80 ①
	Tribute	3.0	1	200@5500	200@4500	3.50x3.13	10.0:1	45 ①
2003	Tribute	2.0	B	127@5500	135@4500	3.34x3.46	9.6:1	54-80 ①
	Tribute	3.0	1	200@5500	200@4500	3.50x3.13	10.0:1	45 ①
2004	Tribute	2.0	B	127@5500	135@4500	3.34x3.46	9.6:1	54-80 ①
	Tribute	3.0	1	200@5500	200@4500	3.50x3.13	10.0:1	45 ①
2005	Tribute	2.3	Z	153@5800	152@4250	3.44x3.70	NA	29-30@2000
	Tribute	3.0	1	200@5500	200@4500	3.50x3.13	10.0:1	11@1500②

SFI: Multi-port Fuel Injection

① The manufacturer does not provide an engine speed specification for oil pump pressure.

② Minimum hot

67162-TRIB-C02

ENGINE TUNE-UP SPECIFICATIONS

Year	Engine Displacement Liters	Engine VIN	Spark Plug Gap (in.)	Ignition Timing (deg.)		Fuel Pump (psi)	Idle Speed (rpm)		Valve Clearance	
				MT	AT		MT	AT	Intake	Exhaust
2001	2.0	B	0.039-0.043	10 BTDC	—	65②	①	—	HYD.	HYD.
	3.0	1	0.052-0.056	10 BTDC	10 BTDC	65②	①	①	HYD.	HYD.
2002	2.0	B	0.051	10 BTDC	—	65②	①	—	HYD.	HYD.
	3.0	1	0.052-0.056	10 BTDC	10 BTDC	65②	①	①	HYD.	HYD.
2003	2.0	B	0.051	10 BTDC	—	65②	①	—	HYD.	HYD.
	3.0	1	0.052-0.056	10 BTDC	10 BTDC	65②	①	①	HYD.	HYD.
2004	2.0	B	0.051	10 BTDC	—	65②	①	—	HYD.	HYD.
	3.0	1	0.052-0.056	10 BTDC	10 BTDC	65②	①	①	HYD.	HYD.
2005	2.3	Z	0.041-0.045	10 BTDC	10 BTDC	39②	①	①	HYD.	HYD.
	3.0	1	0.052-0.056	10 BTDC	10 BTDC	39②	①	①	HYD.	HYD.

BTDC: Before Top Dead Center

HYD: Hydraulic lash adjusters

NA: Information not available

① Refer to Vehicle Emission Control Information Label

② Key on; engine off

67162-TRIB-C03

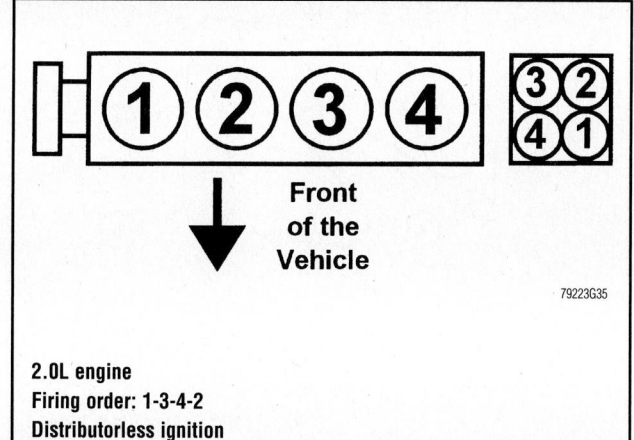

2.0L engine
Firing order: 1-3-4-2
Distributorless ignition

79223G35

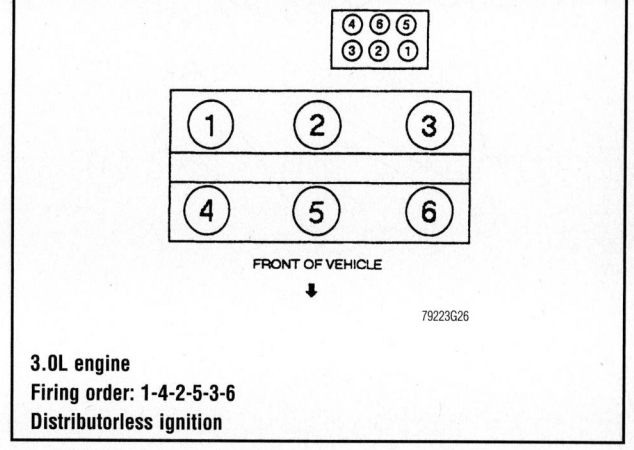

3.0L engine
Firing order: 1-4-2-5-3-6
Distributorless ignition

79223G26

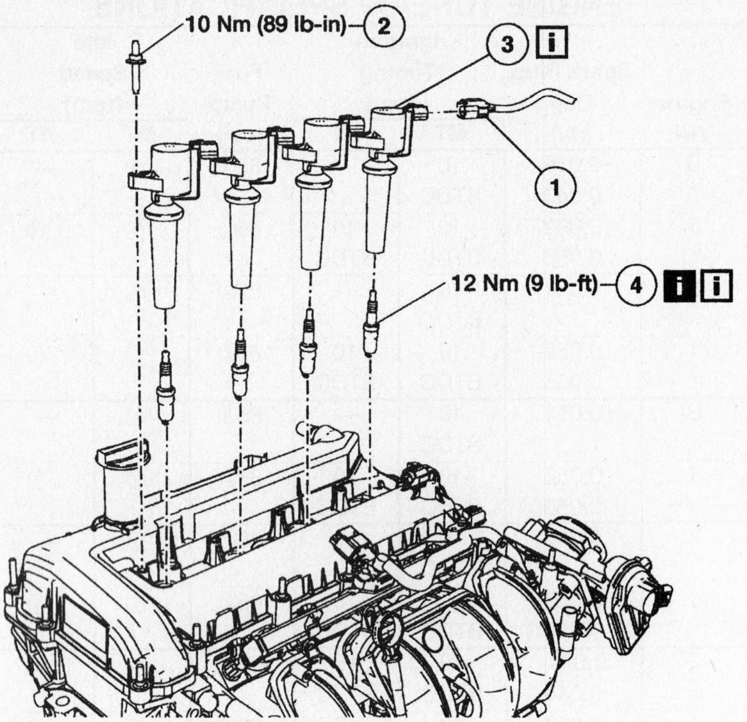

10 Nm (89 lb-in)—(2)

(3) **i**

(1)

12 Nm (9 lb-ft)—(4) **i** **i**

1 Ignition coil-on-plug electrical connectors

2 Ignition coil-to-valve cover bolts

3 Ignition coils

4 Spark plugs

67197-ESCA-G61

Coil and spark plug arrangement—2.3L engine

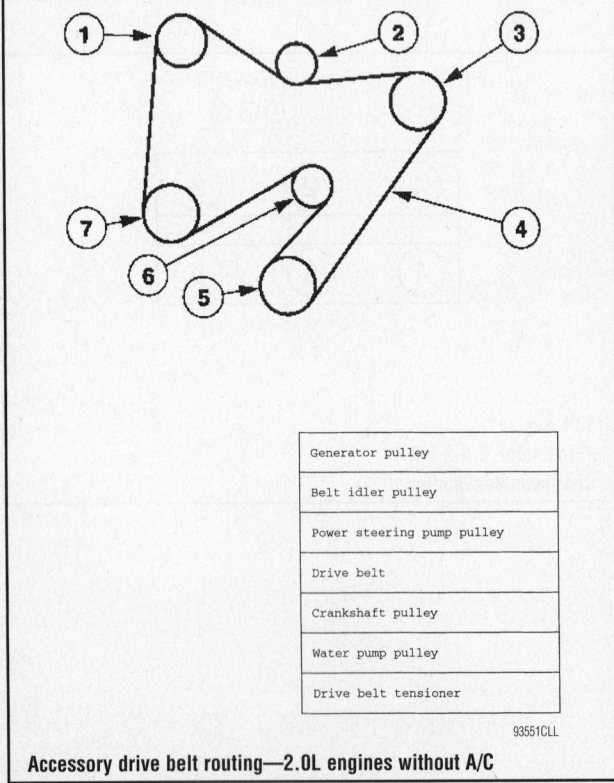

| Generator pulley |
| Belt idler pulley |
| Power steering pump pulley |
| Drive belt |
| Crankshaft pulley |
| Water pump pulley |
| Drive belt tensioner |

93551CLL

Accessory drive belt routing—2.0L engines without A/C

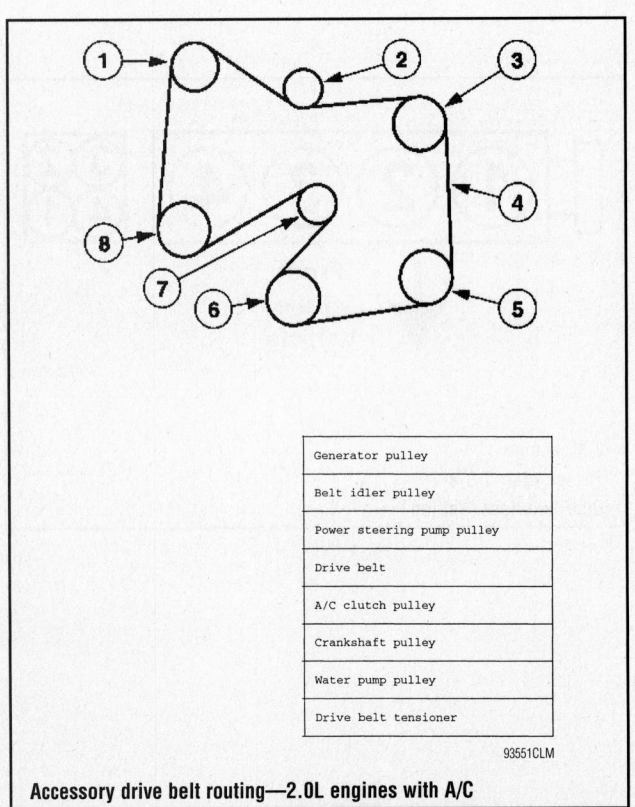

| Generator pulley |
| Belt idler pulley |
| Power steering pump pulley |
| Drive belt |
| A/C clutch pulley |
| Crankshaft pulley |
| Water pump pulley |
| Drive belt tensioner |

93551CLM

Accessory drive belt routing—2.0L engines with A/C

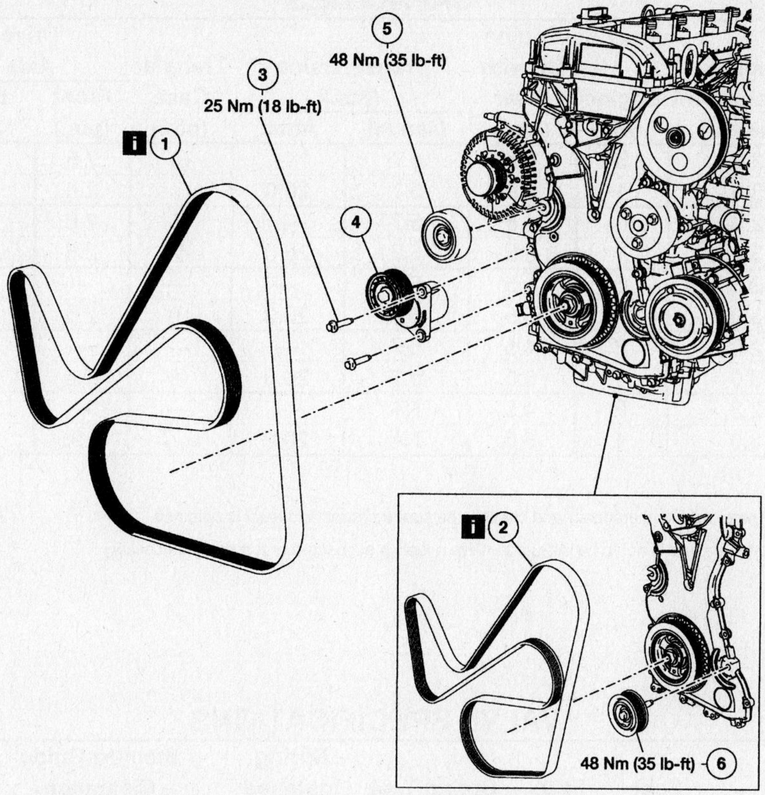

5
48 Nm (35 lb-ft)

3
25 Nm (18 lb-ft)

i 1

4

i 2

48 Nm (35 lb-ft) 6

1 Accessory drive belt (with A/C)   4 Accessory drive belt tensioner
2 Accessory drive belt (without A/C)   5 Accessory drive belt idler pulley
3 Accessory drive belt tensioner bolts   6 Accessory drive belt idler pulley (without A/C only)

67197-ESCA-G62

Accessory drive belt routings—2.3L engine

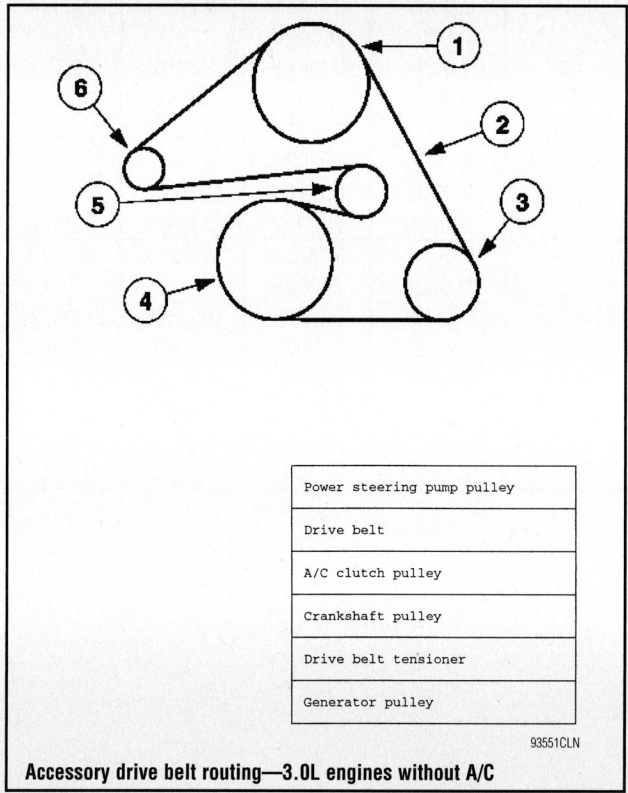

| Power steering pump pulley |
| A/C clutch pulley |
| Drive belt |
| Crankshaft pulley |
| Drive belt tensioner |
| Generator pulley |

93551CLN

Accessory drive belt routing—3.0L engines without A/C

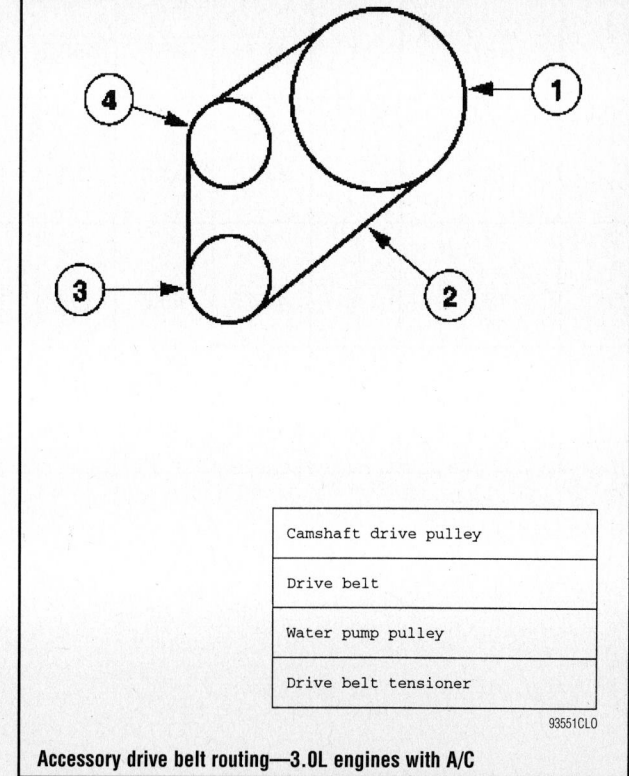

| Camshaft drive pulley |
| Drive belt |
| Water pump pulley |
| Drive belt tensioner |

93551CLO

Accessory drive belt routing—3.0L engines with A/C

CAPACITIES

Year	Model	Engine Displacement Liters	Engine VIN	Engine Oil with Filter (qts.)	Transmission (pts.) Manual	Transmission (pts.) Auto.	Transfer Case (pts.)	Drive Axle Front (pts.)	Drive Axle Rear (pts.)	Fuel Tank (gal.)	Cooling System (qts.)
2001	Tribute	2.0	B	4.5	5.7	—	①	2.6	3.0	15.0	7.0
	Tribute	3.0	1	5.8	5.7	20.0	①	2.6	3.0	16.0	10.5
2002	Tribute	2.0	B	4.5	5.7	—	①	2.6	3.0	15.0	7.0
	Tribute	3.0	1	5.5	5.7	20.0	①	2.6	3.0	16.0	10.5
2003	Tribute	2.0	B	4.5	5.7	—	—	—	3.0	15.0	7.0
	Tribute	3.0	1	5.5	5.7	20.0	①	2.6	3.0	16.0	10.5
2004	Tribute	2.0	B	4.5	5.7	—	—	—	3.0	15.0	7.0
	Tribute	3.0	1	5.5	5.7	20.0	①	2.6	3.0	16.0	10.5
2005	Tribute	2.3	B	4.5	NA	—	—	—	3.0	16.5	7.6
	Tribute	3.0	1	5.5	NA	20.0	0.75	2.95	3.0	16.5	10.5

NA: Information not available

NOTE: All capacities are approximate. Add fluid gradually and check to be sure a proper fluid level is obtained.

① The transfer case is lubricated for life and is not to be checked unless a leak is suspected or a repair is necessary.

67162-TRIB-C04

VALVE SPECIFICATIONS

Year	Engine Displacement Liters	Engine VIN	Seat Angle (deg.)	Face Angle (deg.)	Spring Test Pressure (lbs. @ in.)	Spring Installed Height (in.)	Stem-to-Guide Clearance (in.) Intake	Stem-to-Guide Clearance (in.) Exhaust	Stem Diameter (in.) Intake	Stem Diameter (in.) Exhaust
2001	2.0	B	45	45	①	1.346	0.0007-0.0025	0.0007-0.0025	0.2374	0.2374
	3.0	1	44.75	45.5	153@ 1.18	1.57	0.0008-0.0027	0.0018-0.0037	0.2352-0.2360	0.2343-0.2350
2002	2.0	B	45	45	①	1.346	0.0007-0.0025	0.0007-0.0025	0.2374	0.2374
	3.0	1	44.75	45.5	153@ 1.18	1.57	0.0008-0.0027	0.0018-0.0037	0.2352-0.2360	0.2343-0.2350
2003	2.0	B	45	45	①	1.346	0.0007-0.0025	0.0007-0.0025	0.2374	0.2374
	3.0	1	44.75	45.5	153@ 1.18	1.57	0.0008-0.0027	0.0018-0.0037	0.2352-0.2360	0.2343-0.2350
2004	2.0	B	45	45	①	1.346	0.0007-0.0025	0.0007-0.0025	0.2374	0.2374
	3.0	1	44.75	45.5	153@ 1.18	1.57	0.0008-0.0027	0.0018-0.0037	0.2352-0.2360	0.2343-0.2350
2005	2.3	Z	45	45	38.6@1.49	1.496	0.0009	0.0011	0.2153-0.2159	0.2151-0.2157
	3.0	1	44.75	45.5	153@ 1.18	1.57	0.0008-0.0027	0.0018-0.0037	0.2352-0.2360	0.2343-0.2350

① Intake: 82.1@ 0.988
  Exhaust: 95@ 1.0275

67162-TRIB-C05

CRANKSHAFT AND CONNECTING ROD SPECIFICATIONS

All measurements are given in inches.

Year	Engine Displacement Liters)	Engine VIN	Crankshaft				Connecting Rod		
			Main Brg. Journal Dia.	Main Brg. Oil Clearance	Shaft End-play	Thrust on No.	Journal Diameter	Oil Clearance	Side Clearance
2001	2.0	B	2.2827-2.2835	①	0.0035-0.0102	3	1.8460-1.8468	0.0006-0.0028	0.0040-0.0110
	3.0	1	2.4791-2.4800	0.0010-0.0018	0.0043-0.0091	3	1.9673-1.9681	0.0010-0.0025	0.0039-0.0118
2002	2.0	B	2.2827-2.2835	①	0.0035-0.0102	3	1.8460-1.8468	0.0006-0.0028	0.0040-0.0110
	3.0	1	2.4791-2.4800	0.0010-0.0018	0.0043-0.0091	3	1.9673-1.9681	0.0011-0.0026	0.0039-0.0118
2003	2.0	B	2.2827-2.2835	①	0.0035-0.0102	3	1.8460-1.8468	0.0006-0.0028	0.0040-0.0110
	3.0	1	2.4791-2.4800	0.0010-0.0018	0.0043-0.0091	3	1.9673-1.9681	0.0011-0.0026	0.0039-0.0118
2004	2.0	B	2.2827-2.2835	①	0.0035-0.0102	3	1.8460-1.8468	0.0006-0.0028	0.0040-0.0110
	3.0	1	2.4791-2.4800	0.0010-0.0018	0.0043-0.0091	3	1.9673-1.9681	0.0011-0.0026	0.0039-0.0118
2005	2.3	Z	2.0460-2.0470	0.0007-0.0013	0.0080-0.0160	NA	1.9673-1.9681	0.0011-0.0026	0.0760-0.1200
	3.0	1	2.4791-2.4800	0.0010-0.0018	0.0043-0.0091	3	1.9673-1.9681	0.0011-0.0026	0.0039-0.0118

NA: Information not available

① Journals 1, 2 and 4: 0.0010 - 0.0017 in.
　Journal 3: 0.0012 - 0.0019 in.

67162-TRIB-C06

PISTON AND RING SPECIFICATIONS

All measurements are given in inches.

Year	Engine Displacemen Liters	Engine VIN	Piston Clearance	Ring Gap			Ring Side Clearance		
				Top Compression	Bottom Compression	Oil Control	Top Compression	Bottom Compression	Oil Control
2001	2.0	B	0.0004-0.0012	0.0100-0.0300	0.0100-0.0300	0.0160-0.0660	0.0015-0.0032	0.0015-0.0035	snug
	3.0	1	0.0005-0.0009	0.0039-0.0098	0.0106-0.0165	0.0059-0.0256	0.0016-0.0030	0.0016-0.0033	snug
2002	2.0	B	0.0004-0.0012	0.0100-0.0300	0.0100-0.0300	0.0160-0.0660	0.0015-0.0032	0.0015-0.0035	snug
	3.0	1	0.0005-0.0009	0.0039-0.0098	0.0106-0.0165	0.0059-0.0256	0.0016-0.0030	0.0016-0.0033	snug
2003	2.0	B	0.0004-0.0012	0.0100-0.0300	0.0100-0.0300	0.0160-0.0660	0.0015-0.0032	0.0015-0.0035	snug
	3.0	1	0.0005-0.0009	0.0039-0.0098	0.0106-0.0165	0.0059-0.0256	0.0016-0.0030	0.0016-0.0033	snug
2004	2.0	B	0.0004-0.0012	0.0100-0.0300	0.0100-0.0300	0.0160-0.0660	0.0015-0.0032	0.0015-0.0035	snug
	3.0	1	0.0005-0.0009	0.0039-0.0098	0.0106-0.0165	0.0059-0.0256	0.0016-0.0030	0.0016-0.0033	snug
2005	2.3	Z	0.0009-0.0017	0.0060-0.0120	0.0120-0.0180	0.0070-0.0270	NA	NA	NA
	3.0	1	0.0005-0.0009	0.0039-0.0098	0.0106-0.0165	0.0059-0.0256	0.0016-0.0030	0.0016-0.0033	snug

NA: Information not available

67162-TRIB-C07

TORQUE SPECIFICATIONS
All readings in ft. lbs.

Year	Engine Displacement Liters	Engine VIN	Cylinder Head Bolts	Main Bearing Bolts	Rod Bearing Bolts	Crankshaft Damper Bolts	Flywheel Bolts	Manifold Intake	Manifold Exhaust	Spark Plugs	Oil Pan Drain Plug
2001	2.0	B	①	②	③	80-87	83	13	12	11	18
	3.0	1	④	⑤	⑥	⑦	59	⑧	15	11	NA
2002	2.0	B	①	②	③	80-87	83	13	12	11	18
	3.0	1	④	⑤	⑥	⑦	59	⑧	15	11	NA
2003	2.0	B	①	②	③	80-87	83	13	12	11	18
	3.0	1	④	⑤	⑥	⑦	59	⑧	15	11	NA
2004	2.0	B	①	②	③	85	83	13	12	11	18
	3.0	1	④	⑤	⑥	⑦	59	⑧	15	11	NA
2005	2.3	Z	⑨	NA	NA	⑩	⑪	13	NA	11	21
	3.0	1	④	⑤	⑥	⑦	59	⑧	15	11	NA

NA: Information not available

① Step 1: 15 ft. lbs. (20 Nm).
Step 2: 30 ft. lbs. (40 Nm).
Step 3: Plus an additional 90 degrees.

② Step 1: 18 ft. lbs.
Step 2: +60 degrees

③ Step 1: 26 ft. lbs.
Step 2: +90 degrees

④ Step 1: 30 ft. lbs. (40 Nm).
Step 2: Tighten the bolts 90 degrees.
Step 3: Loosen the bolts one full turn.
Step 4: 30 ft. lbs. (40 Nm).
Step 5: Tighten the bolts 90 degrees.
Step 6: Tighten the bolts 90 degrees.

⑤ Step 1: Fasteners 1-8: 18 ft. lbs.
Step 2: Fasteners 9-19: 30 ft. lbs.
Step 3: Fasteners 1-16: +90 degrees
Step 4: fasteners 17-22: 18 ft. lbs.

⑥ Step 1: 17 ft. lbs.
Step 2: 32 ft. lbs.

⑦ Step 1: 89 ft. lbs.
Step 2: Loosen 1 full turn
Step 3: 37 ft. lbs.
Step 4: 66 ft. lbs.

⑧ 89 inch lbs.

⑨ Step 1: 44 inch lbs.
Step 2: 11 ft. lbs.
Step 3: 33 ft. lbs.
Step 4: +90 degrees
Step 5: Plus 90 degrees

⑩ Step2: 74 ft. lbs.
Step 2: plus 90 degrees

⑪ Step 1: 37 ft. lbs.
Step 2: 50 ft. lbs
Step 3: 83 ft. lbs.

67162-TRIB-C08

WHEEL ALIGNMENT

Year	Model		Caster Range (+/-Deg.)	Caster Preferred Setting (Deg.)	Camber Range (+/-Deg.)	Camber Preferred Setting (Deg.)	Toe-in (in.)
2001	ALL	F	NA	+1.93	NA	-0.84	0.12+/-0.12
		R	NA	NA	NA	-0.04	0.09+/-0.11
2002	ALL	F	NA	+1.93	NA	-0.84	0.12+/-0.12
		R	NA	NA	NA	-0.04	0.09+/-0.11
2003	ALL	F	NA	+1.93	NA	-0.84	0.12+/-0.12
		R	NA	NA	NA	-0.04	0.09+/-0.11
2004①	4-cyl.	F	1.00	+1.72	1.00	-0.48	-0.08+/-0.32
		R	NA	NA	1.00	+0.13	0.04+/-0.17
	6-cyl.	F	1.00	+1.72	1.00	-0.84	0.23+/-0.32
		R	NA	NA	1.00	+0.13	0.10+/-0.17
2005	4-cyl.	F	1.00	+1.70	1.00	-1.00	-0.23+/-0.23
		R	NA	NA	0.75	+0.10	0.10+/-0.20
	6-cyl.	F	1.00	+1.70	1.00	-1.00	0.23+/-0.32
		R	NA	NA	0.75	+0.10	0.10+/-0.20

NA: Information not available

① Assumes 8 gallons of gas

67162-TRIB-C09

TIRE, WHEEL AND BALL JOINT SPECIFICATIONS

| Year | Model | OEM Tires | | Tire Pressures (psi) | | Wheel Size | Ball Joint Inspection | Lug Nuts (ft. lbs.) |
		Standard	Optional	Front	Rear			
2001	Tribute	P225/70SR15	P235/70R16	①	①	NA	0.030 in.	98
2002	Tribute	P215/70R16	P225/70SR15 P235/70R16	①	①	NA	0.030 in.	98
2003	Tribute	P215/70R16	P225/70SR15 P235/70R16	①	①	NA	0.030 in.	98
2004	Tribute	P225/70SR15	P235/70R16	①	①	6.5	0.030 in.	98
2005	Tribute	P225/70SR15	P235/70R16	①	①	7.5	0.030 in.	98

OEM: Original Equipment Manufacturer
PSI: Pounds Per Square Inch
STD: Standard
OPT: Optional
NA: Not Available

67162-TRIB-C10

BRAKE SPECIFICATIONS
All measurements in inches unless noted

| Year | Model | | Brake Disc | | | Brake Drum | | Minimum Lining Thickness | Brake Caliper | |
			Original Thickness	Minimum Thickness	Maximum Run-out	Original Inside Diameter	Maximum Machine Diameter		Bracket Bolts (ft. lbs.)	Mounting Bolts (ft. lbs.)
2001	Tribute		0.940	0.860	0.002	9.00	9.06	0.039	111	26
2002	Tribute		0.940	0.860	0.004	9.00	9.06	0.039	111	26
2003	Tribute		0.940	0.860	0.002	9.00	9.06	0.039	111	26
2004	Tribute		0.940	0.860	0.002	9.00	9.06	0.039	111	26
2005	Tribute	F	NA	①	0.004	NA	9.05	0.118	111	②
		R	NA	0.430	0.004	—	—	0.118	—	26

NA: Information not available
① Base brakes: 0.86 in.
  With 4-wheel discs: 0.95 in.
② With disc/drum: 26 ft. lbs.
  With 4-wheel disc: 33 ft. lbs.

67162-TRIB-C11

SCHEDULED MAINTENANCE INTERVALS
2001-03 Mazda Tribute

TO BE SERVICED	TYPE OF SERVICE	5	10	15	20	25	30	35	40	45	50	55	60	65
Air cleaner filter	R						✓						✓	
Accessory drive belt	S/I	Every 100,000 miles												
Accessory drive belt	R	At 120,000 miles if not previously done so												
Auto. Trans. Fluid level	I		✓				✓			✓			✓	
Auto. Trans. Fluid	②						✓						✓	
Auto. Trans. Fluid	③	Every 150,000 miles												
Brake system ①	S/I			✓			✓			✓			✓	
Cabin air filter	R			✓			✓			✓			✓	
Camshaft belt (2.0L)	R	Every 120,000 miles												
Cooling system hoses and clamps	S/I			✓			✓			✓			✓	
Engine coolant (green coolant)	R	Every 75,000 miles and every 30,000 miles thereafter												
Engine coolant (yellow coolant)	R	At 5 years or 100,000 miles												
Engine oil & filter	R	✓	✓	✓	✓	✓	✓	✓	✓	✓	✓	✓	✓	✓
PCV valve	R	Every 100,000 miles												
Exhaust system & heat shields	S/I						✓						✓	
Fuel filter	R						✓						✓	
Rear axle lubricant	R	Every 150,000 miles												
Rotate tires	S/I	✓	✓	✓	✓	✓	✓	✓	✓	✓	✓	✓	✓	✓
Steering linkage	S/I			✓			✓			✓			✓	
Spark plugs	R	Change at 100,000 miles												
Suspension components	S/I			✓			✓			✓			✓	
Wheels for play and noise	I			✓			✓			✓			✓	

R: Replace   S/I: Inspect and service, if necessary   L: Lubricate   A: Adjust   C: Clean

① Inspect the reservoir fluid level, rotor and or drum, brake lines, hoses, calipers and or wheel cylinders

② Change automatic transmission/transaxle fluid on all vehicles equipped with AX4S, 4F50N, 4R100, 4F27E

Special Operating Condition Requirements

When towing a trailer or using a camper or car-top carrier:

Change engine oil and install a new oil filter every 4,800 km (3,000 miles) or 3 months.

Change transfer case fluid every 96,000 km (60,000 miles).

Change manual transmission fluid as required.

Inspect and lubricate U-joints as required.

During extensive idling and/or low speed driving for long distances, as in heavy commercial use such as delivery, taxi, patrol car or livery:

Change engine oil and install a new oil filter, lube front lower control arm and steering linkage ball joints with zerk fittings (if equipped) every 4,800 km (3,000 miles) or 3 months.

Inspect brake system and check battery electrolyte level (Patrol cars) every 8,000 km (5,000 miles).

Install a new fuel filter every 24,000 km (15,000 miles).

Change automatic transmission fluid, lubricate 4x2 wheel bearings, install new grease seals and adjust bearings every 48,000 km (30,000 miles).

Install new spark plugs and change transfer case fluid every 96,000 km (60,000 miles).

Install a new cabin air filter as required.

When operating in dusty conditions such as unpaved or dusty roads:

Change engine oil and install a new oil filter every 4,800 km (3,000 miles) or 3 months.

Install a new fuel filter every 24,000 km (15,000 miles).

Change automatic transmission fluid every 48,000 km (30,000 miles).

Change transfer case fluid every 96,000 km (60,000 miles).

Install a new engine air filter as required.

Install a new cabin air filter as required.

SCHEDULED MAINTENANCE INTERVALS
2001-03 Mazda Tribute
Footnotes Continued

When operating in off-road conditions:

Change automatic transmission fluid every 48,000 km (30,000 miles).

Change transfer case fluid every 96,000 km (60,000 miles).

Install a new cabin air filter as required.

Inspect and lubricate U-joints.

Inspect and lubricate steering linkage ball joints with zerk fittings.

67162-TRIB-C13

SCHEDULED MAINTENANCE INTERVALS
2004 Mazda Tribute

TO BE SERVICED	TYPE OF SERVICE	VEHICLE MILEAGE INTERVAL (x1000)												
		5	10	15	20	25	30	35	40	45	50	55	60	65
Air cleaner filter	R						✓						✓	
Accessory drive belt	S/I	Every 100,000 miles												
Accessory drive belt	R	At 150,000 miles if not previously done so												
Auto. Trans. Fluid level	I			✓			✓			✓			✓	
Auto. Trans. Fluid	②						✓						✓	
Auto. Trans. Fluid	③	Every 150,000 miles												
Ball joints (2wd)	L			✓			✓			✓			✓	
Brake system ①	S/I			✓			✓			✓			✓	
Cabin air filter	R			✓			✓			✓			✓	
Camshaft belt (2.0L)	R	Every 120,000 miles												
Cooling system hoses and clamps	S/I			✓			✓			✓			✓	
Engine coolant (exc. Premium gold)	R	Every 105,000 miles, then every 3 years or 50,000 miles												
Engine coolant (Premium gold)	R	At 5 years or 100,000 miles												
Engine oil & filter	R	✓	✓	✓	✓	✓	✓	✓	✓	✓	✓	✓	✓	✓
PCV valve (external)	R	Every 100,000 miles												
Exhaust system & heat shields	S/I						✓						✓	
Front wheel bearings and seals	R	Every 150,000 miles, if not previously done so												
Fuel filter	R						✓						✓	
Rear axle lubricant	R	Every 150,000 miles												
Rotate tires	S/I	✓	✓	✓	✓	✓	✓	✓	✓	✓	✓	✓	✓	✓
Steering linkage	S/I			✓			✓			✓			✓	
Spark plugs	R	Change at 100,000 miles												
Suspension components	S/I			✓			✓			✓			✓	
Wheels for play and noise	I			✓			✓			✓			✓	

R: Replace   S/I: Inspect and service, if necessary   L: Lubricate   A: Adjust   C: Clean

① Inspect the reservoir fluid level, rotor and or drum, brake lines, hoses, calipers and or wheel cylinders

② Change automatic transmission/transaxle fluid and filter on all vehicles equipped with AX4S, 4F50N, 4R100, 4F27E.

③ All transaxles

Special Operating Condition Requirements

When towing a trailer or using a camper or car-top carrier:

Change engine oil and install a new oil filter every 4,800 km (3,000 miles) or 3 months.

Change transfer case fluid every 96,000 km (60,000 miles).

Change manual transmission fluid as required.

Inspect and lubricate U-joints as required.

During extensive idling and/or low speed driving for long distances, as in heavy commercial use such as delivery, taxi, patrol car or livery:

Change engine oil and install a new oil filter, lube front lower control arm and steering linkage ball joints with
zerk fittings (if equipped) every 4,800 km (3,000 miles) or 3 months.

Inspect brake system and check battery electrolyte level (Patrol cars) every 8,000 km (5,000 miles).

Install a new fuel filter every 24,000 km (15,000 miles).

Change automatic transmission fluid, lubricate 4x2 wheel bearings,
install new grease seals and adjust bearings every 48,000 km (30,000 miles).

Install new spark plugs and change transfer case fluid every 96,000 km (60,000 miles).

Install a new cabin air filter as required.

When operating in dusty conditions such as unpaved or dusty roads:

Change engine oil and install a new oil filter every 4,800 km (3,000 miles) or 3 months.

Install a new fuel filter every 24,000 km (15,000 miles).

Change automatic transmission fluid every 48,000 km (30,000 miles).

Change transfer case fluid every 96,000 km (60,000 miles).

SCHEDULED MAINTENANCE INTERVALS
2004 Mazda Tribute
Footnotes Continued

Install a new engine air filter as required.

Install a new cabin air filter as required.

When operating in off-road conditions:

Change automatic transmission fluid every 48,000 km (30,000 miles).

Change transfer case fluid every 96,000 km (60,000 miles).

Install a new cabin air filter as required.

Inspect and lubricate U-joints.

Inspect and lubricate steering linkage ball joints with zerk fittings.

67162-TRIB-C15

SCHEDULED MAINTENANCE INTERVALS
2005 Mazda Tribute

TO BE SERVICED	TYPE OF SERVICE	VEHICLE MILEAGE INTERVAL (x1000)												
		5	10	15	20	25	30	35	40	45	50	55	60	65
Air cleaner filter	R						✓						✓	
Accessory drive belt	I ⑤	Every 100,000 miles												
Auto. Trans. fluid level	I			✓			✓			✓			✓	
Auto. Trans. Fluid	③ ④						✓						✓	
Ball joints (2wd)	L			✓			✓			✓			✓	
Brake system ①	S/I			✓			✓			✓			✓	
Cabin air filter	R			✓			✓			✓			✓	
Cooling system hoses and clamps	S/I			✓			✓			✓			✓	
Driveshafts & halfshafts	S/I			✓			✓			✓			✓	
Engine coolant (Premium Gold)	R	Five years or 100,000 miles, then every 3 years or 50,000 miles												
Engine coolant (exc. Premium Gold)	R	Every 105,000 miles												
Engine oil & filter	R	✓	✓	✓	✓	✓	✓	✓	✓	✓	✓	✓	✓	✓
Front wheel bearings and seals (2wd)	R	Every 150,000 miles, if not previously done												
Fuel filter	R						✓						✓	
Man. Trans. Fluid	R	Every 120,000 miles												
PCV valve	S/I	Every 100,000 miles												
Exhaust system & heat shields	S/I						✓						✓	
Rear axle lubricant (4wd)	R	Every 150,000 miles												
Rotate tires	S/I	✓	✓	✓	✓	✓	✓	✓	✓	✓	✓	✓	✓	✓
Steering linkage	S/I			✓			✓			✓			✓	
Spark plugs	R	Change at 100,000 miles												
Suspension components	S/I			✓			✓			✓			✓	
Wheels ②	I			✓			✓			✓			✓	

R: Replace   S/I: Inspect and service, if necessary   L: Lubricate   A: Adjust   C: Clean

① Inspect the reservoir fluid level, rotor and or drum, brake lines, hoses, calipers and or wheel cylinders

② Inspect for end play and noise

③ Change automatic transmission/transaxle fluid and filter on all vehicles equipped with 4F50N, 4R100 and 4F27E.

④ Change every 150,000 miles for all transaxles

⑤ Replace at 150,000 miles, if not previously done

Special Operating Condition Requirements

When towing a trailer or using a camper or car-top carrier:

Change engine oil and install a new oil filter every 4,800 km (3,000 miles) or 3 months.

Change transfer case fluid every 96,000 km (60,000 miles).

Change manual transmission fluid as required.

Inspect and lubricate U-joints as required.

During extensive idling and/or low speed driving for long distances, as in heavy commercial use such as delivery, taxi, patrol car or livery:

Change engine oil and install a new oil filter, lube front lower control arm and steering linkage ball joints with

zerk fittings (if equipped) every 4,800 km (3,000 miles) or 3 months.

Inspect brake system and check battery electrolyte level (Patrol cars) every 8,000 km (5,000 miles).

Install a new fuel filter every 24,000 km (15,000 miles).

Change automatic transmission fluid, lubricate 4x2 wheel bearings,

 install new grease seals and adjust bearings every 48,000 km (30,000 miles).

Install new spark plugs and change transfer case fluid every 96,000 km (60,000 miles).

Install a new cabin air filter as required.

67162-TRIB-C16

SCHEDULED MAINTENANCE INTERVALS
2005 Mazda Tribute
Footnotes Continued

When operating in dusty conditions such as unpaved or dusty roads:

Change engine oil and install a new oil filter every 4,800 km (3,000 miles) or 3 months.

Install a new fuel filter every 24,000 km (15,000 miles).

Change automatic transmission fluid every 48,000 km (30,000 miles).

Change transfer case fluid every 96,000 km (60,000 miles).

Install a new engine air filter as required.

Install a new cabin air filter as required.

When operating in off-road conditions:

Change automatic transmission fluid every 48,000 km (30,000 miles).

Change transfer case fluid every 96,000 km (60,000 miles).

Install a new cabin air filter as required.

Inspect and lubricate U-joints.

Inspect and lubricate steering linkage ball joints with zerk fittings.

67162-TRIB-C17

PRECAUTIONS

Before servicing any vehicle, please be sure to read all of the following precautions, which deal with personal safety, prevention of component damage, and important points to take into consideration when servicing a motor vehicle:

• Never open, service or drain the radiator or cooling system when the engine is hot; serious burns can occur from the steam and hot coolant.

• Observe all applicable safety precautions when working around fuel. Whenever servicing the fuel system, always work in a well-ventilated area. Do not allow fuel spray or vapors to come in contact with a spark, open flame, or excessive heat (a hot drop light, for example). Keep a dry chemical fire extinguisher near the work area. Always keep fuel in a container specifically designed for fuel storage; also, always properly seal fuel containers to avoid the possibility of fire or explosion. Refer to the additional fuel system precautions later in this section.

• Fuel injection systems often remain pressurized, even after the engine has been turned **OFF**. The fuel system pressure must be relieved before disconnecting any fuel lines. Failure to do so may result in fire and/or personal injury.

• Brake fluid often contains polyglycol ethers and polyglycols. Avoid contact with the eyes and wash your hands thoroughly after handling brake fluid. If you do get brake fluid in your eyes, flush your eyes with clean, running water for 15 minutes. If eye irritation persists, or if you have taken

brake fluid internally, IMMEDIATELY seek medical assistance.

• The EPA warns that prolonged contact with used engine oil may cause a number of skin disorders, including cancer! You should make every effort to minimize your exposure to used engine oil. Protective gloves should be worn when changing oil. Wash your hands and any other exposed skin areas as soon as possible after exposure to used engine oil. Soap and water, or waterless hand cleaner should be used.

• All new vehicles are now equipped with an air bag system, often referred to as a Supplemental Restraint System (SRS) or Supplemental Inflatable Restraint (SIR) system. The system must be disabled before performing service on or around system components, steering column, instrument panel components, wiring and sensors. Failure to follow safety and disabling procedures could result in accidental air bag deployment, possible personal injury and unnecessary system repairs.

• Always wear safety goggles when working with, or around, the air bag system. When carrying a non-deployed air bag, be sure the bag and trim cover are pointed away from your body. When placing a non-deployed air bag on a work surface, always face the bag and trim cover upward, away from the surface. This will reduce the motion of the module if it is accidentally deployed. Refer to the additional air bag system precautions later in this section.

• Clean, high quality brake fluid from a sealed container is essential to the safe and

proper operation of the brake system. You should always buy the correct type of brake fluid for your vehicle. If the brake fluid becomes contaminated, completely flush the system with new fluid. Never reuse any brake fluid. Any brake fluid that is removed from the system should be discarded. Also, do not allow any brake fluid to come in contact with a painted surface; it will damage the paint.

• Never operate the engine without the proper amount and type of engine oil; doing so WILL result in severe engine damage.

• Timing belt maintenance is extremely important! Many models utilize an interference-type, non-freewheeling engine. If the timing belt breaks, the valves in the cylinder head may strike the pistons, causing potentially serious (also time-consuming and expensive) engine damage. Refer to the maintenance interval charts in the front of this manual for the recommended replacement interval for the timing belt, and to the timing belt section for belt replacement and inspection.

• Disconnecting the negative battery cable on some vehicles may interfere with the functions of the on-board computer system(s) and may require the computer to undergo a relearning process once the negative battery cable is reconnected.

• When servicing drum brakes, only disassemble and assemble one side at a time, leaving the remaining side intact for reference.

• Only an MVAC-trained, EPA-certified automotive technician should service the air conditioning system or its components.

ENGINE REPAIR

Distributor

The Escape uses a Direct Ignition System (DIS). No distributor is used.

Alternator

REMOVAL & INSTALLATION

2.0L Engine

1. Remove or disconnect the following:
   • Negative battery cable
   • Drive belt
   • Alternator electrical connectors and loosen the upper alternator bolt while moving the alternator to the rear of the engine
   • Alternator. Torque the lower bolt to

35 ft. lbs. (48Nm). Torque the upper bolt to 18 ft. lbs. (25 Nm).

To install:
2. Install or connect the following:
   • Alternator with the upper bolt in the alternator before installation. Torque the bolts to 18 ft. lbs. (25 Nm).
   • Alternator electrical connectors
   • Drive belt
   • Negative battery cable

2.3L Engine

1. Disconnect the battery.
2. Remove the front end accessory drive belt tensioner. Rotate the front end accessory drive belt tensioner counterclockwise to loosen tension on the front end accessory drive belt.

3. Remove the front end accessory drive belt.
4. Remove the alternator B+ terminal.
5. Remove the alternator electrical connector.
6. Remove the alternator lower air duct bolt.
7. Remove the alternator lower air duct. Press the locking tab to release the lower air duct from the upper air duct.
8. Remove the pin-type retainer.
9. Remove the alternator shield.
10. Remove the alternator stud nut.
11. Remove the alternator stud.
12. Remove the alternator bolts.
13. Remove the alternator.
14. To install, reverse the removal procedure. Observe the following torques:
    • Alternator mounting bolts: 35 ft. lbs. (47 Nm)

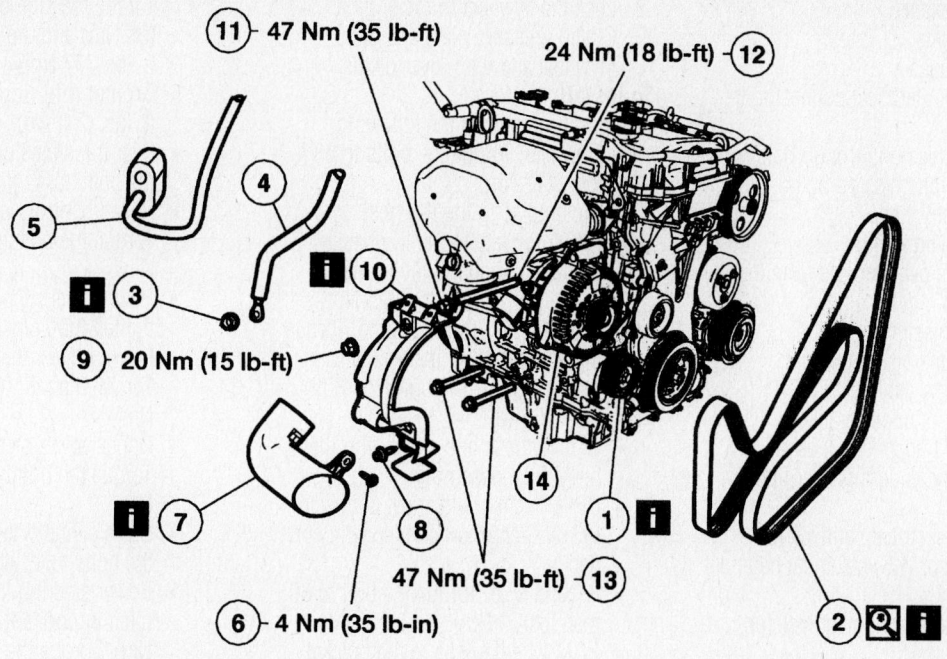

11 - 47 Nm (35 lb-ft)
24 Nm (18 lb-ft) - 12
5
4
3
10
9 - 20 Nm (15 lb-ft)
7
8
14
1
47 Nm (35 lb-ft) - 13
6 - 4 Nm (35 lb-in)
2

1	Front end accessory drive belt tensioner	8 Pin-type retainer
2	Front end accessory drive belt	9 Generator shield nut
3	Generator B+ terminal nut	10 Generator shield
4	Generator B+ terminal	11 Generator stud nut
5	Generator electrical connector	12 Generator stud
6	Generator lower air duct bolt	13 Generator bolts
7	Generator lower air duct	14 Generator

67197-ESCA-G01

Alternator mounting—2.3L Engine

- Alternator stud: 18 ft. lbs. (24 Nm)
- Stud nut: 35 ft. lbs. (47 Nm)
- Shield nut: 15 ft. lbs. (20 Nm)
- Lower air duct bolt: 35 inch lbs. (4 Nm)

3.0L Engine

1. Remove or disconnect the following:
   - Negative battery cable
   - Right side intermediate axle shaft
   - Right side splash shield and retainers
   - Drive belt
   - Alternator electrical connectors
   - Alternator. Torque the mounting and adjusting bolts to 35 ft. lbs. (48Nm).

To install:

2. Install or connect the following:
   - Alternator. Torque the bolts to 35 ft. lbs. (48 Nm).
   - Alternator electrical connectors
   - Drive belt
   - Negative battery cable

Ignition Timing

ADJUSTMENT

Ignition timing is controlled by the Powertrain Control Module (PCM). No adjustment is necessary or possible.

Engine Assembly

REMOVAL & INSTALLATION

2.0L Engine

MANUAL TRANSMISSION

1. Before servicing the vehicle, refer to the precautions in the beginning of this section.
2. Properly recover the air conditioning system refrigerant.
3. Properly relieve the fuel system pressure.
4. Drain the cooling system.
5. Drain the engine oil.

6. Remove or disconnect the following:
   - Hood
   - Battery and battery tray
   - Air cleaner housing
   - Fuel lines
   - Throttle cable and speed control cable, if equipped
   - Exhaust Gas Recirculation (EGR) vacuum valve regulator
   - EGR electrical connectors and vacuum hoses
   - Brake booster vacuum hose
   - Powertrain Control Module (PCM) wire harness and ground
   - Wire harness connector
   - Power distribution board electrical connectors
   - Evaporative emissions (EVAP) canister vacuum lines
   - Upper radiator hose
   - Power steering line bracket
   - Upper power steering pump bolts
   - Coolant hose
   - Heater hoses
   - Speed control unit, if equipped
   - Catalytic converter

- A/C compressor
- Both halfshafts
- Shifter linkages
- Block heater electrical connector, if equipped
- Front transmission through bolt
- Engine-to-transmission bolts
- Lower radiator hose
- Power steering pump
- Clutch slave cylinder line from the bracket and move it aside
- Rear transmission mount
- Left side transmission mount
- Lower ground cable
- Engine mount upper bracket
- Engine and transmission as an assembly by using a proper lifting device
- Alternator electrical connectors
- Knock Sensor (KS) electrical connector
- Oil pressure sender electrical connector
- Starter electrical connector
- Vehicle Speed Sensor (VSS) electrical connector
- Park Neutral Position (PNP) electrical connector
- Fuel charging wire harness electrical connector
- PCM wire harness from the bracket
- PCM ground wire
- Back up lamp switch electrical connector
- Wire harness
- Differential Pressure Feedback (DPFEE) EGR sensor

7. Separate the engine from the transmission.

8. Lock the flywheel to the engine.
9. Clutch pressure plate and disc.
10. Flywheel and rear cover plates.

To install:
11. Install or connect the following:
- Flywheel. Torque the bolts to 83 ft. lbs. (112 Nm).
- Clutch disc to the flywheel
- Pressure plate to the flywheel. Torque the bolts in the proper sequence to 18 ft. lbs. (25 Nm).
- Transmission to the engine. Torque the bolts to 33 ft. lbs. (45 Nm).
- Starter. Torque the bolts to 18 ft. lbs. (25 Nm).
- Wire harness and attach it to the powertrain assembly
- DPFEE sensor electrical connector
- Reverse lamp switch electrical connector
- Ground wire. Torque the bolt to 80 inch lbs. (9 Nm).
- PCM wire harness to the bracket
- Fuel charging wire harness electrical connector
- PNP switch electrical connector
- VSS electrical connector
- KS, Oil pressure sender and starter electrical connector. Torque the fasteners to 9 ft. lbs. (12 Nm).
- Alternator electrical connectors. Torque the fasteners to 71 inch lbs. (8 Nm).
- Powertrain assembly in the vehicle
- Left side transmission mount. Torque the side bolts to 41 ft. lbs. (55 Nm) and the center bolt to 66 ft. lbs. (90 Nm).
- Engine mount upper bracket.

Torque the side bolts to 72 ft. lbs. (98 Nm) and the center bolt to 57 ft. lbs. (77 Nm).
- Ground wire. Torque the bolt to 25 ft. lbs. (34 Nm).
- Rear transmission mount. Torque the bolts to 41 ft. lbs. (55 Nm).
- Speed control unit, if equipped. Torque the bolts to 89 inch lbs. (10 Nm).
- Power steering pump and hand tighten the bolts
- Lower power steering line bracket. Torque the bolt to 89 inch lbs. (10 Nm).
- Upper power steering line bolt. Torque the bolt to 15 ft. lbs. (20 Nm).
- Slave cylinder line and clip. Torque the bolts to 16 ft. lbs. (22 Nm).
- Power steering lines. Torque the retaining bolts to 89 inch lbs. (10 Nm). Torque the power steering pump bolts to 18 ft. lbs. (25 Nm).
- Lower radiator hose
- Engine-to-transmission bolts. Torque the bolts to 33 ft. lbs. (45 Nm).
- Front transmission through bolt. Torque the bolt to 66 ft. lbs. (90 Nm).
- Block heater electrical connector
- Shifter linkages. Torque the upper bolt to 33 ft. lbs. (45 Nm) and the lower bolt to 15 ft. lbs. (20 Nm).
- Coolant hose
- Catalytic converter
- Heater hoses
- Upper radiator hose
- EVAP canister vacuum lines
- Power distribution box electrical connector. Torque the fastener to 9 ft. lbs. (12 Nm).
- Wire harness electrical connector
- Ground wires
- PCM wire harness and ground
- Brake booster vacuum supply hose to the intake manifold
- EGR vacuum regulator valve hoses and electrical connector
- Throttle cable and speed control cable, if equipped
- Fuel lines
- Battery tray and battery
- Air cleaner
- Hood

12. Fill the engine with clean oil.
13. Fill and bleed the cooling system.
14. Recharge the A/C system.
15. Start the vehicle, check for leaks and repair if necessary.

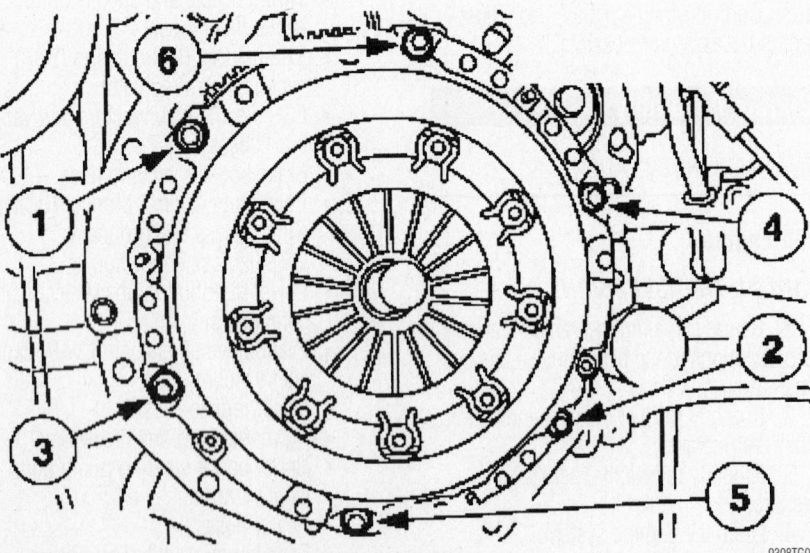

9308TG07

Tighten the pressure plate bolts in the proper sequence–2.0L engine

2.3L Engine

MANUAL TRANSAXLE

All vehicles

1. With the vehicle in NEUTRAL, position it on a hoist.

2. Release the fuel system pressure.

3. Remove the engine air cleaner and air cleaner outlet pipe.

4. Remove the battery tray.

5. Drain the engine oil.

6. Drain the cooling system.

7. Remove the starter.

8. Remove the catalytic converter.

9. Remove the accessory drive belt.

10. Remove the bolts and the lateral support crossmember.

11. Remove the LH front drive half-shaft.

12. Remove the front drive intermediate halfshaft

4x4 vehicles

13. Remove the six bolts holding the driveshaft to the transfer case.

14. Position the driveshaft aside.

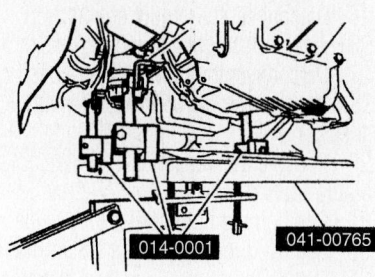

014-0001   041-00765

67197-ESCA-G02

Engine secured to lift table—2.3L engine w/manual transaxle

All vehicles

15. If equipped, remove the bolt and ground eyelet.

16. Remove the power distribution box cover.

17. Remove the nuts and disconnect the cables.

18. Disconnect the electrical connector from the power distribution box.

19. Remove the bolt and disconnect the ground strap. Loosen the bolt and disconnect the 42-pin electrical connector.

20. Detach the wiring harness retainers from the battery tray bracket and position the wiring harness out of the way.

21. Disconnect the clutch hydraulic tube fitting. Detach the tube from the spring clip and position aside.

22. Remove the retaining clips and disconnect the transaxle control cable.

23. Remove the retaining clips and disconnect the transaxle control cable.

24. Disconnect the vehicle speed sensor electrical connector and pin-type retainer.

25. Disconnect the reversing lamp indicator switch and detach the wiring harness retainers.

26. If equipped, disconnect the block heater electrical connector. Detach all the block heater wiring harness retainers and position the wiring harness aside.

27. Disconnect the upper radiator and coolant vent hoses.

28. Remove the nuts and the coolant vent hose brackets. Position the coolant vent hose aside.

29. Detach the heater hose support strap from the stud.

30. Disconnect the heater hoses from the heater core.

31. Remove the retainers and the accelerator cable snow shield.

32. Disconnect the accelerator cable and speed control cable (if equipped).

33. Remove the nut from the accelerator control cable bracket.

34. Remove the nut from the accelerator control cable bracket and position the accelerator control cable and bracket assembly aside.

35. Remove the nut and position the power steering tube and bracket aside.

36. Disconnect the vacuum supply tube and position aside.

37. Disconnect the fuel vapor return tube and position aside.

38. Disconnect the vacuum reservoir tube and position aside.

39. Disconnect the fuel supply tube and retainer and position aside.

40. Detach the electrical connector retainers.

41. Disconnect the powertrain control module (PCM) electrical connectors. Remove the nut and position the harness aside.

42. Remove the bolt and detach the ground wire.

43. Remove the two power steering pump bolts.

44. Disconnect the lower radiator hose from the radiator.

45. Disconnect the A/C compressor electrical connector and remove the four bolts. Position the A/C compressor aside and support the compressor with a length of mechanics wire.

46. Disconnect the power steering pressure (PSP) sensor electrical connector.

➡**The bolt under the power steering pressure tube will remain with the power steering pump.**

47. Remove the bolts and position the power steering pump aside.

48. Remove the front roll restrictor bolt and the two bolts for the engine support crossmember.

49. Remove the rear nut and the engine support crossmember.

➡**The transaxle-to-engine bolts differ in length. Mark the bolts for correct installation.**

50. Remove the two transaxle-to-engine bolts.

➡**The transaxle-to-engine bolts differ in length. Mark the bolts for correct installation.**

51. Remove the two transaxle-to-engine bolts.

52. Using the special tools, secure the engine to the lift table.

53. Remove the engine mount bracket bolt.

54. Remove the nuts and the engine mount bracket.

55. Remove the bolt from the transaxle rear mount.

56. Remove bolt from the LH transaxle mount.

57. Lower the engine and transaxle from the vehicle.

58. Using the engine crane and spreader bar, remove the engine and transaxle from the lift table.

➡**The transaxle-to-engine bolts differ in length. Mark the bolts for correct installation.**

59. Remove the remaining six engine-to-transaxle bolts and separate the engine and transaxle.

To install:
All vehicles

60. Using the engine crane and spreader bar, position the engine and transaxle together. Install the six upper transaxle-to-engine bolts. Torque to 35 ft. lbs. (48 Nm).

61. Using the engine crane and spreader bar, position the engine and transaxle onto the lift table.

62. Using the special tools, secure the engine to the lift table.

63. Raise the engine and transaxle into the vehicle.

64. Install the bolt in the LH transaxle mount. Torque to 76 ft. lbs. (103 Nm).

65. Install the bolt in the rear transaxle mount. Torque to 85 ft. lbs. (115 Nm).

66. Install the engine mount bracket. Torque to 66 ft. lbs. (90 Nm).

67. Install the engine mount bracket bolt. Torque to 66 ft. lbs. (90 Nm).

68. Install the 4 lower transaxle-to-engine bolts. Torque to 35 ft. lbs. (48 Nm).

69. Install the engine support crossmember and nut. Torque to 66 ft. lbs. (90 Nm).

70. Install the two bolts for the engine support crossmember. Torque to 66 ft. lbs. (90 Nm).

71. Install the front roll restrictor bolt. Torque to 85 ft. lbs. (115 Nm).

➡ **The bolt under the power steering pressure tube will remain with the power steering pump.**

72. Position the power steering pump and install the bolts. Torque to 18 ft. lbs. (25 Nm).

73. Connect the power steering pressure (PSP) sensor electrical connector.

74. Install the A/C compressor and connect the A/C compressor electrical connector. Torque to 18 ft. lbs. (25 Nm).

75. Connect the lower radiator hose to the radiator.

76. Install the two lower power steering pump bolts. Torque to 18 ft. lbs. (25 Nm).

77. Install the ground wire and bolt.

78. Connect the powertrain control module (PCM) electrical connectors. Position the harness and install the nut.

79. Attach the electrical connector retainers.

80. Connect the fuel supply tube.

81. Connect the vacuum reservoir tube.

82. Connect the fuel vapor return tube and retainer.

83. Connect the vacuum supply tube.

84. Install the power steering tube and bracket.

85. Position the accelerator control cable and bracket and install the nut.

86. Install the accelerator control cable and bracket and nut.

87. Install the accelerator cable and speed control cable (if equipped).

88. Install the accelerator cable snow shield and retainers.

89. Connect the heater hoses to the heater core.

90. Attach the heater hose support strap to the stud.

91. Position the coolant vent hose and install the coolant vent hose brackets and nuts.

92. Connect the upper radiator and coolant vent hoses.

93. If equipped, route the block heater wiring harness and attach all retainers. Connect the block heater electrical connector.

94. Connect the reversing lamp indicator switch and attach the wiring harness retainers.

95. Connect the vehicle speed sensor

(VSS) electrical connector and pin-type retainer.

96. Connect the transaxle control cable and install the retaining clips.

97. Connect the transaxle control cable and install the retaining clips.

98. Connect the clutch hydraulic tube fitting. Attach the tube to the spring clip.

99. Attach the wiring harness retainers to the battery tray bracket.

100. Connect the 42-pin electrical connector and tighten the bolt. Install the ground strap and bolt.

101. Connect the electrical connector to the power distribution box.

102. Connect the cables and install the nuts.

103. Install the power distribution box cover.

104. If equipped, install the ground eyelet and bolt.

4x4 vehicles
105. Install the driveshaft. Torque to 15 ft. lbs. (20 Nm).
All vehicles
106. Install the front drive intermediate halfshaft.

107. Install the LH front drive halfshaft.

108. Install the lateral support crossmember. Torque to 85 ft. lbs. (115 Nm).

109. Install the accessory drive belt.

110. Install the catalytic converter.

111. Install the starter.

112. Install the battery tray and battery.

113. Install the engine air cleaner and air cleaner outlet pipe.

114. Fill the engine with clean engine oil.

115. Fill and bleed the cooling system.

116. Bleed the clutch system.

AUTOMATIC TRANSAXLE

All vehicles
1. With the vehicle in NEUTRAL, position it on a hoist.

2. Release the fuel system pressure.

3. Remove the engine air cleaner and air cleaner outlet pipe

4. Remove the battery tray.

5. Drain the engine oil.

6. Drain the cooling system.

7. Remove the starter.

8. Remove the catalytic converter.

9. Remove the accessory drive belt.

10. Remove the left front drive halfshafts.

4wd vehicles
11. Remove the transfer case.

2wd vehicles
12. Remove the bolts and the lateral support crossmember.

13. Remove the front drive intermediate halfshaft.

All vehicles
14. If equipped, remove the bolt and ground eyelet.

15. Remove the power distribution box cover.

16. Remove the nuts and disconnect the cables.

17. Disconnect the electrical connector from the power distribution box.

18. Remove the bolt and disconnect the ground strap. Loosen the bolt and disconnect the 42-pin electrical connector.

19. Detach the wiring harness retainers from the battery tray bracket and position the wiring harness out of the way.

20. Disconnect the transaxle electrical connector.

21. Disconnect the shift cable from the transaxle manual lever.

22. Position the transaxle control cable and bracket aside.

23. Disconnect the transmission range (TR) sensor electrical connector.

24. Detach the transaxle control harness from the brackets.

25. Disconnect the fluid cooler tube.

26. Disconnect the output shaft speed (OSS) sensor electrical connector (black).

27. Disconnect the turbine shaft speed (TSS) sensor electrical connector (white connector).

28. Remove the transmission fluid cooler retaining bracket bolt.

29. Position the fluid cooler tube aside.

30. Remove the bolt and the OSS sensor.

31. Detach the transaxle control harness from the retaining clip.

32. If equipped, disconnect the block heater electrical connector. Detach all the block heater wiring harness retainers and position the wiring harness aside.

33. Disconnect the upper radiator and coolant vent hoses.

34. Remove the nuts and the coolant vent hose brackets. Position the coolant vent hose aside.

35. Detach the heater hose support strap from the stud.

36. Disconnect the heater hoses from the heater core.

37. Remove the retainers and the accelerator cable snow shield.

38. Disconnect the accelerator cable and speed control cable (if equipped).

39. Remove the nut from the accelerator control cable bracket.

40. Remove the nut from the accelerator control cable bracket and position the accelerator control cable and bracket assembly aside.

41. Remove the nut and position the power steering tube and bracket aside.

42. Disconnect the vacuum supply tube and position aside.

43. Disconnect the fuel vapor return tube and retainer and position aside.

44. Disconnect the vacuum reservoir tube and position aside.

45. Disconnect the fuel supply tube and position aside.

46. Detach the electrical connector retainers.

47. Disconnect the powertrain control module (PCM) electrical connectors. Remove the nut and position the harness aside.

48. Remove the bolt and detach the ground wire.

49. Remove the two power steering pump bolts.

50. Disconnect the lower radiator hose from the radiator.

51. Disconnect the A/C compressor electrical connector and remove the four bolts. Position the A/C compressor aside and support the compressor with a length of mechanics wire.

52. Disconnect the power steering pressure (PSP) sensor electrical connector.

➡ **The bolt under the power steering pressure tube will remain with the power steering pump.**

53. Remove the bolts and position the power steering pump aside.

54. Remove the front roll restrictor bolt and the two bolts for the engine support crossmember.

55. Remove the rear nut and the engine support crossmember.

➡ **The transaxle-to-engine bolts differ in length. Mark the bolts for correct installation.**

56. Remove the two transaxle-to-engine bolts

➡ **The transaxle-to-engine bolts differ in length. Mark the bolts for correct installation.**

57. Remove the two transaxle-to-engine bolts

58. Using the special tools, secure the engine to the lift table.

59. Remove the engine mount bracket bolt.

60. Remove the nuts and the engine mount bracket.

61. Remove the bolt from the transaxle rear mount.

62. Remove the bolt from the left transaxle mount.

63. Lower the engine and transaxle from the vehicle.

64. Using the engine crane and spreader bar remove the engine and transaxle from the lift table.

65. Remove the starter motor isolator.

66. Remove and discard the four torque converter nuts.

➡ **The transaxle-to-engine bolts differ in length. Mark the bolts for correct installation.**

67. Remove the remaining six engine-to-transaxle bolts and separate the engine and transaxle.

To install:
All vehicles

68. Using the engine crane and spreader bar, position the engine and transaxle together. Install the six upper transaxle-to-engine bolts. Torque to 35 ft. lbs. (48 Nm).

69. Install new torque converter nuts. Torque to 26 ft. lbs. (35 Nm).

70. Install the starter motor isolator.

71. Using the engine crane and spreader bar, position the engine and transaxle onto the lift table.

72. Using the special tools, secure the engine to the lift table.

73. Raise the engine and transaxle into the vehicle.

74. Install the bolt in the left transaxle mount. Torque to 76 ft. lbs. (103 Nm).

75. Install the bolt in the rear transaxle mount. Torque to 85 ft. lbs. (115 Nm).

76. Install the engine mount bracket. Torque to 66 ft. lbs. (90 Nm).

77. Install the engine mount bracket bolt. Torque to 66 ft. lbs. (90 Nm).

78. Install the 4 transaxle-to-engine bolts. Torque to 35 ft. lbs. (48 Nm).

79. Install the engine support crossmember and nut. Torque to 66 ft. lbs. (90 Nm).

80. Install the two bolts for the engine support crossmember. Torque to 66 ft. lbs. (90 Nm).

81. Install the front roll restrictor bolt . Torque to 85 ft. lbs. (115 Nm).

82. Position the power steering pump and install the upper bolts. Torque to 18 ft. lbs. (25 Nm).

83. Connect the power steering pressure (PSP) sensor electrical connector.

84. Install the A/C compressor and connect the A/C compressor electrical connector. Torque to 18 ft. lbs. (25 Nm).

85. Connect the lower radiator hose to the radiator.

86. Install the two lower power steering pump bolts. Torque to 18 ft. lbs. (25 Nm).

87. Install the ground wire and bolt.

88. Connect the powertrain control module (PCM) electrical connectors. Position the harness and install the nut.

89. Attach the electrical connector retainers.

90. Connect the fuel supply tube.

91. Connect the vacuum reservoir tube.

92. Connect the fuel vapor return tube and retainer.

93. Connect the vacuum supply tube.

94. Position the power steering tube and bracket and install the nut.

95. Position the accelerator control cable and bracket and install the nut.

96. Install the accelerator control cable and bracket and nut.

97. Install the accelerator cable and speed control cable (if equipped).

98. Install the accelerator cable snow shield and the retainers.

99. Connect the heater hoses to the heater core.

100. Attach the heater hose support strap to the stud.

101. Position the coolant vent hose and install the coolant vent hose brackets and nuts.

102. Connect the upper radiator and coolant vent hoses.

103. If equipped, route the block heater wiring harness and attach all retainers. Connect the block heater electrical connector.

104. Attach the transaxle control harness to the retaining clip.

105. Install the output shaft speed (OSS) sensor and bolt.

106. Install the fluid cooler tube.

107. Connect the transmission fluid cooler tube.

108. Attach the transaxle control harness to the brackets.

109. Connect the transmission range (TR) sensor electrical connector.

110. Install the transaxle control cable and bracket.

111. Connect the shift cable to the transaxle manual lever.

112. Connect the transaxle electrical connector.

113. Attach the wiring harness retainers to the battery tray bracket.

114. Connect the 42-pin electrical connector and tighten the bolt. Install the ground strap and bolt.

115. Connect the electrical connector to the power distribution box.

116. Connect the cables and install the nuts.

117. Install the power distribution box cover.

118. If equipped, install the ground eyelet and bolt.

2wd vehicles

119. Install the front drive intermediate halfshaft.

120. Install the lateral support cross-member. Torque to 85 ft. lbs. (115 Nm).

4wd vehicles

121. Install the transfer case.

All vehicles

122. Install the left front drive halfshaft.

123. Install the accessory drive belt.

124. Install the catalytic converter.

125. Install the starter.

126. Install the battery tray and battery.

127. Install the engine air cleaner and air cleaner outlet pipe.

128. Fill the engine with clean engine oil.

129. Fill and bleed the cooling system.

3.0L Engine

1. Before servicing the vehicle, refer to the precautions in the beginning of this section.

2. Properly recover the air conditioning system refrigerant.

3. Properly relieve the fuel system pressure.

4. Drain the cooling system.

5. Drain the engine oil.

6. Remove or disconnect the following:
- Hood
- Battery and battery tray
- Air cleaner outlet tube and housing
- Lower radiator air deflectors
- Fuel lines
- Water pump drive belt
- Accelerator cable and speed control cable, if equipped
- Vapor Management Valve (VMV)
- Powertrain Control Module (PCM)
- PCM ground wire
- Thermostat housing and hose assembly and move them aside
- Power distribution box electrical connector
- Power distribution box cover
- Nuts and cables from inside the power distribution box
- Transmission linkage
- Brake booster vacuum hose
- Heater hoses
- Power steering return line
- Power Steering Pressure (PSP) switch electrical connector
- Power steering supply line
- Oil level indicator
- Catalytic converter
- A/C compressor
- Both front wheels
- Intermediate drive shaft, if equipped

7. Separate both side ball joints.

8. Separate both side tie rod ends from the steering knuckles.

9. Separate both sway bar links from the strut mounts.

10. Separate the struts from the steering knuckles.

11. Remove or disconnect the following:
- Both wheel speed sensors, if equipped
- Brake calipers from the steering knuckles and properly support the struts
- Steering shaft from the rack
- Transmission line bracket bolt
- Transmission cooler lines
- Torque converter inspection cover
- Torque converter nuts
- Block heater wiring, if equipped

12. Install a powertrain lifting devise and raise the vehicle.
- Engine support bracket
- Transmission support
- 2 rear subframe bolts
- 2 subframe side bolts
- Motor mount support bolts
- Engine and transmission as an assembly
- Heated Oxygen (HO$_2$S) sensor
- Transmission Range (TR) sensor
- Transmission harness electronic control switch
- Transmission control harness from the bracket
- Starter and wire harness
- Knock Sensor (KS) electrical connector
- Output Shaft Speed (OSS) sensor electrical connector
- HO$_2$S sensor and Exhaust Gas Recirculation (EGR) tube from the exhaust manifold
- Alternator and electrical connectors
- Right side exhaust manifold and gasket
- Halfshaft support bracket and move it aside

13. Separate the engine from the transmission assembly

To install:

14. Install or connect the following:
- Powertrain assembly on the subframe
- Transmission-to-engine bolts. Torque the bolts to 30 ft. lbs. (40 Nm).
- Halfshaft bracket. Torque the bolts to 18 ft. lbs. (25 Nm).
- Right side exhaust manifold and new gasket. Torque the bolts to 15 ft. lbs. (25 Nm).
- Alternator. Torque the larger bolts to 18 ft. lbs. (25 Nm) and smaller bolt to 89 inch lbs. (10 Nm).
- EGR tube and HO$_2$S sensor electrical connectors
- OSS sensor electrical connector
- KS jumper electrical connector
- Starter. Torque the bolts to 18 ft. lbs. (25 Nm).
- Transmission control harness to the bracket. Torque the bolt to 18 ft. lbs. (25 Nm).
- Transmission harness
- Transmission range sensor
- Powertrain assembly
- Motor mount support. Torque the bolts to 66 ft. lbs. (90 Nm).
- Subframe side nuts. Torque the nuts to 76 ft. lbs. (103 Nm). Raise the vehicle and support the powertrain assembly with a lifting device.
- Transmission mount. Torque the bolts to side bolts to 66 ft. lbs. (90 Nm) and the other bolts to 76 ft. lbs. (103 Nm).
- Motor mount. Torque the bolts to side bolts to 66 ft. lbs. (90 Nm) and the other bolts to 76 ft. lbs. (103 Nm). Remove the powertrain lift.
- Block heater electrical connector, if equipped
- Torque converter. Torque the nuts to 27 ft. lbs. (37 Nm).
- Transmission cover plate and plug
- Transmission cooler lines
- Transmission cooler line bracket. Torque the bolt to 15 ft. lbs. (20 Nm).
- Steering shaft to the rack. Torque the bolt to 18 ft. lbs. (25 Nm).
- Struts to the steering knuckles. Torque the bolts to 75 ft. lbs. (102 Nm).
- Brake calipers to the steering knuckles
- Wheel speed sensors, if equipped. Torque the bolts to 89 inch lbs. (10 Nm).
- Sway bar links to the strut mount. Torque the bolts to 41 ft. lbs. (55 Nm).
- Tie rods to the steering knuckles. Torque the bolts to 41 ft. lbs. (55 Nm).
- Ball joints. Torque the bolts to 52 ft. lbs. (70 Nm).
- Intermediate drive shaft, if equipped
- Both front wheels
- A/C compressor
- Lower radiator air deflectors
- Catalytic converter
- Oil level indicator dipstick tube
- Power steering line and bracket. Torque the bolt to 13 ft. lbs. (17 Nm).
- PSP switch electrical connector

- Power steering return line
- Heater hoses
- Vacuum lines
- Transmission linkage
- Wire harness cables and nuts to the power distribution box. Torque the nuts to 89 inch lbs. (10 Nm).
- Power distribution box wire harness
- Thermostat housing and connect the hoses
- Ground wire. Torque the bolt to 89 inch lbs. (10 Nm).
- PCM electrical connector
- VMV electrical connector
- Accelerator cable and speed control cable, if equipped
- Air cleaner assembly
- Water pump drive belt
- Battery and tray

15. Fill and bleed the cooling system.
16. Fill the engine with clean oil.
17. Recharge the A/C system.
18. Inspect and top off the power steering fluid.
19. Start the vehicle, check for leaks and repair if necessary.

Water Pump

REMOVAL & INSTALLATION

2.0L Engine

1. Before servicing the vehicle, refer to the precautions in the beginning of this section.
2. Drain the cooling system.
3. Remove or disconnect the following:
   - Negative battery cable
   - Right front wheel
   - Splash shield
   - Drive belt
   - Water pump pulley
   - Water pump

To install:
4. Install or connect the following:
   - Water pump. Torque the bolts to 89 inch lbs. (10 Nm).

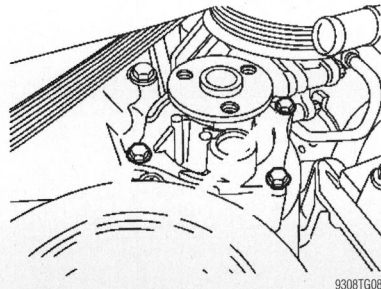

9308TG08

Exploded view of the water pump—2.0L engine

- Water pump pulley. Torque the bolts to 89 inch lbs. (10 Nm).
- Drive belt
- Splash shield
- Right front wheel
- Negative battery cable

5. Refill the cooling system.
6. Start the vehicle and check for leaks, repair if necessary.

2.3L Engine

1. With the vehicle in NEUTRAL, position it on a hoist.
2. Drain the cooling system.
3. Remove the accessory drive belt.
4. Remove the coolant pump pulley bolts.
5. Remove the coolant pump pulley.
6. Remove the coolant pump bolts.
7. Remove the coolant pump.
8. Remove the coolant pump O-ring seal.
9. To install, reverse the removal procedure. Torque the water pump bolts to 89 inch lbs. (10 Nm). Torque the pulley bolts to 18 ft. lbs. (25 Nm).
10. Fill and bleed the cooling system.

3.0L Engine

1. Before servicing the vehicle, refer to the precautions in the beginning of this section.

2. Drain the cooling system.
3. Remove or disconnect the following:
   - Negative battery cable
   - Air cleaner outlet tube
   - Water pump belt tensioner
   - Coolant hoses
   - Water pump
   - Water pump from the housing

To install:
- Water pump to the housing. Torque the bolts to 89 inch lbs. (10 Nm).
- Water pump. Torque the bolts to 89 inch lbs. (10 Nm).
- Coolant hoses
- Water pump belt tensioner
- Air cleaner outlet tube

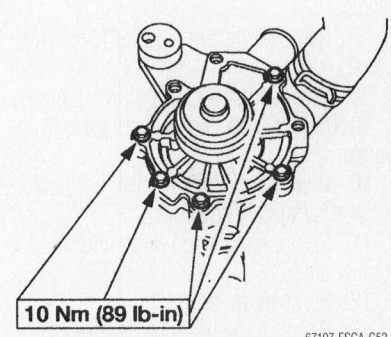

10 Nm (89 lb-in)

67197-ESCA-G52

Water pump mounting—3.0L engine

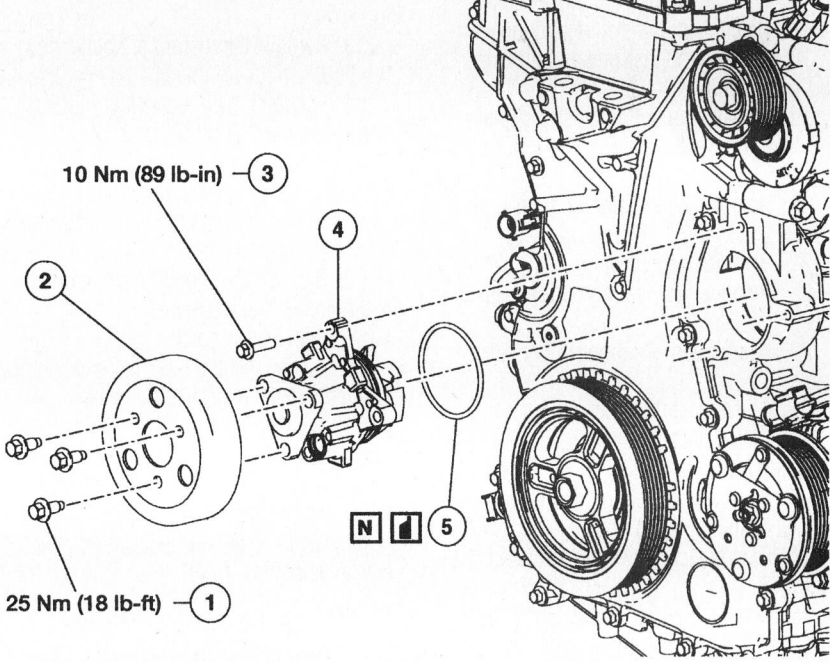

10 Nm (89 lb-in) — 3

2

4

25 Nm (18 lb-ft) — 1

N **5**

1 Coolant pump pulley bolts	4 Coolant pump
2 Coolant pump pulley	5 Coolant pump O-ring seal
3 Coolant pump bolts	

67197-ESCA-G03

Water pump mounting—2.3L engine

- Negative battery cable
4. Refill the cooling system.
5. Start the vehicle and check for leaks, repair if necessary.

Heater Core

REMOVAL & INSTALLATION

1. Drain the engine coolant.
2. Disconnect the heater hoses from the heater core.
3. Remove the driver air bag module.
4. Remove the two front door scuff plates.
5. Remove the four pin-type retainers.
6. Remove the two front door scuff plates.
7. Remove the two A-pillar lower trim panels.
8. Remove the two pin-type retainers.
9. Remove the two A-pillar lower trim panels.
10. Disconnect the electrical connectors located by the LH cowl.
11. Position the hood latch release handle aside.
12. Remove the bolts.
13. Position the hood latch release handle aside.
14. Remove the utility compartment.
15. Remove the four pin-type retainers.
16. Remove the utility compartment.
17. Disconnect the electrical connector.
18. Remove the instrument panel steering column cover.
19. Release the upper clips and rotate the cover outward to release the lower pivot retainers.
20. Remove the steering column lower cover.
21. Remove the screws.
22. Remove the steering column lower cover.
23. If equipped, disconnect the shift cable.
24. Disconnect the shift cable.
25. Disconnect the shift cable from the retaining bracket.
26. Remove the steering column coupler access cover.
27. Disconnect the steering column coupler.
28. Remove the steering column coupler bolt and nut.
29. Disconnect the steering column coupler.
30. Remove the cover panel.
31. Remove the pin-type retainer.
32. Release the retaining clip.
33. Disconnect the electrical connectors.

34. Disconnect the climate control vacuum harness connector.
35. Disconnect the in-line electrical connector.
36. Remove the four instrument panel center brace bolts.
37. Remove the passenger air bag module.
38. Disconnect the vacuum harness connector.
39. Disconnect the temperature control cable.
40. Position the locator pin.
41. Release the locking tab.
42. Disconnect the temperature control cable from the blend door shaft.
43. Close the glove compartment.
44. Press the release tabs inward while raising the glove compartment.
45. Disconnect the electrical connectors at the blower motor.
46. Disconnect the antenna cable in-line connector.
47. Open the four A-pillar passenger assist handle covers.
48. Remove the two A-pillar passenger assist handles.
49. Remove the four bolts.
50. Remove the two A-pillar passenger assist handles.
51. Remove the two windshield side garnish moldings.
52. Remove the instrument panel cowl top cover.
53. Remove the instrument panel cowl top bolt.
54. Loosen the tilt lever (if equipped) and lower steering column.
55. Position the transmission range selector lever (if equipped) down to provide access to the instrument cluster finish panel and instrument cluster.
56. Remove the screws and the instrument cluster finish panel.
57. Remove the screws.
58. Disconnect the electrical connectors and remove the instrument cluster.
59. Through the instrument cluster opening, remove the instrument panel nut.
60. Remove the two instrument panel finish end panels.
61. Remove the four instrument panel cowl side bolts.

➡This step requires an assistant.

62. Remove the instrument panel.
63. Remove the heater blending door levers.
64. Remove the screw for heater blending door.
65. Remove the levers for the blending door.

66. Remove the heater core.
67. Remove the three screws.
68. Remove the cover for the heater core and pull the heater core out of the housing.

➡Before installing the temperature control cable, make sure the blend door, cable and temperature switch are correctly positioned.

69. To install, reverse the removal procedure.

✻✻ CAUTION

Electronic modules are sensitive to static electrical charges. If exposed to these charges, damage may result.

✻✻ CAUTION

Once the new module is installed, it is necessary to download the module configuration information from the scan tool into the new instrument cluster.

Cylinder Head

REMOVAL & INSTALLATION

2.0L Engine

1. Before servicing the vehicle, refer to the precautions in the beginning of this section.
2. Properly relieve the fuel system pressure.
3. Drain the engine oil.
4. Remove or disconnect the following:
- Negative battery cable
- Ignition coil bracket
- Thermostat housing
- Positive Crankcase Ventilation (PCV) tube
- Intake manifold
- Exhaust manifold

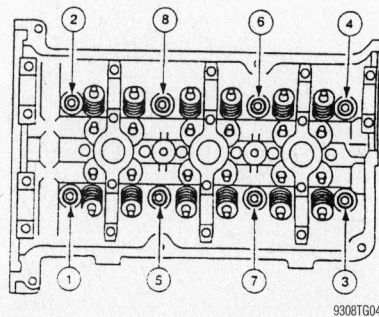

9308TG04

Cylinder head bolt torque sequence 2.0L engine

- Power steering bracket and move it aside
- Valve tappets
- Engine mount lower bracket
- Engine mount upper bracket
- Cylinder head bolts in the proper sequence and discard the gasket

To install:

5. Install a new head gasket and the cylinder head.

6. Lubricate the cylinder head bolt threads.

7. Torque the cylinder head bolts in the proper sequence as follows:

   a. Step 1: 15 ft. lbs. (20 Nm).
   b. Step 2: 30 ft. lbs. (40 Nm).
   c. Step 3: Plus an additional 90 degrees.

8. Install or connect the following:

- Engine mount upper bracket. Torque the 2 upper bolts to 72 ft. lbs. (98 Nm) and the center bolt to 57 ft. lbs. (77 Nm).
- Engine mount lower bracket. Torque the bolts to 37 ft. lbs. (50 Nm).
- Valve tappets
- Power steering pump bracket. Torque the bolts to 20 ft. lbs. (28 Nm).
- Exhaust manifold
- Intake manifold
- PCV tube
- Thermostat housing
- Ignition coil bracket
- Negative battery cable

9. Fill the engine with clean oil and replace the filter.

10. Start the vehicle and check for leaks, repair if necessary.

2.3L Engine

✳✳ WARNING

During engine repair procedures, cleanliness is extremely important. Any foreign material, including any material created while cleaning gasket surfaces, that enters the oil passages, coolant passages or the oil pan can cause engine failure.

1. With the vehicle in NEUTRAL, position it on a hoist.
2. Remove the camshafts.

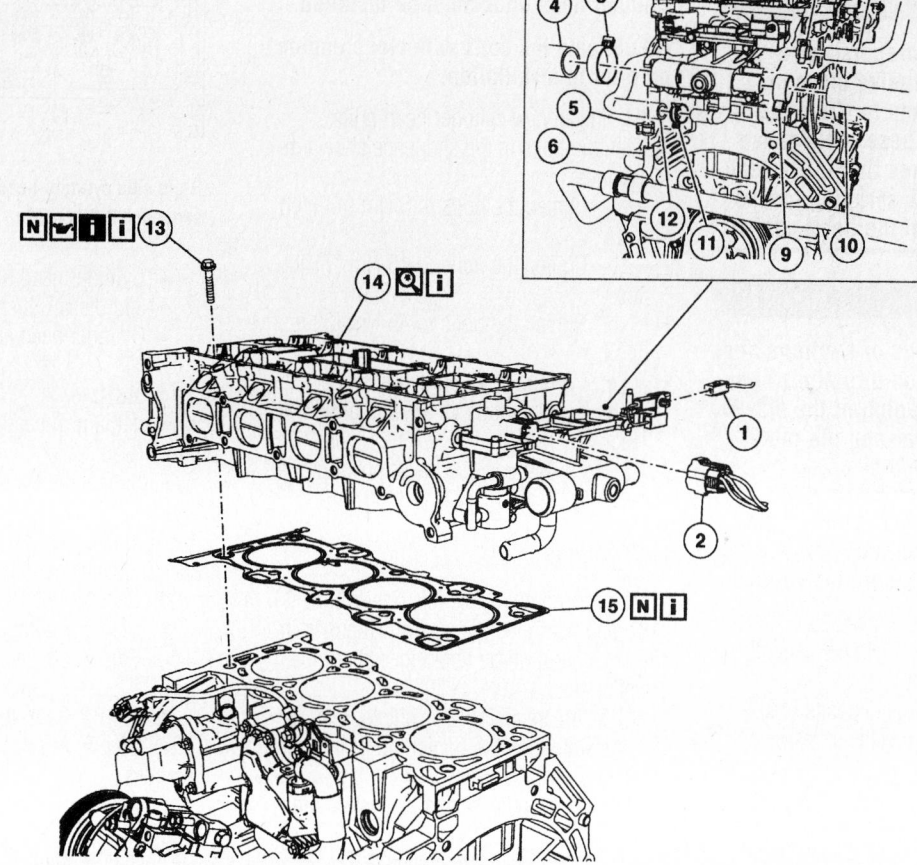

1 Radio ignition interference capacitor electrical connector	9 Heater hose clamp
2 Exhaust gas recirculation (EGR) valve electrical connector	10 Heater hose (position aside)
3 Upper radiator hose clamp	11 Bypass hose clamp
4 Upper radiator hose (position aside)	12 Bypass hose (position aside)
5 EGR coolant tube clamp	13 Cylinder head bolt
6 EGR coolant hose (part of heater hose) (position aside)	14 Cylinder head
7 Engine coolant vent hose clamp	15 Cylinder head gasket
8 Engine coolant vent hose (position aside)	

67197-ESCA-G04

Cylinder head removal—2.3L engine

3. Remove the intake manifold.

4. Remove the catalytic converter.

5. Disconnect the radio ignition interference capacitor electrical connector

6. Disconnect the exhaust gas recirculation (EGR) valve electrical connector

7. Remove the upper radiator hose.

8. Remove the EGR coolant tube clamp.

9. Remove the EGR coolant hose.

10. Remove the engine coolant vent hose.

11. Remove the heater hose.

12. Remove the bypass hose.

13. Remove and discard the cylinder head bolts.

14. Remove the cylinder head.

15. Remove the cylinder head gasket.

16. Inspect the cylinder head for distortion.

✳✳ WARNING

Do not use metal scrapers, wire brushes, power abrasive discs or other abrasive means to clean the sealing surfaces. These tools cause scratches and gouges that make leak paths. Use a plastic scraping tool to remove all traces of the head gasket.

✳✳ WARNING

Observe all warnings or cautions and follow all application directions contained on the packaging of the silicone gasket remover and the metal surface prep.

➡**If there is no residual gasket material present, metal surface prep can be used to clean and prepare the surfaces.**

17. Clean the cylinder head-to-cylinder block mating surface of both the cylinder head and the cylinder block.

18. Remove any large deposits of silicone or gasket material with a plastic scraper.

19. Apply silicone gasket remover, following package directions, and allow to set for several minutes.

20. Remove the silicone gasket remover with a plastic scraper. A second application of silicone gasket remover may be required if residual traces of silicone or gasket material remain.

21. Apply metal surface prep, following package directions, to remove any traces of oil or coolant, and to prepare the surfaces to bond with the new gasket. Do not attempt to make the metal shiny. Some staining of the metal surfaces is normal.

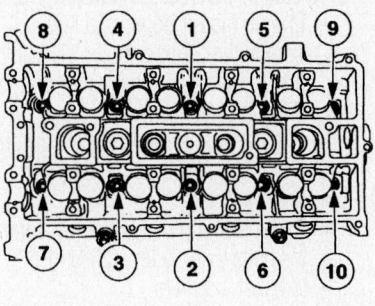

Cylinder head bolt torque sequence—2.3L engine

22. Apply silicone gasket and sealant to the locations shown.

23. Install a new head gasket.

➡**The cylinder head bolts are torque-to-yield and must not be reused. New cylinder head bolts must be installed.**

➡**Lubricate the bolts with clean engine oil prior to installation.**

24. Install new cylinder head bolts. Tighten the bolts in the sequence shown in five stages.

   a. Tighten the bolts to 5 Nm (44 inch lbs.).

   b. Tighten the bolts to 15 Nm (11 ft. lbs.).

   c. Tighten the bolts to 45 Nm (33 ft. lbs.).

   d. Turn the bolts 90 degrees.

   e. Turn the bolts an additional 90 degrees.

25. To install, reverse the removal procedure.

3.0L Engine

The procedure for the left side cylinder head and right side are similar. Changes in the procedure will be noted for either side cylinder head.

1. Before servicing the vehicle, refer to the precautions in the beginning of this section.

2. Properly relieve the fuel system pressure.

   • Drain the cooling system.

3. Remove or disconnect the following:

   • Negative battery cable

   • Camshaft

   • Exhaust Gas Recirculation (EGR) tube, right side only

   • Exhaust manifold

   • Camshaft followers

   • Hydraulic lash adjusters and matchmark them for proper installation

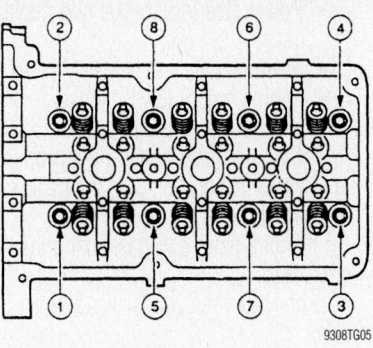

Left side cylinder head bolt torque sequence 3.0L engine

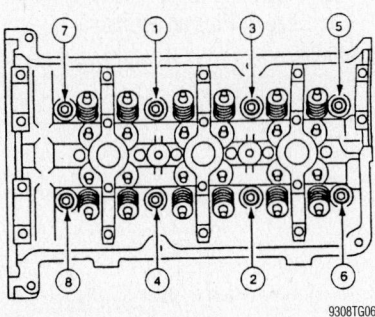

Right side cylinder head bolt torque sequence 3.0L engine

   • Cylinder head bolts in sequence and discard them

   • Cylinder head and discard the gasket

To install:

4. Install a new head gasket and the cylinder head.

5. Lubricate the cylinder head bolt threads.

6. Torque the cylinder head bolts in the proper sequence as follows:

   a. Step 1: 30 ft. lbs. (40 Nm).

   b. Step 2: Additional 90 degrees.

   c. Step 3: Loosen the bolts one full turn.

   d. Step 4: 30 ft. lbs. (40 Nm).

   e. Step 5: Plus an additional 90 degrees.

   f. Step 6: Plus an additional 90 degrees.

7. Install or connect the following:

   • Hydraulic lash adjusters

   • Camshaft followers

   • Camshaft

   • Exhaust manifold. Torque the bolts in sequence to 15 ft. lbs. (20 Nm), right side only

   • EGR tube, right side only

   • Coolant bypass tube

   • Negative battery cable

8. Fill the coolant to the proper level.

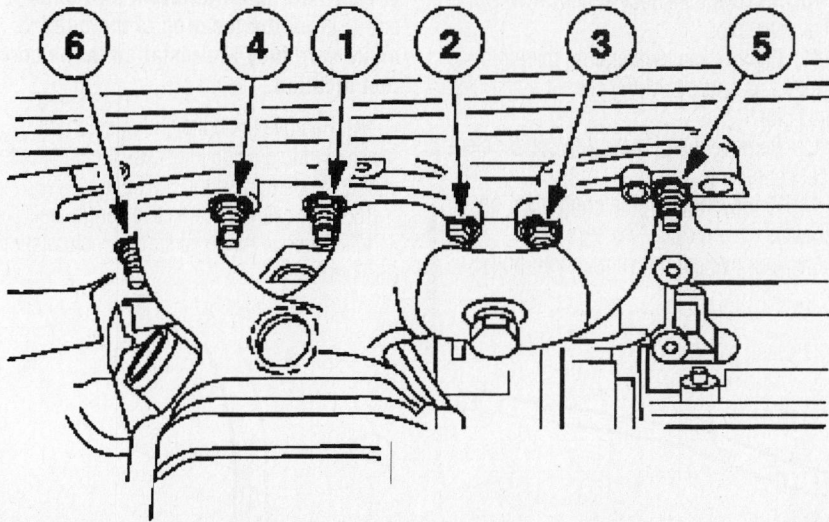

Right side exhaust manifold bolt torque sequence–3.0L engine

9308TG09

9. Start the vehicle and check for leaks, repair if necessary.

Intake Manifold

REMOVAL & INSTALLATION

2.0L Engine

1. Before servicing the vehicle, refer to the precautions in the beginning of this section.

2. Properly relieve the fuel system pressure.

3. Remove or disconnect the following:
   - Negative battery cable
   - Fuel injection supply manifold
   - Throttle Position (TP) sensor electrical connector
   - Idle Air Control (IAC) electrical connector and unclip the harness from the bracket
   - Main engine control sensor wiring
   - Connector from the bracket
   - Powertrain Control Module (PCM) wire harness from the bracket
   - Brake booster vacuum hose
   - 4 additional vacuum lines
   - Positive Crankcase Ventilation (PCV) hose from the intake manifold
   - Knock Sensor (KS) electrical connector
   - Alternator
   - Intake manifold and discard the gasket

4. Clean the mating surfaces.

To install:

5. Install or connect the following:
   - New gasket

- Intake manifold. Torque the bolts, in sequence, to 13 ft. lbs. (18 Nm).
- Alternator
- KS electrical connector
- PCV vacuum line
- 4 vacuum lines
- Brake booster vacuum supply hose
- PCM wire harness to the bracket
- Main engine control sensor wiring
- IAC valve electrical connector and attach the harness to the bracket
- TP sensor electrical connector
- Fuel injection supply manifold
- Negative battery cable

6. Start the vehicle and check for leaks, repair if necessary.

2.3L Engine

1. With the vehicle in NEUTRAL, position it on a hoist.

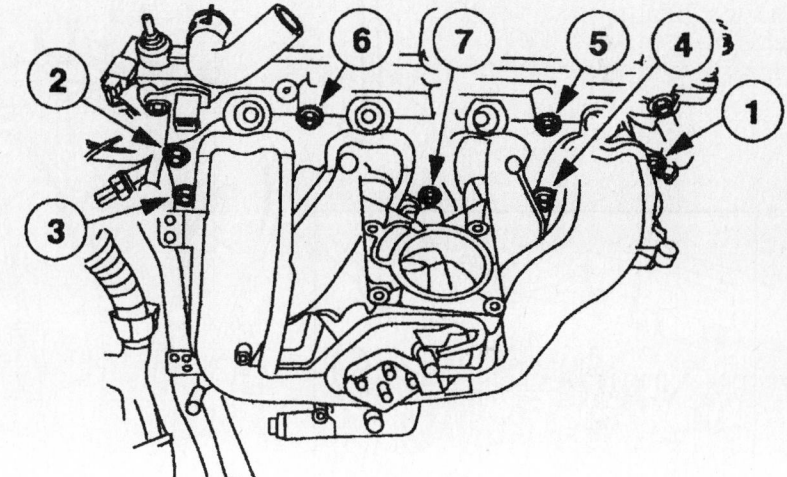

Intake manifold bolt loosening sequence—2.0L engine

67197-ESCA-G64

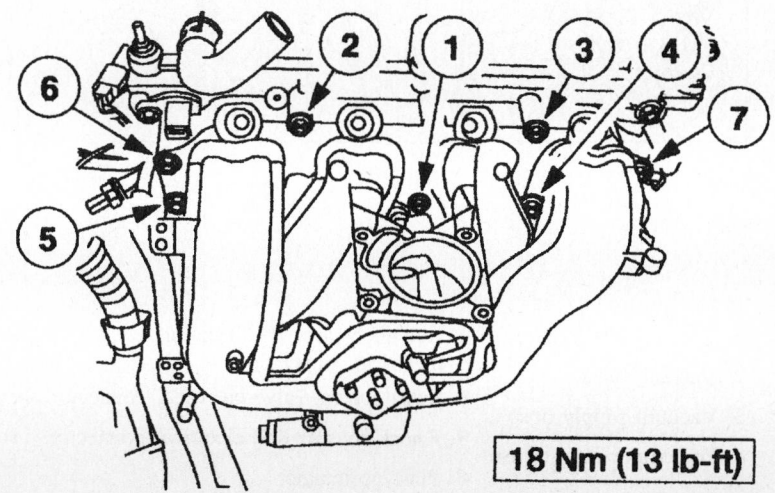

18 Nm (13 lb-ft)

67197-ESCA-G63

Tighten the intake manifold bolts in the sequence shown—2.0L engine

2. Remove the throttle body.

3. Remove the fuel rail.

4. Remove the oil level indicator tube.

5. Remove the vacuum tube.

6. Remove the vacuum supply hose.

7. Remove the fuel vapor return hose.

8. Remove the idle air control (IAC) motor electrical connector.

9. Remove the swirl control valve electrical connector.

10. Remove the knock sensor (KS) electrical connector.

11. Remove the temperature manifold absolute pressure (TMAP) sensor electrical connector.

12. Remove the oil pressure sender electrical connector.

13. Remove the engine control wiring harness.

14. Remove the intake manifold bolts.

➡ **There are three different size bolts used. Mark the location of the bolts to make sure they are installed in the correct location.**

15. Remove the bolts and position the intake manifold aside to access the crankcase vent hose clamp and the EGR tube.

16. Remove the crankcase vent hose.

17. Remove the exhaust gas recirculation (EGR) tube.

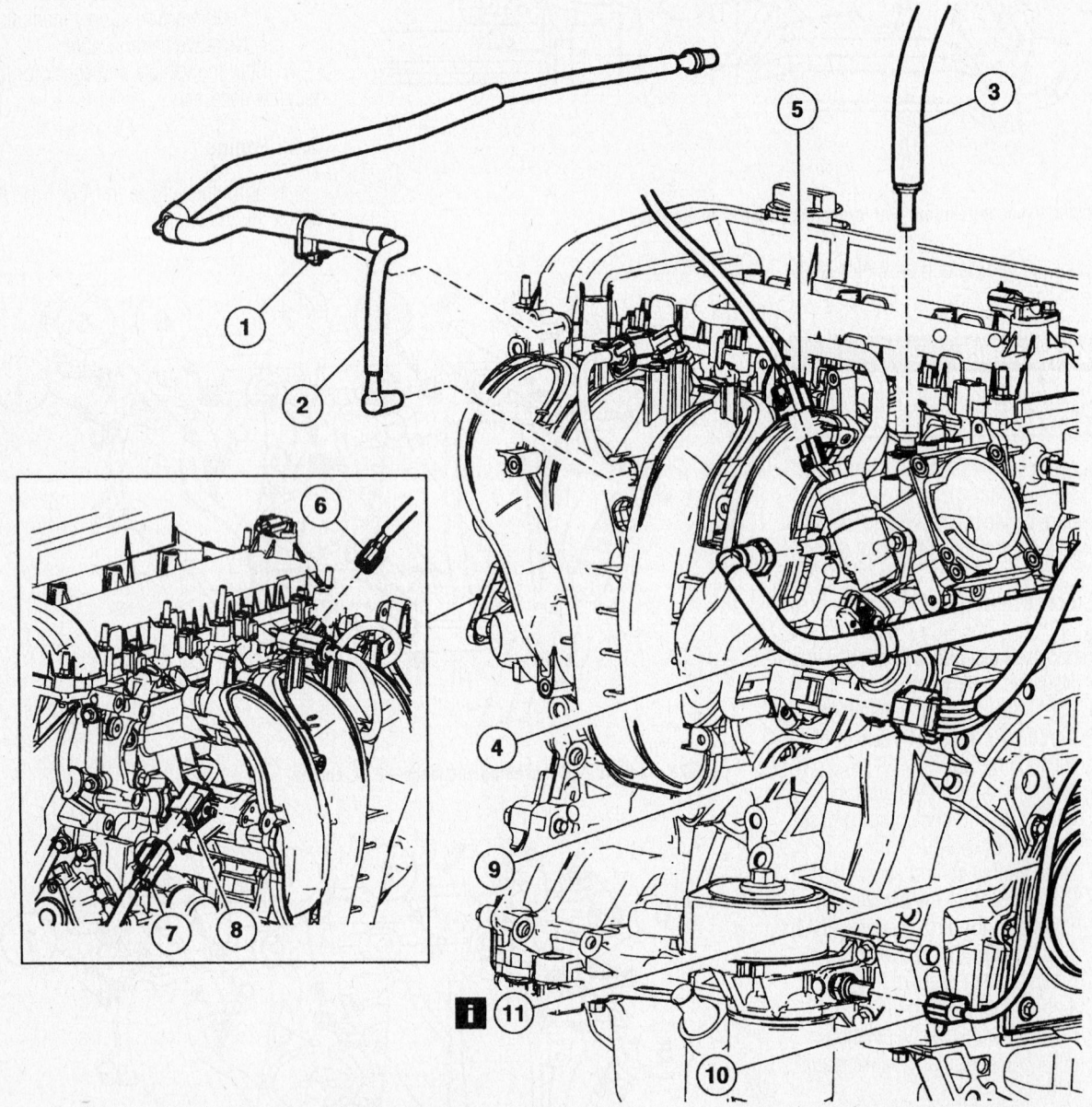

1 Vacuum tube retainer	5 Idle air control (IAC) motor electrical connector	9 Temperature manifold absolute pressure (TMAP) sensor electrical connector
2 Vacuum tube	6 Swirl control valve electrical connector	10 Oil pressure sender electrical connector
3 Vacuum supply hose	7 Knock sensor (KS) electrical connector	11 Engine control wiring harness
4 Fuel vapor return hose	8 Pin-type retainer	

67197-ESCA-G06

Intake manifold and related parts—2.3L engine

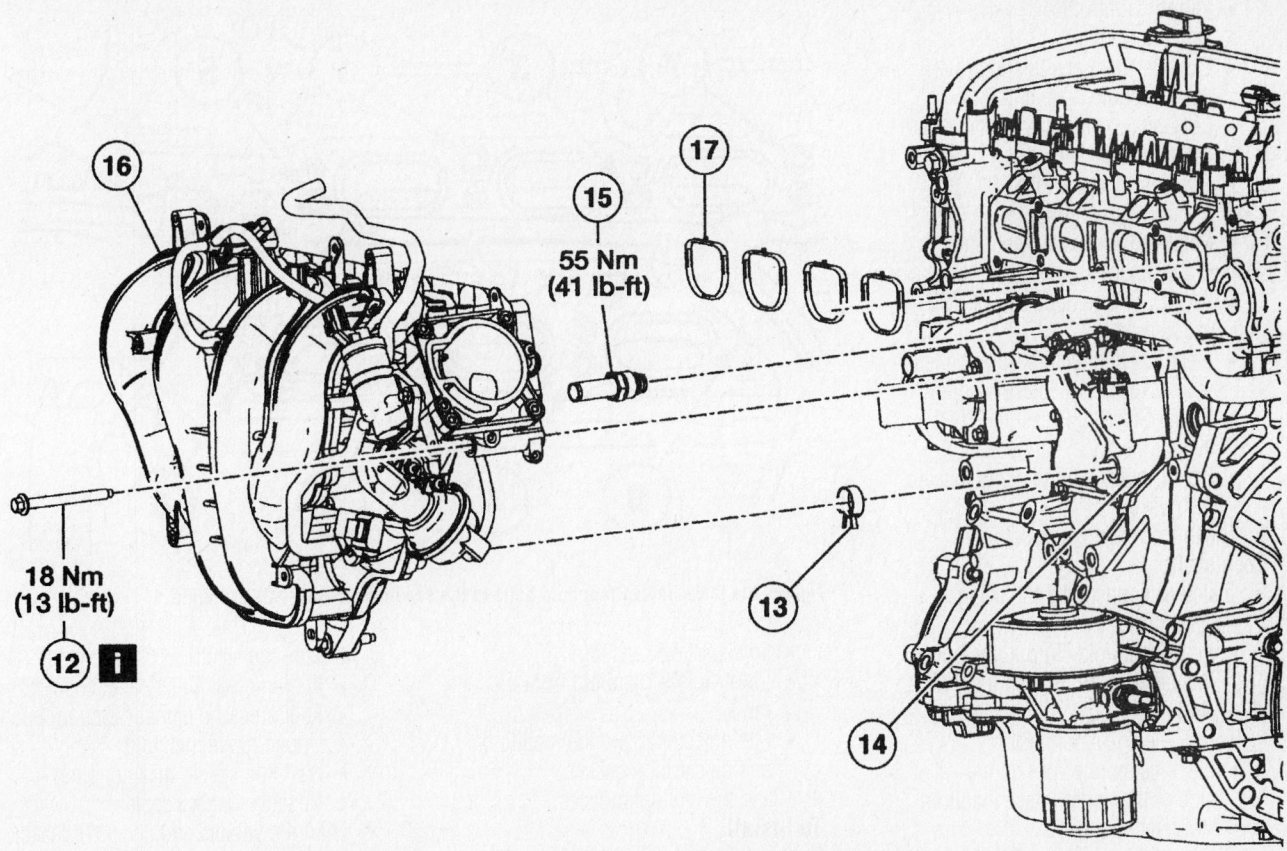

12 Intake manifold bolts

13 Crankcase vent hose clamp

14 Crankcase vent hose (position aside)

15 Exhaust gas recirculation (EGR) tube

16 Intake manifold

17 Intake manifold gasket

67197-ESCA-G07

Intake manifold installation—2.3L engine

18. Remove the intake manifold.
19. Remove the intake manifold gasket.
20. To install, reverse the removal procedure. Torque the intake manifold bolts to 13 ft. lbs. (18 Nm).

3.0L Engine

UPPER

1. Before servicing the vehicle, refer to the precautions in the beginning of this section.
2. Properly relieve the fuel system pressure.
3. Drain the coolant system.
4. Remove or disconnect the following:
   - Negative battery cable
   - Air cleaner outlet tube
   - Engine appearance cover
   - Throttle cable
   - Speed control cable, if equipped
   - Throttle cable bracket
   - Throttle Position (TP) sensor electrical connector

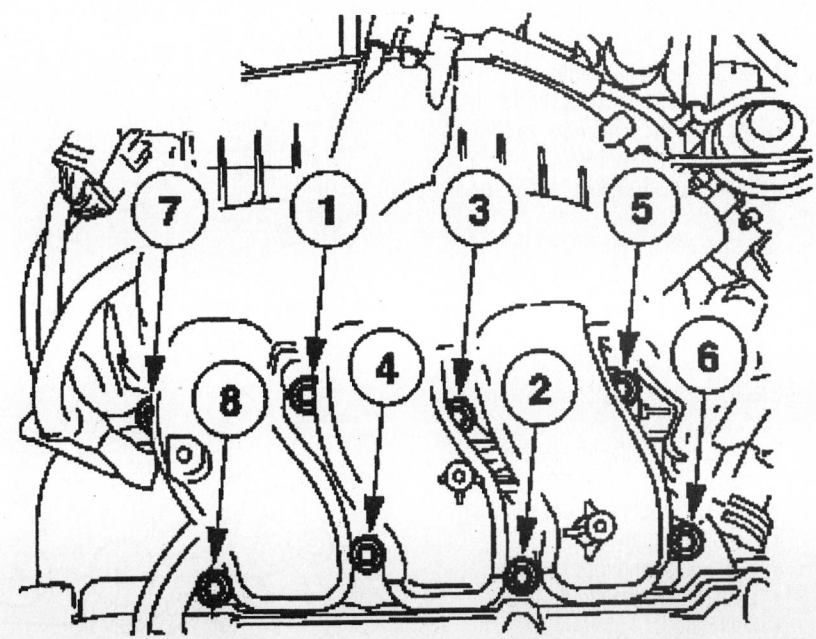

9308TG02

Tighten the upper intake manifold bolts in the sequence shown—3.0L engine

- Idle Air Control (IAC) valve electrical connector
- Exhaust Gas Recirculation (EGR) valve vacuum hose and tube
- EGR vacuum regulator valve electrical connector and hose
- Chassis vacuum hose
- Engine vacuum hose
- Positive Crankcase Ventilation (PCV) hose
- Vapor Management Valve (VMV) vacuum hose
- Electrical connectors from the left side of the upper intake manifold
- Power Steering Pressure (PSP) sensor electrical connector
- Upper intake manifold and discard the gasket

5. Clean the mating surfaces.

To install:

6. Install or connect the following:
- New gasket
- Intake manifold. Torque the bolts, in sequence, to 89 inch lbs. (10 Nm).
- PSP electrical connector
- Electrical connectors on the left side of the upper intake manifold
- VMV vacuum hose
- Chassis, engine and PCV hoses
- EGR valve vacuum regulator
- EGR valve vacuum hose and tube. Torque the nut to 30 ft. lbs. (40 Nm).
- TP sensor electrical connector
- IAC valve electrical connector
- Throttle cable and speed control cable, if equipped. Torque the bracket bolts to 89 inch lbs. (10 Nm).
- Air cleaner outlet tube
- Engine appearance cover. Torque the bolts to 53 inch lbs. (6 Nm).
- Negative battery cable

7. Fill the coolant system to the proper level.

8. Start the vehicle and check for leaks, repair if necessary.

LOWER

1. Before servicing the vehicle, refer to the precautions in the beginning of this section.

2. Properly relieve the fuel system pressure.

3. Remove or disconnect the following:
- Negative battery cable
- Fuel line spring lock coupling
- Upper intake manifold
- Fuel rail
- Fuel injector electrical connectors
- Fuel pressure damper vacuum line

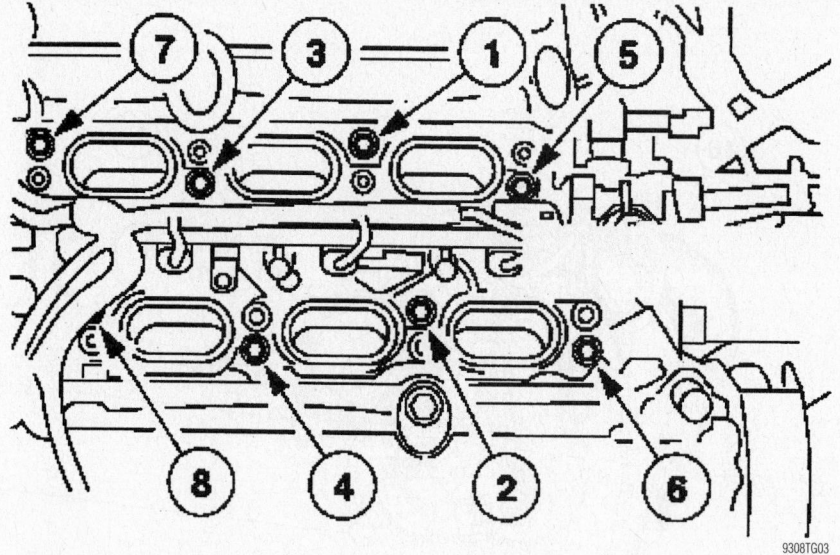

Tighten the lower intake manifold bolts in the sequence shown—3.0L engine

- Lower intake manifold
- Lower intake manifold from the fuel rail
- Fuel injectors from the manifold and discard the gasket

4. Clean the mating surfaces.

To install:

5. Inspect the fuel injector O-rings and replace if necessary.

6. Install or connect the following:
- Fuel injectors into the lower intake manifold
- Fuel rail. Torque the bolts to 89 inch lbs. (10 Nm).
- New gasket
- Intake manifold. Torque the bolts, in sequence, to 89 inch lbs. (10 Nm).

- Fuel rail electrical connectors
- Fuel injector electrical connectors
- Fuel pressure damper vacuum line
- Upper intake manifold
- Fuel line spring lock coupling
- Negative battery cable

7. Start the vehicle and check for leaks, repair if necessary.

Exhaust Manifold

REMOVAL & INSTALLATION

2.0L Engine

1. Before servicing the vehicle, refer to the precautions in the beginning of this section.

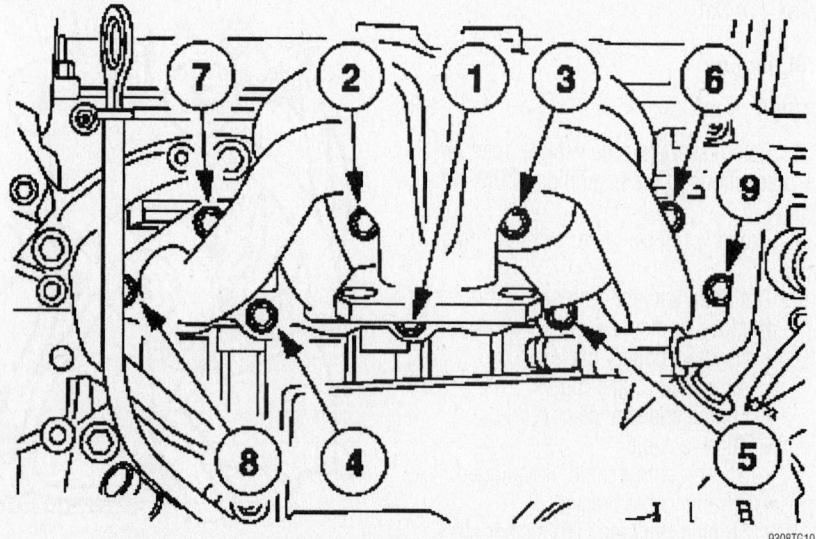

Exhaust manifold bolt torque sequence–2.0L engine

2. Remove or disconnect the following:
- Negative battery cable
- Catalytic converter
- Oil level indicator tube and bracket
- Exhaust manifold and discard the gasket

To install:

3. Clean the sealing surfaces of any old gasket material.

4. Install or connect the following:
- Exhaust manifold and new gasket. Torque the bolts to 12 ft. lbs. (16 Nm).
- Oil level indicator tube and bracket. Torque the bolt to 89 inch lbs. (10 Nm).
- Catalytic converter
- Negative battery cable

5. Start the vehicle and check for leaks, repair if necessary.

2.3L Engine

> ✳✳ **WARNING**
>
> **Do not use oil or grease-based lubricants on the insulators. They may cause deterioration of the rubber.**

> ✳✳ **WARNING**
>
> **Oil or grease-based lubricants on the insulators may cause the exhaust hanger insulator to separate from the exhaust hanger bracket during vehicle operation.**

➡️**Exhaust fasteners are of a torque prevailing design. Use only new fasteners with the same part number as the original. Torque values must be used as specified during reassembly to make sure of correct retention of exhaust components.**

1. Remove the flex pipe nuts.
2. Remove the flex pipe gasket.
3. Remove the manifold bracket bolts.
4. Remove the heat shield.
5. Remove the exhaust manifold nuts.
6. Remove the catalyst monitor sensor.
7. Remove the heated oxygen sensor.
8. Remove the exhaust manifold.
9. To install, reverse the removal procedure. Make sure to apply anti-seize lubricant to the threads of the sensors before installation. Failure to tighten the exhaust manifold nuts to specification before installing the manifold bracket bolts will cause the manifold to develop an exhaust gas leak.
10. Observe the following torques:
- Exhaust manifold-to-head: 35 ft. lbs. (47 Nm)

Left side exhaust manifold bolt torque sequence–3.0L engine

- Flex pipe-to-manifold: 18 ft. lbs. (25 Nm)
- Heated oxygen sensor: 35 ft. lbs. (47 Nm)
- Catalyst monitor sensor: 30 ft. lbs. (40 Nm)

11. Check the exhaust system for proper alignment.

3.0L Engine

LEFT SIDE

1. Before servicing the vehicle, refer to the precautions in the beginning of this section.
2. Remove or disconnect the following:
- Negative battery cable
- Heated Oxygen (HO$_2$S) sensor and catalyst monitor
- Splash shield
- Exhaust crossover pipe
- Drive belt
- A/C compressor and move it aside
- Exhaust manifold and discard the gasket

To install:

3. Clean the sealing surfaces of any old gasket material.

4. Install or connect the following:
- Exhaust manifold and new gasket. Torque the bolts to 15 ft. lbs. (20 Nm).
- A/C compressor. Torque the bolts to 18 ft. lbs. (20 Nm).
- Drive belt
- Exhaust crossover pipe. Torque the bolts to 30 ft. lbs. (40 Nm).
- Splash shield. Torque the bolts to 80 inch lbs. (9 Nm).
- Left side HO$_2$S sensor and catalyst monitor
- Negative battery cable

5. Start the vehicle and check for leaks, repair if necessary.

RIGHT SIDE

1. Before servicing the vehicle, refer to the precautions in the beginning of this section.

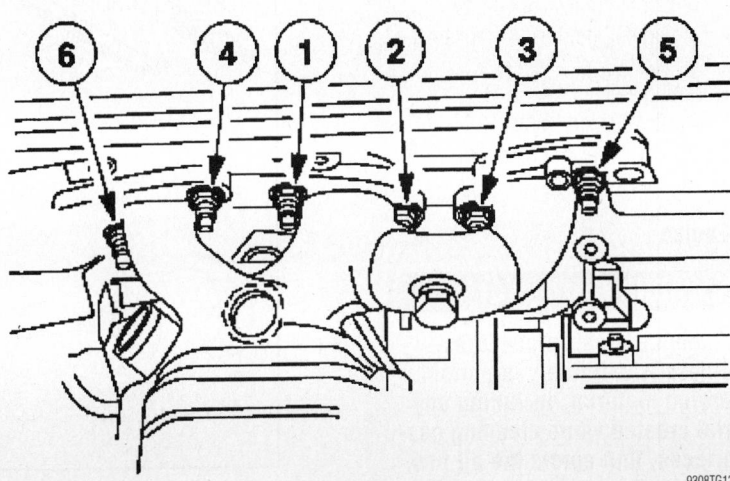

Right side exhaust manifold bolt torque sequence–3.0L engine

2. Remove or disconnect the following:
- Negative battery cable
- Exhaust Gas Recirculation (EGR) tube
- Alternator
- Right side Heated Oxygen (HO2S) sensor
- Right side exhaust manifold and discard the gasket

To install:

3. Clean the sealing surfaces of any old gasket material.

4. Install or connect the following:
- Exhaust manifold and new gasket. Torque the bolts to 15 ft. lbs. (20 Nm).
- Right side HO2S sensor
- Alternator
- EGR tube
- Negative battery cable

5. Start the vehicle and check for leaks, repair if necessary.

Front Crankshaft Seal

REMOVAL & INSTALLATION

2.0L Engine

1. Before servicing the vehicle, refer to the precautions in the beginning of this section.

2. Remove or disconnect the following:
- Negative battery cable
- Timing belt
- Crankshaft sprocket and timing belt guide
- Crankshaft oil seal

➡ **Be careful not to damage the seal surface of the cover.**

To install:

3. Install or connect the following:
- New front crankshaft oil seal
- Timing belt guide and crankshaft sprocket
- Timing belt
- Negative battery cable

4. Start the vehicle and check for leaks, repair if necessary.

2.3L Engine

✸✸ WARNING

During engine repair procedures, cleanliness is extremely important. Any foreign material, including any material created while cleaning gasket surfaces, that enters the oil passages, coolant passages or the oil pan can cause engine failure.

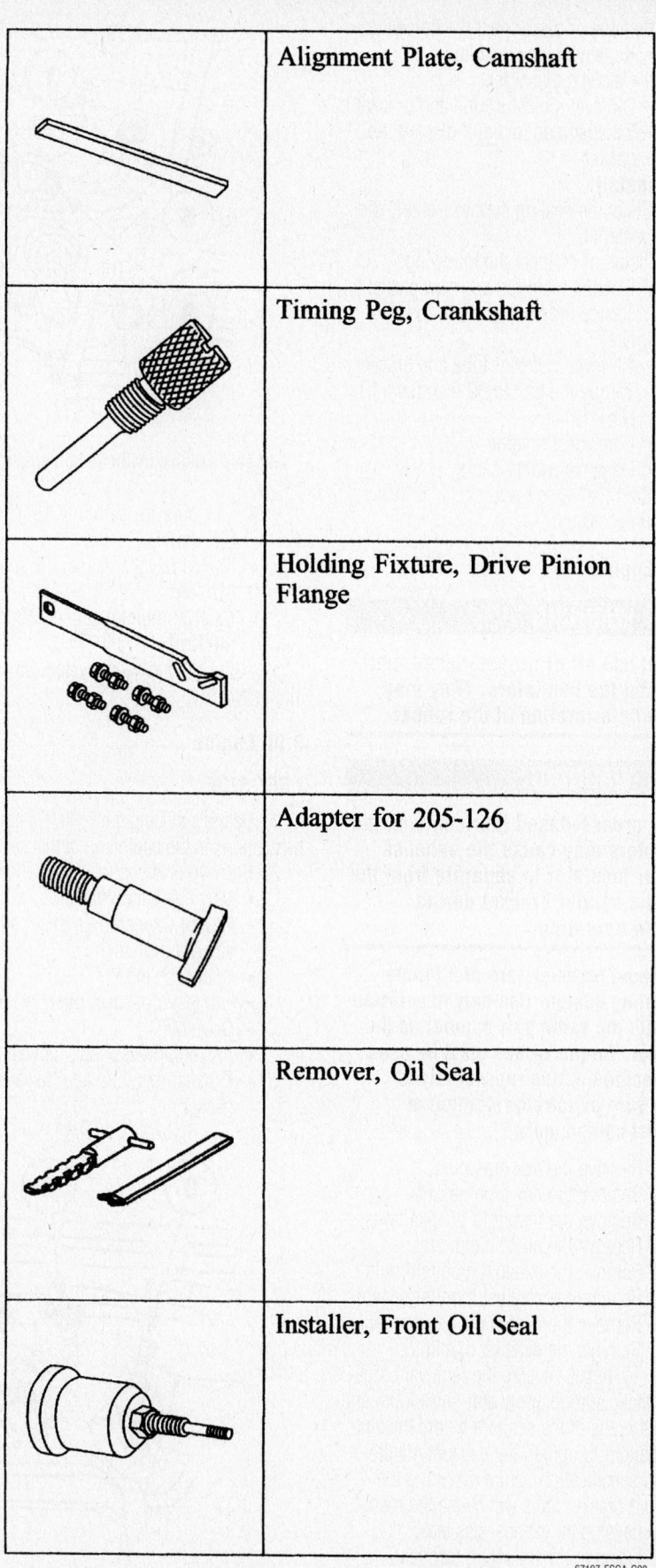

	Alignment Plate, Camshaft
	Timing Peg, Crankshaft
	Holding Fixture, Drive Pinion Flange
	Adapter for 205-126
	Remover, Oil Seal
	Installer, Front Oil Seal

67197-ESCA-G08

Tools necessary for this job—2.3L engine

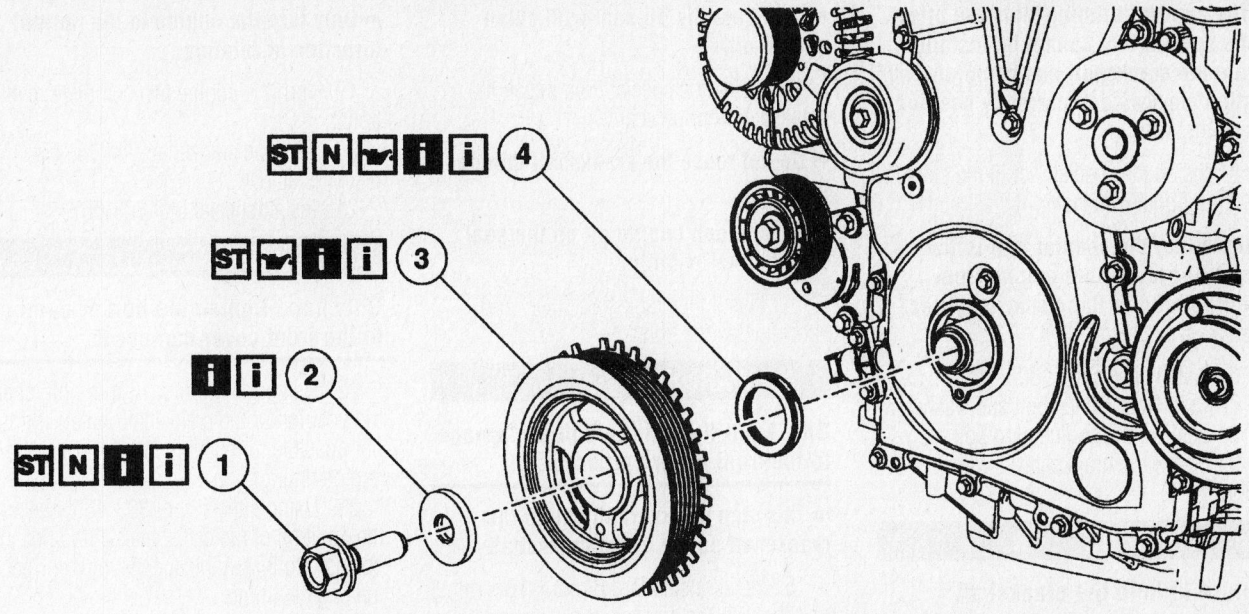

1 Crankshaft pulley bolt
2 Crankshaft pulley washer
3 Crankshaft pulley
4 Crankshaft front seal

67197-ESCA-G09

Crankshaft pulley and seal—2.3L engine

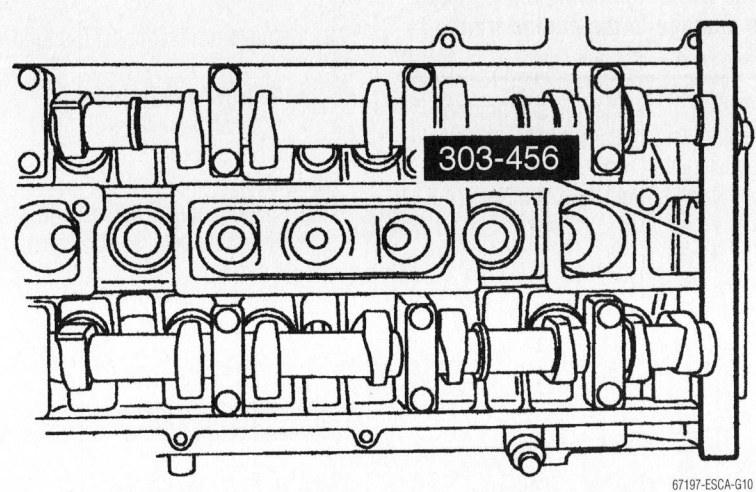

303-456

67197-ESCA-G10

Camshaft holding tool—2.3L engine

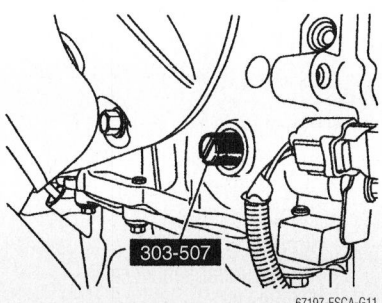

303-507

67197-ESCA-G11

Install special tool 303-507—2.3L engine

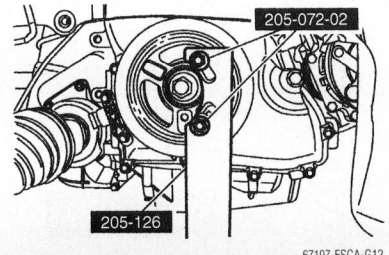

205-072-02

205-126

67197-ESCA-G12

Install the crankshaft holding tools —2.3L engine

✳✳ WARNING

The crankshaft, the crankshaft sprocket and the pulley are fitted together by friction, using diamond washers between the flange faces on each part. For that reason, the crankshaft sprocket is also unfastened if you loosen the pulley. Therefore, the engine must be retimed each time the damper is removed. Otherwise severe engine damage can occur.

1. With the vehicle in NEUTRAL, position it on a hoist.
2. Remove the accessory drive belt.
3. Remove the valve cover.

✳✳ WARNING

Failure to position the No. 1 piston at top dead center (TDC) can result in damage to the engine. Turn the engine in the normal direction of rotation only.

4. Using the crankshaft pulley bolt, turn the crankshaft clockwise to position the No. 1 piston at top dead center (TDC).

✳✳ WARNING

The special tool 303-465 is for camshaft alignment only. Using this tool to prevent engine rotation can result in engine damage.

➡The camshaft timing slots are offset. If the special tool cannot be installed, rotate the crankshaft one complete revolution clockwise to correctly position the camshafts.

5. Install the special tool in the slots on the rear of both camshafts.

➡Installing the special tool in this step will prevent the engine from being rotated in the clockwise direction.

6. Install special tool 303-507.
7. Crankshaft pulley bolt and washer.
8. Remove the engine plug bolt.
9. Install the crankshaft holding tools.

✳✳ WARNING

Failure to hold the crankshaft pulley in place while loosening the bolt can result in damage to the engine.

10. Remove the crankshaft pulley. Remove the crankshaft pulley bolt and washer.
11. Remove the crankshaft pulley.

✳✳ WARNING

Use care not to damage the engine front cover or the crankshaft when removing the seal.

12. Using the special tool, remove the crankshaft front oil seal.

➡Remove the through-bolt from the special tool.

➡Lubricate the oil seal with clean engine oil.

13. Using the special tool, install the crankshaft front oil seal.

➡Do not reuse the crankshaft damper bolt.

➡Apply clean engine oil on the seal area before installing.

14. Install the crankshaft pulley and hand-tighten the bolt.

✳✳ WARNING

Only hand-tighten the bolt or damage to the front cover can occur.

➡This step will correctly align the crankshaft pulley to the crankshaft.

15. Install a standard 6 mm x 18 mm bolt through the crankshaft pulley and thread it into the front cover. Rotate the pulley as necessary to align the bolt holes.

✳✳ WARNING

Failure to hold the crankshaft pulley in place while tightening the bolt can cause damage to the engine front cover.

16. Using the special tools to hold the crankshaft pulley in place, tighten the crankshaft pulley bolt in two stages:
   a. Stage 1: Tighten to 100 Nm (74 ft. lbs.).
   b. Stage 2: Tighten an additional 90 degrees (¼ turn).
17. Remove the 6 mm x 18 mm bolt.
18. Remove the special tools.

➡Only turn the engine in the normal direction of rotation.

19. Turn the engine two complete revolutions.
20. Turn the crankshaft until the No. 1 piston is at TDC.
21. Install special tool 303-507.

✳✳ WARNING

Only hand-tighten the bolt or damage to the front cover can occur.

22. Using the 6 mm x 18 mm bolt, check the position of the crankshaft pulley. If it is not possible to install the bolt, correct the engine timing.
23. Using special tool 303-465, check the position of the camshafts. If it is not possible to install the special tool, correct the engine timing.
24. Remove the 6 mm x 18 mm bolt.
25. Install the engine plug bolt.

3.0L Engine

1. Before servicing the vehicle, refer to the precautions in the beginning of this section.
2. Remove or disconnect the following:

   • Negative battery cable
   • Crankshaft pulley
   • Front oil seal

To install:
3. Install or connect the following:
   • New front crankshaft oil seal
   • Crankshaft pulley
   • Negative battery cable
4. Start the vehicle and check for leaks, repair if necessary.

Camshaft and Lifters

REMOVAL & INSTALLATION

2.0L Engine

1. Before servicing the vehicle, refer to the precautions in the beginning of this section.
2. Remove or disconnect the following:

   • Negative battery cable
   • Camshaft timing sprocket and verify the valve clearance
   • Camshaft journal cap bolts by loosening them in several passes in the proper sequence
   • Camshafts
3. Inspect the camshaft for wear and discard the oil seals

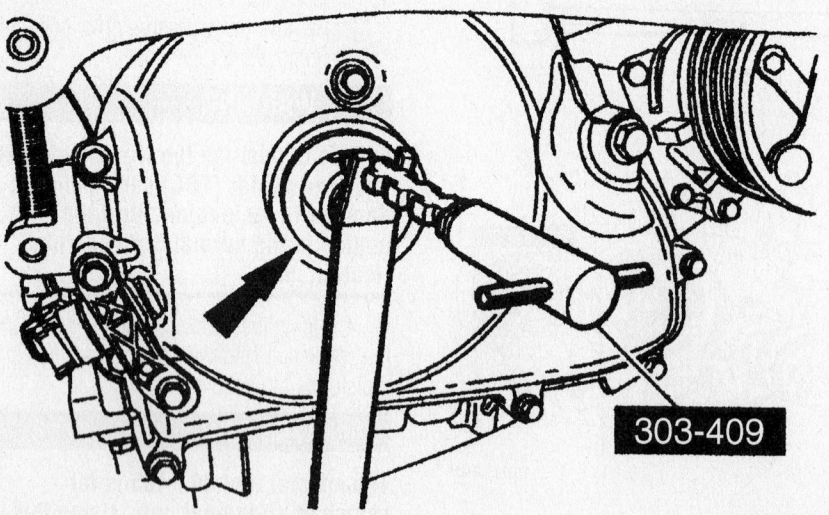

303-409

67197-ESCA-G13

Using the special tool, install the crankshaft front oil seal —2.3L engine

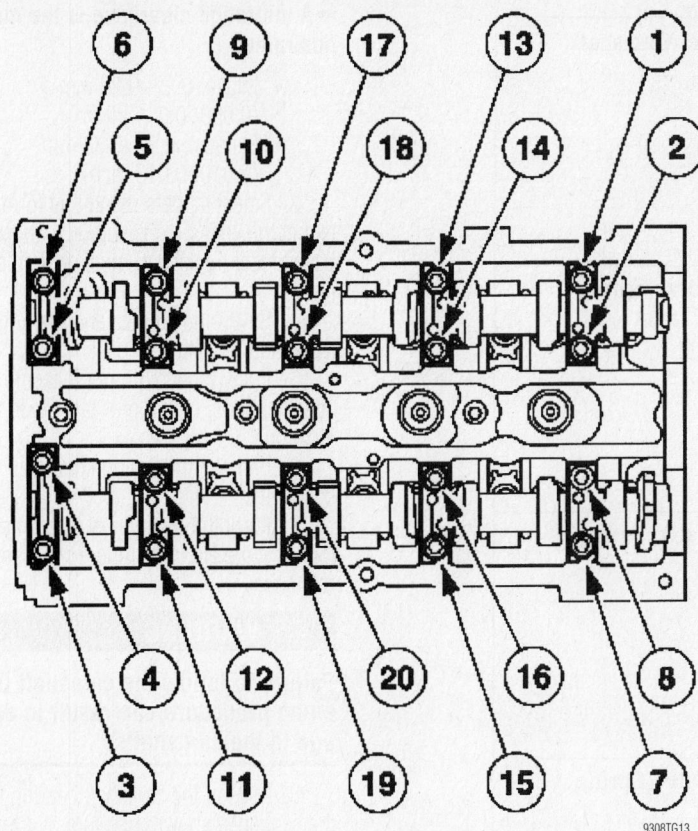

Remove the camshaft bearing caps in sequence–2.0L engine

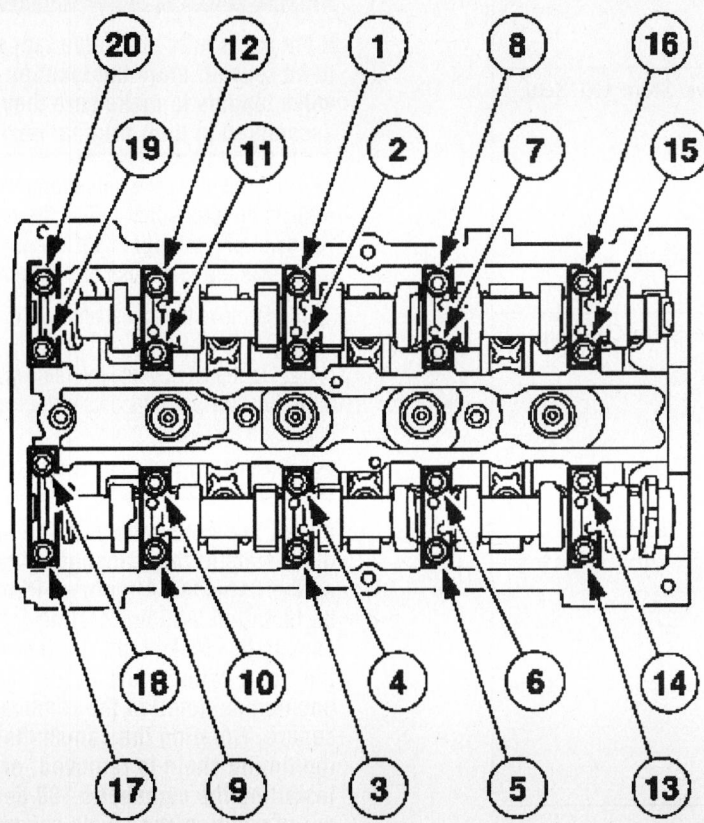

Camshaft bearing cap tightening sequence–2.0L engine

To install:
4. Install or connect the following:
- Camshaft cam followers, lubricate the bearing journals thoroughly. Torque the caps to 14 ft. lbs. (19 Nm).
- Exhaust camshaft oil seal
- Camshaft timing sprocket
- Negative battery cable

2.3L Engine

⁂ **WARNING**

During engine repair procedures, cleanliness is extremely important. Any foreign material, including any material created while cleaning gasket surfaces, that enters the oil passages, coolant passages or the oil pan can cause engine failure.

⁂ **WARNING**

The crankshaft, the crankshaft sprocket and the pulley are fitted together by friction, using diamond washers between the flange faces on each part. For that reason, the crankshaft sprocket is also unfastened if you loosen the pulley. Therefore, the engine must be retimed each time the damper is removed. Otherwise severe engine damage can occur.

1. With the vehicle in NEUTRAL, position it on a hoist.

➡Valve tappets are select fit and the valve clearance must be checked before removing the tappets.

⁂ **WARNING**

Turn the engine clockwise only, and only use the crankshaft bolt.

➡Before removing the camshafts, measure the clearance of each valve at base circle, with the lobe pointed away from the tappet. Failure to measure all clearances prior to removing the camshafts will necessitate repeated removal and installation and wasted labor time.

2. Use a feeler gauge to measure the clearance of each valve and record its location.

➡The number on the valve tappet only reflects the digits that follow the decimal. For example, a tappet with the number 0.650 has the thickness of 3.650 mm.

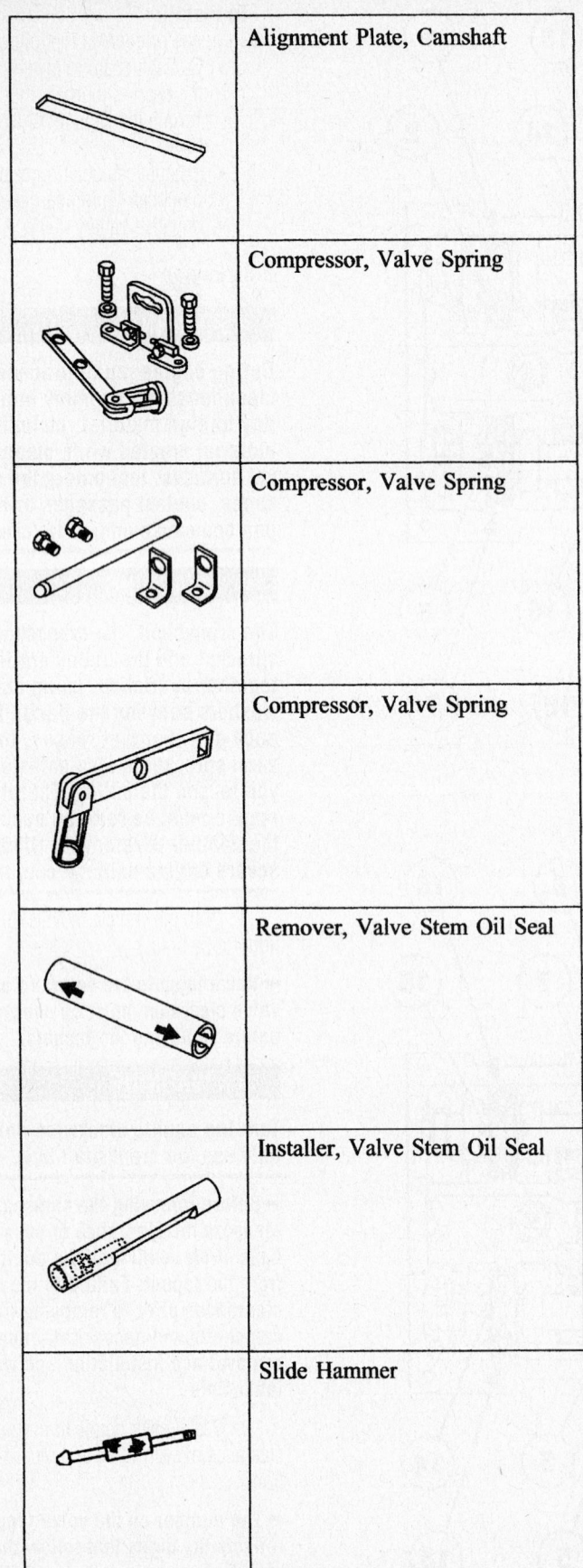	Alignment Plate, Camshaft
	Compressor, Valve Spring
	Compressor, Valve Spring
	Compressor, Valve Spring
	Remover, Valve Stem Oil Seal
	Installer, Valve Stem Oil Seal
	Slide Hammer

67197-ESCA-G15

Tools necessary for camshaft and lifter removal—2.3L engine

➡**A midrange clearance is the most desirable:**

- Intake: 0.22–0.28 mm (0.008–0.011 inch)
- Exhaust: 0.27–0.33 mm (0.010–0.013 inch)

3. Select tappets using this formula: tappet thickness = measured clearance + the base tappet thickness–most desirable thickness.

4. Select the tappets and mark the installation location.

5. If any tappets do not measure within specifications, install new tappets in these locations.

6. Remove the timing chain and sprockets.

7. Mark the position of the camshaft lobes on the No. 1 cylinder for assembly reference.

✳✳ WARNING

Failure to follow the camshaft loosening procedure can result in damage to the camshafts.

8. Loosen the camshaft bearing bolts in the sequence shown, one turn at a time. Repeat until all the tension is released.

9. Remove the camshaft bearing caps.

✳✳ WARNING

If the camshafts and valve tappets are to be reused, mark the location of the valve tappets to make sure they are assembled in their original positions.

➡**The number on the valve tappets only reflects the digits that follow the decimal. For example, a tappet with the number 0.650 has the thickness of 3.650 mm.**

10. Remove the camshafts.

11. Valve tappets.

12. To install, reverse the removal procedure. Coat the valve tappets with clean engine oil and insert them.

✳✳ WARNING

Install the camshafts with the alignment slots in the camshafts lined up so the Camshaft Alignment Plate can be installed without rotating the camshafts. Make sure the lobes on the No. 1 cylinder are in the same position as noted in the removal procedure. Rotating the camshafts when the timing chain is removed, or installing the camshafts 180 degrees out of position can cause severe damage to the valves and pistons.

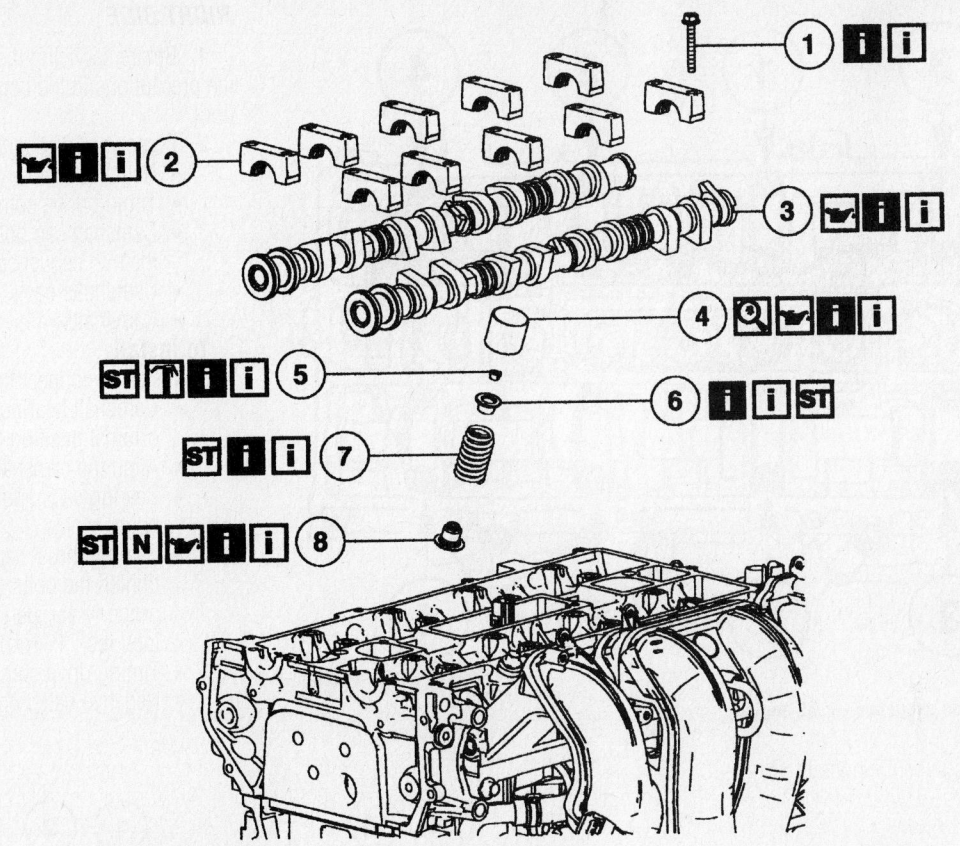

1 Camshaft bearing cap bolt
2 Camshaft bearing cap
3 Camshaft
4 Valve tappet
5 Valve collet
6 Valve spring retainer
7 Valve spring
8 Valve seal

67197-ESCA-G16

Camshafts, lifters and related parts—2.3L engine

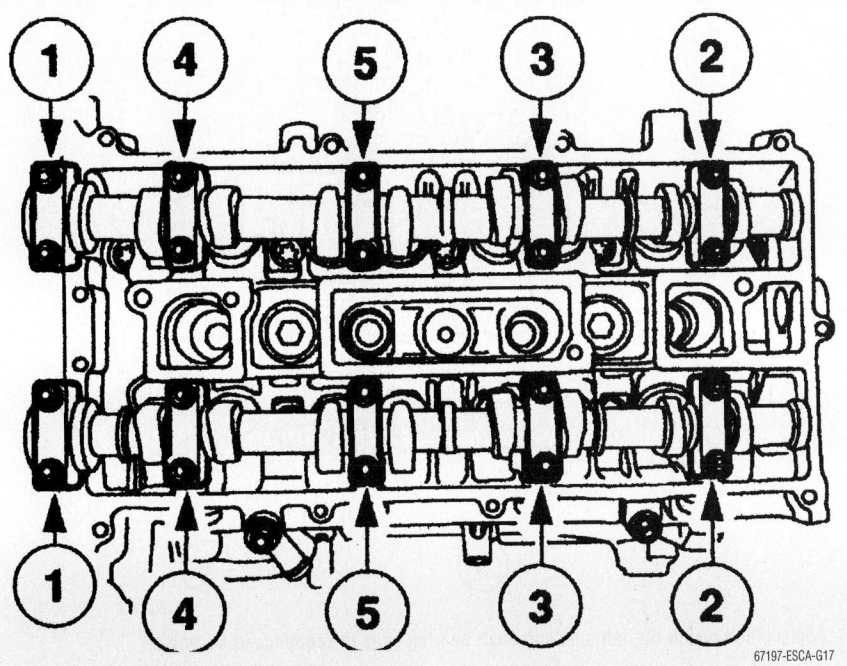

67197-ESCA-G17

Camshaft cap removal sequence—2.3L engine

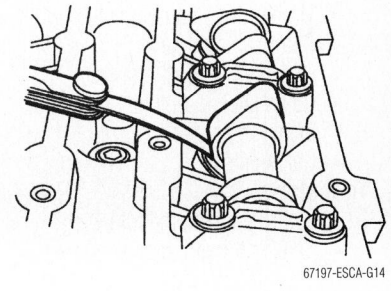

67197-ESCA-G14

Valve clearance check—2.3L engine

➡**Lubricate the camshaft journals and bearing caps with clean engine oil.**

13. Install the camshafts and bearing caps. Tighten the bolts in the sequence shown in three stages.

a. Stage 1: Tighten the camshaft bearing bolt caps one turn at a time until tight.

b. Stage 2: Tighten the bolts to 7 Nm (62 inch lbs.).

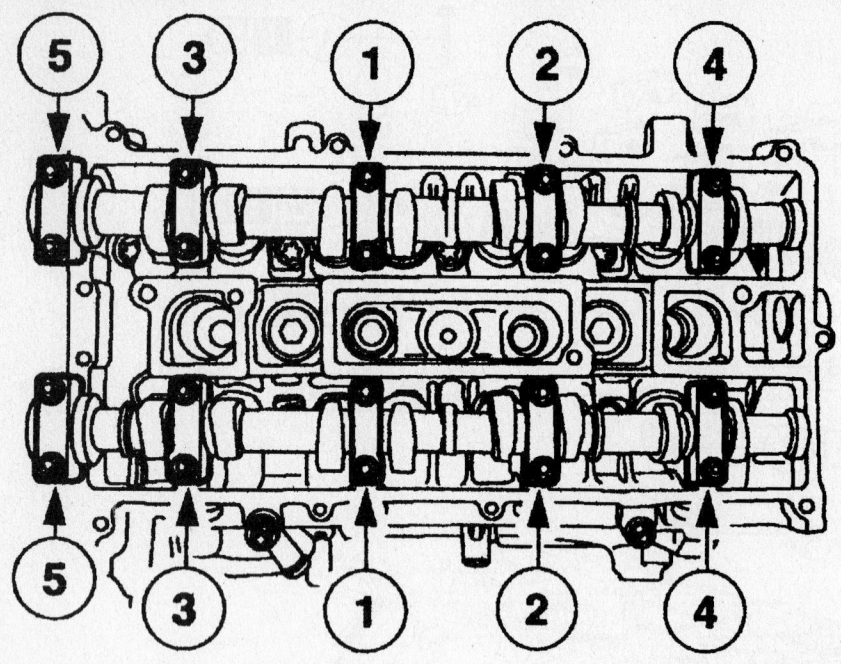

Camshaft cap torque sequence—2.3L engine

67197-ESCA-G18

RIGHT SIDE

1. Before servicing the vehicle, refer to the precautions in the beginning of this section.
2. Remove or disconnect the following:
   - Negative battery cable
   - Timing drive components
   - Camshaft cap bolts by loosening them in sequence
   - Camshafts caps
   - Camshafts

To install:

3. Install or connect the following:
   - Camshaft bearing caps in their original position
   - Align the camshafts
   - Bearing caps and hand tighten the bolts
   - Bearing thrust caps and hand tighten the bolts. When aligned properly, torque the bolts to 89 inch lbs. (10 Nm).
   - Timing drive components
   - Negative battery cable

c. Stage 3: Tighten the bolts to 16 Nm (12 ft. lbs.).

3.0L Engine

LEFT SIDE

1. Before servicing the vehicle, refer to the precautions in the beginning of this section.
2. Remove or disconnect the following:
   - Negative battery cable
   - Water pump belt
   - Timing drive components
   - Camshaft oil seal
   - Camshaft oil seal retainer
   - Camshaft cap bolts by loosening them in sequence
   - Camshafts

To install:

3. Install or connect the following:
   - Camshaft bearing caps in their original position
   - Align the camshafts
   - Bearing thrust caps and hand tighten the bolts. When aligned properly, torque the bolts to 89 inch lbs. (10 Nm).
   - Timing drive components
   - Camshaft oil seal retainer
   - Crankshaft oil seal
   - Water pump drive pulley
   - Water pump belt
   - Negative battery cable

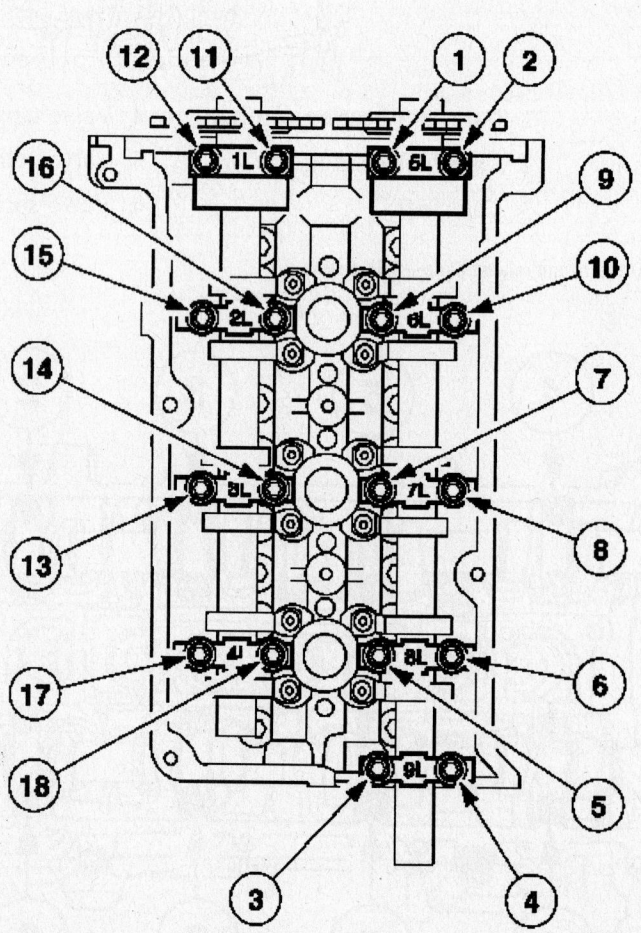

9308TG15

Remove and install the left side camshaft bearing caps in sequence—3.0L engine

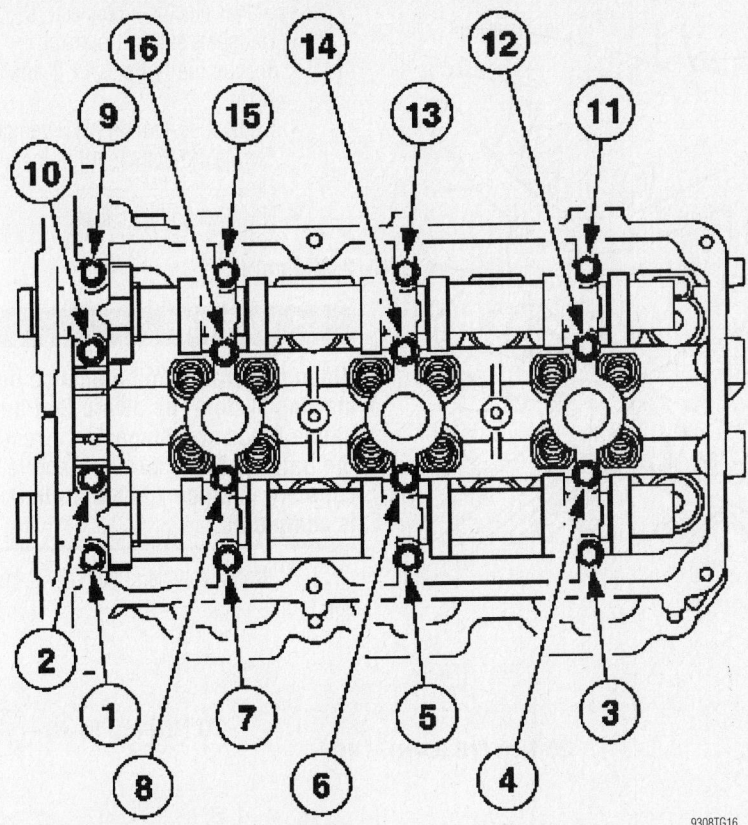

Remove and install the right side camshaft bearing caps in sequence–3.0L engine

9308TG16

Valve Lash

ADJUSTMENT

2.0L Engine

1. Before servicing the vehicle, refer to the precautions in the beginning of this section.
2. Remove or disconnect the following:
   - Negative battery cable
   - Timing belt
3. Measure each valve's clearance at the base circle with the lobe facing away from the tappet.
4. Use a feeler gauge to measure and record each valve's clearance
5. Remove or disconnect the following:
   - Camshafts
   - Valve tappets from the cylinder head
6. A mid range clearance is recommended as follows:
   a. Intake: 0.006 inch (0.15mm).
   b. Exhaust: 0.012 inch (0.3mm).
   To install:
7. Install or connect the following:

   - Valve tappets after lubricating them with clean engine oil
   - Camshafts and verify each valve's clearance at the base circle with the lobe facing away from the tappet
   - Timing belt
   - Negative battery cable

2.3L Engine

➡**Before removing the camshafts, measure the clearance of each valve at base circle, with the lobe pointed away from the tappet. Failure to measure all clearances prior to removing the camshafts will necessitate repeated removal and installation and wasted labor time.**

1. Use a feeler gauge to measure the clearance of each valve and record its location.

➡**The number on the valve tappet only reflects the digits that follow the decimal. For example, a tappet with the number 0.650 has the thickness of 3.650 mm.**

➡**A midrange clearance is the most desirable:**

   - Intake: 0.22–0.28 mm (0.008–0.011 inch)
   - Exhaust: 0.27–0.33 mm (0.010–0.013 inch)
2. Select tappets using this formula: tappet thickness = measured clearance + the base tappet thickness–most desirable thickness.
3. Select the tappets and mark the installation location.
4. If any tappets do not measure within specifications, install new tappets in these locations.

3.0L Engine

1. Before servicing the vehicle, refer to the precautions in the beginning of this section.
2. Remove or disconnect the following:

   - Negative battery cable
   - Camshaft followers
   - Hydraulic lash adjusters

➡**Mark the position of the hydraulic lash adjusters to assure they are assembled in their original position**

3. Inspect the adjusters for scoring or uneven wear in the bore and replace them as required.
   To install:
4. Install or connect the following:
   - Hydraulic lash adjusters after lubricating them with clean engine oil
   - Camshaft followers
   - Negative battery cable

Starter Motor

REMOVAL & INSTALLATION

2.0L Engine

1. Before servicing the vehicle, refer to the precautions in the beginning of this section.
2. Remove or disconnect the following:
   - Negative battery cable
   - Starter bolts
   - Exhaust system, AWD vehicles only
   - Halfshaft support bracket bolts
   - Starter electrical connectors
   - Starter
   To install:
3. Install or connect the following:
   - Starter. Torque bolts to 20 ft. lbs. (27 Nm).

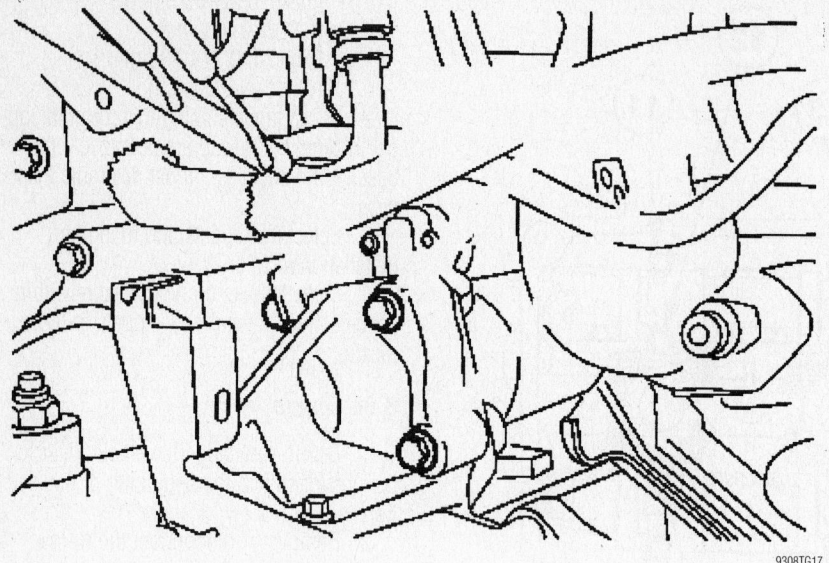

Removal of the starter motor–2.0L engine

9308TG17

- Starter electrical connectors
- Halfshaft support bracket. Torque the bolts to 11 ft. lbs. (15 Nm).
- Exhaust system on AWD vehicles. Torque the bolts to 18 ft. lbs. (25 Nm).
- Negative battery cable

2,3L Engine

❋❋ WARNING

When performing maintenance on the starting system, be aware that heavy gauge leads are connected directly to the battery. Make sure protective caps are in place when maintenance is completed.

1. With the vehicle in NEUTRAL, position it on a hoist

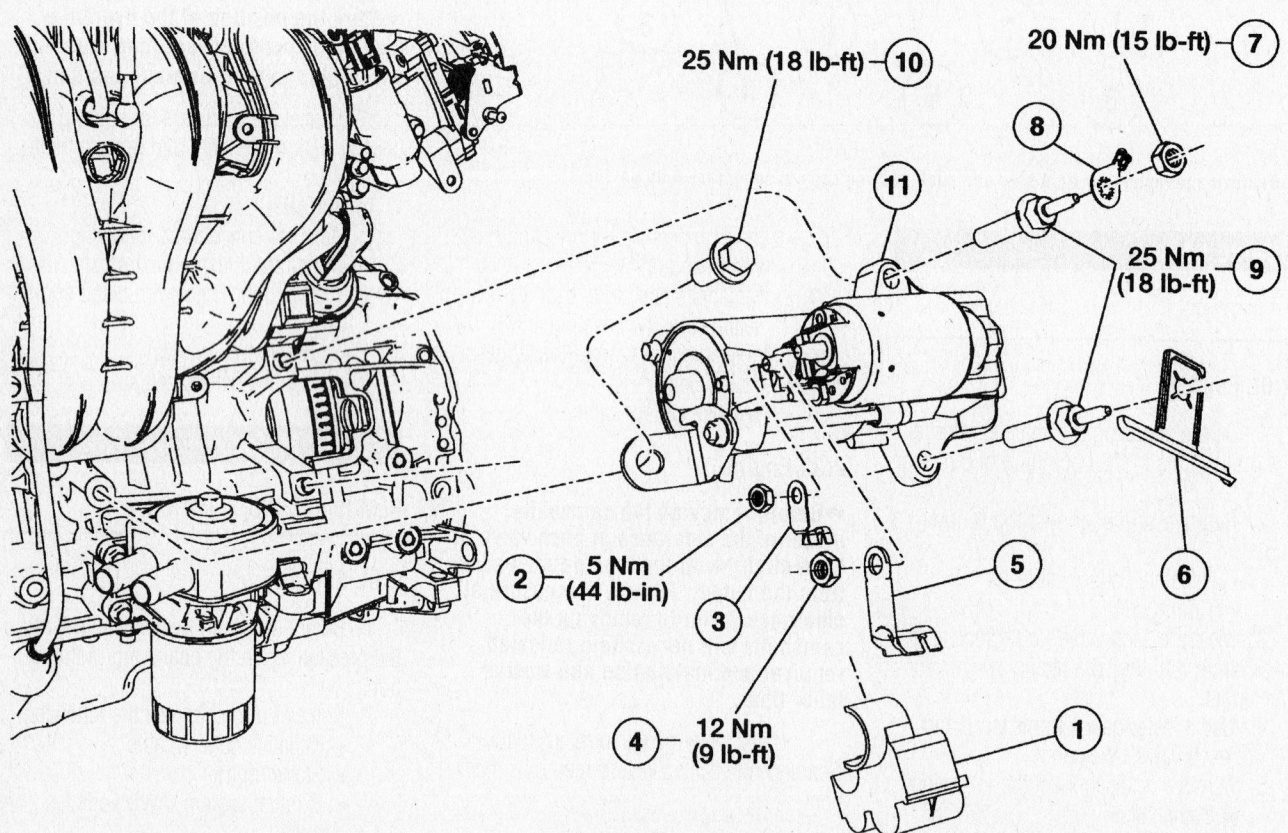

1 Starter motor solenoid terminal cover	5 Starter solenoid battery cable	9 Starter motor stud bolts
2 Starter solenoid wire nut	6 Wiring harness retainer	10 Starter motor bracket bolt
3 Starter solenoid wire	7 Ground strap nut	11 Starter motor
4 Starter solenoid battery cable nut	8 Ground strap	

Starter mounting—2.3L engine

67197-ESCA-G19

2. Disconnect the battery ground cable.

3. Starter motor solenoid terminal cover

4. Starter solenoid wire

5. Starter solenoid battery cable

6. Wiring harness retainer

7. Ground strap

8. Starter motor stud bolts

9. Starter motor bracket bolt

10. Starter motor

11. To install, reverse the removal procedure. Torque the starter and bracket bolts to 18 ft. lbs. (25 Nm).

3.0L Engine

1. Before servicing the vehicle, refer to the precautions in the beginning of this section.

2. Drain the cooling system.

3. Remove or disconnect the following:

- Negative battery cable
- Air cleaner outlet tube
- Coolant hoses and move the thermostat aside
- Starter electrical connectors
- Starter

To install:

4. Install or connect the following:

- Starter. Torque bolts to 20 ft. lbs. (27 Nm).
- Starter electrical connectors and reposition the thermostat
- Connect the 4 coolant hoses
- Air cleaner outlet tube
- Negative battery cable

5. Fill the cooling system to the proper level.

6. Start the vehicle and check for leaks, repair if necessary.

Oil Pan

REMOVAL & INSTALLATION

2.0L Engine

1. Before servicing the vehicle, refer to the precautions in the beginning of this section.

2. Drain the engine oil.

3. Support the powertrain assembly.

4. Remove or disconnect the following:

- Negative battery cable
- Catalytic converter
- Oil pan and gasket

5. Thoroughly clean the gasket mating surfaces.

To install:

6. Apply silicone sealer to the oil pan.

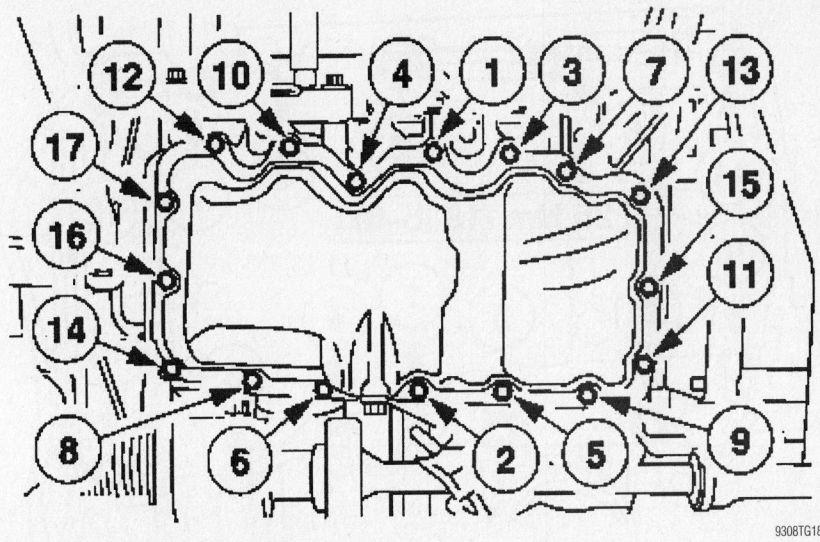

Tighten the oil pan bolts in sequence–2.0L engine

7. Install a new gasket on the oil pan.

8. Oil pan. Torque the bolts in sequence to:

a. Step 1: 53 inch lbs. (6 Nm).

b. Step 2: 106 in lbs. (12 Nm).

9. Install or connect the following:

- Catalytic converter
- Negative battery cable

10. Fill the engine with clean oil.

11. Start the engine and check for leaks, repair if necessary.

2.3L Engine

1. With the vehicle in NEUTRAL, position it on a hoist.

2. Remove the oil level indicator and tube.

3. Drain the oil.

4. Remove the 4 front cover-to-oil pan bolts.

5. Remove the 4 oil pan-to-bell housing bolts.

6. Remove the 13 oil pan-to-block bolts.

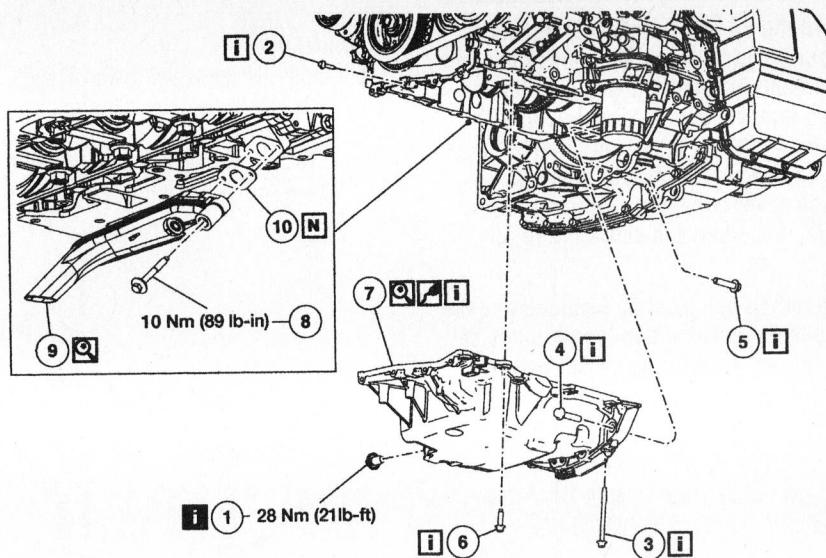

1 Drain plug	6 Oil pan bolt
2 Engine front cover bolt	7 Oil pan
3 Oil pan bolt	8 Oil pump screen and pickup tube bolt
4 Oil pan-to-bell housing bolt	9 Oil pump screen and pickup tube
5 Oil pan-to-bell housing bolt	10 Oil pump screen and pickup tube gasket

Oil pan, pump and related parts—2.3L engine

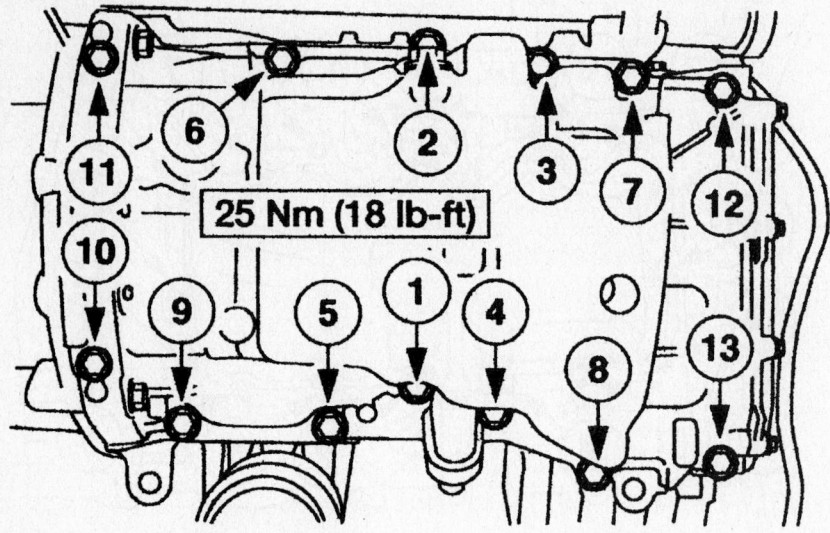

Oil pan bolt torque sequence—2.3L engine

To install:

✷✷ WARNING

CAUTION: Do not use metal scrapers, wire brushes, power abrasive discs or other abrasive means to clean the sealing surfaces. These tools cause scratches and gouges, which make leak paths. Use a plastic scraping tool to remove traces of sealant. Clean and inspect all mating surfaces.

➡ If the oil pan is not secured within four minutes of sealant application the sealant must be removed and the sealing area cleaned with metal surface cleaner. Allow to dry until there is no sign of wetness, or four minutes, whichever is longer. Failure to follow this procedure can cause future oil leakage.

➡ The oil pan must be installed and the bolts tightened within four minutes of applying the silicone gasket and sealant.

7. Apply a 2.5 mm bead of silicone gasket and sealant to the oil pan. Install the oil pan. Install the oil pan-to-bell housing bolts. Torque to 35 ft. lbs. (48 Nm).

8. Install the front cover bolts. Torque to 89 inch lbs. (10 Nm).

9. Install the oil pan-to-bell housing bolts. Torque to 35 ft. lbs. (48 Nm).

10. Install and tighten the oil pan bolts in the sequence shown to 18 ft. lbs. (25 Nm).

11. Fill the engine with clean engine oil.

3.0L Engine

1. Before servicing the vehicle, refer to the precautions in the beginning of this section.

2. Drain the engine oil.

3. Remove or disconnect the following:
   - Negative battery cable
   - Flexible exhaust pipe
   - Downstream catalyst monitor sensor
   - Oil pan and gasket

4. Thoroughly clean the gasket mating surfaces.

To install:

5. Apply silicone sealer to the oil pan.

6. Install or connect the following:
   - New gasket on the oil pan
   - Oil pan. Torque the bolts in sequence to 18 ft. lbs. (25 Nm).

- Flexible exhaust pipe
- Downstream catalyst monitor sensor
- Negative battery cable

7. Fill the engine with clean oil.

8. Start the vehicle and check for leaks, repair if necessary.

Oil Pump

REMOVAL & INSTALLATION

2.0L Engine

1. Before servicing the vehicle, refer to the precautions in the beginning of this section.

2. Drain the engine oil.

3. Remove or disconnect the following:
   - Negative battery cable
   - Oil pan
   - Oil pump screen cover and tube
   - Oil pump and discard the gasket

4. Thoroughly clean the gasket mating surfaces.

To install:

5. Install or connect the following:
   - Oil pump screen cover and tube with a new gasket. Torque the bolts to 89 inch lbs. (10 Nm).
   - Oil pump to the oil pan
   - Oil pan
   - Negative battery cable

6. Refill the engine with clean oil.

7. Start the engine and check for leaks; repair if necessary.

2.3L Engine

1. Before servicing the vehicle, refer to the precautions in the beginning of this section.

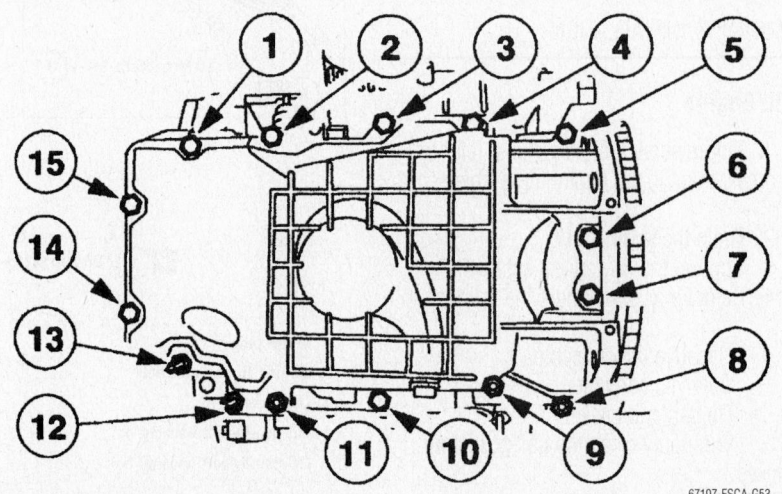

67197-ESCA-G53

Oil pan torque sequence—3.0L engine

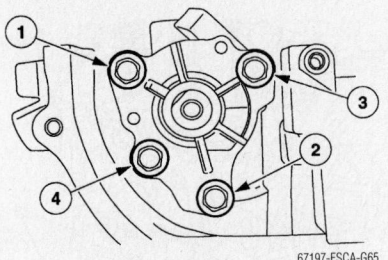

Oil pump torque sequence—2.3L engine

2. Remove the engine from the vehicle and mount it on an engine stand.

3. Remove the oil pan.

4. Remove the oil pump pickup tube and screen.

5. Remove the front cover and the timing chain.

6. Release the tension on the tensioner spring.

7. Remove the tensioner and the shoulder bolt.

8. Remove the guide.

➡ **The oil pump chain sprocket must be held in place.**

9. Remove the oil pump chain and sprockets.

10. Remove the oil pump assembly and gasket.

To install:

11. Install the oil pump with a new gasket. Tighten the bolts in sequence as follows:

   a. Step 1: 89 inch lbs. (10 Nm).

   b. Step 2: 17 ft. lbs. (23 Nm).

12. Install the pump chain and sprockets. Tighten the pump sprocket bolt to 18 ft. lbs. (25 Nm).

13. Install the chain guide, tensioner, and shoulder bolt. Tighten the bolts to 89 inch lbs. (10 Nm).

14. Hook the tensioner spring around the shoulder bolt.

15. Install the oil pump pickup tube and screen with a new gasket. Tighten the bolts to 89 ft. lbs. (10 Nm).

16. Install the oil pan.

17. Install the timing chain and front cover.

18. Install the engine into the vehicle.

3.0L Engine

1. Before servicing the vehicle, refer to the precautions in the beginning of this section.

2. Drain the engine oil.

3. Remove or disconnect the following:

- Negative battery cable

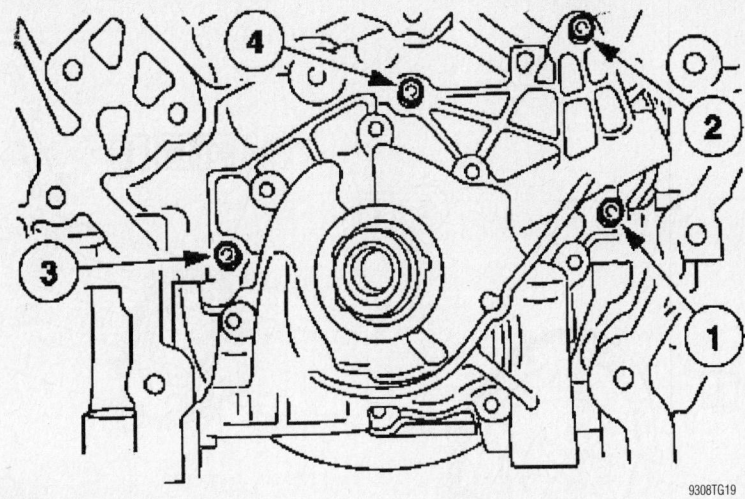

Remove the oil pump bolts in the proper sequence–3.0L engine

Install the oil pump bolts in the proper sequence–3.0L engine

- Timing drive components
- Oil pump screen cover and tube
- Damper bolt and crankshaft sprockets
- Oil pump bolts in the proper sequence

4. Thoroughly clean the gasket mating surfaces.

To install:

5. Install or connect the following:

- Oil pump and bolts in the proper sequence. Torque the bolts to 89 inch lbs. (10 Nm).
- Crankshaft sprockets
- Oil pump screen cover and tube
- Timing drive components
- Negative battery cable

6. Refill the engine with clean oil.

7. Start the engine and check for leaks; repair if necessary.

Rear Main Seal

REMOVAL & INSTALLATION

2.0L Engine

1. Before servicing the vehicle, refer to the precautions in the beginning of this section.

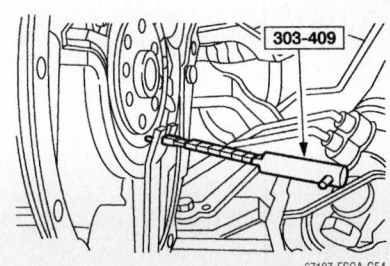

Rear main seal removal—2.0L engine

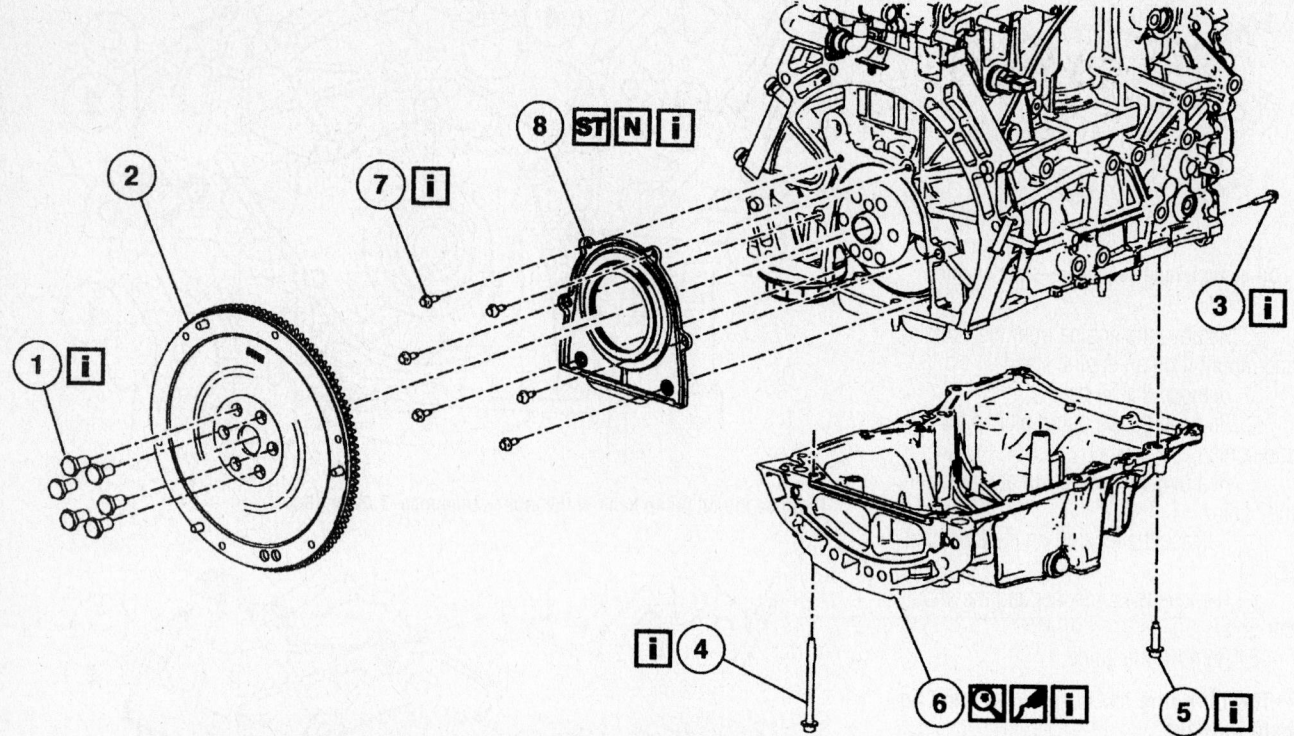

1 Flexplate or flywheel bolt
2 Flexplate or flywheel
3 Engine front cover bolt
4 Oil pan bolt
5 Oil pan bolt
6 Oil pan
7 Crankshaft rear oil seal with retainer plate bolt
8 Crankshaft rear oil seal with retainer plate

67197-ESCA-G22

Rear main seal and related parts—2.3L engine

2. Remove or disconnect the following:
   • Negative battery cable
   • Flywheel
   • Rear main seal
To install:
3. Coat the oil seal with clean engine oil.
4. Install or connect the following:
   • Crankshaft rear oil seal
   • Flywheel
   • Negative battery cable

2.3L Engine

1. With the vehicle in NEUTRAL, position it on a hoist.
2. If equipped, remove the automatic transaxle.
3. If equipped, remove the manual transaxle and clutch.
4. Remove the flexplate or flywheel.
5. Remove the oil pan.
6. Remove the crankshaft rear oil seal with retainer plate
To install:
7. Using a seal installer, position the

crankshaft rear oil seal with retainer plate onto the crankshaft.
8. Install the crankshaft rear oil seal with retainer plate. Tighten the bolts in the sequence shown to 10 Nm (89 inch lbs.).

9. Install the oil pan.

➤ Special bolts are used for installation. Do not use standard bolts.

10. Install the flywheel/flexplate. Tighten

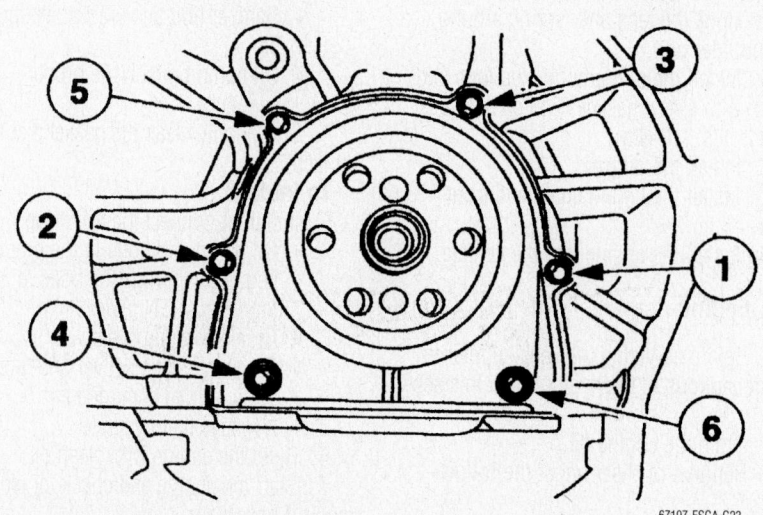

Retainer plate torque sequence—2.3L engine

67197-ESCA-G23

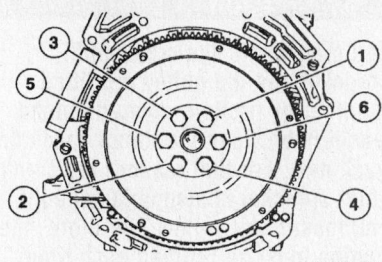

Flywheel torque sequence—2.3L engine

the bolts in the sequence shown in three stages.

    a. Stage 1: Tighten to 50 Nm (37 ft. lbs.).

    b. Stage 2: Tighten to 80 Nm (50 ft. lbs.).

    c. Stage 3: Tighten to 112 Nm (83 ft. lbs.).

3.0L Engine

1. Before servicing the vehicle, refer to the precautions in the beginning of this section.

2. Remove or disconnect the following:

- Negative battery cable
- Flexplate
- Rear main oil seal

To install:

3. Coat the oil seal with clean engine oil.

4. Install or connect the following:

- Crankshaft rear oil seal
- Flywheel
- Negative battery cable

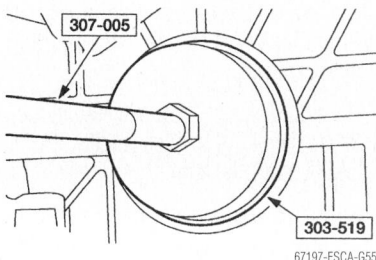

Rear main seal removal—3.0L engine

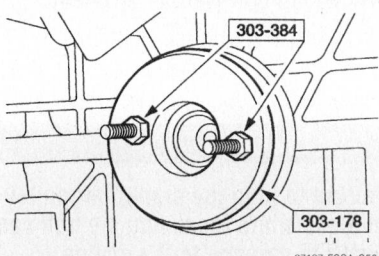

Rear main seal installation—3.0L engine

Timing Belt and Covers

REMOVAL & INSTALLATION

2.0L Engine

1. Remove the valve cover.
2. Remove the spark plugs.
3. Remove the catalytic converter.
4. Remove the bolt, nut, and position the coolant tube aside.
5. Remove the right wheel and tire assembly.
6. Remove the right lower splash shield.
7. Rotate the crankshaft to just before top dead center (TDC) (No. 1 cylinder).
8. Remove the stud.
9. Install the special tool.

➡**Make sure the correct (second) notch in the pulley is indexed to the lower cylinder block.**

10. Rotate the crankshaft clockwise against the peg to bring it to TDC (No. 1 cylinder).
11. Loosen the water pump pulley bolts.
12. Disconnect the battery ground cable.
13. Remove the crankshaft pulley.
14. Remove the bolts and the lower timing belt cover.
15. Lower the vehicle.
16. Install the special tool 303-F072.
17. Remove the ground strap.
18. Remove the engine mount upper bracket.
19. Remove the studs.
20. Remove the knock sensor connector.
21. Remove the bolts and the upper timing belt cover.
22. Remove the water pump pulley.
23. Remove the accessory drive belt idler pulley.

➡**Installation of the alignment tool into the exhaust camshaft may require the camshafts to be rotated clockwise slightly.**

24. Install the special tool and align the camshafts.
25. Raise and support the vehicle.
26. Remove the bolts and the engine mount lower bracket.
27. Loosen the timing belt tensioner pulley and allow to slide down to the bottom of its travel.

✳✳ CAUTION

If the camshaft timing belt is to be reused, mark the direction of the camshaft timing belt to rotation of camshaft prior to removal or premature wear or failure may occur.

28. Slide the timing belt off of the camshaft and crankshaft sprockets. the timing belt for wear. Install a new belt if necessary.

To install:

➡**Make sure the correct (second) notch in the pulley is indexed to the lower cylinder block.**

29. Slide the crankshaft pulley onto the crankshaft and confirm the crankshaft position is at TDC (No. 1 cylinder) by rotating it clockwise against the alignment peg.
30. Remove the crankshaft pulley.
31. Lower the vehicle.
32. Confirm that the timing belt tensioner is installed correctly with the tab positioned in the slot in the inner timing cover.
33. Install the timing belt onto the timing belt sprockets.
34. Adjust the timing belt tensioner.
35. Using a 6 mm Allen wrench, rotate the adjuster counterclockwise and align the marks as shown.
36. Tighten the tensioner pulley bolt.
37. Raise the vehicle.
38. Install the front engine mount lower bracket.
39. Install the accessory drive belt idler pulley.
40. Install the water pump pulley.
41. Hand tighten the bolts.
42. Lower the vehicle.
43. Install the timing belt covers. Torque the cover bolts to 62 inch lbs. (7Nm).
44. Tighten the water pump pulley bolts.
45. Remove the special tool.
46. Install the stud. Torque to 25 ft. lbs. (34Nm).
47. Install the coolant tube.
48. Install the catalytic converter.
49. Remove the special tool.
50. Install the valve cover.
51. Install the spark plugs.

Timing Chain, Gears, Front Cover and Seal

REMOVAL & INSTALLATION

2.3L Engine

✳✳ CAUTION

During engine repair procedures, cleanliness is extremely important. Any foreign material, including any material created while cleaning gasket surfaces, that enters the oil passages, coolant passages or the oil pan can cause engine failure.

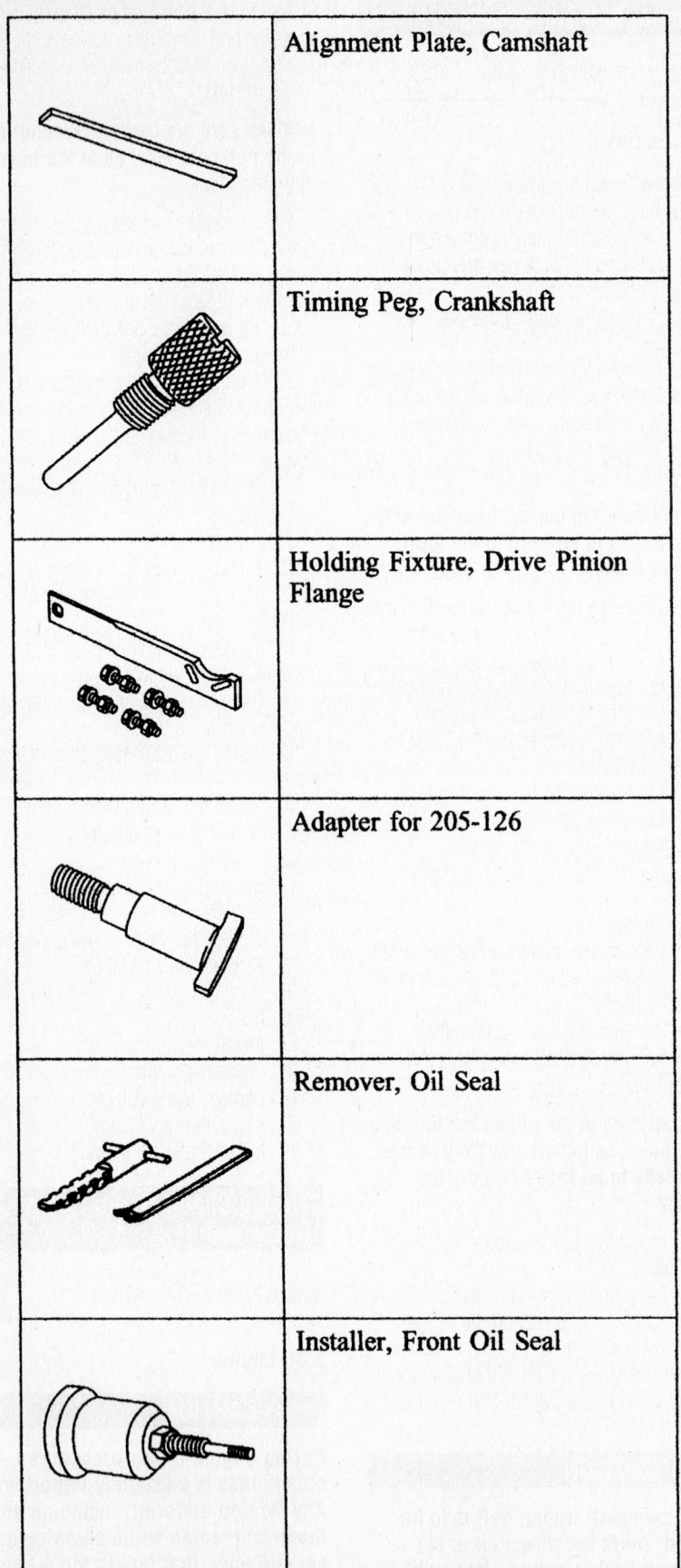

Alignment Plate, Camshaft
Timing Peg, Crankshaft
Holding Fixture, Drive Pinion Flange
Adapter for 205-126
Remover, Oil Seal
Installer, Front Oil Seal

67197-ESCA-G25

Tools needed for timing chain and gears replacement—2.3L engine

✳✳ CAUTION

The crankshaft, the crankshaft sprocket and the pulley are fitted together by friction, using diamond washers between the flange faces on each part. For that reason, the crankshaft sprocket is also unfastened if you loosen the pulley. Therefore, the engine must be retimed each time the damper is removed. Otherwise severe engine damage can occur.

1. With the vehicle in NEUTRAL, position it on a hoist.
2. Remove the accessory drive belt and idler pulleys.
3. Remove the engine mount.
4. Remove the valve cover.

✳✳ CAUTION

Failure to position the No. 1 piston at top dead center (TDC) can result in damage to the engine. Turn the engine in the normal direction of rotation only.

5. Using the crankshaft pulley bolt, turn the crankshaft clockwise to position the No. 1 piston at TDC.

✳✳ CAUTION

The special tool 303-465 is for camshaft alignment only. Using this tool to prevent engine rotation can result in engine damage.

➡The camshaft timing slots are offset. If the special tool cannot be installed, rotate the crankshaft one complete revolution clockwise to correctly position the camshafts.

6. Install special tool 303-465 in the slots on the rear of both camshafts.
7. Remove the engine plug bolt.

➡Only turn the engine in the normal direction of rotation.

➡Installing the special tool in this step will prevent the engine from being rotated in the clockwise direction.

8. Install special tool 303-507.
9. Install the special tools 205-126 and 205-072-02.

✳✳ CAUTION

Failure to hold the crankshaft pulley in place while loosening the bolt can result in damage to the engine.

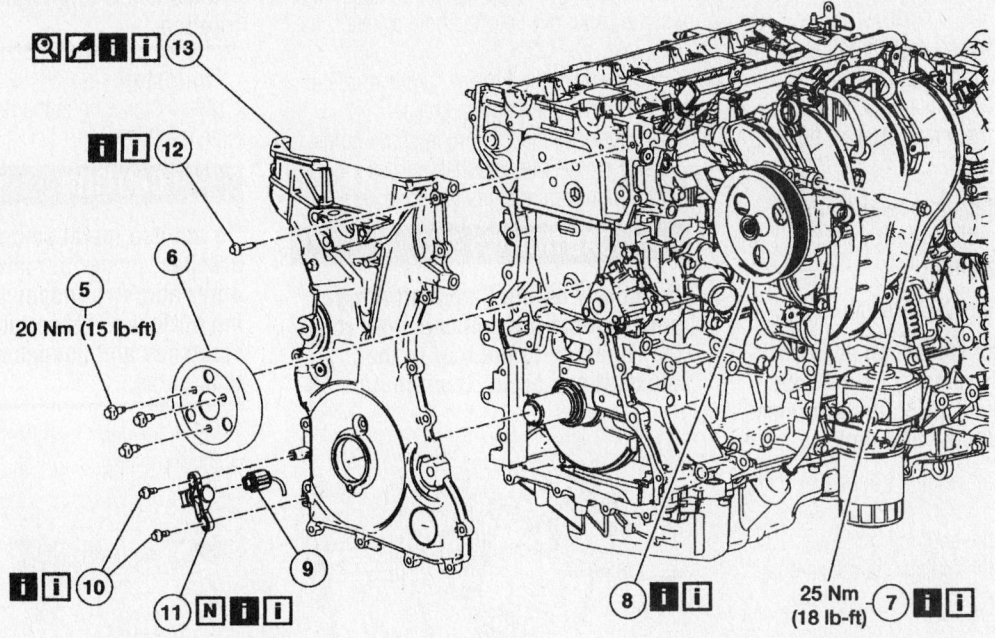

5 Coolant pump pulley bolt

6 Coolant pump pulley

7 Power steering pump bolt

8 Power steering pump (position aside)

9 Crankshaft position (CKP) sensor electrical connector

10 CKP sensor bolts

11 CKP sensor

12 Engine front cover bolt

13 Engine front cover

67197-ESCA-G26

Front cover and related parts—2.3L engine

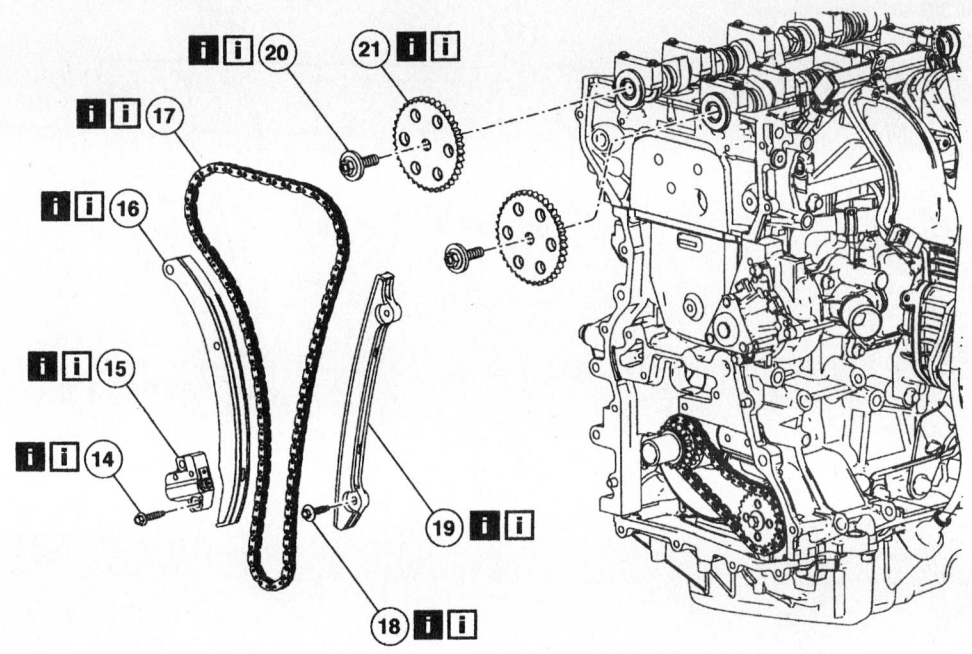

14 Timing chain tensioner bolt

15 Timing chain tensioner

16 RH timing chain guide

17 Timing chain

18 LH timing chain guide bolt

19 LH timing chain guide

20 Camshaft sprocket bolt

21 Camshaft sprocket

67197-ESCA-G27

Timing chain and related parts—2.3L engine

10. Remove the crankshaft pulley bolt and washer.

11. Remove the crankshaft pulley.

12. Remove the crankshaft front seal.

13. Remove the coolant pump pulley.

14. Remove the power steering pump and position it aside.

➡ **The bolt under the power steering pressure tube will remain with the power steering pump.**

15. Remove the CKP sensor.

➡ **Whenever the crankshaft position (CKP) sensor is removed, a new one must be installed, using the alignment jig supplied with the new part.**

16. Remove the engine front cover bolts (there are 22).

17. Remove the engine front cover.

18. Remove the timing chain tensioner.

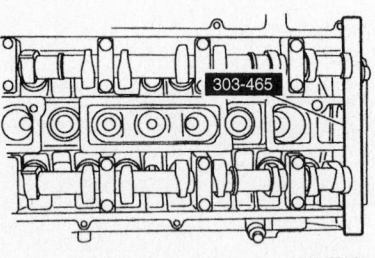

Install special tool 303-465 in the slots on the rear of both camshafts

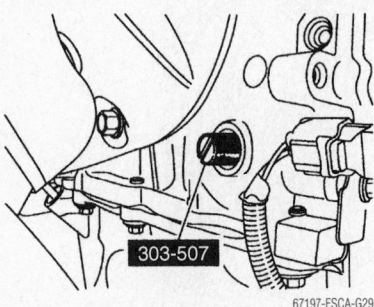

Install special tool 303-507

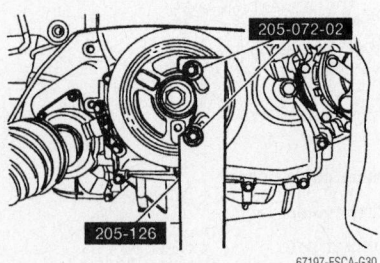

Install the special tools 205-126 and 205-072-02

Compress the timing chain tensioner, and insert a paper clip into the hole to retain the tensioner.

19. Remove the RH timing chain guide.

20. Remove the timing chain.

21. Remove the LH timing chain guide.

22. Remove the camshaft sprocket bolts.

23. Remove the camshaft sprockets.

✳✳ CAUTION

Do not rely on the Camshaft Alignment Plate to prevent camshaft rotation. Damage to the tool or the camshaft can occur. Use the flats on

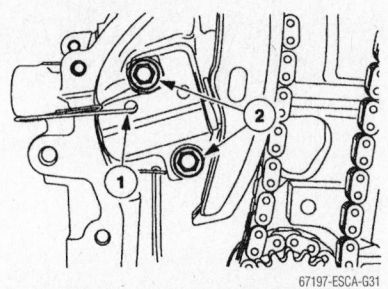

Compress the timing chain tensioner, and insert a paper clip into the hole to retain the tensioner

the camshaft to prevent camshaft rotation.

To install:

24. Installation is the reverse of removal. Note the following:

✳✳ CAUTION

Do not use metal scrapers, wire brushes, power abrasive disks or other abrasive means to clean sealing surfaces. These tools cause scratches and gouges which make leak paths.

25. Clean and inspect the mounting surfaces of the engine and the front cover.

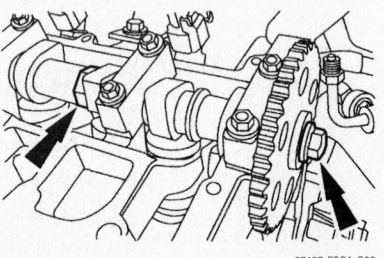

Use the flats on the camshaft to prevent camshaft rotation

Front cover bolt torque sequence—2.3L engine

➡**The engine front cover must be installed and the bolts tightened within four minutes of applying the silicone gasket and sealant.**

26. Apply a 2.5 mm bead of silicone gasket and sealant to the cylinder head and oil pan joint areas. Apply a 2.5 mm bead of silicone gasket and sealant to the front cover.

27. Install the engine front cover. Tighten the bolts in the sequence shown, to the following specifications:
   a. Tighten the 8 mm bolts to 10 Nm (89 inch lbs.).
   b. Tighten the 13 mm bolts to 48 Nm (35 ft. lbs.).

28. Position the power steering pump and install the bolts.

➡**Remove the through-bolt from the special tool.**

➡**Lubricate the oil seal with clean engine oil.**

29. Using a seal driver, install the crankshaft front oil seal.

➡**Do not reuse the crankshaft damper bolt.**

➡**Apply clean engine oil on the seal area before installing.**

30. Install the crankshaft pulley and hand-tighten the bolt.

✳✳ **CAUTION**

Only hand-tighten the bolt or damage to the front cover can occur.

➡**This step will correctly align the crankshaft pulley to the crankshaft.**

31. Install a standard 6 mm x 18 mm bolt through the crankshaft pulley and thread it into the front cover. Rotate the pulley as necessary to align the bolt holes.

✳✳ **CAUTION**

Failure to hold the crankshaft pulley in place while tightening the bolt can cause damage to the engine front cover.

32. Using the special tools to hold the crankshaft pulley in place, tighten the crankshaft pulley bolt in two stages:
   a. Stage 1: Tighten to 100 Nm (74 ft. lbs.).
   b. Stage 2: Tighten an additional 90 degrees (¼ turn).

33. Remove the 6 mm x 18 mm bolt.
34. Remove special tool 303-507.
35. Remove special tool 303-465.

➡**Only turn the engine in the normal direction of rotation.**

36. Turn the engine two complete revolutions.

➡**Only turn the engine in the normal direction of rotation.**

37. Turn the crankshaft until the No. 1 piston is at TDC.
38. Install special tool 303-507.

✳✳ **CAUTION**

Only hand-tighten the bolt or damage to the front cover can occur.

39. Using the 6 mm x 18 mm bolt, check the position of the crankshaft pulley. If it is not possible to install the bolt, correct the engine timing.

40. Using special tool 303-465, check the position of the camshafts. If it is not possible to install the special tool, correct the engine timing.

41. Install the CKP sensor. Do not tighten the bolts at this time.

42. Adjust the CKP sensor alignment jig and tighten the bolts.

43. Remove the 6 mm x 18 mm bolt.
44. Install the engine plug bolt.

3.0L Engine

1. Before servicing the vehicle, refer to the precautions in the beginning of this section.

2. Remove or disconnect the following:
   • Negative battery cable
   • Engine front cover
   • Ignition pulse wheel and install a damper bolt
   • Spark plugs

3. Rotate the crankshaft clockwise to position the keyway at the 11 o'clock position and the camshafts in the correct positions. The No. 1 cylinder will be at Top Dead Center (TDC).

4. Rotate the crankshaft clockwise 120 degrees to the 3 o'clock position to locate the right side camshafts in the neutral position.

5. Remove or disconnect the following:
   • Right side timing chain and tensioner
   • Tensioner arm and timing chain guide

6. Rotate the crankshaft clockwise 2 times to position the keyway at the 11 o'clock position. This will position the left side camshafts in the neutral position.

7. Verify that the left side crankshafts are in the neutral position and mark the link position on the crankshaft sprocket.

8. Remove or disconnect the following:
   • Left side timing chain and tensioner
   • Tensioner arm and timing chain guide
   • Damper bolt and crankshaft sprockets

To install:

9. Install the crankshaft sprockets.

10. Position the timing chain tensioner in a soft jaw vise. Hold the ratchet lock mechanism away from the ratchet stem and slowly compress the timing chain tensioner

11. If the timing marks on the chain are not visible, use a permanent marker to mark the left and right side timing chains. Mark the timing chains in the following sequence:
   a. Mark any link to use as the crankshaft timing mark.

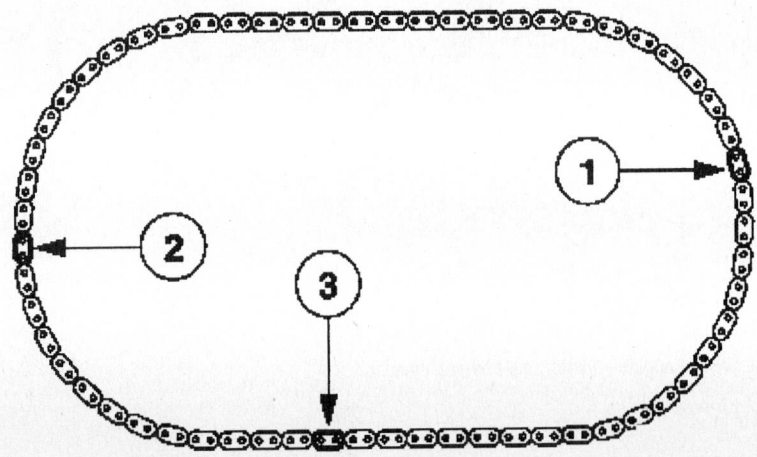

Mark the timing chain in the proper sequence—3.0L engine

9308TG21

b. Count 29 links from the crankshaft timing mark and mark the link as the exhaust cam sprocket timing mark.

c. Continue counting to 42 and mark the link as the intake sprocket timing mark

12. Install the guide. Torque the bolts to 18 ft. lbs. (25 Nm).

13. Install the left side timing chain and align the chain in the following sequence:

a. Mark any link to use as the crankshaft timing mark.

b. Count 29 links from the crankshaft timing mark and mark the link as the exhaust cam sprocket timing mark.

c. Continue counting to 42 and mark the link as the intake sprocket timing mark

14. Install or connect the following:

- Left side timing chain and tensioner arm. Torque the bolts to 18 ft. lbs. (25 Nm).
- Crankshaft damper bolt and rotate the keyway to the 3 o'clock position

15. Verify that the right side camshafts are properly positioned and install the right side timing chain and guide. Torque the bolts to 18 ft. lbs. (25 Nm).

16. Make certain that the timing chain aligns with the marks on the camshaft and crankshaft sprockets

✳✳ CAUTION

Install the pulse wheel with the keyway in the slot stamped 20–25–34Y–30M (Color Blur).

17. Install or connect the following:

- Right side timing chain tensioner and arm. Torque the bolts to 18 ft. lbs. (25 Nm) and remove the damper bolt
- Ignition pulse wheel
- Spark plugs
- Engine front cover
- Negative battery cable

Piston and Ring

POSITIONING

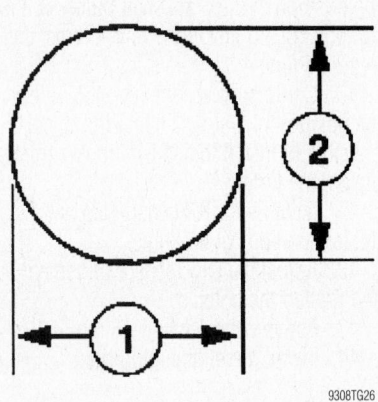

9308TG26

2.0L and 2.3L engine —piston ring end-gap spacing

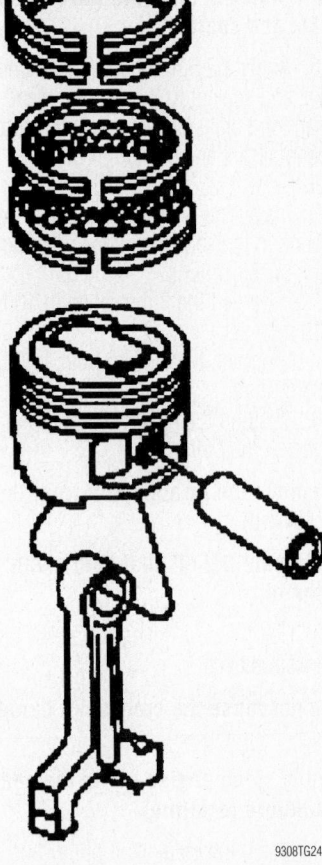

9308TG24

2.0L and 2.3L engine piston and connecting rod positioning ring end-gap spacing

9308TG25

3.0L (VIN 1) engine—piston ring end-gap spacing

FUEL SYSTEM

Fuel System Service Precautions

Safety is the most important factor when performing not only fuel system maintenance but also any type of maintenance. Failure to conduct maintenance and repairs in a safe manner may result in serious personal injury or death. Maintenance and testing of the vehicle's fuel system components can be accomplished safely and effectively by adhering to the following rules and guidelines.

1. To avoid the possibility of fire and personal injury, always disconnect the negative battery cable unless the repair or test procedure requires that battery voltage be applied.

2. Always relieve the fuel system pressure prior to disconnecting any fuel system component (injector, fuel rail, pressure regulator, etc.), fitting or fuel line connection. Exercise extreme caution whenever relieving fuel system pressure, to avoid exposing skin, face and eyes to fuel spray. Please be advised that fuel under pressure may penetrate the skin or any part of the body that it contacts.

3. Always place a shop towel or cloth around the fitting or connection prior to loosening to absorb any excess fuel due to spillage. Ensure that all fuel spillage (should it occur) is quickly removed from engine surfaces. Ensure that all fuel soaked cloths or towels are deposited into a suitable waste container.

4. Always keep a dry chemical (Class B) fire extinguisher near the work area.

5. Do not allow fuel spray or fuel vapors to come into contact with a spark or open flame.

6. Always use a backup wrench when loosening and tightening fuel line connection fittings. This will prevent unnecessary stress and torsion to fuel line piping.

7. Always replace worn fuel fitting O-rings with new. Do not substitute fuel hose or equivalent, where fuel pipe is installed.

Before servicing the vehicle, make sure to refer to the precautions in the beginning of this section as well.

Fuel System Pressure

RELIEVING

2.0L Engine

1. Before servicing the vehicle, refer to the precautions in the beginning of this section.

2. Remove or disconnect the following:

3. Remove the fuel pump relay and start the engine.

4. After the engine stalls, crank the engine 2 more times to be certain that all fuel pressure has been relieved.

5. Turn the ignition switch to the **OFF** position.

6. Install the fuel pump relay.

2.3L Engine and 2005 3.0L Engine

1. Remove the fuel pump relay.

2. Start the engine and allow it to idle until it stalls.

3. After the engine stalls, crank the engine for approximately 5 seconds to make sure the fuel injection supply manifold pressure has been released.

4. Turn the ignition switch to the OFF position.

5. When fuel system service is complete, install the fuel pump relay.

➡**It may take more than one key cycle to pressurize the fuel system.**

6. Cycle the ignition key and wait three seconds to pressurize the fuel system. Check for leaks before starting the engine.

7. Start the vehicle and check the fuel system for leaks.

2001–04 3.0L Engine

1. Before servicing the vehicle, refer to the precautions in the beginning of this section.

2. Remove or disconnect the following:

3. Remove the schrader valve cap at the end of the fuel injection supply manifold and attach a fuel pressure gauge.

4. Open the manual valve slowly and drain the fuel into a suitable container.

5. Continue draining the fuel system to relieve fuel pressure.

Fuel Filter

REMOVAL & INSTALLATION

1. Before servicing the vehicle, refer to the precautions in the beginning of this section.

2. Properly relieve the fuel system pressure.

3. Remove or disconnect the following:
   • Negative battery cable
   • Fuel line to the fuel filter

4. Loosen the clamp and remove the filter

To install:

5. Install or connect the following:
   • New clips to the fuel lines
   • Fuel filter and tighten the clamp
   • Fuel lines to the fuel filter
   • Negative battery cable

6. Start the vehicle and check for leaks, repair if necessary.

Fuel Pump

REMOVAL & INSTALLATION

2001–04

1. Before servicing the vehicle, refer to the precautions in the beginning of this section.

2. Properly relieve the fuel system pressure.

3. Remove or disconnect the following:
   • Negative battery cable
   • Gas cap to relieve any additional fuel pressure
   • Left rear seat cushion and lift the access cover on the scuff plate
   • Pin type retainers and move the carpet aside
   • Screws from the fuel pump module access cover
   • Fuel pump module electrical connectors
   • Fuel and vapor lines from the fuel tank
   • Fuel pump module and discard the gasket

To install:

4. Install or connect the following:
   • New fuel pump module gasket
   • Fuel pump module. Torque the module to 60 ft. lbs. (81 Nm).
   • Fuel and vapor lines to the fuel tank
   • Fuel pump module electrical connectors

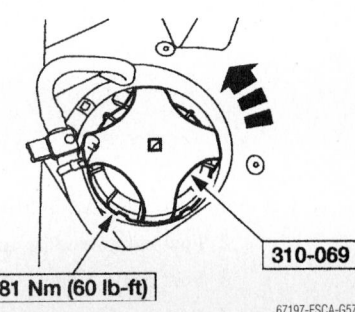

81 Nm (60 lb-ft)

310-069

67197-ESCA-G57

Fuel pump module removal/installation—2001–04 models

- Fuel pump module access cover and tighten the screws securely
- Pin type retainers and reposition the carpet
- Left rear seat cushion
- Gas cap
- Negative battery cable

5. Start the engine and check for leaks, repair if necessary.

2005

1. With the vehicle in NEUTRAL, position it on a hoist.

2. Release the fuel system pressure.

3. Disconnect the battery ground cable.

4. If removing the fuel tank on a four wheel drive vehicle, it is necessary to lower the exhaust system from the catalytic convertor back. Support the exhaust system with a suitable stand, release the three rear exhaust hangers and carefully lower the exhaust system to allow enough clearance to remove the fuel tank.

5. If removing the fuel tank on a four wheel drive vehicle, remove the rear driveshaft.

6. Lift the LH rear seat cushion, posi-

tion the carpet aside and remove the screws and the fuel pump module access cover.

7. Disconnect the fuel pump module electrical connector

8. Fuel vapor control tube assembly valve electrical connector

9. Using a suitable fuel pump lock ring remover, rotate the lock ring counterclockwise and remove.

✱✱ WARNING

The fuel pump module must be handled carefully to avoid damage to the float arm and filter.

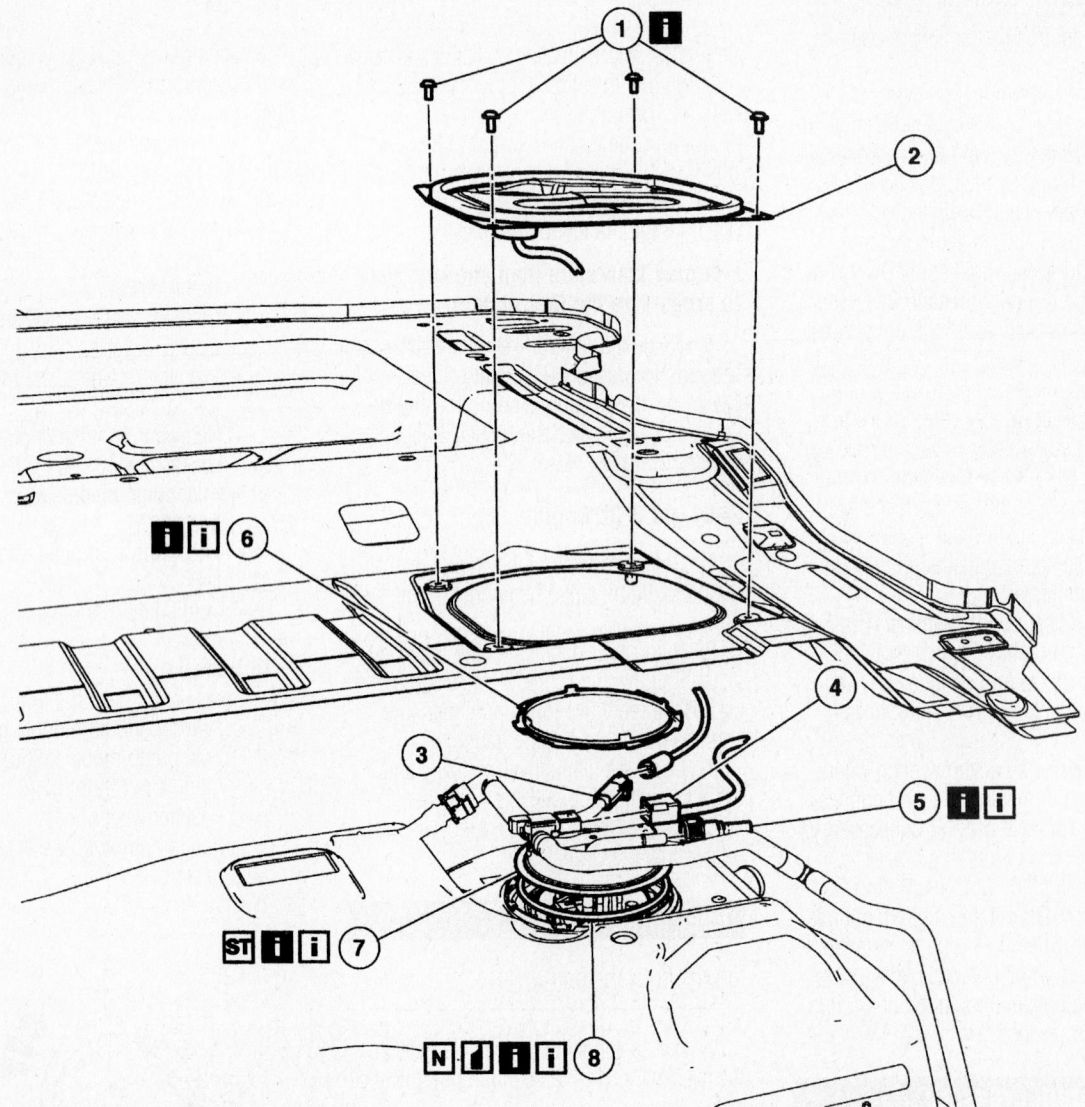

1 Fuel pump module access cover screws
2 Fuel pump module access cover
3 Fuel pump module electrical connector
4 Fuel vapor control tube assembly valve electrical connector

5 Fuel supply tube quick connect coupling
6 Fuel pump module lock ring
7 Fuel pump module
8 Fuel pump module O-ring seal

67197-ESCA-G34

Access to fuel pump module—2005 models

Some fuel will remain in the fuel pump module after draining the fuel tank. Carefully drain the fuel pump module into a suitable container.

10. Prior to completely removing the fuel pump, position it aside and using the special tool and a suitable fuel recovery system, drain the fuel tank.

11. To release the bottom-mounted fuel pump module, reach into the fuel pump module opening and squeeze the retainer tabs on the pump module housing and pull upward.

12. Remove the fuel pump module O-ring seal

To install:

13. Turn the ignition key to the ON position to pressurize the fuel system.

14. Visually inspect the fuel system for leaks.

Make sure the fuel tube clicks into place when installing the tube. To make sure the tube is fully seated, pull on the tube.

➡Apply clean engine oil to the end of the tube before inserting the tube into the connector.

15. Install the fuel tube quick release coupling.

Inspect the surfaces of the fuel pump module flange and fuel tank O-ring contact surfaces. Do not polish or adjust the O-ring contact area of the fuel pump flange or fuel tank. Install a new fuel pump module or fuel tank if the O-ring contact area is bent, scratched or corroded.

Make sure to install a new fuel pump module O-ring and lock ring.

16. Lubricate and install a new fuel pump module O-ring seal upon installing the fuel pump module.

17. When installing the fuel pump module, make sure to align the locator tabs on the fuel tank mounting flange.

➡Be sure the aligning tabs of the fuel pump module unit are positioned in the slot before tightening the lock ring.

18. Holding the fuel pump module O-ring seal in place, rotate the lock ring clockwise until it stops against the retainer tabs.

➡Make sure the collar on the fuel tube is inserted fully into the quick release coupling before the locking tab is locked.

19. Connect the fuel supply quick connect coupling to the fuel supply manifold.

20. Press the fuel supply quick connect coupling locking tab into position.

21. Pull on the fitting to make sure it is fully engaged.

Fuel Injectors

REMOVAL & INSTALLATION

2.0L Engine

1. Before servicing the vehicle, refer to the precautions in the beginning of this section.

2. Release the fuel system pressure.

3. Remove or disconnect the following:
- Negative battery cable
- Fuel injection supply manifold
- Retaining clips and gently twist the fuel injector out of the manifold

4. Check the O-rings and replace if damaged.

To install:

5. Install or connect the following:
- Fuel injector(s) using new O-rings lubricated with clean engine oil
- Fuel injector into the supply manifold
- Retaining clips when the fuel injectors are seated properly
- Fuel injection supply manifold
- Negative battery cable

6. Start the vehicle and check for leaks, repair if necessary.

2.3L Engine

Do not smoke or carry lighted tobacco or open flame of any type when working on or near fuel-related components. Highly flammable vapors are always present and can ignite. Failure to follow these instructions can result in personal injury.

This procedure involves fuel handling. Be prepared for fuel spillage at all times and always observe fuel handling precautions. Failure to follow these instructions can result in personal injury.

1. Before servicing the vehicle, refer to the precautions in the beginning of this section.

2. Disconnect the battery ground cable.

3. Release the fuel pressure.

4. Disconnect the fuel injector electrical connectors.

5. Disconnect the fuel pressure regulator electrical connector and the vacuum hose.

6. Disconnect the fuel injector harness retaining clips from the fuel injection supply manifold.

7. Disconnect the fuel tube.

8. Remove the bolts and the fuel injection supply manifold.

➡Remove and discard the fuel injector O-rings.

9. If necessary, remove the fuel injectors.

10. Install new O-rings and lubricate them with clean engine oil.

11. To install, reverse the removal proce-

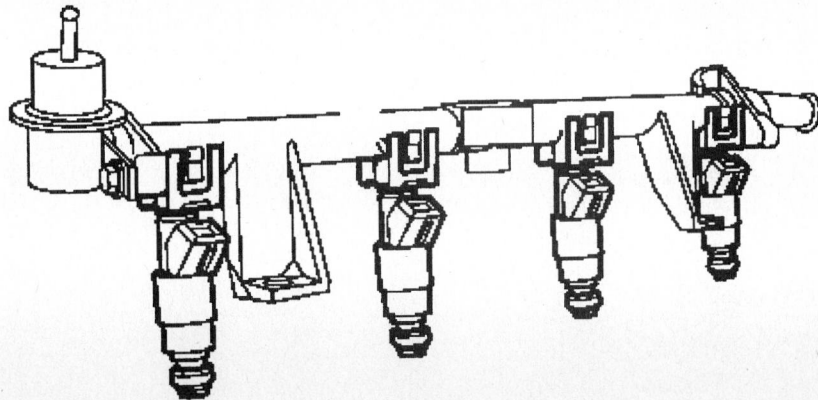

Remove the fuel injectors from the fuel supply manifold—2.0L engine

9308TG22

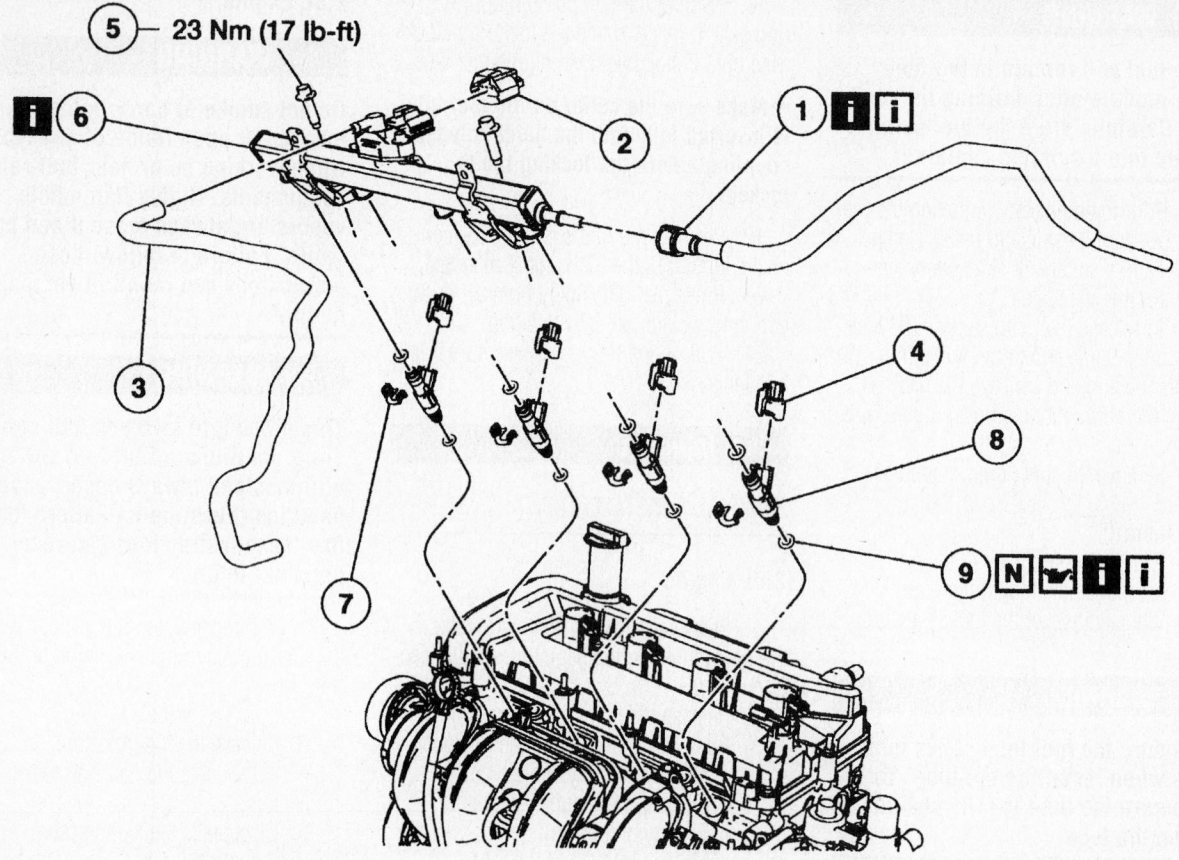

1 Fuel tube quick release coupling (position aside)

2 Fuel rail pressure and temperature sensor electrical connector

3 Fuel rail pressure and temperature sensor vacuum tube (position aside)

4 Fuel injector electrical connectors

5 Fuel rail bolts

6 Fuel rail

7 Fuel injector clips

8 Fuel injectors

9 Fuel injector O-ring seals

67197-ESCA-G35

Fuel rail and injectors—2.3L engine

dure. Tighten the fuel supply manifold bolts to 18 ft. lbs. (25 Nm).

3.0L Engine

1. Release the fuel system pressure
2. Disconnect the battery ground cable.
3. Remove the upper intake manifold.
4. Disconnect the fuel tube quick release coupling.

5. Disconnect the fuel rail pressure and temperature sensor vacuum tube
6. Disconnect the fuel rail pressure and temperature sensor electrical connector.
7. Disconnect the fuel injector electrical connectors.
8. Remove the fuel rail bolts.
9. Remove the fuel rail.
10. Remove the fuel injector.

11. Remove the fuel injector O-ring seals.
12. To install, reverse the removal procedure. Install new fuel injector O-ring seals.
13. Lubricate the new O-ring seals with clean engine oil before installing. Torque the fuel rail bolts to 89 inch lbs. (10 Nm).

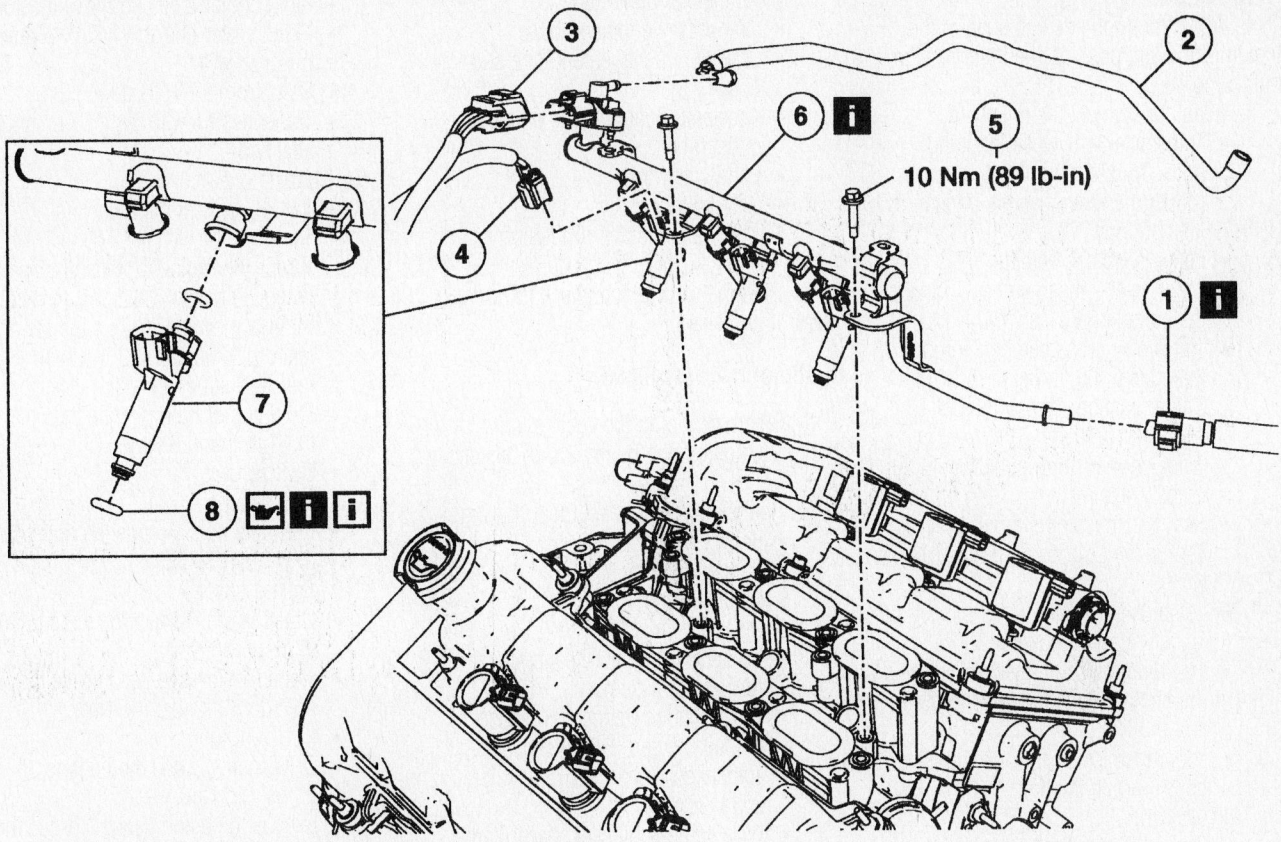

1 Fuel tube quick release coupling (position aside)

2 Fuel rail pressure and temperature sensor vacuum tube (position aside)

3 Fuel rail pressure and temperature sensor electrical connector

4 Fuel injector electrical connector

5 Fuel rail bolt

6 Fuel rail

7 Fuel injector

8 Fuel injector O-ring seals

67197-ESCA-G36

Fuel rail and injectors—3.0L engine

DRIVE TRAIN

Transmission Assembly

REMOVAL & INSTALLATION

Manual Transmission

1. Before servicing the vehicle, refer to the precautions in the beginning of this section.
2. Drain the transmission fluid.
3. Remove or disconnect the following:
   - Battery cables
   - Battery and tray
   - Mass Air Flow (MAF) sensor electrical connector
   - Accelerator cable from the air cleaner outlet tube
   - Emission management tube and hose
   - Crankcase ventilation hose
   - Air cleaner outlet tube

   - Air cleaner housing
   - Back-up lamp switch electrical connector
   - Front wire harness bracket and move it aside
   - Front wire harness bracket spacer
   - Wire harness from the rear harness bracket
   - Park Neutral Position (PNP) electrical connector
   - Rear wire harness bracket and move it aside
   - Vehicle Speed Sensor (VSS) electrical connector
   - Clutch slave cylinder line from the bracket and move it aside while properly supporting the engine
   - Left side transmission support insulator and bracket
   - Rear transmission support insulator

   - Front transmission support insulator and bracket
   - Starter and move it aside
   - Top transmission flywheel housing bolts
   - Front transmission flywheel housing bolts
   - Transfer case, if equipped
   - Left side halfshaft
   - Rear transmission support insulator bracket
   - Shifter linkage and stabilizer bar
   - Transverse crossmember
   - Front to aft crossmember
   - Left side splash shield and properly support the transmission
   - Remaining transmission flywheel housing bolts
   - Transmission and separate the right side halfshaft from the transmission

To install:

4. Align the right side half shaft to the transmission and position the transmission to the engine.

5. Install or connect the following:
- Transmission flywheel housing bolts. Torque the bolts to 33 ft. lbs. (45 Nm) and remove the transmission support
- Left side splash shield
- Front-to-aft crossmember. Torque the bolts to 66 ft. lbs. (90 Nm).
- Transverse crossmember. Torque the bolts to 85 ft. lbs. (115 Nm).
- Shifter linkage. Torque the bolt to 15 ft. lbs. (20 Nm).
- Stabilizer bar. Torque the bolt to 30 ft. lbs. (40 Nm).
- Rear transmission support bracket. Torque the bolts to 66 ft. lbs. (90 Nm).
- Left side halfshaft
- Transfer case, if equipped
- Front transmission flywheel housing bolts. Torque the bolts to 33 ft. lbs. (45 Nm).
- Top transmission flywheel housing bolts. Torque the bolts to 33 ft. lbs. (45 Nm).
- Starter. Torque the bolts to 33 ft. lbs. (45 Nm).
- Front transmission support insulator and bracket. Torque the lower bolt to 66 ft. lbs. (90 Nm) and the 3 upper bolts to 41 ft. lbs. (55 Nm).
- Rear transmission support insulator bolt. Torque the bolt to 66 ft. lbs. (90 Nm).
- Left side transmission support insulator bracket. Torque the bolts to 66 ft. lbs. (90 Nm).
- Left side transmission support insulator. Torque the large bolt to 66 ft. lbs. (90 Nm) and the 3 remaining bolts to 41 ft. lbs. (55 Nm).
- Clutch slave cylinder. Torque the bolt to 15 ft. lbs. (20 Nm).
- Clutch slave cylinder line to the bracket and install the retaining clip
- VSS electrical connector
- Rear wire harness bracket. Torque the bolts to 80 inch lbs. (9 Nm).
- PNP switch electrical connector
- Wire harness to the rear bracket
- Front wire harness bracket spacer and bracket. Torque the bolt to 9 ft. lbs. (12 Nm).
- Back-up lamp switch electrical connector
- Air cleaner housing
- MAF sensor electrical connector

- Air cleaner outlet tube
- Crankcase ventilation hose
- Emission management tube and hose
- Accelerator cable to the air cleaner outlet tube
- Battery and tray
- Both battery cables

6. Fill the transmission to the proper level.

7. Start the vehicle and check for leaks, repair if necessary.

Automatic Transmission

1. Before servicing the vehicle, refer to the precautions in the beginning of this section.

2. Remove or disconnect the following:
- Battery cables
- Battery and tray
- Breather tube
- Mass Air Flow (MAF) sensor
- Intake tube and air cleaner cover
- Air cleaner assembly
- Transmission Range (TR) sensor
- Heated Oxygen (HO2S) sensors
- Transmission harness connector and bracket
- Wire harness bracket spacer and move the bracket aside
- Shift cable
- Shift cable bracket and move the bracket aside
- Starter electrical connectors
- Starter
- Electrical connectors from the valve cover and install an engine support bar
- Upper transmission retaining bolts
- Left side upper transmission mounting plate
- Rear transmission mount
- Right side engine mount bolt and slightly raise the engine
- Both front wheels and splash shields
- Right side halfshaft and intermediate shaft assembly after match-marking them
- Cross brace
- Center exhaust pipe and rubber hanger
- Front exhaust pipe and flange
- Rear exhaust pipe flange
- Driveshaft
- PTU vent tube
- Lower transmission bracket
- Access cover
- Flexplate nuts
- Output Shaft Speed (OSS) sensor
- Turbine Shaft Speed (TSS) sensor

- Fluid cooler tube and move it aside
- Fluid cooler line and install a transmission jack
- Bolts from the PTU unit
- Transmission with the PTU unit attached

To install:

3. Install or connect the following:
- Transmission with the PTU unit. Torque the engine-to-transmission mounting bolts to 30 ft. lbs. (40 Nm).
- Fluid cooler line. Torque the fastener to 17 ft. lbs. (23 Nm) and remove the transmission jack.
- Fluid cooler tube. Torque the bolt to 17 ft. lbs. (23 Nm).
- OSS sensor
- TSS sensor
- Flexplate nuts. Torque the nuts to 27 ft. lbs. (36 Nm).
- Access cover
- Cross brace. Torque the bolts to 96 ft. lbs. (130 Nm).
- Transmission bracket. Torque the bolts to 30 ft. lbs. (40 Nm).
- PTU vent tube
- Driveshaft. Torque the bolts to 15 ft. lbs. (20 Nm).
- Exhaust pipe and flange. Torque the bolts to 21 ft. lbs. (29 Nm).
- Exhaust pipe and rubber hanger. Torque the bolts to 21 ft. lbs. (29 Nm).
- Left side halfshaft assembly
- Right side halfshaft and intermediate shaft assembly by aligning the matchmarks
- Splash shields
- Both front wheels and lower the engine on to the right side engine mount
- Right side engine mount bolt. Torque the bolt to 89 ft. lbs. (120 Nm).
- Rear transmission mount. Torque the upper bolt to 89 ft. lbs. (120 Nm) and the lower bolts to 35 ft. lbs. (45 Nm).
- Transmission mount assemble. Torque the bolts to 30 ft. lbs. (40 Nm) and remove the engine support bar.
- Electrical connectors to the valve cover
- Starter. Torque the bolts to 20 ft. lbs. (27 Nm).
- Starter electrical connectors
- Shifter cable and bracket. Torque the bolt to 14 ft. lbs. (19 Nm) and connect the shifter cable.
- Wire harness and install the harness bracket spacer

- Wire harness bracket. Torque the bolt to 89 inch lbs. (10 Nm).
- HO$_2$S sensor
- TR sensor and make certain it is properly aligned
- Air cleaner assembly
- Intake tube and air cleaner cover
- Breather tube
- MAF sensor
- Battery tray
- Battery and cables

4. Fill the transmission with clean fluid to the proper level.

5. Start the vehicle and check for leaks, repair if necessary

Clutch

ADJUSTMENTS

The clutch is hydraulically driven and therefore no adjustment is required.

REMOVAL & INSTALLATION

1. Before servicing the vehicle, refer to the precautions in the beginning of this section.

2. Remove or disconnect the following:
- Negative battery cable

- Transmission and lock the flywheel to the engine with special tool 303–103
- Pressure plate bolts by loosening them evenly
- Clutch pressure plate and disc

3. Clean the pressure plate and inspect it for burn marks, scores, flatness or ridges, replace if damaged.

4. Inspect the pressure plate diaphragm finger for wear, replace if damaged.

5. Measure the depth of the rivet heads. Minimum depth is 0.012 inch (0.3mm).

6. Inspect the clutch disc for signs of wear and replace if needed.

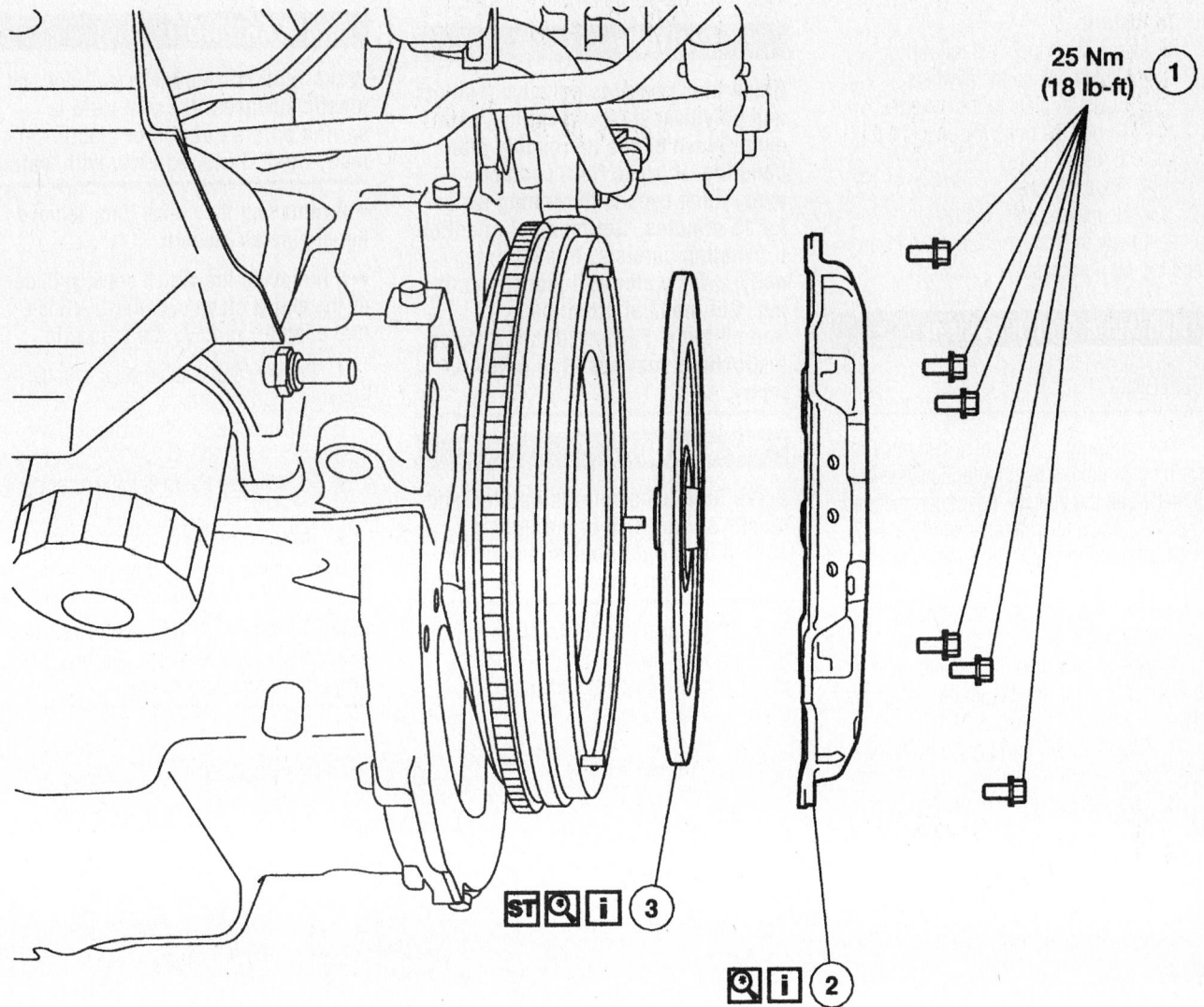

1 Bolt
2 Clutch
3 Clutch disc

67197-ESCA-G58

Clutch components

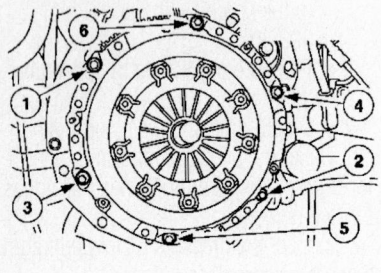

Torque the pressure plate bolts in the proper sequence

7. Check the clutch disc runout. Replace the disc if not with specification: 0.027 inch (0.7mm).

To install:

8. Install or connect the following:
- Clutch disc to the flywheel
- Pressure plate to the flywheel. Torque the bolts in sequence to 21 ft. lbs. (29 Nm).
- Transmission
- Negative battery cable

9. Check the transmission fluid level and top off if necessary.

Hydraulic Clutch System

BLEEDING

The following procedure is recommended for bleeding the clutch hydraulic system installed on the vehicle. It is recommended that the original clutch tube, with quick-connect fitting be replaced when servicing the hydraulic system, because air can be trapped in the quick-connect fitting and prevent complete bleeding of the system.

1. Before servicing the vehicle, refer to the precautions in the beginning of this section.
2. Clean the dirt and grease from the dust cap.
3. Remove the cap and diaphragm and fill the reservoir ¾ of the way with approved brake fluid C6AZ-19542-AB or DOT 3 equivalent fluid (ESA-M6C25-A).
4. Loosen the bleeder screw cover from the slave cylinder and attach a hose to the screw.
5. Place the hose in a container and slowly pump the clutch pedal several times.
6. With the clutch pedal depressed, loosen the bleeder screw to release the fluid and air.
7. Remove the hose and tighten the bleeder screw.
8. Repeat this procedure until all the air is removed from the hydraulic system.

REMOVAL & INSTALLATION

Master Cylinder

2001–04

> **✲✲ CAUTION**
>
> **Note the location and orientation of the clutch push rod retainer. If any damage to the retainer occurs during removal or installation, replace the retainer clip. The retainer must be correctly installed during assembly.**

1. Disconnect the clutch master cylinder rod from the clutch pedal.
2. Remove the clutch pedal bracket nut.

> **✲✲ CAUTION**
>
> **Brake fluid contains polyglycol ethers and polyglycols. Avoid contact with eyes. Wash hands thoroughly after handling. If brake fluid contacts eyes, flush eyes with running water for 15 minutes. Get medical attention if irritation persists. If taken internally, drink water and induce vomiting. Get medical attention immediately. Failure to follow these instructions may result in personal injury.**

> **✲✲ CAUTION**
>
> **Brake fluid is harmful to painted and plastic surfaces. If brake fluid is spilled onto a painted or plastic surface, wash it immediately with water.**

3. Disconnect the clutch master cylinder hose from the clutch master cylinder and plug it to prevent excess fluid loss.

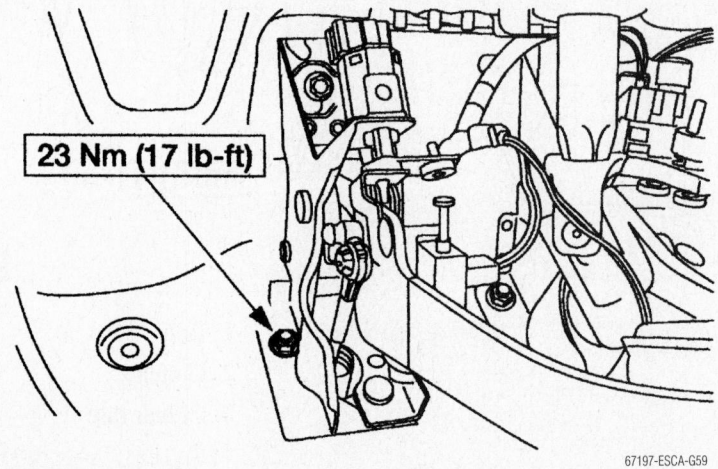

Clutch pedal bracket—2001–04 models

4. Disconnect the clutch slave cylinder tube from the clutch master cylinder.
5. Remove the clutch master cylinder nut.
6. Remove the clutch master cylinder.
7. To install, reverse the removal procedure. Make sure that the push rod retainer is correctly positioned. Observe the following torques:
- Clutch pedal bracket nut: 17 ft. lbs. (23 Nm).
- Slave cylinder tube-to-master cylinder: 22 ft. lbs. (30 Nm).
- Master cylinder mounting nuts: 17 ft. lbs. (23 Nm).

8. Bleed the air from the system.

2005

> **✲✲ WARNING**
>
> **Brake fluid is harmful to painted and plastic surfaces. If brake fluid is spilled onto a painted or plastic surface, wash it immediately with water.**

→ If removing the clutch line, remove the engine air cleaner.

→ If removing the clutch slave cylinder or the clutch slave cylinder-to-clutch line adapter, remove the transaxle.

1. Remove the clutch master cylinder hose
2. Remove the clutch master cylinder push rod
3. Remove the fitting at the master cylinder.
4. Remove the clutch master cylinder

> **✲✲ WARNING**
>
> **Make sure the O-rings are properly positioned on the hydraulic line fittings or leaks may occur.**

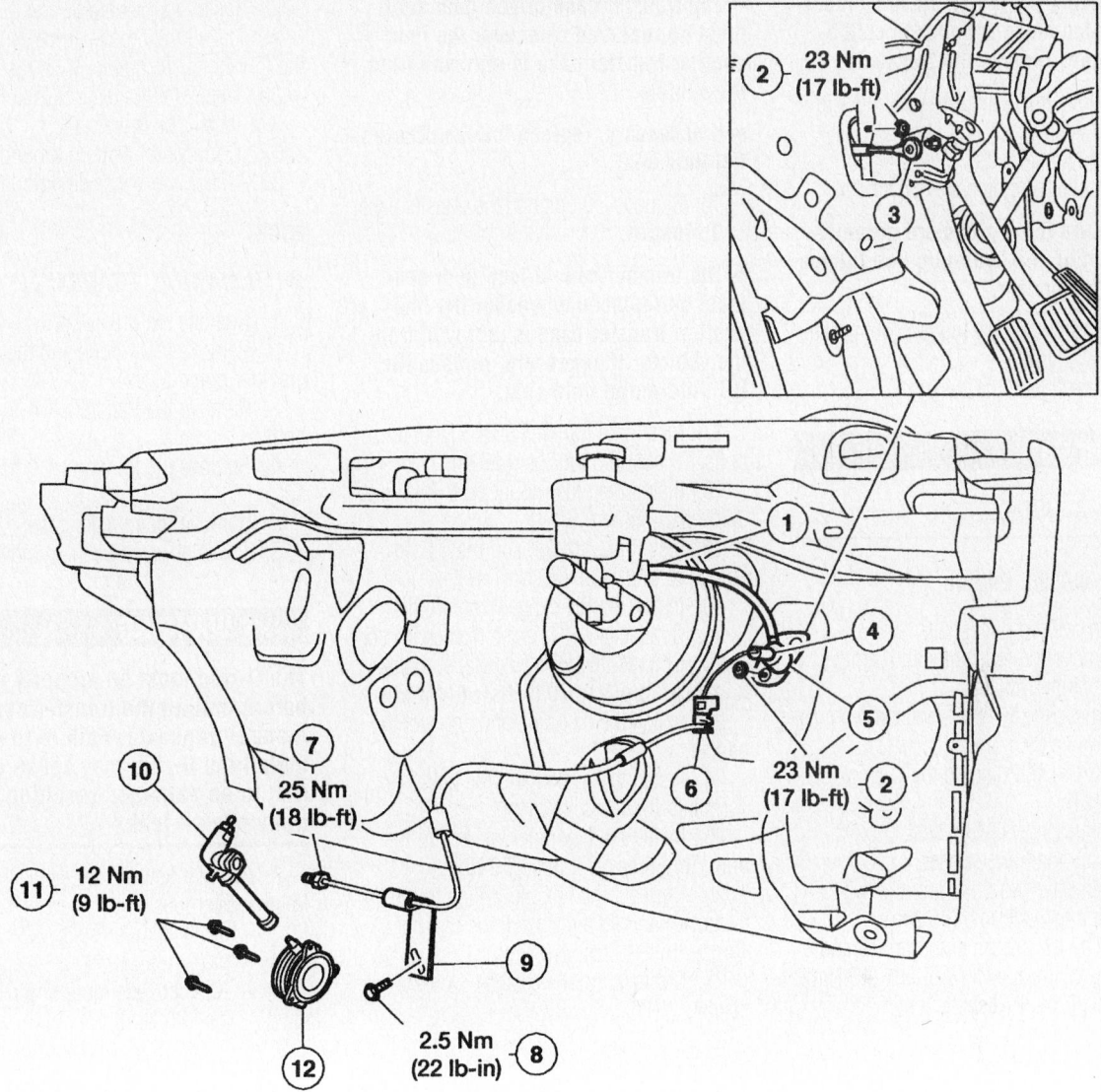

1 Clutch master cylinder hose
2 Nut
3 Clutch master cylinder push rod
4 Fitting

5 Clutch master cylinder
6 Clutch line
7 Fitting
8 Bolt

9 Clutch line bracket
10 Clutch slave cylinder to clutch line adapter
11 Bolt
12 Clutch slave cylinder

67197-ESCA-G60

Hydraulic clutch components—2005 models shown

5. To install, reverse the removal procedure. Torque the mounting nuts to 17 ft. lbs. (23 Nm).

6. Fill and bleed the system.

Slave Cylinder

2001–04

❋❋ **WARNING**

Brake fluid is harmful to painted and plastic surfaces. If brake fluid is

spilled onto a painted or plastic surface, wash it immediately with water.

1. Disconnect the clutch slave cylinder tube and plug it to prevent excess fluid loss.

2. Remove the clutch slave cylinder bolts.

3. Remove the clutch slave cylinder.

4. To install, reverse the removal procedure. Observe the following torques:

- Slave cylinder mounting bolts: 15 ft. lbs. (20 Nm)
- Fluid line-to-slave cylinder: 18 ft. lbs. (25 Nm)

5. Bleed the air from the system.

2005

❋❋ **WARNING**

Brake fluid is harmful to painted and plastic surfaces. If brake fluid is spilled onto a painted or plastic surface, wash it immediately with water.

➡ If removing the clutch line, remove the engine air cleaner.

➡ **If removing the clutch slave cylinder or the clutch slave cylinder to clutch line adapter, remove the transaxle.**

1. Remove the clutch slave cylinder to clutch line adapter.
2. Remove the bolts.
3. Remove the clutch slave cylinder.

➡ **Make sure the O-rings are properly positioned on the hydraulic line fittings or leaks may occur.**

4. To install, reverse the removal procedure.
5. Fill and bleed the system.

Transfer Case

REMOVAL & INSTALLATION

2001–02 w/2.0L Engine

1. Remove the driveshaft.
2. Remove the four bolts and remove the crossmember brace.
3. Remove the two bolts and the two nuts.
4. Remove the transfer case.
5. Remove and discard the O-ring seal.

To install:

6. Install a new O-ring seal.
7. Install the transfer case.
8. Install the two bolts and the two nuts. Torque to 33 ft. lbs. (45 Nm).
9. Install the crossmember brace and the four bolts. Torque to 30 ft. lbs. (40 Nm).
10. Install the driveshaft.

2001–04

1. Disconnect the battery.
2. Drain the transfer case.
3. Remove the driveshaft.
4. Remove the 4 bolts and the crossmember brace.
5. Remove the generator.
6. Disconnect the RH catalyst monitor.
7. Remove and discard the 2 nuts and separate the flexible Y-pipe from the manifold.
8. Remove and discard the nuts. Discard the gasket.
9. Remove and discard the nuts. Discard the gasket.
10. Remove the flexible pipe. Disconnect the hanger.
11. Disconnect the EGR valve to exhaust manifold tube at the manifold.
12. Disconnect the RH heated oxygen sensor.
13. Remove the RH exhaust manifold. Discard the gasket.
14. Remove the support bracket.
15. Remove the 3 transfer case bolts.

➡ **The transfer case driven gear seal must be replaced whenever the link shaft or transfer case is removed from the vehicle.**

➡ **If necessary, replace the RH differential fluid seal.**

16. Remove the bolt and the transfer case.

To install:

➡ **The transfer case driven gear seal must be replaced whenever the link shaft or transfer case is removed from the vehicle. If necessary, replace the RH differential fluid seal.**

17. Install the transfer case. Install the bolts. Torque to 33 ft. lbs. (45 Nm).
18. Install the side mount bolt. Torque to 33 ft. lbs. (45 Nm).
19. Install the bracket and the 6 bolts. Torque to 30 ft. lbs. (40 Nm).
20. Install the RH exhaust manifold.
21. Connect the RH heated oxygen sensor.
22. Connect the EGR to exhaust manifold tube. Torque to 30 ft. lbs. (40 Nm).
23. Install the flex pipe and connect the hanger.
24. Install the gasket and the nuts. Torque to 37 ft. lbs. (50 Nm).
25. Install the flex pipe gasket and the nuts. Torque to 37 ft. lbs. (50 Nm).
26. Connect the flexible Y-pipe to the exhaust manifold and install the two nuts. Torque to 30 ft. lbs. (40 Nm).
27. Connect the RH catalyst monitor sensor.

28. Install the generator.
29. Install the crossmember brace and the four bolts. Torque to 30 ft. lbs. (40 Nm).
30. Check the transfer case fluid level.
31. Install the driveshaft.
32. Connect the battery ground cable.
33. Check the transaxle fluid level.

2005

WITH MANUAL TRANSAXLE

1. Remove the driveshaft.
2. Remove the 4 bolts and the crossmember brace.
3. Remove the transfer case-to-transaxle bolts.
4. Remove the transfer case-to-transaxle nut.
5. Remove the transfer case.
6. To install, reverse the removal procedure.

✳✳ WARNING

The O-ring must be properly installed before mating the transfer case to the manual transaxle. Failure to properly install the O-ring may cause the O-ring to be damaged resulting in transaxle oil leak.

7. Install a new O-ring seal. Observe the following torques:
- Crossmember bolts: 30 ft. lbs. (40 Nm).
- Transfer case mounting nuts: 33 ft. lbs. (45 Nm).

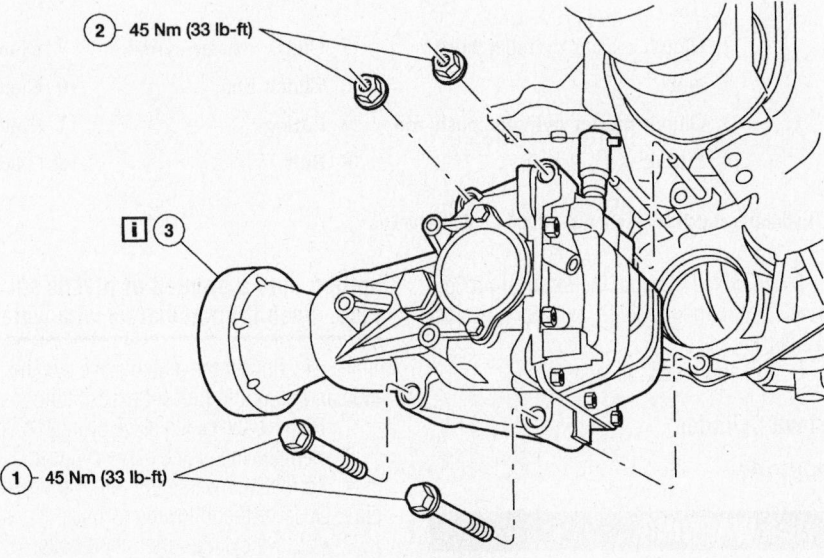

1 Transfer case-to-transaxle bolts
2 Transfer case-to-transaxle nuts
3 Transfer case

67197-ESCA-G37

Transfer case mounting with manual transaxle—2005 model shown

- Transfer case mounting bolts: 33 ft. lbs. (45 Nm).

WITH AUTOMATIC TRANSAXLE

1. Disconnect the battery.
2. Drain the transfer case.
3. Remove the front RH intermediate shaft.
4. Remove the driveshaft.
5. Remove the 4 bolts and the cross-member brace.
6. Remove the generator.
7. Remove the exhaust as required.
8. Remove the heat shield.

9. Remove the transfer case-to-transaxle bolts.
10. Remove the transfer case.

✳✳ WARNING

A new transfer case driven gear seal must be installed whenever the intermediate shaft or transfer case is removed from the vehicle.

➡**If necessary, replace the RH differential fluid seal.**

11. To install, reverse the removal procedure. Observe the following torques:

- Crossmember bolts: 30 ft. lbs. (40 Nm).
- Transfer case-to-transaxle bolts: 33 ft. lbs. (45 Nm).

Front Halfshaft

REMOVAL & INSTALLATION

1. Before servicing the vehicle, refer to the precautions in the beginning of this section.
2. Place the transmission in the PARK position.

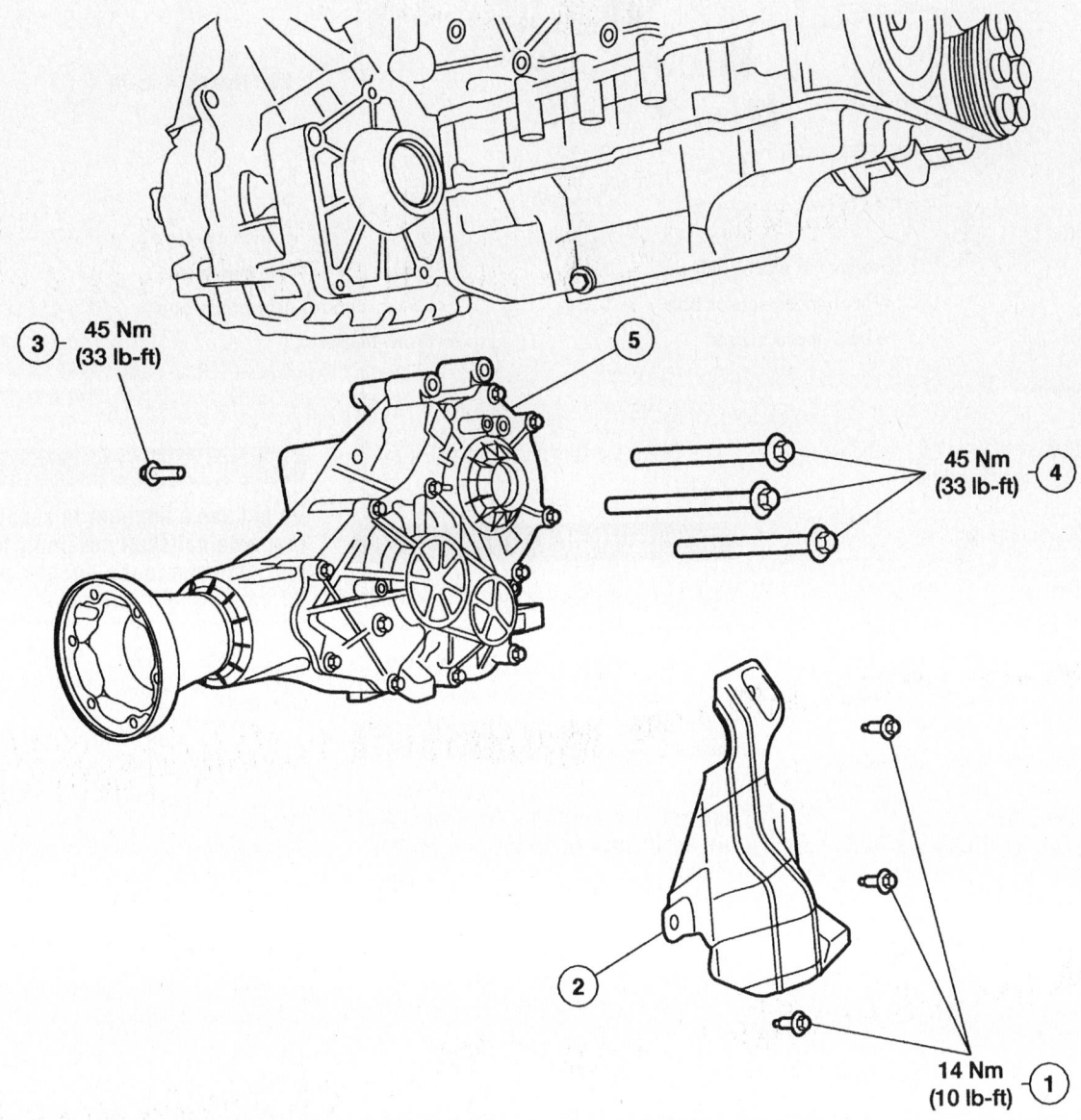

1 Heat shield bolt
2 Heat shield
3 Transfer case-to-transaxle bolts

4 Transfer case-to-transaxle bolts
5 Transfer case

Transfer case mounting with automatic transaxle—2005 model shown

67197-ESCA-G38

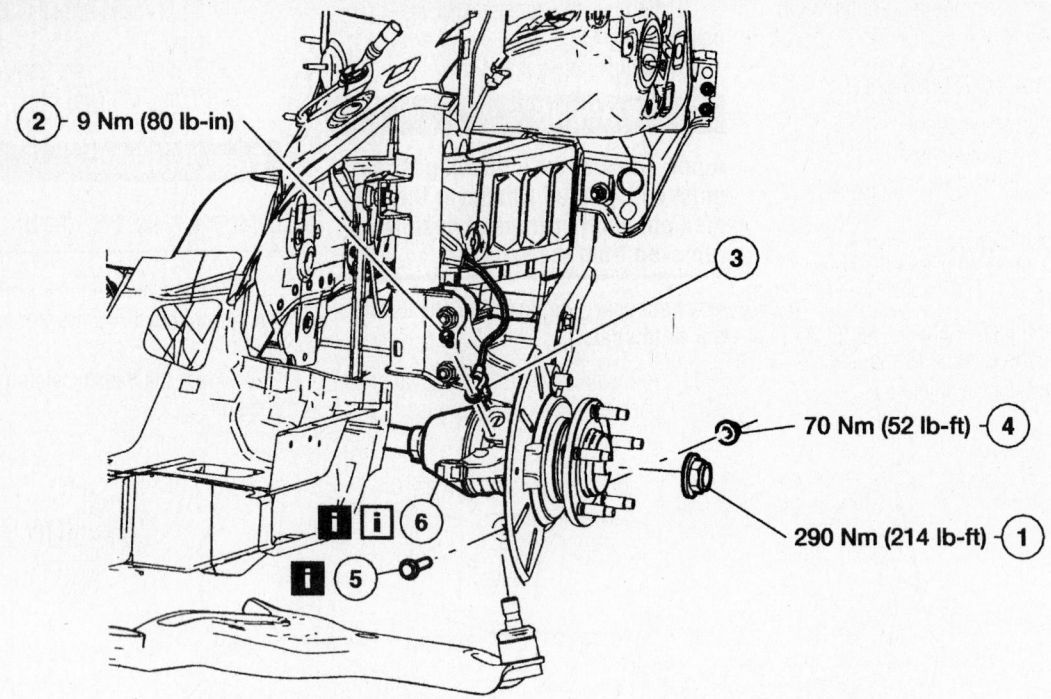

1 Front axle wheel hub nut
2 Wheel speed sensor bolt
3 Wheel speed sensor

4 Lower control arm pinch nut
5 Lower control arm pinch bolt
6 Front halfshaft

67197-ESCA-G39

Front halfshaft

3. Remove or disconnect the following:
 • Negative battery cable
 • Front wheel
 • Front brake disc
 • Front axle wheel hub nut and discard the nut
 • Tie rod end and separate the lower ball from the steering knuckle
 • Halfshaft from the steering knuckle
 • Halfshaft

To install:

4. When seated properly, the halfshaft bearing retainer circlip will snap into the differential side gear groove.

5. Position the halfshaft and joint so that the splines align with differential side gear splines. Push the halfshaft into side gear.

6. Install or connect the following:
 • Halfshaft into the steering knuckle
 • Lower ball joint to steering knuckle. Torque the pinch bolt to 52 ft. lbs. (70 Nm).
 • Tie rod end. Torque the nut to 41 ft. lbs. (55 Nm).
 • New front axle wheel hub nut. Torque the nut to 214 ft. lbs. (290 Nm).
 • Front brake disc
 • Front wheel
 • Negative battery cable

7. Check the fluid level and adjust as needed.

Rear Halfshaft

REMOVAL & INSTALLATION

1. Place the selector lever in NEUTRAL.
2. Raise and support the vehicle.

❋❋ WARNING

Do not loosen the rear axle wheel hub retainer until after the wheel and tire assembly are removed from the vehicle. Wheel bearing damage will occur if the wheel bearing is unloaded with the weight of the vehicle applied.

3. Remove the rear brake drum or brake disc.
4. Remove the rear coil spring.
5. Rear axle wheel hub retainer.
6. Using a puller, press the outboard CV joint until it is loose in the hub.
7. Separate the outboard CV joint from the hub.
8. Remove the rear axle hub assembly.
9. Remove the lower ball joint pinch bolt nut.

❋❋ WARNING

Do not use a hammer to separate the rear axle halfshaft assembly from the hub. Damage to the threads and the internal CV joint can result.

10. Using a prybar, remove the halfshaft.
11. Using a puller, remove the stub and shaft seal.
12. Using a slidehammer and adapter, remove the stub shaft pilot bearing and seal.
13. To install, reverse the removal procedure.
14. Fill the axle with the specified quantity of the specified lubricant.

➡**Lubricate the new stub shaft pilot bearing with rear axle lubricant.**

15. Using suitable drivers, install the stub shaft pilot bearing.

➡**Lubricate the new stub shaft pilot bearing housing seal with grease.**

16. Using the special tools, install the stub shaft pilot bearing housing seal.
17. Install a new circlip on the inboard CV joint.
18. Install the halfshaft end into the hub assembly.
19. Observe the following torques:

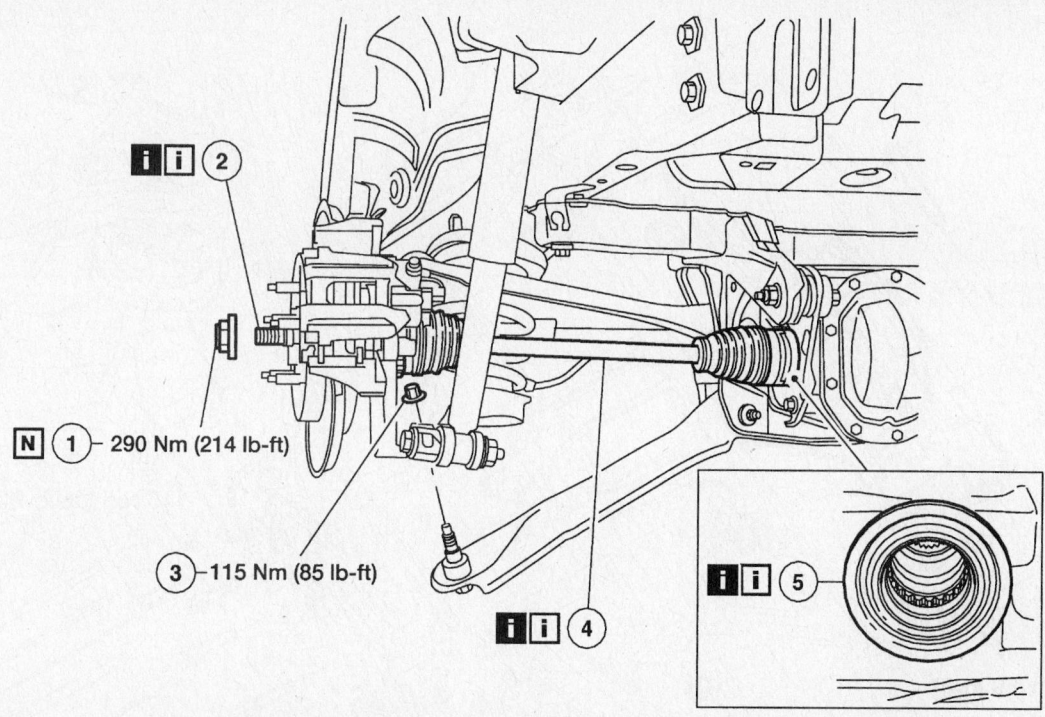

1 Rear axle wheel hub retainer
2 Rear axle hub assembly
3 Lower ball joint pinch bolt nut
4 Rear axle halfshaft assembly
5 Stub shaft seal and bearing

67197-ESCA-G40

Rear halfshaft assembly

- Lower ball joint: 85 ft. lbs. (115 Nm).
- Halfshaft nut: 214 ft. lbs. (290 Nm).

CV-Joints

OVERHAUL

1. Before servicing the vehicle, refer to the precautions in the beginning of this section.
2. Remove or disconnect the following:
   - Negative battery cable
   - Halfshaft and secure it in a soft-jawed vise
   - Inboard halfshaft boot clamp
   - Boot from the inboard CV-joint housing
   - Tripod joint from the CV-joint housing and matchmark the tripod joint to the halfshaft
   - Snapring and boot from the halfshaft
   - Outboard halfshaft boot clamps
   - Outboard boot back to expose the CV-joint and matchmark the joint to the halfshaft
   - Outboard CV-joint from the halfshaft
   - Halfshaft retainer circlip and discard it
   - Boot from the halfshaft

To install:
3. Install or connect the following:
   - Outboard CV-joint and boot
   - New halfshaft bearing circlip
   - Inboard CV-joint to the halfshaft
   - Outboard halfshaft boot forward on to the outboard CV-joint
   - New outboard halfshaft boot clamps
   - Inboard halfshaft boot
   - Tripod joint on the halfshaft by aligning the matchmarks
   - New snapring to the tripod joint and lubricate the needle bearings while filling the housing with CV-joint grease, E43Z–19590–A
   - Inboard halfshaft boot with new clamps
   - Halfshaft
   - Negative battery cable

Rear Drive Axle Housing

REMOVAL & INSTALLATION

2001–04

1. Before servicing the vehicle, refer to the precautions in the beginning of this section.

2. Remove or disconnect the following:
   - Negative battery cable
   - Rotary blade coupling
   - Rear halfshafts
   - Axle assembly-to-front bracket bolts
   - Rear axle-to-side bracket bolts
   - Axle assembly

To install:
3. Install or connect the following:
   - Axle assembly
   - Rear axle-to-side-bearing bolts. Torque the bolts to 59 ft. lbs. (80 Nm).
   - Axle assembly-to-front bracket bolts. Torque the bolts to 59 ft. lbs. (80 Nm).
   - Rotary blade coupling
   - Negative battery cable

2005

1. Remove the spare tire.
2. Remove the rear driveshaft assembly.
3. Remove the rear halfshafts.
4. Position a suitable transmission hydraulic jack to the axle housing. Securely strap the jack to the housing.
5. Remove the electrical connector at the axle.
6. Remove the rear axle differential housing-to-front insulator bracket bolts.

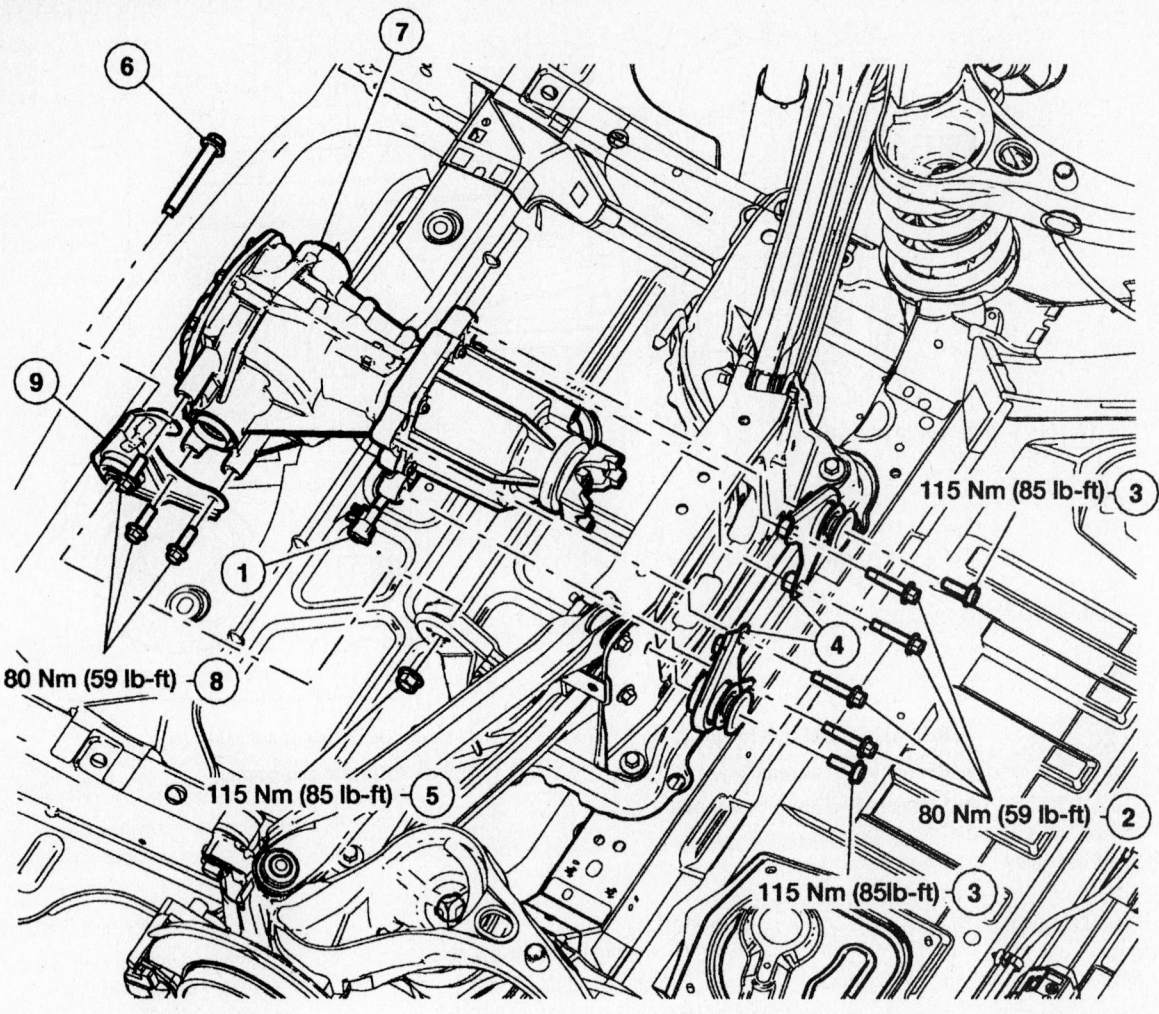

1 Electrical connector

2 Rear axle differential housing-to-front insulator bracket bolts

3 Front insulator-to-bracket subframe bolts

4 Front insulator brackets

5 Side insulator bracket-to-subframe nut

6 Side insulator bracket-to-subframe bolt

7 Rear axle assembly

8 Side insulator bracket-to-rear axle differential bolts

9 Side insulator bracket

67197-ESCA-G41

Rear drive axle removal—2005 models shown

7. Remove the front insulator-to-bracket subframe bolts.

8. Remove the front insulator brackets.

9. Remove the side insulator bracket-to-subframe nut.

10. Remove the side insulator bracket-to-subframe bolt.

11. Remove the rear axle assembly.

12. Remove the side insulator bracket-to-rear axle differential bolts.

13. Remove the side insulator bracket.

14. To install, reverse the removal procedure. Observe the following torques:
- Axle housing-to-front insulator bracket: 59 ft. lbs. (80 Nm)
- Front insulator-to-bracket subframe bolts: 85 ft. lbs. (115 Nm)

- Side insulator bracket-to-subframe nut: 85 ft. lbs. (115 Nm)
- Side insulator bracket-to-differential bolts: 59 ft. lbs. (80 Nm)

Front Driveshaft

REMOVAL & INSTALLATION

1. Place the selector lever in NEUTRAL.
2. Raise and support the vehicle.
3. Remove the front driveshaft-to-transfer case bolts.
4. Remove the universal joint cap bolts. Index-mark the pinion and yoke to the rear driveshaft before removing the bolts.

5. Remove the universal joint cap straps.
6. Remove the front driveshaft.
7. To install, reverse the removal procedure.

Rear Driveshaft

REMOVAL & INSTALLATION

1. Place the selector lever in NEUTRAL.
2. Raise and support the vehicle.
3. Remove the ground strap bolt
4. Remove the universal joint cap bolts. Index-mark the pinion and yoke to the rear driveshaft before removing the bolts.
5. Remove the universal joint cap straps.

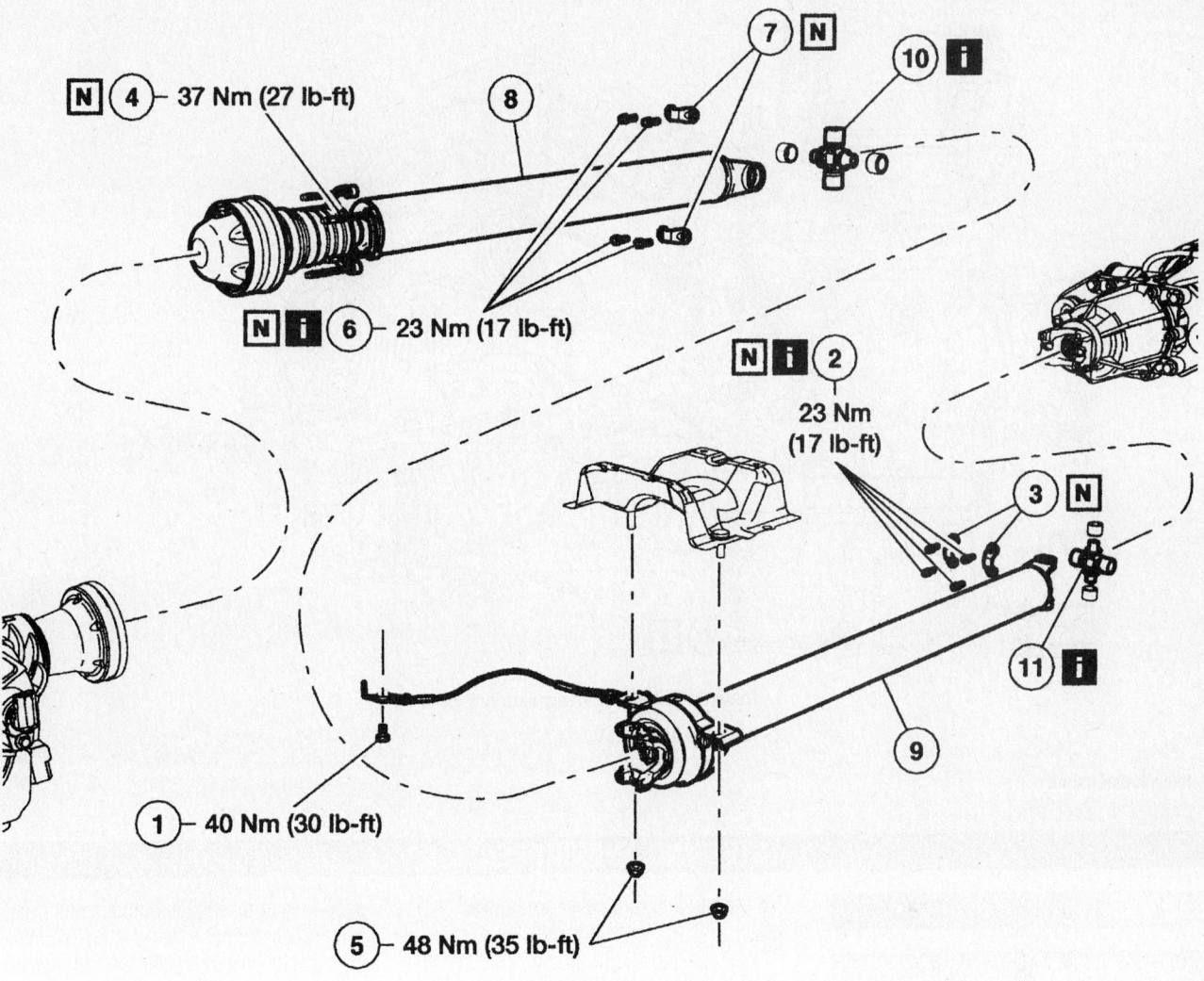

4 — 37 Nm (27 lb-ft)

7

8

10

 6 — 23 Nm (17 lb-ft)

 2
23 Nm
(17 lb-ft)

3

11

1 — 40 Nm (30 lb-ft)

9

5 — 48 Nm (35 lb-ft)

1 Ground strap bolt
2 Universal joint cap bolts
3 Universal joint cap straps
4 Front driveshaft-to-transfer case bolts

5 Center bearing support nuts
6 Universal joint cap bolts
7 Universal joint cap straps
8 Front driveshaft

9 Rear driveshaft
10 Front driveshaft U-joint
11 Rear driveshaft U-joint

67197-ESCA-G42

Rear driveshaft assembly

6. Remove the center bearing support nuts.
7. Remove the universal joint cap bolts.
8. Remove the universal joint cap straps.
9. Remove the rear driveshaft.
10. To install, reverse the removal procedure.

Intermediate Shaft

REMOVAL & INSTALLATION

➡**If removing the intermediate shaft in order to repair a separate component, it should only be removed as an assembly with the RH front halfshaft.**

1. Remove the RH front halfshaft.
2. Remove the inner halfshaft bearing retainer nuts
3. Remove the intermediate shaft
4. To install, reverse the removal procedure. Apply a thin coat of grease to the splines of the intermediate shaft.
5. Verify the front axle lubricant level is to specifications.

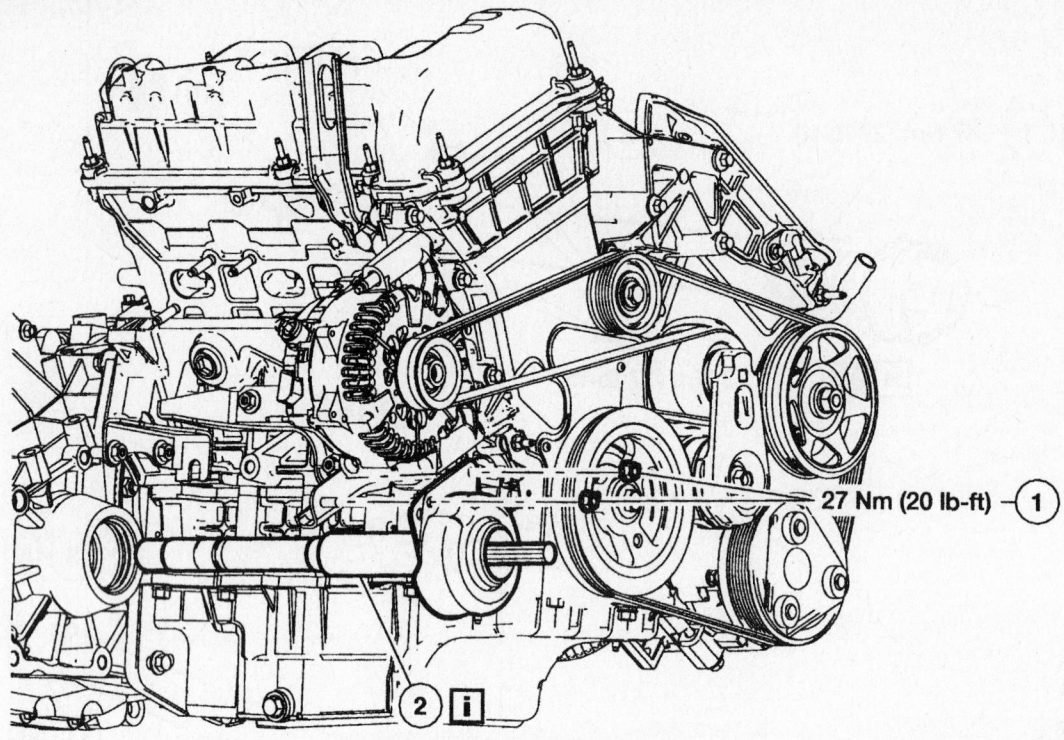

1 Inner halfshaft bearing retainer nuts
2 Intermediate shaft

67197-ESCA-G43

Intermediate shaft

STEERING AND SUSPENSION

Air Bag

※ CAUTION

Some vehicles are equipped with an air bag system. The system MUST BE disabled before performing service on or around system components, steering column, instrument panel components, wiring and sensors. Failure to follow safety and disabling procedures could result in accidental air bag deployment, possible personal injury and unnecessary system repairs.

PRECAUTIONS

Several precautions must be observed when handling the inflator module to avoid accidental deployment and possible personal injury:

1. Never carry the inflator module by the wires or connector on the underside of the module.

2. When carrying a live inflator module, hold securely with both hands and ensure that the bag and trim cover are pointed away.

3. Place the inflator module on a bench or other surface with the bag and trim cover facing up.

4. With the inflator module on the bench, never place anything on or close to the module, which may be thrown in the event of an accidental deployment.

DISARMING

※ CAUTION

The Supplemental Inflatable Restraint (SIR) system must be disarmed before performing service around SIR system components or SIR system wiring. Failure to do so may cause accidental deployment of the air bag, resulting in unnecessary SIR system repairs and/or personal injury.

The positive battery cable must be disconnected for a minimum of 1 minute before beginning any air bag work to de-energize the back-up power supply. It is a good idea to disengage both the positive and negative battery cables to ensure that the Air Bag system is definitely discharged.

ARMING THE SYSTEM

※ WARNING

If the air bag simulators have been used, the air bag simulators must be removed and the air bags reconnected when the system is reactivated to avoid non-deployment in a collision resulting in possible personal injury.

1. Disconnect the positive battery cable.

2. Wait 1 minute, this is required for the back-up power supply in the air bag diagnostic monitor to deplete its stored energy.

3. Remove the air bag simulator from the air bag sliding contact connector at the top of the steering column. Reconnect the driver's side air bag module assembly. Position the driver's air bag module on the steering wheel and secure with the 2 bolts and washers. Tighten the bolt and washer assembly to 8–10 ft. lbs. (10–14 Nm).

4. Connect the positive battery cable.
5. Turn the ignition switch from the **OFF** to **RUN** and visually monitor the air bag warning indicator. The light will illuminate continuously for approximately 6 seconds and then turn off. If a fault occurs, the air bag indicator will either fail to light, remain lighted continuously or flash. The flashing may not occur until approximately 30 sec-

onds after the ignition switch has been turned from **OFF** to **RUN**. This is the time needed for the air bag diagnostic monitor to complete testing the system. If the air bag indicator is inoperative, an air bag system fault exists, a tone will sound in a pattern of 5 sets of 5 beeps. If this occurs, the air bag indicator will need to be serviced before further diagnostics can be done.

Steering Gear

REMOVAL & INSTALLATION

1. Before servicing the vehicle, refer to the precautions in the beginning of this section.
2. Place the steering wheel in the

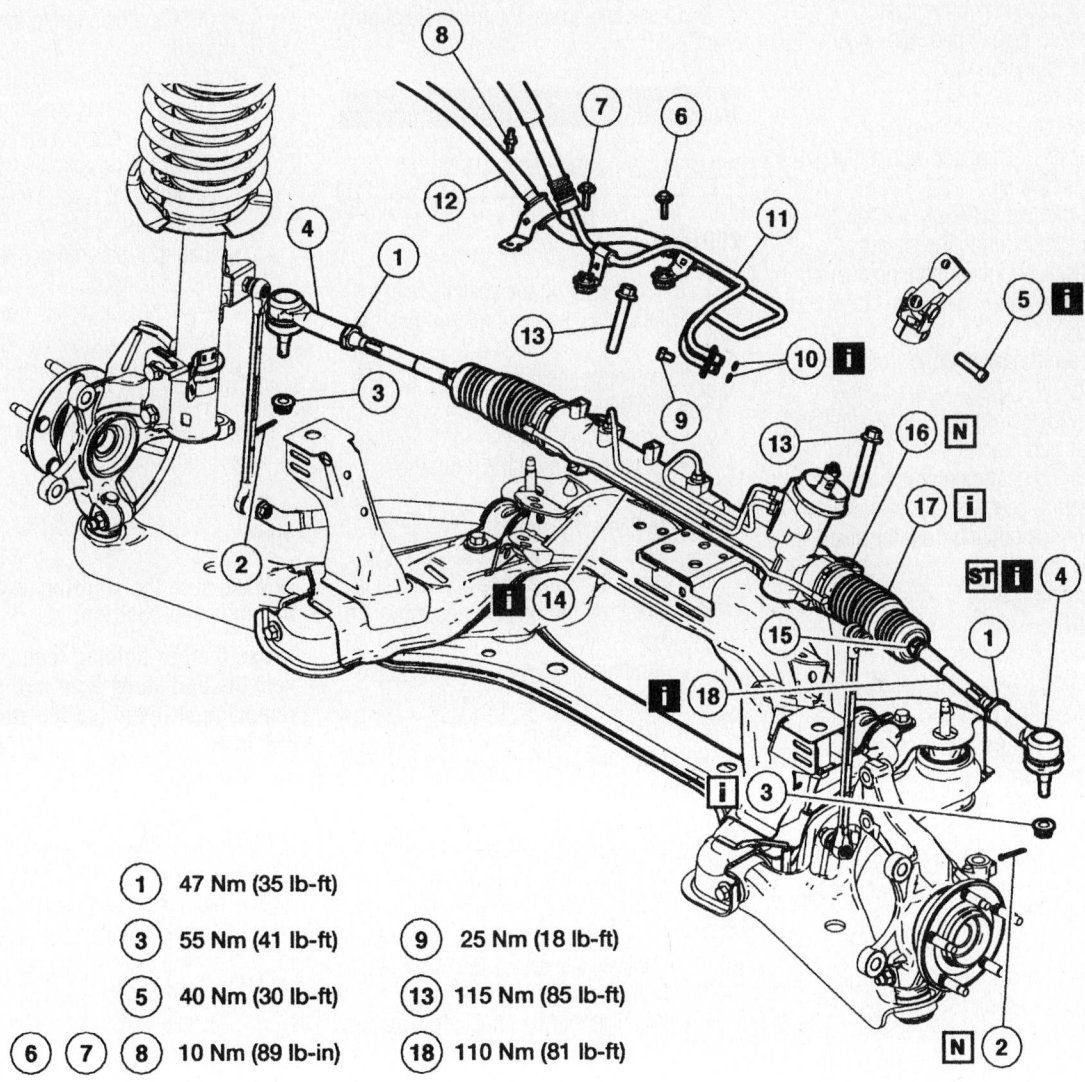

① 47 Nm (35 lb-ft)	
③ 55 Nm (41 lb-ft)	⑨ 25 Nm (18 lb-ft)
⑤ 40 Nm (30 lb-ft)	⑬ 115 Nm (85 lb-ft)
⑥ ⑦ ⑧ 10 Nm (89 lb-in)	⑱ 110 Nm (81 lb-ft)

1 Tie-rod jam nuts (loosen)
2 Cotter pins
3 Tie-rod end nuts
4 Tie-rod end, outer
5 Steering column coupling-to-steering gear pinch bolt
6 Power steering return line bracket-to-subframe bolt
7 Power steering pressure line bracket-to-steering gear bolt
8 Power steering return line bracket-to-steering gear stud
9 Power steering line clamp plate bolt

10 O-rings
11 Power steering pressure line
12 Power steering return line
13 Steering gear mounting bolts
14 Steering gear
15 Tie-rod boot clamp, outer
16 Tie-rod boot clamp, inner
17 Tie-rod boot
18 Tie-rod, inner

67197-ESCA-G44

Steering gear mounting

straight-ahead position. Lock the steering wheel in place, using a steering wheel holder.

➡ **Locking the steering wheel keeps the clockspring in alignment position.**

3. Drain the power steering fluid.
4. Remove or disconnect the following:
   - Negative battery cable
   - Rear transmission insulator
   - Rear transmission insulator bracket, if equipped with an automatic transmission
   - Both front wheels
   - Rear transmission insulator bracket, if equipped with a manual transmission
   - Tie rod end cotter pin and nut
   - Tie rod end from the steering knuckle and record the number of turns required to remove the tie rod end
   - Steering gear coupling pinch bolt
   - Power steering pressure and return lines and bracket
   - Steering gear mounting bolts
   - Steering gear and separate the steering coupling from the steering gear shaft
   - Steering gear

To install:
   - Slide the steering gear rearward to connect the steering coupling to the steering gear shaft

5. Install or connect the following:
   - Steering gear mounting bolts. Torque the bolts to 93 ft. lbs. (126 Nm).
   - Pressure and return lines and bracket. Torque the bracket bolts to 89 inch lbs. (10 Nm).
   - Power steering pressure and return lines to the steering gear. Torque the bolt to 18 ft. lbs. (25 Nm).
   - Steering gear pinch bolt and reposition the boot. Torque the bolt to 18 ft. lbs. (25 Nm).
   - Tie rod end to the tie rod using the number of turns required to remove the tie rod end
   - Jam nuts. Torque the nuts to 35 ft. lbs. (47 Nm).
   - Tie rod end to the steering knuckle. Torque the nut to 41 ft. lbs. (57 Nm) and install a new cotter pin
   - Rear transmission insulator bracket. Torque the bolts to 66 ft. lbs. (90 Nm).

- Both front wheels
- Rear transmission insulator bracket. Torque the bolts to 66 ft. lbs. (90 Nm).
- Rear transmission insulator. Torque the bolts to 66 ft. lbs. (90 Nm).
- Negative battery cable

6. Fill and bleed the power steering system.
7. Start the vehicle and check for leaks, repair if necessary.
8. Check and adjust the front end alignment.

Strut

REMOVAL & INSTALLATION

2001–03

1. Before servicing the vehicle, refer to the precautions in the beginning of this section.

2. Install or connect the following:
   - Negative battery cable
   - Front wheel
   - Brake hose grommet from the bracket
   - Antilock Brake System (ABS) harness from the strut assembly and move the brake hose bracket aside
   - Stabilizer bar link nut and move the bar aside
   - Strut to steering knuckle bolts and support the strut assembly
   - Upper strut nuts
   - Strut and coil spring assembly

To install:
3. Install or connect the following:
   - Strut and spring assembly. Torque the upper nuts to 59 ft. lbs. (80 Nm).
   - Lower strut assembly to the steering knuckle. Torque the lower bolts to 85 ft. lbs. (115 Nm).
   - Stabilizer bar into position. Torque the bolts to 35 ft. lbs. (48 Nm).
   - Brake hose bracket. Torque the bolts to 14 ft. lbs. (18 Nm).
   - ABS harness to the strut assembly, if equipped
   - Brake hose grommet to the bracket
   - Front wheel
   - Negative battery cable

2004

1. Before servicing the vehicle, refer to the precautions in the beginning of this section.

2. Install or connect the following:
   - Negative battery cable

- Front wheel
- Brake hose grommet from the bracket
- Antilock Brake System (ABS) harness from the strut assembly and move the brake hose bracket aside
- Stabilizer bar link nut and move the bar aside
- Strut to steering knuckle bolts and support the strut assembly
- Upper strut nuts
- Strut and coil spring assembly

To install:
3. Install or connect the following:
   - Strut and spring assembly. Torque the upper nuts to 41 ft. lbs. (55 Nm).
   - Lower strut assembly to the steering knuckle. Torque the lower bolts to 85 ft. lbs. (115 Nm).
   - Stabilizer bar into position. Torque the bolts to 35 ft. lbs. (48 Nm).
   - Brake hose bracket. Torque the bolts to 13 ft. lbs. (18 Nm).
   - ABS harness to the strut assembly, if equipped
   - Brake hose grommet to the bracket
   - Front wheel
   - Negative battery cable

2005

➡ **Make sure the steering wheel is in the unlocked position.**

➡ **Use the hex holding feature to prevent the ball studs from turning while removing or installing the stabilizer bar link nuts.**

1. Raise and support the vehicle.
2. Remove the brake jounce hose clip.
3. Remove the brake jounce hose. Pull the brake jounce hose downward slightly to remove the hose from the bracket.
4. Remove the ABS sensor harness bolt.
5. Remove the stabilizer bar link nut.
6. Remove the strut-to-knuckle nuts.
7. Remove the strut-to-knuckle bolts.
8. Remove the strut upper bushing nuts. Reference mark the strut mounting plate nuts.
9. Remove the strut and spring assembly.

✷✷ WARNING

Do not allow the axle shaft to move outboard. Over-extension of the tripod CV joint can result in separation of internal parts, causing failure of the axle shaft.

10. To install, reverse the removal procedure.
11. Align the strut mounting plate nuts to the reference marks.

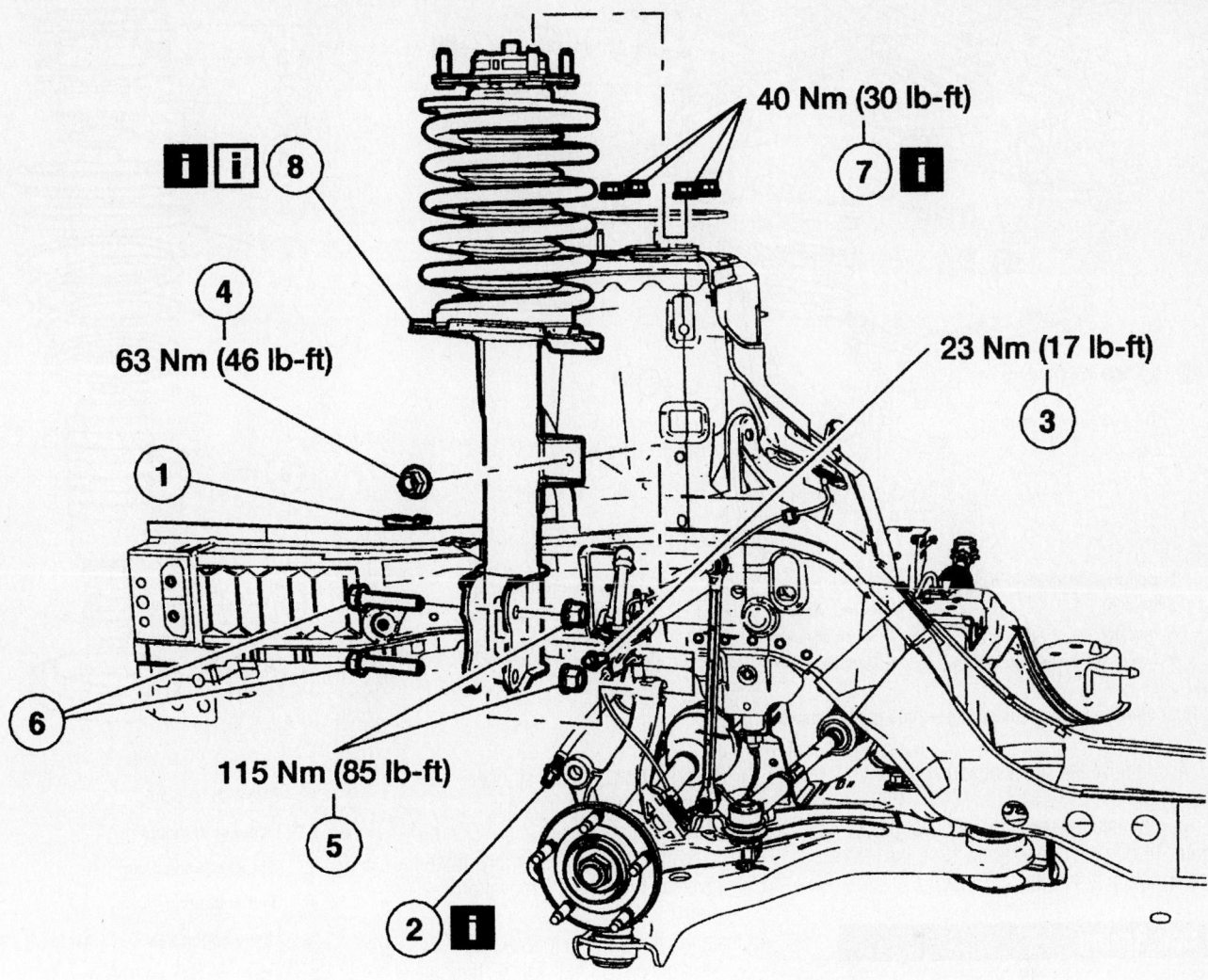

40 Nm (30 lb-ft)

63 Nm (46 lb-ft)

23 Nm (17 lb-ft)

115 Nm (85 lb-ft)

1 Brake jounce hose clip

2 Brake jounce hose (LH/RH)

3 ABS sensor harness bolt

4 Stabilizer bar link nut

5 Strut-to-knuckle nuts

6 Strut-to-knuckle bolts

7 Strut upper bushing nuts

8 Strut and spring assembly

67197-ESCA-G45

Front strut installation—2005 model shown

12. Check the front end alignment and adjust as necessary.

Shock Absorber

REMOVAL & INSTALLATION

Rear

2001–04

1. Before servicing the vehicle, refer to the precautions in the beginning of this section.

2. Remove or disconnect the following:

- Negative battery cable
- Rear quarter trim panel
- Upper shock absorber nut and raise the vehicle enough to relax the suspension
- Lower shock absorber nut
- Shock absorber

To install:

3. Install or connect the following:
- Shock absorber. Torque the lower nut to 85 ft. lbs. (115 Nm).

- Upper shock absorber nut. Torque the nut to 13 ft. lbs. (18 Nm).
- Rear quarter trim panel
- Negative battery cable

200

1. Remove the wheel and tire assemblies.

2. Remove the rear quarter trim panel. Remove the upper shock absorber nut, bushing and washer.

3. Remove the lower shock absorber nut, bolt and washer.

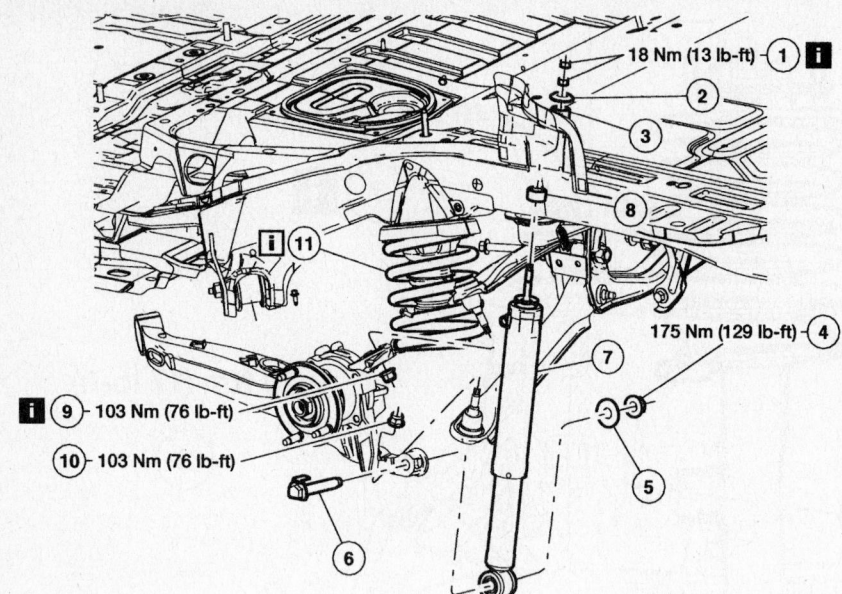

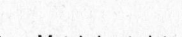

1	Upper shock absorber nuts
2	Washer
3	Bushing
4	Lower shock absorber nut
5	Washer
6	Lower shock absorber bolt
7	Shock absorber
8	Bushing
9	Upper ball joint nut
10	Lower ball joint nut
11	Coil spring

67197-ESCA-G46

Rear shock absorber and spring—2005 model shown

4. Remove the shock absorber and bushing.

5. To install, reverse the removal procedure. Torque the upper nut to 13 ft. lbs.; the lower nut to 129 ft. lbs. (175 Nm).

Coil Spring

REMOVAL & INSTALLATION

Front

1. Before servicing the vehicle, refer to the precautions in the beginning of this section.

2. Install or connect the following:
   - Negative battery cable
   - Front wheel
   - Strut and spring assembly and mount the strut assembly in a holding fixture and compress the coil spring using a suitable tool
   - Strut piston rod nut

3. Coil spring by disassembling the strut in the following sequence:
   a. Step 1: Metal sheet plate.
   b. Step 2: Upper strut mount.
   c. Step 3: Thrust bearing plate.
   d. Step 4: Thrust bearing.
   e. Step 5: Upper spring seat.
   f. Step 6: Upper spring seat isolator.
   g. Step 7: Coil spring.
   h. Step 8: Dust boot.

i. Step 9: Rubber bump stopper.
j. Step 10: Lower spring seat.

To install:
   - Assemble the strut assembly in the reverse order of the removal procedure

4. Install or connect the following:
   - Strut piston rod nut. Torque the nut to 76 ft. lbs. (103 Nm) and remove the assembly from the holding fixture
   - Strut and spring assembly
   - Front wheel
   - Negative battery cable

Rear

2001–04

1. Before servicing the vehicle, refer to the precautions in the beginning of this section.

2. Remove or disconnect the following:
   - Wheel and install 1 lug nut to retain the brake drum
   - Brake line from the wheel cylinder
   - Brake line bracket
   - Bolts from the Antilock Braking System (ABS) sensor bracket and move the sensor aside, if equipped
   - Rear knuckle and loosen the inside upper and lower arm bolts
   - Shock absorber lower nut
   - Spring

To install:
3. Install or connect the following:

1	Metal sheet plate
2	Upper strut mount
3	Thrust bearing plate
4	Thrust bearing
5	Upper spring seat
6	Upper spring seat isolator
7	Spring
8	Dust boot
9	Rubber bump stopper
10	Lower spring seat

9308TG23

Disassemble the strut assembly in the proper sequence

   - Spring to the shock absorber
   - Lower shock absorber nut. Torque the nut to 85 ft. lbs. (115 Nm).
   - Inside upper and lower arm bolts. Torque the bolts to 85 ft. lbs. (115 Nm).
   - ABS sensor bracket, if equipped. Torque the bolts to 80 inch lbs. (9 Nm).
   - Brake line bracket. Torque the bolt to 15 ft. lbs. (20 Nm).
   - Brake line to the wheel cylinder. Torque the fastener to 11 ft. lbs. (15 Nm) and remove the lug nut
   - Wheel
   - Negative battery cable

2005

1. Remove the wheel and tire assemblies.
2. Remove the lower shock absorber nut and washer
3. Remove the lower shock absorber bolt
4. Support the wheel knuckle.
5. Remove the upper ball joint nut.
6. Remove the upper arm inner bolt.
7. Loosen the lower arm inner bolt.

➡**Note the position of the coil spring insulator and coil spring for installation.**

8. Carefully lower the wheel knuckle support.
9. Remove the coil spring
10. To install, reverse the removal procedure.
11. Align the coil spring and coil spring insulator to the previously noted position.
12. Observe the following torques:

- Stabilizer bar link nut: 46 ft. lbs. (63 Nm)
- Strut-to-knuckle nuts: 85 ft. lbs. (115 Nm)
- Upper ball joint nut: 76 ft. lbs. (103 Nm)
- Lower shock absorber nut: 129 ft. lbs. (175 Nm)

Front Lower Control Arm

REMOVAL & INSTALLATION

2001–2003

1. Remove or disconnect the following:
- Negative battery cable
- Front wheel
- Lower ball joint from the knuckle and support the subframe
- Lower control arm

To install:

2. Install or connect the following:
- Lower control arm bolts and hand tighten them
- Pinch bolt to the wheel knuckle. Torque the nut to 52 ft. lbs. (70 Nm) and remove the subframe support
- Front wheel and jounce the vehicle
3. Torque the inner lower control arm bolt to 148 ft. lbs. (200 Nm) and outer bolt 85 ft. lbs. (115 Nm).

2004–05

1. Remove the wheel.
2. Lift the lower arm with a floor jack until the vehicle starts to lift.
3. Record the ride height. It's measure from the center of the halfshaft to the fender lip.
4. Remove the floor jack.
5. Disconnect the ball joint from the knuckle.

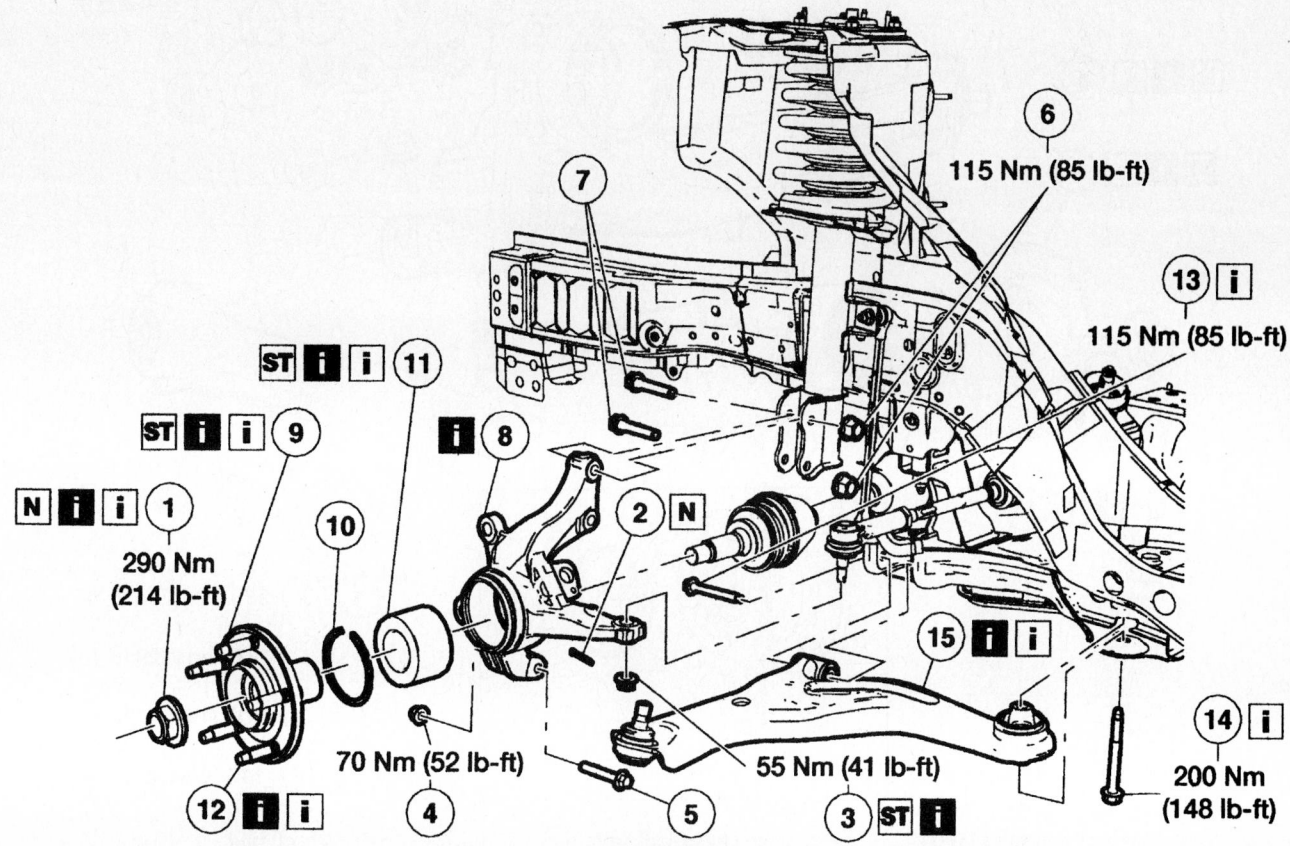

290 Nm (214 lb-ft)

70 Nm (52 lb-ft)

55 Nm (41 lb-ft)

115 Nm (85 lb-ft)

115 Nm (85 lb-ft)

200 Nm (148 lb-ft)

1 Wheel hub nut	6 Strut-to-knuckle nuts	11 Bearing
2 Cotter pin	7 Strut-to-knuckle bolts	12 Wheel stud
3 Tie rod end-to-knuckle nut	8 Wheel knuckle (LH/RH)	13 Lower arm bolt (front)
4 Lower ball joint pinch bolt	9 Wheel hub	14 Lower arm bolt (rear)
5 Lower ball joint pinch bolt	10 Snap ring	15 Lower arm

67197-ESCA-G48

Front lower control arm assembly—2004–05 model shown

6. Support the sub-frame and remove the lower arm.

To install:

7. Install the lower arm, with the bolts loose.

8. Connect the ball joint. Torque the bolt to 52 ft. lbs. (70 Nm).

9. Remove the support.

10. Position the jack under the ball joint and raise the arm to the previously recorded ride height.

11. Tighten the lower arm bolts. Horizontal 85 ft. lbs. (115 Nm); vertical 148 ft. lbs. (200 Nm).

Rear Lower Control Arm

REMOVAL & INSTALLATION

1. Remove or disconnect the follow-ing:
   - Negative battery cable

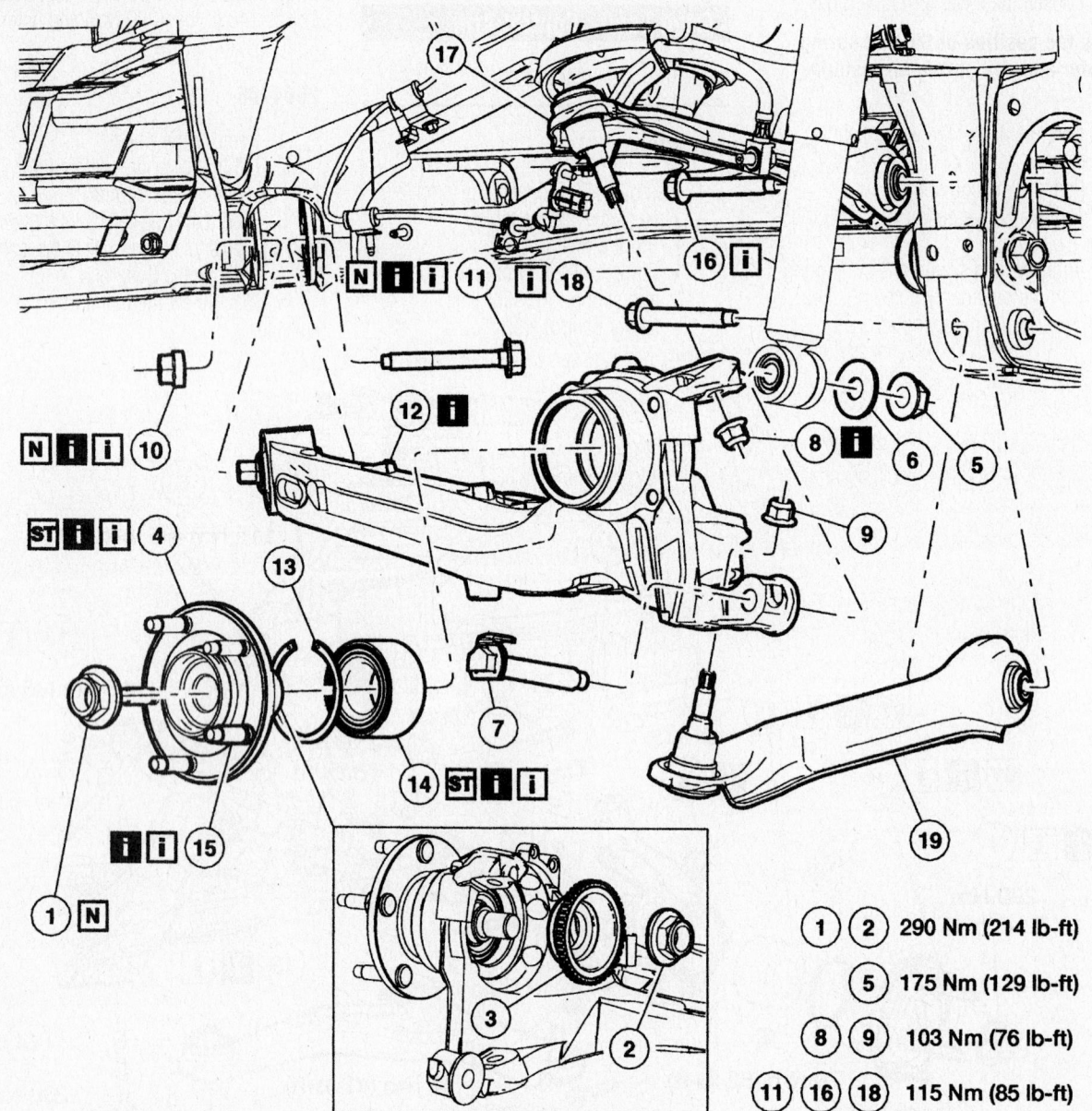

1	2	290 Nm (214 lb-ft)
5		175 Nm (129 lb-ft)
8	9	103 Nm (76 lb-ft)
11	16	18	115 Nm (85 lb-ft)

1 Wheel hub nut (4WD)	8 Upper ball joint nut	15 Wheel stud
2 Wheel hub nut (2WD)	9 Lower ball joint nut	16 Upper arm inner bolt
3 ABS sensor ring (2WD)	10 Cam nut	17 Upper arm
4 Wheel hub	11 Wheel knuckle bolt	18 Lower arm inner bolt
5 Lower shock absorber nut	12 Knuckle assembly (LH/RH)	19 Lower arm
6 Washer	13 Wheel bearing snap ring	
7 Lower shock absorber bolt	14 Wheel bearing	

Rear lower control arm and related parts

- Lower ball joint from the knuckle while holding the ball joint stud from moving
- Lower ball joint nut
- Lower control arm
- Lower control arm inner bolt

To install:

2. Install or connect the following:
- Lower control arm inner bolt
- Lower control arm. Torque the bolts to 85 ft. lbs. (115 Nm).
- Lower ball joint nut
- Lower ball joint the knuckle. Torque the ball joint nut to 76 ft. lbs. (103 Nm).
- Rear wheel
- Negative battery cable

Rear Upper Control Arm

REMOVAL & INSTALLATION

1. Remove the wheel and tire.

➡**It may be necessary to hold the ball joint stud to keep it from turning while removing the nut.**

2. Separate the upper arm from the wheel knuckle. Remove the upper ball joint nut.
3. Remove the upper arm inner bolt.
4. Remove the upper arm.
5. To install, reverse the removal procedure. Observe the following torques:
- Ball joint nut: 76 ft. lbs. (103 Nm)
- Lower arm bolts: 85 ft. lbs. (115 Nm)

Wheel Bearings

REMOVAL & INSTALLATION

Front

1. Before servicing the vehicle, refer to the precautions in the beginning of this section.
2. Remove or disconnect the following:

- Negative battery cable
- Front wheel
- Brake disc
- Wheel hub nut
- Tie rod end cotter pin and nut
- Tie rod end from the knuckle
- Antilock Brake System (ABS) sensor bolt and move the sensor aside, if equipped
- Lower ball joint from the knuckle

- Halfshaft from the knuckle and properly support the halfshaft
- Steering knuckle

3. Press the hub from the wheel bearing and knuckle
4. Press the inner wheel bearing race from the knuckle and remove the snapring

➡**The above step will not be necessary if the inner race remains with the knuckle.**

5. Press the outer wheel bearing race from the knuckle

To install:

6. Install or connect the following:
- Wheel bearing into the steering knuckle
- Snapring
- Wheel hub into the wheel bearing by using a press
- Steering knuckle. Torque the bolts to 85 ft. lbs. (115 Nm).
- Halfshaft into the wheel hub
- Pinch bolt to knuckle. Torque the nut to 52 ft. lbs. (70 Nm).
- Ball joint stud into the knuckle
- ABS sensor. Torque the bolt to 80 inch lbs. (9 Nm), if equipped
- Tie rod end to the knuckle. Torque the nut to 41 ft. lbs. (55 Nm).
- New cotter pin to the tie rod end nut
- Wheel hub. Torque the nut to 214 ft. lbs. (290 Nm).
- Brake disc
- Front wheel
- Negative battery cable

Rear

2WD VEHICLES

1. Before servicing the vehicle, refer to the precautions in the beginning of this section.
2. Remove or disconnect the following:
- Negative battery cable
- Rear wheel
- Rear brake drum
- Wheel hub nut
- Wheel hub
- Inner wheel bearing race from the hub
- Snapring
- Wheel bearing outer race from the knuckle

To install:

3. Install or connect the following:
- Wheel bearing in to the knuckle
- Snapring
- Wheel hub into the wheel bearing

- Wheel hub nut. Torque the nut to 214 ft. lbs. (290 Nm).
- Brake drum
- Rear wheel
- Negative battery cable

4WD VEHICLES

1. Before servicing the vehicle, refer to the precautions in the beginning of this section.
2. Remove or disconnect the following:
- Negative battery cable
- Rear wheel
- Rear brake shoes
- Rear halfshaft nut and loosen the halfshaft from the hub
- Wheel hub and place it in a vise
- Inner wheel bearing race from the hub
- Antilock Brake System (ABS) sensor bracket and move the sensor aside, if equipped
- Parking brake cable from the steering knuckle
- Brake line from the wheel cylinder and support the knuckle
- Lower shock absorber nut
- Lower ball joint by holding the ball joint stud
- Upper ball joint
- Coil spring while noting the location of the insulator
- Steering knuckle cam
- Steering knuckle
- Snapring and press out the outer wheel bearing race from the knuckle

To install:

3. Install or connect the following:
- New wheel bearing into the steering knuckle
- Snapring to the knuckle
- Wheel hub
- Steering knuckle cam and hand tighten the bolt
- Coil spring
- Shock absorber lower nut. Torque the nut to 85 ft. lbs. (115 Nm) for 2001–04 models; 129 ft. lbs. (175 Nm).
- Upper ball joint. Torque the nut to 76 ft. lbs. (103 Nm).
- Lower ball joint. Torque the nut to 76 ft. lbs. (103 Nm). Align the steering knuckle cam and torque the bolt to 85 ft. lbs. (115 Nm).
- Brake line to the wheel cylinder. Torque the brake line bracket bolt to 15 ft. lbs. (20 Nm) and the

brake line fastener to 11 ft. lbs. (15 Nm).
- Parking brake cable to the backing plate. Torque the bolt to 16 ft. lbs. (22 Nm).

- ABS sensor bracket. Torque the bolt to 80 inch lbs. (9 Nm), if equipped
- Halfshaft nut. Torque the nut to 214 ft. lbs. (290 Nm).
- Brake shoes

- Rear wheel
- Negative battery cable
4. Fill and bleed the brake system.
5. Check and adjust the wheel alignment as needed.

BRAKES

Brake Caliper

REMOVAL & INSTALLATION

Front

2001–04

1. Remove the wheel and tire assembly.
2. Remove the brake caliper clip.

3. Remove the brake caliper bolt caps and bolts.
4. Position the caliper aside.
5. Disconnect and cap the brake line from the caliper and remove the caliper. To install, reverse the removal procedure. Torque the mounting bolts to 26 ft. lbs. (35Nm). Torque the brake line to 15 ft. lbs. (20Nm).
6. Bleed the brake system.

2005

1. Remove the wheel and tire assembly.
2. Remove the brake caliper clip.
3. Remove the brake caliper dust boot caps.
4. Remove the brake caliper guide bolts.
5. Remove the brake caliper.

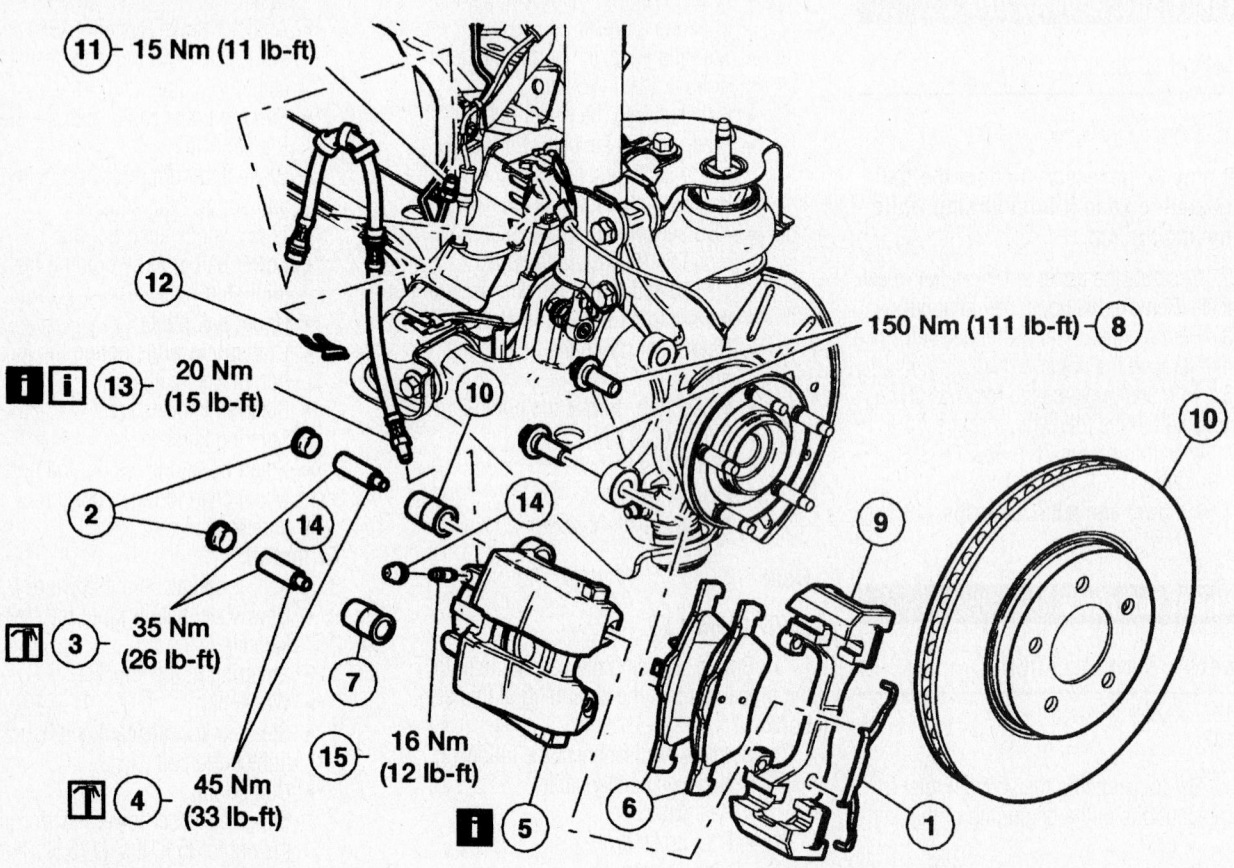

1 Brake caliper clip
2 Brake caliper dust boot caps
3 Brake caliper guide bolts (disc-drum system)
4 Brake caliper guide bolts (4-wheel disc brake system)
5 Brake caliper (RH/LH)
6 Disc brake pads (kit)
7 Brake caliper dust boots
8 Brake caliper anchor plate bolts

9 Brake caliper anchor plate
10 Brake disc
11 Brake line fitting nut
12 Brake caliper jounce hose retaining clip
13 Brake caliper jounce hose
14 Bleeder screw cap
15 Bleeder screw

67197-ESCA-G49

Front caliper installation—2005 model shown

※※ CAUTION

Do not allow the brake caliper to hang by the flexible brake hose.

6. Remove the disc brake pads.

7. Remove the brake caliper jounce hose. Loosen the jounce hose fitting prior to removing the brake caliper.

8. To install, reverse the removal procedure.

9. Torque the caliper pin bolts to 26 ft. lbs. (35 Nm) on disc/drum systems; 33 ft. lbs. (45 Nm) on 4-wheel disc systems.

10. If the hydraulic system was opened, bleed the brake system.

➡ **Thread the brake caliper jounce hose onto the brake caliper before installing the brake caliper.**

➡ **Make sure that the brake caliper jounce hose is not twisted.**

11. Position the brake caliper to the anchor plate and tighten the brake caliper jounce hose.

Rear

1. Remove the wheel and tire assembly.

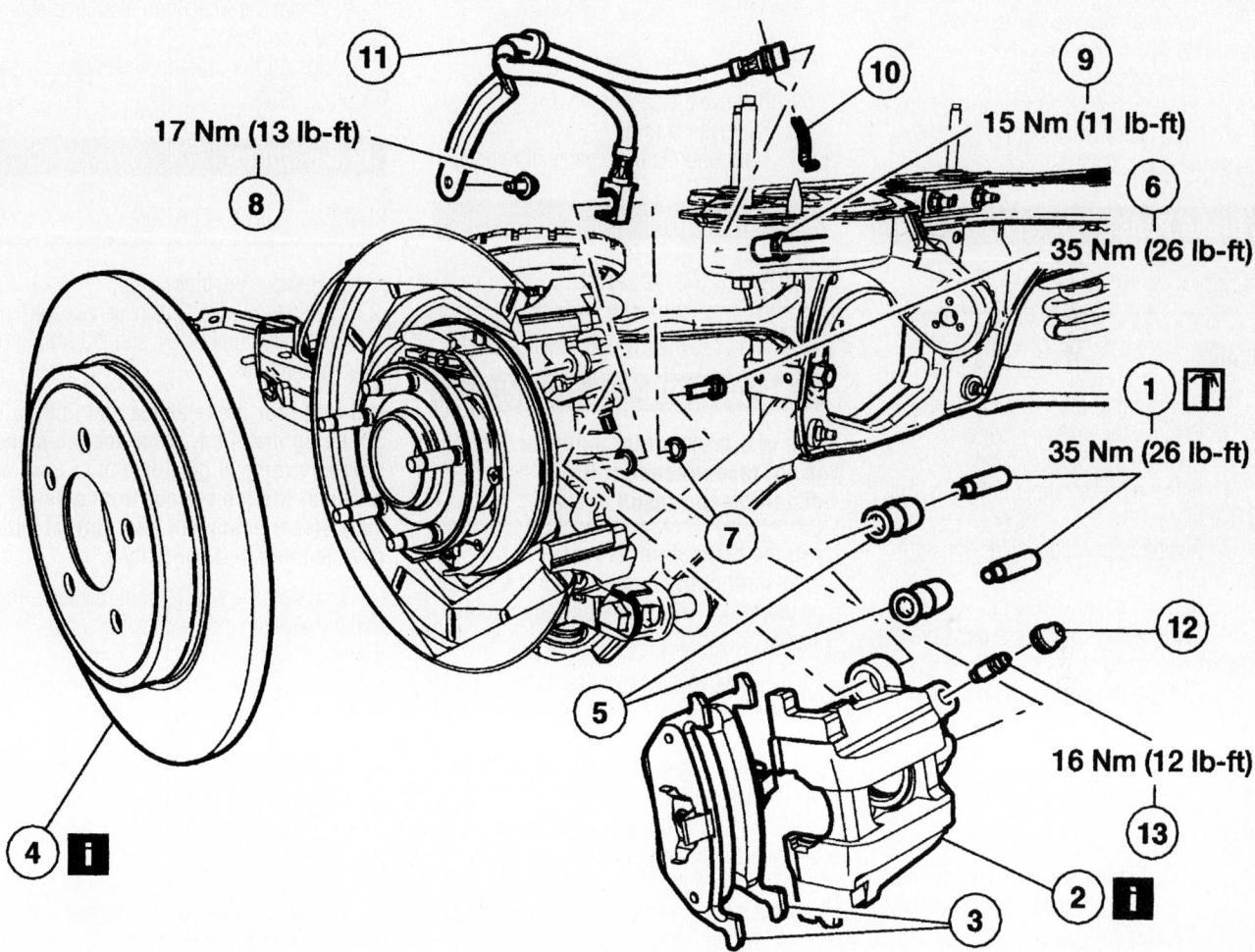

17 Nm (13 lb-ft)
15 Nm (11 lb-ft)
35 Nm (26 lb-ft)
35 Nm (26 lb-ft)
16 Nm (12 lb-ft)

1 Brake caliper guide bolts	8 Brake caliper jounce hose bracket bolt
2 Caliper (RH/LH)	9 Brake line fitting nut
3 Brake disc pads	10 Brake caliper jounce hose retaining clip
4 Brake disc	11 Jounce hose (RH/LH)
5 Brake caliper guide bolt	12 Bleeder screw cap
6 Brake caliper hose flow bolt	13 Bleeder screw
7 Copper washers	

67197-ESCA-G50

Rear caliper installation—2005 model shown

2. Remove the brake caliper guide bolts.

3. Remove the caliper.

✳✳ CAUTION

Do not allow the brake caliper to hang by the flexible brake hose.

4. Remove the brake disc pads.
5. Remove the brake caliper hose flow bolt.
6. Remove and discard the copper washers.
7. To install, reverse the removal procedure. Use new copper washers. Torque the caliper pin bolts to 26 ft. lbs. (35 Nm); torque the flow bolt to 26 ft. lbs. (35 Nm).
8. If the hydraulic system was opened, bleed the brake system.

Brake Pads

REMOVAL & INSTALLATION

Front

1. Remove the wheel and tire assembly.
2. Remove the brake caliper clip.
3. Position the caliper aside.
4. Remove brake caliper bolt caps and the bolts.
5. Position the caliper aside and support.
6. Remove the brake pads.
7. Remove the outer brake pad from the anchor.
8. Remove the inner brake pad from the caliper piston.
9. To install, reverse the removal procedure.

Rear

1. Remove the wheel and tire assembly.
2. Remove the brake caliper guide bolts.
3. Remove the caliper.

✳✳ CAUTION

Do not allow the brake caliper to hang by the flexible brake hose.

4. Remove the brake disc pads.
5. To install, reverse the removal procedure.
6. If the hydraulic system was opened, bleed the brake system.

Brake Rotor

REMOVAL & INSTALLATION

Front

1. Remove the brake caliper anchor plate.
2. Remove the brake disc retaining clips (if equipped) and the brake disc.
3. To install, reverse the removal procedure. Torque the anchor plate bolts to 111 ft. lbs. (150Nm).

Rear

1. Remove the caliper.
2. Remove the rotor.
3. Installation is the reverse of removal.

Brake Drum

REMOVAL & INSTALLATION

1. Remove the tire and wheel assembly.

✳✳ CAUTION

Use of a brake drum puller or a torch is not recommended. Brake drum distortion can result.

➡ If the brake drum is rusted to the axle shaft pilot diameter, tap the center of the brake drum between the wheel studs.

2. Remove the brake drum.
3. If equipped, remove the brake drum retaining clips.
4. If the brake drums will not come off, follow these steps.
5. Move the brake shoe adjusting lever off the brake adjuster screw.
6. Loosen the brake shoe adjuster screw nut by adjusting the nut upward.
7. Using the special tool, 134-R0191, measure the brake drum inside diameter.
8. Install a new brake drum if the maximum inside diameter exceeds the specification shown on the outside of the brake drum.

To install:

✳✳ WARNING

Whenever a wheel is installed, always remove any corrosion, dirt or foreign material present on the mounting surfaces of the wheel or the surface of the wheel hub, brake drum or brake disc that contacts the wheel. Installing wheels without correct metal-to-metal contact at the wheel mounting surfaces can cause the wheel nuts to loosen and the wheel to come off while the vehicle is in motion, causing loss of control. Failure to follow these instructions may result in personal injury.

9. Clean the wheel hub mounting surface and wheel pilot.
10. Install the tire and wheel assembly.

Brake Shoes

REMOVAL & INSTALLATION

1. Remove the brake drum.
2. Use the Brake/Clutch/Service Vacuum to remove brake dust and dirt from the brake assemblies.

➡ If new rear brake shoes and linings are being installed, resurface the brake drums to remove glazing and to ensure an equal friction surface from side-to-side. Resurfacing will also correct out-of-round and bell conditions.

3. Using the special tool, measure the braking surface diameter. If the inside diameter measures more than the maximum specification shown on the outside of the brake drum, install a new brake drum.
4. Remove the parking brake cable from the parking brake cable lever.
5. Remove the hold-down clips and pins.
6. Remove the lower spring.
7. Remove the rear brake shoes.
8. Pull the bottom of the brake shoe forward.
9. Release the upper return spring.
10. Remove both brake shoes together.
11. Remove the self adjuster lever.
12. Remove the self adjuster and spring assembly.
13. Return the self adjuster to the fully seated position.
14. Remove the parking brake lever.
15. Remove the horseshoe clip.

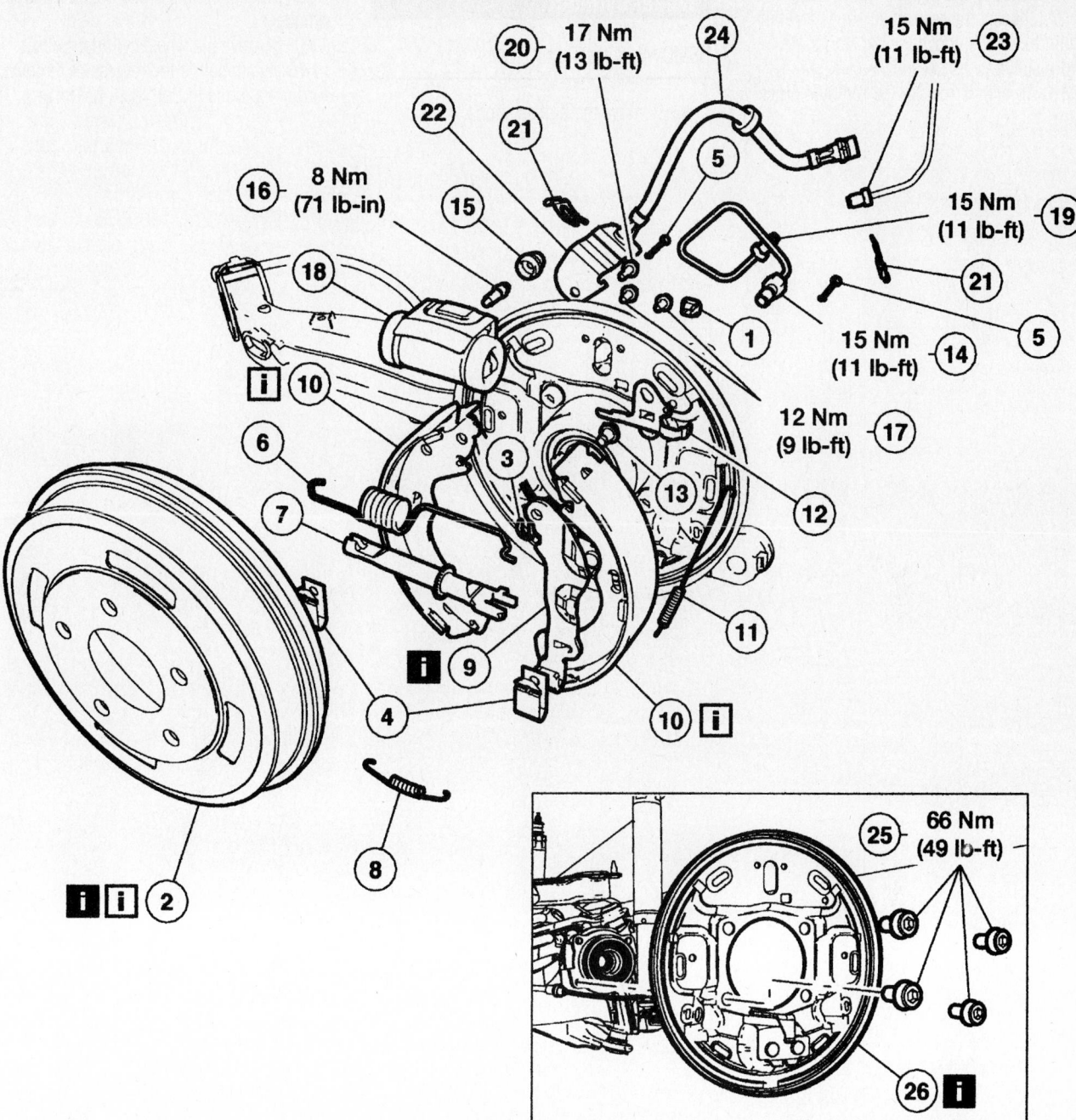

Drum brake exploded view

1 Plug
2 Brake drum
3 Parking brake lever clip
4 Brake shoe retaining clips
5 Brake shoe retaining pins
6 Upper return spring
7 Adjuster assembly (LH/RH)
8 Lower return spring
9 Parking brake actuator lever (LH/RH)

10 Brake shoe (kit)
11 Adjuster lever spring
12 Adjuster lever (LH/RH)
13 Pivot pin (part of 2200)
14 Brake line fitting nut
15 Bleeder screw cap
16 Bleeder screw
17 Wheel cylinder bolts
18 Wheel cylinder

19 Brake line fitting nut
20 Jounce hose bracket bolt
21 Jounce hose retaining clips
22 Jounce hose bracket
23 Brake line fitting nut
24 Jounce hose (LH/RH)
25 Backing plate bolts
26 Backing plate

67197-ESCA-G51

16. Remove the parking brake lever.

17. Inspect the rear brake shoes for minimum thickness above the backing plate, and install new as necessary.

18. To install, reverse the removal procedure.

Wheel Cylinder

REMOVAL & INSTALLATION

1. Remove the brakes shoes.

2. Disconnect the brake line at the wheel cylinder.

3. Remove the wheel cylinder bolts.

4. Installation is the reverse of removal. Torque the bolts to 9 ft. lbs. (12 Nm).